Barr & Stroud

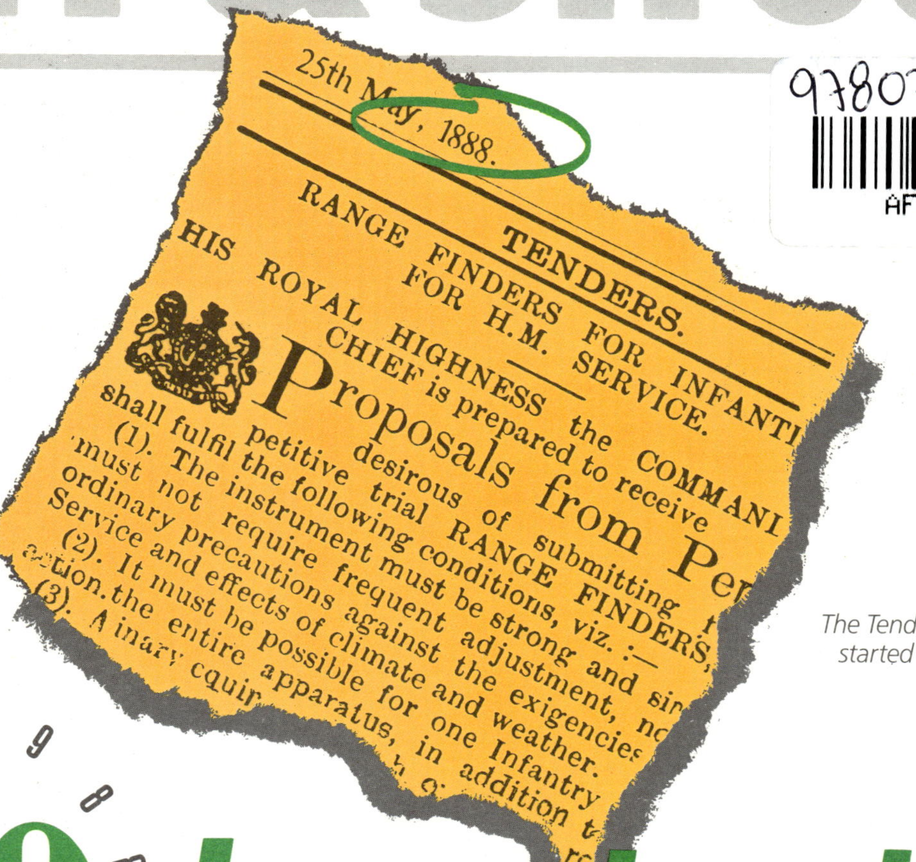
25th May, 1888.
TENDERS.
RANGE FINDERS FOR INFANTI
FOR H.M. SERVICE.
HIS ROYAL HIGHNESS the COMMANI
CHIEF is prepared to receive
Proposals from Per
desirous of submitting
petitive trial RANGE FINDERS,
shall fulfil the following conditions, viz.:—
(1). The instrument must be strong and sin
must not require frequent adjustment, no
ordinary precautions against the exigencies
Service and effects of climate and weather.
(2). It must be possible for one Infantry
action the entire apparatus, in addition t
(3). A inary equir

The Tender that started it all.

1888 | 1988
100
YEARS AHEAD

Innovators then. Innovators now.

In every field, there are those who are best equipped to lead. In the development and manufacture of optical and electro-optical related systems, **Barr & Stroud**, the Defence Systems company of the Pilkington Group, set the pace.

Barr & Stroud's optical and thermal imaging sights, periscopes, laser rangefinders, guided weapon electro-optical systems and fire control systems are proving their effectiveness in service around the world. On land, at sea, in the air.

In the application of new technology, the research and development potential of **Barr & Stroud** keeps them ahead of the competition, reaching ever higher standards of performance.

In a world where only the fit survive, **Barr & Stroud** are fit to lead.

PILKINGTON
◂ Optronics ▸
Barr & Stroud Limited
Caxton Street Anniesland Glasgow Scotland G13 1HZ
Telephone 041-954 9601 Telex 778114 Fax 041-954 2380

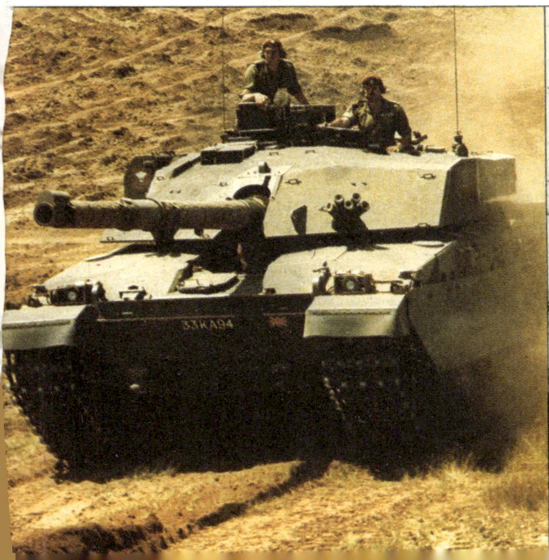

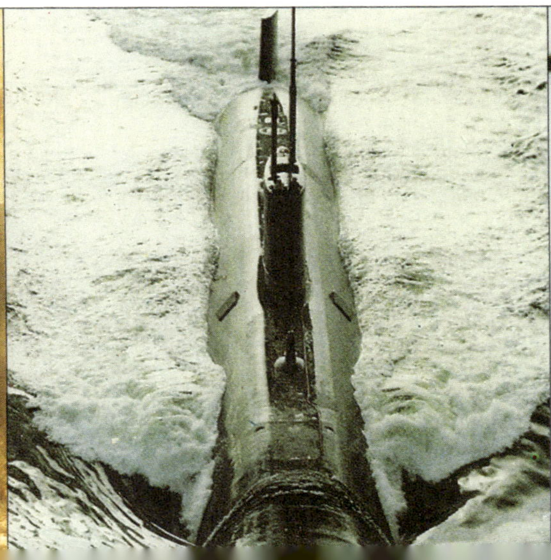

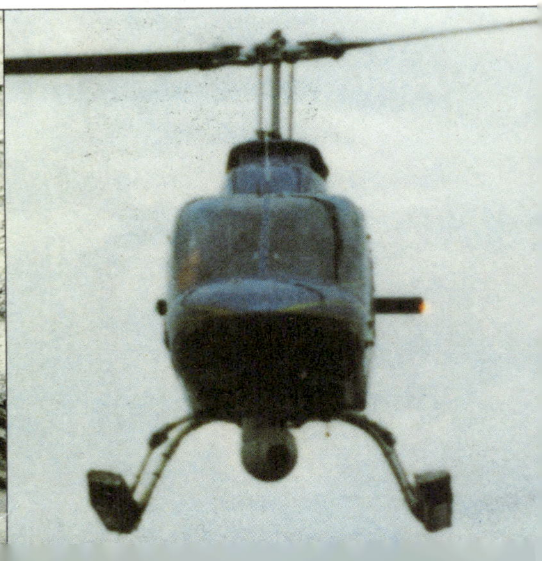

JANE'S WEAPON SYSTEMS 1988-89

Jane's Information Group, 163 Brighton Road, Coulsdon, Surrey CR3 2NX, UK
Jane's Information Group Inc, 1340 Braddock Place, Suite 300, PO Box 1436, Alexandria, Virginia 22313-2036, USA

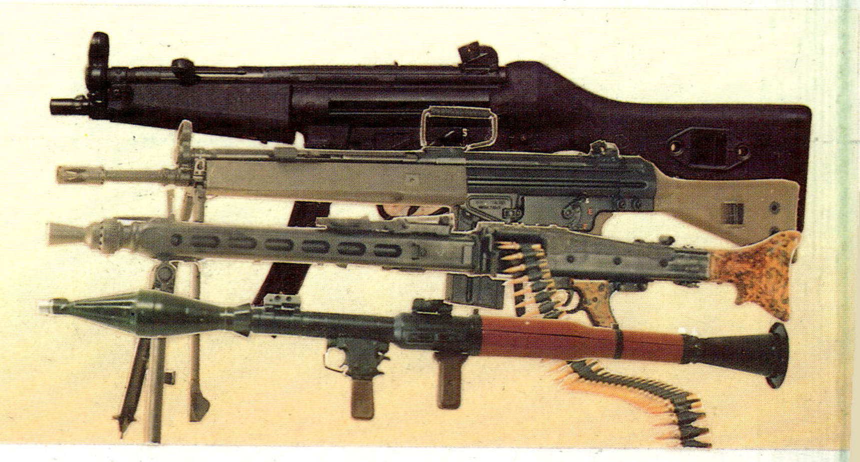

Alphabetical list of advertisers

A

Aérospatiale
Division Engins Tactiques, 2 rue Béranger,
92322 Châtillon Cedex, France .... [14] & [15]

Alkan & Cie
BP 23, Rue du 8 Mai 1945, 94460 Valenton, France...... [30]

Avibras Aerospacial SA
Antiga Estrade de Paraibuna, KM 118, São Jose dos
Campos, São Paulo SP, Brazil .... [38]

B

Barr & Stroud Ltd
Caxton Street, Anniesland, Glasgow G13 1HZ,
Scotland .... *facing inside front cover*

Breda Meccanica Bresciana SpA
Via Lunga 2, 25128 Brescia,
Italy................................ *between main text pages* 512 & 513

Bristol Aerospace (Canada) Ltd
PO Box 874, Winnipeg, Manitoba,
Canada R3C 2S4.......... *between main text pages* 352 & 353

British Aerospace plc
11 Strand, London, England.....................................*bookmark*

C

Consorzio Sistemi Navali Selenia/ELSAG
Via di Sant'Alessandro 28/30, 00131 Rome, Italy .... [48]

Consorzio SMIN
Via Panama 52, 00198 Rome, Italy............ *front endpaper IV*

Creusot-Loire Industrie
Division Mécanique Specialisée, 4 Place de la Pyramide,
Cedex 33, Immeuble Ile-de-France, 92070
Paris-la-Defense, France .... [19] & [29]

D

Direction des Constructions Navales (DCN)
BP 1, 75200 Paris Naval, France...................................... [26]

Defendory International
12 Neofitou Douka Street, GR-106 74 Athens, Greece.... [3]

E

Electro-Optics Industries Ltd (EL-OP)
Advanced Technology Park, Kiryat Weizmann,
PO Box 1165, Rehovot 76110,
Israel.............................. *between main text pages* 832 & 833

Elettronica SpA
Via Tiburtina Km 13 700, 00131 Rome, Italy .... [9]

Eltro GmbH
Postfach 10 21 20, D-6900 Heidelberg 1, Federal Republic
of Germany.. [32]

Euromissile
12 rue de la Redoute, 92260 Fontenay-aux-Roses,
France .... [46]

F

FIAR SpA
Via Montefeltro 8, 20156 Milan, Italy .... [31]

FIAT IVECO
Iveco Defence Vehicle Division, Via Puglia 35,
10156 Turin, Italy .... [21]

H

Hollandse Signaalapparaten BV
PO Box 42, 7550 CD Hengelo, The Netherlands .... [11]

Hughes Aircraft Company
Mail Station C117, 7200 Hughes Terrace, PO Box 45066,
Los Angeles, California 90045-0066, USA .... [13]

I

Industrie Per lo Spazio E Le Comunicazioni SpA
Viale N Macchiavelli 29, 50125 Florence,
Italy ....*between main text pages* 800 & 801

Israel Aircraft Industries
Ben Gurion International Airport
70100 Israel..*back index section*

L

Librascope Division
833 Sonora Avenue, Glendale, California 91201-0279,
USA .... [5]

M

Marconi Command & Control Systems
Chobham Road, Frimley, Camberley, Surrey GU16 5PE,
England......................... *between main text pages* 352 & 353

Matra Defence
Branche Militaire, 37 Avenue Louis-Bréguet, 78146
Vélizy-Villacoublay Cedex, France...................................... [8]

McDonnell Douglas Helicopter Company
Mesa, Arizona 85205, USA .... *back index section*

O

Oerlikon-Buhrle AG
Machine Tool Works, Birchstrasse 155, CH-8050 Zurich,
Switzerland .... [52]

Officine Galileo
Military Systems Division, Via A Einstein 35, 50013 Campi Bisenzio, Florence, Italy .... [23]

Optische Industrie De Oude Delft NV
Van Miereltlaan 9, PO Box 92, Delft, The Netherlands .... *back index section*

OTO Melara SpA
Via Valdilocchi 15, 19100 La Spezia, Italy .... [17]

P

Pakistan Ordnance Factories
Wah Cant, Pakistan .... [2]

Philips International BV
Building BA, PO Box 218, 5600 MD Eindhoven, The Netherlands .... *between main text pages* 352 & 353

Philips Elektronikindustrier AB
S-175 88 Jarfalla, Sweden .... [50]

Philips Usfa BV
Meerenakkerweg 1, Postbus 218, 5600 MD Eindhoven, The Netherlands .... *back index section*

S

Salgad International BV
Stationsplein 45, NL-3001 GC Rotterdam, The Netherlands .... *inside front cover*

SAT
Société Anonyme de Télécommunications, 41 rue Cantagrel, 75013 Paris, France .... *between main text pages* 384 & 385

SATT Communications AB
Telluborgsvagen 90, PO Box 32701, S-126 11 Stockholm, Sweden .... [40]

Selenia Industrie Elettroniche Associate SpA
Defence Systems Division, Via Tiburtina Km 12.400, 00131 Rome, Italy .... *between main text pages* 384 & 385

SEPA
Corso Giulio Cesare 294, 10154 Turin, Italy .... *between main text pages* 832 & 833

SFIM
F-91344 Massy Cedex, France .... [30]

Simrad Optronics AS
PO Box 6114, Etterstad, 0602 Oslo 6, Norway .... [27]

SIPE Nobel SpA
Via Sicilia 162, 00187 Rome, Italy .... *back index section*

SMA SpA
PO Box 200, Florence, Italy .... [32]

SNIA/BPD SpA
Corso Garibaldi 22, 00034 Colleferro, Rome, Italy .... [30]

Sofma
17 Boulevard Malesherbes, 75008 Paris, France .... [25]

T

Tadiran Ltd
11 Ben Gurion Street, PO Box 648, Tel Aviv 61006, Israel .... [44]

TERMA Elektronic AS
Hovmarken 4, DK-8520 Lystrup, Denmark .... *between main text pages* 480 & 481

Thomson Cimsa Sintra
10-12 avenue de l'Europe, BP44, 78140 Vélizy-Villacoublay, France .... *between main text pages* 512 & 513

Thomson Sintra/ASM
Chemin des Travails, BP 53, 06801 Cagnes-sur Mer, Cedex, France .... [36]

V

Vega Precision Laboratories Inc
800 Follin Lane, Vienna, Virginia 22180-4994, USA .... [42]

Vickers Instruments
Haxby Road, York YO3 7SP, England .... *back index section*

Vitroselenia
Via Tiburtina 1020, 00156 Rome, Italy .... [28]

W

Welse Consorzio Sistemi Subacquei
Via G Puccini 2, 16154 Genoa-Sestri, Italy .... [27]

Wild Heerbrugg AG
Segantinisteig 2, CH-8049 Zürich, Switzerland .... [7]

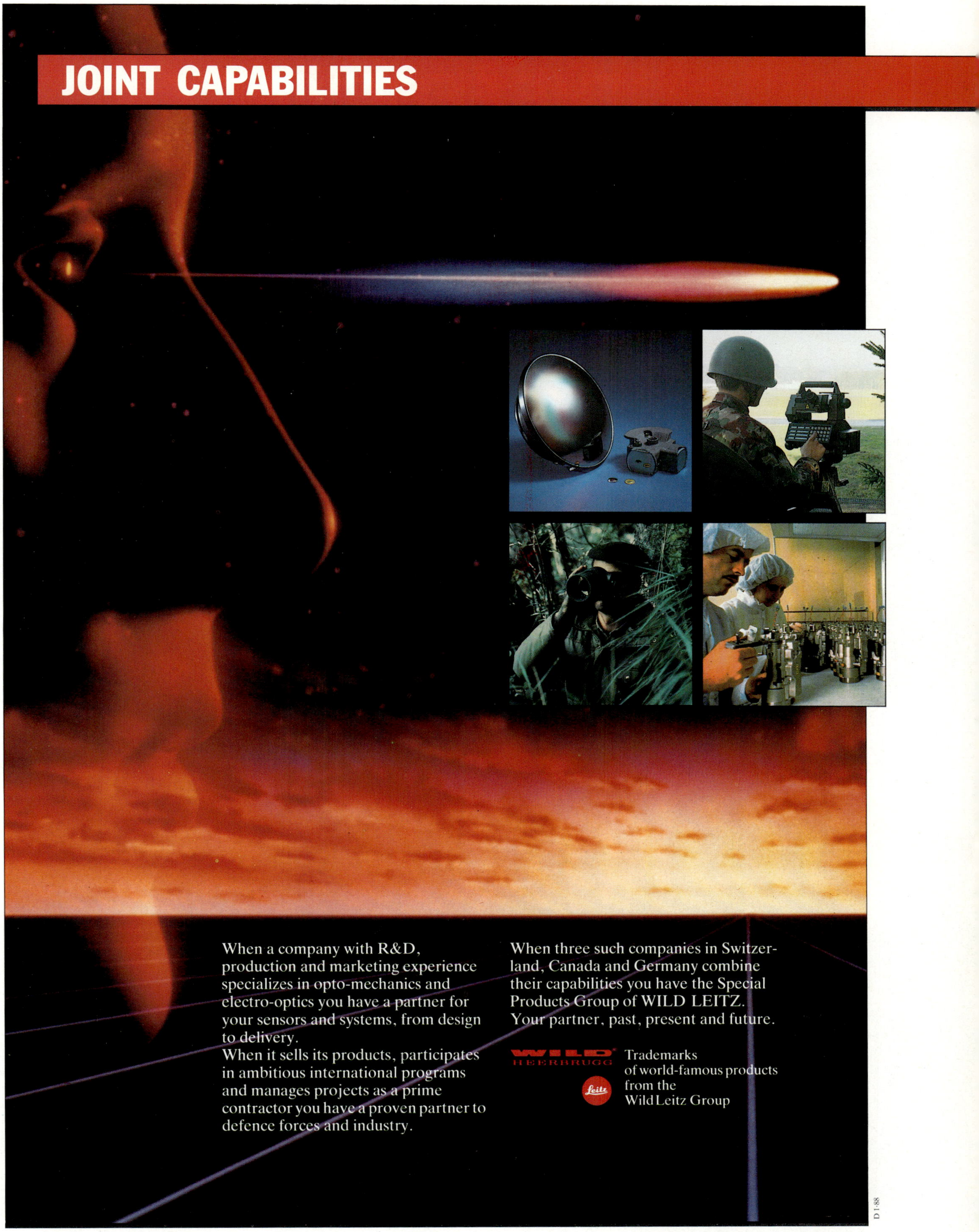
JOINT CAPABILITIES
When a company with R&D, production and marketing experience specializes in opto-mechanics and electro-optics you have a partner for your sensors and systems, from design to delivery.
When it sells its products, participates in ambitious international programs and manages projects as a prime contractor you have a proven partner to defence forces and industry.
When three such companies in Switzerland, Canada and Germany combine their capabilities you have the Special Products Group of WILD LEITZ. Your partner, past, present and future.
WILD HEERBRUGG
Trademarks of world-famous products from the Wild Leitz Group
Leitz
D 1-88

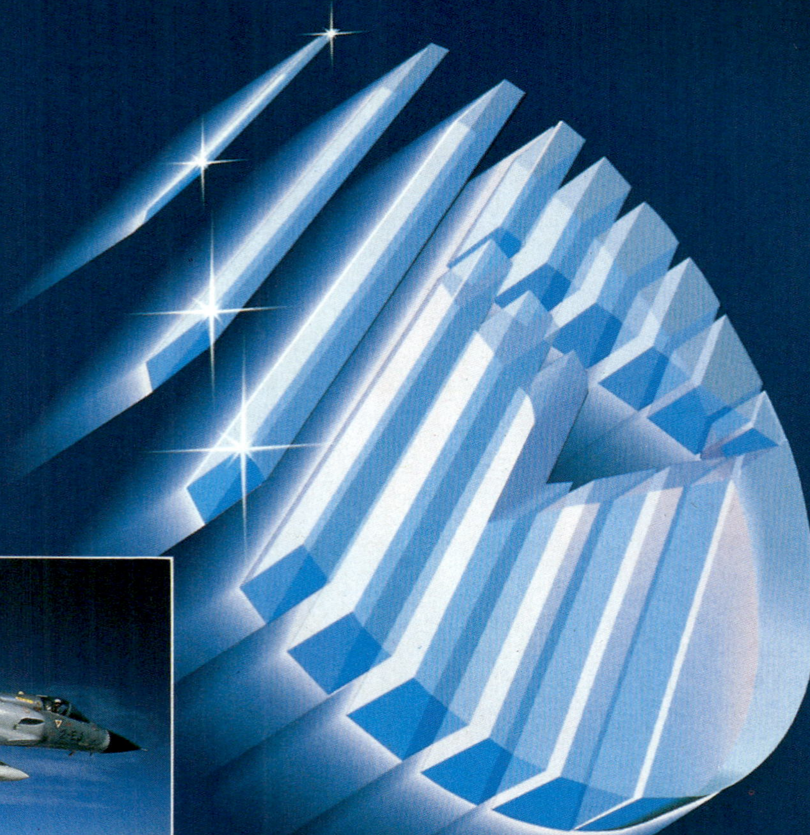

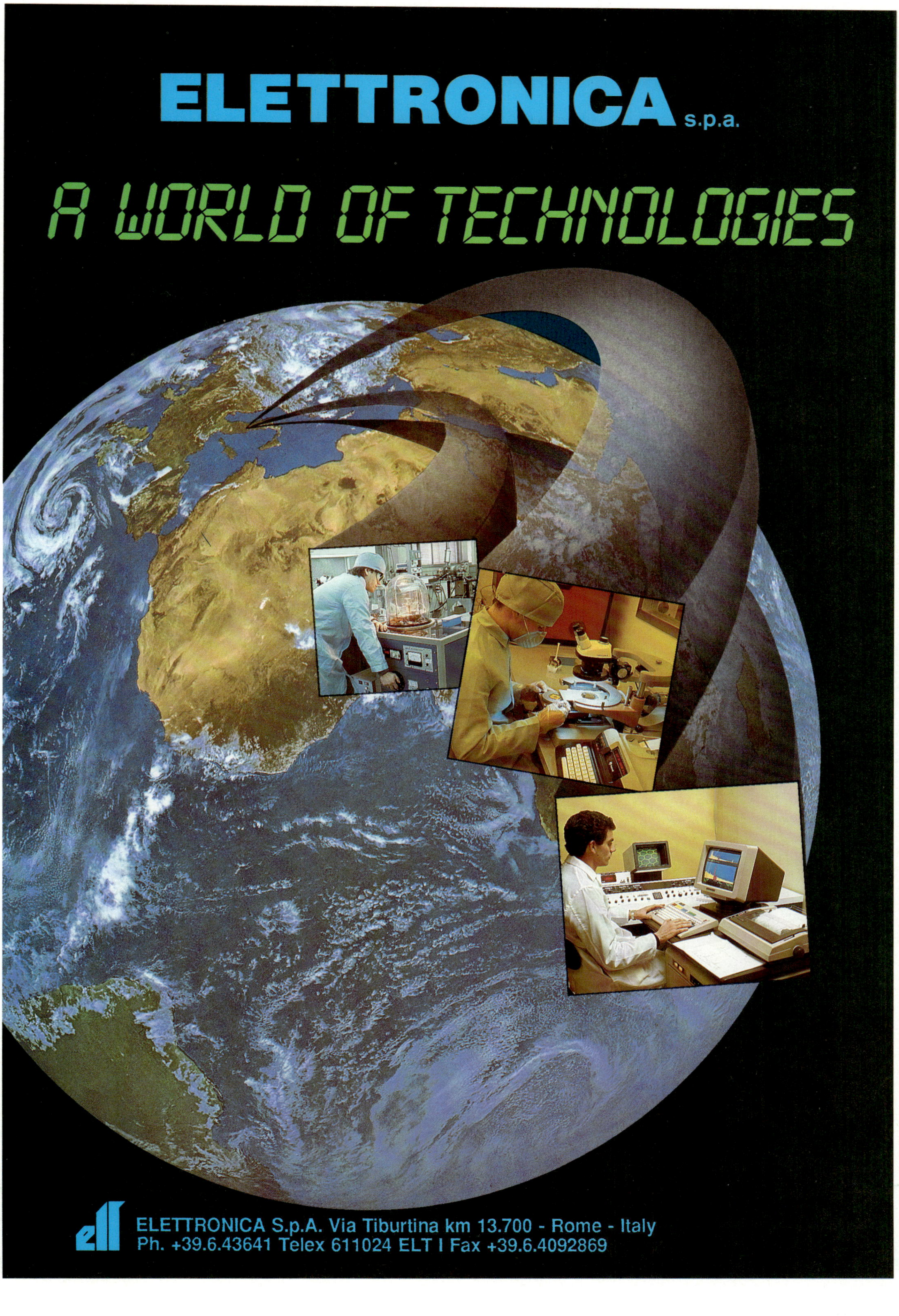
ELETTRONICA s.p.a.
A WORLD OF TECHNOLOGIES
ELETTRONICA S.p.A. Via Tiburtina km 13.700 - Rome - Italy
Ph. +39.6.43641 Telex 611024 ELT I Fax +39.6.4092869

Classified list of advertisers

The companies advertising in this publication have informed us that they are involved in the fields of manufacture indicated below:

Accelerometers
British Aerospace
Matra
SFIM

Acoustic equipment
British Aerospace
Consorzio Sistemi Subacqei
DCN
SEPA
Tadiran
Thomson Sintra/ASM

Acoustic equipment field tracer
Consorzio Sistemi Subacqei
DCN

Aerials/antennas
British Aerospace
Elettronica
Matra
Philips Elektronikindustrier
SMA
Tadiran
Thomson-CSF
Vega Precision Laboratories

AFV sighting systems
Barr & Stroud
British Aerospace
Marconi Command & Control Systems
Simrad Optronics

Air bombs
Matra

Airborne asw detection equipment
British Aerospace
DCN
SMA

Airborne navigation & weapon delivery
British Aerospace
FIAR
SEPA
SFIM
Société R. Alkan & Cie

Air conditioning equipment
British Aerospace
Sofma
Tadiran

Aircraft
British Aerospace
Selenia

Aircraft radar
FIAR
SMA
Thomson-CSF

Air defence systems
Barr & Stroud
British Aerospace
Breda Meccanica Bresciana
DCN
Elettronica
Euromissile
FIAR
Hollandse Signaalapparaten
Hughes Aircraft Company
Marconi Command & Control Systems
Matra
Selenia
Sofma
Thomson-CSF

Alignment equipment
SMA
Wild Heerbrugg (Special Products Division)

Attitude and headings reference systems
British Aerospace
SEPA
SFIM

Ammunition (armoured cars)
SNIA/BPD
Sofma

Ammunition automatic loader
Creusot-Loire Industrie Division Mécanique Specialisée

Ammunition fuzes
DCN
Philips Usfa
Salgad International
SNIA/BPD
Sofma

Ammunition loadings
DCN
SNIA/BPD
Société R. Alkan & Cie

Ammunition loading plants
SNIA/BPD

Ammunition test gun
SNIA/BPD

Anechoic chambers
British Aerospace

Anti-radar weapons
British Aerospace
DCN
Matra
Selenia
SFIM
Sofma

Anti-submarine warfare systems/equipment
British Aerospace
Consorzio Sistemi Navali Selenia/ELSAG
Consorzio Sistemi Subacqei
DCN
FIAR
Hollandse Signaalapparaten
Hughes Aircraft Company
Librascope Division
Matra
Philips Elektronikindustrier
Selenia
SEPA
Société R. Alkan & Cie
Sofma
Thomson-CSF
Thomson Sintra/ASM

Anti-tank weapons
Aérospatiale/Engins Tactiques
Breda Meccanica Bresciana
Bristol Aerospace
Euromissile
Hughes Aircraft Company
Matra
OTO Melara
Pakistan Ordnance Factories
Simrad Optronics

Armament systems
Breda Meccanica Bresciana
British Aerospace
Creusot-Loire Industrie Division Mécanique Specialisée
Matra
OTO Melara
Salgad International
Société R. Alkan & Cie
Sofma

Arming systems
British Aerospace
Marconi Command & Control Systems

Armour
Creusot-Loire Industrie Division Mécanique Specialisée

Armoured fighting vehicles & personnel carriers
Creusot-Loire Industrie Division Mécanique Specialisée
OTO Melara
Simrad Optronics
Sofma

Artillery directing radar
Marconi Command & Control Systems
Sofma
SMA

ASW technology
British Aerospace
Consorzio Sistemi Subacqei
DCN
FIAR
SEPA
Sofma
Thomson Sintra/ASM

ATC systems
Hollandse Signaalapparaten
Selenia
Sofma

Automatic cannons
OTO Melara
Sofma

Automatic flight control systems
SFIM

Automatic rifles
Pakistan Ordnance Factories
Sofma

Automatic test equipment
Barr & Stroud
British Aerospace
FIAR

Automatic testings
Matra
SEPA
SFIM
SMA

Auto-pilot
SEPA
SFIM

Avionics
British Aerospace
DCN
FIAR
Hughes Aircraft Company
SATT Communications
Selenia
SFIM

Azimuth reference equipment
Sofma
Tadiran
Wild Heerbrugg (Special Products Division)

Ballistic measuring equipment
Hughes Aircraft Company
Marconi Command & Control Systems
SNIA/BPD
Sofma
Tadiran
TERMA Elektronic

Ballistic & Tactical missiles
British Aerospace
Hughes Aircraft Company
SNIA/BPD

Batteries
Tadiran

Binoculars
British Aerospace
Officine Galileo
Optische Industrie De Oude Delft

Boards, plotting
SMA

Boats, inflatable
Sofma

Bombs
Hughes Aircraft Company
Matra
Pakistan Ordnance Factories
Sofma

C3
British Aerospace
DCN
FIAR
Hollandse Signaalapparaten
Hughes Aircraft Company
Librascope Division
Matra
Philips Elektronikindustrier
Selenia
SEPA
TERMA Elektronic
Thomson-CSF

Cables, electric & electromechanical
Tadiran

Cameras, tv
FIAR
Thomson-CSF

Chaff dispensing countermeasure launching systems
Matra
SNIA/BPD
Société R. Alkan & Cie

Chaff launchers
Breda Meccanica Bresciana
Matra
Philips Elektronikindustrier
SNIA/BPD
Société R. Alkan & Cie

Chemical warfare agents
Optische Industrie de Oude Delft

Command systems & equipment
British Aerospace
Consorzio Sistemi Subacqei
Hughes Aircraft Company
Marconi Command & Control Systems
Matra
Philips Elektronikindustrier
Selenia
SEPA
SMA

Communication systems
British Aerospace
DCN
Hollandse Signaalapparaten
Hughes Aircraft Company
Librascope Division
Matra
Philips Usfa
SEPA
Tadiran
Thomson-CSF

Communications satellites (military)
Avibras Industria Aerospacial
British Aerospace
FIAR
Hughes Aircraft Company
Matra

Compasses, gyro
British Aerospace
SFIM
Wild Heerbrugg (Special Products Division)

Computerised tank fire control
EL-OP
Hollandse Signaalapparaten
Marconi Command & Control Systems
SEPA
Sofma
SMA

Computers digital
British Aerospace
Hughes Aircraft Company
Tadiran

Consoles
Consorzio Sistemi Subacqei
Hughes Aircraft Company
Selenia
SMA

Control equipment, electronic
SEPA
Tadiran
Thomson-CSF
Vega Precision Laboratories

Converters
Marconi Command & Control Systems

Countermeasure launching systems
Consorzio Sistemi Navali Selenia/ELSAG
DCN
Matra
Philips Elektronikindustrier
Société R. Alkan & Cie

Cyrogenics
British Aerospace
Officine Galileo
Philips Usfa

Data and information processing systems
British Aerospace
EL-OP
FIAR
Hughes Aircraft Company
Librascope Division
Matra
Selenia
SFIM
SMA
Tadiran

Data processing/peripheral equipment
Librascope Division
SFIM
Tadiran

Depth chargers
British Aerospace

Detection and alarm systems
Matra
Sofma
Thomson-CSF

Detectors
SFIM
Simrad Optronics
Sofma

Digital data bus systems, shipborne
Consorzio Sistemi Navali Selenia/ELSAG
DCN
Hollandse Signaalapparaten
SMA

Direct drive servo systems
DCN
Officine Galileo

Display, alpha-numerical large scale
Hughes Aircraft Company
SMA

Display, cathode ray tube
FIAR
Hughes Aircraft Company
Librascope Division
SEPA
SFIM
SMA
Thomson-CSF

Displays, in-cockpit
FIAR
Hughes Airport Company
SMA

Distance measuring equipment
Consorzio Sistemi Subacqei
EL-OP
FIAR
Optische Industrie de Oude Delft
SEPA
Wild Heerbrugg (Special Products Division)

Dopplar sonars
DCN

Drones
DCN
Matra

Ejector release units
Société R. Alkan & Cie

Electronic countermeasures
Elettronica
Hughes Aircraft Company
Librascope Division
Matra
Philips Elektronikindustrier
Selenia
Tadiran
Thomson-CSF

Electronic fuses
Marconi Command & Control Systems
Sofma

Electronic security systems
FIAR
Philips Usfa
Sofma

Electronic warfare systems
British Aerospace
DCN
Elettronica
Hughes Aircraft Company
SATT Communications
Selenia
Sofma
Tadiran
Thomson-CSF

Electronic warning systems
Elettronica
FIAR
SATT Communications
Sofma

Electro-optical systems
Barr & Stroud
British Aerospace
Consorzio Sistemi Navali Selenia/ELSAG
EL-OP
Eltro
FIAR
Hughes Aircraft Company
Matra
Officine Galileo
Optische Industrie de Oude Delft
Philips Usfa
SAT (Société Anonyme de Télécommunications)
Selenia
SFIM
Tadiran

Electro-optical directors (naval)
Barr & Stroud
British Aerospace
EL-OP
FIAR
Officine Galileo
SAT (Société Anonyme de Télécommunications)
Sofma

Encoders/decoders
SAT (Société Anonyme de Télécommunications)
Vega Precision Laboratories

Environmental testing
Barr & Stroud
British Aerospace
Philips Usfa
SMA

EW equipment
British Aerospace
Elettronica
Philips Elektronikindustrier
Selenia
Tadiran

Exhibition organisers
Institute of Industrial Exhibitions (Defendory)

Explosives
Pakistan Ordnance Factories
SNIA/BPD
Sofma

Fibre optics
Hughes Aircraft Company
Matra
Optische Industrie de Oude Delft
SAT (Société Anonyme de Télécommunications)
Selenia
Tadiran

Field artillery
Breda Meccanica Bresciana
OTO Melara
Pakistan Ordnance Factories
Salgad International
Simrad Optronics
SNIA/BPD
Sofma

Fire-control equipment
Avibras Industria Aerospacial
Barr & Stroud
Consorzio Sistemi Navali Selenia/ELSAG
Consorzio Sistemi Subacqei
EL-OP
FIAR
Hollandse Signaalapparaten
Hughes Aircraft Company
Marconi Command & Control Systems
Officine Galileo
Optische Industrie de Oude Delft
Philips Elektronikindustrier
SATT Communications
Selenia
SEPA
SFIM
Simrad Optronics
SMA
SNIA/BPD
Sofma
TERMA Elektronic
Tadiran
Thomson-CSF
Wild Heerbrugg (Special Products Division)

Fire suppression systems
Hughes Aircraft Company

Flexible or rigid disk drives
British Aerospace

Flight data recorders
British Aerospace
SFIM

Grenades (rifle)
SNIA/BPD
Sofma

Ground handling equipment
Société r. alkan & Cie
Sofma

Ground influence mines (sea mines)
British Aerospace
DCN
Sofma

Guidance systems, inertial
Avibras Industria Aerospacial
British Aerospace
Selenia
SFIM

Guidance systems, missile
Aérospatiale/Engins Tactiques
Avibras Industria Aerospacial
British Aerospace
FIAR
Hughes Aircraft Company
Matra
Officine Galileo
SAT (Société Anonyme de Télécommunications)
Selenia
SMA

Guidance systems, optical
British Aerospace
FIAR
Officine Galileo
Tadiran

Guidance systems, radio command
British Aerospace
Hughes Aircraft Company

Gunnery training systems
Sofma
Tadiran

Gun systems
Breda Meccanica Bresciana
Creusot-Loire Industrie Division Mécanique Specialisée
OTO Melara
Pakistan Ordnance Factories
Salgad International
Sofma

Gyroscopes
British Aerospace
Matra
SFIM
Wild Heerbrugg (Special Products Division)

Harbour protection systems (magnetic/acoustic)
DCN

Head-up display
FIAR
Optische Industrie de Oude Delft
Selenia
Thomson-CSF

Heated windows
Barr & Stroud

Hydraulic windows
DCN
Officine Galileo

Hydrofoils
DCN

Hydrophones
DCN
Thomson Sintra/ASM

Iff equipment, shipborne
DCN
SMA

Infantry training equipment
Pakistan Ordnance Factories
Sofma

Infantry weapons
Aérospatiale/Engins Tactiques
Oerlikon-Buhrle
Pakistan Ordance Factories
Salgad International
Sofma

Infra-red detectors for thermal imaging units
EL-OP
SAT (Société Anonyme de Télécommunications)
Sofma

Infra-red equipment
British Aerospace
Eltro
FIAR
Hollandse Signaalapparaten
Hughes Aircraft Company
Officine Galileo
Optische Industrie de Oude Delft
Philips Usfa
SAT (Société Anonyme de Télécommunications)
Selenia
Sofma

Infra-red equipment/materials
Barr & Stroud
FIAR
Officine Galileo
Philips Usfa
SAT (Société Anonyme de Télécommunications)

Infra-red jammers
British Aerospace
SFIM
Société R. Alkan & Cie
Thomson-CSF

Instrument landing systems
SMA
Thomson-CSF

Inverters
Marconi Command & Control Systems

IR thermal imaging systems
Barr & Stroud
British Aerospace
EL-OP
Euromissile
FIAR
Hollandse Signaalapparaten
Hughes Aircraft Company
Marconi Command & Control Systems
Officine Galileo
Optische Industrie de Oude Delft
Philips Usfa
SAT (Société Anonyme de Télécommunications)
Selenia
SFIM
Sofma
Thomson-CSF

IR warning systems
FIAR
SAT (Société Anonyme de Télécommunications)

Laser warning systems
British Aerospace
Eltro
FIAR
Simrad Optronics

Laser rangefinders
Barr & Stroud
EL-OP
Eltro
FIAR
Hollandse Signaalapparaten
Hughes Aircraft Company
Optische Industrie de Oude Delft
Selenia
Simrad Optronics
Sofma
Wild Heerbrugg (Special Products Division)

Lasers
Barr & Stroud
EL-OP
Eltro
FIAR
Hughes Aircraft Company
Optische Industrie de Oude Delft
SAT (Société Anonyme de Télécommunications)
Selenia
Simrad Optronics

Launchers, lightweight rocket
Avibras Industria Aerospacial
Breda Meccanica Bresciana
Bristol Aerospace
Hughes Aircraft Company
Matra
SNIA/BPD
Sofma

Launching equipment
Société R. Alkan & Cie

Law enforcement & riot control equipment
FIAT IVECO
Sofma

Lenses
Barr & Stroud
British Aerospace
EL-OP
Matra
Officine Galileo
Optische Industrie de Oude Delft
Wild Heerbrugg (Special Products Division)

Low frequency sonars
British Aerospace
FIAR

Magnetic immunisation systems
DCN

Magnetometers
DCN

Measuring equipment
SFIM

Message processing equipment
British Aerospace
Marconi Command & Control Systems
SEPA
Tadiran

Meterological equipment
Avibras Industria Aerospacial
FIAR
Marconi Command & Control Systems
Simrad Optronics
Sofma
SMA

Meters
SFIM

Microwave communications equipment
Avibras Industrial Aerospacial
SAT (Société Anonyme de Télécommunications)
Selenia

Microwave landing systems, air defence
FIAR
SFIM
SMA
Thomson-CSF

Mine-hunting sonars
DCN
FIAR
Sofma

Mine neutralisation systems
DCN
Sofma

Mines
British Aerospace
DCN
Marconi Command & Control Systems
Pakistan Ordnance Factories

Missile fuzes
British Aerospace
Marconi Command & Control Systems
Matra
Selenia

Missile launchers & handling systems
British Aerospace
DCN
Matra
OTO Melara
Selenia
SNIA/BPD
Société R. Alkan & Cie

Missile optics
Birtish Aerospace
FIAR
Hughes Aircraft Company
Matra
Officine Galileo
Wild Heerbrugg (Special Products Division)

Missile seekers
British Aerospace
FIAR
Matra
Selenia
SMA

Missiles
Aérospatiale/Engins Tactiques
Avibras Industrial Aerospacial
British Aerospace
Euromissile
Hughes Aircraft Company
OTO Melara
Matra
Selenia
SNIA/BPD

Mobile workshops for maintenance services
Eltro
FIAT IVECO

Mortar locating radar
FIAR
Hughes Aircraft Company

Mortars
Breda Meccanica Bresciana
Sofma

Mortars, canon
Sofma

Motors, hydraulic
Sofma

Mounts, antennas
Avibras Industrial Aerospacial
SMA

Multi sensor collection and correlation systems
DCN
Selenia

Munitions and ordnance
DCN
Pakistan Ordnance Factories
Salgad International
SNIA/BPD

Muzzle boresights
Sofma
Wild Heerbrugg (Special Products Division)

Naval Combat Systems
Consorzio Sistemi Navali Selenia/ELSAG
DCN
Hollandse Signaalapparaten
Marconi Command & Control Systems
Matra
Selenia
SNIA/BPD
Sofma
SMA
TERMA Elektronic

Naval data processing systems
British Aerospace
DCN
Hollandse Signaalapparaten
SEPA
Sofma
SMA
TERMA Elektronic

Naval defence systems
British Aerospace
DCN
Marconi Command & Control Systems
Matra
Selenia
Sofma
SMA

Naval guns
Creusot-Loire Industrie Division Mécanique Specialisée
DCN
OTO Melara
Sofma

Navigation aids
Avibras Industrial Aerospacial
British Aerospace
DCN
FIAR
Selenia
SFIM
TERMA Elektronic
Vega Precision Laboratories

Night vision equipment
Barr & Stroud
British Aerospace
EL-OP
Euromissile
FIAR
Hughes Aircraft Company
Marconi Command & Control Systems
Officine Galileo
Optische Industrie du Oude Delft
Philips Usfa
SFIM
Simrad Optronics
Sofma
Thomson-CSF
Wild Heerbrugg (Special Products Division)

Night vision equipment, components
Barr & Stroud
Officine Galileo
Optische Industrie de Oude Delft
Philips Usfa
Sofma

Noise measurement equipment
DCN

Oceanographic equipment
Creusot-Loire Industrie Division Mécanique Specialisée
DCN

Passive night vision
EL-OP
Hughes Aircraft Company
Marconi Command & Control systems
Officine Galileo
Optische Industrie de Oude Delft
Philips Usfa
SFIM
Sofma
Wild Heerbrugg (Special Products Division)

Personnel carriers
Sofma

Personnel training
Matra
SIPE Nobel

Platforms, rolling
DCN
SMA

Platforms, stable
British Aerospace
SFIM
SMA

Plot extractors
SMA

Portable anti-tank weapons
Breda Meccanica Bresciana
British Aerospace
Euromissile
Hughes Aircraft Company
Matra
Pakistan Ordnance Factories
Sofma

Power equipment
FIAT IVECO
Sofma

Programmable fuses
Sofma

Propellants
Bristol Aerospace
Pakistan Ordnance Factories
SNIA/BPD

Propellant test computers
SNIA/BPD

Propellants
SNIA/BPD

Pyrotechnics
DCN
Pakistan Ordnance Factories

Rocks, bomb
Société R. Alkan & Cie

Radar, airborne
FIAR
Hughes Aircraft Company
SMA
Thomson-CSF

Radar, 3-dimensional (ground mobile)
Avibras Industrial Aerospacial
Hollandse Signaalapparaten
Hughes Aircraft Company
Selenia
Thomson-CSF

Radar and flight systems
Thomson-CSF

Radar, ground
FIAR
Hollandse Signaalapparaten
Hughes Aircraft Company
Marconi Command & Control Systems
Philips Elektronikindustrier
Selenia
SMA
TERMA Elektronic
Thomson-CSF
Vega Precision Laboratories

Radar, ground ship-based
Hollandse Signaalapparaten
Hughes Aircraft Company
Marconi Command & Control Systems
Philips Elektronikindustrier
Selenia
SMA
TERMA Elektronic
Thomson-CSF

Radar high-power early warning (ground)
Hollandse Signaalapparaten
Selenia
·Thomson-CSF

Radar, low coverage (ground)
Hollandse Signaalapparaten
Hughes Aircraft Company
Selenia
SMA
TERMA Elektronic
Thomson-CSF

Radar, phased array
Thomson-CSF

Radar reflective components
British Aerospace

Radar target illumination & tracking (ground, ship)
Hollandse Signaalapparaten
Selenia
SMA
Vega Division

Radio communication equipment
Hollandse Signaalapparaten
Hughes Aircraft Company
Philips Usfa
SMA
Tadiran
Thomson-CSF

Radomes
British Aerospace
DCN
Selenia
SMA

Radomes, air supported
DCN

Reconnaissance equipment
British Aerospace
Tadiran
Vega Precision Laboratories

Reconnaissance vehicles
FIAT IVECO
Sofma

Research and development
Barr & Stroud
Creusot-Loire Industrie Division Mécanique Specialisée
DCN
Hughes Aircraft Company
Optische Industrie de Oude Delft
Tadiran

Retarders bomb
Breda Meccanica Bresciana
Matra

Rocket launchers
Avibras Industrial Aerospacial
Breda Meccanica Bresciana
Consorzio Sistemi Navali Selenia/ELSAG
Creusot-Loire Industrie Division Mécanique Specialisée
Hughes Aircraft Company
Matra
Pakistan Ordnance Factories
SNIA/BPD
Sofma

Rocket motors
Bristol Aerospace
SNIA/BPD

Rockets
Avibras Industrial Aerospacial
Bristol Aerospace
Hughes Aircraft Company
Pakistan Ordnance Factories
SNIA/BPD
Sofma

Secondary surveillance radar
Selenia

Secure communication equipment
Hughes Aircraft Comany
Tadiran
Vitroselenia

Self-propelled anti-aircraft equipment
Creusot-Loire Industrie Division Mécanique Specialisée
Marconi Command & Control Systems
Matra

Self-propelled field artillery
Salgad International

Sensors, transducers
DCN
Hollandse Signaalapparaten
Hughes Aircraft Company
Marconi Command & Control Systems
SFIM

Servomechanics
British Aerospace
DCN
Matra

Servo systems
DCN

Ship & aircraft gun mounts
Sofma

Sights
British Aerospace
Consorzio Sistemi Navali Selenia/ELSAG
EL-OP
Marconi Command & Control Systems
Officine Galileo
Philips Elektronikindustrier
SFIM

Signal processing
British Aerospace
DCN
FIAR
Hollandse Signaalapparaten
SFIM
Thomson-CSF

Simulators
British Aerospace
DCN
FIAR
Hughes Aircraft Company
Sofma
Tadiran
Vega Precision Laboratories

Small arms
Pakistan Ordnance Factories
Sofma

Sonar equipment
Consorzio Sistemi Navali Selenia/ELSAG
Consorzio Sistemi Subacqei
DCN
FIAR
Hollandse Signaalapparaten
Hughes Aircraft Company
Thomson Sintra/ASM

Sonar intercepters
Consorzio Sistemi Subacqei
DCN
Hollandse Signaalapparaten

40-10 WM

OVER 100 Kph

100 HP

INDEPENDENT FRONT SUSPENSIONS

PAYLOAD 1.5 TON

FOUR WHEEL DRIVE

TOWS UP TO 1.5 TONNES

BANKING UP TO 30%

GRADIENT CLIMBABLE: OVER 60%

RANGE 600 Km

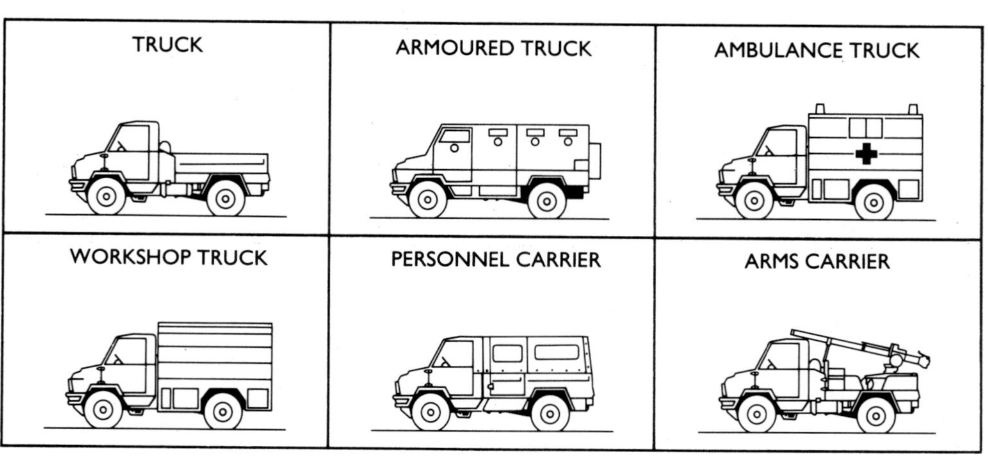

IVECO

IVECO Defence Vehicle Division
Bolzano (Italy) - Ulm (W. Germany)

Sound ranging systems
SATT Communications

Submarine periscopes
Barr & Stroud
Philips Usfa

Submarines & reconnaissance equipment
Hughes Aircraft Company
Marconi Command & Control Systems
SMA

Submarine & systems
DCN
SMA

Submarines & countermeasures
DCN
Tadiran

Switching systems
Hollandse Signaalapparaten
Pakistan Ordnance Factories
Tadiran

Synthesisers, frequency
British Aerospace

Tactical missiles
Aérospatiale/Engins Tactiques
British Aerospace
Euromissile
Hughes Aircraft Company
Matra
Selenia

Tank laser sights
Barr & Stroud
EL-OP
FIAR
Hughes Aircraft Company
Optische Industrie de Oude Delft
SFIM
Sofma

Tanks
Creusot-Loire Industrie Division Mécanique Specialisée
FIAT IVECO
OTO Melara
Sofma

Tank transponders
FIAT IVECO
OTO Melara
Sofma

Telemetry equipment
Hughes Aircraft Company
SFIM
Sofma

Theodolites
Sofma
Wild Heerbrugg (Special Products Division)

Thermal imaging units
Barr & Stroud
EL-OP
Eltro
Hollandse Signaalapparaten
Hughes Aircraft Company
Marconi Command & Control systems
Officine Galileo
Optische Industrie de Oude Delft
Philips Usfa
SFIM
Sofma
Wild Heerbrugg (Special Products Division)

Torpedo homing heads and electronics
Consorzio Sistemi Navali Selenia/ELSAG
DCN
SEPA
Sofma
Thomson Sintra/ASM

Torpedo launching systems
Consorzio Sistemi Navali Selenia/ELSAG
DCN
Librascope Division
Matra
Sofma

Torpedoes
DCN
Hughes Aircraft Company
Sofma

Tow target systems
DCN
Société R. Alkan & Cie

Tracking equipment
Eltro
Marconi Command & Control Systems
Philips Usfa
SFIM
SMA
Vega Division

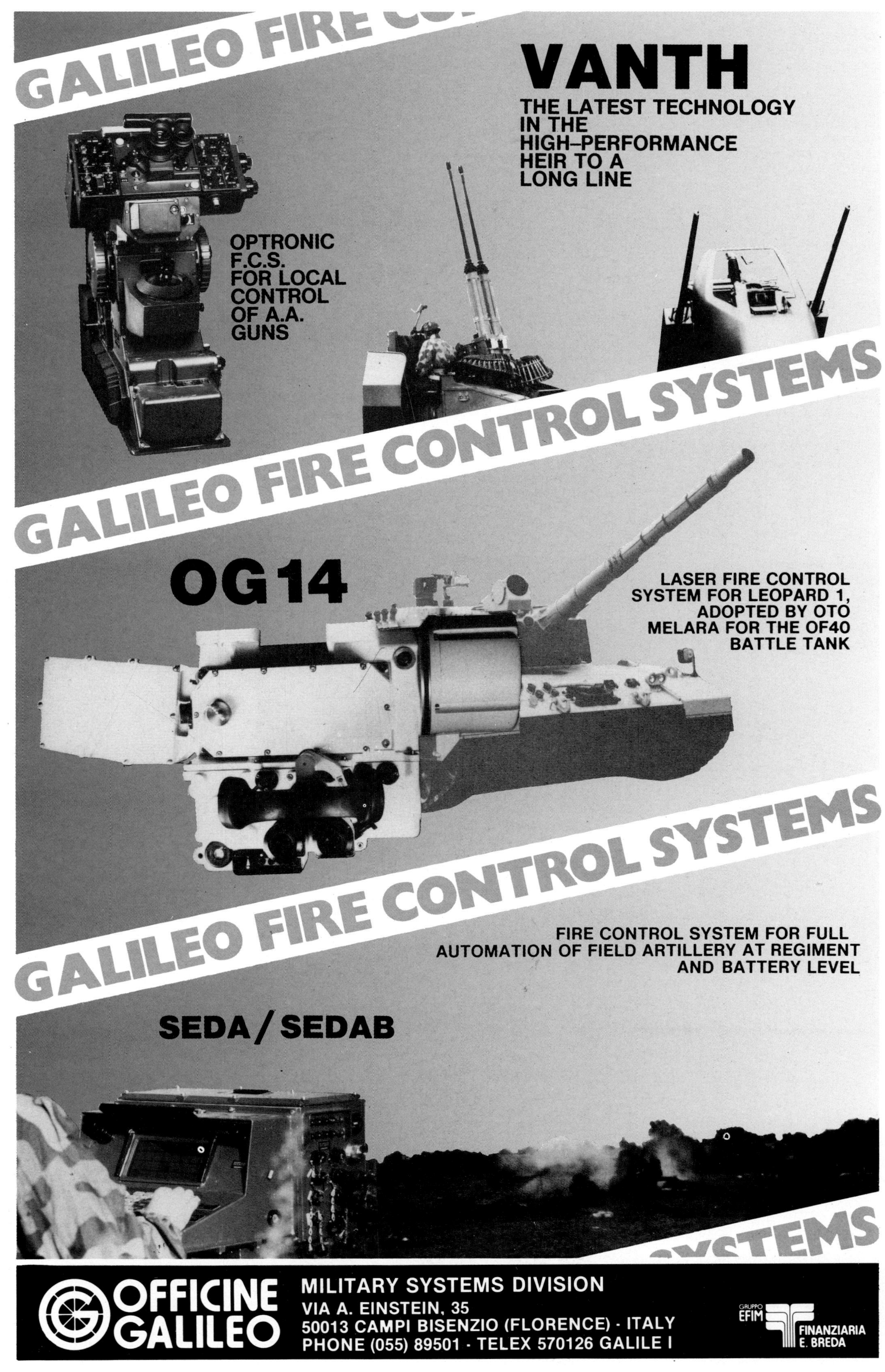
GALILEO FIRE CONTROL SYSTEMS
VANTH
THE LATEST TECHNOLOGY IN THE HIGH–PERFORMANCE HEIR TO A LONG LINE
OPTRONIC F.C.S. FOR LOCAL CONTROL OF A.A. GUNS
GALILEO FIRE CONTROL SYSTEMS
OG14
LASER FIRE CONTROL SYSTEM FOR LEOPARD 1, ADOPTED BY OTO MELARA FOR THE OF40 BATTLE TANK
GALILEO FIRE CONTROL SYSTEMS
FIRE CONTROL SYSTEM FOR FULL AUTOMATION OF FIELD ARTILLERY AT REGIMENT AND BATTERY LEVEL
SEDA/SEDAB
OFFICINE GALILEO
MILITARY SYSTEMS DIVISION
VIA A. EINSTEIN, 35
50013 CAMPI BISENZIO (FLORENCE) - ITALY
PHONE (055) 89501 - TELEX 570126 GALILE I
GRUPPO EFIM
FINANZIARIA E. BREDA

Training
OTO Melara
Singer Libroscope

Training aids and devices
Euromissile
Hughes Aircraft Company
Pakistan Ordnance Factories
SEPA

Training & live mines
British Aerospace
Marconi Command & Control systems
Pakistan Ordnance Factories
Sofma

Training & simulation
British Aerospace
Consorzio Sistemi Navali Selenia/ELSAG
Hughes Aircraft Company
Matra
Pakistan Ordnance Factories
Selenia
SEPA
Sofma
Tadiran

Transponders, radar
Vega Precision Laboratories

Trucks
FIAT IVECO

Turrets for armoured fighting vehicles
Breda Meccanica Bresciana
Creusot-Loire Industries Division Mécanique Specialisée
OTO Melara
Sofma

Underwater combat control equipment
Consorzio Sistemi Navali Selenia/ELSAG
Consorzio Sistemi Subacqei
Hollandse Signaalapparaten

Underwater systems
British Aerospace
Consorzio Sistemi Navali Selenia/ELSAG
Consorzio Sistemi Subacqei
Consorzio SMIN
DCN
Hollandse Signaalapparaten
Hughes Aircraft Company

Underwater telephony
DCN

Universal turrets and tanks
OTO Melara
Sofma

Unmanned aircraft
Tadiran

Up-date, modification & reverse engineering (electronics)
TERMA Elektronic

Vehicles military (track and wheel)
Avibras Industrial Aerospatial
Creusot-Loire Industrie Division Mécanique Specialisée
FIAT IVECO
Sofma

Velocity measuring equipment
British Aerospace
Marconi Command & Control Systems
Sofma

Weapons carriage & release equipment
British Aerospace
Société R. Alkan & Cie

Weapon systems
Aérospatiale/Engins Tactiques
Avibras Industrial Aerospatial
Breda Meccanica Bresciana
Bristol Aerospace
British Aerospace
Consorzio Sistemi Navali Selenia/ELSAG
Consorzio Sistemi Subacqei
Creusot-Loire Industrie Division Mécanique Specialisée
DCN
FIAR
Hughes Aircraft Company
Librascope Division
Marconi Command & Control systems
Matra
OTO Melara
Salgad International
Selenia
SMA
SNIA/BPD
Sofma
WELSE Underwater Systems Consortium

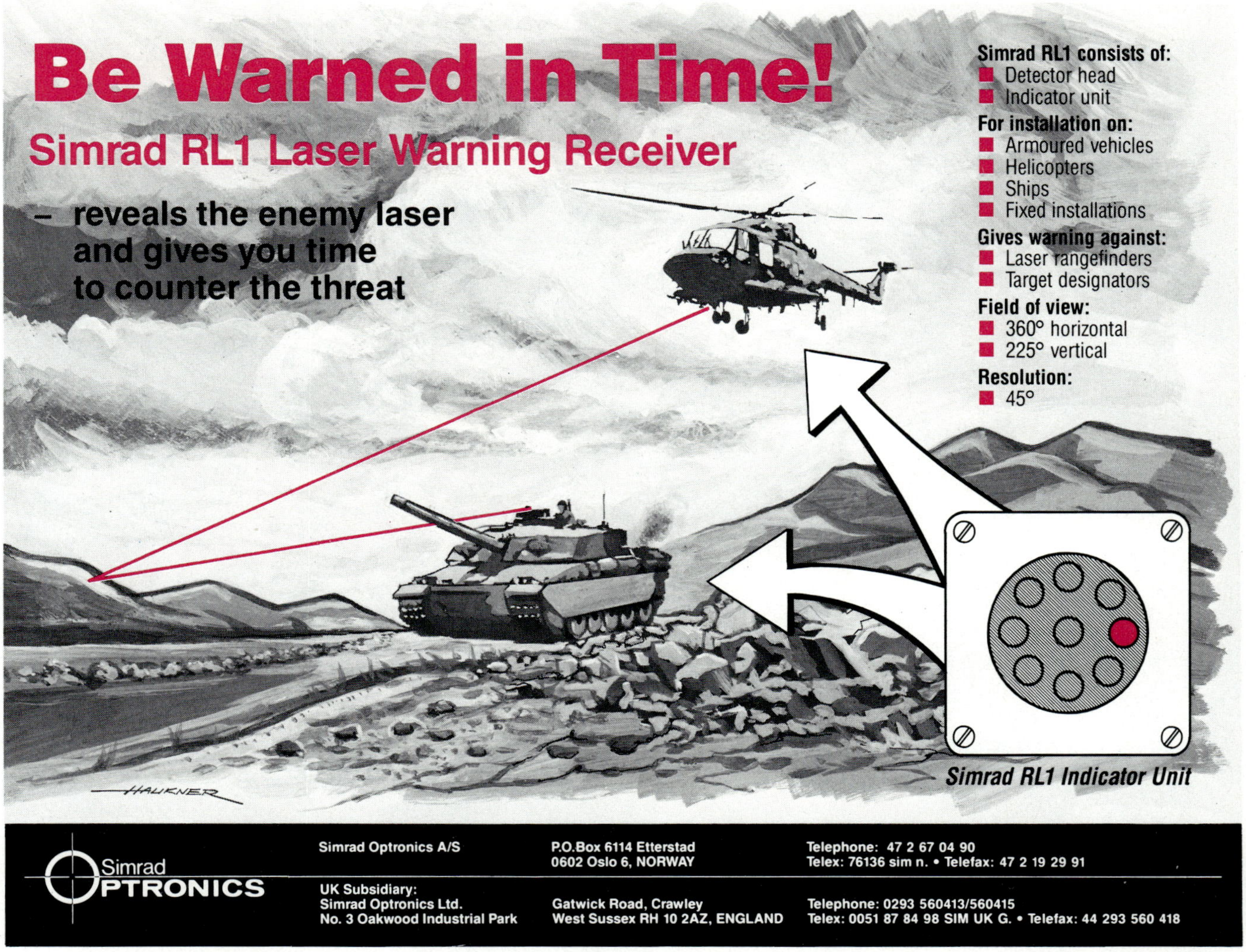

T 20

TL 20 S

TLi 127

TL 2 i

CB 127

TLi A

P 127 A

CP 127 A

CREUSOT-LOIRE INDUSTRIE
TOURELLE T25.

We also make turrets to measure. Please send us your inquiries.

CREUSOT-LOIRE INDUSTRIE

DIVISION MECANIQUE SPECIALISEE

Immeuble Ile-de-France - 4, place de la Pyramide - Cedex 33 - 92070 Paris La Défense
Tél. (1) 49.00.60.50 - Télex : MOTOY 615638 F

CRL 841

WEAPON SYSTEMS

ROME ITALY

GROUND TO GROUND

LAUNCHERS AND ROCKETS
50 mm 122 mm for
FIROS 6 FIROS 25
FIROS 30
SYSTEMS

CONVENTIONAL
AND SUBMUNITION
WARHEADS

AIR TO GROUND

LAUNCHERS AND ROCKETS
for 50mm 2".75 81mm
122 mm SYSTEMS
BOTH FOR HELICOPTERS
AND AIRCRAFTS

CONVENTIONAL
AND SUBMUNITION
WARHEADS

00187 ROMA VIA SICILIA 162 TEL. 06 4680 TLX 610114 BPD RM I

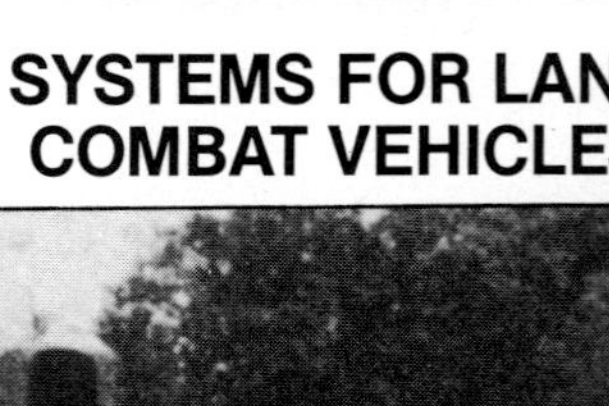

SYSTEMS FOR LAND COMBAT VEHICLES

MULTIPURPOSE SIGHTS

Heading reference systems

Land navigators

Test systems

sfim

F 91344 MASSY CEDEX
Tél.: 33-(1) 69 20 88 90 -
Télex: SFIM 602.164 F

Vendre 040

R. ALKAN & Cie

AIRCRAFT ARMAMENT

- Airborne carriage and release systems.
- Bomb-racks and multi-purpose aircraft pylons. Release and ejector release units. Quick locking anchor units and ejectors. Missile-launcher elements. Multiple ejector racks. Mine-launchers. Heavy load cargo release units and restraining systems.
- Target towing systems and targets. Radar reflectors. Pyrotechnic initiators.
- Cartridge-launchers for ground attack, photographic reconnaissance and illumination.
- Sono-buoy launchers for A.S.W. Torpedo-launchers.
- Weapon control and monitoring systems.

COUNTERMEASURE DISPENSERS

Integrated, scabbed on and pod mounted dispensers with I.R. and E.M. cartridges for aircraft self-protection, with their electronic management and display systems, cockpit control panel.

AIRCRAFT SAFETY OUTFITS

Integrated harness seats. Shock absorber systems for helicopter seats. Parachute release devices. Safety locks for oxygen compartments. Fire spraying buckets.

COMPOSITE MATERIAL STRUCTURES

Pylon structures, fairing elements, radar polarizers, radomes, box-type oil tanks.

GROUND HANDLING AND TEST EQUIPMENT

Portable winches for weapon loading and aircraft servicing applications. Arrest barrier release systems. Armament test equipment. Automatic and programmable cable and circuit testers.

B.P. 23—Rue du 8 mai 1945—94460 VALENTON—France
Tél: (1) 43.89.39.90—Télex: 203876 F.
Télécopie: (1) 43.89.10.61

R. ALKAN (U.K.) LIMITED
Robert House, Shute End
WOKINGHAM
Berkshire RG11 1BJ
Tel: (734) 785.588
Tel: (734) 782.395
Tlx: 846128
Telefax: (734) 774.939

ALKAN U.S.A., INC
1888, Century Park East
Suite 320
LOS ANGELES
California 90067
Tel: (213) 282.8450
Tlx: 651333
Telefax: (213) 232.8530

40 YEARS EXPERIENCE

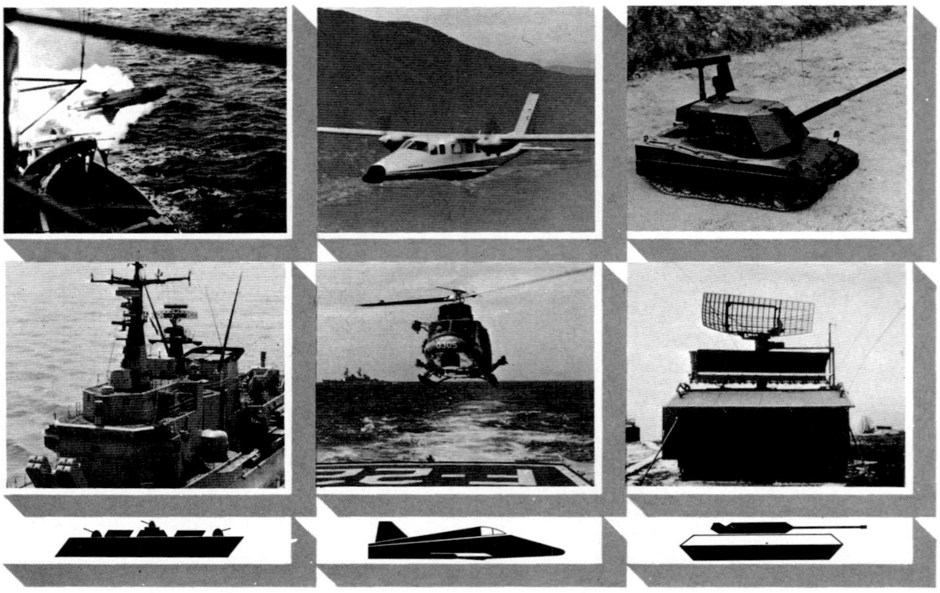

SMA is a flexible company founded in 1943. Since 1948 is active in design, development and production of electronic equipments for defence. Main activities are radars for ground, shipborne, airborne, missile applications and related display systems. The company broad spectrum permits the supply of turnkey systems taylored to the end-user needs.

SMA

SEGNALAMENTO MARITTIMO ED AEREO

SMA's basic product lines and business sectors include: High definition and discrimination **coastal radar control stations**, both for maritime traffic management systems and military application (anti intrusion, coastal defence, mine splash detection); **Shipborne navigation /search radars** devoted to low, very-low air coverage (anti sea skimmers); **Missile active homing heads**, provided with a variety of ECM devices, for SSM (OTOMAT MK2) and ASM (Marte MK2) applications; **360° coverage search and navigation airborne radars** for helicopters and maritime patrol aircraft; **Standard Pod contained search radars** for fixed wing aircraft to be employed in conjuntion with ASM; **Integrated search and tracking radars for A/A defence** to be employed both on specialized trucks (OTOMATIC) or on semi-mobile/fixed station for point defence; applied research aimed toward millimetric waves (up to 100 GHz) in **active and passive sensors** (radiometer); **Duct analisys** and electromagnetic propagation forecast.

P.O. BOX 200-FIRENZE(ITALIA)-TELEPHONE:055/27501-TELEX:SMARAD 570622-CABLE:SMA FIRENZE

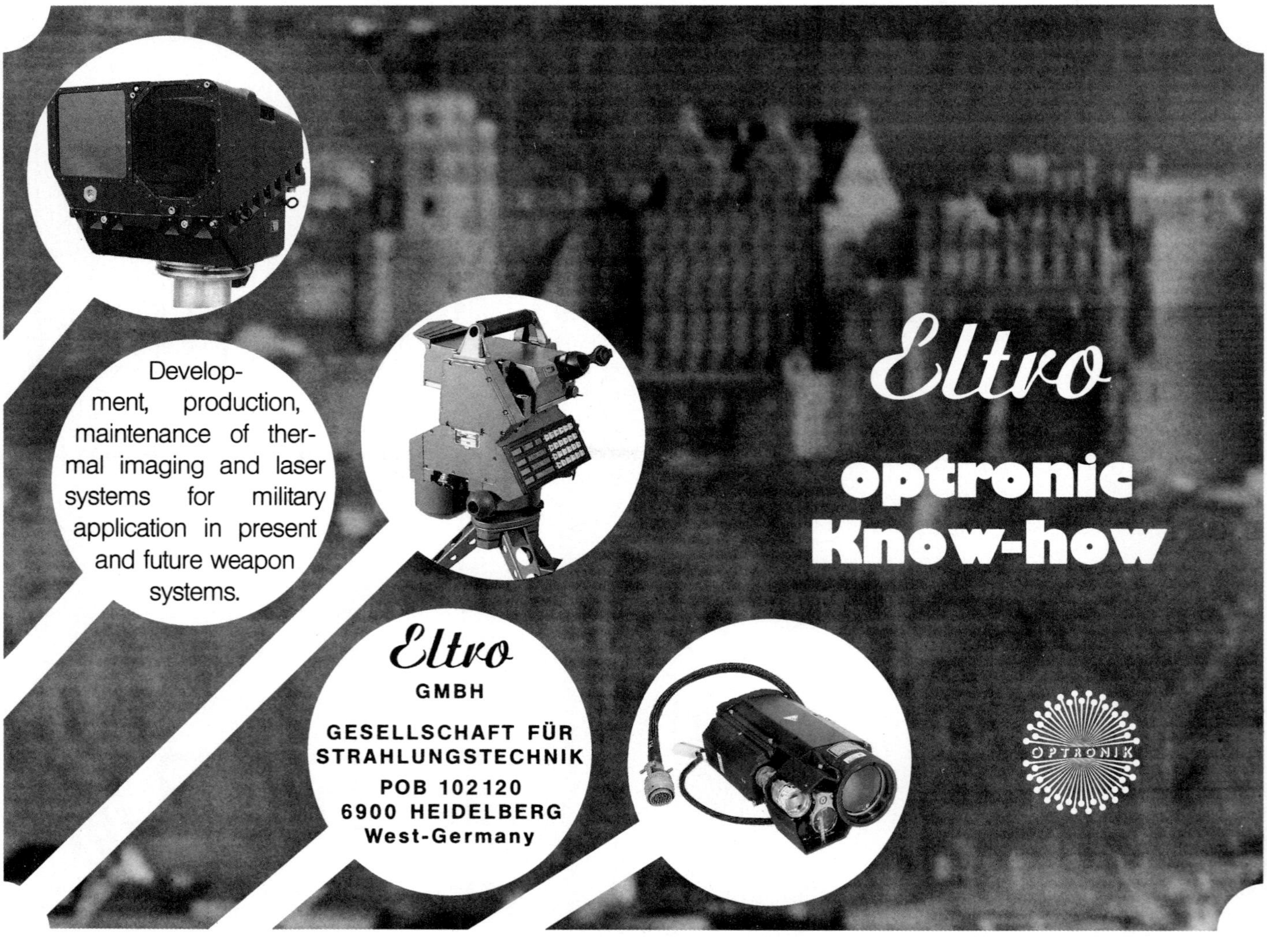

Delta 181, launched by NASA for Strategic Defense Initiative, lifts off from Pad B on Complex 17 at 5:07 pm Eastern time on February 8, 1988
(US DoD)

JANE'S WEAPON SYSTEMS

NINETEENTH EDITION

EDITED BY
BERNARD BLAKE

1988-89

ISBN 0 7106 0855 1
JANE'S YEARBOOKS

In the USA and its dependencies
Jane's Information Group Inc, 1340 Braddock Place, Suite 300, PO Box 1436, Alexandria, Virginia 22313-2036, USA

Contents

FOREWORD	[39]
GLOSSARY	[47]
JANE'S CODE NUMBERS	[51]
STRATEGIC WEAPONS	1-33
LAND-BASED WEAPONS AND SYSTEMS	
Land-based air defence	34-62
Army fire control	63-112
LAND-BASED TACTICAL SURFACE WEAPONS	
Battlefield support weapons	113-135
Anti-tank assault weapons	136-159
Coastal defence weapons	160-105
LAND-MOBILE SURFACE-TO-AIR WEAPONS	
Mobile surface-to-air guided missile systems	167-191
Mobile gun and rocket systems	192-210
Portable anti-aircraft guided missile systems	211-217
Ground radar	218-305
Electronic warfare equipment (land)	306-347
Electro-optical equipment (land)	348-407
SHIPBORNE WEAPONS AND SYSTEMS	
Naval fire control	408-448
Shipborne surface-to-surface weapons	449-461
Shipborne surface-to-air weapons	462-484
Naval guns and bombardment rockets	485-510
Underwater warfare	511-539
Sonar and underwater detection equipment	540-599
Naval radar	600-645
Electronic warfare equipment (sea)	646-676
Electro-optical equipment (sea)	677-690
AIRBORNE WEAPONS AND SYSTEMS	
Air-to-air missiles	691-705
Air-to-surface missiles	706-726
Aircraft armament	727-771
Airborne radar	772-814
Electronic warfare equipment (air)	815-866
Electro-optical equipment (air)	867-903
ADDENDA	904
ANALYSIS	905
NUMERICAL LIST OF ENTRIES	959
INDEX	995

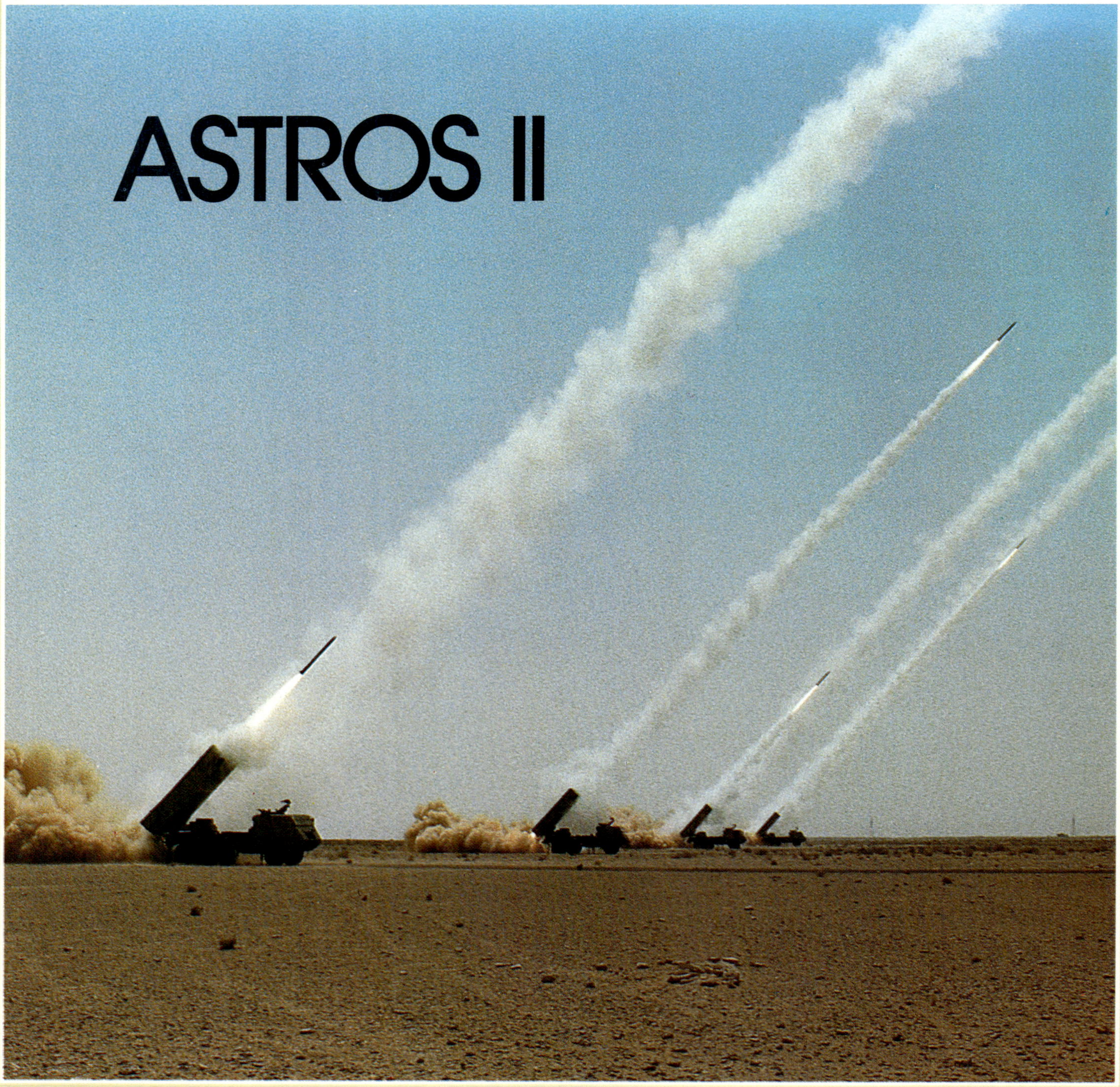

COMBAT PROVEN

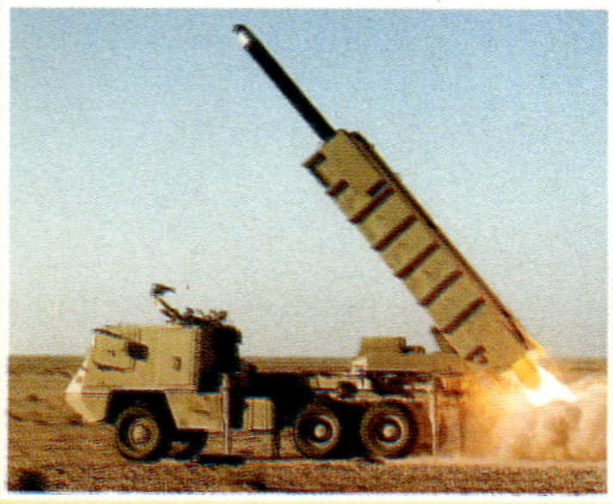

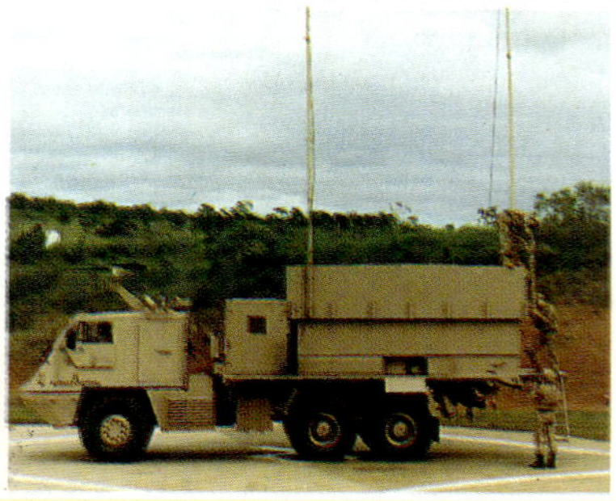

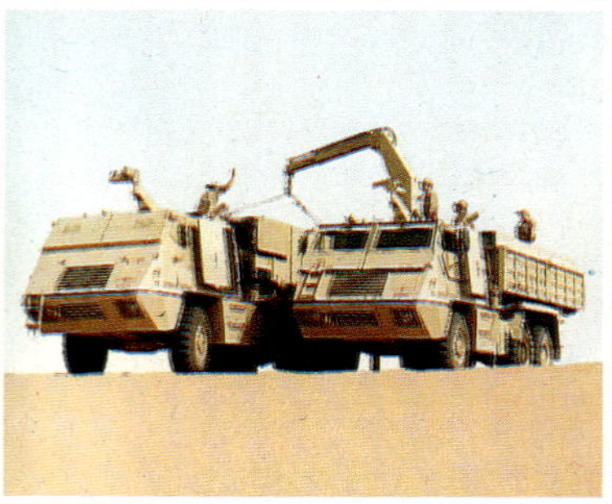

AVIBRAS AEROESPACIAL S.A.

Antiga Estrada de Paraibuna, Km 118 - P.O. Box 229 - 12.200 - São José dos Campos - S.P. - Brazil
Phone (123) 21-7433 - Telex (123) 3493 AIAE BR - FAX (123) 221689

FOREWORD

Although a number of important events have occurred in the weapons scenario during the past year there is little doubt that by far the most important is the INF treaty agreed between Mr Gorbachev and Mr Reagan in December 1987. This treaty aims to ensure that over 2500 United States and Soviet intermediate and shorter range missiles are taken out of operational service and destroyed during the next few years. In addition, another summit has been arranged for the summer of 1988 at which the overall weapon strategy will be discussed and, in particular, methods of reducing strategic weaponry still further. These agreements, which have taken many years of sometimes fruitless negotiations to reach, are the first signs of a new era in worldwide peace since the end of the Second World War in 1945.

At the time of writing (mid-May) the treaty had not been ratified by the US Senate but this appears to be only a matter of time, with the likelihood of ratification before the next summit in June 1988. Although thousands of pages have been written in the international press on this subject, it is worthwhile looking at the hard facts of the agreement and trying to assess its impact and the likelihood of its total implementation.

According to the US Arms Control and Disarmament Agency and the US Natural Resources Defense Council the number of missiles that are to be eliminated are as follows:

(a) Intermediate missiles (those with ranges between 1000 and 5000 kilometres) – 429 US and 470 Soviet deployed missiles, plus a further 260 and 356 respectively which are non-deployed.

(b) Short range missiles (ranges between 500 and 1000 kilometres) – 387 Soviet deployed missiles, plus 170 US and 539 Soviet non-deployed missiles.

Whilst the elimination of 2611 missiles seems a remarkable achievement, it should be remembered that in the overall context of the strategic and non-strategic nuclear weapons that will remain, the total represents only some 6 per cent of the combined US/Soviet stockpile, which includes over 4500 strategic launchers with a total of some 23 500 warheads. Such is the state of modern technology when one considers the latest multiple independently-targeted vehicles (MIRVs), with up to twelve warheads per launcher.

If one then adds in the nuclear weapons owned and operated by the United Kingdom, France and China the worldwide percentage to be eliminated drops to around 5 per cent. It should also be noted that there are no constraints on intermediate submarine-launched ballistic missiles, and intermediate range air or sea-launched cruise missiles.

The intermediate range missiles concerned are the Soviet SS-20, SS-4 and SS-5 types (known in the USSR as RSD-10, R-12 and R-14 respectively), and the United States Pershing II and the BGM-109G (perhaps more popularly known as the Ground-Launched Cruise Missile). The shorter range missiles to be eliminated are the US Pershing 1A and the USSR SS-12 and SS-23 (Soviet nomenclature OTR-22 and OTR-23 respectively).

Article IV of the treaty states that each side shall eliminate all its intermediate range missiles and their launchers, and all support structures and support equipment of the categories listed, so that at the end of three years from the entry into force of the treaty, there shall be no such missiles, launchers, support structures or support equipment possessed by either signatory.

Article V covers the agreement for each side to eliminate all the shorter range missiles designated above, and all associated support equipment, no later than 18 months after entry into force of the treaty and thereafter no such missiles, launchers or support equipment shall be possessed by either party.

Article VI states, with some reservations, that upon entry into force of the treaty neither side shall produce or flight-test any intermediate or shorter range missiles, or produce any stages of such missiles, or launchers of such missiles.

In order to ensure that all aspects of the treaty are carried out, a comprehensive programme of inspection by large numbers of both Soviet and American officials has been agreed. The details of these inspections are quite complex but they cover a period of 13 years of short-notice inspections. This includes the right to visit installations on each side to confirm that information supplied is correct, to conduct special 'close-out' inspections when bases and support installations are eliminated, and to observe the destruction of missiles.

To make it easier to destroy the missiles, each side is allowed to launch up to 100 missiles into oblivion during the first six months after the treaty comes into force. However, none of these are allowed to be used as targets for ABM interceptors to prevent them being used in tests for anti-missile defence systems. The remainder can be destroyed by conventional means at special establishments, although the warheads and guidance systems can be removed before arrival at the site. The nuclear material from the warheads may be recycled for use in either military or civil employment.

So much for the bare bones of the treaty. The question remaining is the willingness of each side to implement the provisions. On the credit side, the treaty has been written much more tightly where constraints are really intended, and it would appear that some of the lessons of the imperfections in previous agreements (SALT 1 and 2) have been learned. It is written in the usual style, aimed at preserving those technical freedoms that the two sides wish to protect and perhaps wish to exploit sometime in the future. However, many people worry about the different attitudes of Western and Soviet nations towards legal treaties.

The Soviet Union views such treaties as legal statements, and is not really concerned with the 'spirit' or 'intent' of the wording which Western governments regard as an essential part of any international agreement. On the Western side the number of freedoms left open makes one wonder how many loopholes the average lawyer could drive through it! It is to be hoped that more of the spirit of agreement is upheld in the future than has been the case with the 1972 and 1979 treaties. If it isn't, then we have all been wasting our time and the status quo will remain as before, with arguments raging back and forth as to who has broken the treaty because of different interpretations placed upon its wording.

One interesting aspect of the treaty's general considerations is that appertaining to test-flying of short range missiles. This appears to let in weapons such as the 'Scud B' battlefield support missile, plus its modernisation and range enhancements, provided that it is not test-flown over ranges greater than 500 kilometres. It is easy to see that the updated version (also known as 'Scud C'), which has a reputed range of some 450 kilometres, could well finish up as a replacement for the eliminated SS-23, particularly if the SS-23 guidance system was incorporated. Rather ironic since the SS-23 was developed as a replacement for the 'Scud' system! It could also mean that the Western powers could match the system with, perhaps, an updated Lance battlefield support missile to replace the Pershing.

Generally speaking, there appears to be no restraint on modernising short range weapons, ie those with ranges that are not currently less than 500 kilometres, providing that they are not test flown beyond the limit. This could be a very dangerous practice and would largely undermine the INF treaty. The Soviet Union has expressed an opinion on several occasions that it would also like to see short range battlefield nuclear weapons eliminated in the foreseeable future. However, NATO is less than happy about an idea which it says would leave Western Europe exposed to the threat of superior Warsaw Pact conventional forces.

The next move is the Moscow summit between Mr Reagan and Mr Gorbachev which, at the time of writing, was scheduled for early June 1988. This meeting is intended to be the forerunner of agreements to limit strategic nuclear weapons. However, the momentum of arms control talks appears to have slowed down considerably since last December and there seems little likelihood of complete agreement between the two sides at the Moscow summit.

One of the problems seems to be deciding whether submarine-launched cruise missiles should be classified as strategic missiles, and sorting out the classifications for air-launched nuclear weapons. With the forthcoming United States Presidential Election getting closer and closer there will need to be a great effort on both sides if an agreement is to be reached before the end of 1988. Even if such an agreement is reached, the general consensus of opinion is that at the most it will reduce the present stockpile of strategic missiles by only 50 per cent. This may sound a lot but it would still leave around 6000 warheads apiece; more than ample to carry out total destruction of either side.

The other main issue dividing the two parties is the SDI programme and what limits can be put on anti-missile tests in space. It is interesting to note that the Soviet Union has now admitted that it has had its own SDI programme for a number of years. This was outlined two years ago in the foreword of *Jane's Weapon Systems*, but was vigorously denied by the Soviet Union.

On the US side, SDI is currently going through rather a rough patch, with the US Senate trying to cut down the programme cost. The result of any substantial cut-back will be to seriously delay the implementation of SDI and will probably mean that a worthwhile space-based anti-ballistic system is unlikely before the end of this century. It is interesting that a recent technical study by the US

The new hearing aid for artillery hunters.

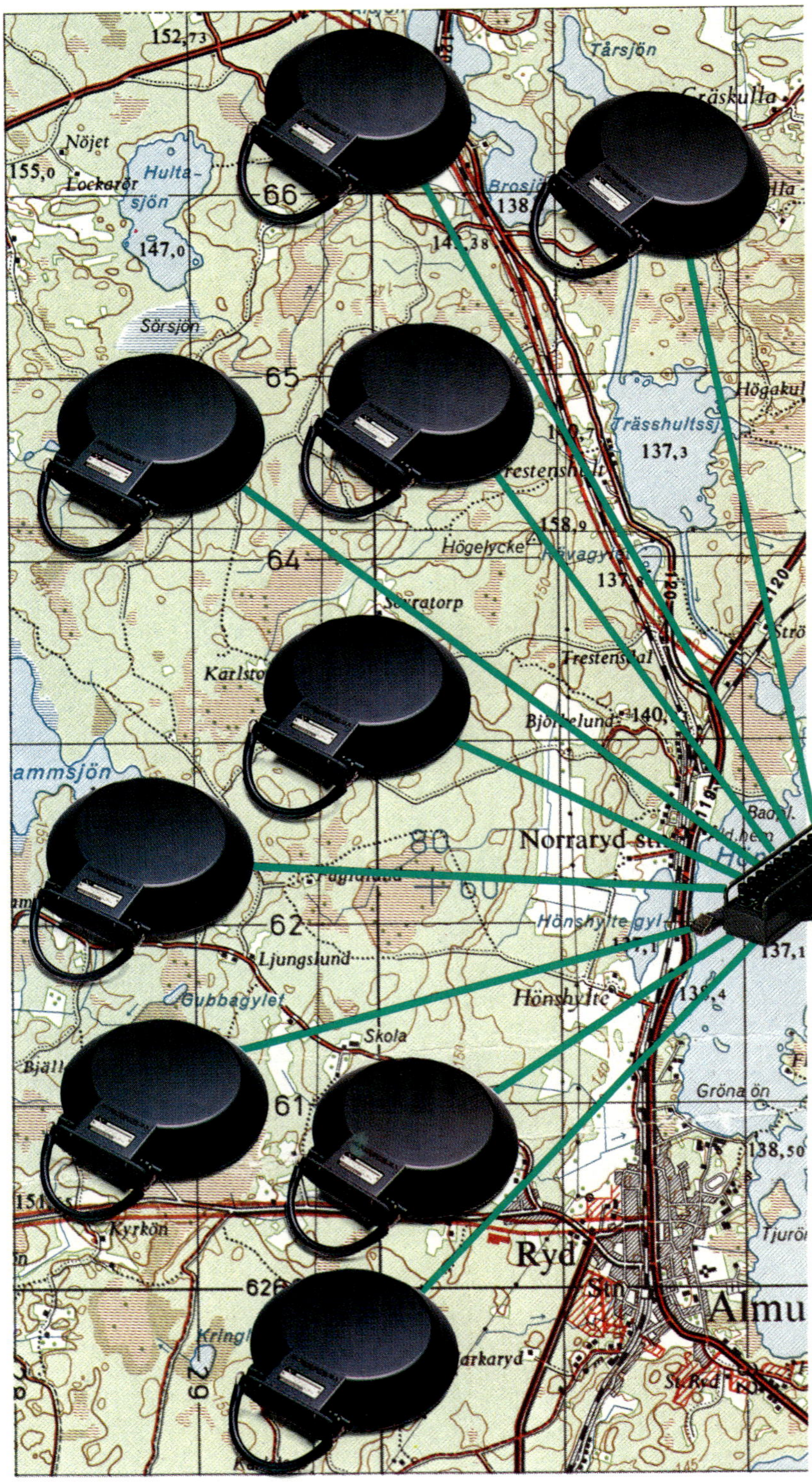

Some people think that artillery ranging with the help of sound is a typical example of a mature technique, which can hardly be improved.

They are wrong. Take a closer look at SORAS 6, a new sound ranging system that we have developed in cooperation with the Swedish Army.

Thanks to new techniques and measuring methods, we can promise a range of up to 30 km under good weather conditions. And an accuracy that varies between 0.5—2% of the measured distance.

A complete system consists of calculator, meteorological sensor and nine microphones. The calculator can be mounted in a vehicle. The microphones are placed on the ground and connected to the calculator via twin cables.

SORAS 6 can be used to measure both muzzle explosions from enemy guns and your own artillery's impact bursts. On the calculator, the measured positions are indicated, either graphically or alphanumerically. The information can also be shown on a paper strip.

The target coordinates are usually transferred to your own artillery in the form of spoken messages, but automatic transfer to target receivers at the gun site is also possible.

When it comes to the price, development has fortunately stood still. An artillery localization radar, with the same performance as SORAS 6, costs around 30 times as much today. Tell your price hunters that.

Congressional Office of Technology Assessment (OTA) stated 'a comprehensive ballistic missile defence is likely to be the most complex system ever constructed. There may always be irresolvable questions about how dependable software would be. The relatively slow rate in improvement in software engineering technology makes it appear unlikely to OTA that this situation will be substantially alleviated in the foreseeable future.' The report also gives an opinion that SDI could not possibly be deployed in the mid-1990s as originally planned since the required technology would not be ready by then.

European defence

Whatever the pros and cons of the INF treaty, and the possibility of a strategic weapons limitation agreement, it has had the effect of concentrating the minds of European governments on an overall co-operation in defence. Admittedly this co-operation already exists in the form of NATO but with the elimination of intermediate and shorter range nuclear weapons, and the rather parlous state of the US economy, it is becoming apparent that Europe as a whole will have to look hard at its own contribution to defence. Apart from the gap left in NATO defences by the elimination of these weapons there has been a growing pressure in Washington for the reduction of US forces in Europe, and a feeling in the US Senate that Europe should do more in its own defence than it has up to now. Overall the most important issue for Europe is self-sufficiency in defence, with leadership coming from West Germany and the two European nuclear powers, the UK and France.

The emphasis in Europe is swinging more and more to conventional forces, of which the Soviet Union has an almost overwhelming number. It should be remembered that France, West Germany and the United Kingdom were far from happy about scrapping the shorter range weapons; in fact West Germany was almost forced to acquiesce for political reasons. The latter would much rather see this type of weapon used to halt any advance before it spreads into the heart of the country. This is quite understandable since in the event of a war between the Warsaw Pact and NATO, West Germany would inevitably finish up as the main battleground. Most German politicians can remember the last few months of the Second World War in Europe and have no wish for a repetition of the devastation, particularly when large quantities of battlefield nuclear weapons would add to it.

However, should hostilities break out in the foreseeable future than it would almost certainly take the form of a surprise attack by the Warsaw Pact forces to counter and prevent the build-up of NATO forces. This attack would probably be launched using multiple spearheads by land and sea with a pre-emptive airborne strike against NATO airfields and submarine bases. Having been pushed back into the heart of West Germany it would then be extremely difficult for NATO to produce a framework of counterattack and to provide supporting forces.

It is much easier for the Warsaw Pact countries to bring their reserves into action than it is for NATO, many of whom would have to be brought across the Atlantic by ship under the constant and considerable threat of submarines. All of which means that the faster the Warsaw Pact countries could strike, the better their chances of winning, or at least of negotiating a settlement on their terms. If NATO had ample warning of a probable attack and could group its forces, the resultant hostilities could well finish up as a long drawn-out stalemate which neither side could afford economically.

The emphasis must therefore be on maintaining a permanent conventional force in Western Europe which is sufficiently strong to deter any possible aggression. This can only be carried out if a large proportion of the actual forces, and their logistical support, is provided by European powers, principally the United Kingdom, France and West Germany. It is gratifying to note that the existing rapport between France and West Germany has been broadened by the new Franco-German defence council and the formation of a Franco-German army brigade in early 1988. There is also a considerable increase in co-operation at high levels of defence planning and budgeting between France and the United Kingdom.

Another reason for increasing conventional forces is the probability that once the USA and USSR have reached agreement in limiting strategic weapons they will start to pressurise the UK and France to reduce theirs. It is essential, therefore, for the NATO countries to modernise their existing forces, both conventional and nuclear. This includes surface ships and submarines, armoured vehicles, aircraft, artillery, battlefield nuclear weapons, communications, radar and electronic warfare systems.

Since 1966 when General de Gaulle took his country out of the military structure of NATO because he feared American domination, France has had a chequered history of its co-operation with that organisation, although it has always been committed to the defence of Europe in the long term. However, during the past few months France has been extending its involvement with NATO and the fact that both Francois Mitterand and Jacques Chirac attended the NATO summit last March leads one to the conclusion that a new era has commenced. In addition, nearly 1000 US Marines spent two weeks training in France in February, the first time that this has happened since 1966 and in January, France agreed to open its territory and ports to NATO reinforcements when necessary. France still feels that there is a likelihood of the United States withdrawing some of its forces in the future but whether this means that France would commit its forces totally to NATO or join the integrated command structure depends very largely on any alterations which would take place within NATO.

There is still a strong element of chauvinism within France, as is evident in the aircraft construction industry, where it withdrew from the EFA programme to build its own Rafale aircraft.

The United Kingdom has a long history of involvement in NATO and is one of the leading exponents of strengthening that organisation. However, the UK insists on maintaining a special relationship with the United States, an insistence that has annoyed France in the past, and to some extent still regards itself as an island race separate from the mainland of Europe. However, this attitude is changing rapidly with increasing involvement with the rest of Europe, and particularly with the European Economic Community, which becomes a completely open market in 1992. The construction of the Channel Tunnel, which is due to open at about the same time, will accelerate the process.

The era of co-operation with the rest of Europe dawned a few decades ago with Concorde, Jaguar and Tornado and has now accelerated with EFA, Trigat, NFR 90 and a number of other projects. As with France, there is still a strong element of chauvinism in the UK's feelings towards other countries. Although the UK was very heavily bombed during the Second World War it has not been overrun by foreign troops for nearly a thousand years, as opposed to France, Germany and Italy, and most of the rest of Europe, during the past century.

West Germany has a particular problem associated with the short range battlefield nuclear weapons. Because any hostilities are likely to be fought out on German soil it would be only too happy to see the short range weapons eliminated to avoid possible wide-scale long-term radiation devastation. This is a fundamentally opposite view to that of France and the UK, who would like to see these weapons modernised to offset the Warsaw Pact numerical superiority in conventional arms. At times West Germany gives the impression that it may be sliding gradually into neutralism, certainly France seems to think so at times, or in the long term perhaps even unification with East Germany at the expense of leaving NATO. However, at the moment West Germany is the linchpin of Western defence strategy and the majority of NATO troops are stationed on its soil.

Research and development

If the Western European countries are to counter the threat of the Warsaw Pact using conventional arms then it is essential that a great deal more money is poured into research and development (R&D) on both a national and international front. R&D is the keystone for the future of industry, whether it be in the defence or in the civil sphere. In the recent past there has always been a marked reluctance in some countries to spend money on this vital aspect of industry and, in the military field, this has meant that many good ideas have never been taken up, or have been left half-finished for want of the necessary funds. This attitude is particularly apparent in industry itself more than in the state-controlled R&D establishments. The latter have always been reasonably well funded, even if perhaps they do not necessarily spend the money wisely, but industry has tended to concentrate upon the commercial side of research, pouring most of its funds into short term development rather than long-term basic research. This is easy to understand, with companies responsible to their shareholders for providing reasonable profit margins and dividends, but many lose sight of the fact that it is the long-term basic research that leads to products being developed in, say, ten or fifteen years time.

Generally speaking, the United States is better than virtually anyone else in this respect, although the Soviet Union state-owned research and development establishments undoubtedly spend vast amounts of money on military projects. The major industrial companies of many European countries are not very good in this matter, mainly because they expect the research work either to be carried out in the government establishments, or to be generously funded by the individual government. It is rather interesting that the UK is considering setting up a wholly government owned company,

Defence Research plc, to take over the present six research establishments.

Once upon a time there was quite a large amount of 'private venture' work carried out by industry but in some companies this has long since been cut back in the interest of short-term profits. Undoubtedly, research and development is a very costly affair, and with new ships such as frigates costing around $200 million each, fighter aircraft costing perhaps $50 million each and main battletanks priced at perhaps $3 million apiece it is easy to understand why industry is loath to commit itself to too much money for the R&D field.

To quote a specific national example, more than 25 per cent of all industrial and government R&D funding in the UK is concerned with various weapons and their associated electronics and mechanics. Of this about £2.5 billion comes out of the defence budget, although this figure has been questioned by the Society of British Aerospace Companies as being too high. Broadly speaking, in the defence field, most basic research is carried out by the government establishments with industry carrying out development on a private venture basis and/or using government funds. With the recently announced cuts in government expenditure on R&D more emphasis may be placed on industrial funding.

In the future it will prove even more necessary for defence research and development, and indeed production, to be funded and carried out on an international basis. With the very large increases in the price of weapons, particularly the platforms such as aircraft, ships and armoured vehicles, it is becoming increasingly difficult for any one country to produce practical results. Even the USA with its vast resources of manpower and technology is feeling the financial strain in this area.

The one area which suffers more than any other is that of true basic research aimed at producing fundamental breakthroughs in technology but associated with a high degree of risk of failure. An example of this type of research is the USA SDI programme, where many projects are being worked on at the very tip of modern knowledge, and sometimes even beyond that. It is no wonder that even with the enormous sums of money being spent on SDI progress is still very slow. It would be interesting to find out how much is being spent by the Soviet Union on defence R&D as a proportion of its GNP.

Weapons

With so much emphasis this year on the elimination of a large number of missiles it is somewhat difficult to get current and future types into perspective. In the USA the ICBM modernisation programme continues, to the extent that some Minuteman silos in Nebraska and Wyoming are being used to house the new Peacekeeper ICBM, with most of the electrical equipment being redesigned to accommodate Peacekeeper.

The first Peacekeepers were in position by December 1986. Most of the flight testing of Peacekeeper has been completed and the missile is in full production. The full planned force of 528 Minuteman III and 450 Minuteman II missiles was operationally deployed by mid-1987.

Development of the Small ICBM (SICBM) continues as part of the overall strategic modernisation programme. The status of this missile under SALT 2 is not clear; since it deviates by more than 5 per cent in dimensions from any predecessor, it is really a new missile. However, since SALT 2 was not ratified by the USA the question is probably irrelevant.

In the field of submarine-launched missiles, Trident II (D-5) is still in full-scale development and the first submarine-launched flight test is scheduled to occur in March 1989. The first operational missiles are scheduled to be at sea in 1989, with 312 deployed in 13 USN submarines by 1998. The UK programme on the Trident II (D-5) missiles for operation in the mid-1990s is underway with the first two Trident submarines being under construction.

In the Soviet Union both official and unofficial sources claim to have evidence of two new ICBMs, in addition to the latest SS-24 and SS-25 types. A cruise missile programme is well underway with the development of a submarine-launched version (SS-N-21), similar to the US Tomahawk. An air-launched version is believed to have achieved initial operational capability in 1984, and has probably formed the basis of the SS-N-21 development programme.

The French SSBS Type S-3D/TN-61 IRBM is operational and deployment began in the early 1980s. A new IRBM, designated Type S-4, has been announced by the French Ministry of Defence, and is intended to replace the S-3 missile in the late 1990s. The latest generation of submarine-launched ballistic missiles, the M-4, became operational in 1985 and is now in service.

In recent years, China has emerged as the fifth nuclear power after the USA, UK, USSR and France. The original missiles were largely based on Soviet types but more recently indigenous types have made an appearance. These include the CSS-4 ICBM and the CSS-N-3 SLBM, plus a reported new SLBM, the CSS-NX-4.

The land-based tactical missile scene has been relatively quiet during the past year, although in the area of International co-operation systems such as the Multi-Launch Rocket System (MLRS) and the Trigat anti-tank assault weapon are going ahead strongly.

Air-launched missiles has been a much more fruitful field, with a considerable number of new weapons in development or about to enter operational service. These latest systems include the MAA-1 air-to-air missile from Brazil, the PL-9 developed by China, and from France the MICA, which is still in relatively early development, and the AATCP Mistral variant due to enter service this year. Two air-to-air missiles which form an important part of the NATO 'family of weapons programme' are the AMRAAM with full-scale deliveries commencing in 1989, and the ASRAAM which is still in the project definition/engineering development stage, with production scheduled to start in 1993. Japan is developing an air-to-air missile known as the XAAM-3 and Taiwan has revealed more details on the Sky Sword weapon.

In the Soviet Union one of the latest air-to-air missiles, designed for use with the MiG-29 and Su-27 aircraft, is the AA-10 'Alamo' and in the United States, the US Navy is about to start the development of the AAAM which will replace the Phoenix missile in the late 1990s.

The air-to-surface scenario has a number of new entrants. Among these is the air-launched version of ANS being developed jointly by France and West Germany and due to enter service in the mid-1990s. The Chinese C-801 anti-ship missile is currently in service in the shipborne role and an airborne version is in late development. The airborne version of the Swedish RBS 15 is in development, with an in-service date expected in 1990. In the Soviet Union, the AS-13 is believed to be a Mach 3.5 missile, which will succeed the AS-4 and AS-6. The US air-launched cruise missile, AGM-129, was due to be in service by 1990 but the latest information indicates that it is one to two years behind schedule, and the SRAM 2 (XAGM-131) is due to be flight tested next year, for eventual deployment on the B-1B and B-2 bombers. Finally, the anti-radiation missile field has had conflicting successes. The USA HARM has been in service with the US Air Force for nearly five years while the next generation, Tacit Rainbow AGM-136 is in development and due into service in the early 1990s. In the UK, ALARM development has been seriously delayed by problems with the rocket motor that proved so serious that a new motor had to be selected. Since the new motor has to be proven, ALARM will not be entering service with the UK Royal Air Force until 1989/90.

Air defence

Like many aspects in this field, air defence is an on-going affair with a number of countries improving their overall air defences. The Soviet Union continues to implement its policy of an umbrella of early warning networks and anti-ballistic missile defences. This includes conventional radars, OTH radars, and airborne forces and missiles of various types. The Moscow ABM system is virtually complete and is equipped with 'Galosh' and 'Gazelle' interceptors and SA-10 and 10B SAMs. The mobile SA-12A is operational and the SA-12B long-range version should be deployed shortly.

In Europe, NATO has a long-term comprehensive update of the NADGE system in the planning stage, which is intended to provide a completely integrated defensive and offensive system by the year 2000. The UK and French national air defence systems are integrated with NADGE, and since both countries are now buying E-3 AWACS aircraft the overall surveillance capability is being vastly improved.

In North America, the USA and Canada are updating the North Warning System (NWS) with a complete chain of new radars and control centres and the BMEWS system is currently being modernised. Part of the OTH eastern segment is now operational and all four Pave Paws radars are in full scale use. The combination of the new NWS, OTH radars, BMEWS sites, Pave Paws, the airborne AWACS aircraft, and a number of other sensors, combined under the Joint Surveillance System in NORAD, makes the North American defensive system the most comprehensive in the world.

A number of other countries have been taking steps to improve their overall air defence. Amongst these is Saudi Arabia, which now has a very comprehensive network of long and short range radar systems, SAMs, AWACS aircraft, interceptors and highly sophisticated control centres. Australia is improving its defences by new interceptor aircraft and the construction of an OTH radar network. The Brazilian integrated air defence/ATC radar system is in the final stages of

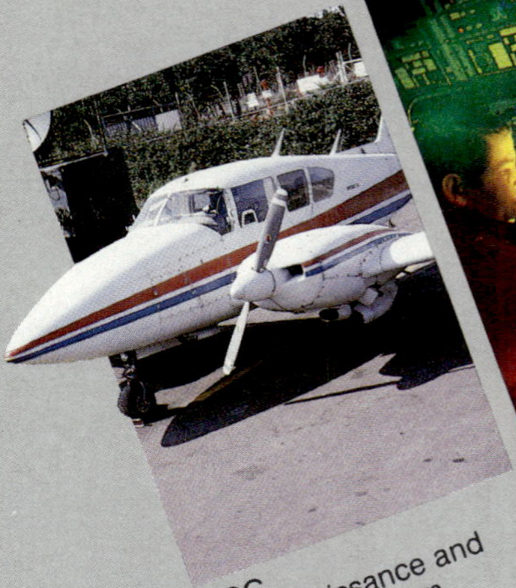

RCC
Reconnaissance and guidance system

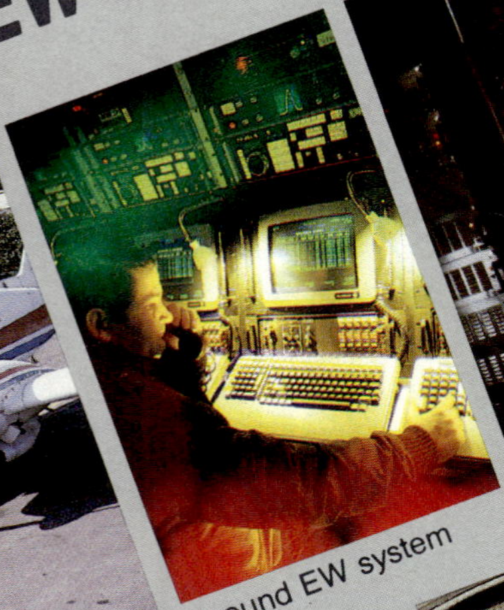

Ground EW system

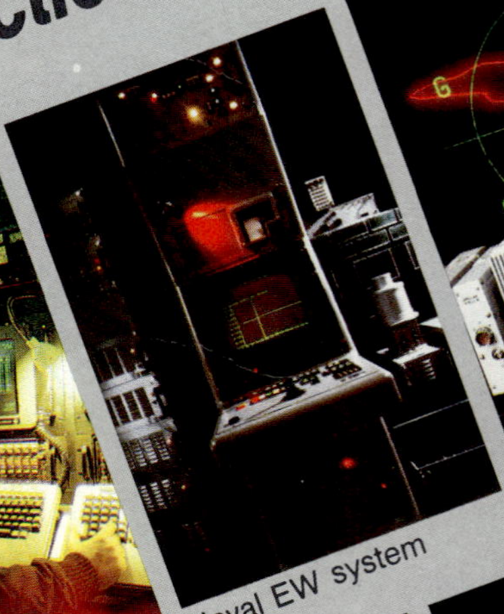

Naval EW system

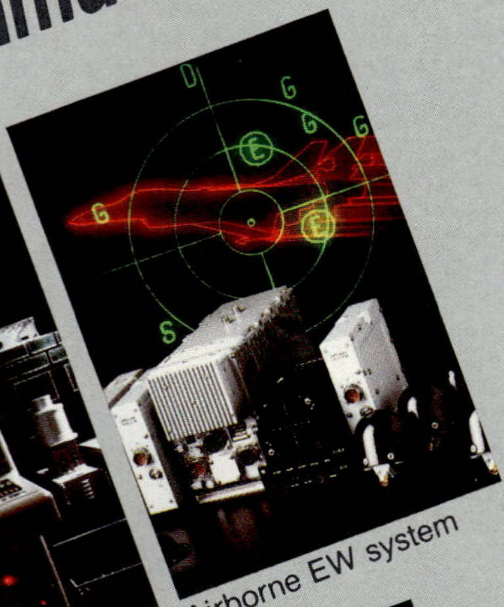

Airborne EW system

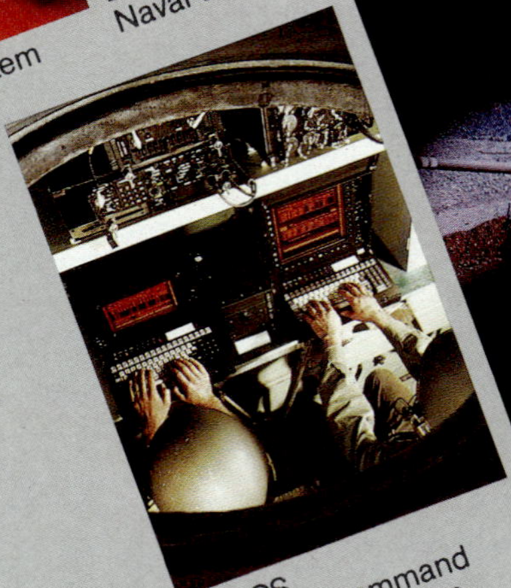

DACCS
Divisional command and control system

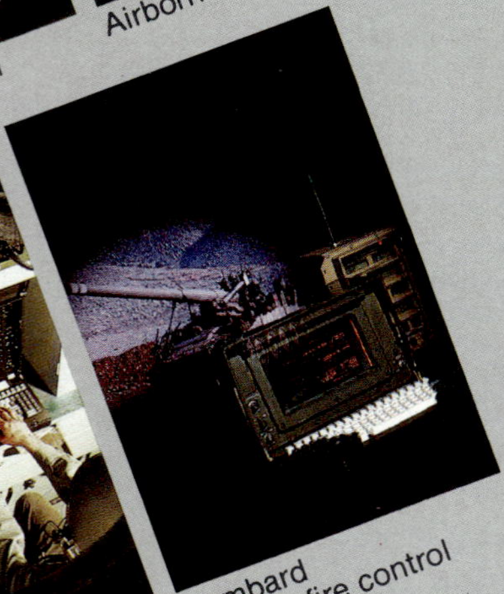

Bombard
Artillery fire control system

completion under a joint programme between Brazilian industry and Thomson-CSF of France.

Electronic systems and sensors

Much of the emphasis during the past year has been in two fields, those of electronic warfare and underwater detection. There is little doubt that one of the major threats to the Western alliance comes from the massive Soviet submarine force. This is particularly significant when one considers two factors; the first being the INF treaty, which would place more emphasis on submarine-launched intermediate-range missiles, and the second being the inevitable fact that any reinforcements from the USA into Europe would have to come across the Atlantic Ocean. With these considerations in mind the Western powers have been carrying out wide-scale design and development into more efficient sonar and underwater detection systems, both surface ship and submarine borne, with a considerable emphasis on towed array systems which are mainly passive in operation. Many of these are either becoming operational or will shortly do so, and combined with the airborne segments using sonobuoys, dipping sonar and sophisticated processors, and static systems such as SOSUS, they are making life much more difficult for the submarine.

Electronic warfare is one of the key elements of the modern battle scenario, protecting one's own forces from attack, denying information to the enemy, and intercepting and disrupting his own voice communications and data links. A widespread network of electronic intelligence stations are operated by both the Western Alliance and the Warsaw Pact countries by land, sea and air, plus jamming systems to deny the other side's communications.

In the airborne role one of the most important considerations is the protection of the aircraft from radar and infra-red guided missiles. A number of very efficient systems are currently in service to warn the pilot and to decoy the incoming missile. Since however the missile guidance designers are constantly producing new ways of eluding decoy and jamming techniques, the electronic warfare designers are having to produce better and better protection systems. Frequently the same manufacturers are doing both! Protection for warships has recently come into the forefront of the naval scenario with the various episodes in the Gulf War, although much of it could have been foretold by the Falklands (Malvinas) campaign. With modern high speed, high penetration missiles carrying conventional high explosive warheads, the destroyer and frigate type of naval vessel are particularly vulnerable to destruction or severe damage. It has become vitally necessary to detect and decoy incoming missiles as well as trying to destroy them with anti-missile missiles and guns.

One could go on for page after page on the relative merits of electronic countermeasures and electronic counter-countermeasures, with each getting the upper hand temporarily, only to be overtaken by the other. Surveillance radars have become vulnerable to anti-radiation missiles and are having to be designed using frequency agility, low sidelobes, decoys and a variety of other anti-missile techniques. For shorter range work more and more emphasis is being placed on passive detection systems, mainly in the infra-red spectrum.

Whatever the sensor system, however, be it radar, sonar, electronic warfare or electro-optical, the heart of the equipment is in the processor and its software programme. The modern digital computer has revolutionised electronics by producing high speed processing so that a vast amount of detail can be extracted from raw sensor data in a very small amount of time. It can be displayed in a variety of formats, graphic or alphanumeric, and the operator can take virtually instantaneous action. Even the latter is becoming automated and the time is not far distant when artificial intelligence will take much of this off the operator's shoulders. One suspects however that the vital decisions will still have to be taken by a human being!

Acknowledgements

One of the pleasures of an editor's tasks is to acknowledge the help and guidance that he has been given over the past year. This year has seen a considerable number of changes in the Jane's organisation with the consolidation of DMS Inc and Interavia SA into the Jane's Information Group.

In particular my thanks go to the two members of the *Jane's Weapon Systems* team who have contributed greatly to the efforts of producing the yearbook on time with the latest information. Duncan Lennox joined the team as Assistant Editor in the middle of 1987 and has been responsible for updating the strategic and airborne weapons sections of the book. With the help of two stalwart contributors, Geoff Loasby and Arthur Rees, he has ensured that these sections have been comprehensively updated. The second member of the team, Ted Hooton, joined at the end of 1987 and has been responsible for large sections of the book, principally those associated with both naval and land-based fire control and weapons. His enthusiasm has meant that many aspects of this side have been unearthed and written up.

Inevitably there are many other people to whom I would like to express my gratitude for help. These include fellow yearbook editors Chris Foss, Terry Gander, Richard Sharpe and Ken Munson. Special thanks are due to the in-house team under Bob Hutchinson, Jane's Publishing Director, and in particular to Anne Horgan who always seems to be able to produce order out of chaos. Unfortunately Anne has decided to leave Jane's to conquer further fields but we have been very lucky in welcoming back Chrissie Richards, who has taken over as Publishing Controller. Thanks are also due to our typesetters, Method Limited, who under Eddie Green have produced the good words under considerable pressure in a short time scale, and also to Michael Badrocke for producing specially drawn diagrams. A number of overseas contributors have provided both information and photographs and I am grateful to Stefan Terzibaschitsch, Wilhelm Donko and Raymond Cheung for their efforts.

Last, but by no means least, my grateful thanks are due to the Jane's Information Systems team under Keith Faulkner who have transcribed all the information provided by the editorial team into database copy and printouts suitable for typesetting. Without the unstinting help of Keith, and his two assistants Ruth Simmance and Julie Barber, life would have been much harder.

Bernard H L Blake May 1988

Glossary of acronyms and abbreviations

AA anti-aircraft
AAA anti-aircraft artillery
AAAW advanced airborne anti-armour weapon
AAH advanced attack helicopter
AAM air-to-air missile
AB assault breaker
ABM anti-ballistic missile
ABRES advanced ballistic re-entry system
ABRV advanced ballistic re-entry vehicle
ACCS air command and control system
ACE Allied Command Europe
ACM anti-armour cluster munitions, or advanced cruise missile
ACMI air combat manoeuvring instrumentation
ACM advanced cruise missile
ACV armoured cannon vehicle
ADCAP advanced capabilities
ADDS Army data distribution system
ADM atomic demolition munitions
ADP automatic data processing
ADPG Air Defence Planning Group
ADS automatic defence system
AE assault echelon
AECB Arms Export Control Board
AEGIS AEW/ground environment integration system
AEW airborne early warning
AEW&C airborne early warning and control audio frequency
AF audio frequency
AFAP artillery-fired atomic projectile
AFATDS advanced field artillery tactical data system
AFB air force base
AFC automatic frequency control
AFSARC Air Force Systems Acquisition Review Council
AFSATCOM Air Force satellite communications system
AFV armoured fighting vehicle
AGC automatic gain control
AGM air-to-ground missile
AIAAM advanced intercept air-to-air missile
AIM air intercept missile
ALARM air-launched anti-radiation missile
ALCC airborne launch control centre
ALCM air-launched cruise missile
ALCS airborne launch control system
ALOC air line of communications
ALWT advanced lightweight torpedo
AM amplitude modulation
AMCM advanced mine counter-measures
AMRAAM advanced medium-range air-to-air missile
ANG Air National Guard
APC armoured personnel carrier
ARINC Aeronautical Radio Incorporated
ARM anti-radiation missile
ARP anti-radiation projectile
ARRADCOM Army Armament Research and Development Command (USA)
ARS automatic reporting system
ARSR air route surveillance radar
ASAT anti-satellite (weapon or system)
ASCM anti-ship cruise missile
ASM air-to-surface missile
ASPJ airborne self-projection jammer
ASR airport surveillance radar or air staff requirement
ASRAAM advanced short-range air-to-air missile
ASROC anti-submarine rocket
ASTAR airborne search target attack radar
ASUW anti-surface ship warfare
ASW anti-submarine warfare
ASW/SOW anti-submarine warfare/stand-off weapon
ATA advanced test accelerator
ATAFCS airborne target acquisition and fire control system
ATB advanced technology bomber
ATC air traffic control
ATE automatic test equipment
ATF advanced tactical fighter
ATGM anti-tank guided missile
ATGW anti-tank guided weapon
AUTODIN automatic digital network
AWACS airborne warning and control system

BADGE base air defence ground equipment
BATES battlefield artillery target engagement
BCC battery control centre
BCP battery control post
BETA battlefield exploitation and target acquisition
BITE built-in test equipment
BMD ballistic missile defence
BMEWS ballistic missile early warning system
BUIC back-up intercept control (for air defence)
BVR beyond visual range

C³ command, control and communications
C³I command, control, communications and intelligence
CAD computer aided design
CADMAT computer aided design, manufacture and test
CANE computer aided navigation equipment
CAS close air support
CASS command active sonobouy system
CASTOR corps airborne stand-off radar
CBU cluster bomb unit
CBW chemical and biological warfare
CCIS command and control information system
CCTV closed-circuit television system
CD civil defence
CEP circular error, probable
CFAR constant false alarm rate
CFV cavalry fighting vehicle
CIA Central Intelligence Agency
CINCEUR Commander-in-chief, European Command
CINCLANT Commander-in-chief, Atlantic Command
CINCPAC Commander-in-chief, Pacific Command
CINCSAC Commander-in-chief, Strategic Air Command
CIR continuous infra-red
CIS Combat identification system
CIWS close-in weapon system
CLOS command to line-of-sight
COMINT communications intelligence
COMSEC communications security
CONUS continental United States
CRAF Civil Reserve Air Fleet
CRP control and reporting posts
CRT cathode ray tube
CSDT control for submarine discharge torpedo
CSEDS combat systems engineering development site
CSLT control for surface launched torpedoes
CSMS corps support missile system
CSOC consolidated space operations centre
CSW conventional stand-off-weapon
CTBT comprehensive test ban treaty
CV aircraft carrier (USN)
CVN aircraft carrier, nuclear-powered (USN)
CVV aircraft carrier, medium-sized (USN)
CW chemical warfare, or continuous wave
CWS container weapon system
CWW cruciform wing weapon

DAR Defence Acquisition Regulation
DARPA Defence Advanced Research Projects Agency
DCS defence communications system
DCU detection and control unit
DDG guided missile destroyer (USN)
DEW distant early warning
DHS data handling system
DIA Defense Intelligence Agency (USA)
DICASS directional command active sonobuoy system
DIFAR directional frequency and ranging (sonobuoy)
DIVAD Division Air Defence
DLA Defence Logistics Agency
DLOS disturbed line-of-sight
DME distance measuring equipment
DMSP defence meteorological satellite programme
DMT deep mobile target
DMTI digital moving target indicator
DMZ demilitarised zone
DNA Defence Nuclear Agency
DoD Department of Defense (USA)
DOT designating optical tracker
DPM digital plotter map
DRTS detecting, ranging and tracking system
DSARC Defence Systems Acquisition Review Council
DSB Defence Science Board
DSCS defence satellite communication system
DSMAC digital scene matching correlation
DSP Defence Support Programme
DTOC divisional tactical operations centre
DTU data transmission unit

ECCM electronic counter—countermeasures
ECIP Energy Conservation Investment Programme
ECM electronic countermeasures
ECR embedded computer resources
EHF extremely high frequency
EJS enhanced JTIDS system
ELF extremely low frequency
ELINT electronic intelligence
EMC electro-magnetic compatability
EMP electro-magnetic pulse
EPC electronic plane conversion
ER enhanced radiation
ER/RB enhanced radiation/reduced blast
ERAM extended range anti-tank mine
ERU ejector release unit
ESM electronic support measures
EW electronic warfare
EXJAM expendable jamming system

FA Soviet frontal aviation
FAA Federal Aviation Agency
FACE field artillery computing equipment
FADAC field artillery digital automatic computer
FAESHED fuel-air explosive helicopter delivered
FASCAM family of scatterable mines
FAST fuze activating static targets
FBM fleet ballistic missile
FCS fire control system
FDC fire direction centres
FEBA forward edge of the battle area
FEMA Federal Emergency Management Agency
FET field effect transistor
FFG guided missile frigate (USN)
FFT fast Fourier transform
FLIR forward looking infra-red
FMF Fleet Marine Force
FMS Foreign Military Sales
FOV field-of-view
FPB fast patrol boat
FPR Federal procurement regulation (US)
FRAS free-rocket anti-submarine
FROG Soviet free-rocket-over-ground for tactical use
FRS Fleet readiness squadron
FSAT full-scale aerial target
FSD full-scale development
FTS flexible turret system
FWE foreign weapons evaluation
FY fiscal year
FYDP five-year defence programme

GAO General Accounting Office
GBU glide bomb unit
GCA ground controlled approach (radar)
GCI ground controlled interception (radar)
GDIP general defence intelligence programme
GDS gun display system
GEADGE German air defence ground environment
GEODSS ground-based electro-optical deep space surveillance
GLCM ground-launched cruise missile
GLLD ground laser locator designator
GMF ground mobile forces
GNP gross national product
GPS Global Position System (formerly NAVSTAR)
GWEN ground wave emergency network

HARM high-speed anti-radiation missile
HDD head-down display
HE high explosive
HEL high energy laser
HF high frequency
HHLR hand-held laser rangefinder
HIPAR high power acquisition radar
HOE homing overlay experiment
HPI high power illuminator
HTKP hard-target kill probability
HUD head-up display
HVM hypervelocity missile

IC integrated circuit
ICAO International Civil Aviation Organisation
ICBM intercontinental ballistic missile
ICC information co-ordination control
ICWAR improved continuous wave acquisition radar
IF intermediate frequency
IFCS improved (integrated) fire control system
IFF identification, friend or foe
IFM instantaneous frequency measurement
IFV infantry fighting vehicle
IIR imaging infra-red
IMINT imagery intelligence
INF intermediate range nuclear forces
INS inertial navigation system
IOC initial operational capability
IONDS integrated operational nuclear detection system
IOS integrated observation system
IPD improved point defence
IR infra-red
IR&D independent research and development
IRBM intermediate range ballistic missile
IRST infra-red search and track
ITSS integrated tactical surveillance system
IUS inertial upper stage
IUSS integrated undersea surveillance system

JCS Joint Chefs of Staff
JINTACS joint interoperability of tactical command and control systems
Joint STARS joint surveillance and target attack radar system
JSC Johnson Space Centre
JSDF Japanese Self-Defence Force
JSS joint surveillance system
JSTARS joint surveillance and target attack radar system
JTACMS joint tactical missile system
JTFP joint tactical fusion programme
JTIDS joint tactical information distribution system

LAAAS low-altitude airfield attack system
LAMPS light airborne multi-purpose system
LANTIRN low-altitude navigation and targeting infra-red night system
LAPADS lightweight acoustic processing and display system
LARS light artillery rocket system
LASR low altitude surveillance radar
LAW light anti-tank weapon
LCAC landing craft, air cushion

LCC launch control centre
LCP launch control post
LCWDS low-cost weapon delivery system
LDS layered defence system
LED light emitting diode
LFICS landing force integrated communications system
LGB laser-guided bomb
LHA amphibious assault ship (USN)
LLL low-level light (television)
LLLGB low-level laser guided bomb
LLTV low-light television
LMTR laser marker and target ranger
LOAD low-altitude defence
LOC lines of communication
LPD landing platform, dock (USN)
LPI low probability of intercept
LRA Soviet long-range aviation
LRAAS long-range airborne ASW system
LRF laser rangefinder
LRINF longer-range intermediate-range nuclear forces
LRSOM Long range stand-off missile
LRTNF long-range theatre nuclear forces
LRU line replaceable unit
LSD landing ship, dock (USN)
LSI large scale integration
LTD laser target designator
LTDP long-term defence programme
LTDS laser target designator system
LUA launch under attack
LWIR long wave infra-red

MAC Military Airlift Command (USAF)
MAD magnetic anomaly detector or mutually assured destruction
MANPADS man-portable air defence system
MAP Military Assistance Programme
MARAD Maritime Administration
MARS mobile automatic reporting system
MBAR multiple beam acquisition radar
MBER mutual and balanced force reductions
MBT main battle tank
MCM mine countermeasures
MCMV mine countermeasures vehicle
MEECN minimum essential emergency communications network
MF medium frequency
MFR multi-function radar
MGT mobile ground terminals
MHSV multi-purpose high speed vehicle
Midgetman see SICBM
MiG Mikoyan aircraft
MILCON military construction
MIRV multiple, independently targetable re-entry vehicle
MLRS multiple-launch rocket system
MMW millimetre wave radar
MNS mine neutralisation system
MoD Ministry of Defence (UK)
MoU memorandum of understanding
MPA maritime patrol aircraft
MRASM medium-range air-to-surface missile
MRV multiple re-entry vehicle
MSC Military Sealift Command
MSI medium-scale integration
MSO ocean-going minesweeper
MTBF mean-time-between-failures
MTCAS Marine Corps Tactical Command and Control System
MTI moving target indication
MTMC Military Traffic Management Command
MTTR multi-target tracking radar of meantime-to-repair
MULE modular universal laser equipment
MX Missile Experiment or Missile-X (advanced ICBM)

NADGE NATO air defence ground equipment
NASA National Aeronautics and Space Administration
NATO North Atlantic Treaty Organisation
NAVSTAR navigation satellite timing and ranging
NBC nuclear, biological and chemical
NCA National Command Authority(ies)
NDRF National Defence Reserve Fleet
NEACP National Emergency Airborne Command Post
NFCS nuclear forces communications satellite
NFIP National Foreign Intelligence Programme
NIS NATO identification system
NMCS National Military Command System
NNK non-nuclear kill
NOD night observation device
NODLR night observation device, long-range
NORAD North American Air Defence Command
NPG Nuclear Planning Group
NSA National Security Agency (US)
NSC National Security Council
NSNF non-strategic nuclear forces
NSR naval staff requirement (UK)
NSWP non-Soviet Warsaw Pact
NTC National Training Centre
NVG night vision goggles
NWS North Warning System

O&M operations and maintenance
OASD Office of the Assistant Secretary of Defence
OCC operational control centre
OED operational evaluation demonstration
OFPP Office of Procurement and Policy
OJCS Office of the Joint Cheifs of Staff
OM operations module
OSD Office of the Secretary of Defence
OTH over-the-horizon
OTHB over-the-horizon backscatter

PADS position and azimuth determining system
PAL permissive action link
PAR precision approach radar
PARCS perimeter acquisition radar characterisation system
PAVE-PAWS phased-array radars
PB particle beam
PCU power control unit
PGM precision guided munitions
PLRS position location reporting system
PLSS precision location strike system
PNVS pilot night vision system
POL petroleum, oil and lubricants
PPI plan position indicator
PRF pulse repetition frequency
PSP programmable signal processor
PWRS pre-positioned war reserve stocks

RAM rolling airframe missile or random access memory
RAP rocket-assisted projectile
RAPIDS radar passive identification system
RB/ER reduced blast/enhanced radiation
RCMAT radar-controlled miniature aerial target
RD&A research development and acquisition
RDF Rapid Deployment Force
RDJTF Rapid Deployment Joint Task Force
RDSS rapidly deployable surveillance system
RDT&E research, development, test and evaluation
REMBASS remotely-monitored battlefield sensor system
RF radio frequency
RFS radio frequency surveillance
RHI range height indicator
ROCC region operations control centre
ROR range only radar
RO/RO roll on/roll off
ROTHR relocatable over-the-horizon radar
RPV remotely piloted vehicle
RRF ready reserve force
RSLS receiver side-lobe suppression
RSVP rotating surveillance vehicle platform
RV re-entry vehicle
RWR radar warning receiver

SAC Strategic Air Command
SACDIN SAC digital network
SACEUR Supreme Allied Commander, Europe
SACLANT Supreme Allied Commander, Atlantic
SACU stand-alone digital communications units
SADARM search and destroy armour munition
SAGE semi-automatic ground environment (air defence)
SAL strategic arms limitation
SALT Strategic Arms Limitation Talks
SAM surface-to-air missile
SAMSO Space and Missile Systems Organisation
SAR synthetic array radar, or search and rescue
SARS static automatic reporting system
SDI Strategic Defense Initiative
SDS satellite data system
SGEMP system-generated EMP
SHAPE Supreme Headquarters Allied Powers, Europe
SHF super-high frequency
SHORAD short-range air defence
SICBM Small ICBM (Midgetman)
SINCGARS single channel ground and airborne (communications) system
SLBM Submarine launched ballistic missile
SLCM submarine (sea) launched cruise missile
SLEP service life extension programme
SLMM sea-launched mobile mine
SLOC sea line of communication
SM Standard missile
SNF short range nuclear force
SNM special nuclear material
SOSUS sound surveillance system
SOTAS stand-off target acquisition system
SP self propelled
SPADOTS spare detection and tracking system
SPAT self-propelled acoustic target
SRAM short-range attack missile
SSB conventional-powered ballistic missile submarine or single side-band
SSBN nuclear-powered ballistic missile submarine
SSKP single shot kill probability
SSN nuclear-powered attack submarine
SSR secondary surveillance radar
SSURADS shipboard surveillance radar systems
START Strategic Arms Reduction Talks
STC satellite test centre
STIR surveillance and target indicating/illuminating radar
STOL short take-off and landing
STP systems technology programme
STR systems technology radar
SUAWACS Soviet Union Airborne Warning and Control System
SUBROC submarine launched rocket
SURTASS surveillance towed array sensor system
SUT surface and underwater target
SXTF satellite X-ray test facility

TAC Tactical Air Command (USAF)
TACS tactical air control system
TACTAS tactical towed array sonar
TADS target acquisition and designation system
T-AKX commercial roll-on/roll-off ship (USN)
TALCM tactical air-launched cruise missile
TAOC tactical air operations centre
TAS target acquisition system
TASM tactical anti-ship missile
TCT target centred tracker
TEL transporter erector launcher (for MX advanced IBM)
TERCOM terran contour matching navigation system
TER terrain following radar
TGSM terminally guided submunitions
TICM thermal imager common module programme
TIR target illuminating radar
TIS thermal imaging system
TISEO target identification system, electro-optical
TLAM Tomahawk land attack missile
TNF theatre nuclear forces or tactical nuclear forces
TNFS³ theatre nuclear forces, survivability, security and safety
TNW tactical nuclear warfare
TOW tube-launched, optically tracked, wire-guided
TRAM target recognition attack multi-sensor
TRI-TAC joint tactical communications programme
TWS track-while-scan (radar)
TWT travelling wave tube

UCS underwater combat system
UHF ultra high frequency
UKADGE UK air defence ground environment
UKADR UK air defence region
USAF United States Air Force
USAFE USAF, Europe
USAFR USAF Reserves
USANG United States Air National Guard
USAR United States Army Reserves
USAREUR United States Army Europe
USMC United States Marine Corps
USMCR USMC Reserves
USN United States Navy
USNR USN Reserves

VDS variable depth sonar
VHF very high frequency
VHSIC very high speed integrated circuit
VLA vertical launch ASROC
VLF very low frequency
VLGM vertical loading gun mount
VLS vertical launch system
VRP visual reporting posts
VSTOL vertical/short take-off and landing

WAAM wide area anti-armour munitions
WP Warsaw Pact
WRM war reserve material
WRS war reserve stocks
WSEP Weapon System Evaluation Programme
WVR within visual range
WWMCCS world-wide military command and control system

Jane's Code Numbers

Jane's entry numbers are usually found to be more convenient than manufacturer's or operator's designation for a system or equipment item. The latter can and frequently do change in the course of the weapon's/system's operational life, with the inevitable difficulties in keeping track of a specific item through several or many editions. By allocating a discrete entry number this confusion is eliminated, and although commercial whims and other vagaries may change the title of an entry with time, its *Jane's Weapon Systems* (JWS) number will always identify it. This is sometimes useful when by chance or design a marketing name is chosen by two contractors for different items/systems, which may in fact be very similar or completely different, 'Hawk' for example: this can either be BAe's strike/advanced trainer aircraft, or Raytheon's Homing All the Way Killer SAM. Giving each its own number resolves the confusion.

For those who prefer not to use numerals, from the second edition of *Jane's Weapon Systems* onward a conventional alphabetical index has been included, giving readers the option of using whichever method they like best to find their way among the admittedly confusingly elaborate variety of subjects that have to comprise the contents. However those who have the patience to familiarise themselves with the JWS numbering system will find that it is capable of use for a number of additional research and collation functions to meet individual readers' needs.

If we briefly summarise how entry numbers are allocated, and the significance of the digits that go to make up the seven-figure number, this will assist in making maximum use of this book. It may be noted that the number consists of two parts: a four-digit number before the 'decimal point', and three more numerals after the point. The first part is discrete to the entry to which it refers, and generally is not changed in subsequent editions, thus identifying that particular item permanently. (Over a period of years, in rare instances, a topic is given a fresh number. This usually occurs if there has been a major alteration to some aspect(s) that materially affects the system or equipment described, calling for new or considerably modified text.)

It is the last three digits that are the most useful. These are code numbers and have no arithmetical significance whatsoever. The last figure of all denotes the section of the book in which a particular entry will be found. Thus it is always either '1', '2', '3' or '4', referring to the Systems, Platforms*, Equipment or Analysis Sections of Jane's Weapons Systems. These appear in that order in the book. This narrows the search for a given entry appreciably by limiting to one of four sections**.

Immediately after the 'decimal point' are two other digits, which, in the case of 'system' entries (ie. entry nos. ending in '1') describe the function of the entry described by means of the following code:

1 – Land
2 – Sea
3 – Air (including space)
4 – Underwater
5 – Data Acquisition
6 – Data Processing
7 – Display or Output
8 – Actuation, Direction or Guidance
9 – Miscellaneous
0 – Not Applicable

Thus, in the case of an air-to-air missile entry, its JWS number would be coded 'XXXX·331', and a submarine-launched ballistic missile 'XXXX·411', the first digit after the point indicating the missile's launch environment (ie. '3' for air-launched, or '4' for submarine-launched), and the next digit denotes the intended target environment ('3' for air, or '1' for land).

Where the system to be described in an entry is something other than a missile, eg. a naval tactical data handling system, the first digit after the 'point' will be used to denote the environment in which the system operates (in this case '2' on board a ship) and one of either '6', '7', or '8' will follow to indicate the predominant function of the system: 'data processing' (6) where computers form the basis of the system, 'display' (7) if a display console is the main element, or (8) if direction is the principal function of the system. A few minutes study of entries and the numbers they have been allocated will soon make the pattern clear.

In the case of Platform Section Entries (ie. those with a final digit '2') only the first digit after the point has any significance, when it indicates either a land vehicle (tank etc) when the figure '1' would be used, '2' for a ship entry, '3' for an aircraft, and '4' for a submarine. The next digit is always a '0'. Thus entry XXXX·302 would describe an aircraft or helicopter.

The coding applied to Equipment Section (final digit '3') is similar to that used for system entries in Section One, with the digits after the 'point' having the same significance as set out in the list above. In equipment entries though, the digit immediately after the 'point' is used to signify the environment in which the equipment operates, and the next figure is used to denote its main function. In practice this means that 'equipment' entries always have '3' as their last figure, and '1', '2', '3' or '4' as the first figure after the 'point'. Thus an airborne radar will be coded XXXX·353, a sonar XXXX·453, and a land-based electronic warfare equipment XXXX·193.

Readers who take the time to familiarise themselves with the system will appreciate the scope for additional uses of the code numbers for their own analysis purposes etc. For example, a list of entries dealing with air-to-surface weapons could be readily compiled by noting all entries with numbers that end '·311', '·321' or '·303'. Similarly, underwater display equipments, for instance, could be extracted by listing those entries ending '·473'.

It is realised that complexity of modern weapons and military equipment, to say nothing of rapidly advancing technology, inevitably throws out new hardware from time to time that does not fall neatly into this coding system, but it is felt that to attempt to embellish the coding system to accommodate such problems would only lead to an unacceptably cumbersome system.

*Each platform category, land, sea and air, now has its own yearbook devoted entirely to that category of vehicle. Consequently, corresponding JWS platform sections were omitted some years ago although the salient data relating to each now appears in tabular form in section 4 of this book.

**Sections now number three only, now that the platform section no longer appears.

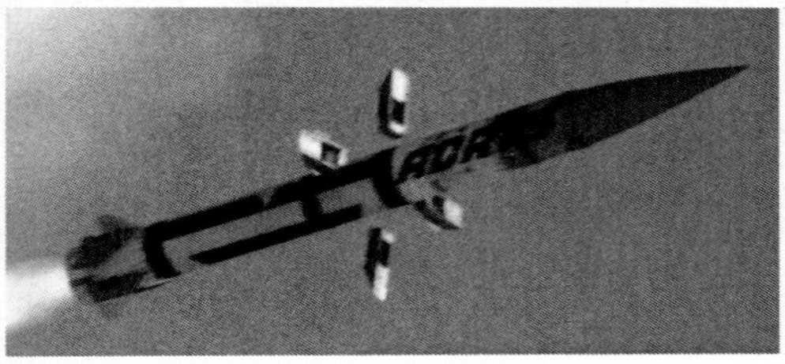

STRATEGIC WEAPON SYSTEMS

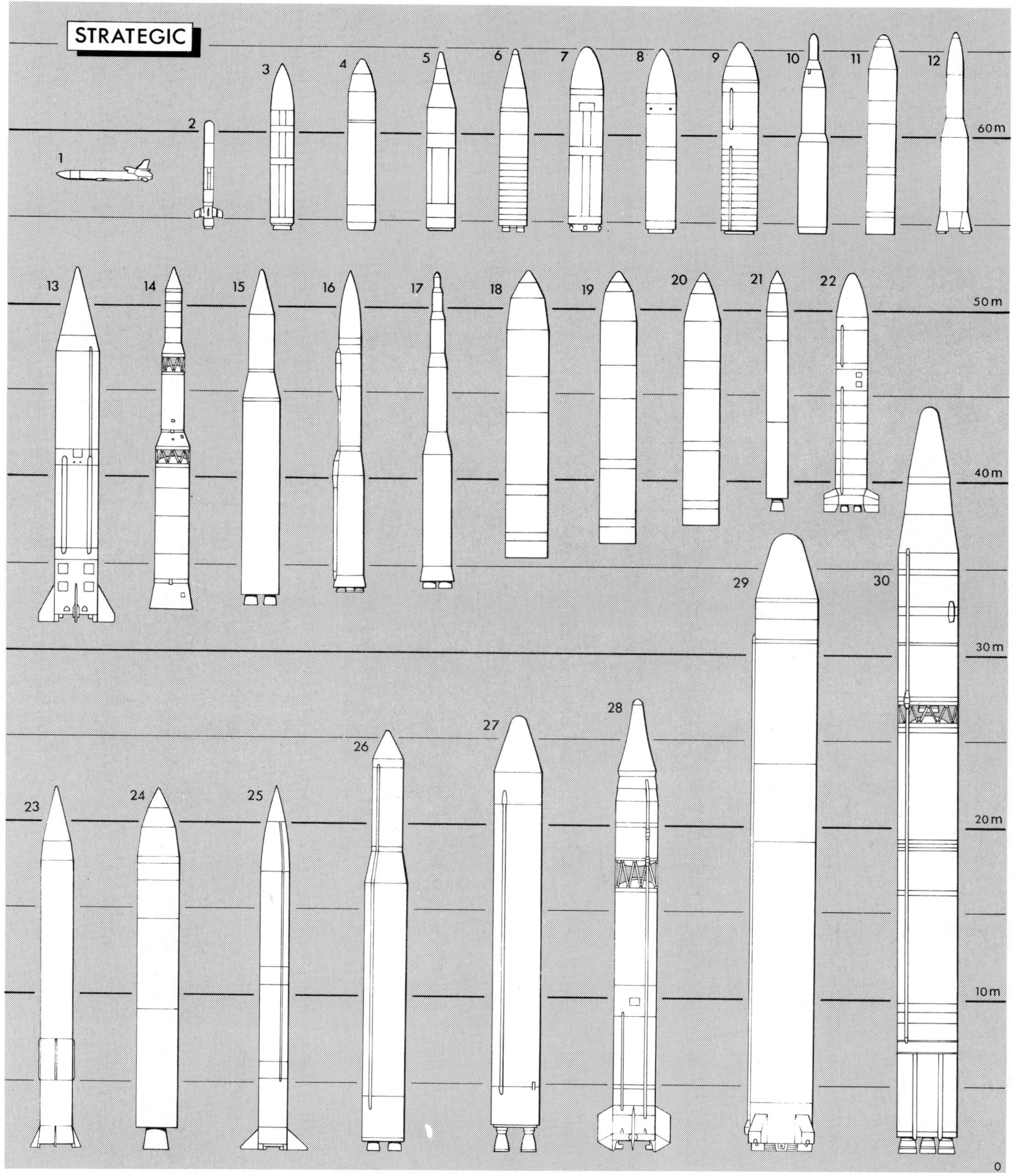

(1) ALCM, **(2)** GLCM, **(3)** Polaris A-3, **(4)** SS-N-6, **(5)** CSS-N-2, **(6)** MSBS (M20), **(7)** Trident C-4, **(8)** Poseidon C-3, **(9)** MSBS (M4), **(10)** SS-N-5, **(11)** SS-N-17, **(12)** SS-20 (provisional), **(13)** CSS-2, **(14)** SS-13 **(15)** SS-11, **(16)** Minuteman III, **(17)** Minuteman II, **(18)** SS-N-20, **(19)** SS-N-18, **(20)** SS-N-8, **(21)** Small ICBM, **(22)** SSBS, **(23)** SS-4, **(24)** Peacekeeper, **(25)** CSS-1, **(26)** SS-17, **(27)** SS-19, **(28)** CSS-3, **(29)** SS-18, **(30)** CSS-4

TERMINOLOGY

The term 'strategic' is subject to different interpretation by different nations. The Soviet Union, for example, tends to regard any weapon that can reach its soil as strategic, regardless of its range capability. Existing treaties categorise strategic weapons as those with a maximum range exceeding 5500 km and intermediate or theatre weapons as those with ranges between 1000 and 5500 km. Systems with ranges less than 1000 km are called short range weapons. The Intermediate Nuclear Force (INF) Treaty, which applies only to land-based systems, covers both intermediate and short range weapon systems with ranges between 500 and 5500 km. *Jane's Weapon Systems* will use these definitions to categorise ballistic missile systems, but we have not altered the location of individual weapon system descriptions from last year's book. As a result whilst SS-20 and SS-4 remain in the Strategic Weapon Systems section, Pershing, GLCM, SS-12 and SS-23 remain as before in the Battlefield Support Weapons section.

CHINA (PEOPLE'S REPUBLIC)

2010.111
CSS-1 (Chinese designation: DF-2)
(*Dong Feng*=East Wind)
Liquid-fuelled intermediate range ballistic missiles similar to the Soviet SS-3 have been operational in north-eastern and north-western China since 1970, although some American reports claim operational use since 1966. Like the SS-3 Shyster, these missiles have a maximum range of 1200 km and it is understood that some 50 were deployed, with production ended in 1969 at 100 missiles. The designation assigned by the Pentagon is CSS-1. The first operational warheads are believed to be smaller than those carried by the Soviet missile; a figure of 15 KT has been suggested but a 20 KT warhead may be fitted to some. In particular it is probable that those which have been most recently deployed carry thermo-nuclear warheads similar to those carried by the longer range systems.

At one stage it was thought that this missile would either be deployed in large numbers or be replaced by solid-fuel missiles. It now seems rather more likely, however, that the Chinese will put the majority of their effort into the ICBM and IRBM programmes and eventually remove the shorter range missiles from service.

CHARACTERISTICS
Length: 21 m
Diameter: 1.6 m
Weight: 26 000 kg
Range: 1200 km
Propulsion: Liquid-fuelled single stage
Warhead: 1 × 15-20 KT

2049.111
CSS-2 (Chinese designation: DF-3)
Although the Chinese intermediate range programmes have been subject to delays similar to those that appear to have affected their ICBM programme, it seems fairly clear that there has been a small-scale (15 to 20) deployment in China of a single-stage system using a storable liquid-propellant. This missile is understood to be similar to the first stage of the rocket which launched the first Chinese satellite in April 1970 and its range has been estimated as lying between 2500 and 3000 km. According to US sources, entry into operational service occurred in 1971 and it has been designated CSS-2.

Basing is understood to be on a basis of deployment at permanent sites from which targets in Central and Eastern Asia can be reached, although recent estimates suggest a capability for relocation of launch facilities without necessarily implying a mobile system. Production is said to total around 50 missiles.

In June 1986, flight tests began of a modernised system with a new delivery system, probably of an advanced MRV type, with some ability to place re-entry vehicles on dispersed targets, according to Chinese sources. This implies that the system now has additional powered stages.

CHARACTERISTICS
Length: 20.62 m
Diameter: 2.46 m
Weight: 27 000 kg
Range: 2700 km
Warhead: (old version) 1-3 MT; (new version) possibly 3 re-entry vehicles, each 50-100 KT

Chinese authorities describe this as an "improved version of a large rocket" - originally thought to be a CSS-3 but later confirmed to be an updated CSS-2 IRBM with modified nosecone

CSS-2 IRBM: Composite picture taken in Beijing. An updated version with MIRVed warheads may be operational shortly (via G. Jacobs)

4210.111
CSS-3 ICBM (Chinese designation: DF-4)
The Chinese CSS-3 strategic ballistic missile is a two-stage weapon believed to have a maximum range of around 7000 km carrying a warhead with an estimated yield of 2 MT. According to American defence analyses, two of these weapons had been deployed in silos in 1978/9, now a maximum total of eight are believed operational and inertial guidance is assumed. Deployment is reported to be concentrated in western China, from where targets in European USSR would be accessible.

Prognostications of Chinese strategic intentions and missile production and deployment should be treated with caution. According to official Chinese reports, the PRC's 2nd Artillery Corps completed a series of tests on its strategic missiles' software in December 1985 which increased the accuracy of its missiles.

Around 30 CSS-3s were produced. A civil variant of this missile is called Long March 2 and has the Chinese designation CZ-2.

CHARACTERISTICS
Length: 32.57 m
Diameter: 3.35 m
Weight: 191 000 kg
Range: <7000 km
Warhead: 1 × 2 MT

CSS-3 ICBM in a ceremonial parade in Beijing (via G. Jacobs)

CSS-2 about 2-dozen

IRBM (~~about 20~~) single-stage liq. prop.

replaced by ICBMs (CSS-3 / DF-4)

4211.111

CSS-4 ICBM (Chinese designation: DF-5)

In May 1980 China launched into an impact area in the Pacific Ocean two test rockets for what is provisionally understood to be a new strategic ICBM weapon designated CSS-4. The two launches achieved ranges of about 7000 km although the figures quoted by different observers vary quite appreciably, some higher (up to 15 000 km) and some lower, and the impact area was within the area bounded by the Solomon Islands, Fiji and the Gilbert Islands. No warheads were carried according to Chinese reports of the event. The similarity of this missile to the Soviet SS-18 implies that the CSS-4 has a payload of 4000 to 7000 kg enabling it to carry any of the nuclear warheads that have been tested by the PRC; the payload is therefore put in the 1-5 MT range. The missile has liquid propulsion and, for a missile of this size and range, three stages can be assumed. US reports claim that CSS-4 missiles were deployed at two sites in hardened silos in central China in 1980/81 with up to five launchers now operational and 20 missiles delivered although Chinese sources maintain the missile became operational in 1984. Given the Soviet concern for the ever present Chinese threat, deployment of this missile may have been one of the reasons for the decision to upgrade the Moscow Anti-Ballistic Missile defence system (**2932.131**).

A civil variant of this missile is called Long March 3 and has the Chinese designation CZ-3.

CHARACTERISTICS

Length: 43.25 m
Diameter: 3.35 m
Weight: 202 000 kg
Range: >10 000 km
Warhead: 1 × 5 MT (payload capability: 4 000 to 7 000 kg)

First stage of CSS-4 showing large nozzles (via G. Jacobs)

Nose-cones and upper stages of CSS-4 ICBM (via G. Jacobs)

4893.411

CSS-N-3 SUBMARINE-LAUNCHED BALLISTIC MISSILE (Chinese designation: JL-1)

The People's Republic of China has been developing a new submarine-launched ballistic missile, for use with the new 'Xia' class of nuclear-powered submarine. The Western designation is CSS-N-3.

It is thought to have been developed, to some extent, from the CSS-2 (**2049.111**). This suggests that the naval weapon will have a similar payload, probably a single 2 MT nuclear warhead and a maximum range of about 2700 km. It is believed to be operational and deployed aboard 'Xia' class submarines with 12 launch tubes in each of the two boats.

Test firings were carried out in 1982 reportedly from a submerged pontoon (30 April 1982) near Huludao, and later on 12 October 1982. In 1985 it was estimated that the CSS-NX-3 system was nearing operational status, and a number of these missiles were ceremonially paraded in Beijing in October 1984.

CHARACTERISTICS

Length: 10 m
Diameter: 1.5 m
Weight: 13 800 kg
Range: 2700 km
Warhead: 1 × 2 MT

'Xia' class SSBN with 12 tubes for CSS-N-3 SLBM (Xinhua Radiophoto/Associated Press)

CSS-N-3 SLBM (via G. Jacobs)

6411.411

CSS-NX-4 SUBMARINE-LAUNCHED BALLISTIC MISSILE

The *Liberation Army Journal* in late 1985 announced the "success" of a SLBM underwater test-firing the previous September that may be associated with this reported programme. Some sources suggest it will also carry a single 2 MT warhead although MIRVing should not be ruled out.

Future Chinese SSBNs may be modified to carry 16 tubes. Up to four 'Xia' class may be planned.

FRANCE

2163.111

SSBS INTERMEDIATE RANGE BALLISTIC MISSILE TYPE S-3D/TN-61

SSBS (*Sol-Sol-Balistique-Stratégique*) is an intermediate range two-stage solid-propellant missile, with thermonuclear warhead, which is stored in and launched from an underground silo.

Launch areas are dispersed and hardened to reduce the effects of an enemy attack. Each area includes the silo in which the missile is maintained in operational readiness, and an annexe housing the automatic launching equipment and support services.

The original S-2 strategic missile force formed the main land-based element of the French '*force de frappe*', now '*force de disuasion*', from the early 1970s.

In 1973 a programme was initiated to develop the second-generation SSBS weapon system, the S-3. This programme entailed the renovation of the first two groups of S-2 silos and replacing the missiles with the S-3 model.

Retaining the same first stage as the S-2, the S-3 has a second stage of higher performance, namely, the P-6 originally developed for the MSBS (**1134.411**). An advanced re-entry vehicle system is also included. This has a thermonuclear warhead and a re-entry vehicle which is hardened against the effects of a high altitude nuclear explosion from an ABM. Other features are a new system of penetration aids designed to counter enemy defences. The complete system is reported to be hardened against the effects of EMP (electro-magnetic pulse).

Improvements to the ground facilities include both equipment modernisation and system modifications to increase reliability and reduce maintenance costs.

CHARACTERISTICS

Type: Intermediate-range silo-launched, ballistic missile

Guidance: Inertial

Propulsion

First stage: SEP Type 902 motor. 4 gimballed nozzles for control

Second stage: SEP Rita II motor containing 6 tonnes of solid propellants. Thrust vector control by 4 freon injectors into a single fixed nozzle

Dimensions

Length: 13.8 m

Diameter: 1.5 m

Weight: 25 800 kg

Range: 3000 km

Warhead: Thermonuclear (1.2 MT)

DEVELOPMENT

Development was covered by five contracts awarded to Aérospatiale spanning the six-year period 1974-80. Four contracts were from the Direction des Engins, and one from the Commissariat à l'Energie Atomique, and their combined value amounted to over $456 million.

STATUS

Development of the S-3 was completed and deployment began in 1980. The first group of nine S-3 missiles and their associated silo installations on the Plateau d'Albion were officially inaugurated in May 1980 and both groups of nine silos were operational by the end of 1982. Reaction time is reported as about three and a half minutes. The French have declared their intention of replacing these increasingly vulnerable fixed based missiles with a new generation of mobile weapons, land and seaborne, over the next decade.

SSBS S-3 IRBM launch from the Landes test site (C.E.L.)

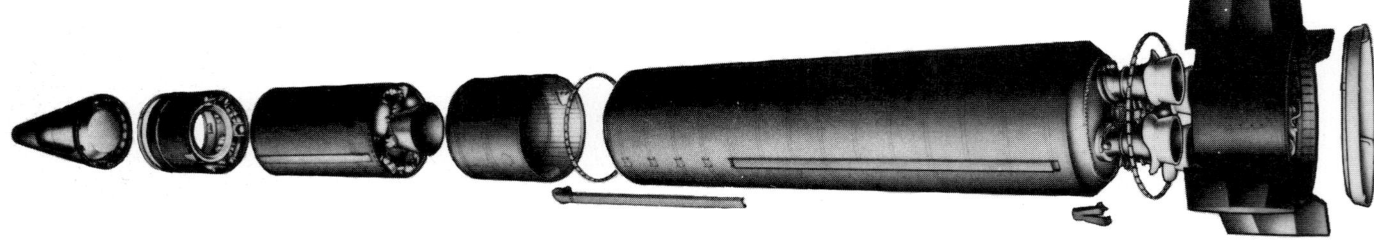

SSBS S-3 constructional layout (Aérospatiale)

6412.111

SSBS INTERMEDIATE RANGE MISSILE TYPE S-4 (SX)

Details have been announced by the French Defence Ministry of a FFr3500 million 10 year programme for the S-4 land-based IRBM to replace the S-3 missiles after 1996.

The S-4 will be a two-stage, solid-propellant, land-mobile system with a range of 4500 km. Present basing plans call for 33 missiles each with 3 targetable and possibly MIRVed warheads in the 20 KT range, capable of random dispersal around France by air or road in a period of tension.

A final decision on the viability of the project is expected in 1988/89 with a possible production start-up in 1990.

1134.411

MSBS SUBMARINE LAUNCHED BALLISTIC MISSILE FAMILY

MSBS (*Mer-Sol Balistique Stratégique*) is an intermediate range, two-stage submarine-launched weapon forming a vital part of the French nuclear deterrent force. In general concept it is similar to the American Polaris missile family.

The M-1 version went into service in 1971 but was phased out in favour of the M-2 in 1974 which by mid-1977 had itself been supplanted by the M-20. Now the M-4 version has entered operational service and there are plans for a M-5 SLBM to arm the seventh French SSBN in the mid 1990s.

The current MSBS force is based on six nuclear-powered submarines (SNLE - *Sous-marines Nucléaire Lanceur d'Engins balistique*), each able to carry and launch 16 missiles. The M-4 is operational in *L'Inflexible* and *Le Tonnant* is undergoing a three year conversion to this configuration. *L'Indomptable, Le Terrible* and *Le Foudroyant* are still carrying the M-20 missile but will be retrofitted (in that order) with the M-4 version. The eldest and fifth M-20 boat, *Le Redoutable*, launched in March 1967, will not be converted.

Three SSBNs are intended to be operational at any time.

Logistical support for the MSBS system is provided by the Ile Longue Naval Base in Brest Bay where there are the assembly and storage facilities necessary to keep the missiles in operational readiness.

M-20 SLBM

The M-20 has an upgraded re-entry vehicle system which includes a thermonuclear charge, hardening

M-4 launch from the conventional test submarine Le Gymnote. *14 development firings were staged; the last in February 1984. Only one was unsuccessful* (C.E.L.)

against the effects of an ABM explosion at high altitude and improved penetration aids.

CHARACTERISTICS - M-20/TN-60 MISSILE
Length: 10.4 m
Diameter: 1.5 m
Weight: 20 000 kg
Guidance: Inertial
Propulsion

First stage: Type 904 motor with control by 4 gimballed nozzles. 10 tonnes of solid propellant

Second stage: Rita II motor containing 6 tonnes of solid propellant. Internal thrust vector system of control using freon injection into single fixed nozzle

Warhead: Thermonuclear TN-60 1.2 MT with penaids
Range: >3000 km

One hundred missiles were manufactured.

M-4 SLBM

A completely new generation of MSBS, designated the M-4, is now operational aboard *l'Inflexible*. The range is significantly increased to over 4000 km and there are six 150 KT thermonuclear warheads probably in a MIRV type configuration. M-4 missiles will be deployed on existing submarines after certain modifications have been carried out. The *Le Tonnant* is now being modified for M-4. Afterwards all the others except *Le Redoutable* will be modified.

A three-stage propulsion system is employed; the first stage (M-401) has a metal casing, while the other two motors (M-402 and M-403) have filament wound casings of Kevlar material produced in the USA.

The experimental diesel-electric submarine *Le Gymnote* was used for M-4 tests with twin tubes. She was paid off 12 October 1986.

CHARACTERISTICS - M-4/TN-70 MISSILE
Length: 11.05 m
Diameter: 1.93 m
Weight: 35 000 kg
Guidance: Inertial
Propulsion: 3-stage solid propellant
Warhead: 6 × 150 KT in bus powered by two SEP/SNPE Type 404 rocket motors
Range: 4500 km

M-5 SLBM

The latest development of the MSBS family of SLBMs is the M-5 version, planned to arm the seventh French SSBN possibly to be laid down in 1988 for service in 1994. The missile will have a range of 11 000 km and sophisticated penetration aids to defeat perceived developments in the Moscow anti-ballistic missile defences. Payload has been variously reported as either 10 or 12 MIRVs.

STATUS

The M-20 is operational and the M-4 became operational in 1985. The first M-4 test launch took place (from the Landes Test Range near Biscarosse) in November 1980. The 14th and final test firing from *Le Gymnote* was on 29 February 1984. Despite its larger diameter, the M-4 missile can be carried in the tubes of first generation SNLEs with certain modifications, and the earlier submarines will be converted from the M-20 to the M-4 one at a time after the M-4 entered service at the end of May 1985 in *l'Inflexible*. It has been reported that an improved version, designated M-4C, is being developed, with slightly extended range to 5000 km, to be carried in *Le Tonnant*. This is effected by the lighter TN-71 warhead being installed.

In March 1986 a French boat launched an M-4 over a 6000 km trajectory.

Underwater launch of M-4 from Le Gymnote

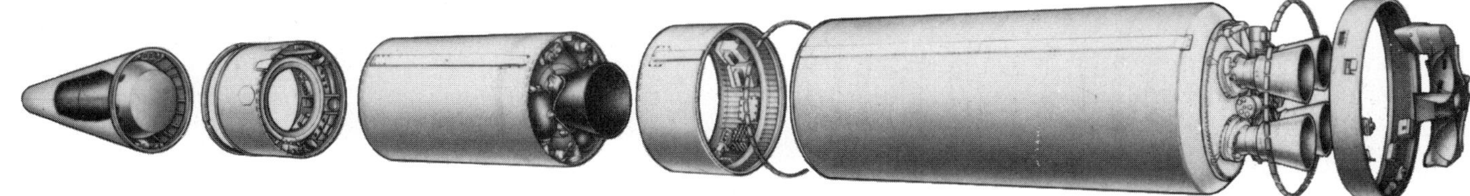

M-20 MSBS still operational in four SSBNs

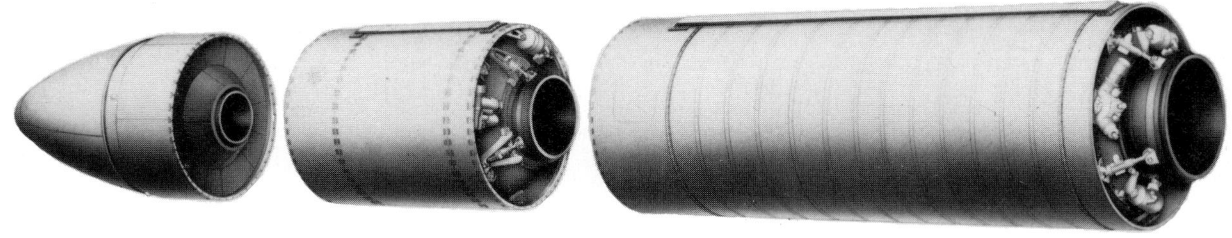

New generation M-4 MSBS at sea in l'Inflexible

UNION OF SOVIET SOCIALIST REPUBLICS

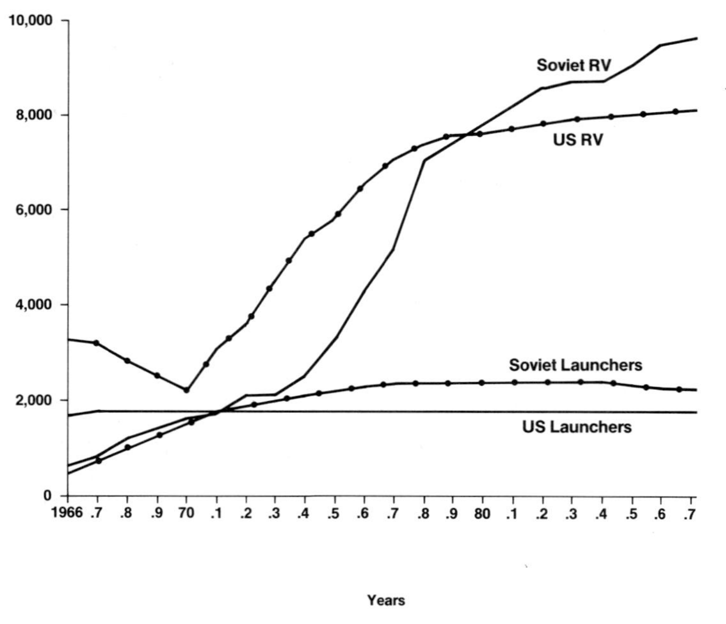

US AND SOVIET STRATEGIC LAUNCHER AND RE-ENTRY VEHICLE DEPLOYMENTS 1966-1987

	Launchers		RV	
	US	**USSR**	**US**	**USSR**
1966	1650	430	3300	600
7	1760	730	3200	800
8	1760	980	2800	1200
9	1760	1250	2500	1400
70	1760	1490	2200	1600
1	1760	1750	3100	1700
2	1760	1880	3600	2100
3	1760	1980	4500	2100
4	1760	2100	5350	2500
5	1760	2170	5800	3300
6	1760	2260	6500	4300
7	1760	2330	7000	5200
8	1760	2360	7350	7000
9	1760	2360	7550	7350
80	1760	2360	7600	7800
1	1760	2360	7700	8200
2	1760	2360	7800	8550
3	1760	2360	7900	8700
4	1760	2360	7950	8700
5	1760	2300	8000	9000
6	1760	2250	8050	9450
7	1760	2250	8050	9550

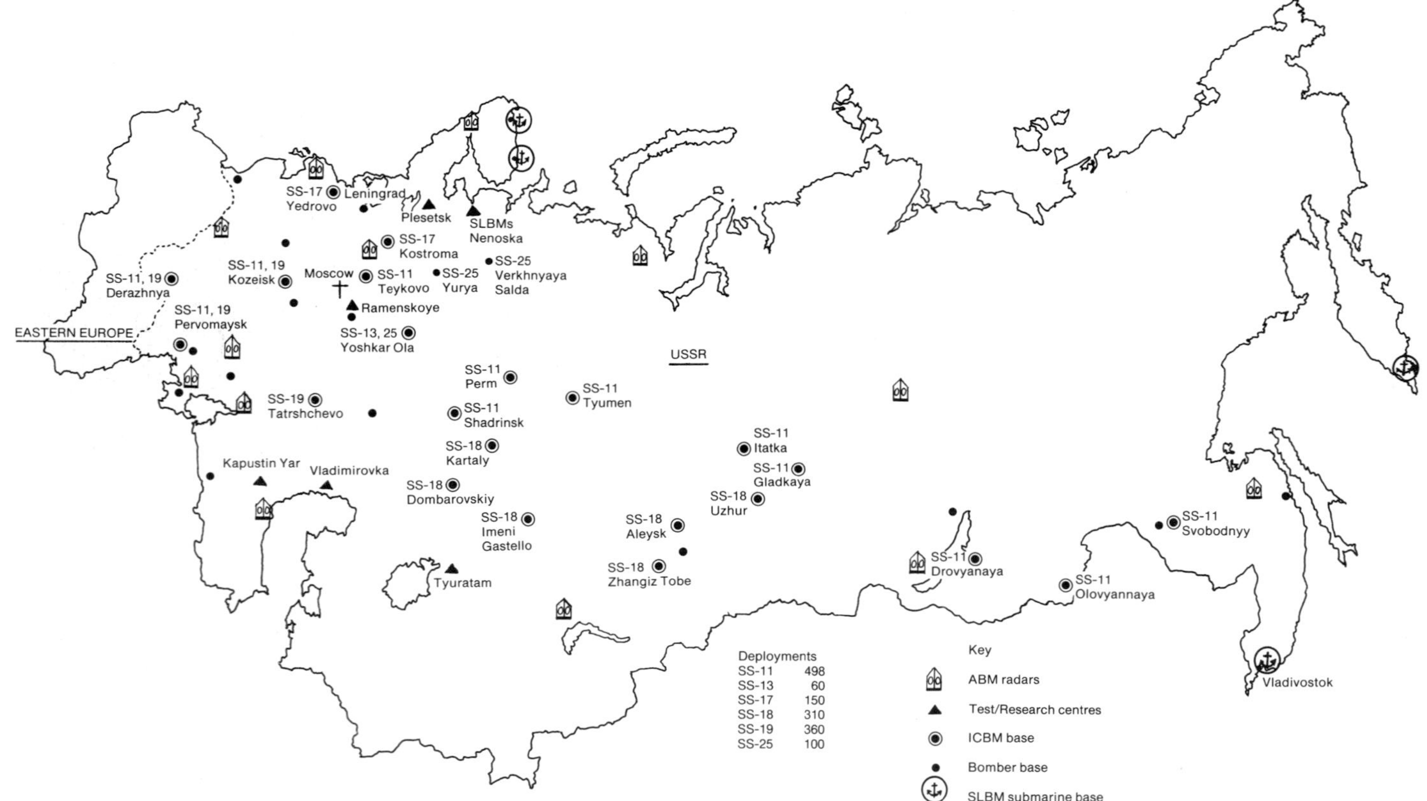

Map summary of principal Soviet strategic assets. Note: details of IRBM force deployments and strength appear in the text

2952.111

SS-4 (SANDAL) (Soviet designation: R-12)

Known also by the NATO code-name Sandal, the SS-4 is an intermediate range ballistic missile which is a developed version of the SS-3 (Shyster).

The Yangel design team completed development in 1957 and the SS-4 was first shown in public in 1961 with IOC in 1958. It became a standard IRBM in the Soviet armed forces. It was also the missile that lay at the root of the Cuban crisis in 1962. Depending on the weight of the warhead the missile has a range of between 1500 and 2000 km.

Initially, like the SS-7 Saddler, (also a product of the Yangel design bureau), the SS-4 used non-storable propellant which therefore required several hours preparation before launch.

The complete weapon system comprises about 12 tractor vehicles with special trailers. Some 20 men are required to erect and launch the missile. Official US sources report that the SS-4 has a reload capacity when fired from soft sites; it is also deployed at hardened sites but without a reload capability. The guidance system employed was originally radio command, operating on guidance vanes in the efflux nozzles, but it was observed at the time of the Cuban crisis that a changeover to inertial guidance had taken place.

CHARACTERISTICS

Guidance: Inertial

Guidance method: By control of elevators and of guidance vanes in the efflux nozzles

Accuracy: 2.4 km CEP

Propulsion: 2-stage liquid propellant

Warhead: Nuclear (1 MT)

Missile length: 22.77 m

Missile diameter: 1.65 m

Launch weight: 27 000 kg

Range: 1500 – 2000 km

STATUS

Since 1977, according to official US estimates, the proportion of SS-4 and SS-5 missiles has slowly declined as the SS-20 successor system deployed, the SS-4 less rapidly than the SS-5. At the signing of the INF Treaty, in December 1987, the Soviet Union declared 65 deployed SS-4 missiles with a further 105 missiles not deployed. There were no SS-5 missiles deployed. The SS-4 missiles are due to be withdrawn within three years, under the terms of the INF Treaty.

4207.111

SS-11 (SEGO) ICBM

The SS-11 Sego ICBM has been reported in four distinct versions at one time known in official American publications as Mods 1, 2, 3 and 4 but now referred to in three variants only. Despite development commencing during 1955-58 and with an initial operational capability achieved in 1966, this ICBM survives as the most numerous in the Soviet strategic armoury. Mods 2/3 were operational in 1973/75.

Peak deployment was in 1972 with nearly 1000 SS-11 silos but six years later the missile began to be replaced by the SS-17 and SS-19 missiles.

The basic SS-11 missile is a two-stage storable liquid propellant rocket with a range of between 10 600 and 13 000 km, which, when it entered service, carried a single warhead payload.

In the case of the Mod 2 there is still a single warhead but this is accompanied by penetration aids. The Mod 3 version, when it was detected soon after the Mod 2, was noted for a different RV and warhead arrangement, in particular an MRV platform equipped with three warheads. No details of the size of these warheads have been discovered but it has been deduced that they are unlikely to be less than the 200 KT of each of the warheads carried by the US Minuteman III, which the SS-11 closely resembles. Many of the modifications to the SS-11 can be attributed to the coincident ABM defence activity in the United States during the late 1960s.

CHARACTERISTICS

Type: Silo-launched intercontinental ballistic missile. Hot launch

Guidance: Inertial

Propulsion: 2-stage storable liquid propellant

Warhead: Single, 950 KT (Mod 1); Single, 1MT, plus penaids (Mod 2); 3 MRVs in range, 100-300 KT (Mod 3)

Dimensions

Length: 19 m

Diameter: 2.4 m

Weight: 48 000 kg

Range: 10 000 km (Mod 1); 13 000 km (Mod 2); 10 600 km (Mod 3)

CEP: 1400 m (Mod 1); 1100 m (Mods 2/3)

STATUS

In early 1981 there were about 500 SS-11s deployed operationally altogether, mostly positioned opposite the Soviet Union's borders with China, Europe, Scandinavia, and the Middle East. The official US view in 1986 was that there were 448, down from 520 two years before, made up of 28 Mod 1 and 420 Mod 2 and 3 versions at eight fields. Other western sources give a total of 470 for all mods.

Sometime after March 1985, the Soviets began dismantling SS-11 silos to compensate for SS-25 (**4779.111**) deployment. This process is expected to continue and by the mid-1990s, all SS-11s will probably be de-activated. There are recent suggestions that the dismantlement process has temporarily been halted.

The silos have not been part of the massive Soviet hardening/refurbishment programme underway since 1972.

SS-11 Sego ICBM in its launch canister which is loaded into the silo with the missile to aid reload and preparation for second launch

---

2958.111

SS-13 (SAVAGE) ICBM (Soviet designation: RS-12/SS-12)

First seen in public in 1965, the year of initial deployment, this missile, known to NATO as Savage, is a three-stage solid-propellant ICBM comparable with the US Minuteman. It was a product of the little known ICBM design bureau headed by Aleksandr D. Nadiradze.

The three stages of the missile shown here are separated by truss structures and each has four nozzles. The upper stages are believed to be identical with the two stages of the SS-14 Scapegoat missile, which was abandoned before deployment.

Although the missile has been in service for some time it has never been deployed statically on a large scale. A total of 60 are based in fields at the Yoshkar Ola base. So far as is known it has never been equipped with other than a single warhead. The Mod 2 version was deployed in 1972. The Soviets claim that the SS-25 (**4779.111**) is a treaty compliant modification of the SS-13; however US experts calculate that the throw-weights of the two missiles differ by considerably more than the 5% allowed for modernised missiles. A problem here, however, is the ambiguity in the agreed definition of 'throw weight' in the SALT 2 Treaty. The Soviets have insisted that no infringement is involved.

SS-13 Savage three-stage solid-propellant ICBM

CHARACTERISTICS

Type: Intercontinental ballistic missile, hot-launch from silo

Guidance principle: Presumed inertial

Propulsion: 3-stage, solid-propellant motors

Warhead: Nuclear, estimated 1 × 750 KT capability

Missile length: Overall 20 m

Missile body diameters: 1st stage 1.7 m. 2nd stage 1.4 m. 3rd stage 1 m. Nose cone and re-entry vehicle 1 m

Launch weight: 33 000 kg

Range: 9400 km

CEP: 1800 m

STATUS

Operational with about 60 deployed in the neighbourhood of Yoshkar Ola, between Gorki and Kazan. The Soviets reportedly fired two SS-13s in 1986 (the first launches in two decades) so the US could judge Soviet claims that the SS-25 is a modernised version of this missile.

6190.111

SS-16 (SINNER) ICBM (Soviet designation: RS-14)

The Soviet Union flight-tested this three-stage, solid-propellant, single warhead ICBM from the Plesetsk Missile Test Centre during 1972-76. It is another product of the Nadiradze design team at Bisk. NATO sources estimate that it was capable of operational deployment in 1977. US analysis indicates that the missile is closely related to the SS-20 IRBM (**3219.111**), with the SS-20 comprising the first two stages of the SS-16.

During negotiation of the SALT II Treaty, the US side was concerned that separating SS-16 and SS-20 deployments could be difficult because of similar characteristics. The Soviets agreed not to produce, test or deploy ICBMs of the SS-16 type and not to manufacture the SS-16s third stage re-entry vehicle or "other appropriate device" (euphemism for post-boost vehicle) for targeting the single re-entry vehicle.

In the period 1976-78, after SS-16 flight tests had been ended, modification and expansion of mobile missile facilities at Plesetsk were staged by the Soviet Union, according to the US Arms Control and Disarmament Agency. The US Government contends that the SS-16 was "probably" deployed at Plesetsk until 1985 and that recent activity at the test centre indicates that the missiles have been removed. Some 50 SS-16s were allegedly deployed in silos, it is understood. At these sites, "there is some evidence that equipment associated with a different ICBM has been introduced".

The US Government "found the (SS-16) activities at Plesetsk to be a probable violation of the USSR's legal obligation and political commitment under SALT II". The 1987 INF Treaty provides for the clarification of the SS-16 deployment status, and this position should therefore soon be resolved.

Pentagon estimates put the SS-16s range at 9000 km. It is (or was) cold-launched. An undetermined number were deployed.

STATUS

Allegedly deployed 1978-85 at Plesetsk. Now dismantled.

FOURTH GENERATION INTERCONTINENTAL BALLISTIC SYSTEMS

The most important characterising feature of this new generation of intercontinental range ballistic missiles was the introduction of an independent targeting capability for their multiple re-entry vehicles, making them genuine 'MIRVed' (Multiple, Independently Targetable Re-entry Vehicle) systems. This capability implies the carrying of guidance computers on the re-entry vehicle platform - the post boost vehicle (PBV or 'bus'). All three missiles comprising this generation, the SS-17, 18 and 19 have also been tested in versions with a single, large, re-entry vehicle which might be intended to compensate for the loss of capability introduced by the use of a post boost vehicle and multiple re-entry vehicles, thus ensuring the maintenance of a hard target capability for at least one of the missile variants. The SS-17 and SS-19 are broadly similar in specification, lending weight to the US suggestion that they were developed in competition.

It is also thought that this generation of missile utilises a different guidance procedure, an on-board computer determining the deviation from the pre-programmed course and directing corrections to that course or plotting a new course depending on the attendant circumstances.

Under the SALT Treaties the SS-17 and 19 are classed as 'light' missiles and the SS-18 as 'heavy' - the only Soviet heavy missile permitted under the treaty.

The SALT Treaties permit a maximum of 820 land-based ICBMs, of which no more than 308 may be of the SS-18 'heavy' type. The full allowance of silos, (820), have been refurbished and hardened for the three new systems since 1972.

2991.111
SS-17 (SPANKER) STRATEGIC ICBM (Soviet designation: RS-16)
Categorised by official US sources as a 'fourth generation' ICBM, the SS-17 two-stage storable liquid propellant missile is one of two Soviet missiles regarded, (with the SS-19), as successors to the widely deployed SS-11, which had long range capabilities but lacked accuracy (**4207.111**).

Both the SS-17 and the SS-19 have improved accuracy, increased throw weight and greater survivability than their predecessor. The SS-17 is slightly longer than the SS-11 and has a larger volume. The cold launch technique is used for ejection from the silo, using a gas generator before the main, first stage, engine is ignited, whereas the SS-19 uses the hot launch procedure, in which the first stage itself propels the missile from the silo. The US DoD states that it is Soviet practise for all liquid propelled ICBMs now deployed to be contained in a canister, within the silo, prior to launch. These canisters serve several purposes, principally those of environmental stabilisation during transportation and protection during silo loading.

CHARACTERISTICS
Type: Intercontinental ballistic missile
Guidance: Inertial. PBV, on-board computer-controlled
Propulsion: 2-stage storable liquid propellant motors, plus PBV. Cold launched
Dimensions: Length appears to be about 24 m and base diameter about 2.5 m but little reliance can be placed on these figures
Weight: Well below the limit permitted for this class of missile (see note on this point below, under the SS-19); estimated to be about 65 000 kg
Warhead: Nuclear; yield not known, but probably not less than 200 KT per warhead for the 4 warhead Mods 1 and 3 versions; the Mod 2 has a single 3.6 MT warhead
Range: 10 000 km (Mods 1 and 3); 11 000 km (Mod 2)

STATUS
Operational. Tests are said to have been observed for the first time during the second half of 1972 and there was intensive testing during 1973 and early 1974. Deployment in converted and modernised SS-11 silos began in 1975. By 1980, 150 SS-17s were deployed in these silos and this was considered to be the total number intended for deployment. This 1982 estimate apparently still holds good, with the additional belief that the deployments at Kostroma and Yedrovo (in the NW of the USSR, some 150 km from Moscow) are all of the Mod 3 version. 150 SS-17s were deployed in 1987.

2992.111
SS-18 (SATAN) STRATEGIC ICBM (Soviet designation: RS-20)
Largest of the 'fourth generation' Soviet intercontinental ballistic missiles and the only 'heavy' missile permitted for the Soviets under the SALT 2 Treaty.

The SS-18 is a large, two-stage, liquid propelled missile and is the functional successor to the SS-9 (**2962.111** in 1980-81 and earlier editions).

By 1983 four different versions of the SS-18 had been identified, each with a different payload/range ratio, as follows:

SS-18	Mod 1	Mod 2	Mod 3	Mod 4
Max range (km)	10 000	11 000	16 000	11 000
Warheads	1	8/10	1	10
Yield	25 MT	500 KT MIRVed	20 MT	500 KT MIRVed
IOC	1974	1976	?	1982

The US describes the Mod 4 as capable of carrying '10 +' RVs, implying that it is capable of exceeding the treaty limit, for land-based missiles, of ten warheads.

Some authorities have reported the commencement of flight testing of a follow-on version of the SS-18. Currently carrying the test range designator TT-09, this missile appears capable of carrying up to 10 RVs and probably is intended to have accuracy and range capability improvements over the earlier versions of the SS-18. Early flight tests of this missile did not prove successful. The first launch, in April 1986, exploded soon after clearing the silo and another launch in September 1986 ended when an explosion occurred during separation of the first and second stages.

The US authorities are sure that increased accuracy has been a major consideration in the development programme of this new modification, presumably with the intention of improving the hard target capability, necessary to counter the silo hardness improvement programme undertaken by the US for the Minuteman and Peacekeeper missiles. Some of the more recent tests, moreover, have been carried out using a single RV, from which it is deduced that the USSR may have a continuing interest in a missile with one very powerful warhead. These single warhead tests may have been part of a comparative test programme designed to establish the best way of countering the improved hardness of modern silos, given that single warhead missiles have a greater accuracy and hard target capability than MIRVed systems.

SS-18 is transported in a canister that is also used to load the missile into its ICBM site as shown in this US sketch

Like other fourth generation ICBMs, the SS-18 is deployed in a launch canister within the silo to provide protection to the missile during transportation and silo loading.

CHARACTERISTICS
Type: Intercontinental ballistic missile
Guidance: Inertial. Computer-controlled post boost vehicle
Propulsion: 2-stage liquid propellant, plus PBV. Cold launched
Dimensions: Length appears to be about 35 m and base diameter about 3 m
Weight: 200 000 kg
Range: 12 000 km (Mod 1); 11 000 km (Mods 2 and 4); 16 000 km (Mod 3)
CEP: Estimated at 250 m (Mod 4)

STATUS
Production estimated at 465, largely of the Mod 4 variant.

Deployed operationally in 1974 in former SS-9 silos and launch complexes converted and improved to accommodate the SS-18s. By 1982 this task was thought to have been completed and the number of SS-18s deployed was then estimated to be 308, the present level of deployment for this system. All are now Mod 4.

In 1983 it was reported that the SS-18 silos had been hardened to enable them to withstand an overpressure of 4200 kg/cm² (6000 psi) presumably to counter improvements in the accuracy of the Minuteman system.

The geographic distribution of the SS-18s within the Soviet Union suggests that they are deployed roughly along the route of the trans-Siberian railway, with missile fields at Dombarovskiy, Imeni Gastello, Kartaly, Aleysk, Zhangiz Tobe and Uzhur.

The Pentagon maintains that the SS-18 Mod 4 was "specifically designed to attack and destroy ICBM silos and other hard targets" in the USA. The Mod 4 is deployed in sufficient numbers to destroy 60-80% of US silos, using two warheads against each silo, with some 1000 warheads remaining available for other targets. At the time of this deployment, however, the US was considering major improvements in the hardness of its silos and these large numbers of accurate RVs may reflect a Soviet concern to maintain their capability against the US force.

2993.111
SS-19 (STILLETO) STRATEGIC ICBM (Soviet designation: RS-18)
The SS-19 is a two-stage, liquid-propellant missile with a computer-controlled post boost vehicle and MIRV warhead system.

Together with the SS-17, the SS-19 appears to be a successor to the SS-11 missile. Although the reasons for the dual development are not clear, there are distinct differences between these two missiles which may suggest a lack of confidence in the two designs, necessitating the development of both. The SS-19 is, for example, hot launched from its silo, whilst the SS-17, like the SS-18, utilises an additional stage to 'cold' launch from the silo.

Although little direct information exists on specification of the SS-19, a great deal can be deduced from its status, under the SALT 2 Treaty, as the heaviest permitted missile in the 'light' category, the category that also contains the Peacekeeper and SS-24 missiles. The constraining characteristics of this category were essentially decreed by the sizing limits on silo modernisation agreed as part of the SALT 2 Treaty. The analogy with Peacekeeper missile would give the SS-19 a throw weight of some 3600 kg, more than three times that of the SS-11 and a launch weight of some 90 000 kg against an estimated 50 000 kg for the SS-11.

The SS-19 is clearly larger than the SS-17 and is also provided with a MIRV capability, reportedly being able to dispense four or six RVs. There is also a single RV version, Mod 2, which appeared after the first MIRV version, the Mod 1. As with the other single RV variants, this may be intended to be a guaranteed hard target capable version, as it combines high accuracy with high yield. The Mod 3 variant, with a payload of six, 500 KT warheads, is said to combine the advantages of the two earlier variants and was reported to have entered service in 1980.

The latest of the 'fourth generation' ICBMs now operational, the SS-19 is also deployed in canisters within the silos, in the same way as the SS-17 and SS-18.

CHARACTERISTICS

Type: Silo launched intercontinental ballistic missile

Guidance: Inertial. Computer-controlled post boost vehicle

Propulsion: 2-stage liquid booster plus PBV

Dimensions: Length about 27 m; base diameter about 2.5 m

Launch weight: 90 000 kg

Throw weight: 3600 kg

Warhead: Nuclear. MIRV system of 4 or 6 warheads. Yield not known but the MIRV versions are probably in the region of 200 KT and the single RV version several megatons

CEP: 300 m

STATUS

Deployed operationally between 1974 and 1982. The first SS-19 test detected took place in April 1973. With the SS-17, this missile is replacing the SS-11 ICBM, the silos being improved at the same time as they are converted to receive the later type of missile. In 1986 it was estimated that there were 360 SS-19 Mod 3 missiles deployed in the major Derazhnya, Kozelsk, Pervomaysk and Tatishchevo missile fields.

3219.111

SS-20 (SABER) MOBILE IRBM (Soviet designation: RSD-10 *Pioner*)

The SS-20 road mobile, solid propellant replacement for the SS-4 and SS-5 intermediate range theatre systems entered development at the A.D. Nadiradze OKB in Biysk in the mid-1960s. It is believed to comprise the first two stages of the intercontinental SS-16, which was also a road mobile system, banned under the SALT 2 agreement. The SS-20 may, therefore, have been more of an opportunistic system than one with a specific role. SS-4, SS-5 and SS-20 deployment between 1978 and 1983 revealed a steady decrease in quantities of the first two types matched by comparable increase in SS-20 deployments.

Whereas quite good photographs of its predecessors have been available for some time, photographic evidence about the SS-20 has until recently been extremely limited and generally of poor quality. Open sources have been mostly forced to rely upon fragmentary details culled from official (principally US) assessments when attempting to estimate the characteristics of this weapon. However, the December 1987 INF Treaty negotiations provided the first real details from an authoritative Soviet source. The SS-20 is a two-stage solid-propellant mobile IRBM 16.49 m long with a multiple warhead capability and a maximum range of 5000 km. Inertial guidance is employed and a figure of 400 m CEP for the Mod 2 version is usually accepted, although inevitably this must rest to a significant extent on the accuracy with which the initial launch site has been surveyed and with range. Some reports claim that there are a significant number of pre-surveyed launch sites in the USSR, but that this is not absolutely essential for operation of the SS-20.

SS-20 Saber (RSD-10 Pioner) on its transporter/erector/launcher vehicle, a photograph released by the Soviet authorities in December 1987

A left-side view of SS-20 on its TEL vehicle

There are 48 regiments in service, mainly of the three-battalion type. Six regiments, at Gezgaly, Petrikov, Vysokaya Pech, Zasimovichi and Zhitkovichi, are the smaller two-battalion type. It would appear that at peak strength, the RVSN could have deployed up to 56 Pioner regiments. A number of Pioner regiments were disbanded prior to signing of the treaty. A division in the Urals was converted to SS-25 Sickle regiments, and their equipment was sent to a holding facility at Sarny in the Ukraine. A Pioner regiment presumably consists of two to three launcher battalions plus command and support units. The regiment has an attached KGB OSNAZ motorised security battalion equipped with BTR-70/80 armoured transporters. A KGB communications alert company is integral to the unit, providing war alert information and weapon use authorisations. Each Pioner battalion consists of a headquarters, a command group, three firing platoons and a technical support platoon.

US sources report SS-20 complexes of up to nine Mod 2 missiles, each with three nuclear warheads and with reloads for each launcher. Artist's impression shows typical deployment at unprepared sites

Pioner regiments, in turn, are believed to be organised into missile divisions. The divisions usually consist of five launcher regiments plus a missile technical regiment, but two of the divisions in the Far East have only four regiments. At present, there appear to be ten divisions, six in the European USSR and four in Asia. At peak strength, there were 11, one having been disbanded in the Urals in 1985. The missile technical regiment consists of a regimental HQ, missile preparation and maintenance battalion, a missile transport battalion and a KGB warhead security battalion. These regiments operate out of the facilities listed as storage bases in the INF Treaty. The missile transport battalion operates ten heavy missile transporters which closely resemble the Pioner launch vehicle. During peacetime, these bases do not stock a large inventory of missiles. Most seem to have only one to two missiles and one to four training missiles. The main peacetime storage facility is at Kolosovo, where 144 missiles are stored, and Zherebkovo, where a further 20 missiles are stored.

The RSD-10 Pioner Transporter/Erector/Launchers (TELs), which are probably derived from the MAZ-7310 tractor, are based in special launch pad shelters which have opening roofs to permit the launch of the missile from within the shelter. According to data released with the treaty, there is only a single such structure per TEL at the operating bases. It would appear that Soviet operational practice is not to launch exclusively from such shelters, but rather to use pre-surveyed entrenchments located in the operating area around the base itself. As a result, the treaty details the operating areas around the base. The regimental operating areas appear to vary considerably in size due to locale. In the case of the division near Drovyanaya, the operating areas vary in size from about 25 to 50 km in length, compressed along the Ingoda River valley. They are supported by a single technical support regiment with ten transporter vehicles.

The SS-20 is said to be capable of refiring after launching the first missile, but considerable heat is involved in each launch so that it may not be possible to prepare for a second launch in less than a few hours, according to some authorities.

The payload of the SS-20 is another feature that has now been clarified. There may have been an early version with a single RV, but as the 3 RV MIRVed version entered service in the same year, 1977, and the US do not designate this version as Mod 2, it must be assumed that the single RV version was not deployed. The Soviets have stated that all SS-20 carry 3 MIRVed RV, with 150 KT nuclear warheads. The US has reported the flight testing of a genuine Mod 2 version, expected to be more accurate than its predecessor. The INF Treaty to remove this category of missile must call into question its original envisaged role; alternatively, its role could be taken over by one of the fifth generation intercontinental missiles.

STATUS

By the end of 1980 it was estimated there were 180 operational SS-20 launchers in use. From January 1981 onward, the pace of SS-20 base construction increased, and in May 1983 official US estimates claimed there were 351 operational SS-20s in the Soviet Union of which 243 were based in the European portion of the USSR west of the Ural Mountains. The remainder were to the east of this range and were presumed to be targeted on China, Japan and South Korea. The total of 243 facing NATO remained the same until early 1984, apparently to substantiate the Soviet moratorium on deployments that lasted until the breakdown of the Geneva arms limitation talks. However, the overall total of SS-20 deployments according to US figures in February 1984 was 378 launchers (with some additional reload missiles). Other sources quoted even higher figures.

Towards mid-1985 SS-20 deployments were quoted as having reached a total of 414 missiles, of which 243 were in the western part of the USSR.

By early 1986, the total had reached 441, with more than 171 based within range of Japan and East Asia including the missile regiment remaining in Central Asia. The Soviet Union declared 405 deployed SS-20 missiles at the signing of the INF Treaty in December 1987, with a further 245 non-deployed missiles. A total of 29 separate locations, some with as many as five operating bases within them, were notified.

Some SS-20 bases like Yurya, are being converted to accommodate the mobile version of the SS-25 ICBM (**4779.111**).

4778.111

SS-24 (SCALPEL) ICBM (Soviet designation: SS-23)

The US code SS-24 has been assigned to a new fifth generation ICBM being developed by the USSR, possibly as an eventual successor to the existing SS-18 ICBM (**2992.111**). Few details beyond those revealed (or leaked) from US Intelligence assessments, are available to open sources. The Soviet Union has indicated that the SS-24 is their one new missile permitted under the SALT 2 agreement. It is in the 'light' category, which could give it a size close to that of the US Peacekeeper missile. The missile has an estimated length of between 21 and 22 m and a payload of 8-10 RVs with yields of 300-500 KT. Range is estimated by various sources as between 8500 and 10 000 km. CEP is reported to be better than 200 m. It has been presumed by official US observers that the SS-24 uses solid propulsion and three stages are assumed. The 'cold' launch technique is used.

Official DoD reports have stated that the SS-24 is probably more accurate than either the SS-18 Mod 4 (11 000 km range version) or the SS-19 Mod 3 (**2993.111**), and it is assumed that both this and the other new ICBM being developed in parallel, the SS-25 (**4779.111**) will have a greater hard target kill probability than their predecessors. If this is so, they can be expected to be able to redress the balance of advantage against the effects of American silo hardening programmes.

Both mobile (transportable) and silo-based versions of the SS-24 are expected to appear.

Mobile versions of the SS-24 are seen as parallel Soviet developments to the US ones during the 1970s in giving their missiles survivability through mobility, the early US mobility experiments covering air, sea and ground launches of the Minuteman missile. All mobile systems will suffer some degradation in accuracy and, presumably, that work led to the preferred options being road or rail borne systems – the latter now again being experimented with by the US, with the Peacekeeper system. The rail-mobile version of the SS-24 is carried on a railway vehicle resembling a freight wagon for movement over the nation's rail network. Flight-testing of this version has continued. Construction of a series of protective tunnels (as MX Peacekeeper) is reported to have been started near Kozelsk and Perm. Another mobile model carried on a tracked chassis was said to be in development.

CHARACTERISTICS

Type: Rail-mobile and silo launched intercontinental MIRV ballistic missile

Propulsion: Multi-stage solid motors plus PBV

Launch weight: 90 000 kg

Throw weight: About 3600 kg

Payload: Up to 10 re-entry vehicles

Range: Greater than 10 000 km

STATUS

The first flight test reported in the West took place in October 1982, and under direction of the Plesetsk Test Centre, testing continued thereafter at a rapid pace so that the 10th test in September 1983 (a failure) narrowly preceded a flight of November 1983 when eight RVs were launched, apparently successfully.

US artist's impression of rail-mobile SS-24

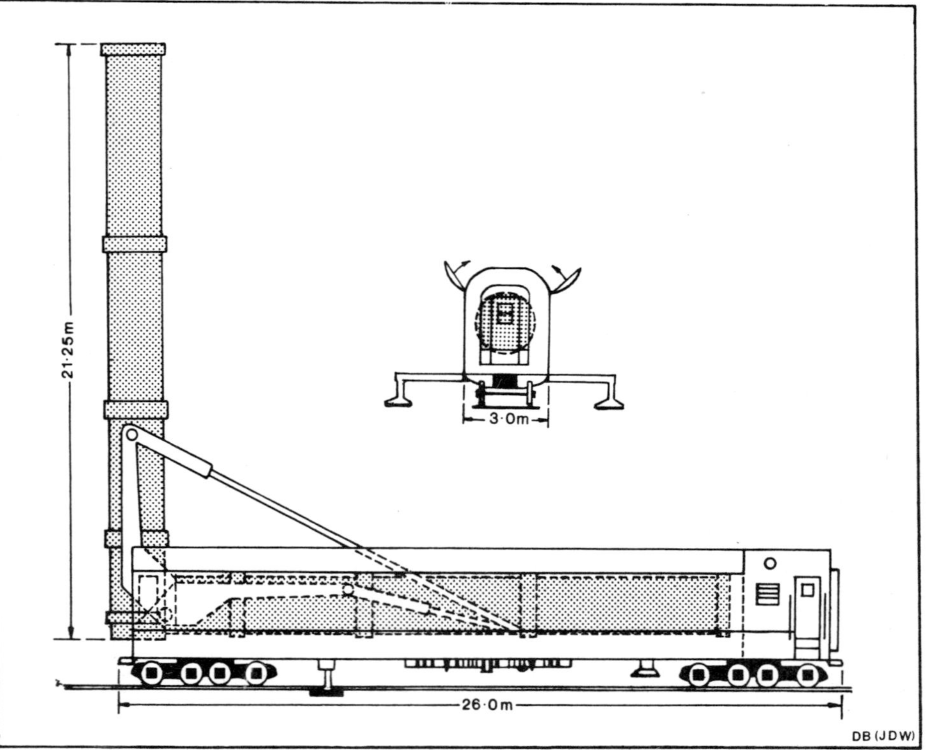

Diagram of rail-mobile SS-24 operation (Jane's Defence Weekly)

The rail-mobile version was deployed in 1987, with silo-based deployment, possibly in existing SS-11 bases, expected to follow in the next two or three years.

4779.111

SS-25 (SICKLE) ICBM (Soviet designation: SS-12M)
The designation SS-25 has been assigned to what is believed to be a new, 'fifth generation' ICBM now being deployed as a possible successor to the existing SS-11 ICBM. US intelligence estimates, upon which most of what has been revealed to date is based, suggest that the SS-25 is of about the same size as the US Minuteman ICBM, with a length of between 18 and 19 m, a diameter of 1.70 m, a weight of 35 000 kg and a range of 10 000 km.

Solid propulsion is used and the missile has three stages. A single re-entry vehicle is thought to be carried with an estimated yield of 550 KT. Like its contemporary, the other new Soviet ICBM SS-24, the SS-25 will be capable of either silo deployment or operation in a mobile form and road mobile in this case. US official drawings of SS-25 deployment (see illustration), which are presumably based on satellite intelligence photographs, reveal a noticeable similarity with the now widely deployed SS-20 mobile IRBM (**3219.111**), a reminder of the widespread Soviet practice of building on experience.

Certainly deployment has been spotted at two SS-20 sites, at Yurya and Verkhnyaya Salda. Here the launcher garages, fitted with sliding roofs for fast-reaction launches, are longer than those sheltering the SS-20 IRBMs and ground-support equipment is different. Another base is in the SS-13 field at Yoshkar Ola. Twenty-eight bases had been identified by early 1987, some in groups of nine on former SS-7 Saddler sites.

The US Arms Control and Disarmament Agency, moreover, alleges that former SS-7 Saddler ICBM sites (destroyed under the SALT I Interim Agreement in the 1970s) are being used "in support of the deployment and operation of the SS-25 mobile" missile.

There has been considerable contention between the US and the Soviet Union regarding the legality of the SS-25 on two technical counts. Both signatories are permitted one new missile under the SALT 2 Treaty; the US claims the SS-25 to be a new missile and hence an infringement, since the SS-24 has already been declared and accepted, as the permitted one new Soviet missile. The Soviet Union claims that the SS-25 is a development of the SS-13, which would make it legitimate under the SALT 2 rules. The second issue relates to the throw weight of this missile, which has also led to considerable debate between the US and the Soviet Union. As a development of the SS-13 the SS-25 would be constrained to a throw weight within 5% of that of its progenitor, which the Soviets claim to be the case.

Gen Yuri Maksimov, Commander-in-Chief of the Soviet Strategic Rocket Forces, in the November 1986 issue of *Soviet Military Review* referred to the SS-25 as the SS-12M, a variant of the SS-12 (called SS-13 in the West). He said the USA omitted elements of the old "SS-12" and included "equipment used only for testing of the modernised SS-12M missile" in the throw weight. This response of Gen Maksimov indicates that, at the source of these problems, is the ambiguity of the treaty terminology, the agreed definitions clearly appearing not to cater for the changes in design that arise between missile generations.

US DoD opinion believes that in any case the SS-25 will possess a greater HTKP than predecessor systems, principally by virtue of better accuracy. Any assessments of this aspect of the SS-25 presumably must have been derived from monitoring telemetry data obtained during test flights. US reports have stated that some of these tests were notable for the fact that telemetry was sent in encrypted form, and the legality of this in respect of the SALT II terms has also been questioned.

CHARACTERISTICS

Type: Single RV, intercontinental ballistic missile. Cold launch

Launch of a road-mobile SS-25 ICBM according to a Pentagon artist. The US says that deployment of this missile violates SALT restrictions

An official US artist's impression of the new SS-25 ICBM system in operation

Basing: From base, using sliding roof garages and road mobile, using TELs from pre-surveyed sites
By Analogy with the SS-13:
Propulsion: 3-stage solid propellant
Launch weight: 33 000 kg
Throw weight: 1000 kg
Length: 18.5 m. 1st Stage diameter 1.7 m
Range: 10 500 km

STATUS

Test flights of the SS-25 that have been reported in the West include: a failure on 27 October 1982; another from Plesetsk to the Kamchatka Peninsula (6436 km) on 8 February 1983; in the period February-May 1983 there was one failure, but one successful flight on 10 May 1983.

The US has announced that 100-plus mobile SS-25 launchers had become operational by early 1987, indicating a fast rate of deployment as only 18 had entered service by Autumn 1985, replacing 20 SS-11s.

Twenty missiles are said by Japanese sources to be based at Spassk Dal'niy, close to the Chinese border. Further deployments were expected on Sakhalin Island.

One US source has stated a belief that a MIRVed version of the SS-25 could be introduced before the end of the decade. Any such introduction would negate the Soviet claim of an SS-13 genealogy for the missile – new developments of a missile must have the same front end design under the treaty conditions. Such a development hence seems unlikely.

4894.111

SOVIET ICBM DEVELOPMENTS

Both official and unofficial/leaked US sources claim to have evidence of development work on two new Soviet ICBMs, additional to those specifically stated to be already operational such as the SS-24 and SS-25. This entry summarises what has been gathered from these various sources regarding those Soviet ICBM programmes that appear to have adequate reference material to justify assigning provisional identities to them.

The Soviet missile development programme has always been characterised by continuity of development, with variants of existing missiles appearing even as new generations are being flight tested. These developments have always been within the strict rules of the SALT Treaties, at least as they seem to be perceived by the Soviets; thus there seems every likelihood that further developments will exploit all the freedoms that the existing and future treaties allow. On this basis, follow on versions of the new, fifth generation, missiles, the SS-24 and 25 can be expected, along with further modifications, as noted above, of fourth generation missiles, such as the SS-18. Given the vulnerability of the fixed based SS-18 missile these latter developments are somewhat surprising, particularly as the SS-24 not only reportedly has the same payload and accuracy as the SS-18, but has the added survivability that mobile basing provides. In view of the impending missile and throw weight reductions likely under the START Treaty, future developments are likely to concentrate on improved yield, accuracy and reliability, which would enable hard target kill probabilities to be improved, with consequent economies in RV requirements against US silos and command and control facilities.

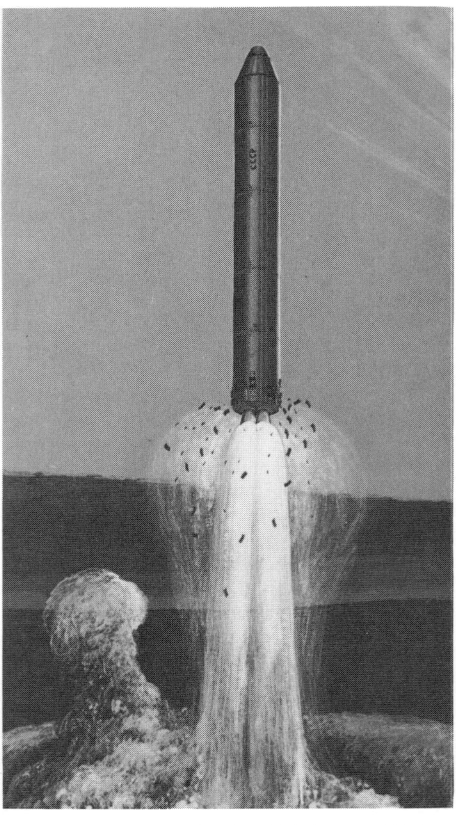

Artist's impression of the SS-18 follow-on ICBM

4819.001

SOVIET CRUISE MISSILE PROGRAMME

The Soviet Union has a long experience of cruise missile operation, admittedly mostly within a naval context, but signs have been detected that a programme to extend the areas of application of this category of weapon in Soviet service is now well advanced. Recent US analysis claims to have evidence of new air-, sea- and ground-launched cruise missiles of several types under development with the air-launched version having reached initial operational capability in 1984.

The first category of cruise missile is believed to be very similar in design concept and physical parameters to the American Tomahawk (**3993.001**, **4197.221** and **4194.111**), and like that weapon, the Soviet cruise missile has appeared in a sea-launched version SS-NX-21 (**4796.411**), as an air-to-surface missile AS-15 (**3206.311**) and as a counterpart to the USAF GLCM in the SSC-X-4 ground-launched version.

Insufficient data had been disclosed to make possible more than a general assessment of this class of Soviet cruise missiles, but the data released by the Soviet authorities at the INF Treaty talks has provided some further information. The SSC-X-4 is 8.09 m long, has a body diameter of 510 mm and a launch weight of 1700 kg. A single nuclear warhead is carried to a range of 3000 km. An air-breathing turbojet is the likely power plant, but the precise form of guidance is more open to doubt at this stage. It can be assumed that the AS-15 and SS-NX-21 are similar.

The Soviets have converted a 'Yankee' class ballistic missile submarine into a test platform for the SS-NX-24 cruise missile

The AS-15, the air-launched, low-altitude, subsonic, cruise missile, is currently deployed on the Bear H aircraft and is expected to be carried on the new swing-wing Blackjack bomber which could be operational in 1988. The sea-launched variant, the SS-NX-21, is small enough to be fired from standard Soviet torpedo tubes. Possible launch platforms include three 'Victor' classes of nuclear-powered attack submarines; a reconfigured 'Yankee' SSN and the new 'Akula', and 'Sierra' classes of boats. The Pentagon believes this cruise missile could become "operational soon" and "could be deployed in submarines off US and allied coasts.

The ground-launched variant, the SSC-X-4, was due to be deployed in 1987 for use in theatre operations in Eurasia. However, with the December 1987 INF Treaty specifically covering US deployed GLCM, it is assumed that the SSC-X-4 will not now be deployed.

In addition to the Soviet 'Tomahawk-equivalent', the USSR is pursuing the development of a much larger class of cruise missile, possibly 13 m long. One, the SS-NX-24, will be flight tested from a specially

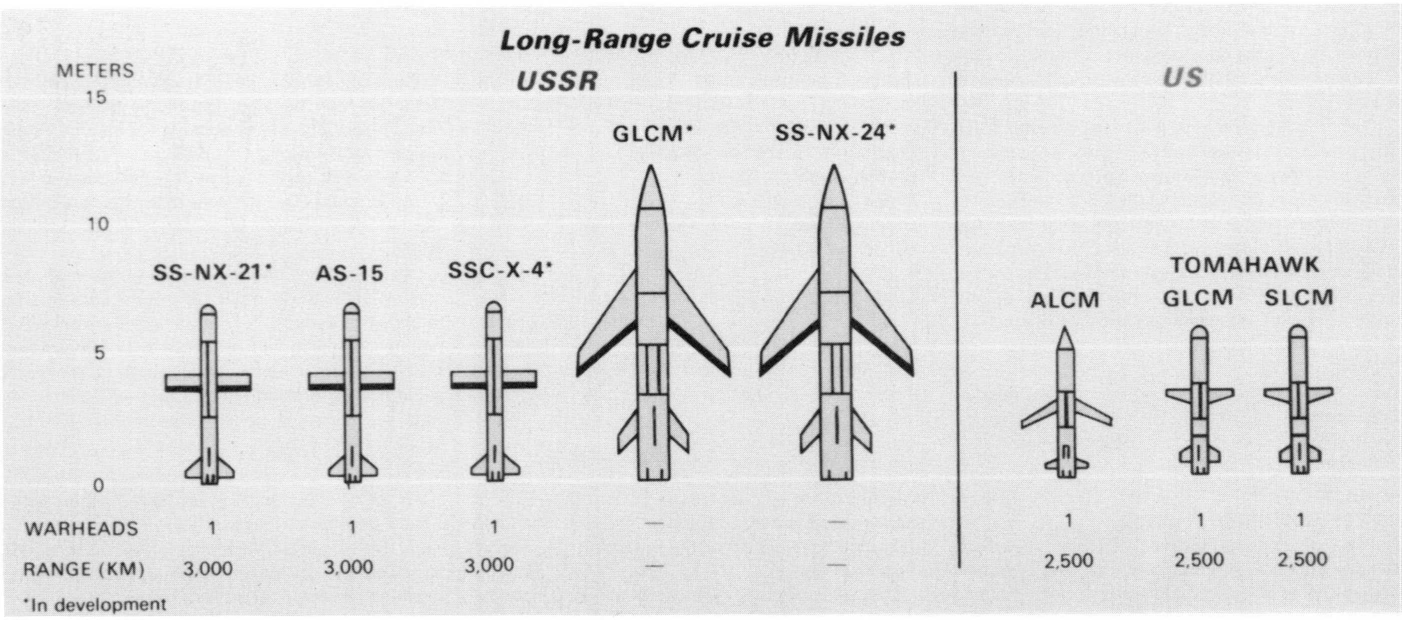

Comparison of the new generation of Soviet cruise missiles with their US counterparts

converted 'Yankee' class submarine and may become operational by 1988. A ground-launched version of this missile may be developed, designated GLCM in the chart above.

All these cruise missiles will probably be equipped with nuclear warheads when first deployed and will be capable of attacking hardened targets.

The decision to provide alternative conventional payloads will depend on the type of guidance system employed, and the capability to attack hard targets that its accuracy provides.

Bear H aircraft, launch platform for the AS-15 cruise missile. Training flights have been intercepted by US and Canadian aircraft near the north-west of the American continent. More than 20 Bear H aircraft are operational with the AS-15

SOVIET SUBMARINE LAUNCHED BALLISTIC MISSILES

The Soviet Union has never enjoyed the easy access to the world's oceans that the US has nor has it had, until recently, the communications and overseas ports that it would need to confidently commit its primary strategic nuclear force into submarine basing. This situation has led it to concentrate its forces in land-based silos whilst building up a force of very long range submarine missiles, which could be deployed, suitably protected, in its home waters and under the Arctic ice cap. The Soviets have now achieved this and they are also now fast approaching the US capability in SLBM throw weight and RV numbers. They can be expected to approach US missile accuracies with their next generations too and it will be interesting to see if they then follow the US and commit more of their missile strength to submarines.

One US direction that the Soviets have never followed with any confidence is the use of solid propellants for their submarine based missiles. They have, with the apparently unsuccessful SS-N-17 and with the SS-N-20, ventured into solid propellants for SLBMs, but apparently with mixed confidence because they have also continued to develop liquids propelled missiles, with the SS-N-23. Western Navies have always feared that liquid propellants, particularly in the restricted confines of a submarine, represented too great a hazard to compensate for their better performance over solids, but the evidence is that the Soviets have not had great problems, the only known liquid propellant related incident being the loss of an old boat in 1986.

1153.411
SS-N-5 (SERB) SUBMARINE LAUNCHED BALLISTIC MISSILE

Serb is the code-name assigned by NATO to the successor to the Sark (**1152.411** in 1980-81 and earlier editions) Soviet submarine-launched ballistic missile. Serb has been described as representing the second generation of this class of missiles in the USSR.

Like its forerunner, Serb is similar in appearance to the Polaris A-2 missile once in service with the USN, and it is understood to be a two-stage vehicle with inertial guidance. In size it conforms more closely with the dimensions of Polaris than does Sark, which is appreciably larger. According to official US information, the SS-N-5 is approximately 12.9 m in

'Golf II' class SSB. A total of 39 are deployed in this class, just under half in the Baltic sea. Three missile tubes are in the submarine fin

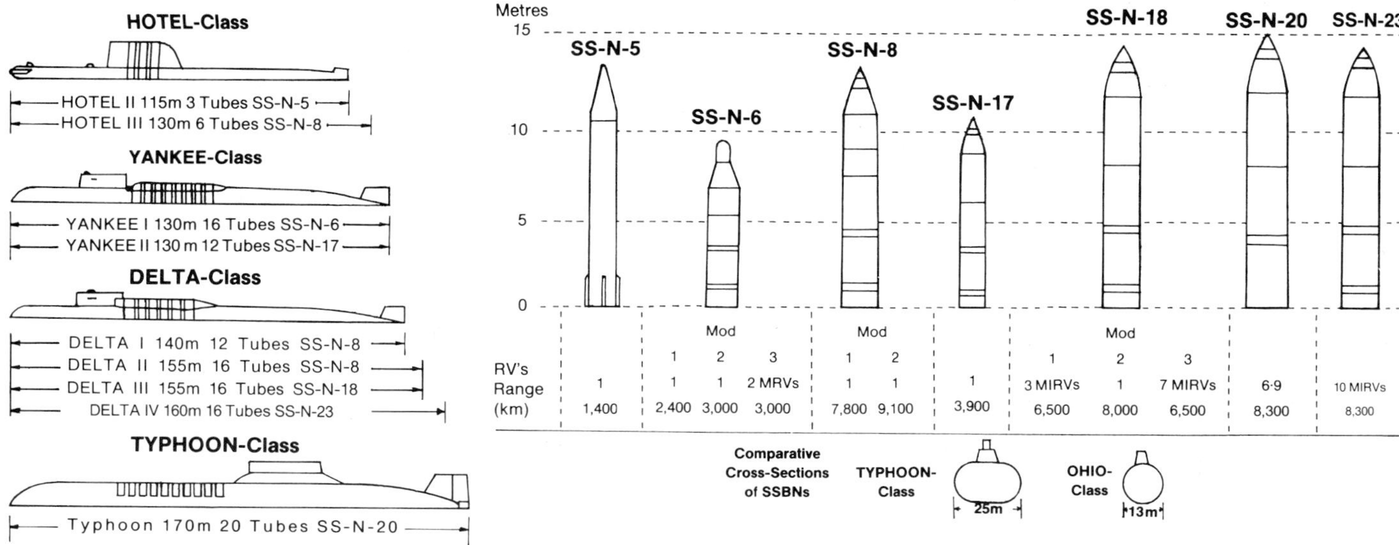

Soviet SLBM capability: Diagram showing composition of Soviet strategic submarine fleet and capabilities of SLBMs they carry. At least four 'Typhoon' class submarines with SS-N-20 missiles are at sea, providing a total of 80 missiles. Production of 'Delta III' class ships armed with SS-N-18 third generation SLBMs is believed to be nearing its end with 14 launched. Each has 16 tubes. 'Delta II' boats have 16 SS-N-8 tubes, while earlier 'Delta I' class boats have 12 tubes. In 1987, 292 SS-N-8 missiles were at sea. One 'Yankee II' class boat is fitted for SS-N-17 (12 tubes) tests, while 17 'Yankee I' class boats are deployed with SS-N-6 Mods 1 or 3. There is one 'Hotel III' trial boat for the SS-N-8, also a 'Golf III' carries the SS-N-8 in six launch tubes. Thirteen 'Golf II's are armed with 39 SS-N-5s; just under half of them based in the Baltic. One 'Hotel II' is also believed operational with the SS-N-5

length and has a maximum diameter of 1.42 m. Launch weight has been estimated at around 18 000 kg. While there is fairly close agreement among the various authorities on these dimensions, there is no agreement on the missile's range capability. Estimates vary from about 1200 km to about 2400 km but the lower figure is more likely. A single re-entry vehicle with a one megaton warhead is carried. CEP is estimated at 2800 m.

When the Serb missile was displayed in the November 1967 parade, what is apparently a different ejection system to that employed in American submarine launcher tubes was revealed. At the base of the missile there is a cluster of 18 small, electrically fired cold-gas nozzles which are used to eject the missile from the sealed launcher tube. On passing clear of the submarine the section carrying these gas motors is jettisoned from the main missile body by explosive bolts fired when the first stage ignites.

STATUS

Operational since 1964 in the later 'Golf' class (conventional) and later 'Hotel' class (nuclear) submarines. Latest US estimate is that 39 remain operational in a theatre role with 'Golf II', half of them deployed in the Baltic for an in-theatre role, the remainder in the Far East. One 'Hotel II' may also be armed with the SS-N-5, now regarded as obsolescent. As a submarine launched ballistic missile, the SS-N-5 is not affected by the terms of the 1987 INF Treaty.

1154.411

SS-N-6 (SAWFLY) SUBMARINE LAUNCHED BALLISTIC MISSILE (Soviet designation: R-21)

All three versions of the SS-N-6 Sawfly submarine-launched ballistic missile are liquid-fuelled, and, based on US information, are 9.65 m long and 1.65 m in diameter.

The SS-N-6 was a product of the design team headed by Viktor P. Makayev, beginning in the mid 1950s.

The Mod 1 is a single 1 MT warhead version with a reputed range of 2400 km. In October 1972 flight tests began of a modified version which became known as the Mod 2 and this was given an increased range of about 3000 km, by virtue of an improved propulsion system. According to the American authorities the SS-N-6 Mod 2 with a 1 MT warhead is now capable of reaching any target in the USA from the 100 fathom contour off the coasts of the USA.

The SS-N-6 Mod 3 is the latest Sawfly development and has a range of some 3000 km, carrying a payload of two 500 KT MRVs. These are not independently targeted.

An American DoD assessment claims that the combination of accuracy and yield of all three SS-N-6 versions is sufficient only for strikes against soft targets. CEP is estimated at 1300 m.

STATUS

The SS-N-6 entered service in 1968; the Mod 2 began operational service in 1973, followed by the Mod 3 in 1974. Current deployments are all on 'Y' class nuclear submarines. It is likely that Mod 1 and 2 missiles have been replaced progressively by Mod 3 types.

Production of 'Y' class submarines ended in 1975, when there were 34 boats of this class. Since 1978, to comply with the terms of SALT I, the Soviet Union withdrew 15 'Yankee I' boats from SLBM duties to allow the overall Soviet ballistic missile submarine fleet to remain within the limit of 62 modern SSBN/950 SLBM limits established in 1972. Some of the laid-up 'Yankees' are thought to have been refitted as attack or cruise missile submarines. A US estimate of early 1987 stated that probably 272 SS-N-6 missiles remained in operational service. Around 1000 missiles were produced. The December 1987 INF Treaty only covers land-based intermediate nuclear forces and the SS-N-6 remains unaffected.

In October 1986, a 'Yankee' SSBN was scuttled in the North Atlantic following an explosion in one of the launch tubes presumably of volatile missile fuel. Her missiles went down with the boat.

'Yankee' class nuclear-powered SLBM submarine armed with SS-N-6 missiles

SS-N-6 Sawfly SLBM on parade in Moscow

3078.411

SS-N-8 SUBMARINE LAUNCHED BALLISTIC MISSILE

For some time past the SS-N-8 was thought to be a possible development of the SS-N-6 (Sawfly), but the US Defense authorities have claimed evidence to justify regarding this weapon to be sufficiently different to warrant another designation.

According to official US data it is about the same length as the SS-N-5 at 12.95 m, but of increased diameter, approximately 1.65 m. It is the same diameter as SS-N-6 so that its greater length gives it an appreciably larger volume. It employs liquid-fuelled propulsion and can carry a single warhead to a range of almost 8000 km. The guidance system is said to incorporate stellar-inertial techniques, capable of CEP values of 0.4 km.

At one stage the possibility of three variants of the

'Delta II' class submarines are armed with 16 SS-N-8 SLBMs compared with the 12 carried by 'Delta I' class boats

SS-N-8 was reported but later information is that two versions entered service: Mod 1 with a single 1 MT RV and a maximum range of 7800 km; and Mod 2 with 2 × 800 KT RVs and an increased range of 9100 km.

STATUS

The SS-N-8 was operational by 1973 and is now deployed in 'Delta' class Soviet submarines. In mid-1983 it was estimated that the Soviet Union had a total of 22 operational 'Delta' class submarines, with the 18 12-tube 'Delta I's and the 4 16-tube 'Delta II's both armed with the SS-N-8 but an official US total of 292 SS-N-8s in service published in 1987 creates a discrepancy in numbers. 'Delta III' class boats are armed with the later SS-N-18 (**3357.411**) and 'Delta IV' class boats with the SS-N-23 SLBMs (**4795.411**).

3356.411

SS-N-17 (SNIPE) SUBMARINE LAUNCHED BALLISTIC MISSILE

The SS-N-17 is one of two second-generation Soviet SLBMs (the other being the SS-N-18, **3357.411**, below) to appear in the mid-1970s and is the USSR's first SLBM to use a solid propellant and to be equipped with a post-boost vehicle (PBV). To date only a single re-entry vehicle payload has been reported but this feature could have implied an original intention to either produce a MIRV modification or to use the PBV to improve the single warhead accuracy.

It began testing in 1976 and is believed to have been engaged in periodic sea trials since 1977 in a converted 'Yankee II' class submarine with 12 missile tubes.

According to American official information, the SS-N-17 is about 11.06 metres in length and 1.65 metres in diameter. Estimated range is 3900 km and inertial guidance is assumed. A CEP of 1400 m has been suggested.

The role of this weapon in the Soviet SLBM programme is far from clear from the available evidence to date. Evidence in the open sources is ambiguous, although the dropping, in official US sources, of the 'X' in the designator, indicates that there is evidence that the missile is now considered deployed.

A single submarine of an ageing class seems an unlikely basis for a new strategic weapon system, and the seeming lack of urgency and/or progress in the programme could be indicative of an essentially R & D oriented effort that may have failed to live up to expectations so far.

Like the SS-N-5 and SS-N-6 missiles the intermediate range SS-N-17 is unaffected by the terms of the 1987 INF Treaty, which only applies to land-based systems.

3357.411

SS-N-18 (STINGRAY) SUBMARINE LAUNCHED BALLISTIC MISSILE (Soviet designation: RSM-50)

The two-stage SS-N-18 SLBM began flight tests shortly after the SS-N-17 (**3356.411**) in 1975. It is similar in some respects to the SS-N-8 (**3078.411**) but somewhat larger, and is believed to be equipped with a more sophisticated guidance system. Liquid propellant is used and US estimates of maximum range give a figure of up to 8000 km.

Like the SS-N-17, it is equipped with a large post-boost vehicle (PBV) and US sources claim to have identified three models:

Mod 1 has a payload of three 200 KT MIRVs and a range of 6500 km (IOC 1977); Mod 2 has a single 450 KT RV and 8000 km range (IOC 1978) and Mod 3 carries seven 200 KT MIRVs with a maximum range of 6500 km. CEPs of Mods 2 and 3 is estimated at 900 m.

Official US information suggests a missile length of about 14.1 m and a diameter of approximately 1.8 m.

'Delta III' class nuclear-powered submarines are armed with 16 tubes each for SS-N-18s, most numerous of the Soviet Union's third generation of SLBMs

STATUS

Before November 1976, the SS-N-18 was tested from land-based launch sites in the USSR, but in early November the first submarine launch took place in the White Sea. There was another test in September 1978 when the SS-N-18 reached an impact area in the Pacific Ocean. In early 1979 the US DoD revealed deployment of the SS-N-18 aboard 'Delta III' class submarines. This had been confirmed by 1983 when it was estimated that there were 14 'Delta III' class submarines, with 16 tubes each, armed with the SS-N-18 SLBM. There was a total of 224 missiles operational in 1987.

4208.411

SS-N-20 (STURGEON) SLBM (Soviet designation: RSM-52)

This submarine launched ballistic missile was first mentioned in the 1981-82 edition of *Jane's Weapon Systems*, when the US designation included an 'X' for experimental or under development, ie SS-NX-20. In 1984 the US authorities deleted the 'X' and stated that with the entry into service of the first of the giant 'Typhoon' class of submarines, the SS-N-20 SLBM system that these ships carry is now considered operational.

Soviet 'Typhoon' class submarine is armed with 20 SS-N-20 SLBMs

According to US Intelligence estimates, the solid-propellant SS-N-20 is marginally larger than the SS-N-18 (**3357.411**), but probably carries a payload that offers more re-entry vehicles (six to nine 100 KT MIRVs compared with the seven of the SS-N-18 Mod 3) while having a greater range than the SS-N-18 Mod 2's 8000 km, at an estimated maximum of 8300 km.

In the 'Typhoon' class submarines, which displace between 25 000 – 30 000 tons, 20 SS-N-20 missiles are stowed in two rows of launch tubes located forward of the boat's sail.

Missile length is estimated at 15 m and diameter at 2.2 m. CEP is believed to be 500 m.

STATUS

Flight tests of the SS-N-20 were first detected in January 1980, reportedly with mostly unsuccessful results. Two more successful tests were reported in 1981, and in October 1982 a simultaneous submerged launch of four missiles took place. The first 'Typhoon' class submarine was launched in September 1980, and the second and third ships of the class by 1984. Four are now operational, providing a total of 80 missiles, with possibly a fifth on sea trials. US authorities believe there may be two or three more of the class now under construction so that by the early 1990s there might be eight of these powerful boats operational.

In 1987 US authorities expected flight testing of a modified SS-N-20 to begin shortly.

A massive explosion at a missile fuel plant at Biysk, 80 km SE of Novosibirsk in early 1986 "seriously" reduced production of SS-N-20s according to Western reports.

4795.411

SS-N-23 (SKIFF) SLBM

This new liquid-propelled submarine launched ballistic missile has been flight tested on board two units of the newest Soviet SSBN, the 'Delta IV'. A fourth 'Delta IV' has been launched, two more are building.

The missile is said to be 14 m long, carry 10 MIRVed warheads, and have a range of 8300 km, similar to the SS-N-20. The number of stages is not known. CEP is estimated at 900 m.

US sources say the SS-N-23 will have greater throw-weight and be more accurate than the SS-N-18 currently deployed in 'Delta III's. There are reports that all 14 of the 'Delta-III's will be modified to deploy the SS-N-23.

A modified version may be developed before the end of the decade and could provide a hard-target capability for Soviet SLBMs.

STATUS

Flight testing is thought to have started in 1983. Forty-eight missiles are at sea. It has been suggested that the SS-N-23 entered operational service in 1986.

'Delta IV' at sea; two units are on sea trials and a third has been launched

4796.411

SS-NX-21 CRUISE MISSILE

The Soviet authorities are believed by the US DoD to be developing a new intermediate range cruise missile similar to the American Tomahawk system (**4197.221**) that can, like the Tomahawk, be launched from submerged submarines. The new Soviet weapon has been given the designation SS-NX-21, the letter 'X' denoting the fact that the system is still at the development stage.

Little is known about the appearance or detailed performance characteristics of the SS-NX-21, but it has been stated that this cruise missile can be launched from a standard torpedo tube so that a diameter of about 53 cm can be assumed. Folding flying surfaces are other likely features, and probably an air-breathing turbojet power plant, possibly coupled with a solid propellant rocket booster motor for broaching the surface of the sea and accelerating the missile to flying speed. These features are illustrated in the artist's impression issued by the DoD (see photograph).

A maximum range of about 3000 km has been suggested, and a single nuclear warhead will probably be carried. Guidance techniques that could be employed include the use of programmed inertial guidance to the target area, based on target designation co-ordinates supplied by the parent submarine, followed by one or other of the various terminal homing options available, according to the nature of the target (eg radar homing, anti-radiation etc).

STATUS

Development programme from AS-15 air launch cruise missile. It has not been confirmed but flight tests are thought to have begun, and US authorities believe the SS-NX-21 could enter operational service "soon", possibly in one or more classes of Soviet nuclear attack submarines if American expectations are realised. These launch platform possibilities include particularly the newer SSNs such as the 'Akula' and 'Sierra' classes as well as existing types such as the 'Victor III' class.

DoD artist's impression of a Soviet SS-NX-21 submarine launched cruise missile attack

New 'Akula' class nuclear attack submarines are seen as possible carriers of SS-NX-21 cruise missiles that could be used for land attack operations

6191.411

SS-NX-24 CRUISE MISSILE

The Soviets have converted a 'Yankee' class submarine into a test platform for a large (13 m long) long-range cruise missile, designated SS-NX-24.

US sources claim the missile has been flight-tested and could become operational by 1988.

SOVIET ANTI-BALLISTIC MISSILE DEFENCES

The 1972 ABM Treaty, modified by the 1974 Protocol, allowed for a single system to be deployed in each of the two signatory countries, the US and the USSR, for the defence of either a missile field or the national capital. The US system, deployed to defend the silo field at Grand Forks, North Dakota, was decommissioned in 1976 and the only system now deployed is the Soviet system, around Moscow. This has, over recent years, been modernised and expanded, in accordance with treaty provisions and, once complete, will take up most of the permitted treaty freedoms.

2932.131
ABM-1B (GALOSH) ANTI-BALLISTIC MISSILE MISSILE (Soviet designation: UR-96)

Galosh is the NATO reporting name of the ballistic missile defence system deployed around Moscow by the Soviet authorities, based on the use of a missile thought to resemble the US Spartan anti-missile missile, (**2811.131** in 1977 and earlier editions), designated ABM-1B under the US system of codes and UR-96 in Soviet nomenclature. In the Moscow ABM complex Galosh is used in conjunction with various radars for target detection, engagement and tracking purposes, as described in the Land-Based Air Defence Systems section of this book (**3214.181**).

Apart from photographs of Galosh in its transport container, which revealed four first stage rocket motor nozzles in a 1974 Moscow parade, the world at large has seen little of this liquid-fuelled, multi-staged missile. However, this enabled the length of the container to be estimated at about 20 m from which some idea of the Galosh dimensions can be gleaned.

The original single layer ABM system for Moscow was initially completed with 64 reloadable above-ground launcher units for the ABM-1B located at four complexes to the north and west of the city, six Try-Add guidance and tracking radars to each complex, and the Dog House and Cat House battle management radars south of Moscow.

At the present time the system is in the process of upgrading and expansion, in the interim there are only 16 Galosh deployed.

When the modernisation and upgrade programme is operational, the new system will consist of a two-layer arrangement comprising: silo-housed modified Galosh missiles designed for exoatmospheric missile interception; hardened silo-based high-acceleration missiles for endoatmospheric interception (see new entry below); associated tracking and guidance radars and a large new radar ('Pill Box') at Pushkino designed to control ABM engagements. ABM-1B silo complexes are at Klin, Novo-Petrovskoye, Verena, Kaliningrad and Aleksandrov. Construction work on new silos is underway or is completed in an inner ring of Schodna, Mervskino and Lytkarino.

By early 1987, there were 98 launchers operational, including those for the new endoatmospheric interceptor. The 1972 ABM Treaty permitted a total of 100 launchers and this figure was expected to be attained during 1987.

US sources maintain that the capability of the launchers is being enhanced by provision of a reload/refire facility. Above-ground test launches of the modified Galosh ABM missile at the Sary Shagan Missile Test Range have demonstrated a reload/refire time of "much less than a day". It is believed that each silo has the capability of one refire.

A network of six new large phased array early warning radars is being built to supplement and upgrade the capability of the Henhouse chain of radars around the borders of the USSR. Five of these meet the treaty requirements of being 'on the periphery (within 500 km, of it, by common consent) and oriented outward'. The sixth, at Krasnoyarsk, appears to be designed to provide early warning of Trident launches from the Pacific and, as it is some 4000 km, from the boundary it faces, would be a treaty infringement if this was its purpose. However, the treaty requirements relating to the functions of these radars are ambiguous and the Soviets dispute the infringement charge.

The US claims that a rapidly deployable ABM system to protect important target areas in the USSR is under development and it is claimed that the production of radars and launchers that would be needed for widespread ABM defences beyond the 100 launcher limit has been initiated. It has been estimated that within ten years the Soviet Union would be in a position to deploy such a system at sites that could be built in a matter of months rather than years.

A typical site might consist of engagement radars, guidance radars, above-ground launchers and the high acceleration interceptor. The peripheral radars together with the Hen House, Dog House, Cat House, and perhaps the Pushkino radar, appear to be designed to fit into a broader ABM system of this type. The overall impression gained by the US authorities from these activities is that the Soviet Union may be preparing a national ABM defence for the whole country.

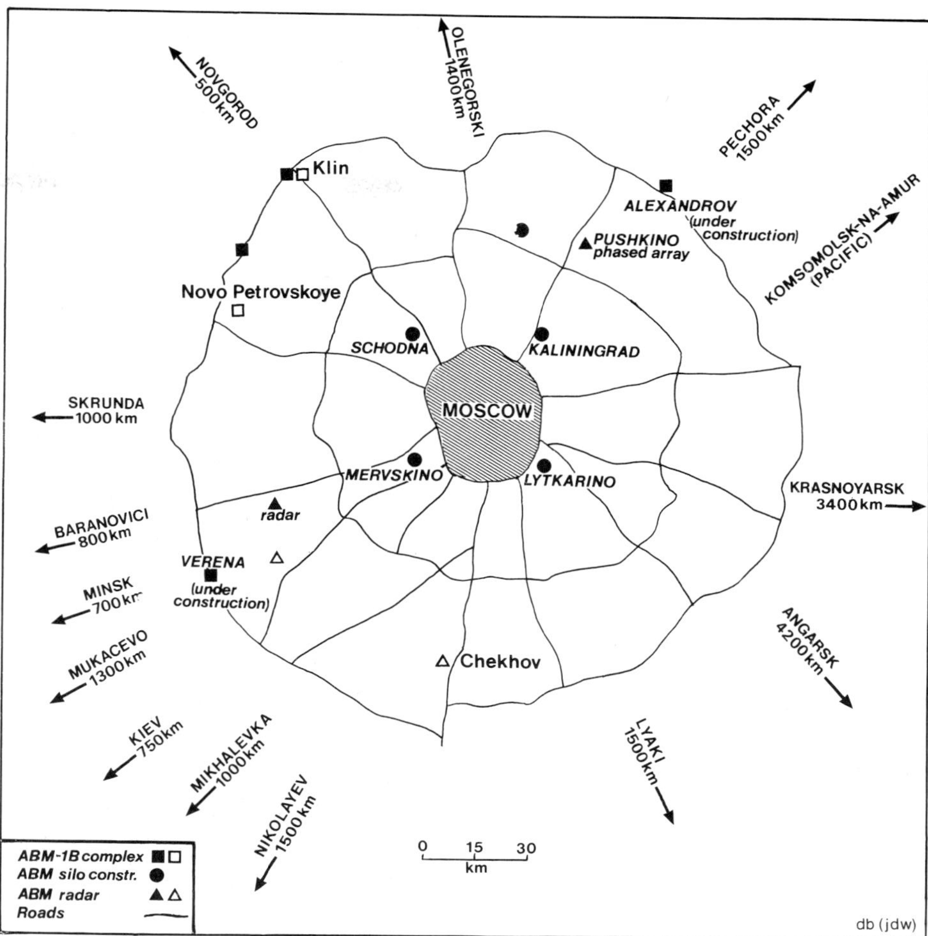

Moscow ABM system map

DoD artist's impression of an ABM-1B Galosh surface launcher launcher of the Moscow ABM system in operation. Compare missile configuration and container with photograph of Galosh in Moscow parade (below)

Additionally, the USSR is deploying the SA-10 Grumble surface-to-air missile system (**3620.131**), and flight testing the mobile SA-X-12B (Giant) (**4774.131**) which is both a tactical SAM and anti-tactical ballistic missile weapon that may have the potential to engage NATO missiles such as Lance, Pershing I and II. It is thought possible that the SA-10 and SA-X-12 might also be able to intercept certain earlier generation strategic ballistic missiles and cruise missiles under some conditions.

The SA-10 Grumble can defend against low altitude targets with small radar cross-sections like cruise missiles. The first site was operational in 1980; now more than 60 are active and work is progressing on at least another 30. Around 50% are located near Moscow suggesting priority for last ditch defence of C[3] installations and key military-industrial complexes.

CHARACTERISTICS

ABM IB Galosh

Type: Surface-to-air anti-ballistic missile

Guidance: Radar command

Propulsion: 3-stage liquid-propellant, with a loiter capability for the third stage

Warhead: >1 MT

Length: 19.8 m

Diameter: 2.57 m

Weight: 33 000 kg

Range: 350 km

STATUS

More than 150 Galosh have been produced, with a modified version operational at four sites around Moscow.

Galosh anti-missile (Novosti)

6192.131

SH-08 (GAZELLE) ANTI-BALLISTIC MISSILE MISSILE

New quick-reaction silo-based high acceleration endoatmospheric ABM interceptor equipped with low yield nuclear (single figure kiloton range) warhead. British sources refer to it as HAV; high acceleration vehicle. Eventually, more than 32 launchers will be operational. This missile, similar to the Sprint developed for the inactive US Safeguard system, was test-launched at Sary Shagan, demonstrating, like Galosh, a reload capability of much less than a day.

CHARACTERISTICS

Range: >80 km

US artist's impression of the SH-08 Gazelle endoatmospheric ABM

4895.131

SOVIET ADVANCED DEFENCE PROJECTS

The Soviets have always devoted a substantial proportion of their military efforts to defence and they set about complementing the very extensive air defence systems with space elements as soon as ballistic missiles became part of the threat. In the early years these were largely passive in nature, comprising launch detection satellites, but they later introduced an anti-satellite weapon based on co-orbiting attack satellites. Western analysts estimate that the proportion of Soviet space launches related to military application has been over 80% since 1984 and it is believed that a good many of these are connected with further experimentation in the use of space for the development of defensive technologies and of other advanced techniques to counter possible Western advances in the military uses of space.

Since the late 1960s, the Soviet Union also has been pursuing a substantial advanced defensive technologies programme - a programme that has been exploring many of the same technologies of interest to the United States in the SDI programme (**4896.131**). In addition to covering a wide range of advanced technologies, including various laser and neutral particle beams, the Soviet programme is much larger than the US effort in terms of the amount of resources invested - plant, capital and manpower.

The high-energy laser programme in the USSR, which dates from the mid-1960s is reported to be much larger than that of the USA and more than half a dozen major R & D facilities and ranges have been built, employing some 10 300 scientists and engineers,

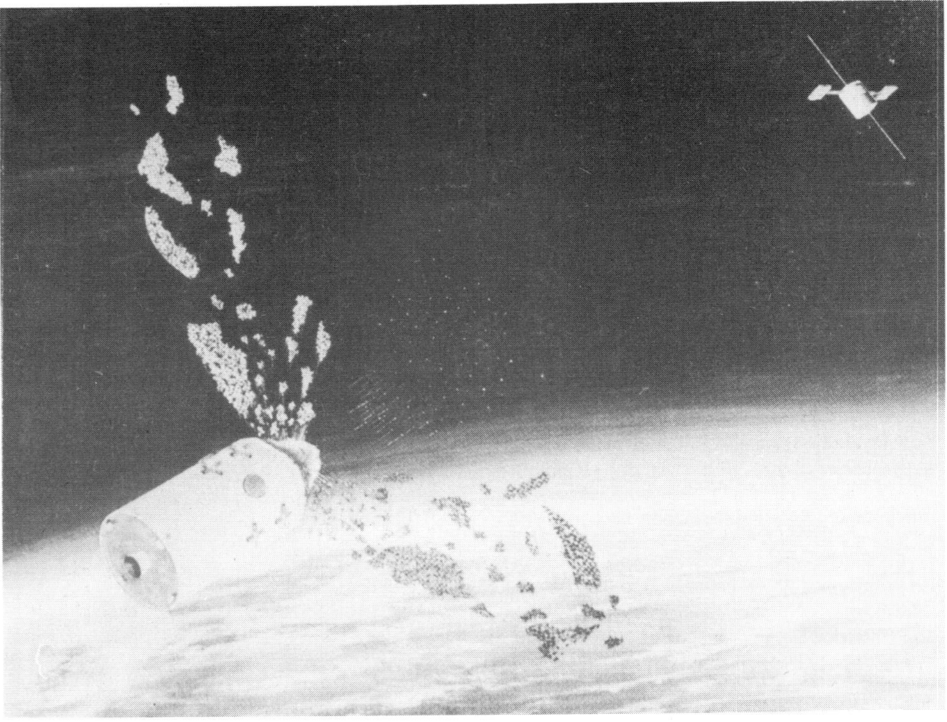

An artist's impression of the satellite interceptor vehicle that the USSR has deployed since the early 1970s; the interceptor destroys the target with a conventional HE warhead dispensing a cloud of pellets

for laser development. A laser weapons programme of the size and scale of the Soviet Union's would cost $1000 million a year in the USA, it is estimated.

These efforts could lead to the testing of space-based ABM systems in the mid-1990s and to their deployment after the turn of the century. The President's initiative, therefore, in adding a major space defence element to the US ABM programme, can be seen in part as a prudent response to Soviet R & D activities in this field and as an insurance against unilateral Soviet efforts to develop and deploy an advanced defensive system.

The US Space Command has noted that Soviet anti-satellite lasers are now capable of disrupting US satellites in geosynchonous orbit and could pose a more serious threat in the next four to five years. Twin ground-based lasers, based at Sary Shagan, are said to have the capability of killing US systems in orbits below 400 km and damaging those up to 1200 km range. The Soviet Union already has two ground-based lasers that have "a limited capability" to attack US satellites. By the late 1980s, believe US authorities, the Soviet Union could have prototypes for ground-based lasers for BMD. Testing of the components for a large-scale deployment system could begin in the early 1990s. The many difficulties in fielding an operational system will require much development time, and initial operational deployment is not likely in this century. However, with high priority and some significant risk of failure, some test stages could be omitted to allow a ground-based laser BMD to be deployed by the early to mid-1990s.

Ground- and space-based particle beam weapons for the same purpose are expected by the US authorities to be more difficult to achieve than laser systems, but it is claimed that the Soviet Union has a 'vigorous' programme under way for particle beam development and could have a prototype space-based system intended to disrupt satellite electronics ready for testing in the late 1990s. A weapon designed to destroy satellites could follow later. Soviet efforts in particle beams - particularly in sources and radio frequency quadrupole accelerators for particle beams - are regarded as "very impressive" by the Americans.

Work is also in progress on technologies that relate to either specific weapon projects, or that might lead to more advanced ASAT systems. These include space-based kinetic energy, ground- and space-based laser, particle beam and radio frequency weapons. The Soviet Union is thought to believe that these techniques offer greater promise for future ASAT application than continued development of ground-based orbital interceptors equipped with conventional warheads. It is also thought that military applications of directed energy technologies hold promise of overcoming weaknesses in the country's conventional air and missile defence.

The USSR has begun development of at least three types of high-energy laser weapons for air defence. This includes research into the gas dynamic laser, the electric discharge laser and the chemical laser. Other types include the excimer and free electron laser, according to British sources. These include lasers intended for defence of high-value strategic targets in the USSR, for point defence of ships at sea, and for air defence of theatre forces. An airborne laser has also been tested which could have ASAT or air defence applications. If past philosophies are pursued, air defence lasers are more likely to be deployed to complement rather than replace existing SAM and interceptor aircraft assets.

The strategic defence laser is probably in at least the prototype stage of development and could be operational by the late 1980s. It is thought most likely to be deployed in conjunction with SAMs in a point defence role. Since the SAM and laser systems would have somewhat different attributes and shortcomings, they would provide mutual support. The shipboard lasers probably will not be operational until the mid-1990s. The theatre force lasers may be operational somewhat earlier and are likely to be capable of structurally damaging aircraft at close ranges and producing electro-optical and eye damage at greater distances.

In arms control discussions in Geneva, the Soviet Union steadfastly denies it has its own co-ordinated SDI programme, although General Secretary M. Gorbachev in a December 1987 TV interview acknowledged that the Soviet Union is pursuing a similar research programme.

Conceptual illustration of a Soviet ground-based laser capable of interfering with some US military satellites (US DoD)

American drawing of the Tyuratam site where the USSR is said to have the capability of launching a series of ASAT weapons

Drawing by a US artist, assumed to be working from satellite intelligence data, of the Soviet directed-energy site at the Sary Shagan proving ground. This and other lasers at the same site are thought capable of use in an ASAT role and possibly a BMD role

The Soviet Union is also developing an airborne laser, and assuming success in development, a limited initial deployment could begin in the early 1990s. If sufficiently accurate, such a laser could have ASAT applications, but its more likely use would be in the protection of high-value airborne assets, and cruise missile defences.

Related weapons technology work in progress is stated to include the means of producing electric power efficiently and high-quality optical component manufacture. A rocket-driven magnetohydrodynamic (MHD) generator that can produce 15 MW of short-term electric power has been developed, and this is a device which it is claimed does not have a counterpart in the West.

A programme to conduct research into the use of strong radio frequency (RF) signals that have the potential to interfere with or destroy components of missiles, satellites, and re-entry vehicles has also been carried out for some time. It is estimated that by the 1990s the USSR could test a ground-based RF weapon capable of damaging satellites.

Soviet programmes for the development and application of directed energy technologies to strategic defence have been pursued very vigorously in the past, according to US sources, and will continue in the future.

In the area of kinetic energy weapons, using the high speed collision of low mass with the target as the kill mechanism, the USSR has a variety of research programmes in progress, which could result in a near-term, short-range, space-based weapon system useful for satellite or space station defence or for close-in attack by a manoeuvring satellite. Longer range, space-based systems probably could not be developed until the mid-1990s or later in the US view.

Soviet Space Programme

Soviet doctrine by no means overlooks the military potential of space and specific areas of military operations where space systems and assets are employed embrace such activities as ASAT operations and systems; intelligence collection; command, control and communications; meteorological support services; navigational facilities; reconnaissance and targeting; and survey and geodesy. Such military support systems are linked to ground, naval and air forces via Earth terminals, and reliance on them is reported to be growing.

Anti-satellite Weapons

Since 1971 the USSR has had the ability to attack satellites in near-Earth orbit with a ground-based orbital interceptor. Using a radar sensor and a pellet-type warhead - the so-called "space shotgun" - the interceptor can attack a target in various orbits during the interceptor's first two orbits. (See **1506.351** *Jane's Weapon Systems, 1971-72.*) An intercept during the first orbit would minimise the time available for the target to take evasive action. The interceptor can reach targets orbiting at altitudes of more than 5000 km, but it is thought likely to be intended for high-priority satellites in lower orbits. The ASAT interceptor is launched from Tyuratam, where launching facilities include a number of launch pads, storage accommodation for interceptors and launch vehicles. Several interceptors per day could be despatched using the facilities that are said to exist at this site. In addition to the orbital interceptor, the Soviet Union has two ground-based, high energy lasers with ASAT capabilities. It is also believed that there is the technological capability to develop electronic warfare against space systems and/or to use existing ABM interceptors in a direct-ascent attack on low-orbit satellites. Significantly, the 1987 INF Treaty approves the development of direct ascent ASATs such as that in development in the US. This could indicate that the Soviets may also be interested in following this route.

Space Reconnaissance Systems

Soviet Military Power, 1987 contains a useful summary of other Soviet space systems dedicated to military missions. These include satellites that perform reconnaissance, missile launch warning detection and attack warning, command and control, and ASAT operations, as noted above. Dual-purpose satellites that carry out some civil functions are used for communications, navigational support, and weather forecasting and monitoring.

The Soviet Union possesses a missile launch detection system based on a satellite-borne infra-red sensor in a highly elliptic orbit, covering US Minuteman ICBM fields. In its present form, it is not capable of detecting submarine launches. An experimental geostationary satellite aimed at improving the performance of the current system has been launched and there is a strong research programme to improve capability in this area. The USA has no equivalent counterparts (as yet) to the satellites used by the Soviet Union for ocean surveillance, the electronic ocean reconnaissance satellite (EORSAT), or the nuclear-powered radar ocean reconnaissance satellite (RORSAT). Their function is to detect, locate, and target US and allied naval forces for Soviet anti-ship weapons. These systems are stated to track naval and merchant shipping, and four such satellites were launched in 1984, two of which were of the same type (RORSAT) that crashed in 1978 spreading debris (with some associated radioactivity) in northern Canada.

Recently the USSR has started using a new radar-equipped satellite system which is designed for mapping ice formations in the polar regions. These satellites are expected to appreciably enhance the Soviet Fleet's ability to operate in icebound waters. It has been noted that Soviet launches of spacecraft into geostationary orbits have increased in recent years. Most are assumed to be for communications purposes but they may not all be for this role, and some doubt remains as to the overall plan for this class of spacecraft. A Soviet equivalent to the American Navstar navigation satellite system for providing accurate position-finding facilities on a world-wide basis is reported to be in the early stages of development under the name of GLONASS. Other developments include improved routine reconnaissance satellites which are probably designed for longer duration sorties than earlier satellites which generally spent much less time in orbit than their US counterparts.

For further details of the Soviet military space programme, see *Jane's Spaceflight Directory.*

UNITED KINGDOM

2390.411

UK POLARIS MISSILES

The UK Polaris force of four submarines, each armed with 16 Polaris A-3 SLBMs have constituted the British nuclear deterrent since 1969. It is planned to continue this role until replaced by the UK Trident II system in the mid-1990s. Originally the standard US UGM-27C Polaris A-3 missile equipped with UK designed and built nuclear payload was used, but to maintain the system's effectiveness a programme to upgrade the payload was instituted under the code-name Chevaline. Missiles equipped with this system are designated Polaris A-3TK and all four RN Polaris submarines are now equipped with this version.

Following retirement of Polaris from US Navy service, the Royal Navy remains the sole operator of this system.

CHARACTERISTICS

US designation: UGM-27C
RN designation: A-3TK
Type: Submarine-launched ballistic missile (SLBM)
Guidance: Inertial
Propulsion: 2-stage, solid propellant
Warhead: 3 × W-58 200 KT warhead, or Chevaline (see text)
Re-entry vehicle: MRV or Chevaline (see text)
Length: 9.55 m
Diameter: 1.37 m
Launch weight: 13 600 kg
Range: 4630 km (approx)
CEP: 900 m

Each standard missile is capable of a payload comprising three RVs with an estimated 200 KT yield each. Under the UK Polaris improvement programme code-named Chevaline, these are replaced by manoeuvrable re-entry vehicles with new penetration aids. Other changes include fire control system modifications.

A-3TK CHEVALINE

In the late 1960s, as the first British SSBN went on patrol, there was concern about Western intelligence assessments of the future capability of Soviet anti-ballistic missile defences around Moscow (**2932.131**). The USA had already decided to follow the MIRV route to counter this threat by using the Poseidon SLBM (**1133.411**). This system was offered to the UK Government in 1970 but the offer was politely declined for a variety of political and technical reasons. A British Cabinet sub-committee then decided secretly to develop a highly complex system of penetration aids which could defeat ABM defences well into the 1990s - codenamed 'Chevaline'. This took research under an abandoned US project, 'Antelope', much further.

Work on Chevaline was mainly at the British Atomic Weapons Research Establishment at Aldermaston. The problem of squeezing a small space craft into the front end of a missile occupied previously by what is euphemistically termed the "equipment section" - but which merely ejected three warheads - proved far more complex than originally thought.

The cost escalated dramatically from the autumn 1972 estimate of í175 million to í495 million at the same price base at the time of the 1976 Defence Review. Then, Chevaline came under control of MoD Procurement Executive and greater industrial participation was sought. In the end, the project cost í1000 million at 1980 prices.

A Chevaline missile comprises a Penetration Aid Carrier (PAC), the British version of the US warhead 'bus' capable of manoeuvring in space, fitted with an onboard computer and a range of sensors to determine orientation. The PAC dispenses warheads and a "variety of penaids" to blind and confuse Soviet ABM radars. As the defences are unable to distinguish between decoy and live warheads, they must expend significant numbers of ABM interceptors in an attempt to eliminate the threat. The Chevaline payload saturates the defences in a similar way to the deployment of multiple warheads. Each warhead has been hardened against electromagnetic pulse and is coated with an ablative substance to give high temperature protection.

Testing began on September 1977 with a series of development flights from flat pads in the USA. In November 1980, launches down the US missile range off the Florida coast revealed unsuspected separation problems. However, these were overcome and the development phase ended in February 1982 with a number of "highly successful" submerged launches from HMS *Renown*, 30 miles off Florida and out of six

HMS Resolution, *one of the Royal Navy's four Polaris submarines. 16 missiles are carried* (Royal Navy)

more launched from HMS *Revenge* in 1983, only one was aborted for range safety reasons.

The Chevaline development is not a MIRV system and does not involve any increase in the yield of the Polaris missiles. However, Chevaline RVs are said to be capable of separation between impact points of a maximum of 70 km.

Complementary to the Chevaline programme was the decision in 1981 to purchase further rocket motors for the missiles themselves because of shelf-life problems. It was decided to recreate the technology of 20 years earlier and produce carbon-copy replacement motors. Lockheed, prime contractor, brought some of its experts out of retirement for the programme, designated A-3R (for replica). Aerojet was responsible for the first stage and Hercules for the second stage motor.

The remotoring programme was completed in 1986 at a cost of í437 million. The first unarmed firings of the remotored missiles were in August 1986. The fourth and final launch, from *Revenge*, was aborted because of a guidance system fault.

It seems probable that not all of the 16 missiles that each Polaris submarine takes on patrol are fitted with the Chevaline payload. As this is specifically designed to penetrate ABM defences (which officially exist only around the Moscow region) missiles which might be targeted elsewhere would carry a normal Polaris payload.

STATUS

The Royal Navy has four nuclear-powered ballistic missile submarines, each carrying 16 Polaris missiles. They are (with launch year in brackets): HMS *Resolution* (1966), *Repulse* (1967), *Renown* (1967) and *Revenge* (1968). The UK Polaris force became fully operational in 1970.

Chevaline is now operational on three Polaris submarines: *Renown, Revenge* and *Resolution*. The final submarine, *Repulse*, is having the system retrofitted in her third major refit at Rosyth.

A three-year gap between the end of the commission of HMS *Resolution* and the first of the Royal Navy Trident submarines on-station date is expected to be compensated by extending the length of commission for the remaining Polaris boats, with longer periods between refits now possible due to the second-generation reactor cores now available. A decision on a fourth refit for *Resolution* does not have to be taken by the UK Ministry of Defence until 1988.

The eventual replacement for the UK Polaris force will be the UK Trident D-5 SLBM system (**4209.411**), a version of the system entering service with the US Navy (**2840.411**). The status of the UK Polaris face is not affected by the 1987 INF Treaty which explicitly applies only to land-based systems. This is an important point, as the UK although not a signatory to the INF Treaty traditionally abides by US defence treaties.

CONTRACTORS

About 800 British firms were associated with the original UK Polaris programme. Some of the major contractors are listed below.

Submarines:
- Vickers
- Cammel Laird

Nuclear reactors:
- Rolls-Royce

Systems:
- George Wimpey and Co
- Vickers
- GEC-Marconi
- British Aerospace
- Thorn EMI Electronics

4209.411

UK TRIDENT SLBM SYSTEM

A UK Government decision to procure the American Trident SLBM system (**2840.411**) as a replacement for the British nuclear deterrent based on the Polaris SLBM system (**2390.411**) was revealed in July 1980. At that time it was intended to acquire the Trident I (C-4) missile from the USA, fit British-built warheads, and deploy it in submarines of British design and construction. The resulting system would be entirely under British ownership and operational control, but the whole force would be committed to NATO in the same way as the current Polaris force. In the event, shortly after the UK decision was made public, the US administration opted to utilise the potential offered by the larger missile tubes fitted in the new 'Ohio' class of SSBNs (Trident I can be carried in the Poseidon SLBM tube, but the tubes for Trident II missiles are both longer and of greater diameter), and procure the more powerful Trident II (D-5) missile in the 1990s. This time-scale coincided with that envisaged for the British Polaris replacement programme and it appeared sensible to adopt the SLBM planned for US service at the same time.

Where the original UK Trident programme was based on building a new class of British SSBN which would use a missile compartment based on that of the USN 640-class SSBN, adoption of the Trident II made this impossible and the solution adopted is that of employing a centre section based on that used in the USN *Ohio* class SSBN. The British version will have only 16 missile tubes, however, compared with the American boats' 24 tubes.

British-built warheads will be carried in a US-designed MIRV bus, and in early 1985 warhead design and development was reported to be progressing well. (A British nuclear test device was detonated in an underground test in the Nevada desert in mid-1983 and this was the 15th in a series of UK tests there since 1980. Though unconfirmed, some if not all, are thought to be linked with the UK Trident programme.)

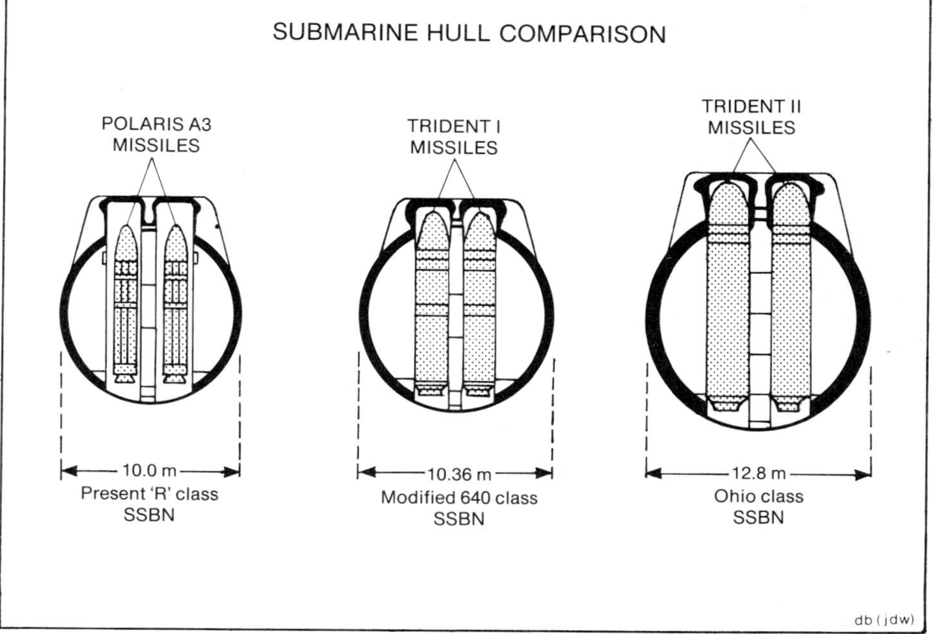

Comparison of submarine cross-sections for Polaris, Trident I and Trident II SSBNs illustrating how hull diameter increases with missile length

In December 1986, George Younger, the British Defence Secretary, disclosed that UK Trident missiles would carry "about two and a half times" the number of warheads carried by Polaris; ie 480.

The British Trident submarines will be powered by a new pressurised water reactor (PWR2) that will also be employed in a new class of British hunter-killer submarines. This reactor will have a longer core-life, so lengthening the period between submarine refits, and will also be appreciably quieter running, making the boat itself harder to detect as well as assisting detection of other submarines by its lower 'own-noise'. The boat will displace more than 15 000 tonnes, and will be 150 m long with a centre section diameter of over 12.8 m.

The Trident II missile is expected to have an 'in-tube' life within the submarine of at least seven years, much longer than Polaris, thereby reducing the on-board maintenance considerably. If the planned fleet of four UK SSBNs is deployed, this should be adequate to provide an 'operational squadron' of three vessels continuously, so that with the fourth submarine undergoing refit there can be one armed Trident submarine on patrol at all times. Refits are planned to take place at intervals of about seven years for each submarine.

STATUS

Planned for entry into service in the mid-1990s to replace the UK Polaris force with a fleet of four SSBNs, to be built by Vickers at Barrow. The Trident II (D-5) missiles will be supplied from US sources under an inter-governmental agreement. Warheads will be a British responsibility.

The UK Minister of Defence, in a statement in September 1982, announced a decision to use US facilities at Kings Bay, Georgia, USA for the initial preparation for service of UK Trident missiles and for

Artist's impression of a UK Trident submarine

their refurbishment at the end of each British submarine's 7 to 8 year commission period. On this basis it will not be necessary to go ahead with previous plans to build the full range of facilities needed for this purpose at Loch Long in Scotland, although some new facilities will need to be constructed.

In January 1988 it was announced that the revised cost of the UK Trident programme had reduced for the second year running and was now at i9043 million. The proportion of the programme to be undertaken in the UK has also risen to 64%.

The keel of the first UK Trident SSBN 06, HMS *Vanguard*, was laid in September 1986, and the order for the second, HMS *Victorious*, was placed in October 1987. The construction of the first two submarines at Barrow, by Vickers Shipbuilding and Engineering Ltd, is going well and both hulls are under assembly. Trials are scheduled to start in the early 1990s.

UNITED STATES OF AMERICA

4560.111

US ICBM MODERNISATION PROGRAMME

In October 1981 President Reagan announced a comprehensive plan for modernising US strategic forces. A key part of that plan was his proposal to remedy growing deficiencies in the US land-based ICBM force. His plan called for deployment of the Peacekeeper (**4561.111**) missile with an initial operational capability in 1986. With this early deployment the President directed an aggressive research programme to identify a more survivable means of basing for the Peacekeeper missile. In November 1982, in response to a Congressional deadline, the Closely Spaced Basing plan was recommended by the Administration as the basing mode that provided the optimal survivability for Peacekeeper. Congress delayed voting on that recommendation pending a thorough review of alternatives, including a new small missile (unofficially dubbed 'Midgetman') in addition to the Peacekeeper.

The review of alternatives conducted by the President's Commission on Strategic Forces reaffirmed the necessity for a new ICBM and recommended that the Peacekeeper missile be deployed in existing silos with supporting R&D programmes to provide alternative forces for the 1990s.

The programme for ICBM modernisation, recommended by the President's Commission on Strategic Forces in 1983 and accepted by the President, calls for deployment of 100 Peacekeeper missiles in existing silos and development of a small missile suitable for either fixed or mobile development. The House-Senate Fiscal Year 86 Authorisation Conference Committee agreed to a limitation of no more than 50 Peacekeepers in existing Minuteman silos. Initial operations of the Peacekeeper system began at the end of 1986.

Peacekeeper

The Peacekeeper is a four-stage ICBM that carries ten independently targetable Mk 21 re-entry vehicles with 335 KT W-78 warheads which has many advantages over missile weapon systems currently in the US inventory (**4561.111,** below). It will be more accurate, able to carry more warheads, with greater range and can be targeted more flexibly than the Minuteman ICBMs. Its greater resistance to nuclear effects and better guidance system provide an improved ability to destroy very hard targets. The Peacekeeper deployment requires a minimum of modifications to the Minuteman silos to accommodate new support equipment, a shock isolation system, and a more sophisticated command and control system. The deployment began in the NW portion of the Minuteman field at FE Warren AFB, Wyoming with 10 in FY86 and 27 in FY87. In addition, the Reagan Administration decided in January 1987 to begin development of a rail-mobile version of Peacekeeper for peacetime deployment. During periods of tension, the missiles could be moved to classified locations on USAF bases.

Minuteman Modernisation and Peacekeeper Deployment

As noted elsewhere in this volume and earlier editions, the existing Minuteman ICBM system (**2716.111** and **2717.111**) and new Peacekeeper ICBM (**4561.111**) are involved in the US ICBM modernisation programme to the extent that silos of the former system are now having the Minuteman missiles they house replaced by newer Peacekeepers, while at the same time there is a general effort to improve US ICBM basing facilities. The most important of these efforts are summarised briefly in the following paragraphs.

A hardened intersite cable system (HICS) will provide a new pressure monitoring system for the cabling between sites at all Minuteman Wings. This was completed in 1986. The integrated command, control and communications (C^3) programme is designed to improve security, reliability, and data handling capability for the National Military Command System. The improved Minuteman physical security system will replace the ageing radar network now used and will reduce the number of nuisance alarms now encountered. Deployment of this system at Wings I and VI, and Squadron 20 started in 1985, and the US Air Force plans to extend this security improvement to all six wings.

A Minuteman extended survival power system (MESP) will substantially extend the launch facility survival period by its greater resistance to the effects of nuclear radiation etc, at a time of conflict. It consists of high-energy lithium batteries, built by GTE, with controls provided for the application of power. Installation of these batteries at 200 Minuteman III launch facilities began in 1984 and this work continued into 1986. The Minuteman transporter/erector fleet of vehicles is to be replaced in the near-term period by 35 new transporter/erectors, under a major upgrading programme. From April 1985, on-site depot teams moved into Minuteman Wings to carry out systematic integrated maintenance to the year 2000 under an Air Force programme called Rivet Mile.

Most of the electrical equipment at Minuteman silos has been re-designed to support Peacekeeper. The existing Minuteman guidance and control cooler will be replaced with a new guidance and control conditioning unit; a new 96 V DC power supply will be added to power the advanced missile guidance system; a new programmer group and UHF transmitter for two-way communication with the Airborne Launch Control Centre (ALCC) will be added; and there will be modifications to the power supply systems and environmental control systems. Software changes for the guidance improvement program were completed in 1984. In 1986 development of a new stage III motor for Minuteman II was due to begin so that replacement of the ageing current motor can be carried out to allow Minuteman II to continue as a credible element of the US strategic ICBM force until the year 2000.

The major facility modification for Peacekeeper deployment involves removing the upper two metres of the launch tube liner to allow on-site assembly of the missile stages and re-entry system. The missile and canister will be suspended by a new shock isolation system that will provide protection from nuclear attack and induced ground shock. EMP protection will also be improved.

The Minuteman launch complex control centre facilities will be modified for C^3 integration. Computer-aided message processing programs will reduce the time from receipt of the launch authentication message to completion of the launch execution command. Hardware modifications will improve visual display and control functions in the launch control console and command message processing group. The HICS will be expanded, resulting in a single command and control system that interconnects ten launch control facilities and 100 launch facilities.

Deployment of Peacekeeper at Minuteman complexes will enhance the capability of the ALCCs by adding a demodulator/decoder and encrypting the existing UHF communications downlink. Remote re-targeting and launch of the Peacekeeper missile from the ALCC will be possible when command access has been permitted. The encrypted uplink at each launch facility allows the ALCC to monitor launch facility and missile status. Other electrical modifications will yield a configuration that supports both Peacekeeper and Minuteman. This entails a re-design of the ALCC power supply, code retaining power supply and other equipment to ensure EMP protection; these changes give the ALCC virtually the same command and monitoring capability as the launch control facility once ALCC access is established.

Funding for some of these efforts is included in the budget allocations for specific missile programmes, such as the Peacekeeper ICBM and the Small ICBM, and readers are directed to the appropriate entries (**4561.111** and **4897.111**) for further details.

Artist's impression of how Peacekeeper advanced ICBMs are deployed in existing Minuteman missile silos. Modifications will be minimal and the silo will not be hardened beyond existing levels

Other areas of modernisation are also being researched. Boeing Aerospace has resumed its role as integrating contractor for the USAF's Re-entry Systems Launch Program (RSLP) at Vandenberg which includes flight tests of Minuteman I missiles with a variety of payloads to test re-entry vehicle shapes, penetration aids and decoys. Rockwell and Acurex Corp have also received a $5.1 million contract for the Advanced Strategic Missile Systems pyrotechnic penetration aids programme.

4561.111

PEACEKEEPER (MGM-118) ICBM

Peacekeeper is the name assigned to the weapon that was formerly known as the Advanced ICBM, or MX, and earlier work was described in **2718.111** in previous editions. It is a major element in the US ICBM modernisation programme (**4560.111**).

The MGM-118 Peacekeeper missile is a four-stage intercontinental ballistic missile (ICBM) which has been designed to deliver 10 Mk 21 re-entry vehicles to independent targets at ranges usually greater than 8000 km. The missile is 21.6 m long, 2.34 m in diameter and weighs 88 450 kg. Three of the four stages use solid propellant materials exhausted through single nozzles. Hydraulically operated thrust vector actuators move the nozzles to guide the missile along its flight path. The fourth stage, called the deployment stage, uses a liquid propellant to power an axial thrust engine and eight small engines which are used for attitude control. The post-boost vehicle includes a guidance and control system, a deployment module and re-entry vehicles. The guidance function is performed by a completely self-contained inertial guidance and navigation system. During flight the missile is completely independent of ground references or commands.

At the time of assembly, each Peacekeeper missile is placed in a support launch canister 23.4 m long. At launch, a solid propellant in the canister base is ignited to eject the missile by gas pressure. The first stage motor ignites automatically after the missile clears the canister. Peacekeeper is the first land-based US ICBM to employ so-called 'cold-launch' techniques.

The first stage solid rocket motor weighs approximately 49 000 kg and is about 9.1 m long. Stage I is equipped with a single movable nozzle which is controlled by signals from the guidance and control system.

The second stage solid rocket motor weighs approximately 27 000 kg and is about 5.5 m long. Both stages use a synthetic rubber-based solid propellant with aluminium and oxidiser additives.

The third stage solid rocket motor, weighing about 7.700 kg, is approximately 2.4 m long. Stage III uses a synthetic polymer plastic-based solid propellant with aluminium and oxidiser additives. Like Stage II, the Stage III single nozzle also has an extendable exit cone to improve performance without increasing the length of the missile. The solid propellants in the first three stages are contained in motor cases made of a kevlar epoxy material.

The fourth stage weighs about 850 kg and is 1.2 m long. Stage IV uses a liquid bi-propellant rocket propulsion system which provides velocity and attitude corrections for this phase of the missile flight. Following the burn-out and separation of Stage III, the post-boost vehicle manoeuvres and each re-entry vehicle is released. The vehicle is then moved by its propulsion systems to new positions where the remaining re-entry vehicles are deployed in sequence. The propulsion system includes the liquid propellants and tanks, an axial engine and its thrust vector actuation system, and attitude control engines. The attitude control engines use the same bi-propellants as the axial engine.

The missile guidance and control (G&C) system is contained in the post-boost vehicle. The G&C system includes the missile guidance and control set containing the inertial measurement unit (IMU) and the missile electronics and computer assembly; ground and flight software; the in-flight cooling subsystem; airborne power supply; missile inter-connection cables; the power and the missile umbilical assembly. Inside the IMU is a floated sphere containing gyros, accelerometers and some associated electronics. The gyros maintain a basic computer commanded frame of reference with which all missile movements are compared. The accelerometers measure all vehicle velocities with respect to that frame of reference. Utilising the position and velocity information generated by the gyros and accelerometers, the missile computer determines all guidance and control functions of the missile. The G&C system also determines the correct release point for re-entry vehicles near the termination of the powered flight path.

The re-entry vehicles, which contain the weapons, are conically shaped and covered with materials to protect the warheads during the flight through the atmosphere to the target. The high speed re-entry will cause extreme heating requiring a surface material to ablate or erode in a controlled manner to protect the weapon throughout its flight.

In addition to the stages and re-entry vehicles, the missile also includes a shroud and the deployment module. The shroud, often called a nosecone, protects the re-entry vehicles during the ascent phases of flight. The shroud is topped with a nose cap made of an alloy called Inconel, and contains a rocket motor to separate it from the missile.

The deployment module provides the structural support for the re-entry vehicles and carries the electronics to activate and deploy them. These vehicles are mechanically attached to the deployment module. The mechanical attachment is severed by explosive units which frees the re-entry vehicles allowing them to separate with minimum disturbances from the deployment module. Each re-entry vehicle is deployed at a position which will allow it to follow a ballistic path to its target.

The Peacekeeper missile can carry 12 Mk 12A warheads or 11 advanced ballistic re-entry vehicles (ABRES), but is limited to ten warheads by the SALT provisions.

The ten warheads currently baselined on Peacekeeper are of the Mk 21 type. According to the US Air Force, Peacekeeper with ten RVs will have the range to reach the most distant planned hard targets.

DEVELOPMENT

The US Air Force Strategic Air Command stated a requirement for a new ICBM in 1972. This weapon should be capable of destroying hardened targets and should itself be based in a survivable manner. After the requirement had been validated, the programme was initiated as the MX (Advanced ICBM), renamed the Peacekeeper in November 1982.

The Peacekeeper programme, which includes a missile and a basing system, began full-scale development in September 1979 (about two years later than intended), but the basing plans then proposed to ensure survivability were not acceptable. On taking office in January 1981, President Reagan initiated a review of the American strategic forces and alternative options for modernising those forces. In October 1981 he proposed, as part of the US deterrent modernisation programme; continuation of Peacekeeper development, with the objective of near-term interim deployment of the missile in Titan or Minuteman silos that had been modified to increase their hardness; cancellation of the MPS (multiple protective shelter) basing system; and deactivation of the Titan II ICBM force.

However, Congress rejected the proposal for interim Peacekeeper deployment, and on 22 November 1982 the President then proposed stationing 100 of these missiles in an array of 100 closely spaced

Seventeenth test of the Peacekeeper ICBM from Vandenberg AFB, California, in March 1987. The flight was the ninth launch from a modified Minuteman silo

Peacekeeper (MGM-118) ICBM is ejected by gas generator at base of canister before first stage motor ignites. Protective pads fall away from the missile as it emerges from the canister

Martin Marietta has developed transportation and handling equipment for the Peacekeeper, including an emplacer for inserting and removing missile stages from silos

'super-hard' silos near FE Warren AFB, Wyoming (the so-called 'Dense Pack' basing mode). Congress also rejected this, allowing funds for Peacekeeper missile development but not for procurement, while restricting funds for basing mode development. Flight testing of Peacekeeper was also forbidden until the basing mode had been approved.

Consequently the President set up a Commission on Strategic Forces in January 1983, to examine basing modes for Peacekeeper. The suggestions of the Commission comprised three main elements:

(1) siting 100 Peacekeeper missiles in existing Minuteman silos
(2) development of a new small ICBM (popularly known as 'Midgetman')
(3) investigation of hardened silos, shelters, and mobile launcher systems.

The President endorsed these recommendations in April 1983 and Congress gave its approval the following month, however, in 1984 Congress limited the deployment to 50 missiles.

Deployment began in August 1986 of the first Peacekeeper missiles in modified Minuteman silos at Warren AFB near Cheyenne, Wyoming. A total of 22 missiles had been turned over to SAC by July 1987, but eight of these were non-operational due to a shortage of inertial measurement units. The total deployment of 50 missiles is expected by the end of 1988.

In December 1986, President Reagan decided to proceed with development of an additional mobile rail garrison basing mode for Peacekeeper, placing the missile on railway cars during peacetime at military installations around the USA. During periods of tension, the missiles would be moved to classified locations. The first contract for rail garrison basing development was awarded in September 1987, to Boeing Aerospace.

The USAF announced in February 1987 that 10 sites had been identified as possible rail garrison bases: Barksdale, Louisiana; Blytheville, Arkansas; Dyess, Texas; Fairchild, Washington; Grand Forks, N. Dakota; Little Rock, Arkansas; Malmstrom, Montana; Minot, N. Dakota; Whiteman, Missouri and Wurtsmith, Michigan.

CHARACTERISTICS – Peacekeeper

Missile length: 21.6 m
Diameter: 2.34 m
Weight: 88 450 kg
Propulsion: 3 solid propellant stages, PBV liquid propellant
Throw weight: 3580 kg
Warheads: 10 × 500 KT, Mk 21 MIRVs
CEP: 40 m
Range: >8000 km
Launch canister:
Length: 23.4 m
Diameter: 2.79 m
Weight (with missile): 131 540 kg

STATUS

After Congressional approval of the Peacekeeper basing mode in May 1983, production of Peacekeeper missiles began with a DoD request for funds for 27 missiles in the 1984 budget. This quantity was not approved however. The House-Senate Fiscal Year 86 Authorisation Conference Committee agreed to limit deployment to no more than 50 missiles in existing Minuteman silos. They agreed to provide $1746 million to procure 12 missiles and 18 basing sets to support silo deployment of 50 Peacekeepers.

The Peacekeeper R & D flight-test programme, consisting of 20 planned launches, was initiated at Vandenberg AFB, California in June 1983. Phase I of the flight test programme, including the first five launches, was completed in June 1984 and was designed to test the missile's functional performance. Phase 2 of the flight test programme was completed with the 17th flight in March 1987. This phase tested missile and RV capabilities and silo integration. The phase 3 flights will concentrate on missile accuracy and testing modifications.

In FY 87 testimony to Congress, the USAF said its requirement would remain at 123 test missiles, whether deployment stays at 50 or the 100 missile level wanted by the DoD. The USAF received $280 million for R & D and $1.114 billion for 12 test missiles in FY87.

The 123 missiles comprise 108 for flight testing and 15 for ageing and surveillance ground tests over the 15 year design life.

Peacekeeper missile deployment in SE Wyoming and SW Nebraska will be supported by the 90th Strategic Missile Wing at FE Warren AFB, near Cheyenne, Wyoming.

Full-scale development for basing began in June 1983, and a draft environmental statement was issued the following October. Public hearings followed and the final statement was issued in January 1984 and construction started on support facilities in mid-1984,

Six Peacekeeper re-entry vehicles splashdown in the target area off Kwajalein Atoll after the ninth test launch of the missile in August 1985

above-ground silo modifications starting in 1985. IOC was reached in December 1986.

In 1984 the Peacekeeper programme moved into production. Primary and associate contracts were awarded for all four missile stages, and in January 1984 eight contracts were placed for G & C production. Another 22 contracts were placed in April 1984 for production of the RV, RV system and fuze, and work was also carried out in support of the Peacekeeper G & C system housed in Stage IV.

In May 1984 it was reported that the USAF had initiated a programme to qualify General Electric as a second source for production to the Mk 21 developer, Avco. GE provides the Mk 21's fuze, with AVCO as a second source for production. In April 1986, Avco Systems Textron completed delivery of the first production Mk 21 RVs.

Actual, planned and projected funding for the Peacekeeper programme is as follows (in $ millions):

Fiscal Year	1985	1986	1987	1988	1989
Procurement quantity	21*	12	12	21	21
Procurement cost	842.8*	1676	1114	1260	1359
Construction cost	95.7	55.8	27.5	9.4	6.5
Total expenditure including spares and RDT&E	2611.1	2528.2	1831.6	N/A	N/A

*$1.5 billion for 21 missiles reapplied from prior programmes.

CONTRACTORS

Aerojet General: Second-stage.
Avco: Re-entry systems integration, Mk 21 vehicle.
Boeing: Basing and deep basing support.
Charles Stark Draper Laboratory: Guidance and control technical support.
General Electric: Mk 21 fuze.
GTE Government Systems: Launch control system.
Hercules: Third-stage systems.
Honeywell: Specific force integrating receiver.
Martin Marietta: Missile assembly test and system support; transportation and handling equipment.
Northrop: Inertial measurement unit; third generation gyro.
Rockwell International: Guidance and control system; flight and ground software; fourth-stage system.
TRW: Targeting and analysis program.
Westinghouse: Launch canister.

This artist's impression of a Peacekeeper train inside a rail garrison igloo, shows some of the cars in a typical configuration. These include, from left to right, the locomotive, security car, missile launch car, launch control car, and finally another missile launch car

4897.111

USAF SMALL ICBM

A small intercontinental ballistic missile (SICBM) is being developed by the US Air Force and the DoD as part of the US President's strategic modernisation programme (see **4560.111**). Following the recommendations of the Schriever commission, the USAF set up a SICBM Programme Office in July 1984 to direct development and acquisition policies for both missile and basing technologies.

Proposed as a 37 000 lb (16 800 kg) missile, the small ICBM would have sufficient accuracy to ensure a high probability of damage to hard targets. A single warhead is to be carried to maximum ranges of approximately 6000 nautical miles (11 000 km). The provisionally accepted characteristics appear below. The re-entry vehicle will be the Mk 21 being developed for the Peacekeeper ICBM with a 500 KT warhead. The SICBM is under consideration for deployment on road mobile radiation-hardened truck launchers to enhance system survivability. During peacetime the launchers will move about at random over large military reservations in a similar fashion to that planned for Peacekeeper when it was still known as the MX. At times of tension, the SICBM launchers would be permitted to travel more extensively to complicate an enemy's targeting task further still.

CHARACTERISTICS

Length: 14.02 m
Diameter: 1.17 m
Weight: 16 800 kg
Payload weight: 453 kg
Propulsion: Solid, three-stage
Guidance: AIRS type or stellar-inertial
Re-entry vehicle: Mk 21 with 500 KT warhead
Range: 11 000 km

The basing mode for the SICBM has not been finally determined, but the missile will be designed to

Full-scale mock-up of so-called 'Midgetman' Small Intercontinental Ballistic Missile (SICBM) at Martin Marietta in Denver, Colorado, USA

accommodate silo launching or the hardened mobile launcher. A full-scale development contract was awarded to Boeing in December 1986, for the Hard Mobile Launcher (HML). The test vehicles are about 30 m long and weigh 90 000 kg. Sites in 15 states are under consideration for basing.

Northrop Corp is supplying guidance units for use in the SICBM programme under a contract awarded by the USAF in December 1984. The units will be Modified Advanced Inertial Reference Spheres (MOD AIRS), based on the guidance systems being produced for Peacekeeper. MOD AIRS has been designated as the base-line guidance unit for the SICBM programme.

STATUS

The recommended programme that is being followed by the USAF includes a competitive system definition phase; a competitive pre-full-scale development phase; and a full-scale development phase. The system definition phase was nearing completion by early 1985, and this built on Peacekeeper technology as well as assessing alternatives to ensure success in meeting the objectives set out for the SICBM. The pre-full-scale development phase began in late 1984 and was expected to continue with competitive contracting for weapon systems integration, propulsion, and the hard mobile launcher. The full-scale development phase for the basing mode(s) began in January 1987, and this programme will lead to flight testing in 1989 with initial operational capability in 1992. Present plans call for a 22 flight test programme at Vandenberg AFB, California.

Eighteen development contracts had been awarded by early 1985 during the SICBM concept definition phase. These included work on the propulsion, hard mobile launcher, guidance and control, command and control, and weapon system concept definition efforts. Hardness validation tests of the mobile launcher began in October 1984 and continued into 1985.

In April 1986, the DoD agreed recommendations by the Defense Science Board Task Force which called for an increase in the proposed weight of the SICBM from 13 600 kg to 17 000 kg to allow "full target coverage, penetration aids and the capability for future payload variations - including a Manoeuvring Re-entry Vehicle (MaRV) or two warheads of a smaller size than the baseline configuration of a single Mk 21". The status of the SICBM under SALT II is unclear; the missile deviates by more than 5% in dimensions from any predecessor and hence, nominally, is a new missile. Since SALT II was not ratified by the US and is considered expired by the current Administration, the missile may be legitimate on deployment.

The US DoD FY89 budget proposals recommended the suspension of this SICBM programme, and its future is uncertain.

CONTRACTORS

Mk 21 Re-entry Vehicle: General Electric Re-entry Systems Division, Philadelphia.

Martin Marietta, Denver: post boost vehicle, assembly test and system support contractor.

Hard mobile launcher: Boeing Aerospace, Seattle.

2716.111

MINUTEMAN II INTERCONTINENTAL BALLISTIC MISSILE

Minuteman II was introduced as a considerably improved version of Minuteman I. The improvements, however, took the form of an upgrading of the missile's capabilities rather than a radically new departure as, in some respects, is Minuteman III (**2717.111**). Like Minuteman I it is a three-stage ICBM carrying a single thermonuclear warhead, but it has increased range and azimuth, providing greater targeting coverage, while carrying a larger payload. A more sophisticated guidance system is capable of pre-storing the locations of a larger number of alternative targets (see **2710.111**) and the over-all accuracy of the missile system is greater than for Minuteman I. The increased payload capability enables the missile to carry a larger thermonuclear warhead together with a number of penetration aids.

The first- and third-stage motors of Minuteman II are believed to be the same as those on Minuteman I but the second-stage motor is new.

CHARACTERISTICS

Service designation: LGM-30F

Type: 3-stage, solid-propellant, intercontinental ballistic missile

Guidance: Inertial, Rockwell International Autonetics

Propulsion: First-stage, Thiokol TU-122 (M-55), approx 91 000 kg st; second-stage, Aerojet SR19-AJ-1, approx 27 500 kg st; third-stage, Hercules, thrust approx 16 000 kg. Thrust vector control on first- and third-stage motors by 4 movable nozzles. Second-stage motor has a single nozzle with secondary liquid injection for thrust vector control

Re-entry vehicle: Avco Type 11B and 11C with Mk 1 and 1A penetration aids, the latter supplied by Tracor Inc

Warhead: Mk IIC RV with thermonuclear single W-56 warhead of a yield unofficially reported as around 2 MT

Dimensions: Length 18.2 m; diameter 1.8 m at first-stage interstage

Launch weight: 31 746 kg

Range: 12 500 km

STATUS

Operational.

The Minuteman ICBM completed 20 years in operational service in October 1982, and seems destined to serve for an appreciable time yet. Additional details of historical background, deployment and development over the years have appeared under entry number **2710.111** in previous editions of *Jane's Weapon Systems.* (See also US ICBM Modernisation Programme, **4560.111**).

CONTRACTORS

Assembly, test and installations: Boeing.

Systems engineering: TRW Defense Systems.

Guidance: Rockwell International.

Propulsion: Aerojet-General, Hercules, Thiokol.

Re-entry vehicles: Avco Systems Textron Division, USA.

Penetration aids: Tracor Aerospace.

2717.111

MINUTEMAN III INTERCONTINENTAL BALLISTIC MISSILE

Minuteman 3 is a three-stage ICBM powered by solid-propellant rocket motors, but it incorporates several features which make it more than a simple improvement on the two earlier members of the series.

Most of the special features relate to the final stage and re-entry system and the most significant, operationally, is the introduction of a MIRV system of three warheads. This MIRV platform is essentially a fourth stage of the missile and is powered by a 143 kg thrust motor and manoeuvred by six small pitch and yaw motors and four smaller roll motors. These motors are controlled by the fourth-stage guidance package which also organises the release of the chaff decoys and the re-entry vehicles. The third stage of the missile has also been considerably improved by the introduction of a new motor using fluid-injection thrust-vector control, which gives a finer control of movement than the earlier arrangement of four movable nozzles and which, with improved guidance, enables the missile to carry its large payload over a greater range and at the same time reduce the missile's CEP to less than 200 m.

CHARACTERISTICS

Service designation: LGM-30G

Type: 3-stage, solid-propellant, intercontinental ballistic missile

Guidance: Inertial

Propulsion: First stage, Thiokol TV-122 (M-55), approx 91 000 kg st; second stage, Aerojet SR19-AJ-1, approx 27 500 kg st; third stage, SR73-AJ-1 Aerojet and Thiokol, approx 15 500 kg st; post-boost propulsion system, Bell Aerospace package comprising 143 kg st bi-propellant engine for fore-and-aft control, 6 × 10 kg st engines for pitch and yaw control and 4 skin-mounted 8 kg st motors for roll control. Propellant tanks are welded shut for long-term silo storage. All motors commanded by fourth-stage guidance package

Re-entry vehicle: General Electric Mk 12. Normal Minuteman III payload is 3 MIRV warheads plus chaff and decoys. 2 or 3 Mk 12 RVs can be carried. A Mk 12A RV is being substituted for the Mk 12 on all Minuteman III missiles

Warheads: Three Mk 12 MIRV with thermonuclear, W-62 170 KT each; or 3 Mk 12A MIRV W-78, 335 KT each

Dimensions: Length 18.2 m; diameter 1.85 m at first stage interstage

Launch weight: 34 467 kg

Range: >11 000 km

CEP: <200 m

STATUS

Operational. By mid-1987, full planned force of 528 Minuteman III and 450 Minuteman II missiles were operationally deployed. Considerable efforts are being sustained to maintain and enhance the operational potential of the Minuteman force, of which the Minuteman III is the principal element. These efforts embrace the Mk 12A, re-entry vehicle, guidance accuracy improvements, silo upgrading and other programmes. In May 1986, GTE Government Systems won a 30-month contract to upgrade the electromagnetic pulse protection plan for this missile.

Replacement of Mk 12 warheads with Mk 12A RVs on some Minuteman III missiles is in progress and all missiles are due to be fitted with the new RV which carries a W-78 warhead. This warhead is about 16 kg heavier than the Mk 12s and there is a consequent slight reduction in the Minuteman III range and MIRV footprint. Deployment of the Mk 12A began in December 1979. In the meantime the NS-20 guidance software improvements are scheduled for incorporation on all Minuteman IIIs.

As noted above, the Minuteman system had served as the main element of the American land-based strategic ICBM force since October 1982. Conversion of Minuteman III silos in Nebraska and Wyoming to house MGM-118 Peacekeeper ICBMs started in 1986. The first MGM-118s were in position by December 1986.

(See also US ICBM Modernisation Programme, **4560.111**)

CONTRACTORS

Assembly, test, and installation: Boeing Aerospace.

Systems engineering: TRW Defense Systems.

Guidance: Rockwell International.

Propulsion: Aerojet-General and Thiokol.

Inertial and microwave systems: Bell Aerospace Textron.

Re-entry vehicle: General Electric Co.

1133.411

POSEIDON C-3 (UGM-73A) FLEET BALLISTIC MISSILE

This is a two-stage, solid-propellant intermediate range ballistic missile designed for launching from submerged submarines of the USN. It was a successor to the Polaris A-2 and A-3 weapons. The range of Poseidon is the same as for the Polaris A-3, 2500 nm (4630 km), but weight is more than doubled at 29 480 kg. Dimensions are correspondingly greater with length 10.36 metres and diameter 1.88 m, but the later missiles were capable of deployment in the launch tubes of Polaris submarines without major modifications to the ship.

A considerably increased nuclear payload of 10 Mk 3 MIRVs carrying W-68 warheads or maximum 14 × 100 KT W-76 warheads enables a single Poseidon to be used against a number of targets. An improved guidance system was also incorporated which provided double the accuracy, it is claimed, of earlier weapons. CEP is estimated at 450 m.

An improved fire control system, Mk 88, was produced for the Poseidon system by the General Electric Company. This interfaces with a later version of the Rockwell SINS (ships inertial navigation system), the missile and the launcher. Mk 88 performs target calculations, insertion of data into the guidance system, test and checkout launch order, and sequence control.

DEVELOPMENT

The first flight model of Poseidon was launched in August 1968 from Cape Kennedy, Florida. The 14th launch, in December 1969, was the first complete test of the weapons system, including the launcher, control, and missile sub-systems. Development flight testing was completed in June 1970.

STATUS

Poseidon became operational in March 1971. An undisclosed number of Poseidon warheads are assigned to NATO targets.

The fiscal year 1984 report by the US Secretary of Defense confirmed that 12 of the original fleet of 31 Poseidon submarines had been converted to operate the later Trident I (C-4) SLBM during the 1970s, and this programme has now been completed. Unless there are any later plans to modify more Poseidon boats to operate Trident I missiles, it is intended to progessively retire the remaining C-3 Poseidon submarines between 1993 and 1999. At 1 July 1986 there were 256 Poseidon C-3 SLBMs operational with the US Navy.

Test launch of Poseidon

CONTRACTORS

Prime contractor: Lockheed Missiles and Space.

Mk 88 fire control system: General Electric Company, Ordnance Systems Division.

SINS: Rockwell International.

3993.001

CRUISE MISSILE PROGRAMME

The American cruise missile programme has been difficult to define satisfactorily for almost a decade, for reasons which have been discussed at some length under the above entry number in previous editions of this book. Summarised, US cruise missiles currently in operation derive from two separate missile programmes each with their own developer and sponsoring service (USAF and USN), which subsequently were co-ordinated under DoD supervision. Since then models and applications have proliferated, thanks to a great extent to the inherent flexibility of the basic cruise missile concept and efforts of the weapons' producers and sponsoring agencies. Thus there have been air-launched versions derived from both the Boeing ALCM, AGM-86 (**1766.311**) and the General Dynamics BGM-109 Tomahawk. The latter has been developed in several versions: the MRASM (AGM-109H) medium-range air-to-surface missile (cancelled by Congress in fiscal year 1984); several ship- or submarine-launched models (**4197.221**); and the BGM-109G ground-launched cruise missile (**4194.111**). Additional details of the ALCM will be found in the Air-to-surface Missiles section of this volume, as will information on the GLCM in battlefield support context in the appropriate pages. Similarly naval applications of Tomahawk are discussed in the Shipborne Weapons section.

The extent to which any particular cruise missile model is 'tactical' or 'strategic' depends on the circumstances and the individual conventions that are adopted. There are no unambiguous definitons of the terms 'strategic' and 'tactical' that have been agreed between the US and the USSR for any weapon system and this situation creates particular definitional problems for the status of cruise missiles, which can only be resolved at the political level. However, arbitrary judgements are frequently based on the range and payload capabilities, so that the three US cruise missiles listed in the following paragraphs normally qualify for consideration under the 'strategic' heading on this basis.

TLAM-N: The nuclear armed version of the land attack Tomahawk (BGM-109A), tactical land attack missile - nuclear (TLAM-N) which provides the USN with a long-range strategic weapon which can be used in support of world-wide objectives, including targets within reach of covert submarines operating in waters where America has neither sea nor air control. The nuclear SLCM is claimed to match a capability that has been deployed by the Soviet Union for a long time.

The original submarine launching mode for the Tomahawk cruise missile employed the submarine's normal torpedo tubes, from which the missile was ejected, but a new vertical launch system is under development by Westinghouse, under contract to the USN. Called the Capsule Launcher System (CLS) it is intended to add firepower to existing and future submarine fleets. Tomahawk missiles are loaded into the capsule at the factory and shipped as a unit ready for use. The vertical launch mode is claimed to offer an increase of up to 50 per cent in weapon loads compared with the present method which is constrained by the torpedo tubes installed. First installations will be in "SSN 688" class submarines, within the bow section but external to the pressure hull and a set of 12 Tomahawk missiles can be carried. These will be available for sequential launch, in addition to the weapons carried in the torpedo room.

ALCM: The AGM-86B air launched cruise missile (**1766.311**) is in production and will form a force of long-range nuclear-armed air-to-ground cruise missiles for use with Strategic Air Command B-52G bombers. Initially, each bomber has 12 ALCMs mounted externally on jettisonable pylons, with the current internal load of gravity bombs and SRAMs. In the mid-1980s, eight ALCMs will replace the internal load on each aircraft. By 1990 all B-52Gs will be armed with 20 ALCMs each. In mid-1987, 98 B-52Gs had ACLMS; a further 46 B-52Hs were armed with ALCMs and SRAM.

GLCM: BGM-109G, the ground launched cruise missile (**4194.111**) was intended as a response to Soviet deployment of intermediate range nuclear weapons such as the SS-20 in the European theatre. The basic combat flight of the GLCM system consists of 16 Tomahawk missiles, four transporter-erector-launchers, and two launch control centres. One flight maintains a quick reaction alert status on each main operating base during normal peacetime conditions, but during periods of increased alert status, the GLCM flights would disperse from the main base to satellite stations to ensure increased survivability if attacked. The initial operational capability (IOC) was achieved in December 1983, and construction of further GLCM basing facilities had already been started in the Federal Republic of Germany, Italy and the UK. Deployment of 48 GLCMs began at Florennes AB, Belgium in 1985 and the Netherlands agreed to deploy a similar number. Original plans were for 464 GLCMs to be based in Europe by 1988, deployed in 29 flights with 116 launchers; Britain and the Federal Republic of Germany agreed to provide bases for about one third of this total each, and the remainder would be distributed between Italy, Belgium and the Netherlands.

The INF Treaty, signed in December 1987, agreed on the removal of all GLCM from Europe within three years. At the time of the treaty the US declared 256 missiles plus 53 spares deployed and a further 133 not deployed.

STATUS

Tomahawk procurement, actual and planned (see following table), reflects the not unusual difficulties

The modernised cruiser USS New Jersey *launches a BGM-109 sea launched cruise missile. USN plans to deploy nuclear tipped Tomahawk cruise missiles on surface ships and submarines for engagement of hard targets*

encountered in fielding a novel weapon system, and also the effects that resolution of such difficulties has on subsequent rates of procurement and deployment. The US Navy budgeted for 249 missiles in fiscal year 1986 and another 324 in the following year, for example. The anti-ship, and nuclear and conventionally armed land-attack versions of the system have already entered fleet service. At present Tomahawks are fitted in battleships and a number of US submarines, and are to be installed in DDG-963 class destroyers. To maximise the number of weapons that can be carried on each ship, vertical launch systems (VLS) are being installed on the new SSN-688 attack submarines, CG-47 cruisers, and DDG-51 destroyers, and this system is to be retrofitted in some DDG-963 ships. The VLS is essentially a cell of launch tubes in which different types of missile and rockets can be clustered.

Actual, planned and projected sea-launched Tomahawk funding and procurement was as follows (in $ millions):

Fiscal year	**1985**	**1986**	**1987**
Quantity	180	249	324
Procurement	553	724.8	790.5

By early 1985, 1080 ALCMs were deployed with about 90 B-52 strategic bombers, and more of these bombers were being prepared to carry this weapon.

Under an agreement signed in late 1982, McDonnell Douglas became a co-producer with General Dynamics of the Tomahawk cruise missile.

CONTRACTORS

AGM-86B ALCM: Boeing Aerospace.

BGM-109 Tomahawk: General Dynamics and McDonnell Douglas Astronautics.

Launch of GLCM during USAF tests

2840.411

TRIDENT C-4 (UGM-93A) STRATEGIC WEAPON SYSTEM

The UGM-93A Trident I (C-4) missile is a three-stage ballistic missile powered by solid-fuel motors, and guided by a self-contained inertial guidance system, with a range approximately twice that of Poseidon (**1133.411**), or about 7000 km.

The guidance system for the C-4 missile, designated Mk 5, is smaller and lighter than that for Poseidon, thereby reducing inert weight and providing more space for missile propulsion. The Mk 5 guidance system is basically a functional equivalent to the all inertial Mk 3 Poseidon system. The most significant difference is the addition of a stellar sensor which takes a star sight during the flight. The post-boost vehicle will correct its flight path based on data derived from the stellar sighting. This enables the Trident missile to meet Poseidon accuracy objectives at longer range. The function of the guidance system is to generate missile steering commands during powered flight that accurately direct the missile to the correct velocity, position, and altitude for re-entry body deployment to the assigned target. Control information is generated by inputs from the shipboard fire control sub-system and flight acceleration history.

The post-boost vehicle (PBV) is designed to permit corrections for errors in launch position data and manoeuvres to deploy re-entry bodies, following third stage separation.

A new Mk 4 re-entry body was developed for Trident I. In addition to the ballistic Mk 4 RV, the Mk 500 Evader manoeuvring RV was being developed as an option but the work is understood to have been discontinued. Instead, if it is decided to equip the Trident II (D-5) missile, development of a completely new Mk 5 re-entry system is more likely.

The launcher sub-system stows and protects the missiles aboard the submarine, pressurises the launch tube and missile before launch, and launches the missile from the submarine. The launch control system is of a new design to provide integrated control and monitoring of 24 missiles and tubes and will be co-located in the missile control centre to improve weapon system control and reduce manning to operate the equipment.

The fire control sub-system prepares the missile guidance system with targeting and launch position data and co-ordinates the preparation of the launcher and missile test and readiness equipment for the missile launch. Two basic fire control configurations were developed: the Mk 88 Mod 2 for retrofitting to Poseidon, and the Mk 98 Mod 0 for the new Trident submarine. A new fire control computer, the Trident Digital Control Computer (TDCC), was required to handle the increased computational work involved in

Submarine Francis Scott Key *(SSBN-657) taking on Trident C-4 missiles from USS* Simon Lake *(AS-33) at Kings Bay, Georgia* (US Navy)

USS Ohio *with Trident C-4 missile tubes opened*

providing pre-settings for the C-4 missile. A digital read-in sub-system (DRISS) will provide a digital guidance interface under the control of the TDCC. The navigation data converter will convert analogue navigation and optical outputs to digital. These systems are common to both fire control configurations.

The Trident navigation sub-system is basically similar in design to that of the 640-class Poseidon SSBNs with the following improvements: (1) the major addition of an electrostatically supported gyro monitor (ESGM) to lengthen the time between navigation fixes and thereby reduce submarine exposure to potential detection; (2) provision for additional closed-loop air cooling to enhance maintenance and reliability; (3) selected modifications to reduce noise, harmonics, and susceptibility to stray electromagnetic interference; and (4) re-engineering of the navigation satellite receiver and the navigation sonar system to ensure economic manufacture.

Development of a Trident missile will utilise the full volume of the Poseidon submarine missile tube and will provide the USN FBM force with the ability to operate beyond the areas possible with the Poseidon 4600 km range. Trident will be able to achieve this without any reduction in payload, whereas any increase in range of the Poseidon is at the cost of a smaller payload.

Two options were considered for Trident II. One was to increase the missile diameter, as well as its length to provide both greater range and increased payload. The second option was to use more of the additional missile volume derived from the increased dimensions for still more payload at some sacrifice in missile range. Some of the enhanced payload capability will be devoted to obtaining increased accuracies. A variety of mid-course updating techniques and terminal sensor systems are under consideration as methods of improving accuracy. Updating by means of the GPS satellite navigation system is one example of the former, and there are a number of terminal guidance techniques which might be applied.

CHARACTERISTICS – Trident 1
Length: 10.39 m
Diameter: 1.88 m
Weight: 29 500 kg
Propulsion: 3-stage solid
Guidance: Stellar-inertial
Payload: Mk 4 RV with 8 × W-76 100 KT MIRVs
Range: Approx 7400 km
CEP: 460 m

OPERATION

Trident may be launched from submerged or surfaced FBM submarines and is ejected by the pressure of expanding gas within the missile launch tube. After the missile has attained sufficient acceleration and reached a specified distance from the submarine, the first-stage motor is ignited, the 'Aerospike' (an extendable spike at the nose of the missile to improve aerodynamic performance) is deployed, and the boost phase begins. When the first-stage motor burns out, it and the interstage are separated from the missile, the second-stage motor is ignited, and the boost phase continues. As acceleration decreases because of second-stage motor burn-out, the second-stage motor is separated and the third-stage motor ignites to continue the boost phase. Third-stage burn-out completes the boost phase, after which the post-boost control system provides equipment section thrust and control until all the RVs have been deployed.

STATUS

In 1987 the major portion of the US sea-based strategic deterrent consisted of 256 Poseidon (C-3) and 384 Trident I (C-4). Eight SSBN-726 *Ohio* class were operational with Trident C-4. This missile is also carried in six retro-fitted SSBN-640 *Franklin* class and six SSBN-627 *Madison* class.

The first eight SSBN-726 Trident submarines are armed initially with Trident I missiles, and Trident II missiles will be deployed on the ninth boat and all subsequent ships of the eventual 24-strong class, as well as being retrofitted in the first eight ships in due course. Initial deployment of Trident II missiles is expected in late 1989.

Trident II (D-5)

Trident II (D-5) is in full-scale development and the first submarine-launched flight test is expected to

Differences in size between Trident D-5 (left) and C-4 are dramatic: the aerospikes emerging from the nosecone extend after the missile clears the water and provides a buffer to streamline airflow

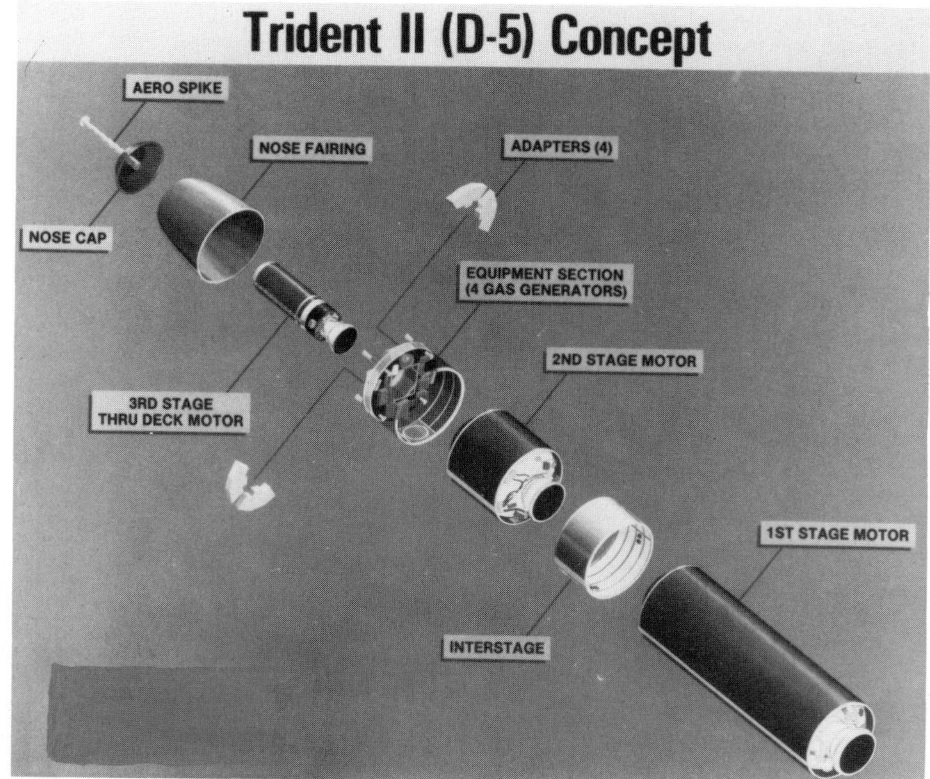

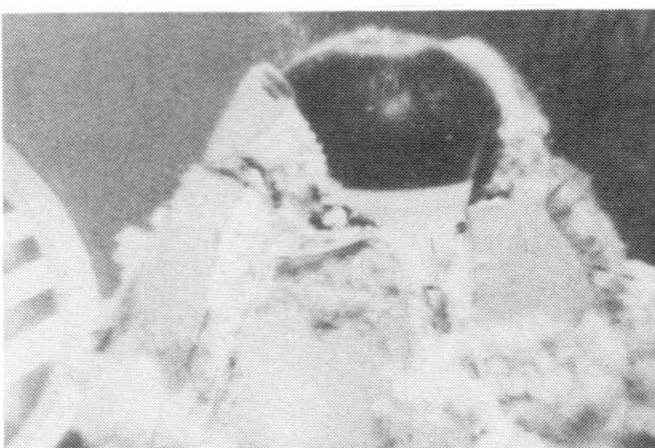
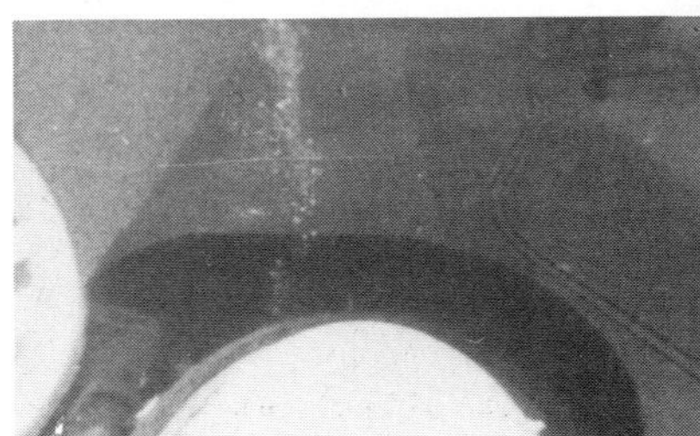

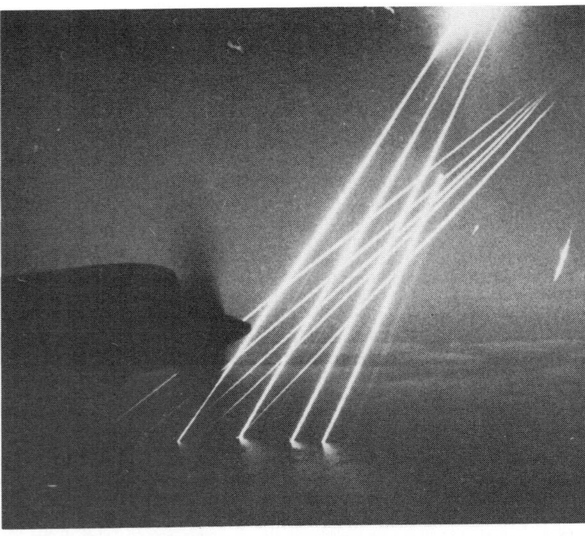

Composite picture showing Trident I C-4 launch. At left, emerging from submerged submarine; centre, ignition of first stage motor after clearing the surface of the sea; right, tracks of dummy re-entry vehicles

occur in March of 1989. Production of the missile is planned to begin in 1987, with 21 missiles; thereafter it is expected to produce 72 Trident II missiles per year from 1988 to 1992, inclusive. The first two stages' casing will be of graphite-epoxy and will have a burn time of around 65 seconds each. The third stage motor, burns for 40 seconds, ending the boost phase. The US DoD has asked Martin Marietta to bid for manufacture of the Trident D-5 missile, to act as a second source to Lockheed. As the total US procurement plans are for 800 missiles, this could result in significant cost reductions.

The first D-5 missiles will be at sea in 1989 with 312 deployed in 13 USN submarines by 1998.

CHARACTERISTICS

Length: 13.42 m
Diameter: 2.11 m
Weight: 59 090 kg
Payload: Mk 5 RV with 8 to 12 RVs each with 300-475 KT warheads.
Range: 12 000 km
Guidance: Mk 6 Stellar-inertial
CEP: 120 m

The US Navy launched the first development D-5 missile from a flat pad at Cape Canaveral in January 1987. This programme will consist of 25-30 flights, of which 20 will be from flat pads. The first submerged launch is planned for the summer of 1989, from the submarine *Tennessee*.

Several years ago it was announced that the British Government had decided to purchase the Trident system for Royal Navy submarines. Originally the Trident I version was to have been procured but after the US decision to discontinue Trident I production in favour of Trident II, the British elected to follow suit and adopt the later system. Details of the UK Trident programme appear in **4209.411**.

CONTRACTORS

FBM submarines: Electric Boat Division, General Dynamics.

Trident missile system manager: Lockheed Missiles and Space.

First and second stage motors: Hercules; Morton-Thiokol.

Third stage motors: Chemical Systems Div of United Technologies.

Post-boost gas generators: Atlantic Research.

Re-entry Body manufacture: General Electric.

Submarine subsystems Launcher: Westinghouse Electric.

Guidance: Charles Stark Draper Laboratory.

Fire Control: General Electric Ordnance Systems.

Navigation: Sperry Systems.

Inertial navigation: Rockwell International.

Instrumentation: Interstate Electronics.

3216.131

US BALLISTIC MISSILE DEFENCE PROGRAMME

Strategic Missile Defence

The United States pursued a very vigorous ballistic missile development programme throughout the 1960s and early 70s, deploying a peripheral early warning radar network and a ballistic missile defence system at Grand Forks N.D. The development of sophisticated penetration aids and, in particular, the emergence of MIRV technology made such defences essentially ineffective and the facility was decommissioned in 1976. The US has, nevertheless, continued to invest in research programmes to recover the effectiveness of anti-ballistic missiles against the evolving threat and it was this US Army work that became the basis of the Strategic Defence Initiative programme in 1983. Many of the advanced technology activities of other DoD Agencies, such as the Defence Advanced Research Projects Agency (DARPA) were subsumed into the SDI Organisation, bringing together all of the potentially useful research work on ballistic missile defence and thus ensuring maximum benefit from the co-ordination of previously independently sponsored projects. Since the advent of the President's Strategic Defence Initiative (SDI, **4896.131**) the BMD programme has assumed a wider significance and assurance, the thrust of the new political support has brought it energy and authority and, as a consequence, many experimental projects and concepts have developed rapidly. For example, the Homing Overlay Experiment (HOE), an important proof of principle step in the development of homing vehicles for use in space, has now become an important part of the development of space interceptor technology. The background and accomplishments of the Army BMD programme have been described under this entry number in earlier editions of this volume. This programme is now managed by the US Army Strategic Defence Command (SDC) and will be described briefly below. The final flight of the Homing Overlay Experiment (HOE) in June 1984 produced the first tangible results for the SDI programme by confirming technology for a non-nuclear capability to home onto and destroy ballistic missile warheads above the atmosphere.

The HOE of June 1984 was referred to as a tactical prototype and the interceptor vehicle employed a Honeywell long wavelength IR sensor and a kinetic energy kill mechanism that consisted of a 15 ft diameter (4.6 m) net with weighted spokes which is wrapped around the interceptor vehicle body. Propulsion was by means of a Rocketdyne bi-propellant rocket motor which could be throttled. The target was an ICBM with a dummy warhead launched from Vandenberg AFB toward an impact point north of the Kwajalein Atoll in the Pacific Ocean. The interceptor was boosted from an island in the Atoll by Minuteman first and second stage motors, on a trajectory that was calculated to miss the target vehicle by over 30 km, so that the interceptor's homing system would have to locate the target, position itself to deploy the kill vehicle, acquire the target, intercept it and unfurl the kill mechanism and destroy the target by the kinetic energy of the resulting collision. This is stated to have been sufficient for the purpose without resorting to nuclear or conventional explosive devices.

Projects in progress at present include concept definition of a missile which would exploit the lessons learned in the HOE project and provide an option for non-nuclear engagement of attacking missiles above the atmosphere.

One major programme of this type is ERIS – the Exoatmospheric Re-entry Vehicle Interceptor System, designed for intercept in the late mid-course phase of the threat vehicle flight (see SDI **4896.131**). Technology development supporting this system includes low cost miniture kill vehicle technology, advanced propellants and structures and guidance, control and missile electronics. The advancements for exoatmospheric systems are focussed on miniaturising kill vehicles, developing ultra high burn rate propellants and demonstrating a two order of magnitude improvement in inertial navigation. Another project at the concept definition stage is a defence system that would use a missile which would operate in the upper atmosphere. A candidate for this role is the High Endoatmospheric Defence Interceptor (HEDI). This vehicle will intercept re-entry vehicles at the beginning of the terminal phase of ICBM flight. Related technologies for this system include advanced seekers, windows and avionics, warhead and fuzes and advanced propulsion systems. Demonstration contracts for the appropriate subsystems were awarded in 1985 and will culminate in an integrated system demonstration in the late 1980s or early 1990s.

In the miniature projectile area, the investigations cover several deployment options for specialised warheads and launchers. Such projectiles might be used in both ground and space based modes, possibly with interceptors of the ERIS and HEDI type. Technology developments in this area include projectile miniaturisation, rapid fire capability and specialised guidance and control. A major current activity under the mini projectile effort is the Lightweight Exoatmospheric Advanced Projectile (LEAP), a concept demanding technology at the limit of capability. Both the US Army and Air Force have LEAP programmes, the basic concept having evolved from the earlier Sagittar/Gremlin concepts; the Air Force approach is being developed by Boeing, with Hughes meeting the Army requirements. The rate of progress in these technologies can be gauged from the fact that the earlier Sagittar requirement, itself demanding at the time, to build a 10 kg projectile, has now evolved into the LEAP requirement, initially for a 5 kg vehicle, with the next phase requiring a 2 kg projectile, qualified for use in space and hardened for use in electromagnetic rail guns. Hardware should be completed during 1988, with fully functional test projectiles expected to be available for test in 1989.

Other activities in the mini-projectile area include the Hypervelocity Launcher Technology programmes, including the TIER I and TIER II programmes, which emphasise increasing projectile mass and velocity, improving launcher efficiencies and the design and fabrication of the guns themselves.

The Flexible Lightweight Agile Guided Experiment (FLAGE) programme is a small, non-nuclear, hit to kill interceptor which uses an active radar and small, multi divert thrusters. This programme was one of the two sponsored hit to kill (KEW) validation experiments conducted during 1986, it achieved three successful intercepts (out of three attempts), with the third flight destroying an air launched re-entry vehicle simulation target.

The second experiment in the series was DELTA 180, launched on 5 September 1986. This complex experiment made observations and a space intercept. The primary purpose of the follow-on flight, DELTA 181, launched on 8 February 1988, was to collect further phenomenology data, particularly that required to support Space-Based Kinetic Kill (SBKKV) flight experiments. Two further Delta flights, numbers 183 and 185, are planned by late 1989, Delta 183 will carry a dual payload, called LACE/RME, for laser communications experiment/remote mirror experiment. Delta 185 will carry HIBREL, or high brightness relay mirror and is due to be launched in late 1989.

A further Army programme is the Extended Range Interceptor (ERINT). The ERINT will modify FLAGE technologies with improved radar seeker performance, the provision of a warhead with a fusing function, larger attitude control motors and a more powerful boost motor.

The picture that emerges from this strategy is one of having a programme – the kinetic energy programme – that covers the requirements of the mid-course and terminal intercept mission. The SDIO purpose, however, is to cover the whole gamut of the global SDI mission, from boost phase to terminal/tactical defence and these wider aspects are examined in the SDI section that follows (**4896.131**). Whilst the US Army is only one of the contracting agencies supporting SDI, its experience in ballistic missile defence has been co-ordinated with the other agencies to provide a key element and in particular to manage the theatre missile defence programmes that will link the technologies of BMD to tactical missile defence and future air defence systems.

In August 1984 a $289 million contract was awarded to Boeing Aerospace to demonstrate the use of airborne IR sensors and data processors in the Anti-Ballistic Missile (ABM) context. A modified Boeing 767 aircraft will be used in the course of a five-year programme to develop what is known as the Airborne Optical Adjunct (AOA) that might be used in a layered ABM system to provide early warning of ICBMs, track RVs, discriminate between warheads and decoys and give tracking information to ABM ground radars and interceptor vehicles. Long-wave IR sensors are being supplied by Hughes Aircraft Co and Honeywell is providing the data processing subsystem.

The LTV Corporation Missiles and Electronics Group is studying a number of SDI concepts for missile defence. LTVs research focuses on kinetic energy weapons that would destroy enemy missiles or warheads by direct impact.

In addition to these and other near-term projects, the BMD organisation is continuing R & D for SDI on advanced technologies for far-term applications, such as a high energy free electron laser and a neutral particle beam device. This organisation is also contributing to realisation of the battle management/command, control and communications (C^3) and non-nuclear lethality aspects of SDI.

The US Army is also pursuing a terminal phase interceptor version of the Patriot surface-to-air (SAM) missile (**2800.131**). On 11 September 1986, a Raytheon Patriot air defence missile was used by the US Army to destroy a Lance tactical ballistic missile. The velocity of the Lance at intercept was near Mach 2 while the Patriot's velocity at impact was greater than Mach 3. The Patriot software was modified versions of that currently used in the missile's computer system.

Theatre Missile Defence

Some of these technology development programmes fall within the Theatre Missile Defence/Foreign Technology Programme, sponsored through the US SDI programme. Electromagnetic Rail Gun Technology is being pursued in the United Kingdom and Combined Chemical and Electrical Propulsion Schemes by Israel.

The SDI Anti-Tactical Ballistic Missile (ATBM) Defence Programme performs research on simulation, component, subsystem and interceptor technologies and arranges for integrating and testing hardware for a theatre defence architecture. These objectives will be achieved under several integrated programmes.

In December 1986 seven international consortia were awarded phase 1 contracts for Theatre Missile Defense Architecture Studies (TMDAS). The US Army Strategic Defense Command then followed these studies with phase 2 contracts for five of the consortia and these will be completed by July 1988. In addition, two further contracts were awarded, to the UK and Israel, for studies into anti-tactical ballistic missile defence architectures. US and allied contractors are to be invited to identify hardware systems for an interim theatre missile defence under a programme called Invite, Show and Test (IST), that will culminate in experimental test firings.

4896.131

US STRATEGIC DEFENCE INITIATIVE (SDI)

American efforts to examine the feasibility of strategic defences against ballistic missiles were reorganised and gathered together under a new strategic defence initiative in response to a speech of March 1983 by President Reagan. The extensive ballistic missile defence activities of both the United States and the Soviet Union over the last 25 years have been summarised in the preceeding section, where it was also noted that the relative ease of sophistication of offensive ballistic missile systems had maintained these forces in a condition of invulnerability. The evolution of defensive capability was inadequate to provide more than a token effectiveness in the defence of either major cities or silo fields, indeed, this was the reason for the decommissioning of the US system at Grand Forks, N.D. in 1976. The new initiative, the Strategic Defence Initiative, arose initially out of a political search for a solution to the growing unease surrounding the continuing expansion of nuclear missile forces and it became established as a revolutionary approach to the missile defence challange once the Presidentially appointed Fletcher commission had reported that the advances in a wide range of technologies, particularly those connected with space utilisation, justified a renewed investment in the subject. The DoD initiated a Defensive Technologies Study to define a long-term research and technology development programme to explore the key technologies needed for an effective multi-layered ballistic missile defence (BMD) system.

To achieve this goal the R & D programmes of the US Army, Navy, Air Force, Defence Nuclear Agency, Department of Energy, and Defence Advanced Research Projects Agency (DARPA) that relate to missile defence were consolidated in 1984 into the Strategic Defence Initiative (SDI). The objective of the research programme is to develop a proven technology base to lead to a defensive system that will destroy Soviet ballistic missiles during the various phases of flight (boost, post-boost, mid-course, and terminal). In the first of these (boost), the first and second-stage rocket motors are burning, and during this phase a unique and intense infra-red signature is produced; in the second phase, the post-boost vehicle (warhead carrier) separates from the main engines and the multiple warheads are deployed,

An artist's impression of the Zenith Star laser spacecraft, designed to be the first US large space laser experiment in the early 1990s. (Martin Marietta)

together with such items as penetration aids, decoys, chaff etc. In the third (mid-course) phase these travel on ballistic trajectories through space, and above the Earth's atmosphere. In the fourth (terminal) phase, warheads and penaids re-enter the atmosphere and become subject to its influences and effects. It is the aim of the SDI programme to find techniques that will permit engagement of attacking missiles in all four phases of flight.

There is an advantage in engaging the target missile in the boost phase because the warheads and penaids etc have not yet separated. There is also a military advantage in engaging the bus during the post-boost phase, since after deployment of the payload elements, it is necessary to discriminate between warheads and decoys so that only the real target(s) are destroyed. The technologies for the terminal defences are also likely to be applicable to defence against shorter range nuclear ballistic missiles that have brief flight times, eg. SLBMs and theatre-range ballistic missiles.

The successful transition from the research phase of this work to development and deployment depends critically on agreement between the US and the USSR on interpretations of the 1972-74 ABM Treaty. There is considerable dispute within the US on the freedoms that exist to allow this work to proceed within the treaty framework and these have to be resolved in addition to the reaching of agreement with the USSR on how new strategic defence studies can proceed.

To manage research efforts, the SDI programme is divided into five major elements.

The first of these, surveillance, acquisition, tracking and kill assessment, is planned to explore the technologies needed to detect, identify, locate, and track ballistic missiles or their components during the boost, mid-course, and terminal phases of their trajectory.

The second element consolidates the directed-energy weapons projects. Here, the research is conducted in the four general classes of directed-energy weapons applicable to missile defence: space-based lasers, ground-based lasers, X-ray lasers, and particle beams.

The third research element is concentrated on kinetic energy weapons that are designed to destroy ballistic missiles or their components by direct impact. Research is focused on 'smart projectiles' that could be fired from the ground or from space. An example of kinetic energy weapons research was the US Army's demonstration in July 1984 of the technologies necessary to intercept and destroy a ballistic missile warhead in space using non-nuclear kill mechanisms. (HOE see **3216.131**). The US Army has named Lockheed Missiles and Space as prime contractor for developing the long range exoatmospheric interceptor for the SDI programme.

The testbed is part of the Exoatmospheric Re-entry Vehicle Interceptor Subsystem (ERIS), a product of the HOE programme. The problem here is one of discrimination between targets and decoys and having enough interceptors to intercept all the real re-entry vehicles.

The final layers of defence would be well down in the atmosphere. The principal system being developed for the upper part of the atmospheric regime is the High Endoatmospheric Defence Interceptor (HEDI), there may be additional defences below this based on, for example, electromagnetic gun principles, described above (see **3216.131**).

The fourth is the critical element and is related to the system architecture and battle management. In this area research is being conducted on how positive command and control might be set up for defensive purposes. The very considerable problem of deciding how to combine these capabilities in the most effective manner to provide a useful and cost effective defence system has been approached through the formation of Defence Architecture Study Teams. The best and most promising features are selected from these and combined within the best of the architectures and it is intended to reduce the number of teams and options as refinement of the architectures becomes possible. Not the least of the difficulties in doing this is the rapid advance of the technologies and the potential performance of weapons, sensors and the battle management systems to operate a system.

The final element is concerned with survivability, lethality, and key technologies, and research is concentrated on the problems of space electrical power sources, launch vehicle requirements, and the critical areas of ballistic missile lethality and defensive system survivability.

SDI – an artist's impression of LTV's proposed space-based system which will use chemically propelled (non-electromagnetic) interceptors to attack missiles during their boost phase. This kinetic energy weapon concept is non-nuclear

By following a co-ordinated programme it is hoped that by the early 1990s it will be possible to permit a decision whether or not to proceed with the development of a strategic defence system for the USA and its allies. In the event of a decision in favour of proceeding with development and deployment, implementation would probably follow this pattern:

Research Phase: This embraces the current phase from the President's speech of 23 March 1983, to the early 1990s, when a decision on whether to go on into systems development would have to be taken.

Systems Development/Full-scale Development Phase: Assuming a decision to proceed in the early 1990s this stage would cover the design and construction of actual system prototypes and components, and their subsequent testing.

Transition Phase: This would be the period of incremental, sequential deployment of defensive systems, with the stated intention for each increment, in conjunction with effective and survivable offensive systems, to increase deterrence and reduce the risk of nuclear war. During this period, as the USA and USSR progressively reduce the value of such (strategic ballistic) missiles, significant reductions in their numbers could be negotiated and implemented.

Final Phase: This would be the period during which deployment of defensive systems would be completed and ballistic missile numbers would reach a minimum.

There is a strong desire in Washington to deploy at least kinetic energy kill weapons, (either ground or space based) within the next decade.

USAF SDI-related programmes

From 1985, several USAF programmes have been integrated into the SDI. The most noteworthy of these are the Advanced Warning System (now known as the Boost Surveillance and Tracking System), the Space-based Space Surveillance System (now called Space Surveillance and Tracking System), and the Strategic Laser Technology Programme (now a part of Directed Energy Weapons). The first two of these, the BSTS and SSTS systems are based on the earlier requirements for warning of attack, to enable aircraft on stand-by to take-off and for submarine warning. Such requirements are modest compared to those needed for the planning of a ballistic missile defence and these systems are being greatly enhanced to enable them to meet this complex need.

The Laser Technology Programme was combined with many of the other directed energy programmes being carried out by the DoD to form a single grouping with greatly expanded objectives, notably the destruction of boosters and post-boost vehicles in the very early part of their flight and the discrimination of warheads from decoys by probing threat objects with neutral particle beams. The achievement of these objectives early in the flight of an attacking missile is considered essential to the overall success of a defence force, because of the importance of reducing the scale of the attack before the less effective terminal weapons are brought into use.

The Directed Energy Weapons Technology Programme addresses four basic concepts; space-based lasers, ground-based lasers, space-based particle beams and nuclear directed energy weapons. Apart from the expected problems relating to surviving attack and of constructing control systems which are proof against jamming, all these systems have to be made with sufficient reserves of energy to cater for unexpected countermeasures that the Soviets may produce.

All ground-based systems need some form of space-based reflector to compensate for the curvature of the earth and also need to be based on a form of energy that can be transmitted through the atmosphere, one candidate for this purpose is the free electron laser, under research at both DoE laboratories and at the Boeing Corp.

Neutral particle beams have attractions of two kinds – they are very penetrating and place a protective burden on the attacker, but they also have the useful secondary value of stimulating the emission of radiation from the attacked vehicle. This can be used by the defence for identification purposes, as the amount of this radiation enables the mass of the object to be determined. As warheads weigh very much more than decoys, this is a crucially important measurement.

The fourth concept, the nuclear directed energy beam, overcomes the principal shortcoming of all other energy sources, that of generating sufficient laser energy for attacking re-entry vehicles. A nuclear explosion is used as a source of a number of X-ray laser beams which can be directed at a large number of targets at once. This concept has to be regarded as very much a contingency approach to the strategic defence requirement for diverse reasons; technically, because its use lowers the nuclear threshold and the generation of nuclear radiation would degrade other sensors and politically because the use of nuclear weapons in space is contrary to the provisions of several treaties and to the original Presidential intention to build a non-nuclear defence system.

These programmes are conducted under the SDI management to meet the broader range of requirements of BMD, while also meeting the present validated tactical warning/attack assessment and space defence requirements. The USAF is also heavily committed to the battle management and command, control and communications aspects, and the supporting technologies programme. During the 1980s experiments and demonstrations will be conducted to assess the feasibility of the applicable technologies and to support development decisions taken in the 1990s.

US Theatre Defence Activities

The original Presidential concept for SDI was to create a global defensive system and one of the early steps taken by the SDI Organisation was to recognise that considerable differences exist between the strategic and theatre environments and to establish a separate set of theatre defence studies in parallel with the strategic ones. There are two approaches to the theatre work, the setting-up of international consortia to produce Theatre Missile Defence Architecture Studies (TMDAS) and separate agreements with the United Kingdom and Israel to produce additional, independent, studies. The more localised nature of the theatre studies makes them very weapon and time sensitive; the European theatre, for example, is dominated by the SS-21, SS-23, SS-12 and SS-20 threats (none of which apply to the North American continent) and this will continue to be the case until the INF Treaty reductions are fully implemented. The Middle East theatre, like that of central Europe, is dominated by very short flight time threats, such as the SS-21; systems like the Israeli 'Arrow' anti-tactical ballistic missile are being studied under SDIO direction.

These various studies are, therefore, concentrating on specialised solutions to local problems.

4186.181

ASAT ANTI-SATELLITE WEAPON

Funded under the heading of Space Defense Operations (PE12450F), the US Air Force is developing a prototype miniature air-launched system (PMALS) to meet a need foreseen for an aircraft-launched anti-satellite (ASAT) capability. This need is justified in the DoD Fiscal Year 1982 Report of January 1981 as being necessary in the absence of agreements limiting the use of space to peaceful purposes but while facing a previously tested Soviet ASAT system.

It is understood that the three main contractors now engaged on this task were already working on this problem, and had been for a considerable period. Boeing for instance is believed to have been occupied in this way since 1975, or earlier, part of the time under contract to Vought. Both these concerns are collaborating with a third, McDonnell Douglas Aircraft, in the development and testing of an advanced technology interceptor missile capable of being launched from an F-15 fighter aircraft. The latter is required to be capable of preparation for the ASAT mission from the standard 'line' configuration in six hours or less.

The weapon envisaged for the PMALS tests comprises a first stage based on the Boeing SRAM (**1107.311**) short-range attack missile, an Altair III second stage provided by Vought and a miniature vehicle warhead terminal stage, also a Vought responsibility. McDonnell is modifying the F-15 for the ASAT role, including provisions for housing the 'carrier aircraft equipment' (CAE) package in the F-15's weapons bay. Guidance algorithms for the aircraft, launch and missile initiating procedure will be stored in the F-15 aircraft computer, and the F-15's head-up display will be employed to provide steering cues to the pilot for launching the ASAT weapon. The SRAM-based first stage is being modified by Boeing to carry the Altair III second stage, the basic SRAM having a range of between 60 and 160 km depending on the flight profile etc, at speeds of Mach 3 or more, under inertial guidance and carrying a very respectable payload. No details have been released on the Vought Altair III second stage vehicle beyond the fact that it is the fourth stage propulsion unit of the Vought-built Scout space launch vehicle.

The prototype vehicle resembles a conventional tactical missile 5.18 m long and with a body diameter of 507 mm. The missile weighs 1180 kg, of which the miniature homing vehicle payload is believed to account for some 16 kg.

The satellite 'kill' mechanism has not been disclosed but the possibility of the ASAT terminal stage making a direct hit on the target has been reported; this is one aspect that may be assumed to be a major aspect of the experimental development programme for ASAT.

F-15 Eagle aircraft carrying ASAT missile

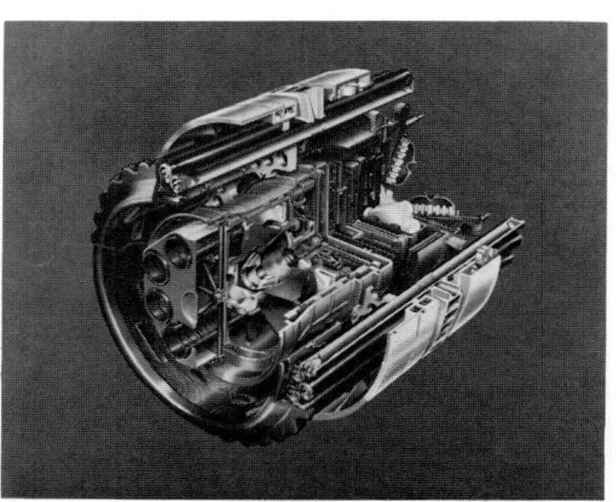

Cut-away drawing of ASAT Miniature Vehicle

CHARACTERISTICS

(Modified Boeing SRAM first stage, modified Vought ALTAIR III second stage and Vought miniature vehicle warhead stage).

Length: 5.18 m
Diameter: 507 mm
Weight: 1180 kg

STATUS

Preliminary work by ASAT contractors began in the 1970s. Vought was awarded a four-year $268 million contract in late 1980 for ASAT development, and Boeing reported its work during 1980 to be worth $6.7 million.

A Space Defense Operations Centre (SPADOC) was established at the NORAD Cheyenne Mountain Complex in 1979/80 to provide command, control and communications to manage space defence operations. This is being modified to handle ASAT operations and these changes are being carried out with hardware and software developments on a schedule to suit ASAT tests.

In June 1983 it was reported that the US Air Force 318th fighter/interceptor squadron, based at McChord AFB, Washington, was being re-equipped with F-15 aircraft (replacing F-106s) to operate ASAT weapons. The first F-15 was handed over in June 1983. The first test flight launch took place in January 1984.

In November 1984 the US Air Force conducted an infra-red probe test of the booster, booster guidance system and miniature vehicle flight sensor using a star as the infra-red source. The ASAT missile was launched from an F-15 which took off from Edwards AFB, California. The test was conducted in air space over the Western Test Range at Vandenburg AFB, California.

A test against a so-called dead satellite was carried out on September 13 1985. The USAF estimates that some 112 ASAT missiles will be required. A total of nearly $500 million has been planned for the research and development programme for FY89 for continuation of the ASAT, but Congress has refused to allow further tests to be carried out and the programme appears to be near to cancellation. This attitude may change in view of the legitimisation of direct ascent ASATs under the 1987 INF Treaty, implying that the Soviets may be interested in pursuing the direct ascent ASAT route. The US DoD FY89 budget proposals recommended the cancellation of this programme, and this may now result in a complete stop as opposed to the uncertainty of the last 2 years.

CONTRACTORS

Vought Corp, Ground Prairie, Texas, USA.

Boeing Aerospace Company, Seattle, Washington, USA.

McDonnell Douglas Aircraft Company, St Louis, Missouri, USA.

LAND-BASED WEAPON SYSTEMS AND ASSOCIATED EQUIPMENT

This main section contains those weapon systems that are land-based, together with their associated equipments. It covers land-based air defence systems, army fire control systems, battlefield support weapons, anti-tank assault weapons, coastal defence weapons, mobile surface-to-air guided missiles, mobile gun and rocket systems, portable anti-aircraft guided missiles, surveillance and tactical radar systems, electronic warfare equipment and electro-optical systems. This alteration in the previous running order of *Jane's Weapon Systems* has been made to enable systems and equipment of each of the four main operational areas, ie strategic, land, sea and air, to be brought together. The numerical system remains as before, thus making it simple to cross-refer to previous editions.

LAND-BASED AIR DEFENCE SYSTEMS

3215.181

AIR DEFENCE SYSTEMS NOTE

This section is concerned principally with national air defence systems, and systems for the command and control of air defence forces comprising a mixture of weapons (eg guns, missiles, and aircraft) intended for deployment in a mobile or semi-permanent fashion in support of an expeditionary force abroad. Within the constraints imposed by what is, by nature, one of the more sensitive areas of defence technology, the aim is to concentrate upon the infrastructure of such systems and the hardware associated with this aspect of air defence. Sensors and weapons are covered elsewhere in this volume.

The distinction between 'air defence command and control systems' above, and certain army fire control systems is clearly a fine one in some instances: the reader is therefore directed also to that sub-section, which follows this one.

As already observed air defence is a particularly sensitive subject and consequently information on some nations' systems and equipment is often severely limited. Political, as well as military reasons frequently are the cause of manufacturers' reluctance to divulge the destination of equipment such as radars or computers, which might otherwise enable assessments of particular countries' air defence provisions. There is an increasing tendency for long- and medium-range radars to be employed for both air defence purposes and air traffic control (ATC) functions. Up to a point, the demands of both services are almost identical and there is much to be said for combining the requirements in terms of sensors, data handling and display equipment, as far as is possible. Where it seems appropriate, therefore, we are giving some details of certain examples of this practice in this section. Such instances will be clearly identified in the text. In other cases air defences are being freshly constructed and may as yet be incompletely defined. In the following paragraphs brief details are given of what is known of some of these programmes.

Abu Dhabi

Recent developments in the Gulf War has initiated an urgent requirement for an air defence system. Radar and telecommunications equipment valued up to $700 million is being sought and discussions with USA and UK have started. The Soviet Union has also indicated a willingness to help. Abu Dhabi is part of the United Arab Emirates (see later) and it is almost certain that any air defence requirements will form part of an overall UAE system.

Angola

For reasons that are not difficult to understand very little is known about the air defence systems operating in Angola. Virtually all military equipment has been supplied by the Soviet Union and much of it is operated by Cuban personnel. It is believed that some $2 billion of Soviet military aid has been supplied to Angola during the mid-1980s. Angola Air Force aircraft known to be in operational service include the MiG-19, MiG-21, MiG-23 and Su-22, as well as a number of helicopters, transports and maritime reconnaissance aircraft.

Ground-based anti-aircraft forces consist of a large number of Soviet SAMs, including SA-2 Guideline, SA-3 Goa, SA-6 Gainful, SA-8 Gecko, SA-9 Gaskin and SA-13 Gopher missiles, plus the necessary missile command and AA radars. Over 20 radar sites have been identified, using a variety of surveillance and search radars, including Tall King, Spoon Rest, Bar Lock, Flat Face, Squat Eye and Side Net systems. Details of these radars are given in the USSR entry of the Ground Radar section. The accompanying diagram gives locations of the radar sites that have been identified, with the coverage up to 20 000 ft.

Argentina

In the early 1970s Argentina was known to have expressed an interest in a new national network, and later the French Thomson-CSF concern undertook a similar programme to that completed for Brazil, namely an integrated ATC/defence network. In 1985, the President of Argentina introduced a new plan to reorganise the defence of the nation with a new National Defence Cabinet. The plan awaits ratification by the government.

In 1978 an $8.79 million US contract for the supply of three AN/TPS-43 air defence radars to Argentina was announced, and later it was reported that a number of Cardion-built AN/TPS-44 radars had been supplied. No status details of either sale was obtained subsequently and no official information has been disclosed, but it is probable that these radars were sited to afford cover for the main naval and air bases near Buenos Aires, Puerto Belgrano, Commodoro Rivadavia, Santa Cruz, Rio Gallegos, and Rio Grande and Ushuaia. In 1982 one of the AN/TPS-43 radars was deployed to the Falkland Islands in support of Argentinian forces landed there. This equipment was captured by British forces when the Islands were reclaimed and after being refurbished in UK, is now based at Port Stanley in the Falklands. It seems likely that early steps were taken to replace this item of lost equipment following Argentina's withdrawal from the Falkland Islands, either with another TPS-43 or a broadly similar type such as the Thomson-CSF TRS 2215 or TRS 2230.

Caribbean Area

To fill a gap in the air defence long-range surveillance radar coverage, the USAF is developing a radar network in the Caribbean basin to provide an enhanced capability to US Atlantic and Southern commands. Eight sites are planned. The USAF is buying the systems and operating them and will share data with the host countries. Westinghouse has been selected to receive a contract for the first of these radars and has received $13.9 million to build the system. It will be installed in Panama at an undisclosed

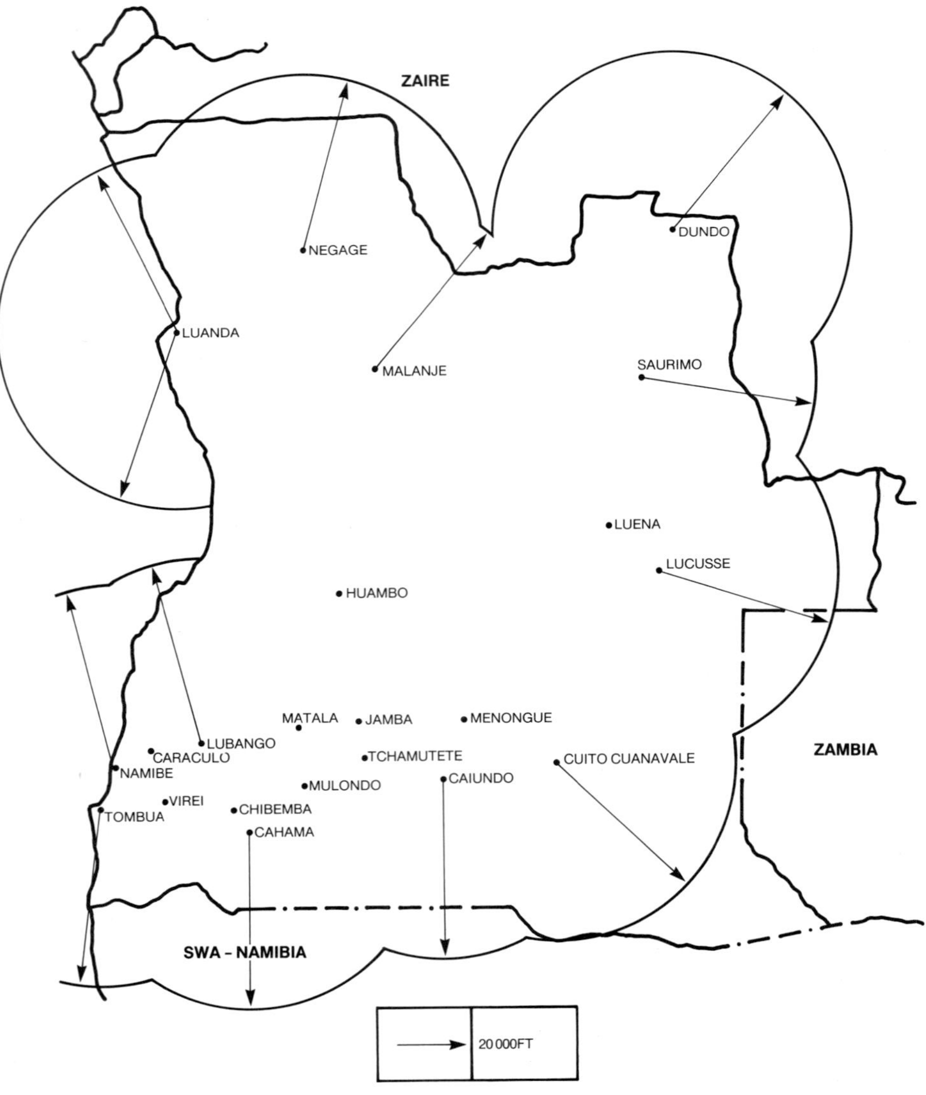

location. Under Project 3159 (Caribbean Basin Radar Network) the USAF plans to erect the remaining seven radars at other locations. No details are available as to the location of the other sites, or the overall time scale although the Panama system was due to be operational by the end of 1987.

Egypt
Egypt is unusual in that it has four armed services, one of which is dedicated to air defence. To what extent this peculiarity is a result of the period when Egypt relied mostly on the USSR for her defence equipment and consequently absorbed much of the military thinking of that state is not known. However, the fact remains that much of the essential hardware for air defence, particularly anti-aircraft artillery, and surface-to-air missiles, was initially of Soviet origin despite the severance of the Egyptian-Soviet link in successive stages in 1972 and 1975. Since that time military equipment procurement has tended to reflect the various phases of Egypt's evolving relationship with differing friendly nations willing (or anxious) to provide arms or to avail themselves of one or other aspect of Egypt's resources or situation.

As a consequence of this somewhat chequered background, the present inventory comprises an extremely varied total of hardware. There are almost certainly still examples of Soviet radars associated with missile systems previously supplied, such as the SA-2 Guideline, SA-3 Goa and SA-6 Gainful, various French types and Egypt has also consistently been reported as one of the unidentified Middle Eastern states said to have been supplied with Plessey's AR-3D three-dimensional radar although this cannot be confirmed.

In 1980 four General Electric AN/TPS-59 3-D radars and, later, eight Westinghouse AN/TPS-63 air defence radars were ordered for Egyptian use at a combined cost of about $154 million. These have now been delivered. In the same period 12 AN/TSQ-143 mobile automated operations centres, which interface with the TPS-59 and TPS-63 radars, were ordered from Westinghouse, for operation of Hawk missile batteries. Each operations centre has four OJ-560 display consoles, two main computers, 12 high-density micro-processors and an audio/video recorder that files all track data, displays and operator actions. Most, if not all, of these 12 centres have been completed and are operational.

In mid-1986 it was announced that Egypt had signed an agreement with Westinghouse Defense and Electronics to co-produce the US company's AN/TPS-63 tactical radar. The deal, stated to be worth $190 million over seven years, will cover joint production of 34 radars by Westinghouse and the Egyptian Benha Electronics plant.

Hughes received a $210 million contract in December 1983 for the first phase of a new Egyptian national air defence system, and this called for integration of all existing radars, missile batteries, air bases, and command centres into an automated command and control system. Also integrated are E-2C Hawkeye airborne early warning aircraft, the first two of which are in service, with three more which were due to be delivered by the end of 1987.

India
In early 1974 an Indian Air Defence Ground Environment System (ADGES) programme was revealed when it was announced that indigenous production of equipment was to start in January 1974 at the Bharat Electronics Ltd plant at Ghaziabad, east of Delhi. The system was installed along the Indian northern and western borders (with Pakistan and China) and a communications system for use in the air defence of a limited area along the northern borders was built with American equipment.

The current air defence system is understood to rely on a miscellaneous assembly of radars of varying origin, reflecting India's changing international associations and allegiances since gaining independence. This includes British, French, Soviet and possibly other types of equipment.

Between 1962 and 1978 at least 24 long- and medium-range radars of Italian design were installed ostensibly for ATC use, but clearly so that target data from some of them at least would be useful for air defence/surveillance. Most of these radars are D-band equipments, eight of them ATCR-2 systems and the remainder ATCR-4Ts (**2248.153**). Licensed production of the latter type was undertaken in India from 1971 onward, more than 10 being provided in this way. It is very probable that the Soviet-built SA-2 and SA-3 air-defence missiles used by the Indian Air Force rely on a number of Soviet radars.

Thomson-CSF revealed in February 1984 a contract to supply four 3-D military radar stations to India, and the construction under licence of five more stations. India is now developing an indigenous air defence radar industry, largely based on co-operation between Thomson-CSF and Bharat Electronics. The TRS-2215D systems being built under licence form part of the ADGES plan.

Indonesia
Following the earlier updating of the previous air defence/ATC network further modernisation has taken place with the acquisition of two TRS-2215 systems (**3272.153**), the transportable version of the TRS-2230 static E/F-band system, which have been operational for some time. More recently, certain equipment of Dutch origin is thought to have been acquired for short-range air defence purposes. Much more recently Indonesia has decided to update its air defence system for the central area and has instituted a competition. The companies bidding include Plessey, Thomson-CSF, Hughes and Westinghouse, but to date no further information has been released. Air defences are also being updated with part licence-built BAe Rapier SAMs. Eight F-16 fighters are on order.

On completion of a contract, announced in June 1984, to provide four new ATC centres, Thomson-CSF will have installed over a period of ten years coverage for most of the Indonesian archipelago with TA 10 or TA 23 primary radars, RS 770 or RS 870 secondary radars, and 10 regional control centres. In mid-1985 Thomson-CSF revealed a further contract to extend the coverage of the eight primary, and 11 secondary surveillance radars then serving Indonesian airspace. This extension to the coverage will mean the supply of several TRS-2215D or 2230D radars.

Jordan
Jordanian air defence is based on F-5 and Mirage F-1 aircraft and Improved Hawk SAMs (**2640.131**), supported by AA guns. The main command and control element is the AN/TSQ-73 air defence missile control and co-ordination system (**2835.181**, later in this section). Following US refusal to supply Jordan with F-16C or F-20 aircraft, or mobile Improved Hawk missiles, it can now be expected that the radars of western origin including Westinghouse AN/TPS-43 equipments which are operational in Jordan, will be joined by Soviet equipment. Jordan has SA-8 Gecko, SA-13 Gopher and SA-14 Gremlin SAMs in operational service, and Shorts Javelins and BAe Rapiers on order. Supply of 20 Mirage 2000 and 8 IDS Tornados has been agreed in principle.

Kenya
In June 1985 Thomson-CSF reported the award of a contract to provide complete air coverage for Kenya by means of a system comprising a central control facility supplied with data from four radar sites. This will probably be an integrated civil/military system with the main centre at Nairobi and with an ancillary control centre at Mombasa in the SE of the country. A number of BAe Rapier SAMs have been ordered.

Kuwait
In recent years Kuwait has significantly increased its air defence forces with acquisitions of Mirage F1 interceptors and Hawk surface-to-air missiles, of which seven or eight battalions have been ordered. Recently acquired sensors include the ITT-Gilfillan TPS-32 mobile three-dimensional radar (**2516.153**). Kuwait is also reported to have been the recipient of one or more Plessey AR-3D air defence radars (**2484.153**) although this has not been confirmed. For command and control of the Hawk missiles two TSQ-73 systems (**2835.181**) were ordered and are thought to have been operational since 1982.

In June 1982, ITT Gilfillan was paid more than $5.3 million by the US Army Missile Command for support of two AN/TPS-32 radars in Kuwait. In November 1983 Thomson-CSF was awarded a contract to modernise the low-altitude air defences of Kuwait and to set up an integrated command system for the entire air defence infrastructure. Gradual implementation involves the supply of TRS 2230 3-D radars. For defence against air attack, Kuwait employs SA-6, SA-7 and SA-8 SAMs.

During 1984 and 1985 various projects were implemented with the aim of further improving Kuwait's air defence. In mid-1984 steps taken with US assistance included: certain radar improvements; AN/TQS-73 Missile Minder system for use with PIP Hawk SAMs; and arrangements for data link facilities to allow Kuwait to share data obtained from Saudi Arabian E-3A AWACS aircraft, and use the information for direction of Hawk missile batteries. In March 1985, Thomson-CSF announced a Fr 1.2 billion contract for the KAFAD command, control and communications programme to modernise Kuwait's air command and control system. This includes a series of radar stations around the country's borders and a central command post.

With the escalation of the Gulf War Kuwait has approached various Western nations to upgrade its air defence network. In particular Kuwait is seeking an anti-missile defence system to protect its oil terminals and tankers from attack by Silkworm missiles. The US Defense Secretary announced in January 1988 that he had discussed Kuwait's air defence needs and teams were in Kuwait helping with improvements. It is reported that Kuwait is interested in buying Contraves Amoun air defence systems from Egypt. Amoun is a version of the Skyguard system manufactured under licence by Egypt.

Libya
For several years plans to expand and modernise the Libyan air defence network have been hampered by political obstacles. US, UK, Italian and French companies and consortia competed for the task, but it is now thought that Libyan air defence is modelled on Soviet principles and heavily reliant on equipment from the same source. A certain amount of air defence radar damage was caused by the use of the Harm anti-radar missile by US aircraft during the Gulf of Sirte incident in April 1986. The extent is not known and was probably repaired or replaced rapidly.

Libya has a separate Air Defence Command as part of the air force. A command and control system known as SENEZH is in existence but no details are available. The Command operates a large number of Soviet missiles, including SA-2, SA-3, SA-5A, SA-6 and SA-8, plus AA guns. The Libyan Army operates SA-7 and French Crotale SAM batteries.

Malaysia
Following the British withdrawal from the Far East in 1971, the Malaysian Government purchased two Marconi S600 mobile radar units (**2426.181** and **1168.153**) to meet the country's immediate air defence needs and to support the two RAAF Mirage squadrons based in Malaysia. About a year later a third S600 system was procured. In early 1983 it was learned that a Hughes air defence radar, HADR (**3721.153**), had been ordered for MADGE (Malaysian Air Defence Ground Environment) for which Hughes Ground Systems Group is the prime contractor.

This system is now operational, providing air surveillance and identification services for civil and military aircraft in airspace of interest to Malaysia. It uses modern data processing equipment, large screen displays, new communications and the Hughes 3-D radar which is probably used in conjunction with earlier existing radars to provide target data for the entire country. Control of radar sensors, aircraft and other air defence assets is exercised from a number of control centres.

Malaysia is very keen to upgrade its air defence capability by the acquisition of AWACS type aircraft, new fighters, low-level air defence radars, various surface-to-air missiles and low-level air defence guns. However, these plans have had to be shelved because of budgetary restraints but will probably be resurrected in the early 1990s.

Morocco
A series of contracts, collectively assessed as worth about $100 million, for the provision of an air defence network was announced in June 1977. The principal contractors named were Westinghouse, Burroughs

and Ford Aerospace. Westinghouse provided the primary radars and integrated the entire system, while Burroughs supplied large-scale B-6700 series computers for processing tracks, ATC data, identification, plot extraction, etc. Ford Aerospace provided microwave communications links under a contract worth about $8 million. Value of the Burroughs contribution was approximately $30 million.

The complete system is based on 16 AN/TPS-43 radars at early warning and coastal sites. A centralised control facility directs Moroccan air defence aircraft. Main air defence forces consist of F-5 and Mirage F-1 aircraft, with SA-7 and M-730 Chaparral SAMs.

Nigeria

In 1984 Nigeria continued its programme of refurbishment and upgrading of the national infrastructure, including the airspace control services. A Selenia RAT-31S has been procured and is in operation for civil/military purposes at Lagos. This radar is also used for control of nearby AA units, consisting of guns and Roland SAM units. Elsewhere in Nigeria, ATC radars that were shut down for many months have been reactivated. These include: Enugu radar installed in 1980 but only commissioned in 1984; Kano and Kaduna airport radars.

Philippines

Although independent since 1946, the Philippines has relied almost entirely on the USA for its defence, and this is reflected in the inventory of the Philippine forces. Under the terms of the treaty between the two countries, the USA enjoys important base facilities in the Philippines, notably the Subic Bay naval base and Clark AFB, and over the years there have been appreciable supplies of equipment of US origin. The Philippine forces operate Hawk surface-to-air missile batteries, and F-5 fighter aircraft are also available for air defence. In addition to the target detection and designation radars associated with the Hawk missile batteries, two Italian-built ATCR-3T radars supplied for ATC purposes at Manila and Mactan in 1972 could also serve a supplementary surveillance role. Four ITT-Gilfillan Series 320 three-dimensional air defence radars have been delivered under the terms of a $13 million contract given to the US company by a Dutch contractor.

During the past few years the Philippines has been preoccupied with internal security and this is likely to be the main consideration in the forseeable future. The US air and naval bases at Clark Field and Subic provide a guarantee against external attack. It should be remembered however that the base agreements expire in 1991 and if not renewed would mean that a substantial increase in the country's armed forces would be needed.

Saudi Arabia

On to an original air defence network based on radars of American and British origin dating from the 1960s, a succession of improvements have been grafted, either for their overall system enhancement potential or as adjuncts or essential elements of a new weapon system acquired for the Saudi Arabian defences. Examples of the latter process are the radars associated with the US Improved Hawk surface-to-air missile batteries or the French Shahine (Crotale). The Saudi Arabian Army has 33 SAM batteries, 16 with 128 Improved Hawk (three missiles each) and 17 with 68 Shahine fire units (six missiles each), with a further 60 Shahine and 70 Improved Hawks operated by the Air Defence Forces, a separate force of equal standing with the Army and Air Force. In 1980, the United States supplied one AN/TPS-43 3-D radar to strengthen the region's defences and at the same time deployed AWACS aircraft to serve as an airborne control post based in Saudi Arabia. Four of these aircraft were used in this way and three additional AN/TPS-43 radars were ordered and supplied by Westinghouse. In 1981 a $75 million contract was signed for additional AN/TPS-43G radars, plus modifying the original TPS-43s to the 43G standard.

Defence ministers from six Arab states in the Gulf area agreed in principle to establish a joint air defence system based on Saudi Arabian airborne warning and control aircraft. The states concerned are Saudi Arabia, Kuwait, United Arab Emirates, Bahrain, Qatar and Oman. The overall project is known as Peace Shield and was finalised to include five E-3A AWACS aircraft; eight Boeing KE-3A tanker aircraft; a command, control and communications complex with its underground command centres, data processing facilities and communications links, and a new ground network of 17 AN/FPS-117 long range radars similar to the North American Seek Igloo radars. Work on the estimated $8.5 billion programme started in 1981 and by 1987 the centralised command operations centre had been constructed and work on building and connecting five sector command and operations centres to be linked to the centralised facility was nearly completed. The first E-3 aircraft was delivered in June 1986 and all 13 E-3 and KE-3A aircraft have now been delivered. The Royal Saudi Arabian Air Force is also receiving 72 Tornado aircraft (48 IDS version and 24 ADV) as well as 30 BAe Hawk aircraft.

Plans to link Saudi Arabian defence systems and those of neighbouring countries with the AWACS aircraft by data link were completed during 1984-85 and are being implemented.

During 1984 an agreement was reached between France and Saudi Arabia to integrate new versions of the Shahine missile with the national air defence network and it has been reported that a contract has been signed for the maintenance of the Shahine network to be undertaken by France. Estimated value was $4 billion. Also in 1984 contracts were awarded to Westinghouse for improvements to the AN/TPS-43 radars, communications and troposcatter links. A new arms sales package from the United States was agreed in November 1987 and included 12 more F-15 aircraft to be delivered individually as replacements for the existing F-15 interceptor force.

Singapore

During 1980 a joint civil/military integrated ATC system based on Signaal LAR II long-range surveillance radar and data handling equipment by the same manufacturer, under the name LORADS (long range radar and display system) was commissioned. Most of the hardware, including the radar antenna, is located at Changi International Airport, and the data stored in this computer-based system is available to both civil and military authorities. In a move, in late 1982, that was seen in some quarters as a preliminary to the formation of a regional air defence system embracing five ASEAN states (Indonesia, Malaysia, the Philippines, Singapore and Thailand), Singapore requested approval for the purchase of four E-2C airborne early warning aircraft from America and contracts were awarded to Grumman in 1984. The four aircraft have now been delivered and are operational with the Singapore Air Force, providing airborne surveillance of most of South-East Asia's airspace.

The island is protected by three layers of surface-to-air missile air defence, consisting of Bloodhound 2, Rapier, Improved Hawk and RBS-70 systems. In line with Singapore's doctrine of the deterrence of threats through the maintenance of a pre-emptive strike capability, the country has a strong air force equipped with F-5E fighter and A-4 attack aircraft. The latter force is being re-engined and equipped with Maverick air-to-surface missiles, with its range considerably extended by the conversion of four C-130s to in-flight refuelling tankers. Eight F-16 aircraft are on order and are due for delivery during 1988-89.

Southern Africa

This area covers Angola, Botswana, Malawi, Mozambique, Namibia, Tanzania, Zambia and Zimbabwe. Angola air defence has been described under a separate entry. In general Southern Africa has been supplied from Soviet sources, although in some cases Chinese equipment is in operational use. Although there appears to be no centralised air defence network the area is well supplied with surveillance radar systems, many of which are Soviet types. The accompanying diagram gives locations of those sites that have been identified, together with their coverage on a 1 m sq target at 20 000 ft.

Most of the surface-to-air missiles deployed are Soviet types, including SA-2, SA-3, SA-6 and SA-7, plus a variety of anti-aircraft guns of various types and calibre. Mozambique operates MiG-17 and 21 aircraft, Zambia has MiG-21 and Chinese J-6s, Tanzania operates J-4, J-6 and J-7s, while Zimbabwe has a variety of BAe Hawks and Hunters, with a J-6/J-7 squadron in formation. Botswana and Malawi do not possess combat aircraft.

South Korea

South Korea is being supplied with Improved Hawk surface-to-air missile systems and is procuring the

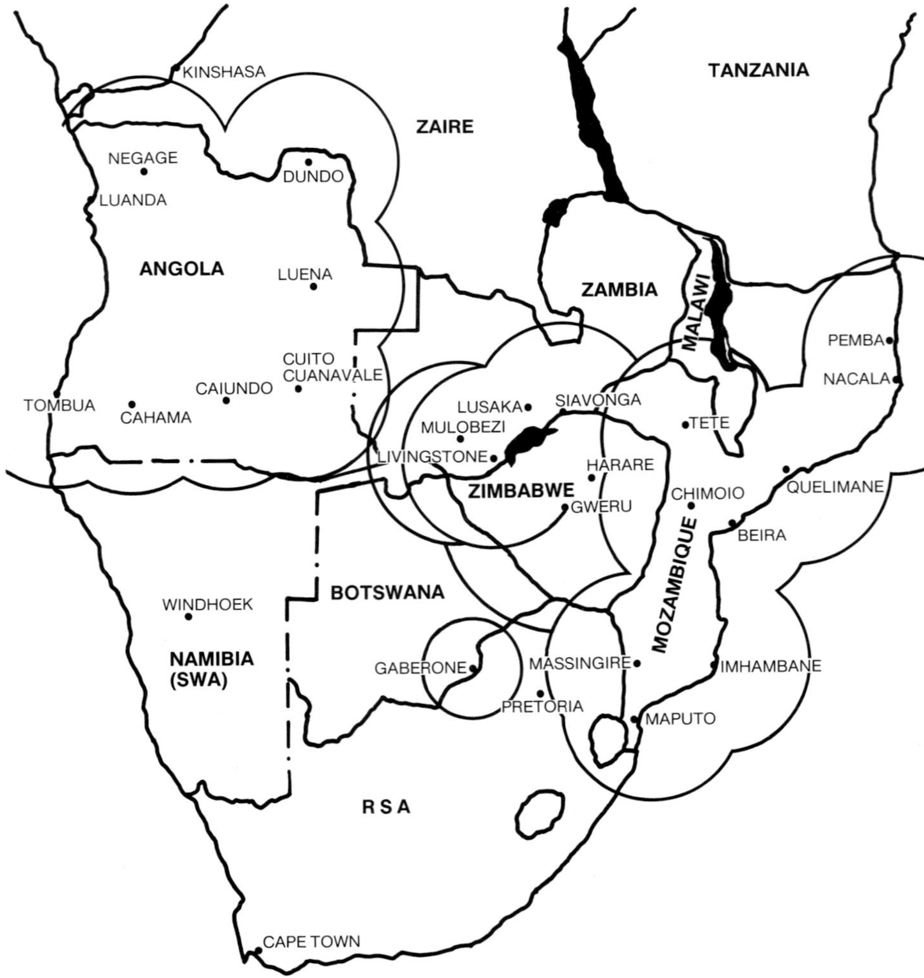

related TSQ-73 system (**2835.181**) for command and control. A new overall air defence system to provide for target assignment, early warning, and co-ordination of other air defence forces such as interceptors is being developed. The main air defence consists of F-4, F-5 and F-16 aircraft, with more of the last type on order.

Taiwan
It is believed that Taiwan is engaged in the development and implementation of an automated air defence network similar to those provided by Hughes for Japan, Switzerland, Spain, and Israel. Taiwan currently possesses over 1600 SAMs, including 800 Hawks, 420 Chaparrals and 400 Nike Hercules, as well as Tien Kung 1 and 2 (Patriot type). On order are large numbers of Improved Hawk, Chaparral and Tien Kung 1 SAMs. These are almost certainly operated by AN/TSQ-73 automated air defence systems supplied by Westinghouse who has also provided TPS-43 radars. Airborne air defence is provided mainly by over 250 F-5E/F aircraft and about 100 F-104s.

Thailand
Air defence assets operated by this state include two AA artillery regiments armed with 40 mm guns and other weapons, two air defence squadrons of F-5 aircraft, and miscellaneous AA weapons operated by other units. The volatility of the potential threats facing Thailand, and its geographic circumstances, pose difficulties in arranging for appropriate air defence command and control.

A programme to automate Thailand's air defence system has begun at an estimated total cost of $207 million. The work, in three phases, will be financed with FMS credits. It involves replacement of the RTAF's present radar network and modernisation of the Central Air Defence Sector command, control and communications system.

The first phase links radars in the Central Plain including areas bordering Kampuchea. This also provides a data link capability for interoperability with US forces. The second and third phases, covering the north and south remain in the balance due to budget problems.

Thailand ordered a new generation AN/TPS-70 series mobile tactical radar designed to track hostile aircraft out to 450 km. The unit, which has now been supplied, is the third modern radar supplied by Westinghouse who delivered two of its AN/TPS-43 series in 1980. All three of these units will be used in the first phase of the automation. A number of Sanders LAADS systems (**3123.153**) were ordered in 1985 and deliveries commenced in 1987.

The Mitre Corp of Bedford, Massachusetts, is providing systems engineering support. On the US side, the overall work is being handled by the USAF's Electronic Systems Division at Hanscomb AFB.

The Contraves Skyguard low-level anti-aircraft air defence system has been ordered. Up to four batteries may be procured. Also on order are the Blowpipe SAM and more Vulcan 24 mm AA guns. 18 F-16 fighters are being procured and delivery was due to commence in August 1988. The F-5E/F fighter force is being upgraded with first flight trials of modified aircraft due in 1988.

United Arab Emirates
It was reported in 1986 that Hughes was negotiating with the UAE for a study contract for an operations centre and system to integrate air defence in the area, including fighters, guns and SAMs. The UAE defence forces currently operate Rapier, Crotale and RBS-70 surface-to-air missile systems, as well as two interceptor squadrons equipped with Mirage 5 aircraft. On order are 36 Mirage 2000s, 24 BAe Hawk aircraft, 42 Improved Hawk missiles and Skyguard air defence systems. There are plans to invest more than $700 million in radar and communications equipment, and in mid-1987 the first of three AN/TPS-70 radars was delivered. The other two are due for delivery in 1988. Westinghouse is also supplying command and control equipment to communicate, correlate and display real-time data from the radars.

Venezuela
The Venezuelan forces include a number of weapons specifically intended for anti-aircraft applications, and although no details of a dedicated air defence network have been received, two Italian-built ATC long-range radars supplied in 1973 do have the useful advantage of being sited at virtually the western and eastern extremities of the country. A Selenia ATCR-2 with computer-based display system is located at Maracaibo, while in the east there is a Selenia ATCR-3. Westinghouse is known to have supplied air defence equipment, which probably includes the TPS-43 air defence radar. A tri-service evaluation committee was set up in 1980 to consider the development of a national air defence system but no further information has been released. As with other oil-producing states the substantial decrease in oil revenues has probably curtailed the defence budget.

Yugoslavia
This communist state, though relying to a marked degree on Soviet and Warsaw Pact sources for much of its military equipment, nevertheless manifests a stoutly independent line in certain areas of its defence procurement. Thus, although there is a marked preponderance of Warsaw Pact systems, such as SA-2, 3, 6, 7 and 9 SAMs, various AA artillery and the ZSU-57-2 self-propelled AA gun in its air defence inventory, Western hardware does find a place in the Yugoslav order of battle. One such instance, reported in 1982, was a Westinghouse contract valued at $39.1 million for six AN/TPS-70 surveillance radars.

AUSTRALIA

3521.181
AUSTRALIAN AIR DEFENCE
A wide ranging defence strategy was outlined in a Defence White Paper produced in March 1987. It commits Australia to high quality intelligence together with flexible, long-range capabilities for maritime surveillance and interdiction, aimed at denying any hostile nation effective use of the sea and air to the north of the country. In September 1987, the Australian Minister of Defence released information on the new defence budget. In a statement at the same time he said: "The Department of Defence will spend $704 million in 1987-88, providing an average real growth in defence spending of 1.7% a year over the five years of the Hawke government." The key elements are the development of Jindalee (see later), the upgrading of the RAAF 20 P-3C maritime surveillance aircraft, the strike force of F/A-18 and F-111 aircraft, plus, of course, naval and ground forces. This strategy, which will update the Hub Cap system described below, will be implemented during the 1980s and 1990s.

Australia's sparse population in conjunction with its continental size combine to make a conventional static air defence organisation along the lines of that of the USA an unrealistic proposition. Instead, an effective compromise has been achieved by the combination of Hub Cap air defence systems (two) (**2016.181**) with the search radars and part of the infrastructure needed for the control of civil and military air traffic serving the main population centres of Australia. In this way the costs of fixed installations are minimised while operational flexibility is maintained.

Hub Cap was described in detail in this section of *Jane's Weapon Systems 1979-80* and in previous editions, but a short summary of the salient features is retained in this entry for convenience.

The main RAAF defence forces available consist of F/A-18, F-111 and Mirage III aircraft, although the Mirages are being phased out as the F/A-18s come into service. The 23 F-111 aircraft will be upgraded with improved avionics, and the operational capability of the 75 F/A-18s, which are either in service or on order, is being improved with the conversion of the RAAF's four Boeing 707s to the in-flight refuelling role. Some of the F-111s, F/A-18s and the P-3C maritime surveillance aircraft are being fitted with Harpoon missiles. In addition the RAAF is considering the acquisition of a number of airborne early warning aircraft, and a request for proposals has already been issued.

The ground-based anti-aircraft forces are equipped with Rapier SAMs (some fitted with DN 181 Blindfire radar), RBS 70 portable SAMs and AA guns. The RBS-70 portable AA missiles are being phased in to replace Redeye.

The Hub Cap system provides three-dimensional airspace surveillance and control of fighters to intercept hostile targets over a wide area from fully mobile air-conditioned shelters designed for deployment by C-130 aircraft.

The use of digital computing permits automatic tracking of targets from the primary radar while plot extraction from the secondary radar returns is used to reinforce the primary radar auto-tracking. For the RAF requirements, the computer is programmed to take account of the aircraft characteristics, perform an assessment of an interception situation when supplied with the relevant data on weapon availability, and provide a number of alternative intercept solutions for selection by a controller. Comprehensive communication facilities are provided with Hub Cap enabling operational personnel to communicate with each other, with aircraft under control, and with units external to the system.

The original Hub Cap radars have now been replaced by two Westinghouse AN/TPS-43 systems (**2517.153**) which were commissioned in 1982. A tactical air defence system, also provided by Westinghouse, is now operational and allows for radar-directed control of defensive fighter aircraft operations wherever the system is deployed.

Construction of a back-scatter over-the-horizon (OTH) radar for the detection of both aircraft and ballistic missiles was first revealed in 1974. Code name for the project is Jindalee and the site is near Alice Springs, close to the US satellite facilities there.

Promising results from the experimental narrow-beam Jindalee OTH radar indicated that future surveillance of the Australian coast and surrounding offshore regions would be undertaken by this technique. By 1978, about $A6 million had been spent on Jindalee research and development. Ranges of more than 1000 miles (1600 km) against civil airliner targets had been obtained with the Jindalee installation, and in late 1978 approval was given for a larger and more powerful version to have a much wider coverage arc (Stage B). Construction started at Alice Springs, with an anticipated total cost of $A24 million. This installation has greater transmitter power and a larger receiving antenna, and will provide for surveillance cover of airspace to the north of Australia. The new Stage B receiving station was constructed at Alice Springs, next to the original Stage A site on Bond Springs Station. The new reception antenna array is 2.8 km long. Much of the large transmitter complex at Harts Range, about 160 km from Alice Springs, had been completed by early 1982 and it was necessary to build a new 3 MW power house and building to house the additional transmitter equipment. A number of advanced items of new equipment developed at the Australian Electronics Research Laboratory, South Australia, was moved to the Stage B sites in 1981.

As a result of highly successful trials it has now been confirmed that a network of three Jindalee OTH radars will be constructed at a cost of some $A500 million. Studies are currently in progress on their precise location, with a probability of two in the north-east of Australia and one in the south-west.

AUSTRIA

3326.181

AUSTRIAN AIR DEFENCE (MRCS-403)

Selenia has supplied a number of mobile and fixed air defence systems for incorporation in the nation's overall defence network. The systems are designated MRCS-403 mobile automatic reporting and control system, and are based on the RAT-31S three-dimensional radar (**1953.153**) in conjunction with a data processing and display sub-system.

The radar's main characteristics are simultaneous range, azimuth and elevation measurements of targets at all altitudes, sophisticated ECCM performance, and high resistance to ground and/or weather clutter.

The MRCS-403 produces an air situation picture which is transmitted back to superior command centres via a narrow-band data link. This is performed together with functions such as automatic track initiation, automatic tracking, automatic and/or manual height measurement and automatic track reporting. The system can process three-dimensional low- and medium-altitude air situations, and is also able to perform autonomous control functions such as identification, interceptor and air defence missile control. In its basic configuration it comprises a shelter-housed RAT-31S radar and data processing and display shelter. For expanded operational capabilities, additional shelters are used on a modular basis to provide more operator and communications facilities.

The operations shelter contains a dual NDC-160 computer complex, three display consoles, a communications control unit, and ground-to-air facilities. The computer is integrated locally with the three-dimensional surveillance radar, and for remote IFF/SSR inputs the system is equipped with a data link interface for receiving plots. The computer processed information is fed to the display sub-system as synthetic/alert track information and to remote centres as track information via a data link.

Typically there are five operator positions: master controller, track controller positions (2), manual input operator, and the radar supervisor. Track capacity is 60 (expansible) and three intercepts can be handled simultaneously. The displays employ synthetic video presentation and there are two video maps for each unit.

STATUS

Fully operational.

CONTRACTOR

Selenia, Industrie Elettroniche Associate SpA, Via Tiburtina 12 400 Km, 00130 Rome, Italy.

Internal view of a Selenia MRCS-403 system operational shelter supplied to the Austrian MoD

The first Austrian site to be equipped with Selenia's MRCS-403/RAT-31S air defence radar

BRAZIL

4898.181

SISCEA INTEGRATED AIRSPACE CONTROL SYSTEM

The integrated civil/military airspace surveillance and control system now being implemented in Brazil is the culmination of a series of predecessor systems representing stages of evolution of the complete system that were known as DACTA I, II and III. These were described under entry number **2166.181** in earlier editions of this book. The overall system is known as Siscea, and is a vast radar and operations centre network covering three-quarters of Brazilian airspace; with Thomson-CSF in partnership with Brazilian industry. The complete system is under the control of CISCEA (Comissão de Implantacção do Sistema de Controle do Espacço Aéreo).

DACTA I was initiated in 1972 and became fully operational in 1976, covering the central part of Brazil. This led to a series of phased extension and upgrading programmes (DACTA II, DACTA III and Amazonia) to expand the coverage and facilities, together with a modernisation at a later stage of the original DACTA I.

The updated DACTA I uses LP-23 M long-range surveillance radars (**3099.153**) and RS 870 SSRs supplying data to the Brasilia regional control centre which provides radar control over the Rio/São Paulo/Brasilia triangle within which most of Brazil's civil air traffic is concentrated. The DACTA II phase relates to the introduction of TRS 2230 3-D radars (**3272.153**) and more RS 870s to provide radar coverage and control over the entire southern part of the country from the Curitiba centre, including the extensive military needs of this area. The completion of DACTA I and II gives unbroken coverage of about four million square kilometres. The continuation, DACTA III extends this cover over the whole north eastern region of Brazil, with a new regional control centre, probably at Recife. Two-dimensional primary surveillance radars with SSR are used in this phase of the programme. In total, the DACTA II and III phases entails the supply of 11 3-D long-range radars and four approach radars, as well as operations centres equipped with between 20 and 30 consoles.

The final stage of the Siscea project is the implementation of the Amazonia phase which in-

A comprehensive radio, telephone and microwave link system co-ordinates the extensive radar and communications network needed for Brazilian ATC/air defence control purposes

cludes the provision of two new autonomous regional control centres with LP-23 M surveillance radars and RS 870 SSRs at Manaus and Belem. The ultimate phase of the programme will result in there being five regional control centres, the main ones being Brasilia, Curitiba and Recife.

Concurrently with, and no doubt closely interconnected with the Siscea system (although to what exact extent is not known in detail) are numerous other radars (civil and military) which have been installed or ordered in recent years. Among such items are a number of Selenia RAT-31 3-D surveillance radars (**1953.153**) and various airfield surveillance and terminal area radars of French and Italian manufacture. A number of French height finding radars were also supplied and installed during the earlier years of the DACTA stages of the programme and it is likely that a significant proportion of these are still operational.

At the Brasilia integrated detection and control centre, target data from search, height finder and SSR sensors are received via microwave links and processed by two automatic data processing systems. The computer programming is fully integrated and its functions include tracking of military and civil aircraft, interception for air defence, and flight plan processing for ATC purposes. The centre consists of separate operations rooms for ATC and for air defence, and there is also an automatic switching centre for message management and communications.

STATUS

Implementation of the Siscea programme began in 1972 and in its DACTA I format the system has been operational since 1976. DACTA II was due to be completed by the end of 1987 and work on the upgrading of DACTA I and construction and installation for the DACTA III extensions is well under way. These phases involve the replacement of some of the originally installed radars, for example, the LP-23 M and Volex 3-D radars at the Tres Marias site have been replaced by a TRS 2230 3-D radar. At least 12 TRS-2230D systems have been procured under DACTA II. The original LP-23 M and Volex radars could be used elsewhere such as the regional control centres at Manaus and Belem in the Amazonia phase of the programme. Completion of the Amazonia phase and implementation of the full five-centre system is estimated to require several more years.

Technology transfers have been such that Brazilian industry is a full partner in the project. The hundred or so consoles, the secondary and approach radars, the telecom and launching aids facilities were manufactured in Brazil with Thomson-CSF partnership. By the end of 1987, radar coverage extended over the centre, south and north-east of Brazil, as well as part of Amazonia.

CONTRACTORS

ESCA Engenharia de Sistemas de Controle e Automacao SA, Brazil.

ELEBRA SA, Eletronica Brasileira, Brazil.

Tecnasa, Eletronica Profissional SA, Brazil.

Thomson-CSF, Detection Control and Communication System Group, Paris, France.

TRS 2230 three-dimensional surveillance radars and RS 870 SSRs like this will cover the entire southern Brazilian airspace when the Siscea ATC/air defence network is completed

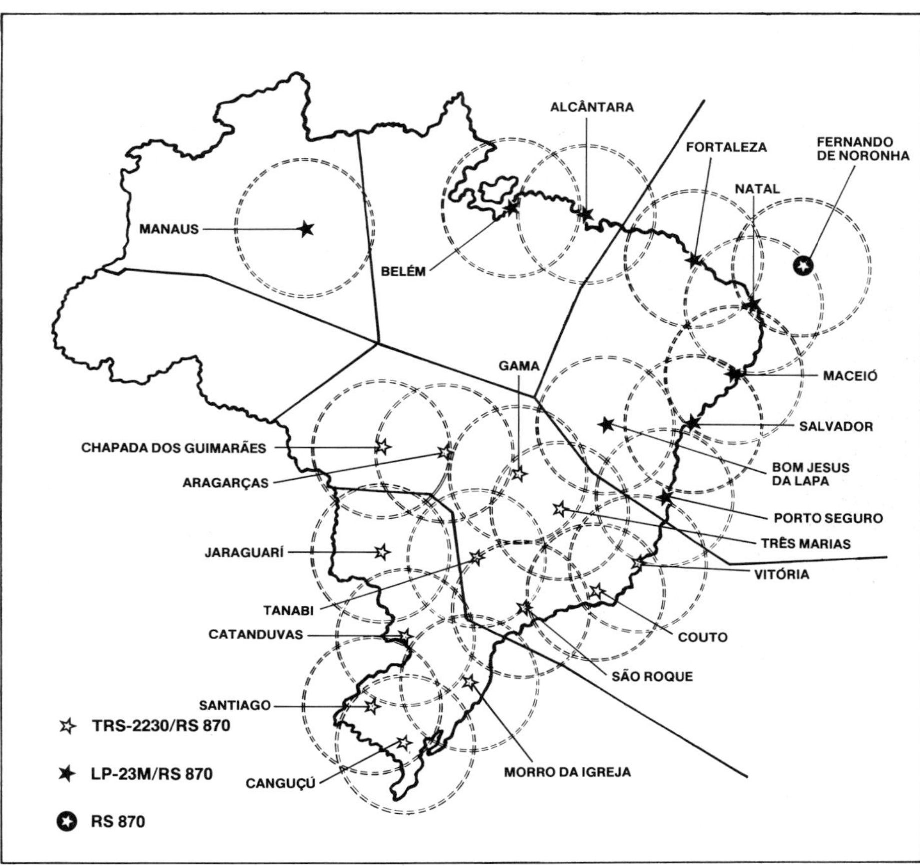

Map showing the full extent of the Brazilian integrated air defence/ATC radar network when the Siscea programme is completed. The six air defence zones, with headquarters at Belem, Recife, Rio de Janeiro, São Paulo, Porto Alegre and Brasilia, are served mostly by LP-23 M surveillance radars in the north and TRS 2230 3-D radars in the south of the country

CANADA

6805.181

CANADIAN AIR DEFENCE

Canadian air defence is inextricably bound up with that of the United States, and the 1985 North American Air Defence Modernization (NAADM) accord between the two countries underlines the joint requirements. A White Paper produced in June 1987, entitled Challenge and Commitment, outlined Canadian defence policy for the rest of the century and emphasised the basic tenet that Canadian defence policy "will continue to be based on a strategy of collective security within the North Atlantic Alliance, including the continental defence partnership with the United States".

The basic NAADM project comprises a new North Warning System (NWS) to replace the elderly DEW radars, including some new sites in Canada; the OTH-B sites in the United States (see entry **5617.153**); and the use of US AWACS aircraft. This latter accord included the setting up of dispersed operating bases for the E-3s and also provided for the provision of forward operating locations for interceptor aircraft. The NAADM programme is funded jointly by the United States and Canada.

The NWS radars are being positioned in Alaska, North Canada and down the Labrador coast, and consist of both long-range and short-range surveillance systems. Details of the system are contained in the United States entry in this section (**2567.181**). Canada is responsible for the project management, system integration, new facility design and construction, operation and maintenance of the system in Canada and is also supplying the complex communications network of the NWS.

Canada is also procuring a new low-level air defence system for the 1990s, the cost of which is likely to exceed $Cdn1 billion. The project is a most ambitious plan to equip the Canadian Armed Forces with the Martin-Marietta/Oerlikon developed Air Defence/Anti-Tank System (ADATS), the Contraves Skyguard fire control system and the Oerlikon GDF-005 twin-barrel 35 mm anti-aircraft gun. The new defence system was chosen after an international competition won by the team led by Oerlikon-Buhrle of Switzerland. Initial deliveries are due to begin in 1988 and are spread over the following two or three years. The LLAD system is scheduled to be fully operational by 1991.

CHINA (PEOPLE'S REPUBLIC)

2052.181

CHINESE AIR DEFENCES

No detailed information on Chinese air defences is available, but there are some known statistics on their magnitude.

Until comparatively recently the Chinese were reported to have in service about 100 surface-to-air missile sites and some 15 000 air defence guns of various types. An interceptor aircraft force of over 4000 aircraft is maintained, consisting of MiG-15, MiG-17, MiG-19, and MiG-21 types. A new all-weather fighter of Chinese design recently entered service. Associated with all of these were about 1500 air defence radars, but little is known of the detailed deployment of these systems or their organisational structure beyond the general information that the radars constitute a network for air surveillance and control and a missile early warning system for the detection of ICBM launches against China. More recent information, while still leaving much open to question, suggests that the country's defences are being substantially modernised; new radars, it seems, have been designed, and improved versions of the SA-2 Guideline missiles (**2942.131**) which the Chinese originally acquired from the Soviet Union are being produced as the HQ-2J. However a recent American assessment still considers the Chinese air defence system to be subject to major weaknesses, such as an outmoded command and control system.

In February 1984 it was reported that a French company has, or will have, supplied six long-range radars, three terminal area radars, and control centres for airspace management of the Shanghai/Peking region. Certainly China lacks a centralised and computerised air defence system and responsibility for airspace control appears to rest largely with local air defence headquarters. There are about 100 surface-to-air missile sites, mainly equipped with HQ-2 (SA-2b) missiles and approximately 16 000 AA guns.

It should also be noted that at least one very large phased-array radar has been constructed in Western China, presumably as part of the missile warning system. Although information on the latest developments is difficult to obtain, there seems little doubt that China is engaged on an indigenous surveillance radar development programme. The extent of this in relation to an ABM system is not known but with such a vast country it seems likely that this will take many years. In late 1986 it was stated in Beijing that an OTH radar programme was being implemented.

In proportion to the size and population of China, its defence budget appears to be very small. The general impression is that defence spending is a relatively low priority compared to other major reforms being carried out.

FRANCE

4571.181

T.ACCS

Thomson-CSF Air Command and Control Systems (T.ACCS) is the designation of the improved MIDAS (**2106.181**) range of systems that the company designed and developed for such purposes. There are three basic types of station.

All these stations are implemented using the same line of versatile modular equipments, based on the latest data processing and display techniques. In addition to the radar and telecommunications equipment, these equipments include radar plot extractors, data processing systems, and display systems. Compared with the original MIDAS systems, increased computer facilities make possible significantly greater track capacity and data handling capabilities with little if any increase in manning requirements.

The company range of radar plot extractors includes models matched to 2- or 3-D radars, taking into account the importance of low-altitude surveillance and the possibility of enemy ECM actions. They may be specific sub-assemblies incorporated with the radars in new equipments, or programmable assemblies which can be matched to previously installed radars.

Thomson-CSF air defence systems use general-purpose computers and peripherals designed and packaged for various environments: standard environment, aboard transportable cabins or aboard armoured vehicles. For the specific air defence software, Thomson-CSF has developed a full set of packages devoted to the various functions: automatic and rate-aided manual tracking, threat evaluation, weapons allocation, automatic interception, missile designation, simulation. Operational software for each station is then built up by linking these various modules.

Display systems are of the autonomous type; each display console includes its own refreshing memory and a programmable processor, which digitally performs various functions connected with the display, such as scaling and off-centring, synthetic information selection, etc or gives tactical assistance to the operators. These local functions performed by the console processor can be adapted to various requirements, for example for rate-aided manual tracking and to remote these tracks, without having a local data-processing system. These autonomous consoles are also used for displaying mixed raw and synthetic video signals.

Three types of station developed by Thomson-CSF using these families of equipments differ mainly in their operational assignments, sizes, and their fitness for mobility.

These three types of station are classified as:

(1) Fixed Operations Centre: air defence operations centre, sector operations centre and control and reporting centre
(2) MCRC: mobile control and reporting centre
(3) LARP/FACP: mobile low altitude reporting post/forward air control post.

T.ACCS operations room

FIXED OPERATIONS CENTRE

Automated functions of a fixed operations centre include:

(1) establishment and updating of the general air situation using mono or multi-radar tracking
(2) identification of detected aircraft
(3) evaluation of the threat
(4) allocation of targets
(5) assignment of weapons
(6) activation and automatic guidance of interceptors
(7) co-ordination between fighters and surface-air missile units
(8) control of military air traffic
(9) complete simulation for the training of military staff.

A centre can be easily adapted to each specific need, by varying the number of autonomous consoles and the central memory size of the computers. If necessary, a second similar computer complex can be added to the system, allowing various degraded modes in case of a computer's failure.

MCRC

This configuration is inherently capable of operating in various roles, either on an autonomous basis, or linked to a fixed operations centre and to other mobile centres. Its operational capacity is less than that of a fixed centre, but it nevertheless affords wide possibilities which make it usable as an addition to an operation centre in territory air defence for an army corps to provide aerial protection. The MCRC is used to complement the fixed air defence organisation and to carry out the control of tactical air defence missions and battle corps cover missions.

It allows the following functions: automatic initiation and automatic following of tracks from local radar data. Both functions can also be performed in rate-aided manual mode; transmission and automatic or manual correlation of tracks in connection with higher and subordinate levels; identification of tracks; automatic control of up to six simultaneous interceptions; recovery of fighter aircraft; co-ordination with surface-to-air missile units; simulation of eight moving targets, of which four are piloted fighter aircraft.

Typical operational capacities are: 90 tracks; automatic two-way communications with six air defence centres; up to six simultaneous automatic interceptions; eight simulated moving targets of which four can be piloted in real time from pilot consoles.

The MCRC includes MITRA computers, adapted for use in a cabin, with a disc for program loading and data recording, and a teleprinter. There are three AMC 800 consoles, also adapted for use in a cabin by a master controller and two interception controllers.

This typical configuration, which can easily be extended with additional modules, meets the requirements of the operations centre normally associated with the Thomson-CSF mobile 3D radar TRS 2215, for example (**3272.153**).

LARP/FACP STATION

This configuration has a dual operational role: it allows an additional radar coverage, in particular at very low altitude, with minimum personnel and material; and it is designed to carry out control of tactical missions, mainly in aerial support and to aid aerial penetration operations. To fulfil these missions the LARP/FACP makes use of data from a radar adapted to low altitude detection, such as a TIGER radar (**2156.153**), the associated secondary IFF radar, and possibly, visual observation posts. Additionally, it is connected through automatic data link with a fixed or mobile centre.

The LARP/FACP is equipped, for processing and display, with two autonomous consoles TVT 3540.

The main functions are detection of targets; building up of tracks by rate-aided manual or automatic tracking of targets; identification (IFF/SIF); introduction of visual observation data into the system; transmission of tracks to the higher level; reception of tracks from the higher level; control of tactical missions.

STATUS

In production and in service with several overseas countries probably including South American, African, Asian and Gulf States, including Saudi Arabia.

CONTRACTOR

Thomson-CSF, Division Systèmes Défense et Contrôle, 40 rue Grange Dame Rose, 92363 Meudon-La-Forêt Cedex, France.

2161.181

STRIDA AIR DEFENCE SYSTEM

STRIDA (Système de Traitement et de Representation des Informations de Défense Aérienne) is the French air defence data handling system. The system consists of a network of stations covering French territory with the following main functions:

(1) detection and identification of aircraft moving in French airspace
(2) threat evaluation and dissemination of early warnings. The air situation is centralised and synthesised in the air defence operational centre (ADOC)
(3) updating of active means (aircraft and missiles) status in every sector operational centre (SOC)
(4) weapons selection, engagement and automatic intercept guidance
(5) aircraft recovery to air bases
(6) control of military operational and training flights
(7) co-ordination with the air traffic control system to ensure identification and spacing of operational military flights with respect to general air traffic
(8) progressive integration of the information reported by air base radars for improving the low altitude coverage.

STRIDA consists of different types of operational centres (ADOC, SOC/CRC, CRC, ARP) which exchange digital messages by a special telecommunication network, the Air 70, associated with electronic switching stations.

At ARP and CRC levels, signals coming from 2-D and 3-D long range radars (Palmier) and height-finding radars (SATRAPE with phased-array antenna) are used for providing the labelled air situations needed at SOC/CRC and ADOC levels. All information and orders to be exchanged between these different centres for command and control purposes are transmitted through the Air 70 network.

The STRIDA network is connected to the NADGE network (**1181.181**), to the 412L (**2533.181**) system and to Combat Grande (**2322.181**) to provide total coverage of western continental Europe. It is also connected to the French ATC CAUTRA system for co-ordination of military and civil air traffic control.

The data handling equipment of a typical latest generation CDCS station mainly consists of:

(1) an EMIR radar data extractor using a programmed extraction concept
(2) a high power processing system using IBM 370 series computers
(3) a display sub-system including from 20 to 30 operational positions (see below).

In some centres, special equipment is found: in the air defence operational centre, a large screen display and in a centre for intercept controller training a piloted aircraft simulator (SACHA) generates a video signal corresponding to up to 12 simulated aircraft. This signal can be mixed with the live video of the radar for a transparent simulation.

Each operator position or console designed for one controller and his assistant is composed of:

(1) a plan view display for presenting raw video (local radar raw video) and data generated by the processing system (synthetic view of the air picture - mainly tracks). A 40 cm diameter screen is used
(2) one or several monochrome or colour screen(s) (diagonal from 13 to 35 cm and capacity up to 4000 characters) for presenting detailed information on certain subjects (tracks, intercepts) or data received in alpha-numeric form (operational status, flight schedules)
(3) several keyboards and a rolling ball (or a joystick) for selecting the data presented and entering functions and data
(4) control keyboards and panels for communications (radio, telephone, interphone) and secondary radar.

A STRIDA air defence reporting centre and communications node co-located with a 3-D radar in the Alps

STATUS

The research and development programme for the STRIDA system began in 1956 under the responsibility of Service Technique des Télécommunications de l'Air (STTA). The first stations were fitted with specialised IBM/CAPAC real-time computers and Sintra VISU II display sub-system. They became operational in 1963. Since this date, there has been a continuous programme of improvements for software and hardware during the implementation of the whole military programme, every new centre having to be interoperable with the others. The medium and high level coverage main centres are now near completion. The last ones are equipped with IBM 370/158 computers and Sintra VISU IV display sub-systems. The next step entailed the extension of the low-level coverage of the system by the integration of recovery and tactical operations radars. In 1978, Sintra was chosen to supply the STRAPP (STRida-APProach) limited to the low-altitude radar at each approach centre. Data to and from STRIDA main centres are automatically processed and transmitted by the STRAPP.

CONTRACTOR

Thomson-CSF, Division Systemes Defense et Controle, 40 rue Grange Dame Rose, 9253 Meudon La Foret Cedex, France (following the merger of Sintra with the Thomson Group).

VISU IV consoles produced by the Sintra subsidiary of Thomson-CSF for the STRIDA air defence system

6501.181

MYGALE AIR DEFENCE SYSTEM

Mygale is a mobile very short range coordinated air defence system based on two guiding principles. Firstly it is designed to make the maximum use of possibilities offered by missile techniques to automate the firing sequence to the fullest extent, and secondly to provide the weapons chief with the means to lead a coordinated, controlled and aggressive air defence. The system consists of two independent and complementary subsystems, the Mygale Warning and Fire Control Station C2 (WFCS), and passive Aspic firing stations. The maximum distance between the WFCS and an Aspic firing station is 3 to 5 kilometres.

The WFCS carries out the following functions:
(a) air space surveillance via a 2-D, solid-state, frequency-hopping Doppler pulse radar
(b) maintaining watch while on the move with the antenna retracted
(c) threat assessment
(d) firing station coordination
(e) IFF
(f) target designation to Aspic firing stations
(g) integration into tactical command systems (C2 and C3).

The passive operation Aspic firing station is equipped with a Daldo target designation helmet aid for autonomous operation, whether linked or not to a warning network and a firing console. This equipment may be used up to 50 metres from the firing station. Optionally IFF, FLIR and laser rangefinder systems can be also provided. All close-in missiles can be fired from Aspic, including Stinger, Mistral, RBS-70, SA7 and SAX. Aspic carries out the following operations:
(a) reception of tracking data and slewing of missiles
(b) automatic target acquisition and tracking via TV camera and/or infra-red
(c) masked time irreversible sequence
(d) immediate re-engagement.

The WFCS and Aspic sub-systems are built on a modular principle and are independent but complementary. They use the close-in missiles and carrier vehicle adopted by the user. Whether or not integrated on a light vehicle or an armoured vehicle the Mygale weapon system can be air-lifted. Reciprocal localisation and suitable automatic data processing allows flexible and spread-out deployment of an operational unit (one WFCS and four to eight firing stations).

CHARACTERISTICS

Search radar

Range: 14 km on moving airborne targets (7.5 km on a hovering helicopter)

Data renewal time: 1.5 s

MTBF: 1000 h

Data processing: IFF, automatic threat assessment, simultaneous target designation capability

Aspic firing station

Movement: Azimuth 360, positive and negative elevation

STATUS

The Mygale system is in operational service. Thomson-CSF has sold the system to one export customer for use with Mistral missiles, and the company is offering a reconfigured Aspic fire unit as part of its bid for the pedestal-mounted Stinger portion of the US Army's Forward Area Air Defence (FAAD) battlefield system.

CONTRACTOR

Thomson-CSF, Division Systèmes Défense et Contrôle, 40 rue Grange Dame Rose, BP 34, 92360 Meudon-La-Forêt, France.

Mygale Warning and Fire Control Station (C2)

Aspic firing station

GERMANY (FEDERAL REPUBLIC)

4275.181

SILLACS LOW-LEVEL AIR DEFENCE SYSTEM

The Siemens low-level air defence control system (SILLACS) evolved from the West German TMLD system (**3607.181**). The latter system was enhanced by adding more functions and improving its effectiveness. In the new system, SILLACS, extensive use is made of automatic data processing, multiple radar tracking and the support of control operations. SILLACS therefore forms a mobile autonomous low-level air defence control system for sensitive areas that can be interfaced with any nation-wide air defence system for the upper airspace.

Low-level radars are available with ranges from 45 to 140 km for optimum coverage configurations. They combine very high clutter suppression with high ECM-resistance. Two versions of the Siemens pulse doppler radar family are specially designed for use in SILLACS:
(1) the fully mobile DR 151/DR 162 with detection ranges of 45 and 60 km, respectively, consisting of one vehicle with a mast erectable up to 18 m and the radar/operator shelter
(2) the transportable DR 172 with a detection range of 140 km, consisting of two shelters: an antenna unit and an operator shelter. The Siemens IFF interrogators DII 211, DII 111 or DII 511 are integrated in the primary radars. Up to eight of these radars may be connected to the SILLACS Control and Reporting Centre (CRC) which contains the data processing and display equipment as well as the communication facilities for interceptor guidance and data exchange to higher ranking command and control centres.

The CRC consists of one equipment shelter, one display shelter and separate power generators mounted on trailers. The equipment shelter houses all sub-systems for automatic data processing, data and voice transmission and switching. The display shelter contains four operator work stations for air situation display and control and two assistant's positions. This equipment is connected to the CRC equipment shelter via cable link.

Incoming target reports from the radars first undergo a validity check. They are then fed into the multiple radar tracker, which generates a consistent

SILLACS CRC operator's position

unambiguous picture of the air situation. New tracks are initiated automatically. Non-automatic initiation areas can be defined by the operator. Processing and display of target reports can be performed independent of track processing. Thus even in case of computer failure an air situation picture can be obtained from the system. Mission data, eg speed, heading, bearing, distance are provided by the computer for the support of intercept control operations. The operator work stations consist of:

(1) 22-inch (56 cm) diameter air situation displays for target reports and tracks, symbols and vectors
(2) alpha-numeric readouts for status information and track data
(3) function keyboards
(4) alpha-numeric keyboards
(5) rolling balls.

The display control panel incorporates the facilities to select the displayed area, ie to adjust its centre or to off-centre it by means of the rolling ball, to choose the proper scale to switch the output channel of interest to the connected air situation display and to connect the display unit to any input channel. By means of additional push-buttons the operator can combine and adjust the display of video maps, grid lines, plots and tracks as required. Each operator position in the CRC display shelter is supplied with an IFF remote-control panel to specify mode and expected reply code of IFF interrogations. At this panel the actual, remotely-controlled status of the IFF interrogators of up to eight associated radar sites can be indicated simultaneously. The operational program system and the video map are stored on tape cartridges.

Exchange of information between the four operators and two assistants in the CRC and several subscribers is provided. By means of a communication system containing radio link and multiplex equipment, external communication paths are set up. Furthermore, connections to the operators in the associated radar sites and remote-controlled ground-to-air radio link facilities are installed.

STATUS

The system is in full operation with an unspecified number of customers.

CONTRACTOR

Siemens AG, Landshuter Str 26, 8044 Unterschleissheim, Federal Republic of Germany.

3607.181

TMLD LOW-LEVEL REPORTING AND CONTROL SYSTEM

The air defence of the Federal Republic of Germany is intimately bound up with the overall NATO air defence ground environment (NADGE) system of radars, communications and data processing and control centres (**1181.181**), and to varying degrees with systems employed by allied nations present in Europe, such as the American 407L air weapons control system (**2533.181**). However, because of limitations in the low-level cover of parts of the former systems and because the AFCENT sector of the Central European NATO area is particularly subject to the threat of low-flying surprise attacks, the Luftwaffe has set up the TMLD (Tiefflieger-Melde-und-Leitdienst) low-level reporting service, additional and supplementary to the other radars.

This is part of the combat operational system of air defence and is designed particularly for the monitoring of low-level airspace. The TMLD radars are located in the area near the border, the individual sites being chosen to ensure continuous surveillance. When the radar detects a target, the information is converted into data messages and immediately transmitted to the low-level reporting centres TMLZ (Tiefflieger-Melde-und-Leitzentralen) and on to the user. The whole TMLD is completely mobile so that changing tactical and operational circumstances can be accommodated. The main tasks of the TMLD are:

(1) monitoring the lower levels of airspace in selected areas
(2) processing the low-level air situation in the control centres (TMLZ) and passing the information on to the air defence control centres (eg NADGE, GEADGE)
(3) preliminary warning and target forecasting for Hawk SAM units
(4) control of air operations.

The TMLD units are combined into two sections, and are employed in the 2 ATAF and 4 ATAF areas. In each case they come under the command of the commander of a Type B signals regiment. It is planned to assign them as a NATO command force. The main elements of such a unit are:

(1) a radar crew to operate the radar equipment
(2) an evaluation crew to man the control centre
(3) a radio relay crew for setting up communications links
(4) a maintenance crew for the maintenance and repair of equipment.

In peacetime the TMLD mainly limits its activities to monitoring the area in the immediate vicinity of the border, in a quasi-static role. At other times, however, monitoring of airspace must be extended to a greater area, making use of the system's mobility and flexibility. In this it will be supported by the Army's AAD command and control systems, which will be employed in monitoring the airspace over the combat area.

A Siemens MPDR-30 radar of the Federal German Republic's low-level air defence system deployed in the field

The principal hardware used in the TMLD system comprises: MRDR 30/1 mobile pulse doppler radars (**3286.153**); TM control centre; CRC TM display unit; radio link set. The radar consists of two 5-ton trucks, each with a 1½-ton trailer containing a generator, with one truck serving as an equipment carrier and the other as the antenna carrier. In the TM centre all the information provided by all the radar equipment within one sector is processed and displayed. Linked users include: the NADGE/GEADGE air defence control centres and Hawk SAM batteries.

OPERATION

Information on air targets supplied by the TM radar is converted into messages for transmission to the TMLZ. The data transmitted contains the positions of the target as X/Y co-ordinates and the target identification. Two operating modes are planned for the preparation of this information: automatic, and semi-automatic.

Automatic Operation: This is regarded as the normal mode for peace time. Targets detected are automatically evaluated on the basis of whether or not they emit a valid IFF/SIF code, and they are reported accordingly. In addition, the operator in the radar truck (acting on instructions from the control centre) can insert further symbols manually in order to give prominence to certain targets of particular significance.

Semi-automatic Operation: This mode is of special importance when automatic operation with the radar signal processor is not possible for technical or tactical reasons. In this mode the operator can mark targets identified on the radar display with the aid of a rolling ball, using symbols that he can superimpose on the target blip. Transmission of targets to the TMLZ or the user is again performed automatically. For this purpose various types of symbol are available, to each of which a particular meaning is assigned to assist the responsible air defence officer in reaching decisions.

STATUS

The system has been fully operational since 1977.

CONTRACTOR

Siemens AG, Radio and Radar Systems Division, Landshuter Str 26, 8044 Unterschleissheim, Federal Republic of Germany.

3945.181

GEADGE

GEADGE is the acronymic title of the air defence network for the southern part of the Federal Republic of Germany, and is derived from 'German Air Defence Ground Environment', The new system, for which Hughes Aircraft Company was nominated the prime contractor in 1979 under a contract worth more than $150 million, replaced the 20-year old 412L radar network (**2533.181**) operated by the West German Air Force. The system integrates new and existing long-range surveillance radars into a single network based on four centralised command centres, and it embraces manned and unmanned fixed and transportable radar systems. GEADGE now receives radar data directly from E-3A AWACS early warning aircraft. The southern portion of West Germany was not included in the original NADGE system (**1181.181**), but GEADGE fills the gap left in that system and connects directly with it. In addition to fixed and transportable gap-filler radars, the new system utilises two of four new permanently located radars known by the manufacturer's name of HADR (Hughes air defence radar) (**3721.153**), which is an advanced three-dimensional, multi-role radar that will automatically detect, classify, and report on targets intruding into their coverage area.

In the GEADGE system, Hughes is also supplying HMP-1116 minicomputers, H-5118M central computers, and HMD-22 display and control consoles, as well as being responsible for software, installation and integration. The HMD-22 consoles are built under licence by AEG and the communications equipment for the system has been supplied by Tekade.

STATUS
The first of two sites became operational in 1983 and 1984, and the final two centres were operational in 1985. Final system acceptance took place early in 1986.
CONTRACTOR
Hughes Aircraft Company, Ground Systems Group, Fullerton, California 92634, USA.

A West German captain monitors the air traffic over the southern portion of West Germany at a GEADGE site

4276.181
SIEMENS BATTERY CO-ORDINATION POST (BCP)
Site defence against low flying aircraft is generally dependent on the use of more or less autonomous short-range AA weapons such as machine guns and surface-to-air missiles. This new system by Siemens is meant to increase the effectiveness of such existing weapon systems. Depending on the topographical situation of the site to be defended, on the available defensive weapons, and particularly the anticipated threat, the Battery Co-ordination Post (BCP) optimises air defence.

Equipped with an acquisition radar and command and control equipment, the commander is able to control up to 12 fire units deployed around the site. An integrated digital computer evaluates threat and status information. Data link and weapon terminals are provided to connect the BCP with the fire units for the transmission of aiming and firing instructions and for the exchange of data.

STATUS
Siemens have conducted system tests which have demonstrated increased weapons effectiveness. This system is integrated in the Artemis 30 AA gun air defence system being produced for the Greek armed forces, and in the CORAD (Co-ordinated Roland Air Defense) system.

CONTRACTOR
Siemens AG, Radio and Radar Systems Division, Landshuter Str 26, 8044 Unterschleissheim, Federal Republic of Germany.

Radar shelter of Siemens BCP system

6802.181
TLVS MEDIUM-RANGE TACTICAL AIR DEFENCE SYSTEM
TLVS (Taktisches Luftverteidigungssystem) is a conceptual study for a new medium-range tactical air defence system with anti-tactical missile capabilities, designed to be in operation from the year 2000 onwards as a successor to the Hawk system in Germany. It is being developed in a joint programme by MBB, AEG and Siemens, as a surface-to-air missile system incorporating radar, a communications network and a comprehensive command and control station.

The system is designed for all-sector, all-weather defence against a variety of airborne threats, including tactical ballistic missiles, with a high simultaneous fire power in dense ECM environments.

The multi-function radar is intended to detect low-flying objects with small radar cross-sections, as well as high-diving tactical ballistic missiles. The surface-to-air missiles will be fitted with an active seeker head to provide for multiple simultaneous engagement of several airborne targets. Mid-course guidance will be provided by the multi-function radar. A 'fire and forget' capability will exist for close-range targets.

The command and control station will incorporate sophisticated weapon systems computers/processors to enable rapid decisions to be made in complex situations. The provision of artificial intelligence methods is also being investigated. The communications network will be jam-resistant and will be backed up by alternative fibre-optic links.

ISRAEL

2226.181
ISRAEL AIR DEFENCE SYSTEM
A computer-controlled air defence system has been installed in Israel. No official details of the equipment are available, and it is believed that no public statement regarding the existence of the system has been made by either the Israeli authorities or the prime contractor. It appears to be generally accepted that Hughes Aircraft Company was awarded the contract after completing a study for the new system and it is also understood that the system is based on main control centres situated near Tel Aviv and in the Western Negev in the southern part of the country. There are possibly subsidiary centres elsewhere, and the existence of a tactical operations centre with a PPI display of the entire Israeli airspace for viewing by the Cabinet has been claimed.

There are thought to be many similarities with the USAF 407L tactical command and control system (**2824.181**), and Hughes 4118 digital computers as employed in this system are reported as providing the data processing facilities for the Israeli air defence system. Sensors are understood to include two Westinghouse AN/TPS-43 three-dimensional radars (**2517.153**), several FPS-100 radars, and there are almost certainly other radars of British, French, and Israeli origin which have been incorporated into the system.

In January 1976 the sale of four Grumman E-2C early warning aircraft to Israel was reported and these were delivered between November 1977 and mid-1978.

The E-2Cs have either APS-120 or APS-125 long-range airborne early warning radar (**1783.353** and **1965.353**) and Grumman had a study contract to make the E-2C data link compatible with the Israeli air defence network to enable the exchange of data between ground and air sensors, as well as integrating the AEW element into the overall system.

The original Israeli defence network is thought to have been completed and handed over by Hughes some time during 1973.

In 1981-82 a major communication control system for air defence sites was supplied to the Israel Defence Forces by the Electronics Corporation of Israel Ltd (ECI) in co-operation with Elbit Computers Ltd, known as MCCS-800.

Alternatively known as ACCESS, the system provides each one of a large group of operators fast and reliable access to up to 1000 communication devices located in the control centre vicinity or in remote communications stations.

From a single operator ACCESS provides:

(1) VHF, UHF and HF channel selection which can be shared by a large number of stations
(2) direct point-point-line intercom and telephone communication between operators, with a built-in break-in capability for use by supervisor
(3) automatic allocation of local, remote and relay stations and automatic routing of desired equipment based on an unblocked switching matrix
(4) central control of all communication devices and rapid reallocation of equipment plus automatic fault detection and device replacement.

The Operator Unit is capable of reception and transmission on six different channels, intercom to at least six other operators, direct access to directional antenna, direct access to remote stations and individual allocation of encryption devices.

The operation stations cover: Technician Station with complete technical command and control of the system; Preset Station covering system planning and presetting of allocation programmes for various prepared missions and supervisor station to provide total system command and control.

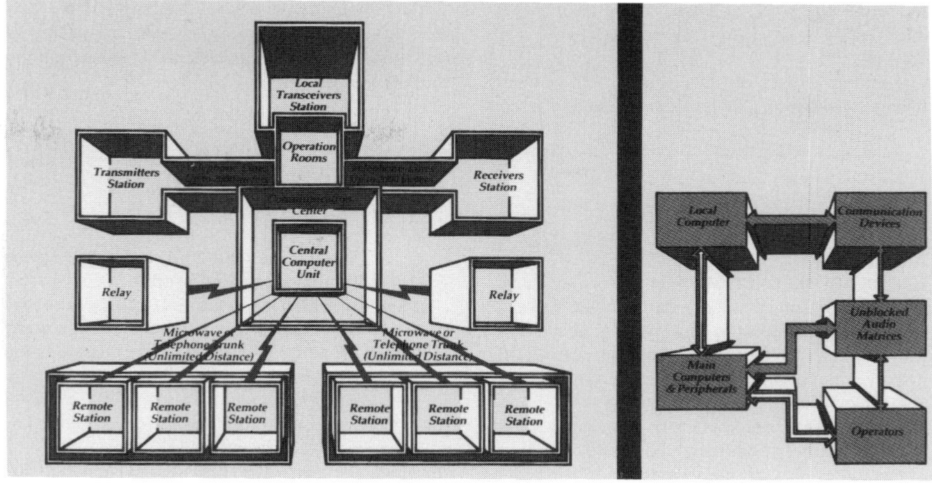

System architecture (left) and block diagram (right) of the ACCESS system

---

ITALY

3324.181

MOBILE AUTOMATIC REPORTING STATION (MARS-402)

The Mobile Automatic Reporting Station MARS-402 is a mobile radar system providing the detection of low flying aircraft, surface targets, generation and reporting of the air/surface situation and controlling function for navigational assistance of home based military aircraft. The system is functionally composed of the following main sections:

(1) operation section
(2) Pluto radar section
(3) power supply.

The Pluto radar section is composed of the radar antenna, transmitter and receiver equipment providing 2-D primary and SSR target detection.

The operation section is composed of the data processing display and communication equipments.

The data processing and display equipments provide tracking, reporting and air controlling functions.

The communication equipment provides internal communications G/G and G/A communications control.

The system is capable of interfacing with the G/A and G/G communication equipment.

The physical components of the MARS-402 in the standard configuration are as follows:

(1) one antenna module
(2) one radar shelter
(2) one ops shelter
(3) mobiliser(s) (optional)
(4) tactical tower (optional).

In the compact version of the MARS-402 the radar and operation section equipments are housed in the same shelter to enhance compactness and mobility.

OPERATION

Operational roles are:

(1) 2-D multiple radar input
(2) early warning surveillance with single or multiple radar input
(3) navigational assistance
(4) fixed or transportable applications
(5) local or remote radar input.

MARS-402 with Pluto SIR and operational control centre cabin

Specific functions include:

(1) automatic target detection and initiation
(2) automatic tracking
(3) multi-radar processing (with optional multiple radar configuration)
(4) rate-aided manual tracking
(5) data link for exchange of track data, command messages, etc with automatic and manual facilities
(6) identification by computer assisted SIF active/passive decoding, automatic data link, or manual input
(7) search radar/IFF video display
(8) synthetic video display.

CHARACTERISTICS

Track capacity: 40 (expansible)
Simultaneous intercept vectoring: 2 (expansible)
Data link: Up to 4 sites
Video maps displayed: 2
Operator orders: 42
Computer: NDC-160, mobile
Displays: MDU-03

CONTRACTOR

Selenia, Industrie Elettroniche Associate SpA, Via Tiburtina 12 400 Km, 00130 Rome, Italy.

6803.181

CATRIN C3I SYSTEM

Catrin is an integrated command, control and telecommunications system for battlefield use. It is in the early development stage with operational deployment scheduled for 1993. From an operational point of view the Catrin system is part of an overall modernisation programme affecting all sectors of the Italian Army, from helicopter gunships to command, control, communications and information systems.

The Catrin system is operational up to army corps level and consists of three subsystems; a telecommunications subsystem (SOTRIN), a surveillance/ground target acquisition and fire control subsystem (SORAO); and a low/very low air surveillance subsystem with tactical command and control of anti-aircraft guns and aircraft (SOATCC).

The SOTRIN subsystem is a mobile network handling voice, data, telegraph and facsimile communications. The necessary redundancy is built-in to ensure continuous operation even in the event of damage or breakdown.

The SARAO subsystem consists of a network of sensors and of data correlation units. The network consists of radars, electro-optical devices, sound ranging equipment and remotely controlled midget aircraft. The correlation network gathers the sensor data for processing and distribution to users.

The SOATCC subsystem comprises a radar network, a command and control network, an information/logistics network, which also locates friendly aircraft. The radar sensor network uses one or more three-dimensional radars and a number of mobile two-dimensional systems, to identify low and very low flying targets. Using the radar data, the command and control centres evaluate the threat, co-ordinate air defence, control the airspace and the missions of friendly aircraft and send alarm signals to all the deployed units. This subsystem also uses the

SOTRIN telecom network to distribute the information.

STATUS

The Catrin project began in June 1987 when the Italian Ministry of Defence commissioned a consortium of companies to produce the system. A six-year programme is envisaged, with pre-prototype testing scheduled to take place three years from the start of development.

CONTRACTORS

The consortium of contractors commissioned by the Italian Defence Ministry consists of Aeritalia, Agusta, Italtel, Marconi Italiana, Selenia and Telettra.

JAPAN

2254.181

BADGE AIR DEFENCE SYSTEM

BADGE (Base Air Defence Ground Environment) is a computerised air defence system designed to provide the information gathering data processing and display functions required for umbrella protection against aerial attack on Japan. The system comprises radars that will automatically detect, track, and identify airborne targets over Japan and a large area of the surrounding ocean, computers to process the radar data and evaluate threats, and other computers to process and furnish data on weapon availability, intercept geometry and related measures, all of which is displayed, together with the processed radar data on complex displays to the appropriate interceptor or missile controllers.

The $56 million system was largely built in Japan for the Japan Self Defence Force. The prime contractor was the Hughes Aircraft Company of the USA and they supplied much of the equipment of the first installation as well as being responsible for system and equipment design. Most of the subsequent manufacture, however, was carried out by Japanese manufacturers.

BADGE sites extend from the northernmost tip of Hokkaido to the southern extremity of Okinawa Island and there are 28 fixed surveillance radar sites. The computers belong to the Hughes H330 series. Japanese airspace is divided into four air defence sectors with a direction centre for each: Western Air Defence Force Control Centre (ADFCC); Central ADFCC, Iruma; Northern ADFCC, Misawa and Okinawa ADFCC. The system became operational in 1969.

BADGE is similar in many respects to ADGE (**2211.181**) and Florida (**2373.181**) and reference may be made to the latter for more detailed information.

In 1982 Japan initiated an extension and upgrading programme called BADGE-X to both increase its coverage and provide for the integration of the Japanese Hawk surface-to-air missile batteries with the existing defence systems.

It was then understood that there were likely to be five phases to this operation, although the precise order was not known. The phases were:

(1) updating and expanding BADGE to improve coverage of the main Japanese islands
(2) linking BADGE to the Nike/Hawk batteries via the JAN/TSQ-51B distribution systems then being built by NEC Corporation under licence from Hughes. In 1985 it was decided to replace the Nike missiles with the Patriot system and this is going ahead. In addition the Hawk batteries are being updated to Improved Hawk
(3) developing an airborne early warning system
(4) further extending BADGE to cover other key areas such as Okinawa
(5) introducing improved (three-dimensional) radars for BADGE to replace the Bendix FPS-20 surveillance radar (**2528.153**) and General Electric FPS-6 heightfinders (see **2523.153**).

STATUS

The BADGE system's computers have all been duplicated, thereby greatly enhancing reliability, ease of maintenance and repair, and improving training facilities. Prime contractor for this work was Nippon Avionics Company Ltd. Additional computers and peripheral equipment, designed to work in parallel with the original hardware have been added, and another feature was the addition of a number of magnetic drum memories upon which all operational data are stored automatically. In the event of a malfunction demanding a computer change-over, the drums have all operational data that were contained in the failed system and they will be available for use in the standby system.

In its fiscal year 1979 budget the Japan Defence Agency announced that after evaluating eight possible airborne early warning aircraft types, it was decided to procure E-2C AEW aircraft to provide the low-flying target capability and extended range facilities not possible with the existing ground-based air defence radars. The first patrol area designated for coverage by the E-2C was operational in 1983 and eight E-2C aircraft are now in service. The Defence White Paper also alluded to other measures to renew and modernise BADGE equipment as well as research and evaluation in support of this.

Older early warning and fighter control radars have been replaced by modern three-dimensional radars. The equipment selected has the JASDF designation J/FPS-2 but is also known as the F3D. Produced by NEC Corporation it is probably either one of the equipments described in entries **2298.153** and **3702.153** or a near derivative. It is understood to have been based on a mobile radar known as the M3D (J/TDS-100) and the first example of the new radar was installed at the Oominata radar site in 1979. The FPS-2 is an electronically scanned array radar which uses mechanical rotation for azimuth scan, other features including dual transmitter/receiver, MTI, digital processing and SIF. The display arrangements employ the same symbols that are used in the BADGE semi-automatic air defence system and six bearing lines and two targets can be displayed simultaneously for GCI. Special ECCM facilities are incorporated.

The JASDF is believed to have at least seven F3D radars installed and the second modernised J/FPS-2 was installed at the Sado radar site, off the coast of Niigata in the Japanese Sea during 1980. Two mobile J/TDS-100 3-D radars were installed in the North and Centre Air Warning Control Force regions, and it was planned to install an improved TDS-100 in the Western Air Defence District.

In March 1982 it was reported that the air defences of Chitose air base and the Tobetsa radar site were to be upgraded and the 1982 budget allowed for three Toshiba Type 81 air defence missile systems. It was also decided to start development of a new long-range air defence radar system, known as the XJ/FPS-3 (**4605.153**), to replace the FPS-20, the new equipment to be an active phased array system with a longer range than the 300 km of the FPS-20.

In early 1983 the Japanese NEC concern was selected as prime contractor to undertake the BADGE-X expansion programme, with Hughes Aircraft Company as a major sub-contractor (for systems design, management and licensing of hardware).

Installation and construction of the BADGE system expansion began in the second quarter of fiscal year 1985. First system was installed at the HQ of the Middle Air Defence Force, followed by Northern ADF, Air Defence Command, Western ADF and SW Composite Air Division. Completion expected in March 1988.

Japan has decided to implement an Over-The-Horizon (OTH) radar system which, it is understood, will be integrated into the overall US OTH network. The transmitter is to be located on Hahajima Island with the receiver on Iwo Jima Island.

CONTRACTORS

Hughes Aircraft Co, Fullerton, California 92634, USA.

Nippon Avionics Company Ltd, NEC Corporation.

NATO

1181.181

NADGE - NATO AIR DEFENCE GROUND ENVIRONMENT

This is a multi-national programme involving 14 NATO countries (so far as funding and contracts are concerned) in the updating and co-ordination of the air defence systems of nine European members of NATO. They are Norway, Denmark, West Germany, Netherlands, Belgium, France, Italy, Greece, and Turkey. The British air defence network interfaces with NADGE. French participation is limited to use of, and contribution to the reporting and control functions, and that country's defence forces will not be directed against hostile targets by NATO.

NADGE, which is now fully operational, was conceived not as an entirely new air defence installation, but rather as an overall plan for the improvement of some existing hardware and the provision of new equipment in certain areas either to fill gaps in the pre-existing systems or to improve their performance. The original cost was an estimated £120 million. The history of the programme and details of the system and elements of the equipment have been described under this entry number in previous editions of *Jane's Weapon Systems*. The following notes summarise the most significant of the latest detail improvements and system enhancements.

Improvements to individual national air defence networks are in train in several NATO countries and these will all be engineered to complement and enhance the overall NADGE capability. Portugal is building a new NATO radar station on the Algarve coast. Transportable or mobile three-dimensional radars have been added to the NADGE network. Norway has acquired three HADRs to enhance the NADGE facilities in that country. Hughes and Kongsberg are producing 18 TPQ-36 derived Low Altitude Surveillance Radars to form a low-level acquisition radar and control system (ARCS) of air defence. The computer-based ARCS is used with the Hawk missile system where it replaces three radars and multiple control centres. The ARCS radar is a mechanically rotated derivative of the phase-scanning mortar locating radar. A network of NATO AWACS bases in West Germany, Italy, Norway, Greece and Turkey is in operation.

The first NATO AWACS E-3A aircraft was delivered to the main operating base at Geilenkirchen, West Germany in February 1982, the last of the 18 being delivered in mid-1985. As noted elsewhere Saudi Arabia received the first of five AWACS in June 1986, with the remainder due to have been delivered by mid-1987. The USAF ordered 35 E-3 AWACS aircraft of which all have now been delivered, and have been, or are being, progressively updated to E-3B or E-3C standard. In December 1986 the United Kingdom decided to buy six E-3 AWACS aircraft, with an option on two more. A seventh aircraft has now been ordered with the eighth remaining as an option. The first will be delivered in 1991. In early 1987 France decided to order three E-3s with two on option. The latter two aircraft have now been ordered to give a total of five and a sixth is on option. NATO has also purchased the first of three second-hand Boeing 707 aircraft, which will be refitted to serve as AWACS flight crew trainers.

The Airborne Early Warning/Ground Environment Integration Segment (AEGIS) programme provides for the controlled exchange of air surveillance data between the NADGE system and the NATO AEW aircraft by augmenting existing ground facilities. This will enable NATO commanders to direct air defence weapons against enemy aircraft intruding into Western European airspace at any altitude. When fully implemented AEGIS will process in-flight AWACS radar data for dissemination to command

and control installations throughout Europe from Scandinavia to Turkey. The award of a $285 million contract in October 1982 for installation of AEGIS equipment at 36 NADGE sites brings the total AEGIS contract value to about $400 million, and the number of equipped sites to 36. Hardware includes the improved terminal version of JTIDS and the Hughes 115118 ME computer. With this facility, ground controllers will be able to watch targets detected and tracked by AEW radars, thus extending the former's operational range. The first two AEGIS ground stations (in Denmark and West Germany) became operational in October and November 1983, and the complete system was due to be operational in 1988.

Norway, Denmark and NATO have awarded contracts to Hughes Aircraft Company to provide an automated command and control information system (CCIS) for NATO's northern Europe command. The system is designed to give NATO commanders current information on the disposition and readiness of their forces in the field and provide information on enemy air activities through the use of high speed computers. Commanders can also use the CCIS to plan deployment of their forces and to speed transmission of orders. Four Scandinavian subcontractors are being used by Hughes and the system is due to be fully operational by the late 1980s. In a similar programme, the UK awarded a contract valued at £37 million in August 1987 to a consortium led by ICL to develop and build the UK AIR CCIS. This project has as a major function linking the C-in-C UKAIR (as local NATO air forces commander) with UK-based air commanders. The CCIS is to be connected to the existing NADGE system. The JTIDS system, currently in late development and operational trials has been ordered by the UK MoD to form part of the British air defence requirements.

To improve detection of low-flying intruders Denmark instituted, in March 1984, the Coastal Radar Integration System (CRIS) which links data from Danish coastal radars that survey the Kattegat and the western Baltic, a corridor that intruders might seek to use. In mid-1986, THORN EMI Electronics installed this coastal radar integration system in Denmark to feed information from the coastal radar network to the Danish elements of NADGE.

In mid-1984 it was reported that 40 LMT RASIT battlefield radars (**2141.153**) had been ordered for use by the West German Army for mounting beneath tethered balloons. These are flown at an altitude of about 800 metres to eliminate shadow zone in low-level radar cover for air defence purposes. Deliveries started in February 1985.

There was a NATO requirement for a number of S-band 3-D surveillance radars to be located at sites in Italy, Greece, Turkey and Portugal to upgrade the cover in the so-called Southern Flank of the NADGE network, and competitive bids were sought from qualified contractors in the early 1980s. The estimated total value is in the region of $300 million and bids were submitted by Hughes, Selenia/Sperry, Thomson-CSF, and Plessey/ITT-Gilfillan. In the Autumn of 1985 it was announced that NATO had selected Hughes to supply eight HR-3000 3-D phased array radars to meet the requirement - three for Turkey, two each for Greece and Italy and one for Portugal. Two additional HR-3000 systems have been ordered subsequently by Portugal.

An air defence commander in a West German NADGE (NATO Air Defence Ground Environment) centre provides instructions to the E-3A AWACS early warning aircraft through the digital manual input station (MIS) under new capabilities provided by AEGIS (Airborne Early Warning/Ground Environment Integration Segment)

Air Command and Control System (ACCS)

NADGE was originally conceived as a purely defensive system against manned aircraft. However, new threats have emerged, and these, together with future threat evolvement, need to be met by a completely integrated defensive and offensive system. An ongoing study has been carried out within NATO for the Air Command and Control System (ACCS) to provide a NADGE update, which will ultimately provide overall air control of all types of air activity within Allied Command Europe (ACE). It is intended that this will be evolutionary in nature, starting with existing facilities and types of organisation for NATO air force control, and proceeding from that point to meet future requirements in a number of stages. By the year 2000, or shortly after, NATO intends to have improved vastly the command and control system of its air forces by implementing a system that will bring together, expand, modernise and automate all relevant subsystems and components of control.

The ACCS programme is currently in the planning phase, and a NATO ACCS team has been established, with the actual creation of ACCS commencing in 1991. The execution of ACCS will be based on a five-part master plan, and in view of the importance of the ACCS programme, both politically and economically, industry has co-operated to form two international consortiums, ACCSCO (ACCS Company) and AMS (Airspace Management System). The first consists of Siemens and AEG-Telefunken (Federal Republic of Germany); Marconi and Plessey (United Kingdom); MBLE (Belgium); Selenia (Italy); Hollandse Signaalapparaten (Netherlands); Thomson-CSF (France); and Hughes Aircraft (USA). The second consortium comprises Boeing and Westinghouse (USA); ESG and SEL (Federal Republic of Germany); Racal Radar and Logica (United Kingdom); Thomson-CSF and ISR (France); and Italtel (Italy). Italtel is also associated with four other Italian companies - Elettronica, FACE, FIAR and Electra. The two consortia differ to some extent in their approach to the ACCS programme in that AMS believes in the early introduction of new technology and operational doctrines while ACCSCO believes in using the existing system as a foundation on which the new equipment and technologies can be built.

RASIT radar mounted on underside of TCOM balloon for low-level gap-filling air defence role in West Germany

To date, four contracts, worth a total of $10 million, have been awarded to the two consortia for the ACCS Multi-sensor Integration Study, and the Communications and Automated Data Processing Design Options Study. In addition AMS has won a contract for the ACCS Data Base Implementation and Automated Data Processing Support Tools Study.

STATUS

The NADGE system is continuously evolving, and in addition to the ACCS programme described above a number of other updating plans are in progress. Turkey has a requirement for an airspace radar control project which probably requires updating or replacing about 12 radar systems and has already asked for proposals from a number of companies. Turkey is also expected to invite bids for a major communications network, known as the Turkish Air Force Integrated Communications System (TAFICS). This latter is reported to be NATO funded. Selenia has won contracts to improve the national air defence network and provided an Argos-10 long-range surveillance in Sicily. 14 RAT-31S E/F transportable 3-D radars have been purchased by Italy, with options for more. The Federal Republic of Germany has developed a low-level radar system to meet national requirements consisting of Siemens MPDR 30/1 radars in a TRMS mobile system. Most of these programmes are national in concept but all contribute to the overall NADGE system.

CONTRACTORS

The members of the original six-nation consortium are listed below. There are also numerous other subcontractors from the other participating countries which are receiving orders in relation to their own country's contribution to the funding of the project.

Hughes Aircraft Company, Fullerton, California, USA.

Marconi Radar Systems, Chelmsford, Essex, England.

AEG-Telefunken, Ulm, Federal Republic of Germany.

Thomson-CSF, Paris, France.

Selenia, SpA, Rome, Italy.

NV Hollandse-Signaalapparaten, Hengelo, Netherlands.

NETHERLANDS

6363.181

WATCHDOG COASTAL DEFENCE SYSTEMS

Watchdog is an integrated coastal surveillance, command and control system. It consists of several types of surveillance sensors and a command, control and communication system for the operations centre. The system is of a modular design and can be easily adapted to changing requirements.

The primary objective of Watchdog is collecting, processing and presenting clear information in order to enable the Command to continuously evaluate the situation and to take proper decisions with respect to tactical operations. Actions are co-ordinated by means of internal and external communication.

In its most extended configuration, the primary sources of information are coastal surveillance radar stations, radio direction finder stations and thermal imaging observation posts. Other sources which add information to the command and control system are maritime patrol aircraft (MPA), offshore patrol vessels (OPV), patrol boats, governmental bodies and privately-owned companies.

Target data and other tactical information are input into the system computer and continuously updated, correlated, stored, distributed to other users, and presented on operator displays together with selectable computer-generated geographical maps of the area concerned. Additional data are displayed on the electronic data displays.

All data of interest and communication traffic are recorded automatically or by operator command for future play-back. After evaluation of the tactical situation, decisions are made with respect to the primary tasks of Maritime Guard, ie maintaining safety, order and health in the EEZ. If required, own forces can be directed to the scene of action.

Operator's console at the harbour operations centre Ymuiden

CONTRACTOR

Hollandse-Signaalapparaten BV, PO Box 42, Hengelo, Netherlands.

NORWAY

6820.181

NORWEGIAN AIR DEFENCE

As part of the NATO organisation, Norwegian air defence systems are integrated into the NADGE system and Norway has recently acquired three Hughes HADR radars to enhance facilities for the NADGE network. In addition, Norway and Denmark are acquiring an automated Command and Control Information System (CCIS) for NATO's northern Europe command. Details of the NADGE system are given under **1181.181**.

For defence against low-level, Norway is introducing the NOAH (Norwegian Adapted Hawk) programme to lengthen the operational life of the Hawk weapon system beyond the year 2000. This includes application of the Acquisition Radar and Control System (ARCS) which is described under entry **6819.153**, a joint venture between Hughes and Norsk Forsvarsteknologi A/S.

A single ARCS is able to replace the three radars and two control units currently needed to operate the three launchers in the Standard Hawk configuration. NOAH has the communication capability to net information from two or more sensors, thus providing an integrated tactical air defence picture giving early warning and battle control.

The NOAH fire conntrol units operate with the Improved Hawk weapons, while a fourth ARCS operates with the L-70 AA guns and their associated fire control system. Each ARCS will also interface with man-portable air defence missiles such as the RBS-70 or Stinger. The NOAH fire control unit consists of the IHawk high-power illuminator, three IHawk launchers carrying three missiles each and one ARCS. Several triad batteries have already been delivered to the Royal Norwegian Air Force.

OMAN

3325.181

OMAN AIR DEFENCE

The air defences of the Sultanate of Oman have undergone a major modernisation and improvement programme instituted in the early 1970s. The ruler engaged BAC, now part of British Aerospace, to undertake the task, with advice and assistance from the British Government. The basic requirement was for a highly mobile and accurate integrated air defence system capable of defending civil or military installations from ground or air attack and able to intercept any intruders anywhere within the Sultanate or its surrounding waters. The hardware selected for these tasks comprised two squadrons of Rapier low-level surface-to-air missiles (**2424.131**) and 12 Jaguar tactical strike and fighter aircraft. The current combat

Marconi Radar S600 series search and heightfinder radars deployed in the desert

aircraft strength amounts to 20 Jaguars, 13 BAe 167 Strikemasters and a number of BAe Hunters. Eight Tornados are on order. To make the most effective use of these forces, a master communication and control network was required urgently to provide the necessary early warning and tactical control and command link to the defence force. Marconi provided the bulk of the radar and communications equipment needed.

The communications system has two main centres, connected by strategically-sited terminal and repeater stations. These link the air defence operations centre with the two sector operations centres, each of which has its own surveillance radar station.

Defence centres in the vicinity of Muscat in the north and along the border with Yemen in the south are linked by a tropospheric scatter system running the length of the country and fed by short haul line-of-sight radio links. This network is used to convey processed data and high priority communications from radar sites and sector operations centre to the main operations centre. One of the Rapier squadrons and the Jaguar aircraft are based at Thumrait, a military airfield 700 km south of Muscat and where there is also a radar and communications installation. There are also other units, including the Rapier Squadron HQ at Seeb on the coast to the west of Muscat.

STATUS

Operational. In mid-1985 it was revealed that Marconi Radar had received a i38 million contract to extend and update the Oman Integrated Air Defence System. The major items of new equipment are two Martello S713 long-range 3-D radars (**3491.153**), their associated MACE display and data handling systems. Part of the contract involves updating and expansion of existing Sector Operations Centres (SOCs) and Control and Reporting Centres (CRCs), as well as the provision of a new CRC. The Rapier SAM squadrons are being updated with Blindfire radar systems.

CONTRACTORS

British Aerospace, Army Weapons, Six Hills Way, Stevenage, Hertfordshire SG1 2DA, England.

Marconi Radar Systems Limited, Writtle Road, Chelmsford, Essex CM1 3BN, England.

SPAIN

2322.181

COMBAT GRANDE AIR DEFENCE SYSTEM

Combat Grande is the name of the programme of automation of the Spanish air defence system by the USAF for the Spanish Government and supplied by a jointly owned Spanish-American company registered in California.

The programme (coded 451D) was designed to automate Spain's manual air defence system by providing a computerised capability for aircraft surveillance and tracking. It includes the development of a combat operations centre and a sector operations centre and modernisation of seven existing long-range radar and ground-to-air transmitter radio sites located throughout the Spanish peninsula. The programme also calls for the improvement and enlargement of an existing microwave system to tie in to the new defence system.

Included in the radar enhancements were the provision of new IFF/SIF facilities, video extractors, and modems. A capability for remote operation of the radar site ground/air communications from sector operations centres for aircraft control was also provided. The existing microwave links between the various stations forming the network will have their capacity increased and redundant routes will be added where alternatives do not exist at present to ensure reliability and security of communications.

The system employs the Hughes H5118M computer, and the Radex system - developed by Hughes in an earlier air defence programme -provides vide extraction and signal processing facilities.

The H5118M is a medium-scale, militarised computer specifically designed for command and control systems. It is smaller, less costly, and more reliable, says the manufacturer, than its predecessors. It has the same maximum capacity as the H3118M, developed 10 years earlier, yet is only one-fifth the size. Software packages developed for a number of other computers may be used interchangeably with the new H5118M, including the H3118M of NATO's air defence ground environment (NADGE) system (**1181.181**), the H4118M used in the USAF 407L system (**2824.18**), and the Mk 158 employed in the

One of the seven long-range radar sites in the Combat Grande Spanish air defence system. Also known as SADA, Sistema Semi-Automatico de Defensa Aerea, it was produced by a jointly-owned Spanish-American company, COMCO

USN improved point defence target acquisition system (IPD/TAS) (**1928.281**).

In its basic configuration, the H5118M consists of a processor, buffered input-output channels, and 16 384 words of core memory, housed in a refrigerator-size cabinet, along with power supplies, maintenance-control panels, and cooling mechanism. It is categorised as a modular, 18-bit, binary, parallel, synchronous system with a 2 microsecond add time and a 4.6 microsecond multiply time. A building-block design enables the H5118M to be expanded to a dual-processor system and a maximum of 131 072 words of core memory, with no increase in cabinet size.

The Radex equipment provides automatic tracking of target returns from primary radars. When other modifications are added, tracking performance will be improved by increasing the number of tracks that can be handled by an individual console operator; track accuracy on non-co-operative tracks; the operator's ability to detect valid targets by providing better recognition than with current narrow pulse video display. Radex was recently bought by the USAF for integration into the 407L system.

STATUS

Under a programme, unofficially known as Combat Grande II, additions have been made to the radar facilities and new microwave communications have been installed to improve air defence network data exchange services. In June 1980 the first delivery of Collins data transmission equipment was made. This was to replace outdated hardware at five of 44 sites being modernised under the programme to improve data transmission from radar sites to Torrejon Air Base, near Madrid, for use in the combat operation centre. This part of the programme was completed by 1982.

In an award of $2.37 million announced in July 1980, General Dynamics Corporation was given the task of modification, test and installation of a search radar at Barbanza in the north-west of the country. The equipment includes an antenna array measuring about 12 × 5.4 m and dual transmitter/receivers. A US-built heightfinder radar is installed at the site. The new Barbanza Mountain radar station was handed over to the Spanish Air Force by USAF Electronic Systems Division (ESD) in April 1984, making a total of eight Spanish air defence radar sites modernised by ESD.

With the formal entry of Spain to NATO, the closer linking of the Combat Grande system with the NADGE air defence system (**1181.181**, above) can be expected. There may also be certain new installations as a result of this development.

CONTRACTORS

Prime contractor is COMCO Electronics Corporation, a company jointly owned by Hughes Aircraft and CESELSA. Each of the two parent companies has a 50 per cent holding in COMCO.

Hughes built the computers and designed the communications equipment for the new system. CESELSA manufactured the communications sub-systems and managed the civil engineering construction.

SWEDEN

2358.181

STRIL AIR DEFENCE SYSTEM

STRIL is a fully automatic air surveillance and operations control system operated by the Swedish Air Force. It is based on four air defence sectors, each having a Sector Operations Centre (SOC) which receives radar data from static and mobile Control and Reporting Centres (CRC).

Inputs to the system come mainly from high and low level air surveillance radars, but a back-up visual reporting service is included to supplement the radar data and to replace it if the radar input is blocked. Information from all these sources is fed into a central data store whence it is extracted for selective presentation to controllers having specific territorial assignments.

OPERATION

When a threat is detected the STRIL controllers, who also have available to them state-of-readiness information on available forces and weapons, decide with what kind of available weapon the threat can best be countered and assign that weapon to the task. In appropriate cases, eg surface-to-air missiles, they will also control the operation. Currently the forces controlled by the system are the SAAB Draken and Viggen interceptor aircraft, AA guns, and Improved Hawk surface-to-air missiles. Most of the control and reporting centres, and the display and computing equipment at the local operations rooms at radar sites have been delivered by Datasaab, which also participated in the overall system development of STRIL.

The STRIL centre is also linked up to the civil defence organisation, and can both alert them and warn industry or the civilian population.

If the threat cannot be countered by the forces under the control of one centre but can be countered by those of another centre, the system provides both voice and transmission links for giving information to other centres. Both narrow- and broad-band microwave links are employed for the exchange of data between centres.

Although primarily an air defence system STRIL is linked to coastal as well as to anti-aircraft artillery.

The original STRIL-60 system has been the subject of sustained development and improvement throughout its operational life and work is now going towards a STRIL-90 standard. The most recent, and so far as is known, the most extensive modernisation programme is now in progress. Recently Datasaab supplied new advanced computerised display systems with such facilities as automatic tracking, etc, and 16 three-dimensional primary radars have been installed to supplement and/or replace older radars.

The older radars are the Thomson-CSF Palmier 3-D types (Swedish designation PS65/66) which were used in the NADGE system. These are static structures, 12 m high with 16 m wide antennas making them vulnerable to attack. The new radars, which are now in operation, are the ITT/Gilfillan AN/TPS-32 (PS806) (see entry **2516.153**) installed on retractable 25 m masts operated from mountain silos. The display equipment, including ECCM control and display processing has been developed and manufactured in Sweden by Datasaab. Some of these sites have chambers which can house mobile control and reporting centres, with multi-processing and display subsystems provided by Marconi Radar.

Low level surveillance is carried out by a chain of PS15 radars, developed and built by Selenia in the late 1960s/early 1970s. These are being supplemented by ITT/Gilfillan systems (PS870), the first few of which have been delivered, with programme completion scheduled for 1993/94.

The two main examples of the Datasaab family of transportable air defence operations centres (TADOC) are the TADOC 311 and TADOC 431. The latter is the more sophisticated with regard to operational functions, computing power etc. These transportable centres can be moved to new locations easily if strategic plans are changed or if it becomes necessary to replace other centres which have been destroyed by enemy action. Another advantage of this technique is that buildings or bunkers in stationary centres can be simplified or even eliminated by housing equipment in TADOC cabins. Up to three TADOC 311s can be collocated, and up to four TADOC 431s. Each cabin can contain from two to four operational consoles, and it is possible to exchange data between cabins within a collocated group.

TADOC 311

Normally, the TADOC 311 interacts with radars in the ASCC system. Data from the radars is sent automatically to the CRC without any operator intervention. The primary radar can be of the 2-D or 3-D type and automatic transmission ensures fast target detection within the entire ASCC system and thus a short system response time (essential in modern air defence systems). However, since systems that may have to be used in wartime must have fall-back modes of operation, the TADOC 311 can exercise a certain amount of control in addition to its normal reporting role. This permits independent operation, should the CRC or its ground-to-ground communication fail. In addition to its 'own' local radar, the TADOC 311 can make use of a remote radar (either a military gap-filler radar or a civil radar). This remote radar delivers extracted radar data via a narrow-band, telephone-type channel such as a radio link.

The TADOC 311 can also serve as a centre for visual reporting posts (VRP). Their data can be filtered by the centre and transmitted to the CRC.

Main operational tasks of the TADOC 311 include:

(1) Producing target information and transmitting it to the CRC. This is normally automated for radar data (transmission of extracted radar data or plot messages). There is, however, a manual back-up function that can be used if the radar or extractor function is degraded. The manual back-up function can also be used to filter and transmit target information obtained from VRPs

(2) ECCM control, which entails evaluation of the ECM situation and the taking of appropriate countermeasures to eliminate or alleviate the consequences of jamming

Secondary operational tasks include:

(3) Autonomous airspace surveillance, which is carried out within the area covered by the connected radar station(s) in the event that ground communication to the CRC or the CRC itself is out of action

(4) Weapons control, primarily, control of fighter aircraft in the situations cited above

(5) Producing target information and transmitting it to nearby SAM/AAA units. This can be performed using the manual back-up function mentioned earlier.

TADOC 431

Normally the TADOC 431 serves as a tactical operations centre such as a CRC in the ASCC system. The TADOC 431 is responsible for all tactical operations within an air defence sector. Sensor information from radars and control and reporting posts (CRP) is transferred automatically to the TADOC 431 which can also receive information from visual reporting posts (VRPs) either directly or via a CRP (in which the information is filtered to some extent). The received airspace information forms the basis for weapons control. The TADOC 431 also reports the airspace situation to a co-ordination centre (CC). This reporting is automatic so that the CC will receive an accurate and up-to-date airspace situation picture of the sector in question. However, since systems that may have to be used in wartime must have fall-back modes of operation, the TADOC 431 can take over some CC functions should the CC or its ground-to-ground communication fail.

Operational console in a TADOC 311 air defence control cabin

The main operational tasks of the TADOC 431 include:

(1) Airspace surveillance: this is carried out utilising information obtained from connected sensors of different kinds and from subordinate centres. An airspace situation picture of the air defence sector is compiled and presented in the TADOC 431 itself and can be reported to a superior centre such as a CC

(2) Weapon control, which involves: evaluation of the threat situation; assignment of the appropriate weapon system (fighter, SAM or AAA); allocation of specific fighter aircraft or SAM/AAA unit and weapon controller; interception and recovery control of fighter aircraft or allocation of targets to SAM/AAA units. Weapons control is exercised throughout the air defence sector and is based upon the airspace situation picture

(3) Civil defence telling and base alerting. This is carried out by issuing warnings to threatened civil targets, such as highly populated areas, industrial areas etc. Warnings are also issued to threatened air bases. Moreover, general airspace situation information is distributed to different 'customers'

(4) Co-ordination of military and civil air traffic control by co-ordinating military and civil air traffic activities in the air defence sector to ensure safety and facilitate identification of civil flights

Secondary tasks include:

(5) Close air support and reconnaissance supervision: for the most part, this entails supervision of reconnaissance flights and supervision and/or direction of close air support aircraft. The objectives are to keep track of these aircraft and be ready to warn them of hostile aircraft

(6) Naval support and co-ordination by co-ordination of naval and air force activities in the air defence sector to minimise the risk of inadvertent firing on friendly units. This can also include the exchange of selected target information between the TADOC 431 and a naval command centre.

STATUS

As was stated earlier, work is in progress to update the STRIL system with a new project entitled STRIL/C 90 which is scheduled to enter full operational service in the mid-1990s. This is intended to be a very flexible, highly automated system, able to operate with current and foreseeable sensors and communication systems. It will certainly entail a complete upgrading of the SOCs and CRCs with new data processing and display systems and their associated software, together with new narrow-band data link communication links. It will also phase in with the introduction of the new Swedish Gripen interceptor aircraft. Requests for tenders have been issued to both indigenous and international companies and the successful bidder was to have been chosen by the end of 1987.

CONTRACTORS

Prime contractor for the STRIL system is Datasaab AB (formerly Stansaab Elektronik AB), 175 86 Järfälla, Sweden. Datasaab is also responsible for the creation of similar systems in Denmark and elsewhere.

SWITZERLAND

2373.181

FLORIDA AIR DEFENCE COMMAND CONTROL SYSTEM

FLORIDA is the name given to a computerised air defence command and control system that was accepted and declared operational by the Swiss military authorities in April 1970. General contractor for the system was the Hughes Aircraft Company.

DEVELOPMENT

The system consists of several military radar stations with 3-D radar and air defence direction centres, including computers, display consoles, and associated equipment. Information from the radar station is fed into conversion equipment in underground air defence direction centres and processed in turn by a high-speed general-purpose computer.

This computer automatically establishes speed, heading, and altitude of an unidentified intruder. Display consoles present a constantly updated picture of the aircraft's flight track as well as information on the various weapons available, their launch ranges, velocities, armament, restrictions, and time-to-kill.

Should the target be identified as an immediate threat, the air defence commander can electronically request all-weather interceptor aircraft or surface-to-air missiles to intercept and destroy, and can alert the civil defence organisation.

SYSTEM COMPONENTS:

The radar used is a long-range three-dimensional radar, with a planar array antenna using the Hughes elevation frequency scanning technique. This provides simultaneous range, bearing, and altitude data. An IFF system is associated with the radar.

The system includes video extractors that analyse radar returns, with the help of a real-time computer, and extract actual aircraft tracks from spurious returns such as clouds and ground returns. Track acquisition and updating is automatic.

This processed information is fed to air defence direction centres where it is accepted by a computer. This takes in information from the missile sites, airfields, and other military installations and stores it for use in selecting defensive measures.

It can also simulate air battles for training and instruction, and can be used as a general-purpose data processing centre.

The computer may be required to provide information on weapons available, their launch ranges, velocities, armament, restrictions, and time-to-kill. Numerals and symbols are used to provide this intelligence to the controller and to depict the battle area, the threat and the defences, the problems, and their solutions. The computer, the Hughes H-3324, is of the same type as those associated with the radars but has a much larger storage capacity. Typically it might be called upon to execute between 50 000 and 550 000 instructions in a second.

In the direction centre the display consoles are arranged to display only that information which is pertinent to the tasks of the operator concerned. The principal functions are:

Identification officer - responsible for designating radar tracks as friendly or hostile aircraft.

Interceptor director - interested in his own controlled interceptor and the assigned hostile aircraft. To him, specific data about this intercept is presented including attack geometry, altitude, speed, time to intercept, etc.

Air traffic co-ordinator - concerned with military aircraft requesting clearance to cross civilian airways.

Missile officer - (surface-to-air) in close contact with the missile batteries and assigns targets to the appropriate site.

Chief of air defence - concentrates on deployment and the threat. Total composite air situation and threat boundaries are important to him.

Aircraft and missile forces currently under control are two Mirage IIIS interceptor squadrons, six F-5E/F squadrons, nine Hunter squadrons and Rapier and Bloodhound SAMs. The Skyguard fire control system is in operational use for the 35 mm AA gun defences.

STATUS

The Florida system officially became operational in April 1970. In the 1979-80 period the Swiss authorities studied competing equipment as a means of providing air surveillance and ground control of interceptor aircraft in the lower airspace. Field tests and evaluations were held of competing systems, among them the Hughes VSTAR (**4080.153**), and the Selenia MRCS-403 mobile air defence system (**3326.181**) and its 3-D radar RAT-31S (**1953.153**). In early 1982 it was announced that under a Swiss air defence programme called Taflir, a version of the Westinghouse AN/TPS-70(V)2 radar, known as Vigilant (**5870.153**) had been ordered and was delivered in 1984 for evaluation. These trials were understood to be successful and it is believed that an order was placed for systems to be delivered, commencing in 1987. However, no official confirmation of this has been made.

CONTRACTOR

System contractor: Hughes Aircraft Company, Fullerton, California 92634, USA.

UNION OF SOVIET SOCIALIST REPUBLICS

3214.181

SOVIET AIR DEFENCE

The Soviet Union maintains and operates an elaborate and extensive air and space defence network consisting of a variety of organisational levels equipped with a multiplicity of sensors, systems and control and command levels which are frequently interdependent and overlap to provide operational redundancy. This ensures maximum protection from equipment failures or other interruptions in service due to hostile action or accident.

Early Warning. What is claimed as the world's most comprehensive early warning system provides separately for detection and alerting to both ballistic missile attack and conventional air attack by missiles and aircraft within the atmosphere. The current ballistic missile early warning system consists of a launch detection satellite network, over-the-horizon (OTH) radars, and a series of large phased array radars on the periphery of the Soviet Union.

The launch detection satellite system can give a warning of any American ICBM launch, as well as determining the area from where it was launched. Two large radars of the OTH variety are in operation facing US ICBM sites and could provide up to 30 minutes warning of attack. Two additional OTH systems are joining them to improve coverage and monitor US aircraft and ship movements in the Pacific areas.

The next layer of ballistic missile early warning facilities consists of 11 Hen House large detection and tracking radars located at six sites on the Soviet periphery, and these can confirm the warning of an attack given by the satellite and OTH radar systems and give an indication of its scale. The Hen House radars are also able to provide certain target tracking data in support of ABM deployments.

Since 1983, an additional large new phased array radar, claimed to be for ballistic missile early warning purposes, is still being built at Krasmoyarsk in Siberia. This brings the total of radars of this type in use or being built to six. This new radar closes the gap in the combined Hen House and new phased array radar early warning and tracking radar. Together, this radar and the five others like it form an arc of coverage extending from the Kola Peninsular in the northwest, around Siberia, to the Caucasus in the southwest. Hen House radar coverage completes the circle. In the US viewpoint, the new phased array radar probably violates the 1972 ABM Treaty in that it is not located on the Soviet periphery nor does it point outward from Soviet territory, as required by the Treaty. At the USSR/USA summit meeting in December 1987 to sign the INF Treaty, the USA arms control delegation was understood to have demanded that the Krasnoyarsk radar be modified or removed before any understanding is reached on the ABM Treaty. The Soviet Union refutes any suggestion that this radar violates the treaty and maintains that it is for space tracking. During the autumn of 1987 the Krasnoyarsk radar site was opened for inspection to a Western delegation which concluded that the half-built system was probably only a conventional early-warning radar with some limited satellite tracking capability. It is likely to be completed by the end of 1989. The complete network provided by these

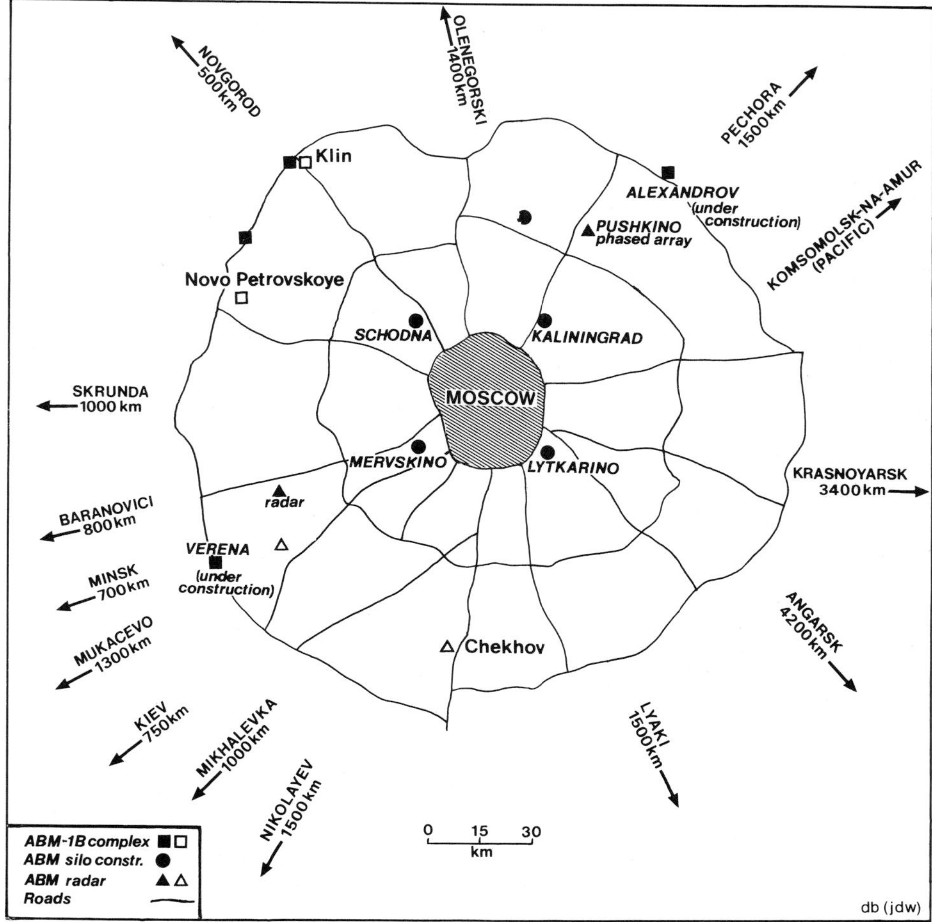

Moscow ABM system map

phased-array radars, which could provide ABM battle management beyond Moscow, is expected to be operational by 1990.

According to United States sources, the Soviet Union is engaged in the construction of three new phased array ballistic warning radars on the periphery of the USSR. One of the new installations is believed to be at Baranovchi near the Polish border. Previously it was believed that five new phased arrays were being built, but sources now believe that nine are engaged in the upgrade of the ballistic missile defences, including the controversial Krasnoyarsk site 3400 km from Moscow. The entire network of large phased array radars will form almost a complete circle around the USSR and the complete system could be operational by the 1990s.

The new phased radars duplicate or supplement the eleven existing Hen House radars at six locations on the USSR periphery. They are believed to be able to track more attacking missiles with greater accuracy than the existing network covering almost all approaches to the Soviet Union. They will check any 30 minute warning of missile launch detected by the Cosmos early warning satellites and the OTH radars directed at the US ICBM fields.

Western analysts believe that the new radar sites will allow the Soviet Union to move quickly to the establishment of a national ABM defence system, rather than one based on Moscow, as at present. It is estimated by the USA that the Soviet Union has spent about $150 billion on strategic defence in the last 10 years.

Ballistic Missile Defence. Throughout the 1980s the Soviet Union has been working to upgrade the Moscow ABM system, which is intended to give a measure of protection for the city itself and for the significant civil and military command installations that exist in and around Moscow. The original single-layer ABM system included 64 reloadable launchers above ground at four complexes for the Galosh ABM-1B, six Try Add guidance and engagement radars at each complex, and the Dog House and Cat House target tracking radars to the south of Moscow. The Soviet authorities are now upgrading and expanding this system to the 100 launchers permitted by the 1972 treaty; in the interim there are only 16 Galosh deployed. When complete, possibly by the end of 1988, it will be a two-layer system consisting of silo-based long-range modified Galosh interceptors designed to engage targets outside the atmosphere; silo-housed high-acceleration Gazelle interceptor missiles designed to engage incoming targets within the atmosphere; associated engagement and guidance radars; and the new Pill Box massive, four-sided, phased array radar at Puskino, north of Moscow, designed to control ABM engagements. It is believed the ABM silos may be reloadable. The SAH-08 (Gazelle) ABM interceptor missile is becoming operational and is being deployed in a number of complexes around Moscow.

The SA-10 (Grumble) SAM has now been deployed at nearly 100 sites, many of which are located near Moscow. A mobile version, the SA-10B, is also being deployed. Coming into operational use is the SA-12A (Gladiator) mobile SAM, whilst a longer range, higher altitude version, the SA-12B (Giant) is being flight tested. This latter SAM is reported to be capable of intercepting aircraft at all altitudes as well as cruise missiles and tactical ballistic missiles.

Air Surveillance. There are an estimated 10 000 air defence radars deployed in the USSR at about 1200 sites. These provide virtually complete coverage at medium to high altitudes of the airspace over the USSR, and in some areas extending hundreds of kilometres beyond Soviet borders. Limited cover against low-altitude targets is concentrated in the west of the Soviet Union and certain high-priority areas elsewhere. From 1983 onward the Soviet authorities began to deploy two new air surveillance radars which will aid the early warning of cruise missile and/or bomber attacks, and will enhance air defence electronic warfare capabilities.

Existing air surveillance radars are also being improved and US officials estimate that more than 20 types are under development. Additionally, the programme to deploy improved air surveillance data networks is continuing, to enhance the exchange of target data throughout the surveillance network to radar outstations to surface-to-air missile sites, and various levels of control posts. This is one aspect of a major overhaul that is thought to be under way, which involves provision of improved capability against low-level threats such as cruise missiles.The programme is understood to embrace the partial integration of strategic and tactical air defences, upgrading of early warning and surveillance facilities, and the installation of the improved data network mentioned above. Other facets are new aircraft and air-to-air missiles, surface-to-air missiles and a Soviet Union AWACS.

Currently the Soviet Union has more than 9000 strategic SAM launchers, about 5000 tactical SAM launchers (excluding hand held) and approximately 2250 air defences interceptor aircraft dedicated to strategic defence. An additional 2000 interceptors could be drawn upon for strategic defence operations.

PVO Strany (*Voiska Protivovzdushnoy Strany*) is the national air defence command of the Soviet Union, with an underground headquarters on the outskirts of Moscow. There are four operations/systems commands:

(1) RV-PVO (*Radio teknicheski Voisk-PVO*) radio-technical troops
(2) ZA-PVO (*Zenitnaya Artilleriya-PVO*) anti-aircraft artillery
(3) ZRV-PVO (*Zenitnaya Raketmye Voiska-PVO*) anti-aircraft missile troops
(4) IA-PVO (*Istrebitel 'naya Aviatsiya-PVO*) fighter aviation forces.

The 10 000 or so air defence radars are the responsibility of the RV-PVO, while missile systems and AA guns are in the hands of the ZA-PVO and ZRV-PVO commands. Fighter aircraft are managed by IA-PVO.

Operationally the Soviet air defence network is divided into four levels of command responsibility: Air Defence (AD) Districts or Rayons, AD Zones, AD Sectors, and 'frontal' AD sub-units defending specific point targets. The rayons are sub-divided into at least two AD zones and these are in turn divided into SAM defence sectors. Each level has a particular command function and operational mission. At rayon level these include search and entry, long-range target identification, designating air threats to lower command echelons, and authorisation of fighter intercepts or missile engagements. Twenty two rayons cover the

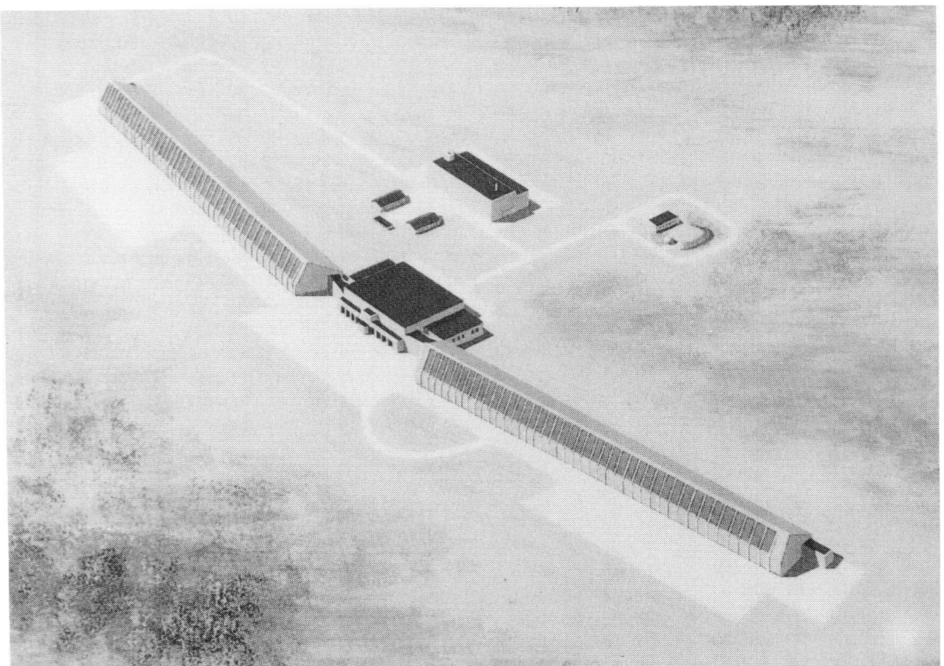

Eleven of the large Hen House ballistic early warning radars are located at six sites around the periphery of the Soviet Union

Artist's impression of the phased array ballistic missile detection array and tracking radar at Pechora in the north of the Soviet Union

US artist's impression of the Soviet Dog House ABM battle management radar

USSR from the eastern border with China to the Warsaw Pact states of Eastern Europe.

The AD districts are divided into AD zones, some districts having as many as four zones and these are under IA-PVO operational command. The command responsibilities at this level embrace target identification friend or foe (IFF), GCI radar operation, and fighter intercept of hostile/unknown aircraft.

An AD sector is a specific, relatively small scale section of airspace under the control of Air Defence Launch Command Centres, and their defence is based on SAMs and their associated target tracking and guidance radars.

Passive defence: The Soviet Ministry of Defence controls the nation-wide civil defence programmes of the USSR. The Chief of Civil Defence is a Deputy Minister of Defence and General of the Army. Full-time civil defence staffs exist at each echelon of the Soviet administrative structure. Civil defence staffs also exist at significant industrial, utility and other installations. In wartime, the civil defence administrative structure, assimilated into an integrated command system, would play a significant role in maintenance of the government and the economy. This goal is supported by the protection provided for the leadership through deep, hard, urban sheltering and an extensive network of hardened relocation sites outside the cities, with redundant communications systems. The programme also provides for continuity of support for the economy in wartime through the protection of the essential workforce by sheltering at work and by the dispersal of off-shift workers to areas away from work sites.

The Soviet Union is deploying the Mainstay AWACS aircraft which is based on the IL-76 Candid transport

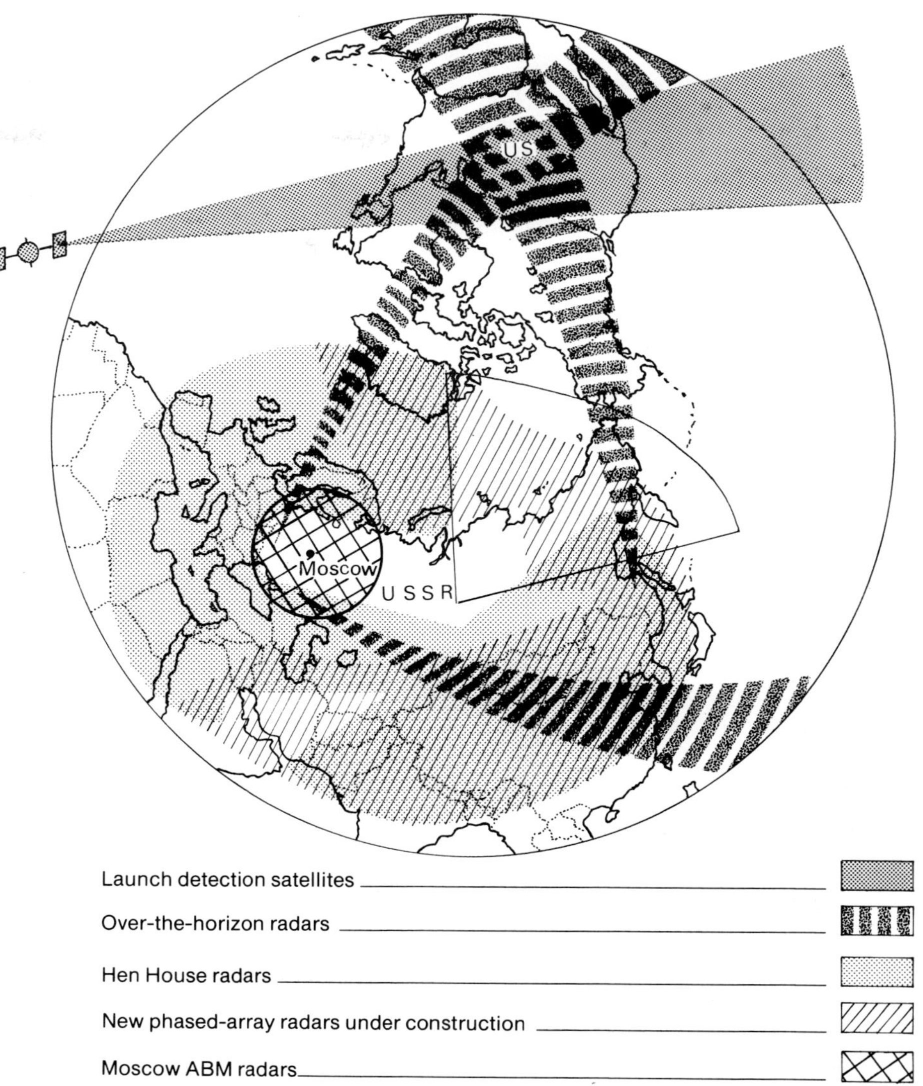

Coverage of ballistic missile detection and tracking systems

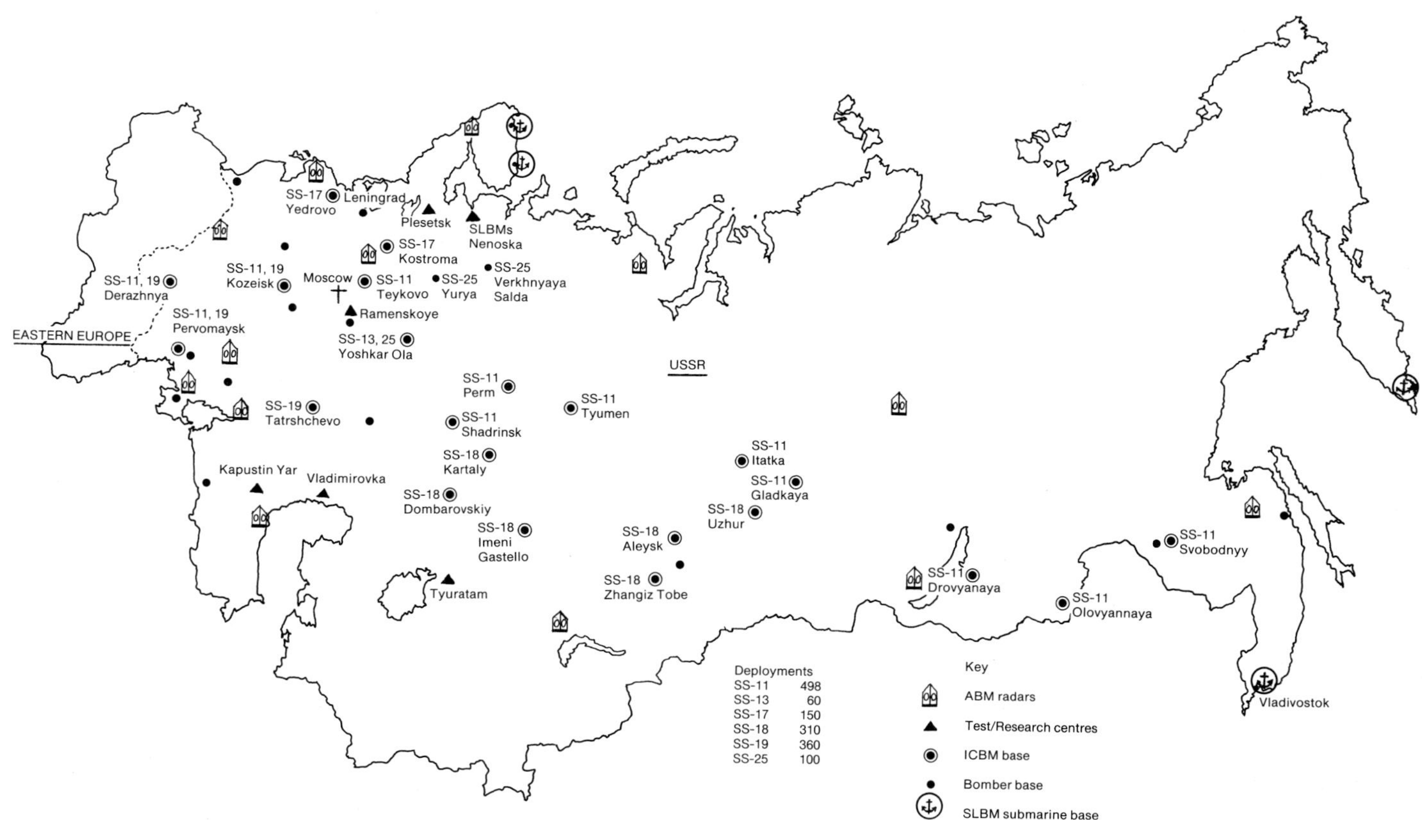

Map summary of principal Soviet strategic assets. Note: details of longer-range IRBM force deployments and strength appear in the Strategic Weapons *section*

UNITED KINGDOM

2444.181

UK AIR DEFENCE SYSTEM

The United Kingdom Air Defence Region (UKADR) forms one of four regional commands of the NATO Integrated Air Defence System and stretches for about 1000 nautical miles in an arc from north of the Faeroe Islands to the western flanks of continental Europe. The systems for warning of attack on the UK are BMEWS (**2525.181**), one station of which is sited in the UK, and the UK Air Defence Ground Environment (UKADGE), the former being concerned with missiles (ballistic) and the latter with aircraft. UKADGE is also linked with NADGE (**1181.181**).

The UK air defence system is coupled into the continental air defence systems by its links with six NADGE stations. The UK data terminal for these links is the West Drayton complex which was originally intended as the heart of Linesman. Incoming data is relayed from West Drayton to the Air Defence Operations Centre (ADOC) at High Wycombe and thence to the sector operations/control reporting centres (SOC) at Buchan, Boulmer, and Neatishead.

Each SOC consists of a radar system and a control and reporting centre, and each is capable of operating independently of other elements of UKADGE.

The 1984 Statement on Defence Estimates stated that a programme of qualitative and quantitative improvements to UK air defences was under way, and this involved the replacement or modernisation of virtually all the RAF's air defence assets. The UKADGE network of ground radars and command and control and communications systems are being totally modernised. In the 1983 Statement it had been confirmed that as part of the programme, 12 transportable 3-D radars had been ordered. This project has become known generally as Improved UKADGE or IUKADGE.

The structure of the new system is illustrated in the map (right). Centralised command of air defence operations in the UK air defence region is exercised from the ADOC at High Wycombe which is linked to four combined sector operations centres and command and reporting centres (SOC/CRC), with, in turn, links to four command and reporting centres (CRP), and there are a number of reporting posts (RP). Other significant installations are the standby ADOC (SADOC) at Bentley Priory, where there is also a software development centre (SDC), and an engineering training facility (ETF). Comprehensive high capacity data links exist or are being established to ensure rapid and secure interchange of data between these sites, as well as to other neighbouring air defence facilities with which interoperability is essential.

The latter include links with the French STRIDA air defence network at Tours and Doullens, NADGE at six sites, NATO and RAF AWACS aircraft, certain naval vessels and various defence assets such as fighter aircraft bases, SAMs etc. Each SOC is responsible for tactical control of operations within its designated region and for the associated CRCs in charge of tracking and interceptor control. The CRPs give a certain degree of back-up fighter control and the RPs represent the detection front line. Each of the RPs has or will have modern 3-D radar(s). Also to be implemented is a supporting network of passive sensors (employing ESM and infra-red detection techniques in all probability).

To manage the vast quantity of data gathered by this array of sensors, ensure its reliable and rapid distribution in a flexible manner, and display it in a better fashion than was possible with the predecessor systems such as Linesman, demands greatly enhanced data processing facilities. Each SOC/CRC incorporates five VAX minicomputers, two for input/output processing, one for radar tracking, one for passive sensor (EW/ESM/IR) tracking, and one for system 'housekeeping'; any single computer failure can be tolerated without impairing system performance.

At the core of the programme is the IUKADGE Command and Control System (ICCS) which has been designed by the UKSL consortium. When completed it will be the biggest software-based system in Europe linking command and control sites with land-based sensors, operational sites, aircraft and warships. The assimilation of data from a wide variety of sources is a complex task and the three main members of the consortium, Plessey, Hughes and Marconi are each responsible for one part of the system. Hughes is providing the large screen colour displays for the ADOC and the SOCs together with the data handling system, Marconi is providing the display and voice communications systems based upon universal consoles and Plessey is providing the communications system or digital data network. The ICCS system should have entered service early in 1987 but delays have put back this date and it is now scheduled for early 1988.

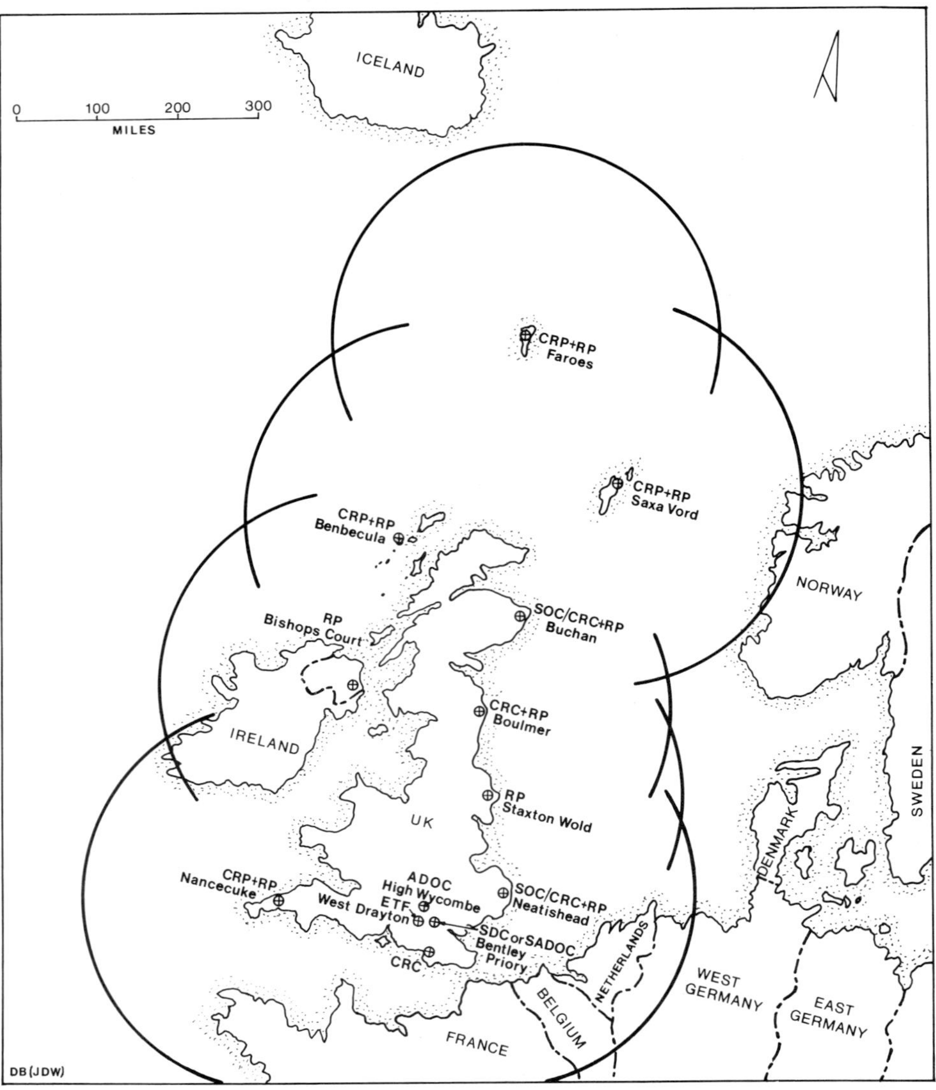

Map of major IUKADGE sites in the British Isles (Courtesy Jane's Defence Weekly)

IUKADGE will be supported by the Uniter communications system designed and produced by GEC telecommunications. Uniter is a secure and survivable integrated network and will link about 120 major sites, including the operational bases, to support both IUKADGE and CCIS (see later). The system is based on the System X telephone system developed by GEC and Plessey and is being carried out in two stages. Stage 1 was completed in 1986 and Stage 2 is scheduled for completion by 1990. Some remote sites will also supplement Uniter by satellite communications through Skynet.

The first of the new ground radar contracts to be awarded was for two D-band American General Electric AN/TPS-592 variants (see below) to meet ASR 887. This requirement called for transportable sets capable of setting up at a pre-surveyed site and going into operation within six hours of arrival. ASR 1585, met soon afterwards, called for E/F-band radars for the UK East Coast chain and this requirement was met by three Plessey-ITT Gilfillan AR-320 equipments, and also included two Marconi Martello D-band radars. A later requirement, ASR 1586 called for another two Martellos and three more AR-320s. The original contracts are reported to have also included certain decoy equipment to divert, confuse or frustrate any Soviet anti-radiation weapons that might be deployed against the radars.

A chain of passive sensors is being established to meet ASR 1584, and in July 1984 a contract was announced by THORN EMI Electronics who were awarded a £5 million contract for the development of a short range air defence alerting device (ADAD). It has not been confirmed at this stage that the THORN EMI contract relates to ASR 1584, but the relevance will be clear to readers of this book.

Both E/F-band and D-band systems are required for the UKADGE updating programme, and a total of seven D-band and six E/F-band radars have been ordered. The first two systems have been installed at Buchan and Benbecula in Scotland and are operational. These are the GE592 (**4321.153**), a variation of the AN/TPS-59, a version of which has been used extensively in the USAF Seek Igloo program me (**2567.181**). This radar has also been supplied, in GE592 form, to Belgium for incorporation in the NATO air defence network of that nation. For the UK, the GE592 has been supplied as two air defence systems, each mounted on five trailers. The antenna is carried on one of these, the other four being for the radar electronics, IFF, simulation, communications and support equipment. This purchase was funded by NATO.

Of the other D-band radars, four are Marconi S723 Martellos (**4798.153**), two of which were ordered under NATO infrastructure funding and two under direct UK MoD funding. All four radars have now been delivered and should be operational in 1988. The remaining D-band radar is a S713 Martello (**3491.153**), an earlier version of the Martello concept. Although no official information has been released on the use of this last system it appears likely that it will be used as a mobile operational spare.

The E/F-band radars are Plessey/ITT-Gilfillan AR320 systems (**3649.153**). Six of these have been ordered, three under NATO funding and three funded by the UK government. Delivery of these radars was scheduled to commence during 1987.

Serious problems experienced with the development of the Mission System Avionics (MSA) for AEW Nimrod continued to delay the proposed introduction of the aircraft and, in December 1986, the UK government decided to abandon the programme completely and has now ordered seven E-3 AWACS aircraft from Boeing, with an option for one more. The first aircraft is due for delivery early in 1991.

Now that a decision has been made in favour of the AWACS aircraft, both the RAF and existing NATO AWACS aircraft will at first use Interim JTIDS before all European AEW aircraft use the full JTIDS format. To enable the UKADGE to play its part in this programme, construction and installation of a network of UK data link stations has been completed, comprising five receiver stations stretching from the Faeroes in the North to Norfolk. THORN EMI Electronics has produced and integrated data link terminals, known as Data Link Buffers (DLB) at these five sites under a programme called Project Cheek to interface with interim JTIDS. The first JTIDS Class 2 terminal was tested in a Tornado aircraft early in 1987.

The DLBs will enable local Command Reporting Centres (CRC), which use a standard NADGE message format, to communicate with the two (later three) message standards of the various types of AEW aircraft and other units.

The withdrawal of Bloodhound from Federal Germany enabled the number of Bloodhound sites in the UK to be increased so that there are now seven: Barkstone Heath, Bawdsey, Coltishall, North Coates, Wattisham, West Raynham and Wyton. This weapon is expected to remain viable until well into the 1990s, and an update programme has been carried out to improve serviceability.

In August 1987, a consortium headed by ICL was awarded an £37 million contract to develop and build a new computer-based command, control and information system (CCIS) for the headquarters of the C-in-C RAF Strike Command (UKAIR). This system is intended to appreciably improve the ability to control the strategy of air operations in NATO's UK air region in times of war. The project will be jointly funded by NATO and UK MoD. CCIS will link UKAIR HQ, and a standby HQ, to major RAF and USAF sites in the UK with facilities to enable UKAIR CCIS to interoperate with other UK and NATO systems including the Central Region CCIS, UK Maritime CCIS and IUKADGE.

Ultimately it is expected that IUKADGE will incorporate a form of Surface Wave Radar (SWR) as part of the defence against low flying aircraft and cruise missiles.

For some years Marconi Radar Systems has been working on SWR, an HF Equipment which, using the sea's salinity, adheres to the surface and follows the earth's curvature. Targets up to 320 km distant can be detected. See **5869.153**.

There have also been unconfirmed reports that the UK MoD is studying the possibility of an OTH radar system, similar to that in use in the USA, based in the Outer Hebrides and Northern Scotland. There has been no official confirmation of this project, although it would tie in well with the CONUS-OTHB system in the USA.

RAF personnel check out computer displays for IUKADGE, undergoing final test at Hughes Aircraft Company, Fullerton, California. Displays shown include Marconi text monitors (left), a Plessey air status display (centre), and a rear-projected HDP-4000 large screen display provided by Hughes. The improved system is produced by UKADGE Systems Ltd

STATUS

As with any air defence system updating UKADGE is an on-going affair and even when the complete IUKADGE system is operational there will be additional upgrading to cope with new technologies and/or threats. The IUKADGE ICCS system is likely to be in operation before this book is published in July 1988 which, considering the size and complexity of the programme, represents an inevitable slippage in the time-scale. The complete system, including the radars, communications networks, full JTIDS, etc will probably become fully operational in 1992-93. The cost of the full updating of the UK's air defences has been about £7 billion, with the cost of IUKADGE itself at around £1 billion.

UKADGE CONTRACTORS

UKSL: Hughes, Plessey and Marconi Radar.

6502.181

MACE COMMAND AND CONTROL SYSTEM

The MACE command and control system fits the Sector Operations, Control and Reporting Centre, and SAM Control Centre roles, operating in either autonomous mode or as part of a fully integrated defence network. Each MACE node in a network can generate a 'recognised air and surface picture' by means of its own automatic track initiation and following programmes using local sensor data, and will automatically crosstell tracks to other nodes for correlation purposes. MACE supports computer-assisted interceptions and recoveries, SAM assignment, and simulation for training and trials purposes.

MACE can be provided either for use on static sites, or fitted in ISO containers for mobility and reduced vulnerability from attack.

Each workstation has a 16 or 22-inch diameter cursively-written PPI display, an interactive touch tabular display, a rolling ball, special function keys, and voice communications facilities. The PPI display presents raw radar from local sensors, digitised and time compressed to allow more time for the addition of computer labelling, graphics and tote data.

Up to three workstations can be driven from a processor set, and MACE operations systems are normally supplied in multiples of three operator positions.

CONTRACTOR
Marconi Radar Systems Ltd, Writtle Road, Chelmsford, Essex CM1 3BN, England.

MACE Command and Control System

6503.181
MARDIS AIR DEFENCE NETWORK

The MARDIS Command and Control System is a powerful air defence network designed for the higher levels of defence management, up to Air Defence Operations Centres. All target inputs are at plot or track level, consequently all data is digitised and can be distributed over the defence network. The tracks generated by all sensors are combined in a full multi-head tracking facility at each centre to give a clear, unambiguous, nationwide recognised air and surface picture. Each centre can perform full threat assessment, weapons allocation, intercept trials, intercept and recovery support, and missile control.

Each MARDIS unit uses distributed processing on the latest 32-bit microprocessor chips, and has a comprehensive diagnostic and fault recovery capability built in. Each workstation has a 16 or 22-inch diameter cursively written PPI display suitable for daylight viewing, an interactive touch tabular display, two rolling balls, special function keys, and voice communications facilities. Hot standby facilities are provided so that any single failure will not affect the operation of the system.

MARDIS can be provided either for use on static sites, or fitted in ISO containers for mobility and reduced vulnerability from attack.

CONTRACTOR
Marconi Radar Systems Ltd, Writtle Road, Chelmsford, Essex, CM1 3BN.

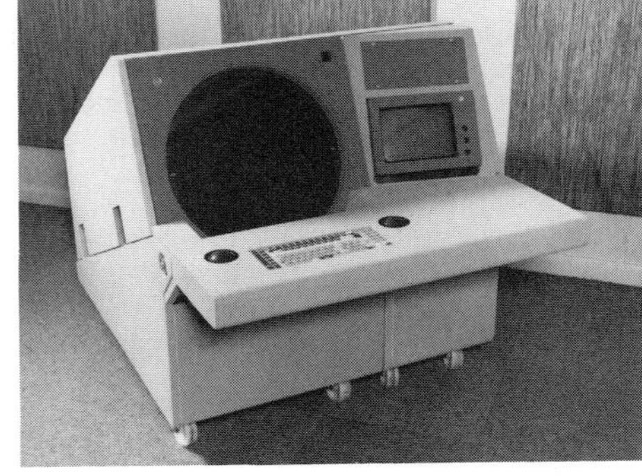

MARDIS Command and Control System Universal Console

UNITED STATES OF AMERICA

2525.181
BALLISTIC MISSILE EARLY WARNING SYSTEM (BMEWS)

The BMEWS (system 474N) comprises a small chain of very large radars for the detection of a ballistic missile attack on North America from the general direction of the USSR. There are three operational sites: Site I in Thule, Greenland, Site II in Clear, Alaska, and Site III on Fylingdales Moor in England.

Three types of radar were used in the original system. At Thule there were four AN/FPS-50 detection radars (**2511.153**) and one AN/FPS-49 tracking radar (**2509.153**), at Flyingdales there are three AN/FPS-49 radars, and at Clear there are three AN/FPS-50 and one AN/FPS-92 radars (the AN/FPS-92 is a more modern version of the AN/FPS-49). Upgrading of the system is currently in progress and includes replacement of the older AN/FPS-49 radars by a version of the AN/FPS-115 Pave Paws (**3174.153**).

Information on BMEWS targets is transferred by a communications network to the North American Air Defense (NORAD) Combat Operations Center.

STATUS
The BMEWS sites in Alaska, Greenland and England have been in operation since 1962 and are reported to have proved highly reliable. In conjunction with early warning satellites the BMEWS radars comprise the principal means of warning of a ballistic missile attack. Improvements are currently being carried out on the radar systems, as described later. Additional warning of SLBM attack was originally provided by six System 474N coastal radar stations (**2810.181**), all but one of which have been replaced by the two Pave Paws phased array radars (Project 259) at Otis AFB, Massachusetts and Beale AFB, California. The AN/FSS-7 radar site in Florida has been retained to provide additional cover for the south-eastern approaches to USA. The SLBM detection system has been further expanded by two additional Pave Paws radars at Robins AFB, Georgia and at Eldorado AFB, Texas. The Robins site became operational in 1986 and the Eldorado site in 1987. The AN/FPS-85 at Elgin AFB, California is still remaining in service.

The upgrade programme for the BMEWS system consists of the incorporation of new computers, new software, increased bandwidth and larger raid tracking capability. New ancillary communications to connect the sites with NORAD will also be installed. The programme also includes replacement of the missile impact predictor sets (MIPS) at all three sites by replacing the original on-site IBM 7090 computers with redundant CDC 170-876 processors.

The system at Thule has now been upgraded with a version of the AN/FPS-115 radar and became operational in late 1986. The UK Ministry of Defence announced in May 1986 that agreement had been reached with the USA for a full update programme at Fylingdales, including the installation of AN/FPS-115 version radars in place of the AN/FPS-49 systems. Contracts will shortly be placed for these radars.

ITT was awarded a contract by the USAF Space Command in late 1987 to operate and maintain the BMEWS system.

The new AN/FPS-115 phased array radars will be more accurate, have a larger track capacity, and provide better prediction facilities. The radars, and the associated processing equipment, will also provide far more accurate detection and analysis of the larger number of smaller warheads in a Multiple Independently Targetable Re-entry Vehicle (MIRV) attack.

In addition to its ICBM-related functions, the latter BMEWS site in the UK will improve warning of missile attacks on NATO territory and of SLBM attacks against the USA, while enhancing surveillance of objects in space.

CONTRACTORS

Main contractors for the BMEWS radar systems were RCA, General Electric, and Western Electric. Raytheon is main contractor for AN/FPS-115.

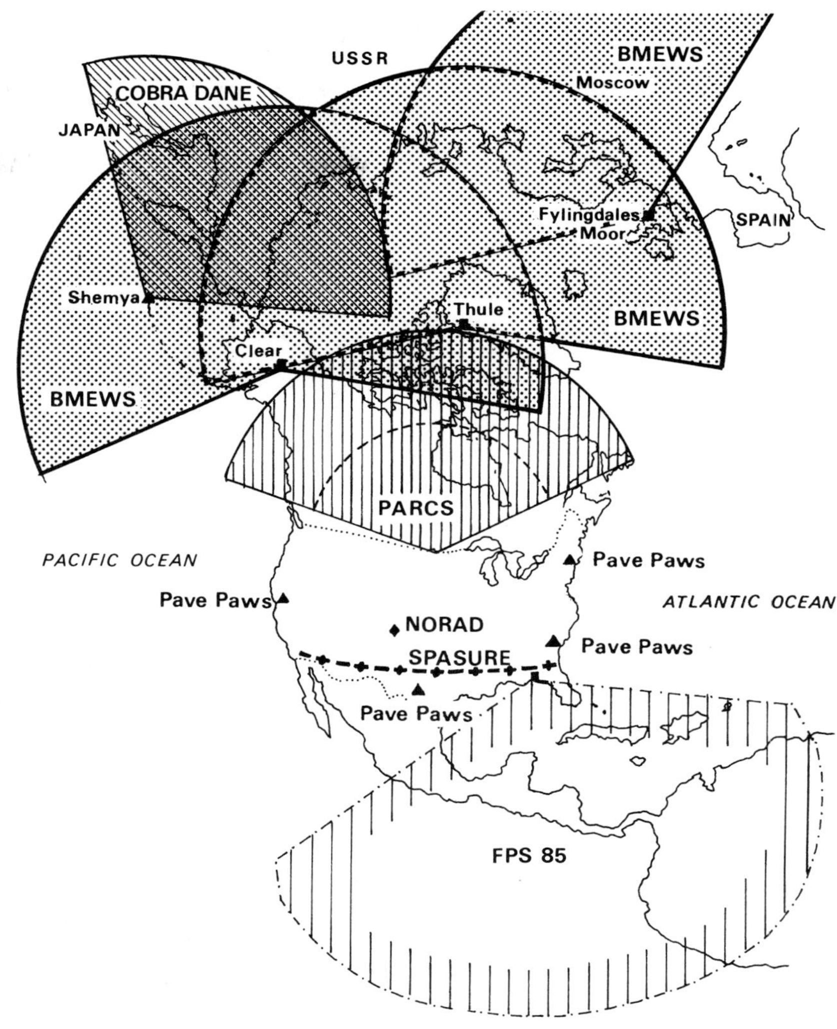

Main elements of the American ICBM warning facilities protecting North America, showing the coverage of BMEWS stations, PARCS (Perimeter Acquisition Radar Attack Characterisation System), and the FPS-85 phased array radar and SPASUR network in the south. All four Pave Paws radars are now operational (see entry **3174.153**)

2750.181
OVER-THE-HORIZON (OTH) RADAR SYSTEMS

The continental over-the-horizon backscatter radar (CONUS OTH-B) system has been developed under US Programme 414L by GE, under the auspices of the USAF Electronic Systems Division. The system is designed primarily as a defence against hostile aircraft and low-flying cruise missiles and will increase the warning of attack by air-breathing air threats by extending US surveillance coverage from the American coasts. It will provide all-altitude capability.

STATUS

The eastern segment of the system achieved limited operational status in late 1987. A detailed description of the design and operational status of the AN/FPS-118 CONUS OTH-B system is given in the Ground Radar section under entry **5617.153**.

CONTRACTOR

General Electric Company, Electronic Systems Division, Syracuse, New York 13201, USA.

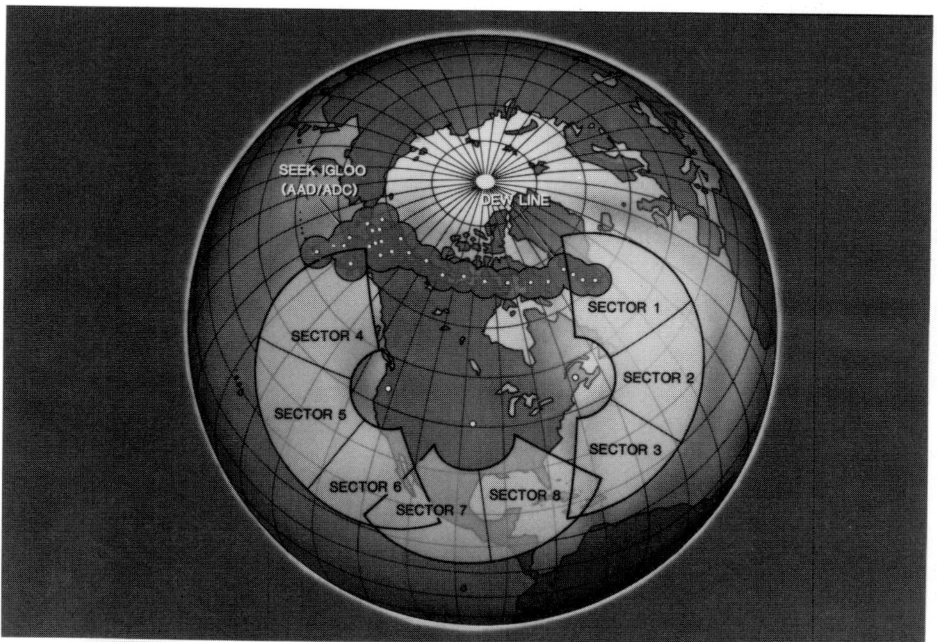

North American air surveillance systems, showing OTH-B sectors for east, west and southern cover and improved DEW Line to north

2567.181
NORTH WARNING SYSTEM (DEW LINE)

The DEW Line is one of the earlier parts of the complex of radar systems designed to warn the US of the approach of hostile aircraft or missiles. Also known as the North Warning System, it is an array of radars that stretches across the northern areas of North America from Alaska to Greenland. The project was originally implemented as Programme 413L.

Main radars of the line have been the AN/FPS-19 and AN/FPS-30. The number of radars in the DEW Line was originally 60 sites but this has been reduced to 31 in Canada and Alaska and four in Greenland. This figure is being maintained while an extensive updating programme is carried out.

The updated North Warning System will consist of 13 AN/FPS-117 long-range radars (11 in Canada and two in Alaska) and 39 short-range, unattended, gap-filler radars. Of the 13 AN/FPS-117 sites, ten will use old DEW Line sites and three will be built in new sites in Canada (two in Labrador and one in Baffin Island). Computer systems, software and communications upgrades in support of the new radar network are also part of the improvement programme.

Sperry is responsible for the overall system engineering, the development of the communications architecture and the development of the short-range radars. General Electric is supplying the AN/FPS-117 radars. The updating programme is proceeding under a joint cost and responsibility arrangement between USA and Canada, as part of the 1985 North American Air Defense Modernization accord between the two countries (see Canadian Air Defence under entry **6805.181**).

STATUS

As of early 1988 a number of the long-range radars had been installed and were being tested. The remaining AN/FPS-117 systems were due to be installed during 1988 with initial operational capability scheduled for the latter part of that year. Satellite communications terminals are being installed at each site and also at the Regional Co-ordinating Centre at North Bay, Ontario. Prototypes of the short-range radars have been built and were due to be installed and tested at North Bay during 1988. Procurement of the short-range systems should commence in 1989. Full operational status of the new North Warning System is currently scheduled for 1992.

CONTRACTORS

For the update programme, Sperry is responsible for the overall system engineering and the short-range radars. General Electric is supplying the long-range radars. In Canada CANAC/Microtel has been selected to co-ordinate and manage the project.

3213.181

JOINT SURVEILLANCE SYSTEM

Under USAF Programme 968H the Joint Surveillance System (JSS) was developed as a successor to the SAGE/BUIC (semi-automatic ground environment/back-up interceptor control) system (**2803.181**) which had been the mainstay of the North American air defence surveillance and warning complex for some 25 years. The expression 'joint' in the title stems from Canadian participation and also by virtue of the fact that in the USA there is joint provision of sensors and data on the parts of the civil and military air authorities (Federal Aviation Administration and the USAF).

Some hardware from SAGE/BUIC is incorporated in JSS but the latter essentially replaces the former system in performing the peacetime air sovereignty mission.

JSS centres (ROCCs) are located at Griffiss, March, McChord and Tyndall AFBs, and they receive data from 46 radar sites. Fourteen more sites feed data to another ROCC in Alaska at Elmendorf AFB, and two radar sites supply data for another ROCC in Hawaii. Twenty-four air surveillance radar networks in Canada feed data to two ROCCs located at North Bay, Ontario.

The function of the ROCCs is to accept data from multiple sensors, automatically process this data, and display data for detection, tracking, and identification of air targets, and the assignment and direction of interceptor aircraft to ensure peacetime air sovereignty. In time of war or emergency, JSS ROCCs serve as the means of transferring the command and control functions to E-3A AWACS aircraft (**1585.353**), and would continue as a back-up to AWACS. In peacetime, six of these aircraft will be assigned to co-operate with the JSS.

The North American Air Defense Command (NORAD) Combat Operations Center (COC), to which ROCC information is passed, is also undergoing modification under PE 12311F, which provides for command control and communications support to HQ NORAD. Within this PE is the NORAD Cheyenne Mountain Complex Improvement Programme (427M), and acquisition programme to update the data processing, display, and communications elements of the NORAD COC (former 425L system), and the Space Defense Center part of the Aerospace Defense Command (ADCOM) Spacetrack system (**2825.181**), the old 496L system.

STATUS

The first region of the JSS to enter operation under the USAF Tactical Air Command was in March/April 1983 and the remaining regions became operational at intervals of a few months until by March 1984 all seven continental US ROCCs were operational with Hawaii, the eighth, following later that year.

Under the SAGE system, 34 radar sites in the USA were operated exclusively by the USAF and 10 jointly with the FAA. With JSS, there are nine solely USAF radar sites and 36 joint sites. Operation and maintenance cost savings amounting to $100 million per year are expected.

CONTRACTOR

Hughes Aircraft Company, Ground Systems Group, Fullerton, California 92634, USA.

Console operators in the JSS Region Operations Control Centres receive information from up to 20 military and civilian radars

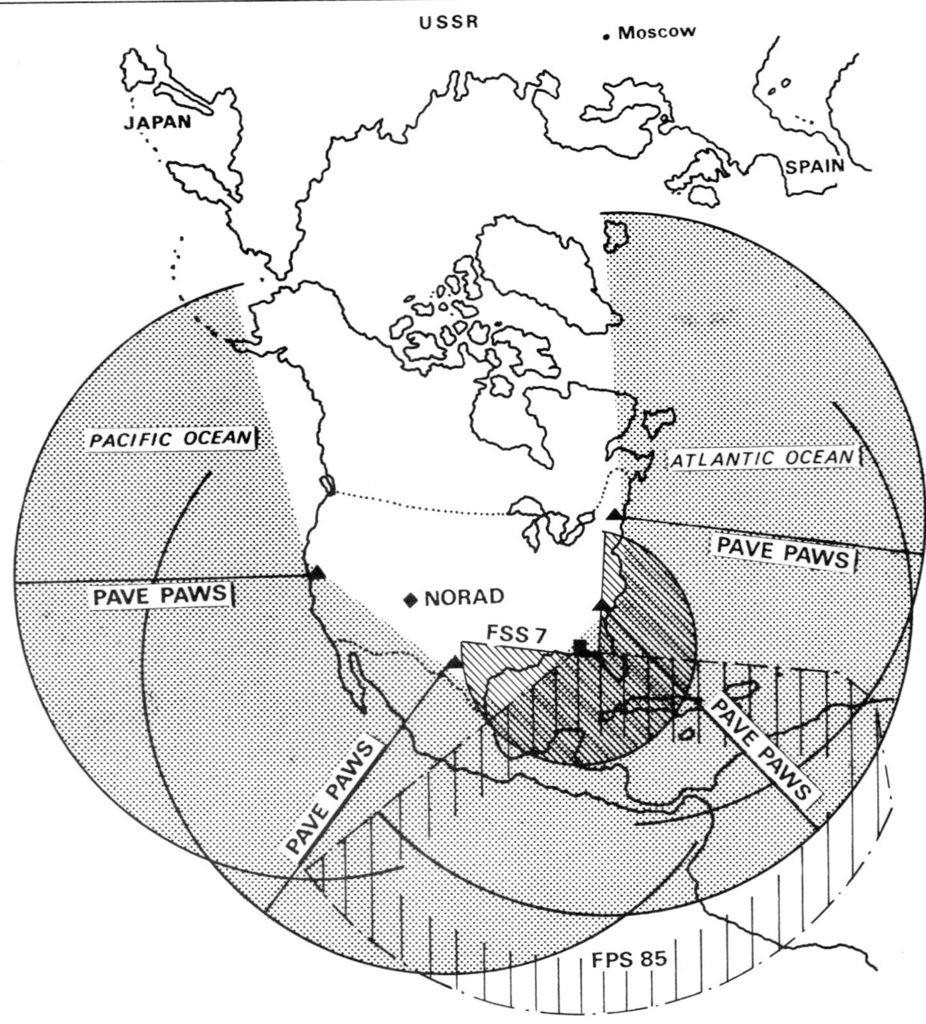

Only one AN/FSS-7 SLBM detection radar needed to be retained after the Pave Paws phased arrays on the East and West Coasts of the USA entered service. Previously there were three FSS-7s on each coast. Although all four Pave Paws radars are now operational, the last FSS-7 and the FPS-85 will be retained for the moment

2810.181

SLBM DETECTION SYSTEM – SYSTEM 474N

This system, which became operational in 1971, is designed to detect missiles launched by submarines operating either in the Atlantic or in the Pacific Ocean. It consisted of seven radars located three on each coast of the USA and one in Texas. The main radar used was the AN/FSS-7 (**2538.153**), made by Avco; in 1972, however, it was decided to supplement these radars with about 20 per cent of the surveillance capacity of the AN/FPS-85 radar (**2546.153**) which is otherwise assigned to the Spacetrack programme

(**2825.181**) and which can provide coverage over most of Central America and the Caribbean.

STATUS

Operational with AN/FSS-7 radars since 1971. Operational with the AN/FPS-85 from mid-1974.

In April 1976, Raytheon was awarded a contract to build the first two AN/FPS-115 Pave Paws radars (**3174.153**) on the East and West Coast of the United States. The two selected sites were Otis AFB, Massachussetts and Beale AFB, California respectively. Each includes a dual faced array about 30 metres high. The Otis system was declared operational in 1980 and the Beale radar followed shortly after. In response to the growing SLBM threat it was decided to build two more Pave Paws systems and the two sites chosen were at Robins AFB, Georgia and Eldorado AFB, Texas. These four, together with the PARCS radar in North Dakota, complete a planned five-site phased-array radar SLBM warning network.

As at mid-1987 all four Pave Paws sites were fully operational. Radar upgrades are in progress for Robins and Beale, and computer upgrades for Otis and Beale. The Eldorado site is also scheduled to support NORAD tracking of space objects. Warning data from all these stations is passed to the NORAD Cheyenne Mountain Complex, Strategic Air Command and the National Command Authorities. Software has been developed by IBM.

2825.181

SPACETRACK

Spacetack is the USAF worldwide system for the detection, tracking and identification of all objects in space. It is composed of large radar, optical and radiometric sensors located around the globe, and its control centre maintains a catalogue of all objects in space. Information from Spacetrack is fed to the Space Defense Operations Center (SPADOC), located at the NORAD Cheyenne Mountain complex and is part of the centrally-managed Space Defense Systems programme involving four functional areas: anti-satellite, space surveillance, space systems survivability and command and control.

Main sensors of the system are the AN/FPS-85 radar (**2456.153**), Cobra Dane (**2546.153**) and the AN/FSR-2 passive optical and radiometric sensors. The AN/FPS-85, of which the AN/FPS-46 was the prototype, is an electronically steerable radar operating in the UHF band and giving three-dimensional information on all satellites and similar objects passing over it.

The complete network of four Spacetrack radar sites, in addition to the AN/FPS-85 at Elgin AFB in the USA and the Cobra Dane radar at Shemya AFB, includes the BMEWS sites at Thule, Greenland; Clear, Alaska; and Fylingdales, England. The Thule site has recently been upgraded with the AN/FPS-115 Pave Paws radar (**3174.153**) and the Fylingdales site is also to be upgraded in the same fashion (see entry **2525.181**). Other Spacetrack radar sites at Pirinclik, Turkey and Kaena Point, Hawaii have also been upgraded.

The current Spacetrack optical system is a four-site, Baker-Nunn camera system with sites at San Vito, Italy; Sand Island in the Pacific; Mount John, New Zealand and Edwards AFB, California. It is capable of tracking and identifying satellites out to synchronous orbit altitudes of 22 300 miles. Improvements continue to be made to Spacetrack sensors. The long term intent is to convert Spacetrack into a near real time space-based system with less dependence on sites overseas.

NAVSPASUR

This system, with the USAF Spacetrack system, forms the NORAD space detection and tracking system (SPADATS) and is the USN contribution to the latter. The NAVSPASUR detects and tracks satellites which pass through a detection 'fence' consisting of a fan-shaped radar beam extending in an east-west direction from San Diego, California, to Fort Stewart, Georgia. A central transmitter is located at Lake Kickapoo, Texas. There are a total of nine stations, six receiving and three transmitting. The beam cannot be steered and detections are made when a satellite passes through a stationary beam. There is a single Baker-Nunn camera at Cold Lake in Canada. The NAVSPASUR HQ and Computation Center is at Dahlgren, Virginia and nine field stations extend across the southern part of the US along a great circle inclined at about 33° to the Equator. The largest transmitter is at Lake Kickapoo, and the two smaller ones at Gila River, Arizona, and Lake Jordon, Atlanta. The six receiver stations are situated at San Diego, California; Elephant Butte, New Mexico; Red River, Arkansas; Silver Lake, Missouri; Hawkinsville, Georgia and Fort Stewart. Other sensors feeding into the system are all three BMEWS radar stations, the USAF Eastern Test Range radar on Ascension Island, the Space and Missile Test and Evaluation Center radar at Vandenberg, and the UK's Malvern radar.

GEODSS

A major segment of the Spacetrack Augmentation programme is the GEODSS (Ground Electro-Optical Deep Space Surveillance) system. An experimental prototype was developed by Lincoln Laboratory at White Sands Missile Range, New Mexco by August 1975 and has subsequently become part of the overall GEODSS system. It consists of two sophisticated telescopes, a 31-inch surveillance and a 14-inch tracker, operating in tandem; associated electro-optics; a computer system with ancillary electronics; and communications equipment. The project consists of five sites at White Sands; Choe San Jong, South Korea; Maui, Hawaii; Diego Garcia; and in Portugal. The first three sites have now been deployed and are operational. The Diego Garcia site is currently under test and evaluation, and the Portuguese site is due to be completed by 1989.

The computer provides for on-line analysis and initial presentation, and it also includes an automatic alarm on detecting unknown space objects. The telescope can detect objects in space which are 10 000 times below the detection threshold of the human eye. A form of MTI is used to control the movement of the telescopes, this being arranged to correspond with either the motion of the target or the star field. The net effect in either case is to isolate the desired target.

STATUS

Spacetrack is a continuing programme and modifications and improvement projects are being carried out on an on-going basis. System upgrades to increase operational effectiveness have been examined and tested and include such features as dynamic or real-time scheduling, an all-sky camera for cloud monitoring, daylight operations and satellite mission assessment improvements. The US Air Force Maui Optical Station (AMOS) is due to become a separate project in its own right in 1988, and is intended to develop techniques for rapid and accurate calibration of Spacetrack radars using US Navy transit satellites. Included in this project is the use of laser technologies to measure and compensate for atmospheric factors that would otherwise distort surveillance data.

SPADOC has evolved in phases from a totally manual system in 1979 to the current semi-automatic capability. A Phase IV updating programme is being carried out by Ford Aerospace under a USAF contract and is being accomplished in three blocks. Block A provides the basic architecture and executive software, and has now been completed, Block B will increase the NORAD Space Surveillance Center capability and Block C will incorporate a number of improvements such as semi-automated space warning, countermeasure management, evasive manoeuvre detection and strike assessment. It is intended that the complete SPADOC system will reach full operating capability by 1992.

CONTRACTORS

Numerous. Some of the principal ones have been Bendix, Cutler-Hammer, General Electric, Ford Aerospace, RCA and SDC.

6806.181

AIR DEFENSE INITIATIVE

The Air Defense Initiative (ADI) project is in the exploratory development stage, and is intended to develop ground, airborne and space-based system concepts, and the necessary technology, for USAF future surveillance missions. Major development technology includes new surveillance radars, counter-ing of low radar signature threats, cruise missile surveillance and ECCM techniques against anti-radar missiles.

Little unclassified information on this project has been released by the USAF, but it is understood that a number of design technologies have been completed, including a lightweight phased array space-based radar antenna using solid-state transmitter modules, a conformal array antenna for an advanced airborne surveillance radar (see **5609.393**), a lightweight surveillance radar decoy and a number of associated hardware systems.

STATUS

Although there are a large number of both contractual and USAF in-house efforts involved in the ADI programme, the project is still in the very early design and development stage. Time scales are unknown but this project appears to be aimed at the mid- to late-1990s.

3073.181

TACTICAL AIR DEFENCE SYSTEM - TADS

The Westinghouse Tactical Air Defense System (TADS) is an integrated, flexible air surveillance and control system for mobile, semi-permanent, or fixed installation. Irrespective of the type of installation, the electronic equipment is identical. TADS can operate autonomously in a manual, semi-automatic, or completely automatic mode, and it can also provide inputs for higher level operations/command centres. Various combinations of equipment can be configured to meet a range of operational requirements in a cost-effective way.

A typical TADS deployment might comprise:

(1) An AN/TPS-43 long-range three-dimensional radar (**2517.153**) or AN/TPS-70 (**4319.153**) complemented by a medium-range 2-D radar such as the AN/TPS-63 (**3475.153**) or Vigilant Tactical Surveillance System (**5870.193**)
(2) One or more operations centres
(3) A communications centre
(4) Cable pallet
(5) Maintenance and storage shelter
(6) Primary power sources
(7) Troposcatter and line-of-sight microwave communications equipment.

A TADS unit can be deployed and fully operational in two hours, and to achieve this mobility, generators, radar antenna, and cable reels are mounted on transportable pallets. All other major units are housed in S-280 shelters. All items can be moved by helicopter, cargo aircraft, or the various land transport techniques.

The communications centre can be adapted to suit a wide variety of air operations and requires only primary electrical power and interfacing with the operation centre to provide ground-air-ground and point-to-point transmission and reception facilities. It typically includes two 28 000-channel HF/SSB transmitter/receivers, interface with line-of-sight microwave equipment, and 24-channel troposcatter interface equipment.

Each operations centre is housed in a standard S-280 shelter, and two or more of these units can be linked to the radar to provide increased operational capability. Equipment in the operations centre includes a radar/IFF signal distribution unit, plotting boards, a video map generator, IFF defruiter, air-conditioning/heater units (operated externally), and associated furnishings and is able to handle up to 100 tracks in a fully automated mode.

STATUS

The configuration was conceived in the early 1970s, and the manufacturers state that TADS is being acquired and deployed by countries in all parts of the world.

Operations centres and communications centres have been custom-manufactured for 13 countries world-wide. They vary in complexity to provide manual, semi-automatic, or fully automatic capabilities. TADS, or related Westinghouse systems, have been ordered for the following countries: Argentina, Australia, Egypt, Federal Republic of Germany, Greece, Iran, Israel, Jordan, South Korea (partial), Morocco, Singapore, Spain, Switzerland (partial), Taiwan, Thailand and Venezuela.

CONTRACTOR

Westinghouse Defense and Electronics Center, Baltimore, Maryland 21203, USA.

4214.181

TAOC TACTICAL AIR OPERATIONS CENTRAL

The TAOC is a transportable automated air command and control system for controlling and coordinating the employment of air defence weapons, interceptors and surface-to-air missiles. Total weight is 6803.8 kg with all equipment, including interconnecting signal and power cables.

The basic system element is the tactical air operations module (TAOM), which has now been designated AN/TYQ-23(V)1. A single TAOM, housed in a standard six-metre ANSI ISO shelter, contains all the air command and control equipment (with the exception of search radar), IFF and prime power requirements needed to perform the air defence function. Four operator consoles are located in each TAOM, each fitted with a multi-colour display and a monochrome alphanumeric display. Full system functional capability is provided by a single shelter.

Operations modules are used at the US Marine Corps Tactical Air Operations Center and provide facilities for accepting inputs from search radar and IFF systems, performing automatic track correlation, acquisition, identification, classification, tracking, threat evaluation and weapon selection, assignment and control; receiving and processing track information orders, command and status data received via digital data link capabilities (these data links are via TADIL A, TADIL B, TADIL C (one- and two-way), ATDL-1, NATO Link 1, 16 Kbps and JTIDS); processing inputs from operator consoles for the entry, deletion or modification of stored information and initiating the appropriate action both within the operations module and for the transmission external to the TAOM; and displaying, on operator consoles, the real-time tactical air situation based on all system inputs, both manual and automatic.

The present system design permits the interconnection of up to five TAOMs through the use of fibre-optic cables. Cables of 500-metre lengths allow the dispersion of TAOMs for tactical considerations or because of terrain constraints. Interfacing radars can be located up to 2 km from the TAOM when connected by fibre-optic cables. Radar/TAOM separation of up to 40 km can be achieved using equipment installed in the TAOM narrow-band secure radio link. A joint development by the US Marine Corps and the US Air Force (see entry **6263.181**), TAOC will replace equipment developed under the Marine Tactical Data System programme. The TAOC equipment completed operational tests in October 1986.

Typical operator console unit (OCU) utilising the finger-on-glass design technique

Typical TAOM shelter

STATUS
A $223 million contract to produce TAOM for the USMC and MCE for the USAF was awarded to Litton in May 1987 and a further contract valued at $153 million was awarded to Litton in February 1988 for nine AN/TYQ-23(V)1 and nine AN/TYQ-23(V)2 systems. Deliveries are scheduled to commence in 1990. Further contracts for these systems are likely over the next few years, with the system due to enter service in 1990.
CONTRACTOR
Litton Data Systems, 8000 Woodley Avenue, PO Box 7601, Van Nuys, California 91409-7601, USA.

TAOC using five operations modules

6263.181
MODULAR CONTROL EQUIPMENT
The Modular Control Equipment (MCE) programme provides the US Air Force with a transportable, modular, automated command control system capable of controlling and coordinating the employment of aircraft in a full range of offensive and defensive air operations. The equipment developed under the MCE programme will replace USAF command and control equipment fielded under the 407L/485L programmes. Expanded automation will be provided at the control and reporting centre and automation will be introduced at the forward air control post as a result of the programme. MCE technology will also provide a ground attack control capability. Each MCE module possesses all of the capabilities included in the USAF message processing centre.

The basic system element is the AN/TYQ-23(V)2 Tactical Air Operations Module (TAOM). A single operations module housed in a standard 6-metre ANSI ISO shelter contains all MCE mission-essential equipment including a full range of tactical digital data links (TADIL-A, TADIL-B, TADIL-C, NATO Link, ATDL-1 and JTIDS). Search radar, IFF and prime power equipments are external to the shelter. Full system functional capability is provided by a single shelter which weighs approximately 5850 kg with all operations equipment, including cables, installed for transport.
STATUS
A joint development with the US Marine Corps TAOC programme, the MCE full-scale engineering development model completed operational testing in January 1987. Production contract award was made in May 1987 with a projected in-service date of 1990 (see entry **4214.181**), and a further contract valued at $153 million was awarded to Litton in February 1988 for nine AN/TYQ-23(V)1 and nine AN/TYQ-23(V)2 systems.
CONTRACTOR
Litton Data Systems, 8000 Woodley Avenue, PO Box 7601, Van Nuys, California 91409-7601, USA.

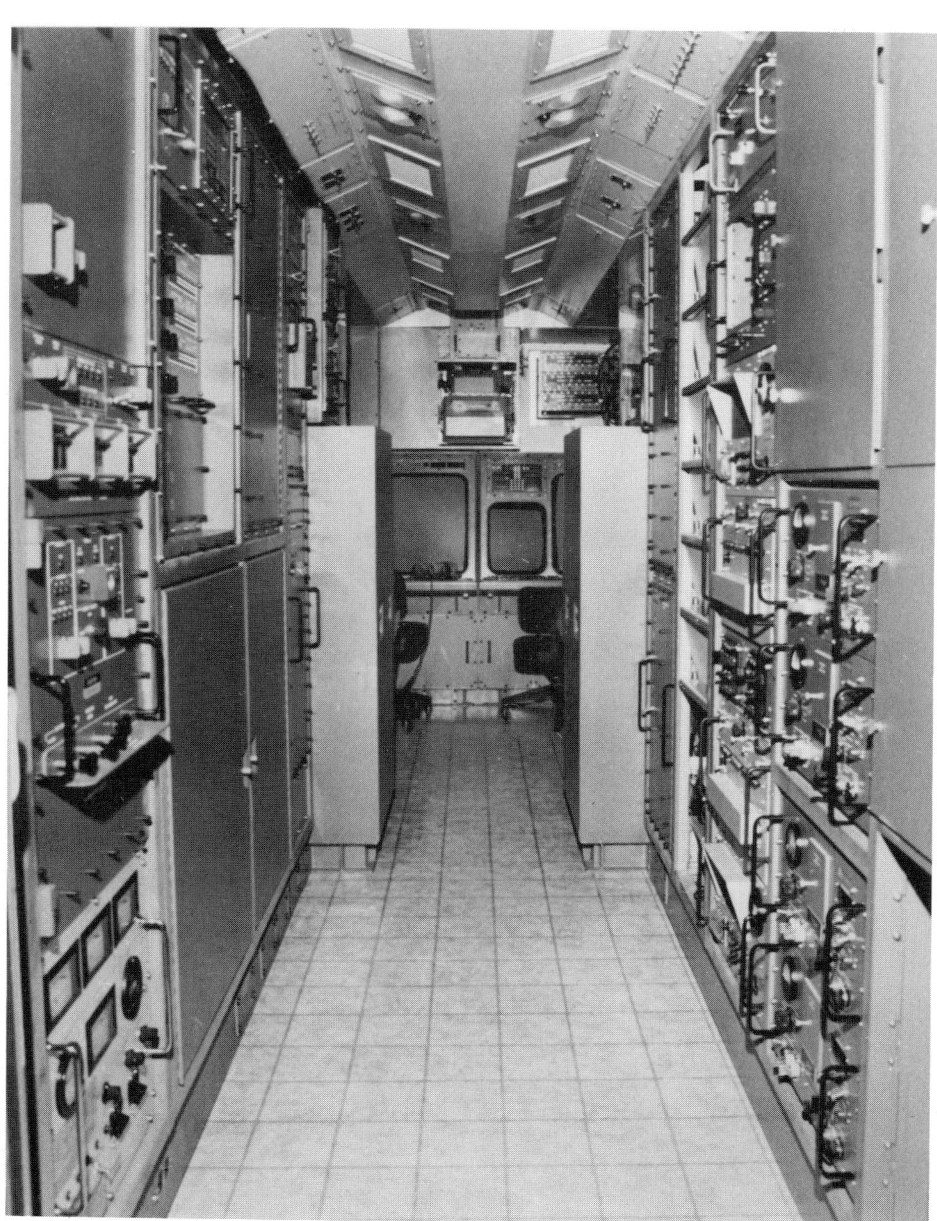

Interior of Modular Control Equipment Shelter

The MCE operator console units provide the man-machine interface system operator with real-time situation and auxiliary display information

2835.181

AN/TSQ-73 AIR DEFENCE MISSILE CONTROL AND CO-ORDINATION SYSTEM

The AN/TSQ-73 is a command and control system for the fire co-ordination of Nike Hercules, Hawk, and Improved Hawk missiles. Developed and produced by Litton Data Systems under contract to the US Army, this 'Missile Minder' system incorporates a dual-processor AN/GYK-12 computer tailored to real-time control and co-ordination requirements. Multipurpose display consoles provide PPI formats for the presentation of radar, track, fire unit, map, and other real-time data. Auxiliary tubular display readouts on the same display surface provide amplifying data on tracks, fire units, operator alerts, and other items. In addition, a light emitting diode (LED) status panel presents fire unit, mission, and system status data.

Radar interface equipment simultaneously and automatically extracts airborne targets from 2-D or 3-D radar and IFF and validates the IFF codes. The integral communications equipment provides for digital communications over missile battery, intra-Army, and TADIL-B data links. Facilities for multiple net voice communications are also provided. Built-in simulation equipment provides realistic training of operators through the simulation of complete air defence scenarios. The AN/TSQ-73 utilises a proven 'repair-by-replacement' maintenance concept consisting of automatic fault detection, semi-automatic fault isolation, and self-contained spares to provide greater system availability and reduced maintenance cost.

In its mobile configuration, the AN/TSQ-73 is housed in a single S-280 type shelter 467 cm long, 203 cm wide and 186 cm high. The system is transportable by cargo aircraft, helicopter, or truck, and is powered by either a 400 Hz diesel generator or a local 50/60 Hz supply.

Processing sensor data provided by a wide variety of currently operational local radars and IFF sets, the two-channel radar interface sub-system automatically detects and acquires the targets sending position reports to the automatic data processor. Special features of the radar interface include an integral IFF beacon decoder and an automatic clutter-mapping function by which extraneous radar returns are removed from consideration as targets.

Litton AN/TSQ-73 interior shelter operation

Each multipurpose console presents situation (PPI) and auxiliary readout (ARO) data on a single cathode ray tube (CRT). Situation information occupies a 37cm diameter circle at the top of the CRT, while auxiliary data is presented in a rectangular area near the bottom. The former includes track symbols position and identification, track number, raid size, height, velocity, extended vectors, and source code. Two area maps, selectable from a list of ten, may be displayed simultaneously, with up to three air-defence-safe corridors. Defence weapons status is also presented for total operational control and includes site position, engagement positions, site number, status, commands, and pairing lines. Using computer-generated alpha-numeric data, the ARO presents more detailed track and fire unit information and permits the console operator to edit the displayed data.

The arrangement of display console controls optimises operator effectiveness and minimises operator training. Operator entry is made by action switches, a full 36-character alpha-numeric keyboard, and a control stick for entering position co-ordinates.

Much of the AN/TSQ-73 equipment is the same as that used by the US Army's TACFIRE artillery command and control system so that there will be significant savings in training, support, and procurement costs. The general-purpose nature of these systems also makes many of them readily adaptable to other electronic command and control systems. The general-purpose nature of the display console controls takes full advantage of an alterable processor which can be easily programmed to meet the specific requirements of other applications, like those of the Air Traffic Management System, control of SHORAD, and SAM-D command and control group.

STATUS

Type classified 'Standard'. Fielded by the US Army. Deliveries include systems for numerous other countries. In 1981 a co-production agreement covering the manufacture of 11 AN/TSQ-73 systems for NATO was established. Under this agreement a contract valued at $103 million was awarded to Litton, with co-production in France by Sintra and in Italy by Litton Italia.

A product improvement to upgrade and expand the memory using CMOS technology was completed in 1987.

CONTRACTOR

Litton Data Systems, 8000 Woodley Avenue, PO Box 7601, Van Nuys, California 91409-7601, USA.

ARMY FIRE CONTROL SYSTEMS

2642.181

INTRODUCTION

This section deals with a number of equipments whose functions are primarily concerned with the control of guns or gun-missile combinations for the engagement of surface or airborne targets.

There are some systems described here that are always associated with the same weapon, simply because they have never been used with any other. The section does not include systems designed specifically for use with a particular weapon. In such instances the control arrangement is generally covered in the relevant entry for the complete weapon system.

However, this section does include some items which are not complete fire control systems but are major adjuncts to such systems – a good example being the British AMETS system (**2400.161**).

Comprehensive control and reporting and similar systems are dealt with in the section on Land-based Air Defence Systems. The main distinction between systems described in that section and anti-aircraft systems described in this section is essentially one of size; the air defence systems cover sensors and weapons deployed over a wide area extending upwards to sub-continental size; those described below are in general associated with only a few weapons – from a single tank gun to an anti-aircraft battery – although major field artillery systems such as TACFIRE (**2827.161**) and BATES (**2644.161**), while providing a service at battery or regimental level, can cover a wider field.

Readers may find it helpful to consult the companion reference book, *Jane's Military Communications*, which deals with closely related topics to those in the following pages of this section, and *Jane's Armour and Artillery* for additional information on some of the weapons controlled by these systems.

BELGIUM

2598.181

SABCA TANK FIRE CONTROL SYSTEMS

SABCA has designed and manufactured over a thousand Laser Tank Fire Control Systems (LTFCS) for Leopard 1, Centurion, Vijayanta, M48, M60 and AMX-30 MBTs and the 90 mm Jagdpanzer self-propelled anti-tank guns of the Belgian Army.

Leopard

The Leopard LTFCS is fully integrated and uses high accuracy computation techniques in the ballistic computer and automatic sensors. The reticle of the laser/optical sight is always centred by a gimballed mirror drive and the laser and optical axes are perfectly aligned. The system is used by the Australian, Belgian and Canadian Armies.

At the end of 1986 the Belgian Government awarded SABCA and OIP-Optics a BFr177 million contract to develop a new fire control system for the Belgian Army's Leopard 1 tanks. A major objective is to improve the night-fighting capability and this will involve integrating a thermal imager with a day/night sight and laser rangefinder.

SABCA will be the prime contractor responsible for modifying the existing fire control system and system integration as well as developing the thermal imaging system using TICM II modules from GEC Avionics. The Belgian company will also modify the existing mirror drive and develop new sight electronics, a muzzle reference sensor for automatic boresight correction and a new gun elevator resolver. It will be assisted by SERE-BEZU of France (mirror drive), Leitz-Canada (muzzle reference sensor) and Wegmann of Germany (gun elevation resolver).

OIP will develop the IR telescope, a new periscope head and the integrated day/night sight.

The prototype system was due to be handed over to the Belgian Ministry of Defence by April 1988. The production is intended to enter service in 1991 and all of Belgium's 334 Leopards will have received it by 1996.

The ATLAS Family for Centurion/M48/M60/AMX-30

The Centurion LTFCS, M48/M60 and AMX-30 ATLAS LTFCS are basically identical. They consist of two main groups:

(1) an integrated day/night laser sight group including a laser visual unit and a second-generation image intensifier equipment (a) (see illustration) or an all weather day-and-night thermal imaging system (b) (see illustration)
(2) a ballistic computer group including controls and ammunition selection units, automatic or manual corrections for standard and non-standard environmental conditions (see illustration). An extensive built-in-test system controls at all times the performance of the system and identifies faults through line replaceable units, thus doing away with the need for tank level test equipment.

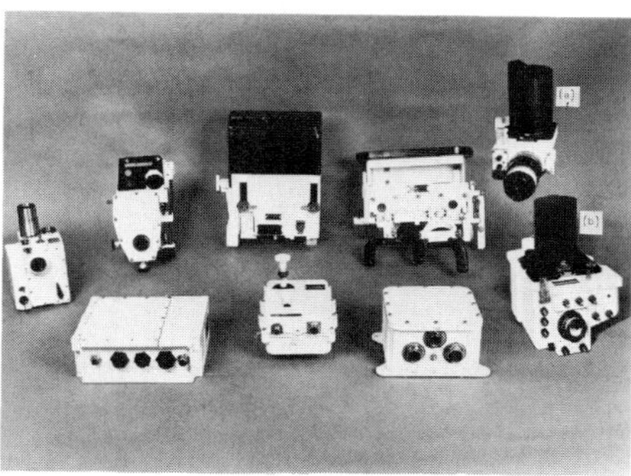

Integrated day/night laser sight sub-system for Centurion, M48, M60 and AMX-30 MBTs (ATLAS)

Ballistic computer sub-system for Centurion, M48, M60 and AMX-30 MBTs (ATLAS)

The ATLAS system is used in Jordan.

UTFCS

The Universal Tank Fire Control System is a full solution fire control system designed to meet current threats and operational requirements as well as those of the next decade (see illustration). The system is equipped with a fully stabilised sight (director system). This approach provides a greater scene stability and continuous tracking of targets. It thus provides a full shoot-on-the-move capability and minimum reaction time. The integrated day/night sight is associated to a muzzle reference mirror to compensate gun droop.

The fire control computer, besides ballistic computation, includes interfaces to gun/turret stabilisation system and the sight head electronics; an extensive built-in-test capability; a gunner training mode whereby the gunner can aim and track simulated targets while his performance is monitored permanently.

The major components of the Universal LTFCS are:

(1) an integrated gunner's sight including a stabilised head mirror unit, a laser visual unit and a passive night sight (image intensifier or thermal imager)
(2) a digital micro-computer based on a high performance 16-bit microprocessor
(3) a vertical sensor unit
(4) a gun position sensor
(5) a wind sensor
(6) gunner and commander control units
(7) a muzzle reference unit
(8) a commander's stabilised sight.

TITAN Mark I

The new T-Tank Laser Fire Control System developed by SABCA upgrades the Soviet built T-54/T-55 and Chinese Type 59 MBTs. This system consists of:

(1) gunner's telescopic day sight including a CRT for in-eyepiece display. This unit also integrates a Neodymium-Yag laser rangefinder.
(2) gunner's periscopic night sight that utilises image intensification in its basic configuration
(3) microcomputer unit that includes a cant angle sensor
(4) data entry control unit that includes a temperature sensor
(5) rate tachometer
(6) muzzle reference unit.

The main advantages claimed by SABCA for this fire control system are improved first round hit probability (doubled at ranges of beyond 1500 m), reduced reaction time, accurate firing against moving targets, capable of operation under both day and night conditions, simple to operate and low maintenance costs.

STATUS

Both UTFCS and TITAN have been demonstrated to potential customers.

CONTRACTOR

SABCA, Chaussée de Haecht 1470, 1130 Brussels, Belgium.

BRAZIL

6346.181

AVIBRAS EDT-FILA ANTI-AIRCRAFT FIRE CONTROL SYSTEM

The EDT-FILA (Fighting Intruders at Low Altitude) has been developed by AVIBRAS to meet the requirements of the Brazilian Army for a mobile fire control system for use with 35 mm and 40 mm anti-aircraft guns and, at a later date, to be used with surface-to-air missile systems. The first prototype was handed over to the Brazilian Army in September 1985.

The basic technology for the FILA was obtained from the Swiss Contraves Skyguard anti-aircraft fire control system (**2377.181**) but with a number of improvements such as an additional Ka band search radar, laser rangefinder and a new C-2001 computer which allows the storage of several types of software such as self-test, diagnosis and combat. The system retains the basic Skyguard X band radars for search and tracking, IFF and TV tracker.

The search radar is of the doppler type with a high detection capability, high clutter suppression, automatic target alarm and an IFF system.

The main X band pulse doppler radar has monopulse evaluation, automated target exchange, ASM detection and alarm and is claimed to be very accurate.

The Ka band auxiliary tracking radar is pulse code and the high precision laser rangefinder is fully integrated. The TV tracker allows for automatic high precision target tracking and has automatic zoom and SIT-Vidicon and quick and accurate parallax measurement.

The third generation digital computer allows real time data processing and has ample capacity core-store, protection against program loss and incorporates a program loader.

The control console has an indicator PPI displaying doppler and raw video as well as symbols and markers for presenting the tactical situation, ECCM control panel, monitor of the TV tracking system, joy stick for manual tracker control and matrix panel for data input and output.

The complete system is installed in a body built of fire-resistant reinforced fibreglass polyester and its cabin is fully air conditioned.

STATUS

Serial units under production.

CONTRACTOR

AVIBRAS Industria Aerospacial SA, Marketing Division, Caixa Postal (P O Box) 229, Sao Jose dos Campos – SP, Brazil.

AVIBRAS EDT-FILA with radars erected

CANADA

6701.181
INTEGRATED FIRE AND MANEUVER CONTROL SYSTEM

The Integrated Fire and Maneuver Control System (IFMCS) is under development and will provide the tank commander with facilities by which he can monitor the operational capability of his vehicle and weapon systems. It can receive and transmit both graphical and textual tactical information and can conduct crew training.

The capabilities will be provided by means of interactive displays capable of displaying map, graphical and textual information. Exchange of information with external sources will be effected through a radio interface using approved message formats.

Data transfer within the tank will be supported on a 1553B dual redundant bus. Overall system performance and maintainability will be enhanced by use of a high level programming language, Ada, and through use of VLSI techniques.

The system comprises:

(1) Computer electronics unit
(2) Commander's display panel
(3) Mass storage device
(4) Navigation system
(5) Improved computer control panel

The computer electronics unit will include two MC68020 microprocessors with 4 Mbytes DRAM. This will be the same volume and size as the Abrams CEU and will feature a hard disc interface and a radio modem. It will interface with a Mil. Std 1553B control bus, a RS422 serial data bus and a RS485 high speed display bus.

The commander's display panel will be an 20.32 × 20.32 cm (8 × 8 inch) EL touch panel. It will be for the display of map, graphical, overlay and textural data as well as for the input of data and for the preparation of messages/orders.

The mass storage device will be a hard disk with 160 Mbytes of formatted memory. It will be used for the storage of digitised terrain data and maintenance procedures. It will be upgradable to an optical disk system.

The navigation system will use the Lear Siegler VNAS and will provide position and heading reference inputs. These will be used to provide a dynamic map display.

The improved computer control panel will feature a microprocessor-controlled 7.62 × 12.70 cm (3 × 5 inch) EL panel which will include numeric and function keys, diagnostics and built-in test equipment. It will be the same size as the Abrams CCP and will be used to input fire control data as well as having displays for graphic and text data.

The IFAMCS will also have a radio interface to enable data transmission over a combat net radio.

CONTRACTOR

Computing Devices Company (a Division of Control Data Canada Ltd), PO Box 8508, Ottawa, Ontario, K1G 3M9, Canada.

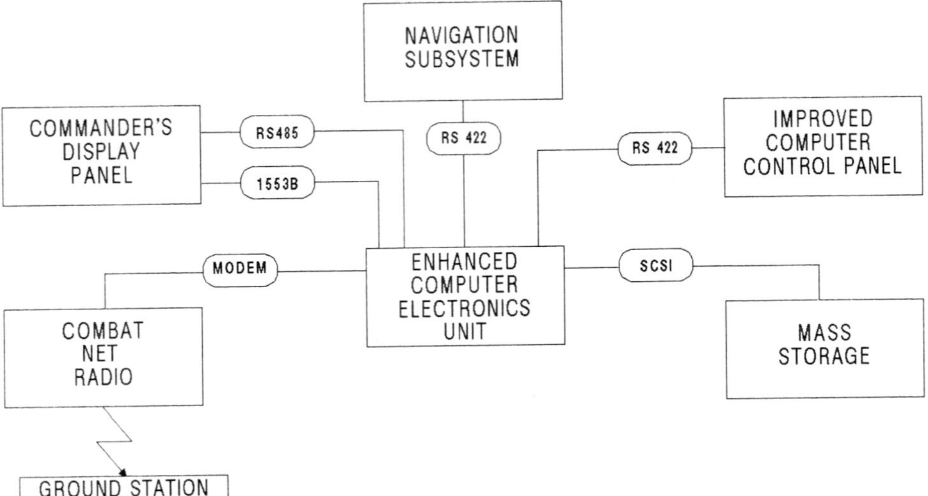

Integrated fire and maneuver control system

6702.161
CDC BALLISTIC COMPUTER SYSTEM

The Ballistic Computer System (BCS) for the M1 and M1A1 Abrams is the first digital fire control computer to be used in large numbers in a production main battle tank. More than 5500 systems have been produced.

The system consists of the Computer Electronics Unit (CEU) and the Computer Control Panel (CCP), the latter serving as the man-machine interface for the gunner.

The CEU comprises a 16-bit microprocessor, analog and digital input/output circuitry and re-programmable, expandable memory to accommodate future changes. The CCP consists of a keyboard containing numeric and function keys, backlit to identify data being entered or displayed, and a 5-character LED display. A significantly improved version of the BCS has been developed in which the CEU incorporates two 32-bit microprocessors and the CCP a 3 × 5 inch electroluminescent panel. The space claim remains the same.

STATUS

In other applications, a modified version of the Abrams BCS is being produced for the Republic of Korea K1 tank and two M48 retrofit customers, with the potential for a number of others. The system has also been successfully integrated with the Close Combat Vehicle Light, the M60A1 tank and the Combat Vehicle Armament System Technology demonstrator. A BCS-based digital retrofit kit for the analog computing system of the M60A3 has also been developed.

CONTRACTOR

Computing Devices Company (a Division of Control Data Canada Ltd), PO Box 8508, Ottawa, Ontario, K1G 3M9, Canada.

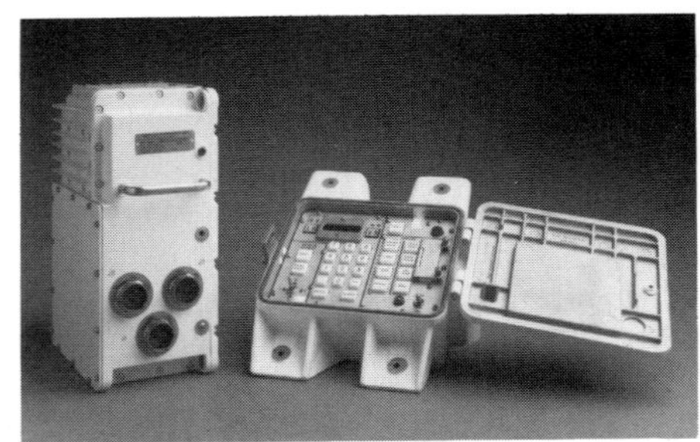

The CDC ballistic computer system (BCS)

6703.161

CDC CONTROL AND DISPLAY UNIT

The Control and Display Unit (CDU) is being developed as a component of the Modular Azimuth and Positioning System (MAPS) programme, and is the man-machine interface of the host system. The functions demanded of the CDU in this MAPS application have produced a design with a wide variety of other applications such as a fire direction calculator, a forward observer data entry device and a gun laying display unit.

Within MAPS, the CDU is a data collection and communication device. Its embedded Intel 8086 16-bit microprocessor enables it to be fully programmable using liquid crystal displays and a keyboard to interface with the operator.

The 12.7 × 10.16 cm (5 × 4 inch) keypad is configured to ensure that existing functions may be easily modified and additional functions incorporated. The keypad is backlit by an electroluminescent strip for operation in darkness.

The liquid crystal display consists of three lines of 12 characters to afford a comprehensive, efficient display of data to the operator. The display is also fully backlit to allow a wide range of ambient light.

The system is programmed in PASCAL to allow for ease of change to Ada for future growth.

The CDU is equipped with one full duplex and one half duplex RS422-compatible serial communication interface, allowing high speed synchronous or asynchronous communications. The communication interface is fully programmable using the Zilog Z8030 Serial Communications Controller.

CHARACTERISTICS

Weight: 1.6 kg
Dimensions: 20.3 × 10.2 × 6.4 cm
Power supply: 24 V DC
Power consumption: 6 W (25 W with display heaters on)
Temperature range: –46° to +65°C

CONTRACTOR

Computing Devices Company (a Division of Control Data Canada Ltd), PO Box 8508, Ottawa, Ontario, K1G 3M9, Canada.

The CDC control and display unit

6704.161

CDC MILITARY COMMUNICATIONS CALCULATOR

The Military Communications Calculator (MCC) is under development by Computing Devices Company as an artillery forward observer (FO) digital message entry device. The unit guides the FO through the composition of various messages related to artillery fire control and allows the FO to review the prepared message before transmitting it to the stated destination. Implementation of a different software suite enables the device to be utilised as a Mortar Data Communications system.

The MCC consists of an embedded Intel 8088 16-bit microprocessor, memory, a MIL-STD-188C serial interface and CCITT V 23 modem. The man-machine interface consists of a 320 × 192 dot matrix electroluminescent display and a touch-sensitive overlay panel. Firmware for the Intel 8088 is programmed in PASCAL.

CHARACTERISTICS

Weight: 2 kg
Dimensions: 24 × 15.5 × 7.5 cm
Input voltage: 16 to 30 V DC
Power consumption: 15 W
Temperature range: –40° to +52°C

CONTRACTOR

Computing Devices Company (a Division of Control Data Canada Ltd), PO Box 8508, Ottawa, Ontario, K1G 3M9, Canada.

The CDC military communications calculator

3937.161

AN/GYG-501 MILIPAC ARTILLERY COMPUTER

The AN/GYG-501 MiliPAC is a portable artillery computer initially designed for the Canadian Department of National Defence to carry out the calculation of firing data for a battery of six guns of 105 mm calibre, as well as a number of related functions such as field and unit survey and fire control.

Since the original development in 1976-79 MiliPAC has been expanded to the battalion level where it is used as a Tactical Fire Control Computer System. This includes the expansion of the data communication capability for the units used at both the battalion and battery Fire Direction Centres and expansion of the capability to receive digital information from the Forward Observers.

The MiliPAC is built as a single unit which is light enough to be carried easily by one man but is meant for installation in a vehicle or other transportable shelter used as the battalion or battery fire control post. The processor used in MiliPAC is the Intel 8086 microprocessor. The memory is a combination of CMOS, RAM and EPROM with battery back-up for the RAM memory. The battery can maintain the contents of the RAM for three months with no recharging. The EPROM memory contains all of the programme code and is non-volatile. The combination of memory used is 12 K words of CMOS RAM and 128 K words of EPROM, expandable to 448 K words. The light emitting diode display used on the Canadian configuration has the capacity of 8 × 48 character lines and provides input and output data, as well as cue messages to assist the operator. An alternative 512 by 256 dot matrix electro-luminescent (EL) panel has been developed to display the Arabic script. In addition to data entry input from the keyboard, provision is made for teleprinter input and output, and gun display system.

The standard software package includes routines for calculation of firing data for up to six guns at up to three positions, a combination of 120 target and observer records, each observer being a protected safety zone, an additional five independent safety zones, five fire support lines, five crest clearance zones, two air corridors, minimum firing elevation for each of the firing positions, 12 non-standard muzzle velocities for each gun/projectile combination, 27 lines of meteorological data, ten sets of registration corrections, and ballistic data for four weapons.

CHARACTERISTICS

Computer: Intel 8086 microprocessor
Memory: 12 K words CMOS RAM and 128 K words of EPROM; expandable to 448 K words
Display: 384 character LED (512 × 256 dot matrix EL panel optional)
Weight: 20 kg
Power supply: 20 to 33 V DC or 10 to 16 V DC
Consumption: 50 W

STATUS

Following trials with three prototypes delivered to the Canadian Armed Forces in 1979 a total of 150 production systems were delivered. In 1986 deliveries began to a Middle East country for a battalion level system. Several other countries are studying this system. These systems include an optional two channel radio/wire data transmission interface.

CONTRACTOR

Computing Devices Company (a Division of Control Data Canada Ltd), PO Box 8508, Ottawa, Ontario K1G 3M9, Canada.

MiliPAC artillery fire direction computer

CHINA

6347.181

AMC-II MORTAR COMPUTER

The AMC-II hand-held mortar computer is a self-contained unit incorporating a computer, keyboard for data entry, batteries, illuminator and a leather carrying case.

It is water-proof and suitable for use under all weather conditions and is easy to operate requiring only one or two days training.

The AMC-II is suitable for all types of mortar and their associated ammunition and in addition to being used for mortar fire computation can also be used for topographical surveying.

It is powered by four Ni-Cd batteries which provide sufficient power for at least 800 operations.

CHARACTERISTICS

Weight with batteries: 1.2 kg
Dimensions: 112 × 243 × 40 mm
Power supply: 5 V DC
Accuracy in survey role: ±1 m
Calculation time in survey role: 1 second
Maximum distance deviation of initial firing data: ±8 m
Maximum azimuth deviation: 1 mil
Firing data calculation time: 4 seconds approx

STATUS

In production.

CONTRACTOR

Chinese state factories. Enquiries to Chinese North Industries Corporation, 7A Yuetan Nanjie, Beijing, People's Republic of China.

6348.181

AVC-101 ARTILLERY FIRE CONTROL COMPUTER

The AVC-101 (Artillery Versatile Computer) has been designed to compute the initial firing data and corrections for individual weapons in an artillery battery with a maximum of eight weapons and three battery reference pieces in a battalion. It can also be used with mortars and multiple rocket launchers.

It can store and recall 30 targets and adjusted data of 10 targets. Targets and points of burst are adjusted by using polar or grid co-ordinates, and data shifted from a known point. In addition the AVC-101 can compute the predicted point data for moving targets. For checking, error codes are provided.

Programmes for the following guns, howitzers, mortars and multiple rocket launchers are already available:

122 mm howitzer Type 54
122 mm gun Type 60
130 mm gun Type 59
152 mm gun-howitzer Type 66
130 mm multiple rocket launcher Type 66
130 mm multiple rocket launcher Type 63
122 mm multiple rocket launcher Type 81
82 mm mortar Type 67
100 mm mortar Type 71
120 mm mortar Type 64

New ballistic programmes for other weapons can however be written.

Variants

AVC-102 Artillery Survey Computer

This is designed to perform all the artillery survey computations and has 20 functions to meet the requirements of survey operations.

AVC-103 Artillery Meteorological Computer

This provides meteorological messages for artillery firing units and sound ranging units. Once the initial data has been entered manually, a meteorological message will be computed and printed out with accuracies of 1 mil in wind direction, 0.1 m per second in wind speed, 0.1° C in temperature and 0.1 mb in pressure.

AVC-104 Artillery Sound Ranging Computer

This replaces plotting equipment and is used with sound ranging equipment to enable the position of a sound source to be obtained in a much shorter time and with greater accuracy. It computes the target co-ordinates and meteorological correction data, and conducts sound ranging adjustments.

CHARACTERISTICS

Weight: 0.8 kg
Size: 220 × 140 × 45 mm
Power consumption: 0.15 W
Power supply: 6 V DC
MTBF: 4000 hours
Operating temperature: –20° to +50° C
Firing data computation for fire direction centre: 10 seconds
Firing data computation for each gun: 4 seconds
Accuracy, gun, range: 1/1000
Accuracy, gun, deflection: 0.6 mils
Accuracy, rocket, range: 1.3/1000
Accuracy, rocket, deflection: 0.8 mils

STATUS

In production.

CONTRACTOR

Poly Technologies Inc, 5/F, Citic Building, 19 Jian Guo Men Wai Street, Beijing, People's Republic of China.

FRANCE

4840.131

BPTSG MODULAR FCS FOR AA GUNS

The BPTSG modular fire control system (FCS) has been designed for twin 20 mm anti-aircraft guns such as the GIAT 76T2, 25 mm twin guns, or 30, 35 and 40 mm systems. The main elements of the BPTSG FCS are (see illustration): a gyro-stabilised sight unit (1) and its associated electronics (2); a mechanical sight for target acquisition (3); a laser rangefinder and a control and display unit (4); a joystick control (5); digital computer (6); voltage chopper/amplifiers for driving the gun (7); electrical DC motors and gears (12); an alternator with a combustion engine for driving the alternator (9); an electrical accumulator group (10) and (11); a simulation box for training, including a TV camera and a miniature viewer (8).

The digital computer performs calculations for accurate computation of lead angles, limitation of firing range etc, and provides an interface for remote control and target designation. There is also an interface for the laser rangefinder.

CHARACTERISTICS

Sight
Field/magnification: 30°/× 2; 15°/× 4
Elevation deflection: –10° to +85°
Azimuth deflection: ±32°
Laser rangefinder
Range: 10 km
PRF: 5 – 20 Hz
Wavelength: 1.06 μm
Performance (on twin 20 mm carriage)
Max azimuth rate: 1.7 rad/s
Max elevation rate: 1.3 rad/s
Max azimuth acceleration: 2.5 rad/s²
Max elevation acceleration: 4 rad/s²
Minimum stable rate: < 0.1 mrad/s²

STATUS

Development.

CONTRACTOR

SAGEM – Département Armement Terrestre, 6 avenue d'Iéna, 75783 Paris Cedex 16, France.

SAGEM BPTSG modular AA FCS equipment locations on typical 20 mm mount

4147.181

COTAC FCS

Tank fire control systems for French vehicles are produced in several versions forming the COTAC (Conduite de Tir Automatique) family of systems. The principal known examples are outlined in the following paragraphs.

COTAC APX M401

COTAC M401 is the automatic fire control equipment developed for the AMX-10RC 6 × 6 reconnaissance vehicle armed with 105 mm gun. It permits daylight engagement of fixed or moving targets. The equipment comprises the APX M504-04 gunner's sight for observation, fire correction data display by M421 optical compensator, laser rangefinding by APX M550 rangefinder, automatic harmonisation of sight and weapon, and emergency firing using a stadiametric ranging scale. Other items are tachometers on both axes; an automatic cant sensor; automatic displays of wind, temperature and altitude; and a computer for calculating fire corrections.

CHARACTERISTICS

Range measurement: 400 – 10000 m
Target speed measurement: 0 – 25 mils/s (azimuth); 0 – 5 mils/s (elevation)
Cant measurement: +0.2 to –0.2 rad
Crosswind setting: +10 to –10 m/s
Outside temperature range: –15 to +45°C
Altitude setting: 0 – 1500 m
Firing correction computation
Bearing: –30 to +30 mils
Elevation: 0 – 50 mils

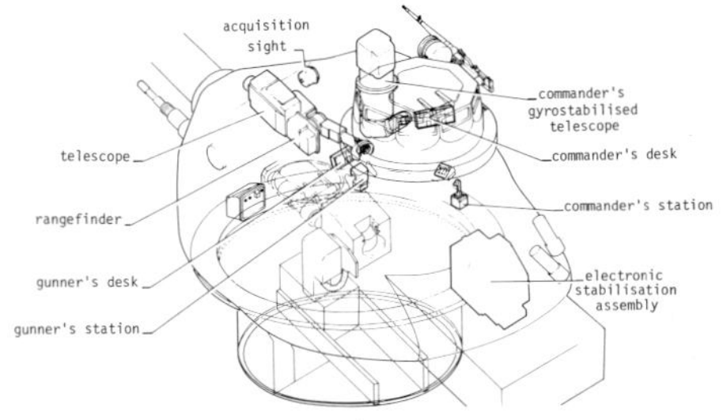

Main components of COTAC APX M581

COTAC APX M581

The COTAC APX M581 is an integrated automatic fire control system developed for the GIAT AMX-30 B2 MBT armed with a 105 mm gun which is used by the French Army. It permits the tank, when stationary, to engage in daylight stationary or moving targets. With the addition of the CASTOR thermal camera, night engagements of stationary or moving targets can also be carried out. It includes the 'tank stationary' fire control system of the original AMX-30 MBT, and, integrated with the same assembly, an APX 550 laser rangefinder, tachometer, slant sensor, analogue computer and various optical devices.

CHARACTERISTICS

Magnification: × 10

Range measurement: 400 – 10000 m

Target speed measurement: 0 – 30 mils/s (both planes)

Slant measurement: + 200 to –200 mils

Crosswind setting: + 10 to –10 m/s

Outside temperature setting: –15 to + 45°C

Altitude setting: 0 – 1500 m

Range of fire correcting computing: 3000 m

STATUS

The APX M401 COTAC has been in series production since 1979 and the APX M581 COTAC since 1981.

CONTRACTORS

GIAT (Groupement Industriel des Armements Terrestres), 10 place Georges Clemenceau, 92211 Saint-Cloud, France.

Laser rangefinder: CILAS.

Gyrometers: SFIM.

Computer: SINTRA.

3979.181

SOPTAC 11 FIRE CONTROL SYSTEM

The SOPTAC 11 fire control system consists of a TJN2 90 telescope, TCV 107 laser rangefinder, computer unit, electronics box for the commander and a gunner's electronics box integrated into the telescope.

The system can be used for observation by both day and night, laser rangefinding and the engagement of both stationary and moving targets. It can be installed in a variety of turrets so long as the telescope can be fitted through the gun mantlet.

Target information fed into the computer includes range to target, speed of target, vertical closing speed, deflection of firing vehicle, temperature, air pressure, type of ammunition being used and side wind.

The TJN2 90 telescope has the following characteristics:

Magnification: × 6 (day path); × 6 (night path)

Field of view: 8° (day path); 6° (night path)

Exit pupil: 4.5 mm (day path); 4.5 mm (night path)

Eyepiece setting: –5 to + 2 dioptres

Laser safety: 50 dB (day path); protected by tube (night path)

The second generation image intensification tube has automatic gain control. The TCV 107 laser rangefinder has a minimum range of 150 m and a maximum range, depending on weather conditions, of 10000 m. It is accurate to ± 10 m

As an option a day telescope with a × 8 magnification and a 6° field of view can be installed with a manually raised micrometer and transfer of the gunner's image to the tank commander's sight.

STATUS

In production. Installed in the GIAT TS 90 two man turrets armed with a 90 mm gun installed on the Renault VBC-90 (6 × 6) armoured cars in service with French Gendarmerie and Oman.

CONTRACTOR

SOPELEM, PO Box 264, 75866 Paris Cedex 18, France.

Elements of SOPTAC tank fire control system showing TJN2 90 day/night sight, commander's control unit, computer and CILAS laser rangefinder

---

6349.181

SOPELEM SOPTAC 36 DAY/NIGHT FIRE CONTROL SYSTEM

The SOPTAC 36 day/night fire control system has been developed as a private venture by SOPELEM for installation in the gunner's station in turrets installed on light armoured vehicles.

The system enables observation and firing by both day and night, laser rangefinding and engaging both stationary and moving targets.

The system is provided with a computer that takes into account the different parameters that govern firing accuracy including distance to the target, its speed in three directions (lateral, vertical and approach/departure), vehicle cant, air temperature and pressure, type of ammunition used for engagement and crosswind.

The SOPTAC 36 fire control system is of modular design and consists of a dual purpose telescopic sight with day and night paths (latter uses same eyepiece as the day path), laser rangefinder module, line of sight projection module of mobile cross hairs, computer unit with micro-processor and control boxes for tank commander and gunner. The sight can also be fitted with a repeater system that transfers the image to the vehicle commander.

Main components of the SOPTAC 36 fire control system with gunner's day/night sight on right

CHARACTERISTICS

Optical equipment

	Magnification	Field
Periscope path	× 1	traverse 42°; elevation 23°
Day path	× 8	7°
Night path	× 8	6°
LI tube		
2nd or 3rd generation micro-channel tube		
Range of path with 3rd generation tube		
	10^{-4} lux (dark night)	10^{-2} lux (quarter moon)
Detection	1400 m	2500 m
Reconnaissance	600 m	1500 m

Laser

Precision: ± 5 m from 150 m to 9995 m

Rate: 1 measurement every 2 seconds, limited to 12 measurements/minute

Fire control

Operating range: 300 to 5000 m

Elevation compensation: 0 to 60 mils

Deflection compensation: ± 30 mils

Cant angle range: ± 20°

Jump compensation: ± 1 mils

Ammunition selection: 4 types

Air pressure range: –300 to + 100 mm Hg

Air temperature range: –40 to + 70° C

Crosswind range: 20 m/s

STATUS

In production.

CONTRACTOR

SOPELEM, 19 Boulevard Ney BP264, 75866 Paris Cedex 18, France.

3843.181

APX M586 TANK FCS UPDATE

The APX M586 is a tank fire control improvement equipment intended for the updating of earlier MBTs such as the M48, AMX-13 etc. The equipment comprises an external unit positioned in front of the original firing sight tube, and a control console for use by the tank commander. There is also a TCV-107 laser rangefinder and an electro-inertial sensor unit which contains all the necessary elements for parameter measurement and fire correction calculations.

By the addition of Thomson-CSF Canasta low light-level television equipment a complete day/night capability can be obtained.

CHARACTERISTICS

Range measurement: 300 – 10000 m

Target speed measurement: 0 – 30 mil/s (both planes)

Cant measurement: + 200 to –200 mils

Crosswind setting: ± 10 m/s

Altitude setting: 0 – 2000 m

Automatic setting of firing corrections: ±30 mils (azimuth); 0 – 60 mils (elevation)

The afocal module is adaptable to the gunner's sight of T-54, T-55 and Type 59 main battle tanks and is fitted at the forepart of the daylight aiming sight. It enables both daylight and night aiming without any mechanical modification on the turret. Night aiming is performed by using a second-generation micro-channel light intensifier tube whereas daylight aiming requires the retraction of a swivel mirror. The module is housed in a sealed water-tight casing and can be removed and reset without having to be boresighted again.

CHARACTERISTICS

Magnification: × 1

Field: ⩾70°

Resolution: >0.5 mrad at 10^{-1} lux

Operating temperature: –40 to +50° C

Performance

Daylight aiming: unchanged

Night aiming: 1000 to 2000 m according to atmospheric conditions

The afocal module incorporates an optical plate for combining the images from the visible light and infra-red beams, a deviation display collimator and a laser rangefinder.

STATUS

Prototype.

CONTRACTORS

GIAT (Groupement Industriel des Armements Terrestres), 10 place Georges Clémenceau, 92211 Saint-Cloud, France.

Sight: SOPELEM.

Laser rangefinder: CILAS.

3935.161

CAC 101 FIELD ARTILLERY COMPUTER

The Thomson Brandt Armements CAC 101 field artillery computer was designed principally for the preparation of firing data for 105 mm and 155 mm guns and howitzers and for heavy mortars. The main operational characteristics of the CAC 101 include storage in the computer memory of six gun positions, 10 targets, 10 observation posts, and 10 protected areas in addition to meteorological data (standard STANAG 4082 message up to line 21). Protected memories are provided. Ballistic data are carried on plug-in modules, each module carrying the data relating to one type of projectile and the weapon firing it. Using this information, the computer calculates the firing data for the main reference piece as well as the subordinate pieces. A built-in test program carries out self-checking of the system.

Data are entered into the computer by means of a keyboard in accordance with the dialogue between the machine and the operator, assisted by an alphanumeric display panel. In addition to the standard computational routines, there are possibilities for optional calculations, such as the ballistic co-ordinates of a given round during fire adjustment, computation of the residual correction at the end of a registration, and calculation of the correction for muzzle velocity. Others are topographical computations such as conversion of rectangular co-ordinates to polar co-ordinates and vice versa, direct and inverse bearings, direct and inverse intersections, triangulations, change of grid system, or goniometric network.

Various peripheral equipments may be connected to the computer, including a teleprinter, telex reader, remote data display etc. The CAC 101 can be rapidly changed from one weapon type to another by the simple expedient of exchanging the plug-in ballistic module.

CHARACTERISTICS

Dimensions: 370 × 355 × 313 mm

Weight: 15 kg

Power supply: 19 – 32 V DC (24 V nominal)

Consumption: 60 W

STATUS

Series production. Believed to have been supplied for export.

Thomson Brandt Armements CAC 101 field artillery computer

CONTRACTOR

Thomson Brandt Armements, 204 Rond-Point du Pont de Sèvres, 92516 Boulogne Billancourt Cedex, France.

4580.161

ATAC ARTILLERY MINI COMPUTER

The ATAC Artillery Mini Computer has been developed by Thomson Brandt as a private venture and is now in production. The system comprises a small hand-held digital computer and an optional console. The computer will perform all calculations for the firing operations of mortars, howitzers and cannon, and is particularly designed for use in autonomous operations where centralised control facilities are not available.

When the lightweight computer unit is fitted to the console, it becomes a system capable of integration into an automated artillery network. The ATAC system can be adapted to any particular weapon (and ammunition type) in the above categories by means of appropriate plug-in modules. Each of these modules carries eight ballistics. The computer has a semi-permanent data memory that can store nine gun positions, 99 target locations, 20 observer locations, ten datum positions, safety data, meteorological information (sounding data STANAG 4082) and prevailing conditions, and other data such as powder temperature and muzzle velocity.

The main functions performed include firing preparation using the arc method, fire control, fire for effect, special firing missions (eg moving targets), and survey. ATAC can be linked to other ATAC computers by cable or radio links, as can the gun remote data display equipment and observers' terminals. Built-in test facilities are incorporated.

CHARACTERISTICS

Dimensions: 245 × 165 × 85 mm (with batteries)

Weight: 3 kg (approx) with batteries

Display: LCD 160-character screen

Power supply: 24 V DC

Endurance: 1 week normal use

Environmental: Military –25 to +55°C

ATAC portable field artillery computer

STATUS

In production.

CONTRACTOR

Thomson Brandt Armements, 204 Rond-Pont du Pont de Sèvres, 92516 Boulogne Billancourt Cedex, France.

3187.161

SIROCCO METEOROLOGICAL RADAR STATION (QR-MX-2)

The Sirocco system comprises an I-band tracking radar for tracking meteorological sounding balloons, and associated telemetry system, processing equipment, and provisions for the transmission of meteorological data for use by artillery units. The last of these may or may not be equipped with an Atila computing system (**3188.161**).

The complete system is built as a self-contained land-mobile unit, the tracking radar being mounted on a two-wheeled trailer, and an air conditioned SH 17 shelter is carried on the towing vehicle. The latter houses the operating crew, the radar and telemetry operating console, a radio-telegraphy system with built-in modem, and stowage for the sondes and radar reflectors. In addition to the radar, the trailer carries the telemetry receiver and two electric generators.

The Sirocco system produces standard met messages, on-line, for transmission by teleprinter or via a radio-telegraphy link for use by artillery units. The messages provide wind speed and direction, and air temperatures at a series of standard altitudes. Maximum range is in excess of 130 km and angular accuracy is better than 0.05°. Range can be measured to within 5 m and windspeed can be determined with an accuracy of better than one knot (0.5 m/s). In its trajectography version, Sirocco establishes the trajectories of a wide range of moving objects; projectiles (radar return), drones and missiles (transponder).

Sirocco mobile meteorological station deployed

CHARACTERISTICS
Radar: I-band
Peak power: 80 kW
Receiver: 2-axis monopulse. Optional transmitted level control
Ranging: Digital in 5 m increments
Angular tracking limits: –8 to +160° elevation; unlimited azimuth
Max range: > 130 km
Angular accuracy: Better than 1 mil
Range accuracy: Better than 5 m
Windspeed accuracy: Better than 1 knot (0.5 m/s)

STATUS
In production and in service with a number of countries including France, Morocco, the Netherlands, Saudi Arabia, Jordan and Venezuela.

CONTRACTOR
Electronique Serge Dassault, 55 quai Marcel Dassault, 92214 Saint-Cloud, France.

3934.161
ATIBA ARTILLERY FIRE CONTROL SYSTEM

The ATIBA system has been derived from the French Atila (**3188.161**) artillery automation system and is designed to provide fire calculations for artillery units at battery level. However there are provisions for integrating ATIBA systems into a larger overall system such as Atila when appropriate and if required.

The ATIBA artillery fire control system provides for control of a battery of up to eight guns of any calibre, performs firing calculations for each gun using data provided by up to ten observation posts, ballistic data for any calibre, charge, shell and fuze, and standard STANAG 4061 meteorological message inputs. Trajectory calculation time is less than five seconds for six guns. The system also performs related functions such as survey calculations and mobile target engagement.

The equipment consists of the central processor and control unit which can be fitted in any type of vehicle, such as a Jeep or truck, a number (up to eight) of gun display units (GDUs) located at each gun position, and a 1200 baud radio link (wire or radio) connecting the central battery unit to the GDUs and observer positions. The observer posts are equipped with radio sets and modems permitting the exchange of data in digital form. The battery unit and the GDUs are provided with display panels for the presentation of firing data, command orders and other relevant data. At the central battery unit these facilities extend to man/machine interface facilities such as a plasma display screen, keyboard and printer. The plasma display screen capacity is 256 characters.

The central processor is based on CIMSA SINTRA military computers and it performs the following functions:
(1) calculation of the firing data for each gun
(2) determination of the charge with respect to the battery target-range
(3) processing of any targets, including point and linear targets, in regard to the best distribution of rounds on the target, and the number of rounds per type of shell required
(4) possible mixes of ammunition types to achieve engagement objectives
(5) fire adjustment
(6) determination of registration firing corrections for each charge
(7) dividing the battery into two sections to engage two different targets.

All the programs needed for the various survey functions exist in the standard software package.

The input data at the battery unit is aided by grids displayed on the screen which list the parameters required appropriate to the function selected. A check is made to indicate to the operator any errors he may have made. By means of the keyboard he is able to insert meteorological data, ballistic data, logistic information, gun co-ordinates, observer and target co-ordinates. Fire requests are entered in a similar fashion. Computed firing data is transmitted to individual guns where the GDUs present data on type of shell and fuze, number of rounds, the charge, bearing(s), elevation(s), fuze time, and triggering mode. Status of guns is returned to the battery unit over the same data link.

A group of ATIBAs can be combined to form one regiment or division Atila without modification other than the addition of a fire co-ordination centre. The latter facility enables the following additional functions to be performed:
(1) centralisation of fire requests at a higher level (regiment or division)
(2) evaluation of the tactical situation and the threat
(3) management of six ATIBA systems from a single co-ordination centre
(4) division of targets among batteries to achieve better concentration of fire on a given target
(5) improved management of ammunition.

CIMSA SINTRA gun display unit (GDU) is used at individual gun positions in ATIBA and ATILA artillery automation systems

STATUS
Adopted by the French Army and in service in six countries.

CONTRACTOR
CIMSA SINTRA, 10-12 avenue de l'Europe, 78140 Vélizy, France.

3188.161
ATILA ARTILLERY AUTOMATION SYSTEM

The Atila (artillery automation) system is a data processing system designed to fulfil three major operational requirements of artillery regiments:
(1) computing: this is concerned with the rapid and automatic calculation of ballistic data taking into account such parameters as topography, meteorological conditions, sizes and locations of targets, and locations of gun batteries
(2) transmission: the communication of calls for fire by individual forward observers to the command post, and the transmission of firing data to guns is performed by the system. Automatic inter-command post links provide for the mutual exchange of information
(3) management: decision making is speeded and enhanced by storage of the tactical situation in the computer memory, and by virtue of the facilities provided for man/machine dialogue.

ATILA system chain of command extends to individual batteries or guns. Fire direction centre vehicle (right) seen with 155 mm GCT self-propelled gun

To meet the needs of different organisational structures, ie from the autonomous battery to the artillery group which combines the fire from 18 batteries, CIMSA SINTRA has organised automation of these units around two types of vehicle both of which are based on the AMX-10 in the French Army system;
(1) fire centre vehicle (fire control post) containing the computer which essentially executes fire calculation functions: this is a technical centre. (AMX-10 PCH SAF)
(2) a command vehicle for the relevant fire unit commander (battery commander, regiment commander, or the group commander): this is a tactical centre. (AMX-10 PCH SAL).

The following elements are linked to these two vehicles:
(A) artillery observers who have artillery tactical terminals (ATT) to send their message coded for fire requests, fire adjustments, intelligence messages. This information is automatically sent to the fire centre vehicle via radio link
(B) the guns receive the fire centre computer-generated fire data on a remote display unit.

Automatic digital data links ensure the reliable and rapid exchange of data between these various elements of the system, using modems and standard radio sets.

If regimental organisation is prevailing, the equipments for a standard regiment are:
(A) 1 Regimental FDC computer
(B) 3 Battery FDC computers
(C) 9 Artillery Tactical Terminals (ATT)
(D) 18 Gun Display Units (GDU).
The regimental FDC computer is intended to provide maximum responsiveness for the employment of fire power. It selects the targets to be attacked; monitors up to six batteries' FDCs; achieves fire planning and allocates fire units and ammunition; provides tactical information; and deals with any type of target.

The battery FDC is intended to provide: rapid and accurate computation of firing data; secure transmission of fire commands; reception of fire request from the observers or the regimental FDC.

The Artillery Tactical Terminal increases the execution task of the fire mission thanks to a simple and accurate description of the fire request, and conducts fire, thanks to an efficient and complete dialogue with the regimental and/or battery FDC.

The Gun Display Unit is intended to display for the gun section chief all the firing data elaborated by the computer and to send back the gun status.

CHARACTERISTICS

Organisation: Battery – Regiment (3 to 6 batteries) – Group (3 to 6 regiments)
System capacity at regiment level is as follows:
Observers: Up to 36 positions
Recorded targets: 500
Recorded fire plans: 4
Safety zones: 40
Muzzle velocity: 540 m/s
Registration points: 20
System reaction time
Transmission of one message: 1s
Battery reaction time: 15s
Regiment reaction time: 35s

Slightly different versions of Atila have been designed for two Middle Eastern customers; variations between these and Atila I include differing organisational structures and the data processors employed at regimental and battery levels.

ATILA II

In the latter half of 1983 CIMSA revealed a new version of the Atila system configured for use with different artillery organisational structures to those of the original system. Atila II incorporates certain design improvements made possible by experience with operation of Atila I and embodies CABA 11 and CABA 21 processors at battery and command posts, but the basic objectives of both systems are essentially the same.

STATUS

In series production and serving in six countries.

Issue of Atila hardware to French Army artillery units began in 1980. By early 1985, 10 of the 16 French artillery regiments due to equip with Atila were reported to have the system, and the remaining six were expected to receive their equipment by the end of 1986. Work is in progress on a later version which will be known as Atila II.

ATILA system battalion fire direction centre computer

A version of Atila has been sold to Saudi Arabia for direction of French-built GIAT 155 mm GCT self-propelled guns, and the designation of this export version is Palmier. Exports of the system are reported to amount to about 30 systems. By February 1986 a total of seven countries had ordered ATILA including France, Saudi Arabia, Thailand (used by Marines and delivered in 1985) and Venezuela (9 systems ordered in 1983 with final deliveries taking place in 1986).

CONTRACTOR

CIMSA SINTRA, 10-12 avenue de l'Europe, 78140 Vélizy, France.

6350.181

ADIVA ARTILLERY DIVISION AUTOMATION SYSTEM

The Brigade Artillery Fire Direction Centre plans and implements the artillery fire of several artillery regiments supporting a large combat force unit.

The main goals of the automatic command/control artillery brigade fire direction centre are to supply the elements necessary to the Artillery Command in the division command post to enable him to carry out his three main missions. Firstly, to advise the commander of the large supported unit concerning the fire, eg the planning function. Secondly, to transmit the orders to the artillery and check their execution, eg the command and control function. Thirdly, to ensure the implementation of the fire resources for which he is responsible.

The essential functions fulfilled by the automatic command and control brigade fire control centre are: acquisition, storage and display of the targets (link with the tactical systems); resource planning (assignment of targets to fire units and authorised consumption within the framework of the manoeuvre; co-ordination of the artillery and fire support units; inventory of the resources and organisation (direct and general support units and available reinforcement unit) and management of concentrated and counter battery fire.

STATUS

Prototype.

CONTRACTOR

CIMSA SINTRA, 10-12 avenue de l'Europe, 78140 Vélizy, France.

CIMSA SINTRA ADIVA artillery division automation system

6706.181

TAD LRM FIRE CONTROL SYSTEM

This system is being developed by CIMSA for the fire direction centres of MLRS units (see **4060.111**). TAD (Transmission Automatique de Données) LRM (Lance Roquettes Multiple) uses hardware from CIMSA's STRATAGEMES (Systèmes de Traitement et de Transmission d'Application et de Transmission d'application Générale pour l'Exploitation Militaire en Environnement Sévère) a modular processing and display system for use in severe electronic environments.

The system is based upon an intelligent terminal incorporating a Motorola 68000 microprocessor. This is linked to a 30 × 30 cm electroluminescent display 768 × 768 pixels, with a power consumption of 10 W or an integral 7 × 11.2 cm graphics display of 200 × 320 pixels with a power consumption of 70 W.

This system, which may be mounted in a VAB wheeled armoured vehicle, will act as both battery and battalion fire direction system and will be linked to ATILA artillery system (see **3188.161**).

CONTRACTOR

CIMSA SINTRA, 10-12 avenue de l'Europe, 78140 Vélizy, France.

GERMANY (FEDERAL REPUBLIC)

4231.181

COMBAT MONITORING SYSTEM

The Combat Monitoring System (CMS) is designed as a tactical command system for target acquisition and identification, threat analysis and target assignment in battlefield combat situations. The equipment can be divided into two functional packages, the weapon system data acquisition unit (WSDA) and the central control and command station (CCS). The data input and output to each participating fire unit is made via a WSDA. These are integrated in each fire unit. The weapon carriers attached to the CMS are monitored and controlled by a central station. This may be installed in a shelter, standard cabin or any armoured vehicle.

The system can be used to monitor or control weapons on the ground or in the air. Data relevant to combat action is transmitted from each weapon system either by FM radio or by cable link, to a central control. After the threat is analysed and the target interrogated by IFF, a target assignment is given to the appropriate weapon system. The tactical situation is presented to the command controller on a colour display unit. A printout of the action is available for analysis.

The CMS can be operated as a tactical training system. Alternatively, it can be used as a range instrumentation system for assessing weapon systems.

CONTRACTOR

Dornier GmbH, PO Box 1420, 8990 Friedrichshafen 1, Federal Republic of Germany.

4144.161

HEROS ARMY COMMAND, CONTROL AND INFORMATION SYSTEM

HEROS (das Heeres-Fuhrungsinformationsystem zur Rechnergestutzten Operationsfuhrung in Staben) supports all general and special staff areas of the Federal German Army.

It supports Army operations, enhances command and control of own forces and deployment of joint reconnaissance systems and provides a fast flow of information to the Army Staff, to the Air Force and Navy as well as to NATO and Allied headquarters.

HEROS is a phased programme which is implemented in the following steps:

Mobile systems:

(1) HEROS-2/1 Corps, Division and Brigade Staffs, Territorial Task Force and Home Defence Brigade Staffs

Static systems:

(1) HEROS-3 Army Staff/Army Staff HQ

(2) HEROS-5 Territorial Command, Military District and Military Region Command

Implementation steps:

(1) HEROS-5-1 Traffic Control System

(2) HEROS-5-2 Engineer Barrier

Planning System

The system is tailored to the command and control process of the Army. Messages are received by the information or communications cell and are distributed to the users after proper recording. This recording also serves as the basis for keeping the war diary. After the users have checked and evaluated the messages, the messages are broken down by the system into their individual elements of information which are incorporated into the actual data base. Thus the complete and up-to-date picture of the actual situation (own as well as enemy forces) is available to the user who can retrieve all data pertaining to his mission from this data base. This retrieved data can be compiled in the form of messages, tables or maps.

Assessment of the situation, operations planning and, if required, staff briefings, can be accomplished in the individual locally dispersed cells.

After decision making by the commander, the operations plan is worked out in detail at the individual work stations. Orders and related material are prepared with automatic data processing (ADP) support in alphanumeric or graphic form and distributed via data communications to the subordinate units.

For the exchange of data, all cells of a command post are interconnected by a command post network. For voice communications, a switchboard is used which also allows party line operation.

The individual command posts are interconnected by the Army's Automated Corps Communications Network (Autoko) or by VHF combat radio network and data communication information exchanged between command posts is encrypted automatically. Situation reports are formatted according to the NATO standards so that interoperability with other national or NATO systems can be realised. Interoperability has been demonstrated in trials with the US Manoeuvre Control System (SIGMA) for example.

HEROS 3 became operational in 1983 while HEROS-2/1-X test-bed systems have been created for the corps, division and brigade levels. One of the purposes of the test-bed system is to prove the feasability of the deployment of ADP supported CCIS in the field. For this purpose hardware is installed in shelters to guarantee mobility within the command posts of units selected for operational testing.

The test-bed system HEROS-2/1-X-2 comprises the following components at corps level: a communications shelter; main computer shelter; graphics shelter and user work station which are located away from the main computer shelter, eg in nearby buildings. Within corps headquarters communication is by means of an Ethernet fibre optic local area network.

The shelters as well as the work stations are tempest protected. All connections between the shelters and the user work stations are accomplished by fibre optic cables. The data communication between the corps communications shelter and an interface shelter at the Army Staff (HEROS-3) is established via fixed circuits of the existing equipment.

The HEROS-2/1-X consists of the following command post cells at division and brigade level: an information cell; tactical operational cell (G2/G3 and S2/G3 for division and brigade respectively); and support cell (artillery). Facilities are provided for simulation of other staff areas of the main command posts.

Each command post cell is equipped with ruggedised commercial equipment including a computer, disc store, two alphanumeric consoles, a printer, the command post network interface, and communication equipment. Additionally provided are a plotter for three colour hard-copies on paper or foil and interfaces for Autoko and the combat radio net in the information cell, and a graphic plasma display unit with slide projection of maps.

The battalion command posts are equipped with I/O devices interconnected with the brigade command post via combat radio.

The system characteristics of the HEROS-2/1 systems include ADP supported assessment and evaluation of the situation; operational planning and issue of orders; employment of an event oriented formatted reporting system based on international standards; automatic processing of message contents into redundant data files; display of command and control data in the form of tables and maps with the possibility of graphic manipulation; hard copy devices for print out of texts and graphics; system operation either by command language or by menu supported dialogue and redundant data and voice communication including radio transmission of graphic information.

The HEROS-2/1-X systems were completed by Siemens AG as main contractor with sub-contractors INFODAS and GFS-METRA and in close co-operation with the responsible procurement agencies especially the Federal German Office for Military Technology for Procurement.

Siemens have also developed the HEROS-2/1-SVK which was employed and successfully tested during the NATO exercise WINTEX 87 with the main emphasis upon the communications aspect. The testbed systems have been mounted in three vehicles; a 10 tonne truck has carried the main computer shelter with 12 terminals and 5 tonne trucks have carried the graphics and communications shelters.

In addition to these shelters there will be a standard set of equipment consisting of work stations, storage devices and printers for each staff cell. There will be nine such cells at corps headquarters, seven at divisional level and six at brigade level.

Field trials of the HEROS-2/1 system with the Second German Corps will begin in 1988 with Siemens AG as the prime contractor. These will use truck-mounted shelters which are likely to be used by corps, division rear (three cells) and brigade rear (three cells) headquarters.

STATUS

Development of both HEROS-3-1 and HEROS-2/1 is continuing. In 1989 further field trials will be held featuring fully-militarised equipment in M577 armoured command vehicles which will be used at forward division (four cells) and brigade (three cells) headquarters. If the trials are successful it is anticipated the first production HEROS-2/1 system will be delivered early in 1990 and the corps will be fully equipped by 1992.

CONTRACTORS

Siemens AG, Radio and Radar System Division, Landshuterstrausse 26, D 8044, Unterschleissheim, Federal Republic of Germany.

GFS-METRA Gesellschaft fur Systementwicklung und Fuhrungssysteme mbH, Siegburgerstrasse 215, D-5000 Cologne, Federal Republic of Germany.

INFODAS Gesellschaft fur Systementwicklung und Informationsverarbeitung mbH, Rhonestrasse 2, D-5000 Cologne, Federal Republic of Germany.

1626.161

FALKE ARTILLERY COMPUTER SYSTEM

Falke is a computerised artillery fire direction system for use at battalion or battery level. The system was developed by AEG-Telefunken for the Federal German Ministry of Defence in close co-operation with the military user. The result of this co-operation is a system which meets the present and most of the future military requirements.

The central unit of the Falke system is the universal military computer AEG 80-20 MARS. This computer is fully militarised, especially designed for mobile operation under rough operating conditions and available in three versions, differing in computing power and memory capacity. It can be run on a 24 V DC power supply. The word-length is 16 bits and the maximum main memory capacity is 1 Mbyte. Many standardised interfaces for different military peripherals are provided. Special peripherals include a hermetically sealed magnetic tape cassette unit (PLG 38) for program loading (soon to be replaced by a magnetic bubble cassette unit) and a data display and transmissiion unit (radio and wire transmission) for use by the fire direction officer, the forward observers and as gun display (OAZ 794).

The application programs are: fire direction for various calibres; simultaneous fire control for different firing units; surveying and reconnaissance procedures; and evaluation of meteorological data.

For operating the computer a flexible dialogue

Military computer AEG 80-20 MARS

principle is used. This means that the computer itself calls for all necessary input data and performs various plausibility checks. Thus the operation of the Falke system is very easy to learn and to execute. Furthermore, since operation is fully programmable, all specific requirements and procedures can easily be realised.

For the acquisition of meteorological data, a fully mobile and militarised meteorological system, based on a 120 km balloon tracking radar, a one-, two- or three-channel radiosonde system and a computerised evaluation system, is available. Radar, radiosonde system and evaluation system (also using the military computer AEG 80-20 MARS) together are installed in a single standard two-wheel trailer, so avoiding any external connections.

STATUS

The Falke system is in production and systems have been delivered to the German Bundeswehr in quantity, as well as to different customers abroad.

CONTRACTOR

AEG-Telefunken, Anlagentechnik, Geschäftsbereich Hochfrequenztechnik, Postfach 1730, 7900 Ulm, Federal Republic of Germany.

Display and transmission unit OAZ 794

4829.181

ARMY COMMAND AND CONTROL SYSTEM PROGRAMMES (FRG)

The German Federal Government has embarked on a comprehensive improvement plan for the Army's command and control facilities. This involves several distinct development programmes for new systems to replace or complement existing C² systems. The major examples of such programmes are briefly summarised in the following paragraphs:

ADLER

Between 1985 and 1988 a new artillery C² system known as Artillerie-daten-lage und Einsatz-Rechner-verbund (ADLER) is planned for introduction, and this will provide for the rapid collection and dissemination of recce/target information between regimental HQs and individual batteries via data link. The objective is to facilitate immediate transfer of information, fire orders, reports etc. to ensure optimum fire concentration with minimum reaction time and expenditure of ammunition.

MR80/20 computers with RDS-80 disc memories will be employed, and other equipment includes: DEA-64 data I/O units and SEM 80/90 radios by SEL, the PLG-39 program loader from ATM Computer, and MT-120 commercial teleprinters from Mannesmann/Tally. The computer (MR80/20) has been selected as the standard machine for German Army C² systems. A main contractor for the ADLER system has yet to be nominated.

ARES

Dornier is developing a new fire control system for use with the Multiple Launch Rocket System (MLRS) (**4060.111**) batteries when this is deployed by the German Army, and it will perform the same function for the 110 mm LARS, Light Artillery Rocket System (**5577.111**). The Artillerie-Raketen-Einsatz-System (ARES) functions will include:

(1) battlefield surveillance
(2) target identification and location
(3) calculation of fire data
(4) production of fire orders
(5) launcher site survey at battery level
(6) data communications management.

ARES will be compatible with the ADLER network (above), and with allied NATO forces operating with MLRS. Troop trials were scheduled to begin in mid-1987.

IFAB

Thyssen Henschel, as main contractor, has developed an integrated fire control system for artillery batteries (Integrierte Feurleitmittel für Artillerie Batterie (IFAB)) which consists of the following parts: the forward observer vehicle based on a modified M113 series armoured personnel carrier chassis fitted with optronic and navigation means for target surveying and data transmission; the fire direction centre vehicle based on a modified M577 series armoured command post vehicle which has a digital ballistic computer and digital data equipment for the calculation and data radio transmission of the fire commands; the data unit installation kits installed in the self-propelled howitzers for receiving the fire commands.

Delivery of the 320 forward observer vehicles to the West German Army was completed in 1985, the 150 fire direction centres are now in production as are the kits for the howitzers.

The principal sub-contractor for the electronics was ESG during the development phase. Computer, navigation system and data units are produced by ATM, BGT Litef and PST.

3797.161

LEMSTAR TANK FIRE CONTROL SYSTEM

The LEMSTAR integrated fire control system is designed to meet requirements for a fire capability on the move and to be equivalent to the fire control systems of the latest Western tanks such as the German Leopard 2 and the American M1 Abrams. Most of the hardware is proven equipment which is being produced in volume.

It was originally designed to meet a West German Army requirement to retrofit the Loepard 1 (the competition was won by the FLP 10, see **4858.161**) but it was also intended for retrofitting in other vehicles such as the M48, M60 and Centurion. The system comprises the following main components:

(1) Stabilised combined laser rangefinder-aiming sight (LV/2 Stab or RZG Stab).
(2) Standard fire control computer (FLER-H or FLER-M).
(3) Weapon stabilisation system (WSS) which is slaved to the primarily stabilised gunner's sight.
(4) Position data transmission between sight and weapon by synchros of high accuracy.

The rangefinder-aiming sight (**3580.193**) acts as the gunner's main sight for observation, aiming and tracking, and ranging the target. It contains an optical periscope with two magnification steps, the integrated Nd-glass laser rangefinder and the stabilisation equipment. The sight is monocular and the range is digitally displayed in the eyepiece. Up to three different ranges can be detected and stored from one laser transmitter pulse.

The fire control computer was originally the Type FLER-H fire control computer (**1619.161**) but later versions of LEMSTAR can use the FLER-M. The computer calculates the super-elevation and the lead angle in azimuth from the ammunition selected, the target's effective range, internal and external ballistic data, trunnion cant and the terrain slope. Correction angles in azimuth and elevation are calculated taking into account the laying speed during target tracking. Input measurement and weapon/sight alignment are carried out automatically.

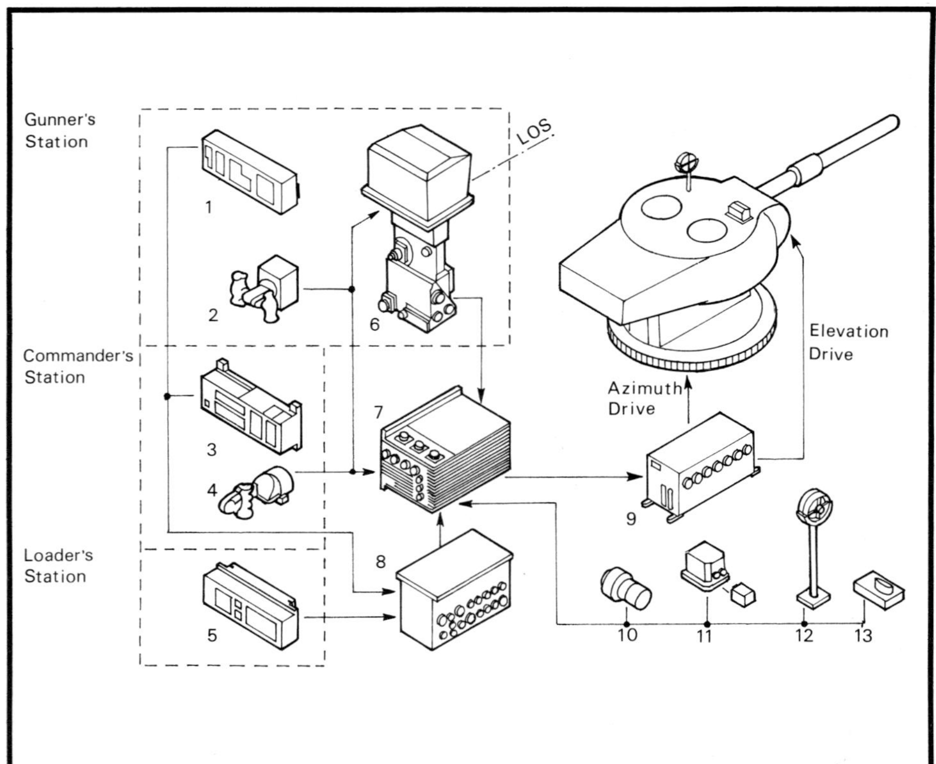

LEMSTAR integrated tank fire control system

(1) *gunner's control panel* **(2)** *gunner's handle* **(3)** *commander's control panel* **(4)** *commander's handle* **(5)** *loader's control panel* **(6)** *gunner's sight primarily stabilised with laser range finder* **(7)** *computer and stabilisation electronics* **(8)** *operation mode logic* **(9)** *weapon stabilisation and slave electronics* **(10)** *vertical gyro* **(11)** *powder temperature sensor* **(12)** *crosswind sensor* **(13)** *air data electronics*

The WSS consists of two subsystems; the hydraulic weapon drive pack, and WSS electronics including the sensors. The hydraulic weapon drive pack is a proven system produced in large numbers for the Leopard 1 MBT. The WSS is slaved to the gunner's stabilised sight and its task is to slave the gun to the Line-Of-Sight (LOS) allowing for the correction angles generated by the computer. It also compensates for gun movement as the vehicle moves.

The position of the gun in elevation and azimuth is

sensed by gyro units. The electrical synchro chain links the stabilised gunner's sight with the gun. It contains position resolvers as input elements at the sight and as reception elements at the weapon and both elements are selected as a pair to ensure the high accuracy required.

The LEMSTAR system philosophy is based upon combining the gunner's stabilised combined sight with the slaved weapon along the LOS for extremely accurate aiming and target tracking even on the move. Accuracy is essentially better compared with the gun because of the sight's lower mass, smaller size and mechanical tolerances with higher hit probabilities being possible by inputs into the computer from specialised sensors.

All inputs are automatically processed to compute the super-elevation and lead angle. The gun is kept in the exact fire position via the synchro chain. The gun fires when the LOS and gun boresight coincide and because this occurs frequently the time gap between pressing the firing button and the commencement of firing is very low. Thus, a high first round hit probability is achieved even when on the move and against moving targets.

The optical/electro-optical equipment for the commander depends upon the tactical requirements of the user. The basic system consists of a panoramic periscope but there are various options including a stabilised one similar to the gunner's sight. If a night fighting capability is required AEG-Telefunken recommend the Type PZB 200 LLL-TV night sight. This is a rugged proven unit for gunshield mouting with separate monitors for gunner and commander.

STATUS

LEMSTAR has been trialled in an Austrian SK 105 tank destroyer as well as in M48 and AMX-30 MBTs.

CONTRACTOR

AEG-Telefunken, Anlagentechnik, Fachbereich Flugwesen und Sondertechnik Steuerungs und Regelungssysteme, Industriestrase 29, 2000 Wedel (Holst), Federal Republic of Germany.

1619.161

FLER-H TANK FIRE CONTROL COMPUTER

The Type FLER-H fire control computer (**1619.161**) consists of the following sub-assemblies: sensors; vertical gyro; control panel; central computer unit; 400 Hz inverter. The computer sensors can be extended to maximum performance by adding a complete line of sensor units such as cross wind, ammunition temperature etc. Later versions of LEMSTAR may incorporate the FLER-M fire control computer.

The computer calculates the super-elevation and the lead angle in azimuth (due to the curved trajectory) from the ammunition selected, the target range, internal and external ballistic data, trunnion cant and the terrrain slope. Correction angles in azimuth and elevation are calculated taking into account the laying speed during target tracking. The computer controls the setting of these angles at the sights and the weapon. The inputs are measured automatically and fed into the computer and aligment of weapon and sight is corrected automatically.

CHARACTERISTICS

Ammunition types: Choice of 4 ballistics
Distance range: 400 – 3000 m
Calculation accuracy: ⩽0.1 mrad
Target velocity: Azimuth 40 mrad/s; elevation 15 mrad/s
Air pressure: 810 – 1215 mbar
Air temperature: –35 to +65°C
Crosswind: –20 to +20 m/s
Powder temperature: –30 to +45°C
Trunnion cant: –12 to +12°
Terrain slope: –20 to +20°
Test circuits: BITE
Power supply: 28 V DC, 90 W
26 V DC, 400 Hz, 10 W
Weight and dimensions
Core: 21 kg, 360 × 272 × 240 mm
Vertical sensor: 4.5 kg, 250 × 190 × 190 mm
Air data sensor: 6.8 kg, 345 × 220 × 178 mm
Powder temperature sensor: 4.1 kg, 205 × 130 × 160 mm + (125 × 65 × 75 mm)
Control unit: 3.2 kg, 180 × 215 × 105 mm

STATUS

The FLER-H computer is in series production for use in Leopard 1A4 tanks. It also forms part of the LEMSTAR tank fire control system proposed for modernising programmes for other types such as the M48, M60 and Centurion (see **3797.161** in this section).

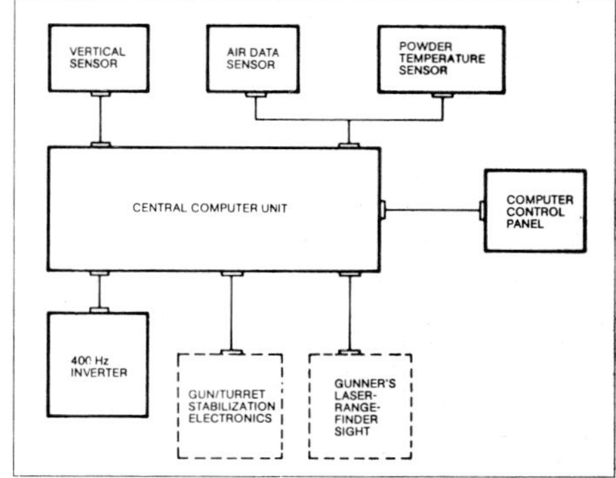

Block diagram of FLER-H tank fire control system

CONTRACTOR

AEG-Telefunken, Anlagentechnik, Fachbereich Flugwesen und Sondertechnik Steuerungs und Regelungssysteme, Industriestrasse 29, 2000 Wedel (Holst), Federal Republic of Germany.

4858.161

FLP 10 TANK FIRE CONTROL SYSTEM

The FLP 10 tank fire control system is designed for the control of main and secondary armament in main battle tanks such as the Leopard 1A1A1, 1A2 and 1A3, providing facilities for the engagement of fixed and moving targets under day or night conditions. After trials with three different fire control systems the FLP 10/EMES 18 combined with a thermal imaging system was accepted for retrofitting to about 1300 Leopard 1 series MBTs of the West German Army in 1984.

The system contains the thermal imaging sight assemblies of the Leopard 2 MBT, with the same performance. The laser rangefinder is also identical with that of Leopard 2, and there are a number of other elements that are common with that vehicle's FCS. Among the major features of the FLP 10 are the following:

(1) Primary stabilised line of sight for the optical and thermal imaging sight.
(2) Integrated thermal imaging sight.
(3) Facilities for the commander to monitor the image presented in the gunner's primary sight, and to input range data obtained passively from the commander's periscope, laser rangefinder or manually.
(4) Six main gun ammunition types, and ballistics for the machine gun.
(5) Dynamic lead angle
(6) No separate parallax processing is needed for the daylight channel, thermal imaging sight or laser.
(7) Facilities to input ballistic parameters.
(8) Common temperature compensated angle transmission mechanism for daylight channel, thermal sight and laser.
(9) Cant angle sensor.

All the principal components of the FLP 10 are housed within the tank turret. They are: (a) gunner's primary sight (GPS), (b) computer control unit, (c) commander's GPS eyepiece assembly, (d) commander's thermal imaging control unit, (e) gun stabilisation electronics, (f) laser electronics, (g) loader's control unit, (h) vertical sensor, (i) gunner's control unit, (j) thermal imaging electronics, (k) interface/test/operating mode control assembly, (l) fire control computer.

In the fire control system, several optical, electro-optical, electronic and mechanical subsystems operate together. A single mirror that is stabilised in azimuth and elevation is used for the daylight visual channel, laser transmission and reception, and thermal imaging. This ensures synchronisation of the optical axes. This mirror enables the gunner and commander to aim at the target using daylight or thermal imaging, and perform laser rangefinding. Identification and tracking of the target in daylight are by means of a high quality day telescope. A reticle, range value and system status information are superimposed on the telescope image. At night or in poor visibility identification and tracking are performed with the thermal imaging sight. Target markers are superimposed on the thermal image which is injected into the daylight channel optical path. After aiming at the target, it is possible to measure range with the laser rangefinder; the line of sight is not affected by changes in the ballistic values.

The gunner, or commander if the latter's GPS eyepiece is being used, controls the line of sight with the aid of hand controls. Utilising the vertical sensor, the effects of cant error are eliminated from the aiming process. The resulting target data are fed automatically into the fire control computer which continuously calculates the superelevation and lead angles according to selected ammunition type and preset parameters. These include: target speed, cross-wind, grain temperature, height above sea level, air temperature, correction values and control signals.

CHARACTERISTICS

Gunner's primary sight: Integrated unit with 4 channels, daylight, laser transmit/receive, night vision (thermal imaging)
Magnification: × 12
Field of view: 5°
Commander's GPS eyepiece
Exit pupil: >5 mm diameter
Transmittance: >10%
Superimposed data: Range, ammunition type, ready to fire
Fire control computer
Ballistics: 6
Max processed range: 4000 m
Rangefinder: Nd:YAG laser
Wavelength: 1.06 μm
Receiver FOV: 0.35 – 0.4 mrad
Measuring range: 200 – 9900 m
Resolution: Less than 20 m
Measurement accuracy: +10 m
Pulse rate: 10 pulses/minute
LOS aiming and stabilisation
Aiming sector: ±3° (azimuth); –10° to +20° (elevation)
Stabilisation precision (channels): Less than 0.1 mrad

STATUS

In production for installation in Leopard 1s of West German Army.

CONTRACTOR

Krupp Atlas Elektronik GmbH, Sebaldsbrücker Herrstrasse, 2800 Bremen 46, PO Box 448545, Federal Republic of Germany.

INTERNATIONAL

3939.161

BC81 (Re 83)/(FARGO) ARTILLERY COMPUTER

The BC81 artillery computer is a battery level digital computer for calculating firing data for guns of various calibres, taking into account the usual meteorological, ballistic and other parameters. Now in service with the Swiss Army, for which the Fargo system was developed, the computer is designated Re 83. The processor is a version of the British Aerospace 1412A, which is used in a variety of other military applications, among them co-ordinate conversion for the Exocet MM38 anti-ship missile system (**1156.221**). The standard software package provides for firing calculations for nine different types of guns and ammunition, as well as a number of targets and observer positions, and has a full survey capability. Input data includes meteorological information, target co-ordinates (including altitude), and ballistic characteristics. In view of the system's intended operating environment (the Swiss Alpine region) elaborate provisions have been made for insertion of crest clearance data, target zones and other restrictions due to the nature of the terrain.

DEVELOPMENT

Began in 1977 at Bracknell, in co-operation with the Swiss company Zellweger Uster AG. Extensive Swiss Army trials and evaluation preceded production decision in 1982/83.

STATUS

Production for Swiss Army confirmed in 1983; large order placed with estimated total value £14 million. In service with Swiss Army. Deliveries were to be completed by the end of 1987.

CONTRACTOR

British Aerospace, Naval and Electronics Systems Division, Downshire Way, Bracknell, Berkshire RG12 1QL, England. (Formerly Sperry Gyroscope.)

BC81 artillery computer incorporates 1412A digital processor. The Swiss Re 83 version differs in minor details

---

6582.181

SUPER-FLEDERMAUS UPDATE

An adaptation of the SATT fire control system update package (**6353.161**) for anti-aircraft fire control systems is being produced by Terma Elektronik A/S together with SATT Communications AB and Philips Elektronikindustrier AB for Denmark.

The Danish package, which is for the Super-Fledermaus, includes:

(1) A new digital fire control computer.
(2) A new Ku-band frequency agility/MTI monopulse tracking radar.
(3) Solid state component modifications to the displays and man-machine interface.
(4) A more accurate tracking servo system.
(5) A built-in trainer.
(6) Interfaces for associated equipment.
(7) Built-in power supply.

By introducing modern components and micro-computers as well as replacing obsoclescent electronics and electro-mechanical parts the updated system substantially enhances target acquisition speed hit probability and overall reliability. There are several optional interfaces which will allow the upgraded system to be used in combination with different types of search radar, laser rangefinder, muzzle velocity measuring equipment and similar devices.

CONTRACTOR

Prime contractor is Terma Elektronik A/S, Hovmarken 4, 8520 Lystrup, Denmark.

ISRAEL

6442.161

EAGLE EYE ANTI-AIRCRAFT FIRE CONTROL SYSTEM

For many years Israel has been using the Swiss Contraves Super-Fledermaus (**2376.181**) anti-aircraft fire control system to control its 40 mm L/70 air defence guns.

To meet the requirements of the Israel Defence Forces (IDF) for a modern electro-optical air defence fire control system MBT Weapon Systems has developed the Eagle Eye air defence fire control system.

Eagle Eye (or Simple Economic Fire Control System) is based on an upgraded Super-Fledermaus system with a new and more powerful tracking computer and an optronic mount comprising a laser rangefinder, optical sight and TV camera and controls. Options for TV tracker and FLIR are also provided for future growth of the system.

Other components of the Eagle Eye system include an EL/2106 search radar and radar display, battery commanders console and communications equipment. The fire control system can control up to six guns or missile launchers of any calibre.

Although the Eagle Eye is based on the Super-Fledermaus it is also applicable to a new weapon system. According to MBT Weapon Systems, the probability of engagement of Super-Fledermaus fitted with the upgrade package is three times better while its kill probability is doubled.

STATUS

In production. In service with Israel Defence Forces.

CONTRACTOR

MBT Weapon Systems, IAI, PO Box 105, 56000 Yehud Industrial Zone, Israel.

4492.161

COMBAT ARTILLERY C³ SYSTEM

Combat is the name given by Elbit Computers Ltd to the artillery command, control and communications (C³) system they have designed to meet the needs of artillery formations at battery and battalion levels. The basic configuration is of a modular design which allows the addition of more units to increase the performance and range of facilities. The basic version of Combat provides for C³ functions necessary for operation at battery level. It consists of a battery C³ unit, a section chief display and communication unit for each gun, and a sight display unit for each sight. By the addition of an artillery computer the system will perform functions such as storing up to 30 positions, 100 targets, 15 simultaneous fire missions and handling of two meteorological telegrams. By adding modems various additional functions can be incorporated to bring the above facilities to forward observers and enable forward observers to undertake tasks such as re-positioning of sights etc.

Combat hardware at the fire direction centre (FDC) is based on an artillery tactical command (ATC) unit, and this includes the ballistic computer and a measurement remote terminal (MRT). The FDC sub-system performs simultaneously on three networks (guns, battalion level, fire control) and displays updated status data on the battery in real-time. It carries the requested artillery computations itself or receives them from the artillery computer when

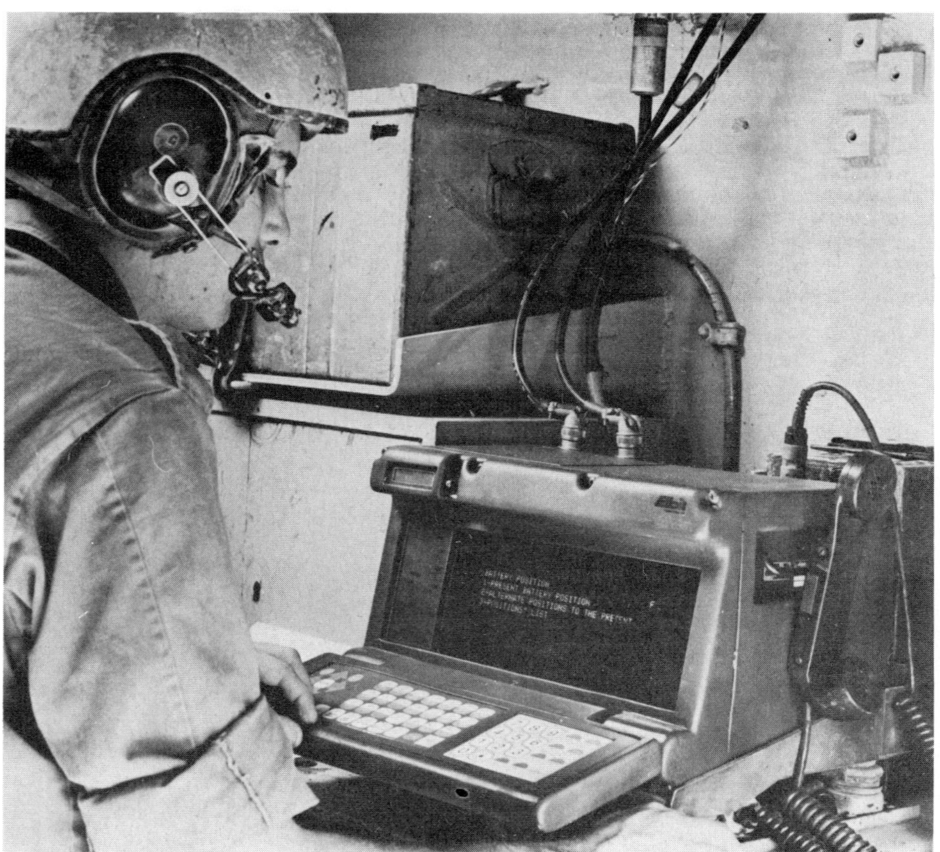

Artillery tactical command unit of Elbit Combat artillery C³ system

incorporated in a larger system. Messages received or transmitted are printed automatically. At the gun level, the hardware consists of the gun central unit (GCU), sight control monitor (SCM), the elevation sight monitor (ESM) and various other sensors and actuators. The gun sub-system receives and displays firing data and other command information sent by the FDC. Elevation and bearing sights are automatically set to the requested firing data. Data from sensors is reported via the gun sub-system to the FDC; for example, the gun sub-system receives measured muzzle velocities and transmits the mean values back to the FDC. Forward observers are equipped with a Mini-SACU (**4278.161**) which interfaces with a digital goniometer, a laser rangefinder, and a gyro north-seeker.

Conventional cable, or digital burst transmission radio links, can be used to interconnect the various elements of the complete Combat artillery C³ system.

STATUS

In production and in service with Israeli Army.

CONTRACTOR

Elbit Computers Ltd, Advanced Technology Centre, PO Box 5390, Haifa 31053, Israel.

4278.161

SACU COMMAND AND CONTROL

Elbit Computers Limited produces a range of stand-alone digital communications units (SACU) for the transmission and management of most kinds of battlefield communications via line (twin conductor) or radio channels. The latter may employ FM and AM in the VHF, UHF or HF bands.

The basic SACU provides simplified, two-way high-speed burst data transmission in standard message format or as free text. The display incorporated in the unit leads the operator through the message entry sequence. Entries are made via the keyboard in accordance with a pre-stored list. Received messages are stored in a memory and displayed on operator request. The system provides for automatic message acknowledgment, and special purpose computations, such as ballistic or navigation calculations are optional. SACU interfaces with standard communications equipment. Optional interfaces include: printer, central processor, tape cassette, and teleprinter. A microprocessor controls the SACU operation and performs data processing functions. The keyboard is a sealed, flat surface, pressure activated device which is self-luminescent for night operation. Characters are formed by 5 × 7 dot arrays in the display which can present up to 256 characters.

CHARACTERISTICS

Data transmission: FSK at 300 or 1200 bits/s; DPSK at 2400 bits/s
Height: 355 mm
Width: 406 mm
Depth: 355 mm
Weight: 24.9 kg

Mini-SACU

This is another member of the SACU family which provides similar facilities to those of SACU (above) but which is packaged for use by forward observers and at observation posts. A touch-panel keyboard is provided for operator access to the system, and the unit has message coding facilities as well as error detection and correction. Other features include: high-speed burst mode transmission; radio network management; automatic message acknowledgement; microprocessor control; long- and short-term memory.

CHARACTERISTICS

Data transmission: FSK at 75, 150, 300 or 1200 bits/s
Height: 220 mm
Width: 105 mm
Depth: 85 mm
Weight: 1.5 kg

Other members of the SACU family of equipment are:

TI-100: an interface unit for linking a teleprinter to SACU

GI-100: an interface for linking special-purpose equipment to SACU

LRCS 100: a long range communication SACU-based system for complex HF networks

PDI 100: a plasma display interface for major command and control centres

SACU-2000: with 2000 character display expanded EPROM and RAM and with a capability for simultaneous transmit/receive on three different networks. Multi-channel digital links between SACU users are established with the use of a high frequency multi-channel adapter.

STATUS

In production and thought to be employed by the Israeli Defence Forces.

CONTRACTOR

Elbit Computers Ltd, Advanced Technology Centre, PO Box 5390, Haifa 31053, Israel.

SACU family with Mini-SACU on left and Micro-SACU, now no-longer available, on right

---

3192.161

DAVID ARTILLERY BATTERY COMPUTER

The DAVID artillery computer was developed to be used by the technical assistant at the battery level as well as the battalion FBC. Using LSI technology, a small and low-cost unit has been designed which performs all the calculations required at this level: firing data, adjustments, fire for effect, corrections, survey, etc, as well as functional checks of data and procedures to prevent operator errors.

Based on experience gained since the system was introduced into IDF service in the 1970s, the David artillery computer has been developed and has now reached the Mk III version. As before, the David performs all the calculations at the battery and battalion level, replacing all the traditional tables and charts. Calculations under fire are much quicker and much more accurate and effective.

The new David replaces the old software with a microprocessor programmed in an Elite language and a 104K memory. This means that it can now store up to 100 targets and handle five simultaneous missions, quicker than ever. Maintenance is simplified through increased modularity. Power consumption is lower. Externally, no change was made in the computer. It is still compact, lightweight and rugged, adaptable to a wide variety of guns. The system is easy to operate, requiring only minimal training.

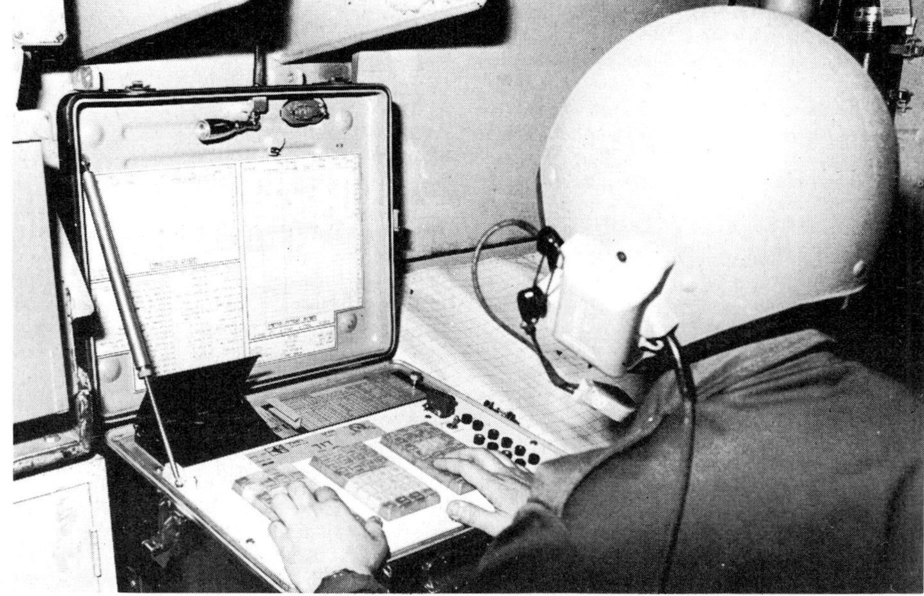

DAVID artillery computer equipment in use

Data can be transmitted from the computer to gun display units and higher level systems such as battalion Field Direction Centres. Direct data transmission from the forward observer to the computer is also provided. Optionally a printer can be connected as a hard copy device. The complete David system includes a David Field Artillery Computer in the battery fire direction centre, and a Gun Display Unit and Slave Unit for each gun. Connections between the display units at the guns and the fire direction centre are via regular field telephone wires.

CHARACTERISTICS
Microprocessor: 8085A
Memory size: 104K (96K EPROM, 8K RAM)
Computation time: 3 – 5 s (fire data); 0.2 s (chart data); 3 – 5 s (corrections)
Guns handled: 6 (in one sequence)
Targets and/or alternate positions stored: 100 (expandable to 200)
Fire missions: 5 (expandable to 15)
Corrections stored: 20 (expandable to 30)
Safety measures: crest clearance; no-fire areas (optional)
Warning measures: visual – error message; aural – buzzer
Peripherals: output to GDU, input/output to/from TDM-326, printer (optional)
Temperature: –30 to +52°C (operation); –40 to +71°C (storage)
Accuracy: 20 m (range); 15 m (deflection); 0.1 s (fuze setting); 1 m or 1 mil (survey calculations)
Dimensions: 500 × 440 × 270 mm
Weight: 27 kg

STATUS
In service with Israeli Defence Forces. Current production model is the Mk III.

CONTRACTORS
RAFAEL, Armament Development Authority, Israel.
Ministry of Defence, Haifa, PO Box 2082, Israel.

3524.161
GUN DISPLAY SYSTEM

The Gun Display System (GDS) is designed to transfer essential data between the DAVID artillery battery computer (**3192.161**) at the fire direction centre (FDC) and individual guns of a battery. The system addresses each gun separately so that the pertinent information is passed only to the relevant gun, with the single exception that in the registration process, all the guns receive the firing data of the ranging gun. Each new command displayed at the gun activates an audio warning to draw the gun crew's attention to the new orders.

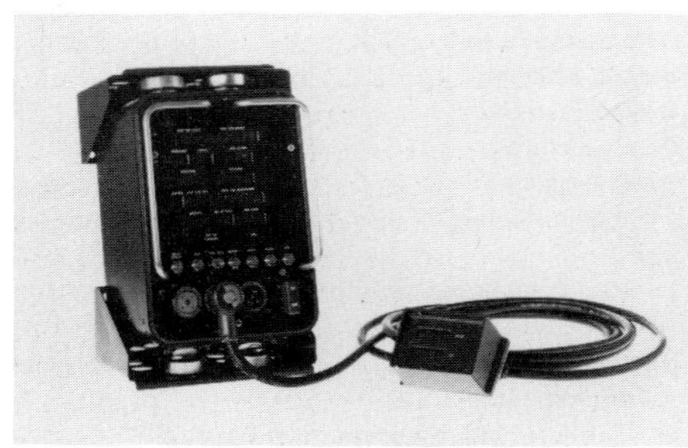

Gun display system

The GDS consists of three main sub-systems:
(1) the display units located at the guns
(2) the message centre for generating and receiving messages in the DAVID artillery computer
(3) a communication system.

A main display unit is mounted in front of each gun crew commander, and an additional display unit can be located near the gun sights. This slave unit will display only the relevant information for the particular gun layer. The GDU receives firing commands issued by the message centre and displays them to the crew commander and individual gun layers. It receives and displays the following data:
(1) preparation for engagement: projectile and fuze types
(2) firing data: deflection, elevation, fuze setting, propellant and charge number
(3) range adjustment data: ranging gun and ammunition type for range adjustment (the ammunition type is displayed only to the ranging gun)
(4) fire for effect data: number of projectiles, firing mode and fire interval.

GDU transmits the following reports back to the message centre:
(1) acknowledgement of a new command (push-button)
(2) transmission error (by means of error detection techniques in the data transfer system)
(3) READY signal (push-button)
(4) FIRED signal (push-button). This also operates the count of rounds fired
(5) MISFIRE signal (push-button)
(6) number of rounds remaining for completion of the mission (automatic)
(7) last round fired signal (operates automatically when round remaining reaches zero).

The numeric data, the projectiles and firing mode data appear on seven-segment indicators. The unit is also provided with a test button and an ON-OFF switch.

The message centre is packaged in the DAVID computer container, and uses spare keys of the DAVID keyboard, and no hardware or software changes are required in the DAVID system to add the gun display system. The communication system is based on normal telephone line communications techniques.

STATUS
Operational in various countries.

CONTRACTORS
RAFAEL, Armament Development Authority, Israel.
Ministry of Defence, Haifa, PO Box 2082, Israel.

3870.161
ELBIT FIRE CONTROL SYSTEM

After a two-year development effort, Elbit Computers Ltd began production of a digital tank fire control system based on a microprocessor central processing unit (CPU) in 1979. Although designed primarily for the Merkava tank, Elbit's concept was to produce a modular system that was flexible enough to permit expansion to meet future requirements and advances, and at the same time to be basically suitable for use in all existing types of tank. Currently Elbit is manufacturing systems (known as the Matador FCS) for M60, Centurion and other MBTs, based on its successful programme for the Merkava. In addition, the company has designed a low-cost system (Lancelot **4786.161**) for retro-fitting in light tanks as well as for use on new AFVs.

In addition to the CPU, the system consists of three operation units (gunner, commander, and loader), control and feedback servo loops, and sensors. The system is capable of operation without power in an emergency. The gunner's unit is the main operation unit and it provides all manual inputs necessary for the ballistic computation. This unit also incorporates a logistics panel which enables bore-sighting and system built-in test as well as display of pre-selected inputs and computed outputs. The gunner's unit also includes controls for the selection of ammunition (any one of six types), and jump compensation for each type of ammunition in elevation and deflection. The commander's unit provides readout of the system's display, range and ammunition inputs and the loader's unit provides ammunition inputs. An integral facility for firing the gun at night is provided, based on the use of a second generation image intensifier.

Elbit Matador FCS installed in Centurion MBT

The Matador system includes the following automatic sensors: laser rangefinder, turret cant angle sensor, target velocity sensor and a meteorological mast with crosswind velocity, air temperature and pressure sensors. The control loop transfers the computed super-elevation information to the hydraulic gun elevation servo and to the ballistic drive. In addition, deflection data is transferred to the moving reticle. The feedback loop ensures that the actual gun super-elevation and reticle deflection data are identical to the computed data and will correct any error accordingly. Thus, the gunner's aiming reticle remains on target (full autolay capability) throughout the entire firing preparation process.

Parameters taken into the elevation and deflection compensation calculation typically include range, type of ammunition, cant angle, parallax, ammunition drift, gun jump, gun wear, moving target angular velocity, crosswind, barometric pressure, air temperature, charge temperature, and options such as muzzle reference system, line-of-sight angle, and gun resolver.

CHARACTERISTICS
Operating range: 400 – 5000 m
Elevation compensation: –1 to +99 mils
Deflection compensation: +25 mils
Cant angle range: ±15°
Jump compensation: ±5 mils (for all ammunition types)
Bore-sighting range: 400 – 5000 m
Ammunition selection: 6 types
Electrically inserted range: 300 – 5000 m
Computer memory: EPROM; 16 K total, 5 K utilised
Dimensions
Gunner's unit: 210 × 220 × 220 mm
Commander's unit: 85 × 115 × 134 mm
Loader's unit: 80 × 72 × 134 mm
Cant angle indicator: 100 × 160 × 190 mm

STATUS
The Matador Mk 1 tank fire control system is installed in the Merkava Mk 1 MBT and is suitable for installation in the M47/M48/M60 and Centurion MBTs. The Matador Mk 2 is installed in the Merkava Mk II and is suitable for installation in the T-54/T-55, AMX-30 and Leopard 1 MBTs. The Matador Mk 3 is installed in the Merkava Mk III MBT and can be installed in other MBTs.

CONTRACTOR
Elbit Computers Ltd, Advanced Technology Centre, POB 5390, Haifa 31053, Israel.

4786.161
LANCELOT TANK FCS

Lancelot is a tank fire control system designed for retrofitting in existing AMX-13 and M41 light tanks, M4 and M47 medium tanks and light armoured vehicles such as the M113 with 60 mm weapons. It has been evolved from experience gained by Elbit in the development of the Matador FCS (**3870.161**) for the Israeli Merkava main battle tank programme.

The Lancelot system consists of a Nd-Yag laser rangefinder, a second generation image intensifier, a digital computer with a built-in cant angle sensor, control units for both commander and gunner and a moving target velocity sensor. Although based on the digital system developed for the Merkava, Lancelot is more compact enabling it to be fitted in the turrets of a wider range of smaller vehicles.

The system corrects the aiming reticle for the same parameters as the MBT FCS (range, ammunition type, cant angle, parallax, ammunition drift, gun jump, gun wear, moving target velocity, crosswind, barometric pressure and air temperature).

STATUS
The Lancelot FCS has been fitted into various armoured vehicles and tanks, and the accuracy and ergonomic features proved. No details of production or series fittings have been released.

CONTRACTOR
Elbit Computers Ltd, Advanced Technology Centre, PO Box 5390, Halfa 31053, Israel.

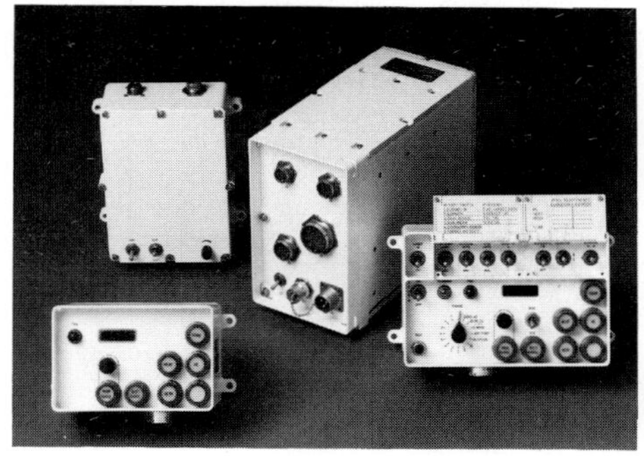

Component units of Lancelot AFV FCS

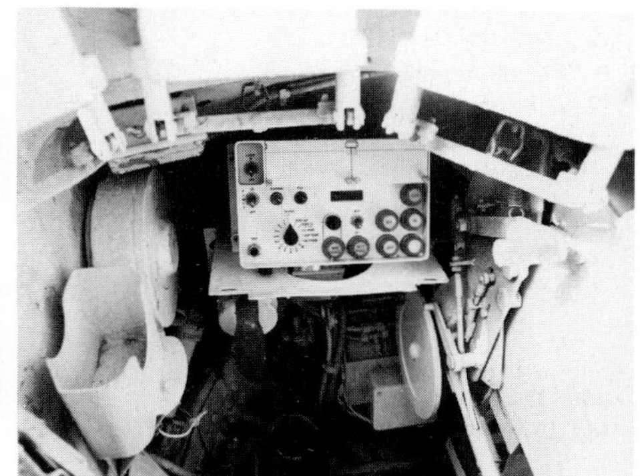

Elbit Lancelot FCS fitted in AMX-13 light tank

ITALY

4490.161

SEDA AND SEDAB ARTILLERY AUTOMATION SYSTEMS

The SEDA is a field artillery automation system which performs the management and process of field artillery information at battalion level and carries out all the procedures needed to compute firing data for the battalion (up to six batteries each composed of up to eight guns).

The most important functions performed automatically by SEDA are:

(1) automatic data transmission and reception among different units of the battalion
(2) fire preparation
(3) evaluation of the fire possibility
(4) evaluation of the targets and the consequent definition of the fire characteristics
(5) firing data computation and fire distribution
(6) fire planning
(7) target recording
(8) ammunition management

The SEDA basically consists of:

(a) one battalion computer at the battalion command post
(b) one battery computer at each battery command post
(c) one man portable terminal (DMT) for each commander and observer
(d) one gun commander display unit (GDC) for each gun

Using the DMT, observers and commanders send target reports, fire requests and adjustments to the battalion and battery computers; the transmission is performed in a fully automatic way through fixed format messages digitally coded by DMT and sent on voice channel of a frequency modulation radio-set. The battalion computer exploits all the information stored in its memory to coordinate the fire among the different batteries. It sends these data digitally coded to the battery computers which compute firing data for each GDC. The battalion computer can be directly linked to various peripheral units of the battalion, such as a met station or a muzzle velocity meter.

Compared with calculation by conventional methods or with other digital systems, SEDA offers the following advantages:

(1) A fast and reliable transmission of information among the different battalion elements, also when the voice communication link becomes particularly weak.
(2) The system can use radio sets already available at the battalion
(3) The computers fulfil fast and accurate calculations of firing data because they perform a numerical integration of the projectile trajectory taking into account point by point all the meteorological and ballistic conditions.
(4) All the information coming from the observers can be introduced also manually into the computers.

The SEDAB, artillery fire control system at battery level, is a part of the SEDA. It can operate autonomously as a separate stand-alone system to provide fire computations and transmission management for a battery of up to eight guns. SEDAB includes: one computer at the battery command post; one man portable terminal (DMT) for each observer; one gun display unit (GDC) for each gun.

STATUS
In production

CONTRACTOR
Officine Galileo SpA, Via Einstein 35, 50013 Campi Bisenzio, Florence, Italy.

Elements of the SEDAB artillery automation system

4855.161

CMB8 MORTAR COMPUTER

The CMB8 is a military, hand-portable computer which automates the procedure needed to compute firing data for a mortar battery composed of six mortars.

It makes use of all the information stored in its memory to compute firing data for the different mortars of the battery, and it can be connected to a Galileo military, hand-portable printer for the recording of the executed procedures. An internal rechargeable battery and the possibility of a connection to an external power supply give the CMB8 a long combat autonomy.

Compared with conventional methods, the CMB8 offers many advantages:

(a) A drastic reduction of time in calculation, together with an increased precision
(b) A reduction in preparation time and crew members, because a menu sequence directs the operator step by step, to achieve the completion of the procedures
(c) The recording, by a printer, of the executed procedures
(d) A possible automatic data transfer through a digital link with other units of the battery (observers) or with upper levels.

Officine Galileo CMB8 mortar computer with optional printer coupled on left

The most important functions performed by CMB8 are:

(1) The fire preparation, including:
  (a) Survey preparation with several programs such as: polar coordinates replot, close traverse replot, shift from known point, intersection replot, reverse intersection.
  (b) Introduction of the position data for all the mortars of the battery.
  (c) Ballistic preparation with the introduction of met message, charge temperature and weight, muzzle velocities, etc.

(2) Evaluation of the fire possibilities taking into account friendly troop positions, (not fire areas and crest clearances) to evaluate the possibility of fire against a target

(3) Firing data computation and fire distribution for efficacy for each mortar (convergence, smoke, parallel, barrage).

STATUS

In production and in service with the Italian Army.

CONTRACTOR

Officine Galileo SpA, Via Einstein 35, 50013 Campi Bisenzio, Florence, Italy.

6537.153

OG14L2A TANK FIRE CONTROL SYSTEM

This is the fire control system for the OTO Melara OF40 battle tank. It includes the C215 gunner's sight (**4216.193**) which incorporates a VAQ-33 laser rangefinder (**3558.193**), the SFIM VS580 day/night sight (**3560.193**), two control units and a number of optional sensors such as TV or infra-red trackers. A lead angle measurement system and a powder-temperature sensor are also included. Information on the individual components is given under the relevant index number.

STATUS

In production for the OF40 tank.

CONTRACTOR

Oficine Galileo SpA, Via Einstein 35, 50013 Campi Bisenzio, Florence, Italy.

4232.161

OG 14 L2B TANK FIRE CONTROL SYSTEM

The OG 14 L2B modular fire control system is designed for use in the 105 mm armed OTO Melara OF 40 MBTs now in service with the United Arab Emirates. It comprises an optical sight with a laser rangefinder, a digital ballistic computer, control panels for the gunner, commander and loader. With the additional sensors the ballistic computer can carry out lead angle computation with the tank and target in motion and integrate data for meteorological conditions, powder temperature and gun wear. The computer is designed also, to accept 120 mm gun ballistic data.

Optional dedicated software permits crew training.

CHARACTERISTICS

Laser rangefinder

Max range: 10000 m

Max pulse rate: 1 pulse/s

Ballistics computer

Max computation distance: 4000 m

Max elevation correction: $\pm 80°$

Max traverse correction: $\pm 30°$

STATUS

In production.

CONTRACTOR

Officine Galileo SpA, Via Einstein 35, 50013 Campi Bisenzio, Florence, Italy.

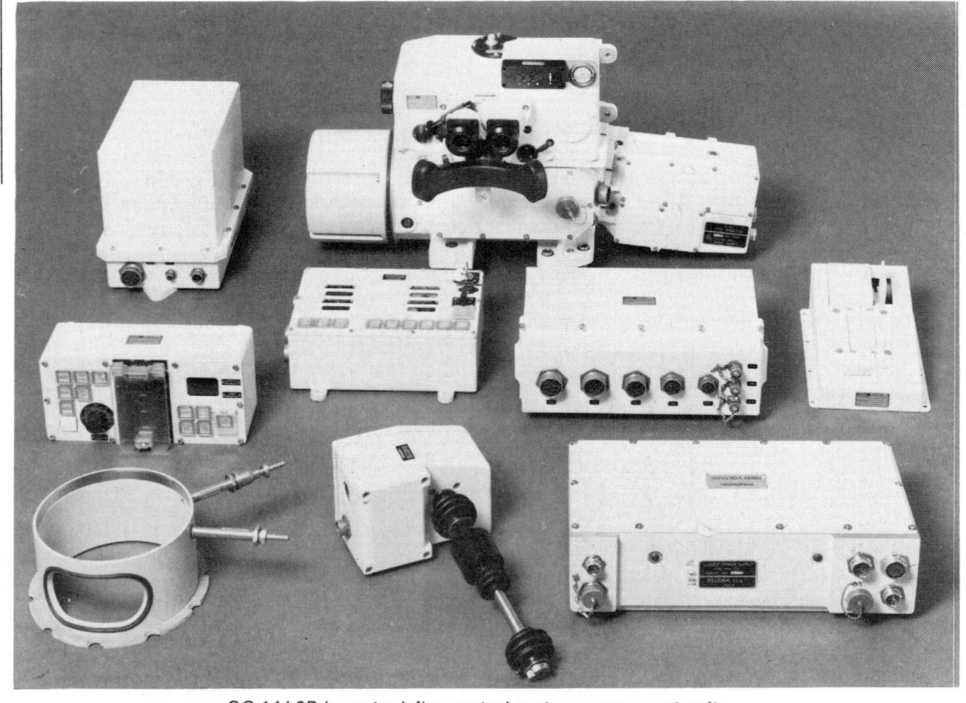

OG 14 L2B laser tank fire control system component units

4856.161

TURMS LASER TANK FIRE CONTROL SYSTEM

The TURMS, formerly, OG14L3 is a third generation LTFCS for battletanks characterised by a fire-on-the move capability, owing to the primary stabilised gunner sight, master with respect to the slaved gun.

TURMS has been conceived to meet the requirements of new main battle tanks but its modular construction makes it suitable for retrofitting. The system consists of a periscope gunner's sight, the commander's SP-T-694 panoramic sight, a digital ballistic computer, sensors, a muzzle reference system and control panels.

The gunner's monocular sight incorporates within a single housing four modules; a primary stabilised head mirror, a visual unit, a laser transceiver and a passive night viewing channel. All visual channels are routed through the common head mirror. The laser is a Selenia MTL-8 Nd-YAG unit while the night vision system can be either an image intensifier with five degree field of view or a Officine Galileo VTG 120 thermal imager with dual field of vision.

The commander's binocular sight is also primarily stabilised and is a development of the VS580 developed by SFIM and adopted for the Officine Gallileo OG14L2 fire control system (see **4232.161**). The original SFIM sight head has been retained but new optics have been developed by Officine Gallileo.

Main components of the Officine Galileo TURMS laser tank fire control system

The system is controlled by a digital 16-bit microprocessor based upon the Texas Instruments 9096. For details see **6528.193** in the Electro-optical (Land) section. This sight can be expanded, if required, to allow computation of both azimuth and elevation lead angles through the fire control computer to provide the commander with a capability for autonomously firing while the vehicle is moving.

Fire control computer options include systems from Marconi Italiana or OTO Melara. The computer performs all ballistic computations taking into account data from the meteorological, tank attitude, gun wear and muzzle reference sensors. It also controls and manages the operation of the sights, rangefinder and servos while additional features include built-in test equipment and a training capability. In the event of failure of other parts of the system the computer will reconfigure its operation for optimum use.

Because of the very high quality and accuracy of line-of-sight stabilisation and firing computation the system has a very high first round hit probability on the move either during the night or during the day.

CHARACTERISTICS

Gunner laser sight
Monocular: 10 ×
Stabilisation accuracy: Better than 0.15°
Laser rangefinder: Nd:YAG ranging up to 10 km
Thermal imaging unit
Spectral band: 8 ÷ 14μ
FOV: 1.1 × 2.2° (narrow); 3.3 × 6.6° (wide)
Commander's sight
Binocular: 2.5 × – 20°; 10 × – 5°
Stabilisation accuracy: Better than 0.5°
Ballistic computer
Ballistic computation range: 300 to 4000 m
Ammunition types: choice of 5 ballistics
Integrated autotest (BITE).

STATUS

Preproduction

CONTRACTOR

Officine Galileo SpA, Via Einstein 35, 50013 Campi Bisenzio, Florence, Italy.

4857.161

ATREOS LASER TANK FIRE CONTROL SYSTEM

ATREOS is an austere LTFCS, specially suitable for armoured vehicles and retrofitting of existing tanks, in a minimal configuration to provide very good performance at a reasonable cost.

The major units, linked as shown in the nearby schematic system layout are:

(1) Day/night gunner sight comprising: laser visual unit, passive night elbow, periscope assembly with × 1 window
(2) Reticle projector unit
(3) Ballistic computer
(4) Gunner control panel
(5) Commander control panel.

The day/night gunner sight is assembled in a self-contained periscope housing, interchangeable with the M32 existing periscope. The laser visual unit and the passive night elbow directly and respectively replace the M32 daylight elbow and the M32 active IR elbow to add day/night range finding and firing capability. Bulk absorption laser filters, one integral with the eyepiece and the other for the × 1 window, provide safety to gunner's eye. The ballistic computer, usually attached to the turret ceiling above the gunner, is a digital computer which generates the traverse and elevation lead angles according to the selected ammunition and the measured range.

CHARACTERISTICS

Telescope
Magnification: × 8
True field of view: 7°
Data and signals displayed in the eyepiece
Laser rangefinder
Passive dye Q-switch Nd YAG
Range: 350 m – 9995 m
Passive night elbow
Intensifier tube: Second generation
Magnification: × 5
True field of view: 10°
Tube gain: × 25000 (min)
Ballistic computer
Target range: 400 to 3500 m
Ammunition choice: 4 types
Computation accuracy: 0.5 mils

STATUS

Preproduction.

CONTRACTOR

Officine Galileo SpA, Via Einstein 35, 50013 Campi Bisenzio, Florence, Italy

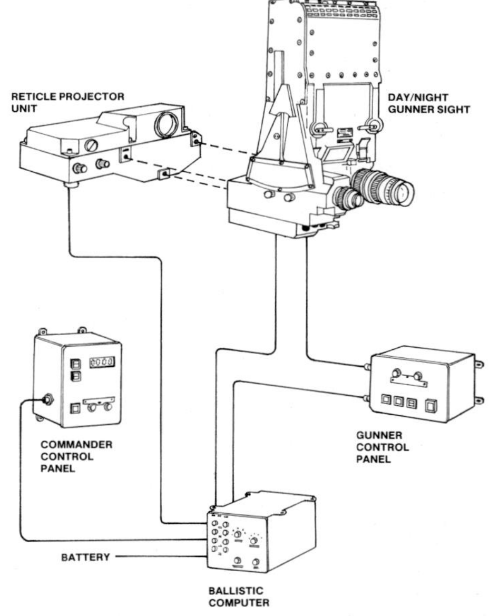

ATREOS system schematic layout

4532.181

CATRIN COMMUNICATIONS SYSTEM

CATRIN (Sistema Campale Trasmissioni Integrato) is an automatic communication system intended for Corps area operations. It functions in the same way as other field communication systems being developed for European deployment such as the US Mobile Subscriber Equipment System.

The CATRIN project is based on a feasibility study prepared between 1970 and 1975 by a consortium made up of Marconi Italiana, Selenia, IBM and Telettra. Technical and operational specifications in line with EUROCOM recommendations were established by the Italian Staff in 1978 and the project is presently carried out by the Italian Army with a consortium of Italian industries. The consortium, composed by Italtel, Face Standard, Telettra, GTE, Fatme and Marconi Italiana, is now working on various components of the system, in particular the central software and hardware.

The CATRIN network consists of trunk-nodes, whose equipment is mounted on shelters for maximum mobility, connected to one another by 32-channel UHF radio-links. Each trunk node may be linked by 16-channel radio relays to Corps and Division access nodes. More mobile users have a radio-access sub-system connected to a radio central.

The grid configuration of CATRIN gives the redundancy necessary to ensure continued operation in the event of damage or failures. A central supervision sub-system continuously checks the operativity of the entire system and gives the Corps signal commanding officer relevant information on the status of the network and instructions for redeployment.

Channels selected for CATRIN are 16 kb/s and 32 kb/s with speech digitally encoded by Delta modulation. Traffic security is obtained through integral ciphering of all communications. A super-cipher is provided for the most sensitive subscribers.

Each user is identified by a seven-digit number which is retained anywhere in the Corps area, no matter how many times he moves.

More mobile users, such as battlegroups or brigade HQs, can use the single-channel radio access (SCRA) sub-system. It gives a subscriber on the move the same local and trunk communication facilities as static users.

The SCRA equipment will be compatible with the new single-channel radio-net family being developed by the Italian Army. This will allow a radio-net user to become a CATRIN subscriber simply by coupling its radio to an SCRA adapter.

A typical Corps network manages 1276 speech users, 248 telegraph and 204 radio users.

The whole system is installed on class 1 and 2 UEO shelters, while terminal equipment is normally mounted on armoured vehicles.

CATRIN network can directly communicate with EUROCOM compatible systems, like the US Mobile Subscriber Equipment System under development by GTE. It can also be connected to NICS, national civil and military networks and integrated with the combat-net radio used on the field by lower echelons units.

STATUS

Operational tests of the full system have started, although many elements of CATRIN will be put into production much earlier.

Radio links and Delta multiplexer, for example, are already in production by Marconi Italiana for use in the HAWK anti-aircraft missile regiments.

6351.181

ARTILLERY FIRE CONTROL SYSTEM

The Artillery Fire Control System (AFCS) is a complete battlefield command, control and communications system that comprises the ARCO 1000 Artillery Computer, GDU 801 Gun Display Unit and the DMG 860 Digital Message Generator.

The ARCO 1000 Artillery Computer is designed to aid the operation of a variety of weapons and can be deployed as a stand-alone system or in a battery environment. It can be used for ballistic, survey and communications roles and all that is required for different weapons and ammunition is to change the software module which is customised to meet the specific ballistics of the weapon/ammunition employed.

A survey software package is normally included in the standard ARCO 1000 which can be expanded at the request of the customer to provide survey units with an effective calculator.

The hardware of ARCO 1000 is provided with modems to drive three radio nets by direct interface to standard UHF/VHF and HF sets. The number of radio nets can be extended to five as an option. The communication software of the ARCO 1000 has been especially designed to make easy, immediate and reliable management of communication traffic between the command post and the peripheral units (forward observers, met stations and gun positions).

ARCO 1000 Artillery Computer

The ARCO 1000 can provide eight gun positions with the relevant firing data to the GDU 801 Gun Display Units, forward observers provide target information with the DMG 860 Digital Message Generator units.

The GDU 801 Gun Display Unit is the peripheral interface and display unit between the ARCO 1000 battery computer and the personnel manning the gun. This rugged unit has been designed to withstand the extreme environmental conditions on the battlefield. It receives data from the ARCO 1000 via radio or wire and displays the following information:

Firing data (deflection, elevation and fuze setting)

Type of ammunition to be used (projectile, charge and fuze)

Fire commands (number of rounds to be fired, firing rate, fire immediate, or on command or fire time)

Fire status (number of projectiles already fired following a fire order, operational status of the gun)

Operational commands (such as cease fire, take cover, alert and so on).

The GDU 801 can transmit back to the ARCO 1000 its own status, for example ready or not available for firing, and to acknowledge the actual firing of the number of rounds requested by the firing order, to allow the battery computer to check for ammunition accounting and fire effect, as requested.

The transmission from the GDU 801 to the ARCO 1000 uses the same line that conveys data from ARCO 1000 to the GDUs.

The GDU 801 reduces the response time of the battery following a fire request, increases the effect of firing and the surprise factor and eliminates human error in the transmission of the firing data from the battery command to the individual guns. This in turn eliminates tedious time consuming checks on the firing data transmitted to the guns.

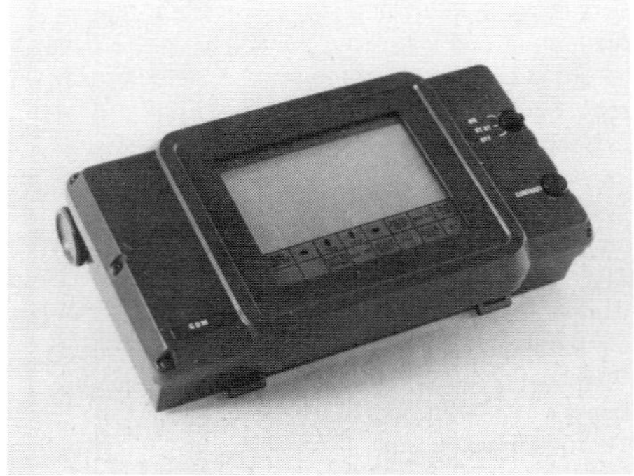

DMG 860 Digital Message Generator

The DMG 860 Digital Message Generator has been designed to allow the operator to transmit and receive data and information under severe communication and environmental conditions. It is fitted with an internal modem which allows data transmission via standard HF/VHF/UHF radio sets. In addition the DMG 860 can be provided with modems to drive telephone lines.

The display area of the DMG 860 acts simultaneously as an output device providing visual display of the data and information stored in the DMG 860, and as an input device, which allows the operator to enter data and information to be stored and/or transmitted by the DMG 860.

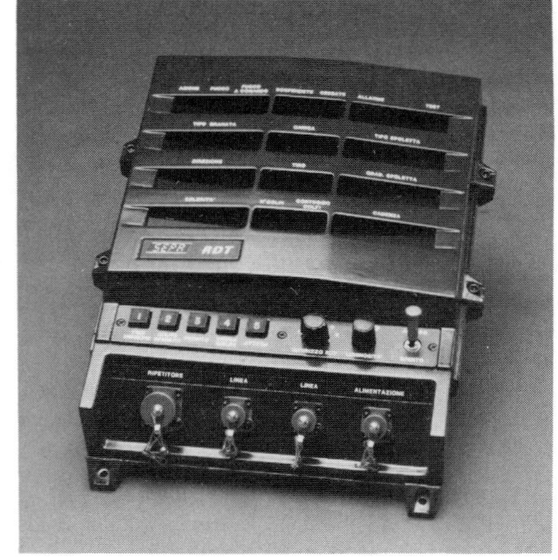

GDU 801 Gun Display unit

CHARACTERISTICS

Component	ARC 1000	GDU 801	DMG 860
Weight	18 kg	5 kg	2 kg
Width	425 mm	250 mm	280 mm
Height	315 mm	350 mm	165 mm
Depth	400 mm	90 mm	90 mm

STATUS

In production.

CONTRACTOR

SEPA – Società di Elettronica per l'Automazione SpA, Corso Giulio Cesare 294, 10154 Turin, Italy.

6443.181

PAC 101 – Portable Artillery Computer

PAC 101 is a Portable Artillery Computer, designed to aid the operation of a variety of weapons, deployed stand-alone or in a battery environment.

The functions provided by PAC 101 satisfy the ballistic, surveying and communication requirements of a tactical operation. To tailor PAC 101 to a specific weapon and ammunition, it is only required to change a software module, customized to meet the specific ballistic of the weapon/ammunition employed and to satisfy the procedures adopted for the specified weapon by the customer.

Quick and accurate computations, such as those obtained by using PAC 101, together with its straightforward use and clear display of the results, allow to greatly increase the effect of firing, since this will improve:

- the surprise factor;
- the promptness of intervention;
- the saving of ammunition.

PAC 101 Portable Artillery Computer

The adoption of PAC 101 will further simplify the personnel training; its use eliminates the need for auxiliary instruments, firing tables and logs required when the calculations are carried on manually or with numerical-graphical methods.

As an option, PAC 101 can be supplied with a plug-in microprinter, to obtain a paper hard-copy of the data, and with a communication module, to enable connections via standard radio or wire links to remote posts. In this situation, PAC 101 can be effectively used as a digital message generator and receiver, via the standard radio sets in use. Optional crypto protection can be provided.

The surveying software package, normally included in the standard PAC 101, can be expanded at the customer request, to provide the Surveying Units with a handy and effective calculator.

PAC 101 is functionally and mechanically composed of several subassemblies:

- Dedicated Keyboard

  Includes full alphanumeric capability, in addition to functional keys used to perform specific functions, as shown on the keytops.

  The keyboard is a rugged, 40-key design, with tactile feedback. It is completely sealed, waterproof and backlighted for night-time use.
- Alphanumeric display

  PAC 101 features a two line, 24 characters per line, alphanumeric LCD display, backlighted for night-time use.
- Computing unit

  Includes a 16-bit CMOS microprocessor and CMOS memory, for low consumption and data retention when switched-off.
- Interfaces

  PAC 101 can be interfaced to external peripherals in a variety of ways, including standard interfaces such as the EIA RS-232 and EIA RS-422.

  Direct interface to a dedicated micro-printer (optional) is provided. A modem provides a simple connection to any standard radio set or to a wire line for remote communications.
- Power supply

  Ni-Cd batteries (rechargeable) provide the prime source of power to PAC 101.

  An external battery-charger and vehicular mount is available to power/recharge PAC 101 from a d.c. source of 12/24 V, and may be supplied separately.

PAC 101 includes the modules described inside a compact and rugged case; the small dimensions of the case allow an easy use of PAC 101 as a hand-held calculator.

CHARACTERISTICS

Weight: 1.9 kg
Width: 292 mm
Height: 147 mm
Depth: 65 mm

ENVIRONMENTAL SPECIFICATIONS

According to MIL-STD-810C, for portable and vehicular equipment.

STATUS

In production for the Italian Army.

CONTRACTOR

SEPA – Società di Elettronica per l'Automazione SpA, Corso Giulio Cesare 294, 10154 Turin, Italy.

KOREA (REPUBLIC OF)

4751.181

PYK-80 FIELD ARTILLERY COMPUTER

The PYK-80 field artillery computer is a digital microprocessor-based system which performs ballistic calculations and related computational tasks for the direction of various field artillery weapons. It is produced in a number of versions appropriate to specific types of weapons, as denoted in the following list:

PYK-80(V)1 – 105 mm howitzer
PYK-80(V)2 – 155 mm howitzer
PYK-80(V)3 – 8 inch howitzer
PYK-80(V)4 – 4.2 inch mortar
PYK-80(V)5 – multiple rocket launchers.

The configuration of the computer can be seen from the illustration which shows the operator keyboard hinged from the front of the unit and the LED display. The unit is adaptable to any type of gun by means of interchangeable plug-in units which carry the necessary program and parameter changes. Calculations performed include firing data, firing adjustment, target data storage and checking, error checks and built-in test. Optional functions include firing data transmission and display using a gun display unit, and target data transmission by means of a data message device.

PYK-80 field artillery computer unit

Input data to the system consists of target information, artillery data, and observer positions, FO information such as target position and adjustment data, meteorological data, and battery information including centre of arc bearing, gun position(s), and minimum elevation. Output data includes: firing data, quadrant elevation, charge number and fuze time.

The PYK-80 computer is connected to the FOs, higher command, a survey team for survey data, a met team and the gun batteries.

CHARACTERISTICS

Type: PYK-80
Width: 390 mm
Depth: 210 mm
Height: 360 mm
Weight: 16 kg
ROM: 16K bytes
RAM: 4K bytes
Cycle time (instruction): 1.2 microsec

STATUS

In production. In service with South Korean Army.

CONTRACTOR

Gold Star Electric Co Ltd, 27-2, Yeoeuido-dong, Yongdunpo-ku, Seoul 150, Republic of Korea.

4752.181

GSQ-250K REMOTELY MONITORED SENSOR SYSTEM

The GSQ-250K remotely monitored sensor system consists of a radio frequency monitor (RFM), miniaturised seismic detector (Minisid), audio add-on unit (AAU), and magnetic intrusion detector (Magid). The Minisid, AAU and Magid are implanted in the area to be observed, where they detect vibration, sounds and magnetic disturbances that might indicate human and/or vehicular activity. This information is sent back to the monitoring site in the form of VHF radio signals. The RFM, which can be installed in a monitoring station, a mobile unit or used as a portable unit, receives, analyses and displays the data gathered.

The main elements of the system are:

(1) USQ-46A radio frequency monitor
(2) USQ-154 miniaturised seismic intrusion detector
(3) DJ-382 audio add-on unit
(4) DJ-509 magnetic intrusion detector
(5) geophone, microphone and auxiliary equipment.

This equipment is combined in various ways to suit the environment and the specific operational requirements of the system. Possible configurations include:

(1) using the Minisid alone, with its built-in transmitter, to monitor vibration
(2) using the Minisid with the AAU to monitor vibration and sound
(3) using the Minisid with the Magid main solenoid only to detect seismic vibration and magnetic field changes
(4) using the Minisid with the Magid's main and auxiliary solenoids to detect magnetic field changes only.

USQ-46A

The USQ-46A RFM consists of a main unit, an antenna and a headset, and is supplied complete with power supply and mounting base. It may be operated in a man-pack, vehicle-mounted, or fixed station configuration. It receives signals on any one of 840 channels. When a signal from a Minisid is received, the RF monitor displays the remote sensor's identity number on its front panel and relays the received audio to the headset.

USQ-46A radio frequency monitor unit

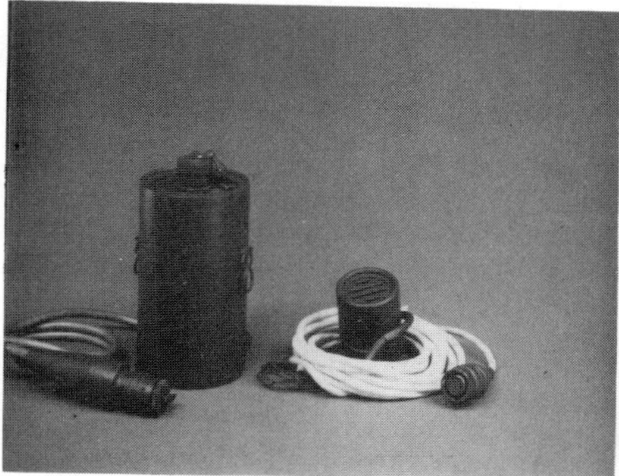

DJ-509 Magid

CHARACTERISTICS

Frequency range: 160 – 176 MHz
Number of channels: 840
Modulation: FM/FSK
Channel spacing: 18.75 kHz
Identity codes: 64 (max)
IF: 21.4 MHz
LO frequency: 3.75 MHz
Dimensions: 333 × 263 × 103 mm
Weight: 8 kg (approx)

USQ-154

The USQ-154 Minisid consists of a main unit, antenna, and a geophone. When the geophone detects vibrations in the ground, such as those caused by troops or vehicles passing within range, it transmits an alarm signal on a frequency suitable for reception by the RFM. The AAU or Magid can be connected to the Minisid for use as supplementary sensors.

CHARACTERISTICS

Transmitter Power: 2 W (min); 4 W (nominal)
Frequency range: 162 – 174 MHz
Transmission: Digital FSK
Spurious emission: 40 dB minimum
Frequency response: 40 – 2000 kHz
Detection range: 50 m dia (troops); 300 m dia (vehicles)
Dimensions: 210 × 210 × 105 mm
Weight: 5 kg (approx)

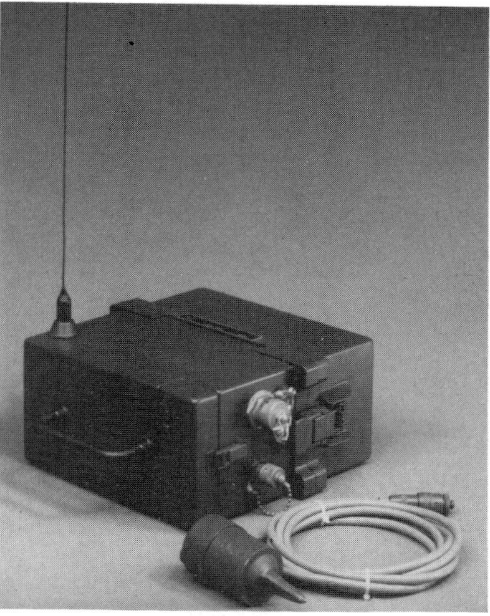

USQ-154 Minisid

DJ-382 audio add-on unit

DJ-382

The DJ-382 audio add-on unit consists of a main assembly containing a microphone and is used as an ancillary device connected to the Minisid, from which it draws its power. The AAU is activated by three successive occurrences of ground vibrations and sends out audio signals within its range of audibility via the Minisid. An optional daylight sensor permits operation at night only if desired.

CHARACTERISTICS

Audio output: 0.3 to 1.0 V rms
Voltage gain: 80 dB
AGC range: 50 dB (keyed AGC)
Detection range: 3 – 4 m dia (troops); 10 – 20 m dia (vehicles)
Dimensions: 135 × 76 mm (dia)
Weight: 1 kg (approx)

DJ-509

The DJ-509 Magid consists of a main solenoid and is used as an ancillary device attached to the Minisid to detect disturbances to the Earth's magnetic field associated with the passage of magnetic material through the area being observed.

CHARACTERISTICS

Voltage gain: 138 to 141 dB
Detection range: 3 m dia (troops); 10 m dia (vehicles)
Dimensions: 310 × 65 mm (dia)
Weight: 1.4 kg (approx)

STATUS

Production complete. In service with South Korean Army.

CONTRACTOR

Gold Star Electric Co Ltd, 27-2, Yeoeuido-dong, Yongdunpo-ku, Seoul, Republic of Korea.

NETHERLANDS

4830.161

FIST ARTILLERY SYSTEM

FIST (Fire Support Information System for Field Artillery) has been developed to replace the US FADAC (Field Artillery Digital Automatic Computer) which has been in use with the Dutch forces since 1967.

FIST is to equip the Dutch corps in West Germany with five major terminals to each battalion (one per battery, and two at battalion level). Each terminal has several microprocessors, a keyboard and a VDU, and is employed to calculate firing data at battery level and to exchange data between battalion and battery via burst transmission (to avoid countermeasures). The system is further composed of an observer terminal and a gun commander's terminal.

Each battery computer is able to store data (such as ammunition levels, gun status and characteristics etc) for up to four batteries, information on up to 150 targets, accept and process information from forward observers and the fire direction officer, carry out aiming calculation for up to eight guns, and the average time between receiving a fire order and displaying firing data is less than 25 seconds for the first seven rounds.

STATUS

A prototype is undergoing field trials. Production is scheduled to start in mid-1988

CONTRACTOR

Hollandse Signaalapparaten BV, PO Box 42, 7550 GD Hengelo, Netherlands.

FIST battery/battalion terminal comprising four modules: processor, display, keyboard and plug-in bubble memory

2301.181

FLYCATCHER AA FIRE CONTROL SYSTEM

Flycatcher is an AA radar fire control system designed to detect, identify, and automatically track aircraft flying at low to medium altitudes. It can be used to control either guns or missiles, or any combination of both.

Main elements of the system are a dual retractable search and track antenna system, two radar transmitter/receivers, a radar display and control assembly, a fire control assembly, and a digital computer. The whole system is housed in a transportable military container. Power consumption is about 11 kVA. Electrical power is provided by its own onboard generator.

With Flycatcher's 3-D tracking channel, three weapon systems can be controlled at a distance of approximately 1 km. Using its 2-D track-while-scan channels Flycatcher can provide target information to even more weapons at a distance of 6-7 km. By providing an optional defence in depth feature Flycatcher is effective against modern medium range weapons as well as being protected against both anti-radiation missiles and stand-off jamming.

The radar operates in the I-band, and features track-while-scan, a high data-renewal rate, and a multiplicity of ECCM and anti-interference facilities. An additional K-band facility provides high tracking accuracy with anti-image performance. The search antenna is a slotted waveguide; the tracking antenna is parabolic with a cassegrain reflector and a monopulse feed horn. TV tracking is incorporated as a complementary system for passive tracking.

The general-purpose computer, which can handle targets with speeds up to Mach 2, performs the ballistic computations for medium calibre guns and data processing for missile guidance and a number of other functions.

Signaal's Flycatcher in operating position showing generator on left side of carriage and antennas in erected position

CHARACTERISTICS

Function: All-weather point and area defence with AA guns and S/A missiles against medium- to low-level air attacks

Sensors: I-band search and track radar, K-band anti-image tracking radar, TV, IFF

Weapons: Up to 3 weapons, using 3-D data plus a number of weapons using 2-D data, employing Reporter TDR (**4469.181**), providing defence in depth.

Search coverage: Up to 20 km (1 m^2 target) track-while-scan

Display: PPI, north-oriented, clutter- and interference-free, scales 10/20 km

Target engagement: Joystick indication, automatic acquisition and tracking, automated weapon control

System design: Container construction, retractable antennas, self-contained diesel-driven power supply, removable wheel-train, micro-miniaturised solid-state electronics

Dimensions: Length 2.73 m; width 2.12 m; height (antenna in) 2.13 m, (antenna out) 3.65 m

System weight: 3000 kg (approx)

Environmental: Operating temperature –30 to +52°C, storage temperature –55 to +71°C, others according to MIL-STD-210, ground equipment, world-wide, vibration and shock capabilities according to MIL-STD-810 proc 13.

STATUS

In production. In service with a number of countries including India and the Netherlands.

CONTRACTOR

Hollandse Signaalapparaten BV, PO Box 42, 7550 GD Hengelo, Netherlands.

4469.181

REPORTER TACTICAL EARLY WARNING SYSTEM

Reporter is a transportable early warning and cueing radar system designed to search and track air targets at altitudes of 15 to 4000 m and ranges up to 40 km. The compact, I-band radar has integrated IFF for target identification, and features a small-size antenna (combined radar/IFF), able to rotate at high speeds. The receiver incorporates digital video-processing (MTI) with double canceller techniques ensuring accurate discrimination between fast and slow targets (eg hovering helicopters).

Up to 20 designated targets are automatically tracked by the built-in video extractor. Target data is transmitted via VHF/FM radio or line communications to an unlimited number of fire-units, equipped with a Reporter target data receiver (TDR). Extensive and successful trials have evaluated the effectiveness of low-level air defence systems consisting of a combination of Reporter with various surface-to-air missiles and medium range guns. Weapons with associated surveillance radars can maintain radar silence by using information from Reporter.

At the weapons site target data from Reporter is processed by the TDR. This is a computer-based, portable, highly intelligent terminal suitable for fire units to evaluate data obtained from Reporter. The TDR is used for various weapon parameters such as position, priority line, and allocated defence sector. With its built-in threat evaluation programme, each TDR displays data of the four highest priority targets with reference to the weapon system position. Various weapon interfaces are available as standard to convert target information into control signals for automatic weapon aiming. The interfaces can be offered separately.

Reporter consists of the radar system with shelterised operator compartment, a number of lightweight portable terminal equipments located at remote fire unit positions, and optional interface equipment for automatic weapon aiming. The radar part is transported on a two-wheel trailer which carries a mast-mounted antenna. Alternatively, the radar can be mounted on a military vehicle. The operator console and associated communications, computer and electronics equipment are housed in a standard shelter to be transported by any suitable military vehicle.

Reporter consists of the radar system with shelterised operator compartment, any number of lightweight portable terminal equipments located at remote fire unit positions, and optional interface equipment for automatic aiming. The radar part is transported on a two-wheel trailer which carries a mast-mounted antenna. Alternatively, the radar can be mounted on a military vehicle. The operator console and associated communications, computer and electronics equipment are housed in a standard shelter to be transported by any suitable military vehicle.

Reporter can be extended with an airspace control function to assist radar operators in discriminating between friend and foe, as an additional means to IFF. In that case, behaviour information from own aircraft, and other air activities, will be displayed on Reporter's PPI, defining airspace information (eg lines, areas and zones) in synthetic form.

STATUS

Over 20 Reporters are in operation in a number of countries. Reporter system elements are incorporated in the WM naval fire-control systems of which over 500 have been sold to 35 countries.

CONTRACTOR

Hollandse Signaalapparaten BV, PO Box 42, 7550 GD Hengelo, Netherlands.

Reporter operator's shelter on light military truck and radar trailer

Reporter terminal equipment (from top): radio set, TDR and optional interface for automatic aiming

6352.161

INTEGRATED TRACKING AND FIRE CONTROL SYSTEM FOR WILDCAT

The private venture Wildcat twin 30 mm self-propelled air defence system (6 × 6) (**3681.131**) has been developed by Krauss Maffei in co-operation with Hollandse Signaalapparaten, Siemens and Mauser. It is fitted with an automatic tracking and fire control system designed and manufactured by Hollandse Signaalapparaten BV.

This system is based on the well proven LIOD and LIROD tracking and fire control systems. The tracking and fire control system is integrated in a turret which can be mounted on a variety of tracked and wheeled vehicles.

Depending on operational requirements, the Wildcat can be equipped with different automatic tracking systems:

TV/laser tracking system for passive tracking

FLIR/laser tracking system for passive day/night applications

Ka-band radar tracking system for all weather applications.

The tracking sensor is mounted on a servo controlled platform in azimuth and elevation at the turret front between the two 30 mm weapons. This ensures a constant alignment with the target during target tracking in combination with weapon aiming and firing.

Wildcat twin 30 mm air defence system with integrated tracking and fire control system by Signaal

The tracking system gives continuously accurate target data to the digital fire control computer that calculates the weapon lead angles. Errors due to canting, meteorological conditions and muzzle velocity variations are accounted for.

A gyro package separates the vehicle movement and target movement and in this way firing on the move is assured.

STATUS

Development of the Wildcat twin 30 mm air defence system is complete.

CONTRACTOR

Hollandse Signaalapparaten BV, PO Box 42, 7500 GD Hengelo, Netherlands.

2302.153

INTEGRATED RADAR SYSTEM FOR 35 mm AA TANK

One version of the 35 mm AA tank systems described in the Systems section (**2370.131** in 1977 and earlier editions) is equipped with an integrated search and tracking radar system designed and manufactured by NV Hollandse Signaalapparaten. This integrated radar system is mounted on a Leopard tank chassis armed with twin, rapid-firing 35 mm Oerlikon AA cannon. The Signaal system is based upon the well-proven advanced integrated radar technology in which the same I-band transmitter is used for both search, for fast acquisition and tracking purposes. A dual-band tracking radar operating in both I- and Ka-bands, with a broad beamwidth compared with that of the search antenna and a very small beamwidth for tracking the extremely low flying targets, reduces reaction time to the barest minimum and allows the tracking of the lowest, tree-top level aircraft. Simultaneous search transmissions are continued through the entire tracking phase, permitting complete coverage against further airborne threats to be maintained. Rapid automatic or manual switching facilities allow the system to cope with a multi-target environment.

Signaal integrated radar system for 35 mm AA tank

The radar system has been designed to detect and identify aircraft at very low to medium altitudes and to track them automatically. It has the following important features:

(1) integrating all-weather search and tracking systems with a very short reaction time
(2) search while tracking through 360°
(3) search on the move, with compensation for vehicle's speed
(4) good performance in clutter and ECM environment
(5) compact rugged design and module concept to facilitate maintenance
(6) dual frequency tracking radar
(7) integral IFF.

CHARACTERISTICS

Antenna system

Search antenna: Slotted waveguide type
Length: 1.5 m
Horizontal beamwidth: 1.4°
Vertical beamwidth: 30°
Polarisation: Horizontal or circular
Rotational speed: 60 rpm
Range: 15 km
MTI improvement factor: 30 dB
Tracking antenna: Parabolic with cassegrain reflector and monopulse feedhorn, plus integrated nozzle motor for Ka-band transmit/receive system
Diameter: 0.6 m
Beamwidth: 4.2°

Transmitter/receiver

Transmitter power: 200 average
Frequency: I-band
Search channel: MTI with double canceller (digital)
Tracking channel: Pulse doppler and built-in anti-image radar in Ka-band
ECCM: Digital video correlator (ISU), pulse-length discriminator (PLD). Passive tracking in I-band, PRF stagger
Noise figures: Search receiver 7 dB, tracking receiver 9 dB

Radar display and control panels

PPI: Diameter 25 cm, north orientated clutter-free picture, compensated for vehicle's speed. Arrangement of controls to high standard of human engineering. IFF completely integrated with the search system
Power consumption: 200 V 380 Hz approx 2 kVA

STATUS

In service with the Royal Netherlands Army.

CONTRACTOR

Hollandse Signaalapparaten BV, Zuidelijke Havenweg 40, PO Box 42, 7550 GD Hengelo, Netherlands.

NEW ZEALAND

6705.161

MERE MORTAR FIRE CONTROL COMPUTER

Mere, a Maori weapon, is a mortar fire control computer designed to meet a New Zealand Army requirement for an 81 mm mortar support system by Physics and Engineering Laboratory of the New Zealand Scientific and Industrial Research Department. It is manufactured by Wormald Vigilant in two versions; military specification and part militarised. The former contains military grade components and parts providing a MTBF of 10200 hours while the latter has some military grade components with the remainder being of high reliability or industrial grade to give an MTBF of 6750 hours.

It is a simple-to-operate, hand-held computer with 24-key pad and a 32 character dot matrix display with transflective LCD. The fire mission data includes range, optimum charge, elevation, bearing and time of flight/fuze. It can store the positions of up to 99 targets, 24 observers/friendly units and nine mortar lines. Positions may be specified by grid or by using a laser rangefinder. Altitudes may be set in feet or metres.

Complete operator prompting makes the system easy to use. When necessary warning messages are displayed to ensure observer safety. If required NATO format meteorological data may be added. If there is no operator action for 45 seconds the system automatically switches itself off.

Ancillary equipment includes an optional four-battery charger and a large character generator for classroom training purposes.

CHARACTERISTICS

Weight: Less than 4 kg
Dimensions: 265 × 80 × 125 mm
Operating temperature range: –31 to +60°C
Target store: 99
Power supply: 6 V rechargeable battery
Battery life: 70 hours (continous use), 1000 hours (intermittent use)

STATUS

The military standard system has been selected by the New Zealand Army. The system has been supplied for evaluation to India and Malaysia as well as being approved for export to the Philippines.

CONTRACTOR

Wormald Vigilant Ltd, PO Box 19545, Christchurch, New Zealand.

NORWAY

6707.181

NORWEGIAN ARMY CCIS

The Army Command, Control and Information System (CCIS) is structured on three levels:

COMLANDNON/COMLANDSONOAR/Area Command

The CCIS for these command, which are located at permanent headquarters, is integrated with the basic NORCCIS concept. Although the current NORCCIS programme is limited to the Air Force, subsequent NORCCIS phases will provide for both Army and Navy requirements at this level.

Division/Brigade

A special CCIS programme for brigade level was launched in 1982. The objective of the first three-year phase was to chart the basic requirements and to establish operational and technical guidelines for future CCIS projects.

The second phase which started in 1987, comprises a mobile test system. Emphasis will be placed on the development of common basic hardware and software for the Army CCIS and ODIN 2 Field Artillery FCS (**2645.161**), as well as specialised software for both projects.

The major functions covered in the mobile test system, in addition to operational functions, are processing of NATO Adat-P3 formatter messages, geographical map presentation combined with tactical information and communication between the workstations in operation at the various operational units in the Army.

Subsequent phases will build upon the mobile test system presently being developed. It is intended to extend this both operationally and technically to achieve a higher level of performance.

Battalion/Battery

This level comprises various units, for example Infantry, Artillery, Intelligence, Engineers, etc.

The general structure of CCIS assumes numerous unit-level CCIS subsystems, each providing automated support to one unit type. All unit-level CCIS subsystems will, however, follow the generalised guidelines (inter-operability, standards, etc) for the Army CCIS.

STATUS

Contract was awarded in October 1987 to NFT A/S Kongsberg, Norway, to provide the mobile test system for limited functions within division/brigade and lower units. The mobile test system is scheduled to be tested in spring 1989.

CONTRACTOR

Norsk Forsvarsteknologi A/S, 3600 Kongsberg, Norway.

3126.161

BODIL COAST ARTILLERY FCS

BODIL is an electro-optical fire control system for coastal artillery applications developed for the Norwegian Coast Artillery by Kongsberg.

Few details have been revealed but the BODIL FCS includes mini-computers, electro-optical sensors, data transmission links, and gun displays. The system requirements are based on reliability and ease of operation. Only one sensor operator and one system operator are required.

The system is flexible, and can easily be expanded to cater for larger batteries and fortifications, for example, by adding more sensors and tactical and tracking displays. The design allows for the transmission and reception of data from other batteries.

STATUS

Operational.

CONTRACTOR

Norsk Forsvarsteknologi A/S, PO Box 1003, N-3601, Kongsberg, Norway.

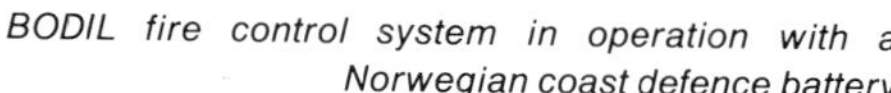

BODIL fire control system in operation with a Norwegian coast defence battery

2645.161

ODIN ARTILLERY FCS

Odin is the name given to a computer-based battalion-level artillery fire control system which is currently in service with the Norwegian Army.

The basis of the system is the Kongsberg NM90 – a special fire-control version of a general-purpose computer. One of these is located at each battery command post, and there is a further computer of identical pattern at battalion command post. Each computer has a field artillery panel, a combined control and display panel for entering information and displaying data. The important feature of this arrangement is that, in association with programs stored in the computer, the panel not only accepts and displays data but also instructs the operator, by various display devices, leading him through the sequence of actions required by the fire control procedure. Firing data are presented at each gun by means of the gun-display unit NO/VGG-20.

Target data inputs will in general be provided by forward observation posts, each of which is equipped with a Simrad LP3 laser rangefinder (**1804.193**) and the NO/VGC-30 data transmission system produced by Siemens A/S in Norway. Other inputs accepted by the system are meteorological data, muzzle velocity measured by, for example, the Nera Bergen doppler radar chronograph, and survey data from standard survey equipment. In addition the computer can store gun ballistic data (up to three types of gun simultaneously) and ammunition calibration data.

The computer comprises a central processing unit, a memory unit, the field artillery panel referred to above, and a power supply unit, all contained in one cabinet. General specification details of the computer are set out below.

CHARACTERISTICS

Type: General purpose, digital, 16-bit word
Organisation: Parallel, negative numbers in two's complement. Floating point arithmetic performed in hardware
Instructions: 96 total (32 basic)
Addressing: Any combination of relative, register modified, and indirect
Main memory: 16 K. Asynchronous operation with CPU. Power failure memory protection. Expansible to 64 K
Memory cycle time: 1.5μs
Input/output: 16 bits in parallel under program control to/from 256 devices. 1 channel for direct memory access
Data words: Single word 16 bits
Double word 32 bits
Floating point number 48 bits
(33 mantissa 15 exponent)
Interrupt: Standard single-level interrupt system
Power requirements: 24 V DC (18 – 32 V)
Power consumption: Approx 400 W (16 K memory)
Dimensions: 630 × 410 × 470 mm
Weight: 70 kg
Environment: DEF-133 Table L-3
Temperature: –40 to +55°C (operating); –55 to +80°C (non-operating)
Vibration: 1*g*
Shock: 50*g* – 10 ms
Humidity: 0 – 100% (with condensation)
RFI: MIL-STD-461A
Peripherals: Interface for paper tape reader digitronics 2540 EP. Other peripherals may be connected via the general I/O channel. There is spare space for introduction of special interfaces according to specifications

DATA CAPACITY

Grid co-ordinates: 9 battery points
18 guns organised in 3 batteries with 6 guns in each battery
36 forward observer locations
100 targets or survey reference points
10 safety zones
2 obstacle points for each of 5 obstacle sectors
Calibration data: 18 guns; 8 ammunition lots
Ballistic data: 3 types of gun at the same time.

Programs are available for the following artillery pieces:
US 105 mm M2A1 or M2A2 howitzer
US 155 mm M1 howitzer
US 155 mm M109G SP howitzer

Other programs are being developed and can be produced for any gun, subject to the availability of ballistic data.

STATUS

In production and in service with Norwegian forces.

CONTRACTOR

Norsk Forsvarsteknologi A/S, PO Box 1003, N-3601 Kongsberg, Norway.

NO/VGG-20 gun intercommunication unit and gun display unit used with the Kongsberg Odin artillery FCS to provide automatic data transmission from computer to individual guns

6708.181

KMT 400 MULTI-PURPOSE MILITARY TERMINAL

The KMT 400 is a powerful and efficient, ruggedised, miniature computer terminal. It is designed to meet a variety of military requirements ranging from message communication to complicated ballistic computations.

With a simple change of application software the terminal can be used for intelligence, logistics, forward observation, missile control, general communications and fire control for both artillery and mortars. The clear, comprehensive and instructive menu permits military personnel to use the KMT 400 with the minimum of training.

The processor used is the 80C88 with an EPROM of 256 Kbytes, a RAM of 96 Kbytes and an EPROM of 8 Kbytes. Alternatively the 80C87 processor can be used. The system features touch matrix overlay on the 72 × 144 mm liquid crystal display screen. The screen has a resolution of 128 × 256 pixels.

In the mortar fire control role the terminal can be

used by forward observation posts linked to the command post by land line or radio link. The terminal is capable of registering 20 targets, nine observation post positions and nine safety zones. Input of observation data may be made in any order because it is automatically sorted and filed.

The terminal also has an area survey calculation function which gives greater accuracy. Communications can be in fixed-format message, free text or fire-order form with optional oral communication. Fire commands to the command post can be transmitted in 0.7 seconds and this, together with error correction coding, gives not only assured transmission but also provides a high degree of ECCM performance.

A command post will use two KMT 400 terminals which fit into a frame with an optional printer. In this configuration the system can deal with:

(1) 50 targets.
(2) 20 observation post positions.
(3) Eight weapons.
(4) 35 circular safety zones with 250 m fixed radius.
(5) One hexagonal safety zone.

The system can transmit messages to all its units either individually or simultaneously.

The ballistic calculations are based upon a mass-point model which takes into account variables such as wind conditions at various heights, temperature variations and different ammunition types.

At mortar unit level a single KMT 400 terminal is used linked to the command post by land line or radio. Contact is established after synchronisation of identification and time simply and rapidly through the menu.

Firing data is presented in large (15 mm) characters which are legible at a distance of 3 m. The electroluminescent backlighting is self-regulating and based upon prevailing visibility. The functions of the terminal at unit level are to present basic target data, the firing order, ammunition status, messages and acknowledgments.

The KMT 400 being used at mortar unit level

CHARACTERISTICS

Size: 70 × 120 × 200 mm
Weight: 2 kg
Operating temperature range: –40 to +55°C
Target store: 20-50
OP store: 9-20
Battery life: 50 hours operational
Power supply: Lithium BA-5052N battery

STATUS

Pre-production systems were to be delivered to the Norwegian Army for field trials and evaluation in February 1988. Full production is scheduled to begin in autumn 1988.

CONTRACTOR

Norsk Forsvarsteknologi A/S, 3600 Kongsberg, Norway.

SOUTH AFRICA

4493.161

AS80 ARTILLERY FIRE CONTROL SYSTEM

The AS80 fire control system has been developed to fulfil the need for a decentralised autonomous system which requires minimal operator training while maintaining fast, reliable and accurate fire control computations.

The system consists of a digital computer; a terminal unit with keyboard, visual display keyboard and visual display facilities; a printer and a gun display system. All units have been fully militarised to ensure survival of the system in vehicles. Low power consumption is achieved by automatically reverting to a standby state during inactive periods. All programs and data are non-volatile, and are retained even when the power is switched off. Communication between the user and the system is achieved by means of an interactive dialogue, using menu selections, which guide the user through the correct sequence of operation. The procedures implemented in the dialogue correspond to the manual artillery procedures for the gun. These procedures may be adapted for other weapons or procedures. Approval for the entry of data into the system may be controlled by the officer in charge by use of the confirmation switch.

Currently available software caters for four simultaneous engagements using fire units of up to eight guns. However, with limitations, a total of up to 32 guns can be accommodated. The main facilities offered by the system are:

(1) entry of data
(2) handling of engagements
(3) performing survey calculations
(4) printing of data
(5) hardware check-out.

On having entered the necessary data for the deployment, engagements may be undertaken using the four phases which are provided. These are:

(1) initial orders for the definition of the fire mission
(2) adjustment of fire on the target
(3) fire for effect
(4) end of mission and registration of targets.

During the adjustment and fire for effect phases gun data is calculated and is immediately available for transmission by either radio or landline to the gun display units, which may be either tripod or gun mounted. In addition to the ability to compute ballistic firing data, the system offers several additional features which include: retention of records; ability to perform survey calculations at battery level; warnings of infringements in 'no fire' areas and crest violations; preparation of fire plans; derivation of meteorological conditions (eg from an S700 met system, **4494.161**); maintenance of target, weapon and ammunition records.

The principal hardware items of the AS80 system include a 16-bit minicomputer, a customised terminal offering protection of the CRT during transportation, a printer using standard teleprinter paper, a gun display unit interface with associated line distribution box and interface to data entry terminals. Up to 32 gun display units may be driven by one interface unit.

AS80 printer unit in its protective case

Inside a fire control post container of the South African AS80 system showing the terminal (with keyboard lowered), digital computer and interface unit on lower shelf

Communication with the self-contained gun display units may be achieved either by the internal radio or by a landline.

CHARACTERISTICS

Computer: 16-bit; 64 K words, expansible to 256 K words; programs and data non-volatile

Terminal keyboard: 56 characters (36 alphanumeric, 4 symbols, 11 specific function keys, 5 unspecified function keys); remote 'confirm' switch for FCO

Printer

Line length: 80 characters/6 lines/in

Speed: 30 cps (max)

Character set: 58 characters from ASCII standard code

Interfacing: 20 mA single current loop

STATUS

In production. In service with South African Army.

CONTRACTOR

Reunert Technology Systems (Pty) Ltd, PO Box 1750, Rivonia, 2128, Republic of South Africa.

4494.161

S700 METEOROLOGICAL STATION

The S700 meteorological system provides a simple method of gathering meteorological data for ballistic computations and related battlefield applications. The system is completely mobile, being fitted in a standard military shelter, and offers automatic tracking of a radiosonde from the launch until termination of a flight. The data received from the radiosonde, together with its position, is fed to the digital computer which calculates wind speed and direction, at the various altitudes, together with a pressure, temperature and humidity profile. The results may be presented in formats suitable for direct use in artillery fire prediction calculations. The deployment of the system is simple, requiring only a single vehicle for transportation and a crew of two for operation.

Tracking and reception unit of S700

The man-machine interface is by means of a dialogue, the first operation undertaken being the calibration of the radiosonde in a ground check chamber included in the system. Calibration is performed for temperature, pressure and humidity. On completion of the calibration process, the radiosonde is attached to a balloon and the antenna is aligned for tracking by means of a CCTV monitor. Ground meteorological conditions obtained from the monitor unit are fed into the computer system. On releasing the balloon, the antenna automatically tracks and data containing temperature, barometric pressure and humidity are received and recorded, being used in conjunction with the tracking antenna position to produce the required meteorological data as a hard copy print-out. The ground receiver antenna automatically tracks the sonde, using the latter's transmissions, thereby eliminating any need for radar tracking of the balloon/sonde. This has the operational benefit of making the ground station passive in use and thus immune to counter-measures.

The hardware consists of a monitor unit, a receiver unit, a tracking unit and a processor unit which offer the following:

MONITOR UNIT

System power conditions, flight clock, CCTV for initial antenna alignment and ground weather conditions.

RECEIVER, TRACKING AND PROCESSOR UNITS

Receiver frequency range: 1680 MHz ± 20 MHz

Tuning: Manual + AFC

IF: 30 MHz

Sensitivity: 100 dBm

Noise figure: 9 dB max

COMPUTER

The digital processor used is a 16-bit mini-computer, understood to be of South African design and manufacture. The system operates from standard lead acid batteries, the supply voltage being 24 V DC. Power consumption is nominally 170 W with a peak requirement of 320 W. Fully militarised, the system conforms to DEF 133 Table L2 (as applicable) and can operate in the temperature range of −10 to +50°C.

DIMENSIONS AND WEIGHTS

Operating console: 825 × 510 × 400 mm (excluding computer terminal)

Radiosonde ground check chamber: 406 × 400 × 360 mm

Antenna unit: Height (max) 1.7 m (excluding raising mechanism); turning radius 0.96 m; disc diameter 1.2 m

STATUS

The system has been extensively deployed under operational conditions and is in regular service with the South African Army.

CONTRACTOR

DIEL (Pty) Ltd, PO Box 228, Plumstead, 7800, Republic of South Africa.

4495.161

EMVA MK 10B MUZZLE VELOCITY ANALYSER

The EMVA Mk 10B muzzle velocity analyser is a doppler radar measuring instrument capable of measuring projectile velocities over the range 30 to 3000 m/s. Results are presented in metres per second for direct application in fire control calculations. It is suitable for use with gun calibres of 20 mm and above. The instrument consists of an antenna head suitable for mounting on the gun, or on a tripod adjacent to the gun, and a processor unit. The processor unit contains a liquid crystal display and printer for data recording. A single 50 m or optional 100 m shielded cable connects the antenna head to the processor. Power can be obtained from standard lead acid batteries. No warm-up time is required and the equipment is ready to operate immediately after the brief automatic self-test sequence has been completed.

Provision of interactive operator dialogue minimises incorrect data entry, and informs the operator of incorrect procedures. It also enables potential operators to be trained in a matter of hours. The dialogue is currently available in two languages.

OPERATION

The operator enters the expected velocity range, and any other relevant data that may be required to be recorded in hard copy form, into the processor via the keyboard. The measurement process is entirely automatic from the time the system commences measurement until the final hard copy print-out. The antenna head ceases to radiate the moment the projectile is beyond instrument range. The instrument is self triggered by the doppler signal. However, an external trigger unit which operates on gun flash or projectile shadow is also provided. When plugged into the antenna head it automatically over-rides the self-trigger circuit.

From the instant the measurement commences until the projectile leaves the range of the instrument approximately 100 individual velocity samples are taken and stored. The exact number depends on the velocity and cross sectional size of the projectile.

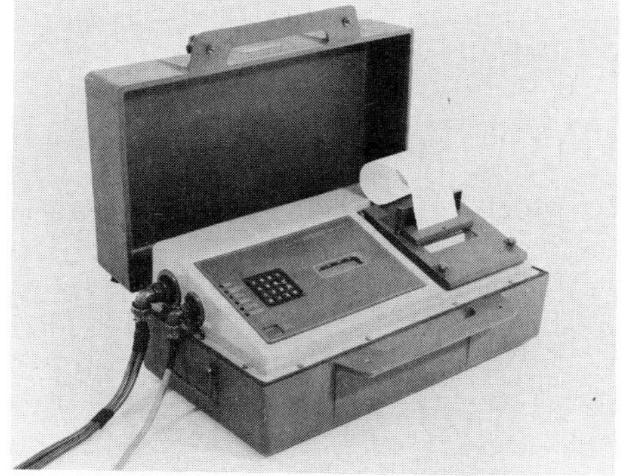

EMVA Mk 10B processor cabinet contains microprocessor, operating controls for system and printer for analysis results

These samples are then evaluated, processed and used to calculate to two decimal places the velocity of the projectile as it left the muzzle of the gun.

This velocity value, which is presented on the liquid crystal display, and printed out within seconds of the round being fired, is the non-standard instrumental velocity. Readout of the standard instrumental velocity may also be obtained provided that the gun tables for any particular gun or guns are known and are pre-programmed by the factory into the processor memory. Similarly the effect of gun wear, elevation and meteorological conditions may also be taken into account. In the rapid fire mode the operator need only press a single key to instruct the processor to assume that all the data entered for the previous round will be the same for the next and to automatically increment the round number.

CHARACTERISTICS

Velocity range: 30 – 3000 m/s

Accuracy: ±0.05%

Power requirements: 24 V DC: 2 A average

Dimensions

Antenna head: 305 × 145 × 165 mm

Doppler radar head for EMVA Mk 10B muzzle velocity measuring analyser

Processor unit: 650 × 380 × 290 mm

Accessory box: 720 × 520 × 270 mm

Tripod: Adjustable height from 600 – 1500 mm

Weight

Antenna head: 9 kg

Processor unit: 30 kg

Accessory box: 20 kg

Tripod: 25 kg

DEVELOPMENT

Developed in the Republic of South Africa.

STATUS

In production and in service with the South African Army.

CONTRACTOR

Global Chemicals (Pty) Ltd, PO Box 5008, Cape Town, 8000, Republic of South Africa.

SPAIN

6709.181

FELIS ANTI-AIRCRAFT FIRE CONTROL SYSTEM

Felis (formerly Linca) has been designed to meet a Spanish Army requirement to modernise the short range air defence system using the Bofors 40/70 gun. Initially the system, as Linca, was developed by CETME assisted by the Defence Ministry's Comision Asosora de Investigacion y Technolgia but Inisel has now taken over full responsibility for it. The objectives are to produce a cost effective fire control system whose features will include quick reaction time and easy maintenance. The manufacturers claim the basic system, without radar, will cost $750,000 which, they claim, will be six times cheaper than comparible equipment.

The system consists of a fire control computer, an electro-optical tracker/designator with laser rangefinder and a radar interface all mounted in a trailer which weighs approximately 2.2 tonnes and can be towed by a Land Rover type vehicle. The fire control computer incorporates two Intel 16-bit microprocessors and one Intel 8-bit microprocessor.

The designator features a TV camera for daytime use, a ND YAG laser rangefinder and an optional thermal imager. It can turn 360° in azimuth at 85° per second. In elevation it can turn from −5° to +85° at a speed of 50° per second. The TV camera can be used either manually or automatically and the picture is shown on a 20.3 cm monitor.

The system has three operating modes; radar acquisition, autonomous electro-optic and pre-determined electro-optic. In the first data from the radar, usually the Contraves LPD-20 radar (see **1529.153**), produced under licence by Inisel, initiates the optical automatic tracking sequence. In the second the electro-optical designator initiates tracking and in the third a pre-determined scanning pattern is carried out by the electro-optical designator until the target is located.

STATUS

Prototype delivered to the Spanish Army in 1987 for technical and tactical evaluation. Inisel anticipates a contract for 100 systems which will be delivered between 1988 and 1993.

CONTRACTOR

Inisel SA, Sor Angela de la Cruz 6-13, Madrid 28020, Spain.

6366.161

SEIMOR MORTAR COMPUTING SYSTEM

The SEIMOR hand-held mortar computing system has been designed to compute firing data for all types of mortar and some artillery fire functions.

For low power consumption and greater reliability CMOS technology has been employed.

SEIMOR uses the NSC-800 micro-processor which has all of the advantages of the Z80 and 8085 so enabling a very compact hardware design and high speed calculations.

It has 64 K of memory (NSC-800) and uses 32 K for the ROM and 8 K for the RAM, with the remainder being available for future applications.

The aluminium case is protected by an epoxy resin outer casing with the touch keypad being made of polycarbonate and is highly resistant to damage, including being dropped from 2 m and immersed in 4 m of water.

Its EPROM (Electrically Programmable Read-Only Memory) can store position data for 80 targets, 12 observers, 42 mortars and nine fire procedures.

The display on top part of the SEIMOR shows type of mortar bomb, range, charge, azimuth and elevation, angle of deflection and time of flight. Two fire missions can be prepared simultaneously.

The batteries can be changed without loss of memory and data is not lost when the computer is switched off. A printer connection facility is provided.

CHARACTERISTICS

Size: 90 mm × 60 mm × 180 mm
Weight: 500 g
Operating temperature range: −20 to +70° C

STATUS

In production. In service with Spanish Army.

CONTRACTOR

Seidef SA, Pso. de la Castellana 140, 28046 Madrid, Spain.

The SEIMOR mortar computing system

SWEDEN

2378.181

BOFI ANTI-AIRCRAFT FIRE CONTROL SYSTEM

BOFI is an all-weather fire control instrument of multi-sensor type developed by Bofors for their 40 mm L/70 gun (**5528.103**) of which large numbers have been produced. The equipment is fully integrated with the gun and each gun then becomes an autonomous firing unit that can combat its targets individually.

The equipment consists of a computer, an operator's panel with joystick, sights for day and night use, laser rangefinder and a tracking radar. The major portion of the equipment is installed in the operator's cab. The main sensor of the multi-sensor fire control is the Ku-band (J-band) radar, using the MTI function during acquisition and frequency agility during automatic tracking to increase the accuracy. Tracking observation can be done on an A-scope or by using the sights. The optronic sensors consist of a 10 Hz neodymium laser rangefinder, a day sight and a night sight with a three-stage light amplifier. The sensors can be used redundantly to give maximum performance and resistance to jamming.

The computer is of hybrid type which calculates the target co-ordinates continuously as well as the target's future position in a cartesian co-ordinate system. External ballistic corrections are set on the operator's panel and are compensated for in the computer's calculation of super-elevation and wind drift etc.

A central search radar is used acting as a combat control centre for a battery of BOFI guns. An optical target indicator is incorporated in the system for designation of targets visually detected at the gun site. The BOFI equipment is designed in such a way that it can also be used on existing 40 mm L/70 field guns.

Bofors 40 mm L/70 anti-aircraft gun with BOFI anti-aircraft fire control system fitted

DEVELOPMENT

Development work and system tests of the all-weather equipment were completed in 1979 and further details will be found in **2349.131** in the Mobile Gun and Rocket Systems section of this volume.

STATUS

The basic version without tracking radar was tested during 1972-73 and put into series production in 1976. The system has been exported to several customers.

CONTRACTOR

AB Bofors, S-691 80 Bofors, Sweden.

6444.181

LINDA C_3I ARTILLERY SYSTEM

LINDA is an integrated C³I system of modular design incorporating both software and hardware modules. This concept offers considerable scope for combining modules and components to create customised systems to meet individual requirements.

The software modules include fire control, intelligence, tactical and fire planning, deployment of units, logistics and data communications. The hardware components include forward observer instruments, fire direction centre (command and control units), battery fire control computers, RIA gun sight or gun display unit (GDU).

The fire direction centre (FDC), whose main operational functions are fire direction and command planning, is the heart of the system. The FDC is mounted either in Lapplander vehicles or mobile shelters with communications by DLT 200 data link terminal. Battalion FDCs will have a single console, brigade FDCs will have two consoles and two brigade FDCs can support a division.

The equipment in the FDC is designed to provide commanding officers with extremely accurate pictures of the tactical situation presented in the most ergonomic manner. The data base for the P857E computer contains information on both friendly and hostile units, logistics, terrain and other relevant factors. A typical brigade FDC will include a monitor with keyboard for map display, a monitor for either large-scale graphics or maps, a graphics tablet and a MIA fire control computer and communication unit (see entry **6445.161**).

The Philips VP 415 Laservision display system is unusual in that it incorporates video discs. Each side of the disc can store and retrieve up to 54000 different maps in a matter of seconds. The maps are displayed on Barco 5137 colour monitors with the locations of military units, targets, area zones and other information superimposed with the aid of a DEC VSV 90 graphics controller.

A version of the system is being developed to provide shelter-mounted command and control units for the Swedish Army these being available in several configurations including intelligence command centre, divisional intelligence unit and a ground station for RPVs. Each features a DEC DRAC (Digital Equipment Ruggedised Army Computer) based upon the MicroVax II minicomputer with Ada software.

The VP 415 Laservision and Barco 5137 system are retained but the units will also feature DEC hardware including the VT 241 text terminal, RD 53 Winchester disc drives, LA 50 printer, LVP 16 plotter and CalComp 2000 graphics tablet. All the subsystems are linked by a DECnet network and PEAB are developing interfaces with the Ericsson TR 8000 combat net radio.

STATUS

The LINDA artillery C[3]I system is under development in two versions; 9FA 400 for the Swedish Army and 9FA 500 for export. Prototype command and control units began a two-year trial with the Swedish Army late in 1987.

CONTRACTOR

Philips Elektronikindustrier AB, 175 88 Jarfalla, Sweden.

9FA 400 LINDA C³I system installed in a 6 × 6 cross-country vehicle

4307.161

9FA 302 FIELD ARTILLERY COMPUTER

This system is developed from the 9FA 301 which is the standard Swedish Army fire control computer in use with the Bofors 155 mm FH-77A field howitzer. The 9FA 302 features software improvements, improved temperature range and higher capacity solid-state memory.

The 9FA 302 is a field artillery computer designed for battery-level operations. Its main application involves computing azimuth, elevation, fuze settings and propellant charges for up to eight guns. Discrete firing parameters are computed for each gun and normally transmitted directly via a data link to the RIA gun sight (**3144.193**). Firing parameters can also be transmitted directly to a gun display unit. All guns receive firing parameters within five seconds of inputting the complete set of target condition data.

For further details see entry **4307.161** in the 1987-88 edition.

CHARACTERISTICS

Width: 580 mm
Height: 420 mm
Depth: 750 mm
Weight: 45 kg
Power: 21-30 V DC approx 200 W
Operating temperature range: –30 to +60°C
Operational data
Battery emplacements: 3
Guns per battery: 8 (max)
Projectile types: 8 (max)
Charges: 10 (max)
Target memory: 100
Firing zone limits: 5 areas

STATUS

Production complete. In service with Swedish Army.

CONTRACTOR

Philips Elektronikindustrier AB, 175 88 Järfälla, Sweden.

PEAB 9FA 302 fire control computer

6445.161

MIA FIRE CONTROL COMPUTER AND COMMUNICATION UNIT

MIA has been developed in close co-operation with the Swedish Army both as a component of the LINDA system (**6444.181**) for use in Fire Direction Centres as a console, as a fire control computer both for batteries (replacing the 9FA 301) and independent guns/rocket launchers. Both in this role and as a forward observer's instrument it is operated by one man.

In the forward observer role MIA can process data for three separate fire unit missions simultaneously. This permits three observers to direct fire against three separate targets or a single observer to call down simultaneous fire on three separate targets.

As a battery fire control computer it can provide ballistic data for a six-gun battery and two more batteries or it can calculate the individual trajectories for up to eight guns. Different kinds of guns and mortars can be included in the battery while ballistic data for up to ten guns and mortars, together with associated ammunition, can be held. The ballistic data base can easily be updated if new weapons or ammunition are introduced.

MIA Fire Control Computer

Fire orders are entered in the MIA and held in a file. Each fire order includes the time, forward observer and the number of guns to be used to provide the operator with a clear picture of the engagement sequence and resources allocated. Azimuth, elevation, fuze setting and charge are then computed with trajectory integration for each gun to ensure both maximum accuracy and flexibility in deploying individual guns. The propellant charge can be selected manually or automatically.

The system has a target data memory with a capacity of 100 targets, it can store two sets of meteorological data plus standard meteorology and it can compute and monitor six firing zone limits.

The data is transmitted to the guns by cable and displayed automatically on the RIA gun sights (**3144.193**). Alternatively the data may be presented on a discrete display at guns equipped with conventional sights or read out in the conventional manner, although the latter should be regarded as a back-up procedure.

MIA may also be used as a communications terminal with the aid of the appropriate external communications system. It can be used to receive up-to-the-minute orders and intelligence data from command centres. Situation, intelligence and logistic reports can be transmitted to command centres or other MIA terminals.

The system is not dependent upon standard formats for message transmission. An incoming message will trigger an alert to advise the operator of its arrival and will then be stored in a file to be called up when the operator is free. Transmitted messages are stored in another file which is usually used for reports which are updated when fresh data becomes available.

The terminal has a large, clear display unit sub-divided into a working area of 16 lines each of 64 characters and a message field with clock and alert signal for incoming messages. Operation is by man-machine dialogue with ballistics and fire control being the main items of the menu. The operator selects the function on the basis of information displayed in the working area and the options available at each level are continuously shown. For speed most dialogue functions are replaced by function keys. A help function is available to provide information on different features and error message codes are also displayed.

CHARACTERISTICS

Size

Width: 340 mm

Depth: 320 mm

Height: 130 mm

Weight: 5 kg

Display: LED

Lines: 16

Characters per line: 64

Keyboard: Diaphragm type with full ASCII character set and dynamically controlled function keys

Voltage: 24 V DC

Power consumption: 10-15W depending on operating conditions

STATUS

Preparations for serial production

CONTRACTOR

Philips Elektronikindustrier AB, 175 88 Järfälla, Sweden.

3313.181

9AA 100 FIRE CONTROL SYSTEM

The 9AA 100, formerly KALLE (**2468.181**), is a light anti-aircraft gun control system developed by Philips Elektronikindustrier AB, as main contractor, in partnership with the Aerospace Division of Saab-Scania who are responsible for the optronic technology. It is intended for fixed or mobile installations controlling typically, two guns of 30 to 40 mm calibre.

The basic system comprises a combined TV director and laser rangefinder assembly, a computer, a display and control unit, and data transmission circuits and adaptors for the guns. The TV director is manually laid for target acquisition, after which it tracks automatically by contrast tracking and with the laser rangefinder provides full target co-ordinates for the computer. The computer is a general-purpose mini-computer and is the same as that used in other PEAB fire control systems, 9 LV 200 Mk 2, 9KA 400 etc.

Controls on the display and control unit include director controls for acquisition and tracking modes and provision for gun supervision and control. Displays include a TV monitor for presentation of the picture and target data in alpha-numeric form. System options include an IR-camera and associated tracking circuits to give night-time capability. It is also produced in a shipboard version, called 9 LV 100 (**3185.281**).

STATUS

In service with Swedish forces.

CONTRACTOR

Philips Elektronikindustrier AB, 175 88 Järfälla, Sweden.

In co-operation with:

Saab Missiles AB, 581 88 Linköping, Sweden.

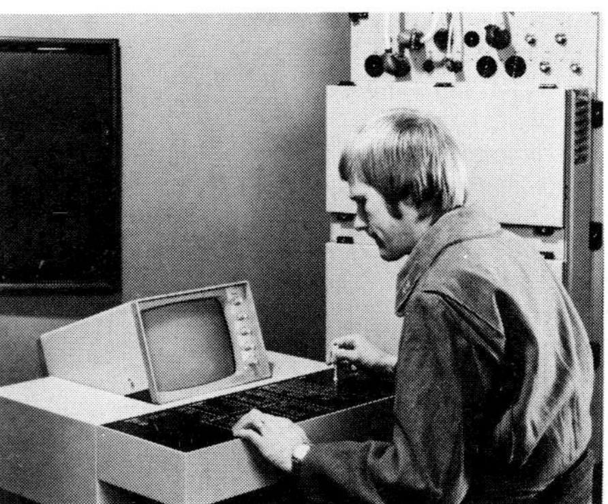

9AA 100 control and display console

4823.181

9AA 200 FIRE CONTROL SYSTEM (TRACKFIRE)

TRACKFIRE is a combination of the optronic system 9AA 100 (**3313.181**) and PEAB's Ku-band tracking radar. It is a frequency agile radar, also capable of operation in MTI-mode. The same radar is also used for the BOFI system and for updating of Super-Fledermaus.

The system is modularly constructed and can be equipped with any combination of following sensors: TV-camera, Laser range-finder, Ku-band tracking radar and IR-camera.

Acquisition of air targets is implemented automatically on designation from a search radar or the optical target designator included in the system.

Air targets are tracked by the tracking radar or the automatic TV-tracker and the laser range finder or different combinations of the sensors. The PEAB developed TV-tracker is using a combination of tracking methods and works independently of the radar tracker. It can also be used for tracking of a second target being within field of view of the TV-camera (IR-camera).

In the basic version the director is mounted on a lightweight cross-country trailer and the control and display units are installed in a shelter.

The system can command up to four guns or missile systems. Gun laying data is transmitted over a simple two-wire data link and converted to synchro signals at the guns.

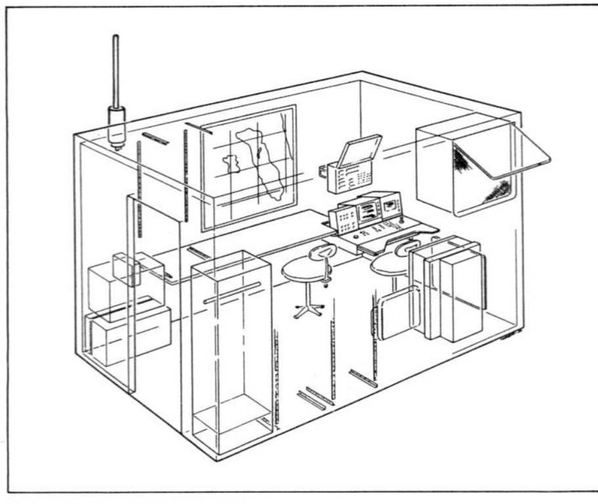

TRACKFIRE fire control centre

STATUS

In production with first deliveries in spring of 1987. This TRACKFIRE is featured as a standard sub-system for the ARTEMIS 30 air defence system (**4511.181**) in production for the Greek Armed Forces.

TRACKFIRE director with tracking radar, TV and laser rangefinder

CONTRACTOR

Philips Elektronikindustrier AB, 175 88 Järfälla, Sweden.

4274.181

BOFORS AFV FIRE CONTROL SYSTEM

The latest Bofors Aerotronics fire control system for armoured fighting vehicles originates from the FCS developed for the Swedish Ikv-91 tank destroyer, described in **2399.181** in earlier editions of *Jane's Weapons Systems*, and evolved through the Type FV FCS series (**3318.181**).

One system comprises a day/night optical sight with a passive image intensifier and a laser rangefinder; and an independent line-of-sight for the gunner and an accurate ballistic and lead-angle digital computer give the system the necessary high technical and performance characteristics needed for the engagement of moving targets from a moving vehicle. This version is suitable for retro-fitting to existing AFVs and can be modified for use in updating many tanks. As an alternative, the gunner's sight can be specified as a day laser sight.

The system has been configured to accept an integrated stabilised commander's sight and provisions for the commander to align the gunner's line-of-sight with his cupola position.

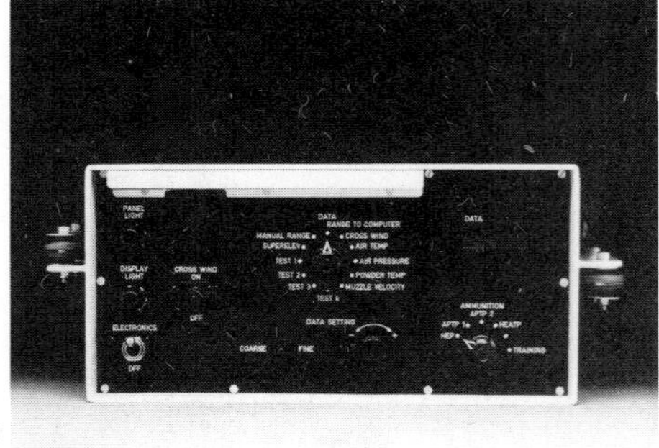

Computer/control panel of Bofors Aerotronics tank FCS

CHARACTERISTICS

Gun sight

Magnification: × 8.5 (night); × 7 (day)

Field of view: 5.3° (night); 9° (day)

Bofors Aerotronics gunner's day/night sight

Input pupil diameter: 150 mm (night); 50 mm (day)
Exit pupil: 7 mm
Computer
Range: 200 – 6000 m (laser and manual)
Number of ballistics: 5
Air pressure range: –300 to +100 mm Hg
MV range: ±50 m/s
Air temperature: –30 to +50°C
Powder temperature: ±40°C
Crosswind: ±20 m/s
Head wind: ±30 m/s
Max superelevation and negative lead angle in elevation: +70 mrad/–10 mrad
Lead angle: ±40 mrad
Max lead angle and wind correction: ±45 mrad
Jump max: ±5 mrad
Trunnion tilt, max: ±14°
Control System
Max azimuth speed: ±300 mrad/s
Max azimuth speed of target: ±80 mrad/s
Max elevation speed: ±80 mrad/s
Max elevation speed of target: ±20 mrad/s
Laser Rangefinder
Wavelength: 1.06 μm
Pulse power: 1.8 MW
Pulse duration: 15 – 20 ns
Time between measurements: 2 s
Effective input pupil of laser receiver: 130 mm
Maximum range: 6 km (typically)
STATUS
In service with Sweden and another country.
CONTRACTOR
Bofors Aerotronics AB, 181 84 Lidingo, Sweden.

6710.181
BOFORS TYPE U FIRE CONTROL SYSTEM
Bofors Aerotronics are developing a modular day/night laser fire control system for use in both light armoured vehicles or anti-aircraft guns. The Type U is intended to be a cost-effective, independent line-of-sight system with high performance and wide coverage.

It will be suitable for retrofitting and its modular stabilisation system will enable it to be added to existing control systems. The Type U will provide a short reaction time between the detection of the target and the beginning of the firing sequence.

CHARACTERISTICS
Anti-aircraft version
Gun sight
Magnification: × 7
Field of view: 8°
Elevation range for line-of-sight: -20° to +60°
Max line-of-sight azimuth deflection: ±20°
Computer
Range: 200-6000 m (laser and manual)
Number of ballistics: 15 (3 quick choice)
Control system
Max azimuth speed: +1.5 rad/s
Max elevation speed: ±1.5 rad/s
Laser Rangefinder
Wavelength: 1.06 μm
Beam divergence: 1.5 mrad
Max range: 6 km (typically)
Tank version
Gun sight
Magnification: × 4,7 (night) × 7 (day)
Field of view: 4.6° (night) 8° (day)
Max line-of-sight azimuth deflection: ±10°
Computer
Range: 200-6000 m (laser and manual)
Number of ballistics: 15 (3 quick choice)
Trunnion tilt max: ±14°
Control system
Max azimuth speed: ±300 mrad/s
Max azimuth speed of target: ±80 mrad/s
Max elevation speed: ±80 mrad/s
Max elevation speed of target: ±20 mrad/s
Laser rangefinder
Wavelength: 1.06 μm
Max range: 6 km (typically)
STATUS
Both versions are at the prototype stage.
CONTRACTOR
Bofors Aerotronics AB, 181 84 Lidingo, Sweden.

Bofors Aerotronics Sight U (tank version)

1548.181
9KA 400 KASTELL COAST FIRE CONTROL SYSTEM
The 9KA 400 is the modern, digital version of the well-proven coast artillery fire control unit MARELD. It is primarily intended for fixed coast artillery or missile batteries. It can also be used as a sensor in the 9CSI 600 STINA naval command system (**4027.281** and **4028.281**).

Sensors are an I/J-band frequency agility radar with a continuously scanning antenna, and a combination TV-camera/laser rangefinder on a separate pedestal. The radar antenna has a very narrow horizontal beam, 0.6°, to give high resolution in azimuth for fall-of-shot observation, and high jamming resistance. Multiple targets are tracked by track-while-scan methods using digital extractor and Kalman filter techniques.

The TV/laser sensor is remotely controlled by the operator and the information is displayed on a TV-monitor and a digital read-out for the rangefinder. The digital computer is of the same basic type as in the 9 LV 200 Mk 2 (**3127.281**) naval fire control system.

Target data, position, and speed vector are generated and fed to ballistic computer units for calculation of gun-laying data. The guns can be dispersed over a large area remote from the fire control unit, and can also receive target data from other fire control units in the vicinity.

The 9KA 400 Kastell can be supplemented with a self-contained electro-optical fire control director using telescopes and a laser for high accuracy computer-aided tracking of one target. Infra-red or image intensifier sensors can be added to obtain all-weather capability.

STATUS
The original 9KA 400 has been operational in large numbers in Sweden, and other countries, since the mid-1960s. The new version entered service in late 1976 and is now in operation and a newer generation Mk 2 version is now in production. At least 70 systems have been ordered for use in Sweden and four other countries including Norway and Spain. Within Spain the 9KA 410 has been produced under licence by Inisel.

CONTRACTOR
Philips Elektronikindustrier AB, 175 88 Järfälla, Sweden.

PEAB 9KA 400 Kastell radar antenna

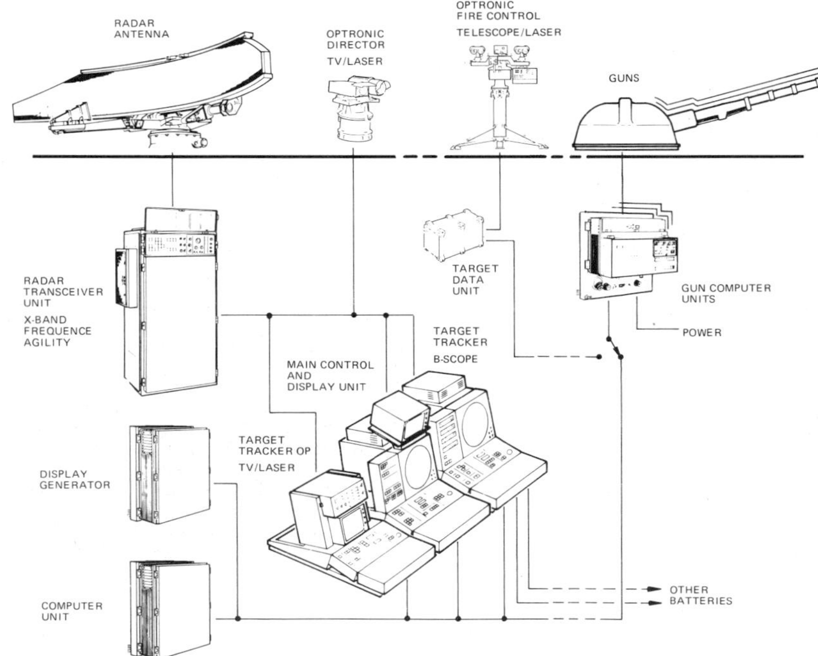

Coast artillery system 9KA 400 Mk 2

4318.281

9KA 500 KARDINAL COASTAL FIRE CONTROL SYSTEM

The 9KA 500 Kardinal mobile coastal artillery fire control system is derived from the well-proven 9KA 400 (**1548.181**) which has been operational in Sweden and other countries since the mid-1960s. All operational functions of the 9KA 400 are retained and many operational and technical features are added. The major new ones are:

(1) higher mobility through reduced weight
(2) greater deployment flexibility through separation into a sensor unit and an operations unit
(3) additional combat control functions, ie improved presentation of the tactical situation and automatic tracking of multiple targets
(4) the 9KA 500 Kardinal can easily be supplemented with functions for engagement with missiles and some of those functions are already implemented.

The sensor unit is a single-axle trailer with a hydraulically-operated mast. The top of the mast houses the radar antenna and, below the antenna, an optronics unit. The optronic sensors are: a TV-camera and a laser rangefinder. The turning gear for the optronics is an 'around the mast' design giving both the radar and the optronic sensor a full 360° field of view. Also mounted on the sensor trailer is the radar T/R unit, a hydraulic pump, and a motor-generator, making the trailer a self-contained unit, remotely controlled from the operations unit. The mechanical design of the trailer is very rigid in order to achieve the stability necessary for fire control purposes, ie fractions of a milliradian even in strong winds.

Two versions of the radar, transmitting in the X- and the Ku-band respectively, are available. In both versions the well-known spin-tuned magnetron is used, giving wide-band frequency agility with the same benefits as in the 9KA 400. The Ku-band version gives the high angular resolution as in the 9KA 400, ie 0.6°, and yet is lightweight and highly mobile. This particularly high resolution is valuable in applications where separation of targets and fall-of-shot observations ('splash-spotting') is required.

The transmitter power at Ku-band (65 kW) is lower than the power at X-band (200 kW) but the shorter wavelength at Ku-band suffers less from the 'beam-lift' due to multipath effects. Thus the detection and tracking ranges against ship targets are the same for the two radars when deployed at such altitudes, and they are both effectively limited by the curvature of the earth.

The sensor unit and the operations unit can be separated by a distance of some kilometres. Transmission of radar and TV video, data, commands, and logic signals between the two units takes place over a lightweight fibre optical link, which greatly facilitates cable laying compared to conventional cables. Furthermore, the fibre can be used in full duplex in both directions by the use of dual wavelength transmission with dichroic prisms in the terminal equipment.

The set-up of displays and control consoles in the operations unit is similar to the 9KA 400. Like that, the number of consoles and their functions can be easily tailored to suit various operational doctrines and functions, and the weapons to be controlled, either guns and/or missiles.

The fixed installation version is called the 9KA 400 Mk 3 Kastell and this has been sold to Norway for use with their coastal artillery units.

9KA 500 Kardinal coastal artillery fire control system deployed

STATUS
In production.

CONTRACTOR
Philips Elektronikindustrier AB, 175 88 Järfälla, Sweden.

6711.181

9KA 100 COASTAL ARTILLERY FIRE CONTROL SYSTEM

The 9KA 100 FCS consists of a well proven self-contained electro-optical fire control sight (KOBRA) and a ballistic computer. The sight uses telescopes, a laser rangefinder and a servo-controlled turntable for high accuracy computer-aided tracking. Infra-red or image intensifier sensors can be added in order to obtain all-weather capability. The ballistic computer can be either a battery ballistic computer serving up to four guns individually or serve each gun in the battery. The fire control sight is used not only to measure the range and bearing to a surface target but also to observe and measure the fall-of-shells. Fire control data (target data and fall-of shell corrections) are automatically transmitted to the ballistic computer unit(s) via a two-way, two-line link. As a supplement or replacement for the KOBRA an extremely lightweight fire control sight (KOLIBRI) can be used for short and medium range coast artillery batteries.

The ballistic computer receives fire control data and performs parallax compensations and ballistic calculations and delivers individually gun laying data as output. The ballistic computations are based on ballistic tables containing projectory and several ballistic parameters. Atmospheric ballistic parameters may be inputted as ballistic values or as a complete altitude forecast with up to ten altitude layers. In the later case the ballistic computer automatically computes the ballistic parameters for pressure, wind and temperature.

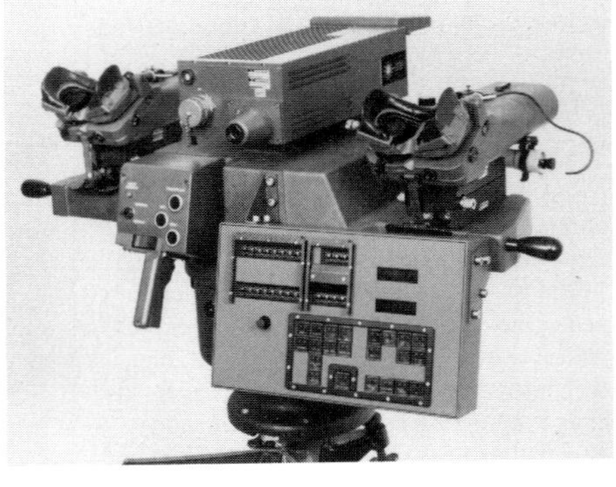

The Kobra electro-optical fire control sight

STATUS
The 9KA 100 system is used as a supplement to the 9KA 400-500 systems but also as initial refitting programme for older coastal artillery batteries, fortified or mobile. At least 20 systems have been ordered for use in Sweden and other countries.

The Kolibri lightweight sight which in Swedish service is designated ARTE 734

CONTRACTOR
Philips Elektronikindustrier AB, 17588 Järfälla, Sweden.

6353.161

SATT UPDATE PACKAGE FOR ANTI-AIRCRAFT FIRE CONTROL SYSTEM

The SATT update package includes software and involves replacing obsolescent electronics and virtually all electro-mechanical components with state-of-the-art technology. These packages are flexible, modular designs featuring micro-processor controlled solid-state electronics.

The packages are adaptable for most fire control systems and have already been used to update the Super-Fledermaus and the Swedish Cig 760. The Super-Fledermaus anti-aircraft fire control system (**2376.181**) was designed over 25 years ago and is still in service all over the world.

Super-Fledermaus anti-aircraft fire control system with SATT update package including Philips radar

Each update package contains a complete fire control computer, interfaces for associated equipment, built-in trainer with 1000 software scenarios, fire control officer's terminal, and gun data receiver with a hand-held gun control terminal for each gun linked to the system.

Because interfaces are available in several options, the capability exists for using the upgrade kit in combination with different types of search and tracking radars, laser rangefinders, muzzle velocity measuring equipment and similar devices. Any type of machine controlled anti-aircraft gun can be connected to the equipment using a two wire link.

The upgraded system will feature the following performance factors: five second reaction time; an 88% hit probability from a 10-round burst at a range of 2000 m; non-linear acceleration prediction capability; improved reliability with an MTBF of more than 500 hours and an MTTR of less than 35 minutes; faster deployment and set up times.

In 1986, prototypes of the updated Cig 760 and Super-Fledermaus respectively were delivered to the Swedish Army and to the Danish and Norwegian Air Forces. In 1988, updated prototypes will be delivered to two other countries. In conjunction with the deliveries, the customers have carried out final acceptance tests with good results. Contracts have already been signed for serial delivery during the period of 1987-1990 to Sweden, Denmark, Norway and one other country.

In addition to the SATT update package, the concepts for the various customers include additional equipment from other suppliers. Philips Elektronikindustrier AB, Sweden, will deliver to the Danish Air Force their KU band fire control radar with frequency agility function. Ericsson Radio Systems AB, Sweden, have updated their fire control radar to incorporate a mono-pulse function for delivery to the Norwegian Air Force and one additional customer.

Ericsson are also manufacturing a laser rangefinder with a range of 8 km for the Swedish Cig 760. Most concepts also include a new muzzle velocity radar from OPOS Electronics AS, Denmark, with an accuracy of 0.2 per cent and a measuring range of 900-1100 m/s.

STATUS

Development of this package started in 1983 in response to customer requirements to upgrade existing air defence systems. Now in production. In service with Danish Air Force.

CONTRACTOR

SATT Communications AB, Box 32701, S-126 11 Stockholm, Sweden.

SWITZERLAND

3314.181
FIELDGUARD ARTILLERY FIRE DIRECTION SYSTEM

Fieldguard is a modern and advanced all-weather Fire Direction System for cannon artillery and rocket artillery. It detects the influences of weather and weapon conditions on the trajectory by radar tracking rounds or rockets in flight. The best firing data is then immediately determined by the Fieldguard computer and transmitted automatically to the weapons.

Fieldguard is compact and modularly designed. This gives a high flexibility to mount Fieldguard on any wheeled or tracked vehicle.

The system is self-contained. It consists of radar tracking unit, digital computer, operator console for man-machine dialogue, digital data transmission with weapon display units, and power supply.

Fieldguard communicates with various kinds of sensors such as drones, battlefield radars, and forward observers.

Fieldguard receives a firing order with target co-ordinates. The computer calculates the firing data for the basic gun or the ground launcher, based on the stored ballistics, and transmits it automatically to that weapon display unit. The so-called pilot firing now starts. Up to three pilot rounds are fired, tracked by radar and destroyed in flight in order not to alert the target.

Fieldguard extrapolates the remainder of the trajectory and computes the firing data of the adjusted trajectory, individually for each weapon.

The fire for effect can be released and the target is under accurate and surprising massed fire.

Without previous pilot firing Fieldguard can check and update best firing data by tracking one round within a volley or one rocket within a ripple of fire for effect.

Forward Observers are equipped with portable target location equipments and connected to Fieldguard by radio or field cable. The target co-ordinates will be located by the laser-goniometer and transmitted to Fieldguard by the Digital Message Device (DMD).

In short, Fieldguard:

- operates under all-weather conditions;
- improves accuracy of fire for effect;
- improves accuracy of unobserved counterfire;
- effects full surprise of enemy;
- renders considerable savings in time;
- minimizes ammunition consumption;
- operates without meteorological information;
- renders considerable savings in personnel.

STATUS

In production and in service with at least six countries including West Germany who deploy it with the 110 mm Light Artillery Rocket System. The system was ordered by Thailand in 1987 and is known to be used with the Avibras ASTROS II (**4146.111**) artillery rocket system.

CONTRACTOR

Contraves AG, Schaffhauserstrasse 580, CH-8052 Zurich, Switzerland.

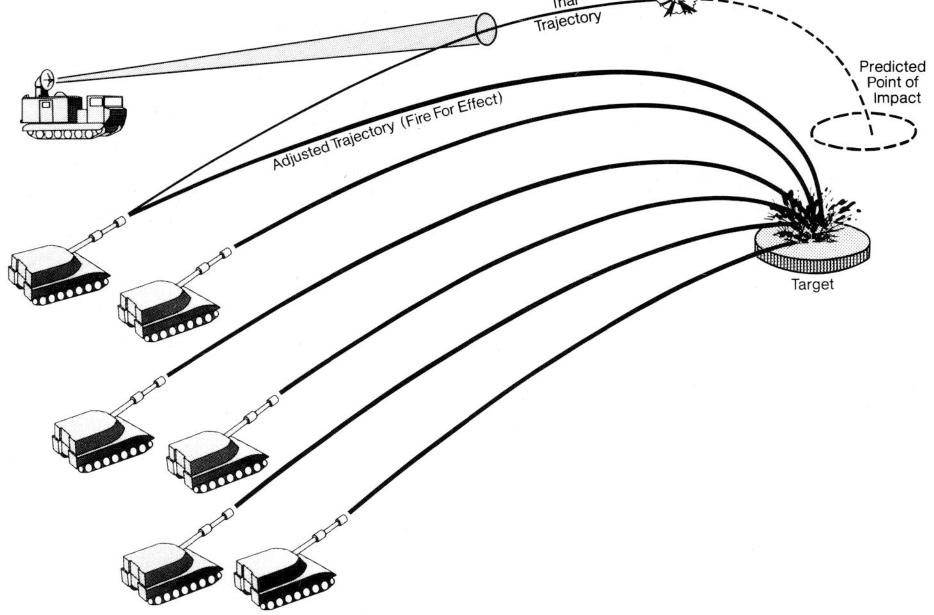

Diagram of Fieldguard pilot firing technique

Fieldguard system deployed operationally

2377.181
SKYGUARD ANTI-AIRCRAFT FIRE CONTROL SYSTEM

Skyguard is the fire control system that Contraves has developed as the successor to the Super Fledermaus (**2376.151**) system that they have marketed so successfully in various versions for many years.

Major differences between Skyguard and its predecessors – apart from a general modernisation of design techniques – are the provision of two radars (one for search and one for track) in place of one and the use of a digital instead of an analogue computer. Another important difference is the use of pulse-doppler type radars in place of pulse radars.

Main system components are an Ericsson UAR 1021 search and tracking pulse-doppler radar (**2357.153**), a TV tracking system, a digital computer, a control console, and a power supply system. The whole is contained in a fire resistant reinforced fibreglass-polyester cabin that can be mounted on a trailer or a wheeled or tracked prime mover, and is air transportable. The power supply system is built into the main equipment but can be removed for external operation and is automatically refuelled directly from cans.

A closed-circuit TV system provides for optical tracking and automatic TV tracking. An optical sight on a rotating chair is provided for visual target acquisition with provision for target indication by means of a flashing strobe on the PPI.

As part of the programme to augment the combat value of the Skyguard, Contraves has designed various technical innovations and developed accessory equipment. These primarily assure optimal acquisition, tracking, and combat of low-flying aircraft under extreme topographical conditions. Besides X-band radar and TV-tracking, Skyguard is now available with fully coherent Ka-band radar. This ensures improved target resolution in terrain where the use of radar is very difficult. At the same time it increases the redundancy of the automatic tracking possibilities. With the additional Ka-band radar Skyguard remains easy to operate. The Ka-data also appears on the Combat Display. The Ka-band radar can be easily and quickly retrofitted to all existing Skyguard fire control units.

The Contraves computer operates in real time to perform threat evaluation functions, calculate ballistic data for guns and command signals for missiles, aid the target tracking operation and monitor, and check-out the entire system.

In addition to the PPI and R-trace displays already mentioned, the control console incorporates a tactical display with numerical read-out, the TV tracking monitor, rolling-ball control for PPI markers, joystick control for manual tracking, and a matrix panel for data input and output.

OPERATION

A search radar data extractor (SRDE) by Contraves AG in Zurich is available and the system can be integrated into Skyguard's fire-control unit. It enhances the degree of automation and the time required for target acquisition and tracking is shortened and threat evaluation automated.

The SRDE system is used for the following main tasks:

(1) Automatic first-target alert coupled with automatic target acquisition.

As soon as the first-target alert is acquired by the search radar, operators are automatically alerted. Thereupon, the target is automatically passed on to the tracking radar which keeps on tracking it. Despite very high clutter suppression of the Skyguard's pulse doppler search radar, alerts will unavoidably be triggered also by false targets in the shape of moving objects (motor cars, ventilating fans, etc). Such false targets are specifically marked and can at any time be wholly or partially erased at operator's discretion.

(2) Automatic threat evaluation regarding targets in the combat area or those already acquired by the search radar.

By the track-while-scan system it is possible to simultaneously track several targets as well as to evaluate automatically the respective degree of threat. Skyguard's digital system computes, among other things, the threat degree for each individual target, based on the angle of the approaching target, velocity, range, etc. Targets are shown simultaneously on the PPI in the order of priority corresponding to threat involved.

As a matter of course, the data supplied by Skyguard's integrated IFF unit are also included in the threat evaluation.

Skyguard is able to control medium calibre guns and/or guided missiles. The Contraves-designed missile launcher uses the proven undercarriage of the Oerlikon 35 mm gun and carries four Sparrow or Aspide missiles in containers, as well as the target tracking and illuminator radar antenna. The launcher is directed either remotely from the Skyguard system or locally by the launcher operator.

For crew training, there are two different simulators available: one is integrated in the system, while the larger one is an additional unit connected to it (Training Simulator 2).

STATUS

Skyguard was developed by Contraves as a private venture project and, after evaluation, it has been adopted by the armed forces of more than 20 countries including Austria, Germany (Federal Republic), Spain and Switzerland, with more than 420 systems having so far been ordered.

Most customers use the system with the Oerlikon GDF twin 35 mm gun (**4537.131**) but it is also used with the Bofors 40 mm L/70. The Italian Army selected Skyguard for use in the Spada point defence missile system (**2250.131**).

CONTRACTOR

Contraves AG, Schaffhauserstrasse 580, CH-8052 Zurich, Switzerland.

Skyguard (left) being used to control two twin 35 mm Oerlikon-Bührle towed anti-aircraft guns with two four-round Sparrow launchers

Contraves Skyguard anti-aircraft fire control system deployed

2376.181

SUPER-FLEDERMAUS

A member of the series of anti-aircraft fire control systems produced by Contraves in Switzerland, the Super-Fledermaus Type D IX with MTI is a radar optical tracking and computing system suitable for use with medium calibre anti-aircraft guns.

The main equipment is compactly mounted on a four-wheel cross-country trailer and comprises a radar/visual tracker for target acquisition and determining the firing data for up to three separate gun emplacements, and a muzzle velocity indicator for measuring the muzzle velocities of up to three guns. A separate power supply unit, mounted on a two-wheel trailer, generates power for the fire control unit. Auxiliary equipment are a signal box for use by the fire control officer and an optical putter-on for rapid visual acquisition of unexpected targets.

A detailed description of the system was published in the 1970-71 edition of *Jane's Weapon Systems*, but systems in use may vary for in recent years Contraves and other firms have offered various improvements as modification kits for rectrospective fitting.

CHARACTERISTICS

Type: Combined radar/visual tracker and computer for remote control of medium calibre anti-aircraft guns or missiles

Brazilian Army Super-Fledermaus fire control system showing tracker's position to left of radar and the controller position under shelter at extreme left of unit

Target detection: Up to a range of 50 km
Target acquisition: All-weather operation with 5 options:
(a) by selectable programmed scanning motion
(b) by optical putter-on
(c) by optical tracker
(d) by tactical radar
(e) by radar/visual combination.
Target tracking: Up to a range of 40 km with 5 options:
(a) automatically by radar
(b) automatically by regenerative control
(c) automatically by memory
(d) visually by tracker operation
(e) automatically in range, visually in angle.
Firing data computation: Continuous computation individually for 3 gun emplacements
Muzzle velocity and meteorological corrections included
Special modes selectable for diving or curved flights
Muzzle velocity measurement: Separate measurement by built-in equipment for each gun
ECCM: High transmitting power; narrow main beam; high sidelobe attenuation; 2 independent transmitting channels; jump changes of frequency over wide band; IAGC; STC; FTC
Clutter suppression: MTI data processor
Mobility: Cross-country trailer
Maintenance: Built-in test and calibration equipment. Use of modular plug-in units

STATUS

Production complete but still in service with more than 20 countries including Brazil, Denmark, India, Israel, Italy, Japan, Norway, Sweden and Switzerland.

CONTRACTORS

Complete system: Contraves AG, Zurich, Switzerland, but was also made under licence in India, Italy and Japan.

4586.181

GUN KING SIGHTING SYSTEM

Gun King is an autonomous sighting system for use with automatic anti-aircraft guns, such as the Oerlikon Diana twin 25 mm AA gun GBF-BOB (**4525.131**) or Oerlikon twin 35 mm AA field guns GDF (**4537.131**) although the system is equally well suited for naval applications or operation of other gun types, and has already been tested on the Soviet 23 mm ZU-23 towed anti-aircraft gun system.

Of modular design, Gun King is based on the use of a new Contraves digital computer coupled with an optical system that incorporates a periscopic sight, laser rangefinder, and includes provision for a TV camera for operator training or for use as a tracking module. There is a common optical path in the sight periscope for the laser beam and the aiming sight-line, and these beams are deflected with high accuracy by a gimballed mirror, thus permitting the use of a fixed eyepiece. The periscope has a multispectral optical system and night operation is possible.

The rangefinding function is performed by the laser rangefinder, which uses a multi-divergence neodymium laser. A novel switch-over device enables the laser beamwidth to be reduced for use against ground targets, eliminating ground clutter.

The digital computer calculates all fire control data at high speed with optimum control and accuracy, using filter and ballistic algorithms. In addition it controls the periscope, laser triggering during the tracking phase and the drive electronics. At the instant of optimum hit probability, the operator is given an audio signal to open fire. Target changes can be effected very rapidly because target data is stored in the computer memory. All data relating to weather conditions and weapon parameters is entered into the computer so that the operator is free to concentrate on precise target tracking. Sight and gun are commanded entirely by the operator's control yoke which provides full hands-on control.

Modern DC power electronics and servo-drives are used for gun control, and the battery capacity alone ensures full operational readiness of the entire system from a passive, silent standby status. Up to five combat cycles can be carried out before the engine of the power supply has to be started to provide electrical power.

When employed in a naval role, the Gun King system can be gyro-stabilised and may also be used as a stand-alone fire control facility for several weapons. It can be slaved to other target designation facilities, such as radar or other fire control systems, in both naval and land-based applications.

STATUS

In series production and ordered by about a dozen customers including Canada.

CONTRACTOR

Contraves AG, Schaffhauserstrasse 580, CH-8052 Zurich, Switzerland.

Contraves Gun King laser/computer controlled sighting system

UNION OF SOVIET SOCIALIST REPUBLICS

3323.181

SOVIET ARMY FIRE CONTROL

Very little reliable information is available for publication concerning the systems and equipment used by the Soviet Army for such purposes as artillery fire direction, survey, meteorological data acquisition and distribution, tank fire control, tactical data automation, command and control.

Artillery Fire Control

Soviet artillery is controlled at battery and battalion level from Command Observation Posts (four per battalion) and Fire Direction Centres-FDC (one per battalion) with the emphasis of control being at battalion rather than battery level as with Western armies. In self-propelled artillery battalions all these vehicles are Armoured Command and Reconnaissance Vehicles (ACRV) with the FDC in an ACRV M-1974-3, while each battery has as ACRV M-1974-3 for each battery's gun position officer. Towed artillery battalions have their FDC in BTR wheeled armoured vehicles or GAZ-66 soft skinned wheeled vehicles.

The Command Observation Posts have only optical and electro-optical equipment such as the DAK-1 laser rangefinder (see **3437.193**), PAB-2 periscope artillery compass, the AST battery commander's telescope and DS-0.9 stereoscopic rangefinder together with radios and telephones to the Fire Direction Centre but not to the gun positions. Each FDC, which also acts as the link between the ground forces and the tactical air force and includes a fire control computer which American sources indicate is similar to BCS (see **3510.161**) and whose features may be similar to the Yugoslavian ARS system (**4584.161**), together with radio and telephone links both to the gun positions and the Command Observation Posts. There is also radio equipment for air controllers.

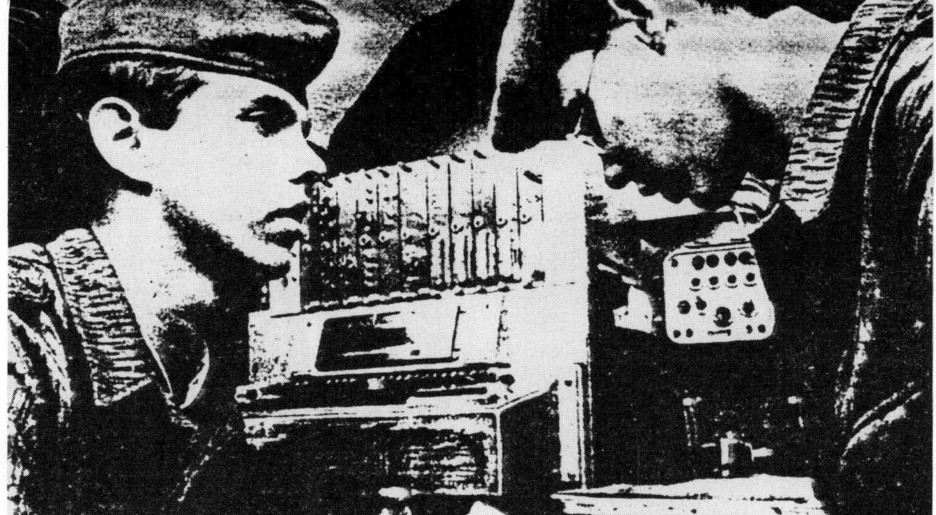

The Soviet artillery fire control computer

The computer, believed to be digital, features an alphanumeric keyboard, a short row of separate keys and a small screen upon which is likely to be displayed alphanumeric co-ordinates. An unusual feature above the display screen, is a plug-in unit with 11 pins. This is likely to be for land-line link to Command Observation Posts, forward observers, regimental and divisional command posts. The computer is fitted on the right hand side of the vehicle while radios, probably the R-108D/M (28-36 MHz FM) transceiver and the R-311 receiver, fitted on the left hand side. The FDC vehicle usually operates in conjunction with a command post vehicle and a radar equipped vehicle.

Artillery units are frequently supported by what is termed Artillery Instrument Reconnaissance (Artilleriyskaya Instrumental'naya Razvedka-AIR) including artillery location radars (Stantsiya Nazemnoy Artelleriyskoy Razvedki-SNAR), mobile ground reconnaissance (battlefield surveillance) radars (Podvizhnaya, Stantsiya Nazemnoy Razvedki-PSNR), mortar location radars (Artilleriyskaya Radiolokatsionnaya Stantsiya Obnaruzheniya Minometov-ARSOM) and "mobile reconnaissance posts" (Podvizhny Razvedyvatel'ny Punkt-PRP). Systems in service include the SNAR-1/2 (Pork Trough) (see **4328.153**), the PSNR-1 (see **4602.153**), the ARSOM-2P (Small Yawn), the PRP-3 (Small Fred) and Big Fred (for both see **4329.153**).

In addition artillery units are supported by an artillery survey vehicle (AT-ST) with land navigation equipment and PNV-57 active IR night driving system. The navigation equipment includes route sensors which measure length in small, straight, sections, a gyroscopic course indicator and a "course plotter". The latter is a computer which adds increments to the co-ordinates of the previous point to calculate the existing position. This is then printed out on a map. A sighting device and an artillery director are used to provide the initial orientation of the survey vehicle or to determine the grid co-ordinates. Accuracy depends upon the length of the route and terrain conditions with the overall deviation 1% of the route travelled on firm ground. Normally it takes the crew 15 minutes to prepare the land navigation system for use but if the system has not been used for some time, or in temperatures below -30°C this preparation time is 25 minutes.

It is likely that the Soviet Army is developing an automated target acquisition system (Reconnaissance-Destruction Complex in Russian terminology). This would be capable of locating and tracking targets at any time of the day, allocating priorities, computing fire data and bringing down fire. American sources indicate the Soviet Army already operates at least three kinds of RPV.

These sources also indicate that the Soviet Army does not appear to have anything comparible to the US Army's Army Data Distribution System (ADDS), formerly PLRS/JTIDS Hybrid (see **3944.161**), the Mobile Subscriber Equipment (MSE), the All Source Analysis System (ASAS) or Maneuver Control System (MCS) (see **3940.161**). However, a Soviet source indicates that regimental, divisional and possibly battalion command vehicles have communications equipment together with "... automation, observation, navigation and other aids".

Tank Fire Control

The T-64/T-72 series of main battle tanks feature an analogue ballistic computer receives laser and coincedence rangefinder inputs which it converts to range and elevation data. Earlier vehicles apparently still rely upon optical acquisition and tracking of targets with lasers used to determine range. Digital ballistic computers are likely to have been introduced into later vehicles but it is not clear if any attempt has been made to retrofit them to the remainder of the tank fleet.

TNA-3 Land Navigation System

This system was introduced with the T-62K command tank in the early 1970s and is apparently used by later models of command tank. It is a fairly accurate system but one which requires 15 minutes to warm up and it must be reset every 1.5 hours.

The system features a GPK-59 gyrocompass, two compass indicators, a latitude correction device, an odometer, a power converter and a calculator. The gyrocompass provides the vehicle azimuth which is relayed to indicators at the driver and commander's position. The odometer, which registers in metres, indicates the distance travelled while the calculator is used to determine and display the grid co-ordinates based on figures entered at the start point. The calculator will provide a continual update of the vehicle location and can also provide the distance and azimuth to a predetermined objective. The system is accurate within 10 m.

UNITED KINGDOM

3320.161

WAVELL ADP SYSTEM

Wavell is the British Army's automatic data processing system designed to assist commanders and their staffs with the conduct of general operations and intelligence work at brigade level and above.

It is designed to use any suitable communications link and to operate with the BATES fire control system (**2644.161**). Stage 1 used the Bruin trunk communications network but Stage 2 uses the Ptarmigan network, each network reserving two channels for Wavell use.

The full system will provide headquarters with automated data processing (ADP) facilities comprising a central processor, bubble memory backing store and two floppy disk drives for loading programmes and data. All these are located in Wavell Processor Installation (WPI) vehicles which are linked to the communications system and the remotely located staff terminals. WPIs are located at the main, alternate and rear corps headquarters, at the main, alternate and rear divisional headquarters as well as brigade headquarters.

The MDP 1000 processor installation includes a GEC 4080M central processor which, with its associated fast store, is in two Air Transportable Racking (ATR) units holding a total of 1 Mbyte of semi-conductor memory. There are three units of bubble memory backing store, each having a capacity of 24 Mbytes. The bubble memory has been selected because it has low volume and low power dissipation. It is also non-volatile if there is power loss and yet can be easily and quickly erased. It is anticipated that during the life of the system advances in bubble memory technology will enable the total system capacity to be increased to 288 Mbytes. Also in the installation are eight full duplex 16 kbits/sec modem channels.

There are two types of installation. The full system will be at the corps and rear division headquarters in a standard 1 tonne container mounted on a Bedford 4 tonne flatbed truck together with a 20 kW generator. The generator provides power for air-conditioning and lighting, as well as providing, through a conditioning unit, a 28 V DC source for the ADP system and back-up batteries. At the remaining division headquarters it will be in a FV 435 armoured command vehicle with power and environmental control equipment on the roof while a more basic system with input/output terminals will be provided for brigade headquarters also in an FV 435.

The staff terminals feature a 21.5 × 21.5 cm alphanumeric plasma panel display together with a detachable keyboard. At present the terminals do not have a graphics capability but it is hoped to add a large screen for this purpose. The total unit weighs 30 kg and they can be located up to 1 km from the WPI.

In operation raw data will be collated with data entered into the system before deployment and with other data generated by the machine itself. It will then be displayed as meaningful subsets of data (or formats) on the staff terminals. These formats will be used by the staff to assist with formulating plans and orders and in the preparation of briefs for higher command. Formations can build up their own data banks which can be switched to other formations or to the central data bank. General purpose free-text facilities will also be provided for messages and orders. All formats can be printed out on a hard copy printer for briefing or fall-back and for staff cells not equipped with Wavell.

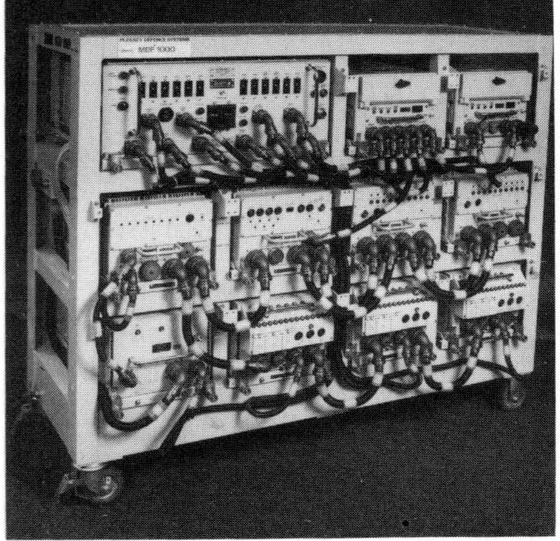

MDP 1000 system of Wavell battlefield C² automation system can be accommodated in a wide range of military vehicles (tracked, wheeled, ships or aircraft) or used free standing as shown here

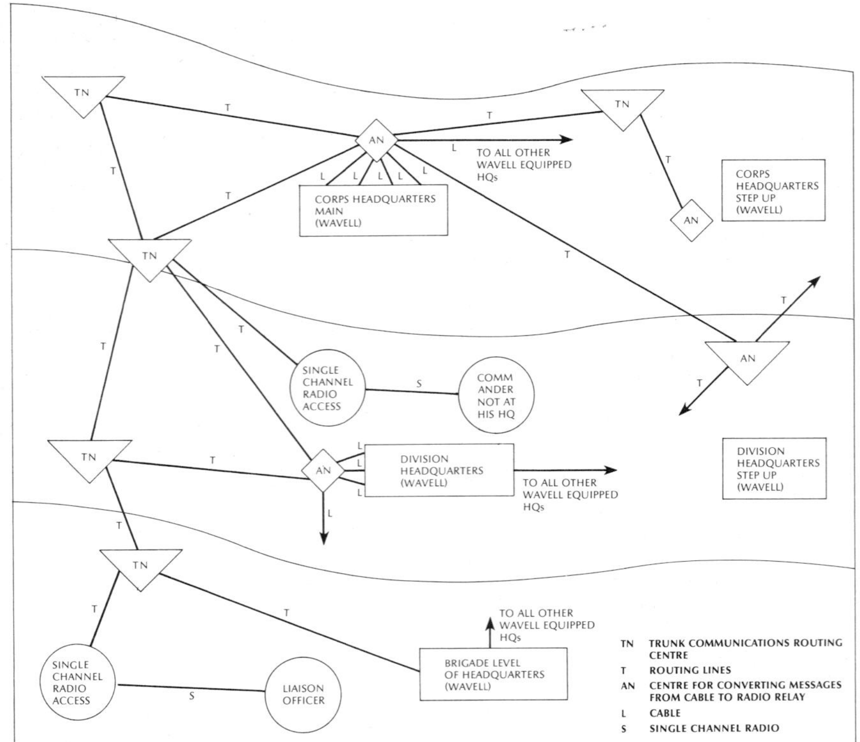

Integration of Wavell system with typical communications network

STATUS

Project Wavell began as a series of staff studies in the early 1970s and in 1976 it was decided to field a minimum system using commercial hardware with minimum ruggedisation at a cost of £6 million. This system was used for operations and staff functions using existing staff proecedures.

Two years later the Stage 1 equipment was fielded with 2nd Armoured Division in Germany using PDP 11/35 processors and Cossor DID 400 VDUs as staff terminals. The system, mounted in hard-topped Land Rovers, was initially to help provide commanders with an overall picture of the battlefield. It proved so successful it was extended for use by artillery, engineer and logistic organisations.

In 1979 the go ahead was given for Stage 2 production and Plessey received a £25 million contract to produce 33 Wavell systems including a Secure Information Management System (SIMS). The system began entering service with the British Army in 1985. In 1980 the Defence White Paper estimated that Wavell would cost £55 million to implement.

CONTRACTOR

Plessey Defence Systems Limited, Grange Road, Christchurch, Dorset BH23 4JE, England.

4753.161

QUICKFIRE ARTILLERY FCS

The Marconi Quickfire artillery fire control system, introduced at the 1984 British Army Equipment Exhibition, is designed for use at battery and battalion level with all types and calibres of indirect fire weapons. It was a private venture development by Marconi Command and Control Systems, and is a successor to the earlier FACE system (**2415.161**).

While many of the FACE features, such as the tactile colour-coded keyboard and customer-language matrix, have been retained in Quickfire, use of later technology (eg CMOS devices) has permitted significant reductions in weight and volume of the equipment. The result is a unit weighing six per cent that of its predecessor, with three per cent of its volume, at less cost than comparable previous generation equipment.

Quickfire uses full ballistic modelling to provide fire control data and the system can be configured as a Battery (serving up to eight guns) or a single gun system. It is based on a common hardware unit that can be configured for use by a forward observer (FO) as a data module, as battery or battery command post modules, and as individual gun display modules. A plug-in software module stores ballistic data to adapt the unit for operation with specific types of weapon. This basic data is available from a library of FACE programs which currently cover most weapons in general use throughout the world.

Entry of mission data into Quickfire units is by means of a menu-led formats of the matrix whereby visual prompts are given at each stage of the operation. Battery fire control data is computed in less than one third of the time of flight, and the solution is presented in the form of a back-lit LCD which is easily visible in all lighting conditions likely to be encountered. There is storage for up to 100 target locations, ten firing zones, and ten friendly force areas. The system provides for two concurrent fire missions.

Data is conveyed between FOs, command post and gun positions by either radio or cable links, and Quickfire modules have three communications channels which allow for burst transmission of digital data.

As used by a forward observer, the Quickfire module matrix provides preformatted messages to allow the FO to send complete orders for target engagement and to report his own position and friendly locations directly to the command post module. It also receives and displays messages in response.

At individual gun positions, or rocket launchers, Quickfire units are operated as display modules to present all incoming data, while providing answer-back facilities. Individual gun firing data such as, ammunition, charge, fuze, bearing elevation, method of fire and number of rounds is presented.

As a battery or battalion computer module, Quickfire carries out all the necessary survey and ballistic calculations and aids command and control of the guns. Each input from FOs is automatically entered and displayed as required, and command post staff have full authority over the input of data and checking of computations. Under operator control, outgoing messages are automatically transferred and transmitted over the radio network.

CHARACTERISTICS

Dimensions: 321 × 291 × 85 mm

Weight: 7.2 kg; 6.5 kg FO and gun display module

Power consumption: 5.0 W; 4/1 W FO and gun display module

Met data message: Up to 26 lines (any standard)

Target records: Up to 100

Communication channels: 3

Facilities: Printer interface, real-time clock, touch keyboard, 10 firing zones, 2 concurrent missions, survey, etc

STATUS

In production. In mid-1986 it was announced that Bofors of Sweden had selected Marconi to supply the complete fire control package for the 155 mm FH-77A howitzer system being sold to Sweden. This consists of the Quickfire modular fire control system for both the observer and battery computing functions and a quantity of GUNCOS hand held fire control calculators.

CONTRACTOR

Marconi Command and Control Systems Ltd, Chobham Road, Frimley, Camberley, Surrey GU16 5PE, England.

Quickfire module used in the forward observer module role

Quickfire artillery FCS uses common hardware. Same basic module can be used for FO, command post positions as well as gun display module, as here

6446.161

MISTRAL PASSIVE METEOROLOGICAL SYSTEM

The MISTRAL passive meteorological system has been developed as a private venture by Marconi Command and Control Systems and was unveiled in 1986.

MISTRAL consists of a mobile station in a single container that can be mounted on a variety of chassis, a 6 × 6 version of the Land Rover being used for the

Marconi MISTRAL passive meteorological system mounted on rear of 6 × 6 Land Rover chassis

prototype system. This can maintain a 24 hour service using only four men.

Inside the container is the operators workstation where all of the data gathering is monitored. The required military meteorological messages are produced using a variant of the Marconi Quickfire (**4753.161**) module which uses a simple and reliable menu-led routine to maintain speed and accuracy and also provides its own interface with the communications system specified by the user.

The module also interfaces with a printer to provide a hard copy if required. Messages can be produced to NATO and Eastern standards as well as civil, or other formats as required.

MISTRAL uses established international NAVAID systems that provide worldwide systems which have been used for many years for both navigational and meteorological purposes.

A balloon carried radio sonde can sample temperature, pressure and humidity and receives radio transmissions from at least three of the world wide ground stations. All of the data, including the ground station signals, are re-transmitted to MISTRAL where the NAVAID signals are processed to determine the relative position of the sonde, and so produce wind and speed.

A meteorological sounding takes an hour to reach a height of 20 km, though partial messages to a lower height can be produced at earlier times during the sounding. Greater heights can be achieved using a larger balloon, though this will take longer than an hour.

In addition to being highly mobile and air transportable, MISTRAL is a passive system with no radar emissions that can render it vulnerable to anti-radiation missiles or radio direction finding.

STATUS

Development complete. Ready for production.

CONTRACTOR

Marconi Command and Control Systems Ltd, Chobham Road, Frimley, Camberley, Surrey GU16 5PE, England.

4754.181

MARCAL GUN CALIBRATION SYSTEM/MOGUL

The Marconi Marcal gun calibration system is an on-gun system that measures the velocity of every round fired, stores this data and produces a statistical basis for gun calibration, offering a more rapid and accurate way of continuously correcting for variables such as bore wear, met conditions, muzzle velocity etc, that affect the accuracy of fire.

Typically, shell velocity, or a corrected quadrant elevation and fuze setting, where applicable are displayed after each round fired, so that effectively the gun performance is continuously monitored. At the same time the gun data base is automatically updated, enabling calculation of gun laying adjustments on a round by round basis. The equipment is designed for permanent fitting on all categories of field guns of 30 mm or above, and consists of two small units: a doppler radar head, and a control unit containing a microprocessor.

The doppler radar head incorporates a microprocessor and provides accurate velocity measurement with low transmitted power. The control unit includes a data analysis and storage system, an operator display, and an automatic data interface for system expansion. A non-volatile store holds the firing data of the last 50 rounds fired and this information can be retrieved on a simple portable recorder.

CHARACTERISTICS

Radar: 5 mW spread spectrum technique
Transmission duration: 1s (max per round)
Beam: 5° (horizontal) × 10° (vertical)
Velocity measurement error: 0.2% (max)

Marcal gun calibration system

Projectile speed range: 100 – 1000 m/s
Processing speed: 30 rounds/minute
Min projectile calibre: 30 mm

MOGUL

The Marcal muzzle velocity measuring radar is also employed in the Marconi's Mogul modular gun laying system which incorporates the Marcal system with additional items to suit it for use with SP and towed field guns. In the Mogul system the central unit is the display and processor unit (DPU) which is provided with a fast two-way data bus for on-board operations. A data link compatible with a variety of artillery fire control systems allows for inputs from:

(1) An electronically enhanced indirect and direct fire sighting system. For towed guns and as a back-up system, a direct and indirect optical sight is available which automatically corrects for trunnion cant and tilt using precision instrumented inclinometers
(2) An inertial gun reference unit mounted on the trunnions provides azimuth and elevation data for MOGUL
(3) Gun control electronics. MOGUL automates the gun laying procedure using the on-board sensors and the micro-processor based control system. A gun attitude reference unit employing a micro-processor and strapdown inertial technology is fixed to the gun mounting to provide direct measurement of gun elevation and azimuth values. A control unit interfaces with the gun laying system and provides rapid autonomous gun laying from the DPU
(4) Gun safety instrumentation. Also available is a shell position sensor to confirm that the shell is correctly rammed to avoid fallback, and a gun tube temperature sensor to monitor the approach of cook off conditions.

STATUS

In September 1985 it was announced that the British Army had ordered MARCAL on a one per gun basis for all guns of the Royal Artillery. MOGUL has also been demonstrated in hardened form with a capability to incorporate position indication and ballistic computation.

CONTRACTOR

Marconi Command and Control Systems Ltd, Chobham Road, Frimley, Camberley, Surrey GU16 5PE, England.

6368.161

GUNZEN ARTILLERY FIRE DATA COMPUTER

The GUNZEN hand held Artillery Fire Data Computer was developed by Zengrange to meet the requirements of the British Army following the successful introduction of the Zengrange MORZEN (mortar fire data computer) into the British Army.

The British Royal Artillery use GUNZEN with all equipments ranging from the Royal Ordnance 105 mm Light Gun up to the Lance surface-to-surface tactical missile system.

Like MORZEN, GUNZEN is mounted in its own ruggedised portable case and is powered by standard commercial batteries which last several months with an early warning of battery failure being provided.

GUNZEN can be easily configured to provide fast, accurate firing data for any gun, howitzer or heavy mortar using any ammunition and any sighting system, in the indirect fire role. All procedures are computer led with comprehensive error reporting and protection against inadvertent or unsafe miskeying by the operator.

All software is permanently etched onto incorruptible plug-in ROM (Read Only Memory) modules. The standard GUNZEN package uses two general purpose map data plotting and procedural modules, and one weapon specific ballistic module. Weapon modules can be quickly and easily changed to accommodate different equipment and ammunition systems.

GUNZEN provides individual data for up to eight guns, stores up to 47 targets by NATO's standard target numbering system and 9 observation posts by callsign.

It computes or accepts external registration data, uses ballistic meteorological data if this is available and performs all of the ballistic calculations required by artillery weapons.

Corrections are made automatically for non-standard muzzle velocities, projectiles, weights, non-rigidity, met and gun displacement in both high and low angle.

Missions are initiated with grid co-ordinates (6, 8 or 10 figs), Polar co-ordinates, a stored target, a correction from a stored target and a correction from grid co-ordinates.

Adjustment of fire is by Polar or target grid corrections with up/down corrections for fuzed projectiles or altitude.

At fire for effect, lines can be parallel, converged, linear or sheaf pattern and two missions can be run simultaneously without fear of confusion.

An integral clock provides real time or time relative to H-hour. Survey routines (including fixation, traverse and change of grid) enable fast deployment and location of battery and individual gun positions.

The GUNZEN Artillery Fire Data Computer has passed the British Army testing for high and low temperature (–30° to +55° C), dust implosion, drop testing, vibration, immersion in water, bounce, nuclear flash burn and electro-magnetic compatibility. It is qualified to DEFSTAN 00-1 and DEFSTAN 07-55.

CHARACTERISTICS

Size: 200 mm × 118 mm × 58 mm
Weight: 625 g
Operating temperature range: –30° to +55° C
Target store: max 47
OP store: max 9
Battery life: 80 hours
Power supply: 4 × 1.5 V dry cell
Timings
Bearing and charge: <6 s
Elevation and time of flight/fuze: < 12 s
Accuracy
Map range: ±1 m

GUNZEN artillery fire data computer

Bearing and elevation: ±1 mil
Fuze setting: ±0.5 units
Time of flight: ±0.5 seconds
Coordinates: ±1 m

STATUS

In production. In service with British Army and other armies.

CONTRACTOR

Zengrange Ltd, Greenfield Road, Leeds LS9 8DB, England.

6367.161

MORZEN II MORTAR COMPUTING SYSTEM

Following the successful introduction of MORZEN into service with the British Army, MORZEN II is the second generation mortar fire data computer which complements the increased sophistication of latest weapon technology and provides fast, accurate fire data for any mortar, using any ammunition and any sighting system, in the indirect-fire role. It is in full service with the British Army, who use it with the Royal Ordnance 81 mm mortar, and an increasing number of other countries around the world.

Lightweight, hand-held, robust and simple to operate, MORZEN II uses standard prompts and procedures familiar to each user. With only a few hours training, MORZEN II can be operated successfully in the field, to produce faster, more accurate mortar fire than any other proven system.

Main features of MORZEN II can be summarised as all computations to 10 digit accuracy from 6, 8 or 10 digit reference points, stores up to 314 target locations and 9 observation post locations plus 14 mortar lines for normal or shoot and scoot operations, polar co-ordinates for use with laser rangefinders, easily used DATUM procedure for accurate location of new mortar line, without maps procedures with improved facilities, on board clock for real time, relative to H-hour, splash and co-ordinated illumination, easily understood error messages generated by miskeying or invalid data input.

The MORZEN II has a battery life in excess of 80 hours of continuous use with advanced warning of battery failure being given. It has been designed for use with NBC or arctic clothing and operates under a wide range of environmental conditions.

The MORZEN II Artillery Fire Data Computer has passed the British Army testing for high and low temperature (–30° to +55° C), dust implosion, drop testing, vibration, immersion in water, bounce, nuclear flash burn and electro-magnetic compatibility. It is qualified to DEFSTAN 00-1 and DEFSTAN 07-55.

CHARACTERISTICS

Size: 200 mm × 118 mm × 58 mm
Weight: 625 g
MFBF: 10 000 hours
Operating temperature range: –30° to +55° C
Storage temperature range: –40° to +75° C
Target store: max 314
OP store: max 10
ML store: max 14
Battery life: 80 hours
Power supply: 4 × 1.5 V dry cell
Timings
Bearing and charge: 6 seconds
Elevation and T of F/Fuze: 12 seconds
Accuracy
Map range: ±1 m
Bearing: ±1 mil
Elevation: ±1 mil
Fuze setting: ±0.5 units
Time of flight: ±0.5 seconds
Coordinates: ±1 m

STATUS

In production. In service with British Army and other armies.

CONTRACTOR

Zengrange Ltd, Greenfield Road, Leeds LS9 8DB, England.

MORZEN II mortar computing system with cover open

3523.161

MORCOS MORTAR COMPUTING SYSTEM

The Marconi Morcos mortar computing system computes firing data for all types of mortar and is intended to replace the existing manual methods of fire prediction calculation which are slow, prone to inaccuracy and wasteful of ammunition.

Morcos is a single self-contained unit incorporating a computer, keyboard for data entry, display and batteries. The case is built of durable polycarbonate and profiled to fit comfortably in the hand. It is built to full military specifications to withstand adverse climatic, environmental and nuclear effects.

A wide range of drills is available and, as an example, the program for the British Army's L16 81 mm mortar, includes drills which produce data for HE (high explosive) smoke and illuminating missions, survey problems, target reduction and three concurrent fire missions. It incorporates inputs which take into account variables such as meteorological conditions, charge temperature, difference in altitude between mortar and target, position of own troops, (for warning when own troops are in the danger area).

The computer in Morcos is a microprocessor with access to a semi-conductor backing store. Each ballistic program can predict for one type of mortar and its attendant ammunition. The program can be changed in seconds by removing a plug-in memory unit (called a ballistic module) and by replacing it by another for a different ammunition system. This can be done, in the field, in all weather conditions because the moisture seal in the equipment is not broken.

The elastomeric type keyboard is both waterproof and reliable and is illuminated for night operation by a variable brightness electroluminescent panel. The keyboard has twenty-four keys, ten of which are for digits 0 to 9. The display has a capacity for eight alpha-numeric characters and uses high efficiency, extra bright, light-emitting diodes, which have a variable lighting control.

Morcos is powered by batteries, which can be either throwaway or rechargeable, or from an external power source. It weighs, including batteries, less than 1.35 kg and measures 23 × 10.1 × 5.4 cm. It is a highly reliable system based on the use of large scale integration and hybrid techniques. Faults can be quickly repaired by easily replaceable sub-assemblies, and no special first line test equipment is needed.

STATUS

Full production and supplied to 11 countries.

CONTRACTOR

Marconi Command and Control Systems Ltd, Chobham Road, Frimley, Camberley, Surrey GU16 5PE, England.

6369.161

MAMBA ARTILLERY AND MORTAR BALLISTIC AID

MAMBA (Marconi Artillery and Mortar Ballistic Aid) has been developed as a private venture by Marconi Command and Control Systems and was announced in June 1986.

It is a hand-held computer producing ballistic data for artillery guns and rockets, or for mortars. In addition the equipment can be configured as a data entry or a display terminal.

It produces gun data in approximately 2.5 seconds (less for a mortar), and to switch weapons within a role, involves only a simple and quick change of a small personality module.

MAMBA has an LCD display with two lines of alpha-numeric characteristics addressed by six function keys and ten dual function keys allowing numeric entry and menu option selection, all with electro-luminescent back lighting.

It uses two 9 V commercial batteries giving up to 700 hours of life with the added ability of using an alternative outside power source in the range 9 to 30 V DC.

It has 32 Kbytes ROM and 4 Kbytes RAM (expandable to 96 Kbytes ROM and 16 Kbytes RAM) and is provided with an uncommitted serial data port (RS232/RS422) for connection to a printer, modem, etc.

MAMBA is designed and tested as a full military specification unit to meet the relevant clauses in DEF-STAN 07-55 for use throughout the world and has an estimated mean time between failure of 6000 hours.

STATUS

Development complete. Ready for production.

CONTRACTOR

Marconi Command and Control Systems Ltd, Chobham Road, Frimley, Camberley, Surrey, GU16 5PE, England.

MAMBA (Marconi Artillery and Mortar Ballistic Aid)

2400.161

AMETS – ARTILLERY METEOROLOGICAL SYSTEM

AMETS is the name given to a self-contained, computer-based, mobile meteorological system developed for the British Army by Marconi Command and Control Systems Ltd and Plessey Radar, for obtaining and processing information on atmospheric conditions automatically. It is capable of processing data from a wind-finding radar and radio sonde complex to an altitude of 20 km every hour, and incorporates a facility to output partial messages within seconds of the radio sonde passing through the relevant height band. A single AMETS will normally provide all meteorological data needed in the area of one army division.

AMETS uses a specially-developed radio sonde with a very accurate temperature sensor. The outputs from radio sonde and radar are fed into a computer which can then calculate the required meteorological information. Measurement of surface pressure, and the average humidity figures which can be provided for the particular zone of operation are stored in the computer memory. The loss of accuracy resulting from this simplification of the radio sonde is negligible for artillery purposes, yet it offers appreciable gains in terms of operational simplicity and reliability in military field conditions.

CONSTRUCTION

The data processing part of the system is based on an MC1800 computer, produced by Marconi Command and Control Systems, and already in use in a wide variety of military applications including FACE. The Plessey WF3M radar is built to full military standards, and is derived from the Plessey WF3 wind-finding radar used for civil meteorology.

AMETS is mounted in a single instrumentation vehicle with a small trailer carrying the radar. In addition there will normally be a command post vehicle, and reconnaissance and stores vehicles to carry hydrogen and other troop stores. The instrumentation vehicle will carry all the data processing equipment, the radar and radio sonde monitoring and display units and a full range of diagnostic test equipment.

STATUS

In service in the United Kingdom and some overseas countries.

CONTRACTORS

Marconi Command and Control Systems Ltd, Chobham Road, Frimley, Camberley, Surrey GU16 5PE, England.

Plessey WF3M mobile wind-finding radar is primary sensor of upper air winds for AMETS. Well over 50 of these radars have been sold

The Plessey Radar Ltd, Oakcroft Road, Chessington, Surrey KT9 1QZ, England.

2644.161

BATES – BATTLEFIELD ARTILLERY TARGET ENGAGEMENT SYSTEM

BATES (Battlefield Artillery Engagement System) is a semi-automatic data processing and fire control system being developed for the British Army. It will feature distributed data processing with each command level having computing power appropriate to its function while information will be on call from other parts of the system using digital communications.

Development of the system began in 1976 with a definition study with full scale development beginning in 1980. At this time Marconi was responsible for the hardware and Scicon was responsible for the software. Spending restrictions and software difficulties meant the system, which was scheduled to enter service about 1985, slipped two years. In October 1985 Marconi Command and Control took over responsibility for both hardware and software when they were awarded a £100 million contract for Phase 1 of the programme. Hunting Hivolt were responsible for vehicle installations while Marconi Secure Radio Systems were responsible for encryption. The first phase involves the delivery of 200 systems against a total British Army requirement of 600-800.

The hardware will include the MCS 2010 digital processor (based upon the Intel 8086 microprocessors), MCS 2030 plasma visual display units, MCS 2040 data entry units, MCS 2050 gun display units and a MCS 2061 communications interface with the Ptarmigan tactical network.

The function of the system is to accept inputs from observers and artillery staff using sensors such as radar, sound ranging systems and reconnaissance devices. The data is then stored, analysed and displayed to assist the artillery commander in decision making. The system differs from TACFIRE (**2827.161**) by retaining the commander in the control loop while the American system is totally automatic.

Net and trunk communications systems will link target acquisition with artillery headquarters and the fire units. The automatic data processor will allow forward observers to feed target data to the command posts and speed up the enagagement process by reducing the human links in the chain.

The system is intended to replace FACE (**2415.161**), AWDATS (**3215.161**) and the data processing elements of AMETS (**2400.161**).

STATUS

In full scale development. During the summer of 1987 the British Ministry of Defence requested tenders for Phase 2 of BATES involving 66 per cent of the programme including computer units, printer control units, truck interface equipment and installation kits with deliveries between 1989 and 1991. It is believed the basic BATES system will now enter service late in 1988 or early in 1989.

CONTRACTOR

Marconi Command and Control Systems Ltd, Chobham Road, Frimley, Camberley, Surrey, GU16 5PE, England.

2415.161

FACE – FIELD ARTILLERY COMPUTER EQUIPMENT

FACE is a field gunnery computer system specifically designed to perform quickly and accurately the lengthy calculations which are involved in producing artillery firing data.

FACE undertakes two specific roles. First, as an artillery survey computer it carries out all the calculations necessary in this task. Second, and more importantly, as a gun data computer it processes all the information required to place a shell accurately on the target and displays this information on a console.

FACE has been modernised by replacing the 920B computer by the MC 1800, which uses the latest LSI sliced technology. This has greatly increased the speed of operation and the memory capacity. This improved capability has resulted in a system that can control the fire of up to 32 weapons, deployed as four batteries, each of up to eight weapons. For greater tactical flexibility, FACE can still be deployed at each battery.

OPERATION

While the guns are deploying for action, the console operator in the command-post vehicle begins to receive setting-up data. This is added to the information already in the computer's store which includes a latitude function allowing for earth rotational effects and the muzzle velocity for each gun for each charge. The first input is the geographical reference of the centre of the battery position – the 'battery grid'; next the position of each gun in relation to battery centre, and the temperature of the ammunition charges. Then meteorological information is received by radio and is also entered into the computer. The equipment is now ready to receive data for shooting in the form of target position and height, the type of ammunition to be used, etc. Once all the relevant information has been fed into the computer, it produces answers which are passed to the guns via the AWDATS (**3215.161**).

The forward observer, noting the fall of shot, radios corrections, determined visually or by the use of a laser rangefinder, back to the command-post. The console operator feeds these straight into FACE which computes updated gun data for relay to the guns.

Frequently gunners are required to deliver fire in support of a pre-arranged battle plan. In practice, battle conditions often call for changes to this fire plan and because FACE has great speed and flexibility, these changes can be easily accommodated.

A wide variety of gun and rocket programs are available for use with FACE.

STATUS

Production complete. Sales of FACE have been made to 18 countries including Australia, Canada, China, Egypt, Ireland, Nigeria and the UK.

CONTRACTOR

Marconi Command and Control Systems Ltd, Chobham Road, Frimley, Camberley, Surrey GU16 5PE, England.

4157.181

PADS POSITION AND AZIMUTH DETERMINING SYSTEM

The Ferranti position and azimuth determining system (PADS Mk II) has evolved from PADS Mk I, now in service with the British Army, and FILS the commercial inertial land surveyor system. It is designed to meet an extended military requirement for a compact vehicle- or helicopter-mounted, accurate survey and navigation system, capable of on-board mission processing in near real-time. The equipment displays continuously its position in geographical or UTM grid co-ordinates, together with orientation relative to true or grid north. Alternatively given the co-ordinates of a destination point, the equipment will display continuously the range and bearing to that point, to assist navigation. The system permits the storage of up to 62 waypoints, which may be a combination of established control and new points being fixed, enabling the accuracy of computed values to be improved as new control data is provided during the traverse. Position information is accurate to within three metres and azimuth to within 0.3 mil after mission processing of an open traverse. For traverses of up 15 km between established control, points may be fixed to a relative accuracy of

Main unit of Ferranti PADS Mk II

better than one metre. The equipment is entirely self-contained and is powered from a vehicle 24 V DC supply.

The main application for PADS is to provide artillery units with position and orientation data for guns, launchers, observation posts and to carry out surveys for target acquisition devices. This greatly improves the effectiveness and flexibility of artillery on the battlefield and provides a saving in survey personnel. PADS Mk II is a single box system in a mounting tray with overall dimensions of 540 × 470 × 500 mm and 51 kg weight, together with a control and display unit which may be mounted in any convenient position in the vehicle. Azimuth is transferred by auto collimation on the systems mirror or by fixing two bearing pickets.

STATUS

PADS Mk I is in service and quantity production for the British Army. PADS Mk II has been developed to meet overseas orders and over 200 have been sold for military use in Scandinavia, Western Europe, Middle East, Far East and Australasia.

CONTRACTOR

Ferranti Defence Systems Ltd, Navigation Systems Department, Silverknowes, Edinburgh EH4 4AD, Scotland.

Control and display unit for PADS Mk II

3125.161

AWDATS – ARTILLERY WEAPON DATA TRANSMISSION SYSTEM

AWDATS is an artillery weapon data transmission system which displays the firing data calculated by the field artillery computer equipment (FACE) (**2415.161**) at the individual guns of a battery. Due to the different geographical location of each gun within the battery and variations in muzzle velocity between guns, there is a small but significant difference between the data for each gun in the battery.

The guns are provided with a data display unit which receives the data from a coding and distribution unit in the command post. The display units can either be installed in the vehicle for self-propelled guns or mounted on a frame for towed guns. Data transmission is by a frequency shift modulation audio frequency signal over existing radio or line channels which are also used for voice communications. Transmissions are audible and take less than three seconds for a complete battery of six guns. AWDATS makes a valuable contribution to the speed of reaction and efficiency of guns within the battery by eliminating the necessity for verbal transmission, checking of gun data and greatly reducing the probability of errors being made.

STATUS

Currently in service with the armed forces of the United Kingdom and a number of other countries.

CONTRACTOR

Marconi Command and Control Systems Ltd, Chobham Road, Frimley, Camberley, Surrey GU16 5PE, England.

4582.181

EFCS 600

EFCS 600 has been developed to apply high accuracy and simplified engagement procedures to any gun turret old or new, large or small. Presently in service and with proven ability, the system has been used to fire over 3000 rounds on vehicles such as T-55/Type 59, Vickers Mk 3, M48, Scorpion, Chieftain, Centurion and Challenger. It is designed to ensure simplicity of installation, ease of operation, ease of maintenance and high reliability. Being modular in construction, it has inherent flexibility in configuration and installation and can be simply adapted to meet the requirements of most customers.

The microprocessor-based computer cubicle is used to store ballistic information and carry out the calculations required to produce the offsets. A simple to operate commander's control panel incorporates switches to manually input ammunition type, barrel wear and if required, range and meteorological data. An alpha-numeric display provides the user with full function information, and a further switch operates a complete system built-in test routine. Sensors supply azimuth and elevation rates, wind speed/direction and air density, and gun trunnion tilt. A sight mounted laser rangefinder supplies target range.

To engage a moving target, ammunition type is selected then the target tracked with the laser button depressed. On releasing the button, the laser is fired, feeding the range and sensor information to the computer, which calculates the offsets required. The resultant aiming mark can then be laid onto the centre of the target ready for firing. When used with a CRT aiming mark system, an Aiming Mark Graphic Unit (AMGU) controls the CRT and provides a dot enclosed by an expanding/contracting circle in the sight's FOV.

The computer is programmed for the particular gun and ammunition natures used, however by simply changing the ROMs the system can be converted for a different gun or ammunition.

Through a continuous process of enhancement, additional facilities have been incorporated to ensure that EFCS 600 provides increased benefits not just through performance, but in the areas of training and interface to other systems.

In addition to direct fire engagement, a semi indirect facility enables the system to be used for accurate engagement of long range targets. The AMGU is a new development that provides not only the CRT offset control, but two further features. To allow engagement sequences to be practised in the vehicle, the AMGU allows graphic representations of targets to be displayed and engaged in the sight. This facility can also be used to display system and ammunition status information in the gunner's primary sight, and if the correct CRT is incorporated to display a thermal picture in the day sights.

MANTIS

To provide a full day/night capability for EFCS 600, Marconi have developed Mantis. This utilises a servo driven mirror in the head to allow the use of an accurate electronic link to the gun, thereby providing a degree of accuracy not normally possible with mechanical links. The sight is designed for installation into T-54/55 as well as Type 59 without the need for armour cutting or turret modification. It can be provided in either a Thermal Imager or Image Intensification form through the use of interchangeable elbows such as the OEC II elbow or the MCCS TIVS TI elbow. With the thermal elbow, the glass lens would be changed to germanium. Mantis has been thoroughly tested on firing trials and has proved to provide a flexible, simple to install, simple to use sight with a high degree of accuracy.

STATUS

Currently in service with two armies and in quantity production, the EFCS 600 is fitted as standard equipment to the Vickers Mk 3 main battle tank.

CONTRACTOR

Marconi Command and Control Systems Ltd, Chobham Road, Frimley, Camberley, Surrey GU16 5PE, England.

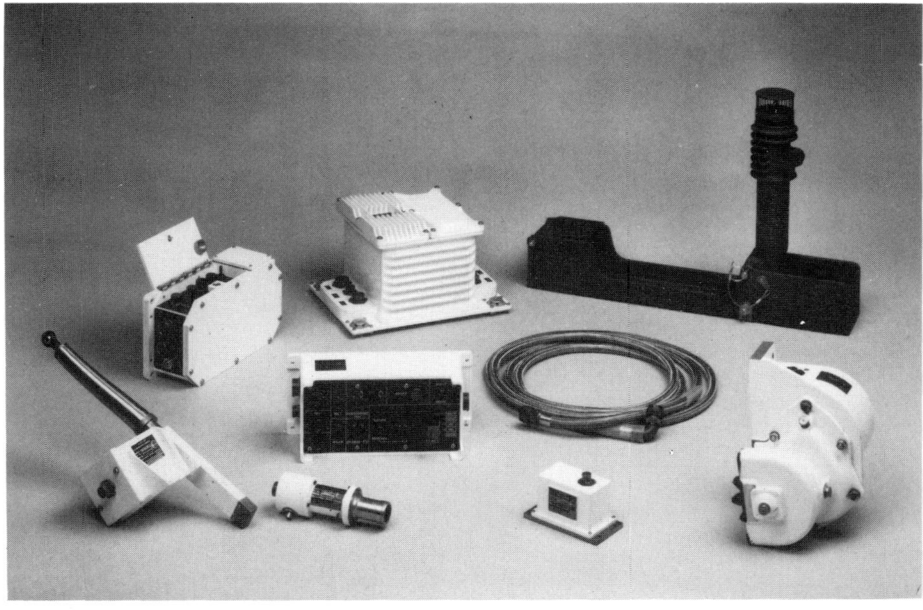

Group photo of EFCS 600 equipment

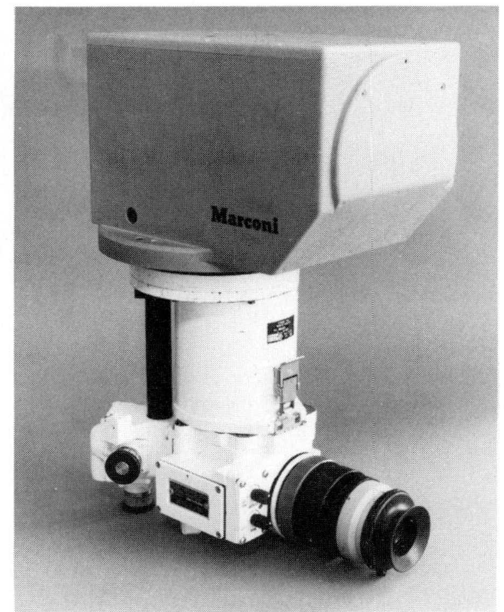

Mantis in II configuration

4831.181

MARCONI GUN CONTROL AND STABILISATION EQUIPMENT

Marconi Command and Control Systems experience in Gun Control dates back to the first electric stabilisation system for Centurion. The considerable expertise built up over this period has provided the background for a range of electric systems designed to meet all sizes of vehicle. All systems supplied by Marconi can be interfaced with fire control systems such as the Marconi EFCS 600.

GCE 628

For main battle tanks, GCE 628 provides an extremely high level of performance previously unattainable with electric gun control systems. The system has been well proven on vehicles such as the Vickers Valiant, Engesa Osorio and Chieftain 900, achieving extremely high slew speeds, excellent low speed tracking and providing excellent stabilisation performance. Although originally designed for modern MBTs, GCE 628 has the flexibility to allow installation into a wide variety of vehicles being suitable for T-tanks, M-tanks and others. The system comprises state-of-the-art processing electronics allied to a unique unit, the solid state power amplifier with an efficiency double that of its predecessor, the rotary metadyne. The solid state power amplifier allows full gun control facilities direct from the vehicle batteries with no additional power required from an APU. GCE 628 gives instant battle readiness and, through the high performance available, provides the ability to rapidly acquire and accurately track targets. Additional benefits of GCE 628 include minimal heat dissipation thus reducing thermal signature and solid state safety, thereby removing the inherent danger associated with hydraulic systems. Furthermore, a discreet surveillance capability is possible through the avoidance of accumulators required for hydraulics and APUs required for other electric systems.

PD700 POWER DRIVES

To meet the requirement for a power drive for light armoured vehicles, PD700 provides an inherently safe, compact, high performance, electric system that can be fitted to new vehicles or simply retrofitted to existing vehicles. Extensively trialled on vehicles such as Scorpion (with both 76 mm and 90 mm turrets) the Engesa Cascavel and on the Cockerill 90 mm turret, PD700 has demonstrated the ability to comply with the needs of the user. It enables the crew to rapidly and accurately acquire and engage targets and to carry out silent watch surveillance using power traverse. In addition, it complies with the requirements for simplicity of operation, reliability, ease of maintenance and integration to a computer based fire control system such as the Marconi EFCS 600. The equipment provided includes controllers for both gunner and commander, providing both with power control of the gun. In addition, a full range of safety switches and firing handles are available to ensure full compatibility with any vehicle. The high efficiency control cubicle is small enough to be unobtrusively installed and allows power traverse and elevation to be used with battery power alone for long periods.

Solid state power amplifier used in GCE 628 and Centaur

Vickers Mk 7 MBT fitted with Centaur weapon control system

CENTAUR

Marconi's expertise in both the fire control and gun control aspects of tank weapon control systems has enabled the development of Centaur, a combined system that ensures a fighting capability in keeping with the designs and requirements of the modern battlefield.

Centaur can be offered at two levels. For modern Main Battle Tanks such as the Engesa Osorio and Vickers Mk 7, where an excellent shoot on the move performance is required, a new high performance fire control system has been designed to operate with the sophisticated independently stabilised sights required for such performance. This has been fully integrated with the GCE 628 gun control and stabilisation system through the use of a combined cubicle containing both the FCS computer and the GCE processing electronics. This facilitates total system BITE, reduces space requirements and allows cost savings. An intelligent control panel with its own micro-processor allows greater control and a choice of operating modes to ensure complete system flexibility. Combined with a full range of dynamic sensors, thus Centaur ensures that the highest performance is available for rapid acquisition and accurate engagement from both static and moving vehicles.

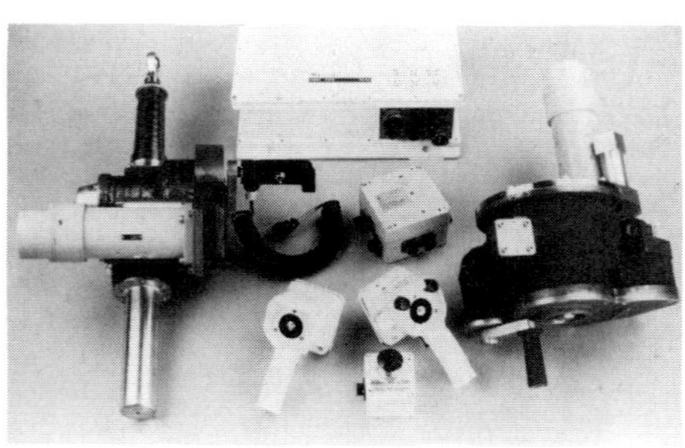

PD700 equipment group

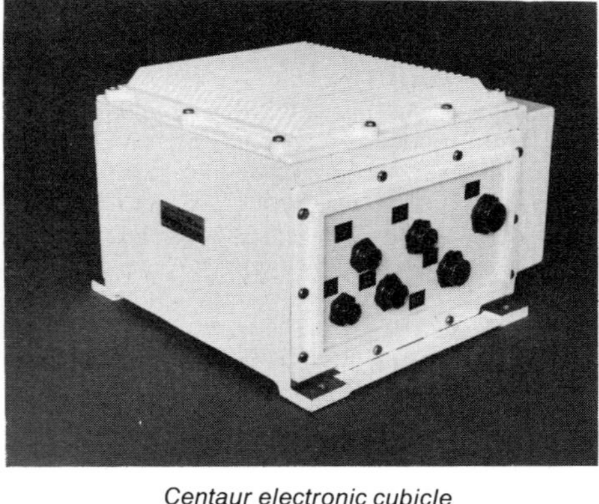

Centaur electronic cubicle

For AFVs employing fixed sighting systems such as T-tanks and light tanks, a simpler version of Centaur can be provided that combines many components from EFCS 600 with a less powerful version of GCE 628. Utilising the same combined processing unit, this system applies the benefits of full integration in a package well suited to the performance and cost requirements of such vehicles. A new development that has now been incorporated is the stabilised aiming mark. This provides a significant improvement in shoot on the move performance through standard fixed sights thereby ensuring a cost effective solution to the requirement for an effective all round capability.

Another development applicable to both versions of Centaur is the onboard training system. A graphics display facility allows simulated targets to be produced by the CRT in the sights. Controlled by the commander to give a variety of engagement types, these can then be engaged by the gunner using his standard engagement procedures.

Centaur is well proven. Systems have been trialled on the Engesa Osorio and Vickers Mk 7 with sophisticated sights, and on T-55, Type 59 and the VFM 5 with fixed sights. In all cases, high levels of performance have been achieved.

CONTRACTOR

Marconi Command and Control Systems Ltd, Chobham Road, Frimley, Camberley, Surrey GU16 5PE, England.

4832.181

COMPUTER SIGHTING SYSTEM (CSS)

Designed primarily for use in the Chieftain and Challenger main battle tanks the CSS is a computer based integrated tank fire control system based on the Marconi MC1800 microprocessor which interfaces to the Barr and Stroud TLS-2 tank laser sight. It is a compact system intended for retrofitting in existing vehicles with minimum effort to give maximum improvement in fire power performance. Sensors mounted on the tank provide the computer with variables such as range, wind speed and direction, air pressure, air temperature, trunnion tilt, angle of sight, turret azimuth position and gun elevation position. Together with other variables and ammunition performance the computer calculates the gun super elevation and lead angle.

The system's design provides simple crew operation, rapid engagement time and high hit accuracy to give good battlefield survivability. CSS was developed from the British Army Integrated Fire Control System IFCS. A full suite of fault diagnosis equipment is available and with the system self test facilities identification of faults to component level is possible.

STATUS

IFCS and CSS are in quantity production for the Chieftain and Challenger main battle tanks for service in the British Army and elsewhere. A development programme is currently under way to further shorten crew tasks and engagement times and to significantly improve the fire on the move capability.

CONTRACTOR

Marconi Command and Control Systems Ltd, Chobham Road, Frimley, Camberley, Surrey GU16 5PE, England.

2647.181

IFCS – IMPROVED (TANK) FIRE CONTROL SYSTEM

Designed primarily for use in the Chieftain main battle tank (**5006.102**) as a replacement for the FV/GCE No 7 Mk 4 equipment, IFCS is a computer-based integrated digital tank fire control system. Based on the Marconi 12-12P general-purpose digital computer and the Barr and Stroud TLS – 2 tank laser sight (No 1 Mk II), it is a compact system intended for retrofitting in existing vehicles with minimum reduction in the quantity of ammunition carried.

The computer, which can accurately simulate ballistic trajectories out to the longest ranges likely to be used in tank engagements, accepts data for the trajectory calculation from various sensors, which provide continuous information on such parameters as wind speed and direction, air pressure, air temperature, trunnion tilt and angle of sight. Muzzle velocity is derived from ammunition to be fired, charge temperature and barrel wear, which is predicted from a count of equivalent full charges fired. All this information may be readily monitored by the tank commander and can, if required, be overridden by means of a manual input. The equipment is suitable for engaging fixed or moving targets from a stationary or moving vehicle.

OPERATION

When the commander has slewed the turret to the approximate azimuth and elevation of the target, and the gunner has confirmed that it is in his sight, the main armament and ammunition are selected on the firing handle – an operation that automatically charges the laser. The gun is then loaded and simultaneously the gunner lays the sight aiming mark on the target, flashes the laser, and continues to track the target. Range is displayed instantly, both as a digital display in the sights and as an elliptical aiming mark surrounding the target in the sight graticule.

Within about one second (depending on the target and the ammunition to be used) after the laser is flashed, the computer calculates the ballistic trajectory and aim-off for target movement. Then, as soon as the gunner presses the autolay switch, the computer offsets the elliptical aiming mark in the sight to give a new point of aim for target range and speed. Using these corrections, the computer also automatically re-lays the gun, and the gunner fires when the target and aiming mark again coincide.

The controls for all the operator-selected functions are mounted on a single firing handle, which is operated by the gunner's left hand: his right hand is free to lay the gun in azimuth and elevation, using a thumb controller. All system tell-back displays used during an engagement are projected in the sight, so that the operator need never take his eyes off the target nor change his hand positions. Firing handles, thumb controllers, and sights are provided for both the commander and the gunner, but with the difference that the former can override the latter.

STATUS

The system is now in quantity production for the UK armed forces. This system is being adapted for fitting to other tanks and fittings have been made in the FV4030/2 (Khalid) MBT for Jordan and the Challenger.

Chieftain tank equipped with the IFCS during trials of the system

CONTRACTOR

Marconi Command and Control Systems Ltd, Chobham Road, Frimley, Camberley, Surrey GU16 5PE, England.

4581.181

MARCONI DFCS

The Marconi Digital Fire Control System (DFCS) was originally designed to cater for both new vehicle and retrofit/modernisation markets where improved fire control facilities are required to meet modern AFV specifications.

The modular design and compact size of DFCS enables it to be fitted into the small turrets of light armoured vehicles and into the restricted space in tanks such as M41, M48 and T series requiring modernisation. Particular attention in the design of DFCS to manufacturing, quality and testing methods has ensured that production costs are compatible with the weapons fit budget of these vehicles. DFCS offers the full capabilities essential to a modern fire control system including moving target capability, automatic gun laying (Autolay) and a built-in gunnery training simulator. It also has a fire on the move capability for MBT applications.

The DFCS 16-bit digital computer unit calculates a ballistic solution including inputs for meteorological data, relative target motion, barrel wear and trunnion tilt. Its performance is compatible with the accuracy of 76, 90, 105 and 120 mm calibre guns even at extended ranges using low velocity HEAT and HESH rounds. Simple engagement procedures are designed to reduce the dependence on human judgement and skill under stressful battlefield conditions. System ergonomics are such that the gunner maintains visual contact with the target over the entire engagement sequence with all relevant information displayed in the gunner's sight. All necessary controls are mounted on the gunner's handles. A feature of DFCS is that it offers an automatic servo-controlled final lay of the gun in readiness to fire which enables DFCS to achieve faster and more accurate engagements than manual fire control systems. The combination of high first round hit probability, extended range and fast engagement ensures a high probability of battlefield survival.

DFCS trials have been carried out in the Cadillac Gage Commando Stingray

The main elements of the DFCS are described in the following paragraphs.

COMPUTER

The computer incorporates self-test software and permanently retains data such as barrel wear and 'shooting-in' adjustments. The main program is held on a plug-in module.

CONTROL PANEL

This provides the following commander's facilities and displays in any language or script:

(1) system and calibration confidence checks
(2) designation and display of ammunition type
(3) entry of ballistic information such as charge temperature, air density, barrel wear and static offset corrections
(4) monitor of sensor inputs
(5) entry of add/drop corrections
(6) manual override of automatic sensors.

CRT CONTROLLER

This produces computer controlled display images in the sight including aiming mark, alpha-numeric range and ammunition type. It also provides simulated target shape, and tracer trajectories in the training mode.

BALLISTIC SENSORS

To increase system accuracy further sensors for trunnion tilt (cant) and meteorological conditions are fitted, plus optical encoders for Traverse and Elevation to provide high quality target tracking and to ensure accurate automatic gun laying.

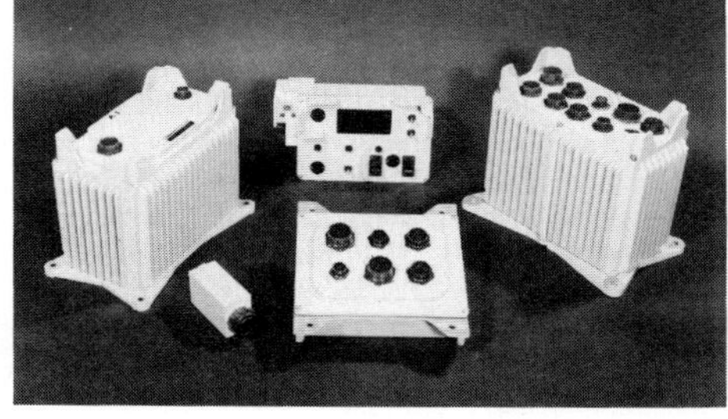

Main DFCS units: power supply (left), CRT controller (back), computer (right), with other elements (front)

CONTROLS

For a power drive turret servo the firing handle and thumb controller provide the gunner with all the controls necessary to conduct an engagement ie weapon selection, ammunition designation, lasing, autolaying and firing on the firing handle, and target tracking on the thumb controller. Controls are available for both electric and hydraulic driven turret servo systems.

STATUS

Preliminary firing trials took place in 1983 in a Cadillac Gage Commando V-150. Since then the DFCS has been successfully tested in the Cadillac Gage Stingray light tank and the V-600 (previously V-300 A1 (6 × 6)) armoured vehicle.

CONTRACTOR

Marconi Command and Control Systems Ltd, Chobham Road, Frimley, Camberley, Surrey GU16 5PE, England.

4148.161

FERRANTI FIRE CONTROL SYSTEM

The Ferranti Fire Control System is a digital system based on the F100-L microprocessor, providing a versatile and cost-effective means of enhancing the combat capability of gun-armed fighting vehicles. Compact dimensions, simplicity of operation and low cost ensure that the system application is not restricted to main battle tanks. The versatile compact computer has been applied successfully to AD platforms, and for the control of main armament on small naval craft. It is suited for both new vehicles and retrofit installations.

The system enhances the combat capability by significantly improving main armament accuracy. It automatically produces an accurate determination of range and a precise compensation for the effects of ballistic and transient variables; it substantially increases the probability of a first round hit. Accuracy is improved against both static and moving targets. The gun-laying sequence is simplified by the injection of a computer-generated aiming mark in the gunner's sight.

The system accepts inputs from a variety of automatic and semi-automatic sensors that enable

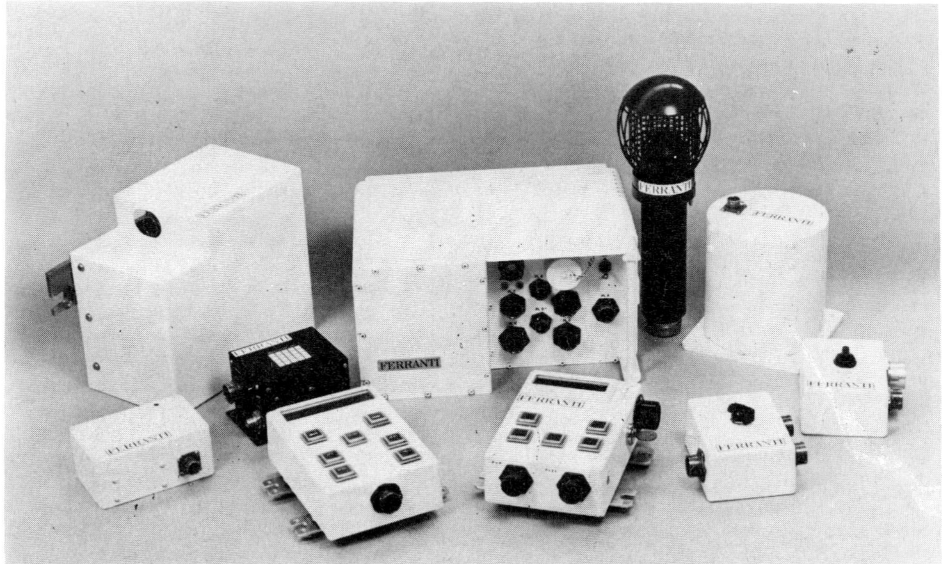

Ferranti fire control system equipment

appropriate fire control parameters, eg meteorological conditions, trunnion tilt, laser rangefinder, etc to be used in aiming ,and gun-laying computations. Accurate determination of target range and the automatic compensation made for crosswind velocity, charge temperature, air temperature and pressure, increases the probability of a first round hit at all ranges. Engagement time is significantly reduced by the production of a ballistically and environmentally corrected aiming mark during the gun-laying sequence, and by increasing the speed at which the crew can respond to events. The vehicle commander can attend to target acquisition and tactical decision making while the gunner can concentrate on the designated target and on computer-generated data presented to him via his sight.

CONTRACTOR

Ferranti Computer Systems Ltd, Bracknell Division, Systems Department, Ty Coch Way, Gwent NP44 7XX, Wales.

4531.181

CLASSIC (RGS 2740)

RGS 2740 CLASSIC (Covert Local Area Sensor System for Intrusion Classification) is an extension of the local area sensor system (Project LASS) carried out by the Royal Signals Research Establishment and Racal Ltd to evolve a system that detects, classifies and remotely displays 'target' information on personnel, wheeled and tracked vehicles by means of a variety of remote ground sensors.

The basic system consists of two main units; the sensor and the monitor. Up to eight sensors can be used with each monitor, and the former are designed to be hand emplaced in suitable tactical locations. Each sensor is coupled to a transducer, either a geophone or an infra-red detector. The unit contains signal processing circuitry which classifies the input and broadcasts a tone-coded message by means of a built-in VHF/FM transmitter. The monitor unit receives this signal, decodes the data and presents the information on an LED display to show sensor identification, type and frequency of intrusion. To extend the range of the sensor transmission, a relay unit is available.

CLASSIC is a modular system with a range of optional accessories, which includes alternative antennas and battery units, transducer/pressure pad switches and a hard copy printer to meet a wide range of ground sensor applications. Standard sensors are the TA 2741 used with the MA 2743 seismic transducer, or the MA 2744 infra-red transducer. The TA 2741 has a switch to select High, Medium or Low Personnel seismic sensitivity, and another switch to select the classification code.

The RTA 2746 monitor comprises a VHF receiver, a tone decoder and an LED display. On receipt of a transmission from a sensor unit, an audible alarm alerts the operator. The sensor's identification and alarm mode setting is displayed on a matrix of three LEDs for each of the eight possible sensors. The display is inhibited after approximately 14 seconds to avoid excessive battery drain, and a pushbutton is provided to enable viewing on demand if required. A user's map panel provides for drawing of a diagram of a tactical deployment.

STATUS

In production. In service with many armed forces including the British Army. In 1987 Australia placed a contract worth £820000 for CLASSIC.

CONTRACTOR

Racal Causec Ltd, Milford Industrial Estate, Tollgate Road, Salisbury, Wiltshire, SP1 2JG, England.

CLASSIC monitor unit gives audio/visual warning and indicates which sensors have been activated

UNITED STATES OF AMERICA

3943.161

MIFASS

Development of MIFASS under a $843.8 million contract ended in 1987. The retiring US Marine Commandant, General P. X. Kelley, commented that the system was deficient "... in terms of configuration, reliability, speed of operation, communications support, user interface and mobility".

CONTRACTOR

Norden Systems Inc, Norwalk, Connecticut 06856, USA.

3510.161

BATTERY COMPUTER SYSTEM

The Battery Computer System (BCS), an advanced fire control system developed for the US Army, has been designed to replace the current field artillery digital automatic computer (FADAC) (**1125.161**) and extend the capability of TACFIRE (**2827.161**), the tactical fire direction system. It provides computerised control of artillery fire at the battery level. A real-time command, control and communications system, it is capable of providing simultaneous computations for up to 12 howitzers or guns and handling three concurrent fire missions.

It is a portable, single unit system containing 256 K of memory capacity. By providing direct digital access for forward observers and to battalion TACFIRE, it extends that system's capability to the firing battery level. It also operates in an autonomous mode and can be used for fire planning and the engagement of moving targets.

The BCS is composed of the battery computer unit (BCU) located at the battery fire direction centre and one gun display unit (GDU) for each weapon in the battery. The BCU is the electronic centre for fire control. It maintains a digital communication link with fire direction officers, fire support officers, forward observers, and weapon section chiefs. It performs all required computations for first-round accuracy, with results appearing on a 1728-character plasma display. The BCU contains a data management system which maintains target, fire unit, ammunition, meteorological, map, target, forward observers, and other system information to simplify the FDC tasks. The central processor within the BCU contains 256 K × 24-bit words of memory, with error detection and correction circuitry for fail-safe reliability. The time taken to compute a fire mission with individual data for 12 weapons is two per cent of the time of flight to the target.

Mission assignments for the BCS are received from forward observers who digitally transmit target information to a battalion TACFIRE or directly to the battery computer unit located at the field artillery level

The battery computer unit, one of two elements of the BCS, acting on the information from the foward observer, automatically computes the firing data and displays it on the unit's screen. Also displayed is the status of the battery's howitzers. The soldier will select a gun for the mission and transmit the firing data to the gun display unit located at the howitzer

The BCU houses a magnetic tape cartridge for software program and data storage, the former permitting central processor programming in under one minute. It interfaces with existing US Army COMSEC equipment and printers. Data are transmitted to the guns in one second. The BCU communicates on the battlefield with WD-1 wire and AM/FM radio.

The communications control device is driven by a microprocessor controller and offers high-speed direct access to the memory module. The BCU contains the main power switches and self-test status panel, so that faults in individual elements can be easily isolated and corrected.

The GDU is the final link in the BCS communications chain. Its place is with each weapon in the firing battery. The GDU consists of three assemblies, a section chief assembly (SCA) and two gun assemblies (GA), plus a signal and power distribution unit. All elements are of solid-state sealed construction.

The SCA is a personal tool for the section chief (weapon commander), the man who commands the weapon crew. Slightly larger than a commercial hand-held calculator, the SCA is connected by cable to the signal and power distribution unit, giving instant access to all gun-related data and commands. The device contains its own memory bank for sequential or direct access to stored fire mission information. It is equipped with audible alarms and visual cueing to alert the section chief to fire missions and check fires. An eight-digit display shows positive indication of gun commands under all ambient conditions. A single button sequences the display of gun orders.

The two identical GAs on the weapon show the gunner and assistant gunner separate elevation and deflection display. They are connected by wire to the signal and power distribution unit and receive the commands directly.

DEVELOPMENT

Norden Systems of Norwalk, Connecticut, USA, was awarded a multi-million dollar contract for development of the Battery Computer System in September 1976. Marconi Space and Defence Systems Limited (now Marconi Command and Control Systems Limited) of England was Norden's principal sub-contractor, and has incorporated much of the experience gained in the design and development of the British FACE (**2415.161**) field artillery computer equipment.

STATUS

A production contract was awarded to Norden in April 1980. Deliveries began in November 1981. The multi-year contract is valued at about $97 million and in June 1980 Norden awarded MSDS Ltd a $20 million contract for the supply of BCS GDUs. These units are based on the MSDS MC1800 microprocessor which is also employed in the latest version of the FACE field artillery computer equipment (**2415.161**). In March 1986 Marconi Command and Control Systems was awarded a further contract worth over £5 million for use with the BCS. This order, for over 1000 GDUs, followed six previous contracts and brought total value of GDU contracts up to over £23.5 million. Deliveries for the latest contract will run from 1986 to 1988. In addition to being deployed by the US Army, BCS is also deployed by the US Marine Corps.

CONTRACTORS

Norden Systems Inc, Norwalk, Connecticut 06856, USA – Prime contractor.

Marconi Command and Control Systems Ltd, Chobham Road, Frimley, Surrey, GU16 5PE, England.

2643.181

INTEGRATED TANK FIRE CONTROL SYSTEM

Intended primarily for installation in the M60 MBT, this system is in the later stages of production. This modern integrated tank fire control system provides full solution capability and can be used for the control of different guns in other tanks.

The system is based on the M21 Ballistic Computer System which includes sensors to detect and measure trunnion cant, crosswind and turret rotation rate. The system provides for the operator selection of ammunition type and the manual input of air temperature, air pressure (altitude) and barrel wear. Range information is provided from an associated laser rangefinder system.

A typical configuration is utilized in the M60A3 tank and is composed of two sub-systems: the AN/VVG-2 laser rangefinder system; and the M21 Ballistic Computer System.

The AN/VVG-2 laser rangefinder system is composed of two main units:

(1) a commander's integrated laser/sight with control unit
(2) a laser electronics unit. Range may be fed automatically to the computer or, if more than one return is received, the commander may select the range return based on his assessment of the situation. Since the gunner's sight is boresighted with the laser rangefinder, either the commander or the gunner can fire the laser and/or the gun.

The computer sub-system includes two ammunition selection units. The commander or gunner may select one of the four basic types of ammunition to be used. Actually, the computer stores solutions for six ammunitions within the four basic types to accommodate ammunitions stored in different geographical locations. The switch at the bottom of the ammunition selection unit is used to set up either tank moving or stationary conditions.

The computer is a solid-state hybrid. It processes all the input data and commands the sight lines for laying the gun and positioning of gunner's reticle.

The top half of the gunner's control unit panel contains three manual inputs: air temperature, pressure altitude (slowly varying inputs) and manual range in the event of laser failure. The row of lights at the top of the unit is for self-test. This self-test system removes the trouble-shooting burden from the tank crew, and replacement of system units in the field is easily accomplished. The bottom half of the panel (normally covered) is used for boresighting, zeroing, manual crosswind input (if desired) and a switch to select the ammunition within the four basic types.

Automatic sensors feed data directly into the computer. They are:

(1) the rate tachometer unit – used when tracking moving targets from a stationary tank. When the tank is moving, this unit (which is most accurate when the tank is stationary) is switched out and replaced by input from the gun stabilisation system gyros, since inertial tracking rates are required
(2) the crosswind sensor – can be stowed horizontally and is spring mounted to prevent damage when the tank encounters low branches. Since this is the only unit not under armour, a circuit in the computer continuously monitors the crosswind sensor output and can be switched to manual input in the event of failure
(3) the cant unit – senses gun trunnion roll.

There is also a gunner's reticle projector that mates with his periscope and provides azimuth deflection of the reticle, and the output unit which drives the periscope head mirror, and the commander's sight in elevation and super-elevation.

CHARACTERISTICS

Operational range: 200 – 5000 m
Range resolution: 2 targets can be resolved at a minimum of 20 m apart
Range accuracy: ± 10 m
Target angular velocity: 0 – 50 mils/s
Max target speed: 0 – 20 m/s
Cant angle: 0 to ± 15°
Crosswind (auto): 0 – 18 m/s
Crosswind (manual): 0 – 23 m/s
Air temperature: –54 to + 52°C
Altitude: –200 to + 3000 m

STATUS

During 1977 Hughes, and Kollsman were awarded quantity production contracts for systems for fitting in M60A3 MBTs. Hughes received a production contract for an additional 1169 systems in 1985. Options for up to 585 systems were also included.

SABCA, Hughes European licensee has produced laser tank fire control systems for Leopard 1 MBTs used by Australia, Belgium and Canada and modified Centurion tanks used by Jordan. The UK licensee, Barr and Stroud, has been delivering laser rangefinders for a number of years for the Chieftain tank in the UK and overseas.

In April 1983, Spain selected a laser tank fire control system based on the M60A3 to modernise its M48 tanks. About 70 per cent of M60A3 components are used in the Spanish Army system and many are produced in Spain by ENOSA/EISA of Madrid, under licence.

Kollsman has completed an in-house product improvement of the AN/VVG-2 Laser Rangefinder System. The improved design employs Neodymium Yag Technology, utilizing AN/GVS-5 hand-held laser rangefinder modules, to provide improved ranging capability in inclement weather conditions, and an order of magnitude improvement in reliability and maintainability. The improved design incorporates a continuous self test feature capable of identifying faults down to the lowest replaceable unit (LRU). The entire rangefinder is packaged within the commander control unit thus eliminating the electronic unit and associated system interconnect cables and resulting in smaller overall packaging of the system.

In 1985 Kollsman supplied a maintenance depot and related test equipment to the Arab Republic of Egypt to support its M60A3 tank fire control systems. Kollsman supports this facility with field service engineering.

CONTRACTORS

Hughes Aircraft Company, El Segundo, California 90245, USA.

Kollsman Military Systems Division, 220 Daniel Webster Highway, Merrimack, New Hampshire 03054, USA.

M60A3 MBT fire control system in use; crewman is using laser rangefinder

6354.171

BRIEFCASE TERMINAL

The Briefcase Terminal (BCT) is a lightweight portable, intelligent communications and display terminal intended for use by commanders at all echelons of the US Air Force, Army and Marine Corps. Designed especially to help commanders co-ordinate troop actions quickly and efficiently, the BCT provides simultaneous communication over six separate radio or wire nets and presents commanders with a fully interactive alpha-numeric and graphic display. Up to three ancillary devices can be connected simultaneously to the BCT. These may include a digitizer, a typewriter keyboard, a printer and a mass storage device such as magnetic type floppy disc.

Key elements are the LED modules, the transparent membrane switch, the RAM memory packages, the hybrid MIL-STD-188C interface, which can transmit at up to 34.8 Kbits/s for each of three channels, and the programmable hybrid modem, which offers FSK rates ranging from 75 bits/s to 2.4 Kbits/s, and digital baseband rates ranging from 75 bits/s to 16 Kbits/s simultaneously, on up to six channels. Net busy sensors automatically prevent transmission when radio nets are being used. The BCT's modems are fully compatible with a range of field radios and secure equipment.

The BCT interoperates with the US Army Battery Computer System (**3510.161**) and TACFIRE systems (**2827.161**). When used with its map digitizer and graphics capability, it provides a graphic picture of the battlefield to aid the commander in making decisions related to technical and tactical fire control at the battery and battalion level.

The entire terminal is housed in a case which is 457 mm wide, 279 mm high and 152 mm deep, and weighs just over 11.3 kg, including a military standard lithium battery which will power the unit for up to 12 hours. A 24-volt vehicular power supply is also an integral part of the unit.

STATUS

In production.

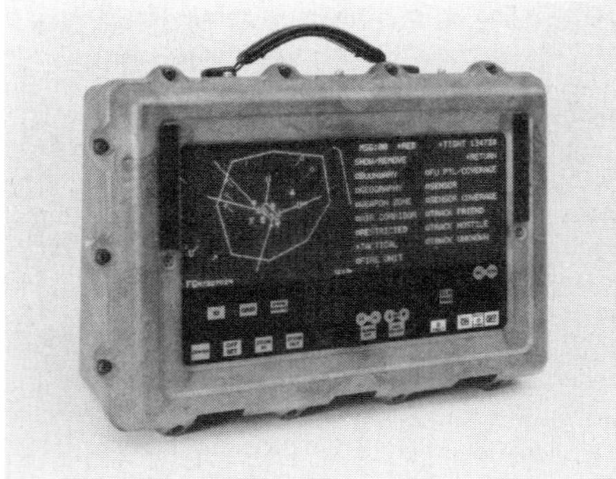

Litton Data Systems Briefcase Terminal (BCT)

CONTRACTOR

Litton Data Systems, 8000 Woodley Avenue, PO Box 7601, Van Nuys, California 91409-7601, USA.

6354.161

AN/PSC-2 DIGITAL COMMUNICATIONS TERMINAL

The AN/PSC-2 Digital Communications Terminal (DCT) is a lightweight hand-held communications message processor providing the user with point-to-point and netted communications over a variety of military radios. It provides user prompting to aid in the composition of messages. The capability to compose, edit, display and process communications messages is provided through a stored program and a micro-processor to augment and speed the composition of messages. A map display is provided to assist in position entry of data. This map is scaled from a field map and can identify terrain and tactical data.

Two of the principal advantages of the DCT over earlier message devices are its extremely small size and weight (2.2 kg) and its increased memory capacity (128 K RAM). Battery life is also significantly improved through the use of low power CMOS components.

AN/PSC-2 Digital Communications Terminal (DCT)

The DCT can receive and transmit multiple messages over tactical nets with operator alerting and automatic storage of specified messages. These messages can be in fixed or variable format. The message processor performs all tasks of format composition, address coding, error control and error checking, as well as net protocol. The operator can specify message data rate, addresses of receiving agency and keying time.

In addition to its direct message capability, the DCT interfaces with the Modular Universal Laser Equipment (MULE) (**3816.193**) and a Ground Locating Laser Designator (GLLD) providing a designate command and accepting a target range, azimuth and height report from the MULE/GLLD. With MULE/GLLD reports, the DCT can generate a target position for message composition.

The DCT provides interaction with the operator by means of the LED display with a transparent overlayed 54-position switch entry. Using this direct input/readout mechanism the function of each switch can be redefined by operator menu selection under program control. The operator can compose from a full alpha-numeric keyboard on selected and formatted messages. To augment this capability and reduce position entry, direct graphic capability is provided with touch or cursor selection and entry of individual map positions.

When used by an artillery Forward Observer (FO), the FO can compose request-for-fire messages and transmit them to higher echelons for action. In this mode, the DCT is programmed to lead the FO through various message types. After all entries have been made, the FO can review the entire message. Corrections are made by touching the display at the field to be changed. This causes the display to return to the keyboard and depict the menu originally used to generate the input. After making the correction, touching the review switch causes the corrected message to be displayed. When a satisfactory message has been composed, the message is transmitted by touching the transmit control on the display.

The DCT compresses and formats the message and adds error correction and detection codes. When messages are received, audible and visual alarms signal the FO. The DCT accepts incoming messages while the FO is composing a message and stores them until the received messages are called up for viewing. The message buffer is capable of holding three composed and five incoming messages. The message buffer size is user program controllable.

STATUS

In production for US Army, Air Force and Marine Corps.

CONTRACTOR

Litton Data Systems, 8000 Woodley Avenue, PO Box 7601, Van Nuys, California 91409-7601, USA.

6356.161

INTELLIGENT TERMINAL

The Lightweight Digital Command Terminal (LDCT) is the heart of the Intelligent Terminal System and is a portable, intelligent communications, computation and display terminal intended for use by commanders at all echelons of the US Air Force, Army and Marine Corps. Designed especially to help commanders co-ordinate troop actions quickly and efficiently, the LDCT provides simultaneous communication over up to six separate radio or wire nets and presents commanders with a fully interactive alpha-numeric and graphic display. Up to three ancillary devices can be connected simultaneously to the terminal. These

Typical Intelligent Terminal (IT) equipment configuration

can include a digitizer, a typewriter keyboard, a printer and a mass storage device such as magnetic tape or floppy disc.

The intelligence of the terminal is provided by up to six micro-processors and 2.4 Mbytes of ram memory. Up to five of the micro-processors provide independent control of the modems, the three MIL-STD-188C serial interfaces, the LED display (which contains more than 55000 addressable picture elements) and the overlay membrane switch (which offers more than 400 programmable and fixed positions). Each of these micro-processors has up to 64 kbytes of memory associated with it. The sixth micro-processor is a 16-bit device which provides up to 2048 Kbytes of ram memory for complex, high-speed processing.

The MIL-STD-188C interface is capable of transmission at up to 38.4 kbits/s for each of three channels, and the programmable hybrid modem offers FSK rates ranging from 75 to 2400 bits/s and digital baseband rates ranging from 75 to 16000 bits/s simultaneously, on all channels. Net busy sensors automatically prevent transmission when radio nets are being used. The LDCT's modems are fully compatible with a full range of field radios and secure equipment.

STATUS

In production.

CONTRACTOR

Litton Data Systems, 8000 Woodley Avenue, PO Box 7601, Van Nuys, California 91409-7601, USA.

2827.161

TACFIRE TACTICAL FIRE DIRECTION SYSTEM (AN/GSG-10)

TACFIRE is a computer-based automated command and control information system for planning and directing artillery fire. It consists of central fire direction centres (FDC) at division artillery and at battalions with a remote device at batteries, the battery display unit (BDU), and another remote device, the variable format message entry device (VFMED) with fire support officers (FSO) and fire support elements (FSE). Digital communications between FDCs and remote devices are by tactical radio or wire and can be clear or encrypted. There are provisions for voice back-up.

The battalion level TACFIRE is housed in a single S-280 shelter and consists of the following major sub-systems:

(1) computer (AN/GYK-12) comprising central processor unit (CPU); input output unit (IOU)
(2) mass core memory unit (MCMU)
(3) power convertor group (PCG)
(4) data terminal unit (DTU)
(5) module test set (MTS)
(6) communications control unit (CCU)
(7) removable media memory unit (RMMU)
(8) artillery control console (ACC)
(9) digital plotter map (DPM)
(10) electronic line printer (ELP)
(11) remote control monitor unit (RCMU).

The division artillery level TACFIRE is housed in two S-280 shelters and consists of the same major sub-systems as shown in the battalion plus one additional MCMU, a dual RMMU, an electronic tactical display (ETD), and an auxiliary DTU.

The two remote equipments are:

(1) Variable format message entry device (VFMED) consisting of:
Remote data terminal (RDT)
Display editor (DE)
Keyboard
Electronic line printer (ELP).
(2) Battery display unit (BDU) consisting of:
Electronic line printer (ELP)
Remote data terminal (RDT).

The AN/GYK-12 computer, CPU, IOU, memory, is the heart of the TACFIRE system. It is a general-purpose micro-electronic computer specifically programmed to facilitate solution of the real-time command and control system problems required in TACFIRE.

TACFIRE artillery control console. Located at battalion and divisional artillery fire direction centres, the ACC is the primary device to control data into and from the computer manually. This control and display device provides a visual presentation on two CRTs of both incoming messages and computer mission solutions. It also provides for control and initiates action on those data by use of an alpha-numeric keyboard which permits real-time entry and interrogation of the computer in normal user language

Currently the divisional artillery and battalion computers are identical, except for one additional memory at the former. However a product improvement programme will replace the AN/GYK-12 with the CP-1822/GSG-10(V) computer into the division artillery level system. This next generation computer is common to the AN/TTC-39 and MSE systems and will eliminate one of the S-280 shelters and associated generators and cables. This improvement will be fielded in 1988. The other major items of hardware of the current system were described in some detail in 1983-84 and earlier editions.

STATUS

In 1981, the US Army procured its fifth option for AN/GSG-10 TACFIRE systems, and production was completed in 1984.

CONTRACTOR

Litton Data Systems, 8000 Woodley Avenue, PO Box 7601, Van Nuys, California 91409-7601, USA.

6357.161

LIGHTWEIGHT LTACFIRE

This system is installed in a High Mobility Multipurpose Wheeled Vehicle (HMMWV), requires no generator, and has all of the software capabilities of TACFIRE, less Technical Fire Control (TFC) which is performed at the Battery Computer System (BCS). The LTACFIRE Battalion Fire Direction Center (BN FDC) provides all the essential capabilities of TACFIRE with numerous significant enhancements, in a small, lightweight, low power, soldier-friendly system. The LTACFIRE:

(1) replaces the TACFIRE BN FDC
(2) replaces the LTACFIRE VFMED utilizing the LTACFIRE Intelligent Terminal (IT)
(3) interoperates with TACFIRE, BCS, Fire Support Team (FIST), Digital Message Devices (DMD), Firefinder, radars, DMDs, DCTs, etc
(4) provides a split and jump capability for BN FDCs
(5) provides a graphics capability
(6) Greatly reduces size, weight and power requirements.

One significant functional enhancement of LTACFIRE is the ability to perform Tactical Fire Control (TFC) and thus co-ordinate 81 mm and 107 mm mortars. The Briefcase Terminal (BCT) (**6354.171**), a part of the LTACFIRE, also offers a tremendous increase in digital transmission rates with its 16000 BPS speed which is valuable in minimizing transmission time and thus avoiding enemy detection. Another significant advantage is that two-man machine interfaces are provided at the Fire Direction Centers (FDCs); one for planning and one for current operations. This, along with the improved soldier-machine interface, should eliminate the TACFIRE Artillery Control Consol (ACC) 'choke point'. The ability to rapidly input Fire Support Co-ordination data into the BCT through the use of the map digitizer is also a tremendous system enhancement.

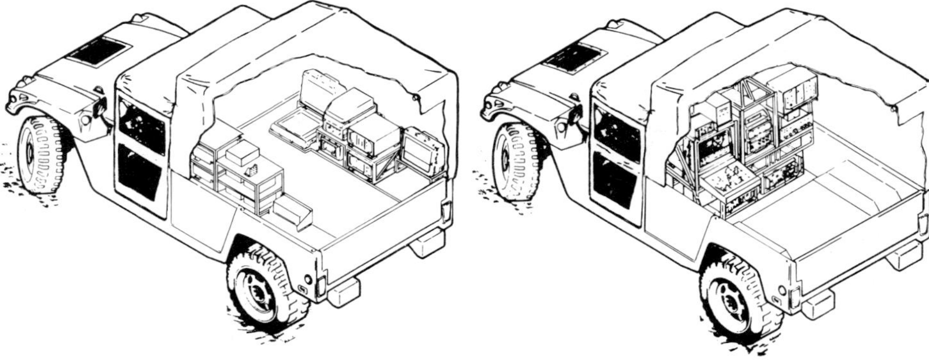

Lightweight TACFIRE installed in the HMMWV. On the left is the battalion FDC (BN FDC) and on the right is the Intelligent Terminal (IT) version

One of the most significant improvements of LTACFIRE is the FDC's ability to perform its mission through the capabilities of the Intelligent Terminal (IT). FDC operators do not have to spend valuable time responding to VFMED requests for data/formats, since the IT has all the message formats resident.

Litton's IT dramatically improves the remote device capability by providing a fully functional device independent of the mainframe computer; thus providing a distributed system architecture rather than a centralized one. The IT has many significantly improved capabilities and for the first time provides Fire Support Section/Fire Support Element (FSS/FSE) personnel with a piece of C³ equipment that allows them to perform their duties in accordance with approved doctrine. The Fire Support Officer (FSO) is truly able to co-ordinate his subordinates utilizing the IT because it is fully interoperable with Fire Support Teams and has its own Tactical Fire Control (TFC) and fire planning capability. The IT operator can call up fire plans, Fire Support Co-ordination Messages, geometry, fire missions, etc. from its data base and does not have to query TACFIRE for message formats since the IT has them all stored. Additionally, the soldier-machine interface eliminates the requirements for VFMED operators to have all valid entries memorized.

STATUS

Fielded at US Army 9th Division. Operational Testing (OT) was completed in May 1987.

CONTRACTOR

Litton Data Systems, 8000 Woodley Avenue, PO Box 7601, Van Nuys, California 91409-7601, USA.

Lightweight TACFIRE Fire Direction Center (FDC)

3317.161

TACTICAL COMPUTER TERMINAL (TCT) AN/UYQ-30

The TCT is a fully militarised tactical computer terminal compatible with existing and planned military communication equipments. The TCT was developed as an intelligent stand-alone terminal capable of communicating digital messages (text and graphics) with other TCTs, and with forward observer equipments, and will be interfaced with the Manoeuvre Control System (MCS) and World-wide Military Communication System (WWMCS).

The tactical computer terminal consists of a plasma display panel microprocessor, a keyboard, communications modems, controllers for external peripheral devices such as a printer, a bulk memory and a central microprocessor. Motorola MC 68000 microprocessors provide complete functional flexibility with capabilities determined by the software resident in the TCT. Read only memory (ROM) and random access memory (RAM) totalling 1 Mbyte optionally expandable to 8 Mbytes, with battery back-up, are packaged in the unit. The latest software packages include higher-level languages. Option for a 32-bit MC 68020 microprocessor will be available.

All of these elements, including the required power supplies, are packaged in a single combination case 38 cm high, 43 cm wide and 59 cm deep. The keyboard is an integral part of the case and folds up to form a watertight cover in the transport configuration. In this configuration, the unit is fully immersible to a depth of 1 metre. Cooling is accomplished entirely without the use of fans, either internal or external. The unit operates directly from 24 ± 6 V DC vehicular power.

The TCT communicates over two independent full or half duplex ports. It also has provisions for controlling a militarised printer/plotter. By simple change of the controller circuitry, other hard copy devices can also be accommodated. An additional port is provided through which the TCT may interface with a bulk memory device such as a magnetic tape, large scale semiconductor memory, bubble memory modules or recorder-reproducer flexible disk RD-508/G, a random access memory providing more than five M-bits of auxiliary storage or interchangeable media. An optional 8 Mbyte magnetic bubble or 100 Mbyte Winchester disk recorder-reproducer is also available. An automatic dialling capability is incoporated.

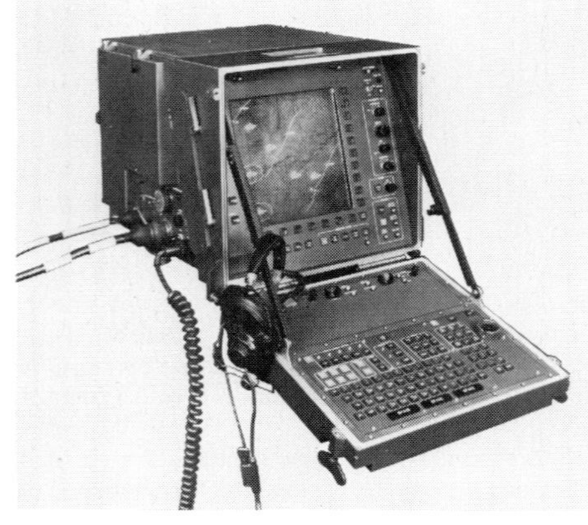

Tactical computer terminal (TCT) is compatible with existing US Army communications equipment

The display consists of a 22 cm square plasma panel which is utilised in both the message text mode and the graphics mode. When used in the graphics mode, tactical information is superimposed over a paper map positioned behind the transparent plasma panel. Larger displays, such as large-screen multi-colour displays at higher command echelons, will interface with the current equipment.

In the graphics mode, the display has the capability of overlaying standard topographic map segments using lines, alpha-numerics and standard or special symbols. It also has the capability of storing these overlays in memory and recalling them at some later time for modification or update. The overlays may be transmitted by wire or radio and hard copies may be obtained by use of an external printer.

The TCT has been packaged to give the maximum mobility possible to operate in the field environment. It meets all appropriate military specifications, including TEMPEST and nuclear survivability requirement.

CHARACTERISTICS

Display medium: Plasma panel
Element resolution: 60 elements per inch 512 × 512 matrix
Character size: 5 elements wide, 7 elements high
Character capacity: 46 lines of 85 characters each
Writing rate: 23 µms per element
Character writing period: 800 µms max
Vector writing rate: 35 µms per element average (computation time included)
Display intensity: 50 ft-lamberts max adjustable to 12.5 ft-lamberts
Refresh: No refresh required

STATUS

During 1976, the TCT was developed from the display/keyboard module of the AN/UYQ-19 tactical computer system, produced for the US Army Communications Research and Development Command. The manufacturer has recently completed a contract for the first 104 production units, and is under a follow-on contract for an additional 130 units.

CONTRACTOR

The Singer Company, Librascope Division, Glendale, California 91201, USA.

3944.161

PLRS (AN/TSQ-129)

The PLRS AN/TSQ-129 is a portable position location and navigation system which uses frequency and time division multiple access (FTDMA) techniques to provide real-time position tracks on several hundred ground and air user units. The characteristics will be found in **3944.161** on Page 103 of the 1987-88 issue.

The AN/TSQ-20 master station is housed in a standard US military S-280 shelter, which is air transportable. This contains the network communications facilities, computers, data displays and related equipment required to support a single independent PLRS network, or to co-operate with other master stations in a multi-PLRS network. An AN/UYK-7 computer is used for position calculation, time processing and display format, an AN-UYK-20A computer receives data from the UYK-7 while a second UYK-20A monitors and controls the PLRS station network. The system uses CMS-2 standard software.

The basic user unit consists of a multi-function transmitter/receiver and processor packaged in a 1-ATR short, low case, and equipped with either of two interchangeable input/output devices. In the AN/PSQ-4 manpack configuration the hand-held user readout module is employed, together with a battery pack giving 24-hour service. The complete manpack weighs 6.8 kg. The other input/output device is the pilot control display panel for use in fixed- and rotary-wing aircraft, tanks, amphibious assault vehicles, APCs or similar mobile units. This is mounted in a suitable place on the panel and is driven by a remotely mounted user unit elsewhere in the vehicle. Vehicles and boats use the AN/GRC-210 while aircraft use the AN/ASQ-117(V).

EPLRS

The Enhanced Position Location Reporting System (EPLRS) is the present designation of the former PJH (PLRS/JTIDS Hybrid) which was renamed in 1986. Like PLRS it is a digital communications network to provide data distribution requirements at divisional level but it has a higher throughput rate of 1200 bits per second. It is designed to support the US Army's five mission areas in a tactical battlefield; maneouvre control, air defence, fire support, intelligence/ electronic warfare and combat service support and will link command and control data systems within each mission area.

The system is designed to support forces in a fluid battlefield even in irregular terrain such as forests or mountains. The EPLRS network can be built up from nothing and will automatically conform to the battlefield conditions allowing for terrain masking and user motion. Each division will have a Net Control Station (NCS) at brigade and division level with another in the division rear. Should an NCS be lost the divisional or a neighbouring NCS automatically assumes control of the effected Enhanced PLRS User Units (EPUU). A division will employ between 500 and 1000 EPUUs.

Each EPUU has a line-of-sight system with automatic relay functions while the system includes such ECCM features as spread spectrum, frequency hopping, error detection and correction as well as automatic re-routing to counter ECM. Locations accurate to within 15 m are automatically displayed on a small hand-held device attached to the user unit. This relieves combat radio networks of up to 75 per cent of voice traffic. The EPLRS system uses time division multiple access technology to enable multiple users to transmit information simultaneously in bursts during pre-determined time slots.

The EPUU, an improved version of the PLRS user unit, features a new NCS 800 digital signal processor, a modified signal message processor, a new multi-level secure data unit and an interface with other battlefield systems. It weighs 11.79 kg (26 lbs) and measures 35.56 × 25.4 × 12.7 cm.

The NCS is a development of the PLRS master station with a third UYK-20A computer to provide additional processing for automatic management of the data communication network. One is allocated to each brigade to manage data distribution as well as providing location, navigation and identification services and it will control up to 250 EPUUs. The data communications requirements specified for each tactical area are entered into the NSC data base. The NSC automatically selects two independent paths for each line of communications and allocates sufficient time slots to accommodate the message and for a response.

PLRS user unit

JTIDS (Joint Tactical Information Distribution System) 2M terminals will be used at centres with very high digital message traffic or where joint service interface is required, for example in the Short Range Air Defense (SHORAD) centre. It is anticipated each division will require 50-70 JTIDS terminals.

Deployment of EPLRS is expected in 1988 and it is reported that the total cost to the US Army by 1990 could be $4 billion.

STATUS

The PLRS has been developed for the US Army and US Marine Corps. A $276 million multi-year contract for 11½ systems including 23 master stations and 3600 user units was awarded to Hughes in July 1983 with first deployment in 1986.

CONTRACTOR

Hughes Aircraft Company, Ground Systems Group, Fullerton, California 92634, USA.

3936.151

REMBASS

The Remotely Monitored Battlefield Sensor System (REMBASS) is an unattended ground sensor system evolved from Remote Sensors (REMS), which is already in service with the US forces. REMBASS will detect and classify both men and vehicles and it is intended to provide forward units and formations with detailed information on the happenings in the zone immediately to their front. It can also be useful in rear areas for the detection of intruders.

The system works by using remotely monitored sensors placed along likely approaches. These sensors respond to a wide variety of influences; infra-red, acoustic, seismic, mechanical energy, magnetic field changes and others. Each sensor produces a signal giving a detection and sends this signal by FM radio incorporated in the body of the sensor. The sensor remains passive when it is not transmitting a detection signal, so it is virtually impossible to locate. The signal may pass direct to the monitor unit, or it may be repeated by emplaced repeaters. With the use of relays/repeaters for ground application, REMBASS is designed to provide targeting data at up to 100 km range. Once at the monitor the signal is demodulated, decoded, displayed and recorded. The system provides up-to-date information over a wide area and, due to the variety of sensors, it is largely self-checking, ie the information from one type of sensor is checked against that from another, virtually eliminating false alarms.

The sensors can be emplaced by hand, 155 mm

REMBASS sensor monitoring set in use. It incorporates a built-in printer to record data as well as serving as a field operations centre

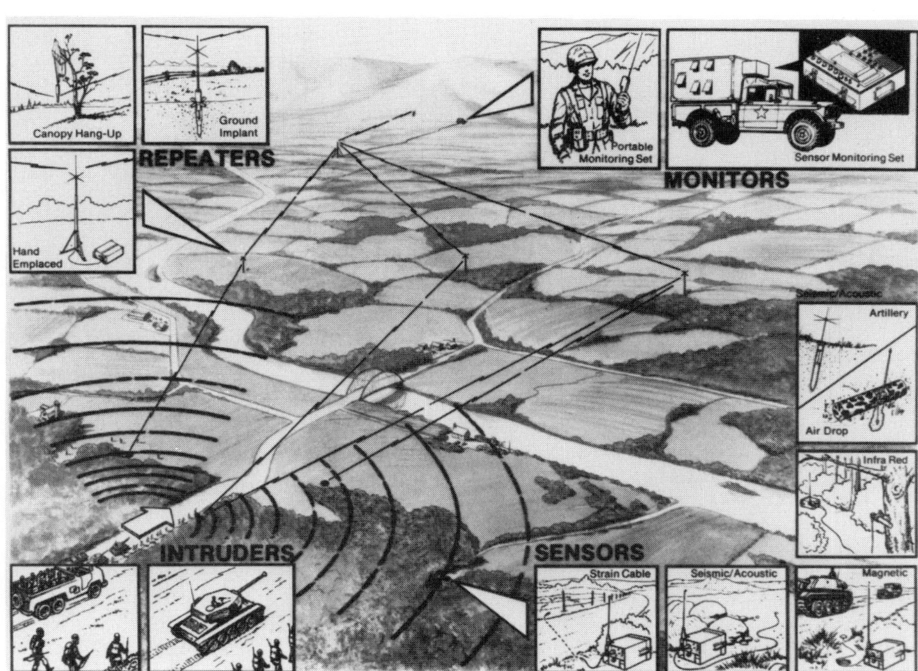

Main components of the REMBASS (Remotely Monitored Battlefield Sensor System) (US Army)

artillery shell or dropped from aircraft. Once in position they remain passive until some activity alerts their detection system. They then detect and identify the target and transmit that information before relapsing into passivity, waiting for the next movement near them. The range over which the information radios work is limited and the repeaters are themselves emplaced by the same means. All sensors and repeaters contain their own batteries and have an operational life of weeks rather than days. They are not re-charged when the batteries run down; another sensor is dropped into place.

The monitors display information as it comes in, but they also store it and can produce it as a hard copy for analysis.

STATUS
The development phase ended in 1983 and full scale production started in 1985. Initial operational capability (IOC) was planned for the second quarter of fiscal year 1987.

CONTRACTOR
RCA Automated Systems Division, PO Box 588, Burlington, Massachusetts 01803, USA.

3940.161

US TACTICAL DATA HANDLING PROJECTS

The US Army and Air Force and the Defense Advanced Research Projects Agency (DARPA) are responsible for a number of projects and programmes, all of which are intended to support one or other aspect of ground forces data collection, processing, dissemination and presentation facilities. The following paragraphs of this entry summarise the main examples of this series of projects.

ACS
Development of the Artillery Computer System (ACS) for the US Marine Corps is continuing. This will be a lightweight, battery-operated computer that will provide rapid, accurate gun-pointing data for individual batteries.

ADDS
The US Army Data Distribution System (ADDS), will modify, combine, and integrate components of Enhanced PLRS (see **3944.161**) and JTIDS in an evolutionary five-phase programme to take advantage of the state of development of these two projects. The ADDS programme will support Army near real-time data distribution requirements at divisional level. It contains three primary elements: Enhanced PLRS user units (EPUU); JTIDS Class 2M terminals; and net control stations (NCS). The first of these are assigned to almost all units in a division area that participate in data communications, identification and position location/navigation. JTIDS terminal will be allocated to those users whose volume of data cannot be handled by the EPUU and who are heavily involved in interservice communications, for example air defence formations which need to exchange information with both ground units and air forces. The NCS provides overall system network management. The ADDS system will support data communications requirements in the five functional areas of fire support, air defence, intelligence/EW, manoeuvre control and combat service support.

The ADDS programme was reviewed in April 1987 to assess its readiness to enhance a low volume of EPLRS receiver with a product improvement. Emerging test information from the PLRS and Enhanced PLRS will be used in making this assessment. Plans are for all PLRS hardware to be modified to the ADDS configuration by the third quarter of 1990.

PLSS
The precision location/strike system (PLSS) will provide a 24-hour all-weather capability for USAF aircraft to attack a variety of ground targets. In particular it is intended for tactical use against air defence systems that depend on very accurate guidance and detection radars. The system has been described in detail (**3228.351**) in previous editions of *Jane's Weapon Systems* (1978 and earlier), but in brief, it relies on the use of specialised distance measuring equipment and other equipment on highly instrumented aircraft to detect electronic emissions from enemy radars and relay that information to a ground-based central processor facility. The Lockheed TR-1 aircraft is the airborne platform for the tactical reconnaissance system (TRS) and PLSS. In the former configuration the TR-1 carries the ASARS II (Advanced Synthetic Aperture Radar System) imaging radar and several other sensors. Information collected by these sensors is transmitted on a digital data link to a TR-1 ground station that processes and sends out reports in less than 15 minutes to users from Army Group or Air Force level down to division and air operations centre levels. The concept is similar for TR-1 PLSS operations. Three TR-1 aircraft carrying PLSS sensors form a triad that transmits threat radar information to a ground station. The PLSS ground station, if necessary, can direct a PLSS-equipped strike aircraft or guided weapon to attack the radar, or any other precisely known target. The TR-1 aircraft relays strike guidance to the attacking aircraft or weapon.

One hardened European ground station and one deployable ground station will support PLSS operations. Two hardened ground stations in Europe will support reconnaissance missions.

Joint STARS
Joint Surveillance and Target Attack Radar System (Joint STARS) is a combined US Army/USAF development programme formed by the merging of USAF's Pave Mover project with the Army's Battlefield Data System (BDS), which, in turn, supplanted SOTAS. (See this entry number in 1983-84 and earlier editions of *Jane's Weapon Systems*). The Joint STARS radar system will locate and track moving targets at extended ranges in the land battle areas of the 2000s. It will give target position information for indirect artillery fire and use of other weapons such as the Joint Tactical Missile System (JTACMS). It also provides the ground commander with real-time moving target data from both Army and USAF platforms via a common secure data link.

Battlefield data will be handled automatically by the US Army in centres similar to this RCA mockup of an ASAS/SEWS signals intelligence control and analysis centre

Airborne tests of the Norden Systems multimode surveillance radar (MSR) for the JSTARS programme have demonstrated detection and location of ground targets in real-time. The ability to perform MTI and synthetic aperture radar (SAR) from an airborne platform has provided essential data on typical tank column manoeuvres. An unprocessed airborne MTI image showed a column of tanks beyond visual range, and viewed against a map overlay this showed the tanks in relation to the road. A time integration display shows ground processed MTI and map underlay with the dots denoting moving targets. Each sweep of the radar cumulatively adds new detections to the displays and in this way high traffic areas can be detected and additional intelligence data derived from it.

Target ranges contemplated extend to more than 100 km into hostile territory, and initially the Army plans to employ the OV-1D Mohawk aircraft as the carrier platform. Grumman was named prime contractor for Joint STARS in September 1985. The Grumman team includes Boeing Military Airplane Company (BMAC) and United Technologies Corporation, Norden Systems Division. Boeing will modify Boeing 707 type airframes to accept the Norden built radar system.

BETA-LOCE
By 1983 BETA had been deployed with US Forces in NATO in Europe, as the Joint Tactical Fusion – Limited Operational Capability Europe (JTF-LOCE), using a computer system developed by TRW under DoD direction. Two more testbed installations for demonstration and evaluation of ASAS (see below) data correlation were also deployed at USAF and Army sites in the USA. In Europe, the LOCE deployment consists of a central automated data processing centre, interface units at sensor ground stations, and local/remote user display terminals. Sensor inputs to the system include: Sigint, imagery intelligence, and human intelligence (Humint).

The European configuration (BETA-LOCE) employs computer software architecture designed to detect, identify and locate target and significant battlefield threat activities by the simultaneous processing of numerous sensor reports through self-correlation, cross-correlation, and aggregation. The last of these

GTE's Government Systems has developed and produced a number of remote tactical data relay sub-stations, similar to this one, and a centralised processing centre for the US Army

is the computer aided process of identification of parent units based on 'sensed' or known data concerning subordinate entities. This process is operator initiated.

In exercises carried out in Europe during 1983 using BETA-LOCE, which was installed in late 1982, it is reported that analysts were able to predict battlefield events significantly earlier than was previously possible.

Experience gained from use in the field of the LOCE system will be applied in the development of more sophisticated systems such as ASAS/ENSCE, below.

JTF

The Joint Tactical Fusion (JTF) system programme is part of a broad US programme of development in the area of new technologies for defeating large scale armoured attacks by means of deep interdiction tactics. JTF is intended to provide automated facilities to process, analyse, and distribute intelligence reports obtained from a wide range of sources. This information will assist battlefield commanders in assessing the status of opposing forces and their dispositions. In the longer term, the authorities are considering a more advanced version that will permit access to direct real-time intelligence.

MCS

The Manoeuvre Control System (MCS) is a command and control system being developed by the US Army to provide an automated information network to assist battle staffs in operational control of the mobile units under their command. Development began in September 1980 with the introduction in the European theatre of a prototype system using microprocessors and minicomputers. From an initial introduction of three systems, the prototype has grown to some 20 systems which are being used in the field. The eventual system will consist of a network of small computers that will enable tactical commanders to have access to information on the status and disposition of their forces and those of their opponent. The European prototype version is being used for additional development.

The first MCS production contract was awarded in 1983 and deployment of this equipment was planned with an IOC in the latter half of 1986. Contractors involved include Singer Librascope, General Dynamics, Datametrics, Ford Aerospace & Communications, and Rolm Corp.

ASAS/ENSCE

The all source analysis system (ASAS) will be the control system for the Intelligence/EW subsystem of the Army Command and Control System, and is being developed together with the US Air Force Enemy Situation Correlation Element, within the Joint Tactical Fusion Programme. Numerous sophisticated sensor systems have been deployed, but the ability to process and distribute their information to battlefield commanders is still restricted in scope and speed by the manual and partially automated processing facilities now available. ASAS will provide modern minicomputer systems to speed up the process and enhance its accuracy and effectiveness. The acquisition plan will provide initial components of baseline systems in the mid-1980s, and systems will evolve via pre-programmed product improvements and software refinement based on the experience gained in the field by users. The ASAS/ENSCE system architecture is modular. Division ASAS, corps ASAS, and ENSCE systems can be configured differently as force structure and mission requirements dictate. Design began in 1984 and field deployments are planned to follow later this decade.

CONTRACTORS

Boeing Military Aircraft Company
Control Data Corporation
E-Systems
General Electric
Grumman
GTE Government Systems
Harris Corporation
Honeywell
Hughes Aircraft Company
IBM
Litton Data Systems
Lockheed Missiles and Space Company
Motorola
Norden Systems
RCA
Rockwell International
Singer's Electronic Systems Division
Sperry Univac
TRW

Army Command and Control System (ACCS)

The ACCS is designed to improve synchronisation of combat forces through efficient information management of the five battlefield functional areas: manoeuvre control, fire support, intelligence/electronic warfare, air defence and combat service support.

In addition all five ACCS systems will be networked for critical information exchange. Information entering the network will be distributed to commanders and their staffs in near real time using three primary means of communication: voice, data and area support.

The voice requirements will be satisfied with the Single Channel Ground and Airborne Radio System. The data system will be the Army Data Distribution System, while Mobile Subscriber Equipment will provide secure, automatic area communications services.

Advanced Field Artillery Tactical Data System (AFATDS)

AFATDS will provide automated support for the planning, coordination and execution of close support, counterfire, interdiction and suppression of enemy fires. It will meet the fire support requirements of the ACCS during the 1990 to 2010 time frame. It will be provided compatibility with all existing and planned US and allied field artillery systems and sensors.

Combat Service Support Control System (CSSCS)

This is the US Army's automated command and control system for the provision of troop services such as supply, maintenance, ammunition, transportation, medical and personnel.

YUGOSLAVIA

4584.161

ARS FIELD ARTILLERY COMPUTER SYSTEM

The ARS field artillery computer system is a digital computer-based system designed to provide user-oriented fire control facilities for the direction of artillery fire at the artillery battalion level. Facilities include automated computation, transmission and display of firing data and commands within the battalion.

The basic configuration of the system comprises:

(1) the main equipment within a computer control sub-system (CCS), and consisting of the artillery computer unit, tape reader, three VHF transmitter/receivers, one radio receiver, a telephone switchboard and power supplies

(2) three battery display sub-systems (BDS). Each BDS is a compact two-man portable equipment designed for use at firing battery command posts. The main items of equipment are carried in a rugged frame and comprise a firing data display, VHF transmitter/receiver, telephone set and battery pack.

The CCS is generally housed in a military shelter carried on a 4 × 4 cross-country vehicle such as the TAM 110 T7 BV.

The main operational features of the ARS system include integrated fire control of an artillery battalion with up to three batteries of guns, howitzers and rocket launchers against static or mobile targets. It can conduct three separate fire missions simultaneously or conduct salvo fire from the whole battalion.

The system can accommodate 200 target locations, including 30 correction points, with ten forward observers. It provides voice communications between the CCS, BDS and forward observers, higher commands and sensors while data transmission can be conducted from the CCS to the BDS. Forward observers can transmit data by radio or cable. Simultaneous data transmission and voice communication can be conducted over the same channel.

The computer is a single unit incorporating a dual Čajavec-52 microprocessor, power supply and an operator's console. Input data includes battery information such as: grid co-ordinates; base direction, deflection and site; muzzle velocity changes or measured value by MV radar; projectile weight variations; powder temperature (for each battery). Ammunition data inserted into the computer includes: shell type, fuze type, low/high trajectory, charge number (manual or automatic selection). Three different types of standard met messages up to 30 lines can be accepted, and target data consists of grid co-ordinates, polar co-ordinates relative to the forward observer, and grid-shift from known point. Additionally, for moving target engagement, there is provision for insertion of time intervals between firings, target speed and direction, or polar co-ordinates relative to FO (artillery radar measurements), or selected impact point co-ordinates. All input data parameter values are automatically checked against acceptable standard limits as a validity test and checked data can be accepted, rejected or left to the operator's decision.

ARS formats firing data output messages and transmits them to the selected battery. An input/output display presents input data as well as computer messages. There is software prompting for the operator in entering data and to ensure the correct operating sequence. Other software facilities include a target file manager, test module, input/output data conversion module, ballistic calculation module, fire transfer/correction module, and a dual processor executive system.

System enhancements available as options are: FO direct data entry facilities; on-line laser rangefinder operation; on-line sound ranging equipment; land navigation sub-system; and an advanced command display unit.

CHARACTERISTICS

Accuracy: 0.001 of range (in metres) in bearing and elevation
Fuze setting: 0.1 s
Firing data computation time: 0.5 to 5 s
Data link speed: 1200 bit/s
Correction code: BCH
Modem: DPSK
Firing data display unit: 300 × 200 × 150 mm
Weight: 12 kg
Power supply: 21 – 29 V DC, 15 W
Computer unit: 680 × 420 × 550 mm
Weight: 60 kg
Power supply: 21 – 29 V DC, <200 W

STATUS

In production and believed operational with Yugoslavian Forces.

CONTRACTOR

SPDR, Federal Directorate of Supply and Procurement, PO Box 308, 9 Nemanjina Street, Belgrade, Yugoslavia.

4585.161

SUVOA FIRE CONTROL SYSTEM

Suvoa is a Yugoslavian coastal artillery fire control system for the direction and control of coast defence gun batteries. Target data is principally obtained from an M70 optical director, supplemented by radar in certain conditions.

Target information from the M70 optical director is fed to a battery distribution unit from where firing data is passed to individual gun batteries. The data is automatically read off and transmitted in a synchronised way. The necessary corrections are set on the rangefinder which calculates them and passes them on to the guns. Transmission is by means of cable data links. Gun crews have only to bring gun traverse and elevation settings into coincidence with the values sent from the director.

In poor visibility or darkness radar range and bearing data may be substituted for the optical target data, but for safety reasons it is recommended that the radar should be separated from the gun battery by a distance of about 1000 metres. The range and bearing information derived from radar must be corrected for parallax before being sent to weapons from the rangefinder.

CHARACTERISTICS

Traverse: 360° unlimited
Range: 1000 – 50000 m
Vertical base variation: 5 – 600 m
Total reading error: 50 m to 10 km range; 200 m to 20 km range; approx 3% to 50 km range
Measuring telescope
Magnification: × 18
Field of view: 4°
Inlet pupil diameter: 54 mm
Exit pupil diameter: 3 mm
Exit pupil distance: 14.8 mm
Adjustment: ±5 dioptres
Reticle: 0 – 35 mils L/R
Periscope: 356 mm
Observation telescope
Magnification: Variable × 3 to × 10
Field of view: Variable 20° to 6°
Inlet pupil diameter: 15 mm to 50 mm
Exit pupil diameter: 5 mm
Exit pupil distance: 15.8 mm
Adjustment: ±5 dioptres
Reticle: 0 – 35 mils L/R
Periscope: 256 mm

DEVELOPMENT

The system has been developed in both stationary and mobile versions. Equipment at each gun comprises a section commander's control box, range data receiver, siting device and bearing indicator. Electrical power for all units (including radar and M70 optical director) is derived from two mobile engine-driven generators, each of which is connected to the system via a transformer unit. The radar is housed in a transportable shelter with the necessary parallax computer.

STATUS

In production and believed operational with Yugoslavian Forces.

CONTRACTOR

SPDR, Federal Directorate of Supply and Procurement, PO Box 308, 9 Nemanjina Street, Belgrade, Yugoslavia.

LAND-BASED TACTICAL SURFACE WEAPONS

2283.111
INTRODUCTION

This section deals with land-based tactical systems whose primary function is to attack surface targets. As such they inevitably include examples of several widely varying types of weapon, often differing considerably in the degrees of complexity involved in their design. Despite this, in general the entries that follow mostly fall within one or other of a few readily recognisable categories. For example anti-tank weapons are a common preoccupation and there are two broad approaches used in designing systems to counter individual tank targets or armoured formations: the small, often infantry-operated, guided missiles; and the more complex systems intended to engage multiple targets meant to combat tanks by indirect fire techniques. As far as practicable such generic divisions of the subject matter of this section are arranged in logical groups.

For the convenience of study and reference, weapons have been divided into three groups and are listed in alphabetical order of country of origin within these groups:

Battlefield Support Weapons
Anti-tank/Assault Weapons
Coastal Defence Weapons

Although the majority of the entries in this section are concerned with guided missile systems, a number of important weapons that do not come into that category strictly speaking, because their significance and operational potential makes it both sensible and convenient to treat both classes collectively. Typical of such blurring of distinctions which were once clearcut but are now hard to distinguish categorically are modern multiple rocket systems and guided mortar rounds, and new types of sophisticated artillery shells with multiple sub-munition payloads. Weapons of this sort will be found among the first two of the above groups: readers whose interests lie more with the artillery pieces which may be employed to deliver such munitions, or for whom the means of transporting and deploying multiple rocket launchers are a primary concern are directed to our companion yearbook, *Jane's Armour and Artillery*, where full mechanical, constructional and other details will be found. Details of mine warfare systems, including both mines and mine-layers, are given in *Jane's Military Logistics*.

Battlefield Support Weapons

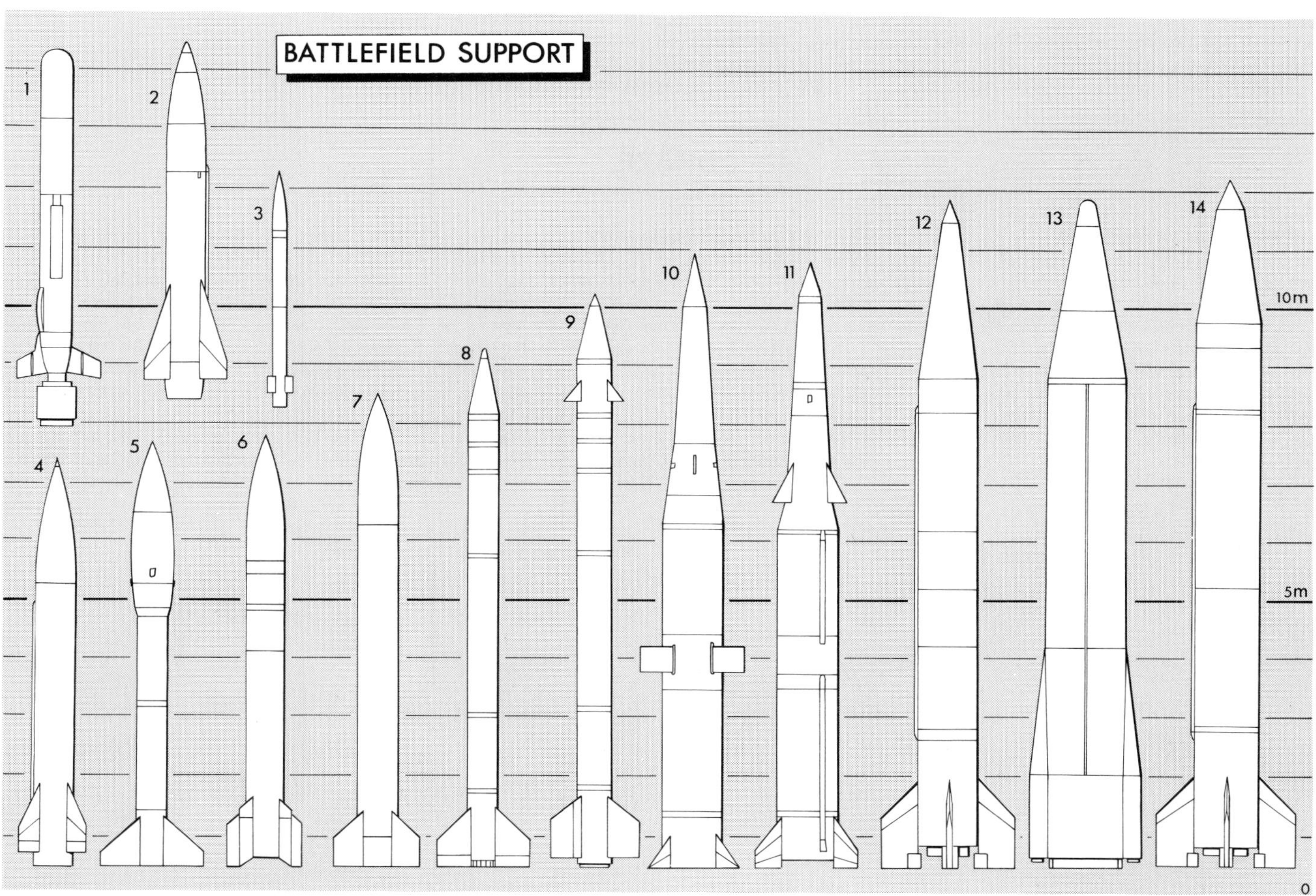

1 *GLCM*, **2** *Lance*, **3** *MLRS*, **4** *Ching Feng*, **5** *Honest John*, **6** *Pluton*, **7** *HADES (provisional)*, **8** *FROG-7*, **9** *SS-21 (provisional)*, **10** *Pershing 1a*, **11** *Pershing II*, **12** *SS-1 Scud 1C*, **13** *SS-12A Scaleboard A (provisional)*, **14** *SS-12B Scaleboard B (provisional)*

ARGENTINA

3674.111
SLAM-PAMPERO 105 mm MULTIPLE ROCKET SYSTEM

The SLAM-Pampero is a multiple artillery rocket system developed in Argentina. A 16-tube launcher is employed for 105 mm calibre rockets which are stabilised by four 'wrap-round' fins at the rear of the missile. A single salvo of 16 rockets will fall within an area measuring 300 × 200 m, with maximum dispersion.

The SLAM-PAMPERO 105 mm multiple rocket launcher has two rows each of eight launcher tubes which are mounted on the rear of a Mercedes-Benz UNIMOG Model 416 (4 × 4) cross-country truck. To provide a more stable firing platform two hydraulic stabilisers are lowered either side at the rear of the vehicle.

CHARACTERISTICS
Calibre: 105 mm
Total weight of rocket: 28 kg

Argentine mobile multiple rocket launcher SLAM-Papmero

Weight of warhead: 10.5 kg
Type of warhead: HE fragmentation with impact fuze
Motor: Single stage solid propellant
Stabilisation: 4 folding fins
Maximum tactical range: 11 000 m
Length of rocket: 1.45 m
Diameter of rocket: 10.5 cm
Time to launch one 16 round salvo: 7.5 s
Angle of elevation: 0 to +52°
Traverse: 90° left and 90° right
CEP: 1%

DEVELOPMENT

Development for the Argentinian Army was undertaken under contract by CITEFA, the Research and Development Institute of the Argentinian Armed Forces.

STATUS

In production. In service with Argentina (Marines).

CONTRACTOR

Instituto de Investigaciones Cientificas y Tecnicas de la Fuerzas Armadas (CITEFA), Zufrategui y Varela, 1603 Villa Martelli, Provincia de Buenos Aires, Argentina.

3675.111

SAPBA-1 127 mm MULTIPLE ROCKET SYSTEM

This system consists of a FIAT 697 (6 × 6) truck chassis with forward control fully enclosed cab to the rear of which is mounted a 36-round 127 mm rocket launcher system.

Each rocket is 2.228 m long and weighs 54 kg of which 18 kg is the warhead, the latter can be of the HE-fragmentation type with a lethal area of 1000 m² or a proximity fuzed anti-personnel type that scatters 4500 steel balls over an area of 6000 m².

When in the firing position the vehicle is stabilised by jacks. The launcher has powered traverse of 90° left and right with elevation being from 0 to +60°. All 36 rockets can be fired singly or at 0.5 second intervals via control box located up to 50 m from the vehicle.

STATUS

Believed to be in production and service with Argentinian Army.

CONTRACTOR

Instituto de Investigaciones Cientificas y Tecnicas de la Fuerzas Armadas (CITEFA), Zufrategui y Varela, 1603 Villa Martelli, Provincia de Buenos Aires, Argentina.

6717.111

CONDOR TACTICAL BATTLEFIELD SUPPORT MISSILE

In 1985 it was revealed that Argentina was developing a missile officially described as for civilian purposes but whose performance seems more suitable for the tactical surface-to-surface role.

Condor is a single-stage missile with solid propellant motor and inertial guidance system. There have been suggestions it may be a derivative of the Martin Pescador (**3678.311**) air-to-surface missile. The range is between 75 km and 150 km with payloads of 475-350 kg respectively. It has been reported that a two-stage development with a range of 650 km was also under development.

In 1987 there were reports that Argentina had supplied 50 Condor missiles to Iran but this was officially denied by the Argentinian government.

CONTRACTOR

Instituto de Investigaciones Aeronauticas y Especiales and FMA.

BRAZIL

4146.111

AVIBRÁS ARTILLERY ROCKETS

For many years, Brazil has been developing unguided rockets. Pioneer work was carried out in the 1960s by the Army Departmento de Estudos e Pesquisas Tecnológicas (Department of Studies and Technological Research). The AVIBRÁS company have designed and developed a wide range of multiple rocket systems, originally many of these were for the Brazilian Army but in recent years most production has been for the export market, especially the Middle East.

ASTROS II (Artillery Saturation Rocket System)

AVIBRÁS has developed the ASTROS (Artillery SaTuration ROcket System) specifically for the export market with first production models being completed late in 1983. Known users include Iraq and Libya.

The ASTROS II system consists of the launching vehicle, ammunition supply vehicles with hydraulic crane and optional fire control units, all of which are based on the same 6 × 6 cross-country truck designed and built by the AVIBRÁS subsidiary company TECTRAN. The launcher can fire three different types of rocket, the AVIBRÁS SS-30 (32-rounds per launcher), SS-40 (16-rounds per launcher) and SS-60 (4-rounds per launcher). The optional fire control system is the Swiss Contraves Fieldguard described in the Army Fire Control Systems section under entry **3314.181**. Ranges of the individual rockets in this system range from 9–60 km, with all rockets having wrap-round fins for improved accuracy.

Sales worth the equivalent of US$1 billion have been made to four countries including Saudi Arabia.

AVIBRÁS ASTROS II artillery saturation rocket system on Tectran (6 × 6) truck chassis. Launcher in firing position

AVIBRÁS ASTROS II artillery saturation rocket system – ammunition supply vehicle (left) and loading launcher (right)

AVIBRÁS ASTROS II fire director unit

AVIBRÁS SS-30 (30 km nominal range) rocket firing from ASTROS II launcher

108R (SS-06)

Designated 108-R by the Brazilian Army, this system consists of a 16-tube launcher for 108 mm, spin-stabilised rockets produced by AVIBRÁS. These carry an HE warhead of about three kg and have a range of about 7000 m. The launcher may be mounted on a two-wheeled trailer for towing by virtually any military vehicle, or may be vehicle-mounted. The launcher crew is said to consist of five men, and the system is fully operational.

SS-07

This system, which has also reached operational status, comprises a 36-tube launcher for AVIBRÁS 70 mm rockets, which are virtually standard FFARs with a four kg warhead and a range of 7.5 km. The launcher may be either trailer- or vehicle-mounted.

SBAT 127

This is an adaptation of the Avibrás SBAT-127 HVAR, with a 20 kg warhead and a range of 15 km. The launcher is a 12-round unit for trailer or vehicle mounting, and the system is operational. Production complete, no longer offered.

X-20

This is the designation of a 180 mm artillery rocket system, known as the X-20 in the Brazilian Army. The rocket carries a 35 kg warhead and has a maximum range of 25 km, being spin-stabilised by cruciform fins. A three-rail launcher assembly is suitable for either mounting on a tracked chassis, or towing by a 6 × 6 vehicle. Production complete, no longer offered.

X-40

Already operational in the Brazilian Army, this system consists of a 300 mm calibre rocket capable of carrying a 146 kg warhead to ranges of 68 km. The launcher is a three-rail assembly mounted on a Bernadini X1A1 light tank known as the XLF-40. Production complete, no longer offered.

SS-300

Early in 1987 it was stated that AVIBRÁS was developing a new medium range tactical missile called the SS-300 which would carry a payload of 1000 kg to a range of 300 km. Mention has also been made of the SS 1000 with a range of 1000 km. The SS-300 will employ the Avibras SIS inertial guidance system which incorporates American technology. The missile is 12 m long, weighs 6 tonnes and has a diameter of 1 m. Tests of the rocket motor were completed by the summer of 1987 and flight tests were reported to begin by the end of 1988. It is hoped to complete development in 1989.

The missile will have a sub-munition payload with anti-amour, anti-personnel and anti-fortification munitions.

CONTRACTOR

AVIBRÁS – Indústria Aeroespacial SA, Caixa Postal 229, São José dos Campos, SP, Brazil.

CHINA (PEOPLE'S REPUBLIC)

6468.111

273 mm (4-ROUND) TYPE 83 MULTIPLE ROCKET SYSTEM

The 273 mm Type 83 multiple rocket system (MRS) is the largest system of its type in service with the Chinese Army with the exception of the specialised Type 74 and Type 762 mine clearing rockets.

The system is based on the unarmoured Type 60-1 full tracked artillery tractor with the five man fully enclosed cab at front and a turntable with the four 273 mm rockets at the very rear. Before launching the rockets two stabilisers are manually lowered at the rear of the chassis to provide a more stable firing platform.

Each rocket has a high explosive warhead, although other types of warhead could easily be accommodated in a rocket of this calibre. The rocket has a maximum velocity of 810 metres a second and all four rockets can be salvo fired in 7.5 seconds with time into action being one minute.

Launcher is traversed and elevated manually with controls being mounted on left side of launcher. Fire control equipment consists of one panoramic sight Type 85 and one indicator Type 85.

CHARACTERISTICS

Calibre: 273 mm
Number of launch tubes: 4
Weight of rocket: 484 kg
Length of rocket: 4.73 m
Maximum range: 40 000 m
Maximum velocity: 810 m/s
Elevation: 5° 30′ to 56°
Traverse: 10° left and right
Crew: 5
Weight of system loaded: 17 542 kg

STATUS

In production. In service with Chinese Army and being offered for export.

CONTRACTOR

Chinese State Factories. Enquiries to NORINCO, China North Industries Corporation, 7A Yuetan Nanjie, PO Box 2137 Beijing, People's Republic of China.

273 mm Type 83 MRS in firing position with launcher elevated

4872.111

130 mm TYPE 70 MULTIPLE ROCKET LAUNCHER

The Type 70 self-propelled rocket launcher is used to fire 130 mm artillery rockets. The 19-tube launcher is mounted on top of a Type 531 tracked APC type vehicle, which gives the system good cross-country capabilities and a certain degree of armour protection.

The spin stabilised HE rocket has an HE warhead (other types are possible but have not been confirmed) and all 19 rounds can be fired in less than 12 seconds. The time to prepare for rocket launch, after coming to a halt, is about two minutes, and half that time to move off again after firing.

130 mm Type 70 multiple rocket system on Type 531 full tracked APC

CHARACTERISTICS
Calibre: 130.65 mm
Number of tubes: 19
Elevation: 0° – 50°
Traverse: 180°
Weight of rocket: 32.8 kg
Weight of warhead: 14.74 kg
Fuze: Jian-1
Max velocity: 437 m/s
Max range: 10 370 m

STATUS
In production. In service with Chinese Army.

CONTRACTOR
NORINCO, China North Industries Corporation, 7A Yuetan Nanjie, PO Box 2137, Beijing, People's Republic of China.

Note: 130 mm (19-round) Type 63 MRS is similar but mounted on rear of NJ-230 (4 × 4) truck chassis.

6172.111
130 mm (30-ROUND) TYPE 82 MULTIPLE ROCKET SYSTEM

This multiple rocket system consists of a 6 × 6 cross-country truck on the very rear of which is mounted a 30-round multiple rocket system with a further 30 rounds being carried in reserve.

The launcher consists of three layers each of ten tubes and two complete salvoes can be launched in five minutes.

The Type 82 multiple rocket system fires two types of rocket, the Type 63 high explosive and the Type 82 high explosive, the latter having an HE filling with an outer layer of steel balls for improved fragmentation effect.

CHARACTERISTICS
Calibre: 130.73 mm
Number of barrels: 30
Length of launcher tubes: 1050 mm
Maximum range: 10 217 m
Elevation: +50° – 0°
Traverse: 170° (32° left, 138° right)
Crew: 7
Weight of system loaded: 9000 kg
Length travelling: 2.28 m
Height travelling: 2.525 m
Ground clearance: 0.285 m
Maximum road speed: 80 km/h

130 mm (30-round) Type 82 MRS in travelling configuration

Range: 550 km
Fording: 0.85 m

STATUS
In production. In service with Chinese Army.

CONTRACTOR
Enquiries to China North Industries Corporation, 7A, Yuetan Nanjie, Beijing, People's Republic of China.

6469.111
122 mm (40-ROUND) SP MRS

During the large military parade held in Beijing in October 1984, a new 122 mm self-propelled gun/howitzer made its first public apperance. It has now been confirmed that its official designation is the Type 83.

NORINCO has recently announced it has developed a new self-propelled 122 mm rocket launcher on the same chassis as the Type 85 self-propelled gun/howitzer. This has been designed to accompany mechanised forces and unlike many other self-propelled multiple rocket systems has a fully armoured chassis.

The engine compartment is at front right with the driver on left side. He is provided with a single piece hatch cover that opens to the rear, heated day periscope WG501 and a TYG-1 night driving periscope.

The other four crew members are seated in the fully enclosed and armoured compartment at the back with bullet proof observation windows in either side and an entry door in the rear.

Suspension is of torsion bar type with each side having six dual rubber tyred road wheels, drive sprocket is at the front, idler at rear and three track return rollers. The first and last road wheel stations have a hydraulic shock absorber.

The 122 mm (40-round) launcher is mounted on a turntable over the rear of the vehicle and is almost identical to the Soviet 122 mm BM-21 (40-round) system introduced in the early 1960s. The launcher can be traversed and elevated under full hydraulic power with manual controls being provided for emergency use. The rockets can be fired one by one or in a salvo with all 40 being fired in 20 seconds.

Mounted over the forward part of the vehicle is a reload pack with a further 40 122 mm rockets which can be reloaded into the launcher automatically. This reload feature is similar to that first used on Czechoslovakian 122 mm RM-70 (40-round) multiple rocket system which made its first appearance in 1972.

122 mm self-propelled multiple rocket system in travelling configuration

The 122 mm rocket has a special shrapnel warhead that can be fitted with either an impact or proximity fuze. The warhead weighs 19.25 kg of which 6 kg comprises the RDX explosive filler with remainder being the casing and some 4100 steel balls which are highly effective against a variety of battlefield targets including vehicles and personnel.

CHARACTERISTICS
Calibre: 122.4 mm
Number of barrels: 40
Unit of fire: 40 + 40 rockets
Length of complete rocket: 2.87 m
Weight of complete rocket: 66.88 kg
Weight of propellant: 20.6 kg
Maximum range: 20 580 m
Rate of fire: 40 rockets in 20 seconds
Elevation: 0 to +55°
Traverse: 66° (right), 102° (left)
Crew: 5
Combat weight: 29 900 kg

STATUS
Development complete, ready for production. Available for export.

CONTRACTOR
Chinese State Factories. Enquiries to NORINCO, China North Industries Corporation, 7A Yuetan Nanjie, PO Box 2137 Beijing, People's Republic of China.

5154.111
107 mm TYPE 63 MULTIPLE ROCKET LAUNCHER

The Type 63 MRL is mounted on a two-wheeled carriage and towed behind a light truck, and in the mountain role. In the latter it is broken down into man-pack sizes and weighs only 281 kg, assembled, when it rests on four trails, ie two to the front and two to the rear. Incendiary and HE warhead rockets are produced for the Type 63 system.

CHARACTERISTICS
Calibre: 106.7 mm
Number of barrels: 12
Elevation: +57°
Depression: –3°
Weight travelling: 602 kg
Weight of rockets: 18.8 kg (HE); 18.74 kg (incendiary)
Length: 841 mm (HE); 915 mm (incendiary)
Warhead weight: 8.33 kg (HE): 7.54 kg (incendiary)
Max velocity: 372 m/s (HE); 383 m/s (incendiary)
Max range: 8500 m (HE); 7885 m (incendiary)
Fuze: Jian-1

STATUS
In service in Albania, China, and Vietnam.
CONTRACTOR
NORINCO, China North Industries Corporation. 7A Yuetan Nanjie, PO Box 2137, Beijing, People's Republic of China.

107 mm (12-round)Type 63 multiple rocket system in travelling configuration (Israeli Ministry of Defence)

4873.111
107 mm TYPE 81 MULTIPLE ROCKET LAUNCHER

The Type 81 multiple rocket launcher is basically the latest production version of the 107 mm Type 63 (12-round) MRS mounted on the rear of a 4 × 4 cross country truck.

The launcher is fitted with magazines so that a group of rockets can be reloaded on the vehicle. There are two firing mechanisms permitting the launch procedure to be carried out either from the vehicle or on the ground. A radio set is provided for crew communications.

The Type 81 can be used with either the HE or the incendiary 107 mm rockets of the Type 63 rocket system, and maximum range is about eight km.

CHARACTERISTICS
Calibre: 107 mm
Number of tubes: 12
Elevation: 0° – 60°
Traverse: 105°
Rate of fire: 10 – 12 s (12 rounds on vehicle); 6 – 9 s (12 rounds on ground)
Total weight in combat: 3615 kg (with 2 sets of rockets)
Rockets: Type 63 HE or incendiary
STATUS
In production. In service with Chinese Army.
CONTRACTOR
NORINCO, North China Industries Corporation. 7A Yuetan Nanjie, PO Box 2137, Beijing, People's Republic of China.

6470.111
CHINESE TACTICAL MISSILES

In addition to the wide range of multiple rocket launchers produced in China for the home and export markets by NORINCO (China North Industries Corporation), CPMIEC (China Precision Machinery Import and Export Corporation/China Great Wall Corporation) is involved in the design and development of longer range missiles for the Chinese Army.

Unconfirmed reports have mentioned a system called the T-5 unguided rocket with range of around 100 km fitted with a tactical nuclear warhead and some of these systems may have been deployed along the Chinese/Soviet border.

In 1986 CPMIEC is understood to have started flight trials of the Model M missile system which has been developed to meet the requirements of the Chinese Army for a highly mobile missile system with a maximum range of 600 km and capable of being fitted with both conventional and nuclear warheads.

The missile is transported and launched from an 8 × 8 launcher with a four door cab at the very front and the missile pivoted at the rear. When travelling the missile lays in the horizontal position and is elevated into the vertical position on arrival at the pre-selected launch site. To provide a more stable firing platform two stabilisers are lowered at the rear of the vehicle.

The missile can be brought into action in about 30 minutes and has a solid propellant rocket motor with four vanes arranged around the exhaust outlet being used to vector the engine thrust with nozzles for the control thrusters being located to rear of the nose and guidance section.

Scale model of the 600 km Chinese M missile on its 8 × 8 TEL chassis (C F Foss)

CHARACTERISTICS
(Model M)
Length: 9.1 m
Diameter: 10 cm
Launch weight: 6200 kg
Range (maximum): 600 km
Guidance: Inertial
Propulsion: Single stage, solid propellant
Warhead: Conventional initial, nuclear possible later
STATUS
Flight tests commenced in 1986 and could enter service with Chinese Army in 1989/1990.
CONTRACTOR
CPMIEC (China Precision Machinery Import and Export Corporation/China Great Wall Corporation).

CZECHOSLOVAKIA

5137.111

122 mm ROCKET LAUNCHER M-1972 (RM-70)

The M-1972 rocket launcher was first seen during a military parade held in Czechoslovakia in 1972. It basically consists of a modified Czechoslovak Tatra 813 (8 × 8) truck chassis, on the very rear of which has been mounted the Soviet BM-21 rocket launcher (**5555.111**). The cab is of the forward control type and is fully armoured. Immediately behind the cab is a rack of 40 rockets, which enable the launcher to be reloaded in five minutes, compared with ten minutes for the Soviet BM-21 launcher. The Czechoslovak vehicle has greater cross-country capability than its Soviet counterpart and is provided with a central tyre pressure regulation system, which enables the driver to adjust the tyre pressures to suit the type of ground encountered.

122 mm rocket launcher (40 round) M-1972

CHARACTERISTICS

	Long Rocket	**Short Rocket**
Calibre	122 mm	122 mm
Length	3.226 m	1.905 m
Weight	77 kg	45.8 kg
Max velocity	690 m/s	450 m/s
Max range	20 500 m	11 000 m

Launcher

Capacity: 40 rounds, plus 40 in reserve
Weight: 33 700 kg (loaded)
Length: 8.8 m
Width: 2.5 m
Height: 2.9 m
Elevation: 0 to +50°
Traverse: 172°

STATUS

In service with Czechoslovakia, East Germany, Libya and Soviet Union. Yugoslavia produces a similar (qv) version of this weapon, mounted on a 12-ton FAP-2220 BS truck. The general arrangement is similar to that of the Czechoslovak system, but with a 32-tube launcher and a slightly different reloading device.

CONTRACTORS

Chassis – Tatra (Czechoslovakia). Launcher – Soviet State Arsenals.

EGYPT

6712.111

SAKR 80

Sakr is developing a battlefield support rocket, Sakr 80, with a range of 80 km and sub-munition payload. It is reported that the Sakr 80 is a single-stage rocket with solid propellant, that it weighs 660 kg and is fired from the same launcher as the FROG-7 missile (**2926.111**). The 200 kg payload consists of 950 M42 dual purpose bomblets or 65 anti-tank mines or high explosive-fragmentation.

STATUS

Under development.

CONTRACTOR

Sakr Factory for Developed Industries, PO Box 33, Heliopolis, Cairo, Arab Republic of Egypt.

4874.111

VAP 80 mm MULTIPLE ROCKET LAUNCHER

The VAP 80 mm 12-tube multiple rocket launcher has been developed by the Egyptian Sakr Factory to provide rapid fire support for light infantry units and similar formations. The system consists of a 12-round launcher mounted at the rear of an Egyptian-built jeep type vehicle, but it is possible to mount it on any similar vehicle of this class. Two stabilisers are employed to give support when firing.

Elevation and traverse of the launch unit are effected manually, and there is provision for remote control of the firing operation at a distance from the vehicle. The 80 mm calibre rockets employed are spin stabilised with four folding fins at the rear end. Two types of warhead are available, illumination and HE fragmentation, and both types use a mechanical impact fuze.

CHARACTERISTICS

Calibre: 80 mm
Length: 1500 mm
Weight: 12 kg
Max range: 8000 m
Rocket burn time: 1 s

STATUS

In production. In service with Egyptian Army.

CONTRACTOR

Sakr Factory for Developed Industries, PO Box 33, Heliopolis, Cairo, Arab Republic of Egypt.

VAP 80 mm (12-round) multiple rocket launcher mounted on rear of an Egyptian-built jeep (4 × 4) light vehicle

4875.111

122 mm MULTIPLE ROCKET LAUNCHER

The 122 mm rocket launcher consists of two banks of three rows of five 122 mm launcher tubes located side-by-side in a single assembly that is mounted on a turntable carried on the rear of a 6 × 6 cross-country truck. Elevation and traverse of launcher is manual.

The system is designed for use against enemy targets spread over areas up to 20 km distant, and the rate of fire enables up to 40 projectiles to be fired in 20 seconds. HE fragmentation warheads, initiated by a mechanical impact fuze with three settings for action against troop concentrations and infantry support weapons, are provided. Two different brake rings are used to control the rocket range and ensure optimum arrival angles.

The Sakr Factory for Developed Industries has also developed a 122 mm calibre rocket with a range of 30 000 metres which can be fired from the existing BM-21 rocket launchers or quadruple ground/mobile launchers.

This rocket is called the Sakr 30 and uses a

Sakr 30 rocket being fired from its multiple rocket launcher

composite propellant and has different types of warheads to increase tactical flexibility including a high explosive pre-formed fragmentation with a proximity fuze and a warhead containing 3 anti-personnel bomblets or 100 dual purpose bomblets or six anti-tank mines.

The fuze uses an electro-mechanical or electronic time fuze for the sub-munition warheads.

CHARACTERISTICS

Type	**HE/PFF**	**Sub-munition**
Calibre	122 mm	122 mm
Length	2580 mm	3180 mm
Overall weight	55 kg	65 kg
Weight of warhead	18.5 kg	27.5 kg
Maximum velocity	1250 m/s	1100 m/s
Maximum range	32 000 m	28 000 m

STATUS

In production. In service with Egyptian Army and other armies.

CONTRACTOR

Sakr Factory for Developed Industries, PO Box 33, Heliopolis, Cairo, Arab Republic of Egypt.

Egyptian-built (30-round) 122 mm multiple rocket system on rear of Soviet ZIL-131 (6 × 6) truck chassis (C F Foss)

FRANCE

4777.111

HADES TACTICAL SURFACE-TO-SURFACE MISSILE

In mid-1983 the French Ministry of Defence decided to go ahead with a new tactical (now designated prestrategic) surface-to-surface missile system to replace the existing Pluton weapon system (**2130.111**).

Prime contractor is the Space and Strategic Systems Division of Aérospatiale under the guidance of the Direction des Engins (DGA). The nuclear warhead will be the responsibility of the Commissariat à l'Energie Atomique (Atomic Energy Commission).

Each Hades system will consist of a semi-trailer towed by a 4 × 4 tractor unit, with the semi-trailer carrying two missiles which are raised into the vertical position for launching. Each truck carries all of the necessary equipment for launching including communications equipment.

On arrival at the selected firing point, no manual operations outside the launch cab are required.

Whereas the current Pluton is deployed at Corps level, the Hades weapons will be concentrated into an autonomous division with a total of four Hades regiments. The Hades division will include the command, communications and support equipment, plus protection units needed to permit autonomous engagements.

Project definition started in August 1983, development in July 1984 with flight testing planned to begin in 1988.

Range of the Hades missile is officially stated to be 480 km, carrying a nuclear warhead of between 10 and 25 KT. After the rocket motor has burned out the aerodynamic control surfaces will enable the mission to change its course. No other details are available at this early stage.

According to the 1983 French Defence Budget, the cost of developing Hades is estimated at between FFr7000 million and FFr10 000 million, and current plans envisage production of a total of 180 missiles. Hades is expected to succeed Pluton in French Army service in 1992.

CONTRACTOR

Aérospatiale, Space and Strategic Systems Division, BP 96 – 78133 Les Mureaux, Cedex, France.

A contractor's impression of the Hades surface-to-surface missile system

2130.111

PLUTON BATTLEFIELD SUPPORT MISSILE

Developed to meet a French Army requirement, Pluton is a surface-to-surface tactical nuclear missile. The weapon, which has a range capability from 20 to 120 km, is installed on and fired from a much modified GIAT AMX-30 tank chassis, the missile container being used as a launching ramp. Missile and warhead are supplied separately to operational units.

In addition to the AMX-30 launching vehicle the system includes command vehicles containing data processing equipment organised round the IRIS 35M computer of the Plan Calcul Militaire. This is a third-generation general purpose computer capable of operation under severe environmental conditions. If operational circumstances require it, target data can be obtained in real-time from an R.20 drone (**2127.351**) equipped with a Cyclope passive IR reconnaissance system (**1070.353**). Missile guidance is by means of a simplified inertial system based on a SFENA semi-strap-down system. Alternative warheads are available: the 25 kiloton yield AN-51 which contains the same MR50 nuclear charge as the AN-52 free-fall bomb carried by French Air Force Mirage III and Jaguar aircraft, or a 15 kiloton nuclear payload. The former is intended for important targets in an enemy's rear, while the smaller yield device would be employed for use nearer to the main battle area.

CHARACTERISTICS

Type: Land-mobile, vehicle-launched, tactical nuclear weapon system

Guidance: Inertial
Propulsion: Dual-thrust solid-propellant rocket motor. Propellant weight 1200 kg
Warhead: Nuclear 25 and 15 KT
Missile length: 7.64 m
Body diameter: 65 cm
Span: 141 cm
Launch weight: 2423 kg
Range: 20-120 km
CEP: Believed to be 150 – 300 m depending on range
Crew (launch vehicle): 4

STATUS

Initial production contracts were awarded in 1972, and deliveries to the French Armed Forces began early in 1974. It is now operational with five French Army regiments (3, 4, 15, 32 and 74). Each regiment has six launchers plus the appropriate support and re-supply units.

CONTRACTORS

Prime contractor, project management, aerodynamic studies, structure, guidance and control system, pre-launch test and control set, vehicle integration: Aérospatiale, Division Engins Tactiques, 2 rue Béranger, 92322 Chantillon Cedex, France.

Others involved are:-
Rocket motor: SEP
Propellants: SNPE
Main body equipment: SFENA, Air-Equipment, Auxilec, Elecma
Warhead equipment: SERAT
Nuclear charge: CEA
Launch vehicle (tank portion): DTAT/AMX/GIAT
Crane: Griffet
Turbine: Microturbo
Vehicle equipment: CII, CIT, SINTRA, SOPELEM
Communication and data processing system: DTAT/GIAT
Command vehicle: SPAIR
Command and control equipment: CII (computer), CIT, SINTRA
Command and control integration: DTAT/SEFT

DESIGN AUTHORITIES

Programme direction: Direction des Engins.
Nuclear warhead: Commissariat à l'Energie Atomique
Command and control system: Direction Technique des Armements Terrestres.

Pluton being launched from its tracked launcher based on AMX-30 tank chassis

GERMANY (FEDERAL REPUBLIC)

5577.111

LARS 110 mm ARTILLERY ROCKET LAUNCHER

The Light Artillery Rocket System (LARS) launcher comprises two groups of 18 launch tubes, which can be mechanically trained and elevated, and from which are launched 110 mm calibre rockets. These rockets can be equipped with a variety of warheads and can be launched singly, in partial salvoes, or salvoes. The rockets are flight stabilised by fins which are folded and protected by a cover for storage and handling.

The complete system was originally mounted on the rear of a Magirus Jupiter (6 × 6) truck chassis which was fitted with an armoured cab, but subsequently the system was removed from these chassis and fitted onto the MAN (6 × 6) truck chassis.

CHARACTERISTICS

Calibre: 110 mm
Propulsion: Single-stage, solid-propellant rocket motor
Warhead: Choice of high explosive, incendiary, training, practice, anti-tank, target, or smoke
Range: 14 000 m (original rocket), 19 000 m or 25 000 m (new rockets)
Fire control: Contraves Fieldguard
Launcher
Capacity: 36 rockets in 2 groups of 18 launch tubes mounted on a 6 × 6 wheeled prime mover
Laying: Mechanical
Traverse limits: +95° (left and right)
Elevation limits: 0 – 55°
Weight (loaded): 17 480 kg
Rate of fire: 36 rockets in 18 s

STATUS

209 supplied to Federal German Army.

LARS 110 mm artillery rocket system on new MAN (6 × 6) truck chassis (Pierre Touzin)

CONTRACTOR

(Launcher)

Wegmann and Co Gmbh, D-3500 Kassel, August-Bode-Strasse, Federal Republic of Germany.

INDIA

6713.111

INDIAN TACTICAL SURFACE-TO-SURFACE MISSILE

India is currently developing a liquid-fuelled tactical surface-to-surface missile designated Prithvi. Few details have been released but it is reported the missile features an inertial navigation system, a range of 240 km and will be capable of delivering a 1 tonne warhead. A second, medium range (150 km), missile designed to carry a sub-munitions warhead is also under development.

STATUS

Test flights of the Prithvi were scheduled to begin in June 1987 and of the medium range missile in 1988. Both missiles are to enter production in 1993.

CONTRACTOR

Indian Defence Research and Development Laboratory.

ISRAEL

6714.111

CHETZ TACTICAL SURFACE-TO-SURFACE MISSILE

Israel is reported to be developing an advanced tactical surface-to-surface missile, Chetz, and during 1987 was asked by the United States to prepare a definitive proposal on the system which might be co-developed. It is reported that the equivalent of $70 million will be required to complete the first stage of development.

4098.111

ISRAELI 290 mm MEDIUM RANGE ROCKET (MAR 290)

The basic 290 mm MAR 290 artillery rocket is 5.45 metres long, with a launch weight of 600 kg of which 320 kg is the warhead. Range is approximately 25 km. These rockets were originally launched from a lattice constructed four-round launcher mounted on the top of a Sherman tank chassis, the vehicle itself being manoeuvred to adjust firing azimuth, with elevation being hydraulically actuated. This model is still in use with the Israel Defence Force (IDF) but a new launch system has been developed based on the use of a Centurion tank hull equipped with a new four-tube elevating and traversing launcher on a superstructure carried on the original turret ring.

Each launcher tube has an internal diameter of 700 mm to accommodate the rocket tail fin span of 570 mm. The overall height of the Centurion with the rocket tubes in the lowered position is 3.6 metres and the vehicle weight is about 50 tons. A crew of four is carried: commander, driver, and two fire control operators. Details of the fire control system employed have not been disclosed. Rockets are launched with all four crew members inside the vehicle. Once all four rockets have been fired, the launch vehicle moves to a fresh position where reload rounds are already emplaced.

The fresh rounds are taken to this location by a truck which lowers them to the ground for the packaging to be removed. For this the truck uses its own handling crane, which is also used to lift the prepared rounds onto a special reloading frame. When the launcher vehicle approaches, it employs a system of two lugs and a bracket on the hull front to engage corresponding items on the loading frame. The frame is then connected to the vehicle's hydraulic system and a ram on the frame lifts the entire frame off the ground into a raised horizontal position until the rockets are in line with the launching tubes. The launch tubes have previously been rotated to point to the launch vehicle's rear. Using hydraulic power the rockets are then pushed into the launch tubes, and the whole sequence is carried out by one man in about ten minutes using a remote control unit. After loading, the frame is lowered to the ground, the hydraulics and mechanical systems are disconnected, and the frame may then be used for more rockets to be loaded by the supply truck.

Earlier lattice construction four-round MAR 290 launcher vehicle based on Sherman tank

CHARACTERISTICS

Type: Unguided, fin stabilised
Rocket diameter: 290 mm
Length: 5.45 m
Fin span: 570 mm
Weight: 600 kg (launch)
Warhead: 320 kg
Propulsion: Solid rocket motor
Range: 25 km (provisional)
Launcher
Tube diameter: 700 mm
Traverse: 360°
Elevation: 0 – 60°

DEVELOPMENT

The Rocket Systems Division of Israel Military Industries started work on the 290 mm artillery rocket in 1965 with the first version to enter service being based on a Sherman tank chassis. Current production model is based on the British built Centurion tank chassis as this is used by Israeli Army in large numbers. Israel Military Industries have also overhauled the Soviet built 240 mm (12-round) BM-24 multiple rocket system on a ZIL-157 chassis, as well as producing identical HE rockets with a range of 10 700 metres. These BM-24 systems are still in service with Israeli Army.

STATUS

The original MAR 290 with the four-round lattice launcher is still operational with the IDF and the new tube-launch version is now in service.

CONTRACTOR

Israel Military Industries, PO Box 1044, Ramat Hasharon 47100, Israel.

Reloading operation of MAR 290 artillery rocket system on Centurion tank chassis (T J Gander)

New tubed launcher for MAR 290 artillery rocket system mounted on Centurion chassis (T J Gander)

4759.111
LAR 160 160 mm LIGHT ARTILLERY ROCKET SYSTEM

The LAR 160 is a multiple launch rocket system designed to direct unguided artillery rockets at battlefield targets at ranges of up to 30 km from tubed launchers mounted on vehicles such as the M548 tracked cargo carrier, M809 6 × 6 5-ton truck, AMX-13 light tank chassis and M47 medium tank chassis. These are typical examples and any vehicle from a standard 6 × 6 military truck to virtually any tank chassis may be used.

The first carrier used during trails was an M548 tracked vehicle carrying 32 rounds, but the later production version used a turretless AMX-13 light tank with two 18-tube LPCs. These vehicles have a three-man crew: driver, commander and signaller/gunner operating the fire control system.

A 30 km maximum range and a warhead payload capability of 50 kg enables a wide range of warheads to be employed, including HE, illuminating, AP, and special purpose payloads such as scatterable mines, and dual-purpose submunitions. The rocket calibre of 160 mm allows virtually any payload that can be accommodated in a conventional 155 mm artillery round to be carried by the LAR 160 rocket. A radar-based fire control system is normally employed to ensure optimum accuracy, and if the Fieldguard pilot round technique is also used, maximum surprise in the target area can be achieved.

M548 tracked cargo carrier with pod of 18 160 mm rockets in rear cargo area

The system consists of the tracked or wheeled carrier vehicle on which is mounted an expendable multiple launcher containing fuzed, sealed-in 160 mm rockets. A typical LAR 160 battery comprises six launchers with up to 50 tubes (M47 medium tank chassis) so that a large quantity of warheads or submunitions can be delivered to the target within about 30 seconds of opening fire. The rockets are fired from launch pod containers (LPCs) made up of varying numbers of tubes, depending on the size and type of carrier vehicle used. The tubes are filament wound in glass fibre and have integral guide rails. There is a polyurethane foam filler between the launch tube and the hexagonal aluminium outer envelope. The hexagonal section permits simple construction of LPCs of different capacities.

The LAR 160 rocket is a free flight, fin stabilised, solid propellant, surface-to-surface round comprising: a low drag aerodynamic nose section, payload section, a short burn-time solid rocket motor, and folding wrap-around stabilising fins that deploy when the rocket leaves the launch tube. The number of rockets in each LPC bundle is adjusted to suit the carrying capacity of the vehicle used, and a suitable turntable with elevating mechanism is normally provided. Reloading is rapidly accomplished by replacing an empty LPC with a freshly loaded one.

Fire control systems suggested for the LAR 160 system are the Contraves Fieldguard (**3314.181**), or the Westinghouse Quickfire system based on the arrangement used in the now cancelled US Army twin 40 mm DIVAD air defence system using a version of the F-16 AN/APG-66 nose radar (**3303.353**). In both cases, precision radar tracks pilot rockets fired to 60 – 75 per cent of their full trajectory before being detonated in mid-air to avoid alerting the target area. The computer then calculates aiming data from the trajectory information derived from the pilot shot(s) and the projected impact point(s).

CHARACTERISTICS

Rocket
Calibre: 160 mm
Length: 3.311 m
Rocket motor length: 2.032 m
Warhead length: 1.279 m
Fin span: 350 mm
Launch weight: 110 kg
Weight at burn-out: 77 kg
Propellant weight: 36 kg
Warhead weight: 50 kg
Range: 12 – 30 km

Launch pod container (18-tubes)
Length: 3.46 m
Width: 1.0 m
Height: 820 mm
Weight: 2522 kg (complete); 540 kg (empty)

Variant
Early in 1986 Israel Military Industries announced a new towed version of the LAR 160. This consists of a two wheeled trailer launching platform and a disposable Launch Pod Container (LPC) with 18 rockets. Once the 18 rockets have been fired the whole LPC is replaced by a new LPC fully loaded and it can then be fired again. A 6 × 6 truck can tow one trailer mounted LPC as well as carrying two additional LPCs with a hydraulic crane being used for loading purposes.

STATUS
In production and operational with IDF. The system has been exported to Venezuela mounted on the French built AMX-13 light tank chassis, and has been offered to the US Army and Marine Corps in association with US defence manufacturers.

CONTRACTOR
Israel Military Industries, PO Box 1044, Ramat Hasharon 47100, Israel.

ITALY

3764.111
FIROS 6 51 mm FIELD ROCKET SYSTEM

FIROS 6 is a field weapon system based on the SNIA BPD two inch (51 mm) rocket, capable of performing an anti-guerrilla and infantry support role within a distance of six km. The launcher of the modular type has 48 launching tubes and is installed on a light vehicle with high manoeuvrability, allowing the system to perform missions on difficult terrain (jungle, forest, desert). Each firing unit can carry the number of rockets required for firing at least two ripples. The fire control unit offers the possibility of firing partial or complete ripples of rockets with pre-selected warheads.

The rocket has high accuracy due to a special folding fin assembly unit and can carry the following types of warhead: HEI (high explosive incendiary), PFF (pre-formed fragmentation), AT-AP (anti-tank anti-personnel), SM (smoke), SP (spotting), illuminating, practice (smoke, flash and inert).

The FIROS 6 system is supplied either in the 'package' version, inclusive of the vehicle, or in a 'kit' version which can be easily installed on vehicles already in service with armed forces. SNIA BPD has developed an installation of the system with a 48-tube launcher on the Land Rover Mod 109. The system has also been successfully installed and demonstrated on the FIAT Campagnola 1107AD and 6614 APC.

FIAT 6614 APC equipped with FIROS 6 rocket launcher system

CHARACTERISTICS
(Typical example installed on Land Rover 109)
Time to bring the system into action: 3 minutes
Ripple time: 5 s (48 rockets)
Disengagement time: 1 minute
Vehicle: Land Rover Mod 109, FIAT Campagnola 1107 AD (4 × 4) light vehicle and FIAT 6614 (4 × 4) APC
Max speed on road: Over 90 km/h (fully loaded)
Loading capability per mission: 2 ripples (96 rockets)
Overall length: 4.46 m
Overall width: 1.65 m
Overall height: 1.95 m

Rocket launcher
Number of tubes: 48
Elevation: –5 to +45°
Azimuth: ±105°
Length: 2.00 m
Width: 0.60 m
Height: 0.91 m

Weights
Empty launcher: (including sighting and firing units) 250 kg
Payload: (96 rockets) 460 kg; empty vehicle 1850 kg; crew (3 men) 240 kg; overall weight of launch unit in combat conditions 2800 kg

STATUS
In production.

CONTRACTOR
SNIA BPD, Setlore Difesa e Spazio, Via Sicilia, 162-00187 Rome, Italy.

5230.111

FIROS 25/30 122 mm FIELD ROCKET SYSTEM

The FIROS 25/30 122 mm field rocket system has been developed as a private venture by SNIA BPD. It is an area saturation weapon system and is based on the 122 mm SNIA BPD unguided rocket. The system is intended for employment either in battery missions or isolated missions and can cover large areas in the 8 to 33 km zone.

A typical battery would consist of a command post equipped with a fire direction system, six firing units and six or more escort units. The firing unit is a modular rocket launcher installed on a Fiat-Iveco 230.38 WM heavy truck (6 × 6). It is available in two different versions, a standard version with a powered/manual system for elevation and traverse which is simple to operate. A fully automatic version with an inertial navigator, a servo control system and a SEPA ARCO 1000 ballistic computer. This automatic firing unit enables the crew to perform topographic and ballistic calculation and aiming procedures through an Automatic Control Unit installed in the cab of the vehicle.

The launcher consists of two removable modules with 20 launching tubes each. The escort unit can carry four launching modules. Loading operations are carried out by removing the empty module by a jib crane fitted on the support vehicle and installing the filled one ready to fire.

The firing unit has a fire control system and a link with the command post to acquire the necessary firing data in real time. The fire control system has been designed to enable selection of the number (from 1 to 40) and type of rockets to be fired.

The rockets, available in two models with different maximum range, can carry a variety of different warheads, from the conventional type (HE, PFF, WO) to the sub-munition types dispensing anti-tank or anti-personnel bomblets and mines.

Initial production systems had an unarmoured cab but current production launchers and command post vehicles have an armoured cab.

FIROS 25 differs from FIROS 30 in having a double base propellant compared with the composite propellant of the later rocket.

CHARACTERISTICS

Time to bring the system into action: 5 minutes (1 minute using automatic firing unit)
Ripple time: 16 s (40 rockets)
Disengagement time: 1 minute (30 s using automatic firing unit)
Reloading time: 5 minutes
Launcher Unit
Number of tubes: 40 (20 per module)
Elevation: 0 to +55°
Azimuth: ±105°
Crew: 3 (2 using automatic firing unit)
Launching Module
Length: 3.7 m
Width: 0.82 m
Height: 0.69 m
Empty weight: 400 kg
Loaded weights: 1470 kg (conventional warhead); 1690 kg (submunition warhead)

STATUS

Rockets and complete system are in production. Six batteries of FIROS 30 have been ordered by the Italian Army.

CONTRACTOR

SNIA BPD, Setlore Difesa e Spazio, Via Sicilia, 162-00187 Rome, Italy.

FIROS 25/30 resupply vehicle with hydraulic crane for loading new pods of rockets

FIROS 25/30 battery command post in latest production configuration with armoured cab

FIROS 25/30 multiple rocket launcher in latest production standard with armoured cab

JAPAN

3673.111

TYPE 75 130 mm MULTIPLE ROCKET LAUNCHER

This weapon consists of a tracked and armoured vehicle with a 30-round launcher for 130 mm calibre fin-and-spin-stabilised artillery rockets. Sequential salvo firing is possible. The 30-round launcher unit can be elevated and traversed by electrical power, and the vehicle can also be equipped with a 12.7 mm machine gun as secondary armament.

CHARACTERISTICS

Rocket
Calibre: 130 mm
Length: 1.856 m
Propulsion: Single-stage, solid-propellant rocket motor
Weight: 43 kg (at launch)
Range: 15 000 m

Launcher vehicle
Weight: 16.5 t
Road speed: 53 km/h
Crew: 3

DEVELOPMENT

The development and production of the Type 75 rocket was carried out by the Nissan Motor Company Ltd under contract from the Japan Defence Agency.

STATUS

Entered service with the Japan Ground Self Defence Force in 1975. By 1985 the Japan Ground Self Defence Force had 66 Type 75 MRS in service.

CONTRACTORS

Aerospace Division, Nissan Motor Co Ltd, 5-1 3-Chome, Momoi, Suginami-ku, Tokyo 167, Japan – Rocket and multiple launcher.

Komatsu Industries Ltd, (and others) – Vehicle.

130 mm (30-round) Type 75 multiple rocket system (K Nogi)

SOUTH AFRICA

4527.111

VALKIRI 127 mm ARTILLERY ROCKET SYSTEM

Valkiri is a highly mobile artillery rocket system capable of producing high rate saturation fire on area targets at ranges of between 8 and 22 km. It can be deployed on its own or in support of more conventional artillery weapons against area targets such as camps, troop concentrations and convoys of soft skinned vehicles.

Standard artillery techniques are applied to sight the launcher; the sight being mounted in the side of the launcher and stored in a special container on the vehicle when not in use. The rockets can be fired singly or in ripples of between two and 24 from within the cab of the vehicle by means of an electronic firing unit mounted inside the cab or remotely via a remote firing unit which can be connected to the launcher via a 50 metre plug-in cable.

A test circuit is incorporated in the firing unit to check the status of the rockets and associated firing circuits as regards possible open or short circuits prior to firing.

The 127 mm calibre rocket has a double base propellant motor, is 2.68 m long and weighs 60 kg. An anti-personnel warhead with 8500 steel balls is carried and both impact and a frequency agile radio proximity fuzes are employed. The latter fuze has been developed by Fuchs and is powered by an internal wind turbine generator. The rocket has a maximum range of 22 km which can be reduced to 15 km by means of a lower launch elevation angle or to 7.5 km by adding steel drag rings to the rocket body.

Valkiri 127 mm (24-round) multiple rocket system in firing position with stabilisers lowered at rear of vehicle (T J Gander)

The Valkiri multiple artillery rocket launcher (MARL) has a total of 24 tubes arranged in three horizontal rows of eight, carried on a traversing and elevating mount which is operated hydraulically. This mount is integrated with a standard light four-ton Unimog 4 × 4 chassis, the only other modification to which is the provision of two hydraulically actuated stabiliser spades which are used when deployed for firing. Hydraulic power for the stabilisers and traverse and elevation of the tube assembly is derived from either the vehicle's power take-off or from the battery. Each launcher vehicle carries the firing crew of two, radio, sight and on-vehicle or remote fire control equipment.

127 mm rocket used with the Valkiri multiple rocket system which has a prefragmentation warhead fitted with a contact or proximity fuze. This model has a practice warhead

Each MARL is supported by an ammunition carrier, based on the SAMIL 5 tonne truck, which carries 48 rockets in containers. Four MARLs and their resupply vehicles comprise a troop, which also has a survey team and a meteorological team, the latter using a Unimog vehicle with a telescopic mast for wind measurement. Two troops form a battery and several batteries form a regiment.

In addition to the mast-mounted sensors, at MARL sites a wind gun firing a small calibre projectile is used to estimate the prevailing local ground wind conditions. Meteorological and upper air wind data obtained by the divisional S700 system (**4494.161**) is transmitted direct to individual troop fire control units of each battery within a division.

A MARL can be ready to fire within five minutes of arriving at a pre-surveyed site so that theoretically a battery can launch a total of 192 rockets in six minutes. Each MARL can be ready to move again in a maximum of two minutes. Alternatively, in ten minutes each could be reloaded with another 24 rockets and ready to fire again.

CHARACTERISTICS

Launcher
Number of launch tubes: 24
Firing rate: 1 per second
Firing mode: Singly or any ripple between 2 and 24
Elevation: 888 mils (max)
Traverse: 1851 mils
Vehicle: 4-ton 4 × 4
Max speed on open roads (loaded): 90 km/h

Crew: 2
Time to load 24 rockets: 10 minutes (max)
Time for deployment of launcher on a surveyed site: 5 minutes
Rocket
Calibre: 127 mm
Length: 2680 mm
Propulsion: Double base solid fuel rocket motor
Warhead
Type: Prefragmentation
Lethal area: 1500 m² average
Fuze: Proximity or contact
Performance
System effectivity (battery of 8 launchers) kill effectiveness on a standard 400 × 500 m
NATO target: 30%
Range: 8 to 22 km (with drag rings)
Precision at maximum range
Standard deviation in azimuth: 270 m
Standard deviation in range: 190 m
Accuracy at maximum range
Standard deviation in azimuth: 290 m
Standard deviation in range: 200 m
STATUS
In production. In service with South African Army.
CONTRACTOR
Armscor Corporation of South Africa Ltd, 0001 Pretoria, Republic of South Africa.

SPAIN

4871.111
TERUEL 140 mm MULTIPLE ROCKET SYSTEM
The 140 mm multiple artillery rocket system used by the Spanish Army is based on mobile launcher vehicles mounted on Pegaso 3050 6 × 6 truck chassis on the rear of which is carried an elevating assembly consisting of two 20-round blocks of launch tubes for the 140 mm calibre rockets. The driving cab of the vehicle can be provided with a roof mounted machine gun for self defence against aircraft or other lightly armoured threats.

The rockets are equipped with a 1.57 second solid fuel propulsion motor, and by the selection of one of three types of aerodynamic air brake three different trajectories can be flown for the same tube elevation angle. Rockets also have flip out stabilising fins. Either an 18.6 kg HE warhead containing 6.6 kg of composition B explosive, or a 21 kg sub-munition warhead can be employed. The latter payload is available in three versions: (a) with 42 anti-personnel grenades, 400 steel pellets and percussion fuzes, (b) 28 hollow charge grenades with contact fuzes and capable of penetrating 100 mm of armour, or (c) six pressure activated anti-tank mines fitted with anti-disturbance devices. The rocket fuzes may be either the PD M34A1 contact type, proximity or timed. Mixed loads of all three types of rocket can be contained in each of the 20-round blocks carried on the launcher vehicle, and may be selected for firing according to the target(s) to be engaged. Reloading is effected manually.

Loading a 140 mm rocket into the Teruel (40-round) multiple rocket system

A re-supply vehicle with four 20-round blocks is allocated to each firing vehicle, as part of the battery, and the system is normally deployed in battalions of three batteries each with six firing vehicles.
CHARACTERISTICS
Calibre: 140 mm
Number of tubes: 40
Length of rocket: 2.044 m (HE); 2.14 m (sub-munition)
Weight: 26 kg (HE); 59 kg (sub-munition)
Max range: 28 000 m
Max velocity: 687 m/s
Reload time: 5 minutes
Max elevation: +55°
Traverse: 240°
Crew: 2 or 3
STATUS
In production. In service with Spanish Army. Gabon has purchased one system for trials.
CONTRACTOR
Empresa Nacional Santa Barbara, Manuel Cortina 2, 28010 Madrid, Spain.

SWITZERLAND

4123.111
OERLIKON 81 mm TYPE RWKO14 ROCKET LAUNCHER
The Type RWKO14 multi-tube rocket launcher is designed for firing Oerlikon 81 mm SNORA unguided solid fuel rockets. This rocket has folding fins and a pure internal burning propulsive element. The RWKO14 comprises two groups of 15 launch tubes mounted symmetrically on a one-man armoured turret, carried on APCs such as the MOWAG Piranha (8 × 8) or full tracked M113 (illustrated). Loading is simple and a typical reloading time is about six minutes. Aiming in traverse and elevation is by manual operated mechanism and laying is accomplished by means of a computer-assisted aiming periscope, Type RLK252. The firing control arrangements provide for single round firing of rockets in pairs and for salvoes, and used for ground-to-ground engagements, ranges of 4 to 10 km are attainable.
CHARACTERISTICS
Calibre: 81 mm
Rate of fire: 2 × 300 rounds/minute (max)
Length of tubes: 2000 mm
Ammunition: 81 mm SNORA rockets, with different warhead types
Length: 1548-1784 mm (depending on type of warhead)

M113 series APC fittted with Oerlikon-Buhrle multiple rocket launcher type RWKO14

Weight of warhead: 15.7 – 19.7 kg (depending on type of warhead)
Max rocket speed: 520 – 670 m/s (depending on type of warhead)
Aiming periscope: Type RLK 252
Magnification: × 7
Field of view: 9°
Slant error correction: 10°
Aiming range
Traverse: Unlimited
Elevation: –10 to +50°
Turret weight: 820 kg (empty)

STATUS
Production complete. No longer offered. In service with undisclosed countries.

CONTRACTOR
Machine Tool Works, Oerlikon Bührle Ltd, CH 8050 Zurich, Switzerland.

TAIWAN

4547.111
CHING FENG (GREEN BEE) SURFACE-TO-SURFACE MISSILE

This weapon made its first public appearance at a National Day parade in Taipei in October 1981 when it appeared preceded by what were reported as associated fire control and radar facilities. Little firm information from sources in Taiwan has been made available but the Ching Feng was described as a medium range surface-to-surface or surface-to-air missile with a maximum range of more than 100 km.

From the photograph the weapon would seem to be carried by a transporter trailer rather than a launcher vehicle, which to some extent leaves the method of deployment and launching procedure details unknown. The size and general arrangement of Ching Feng suggest that the nearest likely relative is the American Lance battlefield support weapon (**2682.111**) and there is a considerable degree of superficial similarity which could well be backed by internal and less visible points of similarity, or even identical features. Co-operation between Israel and Taiwan has been noted in other instances (eg. the Hsiung Feng coastal defence missile, **4442.121**, and its obvious relationship with the Israeli Gabriel, **6019.221**) and Israel is known to operate the Lance SSM.

Lance exists in versions with either conventional HE or nuclear warheads but it seems unlikely at present that either Israeli Lance or the Ching Feng has other than a conventional payload. Therefore the latter probably most closely resembles an earlier version of Lance, with very similar performance and operating characteristics. Significant Lance characteristics in this context include: simplified inertial guidance, range of about 120 km, storable liquid-propulsion, and a launch weight of approximately 1400 kg.

STATUS
Operational with the Republic of China's Army. The Chung Shan Institute of Science and Technology is reported to have been responsible for development of Ching Feng. Production is believed to have ceased.

CONTRACTOR
Taiwan State Arsenals.

A view of Ching Feng SSM reveals similarities to US Lance. Missiles behind Ching Feng are Hsiung Feng coastal defence missiles derived from Israeli Gabriel

A Ching Feng SSM on transporter during National Day parade in Taipei (L J Lamb)

4346.111
126 mm KUNG FENG III/IV MULTIPLE ROCKET SYSTEM

The Kung Feng multiple rocket systems were developed in the early 1970s with the basic Kung Feng (Worker Bee) consisting of two banks of 126 mm calibre rocket launcher tubes arranged in five rows of four.

The rockets are 0.8 m long, weigh about 25 kg and are spin-stabilised. The rockets are fired from launcher tubes over 1.6 m long with the complete load of 40-rounds capable of being fired in 16 seconds.

The Basic Kung Feng IV system is mounted on top of a standard US built M113 series full tracked armoured personnel carrier and used by armoured and mechanised units. Elevation and depression is hydraulic but there is no traversing system, this being achieved by traversing the complete vehicle on its tracks.

For the infantry a trailer mounted version towed by a ¾-ton truck has been developed, this is believed to have the designation of the Kung Feng III.

The Marines use the Kung Feng IV system mounted on top of a LVTP5 series full tracked armoured personnel carrier which is fully amphibious.

The Navy use the Kung Feng Sea on their destroyers for dispensing ECM chaff.

STATUS
In production and operational with Taiwanese forces.

CONTRACTOR
Taiwan State Arsenals.

Kung Feng IV 126 mm (40-round) multiple rocket system mounted on a M113 armoured personnel carrier

4345.111
WORKING BEE-6 117 mm ARTILLERY ROCKET

Under the designation 'Working Bee-6' the Taiwan State Arsenal has developed and produced a 117 mm multiple artillery rocket system which was first revealed in the National Day Parade in October 1981. The weapon consists of a group of launch tubes comprising five rows of nine tubes carried on a trainable platform mounted on a military truck.

The launcher is mounted to rear of fully enclosed three man cab with a traverse of 90° left and right. Before firing two stabilisers are lowered at the rear of the launcher.

The rocket is 1.8 m long and weighs 60 kg at launch, including the warhead. Maximum range is estimated to be 15 000 m.

STATUS
In production and operational with Taiwanese forces.

Taiwanese Working Bee-6 117 mm multiple artillery rocket system (L J Lamb)

CONTRACTOR
Taiwan State Arsenals.

UNION OF SOVIET SOCIALIST REPUBLICS

3327.111

SS-1 (SCUD) BATTLEFIELD SUPPORT MISSILE

The SS-1 family of heavy artillery rockets code-named Scud by NATO has been operated by Soviet forces and some client states for many years. Over the years it has seen a series of steady improvements and modifications (as recorded under this entry number in earlier editions of this volume). The latest model to have been identified by western sources is the SS-1C, which is now, in turn, being progressively replaced by the later SS-23 weapon (**4550.111**). The longer range Scud C was confirmed in a US Armed Services Committee reference in hearings of April 1978 to the KY-3 Scud, when it was stated that this version was first deployed in 1965. It was added that the later version has a longer range than Scud B, but a lower CEP accuracy.

Scud B in travelling configuration

Soviet SS-1 Scud deployed for action. Note crew in CBW clothing

German sources indicate the Scud is designated T-7 by the Soviet Union while American sources state the designation is R-11 for Scud A and R-17 for Scud B.

Scud is a land-mobile system, single missiles being carried on an IS-III vehicle that combines the functions of transporter and erector by means of hydraulically actuated rams which raise the missile to the upright position for launching. Scud A and Scud B have both been deployed on the tracked IS-III chassis, but from 1965 onward Scud B only has been reported on the MAZ-543 eight-wheeled transporter vehicle. As far as it is known the Scud on the original IS-III tracked chassis is no longer in front line service with the Warsaw Pact.

Liquid propulsion is employed, and the launch preparation time is quoted by Soviet sources as about one hour. Associated with Scud batteries, in addition to re-supply and loading vehicles, are a meteorological mobile unit which includes an End Tray weather radar, a ZIL-157V tanker vehicle with a trailerised pump unit, and command and control vehicles. The End Tray is assumed to be for radio-sonde tracking and data collection for use in obtaining upper-air information for ballistics computation prior to launch. A fairly lengthy survey procedure is also entailed on arrival at the firing position, and both tripod-mounted theodolites and optical devices attached to the missile/launcher by special brackets are employed for this. At least two types of theodolite have been noted.

The data below refers to the standard Scud B; Scud A is roughly 0.5 m shorter and has reported range capability of some 130 km. The reputed range of the Scud C model is in the region of 450 km.

CHARACTERISTICS

Type: Surface-to-surface, artillery missile
Configuration: Cylindrical body with conical nose and cruciform cropped delta tail surfaces
Length: 11.25 m
Diameter: 85 cm
Weight: 6370 kg
Propulsion: Storable, liquid
Range: 280 km
Guidance: Simplified inertial
Warhead: Nuclear, optional non-nuclear warhead

STATUS

Operational in Warsaw Pact forces. In addition to the Soviet Forces, Bulgaria, Czechoslovakia, East Germany, Hungary, Poland and Romania have been reported as operating Scud. Non-Warsaw Pact countries to have received this weapon include Egypt, Iraq, Libya and Syria. All but the Soviet missiles are likely to be fitted with non-nuclear warheads. The Scud with a conventional warhead has been used by both Iraq and Iran in the recent Middle East conflict.

Scud missiles in Soviet service are deployed on a scale of missile brigades assigned to armies and fronts, each brigade having 12 to 18 launchers. There is one brigade in the strategic reserve.

From the numbers deployed, their length of service and the successive stages of development and improvement they have undergone, the Scud series of weapons is well regarded within the Warsaw Pact and beyond. The Soviet Union started to replace its Scud missiles with the SS-23 in the early 1980s and this is now being deployed to all of the seven Scud brigades deployed on East German and Czechoslovak territory. This modernisation programme is expected to be completed in 1989 with the number of launchers per brigade being increased from 12 to 18 with each launcher having two missiles.

Scud B mounted on an MAZ-543 (8 × 8) cross country truck chassis

---

2959.111

SS-12A (SCALEBOARD A) BATTLEFIELD SUPPORT MISSILE

The SS-12 Scaleboard surface-to-surface missile is transported and launched from a MAZ-543P (8 × 8) cross-country truck which is also used for the Scud B. Scaleboard is normally found at Front level with a SS-12 Scaleboard brigade having two or three battalions each of which has six launchers.

In 1985 the US Department of Defense stated that a more accurate version of the Scaleboard had been deployed, this is now known to be the SS-22 (**4549.111**) which uses the same launcher as the SS-12 system and has a similar range but is more accurate.

Unlike the Scud missiles, which in transit are exposed to view as they rest in their elevating cradles, the Scaleboard missile is enclosed in a ribbed split metal casing which is elevated with the missile into the firing position.

Scaleboard appears to be approximately the same length as Scud B – rather more than 11 m – but larger in diameter. Its range has been estimated in the region of 700 – 800 km. Nothing is known about its guidance system but it is presumed to be some form of inertial guidance suitable for use with a missile that is elevated from the horizontal to the vertical shortly before firing.

SS-12 Scaleboard missile on the move

CHARACTERISTICS

Type: Surface-to-surface tactical
Configuration: Scaleboard has not been publicly revealed outside the launcher/erector casing which completely envelopes the missile. The following dimensions are based on studies of the container and transporter vehicle (MAZ-543)
Length: 12.0 m
Diameter: 100 cm
Weight: 9700 kg
Propulsion: Presumed storable liquid
Range: 900 km
Guidance: Inertial
Warhead: Nuclear

STATUS

Fully operational in the USSR, but there remains a marked scarcity of reliable information regarding the precise role of this weapon and the scale of its deployment. However it was noted in 1984 that for the first time the Soviet authorities deployed Scaleboard forward to Eastern Europe (Czechoslovakia and East Germany). These missiles are now reported to be deployed normally with Soviet combined arms formations, and current estimates of numbers suggest 60 or more facing European NATO forces, 40 in the Soviet Far East, one battalion in strategic reserve and a similar number facing Turkey.

4548.111

SS-21 (SCARAB) BATTLEFIELD SUPPORT MISSILE

The SS-21 (Scarab) surface-to-surface missile system entered service with the Soviet Army in the mid-1970s as the direct replacement for the FROG-7. The Soviet designation is OTR-21.

The SS-21 is transported and launched from a Type 5937 (6 × 6) amphibious transport vehicle which is also used as the basis for the SA-8 Gecko surface-to-air missile system.

When being transported the SS-21 missile is protected within the hull of the vehicle and only raised for launching. The SS-21 has a solid propellant fuel and is fitted with an inertial navigation system. It has a much longer range and is much more accurate than the FROG-7 system it is replacing. The SS-21, which is called the Tochka (Point) in the Soviet Army, can be reloaded in 15 minutes and can be fitted with different warheads to attack different targets.

STATUS

By late 1985 a total of 700 FROG-7s and SS-21 (Scarab) missiles had been deployed in central and Eastern Europe with each Soviet division normally having one battalion with four launchers plus reloads. The SS-21 has already replaced the FROG-7s in service with the Czechoslovakian Army and some have also been supplied to Syria.

CHARACTERISTICS

Type: Surface-to-surface tactical
Length: 6.0 m
Diameter: 0.85 m
Weight: 3000 kg
Weight of warhead: 450 kg
Range: 120 km
Guidance: Inertial
Warhead: Nuclear, chemical, HE and sub-munition

SS-21 (Scarab) battlefield support missile on its 6 × 6 transporter/erector/launcher vehicle (US DoD)

4549.111

SS-12B (SCALEBOARD B) BATTLEFIELD SUPPORT MISSILE

The SS-12B or OTR-22, formerly designated SS-22, is produced in the Votkinsk Machine Building Plant and is transported and launched from a MAZ-7310 (8 × 8) vehicle. It is believed to have entered service in 1978 and is usually deployed in brigades of 12-18 launchers. There are believed to be six brigades; two in Eastern Europe (36 launchers), one in the Western Soviet Union (14 launchers) and three east of the Urals (43 launchers). In addition there are believed to be four independent battalions with 4-6 launchers within the Soviet Union. The system will be eliminated under the INF Treaty.

CHARACTERISTICS (Provisional)

Type: Surface-to-surface tactical
Length: 12.38 m
Diameter: 1.01 m
Weight: 8.8 tonnes
Weight of warhead: 1 × 500 KT
Range: 900 km
Guidance: Inertial
Fuel type: Solid
Warhead: Nuclear

4550.111

SS-23 (SPIDER) BATTLEFIELD SUPPORT MISSILE

The SS-23 (Spider) battlefield support missile is the replacement for the Scud B (**3327.111**) with a much greater range and accuracy and entered service with the Belorussian Military District in 1985. The Soviet designation is OTR-23.

The missile is transported and launched from a MAZ-543 (8 × 8) chassis and like the SS-21 (Scarab) system the missile is kept within the hull of the vehicle until required for launching.

No detailed characteristics of the SS-23 (Scarab) have been released except that it has an inertial guidance system, solid fuel propellant, nuclear warhead and a range of 500 km.

The SS-23 is reported to have a CEP of less than 200 metres and in addition to the nuclear warhead conventional warheads can be fitted such as HE, sub-munition or chemical. The Soviet Union has stated 82 launchers and 167 missiles are deployed in Eastern Europe.

2926.111

FROG-7 BATTLEFIELD SUPPORT MISSILE

The introduction of FROG-7, first shown to the public in 1967, marked a new departure in carrying vehicles – the vehicle used being a modern, wheeled, erector launcher (ZIL-135). The missile too, had a more modern look about it, reverting to the original single-stage design and having a cylindrical warhead of the same diameter as the missile. This missile is the last of the unguided ballistic free rocket over ground (FROG) series (see **2890.111, 2922.111, 2923.111** and **2924.111** in 1980-81 and earlier editions), some of which were supplied to Warsaw Pact states and other client states by the USSR.

The FROG missiles are called Luna by the Soviet Union. FROG 1 and 3 are reportedly designated Luna 2 (German sources suggest they are also called T-5) while FROG-7 is Luna M.

The main nozzle of the single-stage rocket motor is surrounded by a ring of much smaller nozzles.

Two Soviet FROG-7 missiles ready for launch

CHARACTERISTICS
Type: Surface-to-surface, spin-stabilised unguided tactical missile
Propulsion: Single solid-propellant motor
Warhead: Nuclear or HE 450 kg
Length: 9.1 m
Diameter: 55 cm
Launch weight: 2300 kg
Range: 70 km

STATUS
Operational in the armies of most Warsaw Pact countries. Also supplied in quantity to Egypt, Iraq, Libya, North Korea, and Syria. Some missiles, supplied by the Soviet Union to Syria, were actually fired against the Israelis in the 1973 Arab-Israeli war. They were equipped with high-explosive warheads. It was at one time suggested that these particular missiles may have been a guided version (see **2927.111**) but this has not been confirmed or repeated recently; in any case they appear not to have been very effective as HE weapons.

Each Soviet tank and motorised rifle division has a FROG-7 or SS-21 battalion consisting of battalion HQ, HQ battery and two firing batteries each with two FROG-7 or SS-21 launchers. Each launcher also has one reload missile.

5555.111
ARTILLERY ROCKETS
The Soviet Union has been using multiple rocket launchers since before the Second World War, and since then has continued to develop new systems. Multiple rocket systems are used by all members of the Warsaw Pact and some members have either developed their own systems or modified Soviet equipment to meet their own requirements.

The BM-21 is the most widely used of all of these multiple rocket launchers, and each tank and motor rifle division has a battalion of these, each battalion having three batteries of six launchers. The rockets can be fitted with a variety of warheads including high explosive, chemical, smoke, and biological.

BM-21 122 mm multiple rocket systems ready for action

STATUS
122 mm BM-21: Entered service in 1964 and is used by most members of the Warsaw Pact; it has also been exported to many countries, including Afghanistan, Algeria, Angola, Bulgaria, Chad, China, Cuba, Egypt (improved version also built), Ethiopia, East Germany, Hungary, India, Iran, Iraq, Israel, North Korea, Libya, Morocco, Mozambique, Nicaragua, Pakistan, Poland, Syria, Tanzania, USSR (including naval infantry). Vietnam, North Yemen, South Yemen and Zambia. China has developed at least two 122 mm systems virtually identical to the launcher fitted to the BM-21. These are the 122 mm (30-round) Type 82 system on a 6 × 6 chassis and the 122 mm (24-round) Type 83 system on a 6 × 6 chassis.
122 mm M-1972: Entered service in 1971/72 and consists of a Czechoslovak Tatra-813 8 × 8 truck with a Soviet BM-21 launcher mounted on the rear; a special rack is carried behind the cab which contains a further 40 rockets. This enables the launcher to be quickly reloaded. Full details of this system are given in entry **5137.111** under Czechoslovakia.
132 mm BM-13-16: This is the oldest of all of the Soviet multiple rocket launchers as it entered service in the Second World War. It is no longer in front line service but is still used for training. The rockets are launched from rails rather than from a frame or tube.
140 mm BM-14-16: Entered service in the early 1950s and has been used by Poland and the Soviet Union. It is no longer in front line service.
140 mm BM-14-17: Entered service in 1959 and is a development of the earlier BM-14-16; it is still used by Poland and the Soviet Union.
140 mm RPU-14: This towed multiple rocket launcher is used only by the Soviet airborne divisions and is issued on the scale of 18 per division.
200 mm BMD-20: Entered service in the 1950s but is

Designation	RPU-14	BM-21	BM-13-16	BM-14-16	BM-14-17	BMD-20	BM-24	BM-25	BM-27
Type	Towed	SP	SP	SP	SP	SP	SP	SP	SP
Chassis	—	Ural-375	ZIL-151	ZIL-151	GAZ-63A	ZIL-151	ZIL-157	KrAZ-214	ZIL-135
Number of barrels	16	40	16	16	17	4	12	6	16
Rocket									
Calibre	140 mm	122 mm	132 mm	140 mm	140 mm	200 mm	240 mm	250 mm	220 mm
Length	1.092 m	3.226 m	1.743 m	1.092 m	1.092 m	3.110 m	1.18 m	5.822 m	4.8 m
Weight	39.6 kg	45.8 kg	42.5 kg	39.6 kg	39.6 kg	194 kg	112 kg	455 kg	360 kg
Max MV	400 m/s	699 m/s	350 m/s	402 m/s	402 m/s	N/A	363 m/s	N/A	—
Max range	9810 m	20 500 m	9000 m	9810 m	9810 m	19 500 m	10 300 m	55 900 m	40 000 m
Launcher crew	5	6	6	7	6	6	6	N/A	6
Weight loaded	1200 kg	11 500 kg	6432 kg	6432 kg	5300 kg	8700 kg	9200 kg	18 145 kg	22 750 kg
Length	4.036 m	7.35 m	7.5 m	6.92 m	5.41 m	7.2 m	6.705 m	9.815 m	9.3 m
Height	1.6 m	2.85 m	3.19 m	2.65 m	2.245 m	2.85 m	2.91 m	3.5 m	3.2 m
Road speed	—	75 km/h	60 km/h	60 km/h	65 km/h	60 km/h	65 km/h	55 km/h	65 km/h
Range	—	405 km	600 km	600 km	650 km	600 km	430 km	530 km	500 km
Launcher elevation	0 to +45°	0 to +55°	+15 to +45°	0 to +52°	0 to +52°	+6 to +60°	0 to +52°	0 to +55°	0 to +55°
Traverse	30°	180°	20°	210°	200°	20°	140°	2°	240°
Time to reload (minutes)	4	10	5-10	4	2	10	4	N/A	15-20

BM-25 240 mm rockets in foreground. BM-14-17 rockets on right

no longer used in large numbers. Some are used by North Korea.

240 mm BM-24: Although developed in the 1950s this still remains in service with a number of countries including Algeria, East Germany, Egypt, and Syria. A number were captured by Israel and these are now used by the Israeli Army.

240 mm BM-24T: This is same launcher as BM-24 but mounted on an AT-S medium tracked artillery tractor. It is no longer in service.

BM-25: Entered service in the 1950s. Its rockets are believed to have a liquid, rather than a solid-propellant. The rockets are the heaviest and have the longest range of all the Soviet multiple rocket launchers.

BM-27: The 220 mm (16-round) multiple rocket system was introduced into the Soviet Army in 1977 and is mounted on a ZIL-135 (8 × 8) truck chassis. It is normally supported by a resupply vehicle which enables 16 replacement rockets to be reloaded in between 15 and 20 minutes. The rockets can be fitted with a variety of warheads including chemical, high explosive and bomblet (minelet, fragmentation and incendiary). In addition to being used by the Soviet Union it is also known to be used by Syria.

Notes:

Poland has developed a towed 140 mm (8-round) multiple rocket system designated the WP-8 which fires the same rockets as the Soviet 140 mm (17-round) BM-14-17, (16-round) BM-14-16 and towed (16-round) RPU-14 systems. The WP-8 has a maximum range of 9810 m.

Romania has deployed a number of rocket launchers of different chassis including the 122 mm BM-21 (21-round) system on the rear of a SR-114 (4 × 4) truck and the Czech 130 mm M52 (32-round) system on a ZIL-151 (6 × 6) chassis.

UNITED STATES OF AMERICA

4194.111

GLCM GROUND LAUNCHED CRUISE MISSILE

The GLCM system consists of the Tomahawk missile (BGM-109) integrated on an air-transportable, ground mobile-launcher unit. Together with its launch control van, it is protected in its peacetime location within a hardened shelter. The ground mobility will lessen the risk of destruction by an enemy pre-emptive strike in the event of war. The USAF is implementing plans to deploy GLCMs and Pershing II surface-to-surface missiles (**2767.111**) in various parts of Europe as the major element of the NATO Theatre Nuclear Forces modernisation programme, initiated principally as a response to Soviet deployment of SS-20 IRBMs in Europe.

The GLCM can be used for both selective or general nuclear release options against fixed targets such as lines of communications, logistic facilities, airfields, command posts and stationary tactical targets such as staging and assembly areas.

The basic GLCM combat unit consists of 16 missiles loaded on four TELs (transporter/erector/launcher vehicle) with two mobile launch control centres. As noted above, the missile is the BGM-109 Tomahawk air-breathing cruise missile with a 2500 km range, flying at low altitude to avoid radar detection and AA defences, and carrying a nuclear warhead. An inertial navigation guidance system updated by terrain contour matching (TERCOM) information directs the missile in flight. More details of this system have been published in earlier editions of *Jane's Weapon Systems*. Each TEL provides transport, protection, elevation and launch support for four cruise missiles, and enables rapid movement from main operating bases to remote launch locations. In combat-alert conditions, GLCMs may be moved from protective hardened shelters to areas where natural terrain offers concealment for movement and firing operations. Until launch, each missile rests in an aluminium canister which is loaded within the TEL, the wings, control fins and engine inlet being retracted during storage. To launch the GLCM, a solid-propellant rocket motor boosts the missile to cruising speed. During the boost phase fins unfold, the engine inlet and wings deploy. The turbofan engine then ignites to sustain flight; after approximately 13s the booster motor burns out and is then jettisoned.

The TEL vehicle, produced by General Dynamics also, weighs about 33 055 kg and is 16.5 m long. It is capable of off-road driving and can be carried in C-130, C-141 and C-5 transport aircraft. The design incorporates provisions for nuclear safety and protection against small arms fire.

Ground Launched Cruise Missile BGM-109G climbs away under booster power after leaving TEL tube during a test firing (USAF)

Two launch control centre vehicles (LCCs) accompany each GLCM combat unit. The launch control sub-system consists of two launch consoles with displays and a computer and software for system control, status monitoring, mission data storage and transfer, and missile re-targeting. The launch control sub-system is integrated with the communications sub-system within a shelter carried on the LCCs, and primary and reserve power supplies are incorporated on these vehicles. There will be redundant communications links and a lightweight fibre-optic data link between each TEL and LCC vehicle. Environmental control and protection for the crew from CBR hazards are also included. Each LCC vehicle weighs about 36 000 kg and is approximately 17.3 m long.

Within the secure main operating bases, protected buildings will provide weapon storage, maintenance and inspection. It is planned that the crews will install warheads, load TELs and undertake periodic inspection and maintenance of all GLCM combat unit parts. The missiles will be fuelled in advance and can be stored for months in their aluminium canisters without requiring scheduled maintenance or handling. Hardened concrete shelters designed to house two LCCs and four TELs protect each GLCM combat flight group from conventional attack during storage, including precision air attacks. The GLCM shelter concept divides each structure into three cells which each hold two TELs or LCCs.

GLCM will be deployed in 'flights', each consisting of a total of 26 vehicles (2 LCCs and 4 TELs carrying 16 missiles, and 20 support vehicles) manned by a total complement of 70 personnel. In transit from the main operating base (MOB) to remote launch sites it is planned to move GLCM flights in two parts called the first and second critical elements, respectively. The TELs are towed by West German MAN (8 × 8) high mobility trucks which have a good cross country mobility.

CHARACTERISTICS

Type: BGM-109G GLCM
Length: 6.4 m (with booster)
Max diameter: 53 cm
Wing span: 2.5 m
Weight: 1773 kg (with booster)
Speed: High subsonic
Range: 2500 km
Warhead: Nuclear W84
Propulsion: Williams International Corp turbofan; Atlantic Research Corp solid booster
Guidance: Inertial navigaton plus TERCOM

STATUS
The GLCM system is in production, with initial operating capability achieved in December 1983 and with a total of 560 production missiles planned by 1988. By then 132 TELs should have been delivered, and 74 LCCs.

Under the NATO Theatre Nuclear Force modernisation programme it was planned to deploy 464 GLCMs in Western Europe and the UK at a cost of $3000 million. The missiles will now be removed from Europe and destroyed under the INF Treaty.

CONTRACTOR
General Dynamics Corporation, Convair Division, PO Box 80847, San Diego, California 92138, USA.

2682.111
LANCE BATTLEFIELD SUPPORT MISSILE

Lance is a surface-to-surface guided missile designed to provide general battlefield fire support for an Army corps. In the US Army it has replaced both Honest John and Sergeant missiles. It can carry either a nuclear or conventional high-explosive warhead and its range has been estimated at approximately 130 km. It requires an eight-man crew, half the number required for Honest John (**2652.111**) and Sergeant (**2804.111** in 1980-81 and earlier editions).

This highly mobile missile can be delivered by aircraft or helicopter, has good mobility over ground and can swim inland waterways. Lance uses simplified inertial guidance, and is the first US Army missile to use ready-packaged and storable liquid propellants. Major ground support equipment includes a tracked self-propelled launcher, a fully mobile lightweight launcher, and a transporter/loader.

The Lance missile comprises a warhead section, a guidance package, fuel tankage, and an engine. The guidance package is a rugged low-cost device conceived and developed in the US Army Missile Command's Guidance and Control Laboratory. Over the operational range of the missile it is claimed to have a high degree of accuracy and reliability.

Spin-stabilisation in flight is achieved initially by venting some of the propellant gases through canted vents in the body of the missile. After initial spin has been imparted in this way it is maintained by aerodynamic pressure on the canted fins at the base of the missile.

Propulsion is by a Rocketdyne storable liquid-propellant engine comprising two concentrically-mounted sections. The outer section provides thrust during the boost phase of the missile's flight, and during this phase, which corresponds to a flight of about a mile (1600 m), the missile is under tight control by the control and guidance electronics. When the on-board inertial system detects that the predetermined velocity has been achieved, the boost motor is extinguished and the missile continues on a zero-*g* flight powered by the (inner) sustainer motor. At a predetermined point in the flight this motor too, cuts out, and the missile finishes its journey in free flight.

For land-mobile operations the weapon system is transported on two tracked vehicles (US Army's M113 family) one of which (M752) functions as an erector/launcher while the other (M688) carries two extra missiles and a loading hoist. Both vehicles are on the M667 chassis, with a road speed/range capability of 68 km/h/450 km. For air mobile operations the lightweight launcher will be more appropriate and this and the missile can be taken into the battle zone by helicopter or parachuted from a fixed-wing aircraft.

CHARACTERISTICS
Designation: MGM-52C
Type: Mobile surface-to-surface tactical guided missile
Guidance: Simplified inertial
Propulsion: 2-part, concentric pre-packed storable liquid-propellant motor
Warhead: Nuclear (W70-1/2/3) or HE
Length: 6.14 m
Diameter: 56 cm
Launch weight: 1285 – 1527 kg
Speed: Mach 3
Range: 130 km

DEVELOPMENT
The Lance prime contractor was selected in 1962. The research and development contract was awarded in January 1963. The first test firing was made at White Sands Missile Range in March 1965. The first firing from the engineering model lightweight launcher was in July 1965 and first firing from the self-propelled launch vehicle came the following month. The first production model was delivered to the US Army for service testing in April 1971; the first Army launch took place that August and service testing was completed in March 1972.

During development a total of 156 flight tests were conducted – including 37 Army launches, all but one of which were successful.

Nuclear armed Lance missiles were initially equipped with the W70-1 and later the W70-2 warhead, subsequent development led to initial operational capability for an enhanced radiation/reduced blast (ER/RB) warhead for Lance, the W70-3, which was available from 1982 onward.

STATUS
Lance was type-classified 'standard' early in 1972 and the first training battalion was activated in April. The weapon was subsequently deployed with six US battalions (108 launchers) in Europe and two in the USA as a training base (144 launchers). The US Army has recently decided to redeploy Lance to South Korea. The first operational launch took place in November 1972. Lance has been purchased by Belgium (5), West Germany (26), Italy (6), Israel, Netherlands (6) and the UK (12). Lance is regarded as possibly being the basis of a missile known as Ching Feng (Green Bee) (**4547.111**) in the Republic of China (Taiwan) Army by whom it is operated.

To extend the range of tactical applications of the weapons the US Army has a non-nuclear warhead which employs cluster bomblets and would be used against different targets from those for which the nuclear warhead is required. Lance production was completed in 1980, but an advanced, solid-propellant flight vehicle plays an important part in the Assault Breaker anti-armour programme as the T-22.

Late in 1986 the US Army awarded a $6.5 million contract to LTV for work to maintain the service life of the Lance battlefield support missile. Vought will perform about 110 000 hours of engineering services to support field operations and any necessary hardware refurbishment.

Production of the Lance was completed in 1980 and many missiles are now reaching the end of the 10-year service life guaranteed when delivered. Refurbishment will keep the Lance in active service at least to the mid-1990s.

Plans are under way for an Improved Lance with longer range.

CONTRACTORS
LTV Missiles and Electronics Group, Missiles Division, PO Box 650003, Dallas, Texas 75265-0003, USA. – Prime Contractor.

American Bosch Arma Corp, Arma Division, Roosevelt, New York, USA – Gyroscope.

Systron-Donner Corporation, Donner Division, Concord, California, USA – Guidance Components.

North American Aviation Inc, Rocketdyne Division, Canoga Park, California, USA – Propulsion.

Hawker-Siddeley Canada Ltd, Toronto, Ontario, Canada – Lightweight Launcher.

FMC Corp, Ground Systems Division, San Jose, California, USA – Vehicles.

Lance missile being prepared for launch

2765.111
PERSHING Ia BATTLEFIELD SUPPORT MISSILE

Pershing Ia is an improved version of the Pershing ground-to-ground guided weapon system (**2764.111** in 1973-74 and earlier editions).

The most obvious change that has taken place is the replacement of the XM474 tracked vehicles on which system components were mounted by a set of wheeled vehicles based on the M656 five-ton truck. The system now has an improved erector/launcher, an articulated truck and trailer combination that carries both the missile and its warhead (previously carried on separate vehicles) and capable of both paved road and cross-country travel. The other vehicles are a transporter for the programmer-test and power stations, the firing battery control central truck, and the radio terminal set vehicle with an inflatable antenna. All the equipment is transportable by the C-130, C-141, and C-5A cargo aircraft.

System reaction time of the Pershing Ia has been reduced by improvements in the erector/launcher and by the introduction of automatic countdown. The radio communications system has also been expanded. There is no change in the missile itself.

Most recent of the improvements is an automatic reference system (ARS) which employs gyro compass techniques similar to those used in navigation and which enables the missile to be fired from a previously unsurveyed site. Associated with ARS is a sequential launch adapter (SLA) which permits the Pershing commander to count down and launch up to three missiles from a single control station without uncabling and recabling after each missile is launched. European-based Pershing units received ARS/SLA ground equipment in 1976/77.

DEVELOPMENT

A $66 million development contract was awarded to Martin Marietta in 1966. This was followed by two production contracts, one for $52 million in 1967 and one for $32 million in 1968. Production started in November 1967 and was completed in 1971.

Pershing Ia equipment began deployment with US and West German forces in Germany in the summer of 1969 and was completed in 1971.

CHARACTERISTICS

Military designation: MGM-31A

Type: Land-mobile air transportable surface-to-surface tactical ballistic missile system

Guidance: Inertial

Propulsion: 2-stage solid-propellant rocket motor

Warhead: Nuclear W50: these are believed to be 400KT warheads

Missile length: 10.5 m (approx)

Missile diameter: 1 m (approx)

Launch weight: About 4600 kg

Speed: Mach 8

Range: 160 – 740 km

STATUS

The US Army withdrew its Pershing Ia missiles from Europe by late 1985 with the deployment of the Pershing II systems. The West German Air Force deploys eight squadrons of Pershing Ia missiles although these are scheduled to be dismantled in due course as part of the INF agreement.

CONTRACTORS

Complete system: Martin Marietta Orlando, Orlando, Florida, USA.

Propulsion: Thiokol Chemical Corporation.

Communications: Collins Radio Company.

Inertial navigation: Bendix Corporation, Eclipse-Pioneer Division.

2767.111

PERSHING II BATTLEFIELD SUPPORT MISSILE

The Pershing II system is a modular modernisation of the Pershing Ia (**2765.111**).

A new inertial navigation system developed by Singer-Kearfott, and a new re-entry vehicle (RV) which incorporates a high-accuracy, terminal guidance system developed by Goodyear Aerospace are used in the later version of Pershing. Range is increased to approximately 1800 km. The enhanced accuracy combined with terminal guidance and control facilities enables warheads of appreciably reduced yields to be employed to engage small hardened targets. Earlier Pershing missiles carry W50 warheads, which are thought to be 400 KT, and these have been criticised for being excessively large and destructive for battlefield use. Pershing II carries a W85 nuclear warhead with two selectable yields, 5 to 50 kt or 10 to 20 kt.

The Pershing II terminal guidance system incorporates an all-weather radar correlation unit that compares the live radar returns with a pre-recorded and stored radar image of the target area. Continual automatic comparisons of the two provides control signals to actuate the control vanes which are a distinguishing feature of the Pershing II RV. The RADAG (from radar area guidance) system relies upon inertial navigation system information from the second stage booster for position data prior to separation. Soon after re-entry, the Goodyear RADAG system is activated to sweep the target area below the rapidly descending RV, and provide accurate position updating to the inertial system. The scanner of the RADAG guidance package rotates at 120 rpm, and corrective commands are computed continuously to bring the RV back on course to its designated impact point.

Pershing II missile erected from its erector/launcher vehicle and ready to fire

Under a contract awarded by Goodyear, Norden Systems is delivering computerised systems to be used for the preparation of reference scenes that will be used to guide Pershing II missiles to their targets. Norden will supply six systems: five are miniaturised PDP-11/70Ms (also known as AN/UYK-42(V)), and one a commercial PDP-11/70 for use by Goodyear for software development. The Norden equipment is used in conjunction with digitised computer discs of possible target areas, which are supplied by the US Defense Mapping Agency (DMA). These discs, called the data base, cover such land features as terrain, elevation, rivers and buildings, and the data base is fed into the Norden computer system which predicts what the missile will see on its radar during flight.

The Norden systems are being delivered to Goodyear, apart from the two for the DMA headquarters in Washington and another for the US Army Engineer Topographic Laboratories at Fort Belvoir, Virginia. Those going to Goodyear are for mounting on government-furnished five-ton trucks.

STATUS

The Pershing II advanced development programme was completed in May 1978. Five flights of the missile were conducted successfully to test and evaluate the terminal guidance system. Full-scale development began in February 1979 with the award of a $360 million contract to Martin Marietta. Sub-contractors for Pershing II include Bendix for airborne and ground computers, Goodyear for the radar guidance sub-system, Hercules for propulsion motors, and Singer-Kearfott for the inertial measurement system.

Production deliveries to the US Army began in May 1983, and at 1 January 1984 production contracts awarded to Martin Marietta amounted to $725 million. That month the first Pershing II entered service in West Germany where 108 systems were deployed. These missiles will be removed from Europe and destroyed under the INF Treaty.

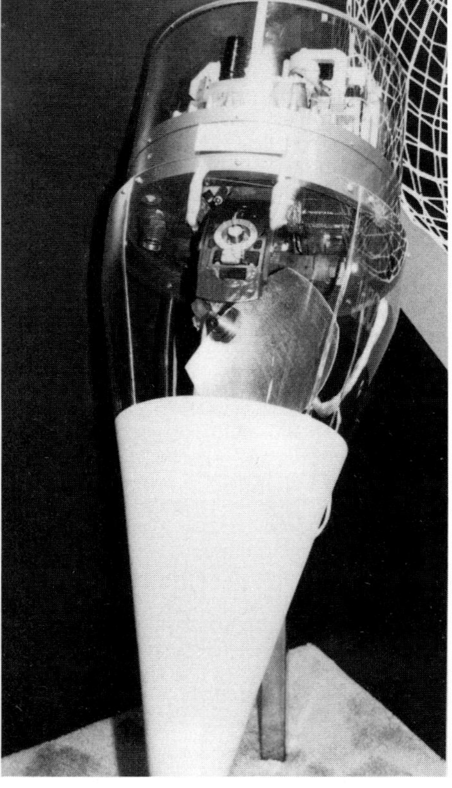

Close-up of Goodyear radar area guidance (RADAG) system in nose of Pershing II showing gimbal-mounted scanner (T J Gander)

CONTRACTOR

Martin Marietta Electronics and Missiles Group, Orlando, Florida, USA.

2652.111

HONEST JOHN BATTLEFIELD SUPPORT MISSILE

Honest John is a simple surface-to-surface free-flight rocket that has the accuracy of standard artillery and considerably better battlefield mobility. The unguided rocket is driven by a single-stage solid-propellant motor and can carry either a nuclear or a high-explosive warhead.

Designed to fire like conventional artillery in battlefield areas, Honest John is now the oldest US-developed missile system still fielded. Studies were begun by US Army Ordnance in 1950, and shortly after that Douglas Aircraft (now McDonnell Douglas) submitted proposals based on the Ordnance specifications and became prime contractor for the system. Firing tests were successfully completed in 1951 at White Sands, New Mexico.

The resultant system comprises a simple rocket and a highly mobile 6 × 6 self-propelled launcher. Operation is simple – there are no electrical controls – and normal crew training and standard fire control techniques are employed.

CHARACTERISTICS

Military designation: MGR-1B

Type: Mobile surface-to-surface unguided tactical missile.

Propulsion: Single-stage solid-propellant rocket motor

Warhead: Nuclear (W31) or high explosive

Length: 7.5 m

Diameter: 76 cm

Launch weight: 2040 kg

Range: 7.5 – 37 km

STATUS

Honest John has seen service with the armed forces of Belgium, Denmark, the Federal German Republic, France, Greece, Netherlands, South Korea, Turkey, the United Kingdom, and the USA. Production ceased some time ago and the missile has been replaced in US Army service by Lance (**2682.111**). Honest John remains in service with Greece, South Korea and Turkey.

6364.111

70 mm RADIRS MULTIPLE ROCKET SYSTEM

The 70 mm RADIRS (Rapid Deployment Integrated Rocket System) multiple rocket system has been developed as a private venture by BEI Defense Systems Company Inc and essentially consists of the standard US Army 70 mm HYDRA 70 air-to-ground rocket system adapted for use in the ground-to-ground role. As the US Army already has the Multiple Launch Rocket System in its inventory the 70 mm RADIRS system is aimed mainly for use by light and medium forces such as the new light and motorised divisions now being formed by the US Army.

Up to six 19-tube replaceable launchers can be mounted on a modified towed M91 chemical rocket launcher or the carriage of the 105 mm M102 howitzer. Other versions include two 7- or 19-tube launchers on 4 × 4 light cross-country vehicles, two 7-tube launchers on a fast attack vehicle, three 19-tube launchers on a HMMWV, six 19-tube launchers on a 6 × 6 2.5-ton truck, four 19-tube launchers on an M42 carriage or an M55 quad 12.7 mm machine gun mount, or a special 4-tube launcher on a tripod for infantry or special forces use.

For mounting on commercial trucks a 5 × 5-round square launcher has been developed. The launchers are reusable 7-tube M260 or 19-tube M261 lightweight aircraft pods or are specially configured reusable launchers as in case of the tripod and commercial utility truck mounted systems.

Up to eight fire units can be controlled either by cable or radio link from one location with the operator able to select single, pair or multiple round bursts at one time from each launcher. The operator may also select the total quantity to be fired and the firing rate for a salvo. Minimum range of the system is 700 m with a maximum range being over 15 000 m.

The rockets can be fitted with a wide range of warheads, most of which are already in the US Army inventory. These include M261 sub-munition, M151 High Explosive Remote Reset, M247 shaped charge anti-armour, M262 illumination, M264 screening smoke, M255 flechette plus training warheads and a number of other types under development.

STATUS

Development complete. Ready for production. All major components for this system have already been type classified for other applications.

HYDRA 70 lightweight multiple rocket system. On left is RADIRS VI launcher with 6 19-tube launchers and on right is a light vehicle with two 7-tube launchers

CONTRACTOR

BEI Defense Systems Co, 11312 South Pipeline Road, Euless, Texas 76040, USA.

4060.111

MULTIPLE LAUNCH ROCKET SYSTEM (MLRS)

The multiple launch rocket system (MLRS) is a highly mobile automatic rocket system developed to enable a firing crew with a minimum amount of training to shoot a complete 12-rocket load, reload rapidly and fire again. Its purpose is to complement cannon artillery in engaging massed attack by enemy formations on an area basis rather than by engaging individual targets in succession. A memorandum of understanding covering the intended adoption of MLRS by Britain, France and Germany has been concluded by these states and the USA (Italy joined the group in 1982).

The MLRS is designed to be operational day and night in all types of weather and to engage and defeat the following targets: tube artillery and rocket counter-batteries, air defence concentrations, trucks, light armour and personnel carriers as well as supportive troop and supply concentrations. Each MLRS launcher has the capacity to neutralise one artillery battery or an equivalent target with a full 12-rocket launcher load.

The armoured self-propelled launcher-loader (SPLL) incorporates the chassis and running gear of the US Army Bradley fighting vehicle and accommodates a crew of three and two launch-pod containers each with six rockets. The containers can be loaded or unloaded singly or simultaneously. The rockets are preloaded and sealed into the container launch tubes.

The SPLL is autonomous with its own fire control system, position determining system/stabilised reference package, launcher drive system and devices for self loading/unloading. The fire control system uses elements of the BCS (**3510.161**) and has four elements: the electronics unit which directs all system activities, the fire control unit which interfaces with the rockets, the operator's fire control panel and a small hand-held unit for loading/unloading the launch pods. There are digital communications with the BCS or TACFIRE (**2827.161**).

The fire control system functions, which can be entered automatically or manually, include semi-automatic processing of fire missions, re-aiming between rockets during ripple fire, ballistic computations and load/unload operations. The system provides automatic retargeting to permit a rapid, accurate volume of fire on numerous individual targets. The alphanumeric display provides simple instructions, with prompts, in any language.

To eliminate the need for surveying sites the Position Determining System (PDS) continuously computes the launcher position by measuring the distance and direction travelled by the vehicle. The positioning facility was added to the Stabilisation Reference Package (SRP) north-seeking gyro compass.

The rockets can be fired singly or ripple fired in units of from 2 to 12. The entire load can be ripple fired within a minute. Each load can place nearly 8000 M77 sub-munitions in an area (typically) 23 hectares at mid range (15 km) against a single target. At maximum range this is approximately doubled. Each M77 has about the same destructive power as a hand grenade and it contains a shaped charge which permits it to penetrate light armour.

MLRS self-propelled launcher/loader in travelling configuration

The dual purpose anti-material/anti-personnel M77 warheads were the first of three developed for the MLRS. Also planned are scatterable anti-tank mines (AT-2) and guided anti-armour sub-missiles (MLRS-TGW).

CHARACTERISTICS

Type: Artillery, surface-to-surface rocket system
Guidance: None (free rocket)
Length: 4 m
Diameter: 227 mm
Range: 30 km +
Propulsion: Solid fuel
Warhead: M77 sub-munition, dispensing
Rockets carried: 12 per launcher
Max launch rate: Rapid ripple, 12 rounds in less than 1 minute

DEVELOPMENT

US Army MLRS concept definition study contracts were let in March 1976, and these were completed later that year. Competitive development by the Boeing and Vought teams began in 1977. Between 1972 and December 1979 the system was known as the General Support Rocket System.

The overall MLRS programme embraces several planned system enhancements and modifications, and there are also a number of other projects relating to it that have varying degrees of official backing. The most significant of all these planned or potential variants are summarised in the following paragraphs.

Lightweight MLRS: This manufacturer's project is an attempt to provide the fire power of a heavy division combined with the mobility of a light division, by mounting a six-round launcher assembly on a modified M113 tracked vehicle. Lightweight MLRS has more than 80 per cent of its parts which are common to the standard MLRS and M113 equipments. The loaded weight of a Lightweight MLRS vehicle is 10 437 kg compared with the 24 276 kg of the basic MLRS. Various six-round trailer mounted versions of the Lightweight MLRS have been proposed for the new US Army light divisions.

AT-2: This is a German project developed by the RTG consortium of MBB and Diehl which will cost the equivalent of $100 million and involves providing a payload of anti-armour mines. The mines, which can penetrate 140 mm of steel plate, are being developed by Dynamit Nobel and fit into seven canisters each of which will have four mines. The rocket, which weighs 257.5 kg, dispenses the canisters which in turn distribute the mines. A single MLRS can scatter 336 mines over an area 1000 × 400 m.

MLRS-TGW: The MLRS Terminal Guidance Warhead programme is the result of an agreement between the USA, United Kingdom, Federal Republic of Germany, France and Italy to provide the MLRS with an active anti-armour capability by the development of terminally guided warheads. The basic concept is as follows:

The MLRS-TGW rocket flies a ballistic trajectory to a predetermined point in space. The TGW dispersal system is then initiated by the rocket fuze to dispense the six sub-missiles near the target area. Each is preprogrammed to a preset trajectory. The sub-missile seeker is activated automatically to search the terrain below for an appropriate target. When this is accomplished, the seeker will guide the warhead to engagement of the target. The TGW is being designed and developed by an international contractor team that includes: Martin Marietta Aerospace (US), Diehl GmbH, (FRG), Thomson Brandt Armements (France), and THORN EMI Electronics (UK), these four national prime contractors jointly owning MDTT Inc, a company which received a $90 million US Army Missile Command contract in late 1984 for TGW development.

Basic MLRS prime contractor, Missiles Division, LTV Missiles and Electronics Group, has associate agreements with the various TGW contractors and expects to play an important part in the integration of TGW with the MLRS system. The fire control system will be updated for MLRS-TGW operation, and a new electrical unit will be incorporated to support MLRS growth over the next 15 years. The latter involves digital data interfaces, software, enlarged computer memory, and multiple processors.

The TGW will consist of three sub-missiles which have external dimensions and ballistic characteristics identical to the AT2 warhead section used for dispensing German scatterable mines in the MLRS Phase II programme. The warhead section will also contain a sub-missile dispersal system activated by the warhead fuze at a designated point in space near the target(s). It is anticipated that millimetric wavelength radar will be used by the TGW seeker.

ABRS: The Assault Ballistic Rocket System is a Vought proposal to employ the MLRS system as a shore bombardment weapon aboard US Navy amphibious assault forces. In this role, it is claimed, ABRS would be 12 times more effective than the 155 mm vertically loaded naval gun and 25 times more effective than the 5-inch/54 gun.

MLRS rocket launch

An artist's impression of the MLRS-TGW dispenser

Binary warhead

A Binary chemical warhead is under development for the MLRS for deployment with the US Army.

SADARM

Late in 1986 two companies (**3988.111**) were each awarded contracts for the development of SADARM (Sense and Destroy Armour) weapons to be dispensed from 155 mm artillery projectiles and a new version of the MLRS.

Army Tactical Missile System: The new US Army Tactical Missile System, covered in the next section, will be launched from a MLRS system.

STATUS

In April 1980 the US Army selected LTV as prime contractor for MLRS. Subsequent funding awarded under four separate contracts (maturation research and development, initial production facilities, low-rate production and multi-year procurement) amounts to a total of approximately $2 billion, calling for delivery of 333 SPLLs and almost 400 000 rockets in launch-pod containers. Initial production began in early 1982 and the first production SPLL was delivered in August 1982. Ultimately the production rate is expected to reach 6000 rockets per month. It is planned to employ MLRS in the NATO forces of Britain, France, West Germany, and Italy as well as those of the USA.

European production of MLRS will be undertaken by the MLRS EPG which has four shareholders; Aérospatiale (France), BPD (Italy), Diehl (Germany), Hunting Engineering (United Kingdom). MLRS EPG contracts out production under a contract from the German BWB. In addition LTV and MLRS EPG have created the MLRS International Corporation (MIC) to market the system.

The contractors for the European MLRS are:

France:

Aérospatiale (launch pod container structure, 42 per cent of SPLL, turret and base, electrics), Creusot Loire (sprocket track), GIAT (rocket igniter, hull), SNPE (65 per cent of rocket motors).

West Germany:

Diehl (launch pod container integration, M77 warhead), Hagen Batterie (batteries), MBB (launch pod container assembly integration and trainer, rocket fins, launch tube, cables, warheads, practice rockets), Rheinmetall (55 per cent of M77 grenades), Thyssen Henschel (carrier integration, final drive, power take-off), Vickers System (launcher drive system), Wegmann (58 per cent of SPLL, cage), Western Gear (azimuth drive).

Italy:

Selenia (fire control system panel, SNIA-BPD (35 per cent of rocket motors).

United Kingdom:

David Brown Gear (transmissions), GEC Avionics (stabilised reference package, gyroscopes), Hughes Micro-Electronics (umbilical clamp head, umbilical housing, internal and external flex printed wiring), Insulating Company & Materials Ltd (rocket nozzle mounts, supports), Lucas Aerospace (rocket motor cases), Marconi Command and Control Systems (fire control system), Royal Ordnance, Chorley (45 per cent of M77 grenades), Royal Ordnance, Birtley (warhead structures), Vickers Defence Systems (cab).

United States:

Cummins (diesel engine), Kerr McGee (ammonium perchlocrate).

In January 1983, Britain became the first foreign country to order MLRS when a $5.2 million contract for four launchers and 108 practice rockets plus training equipment was signed. The Federal Republic of Germany followed suit in March 1983 with a $2.3 million contract for two launchers plus training hardware. Total requirements for NATO countries are as follows:- United Kingdom (71), West Germany (202), France (56), Italy (20). The Netherlands have decided to procure the MLRS direct from the US rather than from the European production line. A total of 21 SPLL will be delivered to the Netherlands with first deliveries planned in 1987. Other potential customers include Turkey.

MLRS was fielded by the US Army in early 1983 and

is now deployed with Army units in the USA, Europe and South Korea. By early 1987 the US Army had deployed 21 MLRS batteries each of which has nine launchers. Three batteries form a battalion with a logistics platoon of 18 Resupply Vehicles (RSV) and a fire control section with a M577A2 vehicle carrying a BCS.

US CONTRACTORS

Self-Propelled Loader Launcher

Sperry-Vickers: Launcher drive system

Norden Systems Inc: Fire control system

Bendix Guidance Systems: Stabilised reference package/position determining system

LTV: Launcher/loader module

Launch Pod Container

Brunswick: Launch tubes

LTV: Launch/pod container integration

Rocket

Atlantic Research Corporation: Propulsion system

LTV: M77 munition warhead

LTV: Rocket integration

Government Furnished Equipment

FMC Corporation: MLRS carrier

US Army: M77 sub-munitions

LTV Missiles and Electronics Group, Missiles Division, PO Box 650003, Dallas, Texas 75265-0003, USA – Prime contractor.

MARKETING

MLRS is marketed world-wide by the MLRS International Corporation (MIC), a US corporation jointly owned by LTV Aerospace and Defense Company and MLRS-Europäische Produktionsgesellschaft mbH (MLRS-EPG). Inquiries to: MLRS International Corporation, 8th Floor, Bowater House East, 68 Knightsbridge, London SW1X 7LT, GB. Tel: (01) 589 6383, Fax: (01) 225 0191, Telex: 8952636 HUNGRO G.

YUGOSLAVIA

2266.111

128 mm MULTIPLE ROCKET LAUNCHER M-63

It comprises a 32-tube launcher for 128 mm spin-stabilised rockets. The launcher is mounted on a two-wheel trailer and the road wheels, together with the spread trails form the platform for the launcher when in operation. The launcher assembly can be elevated or depressed by handwheel; there is also a traverse hand-wheel, but the available traverse is limited to a few degrees for correction of aim – otherwise the launcher box would foul the road wheels. Firing is electrical.

CHARACTERISTICS

Rocket

Calibre: 128 mm

Length: 814 mm

Weight: 23 kg

Warhead weight: 7.55 kg

Launcher

Capacity: 32 rockets

Weight: 2.134 t approx

Elevation: 0 – 48°

Traverse: 30°

Max rocket velocity: 420 m/s

System range: 8.6 km

Rate of fire: 160 rounds/minute

Time to reload: 3 minutes

Crew: 3 – 5

STATUS

In service with the Yugoslav Army.

CONTRACTOR

SPPR, Federal Directorate of Supply and Procurement, PO Box 308, 9 Nemanjina Street, 11001 Belgrad, Yugoslavia.

M-63 rocket launcher

5153.111

YMRL 32 128 mm MULTIPLE ROCKET LAUNCHER M77

The YMRL 32 (M77) was first seen in 1975 and is a modified civilian FAP 2220 BDS 6 × 4 truck chassis with a 32-round rocket launcher mounted on the rear of the body. Behind the cab is a reloading system with a further 32 rockets. In concept the YMRL 32 128 mm system is similar to the Czechoslovak BM-21/M-1972 rocket launcher (**5137.111**). Current production models are mounted on the new FAP 2020BS (6 × 6) truck with much improved mobility and a central tyre pressure regulation system.

The launcher has an electric/hydraulic control system and the rockets can be fired from the cab of the truck, or away from the vehicle with the aid of a remote control system. When the 32 rockets have been fired, the launcher is swung into the horizontal position and the reloading system is raised until it lines up with the launcher. It takes about two minutes to reload the complete launcher.

Launch of Yugoslav YMRL 32 126 mm artillery rocket

CHARACTERISTICS

Calibre: 128 mm

Length: 2.6 m

Weight: 65 kg

Warhead weight: 20 kg (inc 5 kg explosive)

Number of tubes: 32

Range: 20 km

Rate of fire: 18s for 32 rockets (single or ripple)

Elevation: 0 – 50° (auto or manual)

Traverse: 360° (auto or manual)

Time to fire and move (64 rounds): 4 – 5 mins

Dispersion: 0.6% of range (approx)

STATUS

In service with the Yugoslav Army.

CONTRACTOR

SPPR, Federal Directorate of Supply and Procurement, PO Box 308, 9 Nemanjina Street, 11001 Belgrad, Yugoslavia.

Details of the following multiple rocket systems are given in *Jane's Armour and Artillery*:-

Country	System
Austria	130 mm M51 (32-round) on Steyr 630 M3 6 × 6 chassis
Belgium	70 mm LAU97 (40-round) on Streaker, Land Rover or truck chassis
China	284 mm (10-round) Type 74 on truck (6 × 6) chassis
Czechoslovakia	130 mm M51 (32-round) on Praga V3S (6 × 6) chassis
India	122 mm system called LRAR mounted on 5000 kg (6 × 6) chasis is in service with Indian Army
Japan	Type 67 rocket launcher on Hino (6 × 6) chassis
Korea, South	130 mm (36-round) system on KM809A1 (6 × 6) chassis

Anti-Tank/Assault Weapons

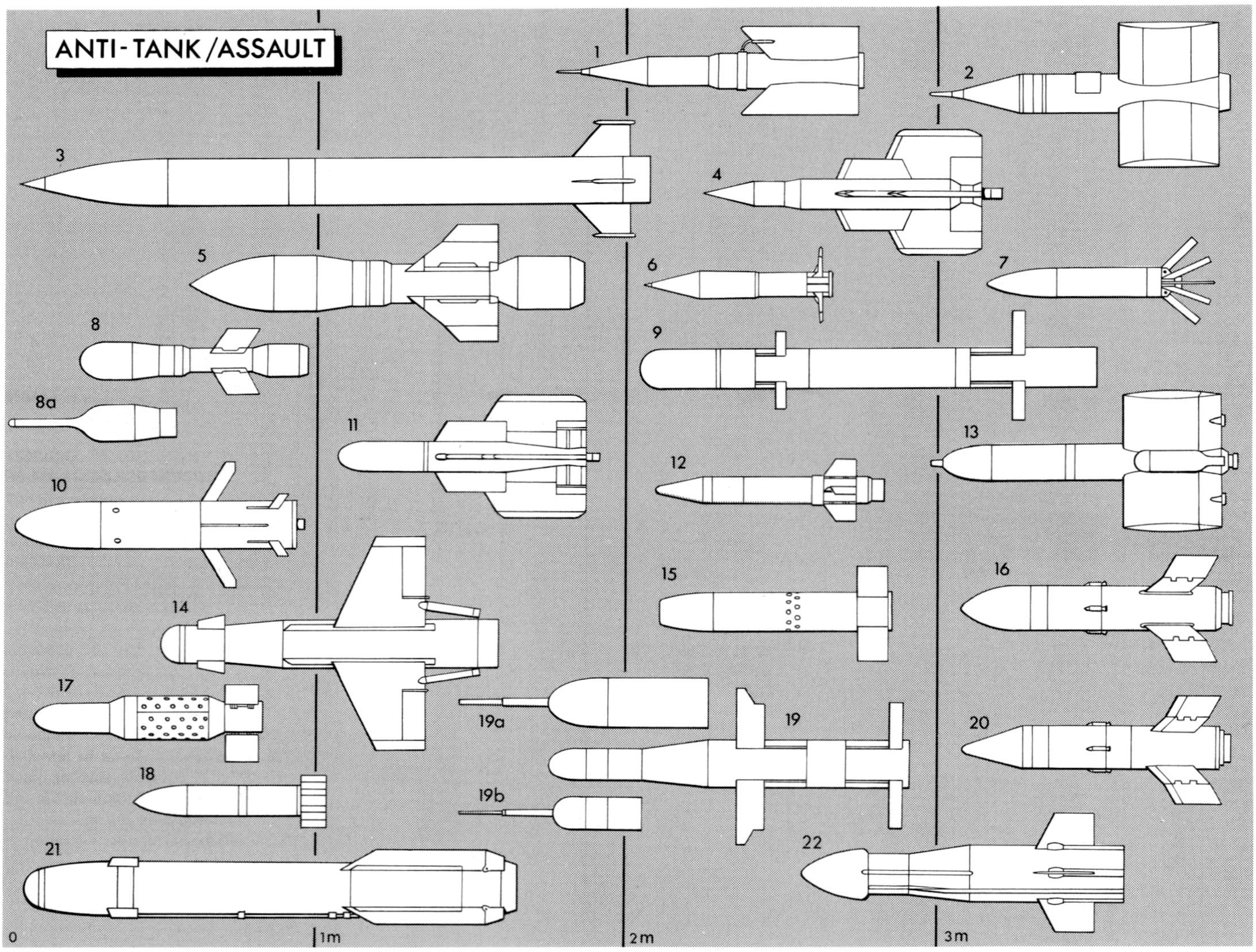

ANTI-TANK/ASSAULT

1 Mamba, **2** Cobra 2000, **3** ADATS, **4** Mathogo, **5** HOT, **6** ACL-STRIM, **7** Folgore, **8** MILAN, **8a** K115 long fuze MILAN version, **9** KAM-9 (T79), **10** RBS 56 Bill, **11** Bantam, **12** Picket, **13** KAM-3D (T64), **14** AT-2 Swatter, **15** Strix, **16** Kun Wu, **17** Dragon, **18** LAW 80, **19** TOW, **19a** TOW 2, **19b** Improved TOW (ITOW), **20** AT-3 Sagger, **21** Hellfire, **22** Swingfire

NOTE: This section of *Jane's Weapon Systems* concentrates mainly on anti-tank guided missiles. Details of additional unguided anti-tank weapons will be found in *Jane's Infantry Weapons* while details of artillery and armoured vehicle projectiles are given in *Jane's Armour and Artillery*.

ARGENTINA

3677.111

MATHOGO ANTI-TANK MISSILE

The Argentinian Mathogo anti-tank missile is a first-generation weapon relying upon wire guidance. It is designed to be carried and operated by an infantry soldier, but can also be launched from vehicles, helicopters or light aircraft. It is fitted with a hollow charge warhead.

The complete system comprises: a water-tight container and launcher with 50 metres of cable, and the control unit and binocular sight. Each control unit can be connected to four missiles and can be augmented by a distributor unit.

CHARACTERISTICS

Type: Surface-to-surface anti-tank guided missile with air-to-surface capability

Guidance principle: Wire-guided command to line-of-sight

Guidance method: By control of spoilers in trailing edge of wings

Gyro: Powered by solid propellant pellet

Propulsion: Booster and solid sustainer

Internal power: Thermal battery

Missile length: 998 mm

Diameter: 102 mm

Wing span: 470 mm

Weight of warhead: 2.80 kg

Penetration: 400 mm

Weight: 11.3 kg (missile); 19.5 kg (missile and launcher)

Effective range: 400 – 2100 m

Cruising speed: 90 m/s

Initial acceleration: 10 *g*

Length of time of powered flight: 23 s

Launcher dimension: 0.35 × 0.35 × 0.75 m

A special simulator for training purposes, using moving or fixed targets, can be connected to the standard control unit. There is also a battery charger that is connected to the control unit.

DEVELOPMENT

Mathogo was designed and developed by CITEFA (the Argentinian Scientific and Technical Research Institute of the Armed Forces) under contract from the Argentinian Army. In 1986 it was revealed that there were two versions of the Mathogo anti-tank missile, one with a range of 2000 metres and the other with a range of 3000 metres, both however used the same firing post and same training simulator.

STATUS

Development was completed in 1978 after a test programme involving more than 100 launches. It is in production.

CITEFA Mathogo anti-tank missile shortly after launch. Note absence of smoke and flame plume

CONTRACTOR

CITEFA (Instituto de Investigaciones Cientificas y Tecnicas de las Fuerzas Armadas), Zufriategui y Varela, 1603 Villa Martelli, Provincia de Buenos Aires, Argentina.

BRAZIL

6715.111

BRAZILIAN ANTI-ARMOUR MISSILE

Work on Brazilian anti-armour missiles began in 1976 when Avibras and the Brazilian Army's Research and Development Institute (Instituto de Pesquisa e Desenvolvimento-IPD) began producing a development of the MBB Cobra (see entry **2181.111**) as the SSX-1.AC1 but the programme was discontinued in 1981.

Currently Orbita are collaborating with OTO Melara on the MSS 1.1 (see entry **6466.111**) while Engesa are reported to be developing two missiles. One will be laser guided and have a range of approximately 5 km while the other will combine laser guidance with an autonomous homing capability and have a range of 8 km. Both missiles are reported to be designed to travel at 1000 kph and to penetrate 800 m of armour plate.

CHINA

6465.111

RED ARROW 8 (HJ-8) ANTI-TANK MISSILE

The Hong Jian (Red Arrow) 8 ATGW was first revealed in the early 1980s and is a second generation system designed by China North Industries Corporation to engage tanks and other armoured fighting vehicles at ranges of 100 to 3000 metres. In addition it can also be used to destroy enemy fortifications and other battlefield targets.

So far at least four versions of the Red Arrow 8 ATGW system have been revealed: mounted in rear of a cross-country truck, tripod mounted, turret mounted with four missiles in ready to launch position on top of Type YW 531 full tracked APC and mounted on a pedestal in a 4 × 4 cross-country vehicle.

Many aspects of the Red Arrow 8 are very similar to earlier Western missile systems, for example the tripod is almost identical to the US TOW while the launcher is similar to MILAN and the actual missile is similar to the British Swingfire.

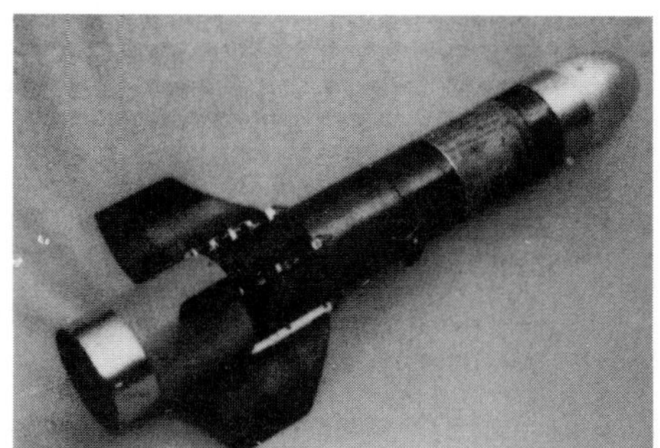

Red Arrow 8 ATGW with fins unfolded

Red Arrow 8 is a tube-launched, optically tracked, wire command-link and semi-automatic infra-red (IR) guided missile and all the operator has to do is to keep his sight hairs on the target until the missile impacts.

The missile is launched from a circular tube, and has four wrap round fins which unfold as the missile leaves the tube. It is fitted with a HEAT (High Explosive Anti-Tank) warhead.

Main components of the system are the missile in its launch tube which also serves as its transport container, tripod, top section of launcher with control box and infra-red goniometer, each of which can be carried by one man.

A training simulator has been developed for Red Arrow 8 as has first and second echelon maintenance equipment.

Red Arrow 8 mounted on 4 × 4 chassis

CHARACTERISTICS

Length of missile: 8.75 cm
Diameter of warhead: 120 mm
Wing span of missile: 3.20 cm
Weight of missile: 11.2 kg
Diameter of container with missile: 2.55 cm
Length of container with missile: 15.66 cm
Weight of container with missile: 24.5 kg
Weight of container with missile ready for firing: 22.5 kg
Weight of top section of launcher with control box: 24 kg
Weight of tripod in carrying condition: 23 kg
Weight of infra-red goniometer in carrying condition: 16 kg
Field of view of infra-red goniometer: 49 mrad (wide); 7 mrad (narrow)
Effective range: 100 to 3000 m
Penetration: 800 mm (static)
Hit probability: 90% approx
Flight velocity: 200 to 240 m/s
Rate of fire: 2 to 3 rounds a minute
Operational environmental temperature: –40 to +50°C
Relative humidity: 95%

STATUS

In production. In service with Chinese Army.

MANUFACTURER

NORINCO, China North Industries Corporation, 7A, Yuetan Nanjie, Beijing, People's Republic of China.

FRANCE

2139.111

SS 11 BATTLEFIELD MISSILE

SS 11 is a line-of-sight wire-guided battlefield missile intended for firing from land vehicles, naval vessels, and slow-moving aircraft. Normally fired from a launching ramp it may also be used with a simplified ground launcher. The designation SS 11 is applied to the surface-to-surface version; the similar air-to-surface version is known as the AS 11 (**1173.311**).

OPERATION

The operator acquires the target by means of a magnifying optical device. As soon as the missile enters his field of vision after launch, the operator commands it to his line of sight by means of a joy-stick. The signals are transmitted over wires trailed from the missile. Tracker flares are installed on the rear of the missile for visual reference.

When installed in a helicopter or ship the simple sighting device used for land vehicles is replaced by a special stabilised sight.

CHARACTERISTICS

Type: Surface-to-surface tactical guided missile
Guidance principle: Command to line-of-sight
Guidance method: Optical tracking and manual wire-guided remote control of missile by vectoring the thrust of sustainer efflux (TVC)
Propulsion: 2-stage solid-propellant rocket motor
Warhead: Various (see text)
Missile length: 120 cm
Missile diameter: 16.4 cm
Span: 50 cm
Launch weight: 29.9 kg
Average cruising speed: 115 – 190 m/s
Minimum turning radius: 1 km
Range: 500 – 3000 m
Armour penetration (140AC warhead): 600 mm minimum

Three Aérospatiale SS 11 ATGWs mounted on rear of a 4 × 4 cross country vehicle

SS 11 B1

Since 1962, the SS 11 B1 version, using transistorised firing equipment, was produced with a variety of different warheads, including an inert type for practice: the Type 140AC anti-tank warhead capable of perforating 60 cm of armour plate, the Type 140 AP02 explosive semi-perforating anti-personnel warhead (2.6 kg of explosive) which will penetrate an armoured steel plate 1 cm thick at a range of 3000 metres and explode about 2.1 metres behind the point of impact, and the Type 140 AP59 high-fragmentation anti-personnel type with contact fuze.

HARPON

A variant of the SS 11 was produced and supplied to French, West German and Saudi Arabian forces under the name Harpon. It was an SS 11 with a modified guidance system and was described in some detail in 1981-82 and earlier editions of this book.

STATUS

Some 180 000 of the SS 11 series of missiles have been delivered, including both ground and airborne versions and including also the Harpon (**2095.111**) version, to more than 30 customer countries. Production has now ceased.

CONTRACTOR

Aérospatiale, Division Engins Tactiques, 2 rue Béranger, 92322 Châtillon Cedex, France.

2055.111

ACL-STRIM ANTI-TANK ROCKET LAUNCHER

This is a man-portable anti-tank weapon that has been adopted by the French Army where it is known as the 89 mm LRAC Model F.I.

Designed primarily for infantry use, the ACL-STRIM is a lightweight rocket launcher that fires an 89 mm rocket with an effective range of 400 metres.

The launching tube is made of glass and resin laminate and its shoulder piece and foregrip can be adjusted to suit the user. An APX M-309 sighting telescope, which is detached for transport, is clipped to the side of the launching tube.

The rocket comprises a warhead and a propulsion system. The warhead has a streamlined plastic cap in which is mounted a piezo-electric generator that detonates the charge on impact with the target. The shaped charge is contained in an aluminium alloy casing streamlined by a plastic fairing. To the rear of this is the base fuze which is at safe in the transport position and is armed in flight by gases tapped off from the propulsion system.

For transport the rocket is carried in a sealed container which in operation forms part of the launching tube and contains the electrical circuits that mate with the firing circuits in the launcher.

At normal incidence the rocket will penetrate 400 mm of armour and the weapon is effective at incidence angles up to 75° from normal.

DEVELOPMENT

Development of the ACL-STRIM was initiated by Luchaire SA and the first design study was made in 1964. The system is now in production and went into service with the French Army in 1969.

CHARACTERISTICS

Weapon

Calibre: 89 mm

ACL-STRIM loaded and ready to fire

Length: During transport, 1.168 m; in firing position, 1.60 m

Mass: During transport, with 0.5 kg sighting telescope, 5.4 kg; in firing position, 8.6 kg

Rocket

Diameter of shaped charge: 80 mm

Total length: 0.600 m

Container length: 0.626 m

Total mass: Without container, 2.2 kg; with container, 3.2 kg

Mass of explosive of the shaped charge: 0.565 kg

Ballistic characteristics

Muzzle velocity: At +20°C, 291 m/s; at +51.5°C, 295 m/s; at −31.5°C, 287.2 m/s

Combat range: For a 1.21 s flight time (max ordinate: 1.8 m): 315 m

For a 2.3 m max ordinate (1.36 s): 360 m

Effective range: 400 – 500 m

Time of flight at 400 m range: 1.56 s

Ammunition: Anti-tank, anti-personnel/light armoured vehicles; smoke; incendiary; practice. For training, a sub-calibre barrel for firing 7.5 mm cartridges at combat ranges is available

Effectiveness of the shaped charge

Thickness of armour plating steel perforated at 0° incidence with T3 warhead: 480 mm

NATO targets 100% perforated: Single target, heavy tank, 120 mm/65°; dual target, heavy tank, 40 + 110 mm/60°

Operating limit incidence: 74 – 75°

STATUS

In production. In service with French and other armies. This weapon has been sold to 21 foreign countries.

CONTRACTOR

The system is being produced jointly by Luchaire SA, 180 Boulevard Haussmann, 75 Paris 8e, France, and by the Manufacture Nationale d'Armes de St Etienne.

4552.111

ERYX ANTI-TANK MISSILE

The Aérospatiale Eryx is a short range anti-tank missile for use by foot soldiers in the battlefield at ranges of up to 600 metres, but principally with fighting at close quarters in urban districts in mind. The Aérospatiale concept calls for a very high penetration missile in preference to the alternative of rockets, the accuracy of which can be adversely affected with increasing range. The ACCP has an effective range of 25 to 600 metres, and a very short flight time enables the use of simple, inexpensive components. The propulsion system also is low-powered so that the weapon can be fired within a confined space.

A simple guidance system is employed that uses an IR beacon on the missile which is tracked throughout its flight by a CCD camera in the firing unit so that corrections can be signalled to command the missiles the line-of-sight to the target via guidance wires. Thrust vector control is used to provide better response to the controls and for high (4 *g*) speed maneouvres.

CHARACTERISTICS

Weight: 14.4 kg (ready to fire)

Missile: 11 kg

Warhead: 3.6 kg

Length: 885 mm

Diameter: 135 mm

Guidance: CLOS

Propulsion: Solid

Range: 50-600 m

Max speed: 275 m/s (terminal)

Penetration: 900 mm steel

Sight: CCD camera

Optical field: 200 mrad

Magnification: × 3

STATUS

Eryx was accepted for the French Army in September 1985 and first deliveries are anticipated in 1990. The French Army plans to buy 2500 firing posts and 50 000 missiles which will replace the ACL-89 STRIM (see entry **2055.111**).

At the end of 1987 it was anticipated that France and Canada would sign an agreement for joint acquisition and production of the system. Canada plans to buy 1000 firing posts and 20 000 missiles at a cost of C$250 million over ten years to replace the Carl Gustav (see entry **5557.111**). It is envisaged that there will be a single production line in France which will accept Canadian sub-assemblies, with industrial participation being equal at least to the value of the Canadian order.

CONTRACTOR

Aérospatiale, Division Engins Tactiques, 2 rue Béranger, 92322 Châtillon Cedex, France.

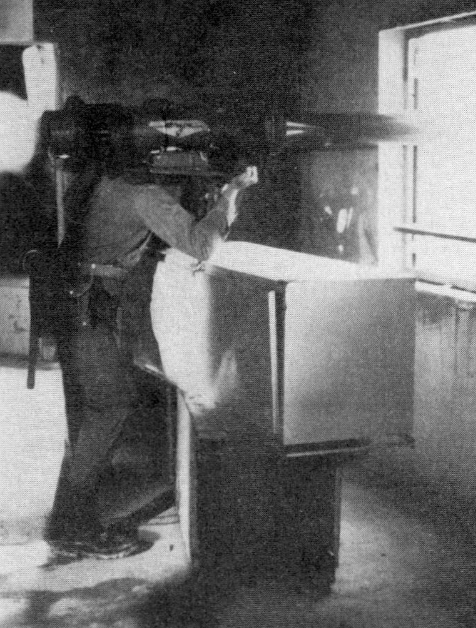

Aérospatiale Eryx short range anti-tank missile system being fire in confined space

---

GERMANY (FEDERAL REPUBLIC)

2181.111

COBRA 2000 ANTI-TANK MISSILE

This weapon system – full designation BO 810 Cobra 2000 – comprises a small wire-guided anti-tank missile, a control box, and cable links. It is a lightweight low-cost weapon suitable for operation by one man, who can readily carry, set up, and fire two missiles.

More details are given under this entry in previous editions of *Jane's Weapon Systems*.

STATUS

In service with the forces of Argentina, Brazil, Denmark, Greece, Israel, Italy, Pakistan, Spain, and Turkey. West German production has ceased.

CONTRACTORS

Originally manufactured by Messerschmitt-Bölkow-Blohm GmbH, in the Federal Republic of Germany. The warhead and sustainer motor were made by Oerlikon-Bührle, Switzerland. Manufacturing licences were granted in Brazil, Italy, Pakistan and Turkey.

2188.111

MAMBA PORTABLE ANTI-TANK WEAPON SYSTEM

Mamba is an anti-tank missile first announced by MBB late in 1972. Of the same general type as Cobra (**2181.111**), it incorporates two major improvements in the missile, in addition to numerous other changes, and MBB has developed a completely new controller for the system. The new missile, however, will accept Cobra warheads and can be used with existing Cobra controllers by users wishing to change to the new weapon, and existing Cobra vehicle launchers can readily be modified to accept Mamba.

One of the major differences between Mamba and Cobra is that, whereas the latter used an oblique-thrust booster (to give a jump start) and an axial sustainer, Mamba has a one-cell dual-purpose oblique-thrust engine which not only retains the jump-start facility but also provides a weight compensation during the cruising phase. This saves the operator from continually having to counteract gravitational droop during the cruising phase. The other major change is the increase in maximum speed from 85 to 140 m/s.

A more detailed description is given under this entry in previous editions of *Jane's Weapon Systems*.

STATUS

Production complete. No longer offered by MBB.

CONTRACTOR

Messerschmitt-Bölkow-Blohm GmbH (MBB), Post Box 801149, 8000 Munich 80, Federal Republic of Germany.

4879.111

RHEINMETALL ANTI-ARMOUR AMMUNITION DEVELOPMENTS

Rheinmetall is developing a range of both 'smart' munitions and so-called 'brilliant' artillery ammunition for use against existing and anticipated armoured formation threats. Known details of some of these projects are summarised below.

The underlying philosophy rests on the view that long range artillery systems firing smart munitions can effectively engage multiple light- and heavy-armoured targets at greater range than main battle tanks using kinetic energy penetrators. However, these new rounds will not replace conventional artillery ammunition used in close support of manoeuvring units. Instead, such ammunition enables approaching armoured units to be engaged before they join first echelon formations, so that the defender has an ammunition mix to meet any tactical situation.

EPHRAM 'brilliant' 155 mm sub-munition scans for target and uses lateral thrusters to steer onto target after lock-on is achieved

ZEPL sub-munitions scan target area. At target detection the formed projectile is fired onto the target

ZEPL 155 mm/203 mm Target Locating Self-forging Projectiles: Rheinmetall is developing the ZEPL 'smart' submunition for 155 and 203 mm artillery. ZEPL is a German acronym for a target locating, self-forging projectile submunition. In concept a thin wall, spin-stabilised carrier shell ballistically transports three ZEPL submunitions to the target area where they are ejected from the carrier by a time fuze. An auto rotating parachute stabilises the ZEPL submunition in a spinning motion during descent causing the body-fixed sensor to search in a tight spiral pattern over the target area. On achieving a target 'lock-on', a charge is ignited so that the submunition forms a projectile which penetrates the target almost vertically and at high velocity. Both the 155 and 203 mm ZEPL shells can be fired at the highest artillery zone charge.

EPHRAM: EPHRAM is the German acronym for a terminal guided, tube-launched projectile, a 155 mm artillery round that Rheinmetall calls 'brilliant', designed to engage light- and heavy-armour at extended ranges in top attack. The method of operation is as follows: Above target area there is a first separation between the carrier shell and the canister with the submunition. Spin brakes and a parachute reduce the canister rotation and velocity. After a second separation, the expelled submunition sensor starts scanning the target area at an altitude of about 800 metres. After identifying and locking on to a target, lateral thrusters are fired which direct the projectile onto the target. A shaped charge warhead then detonates, penetrating the target on impact.

Rheinmetall is producing for the German Army its own models of 155 mm carrier shell and bomblet 'cargo', designated DM 612 (RB 63) and RH 49. The most recently improved bomblet used in the Rheinmetall round has the same modes of operation as the US munitions, but is somewhat larger and heavier. Each bomblet also contains a new fuze and a self destruction mechanism. It is claimed to have a lower failure rate. Numbers following the 'RB' and 'RH' designation denote the number of bomblets carried and the reduced number in the RH 49 provides the space needed for the base bleed feature of the longer range shell. The basic RB carrier shell has a length of 899 mm. When fired with a zone 8 charge, the RB 63 has a maximum range of 22 km and the RH 49 a range of about 30 km.

STATUS

Production DM 612/642. Type classification RH 49.

CONTRACTOR

Rheinmetall Defence Engineering, Ulmenstrasse 125, PO Box 6609, D-4000 Düsseldorf 1, Federal Republic of Germany.

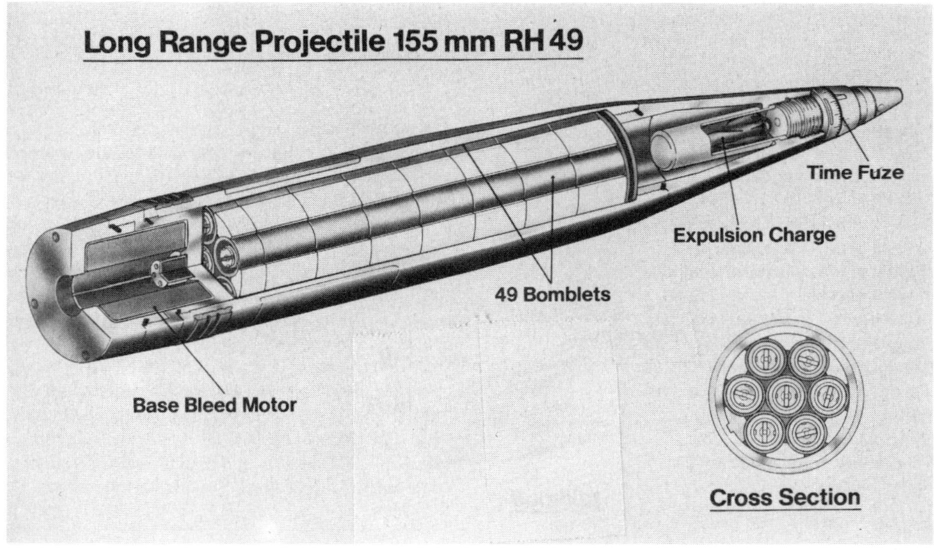

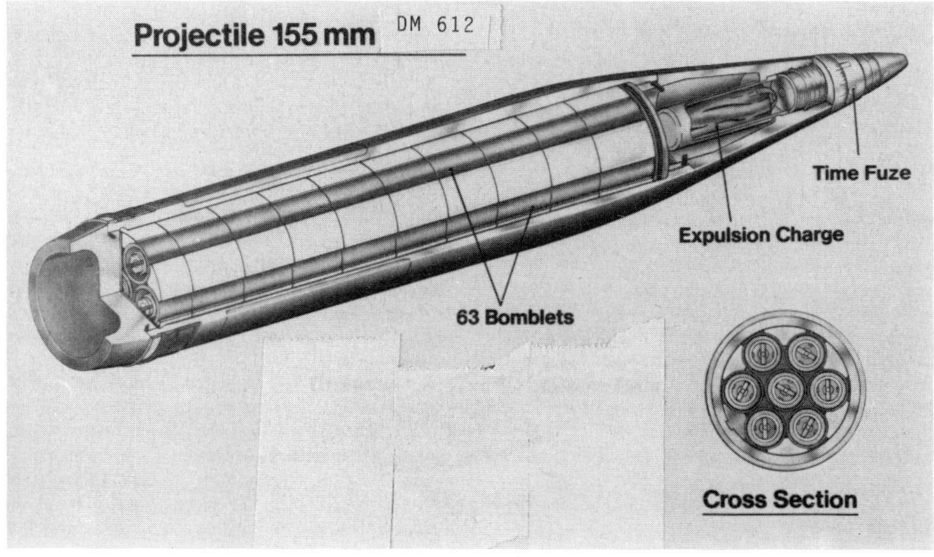

Rheinmetall 155 mm improved conventional munitions (ICM)

INDIA

6718.111
INDIAN ANTI-TANK MISSILE

India is developing an anti-tank missile designated Nag (Snake) but no further details are available. India currently manufactures Milan missiles.

INTERNATIONAL

6466.111
MSS 1.1/MAF ANTI-TANK MISSILE

The MSS 1.1 ATGW is based upon the Italian OTO Melara MAF (Missile Anti-carro della Fanteria) laser guided anti-tank guided weapon. The system is man portable but can also be installed on light vehicles such as Jeeps. Although its primary role is anti-tank it has a secondary role against battlefield fortifications and hovering helicopters.

The complete system consists of the missile in its container which also acts as the launcher and the firing post which includes the adjustable tripod, projector and sighting system.

The missile has a two stage solid propellant rocket motor with the first stage completing its burn inside the launcher and providing sufficient power to get the missile out of the launcher. At a safe distance from the operator the second stage is ignited and boosts the missile to its maximum speed.

The missile configuration is of the canard type (roll free) with conical nose and tail. Folding stabilisers are mounted on rear stakes to provide high manoeuvring capabilities and aerodynamical damping. The HEAT warhead is mounted in the nose of the missile.

The guidance system is of the beam rider type with the beam being obtained by a laser modulating emitter operating in the near infra-red band. The guidance link is generated in the launching post and in the receiver located in the rear of the missile body. Missile guidance is ensured by a loop of servos which correct the missile trajectory as a function of the deviations from the line of sight.

In good weather conditions the sighting method is through an optical sight and by night by a thermal imaging sight.

CHARACTERISTICS

Calibre of missile: 130 mm
Length of missile: 1380 mm
Weight of missile at launch: 14.5 kg
Weight of HEAT warhead: 4.1 kg
Weight of missile and container: 20 kg
Length of missile and container: 1390 mm
Diameter of missile and container: 220 mm
Weight of firing post without round: 23 kg
Maximum range all weather: 2000 m
Maximum range fair weather: 3000 m
Minimum range: 70 m
Time of flight to 3000 m: 16 s
Hit probability (over 400 m): 95%
Reaction time: 2 sec from target acquisition

MSS 1.1 ATGW system

STATUS

The MSS 1.1/MAF was originally developed as a private venture by a team under OTO Melara as prime contractor with Officine Galileo responsible for the guidance, Breda Meccanica responsible for the firing post and SNIA-BPD providing both the warhead and rocket motor. In 1986 ENGESA of Brazil signed an agreement under which the two companies would work together. Early in 1987 a new company, Orbita Sistemas Aeroespacials, was formed to replace ENGESA in the project.

Development of the MSS 1.1/MAF is virtually complete apart from test and training equipment with more than 60 missiles having been launched in Italy. Final development costs of the system will be shared between OTO Melara and Orbita, with the latter's contribution funded by the Brazilian armed forces. By late 1988 one firing post, ten missiles, a test system and a simulator must be delivered to the Brazilian Army for technical evaluation and qualification. The Brazilians reportedly have named their version Leo.

It is envisaged that two production lines for the MSS 1.1 will be established, one in Italy and one in Brazil, with each of these feeding the other with components so there will be a single production source for virtually all parts.

Orbita will have exclusive marketing rights in the Caribbean and South America while OTO Melara will have exclusive rights in NATO countries.

CONTRACTORS

OTO Melara, via Valdilocchi 15, 19100 La Spezia, Italy.

Orbita Sistemas Aeroespaciais SA., commercialised through ENGEXCO, 06400 Barueri SP, Brazil, PO Box 152.

4159.111
ADATS (AIR DEFENCE ANTI-TANK SYSTEM)

ADATS is the title of a battlefield weapon system to acquire, track and engage low-altitude high performance aircraft, helicopters, armoured vehicles and RPVs. The programme is led by Oerlikon and Martin Marietta, with Litton Systems of Canada and Contraves Italiana as partners.

Although ADATS is a unique air defence and anti-armour system initial purchases are primarily intended to meet Canadian and United States air defence requirements with a secondary anti-armour role. For this reason further details will be found in entry **4158.131** in the Mobile Guided Surface-to-Air Missile section.

2212.111
HOT ANTI-TANK MISSILE

HOT (Haut subsonique Optiquement téléguidé tiré d'un Tube) is a heavy anti-tank weapon developed by Aérospatiale and Messerschmitt-Bölkow-Blohm. With low-speed spin-stabilisation like MILAN (**2215.111**), it is a tube-launched, wire-guided missile of larger size and higher performance than MILAN but with the same general principles of operation. HOT was planned as a replacement for the SS 11 (**2139.111**) missile and its mission profile corresponds to a NATO requirement for a missile to operate primarily from armoured or unarmoured vehicles and helicopters.

OPERATION

Like MILAN, the missile has tail fins which fold down against the body in its launch tube and open out in flight. Guidance is automatic command to line-of-sight (CLOS) with an IR tracking system. The latter was developed by SAT and is produced in collaboration with Eltro in West Germany. Several versions are produced: for rotating turrets, light unarmoured vehicles, for vehicles without turrets and for helicopters.

When the missile is launched, IR radiation from its tracer flares is detected by a precision goniometer that is associated with the optical sight and that has its reference axis accurately parallel to the optical axis. Departure of the missile from the optical axis gives rise to an angular error signal which can be combined with an estimate of range (based on the known flight characteristics of the missile) to give a measure of the linear departure of the missile from the line of sight. This measurement is then used to generate command correction signals for the missile whose flight is controlled by means of a jet vane system.

Once the target has been visually acquired, all that the operator has to do is aim carefully at the target, launch the missile, and maintain his aim steadily during the missile's flight. An advantage of this guidance system is that, because the departure of the missile from the line of sight is measured by the system, the amount of flight correction required can be determined to a similar degree of accuracy and, as a result, the initial gathering of the missile on to the line of sight can be accomplished quickly. This has the important effect of giving the system a good short-range performance.

INSTALLATION

The list of vehicle types (land and airborne) to which the HOT system has been fitted is extensive and includes the following:

AFVs:
- M113 series APC – 2 tubes: 11 rounds inside
- AMX 10P APC – 4 tubes
- Panhard VCR APC – 4 tubes; 14 rounds inside
- VAB APC – 4 tubes; 8 rounds inside
- Raketenjagdpanzer 3 – 1 tube; 8 rounds inside

Helicopters:
- B0105M PAH 1 – 6 tubes
- Gazelle SA 341/SA 342L – 4 or 6 tubes
- Alouette III (trials only)
- Dauphin SA 361H – 8 tubes
- Lynx – 8 tubes

Light unarmoured vehicles:
- Land Rover One Ten (single mount)
- Peugeot P4 (single mount)

Mephisto Launcher

This installation consists of four launchers in two armoured boxes. When lowered for travelling or reloading it lies flat on the vehicle roof. For firing the launcher is raised clear of the roof on an elevating mechanism which can rotate it through 360° in azimuth. The launcher is slaved to an AMPX M-509 periscope, stabilised, dual magnification (× 3, × 12) sight which can be used both for surveillance and aiming. The sight can feature an optional Mentor thermal imager which permits firing of HOT missiles up to a range of 4 km. The total weight of the system, with eight reload missiles, is less than 2 tonnes. The Mephisto system installed on the VAB (4 × 4) armoured personnel carrier is in service with the French Army while the Moroccan Army uses the system on a VAB (6 × 6) armoured personnel carrier. Successful trials have also been carried out with the system installed on a Swiss MOWAG Piranha (8 × 8) chassis.

VAB (4 × 4) of French Army with Mephisto system in ready to fire position with four HOT missiles

UTM 800 Turret (Electric)

Here operation of the gear is undertaken by only one man, sitting. The movement in azimuth and elevation direction is achieved electrically. The guidance system in separated elements version is installed within the turret itself. Four missile ramps can be loaded manually. The turret has a relatively low weight and small installation volume. It can be installed on several types of vehicle. The UTM 800 turret is in service with Iraq on the Panhard VCR (6 × 6) vehicle known as the VCR/TT and on VAB (6 × 6) with Qatar. The weight with four missiles is 900 kg.

Panhard VCR/TT with UTM 800 installation as used by Iraq

Jaguar 1 tank destroyer of West German Army with K3S HOT system in ready to fire position

Lancelot Turret (Electric)

The AMX 10 HOT vehicle is equipped with the Lancelot HOT turret, which is operated by two men and is derived from the TH 20. Azimuth and elevation movement is performed electrically by direct command. The guidance system is divided into small boxes and installed in the turret using a special sight for the gunner. Four launching ramps are manually loaded from the rear and the turret with four missiles weighs 1.32 tonnes. The Lancelot/AMX-10P combination are in service with Saudi Arabia. Both the UTM 800 and Lancelot turrets can be fitted with the CASTOR passive infra-red camera.

Launching Gear for Casemate Vehicle (K3S)

This consists of a retractable clamp system and a periscope guidance unit. Empty launch tubes are automatically ejected clear after launching. The reloading of up to eight missiles is automatic on retraction of the clamp system in the vehicle. Operation is entirely under NBC protection. This version is only in service with the West German Army who have modified 316 vehicles to this configuration known as Jaguar 1.

HCT Launching Gear (Electro-Hydraulic)

Due to the absence of a turret basket assembly, the HCT (formerly HAKO) installation has reduced weight and dimensions. The standard-design guidance equipment is located at various points in the vehicle body and linked to the turret assembly by flexible cable. The operator can direct the turret assembly by flexible cable. The operator can direct the turret through ±180° in relation to the vehicle axis for rapid target acquisition; he has the use of a stick control to rotate the turret electro-hydraulically through ±30° in relation to its initial position. This enables precision pointing and target tracking. The stick also controls the angle of elevation of the sighting optics. A stabiliser system may be added for installation on wheeled vehicles. The turret can hold two or four ready-to-fire munitions; reloading is manual.

HOT Commando (ATLAS)

ATLAS (Affut de Tir Leger Au Sol) is the latest development of the HOT ATGW family and was first introduced at the 1986 Farnborough Air Show where it was shown installed on a Land Rover One Ten (4 × 4) light vehicle. It can however be installed on other light vehicles such as the Peugeot P4 (4 × 4) and on this vehicle has been bought by Morocco. It is designed for long-range patrol use where light units could face heavy armour.

The system, weighing 140 kg, consists of a mounting that is installed on the floor of the vehicle rear, a sight with an infra-red tracker (and possibly an infra-red camera), a launching ramp and the standard HOT 2 missile.

Wherever possible, proven components have been used in the HOT Commando, for example the launching ramp from the Gazelle helicopter version of HOT and the Castor night vision equipment from the VAB. A ground tripod has also been developed.

A Land Rover One Ten could carry one HOT Commando ATLAS system and a mix of standard HOT 2 and multi-purpose warhead missiles.

CHARACTERISTICS

Type: Long-range, vehicle mounted, wire-guided, anti-tank weapon with airborne applications
Guidance principle: Command to line-of-sight. Optical aiming with automatic IR tracking (SACLOS).
Guidance method: Wire-guidance control of jet vane system
Propulsion: Solid-propellant booster and sustainer
Warhead: 6.48 kg hollow charge
Missile diameter: Warhead: 136 mm (HOT 1); 150 mm (HOT 2)
Span: 310 mm
Launch tube length: 1.3 m
Launch tube diameter: 175 mm
Weight: Missile and container 32 kg. Missile at launch 23 kg
Max speed: 240 m/s
Flight duration:
2000 m : 9 s
3000 m : 13 s
4000 m : 17.3 s
Range: 75 m – >4 km
Penetration: >800 mm of solid armour (HOT 1); >1300 mm (HOT 2). NATO triple-armour target at 65° incidence. Effective against composite armour
Operating conditions: –40 to +52°C at up to 95% rh

DEVELOPMENT

The first design study for the system was made in 1964. Development is now complete and the new HOT 2 missile with improved warhead entered service during 1985. For air-launched versions see **1771.311**.

In 1986 Euromissile disclosed that it was developing a new multi-purpose warhead for the HOT system. This will have a 150 mm calibre HEAT warhead capable of defeating 350 mm of armour but also containing a thick metallic envelope that breaks up when the charge explodes, dispersing more than 1000 splinters with an effective radius of 25 metres. Development of the private venture multi-purpose warhead was expected to be complete in 1987.

STATUS

The system has been ordered by the French and West German forces following user trials which included the helicopter version. By 1987 a total of 15 countries had ordered HOT in vehicle or helicopter launched configurations and 1434 launchers (732 on vehicles, 702 on helicopters) together with 70 350 missiles have been ordered or produced.

CONTRACTOR

Responsibility for management, sales, and production is vested in Euromissile, 12 rue de la Redoute, 92260 Fontenay-aux-Roses, France.

2215.111

MILAN ANTI-TANK MISSILE

MILAN (Missile d'Infanterie Léger Antichar) is a wire-guided, anti-tank missile system.

The current second-generation system, MILAN 2, incorporates a semi-automatic guidance technique. The only task of the gunner is to keep the cross hairs of his guidance unit on the target until it is hit. The system comprises a missile in a container and a launch and control unit. Before the launch, the container – which also serves as a launch tube – is mounted on the launch and control unit. This in turn can be either mounted on a tripod to be fired from a ground position or, mounted on a pivot, from a vehicle.

MILAN now has a night-firing capability through addition of the MIRA thermal imaging device manufactured by TRT, Siemens and Marconi, and adopted by the French, German and British armies. It consists of an optronic unit weighing approximately nine kilograms which can be mounted on the standard firing post. Target detection is possible at a range of over three kilometres, thus allowing total advantage of the extreme range of the system.

To contend with improved tank armour making its appearance, the three nations collaborating in MILAN and MIRA production and development (France, West Germany and the UK) evolved an improved warhead (K115) which was introduced in 1984. This has an increased diameter of 115 mm compared with 103 mm of the standard item. It also has a different configuration as can be seen from the photograph. The extended nose is arranged to ensure a greater effective stand off distance for the hollow charge. The launcher tube is not modified to accommodate the larger diameter warhead. The increased stand off results in appreciably enhanced armour penetration characteristics.

OPERATION

At launch the missile is ejected from its container by a piston which in turn is propelled by a gas generator located at the rear of the launch tube. The ends of the tube are protected for storage and transit by caps which must be removed before firing.

As the missile emerges from the tube, the tube is disconnected from the launch unit, and the forces in the system are sufficient to throw it backwards from the launcher to a distance of about three metres.

MILAN 2 with K115 warhead in-flight configuration

MILAN firing post equipped with the MIRA night firing device

When the missile is ejected from the tube the wings, which impart a slow spin to the missile, flick open. The missile then coasts forward until, at a sufficient distance from the launcher to avoid harm to the gunner, the warhead is armed.

Built into the launcher/sight unit is an IR TCA guidance system similar in operation to that described above for HOT (**2212.111**). After launching, the missile is gathered to the gunner's line of sight. MBB in West Germany with SAT in France and Eltro in West Germany are jointly responsible for production of the IR guidance system, and Aérospatiale for missiles. BAe and MCCS are responsible for the major portions of the appreciable quantity on order for UK forces' use.

CHARACTERISTICS

Type: Surface-to-surface, man-portable, anti-tank weapon system
Guidance principle: Automatic command wire-guidance (SACLOS)
Guidance method: Optical tracking of target only. Automatic IR tracking of missile and control by vectoring thrust of sustainer efflux
Propulsion: Ejector in tube (gas generator). Solid-propellant boost/sustainer motor
Warhead: Hollow charge. 115 mm
Penetration: Against a solid target at maximum lethality range between 850 and 1000 mm
Missile length: 0.77 m
Body diameter: 0.09 m
Fin span: 0.27 m
Weights: Missile and container: 12 kg; launcher and control unit with tripod: 16.5 kg. Missile: 6.70 kg
Max speed: 200 m/s
Range: 25 – 2000 m
Time of flight: 7.3 s to 1000 m; 12.5 s to 2000 m

DEVELOPMENT

Following the Franco-German agreement, discussion started between Nord-Aviation and Messerschmitt-Bölkow in 1962, and a design study was completed in 1963. Technical evaluation and firing demonstrations by and for the French and West German forces were completed in 1971. User trials in France, Germany, and other countries followed. In June 1972, a successful series of demonstrations was given to representatives of some 30 countries. During the official Franco-German trials, which were designed to test the weapon's efficiency in a wide range of environmental conditions, a hit probability of 88.6% was established in 665 firings.

Before December 1984, more than 50 000 MILAN missiles were fired by various customers. The percentage of satisfactory technological performance of the firings overall was 94%, and among the latter ones, the direct hits scored amounted to more than 92%.

The MILAN ATGW has been used in combat by Britain during the Falklands campaign where it was used to neutralise bunkers and other field fortifications, Iraq has used MILAN in the conflict with Iran and in addition to being used to engage armoured vehicles has also successfully engaged helicopters. Chad has used MILAN against Libyan armoured vehicles in both 1983 and 1987 while Morocco has used MILAN to engage T-54/T-55 tanks and BMP-1s used by the Polisario rebel forces.

MBB are developing a dual purpose warhead with a long telescopic stand-off probe to beat reactive armour. It will be available in the autumn of 1988.

STATUS

Over 210 000 missiles and 7600 firing posts have been ordered by the forces of 37 countries. Current production is running at a rate of 850 missiles and about 20 firing posts per month.

Series production of the MILAN 2 improved K115 warhead began in late 1984.

The MILAN is also manufactured under licence in Italy (by OTO Melara), India (by Bharat Dynamics) and the United Kingdom (British Aerospace, Dynamics Division).

MILAN 2 is being evaluated as a medium range anti-tank weapon to supplement Dragon before AAWS-M is fielded. Operational tests began in February 1988.

CONTRACTORS

Messerschmitt-Bölkow-Blohm GmbH, Munich, and Aérospatiale, Paris, are main contractors. BAe, TRT, Siemens, MCCS, Eltro, Luchaire-SNPE and SAT are also involved.

Management, sales, and responsibility for production: Euromissile, 12 rue de la Redoute, 92260 Fontenay-aux-Roses, France.

4545.111

TRIGAT PROGRAMME

The British, French and German governments have drawn up a memorandum of understanding to cover collaborative development of two anti-tank missiles to meet their national requirements, the so-called Third Generation Anti-Tank (Trigat) missile (also known as PARS-3, ATGW3 and AC3G). The three governments agreed in 1976 to develop joint requirements for the missiles and in 1980 the Euromissile Dynamics Group (EMDG) was created by Aérospatiale, British Aerospace and MBB to produce the systems. Project definition of the system began in 1983 and was completed in 1986.

Two basic versions of the Third Generation Anti-Tank Guided Weapon (ATGW3) missile will be developed; the Medium Range (ATGW3-MR) to replace Milan and the Long Range (ATGW3-LR) to replace HOT, Swingfire and TOW. Aérospatiale will have responsibility for the medium range weapon and MBB responsibility for the long range weapon which is likely to feature a German warhead and electro-optics.

The ATGW3-MR will be an infantry-portable system designed to replace the extremely successful Milan against targets up to 2 km away. The missile will weigh 16 kg and the launcher will be similar in weight to that of Milan (15.5 kg). It will have four folding rear fins which will be canted to spin the missile during flight and will be steered by two lateral exhaust nozzles on the sustainer around the missile's centre of gravity. When the missile is launched from its tube by a low power boost motor the lateral exhaust nozzles face downwards to prevent it striking the ground but as it accelerates to cruise speed (2 km in ten seconds) these swivel to the rear and move independently to steer the missile.

The missile will be an Optical Beam Riding (OBR) weapon, using a carbon dioxide laser projector and rear-mounted detectors on the round, making the system less vulnerable to countermeasures. Thrust vector control will be employed both for greater accuracy and to achieve low launch velocities thus permitting the weapon to be fired from confined spaces. As with the Milan there will be an optional detachable thermal imager sight for use at night or during poor visibility.

The launch unit will be capable of fitting to vehicles using a simple mounting and, as with Milan, it is planned to produce a four-round version which will fit into a compact turret for small wheeled and tracked armoured vehicles. The new missile is planned to be extremely lethal against the projected future main battle tanks using a tandem forward facing hollow charge warhead detonated at the correct stand-off distance by a laser proximity fuze. Like Milan, it may be used against fortifications ('bunker busting') and will have a good capability against attack helicopters.

The long range missile, ATGW3-LR, will be largely of plastic construction and will have a conventional cruciform layout with folding wings and fins. It will be a diving attack system designed to engage targets at ranges around 5 km using an unspecified 'very advanced warhead' which is reported to be of forward facing tandem design.

The missile will also feature an imaging infra-red homing seeker for automatic target tracking and the system will be capable of salvo fire against multiple targets. EMDG have claimed that an operator will be able to acquire a target, begin autotracking and launch a missile in six seconds and can engage four targets simultaneously.

In addition to the missile the system will include a modular stabilised sight (see **5907.193**) which will also be capable of directing other weapon systems, such as guns, as well as acting as a back-up night vision system to helicopter pilots. The system would include a Military Standard 1553B data bus for interfacing with the lock-on-before-launch alignment processor, the fire control computer and launchers permitting salvo firing.

Two versions of the ATGW3-LR are planned, helicopter-launched and ground-launched. MBB is the lead company for the air-launched version which will be based upon two four-round launchers and initially a mast-mounted sight version is planned for the Eurocopter indeed one mock-up missile pod was shown on the Eurocopter mock-up at the Paris Air Show in 1987. However, roof and nose-mounted sights (see **5907.193**) will be available together with direct view units. As with the ATGW3-MR the air-launched long range version will have an anti-helicopter, as well as an anti-armour, role.

The ground-launched version will include a conventional launcher in a compact turret and a more radical model on an expandable 12 m mast, with launchers and sight on the combined mast head, for use in forests or built-up urban areas. Each would feature twin two-round tubes which could be reloaded automatically under armour in NBC conditions.

STATUS

An MoU development of Trigat was signed by the three member nations in November 1987 and development contracts were due to be let early in 1988. Shortly afterwards Belgium, Greece, Italy, the Netherlands and Spain are expected to join the agreement.

Development of the two versions is likely to cost the equivalent of $1.70 billion. When project definition began in 1983 it was anticipated that deliveries of ATGW3-MR to the three initial signatories of the MoU would take place in 1992 and of ATGW3-LR in 1994. It is now anticipated that deliveries of the former will begin in 1995 and of the latter in 1997 with the systems becoming operational two years later.

ATGW3-MR may be purchased by Belgium, France, Greece, Italy, the Netherlands, Spain, the United Kingdom and West Germany.

ATGW3-LR may be purchased by Belgium, France, Italy, Spain, the United Kingdom and West Germany.

DEVELOPMENT

EMDG, with a trilateral programme office in Paris, will receive contracts from governments and will place contracts with its three member companies who will, in turn, award subcontracts. Work-sharing agreements will be divided equally between the three countries on a project basis. It is likely that EMDG will have responsibility through the partners for integrating the missiles and their launch platforms.

Within the agreement there are three 'safeguarded technologies' areas which apply to both versions and for which a lead country is designated and the work is assigned to a company from that country.

The three areas are:

Warheads: Led by MBB assisted by Royal Ordnance and SERAT

Passive homing: Led by British Aerospace assisted by BGT and Thomson-CSF

Electro-optics: Led by SAT assisted by THORN EMI Electronics and Eltro (SATEL).

The present work share of the two versions is as follows:

ATGW3-MR

Programme development: Aérospatiale
Warhead and support equipment: Aérospatiale
Laser beam-riding guidance: TGZ
Lasers: Ferranti Industrial Electronics, SFENA
Firing unit: MBB
Electro-optical system: SATEL
Thermal imager system: SATEL
Laser guidance: TGZ
Fuze: THORN EMI

ATGW3-LR

Programme development: MBB
Air-launched version lead: MBB
Ground-launched version lead: British Aerospace
Modular stabilised sight: Aérospatiale
Mast-mounted sight: SAGEM and SFIM
Thermal imager: SATEL
Warhead: British Aerospace (Royal Ordnance)

When the five new members join the project they will pay an 'entrance fee' and fund areas of specified development. These have still to be allocated and the national contractors will either be nominated by their respective governments or be selected by EMDG.

CONTRACTORS

Prime contractors:

AEG.
Aérospatiale, Division Engins Tactiques, 2 rue Beranger, 92322 Chatillon Cedex, France.
British Aerospace, Dynamics Division, Six Hills Way, Stevenage, Hertfordshire, SG1 2DA, England.
Messerschmitt-Bolkow-Blohm (MBB), Post Box 801149, 8000 Munich, Federal Republic of Germany.

Specialist contractors:

Bodenseewerk Gerate Technik (BGT).
Elektro Spezial (Phillips GmbH).
Eltro.
GEC Avionics/MSA.
Mullard.
Royal Ordnance.
SATEL (consortium of SAT, THORN EMI, Eltro).
Societe Anonyme de Telecommunications (SAT).
SAGEM.
Societe de Fabrication d'Instruments de Mesure (SFIM).
Societe d'Etudes et Applications Techniques (SEAT).
Luchaire.
Societe Nationale des Poudres et Explosifs (SNPE).
SOFRADIR.
Telecommunications Radioelectriques et Telephoniques (TRT).
TGZ (TRT, GEC Avionics, Zeiss).
Thomson-CSF.
THORN EMI Electronics.
Zeiss.

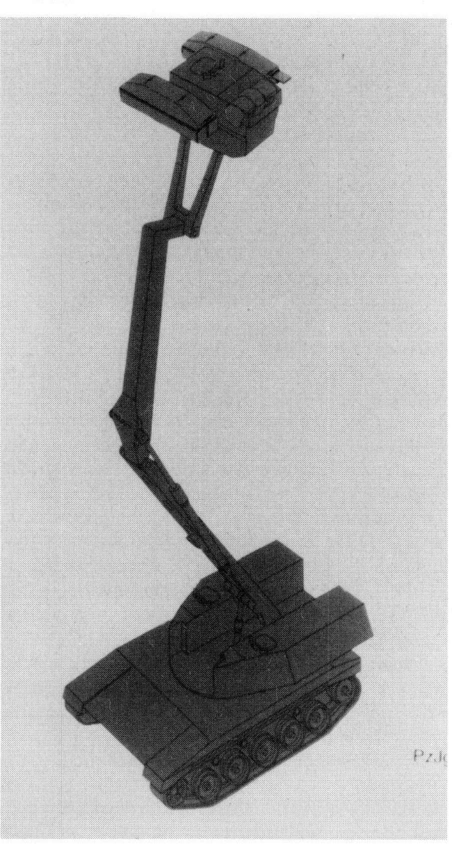

Long range TRIGAT will be a fire and forget missile with a range of over 4 km and will be mounted on a helicopter or ground vehicle. Artist's impression shows system on an elevating arm mounted on a tracked vehicle

Full scale mockup of the medium range TRIGAT

ISRAEL

4859.111

MAPATS ANTI-TANK WEAPON

The Systems Division of Israel Military Industries (IMI), faced with a requirement for a new anti-tank missile capable of challenging existing and anticipated new generation armour, has developed a weapon system designated MAPATS, that was revealed in 1984. It was initially described in the TOW entry (**2830.111**) in the 1984-85 edition of *Jane's Weapon Systems*, and the Israeli weapon does in fact share many features of the US-designed TOW missile.

However, although MAPATS employs similar missile body and control surface arrangements to TOW, it uses an entirely different guidance technique and is more than merely an improved version of TOW. The MAPATS system consists of an infra-red beam-riding missile and a launcher/sight unit. The maximum range is 4500 metres and a hollow-charge armour-piercing warhead capable of penetrating 800 mm of brinel 300 steel is carried. It is launched from a crew-transportable launcher, and the missile's glass-fibre container forms part of the launching system. When fired the missile follows a laser-generated beam which is pointed at the target by the gunner.

The IR beam projector is mounted on the traversing unit of the launcher system and can be steered in azimuth and elevation, thus permitting precise target tracking. Built into the beam projector is a high-power optical sight which is used for target acquisition and tracking. The sight has a crosshair optically aligned with the centre of the beam. In flight the MAPATS missile senses the IR beam via a rear-mounted sensor, and measures its deviation from the beam centre, using this information to correct its trajectory to regain the line-of-sight (LOS). This system is claimed to be immune to jamming and completely automatic. The gunner has only to maintain the sight crosshair on the target until the missile hits.

For protection and storage the missile is packed in an expendable glass-fibre tube, which is loaded into the launcher complete. After firing the empty tube is discarded. The missile is powered by two motors, an ejector and a flight motor. The former burns out and separates from the round within the launch tube. The moment this occurs, the IR detector in the tail and the guidance loop are activated. At a safe distance from the gunner, the flight motor ignites, accelerating the missile to peak velocity. From this point the round coasts until it hits the target. It is stabilised in flight by four wings which flip out and lock into position on leaving the launch tube. Aerodynamic control is by four surfaces that individually respond to the commands of a digital autopilot. There is an on-board gyro for attitude control, and both this and the control surface actuators are cold gas powered.

IMI MAPATS laser beam-rider version of TOW anti-tank missile (T J Gander)

To maximise the effects of the hollow-charge warhead, it is detonated at the optimum stand-off distance by means of an extending nose probe that operates immediately after launch. This essentially mechanical method of ensuring stand-off detonation is preferred to electronic proximity fuzes, some of which can be jammed.

The main components of the launcher unit are: the beam, generator, tripod, traversing unit, launch tube and electronics unit including batteries etc. The combined weight is less than 70 kg and the launching system can be installed on a jeep, M113 APC and a variety of armoured vehicles including tanks.

CHARACTERISTICS

Length: 1.45 m
Diameter: 148 mm
Weights: 29 kg (in container); 18.5 kg (out of tube); 14.7 kg (after motor burn-out)
Hollow charge weight: 3.6 kg
Hollow charge diameter: 144.4 mm
Range: 4500 m
Minimum range: 65 m
Ejection velocity: 70m/s
Peak velocity: 315 m/s
Time to target at 4500 m: 23.5 s
Penetration: 800 mm
Traverse unit
Max tracking rates: 140 mrad/s, azimuth and elevation
Azimuth range: 360°
Elevation range: –20° to +30°
Beam generator/sight
Optical sight magnification: × 13
Optical sight viewing angle: 5.5°
Laser peak pulsed power: Over 20 W
IR beam modulator: Reticle system
IR beam zoom ratio: 1:60
Weight of launcher: 72 kg (including 6 kg night vision system)

STATUS

In production. In service with Israeli Army.

CONTRACTOR

Israel Military Industries, PO Box 1044, Ramat Hasharon 47100, Israel.

ITALY

2257.111

FOLGORE ANTI-TANK WEAPON

Folgore is the name given to a light, recoilless, anti-tank rocket projector developed to meet the requirements of the Italian Army by Breda Meccanica Bresciana with the collaboration of SNIA BPD. It is a two-man system classified by its designers as a short-range system, having a maximum range of 1000 metres.

Folgore can be fired from the shoulder or from a tripod. In the tripod version two men perform the function of aimer and loader and the aimer is aided in his task by a telescopic sight fixed to the launcher which enables him to estimate target range, target speed and elevation angle in a few seconds. In the shoulder version the weapon is provided with a lighter optical device and a bipod, and if necessary, can be carried, loaded and fired by one man. The ammunition, which is the same for both versions, is composed of a rocket and a launching charge; the rocket is fin-stabilised and has a hollow-charge warhead.

In addition to the tripod-mounted and man-portable versions of Folgore, a number of designs for turret installations of this weapon for fitting in various armoured vehicles were revealed in 1982. One such, for the OTO R3 light armoured vehicle introduced by OTO-Melara also mounts a 7.62 mm machine gun and the sighting system. Other Folgore turret designs exist for use on different APCs, light tanks or other AFVs including: a power-operated turret with two Folgore tubes and a 12.7 mm machine gun; a power-operated turret with a single 7.62 mm machine gun; and a manually-operated turret with one Folgore tube and a single 7.62 or 12.7 mm machine gun.

Folgore anti-tank system deployed for action

CHARACTERISTICS

Weapon off tripod
Launcher length: 1.85 m
Launcher calibre: 80 mm
Max range: 1000 m
Minimum range: 50 m
Weapon at shoulder
Data as above except that omission of tripod and electro-optical sighting system gives a light, one-man version, intended for ranges up to 700 m.
Ammunition
Length: 74 cm
Weight: 5.2 kg
Rocket weight: 3 kg
Muzzle velocity: 380 m/s
Max speed: 500 m/s
Max trajectory height: 2.2 m at 500 m range; <3 m at 700 m range

STATUS

Development and field trials complete. Preparation for mass production in progress. The Italian Army has ordered 800 launchers, 37 000 missiles and 100 training systems.

CONTRACTOR

Breda Meccanica Bresciana SpA, Via Lunga 2, 25100 Brescia, Italy

JAPAN

2262.111

KAM-3D ANTI-TANK MISSILE (TYPE 64)

KAM-3D is a command-controlled wire-guided anti-tank missile system which is standard equipment for the Japanese Ground Self-Defence Force (JGSDF) where it is also known as the Type 64 MAT (Missile Anti-Tank).

The KAM-3D missile has an orthodox configuration, with a cylindrical metal body and cruciform metal wings incorporating full-span trailing-edge spoilers for control. Propulsion is by a two-stage Daicel Co solid-propellant rocket motor, the booster stage of which accelerates the missile to its cruising speed in 0.8 s, after which the sustainer stage maintains this speed.

The missile can be fired singly or in multiple units by infantry and is carried by jeeps and helicopters. It is launched at an elevation of 15° and the operator controls it via an optical tracking system, using a flare by day and the sustainer rocket exhaust by night as a visual reference. A gyro-stabilisation system is embodied in the missile. A three-man firing team is required to operate the system. Control is achieved by using a thumb-button control box.

CHARACTERISTICS

Type: Man-portable wire-guided anti-tank missile
Guidance principle: Command to line-of-sight with optical tracking
Guidance method: Wire-guidance control of vibrating spoilers on wing trailing edges
Propulsion: 2-stage solid-propellant rocket motor
Missile length: 100 cm
Missile diameter: Body: 12 cm. Span: 60 cm
Launch weight: 15.7 kg
Cruising speed: 85 m/s
Turning radius: 250 m
Range: 350 – 1800 m

DEVELOPMENT

Following a design study carried out in 1956, development of this missile was started in 1957, under a contract awarded by the Technical Research and Development Institute of the Japan Defence Agency. The KAM-3 weapon system was adopted as standard equipment of the JGSDF in 1964 after several hundred test rounds had been fired, and has the official designation, Type 64 ATM. Trials showed that the missile's velocity control system enabled three out of four unskilled operators to hit a target with their first round after completion of a two week training course with the ATM simulator. Skilled operators could score 19 hits in 20 firings.

A field test set (KAM-3TE) and a simulator (KAM-3TP) are made by the main contractor.

Jeep (4 × 4) cross-country vehicle of Japanese Ground Self-Defence Force two KAM-3D ATGWs (Kensuke Ebata)

STATUS

Production complete. In service only with Japanese Ground Self-Defence Force.

CONTRACTORS

Manufacturers of the complete system are Kawasaki Heavy Industries Ltd, Aircraft Group, 4-1 Hamamatsu-cho 2-chome Minato-ku, Tokyo. Other manufacturers known to have been associated with the system are:

Warhead – Daikin Kogyo Co.
Rocket motor – Daicel Co; Nippon Oil & Fats Co.
Missile guidance – Nippon Electric Co.
Guidance wire – Fujikura Cable Works.

2263.111

KAM-9 (TYPE 79) BATTLEFIELD MISSILE

This is known also as the Type 79, Jyu-MAT, (Jyu = heavy Missile Anti-Tank), and is an extended range SACLOS anti-tank weapon having a higher performance than the KAM-3, which can be used against armoured vehicles on both land and water. The missile is launched from a tubular container, which is used also for transport and storage. A solid-propellant launch motor ejects the missile from the container to a safe distance from the operator; the flight motor then ignites and accelerates the missile to its cruising speed of 200 m/s in a few seconds.

The Type 79 has been designed to meet a requirement for a defensive weapon to engage the landing craft of an amphibious assault force, as well as tank targets. A special warhead was developed to cater for this dual role. It is basically a HEAT round but incorporates an enhanced fragmentation effect, and two types of fuze are employed: for the purely anti-tank role the contact fuze which has a piezo-electric element is used and the other is a variable delay fuze.

The missile is placed in its transport and storage container at the factory, and fired from the container.

Type 79 Heavy-MAT anti-tank weapon system on its Type 73 carrier vehicle

Prior to firing, the container is placed on the launcher which comprises the firing control device, missile checkout, tracking mechanism, and built-in sight unit. The optical sighting device is designed to be operated by one man. During the missile's flight, the operator simply keeps the optical sight trained on the target, and sensors translate the course deviation to electrical signals which are fed into a computer. The computer then feeds course corrections into the missile through the guidance cable.

Instead of a flare for IR missile tracking, an xenon lamp is used as the IR source, this being powered by a thermal battery which also provides electrical power for the missile guidance system.

The launching system has two operational modes: direct and separate firing.

CHARACTERISTICS

Missile
Length: 1565 mm
Diameter: 152 mm
Span: 33 cm
Container: 170 × 20 cm
Weight: 42 kg (in container)
Propulsion: Solid-propellant booster and sustainer
Guidance: SACLOS. Optical sighting, IR tracking; wire command link
Warhead: Hollow charge with piezoelectric fuze or SAP HE with contact/electromagnetic fuze
Range: 4000 m (max)
Speed: 200 m/s

A complete Type 79 firing unit consists of two launcher units, a sight unit, control/guidance electronics unit, and a connecting cable reel. Each launcher is normally mounted on a tripod, and one of them also carries the sight. The other launcher may be sited up to 50 metres away from the sight and is remotely operated via connecting cable. The remote facility coupled with the smokeless rocket motors used offers the operating crew good protection from enemy counter fire and enhanced operational flexibility. The complete system weighs 278 kg and requires a crew of five; it is carried on two Type 73 jeep-type small trucks.

KAM-9 ATGW in foreground with launcher in background (Kensuke Ebata)

STATUS

Operational with the Japanese Ground Self Defence Force. Reports received in 1980 indicated that work started in 1978 on a medium anti-tank missile, KAM-40 (medium-MAT) to follow existing Japanese weapons in this category. The guidance technique will probably be a semi-active laser homing system, and research and trial manufacture of the homing head and laser designator has been carried out with three competitive designs.

In late 1983 it was reported that a prototype batch of eight missiles had been completed, and delivery of a laser designator, launcher and test equipment was scheduled for early 1984. A second batch of 60 rounds with two designators and launchers was scheduled for mid-1985. No further information on this has been released.

CONTRACTOR

Kawasaki Heavy Industries Ltd, World Trade Centre Building, 4-1 Hamamatsu-cho, 2-Chome, Minato-ku, Tokyo, Japan.

SWEDEN

4557.111

STRIX ANTI-TANK MORTAR BOMB

Strix is the name of a joint FFV-Saab Missiles development project aimed at producing an IR homing anti-tank bomb which can be launched from the Swedish Army's m/41c 120 mm calibre mortars. In general the philosophy of Strix is to concentrate upon engaging stationary armoured targets although it can be used for engaging moving ones.

The projectile consists of a warhead and fin assembly, the target seeker, processor unit, thruster rockets assembly, a thermal battery and an impact fuze system. A separate tail unit with increment charges is used to launch the round from the mortar. A sustainer rocket motor can be added to the projectile when ranges in excess of 4.5 km are required.

When a target is located a programming unit will feed details of the propelling charge, size, azimuth angle, elevation and time-to-seeker-activation into the round's programming circuit. If target location and meteorological conditions are known then Strix can be fired like a conventional round although the crew must add the required number, up to eight, of increment charges to the special tail unit. The tail unit is loaded first, followed by the sustainer (where necessary), then the projectile.

The mortar trigger ignites the primary and then the incremental charges and these force the projectile out of the barrel. A plunger mechanism in the tail unit uses the remaining gas pressure to provide sufficient kinetic energy for the tail unit to be ejected when the round is 25-30 m from the mortar. If a sustainer is used this is ignited by a pyrotechnic delay composition once the projectile has cleared the mortar.

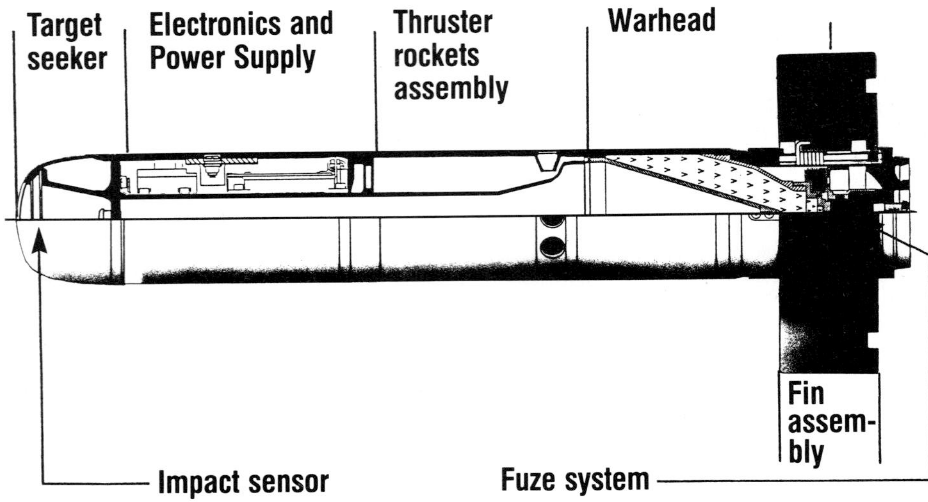

A diagram of the Strix projectile

The Strix 120 mm mortar-launched anti-tank system. Pictured are the projectile, sustainer, tail unit and programming unit

As the round leaves the tube the battery is activated and within two seconds the mechanical arming sequence is complete. Simultaneously the four wrap-round fins unfold to provide spin for accuracy, the fixed cant angle imparting a slow spin with a nominal speed of ten revolutions per second. The round continues to the peak of its trajectory and the initiation circuit of the warhead is charged. If a sustainer is fitted it will be pyrotechnically separated at this stage.

Strix is equipped with an IR seeker which is likely to be a two-colour unit. The seeker is activated at a pre-determined height and the IR signals are amplified and digitised for processing in the electronic unit which includes a 16 bit microprocessor. The signals are compared with real target data and selects the target which most meets the selection data.

The projectile then enters its homing phase firing the 12 thruster rockets arranged peripherally around its centre of gravity. These rockets are fired separately in sequence to correct the trajectory and bring it to the aiming point using predicted position guidance.

Strix will be a top attack system aimed at penetrating the most vulnerable part of an armoured vehicle. A 'tunnel' in the electronics package will increase the warhead lethality which will be increased by unburnt fuel from the thruster rockets. The manufacturers estimate that if three projectiles were fired at a typical armoured formation they would destroy one or two targets.

CHARACTERISTICS

Calibre: 120 mm

Weight: 17.6 kg (projectile), 19.3 kg (with tail unit), 23.2 kg (with tail unit and sustainer)

Length: 81 cm (projectile), 115.4 cm (with tail unit), 133.5 cm (with tail unit and sustainer)

Range: 600-4500 m (without sustainer) or 600-7500 m (with sustainer)

Guidance: IR homing, side thrusters for maneouvring

STATUS

Feasibility studies and project definition began in 1983 and in November 1984 the partners received SEK (Swedish Kronor) 270 million to complete development and begin production. In 1985 it was stated that complete round tests would be carried out in 1987 and production would begin in 1988 but complete rounds will now not be tested until 1988.

The system is likely to be offered for export with a modified tail unit when used with fixed-pin mortars. A slip-ring version for use in rifled mortars is under construction.

CONTRACTORS

FFV Ordnance, S-631 87, Eskilstuna, Sweden.

Saab Missiles AB, Linkoping, Sweden.

5557.111

84 mm RCL CARL-GUSTAV

Two versions of this weapon are described – the 84 mm RCL Carl-Gustav M2 and the more recently developed Carl-Gustav M3.

M2 Weapon

This is a recoilless gun intended primarily for use as an anti-tank weapon but suitable also for other assault roles. It can fire HEAT, HE, smoke, or flare ammunition and has a practical range, with HEAT rounds against moving or stationary targets, of 500 or 700 m, and with HE rounds of 1000 metres. Smoke can be fired up to 1300 metres and flare shells up to 2300 metres. A rate of about 6 rounds/minute is easily achieved.

The weapon can be carried, loaded and fired by one man but a two-man team is standard – one carrying and firing the gun and the other carrying ammunition and assisting in the loading operation.

M3 lightweight Carl-Gustav

The gun is breech loaded and cannot be fired until the venturi has been rotated back into position and locked after the round has been inserted. The round is fired by a percussion-operated side primer, the firing mechanism being contained in a tube on the right-hand side of the barrel. This mechanism is hand-cocked, is operated by a two-pressure trigger and has a safety-catch.

Open or telescopic sights can be used: the telescopic sight has a ×2 magnification and a 17° field of view and is fitted with a temperature correction device. The weapon is normally fired from the shoulder. When fired in the prone position, or over a trench or parapet, a sprung bipod mount provides support just forward of the shoulder pad.

Cartridge cases of ammunition are made of light metal alloy with a plastic blow-out disc at the rear. HE, smoke, and flare shells are spin-stabilised.

The 84 mm FFV 551 HEAT round is rocket-assisted and stabilised by fins which unfold as the round leaves the barrel. The rocket motor increases the shell speed from its muzzle velocity of 254 to 340 m/s maximum. The rocket motor comes into operation when the shell is about 18 m from the muzzle and burns for 1.5 s. The shell will penetrate 400 mm of armour, and the piezoelectric fuze system enables the weapon to function up to 80° incidence.

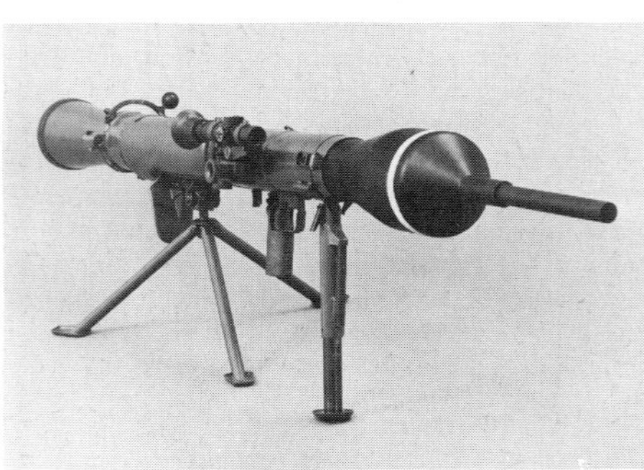

Carl-Gustav with FFV 597 HEAT round loaded

The FFV 551 round is fitted with a Teflon slipping ring to reduce the spin imparted by the rifling of the Carl-Gustav barrel. Practical engagement range with this ammunition is 700 m.

CHARACTERISTICS

	M2	M3
Calibre	84 mm	84 mm
Weapon length	113 cm	106.5 cm
Weight of gun and mount	16 kg	9.6 kg
Weight of HEAT round	3.2 kg	3.2 kg
Anti-tank range	700 m	700 m

Training equipment: Sub-calibre adapters are available for both weapons to enable them to be used with small-arms ammunition for training purposes.

STATUS

In production and in service in several countries.

HEAT FFV 597

The current Carl-Gustav system is designed to defeat present-day combat vehicles, but to enable it to engage the more sophisticated MBTs in the 1990s the system will be upgraded with oversize HEAT round FFV 597. This is rocket assisted and fin stabilised. The warhead is designed to penetrate the front of compound-armour main battle tanks.

The round will be carried in two parts: warhead, and propulsion system (including rocket-motor). Loading will be carried out by inserting the propulsion system as normal Carl-Gustav ammunition and the warhead from the front. The two parts will be joined inside the barrel by a connecting device.

CHARACTERISTICS

Calibre (outer dimension): 132 mm
Weight (complete round): 9.5 kg
Weight (in flight): 8 kg
Length (in weapon): 1.5 m
Length (warhead): 0.75 m
Muzzle velocity: 115 m/s
Maximum velocity: 310 m/s
Range (effective): 300 m
Armour penetration: >900 mm

M3 Weapon

Carl-Gustav M3 is a lightweight version of this weapon. It is built up on a steel liner with the rifling, and covered with carbon-fibre winding. The venturi is made of steel. All external parts are made of aluminium and plastic. All existing types of Carl-Gustav ammunition can be used with the M3 with unchanged performance.

STATUS

In production and in service with undisclosed countries.

CONTRACTOR

FFV Ordnance Division, S-631 87, Eskilstuna, Sweden.

84 mm RCL Carl-Gustav with HEAT 551 round

2363.111

BANTAM ANTI-TANK MISSILE (BS 53)

Bantam is a small wire-guided anti-tank missile, designed for operation by a single infantry soldier, but suitable also for use when mounted in land vehicles or small military aircraft. The missile has a hollow-charge warhead.

The complete system in its simplest configuration comprises the missile in its combined transport and launching container with carrying harness, a control unit, and 20 metres of cable. All this weighs only 20 kg, of which the launching weight of the missile is 7.5 kg. Further details will be found in earlier editions of *Jane's Weapon Systems*.

STATUS

In service with Swedish and Swiss forces but no longer in production.

CONTRACTOR

AB Bofors, 691 80 Bofors, Sweden.

The six Bantam missiles pointing forwards on the Puch-Haflinger vehicle can be fired in two to three seconds after the vehicle has stopped. Six more missiles are carried at the rear of the vehicle

4544.111

RBS 56 BILL ANTI-TANK MISSILE

Bofors, under a contract awarded in mid-1979 by the Swedish Defence Materiel Administration (FMV) has developed a light anti-tank missile called 'Bill' for the Swedish Army, whose designation for it is RBS 56. Development has been carried out in close co-operation with the Swedish Army.

The RBS 56 is a wire-guided command to line-of-sight weapon with an effective range capability extending from 150 to 2000 m, and which is fired from a container/launcher tube that can be either tripod or vehicle mounted. A sight that provides for both day and night operation is carried separately and is attached to the container tube for launching. Setting up for use can be completed in 20 s according to the manufacturer. Propulsion is by a solid-propellant rocket motor which exhausts through nozzles disposed around the circumference of the missile body to the rear of the sustainer motor which is housed in the nose section. Cruciform wings and control surfaces flip out after launch and these are located in the tail section of the missile, where there is also a tracking signal transmitter.

A shaped charge HE warhead is carried which is normally detonated by a delayed proximity fuze. An unusual feature of the warhead is that it is angled to direct its energy at 30° downward from the missile's longitudinal datum. This, and a missile trajectory which is automatically maintained at a height of approximately one m above the gunner's line of sight to the target, are claimed to yield a higher kill probability. The chances of striking the less heavily armoured upper surfaces of a tank are increased, and the angled warhead means that it is less likely to strike a sloping surface at a shallow, glancing angle, thereby enhancing the penetration characteristics. The angled attack also ensures that when sloping surfaces are encountered, the thickness of armour to be penetrated is effectively lessened.

CHARACTERISTICS

Type: RBS 56. Top attack, wire-guided command to line-of-sight
Length: 900 mm
Diameter: 150 mm

Bofors RBS 56 Bill anti-tank missile deployed for action

RBS 56 Bill anti-tank missile demonstrates its 'top attack' capability

Max span: 410 mm
Weights: 20.5 kg (missile and container); 10.7 kg (missile only); 4.2 kg (sight and padding); 11.3 kg (stand)
Warhead: HE shaped charge
Combat range: 150 – 2000 m (stationary targets); 300 – 2000 m (moving targets, 10 m/s crossing speed)
Time of flight: 6 s (to 1200 m); 11 s (to 2000 m)
Setting up time: 20 s
Reloading time: 7 s

STATUS
Development complete and SEK1300 million production order placed by Swedish Army. The US Army has placed a small order for the RBS 56 ATGW for evaluation purposes as a medium range anti-tank weapon to supplement Dragon before AAWS-M is fielded. Operational tests began in February 1988.

CONTRACTOR
AB Bofors, 691 80 Bofors, Sweden.

6716.111
BOFORS 155 mm ANTI-ARMOUR AMMUNITION
In 1986 Bofors Ordnance revealed they have been working on 155 mm anti-armour rounds since 1984 to meet an anticipated Swedish Army requirement. Some subsystems have already been tested.

BONUS is a SADARM-style base-bleed cargo shell with three sub-munitions. The sub-munition will feature an explosion-forged projectile warhead and an IR seeker.

The BONUS sub-munition is deployed in two stages. In the first stage the whole sub-munition package is extracted from the rear of the round by a brake system similar to that used in the company's illumination rounds. In the second stage the sub-munitions are deployed and stabilised by lateral winglets which permit the seeker to locate targets. BONUS could be ready for production in 1989 and be in service by 1991.

Two projects under consideration are BOSS (Bofors Optimised Smart Shell) and Artillery Strix. BOSS is intended to be an intelligent 155 mm shell with either a millimetre wave or IR seeker. BOSS is the subject of a feasibility study. Bofors have also discussed with FFV the possibility of using the Strix anti-armour round (see **4557.111**) as a sub-munition in a 155 mm artillery round, Artillery Strix, which would have a range of 21 km.

STATUS
A development contract was awarded to Bofors for BONUS in 1986 and at that time it was anticipated production would begin late in 1989 with the round entering service in 1991.

CONTRACTOR
AB Bofors, 691 80 Bofors, Sweden.

TAIWAN

4438.111
KUN WU ANTI-TANK MISSILE
Kun Wu is the name given to a wire guided anti-tank missile produced by the Chung Shan Institute of Science and Technology. It was first seen publicly in the 20th Republic of China National Day Parade in Taiwan during October 1981.

The Kun Wu is evidently based upon the Soviet designed Sagger AT-3 (**2950.111**). Reports from Taiwan state that the Chung Shan Institute, which was responsible for Kun Wu development, is known to have studied captured examples of the Sagger said to have been acquired in Vietnam.

The photograph supports such a contention although it is not known if the optical guidance and control system used with the Sagger is retained for the Chinese version of the system or if a different method is used for the direction of the four rounds that are usually deployed on each vehicle. It can be expected that Kun Wu performance will at least equal that of Sagger, including a maximum range of up to 3000 metres with a launch weight of over 11 kg. Other parameters are likely to be similar to those of the Soviet missile, which is described on the next page. Sighting arrangements are not known and it is quite possible that the missiles are fired remotely by the operator some distance from the vehicle. Launch vehicles so far noted all have a prominent radio antenna fixed to one corner of the body. Whether or not this has some special purpose is not known.

Kun Wu anti-tank missiles mounted on Jeep. Four rounds are carried on each vehicle in this configuration (L J Lamb)

STATUS
Production and in service.

CONTRACTOR
State Arsenals.

UNION OF SOVIET SOCIALIST REPUBLICS

2985.111
AT-2 (SWATTER) ANTI-TANK MISSILE
Swatter is the NATO code-name for the anti-tank missile which is known as the AT-2 in the US code and is carried on the BRDM-1 armoured amphibious vehicle. The Soviet designation is PUR-62 (Protivotankovaya Upravlyayemaya Raketa) Falanga.

Of similar size to the AT-1, but heavier (29.4 kg), it is believed to be a more advanced missile – probably in the class of the French SS 11 – and certainly has a different configuraton. Control is by elevons mounted on the trailing edges of the rear-mounted cruciform wings.

The standard mount on the BRDM-1 vehicle carries four missiles mounted on guides. Radio command to line-of-sight guidance is employed, the command link having three frequencies as protection against electronic countermeasures. There is also the possibility of separate terminal homing (probably IR) being provided, and this with radio guidance suits the Swatter for airborne roles such as part of the armament of the Mi-24 Hind attack helicopter.

As deployed on the ground Swatter does not arm until it has flown 500 metres from the launch site. The Swatter launcher on the BRDM-1 has a lateral traverse of 45°, as originally deployed.

The BRDM-1 with four Swatter ATGWs in ready to launch position is the most common version. In 1983 the improved BRDM-2 (4 × 4) amphibious armoured vehicle with a quadruple launcher for the Swatter AT-2 was seen.

In this model, which has not been observed outside of the Soviet Army, the original command-to-line-of-sight guidance has been converted to semi-active infra-red/command guidance. This version of Swatter, sometimes referred to as Swatter-B, weighs 29.84 kg and has a range of 3500 m.

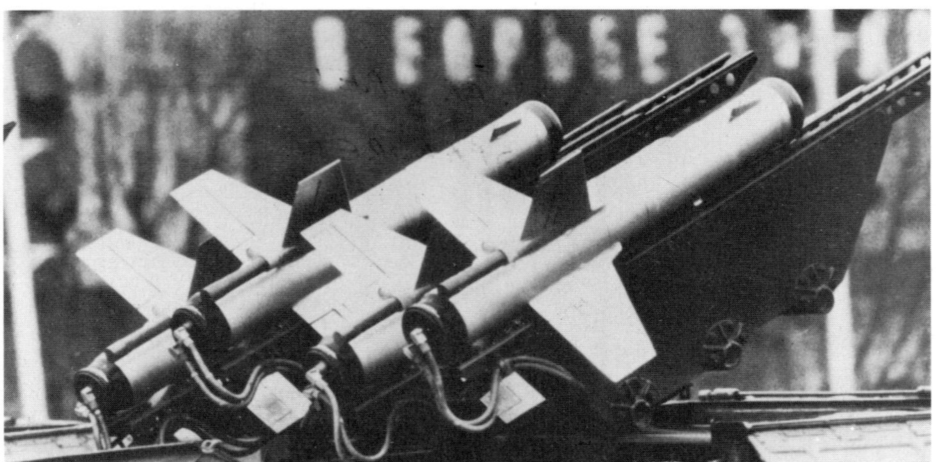

Close-up of AT-2 Swatter anti-tank missile four-round launcher on BRDM-1 reconnaissance vehicle

CHARACTERISTICS
Type: Anti-tank, surface-to-surface and air-to-surface
Configuration: Cylindrical body with rounded nose. Rear set cruciform wings. Two small foreplanes. Ramp launched

Length: 1.16 m
Diameter: 13.2 cm
Span (max): 66 cm
Weight: 29.4 kg
Propulsion: Solid
Range: 500 – 3000 m
Guidance: Radio command to line-of-sight, possible with terminal homing
Warhead: HEAT
Penetration: 500 mm
Average velocity: 150 m/s

STATUS
Believed to be in service only with the Soviet Union, but most probably replaced by more effective AT-5 Spandrel.

2950.111
AT-3 (SAGGER) ANTI-TANK MISSILE

Sagger is the NATO code-name for the small wire-guided anti-tank missile, known by the US code AT-3. More compact than either the AT-1 (Snapper) or AT-2 (Swatter) missiles (**2983.111** in 1984-85 and earlier editions and **2985.111**), but carrying an equally powerful warhead. The Soviet designation is PUR-64 Malyutka.

The AT-3 Sagger ATGW was first observed mounted on the BRDM-1 (4 × 4) amphibious scout car in 1965. This version has six missiles which are under armour protection and when required for action are raised above the top of the vehicle complete with their overhead armour protection. This model has been phased out of front line service and was replaced by the BRDM-2 with an elevating arm with six Sagger ATGWs which for travelling is under armour protection and raised only when required for use. In addition to the six missiles in the ready to launch position a further eight missiles are carried in reserve.

The BMP-1 mechanised infantry combat vehicle, first observed in public during 1967 has a single Sagger ATGW mounted over the 73 mm gun as does the original version of the BMD airborne combat vehicle.

Poland has deployed the OT-64A (8 × 8) armoured personnel carrier with two Sagger ATGWs in ready to launch position above the rear troop compartment. This has been replaced in most units by the much improved OT-64C(1) version which has one Sagger ATGW either side of the machine gun turret.

For training, East Germany uses a modified BTR-40 with three Sagger ATGWs in the ready to launch position.

The Sagger has also been used in the air launched mode from Hind helicopters.

The manpack version is also known as the 'suitcase' Sagger, because the missile is carried in a glass fibre case, with the warhead separated from the rest of the missile. The lid of the case has a rail that enables it to be used as a launcher for the weapon after assembly of the missile. A control unit which incorporates a periscopic sight, a control stick, switches for missile selection, and batteries is carried in a separate container. Sagger is fired from a remote position which enables the operator to be located up to 15 metres from the missile launcher. The periscopic sight has a magnification of × 10. A three-man firing team is normally employed, with a total of four missiles per team. Either one or two men may serve as Sagger controllers, each with a sight and control unit and two missiles, while the third man deploys ahead armed with an RPG-7 for closer range protection if need be. Used in this way the effective engagement range of Sagger is considered to be between 1000 and 3000 metres, and such a team could be deployed with four missiles checked and ready to fire in about 12 to 15 minutes.

As deployed with the BRDM vehicle, the Sagger gunner can operate either within the vehicle itself or move to a position outside up to 80 metres distant. Six missiles can be rapidly fired by use of the selector switch on the control/sight unit which has a × 8 magnification. The BRDM can carry a total of 14 missiles.

Complete AT-3 Sagger man-portable anti-tank weapon with sight on left and missile on carrying case

Client states of the Soviet Union on occasion evolve their own methods of mounting and deploying weapons of Soviet origin, that differ from the system(s) used by the USSR. A recent example of this is the 4 × 4 vehicle mounted version of the Sagger anti-tank missile introduced by Yugoslavia. The armoured vehicle is known as the BOV-1 and it is surmounted by a rotating turret on which are carried six Saggers housed in two rectangular container/launchers with three rounds in each. Sufficient rounds for at least one manual reloading (ie six missiles) are reported to be carried inside the vehicle hull. Yugoslavian forces also employ Saggers in an airborne role, some Gazelle helicopters being fitted with launch facilities for four missiles.

The Yugoslav M-980 mechanised infantry combat vehicle has two Sagger type ATGWs in the ready to launch position, as does the more recent and very similar BVP M80A mechanised infantry combat vehicle.

China North Industries Corporation produces a very similar missile system to the Sagger called the Red Arrow 73 which is also installed on the Chinese version of the BMP-1 called the Wz 501.

CHARACTERISTICS
Type: Surface-to-surface guided anti-tank missile
Guidance: Wire-guided command to line-of-sight, optical tracking
Propulsion: Solid-propellant motor, 2-stage
Warhead: Hollow charge
Missile length: 86 cm
Missile diameter: 11.9 cm
Launch weight: 11.3 kg
Range: 500-3000 m
Penetration: 400 mm
Average velocity: 120 m/s

STATUS
Extensively deployed by USSR and Warsaw Pact forces and by other nations, particularly in the Middle East and elsewhere.

In most members of the Warsaw Pact the AT-3 Sagger has been replaced by the AT-4 Spigot and AT-5 Spandrel systems.

In the early 1980s Taiwan introduced an indigenous anti-tank missile called the Kun Wu which bears some obvious resemblances to Sagger and it has been reported from Taiwan sources that development of Kun Wu (**4438.111**) was based on examples of Sagger, although the warhead could be different.

3638.111
AT-4 (SPIGOT) ANTI-TANK MISSILE

The designation AT-4, Spigot, was at first provisionally assigned to a man-portable, tripod-mounted version of the AT-5, Spandrel, anti-tank missile (**3608.111**). The Soviet designation is Fagot. As employed by infantry, a crew of three men normally constitutes a Spigot section, between them carrying a total of four rounds plus the sight and mount.

Later information however (including an East German photograph showing a BRDM-2 launcher for three AT-4s and two AT-5s) while confirming the similarities between these two missile types makes obvious the larger dimensions of the AT-5 Spandrel. It would therefore appear that while the basic engineering and guidance techniques may be the same for both types, the larger size of the Spandrel is mostly due to the larger propulsion motor needed for the increased range. The Spandrel is recognisable

AT-4 Spigot infantry mounting modified for use on top of BMD-1 airborne fighting vehicle. Spigot anti-tank missile replaces AT-3 Sagger sometimes used in this fashion. This Spigot application first seen on parade in Moscow, October 1983

from the Spigot as the former has a distinct flare shape at rear of launcher tube.

The sight appears to use separate optical paths which are presumed to be employed for tracking of the target (under control of the gunner) and the Spigot missile, corrections to the latter's trajectory being automatic subsequently in response to the gunner following the target.

CHARACTERISTICS

Type: Man-portable surface-to-surface wire-guided anti-tank missile
Guidance: Semi-automatic command to line of sight; optical tracking
Propulsion: Solid
Warhead: HEAT
Launcher length: 1.2 m
Launcher diameter: 13.5 cm
Launch weight: 10-12 kg (estimated)
Range: 2000 m (approx)
Speed: 150-250 m/s (approx)
Penetration: 500 mm

STATUS

In production. Known to be in service with Czechoslovakia, East Germany, Finland, Hungary, Iraq, Libya, Poland, Syria and the USSR.

Preparing AT-4 Spigot anti-tank weapon for firing

AT-4 Spigot anti-tank missile deployed

3608.111

AT-5 (SPANDREL) ANTI-TANK MISSILE

AT-5, Spandrel, is the designation of an anti-tank weapon deployed on Soviet BRDM-2 AFVs. As can be seen from the adjacent illustration, five tubular launchers mounted on a trainable turret are carried on each vehicle. These resemble those of the HOT and MILAN anti-tank missiles (**2212.111** and **2215.111**), and while no official dimensions are available, the AT-5 is probably nearer to the latter in size.

The BRDM-2 has a hatch in the roof immediately behind the AT-5 which is probably for reloading purposes, and, according to official US estimates, the total vehicle load is 10 rounds.

A rotatable optical sighting/tracking head is incorporated in the BRDM-2, mounted in the vehicle roof above the right-hand front seat. The shape of the launch tubes has led some observers to conjecture that some form of gas generator is used to eject the missile before the propulsion motor is ignited, and, also giving cause for speculation, these tubes have always appeared to be unpainted metal of bright or light colour. An East German photograph showed a BRDM-2 vehicle turret on which were mounted a mixed armament of three AT-4 Spigot anti-tank missiles flanked on each side by an AT-5 Spandrel missile. From this photograph is was clear that the Spandrel has a larger diameter launch tube.

The Spandrel ATGW is also fitted as standard on the BMP-2 mechanised infantry fighting vehicle which also has a 30 mm cannon.

CHARACTERISTICS

Type: Surface-to-surface guided anti-tank missile
Guidance: Command to line-of-sight, optical tracking
Propulsion: Solid
Warhead: HEAT
Launcher length: 1.3 m
Launcher diameter: 155 mm
Launch weight: 12-18 kg (estimated)
Range: Approx 4000 m (estimated)
Speed: Approx 150-250 m/s
Penetration: 500 mm

STATUS

In service with Soviet forces and also exported to other countries including Middle East.

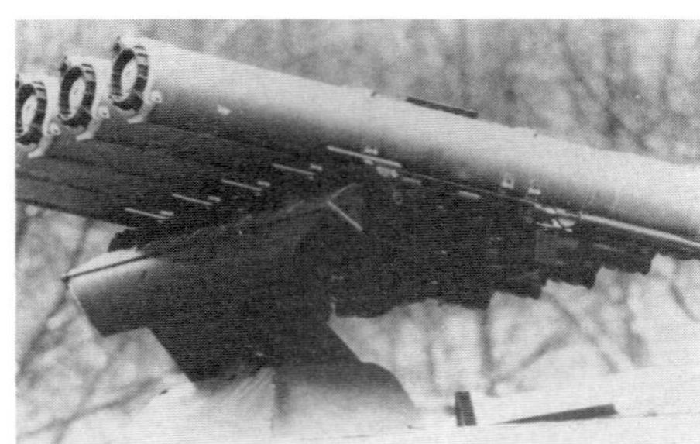

Close up of five-round launcher for AT-5 Spandrel ATGW on BRDM-2 (US Army)

BRDM-2 AFV with five-tube AT-5 spandrel anti-tank missile launcher

---

3639.111

AT-6 (SPIRAL) ANTI-TANK MISSILE

The AT-6 Spiral ATGW was first deployed by the Soviet Union in the early 1980s on the Hind D helicopter and has more recently entered service with East Germany and Poland.

The missile is believed to have a maximum range of about 8000 m, have a HEAT warhead weighing 10 kg and be radio command guided, although some sources have stated that it is IR guided.

STATUS

In service with East Germany, Poland and USSR.

6272.111

AT-7 (SAXHORN) ANTI-TANK MISSILE

This man-portable system is believed to be the Soviet equivalent of the US Dragon anti-tank/assault weapon (**2573.111**) and have a range of about 1000 m. The Soviet designation is Metis.

STATUS

In service with Soviet Army.

6224.111

AT-8 (SONGSTER) ANTI-TANK MISSILE

Early in 1987 it was confirmed that the AT-8, previously known as the Cobra, had been allocated the NATO Codename of the Songster. The Soviet designation is reported to be Kobra. This is launched from the 125 mm gun installed in the T-64B and T-80 Main Battle Tanks. It is believed that there are two versions of the AT-8 Songster, one anti-tank and fitted with a HEAT warhead and the other anti-helicopter and fitted with a high explosive fragmentation warhead. Both are loaded through the breech and not the muzzle as indicated by some sources in 1985/86. The AT-8 Songster has a maximum range of 5000 metres and has laser or radio command guidance.

CHARACTERISTICS

It is reported to have the following characteristics.

Calibre: 120 mm
Range: 4 km
Guidance: Radio command
Weight: 25 kg
Length: 1.2 m
Cruising speed: 500 m/s

STATUS

In production. Only used to date by Soviet Army.

UNITED KINGDOM

2450.111

SWINGFIRE ANTI-TANK MISSILE

Swingfire is a long-range command-controlled anti-tank weapon system capable of engaging and destroying the heaviest armour. Armoured and soft-skinned vehicle variants are available. The missile is wire-commanded, signals being generated by an operator's joystick control, and the commands are interpreted by the missile as demands for a change of heading; thrust vector control is employed. For details see entry **2450.111** in the 1987-88 edition.

The system's missile control equipment can be installed either in a vehicle (AFV variant) or on the launcher pallet.

CHARACTERISTICS

Type: Mobile anti-tank guided missile system. Panclimatic. Missiles prepacked in sealed launcher boxes
Guidance principle: Wire-commanded. Heading demand signals referenced to autopilot. Optical sighting with automatic initial gathering. Manual steering (direct or separated). × 1 or × 10 magnification sights
Guidance method: Thrust deflection of main motor efflux by jetavator
Propulsion: 2-stage boost and sustainer solid-propellant motor
Warhead: Hollow charge powerful enough to defeat all known combinations of armour
Range: Minimum – from less than 150 m at direct fire to 300 m with max separation. Max – 4000 m
Fire arcs: Azimuth ±45° without launcher traverse. Elevation +20°, –10° relative to mounting plane
Special features: Ease of concealment. Immunity to ECM

DEVELOPMENT

British Aerospace Dynamics Division based their design, development, and production programmes on experimental work originally done by Fairey Engineering Ltd. The first design study was made in 1958 and the first public announcement concerning the system was made in 1962. The system went into service with the British Army in 1969.

A new combined sight with visual and thermal channels has been developed which will permit current variants of Swingfire to be used operationally at night and in poor visibility.

Current Swingfire applications are as follows:

(1) **Striker** – the guided weapon vehicle in CVR(T) series. Striker is also in service with the Belgian Army
(2) **Swingfire Pallet** – a crew-portable version which can be transported on and fired from any vehicle or trailer from a long wheel base Land Rover upwards, or fired from the ground. It can be carried over short distances (of the order of 400 metres) and deployed by its three-man crew. Swingfire Pallet is fully established in service with the Egyptian Army and is in production in Egypt. In 1986 it was reported that Egypt had supplied Sudan with some Swingfire missile systems
(3) **Micro-miniaturised (MM) Swingfire** – Development has been completed on an MM standard of programmer and control equipment to replace the existing black boxes, ie the programmer, selector unit etc. The new standard of equipment is compatible with the existing sighting systems including the combined sight. MM electronics will save significant space and weight and will enable Swingfire to be fitted to a greater range of vehicles. Part of this development includes a new lightweight, fixed elevation launcher. There is also a lightweight alternative to the Combined Sight developed by THORN EMI and BAe, which will fit on the separation sight base in place of the present telescope.

British Aerospace have produced 39 000 missiles and an order for 1200 scheduled for delivery by 1990 is likely to be the last.

CONTRACTOR

British Aerospace Army Weapons Division, Six Hills Way, Stevenage, Hertfordshire SG1 2DA, England.

Swingfire Combined Sight enables this long range anti-tank missile to be used at night or in low visibility. This is accomplished by combining thermal imaging with the optical sight using passive IR sensing to detect the target

In its pallet configuration Swingfire anti-tank missiles can be carried and launched from soft-skinned vehicles such as the Egyptian CJ6 jeep

3986.111

LAW 80 LIGHT ANTI-ARMOUR WEAPON

LAW 80 is a one-shot low cost disposable short range anti-tank weapon developed to replace existing weapons which are either light and ineffective or of a size which demands crew operation to achieve lethality. It provides a capability for the individual soldier to engage current and future MBTs at ranges out to 500 m from any aspect including head-on and achieve a high probability of hitting the target.

Much attention has been placed on the need for the firer to obtain a good hit on the target, since the firing of almost all types of anti-tank projectile is likely to reveal the firing position. It is dangerous and time consuming to engage with a second weapon. To achieve this accuracy, a built in 9 mm spotting rifle is used. This spotting rifle contains five pre-loaded rounds, any number of which may be fired without revealing the position of the firer. The ammunition is ballistically matched to the main projectile marked by a tracer with a flash head to record a hit on a hard target. At any time the operator can select and fire the main projectile.

Shock absorbent end caps protect the weapon for carriage and storage. The end caps provide sealing for the tubes to protect the projectile against the effects of water immersion, even though the projectile itself is sealed. The end caps also support the projectile against the effects of mishandling such as end drops. After removal of the end caps, the tube containing the HEAT projectile is extended rearwards from the outer tube. The launch tube is automatically locked into position. This moves the centre of gravity of the weapon from the carrying handle to the shoulder rest to provide an ideal balance suitable for any firing position.

The sight is separately bonded to the front tube and forward movement of a sliding cover allows the sighting prism to erect. The sight is of unit magnification with the graticule projected through the sighting prism into the firer's line of sight, like a head up display. The sight can be used with one or both eyes open and for low light use the graticule can be illuminated with a tritium light source selected by the firer, making the weapon effective down to starlight levels of illumination.

The projectile is initiated with a totally non-electric (and therefore totally EMC proof) system comprising a percussion cap in the launcher connected by a flash tube to the rocket igniter. The rapid burn of the rocket motor which is completed before the projectile leaves the launcher ensures that the efflux is contained and directed rearwards by the launch and blast tubes, leaving the projectile to coast from the muzzle to the target. No debris of afterburn is projected at the firer, the recoil is not discernible, and the peak overpressure at the firer's ear is significantly less than 180 dB.

The forward part of the projectile consists of the HEAT warhead and its fuzing unit, and a double ogive nose switch which also provides the optimum stand-off distance. The fuzing unit generates electrical energy to fire the warhead by means of piezo crystals and contains various safety devices to ensure that the warhead does not arm until safe separation from the firer is achieved. At the rear of the projectile the composite aluminium and filament wound motor case has an extruded vane propellant. Four wrap-around fins are mounted on the rear of the motor. These are spring loaded to erect at muzzle exit to provide stability and to spin the projectile as it coasts to the target.

CHARACTERISTICS
Launcher length: 1.5 m (ready to fire); 1 m (on move)

LAW 80 anti-tank weapon ready to fire

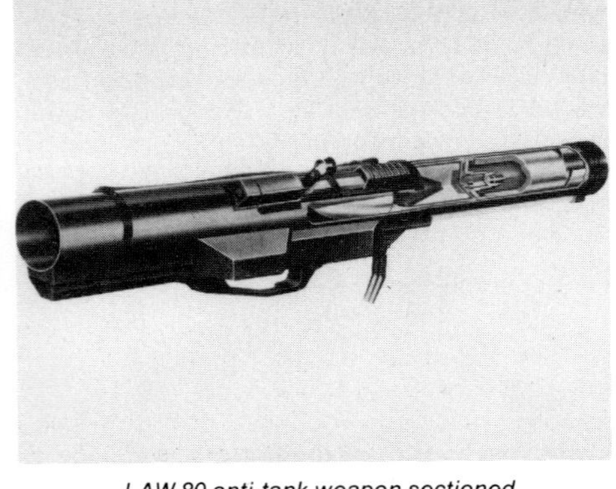
LAW 80 anti-tank weapon sectioned

Shoulder weight: 9 kg
Projectile weight: 4.6 kg (in launcher)
Projectile diameter: 94 mm
Armour penetration: >650 mm
Dispersion: 1 mil approx
Maximum effective range: 500 m
Rear danger area: 20 m approx
Warhead arms at 10 to 20 m
Protected against NBC effects
Inert to RF hazards in battlefield conditions

DEVELOPMENT
It has been developed under the direction of the prime contractor, Hunting Engineering Ltd on behalf of the UK Ministry of Defence. LAW 80 is now in production for the British Army, Royal Marines and RAF Regiment and has also been ordered by Jordan.

CONTRACTORS
Hunting Engineering Ltd, Reddings Wood, Ampthill, Bedford MK45 2HD, England.

4756.111
MERLIN ANTI-TANK MORTAR BOMB

Merlin is an 81 mm mortar bomb compatible with most mortars which, after firing, automatically selects its targets and homes on to it. The system is designed to provide an effective fire and forget capability for infantry confronted by tanks and armoured personnel carriers.

The projectile, which can be augmented by three increment charges, is launched in the conventional manner by dropping down the mortar tube with the shock of the launch initiating both a thermal battery and the safety and arming mechanisms. When the bomb clears the mortar six stabilising rear fins are deployed and as the projectile nears the top of its trajectory four front control fins are deployed and locked. These fins are controlled by an electric motor which twists them to eliminate yaw and roll then adjusts the projectile angle to approximately 45° for tracking.

The fully active millimetric wave seeker becomes activated halfway through the trajectory and initially seeks ground clutter returns. A programmable microprocessor simultaneously processes these returns and data on the projectile's altitude, yaw condition and pitch position. The seeker then examines an area of 300 × 300 m for moving targets and, if none are selected, a 100 × 100 m area is scanned for stationary targets.

When a target is acquired the seeker switches from the surveillance to the terminal phase mode. The control fins then guide the projectile to the geometric centre of the target which it locates by means of the minor change in return from the top of the target. The angle of attack then becomes steeper until upon impact it is nearly vertical. The kinetic energy effect is multiplied by a 'tunnel' in the electronic package which helps develop a penetrating jet. Static firings have shown Merlin can penetrate more than 360 mm of rolled homogenous armour.

The bomb is being designed so its introduction will require no changes either to the mortars or unit organisation. Customers will also not have to purchase specialised test equipment or spares.

Test firing of Merlin 81 mm terminally-guided anti-armour bomb from Royal Ordnance L16 mortar

CHARACTERISTICS
Diameter: 81 mm (on launch)
Length: 90 cm
Weight: 6 kg
Range: 4.5 km
Guidance: Millimetric wave seeker with fin control

DEVELOPMENT
British Aerospace funded development for the Merlin system from September 1981 and this led to Ministry of Defence interest. A general staff target (GST 395) for a mortar anti-tank (MORAT) round appeared in April 1983 and the cost of project definition and pre-development was shared between BAe and the MoD who provided £11 million for the work. However, the MoD subsequently withdrew for financial reasons and the project reverted to being a private venture with an eye on the export market. In late 1985 it was anticiptated that a Merlin round would cost £3700.

STATUS
Company funded development has continued and late in 1985 BAe were aiming for an in service date of 1990. In October 1987 BAe and the System Planning Corporation of Arlington, VA, announced a teaming agreement on Merlin covering development, manufacture and marketing in anticipation of a Defence Department request for analysis, design, development and testing of a millimetric wave, terminally guided anti-tank mortar round.

CONTRACTOR
British Aerospace Dynamics Division, Six Hills Way, Stevenage, Hertfordshire, SG1 2DA, England.

2475.111
VIGILANT ANTI-TANK MISSILE

Vigilant is a man-portable wire-guided anti-tank missile system. In its simplest configuration the system comprises a launcher box containing one missile, a sight controller, a pocket battery, and a length of interconnecting cable. For details see **2475.111** in the 1987-88 edition.

STATUS
Although now regarded as obsolescent, Vigilant remains in limited use with Kuwait and the United Arab Emirates.

6719.111

FURTHER IMPROVEMENTS TO TOW (FITOW)

For several years the British Army has had a general staff requirement for improvements in its helicopter-launched TOW system (see **2830.111**).

In April 1986 Graseby Dynamics and THORN EMI submitted best and final offers for the project known as FITOW (Further Improvements to TOW). In December 1987 it was announced that the Ministry of Defence had selected the THORN EMI contender.

THORN, who are teamed with Royal Ordnance, Hughes Aircraft and Westland Helicopters, are offering a missile which retains the 5-inch warhead but features a new warhead and fuzing system. This is reported to include two downward firing charges with an active infra-red proximity fuze. In addition some modifications will be made to the guidance system to improve ECCM performance.

CONTRACTOR

Prime: THORN EMI Electronics, Defence Systems Division, Feltham, Middlesex TW13 7DZ, England.

UNITED STATES OF AMERICA

4877.111

HELLFIRE MODULAR MISSILE

The Hellfire modular anti-armour weapon, originally developed for use from ground attack helicopters in the anti-tank role **(1391.311)**, has been fired in the ground-launched mode from a ground vehicle during US Army tests. Rockwell International is under contract to the Swedish Government to adapt the weapon to a new shore defence application **(4878.121)**.

The ground vehicle used in the tests carried out by the US Army's 9th Infantry, High Technology Light Division in late 1982 was a one-ton pickup truck on which twin Hellfire launcher rails were pedestal mounted, together with the fire control equipment. Modifications to the vehicle were minimal and consisted of installing the launcher pedestal and its azimuth and elevation controls, missile storage racks for six missiles, and a special firing panel.

As initially operated (in the air-launched mode) Hellfire uses a laser seeker to home onto reflected energy from either an airborne or a ground laser target designator, but combined IR/RF and imaging IR seekers are being developed for the system and this will give the missile a wide range of delivery modes, including engagement from defilade.

CHARACTERISTICS

US Army designation: AGM-114A

Length: 1625 mm (laser); 1727 mm (IR/RF); 1778 mm (IRIS)

Diameter: 178 mm

Fin Span: 326 mm

Weight: 45.7 kg (laser); 47.8 kg (IR/RF and IRIS)

Range: 7000 m

Warhead: Tandem twin shaped charges

STATUS

The air launcher version of Hellfire is in production and in service with the US Army. The anti-shipping version has been tested as has the ground launcher version from a US Army truck.

Congress awarded the Defense Department $5 million to evaluate a ground-launched Hellfire. A 24 month test programme is envisaged with launchers on HMMWV.

CONTRACTOR

Rockwell International, Missile Systems Division, Defense Electronics Operations, 1800 Satellite Boulevard, Duluth, Georgia 30136, USA.

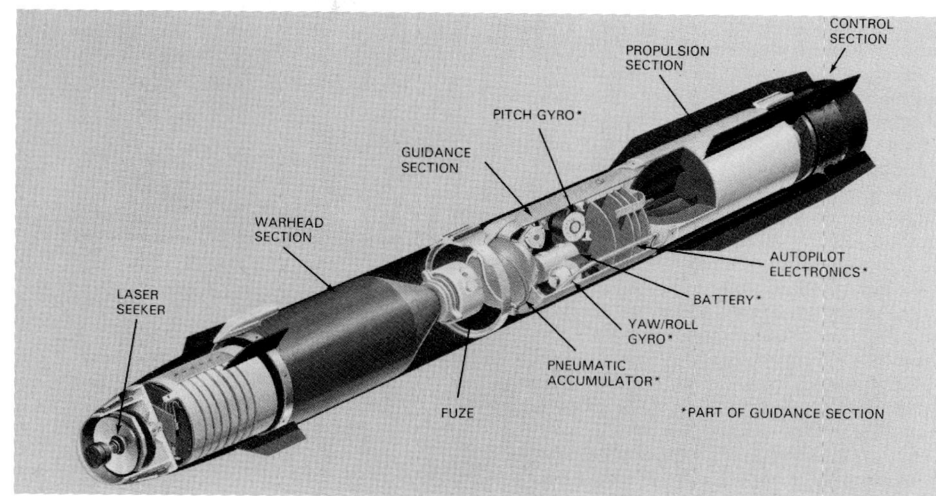

AGM-114A Hellfire anti-armour missile

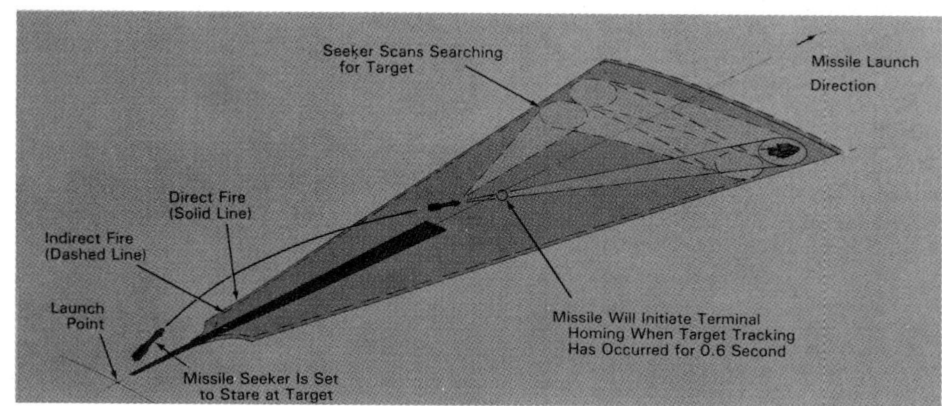

Operating sequence of ground-launched Hellfire anti-tank missile

6225.111

ADVANCED ANTI-TANK WEAPON SYSTEM – MEDIUM (AAWS-M)

The AAWS-M is intended to be a one-man portable weapon weighing no more than 20.4 kg (45 lbs) employed at infantry platoon level to defeat existing and anticipated Soviet tanks day or night through smoke and dust. It is designed to replace the Dragon (see entry **2573.111**) and will have superior lethality, range, as well as a shorter time of flight. The operator will also be less vulnerable and the launch signature will be less than that of Dragon.

Three guidance technologies are under consideration and in August 1986 technical demonstration contracts each worth $30 million were awarded by US Army Missile Command to Ford Aerospace and Communications' Aeronutronic Division (laser beam rider), Hughes Aircraft Missile Systems Group (fibre-optic guidance) and Texas Instruments Defense Systems & Electronics Group (imaging infra-red).

Ford's system is a shoulder-mounted missile, Topkick, in which the operator directs the flight of the missile by sighting directly onto the target. The missile is guided by a low intensity laser beam projected by the projectile and along the operator's line-of-sight. The system will be a top-attack one using two warheads fuzed to detonate when the projectile is in position to do catastrophic damage to the target. The missile will be in a disposable launch tube with a reusable command and launch unit which will be a stabilised sight for both pinpoint accuracy and as a protection from 'gunner flinch'. Ford have ordered five carbon dioxide lasers from Ferranti Electro-optics. Ford are teamed with Hughes Aircraft Company's Missile Systems Group and Raytheon's Missile Systems Division.

A mock-up of the Ford Aerospace and Communications Corporation's contender for the AAWS-M requirement. The system weighs 20.18 kg

The Hughes' system, which will weigh 19.3 kg, consists of an 11.8 kg missile with a medium wave infra-red staring focal plane array seeker infra-red seeker in a disposable launch tube with a reusable command and launch unit. The launch unit will feature an integrated telescope and thermal sight. Although the missile will be a fire-and-forget system the

The Texas Instruments AAWS-M contender, which weighs 18 kg

operator will have an optional use two fibre-optic link by which he will see on the launch unit display the missile seeker's view. This will allow the operator either to guide the missile precisely to the most vulnerable part of the target or to lock it onto the target after launch.

Texas Instruments are proposing a fire-and-forget system using an infra-red seeker. Few details have been released about the missile but it is known it will feature two attack modes; top-attack for use against armour and direct attack against helicopters and fortifications. The company are proposing a system which will have high commonality with their system to meet the AAWS-H requirement.

Each company must deliver one launcher, 18 complete missiles and 60 warheads for a competitive shoot off. Firings were likely to start early in 1988 and each competitor must complete 12 mandatory and six discretionary tests by 1 September 1988. After this competition one company concept will be selected, probably early in 1989, and the winner must be teamed for the follow-on 36 month full scale development contract although once the system reaches the production stage a second source will be established. The selected system will begin limited production in FY 1992 and the anticipated US Army/US Marine Corps requirement is for 7000 launchers and 90 000 missiles. The US Army requires a system weighing less than 45 lbs (20.43 kg) and more than 35 lbs (15.9 kg) with an effective range of more than 1 km.

At the same time as the technology and demonstration phase there is a cost and operational effectiveness analysis on key areas including a product improved Dragon, foreign anti-armour systems, the preferred mix of AAWS-M and AAWS-H (see entry **6467.111**), enemy countermeasures and likely target engagement ranges in urban areas and broken terrain.

The Hughes' Aircraft Company's AAWS-M contender

STATUS

Undergoing a 27 month technology demonstration phase.

CONTRACTORS

Ford Aerospace and Communications Corporation, Aeronutronic Division, Ford Road, Newport Beach, CA 92658-8900.

Hughes Aircraft Company, Missile Systems Group, 8433 Fallbrook Avenue, Canog Park, CA 91304-0445.

Texas Instruments, Defense Systems & Electronics Group, Lewisville, TX.

2573.111

DRAGON ANTI-TANK/ASSAULT WEAPON (FGM-77A/FTM-77A)

Formerly known as MAW (medium anti-tank assault weapon system), Dragon was developed by McDonnell Douglas for the US Army and Marine Corps. It employs a command to line-of-sight guidance system and consists of three main items: a tracker, a recoilless launcher, and a missile. The tracker includes a telescope for the gunner to sight the target, a sensor device, and an electronics package. The tracker is reusable and is attached to the launcher; the missile is never seen by the gunner and after firing the launcher is discarded.

The McDonnell Douglas Kollsman-produced Type SU-36/P IR tracker (**3089.183** in 1979-80 and earlier editions) is the guidance element of the system, and consists of a six-power optical sight aligned with an IR sensor, an electronics package, the trigger, and safety catch. The IR sensor acquires a flare in the tail of the missile as it leaves the launcher. The relative angular positions of the missile and the line of sight to the target are continuously compared with the sight electronics and used to derive correction signals which are transmitted to the missile via the guidance wires.

The missile is ejected from the tube by a gas generator using a recoilless technique. When it emerges folding fins flip open and the missile starts to roll. Thereafter propulsion and control forces are provided by the 60 small sustainers which fire in pairs on demand from the tracker. In operation, the gunner sights the target through the telescopic sight then launches the missile. While he holds his sight on the target, the tracker senses the missile position relative to the gunner's line of sight and sends command signals over wire to the missile.

As commands are sent continually to the missile, the side thrusters apply corrective control forces. The thrusters are fired at appropriate roll angles so that the missile is automatically guided throughout its flight.

CHARACTERISTICS

Designation: M47 (FGM-77A/FTM-77A)

Type: Anti-tank guided missile

Guidance: Wire-guided, command to line-of-sight

Propulsion: Recoilless launched, solid-propellant rocket motor

Warhead: Shaped charge

Launch weight: 13.8 kg (ready to fire)

Range: 1 km

STATUS

Dragon is deployed by US Army and Marine Corps units, but production for this service is now complete.

Late in 1986 McDonnell Douglas Astronautics Company was awarded a $4.7 million contract from the Naval Surface Weapons Center, Dahlgren, Virginia, under the Dragon Product Improvement Programme for the US Marine Corps.

The product improvement programme is in two stages. The first, Generation II Dragon, incorporates a new warhead with 85 per cent improved effectiveness over Generation I Dragon. This warhead has a liner with a non-tapering particle stream to beat reactive armour. It is now in production in Switzerland and has already successfully demonstrated against NATO triple-heavy and composite targets. McDonnell Douglas are currently upgrading 20 000 rounds for the US Army and 12 000 rounds for the US Marine Corps.

McDonnell Douglas have further improved Dragon to produce Dragon III. This features a new warhead with extendable probe which is claimed to have 98 per

Dragon deployed and ready to fire

cent more penetration than Dragon I and a sustainer motor to increase velocity so that the missile can fly 1500 m in 8.6 seconds. More wire has been added to increase range to 1500 m. In addition to the missile a new Day/Night Tracker (DNT) is offered with an imaging infra-red sensor to provide target acquisition to 2 km and recognition at 1.5 km. The tracker electronics have been hardened against counter-measures and thermo-electronic cooler has been added for quieter operation as well as eliminating the need for gas bottles.

Model	Dragon I	Dragon II	Dragon III
Weight carrying	14.6 kg	16.2 kg	22.5 kg
Weight at launch	13.8 kg	14.8 kg	21.5 kg
Length launcher	1154 mm	1154 mm	1154 mm
Length missile	744 mm	846 mm	852 mm
Weight missile	6.2 kg	6.97 kg	15.6 kg
Range	1000 m	1000 m	1500 m

In service with Iran, Israel, Jordan, Morocco, Netherlands, Saudi Arabia, Spain, Switzerland, Thailand and Yugoslavia.

CONTRACTORS

McDonnell Douglas Astronautics Company, Titusville, Florida, USA – developer, initial production, engineering support, and night tracker.

Kollsman Military Systems Division, 220 Daniel Webster Highway, Merrimack, New Hampshire, USA – tracker production.

Raytheon Company, Bristol, Tennessee, USA – round production.

2830.111
TOW ANTI-TANK MISSILE (BGM-71)

TOW is an acronym for tube-launched, optically-tracked, wire-guided, and describes a heavy assault ground-to-ground (or air-to-ground) anti-tank guided weapon system.

The TOW launcher system, for infantry use, is made up of five elements, none of which weighs more than 24 kg, although the complete launcher weighs 78 kg when assembled and ready to fire. Two of these elements are a tripod and a traversing unit, mounted upon the tripod, to which the tripod launch tube and optical sight are attached. The gunner's optical sight is of high magnification and equipped with aiming cross-hairs which when combined with the smooth, stable motion of the traversing unit permit very accurate tracking of moving targets after very little operator training. The fifth individual element is the electronic guidance computer which sends steering commands automatically to the missile in flight.

The missile itself is contained in a sealed storage and transport container which becomes a launch tube extension when placed in the launcher breech. After the breech locks, all electronic contacts to the missile are automatically closed, and TOW is ready to fire.

The missile contains two solid-propellant motors. The launch motor ejects the missile from the launch tube and is burned out by the time the missile has left the tube. Only after the missile has flown several metres does the flight motor ignite, so that no protection is necessary for the gunner against hot exhaust gas and propellant particles. The flight motor is mounted in the centre of the missile with its two exhaust nozzles mounted on either side. This arrangement prevents interference with the guidance wires which are placed at the tail of the fuselage. Steering commands are transmitted by the two wires which uncoil from two separate spools. Short cruciform wings in the centre of the missile and the cruciform rudder surfaces all unfold after leaving the launch tube. Missile manoeuvring is done entirely aerodynamically (without jet vanes) so that TOW maintains good manoeuvrability throughout missile flight. The electronics unit is mounted between the flight motor and the armour-piercing warhead.

After the missile leaves the launch tube, a light source in the tail comes on so that the optical sensor on the launcher, which is bore-sighted with the gunner's telescope, can track the missile along its flight path. The light source is sufficiently strong to allow automatic guidance to the maximum range of the missile under all conditions in which the target is visible to the gunner.

MOBILE INSTALLATIONS

TOW can be installed in most vehicles capable of cross-country travel. Within the US Army it is mounted on Jeeps, the High Mobility Multi-purpose Wheeled Vehicle (HMMWV) as well as the AH-1S helicopter. Armoured vehicles using TOW include the M901 Improved TOW Vehicle (ITV), which has an armoured weapon station featuring two launchers, a day sight and the AN/TAS-4 night and target acquisition sight (see **2830.111** in the 1987-88 edition). The Bradley Fighting Vehicle also has a twin tube retractable installation on the turret side, the tubes being retracted during loading and travelling. The M2 Infantry Fighting Vehicle carries five reload rounds while the M3 Cavalry Fighting Vehicle has ten reloads. The same turret will also be used in 96 anti-armour versions of the Marine Corps' LAV (Light Armoured Vehicle).

The Dutch Army has a similar vehicle to the M901, the YPR 765 PRAT, while Thyssen Henschel converted 162 tank destroyers into Jaguar 2s by removing the 90 mm gun and replacing them with single tube TOW launchers and AN/TAS-4 sights. Since October 1986 Renault Vehicles have an agreement with Hughes who will provide technical assistance in adapting the TOW system to the VAB range of armoured personnel carriers.

TOW is also extremely adaptable to helicopters, as has been demonstrated. For details of helicopter applications see **2831.311**.

CHARACTERISTICS

Military designation: BGM-71

Type: Heavy anti-tank guided weapon system

Guidance principle: Automatic missile tracking and command guidance from optical target tracker

Guidance method: Wire guidance control of gas-operated aerodynamic surfaces

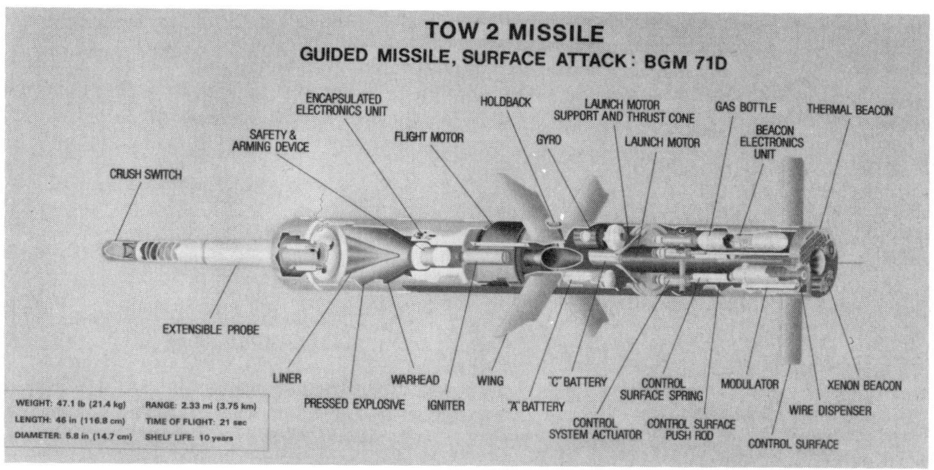

A cutaway drawing of TOW 2

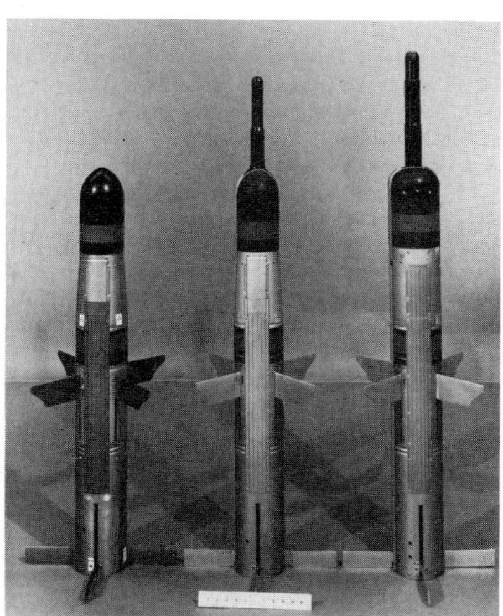

TOW missile family with (l ro r): basic model; improved TOW (ITOW); and TOW 2 with larger warhead, extensible probe and higher impulse flight motor

Infantry version of TOW with gunner's optical sight on left of launcher

Propulsion: Launch motor and a boost coast flight motor which burns for approximately 1.5 seconds
Warhead: HEAT shaped charge
Missile length: 117.4 cm (BGM-71A/C), 140 cm (BGM-71D post launch, pre-launch as BGM-71A/C)
Missile diameter: 15.2 cm
Launch weight: 18.9 kg (BGM-71A), 19.1 kg (BGM-71C), 21.5 kg (BGM-71D)
System weight: 101 kg (BGM-71A), 113.20 kg (BGM-71C), 121.10 kg (BGM-71D)
Speed: 200 m per second
Range: Minimum 65 m, maximum 3750 m
Rate of fire: 3 launches in 90 seconds
Crew: 4

DEVELOPMENT

In 1980 the US Army embarked upon a programme to improve the performance of the TOW system against advanced enemy armour.

The first phase involved an improved 5-inch (127 mm) diameter warhead to increase the armour penetration characteristics. It is of the same size and weight as the standard TOW warhead but of improved design. This version was the Improved TOW (ITOW) and featured a 127 mm long extensible nose probe to provide an optimum stand-off detonation distance to ensure maximum effect from the hollow charge warhead. This system became operational in 1981.

The second phase of the programme, TOW 2, incorporates a heavier 6-inch (152.4 mm) diameter warhead of still greater penetration performance and this occupies the full diameter of the missile body. The missile guidance system has also been improved by the use of a subsystem in which the analogue computer is replaced by dual digital microprocessors to give greater flexibility in guidance programming as well as higher accuracy. To compensate for the added weight the missile motor has an improved propellant to provide a 30 per cent higher impulse. The TAS-4 sight has been modified to function as a totally independent guidance loop and a high intensity thermal beacon has been added to the end of the missile. The missile, which has a unit cost of $10 000, became operational in 1983.

In September 1987 it was revealed that an improved version of TOW 2, TOW 2A, had entered production during the summer. This has a tandem warhead designed to beat reactive armour, a small warhead being added in the missile probe. Externally TOW 2 and TOW 2A are identical and more than 16 000 TOW 2As were scheduled for delivery by April 1987 under FY 1986 orders. The modifications, reportedly, cost only $418 per missile but in the summer of 1987 there were reports that the missile was proving unstable in flight and, therefore, unable to beat reactive armour. However, the US Army denied this stating that the instability problems had occurred ten years earlier and subsequently been solved.

Development of TOW 2B, based on the Over Armor Target Synthesis programme, began in 1978. In this the 127 mm diameter warhead is mounted at right angles to the ground using magnetic with proximity fuzes. It is scheduled to enter service in the early 1990s.

In February 1987 Hughes received a $2.97 million contract to build and test a millimetric wave data link for TOW. By transmitting guidance commands through a secure electronic link the TOW velocity could be increased together with its range. The concept was developed as a private venture and a breadboard system was demonstrated to the US Army Missile Command in 1983. Ground tests began in 1984 using a modified TOW 2 missile and flight tests, using six missiles modified by the US Army's Redstone Arsenal, were scheduled to begin in November 1987.

In mid-1984 Israel revealed it had developed a derivative of TOW using laser beam-riding guidance. Details of MAPATS will be found in **4859.111**.

The United Kingdom is also developing an improved version of TOW for helicopter use as Further Improved TOW (FITOW). Details will be found in **6719.111**.

Basic TOW (BGM-71A) has been in the US Army's inventory since 1970 and 311 000 had been produced by June 1986. The system has been adopted by 36 countries including Bahrain, Canada, Denmark, Egypt, Finland, West Germany, Greece, Iran, Israel, Italy, Japan, Jordan, Kenya, South Korea, Kuwait, Lebanon, Luxembourg, Morocco, the Netherlands, Norway, Oman, Pakistan, Portugal, Saudi Arabia, Somalia, Spain, Sweden (as RB-55), Switzerland, Taiwan, Thailand, Tunisia, Turkey, United Arab Emirates, the United Kingdom, USA, North Yemen. In addition UNITA forces in Angola have received TOW.

Many of these missiles have been converted to ITOW status (BGM-71C) and the AN/TAS-4 night sight has been introduced (see **3824.193**). By June 1986 49 000 dedicated ITOW missiles had been produced. In addition to the United States ITOW is known to have been purchased by Finland (as M-83), Greece, Norway, Italy, Kuwait, Switzerland and Turkey.

Early in 1983 production began of TOW 2 (BGM-71D) missiles and launcher modification kits. In August 1987 Hughes received a $112.47 million contract for 13 755 TOW 2 missiles together with 2731 basic and practice rounds with delivery by March 1989. The following month a $100.79 million contract was awarded to Hughes for 662 TOW 2 systems for Bradley vehicles. Planned TOW 2 procurement in FY 1987 was 21 333 at a cost of $288.7 million. Total deliveries by June 1986 were more than 50 000. In December 1986 the Swiss Parliament approved selection of TOW 2 which will be co-produced under licence. The missiles will be carried by Piranha armoured personnel carriers which will be fitted with Thune-Eureka launcher turrets. In 1987 Congress was asked to approve the sale of 2386 TOW 2 missiles with support equipment, worth a total of $78 million, to Pakistan.

CONTRACTORS

Hughes Aircraft Company, Missile Systems Group, Canoga Park, California 91304, USA, is the prime contractor for the system, having developed it for the US Army Missile Command.

Emerson Electric Co, Electronics and Space Division, 8100 Florissant Avenue, St Louis, Missouri 63136, USA, is the prime supplier of the launcher, missile guidance set, optical sight, and traversing unit.

Kollsman, Military Systems Division, AN/UAS-12C night vision system. Follow on contract awarded for 374 AN/UAS-12C TOW equipment sets was awarded early in 1987 and was a follow on to the contract awarded in May 1986 for 561 sets. Total contract is valued at 53 million. The AN/UAS-12C provides thermal imaging and improved missile guidance to the TOW 2 system.

Texas Instruments Incorporated, PO Box 660246, Dallas, Texas 75266, USA, is the major contractor for the infantry ground launcher TOW 2 retrofit kits which include a digital missile guidance set and the AN/TAS-4A night sight.

6226.111

TACTICAL MISSILE SYSTEM

In March 1986 the US Army selected LTV Aerospace and Defense Company in preference to Boeing Aerospace, to develop the Army Tactical Missile System, a conventional surface-to-surface artillery missile capable of striking targets beyond the range of existing Army artillery, rockets or the Lance missile.

Studies on the programme, originally designated Battle King, began in 1970 and were completed in 1976 with flight tests of the earlier SIG-D in 1976-80 and Assault Breaker between 1978 and 1982. In 1980 preliminary definition studies began of the Army's Corps Support Weapon System (CSWS) and the following year concept definition studies began. In 1982 the Army's CSWS and the USAF CSW programmes were combined to form JTACMS for which pre-full-scale development studies were conducted between 1983 and 1984. In 1984 the US Army selected MLRS as the launcher for JTACMS and decided upon the need for an interim ballistic missile which became TACMS.

TACMS will feature a new solid fuel propulsion system incorporating the Arcadene 360 family of propellants which is being developed by the Atlantic Research Corporation and will give a range in excess of 100 km. Honeywell are developing the H700-3A on-board inertial guidance system with ring laser gyro while Martin Marietta are developing the anti-armour warhead.

A TACMS missile beside an MLRS SPLL which will be capable of carrying two missiles

The initial contract is worth $130.3 million and covers full-scale engineering development, but total value of the whole programme is expected to be $2-3 billion. Under the initial contract LTV is expected to build 50 missiles over a four year period with first production systems following in 1990.

The US Army TACMS will be fired from the current MLRS with each missile being about 3.962 m long, 0.61 m in diameter. It will be controlled by four moveable tail fins. Each MLRS SPLL will carry two pods each with one missile.

The Block I missile will have the M74 anti-personnel/anti-material munitions which will be purchased off-the-shelf to attack enemy troops and material but the Block II missile will feature a warhead, to be developed by Martin Marietta Aerospace, with precision-guided sub-missiles for use against moving armoured vehicles. In the future hard target, mine sub-munitions and runaway attack warheads are proposed.

Following the initial $130.3 million award, the US Army also awarded LTV an $83 million contract to integrate the Army TACMS into the existing MLRS. The changes will principally be in the software of the system's fire control computer so a system can fire either MLRS rockets or TACMS missiles.

STATUS

Full-scale engineering development.

CONTRACTOR

Prime contractor: LTV Missiles and Electronics Group, Missiles Division, PO Box 650003, Dallas, Texas 75265-0003, USA.

6467.111

ADVANCED ANTI-TANK WEAPON SYSTEM – HEAVY (AAWS-H)

During FY 1988 the US Army will inaugurate a programme, AAWS-H, comprising three weapon systems. One is a TOW replacement, the second will be a Kinetic Energy Kill (KEK) weapon and the third will be FOG-M using fibre-optic guidance.

The TOW replacemnet is designated the Advanced Missile System-Heavy (AMS-H) and, like TOW, it will be fired from crew-portable mountings, vehicles and helicopters.

The KEK system will be based upon the LTV High Velocity Missile (HVM), see entry **4448.311** in the Air-to-Surface Missile section. LTV Missiles and Electronics Group have suggested a 284.48 cm long missile with 16.19 cm diameter weighing 77.1 kg which would be fired from four-round launchers on a Bradley vehicle. The missile would travel at 1500 m per second and would be a laser beam rider. The electro-optical fit would include a thermal imager.

The FOG-M system will use the technology developed for FAADS N-LOS. See **6228.131**.

4880.111

US ICM

Under the heading of Improved Conventional Munitions the US Army is implementing a programme to develop a variety of technologies to enhance the capabilities of tubed artillery in battlefield support roles. Ammunition for use against both troops, armoured formations and other important theatre targets are within the scope of this programme which embraces technologies for terminal guidance of projectiles, 'smart' sub-munitions, cargo carrying shells and related technical problems. One such project is the SADARM (sense and destroy armour) programme which is described separately (**3988.111**). Some of the less advanced current programmes being pursued by the US Army are outlined below.

Projectiles: Currently ICM programmes centre on either or both of two standard artillery calibres, 155 mm and 8-inch (203 mm), and as noted the SADARM programme involves the use of cargo-carrying shells of both these calibres. The two standard US cargo-carrying shells are the 155 mm M483A1 and the 8-inch M509A1. The latter dispenses 180 M42 dual-purpose sub-munitions which are effective against both personnel and vehicles. The 155 mm M483A1 carries 88 dual-purpose sub-munitions that offer both light armour penetration and anti-personnel capabilities by including shaped hollow charges and fragmentation payloads.

RAAMS: Also using the 155 mm projectile is the remote anti-armour mine system (RAAMS) where the projectile is used to carry nine M718 or M741 anti-tank mines. The M718 is the 'long life' model, which can remain emplaced for a period of more than 24 hours (reportedly) before self-destruction is automatically initiated, and the M741 has a shorter emplaced life. These shells can be fired to a range of up to 17 km to lay down an anti-tank minefield in the face of an advancing enemy. After the pre-determined active life of the minefield has elapsed it should be safe for friendly forces to advance over the ground. The M781 mines weigh approximately 2.26 kg. A proportion of the mines carried can be equipped with disturbance fuzes to discourage enemy mine clearance activities.

Area Denial Mines: There are long- and short-life versions of area denial artillery munition (ADAM) rounds, M692 and M731, respectively. In both cases each 155 mm shell carries 36 anti-personnel mines which are actuated by a self-deploying trip wire or disturbance fuze. Tests have started on a system whereby a 155 mm shell is used to deploy six expendable communications jammers (EXJAM). These units are ejected from the rear of the shell in flight by a timer/sequencer and then slowed and stabilised in flight by fins and a streamer, to bury themselves on contact with the ground to a depth of a few cm. They then start transmitting to disrupt enemy communications.

XM864 155 mm Projectiles: A base bleed cargo-carrying 155 mm projectile (XM864) has reached the advanced development stage, and this will provide improved range without the impaired accuracy that rocket assisted projectiles sometimes encounter as the price of greater range performance. Ranges of up to 23 km when fired from the M109A2/A3 howitzers, and 26 km from the towed M198 are hoped for.

CGSP: Honeywell has proposed, in answer to the US Army's requirement for a cannon-launched, 155 mm fire-and-forget, anti-armour projectile, the conventional geometry smart projectile (CGSP). This would consist of a 155 mm shell equipped with a ring laser rate sensor providing reference signals to a guidance system that includes a millimetric wavelength seeker which uses the projectile's spin to aid scanning, and fluidic reaction jet control for missile guidance. The CGSP will follow a ballistic trajectory to the target area after firing, a timer will initiate jettison of the nose cone and the radar will begin ranging measurements to the ground. Target search follows shortly after and the target tracking and terminal homing phases are commenced at the appropriate times with corrections to the trajectory by means of the fluidic reaction jets near the rear end of the projectile. This is now known as the Autonomous Precision Guided Munition (APGM) and has become a NATO project. In 1987 it is expected that the US Army Armament Research and Development Command will release an RFP for a 32-month feasibility study for the APGM. This will be developed by members of NATO including the US, Canada, Italy, Turkey, United Kingdom, Netherlands and France.

STAFF: The smart top attack, fire-and-forget (STAFF)

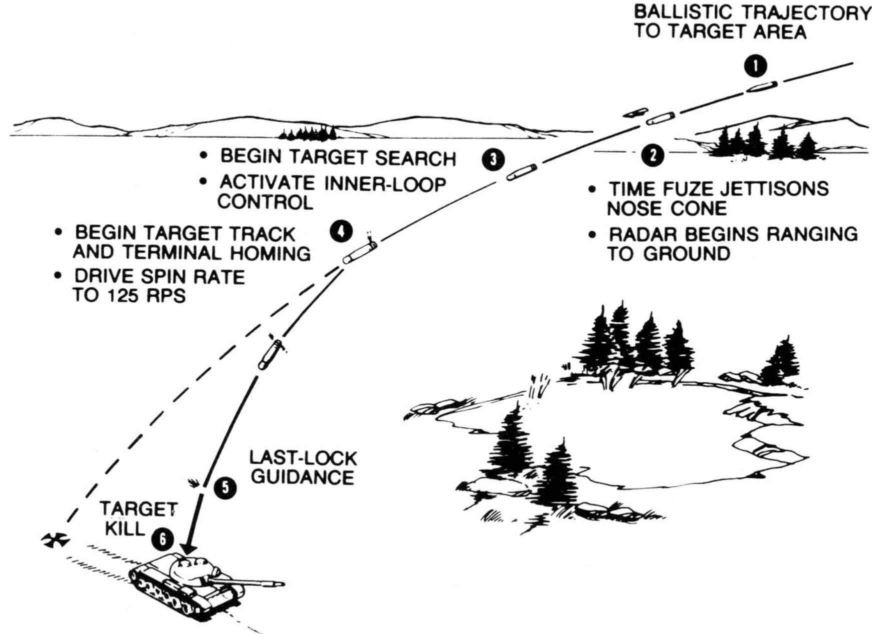

Operating sequence of Honeywell proposed 155 mm CGSP

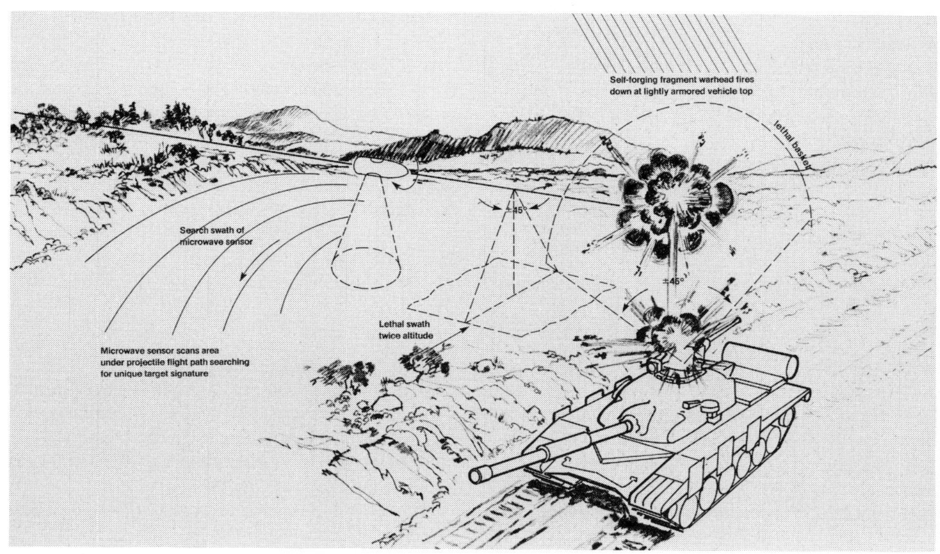

Operating principle of Aerojet STAFF ammunition

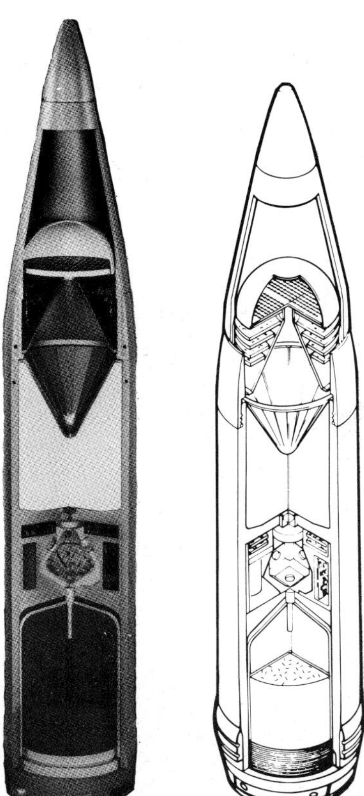

Cutaway drawing of 155 mm CGSP

concept is a novel approach by Aerojet Electro Systems to the engagement of armoured targets by indirect gun fire. Like the Honeywell CGSP (above) the projectile's spin plays an important part in the detection and location of the target, but in the STAFF technique a microwave sensor is used to scan a swath beneath the projectile during its relatively horizontal trajectory over the target area. This is accomplished by pointing the sensor at the right-angles to the projectile's spin axis, which ensures that a belt of terrain parallel to the trajectory is continuously scanned as the projectile spins and passes overhead. Calculations indicate that a swath width equivalent to twice the projectile's height can be scanned. The sensor is programmed to detect specific target signatures, and on encountering one that is within range a self-forming fragment warhead is fired downward to penetrate the top armour of the tank or other target.

CONTRACTORS

Honeywell, Defense Systems Division, 5640 Smetana Drive, Minnetonka, Minnesota 55343, USA.

Aerojet Electro Systems Company, PO Box 296, Azusa, California 91702, USA.

3212.111

COPPERHEAD CLGP (M712)

Copperhead is the name assigned to the M712 cannon launched guided projectile (CLGP), a 155 mm round equipped with a terminal guidance system and launched from conventional howitzers (for example M109 series self-propelled and M198 towed) into an essentially ballistic trajectory. However, by means of a technique known as trajectory shaping, it is possible to use shallow flight path angles which allow Copperhead to be used with cloud ceilings below the specified 3000 feet (914 metres). This provides more guidance time below the clouds and increases projectile acquisition range. During flight the target is illuminated by a laser designator operated by a forward observer. The seeker in the nose of the CLGP acquires the reflected laser signature and uses this signal to operate control surfaces on the projectile to cause it to follow a collision course to the target.

The system was for a US Army requirement to bring indirect fire onto tanks and distant armoured targets with a high kill (preferably first round) probability using a shaped charge warhead. Indirect fire demands the use of a forward observer or a forward air controller who calls for fire in the conventional manner and designates the target during the last seconds of the projectile's flight.

Target designator systems which can be used with Copperhead include the AN/TVQ-2 Ground Laser Locator Designator (**3051.193**), the hand-held AN/PAQ-1 Laser Target Designator (LTD) (**3157.193**), Modular Universal Laser Equipment (MULE) (**3816.193**), the Type 306 Laser Target Marker Ranger (UK), various RPV designator systems, and the AAH advanced attack helicopter target and designation system (TADS) (**3609.363**). The M981/M981A1 Fire Support Team Vehicle (FISTV), now in service with the US Army, can also designate targets for the Copperhead CLGP.

Martin Marietta Copperhead CLGP being unpacked from its transport container (US Army)

The Copperhead round is fired in an inactive state, thus simplifying the precautions needed to protect the guidance system from acceleration shocks. An acceleration-sensitive battery, initiated by firing the projectile, supplies power for guidance and control, as well as initiating and operating a timer which controls the internal operating sequence. Timer adjustment is dependent upon the required range, and therefore is similar to setting a timed fuze.

Associated with a setting of the timing switches is the selection of the trajectory mode for a ballistic trajectory or a fly under/fly out (FUFO) trajectory. The latter is intended for ranges of over 8 km and for bad weather where the projectiles travel at a lower altitude to provide a longer time beneath cloud ceiling for acquisition and guidance. The ballistic trajectory is used for shorter range shots and/or good weather conditions.

Coding of the laser signature which will be accepted by the seeker is also set in at the same time as the above selections are made. This coding will match that being used by the forward operator of the laser designator.

Copperhead is made up of three sections; guidance, payload, and stabilisation and control. The configuration is aerodynamically controlled by cruciform in-line wings and tail fins that provide roll stabilisation and lateral manoeuvrability sufficient to provide an impact 'footprint' adequate for successfully engaging manoeuvring armoured targets.

The guidance section consists of the seeker and electronic assemblies which are housed within the nose dome and steel housing. The seeker uses folded body-mounted optics with a spin stabilised gimballed mirror (seeker gyro). The seeker gyro is spun-up mechanically by a steel spring and sustained and torqued electrically.

The payload section includes a HEAT warhead in a steel structure and a fuze module. The latter incorporates six grazing sensors to detonate the warhead if the nose sensor does not directly hit the target.

The stabilisation and control section includes the flip-out aerodynamic surfaces and associated actuator mechanisms. Deployment of the fins is delayed until after the muzzle exit by the effects of launch acceleration. When deployed at their 20° sweep angle, the tail fins are canted to maintain a six to eighteen revolutions per second projectile roll rate, which aids stability and also allows liberal tolerances in vane manufacture.

CHARACTERISTICS

Type: M712
Weight: 63.5 kg
Length: 137.2 cm
Warhead weight: 22.5 kg
Explosive weight: 6.4 kg
Max range: 16 km
Minimum range: 3 km

A typical impact 'footprint' has been reported as about one kilometre radius from the nominal aim point.

STATUS

Production of Copperhead was authorised in March 1980 with the award of a $62 million contract to Martin Marietta. The US Army expects to procure 31 808 Copperhead Cannon Launched Guided Projectiles of which 16 187 had been procured through FY 1985. The FY 1986 request was for 6900 with the FY 1987 request for 248. The cost of a Copperhead round by the winter of 1987-88 was $32 250 but Martin Marietta were embarking on a programme to trim this by ten per cent. The demonstrated reliability of Copperhead has now reached approximately 92 per cent. It achieved its Initial Operational Capability in Europe in fiscal year 1985 and achieved 10 hits out of 12 shots in an exercise against stationary, moving and multiple targets. By September 1986 Martin Marietta had delivered 12 000 Copperhead CLGP to the US Army. A number of countries have evaluated Copperhead including South Korea, Japan and the UK, but so far no firm export orders have been placed.

Early in 1987 it was stated that a product improvement program had been initiated to enhance the performance of Copperhead and reduce its cost by about ten per cent. The US Army has also evaluated the feasibility of establishing a second source for production of Copperhead so providing an alternative mobilisation producer and inducing competition in future purchases.

In July 1987 the US Army Missile Command sought support for development of a sealed two-colour infra-red hardened seeker for both Copperhead and ATACMS.

CONTRACTOR

Martin Marietta Orlando, PO Box 5837, Orlando, Florida 32855, USA.

3988.111

SADARM SYSTEM

The US Army SADARM (sense and destroy armour) anti-tank sub-munition system is a project to design and develop a weapon for use against massed enemy armour beyond the forward edge of the battle area (FEBA).

It was originally intended that the SADARM would be fired only from the M110 series of 203 mm (8-inch) self-propelled howitzer and much of the development and prototype work up to the early 1980s concentrated on this area.

In 1986 it was decided to develop SADARM for use with 155 mm howitzers such as the M109 series and the Vought Multiple Launch Rocket System (MLRS) which are both, or in the case of the MLRS, will be, available to NATO countries in larger numbers than the M110.

In April 1986 31 contractors were circulated for full scale development of SADARM and three bids were received. In September 1986 the Defense Systems Division of Honeywell and Aerojet Electro Systems were both awarded development contracts for SADARM with development to be completed in June 1991 although it is expected that a production award could be made as early as 1989.

The 155 mm projectile is expected to hold two or three SADARMs while the MLRS will hold six.

Basically the projectile or rocket is fired over the target area and the SADARMs are dispensed from the rear of the projectile by a pre-set time fuze. Once out of the carrier projectile they are stabilised by a parachute and the on-board dual mode sensor searches the ground for potential targets.

Once a target has been detected the sensor will determine the centre of the target and initiate the self forging fragment (SFF) warhead.

The explosive in the SFF warhead forges the metal plate into a slug which travels at a high velocity and will penetrate the at present vulnerable upper surfaces of armoured vehicles.

STATUS

In September 1986 the US Army awarded contracts to Aerojet ($87.2 million) and Honeywell ($95 million) for full scale development of SADARM munitions. The contracts will last 57 months but after 30 months one firm will be selected for production. The ammunition is for MLRS (six sub-munitions) and 155 mm shells (two sub-munitions) and the initial operational capability will by the end of 1990 and 1991 respectively.

CONTRACTORS

Aerojet Electro Systems, PO Box 296, Azusa, California 91702, USA.

Honeywell Incorporated, Defense Systems Division, Minnetonka, Minnesota 55343, USA.

Overall responsibility of the SADARM Programme is with the US Army Armament Research and Development Center, Dover, New Jersey 07801, USA.

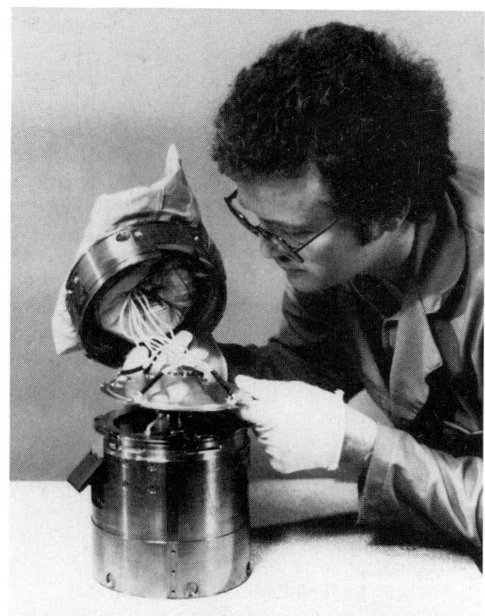

A Honeywell SADARM round is inspected by a technician

Coastal Defence Weapons

BRAZIL

6748.121
SM-70 BARRACUDA SURFACE-TO-SURFACE ANTI-SHIP MISSILE

Three versions are contemplated for this missile, whose development is scheduled for completion at the end of 1989, and which will have a high subsonic speed. At present the emphasis is upon the coast defence role but shipborne and airborne versions are under consideration by the Brazilian Navy and Air Force.

The coast defence version will feature a firing unit mounted on three trucks; a launcher vehicle with four launcher containers, a search acquisition vehicle with retractable antenna and a command vehicle. Each firing unit will have a 12 replacement launcher containers. The vehicle will be the same as used with the Astros rocket system (**4146.111**).

CHARACTERISTICS
Length: 6 m
Diameter: 30 cm
Range: 70 km
CONTRACTOR
Avibras Industria Aerospacial SA, Caixa Postal 229, São José dos Campos, SP, Brazil.

CHINA (PEOPLE'S REPUBLIC)

6723.121
HY-4 COAST DEFENCE MISSILE SYSTEM

The HY-4 appears to be a further development of the HY-2 family but evidently of lighter weight. It is powered by what the manufacturers describe as 'a high-efficiency turbojet' and, like the HY-2, it is fired from a truck-mounted launcher with the assistance of boosters. It is also described as having 'a valid interference-free guidance system' but no further data is available. The guidance system is likely to be similar to the HY-2G (**3984.121**).
STATUS
The missile is under development and will have a range of 130-150 km.
CONTRACTOR
China Precision Machinery Import and Export Corporation, 17 Wenchang Hutong Xidan, Beijing, People's Republic of China.

The HY-4 coast defence missile

3984.121
HY-2 COAST DEFENCE MISSILE SYSTEM

'Hai Ying 2' (Sea Eagle) is the designation reserved for the coast defence version of the CSS-N2 anti-ship missile. SY-1 (Shangyou Yihao-1) is probably the designation for the original CSS-N-2 copied from the Soviet Styx anti-ship missile. The export version is known as Silkworm.

The HY-2G employs monopulse active radar homing, most probably in I-Band (see entry **6581.121** in Addenda). This low frequency dictates the use of a relatively large antenna to achieve a narrow tracking beam-width. The HY-2G is also equipped with a radio altimeter, enabling it to cruise at an altitude of 30 m. During the terminal attack phase the HY-2G can descend even lower, to approximately 15 m.

The HY-2A is an infra-red homing version, using a seeker with a cassegrain lens arrangement, and is thought not to be of the imaging type.

Mobile radar for use with Chinese HY-2 coast defence missile system

HY-2 launch

CHARACTERISTICS
Length: 6.09 m
Weight: 2500 kg
Warhead weight: 450 kg
Max speed: 926 kph
Effective range: 20-95 km
STATUS
Operational and in production in the People's Republic of China. Available for export and thought to have been supplied to some African states and Iran.
CONTRACTOR
China Precision Machinery Import & Export Corporation, 17 Wenchang Hutong Xidan, Beijing, People's Republic of China.

6184.121
C801 COASTAL DEFENCE MISSILE SYSTEM

In late 1984 the Chinese authorities revealed a new anti-ship missile that was claimed to be employed in both ship-launched and coast defence roles. The missile cruises at heights of 20-30 m under automatic control then descends to 5-7 m for the terminal phase. First use of this weapon, the C801 (**4891.221**), in the latter role may well be in mixed batteries with the HY-2 at selected sites.

CHARACTERISTICS
Length: 5.814 m
Diameter: 0.36 m
Max span: 1.18 m
Weight: 815 kg
Warhead: 165 kg
Range: 50 km
STATUS
In operational service.
CONTRACTOR
China Precision Machinery Import & Export Corporation, 17 Wenchang Hutong Xidan, Beijing, People's Republic of China.

FRANCE

2118.121

EXOCET COASTAL DEFENCE MISSILE SYSTEM

The MM-40 coastal defence version of the Exocet anti-ship missile system is now fully operational. It uses the current shipborne MM-40 missile, with the same propulsion arrangements (**1156.221**). Folding aerodynamic surfaces permit the use of small tubular launcher-containers made of wound glass-fibre.

The system is deployed as a mobile battery, all the equipment being carried on military vehicles. The target acquisition and designation radar will usually be located on high ground to ensure adequate range. Three or four launcher vehicles occupy predetermined locations to cover the desired firing area, with supply trucks ready to replenish them, and maintenance vehicles under cover. For engagement of targets beyond the radar horizon, target data supplied by aircraft, helicopter, or ship can be used. The offshore aircraft or ship automatically transmits the target data to the shore radar installation.

CHARACTERISTICS

Type: MM-40
Length: 5.78 m
Launch weight: 850 kg
Range: 70 km

STATUS

The Royal Navy recently equipped Gibraltar with a missile defence based on the MM-38 version of Exocet (see entry **6186.121**). During the South Atlantic conflict a makeshift MM-38 battery caused extensive damage to the DLG *Glamorgan*.

Any type of coastal maritime surveillance radar can be employed for target detection and designation.

Launch of MM-40 from a coastal battery firing vehicle

CONTRACTOR

Aérospatiale, Division des Engins Tactiques, 2 rue Béranger, 92322 Châtillon-Cedex, France.

INTERNATIONAL

1337.121

OTOMAT COASTAL DEFENCE SYSTEM (FRANCE/ITALY)

The Otomat coastal defence system (OCDS) uses the Otomat anti-ship missile, developed by Matra (France) in collaboration with OTO Melara (Italy), and fitted in numerous warships (**1336.221**).

The OCDS is composed of one or more Otomat coastal defence units (OCDUs) carried on cross-country vehicles. Each OCDU comprises a command group, and two to four firing groups. The command group includes one control cabin, a surveillance radar (each carried on its own truck), one command car, and an operational maintenance vehicle. The firing group consists of one firing vehicle with two missiles in containers, and one reload vehicle fitted with a hoist and two spare missiles in containers. One command group can control up to four firing groups. The basic vehicle used is the six wheel drive truck RVI GBD 6 × 6 fitted with sand-duty tyres and having an empty weight of ten tons. Maximum road speed is 75 km/h and range 1000 km.

The command vehicle houses the tactical display console, missile control panel, various communications equipments and the telecommand transmitter. The console enables the tracking of several targets, and the control panel permits the setting of initialisation data into the missiles' computers and authorising their firing by the firing vehicles. The latter vehicles are equipped with a north-seeking gyro-compass to assist initial survey and battery setting up, and an optical sight for reversionary mode use.

Otomat launch from coastal defence launcher vehicle. Ranges up to 180 km can be obtained

VHF radio links are provided between the vehicles permitting flexible dispersion of the OCDU. The problem of over-the-horizon target designation has been solved by employing an external radar, either a relay ship, helicopter or fixed-wing aircraft. The relay platform measures target bearing and range and transmits them via data link to the control cabin. The latter measures range and bearing to the relay plaform and computes the co-ordinates with respect to the firing vehicle. A course correction transmitter allows for in-flight updating of the co-ordinates. Airborne target detection makes it possible to reach extended ranges, thus permitting a single OCDU to cover 360 km of coast line.

An Italian version of the Otomat coastal defence system was revealed at the Mostra Navale exhibition in May 1982. In this version an Otomat coastal battery consists of six OTO C13 tracked vehicles on each of which are two Otomat missiles housed in glass-fibre launcher containers. These are linked by cables or radio to a command post vehicle on another C13 chassis, from where, it is reported, one man can control the entire battery using an SMA MM/SPQ-2F search radar mounted on top of the command post. There can be up to 20 km between the command vehicle and the others of the battery, and it is also possible to employ a naval helicopter such as the Agusta AB 212 equipped with Otomat over-the-horizon guidance facilities for engagement of distant targets. In such a role a range of more than 160 km is claimed.

STATUS

The Otomat coastal defence system is in operational service with the Egyptian Navy, and several others are contemplating acquisition of the system.

CONTRACTORS

Matra SA, 78140 Vélizy, France.

OTO Melara, La Spezia, Italy.

JAPAN

6724.121

SSM-1 COASTAL DEFENCE MISSILE

Development of a ground-launched version of the ASM-1 air-launched anti-ship missile (**1635.311**) known as the SSM-1 is currently under way. During the summer of 1986 full distance firing trials with live warheads were carried out by the Japanese Defence Agency's technical research and Development Institute at the US Navy's Pacific Missile test centre at Point Mugu, California, which was hired at a cost of $6.4 million.

NORWAY

3074.121
PENGUIN COASTAL DEFENCE SYSTEM

The Penguin surface-to-surface anti-ship missile (**1339.221**) is reconfigured to serve in a number of coastal defence roles.

Several alternatives are available such as a truck-mounted self-contained mobile coastal defence unit, fixed coastal battery or as light transportable units for helicopter or truck transport. A missile battery would typically consist of 2 × 4 missile positions and a containerised fire control system including a surveillance radar. The missiles can be launched using data supplied from either the fire control unit or external sources. The missiles can also be launched from sites in-land and set to overfly elevated terrain thus making use of the natural cover.

CHARACTERISTICS

Number of missiles: 4, or more

Missile Data

Weight: 385 kg

Length: 300 cm

Wingspan: 140 cm

Diameter: 28 cm

Speed: High subsonic

Warhead: 120 kg

Fuze: Impact

Trajectory: Variable

Homing: Passive IR terminal

Range: 30 km

DEVELOPMENT

Penguin was originally developed as a ship-to-ship weapon system for the Royal Norwegian Navy by the Norwegian Defence Research Establishment and A/S Kongsberg Vapenfabrikk.

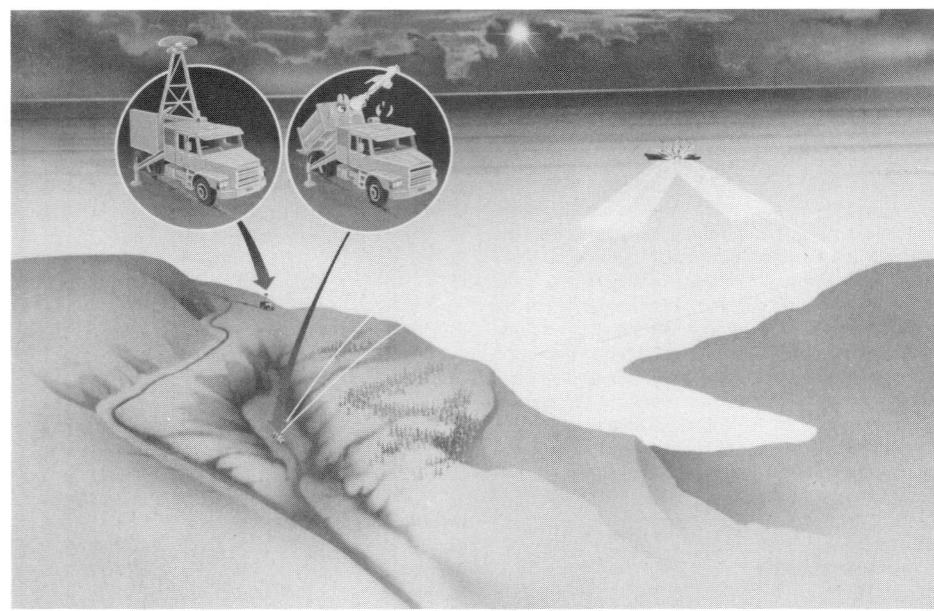

Schematic diagram of coastal defence system based on vehicle-mounted surveillance radar and Penguin launching units

STATUS

Studies of various configurations of coastal defence systems based on the Penguin missile have been conducted for several years by the Norwegian authorities, but no order has been placed.

CONTRACTOR

Norsk Forsvarsteknologi A/S, PO Box 1003, N-3601 Kongsberg, Norway.

SWEDEN

2372.121
RB 08 COASTAL DEFENCE WEAPON

The Swedish RB 08 was the first western anti-ship missile (**2366.221**) is intended for use in coastal defence as well as the anti-ship role. It is based upon the Nord CT 20 target drone.

CONTRACTOR

SAAB Missiles AB, S-58188 Linköping, Sweden.

RB 08A boosted off its launcher

6185.121
RBS 15 COASTAL DEFENCE WEAPON

A truck-mounted coastal defence version of the RBS 15 anti-ship missile (**3976.221**), the RBS 15KA, is being developed by Saab Missiles under a SEK 50 million contract initially to meet a Swedish requirement.

The missile will be identical to the ship-launched version with four launch containers (each weighing 1530 kg) carried on a 6 × 6 truck. Two launch vehicles and a fire control vehicle will form a platoon and two platoons, together with a battery command post and supporting units will form a battery. The fire control equipment will be mounted in a shelter on a 4 × 4 truck and will include a missile fire control panel with alphanumeric keyboard, a target data panel and a communications panel. The battery command post supplies target information through a data link although batteries and platoons can include their own surveillance radar. Launch orders for a pre-programmed target can be issued within ten seconds. Each launcher vehicle can be deployed up to 500 m from the fire control vehicle to which it is linked by land line.

Launch vehicles will normally spend only five minutes at the firing location and they can be reloaded within an hour.

Launch vehicle with four RBS 15 anti-ship missiles

STATUS

Contracts for both the Swedish version (RBS 15KA) and the export version (RBS 15CD) were signed in 1986 and development is well under way. Finland is buying the RBS 15CD version.

CONTRACTOR

SAAB Missiles AB, S-58188 Linköping, Sweden.

4878.121
RBS 17 HELLFIRE SHORE DEFENCE SYSTEM

In late 1984 Rockwell International was awarded a $7.7 million contract by the Swedish Government to adapt the Hellfire modular missile, already developed for helicopter operation (**1391.311**) and battlefield use (**4877.111**), for the shore defence role. It is designed to attack amphibious warfare and mine countermeasures vessels up to 10 km from the coast. Under this two-year adaptation phase, Rockwell was the prime contractor with responsibility for systems integration in collaboration with Swedish industry. Successful development and operational tests have been completed.

Bofors has developed a new blast/fragmentation warhead and post impact delay fuze for the missile, a 23 kg lightweight storage container together with a new 40 kg launcher which can break down into two parts. Rockwell are responsible for system integration and the manufacture of the missile and control electronics in which the guidance section has been modified to adjust the flight trajectory. The modified Hellfire, given the Swedish designation RBS 17, weighs 49.3 kg.

The launcher can be deployed and be operational within five minutes and is set at a fixed elevation of 10° when the missile seeker is locked onto the target before launch or at 20° when fired from a concealed position for a lock-on-launch engagement.

Two launchers, a laser designator and launch control equipment capable of handling the missiles in quick succession make up a battery.

STATUS

Development was completed in 1986 and in 1987 an order for 1000 rounds was placed by the Swedish government. Deliveries will take place between 1989 and 1990.

CONTRACTOR

Rockwell International, Missile Systems Division, Defence Electronics Operations, 1800 Satellite Boulevard, Duluth, Georgia 30136, USA.

A RBS 17 Hellfire being launched

TAIWAN

4442.121

HSIUNG FENG COASTAL DEFENCE MISSILE

The Hsiung Feng missile is only one component within a sophisticated mobile coast defence system. The complete system, thought to be the Honeywell H930 Mod 3, consists of the triple launchers, fire control system, search radar and a Command and Control centre, all carried in semi-trailers. A high degree of mobility enhances the system's chances of survival and also allows a small number of systems to defend a comparatively lengthy coastline. In time of war the Hsiung Feng system would probably be moved to prepared dug-in sites.

There is evidence that locally developed and manufactured subsystems or components are being incorporated into the system. A locally developed I/J-Band surface search radar, the CS/UPS-60(X)C is said to be employed with mobile coast defence units, and can be trailer-mounted. This radar incorporates many ECCM features including side-lobe cancelling, and is therefore ideal as a surface-search and target-indication set for the Hsiung Feng system.

Two mobile coast defence missile units with triple launchers for Hsiung Feng anti-ship missiles

For target tracking and illumination the system uses an RCA R-76 radar. A TV tracker co-mounted with the R-76 antenna, and a separate optical sight provides for back up optical tracking. In this mode an engagement can be carried out in radar silence up until the terminal homing phase. This type of attack may prove particularly effective, given the Hsiung Feng's sea-skimming profile (20 nm cruise and 4 nm terminal attack) and its approach from a land back ground. The victim would probably be unable to detect the attack until too late.

A locally produced fire control radar, the CS/SPG-24 which closely resembles the RCA R-76 (perhaps a licence-built copy) is thought to be associated with the fire control for Hsiung Feng. If this is in fact the case one would expect SPG-24 to replace the US-supplied R-76 as the fire control radar in all future Hsiung Feng systems.

As for the optical tracking systems, it is likely that the H930 Mod 3 incorporates some sort of infra-red night vision device. Local production capability in this field has been demonstrated by the introduction of a thermal imaging coastal surveillance system, CS/KAS-4A.

STATUS

The system is in service with the ROC (Taiwan) Army.

CONTRACTORS

State Arsenals.

Sun Yat Sen Institute (seeker).

UNION OF SOVIET SOCIALIST REPUBLICS

2951.121

SAMLET CRUISE MISSILE (SSC-2B)

Like Salish, the missile known to NATO as Samlet and SSC-2B in the American system (Soviet designations are understood to be 4K-87 and S-2 Komet), is a surface-to-surface version of the air-to-surface jet-powered cruise missile whose NATO code-name is Kennel (AS-1). It has been operational in the USSR, Egypt, Poland, and Cuba as a coastal defence weapon. When shown publicly Samlet has been mounted on what appears to be its launching ramp constructed as a trailer. But other photographs indicate that a more permanent launch emplacement also exists. For launching a JATO bottle is necessary.

CHARACTERISTICS

Length: Approx 7 m

Wing span: Approx 5 m

Weight: Believed about 3 t

Cruising speed: Mach 0.8 – 0.9

Range: 60 – 90 nm with mid-course guidance.

STATUS

An undisclosed number of Samlet missiles was reported to be operated by the coastal frontier force of the German Democratic Republic Navy and the Polish forces. These have been supplanted by the SSC-3 (**4890.121**) coastal defence missile, below. Reported to be no longer operational in Soviet service.

Samlet coastal defence missile preparation

2975.121

SEPAL MOBILE CRUISE MISSILE (SSC-1B)

Sepal is the code-name (SSC-1B in the American system) for the land-based version of the shipboard Shaddock surface-to-surface weapon. (The Soviet designation is reported to be P35). Too little has been seen of this missile on public display for it to be described with confidence. The general view seems to be that it is a ramjet or turbojet-powered cruise missile with hinged wings that open out when the missile leaves its cylindrical launcher. Initial boost is provided by two rocket units under the rear of the fuselage. Length of the missile is estimated at about 10 m and the diameter of the fuselage is probably around 1 m.

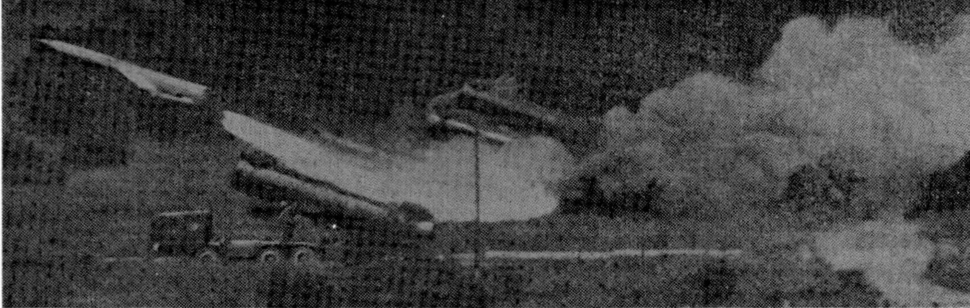

Launch of Sepal coastal defence anti-ship missile

Shaddock is best known as a naval weapon and is described as in entry **2976.221**. It appears, however, that it is used by the land forces of the USSR, and this view is reinforced by the appearance of the container/transporter illustrated here which is clearly capable of elevating the missile to an inclined launch position.

While no doubt it could be used as a ground-to-ground weapon, however, it seems rather more likely that it would be used as a coastal defence weapon, if only because of its ability to cruise at speed at low altitude over water.

Guidance is by radio-command with active radar homing. Range is in the order of 450 km, but mid-course guidance by aircraft or helicopter would be necessary for such a long mission. Each coastal defence battalion is reported to be equipped with between 15 and 18 missiles (inclusive of reloads).

STATUS

Unofficial but reliable sources have suggested that there are probably about 100 Sepal SSMs operated by the Soviet Union Navy's coastal artillery and rocket troops. These are thought to protect principal naval bases and main ports.

4890.121

SSC-3 and SSC-4 COAST DEFENCE MISSILE SYSTEMS

Certain Warsaw Pact countries in 1984-85 began operating a new type of coast defence missile system, but in common with most Pact weapons it appeared in public without any known identification although it is reported to be designated P-15. As yet, no NATO unclassified code-name has been assigned. The only two Pact states known to operate the system are East Germany and Yugoslavia, the latter identifying it as BROM (*Baterije Raketa Obala, More*). In view of the fact that the transport system is based on the MAZ-543 prime mover which is manufactured in the Soviet Union (as probably is the associated radar sub-system) it is likely that this system is used by Soviet Forces also, although no details of deployments have been obtained.

The missile employed is either a standard Styx anti-ship missile (**1155.221**) in its folding-wing SS-N-2B version or the coastal defence model as it is encountered in the Chinese HY-2 system (**3984.121**), or a near derivative of one or all of these (possible Soviet designations applied to missiles of this family are 4K-40 and P-15). From what is known of these systems, the SSC-3 is expected to have a maximum range of 80–90 km at high subsonic speeds, cruising at heights between 15 and 100 metres above the sea surface. Guidance options include active radar homing, with or without a radar altimeter, and passive IR homing. Other details of the missile itself are probably essentially the same as for the parent system(s).

On its own, the choice of a MAZ-543 based vehicle as the means of transporting the SSC-3 system ensured an impressive appearance, with the search and target designation radar equipment, operating/control cabin, and two missiles in individual container/launchers on a single 8 × 8 wheeled chassis. If performance characteristics confirm this, the system's originators will have achieved a notably cost-effective solution.

From photographs of the SSC-3 system deployed and on the move it is possible to speculate on the likely construction and mode of operation. On the eight-wheeled MAZ chassis, immediately behind the driver's cab is a large enclosed housing that contains the radar and electronic equipment and is assumed to accommodate the operations cabin. Recessed into the forward side of this container is the radar scanner mounting consisting of a telescopic mast which is surmounted by the turning gear and the scanner itself. The last of these is a typical coastal radar scanner with a sheet metal reflector about 2.4 metres wide illuminated by an overhanging feed. Operation is presumed to be in the X-band and the principal function surface target detection. In the lowered position for travelling, the scanner faces the front and is positioned over the driver's cab. In this position, the radar only increases the vehicle's overall height by a few centimetres.

The rear portion of the chassis carries a turntable on which is mounted either one large missile container/launcher housing for two Styx-type anti-ship missiles or 'Siamese twin' launcher/container for the missiles. Each missile launch ramp housing has its own doors front and rear which opens upwards to lie flat on top of the container. When deployed for operation, the missile container is slewed by the turntable to roughly 90 degrees of the vehicle's longitudinal axis and elevated by 15 degrees or so for firing. The reloading operation is probably a variation of that employed for Styx missiles fitted in small naval vessels, ie with the aid of a small crane and manhandling into each launcher cell, possibly with some power-assisted mechanical devices.

Apparently the complete operating crew of the SSC-3 does not travel with the carrier vehicle when on the move, as the driver's cab is too small, and it seems likely that an accompanying truck forms part of a complete operational SSC-3 unit. Indications are that in Yugoslavian service a full crew consists of ten men, of which perhaps half their number are naval personnel for management and operation of the missile system, one is the commander, and the others are the soldier vehicle crew. The East German manning arrangements are not known.

CHARACTERISTICS

Length: Approx 7 m

Guidance: Inertial + terminal homing

Range: 80–90 km

Warhead: Nuclear

STATUS

The hardware although undoubtedly fresh to Western eyes, is by no means intrinsically new, though the combination may be. Therefore, this and the evidence of sufficient quantities for appearances in ceremonial parades in East Germany and Yugoslavia, would seem to warrant the assumption that the operational

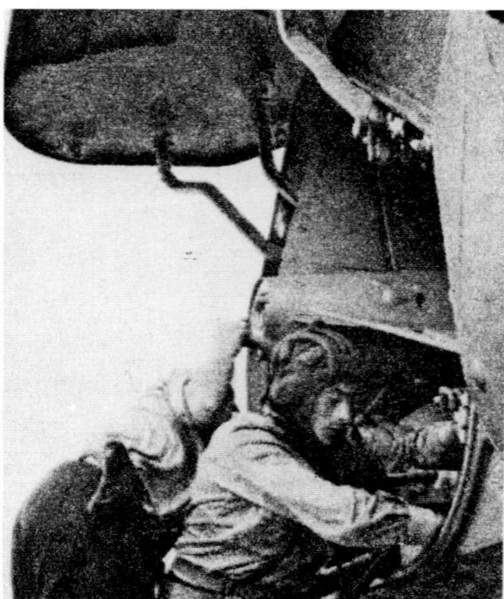

Crew members prepare SSC-3 coast defence missile for firing

SSC-3 coastal defence missile system ready for operation, with target detection radar in elevated position

stage has already been reached. Production lines for the missile employed are known to exist in a number of countries (within the Warsaw Pact and elsewhere), and the radar is not thought likely to incorporate or require any exceptionally advanced technology, so that the only limitation on production numbers is the capacity of the MAZ chassis facilities in Minsk.

Operational deployment with Yugoslavian and East German coastal defence forces can be assumed to have begun in the 1984-85 period, in the latter case possibly as replacements for the SSC-2B Samlet missiles previously employed in that role. The extent of Yugoslav procurements and deployment is not known, and neither is there any indication of Soviet use of the SSC-3 system.

In July 1985 it was reported that deployment of the SSC-3 had started in Syria, where a coastal defence regiment was said to be operating the system in the Latakia region.

Its successor is thought to be the SSC-X-4, a derivative of the SS-N-21 and AS-15 cruise missiles.

Two SSC-3 coastal defence missile units during a ceremonial parade. Front housing contains radar gear and operations cabin; rear housing is on a turntable and contains two Styx anti-ship missiles on separate launch ramps. Note radar scanner in lowered position for travel

UNITED KINGDOM

6186.121

EXCALIBUR COAST DEFENCE MISSILE SYSTEM

Excalibur is based on the MM-38 Exocet mobile missile battery designed and built for the Royal Navy, but can be readily adapted to any new or surplus SSM of the customer's choice. The system comprises a control cabin, a stores and accommodation cabin, two radar/launcher trailers and two mobile diesel generators, making it fully self-contained. It can be transported on its own trailers by land, sea or air to a coastal site.

The control cabin contains the missile fire control system, radar displays for target indication, gyro displays and system interfaces, secure radio facilities and power-supply conversion equipment. The stores and accommodation cabin contains ready-use spares to support the system and limited domestic facilities for the crew during the deployed period. The radar/launcher trailers carry two missiles each, with a hydraulically-operated ramp to permit accurate setting of the firing angle or to compensate for operation on a sloping site. The radar mast is hydraulically extended.

One system, using MM-38 missiles from the stock held for surface warships, was installed by the RN at Gibraltar in 1985, replacing obsolete gun batteries.

Excalibur coastal defence missile system on trial at RAE Aberporth (UK MoD/RAE Aberporth)

CHARACTERISTICS

Radar/launcher trailer

Length: 6.7 m
Width: 2.5 m
Height: 2.4 m
Weight: 10 300 kg (loaded)
Length of cabins: 9.2 m
Width: 3.3 m
Height: 2.4 m
Weight combined: 30 500 kg (max)
Length of generator: 3.3 m
Width: 2.1 m
Height: 2 m
Weight: 2350 kg

STATUS

Operational with Royal Navy (1985).

CONTRACTOR

Vosper Thornycroft (UK) Ltd, Woolston Works, Woolston, Southampton SO9 5GR, UK.

LAND-MOBILE SURFACE-TO-AIR WEAPONS

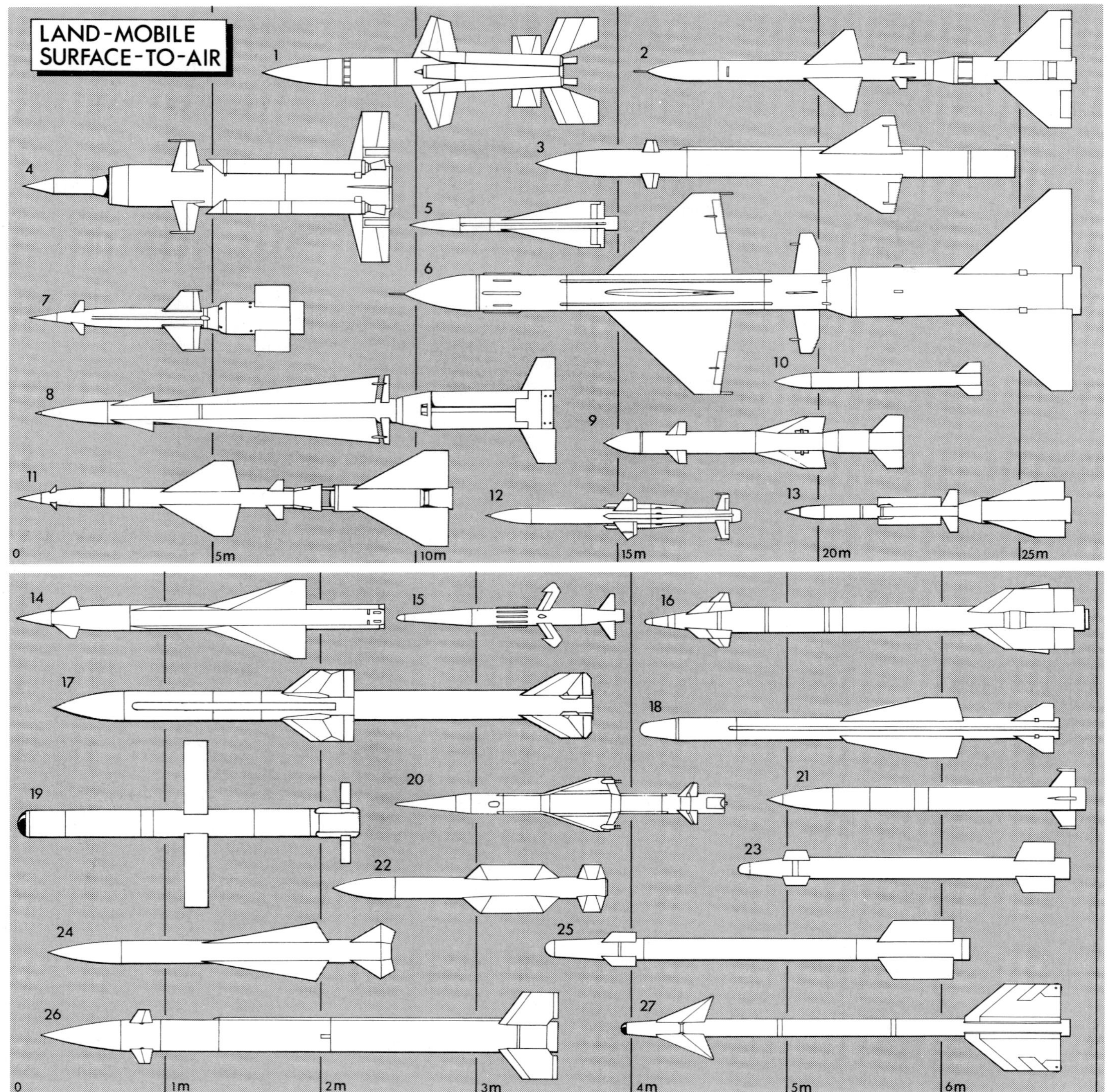

1 Bloodhound, **2** Guideline (SA-2), **3** Guild (SA-1), **4** Ganef (SA-4), **5** I/HAWK, **6** Gammon (SA-5), **7** Goa (SA-3), **8** Nike Hercules, **9** SA-12 Gladiatur, **10** Patriot, **11** HQ-2B, **12** Gainful (SA-6), **13** Aster 30, **14** Roland, **15** RBS 70, **16** Crotale, **17** Aspide (Spada), **18** Tan-SAM (T81), **19** FOG-M, **20** Rapier, **21** ADATS, **22** SA-11 (provisional), **23** SA-9 Gaskin (provisional), **24** SA-10 (provisional), **25** SA-13 (provisional), **26** SA-8 Gecko, **27** Chaparral

2290.131
INTRODUCTION

Described in this section are weapons having a wide range of sizes and performance capabilities. All are suitable for use against hostile aircraft, and a few also have an anti-missile capability. All are to some extent mobile, and some are man-portable, but some of the larger systems take several hours to bring into action and are better suited to a semi-static defence role (eg of airfields or base installations) than to operations in defence of forward units.

Among the more recently developed systems there has been a noticeable emphasis on fast-reaction short-medium range weapons for defence against low-flying aircraft and tactical missiles; there have also been interesting developments in man-portable weapons capable of giving infantry units a degree of self-protection against aircraft attack, which they have never previously known.

For convenience of system comparison, these mobile systems have been divided into three sub-sections – one dealing with mobile (but not portable) guided weapon systems, one dealing with mobile gun and rocket systems, and a third dealing with portable guided missile systems. While the first and third sub-sections are concerned with guided weapons which can be unambiguously distinguished from unguided weapons, there is some difficulty in defining the border between self-contained AA gun systems – which are described in the second sub-section below – and associations of AA guns with separate fire control systems. While we have endeavoured to follow a consistent line here, the reader may find it useful to consult Army Fire Control Systems in the Systems section, as well as the Land-based Air Defence Systems section where incidental references to some specific surface-to-air weapons may be found.

Mobile Surface-to-air Guided Missile Systems

CHINA (PEOPLE'S REPUBLIC)

6720.131

HQ-2B AIR DEFENCE MISSILE

HQ-2B is a mobile surface-to-air air defence missile weapon system derived from the HQ-2J (**4889.131**) which is intended to attack targets at medium and high altitudes. The system has a powerful propulsive engine and strong radio controller power. It adopts anti-jamming code and FM phase comparison fuze. The system has a high effective, all-weather and omni-bearing operation ability in the area of 7-40 km.

The sweep return monopulse system and ranging radar with 3 cm wave length has been adopted in the guidance station and also uses TV tracking system and display device for the manoeuvring targets. There are several guidance modes used in the system so as to fire high speed, manoeuvring targets as well as targets with a jammer in low altitude.

Its launching device which is fitted in a tracked vehicle has a strong manoeuvring and cross-country ability. It carries its own power supply and has the function of automatic level adjustment and function check-up and travels with loaded missiles.

CONTRACTOR

China Precision Machinery Import & Export Corp, 17 Wenchang Hutong Xidan, Beijing, People's Republic of China.

The HQ-2B missile

4889.131

HQ-2J AIR DEFENCE MISSILE

For many years China has used the Soviet SA-2 Guideline (**2942.131**) SAM. The HQ-2J is believed to be a reverse engineered SA-2 although Chinese sources insist it is a new nationally developed missile. In addition to being in service with the Chinese armed forces it is also available for export.

The actual missile comprises two stages in tandem, the sustainer stage (stage II) and the booster stage (stage I). Its effective range is 4-36 km. The warhead is however an improved design with multiple fragments and a large scattering angle.

The complete system comprises the missiles, guidance system, launchers and ground support equipment.

STATUS

Early in 1988 CPMIEC revealed an improved version of the HQ-2 system was under development with the capability of engaging multiple targets. There is no indication when it will enter service.

CONTRACTOR

China Precision Machinery Import & Export Corp, 17 Wenchang Hutong Xidan, Beijing, People's Republic of China.

The Chinese HQ-2J is now offered for export

6418.131

HQ-61 SURFACE-TO-AIR MISSILE SYSTEM

The HQ-61 SAM system has been designed to meet the requirements of both the Chinese Army and Navy and in some respects the missile used by this system is very similar to the US Sparrow air-to-air missile but is larger and heavier.

The system has been designd to engage targets flying at medium and low altitudes with a minimum range of 3 km and a maximum range of 10 km in the horizontal plane extending in an arc up to an altitude of about 8000 metres.

the missile itself has a single-stage solid propellant motor, a chain type high explosive warhead, dual fuze system, homing head using a CW semi-active homing guidance system which is said to have an anti-jamming feature, autopilot, airborne power supply using a combustion turbine generator to supply AC and DC power for all missile components and four hydraulic servo operated canards.

The army version of HQ-61 consists of four 6 × 6 launchers each with two missiles mounted on a turntable at the rear of the vehicle, mobile generators, command post vehicle, tracking and illuminating radar vehicles, and a target indicating radar vehicle also based on a 6 × 6 chassis which is very similar in appearance to the Soviet designed Flat Face target acquisition radar used with the SA-3 Goa system.

A typical target engagement would take place as follows. The target would first be detected by the target indication and radar vehicle and after confirmed as hostile, tracked and illuminated by the rdar vehicle. When the target is within range a missile is launched. It would appear that the system can only engage one target at a time.

HQ-61 missile being launched during trials

CHARACTERISTICS
Length: 3.99 m
Diameter: 28.6 cm
Maximum span: 1.166 m
Maximum speed: Mach 3
Weight at launch: 300 kg

STATUS
The HQ-61 was first revealed in public late in 1986 at which time it had already been under trials for several years. It is expected to enter production in the near future.

CONTRACTOR
China Precision Machinery Import & Export Corp, 17 Wenchang Hutong Xidan, Beijing, People's Republic of China.

Close-up of the HQ-61 launcher showing two missiles in ready to launch position (Christopher F Foss)

FRANCE

4771.131

SAAM SURFACE-TO-AIR MISSILE SYSTEM

Studies have been undertaken by leading French contractors since 1982 into the design and development of a new surface-to-air missile system that could be suitable for use as either a land-based air defence weapon or in a shipboard configuration for naval applications. Provisional designations of the proposed new systems are SAMP (medium range surface-to-air) for the land-based version (formerly SA 90), and SAAM (surface-to-air anti-missile) for the naval model (formerly SAN 90).

The principal collaborators in the study, Aérospatiale and Thomson-CSF, have signed a memorandum of understanding with the French Ministry of Defence in 1983 providing for feasibility studies. Aérospatiale will be responsible for the missiles and the launcher(s), with Thomson-CSF responsible for Arabel phased array, multi-function radar and the guidance system (including fire control).

As noted above, two versions of the system are planned: a future naval surface-to-air anti-missile system (SAAM) using the Aster 15 missile; and a land mobile medium-range surface-to-air system (SAMP) with the Aster 30 missile.

The threat to be addressed includes saturation air attacks with aircraft and missiles, at all altitudes and from multiple azimuth angles. The targets may be expected to be capable of very high manoeuvring capabilities, calling for a defensive missile able to defeat attacking missiles capable of up to 15 *g* and speeds of up to Mach 2.5 or more. Dense and varied ECM and other countermeasures are expected, and high resistance to such challenges is essential, as is maximum autonomy of operation against multiple threats.

The ability to intercept several targets simultaneously and provide a short reaction time will be facilitated by use of the new X-band 3-D phased array radar developed by Thomson-CSF. This will simultaneously carry out surveillance, detection, identification, threat evaluation, designation and tracking of targets, in addition to up-dating the missile inertial navigation system. Rotating at 60 rpm, this radar will incorporate frequency agility, and will be compatible with a missile launching rate of up to one per second. It is possible that electro-optical sensors may be added as a radar back-up.

The Aster missile has been designed for exceptional manoeuvrability; three times that of potential targets is the estimated requirement, ie about 50 *g*. It will have a hybrid control system incorporating aerodynamic and direct-force control elements. Missile guidance will consist of two steps: the initial guidance phase during which the missile heads toward its target under the control of its on-board inertial system, and the terminal phase during which it will use a K-band active radar seeker.

There are two versions of the missile with the same basic missile but with different boosters to meet different range requirements: Aster 15 is for point defence and Aster 30 is for area defence. In each case the Aster basic missile is optimised for engaging maneouvring supersonic missiles and multiple targets. It has four control fins at the rear and lateral thrusters.

An artist's impression of the projected SAMP medium range air defence system including Aribel X-band phased array multi-function radar together with the fire control unit and vertically launched Aster 30 missile

SAMP

Designed to protect high value targets against saturation air attacks, cruise missiles or anti-radiation missiles, the Aster 30 will be a two-stage, vertically-launched missile with an interception range of 25-30 km. It will be designed to engage high speed, manoeuvring targets.

The complete system will consist of a fire control unit, a radar unit with retractable antenna and four mobile fire units each with an eight-round vertical launcher. Each unit will be mouted on a truck capable of cross-country performance and it is intended that a unit should be operational within ten minutes of arriving on a site.

STATUS
Preliminary studies were carried out between 1982 and 1984 leading to a feasibility study contract being awarded to the two partners by the French Defence Ministry in mid-1983. The following year Thomson-CSF were awarded a contract worth the equivalent of US$300 million to develop an Arabel radar and a fire unit while Aérospatiale received a contract for the development of the basic Aster missile.

Missile firing tests began late in 1986 and during 1988-89 there will be an intensive test programme of the Arabel radar and fire unit control at the French CEL test range.

By late 1987 several missiles had been fired leading to a proof-of-principle firing in June 1987. Full scale development is planned to begin by late 1988.

CONTRACTORS
Thomson-CSF, Division Systèmes Electroniques, 1 rue des Mathurins, 92223 Bagneux, France.

Aérospatiale, Division Engins Tactiques, 2 rue Beranger, 92322 Chatillon, France.

6752.131

MISAT SURFACE-TO-AIR MISSILE

At the Paris Air Show in 1987 Aérospatiale and Matra revealed they were developing a 'super velocity' missile, MISAT. MISAT is intended to engage targets which appear briefly and it is intended to be compatible with Crotale/Shahine and Roland using the same command-to-line-of-sight principle used in the Roland system. A new rocket motor is being developed by Aérospatiale which will make it faster than ADATS ie. in excess of Mach 3. MISAT is intended to be available in the late-1990s.

STATUS

Under development.

CONTRACTORS

Aérospatiale, Division Engins Tactiques, 2 rue Beranger, 92322 Chatillon, France.

Matra SA, ave Louis Bréguet, 78140 Vélizy, France.

2074.131

CROTALE LOW-ALTITUDE GROUND-TO-AIR WEAPON SYSTEM (TSE 5000)

Crotale is the basis of a series of low-altitude surface-to-air defence systems designed for the all-weather interception of low-level targets. The system can be mounted on wheeled or tracked vehicles, on transportable shelters or on ships (see also **3972.131, 2111.231**). When vehicle or shelter housed, the complete system comprises an acquisition and co-ordination unit (ACU) linked with two to three fire units. Each fire unit is equipped with a launcher carrying four ready-to-fire containerised missiles and a tracking and guidance radar. The system is intended to counter saturation attacks on strategic or tactical positions or to ensure the protection of battlefield units.

The short reaction time (6.5 seconds to first engagement and 4.5 seconds re-engagement reaction time) required of systems in this category is achieved by means of sophisticated automatic data processing and handling of the operational actions to be performed. The system is broken down into two units to maximise the efficiency of the target designation, target signal processing and detection system by using it in association with several fire units. The automatic processing system is also capable of co-ordinating several air defence systems (eg Crotale, AA guns etc) when saturation attack protection calls for deployment of defence resources in depth, while not increasing risks of either over- or under-kill.

Crotale is installed on wheeled cross-country vehicles that employ electric transmission, and are of two types: the ACUs which control and co-ordinate up to three fire units and form part of an anti-aircraft defence system; and the fire units. The former vehicles are each equipped with a pulse-doppler surveillance radar with fixed echo suppression; a data processing system based on a digital computer which can handle 30 targets, automatically evaluating and tracking the 12 most dangerous targets and designating these to the most appropriate Crotale fire units or to artillery fire control posts; an IFF system; and an inter-ACU co-ordination system. The fire units each have: a turret launcher with four Crotale missiles in containers (ready-to-fire); a monopulse fire control radar that automatically tracks the target and guides one or two missiles, using a very accurate differential angle-error measurement system; a digital computer; TV equipment which is used in clear weather or under jamming conditions for automatic guidance of the missile without radar transmission; an IR angle-error measurement system for aligning the missile on the line of sight to the target immediately after launch (radar beam gathering); and a command link that transmits manoeuvre orders to the missile.

Crotale in operation. Note ACU to right of fire unit launching missile

All the vehicles are provided with an intervehicle communications system using cable and/or radio for the exchange of firing orders and data transmission, and a separate VHF radio system.

The canard-type Matra missile is powered by a powder propulsion unit, and the rounds are housed in watertight containers which are used for transport and launching. The focused-splintering warhead weighs about 14 kg and is activated by a protected IR proximity fuze adjusted to operate according to the intercept geometry computations or on impact. Its lethal range is more than 8 m, which is stated to be greater than the very low clearance distance provided by the accuracy of the angle-error measurement system. The system is therefore claimed to provide a kill probability of greater than 0.8 (0.96 for salvo firing).

Latest production version of the Crotale firing unit with microwave intervehicle link in raised position

CHARACTERISTICS

Type: Land-mobile, automatic, all-weather, surface-to-air guidance weapon system

Guidance principle: Command to line of sight

Guidance method: IR missile gathering; secondary radar missile tracking by target tracking radar; command control of missile control surfaces via autopilot

Propulsion: Single-stage solid propellant motor

Warhead: 14 kg, HE, directed burst, fragmentation type with IR or electro-magnetic proximity fuze

Missile length: 2.94 m

Missile diameter: 16 cm

Span: 54 cm max

Launch weight: Approx 80 kg

Speed: Mach 2.3 reached in 2.3 s. 20 s time of flight to 8 km

Acquisition unit

Radar: E-band, tunable

Sub-clutter visibility: 60 dB

Target capacity: 30 processed per antenna revolution

Antenna rotation rate: 60 rpm

TWS: Automatic, 12 targets

Detection: 1 m^2 fluctuating target within 18.5 km

Intervehicle communications: up to 800 m cable; up to 3 km radio link

Intersector co-ordination: 4 sections separated by up to 10 km link between ACUs

Fire unit

Radar: J-band monopulse in diversity mode

Tracking: 1 m^2 fluctuating target within 16 km

Tracking accuracy: 0.3 mrad

Angle-error measurement accuracy: 0.2 mrad

TV tracking: 14 km range (max); field 3°

Typical target data

Speed: Up to Mach 1.2

Altitude: 15 – 5000 m

Radar cross-section: 1 m^2, fluctuating

Max engagement range: 11 km (50 m/s target); 10 km (250 m/s target)

Minimum range: 500 m

Max crossing range: Up to 6000 m against a 200 m/s target

Reaction time: 6 s from first detection to missile launch

DEVELOPMENT

Work started on the project in 1964 both at Thomson Houston and Engins Matra and a prototype missile was completed in the following year. Overall responsibility for the development lies with the Division Systèmes Electroniques of the Thomson-CSF group while full responsibility for the missile lies with Matra.

STATUS

In production. In service with Chile, Egypt, France, Libya, Pakistan, Saudi Arabia, South Africa, United Arab Emirates and three unidentified countries of which Iraq is believed to be one.

The Crotale weapon system has been developed into at least two specific export versions. Cactus (**2341.131**) was supplied to South Africa some years ago, at an early stage in Crotale's development. More recently, another version was designed and has been supplied to Saudi Arabia as the Shahine. This is described in the following entry (**3972.131**). There is also a naval version (**2111.231**) which appears in the Shipborne Surface-to-air Weapons section of this volume. Crotale in vehicle or shelter mounted configuration has four ready-to-fire missiles. Shahine has six rounds ready-to-fire in vehicle or shelter version, and Naval Crotale has eight ready-to-fire missiles. The shelter mounted version of Crotale has yet to enter production.

CONTRACTORS

Thomson-CSF, Division Systèmes Electroniques, 1 rue des Mathurins, BP 10, 92223 Bagneux, France.

Matra SA, ave Louis Bréguet, 78140 Vélizy, France.

3972.131

SHAHINE AIR DEFENCE SYSTEM (TSE 5100)

Shahine is a later, more powerful version of the Crotale air defence system (**2074.131**) which has been supplied to Saudi Arabia.

The system combines one or two acquisition and co-ordination units (ACUs) with two, three or four firing units each carrying six containerised ready-to-fire missiles with an effective interception range of more than 11 km. Its capability of controlling up to four independent but simultaneous interceptions, and its versatile, modular design, make Shahine an efficient weapon system against saturation air attack.

The Shahine system is a modular system which can be mounted on the AMX-30 type tank chassis or any other tank similar to the ones to be protected; wheeled armoured vehicles; or (air) transportable shelters for the protection of fixed installations. The last version can be moved by C 160 Transall or C-130 Hercules aircraft.

The system consists of:

(1) *ACUs* which can control and co-ordinate up to four firing units each, be co-ordinated with other ACUs and integrated with an air defence network. Each of them is equipped with a pulse-doppler acquisition radar with moving target indication (MTI); a data processing system based on a digital computer providing the simultaneous display of 40 targets, automatic threat evaluation and tracking of the 12 most menacing targets and their designation to the best located firing units; a TV system featuring a TV turret concentric with and independent of the radar turret which provides ground monitoring of moving fire vehicles and optical target reconnaissance and possibly target designation

(2) *firing units* equipped with a turret carrying six containerised ready-to-fire missiles; a monopulse doppler fire control radar providing automatic target tracking and simultaneous guidance of one or two missiles with precise deviation measurement; a digital computer to manage the system; TV equipment designed to perform automatic tracking and missile guidance without any radar transmission, in clear weather or in case of electronic jamming; and an IR localiser to gather the missile on the firing axis after launching.

All vehicles are fitted with a digital link for locating all the units of a Shahine formation and for the automatic exchange of data and orders, and there are radio links between Shahine units and external command posts and air defence centres.

The missile is of essentially the same concept as that used in the Crotale system described above, but in the Shahine application it is slightly heavier at a total weight of 105 kg. A more detailed description of Shahine appeared under this entry number in earlier editions of *Jane's Weapon Systems*.

CHARACTERISTICS

Acquisition range: 19.5 km against 1 m^2 fluctuating target

Tracking range: 17 km

Service range: 0.5 – 11 km against 250 m/s target

Vehicles: Cross-country tracked type or heavy duty wheeled vehicles

Reaction time: 6 s

Missile acquisition distance: 500 m

Crossing range: 8000 m

Warhead: 14 kg

Kill probability: 0.9 (head-on target); 0.99 (two-round salvo)

STATUS

The Shahine for Saudi Arabia successfully completed its trials programme with firing tests which took place in late 1978 and early 1979.

The order from Saudi Arabia was for the supply of eight acquisition and 16 firing units all based on the GIAT AMX-30 MBT chassis, these were all delivered to Saudi Arabia by mid-1982. These are used for the defence of armoured brigades with close in defence being provided by 53 AMX-30 DCA twin 30 mm self-propelled anti-aircraft guns.

The second contract, placed in 1984 and worth some $4 billion, covers the supply of 24 acquisition units and 48 firing units with deliveries to run from mid-1986 through to 1989.

Of the order for 72 systems, two-thirds will be the new Air-Transportable Towed System (ATTS) with remaining one third based on the chassis of the GIAT AMX-30 MBT.

Thomson-CSF is also providing a package to enable the first batch of Shahine systems to be upgraded in Saudi Arabia as well as provision of data links with the Saudi Arabian air defence network (including the Boeing AWACS aircraft), construction of a new defence school, logistics support and training of air defence personnel. By comparison with the Shahine systems which are already operational with the Saudi armed forces, numerous improvements have been embodied in the latest versions, relating to increased detection, guidance accuracy, ECCM performance and integration by data-link with the higher level.

CONTRACTOR

Thomson-CSF, Division Systèmes Electroniques, 1 rue des Mathurins, BP 10, 92222 Bagneux, France.

Shahine firing unit on AMX-30 tank chassis

Shahine on air-transportable shelter firing unit

INDIA

6721.131

INDIAN AIR DEFENCE MISSILES

The Indian armed forces currently operate Russian-made SA-2 Guideline (**2942.131**) and SA-3 Goa (**2938.131**), SA-6 Gainful (**2930.131**), SA-7 Grail (**2941.131**), SA-8 Grail (**3209.131**), SA-9 Gaskin (**3072.131**) together with the British Tigercat (**2465.131**) air defence missiles but a programme of developing indigenous designs is currently under way under the direction of the Defence Research Development Laboratory in Andhra Pradesh. The programme apparently consists of three missiles; Trishul (Trident), Agni (Fire) and Akash (Space).

Trishul is a truck-mounted low-level, quick reaction missile with a range of 9 km which was first test fired on April 29, 1987. Its development will involve 30 to 40 more test firings but it is anticipated the missile will enter production in 1993.

Akash is a medium-ranged surface-to-air missile with a range of some 27 km and it will be supported by a fire control system capable of engaging multiple targets. In addition to the land-based version an anti-ship version for the Indian Navy and an air-launched version are also being considered. Agni, which is also under development, may be a long ranged missile to replace the SA-2 Guideline.

INTERNATIONAL

4158.131
ADATS MISSILE SYSTEM

ADATS (Air Defence, Anti-Tank System) was developed as an international project being jointly undertaken as a private venture by Oerlikon-Bührle of Switzerland and Martin Marietta of the USA. ADATS is a highly accurate, day/night and countermeasure resistant missile system which has a true dual target capability, for engaging air targets or for use in an anti-tank role.

The ADATS configuration consists of an unmanned turret comprised of electro-optical sensor head containing FLIR, TV, laser range-finder and laser guidance sensor, and on which is also mounted a search radar and two four-missile launcher/canisters. The complete assembly can be rotated in azimuth and elevated remotely from operator positions in the vehicle. The Contraves search radar antenna is free to rotate independently for surveillance purposes and can track ten targets simultaneously. Alternative mounting arrangements allow for installation on a variety of wheeled or tracked vehicles and for shelter or container-housed systems. Other chassis envisaged include the MOWAG Shark (8 × 8) and the FMC M3 Bradley CFV. In the M113A2 system the commander and gunner sit side-by-side at the radar and E-O consoles, respectively.

Each ADATS vehicle is regarded as a self-contained, autonomous fire unit, using its own radar for early detection of low-flying aircraft targets, or alternatively being incorporated into a network of larger radars for command and control. On detection of a target, it is handed off to the passive optical tracking system (either TV or thermal-imaging IR may be employed) which is immune to both anti-radiation missiles and ECM. Following passive tracking of the target, the Mach 3+ missile is launched when the target is within range, subsequent guidance being by beam-riding, using a coded carbon dioxide laser beam radiated from the launch vehicle. The dual-purpose warhead, with fragmentation effects against air targets and shaped charge for armoured targets, is detonated by an impact fuze or an electro-optical proximity fuze (aerial targets only). Anti-tank targets are acquired electro-optically and range determined by the laser rangefinder. Remainder of sequence is as for the aerial target.

CHARACTERISTICS

Missile

Length: 2.057 m
Diameter: 152 mm
Weight: 51.4 kg
Warhead: Shaped charge, dual-purpose fragmentation 12+ kg (approx)
Fuze: Impact and laser proximity
Guidance: Laser beam-riding
Propulsion: Solid, smokeless, rocket motor
Max speed: Mach 3+
Range: 8+ km
Ceiling: 5000 m
Search radar: Dual beam, integrated IFF
Search range: Over 20 km
Target acquisition: Radar, FLIR, TV, optical sight
Passive target tracking: FLIR, TV
Crew: 3 (commander, gunner, driver)
Ready to fire missiles: 8 with number of reloads dependent on vehicle type

STATUS

Development of ADATS began in 1979 and the first ballistic flight of an ADATS missile took place at White Sands Missile Range, New Mexico, in June 1981. This was followed by an extensive system and tactical test programme involving 39 test firings, the last of which occurred in August 1984. Since then several missiles have been successfully fired against aircraft, helicopter and ground targets in the United States and Canada.

In 1983 helicopter-mounted applications of ADATS were being studied.

Production of the ADATS has begun led by Oerlikon-Bührle with the following partners:

Litton Systems Canada (electronics), Martin Marietta (electro-optical module) and Contraves Italiana (radar). In mid-1986 Oerlikon opened a new facility, Oerlikon Aerospace Inc near Montreal to develop, produce and integrate the launch and command equipment. Oerlikon-Bührle in Switzerland is responsible for the production and further development of the missile, turret and carbon dioxide laser.

In June 1986 ADATS was selected as part of the solution to the Canadian requirements for a Low Level Air Defense System (LLADS) together with twin 35 mm anti-aircraft guns (**4537.131**) and Skyguard fire control system (**2377.181**). LLADS involves 36 ADATS.

In November 1987 it was announced that ADATS, with a 25 mm Bushmaster cannon, had been selected to meet the US Army Forward Area Air Defense LOS-H requirement (see **6228.131**). The initial contract, worth $97 million, is to produce four vehicles for operational testing in 1989. It is anticipated that contracts worth $1.7 billion will be awarded to Martin Marietta/Oerlikon-Bührle for 166 production units in FY 1990 with deliveries being completed in 1994. Deliveries will be made to heavy divisions in Europe (3rd, 8th Mechanised, 3rd Armoured) and in South Korea (2nd Infantry). There have been reports that the US Army might order a total of 512 at a cost of $5 billion but these figures seem optimistic.

Oerlikon-Bührle spent SFr700 million developing ADATS and will have to sell 300 systems to recoup the costs.

CONTRACTOR

Machine Tool Works, Oerlikon-Bührle Ltd, 8050 Zurich, Switzerland.

An ADATS system on a M113A2 chassis

2218.131
ROLAND MOBILE ANTI-AIRCRAFT WEAPON SYSTEM

The surface-to-air all-weather or clear-weather Roland weapon system is designed for defence against medium, low and very low altitude aircraft or helicopter attack. It is designed for incorporation into various types of armoured vehicles with the aim of including all operational weapon system equipment and functions within a single unit, such as: surveillance radar, tracking radar, optical sight, command computer and command transmitter link. The firing unit can also be installed in a shelter to be carried by a truck or trailer.

FlaRakRad Roland II vehicle for West German Air Force

The weapon system is provided with two launch tubes each housing one missile, and eight missiles for reloading are stored in two magazines. After firing one or both missiles, reloading can be accomplished in about ten seconds.

Roland can be operated in either the clear-weather or the all-weather mode. Target detection is by means of a doppler surveillance radar with a PPI display. The commander sets the turret on to the designated target and then uses the tracking radar to achieve lock-on, or the gunner takes over target tracking and turret control. The target is tracked by optical line-of-sight (LOS) in the clear-weather mode, or by radar in poor visibility. The commander obtains fire authorisation from the fire control computer or passes authorisation to the gunner. Both the tracking radar and the optical sight measure the angle between LOS and the missile. In the radar mode, the missile is tracked via a missile RF beacon, whereas in the optical mode, a flare is tracked using IR sensors. Computer generated missile guidance commands are continuously sent to the missile in flight via a command transmitter.

CHARACTERISTICS

Surveillance radar: L-band pulse doppler
Detection velocity range: 50 – 450 m/s

Roland II on French AMX-30 chassis

PRF: 8.82 kHz
Scan rate: 60 rpm
Detection range: 90% on 1 m² target at 1.5 – 16.5 km
Pulse duration: 5 ms
Power: 200 W (average); 4 kW (peak)
Tracking radar: Ku-band monopulse doppler
Frequency agility: 250 MHz
Power: 10 kW
Beamwidth: 2° (azimuth); 1° (elevation)
PRF: 5 kHz
Pulse duration: 400 ns

Range: 16 km
Optical sight: Monocular
Magnification/aperture angle: × 8/10° – × 12/5° reversible
Missile tracking: IR goniometer, aperture angle reversible
Command transmitter: Ku-band, 50 kW
Roland missile

	Roland 1/2	**Roland 3**
Length	2.4 cm	2.6 cm
Diameter	16 cm	27 cm
Span	50 cm	50 cm
Weight	66.5 kg	85 kg
Speed	500 m/s	600 m/s
Range	6.3 km	8 km
Warhead	6.5 kg HE	9 kg HE

Propulsion: 2-stage solid; 2 s boost, 13.2 s sustain
Altitude: 5500 m (max); 20 m (minimum)
Manoeuvrability: 12 – 16 *g*
Firing unit weight: 6 t
All-up weights: 32 t (AMX-30); 32 t (Marder); 25 t (M812A1)

A version known as US Roland evolved as a result of the American decision to adopt the all-weather system for licence production in the USA. The fire unit can be mounted on an M812A1 (6 × 6) 5-ton cross-country truck which is the version used by a New Mexico Army National Guard unit.

A different configuration has been evolved to meet the needs of the Federal German Republic Air Force and Navy, and US Air Force bases in Europe. Known as the FlaRakRad Roland, this version is characterised by an eight-wheeled military chassis (10 T MAN) on which a special Roland shelter is mounted. This contains all the operational weapon system equipment, and is essentially the same as the Federal German Army version of Roland.

In the following descriptive passages, the Roland I and II versions only are considered.

OPERATION

Target detection in azimuth is performed by a Siemens/Thomson-CSF pulse-doppler surveillance radar on the launch vehicle. Threat evaluation is accomplished within a single antenna scan, and IFF facilities are provided by either a Siemens MSR-400/5 (West Germany) or an LMT (France) interrogator equipment. The former was produced under licence in the USA by Hazeltine for the US Roland.

The weapon system commander is also the operator of the acquisition radar. He selects the target and, in the Roland I configuration, directs the optical sight to the target azimuth. The gunner then searches in elevation and acquires the target, whereupon a missile may be launched. After launch the gunner maintains the LOS to the target and the missile is gathered to the LOS and guided along this track by an IR guidance technique (TCA) which is described in more detail in the entry for HOT (**2212.111**).

In the Roland II configuration a Thomson-CSF tracking radar is used in addition to the periscope sight, otherwise the principle of operation is the same. Since one of the features claimed for the system is that rapid transition from radar to optical or optical to radar tracking is possible even during missile flight, it is reasonable to assume that the radar and optical axes are maintained in alignment.

The tracking radar is a monopulse type with a magnetron transmitting valve and doppler receiver circuits to reduce clutter returns. It is said to have a parabolic cassegrain antenna and to use circular polarisation. The antenna is gyro-stabilised in bearing and elevation – necessary because the radar is required to have a high resolving power and the weapon system is intended to be used, if necessary, on the move.

The radar deals simultaneously with the target and the missile, missile position in relation to the radar beam being established by continuous-wave transmission from the missile beacon.

Command signals are passed to the missile in flight by a radio link and the commands are interpreted as steering movements of vanes in the efflux from the sustainer motor. This command link is a microwave system.

DEVELOPMENT

In Europe a bilateral major improvement programme is in progress. For example: missile performance will be increased (range 6 to 8 km), and speed (500 to 650 m/s), and warhead weight increased by 50% and sub-systems having a direct bearing on either of these three missile parameters are to be up-graded accordingly.

Roland III

In mid-1985 the Euromissile subsidiary of MBB and Aerospatiale revealed the development of an updated version of the system known as Roland III. The principal differences between this model and its predecessors are summarised as follows:

The missile itself has been improved and Roland III rounds have increased range, higher speed and a heavier warhead, but dimensions are virtually unchanged so that Roland III rounds can be handled and fired from existing fire units. A new launcher system has been designed which allows four missiles to be launched without requiring any reloading operation, and this enables each vehicle to carry 12 missiles instead of ten. This increases efficiency against saturation attack by 20 per cent. Other modifications include a new digital processor system that combines the functions of various weapon system sub-assemblies (including self-test) and can be used with Roland II or III missile types, and new interface units that permit connection to a co-ordination centre (CORAD), exchange of target data with other Roland systems, target designation to other attached weapon systems such as AA guns and man-portable SAMs.

A Roland Co-ordination Centre (RCC) has been designed to provide co-ordination for several Roland systems, and possibly guns. AEG are producing 21 RCC for German forces. In a rather similar programme in the USA, tests have been carried out into the use of the Roland fire control system and its surveillance radar to provide target data for friendly air defence systems such as the IR seeking Redeye, Stinger and Chaparral missiles.

STATUS

The first Roland production hardware began initial evaluation trials with live firings in March 1976 and the first fire units began deliveries in November 1977. By the end of March 1987 6441 launchers (231 on AMX-30, 148 on Marder, 234 Shelters, 31 US) and 25 500 missiles had been ordered and delivered.

Euromissile handed over the first production Roland FlaRakPz to the West German Army in June 1981. In the French Army, five regiments are already equipped with Roland. A shelter version has been ordered by Argentina and three other countries.

The Brazilian Army decision to adopt the Roland system was announced in October 1972, and the first system was delivered in July 1977; the partial licence agreement is understood to remain in force.

In March 1981 Iraq was stated to have agreed to purchase a number of Roland systems.

Production of the Roland commenced in the United States in 1979 and was completed in 1985 after 27 Roland fire units and 595 missiles had been built. These have been issued to one battalion of the US Army New Mexico National Guard and earmarked for deployment with Central Command.

It was revealed in April 1984 that Spain has decided to procure the Roland system, the initial purchase comprising 18 fire units and up to 500 missiles. Known users of the Roland system include Argentina (six Shelter mounted systems ordered of which two were deployed in the Falklands), Brazil (four Roland I on Marder tracked chassis), France (AMX-30 chassis), Iraq (Shelter mounted systems ordered on MAN 8 × 8 truck chassis and AMX-30 version), West Germany (Roland II on Marder chassis), Nigeria (16 ordered on AMX-30 chassis), Qatar, Spain (18 ordered on AMX-30 chassis), United States (27 in service on M812A1 6 × 6 truck chassis) and Venezuela (six Shelter units ordered on trailers). The West German Government has ordered 115 FlakRakRad Roland II systems on the MAN (8 × 8) truck chassis, of which 68 are for the West German Air Force, 20 for the West German Navy and 27 for US Air Force bases in Europe, with deliveries to run from 1987 through to 1990.

CONTRACTORS

System: Messerschmitt-Bölkow-Blohm GmbH, Munich, Federal Republic of Germany; Aérospatiale, 92-Chatillon, France.

Radars: Thomson-CSF, Paris, France; Siemens AG, Radio and Radar Division, Landshuterstrasse 26, D-8044, Unterschleissheim.

Management, sales and responsibility for production: Euromissile, 12 rue de la Redoute, 92260 Fontenay aux Roses, France.

Hughes Aircraft Company, Missile Systems Group, 8433 Fallbrook Avenue, Canoga Park, California 91304, USA – (US Roland).

Boeing Aerospace Company, Defense Systems Division, Seattle, Washington 98124, USA – (US Roland).

US Roland leaves fire unit mounted on an Army M812A1 modified (6 × 6) truck at White Sands Missile Range

ISRAEL

4888.131

ADAMS AIR DEFENCE MISSILE SYSTEM

The Israeli company Rafael Armament Development Authority, is developing an anti-missile missile system for air defence purposes under the name of ADAMS (Air Defence Advanced Mobile System). This has already been proposed to both the US Marine Corps and the US Army to meet its air defence requirements.

For the US Marine Corps it is proposed to install the system on the LAV (8 × 8) chassis which is already in service with the US Marine Corps. This would have 10 missiles in the ready to launch position in the rear of the vehicle with the search radar and electro-optical sensor head of the forward part of the vehicle's roof.

For the US Army, a modified Fighting Vehicles Systems Carrier, already used for the Multiple Launch Rocket System (MLRS) would be used. A total of 20 missiles would be installed in the vertical position at the rear of the vehicle.

A Westinghouse AN/APG-66 derivative fire control radar provides all weather, day/night capabilities while the vehicle is stationary or on the move. This radar automatically detects, interrogates, prioritises, acquires and tracks targets and guides the missile to them.

In automatic mode, ADAMS performs the complete engagement decision process that requires only consent to fire by the operator to engage the highest priority threat.

ADAMS also features an electro-optical fire control package which allows engagements on two simultaneous lines of sight and all passive operation when remote target cueing is available or as the tactical situation requires. Details of the naval Barak ship point defence system are given in entry **4200.231** in the Shipborne Surface-to-Air Weapons section.

STATUS

Development project. Vertical launches and transitions had been carried out successfully at Rafael's test range in the Negev Desert by the end of 1984.

CONTRACTORS

Rafael Armament Development Authority, PO Box 2082, Haifa 31021, Israel.

Israel Aircraft Industries Ltd, Ben-Gurion International Airport, 70100, Israel.

Three frames of a filmed record of one of the test vertical launches of the Barak missile used in the ADAMS system now under development

ITALY

2250.131

SPADA POINT DEFENCE MISSILE SYSTEM

Spada is a missile system developed under a contract awarded to Selenia by the Italian MoD (Air Force), for the close-in air-defence of relatively small permanent objectives of strategic importance.

Because of the sophisticated characteristics of the air threat, the extremely short time available for reaction and the high attrition rate required, Spada has been designed so as to provide the following characteristics: very good action co-ordination and very reliable target identification capabilities, relatively short overall reaction time (in the order of ten seconds on first target and about five on the following targets), high fire power with up to four independent reaction centres and sustained action (up to 72 missiles ready to be fired without reloading the launchers), a high degree of automation to optimise action effectiveness, and an adequate degree of mobility for autonomous re-deployment or transfer to different sites.

Aspide missile firng from Spada launcher

Designed for operational integration via links with the national air-defence net, Spada employs the Aspide missile (**1656.131**), which has a high SSKP (better than 70%) and a relatively large coverage which is in the order of 15 km in range and 5 km in altitude, down to a few metres.

The system is modular in concept and its configuration can be adapted to optimise the deployment according to the dimensions of the objective, the topography of the surrounding terrain and the type of threat. Moreover the system has a great flexibility in siting because the various units can be deployed separately, up to five km apart, for the best defence of the objective.

The Spada system is organised in two main modules: the Detection Centre (DC) and the Firing Sections (FS).

The DC consists of the Search and Interrogation Radar (SIR), made up of the SIR antenna pedestal and relevant equipment shelter, and the Operational Control Centre (OCC) shelter.

The tasks of the DC are target search, detection, identification, evaluation and designation for engagement by the FSs. As SIR the Spada utilises the Selenia Pluto F-band radar (**3277.153**), suitably modified for the specific system purpose in its use of the available power. The SIR is a low altitude search radar, particularly suited for operations in dense clutter environment and in presence of severe ECM, which provides data for the OCC. The latter consists of three operational consoles, a data processing system with two NDC-160 interconnected digital computers and a number of communication links (digital data and telephone).

The FSs consist of the Fire Control Centre (FCC), made up of the Tracking and Illumination Radar (TIR) antenna pedestal, inclusive of an on-mounted optronic sensor (TV), the Control Unit (CU) shelter, for TIR and fire control equipment housing, and the Missile Launchers (ML) with six ready-to-fire Aspide missiles each.

The main functions of the OCC operators are those of supervising the correctness of the operations of the automatic systems, setting up some data relevant to the operational situation of the moment, manual intervention to substitute or correct automatic sequences, particularly in the presence of ECM, and selection of battery intervention criteria.

The tasks of the FS, which represents the reaction centre of the system, are the acquisition and destruction, through missile intercept, of the targets assigned by the DC. At the FCC the TIR carries out target acquisition, tracking and illumination for missile guidance. As an additional mode of operation the TIR provides for target search (360 degrees or sectorial scan), detection and self-designation. The TV sensor is used both as a backup to the radar and as an aid for target identification and discrimination as well as for kill assessment. The associated CU supervises these functions which are carried out automatically with the possibility of manual intervention. The MLs provide for missile storing, aiming, selection and preparing to fire as well as for automatic launching. As TIR the Spada uses a pulse doppler monopulse radar derived from the Selenia Orion 30X radar (**1936.253**). It provides, through the same antenna, for target tracking and illumination for missile semi-active guidance, operating in I- and J-bands respectively. The TIR antenna group is available in two configurations: one mounted on a pallet type dedicated structure and the other, more compact, directly installed on the CU shelter roof. The CU consists of an operational console, a data processing system through an NDC-160 digital computer and a number of communication links (digital data and telephone). The MLs consist of a slewable stand which supports a frame type structure on which two rows of three missile canisters each are located. This structure can be moved in elevation both for firing action and for canister loading/unloading operations.

Fire control console inside Spada Firing Section FCC shelter

The main functions of the OCC operator are those of ensuring correctness of the automatic operation, superimposing manual intervention when required, firing the missiles according to the action criteria ordered by the OCC and carrying out kill assessment. A typical Spada battery consists of one DC, with its SIR, linked to two or three FSs, according to the importance of the objective to be defended, with two six-missile launcher units each. The system's highest firepower configuration consists of four FSs under control of the same OCC; a maximum of three MLs can be connected to the same FCC, which allows a very long fire duration without taking any action for reloading the launcher units. The loading process is carried out by means of a small truck-mounted crane which provides for the substitution of the empty canisters with missile loaded ones. The sealed canisters are also used for missile transportation and storage.

STATUS

Field tests of prototype systems were completed in 1977, with production for the Italian Forces following immediately after. Technical and operational evaluation of the first of the series system took place in 1982-83 at the Italian Air Force base of Grosseto, north of Rome.

Two contracts, for a total of twelve systems, were awarded to Selenia between 1983 and early 1985 respectively. The relevant batteries delivered up to now have already been deployed for the defence of the most important selected objectives. Present plans of both the Italian Air Force and Navy indicate the acquisition of additional batteries for a minimum of 20 systems in all at a cost of Lira 459.9 million. In October 1985 the Spanish Government decided to purchase the Skyguard Aspide air defence system with a total of seven batteries to be delivered between 1988 and 1989. The Spanish Army has been using the Skyguard fire control radar system since the late 1970s while the Spanish Navy uses the Aspide missile with the Albatros air defence system. The Spanish name for this system will be the Toledo. Egypt has also bought the system with the combined gun-missile system designated Amoun. There are 18 batteries each with two guns and two missile launchers.

One system will comprise two Aspide launchers, each with four missiles in the ready to launch position, two twin 35 mm Oerlikon-Bührle towed anti-aircraft guns and one Skyguard fire control system.

CONTRACTORS

System: Selenia, Industrie Elettroniche Associate SpA, Via Tiburtin Km 12 400, 00131 Rome, Italy.

Missile (main parts): Selenia, Rome; SNIA BPD SpA (motor, warhead, safe and arming mechanism), Rome; Microtecnica SpA, Turin, Italy.

6727.131

IDRA AIR DEFENCE MISSILE

Idra has been under development since 1983 as a replacement for Aspide but is also being considered as a candidate for the NATO LAAMS (Local Area Missile System) as well as a medium range air-to-air missile. The ground and seaborne versions would be vertically-launched with the support of a new generation multi-function radar.

Idra features a cruciform configuration of wings and control surfaces while propulsion would be by means of a dual-boost reduced smoke solid propellant motor. For extended range versions of 4.62 m long booster would be added.

The on-board guidance system uses an inertial platform with data-link target updating for the mid-course phase. The terminal phase would involve a Selenia J-band pulse doppler radar with solid-state transmitter. Several further guidance options would be available home-on-jam, data-link updating and command to line-of-sight.

CHARACTERISTICS

Length: 3.65 m
Diameter: 2.12 cm
Wing span: 6.44 m
Weight: 210-220 kg

STATUS

Under development.

CONTRACTOR

Selenia Industrie Elettroniche Associate SpA, Via Tiburtin Km 12 400, 00131 Rome, Italy.

JAPAN

3619.131

TAN-SAM SURFACE-TO-AIR MISSILE (TYPE 81)

Toshiba (Tokyo Shibaura Electric) has developed for the Japanese Ground Self Defence Force (JGSDF) an IR homing air defence missile known as Tan-SAM (Tan = short, Surface-to-Air Missile) and officially designated the Type 81. The JGSDF plans to deploy four Tan-SAM platoons of two quadruple launcher units and one fire control vehicle with each division. The quadruple launcher unit, which consists of a pair of launch ramps with a missile rail on the upper and lower surface of each, is carried on a Type 73 3.5-ton 6 × 6 truck. Each missile is loaded onto the launcher by an hydraulic loading mechanism on each side of the vehicle; the missile which is housed in a box container, is placed on the loader manually by the crew after removing the container cover. Four rounds can be loaded in three minutes.

The Tan-SAM missile is similar in appearance to the British Rapier and is powered by a solid-propellant rocket motor giving an effective range of about 7 km. The missile fuselage carries cruciform wings near the mid-point and there are cruciform control surfaces at the tail. The fire control system (FCS) is carried on a modified version of the Type 73 truck and consists of planar phased-array pulse-doppler three-dimensional radar for search and target detection and designation. The commander and radar operator are housed in a cabin equipped with three CRT displays, keyboards and other controls. The radar antenna is slewed mechanically in azimuth and elevation to face the threatened area. Each launcher vehicle can be separated from the FCS vehicle up to 1000 metres, but a distance of 300 metres is more usual; connections between the vehicles for the transmission of target designation, firing orders etc, are by cables.

Close-up of the Type 81 surface-to-air missile launcher showing two of four missiles (K Nogi)

Up to six targets can be tracked by the radar simultaneously, and the radar detection range is estimated at about 20 to 30 km. All targets detected are displayed on the FCS PPI CRT and the commander selects those which are to be tracked. These are then separately displayed together with threat evaluation symbols. Two targets are selected for engagement and the appropriate launcher unit receives target designation data which activates a search pattern by the missile IR homing head and at the same time inserts a pre-programmed autopilot interception course for the missile when lock-on has been achieved. After launch, the missile flies the initial part of its trajectory under automatic pilot control until homing head re-acquisition of the target is accomplished. This in-flight lock-on facility enables dual engagements and rapid engagement of additional targets during the flight of previously launched missiles before they have reached their target(s).

The Type 81 fire control vehicle in travelling configuration with radar lowered (K Nogi)

Each launcher vehicle is equipped with an optical director for use as a standby target designator if ECM or other difficulties rule out radar designation. Special precautions are incorporated in the IF search pattern scanning program to reduce the problems of false acquisition due to the sun and much of this programming adjustment is performed automatically before launch by the computer in the FCS equipment.

CHARACTERISTICS

Type: Surface-to-air, short range
Length: 2.7 m
Diameter: 16 cm
Weight: 100 kg
Speed: Mach 2.4 (approx)
Effective range: 7000 m
Propulsion: Single-stage solid
Warhead: HE, contact and proximity fuzes
Guidance: Autopilot initial phase, in-flight IR lock-on and terminal homing. Radar target designation with optical director back-up

DEVELOPMENT

Development started in 1966 as a replacement anti-aircraft weapon for Japanese Ground Self Defence Force. First complete prototype constructed between 1971-76. Live prototypes produced between 1977-79. Experimental fire control radar built in 1970, with prototypes between 1972 and 1974. Development is estimated to have cost about £21 million.

STATUS

In production. In service with Japanese Self Defence Forces. Recent orders have been eight systems in both fiscal years 1985 and 1986.

CONTRACTOR

Toshiba (Tokyo Shibaura Electric) Co Ltd, 1-6 Uchisaiwacho, 1-Chome, Chiyoda-ku, Tokyo 100, Japan.

SOUTH AFRICA

2341.131
CACTUS

Cactus is the name given to one of the Crotale land-mobile surface-to-air guided weapon systems which is in service with the South African Air Force.

Crotale is described elsewhere in this section (**2074.131**). The Crotale system has undergone some changes in the course of development, and it may well be that the system in service in South Africa is rather different from that which has been supplied elsewhere.

It is interesting to note that the Cactus units in South Africa are linked directly to the early warning radar network which covers the north-east and north-west approaches to the Republic and which has its headquarters at Devon in the Eastern Transvaal. A Mirage interceptor squadron is also linked to the network.

In early 1981 it was reported that six Cactus units no longer required by the South African forces had been re-acquired by Matra and sold to Chile for use there.

South African Cactus fire unit during a military parade

Cactus fire control unit vehicle

SWEDEN

3617.131
VEHICLE-MOUNTED RBS 70 ANTI-AIRCRAFT MISSILE

Bofors has produced new versions of the well-known RBS 70 anti-aircraft laser-guided missile system. One is mounted on any suitable army vehicle and is known as the RBS 70 VLM (Vehicle Launched Missile) and another is mounted on an armoured vehicle and is known as RBS 70 ARMAD.

RBS 70 VL is very similar to the man-portable RBS 70 (**2348.131**). It is mounted on a special stand fixed on a platform of the vehicle. It can be folded down when in transit and covered with a folding hood. Six missiles are normally carried in the vehicle. The system is designed to accompany units on the move and is fitted to receive warning information and target data from mobile search radars. When hostile aircraft are detected by the radar, an alarm is activated in the vehicle and the bearing and range of approaching hostile aircraft are received by a data link from the radar, on a target data receiver mounted in the vehicle cabin.

The crew consists of the commander and the missile operator/driver. When in motion the crew sits in the cab. When alerted by an alarm they drive the

RBS 70 ARMAD version on M113 chassis. Turret carries missile launcher, HARD radar, missiles and crew of 3

vehicle to the nearest suitable site to counter the attack. The operator then erects the stand with its sight attached, loads a missile and takes his position on the seat attached to the stand. The commander orientates the firing unit and then returns to his combat position in the cab and operates the target data receiver and his radio transceiver. The operator launches the missile and keeps his laser beam on the target while the missile automatically finds the centre of that beam. Set-up time for the VL from the moment of a radar warning alarm of an attack can be as short as 30 seconds.

RBS 70 ARMAD consists of an armoured three-man turret mounted on any suitable APC (Bofors have used an M113A1 as prototype). The fire-control system used with ARMAD is an Ericsson solid-state 3-D radar known as HARD, which can detect and discriminate hovering helicopter targets. It provides search, acquisition and track-while-scan functions. The box-shaped antenna can be elevated and is in operation all the time the vehicle is on the alert while on the move, and is electronically stabilised. The HARD radar provides the commander and gunner inside the turret with small displays on which distinct signals are provided for helicopters and strike aircraft. The radar system provides data to a central fire-control computer, and from target acquisition to actually firing the RBS 70 missile the delay can be as little as four seconds. The radar provides ARMAD with a limited all-weather capability, for although the radar can track a target through cloud the gunner must be able to see the target in his sight to provide the missile its laser guidance.

The RBS 70 missiles themselves are stacked vertically inside the right-hand side of the turret and are loaded manually on to the launcher arm which swings up to a horizontal position above the roof passing through an automatically opening and closing armoured door. Once the missile has been fired the spent launcher tube is ejected and the arm lowers for a new round to be loaded. The turret holds seven missiles. The ARMAD crew comprises: commander, gunner and driver. The gunner uses a normal RBS 70 system laser sight. When not in use the antenna arm is lowered on to the turret roof.

RBS 70 APC. In addition to the RBS 70 ARMAD version, a simpler model suitable for use in the open hatch of an APC is also being developed. This consists of a sight and loaded missile container mounted on a stand. When required for action the cover of the missile compartment is folded down and the stand and missile container is raised into the firing position.

RBS 70M Under Swedish government contracts, a night version of the RBS 70 system has reached the full-scale development stage. This has been designated RBS 70M, where M denotes Mörker, the Swedish word for darkness. This version is intended to support the daylight system that will carry the major anti-aircraft burden, since in the Swedish view daylight and good visibility conditions represent the most predominant threat until the advent of the 1990s. From then, however, the incidence of night/adverse weather attacks is likely to increase, and so to meet this threat the RBS 70M version is designed to resolve AA tasks by day and night at ranges up to about six km, when there is an electro-optical line-of-sight to the target.

A local search radar, of new design and the same as that used in ARMAD, will provide target data, enabling targets to be tracked irrespective of visibility so that they can be engaged immediately when they become visible. The launcher is unmanned and remotely controlled by the missile operator. It consists of a stand sight and two ready-to-fire missiles. The sight includes a laser guidance beam transmitter, TV camera and IR camera. The camera image is displayed for the operator on a TV monitor. In daylight, either the TV camera or the IR camera is used while in darkness and poor visibility only the latter is employed.

An improved missile, RBS 70 Mk 2, that has a larger warhead and more powerful rocket motor is also being developed. There will be complete interchangeability between this round and its predecessors.

The details given in the following table relate to the RBS 70 Mk 1.

CHARACTERISTICS
Length: 1.32 m (without booster)
Diameter: 106 mm (without booster)
Guidance: Laser beam-riding
Warhead: HE fragmentation
Fuze: Proximity and impact fuze
Range: 5 km
Altitude coverage: 3000 m
Time of flight to 3 km: 8.5 s
Reaction time: 5 s (RBS 70 VLM deployed); 4.5 s (ARMAD deployed)
ARMAD radar: HARD, 3-D pulse-doppler
Frequency: I-band
Detection range: 9-10 km (helicopters); 10-12 km (aircraft) 90% probability

STATUS
In production.

CONTRACTOR
AB Bofors, Box 500, 691 80 Bofors, Sweden.

Unmanned remotely controlled two-round launcher is a feature of RBS 70M, seen here being set up

TAIWAN

6722.131

TAIWAN AIR DEFENCE MISSILES

Taiwan, which currently operates the Nike Hercules, HAWK/I-HAWK and Chaparral surface-to-air missiles, is reported to be developing two systems which are now designated the Chungcheng 100 series. Formerly known as Tien Kung (Sky Bow) the first, formerly Tien Kung 1, is a rocket-propelled missile similar in appearance to the Patriot while the second, formerly Tien Kung 2, is a ramjet-powered system. No further details are available.

UNION OF SOVIET SOCIALIST REPUBLICS

2944.131

SA-1 (GUILD) SURFACE-TO-AIR MISSILE

Guild is the NATO code-name for an anti-aircraft guided missile system first shown in Moscow in 1960 and identified by the number SA-1 in the US alpha-numeric sequence.

About 12 metres long and 70 cm in diameter, this missile has no separate booster stage and is said by official sources to be liquid-propelled. Movable cruciform foreplane surfaces are another feature of the missile that distinguishes it from Guideline (SA-2, **2942.131**), which has movable tail surfaces but which is otherwise generally of much the same size as Guild. A UK MoD report of April 1976 mentions a range of 32 km for the SA-1 which was confirmed in later official US sources but has since been increased to an estimated 50 km.

STATUS
It is doubtful if this weapon remains operational although it is possible that a small number might be retained on a reserve basis in the Soviet Union.

SA-1 (Guild) missiles in May Day parade in Moscow (Tass)

2942.131

SA-2 (GUIDELINE) MEDIUM-RANGE SURFACE-TO-AIR MISSILE

Guideline is the NATO code-name for a medium-range, anti-aircraft guided weapon system, known also by the US alpha-numeric code as SA-2, which is standard equipment in the Soviet forces and which has been exported in large numbers to many countries outside the Warsaw Pact territory.

The Soviet designation of the system is V-75 Dvina. The system is land-mobile, the missile being transported to the launcher on a trailer towed by a 6 × 6 tractor truck.

Various models of the missile have been observed over the years exhibiting relatively minor changes in external appearance. The details given below relate to a type that was supplied to Egypt, but it is known that later versions exist. In particular, one that was seen in Moscow in 1967 is somewhat longer and has a larger warhead.

CHARACTERISTICS

Type: Surface-to-air tactical guided missile
Guidance principle: Radio command
Guidance method: By control of movable tail surfaces
Propulsion: Solid-propellant booster with liquid propellant sustainer
Warhead: HE 130 kg, proximity fuze
Missile length: 10.7 m
Missile diameter: Booster 70 cm; second stage 50 cm
Launch weight: About 2300 kg
Slant range: 40-50 km
Speed: Mach 3.5 (estimated)
Ceiling: 18 000 m

OPERATION

The radio command guidance system is believed to be quite straightforward. Targets are tracked by radar which feeds data to a computer: this in turn produces signals which modulate the output of the command transmitter. Two sets of four strip antennas are mounted fore and aft of the missile wings to receive these signals.

The radar used in conjunction with these missiles is generally that known to NATO as Fan Song (**2866.153** and **2868.153**). It is said that one problem with the fire control and guidance system is that the missile must be gathered to the radar beam in the first six seconds of flight or it will not be acquired at all.

According to reports published in the American press concerning Israeli experiences with Egyptian Guideline installations, it was at first found that Israeli ECM devices were adequate to deflect the beam-riding missiles. Subsequently, however, improved missiles were introduced having terminal guidance radar with a wider range of frequencies than the Israeli ECM could handle. They were, however, able to obtain improved ECM pods from the USA which they successfully used to jam the missile acquisition, tracking, and guidance systems.

Various warhead and fuze arrangements have been reported. Most of the missiles have high-explosive warheads but contact, proximity, and command fuzes have been reported. Furthermore, the larger version of the missile referred to above as having been first seen in 1967 is believed to have a nuclear warhead.

Recent reports refer to other modifications, including the introduction of optical techniques into the guidance system to give improved ECCM capabilities and low-altitude performance.

STATUS

Operational with the armed forces of the USSR since 1958, and in many other countries. Since the Soviet authorities generally appear reluctant to discard from their inventory any weapon which continues to work reasonably well, it is likely that Guideline will remain in service for some years but the number deployed in the USSR is expected to decline in the coming years. Despite various modifications, however, it is now technologically obsolescent and unlikely to be very effective against an enemy able to deploy reasonably sophisticated ECM equipment.

The SA-2 Guideline remains in service with the following countries: Afghanistan, Albania, Algeria, Bulgaria, Cuba, Czechoslovakia, East Germany, Egypt, Ethiopia, Hungary, India, Iraq, Korea (North), Libya, Pakistan (from China), Poland, Romania, Somalia, Sudan, Syria, USSR, Vietnam, Yemen (North), Yemen (South) and Yugoslavia. There is a separate entry earlier in this section for the Chinese version of the SA-2 known as the HQ-2J (**4889.131**). Egypt is also now producing the SA-2 missile.

SA-2 (Guideline) medium range surface-to-air missile on its launcher (US Department of Defense)

2938.131

SA-3 (GOA) SURFACE-TO-AIR MISSILE

Goa is the NATO code-name for the SA-3 anti-aircraft guided missile used both by the Soviet Army and the Soviet Navy. The Soviet designation of the system is Pechora.

Small enough to be carried in pairs on the vehicle that is used as a tractor for the trailer transporters of both Guideline and Guild, Goa is a two-stage missile that is thought to have been intended for much the same operational role as the US HAWK, that is to say, short-range defence against low-flying targets.

Control of the missile in flight is by movable foreplane surfaces, the rear mounted wings of the second stage being fixed.

In addition to the original twin launcher associated with the SA-3 there is now in use a triple missile launcher, which was first noted at Yugoslav air defence batteries. Still more recently a four-round launcher installation of Soviet origin has made its appearance and is progressively replacing twin launchers in the USSR.

Goa is used in conjunction with an I/J-band fire control radar whose NATO code-name is Low Blow (**2884.153**) and is commonly associated with an acquisition radar code-named Flat Face (**2874.153**).

In addition to its land-based uses Goa is deployed on ships of the Soviet Navy. Further details can be found in the Shipborne Surface-to-air Weapons section (**2939.231**).

CHARACTERISTICS

Type: Shipborne or land-based surface-to-air guided missile
Guidance principle: Command-guided
Guidance method: By control of movable foreplane surfaces
Propulsion: 2-stage solid-propellant
Warhead: 60 kg HE, proximity fuze
Missile length: 6.7 m
Missile diameter: Booster 60 cm; second stage 25-45 cm
Launch weight: 636 kg
Range: 25 km
Speed: Mach 2+ (estimated)
Ceiling: over 13 000 m

STATUS

Introduced into service in 1961 and now widely deployed in the USSR and Warsaw Pact states.

Current users of the SA-3 Goa include Afghanistan, Algeria, Angola, Bulgaria, Cuba, Czechoslovakia, East Germany, Egypt, Ethiopia, Finland (known as SAM-79), Hungary, India, Iraq, Libya, Malawi, Mozambique, Peru, Poland, Somalia, Syria, Tanzania, USSR, Vietnam, Yemen (South), Yugoslavia and Zambia.

SA-3 (Goa) surface-to-air missiles being transported on rear of ZIL-131 (6 × 6) trucks (TASS)

2934.131

SA-4 (GANEF) SURFACE-TO-AIR MISSILE

Ganef is the NATO code-name for a surface-to-air missile launched from a tracked vehicle that was first seen in public in Moscow in 1964. Its US alphanumeric code is SA-4.

Alternative designations associated with this weapon by other sources are the Soviet name 'Kruga' and the description *Zenitniy Upravlyayemiy Raket* (ZUR) or guided anti-aircraft missile. Different models are believed to have the following Soviet numbers, ZUR: 3M8, 3M8M, 3M8M1 and 3M8M2, of which the last two apply to the two versions that appear to be those now in use. The 3M8M1 model is the variant noted by Western observers and dubbed the 'long-nosed' Ganef, and this believed to have a longer range and higher ceiling than the 3M8M2. These notes refer essentially to the obvious external differences and there are likely to be significant variations internally or in performance characteristics to warrant the four designations given here.

Despite its appreciable (by most standards) range capability, the SA-4 is classed as a medium-range air defence weapon by the Soviet Union (ZRK-SD: *Zenitniy Raketniy Kompleks – Srednoye Deistvie*, anti-aircraft missile system – medium range), and is deployed with special air defence missile brigades that are allocated one to each front or army. A brigade consists of three air defence battalions, each of three batteries.

The system comprises five distinct elements, three of which are special to the SA-4 system, and the other two are employed with other Soviet weapons. The latter two are the Long Track search radar (**2937.153**) deployed at brigade level, and Thin Skin nodding height-finder radars (**3292.153**) at the brigade and one per launcher battalion. These radars are also encountered with other air defence systems such as the SA-6 Gainful, for example.

The other (three) elements are: the SPU (*Samokhodnaya puskovaya ustanovka*) mobile launcher unit, TZM (*Transportna-zaryazhyuscha maschina*) transporter-loader vehicle, and the SSNR (*Samokhodnaya stantsiya navedeniya raket*) mobile missile guidance station, known to NATO as Pat Hand (**2936.153**).

The missiles are carried in pairs on the first of these three vehicles, and the SPU/ZRK-SD is a fully tracked, transporter-erector-launcher equipped with a 360-degree trainable mount with elevation mechanism. The two caliper clamps which support the missiles during transit are separately released by manual operation of a worm-screw on the underside of the clamps. Handles for the worm-screws are clipped to the tubular framework carrying the clamps. This frame folds forward into the horizontal position for firing. Reloading of the launcher vehicles is by means of a mobile crane, and there is a special two-point attachment yoke for lifting. Wheeled transporter vehicles are used for re-supply purposes, each carrying one missile.

Each Ganef battery has one TZM transporter-loader vehicle, capable of carrying one reload missile and transferring it to the launcher by means of a crane. Further reload rounds are carried by the brigade, these being transported on semi-trailers which are normally distributed at battalion or brigade headquarters.

The SA-4 missile configuration is fairly typical of the period (early 1960s): a large centre body containing a ramjet main propulsion motor with a prominent nose cone unit, cruciform wings, four wrap-around booster rockets, and four large tail fins. The boosters are solid propellant units, while the ramjet main motor uses kerosene fuel. A transponder to assist missile tracking during an engagement is located on one of the missile tail fins, and there are four interferometer antennas on the leading edges of the wings to enable the missile to home onto reflected energy from the target illuminating radar.

The Pat Hand guidance radar group operates in conjunction with the battery Thin Skin height-finder and Long Track search radar. Once a target is acquired by the latter two equipments, target data is transmitted to the Pat Hand which operates in the H-band and is thought to employ CW illumination for missile guidance as well as target tracking, supplemented by a command link.

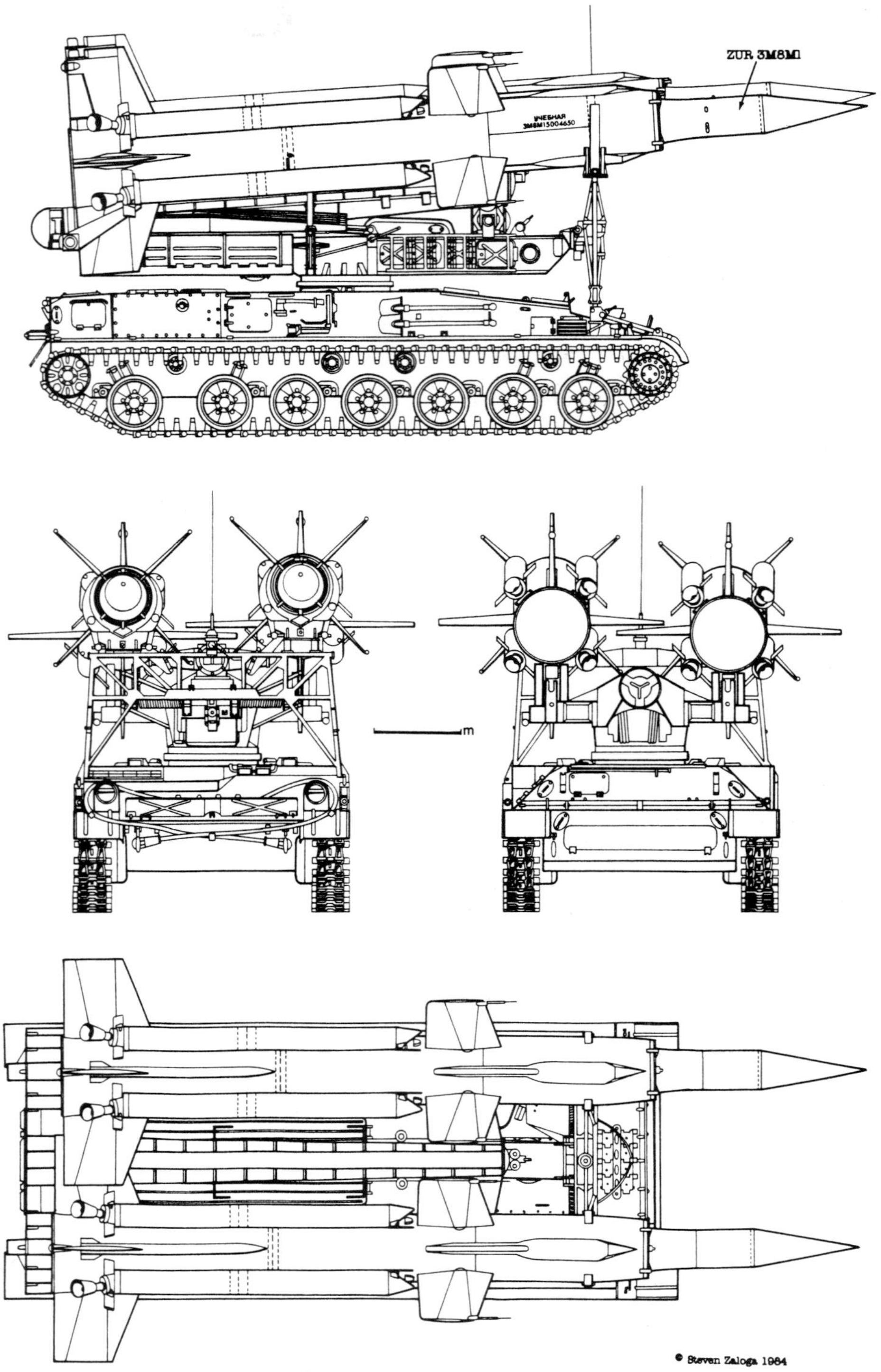

CHARACTERISTICS

Type: Land-mobile surface-to-air tactical guided missile

Guidance principle: Command guidance, plus semi-active radar homing

Guidance method: By control of moving wings on forepart

Propulsion: Ramjet sustainer with four solid-propellant boosters

Warhead: HE, 135 kg proximity fuze

Missile length: 8.8 m

Missile diameter: 0.9 m

Span (tail): 2.6 m

Span (wings): 2.3 m

Launch weight: About 2500 kg

Range: About 70 km

Speed: Mach 2.5

Ceiling: 24 000 m

STATUS

Deployed operationally since the early 1960s, initially with Soviet formations only but later with Egyptian forces, before being withdrawn and replaced by the SA-6 Gainful. Current users of the SA-4 Ganef include Bulgaria, Czechoslovakia, East Germany, Hungary, Libya, Poland and the USSR.

ZUR 3M8M1

m

© Steven Zaloga 1984

General arrangement drawing of SA-4 (Ganef) mobile launcher unit (SPU/ZRK-SD) (Steven Zaloga)

2940.131

SA-5 (GAMMON) LONG-RANGE AIR DEFENCE MISSILE

Griffon was widely accepted as the NATO code-name for a large surface-to-air weapon first displayed in public in 1963 which became associated with the designation SA-5. However in later years little more was seen or heard of Griffon and current practice is to link the designation SA-5 with the NATO code-name Gammon, and a very different looking missile. The system may be designated Volga. According to US assessments it is a weapon about ten metres long, probably with four wrap-around boosters to assist launch and initial acceleration. SA-5 Gammon is estimated to have the longest range of the Soviet SAM inventory with a maximum range of some 300 km. American authorities believe that deployments of the SA-5 have shown a steady increase over the years, as has the improvement and upgrading of the system. Current efforts in this direction are thought to relate to enhancements to suit SA-5 for working in conjunction with the SA-10.

It is understood that there are three versions of SA-5. SA-5A which was first deployed from 1960 and has an HE warhead. The SA-5B was deployed from 1970 onwards and has a nuclear warhead. SA-5C entered service in mid-1970s and can have nuclear or HE warhead and also has improved terminal manoeuvring. Mention has also been made of a version with an anti-radiation seeker.

CHARACTERISTICS
Type: Long-range surface-to-air guided missile
Propulsion: Solid-propellant
Warhead: HE or nuclear
Missile length: Approx 10.6 m
Diameter: 85 cm
Fin span: 290 cm
Range: About 300 km
Speed: Mach 4
Ceiling: About 29 km
STATUS
Operational since 1967, Gammon provides point defence for a number of major areas of strategic importance. The current Soviet deployment plan appears to be based on use of Gammon as the high-altitude missile and the SA-3 Goa (**2938.131**) as the low/medium-altitude weapon for the strategic air defence of the Soviet Union. In addition to being deployed by the USSR within its own borders it is also employed in East Germany and Mongolia. Only known users outside of USSR are Libya and Syria.

2930.131
SA-6 (GAINFUL) SURFACE-TO-AIR MISSILE

Gainful is the NATO code-name for an anti-aircraft missile system known in the US alpha-numeric code as SA-6 and first publicly shown in Moscow in 1967. The Soviet designation is Kub.

Until the 1973 Arab-Israeli war little was known about Gainful in the West. The missile system was, however, used extensively in that war – this being, so far as is known, its first operational use – and scored a number of successes in the early phases of hostilities.

SA-6 (Gainful) tactical surface-to-air missile launcher vehicle deployed

Gainful is both a command guided and a homing missile and is fully mobile, both missile launchers and fire control radar systems being mounted on separate tracked vehicles. The fire control system is both sophisticated and, it would seem, effective. It comprises a primary search and acquisition radar, a target tracking and illuminating radar, a command link with secondary radar response for missile tracking and a missile-borne semi-active homing system. In ECM conditions some of the tracking functions can be performed optically.

The ground equipment of this complex is known by the NATO name Straight Flush (**2885.131**). Its limitation would appear to be restricted search capability when operating without the support of other types of surveillance radar, but it is normal practice to include such longer range radar units in combat formations to obviate this difficulty.

Other radars known to be associated with SA-6 operation include Flat Face and Long Track target acquisition radars (**2874.153** and **2937.153**) and the Thin Skin height-finder (**3292.153**). Straight Flush itself is understood to have a short-range target acquisition capability.

The basic SA-6 unit is a regiment comprising five missile firing batteries. The main elements of such a battery include a Straight Flush radar for target acquisition and fire control and four TELs (transporter-erector-launchers), each of which carries three missiles. Six reload missiles are usually held in reserve on two missile transporter-loader vehicles.

The missile is constructed as a single-stage body containing a dual-thrust integral rocket-ramjet propulsion system. In the boost phase the missile is powered by a solid-propellant rocket motor which accelerates it at about 20 *g* to about Mach 1.5. With the completion of this rocket burn, the tail cone of the missile, which contains the booster nozzle, is jettisoned and the rocket propellant chamber becomes the combustion chamber for the ramjet, ram air being supplied through four intakes disposed symmetrically around the centre section of the missile. The ramjet takes the speed up to almost Mach 2.8.

Two sets of cruciform fins at the centre and tail of the missile provide stability and aerodynamic control, pitch and yaw being controlled at the centre and roll at the tail. The tail fins also carry the command link receiver antenna and the beacon transmitter antenna. The nose of the missile is an ogival radome for the homing head.

CHARACTERISTICS
Type: Surface-to-air tactical guided missile
Guidance: Ground command plus semi-active radar homing. Aerodynamic control
Propulsion: Integral rocket-ramjet
Warhead: HE, probably 80 kg total with 40 kg HE
Fuze: Proximity and impact; possibly also command
Missile length: 5.8 m
Diameter: 33.5 cm
Tail span: 124 cm
Launch weight: 580 kg
Ranges: Max: high altitude possibly 60 km; low altitude probably 30 km. Minimum engagement about 4 km
Speed: Mach 2.8
Ceiling: 18 000 m
STATUS
Operational with Soviet forces and supplied to Algeria, Angola, Bulgaria, Cuba, Czechoslovakia, Egypt, Ethiopia, German Democratic Republic, Guinea, Guinea Bissau, Hungary, India, Iraq, Kuwait, Libya, Mozambique, Poland, Romania, South Yemen, Syria, Tanzania, Vietnam, North Yemen, Somalia and Yugoslavia.

3209.131
SA-8 (GECKO) SURFACE-TO-AIR MISSILE

The SA-8 (Gecko) low-altitude surface-to-air missile is a command guidance system, and in addition to antennas for the search and tracking functions, the radar group (NATO designation, Land Roll) features two separate beacon tracking antennas and two transmitting horns for the command guidance signals. The Soviet designation is Romb.

The main target tracking radar is a cassegrain type array, possibly using conical scan and a pulsed transmission waveform, and the two smaller cassegrain antennas flank the main tracker. They are mounted to provide a limited degree of movement relative to the main array and are assumed to be missile trackers. The command link horns for each are mounted beneath the missile tracker antennas.

The Land Roll radar group uses two frequencies, H-band for target acquisition and J-band for tracking purposes. Both semi-active radar and IR homing techniques have been postulated as the terminal guidance method, and either is feasible.

This arrangement indicates provision for simultaneous firing of two missiles with separate guidance for each. One possible use of this facility would be to enable simultaneous engagement of two separate targets within a formation, and another is to allow 'split engagements' with one target being tracked by the radar and another by the optical tracker which is incorporated in the system.

Both the missile launcher arms and the search radar antenna fold down for reduced transit height.

Control is by four small canard surfaces and there is a cruciform tail assembly of about 60 cm span.

The SA-8 Gecko system is based on a 6 × 6 cross-country truck chassis and is fully amphibious.

SA-8A (Gecko) mobile air defence missile unit on the move. Note folded down search radar

CHARACTERISTICS
Type: Mobile autonomous tactical air defence missile
Guidance: Radar command
Propulsion: Solid
Warhead: HE, proximity fuze, 40 kg
Length: 3.2 m
Diameter: 21 cm
Launch weight: 170 kg (estimated)
Range: 12 km
Speed: Mach 2
Ceiling: 13,000 m

SA-8B

In a military parade in 1980 a new version of the Gecko system appeared in which the two twin missile launcher rails on the vehicle had been replaced by two triple launcher/containers, thereby increasing the ready-to-fire capacity by 50 per cent. This version was later confirmed as the SA-8B Gecko Mod 1 by NATO.

In appearance the six missile containers look slimmer than might be expected from photographs of the original exposed missile installation, and it is possible that the SA-8B missiles are provided with folding tail surfaces to enable their enclosure in the container/launchers. A special vehicle that carries reload missiles accompanies SA-8B fire unit vehicles. Gecko is usually deployed in batteries of four vehicles, each Soviet motor rifle and tank division having five batteries forming an air defence regiment.

STATUS

In service with the Soviet forces since the mid-1970s and operating with armoured and motorised rifle divisions. The level of deployment is reported to be approximately 20 vehicles per division.

Current users of the SA-8 Gecko include Algeria, Angola, German Democratic Republic, Guinea Bissau, India, Iraq, Jordan, Kuwait, Libya, Mozambique, Nicaragua, Poland, Syria and the USSR.

SA-8B (Gecko) mobile surface-to-air missile system on road. Note search radar folded

3072.131

SA-9 (GASKIN) SURFACE-TO-AIR MISSILE

The SA-9 Gaskin designation refers to what at first was generally thought to be an SA-7 Grail (**2941.131**) derivative, but subsequent indications have proved this assumption to have been incorrect. It has been reported that the Soviet designation for the system is Strela 1.

Gaskin is encountered in vehicle-mounted form, on an amphibious BRDM-2 type four-wheeled cross-country vehicle. The usual installation comprises four launcher containers with an aimer's position incorporated in the base. The launcher containers are individually loaded manually on to the tubular construction mounting frame carried by the turret. When operated in an autonomous mode, the method of target acquisition is by visual means with subsequent aiming of the weapons by optical systems. The launcher containers fold down and lie flat on the roof of the vehicle during transit.

When used as part of a larger air defence column, SA-9 vehicles can be linked to search radars to assist in target acquisition, and another method could be that of using radio communications links to 'tell off' targets to individual SA-9 firing units.

CHARACTERISTICS

Type: Mobile short-range tactical air defence missile
Guidance: IR homing
Propulsion: Solid
Warhead: HE, 2.6 kg
Length: 1.8 m
Diameter: 12 cm
Launch weight: 30+ kg
Range: 8 km (approx)
Speed: Mach 1.5+
Ceiling: 4000 – 6000 m

STATUS

US sources claim that the SA-9 was first deployed by Soviet forces in 1968, but since then this weapon has been observed with the forces of Algeria, Angola, Benin, Bulgaria, Cuba, Czechoslovakia, East Germany, Egypt, Guinea, Guinea Bissau, Hungary, India, Iraq, Libya, Madagascar, Mali, Mauritania, Mozambique, Nicaragua, Poland, Syria, Vietnam, Yemen (North), Yugoslavia and the USSR. In Soviet formations, for regimental air defence SA-9s are deployed at a level of 16 per division (four per regiment).

SA-9 (Gaskin) launcher vehicle with four-canister mount in ready-to-fire position

SA-9 (Gaskin) launch vehicle with launcher complete with four missile canisters folded flat for transit

3620.131

SA-10 (GRUMBLE) SURFACE-TO-AIR MISSILE

SA-10 (NATO-code Grumble) is the designation of one of the latest generation of Soviet air defence missiles, but so far no photographs or other information have been made public by Soviet sources.

Over the past few years SA-10 has been described as having been designed specifically to engage low-level targets such as attacking cruise missiles, and as a strategic air defence missile with high performance at both high and low altitudes.

The following paragraphs are based on what constitutes the current official assessments of SA-10 capabilities and estimated characteristics.

The SA-10 is a modern advanced surface-to-air missile system apparently designed for the air defence of the Soviet Union, and is believed to be an 'all-altitude' weapon, which may exist in both semi-permanent and mobile versions. The 1984 edition of the official US publication *Soviet Military Power* states that "the Soviets are developing a mobile version of the SA-10 SAM . . . (which) could be used to support Soviet theatre forces, but, more importantly, if deployed with the territorial defence forces, would allow the Soviets to change the location of those SA-10s in the USSR. The mobile SA-10 could be operational by 1985". Elsewhere in the same document it is stated that the SA-10 is now in series production. The obvious implication is therefore that 'A' and 'B' versions of the SA-10 either exist or are about to.

The SA-10A would be the non-mobile (though probably not fixed, ie transportable or semi-permanent) model which is stated by the US DoD to be operational at 60 sites with another 30 sites under construction in early 1986. Half of these sites are located in the Moscow area. Deployment is said to have been in progress since 1980.

The missile is understood to be about 7 metres long, with a diameter in the region of 45 cm and a maximum range of some 100 km. US drawings of the mobile version of SA-10 suggest that the missile is transported in tubular canisters which elevate to the vertical for launching. The canisters are carried in groups of four on a large 8 × 8 MAZ-7910 wheeled transporter. There is apparently no provision for slewing in azimuth to aim the missile towards the target so that a vertical launch technique is implicit in the US artist's impression (reproduced here). From this drawing it will also be noted that associated with the SA-10 is a planar array guidance radar on a similar wheeled vehicle.

If the above assumptions are correct, and the vertical launch technique is employed, then the semi-permanent SA-10 installations might well utilise a form of silo launcher arrangement with a number of silos installed around important targets for their protection. The silos might or might not be permanently loaded with SA-10 missiles, and similar considerations could apply also to the associated radar(s), these items being deployed to the fixed site only at times of tension or threat, for example. At other times missiles and radars could be deployed elsewhere to meet other threats, for exercise, into hiding, or into storage etc.

A typical SA-10 battery is believed to comprise three TELs and an engagement radar vehicle with perhaps two TZU reload vehicles, all of which are based on an 8 × 8 cross-country chassis.

CHARACTERISTICS

Type: Mobile long range air defence missile
Propulsion: Solid

Warhead: HE, possibly nuclear
Length: 7 m
Wingspan: 1 m
Diameter: 45 cm
Launch weight: 1500 kg
Range: 100 km
Speed: Mach 6
STATUS
Believed to have been deployed in 1980 and operational with PVO Strany (Home Air Defence Command) in the USSR since 1981.

SA-10 (Grumble) mobile air defence missile system includes planar array radar and transporter-erector-launcher units on wheeled chassis according to US analysts (US Department of Defense)

4470.131
SA-11 GADFLY SURFACE-TO-AIR MISSILE

The SA-11 Gadfly mobile missile system has been developed to complement the SA-6 Gainful mobile SAM system. The launcher vehicle is based on the full tracked SA-6b chassis and carries four ready-to-launch missiles on a turntable that can be traversed through a full 360° and a Flap Lid monopulse guidance and tracking radar.

The associated tracked SSNR radar vehicle on a similar chassis carries the 3D Clamshell early warning and acquisition and early warning radars that provide target's height, bearing and range data. Once a target is identified then it is turned over to the SPU launcher vehicle via a data link for tracking and attack.

The Mach 3 semi-active homing missile closely resembles the American Navy's Standard Mk 1 RIM-66 weapon with a maximum range of 28 km and a minimum range of 3 km. It is capable of engaging targets between the altitude limits of 30 and 14 000 m.

An SA-11 Gadfly regiment comprises five batteries each with four SPUs and one SSNR radar vehicle. The SA-11 Gadfly can also be integrated into the SA-6 Gainful air defence network using that system's Straight Flush radar vehicle to provide all the necessary target data in place of its own radar vehicle.

CHARACTERISTICS
Length: 5.6 m
Diameter: 0.4 m
Span: 1.2 m
Launch weight: 650 kg
Propulsion: Solid-fuel rocket motor
Warhead: HE, 90 kg

STATUS
In production. In service with the Soviet Army at divisional level and has also been deployed to Syria.

4774.131
SA-12 GLADIATOR/GIANT SURFACE-TO-AIR MISSILE SYSTEM

The SA-12 Gladiator low- to high-altitude surface-to-air missile system has been developed initially to augment and eventually replace the SA-4 Ganef in the SAM brigades at the Front and Army Level. Two versions apparently exist; the tactical SA-12A Gladiator and the strategic SA-12B Giant with the latter having greater range and unspecified improved capabilities. Giant is still under development.

The SA-12 Gladiator is mounted on a tracked chassis and is 7.2 m long, 0.5 m in diameter, with a wingspan of 1.5 m, launch weight of 2000 kg, fitted with a 150 kg HE warhead and has a maximum speed of Mach 3. It has been designed to counter both high performance combat aircraft and short range tactical ballistic missiles such as Lance, Pluton or the longer range Pershing II. For the former a conventional round is used whilst for the latter a special Anti-tactical Ballistic Missile (ATBM) round is available.

The transporter-erector-launcher (TEL) carries two cylindrical missile container-launchers that can be independently elevated to the vertical for launching and a telescopic missile guidance radar on its rear decking. The TZM missile resupply vehicle with four rounds and the combined phased array multiple target tracking and fire control SSNR radar vehicle also use the same chassis.

There is also an unknown number of KShM command and control shelter vehicles and SSRT search radar vehicles on similar chassis within the unit at brigade, battalion and battery levels.

The minimum and maximum missile engagement altitudes are said to be 90 and 30 000 m respectively. Maximum slant range is 100 km whilst effective range limits are 5.5 and 90 km respectively.

Mobile SA-12 surface-to-air missile system is exepcted to be capable of engaging both high performance aircraft and missiles such as Lance and Pershing. Vertical launch technique is implied in this US artist's impression (US Department of Defense)

CHARACTERISTICS
Length: 7.5 m
Diameter: 50 cm
Max span: 1.35 m
Range: 5-80 km (Gladiator)
5-100 km (Giant)
Altitude: 900-30 000 m
Guidance: Semi-active radar

STATUS
Gladiator is operational in Soviet Union.

4471.131

SA-13 (GOPHER) SURFACE-TO-AIR MISSILE

The SA-13, NATO code-name Gopher, is a low altitude air defence missile system which is intended to replace or supplement the SA-9 Gaskin system (**3072.131**) for the protection of Soviet mechanised formations. Both systems have a very similar configuration, with the SA-9 mounted on a wheeled chassis and the SA-13 on a MT-LB tracked vehicle. It is reportedly designated Strela 10.

Two versions of the SA-13 have been noted, these being referred to as TELAR 1 and TELAR 2, where TELAR stands for transporter-erector-launcher-and-radar. This term embraces several of the main features of the system, namely: mobility; the folding retraction arrangement of the four-canister launcher; and the radar dish mounted between the two pairs of missile canisters.

The radar is a range-only equipment that operates in conjunction with the SA-13's passive IR homing head for missile guidance.

Another function of the radar as a range-only device that has been suggested is that its use would enable the crew to ensure that only targets within lock-on range of the IR seeker would be engaged, thereby avoiding wastage of missiles.

The TELAR vehicle is fully amphibious, and the only significant external difference between the two models is that TELAR 1 carries four small box-shaped units at the corners of its top deck which are antennas for the Hat Box passive radar detection antenna units which are also found on some SA-9 Gaskin SAM systems.

CHARACTERISTICS

Length: 2.2 m
Diameter: 12 cm
Span: 40 cm
Launch weight: 55 kg
Speed: Mach 1.5+
Range: 10 km (max); 500 m (Minimum)
Max altitude: 5000 m
Warhead: HE 4 kg
Guidance: Passive IR homing

STATUS

The SA-13 Gopher became operational with the Soviet Army in 1975 and is now also deployed by Angola, Cuba, Czechoslovakia, East Germany, Iraq, Jordan, Libya, Syria and USSR (Army and Marines).

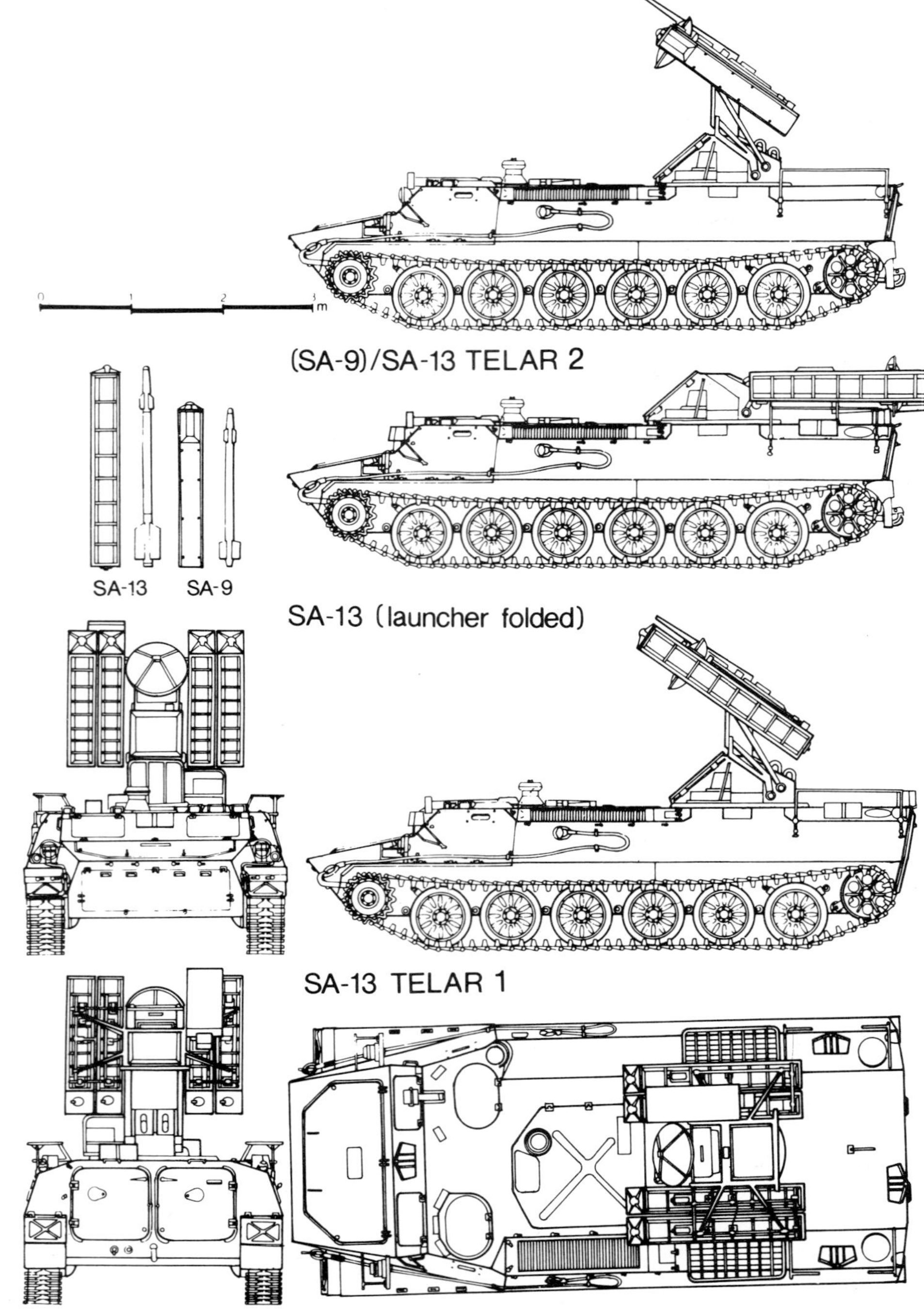

GA drawings of two SA-13 Gopher system versions, with SA-9 and SA-13 missiles and canisters for comparison (Steven Zaloga)

Ceremonial appearance of SA-13 (Gopher) mobile SAM vehicles

6227.131

SA-X-15 LOW-TO-MEDIUM ALTITUDE SURFACE-TO-AIR MISSILE SYSTEM

The Soviet Union is currently developing a new mobile SAM system, designated the SA-X-15 in the NATO designation series, to replace the SA-8 Gecko. No firm details are available at present, but is understood to be based on the naval SA-N-9 system.

STATUS

In development.

UNITED KINGDOM

2406.131
BLOODHOUND Mk 2 SURFACE-TO-AIR WEAPON SYSTEM

Bloodhound Mk 2 is a second-generation mobile anti-aircraft guided weapon sub-system suitable for incorporation into an air defence system. The basic unit, a missile section, typically comprises eight guided missiles and their launchers, a target illuminating radar (TIR), and a launch control post (LCP). The guidance principle is semi-active homing, the missile detecting and homing on the radiation reflected by the target when illuminated by the TIR.

For details of the system operation see **2406.131** in the 1987-88 edition.

In the role for which it was first designed, the Bloodhound missile section is normally equipped with the Firelight TIR (**2413.153**) which is readily mobile and air-transportable. The whole equipment is contained in a single cabin and the antennas are mounted on a retractable pedestal on the roof. For static operations, or where system mobility is not a prime consideration, there is an alternative TIR, the Scorpion (**2428.153**), which is a larger radar giving a longer range. It can be broken down into transportable units but cannot be brought into or taken out of operation as quickly as the Firelight. Both radars are I/J-band CW doppler systems having good performance in the presence of natural or ECM interference.

A Bloodhound Mk 2 being fired during trials

CHARACTERISTICS

Type: Surface-to-air land-based GW system. Land-mobile. Air-transportable
Guidance principle: Semi-active homing. Receiver in missile nose detects and follows radiation reflected by target when illuminated by TIR
Guidance method: Twist-and-steer by pivoting wings
Propulsion: Ramjet with solid-propellant boosters
Warhead: HE, with proximity fuze
Launch control: From LCP by computer using data from TIR
Radar: Choice of 2 TIRs – Firelight or Scorpion
Range: >80 km

DEVELOPMENT

Bloodhound Mk 2 is the successor to Bloodhound Mk 1 which first went into service with the RAF in 1958, development having been started by the Bristol Aeroplane Co Ltd and Ferranti Ltd in 1949. Bloodhound Mk 1 was also exported to Australia and Sweden.

The first design study for the Mk 2 version was started in 1958; the new version went into service with the RAF in 1964. In the same year the equipment also went into service with the defence forces of Sweden and Switzerland (designations RB68 and BL-84, respectively).

In late 1969 BAe was awarded a £10 million contract from the Singapore Government for the refurbishing of the Bloodhound system previously operated by the RAF in that area.

STATUS

The RAF has Bloodhound sites in the UK as part of the NATO and UK Air Defence network. In August 1978 a UK MoD programme drawn up to extend the useful operational life of Bloodhound was announced.

Bloodhound Mk 2 is also in service with the Swiss armed forces who operate about 64 launchers, and the Singapore Air Force with 24 launchers. In early 1984 BAe received a contract from the Swiss Government for a further supply of Bloodhound boost motors, indicating a significant continuation of the system's existing 15-year life in Swiss service.

During 1987 British Aerospace Dynamics (formerly Naval and Electronic Systems) Division completed installation of the first automated Bloodhound operations room which are the first stage of a computer network designed to automate the distribution and presentation of information in all Bloodhound operations rooms. The enhancement of the Bloodhound system began in August 1978 when the UK MoD revealed a programme to extend the missile system's operational life.

Eight automated operations rooms, one for each squadron and for each flight, are being supplied under a £1.2 million contract awarded in July 1986. Each room features a Hewlett Packard HP 1000A computer which are interconnected by landline data highways which are capable of automatic reconfiguration. Data held in store by each computer is common to all and can be constantly updated. In the event of equipment failure or damage the status of the entire British Bloodhound system will be available in all the remaining operations rooms.

Each operations room has a number of work stations, one per flight and four per squadron. The console has two displays with the larger showing the airspace defended by the British Bloodhound system, the positions and tracks of potential targets together with the identities of those launchers (launch control posts) capable of engaging them. Supplementary data in alphanumeric form, including the engineering status data, is presented on the second screen. The menu features more than 100 pages.

In March 1987 the MoD commissioned from British Aerospace a technical evaluation of Bloodhound to assess the changes required to maintain efficiency until the turn of the century. The study was completed later in the year.

CONTRACTORS

The following manufacturers are known to have been associated with the system.

Overall system, missile airframe etc (consortium leader overseas): British Aerospace plc, Naval and Electronic Systems Division, PO Box 5, Filton, Bristol BS12 7QW, England.

Overall electronic system, missile homing system, LCP computer etc, Firelight TIR: Ferranti Ltd, Edinburgh (TIR) and Oldham (remainder).

Ramjet propulsion system: Rolls-Royce Ltd.
Missile fuze: THORN EMI Electronics Ltd.
Scorpion TIR: Marconi Radar Systems Ltd.

2424.131
RAPIER MOBILE LOW-LEVEL ANTI-AIRCRAFT MISSILE SYSTEM

Rapier is a lightweight, mobile, modular air defence weapon system using CLOS guidance. It has been designed by British Aerospace (BAe) to meet a UK Ministry of Defence (MoD) requirement for a fast reaction system for protection against fast (Mach 1+) manoeuvring, low-flying aircraft and helicopters.

It is designed to engage these targets which may appear suddenly at short range from behind terrain screening. It also has the range to destroy aircraft which are either directly approaching or crossing at heights up to 3000 m. The system consists of the missile, the launcher, weighing 1227 kg, a tracker, weighing 119 kg (optical) or 1186 kg (radar), a Tactical Control Unit (TCU) and a Coventry Climax petrol-driven generator with a maximum output of 12.5 kV. The components are interconnected by cable.

The missile, which is of cruciform configuration with wings and control surfaces, is manufactured as a round of ammunition and requires no maintenance once it has left the ordnance depot. It is highly manoeuvrable, being capable of 22 *g* turns, while maintaining a speed of Mach 2 thanks to its two-stage boost motor. Control is by means of a hot gas-driven control surface actuation mechanism while pyrotechnic flares facilitate TV gathering and tracking. The missile is guided by means of commands from the tracker.

The warhead section contains a 0.55 kg semi-armour piercing warhead with the safety-arming unit and a crush fuze. The fuze explodes a small warhead inside the target and this combines with the force of impact of the missile to ensure destruction of even heavily armoured targets.

Towed Blindfire Rapier deployed in field with four-round launcher on left, optical tracker in centre and Blindfire radar on right

The fire unit, which weighs 1227 kg, consists of a surveillance radar with integral IFF, the system computer, the command transmitter and four launcher rails. The command transmitter and the launcher rails are on a 360° turntable.

Either an optical or, in Blindfire, a radar (DN 181 see **2439.153**) tracker is used and when a hostile target is detected the operator is alerted and both launcher and tracker are automatically slewed to the target azimuth. The tracker makes a search in elevation for the target which is tracked manually by the optical system and automatically by the DN 181.

Once the target is within range the operator fires the missile and the engagement system is automatic.

With the optical tracker, which weighs 119 kg, the TV camera automatically tracks flares on the missile and measures any deviations from the Line-Of-Sight (LOS). The Blindfire tracker gathers the missile into the radar beam tracking both missile and target, measuring any divergence from the radar boresight. In both cases these measurements are fed to the fire unit computer which generates and encodes signals for the command transmitter to send to the missile to maintain it on the LOS until impact. The average reaction time from target detection to missile launch is approximately six seconds and re-engagement time is two seconds.

The radar tracker, which has its own generator and a range of 11.5 km, can be added to any optical fire unit to provide all-weather capability 24 hours a day. It operates at a very high frequency and produces an extremely narrow beam which gives excellent discrimination against aircraft flying either in close formation or very close to the ground. Frequency agility is used to reduce the effect of target glint and as an effective ECCM.

Once the four missiles have been loaded (a two-man task) one man can operate the system. When there is a mix of friendly and enemy aircraft, or in ECM conditions, it is usual to use a crew of two (an operator and a tactical controller).

The TCU is used by the detachment commander to define safe lanes for friendly aircraft which are flying in pre-arranged safe lanes. The system's 360° coverage is divided into 32 sectors each with its own switch and by operating them blind arcs can be set, or priority arcs assigned. Should a hostile target be identified in a safe lane, Rapier can revert to its normal mode and engage it. The TCU also includes facilities which are used to neutralise ECM.

The prime production version of the system, Towed Rapier, consists of both Optical and Blindfire fire units which are towed by Land Rover, or similar, vehicles. A single fire unit can cover an area of 130 sq km. The Optical fire unit will have two vehiches and nine reload missiles while the Blindfire fire unit will have three vehicles and 17 reload missiles. Either can be transported by medium lift helicopters as underslung loads or carried internally by heavy lift helicopters.

Customers frequently have a mix of both Optical and Blindfire fire units. The British Army, for example, has a 2:1 mixture to defend units in the field while the RAF Regiment, which defends airfields, uses Blindfire exclusively.

CHARACTERISTICS

Type: Surface-to-air guided weapon system. Land-mobile, air transportable
Guidance: Command to line-of-sight
Length: 2.24 m
Diameter: 13.3 cm
Span: 38.1 cm
Weight: 42.6 kg
Range: 0.5-6.5 km
Time to max range: 15.2 seconds
Speed: Mach 2
Control: Cruciform rear fins
Altitude limits: 0-3000 m
Propulsion: IMI Troy dual thrust
Warhead: 0.55 kg semi-armour piercing with impact fuze

DEVELOPMENT

Development of Rapier began as a private venture, Sightline, in 1961 and in 1963 this became ET.316 whose first test firings took place in 1965. A production contract was issued in 1968 and the system entered service in 1971, the initial systems being to Field Standard A.

Two men can reload the Rapier launcher unit

Tracked Rapier of the Royal Artillery launching a missile during trials

Since then Rapier has been modified to give greater reliability, ease of maintenance and repair, while the performance and ECCM facilities have been adapted to meet the changing air threat now and in the foreseeable future. A continuing UK MoD financed programme to keep Rapier in service at least until the year 2000 has resulted in modifications which have brought the system to Field Standard B at a cost of £500 million. This entered service with the British forces in 1982.

The prime modifications included providing the surveillance radar with a planar array aerial for greater accuracy and better ECCM performance. The IFF system has received Interrogator Side Lobe Suppression (ISLS) to reduce anomalies in the interrogation and identification of targets while an automatic code change allows 48 codes to be pre-set and changed at the required intervals.

In addition a Safe Lane Indicator Computing System (SLICOS) has been added to the Blindfire radar for greater safety to friendly aircraft while microprocessing for self test and diagnostics in the field has been introduced.

Also, in June 1984, BAe announced the development of a battery command post processor, intended primarily for use with Rapier. The microprocessor-based automatic data processing system assists the battery commander in automating the planning, deployment and management duties enabling the optimum position to be chosen for the deployment of tactical weapons.

Other developments include Swiss Rapier (B/L-84), Tracked Rapier, Rapier Laserfire and Rapier Darkfire. General information on these systems is given below but for further details see **2424.131** of the 1987-88 edition.

Swiss Rapier is a version of Towed Rapier specifically configured to meet the requirements of the Swiss Army and the first was officially handed over in June 1984. The modifications included improved acquisition and tracking capability in mountainous terrain, improved ECCM capability, a PPI tactical display, a Swiss IFF system and the use of the standard Pinzgauer (6 × 6) vehicle as the towing and support vehicle.

Tracked Rapier began as a private development in 1974 to enable Rapier to support armoured formations. The system was ordered by Iran in 1974 and cancelled in 1979 but two years later the British Army placed orders initially for 50. Compared with Towed Rapier a Tracked Rapier has a lethality zone of over 140 sq km and a minimum range of 400 m.

Essentially the system consists of an optical fire unit installed in a RCM 748, an armoured version of the M548 cargo carrier, but there are some enhancements. The launcher has eight launcher rails in armoured boxes, while a thermal imager is added to the optical tracker which becomes the Tracker Optical Thermally Enhanced (TOTE, see entry **5906.193**) and is supplemented by the Ferranti Helmet Pointing System (see **4696.093**).

Rapier Laserfire was introduced in 1983 by BAe Dynamics Group, backed by the UK MoD. BAe have declared their intention is to produce a low cost automatic system which is not dependent on operator skills but is capable of day and night operations with a performance comparible to the other Rapier systems.

Laserfire uses the Rapier missile and guidance loop but incorporates an automatic laser tracker and Racal Defence Systems' Possum millimetric surveillance radar (see **5765.153**) for improved ECCM performance and a completely automatic engagement sequence. The system fits on a turntable pallet, weighing 2 tonnes, which can be mounted on any medium size military vehicle capable of a 3 to 4 tonne payload. The pallet may be vehicle or trailer mounted, or can be operated deployed from the ground.

Rapier Darkfire was revealed by BAe in May 1985. This features a six rail launcher, a new remote tactical control console for immediate threat assessment and tactical control and a new, separate, electro-optical tracker with thermal imager. The command guidance system is the same as for other versions, but the surveillance radar is reported to have increased range.

Rapier 2000

Development of Rapier 2000, the next generation (Field Standard C) of the system, began in 1983 and it incorporates many of the features of the Towed Rapier derivatives. In November 1986 the British Government placed an order worth £1 billion with the Army Weapons (now Dynamics) Division of British Aerospace for design, development and initial production of the Rapier 2000 air defence weapon system. In a report to the House of Commons Defence Committee the MoD estimated the total cost of developing and procuring Field Standard C Rapiers at more than £2 billion and said the system would enter service in the early 1990s. BAe literature suggests the system will enter service in the mid-1990s.

Rapier 2000 features a new launcher with eight

TOTE for tracked Rapier is a redesigned optical tracker with added IR capability

missile rails and an automatic thermal imager tracker with manual acquisition and the command guidance transmitter with new command aerial. The tracker, which is similar to that of Rapier Darkfire, will provide optical acquisition and designation facilities in daylight while the thermal imager has a passive scanning mode which can be employed to search for, acquire and track targets by day and night.

Each fire unit will include a Plessey surveillance radar (**6399.153**) to provide for target acquisition in all weathers. The radar, with integral IFF, features a power transmitter to minimise the effectiveness of ECM and separate narrow beams in elevation to minimise the effect of jamming signals. It also has high elevation guard beam which, when an anti-radar missile is detected, inhibits radar transmission.

A new version of the Marconi DN 181 tracking radar can be added to this basic system to provide a fully automatic engagement capability. This generates enough power effectively to break through most jamming systems. Using the M700 processor, frequency management techniques are introduced to evade jamming signals. The radar gathering beam replaces the earlier TV gathering unit and ensures missile acquisition in extreme weather conditions. It has a separate missile guidance transmitter which permits dual simultaneous missile control.

A remote viewing console allows the operator to work from a protected position. The three main elements of Rapier 2000 will be mounted on monocoque two wheel trailers each with its own built-in power supply and having a smoother shape offer a greater degree of protection against nuclear blast as well as a faster wash-down of chemical contamination. The trailer interconnections incorporate fibre-optics and nuclear-hardened chips so offering increased resistance to radiation and EMP effects.

The new Rapier missile, the Mk 2, will enter production early in the next decade and be compatible with earlier Rapier systems. It will have an increased range and a combined armour-piercing and fragmentation warhead. The latter will be used in conjunction with an intelligent proximity fuze against smaller targets.

STATUS

Total Rapier production amounts to 600 launchers, 350 Blindfire trackers and 21 000 missiles. Towed Rapier is in service with, or on order for, 14 countries including Australia (20 launchers), Brunei (12), Indonesia, Iran, Oman (28), Qatar (18), Singapore (10), Switzerland, Turkey (12), United Arab Emirates, United Kingdom (120), United States (24) and Zambia. The US systems are for the protection of bases in England.

Tracked Rapier is operational with the British Army which has ordered 70 units.

Development of Rapier Laserfire and Rapier Darkfire continues. Guided firings of Laserfire were made early in 1987 while troop trials with Darkfire were to take place in mid-1987.

In February 1987 BAe's Naval and Electronics Systems Division announced that Indonesia had placed an order worth up to £500 000 for five battery command post processor systems which is already in service with the British armed forces.

CONTRACTORS

The following manufacturers are known to have been associated with the system.

Weapon system: British Aerospace Dynamics Group, Stevenage, Hertfordshire, England.

Safety and arming system: Marconi Command and Control Systems Ltd.

Surveillance radar and command guidance: Racal Radar Defence Systems Ltd.

Surveillance radar (Rapier 2000): Plessey Radar Ltd.

Secondary radar IFF: Cossor Electronics Ltd.

Servo-optical tracking system: Barr & Stroud Ltd.

Blindfire radar attachment: Marconi Command and Control Systems Ltd.

Missile rocket motor: Imperial Metal Industries Ltd.

Proximity fuze (Rapier 2000): THORN EMI Electronics, Defence Systems Division, Victoria Road, Feltham, Middlesex, TW13 7DZ.

Rapier Darkfire units. This new version features an increased range surveillance radar, six-round launcher, an electro-optical IR tracker and other modifications

Rapier Laserfire deployed ready for action on its modified 4 × 4 Bedford truck chassis

The Rapier 2000 fire unit in ready-to-launch position

2465.131

TIGERCAT CLOSE-RANGE AIR DEFENCE SYSTEM

Tigercat, the land-based version of Short Brothers' naval Seacat, is an autonomous ground-to-air guided missile system for low-level close-range air defence.

In the design of both Seacat and Tigercat the policy was to create a simple guided weapon capable of instant readiness. With any kind of radar warning system there is always the possibility of aircraft approach without detection to a range that permits little time for firing sequences. With Tigercat no prefiring sequences are required on the launcher and auxiliary power services within the missile are activated at launch. The system is thus capable of engaging and destroying at close range even rapidly manoeuvring targets.

Tigercat is compact and mobile and has cross-country capability. The basic fire unit comprises a director trailer, a launcher trailer, and two towing vehicles. These vehicles carry the fire-unit crew of five men, generating equipment, missiles and miscellaneous items. The launcher trailer consists of a two-wheel chassis integral with a three-missile launcher.

Short Brothers' Tigercat deployed in the field

The principle of operation is that after a target is acquired by the aimer the missile is launched into the aimer's field of view. The aimer then directs the missile to fly along the line-of-sight to the target by manipulating a thumb-operated flight controller which applies up, down, left, and right commands to the missile by means of a radio link. Flares at the rear of the missile help the aimer to keep it in sight. As with Seacat the system is also suitable for operation using either radar or TV guidance.

Tigercat has been developed with the facility to accept a 'button-on' fire control radar to provide dark or blind fire in addition to visual operation. The director trailer in this mode incorporates the manually operated pedestal director of the lightweight Seacat system as a combined director and missile aiming sight with Seacat guidance equipment. When in radar control the 'dark fire' aimer's station is in the radar cabin where he is provided with a TV control console. The TV camera is aligned to the radar boresight and provides optical information on the target being tracked. The system comprises a conventional camera and display, and a data processing unit which permits automatic gathering of the missile to the target sightline. It can be integrated with any mobile radar system and also has the facility to receive target information via radio link when employed in the autonomous visual mode.

CHARACTERISTICS

Type: Surface-to-air close-range tactical guided missile

Guidance principle: Command link with optical or radar/TV tracking

Guidance method: Aerodynamic by movable wings

Propulsion: 2-stage solid-propellant motor

Warhead: HE

Missile length: 1.48 m

Missile diameter: 19 cm

Launch weight: 63 kg

STATUS

Tigercat is in production and is in service with several overseas countries including Argentina, India, Iran, Kenya, Qatar and South Africa. At least one foreign Tigercat customer is operating the radar-enhanced version of the system, the Marconi ST850 radar being supplied in 1981.

CONTRACTORS

Short Brothers plc, Belfast, Northern Ireland makes the complete system including special test and training equipment. Other manufacturers known to be associated with the system are Imperial Metal Industries Ltd which makes the rocket motor, THORN EMI Electronics Ltd which makes the fuze and Marconi Radar Systems Ltd which supplies the ST850 radar employed in the radar-enhanced version of Tigercat.

UNITED STATES OF AMERICA

2542.131

CHAPARRAL SURFACE-TO-AIR WEAPON SYSTEM

Chaparral is a ground-to-air IR heat-seeking missile system which was developed in the mid-1960s to meet US Army requirements for a low-altitude air defence weapon.

The system consists of a launch and control assembly, vehicle, and missiles. The missiles are carried on and fired from a turret mounting four launch rails. Eight additional missiles are stored for ready access in the pallet. The gunner in the turret mount aims the missile at the target using optical or electro-optical devices. Once locked-on to the target and fired, the missile guides itself to the target's heat source automatically. The Chaparral missile evolved from the US Navy-developed Sidewinder AIM-9D, which has been modified for ground-to-air rather than air-to-air launch. The Navy procured the original inventory of missiles for Army use.

The launch and control assembly is a self-sustaining unit, capable of operation independent of the carrier vehicle in a ground emplacement mode. For missions requiring full-tracked mobility the fire unit is mounted on an M730 self-propelled tracked vehicle. It is amphibious when a 'swim kit' is used, will carry a crew of five, and can travel at up to 64 km per hour on the road.

One of the operational configurations is the association of Chaparral with the Vulcan air defence gun system (**2850.131**). Forward area alert radar (FAAR – **1526.153**) is deployed with Chaparral/Vulcan to provide early warning and IFF identification within the detection area.

CHARACTERISTICS

US Army designation: M48 A2 (self-propelled); M54 A2 (ground emplaced)

Type: Land-mobile surface-to-air guided missile system

Guidance: IR homing after optical aiming

Missile designation: MIM-72A – MIM-72H

Propulsion: Solid-propellant rocket motor

Warhead: HE, blast fragmentation

Missile length: 2.91 m

Missile diameter: 13 cm

Span: 64 cm

Launch weight: Approx 84 kg

Speed: Supersonic

In addition to the SP and ground emplaced versions of Chaparral (M54 and M48) and improved models of this standard system, two new variants have been added: Night Chaparral and Sea Chaparral.

The former consists of a Chaparral system enhanced by the addition of a forward looking infra-red (FLIR) sensor and an associated display and control panel for the operator to give a night and adverse weather engagement capability.

Chaparral FLIR

This is a thermal imaging device designed as part of the Chaparral improvement programme. It provides considerably improved target acquisition for the operator as well as improved capabilities during daylight/poor-weather conditions in addition to making night operation possible.

The Chaparral FLIR is comprised of US Army standard common modules (**3823.093** and related entries in the Electro-optical Equipment (Land) section contain additional details) specifically the following LRUs: IR receiver; receiver power supply; tracking signal processor; camera/video signal processor; control panel and video display.

The resulting Chaparral FLIR night sight is undergoing installation on all M48 guided missile systems.

The Sea Chaparral employs a slightly modified launch station stabilised to compensate for sea motion and can be operated either with the gunner in the mount or under the control of a remote gunner in the ship's operations room or similar fire control facility.

During 1982 a lightweight towed version of Chaparral emerged which weighs approximately 5445 kg and is capable of being transported by cargo aircraft or helicopter, as well as towing by heavy duty vehicles. The launcher turret on the towed Chaparral has the full 360° azimuth and –10 to +90° elevation launch rail and this assembly is identical to the SP Chaparral. A total of 13 towed Chaparral were built and issued to the 9th Infantry Division at Fort Lewis, Washington.

STATUS

Chaparral is destined to be the US Army's forward area low-level air defence missile up to the end of this century, when it is replaced by the Medium Surface-to-Air Missile (MSAM), according to DoD (see **6419.131**).

By early 1986 Chaparral improvements fielded included an improved infra-red guidance to provide forward engagement capability, an improved ordnance package, IFF system to assist gunner in identifying friendly aircraft, smokeless rocket motor and the FLIR system already described. A new seeker using Rosette Scan Seeker (RSS) and microprocessor technology will start production in FY 1988. Future improvements include adding modifications to improve system performance, reliability and NBC protection. A dual mode (passive RF, passive IR) seeker for performance under all weather conditions and a tail control servo for reduced drag and ease of missile packaging on the launcher are planned improvements for the early 1990s.

MIM-72G with RSS were successfully tested late in 1987. The two colour detector spectral discrimination and scanning spatial discrimination helped distinguish between targets and decoys and six of the seven final firings had a direct hit or were within lethal range of the target. Initial production is scheduled to begin in the second quarter of FY 1988. It is reported the US Army plans to buy 88 fire units and 10 474 missils with RSS at a cost of $1.3 billion.

The United States Army procured a total of 544 tracked Chaparral units in 1985. The fiscal year 1986 procurement was for 64 systems of which 40 were to be reprogrammed from cancelled DIVAD programme with remaining 24 authorised but not appropriated. The total United States Army requirement is for 632 systems, all self-propelled. The additional units are required to equip National Guard battalions. By fiscal year 1990 it is anticipated that the National Guard will have seven battalions of Chaparral systems in the self-propelled configuration.

The Chaparral M48A2 air defence system with day/night and adverse weather capability is now deployed with US forces in Europe and Korea

In addition to being used by the United States Army, tracked Chaparral is also used by Egypt, Israel, Morocco, Portugal, Taiwan and Tunisia. The Sea Chaparral is used only by Taiwan.

CONTRACTORS

Complete system and improved missile: Aeronutronic Division, Ford Aerospace & Communications Corporation, Newport Beach, California 92658-8900, USA.

Vehicle: FMC Corporation, Ground Systems Division, San Jose, California, USA.

6726.131

SABER AIR DEFENCE ANTI-ARMOUR MISSILE

Ford Aerospace are developing a dual purpose air defence, anti-armour missile system called Saber. Two versions are planned, Basic Saber and Long Range Saber. Basic Saber is a shoulder-launched (or vehicle-mounted) system while Long Range Saber will be a vehicle-mounted system which can use TOW launchers or a new launch container. It features a dual purpose shaped charge.

Saber is of conventional design with four fins in cruciform configuration and the missile is attitude stabilised in three axes. It is ejected from the launcher at low speed to reduce the audio and visual signature then when it has travelled a safe distance the motor ignites. The guidance unit features stabilised line-of-sight for target tracking and a guidance laser beam projector to ensure high ECCM performance.

CHARACTERISTICS

	Basic	**Long Range**
Length	1.09 m	1.47 m
Diameter	12 cm	14.7 cm
Weight	10 kg	25 kg

STATUS

Under development.

CONTRACTOR

Ford Aerospace & Communications Corporation, Aeronutronics Division, Ford Road, Newport Beach, California 92658-9983, USA.

4177.131

HAWK SURFACE-TO-AIR MISSILE

HAWK (Homing All-the-Way Killer) has been operational since 1959 and is designed to engage medium and low altitude targets. It uses CW radar homing guidance to discriminate against ground clutter and to achieve effective interception of the lowest flying aircraft.

The system is in service with the US Army, the US Marine Corps and 21 countries (seven NATO) and has been proven in more than 5000 flight tests. The basic system was described in **2640.131** in earlier editions and was followed during the late 1960s by Improved HAWK in which the missile was treated as a round of ammunition without the need for test and repairs. Improved HAWK also saw the addition of a digital automatic data processor (ADP) for target processing, threat ordering and target intercept evaluation.

The HAWK missile has a cruciform configuration with trailing edge control by rear elevons. Its guidance uses proportional navigation with CW, semi-active, all-the-way homing. The guidance system uses state-of-the-art design and an array seeker antenna. The warhead is a blast fragmentation pattern design. It is a certified round requiring no field maintenance or testing. Certification is maintained through periodic batch acceptance testing, annual service firing (there have been 2100 such tests) and periodic batch sampling at special maintenance facilities operated by the contractor.

Improved HAWK has been subject to three upgrade programmes; Phase I (from 1979), Phase II (from 1983) and Phase III. Phases I and II were described in **4177.131** in the 1987-88 edition. Phase III development began in 1983 and is now in production. It involves major modifications both to system deployment configurations and to major systems including the deletion of the AN/MPQ-51 Range Only Radar (ROR) and the AN/MSQ-95 Information Co-ordination Center (ICC) as well as the replacement of the AN/TSW-8 Battery Control Center (BCC) by a Battery Command Post (BCP). The major operational change is the addition of a Low Altitude Simultaneous HAWK Engagement (LASHE) capability against multiple targets. Within the BCP, AN/MSW-11 PCP (Platoon Command Post), AN/MPQ-48 CWAR (CW Acquisition Radar) and AN/MPQ-46 HPI (High Power Illuminator) distributed microcomputers, together with greatly improved software, have been added.

At present the HAWK system batteries have two configurations; Phase II PIP for normal field operations and Triad where increased firepower is required, the US Marine Corps being a major operator. The standard battery has two firing platoons each with a tracking/illuminating radar (HPI) and three triple missile launchers. The Triad battery has three firing platoons; one in standard battery configuration and the others configured as Assault Fire Units (AFU) with a CWAR, a HPI, a PCP and three launchers. The AFUs can be detached from the batteries to operate autonomously where greater mobility or dispersion are required.

Phase III HAWK will be deployed in three configurations; Assault Fire Platoon (AFP), Assault Fire Platoon Plus (AFP+) and Battery. The AFP will have a PCP, a CWAR, a HPI and four to six launchers, the AFP+ having an additional PAR. The battery will have a BCP, a CWAR, a PAR, two HPIs and six launchers.

CHARACTERISTICS

US Army designation: MIM-23B

Type: Mobile surface-to-air guided weapon system

Guidance: Semi-active homing with proportional navigation

Propulsion: 2-stage single chamber solid-propellant motor

Warhead: HE, blast fragmentation

Length: 5.03 m

Span: 1.19 m
Diameter: 0.36 m
Weight: 627.3 kg
Speed: Supersonic
Engagement envelope: Altitude from 30 m to 16 km. Range up to 40 km

MAJOR GROUND EQUIPMENT DESCRIPTION

Descriptions of Phase I and Phase II systems will be found in **4177.131** in the 1987-88 edition. The Phase III systems and modifications now in production are:

Battery Command Post (BCP)

This replaces both the BCC and the ICC and reduces the operating crew from six to four; a Tactical Officer (TO), two Radar Operators (RO) and a Communications Operator. Physically it is similar to the PCP (see below) with the addition of a second radar console, near the Tactical Display and Engagement Control Console (TDECC), and more microcomputers in the digital computer rack.

Platoon Command Post (PCP)

The new PCP uses the same shelter and some of the equipment of the Phase II PCP and is also operated by a three-man crew; TO, RO and Communications Operator. New equipment includes radios, computers, IFF and air-conditioning, with the last cooling both the shelter and the electronic equipment and all have been relocated together with the Communications Operator. NBC protection includes a gas particle filter unit (GPFU) and an entry way air lock.

The ADP has been replaced by microcomputers with higher speeds and more densely packed memories, the TO and RO each have a computer-driven display console while the communications system now has full-duplex serial digital data link communications from the PCP to both the CWAR and the HPI.

In AFP+ configuration another microcomputer is added to the digital computer rack both to process PAR (Pulse Acquisition Radar) data and to interface the PAR with the PCP.

AN/MPQ-48 CW Acquisition Radar (CWAR)

Phase III involves major modifications to the CWAR although its basic function as low-altitude acquisition sensor remains unchanged. However, the transmitted waveform has been changed so the radar can determine both target range and range rate in a single scan. A Digital Signal Processor (DSP) using Fast Fourier Transform (FFT) has been added to process digitally target doppler into detected target data. The DSP provides this data to a new microcomputer within the CWAR and this performs much of the target processing formerly done by the ADP in the ICC. It also transmits processed target track data to the BCP/PCP in serial digital format over a full duplex field telephone wire link replacing the need for a large heavy multi-conductor cable.

AN/MPQ-46 High Power Illuminator (HPI)

There are two major modifications involving the antenna and processing. A wide beam transmitting antenna is added to illuminate a much larger volume for missile guidance during LASHE. A digital microprocessor has been added to process HPI target data and provide full-duplex serial digital communications between the HPI and the PCP.

Product Improvement Programmes (PIPs)

In October 1978 Raytheon announced a series of PIPs to increase operational readiness and system effectiveness. For Phase I and II see **4177.131** in the 1987-88 edition. Phase III streamlines the battery as well as reducing manpower and logistic support requirements and modification kit production has begun for the US Army.

One product improvement programme currently under way is the Norwegian NOAH (Norwegian Adapted Hawk) ARCS (Acquisition Radar and Control System) for which HKV, a joint venture established by Hughes' Ground Systems Group and Kongsberg Vaapenfabrik (now Norsk Forsvarsteknologi A/S), received a contract in January 1984. This involves the delivery of 18 ARCS (one per battery) based upon the Hughes Aircraft Company's LASR (see **4323.153**) while the Norwegians are delivering new fire distribution centres which incorporate the nationally-produced KS-500F computer and KMC-9000 multi-colour displays.

Future Improvements

To ensure HAWK effectiveness into the 1990s a four-step programme is envisaged to improve system mobility. If implemented this would cut the number of vehicles in a firing unit by 50 per cent to seven. The features of this programme are:

(1) Replacing all vacuum-tube circuits on the launcher with a digital microcomputer. This will improve computation of missile launch obstruction avoidance, reduce the number of failed firings and provide full-duplex serial digital data link communications between each launcher and the PCP.
(2) To reduce emplacement and march times mechanical modifications will be made to the launcher to enable it to carry ready-to-fire missiles when it is being moved. At present the missiles have to be removed and reloaded if the launcher is moved.
(3) A new Loader-Transporter will be developed to replace the M-501E3 truck-transported tracked missile loader. The new vehicle will be a flat-bed truck (with hydraulic hoisting crane) which will carry one triple-missile pallet and tow another on a trailer. The crane can load the missiles directly on the launcher.
(4) A North Orientation Module (NOM) will be incorporated to provide rapid indication of the unit's alignment with respect to due North. The NOM would be a north-seeking gyroscope and a digital computer and would be mounted either in a radar or launcher. It will ensure each unit of the system can be independently and correctly aligned day or night in all weathers without either using telescopes or requiring visual contact between units.

A Hawk is launched using targeting data from the LASR (AN/TPQ-36A) radar which will be used in the Norwegian NOAH programme

Agile Continuous Wave Acquisition Radar (ACWAR)

ACWAR (see entry **6400.153**) will improve autonomous detection and if some modifications are made to the missile, data from the ACWAR could be used for missile mid-course control. It may be possible to couple the radar with the HAWK fire control system and use a single command-surveillance vehicle.

Raytheon are suggesting that future medium range air defence requirements might be met by using HAWK as a basis. It would involve introducing ACWAR, modifying HAWK for improved effectiveness against stealth aircraft and cruise missiles, full range simultaneous target engagement, improved ECCM performance and anti-tactical ballistic missile capability. In addition a vertical launched multi-mode missile would be developed.

DEVELOPMENT

Development of basic HAWK began in 1954 with the programme controlled by the US Army Missile Command, Huntsville, Alabama. Engineering of the system was performed by Raytheon which is prime contractor. Production of the US missile and equipment has been principally at Andover and Waltham, Massachusetts.

HAWK was built in Europe by SETEL (Societe

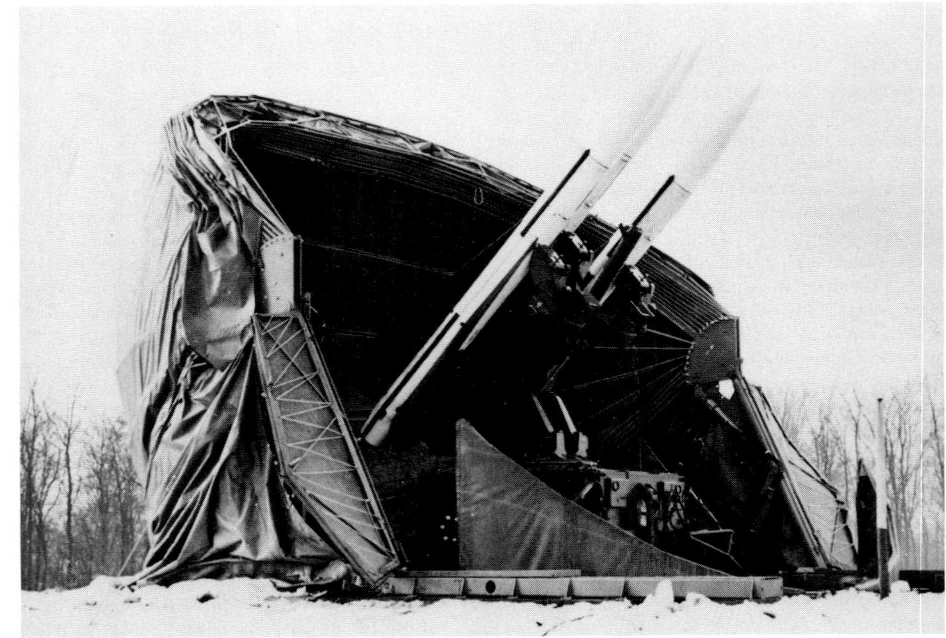

HAWK three-round static launcher in its Koni shelter, in half-open position, at a European NATO site

Europeene de Teleguidage), the initial production programme being completed in 1967 when 4000 missiles were produced. In 1974 the HAWK European Limited Improvement Programme (HELIP) was launched involving Belgium, Denmark, France, Greece, Italy, the Netherlands, and West Germany with Raytheon acting as systems contractor under the NATO HAWK Management Office (NHMO) in Paris. Raytheon produced missile guidance and control units and limited quantities of ground equipment while European contractors (see below) produced ground equipment and some missile parts. They were also responsible for overhaul and conversion of basic HAWK and final assembly of the modified missiles. NHMO's programme management contractor was Raytheon European Management and Systems Company (REMSCO).

American upgrade programmes include LASHE capability and new missile motors. The FY 1988 budget planned HAWK procurement and expenditure for 1988-91 as follows: 1988 – 525 missile units ($161 million), 1989 – 526 units ($152 million), 1990 – 533 units ($96 million), 1991 – 534 units ($88 million).

In addition to the United States the following 21 nations operate HAWK; Belgium, Denmark, Egypt, France, Greece, Iran, Israel, Italy, Japan, Jordan, Kuwait, the Netherlands, Norway, Saudi Arabia, Singapore, South Korea, Spain, Sweden (as Rb-77), Taiwan, United Arab Emirates, West Germany.

CONTRACTORS

Complete system development and US production: Raytheon Company, Lexington, Massachusetts, USA.

Rocket motors: Aerojet General Corporation, El Monte, California, USA.

Wings and elevons: Northrop Corporation, Beverly Hills, California, USA.

Warheads: Iowa Ordnance Depot.

NATO Improved HAWK: GRIP (MBB, Thomson-CSF) and EUROGRIP (MBB, Thomson-CSF, Selenia) are prime contractors under NATO direction together with various subcontractors for the European HAWK user countries.

Japanese production: Mitsubishi Electric Corp and Toshiba.

HAWK battery subject of PIP (product improvement programme) by the European 'GRIP' consortium

2800.131
PATRIOT (MIM-104) TACTICAL AIR DEFENCE SYSTEM

Patriot is a long range, high speed and highly manoeuvrable guided weapon system intended to replace Nike Hercules and some HAWK systems. It uses multiple guidance modes for accurate interception and the best ECCM performance.

The Patriot fire unit (battery), which conducts the engagement, is the smallest combat element of the system and is fully mobile with all units truck or trailer-mounted. It consists of a single radar set, an engagement control station, an antenna mast group vehicle, a AN/MJQ-24 electric power plant with two 150 kW turbine-driven generators and, normally, eight remotely emplaced launching stations although 16 can be used in one fire unit. Each launching station contains four missiles in their firing canisters which also serve as shipping and storage containers.

The fire unit can operate autonomously but up to six can be integrated into a battalion which is controlled by an Information Co-ordination Central (ICC) vehicle. Clusters of fire units are linked to the ICC directly or through a Communication Relay Group vehicle via an Antenna Mast Group to the ICC Antenna Mast Group. Redundancy in communications paths with several fire units means that communications security in both voice and data link modes is very high. Unique address codes in each message ensure the proper unit will receive the message.

Patriot missile test launch from White Sands Missile Range, New Mexico in which a F-102 target was successfully intercepted

The MIM-104 missile has a cruciform configuration with trailing edge control through the four tail-control surfaces. The missile features a single-stage all-boost solid-propellant rocket and, as a certified round, is shipped, stored and fired from its canister without testing or maintenance in the field. Periodic lot sampling of guided missiles on launchers and in storage will provide assurance of their capability. The missile carries a high explosive fragmentation warhead and is capable of outmanoeuvring and destroying any air-breathing threat including conventional and stealth aircraft as well as cruise missiles.

CHARACTERISTICS

Type: Land-mobile, surface-to-air guided weapon system

Guidance principle: Command guidance and semi-active homing. Track-Via-Missile (TVM)

Guidance method: Control of 4 aerodynamic control surfaces

Propulsion: Single-stage solid-propellant TX-486 rocket motor

Missile length: 5.31 m

Missile diameter: 41 cm

Warhead: HE

Speed: Supersonic. In excess of Mach 3

MAJOR GROUND EQUIPMENT DESCRIPTION

Information Co-ordination Central (ICC)

The ICC co-ordinates operations between the individual fire units and interfaces with air defence brigade and higher headquarters. It is similar to the ECS but has fire unit status monitors and no VHF data link equipment. The ICC computer, whose memory has been increased from 196K words to 524K words, processes data from three sources; individual fire units, neighbouring air defence forces and higher echelons to co-ordinate operations of the fire units under its command.

Engagement Control Station (ECS)

The AN/MSQ-104 ECS is mounted on an M-814 truck and acts as the operational command and control centre of the fire unit. It houses the weapons control computer, the man-machine interface and communications terminals. Within the shelter are two display and control consoles. One has a built-in-test equipment panel above it while the other has an environmental panel. Between them are status, generator control and AMG status panels, a hard copy unit and a power distribution unit.

The weapons control computer, whose memory has been expanded from 262K words to 524K words, automatically sequences the system through all tactical engagement operation and monitors operational hardware performance. There are four different groups of software programmes; system initialisation, system operation, diagnostics and training. System initialisation is the basic package and includes specific site data such as radar orientation, coverage boundaries and masked regions. The system operation package provides real-time control of all fire unit elements. The diagnostics enable the computer to detect and locate faults automatically.

The ECS also houses three UHF voice and digital data terminals for communicating between fire units and battalion headquarters. VHF data link terminals provide a secure RF digital data link with launching stations.

Radar

The AN/MPQ-53 radar (see **4478.153**) is a C-band multi-function phased array device which performs the functions of surveillance, target detection and tracking and support of missile guidance which require up to five systems each in the HAWK (**4177.131**) and Nike Hercules (**2723.131**) air defence systems. Mounted on a M-860 semi-trailer operations of the unit are monitored and controlled by the ECS via a digital secure cable link. The antenna is erected to a fixed boresight elevation angle during operation while the entire radar shelter may be rotated in azimuth using a motor-driven pedestal assembly. The antenna unit consists of a 5161 element space fed main array, an IFF antenna for the AN/TPX-46(V)7 interrogator, two ECCM/SLC antennas and a TVM array.

Launching Station

The M-901 launching station is a self-contained, remotely operated unit mounted on a M-860 semi-trailer which includes a 15 kW diesel electric generator, a VHF data link to the ECS, a launcher electronics assembly and four missile canisters. The launcher electronics assembly receives commands from the ECS and reports back to the ECS the status of both the launching station and the missiles. When emplaced the launcher platform is trainable in azimuth and raised to a fixed 38° elevation angle launch position.

DEVELOPMENT

Patriot is the successor to two earlier study programmes managed by the US Army Missile Command's Research and Development Directorate. These studies were concerned with systems known as the Field Army Ballistic Missile Defense System (FABMDS) and the Army Air Defense System for the 1970s (AADS-70s).

In January 1965 the SAM-D requirement was specified and a study was directed towards defence against high performance aircraft at all altitudes as well as short-range missiles. Project definition began in August 1965 and in February 1970 the missile successfully completed its first launch environmental test. The US Army planned to begin engineering development in FY 1971, but was not able to begin until March 1972.

STATUS

In 1974 the programme, now called Patriot, was extended to emphasise development of the Track-Via-Missile (TVM) guidance system. In order not to interfere with the programme it was decided to develop only the minimum necessary hardware.

After a programme involving 14 firings of which 12 were complete successes and one was partially successful authority to resume full scale development was given early in 1976. In August 1976 began a 49 month $425 million contract to complete development and four months later the firing programme was resumed using the first set of prototype ground equipment. When the development programme was completed limited production of Patriot began.

Meanwhile the US Army carried out its own testing and field operational exercises, the latter being completed by the end of 1981. Following parallel confirmation tests on development equipment in 1981 and production hardware in 1982 and 1983, authorisation of increased production was granted.

The stringent 1984 FOE III (Follow-On Evaluation) programme included 16 raids involving a total of 47 sorties against high and low altitude manoeuvring targets but culminated in four hits for four firings in a severe countermeasures environment. These results demonstrated the reliability and troop proficiency levels required for the final deployment decision which occurred in late 1984.

The FY 1986 budget indicated allocations of $78.3 million, $61.3 million, $53.1 million and $38.8 million for continued development in FY 1984-87. Included in this development is the capability to defend against Tactical Ballistic Missiles (TBM) and in September 1986 Patriot successfully destroyed a Lance target in the Anti-Tactical Missile (ATM) mode.

The enhancement of the Patriot system continues with attention focussed upon performance in ECM conditions, against conventionally tipped tactical ballistic missiles, against stealth aircraft as well as conventional manned and unmanned aircraft. By 1987 there were 16 Patriot improvement programmes of which four had completed engineering development and were in production, three had completed development and were entering production and the remainder were in engineering development. Improvements include increasing the memory of the weapon control computer, adding an improved fuze digital processor and improvements both to the software and radar performance. For the last higher density memory modules have been added while the signal processor dynamic range will be extended.

In March 1987 Raytheon was awarded a $3.55 billion multi-year procurement contract beginning in FY 1987 and covering 45 fire units and 491 missiles. Of these 34 fire units and 3862 missiles were for the US while the remainder were for Germany and the Netherlands.

The projected procurement of fire units and missiles for the US Army in FY 1984-87 is given in entry **2800.131** of the 1987-88 edition. The projected procurement and the related budget allocations for FY 1987 and the next five years is as follows:

Fiscal Year	Fire Units	Missiles	Funding ($m)
1987	12	700	1055.2
1988	12	715	985
1989	10	815	907
1990	—	815	742
1991	—	817	567
1992	—	440	615

The US Army plans to purchase 104 fire units and 6077 missiles which it will use to equip nine battalions. By early 1987 it had received 44 fire units and 1161 missiles and the sixth battalion was scheduled to be deployed in December 1987.

In early 1983 Belgium deferred participation in the programme under which European member states of NATO would procure Patriot systems for air defence.

In January 1984 the Netherlands decided to procure four Patriot fire units and 160 missiles through a Foreign Military Sales (FMS) agreement with the USA. Funding for the first fire unit was included in FY 1984 funds. Raytheon is committed in offset to placing $65 million in direct Patriot-related production with Dutch industry and $50 million in Patriot logistics together with $120 million in indirect offsets. By mid-1987 the first Dutch firing unit was operational.

In the same year West Germany also decided to procure the system and agreed to procure 14 fire units and 779 missiles via FMS from the USA. Two fire units and 105 missiles were included in the FY 1985 budget. Through a supplementary compensation agreement, the USA will provide an additional 14 fire untis and 840 missiles from its own inventory. Raytheon is committed to a $500 million direct offset programme with German industry and this will include German trucks, communications and power generation equipment. The Germans have begun their own three-phase project to provide Patriot with an ATM capability. The first phase will involve software improvements which will give the missile a self-defence capability by the early 1990s.

Patriot was also selected by the Japanese Defence Agency in 1984 and in April 1985 the Japanese Defence Agency agreed to a major co-production programme with Mitsubishi Heavy Industry as prime national contractor. It is reported that Japan is seeking a total of 26 Patriot fire units.

There have been reports that Belgium and Italy may begin negotiations for procuring Patriot.

CONTRACTORS

System: Raytheon Company, Andover, Massachusetts, USA (prime).

Principal sub-contractors: Martin Marietta Aerospace, Orlando, Florida, USA (missile airframe and launcher).

Hazeltine Corporation (a subsidiary of Emerson Electric), Greenlawn, New York, 11740, USA (IFF interrogator).

6419.131

MEDIUM SURFACE-TO-AIR MISSILE (MSAM) SYSTEM

MSAM is a conceptual low-to-medium altitude SAM that will replace both HAWK and Chaparral in US Army air defence force in the 1990s. It is envisaged that MSAM will provide greater firepower and tactical mobility with greatly reduced manpower levels over that used by current systems.

MSAM was approved as a major system in August 1986 and is currently a candidate system from cooperative development funded initially under the Nunn Amendment. NATO has an Outline NATO Staff Target, the equivalent of a US ROC (Required Operational Capability) that provides the basis for negotiating a MoU for initial development work in FY 1988.

Raytheon have suggested using a truck-mounted vertically-launched multi-mode missile together with the Agile CW Acquisition Radar (ACWAR) both for division defence, and supplemented by HAWK for area defence.

Hughes Aircraft Company envisage MSAMS as a distributed air defence system with command, control and communications (C3), sensors and vertically-launched missiles in which a battalion could perform 36 simultaneous engagements. The C3 system would provide track distribution to all participants, automated weapon-to-target allocation and secure, jam-resistant, adaptive digital communications. The company appear to be proposing a version of their AIM-120A AMRAAM (see **3340.331**) whose launcher vehicle would have adequate passive sensors for autonomous engagements.

CONTRACTOR

None selected.

2723.131

NIKE HERCULES (MIM-14B) AIR DEFENCE MISSILE

One of the earlier surface-to-air guided weapon systems still in commission, the Nike Hercules missile, successor to the Nike Ajax, was widely deployed in the USA and other countries. It first became operational in the USA in 1958.

Nike Hercules, a second-generation missile, possesses greater destructive ability and has a better performance than Nike Ajax. It has proved successful against high-performance aircraft at a variety of altitudes and has successfully intercepted short-range ballistic missiles and other Nike Hercules missiles in tests.

System units are a low-power acquisition radar, a high-power acquisition radar, a target tracking radar, a missile tracking radar, electronic data-processing equipment and remote-controlled launchers. A later innovation, the high-power acquisition radar (HIPAR – **2499.153**), enables Nike Hercules mobile units to get the same full target detection capability as batteries at fixed sites. The HIPAR's three vans house radar gear transmitter, receiver, and control equipment. One of

Nike Hercules (MIM-14B) surface-to-air missile in S Korean Army service (G Jacobs)

its two semi-trailers hauls the 43 foot (13.1 metre) wide, fan-shaped antenna. The other semi-trailer carries power generators. Before the mobile HIPAR was adopted more than 20 vehicles were required to move the radar system.

When a target is detected by the acquisition radar it is interrogated by the associated AN/TPX-46 IFF Mk XII interrogator and if adjudged hostile its location is transferred to the target tracking radar which pinpoints it for intercept purposes.

The Nike Hercules is launched by remote control, normally at an 85-degree angle. When the booster cluster has separated, the guidance system is activated, programming the missile to roll toward the target and dive into the intercept plane. Steering commands direct the missile to the optimum burst point. The Nike Hercules is equipped with an ECCM capability.

Each Hercules battery can operate as part of a defence network or as an autonomous unit, capable of detecting, tracking, and engaging targets. The system operators are located in a battery control trailer, a tracking radar control trailer, and a launcher control trailer. The Hercules is capable of operating in an ECM environment.

CHARACTERISTICS

Type: Surface-to-air, strategic, guided missile
Guidance: Command
Propulsion: 2-stage solid-propellant rocket motor
Warhead: Nuclear or HE
Missile length: 12.1 m
Diameter: 80 cm
Launch weight: 4858 kg
Speed: Supersonic
Range: $>$ 140 km
Ceiling: $>$ 45 km

STATUS

Major improvements were made to this system in 1961 when new radars and modifications were added to enable the system to remain operational until replaced by the Patriot (**2800.131**) missile. The Nike Hercules was phased out of service in the continental United States some years ago and has now been withdrawn from service in West Germany.

A non-nuclear version of the Nike Hercules missile is being produced by Mitsubishi Heavy Industries Ltd for the Japanese Self Defence Air Force under a licensing agreement with the American manufacturer.

Apart from Japan, Nike Hercules is also widely deployed in NATO and other countries with which the USA has friendly military relationships. Countries believed still to have the missile in strength are Belgium, Denmark, West Germany, Greece, Italy, Norway, Spain, Taiwan and Turkey. The US Nike Hercules batteries in Korea have been transferred to the Republic of Korea, where, it is believed this missile is also operated in an alternative surface-to-surface role.

Due to funding difficulties, from July 1984 the number of Belgian Nike units based in West Germany (eight) was reduced to four, but if an agreement proposed by the USA is concluded these will probably remain in service until replaced by Patriot units in the late-1980s.

In 1981 contracts were placed for a number of improvements to the existing Nike Hercules systems. McDonnell Douglas is refurbishing and modifying NATO Nike Hercules missiles, and the contract also provided for a series of test launches from the NATO test range at Salto di Quirra, Sardinia. A related element of this Nike Hercules updating programme is the digital computer system (DCS) which is being supplied by Norden Systems. The DCS is based on the PDP-11/34M mini-computer and it receives missile and target position inputs from tracking radars, solves the intercept problem and issues guidance commands to the missile. It also performs various other routines, such as fault diagnosis. By replacing the earlier analogue system with the new digital computer, reliability of the Nike Hercules system will be improved and maintenance simplified. DCS deliveries began in late 1981 for NATO batteries and supplies to Far Eastern installations followed. The complete DCS programme could cost $15 million.

CONTRACTORS

Prime: AT & T Technologies, Burlington, North Carolina, USA.

McDonnell Douglas Astronautics Company, 5301 Bolsa Avenue, Huntington Beach, California 92647, USA.

Bell Telephone Laboratories, Burlington, North Carolina, USA.

General Electric Company, Syracuse, New York, USA.

AAI Inc, Cockeysville, Maryland, USA.

Raytheon Company, Wayland, Massachusetts, USA.

6725.131

FIBRE-OPTIC GUIDED MISSILE (FOG-M)

FOG-M is a concept to meet the Forward Air Defense System (FAADS) N-LOS (see **6228.131**) requirement and it involves a 2 m long, 43 kg missile designed by the US Army Missile Command (MICOM) and controlled from a ground station beyond line-of-sight by means of a two-way fibre-optic link. The missile features wings and control fins in cruciform configuration, a booster motor for launching and a sustainer motor. The missile sensor relays battlefield information to the observer and upon target acquisition the missile can either be guided to the target manually or the operator could initiate an automatic tracker. The ground station will provide auto-tracking, land navigation, automatic launch sequence, mission planning and multiple missile monitoring facilities.

The Boeing-Hughes entry for the FOG-M FAADS missile pictured in an HMMVW-mounted launcher

The data link consists of a light-emitting diode or injection laser at the ground station which feeds data through the fibre-optic cable to the missile where it is recovered by a photodiode. A bi-directional coupler allows reception of commands from the launcher as well as transmission of the video image from the missile seeker to the ground station.

MICOM began a demonstration programme of the fibre-optic guidance system in 1979. Tests have already demonstrated that the cable can be paid out at a speed of 600 ft/sec (183 m/sec) which is considerably slower than a TOW missile. The system allows the missile to be launched vertically from behind obstructions before moving to level flight. The operator searches for a target by viewing through the seeker head and once the target has been acquired the missile can be locked on and automatically attack the target. The US Army is seeking a dual-launch capability to exploit the lethality of multiple missile launchers.

Since 1982 the focus of MICOM's attention has been on the potential use of fibre-optic missiles on the battlefield and for four years 15 tests and more than 8500 simulations were carried out. In November 1985, following successful tests against armoured targets, a FOG-M successfully hit a stationary helicopter and three months later a moving helicopter was hit. On 17 December 1986 an rfp was released for an air defence missile. A two-stage programme is envisaged with first maturing the missile designed by the US Army Missile Command. This will have only a TV sensor and will have a range of 10 km. The second stage is to produce a more capable system with the addition of IR sensors.

A request for proposals was anticipated in October 1987 and it was planned to award a contract in June 1988. The missiles will be vertically-launched from 8-32 launch containers mounted on a HMMWV (High Mobility Multipurpose Wheeled Vehicle), tracked vehicles or 5 tonne trucks. Two consortia; Boeing Military Aircraft-Hughes Aircraft-Emerson Electric and Raytheon-Martin Marietta, were competing for the contract with the first named responsible for the missile.

STATUS

Under development.

CONTRACTORS

None selected for the missile or guidance system.

Mobile Gun and Rocket Systems

FRANCE

4502.131

TWIN 20 mm AA WEAPON SYSTEM TA20/RA20S

Electronique Serge Dassault has designed and manufactured the TA20 mobile gun system which, in association with the RA20 S search and target designation radar, provides the means of defending moving formations or vulnerable positions against air attack. Only one RA20S radar is required for a two- or three-vehicle platoon, and this radar may be installed in any one of the vehicles, which then becomes the 'leader', the other one or two vehicles being the 'satellite(s)'. All vehicles are linked by an automatic radio transmission system.

The radar detects low-flying aircraft and assigns a designated target to the leader, to a satellite or both, and automatically transmits the target co-ordinates. This data is fed to the fire control system. Trajectory information is calculated by a processing unit in each vehicle. Separation between vehicles is normally less than 1000 metres. The TA20 system is operated by a crew of two: a radar operator who is also the vehicle commander and platoon commander and a gunner, alone in the turret.

The main components of the system are:

(1) two Oerlikon KAD-B rapid firing 20 mm guns delivering 2000 rounds/minute
(2) a Galileo P56T optical sight
(3) turret with a rotational speed of 80°/s
(4) gyro-stabilisation equipment for guns during firing
(5) control units for turret and gun operation.

This turret can be fitted to a wide range of tracked and wheeled armoured vehicle chassis including the SIBMAS (6 × 6), VAB (6 × 6), EE-11 Urutu, Spartan, M113A2, MOWAG Piranha, AMX-13, AMX VCI and Panhard M3 (see separate entry **2164.131**). It permits engagement of enemy aircraft at a range of 1800 metres, the RA20 S radar detection envelope providing for targets attacking at altitudes between ground level and 2000 metres to be acquired at a range of 12 km.

CHARACTERISTICS

TA20 turret

Track diameter: 1500 mm

VAB (6 × 6) APC chassis fitted with TA20/Rodeo 2 short range anti-aircraft system

Total height: 1200 mm
Combat weight: 1940 kg
Ready use rounds: 2 × 300
Rate of fire: 1000 rounds/minute/gun
RA20 S radar: E-band pulse doppler
TWS: 2 tracks
Antenna unit weight: 60 kg
Range: 12 km

VARIANT

To meet the requirements of the Egyptian Army, Electronique Serge Dassault has developed the TA23E/RAS20S short range air defence system which is basically the TA20/RA20S system but with 20 mm cannon replaced by Soviet 23 mm ZU-23 cannon and the complete system installed on a M113A1 full tracked APC chassis, although it can also be fitted on other chassis such as VAB (6 × 6), Pegaso (6 × 6) or AMX-13 series.

Each of the 23 mm cannon is provided with 300 rounds of ammunition with an effective range of 2000 metres, cyclic rate of fire is 900 rounds per gun per minute. To engage targets at longer range three Sakr Eye surface-to-air missiles are mounted either side of the turret, these are an Egyptian built version of the Soviet SA-7 Grail. The TA23E/RA20S is fitted with the standard P56T Galileo optical sight.

In action one TA23E/RA20S vehicle would act as leader and supply target information to one or two satellite vehicles with the TA23E system but without the radar installed.

STATUS

Production as required, nine systems on the VAB(6 × 6) chassis have been sold to Oman.

CONTRACTOR

ESD – Electronique Serge Dassault, 55 quai Marcel Dassault, 92214 St Cloud, France.

2164.131

M3-VDA 20 mm AA GUN SYSTEM

It comprises a one-man gun turret mounted on the specially adapted M3 and carrying a pair of 20 mm HSS 820 SL guns; a weapon control system with a display for the vehicle commander; and the vehicle itself which is driven by the third member of the crew. The system is also known under the ESD designation TA20, and other variants are the AA 20 system (**3332.131**) which is a transportable version employing towed AA guns, and the VDAA (Véhicule d'Auto-Défense Anti-Aérienne) which is essentially the same as the M3-VDA but mounted on the RVI VAB (6 × 6) chassis with improved cross country mobility. The Panhard M3 (4 × 4) APC chassis has now been replaced in production by the Buffalo which is essentially the M3 with a diesel engine and other improvements so new production systems would use this chassis.

The weapon control system comprises a radar (or alternatively an optical observation system), a computer sight, and high-speed controls for the twin guns and feed system.

The ESD radar is an I/J-band pulse doppler surveillance/acquisition device with a detection range of about 8 km (2 m² target) and a rotation rate of 40 rpm. Targets are displayed on the commander's console, and in target designation the radar feeds target bearing to the optical sighting system and range and target velocity to the Galileo P56 computer sight. The gunner keeps the optical sight trained on the target and the computer calculates lead angle and favourable engagement conditions.

For the simpler (and cheaper) version of the equipment the radar is replaced by two optical observation posts, one for the commander and one for the gunner, with overlapping arcs of observation to give full 360° coverage. When one or other observer is tracking a target the tracking movement generates signals from which the computer calculates lead angle and engagement data as before.

CHARACTERISTICS

Vehicle characteristics: Generally similar to the M3-VTT
Combat weight: 6300 kg
Road speed/range: 90 km/h/1000 km
Armament: 2 HSS 820 SL or similar 20 mm cannon. HSS weapon has MV 1100 m/s, slant range 2000 m, maximum engagement altitude 1500 m. Rate of fire 800-1050 rounds/minute, 600 rounds (2 × 300 belts) carried
Radar: ESD pulse doppler
Computer sight: P56 Galileo

STATUS

In service with Ivory Coast (6), Niger (10) and United Arab Emirates (Abu Dhabi) (42). No longer in production.

CONTRACTORS

System: Société de Constructions Mécaniques Panhard et Levassor, 18 avenue d'Ivry, 75013 Paris, France.

Radar: Electronique Serge Dassault.

Computer: Officine Galileo.

4512.131

SABRE TWIN 30 mm AA GUN TURRET SYSTEM

The Sabre turret system is designed to provide an omni-directional radar search capability and to identify, track, acquire and destroy targets. The turret is of the armoured rotating casemate type. It is equipped with two automatic Oerlikon 30 mm KCB guns, Type 230, with two link belt ammunition bearers and a burst limiter permitting the following firing modes: single-shot firing; 5- or 15-shot bursts.

The ammunition allocation is 300 rounds per weapon, making a total of 600 rounds in the turret. It also supports the Oeil Vert omni-directional search radar used for threat evaluation and acquisition of targets flying at altitudes of not greater than 3000 metres and at radial velocities between 30 and 300 metres/second; the turret can also be equipped with an optional IFF system. Fire control is provided by:

(1) stabilised electro-optical sight which ensures optical tracking of targets
(2) laser rangefinder which ensures highly-accurate target range measurement
(3) automatic TV tracking system
(4) digital computer which receives the data from the radar, TV tracker, rangefinder and calculates firing data and carries out autotest of the equipment.

The Sabre turret is operated by a crew of two: the tank commander who controls the data provided by the radar and orders firing; and a gun-layer/firing operator who controls optical aiming and firing. The turret is autonomous in operation and can be mounted on a variety of tracked and wheeled chassis such as the AMX-30 and Chieftain MBT, TH 400(6 × 6), MOWAG Shark (8 × 8), AMX-10 (6 × 6) and Steyr 4K 7FA full tracked armoured personnel carrier. A trailer version has also been built.

OPERATION

The different operational phases of the Sabre 30 mm twin gun system are as follows:

(1) **Omni-directional search, identification, designation**

Early warning is given by the radar forming part of the system or by radio information from the surveillance net. When the radar is in the 'long-range search' mode, it covers a range 1 to 15 km with Oeil Vert and 1 to 22 km with Oeil Vert Plus, from 0 to 3000 metres in altitude, for a target with a speed varying from 30 to 300 metres/second. As soon as the target is detected by the radar, the computer evaluates the threat and identifies enemy targets (optional IFF system).

(2) **Target acquisition**

In bearing: The tank commander selects the target by overlaying a marker on the echo using a joystick. This operation initialises automatic radar tracking of the target in range and bearing by means of a track-while-scan system. When deemed necessary by the tank commander, at 10 km for example, he slews the turret in the direction of the tracked target.

In elevation: Acquisition is carried out automatically or manually. The gun-laying/firing operator acquires the target in elevation using the stabilised electro-optical sight which includes the TV camera system installed in the turret.

(3) **TV tracking and extrapolation**

Target tracking is by an optical sight and a TV electro-optical system. Automatic or manual system operation is selected by the operator. For automatic operation, all turret movements are slaved by the TV angle-error measurement system which uses the camera video signal to provide angle-error data, ie the position of a target selected by the operator on the TV image with respect to the image centre. The data supplied by the tracking, radar and angle-error measurement system is processed by a digital computer which generates the fire data required to align the two guns on the future target.

(4) **Final engagement**

The tank commander makes the final engagement decisions demanded by the current air situation. As soon as the target comes within the range of the weapon system and firing is authorised, the guns can engage. The guns are continuously supplied with ammunition, thereby permitting several bursts without reload.

Sabre turret on air-transportable trailer/shelter

Thomson-CSF Sabre twin 30 mm defence turret on AMX-30 MBT chassis

CHARACTERISTICS

Weight: 6.5 t (turret)

Surveillance radar

Oeil Vert range: 15 km

Oeil Vert Plus range: 22 km

Sub-clutter visibility: >52 dB

Weapon: 2 × 30 mm KCB guns

Rate of fire: 600 rounds/minute/gun

Burst length: 5 to 20+ rounds

Effective range: 3300 m

MV: 1080 m/s

STATUS

Prototype turrets have been built and tested on a wide range of tracked and wheeled chassis but as of early 1987 no production orders had been placed. The Sabre is a further development of the TG 230A twin 30 mm air defence turret of which 53 were delivered to Saudi Arabia in late 1970s. This in turn is a further development of the turret used on the AMX-13 DCA twin 30 mm self-propelled air defence system of which 60 were supplied to French Army in between 1968 and 1969 and which remain in service.

CONTRACTORS

Prime for system and electronics: Thomson-CSF, Division Systèmes Electroniques, 1 rue de Mathurins, 92223 Bagneux, France.

Chassis and guns: Groupement Industriel des Armements Terrestres (GIAT), 10 place Georges Clémenceau, 92211 St Cloud, France.

Turret: Société d'Applications des Machines Motrices (SAMM), Chemin de la Malmaison, 91570 Bievres, France.

Turret: Creusot Loire Industrie, Division Mécanique Spécialisée, Immeuble Ile de France, 4 Place de la Pyramide, 92070 Paris La Defense Cedex 33, France.

GERMANY (FEDERAL REPUBLIC)

3681.131

WILDCAT TWIN 30 mm AA WEAPON SYSTEM

The Wildcat (formerly AAAT anti-aircraft armoured truck) is a highly mobile, fast road/cross-country vehicle capable of providing anti-aircraft defence for mobile formations, and key installations such as airfields. It is based on the use of a modified chassis of the Transportpanzer 1 (6 x 6) of which 996 have been built for the West German Army by Thyssen Henschel. Five versions have been proposed, among them: one with a surveillance radar and an electro-optical system that includes a TV tracker and laser rangefinder; and an all-weather version with full day/night capability, using surveillance and tracking radars. The crew consists of three men: the commander who is in tactical control of the system, the gunner who conducts engagements, and the driver.

A surveillance radar scans the surrounding airspace at a rate of once per second, whether the vehicle is on the move or stationary. Aircraft and helicopters within range are displayed on the radar PPI and automatically identified by IFF. After a target has been selected on the surveillance radar display and its elevation established, it is tracked automatically by the TV tracker/laser rangefinder unit (daylight/clear weather version) or by tracker radar (all weather, day/night version). The fire control and functional sequences are controlled by a digital computer. This determines the laying of the cannon

Prototype of the Wildcat twin 30 mm air defence system

taking into account target data, vehicle cant and actual meteorological data. When the target enters the engagement zone this is signalled to both the vehicle commander and the gunner. The weapon control system is the responsibility of Hollandse Signaalapparaten. The same concern also provides the LIOD target tracker and SMR micro-computer. Details of the radar, in several applications, will be found in the Flakpanzer entry (**3330.131**) later in this section, the Flycatcher air defence fire control system (**2301.181**), the Goalkeeper naval fire control system (**3516.261**), and in entry **2302.153** in the Ground Radar section of this volume.

CHARACTERISTICS

Vehicle: Transportpanzer 1 (modified)
Combat weight: 18.5 t
Length: 6.88 m
Width: 2.98 m
Overall height: 2.74 m (radar stowed)
Ground clearance: 41 cm
Engine: Mercedes-Benz 8-cylinder, 320 hp
Max speed: 80 km/h
Operating range: 600 km approx
Cannon: 2 × Mauser Mk 30F, automatic belt fed
Calibre: 30 mm
Rate of fire: 1600 rounds/minute (combined)
Ammunition: HEI/SD-T/APDS-T/TP-T
Ammunition capacity: 500 rounds
MV: 1100 m/s (HEI/SD-T)
Traverse: 360° continuous
Elevation: –5 to +85°

Fire control equipment
Surveillance radar: Pulse doppler, digital MTI
Frequency: J-band (about 9 GHz)
Range: 18 km
Data rate: 1 Hz (60 rpm)
IFF: Integrated
Fire control computer: Digital, 24-bit
Tracking radar: All-weather version only
Frequency: Ka-band (about 35 GHz)
Range: 8 km
Antenna: Cassegrain monopulse
E-O director: (Clear weather version)
TV tracker: Militarised daylight television camera, video format to CCIR standard (625 lines, 25 frame/s, 2:1 interlace)
Field of view: 2.4° wide × 1.8° elevation
Laser rangefinder: Neodymium-YAG transmitter, 1.06 micron
Firing rate: 10 Hz
Beam divergence: 1.2 – 1.5 mil
Periscope: Fixed monocular eyepiece
Magnification: × 2, × 8
Field of view: 27°, 7°
Traverse: 360° (continuous)
Elevation: –10 to +70°

STATUS

The first prototype of the Wildcat twin 30 mm air defence system was completed in 1981 and has since been tested in West Germany and overseas. In 1986 a second prototype mounted on a Swiss MOWAG(8 × 8) Shark chassis was completed. This second prototype is for adverse weather operations and has the same surveillance radar as the prototype. For automatic tracking a FLIR/laser rangefinder combination replaces the TV camera.

As of January 1987 no orders for the Wildcat system had been announced.

The two man turret configuration allows the turret to be mounted on a wide range of tracked and wheeled chassis with a turret ring diameter of 2.2 m. The Mauser 30 mm cannon is already in service with two NATO countries and the linkless ammunition feeding system guarantees a shorter loading time and less jamming due to deleted belt links.

CONTRACTORS

System main contractor: Krauss-Maffei Aktiengesellschaft, Ordnance Division, Krauss-Maffei-strasse 2, 8000 Munich 50, Federal Republic of Germany.

Fire control equipment: Hollandse Signaalapparaten BV, Netherlands.

Twin 30 mm cannon: Mauser-Werke Oberndorf GmbH, Federal Republic of Germany.

Surveillance radar: Siemens AG, Federal Republic of Germany.

Ammunition feed system: Heckler & Koch, Kuka Wehrtechnik GmbH, Federal Republic of Germany.

Turret/weapon drives: AEG-Telefunken, Federal Republic of Germany.

Chassis: Krauss-Maffei AG, Federal Republic of Germany.

2182.131

20 mm TWIN GUN ANTI-AIRCRAFT SYSTEM

The AA 20 mm Twin Gun is a light highly mobile automatic system for engaging low-flying aircraft. It can also be used to engage ground targets.

Comprising two automatic rapid-fire guns with an analogue computer sight mounted on a transportable chassis that incorporates the gunner's seat, the system is suitable for rapid deployment in forward areas. The system is operated by a single gunner; to replenish ammunition two men are required and to bring the equipment into or out of action a third man is necessary.

The fire control system is the well-known P56 computing sight and gun laying equipment made by Officine Galileo in Italy. It consists of the following main assemblies: monocular optical sight with swivelling objective prism for laying the gun against air and ground targets; electronic analogue computer for the calculation of kinematic lead values; joystick with two degrees of freedom for speed control of the line of sight; input panel for the target speed and crossing-point distance of air targets or the target distance of ground targets; hydraulic servo drive for the gun traversing and elevating gears; manual traverse and elevation mechanism and a mechanical auxiliary sight.

To engage an air target the gunner first sets estimated target speed and aiming point distance on the input panel then acquires the target in his open sight using the joystick to control the gun. He then changes over to the optical sight and uses the joystick to bring the reticle into coincidence with the nose of the aircraft. Holding the sight in this relative position by means of the joystick he then presses the joystick down to put the guns under control of the computer. Thereafter he keeps the optical sight on the target and the computer points the guns with the degree of lead needed for successful firing. Firing will, however, be inhibited when appropriate by the taboo program which can be inserted into the sight computer.

The taboo facility which can be programmed by the gunner is a safety device to prevent friendly objects from being accidentally engaged. It divides the field of fire into a combat zone and a neutral zone.

Twin 20 mm gun system with MK 20 Rh 202 guns

When engaging ground targets it is necessary only to set the estimated range on the input panel and to bring the target into the reticle of the optical sight before firing.

The guns used in the system are Rheinmetall Rh 202 20 mm rapid-fire (**5600.103** in 1982-83 and earlier editions). These have a very high rate of fire (600 to 1000 rounds per barrel per minute) and a low recoil force. Both gun and ammunition feed are gas operated and independent of weapon and breech movement. In the AA weapon system, cartridges can be fed selectively so that either or both guns can be employed during an engagement.

Twin ammunition boxes containing a total of 550 ready rounds are mounted on either side of the gun assembly and fixed on the upper mount. This ammunition supply enables the weapon to be used for long periods without replenishing the boxes. Ammunition types available are HEI, HEI-T, API-T, API, APDS-T and training and practice TP and TP-T.

Power for the computer and gunlaying systems is provided by an air-cooled engine with a centrifugal governor and a manual starter. This is installed at the rear of the mounting in a cage under the gunner's seat. The complete system is mounted as a two-wheel trailer for towing and can be towed by any vehicle capable of handling an unbraked trailer weighing 2500 kg; the wheels are removed for firing and the mounting is supported on three feet.

CHARACTERISTICS

Cannon: 2 × 20 mm MK 20 Rh 202
Laying system: Hydraulic with emergency manual controls
Elevation: –3.5 to +81.6° (hydraulic); –5.5 to +83.5° (manual)
Traverse: Unlimited
Max laying speed: 80°/s (elevation); 48°/s (traverse)
Max acceleration: 220°/s (elevation); 120°/s (traverse)
Optical system: × 5 magnification; 12° field
Length: 5.035 m (travelling)
Width: 2.36 m (travelling)
Height: 2.075 m (travelling)
Weights: Approx 2160 kg (travelling, less ammunition); 1640 kg (travelling in firing order with ammunition)
Ready rounds: 2 × 280, including 10 on each belt frame
Data input to computer: Target speed 60 – 350 m/s; crossing point distance 100 – 600 m; ground target distance 100 – 2000 m
Generator output: 5.8 kW at 4500 rpm

STATUS

Production as required. In service in Argentina, Chile, West Germany, Greece, Indonesia, Portugal and Turkey. The French Air Force uses the GIAT built Cerbere twin 20 mm which is essentially the Rheinmetall system but with West German MK 20 Rh 202 cannon replaced by the French GIAT M693(F2) weapons. This system has also been tested mounted on the rear of a 6 × 6 cross country truck chassis, but this model has yet to enter production.

CONTRACTOR

Rheinmetall GmbH, 4000 Düsseldorf 1, Federal Republic of Germany.

6728.131

MAUSER ARROW TWIN GUN ANTI-AIRCRAFT SYSTEM

The Arrow system has been designed to defend high value targets such as bridges, airfields, power stations, as well as communications and incorporates two Mauser F 30 mm cannon. The guns, which are capable of firing 800 rounds per second of HEI, API, TP and APDS ammunition, are on a mount produced by Kuka Wehrtechnik and the whole system can be operated in three ways; autonomously through the gunner's electrical control using an optical sight, remotely using a fire control system such as Skyguard or under manual control using a hand crank.

CHARACTERISTICS

Cannon: 2 × 30 mm Mauser F
Elevation: –5 to +85°
Max laying speed: 60°/s (autonomous), 95°/s (remote) traverse, 75°/s (elevation)
Length: 7.88 m (travelling)
Width: 2.38 m
Ready rounds: 2 × 250

STATUS

In serial production and sold to several Asian states including Thailand.

CONTRACTOR

Mauser-Werke Oberndorf GmbH, Federal Republic of Germany.

GREECE

4511.131
ARTEMIS 30 twin 30 mm LIGHT ANTI-AIRCRAFT GUN SYSTEM

The Artemis 30 twin 30 mm air defence system has been designed by EBO S.A. to meet the requirements of the Greek Army and was first demonstrated during the Defendory Exhibition, in Athens in October 1982.

Two main objectives were incorporated from the beginning into the design of the system:

(1) Reliable performance during battle, irrespective of weather and environmental constraints.
(2) User's ability to acquire, operate and maintain the system economically, as well as reliable logistics support, documentation and training with the system.

The Artemis 30 fire unit is a twin 30 mm cannon system towed by a twin-axled split-type carriage. The axle nearest to the towing arm carries the generator which powers the cannon system in action. When deployed this axle/generator assembly is removed and placed some distance away from the mount. The deployed mount is lowered and levelled by means of three hydraulically-operated pads, two on outrigger arms, while the rear axle is power retracted upwards.

The weapons are mounted on each side of a horizontal drum assembly which elevates the cannon. This assembly is placed on a central support on a turntable effecting the weapon traverse movement. The circular ammunition hoppers holding 500 linked rounds (250 for each cannon) are also on the central support. Each cannon receiver has a protective housing that covers the cannon to the base of the barrel. Each of these boxes can be opened to reveal the mechanism to clear jams or for routine servicing. By removing the gun barrel the entire receiver can be swung outwards for more involved repairs and maintenance. The linked ammunition are fed from the hoppers upwards through the central drum and into the feeders from the inside. Spent links and cases are ejected through slots in the cannon housings.

The cannons are the Mk 30 EBO-Mauser Model F, but the barrels are developed entirely by EBO and have some important differences to Mauser's model. They are manufactured according to the cold forging technique since they have a constant twist.

The twin cannon upper mount assembly has an unlimited 360° arc in traverse while the elevation arc is from –5° to +85°.

The angular acceleration of the mount is 200°/s² in azimuth and 166°/s² in elevation while angular speed is 110°/s in azimuth and 75°/s in elevation.

The mount has three different modes of operation:

(1) Operation via a remote fire-control system
(2) Operation via a gunner seated directly behind the central mount support. In this mode the gunner is supplied with all necessary controls including a periscope for ground targets and a gyroscopic angle predicting sight for air target engagement.
(3) Emergency operation via the gunner (no power supplied to the mount). Weapon aiming is accomplished via hand wheels, firing via a foot trigger.

Artemis 30 field AA system deployed in the field

The cannons use 30 mm ammunition especially developed by EBO for the Artemis 30 AA system. They are also compatible with the well-known GAU-8 family of ammunition.

It was established early that Artemis 30 would not be fully dependent on radar fire control. However, it was decided that any fire control method would be of modular form to encompass radar when required.

Under a requirement by the Hellenic Arms Industry (EBO), Philips has contributed a system modified from its original design which uses a compact laser/optronic head mounted on a small single-axle trailer, which presents a video picture to the operator in the fire control centre. The fire control centre is usually housed in an air conditioned shelter carried on a light truck. The shelter also houses the system computer and all necessary communications equipment. The digital computer processes all incoming data and transmits gun-laying data and firing commands to the cannons via a land line. Programs for training and fault finding are handled by the digital computer.

A fire control centre usually controls three or four guns spread over an area about the size of an airfield with each gun operating over a range of 2000 to 3500 metres with a maximum engagement range of 5000 metres.

Diagram of typical Artemis 30 deployment showing six twin 30 mm guns, two fire control centres and E-O tracker units, and battery co-ordination post

Being modular the Artemis 30 fire control system can be expanded and adapted to suit any operational requirement. For use at night an infra-red sensor can be added and for all-weather use an amplitude mono-pulse tracking radar can be installed on the director. The system can also be adapted to handle up to four twin 30 mm cannon carriages and complementary guided missile systems.

Target information and designation is usually supplied to the fire control centre via an acquisition radar system (ARS), combined with a battle co-ordination post (BCP).

The ARS and BCP are integrated into a militarized air conditioned shelter, carried on a military truck, which could be air transported.

The system comprises:

(1) A pulse doppler multifunction radar for air surveillance and target acquisition, IFF interrogator and a command and control tactical display.
(2) The radar operates in the X-band providing optimum performance even in adverse weather conditions and hostile environment.
(3) The radar system can simultaneously track up to 20 targets. Data can be transmitted from the BCP to different FCCs up to a distance of 10 km.
(4) The tactical console can continuously designate hostile targets up to 12 FCCs.

To meet the medium range threat the Hellenic Arms Industry studied a number of missile systems in 1986 and signed an industrial co-operation agreement with Thomson-CSF of France for co-production of the Crotale fire unit and its associated missiles. Hellenic Arms Industry call this the Apollo missile firing unit.

CHARACTERISTICS
Calibre: 30 mm
Barrel length (with muzzle brake): 3.35 m
Operation: gas, automatic
Carriage: 4-wheel, cruciform
Shield: no
Weight (travelling): 7100 kg
Weight (firing, with ammunition): 5600 kg
Length (travelling): 7.85 m
Width (travelling): 2.38 m
Height (travelling): 2.39 m
Track: 1.8 m
Wheelbase: 4.22 m
Ground clearance: 0.32 m
Elevation/Depression: +85°/–5° at 75°/s
Elevation acceleration: 166°/s²
Traverse: 360° at 100°/s
Traverse acceleration: 200°/s²

Rate of fire (both guns): 1600 rpm
Number of rounds (ready use): 500
Overall carriage accuracy: less than 1.5 mrad
Max towing speed: 80 km/h

ARS and BCP
Frequency: X-band
Detection Range: 30 km and digital MTI
Integrated IFF

ECCM package
RF source: Magnetron
Pulse width: 350 nsec
TWS capability: 20 targets
Designation up to: 12 FCCS
Data link: Landline or VHF

VARIANTS

The company has proposed that this system could be mounted on the chassis of the Steyr 4K 7FA armoured personnel carrier which is already being manufactured under licence in Greece for the Greek Army. Other applications are under study. The first system was delivered in June 1987 and the contract will be completed in 1990.

STATUS

In production for Greek Army, Navy and Air Force under a contract signed in October 1984, and first production systems were completed late in 1986.

CONTRACTOR

Hellenic Arms Industry SA, 160 Kifissias Avenue, Athens, Greece.

ISRAEL

4508.131

SPIDER II AIR DEFENCE SYSTEM

The SPIDER II is a ground-based mobile fire control system designed for use against both high- and low-flying targets, aircraft, helicopters or missiles. The system searches for, designates and tracks aircraft and missiles, and controls automatically the aiming and firing of six twin 30 mm cannon turrets. It is claimed to detect low altitude, small radar cross-section targets up to a maximum range of 19 km and will predict ballistics automatically for guns of calibres from 30 to 57 mm.

The basic system consists of a fire control console, tactical control console, search radar, tracking director and a processing unit. These units are normally mounted on a military truck, or housed in a standard shelter/container suitable for transport on a flat-bed vehicle. The search radar is designed primarily for detection of low-flying aircraft and uses coherent pulse doppler techniques. The tracking director is a combination of fire control radar and electro-optical systems with the latter incorporating two TV cameras and a high rate and power laser rangefinder. The fire control console provides the operator with monitoring control of the system by means of a keyboard and a video display. The tactical control console is normally positioned at the commander's station, and displays essential data for fire control monitoring by means of a tactical display of battlefield area and targets detected by the various sensors providing information to the system.

30 mm TCM 30G gun mount used in Spider II air defence system

The system is being offered currently with the 30 mm TCM-30G cannon, but can be used with other weapons such as twin 35 mm Oerlikon or Bofors 40 mm guns.

During 1983 the Spider III version was revealed. This embodies missiles instead of guns to engage air targets and is described elsewhere in this edition.

CONTRACTOR

Israel Aircraft Industries Ltd, Electronics Division, Yehud Industrial Zone, 56000, Israel.

4590.131

TCM Mk 3 LIGHT AA SYSTEM

Based on the considerable experience gained with the 20 mm twin cannon TCM-20 anti-aircraft system, Israel Aircraft Industries has developed a successor system known as the TCM Mk 3. This new mount meets all the requirements of the Israel Defence Forces for mobile air defence, and is capable of deployment in both ground-based and naval applications.

The mount itself has the simple structure, operation, and maintenance demands of the TCM-20 supplemented by advanced sub-systems such as more modern drives and the ability to employ the latest sighting systems. The new gun system can be integrated with the fire control system developed and manufactured by MBT Systems and based on a computer sight with laser rangefinder.

Using suitable adapters, the TCM Mk 3 mount can accept virtually any light 20 – 25 mm AA gun, and aboard ship it may be used as either the main weapon system or as secondary armament to engage air, shore or surface targets.

Command and control of the mount is through a variable speed joystick mechanism. Mechanical movement of the joystick by the gunner is converted into electrical signals that are passed to servo amplifiers for subsequent conversion into drive commands for the training and elevating motors. The TCM Mk 3 has a new drive system consisting of two separate but similar drive units (traverse and elevation), each of which includes a 24 V servo amplifier, torque limiter, gear, and feedback system. This arrangement is very sensitive to joystick demands, enabling the gunner to track fast-moving targets as easily as very slow targets.

The TCM Mk 3 mount can be installed on towed or self-propelled carriages, and a wide range of guns such as the 20 mm HS-404, HS-804, HS-920, Rh 202 and M693, Soviet 23 mm ZU-23 and 25 mm cannon, have been allowed for in the design. A typical system (towed version with 2 × HS-404 guns) weighs 1350 kg ready for action. The basic version is provided with an M18 reflex optical sight, but other options include a × 4 magnification 'starlight' sight for night operation, or the MBT FCS which incorporates a computer sight and integral laser rangefinder.

CHARACTERISTICS

Traverse: Unlimited 360°
Velocity: 0.25°/s, 75 – 80°/s max
Elevation: −10 to +90° (mechanical); −6 to +85° (electrical)
Weight: 1350 kg (2 × HS-404, towed model, ready for combat)
Power supply: 24 V dc
Primary source: 2 × 12 V dc, 80 Ah batteries
Auxiliary source: 28 V dc, 2.8 kW generator

STATUS

In production and operational with Israeli Forces. The original TCM-20 system is in service with Israel and seven other countries. TCM Mk 3 is now in production.

CONTRACTOR

Israel Aircraft Industries Ltd, Ben Gurion International Airport, 70100, Israel.

RAM V1-L combat vehicle with TCM Mk 3 turret with 20 mm HS-404 guns

Towed TCM-20 gun mount for AA defence

ITALY

4535.131

SIDAM 25 AA TURRET

This is a one-man, powered turret mounting four Oerlikon KBA 25 mm automatic cannons, equipped with an electro-optical fire control system. The system is intended to provide effective protection against air attack for troops, armoured columns and other tactically significant targets. Because of the optical sight and the gun calibre, it is also suitable for ground fire purposes.

The turret is constructed of welded aluminium alloy and carries two symmetrical, externally mounted elevating masses, each of which is equipped with two 25 mm automatic guns. Each barrel has a cyclic rate of fire of 600 rounds/minute, and the rate of fire is selectable to provide single shot, 15- to 25-round controlled bursts and full rate firing. Two of the four cannon have dual feed capability for high explosive and armour piercing discarding sabot (APDS) ammunition, and 600 rounds HE and 30 APDS belted ready-to-fire rounds are provided in the turret ammunition magazines. The mode of fire (single shot, 15/25-round bursts, or continuous) can be selected from the commander's control panel, and the system can be operated entirely from within armour protection. Selection of the types of round to be fired is performed mechanically under the control of the commander.

The turret has hydraulically powered traverse through 360° and the elevation limits are –5 to +87°. The design allows easy installation, with minor modifications of the vehicle, in a variety of APCs. The prototypes tested by the Italian Army were mounted on M113A1 armoured personnel carrier chassis but the SIDAM 25 system can be mounted on a wide range of tracked and wheeled chassis, with the former including the OTO C13 private venture.

The system includes an Officine Galileo electro-optical sight that includes a self-stabilised optical dual periscope with a daylight TV camera, low-light TV camera and laser rangefinder, installed in the turret with an IFF device. An electronic tracking unit connected to the sight performs the angular tracking of the target, and target designation from an external source can be accommodated by a target alert display set. The fire control computer is fitted inside the vehicle in the gunner's console, together with an inertial attitude sensor. The fire control system is designed for two operators: the commander sitting in the turret and the gunner in the vehicle hull.

From acquisition by the commander, the target is assigned to the gunner, who carries out automatic or manual tracking by means of a joystick control. All these operations (except firing) can be carried out with the vehicle on the move. Modular design provides for the later enhancement of the system by the addition of such devices as a video compatible FLIR, an inertial navigation system, or passive IR to provide a 24-hour search and detection capability.

CHARACTERISTICS

Turret

Armament: 4 × 25 mm KBA dual-feed cannons
Rate of fire: 2400 rounds/minute (max)
Traverse: 360° continuous
Speed: 90°/s
Acceleration: 150°/s²
Elevation: –5 to +87°
Speed: 90°/s
Acceleration: 120°/s²
Fire control system: Electro-optical with laser rangefinder and computer. Attitude sensor
External designation: Automatic lock-on from external designation via data link
Weight: 3200 kg (combat ready)

Prototype of the SIDAM 25 anti-aircraft turret mounted on a M113A1 APC chassis for trials

DEVELOPMENT

The turret was built by Oto Melara in co-operation with Oerlikon Italiana and Officine Galileo, under an Italian Army programme.

STATUS

Italian Army acceptance tests on the SIDAM 25 were completed in December 1985 and production is now underway.

CONTRACTORS

Turret: Oto Melara SpA, 15 via Valdilocchi 1, 19100 La Spezia, Italy.

Automatic cannon: Oerlikon Italiana, Milan, Italy.

Fire control: Officine Galileo, Florence, Italy.

4594.131

OTOMATIC 76 mm 76/62 SPAAG

Oto Melara has developed a new self-propelled air defence gun based on the Palmaria 155 mm self-propelled howitzer chassis with a modified Oto76 mm 62 calibre naval gun as the weapon. The design also incorporates an all-weather optro-electronic fire control system by Officine Galileo, linked with an S-band surveillance radar and an Ka tracking radar provided by SMA. A major design consideration was an ability to detect and engage aircraft targets (ground attack helicopters in particular) at ranges greater than the anti-tank weapons the aircraft might be carrying, eg 6000 metres or more.

The general arrangement of the OTOMATIC 76/62 SPAAG can be seen from the photograph, from which

Prototype of the Oto Melara OTOMATIC 76/62 self-propelled anti-aircraft gun showing roof mounted search and tracking radars

it will be noted the radar installation is turret mounted. The turret is constructed of welded steel, with the commander and gunner seated on the right and the loader on the left. A 7.62 mm machine gun can be fitted to the turret top for local protection.

Elevation and traverse of the 76 mm gun are hydraulically actuated, with manual mechanical operation for emergency use. Elevation limits are +60 and –5 degrees, while there is unlimited traverse (360 degrees). A total of 28 rounds of ready use ammunition is carried, with a further 36 rounds in reserve in the turret and 26 in the hull. The ready use ammunition consists normally of 25 rounds AA and three anti-tank rounds.

The OTOMATIC self-propelled anti-aircraft gun can fire all the existing 76/62 ammunition at present in use with the standard naval gun systems currently produced by eight countries in four continents. The 76/62 ammunition includes the following types of projectiles:-

HE fitted with point detonating fuze or with proximity fuze.

PFF fitted with point detonating or with proximity fuze. This ammunition is pre-fragmented projectile developed to increase the performance of the conventional HE projectile.

APFSDS projectile is under development by Oto Melara to provide the OTOMATIC system with a good self-defence capability against armoured vehicles.

The 76 mm gun has a cyclic rate of fire of 120 rounds a minute although bursts of up to five rounds would be used against aerial targets and single shots against ground targets. Typical muzzle velocity is 900 metres a second.

The fire control system of the OTOMATIC consists of the following subsystems. There are two independent search subsystems:

(1) Search radar subsystem for all weather operation of both aircraft and helicopter detection, which includes track-while-scan and IFF.
(2) Search optical subsystem for daylight operation of ground target and aircraft/helicopter detection.

There are two independent tracking subsystems:

(1) Tracking radar subsystem for all weather operations including a TV camera.
(2) Tracking optical subsystem for day/night operations including a TV camera, laser and TV tracker.

The turret and weapon subsystem consists of a loading system with turret and weapon servos and stabilisation.

The navigation system consists of a strapdown inertial reference integrating an inertial navigator. The operator's command console consists of two TV displays (one colour and one black and white) for video and data display, command push button panels and command joysticks. The computer system consists of a normal operation computer and a back-up computer.

There are also meteorological sensors and a data link.

STATUS

The first prototype (with radars but without fire control system) was completed in January 1986 while the second prototype (with fire control system) was completed in 1987. Although prototype OTOMATICs use a slightly modified Palmaria chassis the turret can be installed on other chassis.

CONTRACTOR

Oto Melara SpA, 15 via Valdilocchi 1, 19100 La Spezia, Italy.

4589.131

BREDA TWIN 40L/70 AA FIELD MOUNTING

Breda Meccanica Bresciana has developed a field mounting version of its well established twin 40 mm naval anti-aircraft gun for use against low-flying aircraft and missile targets. The gun is fully automatic in operation and there is no need for any on-mount crew when the gun is in action. It is suitable for use with a wide range of fire control and target acquisition systems, for example the Signaalapparaten Flycatcher (**2301.181**).

Two Bofors 40 mm L/70 guns are combined to form a twin elevating mass within a specially designed cradle to comprise the Breda Twin 40L/70. The barrels of standard Bofors guns are located 30 cm apart to reduce elevating mass recoil forces. The mounting consists of a 360-degree traverse training platform supported on a wire race bearing. The platform has parallel vertical lightweight aluminium alloy trunnion supports, and also carries the servo and ammunition feed motors together with the upper feed mechanism, firing mechanism and system junction box. The upper portion of the mounting is completely enclosed within a watertight reinforced glass-fibre cupola which is provided with three access hatches, at the rear and on each side.

Ammunition is stored in a magazine beneath the mounting platform, within the carriage, and rounds are fed to the guns via separate hoisting chains for each gun. The magazine trains with the mounting, and the system is arranged so that if either gun is out of action the other can continue to fire. The forward section of the magazine supplies the left-hand gun and the rear section supplies the right-hand gun. The system is driven by a 400 V ac motor powered from an external generator; fast or slow drive facilities are provided. The slow drive operates the magazine conveyors and the scuttle transferring the rounds to the lower chain hoist; the fast drive operates the ammunition hoists and the scuttle at the top of the hoist. Rounds are fed into the lower magazine manually via the side hatches. Dc electric motors driving epicyclic gearboxes are used for gun elevation.

Towed Breda Twin 40L/70 40 mm AA mount

The fully digital servo-system has the following characteristics and advantages:

(1) Serial data transmission (two wire connection with the fire control system).
(2) Improvement of servo technical performance by a better control.
(3) High tracking accuracy.
(4) Specific diagnostic messages to the gun operator in case of failure, for example "change fuze No 23" or corrective intervention, eg "close the hatches". General automatic test program stored in the computer for a general electric and servo-technical check of the gun under request of the operator. Specific test programs capable of identifying the position of failed electronic components by means of a special test meter.
(5) Easy storage of different obstacle contouring and non firing zone, alignment correction carried out by the local control panel and fully automatic tilt correction.

CHARACTERISTICS

Calibre: 40 mm
Barrel length: 2.8 m
Total weight: 9900 kg (less ammunition)
Weight of ammunition: 1100 kg
Length: 8.05 m (travelling)
Width: 3.2 m (travelling)
Height: 3.65 m (travelling)
Training velocity: 90°/s
Training acceleration: 120°/s²
Elevation velocity: 60°/s
Elevation acceleration: 120°/s²
Magazine capacity: 444 rounds
Rate of fire: 660 rounds/minute (both barrels)

STATUS

Production. In service with Venezuelan forces.

CONTRACTOR

Breda Meccanica Bresciana SpA, 2 Via Lunga, 25128, Brescia, Italy.

4536.131

BREDA LIGHTWEIGHT 30 mm AA FIELD MOUNTING (SENTINEL)

The Breda twin 30 mm mounting is a new lightweight anti-aircraft towed field mounting introduced in 1983 to provide defence against air attack for troops, and other deployed ground forces. It is provided with an autonomous power supply unit and an electro-optical sighting system supplied by Officine Galileo. The latter system incorporates a laser rangefinder and infra-red facilities for night tracking of targets in addition to daylight optical tracking. Aim-off is applied automatically in optical tracking.

The guns fire standard 30 mm GAU-8 high explosive ammunition at a maximum rate of 1600 rounds/minute, but API and APDS ammunition is also available for use against enemy armour and there are frangible rounds for engagement of aircraft targets.

The Breda twin 30 mm AA field mounting is suitable for integration with a variety of gun/missile air defence systems such as the Roland Co-ordination Centre (RCC).

The Sentinel system can, as an option, be supplied with hydraulic road wheel drives to enable it to be manouvred by the operator to a firing site clear of the column or position it after a helicopter drop.

Also provided as an option is fully automatic extension of the out riggers, setting of the jacks and levelling, all of which can be carried out in about 20 seconds.

CHARACTERISTICS

Calibre: 30 mm
Barrel length: 2.458 m
Carriage: 4-wheeled with two outriggers
Shield: Yes
Weight: 5000 kg (with ammunition)
Length: 6.46 m (travelling)
Width: 1.76 m (travelling)
Height: 1.94 m (travelling)

Ground clearance: 0.43 m
Wheelbase: 3.5 m
Elevation/depression: +85°/–5° at 60°/s
Elevation acceleration: 90°/s²
Traverse: 360° at 80°/s
Traverse acceleration: 120°/s²
Flight time to 3000 m: 5.2 s
Rate of fire: 1600 rounds/minute (2 barrels)
Muzzle velocity: 1040 m/s

STATUS

Development complete. Ready for production.

CONTRACTOR

Breda Meccanica Bresciana SpA, Via Lunga 2, 25128 Brescia, Italy.

Breda lightweight twin 30 mm AA towed field mounting deployed

JAPAN

3724.131

AW-X SELF-PROPELLED TWIN 35 MM AA GUN SYSTEM

The AW-X twin 35 mm self-propelled anti-aircraft gun is under development to meet a Japanese Ground Self-Defence Force requirement for a replacement for the old American supplied twin 40 mm M42 self-propelled anti-aircraft guns which are limited to clear weather operations.

The prototype consists of the chassis of the Type 74 MBT fitted with a new power operated turret armed with twin Oerlikon-Bührle 35 mm KDA cannon, and all weather surveillance and tracking radars mounted on the roof at the turret rear.

Mounted externally on left side of the tracking radar is a flat box which is believed to contain a laser rangefinder, optical tracker and possibly a LLLTV. It is believed that a laser device is installed in the forward part of the turret.

STATUS

The first prototype of the AW-X started trials in 1984. Not yet in production or service with Japanese Ground Self-Defence Forces.

CONTRACTORS

Mitsubishi Heavy Industries (chassis), Nippon Seiko-Jyo (armament and ammunition system) with Mitsu-bishi-Denki being prime contractor.

SPAIN

3208.131

MEROKA 20 mm ANTI-AIRCRAFT SYSTEM

Meroka is a novel close-in air defence weapon system that has been under development for a number of years by the Spanish authorities. Both land-based and naval versions have been designed, the latter is described in entry **3176.203** in the naval ordnance section of this volume. Some of the background to the programme has appeared under the above entry number (**3208.131**) in 1983-84 and earlier editions.

The weapon used in both versions is a twelve-barrelled arrangement of 20 mm/120 calibre Oerlikon cannons, grouped together in two horizontal rows of six barrels. A salvo of 12 rounds can be fired in a time of 0.08 seconds, which amounts to a theoretical rate of fire of 9000 rounds/minute. The main purpose of this high rate is less the delivery of a large total of projectiles on to a target but more to meet a need for maximum density of fire when engaging certain categories of air targets, such as low-flying ground attack aircraft, and sea-skimming anti-ship missiles in the naval application. The normal practice is for each salvo to be fired as four groups of three rounds. This minimises the total recoil force applied to the mount as well as having other beneficial operating characteristics.

The group of 12 barrels forms a single rigid unit which is traversed and elevated as a conventional gun, mechanical rigidity being aided by a number of steel bands (two in the army version, four in the naval version) that determine the spacing between the individual barrels. This feature is also used to adjust the dispersion of fire and fire pattern. Ammunition is brought to the 12 guns by two belt feeds which are supplied from a hydraulically powered rotating magazine with a capacity of 720 rounds. Elevation limits are –5 to +85 degrees, and the gun can traverse in azimuth through 360 degrees at a rate of 90 degrees/second.

A simple electro-optical fire control system that incorporates a TV camera, laser rangefinder and digital computer is used, this system being bore-sighted to the gun barrel axis. Aim-off to compensate for target motion is automatically injected when the fire button is pressed after having tracked the target for several seconds.

The prototype land-based Meroka AA gun is mounted on a towed four-wheeled carriage, and incorporates the gunner's position and a self-contained two-stroke engine that drives the hydraulic power unit for gun slewing and elevation, ammunition feed etc.

CETME Mekora 20 mm anti-aircraft system

CHARACTERISTICS

Weapon: 12-barrel 20 mm Oerlikon
Barrel length: 2.4 m (120 cal)
Recoiling mass: 300 kg
Theoretical rate of fire: 9000 rounds/minute
Magazine capacity: 720 rounds
Elevation: –5 to +85°
Traverse: Unlimited 360°
Effective range: 3000+ m
Time into action: Approx 2 minutes
System weight: Approx 5 t, ready for action

STATUS

Prototype(s) completed and initial evaluation trials continuing.

CONTRACTOR

CETME, Centro de Estudios Técnicos de Materiales Especiales, Spain.

SOUTH AFRICA

6588.303

RATTLER WEAPON SYSTEM

The Rattler multi-purpose weapon system has been designed by the Atlas Aircraft Corporation with the assistance of Lyttleton Engineering Works (LEW), for installation in helicopters, tracked and wheeled vehicles, surface craft and fixed installations. Its first application is understood to be in Puma Transport helicopters of the South African Air Force.

The Rattler is of modular design and can accommodate various types of weapon by simply changing the gun cradle and electronics module. When fitted with four 7.62 mm S77 machine guns, the operator is provided with a manual switch to enable him to select either upper or lower twin weapons or fire all four weapons together. When fitted with the 20 mm GAI cannon, ammunition is fed from the left and when fitted with four 7.62 mm S77 machine guns ammunition is fed from an ammunition box either side and below the gunner's seat.

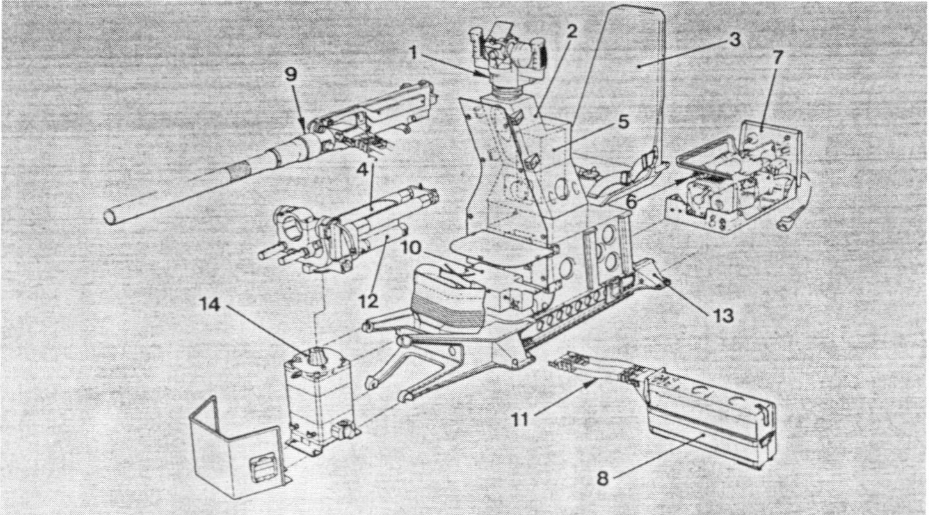

1 *Sighting unit,* **2** *Sight pedestal,* **3** *Gunner's seat,* **4** *Weapon beam,* **5** *Electronics box,* **6** *Accumulator charging point,* **7** *Hydraulic power pack,* **8** *Ammunition box,* **9** *GA1 Automatic gun,* **10** *Relay box,* **11** *Ammunition feedchute,* **12** *Elevation jack,* **13** *Floor structure,* **14** *Servo motor*

Main components of the Rattler multi-purpose servo-operated weapon system fitted with 20 mm GAI cannon

The gunner is seated to the rear of the weapons and is provided with an armour plated seat. He aims the weapons via a Gannet fixed reflector type sight to his front. Traverse and depression is powered via hand grips either side of the sight which also contain the triggers.

The electronics box is mounted below the gunner's triggers while the hydraulic system is located under the gunner's seat. Frontal armour protection is provided for the system.

To compensate for forward velocity and attitude, analogue correction has been incorporated and safety is ensured since the firing trigger can be activated only when the safety and tracking switches are both engaged.

According to ARMSCOR, accuracy has been vastly improved compared with current manual systems, and it is simple to install, operate and maintain.

CHARACTERISTICS

Calibre:	20 mm	7.62 mm
Type of weapon:	1 x GAI cannon	4 x S77 MG
Ready rounds (total):	240	4000
Rate of fire (total):	600 rpm	2000 rpm
Elevation range:	0 to 30°	0 to 38°
Azimuth range:	–40 to +20°	–25 to +25°
Laying speed:	50°/s	50°/s
Length:	2.60 m	1.90 m
Width:	0.50 m	0.80 m
Height:	1 m	1 m
Loaded weight:	310 kg	335 kg
Unloaded weight:	228 kg	300 kg

STATUS

In production. In service with the South African Defence Forces.

CONTRACTOR

ARMSCOR, Private Bag X337, Pretoria 0001, Republic of South Africa.

SWEDEN

4837.131

BOFORS 40 mm ARMOUR TRINITY

Trinity is the name given to a new 40 mm Bofors weapon system designed for use in both land-based and naval applications for close-in air defence purposes. The name derives, according to the manufacturer, from advances made in all three main elements of the system – fire control, the 40 mm gun, and the ammunition. Although initially conceived as an air defence weapon system, in the course of design parallel developments in all three aspects led to it becoming an all-purpose system.

The land-mobile version is known as Armour Trinity. The system is configured to accommodate most of the major elements within an armoured turret which is suitable for mounting on a variety of AFVs such as the MOWAG Shark, TAD, Steyr-Daimler-Puch etc. The gun can be controlled from either the vehicle itself, via a TV viewing the target through the gun sight, or from an armoured cab to the left of the barrel. A tracking radar may be fitted to the right of the barrel and the sight can be equipped with an IR tracker. An on-mount local search radar is regarded as a highly desirable addition.

Trinity has an integrated fire control system in which search and tracking radars can be combined with a digital computer, a Bofors-developed BOF 8016 machine, and other sensors to provide a modular FCS. In addition to primary radar(s) for search and tracking functions, it is possible to integrate IFF. Other sensors that may be incorporated include laser rangefinder, optical day and night channels, and gyro reference and motion sensors for example.

There is an integral sight for all visual lines of sight, daylight, night vision or IR, and laser rangefinder. The fire control computer inhibits fire until it has calculated a sufficiently high kill probability. This results in a need for only short bursts, normally five to 10 rounds, which means that 100 on-mount rounds are normally sufficient for 10-15 engagements.

An important function of the computer complex is the fuze setting program devised by the Swedish Philips organisation. This enables each individual round to have its fuze setting programmed while it is in the final stages of ammunition feed. In this way, the pre-programmed burst of fire can be suited to the nature of each target. Each burst under proximity fuze control can cover 180 m² of airspace with lethal fragments. Using ten-round bursts it is possible to ring an aircraft or missile target to ensure a kill.

An improved version of the L/70 40 mm gun is used in the Trinity system, which improves accuracy by better muzzle energy from the new 40 mm Trinity 3P ammunition that gives reduced ballistic dispersion. The 3P round is all-purpose ammunition, with an adequate anti-armour capability against light armour, thanks to projectile and fuze developments. A penetrator round is projected which will deal with APCs or side-armour of battle tanks.

MOWAG Shark (8 × 8) weapons system carrier being used for Bofors 40 mm Trinity firings in 1985

CHARACTERISTICS

Weapon: Single barrel 40 mm
Rate of fire: 330 rounds/minute
On-mount ammunition: 100 rounds
Fire control: Digital, radar search and tracking, plus optical and E-O lines-of-sight

STATUS

The Bofors 40 mm Trinity system was announced in September 1984 with first firings taking place in 1985 on a MOWAG Shark (8 × 8) weapons system carrier. Expected to commence production in 1987/88 on receipt of orders, but as of January 1987 no firm orders for Trinity had been announced.

CONTRACTOR

AB Bofors, S-691 80 Bofors, Sweden.

2349.131

BOFORS 40 mm BOFI GUN SYSTEM

Based on the Bofors mobile 40 mm L/70 gun (**2360.203**) this new system incorporates an all-weather fire control instrument of the multi-sensor type, BOFI (**2378.181**), integrated with the gun. The system also includes a built-in motor generator and proximity-fuzed prefragmented ammunition. The BOFI gun system is entirely autonomous, each gun forming an individual firing unit able to engage its own targets.

As the main sensor a Ku-band (J-band) radar is used, which, after target designation from a central search radar or optical target indicator, acquires the target fully automatically on fixed frequency with MTI, and thereafter switches to frequency agility during the tracking mode to provide the greatest possible tracking accuracy. By this means the same high precision as achieved with semi-automatic electro-optical tracking using a laser rangefinder and optical sight can be maintained. The tracking radar and the electro-optical sensors are independent of each other and switching between them can be done to match any tactical situation.

A central search radar, such as the Giraffe (**1957.153**) or the Reporter (**4469.181**) system which also acts as a combat control centre for the battery of autonomous guns, contributes to the high effectiveness achieved. To further increase the defence potential a combat control radar can lead a mixed organisation comprised of BOFI guns and RBS 70 missile units.

Latest version of Bofors 40 mm BOFI anti-aircraft gun system incorporates multi-sensor fire control, a computer and power generator

Principal characteristics of the system are:
(1) all-weather capability
(2) automatic acquisition and tracking
(3) high resistance to jamming
(4) capability of individual combat of targets
(5) highly effective against aircraft, helicopters, missiles and ground targets
(6) tactical flexibility and endurance
(7) short deployment and re-deployment times.

CHARACTERISTICS

Weight: 5500 kg
Rate of fire: 300 rounds/minute
Tactical range: 3700 m
Ammunition supply: 118 rounds on gun – sufficient for 12 engagements
System accuracy: 3 mrad
Radar accuracy: Ku-band (J-band) monopulse MTI or frequency agility
Frequency: 15.9 – 17.1 GHz
Frequency agility: 1000 Hz
Lobe width: 1.9°
Laser rangefinder: 10 Hz Nd YAG
Night sight: 3-stage image intensifier
Day sight: Telescope
Ammunition
Type: Proximity-fuzed prefragmented (PFHE)
MV: 1025 m/s
Weight of shell: 880 g
Explosive weight: 120 g
Type: High Explosive Tracer (HE-T)
MV: 1005 m/s
Weight of shell: 960 g
Explosive weight: 103 g

STATUS

The fair weather version of the system has been in production since 1976. The all-weather BOFI gun system completed its design tests during 1979 and has been in production ever since.

Design of proximity-fuzed ammunition was completed during 1974 and is now in large scale series production.

CONTRACTOR

AB Bofors, S-691 80 Bofors, Sweden.

6729.131

UPDATING KIT FOR BOFORS 40 mm L/60 GUN SYSTEM

In view of the large number of 40 mm FAK L/60 anti-aircraft guns still in use Bofors Ordnance have developed a kit for updating these weapons. The kit includes a new sighting and fire control system, an operator console, a power laying gear and various improvements to the gun itself.

The fire control system comprises an optical sight, a laser rangefinder, a fire control computer and operator terminal. The sight presents a magnified target image through a wide field-of-view telescope and provides line-of-sight aiming via a servo-controlled mirror. The computer provides real time computation of gun lead angle and super elevation as functions of target range and predicted kinematics, meteorological data and type of ammunition. The system also has built-in-test equipment. The servo-controlled mirror, optical system and laser range-finder are in a combined housing mounted on the gun. The servo controls make the line-of-sight and laser rangefinder independent of small, unexpected movements of the gun and ensures a constant stream of data to the computer for greater accuracy.

The operator console is mounted on the gun platform and contains the amplifiers and electronics of the laying system servo as well as the controls. The seat has been designed for maximum ergonomic efficiency.

The power laying gear comprises separate traverse and elevation laying units each using a hydraulic motor. The maximum traverse laying speed is a minimum of 80°/s while maximum elevation laying speed is a minimum of 45°/s.

Mechanical improvements are made to the gun to improve the rate of fire to 300 rounds per minute, these include installing new recoil buffers, recuperator springs and rammer rods while a 20-round magazine is fitted to the elevating mass to enable several targets to be engaged without reloading. Other improvements include the installation of a proximity fuze disconnector, the installation of a 24-round ammunition rack on the gun platform and an optical target indicator which is linked to the gun by cable.

A model of the updated Bofors 40 mm L/60 gun

In addition to AP, HE and TP rounds the new gun can use recently introduced PFHE (Proximity Fuzed HE) and APHC-T (Armour-Piercing High Capacity Tracer) rounds.

STATUS

In December 1987 Bofors received a SEK220 million contract from the Dutch Army to modernise 60 guns. The first six will be modernised in Sweden and the remainder will be modernised by RDM in Rotterdam. As a result the guns will now stay in service for some 20 years.

CONTRACTOR

AB Bofors, S-691 80, Bofors, Sweden.

SWITZERLAND

4537.131

OERLIKON TWIN FIELD 35 mm AIR DEFENCE GUN TYPE GDF

The Oerlikon twin 35 mm air defence field gun system has been produced and delivered in considerable numbers over a long period, frequently for use in either the Super-Fledermaus or the Skyguard air defence system. The system has been described in **2374.131** in earlier editions of *Jane's Weapon Systems*, and in related entries. More recently the manufacturer has introduced a range of improvements for the Type GDF 35 mm guns and these are available as kits for updating existing guns, or the modifications may be incorporated in GDF series guns to be delivered. Two levels of modification have been developed, referred to as Kits NDF-A or -C, and at the same time the company has introduced a new generation of 35 mm AD ammunition comprising fragmentation shells with electronic or mechanical impact base fuzes designed to ensure full detonation inside aircraft targets. The full NDF-C kit includes the following items, while omission of certain items results in the NDF-A modification kit:

(1) weapon cover with integrated, automatic weapon lubrication system
(2) optimised automatic cannon function which stabilises rate of fire
(3) automatic ammunition replenishers
(4) integrated power supply unit
(5) Gun-King minisight with laser and digital computer for 3-D target tracking
(6) new gun control system.

The newly developed cover provides an improved, built-in protection of the automatic cannon against dust, sand, snow and water. An automatic weapon lubrication system is built into the cover. New mathematical models studied by the company based on new forms of threat from the air indicate that longer bursts of fire may be necessary in the future. The gas pressure (which is used to unlock the breech block in the 35 mm gas-operated cannon) will therefore increase and cause the rates of fire to increase. This increase is held within reasonable limits by rate-of-fire regulators so that wearing parts and the weapon functions themselves are not unnecessarily overtaxed. In this way automatic fire rate is limited to about 550 rounds/minute.

The automatic units are re-supplied with ammunition manually from the rear by the gun crew. The crew required on the gun is reduced from three to one; the gunner. The rounds are thrust into the ammunition feed system by hydraulic-mechanical devices. The barrel is elevated to 90° for the automatic replenishing cycle. Electrical and mechanical limiters prevent inadvertent firing during the replenishing procedure. The automatic replenisher substantially increases the available ready-for-firing ammunition supply.

Oerlikon-Bührle twin 35 mm GDF-005 air defence gun in firing position

The power supply unit mounted on the gun is quite similar to that of the Skyguard fire control unit. When the gun is in operation, the power supply unit (which is located on the lower mount) is lowered to allow the barrel to turn freely even at the maximum angle of depression. The integrated supply unit simplifies taking up position since the single axle trailer formerly used is no longer required.

A new thyristor gun control system is located on the upper mount. Its coarse and fine transmission system operates through an on-gun microcomputer identical to that of the Gun-King sight. The new gun control system and the integrated power supply result in enhanced combat effectiveness, greater availability, assurance of continued availability etc, reduction in number of vehicles (from five to three per firing unit), and shorter times required to move into position. Diesel or petrol engines can be used (to provide power) as specified.

The minisight, known as Gun-King and developed by Contraves, is an autonomous device that carries out a 3-D electro-optical target tracking for each gun. It can be integrated in place of the XABA or the Ferranti sight. This greatly extends the autonomous operation potential of the gun under fair weather conditions. Sight operation is extremely simple; the gunner only needs to bring the periscope crosshairs into coincidence with the target. An acoustic signal indicates that firing can commence. A built-in laser measures the range of the target which, together with angular data, meteorological data and the continuously registered v_0 values, is processed by an integrated digital computer to give lead data which is then transmitted to the gun control system. The Gun-King sight, as well as featuring in the NDF combat improvement kit for GDF 35 mm AD guns, is also a major element of the Diana 25 mm field AD gun system (**4525.131**). The main features of this new sight are as follows:

There is a common optical path in the periscope for the laser beam and for the aiming line of sight. These beams are deflected with high accuracy by a gimbaled mirror, thus a fixed eye piece is provided. The periscope has an extremely high light transmittance and is fitted with a multi-spectral optical system. Night operation is possible. Range finding is effected by a multi-divergence laser. For ground targets, the laser beam is narrowed by a divergency switch-over device, thus eliminating ground clutter.

The new Contraves high-speed computer for Gun-King calculates not only all fire control data precisely with optimum control, filter and ballistic algorithms, but also controls the periscope, laser triggering during the tracking phase and the gun drive electronics. At the moment of optimum hit probability, the operator receives an acoustic alarm to open fire. Target changes can be effected very rapidly because target data is stored in the memory. Provision is made for the insertion of muzzle velocity and meteorological data. All data relating to weather and weapon parameters is entered into the computer and the operator can therefore immediately concentrate on precise target tracking. Sight and gun are commanded entirely by the control yoke which provides hands-on control.

NDF Series improvements can be carried out on35 mm AD guns as follows:

GDF-001, equipped with the XABA sight. (No longer produced.)

GDF-002, equipped with Ferranti sight.

GDF-001/002, with NDF-A improvement kit integrated. (These guns have already been improved with assemblies that are also part of the NDF-C. Further combat improvements involve the remaining assemblies.) The same applies to the GDF-003 (series production of the 35 mm AD gun with NDF-A assemblies).

In May 1985 the GDF-005 was introduced, which is an overall improvement of the earlier GDF-002, and earlier models can be modified to the GDF-005 standard by use of combat improvement kits supplied by Oerlikon-Bührle. The GDF-005 features a new autonomous gun sighting system, on-board power supply, an automatic loader and other improvements.

STATUS

In production and frequently used by those customers who also use the Skyguard fire control system (**2377.181**). One of the latest customers is Canada which has ordered 20 GDF-005s.

CONTRACTOR

Machine Tool Works Oerlikon-Bührle Ltd, 8050 Zurich, Switzerland.

Oerlikon-Bührle GDF-003 twin 35 mm AD field gun incorporates NDF-C kit improvement features such as Gun-King minisight, new ammunition/reloading provisions, and power supply unit

3723.131

OERLIKON ESCORTER TWIN 35 mm SP AIR DEFENCE SYSTEM

The private venture Oerlikon-Bührle Escorter twin 35 mm air defence system is intended as a self-propelled autonomous air defence system that fills the gap between towed mobile AA guns such as the 35 mm field AA system described in **2374.131** in earlier editions (but which is still widely operated in at least 20 countries), and the more expensive and later AA tanks. Operational applications envisaged include the provision of air defence protection for: mechanised combat support troops (artillery, engineers etc) in action and during deployment; motorised formations; and sites of short-term tactical importance. Operationally, the Escorter is deployed within tactical units and engages targets independently.

The most important sub-systems are:

(1) twin 35 mm Oerlikon gun
(2) drum magazines
(3) high performance sight with FLIR
(4) search radar (consisting of sub-assemblies also employed in the ADATS system (**4158.131**)).

The gun is designed to engage aircraft, helicopters, drones and lightly armoured ground targets at ranges of up to 4000 metres. Three ammunition types with varying ballistics can be selected and fired. With regard to range, accuracy, response time and reliability, the performance of the system in airspace surveillance, target designation and fire control functions is optimised to the deployment spectrum. Full airspace surveillance and IFF functions are maintained even during target tracking.

Effective splinter protection for crew members in the vehicle and gunner's cab is provided. All operating sequences are designed to allow operation of the entire weapon system by a crew consisting of commander, gunner and driver.

CHARACTERISTICS
Calibre: 35 mm
Rate of fire: 2 × 600 rounds/minute
Ammunition: 430 rounds, up to 3 types selectable
MV: 1175 m/s (HEI); 1385 m/s (APDS-T/APFIDS)
Time for ammunition change: 1.5 s
Reload time: 4 minutes
Weapon range: 4000 m
Search radar: Track-while-scan
Range: 20 km
Antenna rotation: 60 rpm
Tracking system
Optical magnification: × 8
Field of view: 8°
FLIR: 5 km range
Laser range: 5 km
PRF: 12.5 Hz

GENERAL DATA
Type: Wheeled
Length: 8.745 m
Width: 2.98 m
Height (minimum): 3.934 m
Wheel base: 4.74 m
Ground clearance (max): 0.78 m
Engine rating: 335 kW
Max speed: 120 km/h
Range: 600 km

Oerlikon-Bührle Escorter twin 35 mm self-propelled air defence system in firing position

Oerlikon-Bührle Escorter twin 35 mm self-propelled air defence system in travelling configuration showing four wheel steering

STATUS
Prototypes of the Escorter (previously called the GDF-DO3) have been built and tested. In January 1986 it was stated that first production vehicles were expected to be completed in 30 months. As of February 1987 no firm sales of the Escorter had been announced although it has been successfully demonstrated in the Middle East. The tracked GDF-CO3 is no longer being offered.

CONTRACTOR
Machine Tool Works Oerlikon-Bührle Ltd, 8050 Zurich, Switzerland.

3330.131

FLAKPANZER 35 mm AA TANKS

The 35 mm AA tank was designed for full combat autonomy featuring its own search radar with integrated IFF for continuous air surveillance, a tracking radar for automatic target acquisition and tracking, and two persicopes combined with a target indicator for optical target acquisition/tracking.

The entire fire control and gun system is integrated in an armoured turret, leaving only the auxiliary power unit and some additional units such as the land navigation equipment to be placed in the hull. This makes the weapon system particularly adaptable to various existing MBT vehicles.

The 5 PFZ-B2 Gepard and 5 PFZ-CA1 Caesar versions – both modified Leopard 1 vehicles – differ mainly in their radar equipment.

5 PFZ-CA1 Caesar system on Leopard 1 chassis

5 PFZ-B2 (Flakpanzer 1 – Gepard)

This model has a Siemens MPDR 12/4 fully coherent pulse-doppler search radar (**3286.153**) with integrated MSR 400 IFF equipment. In addition to its own search data display it can receive and display air situation information from an external surveillance radar. The tracking Siemens-Albis radar is a monopulse-doppler type. The miniaturised and transistorised analogue prediction computer of Contraves design calculates the gun lead angles taking into account the continuously measured muzzle velocity, the meteorological corrections and the vehicle's deck tilt. The computer furthermore gives the open-fire signal and determines the burst length of the gun for high hit probability and most economic use of ammunition.

RADAR CHARACTERISTICS
Search frequency: S-band
Switchable frequencies: 6
Max range: 16 km
Antenna rpm: 60
Sub-clutter visibility: ⩾57 dB
Tracking frequency: Ku-band
Switchable frequencies: 2
Max range: 15 km
Minimum range: 300 m
Sub-clutter visibility: ⩾23 dB
(for further details see **3286.153**)

5 PFZ-CA1 (Caesar)

This model has an integrated search and tracking radar of Hollandse Signaalapparaten manufacture, with a common transmitter for search and tracking. The search section is an MTI system with a digital double canceller. It too uses an integrated MSR 400 IFF system. The tracking section is of the monopulse-doppler type with a cassegrain antenna. There is an additional coaxial anti-image tracking channel. The rest of the fire control and gun system is practically identical to the B2-version.

RADAR CHARACTERISTICS
Search frequency: X-band
Tracking frequencies: X-band; Ka-band
Max range: 15 km
Minimum range for tracking: 300 m
(for further details see **2302.153**)

ATAK-35

Work on this programme has now ceased.

STATUS

The West German Army took delivery of 420 5 PFZ-Bs between 1976 and 1980 while the Belgian Army took delivery of 55 5 PFZ-Bs between 1977 and 1988. The Dutch Army took delivery of 95 CA1 production vehicles between 1977 and 1979. Of the 420 Gepards delivered to the West German Army, the first 195 vehicles were delivered as B2 versions with the remaining 225 as the B2L which are fitted with a laser rangefinder of Siemens design and manufacture.

In order for the Gepard to remain effective into the next decade it will require a mid-life update.

CONTRACTOR

Contraves AG, Schaffhauserstrasse 580, 8052 Zurich, Switzerland.

5 PFZ-B2 Gepard twin 35 mm self-propelled air-defence system with both radars in position

4525.131

OERLIKON 25 mm TWIN FIELD AD GUN (DIANA)

The Oerlikon 25 mm field AA gun (Diana) is a towed high-performance gun intended primarily for the defence of stationary targets (eg airfields, depots etc) against low-level air attack. It is produced in two versions: GBF-AOA with twin KBA 25 mm cannons and Officine Galileo P75 3-D sight; and the GBF-BOB with two KBB 25 mm cannons and a Contraves Gun-King sight. The former version uses NATO standardised 25 mm ammunition and the Italian sight, otherwise the two Diana versions are very similar, apart from the slightly lower rate of fire of the KBA cannon compared with the KBB.

The Diana system has two externally mounted Oerlikon 25 mm automatic belt-fed cannons, a gunner's cab with a sighting system for 3-D target tracking and an integral power supply. The high rate of fire of the twin cannons, the ammunition performance and the effectiveness of the sight together ensure both high hit and high kill probabilities against all hostile targets.

Conversion of the gun from the travelling to the action configuration is rapid: the mounting legs are extended until the two wheels lift clear of the ground and swivel inwards. The location of the gunner's cab in the centre of the rotating structure between the externally mounted cannons enables both the overall length and height of the gun to be minimised, as well as reducing rotational effects on the operator.

Type	NATO Designation	Round Weight	Round Length	MV
Fragmentation HE incendiary	HEI	625 g	288 mm	1160 m/s
Discarding sabot hard core projectile with tracer	APDS-T	550 g	288 mm	1460 m/s
Practice with tracer	TP-T	625 g	288 mm	1160 m/s
Practice with tracer	APP-T	550 g	288 mm	1355 m/s

Target acquisition and designation is carried out either optically and autonomously or by a remote radar unit such as the Skyguard central fire control system (**2377.181**), but in both cases the target is tracked by the gunner using his 3-D sight. This is normally the new Contraves Gun-King laser-computer controlled sight which can be employed for the engagement of both air and ground targets by autonomous gunfire, although the Officine Galileo P75 optical sight may be fitted in the GBF-AOA version. Precise fire control data is automatically determined and updated by a digital microprocessor developed by Contraves, and coupled with a laser rangefinder. The computer controls the combat sequence of the entire gun system.

The system has two 25 mm automatic cannon (KBA or KBB), each with a cyclic rate of fire of 570 or 800 rounds/minute and 250 rounds each of ready-use ammunition. The dual belt feed allows the gunner to select either of two types of ammunition. The mount has an integral power supply for both electric and hydraulic operation and it can also be operated manually if required.

The ammunition employed with the KBB gun has been designed to ensure optimum efficiency against targets of all kinds, and the high muzzle velocities and flat trajectories with short flight times give high hit probabilities. Available ammunition includes four types as shown above:

The fragmentation HE incendiary shell is primarily for use against soft or lightly-armoured targets (such as aircraft, helicopters etc) and is characterised by intense blast and fragmentation effects inside the targets. The discarding sabot ammunition is most suitable for use against armoured targets such as APCs.

CHARACTERISTICS
Gun: Twin cannon (KBA/KBB)
Calibre: 25 mm
Rate of fire: 2 × 570/800 rounds/minute
Barrel length: 2.30 m (92 calibre)
Ammunition: 250 rounds/gun

Diana GBF-AOA with P75 sight and twin KBA 25 mm cannon

Diana GBF-BOB version from rear in firing position

GBF-BOB version of Diana ready for the road

Weight: 1725 kg (less ammunition); 2100 kg (ready to fire)
Length: 4.5 m
Width: 2.0 m
Height: (travelling position) 2.28 m
Elevation: –5 to +85°
Traverse: 360°
Traverse speed: 1.4 rad/s (max)
Traverse acceleration: 1.5 rad/s²
Elevation speed: 1.0 rad/s (max)
Elevation acceleration: 1.2 rad/s²
Sight: Gun-King
Magnification: ×8
Field of view: 8°
Laser rangefinder
Range: ⩾5 km
Accuracy: ±5 m
PRF: 6 Hz
Max lead angle: ±20°

STATUS
Development complete. Ready for production.

CONTRACTOR
Machine Tool Works Oerlikon-Bürhle Ltd, 8050 Zurich, Switzerland.

UNION OF SOVIET SOCIALIST REPUBLICS

5548.131

23 mm ZSU-23-4 SP AA VEHICLE

The ZSU-23-4 has been in service with the Soviet Army since 1965, and subsequently has been built in large numbers. In the Soviet Army the ZSU-23-4 is issued on the scale of four per tank and motorised rifle regiment, giving a total of 16 systems per division. The ZSU-23-4 saw extensive service in the 1973 Middle East campaign when it proved to be one of the most effective of all the low-level anti-aircraft systems employed.

It has a chassis developed from the PT-76 light tank which is almost identical to that used for the SA-6 (**2930.131**). The four ZU-23 cannons are similar to those used on the ZU-23 twin towed anti-aircraft system (**5581.103** in 1982-83 and earlier editions), but are water-cooled rather than air-cooled.

Each barrel has a cyclic rate of fire of 800/1000 rounds/minute, but their practical rate of fire is 200 rounds/minute, and they normally fire in bursts of 50 rounds/barrel. A total of 2000 rounds of 23 mm ammunition is carried, 500 per barrel. Two types of ammunition are used, the HEI projectile weighs 0.19 kg while the API projectile weighs 0.189 kg; both have a muzzle velocity of 970 metres/second. The guns can be used both against ground and air targets and can be laid and fired while on the move.

The turret has full powered traverse through 360°, elevation limits being from –7 to +80°. The B-76 radar (also known in the West as Gun Dish, **2876.153**) is mounted at the rear of the turret and has both target acquisition and fire control capabilities. The radar can pick targets up at a maximum range of 20 km and operates in the J-band. An MTI facility is provided, as is a computer. Optical sights are provided for use in both the ground and air roles. Day and night vision devices are incorporated, and the turret has a TPKU-2 daylight observation periscope plus two BM-190 viewers. The first type, TPKU-2 has an elevation adjustment. For night observation, the TPKU-2 is replaced by a TKN-1T infra-red periscopic sight. A generator is mounted in the forward part of the hull to provide power when the main engine is not being used.

OPERATION
The ZSU-23-4 vehicle is operated by a crew of four: commander, search/surveillance radar operator/ gunner, rangefinder, and driver, the first three being housed in the turret. Either of the first two crew members can control gun firing, stowing and erection, and all three crew members stationed in the turret are separated from the guns by a vertical armour bulkhead which provides protection from fragments, propellant gases and flame. During live firing, guns can be positioned automatically by the radar fire control system or manually by hand-wheels. Radar directed fire is usual against air targets, radar modes providing for search and detection, after acquisition followed by tracking. A conical scan pattern is employed for tracking and this gives angular error signals relative to the target's position from the radar boresight axis. The latter signals are used to calculate drive signals for the turret to bring the stabilised line of sight and the predicted target into coincidence. Two radar CRT displays are provided, one for the search and tracking functions carried out by the radar operator/gunner, and the other for range measurement by the ranging operator. Optical fire direction, with optical target tracking is also possible, possibly allied to radar ranging. To permit firing on the move, stabilisation is provided.

ZSU-23-4 self-propelled anti-aircraft quadruple 23 mm gun vehicle

CHARACTERISTICS
Crew: 4
Weight: 20.5 t
Length: 6.54 m
Width: 2.95 m
Height (radar retracted): 2.25 m
Road speed/range: 44 km/h/260 km
Engine: Model V-6, 6-cylinder water-cooled diesel developing 280 hp at 1800 rpm
Agility
Max step: 1.1 m
Max trench: 2.8 m
Max gradient: 60%
Ground clearance: 0.4 m
Water crossing ability: 1.07 m
Weapon: 4 × 23 mm automatic gun
Rate of fire: 3400 rounds/minute (max combined)
Standard ammunition load: 2000 rounds
Elevation: –4 to +85°
Other equipment: NBC system and infra-red driving lights are provided
Effective anti-aircraft range: 2000/2500 m
Effective ground range: 2000 m

STATUS
In production. In service with Afghanistan, Algeria, Angola, Bulgaria, Cuba, Czechoslovakia, Egypt, Ethiopia, East Germany, Hungary, India, Iran, Iraq, Jordan, Kampuchea, Korea Democratic People's Republic, Libya, Mozambique, Nigeria, Peru, Poland, Somalia, Syria, USSR, Vietnam, Yemen Arab Republic (North), Yemen People's Democratic Republic (South) and Yugoslavia.

In 1983 the ZSU-23-4 self-propelled air defence system started to be replaced in Soviet Army units by the ZSU-30-2 twin 30 mm self-propelled air defence system, but no real photographs of this system have so far been released.

UNITED KINGDOM

4588.131

MARKSMAN TWIN 35 mm ANTI-AIRCRAFT TURRET

The Marksman twin 35 mm all-weather, day/night, anti-aircraft turret has been developed as a private venture by Marconi Command and Control Systems. Development commenced in 1983 with the first prototype, of mild steel construction, being shown for the first time during the 1984 British Army Equipment Exhibition installed on a Vickers Mk 3 MBT chassis.

The second prototype, of all-welded steel armour construction, completed final weapon system trials early in 1986. The shell of the turret was built by Vickers Defence Systems of Newcastle-Upon-Tyne while the 35 mm KDA cannon and the ammunition transfer systems were provided by Oerlikon-Bührle.

The Marksman turret has been designed for installation on a wide range of tank chassis including the Type 59, T-54/T-55/T-72, M47/M48/M60, Chieftain, Centurion, Challenger, AMX-30 and Vickers Mk 3. All that is required to fit Marksman on a tank chassis is an adaptor ring and a simple electrical interface to enable the turret crew to communicate with the driver in the hull. In the case of the Soviet T-series, no adaptor ring is required.

The turret is of all-welded steel construction which provides complete protection against 14.5 mm KPV projectiles over the frontal arc and against 7.62 mm armour piercing projectiles over the remainder of the turret.

The commander sits on the left and the gunner on the right, with observation periscopes to their front, sides and rear. Each crew member also has a roof-mounted gyro-stabilised sight with magnifications of ×3 and ×10 for optical engagement of ground and air targets and the gunner's sight also incorporates a laser rangefinder.

Turret traverse and weapon elevation are electric with manual controls for emergency use. Mounted in the turret rear is an auxiliary diesel generator so obviating the need for a generator in the hull of the tank.

Main armament consists of two Oerlikon 35 mm KDA cannons mounted on either side of the turret on large diameter bearings, with each of these being provided with 230 rounds of anti-aircraft and 20 rounds of anti-tank ammunition. Ammunition is containerised with a reload time of between 5 and 10 minutes.

Mounted on the turret rear is the highly ECM resistant 400 series radar which uses a single antenna for surveillance and tracking. This rotates at 60 revolutions a minute, operates in the X/J band and has a range of 12 km in the surveillance mode.

The digital fire control system consists of the fire control computer, transmitter, synthesizer, signal processing unit, control console, data extraction unit and radar power amplifier.

The fire control system is capable of fully automatic operation, the gunner only needing to press the fire button.

The turret is fitted with an air-conditioning system and a bank of electrically operated smoke dischargers can be mounted on either side. As an option an NBC system can also be installed.

In the summer of 1985 the Marksman turret was taken to Royal Ordnance Leeds, mounted on the Centurion tank chassis used for the original trials. The Marksman turret was then transferred to a Challenger hull which had been fitted with an adaptor ring. This took 1 hour 29 minutes after which the Challenger Marksman was taken out and driven on the Royal Ordnance Leeds test track. During this period the radar was operating on the move and the gun stabilisation system, through which the guns hold their position in space irrespective of hull motion, was successfully tested.

Early in 1986 the second Marksman turret was fitted to a Type 59 tank and was subjected to an extensive series of trials which included automotive and mobility trials as well as system firing trials.

CHARACTERISTICS

Crew: 2 (commander and gunner)
Armament: 2 × 35 mm cannon
Ammunition: 230 (anti-aircraft/barrel); 20 (armour piercing/barrel)
Control: (traverse) 360°, electric at 90°/s slew and 60°/s track; (elevation) –10° to +85°, electric at 60°/s
Weight: 11 000 kg (approx)

STATUS

Development complete. Ready for production. Firing trials of the Marksman have been carried out in the UK (Centurion chassis), Finland (T-55 chassis), Austria (Centurion chassis) and one other country (Type 59 chassis).

Marksman twin 35 mm anti-aircraft gun turret on Vickers Mk 3 MBT chassis

Marksman twin 35 mm anti-aircraft turret installed on Type 59 MBT chassis for trials early in 1986

CONTRACTOR

Marconi Command and Control Systems Ltd, Weapon Systems Division, Leicester, England.

UNITED STATES OF AMERICA

6228.131

FORWARD AREA AIR DEFENSE SYSTEM

Following the cancellation of the M247 DIVAD (Division Air Defense) twin 40 mm self-propelled system in 1985 the US Army decided upon a five point programme to fill the air defence gap as well as denying enemy rotary and fixed wing aircraft sanctuary up to 10 km from the forward line of troops. The programme is known as the Forward Area Air Defense System (FAADS) and it has been estimated that it will cost $11.5 billion to develop and procure by 1996.

The five elements of the programme are:

(1) Improvements in both quantities and sensor types of the current Short Range Air Defense Command and Control (SHORAD C2) which will allow it to distribute data throughout the FAADS with plans for 31 basic sets and 127 ground sensors. The estimated cost of this system is $2.6 billion and it would involve four components; basic, ground, aerial sensors and aircraft identification. The US Army requested $207.8 million in FY 1988 for full scale engineering development of these components. TRW Defense Systems Group has begun development of SHORAD C2 while Westinghouse Defense has proposed a FAADS sensor whose transmitter, receiver and processors would use the same line replacement units as those used in Sgt York and which would also use the IFF, altitude reference unit and

The ADATS which won the FAADS LOS-F requirement contest. Note the 25 mm Bushmaster gun

plasma displays of the same system. It would also have antenna electronics which would be 75 per cent those used in the old gun system. A new elevation monopulse antenna would be developed. Hughes Aircraft are proposing a version of LSAR as a sensor.

(2) The fielding of Non-Line-Of-Sight (N-LOS) missile systems to engage helicopters masked by terrain at ranges up to 10 km at a cost of $2.7 billion. These missiles will be advanced technology systems which will be partly hidden behind tree lines or hills. Two candidates for this are the Fibre-Optic Guided Missile (FOG-M) developed by the US Army's Missile Command at Redstone Arsenal (see **6725.131**) and a version of the US Air Force's AIM-120A Advanced Medium Range Air-to-Air Missile (AMRAAM) which would launched from the ground or from helicopters. The US Army plans a two phase programme to acquire the system. First will come operational evaluation to verify the proof of principle of the basic design. The second, which will require teaming, covers development, production and deployment. This latter phase will include a maturing period of 28 months designed to field operational systems in FY 1991. This will be followed by a 45 month full scale development effort for an extended range version with imaging infra-red seeker supplementing the TV seeker. The US Army wishes to deploy this version by FY 1993.

(3) Deployment of a Line-Of-Sight, Rear (LOS-R) system designed primarily to destroy fixed wing aircraft and deployed no further forward than the rear of a battalion. It will be based upon the Stinger missile and an alternative title for the system is Pedestal Mounted Stinger (PMS) (see following section **2805.131**). The Army wants some 1200 systems at a total cost of $1.35 billion while the US Marine Corps are likely to want a further 300. Late in 1986 three consortia were awarded non-development contracts for missile-gun systems, using eight Stingers and two 0.50 calibre (12.7 mm) machine guns, mounted in the rear of the High Mobility Multi-purpose Wheeled Vehicle. Trials were completed in May 1987. In August 1987 Boeing Aerospace Company won the production contract for this systems with their Avenger (see **6730.131**).

(4) Line-Of-Sight, Forward (LOS-F) is the replacement for the Sgt York and will be an off-the-shelf purchase of an existing system which would combine guns and missiles. The estimated cost is $1.4 billion. From July 1987 four systems were evaluated at the White Sands Missile Range in New Mexico; ADATS (Martin Marietta Aerospace and Oerlikon Bührle), Liberty (LTV Aerospace's Missiles and Electronics Group with Thomson-CSF), Paladin (Hughes and Euromissile) and Tracked Rapier (United Technologies, FMC and British Aerospace). Details of ADATS will be found in **4158.131** and of Tracked Rapier in **2424.131**, although a production version of the latter would have used the Bradley chassis. Liberty was based upon the Shahine missile system upon an AMX-30 tank chassis (production versions would have used the M-1 Abrams chassis) while Paladin used the Roland missile system upon an M-993 tracked vehicle. In November the US Army selected the ADATS system.

(5) The evaluation of combined arms capabilities already in the field with a view to upgrading some to improve their use in the air defence role, the most likely candidate being the Bradley 25 mm cannon which on some vehicles would be given a lead-angle computer, laser rangefinder and increased rate of fire. Also the fielding of helicopter-borne air-to-air missiles such as Stinger and the new Helicopter Team Defense Missile are now in the early stages of development.

An artist's impression of the Westinghouse Defense FAADS sensor mounted in a HMMWV. This would incorporate much of the Sgt York sensor electronics in a vehicle with a variable speed constant frequency generator. However, the FAADS sensor requirement may be deferred during FY 88–89

6417.131
GENERAL ELECTRIC AIR DEFENCE TURRETS

Building on their experience with the 20 mm Vulcan air defence system, the Armament Systems Department of General Electric are proposing new air defence turrets to meet a variety of air defence requirements of the US Army and Marine Corps.

Blazer/US Army

This is being developed as a private venture to meet part of the US Army's Forward Area Air Defense System (FAADS) requirement following the cancellation of DIVAD and will be able to engage aircraft and helicopters out to a range of 6000 m.

General Electric have developed a two man power operated turret that can be installed on the M2 Bradley IFV chassis and be armed with a 25 mm GAU-12/U cannon with two rates of fire (1000 or 2000 rounds a minute cyclic), four Swedish Bofors RBS-70 SAM's and four General Dynamics Stinger SAM's. According to General Electric the 25 mm ammunition/missile load is sufficient to carry out 29 airborne target engagements.

The 25 mm GAU-12/U cannon carries 450 rounds of ready use ammunition with a further 1050 rounds in reserve. Types of ammunition that can be fired include HEI-T and APDS-T. A further four Stinger and

FMC M2 Bradley IFV chassis fitted with Blazer air defence turret

LAV (8 × 8) vehicle fitted with Blazer-25 air defence turret

four RBS-70 SAMs are carried inside of the vehicle for manual reloading. As an alternative to RBS-70 the British Shorts Javelin or Starstreak HVM could be installed.

The SAMs would be used to engage targets out to 6000 m with the 25 mm gun being used out to 2500 m.

The fire control system includes a roof mounted passive electro-optical tracker incorporating a laser rangefinder. As an option an Ericsson HARD or an Electronique Serge Dassault RA20S radar can be mounted on the turret rear for target search. Once the gunner acquires and locks onto the target, autotrack takes over. The Blazer is fitted with a signal processor that receives its FAADS C[2] I data input directly and the crew receive alerting and cueing from platoon and battery control.

Blazer could also be fitted with the passive Infra-red Search and Track (IRST) device in place of the radar, this was originally developed for the US Navy.

Blazer/USMC

The US Marine Corps have a requirement for 125 air defence versions of the Light Armored Vehicle (LAV) and for this General Electric are proposing the Blazer-25 armed with the 25 mm GAU-12/U cannon, 500 rounds of ready use ammunition, four Stinger SAMs in ready to launch position and a passive electro-optical fire control system which has already demonstrated its day-night capability and incorporates an AN/GVS-5 laser rangefinder and automatic lock-on and track with AN/VSG-2 FLIR. As an option, packs of rockets could be mounted on the turret.

STATUS

Development. Prototypes.

CONTRACTOR

General Electric, Armament and Electrical Systems Department, Burlington, Vermont 05402, USA.

2850.131

VULCAN AIR DEFENCE SYSTEM

This is an anti-aircraft weapon system based on the Vulcan 20 mm six-barrel gun. The tracked version, M163, comprises the Vulcan gun, a linked ammunition feed sub-system, and a fire control sub-system, all mounted in an electrically-powered turret and carried on a derivative of the M113 APC. The fire control sub-system consists of a range-only radar and a lead-computing gun sight with its associated current generator. The towed VADS system, M167 A1 (with radar) and Basic (without radar), contains its own batteries and is equipped with a petrol-driven generator for recharging and the whole system is mounted on an M42A1 gun carriage. A total of 1200 gun systems have been produced.

M163 self-propelled Vulcan air defence system on modified M113 APC (M741)

GUN

The Vulcan M168 gun, previously used as aircraft armament at rates as high as 6000 rounds per minute, has been modified for the air defence application to provide alternative firing rates of 1000 and 3000 rounds per minute. There is a burst limiting device, controlled by the gunner, which can be set for 10, 30, 60 or 100 rounds. Three muzzle clamps are available that provide different dispersion patterns for air defence and ground engagement modes. In the former a wide-angle or rectangular pattern is used to provide a high hit probability against low-flying aircraft. A concentrated dispersion pattern is used in the ground role.

In the towed system a conventional belt feed is used for ammunition, the belted rounds being fed from a 500-round container. In other systems where more space is available a linkless feed system can be used and this has a capacity of 1100 rounds.

The turret is electrically driven and its drive is controlled by three solid-state servo amplifiers.

FIRE CONTROL

The fire control system consists of an M61 gyro lead-computing gunsight and a sight current generator. The gunner visually acquires and tracks the target with the gyro lead-computing gunsight. The antenna axis of the radar is servoed to the optical line-of-sight and the radar supplies target range and range-rate data to the sight current generator. These inputs are then processed to provide outputs that are used to control the gunsight.

With inputs of range, range rate, and angular tracking of the optical line-of-sight (measured by a freely gimballed gyro), the sight automatically computes the future target position and adds the required super-elevation to hit the target.

Turret fire control is a disturbed line-of-sight system. The sight case and gun bore are physically fixed in alignment, but the sight reticle, which defines the optical line-of-sight, is positioned by the gyro and is displaced from the gun bore as the gunner tracks the target, thereby establishing the proper lead angle. The amount of optical line-of-sight displacement is dependent on the range and range rate inputs to the sight, and the required tracking time to establish the lead angle is about one second.

The range-only radar, currently supplied by American Electronic Laboratories, is a coherent doppler, moving target indicator (MTI) radar. It will acquire targets up to 5000 m away and is used in both M163 A1 and M167 A1 systems.

The fire control system may be supplemented by an AN/TVS-2B night vision sight and an M134 telescope.

IMPROVED FIRE CONTROL

Lockheed Electronics Company developed an improved fire control system for the Vulcan air defence system which forms the basis of the Product Improved Vulcan Air Defence System (PIVADS), described in entry **4015.131**.

CHARACTERISTICS (TOWED TYPE)

Type: Land-mobile, air-transportable, light anti-aircraft weapon system

Control: Radar ranging, optical aiming, automatic lead computation

Fire power: 6 × 20 mm guns. 3000 rounds/minute rate of fire. 1000 rounds/minute low rate of fire

Dispersion: 12 mrads; 8 × 18 mrads; or 6 mrads (80% 4 rounds fired)

Radar range: 5000 m

Weapon coverage: Azimuth 360°; elevation –5 to +80°

Turret slewing rate: Azimuth 75°/s; elevation 60°/s

Height on wheels: 2.03 m

Height emplaced: 1.73 m

Wheel track: 1.75 m

Weight: 1406 kg

Standby power requirement: 500 W

VARIANT

Saudi Arabia has taken delivery of a quantity of 20 mm Vulcan/Commando mobile air defence systems (**4062.131** in 1986-87 and earlier editions of *Jane's Weapon Systems*) which is essentially the Cadillac Gage V-150 (4 × 4) multi-role armoured vehicle fitted with the complete turret of the towed M167 air defence system. The complete system weighs 9072 kg in combat order and to provide a more stable firing platform outriggers are lowered to the ground before engaging aerial targets.

Vulcan air defence system M167 towed version

STATUS

The M163 A1 Vulcan self-propelled anti-aircraft gun is in service with Israel, Jordan, Saudi Arabia, Morocco, Thailand and USA. The Basic and M167 A1 Vulcan towed anti-aircraft gun is in service with Belgium, South Korea (licensed production), Morocco, Saudi Arabia, Japan (licensed production), Thailand, Uruguay, USA and Yemen Arab Republic (North).

CONTRACTORS

Complete system: Armament and Electrical Systems Department, General Electric Company, Lakeside Avenue, Burlington, Vermont 05402, USA.

Vehicle: FMC Corporation, San Jose, California, USA.

Product Improved Vulcan: Lockheed Electronics Co Inc, 1501 Highway 22 CSI, Plainfield, New Jersey, 07061, USA.

4015.131

PIVADS (PRODUCT IMPROVED VULCAN AIR DEFENCE SYSTEM)

PIVADS, which was developed for the US Army, offers two improvements to the basic Vulcan:

(1) effectiveness: improved probability of hit by replacing the fire control system

(2) availability: reduced system 'down time' by utilising built-in test (BIT) to identify and isolate faults.

This cost-effective improvement of the Vulcan extends the usable service life against current threats and permits a higher degree of readiness. The improvement technology is applicable to new gun systems or to upgrading existing systems and uses the following techniques:

(1) replacement of disturbed line-of-sight (DLOS) gunsight configuration by a stabilised 'director' sight which isolates the sight from turret motion to provide accurate target acquisition and tracking. The tracking recticle is directly controlled by the gunner

(2) computer generated rate-aid signals significantly reduce gunner workload by driving turret to future target position based on historical position data

(3) special servos and turret drive design techniques (rate feed-forward, resonance equalisation, harmonic drives) provide the following characteristics: improved turret stiffness, minimised backlash and high bandwidth turret and sight servos, improved turret transient and steady-state tracking performance

(4) militarised digital microprocessors dramatically enhance system performance and permit adaptive software modifications.

The PIVADS modification kit enhances system effectiveness by the use of a director-type sight, a microprocessor and improved turret drives

DEVELOPMENT

Development was undertaken as a private venture by Lockheed Electronics Company to exploit experience accumulated in the design and development of the Mk 86 GFCS (**1241.281**) and small weapon control systems simulation analysis.

STATUS

The US Army awarded research, development, test and evaluation contracts to Lockheed Electronics in September 1982 for the M163A2 self-propelled PIVADS, and a contract for the M167A2 towed PIVADS in November 1983. In fiscal year 1985, the Army awarded a production contract to Lockheed Electronics for 285 M163A2 and M167A2 PIVADS.

CONTRACTOR

Lockheed Electronics Company Inc, 1501 Highway 22 CSI, Plainfield, New Jersey 07061, USA.

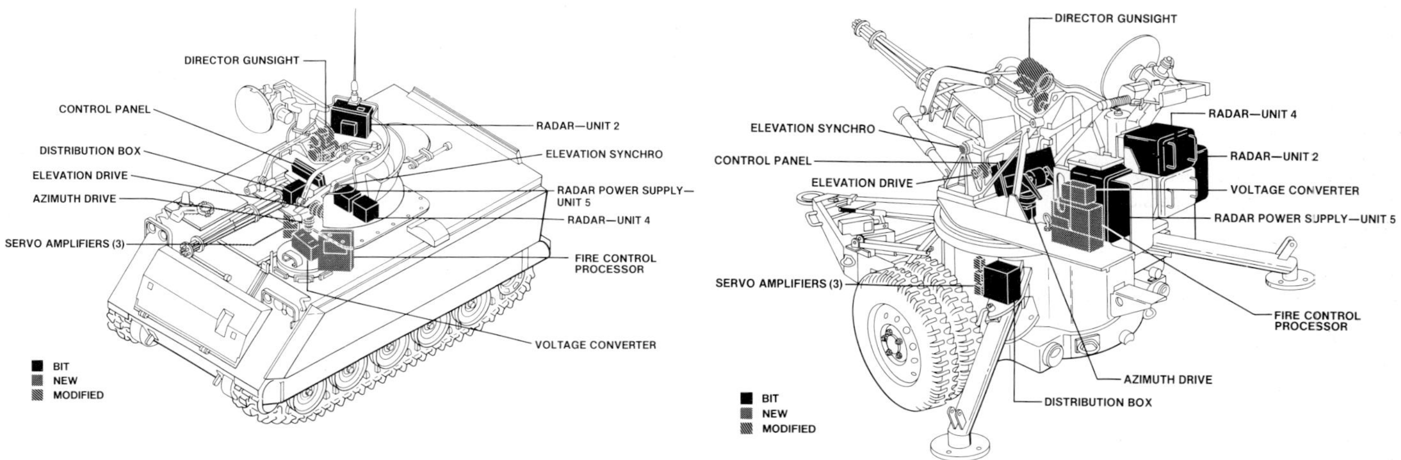

Major elements of the M163A2 self-propelled (left) and M167A2 towed PIVADS (right)

4466.131

GEMAG 25 AA GUN SYSTEM

The GEMAG 25 (General Electric Mobile Air Defence Gun) is a 25 mm towed mobile gun system that has been developed as an air-transportable air defence weapon similar in concept to the Vulcan towed anti-aircraft gun, and of similar size and weight but with a new gun. This is the 25 mm GAU-12/U Bushmaster which is allied to improved gun carriage drives and mountings. The 25 mm Bushmaster rotary multi-barrel gun was selected for the improvements in range and time-of-flight that it offers in comparison with the existing 20 mm Vulcan rounds. The GAU-12/U also has dual-feed facilities which permit the use of either HEI (high explosive incendiary) or APDS (armour piercing discarding sabot) rounds to be fired by the gun without altering ammunition feed arrangements. The gun is fired at rates of either 1000 or 2000 rounds per minute. The carriage has provision for a capacity of 530 rounds.

The fire control system is based on the digital system developed for the PIVADS (**4015.131** Product Improved Vulcan Air Defence System), but with provision for a number of additions including a night vision system based on the AN/TAS-6 forward-looking infra-red night observation device (**3826.193**), and a laser rangefinder. To engage targets at longer range four General Dynamics Stinger missiles are mounted on right side of carriage. At the left rear of the carriage is a French Electronique Serge Dassault surveillance radar which is lowered for travelling.

GEMAG 25 multi-barrel anti-aircraft weapon system deployed

CHARACTERISTICS

Weight: 1814 kg (loaded ready for combat)

Height: 2.057 m (travel configuration); 1.823 m (emplaced)

Length: 3.683 m (travel configuration); 3.589 m (emplaced)

Gun: 25 mm GAU-12/U

Ammunition system: Linked, dual feed, HEI, APDS

Ammunition capacity: 530 rounds
Feed system: Linked hanging rail type
Firing rates: 2000/1000 rounds/minute
FLIR: AN/TAS-6
Radar: Ku-band search and track
Sight: Optical director
Computer: LSI 11M digital
IFF: AN/PPX-3 Mk XII

STATUS
Prototype.

CONTRACTOR
Armament and Electrical Systems Department, General Electric Company, Lakeside Avenue, Burlington, Vermont 05402, USA.

6730.131
AVENGER AIR DEFENCE SYSTEM

The Avenger air defence system has been selected to meet the FAADS Line-Of-Sight, Rear (LOS-R) requirement (see **6228.131**). Developed by the Boeing Aerospace Company and General Electric's Armament System Department, the Avenger is essentially a Pedestal Mounted Stinger system with a General Electric GECAL-50 0.50 calibre (12.7 mm) Gatling machine gun with 200 rounds of ammunition mounted on the back of a HMMWV.

Eight Stinger missiles (see **2805.131**) are mounted on the gyro-stabilised turret which also features an CAI optical sight, a Magnavox Forward Looking Infra-Red (FLIR) sensor (a license built Barr & Stroud IR18), AN/PPX-3B IFF and a Texas Instruments' laser rangefinder. The turret, which is 1.7 m high, 2.13 m long and 2.15 m wide, weighs a total of 1134 kg bringing the total weight of the HMMWV to 3900 kg. The turret can be brought into action in less than ten seconds and the missiles can be activated in less than three seconds. All eight missiles can be reloaded in four minutes.

The operational characteristics include an auto-track to FLIR and missile seeker and automatic lead angle/super-elevation for the missile and gun. During firing trials against ballistic targets at the Yakima Firing Center in 1984 the first Stinger fired when the vehicle was on the move scored a direct hit as did the second fired at night when the vehicle was stationary. The third, fired on the move in rain, was a tactical kill.

Mock-up fits with four Bofors RBS-70 (see **2348.131**) and four Stinger as well as 36 Hyper Velocity Missiles have been made.

The Boeing Aerospace Avenger system mounted on a HMMWV will meet the US Army FAADS LOS-R requirement

STATUS
The US Army placed an initial production contract for 20 systems worth $16.2 million but plans to buy 273 systems worth $189 million by 1995. The total requirement is for 1200 systems while the US Marine Corps is seeking a further 300.

CONTRACTOR
Boeing Aerospace Company, Defense Systems Division, PO Box 3999, Seattle, Washington 98124, USA (prime).

Portable Anti-aircraft Guided Missiles

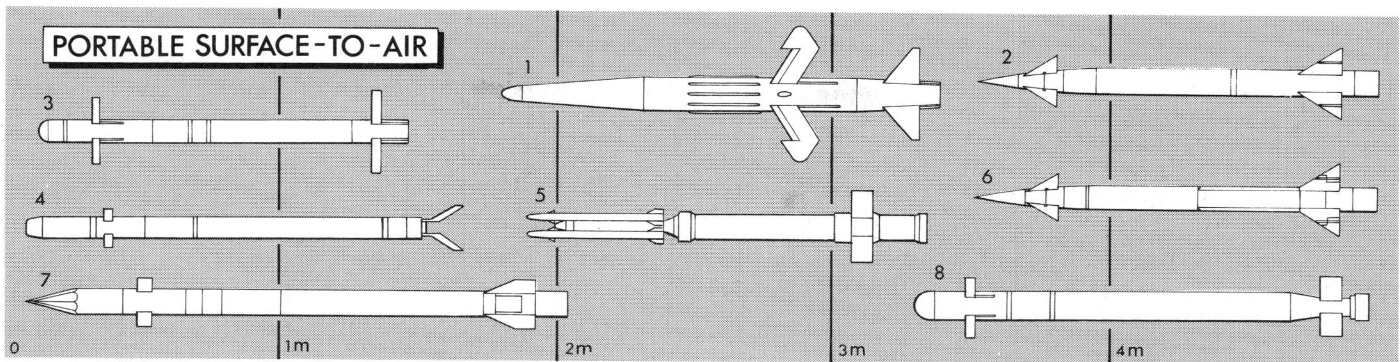

1 RBS 70, **2** Blowpipe, **3** Redeye, **4** SA-7 Grail, **5** Starstreak, **6** Javelin, **7** SATCP (Mistral), **8** Stinger

BRAZIL

6731.131
MSA-3.1 PORTABLE ANTI-AIRCRAFT MISSILE SYSTEM

Currently under development for the Brazilian Army is the Orbita MSA-3.1 man-portable surface-to-air missile. This incorporates the thruster and television guidance technology of the British Aerospace Thunderbolt developed to meet the HVM (High Velocity Missile) requirement of the British Army which was met by the Shorts Starstreak (**6271.131**).

The MSA-3.1 will be a Mach 3 missile fired from a shoulder-mounted launcher which, with the missile, will weigh 25 kg. The missile will have a range of 6 km and will usually have a crew of two. Plans exist for a vehicle-mounted version, MSA-3.2, using a Cascavel AFV and also for a naval version.

CONTRACTOR

Orbita Sistemas Aeroespaciais SA, San Jose dos Campos, Brazil.

CHINA

6479.131
HN-5A PORTABLE ANTI-AIRCRAFT MISSILE SYSTEM

The HN-5A is the current production model of an improved version of the Soviet SA-7 (Grail) (**2941.131**) man portable surface-to-air missile system.

This system is capable of making tail-on engagements against jet aircraft or head-on engagements against propeller-driven aircraft and helicopters under visual aiming conditions.

Main improvements of the HN-5A over the original HN-5 include greater detection range of the IR homing seeker made possible by providing cooling to the IR detector, reduced susceptibility to IR background sources such as bright clouds by incorporating a background noise rejection device into the IR homing seeker and a more powerful HE warhead.

The HN-5A consists of the launch tube which serves as an aiming device and launcher as well as the carrying case, firing unit mounted under the forward part of the launcher which provides launch information and ensures correct firing of the missile, thermal battery mounted on the forward part of the firing unit which provides launch information and ensures correct firing of the missile.

The missile itself is composed of four sections. The infra-red seeker section provided with both cooling and background noise rejection devices. The control actuator section containing a gas generator. The warhead and fuze and the rocket motor section with rear fins attached.

The infra-red seeker detects the thermal radiations emitted from the target aircraft and converts them into missile steering orders to guide the missile in proportional navigation to intercept the target.

The China Precision Machinery Import and Export Corporation also offer the CH-3A integral test and measuring vehicle to complement the HN-5 as well as a training system.

VARIANT

The HN-5C is a HRB-230 (4 × 4) cross-country vehicle with a forward control cab which contains the fire control electronic equipment. To the rear of the cab is a pedestal on either side of which is mounted four HN-5A missiles in the ready to launch position. Between the two backs of missiles is the fire control system which consists of an infra-red tracker, laser rangefinder and TV camera.

Detecting and tracking of the target and launching of the missile can be accomplished either manually or automatically.

Although the prototype system is mounted on an HRB-230 truck, it can be mounted on other chassis as it weighs only 2000 kg. Eight reserve missiles are carried in addition to the eight missiles in the ready to launch position.

Evaluation trials of the HN-5C was completed in June 1986 but it is understood that production has yet to start.

Main roles of the system can be summarised as protection of mechanised units, air defence for forward units and air defence for high value targets such as airfields.

PERFORMANCE CHARACTERISTICS

Maximum operational altitude: 2500 m
Minimum operational altitude: 50 m
Maximum slant range: 4000 m
Minimum slant range: 800 m
Maximum speed of the target in tail-on engagements: 260 m/s (950 km/h)
Maximum speed of the target in head-on engagements: 150 m/s (550 km/h)
Reaction time: Not more than 5 seconds

STATUS

The HN-5A is in production and in service with the Chinese armed forces. HN-5C is still at prototype stage.

CONTRACTOR

China Precision Machinery Import and Export Corporation, Beijing, China.

HN-5A in ready to launch position

Close up of HN-5C system (Christopher F Foss)

EGYPT

4841.131

AYB AL-SAQR PORTABLE ANTI-AIRCRAFT MISSILE

After several years of experience using Soviet-supplied SA-7 Grail man-portable anti-aircraft missile systems, the Egyptian authorities have undertaken production of an indigenous version known as the Ayb al-Saqr at a cost equivalent to $180 million.

The system consists of the missile, housed in a disposable container/launcher, and a grip-stock which includes system electronics together with the sight and firing mechanism. Two complete Sakr Eye rounds are packed into a storage and transport container. The Egyptian version of the SA-7 is understood to be virtually identical to the Soviet item, although better reliability is claimed for the former.

In the absence of an integrated IFF subsystem (French and British add-on IFF equipments are available) a typical engagement follows this sequence: the gunner acquires the target and aligns the weapon using the open sight or the optronic one, and then selects either manual or automatic firing mode. Audio and visual cues are given to the gunner to indicate when the target is within the engagement envelope. The trigger is then squeezed fully and the missile is ejected from the launch tube by a small rocket motor. The booster then accelerates the missile until the sustainer motor is ignited to propel it to the target. If a target is not encountered within 4400 m the Sakr Eye destroys itself.

Homing is by means of a passive infra-red seeker, and the HE warhead is detonated by a fuze that provides for graze initiation of the warhead.

A new state of the art computerised sight can be interfaced with the electronics of the system to give the user the following four advantages: (1) Positive indication of lock-on, loss of thermal target, out of tolerance target or false target. (2) Compensation for missile lead angles. (3) Magnification of target for better target identification when the system is not fitted with IFF. (4) A fire inhibit signal to prevent an invalid missile launch.

Sakr Eye surface-to-air missiles are also fitted to both the Thomson-CSF Nile 23 twin 23 mm self-propelled anti-aircraft gun (two missiles on either side of turret) and Electronique Serge Dassault Sinai 23 twin mm self-propelled anti-aircraft gun (three missiles mounted above each 23 mm cannon). These systems were developed in competition to meet the requirements of the Egyptian Army.

CHARACTERISTICS

Calibre: 72 mm

Weight: 15 kg (ready for use without sight); 18 kg (ready for use with sight)

Max target altitude: 2400 m

Minimum target altitude: 50 m

Max range: 4400 m

Max target speed: 150 m/s (head-on); 280 m/s (pursuit)

STATUS

In production. In service with Egyptian Army. Operational by mid-1987.

CONTRACTOR

Sakr Factory for Developed Industries, PO Box 33, Heliopolis, Cairo, Arab Republic of Egypt.

Close up of turret of Thomson-CSF 23 Nile twin 23 mm self-propelled anti-aircraft gun showing two Sakr Eye missiles mounted either side of the turret

Sakr Eye portable anti-aircraft misile

FRANCE

4198.131

MISTRAL AIR DEFENCE MISSILE

The Mistral/SATCP (sol-air-très courte portée) missile project is a development programme intended to provide a short-range air defence missile system capable of production in several versions ranging from the simplest man-portable type including a variety of vehicle and AFV installations, to airborne and naval variants. The work has been entrusted to Matra, the winner from about five contractors which submitted proposals for the SATCP requirement.

The Matra Mistral SATCP system comprises, in the portable version (basic), two main elements: the missile and the firing station. The former is, of course, expendable whereas the firing unit is not, and the whole assembly can be carried by two men in separate packages.

The missile uses infra-red homing guidance, employing a passive system that is claimed to be very sensitive and capable of true 'fire-and-forget' operation. Impact and proximity fuzing is provided for the 3 kg warhead, the latter being a laser-based device, and an IFF module is available to prevent engagement of friendly aircraft and to minimise reaction time in the face of enemy targets. The system is for defence against both fixed-wing high-speed aircraft and helicopters. The two packages comprising the man-portable version each weigh less than 20 kg. One pack consists of the missile in its launch/transport container tube, and the other includes the stand, seat and sight unit.

Standard version of Matra Mistral with missile being launched

The SAT infra-red homing head, instead of relying on a single-cell IR receiver with consequent risks of unwanted responses from ground objects at low elevation angles etc, employs a multi-element seeker. This and the use of digital signal processing is stated to enhance the Mistral's performance in the face of IR countermeasures, or natural interference from the background. The sensitivity is claimed to be adequate for acquisition of a single jet aircraft target at a range of 6 km or more, or a light helicopter at 4 km, irrespective of the target aspect. Homing-head gyro run-up and head cooling preparatory to launch takes only two seconds, and weapon reaction times of less than five seconds have been achieved in tests without early warning facilities.

The firer has a sighting system that includes a clear collimator and a magnifying telescope. The former allows the display of aiming data enabling the launcher to track the target while following the pre-launch sequence progress and assess azimuth and elevation lead angles. Reload time is quoted as less than 30 seconds.

In addition to the main solid-propellant rocket motor which gives the Mistral a peak speed of Mach 2.6, there is a solid booster which gives the missile a muzzle velocity of 40 m/s, and this is jettisoned when the main motor ignites about 15 m ahead of the firer.

It is intended to employ the same missile in all versions of the system in the interests of logistic simplicity and economy. The homing head is the responsibility of SAT, and SEP and Manhurin undertake the propulsion and warhead manufacture respectively.

CHARACTERISTICS

Guidance principle: Passive IR homing
Missile length: 1.8 m
Missile body diameter: 90 mm
Missile weight: 17 kg
Weight of container and missile: 20 kg
Length of container with missile: 1.85 m
Weight of warhead: 3 kg
Weight of explosive charge: 1 kg
Ejection speed: 40 m/s
Maximum speed: Mach 2.6
Maximum load factor: 8 *g*
Time to 4000 m: 6 s
Maximum engagement range: 6000 m
Minimum engagement range: 500 m

It is proposed to develop thermal sight equipment to provide a night capability for the Mistral weapon system. Five versions are understood to have been considered. The first is meant to fill French Army and Air Force requirements for portable defences for important points in the battle area or near airfields. The French Navy plans to use Mistral for the air defence of small vessels with the SADRAL (Système d'Autodéfense Rapprochée Anti-Aérien Léger), and as a supplement to existing air defence armament of large ships (**4509.231**). The Army envisages its use for air-to-air purposes on anti-tank helicopters (**4510.311**) and it is also seen as suitable for mounting on Jeeps or trucks in the dual missile launcher configuration, and on AFVs (with a six-round turret and radar) for the defence of armoured units and rapid deployment forces. The French Army has selected the Panhard ERC (6 × 6) armoured car chassis for the six-round turret version of the Mistral, although other vehicles can be used such as the Renault VAB.

Matra Mistral/SATCP twin launcher mounted on LOHR FL 400 light cross-country vehicle

STATUS

The portable version is being developed for the French Army and Air Force.

CONTRACTOR

Matra SA, ave Louis Bréguet, 78140 Velizy, France.

JAPAN

4199.131

MAN-PORTABLE SURFACE-TO-AIR MISSILE

A man-portable surface-to-air missile development project was started in fiscal year 1979 by the Japanese Self Defence Force Technical Research Institute. It is understood that the intention is to develop a weapon similar to the existing generation of man-portable SAMs, such as the American Redeye, Soviet SA-7 and the British Blowpipe, but instead of relying on infra-red homing or radar guidance, the Japanese SAM will use an imaging homing system for guidance. This is claimed to be better in that it will have an all-round attack capability and high resistance to jamming. The guidance system will incorporate a CCD (charge-coupled device) to store the target image, and other important factors are the resolution characteristics of CCDs and low cost. It is hoped to achieve a unit cost of less than ten million Yen. Two types of homing head were under evaluation in 1979, and a propellant unit and a wind tunnel test model had been constructed. The former consists of a booster rocket and a sustainer motor. No other details have been made available for publication. In the meantime the Japanese Air Force and Japanese Ground Self Defence Force has been procuring the Stinger man-portable surface-to-air missile system.

STATUS

Development continuing. No schedule for development or production plans have been released.

CONTRACTOR

JSDF Technical Research Institute.

SWEDEN

2348.131

RBS 70 ANTI-AIRCRAFT MISSILE SYSTEM

RBS 70 is a man-portable surface-to-air missile system operating on the laser beam-riding principle, which gives it substantial immunity to jamming. It has a range of about 5 km and is suitable for use against both fighter-bombers and helicopters even at the lowest levels. The engagement of targets is carried out by one man, but to fire several missiles in rapid succession a loader is needed.

The system is suitable for integration with other surface-to-air weapons or as a separate unit. Unlike many other portable anti-aircraft missile systems, RBS 70 can utilise precision target information from a search radar and it is both possible and desirable to deploy several RBS 70 units with one or more search radars to form an anti-aircraft battery. RBS 70 is also designed to utilise IFF signals.

A land-mobile G/H-band (former C-band) radar has been developed by Ericsson Radio Systems for use with the RBS 70. It was introduced as the PS-70R but is now known under the name of Giraffe (**1957.153**). Other search radars, such as Signaal's Reporter, can also be integrated with the RBS 70.

Three packs, stand, sight, and missile in container, make up the complete firing unit which can be assembled and ready to fire in less than 30 seconds. The stand has three legs and forms the suspension device for the telescopic sight which has gyro-stabilised optics and which also contains the guidance laser beam transmitter.

To slew on to the target the operator uses coarse aiming, turning the whole sight housing in the required direction. For tracking, a joystick is used to operate the gyro-stabilised optics.

The missile is launched from a container which also serves for both storage and transport, the missile never being removed from the container except at the moment of firing. A launching motor ejects the missile from this launch container and is jettisoned from it at the end of the tube after which the missile sustainer takes over. The missile carries a receiver for the guidance beam and a small computer for calculation of control signals for the control surfaces. The warhead has both proximity and impact fuzes but if desired the active optical proximity fuze can be switched off before the missile is fired.

Among the items of test and training equipment supplied for use with the missile is a sight simulator for use in training operators. With this equipment an average AA soldier can be fully trained within 15 to 20 hours, and without a need to fire live rounds. Most of the training can be carried out indoors.

Mobile, vehicle-mounted versions of the RBS 70 are described in entry **3617.131**.

Bofors RBS 70 anti-aircraft missile system deployed for firing and fitted with IFF system

CHARACTERISTICS
Type: Man-portable surface-to-air guided missile system
Guidance: Optical beam-riding on laser beam
Propulsion: Booster plus solid propellant sustainer
Warhead: Fragmentation with impact and proximity fuzes
Weight
Firing unit complete: About 80 kg
Missile in container: 24 kg
Container length: 1.6 m
Container diameter: 15 cm
Missile range: 5 km against high speed targets; 6 km against low speed targets
Height coverage: 3 km

STATUS
Development started in 1969 and the system is now in series production and in service with more than 10 countries including Australia, Bahrain, Ireland, Norway, Singapore, Sweden and Tunisia.

CONTRACTOR
AB Bofors, 691 80 Bofors, Sweden.

UNION OF SOVIET SOCIALIST REPUBLICS

2941.131
SA-7 (GRAIL) MAN-PORTABLE ANTI-AIRCRAFT MISSILE

The SA-7 Grail, whose Soviet designation is believed to be Strela 2 (Arrow), is an extensively deployed man-portable infra-red homing light anti-aircraft missile, very similar in concept to the American Redeye (**2784.131**) but preceding the latter into service. It relies upon a tail pursuit interception to engage low-flying aircraft targets and has proved especially effective against helicopters. Two versions exist; Strela 2 and Strela 2M with improved seeker and launcher.

Simple optical sighting and tracking are employed, the IR seeker being activated when the operator has acquired the target. An indicator light denotes seeker acquisition and the operator is then free to fire the missile.

Flares were for a time found to provide an effective countermeasure but it is reported that later models are equipped with filters to combat this tactic. The existence of an improved Mk 2 version of SA-7 has been confirmed, the principal difference lying in a boosted propellant charge to increase range and speed of the missile.

Over the years that SA-7 has been in use – well over 20 – additional forms of deployment have been evolved, apparently by both the USSR and individual client states. In October 1975, for example, a number of Jeep-type vehicles were seen during an Egyptian military parade with a rudimentary platform and support for an SA-7 gunner at the rear of the vehicle. Four reload rounds were attached to the sides of the vehicle. Various ship fittings of this missile have been noted and the designation given to naval applications is SA-N-5.

CHARACTERISTICS
Type: Man-portable, shoulder-fired surface-to-air tactical guided missile
Guidance: Optical aiming, infra-red homing
Propulsion: Solid propellant

	Strela 2	**Strela 2M**
Length	1.34 m	1.34 m
Missile diameter	10 cm	10 cm
Launch weight	9.2 kg	9.97 kg
Warhead weight	1.1 kg	1.1 kg
Max speed	470 m/s	580 m/s
Range	3.6 km	5.5 km
Altitude	3000 m	4500 m
Warhead	HE fragmentation	

STATUS
Although now being replaced by SA-14 Gremlin in Soviet Army, the SA-7 remains in production for both home and export markets and is also produced in China as the Red Tassle Type 5 (qv) and Egypt as the Ayb al-Saqr (qv). In addition to being used by all members of the Warsaw Pact it is also used by Afghanistan, Algeria, Angola, Botswana, Burkin Faso, Cuba, Cyprus, Egypt, Ethiopia, Finland (called SAM-78 Strela), Ghana, Guinea, Guinea-Bissau, India, Iran, Iraq, Kuwait, Laos, Libya, Mauritania, Morocco, Mozambique, Nicaragua, North Korea, Peru, Seychelles, Sudan, Sierra Leone, Syria, Somalia, Tanzania, Uganda, Yugoslavia, Zambia and Zimbabwe.

East German troops training in the use of the SA-7 Grail anti-aircraft missile

6270.131
SA-14 (GREMLIN) MAN-PORTABLE ANTI-AIRCRAFT MISSILE

The Soviet Army has now started to replace its SA-7 (Grail) man-portable surface-to-air missiles with a new system called the SA-14 (Gremlin) by NATO. The Soviet designation is Igla (Needle).

In appearance the Gremlin is very similar to SA-7 but its grip-stock assembly is located farther forward than on the SA-7 and its battery box is located to the rear of the grip-stock.

The SA-14 Gremlin can engage aircraft pulling up to 8 gs and engage targets head-on out to a range of 4000 m, a considerable improvement over the older SA-7 Grail which is limited to engagement of targets from the rear. Laser beam-riding guidance has been reported.

STATUS
In production. In service with the Soviet Army and known to have been exported to Finland (as SAM-86 Igla), Cuba and Syria.

UNITED KINGDOM

2409.131
BLOWPIPE PORTABLE ANTI-AIRCRAFT MISSILE SYSTEM

Blowpipe is a surface-to-air weapon for unit self defence in forward areas against close-range low-level air attack. To carry out this role effectively the equipment is compact, light, and simple so that it can be both carried and operated by one man.

The weapon is suitable for use in extremes of climate with no maintenance needed during long periods in the field. It can be brought into action very rapidly, and reloading time is a few seconds. Blowpipe can be used against both attacking and receding fast aircraft and helicopter targets.

The Blowpipe weapon system is entirely self-contained with no external power requirements, and consists of two main components: the missile sealed within its launching canister, and the aiming unit. The missile/canister combination is treated as a round of ammunition that can be taken out of store and fired with no preparation or testing. The aiming unit houses all the necessary equipment and controls to launch the missile and guide it towards the target.

OPERATION
To prepare the system for action all that is necessary is to clip the aiming unit to the canister, which requires only a few seconds, and the complete system, which weighs about 20 kg, is lifted to the man's shoulder. The aimer acquires his target in his monocular sight, fires the missile, and controls its flight to the target by means of a thumb-operated flight controller.

The act of pulling the trigger activates thermal batteries for power supplies in both the missile and canister (to supply the aiming unit). When these batteries have reached a sufficiently high voltage (in less than one second), the missile first stage motor ejects the missile from the canister. This motor is extinguished before the missile fully emerges and the missile coasts for a safe distance from the aimer before the second stage main motor ignites to boost the missile to supersonic speed.

The missile is fitted with flares which in the early stages of flight are detected by a sensor in the aiming unit to gather the missile to the centre of the aimer's field of view. From then on the aimer guides the missile to the target by means of the controller with up/down and left/right movements. When within lethal distance of the target a proximity fuze in the nose of the missile detonates the warhead.

Commands are transmitted to the missile over a radio link, through the receiver and decoder to the missile control system. This works on 'twist and steer' principles with one pair of nose-mounted control surfaces working differentially to produce roll, and the other pair producing lateral movements.

The missile has a slim, cylindrical body and ogival-nose cone. Cruciform delta-shape canard control surfaces are mounted on the nose cone and a unique type of cruciform delta-shape tail-fin assembly is used. This consists of a sliding ring structure which, in the launching canister, is positioned near the nose of the missile, enabling the diameter of the rear of the canister to be minimised. As the missile is launched, it passes through the tail-fin assembly which finally locks on the rear of the missile. The folded wing tips open to their full span as they emerge from the canister.

Shorts Blowpipe close-range guided missile, although designed for low-level air defence, can also be used against lightly armoured surface targets

CHARACTERISTICS
Type: Man-portable, shoulder-fired surface-to-air (or surface) tactical guided missile
Guidance: Radio command with optical tracking
Guidance method: Twist and steer by nose-mounted control surfaces
Propulsion: 2-stage booster-accelerator solid-propellant rocket motor
Warhead: HE with contact and proximity fuze
Missile length: 1390 mm
Missile diameter: 76 mm
System weight: 20.67 kg

STATUS

In production. In service with 17 forces in 12 countries including Argentina, Canada, Chile, Ecuador, Malawi, Nigeria, Oman, Portugal, Thailand and United Kingdom (Army and Marines). Blowpipe has also seen combat with guerilla units operating in Afghanistan.

CONTRACTORS

The following manufacturers are known to have been associated with the development and manufacture of the system:

Main contractor: Short Brothers plc, Belfast.

IFF system: Cossor Electronics Ltd.

Fuze: Marconi Space and Defence Systems Ltd.

Ignition, safety and arming unit: Royal Ordnance, Blackburn.

Firing circuits: Pye Dynamics Ltd.

Thermal batteries: Mine Safety Appliances Ltd.

The co-ordinating research and development authority is the Royal Radar Establishment.

4533.131

JAVELIN MAN-PORTABLE ANTI-AIRCRAFT MISSILE

Evolved from the very successful Blowpipe man-portable anti-aircraft missile system (**2409.131**), the Javelin shoulder-launched weapon is a surface-to-air missile system that employs a semi-automatic line-of-sight command guidance technique based on a visual system that means that the operator does not have to steer the missile in flight by means of a thumb-stick. Instead a stabilised aiming mark, injected into his field of view, is maintained over the target while a TV camera in the aiming unit picks up the missile flares and generates corrections to maintain the missile in the centre of the aiming mark.

Javelin is a direct development from the Shorts Blowpipe, but has a higher impulse rocket motor and a new warhead. The warhead is in the centre section with the fuzes in the nose. The forward part of the body contains the guidance equipment and the rear part houses the two solid-propellant rocket motors. There are four delta-shaped aerofoil surfaces in the nose for missile aerodynamic control and four at the tail to improve ballistic stability. A feature of the missile is that the nose section is free to rotate independently of the main body, to which it is attached by a low-friction bearing. Twist and steer commands to the control fins guide the missile, resulting in a fast response rate. The new warhead relies on blast/fragmentation effects rather than penetration to effect a kill.

The canister in which the missile is factory sealed is a lightweight environmental container designed to act as a recoilless launcher. The front cap is blown off by gas pressure when the missile gyro is fired, while the laminated rear closure is ejected at launch.

A lightweight multiple launcher (LML) has been developed by the manufacturer to provide a multi-engagement capability for the Javelin system. It is designed to suit a wide range of potential operational deployments, and several versions have been developed for applications other than the man-portable role. All LML configurations use, as clip-on items of equipment: three Javelin canister-housed missiles, and a standard shoulder-launch aiming unit, which are fixed to a collapsible tubular tripod.

In the free-standing application, the main support tube is held vertically by the tripod legs which pivot from an eccentric support collar. Short screw jacks between the top of the sleeve and the legs are used to level the launcher. For trench deployment, it is possible to have one leg as a vertical support within the trench and the two others arranged horizontally to provide lateral support on the trench side(s). There is a version for use with APCs or similar vehicles, the LML(V), in which a pintle mounting provides 60 degrees of freedom, relative to the turret ring, in traverse. The ring is provided with a handgrip and frictional brake, so that the aimer can slew it to the approximate target bearing and then track the target.

A naval version LML (N) has been developed that incorporates an operator's seat mounted on the pillar of the simple pedestal mounting, which provides for manual training and elevation of the three-round aiming and launching head.

The aiming unit is a self-contained firing and control pack with a pistol grip firing handle on the right-hand side. It contains a stabilised sighting system which provides manual tracking and automatic missile guidance through a solid state TV camera. Digital signals from the camera are fed to a microprocessor and the resultant guidance demands are transmitted to the missile by radio link. Controls on the handle include firing trigger, thumb-controlled joystick and system, fuze mode and super-elevation switches. Other controls are channel selector switches for the transmitter and automatic cross-wind correction switch.

VARIANT

Under development for the British Army is the Spartan/Javelin. This consists of an Alvis Spartan full tracked APC fitted with an armoured launcher on the roof with four Javelin surface-to-air missiles in the ready to launch position. A further eight missiles are carried internally as is a lightweight multiple launcher and aiming unit to allow remote engagement.

Shorts LML for Javelin surface-to-air missile in operating configuration

Shorts Javelin man-portable, shoulder launched surface-to-air missile deployed in the field

The Javelin system can be operated from under armour with the controller facing forward and using a panoramic sight for both target search and tracking. Surveillance is provided by the THORN EMI Air Defence Alerting Device (ADAD) mounted on the roof of the vehicle and providing an horizon scanning, passive infra-red detection system.

CHARACTERISTICS

Length: 1.4 m
Guidance: SACLOS
Propulsion: 2-stage solid
Speed: Supersonic
Range: >4 km
Aiming unit: Standard Blowpipe type (as clip-on equipment)
Missiles on launcher: 3
Elevation: –10 to +45°
Traverse: Continuous (stop fitted as required)

Lightweight multiple launchers

Weights: 14 kg (traverse head); 16 kg (tripod support)

LML (V)

Traverse head: 35.5 kg
Turret: 78 kg (with hatch and pintle)

LML (N)

Traverse head: 35.5 kg
Fixed pedestal: 13.5 kg
Seat assembly: 16.1 kg

STATUS

In production. In service with the British Army, Royal Marines and South Korea.

CONTRACTOR

Short Brothers plc, Missile Systems Division, Castlereagh, Belfast BT6 9HN.

6271.131

STARSTREAK HIGH VELOCITY MISSILE SYSTEM

The British Army has a requirement for a High Velocity Missile (HVM) system capable of engaging unmasking close support aircraft and hovering helicopters equipped with anti-tank guided weapons.

To meet this requirement the British Ministry of Defence issued two project definition contracts worth about £3 million, one to the Missile Systems Division of Short Brothers and the second to British Aerospace Dynamics.

In June 1986 the Shorts Starstreak HVM was selected for the British Army. It is expected that three versions of the Starstreak HVM will be developed:

(1) eight HVMs in the ready to launch position on an Alvis Stormer armoured personnel carrier chassis
(2) lightweight multiple launcher with three HVMs in the ready to launch position
(3) shoulder launched by one man.

Starstreak is carried in a sealed canister that serves two purposes; firstly it acts as the launcher and secondly as an environmental container during storage and transit. This completely eliminates the need for testing the round with the only preparation needed to connect the aiming unit.

Starstreak consists of a two-stage rocket motor and three exposed dense metal darts with high explosive warheads. Upon firing the rocket motor accelerates the missile towards the target. As the motor burns out the three metal darts automatically separate from the missile and continue towards the target in formation.

In the Stormer/Starstreak an additional 12 missiles are carried inside the rear of the vehicle for manual reloading. The turret is servo-controlled and collocated with the operator's surveillance, firing and target tracking turret mounted on the forward part of the hull roof. The guidance transmitter is housed within the sight unit and collimated to the target sightline.

A lightweight multiple launcher and aiming unit are carried within the Stormer to enable HVMs to be launched away from the vehicle if required by the tactical situation.

The lightweight multiple launcher consists of three standard Starstreak rounds and a man-portable aiming unit located on a traverse head giving a full 360° arc of fire. This system can stand above ground or be sited in a trench.

STATUS

Under development for British Army. Late in 1986 the UK MoD signed a contract with Short Brothers for the development and initial production of the Starstreak HVM valued at £225 million. Major sub-contractors to Shorts are Alvis (Stormer chassis), RO Westcott (rocket motor) and Avimo (sighting system).

Other applications of the Starstreak HVM are also under active consideration including ship based systems and helicopter launched for use in the air-to-air role. In February 1988 it was announced that Boeing Aerospace will promote Starstreak as a replacement for Stinger.

CONTRACTOR

Short Brothers plc, Missile Systems Division, Castlereagh, Belfast BT6 9HN.

Alvis Stormer with eight Starstreak HVMs in ready to launch position

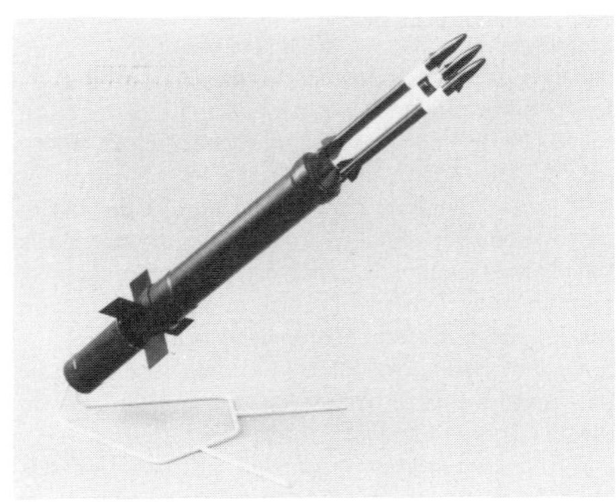

The Shorts Starstreak high velocity missile which contains three highly accurate, extremely manoeuvrable darts each containing a high explosive warhead.

UNITED STATES OF AMERICA

2784.131

REDEYE PORTABLE ANTI-AIRCRAFT MISSILE

Redeye is a shoulder-fired guided missile system designed to give a soldier an effective defence against low-flying aircraft. The US designation is FIM-43 and other users have their own nomenclature as follows: Denmark – Hamlet; West Germany – FLF-1 and Fliegerfaust; Sweden – RB 69. The missile's infra-red sensing device homes on the heat of an aircraft's engines. The Redeye carries a high-explosive conventional warhead and has a two-stage solid-propellant engine. The light launching tube is also a carrying case and can be borne through brush and over rough terrain. A shipping and storage container holds one missile and three battery/coolant units.

On sighting a hostile aircraft, the gunner tracks it in an optical sight. At the same time, he energises the missile guidance system. A buzzer located in the launch tube gripstock informs the gunner when the missile is ready to fire. Upon firing, the booster charge propels the missile out of the launch tube. When the missile has cleared the launch tube muzzle by a distance sufficient to protect the gunner from blast effect (about 6 m), the main rocket ignites and propels the missile the rest of the way to its target. The missile is stabilised in flight by cruciform tail fins and steered by two movable fins close to its nose.

Redeye has no IFF facility and is limited to pursuit course engagements.

CHARACTERISTICS

Designation: FIM-43A
Type: Man-portable, shoulder-fired, surface-to-air guided missile
Guidance: Optical aiming, infra-red homing
Warhead: HE
Missile length: 1.22 m

Practice firing of Redeye portable anti-aircraft missile

Diameter: 70 mm
Propulsion: Solid-propellant booster and sustainer
Weight: 13 kg
Speed: Supersonic
Range: Not disclosed. Probably about 3 km
Crew: Normally 2

Redeye is employed in the forward battle area to protect combat troops against low-level aircraft. In the US Army the weapon is issued to Redeye teams, each made up of a gunner and assistant gunner. From four to six of these teams are assigned to a Redeye section at infantry battalion level.

STATUS

Procurement of Redeye was completed in fiscal year 1970. It has been issued to the USMC as well as the US Army and has been supplied also to Australia, Chad, Denmark (called Hamlet), Greece, Israel, Jordan, Saudi Arabia, South Korea, Sweden, Thailand and West Germany. Replacement by the Stinger system (**2805.131**) is in progress by the US forces.

CONTRACTOR

Prime: General Dynamics Corporation, Pomona Division, Pomona, California, USA.

Rocket motor: Atlantic Research Corporation.

2805.131

STINGER PORTABLE ANTI-AIRCRAFT MISSILE (FIM-92A)

Stinger, Redeye's successor, was in engineering development from July 1972, and is the current US Army and Marine Corps' man-portable air defence system (MANPADS). It is an infra-red seeking missile which enables the soldier to engage effectively low-altitude, high-speed jet, propeller-driven and helicopter aircraft. Designed for the threat beyond the 1980s, Stinger has all-aspect engagement capability, an IFF system, improved range and manoeuvrability and significant countermeasure immunity.

CHARACTERISTICS

Designation: FIM-92A

Type: Man-portable, shoulder-fired, surface-to-air guided missile

Guidance: Passive IR homing

Propulsion: Solid-propellant

Warhead: HE

Length: 1.52 m

Weight: 15.8 kg

An airborne application of Stinger has been trialled, employing a two-round launcher for installation on helicopters to provide such aircraft with an air-to-air self-defence capability and an air-to-ground facility for engaging suitable radiating targets. This project has been named Air Launched Stinger (formerly Multi-purpose Lightweight Missile System, MLMS) in the former role, and ADSM (air defence suppression missile) in the air-to-ground application.

Tests have been carried out where Stinger missiles have been carried and fired from Roland surface-to-air missile system (**2218.131**) fire units. Using a special Stinger container developed by Boeing four Stinger missiles can be housed in the space of a single standard Roland launch tube.

STATUS

Stinger became operational in West Germany in February 1981, and with the US Army 82nd Airborne Division in April 1982.

In addition to the US Army and Marine Corps, it is planned for use by the USAF in small quantities for airfield defence.

The advanced design Stinger-POST version of the weapon entered production in fiscal year 1983. A product improvement, known as Stinger-RMP (Stinger-reprogrammable microprocessor) went into development in September 1984. This improvement is required to retain the ability to up-grade this weapon as the overall threat evolves without having to initiate a costly retrofit programme, and the 1986 budget included $18.6 million to continue Stinger-RMP development. Deliveries of the basic Stinger were completed in January 1987 and initial deliveries of Stinger-POST which has improved performance in an infra-red countermeasures environment began on schedule in September 1986. Stinger POST has an IR search capability and an ultra-violet channel. Deliveries began in September 1986.

Stinger AA missile system

Artist's impression of the General Dynamics/Hughes Aircraft/Thomson-CSF entry in the PMS competition

In September 1987 Raytheon's Missile Systems Division won a $26.4 million contract to become a second source for Stinger. Initially 400 missiles will be supplied with an option, worth $54.4 million for a further 1500. From 1990 General Dynamics and Raytheon will compete for annual contracts.

An agreement for European licensed production was scheduled to be signed in 1988. Some 12 000 missiles are wanted by Greece, the Netherlands, Turkey and West Germany and Dornier were to be the prime contractor. Deliveries are to begin in 1991.

Development of a reprogrammable microprocessor (RMP) is underway. The RMP will provide a capability to facilitate future change to seeker software to counter infra-red countermeasures.

Since its introduction into service in the man-portable role, a number of additional applications of the Stinger weapon system have been designed. Some of these have been developed and tested, others remain proposals; several involve the use of single or multiple Stingers either with novel mountings, or in conjunction with another weapon (gun or missile).

General Dynamics' Valley Systems Division have developed a tripod-mounted version with Marconi FLIR. This was trialled in South Korea during 1987 and in Denmark during 1988.

Setter

The US Army Missile Command (MICOM) began the Setter programme in 1981 with the design and construction of a multiple Stinger launcher prototype, consisting of a power-operated mount with the gunner seated in the centre, flanked by two × four Stinger missiles in pods. Fire control includes a passive RF sensor, day/night vision with TV/FLIR, and the turret is mounted on the HMMWV. Setter also provides for hypervelocity rockets (HVR) in its armament and a total of 54 HVR launch tubes are carried on the turret. A portable gunner fire control console permits dismounted operation remote from the vehicle.

The Stinger provides long-range capability while the HVRs have an effective range of about 1200 m.

Other programmes

The Stinger manufacturers produce as standard items two- and four-round launchers which include system electronics, cooling equipment, and interface, control and display units suitable for use on a variety of platforms. Such launchers feature in numerous other programmes that provide for the application of Stinger to towed systems such as the GEMAG-25 and Vulcan multi-barrel gun systems, powered vehicles such as the LAV, IFV/CFV and deployed with Roland SAM-armed tracked vehicles. Air-to-air use of Stinger has been trialled with several types of helicopter, and naval possibilities include the combination of multiple Stinger launchers on the Mk 15 Phalanx CIWS.

STATUS

In production. In addition to being in service with the United States Army, Air Force and Marine Corps, it is also in service, or has been ordered by a number of other countries including Chad, Italy, Japan, Netherlands, Saudi Arabia, Turkey, United Kingdom (limited use in Falklands) and West Germany. In Europe Stinger will be produced under licence under an agreement concluded in 1983. This consortium is lead by West Germany with the other members being Belgium, Italy, Greece, Netherlands and Turkey.

Early in 1986 the US Army stated that its total requirement for Stinger was 50 664 missiles of which 11 650 had been procured through fiscal year 1985.

CONTRACTOR

General Dynamics Corporation, Valley Systems Division, PO Box 50-800, Ontario, California 91761, USA.

GROUND RADAR

AUSTRALIA

6121.153
JINDALEE OTH RADAR

The Australian Jindalee project is an over-the-horizon backscatter radar, similar in principle to the United States AN/FPS-118 CONUS OTH-B and AN/FPS-71(XN1) ROTHR systems described in detail under entries **5617.153** and **5781.153**. The system operates in the HF band and provides target detection of both aircraft and ships over the 1000 to 4000 km range with a target position resolution of some 20 to 40 km. The main antenna is approximately 2.8 km long, with a transmitter antenna comprising an array of 16 log periodic curtain antennas. Successful trials have been carried out with a receiving station about 40 km north of Alice Springs, in Central Australia, and the transmitter some 100 km north-east of the receiver.

DEVELOPMENT

Development of the Jindalee system started in the 1970s with the Alice Springs station. By 1978, the first stage of the project had been completed by using a narrow fixed beam bounced off the ionospheric 'F' layer giving a detection range of 2000 km. During the early 1980s further development enabled small fixed wing and helicopters targets to be detected, as well as small ships.

The Australian government is now proposing to build an initial network of three systems, costing $A500 million, to cover the northern and western approaches to the country, with the first system expected to enter service in the mid-1990s.

STATUS

In development. The experimental prototype station at Alice Springs is to be upgraded, at a cost of $A57 million, to act as a test bed for the complete system.

Jindalee receiver array at Mount Everard

The main contractor for the whole project is Amalgamated Wireless (Australasia) Ltd.

BRAZIL

Note: Indigenous radar development in Brazil began in 1970, at the instigation of the PEA (Divisão de Eletrônica (Electronics Division) of the IPD – Instituto de Pesquisa e Desenvolvimento (Research and Development Institute)) a branch of the Brazilian Ministry of Aeronautics, the Centro Técnico Aerospacial – CTA (Aerospace Technical Centre). Following preliminary studies and programmes related to the acquisition of basic microwave expertise, and the formation of a nucleus of qualified personnel, two programmes were initiated. As far as is known no other programmes have been initiated and most Brazilian civil and military radars have been obtained from foreign sources; mainly from Thomson-CSF. In addition Selenia has supplied a number of RAT-31S systems for the new air defence system (see entry **4898.181**).

4132.153
S-BAND METEOROLOGICAL RADAR

This radar operates in the 2700 MHz frequency band and has 50, 100, 200 and 400 km. The configuration of the equipment can be seen from the adjacent photograph, from which it will be noted that the large circular reflector can be steered in azimuth and elevation on its tower-mounted pedestal. This presumably allows the radar to be employed for both radiosonde tracking or weather/precipitation surveillance purposes.

Digital techniques are employed in signal processing with presentation as raw radar video. Other facilities include azimuth and range integration, iso-echo and iso-contour presentation with height density levels, colour TV screen display, and remote facsimile presentation. The normal display arrangements provide for sectorial coverage, PPI and RHI displays, and the antenna can be adjusted in elevation from the operating console.

STATUS

It is understood that 32 of these radars have been procured, but no details of installations have been released.

CONTRACTOR

IPD/PEA – Divisão de Eletrônica, 12200 São José dos Campos, SP, Brazil.

Antenna assembly of the Brazilian S-band met radar (Ronaldo S Olive)

CHILE

6383.153
AQUILA MOBILE RADAR

The Chile defence industry has developed a prototype mobile air defence radar system known as Aquila. The equipment has a 50 kW output and will have either eight frequency channels or a frequency synthesiser. Range is given as more than 160 km on airborne targets, even against fixed echo clutter, and either fixed frequency or frequency hopping operation will be available. Pulse duration is stated to be 25 μs, with a pulse compression ratio of 100:1.

The Aquila is modular in design and includes two displays, one with a circular scan for detection, and the other employing a sector scan for tracking airborne targets. An associated powerful processor operates in real time and can be expanded according to requirements. All electronics and displays are housed in a truck-transportable shelter.

STATUS

Prototype. A long-range air surveillance radar is also in development.

CONTRACTOR

Departamento Fabricaciones Militares, Astilleros y Maestranzas de la Armada, PO Box 9-D, Talcahuano, Chile.

CHINA (PEOPLE'S REPUBLIC)

2019.153

CHINESE MILITARY RADAR

During the past few years a considerable amount of detail has been emerging regarding the Chinese radar development programmes to replace the obsolescent Soviet radars that had been in general use. Examples of these are a missile early warning system (**2052.181**), a mobile 3-D tactical radar (**4817.153**) and a number of fire control systems, and with respect to these and other radar developments some code names have emerged from US sources. Cross Legs is the name given to a surveillance radar working in the 1250 MHz band with PRFs of 300 and 600 Hz, suggesting a maximum range capability of something under 500 km. Another name is Thick Skin which is applied to a heightfinder 6500 MHz radar (H-band in the US military frequency band scale).

Information recently obtained from Chinese sources is given in the various entries following this foreword although details of a number of systems are insufficient to provide a meaningful entry. These include the Type 440 two-dimensional surveillance system and its companion heightfinder Type 843A, the Type 797 airport terminal control radar operating in S-band with an effective range of 150 km, the Type 681 secondary radar and the Type 701 meteorological system.

Although China is involved in a substantial indigenous programme, much of the radar is still based on older Soviet equipment, as can be seen from the illustrations which follow. However, as Chinese technology advances, these older systems will be replaced by the more recent developments.

In addition a number of reporting code names, with brief descriptions of the systems to which they refer, are given as follows:

Chop Rest

This is stated to be of Chinese origin and is an early warning radar. In general configuration it is similar to the Soviet Spoon Rest (**2889.153**) VHF equipment, with an array of multiple Yagi antennas, but having a distinctly larger number of elements. Operating frequencies are 162 to 169 MHz and 175 to 189 MHz. Range performance should reach about 250 km.

Cross Legs

Cross Legs is another early warning radar claimed to be deployed throughout China, often in company with heightfinder radar(s), which indicates CGI/air-defence reporting centre installations. The antenna is an elliptical paraboloid with horn feed, supported on a four-legged base.

Cross Slot

Primarily employed for coastal defence, Cross Slot is also an early warning radar, located at fixed sites and often with heightfinder radar(s). Operating frequency is in the E/F-band and the probable range is about 80 km. In addition to installations in China, Cross Slot has been reported in North Korea and Vietnam.

Gin Sling

Fan Song type radars (**2866.153**, **2867.153**, **2868.153**) for surface-to-air missile guidance used by the Chinese armed forces are known by the designation Gin Sling.

Moon Cone

Reported to have been developed in China, Moon Cone is an early warning radar consisting of multiple stacked dipole arrays, located one above the other on a vertical mast. The arrays diminish in area from bottom to top, there being three in a typical installation. VHF operation is almost certain and ranges of 250 km or more should be obtained.

Moon Face

Similar to Moon Cone, but having only one stacked dipole array.

Moon Mat

Similar to Moon Cone, but having only one stacked dipole array.

Moon Plate

Stacked dipole VHF array on trailer mounting.

Slot Rest

Reported to have been developed in China for early warning and target acquisition, Slot Rest is similar to the Soviet Spoon Rest (**2889.153**) radars. Operation is in the A-band (VHF). It is truck-mounted and has been observed in Vietnam as well as China.

Team Work

Team Work is a fire control radar deployed at anti-aircraft artillery sites in China and possibly elsewhere. It was first seen in Vietnam in 1972. Operating frequencies are believed to be in E-band.

4817.153

JY-8 MOBILE TACTICAL 3-D RADAR SYSTEM

The JY-8 is a mobile 3-D air surveillance, target acquisition and interception control radar system operating in C-band. It can be employed as the main radar sensor for an automated tactical defence system, or can be used as an independent radar. The equipment uses advanced signal/data processor techniques and is fully solid-state, with the exception of the magnetrons and thyratrons of the transmitters.

JY-8 3-D mobile tactical radar

The JY-8 consists of four units: an antenna/transmitter-receiver shelter, an operations/reporting shelter, a maintenance shelter and a towing truck. The complete system can be transported by road, rail or sea, and can be assembled or disassembled within one hour.

The system is designed to make use of amplitude comparison for height finding. With the aid of two computers it is able to provide automatically accurate real-time 3-D information on targets. Frequency diversity is used to increase detection probability and enhance ECCM ability. Other facilities such as DMTI, automatic clutter map, automatic residue map, and CFAR are incorporated.

A multi-beam antenna assembly, mounted on the transmitter-receiver shelter and rotating at 3/6 rpm, is employed to form a group of stacked beams. RF power generated by two transmitters is fed to a power-dividing network and then illuminated to the reflector via the waveguide assembly and feed array, creating a group of individual beams stacked vertically to form a wide cosecant squared pattern. In reception, both individual and combined beams are employed, eleven beams formed by a receive network connected with an 11-channel frequency diversity receiver. The received signals are sequentially mixed, amplified, detected and finally sent for data processing and display.

CHARACTERISTICS

Antenna beamwidth: 0.55° (horizontal); 0.9° (vertical) for the focus beam
No. of receive beams: 11
Rotation speed: 3/6 rpm
Power output: 800 kW × 2 (peak)
Pulsewidth: 3 to 3.3 μs
PRF: 365, 500, 600 Hz
Azimuth coverage: 360°
Altitude cover: 25 km
Range: Out to 350 km (maximum)
Elevation: Up to 20°
Accuracy: 500 m (range); 0.3° (azimuth); 600 m (rms) at 185 km range (height)
Target resolution: 1000 m (range); 0.6° (azimuth)
Target capacity: 36 tracks

STATUS

In production and operational service.

CONTRACTOR

ECRIEE (East China Research Institute of Electronic Engineering), PO Box 23, Hefei, Anhui Province, People's Republic of China.

6370.153

JY-9 LOW-ALTITUDE SEARCH RADAR

The JY-9 is a mobile S-band low-altitude search radar intended for use in air defence, gap-filling, airport surveillance and coastal defence. It is designed for the effective detection of targets at low altitude in both ECM and natural clutter environments. It consists of a radar/operations shelter and a small power trailer, and is transportable by air, rail and sea. Setting-up or disassembly time with an experienced crew of four men is 20 minutes.

The JY-9 employs a dual-beam antenna assembly consisting of a deformed parabolic reflector, two horns and two feed channels. One channel is used for transmit, both are used for receive. The antenna rotates at 5/10 rpm to provide 360 degree surveillance in azimuth, and can be mounted on the shelter, on the ground or on a steel tower.

The JY-9 has high anti-jamming and anti-clutter capability due to the use of advanced techniques, including pulse-to-pulse frequency agility, dual channel, JATS, a wide operating band, low sidelobes, DMTI automatic spectrum processing, automatic clutter map, and automatic residue map. MTBF is better than 500 hours, a BITE subsystem is included, and automatic fault detection reaches the level of functional modules.

CHARACTERISTICS

Frequency: S-band
Antenna: Dual-beam with cosecant pattern
Rotation speed: 5, 10 rpm
Beamwidth: Horizontal 1.3°; vertical 40°

Transmitter

Power output: 200 kW peak
Pulse width: 20 μs
PRF: 850 Hz
Coverage: 360° (azimuth); 0-40° (elevation)
Height: 10 km
Range: 150 km (Pd = 80%; Swerling I, 2M²)
Detection accuracy: Range 80 m; azimuth 0.3°
Resolution: 120-140 m; azimuth 1.3°
Tracking capacity: ≥200 targets or groups of targets

STATUS

In operational service.

CONTRACTOR

ECRIEE (East China Research Institute of Electronic Engineering), PO Box 23, Hefei, Anhui Province, People's Republic of China.

6807.153
TYPE 378 BATTLEFIELD RADAR
The Type 378 is a portable battlefield radar, operating in I-band which is used for the detection of vehicles and ships. It is a coherent-pulse doppler system which consists of an antenna/transceiver mounted on a tripod, a rectifier, a signal processor, a generator and a display. After detecting the target a visual and audible alarm is activated, and the position of the target is displayed on the screen. Although claimed to be a portable system, the total weight of 87 kg would appear to require the use of a vehicle to transport it, particularly with the use of a generator for the power source.

CHARACTERISTICS
Frequency: I-band.
Max range: 20 km on vehicles; 6 km on pedestrians
Accuracy: Range 20 m; azimuth 1.5°
Discrimination: Range 60 m; azimuth 3°
Weights: Antenna/transceiver 18 kg; display set 17 kg; rectifier 11 kg; processor 5 kg; generator 30 kg; tripod 6 kg
CONTRACTOR
China Fujian Radio Equipments Factory, Taijung, Sanming, Fujian, People's Republic of China.

Antenna/transceiver of the Type 378 battlefield radar

6127.153
JY-IORIPP RADAR INFORMATION PROCESSING POST
In order to update traditional 2-D radars and heightfinders, and to ensure 3-D coverage, the ECRIEE has developed a mobile 3-D radar information processing post. The intention is to co-ordinate the existing equipments in an automated tactical air defence system to provide a 3-D plan radar/height-finding radar station. The post combines plan extraction with height extraction, incorporates dual microcomputers and microprocessor technology to form a real time target information processing network.
CHARACTERISTICS
Extraction mode: Manual (manual plot keying-in); semi automatic (manual track initiation); fully automatic (hardware/software implementation)
Plan extraction capability: 24 tracks/10 s
Manual extraction capability: 2 × 12 tracks/10 s
Height finding capability: Extraction time for two targets with a relative azimuth of 180° is 15 seconds under the conditions that the height finding radar driving time is 8 seconds and nodding time is 0.5 cycle/s
System error: Range – 500 m; azimuth – 0.5°; height – 300 m
Manual intervention: 68 modes
STATUS
In service
CONTRACTOR
ECRIEE (East China Research Institute of Electronic Engineering), PO Box 23, Hefei, Anhui Province, People's Republic of China

Mobile 3-D information processing consoles

6385.153
MODEL 17-C SURVEILLANCE RADAR
Model 17-C is an intermediate range mobile early warning radar designed for medium and high-altitude surveillance. The Yagi-array type antenna is mounted on a six-wheel truck which also contains the electronics and displays. Alternatively the antenna can be mounted on the ground. The system is small and lightweight and is normally used in the forward area for aircraft detection. The complete system consists of two vehicles, the second truck being a power vehicle containing two diesel generators.
CHARACTERISTICS
Antenna: Yagi array with four rows, two layers and eight units
Frequency: UHF band
Accuracy: 2 km (range); 2° (bearing)
Resolution: 2 km (range); 15° (bearing)
STATUS
In operational use with the Chinese defence forces.
CONTRACTOR
China National Electronics Import and Export Corporation, 49 Fuxing Road, Beijing, People's Republic of China.

4812.153
JLP-40 SURVEILLANCE RADAR
The JLP-40 is a tactical air defence radar designed for use with heightfinder radar(s) for GCI or similar applications. It is similar in design to the Soviet Bar Lock radar (**2861.153**) from which it may well have been derived, and features the same arrangement of two large scanners attached to front and rear sides of a rotating cabin that houses the transmitter/receivers.

Operation is in the S- and L-bands, each parabolic scanner being illuminated by stacked horn feeds to generate families of multiple beams. There are five S-band transmitter/receivers and these are thought to operate with the lower of the two antennas, while the L-band feeds illuminate the upper scanner. An IFF interrogator is incorporated in the system also. Three PPI displays are provided, and one azimuth/range display, plus between two to four additional PPIs. MTI facilities are provided in the L-band transmitter/receiver chains.

The whole system is transportable, though hardly 'mobile' despite the use of wheeled trailers for much of the equipment, and clearly preparing such a large and complex system for operation must occupy several hours from the time of arrival at a surveyed site.

JLP-40 surveillance radar

CHARACTERISTICS
Operating frequencies: L- and S-bands
Range: 270 km
Height: 20 000 m
Azimuth coverage: 360°
Elevation coverage: 0.5 – 30°
Accuracy: 500 m (range); + 0.5° (bearing)
STATUS
In production.
CONTRACTOR
China Electronics Import & Export Corp, 49 Fuxing Road, Beijing, People's Republic of China.

6386.153

HN-401R MOBILE SURVEILLANCE RADAR

The HN-401R is a mobile long range air defence 2-D surveillance radar. It consists of a large horizontally polarised dipole planar array with an aperture of 17 x 8 m, an operations cabin containing all the electronics and two display positions, and a power cabin. The system employs fully-coherent pulse compression techniques and uses a highly stable frequency source. The signal processing system is able to adopt to pulse-to-pulse or burst-to-burst frequency agility, and MTI. The HN-401R comprises three units, an operations shelter, an antenna trailer and a prime power shelter.

CHARACTERISTICS

Operating frequency: UHF

Antenna

Type: Dipole plannar array

Aperture: 17 x 8 m

Gain: 24 dB

Polarisation: Horizontal

Output power: 20 kW (peak); 2 kW (average)

Range: 300 km (1.5 m^2 target)

Resolution: 450 m (range); 9° (azimuth)

STATUS

In service

CONTRACTOR

China National Electronics Import and Export Corporation, 49 Fuxing Road, Beijing, People's Republic of China.

Antenna of the HN-401R surveillance radar

6387.153

TYPE 408-C LONG RANGE AIR WARNING RADAR

The Type 408-C is a mobile long-range air defence warning radar system operating in the VHF band, at 150 to 180 MHz and 100 to 120 MHz. It employs a dipole planar array antenna type, mounted on a pedestal, having two back-to-back dipole arrays which can be operated simultaneously at the high and low bands, giving two transmit/receive channels. The equipment has a wide operating frequency range within each band, enabling it to counter active jamming by rapid change of frequency within the bands. By integrating the video signal it is able to suppress asynchronous pulse jamming, and by modulating the transmitter PRF it counters repeater pulse jamming.

The transmitter system employs air-cooled tower triodes, and uses Hi-Q coaxial cavity tanks to improve frequency stability. The display system consists of one A/R display and 3 PPIs which can scan in opposite directions and display the signals received at the same time from the two antennas within the two bands. The transmitter and receiver can be controlled remotely from the control cabin.

The complete system is mobile with its units installed in eight operation vehicles, two transmitter/receiver vehicles, one display unit vehicle, one antenna unit vehicle, two transport vehicles, and two power supply vehicles (one is a spare).

CHARACTERISTICS

Frequency range: 150 to 180 MHz (high band); 100 to 120 (low band)

Antenna

Type: Two back-to-back dipole planar arrays

Polarisation: Horizontal

Horizontal beamwidth: 8° (high band); 12° (low band)

Rotation speed: Up to 6 rpm

Transmitter

Peak power: 800 kW

Pulse length: 10 µs

PRF: 200 Hz

STATUS

In service.

CONTRACTOR

China National Electronics Import and Export Corporation, PO Box 140, 49 Fuxing Road, Beijing, People's Republic of China.

Antenna of the 408-C radar system

6388.153

TYPE 571 SURVEILLANCE RADAR

The Type 571 is a 2-D L-band radar designed specifically for low altitude air defence warning. Although apparently operating in a different frequency band it bears a striking resemblance to the Soviet Flat Face target acquisition radar (**2874.153**), with its two elliptical paraboloid reflectors mounted one on top of the other, and is almost certainly derived from the latter. The Chinese authorities state that it includes both a frequency hopping and MTI capability.

CHARACTERISTICS

Frequency: L-band

Antenna type: Two elliptical paraboloid net reflectors and horn-shaped radiators

Beamwidth: 4.5°

Rotation speed: 3 or 6 rpm

Peak power: 210 kW

Pulse width: 2 or 4 µs

PRF: 500, 600 or 680 pps

STATUS

In service with the Chinese defence forces.

CONTRACTOR

China National Electronics Import and Export Corporation, 49 Fuxing Road, Beijing, People's Republic of China.

Type 571 surveillance radar

4811.153

TYPE 581 AIR WARNING RADAR

The Type 581 is a medium and low altitude air warning radar for tactical applications. It is a transportable system with the 9.7 metre span scanner mounted on turning gear that is carried on a wheeled trailer, which is stabilised for operational use by large folding legs with screw-jacks at their ends. The antenna rotates at up to 6 rpm. Sector scan is 90 or 30 degrees with selectable scan centres.

The operating frequency is in the L-band, giving a detection range of up to 190 km against a fighter aircraft target. There are two units in a Type 581 mobile convoy, one consisting of the scanner and turning gear (and probably the diesel-electric generator unit also), and an electronics vehicle which includes the transmitter/receiver and operators' cabin.

CHARACTERISTICS

Frequency: L-band

Peak power: 500 W

Pulse width: 4/2 µs

PRF: 300 Hz/600 Hz

Noise figure: 3 dB

Antenna: 9.7 × 3 m

Polarisation: Linear

Gain: 34 dB

Range: 190 km (fighter target)

Accuracy: 1 km (range); 1.2° (azimuth)

Resolution: 1 km (range); 4° (azimuth)

MTI: 20 dB (clutter visibility); 34 dB (cancellation ratio)

Displays: PPI (12-inch); A/R (7-inch)

STATUS

In production.

CONTRACTOR

China Electronics Import & Export Corp, 49 Fuxing Road, Beijing, People's Republic of China.

Type 581 air warning radar

4813.153

JLG-43 HEIGHTFINDING RADAR

The JLG-43 is a nodding heightfinder which follows the pattern of the well-known Soviet 'Cake' series of equipments. It is a mobile system inasmuch as it is readily transportable by military road vehicles, and the complete system can probably be carried in a two-truck load.

Operating frequencies are in the S-band and a 2 MW transmitter provides for coverage out to a range of 200 km and gives a height coverage of up to 25 000 metres.

CHARACTERISTICS

Frequency: S-band
Pulse width: 3 μs
Peak power: 2 MW
Range: 200 km
Height: 25 000 m
Elevation coverage: 0 – 30°
Accuracy: 300 m (height); 2° (azimuth)

STATUS

In production.

CONTRACTOR

China Electronics Import & Export Corp, 49 Fuxing Road, Beijing, People's Republic of China.

JLG-43 heightfinding radar

6384.153

CHINESE OVER-THE-HORIZON RADAR

During the Asian Defence Technology Exposition, Beijing, in November 1986, the Chinese authorities stated that China was carrying out a full-scale development programme into an over-the-horizon radar system. Very little information was provided other than a few parameters which are given below. The system is claimed to have a range of between 700 and 3500 km over an azimuth of 60°, and uses pole antennas similar to those of the Australian Jindalee system.

CHARACTERISTICS

Frequency: HF
Waveform: FM/CW
Transmission beamwidth: 7.5° horizontal; 5 to 36° vertical
Reception beamwidth: 2.5° horizontal; 5 to 36° vertical
Polarisation: Vertical
Output power: 1 MW (20 sets of 50 kW transmitters)
Range: 700 to 3500 km
Capacity: 100 batches of target per 12 to 24 seconds

STATUS

The system is stated to be in already sited. The degree of current operational effectiveness has not been released but it seems likely that the system is in the development phase with trials being carried out.

CONTRACTOR

Hua-Ning Electronics Corporation, PO Box 1315, Nanjing, People's Republic of China.

4814.153

TYPE 404A COASTAL SURVEILLANCE RADAR

The Type 404A is an X-band surveillance radar for use by coastal defence forces. There are two receiver channels operating in frequency diversity and comprehensive signal processing facilities are provided. The elliptical antenna is equipped with a remotely adjustable circular polarising device to reject rain and other clutter. Display arrangements include a PPI, and A- and B-scope presentations.

CHARACTERISTICS

Frequency: X-band
Range: To radar horizon
Antenna gain: 39 dB
Accuracy: 100 m (range); 0.35° (bearing)
Resolution: Better than 200 m (range); 0.8° (bearing)

STATUS

Production.

CONTRACTOR

China Electronics Import & Export Corp, 49 Fuxing Road, Beijing, People's Republic of China.

Type 404A coastal surveillance radar

5761.153

COAST TRACKING RADAR

This is a mobile X-band coastal radar used for surface vessel tracking and the accurate control of coastal weapon systems. Tracking range on a 7500 m target is given as 60 nautical miles (96 km) with an accuracy of 30 m in range and 1.5 mil in azimuth. The rotating antenna is mounted on a flat-bed truck which also carries the operations cabin containing the necessary processing equipment and displays.

CHARACTERISTICS

Frequency: X-band
Antenna beamwidth: 1 × 2°
Peak power: 200 kW
Pulse width: 0.2 μs
Accuracy: 30 m (range); 1.5 mil (azimuth)
Resolution: 60 m (range); 20 mil (azimuth)

STATUS

In production and service.

CONTRACTOR

China Electronics Import & Export Corp, 49 Fuxing Road, Beijing, People's Republic of China.

4810.153

EAGLE COASTAL RADAR

Under the name of 'Eagle' the Chinese authorities are offering for export a Ku-band mobile radar designed for use by coastal artillery units, presumably for target detection and location as well as for fire correction by means of splash-spotting techniques, although this is not categorically stated.

The general arrangement consists of a small (1- to 2-ton) four-wheeled van with a modestly sized mesh antenna unit with its turning gear mounted on the roof of the cab. The latter houses the transmitter/receiver and display equipment in addition to providing working space for the operating crew. Power for the radar is derived from a small portable engine-driven electrical generator that can be located a short distance from the vehicle which is equipped with corner jacks for stability on site.

CHARACTERISTICS
Frequency: Ku-band
Range: 50 km (1000 – 3000 t ship target); 27 km (75 t craft)
Accuracy: 0.09° (bearing); 15 m (range)
Weight: 300 kg (radar only)
STATUS
Production. Available for export.
CONTRACTOR
China Electronics Import & Export Corp, 49 Fuxing Road, Beijing, People's Republic of China.

Eagle coastal radar

4809.153
CHEETAH ARTILLERY RADAR

Cheetah is a Ku-band mobile radar that is stated to be for battlefield use by artillery formations for surveillance and gun fire correction purposes. The complete unit, except for a small engine-driven portable electrical generator located a short distance from the radar unit, is carried on a six-wheeled military truck.

The complete system is probably a Chinese version of the Soviet SNAR-2 radar known to NATO as Pork Trough (**4328.153**), except that in Soviet Army service the equipment is carried on a tracked vehicle.

The Cheetah system is mounted on a flat-bed truck in a cab that contains the electronic equipment and also serves as the operators' cabin. A low-profile rotating antenna is mounted on the roof of the cab with part of the turning gear probably protruding below into the working space. The operational function of the Cheetah radar is battlefield surveillance for targets such as tanks and armoured concentrations, and for correction of artillery gun fire.

Although performance details are sparse, there are claims of some relevance, as indicated in the table below, and references to a two-colour display which shows moving targets in a different colour to the main picture, anti-jamming facilities, and MTI.

CHARACTERISTICS
Frequency: Ku-band
Detection range: 300 m – 16 km (tank target)
Peak power: >70kW
Elevation coverage: ± 10°
Accuracy: 5 m (range); 1 mil (azimuth)
Weight: 7500 kg (including truck)
STATUS
In production. Available for export.
CONTRACTOR
China Electronics Import & Export Corp, 49 Fuxing Road, Beijing, People's Republic of China.

Cheetah artillery radar

4815.153
MW-5 FIRE CONTROL RADAR

The MW-5 is a mobile artillery fire control radar operating on X- and S-band frequencies. It is similar to the Soviet Fire Can radar (**2871.153**) and may well be either a copy or a derivative of that equipment. All the main elements of the equipment are housed in or on a four-wheeled trailer that consists of a container which also serves a cabin for the operating crew.

A circular dish antenna for target search and tracking is mounted on the roof of the cabin. The S-band is employed for target detection and acquisition, possibly with designation by a different long-range search/air warning radar associated with the military unit, and X-band will be used for target tracking. The conical scan technique is probably employed for the latter function. Good ECCM characteristics are claimed for the MW-5 by its manufacturer.

CHARACTERISTICS
Frequencies: S-band (search); X-band (tracking)
Detection range: 55 km (bomber)
Max tracking range: 35 km
Tracking accuracy: 20 m (range); 1.6 mil (azimuth); 1.8 mil (elevation)
STATUS
Production.
CONTRACTOR
China Electronics Import & Export Corp, 49 Fuxing Road, Beijing, People's Republic of China.

MW-5 fire control radar

4816.153

TYPE 311-A FIRE CONTROL RADAR

The Type 311-A fire control radar is for use with anti-aircraft guns and is normally employed with batteries of either 37 mm or 57 mm calibres. It operates on X-band frequencies and is capable of both search and target tracking functions and is generally used with a computer and an optical rangefinder.

The Type 311-A consists of an operations trailer with the main electronic and mechanical elements of the radar, and a towing vehicle in which the operators are carried together with a power generating set, tools and spare parts. The complete system can be set up or dismantled to move within about 15 minutes and the weight of the radar trailer is less than four tons. The towing vehicle with equipment weighs under eight tons.

Detection range on a fighter-size aircraft target in the search mode is at least 30 km, with a maximum tracking range of 25 km. Minimum reliable tracking range is about 500 metres. The radar can be switched to any one of three pre-programmed operating frequencies by the operator, and target position data in the tracking mode can be fed to the computer in either rectangular or spherical co-ordinates. In the search mode, the radar beam is oscillated in the vertical plane at a rate of 4 Hz to broaden the effective beam. A target within this beam that subsequently deviates in either elevation or azimuth by 20 mils or more from the radar boresight axis can then be tracked automatically.

CHARACTERISTICS

Operating frequencies: X-band, 3 switchable
Peak power: 200 KW (0.3 µs pulse); 180 KW (0.9 µs pulse)
Pulse width: 0.3 µs (narrow); 0.9 µs (broad)
PRF: 2500 Hz (narrow pulse); 833 Hz (broad pulse)
Antenna gain: At least 35 dB
Horizontal beamwidth: 2.6°
Vertical beamwidth: 2.4°
Sidelobe: –18 dB (max)
Receiver sensitivity: –92 dB/mW (CW)
Max range: 30 km (detection); 25 km (tracking)

STATUS

Production.

Type 311-A fire control radar

CONTRACTOR

Chuanbei Electronics Company, People's Republic of China.

6389.153

HUNTING DOG RADAR

Hunting Dog is the code name for a moving target search and gun-pointing adjustment radar. It is used to search or track a target manually within 50 km and over the complete 360° in azimuth, and to search for targets automatically within 26°. The system operates in a mode of double-band frequency hopping, using circular polarisation techniques, and is claimed to be very effective against active ECM and natural clutter. Other facilities include a narrow pulse, a narrow beam and electro-mechanical scanning.

The complete system is incorporated in a towed trailer which consists of the main control panel and indicator, with the antenna mounted on top.

CHARACTERISTICS

Frequency: Ka band
Peak power: >30 kW
Range: 100 m to 50 km
Azimuth: 450°; 26° for automatic search
Elevation: –7.5 to +15°
Accuracy: 2.5 m on a stationary target, 5 m on a moving target at a speed of 40 km/h
Weight of system: Approximately 2700 kg

STATUS

In service.

CONTRACTOR

China National Electronics Import and Export Corporation, PO Box 140, 49 Fuxing Road, Beijing, People's Republic of China.

Hunting Dog radar

4807.153

TYPE 791-A PRECISION APPROACH RADAR

The Type 791-A precision approach radar (PAR) is a typical military mobile radar equipment. It is mounted on a six-wheeled truck, although the manufacturers state that both mobile and fixed installations are available.

This Chinese-manufactured system closely resembles the Soviet RSP-7 PAR, known to NATO as Two Spot (**3295.153**), and it could well be a licence-made version although there are slight differences. For example, in the Soviet version the PAR antenna heads are located between the driver's cab and the radar cab, whereas in the Chinese model the antennas are at the rear end of the whole assembly.

However, this could be nothing more significant than the fact that the equipment was designed as a cabin-housed equipment meant to be carried on a flat-bed truck, and capable of being loaded either way round. There are similar slight differences in the scanner outlines.

Operation is in the X-band and dual transmitter/receivers are provided to ensure continuity of service; for the same reason the display indicators are also duplicated.

The two scanners, for search and azimuth guidance and for elevation guidance, are collocated on a mounting at the rear of the vehicle. There is an operators' cab in front of this which also contains communications facilities. Circular polarisation is provided to combat weather clutter and the receiver includes a logarithmic IF amplifier.

CHARACTERISTICS

Frequency: X-band
Range: 35 km (15 km in rain)
Coverage: 1 – 8° (elevation); 20° (azimuth)
Accuracy: 0.35° (elevation); 0.5° (azimuth)
Range accuracy: approx 60 m
Resolution: 200 m (range); 1.2° (angular)

Type 791-A PAR

STATUS

Production.

CONTRACTOR

China Electronics Import & Export Corp, 49 Fuxing Road, Beijing, People's Republic of China.

DENMARK

5629.153
TERMA RADAR AND BALLISTIC INSTRUMENTATION

Terma has been in the field of ballistic instrumentation and the necessary supporting radars since 1961. The company manufactures ballistic instrumentation for internal and external measurements. The instruments are invaluable tools for ballistic research and development projects, and for product quality assurance tests.

Splash Spotting Radar

The splash spotting radar is intended for use on test ranges where firing is made over water and accurately determines the impact co-ordinates of shots. It has two modes, a spotting mode for measurement of slant range and bearing from the weapon to the point of impact, and a surveillance mode to scan the shooting area. It provides digital read-out of range and bearing on an LED display, and a hard copy print-out of measurement results.

The antenna is a 7 ft (2.1 m) slotted waveguide mounted on a gearbox. The gear is driven by a DC servo motor which enables the antenna to either rotate continuously or to scan in preselected areas. The receiver/transmitter and servo-amplifier are mounted inside the antenna supporting structure. The display unit consists of a CRT display, data handling and display controller and the interface and control assembly.

Doppler Radar

The DR 582 doppler radar is the system name for a number of antenna units and a data unit operating with a suitable computer system. They are designed to supply measurement data of rockets, missiles, mortars and other projectiles ranging from the smallest to the largest calibre. They range from small muzzle velocity radars to large scientific systems and supply external ballistic data such as velocity, acceleration, drag coefficient and altitude/distance functions. Two different CW transmitter/receivers are available, one operating in S-band and one in X-band, with power outputs of 20 W at 2.63 MHz and 300 mW at 10.525 MHz respectively. The systems supply continuing on-line analysis. In-depth analysis of the radar data is provided by a specific combination of software and hardware, including fast Fourier transform techniques.

Transient Recorder

This is an instrument specifically designed for the static testing of rockets and in-bore measurement of guns and provides calculations of various parameters.

CONTRACTOR

Terma Elektronik AS, Hovmarken 4, 8520 Lystrup, Denmark.

7° S-band programmed tracking antenna unit

FRANCE

2056.153
ADOUR II INSTRUMENTATION TRACKING RADAR

Adour II instrumentation radar is an improved version of the previous Adour, and is a G-band tracking system using the conical scanning principle. It is intended for acquisition and automatic tracking and is specifically designed for making measurements in:

(1) Flight test centres, for the calibration and evaluation of prototype aircraft, of airborne equipment such as altimeters and autopilots, and of ground equipment such as surveillance radar
(2) Rocket and missile test centres and launching centres for measurements relating to missile, sounding rockets, and satellite launchers.

In addition, the radar is used at such centres for windfinding and similar measurements.

Adour II is available in fixed or mobile form. It has been modified recently to improve its class of accuracy by using more extensive digital techniques and improving the operating mode.

STATUS

Adour is installed on Australian, Brazilian, French, Swiss and other undisclosed test ranges. It is still in production for both French and export needs.

CONTRACTOR

Thomson-CSF, Division Systèmes, Défense et Contrôle, 40 rue Grange Dame Rose, 92363 Meudon-la-Foret Cedex, France.

Note: Adour is a member of an extensive range of French instrumentation tracking radars, other members of this family being:

Artois (**2063.153**)
Atlas (**1943.153**)
Bearn (**2065.153**)
Savoie (**1944.153**)

Details of these radars will be found on succeeding pages of this section of *Jane's Weapon Systems*. More detailed information on Adour is given in previous editions and details of the earlier Aquitaine tracking radar, will be found under entry number **2062.153** in the 1974-75 and previous editions of *Jane's Weapon Systems*.

Adour tracking radar TRS 2504 installed at the Brazilian Army's Marambala Proving Grounds near Rio de Janeiro (Ronaldo S Olive)

2060.153
ANTARES HEIGHTFINDING RADAR

Under the code name ANTARES (for antenna tracking altitude, azimuth, and range by electronic scan) Thomson-CSF has developed a heightfinding radar operating in E/F-band that can be used for air defence and civil air traffic control purposes. This system is designed mainly to measure the altitude of targets tracked by a 2-D search radar which is associated with it, with a very high accuracy and a high data renewal rate. ANTARES is an automated system, and all its input or output data are controlled or generated by an automatic digital processing unit. This feature, together with the utilisation of electronic scanning in elevation, enables the system to satisfy a practically unlimited number of altitude requests.

The ANTARES system consists of a heightfinding radar and a processing unit. Principal units of the radar are an antenna under a radome, a servo system for slaving the rotation of the antenna to that of the associated plan radar antenna, transmitter, receiver, and one angular error measuring receiver. More detailed information is given under this entry number in previous editions of *Jane's Weapon Systems*.

STATUS
No longer in production, but still in operational use.

CONTRACTOR
Thomson-CSF, Division Systèmes, Défense et Contrôle, 40 rue Grange Dame Rose, 92363 Meudon-la-Foret Cedex, France.

ANTARES heightfinding radar antenna. It is normally enclosed within a protective radome

2063.153
ARTOIS INSTRUMENTATION TRACKING RADAR
Artois is an electronic-scan multi-target tracking radar designed to perform highly accurate differential trajectory measurements.

The principal features of this radar are electronic scanning in a cone having a vertex angle of 10° minimum, instantaneous deflection from pulse-to-pulse, and multi-target tracking with controlled variable interlace.

More detailed information is given under this entry number in previous editions of *Jane's Weapon Systems*.

STATUS
No longer in production. One installed at Centre d'Essais des Landes on the French Atlantic coast for ballistic and tactical missile trials and training.

CONTRACTOR
Thomson-CSF, Division Systèmes, Défense et Contrôle, 40 rue Grange Dame Rose, 92363 Meudon-la-Foret Cedex, France.

2065.153
BÉARN INSTRUMENTATION TRACKING RADAR
Béarn is a high-precision G-band automatic tracking radar designed for measuring the trajectory of high-speed missiles at long range.

Facilities offered by the radar include manual or automatic acquisition of a target dynamically or statically designated by the rendezvous method; automatic tracking of radar echoes or transponder signals up to 4000 km; elevation, azimuth, and range co-ordinate read-out as numerical data in real time; automatic changeover to memory tracking in case of signal loss; polarisation switching without interruption of tracking; synchronisation of a chain of radars interrogating the same transponder; and, for shipborne installations, autonomous and automatic stabilisation of the pointing axis.

More detailed information is given under this entry number in previous editions of *Jane's Weapon Systems*.

STATUS
In operational use in France and a number of other countries. No longer in production.

CONTRACTOR
Thomson-CSF, Division Systèmes, Défense et Contrôle, 40 rue Grange Dame Rose, 92363 Meudon-la-Foret Cedex, France.

1943.153
ATLAS INSTRUMENTATION TRACKING RADAR
Atlas is a very high precision tracking radar designed for trajectography and satellite tracking. A monopulse G-band radar, Atlas automatically tracks radar echoes or transponder signals at distances up to about 4000 km. The equipment comprises an antenna turret with which are associated the RF and IF receiver circuits, a 1 MW peak transmitter codable and tunable from 5450 to 5825 MHz, an operating console, an interconnection cabinet, and a mains supply cabinet. The rangefinding unit, in the console, is fitted with a synchronising device that allows the radar to be used in series with other tracking radars of the same or different types.

The Atlas radar is available in fixed or transportable form. Recent improvements include more extensive use of digital techniques and operating mode. Operational optional facilities include TV monitoring, TV tracking and IR tracking.

CHARACTERISTICS
Antenna
Cassegrain feed system:
Diameter: 4 m
Polarisation: Vertical/circular
Gain: 44 dB
Beamwidth (3 dB): 0.9°
Transmitter
1 MW magnetron, codable and tunable: 5450 – 5825 MHz
Pulse length: 0.25 – 1.7 μs
Transmitter frequency controlled by a high-precision standard cavity

STATUS
In production.

CONTRACTOR
Thomson-CSF, Division Systèmes, Défense et Contrôle, 40 rue Grange Dame Rose, 92363 Meudon-la-Foret Cedex, France.

Atlas tracking radar antenna

3275.153
LOUXOR AIR DEFENCE RADAR
Louxor is an experimental prototype electronically scanned radar for short-range air defence applications. It was built as a research and development instrument in the definition and design of a production equipment suitable for use in such systems as Systèmes d'Armes sol-air à Courte Portée (SACP) planned for future close-in air defence applications.

STATUS
Prototype. More detailed information is given under this entry number in previous editions of *Jane's Weapon Systems*.

CONTRACTOR
Thomson-CSF, Division Systèmes, Défense et Contrôle, 40 rue Grange Dame Rose, 92363 Meudon-la-Foret Cedex, France.

3274.153
RAMSA LOW-ALTITUDE SURVEILLANCE RADAR TRS 2140
Ramsa is a mobile short-range, low-altitude surveillance radar station, with the manufacturer's designation TRS 2140. The radar is of the pulse-doppler type, with automatic data processing, and is installed in a cabin. Operational roles are associated with its use in an air defence system for warning, coordination and target designation for gun or missile air defence batteries.

Few details are available for publication other than those main characteristics listed below. An extending tower mounting for the antenna head is an alternative option for the Ramsa system.

TRS 2140 Ramsa low-altitude surveillance radar mounted on container carried by flat-bed truck

CHARACTERISTICS
Frequency: D-band (former L-band)
Peak power: 150 W
Average power: 20 W
PRF: Dual with matching pulse-lengths (optional 3)
Antenna rotation rate: 60 rpm
Detection range: 15 km (fighter aircraft target); 20 km as an option
Associated IFF
STATUS
In production.
CONTRACTOR
Thomson-CSF, Division Systèmes, Défense et Contrôle, 40 rue Grange Dame Rose, 92363 Meudon-la-Foret Cedex, France.

1944.153
SAVOIE INSTRUMENTATION TRACKING RADAR
Savoie is a multi-tracking radar, intended for automatic tracking at long ranges. The equipment is designed to be shipborne or ground based.
CHARACTERISTICS
Type: Monopulse tracking radar
Antenna
Diameter: 8 m
Gain: 28 dB
Beamwidth (3 dB): 6°
Polarisation: Linear
Turret
Weight: 12 t
Azimuth rotation: + 110°
Elevation rotation: 0 – 90°
Speed: 0.25 rad/s
Acceleration: 0.5 rad/s²
Accuracy of analysis axis: 5×10^{-4} rad
Transmitter
Peak power: 150 kW
Mean power: 22 kW
Pulse duration: Pulse compression with several modes
Receiver
Pulse compression: Receiver controlled by computer
Noise figure: 2.5 dB
STATUS
One installed on the trials research ship *Henri Poincaré*. Several improvements have been introduced recently, mainly in the data processing capability.
CONTRACTOR
Thomson-CSF, Division Systèmes, Défense et Contrôle, 40 rue Grange Dame Rose, 92363 Meudon-la-Foret Cedex, France.

Savoie tracking radar antenna

3099.153
TRS 2052/TRS 2053/TRS 2056 D-BAND EARLY WARNING RADAR
The TRS 2052 is a D-band early warning radar which has been developed as a successor to the TRS 2050 (**2100.153** in 1976 and earlier editions). Externally, and in other general characteristics, the later radar is very similar to the TRS 2050. The double-curvature antenna is of the same dimensions (9 × 13 m) and independent high- and low-cover beams are radiated. Cosecant beam shaping in elevation extends to over 40° and sharp ground cut-off of the high-cover lobe produces a strong contrast between air target echoes and ground returns. High-grade circular polarisation ensures rejection of weather returns.

An IFF antenna of the sum/difference type is mounted on the primary radar reflector, as is the side-lobe suppression antenna.

Variations of this successful equipment have led to two more models, the TRS 2053 and TRS 2056. The TRS 2052 air-defence long-range surveillance radar, the medium-range and SRE radars TRS 2053 and TRS 2056, use different antennas with the same electronics. A fully equipped cabin, transportable by air, rail and road, houses a transmitter/receiver unit which comprises mainly: two high-stability transmitter/receivers of the 2 MW/2 kW class, equipped with long-life magnetrons, tunable by single-knob control, operating in dual diversity; an associated IFF interrogator; two reception and processing units comprising a logarithmic receiver and a digital linear MTI filtering assembly.

Optionally, an SLS reception unit to get rid of sidelobes can be included.

In this fashion, a set suitable for both fixed and mobile installations is available to associate with different antennas to constitute the TRS 2052, TRS 2053 and TRS 2056 stations.

With a large antenna (13 × 9 m) featuring high gain and dual coverage, a highly jam-resistant air-defence master station, the TRS 2052, is obtained. This has a detection range capability in excess of 350 km for modern fighter aircraft. With a smaller antenna (9 × 5 m), it becomes the TRS 2053, designed for equipping military airfields as recovery station or as gap-filler radar in an air defence network. These radars are widely distributed around the world. The third one of this family, the TRS 2056, developed for the French Air Force as a movable recovery station, uses an 8.5 × 3.5 m antenna which can be dismounted without special tools. It is known also as Centaure.

Simultaneous use of operationally variable frequency, frequency diversity, constant-false-alarm-rate receivers , of high-performance, compatible MTI, and of sidelobe suppression, endows the radar with excellent anti-jamming capabilities.

All three sets are compatible with fully automatic operational use, and feature built-in test equipment.

CHARACTERISTICS
Type: TRS 2052
Operating frequency range: 100 MHz in D-band
Antenna
Reflector dimensions: 13 × 9 m
Gain (low cover): 36 dB ± 0.5 dB
Gain (high cover): 35.5 dB ± 0.5 dB
Beamwidth in azimuth (3 dB): 1.2°
Cosecant: > 40°
Polarisation: Circular, fixed

TRS 2052 D-band early warning radar

Rotation speed: 6 rpm
Transmitter
Frequency-tunable magnetron
Peak power: ⩾ 2.2 MW
Pulse duration: 3μs
PRF: 350 Hz (3-period staggered)
Receiver
Noise figure of RF amplifier: 3 dB
Noise figure at input to reception chain: 4 dB
Intermediate frequency: 30 MHz
Sub-clutter visibility: ⩾ 25 dB
Digital post-detection integration
Power supply: 3-phase, 220/380 V, 50 Hz
Consumption: 40 kVA (55 kVA for diversity operation)
Performance
Detection range for fluctuating 2 m² target: > 180 nm (330 km)
Detection in altitude: > 60 000 ft (18 300 m)
DEVELOPMENT
The TRS 2052 was developed in the mid-1970s as a successor to the TRS 2050 air defence early warning radar and the other derivatives followed shortly thereafter. The TRS 2053 and TRS 2056 (developed for the French Air Force under the name Centaure), use the same transmitter/receiver as the TRS 2052 but have different antennas (see photographs).
STATUS
Numerous radars of this family have been supplied in France and to overseas customers. The TRS 2056, developed for the French Air Force as the Centaure,

TRS 2056 (Centaure) D-band early warning radar

has been supplied in some quantity, and the TRS 2052 is known to be employed in the DACTA civil/military ATC and surveillance system supplied by Thomson-CSF to Brazil. The TRS 2053, also, is known to have been installed in France and other countries but details are lacking.

CONTRACTOR

Thomson-CSF, Division Systèmes, Défense et Contrôle, 40 rue Grange Dame Rose, 92363 Meudon-la-Foret Cedex, France.

3683.153

TRS 2054 LONG-RANGE SURVEILLANCE RADAR FAMILY

The TRS 2054 is a coherent, klystron-powered, long-range surveillance radar operating in the D-band of the radar spectrum. It has been evolved for air traffic surveillance roles and similar applications, and is thought to have been based upon experience gained with the TRS 2052 series of radars (**3099.153**). It embodies the latest concepts in coherent double-pulse operation and auto-regulating signal processing. This dual-channel 23 cm radar is claimed to be able to detect small aircraft at ranges of up to 200 nautical miles (370 km), and superior clutter see-through capabilities permit detection down to low levels. An integral plot extractor enables the radar to be connected to a data processing system.

The dual beam antenna uses offset feeds for reduced masking of the energy radiated by the reflector, thereby reducing sidelobes, and there is an elongated lower edge to the reflector to intercept radiation from the primary sources so that back lobes are reduced. The feeds are designed to ensure that the reflector edges are only illuminated by low-energy radiation, thus reducing spillover to a level comparable to the background level. A cosecant-squared radiation pattern, boosted at high elevation angles is employed, providing good contrast between short range targets and ground clutter. There is a sharp roll-off at low elevation angles, which also improves the signal-to-clutter contrast, and this is obtained by employing a reflector with a larger-than-usual vertical dimension. Compact stacking of the main and auxiliary feeds, gives good focusing for the high beam and reduced boresight separation between the two beams, while still maintaining a target-to-clutter enhancement of 16 to 18 dB.

TRS 2054 (LP23K) long-range surveillance radar

The transmitter uses a TH 2068 klystron which is stated to have a saturation gain of over 50 dB. It is a pulse-driven tube with five cavities giving a peak output of 5 MW, although in the present application it is deliberately under-driven to increase its service life. It is operated at 3.5 MW peak power and 4.7 kW mean power. The transmitter is crystal-controlled. At each pulse repetition interval, the transmitter generates a coherent pulse train consisting of a long duration pulse followed immediately by a short pulse. This provides a means of obtaining simultaneously in a single radar the benefits of both long- and short-pulse operation. The latter gives a precise radar cell and so improves both target resolution and signal-to-clutter contrast, these characteristics being especially valuable at short and medium ranges. At greater ranges, the longer pulse results in an increased useful detection range.

The radar processing system associated with the TRS 2054 radar is called TVD 1000 and is of the moving target detector (MTD) type. It includes a separate processing channel, called weather channel, and is designed to improve target detection in various forms of clutter, while providing low output false alarm rate. The chosen signal processing algorithms are based on the use of a band of eight Doppler filters, on CFAR thresholding, and on an adaptive clutter map. The weather channel indicates to the user the areas of dangerous precipitation within the radar coverage. Programmable modular processors are used for the digital signal processing and data processing. Their processing enables complex adaptive algorithms to be carried out. This equipment is incorporated in a radar station which can be unmanned, thanks to the automatic built-in-test equipment and the remote maintenance and monitoring system. Results of experiments with the processing system described show clearly that the MTD processing improves the final picture quality. In particular, the detection probability is higher both in clear zones and in clutter, while the false alarm rate remains low.

Other members of the family use the same inherent transmitter/receiver and the antenna of the TRS 2053.

STATUS

In production and service in France and a number of other countries.

CONTRACTOR

Thomson-CSF, Division Systèmes, Défense et Contrôle, 40 rue Grange Dame Rose, 92363 Meudon-la-Foret Cedex, France.

3272.153

TRS 2230/TRS 2215 3-D RADARS

This family of E/F-band air-defence three-dimensional radars, employing electronic phase scanning in elevation, is derived from the very high precision heightfinding radar SATRAPE which has equipped the French Air Force (see **2060.153** ANTARES).

The TRS 2230 is the fixed or relocatable version in this family. The TRS 2215 is the mobile version, transportable by road, rail or aircraft (Transall, C-130, or equivalent), and can be deployed in less than one hour. Both radars employ the same space scanning techniques, the same transmitter, the same reception and processing circuits, and the same antenna components, but make use of two sizes of planar phased array antennas. This commonality, which has particular relevance to maintenance, logistics and training, allows in addition a high flexibility of deployment of the facilities in a system incorporating both versions (eg backing up of a fixed-station antenna by a mobile-station antenna).

The radars analyse the space:

(a) in azimuth by continuous rotation of the mechanically driven antenna system

(b) in elevation by a phase shifter electronic scanning system.

The primary radar antenna is a direct-radiation array of stacked linear sub-arrays fed by two vertical distributors used to form the sum and the elevation difference channels. Each linear sub-array contains elementary sources radiating a circularly polarised wave. The two waveguides allow monopulse operation to perform the heightfinding function. Digital electronic phase-shift networks inserted between the elementary feeds and the corresponding directive couplers control the pointing of the beams.

The transmitter is an amplifying chain transmitter using a high-efficiency crossed-field amplifier as its final tube. It permits the use of the pulse-compression technique and that of pulse-to-pulse frequency agility. It is characterised by high efficiency (greater than 70 per cent) and a modular modulator allows repair operations, if required, to be carried out without shutting down the station.

The receiving section groups the receivers, the MTI unit, the extractor, the plot processing and altitude computing equipment, the control device for the tilting of the antenna according to the mode of operation imposed, and the monitoring desk.

The receiver circuits comprise the anti-jamming chains and three angle error chains. The MTI device processes the signals received from any one of the three lower beams by programming. The extractor performs the extraction of primary and secondary plots and allows them to be associated together. It is followed by a plot processing system which computes altitude, and filters plots so as to transmit only useful data. The device generates the data in a digital form permitting the transmission of tracks to a remote centre. The monitoring desk groups the monitor controls of the station's operation.

CHARACTERISTICS

Detection range: 510 km
Ceiling: >30 500 m
Elevation coverage: –3 to +30°; programmed operating multimode

TRS 2230 long range 3-D radar installed in Brazil

TRS 2215 long-range 3-D mobile radar station

Altitude accuracy:
TRS 2230: ± 250 m at 200 km
TRS 2215: ±450 m at 200 km
Data renewal rate: 10 s
TRS 2215 antenna
Gain: 38.5 dB
Beamwidths (3dB): Azimuth 1.5°; elevation 2 to 4°
TRS 2230 antenna
Gain: 40 dB
Beamwidths (3 dB): Azimuth 1.5°; elevation 1.3 – 3.6°
Polarisation: Circular, fixed
Rotation speed: 6 rpm
Transmission
Peak power: 700 kW
Average power: 10 kW
PRF (average): 380 Hz
Reception: CFAR
MTI improvement factor: 40 dB

STATUS

In production for French and other defence forces. An electronic scanning radar, the TRS 22XX (**6808.153**) which belongs to the same family, has been selected by the French MoD. It complies with NATO standards. The TRS 2215 is being produced in India, under licence, by Bharat Electronics.

CONTRACTOR

Thomson-CSF, Division Systèmes, Défense et Contrôle, 40 rue Grange Dame Rose, 92363 Meudon-la-Foret Cedex, France.

6808.153

TRS 22XX LONG-RANGE 3-D RADAR

The TRS 22XX is a 3-D air defence radar complying with the NATO Class 1 radar specifications. It has been developed primarily to replace the long-range 3-D radars operated by the French Air Force. It is designed for efficient operation in conditions of various and simultaneous clutter, heavy jamming, blast and battlefield conditions. It is transportable by road, rail, sea and air.

The TRS 22XX operates in S-band with a large frequency bandwidth (15%). Its detection coverage is 470 km range, 100 000 ft in altitude and a 20° elevation pattern. It has a number of advantages including a large operating bandwidth, a high grade antenna with very low sidelobes and an electronic phase scanning, non-dispersivity single beam operation. Various transmission modes are programmable by narrow azimuthal sectors, including fixed frequency transmission, pulse-to-pulse or burst-to-burst frequency agility over the complete band, and radar silence. Various operating modes programmable by narrow azimuthal sectors are available in order to optimise detection according to the operational environment. Other advantages include auto-adaptive doppler processing associated with a very short compressed pulse; and anti-jamming techniques such as analysis, automatic extraction of jam strobes, selection of least jammed frequency, and display of jammed sectors.

The TRS 22XX radar station consists primarily of:

(a) the aerial assembly; the azimuth coverage results from the mechanical rotation of the antenna while elevation cover stems from beam and phase scanning techniques. The latter, together with the non-dispersive design of the antenna, allows true frequency agility over the complete bandwidth in all directions
(b) the high average power transmitter, located in a cabin, with its ancillaries container
(c) a reception and signal processing cabin, which houses a local console for the radar station control

The radar may also be operated from a remote position. It is fitted with built-in test equipment which carries out a permanent check of the operational status and delivers diagnostic information to the local or remote monitoring systems.

CHARACTERISTICS

Frequency: S-band
Bandwidth: 15%
Average power: 20 kW
Antenna
Dimensions: 6 m wide × 4.5 m high

TRS 22XX antenna

Rotation speed: 6 rpm
Azimuth beamwidth: 1.2°
Elevation beamwidth: 1.6°
Gain: 40 dB
Range: 470 km
Ceiling: 100 000 ft
Elevation: 20°
Accuracy: Range 50 m; azimuth 0.25°
Resolution: Range 120 m; azimuth 2.5°

STATUS

In production. The first production radar is due for delivery to the French Air Force in late 1988.

CONTRACTOR

Thomson-CSF, Division Systèmes Défense et Contrôle, 40 rue Grange Dame Rose, 92363 Meudon-la-Foret Cedex, France.

3197.153

OEIL VERT RADAR

The Oeil Vert radar is a D-band equipment used in the AMX-30 SA self-propelled 30 mm twin gun anti-aircraft system (**2138.131** in 1982-83 and earlier editions) for surveillance and target detection. It is a pulse doppler system and is mounted on the turret of the AMX-30 carrier vehicle, the indicator and control units being inside and the transmitter/receiver, signal processor, antenna erection mechanism, and junction box being housed in a protective container attached to the rear of the turret.

Two surveillance modes are provided: (a) distant, with a range of 15 km intended for normal use, and (b) close-in, with a range of 6.5 km which is available at the operator's discretion to suit operating conditions. Each mode has several range gates. MTI and track-while-scan facilities are provided and good sub-clutter visibility also is claimed.

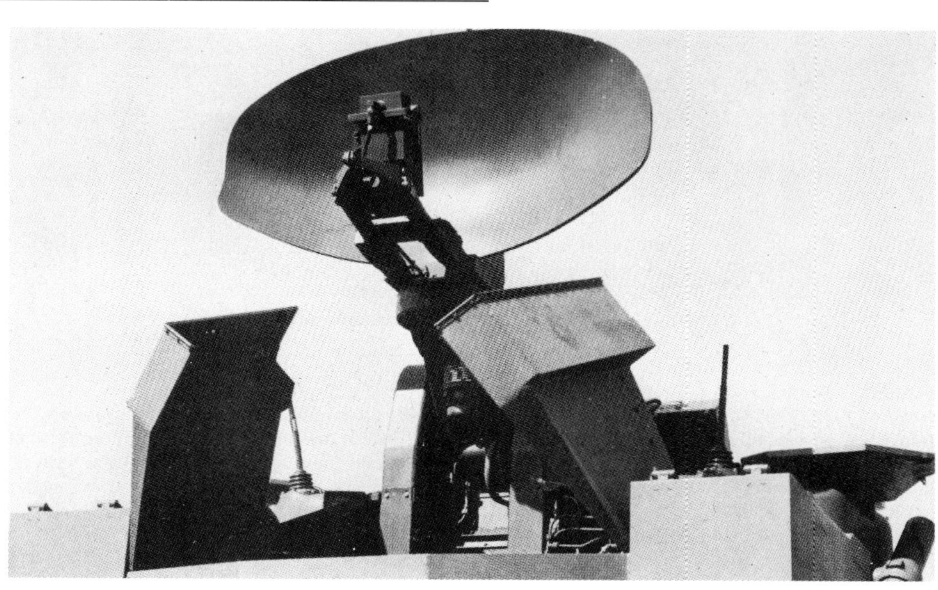

Oeil Vert scanner erected and ready for use

CHARACTERISTICS
Type: Pulse doppler
Frequency: 1710 – 1750 MHz
Power: 120 W (peak)
Max range: 15 km and 20 km as an option
Azimuth accuracy: $\pm 2°$ (approx)
MTI radial speed range: 30 – 160 m/s (slow targets); 160 – 300 m/s (fast targets)
Tracking range: 15 km to almost 0 km

STATUS
In production. Used in the AMX-30 SA tracked AA gun system (**2138.131** in 1982-83 and earlier editions) and the Dragon AA tank project (**3679.131**). The Oeil Vert Plus system has been supplied to the Egyptian armed forces.

CONTRACTOR
Thomson-CSF, Division Systèmes, Défense et Contrôle, 40 rue Grange Dame Rose, 92363 Meudon-la-Foret Cedex, France.

5601.153
RA 20S FIRE CONTROL RADAR

RA 20S is a coherent pulse doppler radar used for the detection and tracking of fixed and rotary wing aircraft flying at low and very low altitudes and is the successor to the RA 20 system described under **3717.153** in previous editions of *Jane's Weapon Systems*. It has been specially developed for integration in anti-aircraft defence systems fitted with 20, 23 or 25 mm twin-gun turrets or carriages, such as the TA20 turret designed by ESD. It can be used as the 'leader' of an anti-aircraft defence unit and operated to serve four vehicles or carriages with automatically slewed weapons.

The radar is discreet, lightweight and small, employing a solid-state transmitter, track-while-scan (TWS) techniques and incorporating a built-in test system. It can be easily operated after a simple training course.

The RA 20S consists of two main sub-assemblies:
(1) an antenna unit comprising a retractable antenna, transmitter/receiver and data extractors
(2) a control console fitted with the controls, processing circuits and a display module.

CHARACTERISTICS
Acquisition envelope
Azimuth: 360°
Elevation: –5 to +45°
Range: 1 – 12 km
Tracking type: TWS
Number of tracks: 2
Range: (to combat aircraft) 12 km; (to tactical helicopter) 12 km

TA20/RA 20S fire control radar mounted on an anti-aircraft defence vehicle

Weight
Antenna unit: 65 kg
Console: 25 kg
STATUS
In full production and more than 50 systems have been delivered to several defence forces.

CONTRACTOR
Electronique Serge Dassault, 55 quai Marcel Dassault, 92214 St Cloud, France.

5602.153
RODEO 2 FIRE CONTROL RADAR

RODEO 2 (Radar d'Observation et de Designation d'Objectifs) is a coherent pulse doppler radar used for the detection, tracking and identification of aircraft flying at low and very low altitudes and is particularly suitable for use against hovering helicopters on watch or at firing position. Identification of the latter is possible from an analysis of its spectrum and comparison with the signature of rotor blades held in a library bank. The basic concept of this equipment is very similar to that of the RA 20S (**5601.153**).

RODEO 2 can be used as the 'leader' of an anti-aircraft defence unit and operated to serve three defence vehicles, fitted with guns or missiles, the weapons of which are automatically slewed. It is IFF compatible and uses frequency agility against electronic countermeasures.

RODEO 2 radar with SATCP SANTAL anti-aircraft system

The equipment is a discreet, lightweight and small radar employing a solid-state transmitter, track-while-scan techniques and incorporating a built-in test system.

The radar consists of two main sub-assemblies:
(1) an antenna unit comprising an antenna, a transmitter/receiver and data extraction circuits
(2) a control console fitted with the controls, processing circuits and display module.

Rodeo 2S is an extended range version of Rodeo 2. It has been designed for the same applications but, in addition, can be used in fixed defence systems. It consists of two main sub-assemblies similar to Rodeo 2. The two versions are interchangeable.

CHARACTERISTICS
Acquisition envelope
Azimuth: 360°
Elevation: –5 to +45°
Beamwidth: 6.5°
Tracking type: TWS
Number of tracks: 2
Range: On aircraft or moving helicopter 12 km (18 km for Rodeo 2S); hovering helicopter at firing position >7 km; helicopter on watch (rotor motion) >5 km
Weight
Antenna unit: 65 kg
Console: 25 kg
STATUS
RODEO 2 is in production and equips the French Army SATCP. A naval version, known as Rodeo Naval, is in late development (**6416.253**).
CONTRACTOR
Electronique Serge Dassault, 55 quai Marcel Dassault, 92214 St Cloud, France.

6823.153
RAPPIST FIRE CONTROL RADAR

Rappist is an electronically-scanned I/J frequency band radar which is being developed for use as a search and fire control system. It is highly resistant to jamming and would normally be vehicle-mounted. The equipment employs a technique for producing a multi-function array, using relatively inexpensive diodes, to form and manipulate beams over an arc to cover 360° in azimuth and 80° in elevation. Rappist uses a cylindrical lens with diode phase shifters, combined with a mechanically rotating illuminator with continuous and discontinuous modes.
STATUS
In development.

CONTRACTOR
Thomson-CSF, Division Systèmes, Défense et Contrôle, 40 rue Grange Dame Rose, 92363 Meudon-la-Foret Cedex, France.

2120.153

SPARTIATE RECOVERY RADAR TRS 2300

Spartiate is a highly mobile helicopter and aircraft recovery radar station developed for the French forces. Its functions include early warning (secondary radar), approach and recovery (primary radar), identification (IFF) and landing. All these functions can be carried out simultaneously.

The station comprises a lightweight trailer supporting the radar transmitter/receiver unit, MTI unit, the IFF interrogator and the antennas, a cabin housing two displays, associated means of telecommunication, and a trailer carrying a power generating set.

The antenna system comprises an azimuth antenna which can rotate continuously and an elevation antenna which scans in the vertical plane. The IFF antenna is integrated with the primary radar antenna.

The transmitter and the receiver operate in I/J-band and are transistorised and miniaturised.

STATUS

15 Spartiates (Systeme Polyvalent d'Atterrissage, Recueil, Telecommunications, Identification de l'Armee de Terre) currently serve the French Army with an excellent performance record. Successful roles include pilot training during tactical exercises and mission support with overseas operations.

Additional details will be found under this entry number (**2120.153**) in earlier editions of *Jane's Weapon Systems*. No longer in production.

CONTRACTOR

Thomson-CSF, Division Systèmes, Défense et Contrôle, 40 rue Grange Dame Rose, 92363 Meudon-la-Foret Cedex, France.

TRS 2300 Spartiate recovery radar

2119.153

PICADOR MOBILE 3-D RADAR

Picador TRS 2200 is a mobile air defence radar station operating in E/F-band. This station combines in a single system the radar functions of 3-D surveillance, heightfinding, and low-altitude surveillance. Designed to be transportable and capable of being deployed rapidly, it can be used for surveillance and interception control either as an autonomous station or as a gapfiller integrated into a large air defence system.

STATUS

No longer in production. More detailed information is given under this entry number in previous editions of *Jane's Weapon Systems*.

CONTRACTOR

Thomson-CSF, Division Systèmes, Défense et Contrôle, 40 rue Grange Dame Rose, BP34, 92360 Meudon-la-Foret, France.

3100.153

COASTAL SURFACE SURVEILLANCE RADAR TRS 3405

This equipment is an I-band radar specially intended for the surveillance of maritime traffic in straits, harbour approaches, access channels, and harbours.

It is designed to provide high performance in range, resolution, accuracy, and protection against the various detection impairments encountered in sea surface surveillance, namely sea and rain clutter, and propagation loss in rain and heavy fog. High resolution is achieved by the use of an antenna with a narrow azimuth beamwidth, and a very short transmission pulse, which results in a very small radar cell. The use of frequency diversity allows decorrelation of sea clutter peaks, reducing the false alarm rate and improving the detection of small ships or buoys.

The radar operates in the I-band (former X-band). To achieve a high gain and avoid unwanted rain echoes which would result from the illumination of clouds at high altitude, the antenna elevation beamwidth is also narrow, and radiation polarisation can be switched from linear to circular. Detection at very close range is ensured by an elevation pattern of the inverted cosecant-squared type down to an angle of −25°.

Round-the-clock operating requirements demand two transmitter/receivers for each radar station. In

TRS 3405 coastal surface surveillance radar

this equipment, they are connected to the antenna through a passive diplexer rather than through a switching device, which enables both units to be operated simultaneously on a frequency diversity basis. As this mode of simultaneous operation is necessary only in heavy weather, the operating cost is not increased significantly. In addition, each reception channel is fitted with an automatic false-alarm regulation control slaved to the mean level of sea clutter.

This equipment is fully compatible with extractors and data remoting.

CHARACTERISTICS

Antenna

Gain: 44 dB

Resolution: 0.24°

Low sidelobes: 25 dB

Inverted cosecant: Squared pattern in elevation down to −25°

Linear and circular polarisation

Rotation rate: 5 or 10 rpm

Transmitter/receiver

Frequency: I-band (formerly X-band)

Peak power: 200 kW

High resolution pulse: 0.05µs

PRF: 1080-570 Hz

Advanced anti-clutter and anti-jamming processing.

STATUS

In production.

CONTRACTOR

Thomson-CSF, Division Systèmes, Défense et Contrôle, 40 rue Grange Dame Rose, 92363 Meudon-la-Foret Cedex, France.

4224.153

TRS 2620 RADAR (GERFAUT)

The TRS 2620 GERFAUT and TRS 2630 GRIFFON (described under entry **6394.153**) family of radar systems is specially designed for battlefield requirements, the self-protection of sites exposed to low and very low altitude air strikes, and close range air defence. The various members of this family use the same modules, the only difference being in the antenna, the number of transmitter modules, and the sequencing and processing programmes.

GERFAUT and GRIFFON are intended for surveillance, early warning and target designation for the deployment of short and very short range weapons. They are designed to suit a variety of configurations, such as 'scattered systems'; combined and co-ordinated systems; and integrated, autonomous and griddable systems.

The TRS 2620 GERFAUT is a surveillance and acquisition radar system which can be used throughout the entire theatre of operations, including forward positions. It acquires and identifies at the first antenna rotation, and locates targets with the accuracy required to guarantee very short reaction time. These performances derive mainly from totally coherent pulse-Doppler, specific 'helicopter' processing, and automatic tracking.

The system has a high anti-jamming capability due to its large bandwidth, the use of burst-to-burst real frequency agility, and by employing antennas optimised for small sidelobes. In addition the TRS 2620 facilities include modular construction which ensures easy integration into any type of vehicle, the antenna under radome being visually discrete, all solid-state equipment (except for the display screen), and built-in-test equipment.

The GERFAUT and the GRIFFON contain the same assemblies, except for the antenna and programming of radar sequencing. GERFAUT and GRIFFON have the same main features in modular architecture.

CHARACTERISTICS

Performance: > 13 km on a combat aircraft; > 8 km on a hovering helicopter

Range resolution: 60 m (175 m on masked helicopter)

Azimuth resolution: 0.8° on moving target; 1° on hovering helicopter; 1° on masked helicopter

Antenna rotation: 40 rpm

TRS 2620 GERFAUT antenna

Number of tracks: 6 tracks (manual initialisation) or optionally 8 tracks (automatic initialisation)

MTBF: 1600 h

CONTRACTOR

Thomson-CSF, Division Systèmes, Défense et Contrôle, 40 rue Grange Dame Rose, 92363 Meudon-la-Foret Cedex, France.

6394.153

TRS 2630 GRIFFON SHORADS RADAR

The TRS 2630 GRIFFON acquisition, early warning and coordination radar is a surveillance system specially designed for short and very short range ground-to-air weapon systems, in both autonomous and decentralised configurations. It is particularly suited to the specific constraints of the battlefield, and is one of a family of battlefield radars, together with the TRS 2620 GERFAUT radar described under entry **4224.153**, and to which reference should be made.

The TRS 2630 GRIFFON and the TRS 2620 GERFAUT contain the same modules except for the antenna (GRIFFON uses a planar antenna) and programming of the radar sequencing. The main features of GERFAUT and GRIFFON are identical.

CHARACTERISTICS

Performance: >18 km on combat aircraft; greater than 8 km on a hovering helicopter
Range resolution: 60 m
Azimuth resolution: 0.6° on moving target; 1° on hovering helicopter; 1° on masked helicopter
Antenna rotation: 40 rpm
Tracking capacity: 6 (manually initialised); optionally 8 with automatic initialisation
MTBF: 1600 h

CONTRACTOR

Thomson-CSF, Division Systèmes, Défense et Contrôle, 40 rue Grange Dame Rose, 92363 Meudon-la-Foret Cedex, France.

TRS 2630 GRIFFON mounted on a cross-country vehicle

3273.153

MOBILE COASTAL SURFACE SURVEILLANCE RADAR TRS 3410

This mobile, cabin-installed station utilises the I-band transmitter/receiver of the TRS 3405 radar (**3100.153**) operating in frequency diversity for decorrelation of sea clutter peaks, coupled with a 4.4 m antenna located on a trailer. Installed at 40 to 50 m above sea level, this station detects small boats up to 25 km away, whereas for larger ships with superstructures at a height of 10 m or more, the detection range reaches 40 km.

The TRS 3410 radar is fitted with a track-while-scan (TWS) device which refines and maintains the data from a surface track, starting from a target designation made by the operator. These data are processed through a computer for the control of various types of anti-ship missiles whose batteries can be located next to the radar station or remotely sited. The compounding of its own accuracy and of the data refinement by the TWS device gives an overall target parameters accuracy which is sufficient for the control of coastal artillery fire against surface targets.

CHARACTERISTICS

Antenna: 4.4 m parabolic reflector
Gain: 39.5 dB
Azimuth beamwidth (3 dB): 0.55°
Elevation beamwidth (3 dB): 4°
Polarisation: Circular
Rotation rate: 11 rpm
Transmitter/receiver
Frequency: I-band
Peak power: 200 kW
High resolution pulse: 0.05 μs
PRF: 1080 – 570 Hz
Advanced anti-clutter and anti-jamming processing

STATUS

In production.

CONTRACTOR

Thomson-CSF, Division Systèmes, Défense et Contrôle, 40 rue Grange Dame Rose, 92363 Meudon-la-Foret Cedex, France.

Thomson-CSF TRS 3410 mobile coast defence radar includes tracking of the co-operating helicopter to provide over-the-horizon target data

2141.153

RASIT E GROUND SURVEILLANCE RADAR

Rasit E is a long range (40 km) ground surveillance radar, for the detection, acquisition, localisation, and recognition of moving targets, either on or near the ground in all weathers. The Rasit E is available in three versions: man-portable and broken down into four 30 kg sections, shelter-mounted and vehicle-mounted. A balloon-borne version is also available (see below).

The equipment is a pulse doppler radar. It operates in I/J-band and uses a coherent receiver and multiple range gates and filters designed for a high probability of detection.

Echoes detected over the surveyed zone are displayed on a B-scope daylight display, connected to an electronic memory.

For each target acquired, the polar and UTM co-ordinates are displayed on the B-scope and the doppler tone is transmitted to the operator either by loudspeaker or by earphone. The operator is thus able to recognise pedestrians, wheeled and tracked vehicles, aircraft, and helicopters.

An automatic system can be arranged to trigger an acoustic alarm as soon as a target enters the surveyed zone. The operator can select the width of the surveyed zone from 30 to 120° and can choose the length surveyed by operating in either of two modes:

(a) Normal: overall radar range of 20 km
(b) Magnifier: any zone 2.5 km long within the total range.

The equipment can be operated by a single inexperienced operator in a variety of conditions. Typical operating times are ten seconds to survey a zone 100° wide and 20 km long, and 30 seconds to acquire a target.

The antenna beam can automatically sweep a sector or can be stopped and then oriented towards a target. The antenna is equipped with a polariser making the radar insensitive to atmospheric perturbations. The radar also has very effective ECCM capabilities.

The RF unit can be mounted on a tripod with the control cabinet 100 m away. Alternatively it can be mounted on any of a variety of military vehicles, such as the APC Steyr, the APC Mowag, the Alvis-British Leyland Spartan, the Panhard M3 or the APC VAB, with the control cabinet inside the vehicle. New improvements include 360° operation, automatic acquisition and tracking, programmable alarm and increased range.

In addition to built-in tests, a tactical test bench can be supplied for third and fourth echelon maintenance. A tactical simulator designed for an efficient and comprehensive training of one or several operators is available.

Rasit battlefield surveillance radar (balloon-borne)

A version of the Rasit radar has been developed by LMT Radio Professionelle, in collaboration with the US company TCOM (a subsidiary of Westinghouse), for installation in balloons as a tactical surveillance system. The radar weighs 55 kg and is suspended from the balloon as a two-axis gyro stabilised package so that the antenna points in the same direction, irrespective of the motion of the balloon. The maximum operating altitude is 750 metres from where the system can, under the best conditions, detect individual persons out to 23 km, light vehicles at 30 km, and heavy vehicles at its maximum range of 40 km (see also entry **3162.353**).

CHARACTERISTICS

Operating frequency: I-band (200 MHz range)
Peak power: 2 kW
Polarisation: Linear or circular
Range (90% probability of detection)
Pedestrian: 23 km
Vehicle: 40 km
Helicopter: 20-30 km
Low-flying aircraft: 40 km
Range reduction: (4 mm/h rain) < 10%
Accuracy: Azimuth ± 10 mrd, range ± 10 m
Sub-clutter visibility: More than 43 dB
Display: B-type daylight oscilloscope, numerical display of polar and UTM co-ordinates, doppler tone, automatic acoustic alarm
Power drain: 150 W (28 V)
Temperature range: –40 to + 55°C
Reliability: Field tested MTBF > 900 h
Overall weight: 90 kg
Number of loads: 4
Weight of the heaviest load: 28 kg
Distance between radar head and operator unit: 100 m max
Provision for target data transmission
Built-in tests
Options: Remote slave TV monitor; plotting table; compatible with Siclamen IFF (**3173.153**).

Tripod mounted version of the Rasit E ground surveillance radar

STATUS

In production. More than 600 equipments have been delivered to 20 armies. A number of Rasit radars were delivered to the People's Republic of China during 1986.

CONTRACTOR

LMT Radio Professionelle, 46-47 quai Alphonse Le Gallo, 92103 Boulogne-Billancourt, France.

6132.153

RB12A (RASSUR) GROUND SURVEILLANCE RADAR

The RB12A is a close range coherent pulse doppler ground surveillance radar, using a solid-state transmitter operating in the Ku band. It is designed to detect targets automatically, to determine bearings and ranges, and to identify targets by means of their doppler characteristics. The equipment consists of a tripod-mounted vertically polarised antenna/transmitter/receiver, a remote hand-held console, a battery pack and a set of headphones. The system is controlled by the remote console over two telephone pair wires. The RB 12A is known as RASSUR in its military version, and Vigile in a civil version.

CHARACTERISTICS

Frequency: Ku band, five switchable frequencies
Antenna beamwidth: 4.5°
Range: 3 km (pedestrian); 5 km (vehicle)
Accuracy: 10 m (range); 1° (bearing)
Surveyed sector: 0 to 360° in sector widths of 20, 30, 60 and 120°
Target velocity: 3 to 100 kph

CONTRACTOR

LMT Radio Professionnelle, 46-47 quai Alphonse Le Gallo, 92103 Boulogne-Billancourt Cedex, France.

RB12A ground surveillance radar

3173.153

SICLAMEN IFF SYSTEM

The Siclamen system has been designed to meet all identification requirements for land vehicles operating in a combat zone, in the same manner as military aviation IFF. Because of the similarity of vehicles, the depth of combat areas and the range of anti-tank missiles, LMT Radio Professionnelle has developed the BIFF (battlefield identification friend or foe) system known as Siclamen (Systeme d'Identification Compatible L-band L'Ami ou Enemi). When installed on land vehicles, this system has a range of 30 km. It discriminates between two targets having a range difference of 100 metres and whose azimuths differ by 2 to 8 degrees, depending on the mode of use.

A Siclamen station consists of an interrogator, a control console and a directional antenna, and interrogates BIFF transponders associated with omni-directional antennas installed on friendly vehicles. The interrogating stations are associated with a primary sensor which may be any field or on-board radar, or even a laser rangefinder or television system, although Siclamen can operate completely autonomously if required. The frequencies used are the same as for IFF, ie 1030 MHz for transmission and 1090 MHz for replies. Information is exchanged by encoded pulse trains.

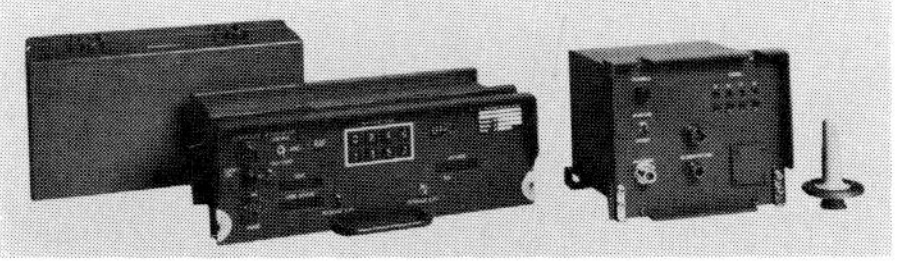

Siclamen IFF equipment

One use of Siclamen is with the Rasit radar (**2141.153**). The latter can operate in:

(a) Mode 1 – surveillance of a sector (apperance of echoes on the display)
(b) Mode 2 – antenna scan interrupt and manual aiming of the antenna at an echo of interest to the operator who can then place a range gate on this echo which generates a doppler sound corresponding to the type of moving object observed (pedestrian, tank, helicopter, etc) in the Rasit loudspeaker.

In mode 1, BIFF interrogations are transmitted continuously. In the case of any friendly vehicle echo, the amber light on the control console comes on when the antenna is pointed at the echo, with a flashing indication next to the echo on the display. It is then possible to check if this friendly vehicle corresponds to a given code entered via a keyboard. If affirmative, a green decoding light comes on and the amber indicator becomes continuous. In mode 2, the Siclamen system indicates on the console, in clear language form, the code of the friendly vehicle at which the radar operator is aiming the antenna. This is known as active decoding. Range of the vehicle is also displayed at the same time as the code.

CHARACTERISTICS

Frequency: 1030 MHz interrogation; 1090 MHz reply
Transmitted power: 10 W peak
Receiver sensitivity: –75 dBm
No of modes: 1 for BIFF
No of codes: 4096
Passive decoder: 1 for BIFF
Range readout: In tens of metres to 30 km
Range resolution: 100 m for two vehicles fitted with BIFF

STATUS

In service.

CONTRACTOR

LMT Radio Professionnelle, 46-47 quai Alphonse Le Gallo, 92103 Boulogne-Billancourt, France.

5782.153

IFF INTERROGATOR NRSI-3

The NRSI-3 IFF interrogator is the latest in a series of these equipments designed and manufactured by LMT Radio Professionnelle. Types of installation are:

(a) a single interrogator with local control, with or without a permanent monitoring system
(b) a dual interrogator with local control installed in a 19 inch (483 mm) rack and consisting of two identical transmitter/receivers, a permanent monitoring system and an automatic switch-over system.

The system is fully solid-state with an optional programmable output power and optional monopulse receiver. Modes of operation are 1, 2, 3/A, B, C, D and compatibility with mode 4. Interlacing facilities are incorporated among the six modes with the possibility of alternating two groups at each antenna revolution. A remote control unit can be fitted which allows all the front panel controls of the dual transmitter/receiver version to be remoted via a 2-wire line, as well as transmission of orders from the decoder.

The permanent monitoring system constantly checks the performance characteristics of the transmitter and receiver. For a dual system operating in the automatic mode, it controls operation on the stand-by unit in the event of the failure of the unit in service. The parameters checked are output power, threshold level, encoding, multiplexing, ISLS switching, SWR and remote control. A naval version of the NRSI-3 is also available.

CHARACTERISTICS

Frequency: Transmit 1030 ±0.2 MHz; receive 1090 ±3 MHz

Peak power: 2 kW

Modes: 1, 2, 3/A, B, C, D and compatible with mode 4

Transmission delay: Adjustable from 28 µs to 926 µs with respect to the synchronisation signal

STATUS

In production.

CONTRACTOR

LMT Radio Professionnelle, 46-47 quai Alphonse Le Gallo, 92103 Boulogne-Billancourt, France.

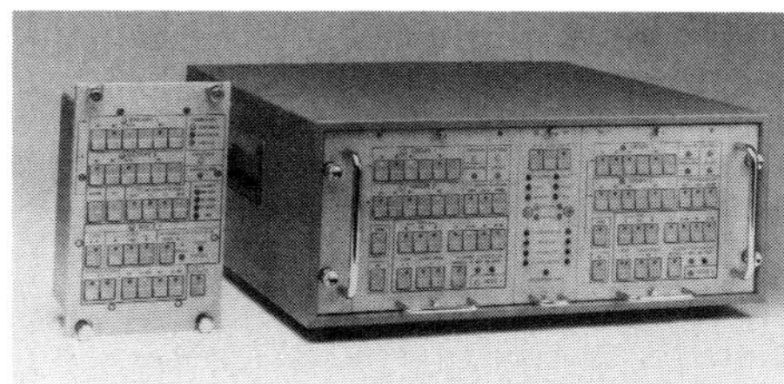

IFF interrogator NRSI-3

6396.153

NR-AI-5A IFF INTERROGATOR-DECODER

The NR-AI-5A IFF interrogator-decoder is designed to operate with the low-altitude short-range radar of a missile system. It consists of an encoder, a transmitter, an antenna switch, two receivers, a discriminator and two passive decoders for 4096 codes. With the exception of one tube the equipment is completely solid-state and makes extensive use of digital integrated circuits, both bi-polar and MOS. All sub-assemblies are plug-in and common to other LMT IFF transmitter/receivers. An internal antenna switch for side lobe suppression during interrogation is fitted.

CHARACTERISTICS

Transmission frequency: 1030±0.5 MHz (optionally ±0.2 MHz)

Receiver frequency: 1090 MHz

Peak radiated power: 400W

Modes: 1, 2 and 3/A, and Mode 4 with appropriate encoder and decoder

STATUS

In service

CONTRACTOR

LMT Radio Professionnelle, 46-47 quai Alphonse Le Gallo, 92103 Boulogne-Billancourt, France.

6397.153

NRAI-6A IFF INTERROGATOR-DECODER

The Type NRAI-6A is an IFF interrogator-decoder designed specifically for the low-altitude, short range radar of the Roland missile system. It consists of an encoder, a transmitter, an antenna switch, two receivers, a defruiter, a degarbler and two passive decoders for 4096 codes. The equipment is completely solid-state and makes extensive use of digital integrated circuits, both bi-polar and MOS. All sub-assemblies are plug-in and are common to other LMT IFF transmitters/receivers. A double receiver is supplied for sidelobe suppression, and an internal switch is fitted for sidelobe suppression during interrogation.

CHARACTERISTICS

Transmission frequency: 1030±0.2 MHz

Receiver frequency: 1090 MHz

Peak radiated power: 100W

Modes: 1,2 and 3/A, and Mode 4 with appropriate encoder and decoder

Weight: 30 kg plus 3 kg for controller

STATUS

In service

CONTRACTOR

LMT Radio Professionnelle, 46-47 quai Alphonse Le Gallo, 92103 Boulogne-Billancourt, France.

6809.153

SB 14 IFF ITERROGATOR

The SB 14 IFF is a small, lightweight, battery operated ground-to-air interrogator for SHORAD weapon systems. A mode 4 internal memory and a reply evaluator model eliminate the need for a cryptographic computer. A programmer loads the modes 1, 2, 3/A, 4 memory with the appropriate code catalogue. The programmer connector and test connector are located on the interrogator. A single cable interfaces the interrogator to the weapon system.

The interrogator includes both ISLS and RSLS circuits. It can be programmed to operate in each mode sequentially, according to the mission requirements. The modes and codes are controlled by the contents of the memory unit, the interrogation procedure being pre-determined at the loading of the memory unit by the programmer. Codes are loaded for a multi-day mission and in the secure mode the interrogation challenges are randomly chosen to avoid compromise.

The IFF challenge is initiated by action of the operator on the weapon trigger before firing and is discrete since a challenge sequence consists of a short burst of interrogations. A positive outcome from a self-test sequence allows the interrogation process to start.

CHARACTERISTICS

Antenna: Three-element dual Yagi

Modes: 1, 2, 3/A and 4

Dimensions: 180 mm high × 250 mm long × 100 mm wide

Weight: 2.5 kg

STATUS

In production.

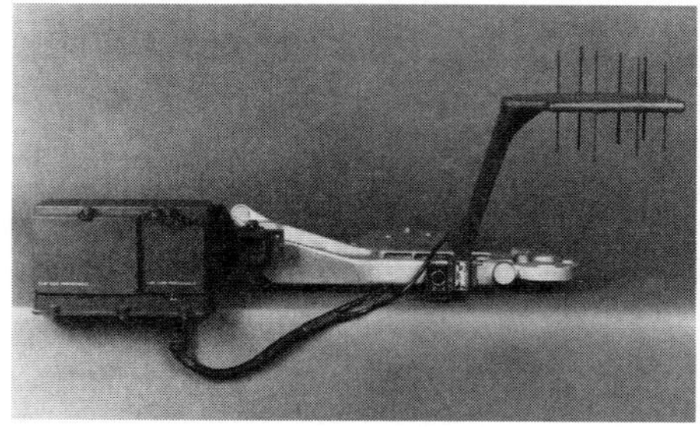

SB 14 IFF interrogater

CONTRACTOR

LMT Radio Professionnelle, 46-47 quai Alphonse Le Gallo, 92103 Boulogne-Billancourt, France.

3287.153

STENTOR BATTLEFIELD RADAR

Stentor is a doppler radar used for ground surveillance of critical areas, such as battlefields and frontiers. It is claimed to have the longest range for ground surface surveillance radars, making possible surveillance of remoter areas and, compared to medium-range radars, giving a significant increase in warning time to aid the interception of hostile units. At shorter ranges the benefits of a powerful signal are evident for high detection probability against very small or intermittent targets, or for minimising the absorption effects of adverse weather conditions. The Stentor ranges are: 30 to 40 km for pedestrian targets; 50 to 60 km for targets such as a Jeep, tank, truck; and 20 to 60 km for boats (depending on sea conditions) and helicopters (a major threat when flying low at low speeds since they are not detectable by classical air defence or low flying aircraft detection radars).

It is a transportable system comprising the radar head and an operator's console which can be remotely sited several hundred metres from the radar head. The radar head can be located on a tower, or on a naturally elevated site, to increase the surveyed area and, thus, distant detection.

The operating frequency is in I/J-band, tunable over 200 MHz with pulse transmission, ground clutter cancellation being by coherent detection and doppler filtering; rain cancellation devices provide for all-weather operation. Moving targets are displayed on a daylight CRT with a digital memory. The surveyed sector can be instantaneously adjusted in aperture and in azimuth.

Stentor battlefield surveillance radar

A tactical data communication system which enables a command post to simultaneously receive and process the data issued from one to ten Stentor radar stations is also optionally available, as are operator training simulators and third echelon bench test units.

Stentor is simple to operate (even by a single inexperienced operator) and to maintain, with high mobility, fast deployment and effective ECCM capability.

CHARACTERISTICS
Frequency: I/J-band, tunable
Peak power: 60 kW
Polarisation: Vertical or circular
Antenna span: 1.6 m
Range: Up to 60 km on tanks and large vehicles
Accuracy: 20 m (range); 4.4 mrad (0.25°) (azimuth)
Weight: 370 kg (radar head and operator's unit)
Power: 1.5 kVA, 220 V, 50/60 Hz
MTBF: > 1000 h

STATUS
New upgraded version in production. Operational with various armies (including NATO) in three continents.

CONTRACTOR
LCTAR (Le Centre Thomson d'Applications Radar), 6 rue Nieuport, 78143 Vélizy-Villacoublay, France (a new subsidiary company of the Thomson-CSF group which handles LCT radar business).

2155.153
TRS 2201 THREE-DIMENSIONAL RADAR

The TRS 2201 is a long-range three-dimensional E/F-band air defence radar. A modern technology version of the ARES radar used in NATO's NADGE network, it ranks among the most powerful radars of its type in the world.

Major features of this radar are its high-power (20 MW peak) transmitter, its elaborate antenna structure, and its range of ECCM facilities that enable it to function satisfactorily in an ECM environment. The equipment is solid-state to the maximum extent possible, is of modular construction, and has built-in test facilities. More detailed information is given under this entry number in previous editions of *Jane's Weapon Systems*.

PERFORMANCE
Range capability of the TRS 2201 is more than 250 nautical miles (460 km) on fighter aircraft, there being no practical altitude limitation. Height measurement accuracy is ±1500 ft (457 m) up to 150 nautical miles (280 km). The data renewal rate is six per minute.

STATUS
No longer in production. Besides France several countries have been equipped with TRS 2201 stations and the equipment was selected by NATO as the main feature of its radar coverage (NADGE plan – see **1181.181**).

TRS 2201 long-range 3-D air surveillance radar

CONTRACTOR
Thomson-CSF, Division Systèmes, Défense et Contrôle, 40 rue Grange Dame Rose, 92363 Meudon-la-Foret Cedex, France.

3271.153
VOLEX IV 3-D SURVEILLANCE RADAR TRS 2206

Volex IV is a fixed, E/F-band, 3-D surveillance radar providing continuously and simultaneously the azimuth, range, and altitude of any target within its cover volume; it may be used either as a 3-D air defence station or as a heightfinder.

STATUS
No longer in production, but still in operational use. More detailed information is given under this entry number in previous editions of *Jane's Weapon Systems*.

CONTRACTOR
Thomson-CSF, Division Systèmes, Défense et Contrôle, 40 rue Grange Dame Rose, 92363 Meudon-la-Foret Cedex, France.

2156.153
TIGER S (TRS 2100 SERIES) RADAR

Under the code name Tiger (terrifically insensitive to ground effect radar), Thomson-CSF has developed a family of lightweight high-performance radars, the first being the TRS 2100 for the detection of low-flying aircraft. A special version is produced for the French Air Force, designated Aladin, and an 'Improved Tiger' version particularly well suited in resistance to high jamming and to non-standard propagation conditions is now being produced. High performance in severe clutter is achieved by the simultaneous use of MTI circuits based on doppler frequency filtering and of pulse compression to limit the volume in which clutter echoes can mix with useful echoes. The radar can be used either as an autonomous detection centre, with local exploitation of radar data, or as a gap-filler station linked to an air defence network. Specific operational functions listed by the manufacturer include: control of tactical operations; low-altitude gap-filler; coastal and maritime surveillance; alert and target designation for SAM batteries; surveillance radar equipment role; and similar mobile or air-transportable applications.

The use of E/F-band makes it possible to have an antenna with a high gain and high angular resolution but with a relatively small reflector (5 × 2.3 m). The reflector profile is C-shaped with a double curvature and the vertical pattern has a steep slope at low elevations and is super-cosecanted at high elevations. An IFF sub-system is incorporated (IFF antenna integral with primary radar antenna) and a BITE facility monitors the main parameters and locates any faulty functions.

The transmitter uses a coherent amplifier chain. The transmission frequency is obtained from a highly stabilised crystal oscillator, and for frequency agility purposes transmission is carried out at different frequencies distributed at random over 250 MHz. The output power is provided by three successive

Improved Tiger S with electronic rack on the antenna trailer

Improved Tiger S with electronics housed in the operations cabin

TRS 2100 Tiger radar on tower

amplifier stages (one transistor stage, one TWT stage, and one CFA stage). The phase stability of this chain is such as to ensure good clutter rejection performance. Pulse compression is achieved by incorporating crystal dispersive networks, the propagation time of which varies with the frequency in the transmit and receive circuits.

The receiver is preceded by a wide-band, low noise RF amplifier. Rejection of fixed echoes is obtained by a digital linear MTI. A constant-false-alarm-rate chain operates with the MTI processing chain.

The radar head can be linked by either a single coaxial cable or microwave link to a remotely sited operations unit. Utilisation can be manual, semi-automatic or fully automatic.

Other members of the family are the Tiger G systems (TRS 2105 and TRS 2106) which are described under entry **6395.153**.

OPERATION

The radar is transportable by helicopter or by conventional truck or cargo aircraft. The transmitter/receiver can be mounted either on a trailer together with the antenna, or in a technical/operations cabin. The antenna can also be mounted on a mast (14 or 54 m).

CHARACTERISTICS

Improved Tiger S

Operational performance: On 2 m² (fluctuating target) with Pd ⩽ 80%

Detection range: 110 km

Altitude detection: > 6000 m

Elevation pattern: Cosecanted up to 45°

Sub-clutter visibility: >40 dB

First blind speed: >1500 knots

Antenna characteristics

Size: 5 × 2.3 m. IFF antenna integrated with primary radar antenna

Gain: 34 dB

Polarisation: Circular, fixed

Azimuth beamwidth: 1.3°

Rotation speed: 12 rpm

Transmitter/receiver characteristics

Frequency range: 600 MHz bandwidth in E/F-band

Mean power: 1 kW

Overall receiver noise figure: <3.5 dB

Frequency-agile using synthesiser; dual-PRF operating mode; linear 3-memory DMTI filter; CFAR; post-detection integration circuits.

STATUS

In production for French Air Force (Aladin) and about 20 other countries. Improved version production commenced in 1986.

CONTRACTOR

Thomson-CSF, Division Systèmes, Défense et Contrôle, 40 rue Grange Dame Rose, 92363 Meudon-la-Foret Cedex, France.

6395.153

TIGER G (TRS 2105/TRS 2106) LOW ALTITUDE RADARS

The Tiger G radar series (TRS 2105 and 2106) is intended for tactical low altitude surveillance. These radars are mainly devoted to low-level coverage and gap-filling in an air defence network, alert and co-ordination of ground-to-air weapons, and control of tactical operations. Both operate in G-band, use coherent transmission, frequency agility, pulse compression, advanced digital I and Q MTI, and CFAR, and benefit by four operating modes. Both are highly resistant to ECM, and are fitted with a processing unit featuring automated radar adaptivity to the environment and automatic initialisation and tracking of local targets.

TRS 2105

The TRS 2105 is housed in a cabin installed on a two-axle trailer. The cabin supports a foldable eight metre mast bearing the 3 × 1.6 metre antenna. The complete system can be carried by road, rail or cargo aircraft (C-130/C-160).

TRS 2106

The TRS 2106 is a derivative of the TRS 2105 and is particularly suited for battlefield operations. It provides detection of low and very low altitude strikes by fixed-wing aircraft, helicopters, RPVs and missiles with the necessary speed, discrimination and accuracy to ensure an efficient assessment of the threat. It also provides early warning and target designation to air weapon systems to optimise the response of anti-aircraft defence systems, and protects friendly aircraft overflying the ground-to-air engagement zone.

The TRS 2106 system is installed on a overland vehicle, together with its operations vehicle. The antenna is secured on top of a mast with an automatic erection and fold-away system which enables very short set-up times, long range coverage at very low altitude, and a low visual profile. The complete system can be airlifted.

The TRS 2106 is a coherent, pulse Doppler radar featuring frequency agility in a wide band and designed to cope with ECM. The main technical features include a wide frequency band (550 MHz in G-band), a planar antenna with very low sidelobe levels, totally coherent transmission chain with pulse compression, several modes of true frequency agility compatible with high performance, MTI processing (transmission by bursts), digital signal processing (transverse/filtering), plot extraction, automatic initialisation and tracking of local targets, operating modes specifically to counter ECM, and IFF compatability. The TRS 2106 is fitted with an operational cabin specially designed for alert and co-ordination of ground-to-air weapons.

CHARACTERISTICS

	TRS 2105	TRS 2106
Frequency band	550 MHz in G-band	
Peak power	70 kW	
Mean power	1 kW	
Detection range	65/100 km	50 km
Rotation rate	5.5 to 11 rpm	30 rpm
Automatic tracking capacity	60 tracks	90 tracks

STATUS

In production.

Tiger G (TRS 2105) mobile low altitude surveillance radar

CONTRACTOR

Thomson-CSF, Division Systèmes, Défense et Contrôle, 40 Rue Grange Dame Rose, 92363 Meudon-la-Foret Cedex, France.

6586.153

SA 90 RADAR SYSTEM

The SA 90 radar system has been designed to work in conjunction with the new SA 90 medium range surface-to-air land-based and shipborne missile systems in development (see entry **4771.131**). It is a multi-function, frequency-agile, X-band 3-D equipment for surveillance, acquisition, tracking and missile guidance. The two axis, electronic scanning antenna is understood to rotate at some 60 rpm and will be compatible with the missile launching rate of up to one per second. It is a joint venture with the French government. On land it will be effective against targets which can include anti-radiation missiles.

STATUS

In development.

CONTRACTOR

Thomson-CSF, Division Systèmes, Défense et Contrôle, 40 rue Grange Dame Rose, 92363 Meudon-la-Foret Cedex, France

---

GERMANY (FEDERAL REPUBLIC)

3286.153

SIEMENS MILITARY RADARS

Siemens-Albis Fire Control Radar

This J-band tracking radar was developed by Siemens-Albis Ltd for use in the Oerlikon-Contraves Type B2 version of the 35 mm AA Gepard tank (**2370.131** in 1977 and earlier editions). A monopulse pulse-doppler equipment, it has an operating range of 15 km and permanent echo rejection of better than 23 dB. In this application it is used in conjunction with the MPDR 12.

MPDR 12

The MPDR 12 is the acquisition radar of the Gepard AA tank Mk3. It is an E-band pulse-doppler radar with high sub-clutter visibility and high data rate (60 rpm). The system is integrated with the MSR 400 interrogator for IFF.

Siemens DR 641 radar in shelter-housed configuration

Siemens MPDR 3002 S low level air surveillance radar used by West German Army

MPDR 16

This is the search radar used in the Roland mobile surface-to-air missile system (**2218.131**) and is a pulse doppler equipment operating in the D-band. The minimum and maximum target acquisition ranges are 1.5 and 16.5 km respectively. The 2 × 1 m antenna rotates at a rate of 60 rpm and is of cosec²-type. An integrated IFF facility is provided, using an MSR 400 interrogator.

The doppler processor eliminates ground clutter. The extractor generates radar plots and feeds the computer of the associated tracking radar with azimuth, range and IFF response.

DR 621/641 (former MPDR 18X/MPDR 30X)

These I-band pulse doppler search radars are multiple application radars for the detection of low and very low flying aircraft and helicopters. Due to their construction, the systems can be installed in shelters, vehicles, tanks (eg Wildcat) and ships. The systems can be used as acquisition and target designation radars (DR 621) in automatic weapon systems, and as radar sensors (DR 641), in site defence systems. They can also be expanded to form command and control centres by adding data processing equipment and display.

DR 151 (former MPDR 45E)

The DR 151 is a mobile low-level air surveillance radar with a range of 45 km. Other improvements include signal processing, ECCM, mechanical design etc. The system is mounted on one truck together with an 18 m extensible mast.

DR 162 (former MPDR 60)

The DR 162 is equivalent to the DR 151, but with range capability extended to 90 km.

DR 172 (former MPDR 90E)

Using a larger reflector and a higher sensitivity receiver in addition to a number of other modifications (eg frequency diversity) the range of the DR 151 can be extended to 90 km (instrumented range to 140 km). This radar can be stowed in two transportable shelters and is particularly suitable for airspace surveillance for airfields and in the SILLACS system (**4275.181**).

MPDR 3002 S

This is a medium-range E-band radar for air surveillance, used by the West German Army. It is equipped with a folding antenna mast mounted on a Marder tank chassis. Range is approximately 30 km.

DII 211 (former MSR 400)

The DII 211 is an IFF-interrogator for use in mobile short- and medium-range radar systems. The equipment employs passive decoding and can be equipped for active decoding. It is compatible with Mk X SIF and Mk XII techniques. Known applications include the Gepard AA tank mentioned above. The DII 211 is compatible with all low level radars of the MPDR series, MPDR 30/1, MPDR 16 (Roland), MPDR 3002 S, DR 151, DR 161 and DR 172.

DII 111/DII 511 (former MSR 200/MSR 2000)

The DII 111 and DII 511 are parts of a new lightweight IFF-equipment family. They employ modern, fully solid-state technology.

The DII 111 is a compact mobile IFF-interrogator for use in short- and medium-range radar systems. The interrogator employs passive decoding (including civil and military emergency decoding as options) in compliance with STANAG 5017, ie it is compatible with Mk X-A and, if applicable, with Mk XII techniques. Applications thus far are in Wildcat (with radar DR 621) and in the mobile radar system Reporter (produced by Hollandse Signaalapparaten BV).

The IFF/SSR-interrogator DII 511 is suitable for all applications in civil and military air traffic control systems. The DII 511 is a more powerful IFF-interrogator for use with larger, long-range radar and weapon systems (ground- or ship-borne).

STATUS

In production.

CONTRACTOR

Siemens AG, Radio and Radar Systems Division, Landshuter Str 26, 8044 Unterschleissheim, Federal Republic of Germany.

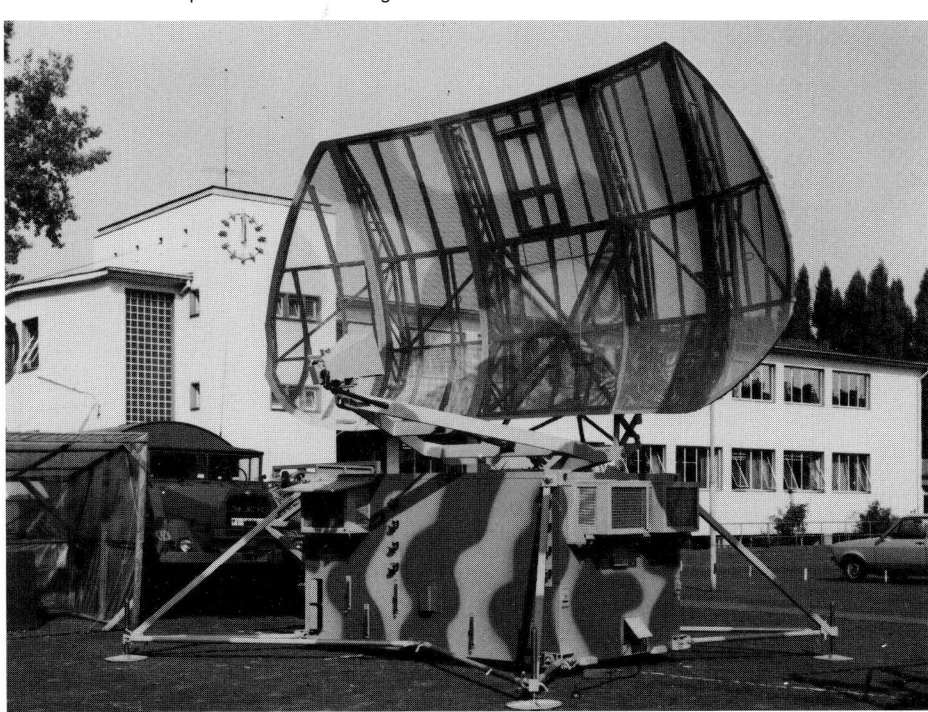

Siemens DR 172 (antenna shelter)

Siemens DR 151/DR 161 on a truck

B2 version of 35 mm Flakpanzer Gepard showing Siemens-Albis fire control radar and MPDR 12 acquisition radar

5762.193

SIEMENS ARTILLERY SURVEILLANCE RADAR

Siemens has signed a co-operation agreement with General Electric of the USA to make a joint bid for the development of a mobile artillery surveillance radar to replace the Green Archer system currently in operational use with the Federal German Army. The equipment is designed to detect, track and locate the source of artillery shells so that the necessary action can be taken, and is described briefly in the International section under COBRA.

The proposed system employs the latest technology, has a C-band, fully phased array antenna and is designed to have a high location rate and accuracy as well as mobility. The radar will detect, track and identify enemy and own shells and from the data obtained it will compute the source and impact of enemy shells and the impact point of own shells. It employs a pencil beam following the horizon to detect the shells, and separate radiators to track them. The complete system is mounted on a 7-ton truck.

CONTRACTOR

Siemens AG, Radio and Radar Systems Division, Landshuter Str 26, 8044 Unterschleissheim, Federal Republic of Germany.

3016.153

TRM/TRS C-BAND RADAR SERIES

AEG has developed a family of mobile and shipborne C-band medium and long-range radar systems. Common to all systems are excellent ECCM features, high modularity and standardised logistics. Although the land-based versions have been designed primarily for mobile configurations they can also be supplied for operation in fixed stations, as coastal radars for sea and air surveillance, and as gap fillers. They are suitable for integration into air defence networks, AA weapon systems, and for integration with other radar equipments.

TRM-S Area Defence System

The TRM-S represents the basic development of a sophisticated mobile C-band 3-D radar. It consists of a remote-controlled sensor and an evaluator shelter which can be separated by up to 300 m, both of which are transported on 15-ton cross-country trucks. On redeployment setting-up by six operators takes less than 30 minutes. The evaluator shelter is NBC proofed and contains up to three operator positions with C3 functions.

A selection of operational modes, pre-programmed to user specific requirements enables the TRM-S to adapt optimally to all feasible tactical conditions. In all modes scanning is performed by a high-precision computer-controlled pencil beam. Up to 4000 targets can be displayed simultaneously; an IFF section is incorporated in the antenna array. Excellent ECCM features are guaranteed by the agility of the main radar parameters transmission, pulse repetition and polarisation.

Phase-coded pulse-compression ensures long-range at low transmit power as well as high resolution at the same time. Additional self-protection can be realised by a stand-off deception jammer, remote-controlled by the TRM-S, and designed with modules of the C-band family.

CHARACTERISTICS

Range: 200 km (radar cross section 2 m, detection probability 60%); 100 km (radar cross section 1 m, detection probability 90%)

Data renewal time: 3 to 15 s (depending on mode)

Max antenna height: 12 m (upper edge)

Max wind speed: 100 km/h (operating); 150 km/h (stand-by)

Power consumption: 45 + 30 kW

STATUS

In service and in production. The West German MoD has ordered the TRM-S as a sensor for the Army mobile AAC3 system to generate target information for the Gepard AA tank and the Roland SAM tank.

TRM-L Point Defence System

The TRM-L is a one-shelter version of a medium-range C-band radar which makes extensive use of TRM-S modules. The NBC shelter is mounted on a 15-ton cross-country truck. After relocation the TRM-L can be set up by three operators in less than 15 minutes.

The TRM-L can be supplied in either 2-D or 3-D versions, later upgrading of the 2-D version is possible. The system possesses high resistance to ECM provided by the agility of the transmission frequency, pulse repetition frequency, polarisation and phase coding of the pulse compression sub-pulses. A high probability of detection of ARMs and low-level aircraft has been achieved by means of sophisticated signal processing procedures. An additional channel ensures the reliable detection of helicopters. Side lobe suppression and IFF functions have been integrated into the antenna.

TRM-L point defence radar

CHARACTERISTICS

Range: 60 km (radar cross section 3 m, detection probability 80%); 46 km (radar cross section 1 m, detection probability 80%)

Data renewal time: 2 to 4 s (depending on mode)

Power consumption: 38 kW

Max antenna height: 12 m (upper edge)

Max wind speed: 100 km/h (operating); 150 km/h (stand-by)

STATUS

In service and in production. TRM-L supplemented by a C3 component has been ordered by the West German MoD as a Command and Control Centre. Within the German point defence system, TRM-L serves as a sensor to control Roland SAM launchers.

TRM-SAM

A one-shelter version of the TRM-S as an air surveillance radar in SAM systems.

TRM-A

A medium-range weapon locating system.

STATUS

In production.

TRS-3D Naval Radar

A 3-D surveillance for small vessels such as fast patrol boats.

TRS-S Naval Radar

A naval version of the TRM-S, adapted for use on board larger vessels such as frigates and destroyers.

TRS-MFR

A multi-function radar for frigates, destroyers or similar size vessels.

TRS-C 2-D Naval Radar

This 2-D system is for use in medium size vessels and is delivered with either a TWT or a magnetron transmitter.

STATUS

The TRS-C has been ordered by the West German MoD for the 'Fleet Service Boats' of the Federal Navy.

CONTRACTOR

AEG Aktiengesellschaft, Radio and Radar Systems, Sedanstrasse 10. 7900 Ulm, Federal Republic of Germany.

6390.153

RATAC-S BATTLEFIELD RADAR

RATAC-S is a highly-mobile coherent pulse-doppler radar with a monopulse antenna configuration used to detect, locate, classify and observe ground targets or low-flying airborne targets. The system was developed by SEL, as a private venture, based on the operational performance of RATAC (see **1528.153** in the International Section). The new system is a compact and easy-to-operate battlefield equipment which uses microprocessor control, digital data processing techniques and integrated components. The use of the latest technology has reduced volume and weight, as well as enhancing reliability with a consequent substantial increase in MTBF.

The equipment combines a medium/long-range battlefield surveillance radar with a high precision artillery fire control system within one unit. Only three cabinets, each weighing about 35 kg, form the complete radar. This consists of two assemblies, an RF Sensor Unit (antenna, transceiver and receiver) and an Operator Unit (operator console, signal-processing, display and computer).

The RATAC-S incorporates a compact high-frequency portion consisting of a rugged magnetron transmitter with modulator, and a highly sensitive, low-noise monopulse Doppler receiver with an IF of 120 MHz which avoids interference with the VHF communication channels. When jamming or interference is experienced, it can usually be overcome easily by selecting an alternative frequency from the built-in, low-phase-noise coherent 10 channel frequency synthesizer.

The entire system operation is controlled by a central computer in the Operator Unit with a co-processor in the RF Sensor Unit. The microprocessor and operational software used enable the elimination of many manual controls and a considerable reduction in the number of components. In addition the system provides an extensive growth potential without the need for hardware change.

RATAC-S provides five modes of operation as follows:

Mode 1 – automatic scanning with a range of up to 38 km and a sector scan width between 30 and 140°

Mode 2 – manual scanning and acoustic target classification up to a range of 38 km
Mode 3 – precise manual target tracking and deviation measurement within an area of ±300 m and ±75 mils, and area observation alarm
Mode 4 – automatic target tracking and deviation measurement
Mode 5 – automatic target tracking and plotting on the display.

The transformation of coordinates from polar to UTM, or vice-versa, is performed by software in the central computer without the need for additional circuitry. All relevant coordinates are automatically presented in allocated text fields on the display. A highly effective moving target indication (MTI) system allows detection and location of slow moving targets with a small radar cross-section.

The use of a graphic display system, consisting of a graphic processor, image storage unit and screen monitor eliminates the need for a storage cathode ray tube. Through digital target image processing the radar map and the target image can be superimposed in semi-brightness. The graphic processor allows display of various markers and cursors which simplify operation. The display storage facility also enables fixed targets to be stored and redisplayed, superimposed on the current display at reduced brightness. Target tracking, accurate target positioning and deviation measurements of artillery shell impacts, acoustic classification of targets, continuous and interruptive BITE testing, and an alarm facility, are also provided by the five operating modes.

CHARACTERISTICS

Ranges available: Two (normal range 0 to 20 km; extended range 18 to 38 km)
Target location accuracy: ±10 m, ±2 mils on a moving target
Ranges (90% detection probability):
Pedestrian: 14 to 18 km
Light vehicle (jeep): 20 to 24 km
Heavy vehicle (tank): 25 to 30 km
Vehicle convoy: Up to 38 km
105 mm shells: Up to 8 km
155 mm shells: Up to 15 km

STATUS

In production.

CONTRACTOR

SEL (Standard Elektrik Lorenz Aktiengesellschaft), Lorenzstrasse 10, 700 Stuttgart 40, Federal Republic of Germany.

RATAC-S battlefield radar

INDIA

6120.153

PSM 33 MOBILE RADAR

The PSM 33 is a transportable E/F-band 3-D air defence control and reporting radar system, which has been developed in cooperation with Thomson-CSF. It consists of an antenna with a tripod type mounting plus two cabins, one housing the transmitter and the other containing the receiving and processing units. It can be transported by road, rail or aircraft and can be deployed in about one hour.

Search in azimuth is by mechanical rotation of the antenna, and electronic scanning in elevation. A stacked beam pattern in elevation is formed by transmitting three frequencies, using monopulse techniques for accurate height measurement. Height finding accuracy is claimed as ±3 mrads. Facilities are incorporated for integration with secondary surveillance radar. Frequency diversity is employed as a protection against ECM and a programme to incorporate an area MTI is underway.

The system has several operating modes:

(a) general surveillance – elevation scanning over 24° with initial elevation adjustable
(b) single or two sector surveillance – elevation scanning over 6 or 12° as required
(c) search/designation: Accurate height measurement possible from –2 to +29.5° while search continues in the lowest sector
(d) pure designation mode: Accurate heightfinding on two closely spaced targets anywhere between –2 and +29.5°.

CHARACTERISTICS

Frequency: S-band
Range: 130 nm on 4 m² target
Range accuracy: 0.125 nm
Height accuracy: Better than ±3 mrads
Antenna:
Beamwidth (3 dB): Azimuth 1.4°, elevation 1.9° per beam
Polarisation: Circular
Rotation: 5 rpm
Transmitter:
Power: 1.8 kW average for each unit
Pulse duration: 5.8 μs
PRF: 500 Hz
Dimensions: Antenna 4.5 × 5.1 m

STATUS

In production.

CONTRACTOR

Bharat Electronics Ltd (a government of India enterprise), Trade Centre, 29/4 Race Course Road, Bangalore 560 001, India.

6205.153

INDRA AIR DEFENCE RADARS

India has developed two air defence radars, one for army use and one for the air force. These are solid-state, low-level mobile systems, the first operating in E/F-band and the second in D-band. Both equipments include computerised data handling and digital MTI facilities. No further details have been made available, although it is understood that the maximum range of the E/F-band system is about 40 km with the ability to detect targets down to 36 m above the ground clutter. The D-band equipment is designed to have a much greater range.

STATUS

In late development. The E/F band equipment has completed trials while the D-band system commenced trials in 1986. No information has been released regarding trials results. Indian R & D facilities are also working on the technology for 3-D electronically scanned, phased array radars with multiple target track-while-scan capability.

CONTRACTOR

Bharat Electronics Ltd (a government of India enterprise), Trade Centre, 29/4 Race Course Road, Bangalore 560 001, India.

INTERNATIONAL

6103.153

COBRA WEAPON LOCATING RADAR

COBRA (COunter Battery RAdar) is the designation of a new weapon locating radar proposed by the governments of France, Germany and UK. It is intended to be a highly mobile equipment to find targets for artillery and rockets, and is planned to enter service in the early 1990s.

A number of multi-national consortia have been formed to bid for the project. The first comprises THORN EMI Electronics, Siemens, Thomson-CSF and General Electric (USA). The second consists of AEG-Telefunken, Marconi Command and Control Systems and a different division of Thomson-CSF. The third comprises SEL, ESG, Ferranti, Marconi Radar and LMT Radio Professionelle (a Thomson-CSF subsidiary). Thomson-CSF is responsible for the teaming arrangements and, consequently, is a member of all three consortia.

STATUS

Definition studies have now been completed and were delivered to the UK, French and West German governments on 1 December 1986. A decision on the competitive bids was due to be followed by a development contract in 1987. At the beginning of 1988 no decision had been made, although it is understood that the first consortium is the preferred contractor. The total cost of development and production is expected to exceed £500 million.

1528.153

RATAC BATTLEFIELD RADAR

RATAC (*radar de tir pour l'artillerie de campagne*) is a lightweight (eight cabinets of 35 kg, or less, each) battlefield radar providing for detection, acquisition, identification, location, and tracking of surface targets such as troops, tanks, vehicles, low-flying helicopters and light aircraft. Other operational functions include artillery direction, surveillance of own forces, and helicopter control. It is comprised of six units and is designed for vehicle mounting. Pulse-doppler and monopulse techniques are employed and the operating frequency is in the 3 cm wavelength region. A highly effective fixed target cancellation system is said to make it possible to locate very slow moving targets with a small radar cross-section, such as a single man. Detection ranges are reported to be better than 15 km against vehicles and 8 km against troops, with an accuracy of 10 m in surveillance. Good all-weather performance is claimed.

The system incorporates its own computer, and target data can be presented in terms of either polar or grid co-ordinates. An optional plotting board unit permits target positions and movements to be recorded. A loudspeaker is provided to aid target identification by listening to the doppler characteristics of targets. Automatic target tracking facilities are included, and there are provisions for the transmission of data to own artillery. Operation with vehicle installed IFF is a possible option.

DEVELOPMENT

Development of RATAC was undertaken by LCT (Laboratoire Central de Télécommunications) under the direction of the French military authority. A subsequent agreement between the French and West German governments provided for joint procurement and production for the armies of both countries. Tests were later carried out by the US Army at Fort Sill.

STATUS

RATAC is in service in France, West Germany, the USA and several other countries. The US nomenclature is AN/TPS-58.

CONTRACTORS

LMT Radio Professionnelle, 46-47 quai Alphonse Le Gallo, 92103 Boulogne-Billancourt, France.

Standard Elektrik Lorenz AG, Lorentstrasse 10, 7 Stuttgart 40, Federal Republic of Germany.

RATAC battlefield radar mounted on an armoured vehicle

ISRAEL

4330.153

EL/M-2121 BATTLEFIELD SURVEILLANCE RADAR

The EL/M-2121 intelligence battlefield surveillance radar has been developed by Elta as an all-weather, mobile, very long range ground surveillance radar capable of detecting, locating and classifying targets moving on or near the ground in any type of weather or time of day/night.

One of the principal uses of the EL/M-2121 is for reconnaissance troops at ground observation posts to provide continuous surveillance and real-time information on activities of enemy forces deep behind the front line. The two operators can choose operational modes as follows:

(a) general continuous search – 360° PPI scan for location and designation of targets
(b) sector scan – angular sector scan to expand selected area
(c) zoom – B-scan display allows greater expansion and finer resolution of target range and azimuth
(d) identification mode – spectrum analysis of target doppler signal.

The complete system can be used as a transportable equipment with the radar and control equipment in a military centre, or divided with a fixed control centre operating a remote radar head. The antenna normally rotates continuously through 360° but can also carry out sector scans, or be pointed at designated targets.

The equipment is of modular construction and employs quadrature (I/Q) digital processing. A strong ECCM capability has been incorporated. It is microprocessor controlled and provides the operator with an eight-colour TV multi-mode display on a 19-inch screen. Built-in expansion options include the ability to be interfaced with other computers to process signals and data.

CHARACTERISTICS

Detection ranges

Pedestrian: 20 km (0.1 m² target)

Small vehicle: 40 km (1 m² target)

Medium size truck: 70 km (10 m² target)

Large truck: 120 km (100 m² target)

Accuracy: Range 15 m; azimuth 0.1°

Resolution: Range 15 m; azimuth 0.3°

EL/M-2121 surveillance radar

STATUS

In production.

CONTRACTOR

Elta Electronics Industries Ltd (a subsidiary of Israel Aircraft Industries Ltd), PO Box 330, Ashdod 77102, Israel.

3708.153

EL/M-2215 SURVEILLANCE RADAR

The Elta EL/M-2215 is a general-purpose surveillance radar, evolved from the successful EL/M-2205 search radar (**1958.153**), and many of the engineering features of the latter radar have been retained in its successor. Improvements made possible by advances in the state-of-the-art, have been incorporated, such as more elaborate signal processing facilities and increased transmitter output power. The transmitter/receiver is a redesigned and repackaged version of the former EL/M-2205. Modular design allows the second-generation EL/M-2215 to be configured in either single or dual transmitter/receiver versions, the latter offering a further choice between a dual-channel arrangement for redundancy or dual-channel with diversity operation. Other features include high resolution; a dual high/low-beam antenna; high clutter suppression; quadrature (I/Q) digital MTI; and fast frequency tuning.

The EL/M-2215 employs the AT-103 single beam antenna (as used in the EL/M-2205) or the AT-104 dual high/low-beam antenna. It may also be used with the AT-102 antenna to form the EL/M-2216 coastal radar for surface and low air cover search applications. (see EL/M-2216V radar).

CHARACTERISTICS

Transmitter

Frequency: 3.1 – 3.3 GHz (2.7 – 2.9 GHz optional)

Output power: 425 kW peak

Pulse width: 0.95 μs (selectable 0.95 or 0.4 μs for EL/M-2216)

PRF: Fixed 1 of 4, or 4 staggered 900 – 1050 pps (1340 – 1560 pps for EL/M-2216 short pulse)

TABLE OF ELTA EL/M-2200 RADARS

Model	**EL/M-2207 Shipborne**	**EL/M-2208 Shipborne**	**EL/M-2215/D Air Traffic Control**	**EL/M-2216 Coastal**
Range	30 nm	30 nm	80 nm	50 nm
Range accuracy	50 m	50 m	50 m	50 m
Range resolution	80 m	50 m	200 m	80 m
Azimuth accuracy	1°	1°	1°	1°
Azimuth resolution	3°	1°	1.5°	1.5°
Power (peak)	425 kW	425 kW (F) 25 kW (I)	425 kW	425 kW
Pulse width	1.4 μs or 0.4 μs	1.4 μs (F) or 0.4 μs (F) 0.25 μs (I)	0.95 μs	0.95 μs or 0.4 μs
PRF (pps)	1000 – 1500 or 500 – 750	1000 – 1500 or 500 – 750	900 – 1050	900 – 1050 or 1340 – 1560
Digital MTI improvement factor	27 dB	27 dB	33 dB	33 dB

Receiver

Type: Superheterodyne

IF frequency: 60 MHz

Noise figure: 2.5 dB (at transistorised amplifier input)

Sensitivity

Normal, –107 dBm

MTI, –104 dBm

Processing

Digital MTI: Quadrature (I/Q) 3-pulse canceller with or without feedback

Improvement factor: 33 dB

Blind speeds: None below Mach 3

Antenna AT-104

Gain: 34 dB low beam; 33 dB high beam

Azimuth beamwidth: 1.4°

Vertical beamwidth: 5°

Cosec squared up to: 42° low beam; 45° high beam

Sidelobes: Better than –24 dB

Polarisation: Linear horizontal or circular, selectable

Rotation rate: 12.5 or 6 rpm, selectable

Reflector dimensions: 5.46 × 3.20 m

Weight: 500 kg

Rotary joints: Dual E/F and D (for IFF)

STATUS

Production.

CONTRACTOR

Elta Electronics Industries Ltd (a subsidiary of Israel Aircraft Industries Ltd), PO Box 330, Ashdod 77102, Israel.

6108.153

EL/M-2216(V) COASTAL SURVEILLANCE SYSTEM

The EL/M 2216(V) is a variant of the EL/M-2215 radar (**3708.153**) which can be supplied as a self-contained shore-based coastal surveillance equipment, or can be custom-tailored for the upgrading of existing systems. It provides long range surveillance of large and medium size surface targets, detection of small sea targets (commando boats, rubber dinghies and fishing boats) in bad weather and rough sea conditions, harbour navigation control, and an optional facility for the detection of low flying air targets. The system employs an AT-102 antenna system consisting of a high resolution dual band (S+X) unit and a high speed X-band antenna. Other details are similar to the EL/M-2215.

CONTRACTOR

Elta Electronics Industries Ltd (a subsidiary of Israel Aircraft Industries Ltd), PO Box 330, Ashdod 77102, Israel.

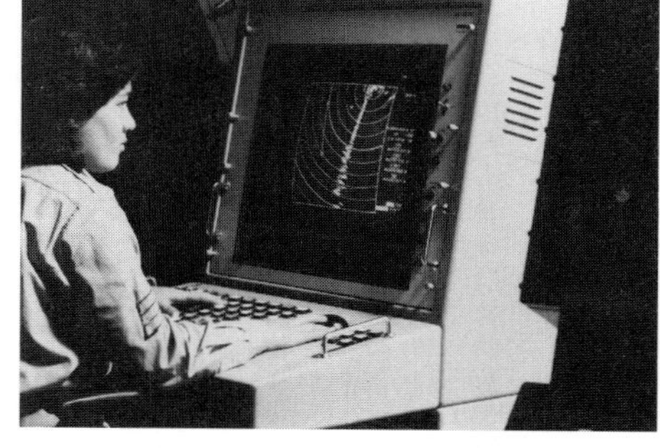

IP-35 bright display unit for the EL/M-2216(V)

4606.153

EL/M-2220 MOBILE SURVEILLANCE RADAR

Another variant of the EL/M-2215 (**3708.153**) is the EL/M-2220, a 2-D mobile air search and surveillance radar designed for tactical aircraft control, AA defence, or as a gap-filler. It features high clutter suppression, quadrature digital MTI, redundancy or dual-channel with diversity and improved signal processing.

Three versions of the system are available:

(a) EL/M-2220F for mobile fast deployment, comprising a S-280 shelter containing an EL/M-2215 transceiver, a 16-inch operator display console, and a small lightweight radar and IFF antenna mounted on top of the shelter. The EL/M-2220F can be deployed within 30 minutes

(b) EL/M-2220S for mobile air surveillance comprising an S-280 shelter containing an EL/M-2215 transceiver and a 16-inch display console, with an AT-103 large antenna on a towable platform

(c) EL/M-2220D for long-range mobile surveillance which consists of two shelters, one containing two EL/M-2215 transceivers (dual-channel), the other housing two consoles; plus an AT-103 antenna on a towable platform.

STATUS

In operational service with the Israeli armed forces.

CONTRACTOR

Elta Electronics Industries Ltd (a subsidiary of Israel Aircraft Industries Ltd), PO Box 330, Ashdod 77102, Israel.

CHARACTERISTICS

Frequency: 3.1 – 3.3 GHz (2.7 – 2.9 GHz optional)

Output power: 425 kW, peak

Pulse width: 0.95/1.4 μs

PRF: 1000/750 Hz, 4 staggered or random variable

Operation	**EL/M-2220D**	**EL/M-2220S**	**EL/M-2220F**
Detection range	80 nm	60 nm	30 nm
Azimuth resolution	1.4°	1.4°	3°
Range resolution	220 m	150 m	150 m
Height coverage	30 000 ft	25 000 ft	18 000 ft

EL/M-2220 mobile air surveillance radar

3646.153

EL/M-2106 POINT DEFENCE ALERT RADAR

The Elta EL/M-2106 point defence alert radar is designed to provide an early warning capability for anti-aircraft emplacements and forward installations. The EL/M-2106 is a portable battlefield radar that can also be used in such applications as target acquisition for surface-to-air missile systems and AA guns, as well as gap-filling roles.

The operating frequency is in the D-band (former L-band) of the spectrum, and typical detection range performance is about 16 km for fixed-wing aircraft and 10 km for helicopters (2 m² targets). Aural and visual warning of aircraft approach is given to the operator, and conventional scanning through 360° is normally provided, but a programmed sector scan with sweep-to-sweep memory is also available. The radar may also be remotely controlled from distances up to 100 metres.

The system is comprised of a transmitter, receiver, antenna assembly and display unit, and the entire installation breaks down into easily carried man-portable units. It can be set up in less than ten minutes and can be operated by one man with only minimal training. A remote control capability is available, up to a distance of 100 m, by the use of a controller weighing about 2 kg.

The EL/M-2106H version is able to detect helicopters in movement or hovering. A processing channel analyses the signal reflected from the rotor blades while the helicopter is hovering and can differentiate between fixed-wing aircraft and helicopters in flight.

The EL/M-2106H is also designed to be fully operational in a moving vehicle such as an armoured car or armoured personnel carrier. A special mast mounted on the vehicle enables different deployment heights according to the geographical and tactical environment. The antenna is a cut-down version measuring 120 × 50 cm and weighing 19 kg.

CHARACTERISTICS
(EL/M-2106H)
Transmitter/receiver frequency: D-band
Peak power: 150 W
Average power: 10 W
Pulsewidth: 6.6 or 13.3 μs
PRF: 8 or 16 kHz
Range gates: 10
MTI improvement factor: 50 dB
Resolution: 1 or 2 km (range); 10° (azimuth)
Accuracy: 1 or 2 km (range); 10° (azimuth)
Height coverage: Up to 8000 ft with antenna setting at +6°
Dimensions: 40 × 26 × 33 cm
Weight: 26 kg
Antenna dimensions: 200 × 80 cm
Weight: 55 kg (including stand and pedestal)
Gain: 23 dB
Horizontal beamwidth: 8°
Vertical beamwidth: 17°
Display: Synthetic with sweep memory. For each range gate, concentric circle of 36 LEDs. Indication of antenna position
Azimuth presentation: LED at 10° intervals
Range presentation: LED range circles spaced at 1 or 2 km intervals, giving max range of 10 or 20 km, selectable

STATUS
In production. An improved version, known as AN/UPS-3 is being produced by GEC Astronics, USA (**5777.153**).

EL/M-2106 point defence alert radar mounted in an armoured personnel carrier

CONTRACTOR
Elta Electronics Industries Ltd (a subsidiary of Israel Aircraft Industries Ltd), PO Box 330, Ashdod 77102, Israel.

1958.153
ELTA SURFACE RADARS

Elta Electronics Ltd produces a range of surface radars for ATC, search and acquisition, and similar applications, with fixed and land-mobile systems for surveillance roles, coastal protection etc.

The base model for most of this equipment, which is built on a modular design concept, is the EL/M-2205, and unless otherwise stated the following text relates to this model. (The current range is now based on the EL/M-2215 (**3708.153**) a modernised and improved version of the EL/M-2205.) Other models are the surface and low air cover search radar, EL/M-2206 and the coastal surveillance radar EL/M-2216 which both employ the AT-102 antenna in place of the AT-103 used in the EL/M-2205, the EL/M-2220 mobile air surveillance radar (**4606.153**), and the naval version EL/M-2207 (**3264.253**) and EL/M-2208. All versions are constructed to conform to the MIL-E-16400 specification and are provided with BITE. Conventional PPI displays or synthetic (raster-type) displays are available for use with all models. Integrated IFF is available as an option and is used in military models. A high performance, compact, and lightweight E/F-band transceiver forms the basic building block, and two of these units may be operated together in dual diversity. Receiver models available include: linear with switchable STC, FTC, IAGC, Log-FTC-Antilog, CFAR. There is single-knob tuning of the transmitter frequency between 3.1 and 3.3 GHz, and a stable synthesiser enabling the whole equipment to cover this band in less than 20 seconds, in the EL/M-2206 version.

Elta EL/M-2205 search radar scanner AT-103 with on-mounted IFF

More details of the characteristics of this radar are given in previous editions of *Jane's Weapon Systems.*

STATUS
Unspecified models of the EL/M-2200 series are stated to be in service in Israel and elsewhere.

CONTRACTOR
Elta Electronics Industries Ltd (a subsidiary of Israel Aircraft Industries Ltd), PO Box 330, Ashdod 77102, Israel.

ITALY

3278.153
MM/SPQ-3 COASTAL RADARS

The SPQ-3 series of radars comprises a variety of equipment combinations, centred on co-mounted I/J-band and E/F-band radars, configured to produce sets appropriate to coastal surveillance and harbour surveillance functions and related applications. The system is transportable by trucks or helicopters, and is generally intended to be integrated in larger defence control networks. The complete equipment is housed within two cabins, plus a trailer motor-generated power supply unit. The first cabin contains the two radar transmitter/receivers, and the associated antennas which are mounted on top can be folded inside for transport. The second cabin contains the operational part of the system, such as displays, processing unit, video recorder, and plotting table. The electronic recording system can be independently connected to the I/J-band or to the E/F-band radar, and is based on processing radar data (range scale, antenna speed, bearing etc) by means of a TV scan converter and connecting these to a magnetic recorder. Restitution of stored data is obtained by connecting the magnetic recorder output data to a TV-type display. Remote and local control facilities are provided.

The principal operational roles are medium to long range detection of air attack or of combined air/seaborne assaults by means of the E/F-band radar; and close to medium range surface surveillance by means of the I/J-band radar to assist defence against surface craft attack, landings etc. In the SPQ-3B version, a special feature is the detection and

Main radar cabin of SPQ-3 system, showing back-to-back scanners for I/J-band spotting and tracking radar and upper early warning E/F-band search radar scanner

location of the splashes caused by air-dropped mines in harbours or confined waters.

Features of the E/F-band radar include: variable-frequency transmission; random PRF; high directivity antenna pattern; Dicke-fix IF receiver; and linear-logarithmic IF receiver. The I/J-band equipment has most of these features with the exception of the Dicke-fix, and two back-to-back antennas as compared with the single horn-fed parabolic reflector of the E/F-band unit. Both are essentially very high image resolution systems with low average emitting power, particularly suitable for operation (by means of frequency and polarisation decorrelation) in high traffic density environments. Sector transmission is possible.

Scan converter displays are provided and there are facilities for video recording and replay. Digital and analogue format remote data transmission outputs are available.

STATUS

Production.

CONTRACTOR

SMA (Segnalamento Marittimo ed Aereo), PO Box 200, Via del Ferrone, 50100 Florence, Italy.

4884.153

HSS-1 HARBOUR SURVEILLANCE RADAR

The HSS-1 harbour surveillance radar system is essentially a very high resolution equipment with a low emiting power, which is particularly suitable for operation (thanks to frequency and polarisation de-correlation techniques) in high traffic density environments (eg. harbours, river estuaries or restricted channels) or as a precaution against infiltration etc.

The system is characterised by a very narrow beam antenna system, X-band frequency diversity operation, a dedicated anti-clutter video processing channel, and a semi-mobile equipment configuration. (See illustration.)

STATUS

Currently in production for undisclosed customers.

CONTRACTOR

SMA (Segnalamento Marittimo ed Aereo), PO Box 200, Via del Ferrone, 50100 Florence, Italy.

HSS-1 operations shelter with high definition scanner mounted on roof

6810.153

TPN-720 MOBILE CONTROL AND LANDING SYSTEM

The TPN-720 is an autonomous mobile radar system for the surveillance and guidance of aircraft during approach and landing, and is particularly applicable to both military and civil applications at temporary landing sites. It operates in I-band (8 to 10 GHz) with MTI and features a highly mobile, shelterised system. Adjustable azimuth glide path coverage up to 21° and elevation coverage up to 21°.

STATUS

In production.

CONTRACTOR

SMA (Segnalamento Maritimo ed Aereo), PO Box 200, Via del Ferrone, 50100 Florence, Italy.

1529.153

LPD-20 SEARCH RADAR

The LPD-20 is a search and acquisition radar developed to counter air threats from low- and very low-level targets in the region below the useful coverage of other radars. Two basic versions have been designed:

(1) Integrated version: The radar can easily be fitted into the Contraves Super Fledermaus family of fire control systems. This simplest arrangement employs the PPI display as the fire control system, so that the search radar is unattended and a single operator is able to control the complete radar system for both search and tracking of targets
(2) Autonomous version: The LPD-20 used as an autonomous radar set is equipped with a separate PPI display, for alerting light AA weapons (guns or very short-range SAM) for different purposes (target designation to light unslaved AA artillery emplacements, low and very low altitude surveillance, etc).

To provide additional operator safety, the PPI display (where remote control of the radar set also takes place) may be sited in a protected and defiladed position, at a distance up to 100 m from the radar site.

Typical deployment of Super Fledermaus fire control system with LPD-20 radar (integrated version)

IFF facilities are also integrated into the radar system. Contraves Italiana SpA in September 1974 concluded an agreement with Cossor Electronics Ltd by which integrated IFF facilities produced by the UK firm can be provided as a customer option for LPD-20 systems.

The main parts of the radar are:

(1) a rotating section, comprising antenna, transmitter/receiver, signal processor, and modulator
(2) turning gear
(3) a cabin containing power supplies and operational controls
(4) trailer chassis.

System weight (trailer version) is 2500 kg.

The LPD-20 is a fully coherent pulse-doppler radar. An MTI improvement factor of better than 50 dB is claimed, together with a detection range of about 20 km on targets travelling at speeds between 20 and 410 m/s at heights up to 4000 m. Angular resolution is of the order of 1.4° and range resolution about 500 m. Rapid reaction to sudden low-level threats is aided by a high data rate, the use of a common PPI for search and fire control, accurate target range and azimuth designation data and transfer from search radar to fire control radar in two to three seconds. Provision is made for the incorporation of IFF Mk 10/SIF facilities into the basic LPD-20 radar. Furthermore, as an optional alternative to the autonomous version, an automatic air target data distribution and display system can be provided, which allows forwarding of target data information to other users. This system consists of a data distribution unit (TD/TU) built into the display unit and a unit for the reception and display of air targets data (TD/DU) which is associated with the user's weapon or weapons.

Numerical data corresponding to target position and identification (polar or cartesian co-ordinates and IFF) is transmitted by the TD/TU (FSK modulator and buffer amplifier) through a two-wire field telephone line or radio link to the user's TD/DU (filter and

Contraves LPD-20 search and acquisition radar with Cossor IFF unit (LPD-20, autonomous version)

Unit for the reception and display of air target data for LPD-20 radar

demodulator), where it is displayed in numerical form. A maximum of three targets can be handled simultaneously by the system. Maximum range of telephone transmission amounts to about 5 km. The TD/DU is provided with autonomous supply (rechargeable DC battery). Data transmission is uni-directional (from display unit to weapons).

STATUS

Although detailed information has not been released, this system is known to be widely used in many parts of the world. Over 150 units have been produced.

Current production of the Super Fledermaus fire control system (**2376.151**) incorporates modifications to enable it to be associated with the LPD-20.

CONTRACTOR

Contraves Italiana SpA, Via Affile 102, Via Tiburtina Km 13150, 00131 Rome, Italy.

5618.153

SHORAR SEARCH RADAR

SHORAR (Short Range Acquisition Radar) is a search, acquisition and target designation system which has been developed for different applications of short range air defence, (ie for the detection of targets flying at medium, low and very low altitudes). The radar has a modular design to allow for easy installation on different types of vehicles (APCs, trailers) and AA turrets, and consists of three modules; an antenna and transceiver assembly, a control console assembly, and an electronics assembly. The antenna/transceiver assembly weighs approximately 400 kg, the antenna being double beam forming in elevation with an IFF antenna integrated on the reflector.

SHORAR is a fully coherent, X-band pulse doppler radar, optimised for the detection of aircraft and helicopters (hovering and moving). It has special features, such as pulse compression, frequency agility, low peak power, low sidelobes, PRF change, and high clutter rejection. Contraves states that as a result of these features, the radar has an outstanding performance in dense ECM environment and heavy clutter.

Computerised and fully automatic procedures, such as track-while-scan, threat evaluation, and target designation, allow for easy operation and multiple target handling. In addition to the alert

Self-propelled version of SHORAR on M-113 APC

target designation function, SHORAR performs the command and control task for associated weapons, which can be MANPADs and/or AA guns, in order to improve their effectiveness. The SHORAR modules are employed for three different applications; as a search radar in the ADATS air defence system; as a self-propelled acquisition and fire coordination unit, installed on APCs such as the M-113, for self-propelled AA systems without a search sensor; and as a search and tactical command post, on a wheeled trailer, for towed AA weapons. The towed version is known as PAGODA, and is described under entry **6391.153**.

SHORAR can, therefore, be employed for air defence of static objectives as well as for escort air defence of mechanised and motorised troops (search-on-move is available as an option).

According to the type and number of AA weapons which receive the target data from a towed or self-propelled SHORAR, it can be used either in a centralised mode (targets are assigned to the best-placed weapon and engagements are controlled by the radar), or in a decentralised mode (weapons receive only target data and then autonomously engage targets according to rules). Target designation can occur both with a direct link to the weapon or through a target data receiver near the weapon.

CHARACTERISTICS

Scan rate: 60 rpm
Frequency: X-band
Azimuth accuracy: 0.5°
Range discrimination: 100 m
Maximum range: 17 to 25 km
Total weight: 580 kg

STATUS

In series production. Successful trials have been carried out with the Egyptian Air Defence Command.

CONTRACTOR

Contraves Italiana SpA, Via Affile 102, Via Tiburtina Km. 13150, 00131 Rome, Italy.

6391.153

PAGODA SEARCH RADAR

Pagoda is the towed version of the SHORAR family of radars (**5618.153**), designed and developed by Contraves Italiana. It is a fully coherent dual-beam X-band pulse doppler radar that is optimised for the detection of RPVs, aircraft and helicopters flying at medium, low and very low altitudes. It can detect a helicopter at 7 km and is claimed to have good

Pagoda search radar (towed version of SHORAR radar)

performance in heavy clutter, as well as high ECM resistance. Other features of the Pagoda include pulse compression, frequency agility, low peak power, and PRF shifting during time on target. The track-while-scan facility for up to 12 targets and threat evaluation is automatic.

There are two main components of the Pagoda system. The single axis trailer contains the antenna unit, transceiver assembly, signal processing unit and IFF equipment. Mounted on the front of the trailer is the self-contained power supply unit and cables to connect the trailer with the second component of the system, the NBC proof shelter. The latter contains a control console, equipped with a PPI for the display of target and tactical information, from which the radar is controlled remotely up to a maximum distance of 100 metres.

The main role of the Pagoda system is target search and acquisition. Weapon coordinating functions are assured by means of the sophisticated control console, incorporating the PPI alphanumeric and graphic display for tactical and air situation information.

The Pagoda can control towed anti-aircraft guns such as four Oerlikon-Buhrle Diana twin 25 mm guns fitted with the Contraves Gun King computerised sight, or man portable surface-to-air missiles such as Stinger.

CHARACTERISTICS

Frequency: X-band
Transmitter: TWT with 2 kW peak power
Maximum range: 25 km
Helicopter detection range: Up to 7 km
Antenna rotation speed: 60/40 rpm
Track-while-scan: 12 tracks with automatic initialisation
Surveillance envelope: 0 to 50° (elevation); 360° (azimuth)
Accuracy: 0.5° (azimuth); 30 m (range)
Total weight: 2600 kg

STATUS

In early production. Successful trials have been carried out with Egyptian Air Defence Command.

CONTRACTOR

Contraves Italiana SpA, Via Affile 102 Km 13150, 00131 Rome, Italy.

3277.153

PLUTO LOW-COVERAGE SURVEILLANCE RADAR

Pluto is a low-coverage surveillance radar for air and coastal defence, designed to detect medium- low- and very low-flying aircraft and small surface vessels under the most adverse environmental conditions (ECM, atmospheric and sea clutter etc). The radar is suitable for fixed and/or mobile installations and can operate either independently or be integrated into a complete defence system. Design features include the following:

Capability to detect and measure the range and azimuth of surface and air targets at medium and low altitudes up to a maximum range of 110 km.

High accuracy and resolution, integral target designation capability and flexibility to interface computer-assisted display and defence systems.

Pulse compression of the radiated waveform with a resultant small scale radar cell size, ie resultant visibility even of very slow moving targets such as small ships (FPB).

Sophisticated ECCM processing with visibility maintained even in presence of heavy active and passive jamming.

Pluto low-coverage radar deployed

Good adaptability and CFAR characteristics under various environmental conditions, ie capability to detect targets and reject interference by automatic evaluation and selection of the best transmission and processing configurations.

Capability of operating in frequency-agility from sweep to sweep, with clutter rejection and sub-clutter visibility of moving targets outgoing or incoming at speeds up to Mach 3 by means of sophisticated MTI processing techniques.

Antenna pattern with high directivity and very low sidelobe levels.

Large transmission bandwidth.

Integrated secondary (IFF) radar.

High MTBF and reduced MTTR.

High reliability, maintainability and supportability.

BITE facilities.

A reduced range version of Pluto is used to supply surveillance data for the Spada air defence system (**2250.131**).

In its basic configuration, Pluto consists of one equipment shelter, one antenna module and a power generator trailer. The transmitter, receiver and processing equipment are all housed in the shelter.

CHARACTERISTICS

Antenna

Type: Reflector, feed

Frequency: E/F-band

Polarisation: Horizontal or circular, selectable

Beamwidth: Horizontal 1.5°; vertical 4°

Beam shape: Horizontal – pencil beam; vertical – super-$cosec^2$ matched to a very high clutter environment

Scan rate: 12/15 rpm

Tilt: –2 to +5° adjustable

IFF antenna: Integrated

Transmitter

Type: Coherent chain

Final stage: TWT

Frequency: E/F-band

Bandwidth: 10%

Transmitted waveform: Phase coded (code agility)

PRF: Prefixed or random stagger

Operating modes: Fixed frequency; random/pre-programmed frequency agility; adaptive frequency selection

STATUS

In production.

CONTRACTOR

Selenia-Industrie Elettroniche Associate SpA, Via Tiburtina Km 12400, 00131 Rome, Italy.

1953.153

RAT-31S TRANSPORTABLE 3-D RADAR

The RAT-31S has been designed as a highly mobile three-dimensional radar capable of performing a wide range of operational functions associated with air defence, tactical air applications, and civil/military air traffic control. The construction method employed makes the system equally suitable for air- or land-mobile applications. The system comprises an antenna group, a transmitter shelter, a receiver and signal processing shelter, and a primary power supply. The antenna group can be moved by a trailer or by a demountable running gear, which is the preferred method when using C-130 aircraft for transportation. The shelters may be moved by any of the methods complying with NATO STANAG standards.

The operating frequency is in the E/F-band and the RAT-31S is intended for aircraft detection and range, elevation and azimuth target measurements at high, medium, and low altitudes. Comprehensive ECCM facilities include the ability to operate in frequency agility from pulse-to-pulse, and/or within pulse-groups, with an automatic adaptive transmission frequency selection. Height measurement of detected aircraft is compatible with frequency agility, and the sub-clutter visibility is sustained even when operating in pulse-to-pulse frequency agility.

The RAT-31S uses a novel antenna system that employs a system of three stacked beams with a phased array control of the elevation of each beam. The configuration is that of a rear-fed planar array, which avoids aperture blocking, minimises sidelobes, and provides good mutual screening characteristics.

Selenia RAT-31S transportable 3-D radar deployed with an MRCS-403 mobile reporting and control system station

A coherent chain transmitter with a TWT final amplification stage is used, which facilitates the frequency agile features of the radar. A staggered position and phase code is used in the transmitted pulse. Features of the receiver include: monopulse operation with Dicke-fix processing; anti-clutter filtering by means of a double canceller; adaptive clutter attenuator; coherent limiter; and moving window azimuth correlator.

The RAT-31S is designed to be an integral element of automated air defence systems. For tactical air defence, Selenia can deploy a mobile command and control centre, the MRCS-403, which is designed to operate with the RAT-31S. This control centre consists of two computers and three control consoles, a computer peripheral set, communications control, and ground-to-air equipment.

CHARACTERISTICS

Antenna

Type: Rear-fed planar array

Polarisation: Linear horizontal

Beamwidth: 1.5° (azimuth and low elevation beams)

Azimuth scan mode: Mechanical rotation over 360°

Scan rate: 5 – 10 rpm

Elevation scan mode: Electronic by phase control from 0 – 21°

Number of beams: 3 different groups of beams are radiated on 3 separate bandwidths

Pattern: Multiple pencil beams

Transmitter

Type: Coherent amplifier chain

Final stage: TWT

Max unambiguous range: >300 km

Stagger sequence: Variable

Transmitted waveform: 3 separate coded waveforms

STATUS

The RAT-31S is in full production. It is operational in the mobile air defence system, MRCS-403, delivered to Austria (see **3326.181**). MRCS-403 systems have been ordered by the Italian Air Force and systems have also been sold to other countries including Brazil.

CONTRACTOR

Selenia Industrie Elettroniche Associate SpA, Via Tiburtina Km12400, 00131 Rome, Italy.

6169.153

RAT-31SL LONG RANGE RADAR

The RAT-31SL is a long range 3-D air defence radar, operating in E/F-band and designed to provide azimuth, range and height data of air targets in heavy clutter and ECM environment. The radar combines stacked pencil beams with phase scanning in elevation to provide volumetric coverage and high accuracy performance. The RAT-31SL is completely adaptive to the operational environment with a multi-channel stacked beam, multi-wave form, mono-pulse antenna system which achieves accurate performance in all modes including truly random pulse frequency agility.

This system is ruggedised to withstand adverse weather, solar radiation, RF, EMI and EMP and is designed for operation during ECM conditions. It draws heavily on the design and technology used in the RAT-31S system (**1953.153**).

STATUS

An engineering development model of the RAT-31SL is being built in co-operation with Sperry Corporation which will be responsible, under subcontract to Selenia, for the manufacture and delivery of the radar transmitter and radar environment simulator. A number of RAT-31SL systems are likely to be procured by the Italian Air Force.

CONTRACTOR

Selenia Industrie Elettroniche Associate SpA, Via Tiburtina Km 12400, 00131 Rome, Italy.

1959.153

ARGOS-10 EARLY WARNING RADAR

The Argos-10 is a high-power, modern early warning search radar, with an effective range in excess of 250 nautical miles (460 km). It incorporates a transmitter designed and manufactured by the Heavy Military Electronics Systems Division of the General Electric Company. The salient features include: a high-power coherent transmitter chain; full adaptability to fixed and/or moving clutter; environmental pattern recognition; high resistance to ECM; reception status matched to the actual environment; fully digital signal processing; integrated primary/secondary video extractor; and modular construction.

The antenna is a large lattice construction with a lattice boom projecting from the lower edge to support the dual horn feeds which provide for dual beam operation. This element of the Argos-10 radar has a marked resemblance to the antenna and feed of the Canadian Raytheon ASR-803 D-band airport surveillance radar, and may be the same design. Provision is made for enclosing the antenna assembly in a protective radome. An auxiliary antenna can be added if required for multiple stand-off jammer

Argos-10 high-power early warning radar in protective radome

cancellation and to provide for unambiguous jam-strobe indication.

The transmitter is water cooled and operates in the D-band. The output has not been stated but is probably several megawatts. A codified output pulse is produced and there is a code agility facility. In accordance with the jamming situation encountered, adaptive frequency selection is available, other operating options being MTI-compatible frequency agility and random frequency agility. The receiver incorporates dual-beam coherent combination; a low-noise RF amplifier; parallel channels for multiple jammer coherent cancellation; and linear with adaptive clutter attenuator facilities.

The digital signal processor has phase and quadrature channels, a three-pulse digital MTI double canceller, and clutter map storage for adaptive clutter cancellation. There is a low velocity clutter filter for reducing or eliminating the effects of rain or chaff. Real-time sensors are provided for moving clutter cancellation and rain contour mapping. Automatic optimum channel selection is incorporated. Measures employed to eliminate second and third time around echoes comprise a pattern recognition technique and multiple stagger codes. There is also a time on target dual threshold integrator in the signal processor section.

Plot extraction with a 10^{-6} false alarm rate in a combined clutter/jamming environment is performed by the video extractor, which also includes a primary/secondary plot combiner, and an interface-buffer to a central computer for narrow-band transmission.

STATUS

Fully operational.

CONTRACTORS

Selenia Industrie Elettroniche Associate SpA, Via Tiburtina Km12400, 00131 Rome, Italy.

General Electric Company, Heavy Military Electronic Systems Division, USA – Transmitter.

6170.153

ARGOS-45 TACTICAL RADAR SYSTEM

The Argos-45 tactical radar system has been designed to provide co-ordination to forward area low altitude air defence systems, to defend battalions and selected combat support units which are vulnerable to high speed, low multi-role aircraft and strike helicopters. The system can take on the role of a low altitude reporting post, as a gap filler in an air defence network, a tactical command post for the control of air support tactical missions, and a designation centre for man portable missiles and anti-aircraft artillery.

Two configurations are available to meet differing operational requirements. In the compact version, Argos-45C, the system consists of a military shelter housing the radar head, cooling unit, two operator consoles, data handling equipment, communications and a power generator. The operators' positions are in a separate compartment within the shelter. The antenna system consists of a lightweight structure mounted on an erectable tower, stowable in transport configuration on the shelter roof, and including an integrated IFF system.

The alternative version, Argos-45T, is suitable for those scenarios where the threat is very high. A new display system and data processing structure (known as Magics and Mara respectively) are used in both versions.

STATUS

In development for the Italian armed forces.

CONTRACTOR

Selenia Industrie Elettroniche Associate SpA, Via Tiburtina Km 12400, 00131 Rome, Italy.

5770.153

SELENIA MOBILE GCA SYSTEM

The Selenia mobile ground controlled approach system consists of an airport surveillance radar (ASR), a precision approach radar (PAR) and an IFF/SSR surveillance radar (SIR). It is suitable for both military and civil purposes and is transportable by helicopter, fixed-wing aircraft, road, rail or ship. The system is operational within two hours of arrival at the site. It is easily adaptable for installation at fixed sites and can be used autonomously or integrated with existing ATC equipment.

The operation shelter is fully equipped with two PPI displays with presentation of aircraft identity and video maps, and two AZ-EL displays for PAR information. VHF/UHF air-ground and ground-ground communications are integrated and autonomous power generators are provided. The complete system is designed to operate over the environmental temperature of –40°C to +70°C and in wind speeds up to 100 km/h.

Selenia mobile GCA deployed

The ASR is a medium-power 10 cm wavelength, solid-state search radar. In the dual channel configuration the equipment works in frequency diversity, providing a range coverage of 75 nautical miles on a target such as a T-33 aircraft. The main feature of the radar is beam switching with beam combination, a system which allows the creation of a third beam with a null on the horizon and adaptive choice of auxiliary or main beam. Also provided are programmed adaptive clutter attenuation, and in-phase and quadrature MTI.

The PAR provides accurate range, height and azimuth information on aircraft within the entire volume of 20° in azimuth and 10° in elevation out to a range of 20 nautical miles. The transmitter operates in X-band and provides coverage to either end of a single runway and to either end of at least two crossing runways by reorientation of the antenna.

CHARACTERISTICS

Airfield surveillance radar

Frequency: 2.7 to 2.9 GHz

Configuration: Single or dual channel

Antenna: Modified cosec; 15 rpm turning speed; beamwidth 1.45°; polarisation continuously variable from linear to circular

Transmitter: Magnetron, 500 kW peak power, 1 µs pulse length, average PRF 1000 Hz typical

Precision approach radar

Frequency: 9 – 9.18 GHz

Configuration: Single or dual channel

Volume coverage: 20° azimuth; –1° to +9° elevation; 20 n miles range

Accuracy: 0.3% horizontal; 0.2% vertical; 1% of the range

Antenna: Double cosec²; beamwidth 0.6° × 2.3° in azimuth and 0.6° × 5.9° in elevation

Transmitter: Magnetron; 200 kW peak power; PRF 3000 Hz average; pulse width 0.25 µs in normal mode, 0.4 µs in MTI mode

IFF/SSR radar

Antenna: 14 ft (4.27 m), on-mounted

System: Dual channel with automatic changeover

Modes: 1, 2, 3/A, B, C.

STATUS

In production.

CONTRACTOR

Selenia-Industrie Elettroniche Associate SpA, Via Tiburtina, Km12400, 00131 Rome, Italy.

3279.153

SIT 431 (AN/TPX-54) IFF INTERROGATOR

The system consists of an interrogator unit, a sum-and-difference antenna and up to six AN/UPA-59A decoders. In SIF operation the primary controls are provided on the UPA-59As. In Mode 4 operation the interrogator is used with an external KIR-1A/TSEC computer. The Mode 4 controls and the secondary SIF controls are contained in a master control box remoted to the display area. An IFF system fault indicator is also included in this unit.

The receiver/transmitter module contains a 1500 W transmitter, a dual-channel receiver and RF interface module. The transmitter module is a completely solid-state unit. A dual-channel receiver provides the receiver sidelobe suppression (RSLS) function when operated with either a sum-and-difference antenna or a sum-and-omni antenna. The use of RSLS provides a means of artificially sharpening and controlling the antenna beamwidth for received signals. One of the unique features of the receiver is the six-stage logarithmic IF amplifier. The output of each stage varies linearly as a logarithmic function of the input. Thus a large dynamic range is obtained without appreciable distortion even at very large signal levels.

CHARACTERISTICS

Frequency: 1030 MHz (transmitter); 1090 MHz (receiver)

Dimensions: 63.5 cm wide × 20 cm high × 36.8 cm deep

Weight: 25kg

STATUS

In production. A more detailed description is given in previous editions of *Jane's Weapon Systems*.

CONTRACTOR

ITALTEL, Defence Telecommunications Division, 66 Via Due Macelli, 00187 Rome, Italy.

JAPAN

4081.153

JAN/MPQ-N1 (TYPE 92) MORTAR LOCATING RADAR

The Type 92 mortar locating radar is apparently a Japanese licence-built version of the American AN/MPQ-4 equipment which is produced by the General Electric Company. The Japanese model, designated JAN/MPQ-N1, is thought to have essentially the same performance as the US-built radar which is described in entry **2497.153**, later in this section of *Jane's Weapon Systems*.

The purpose of these radars is to detect and locate enemy mortar positions so that counter-battery action can be carried out. The two-beam intercept principle is employed, first-shell acquisition is possible and the system can handle multiple targets.

STATUS

Operational with the Japanese Ground Self-Defence Force.

It is usual for the reflector to be folded down when the Type 92 mortar locating radar is being towed (K Ebata)

Japanese Type 92 (JAN/MPQ-N1) mortar locating radar on the move with its reflector erected (K Ebata)

4141.153

J/MPQ-P7 ARTILLERY LOCATING RADAR

This artillery locating radar is apparently of Japanese design, but it is not known if there has been any technical collaboration from America or elsewhere, although there is an obvious similarity with the Hughes AN/TPQ-36 and AN/TPQ-37 mortar and artillery locating radars (**2848.153** and **1976.153**). Like the US-designed radars, the J/MPQ-P7 employs electronic scanning of a planar array antenna to generate the multiple beams necessary for detection and trajectory analysis of incoming projectiles. In this respect the Japanese equipment is so far the only known comparable radar to appear from beyond the American defence industry for operational service. The moderately large antenna array is transported on a wheeled trailer towed by a tracked vehicle which houses items of equipment and the radar crew. The antenna is erected when deployed, but folds flat for transit. Electronic beam steering probably obviates any need for the array to rotate in use. It is likely that a sufficiently broad azimuth coverage arc can be achieved by electronic scanning to provide surveillance of the designated threat direction(s).

Few technical details have been obtained but it is reported that the equipment has a peak output power of 250 kW, operates in the I/J-band (former X-band), and is capable of ranges of more than 30 km.

STATUS

In service with the Japanese Self-Defence Forces.

CONTRACTOR

Toshiba Electric Company, 1-6 Uchisaiwaicho, 1-chome, Chiyoda-ku, Tokyo 100, Japan.

J/MPQ-P7 artillery locating radar equipment in transit configuration (K Ebata)

J/MPQ-P7 artillery locating radar set up for operation

2298.153

NPM-510 MOBILE 3-D RADAR

The NEC NPM-510 3-D radar is a highly mobile system with several novel features. Among these are the antenna system, which is rotated in azimuth but electronically scanned in elevation and which consists of a planar array that can be folded for transport, and a rapidly erected inflatable equipment shelter.

A compact and lightweight system relying extensively on solid-state technology the radar is claimed to be highly reliable. Salient features of the circuitry are a CFA transmitter chain, chirp pulse compression, digital MTI and automatic target detection, tracking, and altitude computation.

STATUS

The radar has been supplied to the Japan Defence Agency.

Inflatable operations centre for the NEC radar

CONTRACTOR

NEC Corporation, 33-1, Shiba 5-chome, Minato-ku, Tokyo 108, Japan.

Planar antenna of the NEC mobile 3-D radar NPM-510

3702.153

NPG-880 3-D AIR DEFENCE RADAR

The NPG-880 is a static three-dimensional air defence radar developed by NEC in Japan; a mobile version, NPM-510, is also produced (**2298.153**, above). Both equipments employ an antenna that embodies a unique phase/frequency scanning technique of a planar phased array in elevation, with mechanical rotation in azimuth. A combination of serpentine feed and ferrite phase shifters is used for scanning a pencil beam in elevation.

The radar embodies energy management concepts and signal processing techniques giving high detection and stable accuracy, with a TWT and CFA transmitter chain, chirp pulse compression and amplitude comparison for height measurement. The advanced digital signal processing techniques used for clutter reduction and rejection (and ECCM) significantly enhance performance for automatic detection and tracking under severe clutter and jamming environments.

The NPG-880 is capable of direct interfacing with any automated air defence network, and it is linked with air defence ground environment and air defence command post systems. The system has dual channel configuration with high reliability design features and substantial built-in-test equipment to minimise system down-time.

The general configuration of the NPG-880 can be seen from the photograph. The associated transmitter/receiver, display, and other electronics can be housed either in permament or temporary buildings.

Planar array of the NPG-880 static 3-D air defence radar

STATUS
Supplied to the Japan Air Self-Defence Force.

CONTRACTOR
NEC Corporation, 33-1, Shiba 5-chome, Minato-ku, Tokyo 108, Japan.

4605.153

XJ/FPS-3 AIR DEFENCE RADAR

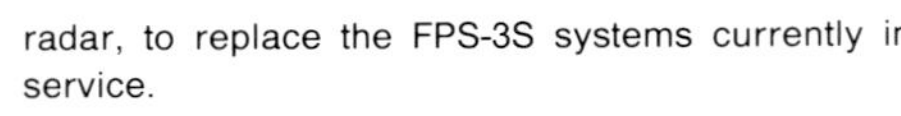

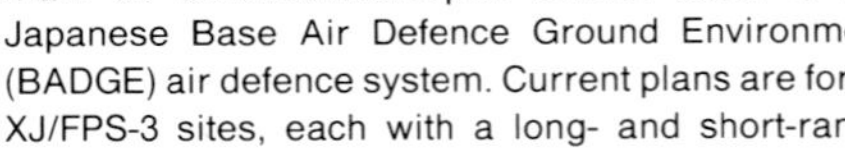

The XJ/FPS-3 air defence radar is a next generation phased array radar project which is intended to be the main air defence/interceptor control radar of the Japanese Base Air Defence Ground Environment (BADGE) air defence system. Current plans are for 28 XJ/FPS-3 sites, each with a long- and short-range radar, to replace the FPS-3S systems currently in service.

STATUS
Three Japanese companies, Mitsubishi Electric, Japan Electric and Tokyo Shibaura Electric have been tendering for the project.

Mitsubishi delivered an advanced phased array radar to the Japan Defense Agency's research and development institute in December 1986. The radar operates in the L- and S-bands and has a decoy system that can transmit up to 400 metres from the radar. Communications with control centres is by optical cables.

The first system is expected to be operational by the end of 1988, following 18 months of research and testing.

6812.153

JAPANESE OTH RADAR

Japan has decided to build an over-the-horizon radar system and information on the siting of the first system has recently been announced. The transmitter will be located on Hahajima Island with the receiver and presumably the control centre, on Iwo Jima Island. The system will be linked into the United States OTH network, in particular that part providing surveillance of the Aleutian Islands chain. No information is available on time scales.

The provision of OTH radar in Japan has been mooted for some time, primarily as an extension of the CONUS-OTH network. Reports in the past have suggested that the complete system will be bought from the United States and will be identical to the American system. Full details of the CONUS-OTH system are given under entry **5617.153**.

NETHERLANDS

3283.153

FLYCATCHER RADAR

The millimetric-wave radar employed in the Signaal Flycatcher low-level anti-aircraft fire control system (**2301.181**) is essentially a repackaged version of that developed for the Leopard 35 mm AA tank CA1 (**2302.153**), allied to an SMR digital computer.

With Flycatcher's 3-D tracking channel three weapon systems can be controlled at a distance of approximately 1 km. Using its 2-D TWS-channels, Flycatcher can provide target information to an extra number of weapons at a distance of 6 or 7 km. By this method, Flycatcher creates and optimises defence in depth. In this way the radar is also effective against carriers of modern medium range weapons, ARMs and stand-off jamming. In addition to this computer-assisted track-while-scan capability, provision is made for the display of targets acquired by other systems.

The general arrangement of the Flycatcher radar employs coaxial mounting of the search and tracking antenna as compared with the separate positions used in the 35 mm AA tank installation. Passive optronic tracking by means of a TV system is incorporated. Change-over to active angle tracking and vice-versa is accomplished automatically if preselected. The main characteristics are otherwise similar for both versions of the radar.

STATUS
Deliveries to the Netherlands forces for defence of air bases and missile batteries began in 1978. The system has also been ordered by Thailand and several other countries. India has ordered the Flycatcher system for its air defence regiments. After initial systems the equipments are now being manufactured under licence by Bharat Electronics and the first Bharat systems have been delivered to the Indian Army.

CONTRACTOR
Hollandse Signaalapparaten BV, PO Box 42, 7550 GD Hengelo, Netherlands.

6167.153

VANGUARD 3-D RADAR

Vanguard is a mobile multi-function 3-D radar based on rotating single plane phased array principles. It is the nucleus of the Signaal Falconer medium surface-to-air missile system programme for the future. This system is designed to be the answer to the threat of fighters, helicopters, tactical ballistic missiles, antiradiation missiles and ECM. Falconer consists of the Vanguard main sensor, an operational command and control centre, plus optional decoys and optronic sensors.

The Vanguard radar is derived from the SMART naval surveillance radar (**5626.253**), and is suitable for both surveillance and tracking functions. The radar is housed in two separate mobile shelters and uses a hydraulically erected antenna array. Vanguard covers a height of 40 km and a range of 120 km. It is guaranteed to track a minimum of 200 targets simultaneously with a refresh rate of two seconds over 360° azimuth and 90° elevation coverage. An IFF radar is integrated and the system is compatible with the NATO Identification System (NIS). Multiple receive beams formed in a digital FFT beamformer and FFT-doppler processing on each beam provides a high degree of resistance to ECM.

STATUS
In development.

CONTRACTOR
Hollandse Signaalapparaten BV, PO Box 42, 7550 GD Hengelo, Netherlands.

6168.153

SIGNAAL COASTAL RADAR

The Signaal Coastal Radar system consists of an I-band transmitter/receiver with an antenna, a variable number of displays, an automatic video extractor with plot processor, and a narrow band microwave-link transmitter/receiver. It can be extended with a helicopter transponder receiver and a secondary video extractor. The system forms the main sensor element of the Watchdog coastal defence system, and can be either locally or remotely controlled.

CHARACTERISTICS

Frequency: I-band (8850 – 9150 MHz) – optionally G-band

Antenna type: Slotted waveguide

Vertical beamshape: Inverted cosec²

Horizontal beamwidth: 0.25, 0.4 or 0.5°

Pulse length: 50 or 200 ns

PRF: Technically adjustable; maximum 4000 Hz for 50 ns pulse; maximum 2000 Hz for 200 ns pulse

Peak power: 40 kW

Signal processing: RF-STC, C-FAR circuit, ISU

STATUS

In service in the Netherlands, Belgium, France, West Germany and Saudi Arabia.

CONTRACTOR

Hollandse Signaalapparaten BV, PO Box 42, 7550 GD Hengelo, Netherlands.

NORWAY

6819.153

NORWEGIAN ACQUISITION RADAR AND CONTROL SYSTEM

In order to meet the challenge of low-altitude and electronic warfare threats against ground forces, a joint venture between Hughes and the former Kongsberg Vapenfabrik is in operation. The result of this co-operation is a high performance, mobile Acquisition Radar and Control System (ARCS). This system represents further development and adaptation of the Hughes Low Altitude Surveillance Radar (LASR) (**4323.153**) and Kongsberg's multi-function console KMC-9000.

The ARCS system provides 3-D search, acquisition and tracking with an integrated IFF system; fully automatic threat evaluation and weapon assignment; centralised, decentralised and autonomous methods of operation; and automatic, semi-automatic and manual modes of operation. Advanced techniques for the rejection of ECM and radar clutter are also included.

The KMC-9000 is designed to meet requirements for military equipment and is based on systems already in use with the Norwegian naval and army forces. Entry **4573.481** gives information on one of these consoles which is in use as a submarine combat system. The basic concept of the KMC-9000 is modular with an intelligent high resolution graphic/alphanumeric terminal as the heart of the system.

The first major application of ARCS is in conjunction with the Improved Hawk SAM batteries at Norwegian Air Force Bases. This combination, known as NOAH (Norwegian Adapted Hawk) is intended to lengthen the operational life beyond the year 2000 (see entry **6820.181**).

STATUS

Several triad Hawk batteries have been delivered to the Royal Norwegian Air Force.

CONTRACTOR

Norsk Forsvarsteknologi A/S, 36000 Kongsberg, Norway.

6392.151

RADAR ELEVATOR SYSTEM

A complete silo system for effective protection of radar antennas has been developed by the Norwegian company Thune-Eureka. Based on retracting the antenna into a protective silo when necessary it enables the radar to be serviced in an environmentally protected area, as well as affording protection in the event of very high winds, bomb blasts or direct shell hits. The silo can be rock-excavated or can be a silo-tower built of concrete above ground. These silos can be used in a variety of applications from long-range air defence radars to air traffic control radars.

Components of the silo installation include a radar antenna elevator, hatches, maintenance platforms located on the internal silo walls, and a programmable controller. The silo elevator and hatches are automatically operated by hydraulical pump, motor and cylinder units. The silo can be operated locally or by remote control to ensure fast and automatic retraction and re-erection. Dimensions of the silo and a lifting capacity of up to 20,000 kg are matched to the particular radar system.

STATUS

In production. The HADR radar systems supplied by Hughes for Norwegian air defence are mounted on elevators (see entry **3721.153**).

CONTRACTOR

Thune-Eureka A/S, Defence Products Department, PO Box 38, N-3401 LIER, Norway (a member of the Kvaerner Group).

POLAND

2340.153

POLISH RADAR

Little information is available on Polish radar developments. It is known, however, that the Industrial Institute of Telecommunications in Warsaw has for many years been working on radar developments and some definite information has been made available on the radars that have been developed for civil ATC purposes.

As an illustration of radar capability, therefore, it may be of interest to note a few details concerning one of these radars, the Avia B D-band airways surveillance radar.

Avia B is a 23 cm wavelength radar using high-power magnetron transmitters and a low-noise parametric amplifier in the receiver. Operating in frequency diversity it has two 1.5 MW peak power transmitters. The parametric amplifier used in the receiver is an original Polish development. The radar has an MTI system, and staggered PRF is used to eliminate blind speeds in the practical range of aircraft performance. The polarisation of the transmitted radiation may be varied from linear to circular.

Maximum range on a 15 m² target with 90 per cent detection probability is 240 km with a ceiling of 26 000 m.

CHARACTERISTICS

Avia B

Operating frequency: 1310 and 1346.2 MHz

Peak power output: 2 × 1.5 MW (frequency diversity)

PRF: 400 Hz mean (recurrence 7: 8)

Pulse length: 3 µs

Noise factor: 3.5 dB

Fixed echo suppression: 30 dB

Antenna rotation speeds: 5 or 10 rpm

Antenna gain: 32 dB

Horizontal beamwidth: 1.3°

Vertical beamwidth: 40°

Polarisation: Variable from linear to circular

6811.153

GAP GATE EARLY WARNING RADAR

Gap Gate is a mobile early warning radar, probably operating in the D/E-band. It appears to consist of a large antenna, mounted on a Czech truck chassis, with the necessary control and data processing facilities contained in the truck. At each end of the antenna are small sub-antennas which are used to reduce sidelobes. One of these may also be used as an IFF interrogator. The truck has a large platform superimposed on the cab which is probably designed to act as a support for the antenna when it is folded down for transportation.

STATUS

In operational service.

Gap Gate early warning radar

SPAIN

5763.153

IRS-10 SECONDARY SURVEILLANCE RADAR

The IRS-10 is a Mark XII secondary surveillance radar (SSR) with Mark XII operating capability. The military version is known as the IRS-10M. The standard configuration of the IRS-10 is dual channel, including supervision units providing automatic changeover of the operational channel in the event of failure. The military version has the capability of generating Mark X SIF modes 1, 2, 3/A and C interrogations and Mark XII mode 4 when associated with a KIR.

The system consists of an antenna, two receiver-transmitters, two coder-extractors, a digital switching unit, an RF switching unit and an antenna control unit. All units other than the antenna are designed for mounting in a standard 19-inch rack. In order to operate the system from a site remotely located from the radar site, a remote control unit is available, and communicates via a telephone line. The CSL-10 antenna is an open array developed according to FAA specifications, and mounted on a pedestal. Three channels provide capabilities for ISLS and RSLS functions and make it suitable for monopulse processors.

CHARACTERISTICS

Frequency: 1030 ±3.5 MHz (transmit); 1090 ±5 MHz (receive)

Polarisation: Vertical

Transmitter peak power: 33 dBW minimum

Weights: Antenna – 350 kg; IRS-10 – 260 kg

STATUS

In military and civil operational service.

CONTRACTOR

CESEL S.A., (CESELSA), Paseo de la Castellana 143, Madrid 29046, Spain.

5764.153

COMPACT IFF/SECONDARY RADAR

The CESELSA COMPACT IFF/Secondary radar is a transportable system, operating at 1030 MHz, designed for a variety of applications, including anti-aircraft artillery, shipboard aircraft identification, helicopter operation and both military and civil air traffic control. The system is equipped with mode 4 for military use. The equipment is of small size and weight and is ruggedised for transport and handling under operational conditions. It provides decoding and processing capability up to 360 km with a power output of more than 230 watts.

STATUS

In production.

CONTRACTOR

CESEL S.A., (CESELSA), Paseo Castellana 143, Madrid 29046, Spain.

6125.193

IRS-20MP MONOPULSE SECONDARY SURVEILLANCE RADAR

The IRS-20MP monopulse secondary surveillance radar (SSR) has been developed for air defence and air traffic control applications. The system is an improved SSR that can be used in a stand-alone surveillance role or in association with a primary/search surveillance radar. It provides accurate plot data on each pulse of an airborne transponder reply. It also provides smoother tracking and better plot data from garbled and defective replies. Because of its inherent lower PRF, less interference (fruit) is generated, and substantially less load to airborne transponders, thereby reducing lock-ups. Its planar open array antenna utilises sum and difference radiation patterns developed to improve azimuthal accuracy. The vertical radiation pattern is designed with a bottom edge that reduces the amount of ground incident energy. No changes to transponders are necessary with the monopulse SSR.

Features of the IRS-20MP include improved accuracy with minimal interrogations, high efficiency with the large vertical aperture antenna, protection against reflections, an all solid-state modular interrogator, a high speed data bus and multiple microprocessors, improved sidelobe suppression, improved degarbling and phantom reply rejection, and programmable output formats.

CHARACTERISTICS

Sum pattern beamwidth: 2.35°

Antenna gain: 28 dB

Sidelobes: –27 dB

Interrogator peak power: 2 kW

GTC: Sector and range programmable

OBA accuracy: Less than 6 minutes of arc

Range: 250 nm

Target capacity: 400 at 7.5 rpm

STATUS

In production.

CONTRACTOR

CESEL S.A., Paseo de la Castellana, 143, 29046 Madrid, Spain.

IRS-20MP monopulse SSR

SWEDEN

1957.153

GIRAFFE SEARCH RADAR

Giraffe is the family designation of a mobile search pulse doppler radar system developed by Ericsson in Sweden. The basic version, PS-70/R, is designed for use with the new Swedish Army/Bofors short-range surface-to-air missile system RBS 70 (**2348.131**) or any short-range missile or AA gun system. The complete radar together with radio communication equipment, tactical control facility and power supply is linked with a number of firing units for precision target designation and combat control.

A prominent feature of this installation is the use of a folding mast for the antenna, giving an effective operating height of 12 m. Together with excellent clutter suppression characteristics (MTI improvement factor is better than 50 dB) this gives the Giraffe very low altitude coverage. The range coverage is 40 km in the surveillance mode and 20 km in the target designation mode.

The radar employs a broadband TWT final amplifier which delivers 15 kW peak, 300 W average, over the G/H-band. The high gain, low sidelobes antenna generates a cosecant-squared beam up to 60° with an azimuth beamwidth of 2°. Digital doppler processing and constant false alarm circuitry are used to automatically extract, detect and present the target of interest. A digital plan position indicator presents video signals and target tracking symbols. The target data link constitutes the interface between the search radar and the firing sites. Information about target geographical co-ordinates, speed and course is derived and transmitted from the target data transmitter, which is located in the search radar. The target data receiver, located at the firing site, performs certain calculations, generates signals for slewing the gun or missile sight towards the target and displays information as to whether the target is engageable or not.

The Giraffe search radar is offered with comprehensive ECCM options and an IFF equipment can be installed. Training equipment, available as an optional accessory, can easily be installed in the cabin without requiring any connections to external units.

The basic Giraffe radar parameters provide for more than adequate range and altitude coverage required by any AA artillery or short-range surface-to-air missile system. The adaptation to different short-range systems is achieved mainly by modifying the interfacing target data receiver. To extend the range coverage required for medium-range surface-to-air missile systems, a more powerful transmitter can be installed. This constitutes the only major modification of the basic design. The Giraffe concept in general and its MTI performance in particular, suit the system for handling the low altitude air surveillance task in early warning and air command systems.

Naval versions (**4299.253**) for installation on board fast patrol boats and up to frigate size ships are available. They combine air surveillance with superior MTI performance, surface surveillance and surface fire control with high resolution capability in one radar only. The MTI improvement factor gives naval versions excellent sea skimmer detection capability.

CHARACTERISTICS

Radar

Frequency: G/H-band (former C-band)

Output power: 15 kW (peak), 300 W (average)

MTI improvement factor: 50 dB

Pulse length: 3 or 6 μs

Range coverage: 20 or 40 km

Target velocity range: 30 – 375 m/s

Antenna revolution rate: 60 rpm

Antenna gain: 29 dB

Beamwidth azimuth: 2°

Elevation pattern: Cosec^2 to 60°

PRF: Either staggered or switched every antenna revolution

Mechanical

Total weight: 6500 kg (excluding vehicle)

Transport width: 2.5 m

Transport height (on vehicle): 3.4 m

Antenna operating height: 12 m

Interior volume: 12 m^3

Interior height: 1.7 m

STATUS

The PS-70/R model has been ordered by the Swedish Defence Materiel Administration for use with the RBS 70 AA missile system and deliveries have commenced. In addition, the Swedish Defence Materiel Administration has ordered the more powerful version of the Giraffe, PS-707/R, for a medium-range surface-to-air missile system, and this is now in production. An AA artillery version has also been ordered and is in production for a number of foreign governments. Yet another version for a different short-range surface-to-air missile is in production. A shipborne version of the Giraffe radar (**4299.253**) has been ordered by the Swedish Government for fitting in 'Spica II' fast patrol boats and by Canada for fitment in the new Canadian Patrol Frigate. A contract worth $9.5 million has been received from the Norwegian Army for the basic Giraffe. Tests on the Norwegian Army Low Level Air Defence System (NALLADS), based on a vehicle-mounted Giraffe were in progress during 1988. The radar is already in service in Norway under previous contracts. More than 250 Giraffe systems have been ordered since 1978 and the radar is in service in 13 countries outside Sweden.

Ericsson Giraffe tactical radar deployed and ready for use

CONTRACTOR

Ericsson Radar Electronics AB, Defence Systems Division, 431 26 Mölndal, Sweden.

2357.153

UAR 1021 COMBINED SEARCH AND TRACK PULSE-DOPPLER RADAR

This combined fully coherent search and track pulse-doppler radar is a successor to the earlier ECSTRA I equipment. It employs a single I-band transmitter which feeds the independent search and track antenna by way of a power splitter. The tracking sub-system uses a monopulse technique with a cassegrain antenna; the search system uses a cheese antenna or a cosec^2 antenna with integrated IFF dipoles.

Important features of the equipment are the high clutter attenuation resulting from the coherent pulse-doppler system; the ECCM capability resulting from the broad-band system (900 MHz), and the possibility of operating either on a fixed frequency (chosen from five with a change time of less than ten milliseconds) in the pulse-doppler mode or with random pulse-to-pulse frequency agility within 900 MHz in the non-doppler mode; the monopulse angular tracking system; and the digital range tracking system.

CHARACTERISTICS

Frequency: 8.6 – 9.5 GHz
MTI improvement factor: 45 dB
Peak power: 26 kW
Noise factor: 12 dB
PRF: 3 different triplets in 4.8 – 8.1 kHz interval with rapid cyclic switching
Pulse width: 1.5/0.5 μs selectable
Tracking antenna gain: 36 dB
Tracking antenna diameter: 1 m
Range coverage: PD-mode 0.3 – 18 km; FA-mode 0.3 – 25 km
Acquisition interval: 1120 m
Velocity window: 22 – 385 m/s
Angle tracking accuracy: 1 mrad

STATUS

In production for the Contraves Skyguard system (**2377.151**). The related UAR 1302 equipment, a coherent-on-receive system, is an integral part of the Marconi S800 series of anti-aircraft radars (**2464.153** and **2466.153**) and has been in production since early 1975.

UAR 1021 combined search and track radar in Contraves Skyguard system

CONTRACTOR

Ericsson Radar Electronics AB, Defence Systems Division, 431 26 Mölndal, Sweden.

5771.153

EAGLE FIRE CONTROL RADAR

The Ericsson Eagle fire control radar is a 'silent' millimetric system intended for use in mobile ground and naval-based air defence systems. The equipment operates in the Ka frequency band (35 GHz), enabling tracking of low flying targets. The Eagle system has been designed with an extremely low radar signature which has been achieved by pulse compression, high antenna gain and almost no sidelobes, in combination with low output peak power. The radiation pattern and a new transmission technique is claimed to make it impossible for escort or stand-off jammers to degrade the radar performance.

The system consists of two units; antenna/transceiver and signal processor, with a total system weight of 130 kg. It is able to track two targets simultaneously, with a multi-path error of less than 0.4 mrad at 10 km, permitting fast lock-over between targets, air-to-surface missile alert or closed-loop fire control. A special operational mode which allows simultaneous pulse-to-pulse frequency agility and MTI is provided. Adaptive MTI provides for automatic wind compensation.

STATUS

In production. Ordered by three customers outside Sweden.

CONTRACTOR

Ericsson Radar Electronics AB, Defence Systems Division, 431 26 Mölndal, Sweden.

Eagle fire control radar

5772.153

HARD 3-D SEARCH RADAR

The Ericsson HARD (helicopter and aircraft radar detection) is a new 3-D local search and acquisition system for ground-based and naval air defence systems. It is a pulse doppler radar having a phase-controlled planar array antenna with electronic elevation scanning. Target data in elevation, azimuth and range suitable for missile and gun air defence systems is provided. A version of this radar will form part of the Swedish Army RBS 70M air defence system and will also include Giraffe search radars (**1957.153**).

The HARD radar is a 'silent radar', so-called because of low peak power and high antenna gain, together with low sidelobes and simultaneous MTI and frequency agility. The manufacturer states that it is not possible for a jammer to detect the radar outside the main lobe or for an anti-radar missile to get enough updating energy unless it is being tracked.

HARD consists of two units; an antenna/transceiver and a signal processor with a total system weight of 110 kg. The transmitter is completely solid-state and is based on pulse compression techniques with a multi-frequency capability. Threat evaluation and separation of helicopter and aircraft/missile targets are carried out automatically. The system also has automatic detection and tracking for several targets simultaneously.

The system is able to operate on a moving vehicle with automatic compensation for any platform movements.

STATUS

In production.

CONTRACTOR

Ericsson Radar Electronics AB, Defence Systems Division, 431 26 Mölndal, Sweden.

HARD 3-D local search and acquisition radar mounted on Bv 206 tracked vehicle

---

4029.153

9KR 400 COASTAL RADAR

The 9KR 400 consists of a 9GR 600 I-band frequency-agile transceiver (**1546.253**), a 9GA 205 antenna and a variable number of displays and/or automatic plot extractors, permitting narrow-band transmissions. It forms the main sensors element of the 9KA 400 Mareld coastal artillery fire control system (**1548.181**) and the 9CSI 600 STINA naval command system (**4028.281**).

The transceiver is completely solid-state with the exception of the magnetrons. The wide-band frequency agility capability offers very high jamming resistance, very long range, and good tracking accuracy. Alternatively, the radar can be run in a fixed-frequency mode for MTI operation.

Other applications are shipborne or ground-based, and further entries can be found in the appropriate sections of this volume.

CHARACTERISTICS
Operating modes: Frequency agility/fixed frequency/MTI
Transmitter tubes (2 off): Magnetron VJ1180/YJ1181
Frequency agility range: Minimum 400 MHz, typically 450 MHz within 8700 – 9500 MHz
Frequency shift: Randomly within frequency agility range from pulse-to-pulse
Pulse peak power: 200 kW
Pulse width/PRF: Typically 1 μs/1000 Hz and 0.25 μs/3000 Hz
Receiver
Frequency range: 8500 – 9600 MHz
IF: 30 MHz
Noise figure: Typically 5.5 dB, max 7 dB with low-noise preamplifier
Receiver characteristics: Lin and log, log/lin, STC
Operational features: Image rejection mixer, Dicke-fix receiver, PLD, IAGC, narrow band jamming suppression, passive ECM mode
Transceiver dimensions: 1200 × 635 × 510 mm
Weight: Approx 225 kg
Antenna 9GA 205
Reflector type: Parabolic
Frequency range: 8.5 – 9.6 GHz
Rotation speed: Up to 40 rpm
Beamwidths
Horizontal: 0.6°
Vertical: 4°
Gain: 40 dB
Sidelobes
Within ± 10° of main lobe: –26 dB
Otherwise: –33 dB
Aperture: 4500 × 625 mm
Weight (including turntable): 260 kg
STATUS
In service. 9KR 400 radar systems form part of the coastal surveillance and defence network for the Yugoslav Navy, integrated with 9LV 200 shipborne tactical systems.

PEAB 9GA 205 antenna

CONTRACTOR
Philips Elektronikindustrier AB, 17588 Järfälla, Sweden.

5773.153
GA 250/K AND GA 250/X ANTENNAS
Philips Elektronikindustrier has designed two antenna systems known as the Types GA 250/K and 250/X. These antennas are designed for a variety of fire control radars, such as the 9KA 500 mobile coastal artillery fire control systems (**4318.281**).
CHARACTERISTICS

	GA 250/K	GA 250/X
Frequency	15.9 – 17.1 GHz	8.5 – 9.6 GHz
Beamwidths		
Horizontal	0.7°	1.2°
Vertical	2.2°	4°
Gain	42 dBi	38 dBi
Polarisation	Vertical	
Type	Parabolic, hornfed	

Reflector: Low drag, air porous cast aluminium
Rotation speed: Two speed ratio 1:2 in interval 6 – 60 rpm plus adjustable sector scan
Aperture: 2.3 × 0.55 m
Weight (including turntable): 160 kg
STATUS
In production.
CONTRACTOR
Philips Elektronikindustrier AB, 17588 Järfälla, Sweden.

GA 250 antenna

TAIWAN

6813.153
TIEN KUNG RADAR
A mobile phased array radar which probably provides target acquisition, tracking and mid-course guidance for the Tien Kung 1 (Sky Bow) missile system has been developed by Taiwan. The radar is carried in a vehicle-mounted carrier with four phased array antennas fitted to each of the four sides of a tower to give 360° all round coverage. The tower is retracted into the container for transport and only elevated into the operating position when the system is emplaced at a Tien Kung battery site. This new system, which has not yet received an official designation, is believed to fulfil much the same function as the AN/MPQ-53 of the US Patriot system, ie search, track-while-scan, simultaneous command guidance of multiple targets and, perhaps, target illumination for terminal guidance.

As yet it is not known whether this radar would provide target illumination for the semi-active version of Tien Kung 1. A missile fire control radar, the CS/MPG-25, has also been developed recently and could conceivably be slaved to the new phased array system to provide CW illumination on a 'time share' basis similar to the US Aegis system.
STATUS
The radar has been developed by the Sun Yat-sen Institute of Science and Technology and was shown at the 1987 celebrations marking the Taiwan National Day. No details of production status are available.
CONTRACTOR
Sun Yat-sen Institute of Science and Technology, Taiwan.

Mobile phased array for the Tien Kung surface-to-air missile

UNION OF SOVIET SOCIALIST REPUBLICS

NOTE
The Soviet Union has carried out a comprehensive upgrading of the Moscow anti-ballistic missile (ABM) system over the past few years. Much of the detail is contained in the Strategic Weapon Systems and Land-based Air Defence sections of this book, but brief details of the radar systems employed are given here to give readers some idea of the types employed.

The first air defence radar associated with the Moscow ABM was the Yo Yo (**2895.153**) which was used in conjunction with the SA-1 missile for anti-aircraft purposes. It became operational in the mid-1950s and lasted until the early 1960s when a series of long-range phased array radars, known as Hen House (**2879.253**) came into operation as part of a missile attack warning system. The first five of these were probably at Angarsk, Lyaki, Skundra, Pechora and Kamchatka. A number of another radar system, known as Square Pair, which is known to be large and powerful (**3290.153**), were constructed to the north-west of Moscow in the early 1960s, in association with SA-5 launchers.

The first real Moscow network, the ABM-1, was constructed in the mid-1960s with a Dog House radar (**2864.153**) positioned to the south of Moscow. Dog House is a medium-range system which received initial warning of incoming missiles from the long-range Hen House radars. Associated with Dog House was a complex of Galosh missiles, and a number of short-range radars, probably the Try Add close-range interception launch control radar. The other radars, Cat House and Chekhov, were also identified with this system, although Chekhov was probably an early name for a 'House' type radar.

A major improvement to the network become apparent in 1980, and has become known to NATO as ABM-3. A battle management phased array radar was built at Pushkino and dwarfed all other radars for sheer size. It is in the form of a chopped-off pyramid, 75 metres high and 150 metres square at the base, with four array faces to give 360° coverage. It is, in fact four times as large as the US Pave Paws antenna array (**3174.153**). Two other radars are also included in the ABM-3 upgrade, known as Flat Twin (**2821.153**) and Pawn Shop (**2822.153**). No detailed information is available on either of these radars except that the former is probably a twin phased-array tracking radar, and the latter is a missile guidance system (with three radomes?)

There are also major improvements to the distant Hen House early warning network. A new radar has been under construction at Pechora (see illustrations at the end of the USSR ground radar section), and nine of the radars are now in operation, at least five of which are adjacent to existing Hen House arrays. In addition the Soviet now has four OTH-B sites (**6135.153**) in operation or under construction.

Details of many Soviet radars are very sketchy and in some cases information is insufficient to justify an individual entry. However, a list of all known Soviet radars is given in the analysis section at the back of this book.

2860.153
BACK NET GCI RADAR
An early Soviet ground control interception (GCI) radar and a predecessor of Barlock (**2861.153**), the radar known to NATO as Back Net has a six-feed antenna system similar to that of Barlock. The antenna consists of back-to-back, parabolic sections, three feeds in each, forming three beams stacked in elevation.

Multiple feeds are widely used in Soviet surveillance and GCI radars. Part of the reason for this is the possibility of increasing the total radiated power within the confines of component limitations; but another reason is the possibility of using such antenna systems to provide some form of height information.

Other predecessors of Barlock include Big Bar A (**2862.153**) and Big Mesh (**2863.153**) both of which also have six-feed antenna systems.

2861.153
BARLOCK GCI RADAR
This large ground control interception and search radar is trailer mounted equipment that has two large truncated paraboloid reflectors with clipped corners.

Operating frequencies of the radar are in the E/F-band. Each of the reflectors is fed by three feeds so that six stacked beams result, each beam having a power output in the region of 1 MW. These beams may enable the radar to be used as a height indicator, but it is also likely that they are part of a system for providing surveillance cover over different altitude bands. Both the reflectors can be adjusted in elevation angle by the operator while the radar is operating to permit manual variations of vertical cover. Individual frequency bands for the stacked beams are:

Barlock GCI radar deployed

2695 – 2715 MHz
2715 – 2750 MHz
2815 – 2835 MHz
2900 – 2990 MHz
2990 – 3025 MHz
3080 – 3125 MHz

Known to the Soviet forces as P-50 this radar has a PRF of 375 pps and pulse widths of 1.83 to 3.1 microseconds for the E/F-band transmission and 2.4 microseconds for a D-band identification (IFF-type) secondary radar transmitter that is often found in association with Barlock.

Range capability of Barlock is about 300 km, range accuracy is about 90 m, and angular accuracy about 0.5°. The entire trailer body rotates for azimuth scanning at rates up to about 12 rpm.

STATUS
Known to have been in operational use since at least the mid-1960s and widely deployed in the USSR and friendly nations.

Inside view of Barlock operations vehicle, showing one of the plotting positions

2862.153
BIG BAR A GCI RADAR
A ground control interception radar operating in the E/F frequency band with a six-feed antenna system similar to that of Barlock (**2861.153**). See also entry for Back Net (**2860.153**) above.

A significant difference is that the two antennas are arranged to provide V-beams in elevation, thereby making possible heightfinding by means of this technique. The rear, upper antenna is displaced so that the major axis of the elliptical reflector is tilted about 25 to 30° from the horizontal.

4329.153
BIG FRED ARTILLERY/MORTAR LOCATING RADAR
The Soviet Union artillery/mortar locating radar (NATO reporting name Big Fred) is a mobile system normally fitted to armoured vehicles. Information on the equipment is scanty but it is believed to be similar to the Cymbeline system manufactured by EMI of the United Kingdom (**1018.153**). The radar is believed to operate in J-band. The antenna folds flat in the forward position for travelling.

OPERATION
The system enables the operator to plot two points in the shell or mortar bomb trajectory and measure the slant range and bearing to each of these positions. The time taken for the projectile to travel between the two points is also measured and the on-board computer uses this information, together with the pre-set elevation angles, to determine the position of the enemy mortar or howitzer. This information is then passed to the field artillery for counter-action.

STATUS
Big Fred is fitted to the multipurpose armoured vehicle MT-LB and the ACRV-2 (armoured command and reconnaissance vehicle) of the Soviet Army. The Yugoslav Army has recently taken delivery of a number of the former vehicles fitted with Big Fred.

The Soviet forces also use a BMP MICV fitted with the Small Fred radar. Small Fred is believed to operate in J-band with a detection range of 39 km and a tracking range of 7 km.

Soviet 'Big Fred' radar is used by Warsaw Pact and other friendly states as well as the USSR

2863.153
BIG MESH GCI RADAR
A ground control interception radar with a six-feed antenna system similar to that of Barlock (**2861.153**). See also entry for Back Net (**2860.153**). Like Big Bar A (**2862.153**), Big Mesh has the rear, upper antenna array tilted to about 25° from the horizontal to permit heightfinding by the V-beam method.

2864.153

DOG HOUSE ABM RADAR

One of several large fixed phased array radars deployed by the Soviet authorities for missile detection. Dog House is the NATO name given to a radar with a very large array antenna.

Complicated signal processing is used in the installations to which these large antenna arrays belong. It is believed that this is an area where Soviet technology is somewhat weaker than, say, that of the USA; and in support of this view it has been pointed out that such signal processing techniques are not much used in smaller Soviet radar installations and that this may be because they have decided to apply limited resources first and foremost to the solution of such critical problems as strategic missile defence.

The principal known use of the Dog House radar is as the intermediate range radar of the Moscow ABM complex. It accepts targets from the long-range Hen House radar (**2879.153**) and hands them on to the Try Adds radars. Dog House operates in the VHF band at around 100 MHz – hence the need for the very large phased array to achieve a sensible beamwidth. It is said to be capable of radiating a peak power up to 20 MW and to have a PRF of 50 pps – which is broadly consistent with its reported range of some 1500 nautical miles (about 2800 km).

Artist's impression of Soviet Dog House radar used for missile detection

STATUS

Operational in the Moscow ABM network.

2866.153

FAN SONG MISSILE CONTROL RADARS

Fan Song is the NATO code-name given to a family of compound radar installations that have been developed for use with the Guideline (SA-2) surface-to-air missile (**2942.131**).

Seven of these radars have so far been identified (by suffix letters A to G) and common to all of them are the detection and tracking of the target, the command guidance of the missile and the tracking of the missile. It is said that these radars can handle up to six targets at once and guide three missiles at a time. It appears, however, that the missile must pass through the guidance beam within a few seconds (reportedly six) of launch if it is to be acquired and steered towards the selected target. A further problem appears to be a limitation in the amount of steering information that the missile is capable of receiving from the ground station. Nevertheless, Soviet forces have kept Guideline (in various versions) in service for many years and it has been used in action successfully by the North Vietnamese. Although the system is probably expensive in operation, as a whole it would appear to be effective.

Operational deployment of Fan Song C

Details of some of the Fan Song series are given below. All members of the series contain as a major element a track-while-scan radar which scans a designated sector with two flapping fan beams radiated from two orthogonal antenna systems. The flapping motion has a sawtooth profile and uses the electromechanical system known as the Lewis scanner. The 'fan' in the code-name refers to these fan beams and the 'song' to the bird-like sound of the demodulated radiation from these radars.

Rear view of Fan Song radar group showing details of antenna feed arrangements

STATUS

Fan Songs of various kinds are in service with Guideline missiles in the USSR, satellite and client countries and are also deployed at Soviet air bases in Afghanistan.

2867.153

FAN SONG B MISSILE CONTROL RADAR

Fan Song B is an E/F-band member of the Fan Song series (see **2866.153** above). Radiation from its two Lewis scanners is at 2965 to 2990 MHz for one and 3025 to 3050 MHz for the other: peak power output is in the region of 600 kW and the equipment has a first-time-round range capability of between 60 and 120 km.

The equipment is trailer-mounted and the scanned sector can be changed by rotating the trailer and tilting the whole superstructure. The sector scanned is about 10° high and 10° wide, these dimensions being the approximate fan beam width, the beam width in the scanning direction being only about 2° for each beam.

A small parabolic antenna, mounted at one end of the horizontal Lewis scanner is used to transmit UHF command guidance signals to the missile.

2868.153

FAN SONG E AND F MISSILE CONTROL RADARS

Fan Song E is a G-band member of the Fan Song radar series (see **2866.153** above). Similar in many respects to Fan Song B (**2867.153**) its main differences are the frequency of the radiation from the Lewis scanners and the fan beam (and hence the sector) dimensions. The beams in Fan Song E are about 7.5° wide in the fan and about 1.5° wide in the scanning direction, carrier frequencies are 4910 to 4990 MHz and 5010 to 5090 MHz and peak power is about 1.5 MW. Unambiguous range is 70 to 145 km.

In addition to the antenna for the command guidance signals, which is similar to that of Fan Song B, Fan Song E has two further parabolic dishes mounted on top of the horizontal Lewis scanner. One of these is horizontally and the other vertically polarised and their purpose is to provide a lobe on receive only (LORO) feature. This is an ECCM technique in which the scanning action of the Lewis scanner is restricted to the receive channel by diverting the transmitted signal, which otherwise would be radiated from the scanner, into a dummy load. This signal is then replaced by a signal from one of the parabolic dishes (the other being for use with the other Lewis scanner) of sufficient power to operate the whole system.

The reason for adding the LORO facility is that by monitoring the Lewis scanner radiation an enemy pilot can tell roughly whether or not he is the prime focus of the attention of the ground forces. If he is, he can employ deceptive electronic countermeasures (DECM) to confuse the radar – such countermeasures typically involving an apparent shift of target angle.

STATUS

G-band Fan Song radars are in widespread use with Guideline missiles. A version that has been dubbed Fan Song F has been reported, identified by a box-like housing on top of the radar shelter. This is said to contain a visual observer's position (with or without optical aids is not known) to assist target and/or missile acquisition. The extent of operational deployment of this modification is not known.

2871.153

FIRE CAN FIRE CONTROL RADAR

This E-band trailer-mounted radar appears to have been derived from the American SCR-584 radar that was extensively used during the latter years of the Second World War and which was supplied to the Soviet forces by the USA. A typical war-time application of the SCR-584 radar was its use on the 'Diver' gun sites that were set up on the east coast of Britain to combat the V 1 attack. The radars then were used with a 3.7 inch (94 mm) AA gun and the electronic predictor. The Soviet radar appears to be used with smaller 57 mm and 85 mm guns.

Fire Can operates at a frequency of 2700 to 2900 MHz, has a pulse width of 0.3 to 0.8 microseconds and a PRF of 1840 to 1900 pps. Peak operating power is believed to be about 300 kW.

Like the SCR-584, the Fire Can antenna is a parabolic dish, perforated to reduce weight and windage, that is mounted on a pedestal on top of the flat roof of the radar trailer. Power is radiated from a rotating dipole feed driven from the rear of the dish and signals from the common T/R system are used, in the tracking mode, to operate a lock-follow system that produces tracking accuracies of about 1.6 minutes of arc in azimuth and 2 minutes in elevation.

The radar can also be operated in a search mode with the antenna rotating continuously in azimuth and nodding in elevation. In this mode the antenna, which has a diameter of about 150 cm, produces a pencil beam with an effective beamwidth of about 5°. In this mode all received signals are displayed on a PPI but the dipole feed is driven all the time so that the operator can switch rapidly from search to track.

It is believed that Fire Can can acquire targets (search mode) at a maximum range of about 80 km and track them from about 35 km. Tracking accuracy in range is said to be about 13 to 15 m.

STATUS

In service. It is reported that some 75 of these radars were at one time deployed for use in North Vietnam together with other versions of the basic system, such as Fire Wheel and Whiff.

2872.153

FIRE WHEEL FIRE CONTROL RADAR

Fire Wheel is the NATO code-name for another SCR-584-derived fire control radar similar to Fire Can and Whiff (**2871.153** and **2893.153**).

2873.153

FLAP WHEEL FIRE CONTROL RADAR

Flap Wheel is the NATO code-name for a conical scan radar which operates in J-band. The antenna is a parabolic dish, horizontally polarised and Yagi array. Very similar to Gun Dish (**2876.153**). It is used to provide fire control data for 57 mm and 130 mm AA guns.

2874.153

FLAT FACE TARGET ACQUISITION RADAR

Flat Face is the NATO code-name of a radar known in the USSR as the P-15. It is a vehicle mounted acquisition radar that is used in conjunction with the Low Blow missile control radar (**2884.153**) and the Goa surface-to-air missile (**2938.131**).

This radar operates in the UHF C-band and radiates its signals from two elliptical paraboloid reflectors each measuring about 11 × 5.5 m mounted one on top of the other. The radar has a range capability of about 250 km with a range accuracy of about 90 m and an angular accuracy of about 0.5°. Frequency bands are about 810 to 850 MHz and 880 to 950 MHz, PRFs are 200 to 800 pps and 600 to 680 pps. Vertical beamwidth is about 5° and horizontal about 2°. Peak power is about 500 kW.

A Flat Face radar was seized in Chad during 1987 and has been subjected to extensive tests at a French research centre. It has proved to be a 'sturdy system with a high level of resistance to countermeasures, and was very good at detecting airborne targets down to low altitudes'.

STATUS

Deployed generally with Goa missiles but also reported with tank and motorised infantry divisions. May also be in use with SA-8 Gecko missiles.

Soviet Flat Face mobile radar unit

2821.153

FLAT TWIN RADAR

Flat Twin is a tracking radar associated with a mobile anti-ballistic missile system and is probably designed for controlling interceptor missiles in their terminal phase. Little is known about this radar although it is probably a flat rectangular faced, phased array type. It is built in modular fashion to enable the various parts to be prefabricated and stored and erected when required. The associated containers or vans would be for receiver signal processing, display electronics and power generation. Various authorities have stated that the antenna would take one or two months to erect and it is assumed from that that the radar is relocatable rather than mobile.

STATUS

Systems have been kept at various test ranges in the USSR and most were believed to have been dismantled in early 1987.

2875.153

GAGE ACQUISITION RADARS

Gage is the NATO code-name for an acquisition radar using stacked elevation beams, that is used in conjunction with the Yo-Yo missile control radar (**2895.153**) which in turn is used with the early Soviet surface-to-air missile Guild (**2944.131**). Operating frequency is in the E/F band and peak power about 2 MW.

2876.153

GUN DISH FIRE CONTROL RADAR

This broad-band fire control radar is used in conjunction with the quadruple 23 mm AA gun mounting on the ZSU-23-4 SP AA vehicle (**5548.103**). This vehicle is used by the AA battalions of the armoured and mechanised divisions of the Red Army and of other Warsaw Pact armies. The NATO code-name Gun Dish is obviously prompted by the appearance of the ZSU-23-4 vehicle.

Designed as it is to oppose low-level aircraft attacks the system of which Gun Dish is a part has to have a short reaction time. The guns can be slaved to the radar in elevation and are mounted, in the ZSU-23-4, in a rotatable turret on which the radar is mounted also and which contains a fire-control computer. A more recent application of this radar that has been reported is its installation on BRDM-2A AFVs mounting SA-9, Gaskin surface-to-air missiles (**3072.131**).

Gun Dish fire control radar with 23 mm AA guns on ZSU-23-4 SP AA vehicle (Novosti)

The Gun Dish radar operates in J-band with a frequency range of 14.6 to 15.6 GHz. The panoramic search radius extends up to 50 kilometres and the target tracking range up to eight kilometres. Antenna diameter is one metre. The radar is normally deployed in batteries of 32 deployed on the front line of the forward edge of battle area.

2878.153

HEIGHTFINDING RADARS

Several basically similar nodding heightfinder radars have been identified and named by NATO. The generic code-name is Cake, and variants on this basic concept are known by such names as Patty Cake, Rock Cake, and Sponge Cake.

These radars, with their large, peel-shaped nodding antennas typically radiate beams with a width of some 3° to 5° horizontal and 1° to 5° in the vertical plane. Nodding frequency is typically about 30 to 40 cycles/minute and range is upwards of 200 km. One such radar, Sponge Cake, has a range in the region of 300 km, with a range accuracy of about 2.5 km at 200 km.

All operate at frequencies in D/E band and have peak powers around 3 MW.

STATUS

Operational.

2877.153

HEN SERIES EARLY WARNING RADARS

Several exceptionally large early warning radars observed by American reconnaissance have been given a series of names containing the word 'Hen'. First to be so named was Hen House (**2879.153**): others whose names have been reported are Hen Egg, Hen Nest and Hen Roost. Few details have been received concerning these radars: the notes below summarise the available information.

Hen Egg

Operating frequency around 2 GHz. Peak power around 3 MW. Pulse width 5 to 15 microseconds and PRF up to 300 pps. These figures suggest the high mean power of about 5 kW per beam.

Hen House

The large antenna array measures 300 × 20 metres and is inclined at 45°. Thought to have an early warning and tracking role in the Soviet ABM system. Probable detection range 3200 nautical miles. (See **2879.153**, below.)

Hen Nest

Operating frequency about 800 MHz. Peak power about 3.5 MW.

Hen Roost

Operating frequency is between 300 and 500 MHz. Peak power similarly believed to be about 5 MW. 2 × 2° beam.

STATUS

All these Hen radars – and probably several others – are believed to be operational although it is quite possible that some of them are only experimental designs (albeit used operationally) that may well not be repeated.

2879.153

HEN HOUSE ABM RADAR

This is the name given by the US authorities to a very large Soviet array that was first detected in the late 1950s and was at that time thought – if indeed it was a radar – to be part of a detection tracking and control network for satellites.

Subsequently, however, it was discovered that these enormous radars perform an important early warning and tracking function in the Soviet ABM weapons system, providing radar coverage comparable with that planned for the US Safeguard (**2798.131** in 1977 and earlier editions) system. No details of the radar are known other than that the antenna array is exceptionally large – perhaps as much as 300 m long and 20 m high. This billboard array is inclined at about 45° from the vertical.

It is reported, however, that the performance of Hen House is comparable to that of the FPS-50 detection radars (**2511.153**) of the American BMEWS system (**2525.181**) – ie a detection range in the region of 3200 nautical miles (about 6000 km) and a track-while-scan capability. It is believed to operate on about 150 MHz and has been reported as having a variety of PRFs from 25 to 100 pps, variable pulse widths, a complicated beam scan pattern with two beams scanning in azimuth, two scanning in elevation and one scanning in a circular pattern. Peak power is believed to exceed 10 MW.

A major improvement to the Hen House network appears to have been made with the construction of nine LPAR systems, most of which are adjacent to Hen House arrays.

STATUS

Hen House radars have been reported as being operationally deployed near Irkutsk, near the Barents Sea and in Latvia not far from the Baltic. See also general notes on the 'Hen' series (**2877.153**).

Artist's impression of the Hen House ballistic early warning radar

---

2880.153

KNIFE REST A EARLY WARNING RADAR

Knife Rest radars, as they are known to NATO, have the Soviet designation P-10. They were the forerunners of the Spoon Rest radars. The array consists of four Yagi antennas in two crossed-brace supported pairs one above the other, each pair mounted on each side of a single tubular support carried on the turning gear. The whole unit is fixed to the top of a tubular mast, braced by guy-ropes.

Knife Rest A has an operating frequency band of 70 to 73 MHz, radiates about 100 kW peak power, has a pulse width somewhere between 4 and 12 microseconds and horizontal and vertical beam-widths in the region of 20 to 25°. Range capability is somewhere around 350 km.

STATUS

Operational – certainly in Vietnam – although the more modern Spoon Rest was deployed in the same theatres.

2881.153

KNIFE REST B AND C EARLY WARNING RADARS

General characteristics of these early warning radars are similar to those of Knife Rest A (**2880.153**) except that the range capability is lower at only about 90 km. Knife Rest B is shown to operate in the frequency band 83 to 93 MHz. Another difference is that Knife Rest B is a vehicle-mounted system, with the Yagi array antenna carried on a rectangular cross-braced sectional mast instead of the tubular pole used in Knife Rest A.

STATUS

Knife Rest B was certainly operational in Vietnam.

3294.153

LONG TALK RECOVERY RADAR

Long Talk is the air search and surveillance radar element of the Soviet standard ground controlled approach and aircraft recovery system used at Soviet air bases. The other elements of the system are the precision azimuth and elevation tracking radars known to NATO as Two Spot (**3295.153**) that form the precision approach radar portion. In some installations all three radars are mounted on one long four-wheeled trailer vehicle, which may also house the two- or three-man operating crew, but the Two Spot installation has been observed mounted on a platform above the driving cab of a standard military truck.

The Long Talk antenna has a moderately large elliptical paraboloid reflector, illuminated by radiating elements carried on a pyramid-shaped strut built up of four tubes. At the horizontal extremities of the reflector, one or two vertically positioned supplementary antennas which could be IFF units are located. Some installations have one such item on each end of the primary radar reflector, others have one only. A radar similar in general appearance to Long Talk and with one vertical supplementary attached to it is reported under the code-name One Eye.

A number of Long Talk radars have been deployed at Soviet air bases in Afghanistan, associated with Two Spot precision approach radar (**3295.153**).

Checking siting of Long Talk air search radar at a Soviet installation. Note associated heightfinder and approach radars

2937.153

LONG TRACK SURVEILLANCE RADAR

Long Track is the NATO code-name for a tactical surveillance radar used with surface-to-air weapon systems as an early warning/putter-on.

Long Track is commonly associated with the Ganef missile system (SA-4 – **2934.131**). It is also associated with the Gainful, SA-6 (**2930.131**) and Gecko SA-8 (**3209.131**) systems. In all cases the radar is mounted on a tracked carrier vehicle. It operates in E-band with a range and altitude in excess of 150 km and 30 000 metres respectively.

STATUS

Operational, certainly with SA-4 missile and possibly elsewhere. Used for SA-4, in conjunction with the Pat Hand fire control radar (**2936.153**).

Soviet 'Long Track' radar is used by client states as well as USSR, this one serves with Yugoslav forces

2884.153

LOW BLOW TRACKING AND MISSILE CONTROL RADAR

Low Blow is the NATO code-name given to a family of I-band radars used with land-based Goa (SA-3) missiles (**2938.131**). These radars are not the same as those used with shipborne Goa missiles which are known as Peel Group radars (**1323.253**). The name Low Blow reflects the ability of the radar to guide the missile towards low-flying targets through heavy clutter.

Like the Fan Song series (**2866.153**) the Low Blow radars use pairs of electromechanically scanning trough antennas mounted orthogonally, but to improve low-angle performance the troughs are mounted at 45° from the horizontal. It appears, too, that the troughs are not Lewis scanners but a form of organ-pipe scanner.

Carrier frequencies of the Low Blow family lie in the 9000 to 9400 MHz band, PRF between 1750 and 3500 pps with unambiguous range correspondingly lying between 40 and 85 km. Radiated beamwidth is in the region of 12° in the fan and 1.5° in the direction of scan. Pulses are between 0.25 and 0.5 microseconds in duration and peak power output is around 250 kW.

STATUS

Operational with Goa missiles. Commonly associated with Flat Face (**2874.153**) or Squint Eye (**2891.153**) acquisition radars.

3296.153

NYSA C LONG-RANGE RADAR

Nysa C is a long-range, early warning air defence radar, according to East German sources, being used in Warsaw Pact states as a part of the overall air defence system to provide target information for interceptor aircraft and surface-to-air missiles.

The antenna array consists of two large parabolic cylinder reflectors mounted one above the other, both facing the same way, carried on a large trailer cabin. The two reflectors are presumably to provide separate high and low cover radiation patterns, and each is fed by a horizontal row of 26 dipole elements. The operating frequency may be in the VHF or UHF band and high power levels are probably employed.

Nysa C long-range early warning radar operated by the Polish air defence forces

2936.153

PAT HAND FIRE CONTROL RADAR

Little information is available on this radar which is used in conjunction with the SA-4 (Ganef – **2934.131**) surface-to-air missile system.

Pat Hand is mounted on a tracked vehicle, and operates in the H-band at 6.44 to 6.88 GHz. Its main tracking beam monitors the target aircraft while a guidance beam from a small antenna on the upper left hand side of the array provides missile guidance. Pat Hand can engage only a single target at one time, but can control up to two missiles against this target. Range is 55-70 km. Each SA-4 Ganef brigade consists of three air defence battalions, each with nine launcher vehicles and three Pat Hand vehicles.

STATUS

Operational with the Air Defence divisions of the Soviet ground forces. Up to the discontinuation of production in the 1970s, about 560 Pat Hands had been delivered with a further 45 to other Warsaw Pact countries.

6822.153

PAWN SHOP RADAR

Pawn Shop is a missile guidance radar which has been in existence since the early 1970s. By virtue of its name it probably consists of three spherical domes to protect the antennas. It is understood to be a relocatable system which is used as a tracking and control radar for anti-ballistic missiles. As with the Flat Twin system, Pawn Shop is prefabricated in sections and erected on site when required. Erection time is of the order of one or two months.

STATUS

The present status is unknown but some of these radars, which were erected at test ranges, have now been dismantled.

4328.153

PORK TROUGH (SNAR-2) COUNTER-BATTERY RADAR

Pork Trough (SNAR-2) is a counter-battery radar designed to locate enemy artillery in order to provide position information to its own guns for counteraction. It operates in J-band and uses a small parabolic cylinder antenna. The antenna and operations cabin are mounted on an AT-LM light tracked artillery tractor. No other information is currently available.

Soviet Pork Trough weapon locator radar is a highly mobile self-contained unit

2974.153

SCORE BOARD IFF RADAR

Score board is the NATO code-name for an IFF-type ground interrogator used by Soviet forces. Presumably very early equipment, it has an antenna array consisting of two horizontal rows of two radiators per element, the two double rows being mounted one above the other to give a four-array of what appears (in some poorly reproduced pictures) to be a total of 32 dipole radiators. The whole is mounted on a rectangular rotatable framework.

STATUS

Apparently operational.

3289.153

SIDE NET HEIGHTFINDER

Side Net is a large transportable nodding heightfinder radar, usually employed in conjunction with early warning surveillance radars such as Barlock, Back Net, and Tall King to provide GCI facilities. Operating frequency is in the range 2560 to 2710 MHz with an effective range of about 180 km at heights up to 32 000 m. The large (8.5 × 3.5 m) narrow elevation elliptical paraboloid antenna is mounted on a pair of lattice trunnions cantilevered from the sides of the cabin containing the associated electronics. The cabin can be rotated in azimuth.

Developed in the early 1960s, Side Net has been exported to countries outside the Soviet Union that have been supplied with Goa and Guideline surface-to-air missile systems.

3290.153

SQUARE PAIR GUIDANCE RADAR

Square Pair is the NATO code-name of a target-tracking and missile guidance radar associated with the SA-5 Gammon long-range air defence missile system (**2940.131**). It is understood to be normally collocated with Back Net (**2860.153**) heightfinder radars, and to be widely deployed within the Soviet Union. No technical or physical characteristics have been obtained to date, although it appears likely to be a dual structure phased array radar. Square Pair is thought to have been operational since the early 1960s.

3291.153

SQUAT EYE ACQUISITION RADAR

Squat Eye is an air search and target acquisition radar, sometimes used instead of Flat Face (**2874.153**), and with Low Blow (**2884.153**) for Goa SA-3 (**2938.131**) air defence missile batteries. Such applications appear to be those where low altitude coverage is of particular importance.

The antenna is of relatively large dimensions and of elliptical paraboloid configuration. It is usually mounted on top of a slender rectangular lattice tower of sectional construction to increase the effective operating height. About 500 kW transmitter power is likely and the radar's role may be essentially that of a gap-filler. A maximum range of about 200 km has been estimated. Squat Eye is reported to employ the same electronics as the Flat Face radar (**2874.153**).

Squat Eye acquisition radar

2891.153

SQUINT EYE ACQUISITION RADAR

This air target acquisition radar is used in place of Flat Face (**2874.153**) and in conjunction with the Low Blow missile control radar (**2884.153**) and the Goa surface-to-air missile (**2938.131**) in circumstances in which improved altitude cover is required.

STATUS

Operational.

2885.153

STRAIGHT FLUSH FIRE CONTROL RADAR

Straight Flush is the NATO name given to the radar and command guidance system used with the SA-6 Gainful (**2930.131**) surface-to-air missile system and possibly with the SA-11 (**4470.131**). It is believed that the system performs the following functions —

Limited search
Low-altitude detection/acquisition
Target tracking and illumination
Missile radar command guidance
Secondary radar missile tracking

Reports of the frequencies used for these functions have not been entirely consistent. It seems, however, that the first three functions are performed in G/H-band, the low altitude function being performed at about 5 GHz and the high-altitude functions at about 6 GHz. Target tracking (and probably illumination) is an I-band function at around 8 GHz and so probably is the command link, while the secondary radar response for missile tracking is probably at a rather lower frequency. Range is believed to be 60 to 90 km, with a maximum altitude coverage of 1000 metres.

The arrangement of the system can be seen in the accompanying picture. The upper, tracking, radar antenna assembly is assumed to be able to rotate independently of the lower antenna and it is further assumed that the two antennas and associated apparatus can be rotated relative to the carrying vehicle on a turntable which is presumably located at the top of the circular turret on which the whole assembly is mounted. Examination of the original photographs suggests that the lower antenna may be pivoted so as to execute a sector scan of some kind and there is some indication that it can be tilted up or down, presumably to compensate for vehicle attitude. These observations are consistent with the assumed functions listed above; the combined radar superstructure is probably too massive for any kind of continuous circular search process and the arrangement probably provides for a slow circular search into an associated sector scan, the circular motion being halted when a target is located and resumed only if the target starts to move out of the sector.

The feed arrangements of both antennas are interesting. The lower antenna appears to have two feeds (which can be seen more clearly in the original picture) the upper one consisting of a single horn, which presumably produces a low angle pencil beam, and the lower feed comprising two or possibly three horns which may well produce a slightly shaped high-cover beam. There could be more sophistication to it than this but if so it is not obvious or obviously likely. The use of two separate feeds, however, is consistent with the suggestion that the low- and high-altitude patterns are radiated at different frequencies.

There is, however, some additional waveguide in this feed which is not explained by what has been said above. It appears to be for a higher radiation frequency (than G/H-band) and could be for one of the other functions (command link or secondary radar – more probably the former). Finally it looks as though both parts of the main feed have a quarter-wave-plate circulariser in front of them. If this is a correct reading of the photographs it tends to date the technology of the equipment: a radar of this sort built in the West in the late 1960s would have used waveguide circularisers. It also looks as though the third (higher-frequency) feed bypasses the plate – which makes sense.

Turning now to the upper antenna, the function of the tripod-mounted device projecting in front of the dish is not clear. The remainder of the assembly is fairly evidently a conical scanning system using a rotating feed driven by a motor in the housing at the rear of the dish. This housing also almost certainly contains the microwave transmitter/receiver stages to avoid the need for multiple rotating waveguide joints.

Two of the struts supporting the projecting device appear to be simple metal rods or tubes secured to the face of the dish: the third (bottom) strut, however, is different, fairly certainly passing through a hole in the dish, and could be a waveguide. The way in which the upper struts meet the 'device' seems a little clumsy if they are no more than supports however, and while it may be that the 'device' is yet another antenna of some sort, we incline to the view that it is either another quarter-wave plate or a beam-spoiler for some operational purpose and that the bottom strut is merely a kind of push rod (hence the need to go through a hole – a waveguide would not) for moving the 'device' into and out of the way.

Apart from the microwave stages of the tracking radar it seems probable that the bulk of the radar electronics is housed in the bin-shaped structure beneath the tracking radar pedestal. The turret may be assumed to house the displays and control gear.

STATUS

Operational. Radars of this type have been used effectively in various Middle East conflicts.

Straight Flush fire control radar. Note pairs of AFV-mounted SA-3 (Goa) missiles in the background (Sava Haery/SIPA Press)

2889.153

SPOON REST A EARLY WARNING RADAR

Known in the USSR as the P-12, the radar known to NATO as Spoon Rest A is an early warning radar that is used sometimes in conjunction with the Fan Song radars (**2866.153**) and the Guideline surface-to-air missile (**2942.131**).

This is a VHF radar working in the 147 to 161 MHz, A-band and radiating at about 350 kW peak from a Yagi array. Range capability is up to about 275 km and antenna beamwidths are around 2.5° in the vertical and something over 1° in the horizontal plane.

The radar is a heavy one and is mobile in two vehicles, one carrying the generator and the other the antenna array and the radar cabinets.

STATUS

Operational, certainly in Vietnam, where 34 sets are believed to have been used operationally, and in Egypt where one was captured by the Israeli forces.

Deployment of Spoon Rest A (P-12) early warning radar

3293.153

TALL KING EARLY WARNING RADAR

Tall King is the name given by NATO to a large early warning radar P-14 for air defence of the Soviet Union and Warsaw Pact states. The origin of this name can be guessed quite easily from the impressive appearance of the structure which incorporates a reflector estimated to measure about 15 m high by 25 to 30 m wide, surmounted by a pylon of perhaps another five metres or so to which are attached the guy-ropes needed to steady this large array.

The method of illuminating the antenna is not clear but operation at metric wavelengths, probably between 150 MHz and 180 MHz, is almost certain. The effective range is probably at least 500 km or 600 km with an altitude range of at least 150 000 feet. Tall King is reported to have been introduced in the 1950s to provide early warning against high-altitude intruding aircraft. Although obviously a fixed-site radar, sectional construction is employed so that, although a lengthy undertaking, Tall King is capable of transportation and re-siting without undue difficulties.

It is believed that Tall King radars have been deployed in Afghanistan to survey activities on the borders with China and Iran.

Side Net (**3289.153**) transportable heightfinder radars have been associated with Tall King installations.

Soviet Tall King early warning radar

Close-up of Tall King early warning radar showing technicians carrying out either final states of erection or routine maintenance

2892.153

TOKEN EARLY WARNING AND GCI RADAR

This E/F-band early warning and ground control interception (GCI) radar is similar to the US CPS-6B radar (the first V-beam radar). Like Barlock (**2861.153**) it has twin truncated parabolic mesh reflectors. Five beams are produced and in the GCI role it has a range of about 150 km and can measure to within 0.5° in azimuth and 300 to 1500 m in elevation. In the early warning role it has a range of 250 to 300 km and a range accuracy of about 1 km.

3292.153

THIN SKIN HEIGHTFINDER RADAR

Thin Skin is the NATO name for an H-band nodding heightfinder radar, generally deployed as part of a ground control interception formation. Both trailer and truck-mounted versions have been observed and its operational role is probably in support of tactical air defence units. In all cases it will be associated with air search and surveillance radars to provide them with complementary height information on targets detected. This radar is reported to be associated with Soviet Ganef, Gainful and Gecko surface-to-air missile regiments. Range is approximately 240 km.

STATUS

First reported in 1965, Thin Skin has since been supplied to nations outside the USSR, both within and beyond the Warsaw Pact states.

Thin Skin heightfinder radar

3295.153

TWO SPOT PAR

Two Spot is the NATO code-name of the radar group employed by Soviet air forces for use in precision approach radar (PAR) aircraft recovery and landing guidance. It is usually associated with the Long Talk air search radar (**3294.153**) which provides aircraft with ATC and homing directions to the airfield, and positioning for the PAR 'talk-down' landing itself. In some installations the Two Spot and Long Talk systems are both mounted on the same trailer vehicle, but Two Spot has also been observed mounted on a platform above the driving cab of a truck.

There are two antennas, both of parabolic section, one for elevation tracking and the other for azimuth. They are carried on a combined mounting, which permits limited rotation of the whole assembly to enable the system to be aligned to the bearing of the runway in use. The antennas and other items can be folded for transit. The operating frequency is in the I/J-band.

Two prominent discone VHF or UHF communications antennas for the ground/air link used for talk-down are carried on a tubular mast, and these may be the origin of the NATO code-name.

Several Two Spot PAR systems have been deployed at Soviet air bases in Afghanistan, associated with Long Talk radars (**3294.153**).

Vehicle-mounted version of Two Spot PAR

2893.153
WHIFF FIRE CONTROL RADAR

This is a van-mounted E/F-band gun fire control radar. Like Fire Can (**2871.153**) it is derived from the American SCR-584 equipment that was supplied to the USSR during the Second World War.

Construction and performance are generally similar to those of Fire Can. The antenna gain of Whiff, however, is somewhat greater.

2895.153
YO-YO MISSILE CONTROL RADAR

An early missile control radar, Yo-Yo is used to control the Guild (**2944.131**) surface-to-air missile (SA-1).

It is reported that this radar can track more than 30 targets at a time. Flapping beam techniques are used for target tracking and the radar uses six rotating antennas to cover a scanning area of some 70° both in azimuth and in elevation. Operating frequency is around 3 GHz and peak power around 2 MW.

4602.153
PSNR-1 BATTLEFIELD RADAR

The PSNR-1 is a portable battlefield radar used in reconnaissance companies of Soviet tank and motorised rifle regiments. It operates in the 9.6 GHz band with a maximum range against vehicles of 10 km and 5 km for personnel. The equipment is tripod mounted, can be carried by three men and takes approximately 15 minutes to prepare for operation. Divisional anti-tank battalions are believed each to have three systems for target acquisition purposes.

6135.153
SOVIET OTH RADAR

The Soviet Union has been engaged in the development and construction of over-the-horizon (OTH) radar for some years, and had its first system operational in the early 1980s. During the early part of 1986 it was reported that an OTH radar was under construction near Nakhodka, in the coastal area of the Sea of Japan. This radar is facing the Pacific Ocean to watch the movement of aircraft, ships and strategic missiles in the area between the Chinese coast and Guam Island. It is the fourth Soviet OTH radar to be constructed, and is assumed to be of the ionospheric backscatter type, similar to the United States CONUS OTH-B system described under **5617.153**. The four systems are sited at Minsk, Nikolayev, Komosolk-na-Amure, and Nakhoda. Performance and capability of the system is unknown but, assuming that the same techniques are used as CONUS OTH-B, the radars will cover the area from about 500 to 1800 nautical miles from the transmitters. Technical characteristics of the system are unknown, although the rapid clicking of 10.5 pps pulses from these radars have earned the nickname of 'Russian Woodpecker'.

In addition to the Dog House and Hen House radars described briefly in this section, the Soviet Union has recently constructed a number of other ballistic missile detection radars. Details of these are virtually non-existent, although the accompanying artist's impressions give some ideas of the installations at Pechora and Krasnoyarsk.

According to United States sources, the Soviet Union is engaged in the construction of three new phased array ballistic warning radars on the periphery of the USSR. One of the new installations is believed to be at Baranovchi near the Polish border. Previously it was believed that five new phased arrays were being built, but sources now believe that nine are engaged in the upgrade of the ballistic missile defences, including the Krasnoyarsk site 3700 km from Moscow.

The new phased array radars duplicate or supplement the eleven existing Hen House radars at six locations on the USSR periphery. They are believed to be able to track more attacking missiles with greater accuracy than the existing network covering almost all approaches to the Soviet Union. They will check any 30 minute warning of missile launch detected by the Cosmos early warning satellites and the OTH radars directed at the US ICBM fields.

Western analysts believe that the new radar sites will allow the Soviet Union to move quickly to the establishment of a national ABM defence system, rather than one based on Moscow, as at present. It is estimated by the USA that the Soviet Union has spent about $150 billion on strategic defence in the last 10 years.

This diagram gives the coverage of ballistic missile detection and tracking systems from the Soviet Union, using early warning radars, over-the-horizon radars and launch detection satellites

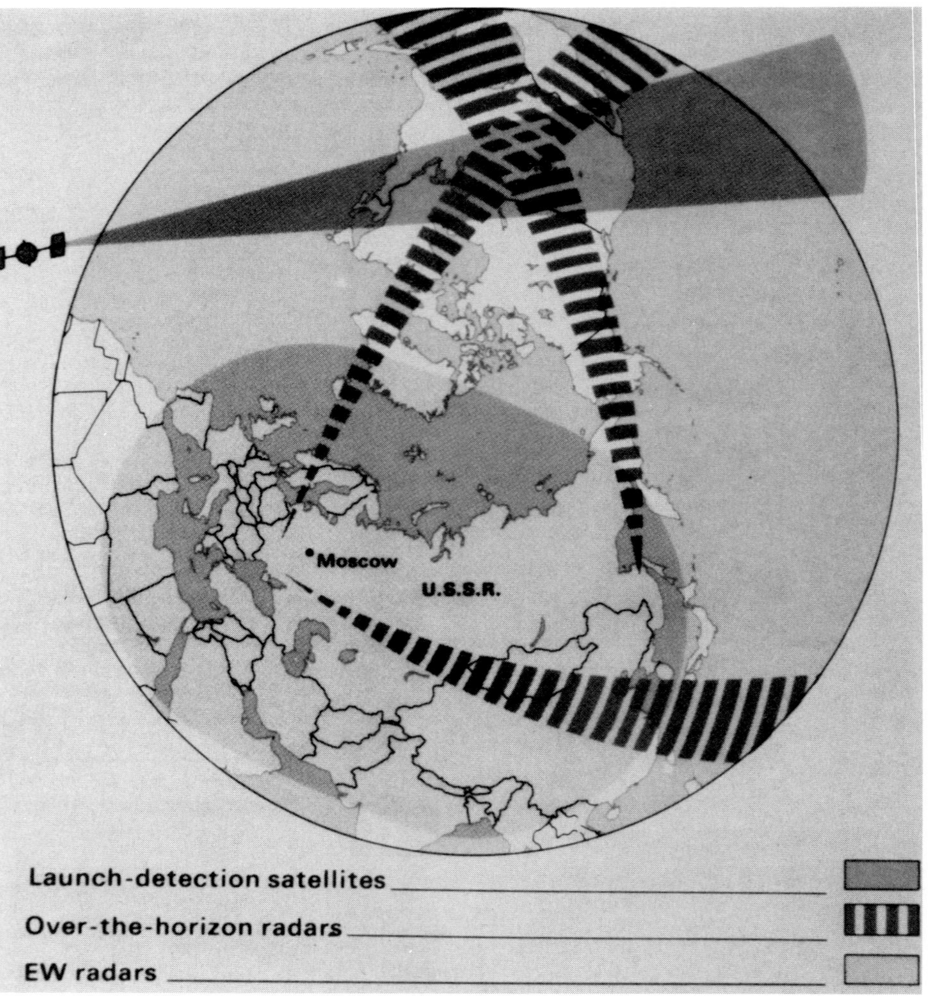

UNITED KINGDOM

1018.153

CYMBELINE WEAPON LOCATING RADAR

Cymbeline is a lightweight, highly mobile self-contained radar (including power supply) with a detachable display unit. The radar is mounted on a four-legged structure supported on screw jacks fitted with hydraulic absorbers. It is produced in three versions of which Mk 1 and 3 are towed on a special two-wheel trailer while Mk 2 is mounted on any suitable AFV or soft-skinned vehicle. The Mk 3 can also be vehicle mounted.

The antenna feed system in Mk 1 and Mk 2 is a mechanical Foster scanner and in Mk 3 an electronically scanned phased array. Both illuminate a parabolic cylinder reflector and produce a pencil beam scanning in azimuth. The complete radar head can be rotated to cover any required sector, for example 3200 mils rotation in 15 seconds. In transit the reflector folds down.

Cymbeline Mk 3 weapon locating radar

Below the antenna is an equipment box which houses the main electronics unit, the power unit, and the display unit during transit.

The main electronics unit contains the transmitter/receiver and the radar timing and computer modules.

The display and co-ordinate indicator units can be removed from the equipment box for remote operation for distances up to 15 m. The display unit consists of a short-persistence 'B' scope on which the radar returns are displayed. It also carries all the controls necessary for the operation of the radar. The mortar position is displayed on the co-ordinate indicator which can be used at distances up to 2 m from the display.

The Mk 3 has a number of special features:

(a) Electronically scanned phased antenna system with increased arc of 1050 mils
(b) Maximum displayed range of 30 km with improved 'B' scope display
(c) Digital computer with LCD numerical readout
(d) Digital data store as standard fit
(e) Key pad man/machine interface
(f) Greater displacement (30 m) from radar to operator

OPERATION

The radar enables the operator to plot two points in the trajectory of the projectile, and to measure the slant range and bearing to each of these positions. The time taken for the projectile to travel between the two points is also measured and the computer uses this information to extrapolate the firing position. This entire process takes place in about half a minute.

Additional facilities have been provided to ensure the maximum accuracy of location and ease of operation over a wide range of operational conditions. For maximum range performance a switched single beam is used, and an additional beam position is available to alert the operator for making the first interception. For short range work a double-beam mode of operation may be selected to reduce operator reaction time errors. This facility also improves the multiple target capability.

Provision has also been made for the internal fitting of an optional digital data storage module. This enables the radar returns to be stored to provide a long-persistence display so that operator concentration can be reduced while improving the marking accuracy. Data storage also improves multiple handling capability.

CHARACTERISTICS

Frequency: I-band
Peak power: 100 kW
Antenna: Mk 1 and 2 – Foster Scanner; Mk 3 – electronically scanned phased array
Sector scan: Mk 1 and 2 – 720 mils; Mk 3 – 1050 mils
Bearing limits: 12 000 mils total rotation
Speed of rotation: 200 mils/s
Presentation: 'B' type display with selected 4000 m and 8000 m zones
Minimum range: 1000 m
Max displayed range: Mk 1 and 2 – 20 km; Mk 3 – 30 km
Weight: 1000 kg (on trailer)
Radar reliability: 200 h MTBF (excluding power source)

STATUS

Total deliveries to the British Army and 16 overseas countries had reached more than 300 sets by 1987 and are continuing. During 1987 a number of additional orders totalling several million pounds were received from undisclosed countries.

An MTI modification has been introduced recently to give a considerable performance improvement in clutter and ECM conditions. British Army systems have been upgraded and New Zealand equipments were due to be upgraded by mid-1988.

CONTRACTOR

THORN EMI Electronics Ltd, Radar Division, Dawley Road, Hayes, Middlesex UB3 1HN, England.

3490.153

RADAR GS20 HOSTILE FIRE LOCATOR CLARIBEL

The Claribel hostile fire locator radar GS20 is designed to provide vehicles of any type or fixed locations with an automatic warning of sniper fire and an indication of the direction of the source of such fire. It is intended to eliminate the problems and disadvantages of locating snipers by traditional methods based on such techniques as the 'crack and thump' acoustic method.

The radar is a CW equipment based on an I/J-band Gunn diode microwave transmitter, and is all solid-state. No mechanical scanning is involved and a number of sensors are arranged around the vehicle (or fixed position) to provide all-round coverage. A circular display unit indicates the direction of hostile fire sources within 30° sectors, the appropriate sector(s) normally being illuminated before the sound of the shot reaches the vehicle. Usually one display is mounted in front of the commander of the vehicle with a second display located in a convenient position for viewing by the other members of the patrol. The GS20 incorporates a second shot capability, so that the location of two snipers firing from different sites can be identified simultaneously. Another useful feature in vehicle installations is an orientation facility which ensures that the information displayed is continuously updated to compensate for any evasive action that the vehicle driver may take after sniper's shot(s).

STATUS

In service with the British Army. More detailed information is given under this entry in previous editions of *Jane's Weapon Systems*.

CONTRACTOR

Racal Radar Defence Systems Ltd, Davis Road, Chessington, Surrey KT9 1TB, England.

3715.153

FALCONER PRECISION APPROACH RADAR

Falconer is a precision approach radar which consists of radar-head equipment installed in a transportable cabin together with ATC display consoles and associated remote control equipment in the operations room. In addition, the radar head cabin is fitted with a display console for engineering use or, with the addition of suitable communications equipment, for use as a stand-alone or fall-back control position. The radar head installation is transportable by road or air and can be quickly and easily reassembled at another site.

The radar-head is a dual-channel system employing solid-state circuit elements, except for the main modulator thyratron and coaxial magnetron. Modular construction is used wherever possible; moreover, the widespread use of small easily replaceable modules ensures relatively inexpensive replacement costs. The use of digital circuitry also maintains drift and significantly reduces warm-up time.

The clear air range of the system on a 1 m² target has been calculated as 33 km; in operation this has reached 42 km so that the operational requirement of 28 km is comfortably exceeded. To enhance target detection in range and in adverse weather conditions, the radar incorporates selectable linear/circular polarisation, a digital MTI and a video correlator to reject uncorrelated clutter returns. A second-time-around rejection circuit and FTC and GTC circuits are also included. In addition, a five-stagger PRF is used to eliminate blind speeds between 20 and 600 knots.

Each system comprises an azimuth and elevation antenna installed in a robust module mounted directly onto the cabin exterior. Each antenna has a beamwidth of 0.5° in its principal plane with a scanning rate of approximately one second. The elevation antenna scans vertically from −1 to +6°, while the azimuth antenna scans from −10 to +10°. Resolution is such that two targets at the same range can be distinguished at 0.6° separation in elevation and 0.8° separation in azimuth. In range, two targets can be distinguished with a separation of 60 m.

Falconer display in control tower at RAF Wyton

STATUS

In operational use at RAF and MoD airfields in UK and elsewhere. More detailed information is given under this entry in previous editions of *Jane's Weapon Systems*.

CONTRACTOR

Cossor Electronics Ltd, The Pinnacles, Harlow, Essex CM19 5BB, England.

3491.153

S713 MARTELLO 3-D SURVEILLANCE RADAR

The Marconi S713 Martello three-dimensional surveillance radar is a mobile D-band long-range equipment designed to provide complete air defence coverage throughout the 1980s and 1990s. An alternative version, the S723, is described under entry **4798.153**.

The system consists of the antenna, a radar container which houses the high power transmitter and signal and data processing systems, a cooling/air conditioning unit and a power generation unit which are all based on the ISO freight container concept so that they can be transported by standard prime movers or carried in transport aircraft such as the C-130. The radar can be ready for action within six hours of arrival at its operational site. The antenna consists of 60 identical horizontal linear array elements, vertically stacked, each with its own receiver. The antenna rotates at six revolutions per minute and each array receives returns which are assembled into a passive IF beam-forming network. This forms eight beam patterns to achieve the required vertical cover and heightfinding accuracy. Cover is provided from the radar horizon up to 30°, with height measurement up to 24° elevation. The array is tilted back at an angle of about 14°, the bottom beam being flat along the ground and all beams being fixed in elevation. The array can be operated in winds of up to 70 knots, and is designed to withstand gusts up to 120 knots.

Range and bearing information is obtained by conventional scanning with elevation data being derived by monopulse processing of the signals in adjacent beams. Since this processing is carried out on a common radiated frequency, accurate height measurements are achieved even when the transmitter is in the frequency agile mode. The three-dimensional data is then processed to deal with clutter, chaff and ECM conditions and correlated with extracted SSR/IFF information before onward transmission to either a local or remote operations centre.

Pulse compression and doppler filtering are also employed to give moving target detection in conditions of clutter and heavy chaff accompanied by significant wind shear.

S713 Marconi Martello mobile 3-D surveillance radar

STATUS
In production. Nine Martello radars have been ordered, five for UKADGE, two for Denmark and two for Oman. Of these, six were of the S723 variant (see **4798.153**), and three were S713s, including the two ordered for Oman. All UK systems have been delivered.

The Royal Jordanian Air Force has placed a £15 million contract with Marconi Radar to supply new air defence radars and to modernise existing radar facilities. The new systems will be either the S713 or S723 Martello.

CONTRACTOR
Marconi Radar Systems Ltd, Writtle Road, Chelmsford, Essex CM1 3BN, England.

4798.153

S723 MARTELLO 3-D SURVEILLANCE RADAR

The S723 Martello radar is a later version of the earlier S713 system (**3491.153**) and differs considerably in both physical appearance and technical characteristics. Physically, the S723 presents a different appearance having a wider horizontal but shorter vertical aperture. Technically, the new planar array antenna incorporates integral solid-state transmitters giving a 'fail-soft' capability, combined with high performance receiver modules. Two versions of the S723 are available, S723A and 723C, the latter having six elevation beams as opposed to eight for the S723A.

The S723 is a 3-D stacked beam radar with a parallel receiving system for heightfinding. Bearing, range and height are available on every target on each revolution of the antenna. The planar antenna is a vertical stack of array elements each fed from solid-state transmitters housed in the antenna spine. Each element has its own high performance receiver. The transmitted RF power is fed to each array with appropriate phase and amplitude relationship to obtain a cosecant squared vertical cover with narrow azimuth beamwidth and low sidelobes.

Returns from the targets are received by all arrays. The individual receiver outputs are then combined in a passive beam-forming network which synthesises either eight or six elevation beams matched to the required elevation cover. All beams have pulse compression and full adaptive signal processing is provided out to maximum range. Target bearing and range are automatically extracted from a series of individual returns by the plot forming system and monopulse measurement of returns in adjacent beams yield corresponding height data.

The trailer-mounted antenna accommodates the solid-state transmitters, the receivers, the beam-forming network, and the IFF/SSR interrogator and its co-mounted antenna. The electronics container houses the eight or six adaptive signal processors, the 3-D plot extraction system, the IFF/SSR data extractor, the plot correlation and digital data link equipment, and a radar management console.

S723 Marconi Martello 3-D radar's new planar array antenna differs noticeably from the earlier S713 Martello

CHARACTERISTICS
Instrumented range: 270 nm
Instrumented height: 200 000 ft
Instrumented azimuth: 360° at 6 rpm
Range accuracy: 525 ft
Height accuracy: 1700 ft (S723A); 2000 ft (S723C)
Azimuth accuracy: 0.13°
Detection range: 270 nm (S723A); 230 nm (S723C) (for small aircraft)

STATUS
Six S723 systems have been ordered, four for UK and two for Denmark. Of these radars, two of the UK systems and the two Danish systems were funded by NATO. All systems have been delivered and have been accepted into service.

Marconi Radar has teamed with Raytheon to bid for a $500 million contract for CONUS FARR (Federal Aviation Agency Radar Replacement), a civil/military radar network of 48 minimally attended radars for the US Joint Surveillance System (JSS), with a radar based on the S723 Martello.

CONTRACTOR
Marconi Radar Systems Ltd, Writtle Road, Chelmsford, Essex CM1 3BN, England.

2464.153

MARCONI 800 SERIES RADARS (LAND-BASED)

The Marconi 800 series of surveillance and tracking radars was introduced in 1972. The early members of the series (type numbers in the range 801-839) are all naval radars and are described in that section of this book. Type numbers in the ranges 840-849 and 850 onwards have been allocated to coastal defence radars and land-based weapon radars respectively. Several radars of these types have so far been announced and are briefly described below.

ST860 surveillance radar on transportable cabin

Type S841 is a coastal defence surveillance radar which provides and continuously updates accurate range and bearing co-ordinates on surface targets for weapon fire control systems.

The radar operates at a fixed frequency, selected in the band 9.3 to 9.5 GHz; it incorporates track-while-scan, digital range, and bearing data extraction and has an autonomous fixed coil display. Several sizes of antenna are employed.

The ST850/M target tracking radar and television system has been designed, in co-operation with Short Brothers and Harland, as an integral part of the Tigercat weapon system (**2465.131**) to improve its performance in conditions of poor visibility by giving a dark fire capability.

The ST850/M is contained in a mobile air conditioned cabin with the radar director on its roof. The system is simple to use, is flexible in operation, and can be speedily deployed. This radar is more fully described in **2466.153**. The ST850/G target tracking radar is a derivative of the ST850 which incorporates a Sperry general-purpose digital computer for the prediction of aim-off to give blind fire gun control. Without the TV system, this unit will control small calibre rapid fire AA guns and the complete equipment gives maximum flexibility to a combined missile/gun point defence system.

ST850 transportable tracking radar

The ST855/856 are range instrumentation radars incorporating full facilities for data recording and analysis. The ST857/858 is similar equipment designed to interface with customers' own instrumentation. A major feature of the ST858 is its ability, when locked on to an aircraft target, to instantaneously change over to track a small object leaving the aircraft. Tracking of projectiles with calibres down to 20 mm is possible.

The S860 is a mobile surveillance radar designed to support the ST850 radars through the provision of target positional data. The S860 is an I/J-band, 200 kW (peak), surveillance radar with a typical range of 60 km against high level targets. Digital MTI is incorporated and there is provision for IFF. The radar is mounted on top of a manned cabin with a dual operator position feeding target co-ordinate data to up to three fire control units. Two PPI displays are fitted, each capable of rate-aided tracking on four targets. Allocation of target data to the selected fire control unit is fully automatic on operation of the appropriate key. Co-ordinate data are parallax corrected to ensure speedy target acquisition and to reduce reaction time.

STATUS
Land-based 800 series radars are in production for the UK MoD and other customers. S860 and ST850 units have been supplied overseas as part of the Short Brothers Radar Enhanced Tigercat 2 air defence missile system. The Royal Aircraft Establishment employs an ST858 as a precision tracker for range instrumentation.

CONTRACTOR
Marconi Command and Control Systems Ltd, New Parks, Leicester, LE3 1UF, England.

2479.153

S613 HEIGHTFINDING RADAR

This is a mobile/transportable heightfinder radar working in the G/H-band. It integrates with any of the surveillance radars in the S600 to S605 range, utilising the same form of construction. The digitally controlled antenna is capable of providing 22 heights a minute and the height is automatically extracted, a unique facility in mobile systems. Modes of operation are normal, sector scan, and volumetric scan.

STATUS

This radar has been supplied to the UK MoD, NATO, and a number of other nations. No longer in production. More detailed information is contained under this entry number in earlier editions of *Jane's Weapon Systems*.

CONTRACTOR

Marconi Radar Systems Ltd, Writtle Road, Chelmsford, Essex CM1 3BN, England.

5615.153

S500 SERIES PRIMARY AIRFIELD SURVEILLANCE RADARS

The Marconi S500 series of radars has been designed primarily for airfield surveillance at military and civil airports. Good sub-clutter visibility in fixed and moving clutter together with excellent ECCM capability make the S500 series also suitable for low cover and gap filling. The series consists of three systems, S511, S511C and S512, operating in the E/F (10 cm) frequency band.

All three systems use the same antenna, signal processing techniques and displays, and differ only in their transmitter/receivers.

The high performance antenna consists of a 5 metre aperture, double cosecant-squared, carbon fibre reflector and a double horn system for dual beam operation. Variable polarisation between linear (vertical) and circular is provided on the lower beam, and fixed circular polarisation on the upper beam.

A modular tower with enclosed upper section housing the turning gear is included, and full remote control and status indication together with remote signal transmission by cable over distances up to 40 km is available. The Marconi Astrid display system is available for use with these radars.

The S511 uses transmitter/receiver type S2022 which incorporates a tunable magnetron driven by a highly stable modulator with fail-soft reliability and a high stability Coho-Stalo receiver. The S511C is a longer range version and uses a high power coaxial magnetron and larger modulator. Frequency diversity using two transmitter-receivers is available with all systems.

The S512 uses a rugged, wide bandwidth cathode modulated TWT transmitter-receiver S2062. This coherent transmitter-receiver uses pulse compression techniques and provides full frequency flexibility (block-by-block and pulse-by-pulse frequency agility).

An advanced three-channel signal processor eliminates both moving and fixed clutter and reduces tangential fading. Three temporal threshold systems (clutter maps) are fitted to ensure a low constant false alarm rate. Optional equipment includes a separate weather processor for controllable weather contours on the display and the S7204 plot extractor which provides the necessary interface between the signal processing and data handling system.

STATUS

In production.

S511 airfield surveillance radar

CONTRACTOR

Marconi Radar Systems Ltd, Writtle Road, Chelmsford, Essex CM1 3BN, England.

4630.153

S700 SERIES MOBILE RADAR SYSTEMS

The S700 series of radars is intended primarily for mobile military use. The series comprises five systems: S704, S705, S706, S711 and S712.

The S704 and S705 are long range radars in D and E/F band respectively, having 25 foot (7.6 metre) single curvature antennas with linear squintless feeds giving cosecant squared cover. The intrinsic sidelobe performance of these antennas is ideal for military operation. Both systems use high power (3.3 MW) wideband transmitter/receivers with pulse compression techniques and full frequency flexibility which, with the S7128 adaptive signal processor and S7204 plot extractor, give good clutter and ECCM performance. The S704 operates with antenna S1061 and transmitter/receiver S2019, and the S705 with antenna S1062 and transmitter/receiver S2018.

The S706 is a medium range equipment operating in E/F band using the same antenna, S1062, as the S705. The transmitter/receiver S2062 uses the rugged wide-bandwidth cathode modulated TWT. Pulse compression and full frequency flexibility is an essential part of the design and with a large dynamic range signal processor and plot extractor provides high quality performance in clutter and ECCM.

The S711 is a short-to-medium range radar for low-level and gap-filling roles. It operates in E/F band and uses a high performance 5 metre aperture double curvature, cosecant squared, carbon fibre reflector, horn-fed antenna, S1074. This antenna can be hydraulically elevated up to 65 feet (20 metres) for visibility over local obstructions. The S2062 transmitter/receiver, S7113 signal processor and S7204 plot extractor provide the necessary high performance.

The S712 is a short-to-medium range radar for low level defence. It operates in the E/F band and has a 2.4 metre aperture, double curvature antenna with a parabolic cover profile mounted on the equipment container. The same transmitter/receiver, signal processor and plot extractor are used as with the S711.

Marconi S711 radar antenna on a 60 ft mast

STATUS

In late development and pre-production.

CONTRACTOR

Marconi Radar Systems Ltd, Writtle Road, Chelmsford, Essex CM1 3BN, England.

5869.153

MARCONI OTH RADAR

Marconi Radar is developing two over-the-horizon medium range HF radar systems for application in over-the-sea early warning systems. One is for land-based equipment, the other for shipborne applications. These employ the well known effect whereby transmissions in the appropriate part of the HF band adhere to the water surface to provide detection well beyond the radar horizon. This is of great value in the detection of surface skimming missiles, low flying strike aircraft, helicopters and ships. Marconi states that depending upon target type, detection can be accomplished up to a distance of 320 km, compared to the 50 km or so by conventional radars. A new signal processing unit has been developed by the company which can continuously handle 10 million floating point complex computations per second. A combination of multiple units provides an OTH radar system capability up to 2000 Mflops.

It is understood that studies and development work on both systems are being funded jointly by the UK MoD and Marconi.

STATUS

In development. It is understood that sea trials have been carried out with an on-board OTH-SW system in conjunction with the UK Admiralty Research Establishment. A significant capability to detect ships and aircraft well beyond the horizon was demonstrated using a prototype equipment. This included detection and automatic extraction of target data (range, bearing and velocity). Marconi claims that a full operational OTH-SW system could be available by mid-1988.

Trials have also been carried out in conjunction with the UK RSRE with a similar but higher powered land-based SW radar providing accurate real time tracking of OTH aircraft targets. Production equipment now in development will use the new processor to give ECCM capability.

CONTRACTOR

Marconi Radar Systems Ltd, Writtle Road, Chelmsford, Essex CM1 3BN, England.

Simulation of a land-based operational display

2484.153

AR-3D AIR DEFENCE RADAR

AR-3D is an E/F-band air defence surveillance radar giving three-dimensional information on targets by mechanically rotating the radar beam in azimuth and electronically scanning in elevation. All elements of the system are trailer-mounted for transport by land, sea or air.

The AR-3D antenna, which combines mechanical scanning in azimuth with electronic scanning in elevation, consists of a compact circular polarised linear array positioned at the focus of a simple parabolic cylinder reflector. The reflector has a height of 4.9 m and a width of 7.1 m. The azimuth sidelobes of the antenna are –36 dB in the 5 to 15° sector and –48 dB in the 15 to 90° sector. An alternative 5.5 m planar array antenna is also available which combines lower sidelobes and accurate beam formation for improved ECCM performance. This latest antenna folds in half for transportation by C-130 aircraft. The electrical characteristics of both antennas are identical to allow retrofit of the new antenna, if required. The transmitter consists of a coherent two-stage E/F-band transmitter. Each of the two stages uses a linear beam pulse microwave tube. Both tubes are designed for excellent phase, noise, and amplitude performance commensurate with the pulse compression and MTI requirements. The transmitter has a peak pulse of 1.11 MW (mean power 10 kW) and a bandwidth of 200 MHz.

AR-3D planar array antenna

The signals received from the antenna are amplified in a wideband (200 MHz) amplifier. After amplification the signals are separated into channels representing elevation bands of approximately 2° (bandwidth 20 MHz per channel). The signals are then time-compressed to 0.1 microsecond and their frequencies measured to enable the fine elevation within each beam to be obtained.

The function of the plot extractor is to detect targets which meet prescribed criteria. This target information is then fed to tracking and display equipment in the form of digital words containing three-dimensional positional information as polar co-ordinates. The on-line processor stores the positional data, carries out azimuth and height calculation, and takes into account various corrections. Plot and track data are then transmitted to the local displays and modems as required.

The varying demands of air defence are met by supplying AR-3D in different operational configurations, as a radar reporting post, control and reporting post, or command and control post. The different requirements are met by variants of the processing and control cabin, with hardware and software organised in modular packages appropriate to the specific function.

As a reporting post the AR-3D provides two operational consoles with automatic plot extraction and reporting to a remote centre of up to 100 combined primary radar and IFF/SSR plots. As a control and reporting post with five operational consoles, automatic initiation and tracking of up to 100 aircraft is provided with automatic reporting of data to a remote operations centre, while up to ten simultaneous computer-aided interceptions can be controlled. The command and control post additionally varies in providing for autonomous computer-aided control of own aircraft and missiles.

CHARACTERISTICS

Coverage: Adjustable. Typically 180 nm on 2 m² target

Radiation: Frequency selective within E/F-band. The operating bandwith is 200 MHz. Independent frequency selective filters for each of 13 elevation channels, each having a bandwidth of approximately 20 MHz.

Antenna 'A'

Type: Linear array and parabolic cylinder

Polarisation: Circular

Beamwidth (azimuth): 1°

Beamwidth (elevation) 2°

Peak sidelobe (azimuth): 25 dB

Peak sidelobe (elevation): 23 dB

Scanning angle: 30°

Gain: 41.3 dB

Adjustment: +4, –2°

Plessey AR-3D radar being deployed with reflector

Rotation rate: 6 rpm

Antenna 'B'

Type: Planar array, horizontal slotted waveguide radiator and serpentine slow wave feed

Polarisation: Linear

Beamwidth (azimuth): 1.5°

Beamwidth (elevation): 1.5°

Peak sidelobe (azimuth): 35 dB typical

Peak sidelobe (elevation): 25 dB

Scanning angle: 20°

Rotation rate: 6 rpm

Transmitter

Peak pulse power: 1.11 MW

Mean power: 10 kW

PRF: 250 pps

Frequency band: E/F-band

Bandwidth (to 1 dB): 200 MHz

Pulse length: 36 μs

PA tube: High power klystron grid modulated

Transmit modes: Continuous or fast on/off

Receiver

Bandwidth: 20 MHz/channel

Compression ratio: 130 :1 (lower channels)

Accuracy

Elevation: 0.15°

Range: 15 m

Range resolution: 40 m

STATUS

In production and in service with the UK MoD and a number of overseas countries. Two AR-3D systems have been deployed in the Falkland Islands. It was announced in December 1986 that Plessey is to supply six AR-3D radars to Iran under a £240 million deal. They are reported to be in a static format and will be deployed on the Afghanistan and Soviet Union borders. The radars are due to be operational in 1990.

CONTRACTOR

Plessey Radar Ltd, Oakcroft Road, Chessington, Surrey KT9 1QZ, England.

1141.153

AR-15M MOBILE SURVEILLANCE RADAR

The AR-15M is a mobile radar system for air defence and ATC medium range surveillance and approach roles. In its air defence role it optimises coverage against low flying targets and operates as a manual reporting post.

The system, providing both primary and secondary radar/IFF facilities, is a mobile unit with a cross country capability which includes the possibility of deployment in desert conditions: the equipment is installed in containers which can be loaded into C-130 transport aircraft when detached from the trailers. The configuration of equipment is as follows:

Antenna Trailer

A special-purpose trailer carries the primary radar reflector antenna, turning gear and IFF/SSR antenna. Facilities are provided to hoist the antenna on to the cabin.

Radar Cabin

The cabin, which is trailer mounted for transportation, contains a dual magnetron transmitter/receiver, the IFF/SSR interrogator/receiver, signal processing and displays. The equipment is of solid-state design and incorporates an ECCM receiver and an advanced digital target indicator. The cabin also contains two operational displays and ground/air communications equipments. It is air conditioned and provides storage for carried spares, test equipment and ancillaries. Jacking gear is fitted to facilitate loading and unloading from the trailer as well as for levelling for operation.

Diesel Generators

Two mobile diesel generators are used for supplying power to the system. One of these is for stand-by purposes. In addition, further cabins can be provided to house the crew and to provide mess facilities if required. The two diesel generators have adequate power to provide the local domestic supply for the crew.

STATUS

In production.

AR-15M mobile surveillance radar

CONTRACTOR

Plessey Radar Ltd, Oakcroft Road, Chessington, Surrey KT9 1QZ, England.

5608.153

WATCHMAN AIR SURVEILLANCE RADAR

Watchman is the latest addition to Plessey's S-band medium-range surveillance radar series. It is designed to offer high performance in terminal area and approach surveillance, fighter recovery, GCA surveillance, approach control, radar sequencing control, outbound control and helicopter surveillance.

New technology in microwave generation and signal processing, and the use of a TWT in the transmitter, has permitted advanced capabilities and considerable flexibility in the frequency and pattern of transmitted pulses. A dual pulse train is used consisting of a very short pulse (0.4 μs) giving high discrimination at short ranges, and a much longer pulse (20 μs) for detecting aircraft which have small echoing areas at long ranges. The long pulse is compressed in the receiver using a surface acoustic wave equaliser. Signal separation is achieved by using different radio frequencies for the long and short pulses. Target detection is further enhanced by interchanging these frequencies every nine pulse repetition intervals.

Watchman air surveillance radar

Watchman thus provides frequency diversity operation with a single transmitter, and the overall stability of the latter contributes significantly to the anti-clutter performance of the system. High receiver sensitivity is provided by low noise field effect transistor (FET) radio frequency amplifiers, one for each beam. A beam switching system ensures that only the selected beam output is passed to the main receiver as required.

In the signal processor, the adaptive moving target detector provides maximum target detection against clutter. Radar returns are processed in parallel through normal radar, ground clutter filter and moving clutter filter channels each of which uses both in-phase and quadrature processing. Target detections from all three processing channels are automatically integrated and passed on for transmission to the display system. This ensures that the operator views a clean video picture, and also provides considerable 'fail-soft' capability. Compatibility with staggered PRF operation ensures good velocity response free from blind speeds and immunity to second-time-around clutter returns.

Transmission of the control and signal data between the radar head and display sites is via a two-core fibre-optic cable link. In addition to the advantages of lightweight and low cost, the complete electrical isolation of this optical link provides absolute immunity to any form of electro-magnetic interference.

An autonomous display is also available to complement the flexibility of Watchman. Designed for high reliability and ease of both operation and installation, each display console incorporates all the requisite hardware and software facilities associated with the accurate display of primary radar signals. A selection of high resolution digital maps may be generated from internally stored data. Supporting ATC data and, where required, SSR plot data and labels, can be selectively displayed with or without primary radar video.

The Watchman system can be used in both civil and military applications. For the military requirement, the equipment is likely to have to operate in a jamming environment in addition to providing a basic control service. Desirable ECCM characteristics such as low antenna sidelobe level, PRF stagger, pulse compression and moving clutter filter, have already been included as standard items in the system but improved ECCM facilities may be required. To further improve the system's resistance to ECM, extensive frequency agility modes can be made available by simple replacement of the driver source by a variant with selected multi-frequency capability.

A number of additional options are also available, these include dual channel operation, primary plot extraction, IFF and secondary surveillance radar.

The Watchman display system is a self-contained, intelligent equipment which allows it to be configured for military and civil uses and for static and tactical air defence.

CHARACTERISTICS

Frequency: 2750 – 3050 MHz

Antenna: Two-beam, horn fed, double curvature reflector

Beamwidth (azimuth): 1.5°

Rotation rate: 15 rpm/12 rpm

Transmitter

Type: Solid-state driver with TWT output power amplifier

Mean power: 1.3 kW

Pulse pattern: Short pulse (0.4 μs) followed by long pulse (20 μs)

Receiver

Type: Low noise FET RF amplifier, double superheterodyne IF with pulse compression

Noise factor: 1.9 dB

Pulse compression ratio: 50:1 (long pulse only)

Signal processor

Type: Adaptive moving target detector

MTI improvement factor: 46 dB

STATUS

In production and in service. Over 60 Watchman radars, including more than 200 displays, have been ordered. The major order for 33 fixed and 6 tactical transportable systems was placed by UK MoD in June 1983, and the first system for the RAF was handed over at RAF Lyneham in September 1986. Other contracts for both single and dual channel installations include the Finnish Air Force, Bahrain, Dubai, China, Spain, Ghana and Oman. These contracts are for both military and civil use. In early 1988, Plessey announced the sale of the first system to India and four further systems to Finland. The Watchman system has also been selected for the UK CAA airport radar replacement programme.

CONTRACTOR

Plessey Radar Ltd, Oakcroft Road, Chessington, Surrey KT9 1QZ, England.

4631.153

TRANSPORTABLE WATCHMAN DEFENCE RADAR

Transportable Watchman is a mobile S-band radar designed to meet a wide range of tactical requirements. Roles include gap-filling, forward area alert, coastal surveillance, missile and gun acquisition, surveillance for a ground approach control system and automatic plot reporting. The basic configuration allows operation as a manual reporting post and can be extended to operate as an automatic post with fall-back for limited control. The system can function autonomously, with its own communications and power supplies, or it can be integrated as a gap filler into a larger air defence system.

Two equipment cabin configurations are available. The first is a single equipment cabin and antenna; both units are transportable by road trailers, heavy lift helicopters or C-130 type aircraft and can be deployed in less than one hour. The antenna is mounted on the cabin during operation. The cabin contains an IFF interrogator/receiver and interference blanker, two operational displays and VHF/UHF communication equipment.

The second configuration (see illustration) consists of two smaller equipment cabins and an antenna mounted on a pallet. All three units are mounted on mobilisers for transport, providing cross-country mobility. The system can be transported by medium lift helicopters or C-130 type aircraft for tactical deployment. During operation the antenna is mounted on the cabin holding the primary radar and TX/RX, and IFF interrogator/receiver, while the other cabin contains two (optionally three) autonomous displays and VHF/UHF communications equipment. The radar cabin and operational display cabin can be separated by up to 4 km, with transmission of control and signal data between them via a two-core fibre-optic cable link.

Two antenna configurations are available, a horn-fed low sidelobe reflector incorporating an integral IFF feed or a high gain horn-fed reflector which provides low sidelobes, with additional high angle enhancement.

The transmitter is the same system as used in the Watchman series (**5608.153**) and uses a solid-state driven, high power TWT amplifier providing wide bandwidth for frequency agile operation.

Options include primary and secondary plot extraction, plot combination, two autonomous displays based on the latest microprocessor technology and data communications to allow operation as an autonomous reporting post.

CHARACTERISTICS

Frequency: S-band

Antenna	**A**	**B**
Beamwidth	1.5° azimuth	1.5° azimuth
Gain	30 dB	33.8 dB
Polarisation	Variable, linear/ circular	Circular
Rotation speed	12/15 rpm	12/15 rpm

Transmitter: Solid-state driver, air-cooled

Peak power: 58 kW

Mean power: 1.3 kW

Pulse pattern: Fixed frequency, pulse-to-pulse agile, burst-to-burst agile

Tactical/transportable Watchman radar

Receiver: Low noise FET amplifier, double superheterodyne IF with pulse compression

Noise factor: 1.9 dB

Pulse width: 0.4 μs (compressed)

Signal processor: Adaptive moving target detector I & Q processing

STATUS

In production for the UK MoD and other countries.

CONTRACTOR

Plessey Radar Ltd, Oakcroft Road, Chessington, Surrey KT9 1QZ, England

5774.153

WATCHMAN COASTAL DEFENCE RADAR

This is a variant of the basic Watchman S-band radar (**5608.153**) with extensive target tracking and data handling facilities for the coastal defence role. The system is designed for operational use in either a static role, in containers or customer supplied buildings, or in cabins for transport by road, rail or air for rapid tactical deployment.

Watchman consists of a frequency agile transmitter/receiver plus signal, MTI and IFF processing units mounted inside a cabin, with a high performance dual-beam primary antenna, incorporating IFF, mounted on top. A second cabin contains the necessary data processing, autonomous display consoles and communications.

Special signal processing techniques are used to ensure optimum detection of surface and slow low level contacts in all conditions, including severe ground, sea and rain clutter, ECM and ECCM. Primary and secondary radar data is processed and tracked to provide a surface and air picture on high definition displays. The displayed information can be used either for local control or can be re-transmitted to a command headquarters.

STATUS

Fully developed.

CONTRACTOR

Plessey Radar Ltd, Oakcroft Road, Chessington, Surrey KT9 1QZ, England.

4607.153

GUARDSMAN DEFENCE RADAR

Guardsman is a two-dimensional mobile radar system designed to meet the requirements of low level surveillance and anti-aircraft artillery alerting. It will also act as a coastal surveillance radar, for both air and surface targets, where reliable detection has to be achieved in the presence of heavy sea clutter.

The system is contained within a single cabin for vehicle mounting which gives cross-country mobility for rapid tactical deployment. An antenna mast allows the antenna to be elevated rapidly to a height of 9.5 metres. The system contains its own prime power generators and is transportable in aircraft of the C-130 type. With a trained crew of two, the system has an in-action time of less than 15 minutes. System 'march-order' action time is less than 5 minutes.

Guardsman features a low sidelobe antenna, a high power wide-bandwidth transmitter and advanced digital signal processing. These provide optimum detection in all forms of ECM and severe clutter likely to be encountered in the forward battle area or in coastal surveillance. At the same time they yield a low constant false alarm rate for maintenance of target tracks.

An associated on-mounted IFF antenna and interrogator/receiver provides secondary plots on the display. Alternatively these can be extracted and combined with primary plots. The combined plots can be transmitted to a track extractor for generation of sensor tracks to an operations centre.

The Guardsman radar can be configured to a variety of uses, for example an autonomous plot/track reporting post or an alerting radar for gun or missile batteries via a command post. It can also be integrated into a long range surveillance air defence system. Alternative communication sub-systems are available for each tactical role.

Guardsman is the C-band variant of the Watchman radar system (**5608.153**) and the two systems are similar in many respects.

CHARACTERISTICS

Frequency: C-band

Antenna: Low sidelobe, line fed parabolic cylindrical reflector with integrated IFF

Azimuth beamwidth: 1.5°

Polarisation: Linear (circular/linear selectable optional)

Rotation rate: 15, 30 or 60 rpm (depending on instrumented range)

Transmitter: Solid-state driven TWT amplifier, fixed frequency, multi-mode operation

Receiver: Low noise FET RF amplifier, pulse compression

CONTRACTOR

Plessey Radar Ltd, Oakcroft Road, Chessington, Surrey KT9 1QZ, England.

Guardsman air and coastal defence radar

5775.153

GUARDSMAN COASTAL DEFENCE RADAR

This is a version of the basic Guardsman C-band radar (**4607.153**) which has been designed for coastal defence operations. The system is highly mobile and is designed for high accuracy detection of surface vessels and low flying aircraft in adverse environmental and ECM conditions. Information obtained is used either for local control or is transmitted to a central headquarters.

Guardsman consists of a single equipment cabin with an antenna mounted on a folding mast. The antenna assembly includes an IFF aerial with the associated equipment contained in the cabin. The latter contains the transmitter/receiver, data processing system, display and communications equipment. The cabin is trailer mounted to provide cross-country mobility and to enable the system to be deployed rapidly. It can be transported by rail or in an aircraft such as a C-130.

STATUS

In production. Two systems have been supplied.

CONTRACTOR

Plessey Radar Ltd, Oakcroft Road, Chessington, Surrey KT9 1QZ, England.

3649.153

AR-320 THREE-DIMENSIONAL AIR DEFENCE RADAR

The AR-320 is an advanced three-dimensional radar system produced jointly by Plessey Radar and ITT-Gilfillan. It meets all aspects of performance stipulated by the NATO Air Defence requirements for radars in the E/F-band.

The system is self-contained in a number of vehicles and is suitable for transport by land, sea or air. It is designed to provide the optimum combination of long range, fine resolution, good performance in ECM and clutter, high reliability, ease of maintenance, and system survivability. The state-of-the-art digital processing and low sidelobe antenna provide immunity to active and passive jamming, inter-jammer visibility, detection probability in the presence of clutter, and constant low false-alarm rate.

The associated radar-mounted SSR/IFF interrogator/receiver provides secondary plots which are extracted and correlated with primary plots. The combined plots are then transmitted to an operations centre for use in the tracking process. These capabilities, and the common module concept of the radar, ensure that it can be configured to meet a variety of tactical or strategic operational needs, and can be used either independently or integrated with a national defence system.

Field Deployment

For tactical operations, the various system elements are installed in cabins mounted on trailers and self-powered vehicles for transport by land, sea or air. The system is designed for rapid deployment and, with a trained crew, can be dismantled, set up, and brought into service again within six hours, transportation time excluded. Typically, the vehicle group will consist of:

(1) an antenna assembly which can be transported on a flat-bed trailer or rail car. In transit the radar antenna is folded down by means of a hydraulically operated mechanism
(2) a trailer-installed transmitter cabin containing primary and secondary radar transmitters
(3) a trailer-mounted processing and display cabin containing signal processing equipment, the radar management suite, operations consoles, and the radar environment simulator
(4) transportable diesel-electric generator(s) able to provide power for the complete system
(5) a trailer-installed cabin containing workshop, human support and storage facilities.

The entire vehicle group can be transported to a prepared or unprepared site and operated as a field group with radio link communications to a reporting post or sector operations centre.

Radome Installation

Where environmental conditions require, the vehicle group can be housed under cover with the radar antenna protected by a rigid radome. In this configuration the antenna is hoisted into the radome and supported on structural cross-members.

Silo Installation

An alternative arrangement allows for installation of the antenna and associated electronic sub-systems within a below-ground shelter or silo. The antenna is permanently mounted on an elevator and, for operational deployment, is raised and unfolded hydraulically. The antenna is capable of rapid retraction into the silo when required. The radar hardware in this configuration is arranged in equipment rooms in conjunction with associated power supplies, heating and ventilation equipment.

CHARACTERISTICS

Long-range high-cover performance within E/F-band
High gain, narrow pencil beam with ultra-low azimuth and elevation sidelobes
Planar phased array utilising within pulse, phase/frequency elevation scan
Wideband frequency agility
SSR/IFF antenna
Wideband transmitter with mean power greater than 24 kW
Variable pulse length and PRF
Multi-channel, pulse compressed signal processing
Digital MTI processing with velocity compensation for good anti-clutter and ECCM performance
Small resolution cell to resolve multiple targets and maximise signal to clutter detectability
Fully automatic, 3-D plot extractor
Accurate 3-D co-ordinates from every plot
High performance Series 9 display system with digital display drive
Bright, synthetic presentation with high quality characters and excellent registration accuracy
Radar environment simulation for operator training
High reliability and availability
Wide environmental tolerance
Transportable configuration for rapid field deployment

DEVELOPMENT

The AR-320 air defence radar is the result of collaboration between Plessey Radar Ltd and ITT-Gilfillan, combining the ITT-Gilfillan Series 320 3-D antenna and transmitter with Plessey AR-3D receiver, signal processor, displays and software. The Series 320 radars represent a family of defence radars based on three-dimensional radars developed for the American Services by Gilfillan, such as the AN/SPS-48 (**1252.253**) and AN/TPS-32 (**2516.153**) which have since been supplied to a number of other countries.

The Plessey AR-3D is a 'within-pulse' frequency scanning radar, in quantity production. The combination of techniques provide an inertialess elevation scan of a single pencil beam which can be accomplished by either frequency or phase control to steer the beam in space. This technique maintains high resolution throughout the coverage volume, high data rate and accurate target location.

STATUS

The AR-320 is in production and six systems were ordered by the UK MoD in early 1983. These are the last major radars in the current improvement programme for the UK air defence network. Three will be NATO funded and three will be funded by the UK government. Delivery was scheduled to commence in late 1987.

CONTRACTORS

Plessey Radar Ltd, Oakcroft Road, Chessington, Surrey KT9 1QZ, England.

ITT-Gilfillan Inc, 7821 Orion Avenue, PO Box 7713, Van Nuys, California 91409, USA.

AR 320 3-D air defence radar system

6376.153

AR-325 COMMANDER AIR DEFENCE RADAR

The AR-325 Commander is a long-range, 3-D, S-band radar, based on the AR-320 system and intended for air defence purposes. The system is capable of detecting targets at ranges up to 470 km and of operating in the most severe electronic warfare environment. It is frequency agile and the transmitter operates with a wide range of pulse lengths and pulse repetition frequencies. This flexibility enables the radar to adapt its transmission pattern to obtain the required coverage. The transmitter provides a high mean power with a relatively low peak power which contributes to an excellent performance in ECM conditions.

Elevation coverage is provided by multiple pencil beams for both transmission and reception. As ground clutter is only illuminated in the lowest beam elevations all other elevations are unaffected, compared with stacked beam radars which transmit a fan beam illuminating all elevations simultaneously. Phase monopulse is employed to measure target elevation independently in each of the beam positions.

The antenna is an ultra-low sidelobe planar array unit giving an azimuth beamwidth of 1.4° and a very narrow vertical beamwidth. These are combined with a compressed pulse length of 0.4 µs to ensure that the clutter cell is extremely small and the detection probability for wanted target returns is consequently increased.

In its tactical configuration the AR-325 Commander provides full facilities for point-to-point and air-ground-air communications, normally contained in a single cabin.

STATUS

Fully developed.

CONTRACTOR

Plessey Radar Ltd, Oakcroft Road, Chessington, Surrey KT9 1QZ, England.

AR-325 Commander air defence radar

2425.153

RAPIER SEARCH AND ACQUISITION RADAR

This system is an integral part of the Rapier anti-aircraft guided weapon system (**2424.131**). Little information has been officially released on this part of the system, and the following is an assessment based on available data.

This equipment comprises a primary surveillance radar working in conjunction with a computer and a secondary radar (IFF) interrogator to locate, identify and, where appropriate, initiate the tracking of enemy targets. The whole equipment is located in the fire unit of the Rapier system.

In the search mode the radar (coherent pulsed doppler equipment) scans continuously through 360° in azimuth. When a target is detected its azimuth range and velocity are measured, and on the same scan it is interrogated by the IFF whose antenna is combined with that of the primary radar. If a 'friendly' response is received the information on that target is cancelled from the system and the search continues without interruption.

If the target is adjudged hostile an alarm signal is sent to the remainder of the equipment and the measured co-ordinates are used to direct the tracker (see description of Rapier) and the launcher towards the target.

It is, of course, possible that a friendly target might fail to respond to the IFF interrogation on a single pass of the surveillance system (eg because of an unfavourable aspect of the airborne antenna) but the system allows for a 'friendly' response to be accepted before the engagement has gone too far.

STATUS

Since the complete equipment forms an integral part of the Rapier system, deployment and similar details given in the description of the weapons system apply equally to the search and acquisition radars.

An autonomous version of this radar, known as Possum, has been developed by Racal for battlefield surveillance against aircraft (**5765.153**).

CONTRACTORS

Primary radar and antennas

Racal Radar Defence Systems Ltd, Davis Road, Chessington, Surrey KT9 1TB, England.

Microminiature IFF equipment

Cossor Electronics Ltd, The Pinnacles, Elizabeth Way, Harlow, Essex CM19 5BB, England.

Cossor IFF transmitter/receiver on the Rapier

6399.153

RAPIER 2000 SURVEILLANCE AND TARGET ACQUISITION RADAR

The Plessey surveillance and target acquisition radar has been designed for the new Rapier 2000 anti-aircraft guided weapon system (**2424.131**). The equipment is constructed in modules to fit the Rapier 2000 trailer, has an advanced planar array, and makes extensive use of VLSI silicon circuits. The system operates in J-band and provides 3-dimensional surveillance, acquisition and tracking of multiple targets, including helicopters, over a wide coverage envelope down to ground level. High resolution ensures prompt precise data on each target, and prevents jamming, interference and clutter.

An integral IFF system is being provided by Cossor Electronics under a separate contract from British Aerospace. It is designed to carry out track validation and primary/secondary correlation as well as its basic task of identification.

STATUS

In development. Much of the design of the Rapier 2000 system has been based on the Grenadier private venture development programme carried out by Plessey. Whether Grenadier as a separate equipment is likely to be marketed by Plessey seems unlikely in the short term.

CONTRACTORS

Primary radar and antennas

Plessey Radar Ltd, Oakcroft Road, Chessington, Surrey KT9 1QZ, England

IFF System

Cossor Electronics Ltd, The Pinnacles, Elizabeth Way, Harlow, Essex CM19 5BB, England.

Rapier 2000 surveillance and target acquisition radar

5765.153

POSSUM AIR DEFENCE SURVEILLANCE RADAR

The Possum surveillance radar provides early warning of the approach of hostile aircraft and passes information to the tracking and fire control systems of up to 32 designated air defence weapons to enable them to lay on and engage targets. Possum can be used in conjunction with both gun and missile systems.

Sited near each radar is a Control/Display Unit (CDU) which enables the unit commander to co-ordinate the operation of all fire units in his group. He is able to control engagements, allocate priorities and monitor the status of weapons under his command. Target information and weapon designation is generated automatically in the threat assessment and tracking computer situated in the radar itself and passed through the CDU to individual Weapon Interface Units (WIU) which are in turn integrated directly the weapon tracking and fire control systems. The type of integration varies with individual requirements from an automatic input of all information direct from the WIU to the tracking system, to a manual or semi-automatic arrangement. Up to eight separate engagements can be conducted simultaneously. The WIU also provides facilities for the commander to feed weapon status information and position co-ordinates into the computer.

An extended range version, Possum (ER), can be provided to control a number of air defence units and to give extra-early warning of aircraft approach, or give added surveillance range to these units. Warning of the approach of targets is passed automatically to the CDUs of subordinate Possum radars. Possum (ER) information is monitored by a CDU (ER) which acts as a command and control centre for the air defence area commander.

Data links between the various units of the Possum system can be either radio or line based.

STATUS

In production. Racal and Officine Galileo are co-operating to offer Possum in combination with the VANTH optronic fire director. This combination is known as Eagle.

CHARACTERISTICS

	Possum	Possum (ER)
Type:	Pulse Doppler	
Frequency:	F-band	
Antenna		
Type:	Planar array	
Rotation speed:	60 rpm	30 rpm
Polarisation:	Horizontal	
Horizontal beamwidth:	8°	2°
Azimuth coverage:	360°	
Elevation coverage:	–5° to +40°	
Detection range		
Fixed wing:	15.5 km	45 km
Helicopter (hovering):	7 km	15 km
Targets tacked/displayed:	Up to 32	Up to 128

Possum system with LWB Land Rover

CONTRACTOR

Racal Radar Defence Systems Ltd, Davis Road, Chessington, Surrey KT9 1TB, England.

5766.153

ZULU MILLIMETRIC SURVEILLANCE RADAR

Zulu is an advanced millimetre wave radar which is designed for both mounted and dismounted roles. It provides automatic target acquisition and accurate cueing data for up to six remotely deployed fire units when dismounted. Operating within M-band and using Surface Acoustic Wave (SAW) signal processing techniques, Zulu provides very high performance against especially difficult targets such as attack helicopters hovering hull-down in heavy ground clutter. Target detection is specifically tailored to close air defence weapon systems and the equipment provides a virtual zero false alarm rate.

A full day/night viewing and alerting display gives the weapon operator initial target engagement geometry and classification. An aural alarm alerts the operator when a hostile target is detected. The alerting display also indicates the selected scan pattern and equipment status, and automatically identifies a faulty line replaceable unit in the event of system malfunction. Integrated within the weapon sight is an Interactive Weapon Display (IWD) which assists the operator to acquire the target without further reference to the sterling display. When the target is within weapon coverage, the IWD will indicate that the engagement may proceed. An MTBF in excess of 500 hours is given and a built-in test system continuously monitors radar performance.

CHARACTERISTICS

Type: Coherent CW doppler

Frequency: M-band

Power: 70 W

Coverage: 6400 mils azimuth; 320 mils elevation

Scan rate: 45 rpm

Scan pattern and data rate: Selectable

Detection range: 7.5 km fighter/ground attack; 5 km helicopter

Detection accuracy: 18 mils azimuth and elevation

Units of the Zulu millimetric surveillance radar

Remote display: Liquid crystal

Facilities: Range, velocity, elevation and azimuth indication; sector blanking and priority are indicated; bite data readout; audio alarm (mutable)

STATUS

In late development.

CONTRACTOR

Racal Radar Defence Systems Ltd, Davis Road, Chessington, Surrey KT9 1TB, England.

6381.153

LAR 200 LOCAL AREA RADAR

The LAR 200 is a low-cost area radar designed for area surveillance, aircraft control and radar vectoring in both military and civil applications. It is a transportable or mobile S-band equipment which allows rapid reaction for emergency purposes, for example as a battle damage radar when the main radar has been eliminated. Overall weight of the equipment is approximately 100 kg and the complete system can be transported by a Land Rover type vehicle.

The LAR 200 system has a 30 kW output with the transceiver integrated into the antenna turning unit. The antenna rotation speed can be varied to give optimum performance for surveillance and radar approach modes. The display uses the Racal-Decca BRIGHT-TRACK colour system to give the optimum use of colour in the display. A variant for naval applications is also available.

Applications of the LAR 200 include local area surveillance, surveillance radar approaches, radar vectoring/sequencing for ILS/MLS or radar approaches, radar monitoring of ILS/MLS approaches, and control of aircraft departing or in transit. A particular role could be the control of VTOL aircraft operating in the battlefield.

CHARACTERISTICS

Frequency: S-band

Antenna

Polarisation: Horizontal

Aperture: 3.6 m

Beamwidth (-3dB): 2 (horizontal); 37 (vertical)

Rotation speed: 22 to 40 rpm

Transmitter

Type: Solid state with pulse forming network driving a magnetron

Peak power: 30 kW nominal

Pulse length: 0.05 μs/1300 Hz; 0.25 μs/1300 Hz; 1 μs/650 Hz

Receiver: Logarithmic with balanced mixer

Weight: Approximately 100 kg for complete system

STATUS

In production and on order by an unidentified customer.

CONTRACTOR

Racal Avionics Ltd, Ground Systems Division, Burlington House, Burlington Road, New Malden, Surrey KT3 4NR, England.

2428.153

SCORPION TARGET ILLUMINATING RADAR

Scorpion is a target illuminating radar used with the Bloodhound surface-to-air guided missile system. Intended primarily for use with static installations of this system, the equipment can nevertheless be dismantled into transportable units.

Basically an I/J-band CW doppler auto-follow radar, the equipment also performs the functions of illuminating the target and thereby providing the reflected radiation to activate the semi-active homing system of the Bloodhound missile. In addition it has a limited target search facility and is equipped for target identification.

More detailed information is given under this entry number in previous editions of *Jane's Weapon Systems*.

CHARACTERISTICS

Type: Static CW doppler auto-follow radar

Frequency: I/J-band

Main antenna: 4.2 m diameter

Facilities: Search and follow. Target identification

Associated system: Bloodhound SAGW system

CONTRACTOR

Marconi Command and Control Systems Ltd, Chobham Road, Frimley, Camberley, Surrey GU16 5PE.

6817.153

MFU 402 FIRE CONTROL RADAR

The Mobile Fire Control Unit (MFU) 402 radar system is designed to give accurate and effective control of anti-aircraft guns or surface-to-air missiles. The lightweight modular construction of the equipment allows easy installation on a tracked/wheeled vehicle, or a trailer. The system is normally offered on a highly mobile trailer, can be rapidly deployed and is readily transported by fixed-wing aircraft or helicopter.

The system main features are fully automatic operation for a fast reaction of less than six seconds, digital MTI processing for good low-level tracking, a proven digital fire control computer and very wide frequency agility for excellent ECCM capability and solution to multi-path problems.

In the remote mode the MFU 402 works in conjunction with an external sensor which provides target range and bearing. After initalisation, the director is parked with the transmitter inhibited. On receipt of target data the transmitter is energised and the target is acquired and tracked. Output orders are passed to the guns or missile at the same time so that a minimum time is required for final aim-off once the predictor has settled. The engagement continues until the external target data is cancelled or the target is destroyed.

Where there is no external target data the local operator takes control. The control position includes a PPI display and with the radar in surveillance, targets can be automatically allocated to the tracking mode by the use of a 'guard zone', or they can be manually allocated by the operator. Once a target has been allocated the automatic sequencing keeps reaction to less than six seconds and the guns can be fired from the control position.

CHARACTERISTICS

Antenna: Offset, front fed parabola, 0.6 m diameter, vertically polarised

Antenna angle control: 60 rpm for surveillance; two axis, gyro stabilised for precision tracking

Transmitter: Frequency agile

Peak power: 2.5 kW

Range: For guns; 12 km self-surveillance, 10 km auto-tracking, increased to 24 km for operation with medium calibre guns

Detection range: 10 km on strike

Tracking accuracy: 1 mrad rms in angle; 5 m rms in range

Elevation coverage: −10 to +85°

System weight: 512 kg excluding trailer

STATUS

Trialled overseas and ready for production.

CONTRACTOR

Marconi Command and Control Systems Ltd, Weapon Systems Division, New Parks, Leicester LE3 1UF, England.

2439.153

RAPIER BLINDFIRE TRACKING RADAR

The DN 181 Blindfire radar was developed by Marconi Space and Defence Systems, with GEC Avionics as a major sub-contractor. On-going production and enhancement development work is now carried out by Marconi Command and Control Systems. This development was carried out on a UK MoD funded programme with BAe as prime system contractor. The radar tracker, with the Rapier SAM (**2424.131**) is designed to meet the threat posed by low-level strike aircraft which can operate at night and in low visibility. Addition of the Blindfire radar gives the system an all-weather/night capability to extend the original 'clear-weather' Rapier devised by BAe with optical tracking to meet this threat. With the radar tracker in the system, the operator has a choice, until the moment of launch, of which system will be the most efficient for the engagement.

Rapier 2000 Blindfire tracking radar

The radar uses differential tracking for simultaneous target and missile tracking. An electronic angle tracking receiver incorporated within the tracker detects boresight errors for accurate positioning of its own antenna and the angular difference between the missile and target sightlines for the generation of command guidance signals. The extremely narrow beamwidth, coupled with precisely positioned individual range gates for target and missile, is the key to the accuracy and very low-level capability of the system, and also reduces the chances of background clutter, such as ground echoes or rainfall, obscuring the target echo. A number of other electronic measures are also designed into the system to minimise the effect of clutter and other unwanted signals, both spurious and deliberate.

The tracking radar is initially alerted to a hostile aircraft by the system surveillance radar, which indicates the approximate direction of the threat. The tracking radar will immediately swing on to this bearing, and rapidly establish the exact bearing, range, and height of the target. When the radar is locked on to the target, the missile is launched and this, too, is tracked by the radar. The difference between the target and missile angles is instantly derived within the system and commands are automatically transmitted to the missile to guide it on to the target.

The radar tracker is a self-contained, 'plug-in' unit, and requires no extra personnel to operate it.

STATUS

The initial UK MoD production contract was awarded in early 1973 and the equipment is now in service. Full production is continuing for the British Army, Royal Air Force and other operators of Rapier. Nearly 400 Blindfire radars have been delivered or ordered up to the end of 1986.

Marconi Blindfire tracker radar

The DN 181 tracking radar also forms part of the RN Type 911 system (**5630.253**) ordered by the UK MoD for the lightweight Seawolf programme.

In January 1987 Marconi Command and Control Systems was awarded a contract of over £100 million for the full development and production of the new dual tracking radar for the Rapier 2000 system see (**2424.131**). This radar will be an updated version of the DN181 equipment to provide all-weather performance in the complex electronic warfare environment of the next century.

CONTRACTOR

Marconi Command and Control Systems Ltd, Chobham Road, Frimley, Camberley, Surrey GU16 5PE.

2466.153

ST850 TIGERCAT RADAR

Designed for use with the Tigercat surface-to-air missile system (**2465.131**) the ST850/M radar with its associated television equipment gives the missile system an all-weather capability as well as enhancing the clear-weather performance through the provision of a television auto-gather facility. In conditions where the missile is visible, it is automatically gathered within a few seconds of launch after which the engagement is completed automatically or manually using the television monitor.

The radar is housed in a mobile cabin and can be integrated with the S860 mobile surveillance radar.

The radar may assume control of either of two Tigercat fire units, providing fully automatic control of the channel whilst the second channel reverts to optical control.

The ST850 is fully autonomous and can carry out its own low-level search in which the antenna is rotated at 40 rpm in a spiral scan. A PPI display unit is provided.

Tigercat deployed showing ST850 radar on right

The antenna is of the twist-cassegrain type. A four-horn monopulse feed and comparator provide three separate RF outputs (viz sum, elevation difference, and azimuth difference signals) for processing in the IF receivers to produce acquisition and auto-follow data. High accuracy of angular data results from the combination of a narrow pencil beam and the monopulse signal processing.

The TV camera is aligned to the radar boresight and provides optical information on the target being tracked by the radar. This system consists not only of a conventional camera and display, but also a data processing system which permits guidance of the Tigercat missile on to the TV line-of-sight.

In a mixed defence environment employing both missiles and guns, the ST850/M fire control unit is extended by the addition of a general-purpose digital computer, to form the ST850/MG. This latter unit provides prediction of aim-off and automatic system control, thus allowing blind-fire gun control, as in the Marconi/British Aerospace Sapphire lightweight gun fire control system (**1828.261**).

In support of the ST850 Series tracking radars, the mobile surveillance radar type S860 is available. The S860 is fully autonomous with its own operators and provides a control designation facility for the larger site. Target data, parallax-corrected, are automatically passed to the selected fire control unit.

STATUS

800 Series radars have been supplied to the UK MoD and other customers. ST850 units have been supplied as part of the Short Brothers Radar Enhanced Tigercat 2 air defence missile system.

CONTRACTOR

Marconi Radar Systems Ltd, Writtle Road, Chelmsford, Essex CM1 3BN, England.

2490.153

ZB 298 SHORT-RANGE GROUND SURVEILLANCE RADAR

The ZB 298 battlefield surveillance radar was developed by GEC Avionics Ltd to meet a British General Staff requirement (which in turn conformed to the NATO basic military requirement) for a portable battlefield radar, and has since been supplied to overseas customers. The radar can be transported by two soldiers and operated by one man to give an all-weather, day and night facility to detect any significant movement on the battlefield. This includes the location and recognition of men singly or in groups, single or multiple vehicles and low-flying helicopters. ZB 298 can also be used in a counter-insurgency role in keeping watch on sensitive areas such as an airfield.

Targets can be detected on land or over water at ranges from 50 metres to 20 km and the radar can be used for observing ground or sea bursts of artillery or mortar fire and for fire correction.

The original 6 km version of ZB 298 is designated by the British Army as Radar GS no 14 Mk 1. The range of the radar has twice been extended and it is now available in a 20 km version. The opportunity was taken while incorporating the 20 km modification, to introduce major engineering changes in the display unit and tripod which utilise the most up-to-date component technology. Supporting equipment for the ZB 298 includes a training simulator, to give realistic training in the classroom, special test equipment for rapid check-out and fault-finding and a telescopic mast to elevate the radar head thereby improving surveillance coverage.

More detailed information on ZB 298 can be found under this entry number in previous editions of *Jane's Weapon Systems*.

STATUS

In service with British, Danish, Dutch, and certain other armies. No longer in production.

CONTRACTOR

GEC Avionics Ltd, Elstree Way, Borehamwood, Hertfordshire WD6 1RX, England.

4600.153

FASTAR BATTLEFIELD SURVEILLANCE RADAR

FASTAR (Forward Area Surveillance and Target Acquisition Radar) is an entirely new system developed by GEC Avionics. It uses advanced technology to give the operator a user friendly display and control unit. The radar head operates in Ku-band.

GEC Avionics states that the key features of FASTAR offer the user an unbeatable capability of detecting moving targets with automatic tracking of designated targets and correction of artillery fall-of-shot. The system is a highly versatile, effective lightweight radar which can be readily mounted on a vehicle, or is man-portable into observation posts which are unapproachable by road transport.

CHARACTERISTICS

Range: minimum 50 m; maximum 15 km
Elevation: $\pm$ 400 mils
Range accuracy: 10 m
Bearing accuracy: 5 mils
Autoscan: 260 to 3200 mils
Weight: 30 kg

CONTRACTOR

GEC Avionics Limited, Elstree Way, Borehamwood, Hertfordshire WD6 1RX.

FASTAR surveillance radar

6398.153

MSTAR MAN-PORTABLE RADAR

MSTAR is a man-portable surveillance and target acquisition radar designed principally for use by forward artillery observers to detect and track targets, and to locate fall of shot for corrections of artillery fire. The radar is easily man-portable, or can be mounted on any AFV or soft-skinned vehicle. It is envisaged that both armoured and infantry reconnaissance units will use this equipment. The system is a completely new design of portable battlefield radar, and is an all-weather, lightweight, solid-state equipment weighing under 25 kg. The microprocessor-based control/display is a single unit with a padded nylon carry case which also acts as a screening hood. The complete system is powered by a standard army field pack battery.

STATUS

After a competitive evaluation, THORN EMI Electronics was awarded a contract, valued at about $20 million, to cover completion of development and the supply of production units for the British Army. Delivery of the first preproduction unit for trials and evaluation is scheduled for late 1988.

CONTRACTOR

THORN EMI Electronics Ltd, Radar Division, Dawley Road, Hayes, Middlesex UB3 1HN, England.

MSTAR man-portable radar

5627.153

PACER MK 2 MUZZLE VELOCITY RADAR

Pacer Mk 2 is a private venture equipment which has been developed to measure muzzle velocities of artillery guns, naval guns and mortars during operational firing missions. It uses doppler radar techniques to measure the velocity in the range from 100 to 1400 metres per second to an accuracy of 0.1 per cent.

Pacer Mk 2 mounted on an FH 70 howitzer

The system consists of only two units; a sensor unit containing all the electronics which can be mounted either on a non-recoiling part of the gun or on a free-standing mount, and a small hand-held terminal display and control unit which is connected by a cable. The radar transmitter is switched on automatically at the moment of firing, by a flash detector, and is automatically switched off after the measurement is complete. This takes between 0.1 and 1 second, depending on muzzle velocity, and the very short measurement period brings detection and ECM risk to a minimum.

The measured muzzle velocity is shown directly in metres per second on the alpha-numeric liquid crystal display, to two decimal places, and the data on up to 16 rounds can be stored in memory. These can be recalled as necessary and any number can be averaged.

Pacer Mk 2 can be connected directly to a fire control computer via its internal interfaces.

CHARACTERISTICS

Frequency: 10.5 GHz
Output power: 500 MW
Transmission time: 0.1 – 1 s
Velocity range: 100 – 1400 m/s
Accuracy: 0.1%
Beamwidth: 10° vertical; 15° horizontal
Calibre: 75 mm and above

STATUS

Pacer equipment has been supplied to eleven countries in Europe, Africa, Middle East and Far East.

CONTRACTOR

Ferranti Computer Systems Ltd, Cheadle Heath Division, Bird Hall Lane, Cheadle Heath, Stockport SK3 0XQ, England.

6204.153

TYPE 282 RANGING RADAR

The Type 282 ranging radar has been developed by Marconi Command and Control Systems to meet the requirements for a highly accurate sensor to form part of a new generation of tracking equipment at UK trials ranges controlled by the Royal Aircraft Establishment. It will be used in conjunction with a television tracker to provide high accuracy range co-ordinates and will replace the kine-theodolites currently in use at these ranges.

The ranging radar is a K-band, short pulse coherent equipment employing digital signal processing for clutter rejection, and filtering for range accuracy. It has achieved static ranging accuracies of better than 0.25 m and dynamic accuracies of 0.4 m. The system consists of two components; a radar head comprising the antenna, microwave injection test equipment, transmitter and receiver; and a signal processing cabinet containing the processing circuits and the electrical interface with the electro-optical equipment.

STATUS

The system is in production for the UK MoD (11 systems), United States (2 systems) and West Germany (1 system).

Type 282 radar fitted to Contraves Goertz electro-optical tracking mount

CONTRACTOR

Marconi Command and Control Systems Limited, Chobham Road, Frimley, Camberley, Surrey GU16 5PE, England.

3692.153

CEL 850/4 IFF INTERROGATOR

The Cossor CEL 850/4 IFF interrogator system is based on the CRS 390 transmitter/receiver which is the standard unit employed in a wide variety of IFF/SSR applications where identification facilities are required in association with long- or medium-range primary radars. These may be fixed or mobile, and one of four standard antennas will normally be employed, either on-mounted with the primary radar scanner, or with its own turning gear. Characteristics of some antennas are given below.

Apart from specially-designed IFF equipments for applications such as the Rapier (**2424.131**) and Bofors RBS 70 missile systems, the Cossor CEL 850/4 IFF system using the CRS 390 transmitter/receiver is normally used. Its compatibility with a wide variety of primary radars, coupled with features that permit ready interfacing, suit it for either slaved, on-mounted or autonomous operation. Small size, lightweight and rugged construction make it suitable for mobile applications, and it is also used at sea. Operating features include: wide range of acceptable trigger pulses; external computer control facility; external modulation facility; flexible mode program and video outputs to match processing system needs. A logarithmic receiver coupled to special video processing circuits make the equipment particularly resistant to ECM interference.

Radar at Stavanger showing the CRS 389 antenna T1

CHARACTERISTICS

Antenna CRS 370/3
Length: 9.65 m
Weight: 305 kg
Gain: 23 dB (minimum)
Half-power beamwidth: 2.2°
Antenna CRS 389
Length: 4.35 m
Weight: 79 kg
Gain: 17.2 dB (minimum)
Half-power beamwidth: 4.5°
Antenna CRS 379
Length: 1.16 m
Weight: 13.8 kg
Gain: 16 dB (minimum)
Half-power beamwidth: 12.5°
Transmitter/receiver CRS 390
Transmitter section
Frequency: 1030 MHz
Power output: 33 dBW
Modes: 1, 2, 3/A, B, C and 4
PRF: 150 – 1000 Hz
Receiver section
Centre frequency: 1090 MHz
Tangential sensitivity: Not less than –85 dBm

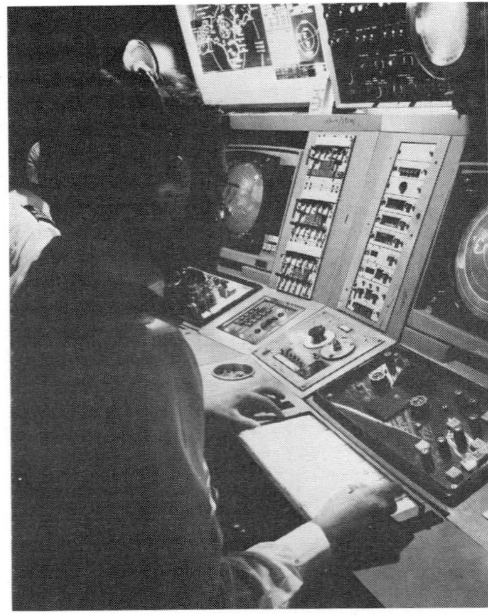

Cossor CVP 190 modular multi-role video processor system in use with a CEL 850/4 IFF system (control panel to left of right-hand display)

Bandwidth: 10 MHz, at –3 dB points
Video outputs: 2 – 5 V, adjustable into 75Ω

STATUS

More than 100 complete systems have been ordered.

CONTRACTOR

Cossor Electronics Ltd, The Pinnacles, Harlow, Essex CM19 5BB, England.

3694.153

880 SERIES IFF INTERROGATORS

The Cossor IFF 880 series of equipments were designed for use in forward battle areas in which there may be a multiplicity of air defence weapons and probable targets. Because of the highly effective nature of many of these weapons, lack of an identification facility with each weapon could lead to the inadvertent destruction of valuable friendly aircraft engaged on strike, interdiction or air defence operations. It was to meet this need that the IFF 880 range of interrogators was introduced with the aim of providing equipments suitable for application to weapons as a stand-alone, self-contained IFF system. Interfaces for integration with towed air defence guns, turret-mounted AA guns, man-portable air defence missiles, and vehicle-mounted air defence missiles have been developed. Outputs from the IFF can be used to inhibit firing, or to provide visual or aural status indications. Integration with primary radars is also possible where required.

Packaging of the transmitter and receiver units is configured to suit the weapon system being equipped. Usually there are requirements for minimum dimensions and weight together with special mounting arrangments. The requirements are easily met by the small size of the equipment. Similarly, a special-to-type antenna is provided in most cases. A typical IFF 880 application is the IFF 880B for the Swedish RBS 70 air defence missile system (**2348.131** and **3617.131**). Variants of the IFF 880B are available for man-portable, Land-Rover mounted and armoured vehicle installations of RBS 70.

See opposite.

IFF 880-5

This variant of the IFF 880 series has been designed as a truly effective identification facility for shoulder launched missiles. The lightweight antenna is simply clipped to the missile canister and connected by a quick release coupling to the transmitter/receiver/processor. The antenna clip houses an interrogation challenge switch and a loudspeaker to indicate target status to the operator.

IFF 880L

This version has been designed for use with the Short lightweight multiple launcher (LML) missile system. It operates as an add-on unit and is self-powered by a battery. The IFF transmitter/receiver/processor unit is attached to the stand of the LML with the antenna secured to the top face of the missile head. Friend or foe status is indicated by audio tones. Weight of the system is 7.5 kg for the interrogator and 2 kg for the antenna.

IFF 880L mounted on Short LML

IFF 880-5 for shoulder-launched missiles

CHARACTERISTICS

Transmitter frequency: 1030 MHz
Power output: 13 dBW (P1, P3), 18 dBW (P2)
Modes: 1, 2, 3/A
PRF: Presettable
Receiver frequency: 1090 MHz
Tangential sensitivity: –76 dBm
Bandwidth: 8 MHz at –3 dB points
Antenna gain: 10 dB
Antenna beamwidth: 20° + 11° – 8° at 3 dB points in azimuth depending upon transponder performance
Dimensions: 355 × 180 × 152 mm, interrogator and antenna but less bracket
Weight: 5.3 kg (less bracket)

STATUS

IFF 880, IFF 880B and IFF 880L are in production for several users.

CONTRACTOR

Cossor Electronics Ltd, The Pinnacles, Harlow, Essex CM19 5BB, England.

6373.193

890 SERIES IFF INTERROGATORS

The 890 series is a new generation of IFF interrogators which has been developed for application to a wide range of weapon systems. Two variants are available and operate on all IFF modes, including Mode 4, with special techniques to narrow the received beamwidth. Operation in Mode 4, in either variant, can be by use of stored codes or by an external cryptographic unit. Extensive use has been made of in-house design VLSI circuits in the signal processing to provide a high, reliable performance in a small volume.

One variant, the IFF 891, has the transmitter, receiver, signal processing and weapon interface contained in one unit. It has been matched to the requirements of the Short Brothers Lightweight Multiple Launcher (IFF 891L) Javelin, Blowpipe or Starstreak. The IFF 891B is matched to the Bofors RBS 70.

The second variant has been designed with the transmitter and receiver in one equipment case for remote operation. This variant, the IFF 892, has the signal processing and weapon system interface contained in a separate unit. It is the selected IFF for the British Aerospace Laserfire and the Marconi Command and Control Systems Marksman. A version, the IFF 892S, has been ordered for the Oerlikon/Contraves Skyguard radar.

IFF 892S fitted to Contraves Skyguard radar

STATUS

In production.

CONTRACTOR

Cossor Electronics Ltd, The Pinnacles, Harlow, Essex CM19 5BB, England.

IFF 891L fitted to Short Brothers' Javelin LML

6422.153

CIRCE NATIONAL SECURE IFF SYSTEMS

For many years countries have been implementing IFF systems in their weapon platforms. The systems, of Western origin, have been generally to the IFF Mark XA standard, or compatible with the IFF Mark XII standard. With experience of use with the Mark XA, it has been found on occasions that it can be easily deceived or exploited. The cryptographic units of the IFF Mark XII standard are controlled by the USA and countries have been unable to operate a secure interrogation mode (Mode 4). The Cossor CIRCE system now provides a nationally secure IFF capability.

CIRCE (Cossor Interrogation and Reply Cryptographic Equipment) is a complete range of modules and units suitable for installation with all types of naval, land-based and airborne IFF systems. This modern, secure IFF capability can be added to existing IFF equipment, from any manufacturer, that is compatible with the IFF Mark XII interface standard. Cossor can supply IFF systems for any requirement and fit CIRCE modules within their equipment, or install them as external applique units.

STATUS

Deliveries are due to commence in early 1988.

CONTRACTOR

Cossor Electronics Ltd, The Pinnacles, Harlow, Essex CM19 5BB, England.

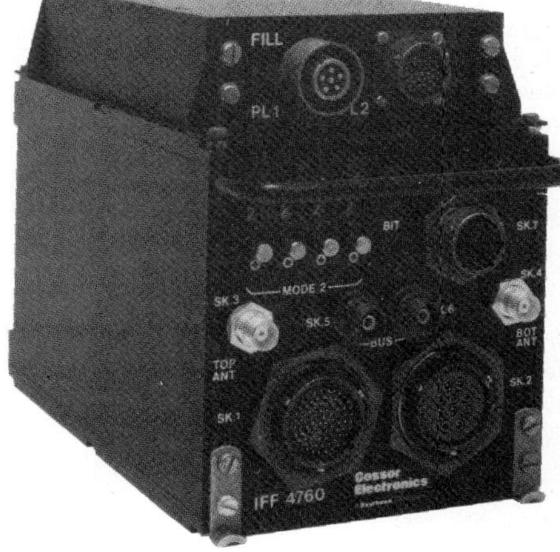

CIRCE TC 100/1 mounted on Cossor IFF 4760 transponder

4604.153

COSSOR MONOPULSE SECONDARY SURVEILLANCE RADAR SYSTEMS

The monopulse technique as applied to secondary surveillance radar (SSR), is a method used to achieve an order of magnitude improvement in target bearing measurement compared to conventional (non-monopulse) SSR. Furthermore, this improvement is readily achievable with a greatly reduced interrogation rate, and has led to the conception of Mode S.

The Cossor monopulse secondary surveillance radar (MSSR) systems have been developed to replace conventional SSR in both terminal movement area and en-route surveillance applications. They can be extended to Mode S operation to accommodate this potential future role. The Cossor MSSR systems achieve the improved accuracy of the monopulse technique, but in addition they incorporate advanced processing techniques to achieve superior decoding and detection performance, and to overcome the other conventional SSR limitations including target capture, military target detection (switched transponders), target code garbling and false target generation (phantoms, ringaround and reflections).

The combination of monopulse and processing techniques provides a superior radar performance in high density traffic and difficult radar environment situations. Cossor MSSR equipments can be successfully deployed at stations which would produce poor conventional SSR performance due to traffic densities, terrain and topography difficulties, or interference from other SSR stations within transponder range.

Two Cossor MSSR standard system types are currently available, the SSR 950 and the Condor 9600. The principal difference between the two is technology and level of system BITE and automatic fault diagnosis. Each type has been developed specifically for monopulse operation and to meet the requirement of ICAO Annex 10 and Stanag 5017. They are available in single or dual configurations, the principal system components being antenna, interrogators, plot extractors, and remote control and monitoring. They include a number of features including interrogator sidelobe suppression, improved interrogator sidelobe suppression, receiver sidelobe suppression, gain time control (R laws), segment gain control (programmable), automatic decoding and plot forming, tracking, false target detection, a high level of BITE, and automatic changeover for dual channel systems.

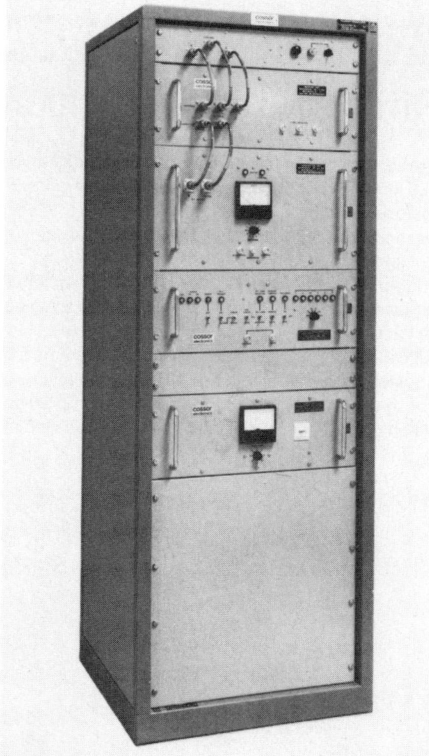

SSR 950/CVP 250 monopulse interrogator

The standard antenna for all Cossor MSSR systems is the large vertical aperture antenna CRS 512 (**5622.153**). It incorporates a controlled underbeam cut off rate to minimise ground radiation, narrow backangle to minimise the antenna cone of silence, and low level sidelobes to achieve good monopulse performance and to reduce the incidence of false targets.

The SSR 950 systems employ interrogators with a tube final transmitter stage, plot extractor using multi-microprocessors and incorporate remote control and monitoring for system configuration and fault reporting.

The Condor 9600 system employs interrogators with completely solid-state transmitters, multi-processor plot extractors, and incorporates automatic fault diagnosis permitting remote control and monitoring for system configuration, fault reporting and fault location.

STATUS

Cossor MSSR systems are currently being deployed in quantity, both in the United Kingdom and worldwide. UK MoD has installed systems at its military airfields throughout UK and West Germany, and 46 of these have now been installed. The UK Civil Aviation Authority has SSR 950 systems in service in the new civil radar network. The Canadian

Department of Transport is deploying 41 dual SSR 955 systems, from which the Condor 9600 system has been evolved, throughout Canada. Cossor MSSR systems are also in service, or are being deployed, in Australia, Greece, Switzerland and several Middle East countries including Saudi Arabia and Bahrain. The Condor 9600 system has been selected by the Swedish Civil Aviation Administration for the country's national air traffic control radar replacement programme.

Future deployments of MSSR equipment will be principally met by the new generation Condor 9600 which will supersede the SSR 950 systems.

CONTRACTOR
Cossor Electronics Ltd, The Pinnacles, Harlow, Essex CM19 5BB, England.

5622.153
CRS 512 SSR ANTENNA
The CRS 512 is a large aperture secondary surveillance antenna designed as a replacement for the familiar linear array which has been the standard antenna since SSR was first introduced. The large vertical aperture, which measures 5 ft (1.53 m), contains a vertical array of radiating elements which give more flexibility in the control of the transmit and receive patterns and provides improved performance. This system overcomes the disadvantages of linear array antennas, which can cause gaps in radar coverage and false replies due to reflections from local terrain and buildings, caused by its narrow aperture.

New constructional techniques have been employed for the radiating elements to minimise weight while retaining strength. This has resulted in keeping the weight to approximately 500 lb (227 kg) for an antenna length of 28 ft (8.25 m). The open array construction also reduces wind resistance and has been designed to withstand a wide range of environmental conditions. The CRS 512 is available for use at airfields, where reflection problems can be most severe, and also for long range surveillance with good coverage at low angles of elevation.

STATUS
In production with more than 75 equipments ordered.

CONTRACTOR
Cossor Electronics Ltd, The Pinnacles, Harlow, Essex CM19 5BB, England.

Cossor large vertical aperture antenna mounted on the radar tower at RAF Scampton

4603.153
MESSENGER SECONDARY SURVEILLANCE RADAR
The Marconi Messenger is a monopulse secondary surveillance radar (SSR) designed specifically to overcome the shortcomings of existing SSR systems. In addition it can be enhanced in the future to full Mode-S standards, with selective addressing and a ground-air data link. Messenger consists of a large vertical aperture (LVA) antenna, an interrogator/responder and a decoder/plot extractor, the electronics being mounted in a standard 19-inch rack.

Marconi Messenger LVA antenna

The LVA antenna produces three azimuth patterns symmetrically disposed about the boresight and forming the means of effecting interrogation sidelobe suppression (ISLS), received sidelobe suppression (RSLS) and monopulse direction finding (MDF). The antenna is manufactured to a computer-optimized electrical design giving a high forward gain of 29 dB and a robust mechanical construction to ensure operation and survival in severe environmental conditions. The LVA antenna reduces the incidence of ground reflections and false targets caused by local buildings, parked aircraft, etc.

The interrogator/responder consists of interrogator, monopulse multiple receiver and a performance monitor. It is completely solid-state, including the final output power stage ranging from 500 W to 2 kW, and is capable of being upgraded to include operation to Mode S standards at a much higher duty cycle than standard SSR. This will enable full SSR data link facilities to be established when Mode-S is introduced on a world-wide basis. The monopulse receivers allow accurate direction finding of targets.

The decoder/plot extractor uses monopulse decoding equipment having special logic included to ensure retention of wanted data in poor or ambiguous signal conditions. The circuits are constructed on printed circuit boards interconnected by a fast multi-highway system. Outputs include range, azimuth, identity and altitude code of all SSR targets in the radar cover.

Data can be output in a number of different formats. Standard modems are used to interface the decoder/extractor output to display and data handling equipment.

STATUS
In production.

CONTRACTOR
Marconi Radar Systems Ltd, Writtle Road, Chelmsford, Essex CM1 3BN, England.

UNITED STATES OF AMERICA

1949.153
COBRA DANE RADAR
Cobra Dane is a large phased array radar at Shemya Air Force Base in the Aleutian Islands under USAF Program 633A. Its purpose is the detection and tracking of ICBMs, SLBMs and satellites. It replaced two radars of the old intelligence variety (AN/FPS-17 and AN/FPS-80), which were themselves pressed into service to undertake Spacetrack duties.

The large fixed phased array of this radar measures about 30 m in diameter and is composed of about 35 000 elements of which some 15 000 are active elements. The system will survey a 2000-mile (3220 km) corridor to collect data on Soviet missile development flights, provide early warning of ICBM launches, detect new satellites, and update known satellite parameters.

The array provides coverage over a 120° sector of the Soviet missile test range, and in its space tracking role the Cobra Dane installation has a range of 25 000 nautical miles (46 000 km). In its data collection role on Soviet missile flights, Cobra Dane has the ability to track up to 100 objects simultaneously with precise data on up to 20 targets. These data are processed at the Shemya site and transmitted to the USAF Foreign Technology Division.

Employed as an early warning system, Cobra Dane has the capacity to track 200 targets and provide detailed information on these to NORAD Headquarters. Among the latter data will be predicted impact locations in the USA.

The installation comprises a total of 12 transmitting groups, each incorporating eight TWTs, with a peak power of 16 MW and an average level of 1 MW.

DEVELOPMENT
The responsible agency is the USAF Electronic Systems Division. The project was authorised in 1971 and Requests for Proposals were issued in 1972, funding also being approved in that year. A contract for $39.6 million was awarded to Raytheon in July 1973. Construction was well advanced at the close of 1974 and the system testing was completed in late 1976.

STATUS
Cobra Dane achieved First Operational Capability in 1977. Total cost was estimated at $68 million. Cobra Dane forms part of the BMEWS system. Display and computer upgrades are planned for the future.

A similar project named Cobra Judy, a shipborne phased array radar for endo-atmospheric tracking to complement Cobra Dane is now in operation in the Pacific (**4301.253**).

CONTRACTOR
Raytheon Company, Equipment Division, Wayland, Massachusetts 01178, USA.

Giant radar at Shemya Air Force Base, near western tip of the Aleutian island chain off Alaska, is a phased array system. Known as Cobra Dane, it is used for tracking Soviet ballistic missile flights. Cobra Dane was constructed by Raytheon Company under contract to USAF Electronic Systems Division

3714.153

AN/FPN-62 PRECISION APPROACH RADAR

The AN/FPN-62 is an updated version of earlier equipment in which new electronics have been substituted for older, more difficult to maintain hardware. The original azimuth and elevation antennas are retained, these being electronically scanned elements providing separate beams for tracking of an approaching aircraft in azimuth and elevation. Guidance signals are normally passed back to the aircraft voice radio link. Operation is in the radar I-band.

STATUS

The AN/FPN-62 was developed by Raytheon in a competitive programme sponsored by the USAF for the updating of fixed PAR facilities based on the AN/CPN-4, AN/MPN-13 and AN/FPN-16 equipments. This resulted in a contract for the upgrading and supply of 40 systems for the USAF, and the prototype was produced in 1976. All 40 systems have now been delivered and are operational.

In 1979, Cossor Electronics, the Raytheon UK subsidiary, won a similar contract for the upgrading of RAF SLA-3 PARs, and is engaged on the production of an Anglicised version of the FPN-62 which is designated the 'Falconer' by Cossor. The UK concern is supplying 43 static and two mobile systems for use at home and overseas RAF bases.

CONTRACTOR

Raytheon Company, Equipment Division, Wayland, Massachusetts 01178, USA.

AN/FPN-62 upgraded PAR

4076.153

SERIES 52 PRECISION APPROACH RADAR

The ITT-Gilfillan Series 52 is a turntable-mounted, precision approach radar (PAR). The radar data is remotely fed to control centres or control towers. Air traffic controllers provide precise approach guidance to aircraft down to the runway threshold through voice communications with the pilot.

The turntable is remotely controlled for aligning to cover multiple runway approaches. Up to six radar displays may be remotely controlled a maximum 3636 m from the equipment shelter. Radar ranges of 10, 15, and 20 nautical miles provide coverage for all types of aircraft and all traffic pattern variations. For transportation, the Series 52 is packaged into two major sub-assemblies: the shelter and turntable pallet.

Radar scan angles of 20° in azimuth and 8° in elevation are developed by the delta-A scanning antennas. These antennas feature minimum sidelobe performance and circular polarisation, coupled with a digital moving target indicator and integrated video. Dual channel radar transmitters and receivers, and high-gain, low-sidelobe antennas, remote control of PAR landing direction, and four sets of preset course and glideslope cursors are operational features of the Series 52 PAR.

CHARACTERISTICS

Frequency: 9000 – 9160 MHz
Peak power: 80 kW
Azimuth antenna gain: 39.7 dB
Elevation antenna gain: 40.3 dB
Max range: 20 nm
Resolution: 0.55° elevation; 0.85° azimuth; 48 m range

STATUS

Production.

CONTRACTOR

ITT Gilfillan, 7821 Orion Avenue, Van Nuys, California 91409, USA.

3174.153

AN/FPS-115 PAVE PAWS RADAR

Pave Paws is the name given to a system of large phased array AN/FPS-115 radars to replace the 474N SLBM detection and warning system radars, AN/FSS – 7 (**2538.153** in 1980-81 and earlier editions). Raytheon was named as contractor to the USAF Electronic Systems Command in April 1976, when a contract valued at $46.6 million was awarded. This covered the first Pave Paws radar at Otis AFB, Massachusetts. Three further sites at Beale AFB, California Eldorado AFB, Texas and Robins AFB, Georgia have been built. All four sites are operational. The ballistic missile early warning stations at Clear, Alaska; Thule, Greenland; and Fylingdales, England are to be updated with a version of the AN/FPS-115 system.

Pave Paws is an important addition to the World Wide Military Command and Control System (WWMCCS), and will be operated and maintained by the USAF Strategic Air Command. The radar's primary mission of detection and warning of SLBM attack also involves the provision of attack characterisation to the NORAD Cheyenne Mountain Complex, the Strategic Air Command and the National Command Authorities. The system's secondary role, in support of the USAF Spacetrack system, feeds in positional and velocity data for the display of all Earth satellites in orbit. Capable of multi-target tracking, it simultaneously detects and discriminates many objects, while providing early warning data, launch, impact, position and velocity information as required.

Automated features of the system include detection, track initiation, and mission decisions. Two standard computers, CYBER 174-12s, which generally serve as the CPU, are programmed for beam steering, the storage and display of data, as well as performing post-mission data reduction and analysis.

Pave Paws operates at UHF frequencies and employs solid-state technology more extensively than Cobra Dane, with transistors driving transmitters to produce radiation. In Cobra Dane TWTs are employed.

Pave Paws consists of a pair of circular planar arrays about 30 m in diameter, each consisting of 5400 elements. The arrays are inclined from the vertical by 20° and mounted in adjacent sides of a building measuring about 32 m high, forming sloping walls on the seaward facing side of the structure. Combined coverage of the electronically steered beams from the two arrays is 85° in elevation and 240° in azimuth. Range is estimated as about 3000 miles (4800 km).

The Otis AFB Pave Paws was declared operational officially in April 1980, with the Beale AFB entering service the following summer.

In mid-1978 it was disclosed that concern over the possibilities of improved Soviet SLBM capabilities required the construction of two more Pave Paws radars to cover additional south-east and south-west facing arcs. These were designated as Robins and Eldorado AFBs, with the latter also scheduled to support NORAD tracking of space objects. The Robins site became fully operational in 1986 and the Eldorado site in 1987.

Pave Paws phased array radar installation at Beale AFB, California

CHARACTERISTICS

Frequency: 420 – 450 MHz
Transmitter: Solid-state
Module peak power: 322 W
No. of active elements/modules (67% filled): 1792
Array type: Corporate feed, density tapered
No. of sub-arrays: 56
Antenna gain: 38.4 dB directive gain
Beamwidth: (transmit/receive): 2°/2.2° at boresight
Polarisation (transmit/receive): Right hand/left hand circular
Array diameter: 72.5 ft (utilised)
Face tilt: 20°
Azimuth: ±60°, 240° with two faces
Elevation: 3 to 85°

STATUS

All four US sites are operational although radar upgrades are currently in progress for Robins and Beale and computer upgrades for Otis and Beale. Work on Thule was completed in 1987, while Fylingdales is scheduled to commence in 1988 with Clear following later.

CONTRACTOR

Raytheon Company, Equipment Division, Wayland, Massachusetts 01178, USA.

Other contractors include CDC, and United Engineers.

2846.153

AN/FPQ-6 AND AN/TPQ-18 PRECISION INSTRUMENTATION RADARS

AN/FPQ-6 and its transportable version AN/TPQ-18 are high-accuracy, amplitude comparison monopulse, C-band instrumentation radars designed primarily for tracking long-range high-velocity targets such as missiles and re-entry vehicles. They can function either in skin-tracking (single-pulse) or beacon tracking (pulse train) modes. The only difference between the two radars relate to the arrangement of their constituent units.

Physically the radars are large, five transmitter cabinets and 60 cabinet racks are required to house the basic electronics. The antenna is 885 cm in diameter, uses a five-horn cassegrain feed and can be rotated continuously in azimuth, and from −2 to +182° in elevation. The transmitter has a peak power of 2.8 MW.

Targets can be tracked out to some 60 000 km and at target speeds up to 18 km/s. Several systems have differing configurations and performance including MW transmitters, pulse compression, 0.03 m/s rms accuracy pulse doppler, 'on-axis' computer aided tracking and low noise high gain feeds. Polarisation is console selectable. The antenna can track targets or be slewed at rates between 0.0006 and 28°/s. One unit was modified recently to track two in-beam targets simultaneously. It incorporates new integrated circuit design which greatly reduces size, power and air conditioning requirements. It is known as the AN/FPQ-19.

CHARACTERISTICS

Azimuth, elevation tracking precision: 0.05 mils rms at 20 dB S/N
Range tracking precision: 4.7 m rms at 20 dB S/N
Antenna beamwidth: 8 mils
Ro, IM^2, 0 dB S/N: 1000 km

STATUS

Eleven of these units are in operation at various USAF, US Army and NASA ranges. A modified version, known as the AN/FPQ-14, is in operation at various sites. The system in Hawaii is being upgraded by ITT.

CONTRACTOR

RCA, Electronic Systems Department, Moorestown, New Jersey 08057, USA.

AN/FPQ-6 precision instrumentation radar

2492.153

AN/FPS-6 HEIGHTFINDING RADAR

For many years the AN/FPS-6 radar has been the principal heightfinder used by US armed forces and many allied nations. At least 450 of these systems have been delivered, including the mobile version, the AN/MPS-14 (**2496.153**).

A high-power E/F-band nodding-beam radar, the AN/FPS-6 is noted for extreme accuracy at long range, and three available versions give it wide versatility under a variety of environmental conditions.

The arctic tower installation consists of a 50 ft (15.25 m) radome (either air-supported or rigid) mounted on a 25 ft (7.6 m), two-storey enclosed tower structure. De-icing is provided by a battery of infra-red lights inside the radome. Radome pressurising (when required), RF, and other electronic equipment is housed in the tower structure.

The temperate tower installation is designed for moderate or tropical climates. The 25ft (7.6 m) supporting structure for the antenna includes an enclosure for RF equipment.

The mobile version (AN/MPS-14) is a six-truck, three-trailer system designed for transport on short notice to new strategic or tactical sites.

The reliability and capability of the basic AN/FPS-6 system has benefited from a continuous improvement programme. Sets with various improvements have been designated as AN/FPS-6A, AN/FPS-6B, AN/FPS-6C, AN/FPS-6D, AN/FPS-89 (**2513.153**) and AN/FPS-90. All these improvements are available in field conversion kit form for the updating of radar sets from earlier production.

CHARACTERISTICS

Peak power: 4.5 MW
Average power: 3.6 kW
Frequency: 2700 – 2900 MHz
PRF: 300 – 405 pps
Pulse width: 2μs
Receiver sensitivity: NF $<$9.3 dB; MDS −108 dBm
Beam characteristics: Pencil beam
Beamwidth: Azimuth 3.2°; elevation 0.9°
Antenna azimuth rate: 180° in 4 s
Elevation rate: 20 or 30 nods/minute
Antenna gain: 38.5 dB

STATUS

Operational.

AN/FPS-6 heightfinder radar, temperate tower and arctic tower versions

CONTRACTOR

General Electric Company, Electronic Systems Division, Syracuse, New York 13201, USA.

2526.153

AN/FPS-7 3-D SEARCH AND GCI RADAR

AN/FPS-7 is a very high power D-band (1250 to 1350 MHz) monopulse 3-D search radar. The equipment uses a stacked-beam heightfinding technique, the stack consisting of seven channels.

Search range is approximately 500 km and height coverage is up to 45 000 m. The transmitter output stage is a high power klystron which delivers 10 MW at any of up to ten preset transmitter frequencies and with a pulse width of six microseconds. A dual-channel transmitting system is provided. When channel A is connected to the antenna through the waveguide switch, channel B is fed to a dummy load. This feature permits full preventive maintenance to be performed without having the equipment off the air.

The basic radar uses several separate but identical receiving systems. Spare receiving channels are available, and may be switched in place of some of the active receivers. The receiver incorporates MTI and anti-jamming circuits.

CHARACTERISTICS

Peak power: 10 MW
Average power: 14.5 kW
Frequency: 1250 – 1350 MHz
PRF: 244 pps
Pulse width: 6μs
Receiver sensitivity: NF $\leqslant$ 9.5 dB
Beam characteristics: Stacked beam, cosec^2
Beamwidth: Azimuth 1.4°; elevation 18°
Antenna azimuth rate: 3.3, 5, 6.6 and 10 rpm

STATUS

Operational. About 30 units produced.

CONTRACTOR

General Electric Company, Electronic Systems Division, Syracuse, New York 13201, USA.

AN/FPS-7 3-D search radar

2493.153

AN/FPS-8 SEARCH RADAR

The AN/FPS-8 is a medium-power D-band search radar designed for aircraft control and early warning and is installed at commercial airports and military bases both in the United States and overseas.

In most installations the antenna is exposed, being mounted on a temperate tower. For severe environmental conditions, the AN/FPS-8 is self-contained in an arctic tower with a protective radome.

Over the years improvements have been made to the basic AN/FPS-8, culminating in the present version whose nomenclature is AN/FPS-88 (V) (**2512.153**).

The AN/FPS-8 also has two mobile versions, the AN/MPS-11 and the AN/MPS-11A (**2495.153**).

CHARACTERISTICS

Peak power: 1 MW peak

Average power: 1.1 kW

Frequency: 1280 – 1380 MHz

PRF: 360 pps

Pulse width: 3μs

Receiver sensitivity: FPS-8: NF 9 dB; FPS-88: NF ⩽ 2.5 dB with parametric amp

Beam characteristics: Cosec²

AN/FPS-8 search radar

Beamwith: FPS-88: azimuth 1.3°, elevation 58°; FPS-8: azimuth 2.5°, elevation 30°

Antenna azimuth rate: FPS-8: 0 – 10 rpm variable; FPS-88: 5, 10 rpm

STATUS

Operational. Over 200 produced.

CONTRACTOR

General Electric Company, Electronic Systems Division, Syracuse, New York 13201, USA.

4327.153

AN/FPS-16 INSTRUMENTATION RADAR

The AN/FPS-16 tracking radar is designed specifically for range instrumentation purposes and is claimed to be the first to be so designed. Its transportable version is the AN/MPS-25. Both systems are C-band units. The antenna pedestal can be rotated continuously in azimuth and from –10 to +190° in elevation. First introduced in the mid-1950s, these units have been extensively modified in the intervening years; almost all now use integrated circuit electronics. Most of the 60 AN/FPS-16 and seven AN/MPS-25 radars have additional modifications including larger diameter antennas, 3 MW transmitters, low noise receivers and pulse dopplers.

CHARACTERISTICS

Angle tracking precision (20 dB S/N): 0.1 mils rms

Range tracking precision (20 dB S/N): 4.7 m rms

Azimuth tracking rate: 750 mils/s

Azimuth tracking acceleration: 1020 mils/s²

Elevation tracking rate: 400 mils/s

Elevation tracking acceleration: 1020 mils/s²

Ro, IM², 0 dB S/N: 235 km

STATUS

In current use. Units are in operation at United States and United Kingdom ranges.

CONTRACTOR

RCA, Electronic Systems Department, Moorestown, New Jersey 08057, USA.

AN/FPS-16 instrumentation radar

2528.153

AN/FPS-20—AN/FPS-100 FAMILY OF LONG-RANGE SEARCH AND GCI RADARS

AN/FPS-20 is one of a family of very long-range dual-channel search and GCI radars. It is widely used partly because it was one of the systems supplied under the MAP programme.

A D-band system, FPS-20 has a search range of some 350 km. The transmitter uses a power klystron with a 2 MW output. Pulse width is six microseconds and the PRF is 360 Hz. The antenna gain is 35 dB and the receiver noise figure 9 dB.

MTI is fitted and the radar has anti-jamming circuits. Data transmission facilities are included.

STATUS

In service in many countries. AN/FPS-20 and subsequent members of the family are used in the SAGE (**2803.181** in 1979-80 and earlier editions) and AWCS (**2533.181** in 1982-83 and earlier editions) systems. Other static radars in the family are AN/FPS-20A, -20B, -64, -65, -66, -67 and -100. AN/GPS-4 and AN/MPS-7 are the mobile versions.

CONTRACTOR

Allied Signal Inc, Bendix Communications Division, Baltimore, Maryland 21204, USA.

2494.153

AN/FPS-24 SEARCH RADAR

The AN/FPS-24 is a high-power low-frequency search radar. Each channel consists of 26 major cabinets, with an additional 11 cabinets that are common to both channels. It has very highly sophisticated anti-jamming features that make it extremely valuable in hostile environments.

The radar is usually installed in a six-storey building with the antenna on top. In severe climatic conditions the antenna is enclosed in a radome which is designed to withstand winds up to 150 knots. The overall antenna system weighs 92.5 tons.

CHARACTERISTICS

Peak power: 5 MW per channel

Average power: 25 kW with 278 pps, 18μs pulse width

Frequency: 214 – 236 MHz

PRF: 278 pps

Pulse width: 6μs without pulse compression; 18μs with pulse compression

Receiver sensitivity: 4.5 dB NF; –113 dBm MDS

Beam characteristics: Cosec²

Beamwidth: Azimuth 2.9°; elevation 30°

Antenna azimuth rate: ¼ rpm CW or CCW; 5 rpm CW only

STATUS

Operational. Eleven systems have been built and are installed at USAF sites in the United States.

CONTRACTOR

General Electric Company, Electronic Systems Division, Syracuse, New York 13201, USA.

AN/FPS-24 search radar

2509.153

AN/FPS-49 EARLY WARNING AND TRACKING RADAR

One of the two basic types of radar used in the ballistic missile early warning system (BMEWS — **2525.181**), the AN/FPS-49 is a very large tracking radar that can also be used in a surveillance mode. It is installed at two of the three BMEWS sites, Thule (Greenland) and Fylingdales Moor (England). At the third site, Clear (Alaska), a slightly different radar, the AN/FPS-92 – which is an improved version of the AN/FPS-49 – is in use. Thus, the tracking radars at each of the three BMEWS stations have different designations, although they are essentially the same: Thule – AN/FPS-49A; Fylingdales Moor – AN/FPS-49; Clear – AN/FPS-92. The description that follows relates in the main to the AN/FPS-49 radars at Fylingdales Moor.

The main reception and transmission element is a 25 m diameter parabolic reflector mounted on a conical pedestal and capable of being rotated in azimuth and elevated and depressed so as either to track a detected target or to search for one continuously. A variety of possible scanning patterns provides for a number of search modes as well as for target acquisition and tracking. Scan rates of at least 10° per second are attainable and the powerful hydraulic drive system employed is said to eliminate backlash.

Dish illumination is by four horns, to provide monopulse operation in both planes, with separate horns for reception. The system radio frequency is in the UHF band and the radar operates on several frequencies in that band. Having regard to this and the dish diameter it is probably reasonable to suppose that the beamwidth, between half-power points, is in the region of 1 to 2°.

AN/FPS-49 radar dome at Fylingdales Moor
(UK Ministry of Defence (Air))

Output power is in the megawatt region, PRF is 27 pps (pulse duration unknown) and maximum range is about 5000 km. Each radar has three power amplifiers and in normal operation two of these are on-line at a time while the third is a 'hot spare' feeding into a dummy load. The receivers incorporate low-noise parametric amplifiers.

The antenna structure is protected from the elements by a 43 m diameter radome. This is of honeycomb-sandwich construction and consists of two high-density skins 1 mm thick with a Kraft-paper honeycomb core 15 cm thick. The honeycomb is made up in hexagonal and pentagonal blocks and 1646 of these blocks make up the radome. The transmission efficiency of the radome is 98 per cent; the average boresight error is 0.1 mil and the maximum error 0.3 mil; the boresight error rate is 0.005 mil.

STATUS

Operational. There was an FPS-49 installation in use at Moorestown, New Jersey, which filled a role in the 474N SLBM detection and warning system (see **2538.153** – FSS-7 radar – in *Jane's Weapon Systems 1980/81*) but this was phased out in December 1974. See note on FPS-50 status (**2511.153**) below.

The BMEWS tracking-stations at Clear, Thule and Fylingdales are being upgraded with the AN/FPS-115 system. Information on this upgrade is given under entry **3174.153**.

CONTRACTOR

RCA, Electronic Systems Department, Moorestown, New Jersey 08057, USA.

2511.153

AN/FPS-50 EARLY WARNING RADAR

AN/FPS-50 is the big static radar of the ballistic missile early warning system (BMEWS). It is installed at two of the three BMEWS sites at Thule (Greenland) and Clear (Alaska); there are three radars at one site, and four at the other.

The radar uses a fixed antenna system comprising a 122 m wide and 50 m high parabolic-torus reflector which is fed by an organpipe scanner. Two narrow beams at two different angles of elevation are produced by this antenna and these two beams are caused to scan simultaneously over a sector in the near-horizontal plane so as to produce two horizontal direction fans, one above the other. As a missile passes through these beams its position and velocity co-ordinates are measured, from which can be calculated the trajectory, impact point, impact time and launch point.

A parabolic-torus reflector was chosen for this radar because it provides an economical measure of scanning the beam from a physically large antenna over a wide scan angle without deterioration of pattern over this arc. The disadvantages of the arrangement are the large size of the total assembly and the large sidelobes that appear in intermediate planes.

Scanning such a reflector can be achieved in various ways, but in the AN/FPS-50 it is accomplished by arranging a series of feeds on the focal points of the torus and switching the transmitter power from one point to the next with an organpipe scanner.

At Clear and Thule the AN/FPS-50 radar is used in conjunction with the AN/FPS-49A or AN/FPS-92 tracking radar. At the third station (Fylingdales Moor, England) only AN/FPS-49 or equivalent radars are used for early warning and tracking.

CHARACTERISTICS

Peak power: 5 MW
Average power: 300 kW
Frequency: 425 MHz range
PRF: 27 Hz
Pulse width: 2 μs

STATUS

The BMEWS systems are still operational but are being upgraded at Clear, Thule and Fylingdales by the AN/FPS-115 Pave Paws radar. Information on this upgrading is given under entry **3174.153.**

CONTRACTOR

General Electric Company, Electronic Systems Division, Syracuse, New York 13201, USA.

AN-FPS-50 early warning radar

AN/FPS-50 early warning radar

2546.153

AN/FPS-85 LONG-RANGE PHASED ARRAY RADAR

This is a very large fixed-array radar located in Florida with its principal axis aligned due south across the Gulf of Mexico and capable of transmitting and receiving over an arc extending 60° on either side of this axis.

A UHF system with a nominal frequency of 442 MHz, the radar uses separate transmit and receive arrays located side by side and both inclined at approximately 45° to the horizontal. Exact dimensions of the arrays are not known but the transmitter array is square, contains 5184 transmitter modules and has a side of something over 30 m. The receiver array is octagonal, contains some 39 000 antennas feeding 4660 receivers, and has a width of some 60 m.

Peak radiated power has been reported as 32 MW with a nominal transmitter beamwidth of 1.4°. Effective receiver beamwidth is said to be 0.8°. As already noted the radar beams are steerable over 120° arc in azimuth: in elevation the coverage is about 105° – from the radar horizon to 15° beyond the zenith. The radar is thus able to track objects northwards at high elevation angles. In the normal tracking mode the receiver complex uses a 3 × 3 matrix of 'beams' (ie receiving directions) in a monopulse mode for optimum resolution. Radiated energy can be varied, by using different modulations, to suit target requirements. In the original design only two energy levels could be selected but additional modulators are being provided to extend the range of choice and hence improve the efficiency of the radar.

This power level selection, together with all other control functions, is performed by IBM 360/65 computers, there being two of these in main and standby roles in the radar complex. These computers are programmed with both routine surveillance and routine and special tracking requirements. They carry in their memory orbital details of space objects for routine tracking so that the radar beam can, at an appropriate time dependent both on its position and the relative priority of the observation, be pointed in the direction in which the target would lie if its orbit had not been perturbed: since both intentional and accidental perturbations are common, the radar will normally have to execute a small search to locate and lock on the target.

Originally designed as the main US-based active sensor of the Spacetrack system (**2825.181**) the AN/FPS-85 is now used also to supplement the AN/FPS-115 Pave Paws (**3174.153**) radars of the SLBM detection system and to operate as an alternative space defence centre (ASDC). The AN/FPS-85 receives its Spacetrack directions from NORAD space defence centre (SDC) which also has as its SPADATS inputs the BMEWS (**2525.181**) and the US Navy SPASUR sensors. At this centre all known data on space objects near the earth are filed in the computer memories and incoming data on a space object can be checked against these memories to determine whether or not it is a known object in approximately two seconds. To guard against computer or other failure at the SDC the same data are to be held on file at ASDC, which has an IBM 360/40 as well as its two IBM 360/65 computers, but in a form requiring more human intervention for a full search than is required at SDC. It is said that a full search at the ASDC may take one minute.

Similarly the SDC processes data from all SPADATS sensors fast enough to issue new orbital data for all stations three times a day, whereas the ASDC will be able to do this only once a day.

AN/FPS-85 phased array early warning radar

Development

The AN/FPS-85 was developed by Bendix, starting in 1962, and was the first large phased-array project in the USA. The first model was destroyed by fire in 1965 but was rebuilt and put into operation in 1969. It is currently operational in its Spacetrack role, dedicating approximately one-third of its time to space surveillance and the remaining two-thirds to routine and special tracking assignments from NORAD. It is also now operational as an ASDC.

It was modified, by the addition of the new modulators referred to above and of suitable software, to perform the SLBM detection function.

Contractors

Prime contractor for the radar: Allied Signal Inc, Bendix Communications Division, East Joppa Road, Baltimore, Maryland 21204, USA.

Computers: IBM.

2512.153

AN/FPS-88 SURVEILLANCE RADAR

AN/FPS-88 is a medium-size D-band static surveillance radar. It is an improved version of the long-established AN/FPS-8 (**2493.153**), the improvements relating to range performance and to signal processing facilities.

The radar features a high-gain antenna, circular polarisation, dual-channel operation, a parametric amplifier receiver, and a radar signal processor with some ECCM capability. There is also provision for using the main antenna as an IFF radiator.

Status

Operational.

Contractor

General Electric Company, Electronic Systems Division, Syracuse, New York 13201, USA.

AN/FPS-88 surveillance radar

2513.153

AN/FPS-89 HEIGHTFINDING RADAR

The AN/FPS-89 is a high-power (4.5 MW, peak) E/F-band nodding-beam radar. Accuracy of heightfinding is ±1000 ft (300 m) at 110 nautical miles (203 km) range. It is an improved version of the AN/FPS-6 radar (**2492.153**), and like the AN/FPS-6 it is available in an arctic tower installation, a temperate tower installation and a mobile version

An Italian licence-built version, AN/FPS-89S, is produced by GE, and among a number of improvements incorporated is a reduction in the receiver noise figure from 9 to 4.5 dB.

Status

Operational.

Contractor

General Electric Company, Electronic Systems Division, Syracuse, New York 13201, USA.

2847.153

AN/FPS-105(V) RANGE INSTRUMENTATION RADAR (CAPRI)

AN/FPS-105(V) is designed for range use to make accurate measurements of target position and velocity for performance evaluation. It can be installed either in a permanent ground station or be packaged for transportable use.

Its standard antenna is a 366 cm diameter solid-surface parabolic dish fed by a monopulse four-horn feed. Polarisation is vertical. The antenna can rotate continuously in azimuth and can be rotated through 200° (–10 to +190°) in elevation, a cable being used to avoid the need for a rotating joint. Microwave circuits are mounted on the rotating section of the antenna mounting. Larger antenna pedestals, 488 or 885 cm, are fitted to some systems. The transmitter will radiate single pulses, pairs, or triplets as required. Peak power is 1 MW and recurrence frequencies and pulse widths can be chosen for 160 and 640 pps and from 0.25, 0.5 and 1 microseconds respectively. Transmitter frequency is in the G/H-band.

The digital range tracker contains a precise

AN/FPS-105 (V) range instrumentation radar

frequency standard and 'nth time round' circuits to enable a relatively high recurrence frequency to be used despite the long measuring ranges. Automatic search and lock-on are provided during target acquisition. Twenty-five bits of range data are supplied by the system at data rates up to 40 times per second.

CHARACTERISTICS

Azimuth tracking rate: 850 mils/s

Azimuth elevation tracking acceleration: 800 mils/s²

Azimuth elevation tracking precision: 0.2 mils rms at 20 dB S/N

Elevation tracking rate: 500 mils/s

Range tracking precision: 4.7 m rms at 20 dB S/N

Unambiguous tracking range: 60 000 km

Ro, IM², 0 dB S/N: 200 km

STATUS

Five units of this configuration are in use at several US ranges.

CONTRACTOR

RCA, Electronic Systems Department, Moorestown, New Jersey 08057, USA.

4326.153

AN/FPS-117 AIR DEFENCE RADAR

The AN/FPS-117 solid-state radar system is an L-band, three-dimensional air defence radar designed to provide long-range accurate aircraft identification and position data for air defence, navigational assistance and tactical control for both close air support and counter air operations. It is essentially a fixed-site derivative of the AN/TPS-59 system (**3282.153**)

The system provides automatic adaptation to, and rejection of, land, sea and weather clutter by using clutter rejection processing. Adaptability in siting the radar is achieved by low-beam coverage enhanced by MTI processing, with doppler filtering for look-down beam positions on elevated sites. Sidelobe nulling is used to eliminate ground clutter for high beam positions.

The planar array measures 7.3 × 7.3 metres and produces a series of pencil beams phase-positioned to scan in elevation up to 20° while the complete antenna rotates in azimuth. The elevation scan consists of 5 to 100 nautical miles short-range beams, and 100 to 250 nautical miles long-range beams. The use of pencil beams is designed to obtain elevation coverage while eliminating clutter problems associated with large transit beamwidths. The transmitter/receiver system consists of 44 row transmitters, each containing eight 100 W RF solid-state power modules, and 44 receiver modules, all mounted on the rear of the array. IFF systems supplied by Hazeltine are incorporated in the radar.

CHARACTERISTICS

Range: 200 nm

Accuracy: 0.25 nm

Altitude: 100 000 ft (30 480 m)

Elevation angle: –6 to +20°

Transmitter power: 24.75 kW peak

Frequency: 1215 – 1400 MHz

Bandwidth: 185 MHz

Agility: 20 frequencies (quasi-random selection, beam-to-beam)

STATUS

Twelve AN/FPS-117 systems have now been delivered and made operational as part of the Alaskan-based air defence system 'Seek Igloo', at a cost of $113 million. The AN/FPS-117 will also form part of the new North Warning System which replaces the old DEW line radar system in Northern Canada and Alaska. GE has been awarded an $89 million contract for the delivery and installation of 15 radars, of which 13 are for the North Warning System. Full operation of the North Warning System is expected by 1992. 16 AN/FPS-117(V) 3 systems are being installed in Saudi Arabia to update the country's air defence system, under the Peace Shield programme. It is understood that a version of the AN/FPS-117 is one of several contenders for the US Federal Aviation Agency Radar Replacement (FARR) joint civil/military programme for minimally attended radars.

In early 1987, GE received a $62 million contract for four long range radars for the NATO Iceland Air Defence System at four sites. The radar type has not been disclosed but it is assumed to be the FPS-117 or one of its variants. The contract is expected to be completed by mid-1990.

CONTRACTOR

General Electric Company, Electronic Systems Division, Syracuse, New York 13201, USA.

3713.153

AN/GPN-22 PRECISION APPROACH RADAR

The AN/GPN-22 precision approach radar (PAR) is intended specifically for use in fixed base high-density ATC system operations under all weather conditions. It operates in X-band and is a development of the high-performance AN/TPN-25 PAR (**3712.153**) used in the USAF AN/TPN-19 landing control system (**2537.153**). Both these radars employ phased array antennas to perform simultaneous scanning and tracking modes over the full radar coverage volume without any mechanical motion of the antenna. In the case of the TPN-25, up to six aircraft targets can be monopulse tracked simultaneously; in the case of GPN-22, the same number of targets can be handled but the coverage pattern measures 20 × 8° compared with the 20 × 15° of the earlier equipment. The range is about 35 km in both cases. Provisions are made for the display of landing aircraft position data in the operations centre up to 3500 m from the radar head.

The only significant difference between the GPN-22 and the TPN-25 is in the physical arrangements of the antenna. The GPN-22 feed assembly is supported by a boom structure projecting over the upper edge of the main reflector, whereas in the TPN-25 it is mounted beside the reflector.

STATUS

Deliveries have been made of 39 sets to the USAF, 2 sets to Austria, 2 sets to South Korea and 11 sets to the Netherlands.

CONTRACTOR

Raytheon Company, Equipment Division, Wayland, Massachusetts 01178, USA.

2497.153

AN/MPQ-4 MORTAR LOCATING RADAR

The AN/MPQ-4 provides quick identification to pinpoint enemy mortar positions in map co-ordinates, enabling artillery units to launch counter-attacks. It uses a two-beam intercept principle of location with positive first-shell acquisition and has the capability for handling multiple targets.

All the equipment is mounted on two two-wheel trailers, each designed for its specific purpose. One trailer carries the primary radar equipment, while the other carries the primary power supply (a petrol engine generator set) and auxiliary equipment such as cable and spare parts.

The AN/MPQ-4 can be operated from virtually any site and can be set up in 15 minutes. Once set up it can be operated by one man. The operating and control console may be used on the equipment trailer or in a remote location.

STATUS

Operational but no longer in production. It has been largely replaced by the Hughes AN/TPQ-36 (**2848.153**). It is produced under licence in Japan as the Type 92 mortar locating radar JAN/MPQ-N1.

CONTRACTOR

General Electric Company, Electronic Systems Division, Syracuse, New York 13201, USA.

4478.153

AN/MPQ-53 GUIDANCE RADAR

The AN/MPQ-53 is a multi-function radar group which performs all the surveillance, IFF, tracking and guidance functions entailed in the Patriot (MIM-104) tactical air defence missile system (**2800.131**). The antenna array is a multi-element phased-array planar configuration carried on a semi-trailer chassis. The antenna unit has separate arrays for target detection and tracking, missile guidance and IFF functions. The last of these tasks is carried out by an AN/TPX-46(V)7 interrogator, using a supplementary array adjacent to the main circular search and track array on the antenna unit. Other supplementary arrays are for command and guidance signal and reception.

STATUS

In service with US forces, planned for NATO use in Europe.

CONTRACTORS

Prime contractor: Raytheon Company, Andover, Massachusetts, USA.

IFF interrogator: Hazeltine Corporation (a member company of Emerson Electric Co), Greenlawn, New York 11740, USA.

Side view of AN/MPQ-53 missile guidance and control radar group for Patriot air defence missile system reveals main planar array and supplementary antennas for other functions on single unit which folds down for transit

5628.153

REAL-TIME VELOCIMETER RADAR (RTVS)

This system is a transportable X-band CW doppler artillery tracking radar designed to measure and record target radial range, velocity, acceleration and angular position. The equipment is carried in two mobile elements; an antenna trailer which contains the transmitter and receiver plus the steerable antenna system and a control cabin which carries the operator control console and data processing units. The RTVS is intended to allow tracking of targets from 100 to 10 000 ft/s and will acquire and track targets down to 0.0032 m². Accuracy at 20 km is stated to be 1 ft/s for velocity, 1 ft/s² for acceleration and 6 m in range. Operating frequency of the radar is pre-tunable between 10 and 10.26 GHz.

Range in the system is determined without modulating the carrier by transmitting three carrier waves simultaneously and measuring the phase shift between transmitted and received beat notes. A high speed array processor is used to perform digital spectral analysis of the returned signal. The fast Fourier transform (FFT) processor extracts target velocity from the raw doppler signal, and spectral details of the transmitted signal resolve the difference.

Unambiguous real time ranging to 75 km is obtained from the FFTs applied to the doppler shifted return signal from the three transmitted tones, where the relative shift is proportional to range. The transmitter uses a stable UHF synthesiser and a low noise master oscillator klystron to drive the final power stage which consists of three klystrons.

STATUS

In operational use at the US Army White Sands Missile Range.

CONTRACTOR

Datron Systems Inc, 20700 Plummer Street, Chatsworth, California 91311, USA.

2495.153

AN/MPS-11 AND AN/MPS-11A SEARCH RADARS

The AN/MPS-11 and AN/MPS-11A are mobile versions of the AN/FPS-8 medium-power D-band search radar (**2493.153**). The difference between the two mobile versions is that the AN/MPS-11 antenna is mounted after arrival at site on a manually erected tower, while the AN/MPS-11A antenna is permanently installed for transport and operation on a 40 mm gun carriage, permitting more rapid initial operating capability.

The AN/MPS-11 may be transported by aircraft or by two trailers and nine M35 trucks. Set-up time, with 15 men, is two hours.

STATUS

Operational.

CONTRACTOR

General Electric Company, Electronic Systems Division, Syracuse, New York 13201, USA.

AN/MPS-11A search radar

AN/MPS-11 search radar

2496.153

AN/MPS-14 HEIGHTFINDING RADAR

The AN/MPS-14 radar is the mobile version of the AN/FPS-6 high-power E/F-band nodding-beam radar (**2492.153**). It is designed for transport by six trucks and three trailers to temporary strategic or tactical sites. Trailers are functionally designed for quick loading and unloading, and assembly and disassembly may be accomplished in rapid and efficient fashion.

The trailers serve the dual purpose of transporting the equipment and supporting certain electronic components during operation. For example, the RF equipment need not be removed from the trailer for operation, and the entire unit is protected by a pre-fabricated shelter assembled at the site. The shelter is equipped with space heaters and lighting fixtures.

The AN/MPS-14 system's electronic components are identical to and interchangeable with those of the AN/FPS-6.

STATUS

Operational.

CONTRACTOR

General Electric Company, Electronic Systems Division, Syracuse, New York 13201, USA.

AN/MPS-14 heightfinder radar

2514.153

AN/MPS-36 INSTRUMENTATION TRACKING RADAR

AN/MPS-36 is a C-band mobile instrumentation radar which can accurately track targets travelling up to 36 000 knots at ranges up to 60 000 km. Data accuracy has been measured at one metre in range and target velocities within 5 cm/s on trajectories of missiles and missile re-entry vehicles.

Advanced integrated circuits are extensively designed into the equipment and pulse doppler capability is integral. A general-purpose computer has been integrated into the radar to provide pulse doppler signal processing. Polarisations are console selectable.

The radar is housed in a 12.4 m electronics trailer and a 366 cm diameter antenna is mounted on an 11 m pedestal trailer. With this arrangement, the radar can be moved from site to site by road, air, or sea and be operational within eight hours after arrival.

CHARACTERISTICS

Transmitter power: 1 MW minimum
Antenna gain: 43 dB minimum
Antenna beamwidth: 1.2° max
Antenna sidelobes: 20 dB minimum
Pedestal coverage (elevation): –5 to –185°
Pedestal coverage (azimuth): Continuous
Azimuth tracking rate: 890 mils/s
Elevation tracking rate: 500 mils/s
Azimuth acceleration rate: 400 mils/s²
Elevation acceleration rate: 400 mils/s²
Azimuth tracking precision at 20 dB S/N: 0.2 mils, rms
Elevation tracking precision at 20 dB S/N: 0.2 mils, rms
Range tracking precision at 20 dB S/N: 2.7 m rms
Range rate tracking precision at 20 dB S/N: <0.03 m/s rms accuracy
Unambiguous range tracking: 60 000 km
Installation time to power on: 8 h max
Receiver noise figure: 4 dB max
Ro, IM², 0 dB S/N: 350 km
Reliability (design goal): MTBF 100 h

STATUS

Fourteen radars are now operational at US test ranges and three have been provided to the Federal Republic of Germany. The radars are used for studies of weapons systems and space programmes.

CONTRACTOR

RCA, Electronic Systems Department, Moorestown, New Jersey 08057, USA.

AN/MPS-36 missile tracking radar

2540.153

AN/PPS-5 COMBAT SURVEILLANCE RADAR

AN/PPS-5 is a portable battlefield radar designed to detect and locate individual men up to 5000 m and groups of men or small vehicles up to 10 000 m. A later model, the AN/PPS-5B, which incorporates advanced technology, is now the standard equipment.

The AN/PPS-5 solves the major problems of personnel detection radars by providing automatic sector azimuth scanning and range coverage with visual display of moving targets over the sectors under line-of-sight surveillance. The radar can be switched from MTI operation to normal operation, permitting the operator to locate fixed targets.

A range-gated filter MTI processor is especially suitable for detecting targets moving as slowly as 1.6 km/h. Operation while scanning is silent.

The control/indicator unit provides remote control of the radar set with the appropriate readout and control devices as well as visual displays of 10 000 m range in 500 m increments on both MTI and normal displays (A-scope and B-scope). Target range is measured by a manually controlled magnetostrictive delay line coupled to a counter; this system was selected instead of the more conventional pulse-delay system because it permits table calibration without having to add calibrators and to make slope, zero, and other adjustments. Target range is measured by a manual range control knob coupled to a direct-reading range counter. A range mark on the display scope shows the position of the range gate. An aural doppler processor consisting of a clutter filter and audio amplifier drives the headphones.

A 50-channel range-gated filter system is used to provide MTI over a 5000 m range. Each range-gated filter channel in the system sequentially covers a 100 m range increment. Each channel is identical in design. The outputs from the 50 filters are added to provide MTI video for the displays.

The selectable scan speeds are held relatively constant by a feedback motor-speed control circuit that supplies only enough power to meet the demand in the presence of varying wind loads; this conserves power. Azimuth co-ordinate readout is obtained remotely from a follow-up servo that repeats the antenna position data. To minimise power consumption, this control is used on demand by the operator only when fine azimuth data are desired.

One version of the radar set has been equipped with a 360° scan and a PPI display kit for the US Army Special Forces. Another version of the radar set has been installed in a vehicle and used for perimeter protection of air bases.

The equipment works in the 16 to 16.5 GHz band (tunable) and has a peak power of 1 kW and a PRF of 4000 pps. An accuracy of ±20 m in range and ±10 mils in azimuth can be achieved.

Physically the equipment consists of three rugged packages weighing about 40 kg. Power consumption is about 57 W.

STATUS

Quantity deliveries of the PPS-5 have continued over a period of some years and some 600 units have been delivered to the US Army. The PPS-5B is now the standard issue in the US Army. In addition to the USA, other unspecified governments have been using the AN/PPS-5A since 1967 and have procured the AN/PPS-5B. The equipment is no longer in production but will continue in operational use for some years.

CONTRACTOR

Eaton Corporation, Command Systems Division, 815 Broad Hollow Road, Farmingdale, Long Island, New York 11735, USA.

6815.153

AN/MPS-39 INSTRUMENTATION TRACKING RADAR

The AN/MPS-39 instrumentation tracking radar has been developed for the US Department of Defense under contract to the US Army's White Sands Missile Range, New Mexico, and is the first MOTR (Multiple Object Tracking Radar) to be produced under this contract. It is a computerised system with an electronically steered phased antenna and a high power transmitter. It can track several targets simultaneously with better accuracy than existing single target radars. These MOTRs will be used by the US DoD to upgrade its test facilities at a number of locations.

STATUS

The first radar was scheduled to be deployed at the White Sands Missile Range in late 1987.

CONTRACTOR

Electronic Systems Department, RCA Aerospace and Defense, General Electric Company, Moorestown, New Jersey 08057, USA.

AN/MPS-39 instrumentation radar under test

5616.153

AN/PPS-11/12 BATTLEFIELD SURVEILLANCE RADARS

The AN/PPS-11 is a hand-held battlefield surveillance radar which weighs only 4.5 kg. It is a modulated CW doppler correlation equipment, operating in Z-band with a range of approximately 1 km against vehicles and an accuracy of 8 m. The operation of the system is extremely simple in that once the equipment has detected a target in the 1 km range, the operator can narrow the options by selecting various range gates until the target is fixed and the range and bearing are available. An audio signal is also available for identification of the type of target.

An improved and expanded version of this equipment, known as the AN/PPS-12, weighs about 7 kg and has an increased range of up to 3 km. A number of refinements, such as automatic scanning, automatic target warning and simultaneous scan and search have also been provided.

STATUS

In operational service.

CONTRACTOR

RCA, Electronic Systems Department, Moorestown, New Jersey 08057, USA.

1947.153

AN/PPS-15A INFANTRY RADAR

The AN/PPS-15A is a J-band, lightweight ground surveillance radar for the detection and location of moving targets such as men, vehicles, or boats. Range and azimuth of a moving target are read directly from solid-state LED displays, and the nature of the target is deduced from its aural signature. The radar set consists of the antenna assembly, control indicator, an antenna drive, a headset, a tripod, remote cable and an internal disposable battery which provides power for 12 hours continuous operation. The AN/PPS-15A can be hand-held, pintle or tripod mounted and remotely controlled. The control indicator and antenna assembly are integral for most operations but may be separated for remote operation at distances up to nine metres. Automatic sector scan can be centred on a selectable bearing, and the scan width also is selectable. Elevation control is manual.

Target returns are fed to an 'all-range' channel and a 'discrete-range' channel. During search, the all-range channel automatically discriminates against clutter and produces both aural and visual alarms when a moving target is detected. During ranging, the discrete-range channel produces a light display and an increased doppler tone when a moving target enters a range gate positioned by an operator-control.

AN/PPS-15A lightweight ground surveillance radar

Both channels are active during the search-range mode, enabling an operator to monitor detected targets on the all-range channel as he ranges a selected target by means of the discrete-range gate control. Five frequencies are available for selection by the operator.

The AN/PPS-15B (**4324.153**) is an advanced version and can be hand-held, pintle-mounted or tripod-mounted on the ground or on a vehicle. It can be remotely controlled.

CHARACTERISTICS

Frequency: Low J-band
Peak power: 50 mW
Antenna: 20 × 30 cm
Mode: Homodyne
Weight: 10.7 kg (remote operation)
Detection ranges: Personnel 1.5 km; vehicles 3 km
Battery: BA4386/PRC-25

STATUS

The AN/PPS-15A radar equipment has been US Army 'type classified' for use to the armed services, and the following have been delivered: 170 to Argentina, 215 to Canada, 25 to Norway, 150 to Spain, 40 to the UN and 1000 to various US military units. Additional sales have recently been made to the UN and Norway, and both Israel and Taiwan have taken delivery of an unspecified number. Advanced versions are being developed and tested, the latest version being in support of the US Air Force ground-launched cruise missile programme.

CONTRACTOR

Amex Systems Inc, Hawthorne, California, USA.

4324.153

AN/PPS-15B BASE SECURITY RADAR

The AN/PPS-15B is a development of the AN/PPS-15A and provides additional surveillance capability in a 350 m range segment, divided into seven 50 m range gates. The system is designed to detect and locate moving targets, eg vehicles and personnel, to provide perimeter surveillance for virtually any type of installation.

The 350 m segment may be positioned by the operator anywhere within the detection range of the system to provide surveillance without interference from, or alarms caused by, traffic moving inside the base perimeter itself. The seven range gates are individually displayed to provide automatic detection. Sector scan width is variable, by the operator, between limits of 22 and 180°. Five transmitter frequencies are available for selection by the operator. Power sources are supplied with the radar which can be operated from either AC (115 or 230 V) or DC (12 or 24 V).

The AN/PPS-15B can be hand-held, pintle or tripod mounted and remotely controlled. Both aural and visual alarms are produced when a moving target is detected, and the LED type display is identical to that of the AN/PPS-15A.

CHARACTERISTICS

Frequency: Low J-band

Detection range: 3000 m

Remoting: 35 m

Power source (external): 115/230 V AC, 12/24 V DC

Weight

Radar: 15 kg (including DC power source and cable)

DC power source: 7 kg

AC power source/battery charger: 9.5 kg

Remote cable and reel: 7 kg

STATUS

In service with the US armed forces. The latest version is being supplied in support of the US Air Force ground-launched cruise missile programme.

CONTRACTOR

Amex Systems Inc, Hawthorne, California, USA.

AN/PPS-15B base security radar

4325.153

TYPE 386 GROUND SURVEILLANCE RADAR

The type 386 radar is the latest in a range of ground surveillance radars produced by General Dynamics, and is a lightweight system with a detection range of 6 km. A number of advanced features have been incorporated, including a hand-held remote control/indicator unit with keyboard push-button controls and a 32 × 16 LED matrix synthetic B-scope display for visual presentation of moving target information. Target detection tracking and alarm warning is automatic and a radar netting capability is incorporated. Remote operation is available out to approximately 1.5 km using a standard military single pair field wire.

Both visual and aural warning are provided, target doppler being fed to the operator's headset for identification. Sector scan width is variable by the operator from 22.5 to 90°.

CHARACTERISTICS

Frequency: Low J-band (previously X-band)

Detection range: To 6 km

Modulation: CW pseudo-random code (PRN)

Power output: 100 mW

Sector scan width: Adjustable 22.5° to 90°

Weight: Basic radar (9 m remote) 12.6 kg; extended remote (optional to 1.5 km) 16 kg

STATUS

The radar has been demonstrated in US Army and US Air Force tests using RF and field wire links for remote operation.

CONTRACTOR

Amex Systems Inc, Hawthorne, California, USA.

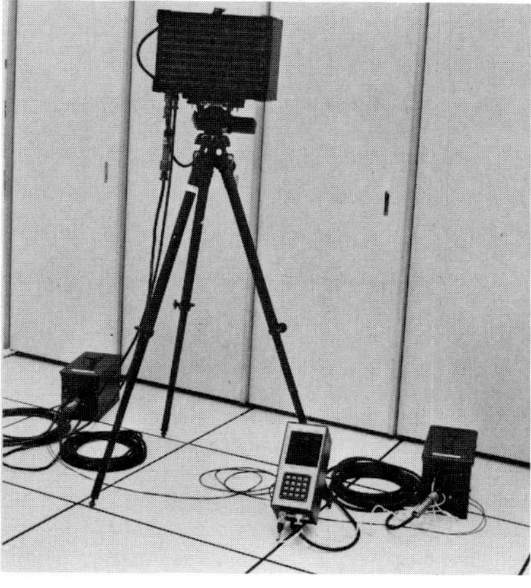

Type 386 lightweight ground surveillance radar

3129.153

AN/TPN-18A/18B RADAR

The AN/TPN-18A ground controlled approach radar system, developed for the US Army, is a lightweight, tactical ground-based radar which provides precise three-function information to airport/heliport controllers. These functions consist of terminal area surveillance, precision approach for control of landings, and heightfinding for aircraft monitoring. This versatile system is readily transported by helicopter, cargo aircraft, or truck, and is suitable for truck/trailer mounting for full mobility in combat-related operations. The earlier TPN-8 version, designed for the US Marine Corps' military all-weather tactical operations, has since been adapted by other military services. Radar set AN/TPN-18A is a component of landing control central AN/TSQ-71A and AN/TSQ-72.

The AN/TPN-18B ground control approach radar system is a modified version of the AN/TPN-18A and features digital MTI (DMTI). The radar receiver-transmitter group has been placed in a shelter to provide improved maintainability. The IFF/SIF equipment used in the AN/TSQ-71B landing control central has been upgraded to a modern solid-state system.

STATUS

The AN/TPN-18A is in US Army service. The AN/TPN-18B is a prototype system which has undergone tests with the US Army. More detailed information is given in previous editions of *Jane's Weapon Systems*.

CONTRACTOR

ITT Gilfillan, 7821 Orion Avenue, Van Nuys, California 91409, USA.

2537.153

AN/TPN-19 LANDING CONTROL CENTRAL

AN/TPN-19 is a terminal ATC system comprising an airport surveillance radar (ASR), a precision approach radar (PAR) (AN/TPN-24 and AN/TPN-25 respectively), and an operations centre. The radar systems are linked to the operations centre by microwave links (10 nautical miles). The system is designed for rapid world-wide deployment as well as fixed base installations. Packaging techniques make the system completely transportable by aircraft, helicopter or road vehicle.

The AN/TPN-24 ASR (**3711.153**) is a dual-channel, 2 × 500 kW E/F-band, MTI radar. Its most noteworthy features are a multiple (12) horn antenna feed which produces lobe-free coverage up to 12 000 metres and 60 nautical miles on a 1 m² target; a low-angle pattern which is electrically variable and programmable in range and azimuth to reduce clutter returns; a digital coherent and non-coherent MTI system to eliminate weather returns, and a staggered PRF to eliminate blind speeds below 1100 knots.

The AN/TPN-25 PAR (**3712.153**) is said to represent a significant advance in the state-of-the-art. Operating in I/J-band, it has a range of 20 nautical miles even with precipitation at 5 cm/hour. This results from the use of the narrow pencil beam (0.75 × 1.4°) produced by the 824-element phased-array antenna of 320 kW. A monopulse system, it can track up to six targets simultaneously. For multiple runway coverage the antenna assembly can be slewed through 270°.

In the GCA configuration there are three displays but through simple modular extension the system provides seven displays for a full RAPCON configuration. All displays are dual mode and can be changed to display either an ASR or PAR presentation simultaneously.

STATUS

Both the complete TPN-19 and separate examples of the two radar sets incorporated in it have been supplied or ordered for US and foreign forces. The US Air Force has 11 TPN-19s; Australia has one TPN-19, six TPN-25 PARs and one TPN-24 ASR; the Netherlands has fixed installations for six TPN-24 ASRs and 11 GPN-22 PARs (the latter being an improved TPN-25); Austria has two GPN-22s; the US Air Force has taken delivery of 39 GPN-22s; and West Germany has been supplied with 44 TPN-24s for fixed base installation.

CONTRACTOR

Raytheon Company, Equipment Division, Wayland, Massachusetts 01178, USA.

4075.153

AN/TPN-22 PRECISION APPROACH AND LANDING RADAR

The AN/TPN-22 is a precision track-while-search radar that uses phase/frequency scanning to provide high data rates for automatic detection and tracking while simultaneously searching a 46 × 8° sector. The radar consists of antenna, transmitter, receiver, digital processor and sub-systems.

The AN/TPN-22 planar array antenna operates at I-band (former X-band) and radiates a computer-controlled pencil beam using phase/frequency scanning. The antenna consists of 94 electronic digitally controlled diode phase shifters feeding 94 serpentine arrays and incorporates circular polarisation for improved performance in bad weather.

The transmitter final stage consists of an air-cooled, high-power TWT using a solid-state regulated high voltage power supply and line-type modulator. The transmitter first stage is also a TWT amplifier.

STATUS

In service with the US Marine Corps. More detailed information is given in previous editions of *Jane's Weapon Systems*.

CONTRACTOR

ITT Gilfillan, 7821 Orion Avenue, Van Nuys, California 91409, USA.

4077.153

PAR-80 PRECISION APPROACH RADAR

The ITT-Gilfillan PAR-80 is a fixed site, turntable-mounted, precision approach radar (PAR) set that features a pencil-beam planar array antenna employing ferrite phase shifters for azimuth beam displacement and mechanical actuator for vertical beam positioning. Radar scan angles of 30° in azimuth and 7° in elevation when combined with the 10 or 20 nautical mile radar range provide large service volume coverage for all aircraft types and traffic pattern variations. The PAR-80 utilises circular polarisation for operation in adverse weather, with optional constant false alarm rate (CFAR) circuitry.

The radar set group (antenna, scan programmer, and the dual receiver/transmitters) are pallet-mounted on a turntable that is remotely controlled for coverage of multiple runway approaches. Control/indicators with an accompanying video processor may be remotely controlled from operations centres or control towers up to 3048 m distant.

The PAR-80 incorporates solid-state design, digital clutter-reduction sub-systems, dual-channel radar transmitters and receivers, high-gain low-sidelobe antenna, remote control of PAR landing direction with four sets of preset course and glideslope cursors and other features.

CHARACTERISTICS

Frequency: 9000 – 9160 MHz
Pulse width: 0.2 µs
PRF: 3450 Hz
Peak power: 150 kW
Beamwidth: 1.1° horizontal × 0.6° vertical
Antenna gain: 45 dB
Sidelobes: –22 dB max
Max range: 37 km
Resolution: 1.1° azimuth; 0.6° elevation; 70 m range
Receiver noise: 11 dB max
Signal processing: Lin/log FTC and linear; log CFAR (optional)

STATUS

The radars have been installed at West German Air Force and Navy bases, and at Belgian Air Force bases.

CONTRACTOR

ITT Gilfillan, 7821 Orion Avenue, Van Nuys, California 91409, USA.

ITT Gilfillan PAR-80 precision approach radar

5623.153

ARSR-3/ASR-30 SURVEILLANCE RADARS

The ARSR-3 and the ASR-30 are minimally attended L-band surveillance radars which are suitable for both civil and military requirements. The systems meet both FAA and military specifications and use the latest technology.

The ARSR-3 provides cover up to 200 nautical miles (370 km) in range and 100 000 feet (30 km) in elevation against a 2 m² target. A number of refinements are used to detect small targets against heavy clutter. These include dual beam elevation pattern to enhance short range detection, an antenna pattern which produces a sharp cut-off underside to reduce ground returns, and RF sensitivity time control to eliminate close-in ground reflections.

The ARSR-3 has complete redundancy in electronics with extensive built-in test equipment, and a demonstrated MTBF of 3000 hours. The system also has a very high weather clutter cancellation feature by the use of circular polarisation.

The ASR-30 is based on the ARSR-3 and is a dual channel L-band surveillance radar that will detect small aircraft (2 m²) out to 120 nautical miles (220 km). Using the circular polarisation mode, targets can be detected throughout severe weather conditions and it is claimed to reject weather with over 40 times the efficiency of S-band surveillance radars.

STATUS

In operational service in USA, Canada and Switzerland.

CONTRACTOR

Westinghouse Electric Corporation, PO Box 746, Baltimore, Maryland, USA.

ASR-30 surveillance radar

5780.153

W-2000 3-D SURVEILLANCE RADAR

The Westinghouse W-2000 L-band radar is a 3-D system designed to meet the current and forseeable future requirements for air defence and civil surveillance systems. It is designed to provide accurate target data while operating in hostile environments caused by interference from active jammers, chaff, ground clutter, heavy precipitation and bird migrations. It is completely solid-state, can be configured for static or mobile applications and may be remotely controlled and monitored if minimally or unattended operation is required. An SIF/IFF system in integrated.

ECCM features include low sidelobes, a wide-band phase-steerable planar antenna, multiple and selectable receive beams, MTI processing, doppler filtering, pulse compression and a method of minimising false alarms. Coupled with these is jam strobe reporting and a jamming frequency avoidance system. This latter includes a signal analyser for clear frequency selection and frequency agility in burst-to-burst and pulse-to-pulse modes.

The spatial volume, up to 500 km in range and 30 000 metres in height, is covered by an optimum search pattern which is achieved by the use of three virtually independent radars operating at different frequencies and elevations. This system triples the available dwell time on all targets.

CHARACTERISTICS

Frequency: L-band
Instrumented range: 270 n miles (500 km)
Height accuracy: ± 1000 ft (305 m) at 100 n miles
Resolution: 1.4° azimuth; 500 ft (152 m) range
Antenna type: Phase steerable in elevation (6 simultaneous beams)
Gain: 39 dB receive; 35 dB transmit
Transmitter: Solid-state, distributed

Antenna of the W-2000 surveillance rader

CONTRACTOR

Westinghouse Electric Corporation, PO Box 1897, Baltimore, Maryland, USA.

3711.153

AN/TPN-24 AIRPORT SURVEILLANCE RADAR

The AN/TPN-24 is the airfield surveillance radar employed in the AN/TPN-19 landing control central system (**2537.153**), and was orginally designed for that system although it has subsequently been widely delivered separately for use on its own or with other equipment. The original version was designed and packaged for use as a highly mobile, tactical radar equipment, but other versions have been configured for different applications, including a fixed site variant, a tower-mounted model (ASR-910), and a semi-static version designated the ASR-909. The antenna and transmitter/receiver units are the same in each version as those of the basic AN/TPN-24 (see photograph on next page).

CHARACTERISTICS

Radar: Dual channel, dual frequency, diversity. Redundant or single channel and spectrum filter options
Frequency: 2700 – 2900 MHz
Transmitter type: Magnetron
Peak power: 500 kW (per channel)
Pulse length: 1 µs
Pulse rate: 1050 pps, average
PRF: 12 staggers
Antenna size: 4.26 × 2.43 m
Azimuth beamwidth: 1.6°
Elevation beamwidth: 4°
Rotation speed: 13 rpm
Beam elevation tilt: + 15 to –2°
Detection range: 110 km (clear weather, 1 m² target)

STATUS

In addition to those supplied for use in AN/TPN-19 systems (a total of about 12) more than 50 separate AN/TPN-24 sets have been supplied, including: 44 fixed base installations for West Germany and six similar installations for the Netherlands. The AN/TPN-24 has also been supplied to Australia and Austria.

CONTRACTOR

Raytheon Company, Equipment Division, Wayland, Massachusetts 01178, USA.

3712.153

AN/TPN-25 PRECISION APPROACH RADAR

The Raytheon AN/TPN-25 is the high-performance precision approach radar (PAR) developed for the AN/TPN-19 landing control central tactical airfield control system (**2537.153**). Designed to fulfil a requirement for a highly mobile system capable of assisting the landing of fixed-wing aircraft and helicopters in adverse weather conditions, the AN/TPN-25 can also be deployed at fixed bases. It employs a unique phased-array antenna that constantly scans the entire service volume of 20 × 15° and 35 km in range, as well as tracking up to six approaching aircraft in elevation and azimuth. No antenna movement is required except when it is necessary to re-orientate it to allow a different runway direction to be used. This can be performed under remote control to permit any one of four runways on an airfield to be served. The TPN-25 is capable of automatic remote feeding of guidance information to the landing aircraft via a data link.

Operation is in the radar I-band and no mechanical scanning is necesary. A four-lobe monopulse feed horn is directed at the main antenna reflector, radiation having to pass through a digitally-controlled microwave 'lens' consisting of an array of 824 phase-shifting elements. The latter are controlled to produce sharp, digitally-steered, step-scanned pencil beams.

CHARACTERISTICS

Radar: Single-channel, dual function (scan/track)
Frequency: 9000 – 9200 MHz
Transmitter type: Solid-state oscillator with TWT and CFA amplifiers
Peak power: 320 kW
Pulse length: 1 μs

AN/TPN-25 PAR

PRF: 3500 pps (average)
Stagger ratio: 3:2
Antenna beamwidth: 1.4° azimuth; 0.75° elevation
Detection range: 37 km (clear weather, trainer aircraft target)

STATUS

Production. Apart from about 12 sets supplied as part of complete US Air Force AN/TPN-19 systems, another eight equipments have been supplied to Australia, both as fixed-base systems and in the tactical mobile mode. A development of the AN/TPN-25, with higher performance characteristics, has been developed as the AN/GPN-22 and this has been ordered and supplied in quantity both to the US Air Force and to foreign customers (**3713.153**).

CONTRACTOR

Raytheon Company, Equipment Division, Wayland, Massachusetts 01178, USA.

2498.153

AN/TPQ-10 RADAR COURSE DIRECTING CENTRAL

The AN/TPQ-10 radar course directing central is designed to provide accurate all-weather guidance to aircraft for performance of close air support missions.

The equipment consists of a tracking radar and an operational shelter containing the radar controls, a computer, displays and communication equipment. Designed specifically for tactical air operations, it is easily transported by helicopter, cargo aircraft or truck.

Further information is given under this entry number in previous editions of *Jane's Weapon Systems*.

STATUS

In operational use but no longer in production.

CONTRACTOR

General Electric Company, Electronic Systems Division, Syracuse, New York 13201, USA.

1526.153

FORWARD AREA ALERTING RADAR AN/TPQ-32/MPQ-49/MPQ-54

The forward area alerting radar (FAAR) was designed to meet a US Army requirement for a lightweight early warning air surveillance radar to increase the effectiveness of air defence weapons, such as Chaparral and Redeye missiles and the Vulcan 20 mm gun system, against low-level aircraft attacks. It has been developed in both transportable and mobile versions, these being designated TPQ-32, MPQ-49 and MPQ-54, the latter two being the mobile versions. The principal system elements and operation are identical or similar in both.

The complete system comprises an integral primary and secondary (IFF) radar antenna assembly, and a truck-mounted shelter containing radar data extraction and processing units, operator's console and VHF data link. The weight of the system is given at 1865 kg.

The 20 km primary radar is a pulse-doppler system using a low-noise stable master oscillator, a grid-modulated TWT transmitter and an all solid-state homodyne (zero IF) receiver. The receiver translates target information to video frequencies which are range gated, doppler filtered, and compared with an alarm threshold. Target data are presented on the operator's console CRT. The antenna measures 183 × 107 cm and is carried on a telescopic mast. The latter is of four sections, pneumatically operated, and capable of extension to a height of about 11 m. 360° coverage is provided. A horn antenna for the IFF is mounted on the upper edge of the main reflector. The IFF system used is the AN/TPX-50, and this provides for either automatic or manual target interrogation.

The operator's console has a 25 cm diameter PPI display which normally presents only processed video which is time-shared with the PPI generator and the symbol generator to show targets. Controls are provided for selection of 'black-scope' or processed video display, sector blanking, threshold adjustment, automatic or manual IFF interrogation, IFF mode selection, radar transmitter/receiver control, self-test functions and data link operation. The console can be removed from the truck or shelter to enable the complete system to be operated from a position up to 50 metres from the radar site.

FAAR AN/MPQ-49

The original FAAR system employed a remote target alert data display set TADDS (GSQ-137) located at each weapon. Each TADDS consisted of a single 6 kg unit comprising a display, audio warning and VHF receiver. The display consists of a replica of 7 × 7 matrix appearing on the operator's PPI display with each square representing a specific area of the total radar coverage. The presence of a target in any one square, as determined by the central operator, is denoted by the appearance of a coloured disc in the appropriate square of the TADDS. Green discs denote friendly aircraft and red hostile. Plotting of target tracks is by means of a wax pencil. Controls on the TADDS unit provide for test functions, data link address code selection, and data link receiver frequency selection between 30 and 76 MHz. The lid of the unit carries a magnetic compass for orientation purposes.

The FAAR today is generally used without the TADDS. Transparent standard grid overlay is used on the PPI display and, voice communication via a VHF link enables the radar operator to alert and cue weapons to engage target aircraft.

One mobile version of FAAR, MPQ-49, is housed in a transportable S-250 shelter, utilising the six-wheel drive M561 vehicle as prime mover. The other mobile version is the MPQ-54 which is mounted on a trailer towed by an M35 truck. The system can also be carried by C-130 transport aircraft, or air positioned in a battlefield environment by a CH-47 helicopter.

DEVELOPMENT

Development was undertaken by Sanders Associates Inc, for the US Army Missile Command under a series of contracts. Prototypes were delivered to the US Army in late 1968 and early 1969, and initial production was ordered in March 1969.

STATUS

The first production model MPQ-49 was delivered to the US Army in January 1972.

CONTRACTOR

Sanders Associates Inc (a Lockheed company), Defense and Information Systems Division, Nashua, New Hampshire 03061-2035, USA.

6129.153
TRACKSTAR LOW ALTITUDE AIR DEFENCE SYSTEM

Trackstar is a self-contained, armoured, fully integrated radar/command and control system. It is a sophisticated AN/MPQ-49 derivative, designed for low altitude air defence and provides radar surveillance and command data links to multiple weapons and gunner displays. The system consists of a radar and a command unit contained in an armoured vehicle, M577, with an antenna system mounted on top of the vehicle.

The radar system employs digital MTI and doppler processing techniques, with multiple ECCM features to counter jamming, and has excellent subclutter visibility. It also provides high order rejection of fixed clutter, ground traffic, weather and chaff. The digital signal processor employs a unique method of detecting and classifying rotary wing aircraft even when homing at ranges up to 30 km. An automated antenna erection, levelling and stowage system is provided, and a fully integrated IFF interrogator and antenna system is incorporated. The Trackstar system gives 360° surveillance out to a range of 60 km.

Dual C and radar displays are contained in the radar/command unit which provides fully automated weapon cueing data composed of near real time target vector and identifying information to ground based weapons to ensure rapid long-range target acquisition. Each Trackstar has a capacity for 64 internal target tracks and can net tracks received by data link from up to three other radars. Target track cueing, fire distribution and IFF data is transmitted via VHF data links to integrated weapon display units which are mounted in or near the associated weapons system, eg Blowpipe, Chaparral, Stinger, Vulcan or similar short-range and man-portable air defence weapons.

STATUS

A number of Trackstar Systems have been sold to Egypt.

CONTRACTOR

Sanders Associates Inc (a Lockheed company), Defense and Information Systems Division, Nashua, New Hampshire 03061-2035, USA.

Trackstar air defence radar

3123.153
LAADS LOW ALTITUDE AIRCRAFT DETECTION SYSTEM

LAADS is a transportable air defence radar system designed for the detection and identification of low-flying targets and for the co-ordination of defensive weapons directed against such targets. It is another AN/MPQ-49 derivative.

The radar employed is a coherent, pulsed doppler D-band system using a TWT transmitter. The 12 operating frequencies are generated by highly stable master oscillators. A zero IF receiver detects moving targets, rejects fixed clutter, and presents analogue data to the digital data processor which drives the local PPI and generates digital data messages for broadcast. The radar location is fed into the processor and the target locations are then computed in map co-ordinates. Target data are corrected on each antenna rotation and transmitted to all associated weapons via the data link. Target track capacity is 64. Instrumented range is 60 km with a 30 km range for hovering helicopters.

An integrated weapon display (IWD) receives digital data messages and displays targets with respect to the weapon location (centre of display). Targets within 16 km from the weapon are displayed in near real-time with 1 km resolution. Targets can be deployed in either a 'B' scan for FLIR or video-equipped weapon systems, or PPI mode for optical systems. Previous positions are scanned to show target direction and to facilitate threat evaluation. The display mounts directly in front of the gunner, with a cursor showing direction of the gun. A hand-held version of the display is available for the unit commander or platoon leader.

LAADS radar deployed

The complete radar equipment, with positions for a crew of three, is housed in a standard US S-280 shelter. This is helicopter transportable or land-mobile on an M36 or similar $2\frac{1}{2}$-ton truck. The system is also available for housing in an H-281 shelter for use on M548 tracked vehicles. The weight of the complete system is 3100 kg.

The 188 × 300 cm scanner is mounted on the shelter and is automatically raised and levelled from its folded stowed position. On-mounted antenna facilities are provided for the integrated IFF TPX-54.

OPERATION

The general concept of operation is for each radar to provide search, IFF and associated facilities for remote weapons positions. Each of the latter is provided with the necessary data link reception equipment and IWD to permit individual target engagement. The IWDs, displays and the VHF data link provides for an unlimited number of weapon sites within line-of-sight radio range (normally in excess of 15 km).

Provisions are made for linking other LAADS radars to form networks to give improved protection against 'nap-of-the-earth' attacks.

LAADS is able to serve and provide target acquisition data to an unlimited variety of weapon types, gun and missile systems, and it embodies the latest signal processing and ECCM techniques.

STATUS

The system has been delivered to overseas governments and is currently in production for several other countries. Thailand ordered a number of systems in 1985 and delivery commenced during 1987.

CONTRACTOR

Sanders Associates Inc (a Lockheed company), Defense and Information Systems Division, Nashua, New Hampshire 03061-2035, USA.

4323.153
HUGHES LOW ALTITUDE SURVEILLANCE RADAR

The Hughes Low Altitude Surveillance Radar (LASR) is designed as a three-dimensional air defence surveillance radar system for the detection and tracking of hovering helicopters and fast, low-flying aircraft. It is a later development of the AN/TPQ-36 system (**2848.153**) and has evolved from this system using 70 per cent common hardware. It is so similar that it is often referred to as the AN/TPQ-36(A).

LASR is an advanced short-range battlefield radar that automatically detects and tracks modern low-flying threats, such as heavily-armed pop-up helicopters and low-flying fixed-wing aircraft unmasking at low altitudes and short ranges. It provides very high three-dimensional accuracy in severe clutter and severe ECM environments.

The equipment features three-dimensional pencil beams and frequency agile phase beam scanning and provides accurate weapons cueing data. High-speed and hovering targets are detected as they appear at extremely low altitudes. LASR target reports are automatic and compatible with current and projected automatic digital data netting systems. Emission control techniques (blinking) and air space management functions are enhanced by the equipment's capabilities which enable radar-to-radar track correlation.

A low sidelobe, electronically phase-scan (elevation), pencil-beam antenna minimises clutter illumination for nap-of-the-earth detection, and reduces enemy ECM effectiveness and vulnerability to anti-radiation missiles. The antenna rotates mechanically in azimuth to provide 360° coverage with a variable rate azimuth scan. Beam scheduling is computer-controlled and adaptive to mission requirements.

LASR is transportable by light vehicles and aircraft and is available with antenna elevator mechanism. Logistics and support are simplified by extensive built-in-test equipment and automatic fault isolation. There are obvious benefits in the area of maintenance and repair because of its commonality with AN/TPQ-36.

STATUS

In production. The system has been selected for the Norwegian Adapted Hawk (NOAH) programme. The contract, awarded in January 1984, calls for 18 systems to be delivered by a joint company, established by Hughes and Norsk Forsvarsteknologi (see also entries **6819.153** and **6820.181**. The first systems were delivered to Norway in 1987. A further six were ordered by Norway in 1986. LASR has been demonstrated successfully against fixed-wing and helicopter targets from 10 to 6000 feet (3 to 1800 m) and from hovering to over 450 mph (724 km/h).

A new target cueing system, allowing shoulder-fired missiles to be fired at aircraft by day or night, uses the LASR to pinpoint incoming targets.

CONTRACTOR

Hughes Aircraft Company, Ground Systems Group, Fullerton, California 92634, USA.

Hughes LASR low-level air defence radar

2848.153

AN/TPQ-36 WEAPON LOCATING RADAR

The AN/TPQ-36 is an artillery, rocket and mortar-locating radar designed as a replacement for the AN/MPQ-4 and older technology weapon locating radars. Although designed to replace the AN/MPQ-4, the AN/TPQ-36 is more than a mortar locating radar and has proven to be very effective in locating artillery and rockets. Location of artillery at ranges beyond the capability of the AN/TPQ-36 is provided by the AN/TPQ-37 (**1976.153**), the other radar that makes up the Firefinder system. Using only a different computer software program, the same operations shelter can be used for either Firefinder radar.

AN/TPQ-36 Firefinder radar during Netherlands Army training exercises

Location of hostile artillery and mortars by the AN/TPQ-36 is completely automatic. The system electronically scans the horizon over a 90° sector several times a second, intercepting and automatically tracking hostile projectiles, then computing back along the trajectory to the origin. The co-ordinates and altitude of the weapon are then presented to the operator. Automatic location is so rapid that the co-ordinates of the weapon are normally with the operator before the enemy round lands.

In addition to its ability to locate artillery, a number of other modifications have been incorporated. The normal 90° sector can be expanded to as much as 360° for use in insurgency operations. The radar can also provide information on the hostile weapon's target by extrapolating the trajectory to the impact point, allowing the information to be used in the priority of return fire.

A very important feature for the modern hostile battlefield is the capability to track and locate weapons firing simultaneously from different locations. Using separate track channels, the radar can track several projectiles at once while continuing to scan for other projectiles in its 90° sector. A further feature allows the radar to ignore subsequent firings from weapon positions already located. As many as 39 different weapon positions simulating an artillery preparation for attack have been used in tests to demonstrate the capability of the system to provide numerous locations in a short period of time.

High mobility allows the system to be used as close as 2 km to the front line. Emplacement takes 15 minutes and displacement five minutes or less. The system can be mounted in a single vehicle plus trailer. Since the operations shelter weighs only 2500 lb (1136 kg), vehicles as small as 1½ tons can carry the shelter and pull the trailer.

The AN/TPQ-36 is a coherent, electronically-scanned, range-gated pulse doppler radar. Beams are moved electronically in the horizontal dimension by phase shifts and in the vertical dimension by frequency shift. Ground, sky and electronic clutter are filtered by a signal processing software system which discriminates hostile projectiles from aircraft, birds, insects or other interference. The three-dimensional radar then tracks the projectiles, computing the firing point and providing the information to the operator within seconds of the firing. The extent of automation allows a single operator to perform all functions necessary to locate the hostile weapon and inform the counterfire systems.

Fault detection sensors constantly monitor performance of each sub-system and provide status information. Should a failure occur, built-in test equipment isolates the problem and informs the operator of the location of the problem, frequently the specific card that must be replaced. These corrective actions should reveal up to 90 per cent of all problems and can normally be accomplished in 15 minutes.

The AN/TPQ-36 has proven to be a reliable system. To date each radar is subjected to not only the normal 'burn-in' at +125°F (+52°C) but is screened for performance at –25°F (–32°C). The result is a reliability of in excess of 125 hours, validating an early field test MTBF of 93 hours. In addition to the high and low temperature screening, the performance of each radar is checked using live artillery and mortar firings.

STATUS

A full-scale three-year production contract was awarded by the US Army to Hughes in August 1978 and since that date over 250 systems have been delivered to the US Army and a number of other defence forces. These have included Australia, Jordan, Saudi Arabia, Pakistan, Netherlands, Thailand, Singapore and Turkey. Five systems were ordered by Greece in November 1987, bringing the total number of systems delivered or ordered to over 250 of which nearly 100 are for export.

Originally configured using the then standard Gamma Goat articulated vehicle, the AN/TPQ-36 now uses a 2½-ton truck to carry the operations shelter and pull the antenna/transceiver trailer. With its versatility of design, the shelter has been configured on the militarised 1½-ton pick-up truck, and will use the Mercedes Unimog for Australia.

An air defence variant, the Low Altitude Surveillance Radar (**4323.153**), has been successfully demonstrated to have the capability to track such different targets as pop-up helicopters and high speed fixed wing aircraft in high clutter environments. The first contract for production of this variant was awarded in January 1984. The high percentage of logistic commonality between the two systems should provide benefits in logistic support and life cycle cost effectiveness.

CONTRACTOR

Hughes Aircraft Company, Ground Systems Group, Fullerton, California 92634, USA.

1976.153

AN/TPQ-37 WEAPON LOCATING RADAR

The AN/TPQ-37 radar provides the US Army's first capability to locate hostile artillery and rocket launchers at their normal firing ranges. It is believed to be the only system with this capability anywhere in the world. Additionally, it is the first tactical radar in large scale production to use phased array techniques for scanning in both azimuth and elevation, by using miniaturised diode phase shifters distributed throughout the antenna.

The AN/TPQ-37 radar programme was initiated in response to the threat of massive concentrations of artillery and rockets in support of both offensive and defensive operations. Prior to this, there had been no effective means to locate and counter indirect fire, with combat troops having to analyse craters to estimate direction and range as a last resort.

The AN/TPQ-37 uses a combination of radar techniques and computer-controlled signal processing for detection, verification, tracking of projectiles, and extrapolating the track-data points to the location from which the projectile was fired. Once the origin of a projectile has been identified, that location can be provided by voice or digital data link to the fire direction centre for initiation of counterfire. These techniques are very similar to those applying to the AN/TPQ-36 weapon locating radar (**2848.153**) described above.

Like the AN/TPQ-36, the AN/TPQ-37 is a coherent, electronic-scanned, range-gated pulse doppler radar. The AN/TPQ-37 incorporates monopulse and its beams are moved in both azimuth and elevation by shifting the inter-element phase using diode phase shifters, the first such application in a tactical system. Significant improvements were incorporated into the AN/TPQ-37 before entering production, ensuring state-of-the-art technology, particularly in the antenna, signal processor and computer. Since then, lightweight kevlar armour plating has been added to protect the antenna.

AN/TPQ-37 weapon locating radar

Operationally the AN/TPQ-36 and AN/TPQ-37 will be complementary, the former between 1 and 4 km from the FLOT to provide for artillery counter-fire against mortars and close-in artillery. Three AN/TPQ-36 radars will be deployed in each division sector. The AN/TPQ-37 will be sited further behind the front line to locate opposing long-range artillery, with two AN/TPQ-37s to each division sector.

Together, the AN/TPQ-36 and AN/TPQ-37 are referred to as the Firefinder System. The modification of the programme to incorporate an operations shelter that can be coupled to and operate either the AN/TPQ-36 antenna or the AN/TPQ-37 antenna provides unique operational and maintenance advantages.

The AN/TPQ-37 consists of two vehicles and an antenna trailer. A five-ton vehicle carries the 60 kW generator and pulls the antenna trailer. The operations shelter is mounted on a ½-ton vehicle. All components are helicopter transportable. Crew size is between 8 and 12, though once emplaced normal operation can be carried out by a single operator.

STATUS

Since 1977 the US Army has ordered 72 systems, and delivery was completed in 1985.

A moratorium on international sales was imposed in 1981 after concern regarding transfer and potential compromise of high technology. After a study of this issue the moratorium was lifted in late 1983 and sales to US allies was authorised. Since that date the system has been ordered by a number of defence forces, including Saudi Arabia and Jordan. At least seven other countries have indicated an intention to procure the AN/TPQ-37, including Pakistan, which wishes to buy four systems at a cost of about $40 million, and Thailand for two systems. The Netherlands has been offered four systems at a cost of $45 million, and the People's Republic of China has shown interest in buying four systems.

A follow-on order for 11 AN/TPQ-37s and an option for 11 more was received in September 1987 from the US government for foreign sales to an unnamed customer. This makes a total of well over 100 systems for US and foreign procurement.

Improvements in mobility have received high priority among the evolutionary changes incorporated during production. In addition to placing the operations shelter on a 2½-ton truck in place of the 1½-ton pick-up, the US Army is replacing the mobilisers initially used for the antenna/transceiver with a tandem-wheel trailer which promises much better stability. Mobility and field tests have shown that the antenna/transceiver can be easily mounted on the FVS carrier, the tracked vehicle chassis used for the Multiple Launch Rocket System in production for the US and several NATO nations.

Since the shelters for the AN/TPQ-36 and AN/TPQ-37 are identical, and operational procedures are similar, those countries that have procured the former will find the transition to the AN/TPQ-37 facilitated.

An air defence radar based on the hardware of the AN/TPQ-37 has been demonstrated in several locations both in the US and overseas. The VSTAR (variable search and track air defence radar) (**4080.153**) has demonstrated a capability to track small aircraft flying through heavy ground clutter and has tracked small sea-going vessels in sea state 4. Most of the VSTAR is common with the AN/TPQ-37 providing considerable logistic and support advantages.

CONTRACTOR

Hughes Aircraft Company, Ground Systems Group, Fullerton, California 92634, USA.

2849.153

AN/TPQ-39(V) INSTRUMENTATION RADAR

AN/TPQ-39(V) is a one-man instrumentation radar of modular design which is available in fixed building and transportable versions. It is intended for range applications involving system evaluation, range safety, vehicle control or other target tracking uses.

Several variants are in use, including the digital instrumentation range (DIR) and Nike digital instrumentation radar (NIDIR). DIR operates in C-band while current units of NIDIR are in X-band. Other variants are also available.

NIDIR variant of the AN/TPQ-39(V) radar

As the equipment's names suggest, solid-state digital techniques are used extensively in the system. A digital computer is used as the basis of the system, with many radar functions (angle and range servos, AFC, AGC, target acquisition and detection, input/output and display control) provided by the radar computer programming.

Many optional extras have been selected for individual range requirements, such as parametric amplifiers, higher power transmitters, data recording facility, plotting boards adjacent to the operator's position, timing equipment and star calibration.

The following characteristics describe the basic DIR variant.

CHARACTERISTICS (DIR Variant)

Azimuth, elevation tracking rates: 500 mils/s
Azimuth, elevation acceleration: 300 mils/s^2
Azimuth, elevation tracking precision: 0.3 mils, rms
Range tracking: 235 km at 640 Hz PRF
Ro, IM^2, 0 dB S/N: 93 km
Range tracking precision: 4.7 m

STATUS

Units are operational at US Army, Navy and Air Forces sites in the USA, and at the West German Aerospace Research Establishment (DFVLR) at Braunschweig. NIDIR variants are operated in the Far East, in the UK and at US Navy sites in the USA.

CONTRACTOR

RCA, Electronic Systems Department, Moorestown, New Jersey 08057, USA.

2529.153

AN/TPS-21 AND AN/TPS-33 PULSE DOPPLER BATTLEFIELD RADARS

These two radars are I/J-band portable manpack systems used to detect persons or vehicles in motion on the ground. Targets are detected at ranges lying between approximate limits of 100 and 18 000 m with a range measurement accuracy of ±23 m or ±1 per cent of range, whichever is the greater. Angular accuracy is ±25 mils.

Scanning is either manual or automatic in range with sector scan. For the TPS-21 radar detection is solely by audible tone in the operator's headphones; in the TPS-33 equipment there is also an A-scope display.

STATUS

Operational with US armed forces.

CONTRACTOR

Admiral Corporation, USA, now incorporated in: La Pointe Industries, 155 West Main Street, Rockville, Connecticut 06066, USA.

2516.153

AN/TPS-32 3-D LONG-RANGE SURVEILLANCE RADAR

AN/TPS-32 is a ground-based, long-range, tactical, lightweight radar which automatically provides precise, three-dimensional position data on multiple targets. The equipment has been designed to operate as the primary 3-D radar sensor for the marine tactical data system. It also provides two positions for operators to vector interceptor aircraft to assigned targets. It is easily transportable by helicopter, cargo aircraft, or by conventional M35 truck. Automatic target detection, clutter elimination, and IFF are included.

Air search coverage of the radar extends to 300 nautical miles (556 km) in range to a ceiling of 100 000 ft (approximately 30 500 m). The search volume is obtained by continuous rotation of the antenna (at six rpm) in the azimuth plane and electronic frequency scanning, –1 to +18° in the elevation plane. The elevation scan is accomplished by changing the frequency of the E/F-band RF energy applied to the planar array, frequency-sensitive antenna. The antenna radiation appears as a series of pencil beams; each beam is 0.84° high and 2.15° wide.

The AN/TPS-32 radar consists of an antenna assembly and three shelters. Two of the shelters contain transmitter stages and the third contains receiver, data processing, including radar and IFF plot extractors, displays and system control equipment. In the transport condition, the radar set consists of two antenna array pallets, one tripod pallet, and three shelters. The complete system is helicopter transportable in six pallets and can be made operational in less than two hours.

To make the most efficient use of transmitter power, the peak power level and inter-pulse period are changed during the complete elevation scan to fit the coverage volume more precisely. This allows more energy to be spent in the longer range portions of the coverage volume at the lower angles of operation.

AN/TPS-32 long-range surveillance radar

CHARACTERISTICS

Coverage
Elevation: 19° within interval –1 to +18° (electronically scanned)
Height: 100 000 ft (30.5 km)
Azimuth: 360° (mechanically scanned)
Displayed range: 2.5 nm (4.6 km) – 300 nm (556 km)
Data rate: 10 s
Probability of detection/range (1 m^2 target): 90% to 200 nm (371 km), 50% at 300 nm (556 km)
False alarm rate: <10 false target messages per azimuth scan
Accuracy (standard deviation)
Range: 750 ft (229 m)
Azimuth: 0.5°
Height: 1200 ft (336 m) at 100 nm (185 km), 3000 ft (914 m) at 300 nm (463 km)
Resolution
Range: 1500 ft (457 m) (manual), 2625 ft (800 m) (automatic)
Azimuth: 2.15°
Elevation: 0.85°
Transmitter
Operating frequency: 2905 – 3080 MHz
Pulsewidth: 30 μs
Interpulse period: 1090 – 3772 μs
Peak power output
Final RF amplifier: 2.2 MW peak (low elevation, long range)
Radar antenna
Elevation beamwidth: 0.84° (nominal)
Azimuth beamwidth: 2.15° (nominal)
Gain: 41 dB (minimum)
Sidelobes (azimuth and elevation): –25 dB (minimum) referenced to peak of main beam
Polarisation: Horizontal
Rotational speeds: 6 rpm
Receiver/processor: Log FTC, IF double canceller MTI and CPAC receivers with CFAR and fast show AGC gain controls; dual nine-channel video processor with 350 kHz 6W per channel (1.3 MHz ind CPAC); automatic 3-D target detection and clutter censoring; automatic IFF/radar correlation

STATUS

The AN/TPS-32 is operational at various US Marine Corps bases. An improvement programme is underway to upgrade the radar for ease of maintenance, improved reliability and improved performance in adverse weather conditions. Versions of the AN/TPS-32, modified to be NADGE compatible (AN/TPS-64) are operational in Kuwait and Turkey.

CONTRACTOR

ITT Gilfillan, 7821 Orion Avenue, Van Nuys, California 91409, USA.

4078.153

SERIES 320 3-D AIR DEFENCE RADARS

The Series 320 family of 3-D air defence radars has been configured to meet various surveillance coverage and siting requirements. The basic system uses a planar array antenna with frequency steered multiple pencil beam groups to scan in elevation, and mechanical rotation to cover 360° of azimuth. Both fixed site and transportable configurations have been produced.

The radar is supplied in either 200 or 300 nm range versions. The 200 nm version has a peak power output of 150 kW and the 300 nm version adds a power amplifier to develop 1.1 MW peak power. Highly stable signal processing techniques maintain target detection and low false alarm rate in high clutter areas. Operation of the radar is highly automatic with extensive on-line monitoring of performance and fault isolation. MTI is employed in areas where ground clutter is present. ECCM capabilities include jammer frequency analysis, jam-strobe reports, and silent sector operation.

Two antenna configurations are available, the 6.4 × 5.18 m antenna pictured provides a nearly circular

1.4° azimuth by 1.6° elevation beam. A 3.65 × 9.75 m antenna, designed for increased elevation accuracy, provides a 2.2° azimuth and a 0.85° elevation beam.

CHARACTERISTICS

Frequency: 2.9 – 3.1 GHz
Peak power: 150 kW/1.1 MW
PRF: Variable
Receiver noise figure: 2.5 dB
Antenna: 6.4 × 5.18 m; optionally 3.65 × 9.75 m
Gain: 40.6 dB (41.5 dB optional)
Azimuth beamwidth: 1.4° (2.2° optional)
Elevation beamwidth: 1.6° (0.84° optional)
Scan rate: 6 rpm
Polarisation: Horizontal
Signal processor: Digital
Improvement factor: 40 dB
Detection performance (2 m² target): 200/300 nm (PD ⩽ 0.5)
Number of false alarms: 5/scan
Azimuth coverage: 360°
Elevation coverage: 0 – 20°
Height coverage: 0 – 30 000 m
Resolution at 100 nm: Range 125 m; azimuth 2.8°; elevation 3.5°
Accuracy (1 sigma at 100 nm): Range 18 m; azimuth 500 m; height 760 m (400 m with option)

STATUS

More than 20 systems are operational throughout the world. An agreement with Plessey, UK, has resulted in a combination of the 320 3-D and the Plessey AR-3D being ordered by the UK MoD from Plessey as the E/F-band radar in the current improvement programme for the UK air defence network. (**3649.153**).

Typical deployment of ITT Gilfillan Series 320 3-D transportable air defence radar system

CONTRACTOR

ITT Gilfillan, 7821 Orion Avenue, Van Nuys, California 91409, USA.

4079.153

AN/MPQ-63 INSTRUMENTATION BALLISTIC RADAR (ARBAT)

The AN/MPQ-63, also known as ARBAT (an acronym for Application of Radar to Ballistic Acceptance Testing of ammunition), is a US Army programme for a range instrumentation radar system that provides ballistic test information on a wide variety of ammunition during live firings. It is a self-contained, autonomous system which performs all the functions of acquisition, track, event detection, impact prediction and associated data reduction. The system collects real-time data on projectiles while they are in flight and provides near real-time readout of this data for accept/reject determination.

The AN/MPQ-63 system is packaged in two major equipment groups; the antenna/transceiver and an instrumentation van. The antenna/transceiver group includes a trailer, antenna, pedestal, transceiver, beam steering buffer and accessories. The antenna is a 10-foot high by 12-foot wide planar array which is steered electronically by phase scan in elevation, and frequency scan in azimuth augmented by mechanical. The transmitter and receiver comprise a fully coherent amplifier chain controlled by a frequency synthesiser tuning over a wide frequency band. This configuration is required for frequency scanning, MTI clutter rejection, and Doppler data extraction. The instrumentation van houses the radar displays, controls, and signal and data processing subsystems.

Information is provided by the AN/MPQ-63 over the full trajectory of the projectile for ballistic acceptance test purposes. The system acquires the projectile rapidly at launch, and transfers from acquisition to track within a nominal 60 milliseconds. It will, therefore, detect and record the location of normal or premature ignition of rocket assisted projectiles. Early burst or parts separation and short rounds data are available in the midflight region. Impact prediction or air burst co-ordinates are recorded at track termination in the target zone. Target data includes position, velocity, acceleration, drag and radar cross section at 20 Hz rate.

Mobile phase-scanned antenna array of the ITT Gilfillan AN/MPQ-63 ARBAT instrumentation radar showing an obvious family resemblance to the 3-D air defence radars produced by this concern

CHARACTERISTICS

Antenna
Beamwidth: 0.55° azimuth; 0.67° elevation
Azimuth scan: 3.8° frequency scan; 170° mechanical scan
Elevation scan: 35° phase scan; 0–25° mechanical tilt
Polarisation: Horizontal
Transmitter
Peak power: 30 kW
Pulse width: 1 μs
Pulse compression: 11 bit phase code
PRF: 1 – 10 kHz
Operating frequency: 9.3 – 10 GHz
Signal processor
FFT: 16 – 128 point
Targets
Types: Mortars, rockets, projectiles
Radar cross section (RCS): –10 to –30 dB m²
Doppler velocity: 20 – 1500 m/s
Elevation rate: 40°/s
Azimuth rate: 12°/s
Range (RCS): 25 km (–30 dB m²); 50 km (–10 dB m²)
Ballistic test reliability: 95% for both event detection and track to impact prediction
Accuracy:
Range: 2 m or 0.05% range
Doppler velocity: 1 m/s
Azimuth: 2 mrads
Elevation: 2 mrads
Event time: 10 μs
Impact prediction: 0.1% of range; 2 mrads in azimuth

STATUS

The Arbat system was accepted by the US Army's Test and Evaluation Command (TECOM) for operation support at the Yuma Proving Grounds, Arizona, USA in January 1985.

CONTRACTOR

ITT Gilfillan, 7821 Orion Avenue, Van Nuys, California 91409, USA

4322.153

FALCON SERIES SURVEILLANCE RADAR

The Falcon (frequency agile low coverage netted) radar is a two-dimensional air/sea surveillance radar providing search, acquisition and tracking of low altitude aircraft and the simultaneous coastal surveillance and tracking of ships, small boats and low-flying helicopters. The automatic tracking of large numbers of aircraft and ship targets can be monitored by operators within the radar control cabin which permits autonomous radar operations. Target data can be transmitted in digital format from a number of Falcon radars, to air and surface control centres at a distant location to provide centralised command and control operations.

The radar equipment can be installed in a protected bunker configuration, in a shelter, or in a mobile configuration, and is designed for unattended operation. Integrated IFF and radar tracks data can be transmitted by digital data link to air and surface operations centres.

The antenna is a dual beam reflector with a cosec² coverage to 25° elevation, and integral IFF. The transceiver is crystal controlled and uses a single-stage high-gain TWT.

The entire system is designed to incorporate strong ECCM capabilities and includes automatic built-in-test equipment to minimise maintenance requirements.

CHARACTERISTICS

Operating frequency: C-band

Antenna type: Reflector, high gain, dual-beam with very low sidelobes
Polarisation is circular, horizontal, operator selectable

Beamwidths: 1.25° (azimuth), cosec² to 25° (elevation)

Search scan: 360° at 6 or 12 rpm

Instrumented range: 100 or 160 km

Peak power: 60 kW

Mean power: 1500 W

Altitude coverage: 0 – 3000 m

Simultaneous tracks: Up to 100, air or surface

Availability: 99.6%

STATUS

The Falcon radar is in production for several countries. Deliveries began in 1986. Falcon coastal defence systems with radars,operations centres and communications links are also in production with deliveries scheduled for 1987.

CONTRACTOR

ITT Gilfillan, 7821 Orion Avenue, Van Nuys, California 91409, USA.

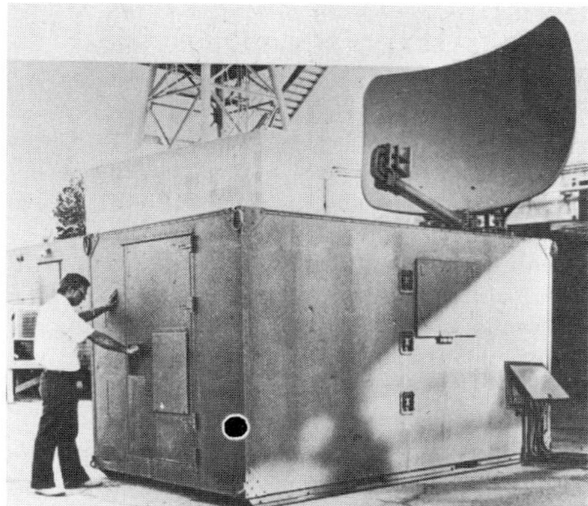

Falcon surveillance radar

2530.153

AN/TPS-34 TRANSPORTABLE AIR DEFENCE RADAR

AN/TPS-34 is a tactical early-warning radar for use in forward air defence systems. It is said to be the first tactical radar to utilise the V-beam 3-D technique.

The radar provides long-range 3-D data on high-speed aircraft and missiles from battlefield sites. It can be transported by helicopter in units averaging less than 200 kg in weight and can be assembled and operating in less than eight hours from touchdown.

STATUS

In service in the British armed forces.

CONTRACTOR

Sperry Corporation, Electronic Systems, Great Neck, New York 11020, USA.

2517.153

AN/TPS-43E TACTICAL 3-D RADAR

AN/TPS-43E is a lightweight air transportable radar designed for use with manned aircraft or SAM batteries in a wide variety of tactical environments. It provides solid 3-D cover to 240 nautical miles on a fighter or fighter-bomber aircraft and measures heights over the full range by signal amplitude comparisons in six channels. Extensive clutter rejection and ECCM features are incorporated in the design, including a digital coherent MTI system, pulse-to-pulse frequency agility, jamming analysis and transmission selection (JATS), coded pulse anti-clutter system (CPACS) and sidelobe blanking. This latest model of the AN/TPS-43 includes extensive equipment refinements and increased operational capability.

For ease of air shipment the equipment divides simply into two pallet loads each of less than 3400 kg. One load comprises the shelter unit, transmitter/receiver and displays; the other consists of the antenna assembly, the feed and ancillary equipment. The entire equipment can be packed into two M35 trucks for road transport.

To minimise weight, light alloys are used wherever possible in the main mechanical structures and micro-miniaturisation techniques are used in the electronics circuits. The transmitter uses a completely solid-state modular modulator to pulse a linear beam klystron.

The feed array features the use of a stripline matrix to form the heightfinding beams. The IFF antenna is a sum and difference type providing an interrogator sidelobe suppression (ISLS) capability. A small printed circuit reference antenna is mounted on the back of the radar feed to act as the radar sidelobe reference antenna; use of this latter antenna during the dead time between transmitter pulses is available for the JATS function.

CHARACTERISTICS

Range: Solid 3-D coverage on fighter-bomber aircraft to 408 km

Data rate: 10 s (6 rpm antenna)

Elevation coverage: 0 – 20°

Accuracy: 107 m range, 0.35° azimuth; ±450 m height

Electrical characteristics

Power output: 4 MW peak, 6.7 kW average

Frequency: 2900 – 3100 MHz in 16 discrete steps (with pulse-to-pulse agility)

Pulse duration: 6.5 μs

Antenna gain: Transmit, 36 dB; receive, 40 dB

Azimuth beamwidth: 1.1°

IFF azimuth beamwidth: 4° (or sum/difference ISLS antenna)

Noise figure: 4.5 dB

Prime power: 400 Hz 3-phase 120/208 V

Mechanical characteristics

Weight: Shelter module 3310 kg; antenna module 2050 kg

Transport: Single C-130 aircraft, 2 M35 trucks, 2 sets of transporters, or 2 helicopter loads

Road speeds: Up to 96 km/h

Air lift altitudes: Up to 15 000 m

Siting requirements: 6 × 10.5 m clear area on slope of 10% or less

Reaction time: 50 minutes with a 6-man team

Wind resistance: Operate to 52 knots, survive 92 knots (tied down)

Operating temperature: –40 to +125°F (–40 to +52°C)

Other

MTBF: 400 h

MTTR: 30 minutes

Outputs to: 3 120 m cables with storage reels

Operations centres PPI: 2 operational positions featuring AN/UPA-62 (40 cm) CRT, digital height readouts, and AN/UPA-59A active/passive IFF decoders. Each position also has access to built-in UHF ground/air communications and HF point-to-point communications facilities.

IFF/SIF equipment: Hazeltine AN/UPX-23 interrogator set and AN/UPA-59A active/passive decoder

Transmitter tube: Linear beam klystron

STATUS

The AN/TPS-43 series of radars has been extensively deployed by the armed forces of the USA and more than 20 other countries. More than 170 systems are deployed worldwide. Continuous improvement in the series continues with the ultra-low sidelobe antenna (ULSA). Westinghouse has been placed under production contract to retrofit the ULSA into the USAF AN/TPS-43 systems, and production deliveries are significantly ahead of schedule. The initial contract for the first production lot was valued at more than $55 million. In addition there has been a major production contract add-on to include advanced signal processing improvements, automatic displays, and numerous reliability and maintainability enhancements. The enhanced TPS-43Es will be front line USAF tactical surveillance radars until at least the year 2010. The enhancements are also applicable to the other AN/TPS-43 configurations with the added benefit of long-term support via the existing USAF logistics pipeline.

CONTRACTOR

Westinghouse Defense and Electronics Center, Baltimore, Maryland 21203, USA.

AN/TPS-43E search radar with ultra-low sidelobe antenna (ULSA)

2518.153

AN/TPS-44 TACTICAL SURVEILLANCE RADAR (ALERT Mk II A/O)

Alert Mk II (AN/TPS-44) is a lightweight, solid-state, transportable D-band air surveillance radar. It was designed for use anywhere in the world as the sensor for the forward air-control post of the US 407L system (TACS).

Alert Mk II is the latest version of the Alert radar series. Information of airborne targets is presented in the form of synchronised radar and IFF video on a 16-inch PPI display console. It also has optional provisions for supplying range and azimuth information to additional tactical operations centres.

The equipment consists of the Alert IIA radar set, a complete IFF/SSR system and the operator display console. All components are housed in two major packages; an equipment shelter houses the electronic components and display console and an antenna pallet carries the folding antenna and feed system, together with the pedestal and antenna drive system. The entire system can be transported by M35 trucks, C-130 aircraft, helicopters or towed on transporters. It can be converted from transport to operational configuration by a team of four people in less than 40 minutes.

AN/TPS-44 tactical surveillance radar deployed

DEVELOPMENT

Prototypes of the equipment were submitted to the US Air Force and successfully passed all tests. Following this the equipment was put into production, and by February 1983 deliveries were known to include 26 systems to the US Air Force, and 13 modified systems to other countries.

CHARACTERISTICS

Antenna

Type

Search: Modified parabolic

IFF: Integral with search radar feed system and reflector

Aperture, search

Horizontal: >45 m

Vertical: >2.7 m

Beamwidth (one-way), search and IFF

Horizontal: 3.8°

Vertical: 8° with cosecant squaring 7 – 27°

Polarisation

Search: Horizontal

IFF: Vertical

Type of feed: Search and IFF: horn

Beam pattern search and IFF

Horizontal: Conventional fan beam (cosine between 3 dB points), sidelobes 25 dB down, back lobes 30 dB down

Vertical: Conventional fan beam with cosecant squaring

Scan

Azimuth: 0 – 15 rpm, clockwise in automatic, or manually searchlighting, clockwise or counter-clockwise. Also available with 6 rpm clockwise fixed-speed option

Elevation: Manually adjustable, –3 to +6°

Transmitter

Power

Peak: 1 MW, or greater

Average: 1.12 kW at 800, 0.745 kW at 533, 1.12 kW at 267

PRF

Manually selectable: 800, 533, 400

Momentary selectable: ±10 µs stagger on 800 PRF. Also available with 4-pulse stagger to eliminate MTI blind-speeds

Frequency: 1.25 – 1.35 GHz, continuously tunable over full range by local or remote tuning control

Tuning rate: 1 MHz/s

Pulse: Width automatically selected when PRF is selected:

1.4 µs for 800 Hz PRF
1.4 µs for 533 Hz PRF
4.2 µs for 267 Hz PRF
2.8 µs for 400 Hz PRF

Rise time: 100 ns, max

Receiver

Noise figure: Typically 3 dB, overall

Dynamic range: 65 dB

Choice of IF: Logarithmic, linear, or wideband limiting

Accuracy

Range: 0.5 nm on 400 Hz PRF; 0.1 nm on 800 Hz and 533 Hz PRF

Azimuth: ±1°

Type of information displays

(a) PPI
(b) A-scope
(c) Output jacks and cabling provided for remote displays

STATUS

In service with the USAF and in other countries.

CONTRACTOR

ISC Cardion Electronics Inc (a member of International Signal and Control/Defense and Space Group), Long Island Expressway, Woodbury, New York 11797, USA.

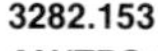

3282.153

AN/TPS-59 3-D TACTICAL RADAR

The AN/TPS-59 is a D-band, long-range 3-D air surveillance phased array radar. The export version is known as the GE 592. Claimed to be unique as a tactical radar, its transmitter is totally solid-state and power is generated by direct amplification at the transmission frequency using D-band power transistors. The antenna is a 9.1 m high by 4.9 m rotating planar array located about two metres above ground level. The three trailer assemblies used as the antenna support structure also carry the array sections during transport. The display and processing equipment, including an AN/UYK-7 general-purpose digital computer, are housed in two air-conditioned cabins (S-280), and these and three 45 kW diesel generators complete the AN/TPS-59 system.

The antenna array is made up of 54 identical row-feed networks and associated row-level transceivers, comprising power supplies, transmitters, pre-amplifiers, phase shifters, duplexers and logic controls. Each set is housed in a single package directly behind the row feed. The vertical, or row-to-row distribution, is accomplished by three column feeds. The antenna rotates mechanically in azimuth, and a pencil beam electronically scans in elevation from 0 to 19° to cover the specified surveillance volume in a raster-scan pattern.

Two basic waveforms are used for the surveillance function: one throughout the short-range interval from 5.5 to 185 km, and the other over the long-range interval from 185 km to the limits of coverage. Eight pencil beams are used to scan the long-range interval in elevation, and 11 are used to cover the short-range. Provisions are made in the lower long-range and short-range beams to counter the effects of multi-path propagation, and a special low-angle heightfinding technique is employed. There is also a special weather mode of operation which automatically optimises energy management and MTI processing according to prevailing weather conditions. This is performed at intervals of five minutes to adjust to weather changes.

AN/TPS-59 3-D phased array tactical radar

In addition to its various functions in beam steering, energy management, and signal processing, the AN/UYK-7 computer stores the position of all detected targets (in three co-ordinates) and correlates the stored positions with fresh data on a scan-to-scan basis. It also performs the special range correlation processing associated with second-time-around returns concerned with MTI operation, and correlates the data from targets in closely spaced beams to avoid the reporting problems associated with multiple detections. All monopulse calculations are performed in the computer.

All controls and indicators for the operation of the AN/TPS-59 are located at the display console, at which there are two operators' positions with three CRTs. Two of these are PPIs and the other is used as either an RHI display or for the presentation of alpha-numeric data. Other facilities include IFF controls, a performance monitoring status display, a clutter gating panel and communications controls.

The AN/TPS-59 is designed primarily to provide long-range surveillance of a tactical airspace to the tactical air operations centre (TAOC). Its detection capabilities (500 targets per ten-second scan), 3-D positional accuracy, and console readouts and controls meet the requirements of an autonomous build-up or back-up ground control intercept role. The system's high data rate capability supports the additional ATC mission. The design specification includes threat criteria such as 1 m² scan-to-scan fluctuating targets flying at speeds up to Mach 4.

CHARACTERISTICS

Radiated peak power: 46 kW

Average duty cycle: 18%

Transmit gain: 38.9 dB

Effective receive area: 14.5 dBm²

Antenna sidelobe levels: –55 dB

System noise temperature: 540 K

Signal processing losses: 2.4 dB

Frequency: 1215 – 1400 MHz

Antenna beamwidth: 3.4° azimuth; 1.7° elevation

IFF subsystem: Modes 1, 2, 3A, C, 4

Surveillance coverage: 360° azimuth; 5.5 – 560 km range; 0 – 19° elevation; 30 500 m altitude

Frame time: 10 or 5 s

Detectability-in-the-clear: PD: 0.7 beyond 200 nm; 0.9 within 200 nm; FAR: 5 per scan

Accuracy: 24 m range; 3 mrad azimuth; 300 m height

Resolution: 60 m range; 3.4° azimuth; 1.7° elevation

Reliability: MTBF 1400 h

Maintainability: 40 minutes MTTR

Weight: 12 700 kg system; 2360 kg max package

Prime power: 89 kW (nominal)

Mobility: 1 h assembly; ½ h disassembly

DEVELOPMENT

General Electric was awarded a contract for the development of the AN/TPS-59 for the US Marine Corps by the Naval Electronic Systems Command in 1972. It began US Marine Corps acceptance testing in 1976-77.

Based upon the modular nature of the antenna

array and its 54 identical row/feed networks, and flexibility inherent in the digital signal processor coupled with extensive use of software control techniques, GE has produced designs for other derivatives of the AN/TPS-59 for a variety of applications. Brief details follow:

Fixed 3-D long-range radar: This has an antenna array measuring 7.3 × 7.3 m (the row/feeds are 50 per cent longer than the AN/TPS-59) and a range capability of about 450 km.

Medium-range transportable 3-D radar: Antenna size 4.3 × 4.9 m, range 225 km, 50-track capabilty.

Long-range 2-D fixed radar: Antenna size 3.7 × 6.1 m with row/feeds disposed vertically instead of horizontally, range 370 km, track-while-scan processing.

Medium-range limited-sector 2-D radar: Antenna size 3.7 × 1.5 m with row/feeds disposed vertically, range 55 km, inertialess azimuth scan of 120°, track-while-scan processing.

A shipboard long-range 2-D radar using an array measuring 4.6 × 2.3 m contained within a 9.2 m diameter radome and with a range of 370 km has also been publicised.

STATUS

15 AN/TPS-59 radars have been delivered, 11 to the US Marine Corps and four to Egypt. The latter are linked to eight Westinghouse AN/TPS-63 systems and 12 operations centres as part of an defence system update in Egypt. A transportable version, the GE 592, is described below (**4321.153**).

CONTRACTOR

General Electric Company, Electronic Systems Division, Syracuse, New York 13201, USA.

4321.153
GE 592 TRANSPORTABLE AIR DEFENCE RADAR

The transportable GE 592 air defence radar system is the export version of the AN/TPS-59 system (**3282.153**). It is a completely mobile equipment and is totally solid-state computer controlled. Operating in L-band and comprising a complete air defence system with IFF or secondary radar for target identification, the equipment is housed in four enclosures and transported on four trailers.

The GE 592 is a versatile system and in its normal transportable configuration is moved by four vehicles which carry the antenna array, data and signal processing units, radar and control management and the power generation equipment. A fifth vehicle is needed to carry the microwave communication equipment, if required. The entire system can also be air-lifted by fixed-wing aircraft or helicopter.

The planar array measures 7.3 × 7.3 metres and phase positions a series of pencil beams to scan in elevation up to 20° while it rotates in azimuth. The transmitter/receiver group is located on the antenna and comprises 44 transmitters, each containing eight 100-watt RF solid-state power modules and 44 receiver modules. The IFF secondary radar antenna is mounted on top of the main array.

The data processor controller provides all radar operational management as well as on-line, automatic performance monitoring to detect possible system failures. The controller provides plot extraction data output and performs sweep-to-sweep and scan-to-scan correlations to reduce false alarms and multiple reports. It also provides automatic radar/IFF correlations as well as monitoring the system for adverse environmental and ECM effects.

Although the GE 592 is completely automatic in operation, it can also be controlled by manual over-ride from operations consoles in the operations and control group trailer.

The secondary radar sub-system (SRS) provides identification, altitude, azimuth and range information on transponder equipped aircraft. The SRS radar coverage is 360° in azimuth, 400 km in range and up to 100 000 feet (30 480 metres) in height.

Composite impression of GE 592 transportable air defence radar systems in operational service as part of the UK air defence network in Scotland

The complete radar system can be housed in a radome/shelter complex, if required. The equipment trailers can be dispersed in any location within 300 metres of the radome. If field dispersal is chosen, the array operates without a radome.

CHARACTERISTICS

Frequency: 1215 – 1400 MHz
Range: Up to 400 km
Azimuth: 360°
Elevation: –4 to +20°
Height: 100 000 ft (30 480 m)
Range accuracy: 100 m rms (5 – 150 nm); 200 m rms (150 – 250 nm)
Height accuracy: 1200 ft (366 m) at 135 nm
Azimuth accuracy: 0.17°
Peak power: 25 kW
Beamwidth: Azimuth 2.2°, elevation 2°

STATUS

Two transportable GE 592 systems are operational at Buchan and Benbecula in Scotland as part of NADGE. A fixed site system is operational in Belgium and one system has been installed in Berlin.

CONTRACTOR

General Electric Company, Electronic Systems Division, Syracuse, New York 13201, USA.

3475.153
AN/TPS-63 TACTICAL SURVEILLANCE RADAR

The AN/TPS-63 is an L-band air surveillance radar used to provide low altitude coverage for tactical air defence operations. This radar, with the companion equipment, AN/TPS-65 (**3476.153**), was built to meet US Marine Corps requirements, and both these radars feature extensive use of micro-electronic and solid-state techniques. An advanced MTI facility provides a 90 per cent probability of detecting a 1 m² target out to a range of 80 nautical miles (approx 150 km). The digital MTI uses a four-pulse canceller with variable time intervals between pulses to distinguish moving aircraft from ground returns while eliminating MTI radar blind speeds. There is an additional digital three-pulse MTI weather canceller and digital CFAR processing. Dual diversity operation is normal and there are built-in test features.

The AN/TPS-63 is designed for transportable or mobile applications, but fixed installations are also possible. The 4.9 × 5.5 m antenna is constructed of sections, which are contained within the shelter during transport, and the system can be operational within one hour of arrival at a new site.

CHARACTERISTICS

Frequency: L-band, dual diversity, 51 selectable frequencies between 1250 – 1350 MHz
Range: 80, 120 and 160 nm (operator selectable)
Data rate: 6, 12, 15 rpm (10, 5, 4 s)
Elevation coverage: 12 000 m shaped coverage to 45°
Power output: 100 kW peak; average 3 kW
Antenna gain: 32.5 dB
Antenna rotation: 6, 12, or 15 rpm
Angular resolution: 2.7°
Range resolution: 152 m
Accuracy: 150 m range; 0.35° azimuth
ECCM: Coded-pulse anti-clutter, frequency agility, PRF stagger
MTBF: 884 h
MTTR: 30 minutes
Weight: 3400 kg
MTI improvement factor: 60 dB

DEVELOPMENT

The AN/TPS-63 was designed and developed under US Navy contracts for use by the US Marine Corps in the latter's Tactical Operations Center to facilitate detection, automatic acquisition and target tracking of aircraft targets. The AN/TPS-65 version serves as the primary airfield surveillance radar sensor for the US Marine Air Traffic Control and Landing System.

STATUS

Over 70 systems have been delivered to the US Marine Corps, US Air Force and seven international customers, including Morocco, Yugoslavia, Korea, Jordan, Venezuela, Taiwan and Saudi Arabia. A version of the AN/TPS-63 has been designed for airship installation to give increased coverage. Entry **3162.353** under the Airborne Radar section gives more details.

Egypt has signed an agreement to co-produce the radar system. A deal valued at $190 million covers an initial batch of 34 radars co-produced by Westinghouse and the Egyptian electronics manufacturer Benha Co. over the period 1985-1992. Eight systems have already been purchased direct from Westinghouse.

AN/TPS-63 tactical air surveillance radar, set up and ready for operation. A four-man team can erect and prepare the radar within an hour

CONTRACTOR

Westinghouse Defense Electronics Center, Baltimore, Maryland 21203, USA.

3476.153

AN/TPS-65 TACTICAL SURVEILLANCE RADAR

The AN/TPS-65 is a two-dimension L-band air search radar developed as the primary airfield surveillance system for the US Marine Air Traffic Control and Landing System. This system provides automated and manual control information for aircraft landings at expeditionary airfields. The radar is similar to the AN/TPS-63, which is designed to provide low altitude coverage for tactical air defence operations, but has a dual channel configuration. Both radars feature extensive use of microelectronic and solid-state techniques.

An advanced moving target indicator (MTI) is claimed to provide a 90% probability of detecting a one square metre target out to the full instrumented range of 80 nm (148 km). When desired, the radar range can be increased to 120 or 160 nm (222 or 296 km). The digital MTI uses a four-pulse canceller with variable time intervals between pulses to distinguish moving aircraft from ground returns while eliminating MTI radar blind speeds. There is an additional digital three-pulse MTI weather canceller, and digital constant false alarm rate processing. Dual diversity operation is normal and there are built-in test facilities. Fixed station configurations of the AN/TPS-65 using sub-systems suitable for commercial airport radar applications are also available.

CHARACTERISTICS

Frequency: L-band, dual diversity, 51 selectable frequencies between 1250 and 1350 MHz
Range: 80 nm (148 km) on fighter-sized aircraft; optionally 120 or 160 nm (222 or 296 km)
Data rate: 6, 12, 15 rpm (10, 5 or 4 s)
Elevation coverage: 12,000 m shaped coverage to 40°
Power output: 100 kW peak; 3 kW mean
Antenna gain: 32.5 dB
Angular resolution: 2.7°
Range resolution: 152 m
ECCM: Coded pulse anti-clutter; frequency agility, PRF stagger
MTBF: 2900 h
MTTR: 20 minutes
Weight: 6200 kg

STATUS

Initial production completed.

CONTRACTOR

Westinghouse Defense and Electronics Center, Baltimore, Maryland 21203, USA.

6401.153

DUAL BAND RADAR

The Westinghouse dual-band radar has been developed out of the AN/TPS-43 and AN/TPS-63 designs to provide a 3-D long-range S-band system that can operate simultaneously with a medium range L-band channel to give high detection capability in extremely heavy clutter and ECM environments. Both S- and L-band elements are contained in one electronics shelter and radiate through one antenna.

The S-band channel operates out to 400 km with adaptive clutter processing and a high power narrow pulse for synchronised, low-range, full volume 3-D cover. The L-band channel operates up to 128 km at a high PRF to obtain the large number of returns necessary to reject clutter and false alarms, and to detect low flying aircraft. The complete equipment is designed to interface with any automated air defence system by simple software adaptations.

STATUS

In operational service.

CONTRACTOR

Westinghouse Defense Electronics Center, Baltimore, Maryland 21203, USA

4319.153

AN/TPS-70(V)-1 TACTICAL RADAR

The AN/TPS-70(V)-1 is a mobile S-band tactical radar designed to detect or track hostile aircraft in a variety of environments at ranges out to 450 km. It incorporates clutter rejection and ECCM features, with a low sidelobe antenna making it very difficult for enemy countermeasures to detect or jam the system. Also incorporated are advanced signal analysis and processing and a digital coherent MTI system.

The antenna, a flat-slotted array measuring 5.5 m wide by 2.5 m high, is combined with programmable doppler signal processing and an automated target extraction/tracking computer to provide the total operational scenario capability for automated and manual command centres. A multiple beam architecture with an entire redundant channel is employed, scanning all elevations and ranges simultaneously for targets by means of six beams. This ensures more 'hits' per target for optimum clutter rejection and target extraction.

The complete system consists of the main antenna, with a secondary antenna mounted on top and an operations/electronics cabin standard shelter unit. For ease of transport the AN/TPS-70(V)-1 is deployed as two modules which can be carried to a forward position by truck, aircraft or helicopter.

The AN/TPS-70(V)-1 is modular in design, the software being flexible to permit fast reprogramming to meet specific operational requirements. The system can be remotely controlled and monitored and can be operated either autonomously or as part of a larger network. The AN/TPS-70(V)-1 incorporates extensive built-in redundancy, self test/diagnosis and solid-state subsystems for virtual 100% system availability. System operation mode and parameters as well as self test are controlled from a centralised touch-sensitive monitoring and fault isolation panel. Should a system fault occur, the faulty LRU (line replaceable unit) is automatically determined and displayed for rapid plug-in replacement. The AN/TPS-70(V)-1 electronics are very similar to those of the USAF AN/TPS-43E enhancement production programme ensuring the benefits of long-term operational and logistic commonality. In addition to the mobile configuration the system is produced in a fixed sight variant, the AN/FPS-700.

CHARACTERISTICS

Frequency: 2900 to 3100 MHz
Peak power: 3MW (mean power 4.9 kW)
Range: 350 km for 1.7 m² target
PRF: Variable
Accuracy: 107 m range; 0.25° azimuth; 450 m height
Elevation coverage: 0 to 20° with option to –3°
Antenna type: Flat-slotted array
Antenna rotation speed: 6 rpm
Azimuth beamwidth: 1.6°

STATUS

The AN/TPS-70 is in full production and has been procured by seven nations with additional procurements underway. It was recently selected for the Caribbean Radar network programme (see entry under **3215.181**).

Three AN/TPS-70 systems have been ordered by the United Arab Emirates for surveillance of the Gulf area. The first of these was delivered in mid-1987 with the remainder due for delivery in 1988. Westinghouse is also supplying command and control systems for these systems. One AN/TPS-70 system has been procured by Thailand to supplement the two AN/TPS-43 radars already in operational service.

AN/TPS-70(V)-1 tactical radar

CONTRACTOR

Westinghouse Defense Electronics Center, Baltimore, Maryland 21203, USA.

5870.153

VIGILANT TACTICAL SURVEILLANCE RADAR

Vigilant is an S-band, medium-range 3-D mobile tactical surveillance radar that has been developed to complement the long range AN/TPS-70 system (**4319.153**). It is a full pulse doppler radar employing a four-pulse modified MTI to provide detection and fast response to low and medium altitude attack out to a range of 110 km (60 nm). It supplies simultaneous range, azimuth and height information of the smallest targets, and is designed to operate in an ECM environment as well as heavy ground and rain clutter.

Vigilant employs a planar antenna, similar to the AN/TPS-70, with a high scanning rate for fast reaction. The system will automatically initiate up to 1000 tracks and update each track on every antenna scan. A 'look-down' beam is provided for use when the equipment is deployed on a mountain top, to ensure detection of low-flying aircraft. An associated Vigilant mobile operations centre can be provided for autonomous or netted command and control. Alternatively it can be integrated with any other type of automated air defence system.

The antenna is a stacked beam type, seven beams giving coverage from –3 to +20°. This gives a fast data rate, a six-second rate being normal, although by operating the antenna at high speed for part or all of the rotation, a 4½ or 3 second data rate can be achieved. This rate allows tracking of targets up to Mach 2 and high aircraft longitudinal and manoeuvre accelerations. Frequency agility, high PRF and pulse coding are employed, as well as doppler techniques for the elimination of chaff. The radar operates at 1250 pps in the pulse agile mode, using eight of the possible 27 frequencies over a 200 MHz bandwidth.

The complete system, including, the operations

Vigilant tactical surveillance radar

centre, is transportable by two vehicles, and can be airlifted by standard cargo aircraft or helicopters. A crew of four can assemble or disassemble the radar within 30 minutes. Vigilant can operate completely unattended if required, all operational functions being controlled remotely via a narrow-band digital link. A built-in monitoring and fault isolation subsystem constantly indicates operational status and displays any fault at the radar and the remote site.

CHARACTERISTICS
Antenna type: Slotted waveguide
Frequency: 2900-3100 MHz
Peak power: 90 kW
Range: 110 km
Accuracy: 40 m range; 0.35° azimuth; ±275 m height
Elevation coverage: –3 to +20°
Data update rate: 6, 4.5 or 3 secs
Azimuth accuracy: 0.35°

STATUS
Following extensive operational tests against small radar cross-section targets in heavy clutter, Vigilant is now in full production for the Swiss government.
CONTRACTOR
Westinghouse Defense Electronics Center, PO Box 1693, Baltimore, Maryland 21203, USA.

6111.153
ADVANCED MISSILE SURVEILLANCE RADAR

The Advanced Missile Surveillance Radar is another variant of the AN/TPS-70(V)-1 tactical radar and is designed for the US Army anti-tactical missile mission. It is intended to detect missiles in intense ECM conditions and has a flexible antenna design which allows the radar to perform normal air defence roles in addition. A unique bi-fold antenna design permits the radar to be configured for either tactical missile detection or typical air defence surveillance. The changeover can be accomplished in less than five minutes.

Up to 1000 targets can be tracked simultaneously, with no degradation in surveillance. Processing capabilities include both impact and launch predictions. For netted operations, the processor design permits transmission of digital track data to a command centre or weapons unit. The radar can operate completely unattended.
CONTRACTOR
Westinghouse Defense Electronics Center, PO Box 1693, Baltimore, Maryland 21203, USA.

6112.153
MULTI-ROLE SURVIVABLE RADAR

The Multi-Role Survivable Radar (MRSR) is a mobile ground-based air defence radar designed to operate and survive in the battlefield environment. It detects and tracks aircraft in intense ECM conditions, and provides netted data with sufficient accuracy to support mid-course guidance of missile systems. The MRSR employs a 'quiet' signal that makes it extremely difficult for anti-radiation missiles to gather the guidance information required to carry out a successful attack on the radar.

The MRSR operates in X-band with a mast-mounted elevated flat-plate phased array antenna. It is contained in a single vehicle, can be rapidly deployed and includes many features to aid survivability. These include low sidelobes, low peak power using phase-coded CW and frequency agility/dual frequency operation, multiple receive beams and a narrow azimuth bandwidth.
STATUS
An advanced development model is due to be delivered to the US Army in 1989 for evaluation.
CONTRACTOR
Westinghouse Defense Electronics Center, PO Box 1693, Baltimore, Maryland 21203, USA.

Multi-role survivable radar

5868.153
ASTAR-3 AND ASTAR-2 AIR DEFENCE RADARS

The ASTAR radars (Advanced Series of Tactical Air Defence Radars) are 3-D and 2-D systems which operate in G-band and have more than 80 per cent commonality of design, with the exception of the antenna. In air defence systems that require both 3-D and 2-D radars, this feature will simplify operations and lower logistics costs of spares, training, documentation and other support requirements.

The ASTAR radars can be used as completely autonomous systems, with the entire radar subsystem, displays and communications equipment mounted in one S-280 shelter. In addition, a small operations shelter with two to four displays can be set up near the radar shelter to form a mini-C³ system. The 3-D planar array antenna and the 2-D reflector antenna can be mounted in several different configurations, ie on a low trailer, together with the radar on one vehicle, on a set of tripod legs, or on top of a low tower. Both radars can also be operated in a completely unattended mode where operational control and radar status/fault defect information is maintained and monitored from a central location.

The ASTAR-3 provides 3-D coverage out to an instrumental range of 200 nm, from –3° to +28° in elevation, and up to 100 000 feet in height. Operationally, the system provides for tactical long-range, 3-D air defence.

The ASTAR-2 provides 2-D coverage out to an instrumented range of 100 nm, from –3° to +32° in elevation, and up to 40 000 feet in height. Operationally, the system is used to provide gap-filler coverage for an overall air defence system or a coastal surveillance radar.

The ASTAR-3 antenna is a G-band planar array which rotates on a pedestal in the azimuth plane, and provides elevation scanning by means of phase shifters in the vertical plane. Dual antenna rotation rates of 6 and 12 rpm are provided. The size of the antenna is approximately 9 × 10 feet so that it does not need to be disassembled or folded for transportation.

CHARACTERISTICS

	ASTAR-3	**ASTAR-2**
Coverage		
Range	200 nm	100 nm
Azimuth	360°	360°
Elevation	100 000 ft	10 000/40 000 ft
Data rate	10/5 s	5 s
Accuracy (at 100 nm)		
Range	50 ft	50 ft
Azimuth	0.2°	0.15°
Height	2500 ft	—
Resolution (PR ⩽ 50 per cent)		
Range	304 ft	304 ft
Azimuth	1.88°	1.57°
Height	2°	—
Plot output rate	100/s	100/s
False plot output rate	0.5/s	0.5/s
MTI improvement factor (surface)	52 dB	52 dB
Circular polarisation (volume clutter)	—	15-20 dB
MTBF	1000 h	1400 h
MTTR	0.4 h	0.4 h
Availability	0.999	0.999

The ASTAR-2 antenna is a G-band cosec² fan beam array which rotates on a pedestal in the azimuth plane, and provides two elevation beams by means of two feed horns. The size of the antenna is approximately 11.5 × 6 feet.

A full range of ECCM features is incorporated in each radar design. Some of these features include very wide operational bandwidth, low antenna sidelobes, anti-chaff MTI processing, jam strobes, sidelobe blanking, CFAR and clutter map processing, coded waveform agility, and pulse-to-pulse frequency agility.

A fully integrated IFF capability is provided on both radars, including radar/IFF correlation. Completely automatic target tracking is also supported which produces target trajectory data for up to 400 targets in the coverage volume. Plot and track data can be displayed locally or transmitted to a remote command and control facility. The communications links can also be used to operate the radar from a remote location.

The ASTAR radar system can take advantage of attended or unattended operation. Extensive microprocessor-based built-in-test and automated operational and maintenance functions achieve total remote capability. This allows a distant command/control site to have full capability of operation, status monitoring, fault detection and isolation.

For local operation, full colour raster scan control/displays are available which provide alphanumeric, graphic, video and real-time data with touch panel system control.
STATUS
In development.
CONTRACTOR
Aydin Radar and EW Division, 32 Great Oaks Boulevard, San Jose, California 95119, USA.

2535.153

AN/TPX-42 PIDP IFF-ATCRBS EQUIPMENT

The PIDP (programmable indicator data processor) is an advanced sophisticated version of the AN/TPX-42 secondary radar system which was built for US Air Force, Navy, civil ATC and military air defence purposes. This latest state-of-the-art version nomenclatured AN/TPX-42A(V)10 is a secondary radar system comprising interrogator/receiver, data processor and display equipment installed in over 100 AN/FPN-47, AN/GPN-12, AN/GPN-20 and AN/FPS-67D airport surveillance radars. The heart of the PIDP, which provides a significant automated extension of existing ATC capabilities, is a newly-developed programmable digital processing system that will provide the controller with more useful information consisting of flight identification, aircraft type, altitude, and ground speed. The PIDP system utilises a NOVA 3 mini-computer (utilising only 26 000 words of programmed storage), new vertical displays (OD-153/T) and/or indicator modification kits (OD-152/T or OD-152A/T) to upgrade existing OD58 and OD58A indicators to accept programmable features.

STATUS

322 basic AN/TPX-42 military and civil systems were built by the Command Systems Division and are presently in place at various sites in the United States and overseas bases. Basic and programmable versions of the AN/TPX-42 have been supplied to the civil/military authorities of Brazil, Bulgaria, Canada, Czechoslovakia, Iran, the Netherlands, Norway, Philippines, Poland, Spain and Taiwan.

CONTRACTOR

Eaton Corporation, Command Systems Division, 815 Broad Hollow Road, Farmingdale, Long Island, New York 11735, USA.

3703.153

AN/TPX-46(V) IFF INTERROGATOR

The Hazeltine interrogator set AN/TPX-46(V) 1-6 is compatible with a wide range of missile systems and associated radars, including the Hawk and Nike-Hercules configurations and most versions of the AN/FPS and AN/TPS radar sets. Interface capabilities for other radar systems are incorporated. The AN/TPX-46(V) interrogates radar targets and, when the aircraft replies correctly, provides indications of aircraft identity for display on the associated radar PPI. The interrogator also has the ability, when its associated radar has been disabled – except for the PPI and its power supplies – to generate its own main trigger, to rotate its own antenna producing antenna synchronising signals, and to display on the radar indicator IFF replies from suitably equipped aircraft. Even when subjected to environmental extremes, the interrogator antenna can synchronise with the radar antenna at speeds up to 25 rpm. In addition, the interrogator transmitter power can be reduced, making it suitable for use with ground control approach radar systems.

The interrogator consists of four basic operating units: the receiver/transmitter, the coder-decoder group, the control box and the antenna group. The antenna consists of an array capable of a beamwidth of 5.5° (nominal). This is achieved by means of sum and difference antenna pattern techniques providing interrogator sidelobe suppression and receiver sidelobe suppression. Until now, production of so narrow a beamwidth has required the use of arrays approximately twice as long.

The AN/TPX-46(V) provides a clean display by means of the defruiting action of its processor unit. Other circuits make possible the automatic countdown of the repetition rate of the radar trigger, when necessary, to a frequency suitable for use by the interrogator set. Also provided is a gain time control circuit.

A solid-state transmitter module is now available as a direct replacement for the original electron tube version, resulting in reduced maintenance costs of labour. The AN/TPX-46(V) operates in Modes 1, 2, 3/A and C or in the Mk XII configuration on Mode 4 in conjunction with a KIR-1A/TSEC computer.

CHARACTERISTICS

Receiver frequency: 1090 MHz
Sensitivity: –80 dBm
Transmitting frequency: 1030 MHz
Peak power: 1000 or 2000 W, selectable; lower power model available
Antenna: 2.23 m long; 22.7 kg weight

STATUS

In service with US Army Improved Hawk missile system, AN/TSQ-38 radars, and AN/TSQ-51 AADCPs; US Marine Corps Improved Hawk missile batteries. Also in service with a number of foreign states. The AN/TPX-46(V)7 configuration is now in production for the US Army Patriot air defence system (**2800.131**).

CONTRACTOR

Hazeltine Corporation (a member company of Emerson Electric Co), Greenlawn, New York 11740, USA.

3705.153

AN/TPX-54 IFF/SSR INTERROGATOR

Using a modular building block approach, the AN/TPX-54(V) provides an interrogator which meets military and civil needs of secondary surveillance radar users in ground and shipborne applications. The modular system provides sufficient flexibility for a single compact interrogator to meet a diverse range of mission requirements, including those of ground and shipborne air defence and military and civil ATC. The AN/TPX-54 is designed to military specifications and operates in up to four SIF modes from Modes 1 or B, 2 or D, 3A and C and also in Mode 4. The unit is capable of operating with up to a five mode interlace when used with a KIR-1A/TSEC computer in Mode 4. The mode selection may be made remotely at an external decoder or at a remote control box. The interrogator front panel also provides mode selection for maintenance purposes.

The basic AN/TPX-54 system consists of five plug-in printed circuit boards to perform the overall functions of transmitting and receiving, synchronization, coding, video processing and BITE. Available as options are 11 additional plug-in circuit boards for defruiting, computer interface, Mode 4 operations, automatic countdown of radar trigger and target data extractor with reply correlation. The target data extractor provides digital target data for use by a radar computer and also allows radar computer control. The AN/TPX-54 predicted MTBF exceeds 2,000 hours.

More detailed information is given under this entry number in previous editions of *Jane's Weapon Systems*.

STATUS

The AN/TPX-54(V) is in use in the United States as well as other countries. It is currently being procured by the USAF as a replacement for all AN/UPX-23, AN/UPX-14 and AN/UPX-21 interrogators. It is being supplied for use in a standard 19 inch (48 cm) rack and meets DoD-AIMS 65-1000, NATO STANAG 5017 and ICAO Annex 10 requirements.

AN/TPX-54(V) IFF receiver/transmitter

CONTRACTOR

Hazeltine Corporation (a member company of Emerson Electric Co), Greenlawn, New York 11740, USA.

4317.153

OX-60/FPS-117 IFF INTERROGATOR

The Hazeltine IFF interrogator OX-60/FPS-117 is the beacon processor group for the AN/FPS-117 minimally attended radar. The OX-60 is a 200 nautical miles system with sidelobe suppression and interrogator sidelobe suppression. It is capable of interlacing Modes 2, 3/A,C, and 4 and parallel processing the replies. The system is all solid-state and to meet the requirements of a minimally attended site, it is designed in a redundant, automatic switch-over configuration. The configuration consists of a control processor, two processors, signal data and two receiver/transmitters which provide an MTBF in excess of 118,000 hours. The automatic fault test and isolation is extensive and under the control of an 8086 microprocessor. Self-test of the beacon system is carried out on a continuous basis and consists of both a local circuits self-check as well as a full test accomplished by the generation of simulated targets. This function occurs in parallel to normal operation and is completely transparent to it.

Once a fault has been detected the microprocessor goes into a fault-isolation function to isolate automatically the fault to a single or group line replaceable unit. A status message is then passed to the computer indicating which module is at fault and needs to be replaced. This automatic fault isolation reduces MTTR and simplifies maintenance procedures.

CHARACTERISTICS

Receiver frequency: 1090 ±3 MHz
Sensitivity: –83 dBm decoding
Transmitter frequency: 1030 ±0.2 MHz
Peak transmitter power: 2 MW
Duty cycle: 1%
Range: 200 nm
Effective beamwidth: 2.75 – 0.5°
Power input: 115/200 V AC 50 – 400 Hz, 3-phase. Single channel 350 W, dual channel 700 W
Specification: Designed to MIL-E-4158
MTBF: 118 765 h (redundant); 2969 h (single channel)
MTTR: 20 min
Dimensions: Signal processor 482 × 266.5 × 520.5 mm; receiver/transmitter 482 × 222 × 609.4 mm; control unit 482 × 133.3 × 381 mm
Weights: Signal processor 29.7 kg; receiver/transmitter 16 kg; control unit 12.3 kg

STATUS

Fitted to the AN/FPS-117 radars which are now operational under the Seek Igloo programme. Also being fitted to the AN/FPS-117 radars of the North Warning System and the Saudi Arabian installations (see **4326.153**). Meets DoD-AIMS 65-1000 and FAA 101051A requirements.

CONTRACTOR

Hazeltine Corporation (a member company of Emerson Electric Co), Greenlawn, New York 11740, USA.

4632.053

MODEL 2679 IFF INTERROGATOR

Model 2679 IFF interrogator is a compact, rugged equipment designed to operate and survive in the battlefield. It provides weapon systems with an all-weather fire capability permitting rapid engagement of enemy aircraft at maximum system range. It is particularly suitable for land mobile systems, helicopters and small ships, where the need is for short range identification, smallness, lightness and low power consumption. The equipment includes both interrogator side lobe suppression and receiver sidelobe suppression capability for good azimuth accuracy while using small antennas. The model 2679 interrogator can be controlled directly from an associated fire control system computer or an associated control box.

The system consists of a receiver/transmitter, video processor, reply processor (sequential observer), timing and control, self-test, memory module and power supply. It has operator controls that allow mode selection, initiation of challenges and target identification. A derivative of the Hazeltine produced AN/MPX-12 interrogator, the Model 2679 can operate in modes 1, 2, 3 or 4.

CHARACTERISTICS

Transmitting frequency: 1030 ± 0.2 MHz

Receiver frequency: 1090 ± 0.2 MHz

Peak power output: 200 W

Sensitivity: –76 dBm

Weight: 28 lb (12.7 kg)

CONTRACTOR

Hazeltine Corporation (a member company of Emerson Electric Co), Greenlawn, New York 11740, USA

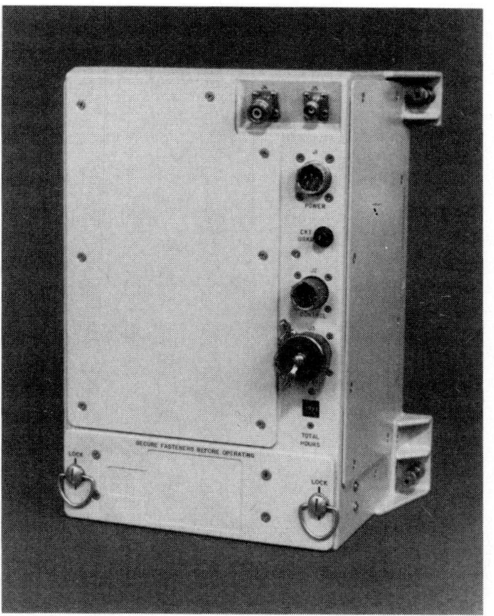

Model 2679 IFF interrogator

6816.153

AN/PPX-3B GROUND INTERROGATOR

The IFF interrogator AN/PPX-3B is a lightweight, battery operated ground-to-air equipment. It is modular in construction with a Mode 4 memory and reply evaluator module to eliminate the need for the KIR-1A computer. A programmer/battery charger loads the Mode 4 memory with the appropriate code catalogue and charges batteries to be used with the interrogator. The antenna/programme connector and test connector are located on the top panel. A single cable interfaces the interrogator to the weapon system during operation. Two cables interface the interrogator to the programmer during IFF coding operations. With slight modifications this equipment can be used for helicopter operations as an airborne air-to-air interrogator.

CONTRACTOR

Teledyne Electronics, 649 Lawrence Drive, Newbury Park, California 91320, USA.

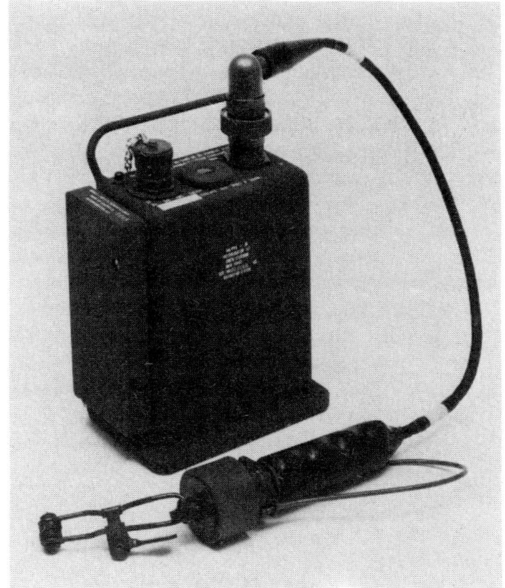

AN/PPX-3B interrogator

5776.153

AN/MSQ-117 IFF TRACKING SYSTEM

The AN/MSQ-117 is a lightweight IFF tracking system which provides advanced target acquisition and tracking for tactical and fast deployment applications. The equipment is completely self-contained and is based on microprocessor technology with a high power solid-state transmitter/receiver.

The system consists of an L-band antenna, a receiver, interrogator/processor and a display. Interrogations generated by the system elicit responses from transponder-equipped aircraft. From these replies, the position, altitude and identity of each aircraft within the system coverage is extracted and displayed on a random access PPI. The equipment also incorporates an adaptive α–β tracker which provides smoothed position and predictions for a selected aircraft. The tracking algorithm supports high speed manoeuvres by tactical aircraft. The digital output of the tracker can be used to position the antennas of other systems, eg ECM/ECCM.

CHARACTERISTICS

Peak power: 1 kW

Range: 60 n miles

Coverage: 360° azimuth; –30° to +85° elevation

Accuracy: ±0.5° azimuth; 300 ft rms range

Operating temperature: –40° to +62°C antenna; 0 to 50°C electronics

Dimensions

Antenna: 62 × 14 inches (157 × 16.5 cm)

Antenna control unit: 20 × 19 × 4 inches (50.8 × 48 × 10 cm)

Coder/processor: 20 × 19 × 5.5 inches (50.8 × 48 × 13.8 cm)

Display processor: (16 × 19 × 8.75 inches (40.6 × 48 × 22.2 cm)

Weights

Antenna: 37 lb (16.8 kg)

Antenna control unit: 36 lb (16.4 kg)

Coder/processor: 25 lb (11.4 kg)

Display/processor: 20 lb (9.1 kg)

STATUS

In production.

CONTRACTOR

Eaton Corporation, Command Systems Division, 815 Broad Hollow Road, Farmingdale, Long Island, New York 11735, USA.

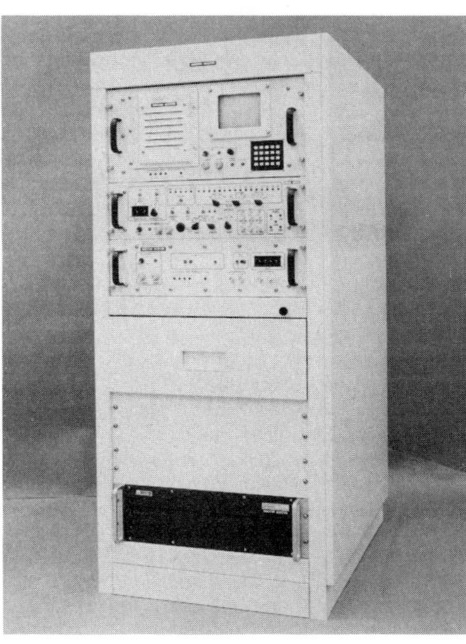

AN/MSQ-117 IFF tracking system

2547.153

AN/VPS-2 RANGING RADAR

This coherent pulse-doppler MTI radar is the range sensor of the Vulcan forward area air defence system (**2850.131**). Targets are acquired and tracked optically in this system but high-speed range and range rate data are required as inputs to the lead computing sub-system.

The axis of the radar antenna is servoed to the optical line of sight. The sight case and gun bore are physically fixed in alignment, but the sight reticle, which defines the optical line of sight, can move relative to the sight cage and is controlled by the gunner (see PIVADS **4015.131**). As the gunner tracks the target, the angular tracking rate and the radar-measured range and range rate are computed to predict future target position and then to control the gun position, introducing between the optical and firing axes the lead angle and super elevation required for successful target engagement.

Although primarily directed by optical means, the AN/VPS-2 radar has an automatic search and lock-on characteristic so that it tracks the target accurately.

CHARACTERISTICS

Frequency: 9200 – 9250 MHz, 6 crystal-controlled channels

Peak power: 1.4 kW

Average power: 10 W

Antenna: 4° conical beam ±20° in azimuth and elevation

Search range: 5000 m
Tracking capability: 15 – 310 m/s
Sub-clutter visibility: 40 dB or better

STATUS

Operational. A modified AN/VPS-2 is being fitted to the Spanish FFG-7 frigates as part of the Merroka CIWS (**3208.131**).

CONTRACTOR

Lockheed Electronics Company, US Highway 22, Plainfield, New Jersey 07061, USA.

AN/VPS-2 range-only radar on a self-propelled multi-barrel gun

---

2499.153

HIPAR (HIGH POWER ACQUISITION RADAR)

HIPAR (high power acquisition radar) is designed specifically for use with the United States Army's improved Nike Hercules surface-to-air missile system. The improved Nike Hercules system is deployed in key areas in the United States and overseas as a defence against high-performance aircraft, air-to-ground missiles, and tactical ballistic missiles. HIPAR was developed as an integral part of the system to provide significantly increased detection capabilities. It supplies long-range coverage and extremely precise azimuth data on small supersonic targets.

In addition to long range, the high-gain narrow-beam antenna used by HIPAR provides high-angle coverage. Its high-altitude coverage is said to be appreciably better than that of most present-day acquisition or surveillance radars.

To permit the full detection capabilities of the high-gain antenna to be realised despite the presence of enemy jamming, what are said to be unique MTI and receiver circuits have been developed for use with HIPAR. These circuits employ anti-jamming techniques that are said to provide capabilities not previously possible with conventional MTI techniques.

Both static and mobile HIPAR systems have been produced for the Army's use in both the United States and Europe. The mobile version (AN/MPQ-43), which uses the same electronics and components as the static version, is said to be the highest-powered mobile radar of its kind in the West. An important feature of the antenna for the mobile version is its ability to be changed quickly from cosec² coverage to fan-beam coverage.

Mobile HIPAR is transported on five flat-bed trailers and can be emplaced by field army personnel after minimum training.

CHARACTERISTICS

Peak power: Classified
Average power: Classified
Frequency: 1350 – 1450 MHz
PRF: 417 – 438 pps
Pulse width: 6 μs
Receiver sensitivity: –113 dBm
Beam characteristics: Cosec² or fan
Beamwidth: Azimuth 1.2°; elevation 0 – 60°
Antenna azimuth rate: 6 or 10 rpm

STATUS

Operational. A total of 91 fixed HIPAR systems and 23 mobile HIPAR systems have been completed.

CONTRACTOR

General Electric Company, Electronic Systems Division, Syracuse, New York 13201, USA.

HIPAR

Mobile HPAR

3721.153

HUGHES AIR DEFENCE RADAR (HADR)

HADR is the acronym by which Hughes refers to an advanced three-dimensional multi-mode radar designed for use in national air defence networks, either to replace ageing equipment, to enhance system performance or to fill gaps in the existing system. One version of the system is known as the H-3000. It is understood that many of the advanced features of this radar are a result of experience gained within the AN/TPQ-36 and AN/TPQ-37 mortar and artillery locating radars (**2848.153** and **1976.153**). Designed to meet new NATO ACE (Allied Command Europe) standard radar specifications, HADR will automatically detect, classify and report on all targets in its area of coverage. Operation is in the E/F-band (former S-band) of the spectrum and a 4.8 × 6 m planar array antenna is used, generally protected within a radome.

The HADR is a member of the same family as the VSTAR (**4080.153**) and there may be commonality of certain parts of equipment as well as operating principles.

The HADR primary radar operates over a bandwidth greater than 12 per cent. The radar achieves a detection range of at least 320 km for a 1 m² target. The instrumented range coverage is 500 km. Versatile computer control enables radar coverage to 30 000 m altitude and up to 24° elevation. False target reports are controlled to prevent ADGE system computer overloads.

Target positional data is provided in three dimensions. Target range is measured to the nearest range cell while target azimuth and elevation are obtained by beam splitting. Performance in clutter is achieved by a digital mean velocity clutter tracker in conjunction with a single or double delay MTI canceller. A range gated pulse doppler waveform and associated processing implementation also is provided. An antenna rotation period of 10 or 12 seconds is consistent with ADGE system data rate requirements. Primary or secondary radar target report rates can exceed 400 per scan at the radar output. However, transmission rates to the remote command and control site may be limited by data link capacity.

The HADR antenna sub-system scans a single pencil beam in the elevation plane under computer control by controlling the phase shifters of the antenna. By properly positioning this beam in sequential steps, complete elevation coverage is provided. The elevation scan is completed before the antenna moves one beamwidth in azimuth, providing complete coverage of the surveillance volume with no holes. Antenna beamwidth and the transmitted waveform are controlled at each beam position to optimise management of the radar time and energy resources. Beamwidth is 1.1° in elevation and 1.7° in azimuth.

The operational modes in the HADR system are defined by tables stored in the computer memory. The computer controls the following elements for each beam in the search raster:

(1) Instrumented range
(2) Peak power
(3) Pulse length
(4) Elevation beamwidth
(5) Waveform
(6) Elevation angle
(7) Frequency
(8) Detection criteria
(9) Plot extraction criteria
(10) Operation changed by software

The secondary surveillance radar (SSR) sub-system incorporates an IFF interrogator and antenna. The HADR antenna makes provisions for mounting the primary IFF antenna as well as the sidelobe blanking IFF antenna and for routeing the RF signals through the rotary joint. The beacon video processor which is part of the multi-function processor provides extraction of IFF video and provides active and passive decoding. It also provides SSR target report outputs to the radar controller. The radar controller provides a correlation of these plot reports with primary radar plot reports for a consolidated output to the command control site.

The HADR primary radar utilises several techniques to reduce the effectiveness of ECM. Sophisticated jammers are thwarted by the radar operational characteristics including frequency agility, coded pulses, multiple pulsewidths and multiple PRFs. The primary radar design has low susceptibility to jamming through the use of pencil beams, a low sidelobe level antenna design, high receiver dynamic range and sharp channel selectivity. Automatic threshold control, target detection correlation and sidelobe blanking are techniques provided to handle any residual jamming which does get into the system. The burn-through mode and automatic frequency selection are special modes available for use in an ECM environment. Single pulse MTI is provided out to 60 nautical miles and two-pulse MTI out to 100 nautical miles. Seven-pulse range-gated pulse doppler is used to suppress rain or chaff clutter.

The system is capable of selecting operating frequency under computer control, and each frequency can be selected in about 40 µs, and may be changed on pulse-to-pulse or dwell-to-dwell basis. When selected by the operator, the system will automatically measure the interference level in each of the available operational channels and select that which is least affected.

The normal surveillance mode can be interrupted, at the operator's discretion, by a special mode which will intensify the energy on a specific target location. Normal surveillance volume is interrupted and all of the energy available is concentrated in a high energy waveform with a long dwell time in a sector automatically centred around the jam strobe which has been measured by the system and which indicates jammer direction. At high elevation angles, full antenna gain will be employed on the target even through the normal surveillance volume might call for a broadened beam.

STATUS

Four HADR systems are now operational in the Federal Republic of Germany, two used in conjunction with the NATO Air Defence Ground Environment (NADGE), and known as GEADGE (German ADGE), where they replace AN/FPS-7 radars of the 412L radar network.

A version of HADR, known as the HR-3000, has been selected for the update of NATO's air defence system on the southern flank. Hughes will deliver three of these radars to Turkey, two to Greece, two to Italy and one to Portugal. Turkey, Italy and Greece will incorporate the equipments into the NADGE system. Portugal exercised an option in late 1986 and has ordered two more. The HR-3000 incorporates a number of modifications to the basic HADR, including increased ECCM capability, wider receiver bandwidth, improved clutter-rejection waveforms and a faster antenna rotation speed to meet the NATO requirement for a higher data rate. The radars will operate on concrete platforms without supporting towers or radomes.

Hughes is also supplying three HADR radars for Norwegian air defence. These systems are used in shelters built into the top of mountains, and are mounted on elevators so that the antennas can be retracted into environmental controlled silos for routine maintenance. The silos are designed and manufactured by Thune-Eureka of Norway. The first system is now operational.

HADR antenna

One HADR is operational in Malaysia as part of the Malaysian Air Defense Ground Environment (MADGE) programme. Hughes is prime contractor on this programme.

CONTRACTOR

Hughes Aircraft Company, Ground Systems Group, Fullerton, California 92634, USA.

4080.153

VSTAR/VSTAR-PT AIR DEFENCE RADAR

Hughes has introduced a new air defence radar based on the operating principle of the company's AN/TPQ-36 and AN/TPQ-37 weapon locating radars (**2848.153** and **1976.153**) and incorporating some elements of the hardware of the latter, thus achieving common logistic support between the two radars. The new radar, VSTAR (variable search and track air defence radar), is an E/F-band (former S-band) equipment designed for rapid deployment and is configured in a variety of transportable forms and is primarily intended as a mobile system and secondly for static use in a gap-filler role. VSTAR has a wide instantaneous field of view (90°) and a 30° elevation field of view. The precision tracking (PT) variant has 90° elevation and 30° azimuthal field of view.

The planar antenna array measures approximately 7 × 11 metres and employs a combination of electronic scanning and mechanical rotation for target search and detection, although multiple targets can be tracked by electronic beam steering with the antenna array stationary. While rapidly rotating, the radar scan covers the entire aircraft penetration envelope, and a precision radar beam searches to a range of 250 km and up to 30 000 metres in altitude. Simultaneously, a beam also scans medium range at a faster rate to provide a quick reaction to low-flying or pop-up targets. The electronically scanned beam can search and track any point within the large azimuth-elevation sector while the antenna is being rotated. This allows electronic back and forward scan for immediate target verification and track initiation, multiple target tracking in high traffic density, variable track update rates, long dwell for non-co-operative identification, and ECM burn-through.

STATUS

VSTAR was demonstrated in three nations in 1980-82 and it is understood that this radar has been submitted for a number of other air defence requirements in NATO and elsewhere. Production contracts have been received but no details are available.

VSTAR air defence radar mounted on a tracked vehicle

CONTRACTOR

Hughes Aircraft Company, Ground Systems Group, Fullerton, California 92634, USA.

5777.153

AN/UPS-3 TACTICAL DEFENCE RADAR

The AN/UPS-3 tactical defence alert radar has been designed to meet the requirement for a short-range air defence command and control sensor for all-weather surveillance and detection of low flying targets. It operates as a tactical alerting and cueing sensor in the extreme forward area of the battlefield and provides automatic identification of helicopters and fixed wing aircraft. Target data can be provided to an adjacent fire unit, or transmitted by digital link to a command location for dissemination to a number of units. The system operates in D-band, is all solid-state, lightweight, portable and highly reliable. It operates autonomously in the field, has a march order time of less than ten minutes and has been air drop qualified. It has also been employed as part of a netted system in the SHORAD C Test Bed by the US Army and can be a low-cost gap filler.

The system consists of a rotating antenna assembly mounted on a pedestal and quadripod, a transceiver/ signal processor also on a quadripod and a display which can be remoted up to a kilometre away. Up to four display units can be operated remotely from the equipment, one being the master by the attachment of

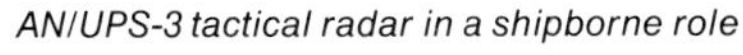

AN/UPS-3 tactical radar in a shipborne role

the remote control unit. The radar can be operated with other displays/terminals through an RS-232 output port. The antenna elevation angle can be adjusted –5° to +10° and provides height coverage from ground level to 10 000 ft. The display presents target range and azimuth on a LED matrix. A digital signal processor which gives significant improvements in accuracy has been developed and tested and this will also provide a track-while-scan capability.

CHARACTERISTICS

Frequency: D-band, eight crystal frequencies selectable
Peak power: 230 W
Average power: 20 W
MTI improvement factor: 55 dB
Antenna
Beamwidth: 8° horizontal; 17° vertical
Rotation speed: 90°/s or 60°/s
Gain: 22 dB
Accuracy/resolution: Range 1 km on 10 km scale, 2 km on 20 km scale; 10° azimuth
Weights
Antenna: 126 lb (57 kg), including antenna, reflector and feeder, pedestal and quadripod
Transceiver: 52 lb (28 kg)
Display: 20 lb (9 kg)
Remote control unit: 2.2 lb (1 kg)

STATUS

In production. Under contract to the US Navy.

CONTRACTOR

GEC Astronics Corp, 3400 Airport Avenue, Santa Monica, California 90406, USA.

5607.153

MULTI-FUNCTION RADAR (MFR)

The Multi-Function Radar (MFR) is a next generation land-based weapon control system that has been developed from the naval FLEXAR programme (**3258.253**). It uses technology derived from this programme and is intended for use with existing I-HAWK and other short-to-medium range missiles to provide a significant reduction in manpower and equipment, improved mobility and transportability, improved operation in ECM conditions and reduced life cycle costs. Further information is given under the FLEXAR entry in the naval radar section.

STATUS

Development is proceeding in parallel with FLEXAR. During 1983, the system successfully completed testing with the Improved Hawk missile and US Navy Standard.

CONTRACTOR

Hughes Aircraft Company, Ground Systems Group, Fullerton, California 92634, USA.

5617.153

AN/FPS-118 OVER-THE-HORIZON RADAR

The CONUS OTH-B (Continental United States over-the-horizon-backscatter) has been developed to provide electronic surveillance of aircraft at extended ranges of 500 to 1800 nautical miles (800 to 2880 km). Operating in the high frequency band, where the radar energy is reflected by the ionosphere, this ground-based radar is designed for over-the-horizon detection and tracking of aircraft and cruise missiles at any altitude. The system operates between 5 and 28 MHz and has two separate (bistatic) transmit and receive sites. On the United States eastern seaboard a prototype system was tested, the transmitter being located at Moscow US Air Force Station, Maine, with the receive site at Columbus Falls US Air Force Station, Maine, 110 miles (177 km) away. To adapt continuously in real-time to the prevailing ionospheric conditions, both the transmit and receive functions are completely computer controlled. Operation of the transmit and receive sites is synchronised in absolute time to better than one microsecond accuracy by Loran-C. The experimental transmissions from the Maine site covered an arc from 16.5 to 76.5° and 500 to 1800 nautical miles (800 to 2880 km) in range. This region was selected to evaluate the radar performance in an area where propagation will be most likely affected by the aurora, and to provide surveillance of the busy North Atlantic routes where aircraft targets are readily available at all times. The entire OTH-B project is due for completion in the early 1990s and will consist of a three-sector eastern seaboard system covering 180°, a similar western seaboard system and a two-sector southward-looking coverage of 120°. A fourth system, based in Alaska and looking southwest along the Aleutian Islands is also to be constructed, plus a number of sites in the Pacific.

CONUS OTH-B E-band transmit antenna section

The transmit antenna array consists of four separate side-by-side 12-element sub-arrays, each optimised to cover a different portion of the total operating range. Together they provide the capability to operate anywhere between 5 and 28 MHz and present low voltage-standing-wave-ratios to the transmitters at all scan angles. The highest frequency sub-array (Band E) uses vertical dipole elements, the other three sub-arrays use canted dipoles. This arrangement provides elevation patterns which match that required for propagation to the desired ranges via ionospheric reflections. The elements are mounted in front of a common backscreen ranging from 45 to 100 feet (14 to 30.5 m) high and approximately 2265 feet (690 m) long. This has been extended to 3630 feet (1106 m). A common groundscreen extends 750 feet (230 m) in front of the arrays.

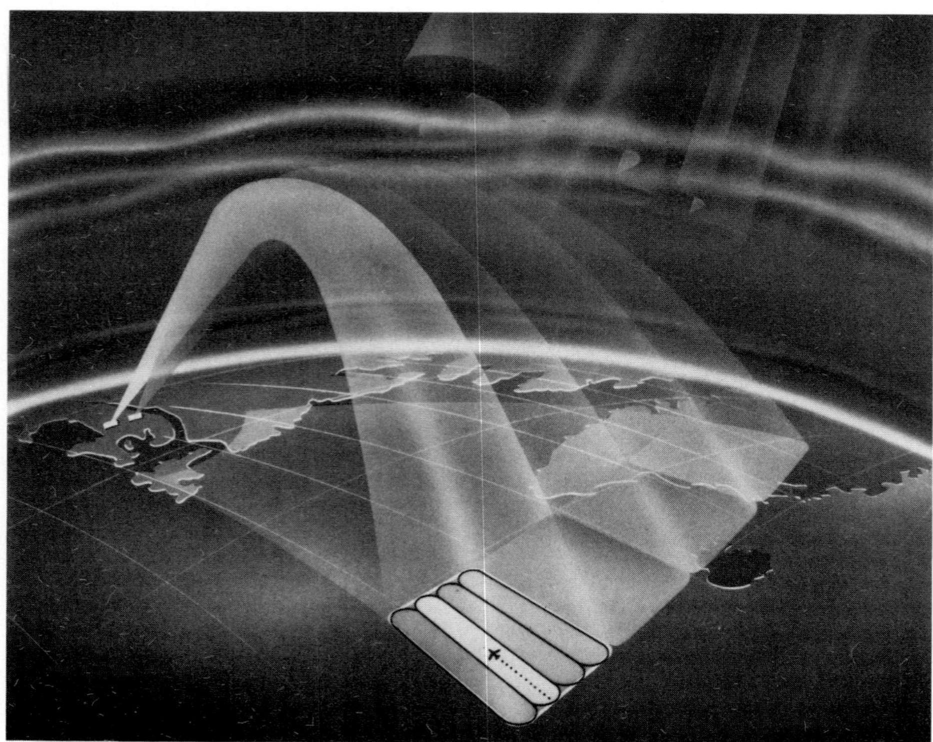
OTH-B range/azimuth sectors

Twelve transmitters operate simultaneously into the 12 elements of a selected sub-array. Each transmitter contains four band-tuned tank circuits which respectively match the four antenna sub-arrays. Each transmitter produces up to 100 kW average power, with very high spectral purity at any operating frequency between 5 and 28 MHz. A 100 kW water-cooled tetrode is used in the final high-power amplifier stage of each transmitter. The 12 transmitters are driven by a beamformer at low power level which causes the transmitter/array combination to collimate the desired 7.5° beam and steer it to the selected positions. In operation the transmit site generates up to 100 MW of effective radiated power.

The receive antenna is a broadside array of 137 fan monopole elements mounted in front of a backscreen which is 3906 feet (1190 m) long and 50 feet (15 m) high. In the operational antenna, the length has been increased to 5230 feet (1594 m). As in the transmit array, a groundscreen extends 750 feet (230 m) in front of the entire receive array. Two bands are used to cover the 5 to 28 MHz range, a low band covering 5 to 11.2 MHz and a high band from 11.2 to 28 MHz. On the low band, every other element is used across the array giving a total of 68 active elements. Dividing the array in this fashion and using variable aperture weights in beamforming results in virtually constant receive beamwidths of 2.75° with no grating lobes over the full 60° azimuth coverage.

The active receive elements are fed via buried coaxial lines to elemental receivers, one for each active element. Each elemental receiver employs 16 RF preselectors to cover the frequency range. In each receiver, the received signals are amplified, filtered and digitised before being passed to the beamformer/signal processor group. The beamformer combines the outputs to form four simultaneous receive beams spatially coincident with the range/azimuth sector. This four-beam cluster scans synchronously with the transmit beam. Beamwidth, beam spacing, pointing angle correction for frequency and real-time measured receive sub-system errors are controlled by the receive control computer. The signal processor processes the four receive beams virtually simultaneously. Functions include MTI, interference suppression, range and doppler resolution processing, non-coherent integration, peak detection and parameter estimation.

At the heart of the system are very high speed computer, data processing and display systems which are used to control the radar, process the returned signals, monitor the HF propagation environment and display the information in a variety of formats to the operator. These displays are both alpha-numeric and graphic types to provide the operator with the maximum amount of information on target speed, track, position, altitude and other special characteristics, and also to adjust the transmission frequency to suit the prevailing HF environmental conditions. When the system is fully operational it is designed to give advanced warning of at least 1 to 1½ hours, even for supersonic aircraft, and, since it uses skywave propagation techniques, it can track very low-flying aircraft and cruise missiles at the same ranges as those flying at higher altitudes.

CONUS OTH-B has three modes of operation:

Normal mode

In this mode a surveillance barrier is established up to 30° wide, anywhere within the 60° azimuth sector, and up to 500 nautical miles (800 km) deep. A step-scan technique is used to illuminate this barrier on a regular, periodic basis by sequential illumination of four contiguous range azimuth sectors. Each sector normally covers a 7.5° by 500 nautical miles (800 km) area, thus four sectors form the 30° wide barrier. Four parallel receive beams, with 2.5° centre spacings, are formed to be coincident with each range azimuth sector and collect the signals reflected from targets in the coverage area. The starting range for each sector in the barrier can be chosen independently.

Interrogate mode

The interrogate mode illuminates one particular range azimuth sector and can be positioned anywhere within the 60° azimuth by 500 to 1800 nautical miles (800 to 2880 km) range capability of the radar. The interrogate range azimuth sector is illuminated by contiguous coherent dwells. As with the normal mode, four parallel receive beams, spaced on 2.5° fixed centres, collect the reflected target energy. The interrogate mode is a special mode intended to provide extra detection energy, contiguous illumination, and high range, velocity and azimuth resolution. It is used to enhance the characterisation of particular targets.

Interleaved mode

This combines the interrogate and normal modes. One interrogate mode dwell is provided after each full scan of the normal mode barrier. Separate radar operating parameters can be selected independently for the interrogate and normal mode portions of the scan. This mode allows the barrier surveillance to be retained while focusing on special target situations anywhere in the coverage region.

DEVELOPMENT

A contract for the prototype system was awarded to General Electric in 1975 and, after the contract was restructured in 1977, an experimental system was developed to demonstrate technical feasibility. Initial testing took place from June 1980 to June 1981 and about one year later General Electric received a contract from the US Air Force to begin full-scale development of the project.

For the western seaboard system, sites have been chosen at Mountain Home Air Force Base, Idaho for the operations centre, Klamath Falls, Oregon for the transmitter and Rimrock Lake, near Alturas, California for the receiver. Work is being carried out on the western coastal sectors and the system is due to be operational in the early 1990s. In December 1986 General Electric was awarded a contract valued at $145 million for the west coast system. Total value of the contract, including options, is expected to be $313 million. An additional contract worth $56 million for the third sector of the west coast system was awarded in November 1987.

Long term planning and site selection for the two southwards-looking sectors has also begun with the operations centre at Grand Forks AFB, North Dakota. A fourth system, based in Alaska, with the transmitter at Gulbana, the receive antennas at Tok and the operations centre at Elmendorf AFB, is also planned and will carry out surveillance down the Aleutian Island chain. In addition, the USAF has commenced construction of OTH radars which will be deployed on Amchitka Island in the Aleutians and at Guam Island in the Marianas. These two radars will be used to look at the area of the Sea of Japan, the coastal area of Siberia and the Sea of Okhotsk. The Japanese government intends to purchase an OTH-B system from the USA, which would be integrated into the overall system and has chosen a site for the transmitter on Hahajima Island, with the receiver and presumably the control centre, on Iwo Jima Island.

Consideration was also given to the construction of a northwards-looking chain but the wide irregularities of the ionosphere and the effects of the aurora preclude a reliable system, although investigations are still going on. The region is already covered by the new Seek Igloo radar chain, with the DEW line being updated into the North Warning System in Northern Canada and Alaska.

The original cost of the complete CONUS OTH-B system has been estimated at $1300 million but escalation of costs and the additional sites in Alaska and the Pacific Ocean will undoubtedly inflate this figure considerably, although there has been a certain amount of cost-cutting by the US Senate.

STATUS

In full development. Two of the three eastern sectors achieved limited operational capability at the end of 1987. Construction of the western sectors has commenced and the system is due to be completed by 1990. Construction of the Alaska system now appears unlikely. (See **5781.153** status)

CONTRACTOR

General Electric Company, Electronic Systems Division, Syracuse, New York 13201, USA.

5781.153

RELOCATABLE OVER-THE-HORIZON RADAR

To complement the CONUS OTH-B system described under **5617,153**, the US Navy requires a relocatable over-the-horizon radar (ROTHR). ROTHR is a tactical land-based, bistatic ionospheric backscatter radar system intended to provide a wide-area over-the-horizon surveillance of both aircraft and ships in support of tactical forces in locations of national interest. At the operator's option the system can be used for overall surveillance and tracking within the coverage area, spotlighting specific regions to handle targets of interest, or the assessment of the size of an attacking force. The ROTHR system is designed to be relocatable to previously prepared sites in support of tactical surveillance missions. Although classified as relocatable the transmitter and receiver antennas are basically similar to those of the CONUS OTH-B system and, therefore, take some days, or even weeks, to erect and dismantle.

Because the ROTHR system is relocatable, it has to adapt to changing ionospheric conditions in various sites. This requirement is met by eliminating the need for external sounder data inputs. The system uses a collocated quasi-vertical incidence (QVI) sounder and a backscatter sounder for propagation management assessment. This approach adapts automatically to the ionospheric conditions of the operational sites.

The system consists of three distinct elements: a transmit site, a receive site and an operations control centre. The transmit and receive sites are separated by about 50 to 100 nautical miles. The control centre is normally collocated with the receive site. The transmit site provides radar illumination in accordance with commands from the control centre. It also provides the transmissions for the backscatter and quasi-vertical incidence soundings used in the propagation management and assessment function. In the receive element, the returned beams are formed digitally in the signal processor which also carries out the range and doppler processing and extracts the target detections. The detections, together with the raw processed data from which they were extracted, and signals from the sounder transmissions are passed to the control centre for further processing.

The operations control centre is the nerve centre of the system and carries out the processing of the returned target information. Within the centre the operators interface with the system via stand-alone displays driven by a data processing subsystem. The system is largely automatic in operation but can be manually overridden if necessary. TRW Defense Systems Group has also received an $18 million subcontract to design and develop an operations control centre for the system. In addition to the design and development work, TRW will provide hardware and software support for the project.

STATUS

Raytheon has received contracts from the US Navy for prototype and engineering models with software support. Systems performance testing began in mid-1987 and is scheduled to be completed in 1988.

The US Navy originally planned to equip seven sites at a cost of nearly $1 billion, but nine are now to be requested for $1084 million. It is intended to carry out trials on a prototype system at Amchitka, one of the Aleutian Islands, during 1987.

CONTRACTOR

Raytheon Company, Equipment Division, Wayland, Massachusetts 01778, USA.

4082.153

ISC CARDION COASTAL RADAR SYSTEM

ISC Cardion has developed and produced a new E/F-band (formerly S-band) coastal defence radar under a contract awarded in 1980 by NATO. A number of these radars are installed around the Danish coast, and the system is also known by the initials DCR (Danish coast radar). It includes an E/F-band primary search radar and a D-band (formerly L-band) secondary surveillance radar (SSR).

The system was designed for use at unattended sites with automatic reporting of data and control signals. Because the equipment is normally unattended, an extensive self-monitoring system has been incorporated into the design. Alerts are sent out immediately upon the first indication of trouble or abnormal situations.

The antenna is a double curvature reflector with integral primary and secondary feed assemblies. An omni-directional antenna mounted on top of the main reflector is used for sidelobe suppression in the SSR system. Since the radar is normally used as a gap filler, the vertical beam shape has been specifically tailored to provide high detection probability of low-level aircraft and surface targets. A tilt mechanism allows adjustment of the beam above and below the horizon.

The transmitter employs a liquid-cooled, coupled cavity TWT, powered by an all solid-state high voltage power supply, to produce a highly stable coded waveform. The transmitter is versatile allowing operation in fixed, burst or full frequency agile operation. All operating modes are sector driven, ie a different mode may be selected for any of the four azimuth sectors, each of adjustable size. As the antenna rotates, the preset operating mode will be automatically turned on in each sector. The transmitter may also be silenced in a sector if desired.

Addition of an automatic detection and tracking unit (ADIT) provides the system with full automation capability. The ADIT features coherent and non-coherent signal processing, primary and secondary plot extraction, automatic track initiation and telephone line interface for data remoting. Video processing concepts allow automatic detection of targets in clutter and ECM environments, with a well regulated false alarm rate. The data processing portion of ADIT extracts plot data on both primary and secondary targets, merging the data into a single report when target correlation can be established. An integrated tracking system provides automatic initiation and continuous track update on all plot data.

One of the main functions performed by the ADIT is the separation and identification of all targets as either surface or air. Either or both types may be selected for transmission. The ADIT also extracts other non-target information such as a clutter map and jamming strobes. This data is transmitted, together with target messages, to the operations centre.

The remote control unit (RCU) is able to control and monitor an entire network of radars using bi-directional communication over telephone lines. Current status information is sent back to the RCU from each of the radar heads where it is displayed. The RCU is equipped with a CRT overlayed with a touch-sensitive screen which serves as the man/machine interface.

CHARACTERISTICS

Antenna

Type: Double curvature, integral with IFF

Polarisation: Circular

Rotation rate: 10 rpm

Size: 16 × 8 ft (4.88 × 2.44 m)

Transmitter

Frequency: E/F-band

Type: MOPA, TWT

Peak power: > 100 kW

Duty cycle: 1%

Waveform: Phase coded pulse for pulse compression

Receiver

Type: Superheterodyne, double conversion with image rejection

MTI: Digital (I&Q)

STATUS

Production and installation. Two further systems were ordered by Denmark in February 1988 making a total of eight systems supplied to, or ordered by, Denmark.

CONTRACTOR

ISC Cardion Electronics Inc (a member of International Signal and Control/Defense and Space Group), Long Island Expressway, Woodbury, New York 11797, USA.

ISC Cardion coastal radar system

6400.153

ACWAR (AGILE CW ACQUISITION RADAR)

The agile continuous wave acquisition radar (ACWAR) is an evolution of the Hawk CW radar technology (**4177.131**). It performs full 3-D target acquisition over a 360° azimuth sector and large elevation angles.

The ACWAR programme was initiated to meet increasingly severe tactical air defence requirements, and the equipment is designed for operation in the 1990s and the 21st century. ACWAR replaces two current Hawk radars to reduce vehicle, manpower and logistics needs. At the same time the system provides full 3-D track information on a large number of targets to accuracies sufficient for cueing and control of other remote weapons as well as data netting. ACWAR is mobile, helicopter and C-130 transportable, and can be vehicle or trailer mounted.

ACWAR consists of an all solid-state exciter/ transmitter, all digital radar control, row board transmit and receive antenna construction for precise sidelobe control and electronic steering in elevation. ACWAR uses a CW frequency agile, phase-coded waveform. Digitised target data are sent from the radar to the Hawk PCP or equivalent.

In addition to its role with Hawk, ACWAR can be deployed autonomously to perform target acquisition and tracking for forward area air defence networks and weapon systems.

CONTRACTOR

Raytheon Company, Missile Systems Division, Bedford, Massachusetts 01730, USA.

5778.153

DR-810 Mk V MUZZLE VELOCIMETER

The DR 810 Mk V muzzle velocimeter is a small one piece CW doppler system that measures the velocity of a projectile at several points along its trajectory. It is intended for use with tactical weapons and can be used with a variety of these, including artillery, small arms and machine guns, air defence weapons, shipborne guns, mortars, and fire control systems. It has been tested with a variety of systems from 5.56 mm to 16-inch calibre. The Mk V is also used in experimental testing range applications where its scientific data output can assist in the development of new munitions.

The equipment is mounted either on a tripod, or directly on the weapon concerned, via a quick-disconnect bracket, and consists of an antenna which contains an integral transmitter/receiver and data processing circuitry. The unit is fully automatic in all its functions. Built-in test equipment, triggering, false triggering logic, anti-jamming frequency agility, and data processing are all conducted without operator input. The system interfaces directly with a fire control computer, or with an optional RS232/422 display. Computed muzzle velocity is fed directly to the user's fire control computer, where it can be used as a real-time fire control input. The system is capable of measuring all types of ammunition, including base-bleed, sabot and rocket assist to an accuracy of ±0.05%. The non-volatile memory will store 1000 shots, which can be filed according to date, time, charge, shell and lot utilising GEC Astronics Mem-Par circuitry. Individual shot data or selected averages can be displayed, together with a variety of scientific data. Velocity output is automatically corrected for non-standard conditions, or is output as an actual velocity.

CHARACTERISTICS

Frequency: 10.525 ± 0.03 GHz

Power output: 130 mW to 3 W selectable

Velocity range: 50 to 1750 m/s

Accuracy: ±0.05%

Dimensions: 315 × 217 × 100 mm

Weight: 8.2 kg

STATUS

In production.

CONTRACTOR

GEC Astronics Corp, Astronics Division, 3400 Airport Avenue, Santa Monica, California 90406, USA.

6109.153

TERMINAL IMAGING RADAR SYSTEM

The Terminal Imaging Radar System (TIRS) is an early design programme, associated with the Strategic Defense Initiative (SDI) programme, to examine technology for a ground-based phased array radar that can discriminate between re-entry vehicles and the many other objects re-entering the earth's atmosphere during an attack. Initial contracts for preliminary design were awarded to Raytheon and Westinghouse and in March 1987 Raytheon was selected to develop, test, manufacture and install a TIRS at Kwajalein Atoll in the Pacific Ocean. Raytheon will also assist in functional tests of the system under a $174 million, 4-year contract from the US Army Strategic Defense Command.

The experiment will use a large phased array, high-resolution microwave radar to detect incoming missile re-entry vehicles in the mid-course and terminal phases of flight.

STATUS

Installation of Kwajalein Atoll is scheduled for the end of 1990 with demonstration/validation testing beginning in October 1991.

CONTRACTOR

Raytheon Company, Missiles Systems Division, Bedford, Massachusetts 01730, USA.

Artist's impression of Terminal Imaging Radar

6110.153

MTAS TARGET ACQUISITION SYSTEM

The Rockwell MTAS multi-sensor target acquisition system is intended to provide a 24-hour, all-weather target acquisition system. The system design is being evolved from the successful demonstration in 1984 that a millimetre wave radar-controlled gun could perform rapid firing against stationary and moving targets, with automatic target detection and priority allocation.

The MTAS is intended to combine a lightweight millimetre wave radar sensor with the capability of handling moving targets, with the existing forward looking infra-red (FLIR) system already in use aboard battle tanks. The signals from the two sensors will be processed in a digital processor and shown on the tank's main display with the radar data superimposed on the FLIR image.

The radar, probably operating at a frequency of 94 GHz, is designed to operate in two modes, coherent moving target indication to detect radially moving targets and area MTI which allows acquisition of creeping and tangential targets. A spread spectrum waveform is used with area MTI to smooth clutter, glint and multi-path effects. The antenna is located in an armoured housing with a 14-inch aperture on the left side of the AFV turret. The system uses an impart oscillator without a limiter.

STATUS

In development. A prototype is likely to be available by the end of 1988 for trials by the US Army.

CONTRACTOR

Rockwell International Corporation, Autonetics Sensors, Aircraft Systems Division, Anaheim, California 92803, USA.

6814.153

RAIDS DETECTION RADAR

RAIDS is a ground-based system intended to detect parachutists, helicopters, hang gliders and other types of ultra-light aircraft. It was designed by Sanders about five years ago by modifying the AN/ALQ-156 airborne missile detection system (**3547.393**). The concept uses a low-beam/horn flare unit to detect the doppler signature of a helicopter rotor or an ultra-light propellor and a high beam to detect parachutists. In the latter case, the system tracks the path of a free-falling parachutist and sounds an alarm when a change in fall velocity indicates that the parachute has been deployed.

STATUS

After the initial development of RAIDS, a lack of interest from the military caused the project to be shelved. However more recently interest has been rekindled and the US Air Force Electronic Systems Division intends to issue requests for proposals to a number of contractors for an airborne intruder type system.

CONTRACTOR

Sanders Associates Inc, Countermeasures Division, Nashua, New Hampshire 03061-2004, USA.

6818.153

ETAS TARGET ACQUISITION SYSTEM

ETAS (multi-target Elevated Target Acquisition System) is a ground surveillance radar being developed by Eaton as part of the US Army's Airland Battle 2000 concept. It is designated 'modular' since the US Army is providing separate segments of the radar as government-furnished equipment. Eaton will integrate this into the modular radar system. The radar consists of a receiver/exciter, low-probability-of-intercept antenna, an ITT-developed TWT, a signal processor developed by the Army and other hardware. The system is designed to detect, classify and track personnel, vehicles and low-flying helicopters.

STATUS

In development.

CONTRACTOR

Eaton Corporation, AIL Division, Commack Road, Deer Park, New York 11729, USA.

ELECTRONIC WARFARE EQUIPMENT (LAND)

CHINA (PEOPLE'S REPUBLIC)

6451.153

MODEL 970 RADAR JAMMING SYSTEM

The Model 970 is a mobile, ground-based radar jammer designed primarily to protect ground targets from air attack. The system operates in X-band and uses noise-modulated 'blanket' jamming to produce a continuous interference sector on the screen of the aircraft surveillance radar. It can also be used to jam airborne guidance radar as a protection against air-launched missiles. The 970 has a variety of jamming types, and can be operated either automatically or manually.

The Model 970 measures the bearing, frequency and antenna rotation speed of the hostile radar, and may be used in conjunction with a pulse analyser to measure other parameters such as pulse width, PRF, radar illumination, etc. The system consists of a number of units including antenna feed, transmitter, coarse and fine receivers, servoes, display and power supply. It is installed in a trailer, for rapid mobility, and has the antenna system mounted on the trailer roof.

CHARACTERISTICS

Frequency: 8 to 12 GHz

Receiver sensitivity: Better than –76 dBW

Power output: Better than 120 W; a 200 W output transmitter is also available.

CONTRACTOR

China National Electronics Export and Import Corporation, 49 Fuxing Road, PO Box 140, Beijing, People's Republic of China.

Type 970 radar jammer trailer showing the antenna system mounted on the roof

FRANCE

3913.193

FRENCH LAND-BASED COMMUNICATIONS EW SYSTEMS

A number of French military EW systems have been identified which are land-based if not necessarily dedicated to the French Army; most are mainly concerned with communications aspects of EW and while there is insufficient detail to warrant individual entries for each, the following paragraphs give what information can be revealed.

BROMURE

This is a vehicle-mounted or static tactical VHF communications jammer and is currently in service with the French Army. It uses an omni-directional antenna array and is believed to have a power output of 1 kW ERP. Thomson-CSF is the manufacturer.

BINOC

Another jammer, covering a different part of the spectrum to Bromure.

ELEBORE

Elebore operates in the HF part of the telecommunications spectrum and performs several functions (eg detection, location, monitoring, etc). It is computer-based and uses a version of the CISMA 15M digital computer.

ELFA

Elfa is the name of an intercept system. The frequency coverage is not known.

ELODEE

Elodee is the name of a DF, emitter locating system. The frequency coverage is not known in detail but it is known that the system operates in the communications bands so that it is probable that the HF, VHF and UHF parts of the spectrum are covered. A CISMA 15M computer is incorporated.

EMERAUDE

Emeraude operates in the VHF band and is a combination system providing several facilities relating to communications monitoring, surveillance, and emitter location. A CISMA 15M computer is used in the system.

STATUS

All are operational with the French Armed Forces. Thomson-CSF is the prime contractor for these systems, although other companies have been involved as sub-contractors. The government DGA (Délégation Générale pour l'Armement) establishment SEFT (Section d'Etudes et Fabrications des Télécommunications) was responsible for development of the above systems.

CONTRACTOR

Thomson-CSF, Division Telecommunications, 68 rue du Fosse Blanc, BP156, 92231 Gennevilliers, France.

3534.053

TMV 200 ELISA ELINT RECEIVER

The Elisa ESM/Elint receiver is designed for search and analysis of pulse or CW signals for use in strategic or tactical operations conducted by land, air or naval forces. It can be deployed in airborne, shipborne or ground based configurations. The equipment consists of a rotatable broadband antenna array, frequency converter, receiver/processor and an operator's console.

The Elisa system was developed to ensure detection of radars having a high pulse compression rate and also radars with fixed, random and coded parameters. The system has a high instantaneous probability of intercept and through the use of reprogrammable software, is able to face changes of the threat.

The digitally controlled receiver features fast frequency commutation, a high sweep rate, modular design and includes a microprocessor unit for data processing and automatic system operation. The frequency converters in association with the basic receiver provide operation over the whole radar spectrum. Automatic measurements are made of the usual parameters, assisted by the use of high accuracy quartz-driven oscillators and low-noise RF amplifiers for high sensitivity. Pulse-by-pulse DF is possible, and the display incorporates a memory device to assist interception of brief signals and frequency-agile radars.

The search pattern is easily programmed by the operator in terms of frequency coverage, disabled frequencies, wide or narrow band selectivity, dwell time, etc, which results in quick adaptation to operational needs. The operator communicates with the system via a keyboard. The console includes a plasma screen showing the panoramic situation, radar parameters and a detailed radar analysis (emission histograms and spectrums).

The system is compatible with peripheral equipments for additional analysis and immediate or delayed operational exploitation.

CHARACTERISTICS

Dimensions	**Width (mm)**	**Height (mm)**	**Depth (mm)**	**Weight (kg)**
Converter	258	194	500	30
Receiver/processor	258	194	500	30
Control unit	410	390	665	30

STATUS

In production and in service. A helicopter version has been designed and has been evaluated by the French Army. Thomson-CSF has also completed work on a ground-based Elint station using Elisa systems.

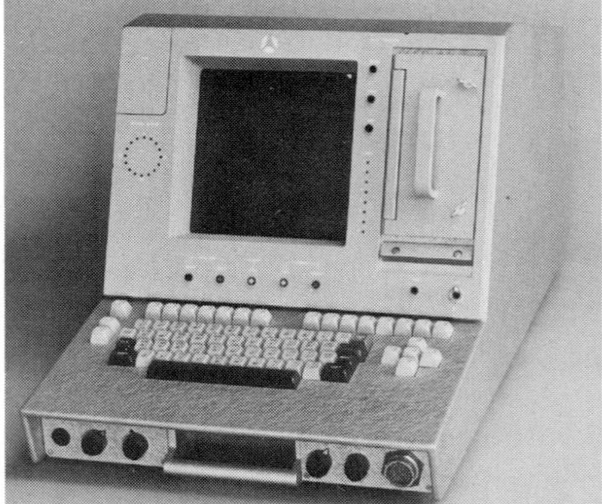

Thomson-CSF Elisa ESM/Elint receiver control/display console

CONTRACTOR

Thomson-CSF – Aerospace Group, 122 Avenue du Général Leclerc, 92105 Boulogne-Billancourt, France.

6452.153

TRC SERIES COMMUNICATIONS INTELLIGENCE EQUIPMENT

Thomson-CSF Telecommunications Division (DTC) has developed a complete family of high performance basic HF and V/UHF radio surveillance equipment compatible with all types of application requirements in land, sea, air, strategic and tactical roles. DTC also advises and assists in the creation of simple or complex EW systems, tailored to the requirements of a customer.

Main examples of the company's EW surveillance systems are described briefly in the following entries.

TRC 195 HF radio direction finder

The TRC 195 radio direction finder provides optimum operational service in the 0.3 to 30 MHz band. Its instrumental features, precision, sensitivity, measuring speed and operational performance make it an essential element of the systems of radio observation and radio monitoring of the spectrum in the HF range.

It enables the specific conditions relating to this range (density of emission and ionospheric scattering) to be taken into account by providing a maximum frequency resolution of ±75 Hz and an antenna adapted to the incidence and polarisation characteristics of the DF signals. It can be operated locally or remotely controlled via standard interfaces.

STATUS

The TRC 195 is in production and has been supplied to French and other defence forces.

TRC 243 HF monitoring receiver

The TRC-243 is a high performance VLF/HF receiver especially designed for radio monitoring applications.

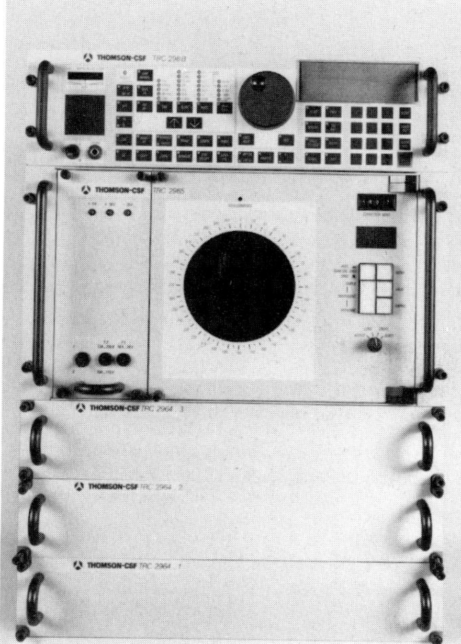

TRC 296 V/UHF radio direction finder

TRC 281 tactical HF jammer with whip antenna

TRC 284 tactical V/UHF jammer with high gain directional antenna (shroud configuration)

When used separately, with no peripherals, the TRC 243 with its capabilities constitutes a complete radio monitoring system. It can scan the discreet channels of all types of modulation, as well as continuous frequency ranges, according to programmable parameters. The equipment can also detect an activity on one channel or on one frequency and stop on it automatically. It can receive all types of modulation with its internal demodulators, including F1B, F3E and high speed data transmissions up to 2400 bits/sec.

The management of all the functions and circuits of the TRC 243 by microprocessors permits new uses which former receiver technologies did not permit.

STATUS

This equipment is in production and has been supplied to French and other defence forces.

TRC 281 tactical HF jammer

Designed to carry out ECM missions within the 2 to 30 MHz range the mobile vehicle-mounted TRC 281 jammer can be integrated in any tactical EW system. It is equipped with the latest ESM features. According to operational requirements it can be operated either locally or by remote control.

The main purpose of this jammer is to attack HF hostile communications, both tactical short- and long-range links. In its omnidirectional antenna configuration, it can be used to jam any links in ground wave propagation while benefitting from high mobility. The same applies in the optional directional antenna configuration; the high power (1 kW PEP) of the amplifier permitting jamming of links in ionospheric wave propagation. It is designed to be a highly automated equipment with a very short reaction time.

STATUS

The TRC 281 is in development and is scheduled to be in production in 1987.

TRC 284 tactical V/UHF jammer

Designed to carry out ECM missions within the 100 to 500 MHz range, the mobile vehicle-mounted TRC 284 jammer can be integrated in any tactical EW system. It is equipped with the latest ESM features and can be either operated locally or by remote control.

The main purpose of this jammer is to jam ground-to-air and air-to-air links. In the standard version, the very high gain directional antenna allows intensive attacks against hostile communications. The TRC 284 in this case has up to 3 to 4 kW ERP power. An optional omnidirectional antenna allows jamming to be carried out while the vehicle is in motion.

STATUS

In development and scheduled for production in 1987.

TRC 285 tactical VHF jammer

The mobile, vehicle-mounted TRC 285 is designed to carry out ECM missions within the 20 to 110 MHz range and can be integrated into any EW system. Other applications and technical details are similar to those of the TRC 284.

STATUS

In development and scheduled for production in 1987.

TRC 296 V/UHF radio direction finder

Including a receiver of the TRC 298 type and a DF processor fully microprocessor-controlled, the TRC 296 is a high operational performance system. It can be operated locally, or remote controlled by means of standard interfaces for integration into a system.

The antenna system allows three different versions, 20 to 160 MHz, 20 to 500 MHz and 20 to 1350 MHz, to be configured with the same operational performance. With its modular design the equipment can be installed in a fixed, semi-fixed or mobile station. A small lightweight antenna version is available.

STATUS

In production and supplied to French and other defence forces.

TRC 296D lightweight antenna mobile direction finder

The TRC 296D is a version of the TRC 296, equipped with a lightweight antenna, and also using the measurement by interferometry principle. Designed for spectrum control missions and radio monitoring, it is especially intended to be fitted on-board any vehicle for mobile missions, or on-board ships or helicopters. It may also be used in a fixed station, when a non-compromising antenna is required. A specific processing of the adequate signal allows different types of transmission used in the V/UHF range to be taken into account, even in severe noise conditions.

It includes a receiver of the TRC 298 family and a fully micro-computerised direction-finding processor which gives it a very high operation flexibility. It can be locally operated, or remote controlled through standard interfaces.

STATUS

In production and supplied to French and other defences forces.

TRC 298A V/UHF intercept receiver

The TRC 298A is an intercept receiver using an ultra-fast frequency synthesiser, intended to constitute the basic element of modern VHF/UHF radio surveillance systems. Fully controlled by two processors, it is highly automated and is provided with an optional remote control unit which enables it to control a radio surveillance mini-system. It provides false processing and performs memory scanning of 1 to 100 frequencies, frequency scanning of 1 to 10 frequency sub-bands, and masking of known frequencies or sub-bands. The equipment is fitted with a sophisticated BITE for fast, accurate fault diagnosis.

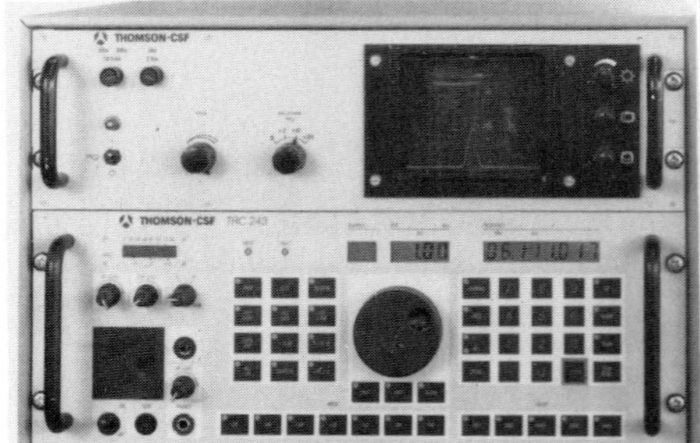

TRC 243 HF monitoring receiver with TRC 3941B panoramic adaptor

STATUS

In production and supplied to French and other defence forces.

TRC 298B V/UHF monitoring receiver

The TRC 298B is a monitoring receiver and forms one of the basic elements of modern V/UHF radio surveillance and electronic warfare. It is microprocessor controlled and highly automated, and is highly suited for integration into systems, due to its optional remote control. Basic functions are similar to the TRC 298A.

STATUS

In production and supplied to French and other defence forces.

TRC 2981 V/UHF panoramic adaptor

The panoramic adaptor TRC 2981, associated with receivers of the TRC 298 family, is the spectrum display equipment essential for operators performing detection, interception, identification, monitoring, direction-finding/location, and jamming missions, as well as spectrum management in the 20 to 1350 MHz range.

STATUS

In production and supplied to French and other defence forces.

TRC 3941B HF panoramic adaptor

The TRC 3941B is a panoramic adaptor designed for association with the receiver TRC 243. It has been designed for radio surveillance and intercept applications on any conventional type of transmission.

The TRC 3941B is operated in IF analysis, associated with the TRC 243, and then permits display of a frequency spectrum around the receiver tuning frequency. The operator can keep simultaneous listening on the receiver.

STATUS

In production and supplied to French and other defence forces.

CONTRACTOR

Thomson-CSF, Division Télécommunications, 66 rue du Fossé Blanc, BP 156, 92231 Gennevillers Cedex, France.

GERMANY (FEDERAL REPUBLIC)

6462.153

HELAS ESM SYSTEM

HELAS is a vehicle-based ESM system developed by Siemens for the West German Army. It is designed for the detection of radar emissions, the data then being passed to a central electronic signal processing station, located to the rear of the battle area. The forward reconnaissance equipment is installed in a standard TPZ-1 armoured vehicle chassis with antenna arrays mounted on hydraulically controlled masts.

The antenna system automatically intercepts, analyses, directionally locates and classifies the emissions in a 100 channel filter bank receiver, before feeding the data to the central processing unit where the signals are further analysed and a display of hostile radar sources is derived from the data. This includes exact position, type, identification in ELINT code, elimination of friendly sources, etc. The evaluation equipment also controls the reconnaissance equipment. Siemens states that the equipment is able to receive all existing and planned radar emissions, including complex emitters that transmit on a number of frequencies simultaneously. The company also states that, because of the HELAS high sensitivity and wide dynamic range, it is able to detect side lobes of remote emitters.

Siemens plans to develop a number of variants of HELAS for use in helicopters and maritime patrol aircraft. This would entail reduction of the volume and weight, and the design of new antennas.

STATUS

Two sets of sensors and one signal processing station have been delivered to the West German Army for field trials. Production was scheduled for mid-1988.

CONTRACTOR

Siemens AG, Radio and Radar Systems Division, Landshuter Str 26, 8044 Unterschleissheim, Federal Republic of Germany.

Antenna head of the HELAS ESM system

4352.053

AEG RECEIVERS AND ANCILLARY SYSTEMS

AEG produces a range of receivers and ancillary equipment for monitoring and surveillance applications. The following paragraphs list the salient features of major examples of this range. All systems listed are in current production.

E 1600 VU VHF/UHF Receiver

The receiver E 1600 VU is designed for radio monitoring in the VHF/UHF range up to 1000 MHz and is used in combination with the WAG 1600 panoramic display. This receiver is designed for operation in both fixed and mobile radio monitoring stations.

The frequency range from 20 to 500 MHz may be extended without increasing equipment dimensions by inserting a module for frequencies up to 1000 MHz. The receiver is characterised by its compact and rugged construction and by simple operation.

For special purposes the receiver is supplied for a frequency resolution of 100 Hz instead of 1 kHz.

General Purpose Receiver E 1700

The E 1700 is characterised by high immunity to large signals, easy operation and compact construction.

E 1700 general purpose receiver

E 1600 VU VHF/UHF receover

With a high quality input circuit and a very low noise synthesiser, the receiver features large signal characteristics not achieved before. The receiver may be complemented with a passive automatically locked preselector which improves its reception properties, particularly for operation on board ships. The equipment is provided with a 30-channel service data memory. Appropriate plug-in modules are supplied for many further types of service, such as ISB, FSK high data rates and diversity operation.

CHARACTERISTICS

Frequency range: 10 kHz – 30 MHz

Operation modes

Standard: A1, A2, A3, A3J, A3A, A3H

Optional: 6A3B, 6A7B, F1, F4, F6

General Purpose Receiver E 1800

The E 1800 is a service monitoring receiver for the frequency range from 10 kHz to 30 MHz. Its technical features include a remote control facility, a microprocessor control and immunity to large signals. Through a serial data interface, the E 1800 may be controlled remotely in all its functions. By means of an addressing system several receivers may be set by one or more central control panels. Depending on the requirement, all types of normal service in the VLF to HF bands may be demodulated by means of integration. In a service data memory with 100 channels all receiver settings may be programmed. The incorporated microprocessor-controlled scanning unit extends the range of uses and allows, for example, the scanning of fixed programmed frequencies or scanning over a programmable frequency range.

VHF/UHF Receiver E 1900

The E 1900 is a search and monitoring receiver for the frequency range from 20 to 500 MHz. The frequency coverage may be extended to 1000 MHz by inserting a plug-in module. Its features are a remote control facility, a microprocessor control and compact construction. Through a serial data interface, the E 1900 may be remotely controlled in all its functions. By means of an addressing system, several receivers may be set by one or more central control panels.

The receiver is designed for the reception of A1, A2, A3, F3, pulse-modulated signals and optionally for A3J signals. In a service data memory with 100 channels all receiver settings may be programmed.

Other technical details are identical with the E 1800 above.

Digital Frequency Measuring Equipment DFM 1500/2

The DFM 1500/2 frequency measuring equipment is used for measuring the frequency of modulated, unmodulated and keyed transmitter signals (CCIR) in the frequency range from 10 kHz to 30 MHz. The measured frequencies are indicated digitally, quickly and accurately. In combination with the type E 1600 receiver and a microwave converter, the frequency coverage may be extended to 1 and 10.5 GHz respectively. Further accessories for digital and analogue registration make it possible to record using suitable printers and recorders and the output of data to computers.

Direction Finder Telegon 7

The Telegon 7 direction finder consists of the VHF/UHF receiver E 1600 and a single-channel DF unit (EP 1650) with a frequency coverage of 20 to 500 MHz. The equipment operates on the principle of the electronically rotating antenna and automatically evaluates the bearing angle. This angle is displayed non-ambiguously in digital and analogue form. In

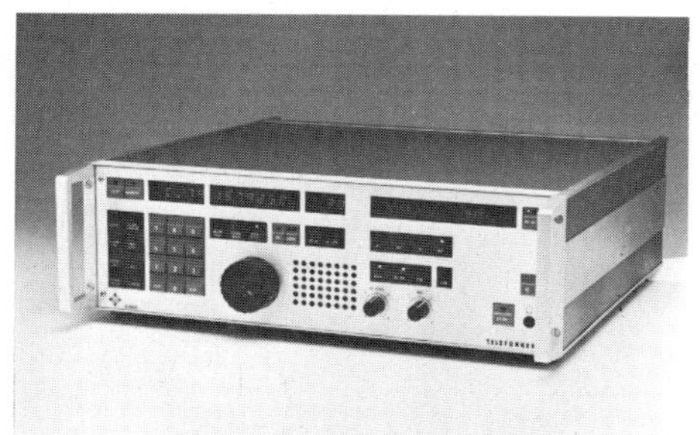

E 1900 VHF/UHF receiver

E 1800 general purpose receiver

addition the relative field strength is indicated, an advantage on mobile operations.

For mobile and camouflaged operation, the manual control unit HB 1650 is supplied and carries the most important control elements for the single-channel direction finder, and also indicates the bearing angle in digital and analogue form.

The Telegon 7 may be installed in fixed, semi-mobile and mobile DF systems. Appropriate data interfaces allow the operation of unattended remotely-controlled DF stations.

Direction Finder Telegon 8

Telegon 8 is a Watson-Watt direction finder with a frequency range from 10 kHz to 30 MHz. It consists of the DF receiver P 1520, control unit BP 1520 and display unit SH 1520. All information important to the operator, ie frequency, digital and analogue bearing, are indicated on a single display.

The more important features of Telegon 8 are:

(1) three-channel DF with non-ambiguous sense indication
(2) microprocessor controlled DF procedure
(3) electronic bearing cursor, digital bearing read-out following setting of cursor
(4) HF stability with ten Hz resolution.

An extensive test programme is integrated into the equipment and the operator can test the entire equipment from antenna input to display, step by step. The digital frequency input makes possible the remote control of the direction finder through other equipment, such as receivers. By the use of additional modules, the Telegon 8 may be used in fully automated DF networks.

Active Adcock DF Antenna AK 1205

The mobile basic omni-directional receiving antenna A 1205 was developed to receive vertically polarised waves in the MF and HF bands. Its horizontal pattern is circular and the vertical pattern is cosine-shaped.

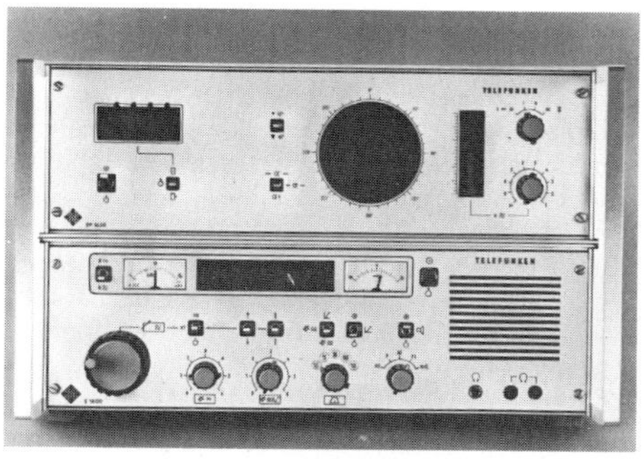

Telegon 7 direction finder

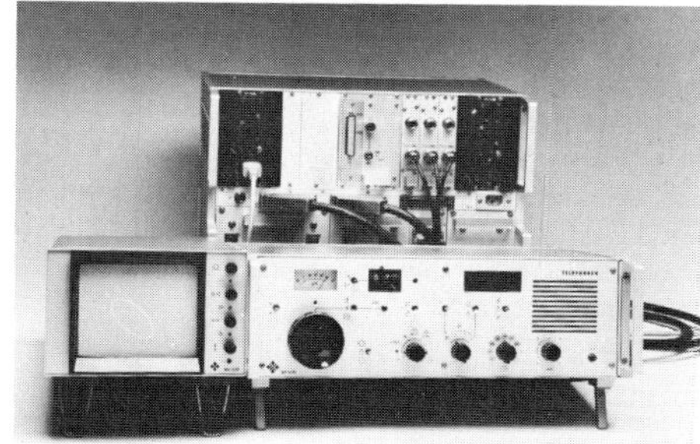

Telegon 8 direction finder

Active receiving antennas of this type are especially suitable for use in antenna groups and Adcock systems since their mutual coupling is negligibly small and the antennas have excellent amplitude and phase synchronism.

A combination, the Adcock System AK 1205, is particularly suitable for use in conjunction with the AEG-Telefunken receiver system Telegon VI or Telegon 8.

Due to light weight, small stowage space and the low constructional height of the antennas, the equipment is highly mobile and is especially suited for use wherever high mobility is demanded and small terrain space is available.

CHARACTERISTICS

Frequency range: 0.25 – 10 MHz (with 20 m base). 2 – 30 MHz (6 m base)
System error: 1° rms
Construction height: 2 m
Weight of each antenna: 7 kg

CONTRACTOR

AEG-Aktiengesellschaft, Postfach 1730, 7900 Ulm, Federal Republic of Germany.

4353.153

HUMMEL VHF JAMMER

The HUMMEL jamming transmitter is an automatically responding VHF jamming transmitter for use against voice and data links. The transmitter works as a multi-channel jamming transmitter in look-through operation in which the frequencies to be jammed may be distributed throughout the entire frequency range without restriction.

A computer controlled transceiver consisting of modules from the 'Baustein-program' is the basic unit of the jamming transmitter. Operating modes are:

(1) transmission and reception with manual control
(2) automatic single-channel operation as responding jamming transmitter
(3) multi-channel fixed frequency operation
(4) search operation with multi-channel jamming transmissions.

The HUMMEL system is normally installed in an armoured station vehicle with integrated power supply which ensures high mobility and effective operation in the forward edge of the battle area.

Through a remote command system, the jamming transmitter may be connected directly to an Elint centre. All settings needed for this operation may be performed directly at this Elint centre without the operator in the vehicle having to make adjustments. One Elint centre can control several transmitters at the same time.

STATUS

In production and in service with the West German Army.

CONTRACTOR

AEG-Aktiengesellschaft, Postfach 1730, 7900 Ulm, Federal Republic of Germany.

6164.193

BLAST JAMMING SYSTEM

Blast is a ground/air acquisition and jamming system which has been developed by AEG. It is a mobile tactical system designed to provide automatic analysis of radar emissions. Although no details have been revealed, it is believed to consist largely of highly sensitive wide band receivers and high power stand-off jammers against airborne radars.

STATUS

Development of the Blast system has been completed. Understood to be intended for use by the West German armed forces.

CONTRACTOR

AEG-Aktiengesellschaft, Theodor-Stern-Kai 1, 600 Frankfurt 70, Federal Republic of Germany.

3536.193

EP 1650 AUTOMATIC VHF/UHF RADIO RECONNAISSANCE SYSTEM

The EP 1650 system consists of a central station, at least three remote DF stations and other equipments. The central station incorporates a computer-controlled search receiver consisting of indicator and analysis receiver for the rapid detection and analysis of radio traffic as well as equipment for the direction of the DF stations and for passing instructions to the monitoring receivers and other equipments.

Technical features include:

(1) recognition of signals present in frequency bands
(2) coarse analysis of new transmissions
(3) DF and position fixing
(4) detailed analysis and monitoring reception
(5) ECM
(6) determination of the times relevant transmitters are on the air.

The search receiver sweeps frequency channels sequentially and in cycles, whereby the reporting accessory detects in a very short time the presence of a transmission and reports the fact to the computer. The computer keeps a file showing the occupancy of the channels and when a new transmission is detected it instructs the analysis receiver to measure the exact frequency. The computer instructs the direction finders in the mobile DF base to take bearings on this frequency. From the bearings determined automatically by at least two DF stations the computer fixes the most likely transmission position.

If the position determined is situated in a definite area then a monitoring receiver is automatically instructed to register or record the contents of the message. The results obtained are available in suitable forms for further processing. The dialogue with the system is performed through a data display station. This dialogue includes both the input of data for the search receivers as well as the printing out of results in formats.

STATUS

In operational service with West German armed forces.

CONTRACTOR

AEG-Aktiengesellschaft, Postfach 1730, 7900 Ulm, Federal Republic of Germany.

Vehicle of the EP 1650 automatic VHF/UHF radio reconnaissance system deployed

1901.053

ROHDE & SCHWARZ DIRECTION FINDING EQUIPMENT

Rohde & Schwarz produces a comprehensive range of radio and electronic DF equipment, embodying a variety of techniques and covering most of the spectrum. Some of this equipment is reviewed in the following paragraphs.

Direction finders operating on the doppler principle are frequently used in the VHF and UHF ranges. In these cases, the DF antenna system consists of a large number of vertical dipoles (16 to 32) arranged in a circle with a diameter of a few wavelengths. To simulate the rotation of a single dipole about a circle, the individual elements are cyclically scanned, thereby producing a frequency modulation of the incoming signal, which can be automatically evaluated. Triangulation systems with unattended direction finders can be realised by using the accessory remote frequency-control or remote bearing-transmission equipment PU 004 or PU 040.

The doppler-type wide-aperture direction finders PA 007 and PA 008 are available for the VHF range and the PA 009 for the UHF range, and are mainly used for air traffic control. For reconnaissance applications, the broadband doppler direction finder PA 055 (see **3532.093**) is recommended for the VHF/UHF range, and model PA 010 for the shortwave range from 0.5 to 30 MHz. The 12-channel VHF PA 022 is designed for coastal use to take bearings of ships and is also suitable for stationary or mobile application against aircraft.

PA 008 VHF direction finder equipment and commutated dipole antenna unit

In addition to the Doppler direction finding systems Rohde & Schwarz produces the PA 2000, operating on the digital Watson-Watt principle and using an Adcock antenna configuration. The system has an integrated fast scanning receiver and covers the frequency of 2 to 512 MHz. It is designed for mobile and fixed applications and has the capability of intercepting and taking bearings of modern military communications systems, including frequency hopping radios.

Operator interaction with the system is menu-driven for both control and display functions. Control is effected via a touch-panel keyboard, with a flat display indicating the control functions available as well as other current system information. Live display of the target bearing information and real-time signal analysis are presented on a colour CRT.

EB 100 Miniport Receiver

The EB 100 Miniport receiver is a lightweight, microprocessor-controlled, portable, battery operated mini receiver designed for ESM purposes. Applications include simultaneous monitoring of several frequencies, assessment of signal strength, and close-range direction finding by means of a manual DF antenna HE 100. Power supply is a 6 V built-in battery which will supply power for four hours continuous operation.

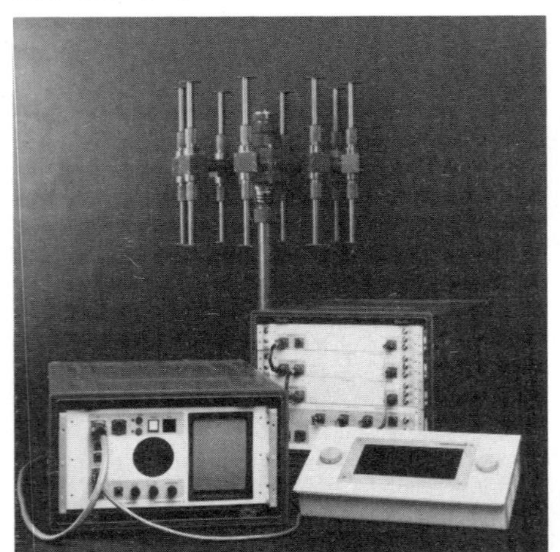

PA 2000 DF system

CHARACTERISTICS

Frequency range: 20 – 1000 MHz
Frequency resolution: 1 KHz (6-digit read-out)
Frequency error: 5 × 10
Demodulation: AM, FM
Memory channels: 18
Dimensions: 186 × 71 × 217 mm
Weight: 3.5 kg

CONTRACTOR

Rohde & Schwarz, Mühldorfstrasse 15, 8000 Munich 80, Federal Republic of Germany.

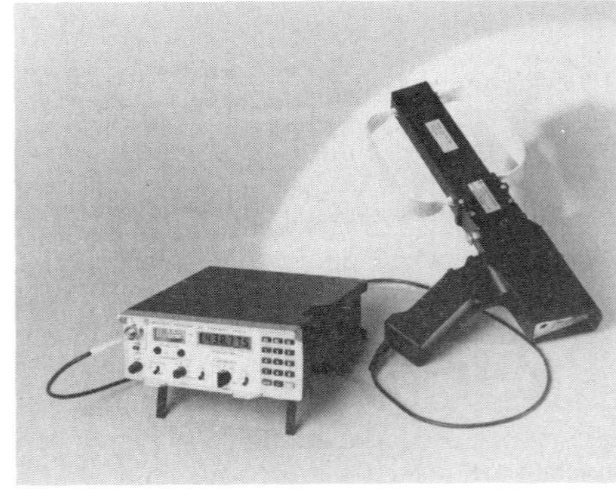

EB 100 Miniport receiver fitted with HE 100 antenna

6453.193

AC 008 DIRECTIONAL ANTENNA

The AC 008 broadband antenna covers the frequency range from 1 to 8 GHz and essentially consists of a special 90 cm reflector with a broadband exciter arranged near its focus and a biaxial adjustment device.

The medium-to-high antenna meets the requirements of broadband radio monitoring for military authorities, particularly for mobile applications. For various modes of polarisation two different exciters are used.

For transmission and reception of signals of linear/circular polarisation, the HL 025/HL 024 A1 log periodic antenna is used. Depending on the mounting position, horizontal, vertical or 45° polarisation can be selected.

With the HL 024 A1 crossed log-periodic antenna, simultaneous reception of horizontally and vertically polarised signals is possible.

An optional polarisation switching network permits the remote selection of any one of the linear and circular polarisation modes (clockwise or anti-clockwise).

For the spatial alignment of the antenna with respect to the incoming wave, the AC 008 incorporates a manual adjusting device. The adjustment range is n × 360° in the azimuth plane and –6 to +36° in the elevation plane. This permits alignment of the antenna for geostationary satellites as well as terrestrial targets.

CONTRACTORS

Rohde & Schwarz, Mühldorfstrasse 15, 8000 Munich 80, Federal Republic of Germany.

AC 008 directional antenna

4138.093

ESM 500 SERIES VHF/UHF RECEIVERS

The ESM 500 is a family of microprocessor-controlled receivers covering VHF and UHF ranges (20 to 1000 MHz) for applications which are primarily in radio detection, radio monitoring and surveillance. It includes broadband desk-top receivers offering a high degree of operating ease and, for use in computerised radio detection systems, hand-off receivers with receive bands tailored to the frequencies of different radio services.

High sensitivity, high overload capability, low oscillator phase noise and tracking input filters provide a wide dynamic range and the high accuracy of frequency tuning is achieved with a built-in synthesiser. Other major features are fast tuning, frequency scanning, non-volatile memory for all the settings associated with up to 99 frequencies, programmed scanning for any of the stored 99 frequencies and BITE for self-testing. The desk-top version, ESM 500A (20 to 1000 MHz), is suitable for use in mobile stations, at manually or remotely controlled observer positions (eg in master-slave operations, one master driving up to ten slaves without a controller), in semi-automatic systems or as a purely remotely controlled receiver in automatically or centrally controlled monitoring and DF systems.

Frequency tuning is either by a keyboard with a buffer memory and check display, by transfer from an internal memory, by a rotary knob with magnetic locking and stepping increments or externally and remotely by a bus or interface. As tuning aids there are two pointer meters for indicating level and offset, plus a built-in panoramic display that makes a signal visible before it is in the IF receiving bandwidth.

For simultaneous monitoring of large numbers of frequencies there are hand-off receivers ESM 508K, ESM 517K and ESM 540K, which, with their facility for centralised control and supply (as many as eight receivers in a group), can be used to form large-scale, multi-channel monitoring systems. The hand-off receivers are driven with sub-addresses, ie only one main address is required for each group. In this way an IEC-bus controller can handle up to 120 receivers.

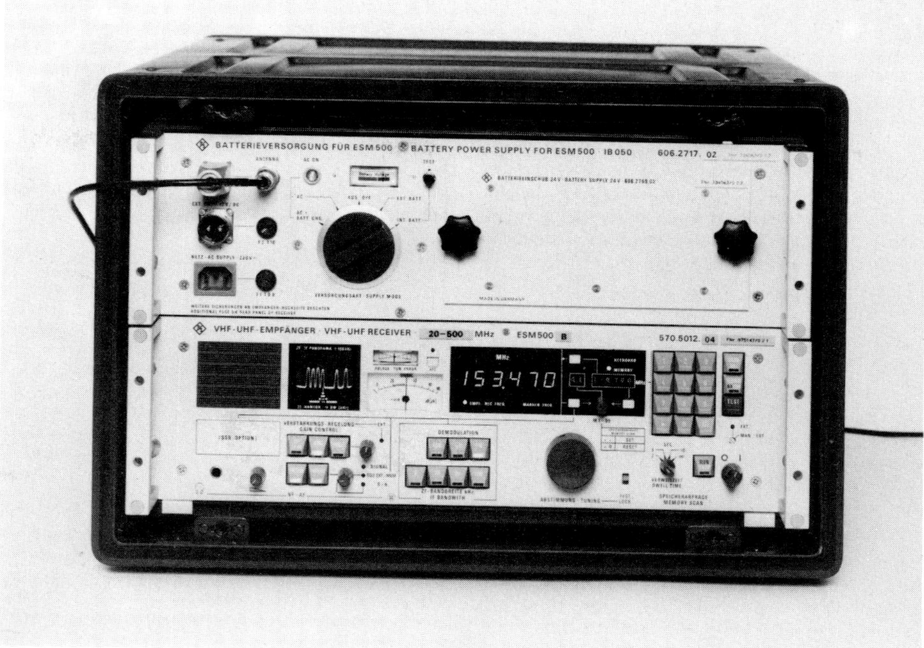

ESM 500B broadband desk-top receiver

STATUS

In production and operational.

CONTRACTOR

Rohde & Schwarz, Mühldorfstrasse 15, 8000 Munich 80, Federal Republic of Germany.

6838.193

ESM 1000 SYSTEM

The ESM 1000 is a family of receivers developed by Rohde & Schwarz as a follow-on to its ESM 500 series (**4138.093**) and covers the same frequency range from 20 to 1000 MHz. Whereas the latter concentrated on the receiver function, the ESM 1000 units have been designed to perform a number of other functions in addition to high speed signal detection. These include step-by-step analysis, task assignment to slave receivers and control of all components of an overall COMINT network.

CONTRACTOR

Rohde & Schwarz, Mühldorfstrasse 15, 8000 Munich 80, Federal Republic of Germany.

4139.093

ESP SCANNING RECEIVER

The ESP automatic receiver 'intelligently' scans the frequency spectrum at a rate of 1000 channels per second in the range from 10 kHz to 1300 MHz. An optional extension to this frequency range is available and extends the range to 2500 MHz.

The receiver is designed for communications intelligence and radio monitoring and, in conjunction with other equipments such as direction finders, monitoring receivers and printers, general-purpose radio monitoring installations can be established and are suitable for computer-control and provide high detection probability.

STATUS

In production and operational.

CONTRACTOR

Rohde & Schwarz, Mühldorfstrasse 15, 8000 Munich 80, Federal Republic of Germany.

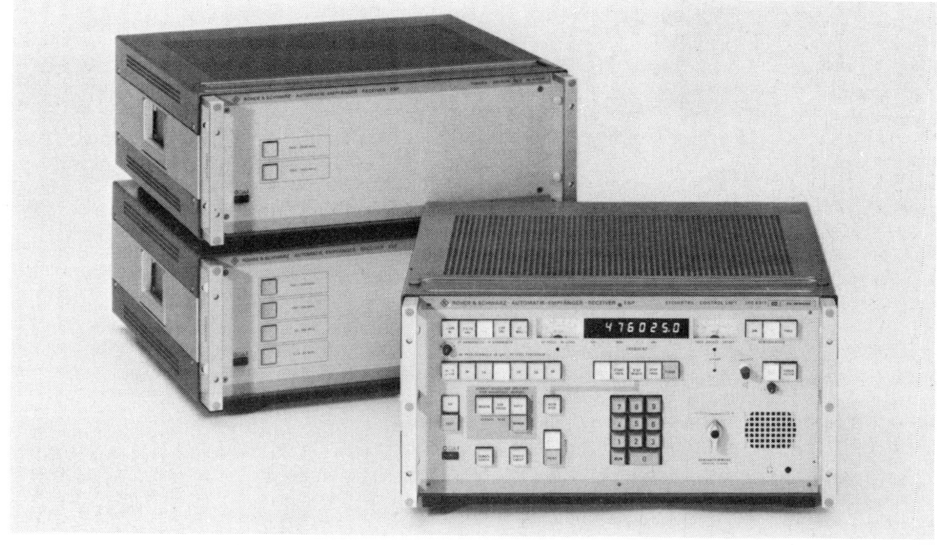

ESP intercept receiver with ESP-T1 and -T2 tuner units

3532.093

PA 055 & PA 010 DOPPLER DIRECTION FINDERS

Rohde & Schwarz has developed the PA 055 broadband doppler direction finder to enable DF and source location as well as detection and signal analysis.

The PA 055 can indicate the direction of a target transmitter to within $\pm 1°$. Frequency coverage is from 20 to 1000 MHz, using two antenna systems, one for the band 20 to 200 MHz and the other for the band 200 to 1000 MHz.

The doppler system employed by Rohde & Schwarz is claimed to offer considerable advantages over conventional Adcock systems, including: wide aperture; low bearing error with high angle of incidence or depolarised waves; the possibility of using normal receivers; simple separation of the antenna system from the actual DF instrumentation; digital bearing readout; fully automatic operation; fully remote operation; high sensitivity through the use of an integrating IF summator and a special method of compensating for unwanted frequency modulation in

PA 010 HF direction finder with circular ground antenna array

the input signal. In addition to the digital bearing readout, there is an analogue display to permit an assessment of bearing quality. Full remote operation is available. The direction finder can already take bearings of signals as short as 10 milliseconds duration.

PA 010

Engineering advances by Rohde & Schwarz have resulted in what is claimed as the first HF doppler direction finder, the PA 010 which covers the range from 0.5 to 30 MHz. The proven technology applied to the shortwave range offers several advantages, for example: the possibility of taking bearings using high-angle signals (sky waves) and indicating their angle of elevation. All the other advantages of the doppler principle, such as uncomplicated remote-control capability, automatic transmission of bearing data, and immunity to multipath propagation due to the use of wide-aperture antennas, are retained with the new HF system.

STATUS

Both the PA 055 and the PA 010 are in production and operation.

CONTRACTOR

Rohde & Schwarz, Mühldorfstrasse 15, 8000 Munich 80, Federal Republic of Germany.

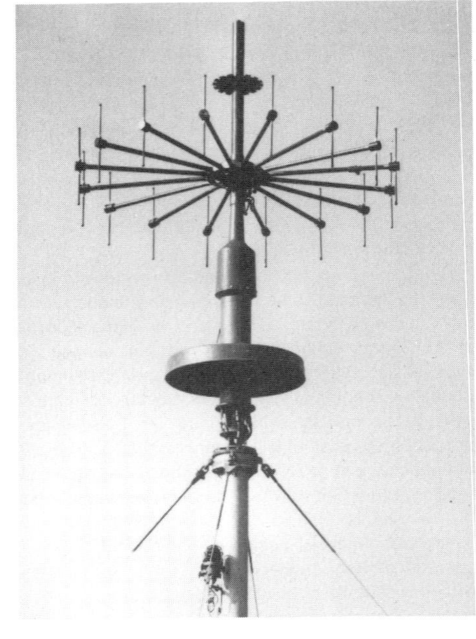

PA 055 DF antenna

4636.193

ROHDE AND SCHWARZ ANTENNA SYSTEMS

In addition to the equipment and systems detailed elsewhere in this section, Rohde and Schwarz design and manufacture a wide range of antenna systems, many of which are used in the ELINT and ESM fields.

HE 010 Active Rod Antenna

The HE 010 has a wide frequency range (10 kHz to 80 MHz), is optimised for maximum dynamic range, has two isolated outputs and is only 1 m high.

HE 013 and HE 015 Active HF Receiving Antennas

The HE 013 active HF antenna consists of the active rod antenna HE 010 and two active dipole antennas HE 002. The HE 015 has the same configuration, including a 90° coupler to form a broadband turnstile dipole. The antennas provide an omnidirectional reception of horizontally and vertically polarised waves in the range 1.5 to 30 MHz (horizontal) and 10 kHz to 80 MHz (vertical).

HE 202 and HE 302 Active Dipoles

The HE 202 and HE 302 active receiving dipoles have a frequency range of 200 MHz to 1 GHz and 200 to 500 MHz respectively and are designed to replace antenna systems of several switch-selected passive elements used for the reception of horizontally or vertically polarised signals. They have comparable electrical characteristics.

Low non-linear distortions and small noise figures are claimed to give a dynamic range seldom attained by passive antennas equipped with quality preamplifiers. In addition the manufacturer claims that almost frequency-independent radiation patterns are obtained over the active frequency range. Input and output of the single-stage, symmetrical push-pull amplifier incorporated in the antennas are protected against the excessive voltage produced by atmospheric discharge or in the immediate vicinity of transmitting antennas.

HK 001 UHF Coaxial Dipole

The HK 001 UHF coaxial dipole is a vertically polarising, omnidirectional antenna for fixed and mobile use. It is particularly suited for installation on ships.

Low wind load and low weight permit the dipole to be mounted at a considerable height above ground, resulting in increased field strength in transmission, higher sensitivity in reception and reduced environmental influences, especially those due to neighbouring parasitic elements.

HK 012 VHF Coastal Dipole

The HK 012 VHF coaxial dipole is a vertically polarising, omnidirectional antenna with high suppresion of current flow on the outside of cables. As a result of its design, it is also suitable for mobile use, particularly on ships.

HK 014 VHF/UHF Coaxial Dipole

The HK 014 VHF/UHF coaxial dipole is a vertically polarised broadband omnidirectional antenna covering the range 100 MHz to 1.3 GHz and is suitable for

Active receiving dipole HE 202

HE 302 active receiving dipole

HE 015 antenna

HK 001 dipole antenna

HK 014 coastal dipole

transmission and reception. As a transmitting antenna the dipole is suitable for CW operation of 400 W maximum up to 400 MHz and of 200 W over the total range.

The VSWR of the coaxial is better than 2 dB over the complete frequency range, the typical maximum value being 1.8 dB. The HK 014 operates as an electrically shortened, unbalanced-fed dipole in the lower frequency range and as a double-cone antenna in the upper range. The selected radiator construction ensures a continuous transition between the modes of radiation, resulting in vertical diagrams with relatively low frequency dependence.

The dipole is particularly useful for ship board fitment.

STATUS

In full production and operation.

CONTRACTOR

Rohde and Schwarz, Mühldorfstrasse 15, 8000 Munich 80, Federal Republic of Germany.

4637.193

PA 510 MOBILE DIRECTION FINDING SYSTEM

The PA 510 is a vehicle mounted mobile DF system. Frequency range is from 1 to 30 MHz with a narrow bandwidth of ± 75 Hz, and a 3-digit readout of instantaneous value or average value plus standard deviation. Excellent results are claimed up to elevation angles of 30 degrees. The antenna can be set up on a tripod and operated remotely to avoid any adverse effects caused by the vehicle.

The vehicle also contains the VHF/UHF DF system, PA 555, which is a mobile version of the PA 055 direction finder (**3532.093**) and has similar characteristics. It has a frequency range of 20 to 1000 MHz.

Only three antennas are required to cover the complete frequency range from 1 to 1000 MHz.

STATUS

In production.

CONTRACTOR

Rohde and Schwarz, Mühldorfstrasse 15, 8000 Munich 80, Federal Republic of Germany.

Vehicle mounted antennas for PA 510 and PA 555

5791.193

GA 082 FSK ANALYSER

The GA 082 FSK analyser is designed to facilitate and speed up the identification of radioteletype signals. Supported by a microprocessor, the analyser automatically analyses, with high accuracy, all parameters of teletype signals of F1B (F1) and F7B (F6) emissions and simultaneously displays the measurements together with all essential information.

The GA 082 can be used as a stand-alone unit for manual control, and as a system unit for computer-aided radio monitoring by military and intelligence authorities.

STATUS

In production.

CONTRACTOR

Rohde and Schwarz, Mühldorfstrasse 15, 8000 Munich 80, Federal Republic of Germany.

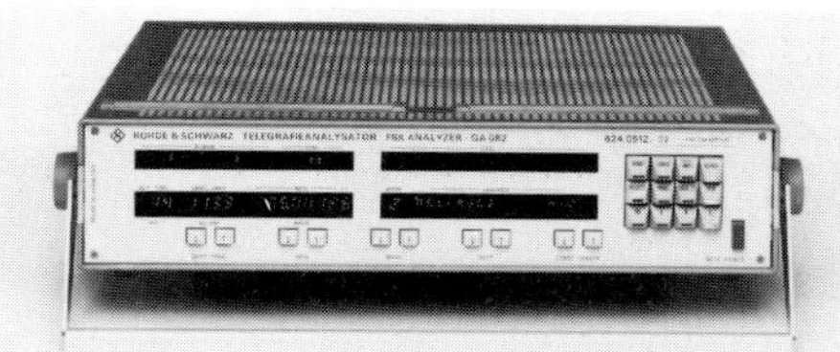

GA 082 FSK analyser

6454.193

PV 017 AZIMUTH ANALYSER

The azimuth analyser PV 017 is intended for use with Rohde & Schwarz HF, VHF and UHF radio monitoring equipment, and other direction finders fitted with a suitable interface. It provides fast and clear real-time display of the DF event in the form of an azimuth histogram, and additionally represents the instantaneous target signal in the form of a waterfall. In this case a point appears at the top edge of the display, vertically above each histogram, and slowly drifts downward in the direction of the histogram. The operator can therefore see the relationship between the incoming signals and their relative frequency of appearance.

These displays permit rapid capture of DF data, simultaneous display of bearing data for several transmitters, determination of the most probable bearing when information fluctuates, and checking of DF data quality. They also facilitate direction finding and increase accuracy and sensitivity for difficult targets.

The operator can determine a bearing by using the peak value of the histogram, and also use the dispersion (standard deviation) of the histogram as a measure of data quality. A further task of the PV 017 is transmitter separation in the case of common-channel operation which is encountered, for instance, with keyed telegraphy. When using Rohde & Schwarz radio monitoring equipment in conjunction with the PV 017, short signal durations, as in the case of frequency hopping, are permissible.

The PV 017 comes as a 19-inch rackmount or a desktop unit.

CONTRACTOR

Rohde & Schwarz, Mühldorfstrasse 15, 8000 Munich 80, Federal Republic of Germany.

6840.193

BABYLON HF JAMMER

Babylon is a jamming system designed primarily for use against HF tactical communication links. Designated MX 502, it is a self-contained mobile system, housed in a shelter and installed on a single-axis trailer. In order to keep costs down, considerable use has been made of commercially available units.

Babylon operates over the frequency range 2 to 30 MHz with an output power deliberately kept down to 1 kW to aid mobility. Fully automatic operation is possible and the system can also be expanded to form part of a computerised net, working to a central station or remote installations. A new high-performance antenna is used which is lightweight and can be deployed and disassembled rapidly.

STATUS

Fully developed.

CONTRACTOR

Rohde & Schwarz, Mühldorfstrasse 15, 8000 Munich 80, Federal Republic of Germany.

6839.193

SIEMENS HF JAMMER

Siemens manufactures three versions of its 1 kW communications jamming system, which are normally installed in a tactical mobile shelter for battlefield use. The family can counter most conventional HF modes, including SSB and USB. The three versions are:

(a) A basic system comprising a cost-effective manually-controlled stand-alone jammer
(b) An intermediate version which is fully computer-controlled and includes ESM functions, plus look-through and signal-initiated jamming modes
(c) A sophisticated version including complete comjam and communications subsystems. The latter provides secure frequency-hopping speech and data links and can be used for control of up to five identical comjam systems.

CONTRACTOR

Siemens AG, Postfach 70074, 8000 Munich 70, Federal Republic of Germany.

5792.193

PLATH DIRECTION FINDING SYSTEMS

C Plath GmbH of Hamburg manufactures a number of automatic and manually operated direction finding (DF) systems. These include DF nets and systems for very long wave to microwave range, stationary and mobile systems, and special receivers and antennas. Three examples with brief details are listed below.

Type 1160 VHF DF System

This is an automatic, remote controlled system for the frequency range 20 to 80 MHz, but extensible to 180 MHz. The complete system consists of three vehicles fully equipped with DF systems and a central station which controls the complete operation, receives the bearings and plots the transmitter position by the normal triangulation method. Each bearing station is equipped with a three-channel DF receiver and associated equipment, with a telescopic mast antenna mounted on top of the vehicle. Although designed primarily for remote operation, each mobile station can be manually operated if required. The system is capable of taking bearings on both very weak and short period signals.

SFP 5200 Visual Direction Finder

The SFP 5200 is a visual direction finder, operating over the frequency range 0.01 to 30 MHz, which is particularly characterised by its high acquisition rate. A maximum of 8 msecs is required from the start of the DF command to indication of the digitally evaluated bearing together with quality criteria. This

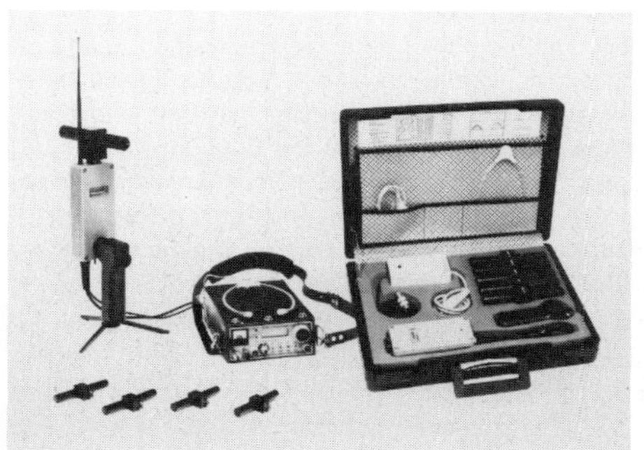

KE 393 miniature DF receiver

time also includes the setting time for the synthesiser at 2 msecs and channel tuning with a new system at 1 msec. With a suitable drive this DF receiver is able to operate against frequency hopping radios and to determine their location with DF networks.

Different bandwidths can be selected for direction finding and audio information. The complete contents of a message can be received with the narrow bandwidths most favourable for direction finding. The digital filters which are used also ensure absolute synchronisation of the receiver during the transient period of tuning or tuning to the filter edge.

SFP 7102 Visual Direction Finder

This is a compact automatic instrument for the VHF frequency range of 20 to 250 MHz. It combines the features of a three-channel Watson-Watt type DF receiver with the small dimensions of a single-channel direction finder. It is especially designed for mobile tasks but also serves as a stationary DF system on airfields, helicopter landing zones, coastal stations, etc. To meet environmental specifications for use in vehicles a ruggedised version is available. Several different Adcock antennas, size and shape according to requirements, are also available.

SFP 7104 Visual Direction Finder

This is an extended range version of the SFP 7102, and is also a three-channel DF receiver, for the HF/VHF/UHF frequency range of 20 to 520 MHz. It is intended primarily for mobile tasks but is equally well suited as a stationary system. To meet environmental specifications a ruggedised version is available.

SFP 7200 Visual Direction Finder

The SFP 7200 radio direction finder consists of a DF receiver and the necessary antenna system, and is designed for automatic DF over the frequency range 10 to 3999.9 KHz. Tuning to the received frequency is achieved by means of a frequency synthesiser in discrete steps of 100 Hz, with crystal-controlled accuracy. The employed frequency is shown as a digital LED display on the front panel. Bearing measurement of the received frequency is automatic in the form of a line on the front panel CRT display whereby the quality of the signal and the presence of possible interfering signals can also be seen easily. The SFP 7200 is normally used in shipborne applications.

SFP 7300 Visual Direction Finder

A compact automatic visual direction finder, the SFP 7300 covers the frequency range 0.1 to 30 MHz. It is a three-channel receiver which is designed particularly for mobile tasks. A ruggedised version is available to meet specifications for use in vehicles.

KE 393 Miniature DF Receiver

This is a miniature lightweight receiver used for taking DF measurements for a variety of purposes, particularly for monitoring in the long, medium and shortwave ranges. Applications include mobile use and with its small size and weight the receiver can, for example, be hidden under the coat of the operator, with a DF antenna placed in the back of the coat. It is supplied complete with a 220 V mains power supply, a battery pack, earphones and a carrying case. Weight of the unit is 2 kg and the dimensions are 140 × 55 × 230 mm.

CONTRACTOR

C Plath GmbH, Gotenstrasse 18, 2 Hamburg 1, Federal Republic of Germany.

Type 1160 VHF/DF system

6843.193

ALBERICH LASER/RADAR WARNING SYSTEM

The Alberich laser/radar warning system is being developed by a consortium comprising Kurt Eichweber, Rheinmetall and Nico Pyrotechnik. The system is designed for use in AFVs to warn of laser or radar threats and protect the vehicle by triggering off the appropriate countermeasures in the threatened quadrant.

Alberich consists of a laser warning receiver, 100 mm high and 130 mm across, which reacts to laser emissions in the 0.66 to 1.1 μm range. It is claimed to be accurate to within 3°. The sensor head also includes a radar warning receiver capable of detecting emissions from search, tracking and fire control radars.

Signals from the sensor head are passed to the analysis unit and after analysis the crew is warned of the type of emission and its direction. The relevant threat information is also sent to a control and monitor unit which can then command the appropriate countermeasures launchers.

STATUS

In development.

CONTRACTOR

Kurt Eichweber Prazisionsgeratewerk, Postfach 50 04 23, 2000 Hamburg 50, Federal Republic of Germany.

INTERNATIONAL

6464.193

COLDS LASER DETECTION SYSTEM

COLDS (Common Opto-electronic Laser Detection System) is a reliable, lightweight, EMI/EMC-resistant, multi-spectral target acquisition system designed for both surface and airborne platforms. The system is able to detect all current and foreseeable laser threats by accurate and precise determination of the threat direction, laser type and coding, so providing the opportunity for selected active and/or passive countermeasures.

CHARACTERISTICS

Wavelength: 0.4 to 2.0 μm (optionally 2.0 to 5.0 μm and 5.0 to 12.0 μm)
Angular coverage: Azimuth 360°; elevation ±45°
Angular resolution: Azimuth ±3°; elevation optional

STATUS

In pre-production and field trials.

CONTRACTORS

Messerschmitt-Bölkow-Blohm (MBB) GmbH, Post Box 801149, 8000 Munich 80, Federal Republic of Germany.

Tracor Inc, 6500 Tracor Lane, Austin, Texas 78725-2070, USA.

ISRAEL

4225.093

ELTA GROUND-BASED EW SYSTEMS

Elta Electronics, part of Israel Aircraft Industries, has disclosed an extensive capability in ground-based electronic warfare systems. These are mainly concerned with intelligence gathering and jamming. Details regarding some of these systems are given below. In addition the EL/L-8300 series of equipments, which are primarily designed for shipborne applications, can also be used in the land-based role. Information on these is given under entry **5841.253**.

COMINT Receivers

For radio monitoring Elta has developed a range of digitally-controlled receivers. The EL/K-1150 is a VHF/UHF equipment operating over the frequency range 20 to 500 MHz and the EL/K-1160 is an HF system covering 0.5 to 30 MHz. Both are synthesised receivers incorporating internal microprocessors to enable local or remote control, and a variety of operational roles such as auto-scan and preset tasks according to remotely controlled programmes.

The EL/K-1250T is a compact dual channel synthesised receiver operating in the 20 to 510 MHz band. Its main use is as a basic building block for larger COMINT or EW systems. The equipment demodulates AM, FM, CW and SSB signals, employing four selectabe IF filters. The frequency synthesiser features fast tuning of less than 500 μs between any selected channels. Receiver operation is by digital remote control. Weight is 5.4 kg.

EL/K-7001 communications EW system

EL/K 7010 VHF communications jammer

EL/K-7001 Communications EW System

The EL/K-7001 is designed to provide close communication intelligence and jamming support at tactical levels. The system consists of two units; the EL/K-7001RC receiving and control unit and the EL/K-7001T transmitter. The EL/K-7001RC incorporates a compact VHF/UHF receiver and microprocessor control circuitry, with a CRT display and keyboard in the front panel permitting convenient data presentation and operation. The EL/K-7001T includes a solid-state wide band transmitter operated by remote control from the EL/K-7001RC. Operating frequency is from 20 to 500 MHz receiving, and 20 to 90 or 100 to 400 MHz for transmission. Normal transmitter output power is 400 W, although up to 1 kW can be provided.

The system is intended for either COMINT operation using only the EL/K-7001RC or both COMINT and COM-JAM activities. A direction finder unit is available to enable more comprehensive COMINT operations. The EL/K-7001 is normally installed in light vehicles, such as jeeps and armoured cars, but is also suitable for installation in light aircraft, helicopters and boats.

EL/K-7010 VHF Communications Jammer

The EL/K-7010 is a family of modular-designed tactical communications jamming systems which provides effective electronic countermeasures support at brigade/division level. The family includes three basic models which cover all the normal frequencies used in tactical military communications. The three basic models are the EL/K-7010, EL/K-7010A and EL/K-7010B which cover the frequency ranges 20 to 90 MHz, 100 to 400 MHz and 1.5 to 30 MHz respectively.

The EL/K-7010 system is integrated in a small S-250

Elta COMINT receivers

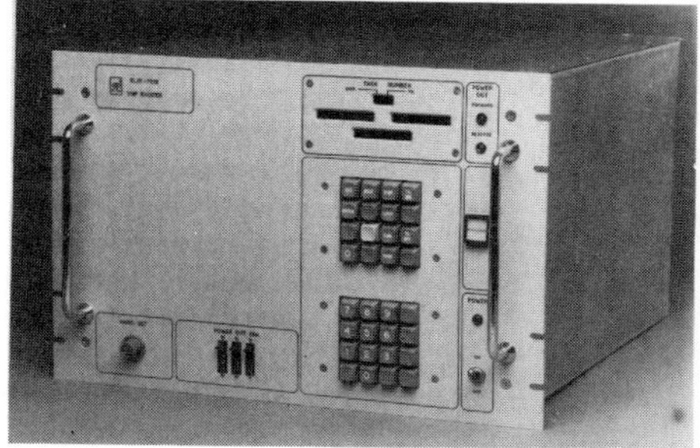

EL/K-7012 jamming exciter

shelter, designed to be mounted on a 1.5 ton military truck or similar vehicle. It includes an exciter, solid state power amplifier, and control unit with CRT display. Log periodic directional antennas on a telescopic mast are used for the VHF and UHF bands. For HF both ground-wave and sky-wave antennas are available.

The system has a stand-alone capability which includes manual or automatic search, acquisition and jamming. Transmit power is 1 kW with selectable COM-JAM dedicated signals transmissions.

EL/K-7012 Communication Jamming Exciters

The EL/K-7012 series of exciters is intended to generate the low level RF signal, further amplified by a power amplifier, for communications jamming purposes. The series consists of three units; EL/K-7012 operating over the frequency range 20 to 100 MHz, EL/K-7012A from 0.5 to 30 MHz, and the EL/K-7012B from 100 to 400 MHz. Each exciter contains two signal generators, a modulation signal generator and an RF synthesiser. Remote control is normally used since the exciter is designed to operate as a basic element of larger ECM computerised systems.

EL/K-7020 EW System

This is a command and control centre deployed at divisional or higher level. Its capabilities consist of automatic search and acquisition, target location and identification, preset tasks monitoring and interception, and countermeasures management by remote control over several EL/K-7010 or equivalent systems. The system is modular and expandable according to requirements.

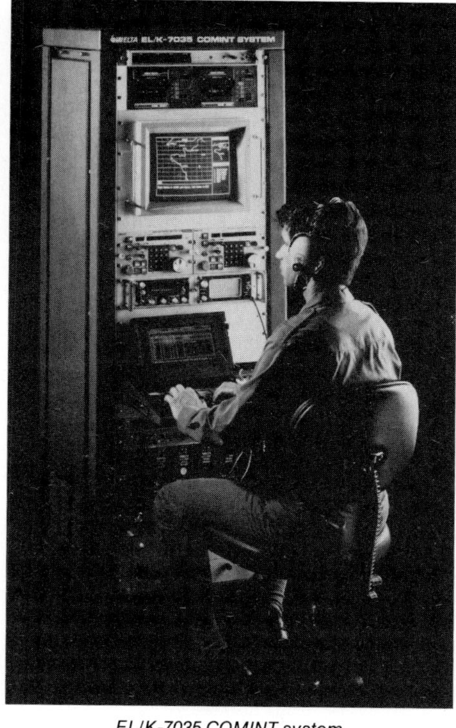

EL/K-7035 COMINT system

EL/K-7035 COMINT System

The EL/K-7035 COMINT system intercepts, monitors, locates and analyses radio communication signals in the 20 to 500 MHz range. The system is based on a multi-task COMINT workstation and can be configured for land, sea or air applications. The basic system consists of two to four receivers, controllers and tape recorders; and a ruggedised computer; plus an optional display. Frequency resolution is 1 kHz.

STATUS

Most, if not all these systems are in service with the Israeli armed forces.

CONTRACTOR

Elta Electronics Industries Ltd (a subsidiary of Israel Aircraft Industries Ltd), Ashdod, Israel.

6824.193

EL/S-8811/2 TACTICAL INTELLIGENT SYSTEM

The EL/S-8811 and 8812 are compact Tactical Video Receiving Systems (TVRS) designed to provide a real-time video battlefield picture for personnel on the battle front. They feature combined video, telemetry and audio reception. The EL/S-8811 is a vehicle-mounted TVRS consisting of a receiver, controller, processor and display unit. The EL/S-8812 is a manpack version which comprises a receiver, controller and display. Both systems provide reception and display of intelligence information transmitted from video sources such as RPVs, reconnaissance aircraft, balloon borne cameras and video binoculars.

The EL/S-8811 is designed for highly mobile units and can be fitted to any armoured or soft-skinned vehicle. The monitor is available in two formats, a fully militarised 5-inch monitor for vehicle installation or a 4-inch flat monitor with a strap for chest carrying. It has processing and graphic capabilities, enabling frame freeze and display of a real-time video picture, complemented by superimposed alpha/graphic information.

The EL/S-8812 is intended for infantry use as a manpack unit with the receiver carried in a back-pack harness. The monitor is the 4-inch flat version designed for chest mounting. It does not include processing and graphic capabilities.

CONTRACTOR

Elta Electronics Industries Ltd (a subsidiary of Israel Aircraft Industries Ltd), Ashdod, Israel.

EL/S-8811 tactical intelligence system

4638.193

RAJ 101 GROUND RADAR JAMMER

The RAJ 101 is a ground radar jammer covering 180 degrees and ranges of up to 30 km. The system uses a high ERP jammer and wide frequency band, and is capable of dealing with three threats simultaneously. In the acquisition mode, RAJ 101 measures frequency, direction, PRF, pulse width and amplitude in order to detect, identify and locate hostile radars. The information collected can also be used for battle analysis.

The high power jammer operates in fully automatic or semi-automatic mode. Frequency and direction are set up to prevent hostile radars from detecting and locating friendly forces. The system is completely self-contained, is vehicle mounted and able to operate and travel in extremely rugged terrain.

STATUS

In production and in service with the Israeli armed forces.

CONTRACTOR

Rafael Armament Development Authority, PO Box 2082, Haifa 31021, Israel.

RAJ 101 ground radar jammer mounted on an APC

6207.193

RATTLER JAMMER

Rattler is a jamming system capable of use in ground, naval and airborne applications, as either part of an overall ECM system, or on its own by means of a local control unit. The system can operate as both stand-off and stand-in and is designed to jam surveillance, search and tracking radar, up to three threats simultaneously. The system consists of four main units, a low power microwave source, a high power wide band amplifier, power supply and a control unit. Output power is believed to be greater than 400 W, on a spot or barrage jamming basis, and the equipment covers the 2-18 GHz frequency band.

The system is ruggedised, designed and tested to MIL-STD-16400 and 5400. Timesharing between the three channels incorporated enables three jamming frequencies through one high power amplifier. Up to 16 Rattler systems can be connected on the same 488 data bus between units and an ESM system, enabling a full system to jam 48 frequencies under the control of a single ESM/ECM suite. Remote control is provided and can control the system from a distance of up to 200 m. Range of the jammer is believed to be 20-30 km.

STATUS

In operational service with the Israeli armed forces.

CONTRACTOR

Rafael Armament Development Authority, PO Box 2082, Haifa 31021, Israel.

4356.093

TIMNEX-4CH AUTOMATIC ELINT/ESM SYSTEM

The Timnex-4CH system is an Elint/ESM receiver intended for intercepting radar signals and for operating in land, sea or airborne environments. It covers the frequency range from 2 to 18 GHz and is designed to intercept simple or complex transmissions and to sort and compare them with a preprogrammed radar library.

The system features a channelised IFM receiver, a sophisticated IDF sub-system and real-time computer processing. Tabular display of the signal parameters with instantaneous spectra and gonio displays are presented to the EW operator. Hard copy print out and data logging on magnetic media are additional features.

Timnex-4CH has been designed to intercept radars at a range of several hundred kilometres. By making several measurements during a known path, computations of the radar location are carried out to provide a full tactical picture of the area. The data acquired may be either used in the vehicle or transmitted to ground/shore control stations by data link. Processed information is displayed on a 19-inch (483 mm) colour raster display in several desired cross-sections.

STATUS

In production for land and sea applications, and in service with the Israeli armed forces.

CONTRACTOR

Elbit Computers Ltd, Military Systems and Products Division, Advanced Technology Center, PO Box 5390, Haifa 31053, Israel.

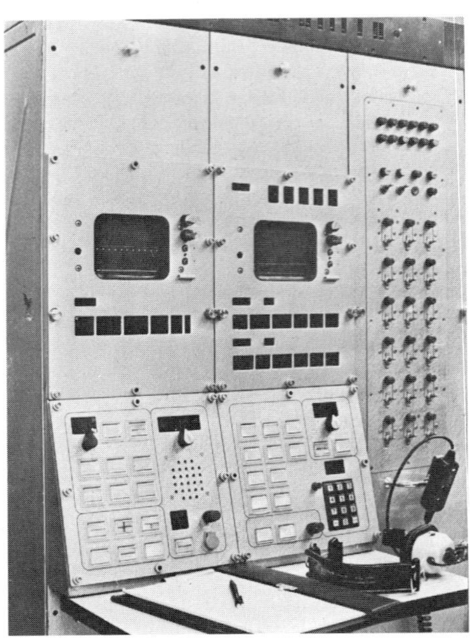

Timnex-4CH automatic Elint system console (submarine version)

4731.153

CR-2740 ELINT SYSTEM

The CR-2740 is a mobile electronic intelligence (ELINT) equipment intended primarily for use in the forward battlefield area. It covers the frequency range from 500 MHz to 18 GHz, and consists of a two-man operations cabin, upon which is mounted a high frequency antenna system, and a trailer carrying a low frequency antenna. The computer-controlled system employs high sensitivity superheterodyne receivers and high gain antennas to carry out automatic signal search and acquisition. After analysis all relevant parameters are presented to the operators on a multi-colour graphics display.

The cabin is a standard S280 shelter, suitable for transport on a variety of vehicles, with two 1.2 metre diameter antennas mounted on top which are used for reception in the 4 to 18 GHz band. A 3 metre diameter antenna for the 500 MHz to 4 GHz range is mounted on the trailer. The two operators, using basically identical equipment, monitor the two bands individually. Most of the data acquired is computer analysed.

Although the system is designed as a mobile forward listening post, the control cabin can be remoted from the antennas, with cable or radio links, for semi-static use.

STATUS

In operational service.

CONTRACTOR

Elisra Electronic Systems Ltd, 48 Mivtza Kadesh Street, Bene Beraq 51203, Israel.

CR-2740 ELINT system

6455.153

CR-2700 MOBILE ELINT SYSTEM

The CR-2700 is a vehicle-mounted mobile electronic intelligence (ELINT) system. Automatic search and acquisition of signals is performed using high gain antennas and superheterodyne receivers which scan the azimuth and frequency range, achieving high sensitivity and dynamic range. The system displays signal parameters, such as frequency, PRF and pulse width, and real-time displays provide in-depth analysis.

The CR-2700 consists of an antenna trailer with receivers for the 0.5 to 4 GHz range, and a shelter with a roof-mounted antenna assembly and receivers for the 4 to 18 GHz range. Two operators are required to work the system, one for each frequency band. The system interfaces with blanking input and provides log video, linear video and outputs. Recording is available for recording of audio and video information.

The system is transportable by truck, helicopter or cargo aircraft, and is quickly deployed and disassembled.

STATUS

In operational service with the Israeli defence forces.

CONTRACTOR

Elisra Electronic Systems Ltd, 48 Mivtza Kadesh Street, Bene Beraq 51203, Israel.

5937.153

AT-4910 COMINT/JAMMING SYSTEM

The AT-4910 is a combat jamming and communications system designed for one-man operation in a number of applications. It is vehicle-mounted to provide high mobility and fast deployment. Frequency coverage is 20 to 100 MHz, 100 to 200 MHz and 200 to 400 MHz, although other bands are also available. The system operates in three modes; communication jamming, a high-power radio station with an output of 400 W, and as a two-way radio relay.

The AT-4910 panoramic receiver covers the frequency range 20 to 500 MHz and provides COMINT surveillance and panoramic monitoring during jamming. The control unit includes all controls, audio and RF switching for the various operation modes, and voice and jamming modulations. The jamming operation is monitored on the panoramic receiver screen where the noise modulation covers the signal. A 'look-through' technique allows search during jamming.

STATUS

In production and in service with the Israeli armed forces.

CONTRACTOR

Elisra Electronics Systems Ltd, 48 Mivtza Kadesh Street, Bene Beraq 51203, Israel.

6165.093

TADIRAN COMINT SYSTEMS

Tadiran manufactures a range of COMINT, SIGINT, COMJAM and DF systems for land-based, naval and airborne applications. Brief descriptions of the land-based systems are as follows.

Series 6000 COMINT System

A microprocessor controlled real-time system operating in the VHF/UHF band (20 to 500 MHz) for the detection of communications signals. The equipment has capabilities for display, monitoring, recording and direction finding.

Series 7000 COMJAM Systems

A family of mobile jamming systems designed for search, identification, monitoring and disrupting hostile communications. The systems cover the frequency range 1.5 to 500 MHz, according to requirements, with high power transmission from 100 W to 10 kW.

Series 9000 COMINT System

A completely automatic system for scanning, detection, recording, direction finding and recognition of voice and data RF signals in the frequency range 2 to 30 MHz, with a selection of modulation modes. The system includes a display unit and has message handling capabilities.

TDF-205 Direction Finder

An automatic direction finder which can operate in a stand-alone capacity or as part of an overall EW system. It can be used on land in fixed or vehicle configuration, at sea or in the air, and can be operated in a remote or manual mode. The frequency range covered is from 20 to 500 MHz with an azimuth accuracy determination stated to be 1.5 rms.

TDF-305 Direction Finder

A low-cost ruggedised version of the TDF-205.

GE-530 ESM Receiver

A fully synthesised microprocessor controlled HF receiver for surveillance applications over the frequency range 5 kHz to 30 MHz. Capable of being controlled locally or remotely it is available in master and slave configurations.

SR-3 Direction Finder

An accurate low cost Adcock single-channel equipment operating from 20 to 500 MHz with up to 100 per cent amplitude and frequency modulation. Fixed, vehicular and shipborne applications.

SR-11 Direction Finder

A dual-channel doppler direction finder operating from 1.5 to 30 MHz. This is a lightweight, ruggedised system for use in both fixed and vehicle-mounted roles.

Fix 500 and 3000

Emitter locating systems operating in VHF/UHF and HF communications bands respectively.

CONTRACTOR

Tadiran Ltd, 11 Ben Gurion Street, Givat-Shmuel, PO Box 648, Tel-Aviv, Israel.

ITALY

4645.093

ELT/128 COMMUNICATIONS DIRECTION FINDER

The ELT/128 is a direction finder, suitable for airborne, ground and naval applications, operating in the HF, VHF and UHF frequency bands. The equipment automatically measures direction of arrival and technical analysis of any kind of signal modulation in the requisite bands. ELT/128 is currently in use on a stand-alone basis in ships and land vehicles, and in integrated ESM systems.

The system consists of an RF unit, control unit, control panel and antenna assembly, the last being tailored to suit the required band coverage, platform and application (fixed or mobile). For the HF band the antenna assembly normally consists of half-dipoles for fixed or transportable applications, and crossed-loop ferrites for mobile work. For VHF/UHF an array of active dipoles for naval or ground applications and active half-dipoles when integrated in ESM/COMINT airborne systems.

Outputs are provided for use in a display unit and a printer for sequential printout.

STATUS

In operational service.

CONTRACTOR

Elettronica SpA, Via Tiburtina Valeria, Km 13,700, 00131 Rome, Italy.

ELT/128 control panel

5953.093

ISIS SIGINT SYSTEMS

ISIS is the name of a series of SIGINT systems developed by Elettronica for ground, naval and airborne applications. They feature the latest technologies and have been designed to intercept and analyse complex and sophisticated radio and radar signals, even in very dense electromagnetic environments. The series consists of ISIS-A for airborne applications, ISIS-G for ground and ISIS-N for naval work, and all systems cover the frequency range from VHF to J band. The systems have a high degree of commonality but have been optimised for the individual operational role.

The SIGINT-G system is a powerful data collection and analysis equipment capable of continuous operation on a 24-hour basis. It is modular in design to allow for expansion, and adaptation of the configuration to an individual operator's requirements. It can be operated in conjunction with radio-link connected mobile slave DF stations to allow emitter location through triangulation.

The SIGINT-A configuration has been designed for installation on medium to large aircraft employed in strategic missions. The system antenna can be integrated easily with the fuselage.

ISIS-N is the naval version and can be integrated with existing EW systems on board practically any warship. It can also be supplied with its own dedicated antenna assembly. It will supplement the ship's normal ESM equipment by providing over-the-horizon targetting features, as well as the data measurement and analysis features.

A number of intercept and analysis modules can be added to the basic systems. These include an extension of frequency covered into the millimetre band, a system to recognise messages of interest in radio-link traffic, and the provision of an HF frequency coverage in the ISIS-G and ISIS-N systems.

STATUS

These SIGINT systems are in operational use with Italian defence forces.

CONTRACTOR

Elettronica SpA, Via Tiburtina Km 13,700, Rome, Italy.

3526.193

IGS-1/IGS-3P MOBILE INTEGRATED ESM/ECM SYSTEM

The IGS-1 is an ECM system mounted on a trailer, and intended to be part of a major EW system. It is capable of autonomous data collection, analysis and reaction, as well as co-ordinated operation with a command and control centre, where intercepted data are received and orders are generated.

The TQN-1 jammer provides the active reaction against the threat. A highly directional antenna is tracked on the target and the jamming is deployed to deny the use of airborne or surface radars.

All equipments composing the system are fitted in an operational console, housed inside an air conditioned shelter. A trailer with wheels suitable for rough and soft terrain is supplied.

STATUS

Operational in several countries but no longer in production.

CONTRACTOR

Selenia Industrie Elettroniche Associate SpA, Special Equipment and Systems Division, Via dei Castelli Romani 2, 00040 Pomezia, Rome, Italy.

3527.153

SELENIA ELECTRONIC SURVEILLANCE SYSTEM

The Selenia Electronic Surveillance System is an ESM system, carried on a Mercedes Unimog type vehicle, and is mainly designed for coastal and battlefield surveillance, as part of a major EW system. Radar emissions are intercepted by RQN-1B radar intercept equipment (**3343.253**), which is a wide open receiver with omni-directional coverage, providing direction of arrival, pulse-width, PRF and frequency band measurement. The same parameters are available in digital form on the PAW-1 pulse analyser and warner, where threat radars can be programmed as such for a prompt evaluation and reaction. A monopulse antenna system is slaved to the RQN-1B handwheel, and feeds an FBR-1 fine bearing receiver for azimuth measurement with high accuracy. At the same time the FR-1 frequency receiver performs spot frequency measurement.

STATUS

No longer in production.

CONTRACTOR

Selenia Industrie Elettroniche Associate SpA, Special Equipment and Systems Division, Via dei Castelli Romani 2, 00040 Pomezia, Rome, Italy.

5664.153

ESS-2 ELECTRONIC SURVEILLANCE SYSTEM

ESS-2 is a mobile integrated surveillance and analysis system designed to operate against ground-based radars, and for coastal and battlefield surveillance against shipborne and airborne radars, as part of a major EW network. It provides high detection probability, high accuracy emitter bearing, optional parameter measurement and identification, and gives instantaneous threat warnings. The system can operate on a stand-alone basis, or it may be integrated into a wider ESM/ECM network.

STATUS

No longer in production.

CONTRACTOR

Selenia Industrie Elettroniche Associate SpA, Special Equipment and Systems Division, Via dei Castelli Romani 2, 00040 Pomezia, Rome, Italy.

JAPAN

6825.093

JAPANESE EW EQUIPMENT

As with many other categories of defence equipment, Japanese requirements for electronic warfare systems have in the past been met mainly from US sources. However, Japan is now designing and manufacturing many defence systems indigenously. Most of these equipments are for airborne and naval applications and very few systems have been identified for land forces use.

An unidentified item of ground-based EW equipment is illustrated and is apparently a transportable battlefield ESM system. However, the flexible waveguides which can be discerned on the original photograph could equally indicate facilities for transmission at microwave frequencies. This would suggest a possible combined ESM/ECM role for the equipment.

Airborne and naval EW equipments are dealt with under their respective sections.

NETHERLANDS

4735.193

SPECTRA ELINT SYSTEM

Signaal has developed a communications intercept and monitoring system covering the LF, MF and HF frequency bands, with an option to cover the VHF and UHF bands. The system is normally operated as a net with several modular designed monitoring stations, tailored to individual requirements, each being equipped to perform specific tasks.

Spectra is intended for emission interception, analysis and storage, as well as emitter location using direction finding techniques. It is completely computer-controlled with operator positions for the monitoring and analysis of voice, telegraphy or facsimile messages. All components are controlled via keyboards and displays from the operator positions.

STATUS

Operational.

CONTRACTOR

Hollandse Signaalapparaten BV, Huizen Division, PO Box 88, 1270 AB Huizen, Netherlands.

NORWAY

3344.153

SIMRAD RL1 LASER WARNING DEVICE

The Simrad RL1 laser warning receiver detects radiation from pulsed laser rangefinders and target markers. The instrument can detect pulsed radiation within the 0.66 to 1.1μm near infra-red band, covering the most common types of pulsed lasers (ruby, GaAs, neodymium) currently being used in rangefinders and target markers. The instrument does not detect continuous radiation.

A detector unit mounted on top of the vehicle has a 360° field-of-view and can in addition detect radiation coming from above. An indicator unit is mounted inside the vehicle. It indicates the approximate direction to the radiating source by means of light-emitting diodes mounted in a circle. Eight 45° sectors are indicated. A ninth emitting diode mounted in the centre indicates radiation received from above. Due to the overlapping fields of view, a total of 17 sectors can be indicated.

The receiver will give an acoustic alarm consisting of a pulsed audio signal in the crew's intercom when a laser pulse is detected. The duration of the alarm is two seconds for a single laser pulse. When more than one pulse is being detected the alarm stays on as long as laser pulses are being received.

CHARACTERISTICS

Detector unit

Number of detectors: Horizontal:4; Vertical:1

Detector field of view: 360°

Type of detector: PIN photodiode
Optical bandwidth: 0.66 – 1.1μm
False alarm rate: $< 10^{-3}$/h
Indicator unit
Resolution: Horizontal: 45°; Vertical: 45°
Number of sectors: Horizontal: 8; Vertical: 1
Display brightness adjustment range: > 100 X
Duration, acoustic alarm: 2 s
Duration, display: 8 s
Operating voltage: 20 – 32 V (24 V nom)

STATUS
Production.

CONTRACTOR
Simrad Optronics A/S, PO Box 6114, Etterstad, Oslo 6, Norway.

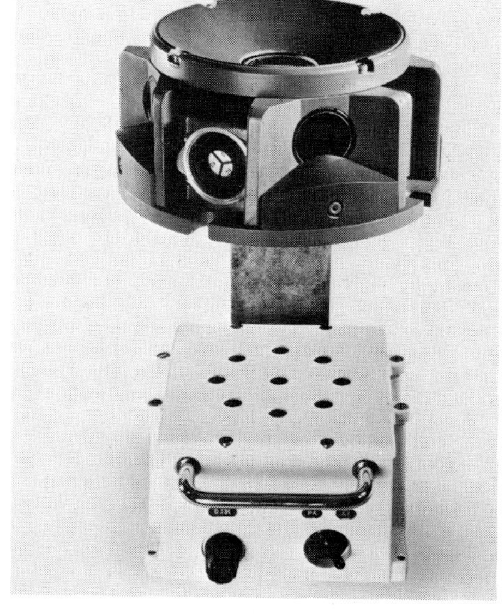

RL1 laser warning receiver detector unit and indicator unit (front)

SPAIN

6826.193

EN/TSD-500 SERIES COMINT SYSTEMS

ENSA of Spain designs and manufactures a range of automatic direction finding equipments which are used in COMINT applications. The systems, known as EN/TSD-501, 502 and 510, are broad and narrow band equipments covering the frequency band from 25 MHz to 1 GHz. The systems can be used in both mobile or fixed applications and are suitable for both land-based and naval use.

The EN/TSD-500 series is modular in construction and consists basically of an antenna array, a twin-receiver/analyser subsystem and a display.

CHARACTERISTICS
Frequency range: 110-140 MHz (EN/TSD-501 and 502); 25 MHz – 1 GHz (EN/TSD-510)
DF accuracy: ±1° (EN/TSD-501 and 510); ± 0-5° (EN/TSD-502)
DF speed: 300 ms (EN/TSD-501 and 502); 10-100 ms (EN/TSD-510)

CONTRACTOR
Electronica Ensa S.A., San Rafael 6, Alcobendas (Madrid), Spain.

6837.193

SEIDAL LASER DETECTOR

SEIDAL is a laser detection device intended for vehicle mounting. It consists of two reception modules, a processor unit and a display unit. The two reception modules can be mounted together, or separately on either side of the vehicle and each contains three photodectors. Indication of illumination by a laser rangefinder or designator is given on a circle of LEDs, accompanied by an aural alarm.

CHARACTERISTICS
Wavelength: 1.06 μm (extendable to 0.5 to 1.1 μm)
Azimuth coverage: 360°
Vertical coverage: –25 to +55°
Azimuth: ±15°

CONTRACTOR
SEIDEF S.A., Paseo de la Castellana 140, 28046 Madrid, Spain.

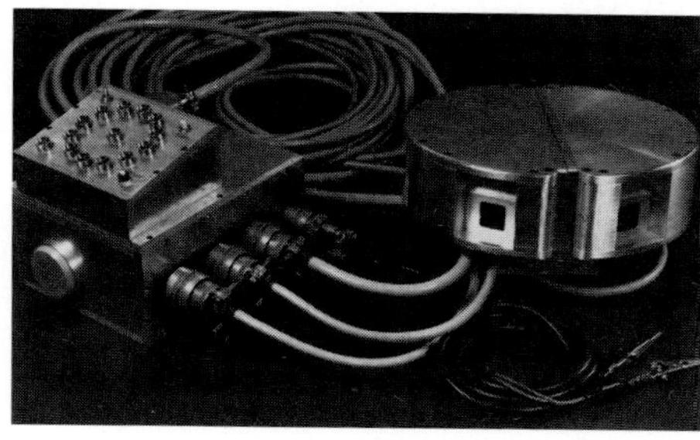

SEIDAL laser detector (two reception modules, processor and display panel)

UNION OF SOVIET SOCIALIST REPUBLICS

6827.193

SOVIET LAND-BASED EW SYSTEMS

The Soviet Union employs large numbers of electronic warfare systems in the ground, airborne and shipborne applications. This covers ESM surveillance systems, radar warning receivers, ECM systems and jammers. In addition, both ships and aircraft have chaff and decoy launchers to deal with radar and heat seeking missiles. Although no definite information has emerged from USSR, there seems little doubt that the Soviets are also developing warning receivers to counter the threats from laser target markers and rangefinders. Very few details of the electronic warfare systems have become available on an unclassified basis, and information on most equipments is insufficient to warrant separate entries in this section. A few details of the Krug and Fix SIGINT systems are given below, together with a list of those land-based EW systems definitely identified. Information on naval and airborne equipments is given under the appropriate Electronic Warfare Equipment section later. The Electronic Warfare Analysis table at the back of this book also gives a list of all Soviet EW systems definitely identified.

The Soviet SIGINT and COMINT organisations are believed to be the largest in the world and figures of 300 000 men have been quoted as their approximate strength.

Fix Series - a family of fixed and semi-mobile DF systems (see below)
Krug - a large HF/VHF DF system (see below)
Loop Series - a family of HF/DF systems, Loop 3 is a trailer-mounted tactical system
Moon - a four-tower fixed HF/DF array
Ramona - a SIGINT system
Ring Two - a trailer-mounted tactical HF/DF system
Small Cross - an Adcock-type antenna DF system
Spaced Loop - a fixed system consisting of a pair of square loop DF antennas
Square Four - an array of four box-shaped DF antennas
Thick Eight - a fixed DF array of vertically polarised HF dipole antennas

Krug SIGINT System

The Krug SIGINT system is a large, land-based HF/VHF DF system which equips many of the large permanently sited Soviet ESM sites. The main purpose of these stations is to monitor tactical radio traffic used by NATO for ground and airborne forces. The early Krug systems, produced in the 1950s from a German World War 2 design, covered frequencies of about 6 to 20 MHz, but the design has been continuously updated since then. The present antenna array is circular, approximately 100 m in diameter and consists of 120 equally spaced broadband vertically polarised antennas. Accuracy of the system is claimed to be better than 0.5°. The system has been further improved by the addition of an inner ring of dipoles that increases coverage into the VHF band. There are large numbers of Krug stations in the Soviet Union and Eastern Europe and sites have also been reported in Cuba and Vietnam.

Fix Series

The Fix family consists of a number of direction finding systems in the HF frequency range. One of the most widely used of these is the Fix-24 which is a circular array of 24 vertical monopoles, placed at 15° intervals in a ring approximately 150 m in diameter. In some stations a second ring of 24 monopoles has been placed inside the outer ring to increase the frequency coverage. Although fulfilling much the same function as Krug it is a much less complex array and the accuracy and sensitivity is reported to be considerably less.

UNITED KINGDOM

4227.193

RAMPART

Rampart is designed to give protection to key military areas such as airfields and missile sites against air attacks and is believed to be the world's first passive defence system for this purpose. It is based on a number of individual firing units which are radio controlled by a compact transmitter from a single central convenient position. Any combination of defensive measures can be activated instantly at any time and up to a range of 4 km.

A complete Rampart system is inexpensive, requires a minimum of personnel and is at immediate readiness for action at all times. It provides protection against manned aircraft, laser and TV guided missiles by the rapid emission of smoke; against heat-seeking missiles by infra-red decoys; and against aircraft and missile radars by chaff decoys. It also offers a unique defence against low-level aircraft attack by the quick release of a mass of 'Skysnare' airborne tethered obstructions which provide a formidable obstruction and which can remain aloft indefinitely. A series of Skysnare balloons is placed around a target area and, because of the kite-like design, will hold station even in the lightest wind. Attacking aircraft will be caused severe weapon aiming and delivery problems and will be forced to climb to an altitude which will expose them to active defence systems.

Firing stations are portable, self-contained, solar-powered units, each capable of firing nine rocket decoys (chaff or infra-red), nine smokes and three 'Skysnare' obstructions. Rapid or slow burning black and white smoke is available to give both immediate area coverage and sustained coverage thereafter. The 'Skysnare' system consists of a balloon, 1000 m of strong line, case, ground anchor, a gas cylinder and pyrotechnic inflation and release. It is at continuous readiness and can be fully deployed within two minutes by radio signal.

A Mk II version is now available designed to meet NATO requirements. This version is man-portable and has an anti-jamming frequency-hopping facility option in addition to a new infra-red screening smoke back-up to complement the existing decoy features.

The system has been further enhanced by the incorporation of a Large Area Smoke Screening system (LASS) which is fully automatic and in response to commands from the Rampart control unit, will generate smoke to counter visual or infra-red sensors for up to 90 minutes.

STATUS

In full production.

Rampart firing station

CONTRACTOR

Wallop Systems Ltd (a subsidiary of ML Holdings plc), Arkay House, Weyhill Road, Andover, Hampshire SP10 3NR, England.

5803.153

SENTRY ESM SYSTEM

Sentry is a family of mobile land-based ESM systems designed for intercept and analysis of the high and low frequency bands used by radars employed in battlefield surveillance, mortar locating, forward SAM and airborne applications. The family is based on a modular design such that systems ranging from simple radar intercept equipments to full ESM systems comprising land mobile individual and master/slave radar information gathering units can be provided, complete with advanced communications facilities.

Sentry I utilises crystal video techniques to provide tactical warning to field commanders, and Sentry II employs swept superheterodyne techniques to gather strategic electronic order-of-battle data for corps and army commander. In the latter role, Sentry would provide a C^3I gathering service.

The Sentry equipment consists of:

(1) radome mounted antennas
(2) ESM receivers, processors and recording facilities in sealed compartments
(3) control and display units
(4) radio links
(5) power generators and air conditioning.

The system is normally installed as a completely self-contained equipment in a Land-Rover type vehicle but can be configured for other vehicles. In the larger systems facilities can be included to feed in radar information collected by aircraft and other sources.

STATUS

Fully developed.

CONTRACTOR

Marconi Defence Systems Ltd, The Grove, Warren Lane, Stanmore, Middlesex HA7 4LY, England.

Sentry ESM system installed in a Land-Rover

3372.193

TYPE 405J MOBILE ECM EQUIPMENT

The ground-to-ground mobile ECM equipment, Type 405J, is a transportable ground jammer which includes an ESM receiving system to detect and analyse the characteristics of a victim radar and an ECM system which provides various types of jamming to combat the victim radar. The equipment is designed to fit into a long-wheelbase Land-Rover which also carries the appropriate receiving and transmitting antennas on a folding mast. The antennas can be adjusted in azimuth and elevation to point towards a chosen victim radar.

The receiver sensitivity, assuming free space propagation conditions and the radiated power output of the jammer, allows the equipment to operate at ranges up to 100 km from a victim radar.

The 405J can be supplied to cover any one of four frequency bands: 1 to 2 GHz, 2 to 4 GHz, 4 to 8 GHz and 8 to 16 GHz. Low jamming-power versions are available to provide a mobile autonomous ECM simulator. The equipment can be located within a few kilometres of an operational radar to give realistic ECM training for radar operators by generating RF signals corresponding to targets and jamming for radiation into the operational radar antenna. High jamming-power versions are available for operational use.

The equipment rack is mounted in the rear of the Land-Rover and there is space for two operators. A portable power supply generator and cables are stored in the rear of the vehicle during transit. Each equipment unit is mounted on slides for easy removal for maintenance. Deployment of the equipment normally takes 10 to 15 minutes and a similar time is taken to dismantle and stow the equipment for transit.

The ESM receiving system, with its processing and display unit, analyses received signals and displays radar parameters in terms of pulse length, PRF, dwell time, radar antenna rotation rate and corrected bearing. The transmitter/receiver control unit receives inputs from the processing unit and generates the selected ECM and target waveforms required by the transmitter. Types of ECM that may be selected include:

(1) CW (spot frequency)
(2) Swept CW (bandwidth controllable)
(3) FM Noise (bandwidth controllable)
(4) Pulse (PRF and pulse length controllable).

The equipment also has the capability to produce false targets in both range and velocity. The Type 405J system is designed on a modular basis and various types of ECM facilities can be included to meet particular customer requirements.

STATUS

In production and supplied to various users.

Type 405J ECM equipment deployed on Land-Rover

CONTRACTOR

Plessey Radar, Oakcroft Road, Chessington, Surrey KT9 1QZ, England.

6830.153

SERIES 500 MOBILE EW SYSTEM

The Series 500 is a transportable EW system, which consists of an ESM receiving system to detect and analyse the characteristics of a target ground radar and an ECM system which provides various types of jamming. The equipment is designed to fit into a cabin which can be transported by a suitable flat bed truck. The antenna can be adjusted in azimuth and elevation to point towards a chosen emitter. The equipment can be supplied to cover any one of five bands: 0.5 to 1 GHz; 1 to 2 GHz; 2 to 4 GHz; 4 to 8 GHz; 8 to 16 GHz. A comparable equipment is available for use against airborne radars.

All equipment except the antenna is mounted in the cabin with space for two operators. A portable supply generator and cables are stored in the cabin during transit. Deployment of the equipment normally takes

10 to 15 minutes and a similar time is required to dismantle prior to transit.

The ESM system, complete with processing and display unit, analyses received signals and displays radar parameters of pulse length, prf, dwell time, antenna rotation rate and bearing. Information from the processor is fed to the transmitter/receiver control unit which then generates the selected ECM and target waveforms required by the transmitter. Types of ECM that may be selected include CW (spot frequency), EM noise (bandwidth controllable) and Pulse (prf and pulse length controllable). The equipment is also able to produce false targets in both range and velocity.

CONTRACTOR

Plessey Radar Ltd, Oakcroft Road, Chessington, Surrey KT9 1QZ, England.

1712.153

PLESSEY HF MONITORING EQUIPMENT

PRS 2282A HF receiver

The PRS 2282A communications receiver is a fast, fully synthesised HF receiver with a frequency range of 10 kHz to 30 MHz, using keypad and tuning control. Designed to meet the needs of communications and monitoring, it is of modular construction with built-in test equipment. In addition to the standard features of an HF receiver it includes ISB with individual AGC on each sideband, synthesiser BFO with displayed and adjustable offsets, carrier derived AGC, diversity operation (master or slave), FM demodulation, and a signal path performance providing excellent reception even in the presence of interfering signals. The equipment has a noise factor of 10 dB or better throughout the frequency range.

PRS 3000 Series

This is a new family of modular EW receivers consisting of three variants, PRS 3830 for band scanning and signal acquisition; PRS 3840 designed for analysis; and the PRS 3850 for extracting information from a signal by manual monitoring. All three are packaged in an identical 3/8 ATR case and can be used for battlefield, naval or airborne applications.

STATUS

In production.

CONTRACTOR

Plessey Radio Systems, Martin Road, West Leigh, Havant, Hampshire PO9 5DH, England.

6829.153

AUTOMATED COMPUTER ASSISTED ELECTRONIC WARFARE SYSTEM (ACEWS)

ACEWS, which was announced by Plessey in February 1988, is a land-based tactical EW intelligence gathering system designed to collect and analyse radio transmissions. It will be used by the British Army in West Germany and will be fully integrated with the existing Vampire transmitter locating stations.

ACEWS receivers acquire RF signals using fast signal processing techniques and data from the receivers is fed into a signal analysis processor where various transmission characteristics are determined. By analysing a number of parameters, a comprehensive picture of the transmission can be built up.

Position fixing of the emitters will be performed by the Vampire direction finding stations, with communications between the intercept complex and the DF stations being carried out by combat net radio. Bearing from each Vampire station is relayed to the intercept complex where the computer system computes the position. Each intercept complex operator can display information on his computer terminal and enter, via a keyboard, his comments. Intelligence analysts at the complex interrogate the data bases holding all the relevant information and the operator's comments and provide information to the various command headquarters on hostile formations and areas of operation.

The EW equipment at the intercept complex includes signal acquisition and analysis using optimised signal processing technology. Within the processing equipment ruggedised plasma displays are being used, together with audio facilities for recording, playback and intercom. The bulk storage device is a ruggedised mass memory disk drive unit with removable Winchester disks. As part of the overall programme the present Vampire stations will be updated and automated.

ACEWS will use the ADA programming language and the system will also employ elements of a new Plessey command, control and intelligence system known as GENERICS.

STATUS

No information is available on delivery dates but ACEWS is a six year development, delivery and field test programme commencing in December 1987.

CONTRACTOR

Plessey Defence Systems Ltd, Grange Road, Somerford, Christchurch, Dorset BH23 4JE, England.

3370.053

CERES COMMUNICATIONS MONITORING SYSTEM

Developed by Rediffusion Radio Systems Ltd in response to a British Government requirement, CERES (computer enhanced radio emission surveillance) provides for communications monitoring in the frequency range 15 kHz to 30 MHz.

A typical system provides monitoring facilities for a team of six operators, each equipped with a console for remote manual and computer control of four receivers and associated equipment, including antenna selection units and two four-channel special purpose tape recorders. Incorporated in the console is a visual tuning aid for each receiver, and audio selection and control circuits for each receiver and tape recorder. The receivers, antenna switching, receiver memories, minicomputer, tape recorders, time-code generator and a disc store, are located as remote equipment in an optimised environment. Automatic control of the remote equipment by the computer is made by means of a visual display unit, which is also used for the transcription of received information. Hard copy is provided by an associated needle printer sited in the console control room.

STATUS

Operational with the UK armed forces. No longer in production.

CONTRACTOR

Rediffusion Radio Systems Ltd, Newton Road, Crawley, West Sussex RH10 2PY, England.

3922.153

RA1792 PROGRAMMABLE RECEIVER

The Racal RA1792 is a programmable synthesised communications and surveillance receiver covering the frequency band 100 kHz to 30 MHz in 10 Hz steps. It was designed as a successor to the RA1770 series of HF receivers. There is an American-built version, designated RA6790, which has considerable commonality with the original UK designed equipment.

The receiver can be controlled directly from its own front panel or remotely using the MA1075 control unit which is visually identical to the RA1792. All controls operate in exactly the same way as those on the receiver front panel and three methods of connection are available:

(1) a simple extension cable, up to 100 m
(2) Racal serial control system (SCORE)
(3) two voice frequency telephone links (with modems).

Frequency setting over the entire range is available from either a keypad or a single, variable rate, flywheel tuning knob. Reception of USB, LSB, AM, CW and FM modes is provided as standard, with ISB operation available as an optional extra. FSK reception is possible by the use of an external modem. It is possible to store, for immediate recall, frequency and mode data on up to 100 channels, and up to 10 of these may be scanned automatically.

Two liquid crystal displays on the front panel indicate channel number, frequency, receiver operating conditions and mode, and include indications of audio line or RF signal level.

STATUS

In operational service but no longer in production.

CONTRACTOR

Racal Communications Ltd, Bracknell, Berkshire RG12 1RG, England.

Racal RA1792 HF receiver embodies novel circuit techniques

3919.153

RA1794 HF/VHF/UHF TACTICAL EW RECEIVER

The Racal RA1794 tactical receiver was designed as a basic block of Racal EW and radio surveillance system (**3918.193**), although it is also ideally suited to other communications EW applications.

The receiver incorporates an LSI synthesiser and microprocessor techniques to provide continuous tuning in 10 Hz steps over the entire coverage of 2 to 512 MHz. The synthesiser embodies a new development by Racal, called the Digiphase technique, whereby the synthesiser generates both the phase and frequency of a signal. It is this feature which enables the entire HF/VHF spectrum to be covered with a single phase-locked loop. The RA1794 is normally controlled by a separate operator control unit, although it may be operated as a stand-alone equipment. Traditional receiver front panel controls have largely been replaced by a multi-function key-pad which provides continuous tuning, at various rates, direct frequency entry, selection and setting of up to 100 frequency and mode channels, filter bandwidth, reception mode and other functions including an in-built test program. All display and metering functions are provided by an analogue/digital LED readout on the front panel.

STATUS

The RA1794 is in production and worldwide operational service.

Racal RA1794 tactical EW receiver

CONTRACTOR

Racal Communications Ltd, Bracknell, Berkshire RG12 1RG, England.

5679.153
RA1795 VHF/UHF RECEIVER

The RA1795 is a programmable synthesised receiver designed primarily for communications intelligence and electronic warfare applications in the frequency range 20 to 1000 MHz, and is suited for use in standard computer-based systems in both fixed and transportable roles. Tuning is in 10 Hz steps across the whole band with reception of AM, LSB, USB, FM, CW and pulse modes. It is fitted in a standard rack mounting unit or in a bench mounting cabinet.

RA1795 VHF/UHF receiver

The receiver uses microprocessor techniques to provide a wide range of operational capabilities combined with a high degree of RF performance. These techniques include automatic scanning of frequencies held in a 100-channel memory, comprehensive display of the receiver operating conditions and provision for external control. A wide selection of filters provides for the interception of all transmissions normally encountered in the VHF/UHF spectrum, including wide band radio relay systems. The 100-channel memory can be loaded with frequencies of interest together with associated receiver control settings ready for immediate access by means of a tuning knob or keypad. Automatic scanning of any of the channels can be carried out, the receiver pausing on each channel for a preset dwell time selected with the keypad.

The receiver can be operated by local, extended or remote control. In local mode, tuning is by means of a multi-purpose tuning knob with fast and slow rates. The fast tune provides an adaptive tuning rate which is varied automatically with the speed of rotation, the slow tune is at a fixed rate in 10 Hz steps. Alternatively, frequency can be entered directly by means of the keypad. The built-in carrier operator relay with adjustable threshold can be set to detect activity in any channel and to activate automatically an external system, eg a tape recorder. All receiver conditions are displayed by either illumination of relevant push-button controls or by an LED display. A compatible signal display unit, type MA1119C, is available and enables a spectrum of up to 8 MHz to be displayed.

STATUS
In production.

CONTRACTOR
Racal Communications Ltd, Bracknell, Berkshire RG12 1RG, England.

4737.053
RA1796 HF/VHF/UHF TACTICAL EW RECEIVER

The RA1796 is a rugged compact synthesised receiver intended for communications EW operation covering the frequency range from 2 to 1000 MHz in 10 Hz steps. This gives complete coverage of the HF, VHF and UHF communication bands. It employs microprocessor techniques to provide a wide range of capabilities, including automatic scanning of frequencies held in a 100 channel memory, automatic search across a specified frequency range and provision for external control.

The equipment is designed to operate under exacting environmental conditions and is suitable for deployment in forward combat areas installed in soft-skinned or armoured vehicles.

The RA1796 incorporates a frequency synthesiser based on a new LSI design of high reliability and performance, with microprocessor control and storage for 100 frequencies and associated control settings. Tuning is in 10 Hz steps across the complete band with reception of AM, LSB, USB, FM, CW and MCW modes.

A compatible signal display unit, the type MA1119, is available and enables a spectrum of up to 3 MHz to be displayed centred on the receiver tuned frequency.

CHARACTERISTICS
Frequency range: 2 – 1000 MHz in 10 Hz steps (2 – 512 MHz optional)
Reception modes: CW, MCW, AM, FM, USB, LSB
Power consumption: 50 W typical
Dimensions: 205 mm high × 390 mm wide × 400 mm deep
Weight: 21 kg

STATUS
In production and in service with a number of NATO and other countries.

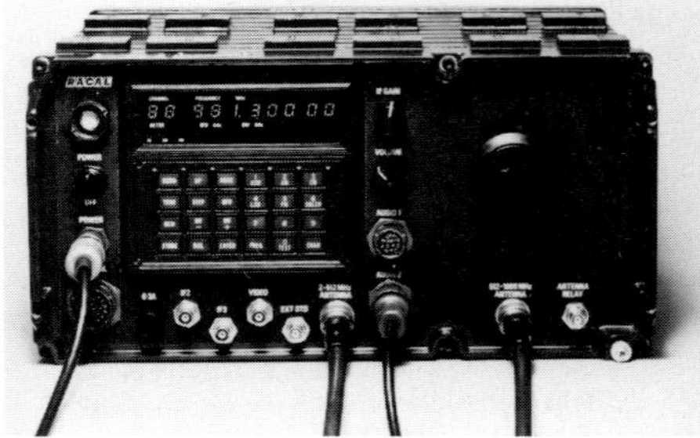

RA1796 receiver

CONTRACTOR
Racal Communications Ltd, Bracknell, Berkshire RG12 1RG, England.

6457.153
RA3700 SERIES MODULAR HF RECEIVERS

The RA3700 series are general purpose, high performance HF receivers covering the frequency range from 15 kHz to 40 MHz in selectable 10 Hz or 1 Hz steps. The range consists of single and dual equipments, the RA3701 and RA3703 being single and the RA3702 and RA3704 dual. The RA3701 and RA3702 have front panel controls while the RA3703 and RA3704 are remotely controlled.

These versatile receivers use microprocessor techniques to provide a wide range of facilities including automatic scanning of channels held in a 100 channel memory, or across a selected frequency range. The basic receivers operate in LSB, USB, AM, FM and CW modes, a synthesiser BFO being provided for CW. The AGC time constant and bandwidth are selected automatically when the mode is changed but may be altered subsequently by using the front panel keyboard. Seven bandwidths are fitted as standard and additional ones may be specified by the user, up to a maximum of 12.

STATUS
In production.

CONTRACTOR
Racal Communications Ltd, Bracknell, Berkshire RG12 1RG, England.

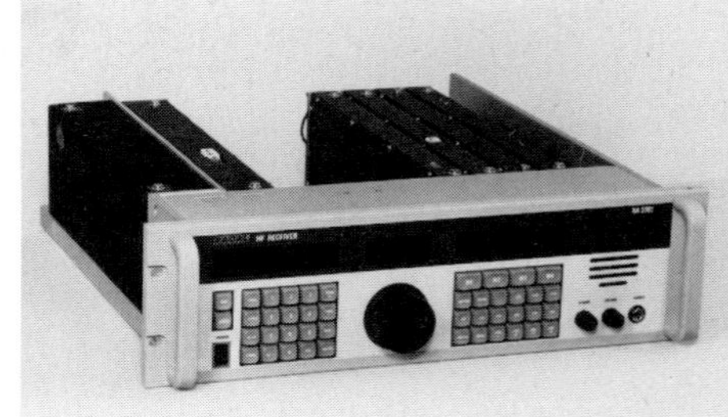

RA3701 HF receiver

3920.153
RACAL TACTICAL DIRECTION FINDING SYSTEMS

Racal radio direction finding systems have been designed principally to meet the needs of EW requirements for emitter location for tactical applications. They cover the HF/VHF/UHF bands up to 1 GHz. Fixed, transportable or mobile versions can be supplied.

The systems use electronically commutated Adcock four-element antenna arrays to generate a signal which is amplitude modulated with azimuth information. After conversion to digital form, this signal is fed to a microprocessor-based bearing evaluation and display unit (MA1110), which derives the bearing information and presents it directly in numerical form. A number of antennas are available covering the segments of the band up to 1 GHz. Up to three can be stacked vertically on a single mast.

Standard Racal communications/surveillance receivers are employed according to the frequency range(s) of interest.

Bearing evaluation and presentation is performed in all systems by the MA1110 processing and display unit. This has a numeric LED readout which is updated five times per second giving a direct indication of bearing. There is also a polar display which shows the bearing as a radial line, and is provided with an electronic cursor. The same display screen may be used for an alternative presentation which includes a dual histogram display in which the horizontal axis represents the bearing angle while the vertical axis shows a continuously accumulating record of bearing data on the selected frequency.

Three or more DF stations can be linked to form a locating net, which can be manually or automatically controlled.

STATUS
In production and in service worldwide.

CONTRACTOR
Racal Communications Ltd, Bracknell, Berkshire RG12 1RG, England.

Tactical direction finding equipment in mobile configuration deployed for operation

3918.193

RS3153 RADIO SURVEILLANCE EQUIPMENT

The RS3153 is a modular tactical ESM system which can be configured to meet various operational requirements.

Generally, an appropriate receiver from the Racal range will be used, both for intercept and DF, to suit the frequency coverage required; RA1792 (100 kHz to 30 MHz), RA1794 (2 to 512 MHz), RA1795 (20 MHz to 1 GHz), RA1796 (2 MHz to 1 GHz).

In the tactical communications intercept role the RS3153 station is built up using two receivers, the MA1115 antenna switching unit, and the MA1116 operator processor and audio switching unit, which are installed in an equipment rack. All facilities are remoted to an operating position, where the operator is provided with the MA1114 operator control unit, a MA1119 panoramic signal display unit and SRA108A cassette recorders. A number of RS3153 stations with intercom and data transfer facilities may be linked together. The MA1114 operator's control unit provides all control and display facilities, including control of receivers in manual or automatic search routines, control of recorders, selection of antennas, and exchange of data with other operators. The RS3153 may be installed in armoured and wheeled vehicles to meet the most demanding tactical environment.

To combine intercept and DF, an RS3153 compatible installation, called Racal Seeker, has been developed. It incorporates the RS3153 station and a DF processor unit (MA1110 or MA1122). Mounted in a Land Rover, it provides the facility to monitor and record transmissions of interest, to produce a single line bearing on target transmitters, and to communicate with a controller. When several Seekers work as a network, target position fixing becomes possible by plotting the intersection of DF bearings. Antenna masts, generators and other ancillary equipment is transported on a trailer. Many Racal Seekers are now in service worldwide, mostly operating in groups to provide a highly mobile EW network for field commanders.

The RS3153 is also upwardly compatible with the Racal Automated Communications EW System (RACEWS), described under entry **5800.153**.

STATUS

System 3000 and Racal Seeker equipments are in production and are in widespread operational service.

CONTRACTOR

Racal Communications Ltd, Bracknell, Berkshire RG12 1RG, England.

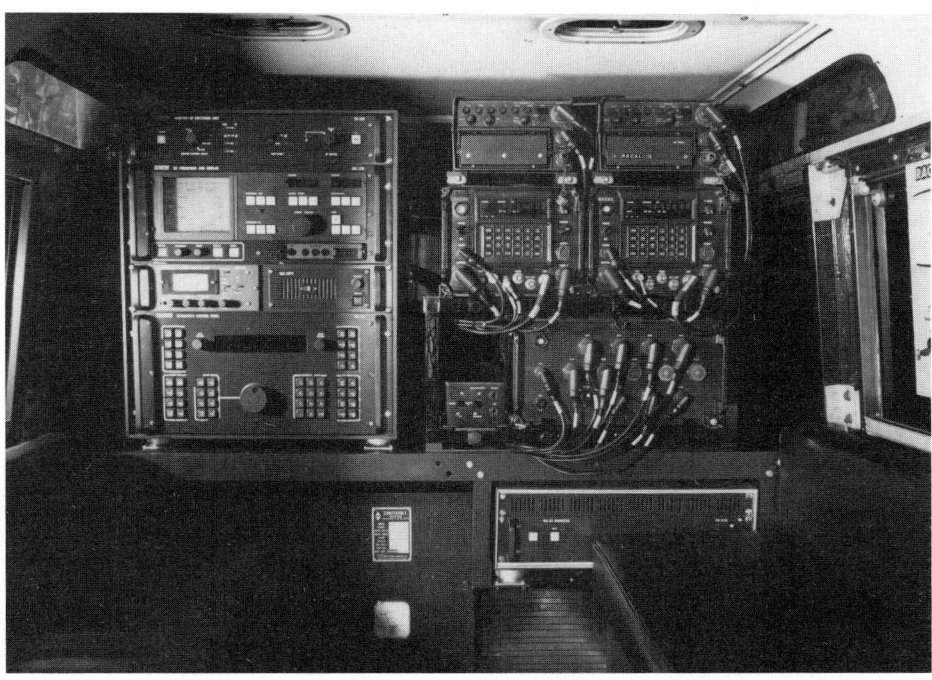

Racal Seeker installation with MA1110 DF processor on left and RS3153 intercept system on right

Racal Seeker with DF antenna

3917.193

RACJAM-V RESPONSIVE VHF JAMMERS

The RACJAM-V series of responsive VHF jammers has been designed to support ground forces in tactical operations. They are capable of rapid deployment and high mobility.

RJS3101

The RJS3101 (US nomenclature AN/TLQ-502) responsive VHF jammer has been designed to meet the requirements for an automatic mobile jammer system covering the 20 to 80 MHz band for ECM applications in the forward combat zone. It can automatically detect and jam signal activity on any one of 16 preselected target channels, on a priority basis. Since jamming only takes place when targets are active, it is possible to disrupt several nets simultaneously depending upon the activity of individual target nets. Alternatively, the RJS3101 can be programmed to scan a specified band of frequencies and disrupt any non-friendly transmissions detected above a preset threshold level.

Integral communications are included in the equipment with facilities for a remote ECM control station to call the ECM operator, if necessary, by interrupting the jamming action. The equipment is designed for mounting in either soft-skinned or armoured vehicles and is fully ruggedised.

The system is controlled from an MA1112 control and display unit which includes a keypad for entering instructions, an internal clock for timing jamming periods and intervals etc, and a comprehensive display for indicating channel information and signal activity. This microprocessor-based unit contains three types of channel store holding data required in the various modes of operation. The jammer channel stores up to 16 target frequencies, each with a priority rating, and a signal level above which jamming is to be initiated; the communications channel stores contain details of up to four frequencies for communications with the control station; and the protected channel stores hold details of up to nine friendly frequencies on which jamming is inhibited when the system is in the 'Band' mode.

RJS3100 100 W jammer installed in a helicopter

Two PRM4090 VHF transceivers are used, for search and jamming, one of them also acting as a communications receiver. In the transmit condition the output of one transceiver is fed into a PA1840 500 W amplifier. The output of the amplifier is then fed via the MA1019 antenna switch unit to a directional antenna which gives a 2 kW ERP signal. An optional whip antenna may be used for receive only.

An MA4280 modem links the jammer system with a printer and keyboard unit, and the MA1017 vehicle interface unit provides for the various interfaces with the other equipments in the system as well as housing the noise generator for modulating the jammer signal.

RJS3100

An alternative to the RJS3101 is provided by the RJS3100 system (US nomenclature AN/ULQ-19). This is a 100 W output system which uses whip antennas for both transmit and receive; otherwise operation is the same as for the 500 W system. The RJS3100 is suitable for mounting in small vehicles as a part load and is ideally suited for operations on the move close to the combat zone or for operations in an urban role. It is also useful as a training aid to give personnel experience of radio operating in the face of ECM. This jammer can also be supplied with a directional antenna to produce 400 W ERP. The RJS3100 is also available as a manpack equipment (3-man load) and is suitable for deployment in back-pack form with paratroops. Each pack weighs 22 kg approximately.

It is also available as a helicopter mounted installation, using a blade antenna and dismountable racks. The use of a helicopter considerably increases the effective range and mobility of the jammer.

RJS3102/RJS3103

These jammers are similar to the RJS3101 but with outputs of 200 W and 1 kW respectively.

RJS3100 RACJAM 100 W jammer in manpack configuration

STATUS

In production and in service in several countries.

CONTRACTOR

Racal Communications Ltd, Bracknell, Berkshire RG12 1RG, England.

4736.193

RJS3140 UNATTENDED EXPENDABLE JAMMER

The RJS3140 is a low power barrage jammer intended for remote deployment against communication nets operating between 20 and 90 MHz. It is a small, easily concealed unit for use against important targets which are out of range of stand-off jammers, and is deployed by hand. A number of them can be grouped together to increase the area of coverage as well as making location by DF more difficult. When used at close range it is claimed to be the simplest available technique for attacking frequency hopping radios.

The RJS3140 is a completely sealed unit which accepts control data via an optical link from the MA1065—a small microprocessor type unit with a 16-character display and keypad for program input.

CHARACTERISTICS

Frequency range: 20 to 90 MHz
Bandwidth: 1 to 31 MHz (set by a programming unit)
Power output: Typically 10W from 14V DC battery
Jamming signal: Pseudo-random continuous noise
Operation time: 2 hrs minimum (–10 to +60° C); 1 hour (–30 to –10° C)
Internal timer delay: 1 minute to 100 hours selectable by programming unit in 1 minute steps
Antenna: Telescope whip adjustable from 1 to 1.8 m
Dimensions: 80 mm high × 133 mm wide × 279 mm deep
Weight: 2.5 kg (including battery)

STATUS

In production.

CONTRACTOR

Racal Communications Ltd, Bracknell, Berkshire RG12 1RG, England.

RJS3140 expendable jammer and programming unit

5680.153

SAT 3311 HF SIGNAL ANALYSIS TERMINAL

The SAT 3311 terminal is a comprehensive aid to search and monitoring operations in the HF spectrum with full provision for technical analysis. It provides the search operator with an ability to detect and examine the structure of unusual or complex signals in addition to the more usual facilities provided by traditional monitor operational terminals. Equally it gives full facilities to the PTT monitor operator for investigating signal usage in conformity with CCIR regulations.

The terminal consists of an RA1792 receiver (**3922.153**) covering the frequency range 100 kHz to 30 MHz, an SA3300/2 digital signal processor unit and an SA3300/1 operator console and display unit mounted in a single standard cabinet or console. An instrumentation recorder can be included if off-line processing is required and an ASCII printer can be provided for hard copy production of teleprinter traffic.

The equipment incorporates a CRT which is used to display up to three traces simultaneously of any intercepted signal, from a selection of time waveforms and instantaneous and averaged spectra. The dual and triple trace displays can also include spectra which are recalled for comparison purposes from a stored library of up to 25 signal patterns relating to known types of signal or to unidentified transmissions still under investigation. In the case of FSK teleprinter transmissions, traffic content is displayed and printed automatically. With these displays the operator can determine target signal parameters including carrier frequency, type of modulation, the nature of the modulating signal and the type of information being transmitted. The spectra or time waveforms are presented together with alphanumeric notation including calculated signal parameters, message text derived from demodulated FSK signals, data concerning the operational status of the equipment and 'menus' for selecting the various operating functions. The maximum bandwidth displayed is 16 kHz and there is a 'zoom' facility to provide a range of bandwidths down to 20 Hz to permit examination of individual transmissions in detail.

STATUS

In production.

CONTRACTOR

Racal Communications Ltd, Bracknell, Berkshire RG12 1RG, England.

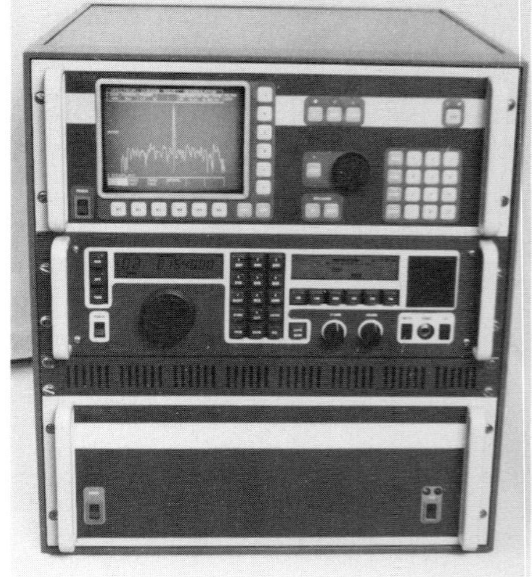

SAT 3311 HF analysis terminal

5800.153

RACEWS ELECTRONIC WARFARE SYSTEM

RACEWS (Racal Automated Communications Electronic Warfare System) has been developed to combine the existing well-established range of tactical EW equipments with small rugged computers. This brings to tactical EW systems the dual benefits of computer assisted information processing and the computer control of equipment, enabling greater exploitation of the target array and the production of more intelligence at faster speeds than is possible in manual operation. This is achieved by computer assistance for signal acquisition, intercept, position fixing, information processing and display, jammer control, and command and control.

The system approach ensures that these computer-assisted subsystems of RACEWS are compatible with the existing range of equipment, creating opportunities for the enhancement of these units where this is appropriate, and for a phased acquisition programme.

Signal acquisition is achieved rapidly and automatically by a computer-driven intercept receiver (RA 1794) which surveys the chosen frequency band rapidly in scan or list modes and reports signal activity. It displays the results of its search on a visual display unit.

Intercept is carried out using the RS3153. In RACEWS, the intercept operator is also equipped with a visual display terminal for access to the EW database.

Position fixing is used in support of intercept operations, and in RACEWS position fixing is automated. An intercept operator tasks DF outstations using his keyboard terminal. The task is passed through the intercept site computer system by radio to the DF outstations. Under computer control the outstation takes bearings automatically and returns the results to the intercept site, where two bearings are computed to fix the target position, enter it into the database, and display it to the intercept operator.

Information processing and display are essential in sorting, storing, retrieving and displaying data to produce intelligence quickly. The RACEWS computer system is designed for this task. Using proven, industry standard equipment, ruggedised and suited to the tactical environment, RACEWS provides considerable computing power at appropriate intercept sites and EW operations centres.

The database is designed for EW operations. The primary retrieval key is a net identifier, preserved over changes of frequency, location, call sign and other characteristics. Secondary keys permit access from other indices. Users have considerable choice of design and layout to suit individual requirements.

The EW database is accessed by the MA1127 visual display terminal, provided to all system operators. It is also used for message handling, the preparation of reports using the processing facility, and the remote control of equipment such as DF sensors or jammer outstations.

A key element in the computer system is the map display, which provides a topographical map of the operational area overlaid with tactical information that has been gathered by the sensors. Position information held in the database can be displayed on the map, permitting the operator to absorb a wide perspective of battlefield information quickly from data that has been flowing into the database.

Jammer control is effected by remotely controlling jammer outstations over net radio.

The command and control system is embedded in RACEWS to provide the following facilities:

(a) data processing and storage
(b) user defined distributed databases
(c) user defined display formats
(d) colour map graphics
(e) message handling
(f) remote equipment control
(g) communications
(h) system management

RACEWS is adaptable to a range of different operational requirements. It has a modular software and hardware architecture to allow flexibility of configuration and deployment. There is choice of EW equipment from the compatible Racal range, and the system can be installed in vehicles of various types and sizes.

STATUS

In production.

CONTRACTOR

Racal Communications Ltd, Western Road, Bracknell, Berkshire RG12 1RG, England.

RACEWS operator using MA1127 visual display terminal and linked map display

5801.153

MA 1122 TACTICAL DF PROCESSING AND DISPLAY UNIT

The MA 1122 has been designed for use in tactical electronic warfare DF systems operating primarily in the VHF/UHF range from 20 to 512 MHz. It can also be used in 'hunting' DF applications in which case the frequency range is extended down to 2 MHz.

Its primary use is with the Racal RA 1794 or 1796 VHF/UHF receivers and a range of DF antennas, including transportable VHF/UHF arrays and special low profile HF and VHF/UHF hunting antennas for mobile operations. Bearing indication is provided by a digital readout to 0.5 degree resolution, and a polar display using LEDs spaced at 10 degree intervals. A quality factor is computed for each bearing and a storage/recall facility allows the last 10 bearings to be displayed after the target signals have gone down. The unit is packaged for tactical use in wheeled or tracked vehicles.

CHARACTERISTICS

Dimensions: 206 × 257 × 412 mm

Weight: 16 kg

STATUS

In production.

CONTRACTOR

Racal Communications Ltd, Western Road, Bracknell, Berkshire RG12 1RG, England.

MA 1122 DF processor and display unit

5802.153

HUNTING DIRECTION FINDING SYSTEM

This system has been specially designed for mobile operations in the HF/VHF/UHF bands up to 512 MHz. It is intended for use by internal security forces or civil authorities to hunt down illegal transmitters and is particularly well suited for operation in urban areas. The system can be installed in a variety of vehicles and full provision has been made for concealment of the type of operation where secrecy is important.

The system is based on the MA1122 processing unit and a suitable Racal receiver, normally the RA 1796. The antenna design has been adapted for mobile applications. The system is used to home in on a target transmission by taking successive bearings while moving. The antennas are concealed for low profile operations. The system can also be used as part of a DF net in conjunction with other mobile vehicles or fixed DF sites.

STATUS

In production.

CONTRACTOR

Racal Communications Ltd, Western Road, Bracknell, Berkshire RG12 1RG, England.

5808.193

JAMCAT JAMMING UNIT

JAMCAT is an EW attachment designed to be used with any combat net radio system to provide a low-cost EW jamming capability. The unit is a dual purpose equipment equally usable for offensive operations or as a training aid. When connected to the radio, via the outlet normally used for the handset, JAMCAT converts it to a unit capable of generating a wide range of jamming signals. Internally generated white noise, single tones, rapidly changing multiple tones or externally generated signals may be used to modulate the transmitter of the radio.

In the offensive role, JAMCAT can be connected into the audio/keying input of most radio transmitters regardless of RF power output. In the training role the equipment can be set up to simulate ECM signals typical of those found on the battlefield.

Four modes of operation are available:

(1) manual mode uses the pressel switch of the associated handset to key the transmitter to produce jamming signals on demand
(2) continuous mode which produces signals without operator intervention

JAMCAT system in manpack configuration

(3) responsive mode which uses an internal signal-activated switch to key the transmitter for a short period when a signal is received during a 'look-through' period. Two rates of responsive jamming are available; 0.1 second delay followed by 2 second jamming, and 0.5 second delay followed by 4 seconds jamming
(4) Fast-look-through mode which simulates the effect of a fast-look-through responsive jammer but is not signal responsive.

The JAMCAT unit can be used in manpack or vehicle-mounted configuration and may be interfaced with any transceiver operated normally through a handset and pressel switch. Power is derived from a clip-on battery pack, or from the transceiver in use if DC is available through the audio interconnect lead.

CHARACTERISTICS

Modulation sources: White noise; continuous tone of 1 kHz sinewave or 500 Hz squarewave; random tones over frequency range 400 to 2400 Hz, changing every 128 milliseconds; external modulation of 0.5 rms into 20 k

Dimensions: 170 mm high × 140 mm wide × 70 mm deep (270 mm high with two Ah Ni-Cad batteries)

Weight: 1 kg (2.1 kg when fitted with batteries)

STATUS

In full production. In service with the Swedish and other defence forces.

CONTRACTOR

Racal-Comsec Ltd, Milford Industrial Estate, Tollgate Road, Salisbury, Wiltshire SP1 2GJ, England.

4162.193

TYPE S373 ECM EQUIPMENT

Under government contract, Marconi Radar has produced the Type S373 ECM system which is a multi-band EW equipment providing high-power jamming. The S373 vehicle also carries an autonomous surveillance system which will detect, locate and analyse the radiations of enemy emitters. It is assumed that S373 ECM vehicles are also employed for such tasks as monitoring the performance of friendly radars under jamming conditions and for training purposes.

The S373 antenna array consists of two horn-fed parabolic dish reflectors which are housed in a double-bubble glass-fibre radome, which is raised to its normal operating elevation by a hydraulic jack system. The two antennas can be turned at rates of up to 200 rpm by means of slab torque motors; these drive the arrays directly, eliminating gear boxes. This arrangement allows scan to be varied from 5 to 180° with positional accuracy of <0.1°, and electrical actuators enable elevation angles of –5 to +8° to be selected remotely. Antenna scan speed and positions in azimuth and elevation are displayed on the operator's console in the cabin. The latter consists of a one-ton container, which houses the electronic equipment and is mounted behind the antenna system on a flat-bed vehicle such as a four-ton Bedford truck.

Each jamming band has an independent high-power TWT (travelling wave tube) transmitter feeding its own antenna, and each transmitter incorporates a digitally controlled variable frequency oscillator (VFO), a driver stage and modulator with an output amplifier stage. A comprehensive range of modulation types and patterns is provided for both transmitters, including AM and wide band FM, which may be sinusoidal, linear swept or noise, pulse modulation over a wide range of pulse widths and duty cycles, and any reasonable combination of these.

Marconi S373 mobile ECM system on its four-ton truck, showing 'double-bubble' antenna housing in raised position

STATUS

No production information or service deployments have been revealed, but it is known to be in service with the British Army. More detailed information is given in previous editions of *Jane's Weapon Systems*.

CONTRACTOR

Marconi Radar Systems Ltd, Writtle Road, Chelmsford, Essex CM1 3BN, England.

4163.193

BARBICAN EW SYSTEM

Barbican is a tactical automated radar intercept, bearing, classification and analysis system for rapid detection of battlefield threat radar emitters.

A typical operational deployment of the Barbican system would consist of three or more ESM stations linked by voice and digital data links to one of the control stations which are the responsibility of Ferranti Computer Systems Ltd. The Ferranti control station of Barbican provides facilities for:

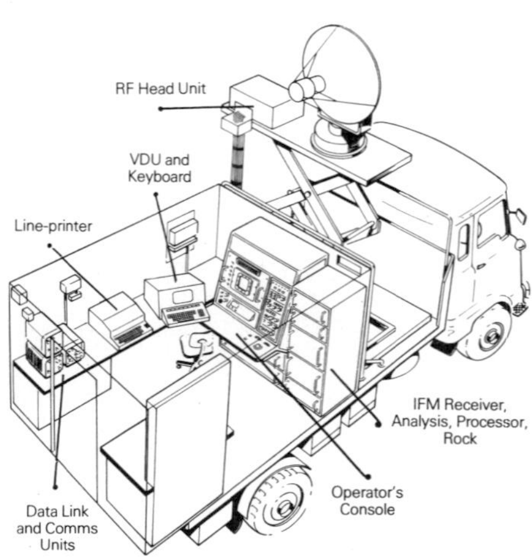

General arrangement drawing of MEL Barbican ESM station vehicle, showing equipment layout

(1) system control and co-ordination
(2) communications control
(3) location and classification of emitters
(4) data management

The arrangement of the MEL Barbican ESM vehicle equipment can be seen from the general arrangement drawing. Barbican system facilities include a high gain DF antenna; an automatic signal logger; a digital instantaneous frequency measuring receiver; digital random access display coupled with multiple display

Barbican EW system mounted on a 4-ton vehicle

options; VDU message formatting; compilation and presentation; hard copy printer; microprocessor control of system switching; video and audio outputs for operator signal analysis; and a digital data link. The frequency coverage is from 1 to 18 GHz with a 360° antenna azimuth cover.

Features of the system compatible with FEBA operations include both operating aspects and performance characteristics. The designed performance provides for high system sensitivity; high intercept probability; accurate DF; frequency measurement on a single pulse; and automatic analysis of radar parameters. One man can operate the Barbican ESM unit due to the comprehensive automatic facilities provided and the equipment is designed for rapid set-up and dismantling for redeployment. The rotating DF antenna, for example, is elevated by a simple scissors mechanism with the turning gear carried on what appears to be a self-levelling platform. Automation ensures rapid processing and interpretation of collected signal data, and system switching is also controlled by the MEL 99 microprocessor, resulting in a compact one-man ESM station providing real-time facilities for the following functions:

(1) signal detection
(2) signal sorting
(3) direction finding
(4) information display
(5) parameter analysis
(6) data formatting
(7) data transmission.

STATUS

The Barbican system is a joint private venture by MEL and Ferranti Computer Systems Ltd. No contract details have been released.

CONTRACTORS

MEL, Manor Royal, Crawley, Sussex RH10 2PZ, England.

Ferranti Computer Systems Ltd, Western Road, Bracknell, Berkshire RG12 1RA, England.

3912.193

MEL MODULAR EW PROGRAMME

MEL is known to be engaged in the development of a range of new technology ESM modules for land-based systems. These modules are designed to enable systems and sensors to be tailored to address a variety of operational requirements.

In one available configuration ESM stations are each able to act as control in a baseline deployment, thus removing the need for a separate control station.

The equipment employs advanced pulse train deinterleaving techniques which can operate automatically or under the operator's direction. A comprehensive data logging capability is also included.

Features of the system include accurate correlation of emitter reports, automatic triangulation, emitter density display for radar deployment analysis, full netting of the system and rapid tasking and steerage.

STATUS

No contract details have been released.

CONTRACTOR

MEL, Manor Royal, Crawley, Sussex RH10 2PZ, England.

5740.093

MEL IFM RECEIVERS

MEL manufactures a range of IFM receivers for use as frequency measuring devices, with digital readout, for use in ESM and ECM equipment. The receivers operate from 0.5 to 18 GHz and are primarily for application in microwave receiving systems where radar analysis pulse train de-interleaving is required on the basis of the frequency of the received signal.

They measure the frequency on a pulse-by-pulse basis for signal durations of 50 ns to CW. This is achieved without prior knowledge of the time of arrival or frequency of the signal; the receivers monitor the whole of the operating band simultaneously. There are two families of these receivers, binary and quaternary. Binary receivers provide high measurement integrity where the signal environment is complex and where high data rates are normal. Quaternary receivers are small, lighter and cheaper and are suitable for applications where measurement integrity requirements are less demanding.

Both families can be made to measure and identify CW signals, or to discriminate against them, while looking at pulsed signals. These pulsed signals can be above or below the simultaneous CW signal being received provided that they exceed the threshold of the receiver.

The receivers are available in various combinations of control functions, data interfaces and frequency range. Sampling facilities cover delayed sample, multi-sample and external sample. A digital coded output is provided with a 10, 11 or 12 bit resolution, the 12 bit versions giving improved frequency resolution and accuracy.

STATUS

In production.

CONTRACTOR

MEL, Manor Royal, Crawley, West Sussex RH10 2PZ, England.

6828.193

CORVUS ELINT STATION

Corvus is a new generation transportable ELINT collection system capable of detecting modern wide band radars in dense electromagnetic environments, even when co-sited with high power emitters. THORN EMI Electronics channelised receiver gives the equipment very high sensitivity so that it can detect and measure the deep sidelobes of emissions out to the radar horizon and beyond.

Channelisation gives a wide instantaneous bandwidth to cope with modern frequency agile emitters, yet the detection bandwidth is narrow enough to give measurement accuracy equivalent to that normally associated with narrow band ELINT receivers.

A typical equipment configuration consists of a tasking receiver and two technical ELINT receivers. All three contain their own channelised receiver and their own separate antenna.

The role of the tasking receiver is to intercept and identify all emitters in the environment, thereby providing the operators with a 'window on the world' from which they can select emitters of interest.

The role of the technical ELINT receivers is to concentrate on particular emitters of interest, as selected by the operators, enabling them to gather detailed information. This data is recorded on magnetic tape for subsequent in-depth analysis.

The complete system provides simultaneous general surveillance and detailed emitter analysis, can characterise jammers and builds its own library. The equipment is contained in a standard NATO RFI screened shelter and can be transported by road, a Chinook helicopter, or a C-130 aircraft.

STATUS

A demonstrator is available.

Cutaway view of Corvus ELINT station

CONTRACTOR

THORN EMI Electronics Ltd, Computer Systems Division, Wells, Somerset BA5 1AA, England.

6216.193

MS SERIES ELINT/ESM SYSTEMS

MSG, a division of THORN EMI Varian offers complete customised ELINT/ESM installations by providing and integrating system modules such as superheterodyne receivers, demodulators, displays, synthesised local oscillators, system controllers, pulse analysers, antennas and peripheral equipment. THORN EMI Varian is a UK company owned jointly by THORN EMI Electronics and Varian Associates of USA.

By using a modular approach with common interfacing, systems can be modified or upgraded as technology advances or as threat scenarios change. ELINT/ESM stations can be configured to operate anywhere in the range 0.1 to 40 GHz with bandwidths of 20 to 1000 MHz, depending on individual requirements.

A feature of the company's ELINT/ESM equipment is that intelligence is distributed amongst key modules to alleviate demand on the system controller, to minimise hardware, and to simplify interface links.

The software design capability ensures development of state-of-the-art microprocessor-based equipment.

As well as producing complete systems, the company has recently developed a complete portable ELINT/ESM system, the MS 3310, for identifying and characterising signals. Combining receiver, synthesiser, demodulator, and integrated display in one unit, it is stable enough to demodulate individual communications channels as well as acting as a general purpose surveillance receiver. Frequency coverage of the MS 3310 is 1 to 18 GHz (50 MHz to 40 GHz with optional converters), with a minimum RF bandwidth of 20 MHz.

STATUS

In production.

CONTRACTOR

THORN EMI Varian Ltd, 248 Blyth Road, Hayes, Middlesex UB3 1HN, England.

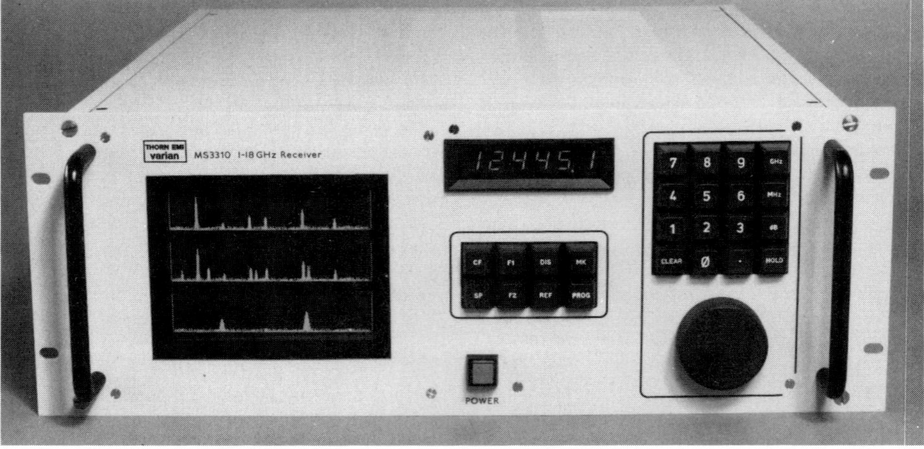

MS 3310 portable microwave receiver system

4411.093

RACAL EW PROGRAMMES

In addition to the various EW systems and equipment described in separate entries in this section, Racal is known to be engaged in a number of other developments in this area. These are mainly the subject of security constraints but the following summaries of certain of these projects can be revealed.

UAF-1

A new ESM system for the Royal Navy is being supplied under a UK MoD contract. The system, known as UAF-1, will ultimately succeed the UAA-1 Abbey Hill equipment in Royal Navy ships. No technical details are available.

ESM/ELINT Systems

Racal is involved in the development of a number of ESM/ELINT systems. These include a ground-based equipment for a central European country, the UAF-1 referred to earlier, a version of Cutlass for the Royal Navy's Type 23 frigates (see **3553.293**), and a submarine system for the Royal Navy Type 2400.

CONTRACTOR

Racal Radar Defence Systems Ltd, Davis Road, Chessington, Surrey KT9 1TB, England.

5669.153

WEASEL ESM AND ELINT COLLECTION SYSTEM

Weasel is a variant of Cutlass (**3553.293**), and is a combined electronic support measures (ESM) and electronic intelligence gathering (ELINT) system operating over the frequency range from 2 to 18 GHz with extension capabilities from 0.5 to 2 GHz and from 18 to 40 GHz. It is an advanced automatic system designed to operate in a dense radar environment and to achieve a low workload for a two-man crew operation.

The ESM part of the system acts as a search receiver with an instantaneous bandwidth equal to the full input range, thereby providing a virtual 100 per cent intercept probability. As a stand-alone equipment it can provide accurate direction finding and identification of intercepted emissions. This latter is achieved by measuring the emitter parameters and comparing them with data stored in a library of known emitter characteristics. This is a fully automatic process and the operator is presented with a tabular display of identification and threat significance. Selected digital information can be recorded, hard copied and transmitted to external stations. An operator can select emitters for close attention and hand over this task to the narrow bandwidth, tunable analysis receiver for attention by the analysis operator. The ESM thereby acts as a filter to reduce the workload on the analysis receiver.

The analysis receiver accepts digital data on any nominated emitter and automatically tunes the receiver to the correct frequency. The higher sensitivity and accuracy of the analysis receiver then permits detailed analysis of the signal. The data obtained can be used to update the library, create new entries and/or be recorded and transmitted to external stations. Both search and analysis receivers are capable of fully independent operation.

The system is designed for installation in a vehicle and for use at both fixed and temporary sites. The equipment is transported and operated in 19 inch (483 mm) special transport housings and is capable of withstanding rough handling and harsh environments.

The latest equipment in the Weasel range is TAC-Weasel which has been designed to meet a requirement for a lightweight, mobile, reliable and cost-effective ESM for use with ground forces, particularly for special task forces, border patrols, special missions, etc. The system covers the frequency range 0.7 to 18 GHz and combines the functions of surveillance and analysis receivers. Technical details of TAC-Weasel are similar to those of the Weasel system.

The TAC-Weasel system is able to be installed in a small vehicle, such as a Land Rover, with a support trailer used to carry the power supply unit. The complete system is air transportable and able to be underslung from helicopters. Although designed for vehicle installation the system can also be used in fixed or temporary sites.

The Cygnus shipborne jammer (**5670.293**) has been developed as a vehicle-mounted jammer to cover the frequency range 4 to 18 GHz.

STATUS

In production for an unspecified customer.

CONTRACTOR

Racal Radar Defence Systems Ltd, Davis Road, Chessington, Surrey KT9 1TB, England.

TAC Weasel installed in a Land Rover (antenna extended)

5674.193

SAVIOUR LASER WARNING SYSTEM

The Saviour system for armoured fighting vehicles reduces vehicle vulnerability to the numerous laser associated weapon threats on modern battlefields by providing the crew with timely and unambiguous threat warning, so enabling appropriate countermeasures to be taken.

Saviour sensor head on a Challenger MBT

The package provides a fighting vehicle with the equipment necessary to enable the commander and crew to receive an immediate aural warning that the vehicle is being illuminated by a laser rangefinder or designator. The commander and driver are also provided with an octantal indication of the direction from which the threat is coming and classification of the emission into rangefinder or designator. As an immediate countermeasure the system includes automatic initiation of the vehicle emergency smoke screen grenades.

The package is provided as an integrated system which includes sensors, power supply, processor and displays. It can be retrofitted easily. A radar warning receiver can be included in the Saviour system if required.

CHARACTERISTICS

Azimuth coverage: 360°, or as required

Elevation coverage: –22.5 to +90°

Wavelength coverage: 0.66 to 1.1 μm

STATUS

In production.

CONTRACTOR

Racal Radar Defence Systems Ltd, Davis Road, Chessington, Surrey KT9 1TB, England.

6372.193

MARCONI UNATTENDED EXPENDABLE JAMMER

The Marconi unattended expendable jammer (UEJ) is a low power barrage jammer designed for remote deployment against hostile communication nets operating in the frequency range 20 to 88 MHz. It is intended for hand-emplacement, either singly or in groups, in areas out of the range of stand-off jammers.

The jammer unit is controlled either locally by a programming unit, or remotely by an HF link from the command centre. Its operating parameters, centre frequency, bandwidth (selectable in 32 steps between 25 and 31 MHz), switch-on time, jam duration and hop mode (up to three centre frequencies and associated bandwidths) may be entered from the programming unit, either directly or indirectly. The jammer radiates a 10 W wide band signal for a minimum of six hours, either continuously or at controlled intervals over a 17 day period.

The HF link for remote control is provided by connecting the programming unit via an AM/HF transmitter to a data link receiver. This is a small HF receiver which demodulates the RF signal and sends the data up a land line, of up to 20 m length, to the jammer. Each jammer has a unique address that allows individual or groups of jammers to be addressed.

CONTRACTOR

Marconi Secure Radio, Brown's Lane, The Airport, Portsmouth, Hampshire PO3 5PH, England.

5736.053

BAe ESM SYSTEM

British Aerospace has designed a J/K band analysis receiving system covering the frequency band from 1 to 18 GHz. The design and development includes receiver and system controllers and a demodulator to handle a wide variety of signals. Both manual and automatic operation of the system will be possible.

STATUS

In batch production.

CONTRACTOR

British Aerospace, Naval and Electronic Systems Division, Downshire Way, Bracknell, Berkshire RG12 1QL, England.

5806.193

SCHERMULY INFRA-RED COUNTERMEASURES

Schermuly multiple-band screening smoke is a vehicle-discharged grenade that produces a large cloud of infra-red radiating material between the vehicle and the enemy position. The emission characteristics of the cloud give an effective screen through visual to the far infra-red frequencies. A salvo consists of twelve 66 mm smoke generators which provide a screen 5 metres in height at 15 to 25 metres from the vehicle, with a spread of 110 degrees giving a front 40 metres long. The screen is established within three seconds from firing using a combination of ground burning grenades and air-bursting submunitions. It lasts for about 40 seconds (IR) and 80 seconds (visual). This screen is claimed to be effective against thermal imagers operating in the 3 to 5 micron and 8 to 14 micron bands, optical and electro-optical devices, and laser rangefinders.

Pains-Wessex Schermuly also manufactures infra-red decoy flares for airborne use. They are available in various forms as listed below:

PW 118 Mk 1	1 × 1 × 8 inch	NSN 1370-99-967-2112
PW 218 Mk 1	2 × 1 × 8 inch	NSN 1370-99-967-1776
PW 218 Mk 2	2 × 1 × 8 inch (complete with impulse cartridge)	NSN 1370-99-967-6825
PW 11	Impulse cartridge	NSN 1377-99-967-2994
PW 21	Impulse cartridge	NSN 1377-99-967-2992

All decoy flares have undergone extensive environmental testing to UK MoD requirements, together with their associated impulse cartridge which are fully Rad Haz protected.

The flares are compatible with the AN/ALE-40 and M130 dispensers or the equivalent.

STATUS

In production and in service with the Royal Air Force and various NATO countries.

CONTRACTOR

Pains-Wessex Ltd, High Post, Salisbury, Wiltshire SP4 6AS, England.

UNITED STATES OF AMERICA

4385.153

AN/ULQ-14 RADAR COUNTERMEASURES

The AN/ULQ-14 is a ground-based radar countermeasures complex operated by the US Army in the 8.5 to 17 GHz frequency band and is designed to be the ground element of the AN/ALQ-143 airborne multi-target electronic warfare system. The system provides electronic countermeasures capability for all army echelons from division upwards for countering ground and air-based threats from radar systems. The primary targets are counter-mortar/counter-artillery radars, surveillance and target acquisition radars and other radar systems in line-of-sight propagation.

The system is housed in a standard US Army S250 shelter for use with a 1½ ton truck and has a data interface with the Bunker Ramo AGTELIS control processor (**4408.153**), the AN/ALQ-143 MULTEWS airborne systems, Quick Look II and the Emerson Electric AN/MSQ-103 Teampack mobile ESM system (**4405.193**). The capabilities of the system include the jamming of four to six signals over a 15 km range.

5834.193

AN/ULQ-19 TACTICAL COMMUNICATION JAMMER

The AN/ULQ-19 system is designed to operate from small armoured or soft-skinned vehicles, or can be installed in a helicopter. It covers the combat net radio frequency band 20 to 80 MHz in 12.5 KHz steps and normally provides 100 W of jamming over this range. Its automatic jamming capabilities include search for and jamming on a priority basis from a list of up to 16 preset channels, or in a specified band of frequencies. It has two variants, a V1 for vehicle installation with a whip antenna, which can operate while on the move, and a V2 which also includes a manually switched amplifier to boost the output to 400 W.

STATUS

In production.

CONTRACTOR

Racal Communications Inc, 5 Research Place, Rockville, Maryland 20850, USA.

1861.053

GTE EW ANTENNAS

The Government Systems Group of GTE has developed and produces a comprehensive range of special antennas for various electronic warfare applications. These are described in some detail under the above entry number in previous editions. Illustrations of typical examples, and summarised information, have been retained for the benefit of readers without access to earlier volumes of *Jane's Weapon Systems*.

AN-10 Broadband Horns

This is a range of microwave horn antennas, usually of the ridged type, and which often are provided with a dielectric phase-correcting lens in the flared section. Individual examples from the range have specific frequency coverage.

GTE AN-14 log periodic array

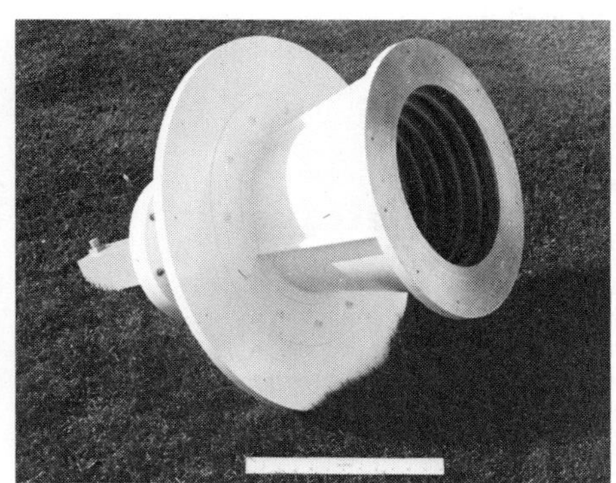

GTE AN-20 corrugated horn antenna

AN-13 High-gain Omni-directional Antennas
This range of omni-directional antennas has been developed for signal acquisition and sidelobe suppression applications. The basic construction is that of a bi-conical horn, with metal grids or meander lines in the aperture, yielding 45° slant linear or circular polarisation, respectively.

AN-14 Log Periodic Antennas
Log periodic antennas are linearly polarised structures covering very broad frequency bands with constant performance characterstics (ie frequency independent). They are directional antennas with medium gain, and are suitable for use with direction finding and electronic reconnaissance systems.

AN-15 Conical Log Spiral Antennas
These are produced by means of a three-dimensional printed-circuit technique for generating precision log spirals on dielectric cones. These conical log spirals maintain essentially constant radiation pattern performance with low polarisation axial ratio over several octaves of bandwidth.

AN-16 Cavity-backed Spiral Antenna
The AN-16 is a precision-backed, planar-spiral antenna covering a frequency band of 2 to 18 GHz.

AN-18 Dual-polarised Horn Antenna
A number of broadband, quadruple-ridged horn antennas have been developed covering frequencies in the range 50 MHz to 18 GHz. This series of antennas was designed for use in reconnaissance and surveillance of signals of unknown polarity.

AN-20 Corrugated Horns
Corrugated horns are used as (cassegrain or prime focus) feeds in reflectors to obtain maximum efficiency and gain. They are ideal as feeds because of their very symmetrical beams and low sidelobes.

The corrugated horn consists of a conical horn with corrugated interior walls. The corrugations act as 'chokes' to cause the current which would exist on the walls of a conventional horn to go to zero. These currents are the cause of high sidelobes and pattern deterioration in conventional antennas.

AN-23 Lindenblad Antennas
GTE has developed printed circuit board circularly polarised omni-directional Lindenblad antennas to cover octave frequency bands. Five different units are available in the 0.5 to 12 GHz range. They are produced as complete sealed units, with integral radome, and are suitable for airborne use in intercept/collection and other applications.

AN-27 Fan-beam Horns
The fan-beam horn design allows independent control of broad azimuth beams and narrow elevation beams, while providing low axial ratios over the entire beam. E- and H-plane beamwidth ratios of approximately 7:1 are characteristic of these horns.

AN-33 Septum Horns
The septum horns are pyramidal horns whose E-plane aperture distribution is compensated to provide sidelobe reduction in that plane.

AN-34 Conical Spiral Array
Under this designation a series of conical-spiral antenna arrays for broadband, high-accuracy tracking and direction finding is produced. Features of such arrays include circular polarisation, low polarisation axial ratio, frequency-independent performance, small bore-sight shift and phase and amplitude tracking among antennas.

AN-121 Rotating DF Antenna Systems
The AN-121 series of rotating DF antenna systems has been developed to meet a wide variety of DF and high gain monitor system applications using GTE Sylvania standard antennas, pedestals, radomes, and control units.

GTE AN-27 fan-beam horns

This monopulse tracking system has a bandwidth of 25:1 and provides both senses of circular polarisation. It consists of a log-periodic array feeding a 30 ft (9.4 m) parabolic reflector. Intercept capabilities are provided by the dually-polarised log-periodic antenna mounted on the side of the dish. A tracking accuracy of ±0.01 beamwidths is claimed

OA-9054(V)4/G quick-erecting antenna masts used with the Patriot air defence system

CHARACTERISTICS

	AN-20A	AN-20C	AN-20D	AN-20F	AN-20H
Frequency (GHz)	8 – 12	8 – 18	1.7 – 2.6	3.95 – 5.85	8.2 – 12.4
Beamwidth	30°	30°	28 – 17°	28 – 18°	30 – 18°
VSWR (max)	1.5:1	1.5:1	1.5:1	1.5:1	1.5:1
Gain (dBi)	15.5	15.5	15.2	15.2	15.2
Aperture (in)	4.75	4.75	27	11.8	5.6
Length (in)	2.4	2.4	37.5	16.4	7.6

Other models covering the bands of 12 – 18, 2.6 – 3.95, and 5.85 – 8.2 GHz are also available.

Magic Mast
GTE Government Systems has developed a family of quick erecting antenna masts to meet the demands of military requirements in the communications and electronic warfare environments. The Magic Mast family consists of a broad range of configurations from 10 – 28 m in height and intended for use mounted on a variety of wheeled and tracked vehicles. The masts are erected either hydraulically or pneumatically and complete operational deployment can be accomplished with one person in less than five minutes.

There are two hydraulic pumps and two air compressors in each system for redundancy. One of each is powered by 115 V 50 – 400 Hz, the other is by 24 V DC. A manual back-up is provided to allow safe lowering of the mast in the event of total AC and DC power loss. Weight of the complete system varies from 136 kg for the 11 m mast to 1418 kg for the 28 m one.

CONTRACTOR
GTE Government Systems, Tactical Electronic Defense Division, 100 Ferguson Drive, Mountain View, California 94039, USA.

4677.153
J-3400/J-3410 MOBILE ECM SYSTEMS
The J-3400 system is a mobile passive/active equipment designed for both signal intercept and jamming in the VHF frequency band. The complete system consists of transmitters, receivers, computers and an antenna system and is normally mounted on a tracked vehicle for mobility in practically any terrain. It is designed for one man operation and can be deployed within three minutes.

In the active mode the J-3400 will automatically jam up to four frequencies in a sequential time-shared mode. There is also a priority jamming mode which jams on a single priority signal whenever it is active. 1 kW (4 kW ERP) jamming power is provided.

In the passive mode the J-3400 functions as a signal intercept system. It continually sweeps all bands, or spot checks preprogrammed frequencies. In either case the receiver pauses on active signals for active analysis.

The J-3410 is an advanced model which offers more capability, including frequency coverage from 20 to 100 MHz and more jamming modes.

STATUS
In production.

CONTRACTOR
GTE Government Systems, 100 Ferguson Drive, Mountain View, California 94039, USA

3032.153
CIRCULARLY DISPOSED ANTENNA ARRAY (CDAA)

A CDAA is an antenna system which consists of a number of monopoles, or dipoles, uniformly spaced in front of a circular reflector that has its axis normal to the earth's surface. The antenna operates as a broadside array and is capable of providing several different types of beams to cover 360° in azimuth.

As the wave front of an emitted radio signal passes over the antenna array, each antenna element is excited at a different time. The output from each element is amplified and distributed by power dividers. The outputs of the element pairs that are symmetrically displaced with respect to the arriving wave front are combined. Each of the combined outputs is then delayed by an amount required to cause the outputs of those elements that are used to form a beam to be coherent in phase.

The summing of these equal phase outputs results in a highly directive beam. The directivity of the beam is a function of how many elements are combined.

During the past 20 years, GTE has designed, developed, and installed many CDAA systems which operate from HF through to the UHF ranges. The physically large antenna systems operate over bandwidths up to 10:1 and achieve DF accuracies that vary between 3 and 5°. These custom-designed systems were developed under government or company-sponsored funding to meet specific requirements. Delivered fully operational, and based on state-of-the-art information, the systems meet or exceed all performance requirements.

Hardware developments over the years have greatly expanded the capabilities of this antenna system. The development by GTE of low NF and high dynamic range preamplifiers led to a large improvement in the system sensitivity. Development of low loss beam forming networks and power combining hybrids provides the capability to form high gain antenna beams simultaneously present over the full 360° azimuth. An arrangement of high dynamic range amplifiers followed by power dividers, patch panels, and selector switches enables the operators to have unrestricted access to each antenna beam.

GTE circularly disposed antenna array (CDAA)

GTE and others have also developed new hardware which permits extension of the frequency range of CDAA systems.

CONTRACTOR
GTE Government Systems, 100 Ferguson Drive, Mountain View, California 94039, USA.

4386.093
SPERRY ANTENNA SYSTEMS EQUIPMENT

Sperry has developed and produces a wide range of special antennas and ancillary equipment for electronic warfare applications. The salient features of major examples are listed below.

Single-axis Antenna Assembly

This is a single-axis antenna designed to operate in severe environmental conditions extending from ocean surface vehicles to high performance aircraft. The primary application is for use as part of an airborne commmand and telemetry link, where maximum gain through beam stabilisation is required to maintain command and control between the aircraft and a ground control station.

Polarisation: Right hand circular
Frequency: 10 – 10.4 GHz
Beamwidth: 18° at 2 dB points

AN/UPQ-3A Steerable Antenna

The AN/UPQ-3A steerable antenna is a versatile directional system designed to operate in severe environments from surface vessels to high altitude airborne vehicles. The primary application is for use as part of an airborne command and telemetry link (as with the single-axis antenna described above).

The antenna is steerable in both azimuth and elevation. Steering is accomplished by the antenna receiving digital pulses, each corresponding to small antenna movements. The antenna will output two three-wire synchro angles representing elevation and azimuth position of the horn relative to its mounting base.

Two-axis Steerable Antenna

This is a directional antenna system with applications identical to the previous item. It consists of two major units: the antenna assembly containing the horn, drive mechanism and position sensing means, and the controller assembly containing programmable circuitry for directional control and three-axis stabilisation compensation. The antenna assembly is normally installed in radomes 12 inches in diameter or less. The mounting base can be supplied with a pressurisation seal for installation through the airframe structure.

Two-axis Corrugated Horn Antenna

This lightweight compact airborne antenna is designed specifically for use where the antenna is located close to other radars or other microwave systems. The corrugated horn design provides low backlobes and minimises interference with other sensors. It is also ideal for installation in pods or mounted in remote locations on the aircraft. Used in tandem with suitable controller and RF switching, this unit can maintain constant link-up with directional ground trackers even when mounted on high performance aircraft.

The antenna is steerable in both azimuth and elevation. Steering is accomplished by the antenna receiving differential TTL digital pulses corresponding to rates up to 60°/s in azimuth and 30°/s in elevation, each corresponding to small antenna movements. The antenna outputs two three-wire synchro angles representing elevation and azimuth position of the horn relative to its mounting base.

Frequency range: 16.7 – 17.7 GHz
Beamwidth: 20° horizontal; 22° elevation
Polarisation: Linear. Circular, right hand or left hand
Azimuth: 360° steerable continuously
Elevation: +45 to –85°
Rate: 30°/s, either axis
Position accuracy: 1.75° at max slew rate in either axis

Multi-band Tracking Antenna

The multi-band tracking antenna is designed for line-of-sight aircraft tracking and control. It will operate in severe environmental conditions and has been tested to military specifications. Two electronic servo control packages are available to operate the antenna. One requires operator presence within 100 feet of the antenna; the other allows remote antenna control from up to 1320 feet away. The servo motor drive uses high-efficiency, pulse-width modulation techniques. The small power ampifiers require no external cooling.

A single-axis tracking feature provides for tracking in areas where multi-path has a tendency to distort the elevation components of spin modulation. The system design permits mobile, fixed or shipborne installations.

Mobile Ground Tracker Terminal

The mobile ground tracker terminal is a rugged multi-band antenna system designed for operation in severe environments. It employs a servo drive using high-efficiency pulse-width modulation techniques to ensure fast slew tracking. Integrated levelling jacks allow rapid set-up in difficult terrain. It can be located up to 1320 feet away from the ground control system to safeguard operator personnel from attacks. The parabolic antenna dish folds down into a stow position and interconnect cables are stowed in the fixed reels to permit air transport.

The ground mobile tracker terminal consists of three major assemblies: the M-390 trailer with levelling system and stowage equipment, the pedestal antenna system, and the electronics enclosure with the antenna control and RF interface electronics.

Frequency range: C, X or Ku
Beamwidth: 1 – 2.25°
Polarisation: Circular, right hand
Slew rates: 30°/s (azimuth and elevation)

125-watt TWT Amplifier

The 125-watt TWT amplifier assembly is a single-unit TWT and power supply designed to withstand the severe environments associated with military aircraft and tactical ground installations. The primary application is a high power, airborne or ground digital data or imagery transmitter.

The assembly consists of three major units: the TWT amplifier providing 150 W (X-band) of CW output power, a solid-state power supply and a built-in test system.

Frequency: X or Ku band
Output power: 150 W

70-watt TWT Amplifier

This consists of two major assemblies: the TWT amplifier providing 70 W (X-band) of output power and the solid-state power supply. As an optional feature this unit is also available with 35 or 50 W power output, and in C or Ku band frequencies with the power output varying slightly with frequency.

Spread Spectrum Modem

The Spread Spectrum Modem has applications as an interface compatible with the AN/UPQ-3(A) data link, as part of an airborne command system, as part of ground-to-air/ground-to-ground/air-to-air data link and is also applicable to reconnaissance use in a jamming environment.

The equipment uses a direct sequence modulated FM carrier at the specified signal levels. The received signal is collapsed to an FM modulated IF by code tracking the PN sequence and multiplying the received signals by the code estimate. The resultant FM signal is phase-lock loop detected and the baseband data is provided at two output ports of the receiver.

Frequency: X or Ku band
Modulation type: FM

CONTRACTOR
Sperry Corporation, 640 North 2200 West, Salt Lake City, Utah 84116, USA.

4387.153

TACTICAL ELECTRONIC RECONNAISSANCE (TEREC) REMOTE TERMINAL (TRT)

The Tactical Electronic Reconnaissance (TEREC) Remote Terminal (TRT) is a collection of equipment, computers and procedures which forms a simple-to-operate portable terminal for tactical operations centres. The system is able to receive, process and report TEREC radar emitter location reports in near-real-time without reliance on ground communications. The TRT accepts voice and data information from the TEREC (AN/ALQ-125) equipped airborne sensor platform and consists of a computer, decryption equipment, a data demodulator, HF and UHF receivers, antennas, and a keyboard and printer.

The radio communications, data processing, keyboard and printing equipment are contained in portable containers which can be transported rapidly (via land, sea or air) and deployed quickly into full operational capability at the operational site. The equipment is fully ruggedised for military use, and can operate from a variety of power sources, including 120 V, 60/400 Hz single phase and 220/240 V, 50 Hz single phase.

STATUS

In operational service.

CONTRACTOR

Texas Instruments Inc, PO Box 405, M/S 3400, Lewisville, Texas 75067, USA.

3031.053

GTE COMPUTER CONTROLLED RECEIVING SYSTEMS

Since mid-1960, GTE, California, has been developing receiving systems designed around modern computers to meet these complex challenges; many of these systems are now in field operation. These systems, deployed in ground based, air, naval, and mobile applications, are divided into two areas: radar receiving systems and communication receiving systems.

Radar Receiving Systems

Typical radar receiving systems consist of five sub-systems: antennas, receivers, displays, signal digitisers, and the computer and its peripherals. In operation the system performs five major functions: monitor, signal acquisition, identification, cataloguing, and reaction.

The computer controls the system's antennas and receivers in a programmed pattern as directed by the operator. Incoming signals are digitised and routed to the computer where they are sorted, identified, and catalogued according to their measured parameters.

Communication Receiving System

Communication receiving systems are divided into eight major sub-systems including: antennas, receivers, demultiplexers, mode analysers, activity detectors, recorders, distribution equipment, and the computer and its peripherals. Incoming communications signals are separated by the system's demultiplexers. The mode analyser determines signal modulation, and passes this information to the computer. The activity detector informs the computer when communications activity takes place. The computer uses this information to control the communications system, routeing signals via the distribution equipment to various recorders and operators, and to maintain activity files.

GTE computer-controlled receiving system

STATUS

Operational.

CONTRACTOR

GTE Government Systems, Western Division, 100 Ferguson Drive, Mountain View, California 94039, USA.

6844.193

RWR 1000 HAND-HELD RADAR DETECTOR

The RWR 1000 radar warning equipment is designed to alert combat troops before they can be detected by anti-personnel battlefield radar. It is a lightweight unit (3.8 kg) and operates over the frequency range 8 to 19 GHz. The unit analyses the pulsed radar signals and presents them in digital format on a LED display, which indicates reception of CW or pulsed radar signals, the direction and polarisation, pulse interval time and frequency band. An audio alarm is also activated.

The complete RWR 1000 consists of a wide band receiver, a signal processing unit, a display unit and a control panel. It will operate continuously for four hours with a 12-volt battery.

CONTRACTOR

GTE Government Systems, Western Division, 100 Ferguson Drive, Mountain View, California 94039, USA.

3105.093

MPI-5A MINIATURE PULSE ANALYSER

The MPI-5A is designed as a militarised display to allow operator analysis of signals as detected in radar environments. The unit is a reliable, miniaturised, solid-state, modular replacement for the AN/ULA-2 with improved characteristics and additional features. It can be used to perform the following:

(1) pulse width measurement
(2) pulse repetition frequency measurement
(3) pulse risetime measurement
(4) signal envelope analysis
(5) pulse signature analysis
(6) signal sorting
(7) frequency band activity monitor
(8) audio monitoring of video signal.

The MPI-5A is a five-gun CRT display, each successive sweep being 10 times slower than the preceding sweep. The sweeps are triggered when the incoming pulse exceeds a threshold. With the use of delay lines, the entire incoming pulse is displayed. The fifth trace can be switched to a panoramic mode which provides a built-in frequency band monitor. An internal calibrator is provided to permit accurate calibration of the display to its scale. An audio output is also provided for audio monitoring or recording of the video signal.

STATUS

No details of applications or deployment have been disclosed.

CONTRACTOR

American Electronic Laboratories Inc, 305 Richardson Road, Lansdale, Philadelphia 19446, USA.

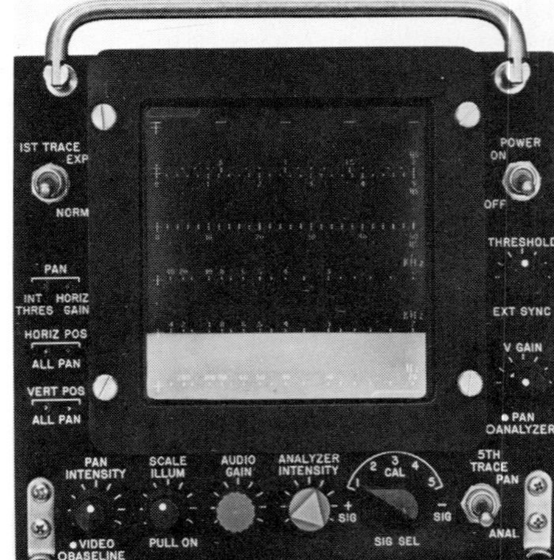

Operating controls and display of MPI-5A miniature pulse analyser

5816.053

R-3030/3050/3080 SURVEILLANCE RECEIVERS

The R-3030 is a new series of lightweight surveillance receivers featuring two completely independent receivers in a rack mount which would normally hold a single unit. Features of the R-3030 including tuning in less than eight milliseconds over the full 5 kHz to 30 MHz range and a modular building block design which makes it easier to meet special requirements for data bus connectors, bandwidths, AGC settings, etc.

The R-3030 provides 100 memory channels, IEEE-488 or RS-232 bus connector modules, EMI/EMC shielding, five bandwidths, five operating modes, versatile scan and sweep modes and a low power requirement.

The R-3050 is similar to the R-3030 in most characteristics, but consists of a single receiver with independent sideband capability. The receiver also includes a built-in front panel speaker.

The R-3080 is a single receiver in a half-rack mount configuration. Otherwise specifications and operation are identical to the R-3030.

CHARACTERISTICS
Frequency range: 5 kHz to 30 MHz
Operating modes: LSB, USB, AM, FM and CW
Bandwidth: 5 bandwidths (0.5 – 8 kHz)
Power input: Approximately 35 W
Dimensions: The dual system fits into a 25 × 19 inch (635 × 482 mm) rack
Weight: 18.1 kg

STATUS
In service with the US Navy.

CONTRACTOR
Cubic Communications Inc, Oceanside, California, USA.

3664.093
WJ-945 RECEIVING/DF SYSTEM

The WJ-945 receiving/DF system is a sophisticated microwave receiving system covering the entire 0.5 to 18 GHz frequency range. The system provides combined control and display functions for a high sensitivity superheterodyne receiver and a servo-controlled rotary DF antenna. The WJ-945 system is configured to minimise the total number of components, offering maximum operator control and display flexibility in the minimum possible equipment rack space.

The system configuration provides each operator with a receiver/DF control unit and two storage type CRT display monitors. Both CRTs in the standard system design are installed directly above the receiver/DF control unit to optimise operation and observation of the system. The microwave tuners used in the WJ-945 system are miniature WJ-940 tuners (described in **3352.053** in 1979-80 and earlier editions). These are mounted in groups of three in tuner enclosure/junction units. Two tuner enclosure/junction units, housing six WJ-940 tuners, allow coverage of the 0.5 to 18 GHz frequency range.

STATUS
Production.

CONTRACTOR
Watkins-Johnson Company, 2525 North First Street, San Jose, California 95131, USA.

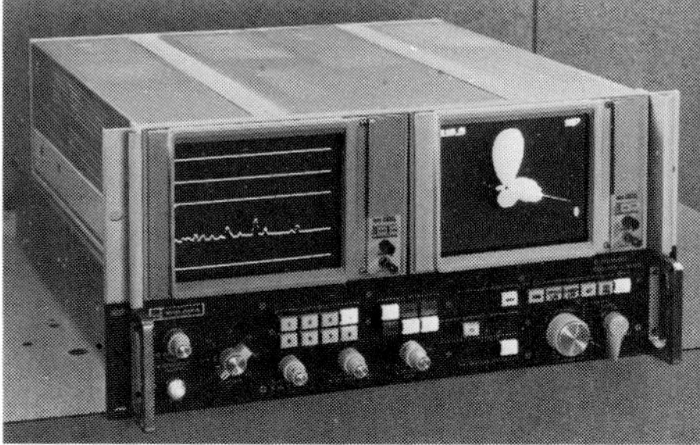

Watkins-Johnson WJ-945 receiving/DF system

4391.053
C-204/WJ-1140 ANALYSIS CONTROL UNIT

The C-204/WJ-1140 analysis control unit (ACU) performs three major receiving system functions: operator interface and system control, automatic signal parameter measurement and automatic signal identification. The ACU is a compact unit that replaces four separate units required in older receiving systems (receiver control, video processing, digital processing and alphanumeric display units). The ACU's automatic operating modes decrease signal identification time and operator training requirements.

The ACU's electronics contain a high-speed microprocessor that controls the unit and performs signal analysis. Under firmware control, the microprocessor compares received signals to a user provided signal library and outputs an alphanumeric code representing the signal source. Signal parameters are derived from time of arrival (TOA), amplitude, pulse width, and angle of arrival (AOA) data.

STATUS
In production.

CONTRACTOR
Watkins-Johnson Company, 2525 North First Street, San Jose, California 95131, USA.

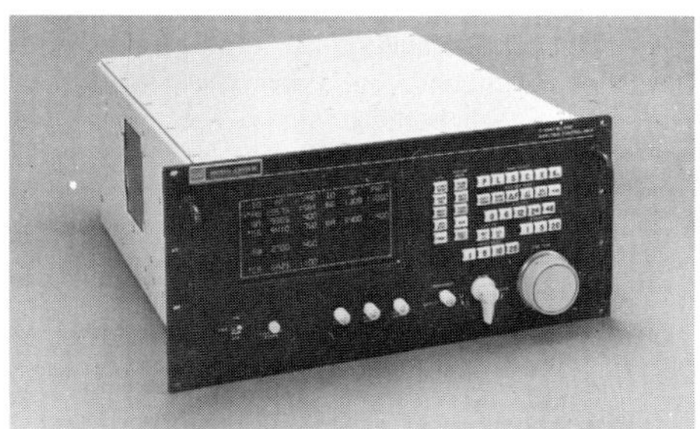

C-204/WJ-1140 analysis control unit

4392.093
WJ-1920 MULTIPARAMETER DISTRIBUTED PROCESSING SYSTEM

The WJ-1920 multiparameter distributed processing (MDP) system, combines superhet and wide band technologies in a unique way to build a wide open receiver that provides instantaneous intercept and recognition of signals in the 0.5 to 40 GHz range. The WJ-1920 system is built upon three functional units: the IFM, the superhet, and the display/filter sub-systems. In the instantaneous frequency measurement unit, signals from an IFM or other broadband receiver channel are processed through a WJ-1921 pulse interval processor (PIP) and tagged in real-time according to their PRI characteristics. This PIP is central to the entire multiparameter distributed processing concept and uses a newly developed PRI time domain transform algorithm which allows a receiver to function in the PRI domain as a channelised receiver does in the frequency domain.

In addition to signal parameters, the WJ-1920 system can incorporate other parameters into its sorting matrix to enhance monopulse characterisation and measurement capabilities (eg, pulse width, pulse amplitude, pulse direction of arrival) and even more subtle characteristics, such as pulse risetime, or pulse shape characteristics. This additional data can then be screened by the WJ-1234 (ACU) channel, displayed on the multiparameter display, or diverted for further automatic processing.

STATUS
In production.

CONTRACTOR
Watkins-Johnson Company, 2525 North First Street, San Jose, California 95131, USA.

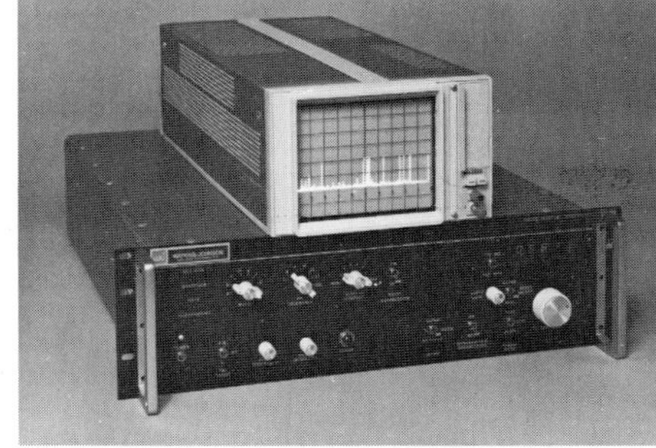

WJ-1921 pulse interval procesor

3665.093
WJ-1205B VIDEO DIGITISER UNIT (VDU)

The WJ-1205B video digitiser unit (VDU) is a microprocessor-controlled, high-speed pulse train analyser with a pulse sample capacity of up to 512 pulses. These pulses may be digitised for analysis of pulse characteristics – pulsewidth, pulse repetition interval (PRI), and average pulse amplitude. The WJ-1205B furnishes average PRI screening for interpreting complex pulse trains. A visual indication of these pulse rates is provided on the front panel of the VDU, along with operator controls. A self-test capability for unit checkout is also provided. Auxiliary equipment, such as an external computer, or line printer may be connected through a single rear-panel connector.

The WJ-1205B may be operated with its own internal, temperature-controlled crystal oscillator or phase-locked to a reference frequency provided by an external signal source (video tape recorder). Front end signal sampling may be gated internally by program control or by an external signal source (video disc). The VDU permits precise pulse-by-pulse analysis of modulated signals as well as other complex-grouped signals. To facilitate the de-interleaving of multiple pulse trains, the unit analyses all PRIs present in the video input stream and can compute up to eight independent PRIs that might exist. Pulse amplitude measurements permit additional analysis of the received signals.

STATUS
Production.

CONTRACTOR
Watkins-Johnson Company, 2525 North First Street, San Jose, California 95131, USA.

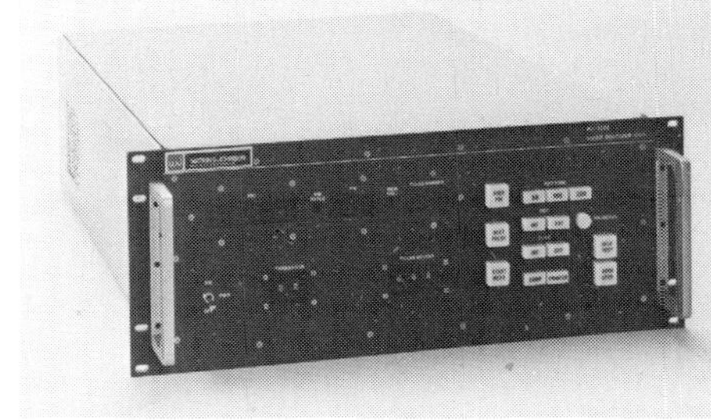

Watkins-Johnson WJ-1205B video digitiser unit (VDU)

4390.053
WJ-927 ESM EQUIPMENT

The WJ-927 system is a highly versatile microwave DF and signal analysis system which provides carry-on ESM capability for airborne, as well as shipboard, mobile, and fixed-site applications. It is small and lightweight, and consumes little power, while attaining high sensitivity and overall exceptional performance characteristics (5 to 18 GHz; expansible to 40 GHz coverage).

The WJ-1234 analysis control unit (ACU), the heart of the WJ-927 system, provides both flexibility and data handling capability previously available only in larger and more complex units. Highly specialised control and analysis functions are performed automatically by the ACU through the use of its solid state base and microprocessor control technology. Specific parameters of signals of known interest can be programmed into the unit. Used as a basis of comparison, this data provides the operator with quick identification of detected signals that match these parameters. In addition, its automatic signal analysis techniques allow the ACU to perform specific or directed search processes. In this mode of operation, the ACU enables only information on detected signals that constitute a match to be displayed to the operator.

In addition, the ACU offers disc and CRT options which allow on-line modifications of the user's library file. Several other interface options are also available to allow integration of the ACU with other on-board systems or units such as the line printer, magnetic tape recorders, and time code reader interfaces, as well as video and IF outputs for analogue recording of signals of interest.

The antenna sub-system of the WJ-927 consists of a direction-finding antenna, an omni-directional antenna, a control/display unit, and an antenna pedestal.

Watkins-Johnson has developed a DF antenna with a fan-beam pattern that yields narrow azimuth beamwidths for accurate direction-finding and broad elevation beamwidth to enhance the probability of detecting signals independent of aircraft altitude. Antenna DF accuracy is $\pm 1°$ rms over the full band and not greater than $\pm 5°$ absolute at the low end of the band.

The WJ-8535-11 omni-directional antenna which has been chosen for this application gives excellent performance in the 0.5 to 18 GHz frequency range with a deviation from omni-directional of ± 3 dB over 90 per cent of the band. The antenna is available with vertical or slant linear polarisation. The antenna itself is 14 cm high with a 14.7 cm diameter base and is mounted inside a low-loss dielectric radome.

The EC-60 control/display unit is the only unit available which provides all control functions (including sector scan) and all display functions, while also containing all of the servo modules necessary for the pedestal.

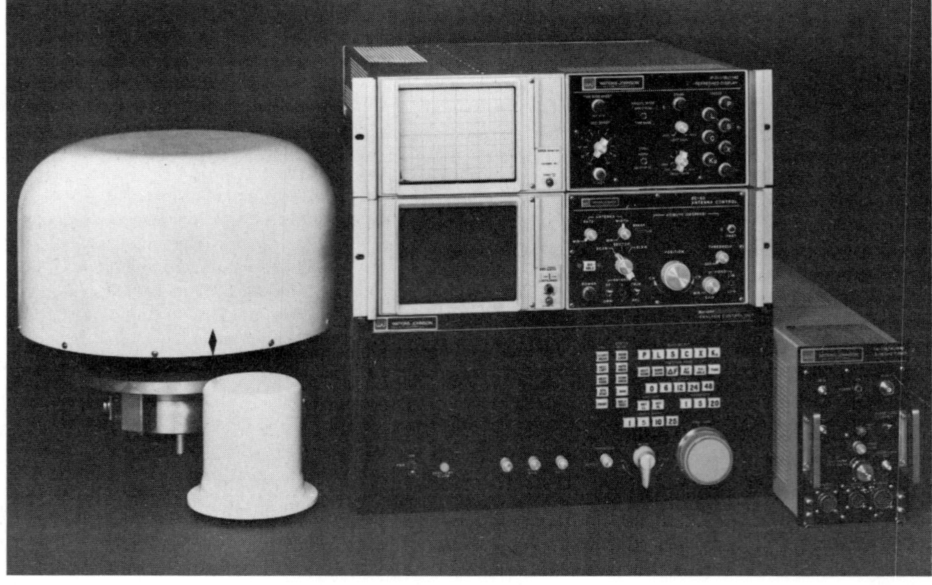

WJ-927 signal analysis equipment

STATUS
In production.

CONTRACTOR
Watkins-Johnson Company, 2525 North First Street, San Jose, California 95131, USA.

4393.053
WJ-1988 AMPLITUDE MONOPULSE DF PROCESSOR

The WJ-1988 amplitude monopulse DF processor is an economical direction finding sub-system designed to be used with various types of static antenna arrays within tactical and strategic ESM applications. By enabling a single wide band tuner to simultaneously process the DF outputs of each antenna within an array, the WJ-1988 eliminates the need for multiple tuners/receivers.

WJ-1988 amplitude monopulse DF processor

The WJ-1988 equipment provides processing of five channels of simultaneous RF information (one omni-antenna and four sector antennas) over the frequency range of 2 to 18 GHz. It consists of an active multiplexer (a small set of components located directly at the antennas), an MD-200 DF processor unit, and a C-100 control and display unit. Antennas and the receiver are separately required and are dependent on the user requirements. Because the composite signal after encoding is a wide band signal (500 MHz bandwidth), a wide band receiver such as the Watkins-Johnson Company TN-1000 MINCEIVER is recommended.

When used in combination with standard wide band receivers, the WJ-1988 develops a monochannel processing technique, which is made possible by a special bi-phase encoding method which the WJ-1988 uses to 'tag' or encode antenna signals. Bi-phase encoding is accomplished by spreading the original spectral power to produce a null at the original signal frequency. This encoding procedure prevents interferometry effects from occurring in signal processing when signals are combined for RF summation, and it also results in an improvement upon the maintenance of original signal integrity during signal processing.

CHARACTERISTICS
Number of channels: 5 (4 DF and 1 omni-channel)
Frequency range: 2 – 18 GHz
DF sensitivity: –78 dBm (typical antennas)
Omni sensitivity: –83 dBm (typical antenna)
Dynamic range: 60 dB minimum
DF accuracy: 4° rms
Instantaneous bandwidth: 5 MHz
Weight: MD-200 16 kg; C-100 and display 13.6 kg

STATUS
In production.

CONTRACTOR
Watkins-Johnson Company, 2525 North First Street, San Jose, California 95131, USA.

1381.053
L-6/A-WJ-1140 MICROWAVE DIRECTION FINDING SYSTEM

This is an integrated microwave direction finding (DF) system, produced by the combination of the L-6/A broadband DF antenna system and the WJ-1140 modular microwave receiving system.

The broadband DF antenna system (L-6/A) is designed to function as a component of an electro-mechanical reconnaissance system when used in conjunction with the WJ-1140 receiver system. The L-6/A provides a visual display of the direction of arrival (DOA) of incoming RF signals that have been processed through the receiving portion of the overall DF system.

Lightweight and compact mechanical packaging makes the L-6/A suitable for use in airborne configurations as well as shipboard and fixed/mobile ground installations. The frequency coverage of this antenna system is 500 MHz to 18 GHz in six bands. Variable azimuth rotation sweep from 0 to 2000 rpm is accomplished through a control on the console.

In addition, the antenna can be steered to any desired azimuth point by a manual control knob located on the indicator. The radiation patterns of the antenna are relatively broad, with circular polarisation. This offers an optimum compromise between probability of signal intercept, direction-finding accuracy and gain. The RF output of the L-6/A system is entirely compatible with the WJ-1140 and the interface is accomplished through the use of standard RF coaxial cabling and connectors.

STATUS
In production.

CONTRACTOR
Watkins-Johnson Company, 2525 North First Street, San Jose, California 95131, USA.

1716.053
WJ-1240 RECEIVING SYSTEM

The WJ-1240 is a computer-controlled receiving system for reconnaissance and surveillance applications over the radio frequency band 0.03 to 40 GHz, and it also has independent digital tuner control and parallel video display. The equipment consists of a modular integrated electronics package comprising the computerised WJ-1240 receiver itself, display and controls, and a direction-finding antenna system.

Independent tuner control permits multiple situation surveillance and provides independent video outputs for each tuner. This control operates in the absence of the computer for manual systems or as a standby mode of operation in computer-controlled versions. A digitally refreshed display with adjustable decay time retains single pulse signals for detailed visual observation. Display traces are automatically scaled during sector scan.

Several antenna assemblies may be used with the WJ-1240 depending upon the total frequency coverage demanded. The WJ-8312 provides direction-finding facilities over the range 0.1 to 18 GHz. Other Watkins-Johnson arrays which can be employed are the L-4/A (40 to 550 MHz), L-5/A (12 to 40 GHz), and the L-6/A.

The WJ-8312 consists of a rigid box structure supported on the turning gear pedestal and carrying a series of microwave receiver horns. Intercepted signals are directed into the horns (which are of dimensions appropriate to the frequency bands covered) by a curved plate reflector which rotates with the whole assembly. This reflector carries at its upper edge a log-periodic array for reception of the lower frequencies.

STATUS
In production.
CONTRACTOR
Watkins-Johnson Company, 2525 North First Street, San Jose, California 95131, USA.

WJ-8312 direction finding antenna

3666.193
WJ-1424 MICROWAVE RECEIVER (TLR-31)
The WJ-1424 microwave receiver is a small, rugged, special-purpose system covering the frequency range from 10 to 40 GHz (the lower frequency limit may be extended down to 1 GHz as an option). Its compact, modular design lends itself well to applications requiring minimum weight and space consumption. Many functions normally designed into individual units have been integrated to provide multi-functional control and display units. The system includes all the elements necessary for a completely operational receiver. The system is comprised of a spinning directional antenna with radome, RF distribution, receiver control, and signal analysis units. The prime power requirement is 110 V AC, 9.2 amps, 37 to 400 Hz.

In its vehicle mounted form, it is believed to carry the designation AN/TLR-31. The receiver control unit provides all system operating control functions including antenna scan control. It also contains pulse analysis circuitry to perform fundamental signal analysis. The receiver operates in three modes: Spectrum, Pan and Analysis/DF. A single-frequency amplitude display is presented to the operator in the Pan mode. When the operator selects DF at the control, the CRT presents signal amplitude in a polar format for bearing information. Signal RF frequency and azimuth are displayed to the operator in digital form via LEDs on the control unit front panel. When utilising the pulse analysis function, the same LEDs provide digital readouts of pulse repetition interval or pulse repetition frequency.

The stand-alone system has been integrated with the AN/MLQ-24 countermeasures receiving set to extend the frequency range coverage into the millimetre wave region.

STATUS
In production and operational.
CONTRACTOR
Watkins-Johnson Company, 2525 North First Street, San Jose, California 95131, USA.

4394.053
WJ-1440 MICROWAVE RECEIVER SYSTEM
Derived from the fully-militarised WJ-1740 microwave receiver system (**3354.053**), the WJ-1440 incorporates added features that allow it to achieve a wider range of application with fewer optional requirements. The microprocessor controlled memory scan, for example, which is an option in the WJ-1740, has been standardised in the WJ-1440 system. This feature permits the operator to store up to 16 scan control blocks for recall. By selecting the memory scan mode, the operator can cause the receiver to automatically scan through any pre-selected number of scan control blocks, causing them to be sequentially displayed on an analysis display, which can be used to display DF and pulse analysis as well as normal receiver functions. Another WJ-1740 option that has been incorporated into the WJ-1440 standard design is a tuner integrated local oscillator synchroniser (TILOS). TILOS increases overall stability and frequency accuracy to 1 part in 10^7 per day standard.

The WJ-1440 also has analysis alternate mode selection and control to enable expansion of a sector within a band without compromising operator contact with a full-band scan display. In alternate mode, the controller alternately sweeps a tuner through a sector and an entire band, which are displayed on analysis display as two distinct traces, with full-band activity displayed above sector activity display.

STATUS
In production.
CONTRACTOR
Watkins-Johnson Company, 2525 North First Street, San Jose, California 95131, USA.

WJ-1440 microwave receiver system

3355.053
WJ-1840 RECEIVING SYSTEM
The WJ-1840 microwave receiver system is a broadband equipment that covers the frequency range 0.5 to 18 GHz with a single two-segment superheterodyne RF tuner contained within an A1D enclosure. The 0.5 to 8 GHz segment is an up-converting dual superhet, while a standard pre-selected single-conversion superhet is used for the 8 to 18 GHz segment. Digital control circuitry permits band scanning, sector scan, and manual modes of operation as well as special dependent, dwell, and intercept modes.

A two-trace display shows signal activity over the full frequency range, over two sectors, or over one sector and a dependent sub-sector. Frequency data is written directly on the face of the CRT in alphanumeric form. A five-trace display is available as an option.

An optional analysis display is available for time base analysis of pulses as short as 10 nanoseconds. Additionally, IF pan analysis with 1 MHz resolution and 200 MHz bandwidth polar display are available within the analysis display. The optional WJ-8535-11 antenna is a slant-linear polarised bi-conical omni-directional antenna covering the whole of the receiver band.

WJ-8535 antenna unit

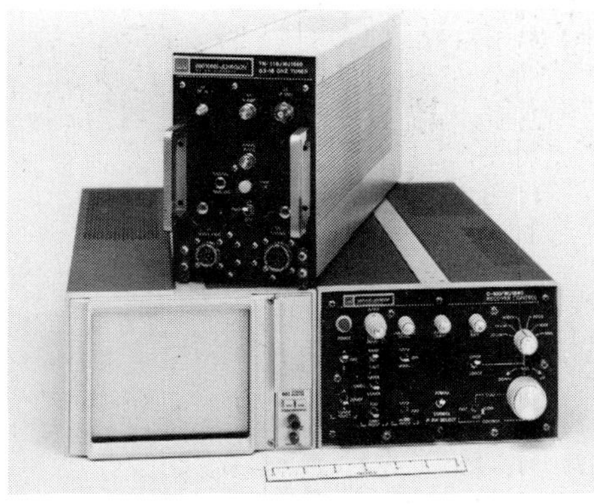

Watkins-Johnson WJ-1840 broadband receiving system

STATUS
In production.
CONTRACTOR
Watkins-Johnson Company, 2525 North First Street, San Jose, California 95131, USA.

4396.053
QRC-259 RECEIVER SYSTEM
The QRC-259 receiver system is a solid-state, electronically-tuned superheterodyne receiver system providing high frequency accuracy, high stability and high sensitivity. It is qualified to meet severe environmental and EMI specifications, and has been deployed under both field and airborne conditions. The QRC-259 also permits unmanned automatic/semi-automatic or fully-integrated operator computer operations.

Up to ten tuners, in any combination of types, can be connected in any QRC-259 receiver system. This flexibility enhances the system's frequency stretching capability above or below typically used frequency limits. All tuners in the system are connected to a tuner relay, which can select one of the tuners for direct operator control by switching tuning commands to the tuner and by switching the tuner's IF output to an IF processor. Tuners not under direct operator control can be controlled independently, as separate tuners, or by auxiliary digital-control inputs applied to the tuner relay.

The QRC-259 system can be operated manually or by automatic band and limited frequency scanning. It can also be operated in either scan-lock or tuner-hold mode. When the receiver is in scan-lock, it intercepts signals exceeding an operator-set threshold level. If an intercepted signal meets criteria for qualification as a true signal, the receiver locks on to the signal by switching to the manual mode at the signal frequency.

In hold mode, the tuner remains tuned to the frequency it was tuned to when the operator put it in hold, and its IF output ·is made available for processing by peripheral equipment. During this time, the other units of the receiver remain under normal operator control.

Another QRC-259 feature is a pulse coincidence detector function that determines the time coincidence between a received pulsed signal and an externally-applied video pulse. The reference pulse used for this purpose can be derived from another receiver or by local generation. When the system is used in a dual-channel configuration (using two operators), two channels can be connected together to examine the environment for coincident emitters. (Coincidence is defined as existing when two pulses are within five microseconds of each other.)

STATUS

Production.

CONTRACTOR

Watkins-Johnson Company, 2525 North First Street, San Jose, California 95131, USA.

6459.093

WJ-1996 ESM RECEIVER

The WJ-1996 tunable IF receiver is designed to operate as either an ESM or a threat warning equipment over the frequency range 0.5 to 18 GHz. Both applications require the ability to automatically search, intercept, analyse, and characterise electromagnetic emissions for signal identification.

As an ESM receiver the WJ-1996 combines the strengths of both IFM and superhet receiver processing. The use of a fast scanning IFM allows rapid signal acquisition and identification, and a 3-D display (frequency/amplitude/time) in a selected 500 MHz band is provided for observation and analysis of signals from frequency agile, multibeam and widebeam CHIRP emitters.

As a threat warning receiver, the 500 MHz wide band tuner is scanned rapidly across the 0.5 to 18 GHz band. The scan rate is chosen to ensure reception of at least two pulses for any signal of interest while the signal is within the 500 MHz window. The system architecture allows operation in dense signal environments.

STATUS

In production.

CONTRACTOR

Watkins-Johnson Company, 2525 North First Street, San Jose, California 95131, USA.

6834.193

WJ-3600 EW ALERT AND COLLECTION SYSTEM

The WJ-3600 EW Alert and Collection System (EWACS) combines narrow and wide band technologies into a receiving and analysis system able to accurately measure frequency-agile radar parameters, including standard signal, staggers, pop-ups, linear and random hoppers and agile/agile emitters. EWACS combines the strengths of narrow and wide band technologies in a parallel architecture where each asset contributes to the area in which it is best suited to operate. The system can execute three tasks simultaneously; automated wide band search, automated narrow band search and manual (operator) analysis.

The WJ-3600 operates over the 0.5 to 18 GHz frequency range, with the capability of extension to 40 GHz. It consists of an integrated system with a superheterodyne receiver, a tunable 2 GHz wide instantaneous frequency measurement receiver and a display/filter.

CONTRACTOR

Watkins-Johnson Company, ESM Division, 2525 North First Street, San Jose, California 95131, USA.

6833.193

WJ-4810 COMMUNICATIONS JAMMING SYSTEM

The WJ-4810 (AN/TLQ-504) communications jamming system provides a minimum of 100 W of RF power to the antenna port over a frequency range from 88 to 500 MHz. Included in the system are a controller, intercept receiver, exciter, power amplifier and RF switching and harmonic suppression capabilities. Threats with accompanying priority or lockout can be stored in a 100-channel memory, either manually or automatically.

The jamming spectrum can be operator tailored or can automatically match the threat signal in both modulation type and bandwidth. A variety of internal modulation sources are available as well as external input. System capabilities include automatic random look-through, simultaneous jamming of up to six targets, syllabic-rate on/off keying of the modulation and adjustable power output.

CONTRACTOR

Watkins-Johnson Company, CEI Division, 700 Quince Orchard Road, Gaithersburg, Maryland 20878, USA.

6461.093

WJ-8976 DIRECTION FINDING SYSTEM

The WJ-8976 is a three-channel direction finder capable of providing accurate azimuth and elevation bearing information for many types of signals, both monopulse and continuous, as well as noise-like signals such as spread spectrum.

The system consists of a three channel slave receiver, a high speed digital processor, an antenna unit, and a WJ-8617B-21 receiver which serves as the master tuner. Several hardware options are available to extend the frequency coverage and enhance the system operating capabilities.

CONTRACTOR

Watkins-Johnson Company, 2525 North First Street, San Jose, California 95131, USA.

6832.193

RESPONSIVE SURVEILLANCE SUBSYSTEM

Responsive surveillance subsystems are being created by Watkins-Johnson to meet the ever increasing complexity and demand in the strategic and tactical communications environment. The subsystems are centred around the WJ-8610A controller and they can include up to 14 VHF/UHF receivers (typically WJ-8615D compact receivers) on a IEEE-4888 bus. Signal monitors, multicouplers, tape recorders and printers are a few of the ancillary components that can be incorporated into the subsystems according to the user requirements. Versatility and flexibility are shown when a simple subsystem is combined with a user's equipment to expand the capabilities and enhance the performance of a programme.

CONTRACTOR

Watkins-Johnson Company, Special Projects Division, 700 Quince Orchard Road, Gaithersburg, Maryland 20878-1794, USA.

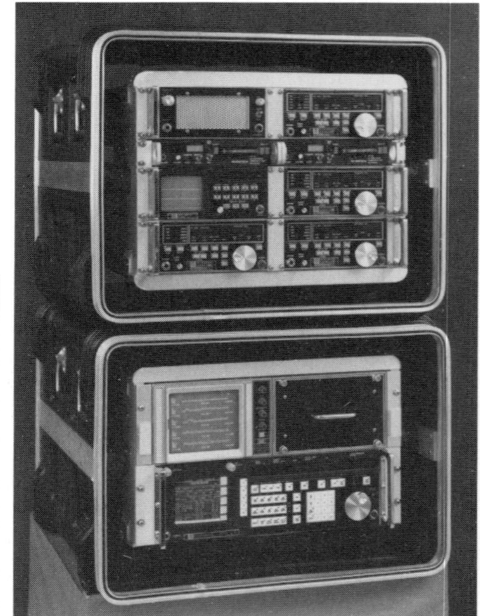

Typical responsive surveillance subsystem

6831.193

WJ-8990 MANPACK TACTICAL INTELLIGENCE SYSTEM (MANTIS)

The WJ-8990 MANTIS is designed specifically for fast moving, forward deployed operations where high mobility is the criterion. It can be transported by a two-man team with minimum set up and take down time. The system comprises a WJ-8972 receiver/DF processor and a WJ-9881 antenna. The former is an extremely compact unit designed to fit in the radio pouch of a standard backpack and provides RF intercept and DF capability over the frequency range 20 to 500 MHz. Through use of HF and UHF expansion modules the coverage can be extended to provide continuous intercept from 0.5 to 1200 MHz and DF coverage from 2 to 1200 MHz.

WJ-8990 MANTIS

The WJ-8972 has provision for manual, hands-on control or automatic search and DF operations, which can be programmed via front panel controls. In addition to conventional scanning, provisions exist for assignable priority channels and lockouts.

The MANTIS system can operate as a stand-alone line-of-bearing system, or as an element of a DF net to conduct emitter location operations. The system is also designed to operate with the WJ-4800 communications jammer. In this configuration, the user has a complete, compact intercept, DF and jamming capability in one simple system.

STATUS

In production.

CONTRACTOR

Watkins-Johnson Company, 700 Quince Orchard Road, Gaithersburg, Maryland 20878-1794, USA.

6458.093

AN/PRD-11 MAN-PORTABLE DF EQUIPMENT

The AN/PRD-11 is a man-portable direction finder operating in the 20 to 500 MHz frequency range. It consists of the WJ-8640-1 Manpack Receiver, the WJ-9180-1 Signal Monitor and the WJ-8975A Manpack DF Processor with a WJ-9880A Antenna. The receiver is a ruggedised unit designed to operate under extreme environmental conditions. It features wide range tuning using interchangeable drop-in units. The antenna system is a lightweight unit which folds up into a cylindrical shape of approximately 91 × 20 cm. It comes with its own tripod to allow the antenna to be elevated to 2.85 metres.

Derived bearing data is displayed digitally by a three-digit LED display, supplemented by a circular array of LEDs.

CONTRACTOR

Watkins-Johnson Company, 2525 North First Street, San Jose, California 95131, USA.

4397.053

CMR-610A WIDE BAND DF RECEIVER

The CMR-610A is a wide band DF receiver designed to provide both visual and aural signal indication of microwave and millimetric activity occurring in the 1.5 to 40 GHz frequency range. The CMR-610A will detect, locate and assign direction to CW, AM, FM and pulse-modulated signals. The receiver has the provision for demodulating audio-modulated signals and has an integral recorder output.

The CMR-610A is a miniature, hand-held, solid-state, battery powered, five-band, crystal video receiver with integral antennas and battery pack. The frequency band to be monitored is selected by the band switch which connects the selected antenna/detector combination to either the video or audio amplifier, according to the mode switch setting in the AM mode of operation. The selected detector output is connected to the video amplifier whose output is routed to the operator's headset, the signal level meter and the recorder output jack. In CW and audio operation the selected detector output is connected to the audio amplifier.

CHARACTERISTICS

Frequency bands

Band 1: 1.5 – 8 GHz; Band 2: 8 – 12.4 GHz; Band 3: 12.4 – 18 GHz; Band 4: 18 – 26.5 GHz; Band 5: 26.5 – 40 GHz

Dimensions: 89 mm high × 89 mm wide × 197 mm deep

Weight: 1.6 kg

STATUS

In production.

CONTRACTOR

American Electronic Laboratories Inc, 305 Richardson Road, Lansdale, Pennsylvania 19446, USA.

4673.093

RG-5202A VHF SCANNING RECEIVER

The RG-5202A is a scanning receiver designed to scan and monitor up to 16 RF channels in the frequency range 20 to 512 MHz. Each channel can be set to receive either AM or FM signals with a choice of IF bandwidths of 10, 20, 40 or 60 kHz. The receiver can be operated by the front panel controls or remotely through control lines. A memory bank stores the tuned frequency, the IF bandwidth, the demodulator mode and the manual/automatic gain choice for each of 16 channels. Frequency information to the nearest 10 Hz is entered into memory by a keyboard.

The time devoted to each channel can be set from 0.1 to 10 seconds. The channel being monitored is indicated by a tone encoded output for remote indication.

CONTRACTOR

Racal Communications Inc, 5 Research Place, Rockville, Maryland 20850, USA.

4398.053

RG-5540 VHF/UHF MONITORING RECEIVER

The RG-5540 is a high performance VHF/UHF receiver designed for reception of signals in the 20 to 500 MHz band with 10 Hz tuning resolution. It can detect AM, FM, CW, USB, LSB and ISB signals. The receiver includes four IF modules with bandwidths from 10 to 500 kHz; 10, 20, 50 and 100 kHz bandwidths are standard.

The RG-5540 has three operating modes; Manual, Remote and Fast Scan. In the manual mode the receiver is a conventional 'hands-off' receiver with operator control of all functions. Tuning is by a large tuning knob or up/down slew key (with three rates), direct numeric keypad entry or an internal scan mode. The scan mode continuously tunes the receiver up and down at nine selectable rates between 10 Hz/second and 1000 MHz/second. Frequency is indicated by an eight-digit LED display; tuning and level meters located under the display aid operator tuning. LED-illuminated keys select IF bandwidth, detection mode, gain control mode, and internal or external reference. A dual (stereo) headphone level control, remote/local switch, RF gain slew key and power switch complete the front panel operator controls. A non-volatile memory preserves all operating parameters when power is removed.

In the remote mode, the receiver functions are controlled via an IEEE-4888 interface. These functions include tuned frequency, IF bandwidth, detection mode, RF gain mode, RF gain frequency scan and internal/external reference. Commands are provided for full set-up or the setting of individual parameters. The receiver status can also be requested, in either remote or local mode. The receiver front panel displays the operating conditions at all times.

In the fast scan mode, the receiver operates under the control of a spectrum surveillance controller to provide a high-speed spectrum scan with digitally refreshed display. A discrete scan of up to 15 channels is also provided.

An optional front panel plug-in spectrum display is available. When installed, this unit, the RG-1320A spectrum display unit, provides a visual display of signal activity about the tuned frequency. The sweep width is variable from 0 to 400 kHz, with 1 kHz resolution.

CHARACTERISTICS

Frequency range: 20 – 500 MHz

Frequency resolution: 10 Hz

Demodulation modes: Standard: AM, FM, CW; Optional: USB, LSB, ISB

Dimensions: 48.3 cm wide × 13.3 cm high × 51 cm deep

Weight: 27.3 kg

CONTRACTOR

Racal Communications Inc, 5 Research Place, Rockville, Maryland 20850, USA.

4674.093

RG-5545A VHF/UHF RECEIVER

The RG-5545A is a high-performance VHF/UHF receiver that has been specifically designed to allow versatile configured features by means of performance options to the basic unit and is ideally suited for monitoring purposes. The basic unit is a 20 to 500 MHz receiver with AM and FM detectors and four selectable IF bandwidths. Tuning is accomplished to 10 Hz resolution by means of a synthesiser. All local oscillators are phase-locked to a single master reference oscillator, either internal or external, for low phase noise and maximum frequency stability. The RG-5545A has been designed for signal handling in dense electromagnetic environments.

A frequency extension option extends the range to 20 to 1000 MHz, 20 to 1400 MHz or 20 to 1800 MHz. These options include preselection above 500 MHz for a +50 dBm 2nd order intermodulation intercept point.

CONTRACTOR

Racal Communications Inc, 5 Research Place, Rockville, Maryland 20850, USA.

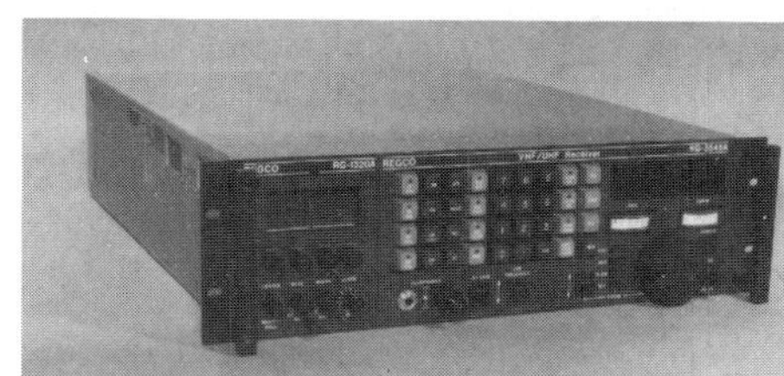

RG-5545A receiver

4676.093

RA6778-C RECEIVER/RA6778-Q CONVERTER

The RA6778-C receiver is a fully synthesised microprocessor based equipment for communication or monitoring over the frequency range from 10 kHz to 30 MHz. The equipment can be controlled manually or by computer and has an interactive man/computer interface. Control capability is hardware adaptable to conform to system requirements.

The RA6778-Q converter is a microprocessor controlled tuneable equipment for communication or monitoring over the band 50 kHz to 30 MHz. It provides a 70 kHz IF signal 100 kHz wide which may be used in applications for wide band spectral analysis or demodulation of HF signal.

Both equipments are of modular construction and have LED tuned frequency indicators. Frequency stability of ±1 part in 10^6 is provided as standard with ±1 part in 10^7 optionally available.

CONTRACTOR

Racal Communications Inc, 5 Research Place, Rockville, Maryland 20850, USA.

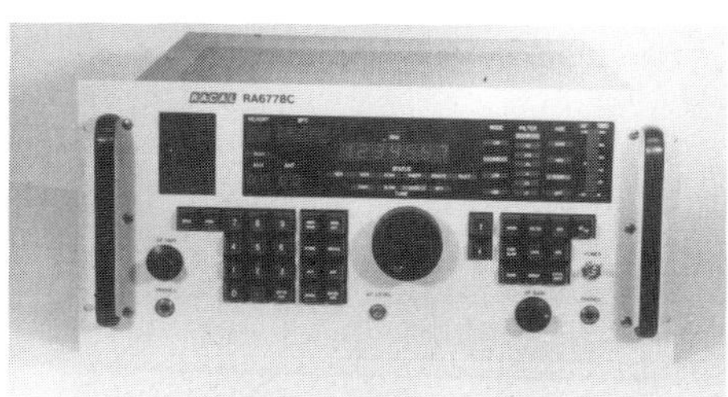

RA6778-C receiver

4675.093

RA-6793A HF RECEIVER

The RA-6793A is an HF receiver covering the range 0.5 to 30 MHz and used for monitoring purposes in the communications field. Two of the features of this system are its ability to provide full scan and sweep facilities under both local and remote control. The standard scan provides a means to scan up to 100 channels actively with a fixed but selectable dwell between channels for operation interaction. A secondary scan mode allows the operator to set different activity recognition thresholds to each scanned channel, whereupon the receiver will halt on activity.

Sweep provides the capability to set start and stop frequencies and step increment. After sweep initiation, the operator can control the incremental speed and halt the sweep if required. The equipment operates in AM, FM, CW, USB and LSB reception modes. Frequency display is by easy-to-read illuminated liquid crystal display on the front panel.

CONTRACTOR

Racal Communications Inc, 5 Research Place, Rockville, Maryland 20850, USA.

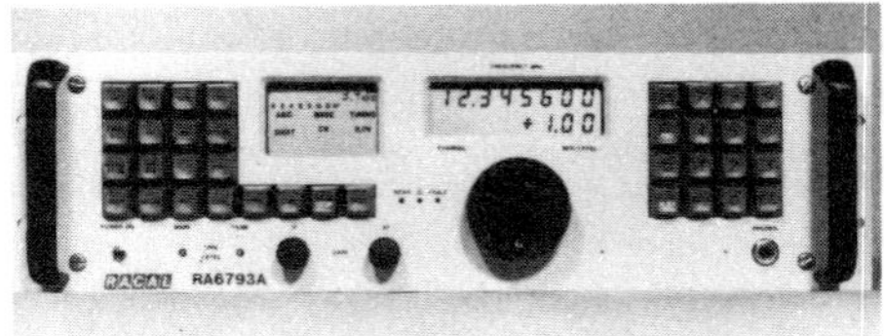

RA-6793A HF receiver

4399.093

RG-1340 SPECTRUM SURVEILLANCE CONTROLLER

The RG-1340 spectrum surveillance controller is designed to work in conjunction with the RG-5540 (**4398.053**) receiver and an external cathode ray tube display forming a surveillance group for monitoring any segment of, or 15 discrete frequencies in, the 20 to 500 MHz band. The RG-1340 provides external scan control to the RG-5540 receiver and X, Y and Z axis information to the display. The spectrum surveillance group may be controlled locally using its own RG-1340 controls or remotely using an IEEE-4888 interface. Alphanumeric information on the display provides status and operator prompting, and the display is refreshed 50 times per second.

The controller has three operating modes: band scan, discrete scan and manual.

In band scan mode, the controller steps the receiver between two programmed frequency limits. The controller also controls the receiver bandwidth. A signal averaging feature enables an operator to integrate noise and make signals more visible. The operator-controlled cursor can be positioned on a signal of interest; alphanumerics display the cursor frequency.

In the discrete scan mode, up to 15 different frequencies can be monitored. Frequency, IF bandwidth, dwell time, threshold level and actual level for each channel are displayed in alphanumeric form on the display. Channel set-ups are automatically arranged from top to bottom in order of ascending frequency, and any channel can be skipped during the scan.

In manual mode, the controller returns control of the system to the receiver front panel. At this time, the display indicates an IF panoramic scan identical to that of an RG-1320A spectrum display unit installed in the left side of the receiver.

The controller front panel consists of two keyboards (functional and numeric), a cursor control, a display brightness control, and local/remote and power switches. To set parameters, the operator selects the operating mode, the prompt function and four keys which can position the prompted parameter. The numeric keyboard is then used to update the parameter.

CONTRACTOR

Racal Communications Inc, 5 Research Place, Rockville, Maryland 20850, USA.

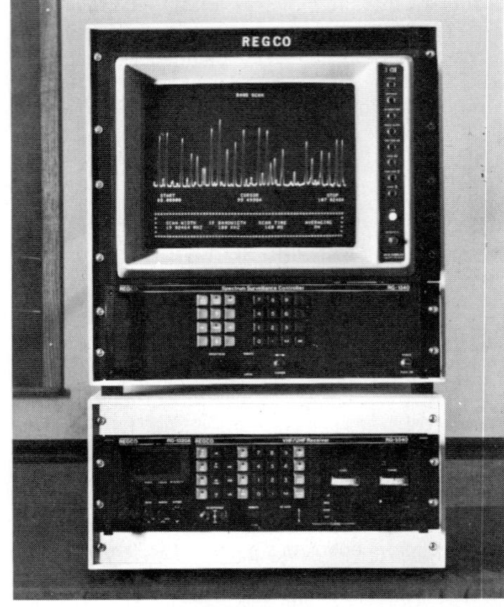

Spectrum surveillance system comprising RG-5540 receiver, RG-1340 controller and display

---

4400.093

RG-1320A SPECTRUM DISPLAY UNIT

The RG-1320A spectrum display unit provides a visual display of the signal activity present around the tuned frequency of an associated receiver. The display is most useful in tuning and analysing signals intercepted by the receiver. The unit plugs into an RG-5540 or similar receiver and derives its power from that source.

The equipment presents a 60 dB logarithmic display of a band up to 400 kHz wide. A 1 kHz bandwidth Gaussian filter provides high resolution and the sweep rate is adjustable from 2.5 to 25 Hz to allow full resolution at maximum sweep widths.

The one- by three-inch cathode ray tube display uses a bright P31 green phosphor with grey contrast and EMI screen. The X and Y deflection signals are available for an external display.

CONTRACTOR

Racal Communications Inc, 5 Research Place, Rockville, Maryland 20850, USA.

4401.193

AN/GLQ-3B COMMUNICATION COUNTERMEASURES SET

The AN/GLQ-3B is a high-power, shelter-mounted communications jamming system designed to intercept and jam voice and data links in the 20 to 230 MHz frequency range. The system incorporates a microprocessor controller, a 1.5 kW solid-state transmitter and a digitally tuned receiver/exciter. Integrated into a standard S-250 shelters, the AN/GLQ-3B can be mounted on any 1¼ ton truck.

STATUS

In current service with the US Army. Eight AN/GLQ-3A systems sold to Turkey.

CONTRACTOR

Fairchild Weston Systems Inc, 300 Robbins Lane, Syosset, New York 11791, USA.

AN/GLQ-3B deployed in the field

4672.093

RG SERIES SPECTRUM MONITORS

Racal Communications Inc manufactures a range of spectrum monitors designed to provide a visual display of the signal activity present and around the tuned frequency of an associated ESM monitoring receiver. Data is displayed on a CRT display forming part of the monitor.

The RG-1307A features a sweep width of 0 to 40 MHz, manual gain control and an accurate frequency marker. Resolution bandwidth is 250 kHz nominal. The RG-1308A provides a variable sweep width of 0 to 8 MHz, a variable sweep rate of 5 to 25 Hz and a resolution bandwidth of 10 kHz.

In group monitor equipments the RG-1309A and RG-1309A-2 provide visual displays of the individual voice frequency channels in the telephone basic group 'B' of 60 to 108 kHz. The RG-1317A and RG-1317A-1 give four independent visual displays of the individual voice frequency channels in the telephone basic group 'B' and also have provision for displaying channels in basic group 'A' (12 to 60 kHz).

CONTRACTOR

Racal Communications Inc, 5 Research Place, Rockville, Maryland 20850, USA.

6140.153

COMMUNICATIONS EW SYSTEM (CEWS)

CEWS is an operational tactical/strategic EW system that is modular in design to allow customised applications. It operates in the HF to UHF bands providing automatic search and surveillance, DF, jamming and analytical facilities against both line-of-sight VHF/UHF and long-range HF links.

CEWS modules in a typical configuration include a command centre, slave DF unit and ECM unit.

Features include computer-controlled automatic search, surveillance, DF and analysis of hostile HF/VHF/UHF communications, priority countermeasures jamming against HF/VHF/UHF links, responsive look-through jamming of multiple emitters and multi-mode automatic and manual jamming.

STATUS

Operational.

CONTRACTOR

Fairchild Weston Systems Inc, 300 Robbins Lane, Syosset, New York 11791, USA.

3037.053

SR SERIES SURVEILLANCE RECEIVERS

Norlin Communications Inc produces a wide range of surveillance and monitoring receivers covering the frequency bands of interest for Comint and Sigint operations. A typical receiver is described in the following paragraphs.

SR-209

The type SR-209 solid-state, modular receiving system is capable of reception of AM, FM, CW, and pulse signals in the frequency range of 20 MHz to 12 GHz. It will accommodate a signal display unit to provide a visual display of signals in a band around the received signal.

Three IF amplifier FM demodulator boards may be used at one time with operational selection by front panel switch. For VHF/UHF, selection of IF bandwidths from 10 kHz to 8 MHz are available. In HF there are three bandwidths.

The wide frequency range of 20 MHz to 12 GHz for a manually tuned receiver is provided through the use of ten plug-in RF tuning heads.

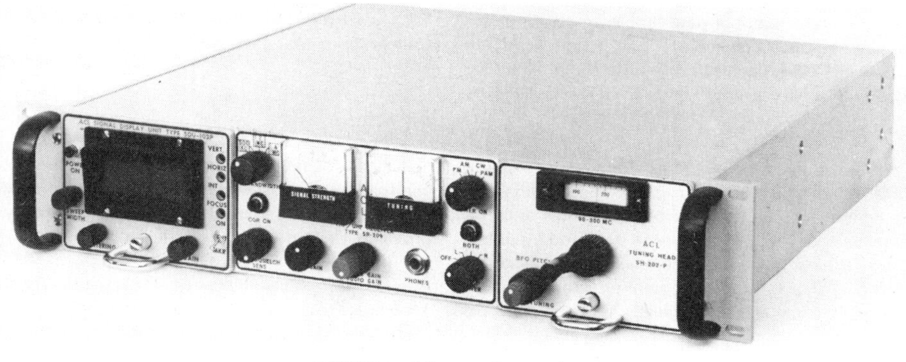

SR-209 modular receiver system

To make the receiver completely self-sufficient for field use a battery pack plug-in unit may be used in lieu of one tuner. No adjustments or changes in the receiver are necessary when this nickel cadmium battery pack is installed. A built-in charger is provided in the battery back unit.

The front ends employ at least two section pre-selectors at the RF input to provide maximum reduction of cross modulation and intermodulation interferences.

The SR-209 basic receiver contains a carrier operated relay to control accessory equipment such as recorders. All operating controls are located on the front panel of the SR-209 and except for the phone jack, all inputs and outputs are located on the rear panel. An exception to this is the optional first local oscillator output which is provided on the plug-in tuning head panel. Two meters, one for tuning and one for signal strength, are on the receiver front panel.

STATUS

Versions of the SR-209 have been supplied for signal intercept, search and DF applications at fixed sites, mobile monitoring stations, airborne, shipborne and submarine collection systems.

CONTRACTOR

Applied Communications (a division of Amstar Technical Products Company Inc), 9125 Gaither Road, Gaithersburg, Maryland 20877, USA.

3903.053

SERIES 2000 RECEIVER SYSTEM

The Norlin Communications Series 2000 range of equipment comprises a number of VHF/UHF receivers, several display units, and a variety of peripheral units from which can be assembled systems for radio surveillance and communications. Modular design permits the combination of units to form equipments appropriate to typical ESM, Elint, Sigint and related functions in both fixed and mobile installations. The basic standard receiver frequency range is 20 to 500 MHz, with a 1200 MHz upper limit as an option.

Six receiver models have been identified:

- SR-2080: Directly controlled 20 – 1200 MHz
- SR-2082: Slave version of SR-2080
- SR-2090: Directly controlled 20 – 500 MHz (option to 1200 MHz) with liquid crystal display status indicator (see below)
- SR-2091: As SR-2090 with 8 memory slots
- SR-2092: Slave version of SR-2090
- SR-2100: Directly controlled version of SR-2090 plus solid-state IF/RF panoramic facility

Norlin Series 2000 VHF/UHF surveillance receiver (above) with a DU-2090 display unit

A three-model series, SR-2070/1/2, with a frequency coverage of 0.05 to 30 MHz has been designed but is not in production.

There are three display units:

- DU-2081: RF/IF panoramic facility, autoscan mode
- DU-2090: RF/IF panoramic facility, 8 memory slots; hand-off to 8 receivers
- DU-2091: RF/IF panoramic facility

Typical specimens from this range are the SR-2090 and DU-2090 receiver and display units, and these are described more fully in the following paragraphs.

SR-2090

The basic receiver tunes the 20 to 500 MHz frequency range, expansible to 1200 MHz. The microprocessor control enables single knob tuning and a single direct-reading display of tuned frequency, even with the frequency expansion option installed. The microprocessor also enables the selection of a wide range of IF bandwidths and detection modes. The receiver is prewired to accept modules meeting all standard digital remote control interface specifications. A liquid crystal display (LCD) presents complete receiver tuning information at one focal point and improves the shock and vibration resistance by elimination of conventional signal strength and tuning meters. The LCD is uniformly backlit with an electro-luminescent panel for display under all ambient lighting conditions.

All tuning commands to the receiver can be applied from a remote location through a digital data interface. The interface translates the remote control data format to digital commands that are accepted by the receiver's microprocessor. A wide variety of digital interface format specifications are accommodated by a selection of the proper interface module. The receiver is prewired to accept all interface modules. Within the command the tuned frequency is in BCD format and the IF bandwidths and detection modes have pre-assigned digital codes. A total of 64 receivers can be addressed by the remote controller. Special command formats and functions can be developed upon request.

DU-2090

The DU-2090 display unit is designed to complement Series 2000 VHF/UHF electronic controlled receivers. The DU-2090 enables the system to provide master/slave operation of up to eight receivers from a central control point. An additional eight signals can be stored in memory for operator convenience. The DU-2090 features a dual trace, digitally refreshed, flicker free, panoramic display of the spectrum in scan operation. Scanning is accomplished by rapidly stepping the synthesiser receiver LO. When Receiver 1 is not swept the CRT becomes an IF Pan display for that receiver. Memory update can be inhibited to freeze the display for elusive push-to-talk targets, because the display is digitally refreshed. Since the stepped synthesiser sweep provides extreme linearity, the marker on the display can be exactly placed over the desired signal on the frozen display. Then, returning the receiver to the manual mode, a 100 per cent intercept probability results on the next target transmission.

STATUS

Production.

CONTRACTOR

Applied Communications (a division of Amstar Technical Products Company Inc), 9125 Gaither Road, Gaithersburg, Maryland 20877, USA.

4659.093

SR-2020 TACTICAL ESM SYSTEM

The SR-2020 is a remote controlled ESM receiver covering the 0.5 to 500 MHz frequency range in a single package. The range can be extended to 1200 MHz. It meets the need for effective, reliable and maintainable mobile tactical EW systems to satisfy Comint, Sigint, ESM and direction finding requirements. It is designed for independent operation or as an integral part of a larger system where a companion power supply will service groups of four receivers. Specifically designed to meet the needs of mobile, airborne or shipboard applications, the equipment is suited for tactical deployment in either wheeled or tracked vehicles under extreme environmental conditions.

The receiver control unit, RCU-2020, provides comprehensive control and status display for up to eight SR-2020 tactical receivers. All receiver functions are controlled by the internal microprocessor and a variety of proven hardware routines. In addition to receiver control, the unit forms the nucleus of a collection system by providing control and data interface ports for system ancillary equipments.

STATUS

In production.

CONTRACTOR

Applied Communications (a division of Amstar Technical Products Company Inc), 9125 Gaither Road, Gaithersburg, Maryland 20877, USA.

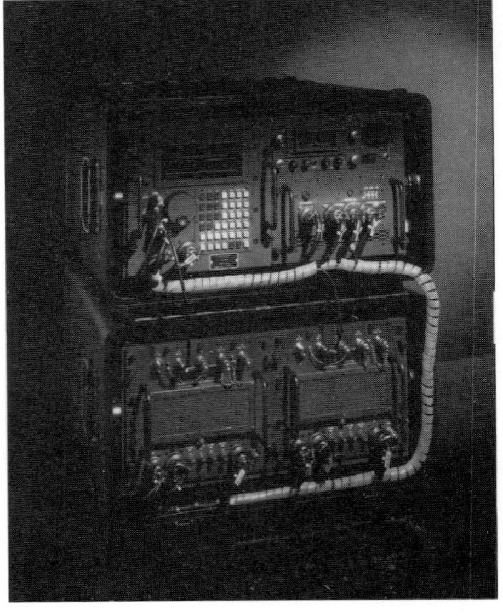

SR-2020 EMS system

4660.193

SRS-2500 VHF/UHF RECEIVING SYSTEM

The SRS-2500 VHF/UHF receiving system is an adaptive and comprehensive collection of units for strategic signal intercept, acquisition, direction finding and analysis applications. It is designed for independent or computer-controlled operation for the reception of AM, FM, CW, SSB and pulse signals over the frequency range 20 to 500 MHz. The coverage may be extended down to 0.5 MHz and up to 1200 MHz if required. System architecture enables additional subsystem groups to be added where each group communicates via a serial multi-address data bus.

The system group consists of a control address unit CA-2100, a digital distribution unit, and IF switch and from one to sixteen SR-2154 surveillance receivers. The heart of the system is the CA-2100 which carries out all the selection, tuning and control functions. It also alters and lists receiver memory content as well as displaying signal parameters of the addressed receiver to be handed off to another receiver. A built-in gas-plasma IF spectrum display and alphanumeric read-out panel provides the operator with a read-out of parameters.

The SR-2154 receiver covers the range 20 to 500 MHz with optional down and up converter units to extend this range to 0.5 to 20 and 500 to 1200 MHz. Each receiver in the system has audio and carrier operator relay outputs for recorder interface or separate operator monitoring.

Fibre optic data links are used so that the receivers, digital distribution unit and the IF switch unit can be remotely controlled by the control address unit.

CONTRACTOR

Applied Communications (a division of Amstar Technical Products Company Inc), 9125 Gaither Road, Gaithersburg, Maryland 20877, USA.

4661.093

SR-2152 VHF/UHF RECEIVER

The SR-2152 receiver is a high-speed, low-phase noise synthesiser applicable to real world Comint requirements. The receiver tunes the 20 to 500 MHz frequency range and is expandable to 1200 MHz. It is fully microprocessor controlled and has a large screen gas-plasma display for IF pan and scanning presentations. The display also presents frequency read-out, signal strength, f1 – f2 frequency band and BFO offset alphanumeric data. The receiver is designed for signal, intercept, search and DF applications. The system may be deployed for use in fixed station sites, mobile monitoring systems, airborne, shipboard and submarine systems.

The receiver accepts a wide variety of standard modules that enable growth and adaptability. Due to this modularity, the receiver is easily configured and specifically designed to form the nucleus of computer controlled and master/slave hand-off systems. AM, FM, CW, LSB, USB and pulse modulation capability is provided. Up to six IF bandwidths can be installed.

CONTRACTOR

Applied Communications (a division of Amstar Technical Products Company Inc), 9125 Gaither Road, Gaithersburg, Maryland 20877, USA.

4664.193

AN/MLQ-T6 COMMUNICATIONS DATA LINK JAMMER

The AN/MLQ-T6 is a mobile, general purpose, ground-based communications and data link jammer, operating in the frequency band of 960 to 1850 MHz, at a power level of 1 kW CW or 2 kW peak. It is capable of simulating the projected threat jammers in the 1980 to 1990 period and is employed during the development, test and evaluation of future communications and data link systems. Major subsystems are a transmitter, receiver, antennas/pedestal, processor/controller, automatic tracker, control and display console and the shelter.

STATUS

In production.

CONTRACTOR

Cincinnati Electronics, 2630 Glendale-Milford Road, Cincinnati, Ohio 45241, USA.

AN/MLQ-TC communications data link jammer

3895.193

AN/MLQ-34 TACJAM

The AN/MLQ-34 TACJAM is a highly mobile tactical communication countermeasures system developed and initially produced by GTE and now being produced by AEL. It is the jamming element of the US Army TACOM EWS (Tactical Communications Electronic Warfare System), and normally operates in conjunction with the AN/TSQ-112 (**4409.153**). TACJAM can disrupt and deceive communication links over a broad, multi-octave VHF frequency range with an effective radiated power (ERP) of 3 to 4 kW. The AN/MLQ-34 has multiple receivers, exciters, and high power transmitters to provide simultaneous multiple threat capability. An upgraded version, TACJAM-A, is currently in development.

The TACJAM equipment operates under computer control to permit quick reaction to changes in the electromagnetic environment. Full manual override is provided such that the operator can cancel or adjust computer activities. A wide range of modulation formats, including FM, CW, FSK, AM and noise, permits effective jamming of virtually any threat emitter present in the equipment's frequency range.

Each multi-kilowatt solid-state broadband transmitter is constructed of basic power modules (BPMs). These are interchangeable and replaceable in the field. All system electronics are programmed for automatic checkout and fault isolation by the AN/USM-410 EQUATE automatic test equipment.

The complete system is mounted on an M-1015 tracked cargo carrier which is highly mobile. Rapid deployment is achieved by using an automatic ground rod driver and an unfolding log-periodic antenna mounted on a pneumatically operated telescopic mast.

All TACJAM electronic and cooling equipments are enclosed in an S-595 shelter which has ballistic protection.

A ground jammer threat simulator, known as the AN/MLQ-T4, is available and gives aircrews an actual experience of specific jamming.

STATUS

Production of 29 systems by GTE was completed by 1984. A contract worth over $44 million was awarded to American Electronic Laboratories (AEL) in June 1983 for 31 TACJAM systems plus spares and engineering support, and another contract in March 1985 for a further 12 systems. In September 1985, AEL received a $135 million contract increase to provide 96 biological/chemical warfare protection kits for AN/MLQ-34 systems. As at November 1987 the contract value for the second production of the AN/MLQ-34 is $146 million, the third production award is $28 million.

An upgraded version, designated TACJAM-A is now being developed and two teams have been chosen by the US Army, AEL/Saunders and Hughes/Argo/E-Systems/Lockheed. The teams have a development contract to produce a prototype TACJAM-A and one will be selected in 1989 for full scale engineering development.

CONTRACTOR

American Electronic Laboratories Inc, 305 Richardson Road, Lansdale, Pennsylvania 19446, USA.

TACJAM (AN/MLQ-34) mobile tactical jammer unit deployed in the field

4405.193

AN/MSQ-103 TEAMPACK ESM SYSTEM

The AN/MSQ-103 ESM system is a ground-based transportable equipment designed to collect, identify and provide the direction of signals from ground-based transmitters. It operates in the 500 MHz to 40 GHz frequency range and is intended for use in the field at division level. The system is housed in a ballistically protected shelter and then mounted on an M-1015 tracked vehicle. It can also be mounted on an M-35 utility truck, a light armoured vehicle, jeep and other types of combat vehicles. It has replaced the US Army AN/MLQ-24 as the basic countermeasures receiving equipment and includes built-in computer processing from a ruggedised mini-computer installed in the vehicle.

The system is also known under the name of Teampack and contains secure voice and wide band data link to a forward control and analysis centre.

A contract for engineering development and enhancement has been completed by Emerson. The enhancement programme, the AN/MSQ-103A, includes improvements to reliability, performance and crew protection, as well as the development of a netted system.

STATUS

In current service with the US Army.

CONTRACTOR

Emerson Electric Co, 8100 W. Florissant, St. Louis, Missouri 63136, USA.

Netted AN/MSQ-103 Teampack system

4403.093

AEL TYPE 6040 UNIVERSAL JAMMER

The Type 6040 ECM universal jammer is a signal source for use in testing system jamming vulnerability. Three classes of jamming signals are generated: barrage noise jamming, spot noise jamming and deception jamming.

Operation is controlled from a digital control unit. The 'dial-a-jam' capabilities of the control unit allow complete operation of the system in a simple and straightforward manner. The system includes a receiver section, a modulation source section and a transmitter.

Depending upon the operator selected mode of operation, the receiver section affects the operation in three different ways. In the transponder modes, the output of the receiver triggers the AEM signal transmissions for which internal RF carrier sources are used; in repeater modes the receiver performs as the front end of the repeater/transmitter configuration; and in the manually actuated modes, the receiver is a passive indicator of the signals which are present in the band of interest. A multi-frequency determining unit and an automatic signal recognition unit are included as part of the receiver section. These signal sorting units identify the class of signal being received, display the information on the control panel, and program ECM response in transponder modes of operation.

CONTRACTOR

American Electronic Laboratories Inc, 305 Richardson Road, Lansdale, Pennsylvania 19446, USA.

3367.193

AN/TLQ-15 COMMUNICATIONS AND COUNTERMEASURES SET

The AN/TLQ-15 communications and countermeasures equipment is a transportable system for tactical use either as a truck-mounted installation or as a deck-mounted unit aboard ship. Fixed or semi-permanent ground-based applications also exist, and the TLQ-15 is capable of airborne use by means of a trailing wire antenna. All units of the equipment are air transportable by fixed- or rotary-winged aircraft.

At the centre of the system is an HF communications transmitter/receiver covering the 1.5 to 20 MHz range.

Associated with this unit are: a panoramic indicator unit; frequency counter; modulation generator; RF amplifier; low-pass filter assembly; and an antenna coupler. In addition to the main antenna (a 35-foot telescopic whip, with counterpoise) provisions are made for feeding into a dummy load and for connection to an auxiliary antenna.

The transmitter/receiver is a solid-state design with synthesised receiver/transmitter tracking, and providing the following operational facilities:

(1) continuous transmission
(2) look-through transmission (90 per cent transmit/ 10 per cent receive)
(3) signal-initiated transmission (signals of 5 µV or above put system into look-through mode)
(4) receive/transmit, AM, FM or DSB voice.

In the first three of these modes, modulation available includes: CW, AM, DSB, FM, AM and FM and FSK.

Automatic frequency control (AFC) tracking of received signals is provided, and there are facilities for signal analysis, various modulations are available for countermeasures operation, and an auxiliary communications link using the AN/VRC-47 can be provided as an option. Additional system potential includes the possibility of using a phased array of two TLQ-15 systems to produce 8 kW of RF power in any desired direction, and continuous reception in the pattern null for a monitoring receiver. The basic power output is 2 kW CW.

STATUS
Operational with US Forces and other countries.

CONTRACTOR
American Electronic Laboratories Inc, 305 Richardson Road, Lansdale, Pennsylvania 19446, USA.

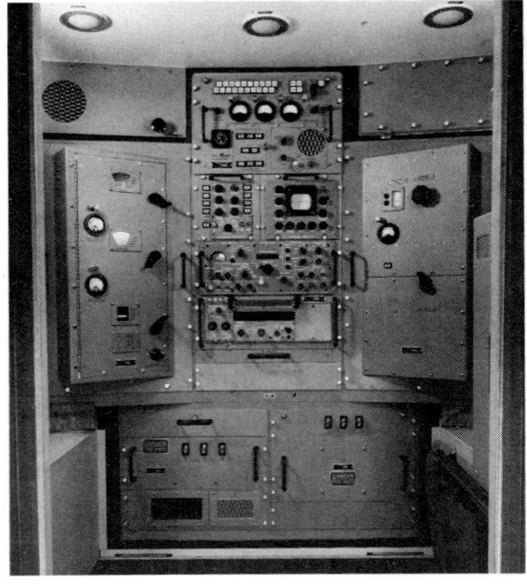

AEL AN/TLQ-15 communications and countermeasures set

6212.153
TACTICAL COMMUNICATIONS SURVEILLANCE SYSTEM

This is a tactical ESM system which intercepts, locates and monitors communication transmissions. Designed for one-man operation, each station covers the HF, VHF and UHF frequencies with automatic environment catalogue, surveillance, monitoring and direction finding modes.

Repetitive functions are automated and an interactive touch entry display screen simplifies control of automated features through menu-driven operation. Integration of the DF and monitoring functions eliminates time delays and errors in associating position fixes with intercept reports.

Multiple stations can be connected by wire or radio links to form a co-ordinated automatic DF and emitter position fixing network.

The self aligning VHF/UHF DF antenna is erected by a pneumatic mast which is integral to the shelter; the lightweight foldable HF/DF antenna is deployed in the field. The S-250 shelter which houses the electronics is environmentally controlled.

Four levels of built-in-test provide rapid fault isolation and operational checkout of major system components. A complete power panel provides power management and protection for all system equipment.

STATUS
In volume production.

CONTRACTOR
Sanders Associates Inc, Surveillance Systems Division, Hudson, New Hampshire 03051-907, USA.

A tactical communications surveillance system deployed in the field

6213.193
TACTICAL COMMUNICATIONS JAMMER SYSTEM

This is a transportable equipment designed for automatic jamming of multiple signals simultaneously on a priority basis. A single station with one operator can search the HF and VHF frequency bands, analyse acquired signals and jam multiple signals in each band.

Automatic features include scanning of preassigned threats, exclusion of friendly or no-interest frequencies and priority jamming. Data can be updated with commands to reflect the changing tactical situation or to take on new tasks and can be stored in a non-volatile memory or loaded from magnetic tape cartridges.

An interactive touch entry display screen simplifies control of automated features through menu-driven operation.

In the active mode, a look-through feature suspends jamming when a target transmission has stopped, immediately directing the system to other frequencies selected for jamming.

The system is rapidly deployable, the log-periodic VHF transceiver antenna is erected by a pneumatic mast which is integral to the shelter. The direction HF transmitting antenna is deployed on the ground. The S-250 shelter which houses the systems electronics is environmentally controlled.

Built-in features and modular design of the system's rack-mounted electronics assure rapid fault isolation and replacement. A complete power panel monitors voltages and protects against transients.

STATUS
In production.

CONTRACTOR
Sanders Associates Inc, Surveillance Systems Division, Hudson, New Hampshire 03051-907, USA.

Tactical communications jammer system deployed in the field

4406.093
AN/TLQ-17A COUNTERMEASURES SET

The AN/TLQ-17A countermeasures jamming and monitoring set has been designed to cover the HF/VHF portion of the frequency band.

Additional modules are available to extend the system capability into the UHF band. Using a modular packaging technique it is available for jeep, truck, shelter or helicopter applications.

Features include continuous HF/VHF coverage, stable synthesised receiver, wide band solid-state transmitter, complete remote capability, microprocessor controlled system operation, synthetic interactive CRT display pan indicator and alphanumeric display and complete built-in-test facilities with CRT readout.

CHARACTERISTICS
Frequency range: Band 1: 1.5 – 20 MHz; Band 2: 20 – 80 MHz
Effective radiated power: 550 W
Receiver tuning time: 1 s
Preselected frequencies: 256
MTBF: In excess of 400 h

STATUS
The AN/TLQ-17A is in production and is currently in service in both the Jeep mounted Traffic Jam configuration and the helicopter Quickfix system. Over 200 systems are in operational service. A reconfigured version to fit in an armoured shelter mounted on tracks has now been developed.

CONTRACTOR
Fairchild Weston Systems Inc, 300 Robbins Lane, Syosset, New York, USA.

AN/TLQ-17A deployed in the field

4409.153

AN/TSQ-112 TACELIS ESM SYSTEM

The AN/TSQ-112 is a location and identification ECM system operating in the frequency range 500 kHz to 500 MHz. The system is fully automatic and performs communication collection, transmitter location and processing functions. It is designed to combat hostile tactical communications radio by generating early warning, target and decision information, and jamming.

TACELIS (tactical automated communications emitter location and identification system) is a major component of the US Army TACOM-EWS (Tactical Communications Electronic Warfare System), and consists of two remote master stations and four slave stations. It is deployed with each operations company forward in one ten-ton truck and three six-ton vans plus a truck tractor. Remote master stations are deployed in three five-ton trucks and each remote slave station is carried by one M113 truck. The component elements of TACELIS include the various stations plus one AN/UYK-7 computer, 13 AN/UYK-19 mini-computers and AN/ULR-17 receivers. The remote slave stations have a direction finding capability only while each master station has 14 receivers and two search/acquisition receivers. The TACELIS system is inter-operable with forward control and analysis centres, the AN/MLQ-34 TACJAM (**3895.193**) via the AN/GRC-103 communications link directly out of the remote master stations, and the Cefly Lancer aircraft.

STATUS

This equipment is highly classified and is in operational service with the US Army. No recent contracts have been announced which indicates that the programme is now complete.

CONTRACTOR

GTE Government Systems, 100 Ferguson Drive, Mountain View, California 94039, USA.

6209.153

AN/TSQ-114 INTERCEPT AND DF SYSTEM

The AN/TSQ-114, known also as Trailblazer, is a local or remotely-controlled tactical, ground-based communications intercept and direction finder system. It is designed to provide real-time intelligence and combat information to commanders at division echelon and below. Frequency range is reported to be 0.5 to 150 MHz with direction finding accomplished between 20 and 80 MHz. The latest version is designated AN/TSQ-114B(V)2.

The system consists of two master control sets (MCS) and three remote slave sets (RSS) which are tied with a UHF data link. The MCS are also linked to a reporting facility. Each MCS is equipped to provide two operators with independent and duplicate capabilities. System-to-operator interface provides the system test, voice link, data storage, intercept, DF display and reporting functions. Additional features provide the operator with a system status monitoring via a caution and warning subsystem that gives both visual and aural alerts.

The Trailblazer system incorporates the following US Army standard equipment: AN/UYK-19B computer, RT-524A/VRC receiver/transmitter, R-442A/VRC radio receiver, RT-1167/ARC-164(V) receiver/transmitter, AN/UNH-17A recorder-reproducer set, CRC/1 recorder/reproducer, AN/UYQ-10 plasma display set and R-1444/UR radio receiver.

The electronics and operator positions are contained in a modified, ballistically protected S-280 shelter, mounted on a standard SIGINT/EW XM1015 tracked vehicle. The antenna group is attached to a 50 ft hydraulic-pneumatic quick-erecting mast group which is attached to the shelter roof, allowing rapid system deployment. A support trailer contains a 30 kW generator for power supplies.

STATUS

In operational service with the US Army as the primary communications and location system for heavy divisions. A product improvement programme has been completed and further upgrading is ongoing.

CONTRACTOR

ESL (a subsidiary of TRW Inc), 495 Java Drive, PO Box 3510, Sunnyvale, California 94088, USA.

Trailblazer vehicle deployed in the field

6841.153

COMMON JAMMER PROGRAMME

The US Army has instituted a programme to develop a common jammer (COMJAM) for use in the later part of the 1990s. The intention is to develop a jamming systems to replace the AN/MLQ-34 and AN/TLQ-17A systems and to ensure that the system is also capable of use in the airborne role for helicopters. The programme envisages a system which would replace existing systems with a common adaptive jammer suitable for use in both light ground vehicles and helicopters.

STATUS

Early conceptual design studies. It is likely that this system will not be available for operational use until the mid- to late-1990s.

6158.193

PIRANHA ADD-ON JAMMER

Piranha (OG-181/VRC) is a low-cost applique jammer that converts any VHF/FM communications equipment to a mobile jamming system. It consists of a power amplifier/control unit, a panoramic display and an antenna/mast assembly, all mounted in a compact rack designed for quick and simple set-up in a vehicle such as a Jeep or Land Rover. Once installed the Piranha system operates with the normal transceiver to generate high power jamming signals capable of disrupting communications in the VHF range. It can be easily transferred to another vehicle at any time.

The power amplifier/control homes in on hostile signals and provides modulation and power for the jamming signal. The power amplifier generates RF power of up to 1 kW ERP which is fed to the high gain antenna. The control unit tunes the jamming frequency and selects the modulation mode (noise, tone or voice). Either continuous or step tuning is possible. An LED display provides a rapid read-out of frequency.

The antenna can be aimed quickly and accurately in any direction. It is of log periodic design and can be rotated completely through 360° in azimuth, in horizontal, vertical and slant linear polarisation. A quick-erect allows the antenna to be raised to a height of 18 ft above ground.

STATUS

Operational with US forces.

CONTRACTOR

American Electronic Laboratories Inc, 305 Richardson Road, Lansdale, Philadelphia, Pennsylvania 19446, USA.

Piranha add-on jammer with antenna deployed

5810.153

D1800 DF AND MONITORING SYSTEMS

The D1800 series of DF systems provides coverage and signal monitoring over the frequency range from 100 MHz to 40 GHz, and is employed in ground-based, naval and airborne applications. The series consists of a rotating antenna, a pedestal, a control unit and a display. A common servo, control and display unit are used with a wide range of antennas, with sizes ranging from six inches to 15 feet. Single antennas are available covering wide bandwidths such as 0.5 to 18 GHz or 8 to 40 GHz with a single feed. Narrow bandwidths are also available. Omni-directional antennas, receivers, signal processors and receiver control/display units can also be supplied.

At least eight different antenna systems are available for ground fixed and mobile, shipborne and airborne applications. They are wide band and cover various frequencies in the range 0.1 to 40 GHz with weights varying from 14.5 to 255 kg.

STATUS

In production and world-wide operational use in aircraft, ships and ground-based installations.

CONTRACTOR

EM Systems Inc, 45757 W. Northport Loop, Fremont, California 94538, USA.

6144.193

S500 ESM SYSTEM

The S500 incorporates two receiver technologies; IFM receiver for high probability signal acquisition and wide band signal analysis, and superheterodynes for sensitivities up to –11 dBm. The frequency coverage is from 500 MHz to 18 GHz. High gain, narrow beam antennas in the microwave band provide high sensitivity and accurate direction finding.

The S500 can function as either a stand-alone system or can be integrated into a signal collection system consisting of three or more collection stations. It can be utilised in ground, shipborne and airborne applications, has performed with remote computer interface, and is provided with auxiliary outputs for fine grain analysis. Intercept is stored digitally for further analysis by magnetic tape recorders or by hard copy printout. Frequency hopping and chirp emitter can be displayed on a unique signal display.

The equipment is completely modular for easy configuration to meet specific requirements. Various system building blocks and capabilities are available, including tuners, analysis demodulators, IF pan displays, RF spectrum displays, time base displays, and parameter displays for any part of the band coverage.

CONTRACTOR

EM Systems Inc, 45757 W. Northport Loop, Fremont, California 94538, USA.

6145.193

R4000 SUPERHETERODYNE RECEIVER

The R4000 receiving system is a modular equipment using the latest microwave and digital technology. It is ideally suited for ESM surveillance, reconnaissance and frequency control applications. Performance from 20 – 40 GHz is achieved by microprocessor controlled tuners, automatic signal analysis capability, and logical display formats.

The system provides optimum performance and reliability, real-time signal processing and computer control capability, graphic and alphanumeric display of data, automatic/semi-automatic/manual operation, RS232 or IEEE-488 digital interfaces, plus provisions for hard copy printout as well as data storage on disk and tape.

Up to 16 low-noise phase-locked RF tuners, octave or wide band, in any combination may be used to monitor the frequency bands of interest. Sixteen frequency monitoring routines may be programmed for the system, all available for instant call-up. Signal analysis using selectable IF bandwidths can be accomplished automatically, operator assisted or completely manually depending on the configuration of the R4000 selected. Displays include RF spectrum analysis, IF pan, signal list, library list, signal data, and a polar display when a DF capability is chosen.

CONTRACTOR

EM Systems Inc, 45757 W. Northport Loop, Fremont, California 94538, USA.

6146.193

OE-320/A DF SYSTEM

The OE-320/A DF system provides frequency coverage over the range 0.5 – 18 GHz. The system is microprocessor controlled and can be operated through an operator/front panel interface, or through digital remote control. The system is completely self-contained and can be used in conjunction with any receiver system. The rotation speed is variable from 0 – 200 rpm. Three control modes, Scan, Sector and Track, are provided. Extensive automatic built-in test has been incorporated.

CONTRACTOR

EM Systems Inc, 45757 W. Northport Loop, Fremont, California 945538, USA.

5819.093

TC-5100 SERIES DF SYSTEMS

The TC-5100 series of direction finders has been developed to conform with the US Army requirements for the 'AirLand Battle 2000' plan. The equipment provides monitoring, intercept and bearing information of signals over the frequency range from 1.5 MHz to 1 GHz. Information is displayed as a three-digit LED display and a circular compass point.

The system employs a small broadband slot antenna with a shape and size that allows for rapid deployment and redeployment without the need to assemble and re-assemble any of the elements. The antenna is designed to be mast-mounted on a vehicle and lowered during movement to a height of only a few inches above the vehicle or shelter roof-line. In certain types of shelter, the antenna can be lowered below the roof-line and still provide bearing information while the vehicle is moving. The antennas are small discs, varying in diameter from 22 to 34 inches (56 to 86 cm) and measuring less than six inches in height. They can be easily mounted or concealed on a variety of mobile ground, shipborne or airborne platforms.

STATUS

In production and in service with the US Army.

CONTRACTOR

Tech-Comm Inc, Fort Lauderdale, Florida, USA.

TC-5100 DF system

6159.193

SCR SERIES RECEIVING AND MONITORING EQUIPMENT

Scientific Communications designs and manufactures a wide range of receiving systems used for surveillance and monitoring purposes. Over four generations of microwave surveillance receiver and pulse analysis products have been developed. Information on a selection of the company's receiver products is given in the following paragraphs.

SCR-2100M Surveillance Receiver

This is a completely solid-state receiver that provides spectrum surveillance, acquisition and analysis capability in the 0.1 – 18 GHz range. The system, also known as the AN/ALR-75(V), can be used in ground or airborne installations and consists of either a 19-inch rack mount, or aircraft style control panels with several remote controlled units in ATR configurations.

The use of independent octave band tuners allows simultaneous panoramic reception over the microwave spectrum, or a mix of multiple tuners on a given band. Any combination of eight tuners can be used. Phase lock frequency accuracy is achieved with a frequency synchroniser and other features include digital control, scratch pad memories, optimised search and acquisition capability, signal identification from memory comparison and digital refreshed displays.

SCR-2100 (AN/ALR-75) in a portable shelter

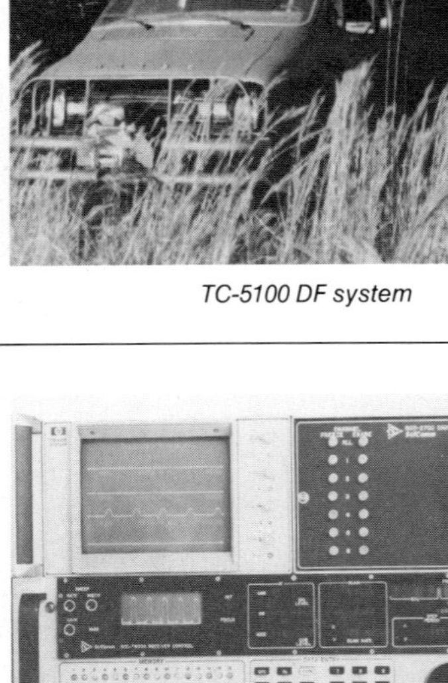

SCR-2700 intercept receiver system

SCR-2700 Intercept Receiver System

This is a semi-automatic search and analysis system with a range of bandwidth tuner configurations which enable the system to cover various bands of frequencies over the range 0.5 – 40 GHz. System components include the RT series of remote tuners, the SCJ-2700 IF matrix, SCC-2700S Slave Remote Control Unit, SCD-2700 Digital Refreshed Display, SCC-7803A Receiver Controller, SCP-2160 Pulse Analyser, and SCP-2700 Dedicated Demodulator.

SCR-2725/2726 Pan Systems

These two SIGINT systems are designed for very fast frequency scan rates and use many components of the SCR-2700. The SCR-2725 covers the range 20 to 1200 MHz and the SCR-2726 from 0.5 to 18 GHz.

SCR-7201 VHF Receiver

The SCR-7201 is a compact synthesiser-tuned receiver designed for both set-on monitoring and limited search. It is available in different versions to provide coverage of various frequency bands, and with a wide range of IF bandwidths. Frequency ranges are 30 – 80 MHz, 60 – 110 MHz, 100 – 210 MHz, 120 – 170 MHz, 160 – 210 MHz and 200 – 250 MHz.

SCR-7202 VHF/UHF Receiver

The SCR-7202 is a full-function, high performance receiver, covering the frequency range 20 – 500 MHz, complete with preselection, and packaged in a $1\frac{3}{4}$-inch (44.5 mm) high by $8\frac{1}{2}$-inch (216 mm) wide chassis. A companion unit, the SCC-7211 frequency extender, works together with the SCR-7202 to provide coverage out to 1100 MHz.

SCR-7204 VHF Receiver

This is a compact synthesiser-tuned radio designed for both set-on monitoring and search of communications signals over the frequency range 20 – 500 MHz. It provides tuning in 1 kHz steps, and simultaneous AM and FM (SSB optional).

CONTRACTOR

Scientific Communications Inc, 2908 National Drive, Garland, Texas 75041, USA.

4008.193

SNAP-1 STEERABLE NULL ANTENNA PROCESSOR

SNAP provides interference protection for VHF/FM tactical communication radios. This equipment was developed jointly by the Sincgars Project Manager and the US Army Electronic Warfare Laboratory for use with the AN/VRC-12 family of radios.

The SNAP-1 processing system consists of two antennas, the SNAP processor and the VHF/FM transceiver to be protected. The processor automatically causes a strong interfering signal to be degraded to a level below a weaker desired signal by pointing a spatial null in the direction of interference. The spatial null is achieved by insertion of the appropriate amplitude and phase corrections in the antenna paths so that a signal subtraction occurs at the arrival angle of the jamming signal.

The SNAP system provides ECCM protection of at least 35 dB (3100 to 1) and is effective against stationary and moving jammers. Operation is automatic on the move or at a halt. It discriminates between friendly communication signals and undesired interference, is transparent to communications channels and requires no special signals. The nulling process occurs during reception of signals only, therefore SNAP protection cannot be detected by enemy radio frequency DF equipment and it uses standard antennas and hardware.

In a typical SNAP installation, a standard AN/VRC-12 receiver/transmitter is used together with its whip antenna. A second standard antenna is added, separated from the first antenna by about 1.5 m. SNAP is installed at any convenient location and cable interconnected with the antennas and the AN/VRC-12. There are no modifications to the radio or the antennas.

STATUS

SNAP-1 is in operational service with US Army and Marine Corps units.

CONTRACTOR

Hazeltine Corporation (an Emerson Electric Company), Greenlawn, New York 11740, USA.

CP-1380/VRC (SNAP) system

5951.153

CHIEF EW COMBAT SYSTEM

Chief is a vehicle-mounted integrated EW combat system that incorporates ESM, ECM, DF, computer and communication subsystems. It is packaged into a commercial or military $1\frac{1}{2}$ ton truck and is modularly designed to allow for reconfiguration or expansion. In addition to the basic EW equipment, the vehicle contains a roof-mounted antenna subsystem, automatic microprocessor control, a power supply, and a software control system for integration of the various equipments. Although designed primarily for vehicle applications, the system can be reconfigured for naval, airborne or fixed ground station use.

The ESM subsystem detects, monitors, characterises and records fixed frequency communication transmissions. This subsystem provides continuous coverage from 0.5 to 500 MHz (expandable to 1200 MHz), of AM, FM, LSB, USB and CW signals, and provides multiple bandwidths. It incorporates automated, panoramic spectrum displays of sectors being searched. Operator search may be manual or automated, and 100 frequencies may be prioritised for lock-in, lock-out or a combination of both.

The ESM subsystem also uses four audio cassette recorders. The recorders provide two or four channel operation with automatic start/stop. In addition, the Chief intercept capability is enhanced through the coordinated use of receivers in the ECM, DF and communications subsystems.

The Chief DF subsystem provides the capability to sort out emitters by frequency, line of bearing, and location (by triangulation from multiple bearings). Multiple bearings may be obtained from a single Chief by changing location or by individual intercepts by multiple units. The DF subsystem consists of two receivers with frequency ranges of 0.5 to 80 MHz and 20 to 520 MHz. A frequency translater option can extend the DF band to 1200 MHz.

The ECM subsystem detects, monitors and disrupts voice communications covering the frequency range of 2 to 500 MHz by combining manual HF and automated VHF/UHF equipment. Both use high efficiency antennas. In HF Chief will jam using a manual transceiver although this may be upgraded and automated if required. In the VHF/UHF scanning mode, the automated jammer scans the frequency band of interest to locate and jam targets. It features a large target table capable of concurrent jamming of up to 12 targets with a 120° radius.

The ECM subsystem transceiver also provides demodulated audio output. In addition, the ECM transceivers can be used for very high powered voice communications covering the frequency 2 to 500 MHz. The Chief system can break through enemy jamming or screen friendly communications using this capability. The standard ECM subsystem has an effective radiated power of up to 2 kW.

The control subsystem consists of computer equipment and software which directs, monitors and controls all system functions. Each individual component may be operated manually, by its own processor, or through the ruggedised MICRO VAX. The latter provides 4 Mbits of virtual memory and 32 Mbits fixed disk. The control subsystem also includes a colour graphics display monitor with alphanumeric keyboard. The control subsystem interfaces to several other Chief subsystems to collect status, send commands and monitor tasking.

The software aims at providing maximum display capability. It consists of menu and graphic software. This provides operational capability including data base functions, computational support, graphics display, man-machine interface support and interface with the DF and location hardware. It also includes hardware control of components (eg receivers, jammers, etc).

STATUS

Integration of the complete system, and development of the control system and software is being carried out by HRB-Singer. The first prototype has been completed and is on trials.

CONTRACTOR

HRB-Singer Inc, PO Box 60, Science Park Road, State College, Pennsylvania 16804, USA – prime contractor.

Chief vehicle-mounted EW combat system

3368.193
COMBAT VEHICLE SACS

The US Army has developed a Semi-Automatic Counterfire System (SACS) for combat vehicles. It has been tested by the US Army in an M60 tank and in the High Mobility Multipurpose Wheeled Vehicle (HMMWV). SACS is produced by the Missiles Division of LTV Missiles and Electronics Group.

SACS uses microprocessor electronics to detect and counter ground-based and airborne threats. Integration of electro-optical and other sensors with the fire control system allows timely response to threats and improves vehicle survival. In addition to providing a warning to the whole crew, the system simultaneously displays the threat information to the commander and semi-automatically initiates a response. Manual override can be activated when required.

In addition to counterfire and evasive manoeuvres, flare, smoke and jammer countermeasures can be used. Current development work is on the integration of alternate passive warning sensors, information transfer through secure data links and unit cost reduction measures to make system costs attractive for potential high volume production.

STATUS
Pre-production for the US Army.

CONTRACTOR
LTV Missiles and Electronics Group, PO Box 650003, Dallas 75265-0003, Texas, USA.

5741.193
HAND-EMPLACED EXPENDABLE JAMMER

The hand-emplaced expendable jammer (HEXJAM) is a solid-state device designed for use against communication links in the HF and VHF bands. It uses wide band frequency barrage jamming that operates continuously against threat signals. It can be used for screening and protecting friendly communications and for supporting countermobility operations by disrupting enemy communications. It is small and lightweight, easily man-portable, and the rugged design allows assured operation. Panel switches allow the user to select the desired function modes and delay times. A self-disable feature will render the unit inoperable.

HEXJAM is re-usable and capable of multimission operational and training use. It is claimed to be effective against both spot jammers and frequency-hopping radios. It has a selectable activation time delay and can be turned on from a remote location using ordinary bell wire.

STATUS
Produced for and delivered to the US Army.

CONTRACTOR
Fairchild Weston Systems Inc, 300 Robbins Lane, Syosset, New York 11791, USA.

6139.193
ARTILLERY DELIVERED EXPENDABLE JAMMER (AD/EXJAM)

The XM867 AD/EXJAM artillery delivered expendable countermeasures equipment is being developed by Fairchild Weston Systems, in conjunction with the US Army Communication Electronics Command. It is a low cost, gun rugged, broadband barrage transmitter capable of disrupting enemy radio communications.

The equipment consists of a standard M483A1 cargo round containing five ECM devices. The rounds are delivered from normal 155 mm howitzers into preplanned target areas, the five devices being ejected as the projectile travels along its flight trajectory. Each device contains a separation system, gun activated power supply, digital timing and RF electronics, self erecting antenna, self deploying ground radials, and retardation and self-erecting system and despin fins.

A complete system consists of the five devices stacked within the projectile in the order of preset ejection times, a nose switch assembly with a reserve battery which is activated during firing, a flex cable connecting the nose switch assembly to a base plug ejection system, an ejection system consisting of an electrically initiated gas generator and a black powder bag, and a standard M577 time fuse which is installed at the time of firing.

To deliver to a preplanned area, the fuse air burst time is selected and set, and the fuse is installed in the projectile. Gun elevation and propelling charge are chosen from standard firing tables. Once fired, the fuse set time starts a series of events that results in the sequential ejection and precise placement of five ECM devices on the ground with antenna erected, ground radials deployed, and the transmitter switched on.

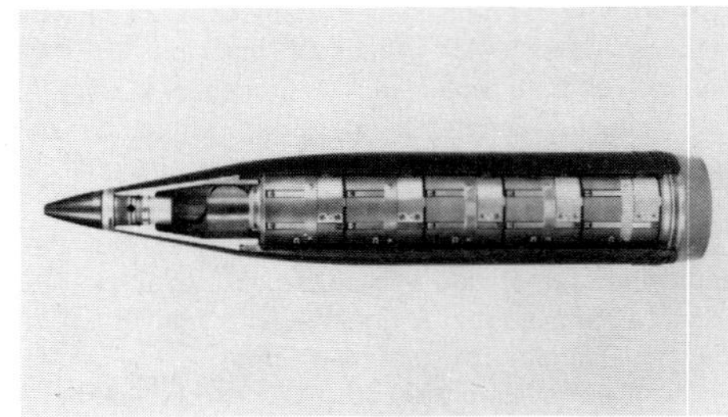

Cutaway view of an artillery delivered expendable jammer

STATUS
In engineering development and evaluation under sponsorship of the Signal Warfare Laboratories of the US Army Communication Electronics Command.

CONTRACTOR
Fairchild Weston Systems Inc, 300 Robbins Lane, Syosset, New York 11791, USA.

4663.193
LQ-102A HAND-EMPLACED EXPENDABLE JAMMER

The LQ-102A is a low-power continuous FM-by-noise jammer intended for hand-emplacement near the intended target's receiver. Two VHF bands are selectable for 'threat' and 'training' functions. The unit is of small size and can be easily transported and deployed.

The jammer features ON/OFF remote control, via a wire pair, up to at least 10 km. Delay of jamming operation is operator selectable in 6 minute increments up to 30 hours. Continuous jamming in excess of five hours is claimed. An electrical destruct mechanism is fitted to the unit.

CHARACTERISTICS
Dimensions: 27.3 × 11.4 × 6.4 cm
Weight: 2.3 kg with battery
Antenna: Folding vertical monopole
Output: 10 W minimum ERP

CONTRACTOR
Cincinnati Electronics, 2630 Glendale-Milford Road, Cincinnati, Ohio 45241, USA.

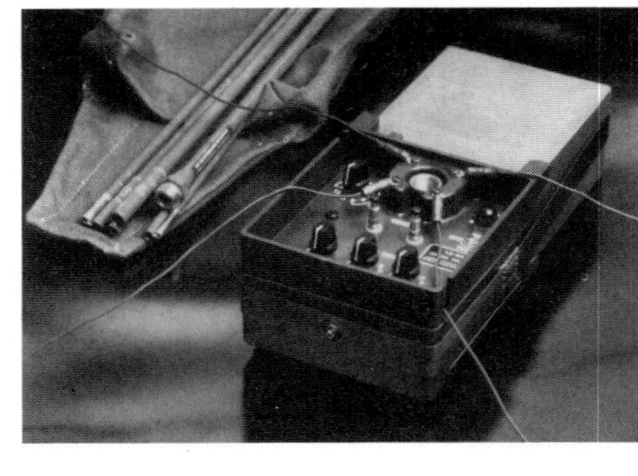

LQ-102A expendable jammer

4658.053
MULTIPLE RADAR EMITTER SYSTEM

The Multiple Radar Emitter System (MRES) is a flexible set of ECM and C³CM threat generators that support personnel training and equipment tests and evaluations in dense electronic operational environments. HF, VHF and UHF communications jammers are available with either high power or moderate power transmitters. Radar, navigation and IFF jammers are also available over the frequency range from 500 MHz to 18 GHz. The jammers can be in either manned or unmanned airborne or field deployable configurations. These threat generators simulate threat signal electromagnetic radiation patterns, modulation techniques and tactics of known or possible threats.

STATUS
In production.

CONTRACTOR
Flight Systems Inc (Tracor), PO Box 2400, Newport Beach, California 92660, USA.

6463.153
MOBILE EW SUPPORT SYSTEM (MEWSS)

The Mobile Electronic Warfare Support System (MEWSS) is a tactical system under development for the US Marine Corps. It is a EW suite of equipment, mainly chosen from 'off-the-shelf' hardware, and designed for installation in the new light armoured vehicles currently entering service. It is intended to provide the ground commander with a mobile EW system capable of operating in a variety of tactical situations.

MEWSS will offer three capabilities, direction finding, communications and jamming, and is designed for operation in the most severe electronic environments. The complete system is likely to consist of the WJ-8618 receiver, AN/PRD-10 direction finder and the AN/VLQ-19 jammer.

STATUS
In development.

6836.193
AN//VLQ-5(V)1 COMFY SWORD

Comfy Sword forms part of the US ADEWS/Comfy Challenge project (**6835.153**) to counter large scale conventional forces attack. Comfy Sword is a series of highly mobile self-contained jamming and deception vans and Comfy Sword III is being developed as a completely modular system to complement Comfy Challenge. It will be used in USAF and combined exercises to inject a C³ countermeasures jamming element to counter interference with USAF air-to-ground communication links.

As a stand-alone equipment, Comfy Sword III can be deployed on a variety of surface and airborne platforms near the forward edge of the battle or exercise area to provide effective jamming using directional and/or omnidirectional antennas.

STATUS
The first phase under contract to Fairchild Weston involves the production of 12 jeep-mounted jammers operating in the VHF and UHF frequency bands.

CONTRACTOR
Fairchild Weston Systems Inc, Electronics Systems Division, 300 Robbins Lane, Syosset, New York 11791, USA.

6835.153

ADEWS (AIR DEFENSE EW SYSTEM)

ADEWS is a mobile ground-based system which is a joint development by the US Army and Air Force and includes the original Comfy Challenge programme. It is intended to counter large scale conventional attack by intercepting and disrupting hostile C³ links and will complement the fire control radars of SAM systems such as Patriot, I-Hawk and the new FAADS weapon. It is a passive and active system to provide detection and jamming of enemy aircraft attempting to penetrate at low altitude.

In its passive mode ADEWS allows air defence systems to detect, locate, identify and track penetrating aircraft without use of surveillance radars. This enables the SAM fire control radar to be used at the last moment and, consequently, largely negates any attempt by the intruding aircraft to use anti-radiation missiles. In an active mode ADEWS uses high power ECM equipment to jam and deny the use of radar frequency aids by hostile aircraft.

The ADEWS ESM capability draws a great deal of technology from the multibeam lens array of the Raytheon AN/SLQ-32 which is in widespread use by the US Navy. The ADEWS operator monitors all activity in a 90° sector of operation, steerable through 360°. Display interface and an on-board threat library provides immediate emitter identification, location, bearing and aircraft classification. The operator may initiate jamming response manually or it may be automatically programmed. The multibeam ECM transmitter provides very high effective radiated power and is capable of addressing multiple threats simultaneously.

ADEWS is designed for tactical mobility and rapid deployment and is installed in a S-280 shelter mounted on a M-927 vehicle.

ADEWS system

STATUS

The ADEWS system is a highly classified project being developed for US Army and Air Force use. The original Comfy Challenge system, designed for the USAF, has now been incorporated in the overall ADEWS programme. Two advanced development models were scheduled to be delivered for testing during 1987 and, as a result of this testing, decisions as to how the programme should proceed were due to be taken in early 1988. See also Comfy Sword (**6836.193**).

CONTRACTORS

Raytheon Company, Electromagnetic Systems Division, 6380 Hollister Avenue, Goleta, California 93117-3197, USA.

GTE Government Systems, Electronic Systems Division, Goleta, California, USA.

6842.153

ASAS/ENSCE

The All-Source Analysis System (ASAS) programme for the US Army is intended to provide the tactical commander with a highly automated capability at division, corps and higher echelons to analyse, correlate, integrate and report intelligence data gathered from field EW units and from numerous strategic and tactical sensor systems. ASAS is a key element in the US Army's Command and Control System. The US Air Force's Enemy Situation Correlation Element (ENSCE) will share information with the ASAS system, will correlate and aggregate multi-sensor data, provide precise location of hostile force structures and provide ground battle situation displays to support the Tactical Air Control Centres.

ASAS/ENSCE is a joint US Army/Air Force development and forms part of the Joint Tactical Fusion programme. The programme innvolves a large number of contractors with two types of prime contracts; Type A for communications and module integration and Type B for an intelligence system. Parts of the system are the AN/TSQ-130(V)1 SIGINT subsystems (SEWS) which will be used at corps level and the AN/TSQ-130(V)2 for divisional level.

STATUS

Although parts of ASAS have been used in exercises the complete system is still in development. Planned production is likely to commence in 1992.

YUGOSLAVIA

6371.193

LASER IRRADIATION DETECTOR

The Laser Irradiation Detector (LID) is a laser warning equipment designed primarily for use in mobile land and seaborne combat and transport vehicles. It responds to ruby.GaAs and Nd.YAG laser systems. The LID set consists of a detector head and a display box. The detector head includes five detectors with corresponding optics, videos and buffer amplifiers, and pulse forming and stretching networks. The optical part of the detector head is constructed to divide the hemisphere above the vehicle into five equal fields of view of approximately 135° × 135°. The head incorporates a circuit that minimises errors caused by laser beam reflections from nearby objects.

The display unit contains the warning signal circuitry, a nine-element LED display, acoustic warning signal generator and power supplies. The LID can display 17 directions, each indicating a conical 46° sector, and also provides a two second aural warning which is injected into the crew headset.

CONTRACTOR

Federal Directorate of Supply and Procurement, 9 Nemanjina Street, 11001 Beograd 9, Yugoslavia.

LID detctor head

ELECTRO-OPTICAL EQUIPMENT

Note: This contains details of electro-optical (also referred to as opto-electronic and/or optronic) equipments. For the convenience of users of *Jane's Weapon Systems* electro-optical equipments have been divided into three sub-sections, devoted to land, sea and air applications. The sea and air sub-sections will be found later in this book in the shipborne and airborne sections respectively. In those cases where an item may be employed in more than one role, it will be found, as a general rule, within either the sub-section relating to its initial application, or the sub-section appropriate to the most common use of the equipment.

Within the complete section readers will find examples of purely optical devices, such as sights and sighting equipment, as well as others which involve the use of different parts of the spectrum beyond the visible region. As a matter of convenience, a number of items will be found in which the use of optical components is incidental (albeit essential). Such entries are mostly related to equipment which has as its main function some role such as weapon delivery, which is a prime objective of most of the entries in this section.

More specifically, the section contains entries concerned with optical sights for land, sea and air use; laser equipment for rangefinding and target designation or marking purposes; infra-red techniques applied to such functions as passive surveillance and detection, thermal imaging sights, night driving periscopes, and forward looking infra-red for aircraft and other uses. From this brief summary it will be appreciated that most of these applications are related closely with one or more of the other aspects of military technology which are the subject matter of this volume, and that certain of them (elaborated to a greater or lesser degree) could equally well be assigned to another part of the book. For example, many of the newer electro-optical (EO) naval tracking equipments are capable of being expanded, with little additional equipment, to form a naval fire control system for small vessels. Similarly, many of the infra-red techniques employed in equipment in this section have applications in the fields of reconnaissance and countermeasures. For such reasons as these, users with specific interests in such areas should also consult related sections such as Army and Naval Fire Control Systems, Electronic Warfare, etc.

Generally speaking, those electro-optical sights specifically designed for infantry small arms, eg infra-red sights and laser rangefinders for rifles and machine guns have not been included. The reader is referred to a companion year book, *Jane's Infantry Weapons*, which carries details of these equipments.

ELECTRO-OPTICAL EQUIPMENT (LAND)

BELGIUM

4743.193

LRS 4 Mk II FIRE CONTROL SYSTEM

This is a portable battery-operated laser rangefinder with a built-in ballistic computer. It is claimed to be one of the smallest fire control systems available and is intended for day and night operation. It is designed for direct firing on weapons such as small turrets on armoured vehicles, recoilless guns, transportable light anti-tank guns and portable anti-tank weapons. Depending on the application, the system can be operated by the built-in rechargeable battery or by an independent battery, eg the vehicle supply.

The LRS 4 Mk II consists of a ×7 day channel, a ×7 image intensifier night sight, an Nd:YAG laser rangefinder, a digital ballistic computer programmed for two different types of ammunition, a computer-driven ballistic aiming mark, a ballistic reticle for emergency firing and a rechargeable battery.

The reticle configuration is identical for night and day observation. There are three separate groups of reticles: a laser cross with moving target marks, a red cross ballistic aiming mark and a ballistic reticle. Most of the controls for the gunner are push-button types, integrated into the keyboard, either on the instrument itself or a remote control unit.

CHARACTERISTICS

Day and night sights

Magnification: ×7

Field of view: ×5.5°

Dioptre adjustment: –3 to +3 dioptre

Laser

Type: Nd:YAG

Wavelength: 1.064 μm

Resolution: 5 m

Accuracy: ±10 m

Range: 200 to 5000 m

Minimum range gating: 200 to 3000 m

STATUS

In production.

LRS 4 Mk II fire control system

CONTRACTOR

OIP Optics, Meersstraat 138, 9000 Ghent, Belgium.

4744.193

LRS 5 FIRE CONTROL SYSTEM

The LRS 5 is a compact day/night laser rangefinder with a built-in ballistic computer. It is designed for use in most types of battle tanks and other armoured vehicles. The LRS 5 is a monobloc, fully integrated system with no separate boxes except for a small overriding control box for the vehicle commander.

The main components of the system are: a unit vision eyepiece, a ×8 power day sight, a ×4 power second generation image intensifier night sight, an Nd:YAG laser rangefinder, a digital ballistic computer able to be programmed for 4 types of ammunition, a moving target tracking system, a computer-driven ballistic aiming mark, a manual range input as the first emergency mode and a ballistic reticle as the second emergency mode. The system provides a firing sequence identical for both day and night modes of operation, a manual correction for secondary parameters such as temperature and humidity and a visual warning to both commander and gunner for every control being operated.

A number of optional items are also available including an automatic computer-controlled correction for cant angles and a slaved commander's sight allowing 100 per cent override facility. In addition, an agreement between OIP and THORN EMI (UK) has provided for the integration of a thermal imager night sight. This is based on the latter company's Multi-Role Thermal Imager (**4260.193**) which is in operational service. The thermal imager can replace the existing image intensifier night elbow, and will also be compatible with the fire control computer, thus offering identical firing sequences for both day and thermal sights.

CHARACTERISTICS

Magnification: ×8 (day sight); ×6 (night sight)

Field of view: 7° (day sight); ×4.8° (night sight)

Laser

Type: Nd:YAG

Wavelength: 1.064 μm

Radiant energy: 8 mJ (typical)

Ranging resolution: 5 m

Ranging accuracy: ±10 m

Range: 200 to 9995 m

Minimum range gating: 200 to 4000 m

STATUS

Over 700 systems sold to six customers, including 200 to Malaysia. The thermal imager module has now been incorporated in a version of the LRS 5.

LRS 5 fire control system with tank thermal sight

CONTRACTOR

OIP Optics, Meersstraat 138, 9000 Ghent, Belgium.

6521.193

LRS 7 FIRE CONTROL SYSTEM

The LRS 7 is a monobloc laser rangefinder for armoured vehicles. The instrument is intended for day and night operation, detection, recognition, identification and aiming of targets. With its anti-airstrike capabilities, the LRS 7 offers an efficient solution for smaller turrets. In order to improve accuracy and reaction time of gun firing a number of main functions are incorporated. These include a ×7 magnifying day sight, a ×7 magnifying night sight, an output mirror mechanically slaved to the gun, and a episcope. The reticle configuration for both day and night observation is identical.

Options include a Nd:YAG laser rangefinder and a ballistic computer, both fully integrated.

CHARACTERISTICS

Day and night sight (image intensifier)

Magnification: ×7

Line of sight angles: –10 to +60°

Laser

Type: Nd:YAG

Wavelength: 1.064 μm

Radiant energy: 8 mJ typical

Ranging resolution: 5 m
Accuracy: ±10 m
Total range: 200 to 9995 m
Minimum range gating: 200 to 3400 m

STATUS
In production.
CONTRACTOR
OIP Optics, Meersstraat 138, 9000 Ghent, Belgium.

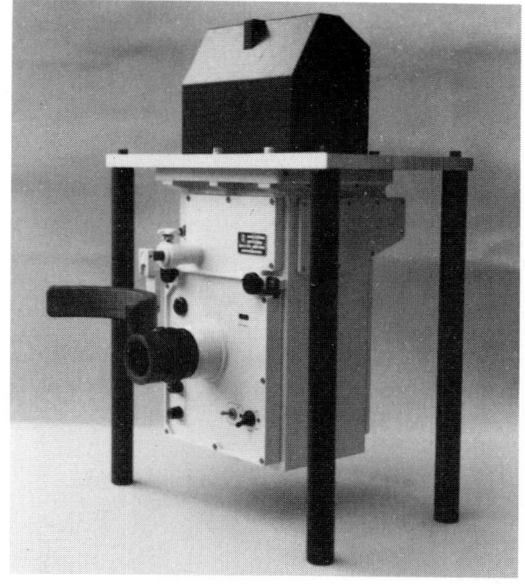

LRS 7 fire control system

CHINA (PEOPLE'S REPUBLIC)

5890.153
CHINESE ELECTRO-OPTICAL SYSTEMS

Until comparatively recently virtually nothing was known about the optical and electro-optical systems used by the Chinese armed forces. However over the past year or so a few details have emerged, particularly since China has appeared as an exhibitor at various exhibitions, albeit in a very small way. Much of the equipment used owes a great deal to earlier Russian design, and nothing is available on any research and development work being carried out in this field. It is known that China has an indigenous radar development programme under way and it seems likely that the same would apply to electro-optics, particularly laser, thermal imaging and TV based systems. Information recently obtained from China is given below, although details are somewhat limited and the only modern equipment is the LR1 laser rangefinder.

LR1 laser rangefinder

LR1 Short-range Artillery Laser Rangefinder
This is a tripod-mounted instrument used by the artillery's reconnaissance and command detachment. It is used for observation, rangefinding, angle measurement and direction. Power is supplied by a silver-zinc battery set. Range is given as 200 to 7000 metres, with an accuracy in range of ±10 metres and in azimuth of ±3 mils. Weight is 36 kg including the silver-zinc battery power supply.

Type 82 laser rangefinder
This is a Nd:YAG laser rangefinder for tanks with a ranging capability from 300 to 3000 metres and an accuracy of ±10 metres. The measured range is indicated by a three-digit display. Radiant energy output is 15 mJ on a 1.06 μm wavelength. Magnification is ×8.

Type SC-83-II hand-held laser rangefinder
The hand-held laser rangefinder model SC-83-II provides target observation and ranging, and is intended primarily for use by artillery forward observation posts and reconnaissance groups. It has a ranging capability of greater than 10 kilometres and an accuracy of better than ± 5 metres. Weight of the complete unit is 2.4 kg.

SR 0.5 0.5m Artillery Rangefinder
This is a conventional tripod-mounted optical rangefinder used by the artillery. It has a magnification of × 10 and a field of view of 6 degrees with a measuring range of 300 to 4000 metres. The weight of the rangefinder itself is 3.6 kg.

SR 0.8 0.8m Artillery Rangefinder
This is another conventional tripod-mounted optical rangefinder. It has a × 12 magnification, field of view of 6 degrees and a measuring range of 400 to 14 000 metres. Total weight including tripod, angle measuring mechanism, and carrying boxes is 29.3 kg.

ASR 1m Anti-aircraft Rangefinder
This hand-held stereoscopic rangefinder is used to measure the slant range of air targets from either land or sea. Magnification is × 10 with a 6 degree field of view. The measuring range is 500 to 6000 metres. The rangefinder weighs 31 kg.

ASR-3-2 3m Anti-aircraft Rangefinder
This rangefinder is employed in conjunction with the fire control director described below. It has a measuring range of 750 metres to 100 km with a magnification of × 32. The field of view is 1.8 degrees, and the weight of the complete rangefinder 252 kg.

GM-03 Anti-aircraft Fire Control Director
This is a precise tracking and computing equipment for 57 mm guns. Using information provided by a fire control radar or a rangefinder the GM-03 calculates the target lead through servo systems and evaluates the firing data for the guns. The complete system consists of an analogue resolver, rangefinder ASR 3-2, and a trailer for carrying and transporting the two units. It is designed to work with one fire control radar and one to eight 57 mm guns.

AS 8 Battery Commander's Telescope
This is a tripod-mounted field telescope used for observation, and measurement of angles and target distance. The magnification is × 8 and the field of view 7 degrees.

AS 16 Battery Commander's Telescope
Similar to the AS 8 but has a magnification of × 16 and a field of view of 4.5 degrees.

AC1 Aiming Circle
Tripod-mounted and used for measurement of magnetic bearings, azimuth and elevation. Also used for observation and, when fitted with a range reticle, to measure distance to target.

Q1 Mortar Quadrant
Used to provide mortars with a firing angle.

CONTRACTOR
China North Industries Corporation, 7A Yuetan Nanje, Beijing, People's Republic of China.

DENMARK

4237.193
TYPE 772 SIGHTING SYSTEM

This tank sighting system employs the JAI Type 772 low light-level television camera which is similar to the Type 771 produced by the same company for naval applications (**3857.293**), in this case employed with an automatic ballistic calculator and an input from a laser rangefinder to provide a day/night sighting system for use in main battle tanks. The data generated by the system is presented on a TV monitor screen together with the camera picture of the outside scene. These facilities are available to both the gunner and the tank commander, each having his own monitor display.

CHARACTERISTICS
Field of view: 64 × 48 mrad
Focus: 30 m to infinity
Range: Dependent on guns and ammunition (typically 0 – 3995 m)
Sensitivity: 2 × 10^{-6} lux, fp (100% contrast, 200 lines)
Recognition limit: 1.5 m target at 1000 m with 10^{-4} lux illumination (target reflection 30%, contrast 50%)
Automatic fire flash protection

STATUS
In production. JAI also manufactures other low-light TV systems for military purposes, including man portable types.

CONTRACTOR
JAI – Jorgen Anderson Ingeniofirma AS, Produktionsvej 1, 2600 Glostrup, Copenhagen, Denmark.

JAI Type 772 low light-level television camera for tank fire control

FINLAND

6523.093

NOKIA ELECTRO-OPTICAL EQUIPMENT

Nokia Electronics is engaged in coastal defence and battlefield surveillance systems and manufactures a number of electro-optical equipment for applications in these fields, and for airborne applications. Examples of this are as follows.

MA 8560 hand-held laser rangefinder

This is designed for a number of applications, such as artillery observation, infantry use, naval fire control and airborne applications. It employs an Nd:YAG transmitter with a radiant output energy of 5 mJ and a beamwidth of 1 mrad. Range of the instrument is given as 150 to 9995 metres, with an accuracy of ±5 metres and a range discrimination of 30 metres. Weight of the unit is 2.2 kg. The MA 8560 includes a number of features, including an auxiliary digital display unit showing the five latest range measurements, enhanced accuracy with an averaging function, auxiliary remote triggering, minimum range controller enabling all targets in the aiming direction to be ranged with only one measurement, two separate displays in the rangefinder and fall-of-shot measurement.

MA 8530 laser rangefinder

The MA 8530 is a tripod-mounted Nd:YAG laser rangefinder designed primarily for coastal defence applications. It can be connected to the control centre computer via telephone or radio links to provide the system with automatic target data in real-time.

MAF 8452 electronic goniometer unit

This is a compact self-contained rangefinder consisting of the MA 8560 hand-held laser rangefinder, an auxiliary display, and a goniometer and tripod.

STATUS

In production.

CONTRACTOR

Nokia Electronics, PO Box 780, SF-00101 Helsinki, Finland.

FRANCE

3557.193

TCV 107 TANK LASER RANGEFINDER

The TCV 107 is a miniature laser rangefinder for use on any type of tank. Small size and weight permit its installation outside the vehicle, or inside where it may be associated with an optical sight. It is of modular construction and consists of two parts:

(1) a sealed transmitter/receiver power unit. A removable optical sight is used for aligning this unit with the gun boresight axis, by means of an integrated beam deviator.

(2) a control box for operating the equipment and displaying the results.

The digital readout of range measurement is displayed on the control box. A warning light indicates when there are several echoes and the distance of the furthest echo is automatically displayed; the range of the nearest one can then be obtained by pressing a button. For installation outside the vehicle, an armoured container with an automatic shutter for the protection of the optics is available.

CHARACTERISTICS

Wavelength: 1.06 μm
Pulse duration: 20 × 10^{-9} s
Peak power: 1 MW
Detector: Avalanche photodiode
Output to vehicle computer (option)
Dimensions: 290 × 140 × 100 mm
Mass: 6.3 kg
Measurement range: 150 – 9500 m
Accuracy: ±5 m
Angular resolution: Better than 0.5 mrad
Repetition rate: 1 measurement every 2 s for 3 successive shots; 12 measurements per minute

STATUS

Series production.

CONTRACTORS

CILAS (Compagnie Industrielle des Lasers), Route de Nozay, 91460 Marcoussis, France.

SOPELEM, 102 rue Chaptal, 92306 Levallois-Perret, France.

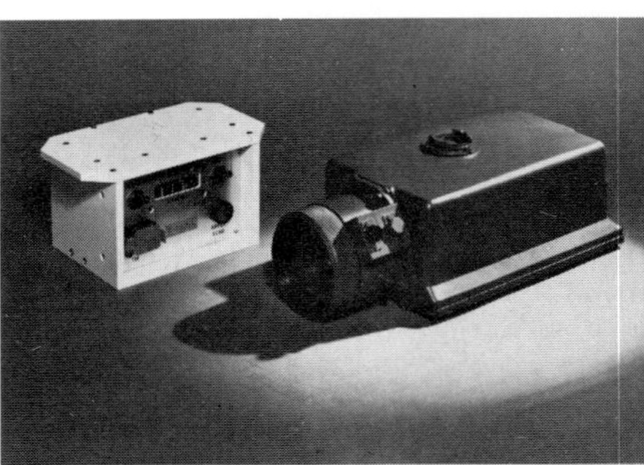

TCV 107 tank laser rangefinder with its associated control box

5897.193

TCV 186 LASER RANGEFINDER

The TCV 186 is a small, lightweight laser rangefinder for integration into the optical sight of the AMX 40 tank. It can also be integrated into the gyro-stabilised sights of helicopters.

The rangefinder is a self-contained unit housing the laser transmitter, receiver, power supply and electronic counter. The measured distance of two targets is transmitted in serial binary form to the fire computer.

CHARACTERISTICS

Wavelength: 1.06 micron
Pulse duration: 25 × 10^{-9} s
Peak power: 1 MW
Detector: Avalanche photodiode
Dimensions: 150 × 110 × 90 mm
Mass: 2 kg

CONTRACTOR

CILAS (Compagnie Industrielle des Lasers), Route de Nozay, 91460 Marcoussis, France.

3040.193

TM 17 (TM 12) ARTILLERY SIGHT/LASER RANGEFINDER

The TM 17 (formerly designated TM 12) is a portable equipment that combines in a single unit an optical sight and a laser rangefinder. The former provides target elevation and azimuth information, and the latter gives range data. Arrangements are made for this information to be fed into an automatic fire control system.

It is also possible to use the TM 17 for taking topographic measurements and survey work by provision of a suitable north seeking system.

CHARACTERISTICS

Laser material: Nd-doped glass
Wavelength: 1.06μm
Pulse duration: Approx 25 × 10^{-9} s
Energy: Over 100 mJ
Peak power: Over 4 MW
Divergence: 0.5 mrad
Measurement range: 150 – 19 900 m. Simultaneous display of 2 echoes
Range blanking: Adjustable between 150 and 5000 m
Measurement accuracy: ±5 m
Repetition rate: 1 measurement every 2 s for 3 successive shots; 12 measurements/minute

Optical sight

Azimuth range: 360°
Azimuth accuracy: 1 mil
Elevation range: –20 to +40°

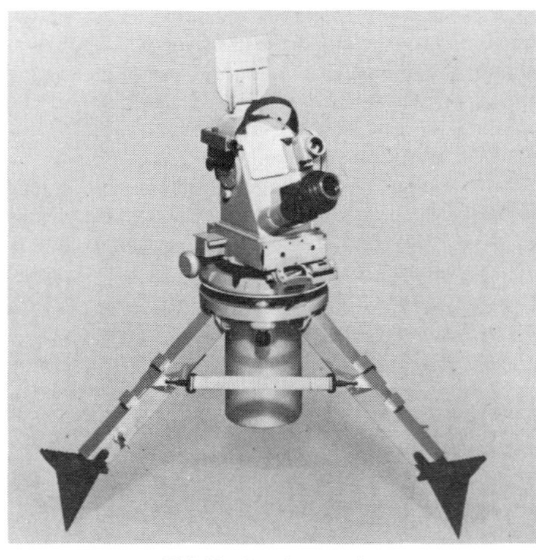

TM 17 tripod mounting

Elevation accuracy: 2 mils
Orientation: By trough compass
Power supply: 24 V Ni-Cd rechargeable battery allowing over 400 measurements
Weight: < 15 kg complete

STATUS

Series production.

TM 17 for vehicle mounting

CONTRACTORS

CILAS (Compagnie Industrielle des Lasers), Route de Nozay, 91460 Marcoussis, France.

SOPELEM, 19 boulevard Ney, 75018 Paris, France.

3416.193

TM 18 (TPV 89) LASER RANGEFINDER

The TPV 89 (French Army designation TM 18) is a hand-held laser rangefinder, similar in size and appearance to a pair of field binoculars. The instrument was designed for observation and rapid and accurate range measurement of targets or other items of interest. In use, the operator centres the cross-hairs on the target and triggers the laser by means of a push-button; the distance in metres is displayed immediately in the eyepiece. Accuracy is ±10 m whatever the target distance, within the range limits of the sight. An optical device provides absolute protection of the operator's eyes during laser emission. Other features include a device to prevent accidental triggering of the laser, a multiple echo indicator, a blanking control adjustable between 150 and 2300 m, and an internal power supply. A tripod mounting with elevation and azimuth control, charger for the nickel-cadmium battery, remote trigger, lithium battery and a removable attenuator are also available.

CILAS TPV 89 laser rangefinder

CHARACTERISTICS

Monocular sight

Magnification: ×6

Field of view: 7° (+120 mrad)

Ocular ring: 5 mm

Laser rangefinder

Wavelength: 1.06µm

Beam expansion ratio: 1:4

Energy: 30 mJ

Repetition rate: 12 shots/minute

Time to range: 0.1 s

Detector: PIN photodiode

Measurement range: 150 – 9990 m

Accuracy: ±10 m

Range blanking: Adjustable 150 – 2300 m

Dimensions: 180 × 150 × 75 mm

Weight: 1.9 kg

Power supply: 12 V 0.5 Ah Ni-Cd rechargeable battery giving 600 shots at 25°C

STATUS

Series production.

CONTRACTORS

CILAS (Compagnie Industrielle des Lasers), Route de Nozay, 91460 Marcoussis, France.

SOPELEM, 19 boulevard Ney, 75018 Paris, France.

3146.193

APX M504 OPTICAL SIGHT AND RANGEFINDER

The APX M504 equipment consisting of a laser rangefinder, an aiming optical system, and an optical deflection device, is part of the COTAC fire control system of the AMX-10RC tank. The equipment provides an optical sight for the gunner, with ×10 magnification and a 120 mil field of view (SOPELEM), a laser rangefinder by CILAS, and a line-of-sight deflector by GIAT (Groupement Industriel des Armements Terrestres). The last of these is used to introduce automatically fire corrections into the optical sight.

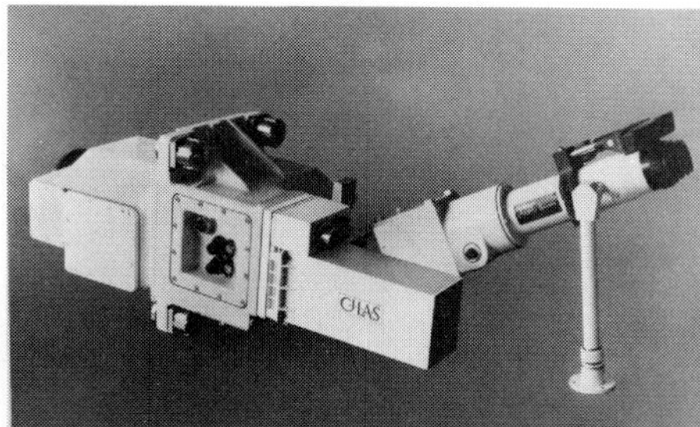

APX M504 sight and rangefinder

In addition to its observation and aiming functions, it provides information automatically on the distance of the target to the firing control system. Its accuracy is ±5 m regardless of the distance of the target within the limits of the range. The maximum measurable distance depends on the visibility, and can reach 10 km. The unit comprises a pilot light indicating the possible presence of two echoes, in which case the measurement of the second echo can be selected.

CHARACTERISTICS

Rangefinder

Laser material: Nd-glass

Wavelength: 1.06 µm

Pulse duration: 25×10^{-9} s

Energy: 100 mJ

Peak power: 4 MW

Divergence: 0.5 mrad

Repetition rate: 1 measurement every 2 s for 3 successive firings, 12 measurements per minute

Receiver

Detector: Avalanche photodiode

Field angle: 0.5 mrad

Dimensions: 250 × 110 × 150 mm. 3.5 kg (rangefinder only)

OPERATION

The gun-layer's function is simply to aim at the target. As soon as this is correctly performed, he triggers the laser rangefinder, and then inserts the corrections which have been computed automatically. A number of sub-systems developed by GIAT enhance the performance of the overall COTAC system: a device for measuring the angular speed of the target in two axes; an automatic cant angle sensor; meteorological sensors; and a computer for producing fire corrections from the measured parameters.

STATUS

The APX M504 was designed under DTAT contract and is now in series production.

CONTRACTORS

CILAS (Compagnie Industrielle des Lasers), Route de Nozay, 91460 Marcoussis, France.

GIAT, 10 place Georges Clémenceau, 92211 Saint Cloud, France.

SOPELEM, 19 boulevard Ney, 75018 Paris, France.

3414.193

APX M550/TCV 80 LASER RANGEFINDER

The APX M550/TCV 80 laser rangefinder was specially designed for mounting on the optical sight of AMX-10RC and AMX-30 tanks, and on some APX gyro-stabilised sights for helicopter applications. A single unit of small dimensions houses the laser emitter, the receiver, power supply and the electronic clock. In addition to the above applications, this rangefinder can be adapted to any fire control system for accurate distance measuring.

CHARACTERISTICS

Laser type: Nd-glass

Wavelength: 1.06 micron

Pulse duration: 25×10^{-9} s

Energy output: 100 mJ

Peak power: 4 MW

Repetition rate: 1 measurement every 2 s for 3 successive shots; 12 shots/minute

Detector: Avalanche photodiode

Minimum detectable power: $\leqslant 2 \times 10^{-8}$ W

Measurement range: 320 – 9995 m. Output to fire control computer

Range accuracy: ±5 m

Dimensions: 260 × 110 × 160/110 mm (height)

Weight: 3.5 kg

Power supply: Vehicle, 19 – 28 V DC

STATUS

Series production.

CONTRACTOR

CILAS (Compagnie Industrielle des Lasers), Route de Nozay, 91460 Marcoussis, France.

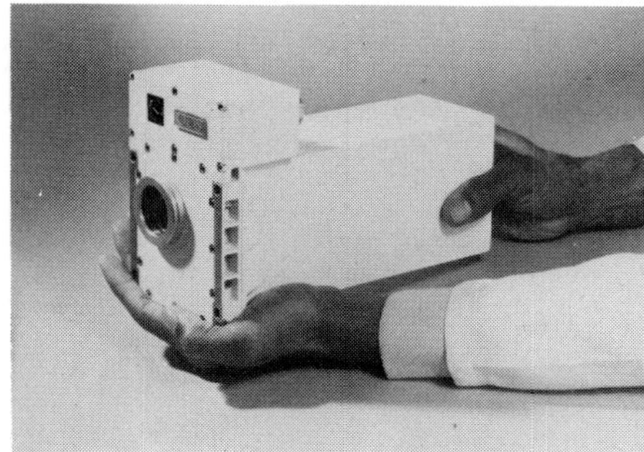

CILAS TCV 80 laser rangefinder

3663.193

TCY 133 LASER RANGEFINDER/ILLUMINATOR

The high repetition rate laser rangefinder TCY 133 was designed for integration into gyro-stabilised sights used with anti-aircraft weapon optical sights. The equipment is provided with a device that projects cross-hairs into the aiming sight of the weapon system, thus aiding alignment of the rangefinder channel, since the rangefinder channel employs the observation optical channel. Rangefinder, electronic clock, and power supply are all contained in the same casing.

CHARACTERISTICS

Laser material: Nd-YAG

Wavelength: 1.06 micron

Pulse duration: 20×10^{-9} s

Peak power: Over 5 MW

Detector: Avalanche photodiode

Detector field: 2 mrad

Dimensions: 470 × 230 × 220 mm

Weight: 15 kg

Power supply: 115 V/400 Hz 3-phase

Measurement accuracy: ±5 m

Repetition rate: 5, 10, or 20 pps (as specified)

STATUS

Series production.

CONTRACTOR

CILAS (Compagnie Industrielle des Lasers), Route de Nozay, 91460 Marcoussis, France.

CILAS TCY 133 laser rangefinder/target illuminator for anti-aircraft weapons

5898.193

TCY 167 MINIATURE LASER RANGEFINDER

The TCY 167 is a miniaturised laser rangefinder designed for integration into the fire directors of individual anti-tank weapons. It is small and lightweight, and can be fitted to a variety of weapons.

The rangefinder is a hermetically sealed self-contained unit which contains the laser transmitter, receiver, power supply and electronic counter. The equipment is controlled from the fire director and passes data on two targets to the computer in serial binary form.

CHARACTERISTICS

Wavelength: 1.06 micron
Pulse duration: 10×10^{-9} s
Detector: Avalanche photodiode
Dimensions: 145 × 105 × 37 mm
Angular discrimination: 2 mrad
Accuracy: ±5 m
Range: 80 to 9800 m
Mass: 0.65 kg

CONTRACTOR

CILAS (Compagnie Industrielle des Lasers), Route de Nozay, 91460 Marcoussis, France.

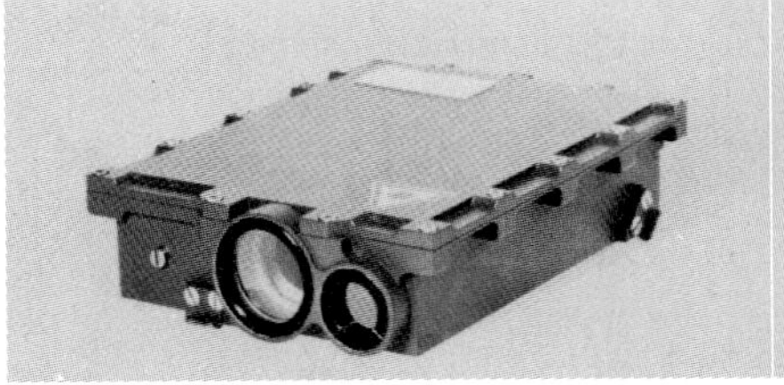

TCY 167 miniature laser rangefinder

3752.193

OB-47 NIGHT SIGHT

The OB-47 night sight is a light intensification monocular sight which enables the gunner of an AMX-30 main battle tank to aim the tank's main armament in darkness. The image intensifier tube employed has automatic gain control and is of the microchannel inverter type. In the AMX-30, the OB-47 is associated with OB-44 observation binoculars, and an M 427-02 laser rangefinder sight. The driver can be provided with a CN2 516 night driving periscope.

CHARACTERISTICS

Magnification: × 5.5
Field of view: 7°
Elevation travel: –8 to +20°
Dioptre adjustment: –5 to +2 dioptre
Range: 1000 m with 10^{-3} lux illumination

STATUS

Production and in service with an unidentified defence force.

CONTRACTOR

SOPELEM, 19 boulevard Ney, 75018 Paris, France.

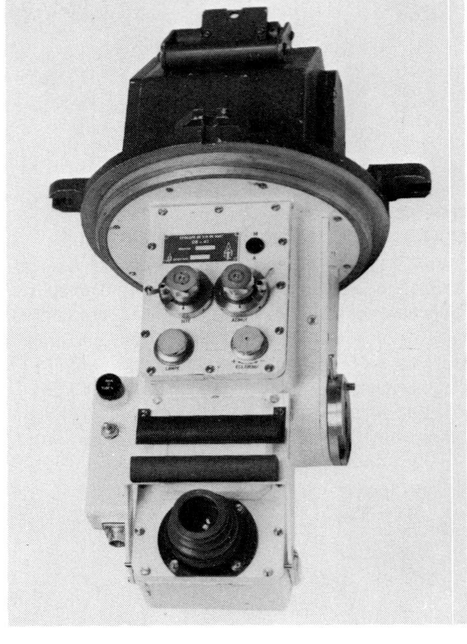

SOPELEM OB-47 night sight

6584.193

CAPITOLE FIRE CONTROL SYSTEM

Capitole is the latest version of the Sopelem SOPTAC 11 anti-tank fire control system (**6286.193**). The light intensification channel of the SOPTAC 11 is replaced, in the Capitole system, by a thermal channel, with the infra-red module being supplied by TRT. The new system consists of a daylight/IR source, a computer, a tank commander's electronic module, a laser rangefinder, and a gunner's electronics module integrated in the sight. Operation of the system is identical to the SOPTAC 11 but with improved performance (extended infra-red camera range under varying meteorological conditions).

CHARACTERISTICS

Sight magnification: × 8
Field: 6° daylight channel; 6° x 3° IR channel
Resolution: 6 seconds of arc

IR Channel

Wavelength: 8 to 12 microns
Detector: HgCdTe photoconductors operating at 77°K
Cooling system: Refrigeration circuit
Typical range: 5 to 6 km detection; 3 km recognition

STATUS

In production.

CONTRACTOR

SOPELEM, 19 boulevard Ney, BP 264, 75866 Paris Cedex 18, France.

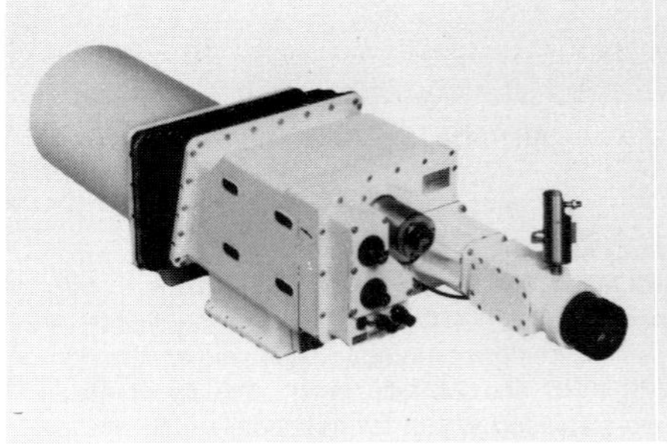

Capitole fire control system

6289.193

TJN2 71 MIXED DAY/NIGHT SIGHT

This sight is intended to equip cupolas with small calibre weapons, and also to equip large scale commander stations. It has an integrated prism head and includes three paths, a day sight, a night sight with a second-generation image intensifier, and a day periscope. The day and night sight paths have a common eyepiece.

It provides for day observation with a wide field, target designation, and day and night firing. A number of versions of the TJN2 71 sight, with different installation heights and elevation interfaces, are available. This ensures that it can be fitted to a wide range of turrets.

CHARACTERISTICS

Magnification: ×7 (day sight); ×4 (night path); ×1 (periscopic)
Field: 7° (day path); 10° (night path); vertical 20°, horizontal 43° (periscopic)

STATUS

In volume production.

CONTRACTOR

SOPELEM, 19 boulevard Ney, 75018 Paris, France.

4269.183

SOPTAL 29 FIRE CONTROL SYSTEM

The SOPTAL 29 fire control equipment is an optical director designed for use with the ACL-STRIM anti-tank missile weapon (**2055.111**). It is used to generate a line-of-sight based on target range (determined by laser rangefinder), weapon cant (pendular graticule), type of ammunition, crosswind (manually inserted), target lateral velocity (transducer between weapon launcher and tripod), temperature and pressure. A detachable afocal unit for night firing is an optional addition.

The equipment comprises three interchangeable basic sub-assemblies: sight, processing electronics and laser module; sight aiming module (afocal system with light intensifier); and base, comprising weapon support and tacho bearing data encoder.

CHARACTERISTICS

Optical

Magnification: × 4
Eyepiece aperture: 4 mm
Focal length: 40 mm
Field of view: 8°

Aiming

Elevation setting accuracy: 0.3 mil
Bearing offset accuracy: 1 mil
Rangefinder: 150 – 2000 m
Elevation correction: Between 150 and 800 m for two types of ammunition
Lateral target velocity limit: 20 m/s
Autonomy: 300 rounds

STATUS

Production.

CONTRACTOR

SOPELEM, 102 rue Chaptal, 92306 Levallois-Perret, France.

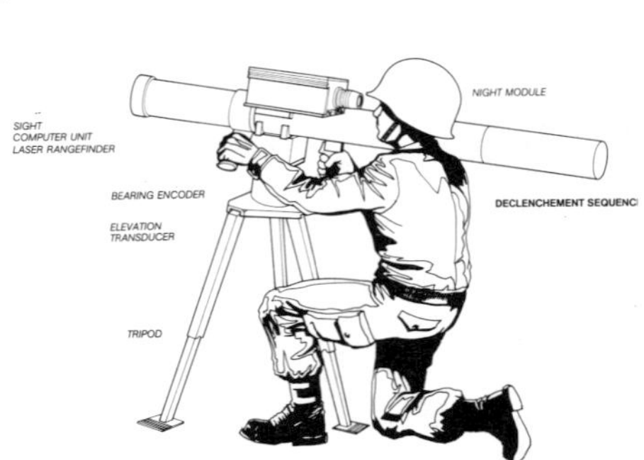

SOPTAL 29 fire control equipment as used with ACL-STRIM anti-tank rocket launcher

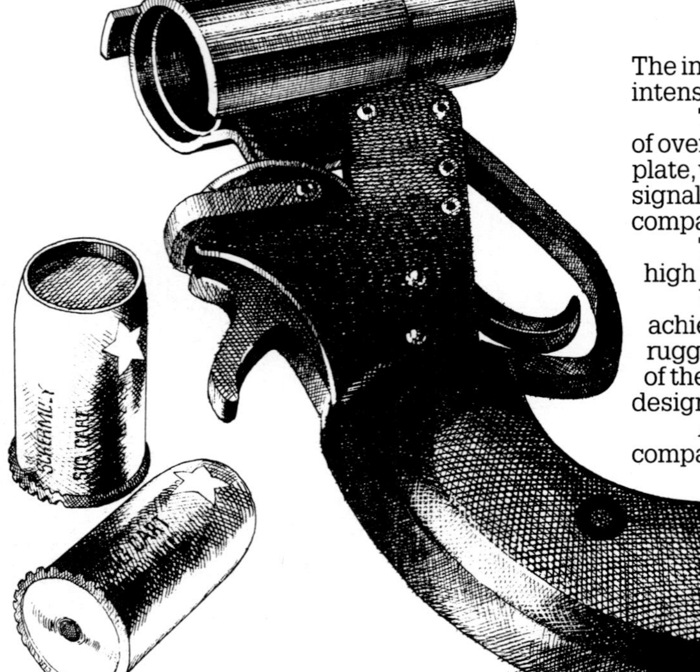

PHILIPS

6285.193

SOPTAC 36 FIRE CONTROL SYSTEM

The SOPTAC 36 day/night periscopic fire control system has been developed to equip the gunner's stations in the turrets of light armoured vehicles. The equipment provides observation and firing by day and night, laser range firing, and firing control function on fixed and moving targets. The system is derived directly from the SOPTAC 11 (**6286.193**), the main differences being the provision of a third-generation image intensifier, instead of a second generation, and is periscopic rather than telescopic.

The system consists of a dual periscope telescopic sight which includes a day periscope/day path/night path using the same eyepiece as the day path, a laser rangefinder module, a line-of-sight projection module of mobile cross hairs, a computer unit with microprocessor, and control boxes for commander and gunner. The sight can also be fitted with a repeater system which transfers the image to the tank commander.

CHARACTERISTICS

Magnification: ×1 (periscopic path); ×8 (day and night paths)

Laser accuracy: ±5 metres from 150 to 9995 metres

Fire control

Operating range: 300 to 5000 m

STATUS

In production.

CONTRACTOR

SOPELEM, 19 boulevard Ney, 75018 Paris, France.

6286.193

SOPTAC 11 FIRE CONTROL SYSTEM

The SOPTAC 11 fire control system provides observation and firing by day and by night, laser telemetry, and a fire control function on fixed and moving targets. It can equip all turrets capable of housing a shield-firing telescope. An integrated computer is fed with data covering the various parameters which condition firing (eg distance, speed of target, vertical closing speed, temperature, pressure, type of ammunition, crosswind, etc).

The SOPTAC 11 consists of five elements, a telescope, laser rangefinder TVC 107, computer unit, an electronics box for the commander, and a gunner's electronics unit integrated with the telescope.

CHARACTERISTICS

Optical

Magnification: ×6

Field of view: 8° (day path; 6° (night path))

Laser

Waveband: 1.06μm

Range: 10 km in good visibility

Accuracy: ±10 metres

STATUS

In production and in service. Nearly 100 systems have been sold.

CONTRACTOR

SOPELEM, 102 rue Chaptal, BP 223, 92306 Levallois-Perret, France.

6287.193

SOPTAC 18 FIRE CONTROL SYSTEM

The SOPTAC 18 fire control system is intended for modernisation of the sighting systems in the FL 12 turret, without modification of the turret. The system allows observation and daytime firing on stationary or moving targets, and laser rangefinding. It consists of a rigid frame fixed inside the turret which holds a calculator case, a tank leader case and a gunner's case. The SOPTAC 18 system takes into account distance, speed of target, cant, temperature, pressure, ammunition type and crosswind.

A night firing sight, $TN_2 1$, can be incorporated to provide night firing by a second-generation image intensifier and observation magnified by ×1 in daytime. This sight is mounted on the left of the day system and is independent of the latter.

CHARACTERISTICS

Operating range: 300 to 4000 metres

Laser accuracy: ±10 metres

System accuracy: ±0.3 metres

Night path ($TN_2 1$): ×6 magnification; 6° field of view

STATUS

In production.

CONTRACTOR

SOPELEM, 19 boulevard Ney, 75018 Paris, France.

3751.193

CN2-500 SERIES PASSIVE VEHICLE PERISCOPE

The CN2-500 series is a range of second generation passive driving periscopes which enable AFVs to be driven at night without the aid of any artificial lighting. Daylight driving is possible also with the same equipment. The second generation light intensification tube employed is stated to eliminate the effects of sudden glare resulting from unexpected exposure to light sources when operating in darkness.

The optical module is fitted with a bi-ocular eyepiece enabling simultaneous viewing with both eyes without adjustment. Models have been designed within the CN2-500 series for fitting in a range of vehicles, as follows:

CN2-508: Panhard, AMX-13T and RV1 vehicles

CN2-516: AMX tanks, Saviem vehicles and RV1

CN2-540: AMX 40

CN2-548: M48 tanks and derivatives

CN2-555: T55 tanks and derivatives

CHARACTERISTICS

Magnification: ×1

Resolution: 1.5 mrad at 10^{-1} lux

Objective lens aperture: 1:1

Horizontal field: 50° bi-ocular vision

Image tube: TH 1313 with AGC

Tube gain: At least 30 000 FL/FC

Luminance gain: >400

STATUS

Production.

CONTRACTOR

SOPELEM, 19 boulevard Ney, 75018 Paris, France.

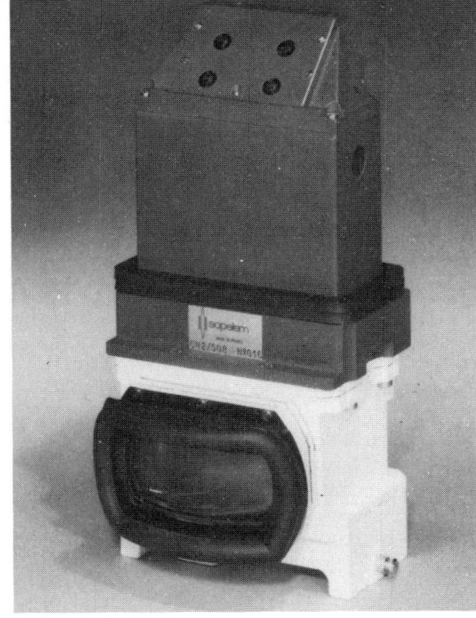

CN2-508 passive driving periscope for Panhard and AMX-13T vehicles

---

6288.193

M 391 EPISCOPIC SIGHT

The M 391 episcope sight has been developed to allow the tank commander station in the TH 20 turret to be equipped with a 20 mm gun. The sight provides day observation, target designation, a firing control function on helicopters and slow moving aircraft, and emergency firing.

The sight is of the periscopic type and includes two optical paths. On the right is a sight path designed for firing on ground targets, while on the left is the episcopic path intended for observation, target designation and firing against aircraft.

CHARACTERISTICS

Angle of elevation: −15° to +55°

Magnification: ×6 (day sight path); ×1 episcopic path

Field (day path): 10°

Field (episcopic path): vertical 12°, horizontal 38° (one eye to 40 mm); vertical 32°, horizontal 77° (both eyes)

Weight: 10 kg

STATUS

In volume production.

CONTRACTOR

SOPELEM, 19 boulevard Ney, 75018 Paris, France.

6536.153

MITHRIDAT TANK FIRE CONTROL SYSTEM

Mithridat is a periscopic tank fire control system being jointly developed by Sopelem and TRT. This is a fully integrated system employing a main periscopic sight module, with a ×1 magnification, a day channel and an infra-red night channel, both of which use a ×8 magnification, a laser rangefinder module, a moving-reticle injection module with an associated digital computer, and the necessary commander's and gunner's control units. The night vision channel is supplied by TRT and is a variant of the Mira night sight in current production for the Milan anti-tank missile. Mithridat is claimed to have an accuracy of about 15 mils.

STATUS

In production.

CONTRACTORS
SOPELEM, 19 boulevard Ney, 75018 Paris, France.
TRT (Télécommunications Radioélectriques et Téléphoniques, 88 rue Brillat-Savarin, 75640 Paris Cedex 13, France.

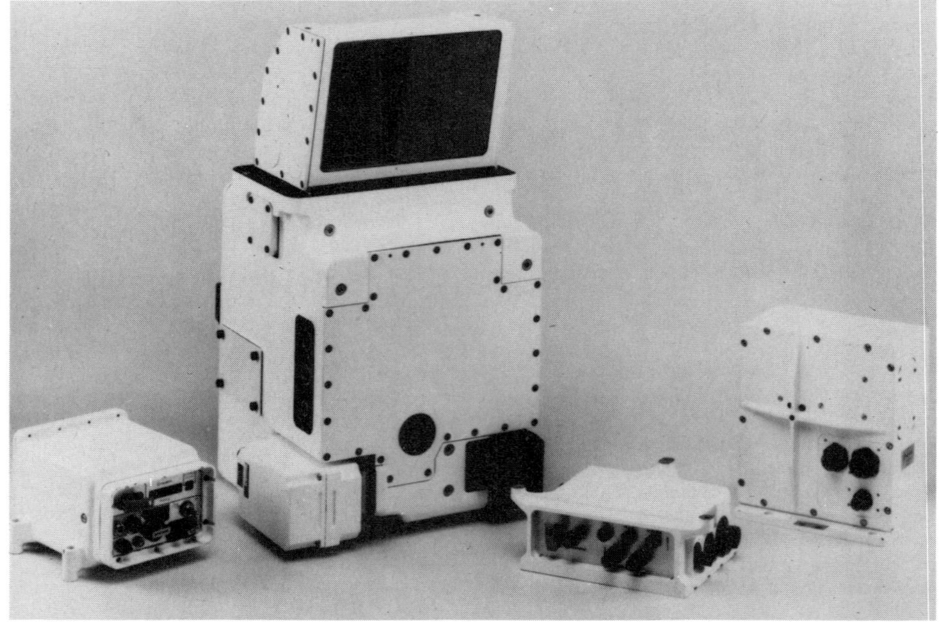

Mithridat fire control system

4245.193
OB-31 NIGHT-DRIVING PERISCOPE
The OB-31 passive periscope is designed to equip AMX-10 and AMX-30 tanks and various other AFVs and may be interchanged with the daylight driving periscope installed in these vehicles to provide combined day/night driving facilities from broad daylight to starlight, without headlights. Two light intensification tubes and suitable optical arrangements provide two separate channels and allow stereoscopic vision. A manually-operated diaphragm permits limiting of image brightness in high ambient light levels and there is a device to protect the tubes from excessive brightness. When the unit is switched off the diaphragm is automatically closed. The use of modern single-stage tubes (XX 1080 type) makes it possible to produce a compact periscope of high performance with simple operation. It can be operated in daylight levels up to about 10 000 lux.
CHARACTERISTICS
Magnification: × 0.9 (approx)
Horizontal field: 48°
Vertical field: 40°
Eyepiece diameter: 20 mm
Lens diameter: 45 mm
Distance between lens axes: 65 mm
Illumination levels: (minimum) classified; (max) 10 000 lux
Dimensions: 327 × 193 × 163 mm
Weight: 11.85 kg (with transport case); 7.85 kg (less case)
STATUS
Production for French and foreign armies. Officially approved by French Army for use in AMX-10, AMX-30 and VAB.
CONTRACTOR
TRT (Télécommunications Radioélectriques et Téléphoniques), 88 rue Brillat-Savarin, 75640 Paris Cedex 13, France.

OB-31 night-driving periscope

4746.193
CASTOR THERMAL IMAGING SIGHT
The Castor thermal imaging camera is designed for use on armoured vehicles in conjunction with observation and fire control systems. It detects and identifies targets by converting their thermal radiation in the IR spectrum into a visible image which can be displayed on a TV monitor. The system is intended for observation and firing under all conditions of daylight, poor visibility and at night. It can be installed in any armoured vehicle, observation vehicle and various turrets, and can be used for artillery observation.
CHARACTERISTICS
Wavelength: 8 to 12 microns
Analysis: Series parallel scanning
Detectors: Photovoltaic HgCdTe cooled by Split-Stirling cycle system
Field of view: 6° × 9° or 5.4° × 8.1° (wide); 2.7° × 4° or 2° × 3° or 1.8° × 2.7° (narrow)
Range: Up to 4000 m, depending on lens
STATUS
In production for French Army AMX30B2 tanks and for artillery vehicles.
CONTRACTOR
TRT (Télécommunications Radioélectriques et Téléphoniques), 88 rue Brillat-Savarin, 75640 Paris Cedex 13, France.

Units of the Castor thermal imager

3845.193
MIRA THERMAL IMAGER
The Mira thermal imaging sight was developed for use with the Milan anti-tank missile system (**2215.111**). The equipment projects an IR image of the landscape into the observer's day-sight eyepiece permitting use of the normal eyepiece for both day and night operation. Automatic harmonisation of the sight and localiser axes ensures that no adjustment is required by the operator. The equipment includes the Mira sight itself, to which is attached a removable battery pack and a compressed gas bottle, plus an interface which allows the sight to be mounted on the Milan firing post by a single person in a few seconds.

The Mira imager is a compact, robust, sealed metal instrument comprising the following:
(1) two germanium lens
(2) a series/parallel mechanical scanner
(3) a detection/display block consisting of a photo-conducting CMT detector array and LED array
(4) associated electronics
(5) a cryogenic cooling sub-system (JT self-regulating minicooler) or cooling machine.

Milan anti-tank weapon with Mira night sight fitted

CHARACTERISTICS
Wavelength: 8 – 13 μm
Processing: Series-parallel
Display: LED
Target detection range: 4000 m (approx)
Target recognition range: 2000 m (approx)
Target identification range: 1500 m (approx)
Field of view: 6 × 3°
Power and cooling supply: J-T minicooler with maximum duration of 2 hours depending upon battery size
Weight: < 8 kg
STATUS
Series production, carried out under the general management of TRT and their contractors, started in 1983. By December 1987 3000 had been produced.
CONTRACTORS
TRT (Télécommunications Radioélectriques et Téléphoniques), 88 rue Brillat-Savarin, 75640 Paris Cedex 13, France.
Siemens AG (Federal Republic of Germany)
Marconi Command and Control Systems Ltd. (UK).

6525.093
CALIOPE THERMAL IMAGER

CALIOPE is a portable thermal imager designed for a variety of applications, including battlefield surveillance, night capability for weapon systems and protection for naval vessels. The imager is easy to use and provides a direct image either by a monocular or binocular, or an output to a TV display. A mechanical interface is designed for quick fitting on a standard tripod or turret. Cooling of the detector is by integrated compressor (Split-Stirling) or by high pressure gas bottle (Joule-Thomson).

CHARACTERISTICS
Spectral bandwidth: 8 to 13 μm
Field of view: 6° × 3°
Magnification: × 7
Scanning: Serial-parallel
Dimensions: 520 × 220 × 190 mm
Weight: 8 kg
CONTRACTOR
TRT (Télécommunications Radioélectriques et Téléphoniques, 88 rue Brillat-Savarin, 75640 Paris Cedex 13, France.

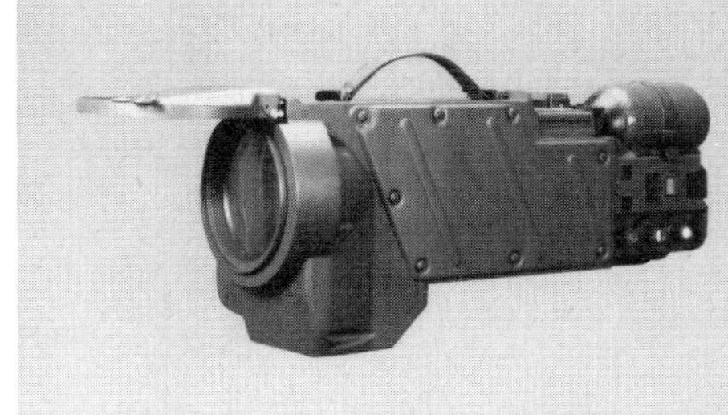
Caliope thermal imager

3560.193
VS 580 GYRO-STABILISED VEHICLE SIGHT

The VS 580 sight is intended for the detection, recognition, identification and acquisition of fixed or moving targets from a moving combat vehicle, with accurate line-of-sight stabilisation.

The VS 580 is a family of models which includes:
(1) VS 580 Observation, a panoramic sight for observation and identification
(2) VS 580 Standard, a panoramic sight with a target acquisition aid
(3) VS 580 Multirole, a panoramic sight with complete target processing capability (stabilised laser rangefinder)

The VS 580 sight can also be used by night, either as a driving aid, for close defence, or to assist night firing of the main gun at a combat firing range. Night vision is through the same eyepiece as daytime vision.

The VS 580 sight is compatible with a large number of existing vehicles and fire control systems.

CHARACTERISTICS
Warm-up time: 6 s
Magnifications: × 3 and × 10.5
Angular freedom: 360° in azimuth and 70° in elevation
Line-of-sight displacement speeds: 0.02 mils/s to 900 or 2000 mils/s in azimuth and elevation
Line-of-sight stabilisation: Aiming error <0.1 mil in normal conditions of use in combat vehicles in motion
Pointing accuracy error: <0.15 mrad
Night vision
Magnification: × 3.5 or × 6
Angular freedom: 360° in azimuth and 70° in elevation
Laser rangefinder
Range: 400 – 10 000 m
Measurement accuracy: ≤5 m
Angular discrimination: <0.3 mil
STATUS
In production. Fitted to 18 different vehicles in various versions.

VS 580-30 sight with thermal imaging
This sight is part of the VS 580 stabilised panoramic family used to detect, identify and fire at thermally imaged targets. It can be fitted with several FLIR types (Castor or Caliope from TRT), depending on user requirements and the type of armament to be slaved. The VS 580-30 is integrated into the fire control system so that the commander and gunner can share operations.

VS 580 gyro-stabilised vehicle sight

This thermal sight may be permanently mounted on the turret roof or, when fully armoured, attached to the outside of the turret for temporary missions (automatic or semi-automatic bore sighting, and fire control system connection to display and command the lines-of-sight). Diffused display is to CCIR standards.

CHARACTERISTICS
One or two fields
Line of stabilisation: <0.1 mil
Accuracy of line-of-sight angular definition: 0.15 mil
Line-of-sight maximum displacement speeds: 900 or 2000 mils/s
Exploration field: –10° to +60° or ±35°
STATUS
In full production.
CONTRACTOR
SFIM (Société de Fabrication d'Instruments de Mesure), 91344 Massy Cedex, France.

VS 580-30 sight with thermal imager

5896.193
VISAA ANTI-AIRCRAFT SIGHT

The VISAA anti-aircraft sight has been developed from the VS 580 range (**3560.193**) and is specially adapted for anti-aircraft engagement from land vehicles. It provides target detection and observation at very long range, target recognition and identification up to 5 km, sight for firing, and target parameter acquisition. The main differences between the VISAA and the VS 580 are an improved travel in elevation and increased tracking speed.

Three modes are available for operational use: an independent mode for observation, a slave mode for alignment on the radar plot, and a master mode for the transmission of target data to the vehicle fire control system. The VISAA sight measures distance and target angular rates. The sight may also be fitted with a built-in tracking system.

STATUS
In production. The sight has been ordered and delivered for several anti-aircraft vehicles.
CONTRACTOR
SFIM (Société de Fabrication d'Instruments de Mesure), 91344 Massy Cedex, France.

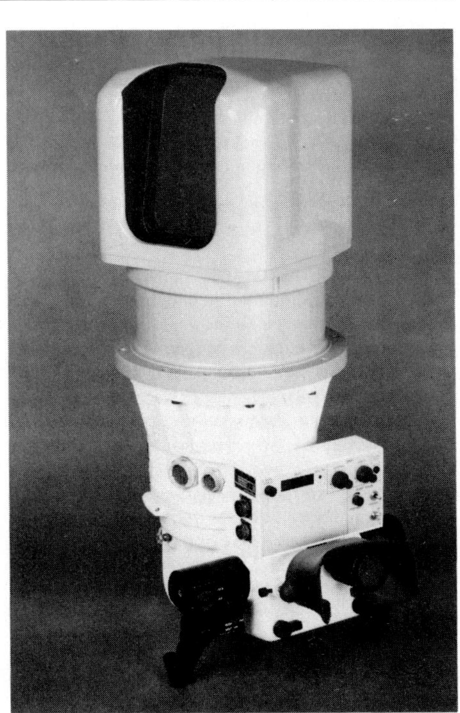
VISAA anti-aircraft sight

4249.193

CDN 22 NORTH SEEKER

The CDN 22 north seeker is an instrument for determining the azimuth for the alignment and orientation of military systems such as: autonomous navigation systems fitted in aircraft, land vehicles or missiles; long range guns; missile launch ramps; and mobile or semi-mobile radars. It is completely automatic in operation and self-contained.

CHARACTERISTICS

Accuracy: Better than 0.5 mil

Run-up and measurement time: 4 minutes

Overall weight: 34 kg

Digital output.

STATUS

In production.

CONTRACTOR

SFIM (Société de Fabrication d'Instruments de Mesure), 91344 Massy Cedex, France.

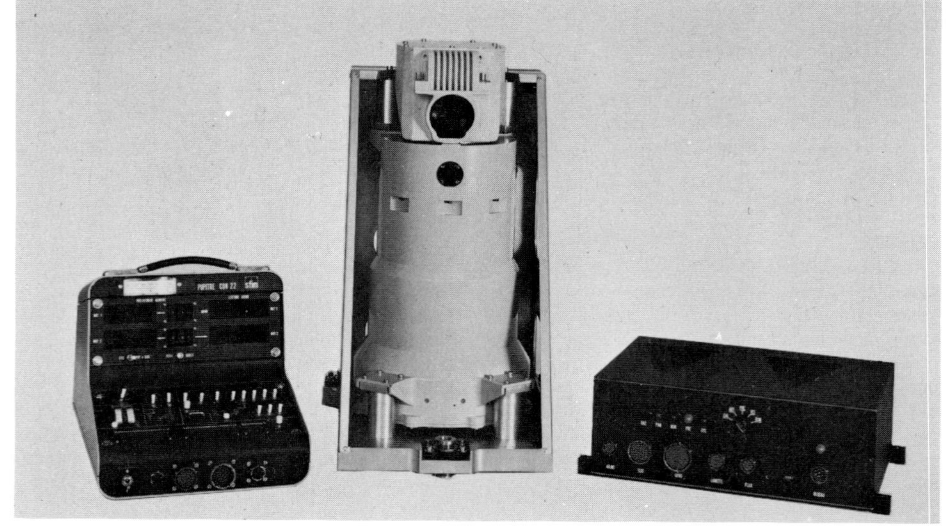

Three units comprising the CDN 22 north seeker equipment

3838.193

APX M363 FIRE CONTROL

The APX M363 is the fire control unit for the French 155 mm GCT self-propelled gun, providing day and night engagements with the following facilities:

(1) indirect fire from an external directional reference by the 'bearing firing method'
(2) direct fire
(3) elevation angle measurement of obstacles
(4) transfer of a reference bearing to another unit.

The equipment weighs, with vertical sensor plate and computer, 160 kg and has an overall height of 1.28 m.

CHARACTERISTICS

Direction finder

Field: 175 mrad

Magnification: × 5

Eyepiece diameter: 4 mm

Elevation: –90 to + 1200 mils

Panoramic observation field

Automatic tilt correction: ± 10°

Max tilt: 10°

Aiming accuracy: 0.5 mil

Direct fire sight

Field: 120 mils

Magnification: × 8

Eyepiece diameter: 5 mm

Aiming (elevation): –90 to + 360 mils

Observation (elevation): –300 to + 600 mils (turret stationary)

Accuracy: 0.5 mil

STATUS

Derived from the APX M589-01, the M363-01 has been adopted by the French Army and is in series production for export.

CONTRACTORS

GIAT (Groupement Industriel des Armements Terrestres), 10 Place Georges Clémenceau, 92211 Saint Cloud, France.

SFIM (Société de Fabrication d'Instruments de Mesure), 13 av Marcel Ramolfo-Garnier, Massy, Essonne, France.

3840.193

APX M520 PANORAMIC SIGHT

The APX M520 is the panoramic sight fitted on the VADAR anti-aircraft vehicle (**3333.131** in 1980-81 and earlier editions) turret and is one of the main elements of that system's fire control. For simultaneous use by the tank commander and the gunner (separate eyepieces), the unit provides panoramic search at low magnification, and target acquisition, allocation, tracking and engagement at high magnification, against low-level aircraft. The sight is provided with means of coupling a tracker television camera to allow automatic target tracking.

The sight is rate controlled from the cross-bar of either of the two operators, for the search or tracking modes.

CHARACTERISTICS

Magnification: × 2 (600 mrad field); × 4 (300 mrad field)

Camera field depth: ∞

Magnification: × 1 (100 mrad field)

Sighting range: n × 360° (azimuth); –8 to + 73° (elevation)

Overall dimensions: 730 × 676 mm

Weight: 115 kg

STATUS

Prototype.

CONTRACTOR

GIAT (Groupement Industriel des Armements Terrestres), 10 place Georges Clémenceau, 92211 Saint Cloud, France.

3841.193

APX M527 TANK SIGHT

The APX M527 multipurpose sight has been designed for installation in the update of the AMX-40 and AMX-32 MBTs. Of modular design it provides the tank commander with stabilised day vision (two magnifications), stabilised image intensifier (optional) and a HL 10 type laser rangefinder (optional). It enables the tank commander, with his vehicle either stationary or on the move, to observe, detect and identify fixed or moving targets and, when coupled to a gun stabilisation system and a 'tank-on-the-move' fire control system, to engage such targets.

CHARACTERISTICS

Day

Magnification: × 8, × 2

Field of view: 100, 400 mils

Exit pupil diameter: 4 mm

Elevation limits: –12 to + 24° (× 2); –20 to + 40° (× 8)

Night

Magnification: × 3.5

Field of view: 580 mils

Elevation limits: –20 to + 40°

Range: 600 m (typically)

STATUS

In production.

CONTRACTOR

GIAT (Groupement Industriel des Armements Terrestres), 10 place Georges Clémenceau, 92211 Saint Cloud, France.

SFIM (Société de Fabrication d'Instruments de Mesure), 91344 Massy Cedex, France.

4107.093

SMT MODULAR THERMAL IMAGING SYSTEM

Under a common module development programme for the French Defence Ministry, TRT in collaboration with SAT has developed a range of thermal-imaging modules to meet the requirements of the three armed services. Under an inter-company agreement each is responsible for the manufacture of specific modules and either may be the prime contractor for system application, depending on the particular project requirements and current workload. Design co-ordination authority, however, is vested in TRT.

The main characteristics of the SMT are: eight to twelve micron spectral bandwidth, full TV display compatibility on a standard CCIR 625-line, a limited number of detector elements, the possibility of application to LED displays, ease of coupling with image processing systems, open or closed-circuit cryogenic cooling, highly stabilised optic axis, and reliable behaviour in the presence of heat sources such as missile traces or fires. The modules include a built-in test system delivering information for the operator and for maintenance purposes.

The modules are assembled in various configurations depending on the particular requirements. They fulfil the basic functions of a thermal imager while specific elements are added to tailor the unit to particular applications. For example, a larger or smaller field of view would require a new optical train, or a TV monitor display would be added should remote viewing of the image be needed.

Current SMT applications are as follows (some are described in more detail in other entries):

Tango

An observation system designed by TRT for the Atlantique 2 maritime reconnaissance aircraft. The thermal camera is fitted in a gyro gimballed platform interfaced with the aircraft. Now in series production.

Mephisto (2212.111)

Firing of a HOT missile from an armoured vehicle.

Castor (4746.193)

A special TRT development for tank applications with all capabilities of a thermal sight in a modern fire control system for guns and missiles. In series production.

Visir

An experimental thermal sight for main battle tanks. This is a very long term programme with production not expected until the early 1990s.

Aphrodite

A night flying aid for helicopter use. The system is mounted in a stabilised turret and can be used with either a helmet type sight or displayed on a TV monitor. This is probably another long term programme.

Victor

A thermal camera associated with a gyro-stabilised mirror designed by TRT enabling a HOT missile to be fired from a Gazelle type helicopter. At the pre-production stage.

Murene
A turret-mounted version of the SMT system for naval gun-laying action. It is in production for the French Navy fire control director associated with the Sadral mount and TTAD turret.

Athos
An SAT camera adopted for the new French tank Leclerc. The camera consists of two blocks, integrated with the day-night gyro-stabilised sight. Two optical fields of view are forecast: 1.9 × 2.9° and 5.7 × 8.6°. The images are displayed on CRT screens to the gunner and to the tank commander.

CONTRACTORS
SAT (Société Anonyme de Télécommunications), 41 rue Cantagrel, 75624 Paris Cedex 13, France.

TRT (Télécommunications Radioélectriques et Téléphoniques), 88 rue Brillat-Savarin, 75640 Paris Cedex 13, France.

SMT thermal imaging modules

4283.193
VIPERE INFRA-RED SURVEILLANCE EQUIPMENT

The Vipere equipment is an infra-red panoramic surveillance system developed by SAT for the Section d'Etudes et de Fabrications de Telecommunications (SEFT) of the French Army Technique des Armements Terrestres. It is designed to detect and locate aircraft and helicopters at low and medium elevation angles in an entirely passive manner. When employed in an anti-aircraft defence system, Vipere will provide target designation data. A very low false alarm rate is claimed for the specially adapted data processing system manufactured in co-operation with Thomson-CSF.

The optical sensor head can be mounted as a fixed installation on the ground or carried on a vehicle with the electronic data processing system. The former unit incorporates a number of mosaic arrays consisting of multiple elements, cooling being by means of a closed-circuit system built into the head with its own compressor.

CHARACTERISTICS
Weight of optical head: <70 kg
Azimuth coverage: 360°
Elevation coverage: >20°
Detection range: >10 km (aircraft target)
Target designation accuracy: 1 mrad (approx)

STATUS
Development of a new field version for French Army will begin in 1988.

CONTRACTOR
SAT (Société Anonyme de Télécommunications), 41 rue Cantagrel, 75631 Paris Cedex 13, France.

6524.193
IRGO HAND-HELD THERMAL IMAGER

This is a family of thermal imagers being developed by SAT for the French Army's Direction des Armements Terrestres, Section d'Etudes et Fabrications. In addition to the hand-held thermal imager the family includes the Lisbeth hand-held aiming unit for use with the Mistral surface-to-air missile (see **4198.131**) and a mini-camera version for remote observation.

The field of view depends on the lens used but the system scans at 25 images per second with 120 lines per image and 200 pixels per line. A Cadmium Mercury Telluride detector is used and the system covers the 3-5 µm waveband.

CHARACTERISTICS
Field of view: × 5.8° (azimuth) (optional 8.7°); × 3.9° (elevation) (optional 5.85°)
Recognition range: >1000 m (NATO tank-size targets)
Focus range: 10 m to infinity
Weight: 2.9 kg (without power supply)
Power consumption: <10 W

CONTRACTOR
SAT (Société Anonyme de Télécommunications), 41 rue Cantagrel, 75631 Paris Cedex 13, France.

Irgo thermal imager

3134.193
THOMSON-CSF CANASTA DAY/NIGHT AIMING EQUIPMENT (DIVT 13)

Thomson-CSF produces a day/night aiming equipment based on low light-level television, and several versions have been developed. The French Army has adopted the equipment for fitting to its AMX-10RC, AMX-30B2 (modernised) and AMX-32 AFVs where it will be integrated with the tanks' IFCS. In French Army service the equipment is designated DIVT 13, while the name employed by Thomson-CSF for export models is Canasta. Different versions were previously referred to as the TMV560/562, but these designations are no longer employed.

The complete equipment includes a night and day TV camera with its control unit and one or two display units for the gunner and/or the tank commander. Its use permits full passive detection, identification and engagement of static and moving targets by main battle tanks and other classes of AFV. A moving electronic reticule, driven either manually from the control unit in accordance with target distance and selected ammunition, or automatically from a ballistic computer in accordance with firing data, gives the gunner aiming information and firing corrections.

The Canasta system is compatible with any gun ballistic and fire control system and is reported to be capable of installation in any type of MBT without turret modification.

CHARACTERISTICS
Lens assembly: Dioptric
Focal length: 210 mm
Aperture: f 1.5
Angular field: 4.1 × 5.5°
Camera tube: Super nocticon – silicon
Illumination range: 10^{-4} – 10^{5} lux
Alignment accuracy: Better than 0.25 mrad
Video output: ITCC standard
Power supply: 20 – 30 V DC

STATUS
Series production. The Canasta system has been installed in M48, Centurion, Leopard, T55, T62, AMX-13, AML 90, ERC 90 Sagaie, and PAC90 vehicles.

CONTRACTOR
Thomson-CSF/Aerospace Group, 122 Avenue du Général Leclerc, 92105 Boulogne-Billancourt, France.

Components of Thomson-CSF Canasta: control unit (top right), TMV 562 camera (centre), TMV 563 monitor (left)

6583.163

TARGET ACQUISITION AID SYSTEM

This is a helmet-mounted target designation aid system designed primarily for anti-aircraft gun defence applications, although it is adaptable to other fire control systems. The basic principle is the slaving of the gun line-of-sight with the observer's line-of-sight, the direction of this system sightline being referenced in an axis system based on the earth's magnetic and gravitational fields. Data is supplied by an magnetometer which detects the components of the magnetic field along the sightline reference axis system, and two accelerometers which give the vertical reference. The target azimuth and elevation angles are then calculated continuously by the system.

The basic equipment consists of a helmet which contains the sensors and a sight system, and a portable computer/control console mounted on the gunner's chest that provides the computation of azimuth and elevation parameters delivered to slave the weapon system in the direction of the target.

CHARACTERISTICS

Accuracy: 1.5° in both azimuth and elevation

Field of use: Azimuth 360°; elevation –80 to +80°

STATUS

Developed for the French Army 76 T2 anti-aircraft twin gun system. Adopted for the Thomson-CSF ASPIC short range missile system and Oerlikon Diana and 35 mm guns.

CONTRACTOR

Crouzet SA, 25 rue Jules Védrines, 26027 Valence Cedex, France.

5699.193

DALDO TARGET ACQUISITION AID SYSTEM

Daldo is a helmet-mounted target designation aid system designed primarily for anti-aircraft defence applications although it is adaptable to other fire control systems. The basic principle is the slaving of the gun line-of-sight with the observer's line-of-sight, the direction of the Daldo sightline being referenced in an axis system based on the earth's magnetic and gravitational fields. Data is supplied by a magnetometer which detects the components of the magnetic field along the helmet reference axis system and therefore along the sightline reference axis system, and two accelerometers which give the vertical reference. The target azimuth and elevation angles are then calculated continuously by the system.

The basic system consists of a helmet cover which contains the sensors and a sight system, a portable control console mounted on the gunner's chest and a processor unit to provide the computation of azimuth and elevation parameters for each gun battery.

A version known as DATAC (device for alignment and target acquisition) has been developed. This variant includes most of the sub-assemblies of Daldo but the headpiece is replaced by a pseudo rifle supporting the line of sight. It provides an accuracy of ±1 degree in both azimuth and elevation.

CHARACTERISTICS

Accuracy: ±3° in both azimuth and elevation (±1° for DATAC variant)

Field of use: Azimuth 360°; elevation –5 to +85°

STATUS

Developed for the French Army 76 T2 anti-aircraft twin gun system.

CONTRACTOR

Crouzet SA, 25 rue Jules Védrines, 26027 Valence Cedex, France.

Daldo target acquisition aid

6732.193

ARES HELMET-MOUNTED TARGET DESIGNATOR

The Ares helmet-mounted target designator system is a development of the Daldo system (**5699.193**) and has been designed for use primarily for air defence applications. It can be used with automatic gun systems and battlefield support rocket launchers. Essentially the gun line-of-sight is slaved to the observer's line-of-sight with the Ares sightline being referenced to an axis system based upon the earth's magnetic and gravitational fields.

The system consists of a helmet containing sensors, a sight system in which the observer can see a reticle collimated to infinity, earphones and a microphone, a portable control console carried on the gunner's chest and a processor unit to compute azimuth and elevation parameters for each gun battery. Data is supplied by a magnetometer which detects the components of the magnetic field along the helmet's reference axis system to provide the sightline reference while two accelerometers provide vertical references.

The system continuously calculates target azimuth and elevation angles. The system is linked to guns by a full duplex serial digital link with a transmission rate of 4800 bauds.

CHARACTERISTICS

Weight: 2 kg (helmet), 1.2 kg (control console)

Dimensions: 30 × 25 × 25 cm (helmet), 19.2 × 18 × 4 cm (control console)

Field of view: 360° (azimuth), -80 to +80° (elevation)

Accuracy: Better than 1.5° (azimuth), -45 to +45° (elevation)

Acquisition time: < 3 s

Power supply: 28 V dc

Power consumption: 100 W

STATUS

Development is complete and the system has been successfully tested with the Thomson-CSF Mygale/Aspic and the Oerlikon Diana systems.

CONTRACTOR

Crouzet SA, Division Aérospatiale, 25 rue Jules-Védrines, 26027 Valence Cedex, France.

Ares helmet-mounted designator

5893.193

VASSIR 85 ANTI-AIRCRAFT OPTRONIC SIGHT

The Vassir 85 stabilised optronic sight is designed for anti-aircraft defence units. It provides automatic tracking and aiming by gyro-stabilisation of the line-of-sight, associated with automatic deviation-measuring type tracking using TV or thermal camera sensors (FLIR). The sight integrates a day observation and tracking camera, a thermal camera which enables observation and tracking by day and night, and a high frequency laser rangefinder for instantaneous target distance measurement. Information on angular rate and distance is passed to the fire control computer.

The Vassir 85 consists mainly of two units:

(1) a stabilised aiming unit using a 2-degree of freedom, dry, dynamically-tuned SAGEM gyroscope

(2) an optical system providing separation of the different channels, integration of the TV field changing functions, laser functions, laser transmission and reception lenses

The aiming head can be fitted with an observation TV camera, a tracking TV camera, a thermal camera or a laser rangefinder in an interchangeable manner.

CONTRACTOR

SAGEM, Departement Armement Terrestre, 6 avenue d'Iena, 75783 Paris Cedex 16, France.

Vassir 85 optronic sight

5892.193

M 389 PANORAMIC SIGHT

The M 389 panoramic sight has been designed for use on the TK 105 turret of the AMX 10C tank. Its function is to provide the tank commander with facilities for all-round observation (magnification ×2 and ×8), acquisition of the target designated by the gunner, designation of a target to the gunner and firing of the auxiliary armament. In case of emergency it provides facilities for firing the 105 mm gun, the axis of which appears within the field of sight. Weight of the sight is 31.5 kg.

STATUS

In production.

CONTRACTOR

SAGEM, Departement Armement Terrestre, 6 avenue d'lena, 75783 Paris Cedex 16, France.

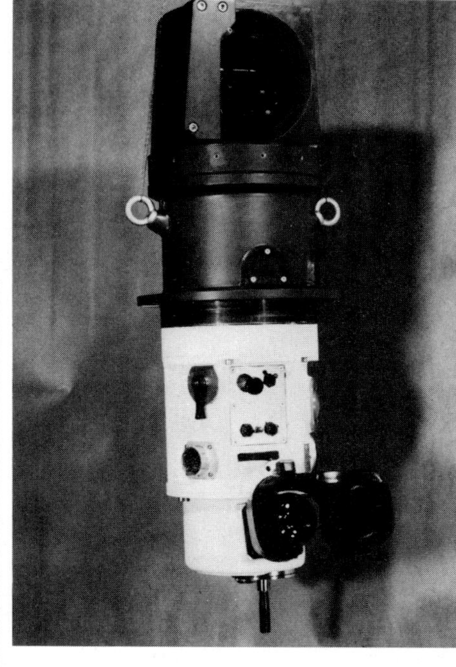

M 389 panoramic sight

5894.193

LGA 389 ARTILLERY GONIOMETER SIGHT

The LGA 389 goniometer sight is used to aim artillery guns. It provides ×2 and ×8 dual-magnification panoramic observation, indirect gun-aiming and direct fire control. The sight is of modular design and can also be fitted with a laser rangefinder or a light intensifier tube for use at night. A digital interface with the Type RS 422 fire control system is provided. Weight of the complete sight and associated electronics is 43 kg.

STATUS

In production.

CONTRACTOR

SAGEM, Departement Armement Terrestre, 6 avenue d'lena, 75783 Paris Cedex 16, France.

5895.193

M 590 GYROSTABILISED SIGHT

The M 590 sight is designed for use with the HOT missile Mephisto firing unit (**2212.111**) installed on a vehicle. It provides all-round observation while the vehicle is moving, and HOT missile firing by day and night.

The sight is a dual magnification type (×3 and ×12) and consists of:

(1) a panoramic head equipped with a platform supporting a FLIR camera, a stabilising unit, and a Gallileic magnification changing system
(2) a lower body grouping the SAT missile guidance unit, a reticle, an image rectifier and the eyepiece
(3) an electronics unit
(4) a bi-directional control stick
(5) a line-of-sight direction indicator.

STATUS

In production.

CONTRACTOR

SAGEM, Departement Armement Terrestre, 6 avenue d'lena, 75783 Paris Cedex 16, France.

GERMANY (FEDERAL REPUBLIC)

3848.193

B171-II IR AIMING UNIT

The Type B171-II infra-red aiming unit is installed in armoured vehicles where it serves for observation, identification and aiming at night.

Both the security area and the target are illuminated by an IR searchlight. The reflected IR radiation is converted to a visible image by the built-in image converter tube. The B171-II IR-aiming unit is part of the Leopard 1 tank standard equipment.

STATUS

Production.

CONTRACTOR

Eltro GmbH, Postfach 102120, 69 Heidelberg 1, Federal Republic of Germany.

3849.193

CE632 LASER RANGEFINDER

The CE632 laser rangefinder provides accurate rangefinding and operates according to the pulse time-delay method with a repetition frequency of 1Hz.

Both assemblies, transmitter-receiver and display control with built-in power supply, are of compact and rugged design. The unit is designed to operate self-sufficiently as well as in connection with vehicle-integrated electro-optical fire control systems.

STATUS

In production and in service with West German Army AFVs.

CONTRACTOR

Eltro GmbH, Postfach 102120, 69 Heidelberg 1, Federal Republic of Germany.

Eltro CE632 laser rangefinder

3852.093

CE618 LASER RANGEFINDER

The CE618 laser rangefinder measures accurately the range to moving and fixed targets by transmitting ten pulses per second by the pulse time-delay method. It consists of a Nd:YAG laser transceiver unit and the power supply with electronics and control box.

Apart from other application fields the unit was designed for use in aircraft for target illumination and rangefinding, in electro-optical fire control systems on ground and on ships, as well as a redundant rangefinder for radar anti-aircraft tracking systems.

CONTRACTOR

Eltro GmbH, Postfach 102120, 69 Heidelberg 1, Federal Republic of Germany.

3850.193

CE624 LASER RANGEFINDER/SIGHT

When used with gated viewing, this rangefinder is a combined gunner's sight for use in battle tanks. The range to moving or fixed targets is continuously measured by a laser transmitter and a gated receiver by means of the pulse time-delay method. Range gating enables the observer to determine which target out of several within view was measured by the laser rangefinder. Thus a selection of single targets is provided.

CONTRACTOR

Eltro GmbH, Postfach 102120, 69 Heidelberg 1, Federal Republic of Germany.

3851.193

CE628 LASER RANGEFINDER

This Nd:YAG laser rangefinder is integrated into the fire control system of the Leopard 2 battle tank and is capable of providing three accurate rangefinding values within four seconds. It consists of laser transmitter, receiver and processing electronics with the related supply unit.

Because of its integration into the observation and aiming equipment, the observer may at all times get an exact digital range readout, the computer simultaneously processing the range data for accurate firing.

STATUS

In production for Leopard 2 MBTs.

CONTRACTOR

Eltro GmbH, Postfach 102120, 69 Heidelberg 1, Federal Republic of Germany.

Eltro CE628 laser rangefinder integrated in periscope control unit, seen here with laser electronics unit

3096.193

RZ1001 THERMAL IMAGING EQUIPMENT

The RZ1001 thermal imaging equipment is the successor to the RZ502 (**1802.193** in 1976 and earlier editions) from which it has been developed. A feature of the RZ1001 is its high geometrical resolution which particularly suits it for the detection of targets at long distances. Operation is possible in day and night conditions or in adverse weather. Due to the equipment's high temperature resolution, the objects of interest are displayed as a flicker-free image on the monitor and with good contrast against their background.

RZ1001 thermal imaging equipment

The equipment is of modular design and is suitable for applications such as target recognition in armoured or 'soft' vehicles, aircraft, helicopters, or ships.

CHARACTERISTICS

Field of view (azimuth): Approx 13° (if desired, variable from 6 – 13°)
Field of view (elevation): Approx 4.5°
Entrance pupil: Approx 130 mm
Focal length: 260 mm
Number of lines: 400
Line interlacing: 4:1
Geometrical resolution: 0.2 mrad
Frame rate: 25 frames/s
Detector: CdHgTe (100 elements, 4 times line displaced)
Waveband: 8.5 – 12.5 μm
Cooling: Joule-Thomson cooler
Monitor: screen diagonal
Power supply voltage: 24 ± 6 V
Power consumption: Camera with Joule-Thomson cooler 200 W approx
Monitor 50 W approx

RZ1002 thermal imaging unit with small monitor unit

CONTRACTOR

Eltro GmbH, Postfach 102120, 69 Heidelberg 1, Federal Republic of Germany.

3434.093

ELTRO ELECTRO-OPTICAL EQUIPMENT

The Eltro organisation produces a wide range of military electro-optical equipment for use both as independent items or as part of a complete weapon system in addition to those items described above. Full details are not available but the following paragraphs briefly describe some significant items.

HOT tracker

This is part of the anti-tank weapon system HOT (**2212.111**) which is for long-range use. The missile guidance via wire occurs semi-automatically. The system was designed as a high-resolution stabilised periscope. Target acquisition and tracking is accomplished via visual optics designed as a periscope. The determination of the missile position is accomplished by an integrated IR-goniometer, whose deviation values are used for missile guidance.

Mini-laser rangefinder

RAMAN laser rangefinder

Milan tracker

This unit is part of the lightweight, semi-automatic anti-tank guidance system Milan (**2215.111**) which is for medium-range use. Target acquisition and tracking is accomplished via the optical sight designed as a periscope. The determination of the missile position is by means of an IR-goniometer, and the combination of the latter with the line of sight gives the guidance function. The Milan anti-aircraft weapon can be fired from vehicles as well as from ground level.

Roland sight

This is part of the low-level air defence system Roland (**2218.131**) which is equipped with guided missiles. The sight serves for target acquisition and tracking. A built-in IR-goniometer supplies the deviation signals of the missile which are used for determining the guidance command signals.

Thermal pointer WOG 3

This equipment serves for passive reconnaissance and observation of heat-radiating objects. It was designed for use in armoured and non-armoured vehicles. In combination with observation and aiming units the heavy contrast thermal signature is superimposed congruently on the image. Thus searching, observing and aiming processes are considerably reduced and facilitated. This is a joint development by Eltro and Zeiss.

CO_2 laser rangefinder

CO_2 laser rangefinder

Eltro is developing a CO_2 laser rangefinder operating in the 10.6 μm waveband. The equipment is in the prototype stage.

Mini-laser rangefinder

This is a device in early development and designed for infantry use. It employs a semi-conductor laser, probably an Nd:YAG, and offers low power transmission and a special measuring system.

RAMAN laser rangefinder

Another development consisting of a Nd: YAG laser and a RAMAN cell. The 1.06 μm wavelength of the Nd:YAG laser is shifted to 1.54 μm in the RAMAN cell for eye-safe operation.

CONTRACTOR

Eltro GmbH, Postfach 102120, 69 Heidelberg 1, Federal Republic of Germany.

4685.193

OFRIS TARGET TRACKER

OFRIS (Optronische-Feuerleit-RIcht-Saule) is a target tracker consisting of a day sight, integrated thermal imager, CO_2 laser rangefinder and IR tracking electronics. To ensure flexibility in the design a modular approach has been adopted to provide various expansion stages and operating modes.

In its basic configuration OFRIS includes a viewing optics block, a drive block and an ocular block with day sight (periscope adaptable to 1.5 ×, 6 × and 12 × magnification), and a thermal imager composed of US common modules. The path of the light rays in the day sight and night sight is deflected at right angles into a common broadband optical system. This is accomplished through a window and by way of a tilting mirror, prior to separate processing in the ocular block. The IR image is displayed on an array of LEDs and can be injected into the periscope for night operations. This version makes use of range data supplied by existing surveillance sensors, such as radar and laser ranger, for lead angle calculation and tracking. Manual tracking, with the aid of a joystick, can also be provided.

Integration of a CO_2 laser and IR tracker, with external monitor where applicable, will afford various additional modes up to a version constituting an autonomous, passive sensor for automatic target tracking. This sensor allows redundant second target tracking and unrestricted night combat capability even in an ECM environment. An autonomous tracking computer, using distributed digital processing can be incorporated.

DEVELOPMENT

The OFRIS concept has been developed for the West German authorities by Carl Zeiss (optics and thermal imager) and Texas Instruments (tracking electronics). Possible uses include the Gepard anti-aircraft tank, the Roland air defence system and the proposed battle tank of the 90s.

STATUS

In development. Production was due to start in the second half of 1987.

CONTRACTOR

Carl Zeiss GmbH, 7082 Oberkochen, Postfach 1369, Federal Republic of Germany.

1727.193

DRIVER'S PERISCOPE TYPE BM 8005

The BM 8005 has been designed for the Marder personnel carrier, Leopard main battle tank, and other armoured vehicles having a standard vision port or the M113 and vehicles with an adapter. It is fitted extensively in NATO forces.

The driver's passive periscope is a binocular night vision device (one optical path for each eye), operated on the principle of image intensification and particularly developed for armoured vehicles provided with the standard vision block port of 65 by 130 mm.

The periscope serves mainly as a driver's vision device enabling night driving, without the assistance of any artificial light sources, down to a minimum illumination level of about 5.4 × 10^{-4} lux.

In addition the periscope can be utilised for surveillance purposes. There is an individual dioptre control for each eye and eyepiece heating prevents the accumulation of moisture.

CHARACTERISTICS

Magnification: × 1
Field of view: 55 × 40° (horizontal × vertical)
Image intensifier tube: Philips XX 1080
Dioptre adjustment: ±4 dioptres
Angular resolution: 0.6 mrad

STATUS

Over 12 000 units have been supplied to armies throughout the world.

CONTRACTOR

Philips GmbH, Unternehmensbereich Systeme und Sondertechnik, Hans-Bredow-Strasse 20, 2800 Bremen 44, Federal Republic of Germany.

BM 8005 driver's passive periscope

3810.193

BM 8025 PASSIVE NIGHT SIGHT

The BM 8025 passive night sight and observation equipment was developed by Philips for the West German armoured personnel carrier Marder and the armoured scout car Luchs. Vehicles equipped with the BM 8025 are qualified for night fighting and do not need any searchlight (IR or white).

In the BM 8025 system, the combination of low light-level imaging and thermal radiation detection offers high tactical efficiency; heat-dissipating targets, like tanks and vehicles, even when camouflaged are rapidly located in the battlefield, and the time for observation, recognition and aiming is shortened to a minimum.

The full field of view of the image-intensifying sight is covered by the scanning thermal target locator. The true position of a detected target is indicated to the operator by a red flashing spot within the light-intensified image of the scene.

CHARACTERISTICS

Magnification: × 6
Field of view: 5.6°
Dioptre focusing range: ±5 dioptres
Thermal target locator
Spectral range: 3 – 5 μm
Field of view: 6 × 2° (horizontal × vertical)
Resolution: 2 mrad

STATUS

The BM 8025 is in service with the West German Army.

CONTRACTOR

Philips GmbH, Unternehmensbereich Systeme und Sondertechnik, Hans-Bredow-Strasse 20, 2800 Bremen 44, Federal Republic of Germany.

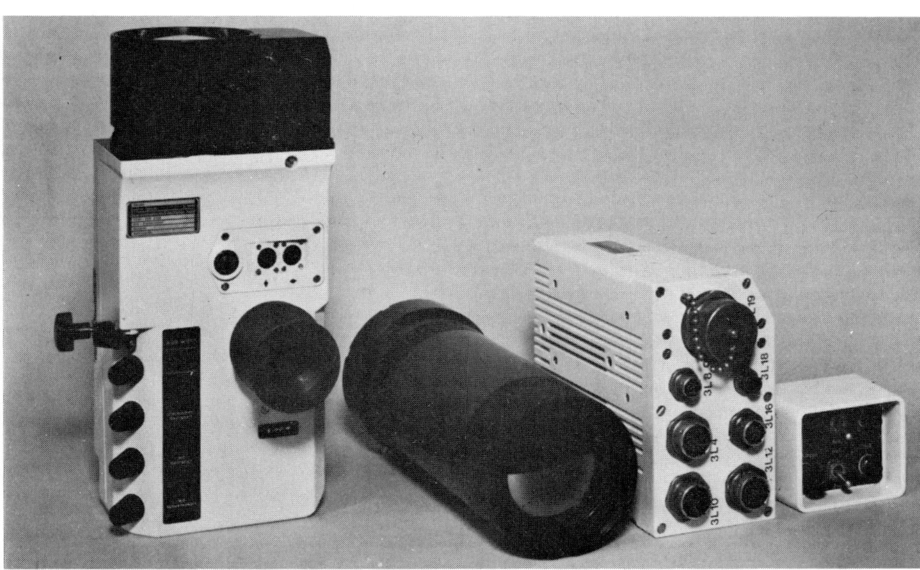

Units of BM 8025 night sight (left to right): image intensifier sight, thermal sensor, power supply unit, and control unit

INTERNATIONAL

5907.193

TRIGAT SIGHTING SYSTEM

Long range TRIGAT sighting system

An international consortium, known as SATEL, has been formed by SAT of France, THORN EMI Electronics of the United Kingdom and Eltro of West Germany, to develop a thermal imager and associated CO_2 laser rangefinder for the sighting system of the Euro-Missile Dynamics Group (EMDG) long-range, third generation, anti-tank guided weapon system (TRIGAT). The consortium has been proclaimed the winner in the competition which was organised in 1985 for the TRIGAT programme.

The work content has been divided between the three companies. SAT is the lead contractor and has the responsibility for the detector, associated electronics and equipment integration. THORN EMI is developing the multiple fields-of-view telescope and the processing electronics. Eltro is responsible for the scanner, laser rangefinder and detector cooling unit.

The thermal imager, CONDOR I, has multiple field-of-view and employs parallel scanning with a mercury cadmium telluride detector mosaic of more than one thousand elements, hybridised with charge-coupled device silicon circuits. It will be used in the sights of the TRIGAT system on both armoured fighting vehicles and helicopters. The Eltro CO_2 laser rangefinder has been in development for some time and operates in the 10.6 micron band.

STATUS

A full scale development programme was expected before the end of 1987.

Medium range TRIGAT sighting system

As for the long range TRIGAT system, the SATEL consortium formed by SAT of France, THORN EMI of the United Kingdom, and Eltro of West Germany was declared the winner in the competition held in 1985 for the medium range TRIGAT sighting system. The TIGER thermal imager uses a serial parallel scanning in conjunction with a MCT detector array of approximately 100 elements. This detector array is fixed on an hybrid circuit including silicon CCD integrated circuits as well. TG2 awarded SFENA and Ferranti a contract to supply lasers for this system.

STATUS

Full scale development programme was expected before the end of 1987 but no further information has been received.

CONTRACTORS

SAT (Société Anonyme de Télécommunications), 41 rue Cantagrel, 75624 Paris Cedex 13, France (lead contractor).

Eltro GmbH, Postfach 102120, 6900 Heidelberg 1, Federal Republic of Germany.

THORN EMI Electronics Ltd, Electro-Optics Division, Forest Road, Feltham, Middlesex TW13 7HE, England.

6535.193

SDFOV THERMAL IMAGING SYSTEM

SDFOV (Simultaneous Dual Field Of View) is a passive, real-time thermal imaging system which uses what the manufacturers state is a unique optical technique to provide both a narrow and a wide field of view of a target scene on a single, split-screen display. The system has been developed by Hughes Aircraft Company of USA and Eltro GmbH of West Germany to enable fire control weapon teams to detect and engage hostile targets more rapidly. The dual field-of-view would, for example, enable tank

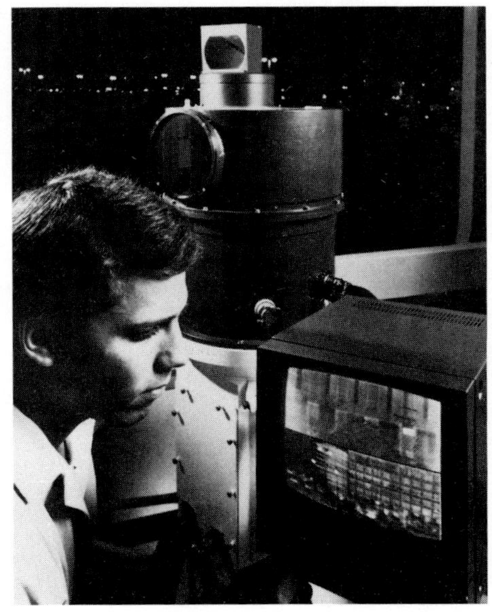

Dual field of view thermal imaging system

crews to respond quickly to battlefield situations, in that the tank commander would search for and identify targets while the gunner could aim and fire the vehicle's weapon systems independently.

The split-screen display is generated by two line-of-sight telescopes focusing infra-red radiation on the same infra-red detector array. The infra-red energy is converted to electronic signals which are eventually formatted for either split or separate screen viewing. The system is based on a modified set of US Army common modules, and can be adapted for airborne and shipborne applications. It could be upgraded in the future to include advanced focal plane array technologies as they become available.

STATUS

In development.

CONTRACTORS

Hughes Aircraft Company, Electro-Optical and Data Systems Group, El Segundo, California 90245, USA.

Eltro GmbH, Postfach 102120, 69 Heidelberg 1, Federal Republic of Germany.

ISRAEL

4244.193

TD22 THERMAL IMAGING SYSTEM

The Rafael TD22 is a compact high performance thermal imaging system for observation at all ranges. It is built to meet the requirements of a battlefield environment with emphasis on simplicity of operation and maintenance, low cost and long life.

The equipment is based on a parallel scanning mechanism operating in the three to five micron band of the infra-red spectrum and the detectors employed are indium antimonide (In-Sb) devices cooled to 77 K by a Joule-Thompson system which uses pressurised nitrogen. The TD22 has a standard television-type video output enabling remote observation as well as viewing on a built-in direct monitor. It is man-portable, and both portable and vehicle-mounted versions of the equipment have been developed. Company literature suggests that various telescopes can be coupled to to the infra-red camera for different applications.

CHARACTERISTICS

Waveband: 3 to 5 μm

Detector: 100 element parallel-scan indium antimonide

Detection range: 6.5 km (tanks); 2.5 km (personnel)

Recognition range: 2 km (tanks); 1.5 km (personnel)

Dimensions: 350 × 200 × 150 mm

Weight: 9.5 kg

STATUS

Production.

CONTRACTOR

Rafael, Ministry of Defence, PO Box 2082, 31021 Haifa, Israel.

4686.193

RAINBOW LASER RANGEFINDER

Rainbow is a CO_2 laser rangefinder capable of ranging on targets out to seven km with a 50 metre separation discrimination. The instrument is of modular design, to facilitate maintenance and repair, and consists of a TEA CO_2 laser and beam expander, a receiver, an optical telescope with ×8 magnification, power supply and an associated display unit. Rainbow weighs approximately 22 kg and measures 200 × 200 × 350 mm. No other details are available.

CONTRACTOR

Rafael, Ministry of Defence, PO Box 2082, 31021 Haifa, Israel.

3802.093

GYRO-STABILISED SIGHTING SYSTEM (GSSS)

The Tamam gyro-stabilised sighting system (GSSS) is an essential aid for observation, identification and tracking during aerial, naval, and ground-based missions. Versions are produced suitable for use in helicopters and light aircraft, ships, and armoured vehicles.

The GSSS neutralises motion-produced disturbances of the vehicle, vibration and environmental conditions affecting visual image and alignment accuracy. The system consists of a dual-magnification periscope/telescope the upper folding element (mirror) of which is gyro-stabilised. The mirror coating is effective over a wide spectral range, from the visual to the near infra-red, and permits optional use of laser or IR tracker in parallel with the visual optical path.

High magnification of the stabilised optics enables the observer to identify and lock on to the target at distances greater than the missile's or gun's range, and to distinguish details sufficiently for weapon guidance to the most vulnerable point.

A TV camera permits bi-ocular conventional sighting. TV enables better sighting conditions, and enhances haze penetration. The recording system allows storage and subsequent retrieval of information received during combat patrols, surveillance sorties and training exercises. It is possible to adapt an automatic contrast lock-on target using the TV option. A CRT and simulator assembly enables the GSSS to function as a training aid during simulated and real field firing conditions.

The high precision goniometer serves as an aiming facility for accurate launching and automatic guidance of missiles against armoured targets. The goniometer is mounted on the gyro-stabilised system optical frame and may be aligned with the optical line-of-sight to within 0.1 mrad or better. Alignment can be checked and adjusted by means of an internal collimator. A laser rangefinding system enables accurate target range information for successful launching.

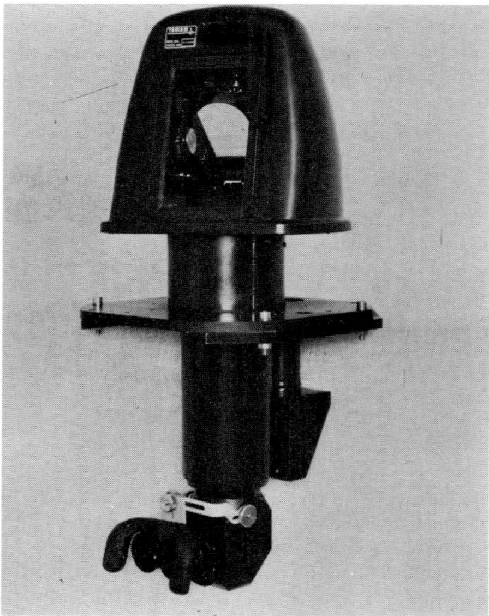

Tamam gyro-stabilised sighting system (GSSS)

CHARACTERISTICS

	Helicopters/ light aircraft	Long range airborne	Ships	Armoured vehicles
Dynamic resolution (mrad)	0.05	0.02	0.06	0.05 – 0.1
Optical resolution (static) (mrad)	0.04	0.01	0.05	0.04
Stabilisation accuracy (vibration) (mrad)	0.02	0.015	0.02 – 0.1	0.02 – 0.1
Surveillance field (*)				
Azimuth	±120 – 360°	360°	360°	360°
Elevation	–40°/+20°	–60°/+30°	–40°/+60°	–30°/+40°

(*) NOTE: Sight surveillance field can be modified to fit customer's specifications.

STATUS

Production.

CONTRACTOR

Israel Aircraft Industries Ltd, Yehud Industrial Zone 56000, Israel.

4287.183

TAMAM PADS

The Tamam position and azimuth determining system (PADS) is a precise ground-based inertial digital navigation equipment supplying instant readout of position, elevation and azimuth for artillery units. The equipment provides firepower co-ordination and orientation for mobile artillery units. It can be mounted on APCs, Jeeps, SP guns or other types of vehicle and is a self-contained, autonomous, all-weather, day/night equipment. Continuous display of co-ordinates and track is available on a moving map and remote display facilities are available. The equipment can be integrated with other sub-systems such as a laser rangefinder or an infra-red sensor.

The main items of the PADS equipment (see illustration) are:

(1) the inertial measurement unit (IMU), which is based on the Singer-Kearfott KT-70 unit operated by the USAF, USN, Israeli and other forces, and its associated power unit

(2) a digital computer, which is software compatible with the Data General Nova and Eclipse families

(3) the control and display unit, which enables the operator to communicate with the computer and accommodates entry of data such as initial position and position updates

(4) a display unit that informs the driver of system status, and particularly any need to stop for velocity updates

(5) an azimuth transfer unit (ATU) which provides remote users with accurate azimuth readings generated by the IMU resolver.

Operation of the ATU is based on a rigid linkage between the optical element and the inertial platform. The gun sight is boresighted and rests with the platform on a common base plate, the whole being gimbal-mounted to permit movement in two axes.

Tamam position and azimuth determining system (PADS)

CHARACTERISTICS

Position accuracy: 27 m (CEP at 50° lat)
Elevation error: 10 m (1 sigma); 33 m (max)
Azimuth error: 0.55 mrad (1 sigma at 50° lat)
Mission duration: 6 h
Latitude limits: 75°N – 75°S
Velocity updates: At 10-minute intervals for 20 s stops

STATUS

Production.

CONTRACTOR

Israel Aircraft Industries Ltd, Yehud Industrial Zone 56000, Israel.

6734.193

YUVAL LONG RANGE SURVEILLANCE SYSTEM

Yuval is an advanced family of cost-effective systems for long range observation which are designed to operate under all ambient light conditions 24 hours a day.

It is based upon three modules; an electro-optical head, a motorised pan and tilt unit mounted upon a collapsible tripod or other platform, and a portable control unit. The system features a dynamic range of ten, very high resolution (-1000 TVL) with the TV picture being transmitted in real-time to many users by land line or data link.

CONTRACTOR

Elbit Computers Ltd, Advanced Technology Center, PO Box 5390, Haifa 31053, Israel.

6735.193

REVEAL VIDEO ENHANCEMENT SYSTEM

To improve the capabilities of long range observation and surveillance systems Elbit have developed their REVEAL (REal-time Video Enhancement At Long range) system which was officially announced in mid-1987. The system combines the process of video stretching (to increase modulation and dynamic range), time domain filters (to enhance high frequency spectra of video signals) and space domain filters (to improve these same video signals). The system can be used in an AFV's FLIR system for target recognition or aboard an aircraft for IFF. The manufacturer claims it doubles the range of most electro-optical observation systems by enhancing, stabilising and sharpening the images.

CONTRACTOR

Elbit Computers Ltd, Advanced Technology Center, PO Box 5390, Haifa 31053, Israel.

Untreated video picture and a video picture enhanced by Elbit's Reveal

---

5699.193

MSZ2 DAY/NIGHT GUNNER PERITELESCOPE

The ELOP Day/Night Gunner's Peritelescope, known as the MSZ2, is a simplified modular integrated electro-optical fire control system for tanks and armoured fighting vehicles.

Its main features are day and night laser rangefinding incorporating enhanced magnification, passive night vision and small size designed for upgrading the ranging and aiming capabilities of tanks and AFVs for a greatly improved first round hit capability. Due to its small size, it can be adapted to any tank or turret where space is limited, and has already been introduced in the M24, M41, M47, M48, M50, M51, Centurion, Saladin and Scorpion vehicles.

Two basic versions are available. The first consists of a mini-laser rangefinder and a × 1 prism, both of which permit observation aiming and rangefinding during daylight, and a night elbow. The ×1 prism is fully interchangeable with the passive night elbow so that observation aiming and rangefinding are possible at night. The elbow contains a standard NATO reticle (or any other required reticle), and a laser aiming mark for night rangefinding. The reticles are projected and collimated on the image intensifier tube giving accurate aiming and rangefinding in the dark.

The second version is a fully computerised model with the addition of a ballistic computer, sensors, and a rotating wedges module for automatic insertion of ballistic compensations, providing a single reticle high accuracy, full solution fire control sytem with fast reaction times.

A number of intermediate versions, between the non-computerised and the fully computerised models are also available.

Day/Night Peritelescopes for commanders are identical to the gunner's model but designed to function with a ×8 day elbow replacing the gunner's mini-laser.

STATUS

The Day/Night Peritelescope is in production and is battle-proven in service with the Israeli Defence Forces.

CONTRACTOR

ELOP-Electro-Optics Industries Ltd, Advanced Technology Park, Kiryat Weizmann, PO Box 1165, Rehovot 76110, Israel.

ELOP day/night gunner's peritelescope

5900.193

MT-18 AND MT-19 LASER RANGEFINDERS

The MT-18 is a lightweight, self-contained, hand-held laser rangefinder specifically designed for infantry use, although it can be used in a variety of applications with specially designed accessories, for artillery, armoured fighting vehicles, and with remote control. Range to the target is provided as a four-digit display output in the sighting eyepiece.

As an artillery rangefinder, the equipment is known as the MT-19. It is tripod-mounted and combined with a goniometer which is electrically and mechanically connected. A rapid target-co-ordinates display is available.

CHARACTERISTICS

Range: 200 m to 20 km

Accuracy: ±10 m

Dimensions: 190 × 210 × 80 mm (320 × 220 × 230 mm with goniometer)

Weight: 2.9 kg (5.9 kg with goniometer)

STATUS

In production.

CONTRACTOR

ELOP – Electro-Optics Industries Ltd, Advanced Technology Park, Kiryat Weizmann, PO Box 1165, Rehovot 76110, Israel.

MT-18 laser rangefinder

5901.193

ELOP TANK LASER RANGEFINDER

The ELOP tank laser rangefinder is designed to provide high accuracy range measurements for the vehicle fire control system. It comprises four units, a commander's control, gunner's control, electronics unit and a laser peritelescope/sight. The equipment allows either the commander or the gunner to measure ranges up to 4995 metres.

CHARACTERISTICS

Operating range: 400 – 9995 m (valid range accepted by computer 400 – 4995 m)

Range accuracy: ±10 m

Range resolution: Capable of resolving 2 targets a minimum of 20 m apart

Dimensions: Control 125 × 130 × 190 mm; electronics unit 330 × 260 × 180 mm; laser peritelescope/sight 325 × 240 × 280 mm

Weight: Control 5.5 kg; electronics unit 11 kg; laser peritelescope/sight 15 kg

CONTRACTOR

ELOP – Electro-Optics Industries Ltd, Advanced Technology Park, Kiryat Weizmann, PO Box 1165, Rehovot 76110, Israel.

6290.193

ELOP MINI-LASER TANK RANGEFINDER

The ELOP mini-laser tank rangefinder is designed to provide high accuracy range measurements for the vehicle fire control system. It consists of two units, an electronics unit and a laser sight. The equipment fully matches the periscopes on the M-32 and M-36, replacing the existing day elbow. With its compact configuration, it can be easily integrated into most tanks and armoured fighting vehicles. A variety of options are provided so that it can be integrated into fire control system replacements in such vehicles as the M41, M47, M48, M60, AMX-13, Centurion, etc.

CHARACTERISTICS

Operating range: 300 to 9990 metres

Accuracy: ±10 metres (optionally ±5 metres)

Range resolution: 30 metres

Sight: ×8 magnification with 8° field of view

Dimensions: 150 × 130 × 90 mm (electronics unit); 310 × 26 × 10 mm (rangefinder unit)

STATUS

In production.

CONTRACTOR

ELOP – Electro-Optics Industries Ltd, Advanced Technology Park, Kiryat Weizmann, PO Box 1165, Rehovot 76110, Israel.

ELOP mini-laser rangefinder

6733.153

SADOT FIRE CONTROL SYSTEM

The SADOT system is a modular, fully computerised, fire control system which can be adapted to fit any AFV. The system is compact, easy to install and features a single reticle for day/night lasing and aiming. It has mechanical back-up and is claimed to be extremely accurate.

The system components comprise the gunner's periscope sight, a ballistic computer and the sensors. The sight features a head mirror, the main body assembly, a mini laser rangefinder (which can be replaced by a ×8 elbow), an image intensifier and a unity window. The sensors can include a cant angle, meteorological mast and target angle velocity sensor.

CHARACTERISTICS

Field of view: ×8 day channel (8° FOV), ×1 day channel (14° × 40° FOV), ×7 night channel (7.5° FOV)

Laser range: 300-9900 m

Defelection: ±30 mils (day), ±20 mils (night)

Elevation: 0-70 mils

CONTRACTOR

ELOP – Electro-Optics Industries Ltd, Advanced Technology Park, Kiryat Weizmann, PO Box 1165, Rehovot 76110, Israel.

6291.193

HRLR-04 AAC LASER RANGEFINDER

The ELOP HRLR-04 anti-aircraft high repetition laser rangefinder is a compact high performance 20 pulses per second solution for AAC upgraded systems. The internal modular structure of the equipment contains highly miniaturised subsystems which enable the modules (3.5 kg version) to be implemented in systems restricted by volume and/or weight.

CHARACTERISTICS

Operating range: 300 – 19 995 metres
Accuracy: ±5 metres
Range resolution: 30 metres
Repetition rate: Single shot to 20 pps continuously
Laser safety: 1.06 μm version – 40 dB minimum eye protection; 1.54 μm option – full protection with reduced sensitivity
Dimensions: 230 × 215 × 430 mm
Weight: 12 kg

STATUS

In production.

CONTRACTOR

ELOP – Electro-Optics Industries Ltd, Advanced Technology Park, Kiryat Weizmann, PO Box 1165, Rehovot 76110, Israel.

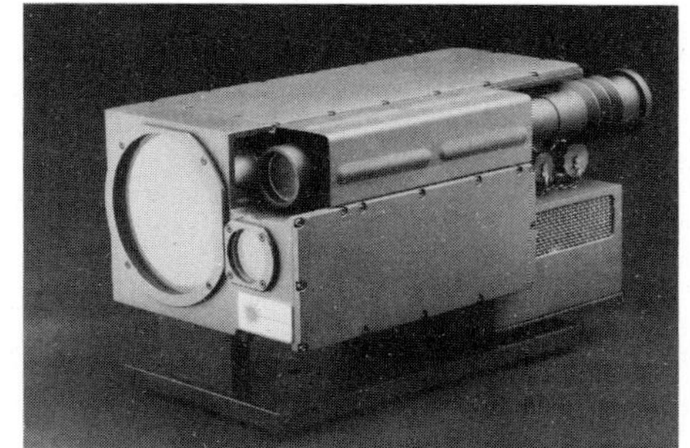

HRLR-04 AAC laser rangefinder

6292.193

BAT-30 FIRE CONTROL SYSTEM

The BAT-30 fire control system is a fully computerised modular system having an electrical interface with the gun and turret. The system comprises a body assembly which is mounted on the turret, with a head mirror mounted on top of the body assembly and bolted to it. A ×8 day sight elbow, incorporating a laser rangefinder and a passive image intensifier, are inserted in the body assembly from below. The peritelescope consisting of these individually sealed items also has a ×1 window to enable large field of view observation.

A ballistic computer group consisting of a ballistic computer, sensors and control boxes complements the control system. Unlike many conventional fire control systems, the BAT-30 has an electrical interface between the system and the tank.

A resolver mounted on the gun trunnion measures the gun elevation angle, and another resolver mounted on the head mirror shaft measures the mirror elevation angle. The difference between the two angles is nulled by the computer. The BAT-30 is, therefore, particularly well suited for upgrading tanks that have no ballistic drive, eg Soviet T-tanks, AMX-30, Centurion, etc.

Two optional additions to the system are a commander's observation device consisting of a head mirror and interchangeable day and night elbows, and a passive image intensifier driver's night viewer.

The BAT-30 has a single reticle, for both ranging and aiming, that always remains at the centre of the field of view. A rotating wedges module which is part of the body assembly moves the scenery and not the reticle. Thus the gunner always ranges, aims and fires with the reticle at the centre of the field of view where resolution is best.

STATUS

In full production.

CONTRACTOR

ELOP – Electro-Optics Industries Ltd, Advanced Technology Park, Kiryat Weizmann, PO Box 1165, Rehovot 76110, Israel.

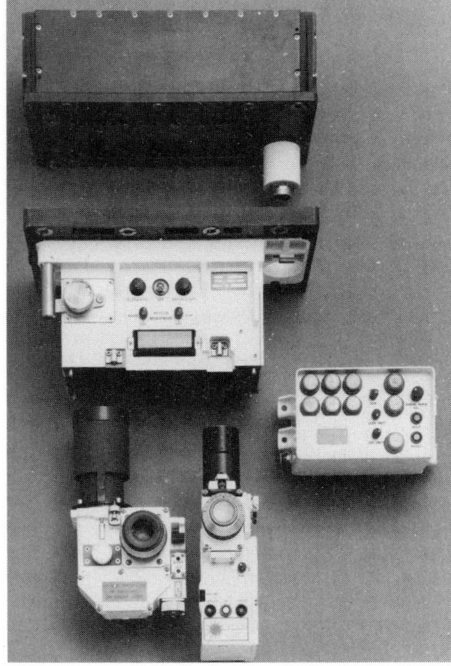

BAT-30 fire control system

6293.193

MATADOR FIRE CONTROL SYSTEM

Matador is a fully automatic fire control system. It is based on an improved M32 periscope with optics that can withstand laser radiation damage, a rotating wedges module for automatic insertion of ballistic compensations in deflection, and higher accuracy rotating mirror servo mechanism.

The peritelescope has three observation channels; a ×8 magnification day sight incorporating a laser rangefinder, a ×1 unity window, and a night elbow with a second generation image intensifier tube. It has a single reticle for both ranging and aiming that always remains at the centre of the field of view since the rotating wedges module moves the scenery and not the reticle. Thus the gunner always ranges, aims and fires with the reticle at the centre of the field of view where resolution is the best.

In addition the system includes a ballistic group containing a ballistic computer, control boxes and sensors. There is an optional optical relay to the commander, or a conversion kit for the commander's existing optical rangefinder, including a range display and a rotating wedges module for automatic insertion of ballistic compensation to enable firing from the commander's position.

STATUS

In production.

CONTRACTOR

ELOP-Electro-Optics Industries Ltd, Advanced Technology Park, Kiryat Weizmann, PO Box 1165, Rehovot 76110, Israel.

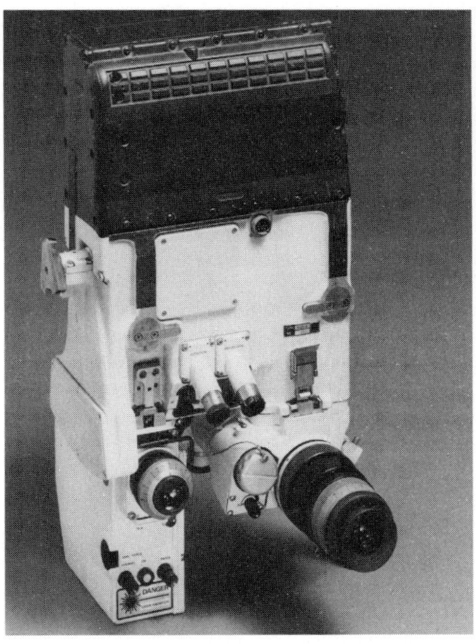

Matador fire control system

ITALY

4216.193

C 215 ARTICULATED TELESCOPIC SIGHT

The C 215 articulated telescopic sight, designed by Aeritalia, is part of the secondary direct fire control system of the new Italian-designed OF 40 main battle tank. It is mounted coaxially with the weapon and is used in conjunction with the M114 telescope mount. The instrument consists of a head assembly which includes an eyepiece, support arm and adjustable head-rest, an articulated body in which the prisms are located, and an objective tube assembly.

CHARACTERISTICS

Magnification: ×8
Field of view: 7.5°
Objective focal length: 305 mm
Eyepiece focal length: 38 mm
Exit pupil: 5 mm diameter
Eye relief: 24.5 mm
Eyepiece setting: –4 to +4 dioptres
Dimensions: 1200 × 178 mm
Weight: 25 kg
Operating temperature range: –40 to +65°C
Storage temperature range: –60 to +70°C

STATUS

In production for OF 40 main battle tank.

CONTRACTOR

Aeritalia (Società Aerospaziale Italiana) p.A., Gruppo Sistemi Avionici ed Equipaggiamenti, Viale Europa, 20014 Nerviano, Italy.

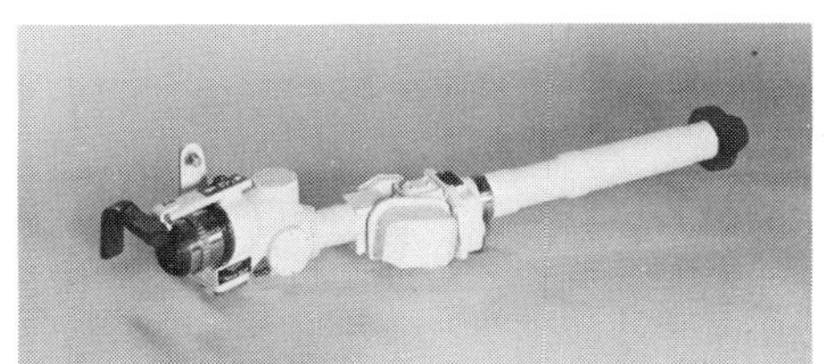

Aeritalia C 215 articulated telescope for new Italian OF 40 tank

3558.193

VAQ-3/VAQ-33 TANK LASER RANGEFINDERS

The VAQ-3 neodymium-glass laser rangefinder for battle tanks has been designed for integration with a computer fire control system. Long-range firing with a very high first round hit probability is possible, even against moving targets. Furthermore, a much shorter reaction time is achieved. The VAQ-3 is now the standard laser rangefinder fitted in the Leopard main battle tank of several armies. An updated version, the VAQ-33, has been produced, and is the standard fit in the OF 40 tank.

CHARACTERISTICS

Type: VAQ-3
Laser material: Nd-glass
Mode of operation: Q-switched
Wavelength: 1.06 μm
Peak power: 6 MW
Energy per pulse: 150 mJ
Pulse repetition rate: 12 pulses per minute or bursts of 3 pulses in 4 s with an interval of 11 s between bursts
Detector: Avalanche photodiode
Range capability: 400 m – 10 km
Range accuracy: ± 5 m
Distance discrimination: 30 m
Azimuth discrimination: 0.25 mrad
Signal to computer: Digital in BCD code with DC logic level of TTL circuits

System composition

Laser optical head (transmitter/receiver)
Gunner's laser control unit (GLCU)
Commander's laser control unit (CLCU)
Laser power supply unit (LPSU)

Laser Optical Head
Weight: 6.8 kg
Dimensions: 280 × 116 × 180 mm

GLCU
Weight: 1.9 kg
Dimensions: 280 × 170 × 75 mm

CLCU
Weight: 0.8 kg
Dimensions: 218 × 131 × 91 mm

LPSU
Weight: 13.9 kg

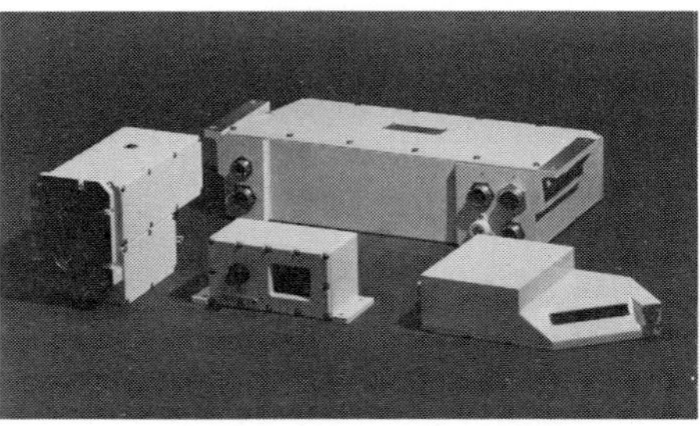

Selenia VAQ-3 tank laser rangefinder

Dimensions: 511 × 126 × 257 mm
Power supply: 24 V DC (vehicle supply)

STATUS

In quantity production. Several hundred units have been produced for the Leopard tank.

CONTRACTOR

Selenia Defence Systems Division, Tiburtina Km 12400, Rome, Italy.

4253.193

VTG 120 LIGHT THERMAL IMAGER

VTG 120 is a man-portable thermal imaging equipment for night surveillance and aiming of medium-range weapons, with particular application to the TOW anti-tank missile. It operates in the 8 to 14 micron band and is provided with an eyepiece for direct viewing. It can also be connected to an external CRT for remote monitoring of the observed scene. An electrically generated reticle is correctly positioned for aiming purposes.

The system is completely autonomous with a battery and air bottle which allows more than two hours of continuous operation. The empty bottles and battery can be easily replaced. A range of ancillary equipment is available including a battery charger, boresight collimator for reticle alignment, gas bottle charger and a tripod mounting interface. The VTG 120 can be supplied with a closed-cycle cooler if required. Standard performances on a tank target are 3000 metres detection and 2000 metres identification in normal meteorological conditions.

CHARACTERISTICS

Wavelength: 8 – 14 microns
Field-of-view: Narrow 20 × 40 mrad; wide 60 × 120 mrad
Magnification: × 12 (narrow FOV); × 4 (wide FOV)
Resolution: Narrow 0.16 mrad; wide 0.5 mrad
Detector: CMT 60 element array
Cooling system: Joule Thomson minicooler
Weight: 11 kg

STATUS

Production.

CONTRACTOR

Officine Galileo SpA, Via Einstein 35, 50013 Campi Bisenzio, Florence, Italy.

VTG 120 thermal imaging night sight fitted to TOW anti-tank missile launcher

---

6736.193

OG C-102 COAXIAL TELESCOPE

The OG C-102 coaxial telescope is an aiming device for direct fire operations of the battle tank main weapon and the coaxial machine gun when mounted. The telescope follows the movements of the main weapon.

The principal parts of the equipment are as follows:

(a) Objective lens assembly which includes the focusing objective and the aiming reticle. Externally two spherical reference collars are available of which one is directly interfaced with the vehicle and the other by means of an alignment mount
(b) Articulated body which permits telescope movements and includes the aiming reticle selector and the laser filter
(c) Eyepiece assembly, including the eyepiece and headrest
(d) Alignment mount and support arm. By means of two orthogonal movements the mount allows alignment of the telescope with the main weapon. The support arm is connected to the eyepiece assembly and minimises eyepiece movement when the weapon is operating in elevation.

Galileo OG C-102 coaxial telescope

CHARACTERISTICS

Magnification: × 8
Field of view: 7.5
Elevation field: -9 to +20°
Height: 264 mm
Length: 927 mm
Depth: 245 mm
Weight: Telescope 22 kg; mount 5 kg

STATUS

Adopted by the Italian Army for its new C1 MBT (Meduno).

CONTRACTOR

Officine Galileo SpA, Via Einstein 35, 50013 Campi Bisenzio, Florence, Italy.

3589.193

P56 SERIES COMPUTING SIGHT AND GUN DRIVE

The basic P56 equipment is a self-powered computing sight and gun laying equipment for local control of light calibre guns against low- and very low-flying targets. It may also be used for the engagement of ground targets. Suitable for fitting on any existing gun, it has been chosen for use with guns of 20 to 40 mm calibres by several countries.

The sight consists of one panoramic articulated telescope with fixed eyepiece for aiming at both air and ground targets. The computer is a solid-state electronic machine. The gun drive consists of an axial pump and hydraulic motor, with prime power being provided by an internal combustion engine or by an electric motor. This motor also drives a generator to provide electrical power for the computer. Tracker training equipment is also available

Sights of this series are encountered frequently on 20 mm guns produced by Rheinmetall in West Germany, GIAT in France and Swiss Oerlikon weapons. The P56/40 version was specially designed for upgrading the Bofors 40/60 weapon.

CHARACTERISTICS

Sight

Elevation: –7 to +85°

Magnification: ×5

True field: 12°

Target speed: 60 – 350 m/s

Crossover point range: 100 – 600 m

Total lead angle: Up to 21°

P56T

The P56T is a turret computing sight and gun laying equipment based on the standard P56 and specially designed for installation on the turret of any wheeled or tracked vehicle intended for AA defence with light calibre guns.

P56/40

This is a special version of the P56 designed for retrofitting to existing 40 mm AA guns, such as the 40/60, to provide an autonomous capability and to enhance performance.

Specific versions for upgrading other weapons are available.

STATUS

Several thousands of the P56 equipment have been supplied for many types of guns of 20 to 40 mm calibre for European and other armed forces (20, 30, 37 and 40 mm calibre). Over 1500 units are in operation in NATO countries and some hundreds in other countries. A series (P56T version) has been produced for use in turrets manufactured by Electronique Serge Dassault.

CONTRACTOR

Officine Galileo SpA, Via Einstein 35, 50013 Campi Bisenzio, Florence, Italy.

P56 computing sight and gun drive for fitting to AA guns

5902.193

VANTH OPTRONIC FIRE CONTROL SYSTEM

This is a family of optronic fire control systems for local control of AA guns, and is the latest generation of an original Galileo concept evolved through the P36, P56 and P75 models. The systems can operate autonomously, can be integrated into a SHORAD system fitted with a co-ordination and control centre and a detection radar, or can be remotely controlled for unmanned operation in association with a fire control unit.

The Vanth family, which incorporates the P75, is characterised by:

(1) up-to-date technology, including a microprocessor-based computer, laser rangefinder and, optionally, optronic sensors for twilight and night operation
(2) modularity and growth capability
(3) direct interchangeability with previous P models on any weapon
(4) adaptability to virtually any AA mount from 20 mm calibre upwards
(5) suitability for retrofit and upgrading of existing guns, with or without own servo systems.

In addition it is possible to upgrade any P56 system into a Vanth model.

The basic model of the Vanth family features fully automatic computation based on laser rangefinding, local control with remote designation (radar or optical), and integrated autotest. This model can be expanded with any one or more additional options, such as remote control by an external fire control system, firing taboo arcs setting, parallax and meteoballistic introduction, integrated training system, TV automatic tracking, light intensifier channel, and a thermal imaging channel.

The basic sub-assemblies of the system are:

(1) an optical head with acquisition open sight, telescopic line-of-sight, laser rangefinder, optional night sight, and control panels
(2) main structure with control joystick, manual manoeuvres and digital computer
(3) hydraulic servo systems, drive gears, prime mover and electrical generator.

The optical head integrates the visual, laser and light intensifier channels, and the TV, into a single optical path. This multiple line-of-sight is independent of the line-of-fire so that target aiming is always at the centre of the reticle, even when the gun is given lead angle corrections. The laser rangefinder is an Nd:YAG type with high repetition and accuracy.

The digital computer performs all the firing computations at high speed, controls and manages the operation of the optical sights, laser rangefinder, servos and accessory functions. It also provides facilities for the reconfiguration of the basic system by the addition of optional modules.

Sight and gun control is performed by the operator through a two-degree of freedom joystick. A back-up manual system is also available. The servo system is electro-hydraulic and provides high speed and acceleration with a very small dynamic error.

STATUS

In production.

Vanth fire control system

CONTRACTOR

Officine Galileo SpA, Via Einstein 35, 50013 Campi Bisenzio, Florence, Italy.

5903.103

MADIS SIGHTING AND DRIVE SYSTEM

Madis is a sighting and turret drive system originally designed for the Oto Melara Quad 25 mm self-propelled gun system. It consists of a daylight optical head, a night optical head, and a hydraulic servo system and control unit. The three units are compact and modular in construction and can be installed in other AA armoured vehicles, either in total or in part.

The daylight optical head contains the visual telescope, self-stabilised by a gyro directly coupled to the line of sight (LOS) to allow operation on the move. It can also operate without the gyro as a servoed optical sight. Provision is made for compensation of image rotation due to the azimuth angle. The head can also accommodate a laser rangefinder and a TV camera, with integration into a single output path of the visual, laser and TV channels to ensure the necessary parallelism.

The night optical head is an optical device designed for integration with a low-light level TV camera in a compact and rugged housing. The LOS can be servo-stabilised to enable use on the move.

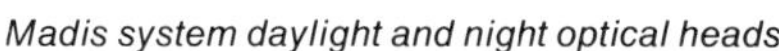

Madis system daylight and night optical heads

The power servo system of the turret and weapon consists of an axial pump and hydraulic motor with an asynchronous motor as prime mover. This assembly also includes the first stage of the reduction gearing and, on the upper part, the computer and control panel.

CHARACTERISTICS
Daylight optical head
Field of movement: Azimuth 45°; elevation –10° to +85°
Telescope magnification: × 5
Telescope field-of-view: 12°
Night optical head
Field of movement: Azimuth 45°; elevation –10° to +75°
Field-of-view: 4.76°

STATUS
In production.

CONTRACTOR
Officine Galileo SpA, Via Einstein 35, 50013 Campi Bisenzio, Florence, Italy.

4220.193
V 200 VISION BLOCK
The V 200 vision block is designed for observation from the inside of the M113 APC and other armoured vehicles, giving the observer protection from small arms fire. The V 200 is capable of withstanding direct hits from 7.62 mm ammunition.

CHARACTERISTICS
Magnification: × 1
Field of view: 90° horizontal; 30° vertical
Transmission: 64%
Dimensions: 291 × 187.5 × 114 mm
Weight: 10 kg
Operating and storage temperature range: –54 to +68°C

CONTRACTOR
Aeritalia (Società Aerospaziale Italiana), Gruppo Sistemi Avionici ed Equipaggiamenti, Viale Europa, 20014 Nerviano, Italy.

4217.193
P 186 TELESCOPIC SIGHT
The P 186 telescopic sight is designed for use as part of the primary fire control system of the Palmaria 155 mm self-propelled howitzer. Fitted on the Bofors telescopic mount, the device is used for laying the weapon in azimuth for indirect fire control. The instrument consists of a head assembly with a right angle prism, top and bottom body assemblies, and an eyepiece which contains an Amici prism.

CHARACTERISTICS
Magnification: × 4
Field of view: 10° (177 mils)
Objective focal length: 110 mm
Eyepiece focal length: 28 mm
Entrance pupil: 19 mm diameter
Exit pupil: 4.75 mm diameter
Eye relief: 19.5 mm, 435 mm
Azimuth range: 6400 mils
Elevation limits: –645 to +345 mils
Dimensions: 525 × 162 × 240 mm
Weight: 8 kg
Operating temperature range: –25 to +70°C

STATUS
In production for the Italian-built Palmaria 155 mm SP howitzer.

CONTRACTOR
Aeritalia (Società Aerospaziale Italiana), Gruppo Sistemi Avionici ed Equipaggiamenti, Viale Europa, 20014 Nerviano, Italy.

Aeritalia P 186 telescopic sight for the Palmaria SP howitzer (right), with the P 204 periscope (see next page) on the left

---

4218.193
P 192 NIGHT DRIVING PERISCOPE
The P 192 is a passive night vision device for drivers of AFVs using a second generation light intensifier tube. With this periscope, it is claimed to be possible to drive vehicles at a speed of more than 40 km/h in starlight conditions without any artificial illumination. It consists of a head prism, S 25 micro-channel light intensifier and a bi-ocular eyepiece. A head prism was selected for the instrument instead of a mirror, because of its smaller size for a similar field of view, and greater strength. The light intensifier is fitted with a gain control which is automatically switched off when the photocathode illumination level becomes too high. A filter allows a correct observation within a luminance range from 10^{-3} lux to 10^{4} lux. The instrument is equipped with a bi-ocular display which enables the driver to view the scene with both eyes through one large diameter eyepiece. The equipment is designed to be fitted on various types of vehicle including the 6616, M113, VCC 1, Marder, M41, M47, M48, M60, Leopard and the TAM.

CHARACTERISTICS
Magnification: × 0.9
Field of view: 48° horizontal, 40° vertical
Objective lens focus: 25 mm
Relative objective aperture: 1:1.3
Eyepiece lens focus: 43 mm
Eyepiece aperture: 1:0.54
Focus range: 10 m – ∞
Eyepiece pupils distance: 35 mm

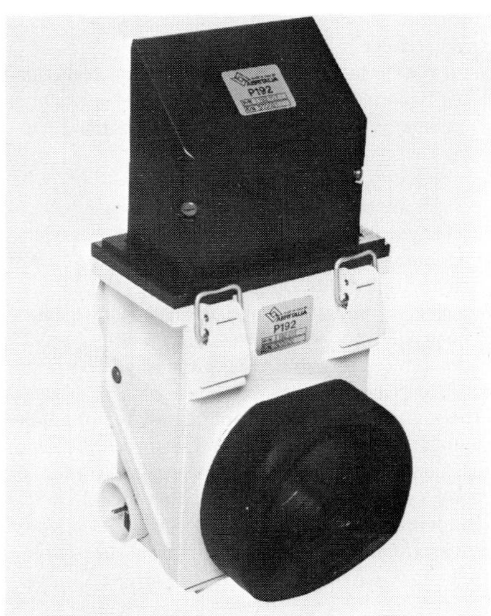

Aeritalia P 192 night driving periscope

Transmission: >75%
Weight: 7 kg
Operating temperature range: –30 to +45°C

STATUS
In production.

CONTRACTOR
Aeritalia (Società Aerospaziale Italiana), Gruppo Sistemi Avionici ed Equipaggiamenti, Viale Europa, 20014 Nerviano, Italy.

4689.193
P 233 NIGHT VISION PERISCOPE
This passive night vision device is intended to satisfy the technical requirements of the commander on various types of armoured vehicles (armoured personnel carriers, infantry fighting vehicles, etc). It can also be retrofitted to earlier types of battle tanks. The instrument consists of the head assembly, including the right angle prism, and the body assembly which contains the objective lens, the second generation image intensifier tube, the reticle, the binocular eyepiece and the electronics.

The periscope has an operational illumination range from 10^{-3} to 10 lux. Range in starlight is up to 800 metres. The instrument is interchangeable with all European and US daylight periscopes.

CHARACTERISTICS
Magnification: × 3
Field of view: 16° horizontal, 15° vertical
Weight: 7.5 kg

CONTRACTOR
Aeritalia (Società Aerospaziale Italiana), Gruppo Sistemi Avionici Equipaggiamenti, Viale Europa, 20014 Nerviano, Italy.

4219.193

P 204 PERISCOPE

The P 204 periscope is designed for day and night operations as part of the primary fire control system of the BM 6616 armoured car. It can be adapted to replace M32, M34, and M36 periscopes installed in other vehicles. The instrument consists of the body and either a P 265 IL daylight elbow sighting telescope or a night elbow sight with a second generation image intensifier. The two sights are interchangeable. The body contains the head prism, a unity power optical system and includes lever linkage to the weapon.

The P 204L incorporates a laser rangefinder module which is built into the instrument. This is a Nd:YAG rangefinder with an accuracy of ±5 m and a maximum range of 9995 m. Output power is 24 mJ.

(A photograph of this instrument appears in entry **4217.193** for the P 186 telescopic sight, above.)

STATUS

In production.

CHARACTERISTICS:

	day	night
Magnification	×8	×7
Field of view	9° (160 mils)	7.9°
Resolution (contrast 1 lux)		0.3 mil
(contrast 10^{-3} lux)		0.4 mil
Reticle setting	–7 to +7 mil	–7 to +7 mil
Objective focal length	240 mm	173 mm
Eyepiece focal length	30 mm	25 mm
Relative aperture		1.5
Entrance pupil	40 mm	96 mm
Exit pupil	5 mm	
Eyepiece setting	–3 to +3 dioptres	–5 to +5 dioptres
Dimensions	466 × 295 × 202 mm	
Weight	12 kg	
Operating temperature range	–40 to +70°C	

CONTRACTOR

Aeritalia (Società Aerospaziale Italiana), Gruppo Sistemi Avionici ed Equipaggiamenti, Viale Europa, 20014 Nerviano, Italy.

6361.193

OG-PO7B PERISCOPE

The periscope OG-PO7B is a binocular optical sight designed to satisfy the technical requirements of observation and sighting of light armoured vehicles in which the cab does not rotate with the turret. The scanning movements of the line of sight, rendered integral with the line of fire during assembly, are given in traverse by the rotation of the turret, and in elevation by a mechanical connection which transmits the angle assumed by the weapon to the prism of the periscope. A monocular version, the OG-PO7 is also available.

CHARACTERISTICS

Magnification: ×1

Field of View: 44°

Elevation scan: –10° to +50°

Azimuth scan: 360°

Dimensions: 550 × 290 × 246 mm

Weight: 19 kg

CONTRACTOR

Officine Galileo SpA, Via Einstein 35, 50013 Campi Bisenzio, Florence, Italy.

OG-PO7B periscopic sight

4688.193

OG-P20 PERISCOPE

The OG-P20 is an optical periscope designed for use in light armoured vehicles. It is primarily intended to fulfil the technical and operational requirements for observation and aiming in vehicles where the crew accommodation does not rotate in concert with the turret.

CHARACTERISTICS

Magnification: × 1.5 and × 4.5

Field of view: 28° (×1.5) and 10° (×4.5)

Elevation: –10 to +60°

Azimuth training: 360°

Transparence: 42% (×1.5) and 40% (×4.5)

Dimensions: 680 × 320 × 340 mm

STATUS

In production.

CONTRACTOR

Officine Galileo SpA, Via Einstein 35, 50013 Campi Bisenzio, Florence, Italy.

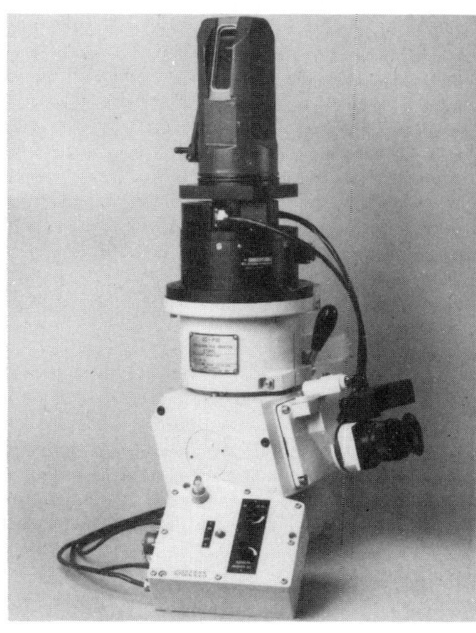

OG-P20 periscope

6294.193

OG-P101 PERISCOPE SIGHT

The OG-P101 optical sight is intended to fulfil the technical requirements of observation and aiming on armoured fighting vehicles. The sight can replace, with suitable modifications, the M 32, M 34 and M 36 periscopes. A second generation light intensifier night vision elbow allows night observation and aiming.

CHARACTERISTICS

Magnification: ×1 and ×8

Field of view: 26° × 10° (×1) and 9° (×8)

Elevation: –10° to +70°

Dimensions: 470 × 205 × 295 mm

STATUS

In production.

CONTRACTOR

Officine Galileo SpA, Via Einstein 35, 50013 Campi Bisenzio, Florence, Italy.

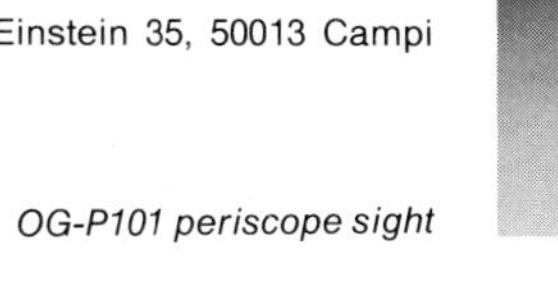

OG-P101 periscope sight

6528.193

SP-T-694 STABILISED PANORAMIC SIGHT

The SP-T-694 AFV panoramic day/night sight allows the tank commander to carry out panoramic survey of the battlefield, supervise the gunner's operation, slave his line of sight to the gunner's, and have autonomous control and firing of the main weapon. A good transparency, together with a large objective aperture and 2nd generation image intensifier tube, provides good performance during night operations.

All the operations are controlled by a 16-bit microprocessor, whose functions can be expanded up to the computation of lead angles (azimuth and elevation) as an alternative to the gunner's fire control system. This enables the commander to carry out control and firing on the move. The interface is digital as well as analogue and allows the sight to be integrated with most modern fire control systems.

CHARACTERISTICS
Elevation: –10 to +10°
Bearing: n × 360°
Number of oculars: 2
Magnification: × 2.5; × 10
Night channel: 2nd generation light intensification tube
Magnification: × 6
Field of view: 8°

STATUS
Pre-serial production for the Italian MBT, MICV, OTOMATIC programmes, with SFIM (France) as the main supplier of the stabilised mirror.

CONTRACTOR
Officine Galileo SpA, Via Einstein No 35, 50013 Campi Bisenzio, Florence, Italy.

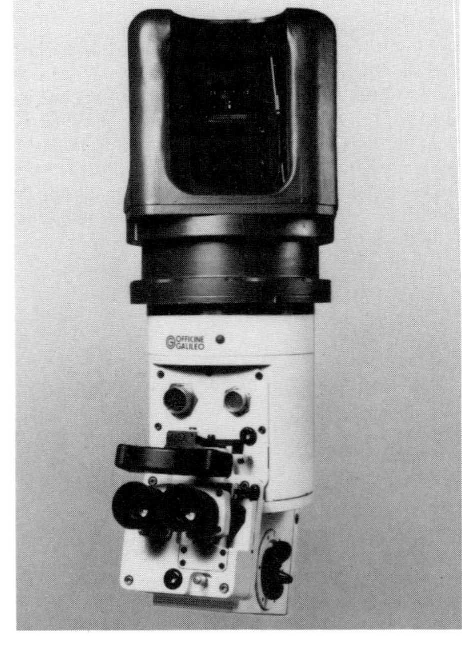

SP-T-694 stabilised panoramic sight

4221.193

T178/278 GONIOMETER

The T178/278 goniometer is designed by Aeritalia to meet artillery requirements. It consists of an elbow sighting telescope, reticle for distance measurements, a reading microscope, and a built-in magnetic declinator device. Two graduated circles provide measurement of horizontal and vertical angles. The instrument is equipped with an optical plumb to aid positioning. Two versions are produced with different illumination systems. The T178 uses a tritium self-illumination system. The T278, designed for the Italian Army, is equipped with a diode illumination system.

CONTRACTOR
Aeritalia (Società Aerospaziale Italiana), Gruppo Sistemi Avionici ed Equipaggiamenti, Viale Europa, 20014 Nerviano, Italy.

CHARACTERISTICS
Magnification: × 8
Field of view: 7° (120 mils)
Objective focal length: 128 mm
Eyepiece focal length: 16 mm
Focusing range: 10 m – ∞
Entrance pupil: 24 mm diameter
Exit pupil: 3 mm diameter
Eye relief: 12.2 mm
Eyepiece setting –5 to +5 dioptres

	Horizontal circle	Elevation circle	Declinator
Graduation	6400 mils	–800 to +1200 mils	553 mils E – 553 mils W
Direct reading	0.5 mil	0.5 mil	1 mil
Estimated reading	0.25 mil	–	0.5 mil

Weight: 5 kg
Operating and storage temperature range: –40 to +55°C

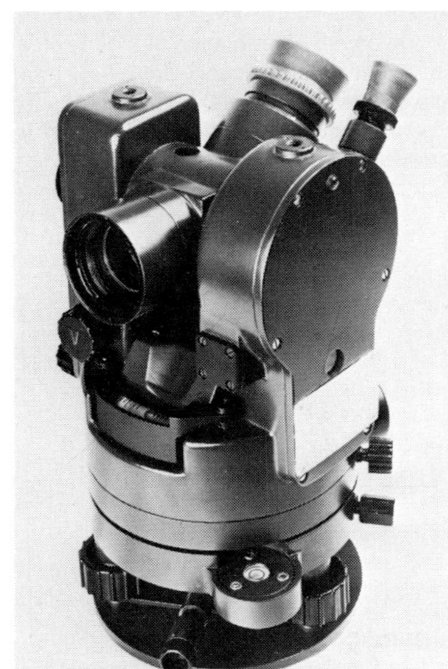
Aeritalia T178 artillery goniometer

NETHERLANDS

3578.193

TYPE TS7TS PASSIVE AIMING SIGHT

The passive aiming sight TS7TS has been designed to replace the infra-red sight on various types of tanks. It enables gunners to locate, identify and engage targets at night, without the use of any artificial illumination. The combination of high-speed entrance optics, a 25 mm image-intensifier tube and an optimally shaped eyepiece ensures a clear view of the field of fire. A gun-following mirror links the two parts (gun and sight).

CHARACTERISTICS
Total magnification: × 7
Field of view: 6°
Focus: 100 m – ∞
Eyepiece adjustment: ±5 dioptres
Image-intensifier tube: 3-stage with ABC, DC and FP
Power supply: 21-30 V DC

STATUS
Production.

CONTRACTOR
Oldelft (BV Optische Industrie De Oude Delft), PO Box 72, 2600 MD Delft, Netherlands.

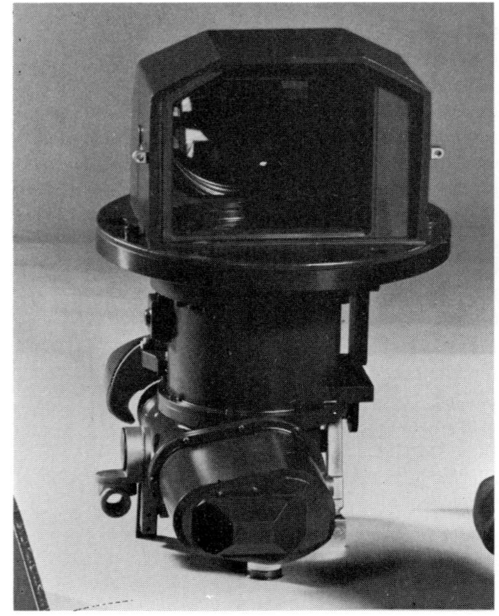
Type TS7TS passive tank sight head

5904.193

SAMTOR THERMAL OBSERVATION SYSTEM

Oldelft has developed a thermal observation system for coastal and frontier surveillance. The system is completely mobile and can either be transported by a Land-Rover type vehicle, or is man-portable by four people. The equipment can be assembled and be operational on site in less than one hour. It can also be mounted on a vehicle for instantaneous use while stationary or on the move.

The SAMTOR system consists of:

(1) a thermal imaging camera
(2) a head assembly to allow movement in azimuth and elevation
(3) tripod mounting
(4) an electronics unit
(5) a control and video display unit
(6) an optical laser rangefinder
(7) an optional 220 V generator.

The system is modular based with an afocal telescope providing a dual field of view, scanning module, IR imaging system, closed-cycle cooling system and an electronics module. The last provides system control, processing of the image and multiplexing of the video system. The optional laser rangefinder is fitted in parallel with the camera reticle.

The control/display unit controls the camera position in azimuth and elevation, adjusts brilliance and contrast, changes the image positive/negative, adjusts focus, selects the field of view and controls the rangefinder when fitted. It also provides a read-out of azimuth and elevation of the camera, plus distance measured by the rangefinder. The thermal image is displayed on a 9-inch (229 mm) high-resolution video screen and can be relayed to outside monitors. Camera azimuth and elevation, and distance from the rangefinder, can also be displayed on the screen.

CHARACTERISTICS

Thermal camera

Field of view: Narrow 1.7° × 4° (optionally a second narrow field 0.85° × 2°); wide 5.1° × 12°

Detection range: Tank 6 – 8 km; person 1.8 km

Reconnaissance range: Tank 2.5 km; person 1 km

Wavelength: 7.5 – 11.7 microns

Cooling system: Closed-cycle Stirling

Time into operation: 10 minutes from starting cooling system

Head assembly

Scanning angle: Azimuth 360° (scan sector is adjustable); elevation –90° to +90°

Accuracy: 3 mils

Laser rangefinder

Type: Nd:YAG

Wavelength: 1.06 microns

Power output: 40 mJ

Range: 200 – 9990 m

Accuracy: 10 m

Field of view: 7°

STATUS

In production.

CONTRACTOR

Oldelft (BV Optische Industrie De Oude Delft), PO Box 72, 2600 MD Delft, Netherlands.

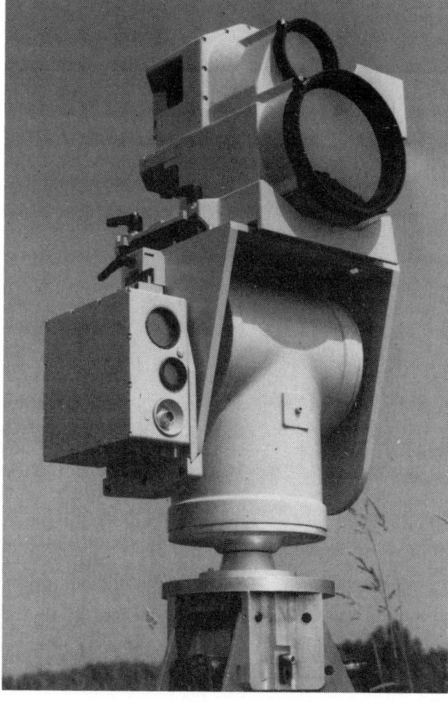

SAMTOR thermal observation system

3576.193

MINILASER RANGEFINDER

The Oldelft Minilaser is a small lightweight laser rangefinder suitable for use in either tripod or vehicle-mounted form. Adapters are available for the latter applications. The Minilaser is of modular construction and has the following sub-assemblies: rangefinder, goniometer, tripod, battery unit, and vehicle adapter (if required). The rangefinder and goniometer have separate housings, enabling each unit to be used independently. A north-seeker incorporated in the instrument facilitates orientation of the observer in respect of magnetic north. Instant digital readouts are provided for distance, azimuth, and elevation angles.

CHARACTERISTICS

Transmitter

Type: Nd:YAG

Wavelength: 1.06 micron

Output energy: 40 mJ

Pulse width: 7 μ sec

Power output: 6 MW

Beamwidth: 0.7 mrad

Pulse rate: 10 per minute

Receiver

Field of view: 0.3 mils

Aperture: 45 mm

Range accuracy: ± 10 m

Range (minimum/max): 200/9900 m

Target selection: First or last reply mode

Sight

Magnification: × 7

Field of view: 7°

Eyepiece adjustment: + 5 to –5 dioptres

Goniometer

Azimuth: 0 – 6399 mils

Accuracy: ±2 mils

Elevation: ±300 mils

Resolution: ±2 mils

Dimensions

Rangefinder: 184 × 184 × 85 mm

Goniometer: 250 × 225 × 180 mm

Battery: 220 × 110 × 100 mm

Weight: 7.8 kg (rangefinder and goniometer)

STATUS

In production for various armies.

CONTRACTOR

Oldelft (BV Optische Industrie De Oude Delft), PO Box 72, 2600 MD Delft, Netherlands.

Oldelft Minilaser

5702.193

RSI TANK LASER RANGEFINDER

The RSI (Rationalisation, Standardisation, Interoperability) laser rangefinder is specifically designed for easy retrofitting to existing battle tanks provided with the M 35 E1 periscope (eg Centurion, M 48 and M60). Daylight aiming sight and laser rangefinder are combined in one unit to replace the existing daylight elbow in the M35 periscope. For use as part of a fire control system, a projected reticle channel is provided to display reticle offsets in the eyepiece. To minimise volume at the gunner's station, the electronic elements are contained in one unit.

The field of view of the instrument is 7° with a magnification of × 8. A minus blue filter and a neutral density filter are included, and can be actuated with a lever on the housing of the optical unit. The aiming reticle is illuminated to stand out against a dark background. The display in the eypiece field of view gives measured range, first/last return mode indication, multiple return indication, master failure indication and projected reticle (azimuth offset angle).

STATUS

In production for various armies.

CONTRACTOR

Oldelft (BV Optische Industrie De Oude Delft), PO Box 72, 2600 MD Delft, Netherlands.

RSI tank laser rangefinder

6531.193

TILAS TANK LASER SIGHT

TILAS (Tank Integrated LAser Sight) is a compact, modular all-in-one unit incorporating an armoured two axes servo-controlled head mirror unit with the following provisions:

(a) laser rangefinder
(b) day sight
(c) thermal sight for day/night, interchangeable with an image intensifier night sight
(d) gunner's thermal display and control unit
(e) muzzle reference capability.

The sight is mounted through the roof of armoured vehicles such as MBTs and AIFVs. It features a servo-controlled armoured periscopic mirrorhead which can be provided with gyro stabilisation for shoot-on-the-move capability. In addition to the gunner's station the TILAS system comprises a remote commander's thermal display and control unit with override functions, a laser rangefinder display and control unit, a servo/gyro control unit, a thermal electronics unit, a gun resolver unit and a muzzle reference mirror.

CHARACTERISTICS

Laser rangefinder
Type: Nd:YAG
Wavelength: 1.064 μm
Range: 200 to 9900 m
Resolution: 10 m
Thermal sight
Field of view: 4 × 1.7° (narrow); 12 × 5.1° (wide)
Tank detection: 8 km approx
Image intensifier
Type: Second generation tube
Magnification: × 5.6
Field of view: 6.6°
Daysight
Magnification: × 8 and × 3
Field of view: 7° and 18°

CONTRACTOR

Oldelft (BV Optische Industrie De Oude Delft), PO Box 72, 2600 MD Delft, The Netherlands.

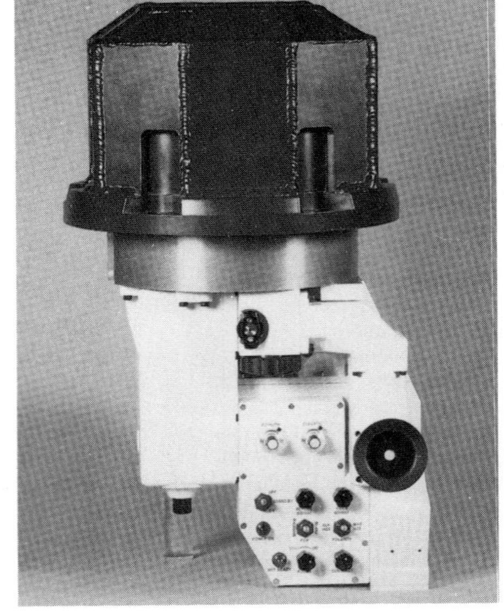

TILAS with thermal sight option

3429.193

UA 9124/9126 DAY/NIGHT PERISCOPE SYSTEM

This periscope system has been designed for both day and night observation and gun-laying, either in armoured fighting vehicles equipped with medium calibre main armament, or in main battle tanks. The modular construction permits independent operational use of the day and night systems. Apart from the day and night periscopes a vision block of unity magnification has been incorporated. The entry optics of both periscopes are coupled to the aiming mechanism of the main armament providing for the elevation of the line-of-sight in collimation with the gun. The vision block has a steady line-of-sight to create a continuous forward observation capability for the gunner, independent of the gun elevation.

The night system is provided with: a second-generation image-intensifier tube with automatic brightness control and point highlight suppression; a built-in and adjustable sky-screen; a reticle for ground targets; and an adjustable iris-diaphragm.

The daylight system is provided with: switchable magnification; a built-in and switchable filter for excessive brightness; and reticles for ground and air targets.

The reticles of both systems are adjustable for both azimuth and elevation (±5 mrad) and can be illuminated.

The system can be extended with a ballistic compensator proportionally designed to fit within the system dimensions.

CHARACTERISTICS

Night system
Magnification: × 7
Field of view: 7°
Eyepiece adjustment: –3 to +3 dioptres
Elevation: –10 to +20°
Daylight system
Magnification: × 6 or × 2
Field of view: 170 and 510 mils
Eyepiece adjustment: –3 to +3 dioptres
Elevation: –10 to +60°
Vision block
Magnification: × 1
Horizontal field of view: 75°
Vertical field of view: 30°
Power supply: 20 – 30 V DC 20 W

STATUS

In production.

Type UA 9124/9126 day/night periscope

CONTRACTOR

Philips Usfa BV, Meerenakkerweg 1, PO Box 218, 5600 MD Eindhoven, Netherlands.

4235.193

UA 9092 TANK THERMAL IMAGER

The UA 9092 tank sight is a thermal imaging system designed for target aiming and observation where a target-to-background temperature difference exists in the infra-red wavelength 'window' between 8 and 12 microns. The thermal radiation received from the scene is converted into a CCIR video signal and is presented on video monitors. A controllable aiming mark is added to the video signal for aiming orientation during gun-laying. The system features a stabilised line-of-sight for mobile observation as a search mode. It is supplied ready to interface with a fire control computer and is fitted with BITE. Positioning of the thermal sensor in a vehicle is not restricted by the location of the gunner's or commander's station.

CHARACTERISTICS

Field of view: Wide 265 mil horizontal × 160 mil vertical; narrow 85 mil horizontal × 50 mil vertical
Spectral range: 8 – 12 microns
Focal adjustment: 50 m-infinity
Cooling system: Philips Usfa UA 7011 Stirling military mini-cooler
Time to reach camera operational status: 10 minutes
Range of elevation: –190 to +370 mil

STATUS

In production.

CONTRACTOR

Philips Usfa BV, Meerenakkerweg 1, PO Box 218, 5600 MD Eindhoven, Netherlands.

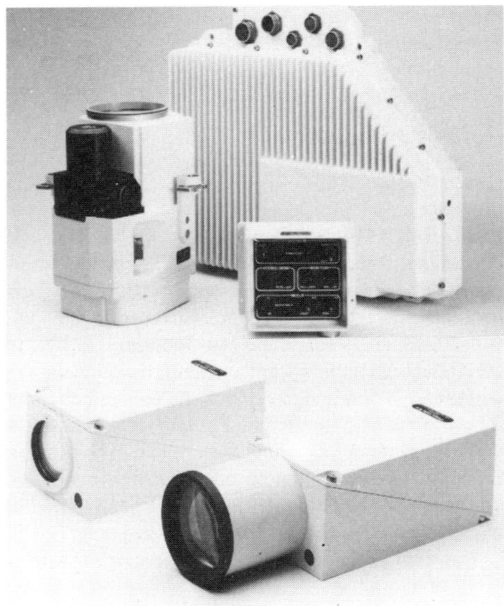

UA 9092 thermal imaging system

3430.193

UA 9630 SERIES UNIVERSAL PASSIVE PERISCOPE

The universal passive periscope has been designed to enable armoured vehicles to be driven at considerable speeds under starlight conditions. Although it has been optimised for driving under starlight conditions it can be employed at higher illumination levels up to daylight. This leads to uninterrupted operation and avoids disruption of the vehicle's sealing under severe conditions such as an NBC warfare environment. The image brightness can be adjusted to suit the user by means of a manually controlled iris-diaphragm. Automatic brightness control (ABC) has also been incorporated and the periscope is fitted with a device providing point highlight suppression, automatically reverting to normal use afterwards.

The periscope is of full binocular design and contains two NATO-standard tubes XX 1080 with a long operational life. The required input power is drawn from the vehicle's battery. Built-in electrical heating of the eyepieces can be switched on to prevent misting and condensation. The equipment meets the optical, mechanical and climatic requirements of the relevant military specifications.

The periscope consists of a main instrument, type UA 1624, and an exchangeable top prism. Changing of the top prism does not affect the sealing of the main instrument. A range of top prisms is available facilitating the fitting of the periscope to all types of armoured vehicle. Among others, top prisms are available for main battle tanks such as Leopard, Centurion and M60 and armoured vehicles like M113, AIVF and Scorpion.

CHARACTERISTICS

Magnification: × 0.9
Horizontal field of view: 50°
Vertical field of view: 40°
Illumination range: 10^{-3} to 10^{4} lux
Range: 5 m – ∞
Tubes: Philips Type XX 1080
Power supply: 20 – 30 V DC
Weight of main instrument: Approx 6 kg
Weight of top prism: 1.5 – 6 kg depending on type

STATUS

In production.

CONTRACTOR

Philips Usfa BV, Meerenakkerweg 1, PO Box 218, 5600 MD Eindhoven, Netherlands.

Type UA 9630 series universal periscopes, for Leopard (left) and M113 (right)

6283.193

UA 9127 THERMAL IMAGING/DAYLIGHT AIMING AND OBSERVATION SYSTEM

The UA 9127 is a combined thermal imager and daylight aiming and observation system designed specifically for use in armoured fighting vehicles. In this system the thermal camera, gunner's monitor and control unit are built into a dual purpose periscope unit, together with an advanced daylight subsystem. The thermal camera can also be used by the vehicle commander.

Both subsystems look out through a single periscope head which follows the gun in azimuth and elevation. A ballistics compensator is incorporated in the linkage.

CHARACTERISTICS

Total field of view: 85 × 50 mrad; 260 × 140 mrad
Spectral bandwidth: 8 – 12 μm
Detector cooling: Philips UA 7044 military mini-cooler
Time to reach camera operational status: 10 min
Time to reach optimum performance: 15 min
Power consumption: 400 W
Weight: 70 kg

STATUS

In production.

CONTRACTOR

Philips Usfa BV, Meerenakkerweg 1, PO Box 218, 5600 MD Eindhoven, The Netherlands.

UA 9127 thermal imaging/daylight aiming and observation system built into an armoured vehicle

1801.193

TYPE TP1MS NIGHT PERISCOPE

The TP1MS is a passive night periscope for combat vehicles. It is a small, binocular, wide angle, passive night vision image intensifier system for direct visual observation by the driver of the road and terrain in front of the vehicle under ambient night illumination conditions. The system will enable movement of combat vehicles at a speed of approximately 40 km/h under starlight conditions without the aid of artificial illumination, and with maximum security from detection.

The field of view of 55° in the horizontal plane gives the driver a clear view of the roadside.

The high-tension power packs are stabilised and will automatically protect the tube against high photo-cathode illumination, and automatically restore normal use afterwards.

CHARACTERISTICS

Magnification: × 1
Field of view: Horizontal 55°; vertical 40°
Centre resolution: Under starlight conditions 10^{-3} lux, 2.5 mrad
Weight: Approx 7 kg
Number of tubes: 2
Tube type: Special diode, photocathode S25, phosphor screen P20
Power supply: 20 – 30 V DC
Entrance optics: Focal length 45 mm relative aperture 1:1.1. Fixed focus 15 m
Ocular system: Binocular and adjustable between +5 and –5 dioptres
Operating temperature range: –20 to +50°C
Storage temperature range: –30 to +60°C

STATUS

Series production.

CONTRACTOR

Oldelft (BV Optische Industrie De Oude Delft), PO Box 72, 2600 MD Delft, Netherlands.

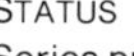

Type TP1MS night periscope

4691.193

TP1MC NIGHT PERISCOPE

The passive driving periscope TP1MC is a small binocular wide-angle night viewing system for direct observation of the road and terrain in front of the vehicle. It consists of two independent viewing channels, each comprising one objective lens system, one second generation flash-protected image intensifier tube, one high voltage unit and an eye piece. The TP1MC can easily be adapted for use in a great variety of armoured combat and personnel carrying vehicles.

CHARACTERISTICS

Magnification: × 1
Field of view: 50° horizontal; 26° vertical
Centre resolution: 2.3 mrad at 3 × 10^{-2} lux; 3.2 mrad at 10^{-3} lux; 4.2 mrad at 10^{-4} lux
Focal length: 20 mm (approx)
Relative aperture: 1:1.1
Focus setting: Fixed 15 m
Dioptre setting: –3 to +3 dioptres
Tube type: Second generation 18 mm microchannel wafer with fibre-optic input and twisted fibre output window
Weight: 6 kg (approx)

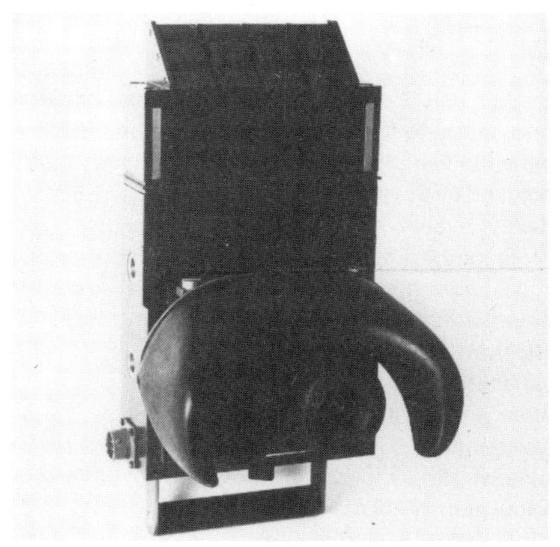

TP1MC night periscope

STATUS

Series production for various armies.

CONTRACTOR

Oldelft (BV Optische Industrie De Oude Delft), PO Box 72, 2600 MD Delft, Netherlands.

6379.193

OLDELFT AIFV THERMAL SIGHTS

Oldelft has developed a range of thermal sights for armoured infantry fighting vehicles (AIFVs) consisting of a thermal camera with gunner's display, day sight, armoured mirror head, commander's display, junction box and laser rangefinder. The range is compact and modular and is suitable for adaptation to different types of AIFVs and tanks. The thermal camera can be provided with 60- or 120-element detectors. The cooling is performed by a closed-cycle Sterling type cooling engine.

STATUS

In production for the YPR-765 AIFV of the Royal Netherlands Army.

CONTRACTOR

Oldelft (NV Optische Industrie de Oude Delft), PO Box 72, 2600 MD Delft, Netherlands.

Odelft AIFV thermal sight

NORWAY

1804.193

LP3 LASER RANGEFINDER

The SIMRAD LP3 laser rangefinder is a one-man portable, tripod-mounted combined sight and laser rangefinder. It enables range, bearing, and elevation data for targets to be rapidly acquired and used directly by weapons or fed into computerised artillery fire control systems. A quick-aiming sight is provided to permit rapid orientation towards targets appearing only briefly, such as shell bursts or gun flashes. A Q-switched neodymium-doped glass laser is used, providing a peak power of 1.5 MW. With a pulse length of 30 nanoseconds this amounts to an equivalent radiated energy of about 45 mJ. A range display shows the ranges to two targets on LED numeric indicators with a resolution of five metres and range discrimination of 30 metres.

CHARACTERISTICS

Transmitter
Laser: Q-switched Nd-doped glass
Wavelength: 1.06 μm
Power (peak): 1.5 MW
Pulse length: 30×10^{-9} s
Pulses per minute: 12 continuous; 30 intermittent
Beamwidth: (90% energy) 1 mrad
Receiver
Field of view: 1 mrad
Detector: Photodiode optimised for 1.06 μm
Effective aperture: 70 mm
Sensitivity: 0.5 A/W
Minimum/max range: 200/20 000 m
Resolution: 5 m
Range discrimination: 30 m
Minimum-range setting: 200 – 6000 m continuously adjustable
Sight
Magnification: ×7
Field of view: 7°
Effective aperture: 50 mm
Eye protection: Dichroic beamsplitter and absorption glass in eyepiece
Goniometer
Elevation: ±350 mil
Azimuth: 6400 mil
Resolution: 1 mil
Dimensions and weights
Rangefinder with goniometer: 260 × 280 × 260 mm; 7.9 kg
Battery pack: 65 × 85 × 230 mm; 1.5 kg
Tripod, folded: 370 × 140 mm diameter; 2.2 kg
Weight, total: 11.6 kg
Transit case: 350 × 350 × 350 mm; 7.8 kg

DEVELOPMENT

The SIMRAD LP3 laser rangefinder was developed for the Norwegian Army under government contract. The programme was supported by the Norwegian Defence Research Establishment.

STATUS

In production for the Norwegian Army and the forces of unspecified countries. It is also in production for the British Army, where it is known as the LP6, in this form being supplied without a tripod, goniometer, and battery.

SIMRAD LP3 laser rangefinder

CONTRACTOR

SIMRAD Optronics A/S, PO Box 6114, Etterstad, Oslo 6, Norway.

3579.193

LP8 LASER RANGEFINDER

The SIMRAD LP8 laser rangefinder is a very small lightweight unit primarily intended for mounting directly on weapons, night observation devices, theodolites and anti-tank weapon systems. The rangefinder contains all electronic and optical units required to enable the operator to lay the instrument on to a target, make measurements and read the range to the target from a digital display.

SIMRAD LP8 laser rangefinder (with battery and control unit)

The transmitter is a miniaturised Q-switched Nd:YAG laser. The receiver uses an avalanche photodiode giving an operational range to typical targets of about 5 to 6 km when optical visibility is 10 km. Unwanted nearby echoes may be gated out from the display by a minimum range control on the operator's side of the instrument.

The range display shows range to the target to the nearest five metres. In addition there are two indicators to warn the operator either that more than one target has been detected or that one or more targets have been gated out.

The 7-segment LED display lights up when a range has been measured and switches off after three seconds. The rangefinder has been designed on modular principles to simplify repair and maintenance and is contained in a sturdy waterproof housing. Mechanical interfaces suitable for weapons, theodolites and night observation devices are available.

CHARACTERISTICS

Transmitter
Laser type: Nd:YAG
Wavelength: 1.064 micron
Output energy: 5 mJ
Pulse width: 10×10^{-9} s
Pulse rate: 12/minute (30 intermittent)
Beamwidth: 1.5 mrad (90% energy)
Receiver
Field of view: 1.3 mrad
Detector: Avalanche diode
Aperture: 45 mm
Clock frequency: 29.97 MHz
Range resolution: 5 m
Display: 7-segment LED
Minimum/max range: 150/10 000 m
Telescope
Magnification: ×7
Field of view: Approx 120 mils
Aperture: 45 mm
Eye protection: Dichroic beamsplitter and absorption glass in eyepiece

STATUS

Production.

CONTRACTOR

SIMRAD Optronics A/S, PO Box 6114, Etterstad, Oslo 6, Norway.

3755.193

LV5 LASER RANGEFINDER

The SIMRAD LV5 laser rangefinder is designed for external mounting on armoured fighting vehicles to improve their combat effectiveness by providing them with accurate range information for weapon aiming.

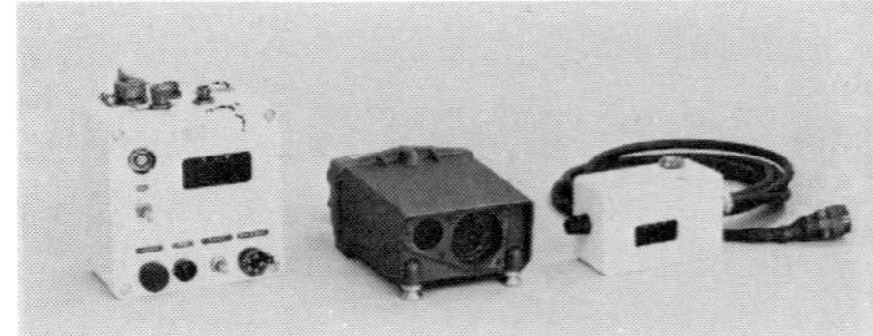

SIMRAD LV5 laser rangefinder outfit for AFV installation, showing (left to right) control/display unit, transmitter/receiver, and optional range display unit

The compact and low profile laser transmitter/receiver unit is mounted on the gun and accurately boresighted to it. The control/display unit is located inside the vehicle in a position where it may be seen and used by the gunner or commander; an LED range display shows the ranges to two targets with a resolution of 5 metres. An auxiliary output provides range information either to an external fire control computer or to an optional range display unit.

CHARACTERISTICS

Transmitter
Laser: Nd:YAG
Wavelength: 1.064 micron
Pulse length: 10×10^{-9} s
Output energy: 10 mJ
Firing rate: 12 continuous; 30 intermittent pulses/minute
Beamwidth: 2 mrad
Receiver
Field of view: 1.3 mrad
Detector: Avalanche photodiode
Effective aperture: 45 mm
Minimum/max range: 150 m/10 000 m
Range resolution: 5 m
Range discrimination: 30 m
Minimum range setting: 150 – 400 m continuously variable
Sight
Magnification: ×7
Weight: 2.2 kg (transmitter/receiver); 2 kg (control unit)

STATUS

In production.

CONTRACTOR

SIMRAD Optronics A/S, PO Box 6114, Etterstad, Oslo 6, Norway.

5703.193

LP160 LASER RANGEFINDER

The LP160 lightweight, one-man portable laser rangefinder is for use by artillery and mortar forward observers. It measures range, azimuth and elevation angles to targets and may be used for fixing own position, target position and for adjustment of fire. Azimuth and elevation angles to the target are measured within 1 mil.

The transmitter is a miniaturised Q-switched Nd:YAG laser with an output energy of typically 10 mJoules. The receiver uses a silicon avalanche photodiode giving a range of up to 20 km. Unwanted echoes can be gated out from the display by a minimum range control which can be set at any range from 150 to 6000 metres.

The operator can select between measuring the range to the first target detected beyond the minimum range setting or the last target within the maximum range. The range display shows the range to the nearest 5 metres. In addition there are indicators to warn the operator if more than one target has been detected, one or more targets have been gated out, battery voltage is too low, laser output is too low or the receiver electronics have failed. Display time is limited to three seconds to save the batteries.

CHARACTERISTICS

Dimensions and weights

Rangefinder and goniometer: 270 × 240 × 230 mm; 5.8 kg

Battery and control unit: 45 × 80 × 220 mm; 0.5 kg

Tripod, folded: 370 × 140 (dia) mm; 1.6 kg

Minimum/max range: 150 – 20 000 m

Transmitter

Laser: Nd:YAG

Wavelength: 1.064 μm

Output energy: 10 mJ (nominal)

Pulse width: 8 ns

Pulse rate: 12 pulses/m (30 intermittent)

Beamwidth: 2 mrad (90% energy)

Receiver

Field of view: 1.3 mrad (nominal)

Aperture: 45 mm

Range resolution: 5 m

Discrimination between close targets: 50 m

Display: 5-digit LED

Telescope

Magnification: × 7

Field of view: 120 mils

Goniometer

Elevation: ± 350 mils

Azimuth: 640 mils

Resolution: 1 mil

STATUS

In production.

CONTRACTOR

SIMRAD Optronics A/S, PO Box 6114, Etterstad, Oslo 6, Norway.

LP160 laser rangefinder

4692.193

LP100 LASER RANGEFINDER

The LP100 is a miniaturised combined weapon sight and laser rangefinder for mounting on various types of anti-tank and other direct firing weapons. It replaces the traditional sights and provides instant target range and aiming information. A built-in ballistic computer, pre-programmed for the type of weapon and ammunition used, calculates the elevation angle and the time of flight for the projectile.

The transmitter is a miniaturised Q-switched Nd:YAG laser with a nominal output energy of 3 mJoules. The receiver uses an avalanche photodiode giving an operational range of up to 500 metres.

LP100 laser rangefinder mounted on a 88 mm Carl Gustav

Unwanted nearby echoes may be gated out by a minimum range control which can be set to any range between 50 and 2000 metres. First/last target logic is also included.

The operation of the LP100 has been made as simple as possible with the number of controls reduced to a minimum. For ergonomic reasons the laser trigger button is mounted on the weapon. During operation of the rangefinder, all information is displayed in the sighting eyepiece. The range display shows the range to the target to the nearest five metres and there are other indicators to warn the operator if more than one target is detected, battery voltage is too low, start pulse is not present and no target is detected.

The reticle pattern can be designed to comply with various types of weapons and ammunition. Triggering the laser yields a range reading to the target. Simultaneously this range information is fed to the ballistic computer which controls an array of 36 LEDs along the vertical axis. The correct superelevation angle corresponding to this range is indicated by lighting up one of the diodes. This diode is then used for aiming when the weapon is fired. The ballistic computer also activates an indicator when the calculated time of flight for the projectile has elapsed. This information is used to estimate the lead angle applied when firing.

CHARACTERISTICS

Dimensions and weight: 75 × 105 × 240 mm; 1.2 kg

Minimum/maximum range: 50 to 5000 m

Transmitter: Nd:YAG

Wavelength: 1.064 μm

Output energy: 3 mJ (nominal)

Pulse rate: 12 pulses/m (30 intermittent)

Beam divergence: < 2.0 mrad w/6x beam expander (90% energy)

Receiver

Field of view: 1 mrad

Aperture: 21 mm

Range resolution: 5 m

Telescope

Magnification: × 3

Field of view: 200 mils

Aperture: 21 mm

Reticle pattern: As required

Eye protection: Dichroic beamsplitter and absorption glass in eyepiece

STATUS

In production.

CONTRACTOR

SIMRAD Optronics A/S, PO Box 6114, Etterstad, Oslo 6, Norway.

6280.193

LV350 SERIES LASER RANGEFINDERS

The LV350 series of laser rangefinders have been designed for easy integration with a large range of vehicle sights. The series consists of the LV352, LV353, LV354 and LV355, all of which are based on a modular design. In each case, the complete instrument consists of two separate sub-units, a transceiver unit and an electronics unit.

The rangefinders can measure ranges between 150 m and 10 km with a resolution of 5 m. As an option, the LV353, LV354 and LV355 models can be delivered with a range counter for measurements up to 20 km. Data output for all models is in serial binary form, either to an external microprocessor (the LV352 model), or to fire control equipment via optocouplers.

The transceiver unit is designed to be used together with an optical sight comprising a Galileian telescope with a magnification of typically × 10. For the LV352, LV353 and LV354 models the same telescope is used for both transmitter and receiver channels. The LV355 requires two separate channels through the sight.

CHARACTERISTICS

Range: 150 to 9995 m (20 000 m optional for LV353, LV354 and LV355 models)

Transmitter: Nd:YAG Q-switched

Wavelength: 1.064 μm

Output energy: 20 mJ

Pulse length: 8 ns

Receiver: Silicon avalanche photodiode detector

Receiver field of view: 8 mrads for LV352, LV354 and LV355 models; 2.7 mrads for LV353

Weights: Transceiver 1.9 kg; Electronics unit 0.35 kg

CONTRACTOR

Simrad Optronics A/S, PO Box 6114, Etterstad, Oslo 6, Norway.

4693.193

TC10 THERMAL IMAGER

The TC10 thermal imager is intended specifically for naval and coastal artillery use to provide high penetration of smoke, haze and camouflage. It is a passive system giving high resolution with a minimum resolvable temperature difference of 0.08°C for long-range detection and identification. Detection is independent of ambient light and the system operates by day and night. The thermal picture is displayed on a TV monitor.

The equipment consists of a thermal camera, signal processing unit and a TV monitor. In a typical shipborne application the thermal camera is mounted on a stabilised platform. The thermal image is scanned across the high sensitivity detector array by a moving mirror. Cooling is by compressed gas.

CHARACTERISTICS

Field of view: 3° horizontal, 1° vertical

Resolution: 0.22 mrad

Focusing range: 25 m – ∞

Detector: Cadmium Mercury Telluride

Spectral band: 8 – 12 μm

Detector cooling: Compressed gas

Mission time: 20 h with a 5 litre cooling bottle at 300 atmospheres

Weight: 8 kg (camera); 7.5 kg (signal processor)

CONTRACTOR

SIMRAD Optronics A/S, PO Box 6114, Etterstad, Oslo 6, Norway.

SOUTH AFRICA

5705.193

E5010 LASER RANGEFINDER

The E5010 is a ruggedised laser rangefinder with up to 8 km ranging capability. Its low mass and small size make the instrument ideal for hand-held and support mounted ranging applications such as:

(1) fire control for mortars, artillery and armour

(2) target ranging for missiles and light anti-tank weapons

(3) air-to-ground ranging.

The design is based on a combined transmitter/receiver and sighting telescope optics. This ensures a compact and convenient-to-use instrument. Range is displayed instantaneously in the eypiece on pressing the firing button with the telescope reticle laid on the target. A continuously adjustable minimum range gate is provided to gate out unwanted echoes. Multiple echo with range-to-first echo, and no-echo indications are also provided.

CHARACTERISTICS

Transmitter: Nd:YAG, beam divergence 0.7 mrad

Monocular sight: × 6 magnification, 6° field of view

Ocular adjustment: –4 to +4 diopters

Accuracy: ± 5 m

Minimum range adjustment: 200 – 3200 m continuous

Pulse energy: 8 mJ nominal

Dimensions: 125 × 160 × 70 mm nominal

Weight: 1.5 kg

STATUS

In production.

CONTRACTOR

Armscor, 224 Visagie Street, Pretoria, Republic of South Africa.

5706.193

E2030 NIGHT DRIVERSCOPE

The E2030 passive night driverscope is a bi-ocular viewing wide-angle night periscope employing a second-generation image intensifier tube. The instrument enables vehicles to be driven by night at normal speed, without artificial lighting.

The special fixed focus bi-ocular eyepiece enables simultaneous viewing with both eyes to provide ease of observation and free head movement for instrument viewing. A fixed focus objective lens is used and adjustable gain setting of the image intensifier tube provides brightness control for high light levels.

The head prism is designed to be removed easily and fits into a standard NATO driver's day periscope mount, thereby making the periscope adaptable to a wide variety of armoured vehicles. A horizontal scanning mechanism provides manual scanning capability of 30° total in the horizontal plane. Instantaneous field of view is > 40° in both horizontal and vertical planes.

CHARACTERISTICS

Magnification: × 0.92 ± 0.05

Field of view: > 40° horizontal and vertical

Horizontal scanning: ± 15°

Focus: Fixed 15 m

Periscopic height: 227 mm max

Exit pupil: >78 mm

Eye relief: 55 mm

Dioptre setting: –2 ±0.5 dioptres

Dimensions: 360 × 200 × 220 mm

Weight: 6.95 kg; 18 kg with case and accessories

STATUS

In production.

CONTRACTOR

Armscor, 224 Visagie Street, Pretoria, Republic of South Africa.

E2030 night driverscope

SPAIN

5707.193

ENOSA ELECTRO-OPTICAL SYSTEMS

A range of aiming sights and periscopes is manufactured by Empresa Nacional de Optica SA (ENOSA) of Madrid. These electro-optical instruments are intended primarily for the Spanish armed forces. In addition, as a result of a Franco-Spanish agreement, ENOSA manufactures the optical equipment for those AMX-30 tanks produced in Spain.

Brief details of the equipment range are as follows:

PP-01 Aiming Periscope

The PP-01 has been designed and developed to equip CETME TC-3 turrets mounted in armoured vehicles such as the BMR and M-113. It consists of a prism head and sight on a mounting stand. A reticle with cross-hairs in the display sight facilitates accurate aiming. Field of view is 43° horizontal and 11° vertical.

PP-02 Observation and Aiming Periscope

This sight consists of a prism head, binocular observation sight and a monocular telescope with a magnifying power of × 2 and × 6 respectively for land and air aiming. It has been designed for the BMR armoured vehicle turret but can be adapted to other automatic gun or machine-gun mounts.

M-92D Elbow Telescope

The M-92D is an optical aiming telescope for the CSR 106 mm gun. It has a × 3 magnifying power and can be used both by day and by night. Aiming by night can be carried out with illumination by either a dry battery or by photoluminiscence using tritium-phosphorus gas capsules.

106 mm Gun Night Aiming Sight

This is an elbow night sight using image intensifier techniques. Magnifying power is × 2.5 with a real field of view of 9.5°.

AA 40/70 Bofors Gun Collimator Sight

A manually operated bi-ocular sight to cater for fire prediction against low-level aircraft attack.

Goniometric Aiming Devices

Goniometric aiming devices are available for 60, 81-105 and 120 mm mortars. They are based on a 'plumb-line' system whereby the aiming device, freely suspended from a tilted axle and on an assembly of high sensitivity balls, automatically and accurately guarantees the alignment in the vertical plane with the mortar bore. The optical system is an elbow telescope, and reticle lighting is provided by means of a dry battery.

AMX-30E Tank Optics

As a result of agreements with a number of French companies, ENOSA manufactures the optical systems for the AMX-30E tanks being built in Spain. These systems include telescopes, rangefinders, periscopes and night driving equipment for aiming and observation under both day and night conditions.

CONTRACTOR

Empresa Nacional de Optica SA, Avenida de San Luis 91, Madrid 33, Spain.

SWEDEN

3144.193

RIA GUNSIGHT

For modern guns and howitzers Philips Elektronikindustrier AB has developed an automatic electronic gunsight called RIA, which can be used both for indirect and direct firing. It has a very simple interface to the gun and can therefore easily be mounted on various types of guns.

RIA comprises the following main units:

Control and Display Unit: The panel has displays for presentation of firing data received from a fire control computer, knobs for different operating modes, and preset values. The unit also comprises electronics for necessary computations including compensation for tilt of the gun and servo-control circuitry for the sight.

Sighting Unit: This unit comprises a panoramic sight and an automatic levelling device. It includes a zero setting instrument for laying the gun in elevation, and also includes, as a reserve mode, mechanics and

RIA automatic gun sight for field artillery, and associated control and display unit

levelling bubbles for manual laying. The panoramic sight has a servo-driven setting of bearing.

Direct Sight: For laying the gun by direct firing.

The equipment receives firing data (bearing, elevation, fuze setting, and charge) from a fire control computer via a data link. Based on these data and the tilt of the gun, etc, RIA automatically calculates relative bearing and elevation.

In azimuth this information controls a servo-driven top-prism in the panoramic sight and thereby the line of sight. The reference for aiming in traverse can be sticks, plates, or a collimator. Gun laying in elevation is done with the aid of a zero-setting instrument. Setting in traverse and elevation can take place simultaneously.

RIA has the following tactical functions:

(1) automatic reception of firing data
(2) presentation of firing data on the control and display unit
(3) settings of gun corrections in azimuth and elevation
(4) automatic computation of relative bearing and elevation, including automatic correction for the tilt of the gun
(5) automatic levelling of the sighting unit
(6) automatic control of the sight in azimuth including automatic setting of bearing
(7) reserve modes for use when the data transmission or the computation is out of order
(8) supervision of tactical and electrical functions by built-in test circuits.

STATUS

In service with the FH 77A in Sweden and the FH 77B gun in other countries.

CONTRACTOR

Philips Elektronikindustrier AB, 175 88 Järfälla, Sweden.

3042.193

ERICSSON ANTI-AIRCRAFT LASER RANGEFINDER

The Ericsson high repetition rate anti-aircraft laser rangefinder is designed to be the ranging sensor in electro-optical fire control systems where the tracking is performed either automatically, by a TV or infra-red sensor, or manually. It can also be used in radar fire control systems to give range information under jamming conditions.

Ericsson AA laser rangefinder (left) mounted on director with Saab-Scania TV camera

The laser rangefinder consists of a transceiver unit, a power module and a range counter, which can easily be integrated into existing systems. The transceiver unit contains the YAG-laser transmitter with its air-cooling system and the optical receiver.

The YAG-transmitter of a high repetition rate laser rangefinder dissipates a considerable amount of heat and must be effectively cooled. The Ericsson anti-aircraft laser rangefinder uses forced air cooling, which is a simple and effective method. The transceiver unit is completely sealed. Via the fan, the air in the unit is circulated through the transmitter. The heated air is then cooled on the case by the cooling fins. This forced air cooling gives simplicity of construction, high reliability and easy maintenance.

CHARACTERISTICS

Range capability: 8-14 km depending on target size. Extinction value 50 dB

Coverage: 280-20 475 m (12-bit parallel form, TTL-level)

Accuracy: 4 m

Adjustable minimum range: 280 – 20 440 m (9-bit parallel form, TTL-level)

Range gate: Length 500 m (other values optional) or to the end of the range coverage

Transmitter

Laser: Nd:YAG

Wavelength: 1.06 μm

Beamwidth: 1.2 – 1.5 mrad

Repetition rate: 10-20 pulses/s

Receiver

Detector: Silicon avalanche diode

Area of optics: 80 cm²

Field of view: 1.5 mrad

Power

Supply: Different power supplies are available for both indoor and outdoor mountings

Max consumption: 300 to 400 VA

Weights

Transceiver unit: 13 kg

Power module and range counter: 12 kg

STATUS

In production. The Ericsson anti-aircraft laser rangefinder has been ordered by a large number of system manufacturers both in and outside Sweden. The total number of lasers ordered exceeds 300 systems.

CONTRACTOR

Ericsson Radar Electronics AB, Defence Systems Division, 431 26 Mölndal, Sweden.

3043.193

ERICSSON ARTILLERY LASER RANGEFINDER

The Ericsson artillery laser rangefinder is used by forward observers for range determination of stationary and mobile targets as well as ranging airbursts up to a distance of 5 to 6 km. The rangefinder, weighing less than 10 kg, consists of transceiver unit, goniometer, tripod and battery, and collapses easily for transport in a heavy-duty glass-fibre container. The transceiver unit uses a neodymium-YAG transmitter enabling instant use and no limitation with respect to range measurement rate.

Manoeuvring controls are kept to a minimum in order to secure easy and rapid use which results in controls for three functions only:

(1) trigger button
(2) a button to present blocking range in display and a dial for continuous adjustment of blocking range
(3) a dial for adjustment of reticle illumination.

The rangefinder is equipped with two counters enabling two ranges to be measured simultaneously in the same line of sight. The goniometer is a precision instrument for measuring azimuth and elevation angles within an accuracy of one mrad.

STATUS

In production.

CONTRACTOR

Ericsson Radar Electronics AB, Defence Systems Division, 431 26 Mölndal, Sweden.

Ericsson artillery laser rangefinder

3044.193

INTEGRATED LASER FOR MAIN BATTLE TANKS

This integrated laser tank sight uses an advanced miniaturised neodymium-YAG transmitter for maximum repetition rate and temperature independence. Boresighting of the laser to the gun bore is facilitated by the adjustable inner chassis of the laser sight. For operator's convenience the laser sight is equipped with a continuous blocking range. For night fighting purposes a mechanical filter, protecting the operator's eyes from light, is built in and is operated when the laser as well as the gun is triggered.

In addition to the normal eyepiece for gun firing, there is a large field of view sight for observation purposes. The laser rangefinder is designed to operate in co-operation with a ballistic computer. The result from the computer, the aiming mark, is presented in the gunner's sight as a red dot on the target (spot injection).

The laser trigger button is placed on the gunner's control handle.

The Ericsson laser rangefinder sight features:

(1) gunner's sight and laser rangefinder in one unit only
(2) spot injection for computer co-operation
(3) simple alignment of gunner's sight to the main gun of the vehicle
(4) easy and rapid installation in the vehicle.

STATUS

Initially developed for modernisation of the Centurion MBT, this sight is now in production and available for fitting in other tanks requiring updating. Orders have been received from two countries, one being Denmark for retrofit to Centurions.

CONTRACTOR

Ericsson Radar Electronics AB, Defence Systems Division, 431 26 Mölndal, Sweden.

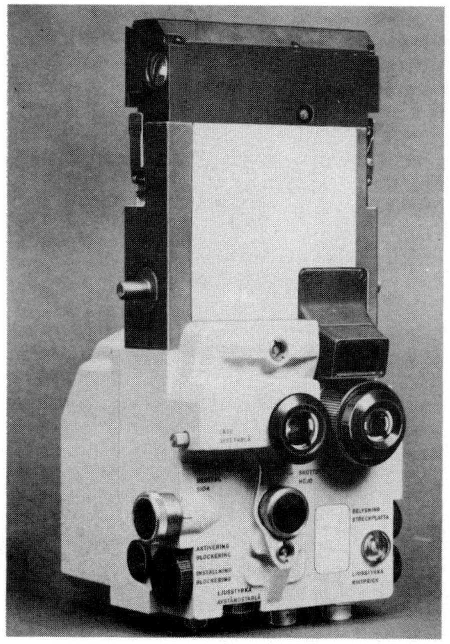

Ericsson gunner's sight for main battle tanks, with integrated laser rangefinder and spot injection

6737.193

UAL 11636 LIGHT ANTI-AIRCRAFT LASER RANGEFINDER

The UAL 11636 is the latest laser rangefinder from Ericsson Radar Electronics and is designed for use with light anti-aircraft guns. It uses a Neodymium-YAG laser with a variable pulse repetition rate up to 25 Hz and a beamwidth of 1.5 or 3 mrad. The system features internal forced air cooling and for naval applications the laser is equipped with a front-end wiper.

CHARACTERISTICS

Weight: 15.5 kg
Wavelength: 1.06 μm
Field of view: 3 mrad
Instrumented range: 200-20 475 m
Range accuracy: 5 m
Power: 115-380 V ac

STATUS

The UAL 11636 has been selected by the Canadians for incorporation in the Skyguard Mk II radar.

CONTRACTOR

Ericsson Radar Electronics AB, Defence Systems Division, 431 26 Mölndal, Sweden.

Ericsson UAL 11636 light anti-aircraft laser rangefinder

3836.093

ERICSSON LASER TRACKER

The Ericsson three-co-ordinate laser tracker has been designed for automatic target tracking and prediction. It is a neodymium-YAG type, with a wavelength of 1.06 microns, a beamwidth of 3 milliradians and a PRF of 25 Hz. It can be integrated into a wide variety of fire control systems (FCS), from relatively simple optical units to complex radar systems. The tracker is a further development of the anti-aircraft rangefinder (described in entry **3042.193**) and uses the same air-cooled transmitter and receiver range measuring circuits. It consists of three sub-units: transceiver, power supply and interface electronics. The interface is a communication link between the transceiver unit and the FCS.

The tracker simultaneously measures target range and angular offset in azimuth and elevation with high accuracy. The narrow beam and short wavelength of the tracker ensure that it is not affected by multipath effects.

STATUS

In development.

CONTRACTOR

Ericsson Radar Electronics AB, Defence Systems Division, 431 26 Mölndal, Sweden.

Ericsson laser tracker

3041.153

ERICSSON COAST ARTILLERY LASER RANGEFINDER

A successor to the UAL 10102 marine laser rangefinder (**1213.253** in 1979-80 and earlier editions), it is now in operational service with the Swedish and Norwegian naval forces. Few details concerning the system have been revealed apart from the following brief description.

The laser rangefinder consists of three units: the transceiver unit, the display unit, and the trigger pedal. The transceiver uses a high PRF neodymium-YAG air-cooled laser transmitter to permit rapid fall-of-shot measurements. The display unit, in addition to the main range display, has a memory display to assist in correction of gun fire. This coast laser rangefinder is designed to interface both electrically and mechanically with the UAL 10102.

STATUS

In production.

CONTRACTOR

Ericsson Radar Electronics AB, Defence Systems Division, 431 26 Mölndal, Sweden.

Ericsson coast artillery laser rangefinder

4694.093

ERICSSON THERMAL IMAGER

This is a family of thermal imaging systems which is based on a number of common modules, and is suitable for use in a variety of land, sea and air applications for surveillance purposes, or can be integrated into fire control systems. The sensor can be equipped with one or two fields-of-view and has electronic magnification. A reticle is electronically generated in the image which is presented on a TV monitor. External synchronisation provides for direct picture changing on a monitor between the imager and other sensors.

The imager uses local, dynamic compression techniques which enhance details of the observed scene so that small objects are not lost against complex background variations. This enables early detection of targets. The system is TV compatible and can be easily integrated with a TV tracker.

CHARACTERISTICS

Spectral bandwidth: 8 – 12 μm
Total field of view: 4° × 2.6° (2° × 1.3° with electronic magnification)
Instantaneous field-of-view: 0.15 × 0.15 mrad
Detector: 8-element Sprite
Cooling: Stirling mini-cooler
Dimensions: 550 × 360 × 220 mm
Weight: 30 kg

STATUS

In production. The Swedish armed forces have ordered systems based on these sub-units for the RBS 70M air defence system, the army helicopter sight system (Helitow) and the naval fire control system Arte 726. The last named, known as the UAM 11103, is a complete all-in-one imaging system for detection and tracking.

CONTRACTOR

Ericsson Radar Electronics AB, Defence Systems Division, 431 26 Mölndal, Sweden.

Ericsson thermal imager

3575.083

VTS-800 VIDEO TRACKING EQUIPMENT

The Saab VTS-800 is an advanced digital correlation tracking equipment introduced by Saab-Scania for various target tracking applications. The tracker embodies the latest technology featuring TV/IR compatibility, centroid tracking and adaptive tracking gate size. It is used for tracking against target backgrounds of a complexity that goes beyond the ability of conventional contrast trackers.

No technical details have been cleared for publication.

STATUS

In operational service.

CONTRACTOR

Saab Missiles AB, 581 88, Linköping, Sweden.

Monitor picture of VTS-800 target tracker display, showing performance against a surface target

3435.193

SAAB IRS-700 IR SURVEILLANCE EQUIPMENT

The Saab IRS-700 is a passive infra-red surveillance equipment developed principally for anti-aircraft defence applications, although also suitable for other roles where an alternative to radar is required. Features emphasised by the manufacturers include passive operation, resistance to jamming, lightweight and small dimensions, high accuracy in bearing and elevation, and lower cost than comparable radar equipment. The sensor head and display/control unit can be separated by up to 1000 m. The main characteristics are tabulated below. No other technical details have been released.

CHARACTERISTICS

Elevation scan: 27° within –5 and +45°
Azimuth scan: 360°
Rotation rate: 30 rpm
Wavelength: 8 – 13 micron
Weight: Approx 65 kg (sensor head)

STATUS

Prototype developed and tested under a contract from the Swedish Defence Materiel Administration. No information on production has been made available.

CONTRACTOR

Saab Missiles AB, 581 88, Linköping, Sweden.

5708.193

PKD-8/KOL-2 SIGHTING SYSTEM FOR ARTILLERY

PKD-8/KOL-2 is an optical sighting system designed to increase the mobility of field artillery by reducing the deployment time for indirect fire. The system combines a PKD-8 panoramic sight on the gun with an indpendent optical collimator (KOL-2) which provides the fixed reference needed for initial alignment and subsequent indirect firing.

The PKD-8 sight is a monocular panoramic instrument for use with either the KOL-2 collimator or conventional aiming references such as pre-surveyed posts or stakes. It has been designed for mounting on the elevation sighting equipment of new guns or retrofitting to existing weapons. Engraved glass scales operated through worm-screws, spring-loaded to eliminate backlash, ensure long-term accuracy.

The KOL-2 is a collimator used in combination with the PKD-8, or other traverse sights, and provides the aiming reference necessary for the initial alignment procedure. The instrument has a built-in traverse scale to allow its use for this purpose and also to allow the gun to be positioned anywhere in a 360° arc around the collimator to simplify deployment. A reticle designed as a symbol plate corresponding to the reticle of the PKD-8 sight is in the objective system of the collimator and is used in the alignment procedure. When the two instruments are used together, the same symbols are seen in the two reticles.

STATUS

In series production.

CONTRACTOR

Bofors Aerotronics AB, S-18184 Lidingo, Sweden.

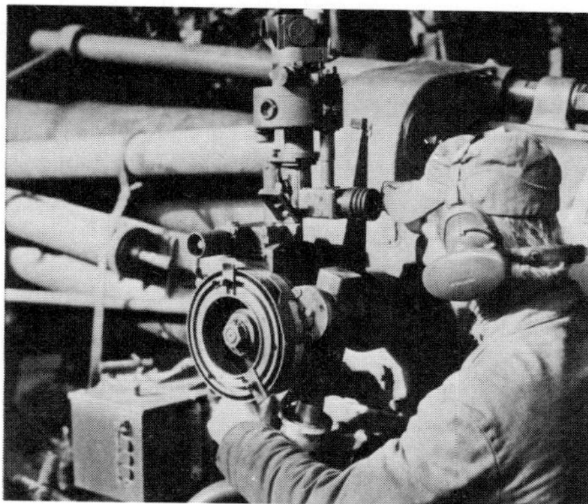

PKD-8 sight in operational use

5709.193

SRS-5 REFLECTING SIGHT

The SRS-5 instrument is an all-purpose reflecting sight for use with automatic guns against air and surface targets and can be used with both land-based and naval anti-aircraft guns. The field of view is 600 mils in azimuth and 300 mils in elevation. A large objective diameter optical system of 96 mm allows the operator to use both eyes for observation. The reticle is projected directly on to the target and is free from parallax.

STATUS

In series production.

CONTRACTOR

Bofors Aerotronics AB, S-18184 Lidingo, Sweden.

SRS-5 reflecting sight

UNION OF SOVIET SOCIALIST REPUBLICS

3437.193

SOVIET SIGHTS—NOTE

The Soviet Union has developed and deployed a wide range of night vision equipment for all types of equipment ranging from rifles, machine guns, anti-tank launchers, anti-tank guns to armoured vehicles. Most of this equipment is of the infra-red active type, although the APN-2 and NSP-2 can be used both actively and passively. Night vision equipment on the T-62 MBT includes an infra-red searchlight to the right of the main armament (maximum range 800 m), infra-red spotlight on the commander's cupola (maximum range 200 m), and an infra-red driving light for the driver (maximum range 60 m). It is thought that the T-64 (or T-72) MBT (**5112.102** in 1978 and earlier editions) is provided with passive night vision equipment. The USSR places great emphasis on night actions and spends between 30 and 40 per cent of exercise time on night training. Brief notes on specific types follow:

APN-2

This is mounted on anti-tank guns and has a maximum range of 900 m; it can be used both actively and passively.

APN-3

This infra-red sight is mounted on anti-tank guns and has a maximum range of 2000 m.

APN-57

This infra-red sight is mounted on anti-tank guns and has a maximum range of 700 m.

TVN-1

This is an infra-red driving aid and is mounted on the glacis plate of MBTs such as the T-55. The driver is provided with a monocular observation piece.

TVN-2

This is an infra-red driving aid and is mounted on the glacis plate of MBTs such as the T-55. The driver is provided with a bi-ocular observation piece.

Infra-red Detection Equipment

The 7.62 mm SVD semi-automatic sniper's rifle and some field binoculars are provided with an infra-red detection system.

Artillery Laser Rangefinders

Soviet artillery has been using both tripod- and vehicle-mounted laser rangefinders for a number of years. The DAK-1 is one such system. It is a battery-powered unit with a maximum range of 20 km. Its practical range is about 10 km. Approximate dimensions are 600 mm long × 300 mm wide × 30 mm deep. It is normally tripod-mounted.

UNITED KINGDOM

4784.193
AV 62 COMMANDER'S SIGHT
The AV 62 is a surveillance sight designed for fitment to lightweight cupolas in light armoured vehicles such as the Spartan and Striker. It is a daylight viewing monocular instrument with provision for a separate periscopic viewer for daytime which can be easily and rapidly replaced with an image intensifier device for night surveillance. The main day sight provides × 1 bi-ocular vision and, optionally, by means of a switch, × 10 magnification through the left ocular. A night vision channel, Type L7A1 is a second generation modular unit designed specifically for use with the AV 62. It provides a × 1 magnification through a monocular eyepiece or, alternatively, through a bi-ocular magnifier.

CHARACTERISTICS
Magnification: × 10 or × 1
Field angle: 7° 45'
CONTRACTOR
United Scientific Instruments Ltd, 215 Vauxhall Bridge Road, London SW1V 1EN, England.

3091.193
LP3 LASER RANGEFINDER
The LP3 laser rangefinder is a lightweight, man-portable instrument designed for use by artillery forward observers to obtain instantaneous measurements of range, azimuth, and elevation angles to targets. It can range at targets from 200 to 20 000 m with a resolution of 5 m. Azimuth and elevation angles can be determined to an accuracy of one mil.

The equipment consists of a laser transmitter and receiver system which are mounted as a unit with the associated control and counting circuits, and sighting telescope on a goniometer mounting which is supported in use by a tripod. The necessary battery power supply pack is carried on one of the tripod legs.

CHARACTERISTICS
Transmitter
Type: Q-switched Nd-doped glass laser
Wavelength: 1.06 microns
Peak power: 1.5 MW; 45 mJ
Pulse rate: 12/minute (continuous); 30/minute (intermittent)
Beamwidth (90% energy): 1 mrad
Receiver
Field of view: 1 mrad
Detector: PIN diode
Effective aperture: 70 mm
Range: 200 – 20 000 m
Resolution: 5 m
Range discrimination: 30 m
Range presentation: Numeric (LED) displays
Multiple target presentation: 2 ranges displayed
Time variable gain: Included
Optical
Magnification: × 7
Field of view: 6.5°
Effective aperture: 50 mm
Eye protection: Dichroic beamsplitter plus absorption glass
Quick aiming sight: Included
Total weight: 19.4 kg (with case)
STATUS
In production.
CONTRACTOR
United Scientific Instruments Ltd, 215 Vauxhall Bridge Road, London SW1V 1EN, England.

4103.193
LP6 LASER RANGEFINDER
The LP6 laser rangefinder can be used for the rapid and accurate determination of target range by forward observers. It has been designed to interface with a variety of tripods, angulation heads and night observation devices. The LP6, used in conjunction with these devices, may be used to determine observer and target positions and to provide data for adjustment of fire even under low light conditions. It can also interface with existing field artillery computing systems. Target ranges can be measured from 200 to 20 000 m with a resolution of ±5 m.

CHARACTERISTICS
Operating temperature: –40 to +55° C
Magnification: × 7
Field of view: 6.5°
Effective aperture: 50 mm
Eye protection: Dichroic beamsplitter and absorption glass in eyepiece
Quick aiming sight: Included
Transmitter
Laser: Q-switched (rotary prism) Nd-doped glass
Wavelength: 1.06 μm
Radiant power: 1.5 MW
Pulse length (half width): 30 × 10^{-9} s, nominal
Quantity of radiant energy: 45 mJ
Pump energy (electrical): 15 J
Pulse rate: 12/minute (continuous); 30/minute (intermittent)
Beamwidth (90% energy): 1 mrad
Receiver
Field of view: 1 mrad
Type of detector: PIN photodiode
Effective aperture: 70 mm
Clock frequency of counter: 29.973 MHz
Minimum/max range: 200/20 000 m (determined by range counter)
Resolution: ±5 m
Range discrimination: 30 m
Minimum range setting: 200 – 6000 m continuously
Range presentation: LED displays
Power supply: 24 V Ni-Cd battery
Overall depth: 230 mm
Overall height: 170 mm
Overall width: 120 mm
Weight (total): 15.3 kg (including transit case)
STATUS
In production.
CONTRACTOR
United Scientific Instruments Ltd, 215 Vauxhall Bridge Road, London SW1V 1EN, England.

LP6 laser rangefinder on angulation head of British Army common mounting system

---

3092.193
LV2 LASER RANGEFINDER
The Avimo-designed LV2 laser rangefinder is fully interchangeable with the existing Centurion tank gunner's sight and can therefore be readily fitted without modification to this range of armoured vehicle. The instrument is fitted with dual magnification to facilitate target acquisition and gun aiming, in addition to the integrated rangefinder capability. It is a completely self-contained unit, and among the design features of the LV2 are the following:

(1) time variable gain regulator to avoid false registration of unwanted natural phenomena
(2) inbuilt eye protection for the user
(3) remote display facility for use by the tank commander
(4) laser transmitter malfunction, and minimum range setting indicators
(5) additional targets other than first and second echoes indicator.

CHARACTERISTICS
Transmitter
Type: Q-switched Nd-doped glass laser
Wavelength: 1.06 microns
Peak power: 1.5 MW; 45 mJ
Pulse rate: 12/minute (continuous); 30/minute (intermittent)
Exit pupil diameter: 10 mm
Receiver
Field of view: 1 mrad
Detector: PIN diode
Effective aperture: 40 mm
Range: 200 – 20 000 m
Range accuracy: ±5 m up to max range
Range discrimination: 30 m
Range presentation: Numeric (LED) displays
Multiple target presentation: 2 ranges displayed
Time variable gain: Included
Optical
Magnification: × 8 and × 1.2
Field of view: 9° (× 8), 45° (× 1.2)
Exit pupil diameter: 5 mm
Total weight: 10.2 kg
CONTRACTOR
United Scientific Instruments Ltd, 215 Vauxhall Bridge Road, London SW1V 1EN, England.

3756.193
LV3 LASER RANGEFINDER
This is a laser rangefinder developed to meet Army requirements for a unit capable of being mounted externally on AFVs. A boresight is built into the LV3 to facilitate accurate boresighting to the carrier vehicle's armament. The control box and display unit is sited in a convenient position for use by the commander or gunner. Ranges on targets between 200 and 20 000 m can be measured with a resolution of 5 m. A continuously variable minimum range gate operative between 200 and 3000 m is incorporated. An additional display can be provided to give the commander and gunner individual information.

CHARACTERISTICS
Transmitter
Laser: Nd-glass
Wavelength: 1.06 micron
Output: 1.5 MW nominal
Pulse length: 30 × 10^{-9} s nominal
Pulses per minute: 12 (continuous), 30 (intermittent)
Beamwidth (90% energy): 1 mrad
Receiver
Field of view: 0.7 mrad
Detector: PIN diode
Effective aperture: 70 mm
Sensitivity: 120 nW
Clock frequency: 29.973 MHz
Range discrimination: 30 m

Sight
Magnification: × 7
Field of view: 6.5°
Dimensions: 290 × 180 × 140 mm
Weight: 10 kg
Operating temperature range: –40 to +55° C

STATUS
Production.
CONTRACTOR
United Scientific Instruments Ltd, 215 Vauxhall Bridge Road, London SW1V 1EN, England.

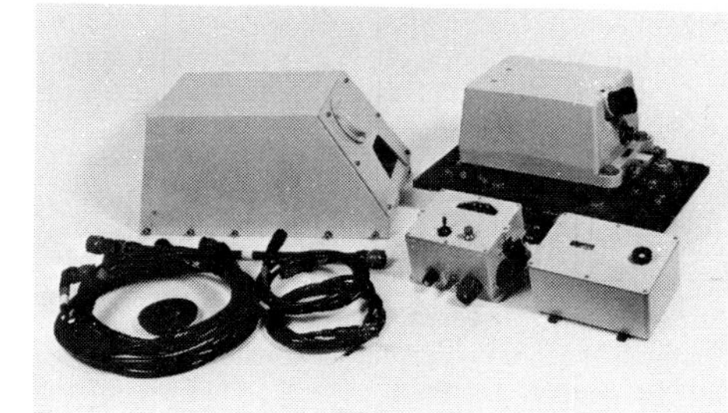

LV3 laser rangefinder component units

5714.193
NVL53 DAY/NIGHT SIGHT AND RANGEFINDER
The NVL53 is a combined day and night sighting periscope with an integrated laser rangefinder facility. It is derived from the M32E1 periscope (**4155.193**) used on M48 and M60 battle tanks but is a more compact system suited for light armoured vehicles while still being capable of fitment to larger fighting vehicles. The NVL53 can be installed in both the gunner's and commander's positions, although the commander's sight normally has a daylight elbow fitted in place of the laser rangefinder and is designated NV53.

The instrument provides both visual readout and ballistic computer data outputs to interface with any desired level of fire control ranging from manual super-elevation of the gun to a fully computerised fire control system. The sight may additionally be supplied with a spot injection unit which provides an illuminated aiming mark when used in conjunction with a fire control computer.

The NVL53 is a monocular type instrument comprising three independent modules; a head assembly, a daylight elbow/laser rangefinder assembly and a passive night vision elbow assembly which incorporates a second-generation image intensifier tube. The laser rangefinder may be used with either the daylight elbow or the night vision system.

The daylight elbow and laser rangefinder is a tri-axial optical system to enable visible light, laser transmitter and laser receiver to use the same basic optical path. Laser light is separated from visual by dichroic beam splitters and absorption filters. A four-digit LED display is projected into the sighting eyepiece and may be duplicated on a remote range readout. Target ranges from 200 up to 9995 m can be measured with a resolution of 5 m.

CHARACTERISTICS
Head assembly
Magnification: × 1
Field of view: 25° horizontal; 8° vertical
Daylight elbow
Magnification: × 8
Field of view: 8°
Entrance pupil: 35 mm diameter
Exit pupil: 4.375 mm diameter
Laser transceiver
Type: Nd:YAG (dye Q switched)
Wavelength: 1.064 μm
Output: 1 MW typical
Range: 200 to 9995 m
Range accuracy: ± 5 m
Range discrimination: 30 m
Night vision module
Magnification: × 7.1
Field of view: 7.3°
Entrance pupil: 198 mm diameter
Exit pupil: 14 mm diameter

NVL53 day/night sight and rangefinder

CONTRACTOR
United Scientific Instruments Ltd, 215 Vauxhall Bridge Road, London SW1V 1EN, England.

5712.193
TL10-T ARTICULATED TELESCOPE LASER RANGEFINDER
The TL10-T laser rangefinder, when integrated into the 'T' series gunner's articulated telescope provides a compact neodymium-YAG laser ranging facility. It replaces the forward objective lens assembly on the existing telescope, thus providing accurate range measurement from under armour without mechanical modification to the vehicle.

The laser assembly incorporates an aiming mark injection facility to enable integration with a computerised fire control system. Range and aiming mark data are displayed in the secondary eyepiece assembly also provided. The basic operating controls are located on a remote panel assembly mounted on the body of the telescope adjacent to the eyepiece. The modified gunner's articulated telescope can then be directly installed in the 'T' series vehicles.

CONTRACTOR
United Scientific Instruments Ltd, 215 Vauxhall Bridge Road, London SW1V 1EN, England.

TL10-T laser rangefinder

6281.193
VML23 VEHICLE LASER RANGEFINDER
The VML23 laser rangefinder is designed for external mounting on armoured fighting vehicles as a solution where various considerations prohibit the installation of a periscopic type. It is normally mounted on the gun mantlet and protected by an armoured cover. Alignment with the gun bore allows use of the vehicle's standard gunner sight for aiming the laser.

The basic optical system of the VML23 is of the monocular type and incorporates a simple reticle for boresighting purposes. A laser protection filter is provided within the eyepiece for eyesight protection during boresighting. Range display is presented in digital format on a remote control box. The information remains displayed until another firing, or the power is switched off.

CHARACTERISTICS
Range: 200 – 9995 m
Range accurary: ±5 m
Transmitter: Dye Q-switched Nd:YAG
Wavelength: 1.064 μm
Radiant power: 1 MW
Pulse length: 8 ns nominal
Pulses: 12 per minute continuous
Beam width: Not greater than 1.5 mrad

CONTRACTOR
United Scientific Instruments, 215 Vauxhall Bridge Road, London SW1V 1EN, England.

3834.093
IR18 MkII THERMAL SCANNER
The IR18 is a compact, lightweight, high performance thermal imager with television CCIR compatible video output of 625 or 525 lines. This imager, capable of seeing by day or night to a much greater degree than image intensifiers or low light television systems, is wholly passive, independent of ambient light levels, sees through mist, smoke or shadow and is not blinded by sun glare, flares or searchlights.

It is modular in design and consists of a scan head, electronic processor, power supply unit, cooling system and auxiliary telescope.

The scan head contains the line and frame mechanisms, drive circuitry, syncs, detector, pre-amplifiers and bias circuits. It can be separated from the electronic unit. Without an auxiliary telescope it has a field of view of 38° × 25.5° with a 1.73 mrad resolution, and has an afocal optical input which allows it to be coupled to a wide range of telescopes and short periscopes.

IR18 thermal imager with a × 9/ × 2.5 telescope

The electronic processor houses all the necessary circuitry to provide an output giving a 625 or 525 line display on a TV monitor of any size. Facilities are included to enable alpha-numerics, symbols or graticules to be injected into the output video signal.

The standard system operates with a Joule-Thompson minicooler normally supplied from a remotely located pure air supply unit like the Barr & Stroud QK4 compressor or air bottle. Alternatively the imager can be run using a split Stirling cooling engine. The power supply unit provides all the required IR18 voltage lines when supplied with a nominal 24 volt DC input.

Telescopes have been manufactured with magnifications from × 1.5 to × 14, including dual field of view (typically × 9/ × 2.5 and × 14/ × 4). They can have focal athermalisation, remote focus control, an engraved graticule for sighting purposes (an alternative to an electronic graticule), and the external lens surfaces can be given a very hard coating (ARG4) suitable for use with a wiper in sand, mud or salt water conditions. A temperature reference can also be provided. Where an application requires a special telescope, the company will undertake design to meet the specific requirements.

The IR18 can be used in two formats: assembled with a suitable telescope it is a thermal imaging scanner producing a TV compatible video output, or as a series of modules it can be incorporated into a large variety of weapon aiming and surveillance systems which employ thermal sensing. It is particularly suited for integration into sighting systems in armoured vehicles, helicopters and ships. It can be used to sight direct-line guns, guided weapons, or to control indirect artillery or mortar fire. The IR18 can also be used for identification at night or in poor visibility to help navigation of ships, hovercraft, aircraft, helicopter and land vehicles.

CHARACTERISTICS

Basic scanner

Field of view: 38° × 25.5°

Resolution: 1.73 mrad

Pupil diameter: 14.5 mm

Detector: Mullard Sprite

MRT (at 0.289 c/mrad): Typically 0.17 – 0.35°C

Spectral bandwidth: 8 – 13 µm

Video output: CCIR compatible (or ETA composite video)

Display: 625/525 lines/frame

Field rate: 50/60 Hz

Frame rate: 25/30 Hz

Weight: 7.4 kg

STATUS

The IR18 has been in serial production since 1981. It is in service with UK MoD and has been sold worldwide. A helicopter version is described under **4783.393**. A version has been developed for installation in the Optica observation aircraft. See also **6592.093** below.

CONTRACTOR

Barr & Stroud Ltd, Caxton Street, Anniesland, Glasgow G13 1HZ, Scotland.

6591.193

LF 19 LASER RANGEFINDER

The LF 19 has been designed around the module used on the tank laser sight for the British Army Challenger MBT. It is intended for external mounting on an AFV and, when boresighted to existing equipment, can be adapted to suit a variety of applications in the military field.

The rangefinder is in two main components; the optical and electronic units. The later unit contains all the rangefinder processing electronics and can be interfaced directly to a fire control system computer for range display and ballistic calculations. For applications where there is no fire control computer a separate control and display unit will be required.

CHARACTERISTICS

Wavelength: 1.06 µm

Output energy: 30 mJ

Pulse width: 6 ns

Beam divergence: 0.6 mrad

Receiver field-of-view: 0.75 mrad

Minimum/maximum range: 300/9990 m

Range accuracy: 10 m

STATUS

In pre-production.

CONTRACTOR

Barr & Stroud Ltd, Caxton Street, Anniesland, Glasgow G13 1HZ, Scotland.

6592.093

IR26 THERMAL IMAGING SENSOR

The IR26 thermal imaging sensor head is the latest member of the IR18 family, and has been designed and built around the latter's scan head with its processing electronics on plug-in printed circuit boards and housed in the thermal imaging processing unit.

The IR26 is a rugged sensor using cryogenically cooled SPRITE detector technology and operating in the 8 to 13 µm waveband. It is designed primarily for AFV and naval applications. The equipment uses a dual field of view telescope giving a wide field of view suitable for surveillance and target detection, and a narrow field of view suitable for target recognition/identification and fire control/missile guidance.

The equipment is packaged in two units to ease installation and maintenance. The TI sensor head contains the optics, detector and essential processing electronics, while the processing unit contains the remainder of the electronics and power supplies. The processing unit can be mounted with separation up to 10 m.

CHARACTERISTICS

Wavelength: 8 to 13 µm

Fields of view: Horizontal 13.6° and 4.75°, vertical 9.1° and 3.18°

Weights: Sensor head 19.9 kg; processing unit 12 kg

STATUS

In production.

CONTRACTOR

Barr & Stroud Ltd, Caxton Street, Anniesland, Glasgow G13 1HZ, Scotland.

IR26 thermal imaging sensor head

4750.193

THERMAL OBSERVATION GUNNERY SIGHT (TOGS)

TOGS is a thermal imaging system, operating in the 8 to 11.75 µm waveband, which completes the day/night gunnery capability of the UK Chieftain and Challenger main battle tanks. It is, however, sufficiently versatile to be fitted to other tanks. The system enables the operator to detect, track and engage potential targets by day and night, and in adverse weather. It provides both the commander and the gunner with a picture of the scene, enabling either to engage a target. A dual field of view telescope provides a target acquisition and vehicle navigation facility in the wide angle, and allows target recognition and gun aiming in the narrow angle. It has been designed for easy retrofit to all tanks and will operate in conjunction with the existing visual sighting system if this is retained.

The TOGS equipment comprises three separate elements; a Thermal Surveillance System (TSS), a Gunnery Sighting System (GSS) and a Test System. The TSS comprises a thermal imager sensor, a processor unit and TV standard displays for both commander and gunner. The thermal sensor is mounted externally on the tank in an armoured barbette and is servo-driven in elevation. When operating in the thermal mode the gunner acquires and engages the target using the display. The GSS element integrates the TSS into the tank fire control system and interfaces with the existing IFCS provision in Chieftain and Challenger. An important component of the GSS is the symbology processor unit which accepts information from the fire control system and generates aiming marks and other symbols for superimposition on the displays.

STATUS

TOGS is in full production and is being fitted to Challenger and retrofitted to Chieftain main battle tanks for the British Army. The sight was designed by Barr & Stroud, but the latest production order has been split between Barr & Stroud and the Avimo subsidiary of United Scientific Instruments. The contracts are worth £17 million to each company over the period 1988 to 1990.

TOGS in servo trunnion unit

CONTRACTOR

Barr & Stroud Ltd, Caxton Street, Anniesland, Glasgow G13 1HZ, Scotland.

3137.193

BARR & STROUD TANK LASER SIGHT

The tank laser sight, now with Nd:YAG laser rangefinder, has its origin in the Chieftain tank laser sight. Simplicity of design and replacement of off-mounted mini-modules by internally located ones has substantially reduced the mechanical and electrical problems of installation and made it suitable for a wide range of AFVs. The sight can also be used with thermal imaging systems.

A fundamental feature of the design is the in-built capability to accept a conversion kit which provides a graticule aiming mark projection system for integration within a computer-driven fire control system.

The tank laser sight unit houses the laser transmitter, the receiver system, and the optical sight. The line of sight in elevation is aligned to the axis of the gun by means of a precision parallel linkage, and in azimuth by coincident turret mounting. Boresighting is achieved using controls on the sight unit in conjunction with either a muzzle boresight or the muzzle reference system. Gun laying is achieved through a ballistic graticule or a laying mark provided by a fire control computer.

Laser ranging can be initiated by the gunner or remotely by the tank commander. The range is displayed in the left eyepiece of the sight unit, and remotely at the commander's station on the commander's range read-out unit.

When a target is partially obscured by smoke or other obstacles, the selection of 'last range' on the range unit allows the true range to be obtained by eliminating all received echoes other than the last echo, which comes from the target.

CHARACTERISTICS

System performance

Operating range: 300 – 10 000 m

Accuracy: ±20 m max error (±10 m for 90% of shots)

Read-out: First or last range, display logic, multiple station

Range gate: Preset

Firing rate: 1 pps

Capacitor charging time: <1 s

Optical system

Sighting telescope: × 10

Field of view: 8.5°

Exit pupil: 6 mm

Graticule: Combining laser aiming mark and ballistic data
Acquisition sight: × 1
Transmitter
Power output: 2 MW (nominal)
Pulse width: 6 ns
Beamwidth: 0.5 mrad
Wavelength: 1.064 micron
Receiver: Swept gain
Dimensions
Sight unit: 333 × 289 × 518 mm
Commander's range read-out: 152 × 54 × 60 mm
Weight
Sight unit: 21 kg
Commander's range read-out: 0.68 kg
AMEU: 7 kg
Power supplies
Nominal input voltage: 28 V DC

STATUS

The Barr & Stroud tank laser rangefinder equipment was developed under MoD contract as the standard gunner's laser sight for the British Army Chieftain MBT. It has since been fitted in Centurion, Leopard and Scorpion AFVs.

CONTRACTOR

Barr & Stroud Limited, Caxton Street, Anniesland, Glasgow G13 1HZ, Scotland.

Barr & Stroud tank laser sight

4284.193

PASSIVE INFRA-RED SURVEILLANCE EQUIPMENT

British Aerospace was selected to develop passive infra-red surveillance equipment for use in low-level air defence systems to complement active detection systems such as radar. Radiating active surveillance systems will need to become increasingly more complex to counter the effects of jamming and other countermeasures likely to be used by an adversary in the future, and the use of radiating systems also exposes the defender to the risk of detection and location. One low cost solution to these problems has been the development of a passive surveillance system with no detectable emissions, and since 1975 the British Aerospace has been researching into the use of infra-red analogues of microwave radar systems.

To undertake the numerous field experiments to prove this concept, the equipment has been installed in a trailer and in the course of a two-week trial to prove its effectiveness, 96 per cent of engageable aircraft targets were observed and located passively at ranges which would have made engagement by a surface-to-air missile system entirely practicable. It was estimated that most of these targets would have been destroyed.

DEVELOPMENT

The company has spent about £3 million on this programme since 1975. The work has been carried out mostly in the Electro-optics Department at Stevenage with support from the company's Bristol facilities.

BAe passive infra-red surveillance equipment

STATUS

Development project. Hardware for the first model of the military version of the equipment was available in 1981 and was compact enough to be mounted on a Land-Rover vehicle. The company has also developed a stabiliser, for hand-held laser range-finders, consisting of a gyro-controlled mirror system.

CONTRACTOR

British Aerospace, Dynamics Division, Six Hills Way, Stevenage, Hertfordshire SG1 2DA, England.

5906.193

TOTE MISSILE SIGHT

TOTE (Tracked Optical Thermally Enhanced) is designed to give the British Aerospace Tracked Rapier system (**2424.131**) a day and night capability. It allows the system to operate in an entirely passive mode by providing an automatic search with the surveillance radar switched off. The equipment combines the optical tracking facilities available on Tracked Rapier with a thermal imaging system using TICM II modules (**4280.193**). The optical tracker has been redesigned to accommodate three new units; a passive IR telescope scanner, an electronics pack to process the IR signals, and a monitor unit on which the target is displayed to the operator.

The operator can select optical or IR mode. In the IR mode the target is acquired either from the surveillance radar which puts the TOTE onto the target azimuith, or by the TOTE carrying out a programmed search pattern. The operator acquires the target and tracks it using the crosswires as in the optical system.

STATUS

Is now fitted to British Army Tracked Rapiers being delivered. TOTE will be retrofitted to existing Tracked Rapiers already in service.

CONTRACTOR

British Aerospace, Dynamics Division, Six Hills Way, Stevenage, Hertfordshire SG1 2DA, England.

TOTE missile sight

5713.193

TYPE 312 LASER BEACON AND TARGET SIMULATOR

A ground-based laser target simulator and navigation and recovery beacon for use with laser seeker equipped aircraft, the Type 312 employs a transmitter module identical with that of the Ferranti Type 306 laser designator/ranger (**1742.193**). Simple optics, which are easily attached to the unit, provide the differing fields of illumination required for the three main applications.

The equipment simulates a laser-designated target giving realistic ranges for airborne seekers operating on a wavelength of 1.06 μm. It may also be used as a navigation beacon to provide accurate updates to an aircraft navigational system and as an aircraft recovery aid to an airfield or landing strip where no navigation aid is available. The Type 312 is rugged, lightweight and man-portable and can be operated from standard army batteries or suitable vehicle power supplies. A data transfer module can be added to enable ground-to-air messages to be passed to the aircraft using the laser receiver in the aircraft.

CONTRACTOR

Ferranti Defence Systems Ltd, St. Andrew's Works, Robertson Avenue, Edinburgh EH11 1PX, Scotland.

Type 312 laser beacon and target simulator

5757.193

TYPE 520 LASER RANGEFINDER

The Type 520 laser system is intended for use in armoured fighting vehicles (AFV) and provides an accurate range to the target. Being a transceiver it can be retrofitted to existing AFV sights. The electronics unit provides the control logic for the power supply and converts return pulses to range data for display on either a commander's or gunner's display.

An integrated optical test system (IOTS) can also be supplied. This set can be used to test the Type 520 completely and to locate any sub-fault. The set can also perform a full serviceability check on the laser when fitted in an AFV sight.

CHARACTERISTICS

Laser type: Nd:YAG
Wavelength: 1.064 μm
Radiant peak power: 2 MW minimum
Emitted energy: 24 mJ
PRF: 1 pulse/s continuous
Beam divergence (90% energy): 4 mrad maximum (giving 0.4 mrad with × 10 telescope)
Q-switch: Lithium Niobate Pockels cell
Ranging: 300 to 9995 m
Ranging accuracy: ±5m
Power supply: 19 to 32 V DC (vehicle or battery)
Dimensions: 148 × 90 × 86 mm
Weight: 2.5 kg

STATUS

In production. A substantial order was received in October 1986 for incorporation into the Vickers L20 gunner's sight (**4705.193**). The Type 520D has also been selected as a key element in the integrated weapon system of the Sikorsky H-76 'Eagle' helicopter.

CONTRACTOR

Ferranti Defence Systems Ltd, St. Andrew's Works, Robertson Avenue, Edinburgh EH11 1PX, Scotland.

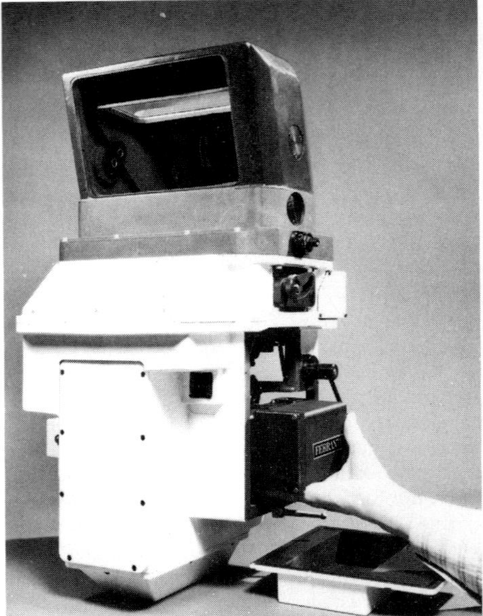

Type 520 in Argus day/night sight for AFVs

4787.193

TYPE 518 AND TYPE 531 LASER SYSTEMS

The Type 518 laser rangefinder is a 1.06 μm Nd:YAG equipment, developed for fitting to kinetheodolites used on military test ranges to track the flight path of missiles and aircraft. It has a PRF of 30 Hz which enables the laser to match the very high accuracy of data provided by the equipment.

The Type 518 can range out to 99.9 km with an accuracy of 2 metres. Such ranges are made possible by the sensitivity of the receiver, the very high clock frequency (74.93 MHz), and by the wide aperture receiver optics. The transmitter optics have a variable beamwidth to allow close targets with a high angular movement past the kinetheodolite to be illuminated continually, while relatively slow distant targets can be tracked using a much narrower beam.

The Type 531 is another 1.06 μm Nd:YAG laser developed for fitment to the auto tracker in Rapier Laserfire. The very narrow beam provides excellent laser line-of-sight for command and guidance, and is proof against ECM.

STATUS

The Type 518 laser is in operational service. The Type 629D is a variant of the Type 629 and is in the post development stage. An order for 38 Type 629G lasers was placed in March 1987 for use in the Canadian Forces Low-Level Air Defence System.

CONTRACTOR

Ferranti Defence Systems Ltd, St. Andrew's Works, Robertson Avenue, Edinburgh EH11 1PX, Scotland.

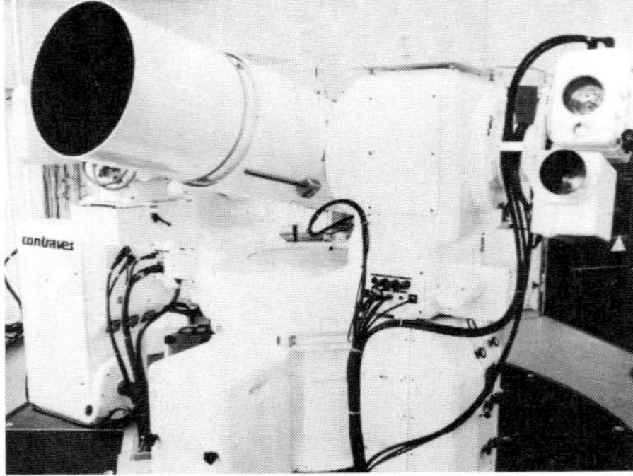

Type 518 laser rangefinder

Type 531 laser transceiver

1742.193

TYPE 306 LASER DESIGNATOR/RANGER

The Type 306 laser designator/ranger has been designed for forward air controllers to designate ground targets for either co-operating laser-seeker equipped ground attack aircraft, or for laser guided weapons. Aircraft using the airborne laser seeker counterpart are able to carry out low-level, high speed, single pass attacks without the pilot having to see the target. This concept greatly reduces exposure to ground defences and, at the same time, considerably improves weapon delivery accuracy since pilot tracking errors are eliminated.

The equipment also has built-in laser ranging for artillery and other purposes. When co-mounted with a thermal imager, a night observation capability is provided.

The system consists of a main unit weighing approximately 9.5 kg, including cables, and a remote firing control. Overall lightness, compactness, ruggedness and reliability, coupled with high performance, are key design features.

The liquid-cooled Nd:YAG laser target marker and ranger will range out to 10 km with an accuracy of ±5 m, and has an endurance to permit target marking for approximately 35 minutes (within a specified duty cycle) before battery recharging is required. The PRF can be either 10 or 20 pps.

OPERATION

The Type 306 is the ground element of the close air support system. The forward air controller aims the laser at the chosen target and calls in ground support aircraft by radio. When the attacking aircraft approaches, the laser is fired and the airborne marked target seeker acquires and tracks the scattered energy, driving the pilot's head-up display to indicate target position. The pilot may then carry out the attack without ever having seen the target. If laser guided weapons are released, the bomb or missile homes precisely on to the marked target.

Alternatively, the Type 306 can be used to provide designation for all the current range of NATO laser

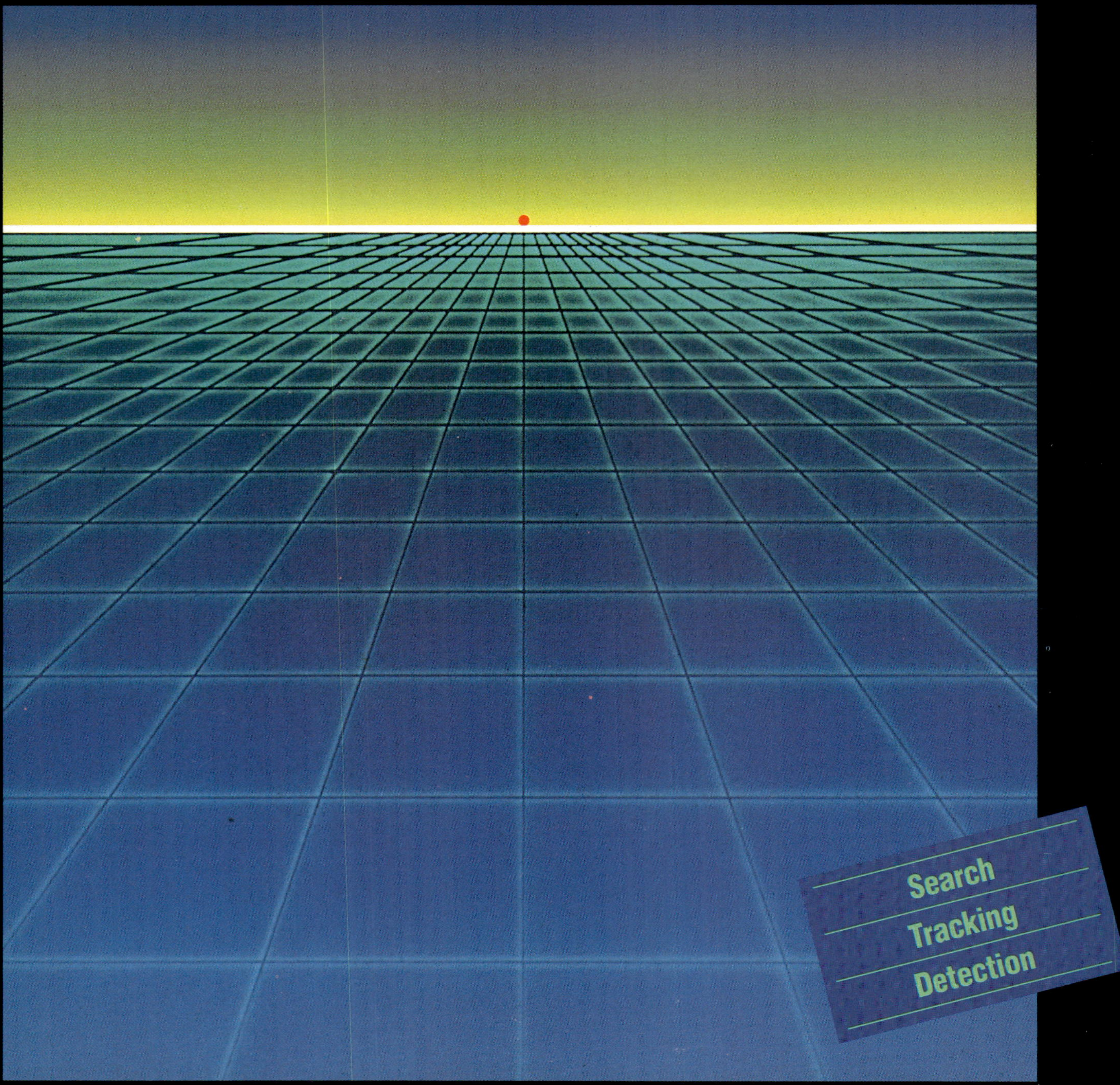

guided weapons, including laser guided bombs, the Copperhead 155 mm shell and the Hellfire missile.

STATUS

The Type 306 was developed under a UK MoD contract and is in service with the British Army and several overseas forces. New Zealand became the sixth customer in 1987.

CONTRACTOR

Ferranti Defence Systems Ltd, St. Andrew's Works, Robertson Avenue, Edinburgh EH11 1PX, Scotland.

Type 306 laser designator/ranger in operation

5755.193

FORWARD OBSERVER LASER DESIGNATOR (FOLD)

With the emergence of weapons such as the Hellfire air-to-surface anti-armour missile and the Copperhead cannon-launched guided projectile, there is an increasing emphasis on moving targets such as battle tanks. To meet this requirement, Ferranti has developed a version of the Type 306 designator (**1742.193**) known as FOLD. The system uses a Type 1906 tracking mount which enables the designator to track and designate armoured fighting vehicles. For night use, a thermal imager is added, co-mounted with the designator. See also BOLD (**5756.193**) and HELD (**5747.393**).

STATUS

In production.

CONTRACTOR

Ferranti Defence Systems Ltd, St. Andrew's Works, Robertson Avenue, Edinburgh EH11 1PX, Scotland.

FOLD system using Ferranti Type 306 designator, Vinten Type 1905 mount and Thorn EMI multi-role thermal imager

---

5756.193

BATTLEFIELD OPERATIONS LASER DESIGNATOR (BOLD)

BOLD is a lightweight laser designator system. The experience gained during the Falklands campaign highlighted the need for a lightweight, man-portable equipment. As with FOLD (**5755.193**), the system is based on the Type 306 laser designator (**1742.193**) mounted on a lightweight, low-cost tracking mount. A third system, known as HELD (**5747.393**) is a helicopter mounted version of BOLD and the basic Type 306 can be interchanged easily between ground and air use by unbolting from one mount on to the other.

STATUS

In production.

CONTRACTOR

Ferranti Defence Systems Ltd, St. Andrew's Works, Robertson Avenue, Edinburgh, EH11 1PX, Scotland.

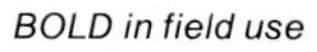

BOLD in field use

3411.193

GSA SERIES GUNSIGHTS

The GSA series of gunsights are designed to replace older fixed reticle gunsights on close-range anti-aircraft guns where the gunner had to judge by eye where to point his gun in order to hit a moving target.

The aiming allowances which have to be considered are for target motion, gravity drop of the ammunition and the effect of crosswind, but by far the most important is that for target motion. This allowance may be formulated in many ways using different parameters, but one that is particularly useful, in the absence of information about the target's speed and direction, is in terms of rate of rotation of the sight line which may be measured by a rate gyro within the gunsight. There is uncertainty about the time of flight of the shell, because of the gunner's limited ability to assess range, but rather than attempt to adjust the range input to the gunsight continuously throughout the attack, the principle adopted is to set into the gunsight a range through which the target must fly and so obtain, with sustained firing, hits at some period of attack – approximately at the time when the actual range equals that set-in. In the case of the GSA 400 series, any one of four ranges may be selected during the course of an attack and that selected is indicated to the gunner as a number in the collimated reticle pattern. For the GSA 200 series, two ranges are available, a maximum and minimum preset by the operator.

Range rate also appears in the aiming condition, but, in the formulation with sight-line rate of turn as a primary parameter, it does so only in a secondary way. The effect of actual range rate being different from a mean assumed value is merely to alter by a few per cent the range at which hits will be obtained.

The principle of generating the aim-off angle as embodied in gyroscopic gunsights is shown diagrammatically (left). The gyro rotor is free to deflect in both the horizontal and vertical planes about its suspension and the deflection from the symmetrical position is proportional to the rate of turn (precession) of the rotor axis in space. By attaching a mirror to one end of the rotor and reflecting a reticle from it, the optical system presents this deflection to the gunner as the required aim-off. The other end of the rotor is in the shape of a dome which rotates in a magnetic field produced by a current through the sensitivity windings and the interaction between the induced eddy currents and field provides the torque necessary to precess the gyro rotor in the direction of the target as the gunner tracks the target. By making the current in the sensitivity winding a suitable function of range, the rotor deflection from its position of symmetry as the gunner tracks the target will give a sight-line deflection equal to the aim-off requirement.

The gyro also has gravity windings which enable the rotor to be deflected statically so that the rest position of the sight-line can be compensated for the gravity drop of the ammunition as a function of range.

Ferranti GSA Mk 3 sight

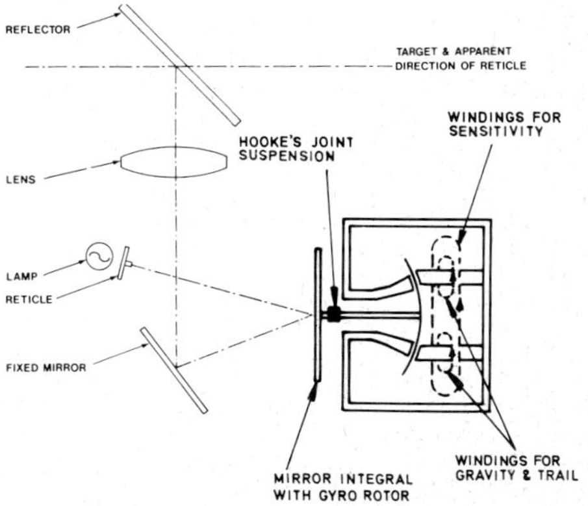

General arrangement of GSA gunsight

A lens in the optical path focuses the aiming reticle at infinity making a 'backsight and foresight' unnecessary. The reticle is reflected into the gunner's eye by a semi-reflecting glass through which the target may also be seen and so aligned with the reticle by rotating the gunsight.

The type of prediction embodied in the gunsight is known as 'disturbed sight-line' because the sight-line is not fixed in relation to the body of the gunsight or gun but lags behind it by a variable amount. It can only be moved, however, through the medium of the gun itself, which may create some difficulty for the gunner whose ability to track the target is clearly limited by the combined effect of the sight-line response and the servos which move the gun. Practical trials have shown that with modern gun servos the gunner can quickly learn to cope with the combined response and achieve a good standard of aiming. Moreover, there is the advantage that any errors in the servos do not affect the accuracy of prediction since they are within the aiming control loop.

The GSA is compatible with any gun with a calibre of 20 to 40 mm and the necessary ballistic data is inserted into the computation by means of an adjustment of trim potentiometers on the electronic card module of the unit. This gives the user the flexibility of having one gunsight for a number of guns of different calibre.

STATUS

Oerlikon-Bührle have adopted two variants of this gunsight as standard fit for their equipment – the GSA Mk 3 for field use with the GDF-B and the GSA Mk 3M for their naval GDM-A twin 35 mm AA guns. One African state has adapted the Bofors L40/60 guns in all its navy ships to accept the GSA 302 system. A South American state fits the GSA 32-JSW (a GSA 200 series type) to its twin 20 mm AA guns. The UK BMARC, a subsidiary of Oerlikon-Bührle, uses the GSA 201 on its GDM-A003 30 mm twin gun mountings. The equipment is now in quantity production for fitment to the Oerlikon-Bührle Type GDF-003 35 mm twin AA field gun and Type GDM-A 35 mm twin AA naval gun. A number of smaller orders have been received for use with optical fire directors and to update Bofors 40/60 40 mm guns for both field and naval applications.

CONTRACTOR

Ferranti Defence Systems Ltd, St. Andrew's Works, Robertson Avenue, Edinburgh EH11 1PX, Scotland.

4696.093

FERRANTI HELMET POINTING SYSTEM

The Ferranti helmet pointing system (HPS) is a target sighting equipment which directs weapons and sensors towards any target at which the wearer is looking. The HPS provides existing weapon systems with a high-speed visual acquisition mode which can be used when surveillance radar data is unavailable or when passive operation is required.

The HPS is a lightweight, easy-to-use and simple system consisting of the helmet-mounted sight and sensor, a fixed radiator, a signal processing unit, a control unit and a power supply. The observer's sight is light, unobtrusive and compatible with face masks.

When used with Tracked Rapier, the commander (or observer) searches for possible targets. The sight has an illuminated aiming mark focused at infinity representing line-of-sight. A tiny sensor, integrated and pre-aligned with the optics, continuously monitors the sightline relative to a radiator fixed to the vehicle. The commander may operate the sight using either right or left eye, or can retract the sight when using binoculars. Once the commander has spotted a target he overlays it with the aiming mark as a means of designating his target. By pressing a button the sightline is transferred to the optical tracker instantaneously, directing it in azimuth and elevation to the target. The tracker operator then follows his normal procedure. The whole procedure takes from two to five seconds. In a more general case, the HPS may be adopted to transfer an observer's sightline to a weapon or sensor which is then instantly directed at the target. A variant of the system can be supplied for simultaneous dual observer operation using a common electronics unit.

Ferranti helmet pointing sight

DEVELOPMENT

The HPS was originally conceived as an aircraft aid and provides unlimited angular coverage and measures line-of-sight angles in azimuth, elevation and roll to an accuracy of 0.5° CEP. It is, however, as a land-based system that it gained its first contract as part of the BAe Tracked Rapier missile system for the British Army.

STATUS

The Ferranti HPS is standard equipment on the British Aerospace Tracked Rapier missile system and is in quantity production for that purpose. The system also has considerable potential in air and sea operations, and sea trials by the Royal Navy took place in mid-1986.

CONTRACTOR

Ferranti Defence Systems Ltd, St. Andrews Works, Robertson Avenue, Edinburgh EH11 1PX, Scotland.

4131.193

L1A2 ARTILLERY DIRECTOR

Based on the Hall and Watts Microptic theodolite ST.190, the L1A2 artillery director is used by the British Army and other forces for artillery survey and gun direction purposes. The director's position is established by normal survey techniques using the instrument to measure angles of prominent features of the landscape, or by reference to the magnetic meridian or astronomical observations. The director then acts as a reference point from which the individual guns of a battery can use their dial sights to set off their bearings to engage a target.

A version of the L1A2, known as the L1A2B, has the basic characteristics of the former director plus an auto-collimating system specifically designed for use with inertial platforms. Also incorporated are improved one mil circles which will decrease training time, enhancing survey and gun orientation.

CHARACTERISTICS

Magnification: × 4.2

Field of view: 213 mils

Eyepiece adjustment: –0.5 to 1 dioptre

Graticule: Vertical and horizontal scales divided every 2 mils to ±100 mils figured every 20 mils

Eye relief: 6.6 mm

Weight: 3.63 kg (instrument only); 7.6 kg (instrument in case)

Case dimensions: 317 × 190 × 178 mm

STATUS

The L1A2 is in service with the British Army. The L1A2B is in service with the Belgian Army.

CONTRACTOR

Hall and Watts Ltd, 266 Hatfield Road, St Albans, Hertfordshire AL1 4UN, England.

L1A2B artillery director

4280.193

TICM II THERMAL IMAGING SYSTEMS

TICM II thermal imaging systems are based on modules developed for the UK MoD under the Thermal Imaging Common Module programme. They are built to full military standard and use advanced optical and electronic components, including the Mullard 'SPRITE' detector, to give high resolution and sensitivity even at long range. The modular basis gives flexibility in system design coupled with ease of maintenance and repair for cost-effective use in both new weapons and for retrofit. Telescopes and displays can be selected by system designers to meet precise operational requirements.

These indirect view systems are fully television compatible and produce a high-quality video image which can be displayed on television monitors or head-up and head-down displays. Unique among thermal imagers, they are fully automatic in operation and always provide the best possible image, as well as allowing module replacement without any adjustment or set-up time. The systems are particularly suited to applications demanding the highest resolution and reliability such as pilot night vision, target detection and tracking, weapon aiming and surveillance.

CHARACTERISTICS

Spectral bandwidth: 8 – 13 μm

Field of view: 60° × 40° (625-line 50 Hz); 48.4° × 32.5° (525-line 60 Hz)

Resolution: 2.27 mrad (60° field of view)

Pupil diameter: 10 mm

MRTD: Typically better than 0.1°C

Detector: 8 parallel CMT 'SPRITE'

TICM II thermal imaging system fitted with a rugged telescope

Video output: CCIR Systems I/EIA-RS170, compatible 625/525-line 50/60 Hz
Weights (modules in enclosures)
Scanner head: 9.5 kg
Processing electronics: 6.5 kg
System control panel: 1.03 kg
X20 telescope: 10 kg
STATUS
GEC Avionics and Rank Taylor Hobson are in large scale production for the UK MoD supplying modules for a wide variety of roles in the British armed forces.

TICM II based imagers are being supplied as the standard thermal equipment for the British Army Chieftain and Challenger Thermal Observation and Gunnery System (TOGS). TICM II also forms part of tracked Rapier's new night firing system 'TOTE' (**5906.193**).

CONTRACTORS
GEC Avionics Ltd, Electro-Optical Surveillance Division, Christopher Martin Road, Basildon, Essex SS14 3EL, England.

Rank Taylor Hobson, PO Box 36, 2 New Star Road, Leicester LE4 7JQ, England.

4701.193
HHI-8 HAND-HELD THERMAL IMAGER
The HHI-8 is a lightweight, high performance, self-contained thermal imager with a dual field of view objective lens system. The flexibility of design is such that the imager can be used hand-held and, by separating the display lower case from the main housing, offers a remote viewing mode enabling the operator to observe with greater protection.

As a compact general purpose surveillance aid it may be used for reconnaissance, patrolling or, when statically mounted, for other roles such as harbour/border security surveillance or weapon fire control. The HHI-8 is designed for simplicity of operation and ease of maintenance, and is ready for use in less than 30 seconds.

Coolant is provided from a standard 0.6 litre rechargeable pressure vessel and electrical supplies from a battery pack contained within the handle, both being charged within seconds and giving up to 5 hours operating time. Alternative batteries and compressed air bottles can be provided for extended operation.

A scan converter unit, offered as an option, produces CCIR video output to standard monitors. This offers the ability for indirect viewing, video recording and the transmission of the video signal via communication links.

CHARACTERISTICS
Spectral band: 8 – 12 μm
Field of view
Fixed: 90 × 150 mils (5° × 3°); magnification ×3
Switched (wide): 170 × 100 mils (9.6° × 5.6°); magnification ×2
Switched (narrow): 70 × 40 mils (4° × 2.3°); magnification ×5
Display: 1-inch CRT
Detector cooling: Joule-Thompson minicooler using compressed air
Battery endurance: Battery pack size C lithium – 5 hours; battery pack size D NiCad – 3 hours
Compressed air endurance: Bottle 0.6 litre – 5 hours continuous
Dimensions: 410 × 140 × 140 mm (including bottle and battery pack)
Weight: 4.8 kg (including battery and air bottles)

HHI-8 hand-held thermal imager

CONTRACTOR
Marconi Command and Control Systems Ltd, Chobham Road, Frimley, Camberley, Surrey GU16 5PE, England.

5715.193
THERMAL IMAGER VEHICLE SIGHT
The Thermal Imaging Vehicle Sight (TIVS) is a gunner's sight designed to provide day and night observation and weapon aiming based on the normal thermal energy radiation. It is a plug-in unit suitable for the direct replacement of image intensifier night sights in a range of sight housings currently fitted to a large number of vehicles. For most vehicles, including M-series and T-series battle tanks, armour modifications are not necessary

The TIVS operates in the 8 to 12 μm band where high energy radiation from an object will penetrate smoke, mist and most fog, and will provide a high thermal contrast picture. The display may be optimised by adjusting the focus, brightness and contrast controls, and the image polarity may be changed from black-on-white to white-on-black. Incorporated within the sight are presettable controls for graticule and aiming marks. The magnification may be changed from ×2 to ×4.9. Suitable periscopic housings can be provided for a range of vehicles.

The display is presented on a CRT, allowing flexibility in layout within the vehicle. A parallel display for the commander can also be fitted.

CHARACTERISTICS
Waveband: 8-12 μ
Optics: Coated lens assembly with two fields of view
Field of view: Wide 9.6° × 5.6° magnification ×2; narrow 3.85 × 2.26° magnification ×4.9
Cooling system: Joule-Thompson mini-cooler, compressed air

CONTRACTOR
Marconi Command and Control Systems Ltd, Chobham Road, Frimley, Camberley, Surrey GU16 5PE, England.

Marconi thermal imager vehicle sight

3410.193
PPE PASSIVE DRIVING PERISCOPE
Pilkington PE Limited is now producing a new passive driving periscope under contract to the UK MoD. The device enables drivers of armoured fighting vehicles to drive efficiently, hatch down, without the need of artificial illumination. The periscope incorporates a micro-channel plate intensifier tube manufactured in the UK by Mullard. The tube provides the high gain necessary to achieve good contrast image, particularly in low light-level conditions. In addition it is far less affected by bright lights (eg gun flash, street lights, etc) in the field of view than first generation (cascade tube) systems.

The optical design features a bi-ocular eyepiece which affords the driver considerable eye relief, minimising the risk of facial injury and reducing eye fatigue over a period of driving. The modular design ensures that the basic periscope fits a variety of armoured fighting vehicles. Assembly is effected by special vehicle mounting brackets and upper casings. The sight is interchangeable with existing day periscopes in a variety of vehicles.

The device is designed to fit the following range of armoured fighting vehicles, without modification: Leopard, Marder, Chieftain (ARV, AVRV, MBT versions), CVRT, CVRW (Scorpion, Scimitar, Striker, Spartan, Samson, Samaritan), 430 series, and all M series (M107, M109, M110, M113, M60) vehicles. In addition, with some modification to the vehicle, it will fit the French AMX-30 and Panhard AML series.

PPE passive driving periscope, universal model

PPE night driving periscope installed in Scorpion light tank

CHARACTERISTICS
Magnification: × 1 bi-ocular
Field of view: 50° horizontal, 40° vertical
Eye relief: 100 mm
Image tube: Mullard 50/40 mm micro-channel plate
Weight: 7.5 kg

STATUS
In service with the British Army and overseas forces.

CONTRACTOR
Pilkington PE Ltd, Glascoed Road, St Asaph, Clwyd LL17 0LL, Wales.

5710.193

CONDOR COMMANDER'S DAY/NIGHT SIGHT

This was designed originally for the Royal Ordnance Factory's main battle tank and provides the tank commander with a 24-hour capability for surveillance, target acquisition and firing of the main armament. It incorporates independent channels for day and night use and interfaces with the main armament via a reticle image projector (PRI) unit. The PRI injects into the sight in line-up a collimated ballistic graticule to enable the commander to engage the target. The unit will also inject the optimum target engagement data derived from the fire control system, if fitted, to improve probability of first round hit by day and night.

Condor is designed to fit the Vickers Mk III main battle tank and Centurion, and the ROF main battle tank. Fittings for other vehicles such as the Vickers Valiant, M47, M48, T54 etc, can be made available.

CHARACTERISTICS

Above armour height: 330 × 330 mm

Elevation range: –10 to +35°

Magnification: × 10 day channel; × 4 night channel; × 1 periscope

Field of view: 5° day channel; 8.9° night channel; 56° horizontal/11° vertical (× 1 channel)

Weight: 56 kg

STATUS

In production for Vickers and ROF main battle tanks. In December 1980 a contract was awarded for over £6 million for the supply of units to fit in ROF main battle tanks. In March 1983 a further contract worth over $4 million was awarded by the UK MoD for the supply of nearly 1000 PRI units for the Chieftain main battle tank.

CONTRACTOR

Pilkington PE Ltd, Glascoed Road, St Asaph, Clwyd LL17 0LL, Wales.

Condor day/night sight

3403.193

No 63 PASSIVE NIGHT SIGHT (L3A1)

The No 63 sight was designed by PPE for use with the No 62 day sight in the No 16 cupola of the Spartan APC and other vehicles. It is also suitable for use in vehicles and other installations as a straight telescopic sight or as a periscopic sight, as on the 155 mm self-propelled gun.

The sight can be quickly detached from the No 62 sight and fitted to a tripod and periscopic mirror attachment for surveillance on foot.

The wide field of view and excellent range provide very effective surveillance capability coupled with accurate firing of the 7.62 mm cupola machine gun.

CHARACTERISTICS

Magnification: × 1.4 monocular; × 1.1 bi-ocular

Field of view: 20°

Image tube: 25 mm 3-stage cascade

Weight: 4.5 kg (7.5 kg with tripod)

STATUS

In production for the British Army.

CONTRACTOR

Pilkington PE Ltd, Glascoed Road, St Asaph, Clwyd LL17 0LL, Wales.

PPE passive night sight for Spartan shown in dismounted role

5908.193

RAVEN DAY/NIGHT SIGHT

Raven is a combined day/night sight and is intended primarily for the British Army's Warrior vehicle, but can be fitted to most AFVs. Each Warrior will carry two sights, one for the commander and the other for the gunner, with the gunner's sight interfaced to the gun linkages. The sight will provide × 1 and × 8 day vision and × 2 and × 6 by night using image intensifier techniques. Ballistic information and all sight functions are operable over an angular range of +47° to –12° with respect to the line of sight.

Raven has been designed for installation from under or above armour. A lifting frame is supplied to aid installation and removal. The eyepieces have been designed to accommodate lateral offset arising from sight/seat centreline differences.

CHARACTERISTICS

Line of sight: –12 to +47°

Magnification: × 1 and × 8 (day sight); × 2 and × 6 (night sight)

Target recognition distance: 3000 m (day); 1000 m (night)

Detection range: 5 km (day); 2.5 km (night)

Dimensions: 560 × 350 × 450 mm

Weight: 50 kg

STATUS

The Raven sight was originally designed as a private venture. It has been ordered by the British Army and production commenced in June 1986. 422 sights have been ordered for installation in the first batch of Warrior vehicles. The second batch contract for approximately 1600 sights has been won by Avimo, a subsidiary of United Scientific Instruments.

CONTRACTORS

Pilkington PE Ltd, Glascoed Road, St. Asaph, Clwyd LL17 0LL, Wales.

United Scientific Instruments Ltd, 215 Vauxhall Bridge Road, London SW1V 1EN, England.

Raven day/night sight

6306.193

OSPREY DAY/NIGHT AFV SIGHT

Osprey is a combined day/night laser fighting vehicle sight and has been developed as the next generation after the Raven sight (**5908.193**). The sight is modular in design offering scope for easy maintenance and replacement of modules at first line. The modular construction enables the sight to be offered with a wide choice of facilities, including thermal imagers, image intensifiers, laser rangefinders, stabilisation and interfaces to fire control systems, and training simulator systems.

A typical Osprey configuration for a light armoured vehicle would offer a medium elevation head, day channel, thermal imager channel and a single shot laser rangefinder, with built-in displays of thermal scene and laser range, viewed through the twin day channel eyepiece. Remote displays for the thermal and laser information can also be provided. A high power repetition rate laser rangefinder can replace the single shot system for use in the anti-aircraft role. A variety of other configurations can be provided.

CHARACTERISTICS

Elevation range: –10° to +20° or –10° to +45° or –15° to +68°

Magnification: × 8 (day channel)

Image intensifier: 2nd or 3rd generation type as required

Thermal imager: Type as required

Laser

Type: Nd:YAG

Wavelength: 1.064 μm

Repetition rate: Single shot or 6 Hz continuous

Beam divergence: 0.5 mrad (single shot); 2 mrad (continuous)

STATUS

The typical Osprey configuration described above has been selected for use in the British Army's GKN Warrior MAOV variant. First prototypes were due to be completed in 1987 with production units in 1989.

CONTRACTOR

Pilkington PE Ltd, Glascoed Road, St. Asaph, Clwyd LL17 0LL, Wales.

6307.193

SABRE DAY/NIGHT VEHICLE SIGHTS

The Sabre system is a day/day and night vision vehicle sight which has been designed by Pilkington PE as an inexpensive, easy to operate compact equipment. It is intended for use by the AFV commander or gunner for surveillance and target acquisition and can be retro-fitted either under or above armour in existing vehicles.

Sabre has been designed in two variants:

(a) a day sight with a unity periscope channel and a × 6 magnified channel

(b) a day/night sight with × 1.5 and × 6 magnified day channels, and a × 4 image intensifier night channel. Either second or third image intensifier tubes can be incorporated.

CHARACTERISTICS

Field of view: 6°

Target recognition distance: <4000 m

Target identification distance: <2500 m

Elevation: –15° to +50°

STATUS

Sabre is a private venture system and is in development. Prototype systems have been built.

CONTRACTOR

Pilkington PE Ltd, Glascoed Road, St. Asaph, Clwyd LL17 0LL, Wales.

3796.093

ARGUS DAY/NIGHT SIGHT

The Argus combined day/night vehicle sight in its various configurations (SS122 to SS126) provides a full 24-hour surveillance and gunnery capability for armoured fighting vehicles. The sight fits into the gunner's and commander's positions of many current armoured vehicles and is mounted in the roof of the turret giving a very low above-armour profile of 230 mm. The sight consists of an optical system which provides the operator with a × 9.3 fixed focus day system, a × 9 variable focus night system, and a unity magnification fixed focus daylight facility. Changeover from day to night systems is effected by a hand-operated lever.

Elevation of the line-of-sight is accomplished by the movement of a tilting head mirror. With the gunner's sight this is mechanically linked to the armament, whereas with the commander's sight it can be supplied to operate either linked with the main armament or manually operated fully independent of the gunner's sight. High elevation versions of the tilting head mirror are also available. Each sight can be fitted with an integral laser rangefinder as an optional extra. When fitted to the gunner's sight only, range information can be supplied either to the commander's sight or to a remote commander's display unit.

The SS122 series can be provided with an optional

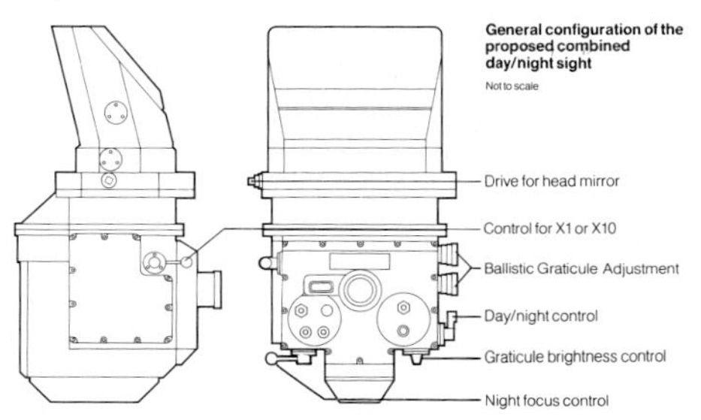

Rank day/night sight

Commander's sight, rear view

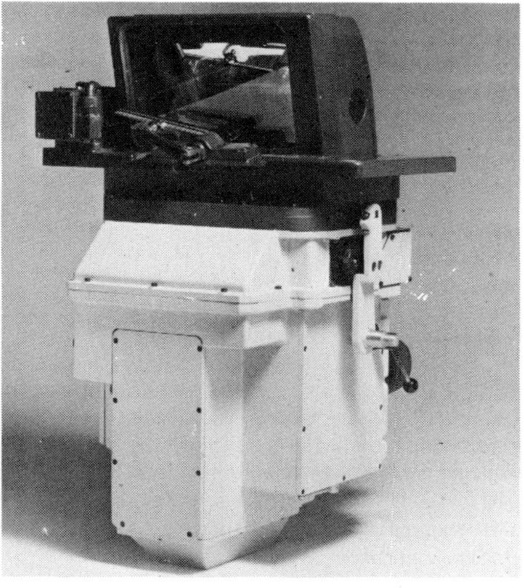

Gunner's sight, front view

Cathode Ray Tube (CRT) based injection unit which permits the sight to interface with sophisticated fire control computer systems thus permitting accurate target engagement by day or night. In conjunction with the CRT injection unit a simple fire control computer is available. The system is designed to simplify the gunner's task by eliminating the use of complex ballistic reticles, thereby preventing a misinterpretation, or incorrect interpretation, when using the latter.

The computer takes into account four main parameters: range of target from the laser rangefinder, trunnion tilt of the main armament, type of ammunition selected and crossing speed of the target. Additional manually input parameters, such as altitude and temperature, are available options.

The injected ballistic reticle is retained and may be switched in for use in either reversionary or callibration mode.

CHARACTERISTICS

Dimensions: 625 × 360 × 350 mm

Weight: 44 kg (sight); 4 kg (control box); 3 kg (laser rangefinder)

Day sight

Magnification: × 9.3 or × 1

Field of view: 5.5° or 25° horizontal × 10° vertical

Eyepiece: Monocular (high magnification) or biocular viewing window (low)

Night system

Magnification: × 9

Field of view: 6.5°

Image intensifier: 25 mm 3-stage Cascade

Eyepiece: Monocular

STATUS

In operational service in 11 countries.

CONTRACTOR

Rank Pullin Controls Ltd, Langston Road, Debden, Loughton, Essex, IG10 3TW England.

3063.193

SS100/110 NIGHT SIGHT

The designation SS100/110 refers to the two versions of the image intensifier night sight produced for the Fox and Scorpion AFVs. They have the British Army designations of SPAV L2A1 and SPAV L3A1, respectively.

The equipment is turret-mounted in front of the gunner and alongside the main armament. It enables gunners to locate, identify, and engage targets at night without the use of any artificial lighting, and it is also used for general surveillance. A dual role capability is achieved by the use of two objectives, mounted one within the other. The outer objective gives a magnification of × 5.8, while the inner, for surveillance, has a magnification of × 1.6. When high magnification is in use, a shutter isolates the low magnification objective. When the other magnification is selected, an iris diaphragm isolates the high magnification objective.

The image intensifier tube is protected from the effects of gun muzzle flash by a shutter which is operated electrically from the gun firing circuit. An illuminated ballistic graticule with brightness control is injected automatically into the optical system when high magnification is selected, and this is used for laying the main armament. Automatic brightness control for the image intensifier tube is provided.

CHARACTERISTICS

Fox night sight

Magnification: × 5.8 and × 1.6

Field of view: 142 mils (8°) and 498 mils (28°)

Objective focus: Variable 50 m – ∞ and fixed at 30 m

Length: 112 cm

Height: 355 mm

Width: 305 mm

Weight: 59 kg

Scorpion night sight

Magnification: × 5.8 and × 1.6

Field of view: 142 mils (8°) and 498 mils (28°)

Objective focus: Variable 50 m – ∞ and fixed at 30 m

Length: 112 cm

Height: 400 mm

Width: 355 mm

Weight: 59 kg

CONTRACTOR

Rank Pullin Controls Ltd, Langston Road, Debden, Loughton, Essex, IG10 3TW England.

4791.193

SS120 NIGHT SIGHT

This night sight, No 37N, was designed and developed for the UK MoD for use by the commander of the Chieftain tank. The specification called for a surveillance, acquisition and gunlaying capability. The sight can be fitted with a 2nd or 3rd generation image intensifier tube, and filters allow it to be used as a stand-by sight in daylight.

The sight is interchangeable with the periscopic sight AV No 37 Mk 4 and, if used with the PRI AV No 22 or No 24, allows the commander to lay the vehicle main armament.

A tilting head mirror is connected by a half-speed linkage to a lever arm actuated by the machine gun mountings AC No 8 and 10. The design is based on discrete assemblies, which break into four main sub-assemblies to ease replacement and repair.

CHARACTERISTICS

Field of view: 8° (night); 35° (day)

Magnification: × 5 (night); × 1 (day)

Fitting aperture: 236 × 115 mm

Weight: 24.8 kg

STATUS

In operational service with the British Army.

CONTRACTOR

Rank Pullin Controls Ltd, Langston Road, Debden, Loughton, Essex IG10 3TW, England.

3817.193

SS130 NIGHT DRIVING PERISCOPE

The SS130 night driving periscope has been developed to fit into many vehicles in service with the UK and other armed forces. It can also be adapted to viewing through the windscreen of conventional vehicles such as ships, boats, Land Rovers and most other vehicles.

The sight has been designed to be fitted easily from inside the vehicle. It is very compact, comprising an objective, channel tube and biocular eyepiece. The fixed-focus eyepiece enables the user to drive the vehicle using both eyes. It is also fitted with a thermostatically controlled heater to prevent condensation.

The periscope is fitted with an iris diaphragm which permits the use of the night sight during daylight hours in an emergency. The power supply is from the vehicle battery. A trickle charged battery is provided in the sight to allow it to be used when the vehicle power supply is not operating.

CHARACTERISTICS

Field of view: 48.7° horizontal; 40° vertical

Fitting aperture: 119 × 65 mm

Weight: 7 kg

STATUS

In service with 11 countries.

CONTRACTOR

Rank Pullin Controls Ltd, Langston Road, Debden, Loughton, Essex, IG10 3TW England.

SS130 driving periscope

4792.193

SS141, SS142 AND SS162 COMMANDER PERISCOPES

The SS141 is a ×2.7 magnification commander's night periscope which can be fitted into most turreted vehicles. It uses a 2nd generation image intensifier tube and has integrated filters which allow it to be used during dawn, dusk and daylight. The SS141 can also be fitted with a reticle to suit a cupola or a turret-mounted light machine gun, permitting engagements under armour.

The SS142 is a variant of the SS141 fitted with an elevating-head prism to provide an increased vertical field of view to allow engagement of low flying helicopters.

The SS162 has a ×3 and ×1 magnification day periscope designed to interchange with Rank night periscopes. It can have either an elevating or fixed head prism.

STATUS

In production.

CONTRACTOR

Rank Pullin Controls Ltd, Langston Road, Debden, Loughton, Essex IG10 3TW, England.

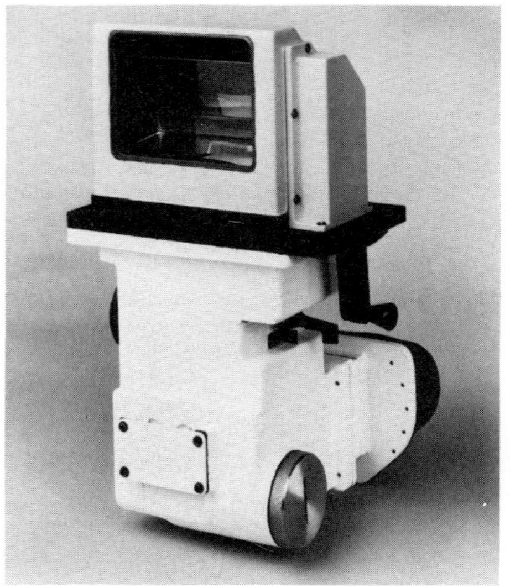

SS142 commander's night periscope

6533.193

SS170 DAY/NIGHT SIGHT

The SS170 combined day/night sight was designed to provide an instantly switchable vision capability to either the commander or gunner of an armoured vehicle. It differs from the Argus range of sights (**3796.093**) in being more compact and more readily adaptable for use in light armoured vehicles and armoured personnel carriers. It has a head prism of exceptionally high elevation for directing small-calibre automatic weapons, from under armour, against both airborne and ground targets. The sight is provided with illuminated integrated ballistic reticles which are boresighted from within the vehicle.

The optical system provides the operator with ×5 and unity magnification day systems, using monocular and biocular viewing respectively, and a ×3 night system. The night channel uses a 2nd generation image intensifier tube and switching between day and night channels is effected instantly by a hand-operated lever.

CHARACTERISTICS

Day system

Head mirror: -35 to -10°

Magnification: × 5 and × 1

Field of view: 8°; 34° horizontal/14°vertical

Night System

Magnification: × 3

Field of view: 13.3°

Focus: 20 m – ∞

STATUS

In service.

CONTRACTOR

Rank Pullin Controls Ltd, Langston Road, Debden, Loughton, Essex IG10 3TW, England.

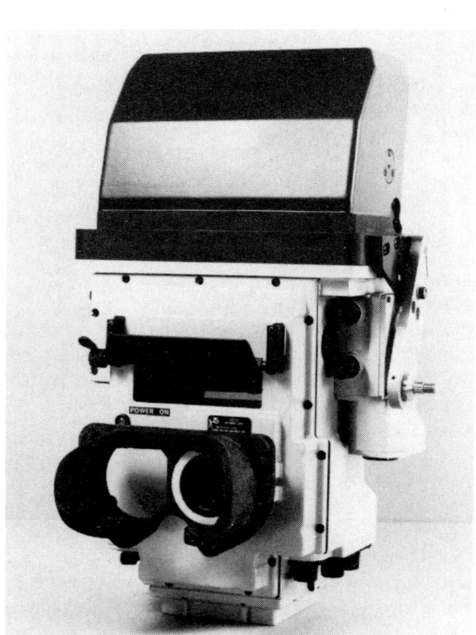

SS170 day/night sight

4788.193

SS200 FIRE CONTROL SYSTEM

The SS200 is a computerised fire control system based on the various versions of the Rank Argus day/night sight. It is intended to enhance the fighting capability of armoured vehicles and can be designed to meet customer requirements.

The system incorporates a laser rangefinder and a fire control computer, designed by Rank Pullin, which is coupled to a moving aiming mark within the sight. It takes into account four main parameters: range of target from the laser rangefinder, trunnion tilt of the gun, type of ammunition selected and the crossing speed of the target. The sighting system avoids the use of a ballistic reticle, thereby preventing a misinterpretation or incorrect interpolation when using the latter. The cross-shaped aiming mark is automatically adjusted by the computer during the complete tracking and firing sequence. The reticle is retained as a reversionary mode.

The fire control system has five elements:

(1) the computer which is the ballistic calculator and provides interface between system, gun, laser, sensors, position encoder, sight displays and crew
(2) the commander's control panel which includes a range display panel and position switches for designation of ammunition type
(3) the trunnion tilt sensor
(4) turret angle sensor which measures the angle turned by the turret over a measured time
(5) the gunner's sight which has a day/night capability with an integral laser rangefinder.

The SS200 is intended primarily for the retrofit market, for both medium and small battle tanks, and also for vehicles such as armoured personnel carriers which mount a gun.

CONTRACTOR

Rank Pullin Controls Ltd, Langston Road, Debden, Loughton, Essex IG10 3TW, England.

4790.193

SS500 ARMOURED VEHICLE SIGHT

The SS500 day and thermal vehicle sight is a complete sighting system incorporating day and thermal channels, processing electronics, CRT display, and laser rangefinder if required, all within one compact sight housing. The high magnification day channel and the dual field of the thermal channel are viewed through a common eyepiece and are instantly switchable. This allows engagement of targets throughout 24 hours. Provision is made for an additional remote display unit to be connected to the system which allows both the commander and the gunner to view the scene simultaneously.

The head unit has been designed to keep the above-armour profile to a minimum. The space envelope dimensions required are small enough to allow fitment in a wide variety of armoured vehicles ranging from armoured personnel carriers and light armoured cars to main battle tanks. Other versions in the range provide options such as an integral laser rangefinder, fire control interfacing, and thermal channel only.

CHARACTERISTICS

Magnification: ×9 and ×1 (day channel); ×10 and ×3 (thermal imager)

Field of view: 4.67° or 23° horizontal × 20° vertical (day channel); 4° horizontal × 2.6° vertical or 13.3° horizontal × 8.5° vertical (thermal imager)

Fitting aperture: 278 × 176 mm

Weight: 40 kg

STATUS

In production.

CONTRACTOR

Rank Pullin Controls Ltd, Langston Road, Debden, Loughton, Essex IG10 3TW, England.

6534.193

SS550 THERMAL ELBOW

The SS550 thermal elbow, in various configurations, is a cost-effective means of providing thermal imaging capability to a range of armoured vehicles, in particular the M and T series tanks. The elbow, which is a thermal-only system, is designed for fitment to a range of head units and, with some modifications, can replace image intensified elbows fitted to many of these vehicles.

The compact scanner and the modular design of the elbow together result in a sighting system which makes maximum use of the limited space available in armoured vehicle turrets. When a servo-driven head unit is fitted, the sight unit can be mounted in any suitable position within the turret; the crew is then provided with remote display and control modules.

The system is designed to interface with fire control systems and laser rangefinders to provide an effective 24-hour combat capability.

CHARACTERISTICS

Magnification: ×10 or ×3

Field of view: 4° horizontal × 2.6° vertical; 13.3° horizontal × 8.5° vertical

Eyepiece: Monocular

Display: 25 mm CRT

Weight: 13.5 kg (sight); 9 kg (electronics unit)

STATUS

In production.

CONTRACTOR

Rank Pullin Controls Ltd, Langston Road, Debden, Loughton, Essex IG10 3TW, England.

SS550 thermal elbow

4789.193

RANK THERMAL IMAGER

The Rank thermal imager is a compact, lightweight, high performance unit of modular design. It operates in the 8 to 13 μm band and produces high thermal and spatial resolution by the use of advanced SPRITE detector technology. The system has been designed for versatility and provides users with a choice of components which can be tailored to suit requirements. These include a range of telescopes with single or dual field of view, direct or remote view display units, remote controls, auto-control which provides automatic optimisation of gain and window controls, and an image processing unit. Cooling is provided by high-pressure gas bottles but compressors or cooling engines are also available.

The unit has been configured for use in a variety of roles, including land and sea surveillance, armoured vehicle sighting systems and airborne systems. It can be used as a surveillance tracking aid for missile systems and a variant is being supplied for the TOW roof sight on British Army Lynx helicopters.

Integrated thermal imager

The SS600 integrated thermal imager is designed for long-range surveillance applications. The system combines the telescope, scanner and processing electronics unit in one environmentally sealed body.

The standard display unit is a 230 mm video monitor, but a 25 mm module, viewed through an eyepiece or relaxed viewer, is available. This can be fixed to the camera for direct viewing or used remotely.

Ancillary equipments available include a range of rugged tripod mounts, manual and electrically driven pan and tilt heads and telemetry interfaces.

STATUS

In production.

CONTRACTOR

Rank Pullin Controls Ltd, Langston Road, Debden, Loughton, Essex IG10 3TW, England.

SS600 integrated thermal imager

4260.193

THORN EMI MULTI-ROLE THERMAL IMAGER

The Multi-Role Thermal Imager (MRTI) is one of a number of thermal imagers developed by THORN EMI under the UK MoD Thermal Imaging Common Modules (TICM) programme. The MRTI uses Class 1 modules from this and has been developed as direct and indirect view systems. Both systems are now in use with the British Army and the Royal Navy. MRTI is suited to a wide range of applications on the battlefield, for offshore patrol, airborne surveillance, weapon aiming and observation of mortar and artillery fire.

Thermal radiation is detected by the cadmium mercury telluride detector array which is scanned across the scene in a series-parallel mode. The detector array operates at 80 K and can be cooled to its working temperature by either a Joule-Thompson cooler using compressed air or by a closed-cycle cooling engine. MRTI is configured to use a 0.6 litre air bottle which, when charged to a pressure of 300 atmospheres, will cool the detector for up to 4.5 hours. The alternative closed cycle cooler has a power consumption of about 40 watts. Power to the rest of the imager can be supplied at voltages between 10 and 30 volts, with a consumption of less than 7 watts.

MRTI in use in the field

The system operates in the 8 to 13 μm band and two telescope fields of view are provided. Alternative telescopes are available to give enhanced performances. The equipment maintains focus when switched from one field of view to another. Controls for temperature offset and temperature window are provided and enable the operator to emphasise different temperature ranges within the scene under view. Hot-white or cold-white is switch selectable.

When used in the direct role the thermal scene is reconstructed using a scanned LED array. By adding another module to the direct view system a TV Compatible (CCIR) output to a remote display can be obtained to form the indirect view system. The TV monitor can be either mounted on the imager or remoted.

DEVELOPMENT

A number of variants of the MRTI have been developed. The OTIS equipment is described under entry **4703.193** and the hand-held version under entry **4704.193**. Four other variants are in various stages of development, and are described under separate entries. The Surveillance Thermal Imager (STI) (**6298.193**) is intended for perimeter surveillance (airfields, harbours, vital installations, etc) and is an indirect view configuration combined with a TV camera. The Tank Thermal Sight (TTS) (**6297.193**) is a compact periscope type for integration with an AFV fire control system, or as a main armament sight. The Naval Thermal Imager System (NTIS) is intended for surveillance, ship detection, target tracking, weapon control, etc. It is configured for continuous operation and provides a TV video output to below-decks displays. The fourth system is a Thermal Imaging Gun Sight (TIGS) which is suitable for use on 20 mm, 25 mm and 30 mm gun mountings to provide a day and night aiming capability. Three TIGS equipments have already been delivered to the Royal Navy for evaluation.

CHARACTERISTICS

Waveband: 8 – 13 µm

Temperature sensitivity: 0.1 K

Field of view: 4.9° × 3.2° narrow; 12.9° × 8° (wide)

Focusing range: 10 m – ∞ (wide); 30 m – ∞ (narrow)

Display: Red LED raster with illuminated graticule and warning symbols for low battery and low air supply.

Dimensions: 480 × 280 × 190 mm

Weight: Thermal imager 10.7 kg (including bottle and battery); remote viewing unit 1.3 kg

STATUS

In production. The equipment is also being developed as a weapon fire control system on helicopters. Trials have been conducted on an Army Air Corps Scout helicopter. British Aerospace are offering the MRTI for use on the Swingfire missile to allow use at night or in poor visibility. THORN EMI is co-operating with OIP Optics, Belgium in the production of a thermal imager, based on the MRTI, for integration with the LRS-5 fire control system (**4744.193**).

CONTRACTOR

THORN EMI Electronics Ltd, Electro-optics Division, Forest Road, Feltham, Middlesex TW13 7HE, England.

6738.193

AIR DEFENCE ALERTING DEVICE (ADAD)

The Air Defence Alerting Device (ADAD) has been developed by THORN EMI Electronics as a lightweight, compact, autonomous system to detect rotary and fixed wing aircraft through their infra-red emissions.

It comprises a scanner head unit, a remote display unit and an electronics processor unit. The scanner head unit provides coverage over a wide field of view in azimuth with an elevation baseline adjustment.

The remote display unit alerts the operator to the presence of the target and gives its bearing. Data and tactical information is entered by a simple alphanumeric keyboard which permits the entry of primary and secondary arcs of responsibility, blind arcs and restricted zones. Each ADAD system can service up to four display units at distances up to 500 m from the processor. The electronics processor unit interprets signals from the scanner head unit and transmits target alerting data to the remote display unit through a standard field telephone cable.

The system, which is powered by a 300 W/28 V portable generator, can be brought into action within five minutes and can be dismantled in half this time.

STATUS

ADAD has been ordered by the British MoD under a £70 million contract.

CONTRACTOR

THORN EMI Electronics Ltd, Electro-Optics Division, 1 Forest Road, Feltham, Middlesex TW13 7HE, England.

THORN EMI ADAD

4703.193

OBSERVER THERMAL IMAGING SYSTEM (OTIS)

OTIS is an adaption of the multi-role thermal imager (MRTI) (**4260.193**) together with a laser rangefinder and angulation head. The MRTI and the laser rangefinder are mutually boresighted. The equipment is man-portable and is intended for application in battlefield surveillance, observation and control of artillery and mortar fire and target tracking for laser guided weapons.

Azimuth and elevation readings are available as mechanical slipping scales or digital readouts. All other technical characteristics are as given for the MRTI.

STATUS

In production for the UK MoD under a contract worth almost £15 million.

CONTRACTOR

THORN EMI Electronics Ltd, Electro-optics Division, Forest Road, Feltham, Middlesex TW13 7HE, England.

OTIS system in use in the field

4704.193

THORN EMI HAND-HELD THERMAL IMAGER

The THORN EMI hand-held thermal imager is a compact lightweight unit intended for use by reconnaissance troops and forward observation officers, as well as a wide range of civil applications.

THORN EMI hand held thermal imager

The equipment uses modules developed from the UK Class 1 thermal imaging common module programme and employs a great deal of the technology which led to the MRTI (**4260.193**) and OTIS (**4703.193**) thermal imagers.

It is, however, considerably lighter than either of those equipments, with the electronics having been hybridised to reduce the weight, but maintains considerable commonality to reduce spares holdings and ease maintenance.

CHARACTERISTICS

Waveband: 8 – 12 µm

Detector: Series/parallel Cadmium Mercury Telluride

Field of view: Wide 20° × 8.6° (× 2 magnification); Narrow 8° × 3.4° (× 5 magnification)

Focusing range: 5m – ∞

Display: Red LED raster with illuminated graticule and warning symbols for low battery and low air bottle supply

Minimum resolvable temperature difference: 0.3 K

Dimensions: 470 × 210 × 160 mm

Weight: 5 kg (including cooling air bottle and battery)

STATUS

The imager is in service with the British Army, the Royal Navy, the Royal Marines, The Royal Air Force, the United States Air Force, the United States Navy, the United States Marine Corps, the Royal Danish Air Force, together with Middle Eastern and Far Eastern countries.

CONTRACTOR

THORN EMI Electronics Ltd, Electro-optics Division, Forest Road, Feltham, Middlesex TW13 7HE, England.

6297.193

TANK THERMAL SIGHT

The THORN EMI Tank Thermal Sight (TTS) is a compact, periscopic configuration thermal imaging equipment based on the UK MoD Class 1 common modules. It is another development from the MRTI (**4260.193**) and has applications for integration into fire control systems and as a main armament sight.

The TTS is of small size and low power consumption and has a dual field of view, the wide field being 17.5 × 11.4°, and the narrow field 7 × 4.6°. Display is standard CCIR (625 lines, 50 Hz) built into the sight, monocular and binocular options are available. Separate display outputs can be provided for commander/training purposes. Weight of the equipment is 10 kg (excluding detector cooling services and remote control).

CONTRACTOR

THORN EMI Electronics Ltd, Electro-optics Division, Forest Road, Feltham, Middlesex TW13 7HE, England.

6298.193

SURVEILLANCE THERMAL IMAGER

This is another development of the MRTI (**4260.193**) and is primarily intended for perimeter surveillance around high security areas, harbours, airfields, depots, etc. It can also be used for range control and as a thermal imaging training aid.

The configuration consists of an indirect view MRTI and a TV camera with 200 mm lens mounted on a Dennard pan tilt head and Vinten tripod. Space is also available for additional payload, eg a laser rangefinder. All functions are controlled remotely with monitors displaying either composite TI/TV or separate TI and TV via a video mixer. The thermal imager has two fields of view, the wide field being 11.1 × 7.1 degrees and the narrow field 4.3 × 2.8 degrees. The system weight, including pan tilt head and tripod is 70 kg. Control unit, power supply and cables weigh another 20 kg.

CONTRACTOR

THORN EMI Electronics Ltd, Electro-optics Division, Forest Road, Feltham, Middlesex TW13 7HE, England.

4783.193

VM SERIES AFV EQUIPMENT

In addition to other entries in this section, United Scientific Instruments manufactures a range of periscopes and viewers for armoured fighting vehicles. Brief details of these are given below.

NV39Q Driver's Night Vision Viewer

The NV39Q provides the 'T' series battle tanks with a complete night driving capability. It has been designed to mount in the vehicle as a direct replacement for the daylight periscope and uses a second generation tube.

NV40 Passive Night Vision Periscope

The NV40 night vision periscope comprises a night vision elbow, a periscope head and a mount. The night vision elbow is a modified and improved version of the standard US Army elbow. The persicope head is similar to the existing IR sight but all the optics have high efficiency coatings to prevent glare and improve image contrast.

NV44 Tank Periscope

The NV44 periscope is intended as a replacement for the M20 daylight periscope on the M41 and M47 tanks and provides both day and night fire control by adapting the fire control module of the M60A1 tank. The day body contains a ×9 power telescope with a unit vision window. The passive night elbow mounts onto the day body and uses a second generation image intensifier.

NV46S Passive Night Vision Periscope

The NV46S provides passive night vision for the commander in the T54, T55 and T59 battle tanks. The instrument employs a second generation image intensifier and is directly interchangeable with the existing IR sight. It provides a magnification of ×4.4 and a field of view of 11 degrees.

CONTRACTOR

United Scientific Instruments Ltd, 215 Vauxhall Bridge Road, London SW1V 1EN, England.

5725.193

VICKERS/GEC CO_2 TANK LASER RANGEFINDER

This is a development, by Vickers Instruments and GEC Avionics, as part of a programme to prove the advantages of CO_2 lasers in the penetration of smoke and fog. It is based on the CO_2 TEA laser (transversely-excited laser) which operates near atmospheric pressure and transmits very short pulses of radiation at high power. The programme is intended to provide a laser rangefinder as part of a new battle tank sighting system.

The 10 micron wavelength of the laser rangefinder is compatible with that of thermal imagers and consequently offers advantages in operational use and in optical design. The CO_2 laser is also intrinsically superior in the penetration of smoke, mist and fog compared with the currently used ruby and neodymium-YAG lasers.

The periscopic upper sight has been developed by Vickers Instruments and the laser rangefinder by GEC Avionics.

DEVELOPMENT

Early research into CO_2 lasers was carried out in the UK by the Royal Signals and Radar Establishment. This latest development is part of a contract placed on Vickers Instruments by the UK MoD for the design of a prototype battle tank sight for the Chieftain and Challenger.

STATUS

In development. Successful trials have been carried out in a main battle tank, and it is understood that the smoke and mist penetration of the CO_2 laser is being confirmed as significantly superior to that of ruby and YAG lasers.

CONTRACTORS

Vickers Instruments, Haxby Road, York YO3 7SD, England.

GEC Avionics Ltd, Applied Physics Division, Elstree Way, Borehamwood, Hertfordshire, England.

Vickers/GEC CO_2 tank laser rangefinder

4705.193

L20 LASER RANGEFINDER SIGHT

The L20 laser rangefinder sight provides a modular system periscope monocular sight which can be fitted to most main battle tanks. Modules are available for new installation and for retrofit on Centurion, Vijayanta, Chieftain, Challenger, Vickers Mk3, M41 MBT and CVR(T) Scorpion tanks.

The L20 consists of five interchangeable modules: optics, optical head, optics/head adaptor, laser and electronics. The laser and electronics modules can be removed without dismantling the sight allowing either module replacement at first-line or reversionary use of the sight. Substitution of optical head and adaptor modules facilitates installation in different tanks.

CHARACTERISTICS

Type: 2nd generation Nd:YAG

Wavelength: 1.064 µm

Q-switch: Electro-optic
Transmitted energy: 20 mJ
Clock frequency: 30 MHz
Beam divergence: 0.4 mils
Field of view: 0.8 mils
Range limits: 300 to 9995 m
Range accuracy: 5 m
PRF: 1 pps

STATUS
The L20 is installed in Chieftain, CVR(T), Scorpion, and Vickers Mk 3 MBTs. Significant orders have been obtained from the Far East, Middle East and Africa, and the sight has been trialled successfully in all other MBTs for which it is designed.
CONTRACTOR
Vickers Instruments, Haxby Road, York YO3 7SD, England.

L20 laser rangefinder

4706.193
L50 CONVERSION UNIT
The L50 is a Sight Laser Rangefinder Conversion Unit, designed and developed for the MBTs T54, T55 and T59. The L50 converts the existing gunner's telescope into a true laser rangefinder telescope. No cutting of armour is required, the existing sight aperture is used. The operation of the converted telescope is identical in most respects with that of the original telescope.

The L50 is available in two build standards; providing laser rangefinding and aiming-mark injection for available computer-based fire control systems, or laser rangefinding only for conventional mechanical fire control systems. The L50 is modular in construction and comprises an optics module, laser module, gunner's control module, CRT module (for injection of computer-generated aiming-mark), foot firing pedal and commander's display module.
CHARACTERISTICS
Type: 2nd generation Nd:YAG
Wavelength: 1.064 μm
Beam divergence: 0.4 mils

L50 laser rangefinder

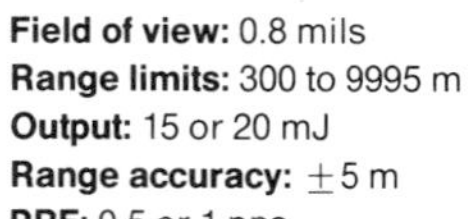

Field of view: 0.8 mils
Range limits: 300 to 9995 m
Output: 15 or 20 mJ
Range accuracy: ±5 m
PRF: 0.5 or 1 pps

STATUS
Orders have been placed by two unspecified countries and successful trials have been carried out in many others.
CONTRACTOR
Vickers Instruments, Haxby Road, York YO3 7SD, England.

6532.193
ARTILLERY SIGHTING SYSTEM
The Artillery Sighting System is a highly accurate, reliable and battle-proven sighting system suitable for artillery and other applications. It is designed to be easy to use, maintain and repair. Designed originally for service on the Royal Ordnance 105 mm Light Gun, the system is fitted to the export version of the VSEL FH70 155 mm Towed Howitzer and to other indirect-fire weapons.

The Artillery Sighting System consists of five modules: dial sight carrier, dial sight, dial sight beacon, daylight direct fire/anti-tank sight and image-intensifying direct fire/anti-tank sight. All vials and scales are illuminated by nuclear light sources, eliminating the need for external power supplies.

STATUS
The Artillery Sighting System is in service with many countries throughout the world.
CONTRACTOR
Vickers Instruments, Haxby Road, York YO3 7SD, England.

6360.183
INFRA-RED SCANNING SEEKER HEAD
GEC Sensors is developing a infra-red scanning head to provide a compact seeker design which offers passive, autonomous guidance for the next generation of guided weapons. The design is based on a 64-element, linear CMT array cooled to 77 K, which is scanned across the field of view by a servo controlled mirror. The signal outputs from the individual elements are fed into a RAM store from where a video output can be provided if required as well as data for further processing.

The design is such as to allow a change of field-of-view and scan rates, eg from an image size of 128 × 64 pixels at a scanning rate of 20 Hz to 52 × 64 pixels at 110 Hz. In addition, the scanning mirror servo loop can be controlled to give a non-linear scan if required, with positional information output to processing electronics. The seeker head incorporates a signal processing unit capable of acquiring targets in the presence of battlefield clutter. Mid-flight target tracking and aim point refinement procedures are also included.
STATUS
In early development leading to evaluation flight trials.
CONTRACTOR
GEC Sensors Ltd, Electro-optical Advanced Systems Division, Christopher Martin Road, Basildon, Essex SS14 3EL, England.

6300.193
RADAMEC ELECTRO-OPTICAL DIRECTORS
Radamec Defence Systems designs and manufactures a wide range of directors and mounts for electro-optical applications in land, sea and air environments, as well as for electronic warfare and radar systems (see appropriate sections). The Series 2000 is a series of eight system packages designed for fire control, surveillance and tracking. Also included in the overall Radamec range are the Type 419 single-axis mount, Types 409, 429 and 449 two-axis, Type 413 three-axis, Type 459 direct drive dual-axis yoke mount, Type 460 target designator, and Type 469 mirror mount. All mountings provide fast, high precision stabilisation and directional capabilities. Brief details of some of the latest systems are given below.
Type 449 Director
This is a two-axis director featuring direct drive of azimuth and elevation by means of samarium-cobalt DC torque motors. Precision encoders are fitted for accuracies to 0.5 arc second for special tracking applications. Dual axis stabilisation is by rate integrating gyros.
Type 460 Target Designator
The Type 460 target designator is used for the visual acquisition and tracking of surface or airborne targets, the bearing and elevation of which are transmitted in synchro or digital format to the weapons control centre. Two versions are available, the shipborne Type 460-001 with handgrip control of azimuth and elevation, and a land-based Type 460-002 with separate handwheels for each axis. Either type may be customised with a binocular sight, monocular sight with superimposed data-on-target, intensified monocular sight, optical ring sight or mechanical pointer.
Type 469 Mirror Mount
The Type 469 is a mirror mount designed for electro-optical sight-line tracking. Directional control and stabilisation are achieved by mounting the mirror on two orthogonal axis, driven by samarium-cobalt direct-drive torque motors. The mirror directs the optical path through a de-rotation prism to the sensors, of which a wide range of types and configurations may be fitted. It has two switchable lens, positioned in front of the mirror, for different fields of view (42.1° wide field and 2.7° narrow field) in the acquisition or tracking mode. The mirror surface is suitable for visual and thermal imaging wavelengths, thermal imaging data being transmitted using germanium optical components. Weight is 25 kg.
STATUS
In production for UK MoD and a number of overseas navies and defence agencies.
CONTRACTOR
Radamec Defence Systems Ltd, Bridge Road, Chertsey, Surrey KT16 8LJ, England.

6739.193
RADAMEC ELECTRO-OPTICAL POD TRACKING SYSTEM (E-OPTS)
The Radamec E-O POD Tracker is a precise, compact, high performance electro-optical tracking system for airborne and surface targets and is one of the wide range modular product family Radamec 2000 Series of electro-optical tracking and fire control systems. It is a versatile and adaptive system which is ideally suited for a wide range of AFV applications.

The E-O PTS contains three field proven high performance electro-optical sensors, including a sprite-based thermal imager, a CCD camera and a high pulse repetition frequency laser rangefinder. All the sensors are precisely boresighted together on a stabilised precision tracking mount.

The mount utilises direct drive torque motors, dc tachomotors and precision read-out synchros which provides a lightweight director with exceptional power and positional accuracy for precision electro-optical tracking and fire control applications.

This E-OPTS is environmentally sealed and fully ruggedised to high shock and vibration levels for exposed installations on-board ship, tracked vehicles and directly 'on-weapons'.

The E-OPTS is available as a stand-alone or as an integrated complete tracking and fire control system.

STATUS

Development complete. Undergoing customer trials analysis.

CONTRACTOR

Radamec Defence Systems Ltd, Bridge Road, Chertsey, Surrey KT16 8LJ, England.

Radamec E-OPTS

3765.093

VINTEN TRACKING MOUNT AND TRIPODS

Tracking heads

The Type 1906-3A head is designed to mount one or two payloads. It has boresighting and digital azimuth and elevation display facilities, and is designed to support a wide variety of surveillance and target acquisition equipments. It is in service with the British Army as a fire control system, carrying simultaneously a thermal imager with a laser target designator or laser rangefinder. It is also supplied for ground reconnaissance surveillance as a mechanical tracking system without electronics (known as Type 1906-3B). Other models are supplied with raised or repositioned elevation units to carry special payloads and to provide encoded azimuth and elevation data at a rate of 50 times a second.

The 1906 heads provide a manual means of smoothly, rapidly and precisely acquiring, tracking and designating targets for precision fire control. Azimuth and elevation data is displayed at 1 mil intervals. An innovative payload interface allows many types of equipment to be mounted securely and dismounted rapidly without impairing their use in other operational roles.

A simple facility enables operators to boresight the sightline of one payload to another to an accuracy in the order of ± 0.1 mil. Boresighting is retained between moves, which makes it possible to use laser rangefinders and designators by night under the control of a night observation device without further calibration.

Other innovative features include variable drag adjustment so that operators can select preferred tracking feel. Counterbalance is achieved by mounting equipments with their centre of gravity on the elevation axis.

The 7.7 kg head supports two payloads weighing up to 25 kg. Azimuth and elevation rotation is continuous. The azimuth read-out is displayed to four digits and elevation to three digits. The head consumes 0.09 A with the display off and 0.25 A when the display is commanded to operate. A single 20-30 V DC battery supplies power for up to two payloads and data display.

The Type 1906 head is designed to operate in temperatures ranging from −36°C to +70°C. The head is tested to Defence Standard 07-55.

Tripods

The 1926 series tripods is in service with Type 1906 heads in the British Army and can be used to support other mounting systems where light weight and tracking performance are important.

Each tripod can support a maximum load of 35 kg. They are constructed entirely from non-ferrous, lightweight rigid materials for strength, stability and high torsional rigidity. This set of tripods weighs 1.8, 3.1 or 4.4 kg and provides a range of height adjustments from 80 to 1090 mm.

Leg adjustment is controlled by a single lever which simplifies leg and levelling adjustment on uneven ground. A Betalight-illuminated 1/2 mm run bubble is used for levelling. Tripod feet can be set to varying angles and are designed to act as rigid pads. Anchor holes are provided. An interface unit is available to facilitate the mounting of other systems on Vinten tripods.

STATUS

In production, accepted into service by the British Army.

Vinten tracking mount with typical payload

CONTRACTOR

W. Vinten Limited, Military Systems Division, Western Way, Bury St. Edmunds, Suffolk IP33 3TB, England.

UNITED STATES OF AMERICA

3200.193

AN/GVS-5 LASER RANGEFINDER

The AN/GVS-5 is a portable, hand-held laser rangefinder resembling, and about the same size as, a pair of binoculars. The unit is completely self-contained and incorporates an internal power supply in the form of a Ni-Cd battery. A neodymium-YAG laser is employed, Q-switched by a chemical Q-switch wafer. The latter consists of a saturable absorbing dye in acrylic plastic which becomes transparent at the correct power level to produce the laser pulse.

The transmitter and receiver optics are combined into a single group comprising the housing, transmitter telescope, and receiver/sighting telescope. This is essentially like one half of a normal binocular assembly with an objective lens and an eyepiece. However, a beam splitter in the optical path provides a means for projection of a reticle and LED range display to the eyepiece, and for providing an optical path to the photodiode detector for the returned laser energy.

CHARACTERISTICS

Type: AN/GVS-5

Laser: Nd:YAG. Q-switched

Wavelength: 1.06 microns

Energy output: $<$15 mJ

Max range: Classified

Range gate: Continuously adjustable to 5 km

Beam divergence: $<$1 mrad

Time to range: 1 s

Ranges per battery charge: 700 at 24°C; 350 at 71°C; 200 at −32°C

Operating temperature range: −46 to +71°C

Weight: 2.27 kg

STATUS

RCA has been awarded several contracts since 1977 which have totalled over 5000 units. The AN/GVS-5 is in widespread service with the US Army and Marine Corps. It has also been sold to Argentina, Canada, Portugal and Saudi Arabia.

CONTRACTOR

RCA Automated Systems, Burlington, Massachusetts, USA.

AN/GVS-5 laser rangefinders undergo final inspection

4711.093

TK-640 LASER RANGEFINDER*

The TK-640 is a small rangefinder readily adaptable for use as an onboard fire control system for battle tanks, assault vehicles and landing craft. Range data is provided as a parallel digital output.

CHARACTERISTICS

Type: Nd:YAG
Wavelength: 1.064 μm
Output energy: 12 mJ nominal
Beam divergence: 0.5 mrad
Pulse rate: 60 per minute
Aperture: 38 mm
Maximum range: 9995 m
Range resolution: 5 m
Dimensions: 190 × 56 × 102 mm
Weight: 2.3 kg

CONTRACTOR

Litton Laser Systems, PO Box 7300, Orlando, Florida 32854, USA.

TK-640 laser rangefinder

4712.093

N-90 SERIES LASER

The N-90 laser was originally designed as a laboratory or industrial system but various options and combinations are available for use in military applications. When combined with an LLS high or medium sensitivity receiver and range counter, the NT-90 becomes a high performance rangefinder with range capabilities out to 20 to 40 km. In this option, the high sensitivity receiver is intended for use against small, long-range targets or when the use of a low laser power is required. The medium sensitivity receiver provides for normal ranging applications.

The addition of an LLS Simulator Module to the basic laser results in a low-cost target simulator for use in verifying aircraft sensor systems and pilot training.

A number of other models are available for specific purposes and various power output options.

CHARACTERISTICS

Type: Nd:YAG
Wavelength: 1.064 μm
Max energy: 90 mJ
Max repetition rate: 20 pps
Pulse width: 18 ±4 ns
Beam divergence: 2.5 mrads raw beam; 0.5 mrad collimated
Beam diameter: 6.3 mm raw beam; 38 mm collimated
Dimensions: 457 × 254 × 152 mm
Weight: 10 kg

CONTRACTOR

Litton Laser Systems, PO Box 7300, Orlando, Florida 32854, USA.

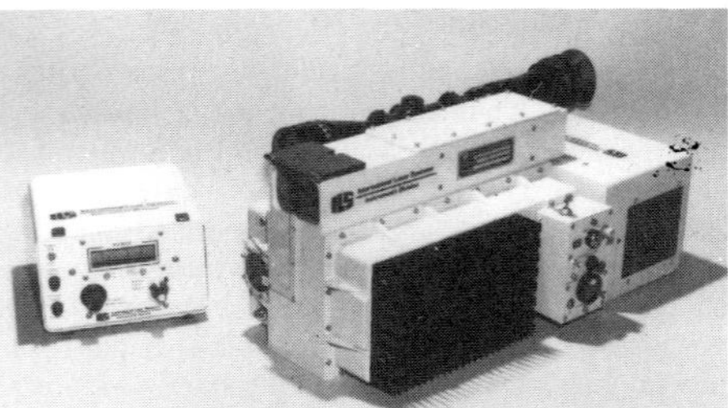

NT-90 MSR receiver and HRC counter

6295.193

LWDR LIGHTWEIGHT DESIGNATOR/RANGEFINDER

The LWDR was originally designed to be a self-contained, lightweight target marker that could be easily transported by one man and used to mark targets for delivery of laser guided munitions. Later a rangefinding capability was incorporated to create the LWDR, and an optional narrow beam divergence was offered for smaller targets and/or longer range capability.

The LWDR is designed for either handheld use or operation on a tripod. An optional lightweight viscous-damped tripod is available for accurate designation of moving targets.

In operation the procedure is simply to switch on the power, aim and press the thumb-operated fire button. Range is displayed in the eyepiece. Standard NATO pulse coding is set by three push button digital switches on the unit. The LWDR is fully compatible with all NATO coded munitions and spot trackers. The equipment is completely self-contained, including a lithium battery pack. A remote fire connector allows 'heads-down' firing or remote control firing against stationary targets.

The LWDR has been modified to provide operation at switch-selected 1.064 and 1.543 μm wavelengths. The latter provides eyesafe operation for training and battlefield ranging.

CHARACTERISTICS

Type: Nd:YAG; optional Raman converter
Wavelength: 1.064 μm; optional 1.543 μm
Output energy: 80 mJ minimum at 1.064 μm
Pulse rate: All standard NATO codes
Beam divergence: 0.5 mrad standard; 0.25 optional
Maximum range: 20 km
Range resolution: 5 m
Sight: 7 × 50
Dimensions: 9 × 7.6 × 14.8 inches (228 × 193 × 376 mm)
Weight: 18 lb (8.2 kg); optional tripod 12 lb (5.45 kg)

CONTRACTOR

Litton Laser Systems, PO Box 7300, Orlando, Florida 32854, USA.

Litton lightweight designator/rangefinder in the tripod-mounted configuration

4112.193

AN/VSG-2 TANK THERMAL SIGHT

The AN/VSG-2 is a thermal imaging infra-red device designed for fitting in US Army M60 tanks as part of an upgrading programme for that series of vehicles. New production M60 tanks have been designated M60A3 since February 1978 when the AN/VVG-2 laser rangefinder and a solid-state ballistics computer were introduced to the tank's fire control system, and the AN/VSG-2 thermal imaging equipment replaced the gunner's passive night vision periscope from mid-1979 onward.

Technical details and performance data have not been released but the latter may reasonably be expected to be no less than those achieved by other land battlefield IR systems based on the range of infra-red 'common modules' developed by Texas Instruments (eg AN/TAS-4, AN/TAS-5, AN/TAS-6).

STATUS

In production with nearly 4000 sights delivered. The US Army plans to convert all M60A3 passive tank sights to thermal imaging and to upgrade a large number of M60A1 vehicles to the same system. The system is also used by Egypt.

CONTRACTOR

Texas Instruments Incorporated, PO Box 660246, Dallas, Texas 75266, USA.

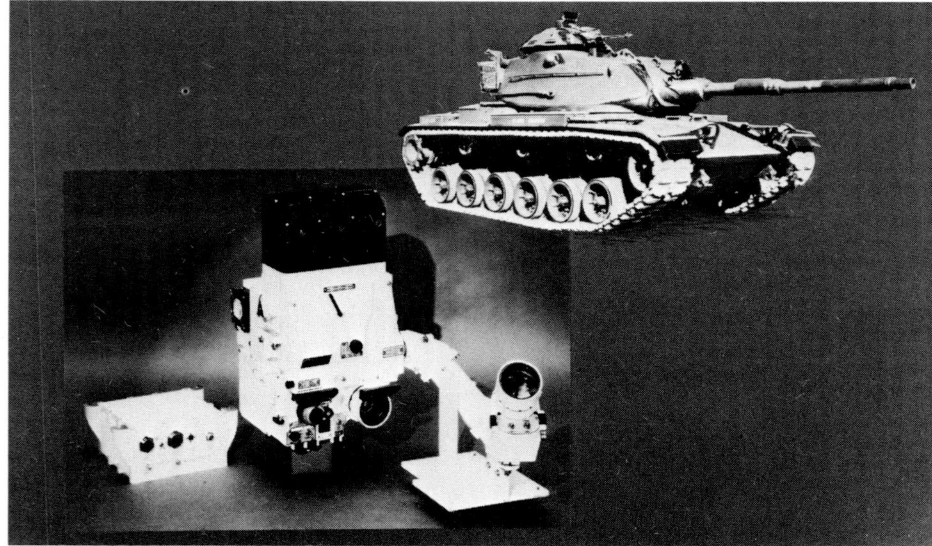

Texas Instruments AN/VSG-2 thermal sight for M60 series

1803.193

AN/VVS-1 TANK LASER SYSTEM

This system was developed by Hughes for the US Army M60A1 main battle tank. It consists of a laser rangefinder integrated with a complete fire control system which includes its own computer and the relevant sensors for ballistic calculations. The project was undertaken in the early 1970s to increase the first-round hit capability of the M60A1's 105 mm gun, against both standing and moving targets. Tanks converted by addition of the new fire control system are designated M60A3.

The target range data is processed in the solid-state computer together with crosswind component, gun trunnion cant, bore wear, air temperature, altitude, and tracking rate to provide the correct azimuth and elevation gun pointing commands values. The computer can provide computation for six types of ammunition, and the system incorporates a self-test mode of operation selectable by the operator.

Kollsman has completed an in-house product improvement of the AN/VVG-2 Laser Rangefinder system. The improved design employs Neodymium Yag technology, utilizing AN/GVS-5 hand-held laser rangefinder modules, to provide improved ranging capability in inclement weather conditions, and an order of magnitude improvement in reliability and maintainability. The improved design incorporates a continuous self-test feature capable of identifying faults down to the lowest replacable unit (LRU). The entire system is packaged entirely within the commander control unit thus eliminating the electronic unit and associated system interconnect cables and resulting in smaller overall packaging of the system.

STATUS

In production with several thousand systems supplied by both Hughes and Kollsman.

In 1985/86 Kollsman supplied a maintenance depot and related test equipment to the Arab Republic of Egypt to support its M60A3 tank fire control systems. Kollsman supports this facility with field service engineering.

Hughes received a production contract from the US Army for an additional 1169 systems in 1985. Options for up to 585 more systems were also included.

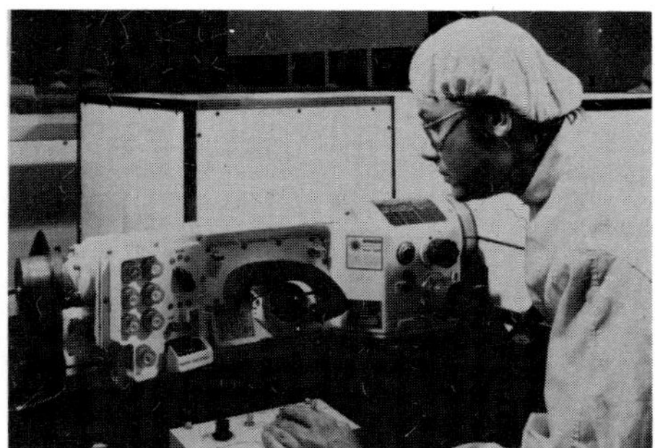

AN/VVS-1 laser rangefinder for M60A1 tank FCS under test at Hughes Aircraft Company laboratory

CONTRACTORS

Hughes Aircraft Company, PO Box 902, El Segundo, California 90245, USA.

Kollsman Military Systems Division, 220 Daniel Webster Highway, Merrimack, New Hampshire 03054, USA.

3048.193

AN/VVS-2 DRIVER'S NIGHT VISION VIEWER

The VVS-2 is a passive night vision device for drivers of AFVs and has been developed for use by the US Army, initially in M60 tanks. Unlike systems such as the active infra-red M19 and M24 drivers' viewers, the VVS-2 image-intensifier equipment is a totally passive system requiring no active illumination, such as IR searchlights, or headlights. The new system is also claimed to offer a 65 per cent greater field of view than earlier devices. Another advantage is the 'bi-ocular' display which allows the driver to view the scene with both eyes through one, large diameter, eyepiece from a distance of several inches. The oval-shaped field of view of 45° horizontal by 38° vertical is presented with unity magnification. All objects located more than four metres from the tank are presented in focus. The head prism protruding from the tank hull is tilted to prevent reflection of enemy searchlights. There are different versions of the AN/VVS-2 viewer for specific vehicles: the AN/VVS-2(V)1A is for the M60A3, the AN/VVS-2(V)2A is for the M1 Abrams tank, and the AN/VVS-2(V)3 is for the M2 and M3 Bradley troop carrier/reconnaissance vehicles. These viewers have the 'internal azimuth rotation mount.' This mount is engineered to provide a water-tight seal, easy viewer installation and removal, and a reliable azimuth rotation. These versions are specified by the US Army as standard items for the designated vehicles and fit directly without modifications.

For other US armoured vehicles such as M60A1, M60A2, M48, M47, M41, M113, and LVTP-7, Baird developed the NDS-2 Model 1924. This viewer is an exact replacement for the M19 and M24 active IR viewers. No vehicle modification is required and the viewer installation can be accomplished in the field. All vehicles that use the NDS-2 model 1924 mount can be field retrofitted with a modified tank hatch insert that will accept the 'internal azimuth rotation mount'.

NDS-2

This is the designation given to an export package version of the VVS-2, of which over 200 have been sold to Israel. It provides a rotatable 45° field of view and incorporates a second generation image intensifier. NDS-2 night driving system packages are configured to suit specific AFV types and consist of a vehicle modification kit and test equipment to support the viewer. Baird has designed NDS-2 packages for both US and foreign vehicles.

CHARACTERISTICS

Field of view: 45 × 38°
Total field coverage: 135° horizontal × 38° vertical
Depth of focus: 4 m – ∞
System resolution: 1.2 mils
System magnification: × 1
Linear distortion: <4%
Image brightness: Variable
Objective lens: f1.09, 33.5 mm focal length
Image intensifier: 25 mm with automatic gain control and bright source protection
Image tube gain: 25 000 minimum
Bi-ocular magnifier: 45°, fixed focus

STATUS

In production. In July 1987 a further 1895 VVS-2 systems were ordered for delivery by June 1991.

CONTRACTOR

Baird Corporation, 125 Middlesex Turnpike, Bedford, Massachusetts 01730, USA.

Three current production versions of Baird Corporation AN/VVS-2(V) night vision driving equipment (left to right): Model AN/VVS-2(V)2A for M1 tank, Model AN/VVS-2(V)1A for M60A3, and Model 1924 which can be mounted on M60A1/2, M48, M47, M41, M113 and LVTP-7 vehicles

3571.193

AN/VVS-501 NIGHT DRIVING/OBSERVATION VIEWER

The VVS-501 has been developed for the Canadian Defence Forces for use by the drivers and commanders of AFVs. Similar to the US Army VVS-2 (**3048.193**), the VVS-501 is a passive night vision device employing the 25 mm micro-channel intensifier. Thus it requires no supplementary illumination such as for active IR viewers and provides variable gain within the field not available in viewers based on first-generation tubes. This latter fact is most important when lights or fires appear in the field of view.

The VVS-501 is mounted by replacing the day vision block of the driver's and commander's stations of the Leopard tank and the identical day vision block of the driver's station of the General Motors Canada AVGP. The entrance assembly interfaces with the Leopard windshield wiper and the eyepiece assembly is identical to the VVS-2. A stowage case is provided for the viewer or day vision block when not in use. Vehicle power or batteries may be used.

CHARACTERISTICS

Field of view: 45 × 38°
Total field coverage: 135° horizontal × 38° vertical
Depth of focus: 4 m – ∞
System magnification: × 1
Linear distortion: <4%
Image brightness: Variable
Objective lens: f1.08, 33.6 mm focal length
Image intensifier: 25 mm with automatic gain control and bright source protection
Image tube gain: 25 000 minimum
Bi-ocular magnifier: 45°, fixed focus

STATUS

Contracts with GLS Division of Krause-Maffei, Munich, West Germany for Leopard 1 and 2 tanks and DSS Canada for AVGP. Item fully integrated into NATO logistic systems.

CONTRACTOR

Baird Corporation, 125 Middlesex Turnpike, Bedford, Massachusetts 01730, USA.

Baird Corporation AN/VVS-501 night driving observation viewer for Leopard 2 tank

---

4155.193

M32E1 MODEL 9880 NIGHT VISION ELBOW

The M32E1 Model 9880 passive elbow provides a visually secure night fire control capability for a number of armoured vehicle types. It is designed as an interchangeable replacement for the infra-red M32 and M36 elbows with no vehicle modification required. The configuration and performance of the elbow is in accordance with the current revision of the US Army specification MIL-E-48441.

The equipment uses a 25 mm second-generation image intensifier tube also produced by Varo. Image tube gain and reticle brightness are manually adjustable to compensate for different levels of ambient lighting. Automatic gain control circuitry is employed to maintain automatically the viewed scene illumination at a constant level during periods of changing light level conditions such as the period from sunset to full darkness. This allows the operator of the sight to use the sight without having to re-adjust the tube gain control every few minutes during this period. The tube features muzzle-flash protection which prevents the tube from being damaged by high intensity short duration flashes of light. The flash protection circuit is designed to recover in time for the observer to see the round hit the target.

CHARACTERISTICS

Optical
Magnification: × 7.2
Field of view: 7.1°
System resolution
At 3 × 10^{-3} ft L: 5.6 lp/mrad
At 3 × 10^{-5} ft L: 2.7 lp/mrad
Objective focal length: 200 mm
Focus range: 50 m – ∞
Eyepiece focal length: 27.9 mm
Eyepiece dioptre focus: + 4 to –4

CONTRACTOR

Varo Inc, Integrated Systems Division, 2201 W Walnut Street, PO Box 461426, Garland, Texas 75046, USA.

4154.193

MODEL 9891 COMMANDER'S PERISCOPE

The Model 9891 commander's periscope was developed as a direct replacement for the TKN-1 used by the commander in the Soviet-designed T55 and T62 tanks to provide both day and night viewing capability. The image intensifier tube features muzzle-flash protection which prevents the tube from being damaged by high intensity short duration flashes of light. The flash protection circuit is designed to recover in time for the observer to see the round hit the target. Automatic brightness control is a standard feature, and permits viewing during the twilight hours of dusk and dawn and protects the tube against bright flashes of light.

The periscope features a stabilised head mirror to provide the commander with improved on-the-move viewing capability. An illuminated projected reticle permits range information from Varo Model 9897 laser rangefinder (**4151.193**) to be displayed in the field of view.

CHARACTERISTICS

Optical
Magnification: × 4
Field of view: 10.5°
Objective focal length: 135 mm
Objective f number: 1.5
Focus range: 25 m – ∞
Eyepiece dioptre focus: + 4 to –4
Eyepiece interpupillary adjustment: 58 ± 5 mm
Electrical power: 18 – 30 V DC
Image intensifier tube
Type: 25 mm micro-channel inverter
Typical resolution: 28 lp/mm
Photocathode response: S-20VR
ABC: Standard

Performance (detection)*	Light Level	
	10^{-1} lux	10^{-3} lux
Man	763 m	448 m
Jeep	1215 m	715 m
Tank	1863 m	1096 m

*Range is calculated based on a 30% contrast target and 3 resolution elements required for detection

System resolution (100% contrast)
0.001 lux: 2.70 lp/mrad
0.1 lux: 3.51 lp/mrad
(30% contrast)
0.001 lux: 1.35 lp/mrad
0.1 lux: 2.43 lp/mrad

CONTRACTOR

Varo Inc, Integrated Systems Division, 2201 W Walnut Street, PO Box 461426, Garland, Texas 75046, USA.

4153.193

MODEL 9892 DRIVER'S VIEWER

The Varo Model 9892 driver's viewer was developed to provide night-time operational capability for a wide range of Soviet-designed armoured vehicles such as the T54, T55 and T62. It complements Varo's Model 9894 gunner's periscope to provide an effective night fire control system capability. The periscope was developed as a direct replacement for the TVN-2 IR driver's viewer and mounts interchangeably with the existing daylight vision block. Two-channel binocular viewing provides maximum stereoscopic viewing for added depth perception and increased user comfort.

CHARACTERISTICS

Optical
Magnification: × 1
Field of view: 40°
Objective focal length: 27 mm
Objective f number: 1.2
Focus adjustment: Fixed (4 m – ∞)
Eyepiece dioptre focus: + 4 to –4
Eyepiece interpupillary adjustment: 58 ± 5 mm
Image intensifier tube
Type: 18 mm micro-channel
Typical resolution: 28 lp/mm
Photocathode response: S-20VR

Performance (detection)*	Light Level	
	10^{-1} lux	10^{-3} lux
Man	200 m	117 m
Jeep	295 m	187 m
Tank	452 m	395 m

*Range is calculated based on a 30% contrast target and 3 resolution elements required for detection

System resolution (100% contrast)
0.001 lux: 0.39 lp/mrad
0.1 lux: 0.59 lp/mrad
(30% contrast)
0.001 lux: 0.25 lp/mrad
0.1 lux: 0.49 lp/mrad

CONTRACTOR

Varo Inc, Integrated Systems Division, 2201 W Walnut Street, PO Box 461426, Garland, Texas 75046, USA.

4150.193

MODEL 9894 GUNNER'S PERISCOPE

The Model 9894 integrated gunner's day/night periscope was developed as a direct replacement for the TPN-1 gunner's periscope used in Soviet-designed T55 and T62 tanks. Installation in these tanks can be accomplished without modification. When integrated with Varo's Model 9895 ballistic computer system (**4156.181**), it provides a full solution fire control system. The major components of the Model 9894 gunner's periscope include: a head/mount assembly with stabilised head mirror for improved viewing and increased ranging accuracy while moving; Model 9880 passive night vision elbow; and a Model 9897 daylight elbow with an integrated laser rangefinder for fast, accurate (±5 m) ranging. Range data is displayed in both day and night fields of view. Apart from the mechanical changes necessary to accommodate the Soviet tank design, this equipment is virtually the same as the Model 9898 (**4149.193**).

CHARACTERISTICS

Magnification: × 7 (night); × 8 (day)
Field of view: 7.5° (night); 8° (day)
Objective focal length: 200 mm (night)
Focus range: 50 m – ∞ (night); fixed (day)
Eyepiece dioptre focus: ±4 (night/day)
Image intensifier tube: 25 mm MCP

Performance (night elbow)*	Light Level	
	10^{-1} lux	10^{-3} lux
Man	1316 m	626 m
Jeep	2100 m	1000 m
Tank	3220 m	1533 m

*Range is calculated based on a 30% contrast target and 3 resolution elements required for detection

CONTRACTOR

Varo Inc, Integrated Systems Division, 2201 W Walnut Street, PO Box 461426, Garland, Texas 75046, USA.

4152.193

MODEL 9896 NIGHT VISION ELBOW

The Model 9896 passive elbow provides fire control capability for armoured vehicles where space is limited and a small compact periscope with high performance is required. It was specifically developed to fulfil the night vision requirements of the M41, M47, and M48 tanks currently using the M20 daylight periscope. The Model 9896 passive elbow can be used with the existing M20 daylight periscope to provide both day and night viewing capability.

The passive elbow features an illuminated projected reticle that is fully adjustable to compensate for varying scene brightness and target contrasts. The standard NATO reticle is provided unless otherwise specified. The tube features muzzle-flash protection which prevents the tube from being damaged by high intensity short duration flashes of light. The flash protection circuit is designed to recover in time for the observer to see the round hit the target. Automatic brightness control is a standard feature. This feature permits viewing during the twilight hours of dusk and dawn and protects the tube against bright flashes of light. The equipment is designed to operate from vehicle power under normal conditions or from a single 2.7 V mercury battery under emergency conditions.

CHARACTERISTICS

Optical

Magnification: × 5.2

Field of view: 10.5°

Objective focal length: 135 mm

Objective f number: 1.5

Focus range: 25 m – ∞

Eyepiece dioptre focus: 2 to +5

Image intensifier tube

Type: 25 mm micro-channel inverter

Typical resolution: 28 lp/mm

Photocathode response: S-20VR

Performance (detection)*	Light Level	
	10^{-1} lux	10^{-3} lux
Man	763 m	448 m
Jeep	1215 m	715 m
Tank	1863 m	1096 m

*Range is calculated based on a 30% contrast target and 3 resolution elements required for detection

CONTRACTOR

Varo Inc, Integrated Systems Division, 2201 W Walnut Street, PO Box 461426, Garland, Texas 75046, USA.

4151.193

MODEL 9897 LASER RANGEFINDER

The Model 9897 daylight elbow/laser rangefinder was developed as a major component of Models 9894 and 9898 (**4150.193** and **4149.193**) gunner's periscopes and is a direct replacement in the US Army's M32 periscope. The elbow features a ×8 visual optical system and a tri-axial laser rangefinder. Range data from the laser rangefinder is automatically displayed in both the day and night elbow fields of view and can be remotely displayed to the commander's periscope.

CHARACTERISTICS

Optical

Magnification: × 8

Field of view: 8°

Range focus: Fixed

Eyepiece dioptre focus: ±4

Laser rangefinder transmitter

Type: Nd:YAG

Beam divergence: ⩽5 mil

Logic: First/last pulse

Power: 1 MW, 6 mJ

Laser rangefinder receiver

Field of view: 1 mil

Minimum range: 400 m

Max range: 10 000 m

Range accuracy: ±5 m

CONTRACTOR

Varo Inc, Integrated Systems Division, 2201 W Walnut Street, PO Box 461426, Garland, Texas 75046, USA.

4149.193

MODEL 9898 GUNNER'S PERISCOPE

The Model 9898 integrated gunner's day/night periscope was developed as a direct replacement for the M32 gunner's periscope used in the M60 and M48 series tanks. It can be used in many other armoured vehicles, such as the M47 and Centurian tanks, by minor modification to the tank. When integrated with Varo's Model 9895 ballistic computer system, it provides a full solution fire control system (**4156.181**). The major components of the Model 9898 gunner's periscope include: the head/mount assembly with stabilised head mirror for improved viewing and increased ranging accuracy while moving; passive night vision elbow (Model 9880); and a Model 9897 daylight elbow with an integrated laser rangefinder for fast, accurate (±5 m) ranging. Range data is displayed in both day and night fields of view.

CHARACTERISTICS

Magnification: × 7 (night); × 8 (day)

Field of view: 7.5° (night); 8° (day)

Objective focal length: 200 mm (night)

Focus range: 50 m – ∞ (night); fixed (day)

Eyepiece dioptre focus: ±4 (night/day)

Image intensifier tube: 25 MCP

Performance (night elbow)*	Light Level	
	10^{-1} lux	10^{-3} lux
Man	1316 m	626 m
Jeep	2100 m	1000 m
Tank	3220 m	1533 m

*Range is calculated based on a 30% contrast target and 3 resolution elements required for detection

CONTRACTOR

Varo Inc, Integrated Systems Division, 2201 W Walnut Street, PO Box 461426, Garland, Texas 75046, USA.

5719.193

KOLLMORGEN ELECTRO-OPTICAL SYSTEMS

The Kollmorgen Corporation manufactures a wide variety of land-based electro-optical systems for infantry and armoured vehicle use. Brief details of the main equipments are as follows:

Model 938 Commander's Weapon Station Sight

This is a ×3 power monocular daylight periscope used as a sighting device for the commander's weapon. The instrument is mounted near, and mechanically linked to, the weapon. The line of sight is capable of movement in the vertical plane from −10 to +65°, via a mechanical link to the weapon. A reticle pattern is presented in the field of view.

CHARACTERISTICS

Magnification: × 3

Field of view: 21°

Entrance pupil: 18 mm

Exit pupil: 6 mm

Model 939 Gunner's Auxiliary Sight

The gunner's auxiliary sight is a sealed and pressurised ×8 power optical device that can be used as a primary or back-up sight for the main coaxial armament. The instrument is mounted coaxially with the main gun and is articulated to minimise eyepiece movement when the main/coax weapon is elevated or depressed. A dioptre adjustment is provided to accommodate focal variations between individual observers. Individual reticle patterns may be presented in the field of view. A switchable ND filter is provided to allow use of the sight during high brightness conditions.

Model 939 gunner's auxiliary sight

CHARACTERISTICS

Magnification: × 8

Field of view: 8°

Entrance pupil: 48 mm

Exit pupil: 6 mm

Model 957 Squad Leader's Search Periscope

This instrument is a sealed single magnification optical device that is used as a target acquisition and vehicle protective surveillance sight for the armoured personnel carrier squad commander. The instrument is mounted on the inside of the top deck of the APC with viewing and control functions internal to the vehicle.

CHARACTERISTICS

Magnification: × 4 minimum

Field of view: 12.5°

Exit pupil: 4.5 mm minimum

Model 998 Fire Control Back-up Sight

This is a ballistic periscope fixed-power daylight monocular sight with a range estimating capability. The sight was designed for sighting and firing the primary 25 mm weapon from the electrically stabilised turret of an M2/M3 tracked vehicle. The sight is purely mechanically linked and will allow either the gunner or the commander secondary gun sighting capability in the event of failure of the primary. The sight has laser protection and an armoured head.

STATUS

All equipments are in production.

CONTRACTOR

Kollmorgen Corporation, Electro-Optical Division, 347 King Street, Northampton, Massachusetts 01060, USA.

6740.193

LIGHTWEIGHT TARGET ACQUISITION/SURVEILLANCE SYSTEM (LTASS)

Designed for use by light ground forces LTASS is to be a low-cost, advanced, line-of-sight surveillance system. It consists of a rapidly erectable mast, the Lockheed Adaptive Modular Payload (LAMP), an electronic sensor package, display and control electronics, a communication system and an auxiliary power unit all mounted in a vehicle.

The sensor package is similar to that designed for the Aquila remotely piloted vehicle with several daylight TV cameras or infra-red imagers. The IR imager has three fields of view; 3.4, 5.1 and 30.6°. The TV camera has a field of view from 5 to 50°. The sensor package is designed so that it can be removed to a place the vehicle cannot reach.

The platform and sensor package weigh 31.75 kg and the auxiliary power unit needs to generate only 28 V.

STATUS

Initial field appraisals have been carried out by the US Army using a HMMWV (High Mobility Multi-purpose Wheeled Vehicle).

CONTRACTOR

Lockheed-Georgia Company, 86 South Cobb Drive, Marietta, Georgia 30063, USA.

4259.193

M61/M61A1 FIRE CONTROL SIGHT

The M61 is a visually-aimed lead computing gyro sight for weapon aiming. The sight is a sealed unit containing a lead computing gyro, range magnet, optical system, synchro transmitters and caging device. The range magnet applies the lead computing precession to the gyro while a separate torque motor applies gravity drop corrections. Synchro transmitters provide angular signals to the fire control computer and range-only radar. The fire control computer also supplies signals to the sight to enable it to generate a lead angle and super-elevation.

The main application of the M61 is as part of the Vulcan air defence system (**2850.131**) and the contractor has developed a similar sight, known as the M61A1, for the PIVADS (**4015.131**) product improved Vulcan air defence system.

CHARACTERISTICS

Magnification: × 1

Lead angle: ±25°

Weight: 15.9 kg

Power: 115 V, 400 Hz; 28 V DC

STATUS

In service with US Army and in the Republic of Korea. More than 800 M61 sights have been manufactured for use in Vulcan and PIVADS. The M61A1 is in use by the US Army on M163A2 self-propelled and M167A2 towed PIVADS.

CONTRACTOR

Fraser-Volpe Corporation, 1025 Thomas Drive, Warminster Industrial Park, Warminster, Pennsylvania 18974, USA.

5717.193

M19A1 DRIVER'S NIGHT PERISCOPE

The M19A1 periscope sight is a sealed unit for night driving. The vehicle is equipped with infra-red or spot lights and the reflected rays are picked up by the periscope. Within the instrument an image converter tube changes the infra-red rays to visible light images which can be seen through the two eyepieces.

The instrument consists of two main parts: the head assembly projecting through the body armour of the vehicle, and the body assembly. The former contains a large prism which directs the incoming infra-red rays down to the body assembly where the objective lens and image converter are situated. The M19A1 has been adapted to all types of vehicle in service with the US Army, including combat tanks, personnel carriers etc. Recent modifications have considerably increased the reliability of the converter tubes.

CHARACTERISTICS

Magnification: × 1

Eyepiece focus: –0.75 to –1 dioptre

Objective focus: ∞

Resolution: 4 minutes arc

Weight: 6.35 kg including head assembly

STATUS

In operational service with the US Army.

CONTRACTOR

Fraser-Volpe Corporation, 1025 Thomas Drive, Warminster Industrial Park, Warminster, Pennsylvania 18974, USA.

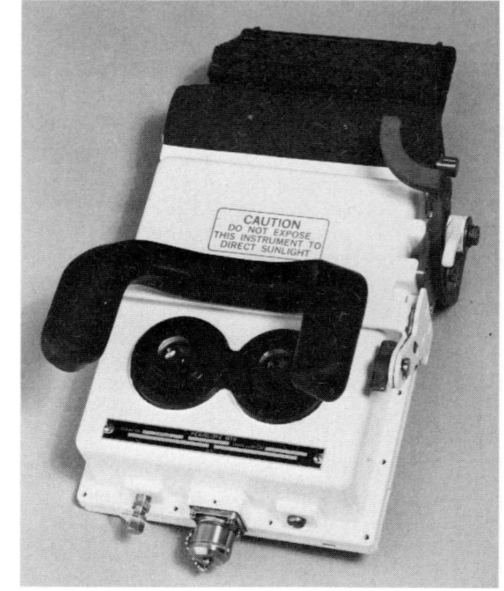

M19A1 night periscope

6359.193

M6043 FIRE CONTROL DIRECTOR SIGHT

The M6043 director sight forms part of an air defence system to acquire and track targets. It is a sealed unit housed in a cast aluminium case that contains gimbal mounted rate integrating gyros and associated circuitry. The gyros provide output signals for rate in two axes. Synchro transmitters located on the gimbal axes provide angular signals, indicating line-of-sight relative to gun line, which are transmitted, to the ballistics computer.

The optical system produces a collimated reticle with a 5 mil centre circle for airborne targets, a concentric 15 mil circle to provide 7.5 mil super-elevation for the ground mode, and an outer 60 mil circle corresponding to radar beam width. A range meter indicates the distance to the target. A ready-to-fire and a radar track/jam light are incorporated.

CHARACTERISTICS

Lead angle: ±25°

Magnification: × 1

Dimensions: 380 × 320 × 290 mm

Weight: 15.9 kg

STATUS

The M6043 sight is standard equipment on the PIVADS air defence sysem (see entry **4015.131**).

CONTRACTOR

Fraser-Volpe Corporation, 1025 Thomas Drive, Warminster Industrial Park, Warminster, Pennsylvania 18974, USA.

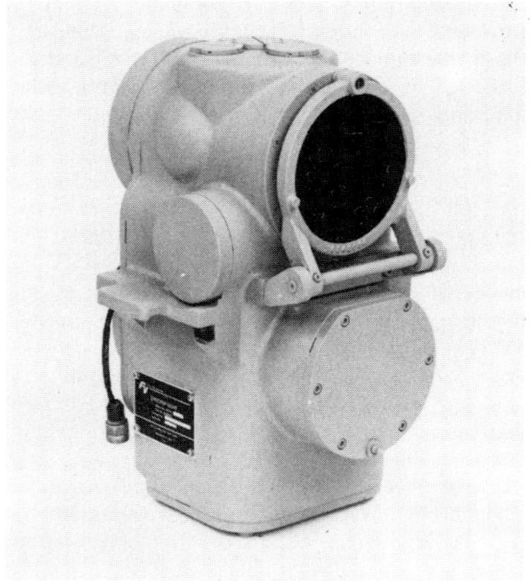

M6043 fire control director sight

3814.193

M1 LASER RANGEFINDER

Hughes Aircraft Company has developed a laser rangefinder for the US Army's M1 Abrams main battle tank, this work proceeding in conjunction with the company's development of a thermal imaging system (**3570.193**) for the M1. Both equipments are integrated into the tank's fire control system. The laser is of the Nd:YAG type but development is proceeding to provide a CO_2 laser in the late 1980s (see under DEVELOPMENT).

The laser rangefinder consists of a transmitter, receiver, power supply, timing and logic circuitry. It includes a 'safe-switch' that allows the entire fire control system to operate without firing the laser, to guarantee safety during training exercises. Built-in test equipment is used to isolate faults to individual sub-assemblies, which speeds and simplifies maintenance. Designed for extreme reliability, the M1 laser system exceeds 1800 hours mean time between failure.

In action, a tank crew member aims at a target and fires the laser. Its beam travels to the target which reflects the beam back to a receiving telescope. Elapsed time of beam travel to and from the target provides accurate range information to the tank fire control computer almost instantaneously. The computer processes this information in milliseconds, together with other data such as crosswind velocity, air temperature, gun trunnion position, air density, target tracking rate and ammunition ballistics. The computer then delivers correct azimuth and elevation firing commands to the target turret and main gun.

A significant development is a change from a pink ruby as the lasing medium to the more recently developed neodymium-doped yttrium aluminium garnet (Nd:YAG) rods. Not only are the latter significantly smaller and lighter, but their power consumption is lower than ruby lasers because they require less energy to produce a laser beam.

The system is capable of distinguishing between two targets separated by as little as 15 m in range. It also allows the fire control system to differentiate between close-up and far-off targets, by allowing the gunner to select the first reflection (close target) or last reflection (far target).

DEVELOPMENT

Hughes has been funded to carry out development work on a CO_2 laser system for the M1 tank. The first system, using a GEC Avionics Mk III TEA laser was delivered to the US Army in 1979. Development continued and in 1984, advanced development

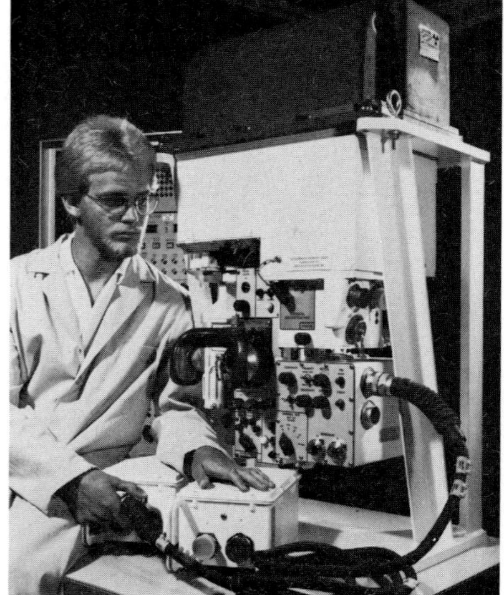

M1 Abrams tank gunner's primary sight incorporating integrated laser rangefinder and thermal imaging systems produced by Hughes

models were produced and evaluated. Full-scale development was initiated in 1985 and development has continued through 1986 and 1987. The CO_2 rangefinder is being developed as part of an overall improvement package for the M1 tank. Production was due to be initiated in late 1987.

CHARACTERISTICS

Laser: Nd:YAG
Range: 200 – 7990 m
Range accuracy: ±10 m
Firing rate: 30 pulses/minute
Power supply: 18 – 30 V DC
Weight: 11.3 kg

STATUS

The laser rangefinder programme started in February 1978 and deliveries began in the latter half of 1979. Several thousand rangefinders and thermal imagers had been delivered by early 1987, and both are still in full production.

CONTRACTOR

Hughes Aircraft Company, Electro-Optical & Data Systems Group, PO Box 902, El Segundo, California 90245, USA.

3570.193

AN/VSG-X(TIS) THERMAL IMAGING SYSTEM

Hughes Aircraft Company developed and is producing the thermal imaging system (TIS) to enable the crews of the new US Army M1 Abrams tank to see through darkness, smoke or haze. It produces an image by sensing the small differences in infra-red heat radiated by objects within view and converting the detected energy into electrical signals which can be displayed on a cathode ray tube similar to a normal television picture. The image displayed is projected into the eyepiece of the M1 tank's gunner.

In addition, the sight gives the gunner target range information and indicates the existence of more than one target – data which is received from the M1 laser rangefinder (also a Hughes product). Ready-to-fire indication and confirmation that systems are working properly are also provided.

The thermal imaging system generates cross-hairs bore-sighted to the laser rangefinder, as well as computer symbology. This allows the gunner to operate the TIS just as he would the day sight.

The infra-red sight will be housed under armour on the M1 to help it withstand the rigours of the tank environment. However the same system can be packaged for use in lighter vehicles, due to flexibility which allows use of multiple or remote displays. This feature results from replacement of the LED direct-viewing display, commonly used in infra-red imaging systems, with a cathode ray tube.

STATUS

M1 sight development began in late 1976. The first prototype system was delivered nine months after award of the contract. The system is in full production and, by early 1987, several thousand M1 laser rangefinders and thermal imagers had been delivered under contracts with the prime contractor, General Dynamics Corporation and the US Army. A five year follow-on production contract for 3299 TIS was awarded to Hughes Aircraft in 1987.

CONTRACTOR

Hughes Aircraft Company, Electro-Optical & Data Systems Group, PO Box 902, El Segundo, California 90245, USA.

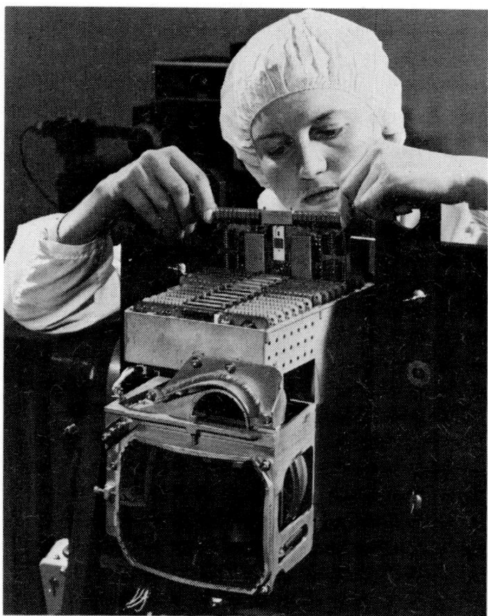

Replacing electronic circuit card of receiver unit of Hughes thermal imaging system for M1 Abrams tank

---

6277.153

M60A3 LASER TANK FIRE CONTROL SYSTEM

The M60A3 laser tank fire control system has been developed by Hughes to improve the effectiveness of the M60A3 battle tank both by day and by night. It consists of two systems, a laser/sight and a computer, and has a dual capability for both commander and gunner.

The laser/sight subsystem is composed of two main units, a commander's integrated laser/sight with control unit, and laser electronics unit. Range may be fed automatically to the computer or, if more than one return is received, the commander may select the range return based on his assessment of the situation. Since the gunner's sight is boresighted with the laser rangefinder, either the gunner or the commander can fire the laser and/or the gun.

The computer subsystem includes the ammunition select sensors. The commander or gunner may select one of the four basic types of ammunition to be used. The computer stores solutions for six ammunitions within the four basic types, to accommodate ammunitions stored in different geographical locations. A switch on the ammunition select unit is used to set up either tank moving or stationary conditions. Automatic sensors feed data directly into the computer. These are a rate unit for use against moving targets, a crosswind sensor, and a gun trunnion cant angle unit.

CONTRACTOR

Hughes Aircraft Company, Electro-Optical & Data Systems Group, PO Box 902, El Segundo, California 90245, USA.

5912.193

ROKIT TANK SIGHT

A new tank gunner's sight which is based largely on the thermal imager (**3570.193**) and laser rangefinder (**3814.193**) designed for the M1 Abrams tank and the West German Leopard 2 has been developed by Hughes for the XK-1 Republic of Korea Indigenous Tank (ROKIT). The sight is stabilised in both azimuth and elevation to enable the gunner to shoot on the move. Also included in the sight are the associated electronics and a vertical sensor. The commander's sight is supplied by SFIM, France.

STATUS

The XK-1 tank is in production and in service.

CONTRACTOR

Hughes Aircraft Company, Electro-Optical & Data Systems Group, PO Box 902, El Segundo, California 90245, USA.

3051.193

AN/TVQ-2 GROUND/VEHICULAR LASER LOCATOR DESIGNATOR (G/VLLD)

Hughes Aircraft Company is producing the AN/TVQ-2 ground/vehicular laser locator designator (G/VLLD) for the US Army Missile Command. It is designed for use by ground-based forward observers to enable them to direct laser-guided missiles, bombs, or artillery shells on to targets designated by the ground forces.

The equipment is a lightweight, tripod-mounted unit containing a laser designator, a laser rangefinder, batteries, a tripod and a viscous damped tracking unit. The rangefinder/designator weighs 12.7 kg, the total system weight is 23.6 kg. It is man-portable in the field by a crew of two, only one of whom is needed to operate the equipment. It can also be vehicle-mounted.

The operational procedure is to use the telescopic sight to scan and search for 'hard' targets such as tanks or bunkers. When a suitable target has been identified, the laser is used to obtain range and bearing data. In co-operation with units deploying laser guided weapons to whom the target has been assigned, the laser is then used to mark the target at the desired impact point to provide guidance for the bomb, missile, or other munition. Provision is made for coding the marking laser energy in wavelength and PRF to permit weapons to differentiate between G/VLLDs. The system is designed to operate for a period exceeding 600 hours.

STATUS

In production since spring 1979 under an $83 million contract and in service with the US Army with first deliveries late in 1981. All systems are due to be delivered to the US Army by 1988.

CONTRACTOR

Hughes Aircraft Company, Electro-Optical & Data Systems Group, PO Box 902, El Segundo, California 90245, USA.

AN/TVQ-2 ground/vehicular laser locator designator. Picture shows tripod-mounted version but combined laser designator and rangefinder can also be mounted on M113 fire support team armoured vehicles

3157.193

AN/PAQ-1 LASER TARGET DESIGNATOR

The AN/PAQ-1 laser target designator (LTD) is a laser target marker developed as a hand-held instrument for use by ground troops. In addition to its use for target designation to supporting ground attack aircraft and helicopters, the LTD is seen as suitable for such applications as a means of making their position known to rescue or supply aircraft by troops who may have been cut off by opposing ground forces.

The equipment is also compatible with the AN/TVS-5 night vision sight, and so may be used in darkness as well as by day.

The LTD consists of three easily-replaceable modules designed to withstand rough field handling. The transmitter module is a sealed unit which is dry nitrogen purged to provide a clean environment for the laser and optical elements. It can be easily detached without disturbing the optical alignment or the seal.

The electronics module houses the power supply and control circuitry, code-set panel, external cooling air blower and ducting, heat exchanger and trigger assembly.

A quick detachable battery module completes the basic unit. It is completely sealed and shaped to function as part of the rifle stock, forming the butt end.

When the LTD is operated from an auxiliary power source, the power is supplied through the battery interface connectors, terminated in a module identical to the battery module so that the overall LTD configuration is maintained.

No technical details of the LTD have been released.

STATUS

Production is now complete and 377 sets have been delivered to the US armed forces (352 to the Army and 25 to the USAF).

CONTRACTOR

Hughes Aircraft Company, Electro-Optical & Data Systems Group, PO Box 902, El Segundo, California 90245, USA.

AN/PAQ-1 laser target designator marking target for aircraft equipped with Pave Penny laser tracker during USAF tests

3816.193

AN/PAQ-3 MODULAR UNIVERSAL LASER EQUIPMENT (MULE)

The AN/PAQ-3 MULE is based on the AN/PAQ-1 laser target designator (**3157.193**), AN/GVS-5 hand-held laser rangefinder (**3200.193**), plus tripod design concepts from the AN/TVQ-2 ground laser locator designator (**3051.193**). This results in significant cost of ownership savings over the life of the equipment, since a large percentage of MULE components are interchangeable, directly, with items from the other programmes. The laser transmitter and electronics from the AN/PAQ-1 are used, with a change in output optics to meet different beam characteristic needs. Rangefinder components carried over from the AN/GVS-5 include detector, video amplifier, range counter/display and low voltage power supply. MULE features built-in test circuits that monitor its own operation automatically. They warn the operator when the battery pack voltage is getting low, when the laser output falls below the laser threshold and certifies other functions. Weighing 17.23 kg, the system consists of laser designator/rangefinder and stabilised tracking tripod modules. MULE can be carried readily by two-man forward observer teams over rough terrain into battle areas and set up for use in less than five minutes. For instant target ranging, an observer trains MULE on a target manually, using its telescopic sight for precise aim. He fires a laser beam and MULE measures the time it takes the beam to reach the target and return. The system's computer multiplies half the elapsed time by the speed of light and displays the range immediately on a digital readout. The tripod automatically determines and displays azimuth and elevation co-ordinates. This information, along with range, can be sent quickly in digital form to a fire control centre via a digital communication terminal, or relayed to an artillery battery by voice channel. To illuminate targets for laser guided weapons, the observer simply trains his sights on the target and triggers the laser beam. Laser-homing missiles, projectiles (and aircraft with laser spot trackers) follow the reflected beam to the target.

Hughes modular universal laser equipment (MULE) deployed in the field

Sperry has developed a new lightweight, low cost gyrocompass for use in the MULE system, and known as a north finding module (NFM). It is an automatic gyrocompass with digital readout which uses a micro-computer to determine automatically the angle from true north without pre-orientation or insertion of latitude. The NFM will provide precise azimuth data on targets and this unit is also being evaluated for use with the US Army's AN/TVQ-2 ground/vehicular laser locator designator (G/VLLD) (**3051.193**), the Hughes Firefinder radars (**2848.153** and **1976.153**), and other systems.

The NFM weighs 1.8 kg and provides an accuracy of two mils within two minutes and 0.3 mils within ten minutes. There are LED and digital outputs.

STATUS

MULE is in production under a contract for the US Marine Corps. The contract was completed in 1987. Hughes are to deliver 37 transmitters for MULE by September 1989.

CONTRACTOR

Hughes Aircraft Company, Electro-Optical & Data Systems Group, PO Box 902, El Segundo, California 90245, USA.

5910.193

AN/VAS-3 DRIVER'S THERMAL VIEWER

The AN/VAS-3 is a periscopic-based thermal imaging system operating in the 7.5 to 12 μm spectrum, configured as a bi-ocular display driver's viewer. It is designed for closed-hatch combat vehicle driving and provides the driver with the means to continue during darkness and battlefield conditions of degraded visibility, such as smoke, haze, fog, etc. IR energy entering through the IR optics of the periscopic sight is mechanically scanned to obtain a rectangular field-of-view. An array of cooled mercury cadmium telluride detectors photoconductively transduce the scanned IR radiation into electrical signals. The signals are multiplexed and processed into visible light by means of a CRT, and then presented into the bi-ocular eyepiece in real-time.

CHARACTERISTICS

Spectral band: 7.5 – 12 μm

Field-of-view: Azimuth 40°, elevation 20°

Magnification: × 1

Size: 450 inches^3 (7374.18 cm^3)

Weight: 11.4 kg

STATUS

Three proof-of-design models and 17 engineering development models are being delivered, under a contract awarded by the US Army, for evaluation and trials. A production decision was anticipated late in FY 1987-88.

CONTRACTOR

Hughes Aircraft Company, Electro-Optical & Data Systems Group, PO Box 902, El Segundo, California 90245, USA.

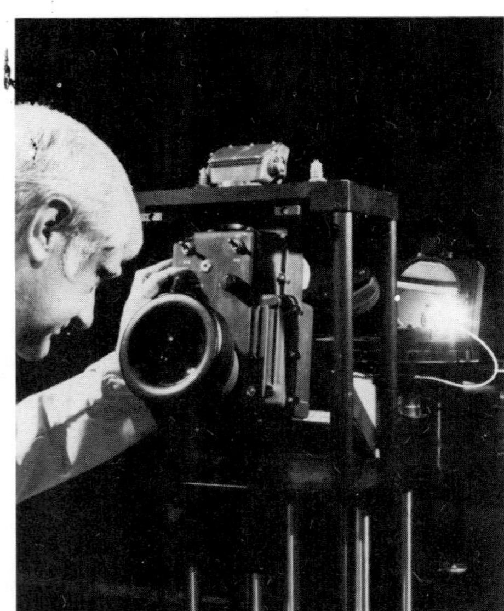

Prototype model of the AN/VAS-3 driver's thermal viewer

5911.193
AN/PVS-6 MELIOS HAND-HELD LASER RANGEFINDER

Melios is a hand-held, eye-safe Raman shifted neodymium-YAG laser rangefinder operating at a wavelength of 1.54 microns, and is approximately the size of a pair of binoculars. It is designed as a low cost unit to determine ranges from 80 to 4000 metres and has a field of view of 7.5°. Melios is capable of continuous operation at 20 pulses per minute, with higher burst availability, and will operate in both first and last reply modes. It is battery powered and will fire 400 rangefinding pulses before needing to be recharged. The complete system consists of the rangefinder, a tripod, battery and a carrying case.

STATUS

Advanced development contracts were issued to Hughes Aircraft Company and Optic Electronic to produce eight models each. A production decision is scheduled for 1988. Development testing and the second phase of operational testing is scheduled to begin in FY 1988 and to be completed by FY 1989.

CONTRACTORS

Hughes Aircraft Company, Electro-Optical & Data Systems Group, PO Box 902, El Segundo, California 90245, USA.

Optic Electronic, Dallas, Texas, USA.

3823.093
MCTNS MAN-PORTABLE COMMON THERMAL NIGHT SIGHTS

Under US Army sponsorship, American manufacturers have evolved a range of basic elements for night sight equipments used with man-portable weapon systems, such as TOW and Dragon. This system of modular equipment is formally known as the man-portable common thermal night sights family of units. Texas Instruments has designed the basic receiver, based on common modules.

Man-portable common modules are assembled into a chassis to become the basic man-portable receiver. This receiver provides the essential FLIR functions for all of the man-portable night sight equipments.

The AN/TAS-4 sight (**3824.193**) is used on the TOW weapon system and the ground/vehicular laser locator designator (G/VLLD) (**3051.193**) system. For these applications, the basic receiver is inserted into a combination case, with a focal telescope and a mount. The basic receiver is inverted and used with a different case (afocal telescope) and mount to provide an AN/TAS-5 (Dragon) night sight (**3825.193**). With its optical display removed, the basic receiver is used with an afocal telescope, cover, mount, and bi-ocular/image intensifier to provide an AN/TAS-6 night observation device, long range (NODLR) (**3826.193**).

The FLIR common modules are the foundation of the commonality concept exemplified by the man-portable sights, but they are also applicable to many other uses for army, air force and navy systems and equipment. The common modules are assembled into the receiver to convert an image formed in the infra-red (IR) spectral band into a visible image displayed to the operator. The modules are:

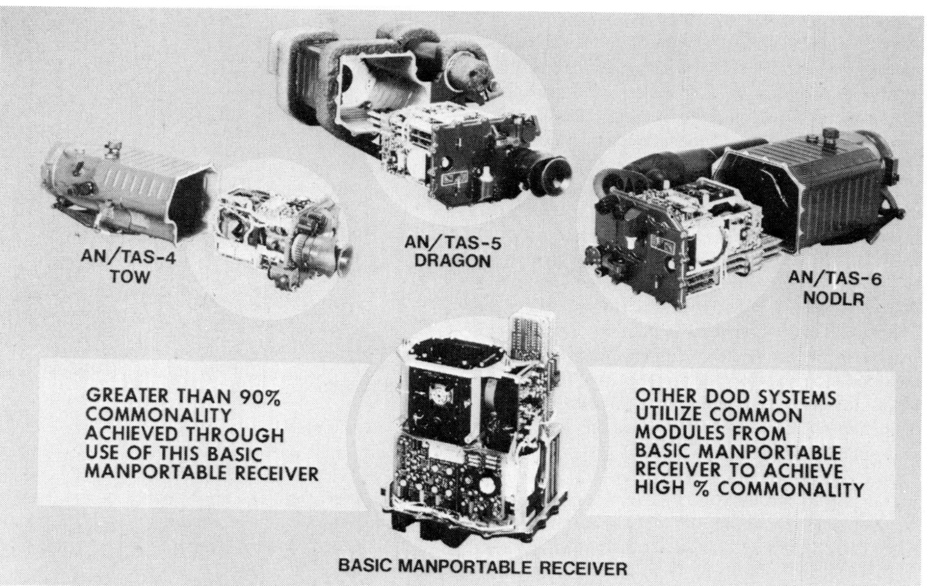

Examples of US Army's range of man-portable common modules employed in forming these thermal imaging night sights

Mechanical scanner: collects IR energy emitted from a scene or object, and displays visible energy from the LED array to the operator.

IR imager: focuses energy collected by the scanner and transfers it to the detector/dewar module.

Detector/dewar: detects IR energy from the imager and processes the signals into electronic energy for the amplifiers. This is a 60-element array, 120 or 180 element arrays being available for other applications.

Preamplifiers/postamplifiers: boost the low level electrical signals from the detector array to a level sufficient to drive the LEDs.

LED array: changes the amplified detector signals into a visible image.

Visual collimator: projects the light from the LED display to the operator.

Video auxiliary control: provides the interface function between the controls and the postamplifiers for such functions as brightness and contrast.

30 Hz scan/interface: provides electronic control for the scanner.

Among the non-man-portable sight applications is the AN/VGS-2 tank thermal sight for the M60 MBT which is now in series production by Texas Instruments; for the USN, the AN/AAS-36 IR detecting set uses TI IR technology, as does the USAF's AN/AAQ-9 FLIR equipment. (Both of the latter are airborne equipments.)

STATUS

Series production by at least two sources (see below) and probably licensed overseas. In February 1983, a contract worth over $24 million was awarded to Kollsman for 300 SU-108 man-portable receivers and 1000 AN/TAS-4 sights for the US Army. Delivery commenced in early 1984. By August 1987 Texas Instruments revealed they had delivered more than 20 000 common module systems to domestic and foreign customers.

CONTRACTORS

Texas Instruments Inc, Electro-Optical Division, PO Box 226015, Dallas, Texas 75266, USA.

Kollsman Military Systems Division, 220 Daniel Webster Highway South, Merrimack, New Hampshire 03054, USA.

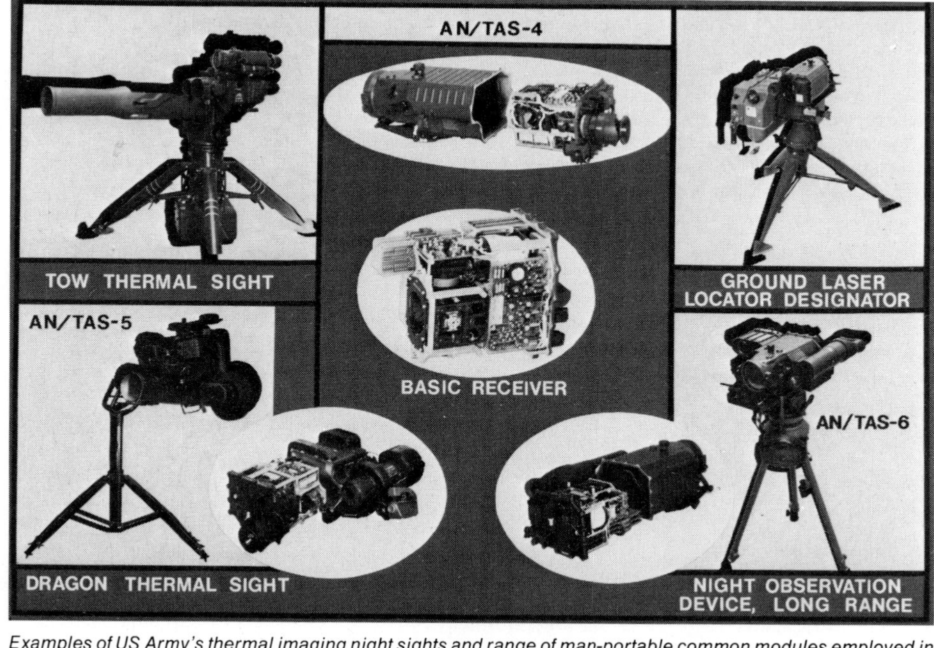

Examples of US Army's thermal imaging night sights and range of man-portable common modules employed in forming these equipments

3824.193
AN/TAS-4 TOW NIGHT SIGHT

The AN/TAS-4 night sight provides a night operating capability for the TOW weapon system. The night sight can be rapidly mounted on and removed from the TOW weapon system by simple movement of the interface locking handle. This quick mount/dismount capability allows the soldier to adapt his TOW weapon system to night and degraded visibility conditions rapidly and without modification to existing hardware. The sight is based on the basic man-portable IR receiver (see **3823.093**, US Army designation SU-108/TAS).

A gas bottle provides cryogenic cooling of the detector. The use of compressed air eliminates hazards associated with more exotic coolants. This bottle will provide two hours of system operation and can be replaced rapidly.

System electrical power is provided by a rechargeable battery package which will power the night sight for two hours and is also quickly replaceable. Combining the consumable elements of the AN/TAS-4 into small self-contained, replenishable components allows the night sight to be truly man-portable and virtually self-contained within a weight of less than 7.2 kg.

The AN/TAS-4 when mounted, provides an additional eyeport for the gunner to detect, identify, track and classify targets directly. Since the night sight detects heat energy emissions, it does not depend upon visible light and can 'see' in degraded visual conditions such as smoke, haze, darkness, camouflage, and foliage.

The AN/TAS-4 has been designed and tested to meet the environments encountered by the TOW weapon system and places no special operating constraints upon the use of the TOW.

DEVELOPMENT

Texas Instruments (TI) received an Engineering Development Programme (Phase 1) contract from the Night Vision Laboratory in 1973, to provide a FLIR for the TOW weapon system. This equipment was to extend the operating capability of the system to the night environment. Using the 'common-module concept' (developed by TI in conjunction with the Department of Defense), a lightweight (man-portable) self-contained far IR device was developed. Phase 1 of the programme culminated in a Research Development Acceptance Test in which the TI system was field tested in competition. Success in these tests resulted in the Phase II contract award to TI in 1975.

In 1981, Texas Instruments was placed under contract by the US Army to develop improved operational capabilities to accommodate the TOW 2 missile system. This initially involved developing the revisions necessary to provide the night sight to the TOW 2 missile. A subsequent improvement was the elimination of the gas bottle cooling through use of a closed cycle cooler. This improvement has been included in the same configuration. The TOW 2 night sight, designated AN/TAS-4A, is in production with Kollsman Military Systems Division and Texas Instruments.

AN/TAS-4A night sight for TOW 2 missile system

STATUS

Production. In February 1983, Kollsman was awarded a $24 million contract by the US Army for the production of almost 1000 AN/TAS-4 sights and over 300 SU-108 receivers. Delivery commenced early 1984. In September 1985, Kollsman was awarded a sole source contract by the US Army Missile Command for 775 AN/TAS-4A TOW 2 night vision sights. A new contract and options in 1986 raised the total quantity to 957. These sets include the AN/TAS-4A IR sight and ancillary equipment. In 1986 Kollsman was awarded contracts totalling $55 million for production of the most recent version of the TOW night sight, the AN/TAS-4C.

Early in 1987 Kollsman announced an improved TAS-4 with Kollsman 60-element digital scan convertor. This generates a sharper image than the standard CCD video FLIR units as well as showing greater sensitivity to lower heat levels.

CONTRACTORS

Texas Instruments Inc, Electro-Optical Division, PO Box 226015, Dallas, Texas 75266, USA.

Kollsman Military Systems Division, 220 Daniel Webster Highway South, Merrimack, New Hampshire 03054, USA.

3826.193
AN/TAS-6 (AN/UAS-11) NIGHT OBSERVATION DEVICE, LONG RANGE

The AN/TAS-6 night observation device, long range (NODLR) provides night observation post capability to the soldier on the battlefield. This night sight is provided with a tripod, with no need for tools for mounting and dismounting. When used in conjunction with the AN/GVS-5 hand-held laser rangefinder (**3200.193**) this night sight can provide information to supporting artillery in degraded battlefield conditions. The complete system is known as the AN/UAS-11.

The AN/TAS-6, in conjunction with its tripod, the AN/GVS-5 hand-held laser rangefinder, and a boresight mechanism, can provide a two-man team with a versatile man-portable forward observation post, which allows operating capabilities in the presence of degraded conditions such as smoke, haze, darkness, camouflage, and foliage.

The AN/TAS-6 is based on the basic man-portable IR receiver (**3823.093**) and uses the same compressed air gas bottle and rechargeable battery pack as the other man-portable systems (AN/TAS-4, AN/TAS-5). This allows the use of the same logistics supply facilities, thereby increasing support flexibility.

The AN/TAS-6 and the AN/GVS-5 are mounted together with the AN/TAS-6 at the left. The bi-ocular eyepiece is covered with a dual eye shroud which provides visual security for the display. This display is green in colour and is produced by an image intensifier tube.

STATUS

Approximately 7000 sights have been procured over a 10 year programme. Series production for the US Army has probably been completed.

CONTRACTORS

Texas Instruments Inc, Electro-Optical Division, PO Box 226015, Dallas, Texas 75266, USA. (AN/TAS-6).

Kollsman Military Systems Division, 220 Daniel Webster Highway South, Merrimack, New Hampshire 03054, USA.

6362.193
TIMS

Texas Instruments have introduced a new thermal imaging multisensor (TIMS) stabilised sight suitable for surveillance and fire control used on armoured fighting vehicles or naval craft. The system provides passive day or night automatic tracking and fire control solutions for a variety of applications including air defence, surveillance, and target ranging and tracking. TIMS is fully stabilised and integrates a high-resolution thermal imaging FLIR with CCD television, a carbon-dioxide laser, and an automatic tracker, and this sensor package can be turret-mounted or remotely located aboard ship. Stabilisation is in two axes and the sensor provides hemispherical coverage. A digital processor is used for automatic tracking. A remote control/display console is provided for the operator.

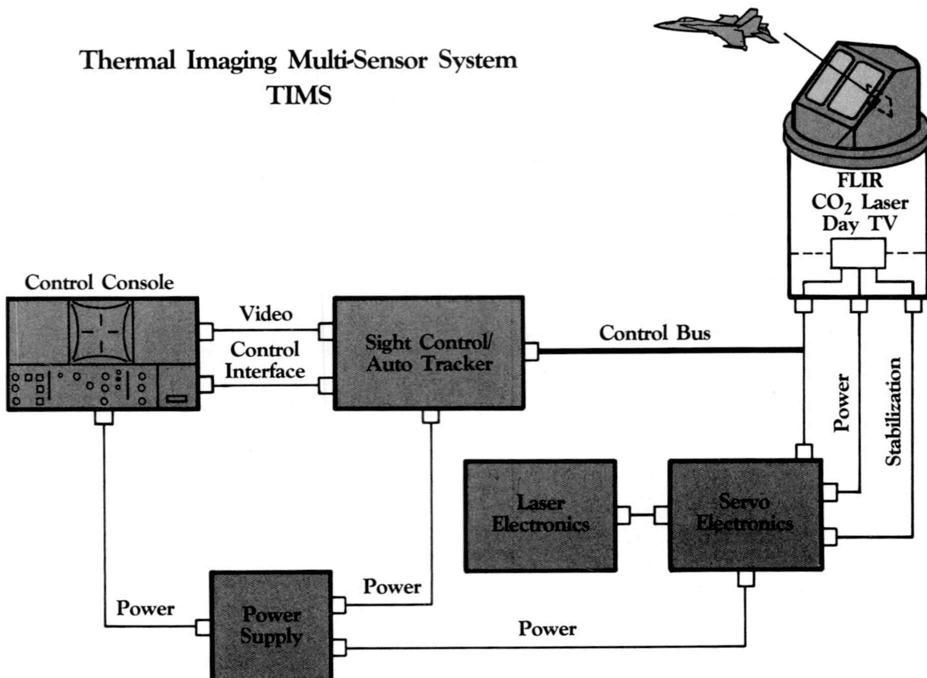

Texas Instruments TIMS sensor unit assembly

CHARACTERISTICS
FLIR
Wide field of view: 1/2 mrad, 2.3° × 3.1°, f/2.0
Narrow field of view: 1/6 mrad, 6.9° × 9.2°, f/1.9
Field of regard: –20° to +75° elevation; 360° azimuth
TV: High resolution CCD (245 × 780 pixels)
Field of view: Approximates to that of FLIR
Laser: CO_2, FLIR integrated
Range: 10 km
CONTRACTOR
Texas Instruments Inc, Electro-Optical Division, PO Box 226015, Dallas, Texas 75266, USA.

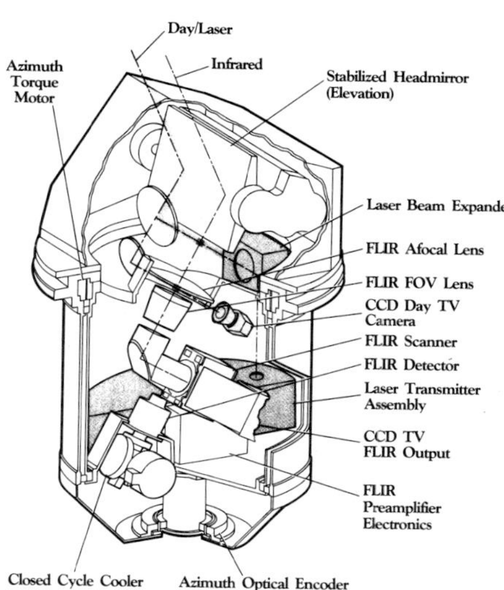

Schematic diagram of Texas Instruments TIMS system

4240.183
TAS TRACKING ADJUNCT SYSTEM (HAWK)

The Tracking Adjunct System (TAS) is an electro-optical tracking system developed as a product improvement for the Hawk surface-to-air missile (**4177.131**) air defence fire control system. It was derived from the TISEO (target identification system, electro-optical) equipment (**1806.363**) which is currently in service and being used by the USAF and a number of foreign countries. TAS provides a passive tracking capability to the Hawk fire control system radar with remote real-time video presentation. It is designed as a complement to the tracking radar. Optical tracking can be coincident or independent of the radar line-of-sight. Manual, automatic acquisition and tracking, rate memory and preferential illumination are key features of the system.

The system also enhances the tracking radar by providing facilities for positive identification, threat assessment, kill assessment, spectre recognition, tracking with minimum doppler, silent tracking and independent tracking. These capabilities represent a positive advantage in a countermeasures environment. More details of the Improved Hawk missile system and its relationship with TAS, along with illustrations of items of TAS hardware, will be found in the entry for that system (**4177.131**) in the Systems Section of this volume. Additional information on the basic TISEO equipment is contained in the Electro-Optical (Air) pages of this section, which follow.

STATUS
Development of the Tracking Adjunct System began in 1976 and production for the US Army and Marine Corps was started in March 1980. The system is being delivered to Hawk users under the Phase II product improvement programme.

CONTRACTOR
Northrop Corporation, Electro-Mechanical Division, 500 East Orangethorpe Avenue, Anaheim, California 92801, USA.

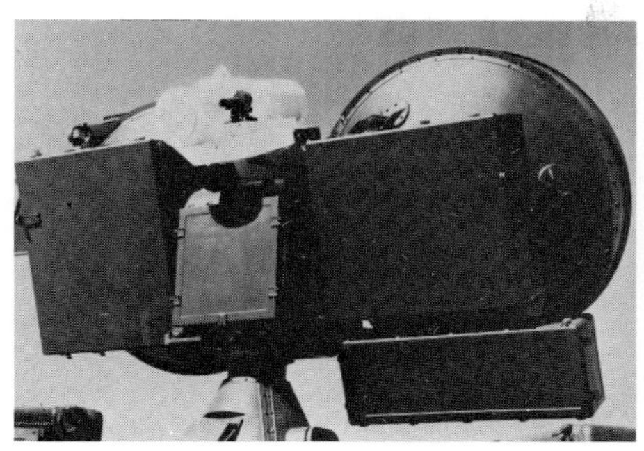

TAS sensor unit mounted with associated electronics unit on Hawk high power illuminator radar, between the two dishes

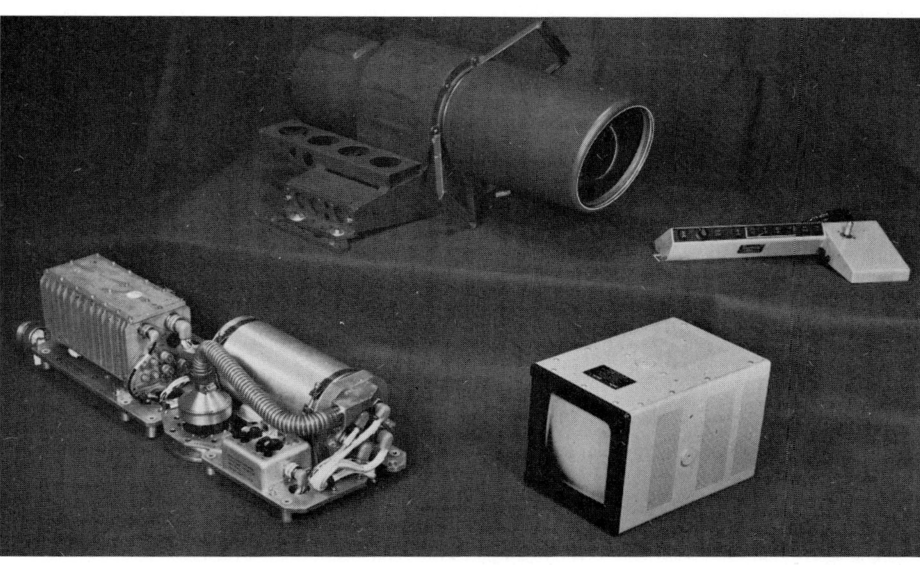

Hardware of Northrop Tracking Adjunct System used with Hawk surface-to-air missile system

6358.193
HONEYWELL AFV THERMAL SIGHTS

Honeywell manufactures a number of thermal imaging systems for a variety of combat vehicles, including M60A1, M60A3, M48A5, Centurion and T-tanks. Principal products include:

Gunner's thermal sight
This is a day/night thermal imaging system operating in the 8 to 12μm waveband. The sight is compatible with mechanical, digital or analogue fire control systems, and will fit within M32 or M35 periscopic envelopes without modification to the tank armour. It provides high performance in the field and is lightweight and ruggedly constructed for use in the combat vehicle environment. The sight can be installed on a variety of vehicles, including M60A1, M60A3, M48A5 and Centurion tanks. It is in production for international applications.

Honeywell International Thermal Sight (HITS)
HITS is a low cost, modular system for 24-hour target acquisition and surveillance on the battlefield. It consists of an 8 to 12 μm thermal imaging system for retrofitting and upgrading older tanks, and may be used as either a commander's or gunner's sight. The sight has been successfully installed and field-tested on Leopard 1 and T-type tanks.

CONTRACTOR
Honeywell Aerospace and Defense, Electro-Optics Division, 2 Forbes Road, Lexington, Massachusetts 02173, USA.

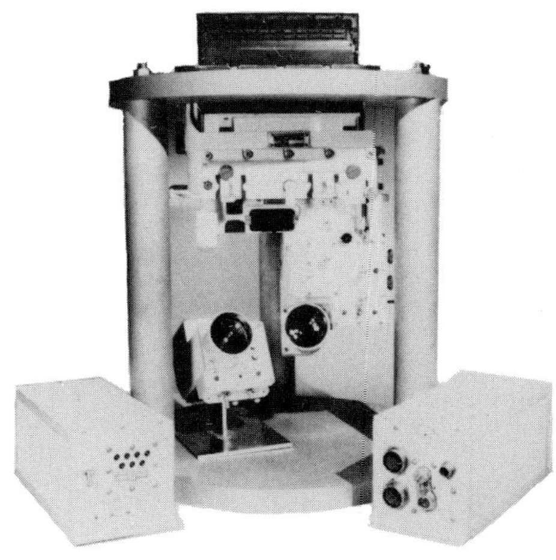

Honeywell gunner's thermal sight

6539.193
FOG-M MISSILE GUIDANCE SYSTEM

FOG-M is a fibre-optic cable which performs much the same function as more conventional wire guidance, but is employed from a position outside the line-of-sight. A TV sensor in the missile relays battlefield information to an unexposed observer through the fibre-optic cable at distances up to 10 kms. Upon target acquisition the missile can be guided manually or through initiation of an automatic tracker.

The data link consists of a light emitting diode or injection laser at the launch end which feeds data through the fibre-optic cable to the missile, to be recovered by a photodiode. A bi-directional coupler allows reception of commands from the launcher as well as transmission of the video image from the missile seeker to the launch station. Tests so far have established that the cable can be paid out at a speed of about 600 ft/sec, considerably faster than the speed of a TOW missile. The system would allow the missile to be launched vertically from behind obstructions before moving to level flight. The operator could search for a target by viewing through the seeker head and, once the target had been acquired, the missile could be locked on and automatically move in on its target.

A particularly effective mode is incorporation of a feature so that the missile goes into a vertical dive while immediately over its target, thereby striking the thin roof armour. Such a system would allow a terminal kill assessment immediately prior to impact to confirm whether another missile is required.

STATUS
In advanced development.

CONTRACTORS
A number of manufacturers and US Army agencies are involved in this project. The manufacturers include the Southern Research Institute, Hughes Aircraft Company, ITT, AT&T, Corbett Technology, Nichols Research and Optelcom.

5721.193

SECO ELECTRO-OPTICAL SYSTEMS

The Standard Equipment Company (SECO) produces a wide range of laser, thermal imaging and electro-optical devices for battlefield use. Brief details of the company's systems are as follows:

SECO Vision 220

This is a portable thermal imaging system comprising an infra-red thermal viewer and a video interface unit. It features four-stage thermo-electric cooling and provides real-time two dimensional images to a standard TV or composite video in 525 or 625 lines.

AD-NDV-2 (AN/VVS-2) Night Vision Driver's Viewer

This is a multi-vehicle passive night vision surveillance and reconnaissance sight which can be mounted in most types of armoured vehicles. It includes a 25 mm second-generation image intensifier with a large diameter binocular display.

CHARACTERISTICS

Magnification: × 1

Field of view: 44° horizontal; 35° vertical

Field coverage: 135° horizontal; 35° vertical

V-NOD-2 Night Observation Device

The V-NOD-2 is a tripod-mounted night vision sight used for long-range surveillance. It uses a 25 mm micro-channel plate image intensifier tube to allow viewing during the twilight hours of dawn and dusk such that a man can be distinguished at ranges in excess of 1000 m in light levels of 10^{-3} lux. A laser rangefinder can be supplied as an optional extra.

CHARACTERISTICS

Magnification (optical): × 9.4

Field of view: 5.6°

Performance: At 10^{-3} lux a man can be distinguished at more than 100 m, a Jeep at more than 1700 m and a tank at 4000 m.

M1A1 Collimator

The M1A1 collimator is used as an infinity reference by artillery gunners to replace aiming posts and lights. It employs a tritium activated light source which requires no batteries or incandescent lamps.

CONTRACTOR

Standard Equipment Company, 9240 No. 107th Street, PO Box 23060, Milwaukee, Wisconsin 53224, USA.

YUGOSLAVIA

3581.093

RLD-2 HAND-HELD LASER RANGEFINDER

The RLD-2 hand-held laser rangefinder can also be goniometer-and/or tripod-mounted, and is intended for forward artillery observation, anti-tank weapons and for infantry use. In use, the operator centres the cross-hair on the target through the right-hand ocular. By pressing the trigger button the laser is fired and the distance is displayed in the left-hand ocular. The display also has three indicator lamps: one shows if there is more than one target in the laser beam; and another shows the existence of targets within the blocked range. The third lamp indicates low battery level. The display is switched off automatically five seconds after laser firing. Continuously adjustable minimum range between 200 and 3000 metres permits selection between various targets.

CHARACTERISTICS

Laser type: Nd:YAG

Peak power: 0.6 MW

Beam divergence: <1 mrad

Pulse rate: 30/minute

Range (minimum/max): 200/19 995 m

Accuracy: ± 5 m

Resolution: Better than 40 m

Blocking: Continuous from 200 to 3000 m

Detector: Silicon avalanche diode

Receiver field of view: Approx 1 mrad

Telescope

Magnification: × 7

Field of view: 7°

Eyepiece adjustment: ± 5 dioptres

Power supply: 12 V DC

Weight: Approx 2.5 kg

STATUS

Production.

CONTRACTOR

Federal Directorate of Supply and Procurement, PO Box 308, 9 Nemanjina Street, 11105 Beograd, Yugoslavia.

RLD-2 hand-held laser rangefinder

3582.193

TLMD SERIES TANK LASER RANGEFINDERS

The TLMD series laser rangefinders were specially designed for Soviet T-type battle tanks (T-34, T-54, T-55 and T-62 for the -2 version, and T-54, T-55, T-62 and T-72 for the -3 and -6), and considerably enhance the fire control capability of these vehicles. The equipments consist of a single unit.

The optical periscope sight is integrated in the laser which is powered from the tank battery. The TLMD-2, -3 or -6 is mounted in the commander's cupola and the existing commander's periscope is removed from its pivot suspension and replaced by the laser. Connection of a suitable power supply is the only other change necessary. A remote display for the gunner is provided.

CHARACTERISTICS

Laser: Nd:YAG

Max range: Up to 20 000 m, depending upon weather conditions and target.

Range accuracy: ± 5 m

Radial resolution: 30 m

Number of targets measured: 2

Number of targets displayed: 1

Blocking range: Continuous from 200 to 3000 m

Pulse rate: 30/minute

Telescope

Magnification: × 7

Field of view: 5°

Eyepiece adjustment: ± 5 dioptres

Power supply: 18 – 29 V DC, approx 20 W

Computer output: BCD code

Weight: Approx 10 kg

CONTRACTOR

Federal Directorate of Supply and Procurement, PO Box 308, 9 Nemanjina Street, 11105 Beograd, Yugoslavia.

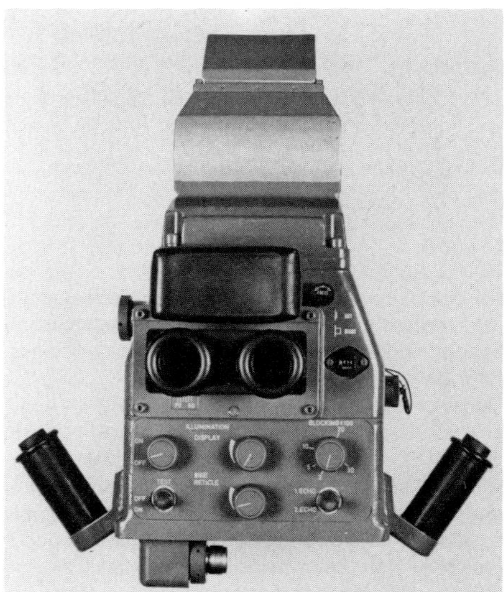

TLMD-6 tank laser rangefinder

6301.193

SGS-55 TANK DAY/NIGHT SIGHT

The SGS-55 gyro-stabilised sight is an electro-optical equipment designed for further improvement in tank sighting systems. Three versions of a basic system are either available or in development: the SGS-55-A which is normally used in conjunction with one of the TLMD series laser rangefinders; the SGS-55-P which is a passive sight using an image intensifier; and the SGS-55-T in which the passive night channel is replaced by a thermal imaging system.

The basic SGS-55 system uses stabilised mirrors for observation of the battlefield while the tank is on the move. This enables the commander to gather tactical information, detect and acquire stationary or moving targets, measure ranges up to 3000 m, bring the turret in the direction of the target by using an acquisition mode, and rotate the turret using a turret mode. The field of view is $\pm 10°$ in azimuth, and –15 to +20° in elevation.

STATUS

The SGS-55-A is in production, the SGS-55-P is in prototype form, and the SGS-55-T is in development.

CONTRACTOR

Federal Directorate of Supply and Procurement, PO Box 308, 9 Nemanjina Street, 11105 Beograd, Yugoslavia.

SGS-55-A gyro-stabilised sight

5723.193

SUVOA ARTILLERY FIRE CONTROL SYSTEM

SUVOA is a tripod-mounted optical fire control system for use with coastal or field gun batteries. It is designed to give range to the target from 1000 to 50 000 metres, depending upon the altitude of the observation post. Information on ranges is transmitted to the gun by a transmission system. A radar equipment is co-sited with the instrument in order to provide range and bearing information in poor visibility or by night.

CHARACTERISTICS

Range: 1000 – 50 000 m

Accuracy: 50 m up to 10 000 m; 200 m up to 20 000 m; 3% up to 50 000 m

Measuring telescope

Magnification: × 18

Field of view: 4°

Observation telescope

Magnification: Variable from × 3 to × 10

Field of view: Variable from 6 – 20°

CONTRACTOR

Federal Directorate of Supply and Procurement, PO Box 308, 9 Nemanjina Street, 11105 Beograd, Yugoslavia.

6303.193

EFCS-2 TANK FIRE CONTROL SYSTEM

The EFCS-2 is an expandable fire control system for T-55 and T-62 battle tanks. It consists of a number of sub-units, a commander's sight, commander's panel, gunner's sight with left eyepiece, a vertical gyro to measure tilt angle and pitch and roll when the vehicle is moving, a computer and an interconnection box. Optional units include an ammunition temperature sensor, and a meteorological sensor to measure crosswind, air pressure and air temperature.

The commander's sight is a gyro-stabilised periscope equipped with a laser rangefinder and situated in the cupola. The associated control panel provides for entry of ballistic information, selection of ammunition, a display of sensor information, and controls for self-test and calibration and manual entry of data. The computer performs all the ballistic computations required by use of the various sensor inputs.

STATUS

In production.

CONTRACTOR

Federal Directorate of Supply and Procurement, PO Box 308, 9 Nemanjina Street, 11105 Beograd, Yugoslavia.

EFCS-2 commander's control panel

5914.193

OMU-2 ARTILLERY LASER RANGEFINDER

The OMU-2 is a binocular periscopic laser rangefinder and integrated electronic goniometer, with a tubular compass, fire control facilities and a tripod. It forms part of the SODA (System for Observation and artillery Data Acquisition) equipment to enable a forward artillery observer to acquire and transmit target distance, target co-ordinates and fire correction data. The SODA system consists of the OMU-2, the TM-2 transmitting and receiving modem and the RDC-2 remote display system.

CHARACTERISTICS

Binocular periscope

Magnification: × 10 and × 20

Field of view: In the right channel 7° and 3.5°; in the left channel 7° and 3.5° horizontal, 5° 17' and 2° 38' vertical

Laser

Type: Nd:YAG, passive Q-switch

Peak power: 2 – 2.5 MW

Range: 190 – 29 995 m

STATUS

The SODA system is in the prototype stage.

CONTRACTOR

Federal Directorate of Supply and Procurement, PO Box 308, 9 Nemanjina Street, 11105 Beograd, Yugoslavia.

OMU-2 artillery laser rangefinder

6302.193

TS-M THERMAL SIGHT

The TS-M is a tripod-mounted thermal imaging sight intended primarily to supply increased operational capability for manually controlled anti-tank guided missiles. The sight can also be used as an independent thermal imager for general purpose surveillance and target detection. In normal night conditions, the TS-M provides detection of a moving tank target at 3000 m, with recognition of the target at 1800 m.

The TS-M can be brought into operation in less than two minutes and provides a minimum of two hours operation by a single set of batteries and one bottle for compressed air cooling.

CHARACTERISTICS

Field of view: 3.5° × 2° (narrow); 1.7° × 10° (wide)

Magnification: × 9.5 (narrow); × 1.9 (wide)

Display type: CRT

Frame rate: 25 Hz

Number of lines: 200

Detector format: Linear array

Weight: Approx 10 kg

STATUS

Prototype

CONTRACTOR

Federal Directorate of Supply and Procurement, PO Box 308, 9 Nemanjina Street, 11105 Beograd, Yugoslavia.

6304.193

10 Hz LASER FOR ANTI-AIRCRAFT SYSTEMS

The Iskra 10 Hz laser has been designed for use in anti-aircraft defence systems and consists of a laser head and complete power and control electronics contained in an aluminium housing. The modular design enables simple upgrading into a laser rangefinder. The laser is provided with connectors which enable it to be coupled to the power supply source, and to other parts of the air defence system.

CHARACTERISTICS

Laser: Nd:YAG

Wavelength: 1.064 μm

Energy per pulse: 100 mJ

Pulse duration: 20 ns

Repetition rate: 10 pps

Cooling: Compressed nitrogen

Dimensions: 320 × 270 × 165 mm

Weight: 18 kg

STATUS

Prototype. A number of 10 Hz lasers have been tested successfully in various anti-aircraft defence systems.

CONTRACTOR

Federal Directorate of Supply and Procurement, PO Box 308, 9 Nemanjina Street, 11105 Beograd, Yugoslavia.

Iskra 10 Hz laser

SHIPBORNE WEAPON SYSTEMS AND ASSOCIATED EQUIPMENT

This main section contains shipborne weapons and their associated equipments. It covers surface-to-surface and surface-to-air missiles, guns, fire control systems, torpedoes, mines, underwater warfare systems, sonar, radar, electronic warfare systems and electro-optical equipment, for both surface vessels and submersibles of all types. This alteration in the running order of *Jane's Weapon Systems* has been made to enable systems and equipments of each of the four main operational areas, ie strategic, land, sea and air, to be brought together. The numerical system remains identical, thus making it simple to cross-refer to previous editions.

NAVAL FIRE CONTROL AND ACTION DATA AUTOMATION SYSTEMS

CANADA

4046.261

ADLIPS

ADLIPS (Automatic Data Link Plotting System) is a modular low-cost, command, control and tactical data communication system designed to replace electro-mechanical plotting tables and associated radar displays in non-TDS (tactical data system) equipped ships. The system provides transmission, reception, filtering and display of target track data from Link 11 and Link 14 data communication networks. ADLIPS also collects and processes own-ship sensor data to compile rapidly a composite tactical display and interchanges target track, weapon assignment and status data with own ship's weapon systems. When operating in a TDS environment, ships fitted with ADLIPS can share tactical data with each other as well as aircraft and other ships that are data link equipped.

An HF Link 11 radio transmitter/receiver provides for communications over this data link, and a Link 11 modem (AN/USQ-76) allows operation in either picket or net controlled modes. This unit generates and recognises Link 11 control codes, preambles and data formats and provides error detection and correction. For Link 14 a CV-2757/SGC interface converts the low-level signals from the computer to the high-level signals required by the cryptographic equipment and the inverse conversion.

The complete ADLIPS has a capacity of up to 140 air, surface and sub-surface tracks, and has a full NTDS repertoire.

STATUS

Litton has built 16 ADLIPS sets for the Royal Canadian Navy.

CONTRACTOR

Litton Systems Canada Limited, 25 Cityview Drive, Rexdale, Ontario M9W 5A7, Canada.

2046.261

CCS-280 COMMAND AND CONTROL SYSTEMS (DDH-280) CLASS

This equipment is a shipborne automatic data handling and display system based on the Litton L-304 single processor version digital computer. Designated AN/USQ-501(V) data processing set, it comprises one digital computer, eight multi-function displays, and other peripherals.

The computer is fitted with a 40K 32-bit word memory capable of expansion to 80K words by addition of plug-in modules to the existing processor mainframe. Increased I/O capability can be achieved by adding units to the existing data bus using available minor channels in the I/O modules. The system includes radar signal simulation, which provides superimposed or separate radar targets that are variable in size and brightness for training. Displays are multi-purpose and identical. Quick action buttons provide mode selection and independent selection of up to seven separate radar videos with independent offset and range scales of 1 to 152 miles (1.6 to 244.6 km) in binary increments. Category selection of symbology is programmable on-line, independently, at each display.

Basic operating software has been provided by the manufacturer. In the case of the CCS-280 the Canadian Armed Forces chose to write all the operational software in a program generation facility provided by the contractor. The overall system provides real-time co-ordination of sensors and weapons by the command at the unit and force level to:

(1) collect, process, and display tactical information from all sensors
(2) control aircraft and other remote surface units
(3) direct own ship's and force weapons
(4) control own ship manoeuvres
(5) exchange information and orders within own ship and force.

STATUS

The system is currently operational at sea in HMCS *Iroquois*, *Athabaskan*, *Huron*, and *Algonquin*. A shore installation for training and program maintenance and generation is fitted in the Combined Support Divisions, Halifax.

CONTRACTORS

Litton Systems (Canada) Limited.

Collins Radio (Automatic Link Equipment).

2047.261

UNDERWATER COMBAT SYSTEM (UCS) 257/280

The UCS 257 and/or UCS 280 comprises two AN/SQS-505 sonar sets (**1792.453**), an anti-submarine warfare data system (ASWDS), and ancillary equipment for both hull-mounted (HMS) and variable depth sets (VDS). Both AN/SQS-505 sonars are electronically identical and comprise a transmitter group, a receiver group, a control indicator group, and a transducer. The ancillary equipment for the hull-mounted set consists of a retractable transducer with fixed faired dome. The ancillary equipment for the variable depth set consists of the AN/SQA-502 hoist group (including tow cable) and a variable depth body (including gyro compass).

The ASWDS is an action data system designed to accept primary target information from hull-mounted and/or variable depth sonars and secondary target information from radars. This information is processed by a high speed general purpose computer to provide action information to the command, computation of fire control solutions, and a weapon control.

A prototype of the system was installed and evaluated in *Terra Nova*. Production models are fitted in *Terra Nova*, *Gatineau*, *Kootenay*, and *Restigouche*. The AN/SQS-505 sonar system (hull-mounted (HM) and VDS) is installed in the four 'Iroquois' (DDH-280) class anti-submarine destroyers.

STATUS

In production. The latest AN/SQS-505(V)6 sonar system (HM) is to be installed in the six new Canadian Patrol Frigates ('Halifax' FFH-330 class).

CONTRACTORS

Westinghouse Canada Inc, PO Box 5009, Burlington, Ontario L7R 4B3, Canada – Prime contractor, AN/SQS-505.

Hollandse Signaalapparaten BV, PO Box 42, 7550 GD Hengelo, Netherlands – Prime contractor, ASWDS.

FRANCE

4049.261

RADOP SERIES NAVAL FIRE CONTROL SYSTEMS

The Defence Division of CSEE offers a range of naval fire control systems suitable for the smallest patrol craft to vessels of more than 200 t, for directing guns against air and surface targets. In addition, these systems provide for surface and low altitude radar surveillance, optical surveillance, and target identification. The range of systems consists of differing combinations of CSEE optical or electro-optical directors, and one or other of the Racal/Decca marine radars.

Four basic systems have been developed in this way:

(1) RADOP 10 series with the CSEE Lynx (**3426.293**) director
(2) RADOP 20 series with the CSEE Panda Mk 2 (**3425.293**) director
(3) RADOP 30 series with the CSEE Naja (**3561.293**) director
(4) RADOP 40 series with the CSEE Najir (**4255.293**) director.

Essentially, all four have the same optical surveillance capability, but they differ in their weapon control performance. For each RADOP series, all the Racal/Decca marine radars of group 12 (1226, 1229, 1230 and 2459) are suitable, so that it is possible to select between 3 or 10 cm wavelength radar, 12- or 16-inch (30 or 40 cm) diameter CRT display, and relative motion, true motion or anti-collision presentation.

Each RADOP version can be optionally complemented by an automatic plotter, radio navigation aids, and an air surveillance radar.

Additional details of the CSEE E-O directors are given in the appropriate pages of the Equipment

Section of this volume, under the reference numbers listed above, and in the following paragraphs the characteristics of individual systems of the RADOP series are summarised. (The Racal/Decca 1229 radar has been employed as an example in all cases, and obviously use of a different model would modify weights and/or dimensions.)

STATUS

In production for various navies including those of Cameroon, Saudi Arabia, Senegal and Uruguay.

CONTRACTOR

CSEE (Compagnie de Signaux et d'Entreprises Electriques) Division Défense, 17 place Etienne-Pernet, 75738 Paris Cedex 15, France.

3102.281

CANOPUS NAVAL WEAPON CONTROL SYSTEM

Canopus is a lightweight naval tracking system capable of directing and controlling anti-air and anti-surface weapons and also providing optical target designation facilities. The system is designed for all-weather operation and is therefore able to deal with a wider range of operational requirements than an entirely optical or electro-optical tracking unit. Included in the system is an operational fire control console with its associated gun fire computer.

The Canopus system performs the following functions:

(1) autonomous optical surveillance over the full horizon
(2) automatic or semi-automatic target acquisition on radar target designations
(3) tracking of air targets including sea-skimming missiles and surface targets with the Ka-band radar, the TV tracker in bearing, elevation and radar range, and, optionally, the infra-red tracker
(4) target designation to or from other fire control systems.

Target data is sent to the gunnery computer which performs ballistic computations for gun fire control. Association of the Canopus tracker unit with a gunnery computer constitutes the Vega Canopus fire control system.

Thomson-CSF Canopus lightweight E-O fire control tracker

STATUS

In production and delivered to several navies.

CONTRACTOR

Thomson-CSF, Division Systèmes Défense et Contrôle, 40 rue Grange Dame Rose, 92363 Meudon-La-Forêt Cedex, France.

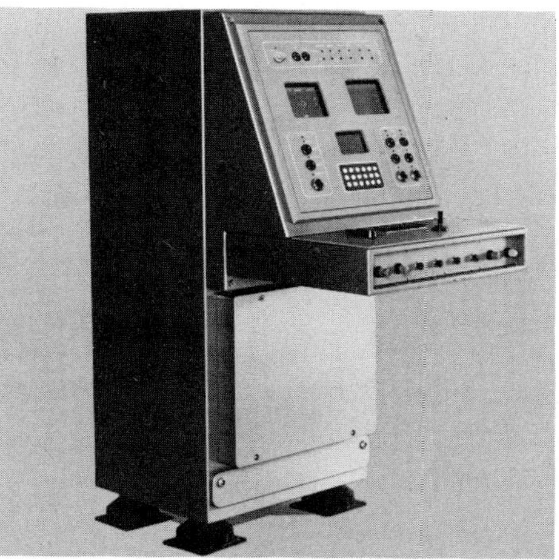

Typical Canopus console

4572.281

THOMSEA NAVAL COMBAT SYSTEMS

Thomson-CSF now produces the former VEGA series in their latest versions as the Thomsea naval combat systems, a range of modular integrated naval systems. The basic components of a Thomsea NCS are as follows:

(1) a combined (air and surface) surveillance and target designation radar (Triton G, Triton II MTI, Triton S or Sea Tiger)
(2) one or more fire control radars, Castor II fitted with television tracking
(3) one or more E-O directors fitted with television tracking, infra-red tracking, laser rangefinder or K-band radar
(4) a computer, or several co-operating computers
(5) tactical information unit: console and tactical table
(6) a track-while-scan (TWS) system, incorporated with the tactical table or the console; it allows simultaneous tracking of 16 to 64 air or surface targets detected by the surveillance radar.

Various other equipments may be associated with the system to augment the ship's capabilities in performing the following functions: designation and tracking (optical sights and directors), radar identification and analyses (IFF, ESM), electronic countermeasures (jammers, chaff-launcher), submarine detection (sonar), navigation (radar, log, heading and vertical reference system). The following weapons are controlled or are fed target co-ordinates:

(1) guns of any calibre
(2) surface-to-surface missiles
(3) surface-to-air missiles
(4) surface torpedoes, conventional or wire-guided
(5) ASW weapons (torpedoes or rockets).

The number and type of fire control systems as well as computer capacity are essentially dependent on the ship's armament. The choice of surveillance and ASW equipment is more dependent on ship's tonnage and missions. Three basic systems are available:

Castor IIB for corvettes and offshore patrol vessels

Castor IIC for destroyers, anti-submarine frigates and missile fast attack craft

Castro IIJ for aircraft carriers, cruisers, destroyers and air defence frigates.

Tactical information units are designed to take into account diverse operational criteria defined by users. The basic unit is constituted of the operational consoles for surveillance and fire control radars, to which is added the tactical table. On the surveillance console and tactical table PPIs, analogue video is complemented with synthetic marks, alpha-numeric labels relating to followed tracks, vectors, symbols and track plotting. The data fed by associated equipments – fire control and navigation radars, radar analyser (ESM), optical devices, sonar – are displayed on the PPIs. Complementary devices added to the display unit allow for: fire correction for surface targets, from splash plotting; data display; inter-ship exchange of track data, from stations equipped with a data link (received tracks are displayed on the PPIs); and recording of simulation or tracks for exploitation or operator training.

Tactical problems are solved by calculations inside the system providing: true-motion presentation on PPI; display of relative bearing and range between two tracks, one of which may be the radar carrier (RBL); display of the collision course and of the time required to intercept a track (CTS); the closest point of approach (CPA); extrapolation of tracks for a given time (time to go); generation of fixed points and measurement of the time elapsed since their initiation (datum).

The operator communicates with the system easily and rapidly through a keyboard. This easy reading of decision-making elements by the operator and technical problem solving reduces to a minimum the reaction time between detection and fire command.

The fire control radar console, which includes the computer, provides the operator with:

(1) acquisition of the designated air or surface target
(2) choice of the operating mode
(3) monitoring of the quality of tracking, using radar or television screens
(4) gun control and fire correction.

Interfacing with associated equipments, whether analogue or digital is achieved with specific couplers. In the tactical table and in the console, digital data flow over a bus-type line. A microprocessor controls display and operational functions.

STATUS

Development of these naval combat systems started in 1967. The first system manufactured, a Vega type I-43, controls two guns, Exocet missiles and AEG SSI4 torpedoes. It has been operational since the end of 1971. From the early systems, mostly adapted to small tonnage ships, to the systems currently available, the evolution has been towards an enlargement in the choice of basic equipment to meet the requirements of a broad range of tonnage and missions, with a particular emphasis on the self defence capability against sea-skimmer missiles.

CONTRACTOR

Thomson-CSF, Division Systèmes Défense et Contrôle, 40 rue Grange Dame Rose, 92363 Meudon-La-Forêt Cedex, France.

4053.261

TAVITAC TACTICAL DATA HANDLING SYSTEMS

The former Vega 3C tactical data handling systems, now designated TAVITAC (Traitement Automatique et Visualisation Tactique) are of modular design for both hardware and software. This makes them highly flexible, fit for any type of ship, and capable of meeting any particular operational need. The display consoles are vertical and horizontal. They can be fitted with different types of screen: monochrome or colour, circular or rectangular. Consoles can receive raw video from several radars, superimposed on synthetic video. Display capacity can be adapted to needs by means of a video compression device associated with optimised processing for picture generation. The consoles can be associated with various data processing computers which can be integral with the consoles or set in separate cabinets.

Software can be adapted to associated sensors and operational requirements, and can be produced on civilian computers which are compatible with the Thomson-CSF computers employed in TAVITAC systems. Standard functions are:

(1) automatic tracking of air and surface tracks, associated with an extractor (TWS), with manual or automatic initialisation, allowing anti-surface gunfire
(2) rate-aided manual tracking
(3) manual entry of tracks and special points
(4) track identification

TAVITAC operation centre

(5) threat evaluation
(6) display of local and external tracks received via data link, with associated symbols, speed vectors and alpha-numeric data
(7) display of optical and ESM bearings, of positions of targets tracked by fire-control directors, of circles, vectors, launching figures
(8) transmission of local tracks via data link
(9) dispatch of target designations to weapons
(10) aids to tactical navigation.

Extended capabilities are:

(11) fully automatic process of engagement of air targets, including track initialisation, threat evaluation, timely target designation, weapons coupling, fire orders, optimised management of directors and weapons in case of simultaneous attacks
(12) tactical situation recording and replay in real- or accelerated-time

TAVITAC 2000 console

(13) training facilities in association with a video simulator
(14) coastline storage and display on PPIs on request
(15) tracking of submarine positions, processing of ASW weapon parameters, guidance of attacks by ASW helicopters or other aircraft.

The display system configurations can be adapted to suit particular ships and their weapon systems. Typical systems might include one console only for the smaller systems, with automatic tracking of 16 air or surface tracks, two to four consoles for larger systems, suitable for modern medium-tonnage ships, with automatic tracking for 16 to 32 air or surface tracks, or up to ten consoles, or more, for larger systems designed for heavy tonnage ships, allowing automatic tracking of up to 200 tracks, with vectors, circles, circle sectors, coastlines, etc.

For all proposed structures, the number of computers devoted to processing of tactical information is defined according to the needs and to the degree of redundancy required.

Different types of console are available: vertical consoles for incorporation with a fire control console, modular vertical consoles, tactical tables with one or two operator positions, and large horizontal consoles with two or three operator positions.

A console is made up of the following modules: a combined radar-synthetic image generator module including a renewal memory of the associated computer, a PPI display module, input devices (order keyboard(s)), video control panel, rolling-ball control unit(s)), and alpha-numeric display on a TV or a plasma screen (alpha-numeric information can also be displayed on the PPI edges). Display capabilities are adjustable to customer requirements up to 200 tracks and 100 different symbols.

In the French Navy such systems are known under the name SENIT (Systeme d'Exploitation Navale des Informations Tactiques). Four 'Madina' class frigates that France has built for Saudi Arabia are fitted with SENIT naval tactical display systems produced by Thomson-CSF.

TAVITAC 2000 – Tactical Data Handling Systems

The new TAVITAC 2000 tactical data systems, like TAVITAC, are of modular design and highly flexible. The raster scan display consoles use high resolution (1024 × 1280) colour monitors.

The powerful radar scan converter allows the radar video display to be superimposed on the graphic display. The consoles are associated with data processing computers of the same range as the display processor, all based on the powerful 68020 microprocessor. Programming is modular and is adapted to the requirements using the ADA language.

The system is linked by a local area network based on the Ethernet standard. Associated sensors and weapons are either connected directly to the network or via a network interface unit.

The system performs all the necessary tactical functions for any size of vessel and in addition, using ruggedised disk storage, offers a data base management capabiltiy for map display, ship resources and management.

STATUS

Production.

CONTRACTOR

Thomson-CSF, Division Systèmes Défense et Contrôle, 40 rue Grange Dame Rose, 92363 Meudon-La-Forêt Cedex, France.

3743.481

DLA ANTI-SUBMARINE FIRE CONTROL SYSTEM

The DLA anti-submarine fire control system is designed on a modular basis to provide the necessary flexibility to permit its use aboard various classes of surface ships, to accommodate various types of weapons, and to enable it to be completely integrated within a ship's combat system. This is achieved by assembling the DLA system from two basic modules: a 'target' module and a 'weapon' module.

The former is built around a large CRT for the display of tactical information, and a microprocessor. The microprocessor is of similar or identical technology to that employed in the weapon module and its function is to accept target data from sonar systems (eg bearing, range, and doppler), and from these to compute the target parameters required by the weapon module (course and speed).

Weapon modules are specific to the weapon(s) fitted in the ship (eg torpedo, A/S missile, rockets, or depth-charges), and current 'standard' weapon modules are available for all French torpedoes as well as the Mk 44, Mk 46 and A 244 types, and for the Creusot Loire/Bofors 375 A/S rocket launcher. Other types of weapon can be accommodated by the designers on request. The weapon module computer accepts data on target bearing, range, course and speed, and processes this information to produce the necessary parameters for the sequencing, setting, and launching or firing of the A/S weapon.

The DLA is supplied either as a completely self-contained system, with both types of module, or as a weapon module (appropriate to the ship's weapon fit) with a suitable interface for integration into a vessel's combat system performing target designation and computation.

STATUS

In series production for the French Navy and one other navy.

CONTRACTOR

Thomson Sintra, Activités Sous-Marines, 1 avenue Aristide Briand, 94117 Arcueil, France.

Tactical weapon control console

1522.261

SENIT NAVAL TACTICAL DATA HANDLING SYSTEMS

The French Navy's naval tactical information systems for surface vessels are identified under the acronymic designation SENIT (Système d'Exploitation Navale des Informations Tactiques), and five systems of this kind have been identified. All five have the broad functions of gathering, co-ordinating, and distributing sensor data and other information, presentation of information on appropriate displays, and certain weapon system control functions. A digital computer, or computers, provides the central processing facilities. The five systems differ in accordance with the nature and operational role of the vessel fitted and the sensors and armament installed. Details of the five versions identified to date appeared in the 1979-80 edition of *Jane's Weapon Systems*, and in preceding editions. No information has been received concerning any successor systems that may be employed aboard new generation French warships.

STATUS

In addition to the SENIT systems produced for French Navy vessels, a certain number of systems have been supplied to foreign ships according to the manufacturers, but no details have been obtained. Four 'Madina' class frigates for Saudi Arabia are equipped, as noted in **4053.261**.

CONTRACTOR

Thomson-CSF, Division Systèmes Défense et Contrôle, 40 rue Grange Dame Rose, 92363 Meudon-La-Forêt Cedex, France.

1739.461

DLT-D3 SUBMARINE TORPEDO FIRE CONTROL SYSTEM

The DLT-D3 torpedo fire control system is used in French Navy submarines. All types of torpedo employed by the French Navy can be launched, including wire-guided. The system may also be expanded for anti-surface missile applications.

Target data is fed to the system from on-board sensors, which comprise fore and aft sonars, acoustic rangefinder and attack and surveillance periscopes.

The system employs a general purpose digital computer, associated with a CRT display terminal. The following functions are performed by the system:

(1) updating of the tactical situation from the data delivered by the various sensor systems fitted and the navigation equipment
(2) assistance in calculating target components by the use of special recorded programs
(3) weapon control, ie computation of the firing path, remote setting of torpedoes, and firing sequence control. The system is designed for launching any type of torpedo
(4) maintaining a chronological record
(5) maintenance assistance by means of test programs.

The DLT-D3 operating programs have been compiled to permit tracking of eight targets, simultaneous guidance of two wire-guided torpedoes and preparation of a third for launching. Each of the three displays of the terminal is dedicated to the presentation of the following data, in accordance with the program implemented: tactical situation, firing path, and alpha-numeric display of parameters (tote) and decoding of the designations and functions of the two common keyboards. Conversation between the operator(s) and the computer are by means of these keyboards. Emergency launch of torpedoes is possible from either bow or stern station.

The system is designed to be served by one or two operators at the operations' centre, and one operator at the bow station with possibly another at the stern station. The equipment arrangement, typically, is one CIMSA 15M125 digital computer (QTD), a monitoring and control console (VIC), an azimuth relay (RZ), a true-bearing diagram (GZ) in the operations' centre, a tube selection panel (PAT), and tube servicing station (PST) in the bow and/or aft torpedo tube compartments.

STATUS

This system is in service in 'Daphné', 'Agosta' and 'Rubis' class SSNs of the French Navy, and is in production for Spanish and Pakistan 'Agosta' class submarines.

CONTRACTOR

Thomson Sintra, Activités Sous-Marines, Chemin des Travails, BP 53, 06801 Cagnes-sur-Mer, France.

DLT-D3 submarine torpedo fire control system

4501.481

SUBMARINE INTEGRATED COMBAT SYSTEM

Based on experience in producing various types of sonars and fire control systems, Thomson Sintra ASM is developing an integrated combat system for submarines.

This combat system enables the tactical situation to be established and targets to be engaged on the basis of information supplied by the various sonars and by non-acoustic sensors.

A high degree of integration and flexibility, coupled with the use of colour consoles, allows effective operation of all functions. Redundancy is provided through the use of reconfigurable standard modular computers and multi-function displays.

The principal aims of the designers are to increase range and accuracy of the passive sonars so as to make the entire detection system consistent with medium- and long-range weapons and to reduce the number of operators while still retaining the ability to deal virtually with all the information at both search and action stations.

STATUS

Development.

CONTRACTOR

Thomson Sintra, Activités Sous-Marines, Chemin des Travails, BP 53, 06801 Cagnes-sur-Mer, France.

Thomson Sintra integrated combat system in a mock-up submarine control room

GERMANY (FEDERAL REPUBLIC)

1824.281

AGIS COMBAT INFORMATION SYSTEM

AGIS (Automatisiertes Gefechts- und Informationssystem für Schnellboote) is a fully integrated command and fire control system which has been developed for the West German Bundesmarine's Type 143 fast patrol boats.

AGIS includes the areas of command and fire control system WM 27 (HSA) (**1259.281**), the optical auxiliary fire control system OG.TR 7/3 (Officine Galileo), weapons, data transmission system LINK 11 (Collins), log system SAL 59 (Jungner), gyro compass system PL-41 (LITEF), IFF system (Siemens) and ESM system. The weapons associated with the system are two 76/62 OTO C guns (OTO Melara) both against air and surface targets (**5533.203**), two wire-guided torpedoes Seal (AEG-Telefunken) (**2178.441**) and four surface target missiles MM 38 Exocet on two double launchers (**1156.221**).

In addition to AGIS, the RF, VHF and UHF transceiver units and their associated communications' operator stations serve for external communication in the radio-telephone, radio teletyping and radio keying modes – unencrypted and encrypted. Internally, communication is ensured via the simplex/duplex system (Philips). Furthermore, navigation and direction finding aids include an echo sounder, magnetic compass, Decca navigator and navigation radar.

The AGIS mission centre is the combat information centre with the two control units for torpedo direction, the artillery control system for guns and missiles, the horizontal tactical display and the vertical plot. From here, the S143 can operate as lead boat for other units (eg S142 or S148). The system is controlled by two freely programmable digital computers processing the sensor data for displaying on the different displays or for weapon control. One of the computers is for generation and utilisation of the tactical situation and for target assignment to the command units, and the other one operates as fire control computer for target tracking, determination of firing parameters as well as for control of weapons and weapon consoles.

In addition to own ship information acquisition, the automatic data transmission system, LINK 11, permits the display of data on targets outside own ship detection range and to transmit tactical information to other units.

The AGIS is also employed on the new class 143A fast patrol boats. A successor to the class 143 combat system, it mainly differs in having the ASMD (RAM) system in place of the stern gun and the torpedo system is replaced by a mine launcher. The class 143A is equipped with an FL1800 ESM/ECM system.

STATUS

Fitted in Type 143/143A fast attack craft.

CONTRACTOR

AEG-Telefunken, 7900 Ulm, Elisabethenstrasse 3, Federal Republic of Germany – Main contractor.

1825.281

SATIR ACTION INFORMATION SYSTEM

SATIR (System zur Auswertung Taktischer Informationen auf Raketenzerstören) is the tactical data automation and display system fitted in the three 'Lütjens' class guided missile destroyers of the West German Bundesmarine. These ships are a modified version of the American 'Charles F Adams' type and are fitted with US radars such as the SPS-52 three-dimensional search and target designator, SPS-10 surface warning set, SPS-40 air surveillance, and SPG-51 fire control radars for the Tartar surface-to-air missiles. Weight and space considerations decided the Bundesmarine against the installation of the US NTDS (Naval Tactical Data System) and SATIR was developed in its stead to meet the German requirement and in general compliance with the B-2 Concept, a NATO standardised system for destroyers and above. Beyond the participation of Univac, no other reliable details have been obtained.

A version of the SATIR system is fitted in the six vessels of the Type 122 'Bremen' class; this is described in detail in 1982-83 and earlier editions (**3504.281**).

3504.281

CLASS 122 FRIGATE COMBAT SYSTEM

The combat system of the 'Bremen' class frigates for the German Federal Navy permits the engagement and destruction of air, surface and underwater targets as well as communication with other units at sea, in the air and ashore. For the performance of these tasks the combat system consists of a central computer system for processing and correlation of the incoming information which is acquired by means of sensors and data transmission, as well as of display consoles with the synthetic scenario for dialogue during decision-making, and for any resultant missions.

All major functions can be monitored from the combat information centre with the aid of an optimised number of personnel in a reliable, co-ordinated and effective manner and can be controlled directly or indirectly. The high processing speed and the large storage capacity of the computers as well as the very fast peripheral equipment and a considerable redundancy permit measures for equipment maintenance and repair in parallel with the major tasks outlined above. A suitable computer set-up, special system instructions as well as a modular software structure permit effective real-time processing. As far as software programs are concerned, a distinction is made between the mission software, maintenance software, support software and general software for administrative and other purposes.

This system, which has been deployed with the West German fleet, is described in detail in 1982-83 and earlier editions of *Jane's Weapon Systems*, together with a list of sensor equipment and weapons fitted in these ships.

CONTRACTORS

AEG-Telefunken, 7900 Ulm, Elisabethenstrasse 3, Federal Republic of Germany – Main contractor for weapon system and integrated logistic support.

Bremer Vulkan – Prime contractor.

INTERNATIONAL

6754.281

KANARIS FIRE CONTROL SYSTEM

This system is being developed by Unisys in collaboration with the Hellenic Navy's Research and Development Centre (GETEN) for half the Greek Navy's 'Glavkos' class (Type 209) submarines.

Each system will consist of a two-cabinet firing distribution unit and the main control unit with two consoles. The latter are the Unisys Tactical Modular Displays (TMD) which feature embedded AN/UYK-44 computers permitting the system to control six weapons simultaneously.

STATUS

Five systems have been ordered, four for shipborne use and the fifth as a land-based training unit.

CONTRACTORS

Unisys Defense Systems, Computer Systems Division, Great Neck, New York, 11020, USA.

GETEN (Hellenic Navy Research and Development Centre).

3509.281

SEAGUARD NAVAL COMBAT SYSTEM

Seaguard is a range of modules which can be integrated in a variety of configurations. This entry is concerned with the general combat system configuration. The specific Seaguard CIWS configuration is described in the Shipborne Surface-to-Air Weapons Section under entry **4016.231**.

The Seaguard equipment modules comprise the following:

(1) Dolphin Search Radar Module (SRM) (**5620.253**)
(2) Command and Control Module (CCM)
(3) Command Support Module (CSM)
(4) Ku-band Tracking Radar Module (TMK)
(5) X-band Tracking Radar Module (TMX) (**6119.253**)
(6) 25 mm Gun Module (25 mm GM) (**2395.20**)
(7) Weapon Control Module (WCM)

Sensors such as long range radar, navigation radar, data link, sonar, ESM, Helo data link, etc., can be integrated in the combat system, as can weapons such as SSM, SAM, CHAFF and ASW.

The SRM is a dedicated high data rate, high accuracy, dual beam C-band radar that provides target detection and indication in severe environmental and ECM conditions of sea-skimming and high diving missiles, as well as a near hemispherical general air and surface surveillance.

The CCM consists of multiple autonomous Nautic consoles with replicated data bases linked through a command system highway. The action-information data base management is organised through the combat system highway which provides very high flexibility in operation with a high level of redundancy. The versatility of the CCM permits interfacing of all sensors and the integration of fire control channels and weapons for tactical command and control.

The CSM is the ship interface locus for all information to and from Seaguard CCM, sensors, tracking and weapon modules. It provides the processor capacity for the Seaguard SRM data processing and track-while-scan of air and surface targets. In automatic engagement modes it provides correlation, threat evaluation and weapon assignment of air targets.

The CSM's floppy disk controller stores the software program related to logistics (diagnostic test) and training programs (training scenario generator).

The TMK is a unique, high dynamic, three axis multi-sensor (Ku-band radar, FLIR and laser), all weather, acquisition and tracking system (stabilised with a local strap down device), impervious to the multi-path effect. The TMK enables tracking down to close-in-ranges of high speed manoeuvring missiles throughout the hemisphere including sea-skimming and zenith diving missiles in a dense electronic warfare environment. The multi sensors (radar and FLIR/laser) are intimately combined with specialised and innovative signal processing and advanced control algorithms into a very modern, high performance, missile/aircraft tracking system.

Detailed monitoring and mode selection of the TMK tracking module are performed by a single operator at the tracker/weapon console (T/WC).

The TMX is a general purpose tracking module for the control of medium calibre guns or ship-to-air missiles consisting of a unique, high dynamic, three axis multi-sensor, (X-band radar, FLIR and laser), acquisition and tracking system enabling long range accurate tracking throughout the hemisphere. Its tracker/weapon console (T/WC) provides the detailed monitoring and mode selection of all TMX engagement sequences.

The 25 mm GM is a high dynamic, canted, two-axis gun mount carrying four independant KBB 25 mm cannon having a flexible and combined rate of fire of 3400 rds/mn, permitting engagement of high speed manoeuvring missiles down to close-in-ranges throughout the hemisphere (–14° to +127°) including sea-skimming and zenith diving missiles, in all conditions of ship motion. The below-deck ammunition feed system, protected from environmental and NBC conditions, has a ready-use capacity of 1660 rds (18 engagements) and permits topping up during firing. The high energy 25 mm KBB/AMDS ammunition with subcalibre penetrator has proven capability against all types of anti-ship missiles and is also effective against aircraft, helicopters and RPVs.

The WCM is a data processor providing optimum gun control, ballistic computation and gun aim point prediction for any type of medium calibre gun in anti-air/surface modes.

Each Seaguard module has its own dedicated data processing facility inter-connected by data highways. The Seaguard combat system links all elements of the overall ship fit by using a modern data handling concept having a number of data highways:

(a) the combat system highway for the exchange of sensor and tactical data between users
(b) the command highway for the exchange of tactical data between command display users

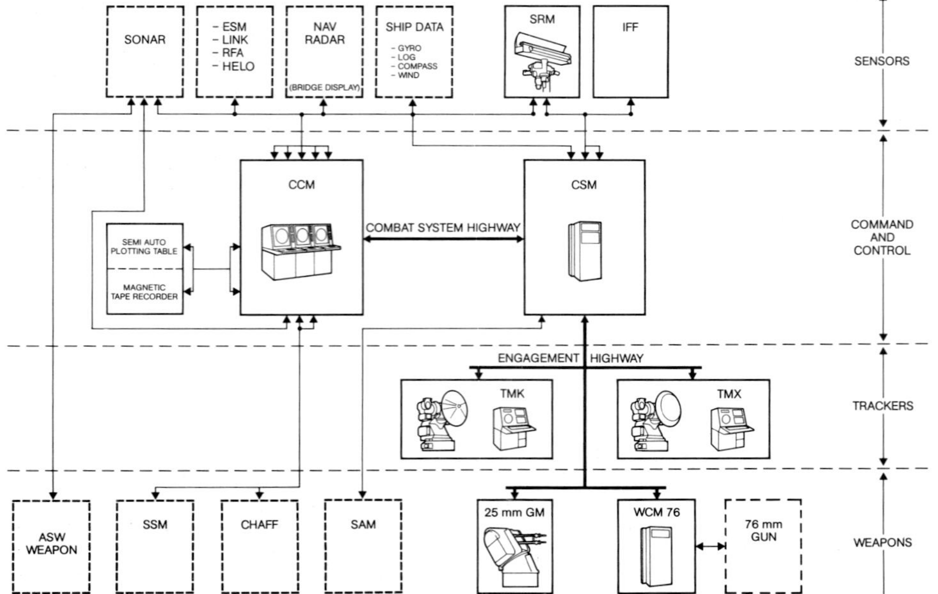

Example of the layout of a Seaguard combat systems architecture for a corvette

(c) the engagement highway controlling the selected target tracking and weapon channels.

The modular concept of the Seaguard combat system provides the integration flexibility necessary for configurations ranging from simple fire control channels up to systems providing surface and shore offence with layered air and sub-surface self-defence.

STATUS

In series production.

CONTRACTORS

Contraves AG, Schaffhauserstrasse 580, 8052 Zurich, Switzerland.

Contraves Italiana SpA, Via Affile 102, I-00131 Rome, Italy.

Machine Tool Works, Oerlikon-Bührle Ltd, Birchstrasse 155, 8050 Zurich, Switzerland.

Plessey Radar Ltd, Oakcroft Road, Chessington, Surrey KT9 1QZ, England.

Plessey Naval Systems, Templecombe, Somerset BA8 0DH, England.

British Manufacture & Research Company Ltd, Springfield Road, Grantham, Lincolnshire NG31 7JB, England.

4489.281

SEA TIGER

Sea Tiger is a derivative of the naval electro-optical fire director system originally developed for the US Navy as the Seafire project (**3727.281**) by two of the participants in that programme, Texas Instruments and Ferranti. The Sea Tiger system is designed to provide passive surveillance using high resolution daylight television and high resolution forward looking infra-red (FLIR). Typical applications include navigation, search and rescue, aircraft and missile detection, and detection of surface craft (ships etc) at the horizon. For fire control purposes three-dimensional tracking is provided by an automatic video tracker used with a laser rangefinder. The system processor carries out calculation of gun or missile orders using three-dimensional track information, ship's sensor inputs and stored ballistic data. Other facilities include handing-off of targets and line-of-sight designation from other ship systems and there is spare computer capacity for distributed processing. Laser target illumination facilities are provided for guidance of laser seeking ordnance such as the US Navy's Deadeye 5-inch laser guided shell and air-launched weapons such as LGBs.

The Sea Tiger equipment comprises an E-O director mounted above deck and incorporating a daylight TV camera, thermal imaging IR sensor, and a Ferranti Type 905 multi-role laser for designation and laser rangefinding. The E-O director is connected by a data link and video link with a below-deck console that contains the computer for the automatic video tracker and fire control calculations. Provisions are included for interfacing the Sea Tiger system with other ship's sensors and data processing facilities.

Operational applications of the system include passive surveillance by day and night, target detection and designation for defensive and attack purposes, close-in point defence capability, and laser target marking for laser guided weapons such as naval ordnance or air-launched bombs.

STATUS

Joint marketing programme by Texas Instruments and Ferranti world-wide, excluding the USA where it is understood that a US Navy contract for the Seafire system (which contains a number of system elements common to the Sea Tiger system) was awarded to Texas Instruments in November 1983.

CONTRACTORS

Texas Instruments Inc, PO Box 660246, Dallas, Texas 75266, USA.

Ferranti Defence Systems, Robertson Avenue, Edinburgh EH11 1PX, Scotland.

6197.281

SEA COBRA FIRE CONTROL SYSTEM

The Sea Cobra fire control system is a Breda 30mm Compact Naval Mounting with the director of a Marconi 400 Series tracking radar mounted on top.

Working in close collaboration, Marconi and Breda Meccanica Bresciana have evolved a system to provide an effective, but affordable, close-in weapon system (CIWS). The combination of the new Breda compact 30 mm gun and the radar fire control system is the optimum in terms of rate of fire and calibre of gun for effective defence against missiles.

Reaction time is of paramount importance in a point defence system especially when countering sea-skimming missiles. The operation of Sea Cobra is fully automatic from the detection of the threat by the radar in its surveillance mode, through 'lock-on and tracking', to the point where the target is in range and the operator fires the gun and destroys the target.

Technical Details

The ship fit comprises two parts, the armament and the control console. The radar surveillance/director transmitter, synthesiser, signal processor and data extraction unit are all mounted on a frame and are housed within the glass fibre weather protection housing of the armament; the control panels, monitors and predictor are mounted in the control console.

Fitted with 30 mm calibre 2458 mm long barrels, the Breda Mauser Compact Naval Mounting is fully automatic and has a rate of fire of 1600 rounds/minute with a high muzzle velocity.

The radar is a modern advanced equipment employing a comprehensive range of ECCM techniques, including frequency agility, and provides both the surveillance and tracking functions within the one unit. Of particular importance is its ability to detect small missile targets and the ability to minimise the angular errors caused by multi-path effects.

To match the range of the gun the radar has an instrumented range of 12 km in the surveillance mode and 10 km in the auto-tracking mode.

The combined precision of the radar fire control system and the low dispersions on the Breda gun leads to an effective system to destroy sea-skimmers in the range 100 to 1000 m using APDS ammunition. In addition to effectiveness against sea-skimming missiles by penetrating the missile warhead, the system provides excellent defence against strike aircraft when high explosive (HE) ammunition is used. The 30 mm Breda gun fires GAU-8 ammunition.

CONTRACTORS

Marconi Radar Systems Ltd, Writtle Road, Chelmsford, Essex, CM1 3BN.

Breda Meccanica Bresciania SpA, 2 via Lunga, 25128 Brescia, Italy.

ISRAEL

3460.281

ELBIT ISRAELI ACTION INFORMATION SYSTEM

Elbit's AIO is designed to offer today's ship commander increased speed and accuracy in tackling the problems of target acquisition, joint forces' organisation and weapon systems' operation. Using inputs from the ship's own sensors and other force participants the system processes the data and presents the tactical situation in real time.

The AIO also performs navigational computations, message exchanges between task force units and shore-based headquarters, and records tactical data for future debriefing and off-line analysis.

The system comprises: AIO cabinet, Graphic Display Console, Tabular Display Console, SACU (Elbit's stand-alone communication unit as a data link), and the ship interfaces.

The AIO is functionally divided into the following subsystems: tactical computer subsystem (with floppy disk), conversion subsystem, extraction subsystem, graphic display subsystem, tabular display subsystem and data link subsystem.

System functions include:

(1) compilation and presentation of tactical situation pictures
(2) management of tactical data exchange between task force units
(3) detection, tracking and processing of target data
(4) correlation of EW/ESM and radar data
(5) computation of navigation and manoeuvring data
(6) computation of EW racket area
(7) presentation of correlating charts from computer
(8) data recording

Elbit's AIO has been designed to meet the most demanding operational, tactical and environmental requirements of naval warfare.

It may be specified for newly-constructed ships or installed as part of retrofit programmes.

CONTRACTOR

Elbit, POB 5390, Haifa, Israel 31053.

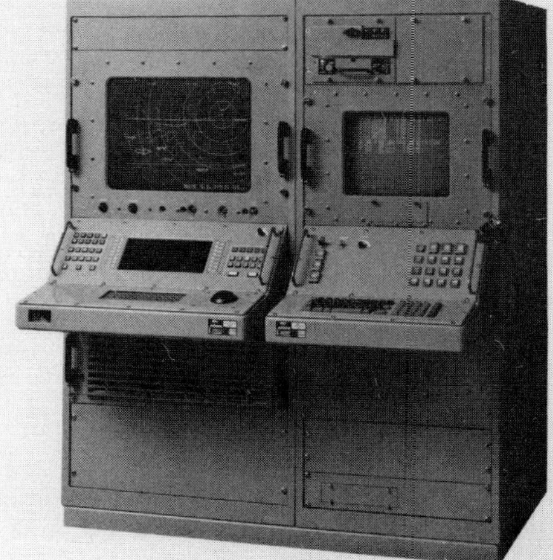

Elbit AIO colour tactical display console

4579.281

RESHET NAVAL COMMAND AND CONTROL SYSTEM

Reshet is a shipborne tactical command and control system for vessels of all sizes. It provides the command with an updated and comprehensive picture of the overall tactical situation, based on data from own ship's sensors and information derived from consorts and/or shore-based facilities, via data links.

Several versions have been designed (**3800.261** in 1983-84 and earlier editions), but all are composed of one or more tactical display consoles (TDCs), a central data processing unit, and in certain cases a data management terminal, as well as the essential interface equipment needed to connect the system with sensors and weapon systems.

Operational requirements met by a typical Reshet system include: real-time assembly, assessment, display and application of all available data; co-ordination of task force actions; over-the-horizon targeting; prevention of 'data saturation'; and modernisation of existing CIC facilities.

To fulfil these functions the following facilities are required:

(1) automatic tracking of multiple targets from search and fire control radars' video data
(2) exchange of tactical information and messages within data link network stations
(3) threat evaluation by classification, correlation, execution of triangulations and selective data display according to operator-defined categories

(4) navigational calculations and other computations for operational purposes
(5) tactical display in colour, with maps, grid, targets and bearing lines, datum points, danger zones, future positions, track histories and other tactical data
(6) comprehensive display, editing and data manipulation facilities such as zoom, off-centring, true/relative motion display, declutter, etc
(7) anti-submarine warfare calculations and display
(8) alpha-numeric presentation of target data, cursor position, system, messages and other information
(9) designation of tracked target to weapon systems and other units
(10) hard copying of tactical scenario for analysis.

All these facilities are provided by a typical Reshet multi-functional system, which might consist of a central processor unit, a number of consoles used as the target acquisition and designation (TAD) position, tactical editor console, commander's tactical console, and data link equipment. The Reshet system's modularity allows two alternative simplified versions to be configured for smaller ships. These differ from the complete system mainly in the display capabilities. The single console configuration replaces the tactical editor console with a data management terminal that provides the tactical editor with data in tabular form, and supplies the commander with a full picture on the tactical console. The minimal system configuration, Reshet TADS/Data Link includes the target acquisition and designation system console and data management terminal, with the central processor unit. This configuration is sufficient to provide: display of data from ship's own radar(s), automatic tracking, synthetic symbols, data link management and display, and target designation capability.

The tactical display console (TDC) component of the Reshet system can be used as an independent fully computerised multi-function unit. In such a role it can provide: local data processing, display of maps, grids and zones, navigational solutions and display, automatic updating of own position, local processing of operator requests, definition of targets and bearing lines, computation and presentation of targets and their track histories, etc.

STATUS

In production and operational with Israeli and other naval forces.

CONTRACTOR

IAI, Israel Aircraft Industries Ltd, Ben Gurion International Airport, 70100, Israel.

4057.261

DG NAVAL FIRE CONTROL SYSTEM

The DG (Dvora/Gabriel) weapon control system is based on the FCS developed for the Gabriel anti-ship missile system (**6019.221**) as fitted in the 'Dvora' fast missile boat. It is especially adapted for smaller naval craft like 'Dvora' class vessels, enabling boats equipped with this system to enjoy a fire power normally available only on larger ships.

A variant of the basic Gabriel integrated fire control system, the DG system has been designed for missile control and guidance only. The system handles search, acquisition and tracking of surface targets, target designation from one unit to another and Gabriel missile trajectory precision guidance.

The main sub-units of the DG system are: fire control tracking radar (FCR) components, comprised of antenna (FCRA), transmitter (FCRT), and receiver (FCRR); a standard search and navigation radar (adapted by the Electronics Division of MBT Weapon Systems to designate targets to the weapon system); a stabilised periscopic optical sight (OS); a display control and processing unit (DCPU); a vertical reference unit (VRU); and Gabriel missile fixed single launchers.

Only two operators are required to operate the system. One, the optical sight operator, handles search and tracking by means of the periscopic OS from the closed bridge, while the other, the DCPU operator, handles FCR operation and missile control and guidance.

STATUS

No information released.

CONTRACTOR

MBT, Israel Aircraft Industries Ltd, Yehud Industrial Zone 56000, Israel.

ITALY

4058.281

NA 18 FIRE CONTROL SYSTEM

The NA 18 is an optronic fire control system (FCS) developed as a lightweight autonomous system for small and medium tonnage ships. In larger ships it serves as a back-up to the existing radar FCS, especially in the presence of heavy ECM.

Fitted with TV, IR and laser sensors, the system is capable of detecting and tracking targets out to a range compatible with the efficient use of a 76 mm gun.

The NA 18 can provide small vessels with a FCS which incorporates command and control and navigation facilities in addition to its FCS function.

Operational features include presentation of search radar video, autonomous search, rate-aided tracking of radar targets, automatic optronic tracking with target-matched tracking prediction models, line-of-sight, line-of-fire stabilisation, control of two guns with ballistic calculations for two different calibres.

CHARACTERISTICS

TV camera
Type: Silicon target
Focal length: 30 – 360 mm

IR camera
Type: Spectral band 8-12 µm
Field of view: 4 × 2.6° (wide) 2 × 1.3° (narrow)

Laser rangefinder
Type: Nd: YAG^3
Peak power: 18 MW

STATUS

In current production. In service with the Italian and several other navies.

CONTRACTOR

Selenia-Elsag Naval Systems Consortium, Via di S. Alessandro 28/30, 00131 Rome, Italy.

NA 18 E-O director equipped with TV camera, laser and infra-red thermal camera

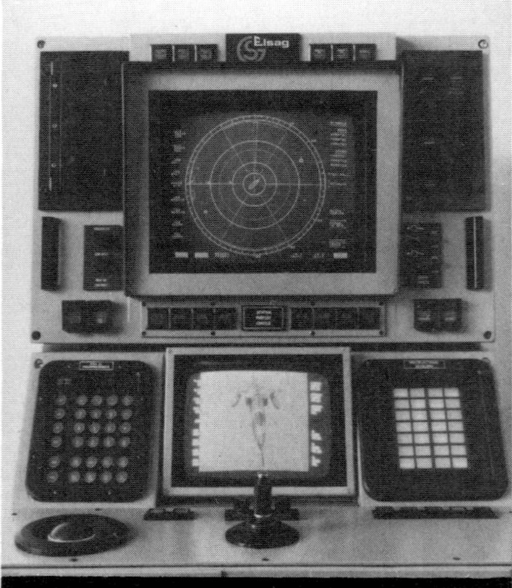

NA 18 operator's console with radar and TV image presentation

4059.281

NA 21 FIRE CONTROL SYSTEM

A development of the well proven and highly successful NA 10 series. Using digital technology with Elsag ESA 24 microprocessors, the NA 21 is capable of controlling three guns of two different calibres and a surface-to-air missile system as well as designating targets for a surface-to-surface missile system.

Other operational features of the system include, programmed autonomous tracker search (panoramic/sectorial), tracking of missile/aircraft and surface targets with target-matched tracking prediction models and presentation of the search radar video.

The fire control radar associated with the NA 21 system is the Orion RTN-10X operating in the I-band (X-band). The RTN-10X features tri-port conical scanning, high tracking accuracy, frequency agility and good ECCM capability.

A TV camera is mounted on the radar director for monitoring and visual control.

NA 21 FC radars as installed aboard and Ecuadorean corvette. Search radar is the RAN-10S

CHARACTERISTICS
Radar
Frequency: I-band (X-band)
Antenna type: Slotted parabolic reflector, high gain, tri-port feed
Tracking method: Conical scan
Transmitter: Frequency changeable high power magnetron
TV
Type: Silicon target
Focal length: 300 mm
STATUS
In current production. In service with the Italian Navy and numerous other navies.
CONTRACTOR
Selenia-Elsag Naval Systems Consortium, Via di S. Alessandro 28/30, 00131 Rome, Italy.

6740.281
NA 25 WEAPON CONTROL SYSTEM

The NA 25 is a modular weapon control system to which a command and control capability can be associated by using a high power multi-processing system together with a new generation of display systems featuring a very high graphic capability.

The fire control section of the NA 25 is provided with radar and electro-optical sensors and is capable of controlling weapons of medium calibre in the anti-aircraft and anti-surface roles as well as small calibre weapons in the CIWS role. Up to two guns of different calibre can be controlled simultaneously.

The fire control radar associated to the system is the Orion RTN-25X, a fully coherent J-band (X-band) radar which is characterised by anti-nodding, ECCM and anti-clutter features together with high tracking accuracy.

An electro-optic system (IR/laser) can be mounted on the radar director to provide an alternative line-of-sight on the same target while a TV camera is used for monitoring and visual control as well as an alternative tracker.

In order to enable the NA 25 to perform command and control functions in addition to fire control, the system makes use of a MARA 286 multi-processor and the newly developed 'MAGICS' console, provided with two high resolution colour monitors capable of displaying raw video from different sensors as well as the tactical situation processed by the computing system. A second display console can be added to enhance the command and control capability of the system.

Operational features of the NA 25 include, programmed autonomous tracker search (panoramic or sectorial), surveillance based on the ship's search radar video, threat evaluation and automatic engagement of the priority target, tracking of missile/aircraft and surface targets with target-matched tracking prediction models, control of two guns of different calibre, automatic target engagement for CIWS functions, operational command and control functions (such as tracking, identification, situation display, data link management, aircraft control, etc).

CHARACTERISTICS
Radar
Frequency: J-band (X-band)
Antenna type: Horn cassegrain
Tracking method: Monopulse
Transmitter: Coherent with TWT. Coded waveform
IR camera
Type: Thermal imaging TINA (8-12 µm)
TV camera
Type: Silicon target
Laser rangefinder
Type: Nd: YAG^3
Peak power: 18 MW
STATUS
Development.
CONTRACTOR
Selenia-Elsag Naval Systems Consortium, Via di S. Alessandro 28/30, 00131 Rome, Italy.

3861.281
NA 30 SERIES FIRE CONTROL SYSTEM

The NA 30 is a multi-purpose naval weapon control system developed for use with a surface-to-air missile system and medium and close range guns (up to three gun outputs and two ballistic calculations). The NA 30 can provide ships of any size with a complete weapon control system for defence against threats from medium to close range.

The fire control radar associated with the NA 30 system is the Orion RTN-30X. This radar is a monopulse fully coherent I-band (X-band) tracker radiating a coded waveform. The Orion RTN-30X has many common components with the RAN-30X surveillance radar.

An electro-optic system (IR/laser) can be mounted on the radar director to provide an alternative line-of-sight on the same target, while a TV camera is used for monitoring and visual control as well as alternative tracker.

Operational features include, automatic engagement of targets initiated by search radar, evaluation of the priority target in a prefixed guard ring, automatic acquisition and tracking of the priority target, target correlation for reaction co-ordinating with other fire control systems, specialised tracking prediction model for the missile threat, target-matched models for air/surface targets, automatic combined reaction with surface-to-air missiles and guns with co-ordination of the firing of the different weapons, control of the surface-to-air missile system and of the guns (up to three guns of two different calibres).

CHARACTERISTICS
Radar
Frequency: I-band (X-band)

Electro-optical director for NA 30

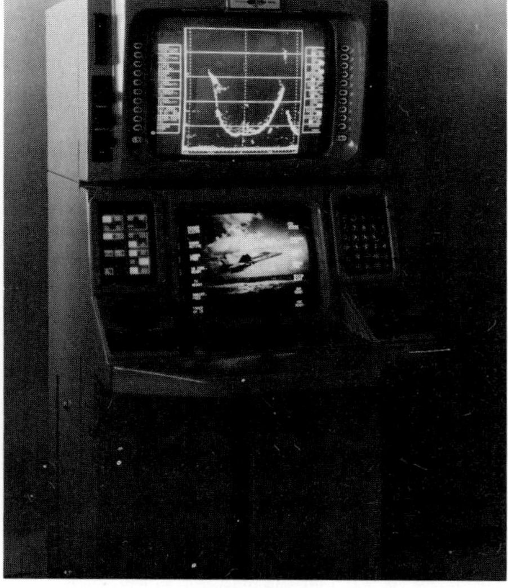
NA 30 multi-function supervision console

Transmitter: Coherent with TWT. Coded waveform
Antenna type: Cassegrain
Tracking method: Monopulse
IR camera
Type: IR pointer (3-5 µm) or IR imaging (SPRITE 8-12 µm)
TV camera
Type: Silicon target
Focal length: 300 mm
Laser rangefinder
Type: Nd: YAG^3
Peak power: 18 MW
STATUS
In current production. In service with the Italian Navy in helicopter carrier *Garibaldi* and 'Maestrale' class frigates.
CONTRACTOR
Selenia-Elsag Naval Systems Consortium, Via di S. Alessandro 28/30, 00131 Rome, Italy.

1920.281
DARDO FIRE CONTROL SYSTEM

Dardo is a Close-In Weapon System (CIWS) developed specifically for 'Lupo' and 'Maestrale' class frigates and the helicopter carrier *Garibaldi*. This system is for use against sea-skimming missiles, diving attack missiles and aircraft-launched rockets and missiles. The Dardo system controls one Breda twin 40L70 automatic gun mount.

The fire control radar associated with the Dardo system is the Orion RTN-20X operating in the J-band (X-band). This monopulse, fully coherent radar has been specifically developed for the CIWS task, one of its principal features being the effective reduction of errors caused by image effect. Other features of the radar include high sub-clutter visibility and good ECCM capability, nodding free self adaptive tracking as well as automatic spotting of rounds.

The operational features of the Dardo system include, automatic detection of targets based on search radar video extraction, the evaluation of the

Dardo supervision consoles

priority target within a preset guard ring and automatic engagement, a specialised tracking prediction model for the missile threat, automatic acquisition and tracking of a missile launched by the tracked

Orion 20X tracking radar with TV camera used in Dardo system

platform, automatic firing action, fast target shifting in case of multiple threat, alternative semi-automatic or manual operational modes and a TV camera mounted on the radar director for monitoring and visual control.

CHARACTERISTICS

Radar

Frequency: J-band (X-band)

Antenna type: Cassegrain

Transmitter: Coherent chain with TWT amplifier

Tracking method: Monopulse

TV

Type: Silicon target

Focal length: 300 mm

STATUS

In current production. In service with the Italian Navy and several other navies.

CONTRACTOR

Selenia-Elsag Naval Systems Consortium, Via di S. Alessandro 28/30, 00131 Rome. Italy.

6276.281

PEGASO FIRE CONTROL SYSTEM

The Pegaso system is an optronic fire control system (FCS) designed to perform two distinct functions – as the main FCS, for vessels whose use for combat is not foreseen as the primary task (landing craft, offshore patrol vessels, etc), and as a secondary FCS on vessels of any tonnage as an adjunct to a main radar FCS.

The system sensor fit can be configured in various ways dependent upon the requirement.These range from only a TV sensor for angular tracking when a simple optronic aid to a radar FCS is required, to a full complement of TV, IR and laser sensors when acting as an autonomous system. In the full configuration Pegaso is capable of receiving designations from external sources and controlling weapons of up to and including 76 mm calibre.

The operational features include, autonomous search, target designation from ship's detection sources, special acquisition patterns, automatic optronic tracking with target-matched tracking prediction models, line-of-sight, line-of-fire stabilisation and control of one gun.

CHARACTERISTICS

TV camera

Type: Silicon target

Focal length: 30 – 360 mm

Lens: Zoom type or fixed

IR camera

Type: Spectral band 8-12 μm

Field of view: 4 × 2.6° (wide) 2 × 1.3° (narrow)

Laser rangefinder

Type: Nd: YAG[3]

Peak power: 18 MW

STATUS

In current production. In service with the Italian Navy.

CONTRACTOR

Selenia-Elsag Naval Systems Consortium, Via di S. Alessandro 28/30, 00131 Rome, Italy.

4824.281

SEN SERIES COMBAT SYSTEMS

With the aim of meeting the weapon control requirements of medium to low tonnage warships in which there are usually space and resource limitations, Selenia-Elsag have developed a family of combat systems known as the SEN series.

The philosophy of the SEN series configuration has been to allow systems to respond adequately to the two basic operational requirements of a naval combat system, ie command and control and weapon control.

The SEN series includes a range of configurations to meet customer requirements. The components are drawn from the Selenia-Elsag product range.

The SEN series has a 'low end' and a 'high end' with a range of intermediate possibilities. An example of the 'low end' being a system comprising a search or navigation radar, a multi-functional console and an optronic fire control system. Under the management of a single operator this would give a small ship such as an OPV, Hydrofoil or small FPB, the following capabilities:

(1) radar surveillance
(2) electro-optical surveillance
(3) radar tracking of surface target in TWS mode
(4) optronic tracking
(5) gun control
(6) surface-to-surface missile management.

At the 'high end' of the SEN series a system can be configured which is suitable for vessels of around 400 to 600 tons operating in a heavy threat environment, both surface and air. Such a system would include a search radar and appropriate number of multi-functional consoles, a radar fire control system, additional optronic lines-of-sight and active and passive EW systems. Under the management of a minimum of two operators the system enables the ship to undertake the following tasks:

(1) extended radar and IR surveillance
(2) tactical evaluation, command and control
(3) auto-tracking and tactical picture compilation (air, surface)
(4) data link management
(5) target tracking by own tracking radar or optronics
(6) gun control
(7) SSM management
(8) ESC surveillance
(9) ECM control.

STATUS

The SEN series combat systems are assembled from components in the existing Selenia-Elsag product range. In series production.

CONTRACTOR

Selenia-Elsag Naval Systems Consortium, Via di S. Alessandro 28/30, 00131 Rome, Italy.

4263.281

OG 30A NAVAL MULTI-SENSOR DIRECTOR

The OG 30A is a naval multi-sensor director which is composed of the Officine Galileo thermal imaging camera, the FIAR P0700 laser rangfinder and the P4651 TV camera, mounted on a servo-controlled two-axis stabilised pedestal. This multi-sensor with its related video converter, TV tracker and displays is integrated to provide the electro-optical (E-O) line-of-sight in the NA 30 FCS (**3861.281**), but is equally suitable for use with other FCS or alone as an autonomous FCS sensor system.

It performs the following functions: panoramic surveillance; observation and identification; acquisition and automatic tracking of surface and aerial targets.

CHARACTERISTICS

Daylight TV camera: P4651

Field of view: 1.8 × 2.4°

Image format: 3:4

Scanning standard: TV CCIR

Sensor: Vidicon tube (S18 Phosphor)

Minimum sensor resolution: 700 TV lines per picture height

Galileo thermal camera

Field of view: 3 × 6°

Image format: 1:2

Instantaneous field of view: 0.25 mrad

Frame rate: 25 Hz

Lines per frame: 200

IR detector type: CMT linear array

Wavelength: 8 – 14 microns

Operating temperature: 77 K

Laser rangefinder P0700

Wavelength: 1.06 microns

Output: 20 – 25 MW

PRF: 15 Hz

Range: 400 m – 15 km

TV tracker P4700

Pedestal

Elevation: –25 to +85°

Azimuth: 360° continuous

Velocity: 2.2 rad/sec (elevation); 1.5 rad/sec (azimuth)

STATUS

In service with the Italian Navy.

CONTRACTOR

Officine Galileo SpA, Via Einstein 35, 50013 Campi Bisenzio, Florence, Italy.

OG 30A naval multi-sensor director

4498.461

SACTIS COMBAT INFORMATION SYSTEM

SACTIS is the abbreviated title given to the submarine action information system being developed for Italian Navy submarines of the Sauro class by SMA in collaboration with Datamat and other contractors. The former concern was selected as prime contractor for this project in 1981 when it was decided to equip this class of submarine with a new AIO system. Two versions are to be produced known as SACTIS 1 and 2, or by the Italian Navy designations MM/BSN-716(V)1 and MM/BSN-716(V)2, respectively. The two versions are intended for fitting in the second and projected third batches of Sauro class boats. In both cases the principal functions include:

(1) automatic data acquisition from all sensors (eg hydrophone, sonar, passive ranging system, search and attack periscopes, radar ESM, navigation system, depth sounder etc)
(2) manual data input
(3) real-time computation of ship's position and target position (from bearing data only)
(4) data display (raw and processed) with four different presentations: unfiltered situation; tactical situation; time/bearing display; tactical operation tabular evaluation
(5) calculation and display of data for typical manoeuvres such as screen or barrage penetration, evasion, target interception or collision course, approach and divergence routes, CPA
(6) data and event recording
(7) playback for training and analysis.

The hardware comprises two display groups which have vertical CRT displays and keyboards for data input and system management, with separate alpha-numeric 'tote' displays above the situation display CRTs; a central processor unit based on a Rolm MSE 14 digital computer; disk memories; a printer; a specially designed interface unit for input/output between SACTIS and the various sensors. The two displays are arranged side-by-side to form a two-man operating console and up to 30 separate targets can be displayed, from which individual track histories can be selected for examination and ten of them presented as filtered targets. Main sensors connected to the system include the BPS-704 radar, IPD-70 sonar system, and the Thetis electronic warfare system.

The later SACTIS 2 version includes provision for connection with the submarine's A184 torpedo fire control system (**3252.481**), and inclusion of the ELT/810 sonar prediction system. There is also extended computer capacity and a third operator position at the SACTIS console.

STATUS

Italian 'Sauro' Class submarines are fitted with the MM/BSN-716 system.

CONTRACTORS

SMA – Segnalamento Marittimo ed Aereo SpA, Via del Ferrone-Soffiano, 50100 Florence, Italy. (Main)

Datamat Ingegneria dei Sistemi SpA, Via Simone Martini 126, 00143 Rome, Italy. (Software)

Three-console SACTIS system

4827.481

SUBMARINE INTEGRATED COMBAT SYSTEM (SICS)

The Italian submarine integrated combat system (SICS) has the following functions:

(1) sensor and tactical data handling
(2) fire control
(3) command
(4) common services

The system hardware is designed in a compact, modular fashion to permit adaptation or expansion as required. Other features include data and video distribution network, advanced computer technology and sensor data processing for optimum weapons utilisation. Growth potential was considered essential to permit changes in sensor and weapons outfit and modifications to operational procedures are unavoidable during the life of a modern submarine.

The system uses general purpose consoles which are completely interchangeable for operational functions. This allows consoles to be reconfigured simply to accommodate changing role requirements, and also permits consoles to change their roles while the boat is 'on patrol' according to varying operational circumstances to permit a reduction in manning requirements whenever possible.

The system's sensor and tactical data handling controls the processing of inputs from sonar, periscope, ESM, radar, log and compass, and other navigational equipment.

The FCS function provides for weapon status monitoring, computation of optimum target engagement and the computation of weapon pre-settings before launch.

The co-ordination, evaluation, designation and other command functions are carried out by the system together with comprehensive monitoring and display facilities, using data directly from sensors to produce a filtered situation picture.

It is possible that different versions of SICS will appear to comply with the different weapons fits of various classes of submarine. No confirmed details of hardware have been received, but contractor data reaching *Jane's Weapon Systems* indicates individual weapon control sub-system configurations from different manufacturers.

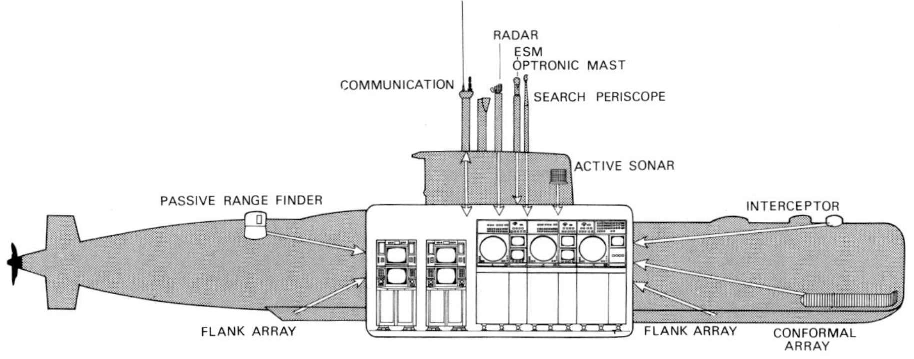

Outline of typical SICS arrangement

STATUS

Development.

CONTRACTORS

SICS is manufactured by Elsag in collaboration with two companies of the FIAT Group, Whitehead-Motofides, and SEPA, with the participation of USEA in the sonar sensor element. This group of Italian underwater warfare interests have since formalised the relationship with the establishment in November 1984 of the WELSE consortium, in which Elsag has a 50 per cent holding, with Whitehead and SEPA having 25 per cent interest each. The WELSE consortium is based at the Elsag headquarters in Genoa, and future SICS activities will probably be directed from this establishment.

4828.281

SEPA NAVAL FIRE CONTROL SYSTEM

The Società di Elettronica per l'Automazione (SEPA) member of the FIAT Group, in addition to participating in the WELSE underwater warfare consortium and contributing to the SICS programme (**4827.481**), offers naval fire control systems (FCSs) for various other applications independently. Such digital systems are mostly based on the use of SEPA militarised mini and microprocessors of the company's ULP Series.

For example a developed version of the torpedo and weapons control sub-system for SICS is available separately as the FCS Mk 3, for submarine command and control which will be used in the 'Improved Sauro' class submarines, and another version has been developed for 'Maestrale' class surface ships as the FCS Mk 2.

FCS Mk 2

The surface ship FCS Mk 2 includes provision for full integration with the ship's combat system (SCS). It incorporates SEPA digital processor, and the main functions of the FCS Mk 2 are:

(1) evaluation and display of the tactical situation
(2) weapons control
(3) acoustic sensor prediction
(4) tracking, classification and evaluation of surface and underwater targets
(5) threat evaluation
(6) countermeasures control.

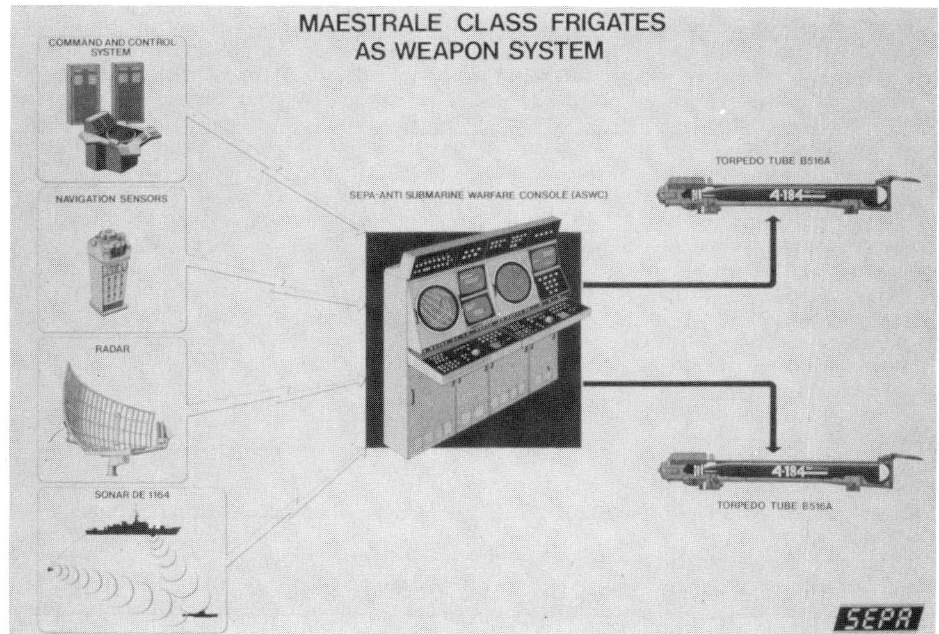

Schematic diagram of FCS Mk 2

The main sensor inputs will include navigational data, search radar information, and DE 1164 sonar signals, and the FCS Mk 2 interfaces with the ship's command and control system. The principal weapons controlled are the port and starboard B516A torpedo tubes.

FCS Mk 3

This is a fully integrated submarine command and control system having as its main functions:

(1) acoustic sensor performance prediction
(2) prediction of counter-detection range
(3) target motion analysis
(4) threat evaluation
(5) target designation for attack
(6) weapons control (torpedoes and missiles)
(7) countermeasures control.

Sensor inputs to the system include radar, navigation and attitude information, periscope and acoustic data from the integrated active/passive sonar system.

Target and ship's own data from these sources are used to calculate and display target positions and target vectors, impact point predictions, tactical situation, a launch and guidance display, and wire guidance signals. The local control and switching box connects the signals from the FCS to the selected torpedoes, provides lauching tube control, and feeds electric power to torpedoes in the tubes. Within the torpedo, the electronic control unit carries out wire signal interfacing, torpedo steering in accordance with FCS commands, computation of target data from the torpedo acoustic head, and torpedo homing.

Inputs to the system include: active and passive sonar data, navigational information, radar and ESM/EW sensors, and periscope. Weapons include A184 torpedo tubes, which are connected via a local control and switching box.

STATUS

In production.

CONTRACTOR

SEPA – Società di Elettronica per l'Automazione SpA, Corso Giulio Cesare 294, 10154 Turin, Italy.

Consoles for the SEPA FCS Mk 3

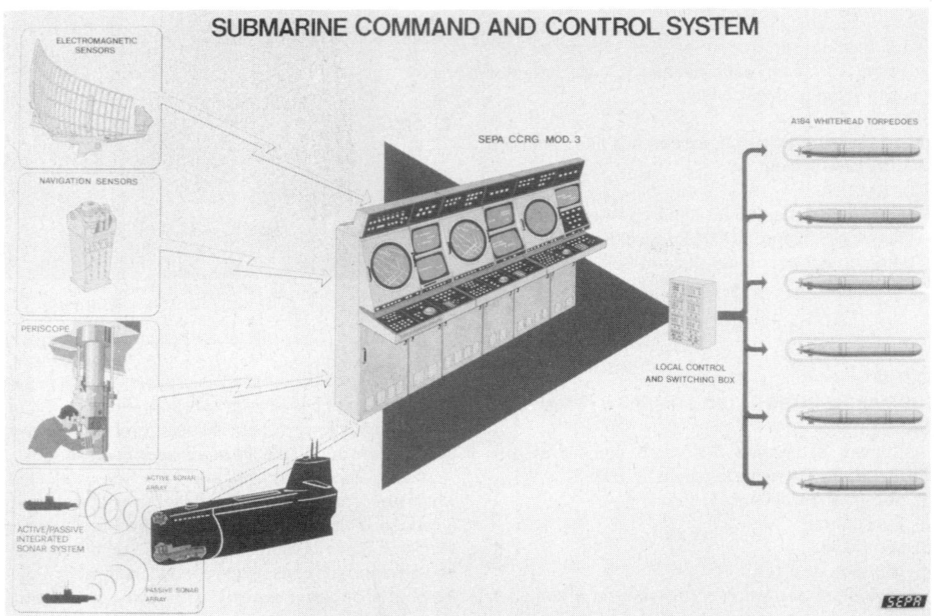

Schematic diagram of FCS Mk 3 for submarines

4826.481

IMICS (INTEGRATED MINE-HUNTING COMBAT SYSTEM)

The integrated mine-hunting combat system (IMICS) is a combination of the Datamat/SMA MM/SSN-714 mine-hunting action information sub-system (MACTIS, **4499.261**), the FIAR AN/SQQ-14A sonar (**3234.453**), SMA MM/SPN-703 navigation radar (**3482.253**), an Elmer telecoms sub-system, and the Datamat MM/SSN-715 navigation sub-system.

Its functions are:

(1) to plan mine-hunting operations
(2) search, detection and classification of submerged targets
(3) analysis of target characteristics
(4) generation, display and storage of action information for the co-ordination and control of mine-hunting operations
(5) navigation
(6) external telecommunications and data link for the automatic exchange of data among fleet units
(7) guidance of surface or submersible vehicles for mine neutralisation
(8) recording of operational data and events.

The MACTIS element is located between the sensor sub-systems, the neutralisation sub-system and the platform. The following units comprise MACTIS: an operator station with one 16 inch graphic video screen, a functional keyboard and a tracker ball, two display units (bridge and CIC/ops room) for data display and navigation, processing signal distribution and power supply units.

The AN/SQQ-14A is a solid-state mine detection and classification sonar, with a two frequency variable depth beam, which enables two operators to simultaneously detect, memorise and classify mines. Separate search, classification and memory display consoles are provided. Comprehensive navigation equipment, in addition to the MM/SPN-703 navigation radar, is provided to ensure precise positioning of the ships and accurate plotting of underwater objects encountered.

STATUS

Fitting in Italian Navy 'Lerici' class minehunters.

CONTRACTORS

Datamat Ingegneria Sistemi SpA, Via del Simone Martini 126, 00143 Rome, Italy.

Elmer, Viale dell' Industria 4, 00040 Pomezia (Rome), Italy.

FIAR, Via Montefeltro 8, 20156 Milan, Italy.

SMA Segnalamento Marittimo ed Aero SpA, Via del Ferrone-Soffiano, PO Box 200, 50100 Florence, Italy.

4499.261

MM/SSN-714 MINESWEEPER DATA PROCESSING SYSTEM

Under the designation MM/SSN-714, SMA and Datamat have developed a digital navigation and plotting system for the Italian Navy's new Lerici class of mine countermeasures ships. The main functions of the SSN-714 MACTIS (mine-hunting action information sub-system) are:

(1) automatic computation and presentation of the ship's current position
(2) display of the tactical situation
(3) analysis and presentation of target characteristics
(4) location of surface targets
(5) event recording
(6) operations planning
(7) guidance of surface and underwater craft.

The system is based on a computer of the same or similar type to that employed in the SACTIS submarine AIO system (**4498.461**), namely a Rolm MSE 14 machine. In the SSN-714 system this is interfaced with recording units, display units, controls, printer and ship's sensors. The principal items in the last of these categories are radar, sonar, compass, log and various navigation aids. The operator's display has a vertical screen CRT and a keyboard for communications with the system, and there are supplementary data readout display units and associated input controls on the bridge and in the operations room.

MACTIS is the central sub-system of the Integrated Mine-hunting Combat System (IMICS, **4826.481**).

STATUS

Production.

CONTRACTORS

SMA – Segnalamento Marittimo ed Aereo SpA, Via del Ferrone-Soffiano, 50100 Florence, Italy. (Main)

Datamat Ingegneria dei Sistemi SpA, Via Simone Martini 126, 00143 Rome, Italy. (Software)

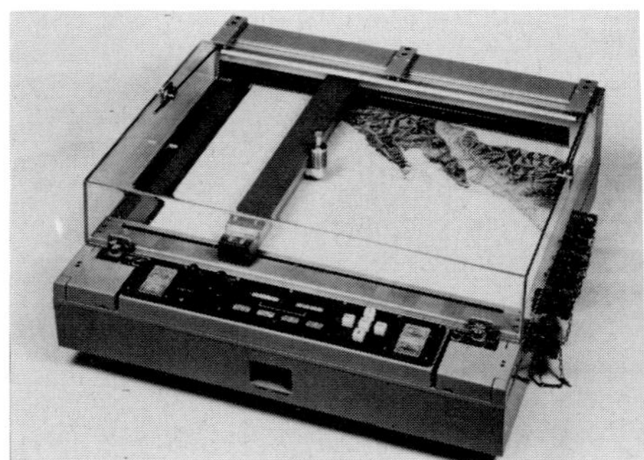

Digital plotter unit of MM/SSN-715 system

4500.261

MM/SSN-715 NAVIGATION AND TRACKING SYSTEM

This system was developed for the Italian Navy's Sparviero class of military hydrofoils and it performs the following functions:

(1) mission planning
(2) navigation and tactical plotting
(3) navigational calculations for own ship and various categories of other targets
(4) vector computation and presentation.

The system is based on the use of a Rolm micro-computer which is interfaced with a hand-held input/output terminal device, a video extractor unit, and an X/Y plotter display unit. Inputs to the system include ship's radar, gyro and log.

The SSN-715 is used as an auxiliary system for navigation purposes to provide tracking of own ship's position as well as that of a target tracked by the navigation radar, presented on a chart by means of the X/Y plotter. Use of the latter means of presentation ensures rapid interpretation of the data and removes possible ambiguities, and the provision of a portable hand-held input/ouput control for operation of the system materially assists the tasks of the ship's command team personnel.

It is also employed as an important sub-system of the IMICS installation (**4826.481**).

STATUS

Production.

CONTRACTORS

SMA – Segnalamento Marittimo ed Aereo SpA, Via del Ferrone-Soffiano, 50100 Florence, Italy. (Main)

Datamat Ingegneria dei Sistemi SpA, Via Simone Martini 126, 00143 Rome, Italy. (Software)

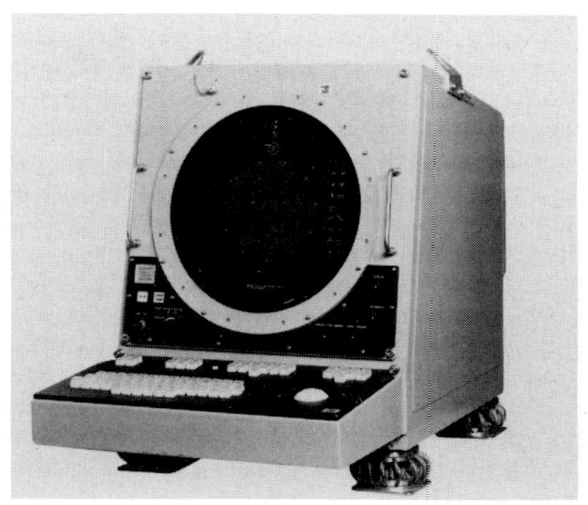

Graphic display for MM/SSN-714 system

1551.281

ALBATROS WEAPON SYSTEM

Albatros is a naval all-weather missile and gun weapon system. The missile used is the Aspide (**1656.331**), a new multi-role weapon developed by Selenia under an Italian MoD contract. Missile control is provided by the Albatros Missile Section (**2228.231**), which is integrated with the gun fire control system to achieve a total weapon system. The naval guns can be of any type, provided the required performance matching with the relevant gun fire control system is ensured.

The original development, referred to as Albatros Mk 1 Mod 1, was a system for integration with an Elsag gun fire control system and it evolved in various production configurations with the common designation of Albatros Mk 2. For details see the 1987-88 edition.

For each configuration the system is available in single or double headed version (type 1/1 or 2/1), the latter having the ability to engage two targets simultaneously. Missile guidance is provided by radar illumination of the tracked target through the director antenna. Pre-flight information and launching orders are produced by the Albatros Missile Section and the relevant consoles are contained in a panel fitted in the weapon control console so as to enable full fire control by a single operator, both for missiles and guns. The two weapons can be employed simultaneously; however missile engagements are normally performed at long ranges (out to 15 km), permitting target shifting or reiteration of firing, while anti-air gunfire actions are initiated at shorter ranges or the guns are used in the surface fire or shore bombardment roles. Engagement of surface targets can also be accomplished by means of the Aspide missile.

Albatros dual-channel tracking and illumination radar

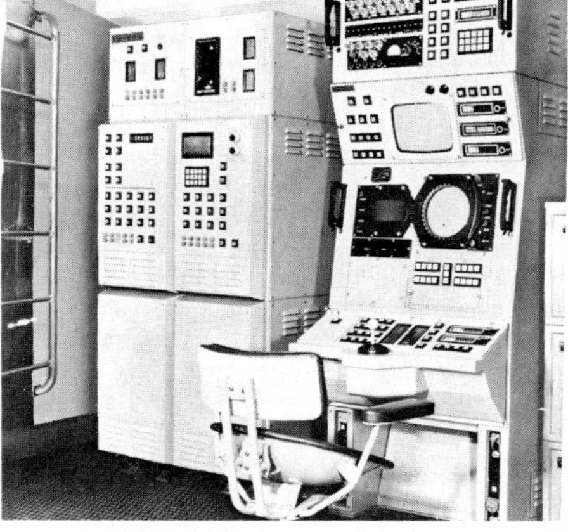

Albatros Mod 3 operator's control console

STATUS

At the end of 1987 more than 100 systems were in service. These include: 10 Lupo class frigates, 7 Blohm & Voss frigates, 8 Descubierta class, Libyan 'Dat Assawari', the carrier 'Giuseppe Garibaldi' and the Thai Sukhodhai class corvettes.

CONTRACTOR

Consorzio Sistemi Navali Selenia-Elsag, Via Panama 52, 00198 Rome, Italy.

3184.261

IPN TACTICAL DATA DISPLAY SYSTEMS

IPN is a series of command and control systems for naval operations. The systems are able to gather, correlate and filter the information coming from ship sensors, communications and data networks and to display clearly the tactical situation.

In addition to this, the systems, which can be configured according to the operational requirements of the user, provide assistance to the command in the following areas:

(1) evaluation of the situation
(2) threat evaluation and control of the weapons
(3) management of electronic countermeasures
(4) aircraft and helicopter control
(5) conduct of ASW operations
(6) management of data links
(7) management of data base.

Single operator vertical console of IPN-10 command and control system

The IPN-10 systems are configured in a series of central units, engineered in small cabinets and a number of display consoles, according to the function required.

The IPN systems, as originally conceived, are based on the NDC-160 and CDG-3032 computers and on display systems consisting of the three-operator horizontal console and the single-operator vertical console. Such systems have now been considerably enhanced by the introduction of the new computers of the MARA 286 type and MAGICS display consoles.

The MARA 286 computers are characterised by advanced hardware and software technology, very high modularity which enables the use of a small range of hardware module types to construct systems with a wide range of processing power, memory and interface capacity, strong support for fault detection, fault tolerance and maintainability.

The MARA 286 computer uses the same software which runs on a Digital VAX commercial computer and uses the ADA language.

The MAGICS display consoles are characterised by high modularity with the scope of enabling the composition of differing configurations to fulfill the requirements of the combat system and platform control, use of the raster scan technique and bit map concept associated with high resolution colour monitors, and the use of MARA 286 computers as display processor.

Multiple operator horizontal console for IPN-10 system

STATUS

IPN systems are in production and are in service with the Italian and several other navies. The IPN-10 is installed in the 'Saettia' class patrol boat and 'Minerva' class corvettes while IPN-20 is installed in 'Lupo' class frigates.

CONTRACTOR

Selenia-Elsag Naval Systems Consortium, Via di S. Alessandro 28/30, 00131 Rome, Italy.

NETHERLANDS

4271.281

GEMINI NAVAL FIRE CONTROL SYSTEM

The Gemini weapon control system has been designed for use aboard smaller naval vessels to provide one-man fire control facilities for guns and/or missile armament for low-level self-defence purposes. It is engineered on a modular basis so that extended versions can be configured to meet differing operational requirements and to enable its use on various classes of ship from corvette to frigate size.

The system consists of a two-axis stabilised I-band search antenna mounted coaxially with a K-band tracking antenna. The most important benefits of the two antennas in one assembly are: all round unobscured coverage for both search and track; both antennas can be located in the most favourable position, ie on the masthead. Surface targets are detected by the I-band search radar to the radar horizon while fighter-sized aircraft can be detected at approximately 25 km.

Full 3D tracking facilities are provided by the K-band radar which has the same processing system as the Goalkeeper CIWS. Fast Fourier analysis techniques are used to overcome sea clutter and maintain contact with targets having small radar reflecting targets. The K-band tracker permits accurate, pinpoint tracking down to sea-skimmer levels and considerably reduces the mirror effects. The narrow K-band pencil beam is also very difficult to jam.

Data from the sensors is fed into the system's control console which is below deck. A labelled position display presents the search radar video and pertinent synthetic data for surveillance and target indication. A monitor and control display shows the tracking information (A-scope) and alpha-numerics and also displays the lightpen operated controls. Lightpen operation with associated microprocessing minimises the number of hardware controls. Functional changes can easily be carried out since, in most cases, only software is affected.

While the basic configuration will serve the needs of numerous vessels, a number of options are available to expand the system and adapt it to special requirements. These include:

(1) radar: more powerful transmitters, travelling wave tube, larger search antenna, combined I- and K-band tracking; transmitters/receivers (and display) for navigation, circular polarisation, lin/log/MTI/DF
(2) vertical reference platform
(3) optronics: TV camera/IR camera (combined with autotrack), laser rangefinder
(4) target tracking facilities: track-while-scan; video extractor
(5) IFF
(6) shock absorbing mounting for director
(7) radar data handling.

STATUS

In production.

CONTRACTOR

Hollandse Signaalapparaten BV, PO Box 42, 7550 GD Hengelo, Netherlands.

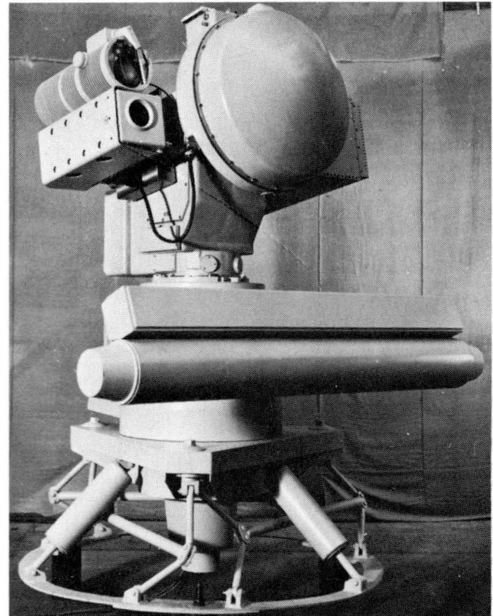

Gemini naval FCS sensor assembly on shock mounts, with integrated radar and E-O sensors

---

3767.281

LIROD 8 NAVAL DIRECTOR

LIROD (Lightweight Radar Optronic Director) is an upgrade of the LIOD optronic fire control system (**3143.293**). It combines an 8 mm wavelength pulse doppler radar/electro-optical director. It is particularly suited to all-weather applications where high accuracy combined with low-level capability and high resistance to jamming are required. Its light weight makes the system appropriate for stand-alone gun direction on small ships, while it may also be used as an additional tracking and monitoring channel within a WM20 Series weapon control system (**1259.281**).

LIROD 8 Mk 2 tracker

A feature of the LIROD system is the optional multi-sensor logic which will automatically select the most reliable signal from the sensor array. The standard LIROD set of sensors (K-band radar and TV camera) can be extended with a laser rangefinder, laser tracker or an IR camera, allowing LIROD to be deployed as an active as well as passive system.

Operational functions that the system can perform autonomously include: optical and radar surveillance for air and surface targets; target designation for anti-air and anti-surface engagements; air or surface fire control; automatic radar/electro-optical tracking or rate-aided tracking of one air or surface target; air or surface target gun prediction calculation; simultaneous control of two dual-purpose guns of the same or different calibre; engagement monitoring, kill assessment and gunfire correction; and system status monitoring.

CHARACTERISTICS

Basic version

Radar system

Antenna type: Parabolic with monopulse cluster
Diameter: 0.6 m
Beam width: 1°
Polarisation: Vertical
Frequency band: 35 GHz
IF frequency: 60 MHz

Optronic system

TV camera: Militarised daylight television camera, remotely controlled field of view ranging from 3 to 30°, automatically selected lens filters, video format to CCIR standard, (625 lines, 25 pictures/s, 2:1 interlace)

Extended version

IR camera: Militarised thermal imaging system, long focal length lens system (germanium) field of view 3°, closed cycle cooling system, video format CCIR standard (625 lines, 25 pictures/s, 2:1 interlace)
Laser rangefinder: Hand or computer-triggered laser range measuring unit, beam divergency 1.2-1.5 mrad, measuring range 200-32766 m, solid-state Neodymium-Yag laser transmitter with 106 µm wavelength and 20 ns pulse length, maximum repetition frequency 10 Hz
Laser tracker: Beam divergency 1.5 and 6 mrad, transmitter identical to laser rangefinder pulse rate 25 Hz
TDS: Manned, fast reaction, pedestal type sight, telescoping sighthead with 7 × 50 binocular and coarse sight, 7-inch TV monitor for presentation of TV picture and alphanumeric data, training movements -170 to +170°, elevation movements –15 to +65°.

STATUS

Both standard and extended versions are in production.

CONTRACTOR

Hollandse Signaalapparaten BV, PO Box 42, 7550 GD Hengelo, Netherlands.

1259.281

M20 SERIES FIRE CONTROL SYSTEMS

The M20 Series comprises a range of integrated ships' fire control systems for use with guns, missiles and torpedoes, and designed for fitting in vessels ranging in size from fast patrol boats to destroyers. Principal applications are medium/close-range air defence and close-range surface defence. Simultaneous engagement of several targets is possible. A further use is for shore bombardment.

Computation and control is based upon a central digital computer, and the range of available sensors includes separate search and tracking radars, and optical trackers. The use of TV trackers is a further option.

The search and target tracking radar sub-system comprises separate search and track antennas, mounted on a common stabilised platform, the whole being enclosed in a weather-proof radome. Both antennas are fed from a common I-band transmitter. During search 100 per cent of the output power of the transmitter is supplied to the search antenna. After target detection and designation to the track radar, the output power is divided, by means of the waveguide switching, so that a certain percentage is supplied to the track antenna for target tracking. A dual arrangement is available with separate search and tracking radars.

In one version of the system (the WM26), only the search antenna is fitted. In this case it is mounted on top of the stabilised platform, and enclosed in a hemispherical radome. In all other versions the track antenna is fitted above and the search antenna below.

Individual 'M' and 'WM' series versions of this system have been described in greater detail in previous editions (**1259.281** 1983-84 and earlier).

STATUS

The total number of delivered or contracted M20 Series systems of all versions exceeds 300. Many versions of the system are fitted in numerous classes of vessel.

CONTRACTOR

Hollandse Signaalapparaten BV, PO Box 42, 7550 GD Hengelo, Netherlands.

1663.261

CIDIS – COMBAT INFORMATION DISPLAY SYSTEM

CIDIS is a modular display system designed for use in ships' combat information centres, where data from a number of external sources such as search radars, IFF and data link have to be processed by digital computer and organised and presented for use by the command team. The normal computer is a Signaal SMR family machine, but other types may be employed. The combination of CIDIS and digital computer constitutes DAISY, a digital action information system (**4496.261**).

The equipment comprising a CIDIS installation consists of a display central unit and a number of display consoles appropriate to the operational functions of the system. The former unit can contain a sensor data distribution unit which acts as the interface between the various sensors and the displays, a video extractor which converts the sensor signals into digital form used in the computer for target detection and tracking, and a computer data distribution unit which interfaces the displays to the computer.

There are two types of display console, vertical and horizontal or 'conference' type. The former is produced in two versions, one for the presentation of radar video and one for sonar video; in each case the console has two indicators, a 40 cm labelled position display (LPD) and a 7-inch (178 mm) TV-type alpha-numeric display (tote). The horizontal console, with a 40 cm LPD for radar presentation, has two operator positions and a third supervisory position. One or two 7- or 15-inch (178 or 381 mm) TV-type totes can be associated with each horizontal display for the presentation of either supplementary alpha-numeric information or TV video.

Instead of the 40 cm LPD unit, a 40/50 cm PVD unit can be provided for bright synthetic presentation.

Functions of the CIDIS include:

Display of raw video from ship's sensors on the LPD.

Display of synthetic data from the computer in two forms: (1) as synthetic data consisting of symbols superimposed on raw radar video on the LPD, with or without track labels, at the operator's discretion or (2) as alpha-numeric information on the tote display.

Communication with the computer by means of manual input keyboards. Quick-entry keyboards, handwheels and a rolling ball, and as special options a light-pen and touch-input device, can be connected.

STATUS

CIDIS and DAISY systems have been delivered for the navies of: Argentina, Belgium, India, Malaysia, Netherlands, Nigeria and Thailand.

CONTRACTOR

Hollandse Signaalapparaten BV, PO Box 42, 7550 GD Hengelo, Netherlands.

4496.261

DAISY AND FORESEE SYSTEMS

DAISY is a modular digital action information system designed to be used in combat information centres on board ships where data received from several sources, such as radars, secondary radars, sonar, data links, etc, must be processed and presented to the command and control team. DAISY forms the heart of a sensor, weapon control and command (SEWACO) system (**1664.281**).

The DAISY/Foresee tasks include:

(1) presentation of raw or digitalised information from primary and secondary radars
(2) compilation and presentation of tactical air, surface and sub-surface pictures
(3) data link operation
(4) designation of targets to weapon systems
(5) assistance in operations, including ASW helicopter direction and tactical navigation.

A DAISY system consists of a data handling cabinet and a number of display consoles appropriate to the operational functions of the system. The data handling cabinet contains the data handling computer, a sensor data distribution unit and one or more video extractors. The data handling computer normally used is a Signaal SMR family machine, but other types may be employed. The sensor data distribution unit performs the interface between the various sensors and the displays, while a video extractor converts the video from either a primary or secondary radar into a digital form used in the computer for target detection and tracking. The combination of DAISY hardware and system software constitutes the Foresee action information system for command and control.

For the display subsystem two types of display consoles are available: vertical (VDC) and horizontal (HDC) or "conference" type. For the presentation of either alphanumeric data in a tabular format or TV video the VDC is equipped with one 7-inch (178mm) alphanumeric display and the HDC with one or two 7- or 15-inch (178 or 381 mm) alphanumeric displays. The main display unit for VDC and HDC is a 40 cm labelled position display (LPD) or a 40/50 cm display unit of the plan view display (PVD) family.

Communication with the data handling computer is performed by means of manual input keyboards, programmable quick-entry keyboards, a rolling ball and handwheels. Other special input and designation devices such as a light-pen and touch input device can be integrated optionally.

The architecture of the latest generation of the DAISY/Foresee system and its components represent a major step ahead in system design with a high degree of integration and automation, and constitutes a highly improved man-machine interface.

In order to achieve this, sets of system tasks are defined as "system functions" (logical subsystems). Every system function can be independently accessed from any workstation in the system. Data communication between system function and workstation or other system function is handled by a data communication network, including high-speed data busses and/or direct connections and/or intelligent communications handling package (COMPASS). The computer complex in the system consists of Signaal's SMR-family computers. The complex provides fail-safe and fail-soft facilities through duplication of critical parts. Parts of the complex are decentralised to achieve optimum damage control.

General Purpose Workstations

To optimise operator flexibility the display subsystem consists of SIGHT (Signaal General-purpose High-resolution Tactical) workstations. These workstations are of the colour raster scan type. Simultaneously different types of data sources for display can be handled: graphic data (characters, vectors, etc), radar data, image data (maps, sonar images).

Storage for maps, manuals, procedures, etc, can be provided using laser optical disc devices. The workstation enables multiple system access (up to the four system functions may be linked functionally to the workstation), facilitating operator system access through modern windowing techniques.

Each workstation can be deployed for any of the system functions within the security limits of any given operator.

The data bus system provides the data highway between the major components of the system. It comprises a multiple data bus based on Ethernet. The busses (up to 4) may be applied in parallel (capacity up to 40 Mbits/sec) or in a redundant fail-safe configuration.

SINCOS (Ship's Integrated Communication System)

Any SIGHT workstation can also control the integrated external and internal communication system. A new communication system, SINCOS 1200, has been developed by Signaal. This system is specifically suitable for the use of GPW, since it adapts itself automatically to the functions required by a given GPW operator. The external part covers all military frequency bands from VLF up to UHF. The HF and VHF/UHF bands have full ECCM capability based on frequency hopping.

The internal part of SINCOS uses the digital FOCON-A fibre optical bus for the distribution of voice and equipment control signals. SINCOS has built in automatic reconfiguration facilities for fail-safe/fail-soft operation.

The latest generation of DAISY/Foresee provides a high degree of flexibility, and reaction speed is increased by drastic automation.

DAISY is a designation of the Royal Netherlands Navy.

STATUS

DAISY and Foresee systems have been delivered for the navies of Argentina, Belgium, Canada, Greece, India, Indonesia, Malaysia, Netherlands, Nigeria, Thailand, Turkey and Peru. SINCOS 1200 is in production for the Royal Netherlands Navy.

CONTRACTOR

Hollandse Signaalapparaten BV, PO Box 42, 7550 GD Hengelo, Netherlands.

1664.281

SEWACO – SENSOR, WEAPON CONTROL AND COMMAND SYSTEM

Under the SEWACO designation Hollandse Signaalapparaten produces integrated sensor, weapon and command systems of varying configurations intended for corvettes, frigates and higher-level combat ships. The same designation is now used also for integrated submarine systems. Full advantage is taken of the integration of the various sub-systems. System functions can be carried out using centralised or decentralised data processing, depending on the operational status of the combat system components.

A SEWACO system in general comprises primary radar sensors for long, medium and short range air and surface warning, and the search part of the combined search and tracking radar of the WM Series weapon control system, or CIWS search radars, secondary radar sensors such as IFF and helicopter transponders, sonar, ESM, and IR sensors. Furthermore, the system includes an action information system of the DAISY or Foresee type, a weapon control system composed of one or more separate tracking and illumination radars (STIR), radar/optronic (LIROD) systems, electro-optical (LIOD) system and optical target designation sights. Communication equipment including automatic data links, one or more primary and secondary video extractors and/or plot history generators for automatic initiation and tracking of air and surface targets, an ECM system and a newly developed sophisticated very short range air defence system against sea-skimmers completes the system. As the system components are of modular concept, the SEWACO system's configuration can be matched to specific requirements.

In some configurations not all the above system components will be of Signaal manufacture. In these cases, however, Signaal takes the responsibility for system integration.

The first integrated SEWACO system was developed for the Royal Netherlands Navy frigates of the 'Tromp' class. This system has been followed by new generations of SEWACO systems in which new technologies have been incorporated, eg for the sensors, computers and displays.

Note: SEWACO sequence numbers are used only for the Royal Netherlands Navy. All other SEWACO systems are assigned a specific customer-related number. Software for most SEWACO systems and sub-systems is also produced by Signaal.

SEWACO I

The principal sensors of the SEWACO I system in the new Royal Netherlands Navy guided missile frigates are the Signaal 3D multi-target tracking search and target designation radar (**1589.253**) and the combined search and tracking radars of the WM25 fire control system (**1359.281**) which is integrated into SEWACO I and the Signaal ESM system. The 3D radar also incorporates a slotted wave-guide antenna for IFF/SIF facilities. Among the weapons associated with SEWACO I are Tartar, Standard SM2 MR, the Harpoon SSM, the NATO Sea Sparrow point defence system, and a twin 4.7 inch (109 mm) gun turret.

SEWACO II

This system is for ten Royal Netherlands Navy standard frigates and for two Hellenic Navy S frigates. Sensors include the Signaal LW.08, and ZW.06 radars, the combined search and tracking radar of the WM25 weapon control system and ESM system, two optical target designation sights and an automatic data link. Non-Signaal sensors include IFF, a helicopter transponder system and the AN/SQS-505 sonar set. The SEWACO system consists of Signaal's DAISY digital action information system, WM25 weapon control system, including a separate tracking and illumination radar (STIR) on which a TV-camera is mounted, while the weapons fit comprises the OTO Melara 76/62 dual-purpose gun, Harpoon SSM, Nato Sea Sparrow, ASW torpedoes and a Lynx ASW helicopter.

SEWACO III

Discontinued programme.

SEWACO IV

This is the designation of the SEWACO system supplied for the 'Westhinder' class of escorts for the Belgian Navy. No official details have been revealed but the following data derived from published sources correspond with what is known of these new vessels and their weapons and equipment fits. Functions of the SEWACO system include:

(1) air, surface, and sub-surface warning
(2) compilation and display of tactical air, surface and sub-surface situations
(3) threat evaluation and automatic air target selection for engagement by the appropriate weapon system

Weapon assignment and fire control by the WM 25 FCS against air, surface, sub-surface, and shore targets

(4) anti-submarine warfare
(5) electronic warfare
(6) data link operation
(7) assistance in tactical operations, including ASW-helicopter direction and navigation
(8) target simulation for training purposes.

Sensors comprise a DA.05 air and surface warning and target indication radar, WM25 combined search and tracking radars, a commercial Raytheon surface search/navigation/helicopter control radar, IFF/SIF, helicopter transponder, a French-made ESM system, data link, optical director, and sonar. Weapons controlled by the SEWACO system include NATO Sea Sparrow surface-to-air missiles, Exocet surface-to-surface missiles, dual-purpose gun, anti-submarine rocket launcher, ASW torpedoes, and ECM systems. The weapon control and combat information equipment consists of the WM25 FCS for missiles, gun, and A/S rockets, three horizontal displays for combat information functions and each with its own separate alpha-numeric 'tote' display, a sonar display console, radar plotting table with true-motion computer, and one video extractor for automatic air and surface tracking.

SEWACO V

This designation is given to the SEWACO system for the mid-life conversion of the six 'Van Speyk' class frigates. Sensors include an LW.02 air warning radar, a DA.05 air and surface warning and target indication radar, a Decca navigation radar with two displays, IFF and helicopter transponder system, two optical target designation sights, two sonar sets, an ESM system and an automatic data link. The action information system is of the DAISY type and includes four horizontal tactical display consoles with alpha-numeric displays and a plotting table. The weapons fit comprises an OTO Melara 76/62 dual-purpose gun, controlled by an M45 weapon control system, Seacat launchers controlled by an M44 system, a Harpoon SSM system, ASW torpedoes and ECM equipment. The ship will be equipped with an ASW helicopter.

SEWACO VI

The first of two Royal Netherlands Navy 'Jacob van Heemskerck' class frigates was commissioned on 16 January 1986. The SEWACO system for these vessels comprises: Signaal LW.08 extended range and elevation cover version, a stabilised DA.05, ZW06 and Goalkeeper search radar sensors; multiple video extractor systems; a Sphinx ESM system; Ramses ECM system; two optical target designation sights; and guided missile control system, SMCS, consisting of three STIRs (one 1.8 m and two 2.4 m scanners) with high power coherent transmitter chains. CW illumination for Standard SM1 MR and for Sea Sparrow missiles is provided by each STIR. The system provides a high level of autonomy by means of automatic threat evaluation and STIR and weapon selection for example. The Goalkeeper close-in weapon system will be installed. The DAISY system includes six vertical consoles, two horizontal consoles and a sonar display console, primary and secondary video extractors. In addition to the normal DAISY functions it has extensive air defence software facilities. IFF, VESTA and data link are also integrated in the SEWACO system.

The weapons are the SM1 MR to be launched by Mk 13/4 launcher, Sea Sparrow from eight-cell launcher, Harpoon surface-to-surface missiles, RBOC countermeasures, Mk 46 torpedoes and the EX 83 30 mm gun portion of the Goalkeeper CIWS.

SEWACO VII

This version has been nominated for the Royal Netherlands Navy's 'Karel Doorman' class frigate SEWACO system.

SEWACO VII is a member of a new generation SEWACO systems with an increased level of integration and automation. In mid 1985 the contract was signed for the nucleus of the M frigate system, being the command and control equipment including the central computer system, the integrated external and internal communication system SINCOS 1200, the high speed data bus and the standard colour workstations.

STATUS

In early 1986 a contract was signed for the multiple weapon control system (a system very similar to the SEWACO VI GMCS, for control of the VLSS and 76 mm gun), SMART the new 3-D F-band radar, the LW.08 long range D-band radar with lightweight pedestal, a multiple radar extraction and tracking system and the VESTA helicopter transponder system. Later in 1986 other SEWACO VII system components were due to be ordered, such as the CIWS Goalkeeper, ESM/ECM equipment, hull mounted sonar, long range towed array sonar and a database management computer complex. The architecture of the SEWACO VII system represents a major step ahead in system design with emphasis on integration, automation and highly improved man-machine interface. The subsystem is connected by either a databus or point-to-point connectors.

In July 1987 Portugal awarded Signaal a contract for three SEWACO systems for Meko Type 200 frigates. The Turkish Navy has ordered SEWACO systems for its four Meko Type 200 frigates.

CONTRACTOR

Hollandse Signaalapparaten BV, PO Box 42, 7550 GD Hengelo, Netherlands.

Vertical display consoles for Signaal SEWACO system

3516.261

GOALKEEPER NAVAL FCS

The Goalkeeper system is based on an integration of the GAU-8/A 30 mm multi-barrel Gatling gun produced by General Electric in the USA (**4042.203**) and a new Signaal radar and fire control equipment to form an autonomous, rapid-reaction close-in weapon system (CIWS) for ship defence (**3616.231**). The design is especially directed toward the anti-ship missile threat.

Rapid target detection is achieved by the I-band search radar with its fully stabilised antenna rotating at high speed. Threat evaluation is accomplished through an integrated video processor and threat evaluation program. Target designation and acquisition are rapidly effected because of the close integration of all the system elements on a single mount. A combined I-band and K-band tracking radar enables the system to track accurately all types of target, including sea-skimmers, under all sea conditions. Anti-clutter and ECCM features are included to maintain this capability in adverse conditions.

Enhanced track processing, including curved flight prediction is inserted to ensure a high target prediction accuracy. Firing accuracy is achieved through the use of a low-dispersion gun coupled with appropriate ammunition and a rigid, stable mount. Residual bias errors in the system are eliminated by a closed-loop spotting control process.

The kill mechanism is warhead detonation (assured kill); although 'control kills' may be expected these are considered as a bonus if they occur.

The main operational tasks of the Goalkeeper system are to counter the following threats:

(1) priority 1 threats such as anti-ship missiles like Exocet (**1156.221**) characterised by a more or less constant low-level trajectory and speeds of up to Mach 2.5
(2) priority 2 threats such as sea-skimming anti-ship missiles which may execute in the final attack phase a pull-up manoeuvre followed by a dive onto its target ship at supersonic speeds of up to Mach 2.5.

Additionally, Goalkeeper can engage effectively manned aircraft, or small patrol craft at considerable ranges.

STATUS

Goalkeeper systems have been ordered for the Royal Netherlands Navy, and for the Royal Navy.

CONTRACTOR

Hollandse Signaalapparaten BV, PO Box 42, 7550 GD Hengelo, Netherlands.

1261.281

M40 SERIES FIRE CONTROL SYSTEMS

The M40 Series of fire control systems comprises a range of combined radar and optical directors for the control of short-range missiles and gunfire against air and surface targets. Target designation is performed by one of the ship's search radars. The M44 is used with the Seacat missile and the M45 is for the control of medium and light calibre guns.

The M44 is a fire control system for ships of frigate or destroyer size. The radar/visual director allows radar-automatic target tracking and visual detection and tracking in bearing and elevation. The stabilisation equipment is embodied in the director, so that no axis conversion is needed and the system is lightweight. It supplies accurate fire control data to the short-range Seacat.

The computer is programmed for: tracking of an air target, tracking of a surface target, firing bracket computation, launcher control.

After the Seacat has been launched, it is gathered by the director operator by means of a flight controller. This is fitted on the aiming bar, which is used for visual target acquisition and tracking. The guidance process is also visual, but provision is made for blind guidance by means of a second range gate and an F-scope on the computer display panel.

For target indication the system can be connected to a TI unit, thus providing target bearing and range to the M44 director and the radar. The lock-on and tracking procedure is fully automatic.

The radar is a conventional I/J-band fire control radar and includes various ECCM devices and special provisions so that it can operate under adverse weather conditions. For training purposes the computer can be provided with a special program to control an electronic marker in the field of view of the binoculars, thus simulating a Seacat flight.

STATUS

M40 Series FCS have been fitted or specified for: Netherlands 'Van Speyk' class frigates; Indian 'Leander' class frigates; the Malaysian frigate 'Hang Jebat'; the Thai frigate 'Makut Rajakumara'; Australian 'River' class anti-submarine frigates. Licence production in India for 'Leander' class frigates.

CONTRACTOR

Hollandse Signaalapparaten BV, PO Box 42, 7550 GD Hengelo, Netherlands.

3766.481

GIPSY SUBMARINE DATA HANDLING AND WEAPON CONTROL SYSTEM

Gipsy is an automated data handling and weapon control system designed for the Walrus II class submarines of the Royal Netherlands Navy. The system forms the link between the submarine's sensors and its weapons. The system consists of seven identical display and computer consoles (DaCCs). The built-in computer is of the SMR-MU type; the 16-inch (406 mm) plan view display (PVD) provides a high-load-synthetic picture together with compressed radar video or sonar video and the control panel is a multi-purpose unit. The use of identical DaCCs offers maximum flexibility.

A central control unit (CCU) is used to regulate the mutual data transfer between the DaCCs and the sensors and weapons. For this function two SMR-MU computers are provided, one active and the other a hot standby machine. The large amount of data from the sonars is handled by two extra SMR-MUs, also housed in the CCU. Three types of sonar are fitted (towed array, flank array and circular array), and other sensors and data sources include a noise analyser, ESM facilities, radar, periscopes, and position finding equipment.

The weapon control system can control a mixed load of weapons for sub-surface and surface engagements, and the ship's launching system consists of two mutually independent sections, each of which is controlled from a launching system control panel (LSCP). The interface with the weapons (modern sub-surface missile) is formed by two identical distribution cabinets. All hardware necessary for the integration of these weapons is included in the distribution cabinet, avoiding the necessity for additional equipment.

The main functions of Gipsy are:

(1) sonar display and control. This makes it possible to omit the original sonar displays and controls
(2) contact evaluation. This entails displaying information obtained from all sensors in a time/bearing format
(3) classification. Where ESM, ASM and noise information is compared with cassette tape stored libraries to provide a rapid classification of the target
(4) contact motion analysis. Modern tracking filters are used for the automatic determination of target movements
(5) tactical plot and general plot functions giving an up-to-date survey of the tactical and/or navigational situation or an historical situation survey
(6) weapon control. Whereby the weapon systems are provided with the requisite aiming data
(7) simulation and test to verify the system's operability and to aid in operator training for the functions (1) to (6).

Additional functions can be inserted easily due to the fact that the system is highly software-oriented.

STATUS

Two systems have been delivered for fitting in the first two boats of the 'Walrus II' class for the Royal Netherlands Navy, together with a reduced system for shore-based training.

CONTRACTOR

Hollandse Signaalapparaten BV, PO Box 42, 7550 GD Hengelo, Netherlands.

1662.481

SINBADS – SUBMARINE INTEGRATED BATTLE AND DATA SYSTEM

SINBADS is a compact data handling and weapon control system suitable for use in submarines of the small coastal type and up to the larger, ocean-going type. It succeeds the Signaal M8 Series of torpedo FCS (**1375.481**) for submarines. The computer used is the Signaal fourth-generation general purpose machine SMR-MU, and the complete system combines the weapon control and data handling functions.

The data handling function covers sensor display and sensor selection for track initiation. The raw sensor data of all sensors can be displayed. Each individual sensor is indicated with a unique label. The development of SINBADS is based on the application of modern passive sonars, capable of automatic tracking of targets. The tracking algorithm is based on modern filtering techniques and the tracking system basically includes as its main mode of operation a 'bearing-only' analysis which, however, also accepts other inputs of target information. The complete tracking system functions as an interactive system in which the command can intervene in the tracking process. For this purpose there are four display formats available.

SINBADS can handle five targets and three torpedoes simultaneously. These can be guided, unguided or a mixture thereof. The system also includes data recording from on- and off-line use, weapon simulation for training, on-line failure monitoring, and an emergency mode for which hand controls are provided to enable torpedoes to be set, fired and controlled in case of a computer or display failure.

STATUS

Produced for the following navies: Argentina, four for 'Santa Cruz' (TR-1700) class; Greece, four for 'Glavkos' (Type 209) class; Indonesia, two; Peru, four for 'Casma' (Type 209) class; Taiwan, two for 'Sea Dragon' class; Turkey, four for 'Atilay' (Type 209) class.

CONTRACTOR

Hollandse Signaalapparaten BV, PO Box 42, 7550 GD Hengelo, Netherlands.

1375.481

M8 SUBMARINE TORPEDO FIRE CONTROL SYSTEM

The M8 is a digital computer-based fire control system for use in submarines for the direction of torpedoes against either surface shipping or submerged targets. It was produced in several versions with designations ranging from the M8/0 of the mid-1950s prototype to the latest. M8 is now out of production and succeeded by SINBADS (**1662.481**), Gipsy (**3766.481**) and Submarine SEWACO (**4587.481**). The basic system comprises a torpedo display control and computer console, a sound path display unit, amplifier and supply unit, distribution box, local control panel(s), and gyro angle setting units. Complete system weight is about 900 kg. The system may be operated by one man, or two men if the submarine is operating with consorts.

The system will accept target data inputs from a range of sensors which includes radar, sonar, passive sound detection systems, periscope observation and consort reports. Ship's own navigational data is also fed into the M8 computer. The display, which has range scale settings for 20, 10, and 5 km, presents the positions of all contacts from all sensors, simultaneously. One or more sensors may be connected to the computer for torpedo engagement, and up to three targets may be attacked simultaneously. The computer is programmed to provide firing data for wire-guided, programmed, conventional, and other types of torpedo, and performs automatic calculation of target position, course, and speed. The CRT display can give true motion, relative motion, or off-centred presentation of the tactical situation.

DEVELOPMENT

The M8 Series originated in a Royal Netherlands Navy contract awarded in 1955 for the development of a torpedo FCS for the 'Dolfijn' class submarines. West German interest in the system gave additional impetus to subsequent development and led to the successful development of systems for use in both submarines and surface ships. Two types of torpedo were involved, the AEG Seal and Seeschlange, for use against surface vessels and submarines respectively. This successful collaboration with the West German industry continued and M8 Series systems are standard fitting on all submarines produced by Howaldswerke Deutsche Werft in Kiel. The ultimate development is the SINBADS system described in **1662.481**, but other related developments are the M9 systems in the West German 'Köln' and 'Thetis' class vessels, and the M11 systems for two new Argentinian fast patrol boats.

STATUS

The following list is believed to record accurately the known installations.

Argentina	Type 209 (2)
Colombia	Type 209 (2)
Equador	Type 209 (2)
Germany (Federal)	Type 205 (5)
	Type 206 (18)
Greece	Type 209 (4)
Netherlands	'Dolfijn' class (2)
	'Potvis' class (2)
	'Zwaardvis' class (2)
Norway	Modified Type 205 (14)
Peru	Type 209 (2)
Turkey	Type 209 (2)
Venezuela	Type 209 (2)

CONTRACTOR

Hollandse Signaalapparaten BV, PO Box 42, 7550 GD Hengelo, Netherlands.

4587.481

SPECTRUM I AND II SUBMARINE SEWACO SYSTEM

SPECTRUM I is a fully integrated combat system designed to carry out all tactical functions for small displacement and larger patrol submarines. A particular feature is the full Fourier frequency analysis on all preformed beams and the use of this information for spectral and spatial normalisation of sonar information before narrow and broadband detection to provide a digital system that automatically optimises detection, analysis and contact tracking under all conditions.

In its full configuration, SPECTRUM I includes a multiple array broad and narrow band sonar suite (SIASS), a command system including interchangeable multi-function display control consoles, a ships integrated communication system (SINCOS) (**4496.261**) and a weapon control system compatible with modern wire-guided torpedoes and air-flight weapons.

A navigation radar (ZW07) also forms part of the complete system configuration. SPECTRUM I subsystems are fully integrated and incorporate a software-based multi-processor modular design.

Data collected by SPECTRUM's sonar suite and other sensors (eg ESM and radar) is processed to localise and identify contacts. The search results from the multiple sensor fit are correlated and displayed on the interchangeable consoles to generate a concise tactical picture. From this information decisions are made about target designation and priorities, threat assessment, weapon assignment and tactical manoeuvring of own ship. Weapon employment from warm-up, presets through launch and post-launch guidance are accomplished by the weapon control function. SPECTRUM I also incorporates a tactical data handling function and functions for navigation, data logging and crew and operator training and is capable of integration with electro-optical sensors (periscopes) and electronic navaids.

Sonar suite (SIASS)

The sonar suite consists of:

(a) a Cylindrical Array Sonar acting as the primary attack sonar. The combination of narrow and broadband processing is to extend detection ranges, to enhance tracking and to provide quick and accurate classification of contacts

(b) a Flank Array Sonar, using 24 acceleration cancelling type hydrophones on either side of the boat, serves as long range detection and tracking sonar especially for the lower frequencies

(c) a Passive Ranging Sonar with three arrays on either side of the boat determining target ranges. The passive ranger's signal processing uses a method of frequency-domain correlation to measure wavefront curvature and thus provide range information. The use of variable tracking bandwidths reduces the influence of unwanted noises and maximises the signal-to-noise ratio

(d) an Intercept Sonar providing coverage up to 90 kHz and employing a fully digital processing utilising fast Fourier transform techniques to detect, correlate and classify intercepts of sonars and torpedoes. Automatic alarms are given for a number of preset threats.

Contacts in automatic track are automatically subjected to passive bearings-only Contact Motion Analysis (CMA). A new generation CMA algorithm is employed which has demonstrated speed and accuracy at sea trials. An electronic classification library is included with a capacity of up to 300 specific platforms.

Weapon control

Two identical weapon interface cabinets are incorporated in the combat system, each containing interfacing and processing equipment required for the control of the weapons in one half of the launching system. Normal and emergency modes are provided to ensure high operational availability.

The weapon control function is compatible with various submarine launching systems and submarine-launched weapons with three firing modes provided. The system is capable of simultaneous launch and control of four modern acoustic wire-guided torpedoes and four submarine-launched missiles.

Tactical data handling

The tactical data handling complex of the SPECTRUM I comprises two redundant computers, interfaces to own ships sensors and a data logging capability. The complex gathers data from all subsystems to compile a comprehensive picture of the tactical situation, correlates data from the various subsystems, supports threat assessment, threat priorities, weapon assignment and tactical manoeuvring, provides aids to navigation such as CPA calculations, stationing and position fixing, carries out environmental predict and system control and monitoring.

SPECTRUM II is an advanced submarine combat system based on SPECTRUM I. The system includes all of SPECTRUM I system functions using standardised processing and display components recently developed for the 'Karel Doorman' class frigate, programme of the Netherlands. This has resulted in the following major upgrades:

(a) the multi-function operator consoles are equipped with high resolution (1376 × 1024 pixels) colour raster scan displays capable of showing brightness modulated sonar pictures and radar video

(b) system data processing is distributed between several 68020 family-based general purpose processors. Distribution of processing tasks and resources is shared co-operatively

(c) an Ethernet-based data bus is used to interconnect the various processors within the system. It transmits information over two mutual independent triaxial cables at a rate of 10 Mbit/s. Distributed control of the network is fundamental to the SIGNAAL data bus architecture

(d) for storage and retrieval of program, library and mission data use is made of two digital optical recorders capable of storing up to 250 Mbits data on compact disk.

All major processing electronics are housed in a newly developed standard 19-inch cabinet. Features of the new design include Eurocard construction format, standardised power supplies and a flexible cooling system. The cabinets can be delivered in three different heights for optional accommodation on board submarines.

STATUS

SPECTRUM I is operated in the 'Sea Dragon' class submarine. SPECTRUM II is on offer to various navies.

CONTRACTOR

Hollandse Signaalapparaten BV, PO Box 42, 7550 GD Hengelo, Netherlands.

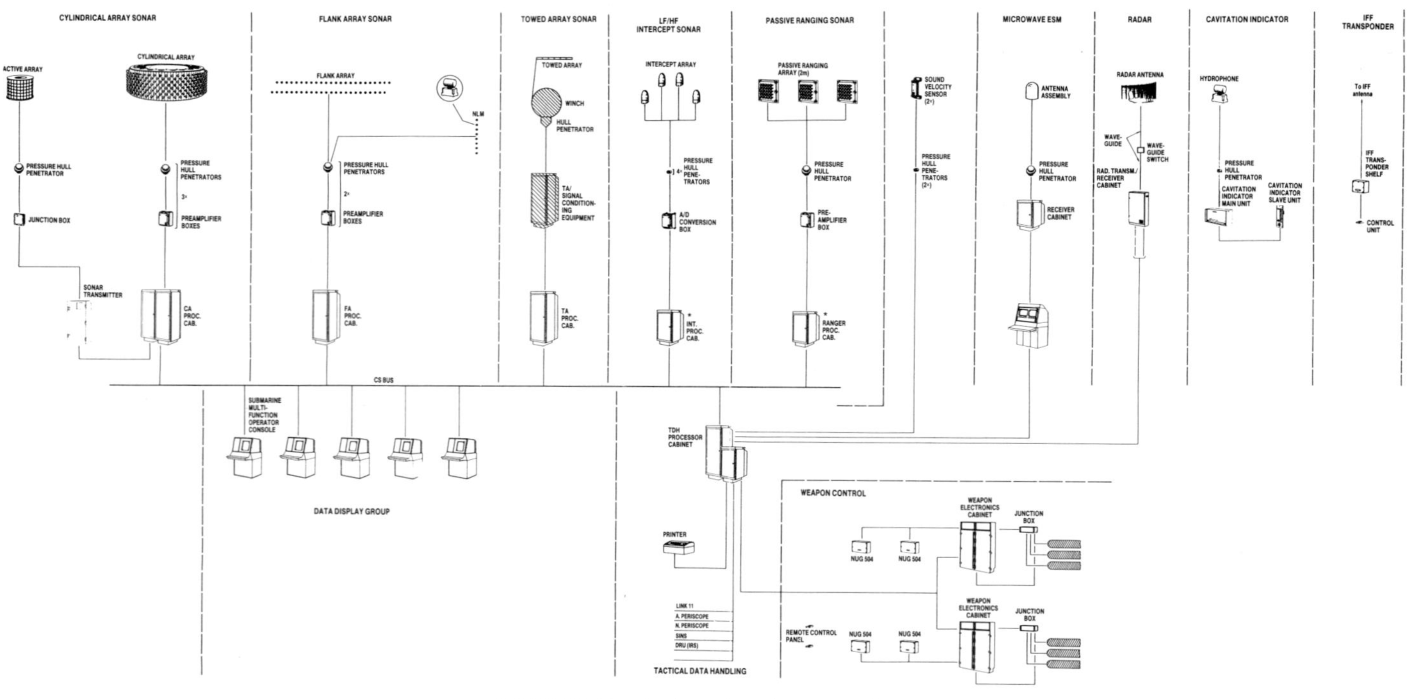

SPECTRUM II main elements and sensor systems

NORWAY

3744.281

NAVKIS

NAVKIS is the navigation, command, control and information system designed and developed for 2800-ton coast guard vessels in service with the Royal Norwegian Navy. When the development and production contract was awarded in 1978, three main priorities for the NAVKIS system were established: navigation, command and control, and air surveillance and gun control. To satisfy these priorities, while meeting a requirement for the use of equipment of differing levels of sophistication, it was decided to base the design on a number of autonomous sub-systems. Data transfer between the sub-systems and most of the sensors is provided by an autonomous multiplex data transmission system known as BUDOS.

Sensors providing navigation or target information to the system include a Plessey AWS-4 air and surface surveillance radar (**3171.253**), two Decca I-band navigation radars (**3127.281**), a PEAB 9LV 200 Mk 2 autonomous fire control system (which incorporates tracking radar and low light-level television), two optical target designation sights, and navigational inputs such as Decca Navigator, Loran C, navigation satellite receiver, echo-sounder, compasses, gyros, ship's log etc.

A navigation computer, KS-500 with a 64K store, continuously monitors the radio navigation receivers and ship's reference sensors to calculate 'own' position and velocity through Kalman filters. All relevant material is recorded on a cassette. In addition, the computer calculates factors such as sea currents, log errors, gyro drifts etc. Display/operator consoles for the navigation facilities are provided in the operations room and on the bridge.

The command, control and information system (CCIS) is the main operational unit within the overall NAVKIS system, and its main functions are those of surveillance and tracking of all targets. It is also employed for control of the ship's helicopter. This part of the system has much in common with the Kongsberg MSI-80S naval AIO system (**1975.281**) and the main operating console is situated in the operations room, together with the AS/GC (air surveillance and gun control) console.

Functions of the CCIS in addition to target tracking and surveillance include: presentation of radar data from any of the ship's radars on the main tactical situation display; identification and continuous automatic tracking of 16 targets; identification and automatic plotting of up to 64 other targets; display of targets from the air surveillance console; video map display of geographical features and/or reference grids etc.

The radar sub-system provides distribution of radar data and integrates information from the two Decca navigation radars and the Plessey AWS-4 (which has digital MTI), as well as providing an interface with the 9LV 200 Mk 2 tracker radar.

STATUS

In production and operational.

CONTRACTOR

Norsk Forsvarsteknologi A/S, PO Box 1003, N-3601 Kongsberg, Norway.

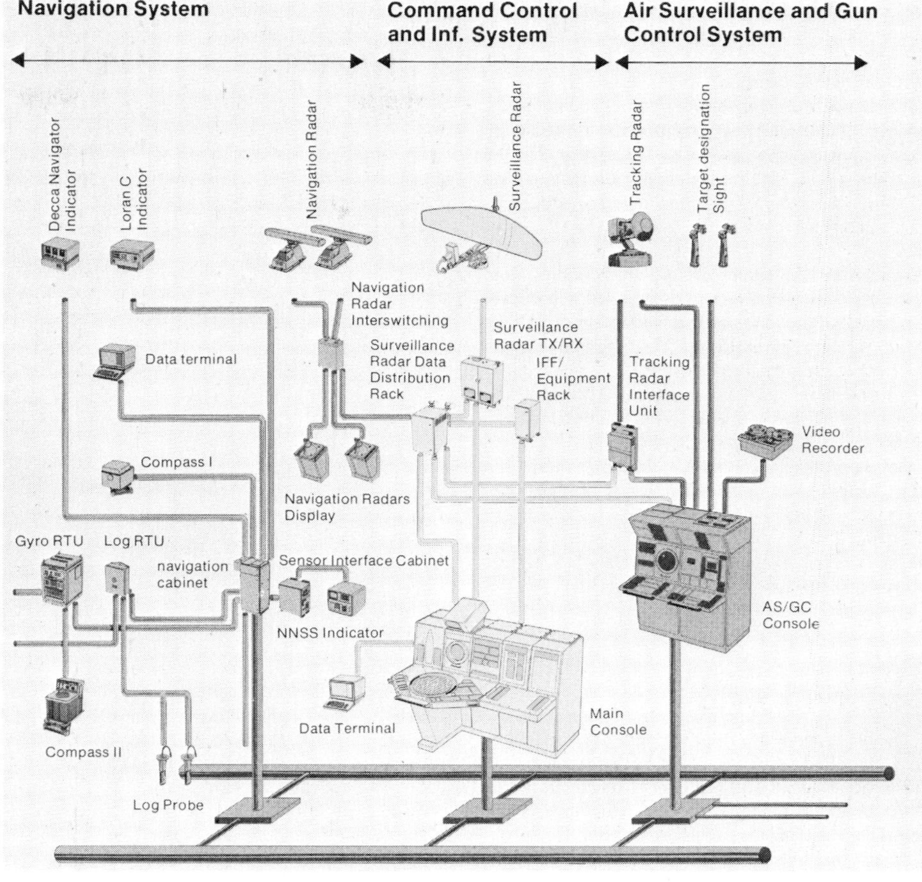

Diagram of Kongsberg NAVKIS system for Norwegian Coast Guard ships

---

1975.281

MSI-80S FIRE CONTROL SYSTEM

The MSI-80S is a weapon control system for small ships and was initially specified and developed for the Royal Norwegian Navy's new 'Hawk' class fast patrol boats. The system is equally suitable for tactical data handling and fire control functions in larger ships such as corvettes or frigates. The system provides a multi-target capability using low-cost conventional navigational radars, stabilised electro-optical sensors for night/passive operations, and target filtering/tracking techniques. Homing surface-to-surface missiles, wire-guided torpedoes, and an AA gun can be controlled individually or simultaneously.

Typically, the armament consists of six Penguin SSMs, four torpedoes, and a 40 mm gun. Inputs to the MSI-80S include the twin Decca TM1226 radars, an electro-optical tracker system which incorporates a TV tracker and laser rangefinder, infra-red scanner, optical sights, and electronic warfare sensors. The low-light television camera channels and displays supplied by GEC Sensors Ltd comprise a GEC low-light camera, incorporating an automatic light control system enabling both daylight and low-light operation. Line-of-sight stability is assured by mounting the camera on a stabilised sensor platform. This platform, manufactured by the former A/S Kongsberg Vapenfabrikk, also carries the other sensors, both active and passive, which are collimated with the television camera. To preserve the advantages of concealment and surprise attack, the basic philosophy is to employ the various passive sensors. The efficacy of this method rests largely upon the tracking and other software of the fire control computer. In the MSI-80S reliance is placed upon programs which have already been proved in the MSI-70U submarine system. The tracking and guidance problems are solved either automatically or semi-manually by man/machine integration and in the latter case the computer serves as a data bank. Data from all available sensors can be used at any time and in any order.

A three-man operating console is provided, this being crewed by:

(1) tactical operator, who initiates and controls target data calculations and the overall tactical situation plot
(2) weapon control operator, who selects, controls, and designates the guidance mode(s) of the relevant weapons
(3) passive sensor operator, who operates the passive sensors and feeds relevant data to the target tracking system.

A fourth position is provided at the console for the commanding officer.

A horizontal 23-inch (584 mm) CRT is used for the presentation of raw radar and computer-generated data, and a 12-inch (305 mm) CRT is provided for the display of alpha-numeric data. The latter can also serve as a back-up for the main tactical display (23-inch).

STATUS

In production.

CONTRACTOR

Norsk Forsvarsteknologi A/S, PO Box 1003, N-3601 Kongsberg, Norway.

4573.481

MSI-90U SUBMARINE AND WEAPON CONTROL SYSTEM

In 1982 the former A/S Kongsberg Våpenfabrikk presented a basic command and weapons control study as the result of a concept and definition phase for the new German and Norwegian (P-6071 or 'Ula' class) submarine projects.

The defined basic command and weapons control system, also called MSI-90U, is a software-based system which uses distributed processing, a high capacity serial data transmission system and multi-function common consoles. The MSI-90U system is the common core of identical hardware and software for both projects (Norway and Germany), covering the following areas:

(a) data transmission system
(b) central computing area
(c) display and operating controls related to the situation processing and to command and weapons control
(d) the torpedo/ship interface.

Consequently the two national Integrated Combat Systems will consist of the MSI-90U Basic Command and Weapons Control System, the actual sensor system chosen by the respective navies and the weapons.

System architecture

MSI-90U has a highly standardised and modularised design. The number of building blocks have been kept as low as possible within the given requirements. MSI-90U consists of a microprocessor-based computer system, KS-900F; a standardised multi-purpose operator console, KMC-100 (from the KMC-9000 family); a Kongsberg BUDOS data bus system; and a torpedo board interface, developed by AEG.

Computer tasks

The MSI-90U is a multi-computer-based system containing software tools for sensor integration and weapons assignment/employment. In addition to the main tasks a number of supplementary facilities are provided by the system.

Man/machine interaction

Each console is equipped with a console computer for operator input handling, picture/data presentation, local computations and data exchange system communication. Picture and data are presented on two equivalent high resolution colour raster scan displays on each console.

STATUS

Development of main system components and special software modules started in 1980. A full scale system development contract was signed in 1982. Demonstration of prototype of the multi-purpose operator console took place in 1983.

An evaluation system of the MSI-90U was installed in a factory-based test stand and testing started in 1986.

Series production of MSI-90U has started and the first system was scheduled ready for onboard installation at the beginning of 1988.

CONTRACTOR

Norsk Forsvarsteknologi A/S, PO Box 1003, N-3601 Kongsberg, Norway.

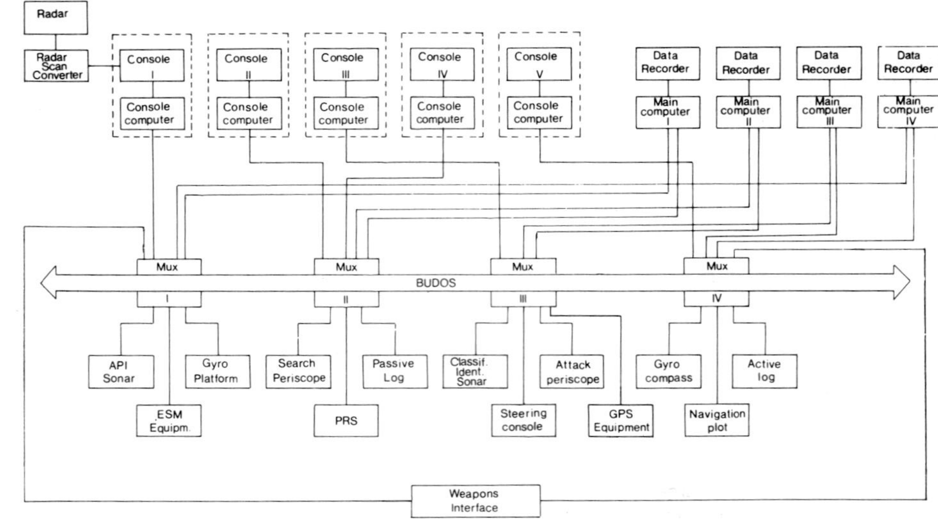

Block diagram of a typical MSI-90U configuration

6741.281

MSI-340/350 FIRE CONTROL SYSTEM

The MSI-340 is a compact fire control system specially designed to meet the requirements of small ships. This integrated system is capable of performing both the target tracking and fire control functions in fast patrol boats or similar warships. In addition, a fair selection of tactical support is implemented.

The system provides automatic target detection and tracking and automatic weapon control as well as extensive picture compilation. Target data is supplied to two different weapon systems either separately or simultaneously. A flexible design allows the customer to select weapons and sensors.

The MSI-340 requires one operator only. The console has a 20-inch high resolution raster scan colour display vertically mounted. The multi-colour display provides scan converted radar pictures and synthetic overlay. Operator inputs are mainly performed via a programmable entry panel.

The main functions of the MSI-340 are multi-target tracking, tactical support and weapon control.

The target tracking system is based on the radar extractor REX-2 which performs automatic target detection, acquisition and tracking. The extractor can interface a wide range of different radars. The tracking system can also combine data from radar and electro-optical sensors.

Target data is transferred to a surface-to-surface missile system and a gun either individually or simultaneously. The operator selects targets and assigns weapons. He also decides the firing mode and fires the weapons. The missile flight trajectory, gun bearing and other tactical data will be shown on the display.

The MSI-350 is an extension of the MSI-340 with added tactical command and control facilities to provide an integrated Combat Control System. The MSI-350 system is designed to accept inputs from a wide variety of surveillance and tracking sensors and supply outputs for control of surface-to-surface missiles and two independent guns.

The system performs automatic target detection and tracking and automatic weapon control as well as extensive picture compilation. The tactical data handling controls the processing of inputs from ESM, navigational receiver and data link, if fitted.

The combat control system requires two operators. One weapon operator will handle the weapon control and one tactical operator will exercise command and control.

The double console has two 20-inch high resolution raster scan colour displays vertically mounted. The multi-colour displays provide scan converted radar pictures and synthetic overlay. Operator inputs are mainly performed via programmable entry panels.

STATUS

The multi-function console standard modules in the KMC-9000 series are in production for several current projects.

CONTRACTOR

Norsk Forsvarsteknologi A/S, PO Box 1003, N-3601 Kongsberg, Norway.

6742.281

MSI-3100 MULTIPLE INTEGRATION SYSTEM

MSI-3100 is a fully integrated and compact system originally built for the Combat Information Centre in Norwegian frigates. It is designed to exercise command and control in an efficient and intelligible manner.

The flexible nature of the system makes it suitable for a large variety of applications on board small and medium-sized naval ships.

The system conducts action information and enables the ship to counter various threats simultaneously. The most important weapons and sensors assigned to the AAW, AWW and ASW warfare areas are computer controlled. A distributed computer arrangement provides simultaneous control of a sufficicent number of sensors and weapons.

The de-centralised arrangement also permits independent management of each warfare area.

The principal functions peformed by MSI-3100 are surveillance, target detection and tracking, threat evaluation and weapon and track sensor allocation.

Functions are carried out automatically taking into account manual override or support at any stage.

The auxiliary functions include Electronic Support Measures (ESM), manoeuvring aid and tactical co-ordination and support.

The concept of full sensor and weapon integration alleviates decision-making and reduces reaction times. From the moment of target detection, 'gun settled and ready to fire' is accomplished in less than ten seconds.

The central part of the system is made from two double operator KMC-9000 consoles. The consoles are function-oriented of the multi-role type. They are manned by four operators altogether. Tactical pictures are shown on high resolution raster scan displays. Radar video is digitised. Operator entries are applied from standard keyboards and programmable touch panels.

STATUS

MSI-3100 is in operation onboard the first updated vessel in the refurbishing programme of the 'Oslo' class frigates.

CONTRACTOR

Norsk Forsvarsteknologi A/S, PO Box 1003, N-3601 Kongsberg, Norway.

SPAIN

6743.281

BAZAN SCPVZ COMBAT SYSTEM

The SCPVZ (Sistema de Combate para Patrullero de Vigilancia de Zona) fast patrol boat combat system has been developed for fitment to the 'Cormoran' class. Unlike the traditional centralised combat system SCPVZ provides a higher survivability, soft system degradation and a modular architecture which allows system growth, use of common hardware and software functions, and low weight and compactness for easy onboard installation.

The main missions of this combat system are sea traffic control, surface surveillance, surface warfare and anti-aircraft defence. The basic elements of the system comprise an X-band surveillance radar system, a tactical console, a data multiplexer, ballistic computer, two fire directors, a 76/62 Bazan OTO Melara gun and a 40L/70 Bazan/Bofors gun. Optional equipment includes air search radar, a FLIR, a tracking radar, IFF, a data link module, an ECM unit, a surface-to-surface missile unit and an EW module.

The heart of the system is the Cormonet data bus digital communications network. This features high transmission data rates, high immunity to noise, high survivability and the ability to link with almost any other equipment.

STATUS

In development. Functional tests of the gun fire director were performed in 1986. Overall sea trials were scheduled for 1987.

CONTRACTOR

Empresa Nacional Bazan, San Fernando apdo 18, Cadiz, Spain.

SWEDEN

3512.281
ERICSSON COMMAND AND CONTROL INFORMATION SYSTEM (CCIS)

The Ericsson CCIS is a multi-display, computer-controlled system for the collection, storage, evaluation and transmission of naval tactical data, commands and messages. Although specially adapted to the handling of action data on modern frigates and corvettes, the basic principle of the system also applies to smaller or larger vessels.

Typical configuration of the modular equipment includes four horizontal and two vertical displays. Horizontal displays include a local air plot to present the tactical air space situation within the ship's operating area, a surface plot depicting all surface movements within a given area, a sub-surface plot for ASW purposes, and a weapon co-ordination and control station. Operational and tactical area situation is displayed vertically on two large screens.

A horizontal control position consists of a console for three operators with a 16-inch (406 mm) PPI and, for each operator, identical input devices (rolling ball and alpha-numeric keyboard). The PPI presents true-motion raw radar picture, track information, target designation and tell-back symbols, vector lines and intercept solutions. Using the input devices on the console, the operators have control of radar selection, tracking, data link communication, weapon co-ordination display selection and recording.

Significant operational features of the system include the following:

(1) presentation of data from ship's sensors – radar, ESM, sonar etc
(2) automatic and rate-aided tracking
(3) presentation of geographical reference information
(4) weapon co-ordination and control
(5) data link communication
(6) on-line recording of tactical information
(7) tactical navigation
(8) intercept calculation
(9) ship data library
(10) simulation and training.

STATUS
A version of this system has been delivered to the Danish Navy for installation in frigates and corvettes.

CONTRACTOR
Ericsson Radio Systems, 17586 Järfälla, Sweden.

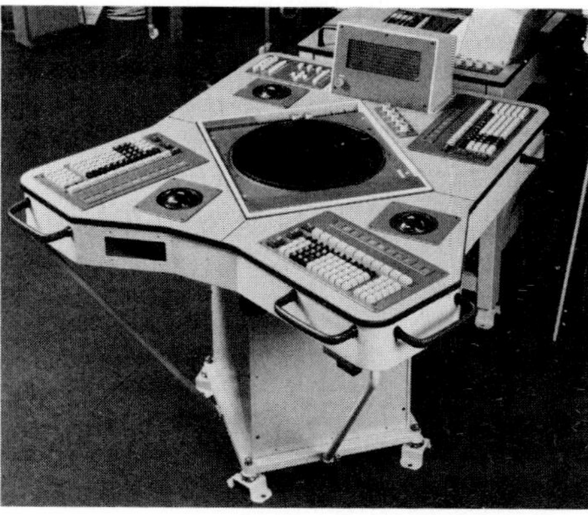

Ericsson CCIS horizontal display with PPI designed for three operators with alpha-numeric keyboard and rolling ball for target designation

---

3186.261
EPLO TACTICAL DATA SYSTEM

EPLO (Electronic Plotting System) is designed to overcome the inherent shortcomings of manual techniques for handling tactical information on board small ships such as fast patrol boats and corvettes. It is an automated system in which the level of sophistication has been governed by cost-effectiveness and space restrictions. The system compiles a tactical picture of an area of operations, and allows data exchange with other ships and with naval radar control centres equipped with a compatible system.

Operational features include the following:

(1) target selection and semi-automatic tracking by means of track symbols on a true motion stabilised raw radar picture
(2) exchange of tactical data via a data link with other ships and shore radar stations
(3) target assignment to on-board fire control system by special assignment symbols
(4) synthetic presentation of the plotted tactical situation, including reference information such as maps, geographical grids, and fixed points
(5) calculation and numerical presentation of range and bearing to target as well as speed and course of target
(6) continuous recording on magnetic tape of the tactical situation.

Targets are selected from the raw radar information displayed on a 16-inch (406 mm) PPI by positioning a free target symbol over the echo. There is a repertoire of 13 different symbols and 24 selectable numbers are available for the labelling of tracks. The symbol co-ordinates are stored and the symbol is presented on the radar PPI and the 24-inch (610 mm) plot indicator. Updating of target position is achieved in cyclic fashion, with either a slow cycle time of two minutes for updating 24 tracks or a faster 30-second cycle time for updating six tracks. Selection of a particular mode is made from the keyboard and depends upon the number of tracks involved.

To correct deviations when tracking in the semi-automatic mode, the target may be moved to the new echo position by means of a rolling ball tracker control. The PPI displays only the new position together with the track history consisting of up to 15 symbols in reduced size. All selected targets in either cycle-time mode are described by their course and speed as well as range and bearing.

STATUS
The system has been supplied to the Royal Swedish Navy and to the Danish Navy. No other fitting details have been released.

CONTRACTOR
Ericsson Radio Systems, 17586 Järfälla, Sweden.

3248.281
9 LV 100 FIRE CONTROL SYSTEM

The 9 LV 100 is a lightweight, inexpensive ship-board system derived from the 9 AA 100 (**2648.181** in 1977 and earlier editions) land-based system. It is intended to control one, or possibly two, 30 to 76 mm guns against air and surface targets, and surface-to-surface missiles. Because of its light weight, it can be used on very small boats, from 50 tons up, but it is also well suited for larger ships, eg 300 to 400 tons, with a limited armament. It is also seen as a secondary, independent director-computer sub-system to supplement more comprehensive weapon control systems on board heavily armed ships, when employed with a 9 LV 200 Mk 2, for example. This combination is designated 9 LV 300 (**4026.281**).

The main units of a 9 LV 100 are:

(1) an optronic director with TV and laser, with provision for the addition of an IR-tracker
(2) a control and display console, with TV-monitor and gun and director controls
(3) two bridge pointers for designation
(4) computer and associated units.

The 9 LV 100 systems allow for:

(1) tracking of one/two air or surface target with the director in fair weather and daylight; and with IR at night
(2) gun, or missile, fire control using director or track-while-scan data
(3) designation from other radars or designators
(4) slaving to other types of directors.

STATUS
In production.

CONTRACTOR
Philips Elektronikindustrier AB, 17588 Järfälla, Sweden.

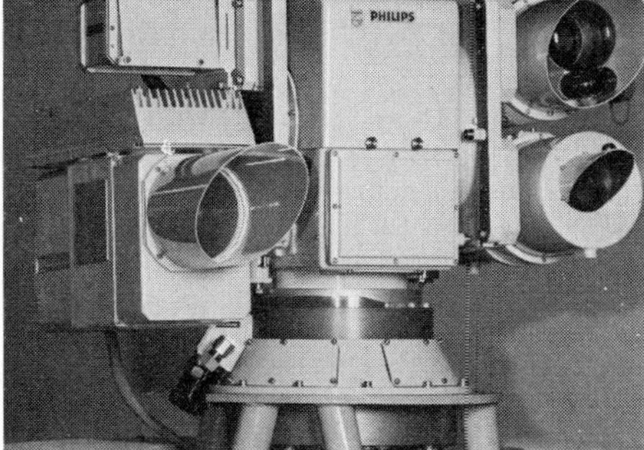

9 LV 100 optronic director equipped with IR camera, TV camera, and laser

---

1542.281
9 LV 200 FIRE CONTROL SYSTEM

The 9 LV 200 has been developed as an analogue fire control system for use on small ships from 80 tonnes and upward, for the direction of dual-purpose guns (40 to 120 mm) and other weapons such as anti-ship missiles (Exocet, Penguin, Otomat, etc) or wire-guided torpedoes. In its basic form the system has the capability for directing one or several guns against one or more targets simultaneously.

Target data are obtained from an I-band search radar and a J-band tracking radar, and there is provision for D-band air warning radar with IFF to be incorporated also. The I-band equipment in the 9 LV 200 is used for both search and tracking of surface targets and for air search. If the D-band pulse-doppler radar is included, a combined antenna system for I-band, D-band, and IFF Mk 10 is used. The

Danish Navy Willemoes *class fast patrol boat fitted with a 9 LV 200 weapon control system which incorporates television tracker based on JAI 771 low light-level TV camera by JAI*

J-band tracking radar can also be modified with a pulse-doppler function to improve performance in adverse clutter conditions against small low-flying targets.

The search radar is of the frequency-agile type, incorporates comprehensive ECCM facilities, and is provided with a stabilised mounting. A helical scan pattern is used. The director radar is a monopulse equipment, with frequency agility, and the pedestal also carries a TV camera.

A full description of operational functions appeared in the 1976 edition of *Jane's Weapon Systems*.

STATUS

The 9 LV 200 has been operational with the Swedish Navy since 1973 and has been delivered to other navies as well, including the Danish Navy. The original 9 LV 200 is now being followed by the 9 LV 200 Mk 2 (**3127.281**) and Mk 3 (**6449.281**).

CONTRACTOR

Philips Elektronikindustrier AB, 17588 Järfälla, Sweden.

3127.281

9 LV 200 Mk 2 WEAPON CONTROL SYSTEM

The 9 LV 200 Mk 2 combat information and weapon control system is the digital successor to the analogue 9 LV 200 and is suitable for use in all types of naval and offshore patrol vessels of 100 tons and upwards. It is optimised for data gathering, compilation and tactical evaluation for the control of one dual-purpose gun (57, 76 mm or larger), one anti-air threat gun (single or twin-mounted 20-40 mm), and surface-to-surface missiles. There are also options for torpedo control. Simultaneous engagement of separate targets is possible, one of which can be an air target. Modular design enables the system to operate independently or as part of a Philips C³I system aboard larger ships.

A typical 9 LV 200 Mk 2 installation consists of:

(1) an X-band search and surveillance radar with frequency agility or MTI, with or without on-mounted IFF
(2) a combined Ku-band electro-optical director for gun fire control (with or without TV tracker). IR and laser trackers are options
(3) tactical and missile control console for overall viewing, threat evaluation and analysis, data link control and target designation to guns, missiles, or torpedoes
(4) a gun control console, mainly for air surveillance and target designation to the director and fall of shot observation; this console can also be used for fire control against surface targets
(5) a surface gun console primarily for gun fire control against surface targets tracked by the surveillance radar; this may also be used as an air surveillance back-up console
(6) two 'bridge pointer' target designators for use by look outs for direct optical target designation or acquisition, typically against surface-to-surface missiles
(7) wind sensor.

KD Handalan *of the Royal Malaysian Navy, with Bofors 57 mm and 40 mm guns and Exocet MM 38, all controlled by a version of 9 LV 200 Mk 2*

OPERATION

Surface targets: a three position console with 16-inch tactical display provides for presentation of search and surveillance data from the I/J-band frequency agile search radar to the Plotter, Combat Information Officer and Missile Officer. Automatic TWS plotting of up to 16 multiple surface or air targets is possible, with presentation as numbered symbols and with full interchange of designated target data with weapon systems. Target data can also be exchanged with other ships via a VHF/UHF voice channel and short burst transmission.

Air targets: search radar video is presented on a 12-inch PPI on the AA gun control console. Pulse doppler processing for MTI is available. Target designation to the director for automatic acquisition of air or surface targets is provided and TWS facilities for air targets can be provided. The Ku-band monopulse tracking radar is mounted on the director and has automatic acquisition with CFAR control. Facilities include switchable frequency agility. Electro-optical sensors can include TV, infra-red camera and laser with automatic tracking capability. CW illumination for operation of semi-active radar guided missiles is also available.

Computation: this is made by means of a digital processor and simultaneous engagements of air or surface targets are possible. An additional console is provided if the torpedo fire control is exercised.

STATUS

In production. In service since 1978 with the Royal Malaysian Navy.

CONTRACTOR

Philips Elektronikindustrier AB, 17588 Järfälla, Sweden.

6449.281

9 LV MK 3 WEAPON CONTROL SYSTEM

The 9 LV Mk 3 series of systems has been developed from the well-known 9LV 200 Mk 2 weapon control system. It has the capacity to correlate data from various surveillance sensors, including a new generation search radar and an advanced ESM surveillance system. The design also allows operation in a multiple threat environment by integrated air defence functions employing guns and ECM.

The installation in the RSwN 'Göteborg' class coastal corvettes exemplifies a total system with a full range of operational functions. The modular design permits simple breakdown into smaller systems to fit any size of ship. This is achieved by local area network internal communication, a distributed processor concept, and the modular hardware and software block design.

A typical 9 LV Mk 3 system comprises:

(1) a radar surveillance system with a dual band (X and C) search radar and target extractors for automatic plotting of air and surface targets
(2) a command and control system with powerful computers, with functions for tactical analysis, target management (correlation threat evaluation, target designation a.o.), situation picture compilation, and presentation
(3) a duplicated data storage system where all the subsystems have access to the latest data
(4) a gun fire control system with the 9 LV Mk 3 TWT tracking radar, and a choice of electro-optical sensors as options.

The system can be extended by adding one or more of the following options:

(1) additional gun fire control channels as above
(2) EW systems comprising sensors and countermeasures, and which can be integrated with the gun fire control system to achieve optimised counter-action in a multiple threat situation
(3) interfaces to a range of weapon systems including SSM, SAM and torpedo systems
(4) data link interfaces.

Data processing capacity is distributed to the various subsystems by Motorola 68020 processors.

OPERATION

The system will be fitted with workstations corresponding to the amount of subsystems involved. The workstations are of general design with software-controlled man-machine interfaces to enable reallocation of functions between different workstations. The displays can be vector scanned or raster scanned. Functions are controlled by menus on touch-input displays.

STATUS

In production for four Swedish 'Göteborg' class, seven Danish 'Flyvefisken' (Stanflex-300) class and four Finnish 'Helsinki' class fast attack craft.

CONTRACTOR

Philips Elektronikindustrier AB, S-17588 Järfälla, Sweden.

4026.281

9 LV 300 NAVAL WEAPON CONTROL AND C³ SYSTEM

The 9 LV 300 consists of two autonomous AA channels, normally a 9 LV 200 fire control system and a 9 LV 100 fire control system. All the capabilities of these two systems are inherent in the 9 LV 300 enabling simultaneous engagement of two air targets, surface target engagement using TWS and firing of SSMs.

STATUS

In service.

CONTRACTOR

Philips Elektronikindustrier AB, 17588 Järfälla, Sweden.

Swedish mine warfare vessel Carlskrona *is fitted with a version of the 9 LV 300 command and control system for operational as well as training purposes*

4025.281

9 LV 400 SERIES C³I SYSTEM

The 9 LV 400 Command, Control, Communication and Intelligence (C³I) system is designed for tactical command from flotilla leaders of frigate size and upwards. It has greater surveillance capabilities than the 9 LV 300 providing a separate air search radar and integrated ESM/ECM which permits automatic tracking and threat evaluation of a large number of air and surface targets.

The command and control functions are also an improvement on the 9 LV family. The 23-inch tactical electronic plot is included as standard. Separate consoles are used for air and surface surveillance.

The command capability is further augmented by a high capacity data link using the time division multiple access principle. This data link allows for exchange of target data and commands between ships and also between ships and command centres ashore. All transmissions are encrypted to a very high degree of security.

In its standard configuration the 9LV 400 WCS provides simultaneous control of four separate guns of different calibres, SSM and various kinds of ECM. Optional control of a SAM system is also available.

STATUS

In service.

CONTRACTOR

Philips Elektronikindustrier AB, 17588 Järfälla, Sweden.

4568.281

9 LV + 50 ASW SYSTEM

For surface anti-submarine vessels the 9 LV + 50 ASW sub-system may be added to the 9 LV fire control systems raising the designation to 9 LV 350 and 9 LV 450 respectively. The ASW sub-system contains its own computer hardware and software. The ASW plot contains one PPI, an alpha-numeric display, officers' and operators' panels. The same computer and display techniques are employed as in the 9 LV systems (**4025/6.281**).

The main tactical functions of the system are:

(1) search and target tracking in association with the ship's sonar system
(2) presentation of sound path propagation on a sound path indicator presentation, integrated with the ASW plot (ie on the same CRT display as the PPI)
(3) continuous presentation of the combat situation
(4) evaluation of target data and prediction of future target position
(5) calculation of control signals for A/S rocket launcher
(6) calculation of own ship's course and time to fire charges for depth charge attack. The last facility is optional. The basic system is designed for control of the Bofors 375 mm twin A/S rocket system with all available types of A/S rockets including Erika and Nelli. Any type of A/S homing torpedo can be accommodated if this option is specified. Sonar is the principal sensor and various models of different manufacture may be employed. Other inputs to the system are the ship's gyro reference system, and the ship's log
(7) Calculation of control data for A/S homing torpedoes.

STATUS

In production.

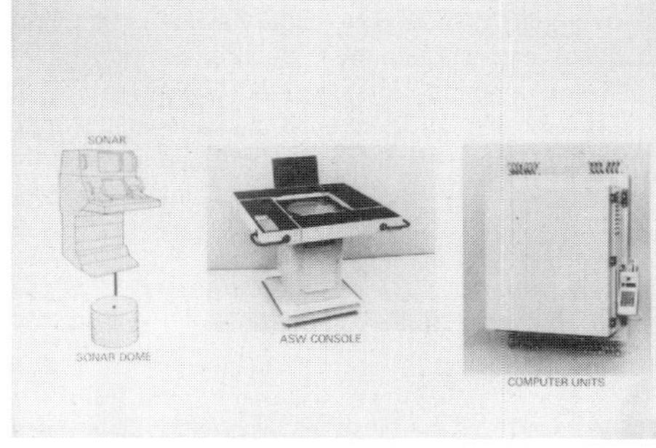

Main elements of typical 9 LV + 50 ASW FCS

CONTRACTOR

Philips Elektronikindustrier AB, 17588 Järfälla, Sweden.

3747.281

EOS-400 FIRE CONTROL SYSTEM

The EOS-400 is a TV tracker and laser rangefinder fire director system. Accurate target position is obtained using the well-known Saab television tracker for automatic tracking in conjunction with a laser rangefinder, both these units being mounted on a common training and elevating stabilised platform. This unit is the only above-deck item of equipment, apart from the weapon being controlled (usually a gun or guns). Below deck, the control/display unit is usually located in the operations room. The only other major item, the electronics cabinet, can be sited in any convenient position in the ship.

The multi-sensor head of the Saab EOS-400 fire control system

The control/display unit contains a TV monitor and a joystick for sensor platform operation, a mode selector, and target data displays. The electronics cabinet houses the video tracking electronics and other control electronics. The video output is connected to the platform and the video output to the control/display unit.

The system is capable of tracking one target among several in close formation, tracking when radar tracking is not possible, and either surface or air targets may be tracked. Inputs from other systems include those from a ship's surveillance radar, and reference data from ship's compass. The system calculates ballistics for several guns and gives gun-laying data.

CHARACTERISTICS

Tracking parameters
Training: Unlimited
Elevation: –30° to +85°
Platform
Angular speed: ±175°/s in azimuth and elevation
Angular transducers: Multi-speed synchros
Stabilisation: Rate gyros in 2 axes
Tracking: Digital correlation
Contrast selection: Automatic black/white
Lock-on area: Whole field
Daylight camera
Scanning standard: 625 lines
Camera tube: Si-vidicon
Night camera: IR camera optional
Laser rangefinder
Wavelength: 1.06 μm

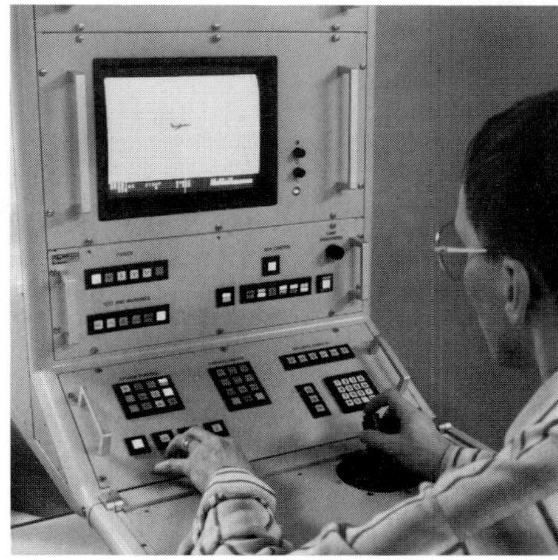

Operator's console of the Saab EOS-400 fire control system

Repetition rate: 10 Hz
Beamwidth: 1.5 mrad

STATUS

Introduced in September 1979. A new version entered the market in 1983, and is installed in Finnish Navy 'Turunmaa' class corvettes and for one additional export customer.

CONTRACTOR

Saab Missiles AB, 58188 Linköping, Sweden.

4027.281

9CSI 500 STINA

The 9CSI 500 STINA (coastal/offshore surveillance and information system) is primarily designed to support a number of peacetime duties, including, but not limited to: maritime traffic control; monitoring of ships with dangerous cargoes; supervision of warships; surveillance of territorial limits for illegal immigration, smuggling etc; surveillance of economic zones, eg for the prevention of illegal fishing; and command of sea rescue operations.

A complete system, eg for nation-wide coverage, normally comprises more than one centre. The concept is modular. A complete system furthermore comprises sensors of various kinds, mainly coastal radars, and communication equipment for data and voice.

Although primarily a peacetime system, the 9CSI 500 can also be used as a naval headquarters in wartime. For mainly wartime use, however, the more comprehensive 9CSI STINA is available (see **4028.281**, below).

For further details see Page 456 in the 1987-88 edition.

STATUS

In service with the Swedish Navy.

CONTRACTOR

Philips Elektronikindustrier AB, 17588 Järfälla, Sweden.

4028.281

9CSI STINA

The 9CSI STINA (naval command, control, communication and intelligence system) system concept designed to automate many of the functions involved in information gathering, filtering and compilation of data, presentation, and transmission of orders and commands with high speed and accuracy. These systems are designed as general purpose equipment but in particular as supplements to the extensive 9 LV family (above) of shipboard combat information and weapon control systems. Like 9 LV systems, 9CSI systems are highly modular using standardised building blocks. The STINA system concept is gradually being updated by stages using new modules developed under the extensive development programme for BASSYSTEM 2000. The latter features a new generation of computers, a local area network, standardised consoles and associated software modules.

A STINA system comprises not only one or more command centres, but also sensors, mainly radars, for gathering input data, and data link communication equipment for exchange of data ship-to-shore and vice versa and for exchange of orders and commands. The 9CSI is primarily designed as a military system for use during a general alert and in wartime. However, many of the functions implemented are equally useful in peacetime for maritime traffic control, coast guard prevention of smuggling and illegal immigration, co-ordination of sea rescue operations etc.

Designed as a modern military system, the STINA concept incorporates fast, secure communication with anti-jamming capability and ECCM features in the radars; all in order to enhance survivability of the functions. Speed of communication is achieved in the system by using a mesh of point-to-point links and the time division multiple access (TDMA) principle over radio links to exchange data and command messages. Security is obtained by encryption of all messages. Anti-jamming and low probability of intercept are inherent in the TDMA principle and can be further enhanced by using spread-spectrum or frequency-hopping techniques.

Other features of the communications are a high degree of automation and routeing of messages adapted to a changing network with priorities and error correction. The speed of transmission is adapted to the medium used for different routes.

Provisions have been made in the 9CSI STINA centres for storage of intelligence in a data base, which can also contain information about own forces, their status etc. In order to assure proper secrecy for such information, a part of the data base is intended to be 'user programmed'.

A major sensor in STINA systems is the Philips coastal surveillance radar (**1549.153**). This is available for fixed or mobile installations and normally provided with an automatic plot and/or target extractor to permit unattended operation and data transfer over narrow band channels to the centres. The plot and target extractors can work not only with the Philips but also with other radars to allow upgrading and automatic inputs. The centres will also accept semi-automatic and manual inputs.

The design is highly modular, which permits the building of command centres of various sizes for different tasks, eg central and regional headquarters. The number of operators' consoles and commanders' consoles is variable, as are computer capacity and the exact display configuration. A standard building block is the operator's console shown in the photograph. It consists of a 41 or 51 cm monochrome or colour display (PPI), an alpha-numeric display terminal with a keyboard, and a communications panel. The latter permits fast access to radio channels and telephone lines.

The computers for storage and processing of data receive inputs in different ways. Inputs from extractor equipped radars are entered automatically, as are data link reports. Other inputs can also be entered automatically if received in standardised formats. Non-standardised messages, eg verbal reports over radio or telephone, are entered by the operators using the keyboard of the alpha-numeric terminal. The operators compile a picture of a situation which is continuously updated. The commander and his assistants can now be assisted by the computer to assess the situation and plan the tactics by using prediction and simulation of alternatives. After decisions, commands are transmitted over the data link using a library of standard orders. A situation picture compiled in one centre can be transferred in full or in selected parts to other centres and to ships. In this way, commanders at all levels have continuous access to up-to-date, accurate information, including relevant parts of the intelligence data base.

STATUS

In production and in service.

CONTRACTOR

Philips Elektronikindustrier AB, 17588 Järfälla, Sweden.

Typical 9CSI STINA operator's consoles

3511.481

ERICSSON SUBMARINE ACTION INFORMATION AND FIRE CONTROL SYSTEM

This system, or a specific version of it, is also known by the initials NEDPS, from Näcken Electronic Data Processing System, because of its use in the 'Näcken' class of submarines of the Swedish Navy.

The AI FCS is a fully integrated system to acquire, process and display information for tactical evaluation and to form a basis for decisions regarding selected targets and torpedo guidance. The system includes complete hardware and software for controlling wire-guided homing torpedoes. Also included is computer control and monitoring of ship's function, such as propulsion, steering, depth keeping, trim, storage battery condition, etc.

The system has a dual computer configuration, the information being gathered through an extensive data collecting system from the different surveillance and weapon systems as well as from the navigation system. The information is presented to two operators on two separate displays, each of which is equipped with input facilities in the form of special keyboards and rolling balls. After target acquisition the tracking is carried out automatically and relevant fire control data are calculated continuously. Upon firing, the torpedoes are normally controlled fully automatically with graphic and alpha-numeric presentation of all relevant information. The operators at the tactical and fire control displays can at any time take over or adjust the procedure.

Ericsson submarine action information and fire control system operators' consoles

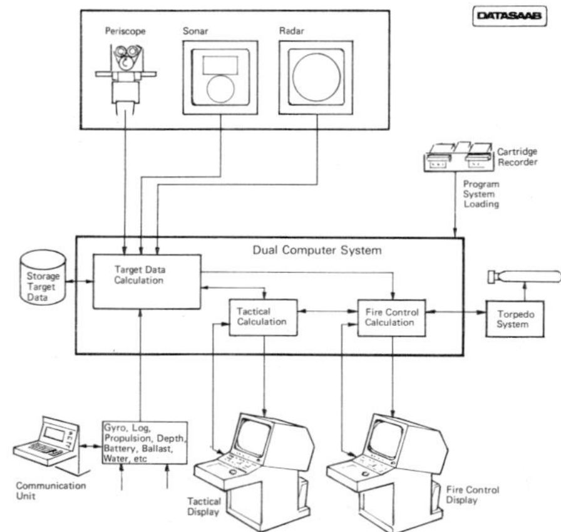

Block diagram of Ericsson submarine action information and fire control system

STATUS

The Mk 1, or NEDPS, is in service with the three 'Näcken' class submarines while the Mk 2 which uses software developed from the Mk 1 with a new single multi-function console, is in production for four 'Västergötland' class submarines and three 'Sjöormen' class submarines. Ericsson have studied retrofits for HDW Type 205 and Type 209 submarines. A Mk 1 has also been installed in a training simulator at Bergas Naval College.

CONTRACTOR

Ericsson Radio Systems, 17586 Järfälla, Sweden.

UNITED KINGDOM

4045.261

CACS 1

CACS 1 was the first of a family of new command systems for the Royal Navy, incorporating modern computers and a new operator interface. It provides a substantial increase in system capability which can be directly employed to enhance the command functions necessary in a modern warship. Both the hardware and software are constructed in modular fashion so that further system development and expansion can be implemented after installation on board and experience at sea, enabling the system performance to keep pace with changing and increasing requirements.

The CACS system architecture is based on a powerful computing facility consisting of two FM1600E computers. These are connected by a distributed highway to several display positions, each of which contains its own Argus M700/20 microprocessor system to optimise operator effectiveness.

General view of HMS Boxer*'s operations room*

The consoles are based on the new Plessey Series 9 display system which, with greatly increased quality and capacity, supersedes the Mk 8 display technology, used in ADAWS. There is extensive use of video re-timing and scan compression techniques. The principal input device is the light pen, which replaces most of the rolling ball and keyboard functions.

CACS employs an all-vertical display concept which, besides offering the operator a more comfortable and efficient working position, overcomes severe shipfitting constraints and the problems associated with conventional horizontal dual-operator displays.

The most significant advance from the user's viewpoint will be the implementation of the 'prompt' interface for operator communication with the system. For CACS 1 the interface has been further developed to include the use of light pens and two TV monitors, one for tote data and one for handling operator inputs. With this interface the tote displays information relevant to the current task in hand and indicates 'prompts' or choices, which when used in conjunction with the light pen and/or keyboard allows the operator to input data and control the system in a logical sequence. The need to remember complicated manual injection sequences has been eliminated with a consequential reduction of operator training requirements.

STATUS

CACS 1 is now fully operational in Type 22 Batch 2 frigates. Ferranti supply the FM1600E, Argus M700, and F100-L computers as well as acting as prime contractor and design authority. Plessey Naval Systems is the main display contractor. Further feasibility studies have been carried out for the Royal Navy, in which the application of the CACS family to other ships has been assessed.

CONTRACTORS

Ferranti Computer Systems Ltd, Western Road, Bracknell, Berkshire RG12 1RA, England.

Plessey Displays Ltd, Station Road, Addlestone, Weybridge, Surrey KT15 2PW, England.

6747.281

'DUKE' CLASS (TYPE 23) FRIGATE COMMAND SYSTEM

The Royal Navy's new 'Duke' class (Type 23) frigates were originally scheduled to receive the CACS 4 while the 'Cornwall' class (Type 22 Batch 3) were scheduled to receive the CACS 5 system. However, in 1986 the Ministry of Defence decided that neither system fully met the requirements of surface ships in the 1990s and requested fresh bids.

Some 19 companies were invited to tender but in April 1987 Plessey Naval Systems with Software Sciences and Ferranti Computer Systems together with Logica and Systems Designers were awarded the contract. However, Plessey later announced they were unable to meet the terms of the contract and in August new bids were requested.

Two consortia were awarded year-long project definition contracts in November 1987. Ferranti, in association with Logica and Systems Designers, are proposing the Type 23 Command System based upon the 500 Series (see **6595.281**). Surface Ship Command System, consisting of Gresham-CAP and Racal, are offering a system similar to Successor.

3505.281

WSA-420 AND CAAIS 450 SERIES FIRE CONTROL AND ACTION INFORMATION SYSTEMS

These systems are based on and incorporate many of the proven features of WSA-4 (**1496.281**), WSA-400 Series (**1524.281**) and CAAIS (**1046.261** in 1980-81 and earlier editions) now in service with the Royal Navy and other navies.

The main differences between these new systems and their proven predecessors is the computer and associated hardware architecture, where a modern generation of computer and display technology is used.

Bearing in mind the operational requirements, Ferranti has designed a range of new systems which embody a combination of the latest technology and proven techniques, to operate with modern sensors and weapons. These include:

(1) WSA-421 – fire control system (receiving target indication from an external source)
(2) WSA-422 – fire control system with limited AIO facilities
(3) WSA-423 – integrated fire control and AIO system
(4) CAAIS 450 Series – AIO systems.

All systems can include on-board training aids, such as radar echo generators, simulated weapon tellbacks etc.

WSA-421 Fire Control System

This is designed primarily for frigates and corvettes fitted with an independent, but linked, AIO/CIC system

Ferranti combat system WSA-422 system console

such as CAAIS 450. For such vessels the main armament would be a 76 mm-calibre gun or a heavier weapon such as a 4.5- or 5-inch (114 or 127 mm) gun. The secondary armament could be a similar or smaller calibre weapon, or a surface-to-air missile. Two tracking heads provide the WSA-421 with two independent channels of fire. The fire control prediction is performed by a Ferranti FM1600E computer. The system accepts target indication from the ship's AIO system.

WSA-422 Fire Control System with Limited AIO

Designed primarily for vessels of patrol craft size to provide weapon control and surveillance tracking for deployment of 35 mm to 127 mm-calibre guns and/or surface-to-air missiles and optionally, surface-to-surface missiles, the system uses the FM1600E computer.

The main sensor is the ship's surveillance and/or navigation radar. A tracker radar and OFD which can mount a laser rangefinder in addition to the E-O or CCTV is also fitted, thus providing two independent channels of fire and enabling two gun mountings to be controlled. Although the tracker radar will normally be paired with the main gun, the flexibility of the system enables any sensor-target pairing to be achieved. The composite weapon system console is manned normally by a single operator with provision for up to three operators in active conditions, and houses all displays, indicators and controls for surveillance tracking, target designation, control of weapon sensors and of the armament. Manual rate-aided tracking, with computer-generated symbology, is available on up to 60 targets. The power of the FM1600E computer is such that additional facilities may be provided if required, such as a remote-controlled, stabilised video tracker, comprising CCTV and laser rangefinder, to assist tracking in poor ECM environments.

WSA-423 Integrated Fire Control and AIO System

This system is designed for vessels of patrol craft and

corvette size to provide weapon control and AIO/CIC facilities for the deployment of guns (35 mm to 130 mm), surface-to-air and surface-to-surface missiles.

WSA-423 is an extension of the WSA-422 system with added AIO facilities to provide an integrated system.

CAAIS 450 Series AIO Systems

This series is designed to provide AIO facilities for a range of vessels of FPB size or larger. Provision is made for passing target indication data to associated fire control systems, which could be a sub-set of WSA-422.

CAAIS 450 can accept tactical data from two surveillance radars, from sonar and ESM equipments. Auto-tracking of targets is provided on both radars. Although the FM1600E computer's capability includes driving up to eight 12- or 16-inch (305 or 406 mm) LRDs, the exact display fit will depend upon the particular application. A basic configuration would have three displays: two for picture compilation and one for command appreciation. Additional display requirements can be envisaged for such functions as ASW, helicopter control, ESM/ECM management or data link management. The software provided is generally derived from the CAAIS programs. However, due to the versatility of this configuration, which results from the use of a powerful computer, specific applications may require that some special software is needed so that the full capability of associated weapons and sensors can be realised.

STATUS

Production began in 1980. The first systems of this series are now at sea. WSA-421 has been selected by the Royal Navy for firing systems. WSA-420, CAAIS 450 and WSA 423 series systems are in production for a number of overseas navies, including Brazil.

CONTRACTOR

Ferranti Computer Systems Ltd, Western Road, Bracknell, Berkshire RG12 1RA, England.

6745.481

OUTFIT DCB SUBMARINE TDHS

Outfit DCB is a well proven tactical data handling system with an integrated tactical fire control system. The equipment is in service with Royal Navy submarines to provide computer assistance to the Action Information and Fire Control organisation and to reduce the manpower requirements of the attack team. Rationalised displays to facilitiate command appreciation are also provided.

The DCB system typically comprises equipment installed in the control room (fire control and AIO consoles), in the computer room (two 1600B computers) and in the weapons compartment (weapons interface, service supply modules and battery monitoring equipment). Synchro and resolver outputs from the submarine's navigation equipment, periscopes and sonars are converted in computer peripheral units and input as digital data to the AIO computer. Classification and intercept data are input in coded digital form from keyboards. Tracks of all contacts are computed and shown in plan and time/bearing form on two displays on the AIO console. Control of track association can be exercised by the AIO operator. The AIO computer is programmed to perform tactical calculation when ordered by the operator. The computer also outputs contact data for printing on an automatic Contact Evaluation Plotter (CEP).

Digital data for potential targets can be passed through a transfer organiser from the AIO computer to the fire control computer. The computer fire control solutions are shown in plan form on two CRT displays at the fire control console. The weapon settings are computed and shown in tabular form on the CRT displays. When a wire-guided torpedo is fired, data feedback from the weapon are returned to the computer so that weapon settings can be updated and steer commands generated.

Weapon settings data, tube orders, weapon orders and training data are transmitted to the weapons compartment as digital words multiplexed on duplicated serial highways. Response words carrying data from the weapons and weapon compartment are transmitted to the control room on two return serial highways. Weapon compartment equipment decodes the digital data from the control room to provide pre-fire supplies, guidance commands and discharge gear drive pulses to the weapons and tubes as necessary.

Duplication of data highways and cross-linking between units provides a degree of redundancy and a choice of fallback modes. Modular concepts are employed to improve maintenance and reduce down-time for fault rectification.

STATUS

In service with the Royal Navy.

CONTRACTORS

Dowty Maritime Systems Ltd, Gresham Division, Twickenham Road, Feltham, Middlesex TW13 6HA, England (prime contractor).

Ferranti Computer Systems Ltd, Western Road, Bracknell, Berks RG12 1RA, England (subcontractor – computers).

3746.481

OUTFIT DCC SUBMARINE TDHS

Outfit DCC is the action information organisation (AIO) and fire control system designed for the new class of SSK the 'Upholder' class. It has been given the type designation DCC and is derived from DCA/DCB systems in use in British nuclear submarines.

The weapons fit for the 'Upholder' includes dual-purpose wire-guided torpedoes (anti-submarine/anti-ship), anti-ship air-flight missiles, and submarine mines. The fire control for weapons is integrated in the tactical system, which comprises: long-range passive sonar, a medium-range passive/ active sonar with a cylindrical array located in the bow, sonar intercept, analysis and classification facilities, passive ranging using three aligned arrays on each side of the vessel, underwater communications, bathythermograph, and a sound speed-measuring system.

Other sensors include search and attack periscopes, and both are provided with facilities for fitting a variety of electronic and electro-optical devices. A Type 1006 surface surveillance and navigation radar, and omni-directional and directional warning antennas for the boat's ESM system are carried on a separate mast or incorporated in the search periscope.

Digital data transmission and processing is employed and the sensors and DCC are closely integrated; target bearing and range and velocity data are processed to provide accurate target motion analysis. The tactical system consists of a three-position console with electronics in the lower compartments and six 12-inch (305 mm) circular cursive displays, light-pen units, fire control order and status units in the upper positions. DCC receives tactical and environmental data from the submarine's sensors via a digital data bus. The data is collated, processed, evaluated and presented on the circular displays for final assessment and decision.

The system also provides all necessary weapon control functions. The system computer is the Ferranti FM 1600E: a second fully redundant computer is provided to allow for fallback modes and future software expansion. Light pens provide the main man/machine interface, and all selection and switching facilities are provided at the operator console positions.

STATUS

First systems delivered for four 'Upholder' class (RN). To go to sea late 1988.

CONTRACTORS

Ferranti Computer Systems Ltd, Western Road, Bracknell, Berkshire RG12 1RA, England (prime contractor).

Dowty Maritime Systems Ltd, Gresham Division, Twickenham Road, Feltham, Middlesex TW13 6HA, England.

4272.481

KAFS SUBMARINE AIO AND FCS

KAFS is suitable for conventional submarines of all sizes. The system uses the advanced computer technology, latest processing techniques and software algorithms incorporated in similar systems in RN nuclear submarines and the new 2400 tonne class 'SSK'; these provide operational facilities such as picture compilation, target motion analysis and fire control so that the commanding officer may carry out the wide variety of tasks on which such a submarine can be employed. These include area surveillance, attacks on surface and underwater targets, operations in inshore waters (such as minelaying or periscope reconnaissance), intelligence gathering and participation in ASW exercises.

The system hardware and software has been specifically designed in a compact, modular manner to provide a wide range of operator facilities which can be adapted or expanded as necessary to meet individual system requirements.

The operator console is located in the control room and is linked by a MIL-STD-1553 data bus to the weapon control equipment in the torpedo room. The base of the operator console houses the FM1600E central processor, the input/output control unit, 256 K semiconductor store, display drive unit, magnetic cartridge unit and PCU, interface units and power supply units. The processor is one of a series of 24-bit F1600 computers designed by Ferranti and supplied for RN surface ship and submarine systems and a wide variety of overseas systems.

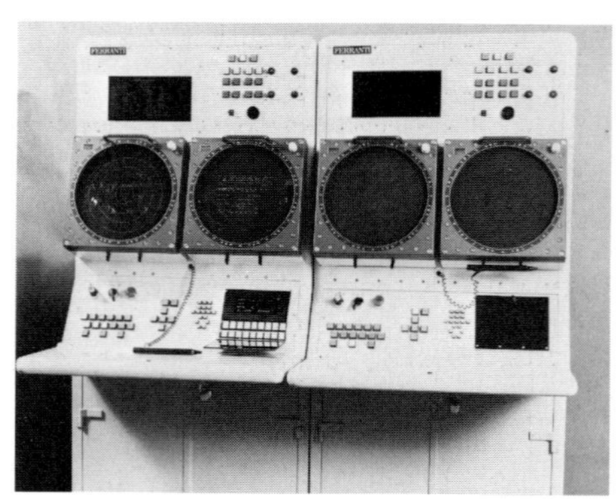

Double KAFS console

The top half of the console has two identical operator positions each having two display areas, a light pen, control panel and a command data panel. Either operator position can control the whole system thereby allowing one position to be shut down in patrol states. The display areas provide the operator with a labelled plan or time/bearing display according to his selection, and a tote providing pages of data and instructions to enable the operator to interact with system control, picture compilation, navigation modes, target motion analysis and weapon fire control.

Training can be carried out on AIO, fire control and weapon control procedures while the system is in the operational mode. Facilities are provided for emergency control of weapon discharge and guidance independent of the central processor. Tube status, weapon status and system status are displayed on a local control panel which also provides weapon discharge and control facilities in fallback mode; maintenance and fault-finding facilities; and training facilities. Microprocessor-based port and starboard weapon interface and tube switching units are provided on a modular basis according to the number of tubes and weapon types. Facilities are provided for weapon pre-heating in stowage and battery/gas monitoring to detect hazardous conditions.

DEVELOPMENT

Studies of an AIO and FCS for RN nuclear submarines carried out by Ferranti in the 1960s led to the Tactical Data Handling System (**3746.481**) which was built under MoD contract for the 'Swiftsure' class SSNs. A production order has been placed for the first four RN 'Upholder' class SSK systems and the Brazilian Navy has placed an order for at least three systems. In 1987 Brazil ordered another two systems for Type 209 submarines under a £4 million contract. One will be licence built by SFB Informatica in Rio de Janiero.

CONTRACTOR

Ferranti Computer Systems Ltd, Western Road, Bracknell, Berkshire RG12 1RA, England.

6746.481

OUTFIT DCG SUBMARINE TDHS

DCG is a tactical data processing system which has been designed, developed and built by Gresham-CAP to support the command and control system in UK submarines. The system exploits commercially available computing units to achieve high performance and high reliability within a low life time cost and short procurement lead times.

System DCG is linked to the main submarine command (action information) computer from which it derives data describing the submarine's own movement and the perceived environment. System DCG is able to subject this data to more vigorous and intensive analysis than the command computer to produce a considerably more accurate picture of the tactical situation.

While the system was developed originally for UK submarines, connections to any automated command system may readily be achieved using standard interfaces. Alternatively, the manual data entry capability enables System DCG to be used completely stand-alone.

Outfit DCG consists of:

(1) an array processor which is able to perform high speed matrix mathematics. It executes up to eight million floating point operations per second; about one hundred times the speed of a conventional minicomputer
(2) a mass memory device to hold long timespans of data necessary to detect trend in slow moving events
(3) sophisticated graphics terminals to present the data in easily assimilated picture form
(4) lightweight keypads with an ergonomic layout specifically designed for DCG interaction
(5) a general purpose minicomputer to organise the data and control the devices
(6) a high resolution colour graphics printer.

The skill of the system integrator consists in selecting matched units which, when combined, deliver the system capability and the system reliability demanded.

System DCG supports an extensive catalogue of tactical software, and the operator terminals are able to perform different functions simultaneously.

These command aids, tracking and combat functions are accessed through a menu driven operator interface, with emphasis on a common style of interaction, data compatibility between functions and ease of extension.

STATUS

In service with the Royal Navy.

CONTRACTOR

Gresham-CAP Ltd, Gresham House, Twickenham Road, Feltham, Middlesex TW13 6HA, England.

6275.481

COMKAFS (OUTFIT DCH) SUBMARINE COMBAT SYSTEM

COMKAFS is one of a range of compact submarine combat systems and is a variant of the KAFS systems (**4272.281**) fitted in a number of submarines. It is suitable for all classes of conventional submarines but has been specifically designed for easy installation in smaller boats. The Royal Navy version of COMKAFS is known as DCH and is being supplied for the 'Oberon' class update programme.

The system uses advanced computer technology, the latest distributed processing techniques and modular software. Operational facilities include picture compilation, tactical evaluation, target indication, fire control, navigation and sonar ray path analysis. COMKAFS can engage surface and subsurface targets, and assists the command during participation in ASW exercises, and in covert operations such as minelaying, periscope reconnaissance and intelligence gathering.

The AIO screen formats include labelled plan display, bearing display, vertical section display, and extended SITREP display. Plasma panels with touch sensitive overlays provide the main operator interaction medium using a menu prompt system.

Information from the submarine's sensors is received via a data bus. COMKAFS employs the Military Argus Compact (MAC) computer, incorporating Argus M700 series processors which provide self-contained computing facilities. Onboard training facilities, built-in status monitoring and fault finding, and two fall-back modes of operation are included. Any number of display positions can be provided, depending on the class of submarine. Colour raster is available as a customer option. Further options include auto guidance of more than two standard torpedoes, an automatic plotting table and an automatic contact evaluation plot. COMKAFS is also available fully integrated with the Triton submarine sonar system which consists of a coupled federation of independently designed and developed sensors from the three major UK sonar contractors, Ferranti, Plessey and Marconi.

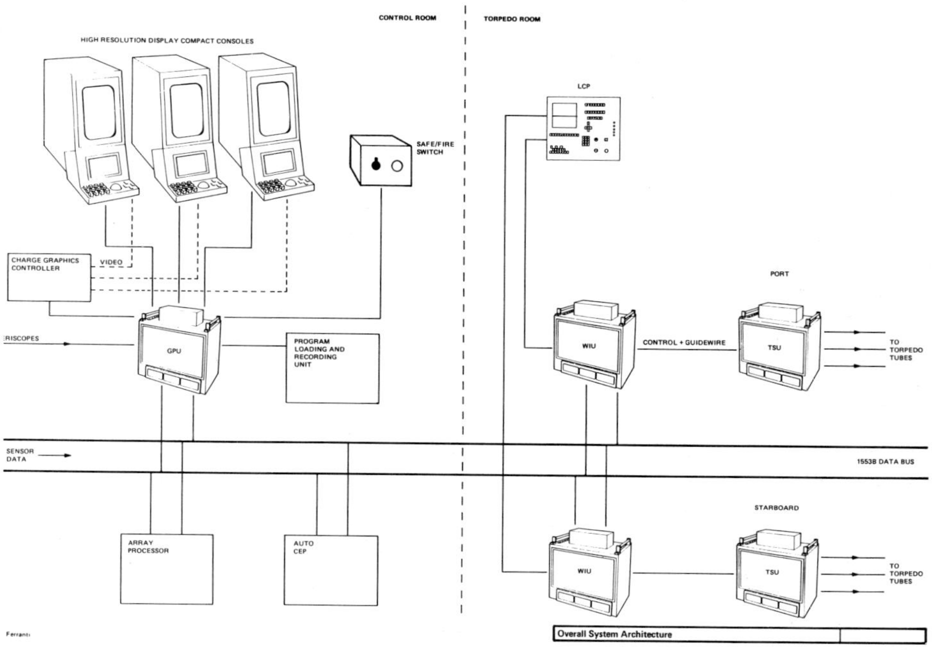

As with KAFS, the high level of automation minimises the need for operator intervention so that one operator can carry out urgent attacks controlling the complete system.

Overall system architecture of COMKAFS

STATUS

In production for ten Royal Navy updated 'Oberon' class SSKs. The first two of 12 systems had been delivered by December 1987.

CONTRACTOR

Ferranti Computer Systems Limited, Western Road, Bracknell, Berkshire RG12 1RA, England.

6594.281

FERRANTI INTEGRATED COMBAT SUITE

The Ferranti Integrated Submarine Combat Suite (FISCS) is the latest system of this type to be produced by the company. It incorporates the necessary system architecture, processors and display technology to provide a combat suite combining state-of-the-art sonars with advanced data handling and weapon control facilities.

FISCS accepts multi-sensor inputs and presents a comprehensive tactical picture by the integration of target data from all sensors. Tactical evaluation facilities include relative velocity calculations, background data manipulation, and command advice. Weapon control capabilities extend to salvo fire of wire-guided torpedoes and submarine-launched missiles. Other features include comprehensive onboard data link facilities, a minelaying package, and automatic deployment of decoys and countermeasures. Periscope and radar video may be fully integrated within the combat suite.

FISCS comprises a six-position operator console with 17 shelves of electronic equipment in the control room, a control panel and seven shelves of electronics equipment in the torpedo room, and the submarine's complete fit of sonar arrays and equipment.

The system draws its information from a variety of sensors, but the heart of FISCS is the advanced sonar capability which provides the acoustic information for subsequent operational activities by the submarine, both inshore and in deep ocean environments. It is optimised against surface vessels, submarines, ASW aircraft and their associated weapon systems. Acoustic detection, signal processing and analysis are employed to maximise detection ranges, provide an advanced warning capability, and capitalise on the latest methods of target analysis and classification. A comprehensive transducer suite facilitates all-round coverage over selectable or fixed frequency bandwidths, and includes towed, flank, bow, intercept and specialist array systems such as passive ranging array groups.

The received information is processed to provide wideband, broadband, narrowband and specialist data for subsequent analysis and use. Nine-core elements operate either individually or in conjunction with other elements, and are integrated together for presentation on any or all of the display screens. Information is readily interchangeable between sonar elements or with the tactical data handling and weapon control subsystems, and the latest ergonomic and man-machine interface designs are incorporated.

CONTRACTOR

Ferranti Computer Systems Ltd, Cheadle Heath Division, Bird Hall Lane, Cheadle Heath, Stockport SK3 0XQ, England.

Ferranti Integrated Submarine Combat System

6595.281

FERRANTI 500 SERIES COMMAND AND CONTROL SYSTEMS

Ferranti Computer Systems has developed a new, integrated family of shipboard systems, primarily intended for the export market. The series consists of three main system elements, the WS500 weapon control system, CS500 command system, and MC500 mine warfare command system. These elements enable systems to be assembled rapidly for a wide range of warships from patrol craft and minehunters to frigates and destroyers.

The 500 series has been developed as a highly flexible system family which can be tailored to varying operational requirements, installed in a wide variety of ships, interfaced with a broad spectrum of sensors and weapons, and implemented with a variety of man-machine interface/operational features. The development of the series involved:

(a) definition of a flexible, distributed, hardware-independent system architecture as the common basis of all members of the family

(b) provision of an infrastructure to support the application software. The key element is a virtual system environment suite (VSE500)

(c) creation of a method, supported by tools, to carry out system software development. Most of this supports the real-time application which is customer-specific. ADA was the only realistic language choice in this area

(d) adoption of a design process which exploited widely-used commercial interfaces and standards wherever applicable.

This approach simplifies the task of optimising systems for individual customers and provides a constant basis of system services which can be layered on the chosen hardware architecture/configuration.

STATUS

A dedicated version of the CS500 for the Royal Navy has been offered for the Type 23 command system competition. With the short timescale involved, Ferranti teamed with two software houses, Logica and Systems Designers, to provide a design to meet the Royal Navy specification.

CONTRACTOR

Ferranti Computer Systems Ltd, Bracknell Division, Western Road, Bracknell, Berkshire RG12 1RA, England.

Ferranti MC500 console

4577.481

VSEL COMBAT SYSTEM DATA BUS

Developed from the interfacing expertise gained as a result of the WSDB development VSEL has designed, developed and produced a range of Data Distribution Units (DDU) for naval vessels. These units are based on a highly flexible digital architecture, which enables a wide range of requirements to be met. The VSEL DDUs are capable of meeting a wide range of digital and analogue interface standards, making them ideally suited to both refit and new build applications. The highly flexible architecture allows a mixture of analogue and digital interfaces in the same unit. Synchro, resolver and most of the common range of digital interface standards (including data highway interfaces) are available from the current range. The processing capability of the units allows for the conversion of interface standards within the unit. The processing capacity of the units allow dynamic simulation to be carried out on board without any additional equipment. The built in self test capability ensures the rapid detection and location of faults.

The units are configured from a range of high reliability, proven sub-assemblies to meet customer requirements in a very cost effective package. The units are very compact and are suited to both small and large naval vessels. The units are designed to ensure easy ship fitting.

Typical applications of the VSEL range of DDUs are ship data distribution in synchro and digital formats; and synchro retransmission units.

STATUS

In production.

CONTRACTOR

Vickers Shipbuilding & Engineering Ltd, Barrow-in-Furness, Cumbria LA14 1AF, England.

Remote terminal of VSEL CSDB

3508.281

CONTROL FOR SURFACE-LAUNCHED TORPEDOES (CSLT)

CSLT is an independent miniaturised torpedo fire control developed by Vickers for operation on any ship having Mk 32 torpedo tubes armed with either MK 44, Mk 46 or other tube-compatible torpedoes. The main features of the system are: small size (control room equipment about 0.1 m^3); micro-processor computation; sensor smoothing and target prediction software derived from proven in-service systems; modular construction; flexible interfaces. The equipment is of rugged construction designed for operation in severe naval environments and comprises:

(1) calculator and display module
(2) torpedo selection and setting module
(3) shipboard indicators
(4) local control and display modules (port and starboard)
(5) power supply and switching unit.

The calculator and display module is supplied either as a bulkhead-mounted unit or as a VCS module for incorporation into a user console. Its interfaces accept data from the ship's sensors and a manual keyboard input, and send processed information on relative target motion to the torpedo selection setting module, shipboard indicators and the AIO.

The torpedo selection and setting module performs these functions in either a semi-automatic or a manual override mode. For certain torpedoes, an ancillary setting module is required. The shipboard indicators display course-to-steer and up to six in number may be sited as appropriate. Two local control and display modules are situated above decks, one each in the vicinty of the port and starboard mountings, to provide monitoring, local control and override facilities.

STATUS

In development.

CONTRACTOR

Vickers Shipbuilding and Engineering Ltd, Barrow-in-Furness, Cumbria LA14 1AF, England.

6748.481

SUCCESSOR SUBMARINE COMMAND SYSTEM

Successor has been selected as the new command system for Royal Navy nuclear-powered submarines beginning with the 'Vanguard' class ballistic missile submarines.

The system features Intel 80386 32-bit general purpose processors enhanced with special purpose Inmos transputer cards for tasks requiring fast parallel processing. Most of the programming uses ADA, although the transputers use OCCAM. Communication is via a high capacity fibre-optic network with a self-healing capability.

Successor also features multi-function consoles with high resolution colour graphics displays. The manufacturers have paid special attention to operator interaction with special features provided for different levels of operation to assist concentration on operational tasks.

STATUS

Ordered for the Royal Navy. Export variants are also available.

CONTRACTOR

Gresham-CAP Ltd, Scientific House, 40-44 Coombe Road, New Malden, Surrey KT3 4QF, England.

4881.281

THORN EMI TORPEDO FIRE CONTROL EQUIPMENT

The THORN EMI torpedo fire control equipment is a lightweight system for deck launched torpedoes and is currently compatible with Mk 44 and Mk 46 lightweight torpedoes.

The equipment may be used as an independent system with full manual setting and firing capability which could be appropriate to smaller vessels. Alternatively it can be interfaced with conventional AIO and weapon system computers found in a wide range of warships for automatic or semi-automatic monitoring and fire control with full manual fall-back, or stand-by facilities.

The equipment will set and monitor the torpedo operating parameters, state of readiness and torpedo tube states and operation for up to two triple barrel launchers. Local displays indicate ships head, optimum, launch bearing, sonar contact and allow optimum launch conditions to be established. It includes test facilities for system checking and which allow simulated firings to be practised.

The equipment is supplied in modular form to ease installation and is VCS compatible.

STATUS

A number of these systems have been supplied and have been in service for several years.

CONTRACTOR

THORN EMI Electronics Ltd, Naval Systems Division, PO Box 4, Rugeley, Staffs WS15 1DR, England.

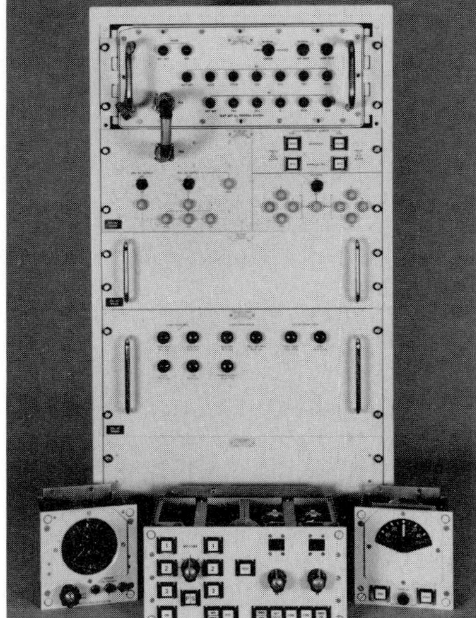

Thorn EMI torpedo FC equipment hardware

6753.261

THORN EMI SHIP'S WEAPON DATA DISTRIBUTION

An essential element in the ship's data flow is the distribution system which accepts data from ship's sensors eg heading, roll and pitch, log, depth or wind data; performs any necessary processing eg format changing; and distributes them to the ship's user equipment.

THORN EMI Electronics has more than 20 years experience of designing and building ship's data distribution and interface system for weapon and navigation requirements. The latest generation of these systems which uses microprocessor and software techniques is known as the THORN EMI D3 system.

The D3 concept enables input/output displays and diagnostic modules to be configured to suit a customer's exact requirements. This is achieved by selecting standard software modules for the processor which, tailored to system requirements, enable interconnection of data between the hardware elements of the system.

As well as all commonly encountered synchro, analogue and digital interfaces, the D3 system can accommodate an embedded data bus remote terminal or bus controller. It can also selectively combine and repackage data from a number of sources for onward transmissions.

STATUS

More than 400 data integration and distribution systems have been supplied for operation in ships with the navies of 18 countries.

Currently D3 systems are being supplied for all the major Royal Navy surface ships and submarines, the 'Kobben' class submarines and for the Canadian TRUMP project.

CONTRACTOR

THORN EMI Electronics Ltd, Naval Systems Division, PO Box 4, Rugeley, Staffs WS15 1DR, England.

Thorn EMI data distribution unit in service

6198.281

LAUNCHING SYSTEMS FOR LIGHTWEIGHT TORPEDOES

Plessey Naval Systems has supplied the Royal Navy with two generations of shipborne torpedo weapon systems (STWS), first with STWS 1, superseded by STWS 2 and latterly with a commercial variant, the PMW 49A.

STWS 1

STWS 1, which was fitted in a large number of ships, was designed to interface effectively with the Royal Navy's fleet sonars Types 177 and 184 and to launch and control the Mk 44 and Mk 46 torpedoes.

STWS 2

STWS 2 is a second generation system developed from STWS 1, suitable for use against any sub-surface target held by the AIO/CIC of the ship. It is a quick reaction anti-submarine weapon system, capable of pre-setting and firing the new Stingray and Mk 46 torpedoes.

STWS 2 launcher fitted to a Type 22 frigate

The system basically consists of triple torpedo tubes fitted to port and starboard of the ship and a microprocessor-based tracker and predictor working in non-real time. The system accepts target track data on its information channel simultaneously from one of three sources:

(1) from the AIO/CIC either automatically or manually, as the primary mode of operation
(2) from the ship's sonars, as the secondary mode
(3) by manual injection, as the fallback mode.

STATUS

STWS 2 is fitted in 'Leander' class frigates, all 'Broadsword' and 'Boxer' class frigates, and 'Birmingham' class destroyers. Many overseas navies also operate Plessey Marine launch systems, from simple manual variants to sophisticated systems which interface with the ship's command and control system to provide fully automatic fire control systems.

PMW 49A DECK LAUNCHER SYSTEM

The export version of the STWS 2, PMW 49A consists of twin triple torpedo tube launchers plus a launch control presetter unit coupled to the launchers via local control panels mounted adjacent to the launchers on each side of the ship. The basic system can preset and discharge most current lightweight torpedo types.

Each triple torpedo tube launcher comprises three glass reinforced plastic (GRP) torpedo tubes arranged with their axes parallel in a triangular configuration, cradled on a training mechanism which in turn is supported by a stationary base bolted to the ship's deck.

Muzzle covers and anti-condensation heaters provide an all-weather storage capacity for the torpedoes. Torpedoes are fired by the release of compressed air stored in the breach assembly air reservoir. Air reservoir charging terminals are installed at each launcher position and connected to the ship's high pressure air supply.

CONTRACTOR

Plessey Naval Systems, Templecombe, Somerset BA8 0DH, England.

1044.261

ADAWS 2 ACTION DATA AUTOMATION WEAPONS SYSTEM

This was the first of a new series of action data automation systems developed for the RN and based upon Ferranti FM1600 series digital computers, and using Plessey display equipment, to provide improved performance and facilities over those of the ADAWS 1 which used Poseidon computers. The important operational difference is that ADAWS 2 and later systems provide, in addition to the action data automation functions, fire control facilities for the various weapons carried by the vessel.

The ADAWS 2 was designed for the Type 82 guided missile destroyer HMS *Bristol*, and accepts inputs from multiple sensors including radars, sonar, passive devices, and data link. The FM1600 micro-circuit computer suite performs the co-ordination, processing, distribution, and display of tactical information derived from these sources. Fire control calculations and co-ordination are also performed by the computers for the armament of Sea Dart SAMs, the Ikara ASW weapon, 4.5-inch (114 mm) gun and other weapons fitted.

Inputs to the system comprise the search radars Types 1022, 992 and 1006, two Type 909 Sea Dart tracker radars, data links, Types 170 and 184 sonars, unspecified EW and passive direction finding equipment, and various manual input facilities.

Among the conventional weapons linked to the ADAWS 2 are included facilities for the direction of various categories of co-operating aircraft such as fighters and ASW helicopters.

STATUS

The ADAWS 2 equipment for HMS *Bristol* was installed in 1970; further ships of this class were cancelled for economy reasons. More details of ADAWS 2 are given in previous editions of *Jane's Weapon Systems*.

CONTRACTORS

Ferranti Computer Systems Ltd, Western Road, Bracknell, Berkshire RG12 1RA, England – main contractors, digital equipment, and software.

Plessey Naval Systems, Templecombe, Somerset BA8 0DH, England – displays, interface, and drive equipment.

1234.261

ADAWS 4 ACTION DATA AUTOMATION WEAPON SYSTEM

This system is essentially a smaller version of the ADAWS 2 (**1044.261**), using the same hardware, but with variations to accommodate the requirements of the 'Birmingham' (Type 42) guided missile destroyer class of vessels. ADAWS 3 was designed for the CVA 01 carrier which was subsequently cancelled.

The ADAWS 4 is produced in two slightly different versions for the RN 'Birmingham' class and the 'Hercules' class supplied to the Argentine Navy. The systems fitted in the Batch 2 and Batch 3 ships have minor differences to reflect the changes to weapon and sensor fits. They are designated ADAWS 7 and 8 respectively.

The ADAWS family of systems also includes ADAWS 5 (**1235.261**) and ADAWS 6 (**1830.281**). Specific benefits of the modular nature of these systems are reductions in the development effort required and in the time necessary for system proving.

The central data processing complex includes two Ferranti FM1600 micro-circuit digital computers, with a Plessey Mk 8 digital display system for the presentation of tactical and other information. The computers provide up to a total of 2 × 192 K words of core store, and they will gather data from all the ship's sensors, assemble and correlate this information and that from other vessels obtained via data link, and present it to the command teams on appropriate display consoles.

All information from 'own ship' and consorts in data link contact will be accessible at any console, and on a selective basis, eg either air situation or underwater situation. The computers also include programs to assist in the evaluation of the relative threat posed by different targets, thus assisting the command in the selection of targets and assigning weapons, particularly the twin-headed Sea Dart missile system. When action is joined, the computers also provide control of ship's weapons. In addition to providing all the aiming and fire control computations for the Type 42's Sea Dart missiles, the ADAWS 4 also includes fire control for the 4.5-inch (114 mm) automatic gun mounting, and the calculation of intercept instructions to be passed by radio to fighter aircraft or anti-submarine helicopters.

STATUS

In service with the Royal Navy and Argentinian Navy.

CONTRACTORS

Ferranti Computer Systems Ltd, Western Road, Bracknell, Berkshire RG12 1RA, England.

Plessey Naval Systems, Templecombe, Somerset BA8 0DH, England.

1235.261

ADAWS 5 ACTION DATA AUTOMATION WEAPON SYSTEM

This system is essentially a further version of the ADAWS 2 system that is fitted to the Type 82 guided missile destroyer, HMS *Bristol*, and uses the same basic hardware and a large proportion of the software of that system, but configured to meet the requirements of vessels of the 'Leander' class of frigates equipped with the Ikara ASW weapon system. A major difference is that only a single computer is employed. The requisite software for all the ADAWS family of systems is generated and maintained by a Ferranti programme team working in collaboration with the ARE, Portsdown. The sensor, computer programming, and display arrangements for the ADAWS 5 are optimised for the principal role of this class of vessel (ASW) and the Ikara anti-submarine weapon which is the main armament.

The system receives inputs from the search radars Types 994 and 1006, and provides auto-tracking of targets by LAX (limited area auto-extraction), including the special processing of transponder returns from helicopters engaged in anti-submarine warfare operation and Vectas. The IFF associated with the Type 994 radar is also controlled by the computer in association with its integral auto-decoder. On the sonar side, the system receives inputs from Type 184 (main search sonar), Type 199 (variable depth sonar) and Type 170 (attack sonar) which also controls the A/S Mortar Mk 10. Target data are processed, correlated and formed into tracks by the computer, which also controls the firing of the Ikara weapon and provides assistance to the helicopter controller during Vectas. The AS 1077 sonar trainer is also fitted to provide on-board training, both for the sonar operators and the AIO team. Other facilities include data banks and Seacat missiles.

The system hardware is similar to that used in other ADAWS but on a reduced scale, namely, one FM1600 computer, one display central equipment, two tactical displays and six LPDs/totes.

DEVELOPMENT

Development has been carried out in conjunction with the other ADAWS programmes by the same companies and government agencies, and originated in 1967.

STATUS

Eight systems have been manufactured by Ferranti for HMS *Leander* and other ships of the first batch of 'Leander' class frigates.

Software and hardware modifications, including increasing the core store to 128 K words, have been carried out in order to keep the system up-to-date with the latest operational requirements.

CONTRACTORS

Ferranti Computer Systems Ltd, Western Road, Bracknell, Berkshire RG12 1RA, England.

Plessey Displays Ltd, Station Road, Addlestone, Weybridge, Surrey KT15 2PW, England.

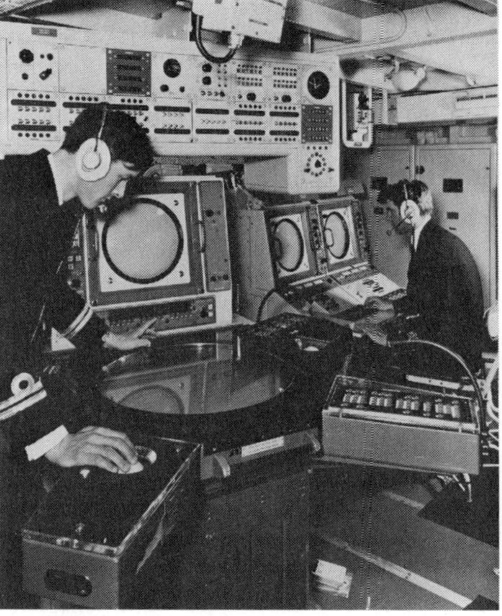

Part of the digital display system supplied by Plessey for ADAWS 5 in HMS Leander

---

1830.281

ADAWS 6 AND 10 ACTION DATA AUTOMATION WEAPON SYSTEMS

ADAWS 6 is the designation of the system for the RN anti-submarine cruisers HMS *Invincible* and HMS *Illustrious*. A slightly different version, designated ADAWS 10 has been fitted in HMS *Ark Royal*. Ferranti and Plessey, who in partnership have produced ADAWS for preceding RN ships, were contractors for the computer complex and display systems for this class of ships.

HMS *Invincible*, the largest ship to be built for the RN since the Second World War, is claimed to have a more comprehensive automated AIO than any previous vessel, and its functions include deployment and control of the ship's own weapons and aircraft as well as command and control of co-operating forces.

Invincible's armament includes the Sea Dart missile (twin launcher), Sea King helicopters, and Harrier V/STOL strike aircraft. Sensors include a Type 1022 long-range surveillance radar, IFF/SIF, Type 992Q search and target indication radar, two Type 909 radars for the Sea Dart system, Type 1006 navigation radar, data links, EW and sonar.

There is a significant area of commonality in both armament and sensors with other RN ships (Type 42 'Birmingham' class and HMS *Bristol*) that have Ferranti/Plessey ADAWS, and as would be expected the new ADAWS for HMS *Invincible* embodies parts of the systems developed for these classes. In ADAWS 6 the two computers are employed in a very similar fashion to those in the ADAWS 2, including the ability to operate in a range of reduced modes in the event of malfunctions. Additional software to meet the special requirements of the *Invincible*'s new operational role were generated by a Ferranti team in collaboration with the ARE, Portsdown.

STATUS

The first ADAWS 6 system was delivered to *Invincible* in 1977 and the system for *Illustrious*, the second ship of the class, in 1980. ADAWS 10 for *Ark Royal* was delivered in 1983.

CONTRACTORS

Ferranti Computer Systems Ltd, Western Road, Bracknell, Berkshire RG12 1RA, England.

Plessey Displays Ltd, Station Road, Addlestone, Weybridge, Surrey KT15 2PW, England.

4578.281

NAUTIS COMMAND AND CONTROL SYSTEMS

NAUTIS is a new-technology family of integrated naval command and control systems from Plessey Naval Systems (formerly Plessey Displays).

NAUTIS systems are modular. They range from a single console, capable of meeting the action information and navigation system requirements of small patrol boats and mine countermeasure vessels, to multi-console systems for more complex requirements of larger warships.

All NAUTIS systems are based on the Plessey naval autonomous intelligent console, NAUTIC. NAUTIC employs the latest electronic design techniques to give a multi-purpose workstation module. A large, high resolution display, together with operator controls, system interfaces and powerful micro-processors with large capacity memory, are all contained within a compact console enclosure. Each NAUTIC console has the computing power to handle the database of a major warship.

The function of individual NAUTIC consoles is determined mainly by the application software and in-built selection of input/output interfaces. Typical functions include:

(1) Air and surface surveillance with auto-tracking from one or more radar inputs.
(2) ESM/ECM management.
(3) ASW tactical display integrating sonar, radar and torpedo systems.
(4) Aircraft control.
(5) Mine countermeasures.
(6) Command display.
(7) SAM and SSM weapon control.
(8) Information transfer by data link.

The console computer uses the internationally available Intel range of micro-processors in multi-bus configuration. An extensive random access memory, backed-up by non-volatile magnetic bubble memory, has the capacity for a command system database that exceeds 1000 × 1000 n miles. This high-resolution database is fully available to the console operator and is preserved by the bubble memory in the absence of power supplies.

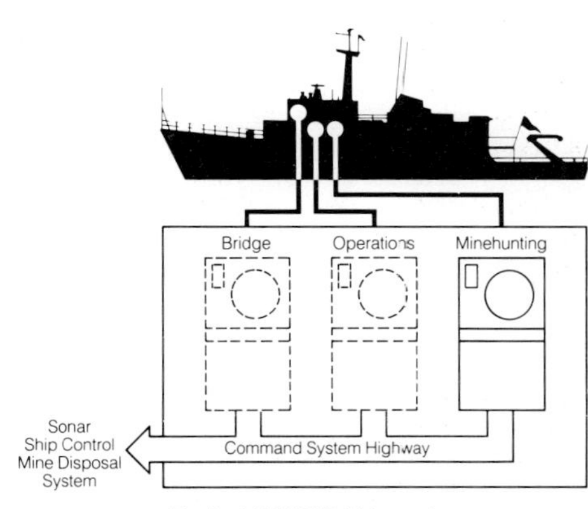

Typical NAUTIS-M layout

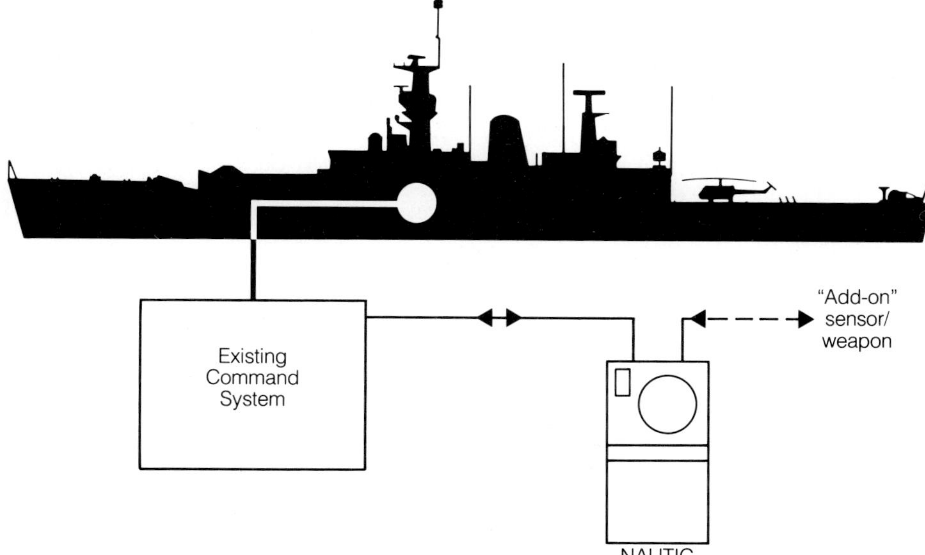

NAUTIS for system updates

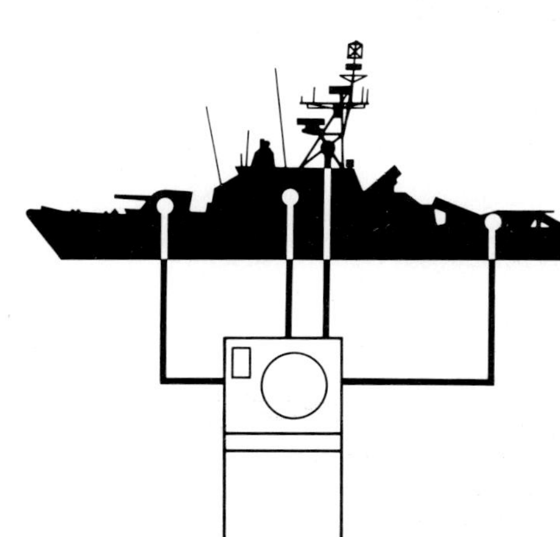

NAUTIS-P for patrol boats and light forces

In larger NAUTIS systems, comprising multiple NAUTIC consoles, intercommunication is effected by means of a dual-redundant command system highway. In such a system a replica of the whole command system database is maintained in each NAUTIC workstation. This feature of database replication, together with the full distribution of data processing in each console, gives NAUTIS systems outstanding advantages in system redundancy, flexibility in task allocation between command system workstations, fewer workstations, reduced manning and high system availability.

NAUTIS systems are designed for use by all navies and maritime policing authorities. The latest man/machine interface techniques are employed, giving an easily understood display with simplified control procedures that can be readily used without extensive operator training.

The needs of the maintainer are also fully considered. Comprehensive built-in test facilities indicate faults to line replaceable modules. This in-depth fault diagnosis, together with minimum routine servicing, enables NAUTIS systems to be maintained on board with minimum specialised skills and training.

The versatility of the NAUTIC console and its various applications in NAUTIS systems, enables navies to achieve unprecedented standardisation of command system equipments. For example:

NAUTIS-A: Auxiliaries
NAUTIS-F: Frigates and large corvettes
NAUTIS-L: Amphibious warfare ships
NAUTIS-M: Minehunters and mine sweepers
NAUTIS-P: Patrol vessels, strike craft and corvettes.

A low-cost update to the command system of an existing ship is often necessary to utilise fully a new sensor or weapon, or to enable a ship to take on a different role. This can now be achieved by adding to, or replacing, an existing action information system with a NAUTIC console. Such a console can be integrated into the ship with minimum changes to the ship's equipment layout. Further NAUTIC consoles can be added, and integrated via the command system highway, for a stage-by-stage modernisation programme.

STATUS

NAUTIS-M is in production for the Royal Navy's new minehunters. NAUTIS-L was ordered by the UK MoD in June 1986 for the amphibious warfare ship HMS *Fearless*. Under consideration for various warships in UK and overseas including patrol vessels, corvettes, MCM vessels, ASW frigates.

CONTRACTOR

Plessey Naval Systems, Templecombe, Somerset BA8 0DH, England.

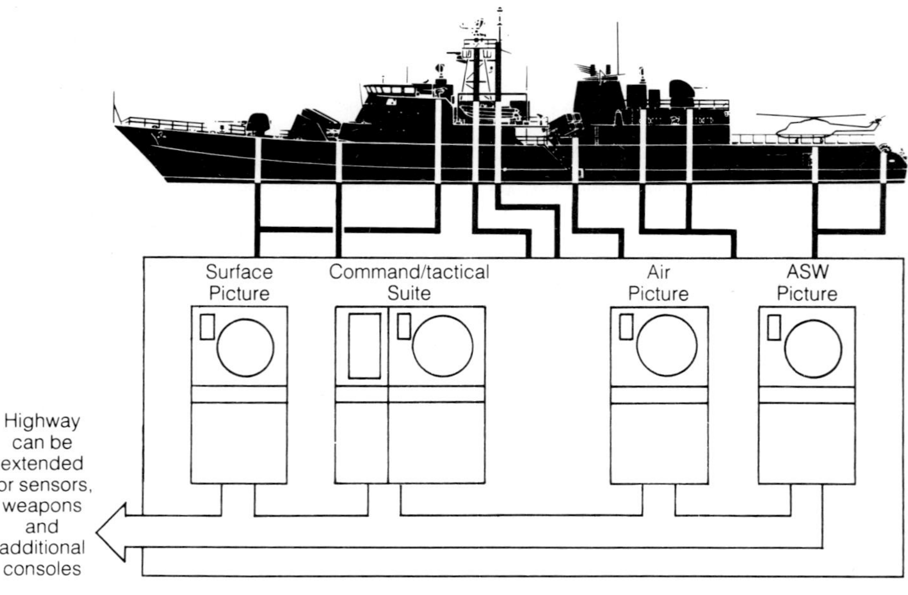

NAUTIS-F for frigates and large corvettes

Plessey's naval autonomous intelligent console (NAUTIC)

6199.281

CANE/MAINS

The CANE series of command systems are designed specifically for use in light frigates, corvettes, fast attack craft, patrol craft and auxiliaries. It is in widespread use with the Royal Navy and overseas navies and defence forces.

Its modular design enables the system to be configured to meet the specific command and control requirements of individual vessels and it can interface with a wide range of sensors and weapon systems to provide a completely integrated command AIO/CIC system.

Racal Marine Systems CANE action information system

CANE systems are available in single or multiple intelligent tactical command console configurations linked by dual redundant digital data highways. Each console is equipped with a high performance monochrome or colour command/operations display.

The system interfaces with a wide range of navigation systems, logs, gyros and air and surface surveillance radars to provide precise positional information and tracking of both surface and airborne targets. Target information includes: target position, course and speed; closest point of approach and relative velocity solutions.

When interfaced to a wide range of tactical sensors, including sonar, ESM, IFF, optronics and intership data links, CANE provides comprehensive tactical picture compilation and threat evaluation. Outputs include target indication information to weapon fire control systems encompassing guns, missiles and ASW.

A version of CANE is available to meet the specialised command system requirements of mine countermeasures vessels in terms of both minesweeping and minehunting roles.

Racal MAINS – Minehunting Action Information and Navigation System

Racal MAINS is an advanced minehunting integrated action information and navigation system with low magnetic signature for minehunting and mine disposal vessels. A derivative of CANE, it is compatible with hull-mounted and variable depth sonars. Designed for the control of sonar, mine disposal and neutralisation operations, an automatic chart table receives inputs from various radio/microwave positioning systems, including Hifix 6/Hyper-Fix, Trisponder/Microfix, and provides auto tracking of targets, steering guidance in MCM operations, with magnetic tape and electronic plot recording facilities.

STATUS

The CANE series has been ordered by the United Kingdom (Aviation Training Ship RFA *Argus*, RFA *Sir Galahad* and *Sir Tristram*), Sweden ('Landsort' class MCMV) Finland (two systems operated by Frontier Defence Force), Germany (three Navy Surveillance Vessels), Hong Kong (two systems operated by Marine Police), South Korea (one system in 'SK 5000' class MCMV).

CONTRACTOR

Racal Marine Radar Ltd, Naval and Harbour Systems Division, Burlington House, 118 Burlington Road, New Malden, Surrey KT3 4NW, England.

6744.281

880 SERIES MINE COUNTERMEASURES SYSTEMS

The 880 series is a range of systems which provide all the command, control and navigation facilities required in mine countermeasures operations. They are modular in design to enable a system to be configured to meet particular operational requirements, or financial considerations.

The range includes a low cost MCM system which integrates the inputs from a number of precise positioning systems such as Racal Hyper-Fix with the ship's log and gyro to provide accurate navigation and track guidance. This can either be automatic through an auto pilot system, or manual via a helmsman's guidance display providing speed over the ground, course made good, along and cross track error etc. A constant check on the quality of the position fix is carried out. Other facilities include task planning which can be prepared in advance to include routes and known sonar contacts etc, and recorded on disk for entry into the system. A hard copy navigation plot provides an historical record of planned and actual track, sonar contacts etc, and a VDU display provides an easy operator interface for control of the system.

This system also forms the basis of an advanced MCM system which provides the additional command and control facilities required in minehunting.

The additional facilities are centred around a colour graphics tactical display and data base which provides an accurate presentation of current operational data including planning data, known classified sonar contacts, new unclassified sonar contacts and

ROV position and track. Other features include the interfacing to and integration of sonar allowing the range and bearing of sonar contacts to be recorded in the system's data base. It also provides the facilities for the control of the sonar during classification. A radar interface allows the range and bearing of radar contacts to be accepted and displayed on the tactical display. The system can accept the input of data from an underwater acoustic positioning system to facilitate the safe navigation of ROVs used for classification and disposal. A data base is maintained on disk and can be passed between vessels assigned to the task area, with specific task plans and operational records added.

STATUS

In operational service.

CONTRACTOR

Racal Marine Systems Ltd, Burlington House, 118 Burlington Road, New Malden, Surrey KT3 4NW, England.

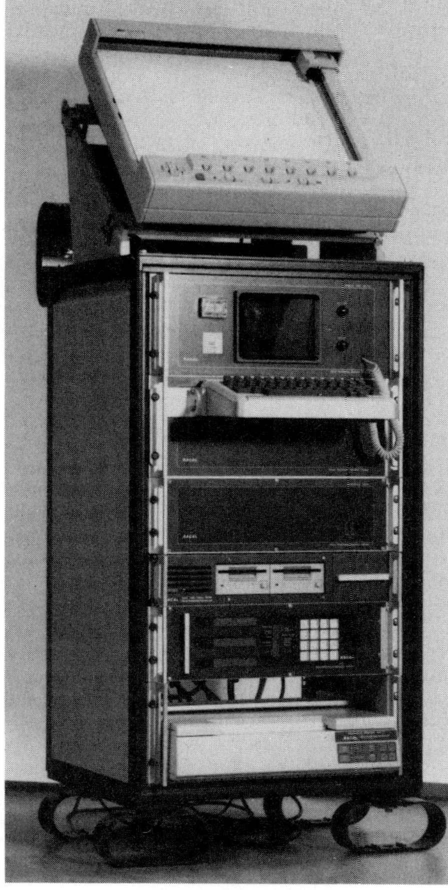

880 series MCM system

3506.281

FERRANTI MINE COUNTERMEASURES CONTROL SYSTEM (MCCS)

In anticipation of future requirements both within the UK and overseas, Ferranti Computer Systems Ltd (FCSL) is developing a new generation of system to fulfil the MCMV task for the 1990s and beyond. The system uses a single high reliability multi-function console, the design of which has paid particular attention to current ergonomic requirements and which employs a full colour raster graphic display screen and a combination keyboard and joystick user friendly man/machine interface.

The architecture employs state-of-the-art technology of proven reliability configured to produce the optimum economic solution to the mine countermeasures problem. The operational software currently written in the UK MoD specified Coral 66 high level language and proven both through exhaustive trials and on active service, is being mapped into the NATO military specified ADA language to meet the common requirements of the major NATO forces.

The Mine Countermeasures Control System (MCCS) may be employed as an autonomous outfit connected directly to all necessary navigation sensors, with interfaces to a minehunting sonar if required as well as sweep, monitoring or minelaying equipment. Alternatively the console may be interfaced to a ship's data bus taking all necessary information from and transmitting instructions via the bus.

The MCCS provides computer assistance and display in the mine countermeasures task which provides for the automation of tasks which have previously been carried out manually. Hence manning levels are reduced. By this means the Mine Warfare Officer is able to control the operation effectively while the MCCS receives, processes and displays navigational and operational data, provides various interactive operator commands and controls outputs to the mine countermeasures and ship control equipments.

Although designed primarily as a Mine Countermeasures System all the functions of MCCS may be applied to a minelaying scenario, as the requirement for accurate navigation, data handling and ship control are similar.

The integration of navigational sensors with command functions and mine countermeasures equipment not only provides a high degree of automation, but also enables a further integration with ship control functions to be readily achieved. The flexibility for subsequent system enhancement is inherent in the system design, for example, the addition of interfaces and command functions for ESM, surveillance radar, gun, missile and data link subsystems are all within the MCCS capabilities.

STATUS

At an advanced stage of development.

CONTRACTOR

Ferranti Computer Systems Ltd, Western Road, Bracknell, Berkshire RG12 1RA, England.

4273.281

F100-L FIRE CONTROL PREDICTOR

For naval optical fire directors (OFD) and on-mount directed gun systems, Ferranti Computer Systems are offering a gun predictor based on the F100-L microprocessor. The predictor calculates the required aim-off angles that are applied to correct the gun aimer or the operator's sight line, enabling the gun to be accurately repointed.

The principal sensors are binocular or IR sights and laser rangefinder, and the system can accept additional inputs for target indication. Alternatively, manual range input is available for surface mode, and fixed fly-through range for an air engagement.

In action the operator tracks the target while the computer monitors tracking rates and calculates the necessary aim-off angles.

STATUS

An order for the predictor from Marinha do Brasil has been received for the supply and fitting of the Defence Equipment & Systems Ltd Mk 5 OFD, in new construction Brazilian corvettes for emergency fire control of Vickers Mk 8 4.5-inch and Bofors 40 L70 40 mm guns. These predictors underwent joint integration trials prior to delivery in July 1985.

Land and sea trials of the DE&S Ltd LS30R (Rarden) mounting, incorporating a prototype predictor, were carried out by the Royal Navy in 1981, following which an initial order for 27 LS30(R) mountings using an Oerlikon KCB cannon was placed with DE&S Ltd.

The F100-L Predictor is also offered as part of Ferranti's Above Decks Defence System (ADDS) which is a low cost, flexible, small ship system which can be installed without deck penetration. ADDS may use a variety of close range guns and Ferranti equipment as required, such as laser rangefinders, night goggles and helmet sights.

CONTRACTORS

Defence Equipment & Systems Ltd, PO Box 25, Norwich NR1 1JD, England.

Ferranti Computer Systems Ltd, Western Road, Bracknell, Berkshire RG12 1RA, England.

Ferranti F100-L microprocessor-based predictor for naval fire control

1828.261

MARCONI/BAe LIGHTWEIGHT GUN FIRE CONTROL SYSTEM (SAPPHIRE)

The Dynamics Division of British Aerospace and Marconi Radar Systems have collaborated to create a cost effective lightweight radar gun fire control system. Designed for any size of warship from fast patrol boats upwards, the system is capable of maintaining rapid and accurate control over small and medium calibre guns against air, surface, and shore targets. The system is fully automatic, thus keeping manning requirements to a minimum: only one man, in an essentially supervisory role, is required to operate the system.

Egyptian Navy 'Ramadan' class fast missile craft with two Sapphire fire control systems

The main components of the system are:

(1) the Marconi 1800 series autonomous tracking radar ST1802 (**1508.253**).

(2) The BAe Sea Archer predictor.

(3) The control console, incorporating the display and control units for one man to supervise the radar, predictor, and gun.

The total package is designed to be compact and lightweight, thus being suitable for installation in FPBs, corvettes, frigates etc. The predictor uses a dedicated computer leading to operational simplicity and autonomous operation.

The radar is fully coherent and incorporates random frequency agility, monopulse tracking, and MTI processing, essential for accurate tracking of small low targets. The system will control servo-operated guns with calibres from 20 mm upwards.

The system provides fire control for all three modes of naval gunnery:

(1) Anti-aircraft (AA) for defence against high- and low-level air targets.

(2) Surface (SU) for use against surface vessels moving at speeds up to 60 knots.

(3) Naval gunfire support (NGS).

In SU and NGS modes, fuze computation is provided for use with star-shell and window. The system can also be used to provide surveillance by means of horizon search or sector scan.

CHARACTERISTICS

Radar

Radar beamwidth: 2.4°

Frequency: Band 8.6 – 9.5 GHz

Peak power: 50 kW (typical)

Mean power: 150 W (typical)

Pulse length: 0.67, 0.3 and 1.0 μs

PRF: Non-MTI 3000 Hz; MTI 4400 Hz

Receiver noise figure: 9.0 dB

MTI: Improvement factor: typically 37 dB

Predictor

Type: BAe Sea Archer

Computer: Dedicated

Gun/AIO interfaces: Synchro or Digital

Program: Semi Conductor (EProm)

STATUS

Evaluated by Royal Navy 1975, introduced in 1976 and improved in 1985. More than 18 systems have been sold and the system is still in production.

CONTRACTORS

British Aerospace, Naval and Electronics Systems Division, Downshire Way, Bracknell, Berkshire RG12 1QL, England.

Marconi Radar Systems, Writtle Road, Chelmsford, Essex CM1 3BN, England.

3246.281

SEA ARCHER 1 FCS

The British Aerospace Sea Archer 1 family of naval fire control systems now embraces several versions, ranging from the purely optical Sea Archer 1 to models with multiple sensor capability and the correspondingly greater operational facilities. The various versions which have been identified so far are described in the following paragraphs.

Sea Archer 1

The Sea Archer 1 optical fire control system has been designed using a dedicated digital computer to provide comprehensive facilities while maintaining ease of operation. The predictor itself is similar to that used in the Marconi/British Aerospace Sapphire system (**1828.261**), a full blind fire radar system. Sea Archer controls servo-operated guns with calibres from 30 mm upwards against aircraft, missiles and surface vessels. While designed as the prime gun controller for FPBs, the system is equally suitable as a secondary weapon controller for larger ships.

The system is comprised of two basic modules, an optical fire director (OFD) with a combination of EO tracking sensors and a laser rangefinder: a gun control console containing a predictor operator control panel and system communications. In the simplest configuration, an aimer uses binoculars fitted to the stabilised OFD to acquire and track the target. To assist in target acquisition the system can accept information derived from the vessel's surveillance radar to slew the OFD on to the target bearing. Optronic sensors, IR or television tracking cameras are fitted in other configurations to give below-decks operation. A TV auto-tracking unit is available if required.

Information regarding the target range, training and elevation is automatically transmitted to the weapon control console. The computer, utilising ballistic and meteorological data already entered into the system, then calculates gun aiming order. The gun is controlled from the control console. To reduce reaction time to a minimum, gun slew occurs as soon as the target is acquired and offset adjustments are then made when a valid prediction solution is

Sea Archer 1A gunfire control system

computed. Typically this process takes two seconds. The gun is fired by means of the console foot push when a valid prediction is achieved and the future range display indicates that the target is within effective gun range.

The computation techniques and electronics interfaces have been successfully proved with 30 mm, 40 mm, 76 mm, and Vickers 4.5 inch (114 mm) Mk 8 guns.

The system is capable of three modes of gun fire control as follows:

(1) AA

In this mode the system is designed to achieve minimum reaction time against fast moving air targets. The director aimer acquires the target and activates the rangefinder. Target data are fed to the control console where the digital weapon controller calculates gun aiming signals. The gun slews into line with the computed data and when the target is within gun range the digital weapon controller indicates that the system is ready to engage the target. Firing can be achieved from the director or the control console.

(2) Surface

In the surface mode the optical director is aimed at the target and the rangefinder is activated to establish the target range. Gun aiming orders are then computed from the target data transmitted by the director. Splash spotting corrections, observed by the director aimer, are applied via the control console.

(3) Naval Gunfire Support

In this mode the gun is used to bombard shore targets. The target co-ordinates are inserted manually at the control console and gun orders calculated and corrected for ship movement by dead reckoning. Spotting corrections, observed by a shore or ship based observer, are applied at the control console.

Sea Archer 1A

This version is an enhancement of Sea Archer 1 which provides complete integration of a TV sub-system with below-decks control by joystick giving an improved surveillance capability and the option of manned or unmanned director on deck. It also has provision for controlling two different types of gun-mounting (alternately). The ballistics for both gun types can be carried in the 1412A computer. Infra-red sensors may also be integrated in the Sea Archer 1A system.

STATUS

There are 21 Sea Archer 1/1A systems in service with four navies in six classes of ship; Brunei ('Waspada' class FAC), Oman ('Province' class, *Al Mansur* FAC, 'Al Wafi' class FPB, *Al Munassir* logistic support vessel), Thailand, United Kingdom ('Peacock' class OPV). Five systems are on order including two 1A Mod 1 for Thai amphibious support ships.

CONTRACTOR

British Aerospace, Naval and Electronic Systems and Equipment Division, Downshire Way, Bracknell, Berkshire RG12 1QL, England.

6450.281

SEA ARCHER 30 FIRE CONTROL SYSTEM

The Sea Archer 30 family of fire control systems uses a very light remotely controlled electro-optic director incorporating the most advanced passive sensors including the British Aerospace UK class II thermal imaging camera, a TV camera and laser rangefinder. The director not only provides extremely accurate line-of-sight target data for gunnery and missile engagements, but also provides the ship with a long range, 24 hour, 360° passive surveillance and observation facility.

The below decks equipment which consists of the gun control console, the predictor and the director electronics unit and is based on the latest single board digital processors to ensure the equipment can be easily integrated with other sensors and weapon systems. From the single operator's control, the system offers full autonomous control of either one or two guns against air or surface targets. Sea Archer 30 will accept TI information from radar trackers or from the ship's AIO and offers the facility of both auto-acquisition and auto-tracking as required.

Should a radar controlled variant of Sea Archer 30 be required a K-band radar version of the director is available.

A variant of the Sea Archer 30 system is designated GSA8 by the MoD(N).

STATUS

15 GSA8 systems have been ordered by the Royal Navy for the four 'Cornwall' class (Type 22 Batch 3 frigates) and the new 'Norfolk' class (Type 23 frigates) to control the Vickers 4.5-inch Mk 8 gun.

CONTRACTOR

British Aerospace, Naval and Electronic Systems Division, Downshire Way, Bracknell, Berks RG12 1QL, England.

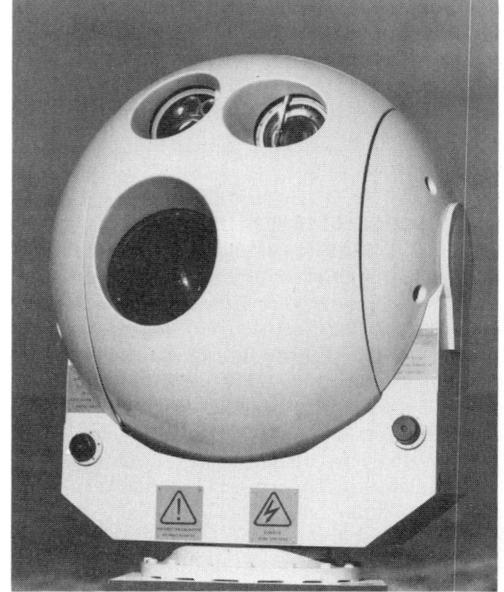

Sea Archer 30 electro-optical director

UNITED STATES OF AMERICA

1928.281

MK 23 TARGET ACQUISITION SYSTEM (TAS)

The Improved Point Defense Missile System (IPDMS) is a surface-to-air system that will provide self-defence for a variety of ships against the anti-ship missile threat through the 1990s. It can react automatically in seconds to threats in any type of weather. It is a highly reliable system that can be maintained through the use of on-line monitoring, built-in tests and equipment, daily system operability tests, and computer-aided fault isolation. It also has been designed to withstand the adverse effects of physical environment and electro-magnetic interference.

IPDMS is formed by the integration of a target acquisition system (TAS) and a fire control and launcher system, the NATO Sea Sparrow Missile System (NSSMS). These two systems are under two separate projects, with TAS under contract from the USN with Hughes Aircraft Company in Fullerton, California. Installation of TAS in DD-963 'Spruance' class destroyers and in high-value auxiliary ships began in early 1980.

Although developed as a sub-system to IPDMS, TAS is designed to designate targets to other weapon systems, such as guns. It is an automatic system composed of a radar, an identification friend or foe (IFF) system, an AN/UYA-4 console, and an AN/UYK-20 digital computer. The system is designed so that the functions of threat detection, track, IFF, evaluation, and designation can be carried out automatically with the operator only monitoring system operation.

As originally conceived, TAS comprised a family of five versions, of varying complexity, and these were designated from Mk 20 to Mk 24 inclusive. The differences between the various TAS configurations were described under this entry number (**1928.281**) in earlier editions of *Jane's Weapon Systems*. The model currently being produced is that known as Mk 23, this being the Radar Automatic version, and as such represents the most sophisticated model except that the infra-red search system has been deleted.

The lightweight TAS Mk 23 antenna sits on a roll-stabilised pedestal and rotates at a rapid two-second scan rate. It searches in two regions: in excess of 20 nautical miles for point defence target designation; and more than 90 nautical miles for surveillance and air control. A third, mixed, mode is a combination of normal and long-range modes.

TAS provides optimum radar performance against close-in and distant targets in detecting moving targets in clutter, and discriminating targets in jamming environments.

NSSMS automatically accepts target designation from TAS and assigns a tracking illuminator radar to the target. The illuminator slews to the location designated by TAS, acquires the target, and locks on. The computer assesses engageability, assigns the launcher, and transmits the firing order to one or more of the missiles. When more than one target is designated, a second tracking illuminator is assigned to the second target and the procedure repeated. Additional targets may be sequentially engaged as each tracking illuminator becomes available.

The firing officer's console contains the controls and indicators for assignment of the launcher system, missile frequency assignment, missile run-up, firing orders, and status monitoring. The controls and indicators provide him with the means to override the automatic operation throughout the engagement sequence.

The main IPDMS configuration, as in the 'Spruance' class destroyers, consists of a TAS and an NSSMS fire control radar set. However, IPDMS may be installed in larger or high-value ships with several NSSMSs interfaced with one TAS.

STATUS

In operational service with four aircraft carriers.

A 3-D version, M-TAS (multi-mission) is being developed by Hughes for the US Navy's new 'fire and forget' missiles.

CONTRACTOR

Hughes Aircraft Company, Fullerton, California 92634, USA.

The USS Ranger *is fitted with Mk 23 TAS, the first carrier to receive this equipment*

3748.281

R-76 FIRE CONTROL SYSTEM

The R-76 fire control system is a naval director system which operates on both surveillance and tracking modes. It is based on a multi-mode radar which can be mounted high on a ship's mast, thereby assisting overall performance, particularly early target detection. Gyro-stabilisation is provided, and in the HR-76 version dual monopulse tracking permits simultaneous missile and target tracking, displayed on the dual A-scopes. There is also provision for instantaneous action data entry for weapon system control.

In addition to the radar, there is an alternative director configuration having a TV contrast tracker unit mounted on the radar sensor platform for low-level target tracking and operation in the face of countermeasures.

Operational functions performed by the R-76 include search for air and surface targets, automatic detection, precision tracking, and direction of weapons which may include guns and missiles. The radar is capable of operating in several modes, among them 'track on jam'; simultaneous target tracking and intercept of air targets; surface target tracking and illumination; detection of enemy missile launch; continuous surface search. In addition to control of fire against surface and air targets, the system provides direct and indirect naval gun fire support and illumination fire. All data processing is accomplished by distributed dedicated micro-computers.

The below-deck operator console has two A-scope displays for simultaneous tracking of two targets, with independent control handles for each and instantaneous action entry for weapon system control. Using this system it is possible to track one target on one display and control a missile on the other. There is also a tactical console with a normal PPI display and operator controls for management of the overall system.

STATUS

The HR-76 made its first appearance as a major element of the H-930 combat control system marketed by Honeywell (**3749.281** in 1982-83 and earlier editions). In September 1979, the manufacturer of HR-76, RCA, announced that this system is to be available on its own with the designation R-76, for fitting aboard all classes of vessel from patrol boat size craft upward, and also for shore coastal defence applications. R-76C5 systems have been installed in two RNZAF 'leander' class frigates.

CONTRACTOR

RCA, Aerospace and Defence, Missile and Surface Radar Division, Moorestown, New Jersey 08057, USA.

2633.281

GUN FIRE CONTROL SYSTEM Mk 56

Gun fire control system (GFCS) Mk 56 is a ship-board, intermediate range anti-aircraft gunfire control system designed for use against subsonic aircraft and surface targets.

Its primary functions are first to track aircraft targets and secondly to compute train, elevation, parallax, and fuze orders for shipboard guns. The system can also be used to control gunfire against surface targets.

To accomplish its purpose, the system uses a radar sub-system for automatic tracking of either visible or obscured targets, an optical sub-system for tracking visible targets, an electromechanical computer sub-system for computing gun orders, and servo-mechanisms for operating the director power drives.

GFCS Mk 56 may be used against different types of target under different operational modes. The computing sub-system is designed for optimum performance against the types of target most dangerous to own ship, which are incoming high-speed torpedo planes, dive bombers, and medium altitude bombers. Slightly less effectively, the system can also be used against all types of surface vessels.

STATUS

This classic GFCS is still maintained as operational equipment aboard many USN vessels. Other navies known to have Mk 56 are those of Argentina, Brazil, Chile, Colombia, Germany (Federal Republic), Greece, Japan, Peru, Spain, Taiwan, and Turkey. More detailed information is given in previous editions of *Jane's Weapon Systems*.

CONTRACTORS

Complete system: General Electric Company, Ordnance Systems Division, 100 Plastics Avenue, Pittsfield, Massachusetts 01201, USA.

Computers: Librascope Inc; Reeves Instrument Co; Ford Instrument Co; Bosch-Avma Co.

Radar: General Electric Co, Syracuse, New York.

Wind Transmitter and Parallax Corrector: Reeves Instrument Co; Belock Co.

2632.281

GUN AND GUIDED MISSILE DIRECTOR Mk 73

This equipment is a major component of the gun and guided missile fire control system Mk 74, one application of which is to the Tartar surface-to-air guided weapon system, and in later models, the Standard naval air defence missile.

Gun and guided missile director Mk 73 is an unmanned, direct-drive motor-driven two-axis unit. It serves as a mount for the antenna of another major component of the system – radar set AN/SPG-51. Three gyroscopes with on-mount electronics provide rate stabilisation. It also provides a platform for the low light-level TV camera Mk 15. The radar line of sight can be rotated continuously in train and can be elevated from –30 to +83° relative to the deck plane. The Mk 73 director mount motors are controlled by the Mk 26 controller developed by General Electric as a more modern, solid-state power drive system, of lighter weight and more compact dimensions than earlier amplidyne systems. The Mk 26 controller contains the controls, servo electronics, and power conversion amplifiers to power the direct-drive motors in the Mk 73 director's elevation and train axes.

STATUS

Three USN ships of the DDG 15 ('Berkeley') class were the subject of a modernisation programme which entailed refurbishment and modernisation of the Mk 73 gun and guided missile directors. General Electric was awarded a major contract for this work in late 1979 and the first hardware for shore tests was delivered in September 1979, followed by a second refurbished Mk 73 in December of the same year. The modernised equipment is designated Mk 73 Mod 6. General Electric is also under contract to produce two Mk 73 directors and associated hardware for a Japanese guided missile destroyer and to supply Mk 73 kits and parts for the Federal Republic of Germany. Under 1984 funding, three Australian ships are being modernised with two Mod 6 directors each.

Since the late 1950s General Electric has supplied more than 150 Mk 73 directors.

CONTRACTOR

General Electric Company, Ordnance Systems Division, 100 Plastics Avenue, Pittsfield, Massachusetts 01201, USA.

New Mk 73 Mod 6 director prior to delivery for USN DDG-15 modernisation programme

---

4014.281

Mk 82 (AEGIS) GUN & MISSILE DIRECTOR

The director Mk 82 and the associated director control Mk 200 are components of the Director Group Mk 81 used aboard new Aegis-fitted ships of the USN. The Mk 82 evolved from the Mk 73 family of directors (**2632.281**) employed for Tartar (**6006.231**) and Standard (**1122.231**) naval surface-to-air missile guidance and control. It incorporates the best features of the earlier system but with enhanced search and slew capabilities and a considerable reduction in weight. Other improvements are a new DC torquer gyro and the elimination of hydraulic brakes. These and other improvements are stated to bring a reduction of about 50 per cent in peak power requirement, a calculated MTBF of 2000 hours, and a reduction in weight from approximately 3450 kg to 1217 kg. Improved search and slew capabilities were required to match the SM-2 version of Standard employed in the Aegis weapon system, the quoted characteristics being a horizon search rate of 72°/s, with a rate accuracy of 0.5 mrad/s and an acceleration of 2.5 rad/s².

The Mk 82 director carries the AS-3156/SPG antenna assembly for the Raytheon Mk 99 Mod 1 FCS missile guidance radar (**1247.253**) on an elevation-over-train pedestal and provides stabilisation for the radar line-of-sight. The director provides search pattern and tracking by responding to rate commands from the fire control system. There are no limitations in train or elevation and the system provides train and elevation data and radar line-of-sight rates in traverse, elevation, and cross-traverse for the fire control system computer.

Compared with the Mk 73 director, the Mk 82 has a much lower profile, in addition to being much lighter.

DEVELOPMENT

The Mk 82 director was designed and developed by General Electric Company, Ordnance Systems Division for the USN Aegis fleet air defence system (**2507.231**). It is developed from the Mk 73 director, also produced by General Electric, and some of the Mk 82 modifications and improvements have been incorporated in the Mod 2 and Mod 5 versions of the earlier system.

STATUS

In production for the USN Aegis air defence system, under contract to Raytheon Equipment Division, who produce the Mk 99 Mod 1 FCS guidance radar for Aegis under contract to RCA. General Electric is building Aegis directors for the USS *Ticonderoga* (CG 47) class of ships. Ten ships (40 directors) are included under present funding. Twelve more directors are included in the 1984 budget (for three ships, CG 57/58/59, with four directors each). At least 24 ships of this class are planned for construction in the next five years.

CONTRACTOR

General Electric Company, Ordnance Systems Division, 100 Plastics Avenue, Pittsfield, Massachusetts 01201, USA.

Aegis Mk 82 directors in final assembly and testing. Mk 200 control is partially visible (right centre) in system test area

---

1241.281

GUN FIRE CONTROL SYSTEM Mk 86

The Mk 86 GFCS was designed as the major fire control system for the gun batteries being installed in the five most recent classes of USN ships. The Mk 86 also provides the third Standard missile tracking/illumination channel in the CGN 38, DDG 993 and DDG 19 (upgrade) classes. The main elements of the system are a high data-rate I-band surface search radar (AN/SPQ-9), an I-band air target tracking and horizon search radar (AN/SPG-60), an optical sensor system, a computing system based on the Mk 152, AN/UYK-7 or AN/UYK-43, and the operator display consoles. Surface, shore, and air targets can be engaged, and the computer is stated to be capable of tracking automatically more than 100 targets using a track-while-scan routine.

Expanded operational capability can be provided by the integration of the USN's new Deadeye 5-inch (127 mm) guided projectile (**3727.281**).

OPERATION

The Mk 86 system handles multiple radar sensor inputs, with each radar operating independently and linked via the computer. The SPQ-9 provides surface search as well as low-level air coverage for defence against low-flying threats and for helicopter tracking. The SPG-60 is an automatic scan-to-acquire and tracking radar that can also operate in a horizon search mode. The optical sensor system can range from a simple CCTV camera chain to a comprehensive electro-optical sensor system. The complete Mk 86 can be interfaced with other ship's systems (NTDS etc) and can receive data from other onboard sensors as well as the SPG-60 and SPQ-9 radars.

The entire Mk 86 operation is controlled and monitored by a control officer and weapon controllers. The control officer's console displays all search radar returns on a PPI presentation. This is used for threat evaluation and selection of targets to be automatically tracked. When a target track has been established, the control officer assigns a specific target and weapon, or combination of weapons, to a weapon control console.

A B-scan and TV display are provided on the weapon control console, and laser and IR displays may be added if the appropriate sensors are fitted. The B-scan gives the weapon controller an expanded radar presentation of his assigned target area, and by examining the display gate around the target on the B-scan he can assess the quality of the track, and for surface engagements, use it to spot shell-splashes. During an air engagement, the B-scan monitor displays an SPG-60 train and elevation error,

SPQ-9 surface search radar (in radome) and the SPG-60 air track radar are major components in the Mk 86 weapon control system

enabling the weapon controller to assess track quality. The TV monitor can be used for splash spotting and other functions. Each weapon control console provides a digital readout of target parameters such as bearing, range, course, speed and elevation, and a keyboard is provided for inserting ballistic, target and other data into the computer.

Several operational modes are available:

Radar Surface Fire

Upon target acquisition, a track-while-scan routine is used to track simultaneously multiple designated targets. Smoothed radar data, with appropriate ballistics data, are used to compute weapon orders.

Air Action

The computer automatically and adaptively controls search patterns to acquire and track air targets. Target data, plus a TV view along the SPG-60 boresight axis, are displayed on a weapon control console. The engagement with SM-1 and SM-2 missiles by ships so fitted is facilitated by virtue of continuous wave illumination injection to the SPG-60 tracking radar.

A pre-action calibration of gun weapons can be initiated from the weapon control consoles.

Anti-Ship Missile Defense (ASMD)

The track-while-scan acquisition gate is expanded to a range sufficient for threat detection. (Full 360° coverage, or over a number of segments to provide surveillance of selected areas, can be chosen.) All targets passing through the acquisition gate are then automatically detected by the system. After target detection, the following functions are performed automatically:

(1) threat evaluation
(2) weapon assignment
(3) assignment of targets to tracking channels
(4) slewing of AA director and a weapon to the target
(5) scan of AA director and 3-D track of targets
(6) initiation of firing; transfer of fire to second target.

Visual Surface Fire

A future facility will allow the weapon controller to track the target visually with the TV system and fire the laser to obtain range data. Target range and angle data are entered into the system automatically, and ballistics are computed.

Indirect Shore Bombardment

For shore fire support missions or engagement of targets that are not within view of either radar or optical sensors, target and own-ship co-ordinates are entered at a weapon control console. The system automatically updates these with inputs from the ship's compass and log, or by means of a beacon or other navigational aids. Spotting data is entered at the keyboard. Automatic compensation for orientation of the target line by an on-shore observer is provided.

DEVELOPMENT

Initial in-house studies were started by Lockheed in 1961, and the first prototype was completed in 1966.

STATUS

Deliveries by the end of 1987 totalled nearly 100. The system is on board the US Navy's most powerful ships, including the 'Ticonderoga' (CG 47) class Aegis cruisers, the 'California' (CGN 36) and 'Virginia' (CGN 38) class nuclear cruisers, the 'Kidd' (DD 993) and 'Spruance' (DD 963) class destroyers and the 'Tarawa' (LHA 1) class assault ships. The Mk 86 has also been retrofitted to modernised 'Charles F Adams' (DDG 2) class destroyers and their German equivalents the 'Lütjens' class.

CONTRACTOR

Lockheed Electronics Company Inc, US Highway 22, Plainfield, New Jersey 07061, USA.

4538.261

AN/USQ-81(V) COMMAND CONTROL SYSTEM

The AN/USQ-81(V) is a computer driven real-time tactical data display system which receives, processes, correlates, displays and prepares surveillance reports for transmission to similarly equipped units. Display systems are installed in submarines, surface ships, aircraft and command centres ashore and are designed for the requirements of each site.

The AN/USQ-81(V) collects and processes surveillance data from a variety of sensors and sources and converts this into near real-time information which is continuously updated. The information is designed to give the overall commander the maximum amount of information on the tactical situation on land, sea and air, as well as up-to-date data on the location and operational disposition of both his own forces and those that he is opposing.

The system operates in interactive real-time mode with an established data base that can be searched both forward and back in time. Manual data control is achieved by programme function and alpha-numeric keyboards in the operator's console. The operator can select a variety of reports, graphics and calculations to be displayed for briefing, editing, comparing documentation and transmitting. The information can be displayed in great detail or it can be filtered to display the specific data a commander wants to see.

A typical AN/USQ-81(V) configuration includes a central processing unit, an operator interface terminal containing the keyboards and display, the computer control panel, graphic display unit, fixed head memory disc, tape unit, paper-tape reader and a hard-copy unit.

CONTRACTOR

Lockheed Missiles and Space Company, 1111 Lockheed Way, Sunnyvale, California 94086, USA.

1835.281

FIRE CONTROL SYSTEM Mk 92

The Mk 92 FCS has been developed for fitting in two new classes of USN ships, the guided missile frigate (FFG) and patrol combatants missile (hydrofoil) (PHM), and for the US Coast Guard medium endurance cutter (WMEC). The PHM is also known as the NATO hydrofoil, since the overall programme of development and construction is being carried out in collaboration with West Germany and Italy, who planned to have essentially the same craft but differing in some aspects of weapon and sensor fits. One such variation is in respect of the fire control system: the Italian Navy will employ an Argo FCS, while the USN and German Bundesmarine will have virtually the same system, namely the Hollandse Signaalapparaten M28 which has been given the American designation Mk 92 Mod 1, produced under licence in the USA by the Sperry Corporation. Details of the M20 Series of HSA fire control systems will be found in **1259.281**, earlier in this section.

Mk 92 FCS radar unit (US Navy)

The Mk 92 is very similar to the M25 and M27 systems built by HSA for the Belgian, West German, and the Netherlands navies, the former providing for control of air and surface engagements by guns and missiles, while the M27 additionally has provision for torpedo control. The Mk 92 is used for air and surface actions using guns and missiles, and there are a number of other changes from the basic M25/M27 configuration to suit USN requirements. In compliance with the differing sensor arrangements and armament in the three classes of US vessels fitted with the Mk 92, this system is produced in different versions, the Mk 92 Mod 1 for the PHM and WMEC, and the Mk 92 Mod 2 for the FFG. A further version, Mk 92 Mod 5, is in production for ships of the Saudi Arabian Navy. All versions have been built, tested and are in serial production.

Patrol Hydrofoil Missile

The USN variant of the PHM which employs a Mk 92 Mod 1, has as its armament eight Harpoon anti-ship missiles (**2641.221**), and a US-built OTO-Melara 76 mm dual-purpose gun (Mk 75). In addition to the collocated search and tracking radars which are a feature of the M20 Series fire control system, other sensors include an I/J-band navigation radar, inertial navigation system, and an electronic warfare suite.

Guided Missile Frigate

The new class of guided missile frigate (FFG) for the USN has as its principal operational roles anti-submarine warfare, anti-air warfare including both anti-aircraft and anti-missile capability, and anti-ship missions. These ships have a dual-purpose launcher able to launch either Harpoon surface-to-surface or Standard surface-to-air missiles. Other armaments include a US-built OTO-Melara 76 mm dual-purpose gun (Mk 75).

The Mk 92 Mod 2 FCS for this class of vessel will operate within a more elaborate environment than the Mod 1 version. In addition to its own search and tracking radars (with integral IFF), other sensor equipment includes an AN/SPS-49 search radar, AN/SPS-55 I/J-band navigation radar, and an AN/SPG-60 STIR (standard target illuminating radar). The complete Mk 92 Mod 2 is integrated with the ship's digital rapid reaction command and control system, Sperry Univac AN/UYK-7 computers being used for this and the Mk 92.

Medium Endurance Cutter

The WMEC employs a Mk 92 Mod 1 FCS which controls a US-built OTO-Melara 76 mm dual-purpose gun (Mk 75) and an optical sight. The ship also contains a Sperry collision avoidance system and command display and control system (COMDAC). This is a shipboard command display and control system which combines both weapons control and ship control. The COMDAC utilises the ship's radar and other sensors to detect and create a display of airborne and seaborne objects and land coastlines. The displays are used by operators to perform navigation/collision avoidance and mission requirements including search and rescue, enforcement of law and treaties, and military operations.

Hazeltine Corporation will supply three types of electronic equipments in various quantities for the COMDAC system: cue generators, digital scan converters, and a video sync generator for each installation. The cue generator and digital scan converter will generate composite, television-compatible data for display of radar and computer-generated information; the third unit provides the timing and synchronisation for all the electronic equipments in the COMDAC system.

CONTRACTOR

Sperry Corporation, Electronic Systems, Great Neck, New York 11020, USA.

4491.281

H-930 SERIES COMBAT/WEAPON CONTROL SYSTEMS

The H-930 Series of combat weapon control systems is designed to accept inputs from a wide variety of available and future surveillance and tracking sensors and provide outputs to a similar variety of weapon sub-systems for ships ranging from fast patrol boats to destroyers. Significant operational features of the H-930 Series include:

(1) automatic detection, tracking, threat analysis and display of targets from an air search radar (ASR)
(2) semi-automatic detection, and automatic tracking and display of targets from a surface search radar (TWS)
(3) precision tracking from up to two track radars (TR) and an optronic device (OD)
(4) split mode tracking utilising best combination of angle and range data from any pair of track data sources from the same target
(5) assignment of any of a ship's guns to any system track data sources simultaneously or all guns on a single track data source or any combination thereof
(6) assignment of a tracking and illuminating radar (TIR) to Standard missile (SM-1)
(7) assignment of a tracking radar to a surface-to-surface missile (SSM) system
(8) integration with IFF, EW and data link in various levels of intimacy
(9) integration with Mk 114 UBFC system (sonar, ASROC and OTS) (**3222.281**)
(10) tactical navigation aids such as intercept/manoeuvre calculations, CPA and time to CPA calculations, true or relative motion display, maintenance on earth grid, and manual or automatic compensation for set and drift
(11) manual, semi-automatic or automatic setting of shell splash spotting data
(12) automatic recommendation of tracker designations and weapon assignments including a means of quick operator acceptance of same
(13) automatic sequencing of tracker designation and weapon assignment as preset by the operator. Such sequences are triggered by an event pre-selected by the operator
(14) demonstrable gun pointing accuracy under realistic test conditions.

The various models within the H-930 Series include all or combinations of the above features. To achieve flexibility and still retain ease of operation, the H-930 operator interface features a common command and control syntax and, in most models, employs a plasma panel that is programmed to accommodate many operator actions and displays, yet preclude illegal actions.

The H-930 Series began with the production of the EX-93 gun fire control system for the USN's Coastal Patrol Interdiction Craft (CPIC) programme. This system was successfully evaluated by the USN and subsequently transferred to South Korea in 1974. Since then the H-930 Series has evolved and continually been improved over the years.

During the development of the Mod 0 system, it became apparent that the H-930 Series must be truly modular in order to be compatible with future technology developments as well as future applications. Accordingly, in the mid-1970s, Honeywell merged the benefits of a distributed computing development programme that had progressed for several years for the USAF into the H-930 Series. As a result, the Mod 1, 2, and 3 systems are the first operational modular ship-board control systems based on a distributed computing architecture for fast patrol boats, destroyer modernisation, and land-based coastal defence installations.

These systems are functionally modular and physically distributed. They have been installed, checked out, and formally accepted by a foreign navy after extensive, rigorous live firing and operational test exercises at sea. The H-930 programme has involved the integration of a wide variety of new and existing weapons and sensors from US and world-wide suppliers.

The most recent modular combat system (H-930 MCS) incorporates the latest technological improvements to realise the full potential of total physical and functional modularity for both new ship classes and modernisation programmes. This new system draws on Honeywell's experience in systems for foreign navies, as well as on research conducted and demonstrated for the USN.

Under the MCS concept, modularity is achieved through a module that is defined as a distributed computing unit (DCU) and the attached system devices that it serves, combined with an equally modular software architecture. The DCU consists of a set of circuit card assemblies, including processors and memory, that perform all of the processing, data routeing, signal conditioning, and protocol functions to allow the attached device(s) to perform effectively as a unit of the total system. Each DCU has an embedded resource computer, either a Rolm AN/UYK-19 or a Sperry AN//UYK-44. These circuit cards, along with special power supplies and cabling connectors, are either mounted in a stand-alone, drip-proof, air-cooled module or are embedded in the cabinets of system control display consoles.

System data is routed as digital messages via the combination of a powerful operating system, local input/output control (IOC) processors, and a multiple redundant data bus system. All addressing and local data transfers are accomplished by the IOC and the executive function of the resource processor. All inter-modular tranfers on the data bus system are effected by the global bus interfaces (GBI). The total data routeing system is not dependent on a central master control point that can fail and bring down the entire system. It is this degree of controlled autonomy that allows the system to tolerate the addition or deletion of a given module with the only immediate effect being the presence or absence of that module's function. Access to the global bus is allocated on a rotating basis such that addition of a module will require a slight adjustment, but the inherent intelligence of the bus system can recognise and recover from a non-responding module. Normally, the designed removal of a module will be accompanied by a compensatory change to the allocation sequence.

An area of significance to the shipbuilder is that of long-run cabling. In a recent study for the US Navy, it was determined that the implementation of a data bus in an FFG-7 class ship could eliminate almost 12000 m of complex signal cables, or more than half of the total cable length. The weight saving was determined to be five tons.

STATUS

In service. Ten systems, each with six consoles, have been supplied for destroyers, seven for FAC in two navies and five for coast defence purposes.

CONTRACTOR

Honeywell Inc, Training and Control Systems Operations, 1200 East San Bernardino Road, West Covina, California 91790, USA.

4270.281

CHALLENGER SA-2

This electro-optronic fire control system primarily designed for fast patrol boat installation provides precise gun orders or missile guidance functions for a large inventory of weapons. (It was originally introduced as Sea Archer 2 and re-named in late 1982.) Gun calibres accommodated range in size from 20 mm to 76 mm and surface-to-surface missile control includes both inertial and laser guided types. Having both surface and air target capability, SA-2 utilises infra-red (three to five micron and eight to twelve micron common module), TV, and laser sensors mounted on a remotely controlled director. An optional M-band (95 GHz) radar tracker module is also available for mounting on the director. Associated with the one-man total system operational concept is a below-deck control console which contains TV/IR radar and alpha-numeric displays as well as system control functions including weapon designation information and tellback information. Automatic operation is provided through an AN/UYK-502 militarised 16-bit computer utilising standard USN CMS-2 software compilation.

Since the line-of-sight stabilised director can support over 150 kg of payload and is trainable through 360° with elevation angle coverage of –30 to +85°, a wide variety of target sensors can be accommodated. Currently in use are a 3.4 to 4.2 micron thermo-electrically cooled imaging and tracking sensor, 7.75 to 11.75 micron common module IR with a new long-life Stirling cycle cooler, a low light-level TV with 10:1 zoom lens elements, and a 1.06 micron high repetition rate rangefinder laser. When laser illumination is required for missile guidance functions a high power coded laser with variable beam divergence is employed. In addition the 95 GHz radar tracker, which is mounted integrally with the E-O director, uses a narrow 0.5° beamwidth to minimise specular multi-path reflections and sea clutter returns. A power output in excess of 1 kW peak provides range and angle information compatible with the detection ranges and performance of the IR and TV sensors.

Processing of data from the available sensors utilises algorithms for edge, centroid, and correlation target tracking to generate angle error data for servo control. Automatically adjusted gate size corresponding to target size and dynamics reduces track jitter and aids the many signal enhancement techniques used to reduce background clutter effects. Images from the TV and IR sensors as well as the radar module can be displayed simultaneously on the control console for operator interpretation.

Target designation for SA-2 can be accomplished from any of the ship sources such as search radar, other FCS equipment, CIC, or optical director. When initial designation is received, the FCS is programmed in automatic acquisition patterns dependent upon accuracy of the input data. Multiple input/output data channels provide the interchange of information between SA-2 and the controlled weapons.

The SA-2 above-deck equipment weighs 305 kg and below-deck equipment totals 422 kg. Other ship inputs required, dependent upon weapon order generation, include heading, roll, pitch and speed in either synchro or digital format.

STATUS

In production.

CONTRACTOR

Sperry Corporation, Electronic Systems, Great Neck, New York 11020, USA.

1922.481

TRIDENT FIRE CONTROL SYSTEM Mk 98

General Electric Ordnance Systems Division has received a contract award for the development of US Navy fire control systems and guidance support equipment. The contract covers work extending over a four-year period for design and initial manufacture of fire control systems for the Trident submarine and for the modifications required for Poseidon fire control systems currently deployed in order to accommodate a larger C-4 missile (**2840.411**).

As a key sub-system of the Trident strategic weapons system, the GE fire control technology will serve the primary purpose of preparing the missile guidance system for flight and controlling the missile launch sequence.

Support equipment will consist of: guidance system test equipment for shore-based activities and nuclear submarine strategic weapons facilities; guidance system containers and handling equipment; and equipment to support fire control and guidance operations at all maintenance levels. Deliveries of fire control systems and support equipment are continuing. No further details were available at the time of publication.

CONTRACTOR

General Electric Company, Ordnance Systems Division, 100 Plastics Avenue, Pittsfield, Massachusetts 01201, USA.

1925.281

PASSIVE FIRE CONTROL SYSTEM Mk 105 (TAC)

The Mk 105 target acquisition console is a passive fire control system for the detection of targets radiating anywhere in the microwave spectrum and the direction of appropriate ship's weapons against those which are identified as threats. The main elements of the system are a broadband microwave antenna array with direction finding facilities, a countermeasures receiver, signal data processor, and an indicator control unit.

The antenna array consists of eight cavity-backed planar spiral elements with broadband balun feeds, these being disposed on a masthead mounting to permit direction finding by the monopulse amplitude technique.

The countermeasures receiver, derived from the AN/ALR-47 radar homing and warning system developed by IBM for the S-3A, Viking ASW aircraft (**1862.353**), is a four-channel, digitally tuned superheterodyne receiver with a fast-acting diode matrix switching head.

The signal data processor measures the characteristics of the video outputs of the receiver and converts the video pulses to digitised, monopulse, parametric data. All signal processing and system controls are handled by software that can be modified by the operator or by selection in advance of a mission. The processor performs all input/output signal conversions and storage, processes and filters the data, and transmits information to the display for use by the operator. All communication with external displays or interfaces with other of the ship's systems are handled by the signal data processor, which has two memories, the combined capacity of which may be as much as 24 K words. The computers are machines of IBM's System/4 Pi range, model SP-OA.

The indicator control unit is provided with an advanced CRT display for the presentation of emitter bearings, and the associated emitter signal characteristics in alpha-numeric form. Its operation and character generation functions are controlled by the signal processor computer. The operator is provided with a control panel on which are push-buttons and thumb-wheels that serve as data insertion devices. The buttons are illuminated internally and also act as indicators. System software can revise the function of each of the 28 push-buttons without affecting the hardware.

STATUS

The first of the production systems was delivered in spring 1974 and in 1976 delivery was completed for guided missile destroyers. More detailed information is given in previous editions of *Jane's Weapon Systems*.

CONTRACTOR

IBM Federal Systems Division, Owego, New York 13827, USA.

3221.481

FIRE CONTROL SYSTEM Mk 113

The FCS Mk 113 Mods 6, 8 and 10 are used in SSN attack class submarines of the US Fleet. Target bearings from the BQQ-5 sonar suite are pre-filtered, and together with own ship motion data, are processed in the Mk 130 digital computer, the first electronic digital computer to be used aboard attack class submarines. With input of own ship state vector information, the target range, course and speed can be computed from passive bearings only. The operator interacts with the target motion analysis (TMA) function at the analyser console Mk 51.

The FCS Mk 113 consists of the following equipments:

(1) the attack director Mk 75 receives own-ship and target information to compute torpedo ballistics and wire-guide controls, and performs position keeping on both target and torpedo. The Mk 75 also supplies necessary data to levelling computer Mk 129 for SUBROC ballistic computations

(2) the attack control console Mk 50 is the system firing panel, tactical display, and torpedo room status panel

Attack control console of Mk 113 fire control system

(3) the levelling computer Mk 129 performs SUBROC ballistic computations, and in conjunction with the reference sensing element Mk 1, provides signals for levelling the SUBROC inertial platform.

Addition of two torpedo control panels Mk 66, two tone signal generators Mk 47, together with major modifications to existing Mod 2 equipments, resulted in FCS Mk 113 Mod 6 and Mod 8 which accommodate the Mk 48 torpedo. Torpedo control panel Mk 66 (which represented the first application of MSI/LSI circuit technology on submarines) acts as a preset panel while TSG Mk 47 generates the signals for transmission of data to the torpedo. The Mod 6 is a field-modified Mod 2 while the Mod 8 comes directly from the factory with the modifications built in.

STATUS

The FCS Mk 113 has been operational since 1962. It is the weapon control system deployed aboard the US Navy nuclear attack class, and with the addition an analyser console Mk 78 (FCS Mk 113 Mod 9), SSBN fleet ballistic missile submarines. In the late 1960s additional equipment required for the Mk 48 torpedo was developed. During the past 10 years, the attack class submarines have been upgraded with the FCS Mk 117 and Combat Control System (CCS) Mk 1 equipments. Only four FCS Mk 113 systems are currently operational on SSN attack class submarines. However, approximately 30 FCS Mk 113 Mod 9 systems remain operational on SSBNs.

CONTRACTOR

The Singer Company, Librascope Division, 833 Sonora Avenue, Glendale, California 91201, USA.

3222.281

ASROC Mk 114 FIRE CONTROL SYSTEM

Fire control system Mk 114 provides the means of solving the shipboard anti-submarine attack problem, generates launching orders, prepares the weapons for firing, and generates designation data for ASROC (**6001.241**) missile tracking by ship's radar. The ASW weapons controlled by FCS Mk 114 range from Hedgehog projectiles and older Mk 44 torpedoes to the newer Mk 46 torpedoes as well as the ASROC missile.

The FCS Mk 114 consists of the following equipments:

(1) the attack console Mk 53 consisting of an attack plotter and a ballistic computer

(2) a stabilisation computer Mk 134 which generates stabilised azimuth and depression orders for the ship's sonar

(3) the position indicator Mk 78 provides a continuous display of own ship's target and weapon tactical information including firing control to the commanding officer

(4) the relay transmitter Mk 43 tests ASROC readiness, and programmes and monitors the ignition and separation assembly of the selected ASROC missiles

(5) a relay transmitter Mk 44 which transmits train and elevation orders to the selected gun or missile fire control system for radar tracking of the ASROC missile.

These equipments employ a firing panel and transistorised, electromechanical analogue computer modules which combine automatic computation with graphic plotting on the Mk 53 console for target motion analysis, and ballistic computations.

STATUS

The Mk 114 FCS, and its predecessor the Mk 111 fire control group have been installed on US Navy destroyers and frigates, and on ASW ships in service with navies of Brazil, Germany (FR), Greece, Italy, Japan, Korea (S), Pakistan, Spain, Taiwan, and Turkey. Originally developed in the early 1960s, the USN maintains a refurbishment and repair facility for the Mk 114 at NTS, Keyport, Washington.

CONTRACTOR

The Singer Company, Librascope Division, 833 Sonora Avenue, Glendale, California 91201, USA.

3223.281

Mk 116 FIRE CONTROL SYSTEM

The Mk 116 Mod 4 underwater fire control system (UFCS) is fitted in USN anti-submarine warfare ships. The system has evolved during the period it has been in operation and the Mod 4 uses an AN/UYK-7 computer as the system central processor. It interfaces with the various guided missile launch systems to launch ASROC missiles against submerged targets at stand-off ranges, and with trainable Mk 32 torpedo tubes to launch the Mk 46 torpedo against submarines at close range.

The principal sensor is the AN/SQS-53 sonar set, with environmental inputs from the electro-magnetic log, wind indicator Type F, advanced design stabilised gyro, AN/SQQ-61 bathythermograph, AN/UQN-4 fathometer, and data from ship's command and decision system. The Mk 116 Mod 4 UFCS includes standard NTDS computers, displays, and interface equipment. It consists of two major sub-systems, the computer processing sub-system and the weapon control and setting sub-system (WCSS).

The Mk 116 Mod 4 comprises:

(1) computer processing sub-system: AN/UYK-7 digital computer with advanced ASW operational program; RD-358 magnetic tape unit; OA-7984 input/output console (paper tape and teletype)

(2) WCSS: weapon control panel Mk 329, which is normally the only manned unit; missile setting panel Mk 330, one for each launch system; torpedo setting panel Mk 331; weapon status and approved panel Mk 332, installed at the bridge and combat information centre; bridge display panel Mk 333; and interface control panel Mk 377, which provides power distribution and signal routeing for missile/torpedo functions.

OPERATION

The underwater battery supervisor, stationed in the combat information centre, operates the UFCS from the weapon control panel and an adjacent display console. ASW team members are also located at the torpedo setting panel to facilitate local operations if required, and at the weapon status and approval panels to provide ASROC launch approval.

DEVELOPMENT

The Mk 116 Mods 1-4 WCSS were developed and produced by Singer-Librascope under contract to the Naval Ocean Systems Center, San Diego.

STATUS

The Mk 116 Mod 4 and its predecessor systems Mk 116 Mods 1, 2 and 3 have been installed in several classes of surface ships: Mod 1 on CGN 38 class cruisers, Mod 2 on DDG 993 class destroyers, and Mod 3 on CG 26 with a single launcher, as part of its modernisation programme. Mk 116 Mod 4 is currently installed on CG 47 'Aegis' class cruisers.

The Mk 116 Mod 5 is an upgrade of the Mod 0 system. The latest version, Mod 6, provides the capability to launch ASW weapons from the Vertical Launch System (VLS) installed in 'CG 56' class cruisers. Singer have developed a new torpedo setting panel to support the Mod 6 upgrade.

CONTRACTOR

The Singer Company, Librascope Division, 833 Sonora Avenue, Glendale, California 91201, USA.

4822.481

SUBMARINE COMBAT CONTROL SYSTEM Mk 2

This is an improved version of the submarine fire control system (SFCS) Mk 1, that was initially produced for the Royal Australian Navy's modernised 'Oberon' class submarines and subsequently fitted in boats of the Canadian and Indian navies. The new system is mainly for export applications, although its use in certain classes of US ship, or surface vessel fittings, should not be ruled out. The original Mk 1 SFCS Mod 0 was described in the Australian section of this part of *Jane's Weapon Systems* in previous editions under entry number **3220.481**.

The Mk 2 combat control system currently in production is a completely integrated action-information system for submarines, corvettes, and frigate-type ships. It provides comprehensive processing and evaluation of all combat data and co-ordinated operational control of all sensors, weapons, and combat manoeuvres. The system's unique and innovative architecture has resulted in lower costs, higher reliability, and easier operation and maintenance than its predecessors.

The system provides for surveillance of a large number of contacts and the simultaneous engagement of multiple threats. The surveillance process is highly automated due to the large amount of navigation, contact tracking, communication link, and weapon telemetry surveillance data that must be processed and evaluated. The results of the automatic processing are presented to system operators in user-friendly pictorial and natural language forms. The highly automated approach frees the operators from perfunctory tasks and allows them to manage by exception. This results in a reduction in the number of operators required for surveillance and allows the assignment of additional operators to the prosecution of contacts which are a threat to the ship's mission. The contact prosecution process, while operator intensive, includes the display of tactical options and recommendations to aid operators in attacking threats using torpedoes and missiles. In addition, the system provides support for performance monitoring and fault localisation, combat data recording and reconstruction, and operation and maintenance training.

The system hardware is completely redundant and can therefore sustain single failures of any type without any loss of tactical capability. The equipment is very compact and will fit down a 650 mm hatch. Central to the system are a set of distributed computers which are embedded in each of the hardware elements. The computer design is modular and therefore expandable by inserting additional processing and memory cards. The system is comprised of the following hardware elements: a sensor data converter/processor with two computers, multiple combat control consoles with one computer each, and a weapon data converter/processor with two computers, all interconnected by a dual combat data bus.

The sensor data converter/processor provides:

(1) interfaces with the navigation sensors, the tracking sensors (including passive/intercept/active sonar, radar, electronic support measures, TV, and periscopes), and the communication link
(2) automatic surveillance and support processing
(3) combat data recording.

Singer submarine combat control system Mk 2 displays

Combat control consoles provide a truly multi-function display of sensor video and processor generated graphics and alpha-numerics. The main display surface is a 45 Hz non-interlaced raster scan colour CRT with 1280 × 1024 picture element resolution. Each console independently provides for the simultaneous display of video from any two sensor sources plus the overlay of graphic and alpha-numeric annotation. Radar, navigation charts, and annotation may be combined for use in combat navigation. The console contains a 17 × 8.5 inch (430 × 22 mm) touch interactive plasma display which is used as a tactical summary display and for interaction with the weapon launch compartment personnel during preparation for a launch.

The weapon data converter/processor provides:

(1) automatic contact prosecution and support processing
(2) interfaces with the launch tubes and weapons
(3) emergency weapon setting, launching, and control.

A plasma display like the one in the combat control console is included and is used for interaction with the console operators during preparation for a normal launch and for local control during emergency operations.

The combat data bus provides communication between the other elements of the system at a one million bit-per-second serial rate over each half of a dual bus in the Mil Std 1553B format. Data is broadcast on the bus from each computer to all other computers simultaneously so that all computers receive the data simultaneously. Each half of the bus has the capacity to provide all of the required system communication. The bus is available in both copper and fibre-optic forms.

The sensor and weapon data converter/processor performance monitoring tasks combine to provide a double check on system failures and in conjunction initiate automatic switching to redundant hardware/software in the event of a single point of failure. Switching processors is simplified as program loading and data base updates are not required. Also, switching halves of the data bus is simplified by switching all communication from one half to the other in the event of any failure on the half that is providing system communication.

The Mk 2 combat control system is a completely integrated action-information system which can survive any single point of failure without any loss of tactical functions. It centres around multi-function colour display consoles, and the cost is low due to the simple and innovative design.

STATUS

The preceeding Mk 1 version was initially supplied to the Royal Australian Navy for fitting in modernised 'Oberon' submarines, and was later adopted for Canadian Navy boats of the same class, as well as being procured for West German-built 1500-ton submarines for the Indian Navy. The Mk 2 SCCS is currently in production. The Mk 3 is also scheduled for retrofitting in Egyptian 'Romeo' class submarines.

CONTRACTOR

The Singer Company, Librascope Division, 833 Sonora Avenue, Glendale, California 91201, USA.

1493.261

US NAVAL TACTICAL DATA SYSTEM

NTDS (naval tactical data system) is the designation given to the combinations of digital computers, displays of various types, and data links which are installed on USN ships for the on-line collection, processing, storage, and presentation of information from sensors such as sonar, radar, optical, and aircraft or ship consorts, via data link. NTDS also is engineered to interface with ATDS (airborne tactical data system) and MTDS (marine tactical data system). All three systems have the prime function of providing automated organisation and display of information for command and control teams for such purposes as threat detection and assessment, and weapon-target allocation.

Although designed to a common concept, NTDS exists in a number of versions, varying in size and equipment complement according to the age of installation and the size of vessel fitted. The overall system has also been subject to a continuous process of updating since its introduction into service in the late 1950s. By 1969, for instance, three generations of display devices had been employed. The basic digital computer for the system, the AN/UYK-20, was preceded in some installations by other types, and later NTDS employ the Univac AN/UYK-7.

STATUS

In addition to USN fleet fittings of NTDS, installations aboard ships of the Australian, French, Italian, Japanese and West German navies now total more than 100.

In the early 1980s the USN selected the AN/UYK-43 and AN/UYK-44 computers to replace the AN/UYK-7 and AN/UYK-20 as 'standard' digital computers. It was also decided to replace the older machines progressively by UYK-43 and UYK-44 processors as requirements expand beyond the capabilities of the former processors, or other circumstances make appropriate. It was also decided to adopt the 'embedded' computer philosophy which provides distributed processing facilities with benefits in survivability, maintainability, flexibility etc. The NTDS has been the object of a programme of up-dating involving replacement of obsolete computers, displays and other equipment, and software modifications to reflect the integration of new sensors and weapon systems into the Fleet.

Hughes has delivered about 2500 displays (UYA-4 and UYQ-21) for these systems. As examples, in the USN, the AN/UYA-4 standard NTDS display system in a 'Perry' class ship includes seven consoles, while in 'Ticonderoga' class vessels 17 consoles and the AN/UYQ-21 large screen are fitted. From 1973 to 1982, 87 UYA-4 systems were procured for the USN at costs that varied from $3.5 million for an FFG 7 to $9 million for the CG 47, with 42 AN/UYQ-21 systems procured in the period 1979 to 1982.

CONTRACTOR

Hughes Aircraft Company, Fullerton, California 92634, USA

Combat information centre of the USS John Young *(DD 973), showing part of the ship's NTDS*

3514.261

W-1200 SERIES FIRE CONTROL SYSTEMS

The W-1200 Series fire control systems are designed to control guns, missiles and defensive measures for fast patrol boats (FPBs). Additional capability can be added to handle fire control requirements up to destroyer size ships.

The W-1200 Series systems have been optimised for FPB's offensive and defensive missions against air, surface and shore targets. Each system of the series is an integration of matched sensors, a versatile fire control computer and a display control console that has been organised for FPB operators. The elements of these systems are fully developed, off-the-shelf hardware. Each hardware elements has been selected and integrated with interfacing to provide fire control based on boat size and requirements, enhanced target detection and acquisition, accurate tracking, and weapons direction. Each system provides operators with full and relevant target information. The systems feature quality performance, rapid data processing, high reliability, ease of maintenance, lightweight and rugged construction. Operation is straightforward, minimising the number of dedicated crew members required while sharply reducing the complexity and load of duties or tasks performed by operators. The W-1200 and W-1210 systems for FPBs up to 100 plus tons require one operator while the W-1220, -1230 and -1240 systems only require two operators.

STATUS

All components of the systems are currently operational on FPBs.

In late 1982, the W-160 pulse doppler monopulse radar replaced the W-120 as the principal tracking radar in the W-1200 Series of fire control systems.

CONTRACTOR

Westinghouse Electric Corporation, PO Box 1897, Baltimore, Maryland 21203, USA.

Systems of the Westinghouse W-1200 Fire Control Series

System	Sensors	Weapon Control	Application
W-1200	Navigation/search radar Optical director	One or more weapons, under computer-supported optical director or radar control, against an air, surface, or shore target.	FPBs from 30 to 60 tons with one or two powered gun mounts. Optical director is the primary fire control director.
W-1210	Navigation/search radar W-120 fire control radar	One or more weapons controlled by the W-120 against an air, surface, or shore target. One or more weapons controlled by search radar against a surface or shore target. System also directs chaff/flare launch.	Applies to FPBs from 60 to 100 plus tons with one to three powered gun mounts, and chaff-flare launchers. W-120 can direct weapons against a target of any type, particularly effective against small, fast, low fliers.
W-1220	Navigation/search radar W-120 fire control radar	One or more weapons under W-120 control, and one or two weapons engaging two targets – air, surface, or shore. One or more weapons controlled by the search radar against surface or shore target. System also directs chaff/flare launch.	Applies to FPBs from 100 to 180 tons with one to three powered gun mounts with a primary fire control director. Optical director supplements W-120 or enables divided fire.
W-1230	Navigation/search radar SPS-58 air search radar W-120 fire control radar Optical director	One or more weapons under W-120 or W-160 control, and one or two weapons under optical director control, thereby engaging two targets – air, surface, or shore. One or more weapons controlled by the search radar against surface or shore target. System also directs chaff flare launch, and missile launch.	FPBs from 180 tons upward with one or more powered gun mounts, chaff flare launchers, and missile launchers. W-120 or W-160 and optical director split the fire control responsibility. SPS-58 in MTI mode enables point defence against quick appearing, fast closing hostiles (pop-up threats).
W-1240	Navigation/search radar SPS-58 air search radar W-120 fire control radar Optical director ESM equipment ECM equipment IFF	Same as W-1230 above, plus gun cueing during radar silence using ESM bearings, plus ECM jamming capability, plus IFF	Same as W-1230 above.

NOTE: Later versions employ the pulse doppler monopulse W-160 tracking radar instead of the W-120.

1834.281

AEGIS COMMAND AND CONTROL SYSTEM

Aegis, formerly designated the Advanced Surface Missile System (ASMS), is a surface-to-air weapons system primarily designed to defend against anti-ship missiles. Major components of the Aegis system are the missile, its launching system, the fire control and weapons control systems which include standard AN/UYA-4 and AN/UYQ-21 weapon control displays and four two-colour rear projection computer-controlled displays, the multi-function array radar system with its computer control, the system command and decision, and the operational readiness test system. See **2507.231**.

The system accepts general operational commands from the Mk 1 Aegis command and decision system and mid-course guidance commands from the Mk 1 control system; and its operational outputs are processed signals giving target detection data to the Mk 1 command system and target and missile track data to the Mk 1 control plus mid-course guidance commands to one or more missiles.

Together with the AN/SPY-1A radar system and the Mk 1 control, the Mk 1 command and decision system make up the detection and decision loops of the Aegis system. The Mk 1 system is the ship's command and control centre and targets enter the detection and decision loop from the AN/SPY-1A radar, from other ship's own sensors or from data supplied by other ships or aircraft. What happens next depends on which of the four Aegis operating modes has been selected. The modes are automatic special, automatic, semi-automatic, and casualty; in the automatic special mode targets meeting certain predetermined threat criteria are automatically fired upon unless manual override is invoked; in all other modes positive human action to initiate firing is needed.

In these three modes the Mk 1 weapons control system inserts targets into the engagement queue and schedules equipment for launching and terminal illumination. Trial intercepts are computed and a time to fire predicted. Resulting data are fed back to the Mk 1 command and decision centre which is also receiving target detection data from the radar control system, operational readiness data from the Mk 1 operational readiness test system and electronic warfare and other data from the command and decision centre. Incorporated in the Mk 1 system is another four-bay AN/UYK-7 computer and a comprehensive display system, the AN/UYA-4. On the basis of all this data, threat evaluation and weapon assignment processes are carried out so that the engagement decision can be taken.

STATUS

Operational in 'Ticonderoga' (CG 47) class cruisers. Modified version to be fitted in 'Arleigh Burke' (DDG 51) class destroyers and new Japanese DDGs.

CONTRACTORS

Prime Contractor: RCA Government Systems, Moorestown (three major RCA divisions are engaged in the project).

Weapon control displays: Hughes Aircraft Company.

Computer software: Computer Sciences Corporation.

Technical advice to USN: The Johns Hopkins University, Applied Physics Laboratory.

Also more than 600 other suppliers.

Shipborne Weapons

INTRODUCTION

This section deals with shipborne surface-to-surface weapons for anti-ship and shore bombardment purposes. In certain cases it will be found that the missile employed in such systems has other roles, and the practice in these instances is to cross refer to the appropriate entries for other applications of a given missile. In general it will be found that the most comprehensive description of the missile as a 'round' will appear under the entry for its principal function, but adequate data for an appreciation of a missile's performance is given in the entry for each system in which it is employed.

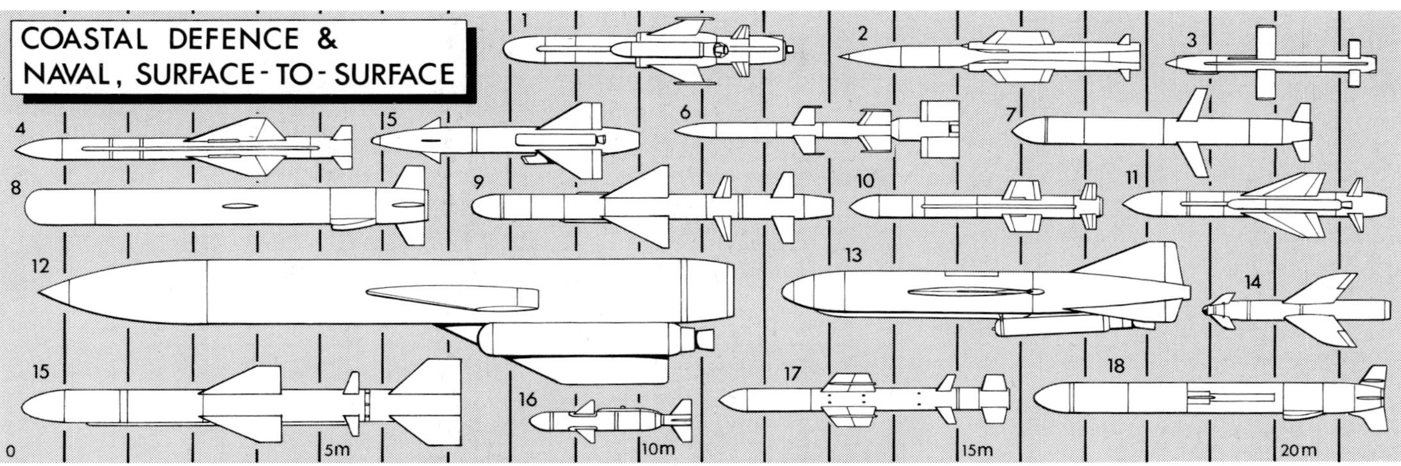

1 Otomat, **2** ANS, **3** Gabriel, **4** Exocet MM 38/9-40, **5** RBS 15, **6** Sea Killer II, **7** Gabriel long range, **8** SS-N-21X (provisional), **9** Hsiung Feng, **10** SSM-1, **11** Shaddock/Sepal (SS-N-3 etc), **12** Sea Eagle, **13** Styx family (HY-2, SS-N-2 etc), **14** Penguin, **15** C801 (provisional), **16** Sea Skua, **17** Harpoon, **18** SLCM Tomahawk

CHINA (PEOPLE'S REPUBLIC)

4891.221

C801 SURFACE-TO-SURFACE ANTI-SHIP MISSILE

In late 1984 the People's Republic of China revealed the existence of an hitherto unknown anti-ship missile of a new design, employed in both the coastal defence role and aboard ships.

The new missile was first known as C.801. It employs a high frequency (probably J-band) active radar seeker, has a sea-skimming attack profile, and terminal altitude can be as low as 4 m, depending on sea-state.

In addition to shipborne and coastal defence applications the C801 can also be launched from submarines and aircraft. The ES5G 'Wuhan' class SSK carries six single C801 launchers in a casing outside the pressure hull. To fire the missiles the submarine must surface and elevate the launchers to 45°, an arrangement similar to the Soviet 'Whiskey Long Bin' and 'Juliett' class SSGs.

Initial target acquisition is probably by passive sonar, but more precise data such as range and bearing fed to the missile immediately before launch is probably supplied by the surface search radar. The air launched version would dispense with the booster.

China's new anti-ship missile, the C801. Note the similarity to the French Exocet weapon, the large booster motor, and the ribbed-box launcher. It is employed in both shipborne and coastal defence roles

CHARACTERISTICS

Length: 5.2 m
Span: 1 m
Weight: 1000 kg
Warhead: 165 kg
Range: 70 km

STATUS

The first installation observed is aboard the prototype 'Jianghu V' type frigate No. 535, which was running trials in 1986 at Shanghai. She carries eight C801 missiles. The weapon is also likely to form the main armament of two new CODOG destroyers, two of which will be completed late in 1987 or early 1988. Several designs for frigates, destroyers and FPBs armed with the C801 have been exhibited. A coastal defence version also exists (**6184.121**).

CONTRACTOR

China Precision Machinery Import & Export Corporation, 17 Wenchang Hutong Xidan, Beijing, People's Republic of China.

6447.221

C.101 SUPERSONIC ANTI-SHIP MISSILE

A prototype supersonic anti-ship missile was displayed by CPMIEC at the Paris Air Show in 1985. Range, dimensions and most other details are not known.

CHARACTERISTICS

Speed: >Mach 1
Powerplant: Two solid-fuel (ramjet?) plus solid fuel sustainer
Homing: Active radar

STATUS

Prototype flew by June 1985. No details have been published.

3973.221

FL-1(CSS-N-2) NAVAL SURFACE-TO-SURFACE MISSILE

As a former client of the Soviet Union, the People's Republic of China received a number of SS-N-2 Styx anti-ship missiles (**1155.221**), and this weapon now forms a major element of the Chinese inventory. It is extensively employed in both the shipborne surface-to-surface role and for coastal defence (**3984.121**).

So far as can be ascertained, few if any changes have been made to the original Soviet design, although indigenous production is now in hand and the possibility of internal detail changes can hardly be ruled out. Chinese-built missiles of this type are designated Fei Lung-1 (Flying Dragon) and they were offered for export in 1984. In the 'Luda' class and 'Anshan' class Chinese vessels on which the FL-2 is currently deployed, the launcher arrangements are obviously of local origin. The 'Luda' class has two triple trainable launchers, while the 'Anshan' has two twin trainable launchers.

The FL-2 is produced in three versions, differing in their guidance sub-systems. Cruise altitudes vary between the models: 100 m (FL-2), 50 m (FL-2G) or 30 m (FL-2A).

The original CSS-N-2 is probably referred to by the Chinese as SY-1 (see **3894.221**). At Farnborough in 1986 CATIC provided information on the later CSS-N-2 version, the Fei-Lung (Flying Dragon-1). This missile uses a new monopulse active radar homer. A substantially smaller antenna suggests a shift in operating frequency to J-band from the original I-band, as this higher operating frequency means that a smaller antenna suffices to achieve the required tracking beamwidth. The FL-1 also employs a radio altimeter to achieve a 30 m cruise altitude and a 15 m terminal attack altitude.

CHARACTERISTICS

Length: 7.36 m

Range: 95 km (higher for dynamic range)

Warhead: 1100 lbs

Guidance: Active radar (HY-2) + radio altimeter (HY-2G), passive IR homing (HY-2A)

STATUS

In operational use and production continues. 'Luda' class destroyers have two triple launchers for FL-2 missiles and converted 'Anshan' class destroyers have two twin launchers although there have been reports of some of these ships having been fitted with triple launchers. Chinese 'Jianghu' class frigates have two twin launchers and frigates of the 'Chengdu' class have one twin launcher. Two main classes of fast missile boat are operated by the Chinese Navy, both derived from their Soviet 'Osa' and 'Komar' class prototypes. The 'Hola' class, of which there are over 100, have four single FL-2 launcher/containers while the 'Hegu' class boats, numbering more than 80, have two single launchers.

CONTRACTOR

China Precision Machinery Import & Export Corporation, 17 Wenchang Hutong Xidan, Beijing, People's Republic of China.

Chinese 'Luda'-class destroyer with triple launcher for FL-2 missiles (G. Jacobs)

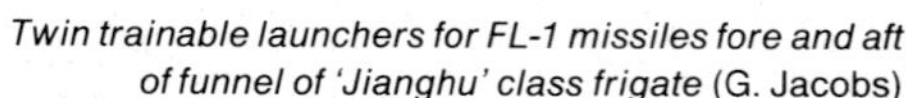

Twin trainable launchers for FL-1 missiles fore and aft of funnel of 'Jianghu' class frigate (G. Jacobs)

6448.221

FL-7 ANTI-SHIP MISSILE

A new missile developed by CATIC is being marketed as an alternative to the C.801 from CPMIEC. It is compatible with existing CSS-N-2 Styx shipboard systems, an advantage claimed by the manufacturers over the rival C.801.

CHARACTERISTICS

Speed: Mach 1.4

Powerplant: Single solid-fuel booster plus liquid sustainer

Cruise altitude: 50 m

Dimensions: 6.6 m long × 54 cm diameter

Weight: 1800 kg

FRANCE

1156.221

MM-38 EXOCET SURFACE-TO-SURFACE MISSILE

Exocet MM 38 is a surface-to-surface tactical missile designed to provide surface warships with all-weather attack capability against other surface vessels. It can be fitted in major and minor warships including fast patrol boats and hydrofoils.

The missile has a streamlined body equipped with four cruciform wings and four tail control surfaces in the same planes as the wings.

Propulsion is provided by a two-stage, solid-propellant motor.

The MM 38 Exocet's range is about 23 nm (42 km), flying at very low altitude. Its cruising speed is high subsonic and it carries a high explosive warhead.

Missiles are stored in box-type launcher-containers, which are mounted on deck.

The missile flight consists of a pre-guidance phase during which it flies towards the target. Range and bearing have been determined by the fire control computer and set up in the missile pre-guidance circuits before firing. In the final guidance phase the missile flies directly towards the target under the control of its active homing head. Throughout the whole flight the missile is kept at very low altitude by a closed-loop control system comprising a radio altimeter and the inertial guidance platform.

The terminal guidance phase (approximately the last 10 km) is by means of the ADAC radar homing head.

MM 40 Exocet anti-ship missile leaving portside canister of a French Navy 'aviso'

Subsequent to the introduction of the original MM 38 model, a number of derivative versions evolved, three of which have been produced in quantity. The air-launched AM 39 model is described in **1770.321** which appears in the section on air-to-surface tactical weapons later in this book. The surface-launched MM 40 variant has a longer steel-cased sustainer motor than the MM 38, resulting in a further increase in range. The MM 40 homing head also has a wider angle target search and acquisition 'gate' commensurate with its greater range (more than 70 km). It is carried in a tubular glass-fibre container/launcher. Four of these launchers can be mounted in the space required for one MM 38 housing.

CHARACTERISTICS
Length: 5.21 m
Diameter: 35 cm
Max span: 100.4 cm
Weight: 735 kg
Range: 42 km

To launch Exocet the ship's fire control system is normally linked to the ITS (*Installation de Tir Standard* or standard firing system). The ITS is assembled by ECAN Ruelle (part of the DCN organisation) under Aérospatiale as prime contractor, with several British sub-contractors. It is fitted in all ships armed with the original MM 38 version.

An updated digital version, known as ITL (*Installation de Tir Légère*) is in production for MM 40 installations, but can be used with MM 38 as well.

The French Navy uses a submarine-launched version, SM 39, which is similar to the air-launched AM 39.

DEVELOPMENT
Development started in 1967 by Aérospatiale in collaboration with French military research establishments. Although designed to meet French naval requirements, the first customer was the Royal Navy, which ordered 300 rounds in 1970, followed soon after by the French Navy. As a result of close collaboration between the two main customers some 17% of component manufacture was given to British companies.

STATUS
Development of MM-38 was completed in 1974.

By the end of 1987 more than 2600 rounds of Exocet in all its versions had been ordered by some 27 countries, among them being: MM-38 – Argentina, Belgium, Brazil, Brunei, Chile, France, West Germany, Greece, Iraq, Malaysia, Morocco, Nigeria, Oman, Peru and the UK. MM-40 – Argentina, Brazil, Colombia, Ecuador, Egypt, Malaysia, Morocco, Tunisia.

CONTRACTORS
System manufacturer: Aérospatiale, Division Engins Tactiques, 2 rue Béranger, 92322 Châtillon Cédex, France.

ADAC homing head: Electronique Serge Dassault.
Altimeter: TRT.
Radomes: British Aerospace.

INTERNATIONAL

4446.221
ANS ANTI-SHIP MISSILE

The ANS (Anti-Navire Supersonique) anti-ship missile project was initiated by Aérospatiale (prime contractor) and MBB as a replacement for the abandoned ASSM/ASEM programme (**3331.221**) to develop a successor within NATO for the Exocet, Kormoran and other anti-shipping weapons. The Franco-German programme is intended to result in a missile which will rely constantly on high speed (Mach 2+ during the whole flight) and high manoeuvrability (+10 *g*) to penetrate ships' defence systems in the future. It will have a range of 200 km. Other design requirements include long-range engagement without requiring monitoring or guidance participation by the operator. Programmable microprocessors will permit adaptation to evolving operational procedures and techniques, and the ability to be used with various types of ship or aircraft launch-platforms.

The ANS is an airbreathing ramjet-powered missile. The air-launched version will be accelerated by a boost motor integrated into the afterburner chamber of the ramjet. The ship-launched version differs solely in the laterally-mounted boosters. Ship-based ANS missiles will be launched from tactical containers arranged in clusters.

Aérospatiale as the prime contractor is responsible for the ANS system. On the basis of the industrial contract, MBB as the cooperating partner is producing the propulsion system with integral booster and a semi-armour piercing warhead. A lightweight helicopter-launched version, ANL (see **4450.321**), is under development as a joint private venture.

STATUS
The first flight took place in 1987 and a development contract was expected in 1988 with the intention of the system entering service in 1994.

An ANS being launched at a test range

CONTRACTORS
Aérospatiale, Divisions Engins Tactiques, 2 rue Béranger, 92322 Châtillon Cédex, France.

Messerschmitt-Bölkow-Blohm GmbH (MBB), Post Box 801149, 8000 Munich 80, Federal Republic of Germany.

1336.221
OTOMAT ANTI-SHIP MISSILE (FRANCE/ITALY)

Otomat is an anti-ship missile intended initially for launching from naval platforms of any size from fast patrol boat upwards, but also capable of land deployment (fixed or mobile). Weight is identical in both applications.

CHARACTERISTICS
Length: 4.46 m
Diameter: 46 cm
Wingspan: 1.35 m
Weight: 770 kg
Speed: Mach 0.9
Range: 60 km (effective); 180 km (max)
Warhead: 210 kg including 60 kg of explosive
Fuzes: Impact and proximity
Guidance: Inertial cruise, active radar homing
Propulsion: Solid boosters; turbojet cruise

The use of the 'fire-and-update' technique permits operational ranges of more than 160 km. Salvo firing achieves a similar improvement to the effective range, with no more than initial target designation ('fire-and-forget').

Otomat missiles are delivered in containers which also serve as launchers, the containers being mounted on prepared locations. Propulsion during cruise is by a Turboméca turbojet engine, the four air inlets for which are in the wing roots. Two lateral jettisonable boosters are attached for the launch phase. Target information can be derived from radar, active or passive search systems, or external data sources such as data link. The ability to gyro-angle the missile through $\pm 200°$ obviates the need for any ship manoeuvres at launch.

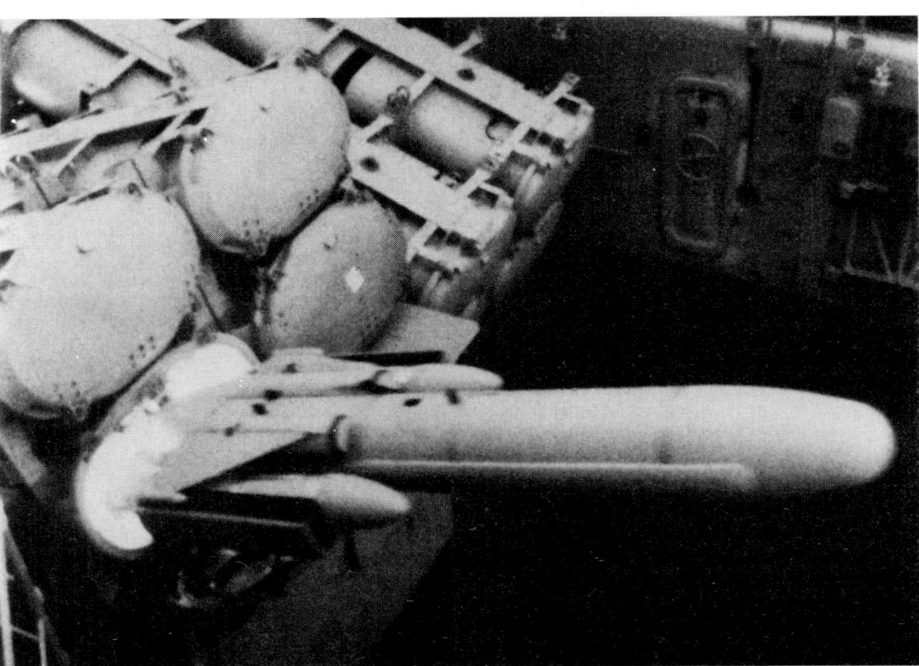
Firing of a Matra Otomat Compact from an 8-cell launcher installation

Since stemming from a common origin in a joint Franco/Italian collaborative development programme begun in 1969 (see below), the Otomat system has evolved in several distinct Italian and French versions. However, considerable commonality exists between the OTO Melara and Matra models. The following paragraphs summarise the main differences.

A new homing head produced by the Italian SMA (Segnalamento Marittimo ed Aero) company allows Otomat to follow a sea-skimming flight path up to the point of impact with the target.

A version of the Italian Otomat has been developed by OTO Melara, the principal feature of which is the provision for mid-course guidance up-dating. This version has been adopted by the Italian Navy in which Service it is designated Teseo (see below).

Subsequent to launching, the Otomat follows a cruise phase towards the target's predicted position, flying at low-level under radio altimeter control and autopilot guidance. An active homing head is used for the terminal phase.

The Otomat is compatible with any normal target designation radar and fire control system.

The Otomat as used by the Italian Navy employs a command system that provides over-the-horizon operation. The complete sub-system comprises a shipboard guidance set (TG1), a missile transponder and an aircraft transmitter (TG2). Both the shipboard and airborne elements are comprised of items from the Marconi Italiana PRT 400 Series mid-course guidance system.

The PRT401 comprises a high-gain directional antenna, stabilised in pitch and roll, and below-deck equipment. The latter includes all control command and tracking units, and a high-power microwave transmitter, a high-sensitivity receiver and a four-channel position plot data extractor, all interfaced with the system main computer. In the missile, the PRT402 equipment consists of receiver and transmitter antenna heads and a transceiver with RF sections and signal processor facilities. The PRT403 consists of two units, a high-gain rotating microwave antenna, and a transmitter, for mid-course missile guidance purposes.

There are several additional items in the PRT 400 series to extend the Otomat range of operational facilities. A lightweight shipborne guidance system, PRT 404 consists of a solid-state transceiver unit and a two-axis stabilised high gain antenna. The former contains a receiver, four-channel position data plot extractor and a klystron-powered transmitter, all interfaced with the system main computer. The PRT 404 is suitable for Otomat installations in hydrofoils, such as the Italian *Sparviero*, or fast patrol craft.

The PRT 405 is a helicopter tracking and two-way data link for use in Otomat midcourse guidance beyond the horizon. There are three separate units: transceiver, omni-directional antenna, and a display/keyboard panel. Total weight is under 14 kg and the system is thought to have been fitted in Italian AB212 helicopters.

A new Mk 2 folding-wing version of the Italian Otomat missile, and new smaller lightweight launchers were revealed publicly in 1984. At the same event (Mostra Navale, Genoa) a project to develop a system that will combine Otomat with the Ikara ASW missile system was disclosed.

The French Otomat Compact system also employs folding wings and a lightweight launcher, and this Mk 2 system uses the ERATO (extended range automatic targeting of Otomat) guidance system that can control up to eight or 16 missiles on four or eight launcher deck mountings. The main feature of this system is its value in utilising the full operational range of the Otomat missile by virtue of the update procedure. It allows salvo firing, with two successive missiles fired at three second intervals; several missiles in flight can be handled at the same time, attacking one or several targets.

In this technique the helicopter or maritime patrol aircraft measures the bearing and range of the surface target and transmits them by data link to the launch ship. Meanwhile the latter measures the bearing and range of the helicopter and computes the co-ordinates with respect to its own position. After firing, in the midcourse phase of missile flight, a fresh complete detection and calculation process is carried out and correction commands are sent to the missile by a telecommand transmitter. Two-axis active radar terminal homing is an advantage, permitting choice of sea-skimming or climb-and-dive attack profile.

DEVELOPMENT

Development was started in 1969 with OTO Melara and Matra working in collaboration. Manufacturers' tests were carried out from 1970 to 1974. The production of missiles began in 1975-76 for the Italian Navy and several other customers.

Milestones in the French development include:

1979: Qualification at sea on board the frigate *Le Basque*.

1980: First firing (in May) beyond 100 km in the 'fire and update' mode.

1983: In November final qualification tests at sea of Otomat Compact with the latest version of the ERATO ship-to-ship FCS for export sales completed with successful 100+ km firing from *Ile d'Oleron*.

STATUS

In series production in France and Italy. Operational with a number of navies.

It is not possible to identify the respective Italian and French production or sales figures, but the total number of missiles ordered by January 1987 was approaching 1000 rounds which will be deployed on more than 80 ships of various classes belonging to at least seven navies. There are also coastal batteries which use Otomat (**1337.121**). The Teseo/Otomat system is being procured for Egyptian, Iraqi, Italian, Kenyan, Libyan, Nigerian, Peruvian, Saudi Arabian, and Venezuelan forces.

In April 1985 a French Otomat Compact missile was given an offset launch under Saudi Arabian command at the French Mediterrannean Test Centre from the frigate *Madina*. The ERATO guidance system was employed with target designation from the ship's Dauphin helicopter and a moving target was hit at a range of 107 km.

CONTRACTORS

OTO Melara, La Spezia, Italy.

Engins Matra, 78-Vélizy, France.

Otomat launchers (top to bottom): standard Italian pattern, French Otomat Compact, Italian Mk 2 lightweight

ISRAEL

6019.221

GABRIEL SHIPBORNE SURFACE-TO-SURFACE MISSILE

Gabriel is a shipborne anti-ship missile developed in Israel and designed for installation in ships from about 50 tons upwards.

A sea-skimmer, the missile is transported in, stored in, and launched from hermetically sealed, reinforced fibreglass containers which also hold the missile launching rail and the hydraulically-operated container door. The missile is pre-adjusted and tested and requires no further tests or adjustments after installation. The standard launcher is a single fixed cell. An optional configuration is three missile cells mounted on a single, rotatable pedestal. The missile, is powered by a solid-propellant rocket motor.

A sophisticated guidance and homing electronics system is carried by the missile, which was developed in Israel.

After initialisation, using target data obtained from the launching vessel's own sensors, Gabriel is launched on a cruise flight path to the target area under twin-gyro platform plus radio altimeter control. The first part of the trajectory is flown at about 100 m followed by a descent to about 20 m for the major part of the flight. After target acquisition, and depending on sea state, the missile may descend to a still lower height for the terminal homing phase. Despite statements that semi-active radar homing is employed, it is thought that the basic terminal guidance mode is I/J-band (former X-band) active radar, together with additional optional modes such as 'home-on-jam', and optical plus command link. A frequency-agility capability in the radar homing system has been reported.

Latest Gabriel anti-ship missile seeker uses planar scanner and modular electronics (T. J. Gander)

A manual/optical alternative control system is available for use in conditions where radar guidance is not practical, provided the ship is fitted with the requisite sensors.

Three models of Gabriel have been produced, the Mk II having approximately double the range of the Mk I model (ie about 36 km) but in most other characteristics being the same or very similar. The Mk III model, however, has this longer range but also is larger and heavier, using active radar guidance, and has enhanced performance.

Three modes of operation have been outlined for Gabriel Mk III, known to the manufacturers as:

(1) Fire and forget (F & F)
(2) Fire and update (F & U)
(3) Fire and control (F & C)

In the F & F mode, target detection is by means of the launching ship's search radar and the missile is despatched to the target area to acquire and engage a designated target without additional data or actions after launch. Using the F & U mode, target data is updated by data link during the missile flight, this reducing uncertainty of target location, increasing hit probability and decreasing the effects of enemy ECM. For the final mode, F & C, the ship's fire control radar is used to provide more precise target data for use in guiding the Gabriel missile. Other ship's systems, eg optical fire directors, data automation systems, etc, may be linked with the Gabriel system in the F & C mode.

The missile control console is manned by one operator and incorporates: a PPI display of targets designated for engagement by Gabriel; search radar raw video PPI presentation; alphanumeric data display; target designation to missile facilities; and launch controls. A missile control unit (MCU) computes all pre-launch data and controls the firing sequence, and the automatic tracking computer (ATC) in the same unit computes track-while-scan data and designates targets to the FCS, using video signals from various sensors.

CHARACTERISTICS

Model	Mk I	Mk II	Mk III
Length	3.35 m	3.41 m	3.81 m
Diameter	34 cm	34 cm	34 cm
Wingspan	1.35 m	1.35 m	1.35 m
Launch weight	430 kg	520 kg	560 kg
Warhead	*c* 100 kg	*c* 100 kg	*c* 150 kg
Propulsion	Solid, 2-stage rocket		
Speed	Mach 0.7		
Range	20 km	36 km	36 km

Launch of Gabriel III anti-ship missile

STATUS

Developed in Israel during the 1960s, the missile is now operational in fast patrol boats of the Israeli Navy, and is in service with at least five other countries, according to Israeli sources.

Sales and fittings of Gabriel are not widely publicised but it is no secret that the Israeli Navy operates the system aboard the 'Sa'ar' and 'Reshef' class missile boats. Other operators (some officially unconfirmed) are Argentina, Malaysia, Singapore, South Africa and Taiwan. One of the latest foreign users of the system to be identified was Brazil, whose new corvettes are fitted.

The Mk II Gabriel became operational in the mid-1970s and the improved Mk III version entered service in 1979/80. In the course of 1982 an air-launched version was introduced and this is illustrated and described in entry **4529.321**.

Military parades in the city of Taipei in Taiwan have revealed a number of missiles having a marked similarity to Gabriel, mounted on the backs of military trucks. These are of Taiwanese manufacture, and were named Hsiung Feng (male bee). It has been stated that these were for coastal defence applications, but a number have been fitted to fast attack craft and retrofitted to old destroyers. The basic Gabriel design has been followed, but with a new seeker developed by the Sun Yat Sen Institute.

South Africa's Skorpioen (scorpion) system is a locally-manufactured version of Gabriel II. (see **4444.221**). Further details appear in entry **4442.121**.

In June 1985 it was disclosed that a long-range version of Gabriel was being developed to have an anticipated maximum range of 200 km. See entry **6187.221**.

CONTRACTOR

Israel Aircraft Industries Ltd, Ben Gurion International Airport, Israel.

6187.221

GABRIEL LONG RANGE ANTI-SHIP MISSILE

In June 1985 it was revealed that a long-range version of Gabriel is being developed. To achieve a maximum range of 200 km inertial mid-course guidance will be necessary, with updating through a data-link and active radar for terminal homing.

CHARACTERISTICS

Dimensions: 4.7 m × 440 mm
Wingspan: 1.6 m
Propulsion: turbojet cruise motor, solid-fuel booster
Warhead: HE

STATUS

Under early development, to be ready in the 1990s.

CONTRACTOR

Israeli Aircraft Industries Ltd, Ben Gurion International Airport, Israel.

NORWAY

1339.221

PENGUIN Mk 2 ANTI-SHIP MISSILE

Penguin is a surface-launched anti-ship missile primarily intended to provide small, fast naval craft with a powerful striking capability against larger vessels. It carries a 120 kg Bullpup ASM-N-7A warhead at Mach 0.8 over a range of 30 km. An impact fuze is fitted and the boost and sustainer motors are solid-propellant powered.

The missile is mounted in a simple box-launcher, delivered as a sealed and tested item from the base to the ship. Combined weight of box-launcher and missile is about 650 kg. Typical ship installations comprise six or more deck-mounted box-launchers. These are located on prepared mounts which incorporate an umbilical connection. When fitted, the missiles are ready for immediate firing and no onboard service or repair is performed. Automatic testing can be effected from the missile control panel in the operations room.

OPERATION

The Penguin missile employs inertial guidance for the cruise phase of its flight path, with passive infra-red homing to the target for the terminal phase. After launch the missile is thus independent of the parent vessel which is thereby freed to take evasive action or engage a further target.

Design requirements include missile range comparable to the maximum effective radar range for fast patrol boats and a passive mode of operation. Target detection, acquisition, and designation is therefore by means of the launch vessel's radar and fire control

A Penguin being launched by the Norwegian Storm Class FAC HNorMS Skudd

system. Upon target acquisition, the fire control computer calculates the bearing to the predicted point of impact and the missile inertial guidance system is automatically slaved to the Kongsberg SM-3 computer data. The missile is then fired in the general direction of the target. After launch, the missile follows a programmed trajectory toward the predicted impact area. This program can be varied.

Prior to acquisition of the target by the missile's infra-red homing system this operates in a search mode, scanning a sector ahead of the flight path. Upon detection of the target, the homing system tracks the target and generates signals which are used to direct the missile to the target.

The Norwegian operational requirements relating to the Penguin system call for a warhead capable of inflicting serious damage on a destroyer.

CHARACTERISTICS

Propulsion: 2-stage solid-propellant motor
Warhead: 120 kg semi-armour-piercing. Impact fuze
Length: 3.0 m
Diameter: 28 cm
Span: 1.42 m
Launch weight: 360 kg
Weight with box-launcher: 700 kg
Cruising speed: Mach 0.8
Range: Mk 1, 20 km; Mk 2, 30 km (max); 2.5 km (minimum)

DEVELOPMENT

Development was initiated in the early 1960s at the Norwegian Defence Research Establishment as a response to an assessment of Royal Norwegian Navy requirements. These two bodies and A/S Kongsberg Våpenfabrikk have worked in close co-operation since the inception of the programme. All Mk 1 missiles now brought up to Mk 2 standard.

Although originally designed for deployment in smaller naval craft, the Penguin system is well suited to use in larger vessels. It is also available for use in coastal defence and from helicopters. In the latter case, only minor modifications are stated to be necessary. Additionally, a redesigned Mk 3 version without the booster motor and with a smaller wingspan are in production for use from jet fighters such as the F-16.

STATUS

Operational in 'Storm', 'Snögg', and 'Hauk' class patrol boats and 'Oslo' class frigates of the Norwegian Navy, and in 'Kartal' class patrol boats of the Turkish Navy. Penguin is also fitted in Royal Swedish Navy 'Hugin' class patrol boats as the RB 12, and in 'Combattante IIIB' fast patrol boats of the Hellenic Navy.

The USN has completed a full scale technical and operational evaluation at the Naval Weapons Centre, Dahlgren Laboratory and at Noto, Cape Canaveral. Penguin installations are being tested for US Navy ships and helicopters as well as for the US Coast Guard.

CONTRACTOR

Norsk Forsvarsteknologi A/S, PO Box 1003, N-3601 Kongsberg, Norway.

A Penguin missile in its launch-container

SOUTH AFRICA

4444.221
SKORPIOEN ANTI-SHIP MISSILE

Skorpioen is the name under which the South African Navy operates the Israeli Gabriel Mk II anti-ship missiles (**6019.221**) which arm the 'Minister' class fast attack craft serving with the SAN. These vessels are similar to the Israeli 'Sa'ar' class and three of them were constructed in Haifa, reaching South Africa in July 1978. Launchers are made in South Africa, and a number of components, but the extent to which the SAN is independent of Israeli sources is not known. For details of operation, performance etc refer to the entry for Gabriel.

STATUS

Successful trials were carried out in February 1987 from a 'Minister' class FAC.

SWEDEN

3976.221
SAAB RBS 15 ANTI-SHIP MISSILE

The RBS 15 is a new-generation anti-ship missile system suitable for use by aircraft, coast defence batteries and naval vessels. The first installation has been completed aboard a Royal Swedish Navy *Spica* class fast patrol boat. They are housed in launcher/containers. The air-launch model (**3975.321**) programme is following.

An air-breathing propulsion system provides long range, with a high subsonic cruise speed at sea-skimming height. Two solid-propellant booster motors aid launch, but these will not be required in the air-launch version. The mission profile includes a programmable trajectory with a sea-skimming final phase.

The RBS 15 has a fire-and-forget capability and is provided with an ECM-resistant homing head made by Philips Elektronikindustrier (PEAB). Pre-launch operations are minimal and all calculations are performed automatically by an on-board computer. A radar altimeter ensures safe cruising at low level.

Eight launcher/containers for RBS 15 anti-ship missiles are fitted on the stern of the patrol boat Halmstad

Launch of RBS 15 missile from Swedish patrol boat HSwMS Piteå

CHARACTERISTICS

Length: 4.35 m
Diameter: 50 cm
Wing span: 1.4 m; 85 cm (folded)
Launch weight: 770 kg
Weight: 600 kg (without boosters)
Guidance: Probably programmed autopilot cruise with height hold, followed by active radar homing
Range: 150 km (estimated)
Cruising speed: >Mach 0.8

STATUS

The system is operational aboard 12 Swedish *Spica* type missile boats and two *Stockholm* class corvettes. The initial contract worth SK 600 million for the development and production of the naval version was awarded by the Swedish Defence Material Administration in July 1979, and in June 1982 a contract was awarded for the air-launched version.

The first export order for RBS15 was placed in March 1983 when a contract for the system to be fitted in new *Helsinki* class fast patrol boats for the Finnish Naval Forces was signed. Additional orders have been made by both the Finnish and the Swedish Navies.

CONTRACTORS

SAAB Missiles AB, 58188 Linköping, Sweden.

TAIWAN

4443.221
HSIUNG FENG ANTI-SHIP MISSILE

This is a licence-built Taiwanese version of the Israeli Gabriel anti-ship missile. As described in the entry for the coastal defence application of this weapon (**4442.121**) the many similarities between the Gabriel and the Hsiung Feng include the triple launcher which originally was a noteworthy feature of the Israeli system, and therefore the main performance characteristics and other details are assumed to be essentially identical. The major difference is in the guidance system. This has been replaced by a new seeker from the Sun Yat Sen Institute, probably redesigned to suit Taiwanese manufacturing methods.

STATUS

In production since 1980. Hsiung Feng anti-ship missiles are fitted aboard Taiwanese destroyers of the US 'Gearing' and 'Allen M Sumner' classes, in triple launcher configuration, and in single launchers aboard two classes of fast attack craft. In many installations the missiles are associated with Selenia Orion RTN-10X fire control radars. A Hsiung Feng II with 80 km range has been reported.

CONTRACTOR

Chung Shan Institute of Research and Technology.

Hsiung Feng anti-ship missile containers aboard a Taiwanese destroyer. Note 'Fletcher' *class destroyer in background with quadruple Sea Chaparral air defence missile launcher*

UNION OF SOVIET SOCIALIST REPUBLICS

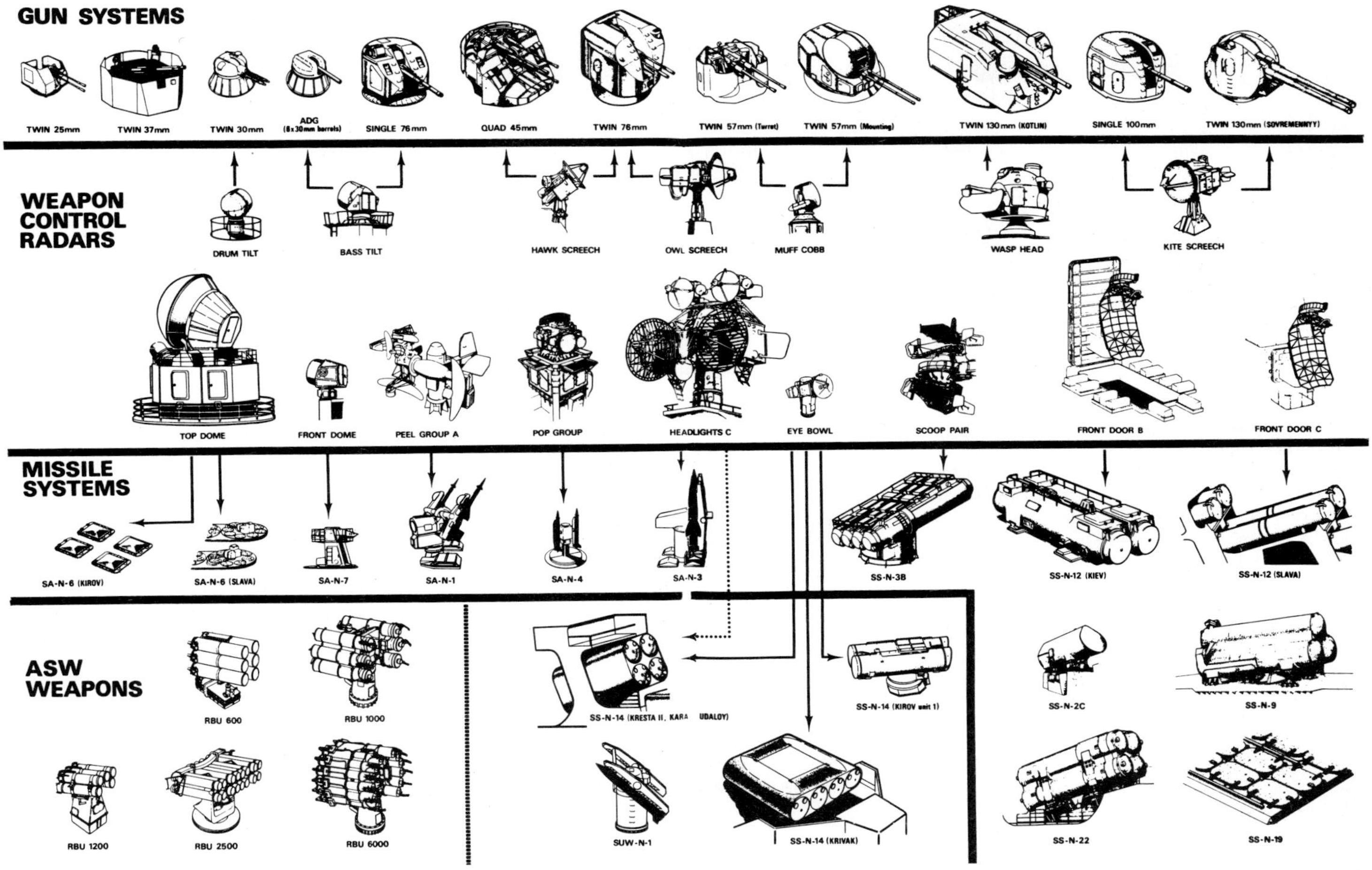

GUN SYSTEMS

TWIN 25mm, TWIN 37mm, TWIN 30mm, ADG (6x30mm barrels), SINGLE 76mm, QUAD 45mm, TWIN 76mm, TWIN 57mm (Turret), TWIN 57mm (Mounting), TWIN 130mm (KOTLIN), SINGLE 100mm, TWIN 130mm (SOVREMENNYY)

WEAPON CONTROL RADARS

DRUM TILT, BASS TILT, HAWK SCREECH, OWL SCREECH, MUFF COBB, WASP HEAD, KITE SCREECH, TOP DOME, FRONT DOME, PEEL GROUP A, POP GROUP, HEADLIGHTS C, EYE BOWL, SCOOP PAIR, FRONT DOOR B, FRONT DOOR C

MISSILE SYSTEMS

SA-N-6 (KIROV), SA-N-6 (SLAVA), SA-N-7, SA-N-1, SA-N-4, SA-N-3, SS-N-3B, SS-N-12 (KIEV), SS-N-12 (SLAVA)

ASW WEAPONS

RBU 600, RBU 1000, SS-N-14 (KRESTA II, KARA UDALOY), SS-N-14 (KIROV unit 1), SS-N-2C, SS-N-9, RBU 1200, RBU 2500, RBU 6000, SUW-N-1, SS-N-14 (KRIVAK), SS-N-22, SS-N-19

Soviet warship weapon systems

1155.221
SS-N-2 (STYX) SHIPBORNE SURFACE-TO-SURFACE MISSILE

The SS-N-2 Styx anti-ship missile, whose Soviet designation is reported to be P-15, is a short/medium range weapon which is very widely deployed by the Soviet Fleet and the navies of many of its allies. Entry into service was in 1959 or 1960, but despite the length of service this indicates, new Styx installations are apparently still being made. This weapon also has the distinction of being the first of its type to have been used operationally, in the 1967 war between Israel and Egypt, and later by Indian ships against Pakistan vessels in December 1971. More recently it has apparently been given a new lease of life by its employment in the coastal defence role with an entirely new ground system (**4890.121**) with East German, Syrian and Yugoslav forces.

The general configuration is that of a small aircraft, with a delta planform wing and a triple tail-surface arrangement. A jettisonable booster rocket is used for the launch and acceleration phases, after which an internal motor sustains a cruising speed of about Mach 0.9. Range of early models is estimated as about 20 nm (40 km) maximum.

Three distinct versions have been reported, identified as the SS-N-2a, b and c models, respectively. Soviet designations applied to this series of Styx missiles include 4K-40 and P-15. The main known difference between the first two is that the SS-N-2a has fixed wings while those of the SS-N-2b fold, and this variation is reflected in the housing for the two different models. Less is known of the SS-N-2c apart from its performance characteristics, which are thought to include an increased range of some 80 km and a sea-skimming terminal trajectory. The latter can be preset to a limit of 300 m, and a third party (eg helicopter or fixed-wing aircraft) may be employed for guidance updating at extended ranges. The SS-N-2c is also credited with a heavier warhead.

During the relatively long service life of Styx it is probable that several guidance modes have been employed, and references probably relate to various combinations of the following alternatives: (1) the cruise phase could be carried out under either autopilot or radio command guidance; (2) the terminal phase could rely upon continuation of command guidance, active radar, or infra-red homing. Over the years, the launcher/hangars associated with Styx have undergone progressive changes and it seems reasonable to assume that the guidance apparatus also has received periodic updating, in which case all of the above methods may be employed in different installations.

The three types available with the Chinese-built version (HY-2, **3973.221** and **3984.121**) probably represent the methods most favoured by the originators and other operators of the system, and readers are referred to those entries for additional data.

CHARACTERISTICS

	SS-N-2a	SS-N-2b	SS-N-2c
Length	5.8 m	5.8 m	6.55 m
Diameter	750 mm	750 mm	750 mm
Weight	2300 kg	?	?
Warhead	400 kg HE	400 kg HE	450 kg HE
Range	40 km	40 km	80 km
Speed	high subsonic	high subsonic	high subsonic

STATUS

Styx has been in operational service in its SS-N-2a version since 1959-1960 and is now widely deployed with Soviet Navy forces and those of numerous friendly states. Most widespread are the fittings on board the 'Osa' and 'Komar' class missile boats. The SS-N-2b model appeared in 1965.

Since the appearance of SS-N-2 missiles on 'Osa' and 'Komar' boats, there have been fittings on a number of other classes of ship, both Soviet and foreign. A later class of 'Osa' missile boat also appeared ('Osa II') on which a new missile was deployed. This was known as the SS-N-11 but is now designated SS-N-2b.

A total of more than 150 'Osa' class missile boats have been transferred to foreign navies, including those of: Algeria, Angola, Bulgaria, Cuba, Egypt, East Germany, Ethiopia, Finland, India (MR 301 fire control system), Iraq, North Korea, Libya, Poland, Romania, Somalia, Syria, South Yemen, Vietnam and Yugoslavia.

The People's Republic of China is another important user of SS-N-2 where it is produced as the HY-2 after having been supplied with initial quantities of Styx missiles and 'Komar' and 'Osa' missile boats. In addition, China seems to have been the first nation to add Styx to the armament of larger vessels, the 'Luda' class of destroyers. SS-N-2c has since been retrofitted to several older Soviet destroyers. (See also **3984.121** and **3973.221** for Chinese applications of the SS-N-2.)

Yugoslav Styx installation/maintenance in progress reveals details of missile handling and wing folding points

'Osa II' class missile boat of the Soviet Fleet showing the four SS-N-2b container/launchers (Novosti)

2976.221
SS-N-3 (SHADDOCK) SHIPBORNE SURFACE-TO-SURFACE MISSILE

Shaddock, whose Soviet designation is reported to be P-6 (SS-N-3a) or P-7 (SS-N-3c), is the largest of the Soviet cruise missiles. It is not known to what extent, if any, the missile that is installed in many surface vessels and submarines of the Soviet Navy differs from the Shaddock missile displayed (in its container) at Moscow parades and used by Soviet land forces (NATO code-name Sepal, **2975.121**), whose Soviet designation is reported to be P-35 but it is generally assumed that there are some differences. Moreover it appears that there are several different missiles of the Shaddock type in naval use. These have been designated SS-N-3a, SS-N-3b and SS-N-3c, while versions or derivatives fitted in later classes of ship have been numbered SS-N-12 in the American sequence and given the NATO name 'Sandbox'.

Command guidance is used, and for surface vessel installations the missile is tracked by Scoop Pair radar (**1324.253**) and course corrections transmitted to it by radio. Soviet technical literature has described methods of terrain following for cruise missiles, without referring to any specific missile but it is quite probable that a measure of terrain following capability is provided for Shaddock in its overland role. This relies upon a radio altimeter which would also be useful in the anti-ship uses of Shaddock. For the terminal phase it is believed that infra-red homing was used in earlier models, but later versions use an active radar homing head.

Missile speed is believed to be transonic and range is limited mainly by radio/radar horizons. With inertial mid-course guidance the maximum achievable range is believed to be at least 450 km, but practical ranges are nearer to 180 km for missiles launched from cruisers and less than that for submarine-launched missiles.

One of the most interesting aspects of this weapon system is the extent to which it has been installed in submarines, and the length to which Soviet engineers have gone to find the most satisfactory form of installation. The most elaborate is the four-tube arrangement in the 'Whiskey Long Bin' class submarine. Other boats of this class have what is known as the 'Twin Cylinder' fitting, with two missile containers attached to the outer casting.

Close-up reveals massive proportions of quad Shaddock launcher aboard Kynda-class cruiser 'rozGrozny' (G. Jacobs)

CHARACTERISTICS

	SS-N-3a	SS-N-3b	SS-N-3c
Length	10.2 m	10.2 m	11.75 m
Range	460 km	460 km	735 km
Propulsion	turbojet + two solid-propellant boosters		
Warhead	1200 lbs conventional or nuclear		
Guidance	inertial		

STATUS

Shaddock-type missiles are operational in 'Kresta I' and 'Kynda' class missile cruisers and in submarines. The nuclear-powered 'Echo II' boats carry eight missiles in pairs, the launchers being faired into the casing so as to present a smooth surface. The 'Juliett' class diesel-electric boats have similar arrangements for four missiles.

The 'Kynda' class cruisers carry quadruple launchers fore and aft, with additional missiles reloadable from superstructure magazines. The 'Kresta I' launchers and all the submarine launchers are elevated for firing, but cannot be trained; only the 'Kynda' launchers can be trained through about 250° and elevated to about 30°.

Only 19 of the 29 'Echo II' class, as well as the 'Juliett' and 'Whiskey' class submarines are believed to be armed with the 'A' version, while the 'Kresta I' and 'Kynda' classes have the 'B' model. The SS-N-3c strategic variant was deployed in the five 'Echo I' SSGNs.

Shaddock is also deployed in a land-mobile role as a coastal defence weapon under the NATO designation 'Sepal' (**2975.121**).

Shaddock launchers on a 'Kresta I' guided missile armed destroyer (USN)

2987.411
SS-N-7 (STAR BRIGHT) SUBMARINE-LAUNCHED CRUISE MISSILE

The SS-N-7 was the first Soviet under water launched anti-ship missile when it became operational in 1971. It is said to have a speed of Mach 0.9 and to cruise on autopilot after being launched from angled tubes housed in the external casing of the submarine. Target data is presumably supplied from other sources, either ships or aircraft.

CHARACTERISTICS
Weight: Approx 3375 kg
Length: 7 m
Propulsion: Solid-fuel rocket
Range: 55-64 km
Guidance: J-Band radar homing
Warhead: 500 kg (nuclear or HE)
Max speed: Approx 0.9 Mach

STATUS
Embarked in the 12 'Charlie I' class missile submarines (eight per boat) but now superseded in the inventory by the SS-N-9.

1760.221
SS-N-9 (SIREN) SHIPBORNE AND SUBMARINE LAUNCHED SURFACE-TO-SURFACE MISSILE

Some confusion exists between this missile and the SS-N-7. It first appeared in 1968-69 in the 'Nanuchka' class corvettes, associated with the Band Stand radar, but was then reported to have replaced the SS-N-7 in later 'Charlie' class submarines, and the prototype 'Papa' boat.

The original estimate of range has been revised downward from about 275 km to roughly 110 km, but a normal operating range of about 75 km seems likely.

CHARACTERISTICS
Weight: Approx 2950 kg
Length: 8.84 m
Propulsion: Solid-fuel rocket
Range: 110 km
Guidance: IR + active radar homing (mid-course guidance required for maximum range)
Warhead: 500 kg (nuclear or conventional)

STATUS
'Nanuchka' class corvettes, 'Sarancha' class hydrofoils, 'Charlie II', 'Charlie III' and 'Papa' class SSGNs are reported to have versions of the SS-N-9, but 'Nanuchkas' supplied to India and other customers are armed with the SS-N-2c instead.

Soviet 'Nanuchka' class missile corvette. Triple launchers on each side of bridge structure house SS-N-9 surface-to-surface missiles. Also to be seen is the lid of the retractable launcher for the SA-N-4 anti-aircraft missile system on the foredeck (Novosti)

4445.221
SS-N-12 (SANDBOX) ANTI-SHIP MISSILE

The designation SS-N-12 (NATO code name Sandbox) has been applied to the missile system that is housed in the new style of launcher noted aboard the Soviet 'Kiev' class vessels. This is generally presumed to be an improved version of the SS-N-3 Shaddock (**2976.221**), and to have a similar purpose, namely long-range anti-ship engagements. An estimated maximum range of 550 km has been suggested. So far there are no known publicly released photographs of the actual weapon housed in the SS-N-12 launcher/container so that there is little firm evidence on which to base speculative assessments of the performance. However, its considerable size can be gauged from the external appearance of the launchers and from this it seems reasonable to infer considerable payload/range characteristics. Autopilot/inertial guidance, probably aided by a degree of command updating at intervals and active radar terminal guidance would appear to be the most likely form of operation. The relatively long range assumed would also suggest that facilities for mid-course guidance by a co-operating aircraft would be a useful capability.

This last feature would be particularly valuable in the case of Sandbox engagements which involve launches of SS-N-12 missiles from submarines, which would be less well equipped than major surface vessels to provide target designation.

Four twin SS-N-12 launchers aboard Soviet carrier Minsk. Note missile handling crane (G. Jacobs)

CHARACTERISTICS
Length: 11.7 m
Propulsion: Turbojet
Range: 550 km
Guidance: Radio command and active radar homing
Warhead: 1000 kg nuclear or conventional

STATUS
The SS-N-12 Sandbox is deployed aboard aircraft carriers of the 'Kiev' class, with eight launcher containers on the foredeck. The system has also been reported in at least 10 'Echo II' class submarines. The 'Slava' class missile cruiser was added to the list when the first of class appeared with 16 launchers in 1982.

4196.221

SS-N-19 (SHIPWRECK) ANTI-SHIP MISSILE

SS-N-19 is the designation of the vertically-launched anti-ship missile system deployed in the Soviet battle cruisers *Kirov* and *Frunze* and also in 'Oscar' class submarines. Both classes first appeared in 1980, and the former type has 20 launch tubes, while the submarines are able to launch 24 missiles. They could well be intended to operate together (with other vessels) as the nucleus of an 'ocean superiority' task force able to challenge western fleets over wide areas. The *Kirov* and the 'Oscar' class submarines are nuclear powered.

Very few details of the SS-N-19 missile system have been disclosed by official sources in the West (still less in Warsaw Pact states) but estimates that have been made are summarised here.

The main role is that of an anti-ship weapon, with a range of up to 550 km flying at supersonic speed (at least Mach 2.5 has been quoted, but Mach 1+ is more likely). Initially it was pointed out that no radar system associated with the SS-N-19 has been identified (on either 'Oscar' submarines or the *Kirov*), and with a range of 550 km such a sensor could be of only partial relevance, due to the line-of-sight range limitations of radar. However, later analyses have claimed an association between the SS-N-19 Sandbox and a radar code-named Trap Door, which is located in the deckhouse immediately aft of the RBU 6000 anti-submarine rocket launcher in the bow of the ship. The precise functioning of this radar is not clear as it is certainly not sited for maximum line-of-sight range performance. Possibly some form of initial 'missile' capture after vertical launch is received to ensure the missile adopts the proper approach heading to the target. Thereafter it seems probable that inertial guidance, allied to an external target detection and designation system (satellite or aircraft, for instance), with self-contained homing is the likely guidance technique. Among the imposing array of sensors and antennas displayed by the two *Kirov* class, it is more than likely that one or other is capable of providing data link or command link facilities for an outgoing SS-N-19 missile during its mid-course phase, for such purposes as target location updating or missile course correction. Clearly submarines have much less capability in this respect, but if the *Kirov* and *Frunze* and 'Oscar' class submarines are intended to operate together, there is no reason why the submarines should not be employed merely as launch platforms with the surface ship (and its better sensor and communications facilities) acting as the directing and controlling element. It should be noted that the *Kirov* and *Frunze* are well provided for helicopter operations; it is estimated that four or five Ka-25 helicopters could be accomodated, possibly including both 'A' anti-submarine and 'B' missile guidance versions.

The Soviet battle-cruiser Kirov, *despite an impressive multiplicity of sensors of all sorts appears at first glance to have little in the way of armament. This impression is aided by the fact that two of the latest Soviet weapon systems fitted in this ship (SS-N-19 and SA-N-6) are vertical launch systems and spend most of their time below decks* (RN)

CHARACTERISTICS

Propulsion: Turbojet

Range: 445 km+

Guidance: Intertial with terminal radar homing

Warhead: Conventional or nuclear

STATUS

Main armament of large cruisers *Kirov* and *Frunze*, 'Slava' class cruisers and 'Oscar' class SSGNs.

4761.221

SS-N-22 (SUNBURN) ANTI-SHIP MISSILE

The designation SS-N-22 has been assigned to the missiles housed in the container/launchers noted aboard the recent Soviet *Sovremenny* and 'Tarantul II' class ships which made their appearance since 1981. The former carries a total of eight launchers (in two groups of four near the bridge), and the 'Tarantul II' missile corvettes are fitted with two twin launchers.

The assumed role is that of a naval surface-to-surface weapon with a main operational mission of anti-shipping engagements, and until more definite evidence is forthcoming it seems reasonable to accept the general conjecture that the missile itself is a developed version of the SS-N-9 Siren (**1760.221**) first noted in 'Nanuchka' class ships. The first of the two new types to make its appearance (*Sovremenny*) is also fitted with the Band Stand radar associated with the SS-N-9 on the 'Nanuchka' ships, which tends to support the suggestion of a common lineage for the SS-N-9 and the SS-N-22.

The presumed range of the SS-N-22 is in the 100 km category, but this suggestion probably relates to the missile's autonomous range using own guidance and the shipboard equipment. Sovremenny class ships, however, have an embarked helicopter that could be employed to provide extended range and this might also be used to facilitate multiple-target engagements.

No estimates of missile speed or payload are possible from the extremely limited information so far available, and for similar reasons conjecture about its appearance/configuration and other characteristics is an equally unrewarding exercise.

CHARACTERISTICS

Propulsion: Solid-propellant rocket

Range: Approx 110 km

Guidance: Mid-course update with terminal radar homing

Warhead: Conventional or nuclear

IOC: 1981

STATUS

Believed operational in at least four 'Sovremenny' class and one 'Tarantul II' class ship.

UNITED KINGDOM

4447.221

SEA EAGLE SL ANTI-SHIP MISSILE

Formerly known as the P5T the Sea Eagle SL is a proposed ship-launched version of the Sea Eagle aircraft-launched anti-ship missile for India (for details see entry **3630.321** in the Air-to-Surface Missiles section of this book). It is intended for the arming of ships from 200 tonnes upwards and is capable of engaging a target well beyond the horizon. The system comprises missiles, launch boxes and fire control equipment.

Although not chosen by the RN as the successor to the Exocet MM-38, the system is still available for export. Only slight modification is needed to produce a helicopter version.

STATUS

Development only (private funding). The first firing of a ship-launched version of Sea Eagle was successfully carried out from a box launcher on the Aberporth, UK range in March 1987.

CONTRACTOR

British Aerospace plc, Dynamics Division, PO Box 5, Filton, Bristol BS12 7QW, England.

A ship-launched Sea Eagle being launched from a box launcher in March 1987

4553.221

SEA SKUA SL ANTI-SHIP MISSILE SYSTEM

The Sea Skua ship launched (SL) anti-ship missile is a variant of the Sea Skua helicopter air-to-surface missile (see entry **1530.321** in the Air-to-Surface Missiles section) intended for use on naval vessels as small as 20 metres long. Sea Skua SL comprises the Ferranti Seaspray surveillance and illuminating radar or equivalent, a control console and deck-mounted launch boxes for the missiles.

The control console is operated by one man and would normally be located in the vessel's tactical control centre or on the bridge. The deck launcher is a lightweight unit mounted on the deck and holding one or more launcher boxes for firing. The launcher boxes can be treated as conventional ammunition rounds and double as transport containers. A two-box launcher, with missiles, weighs 850 kg and would be 2.5 metres long and 1.1 metres wide. The control console weighs 140 kg.

In operation Sea Skua SL uses the radar to locate and later illuminate the target. To commence an engagement the console operator selects the missile to be fired (ripple firing is possible) and sets the missiles terminal sea skimming height. When armed and ready the missile can be launched. During the boost phase the azimuth and height guidance loops are closed and the homing head locks on to the target illumination radar returns. When the boost phase is complete the sustainer motor maintains flight and the missile descends to an interim sea skimming height under radio altimeter control. Later in the flight path the missile descends still further to the terminal sea skimming height as selected before release. Four heights can be selected by the operator. The homing head then guides the missile in azimuth to target impact.

CHARACTERISTICS

Length: 2.5 m
Body diameter: 0.25 m
Wing span: 0.72 m
Speed: High subsonic
Range: Up to 15 km

STATUS

Possible procurement of the system by India has been reported.

CONTRACTOR

British Aerospace plc, Dynamics Division, PO Box 5, Filton, Bristol BS12 7QW, England.

Artist's impression of ship-launched version of Sea Skua

UNITED STATES OF AMERICA

4197.221
TOMAHAWK SEA-LAUNCHED CRUISE MISSILE (BGM-109)

The General Dynamics BGM-109 Tomahawk is the USN's long-range cruise missile. A predominantly tactical role is currently envisaged for Tomahawk by the USN, though sight should not be lost of the fact that Tomahawk has been selected for use by the USAF in NATO for the Ground-Launched Cruise Missile (GLCM) role (**4194.111**) as a major element in the NATO Theatre Nuclear Force modernisation programme.

Studies for the USN Tomahawk Sea Launched Cruise Missile (SLCM), designed for launch from submerged submarines and surface ships, were started in 1972. For submarine use the missile can be fired from a torpedo tube, while for surface ship applications armoured box launchers on deck or vertical launching systems can be employed. The Tomahawk family of cruise missiles incorporates a high degree of commonality in the basic airframe and major system components. Ahead of the mid-section on which are mounted the 'flip out' wings, a forward body contains the warhead and the guidance section, while the rear section houses the turbofan jet engine. A rocket booster motor is attached to the tail cone in the case of the SLCM and GLCM versions to propel the missiles from their tubes and accelerate them to flying speed and altitude where the turbofan can be started to sustain flight.

The SLCM is designed to be able to carry out both land attack and anti-ship missions using either nuclear or conventional high explosive warheads, and there are alternative guidance systems for each mission. These consist of the original advanced TERCOM (terrain contour matching) system which is used for both the land attack SLCM and the USAF GLCM versions of Tomahawk, and a specially modified version of the Harpoon (**2641.221**) missile radar guidance system for use in the anti-ship version of the SLCM. The latter enables Tomahawk to be launched in the general direction of the target area to which it flies at low altitude to avoid detection by hostile radars, and at a pre-programmed distance the missile's active radar is switched on to detect and acquire the target ship. The TERCOM system used in the land attack version of Tomahawk uses inertial navigation to guide the missile to the target using the known location of the launch vehicle and that of the target, this being inserted immediately prior to launch. A pre-programmed flight path is followed to the target, with the terrain profile characteristics encountered by the missile along its flight path being compared automatically by the TERCOM equipment with computer-stored terrain data covering pre-selected segments of the route to confirm and adjust the Tomahawk's flight path.

For submarine launching, Tomahawk is loaded into a stainless steel capsule which provides protection during handling and underwater launch. At the moment of firing, the SLCM is ejected from the capsule and the torpedo tube. The boost motor propels the missile through the water and to the surface, and once in the air the boost motor burns out and is jettisoned. The turbofan cruise motor then sustains flight for the remainder of the mission.

CHARACTERISTICS

Length: 6.4 m
Diameter: 53 cm
Wing span: 2.61 m
Cruise engine: Williams International Corp 275 kg thrust turbofan
Boost motor: Atlantic Research Corp 3170 kg thrust solid-propellant rocket

A US Navy Tomahawk missile approaches a concrete building, penetrates the building and explodes. The missile was launched from a submerged submarine and flew nearly 400 miles to its target

Range: 2500 km (land attack version); 450 km (anti-ship version)
Cruising speed: 885 km/h
Guidance: TERCOM-aided inertial (land attack); modified Harpoon active radar homing (anti-ship)
Warhead: HE or nuclear (land attack); HE (anti-ship)

STATUS

In January 1981 the DoD stated that full-scale engineering development of the Tomahawk land attack missile with conventional HE warhead (TLAM/C) was going ahead for use aboard USN nuclear attack submarines, and the potential of digital scene matching area correlation terminal guidance techniques, allied to a conventional warhead, offered a promising method of attacking fixed targets on land. Operational objectives for this variant, which will be deployed on nuclear attack submarines and surface ships, were to provide naval forces with a long-range cruise missile capable of attacking and neutralising enemy facilities and to degrade their defensive capabilities without resorting to nuclear weapons. The TLAM/C has a CEP of 72 m over a range of 1300 km.

In July 1981 a BGM-109C Tomahawk cruise missile completed the first submarine-launched strike against a land target, located about 480 km from the launch position off the Californian coast. The target was at the Tonopah Test Range in Nevada, and it was reported to have been hit by the inert warhead carried by the Tomahawk, which was fully guided throughout the flight. Well over 50 previous test and evaluation flights of Tomahawk missiles, of various models, had been carried out successfully prior to this flight.

Readers are also directed to entry **3993.001** in the Strategic Weapon Systems section for further information and background on US cruise missile activities.

DoD plans to procure and deploy the missile in large numbers. The programme for fiscal years 1984-88 requested funding for procurement of 1861 missiles. Projected procurement for 1984 and 1985 was 124 and 353 missiles, respectively.

In 1982 McDonnell Douglas became a co-producer of the Tomahawk, when the Joint Cruise Missiles Project Office awarded McDonnell Douglas and General Dynamics contracts to exchange Tomahawk technology, in March 1982.

In November 1985 the USS *Pittsburgh* (SSN-720) became the first submarine in commission with VLS Tomahawk, followed shortly afterwards by the *Providence* (SSN-719).

Eight Armoured Box Launchers (ABLs) with four Tomahawks each have been installed aboard the battleships *New Jersey*, *Iowa*, and *Missouri* and *Wisconsin* will follow. Two ABLs each are being installed in the seven nuclear cruisers of the 'Virginia', 'California' and 'Long Beach' classes, along with seven of the 'Spruance' class destroyers. VLS Tomahawk will be installed in the remaining 24 'Spruance' class (37 missiles per ship), the later 'Ticonderoga' class cruisers (19 missiles), the new 'Arleigh Burke' class destroyers (16 missiles), and the later 'Los Angeles' class attack submarines (12 missiles). Older submarines will receive four tube-launched Tomahawks each: the 'Narwhal' (SSN-671), 'Glenard P Lipscomb' (SSN-685), 37 'Sturgeon' (SSN-637) class and the 31 earlier 'Los Angeles' (SSN-688) class.

The first five 'Aegis' cruisers *Ticonderoga*, *Yorktown*, *Vincennes*, *Valley Forge* and *Thomas S Gates* (CG-47 to 51) cannot launch Tomahawk from their Mk 26 launchers, so consideration is being given to retrofit VLS at some future date.

In September 1987 a submarine-launched Tomahawk successfully demonstrated the terminal dive mode of attack against a reinforced concrete structure after a 640 km (400 nm) flight.

CONTRACTORS

General Dynamics Corporation, Convair Division, 5001, Kearny Villa Road, San Diego, California 92123, USA.

McDonnell Douglas Corporation, Saint Louis, Missouri 63166, USA.

2641.221

HARPOON SHIPBORNE SURFACE-TO-SURFACE MISSILE (RGM-84A)

Harpoon (RGM-84A) is a high-subsonic anti-ship tactical cruise missile operational with the USN.

This missile is officially regarded as the principal anti-ship weapon and has been designated for launch from P-3 and A-6 aircraft, 'Iowa' class battleships, 'Knox' class frigates, all destroyers and guided missile cruisers, 'Perry' class frigates, PHM patrol hydrofoils, and nuclear attack submarines. The airborne version is described elsewhere in this book (**1301.221**) and the following description relates primarily to the shipborne version. The submarine version was described in **2441.421** in 1976 and earlier editions, where details of the alternative launching arrangements will be found.

An all-weather system, Harpoon was designed to achieve operational flexibility and adaptability to existing systems. The main body of the missile with its cruise-phase propulsion and guidance systems, homing and terminal manoeuvring systems, ECCM facilities and warhead is common to all applications. The aerodynamic surfaces of this part of the system exist in several forms for compatibility with various aircraft, shipboard, and submarine launchers. All surfaces are designed for quick attachment, are interchangeable, and in some cases are folded for storage.

For all other than airborne launchings an additional boost section contains a solid-propellant boost motor. This propels the missile on a ballistic trajectory which it follows until the booster separates; after which the missile descends to a low cruise altitude, determined by its altimeter, and flies to the target under the power of its turbojet cruise engine. In the terminal phase the missile may execute manoeuvres to evade close-in enemy defences and enhance the effectiveness of its warhead. Aerodynamic control throughout is by four movable fins on the main missile body.

Harpoon four-tube Mk 141 launcher (W. Donko)

A new weapon control system for Harpoon AN/SWG-IA(V), is produced by McDonnell Douglas, with Sperry as subcontractor. Targeting data provided by shipboard systems are interfaced with the missile through this command and launch system. The Harpoon data processor, a general-purpose digital computer, receives targeting and attitude data from standard shipborne equipment (or from a third party for over-the-horizon operations) and computes the necessary missile and launcher orders. After launch, en route guidance is provided by a missile-borne system consisting of a strap-down attitude reference assembly and a digital computer; no data inputs from the ship are required by the missile after it has been launched. Cruise altitude is monitored by a radar altimeter and flight control commands are interpreted by electromechanical actuators operating on the fins. This guidance system is designed to operate satisfactorily when the missile is launched at any angle up to 180° from the required course line.

Terminal guidance is achieved by means of a Texas Instruments DSQ-28 active radar homing system which maintains its lock until final impact. Included in the terminal guidance system is provision for the terminal manoeuvre referred to above. It should be noted, also, that the high degree of manoeuvrability implicit in this terminal pattern also means that the missile is capable of engaging targets that are taking high-speed evasive action. The radar homing system is frequency-agile and this facility, coupled with extensive on-board computer logic circuitry, provides considerable ECCM capability to the missile in the terminal phase.

CHARACTERISTICS

Type: Shipborne surface-to-surface tactical guided missile
Guidance principle: Pre-programmed attitude reference plus radar altimeter in cruise phase; active radar homing
Guidance method: by electromechanical control of 4 moving fins ($\pm 30°$ travel)
Propulsion: Solid-propellant booster. Teledyne CAE J402 turbojet cruise engine
Warhead: HE penetrating blast type
Missile length: With booster 4.58 m; without booster 3.84 m
Missile diameter: 34 cm
Launch weight: With booster 681 kg; without booster 519 kg
Range: Over 50 nm (90 km)

STATUS

The USN operational evaluation was completed successfully in March 1977, and Fleet introduction was started immediately on surface ships and submarines. The airborne P-3 and A-6 applications were implemented in the autumn of 1979 and 1981, respectively.

Launch of RGM-84 Harpoon missile from USS Fletcher

In June 1982 the USN took delivery of the first of an advanced version of Harpoon, designated Block 1B. Based on the advanced guidance design developed for the RN, the Block 1B Harpoon does away with the terminal "climb and dive" manoeuvre, making detection and defence more difficult for the target. In late 1982 flight test development for a still more advanced version, Block 1C model RGM-84D, was completed. This model provides increased range, programmable waypoints, and selectable terminal trajectories.

The Royal Navy's 'Cornwall' and 'Norfolk' class frigates are to be armed with Block 1C. Submarine and aircraft launched versions of Block 1B already serve with British forces. Canada's new patrol frigates will have Block 1C missiles.

Harpoon is used by the navies of Australia, Denmark, West Germany, Greece, Iran, Israel, Japan, South Korea, Netherlands, New Zealand, Saudi Arabia, Singapore, Spain, Turkey and the United States.

Sub-Harpoon is used by the United Kingdom and the United States.

In August 1987 McDonnell Douglas received a $200 million contract to provide 236 Harpoon missiles in FY 1987 for the United States (80), Japan (38), Spain (16),

Turkey (2), Thailand (6), Australia (36), Greece (32), Netherlands (24), Pakistan (1) and Denmark (1). Deliveries are to be completed in October 1989.

CONTRACTORS

Programme Direction: US Naval Air Systems Command supported by the Naval Sea Systems Command.

Systems Contractor: McDonnell Douglas Astronautics Company, PO Box 516, St Louis, Missouri 63166, USA.

Propulsion: Teledyne CAE (turbojet), Aerojet Tactical Systems and Thiokol (booster).

Guidance: IBM, Lear Siegler, Texas Instruments, Northrop.

Radar Altimeter: Honeywell, Kollsman.

4213.211

DEADEYE 5-INCH GUIDED PROJECTILE

The Deadeye 5-inch guided projectile represents the initial implementation of a potential family of laser homing projectiles for use with standard Deadeye ordnance pieces of various calibres from the 5-inch upward. The original operational requirement (OR-S0957-SH) was dated 15 April 1977 and a four to six year programme was envisaged. The operational objectives were to improve naval gunfire systems when used in support of army and marine corps amphibious assault operations by effectively improving accuracy against fixed and certain moving targets designated by laser target marking systems deployed on ships, aircraft or by ground forces. This would also yield considerable reductions in the number of rounds of ammunition consumed. Precision guided artillery missiles of this type will also be available for use against surface vessels, and to supplement ships' point defence anti-aircraft weapons against air targets. A range of munitions is foreseen which includes the 5-inch projectile with which this entry is concerned, and later 8-inch and 16-inch versions.

The Deadeye guided projectile will be employed with the Mk 45 5-inch/54 calibre naval gun, and it features a semi-active laser seeker used in conjunction with laser designators such as the Modular Laser Equipment (MULE) or airborne equipments such as the A-6 target recognition attack multisensor (TRAM) or laser designator-equipped OV-10 aircraft. The projectile is fired from the ship into an acquisition 'basket' where its sensor acquires the reflected energy from the laser designator and automatically homes onto the target.

The projectile consists of four main sections: guidance, control, warhead, and rocket motor. A ramming sheath is provided to protect the tail fins and to assist the loading process. The guidance section houses the laser seeker head and processing electronics necessary to detect, acquire, process, track, and follow the reflected laser energy to the target. The control section of the projectile contains a cold gas control actuation system, and batteries for the main system and timer. The payload section normally contains a 13.6 kg fragmentation/shaped charge warhead, but this space can be used for a telemetry package during tests or trials.

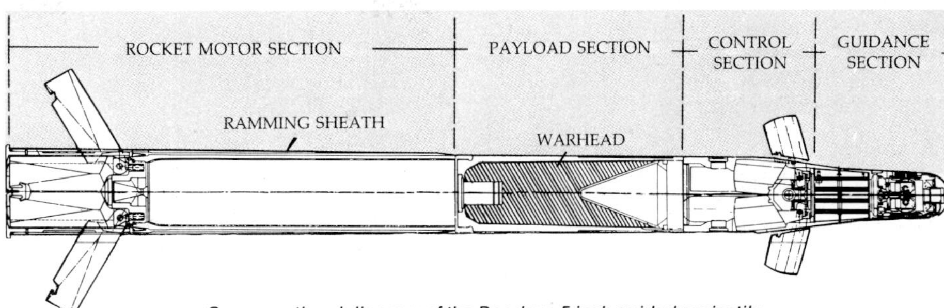

Cross-sectional diagram of the Deadeye 5 inch guided projectile

The rocket motor section consists of a nozzle/fin assembly, a motor case tube loaded with liner and propellant, and a slip obturator. The rocket motor is ignited after launch from the gun barrel and provides a significant extension of the typical range of some 24 km that can be obtained with conventional 5-inch shells and Mk 45 guns.

Technical advances that were instrumental in achieving this performance included the design and development of miniaturised and shock-hardened electronics able to withstand the high loadings imposed by the firing of these shells (up to 10000 *g* acceleration and 25 000 *g* shock) and the production of a new hardened gyro. Microminiaturisation of the seeker electronics achieved a packaging density roughly 17 per cent of the Copperhead land-based artillery laser guided shell, to fit into the 76 mm diameter nose required for automated loading into the gun aboard ship.

CHARACTERISTICS

Length: 1.54 m
Calibre: 5 in (127 mm)
Weight: 47.5 kg
Guidance: Semi-active laser homing, proportional navigation
Control actuation: Cold gas pneumatic
Warhead: 13.6 kg dual mode, fragmentation/shaped charge
Range: >24 km

STATUS

The Deadeye 5-inch guided projectile (GP) programme was started in 1977 and the planned initial operating capability (IOC) in July 1981. The latter date was allowed to slip by a few months as a result of reprogramming the funding in 1980. The 1982 DoD budget allocated funds for manufacture, test and procurement of 5-inch (127 mm) and 3-inch (76 mm) rounds. Operational test and evaluation of Deadeye was completed in 1981 in the destroyer *Briscoe* when test firings showed improvements in accuracy of more than 80 per cent. In November 1982 approval was given for its use by the US Navy.

In December 1983 Martin Marietta received a contract for pre-production tooling, test equipment and other facilities. In January 1986 the Defense Department awarded a $73 million contract to Martin Marietta for 500 rounds but at the end of the year the DoD ordered the company to stop work and this order was lifted only in February 1987.

The programme to build the Block 0 design was cancelled and the electronics were redesigned to the improved Block 1 configuration. Production was resumed with Martin Marietta being awarded a $61.6 million contract in September 1987.

Earlier, in April, the US Navy sought to develop and qualify two second sources for the programme and Martin Marietta was asked to prepare 30 sets of Deadeye hardware for use by potential second source contractors as qualification rounds. In September three contracts for the modification of rounds were awarded to Martin Marietta ($44.2 million), AAI ($11.3 million) and General Dynamics ($13.9 million).

It is anticipated the first production rounds would be delivered late in 1989 but early in 1988 the programme again came under scrutiny as part of the scheme to reduce the DoD budget.

The Deadeye 5 inch semi-active laser homing guided projectile

CONTRACTOR

Martin Marietta Orlando PO Box 5837, Orlando, Florida 32855, USA.

SHIPBORNE SURFACE-TO-AIR WEAPONS

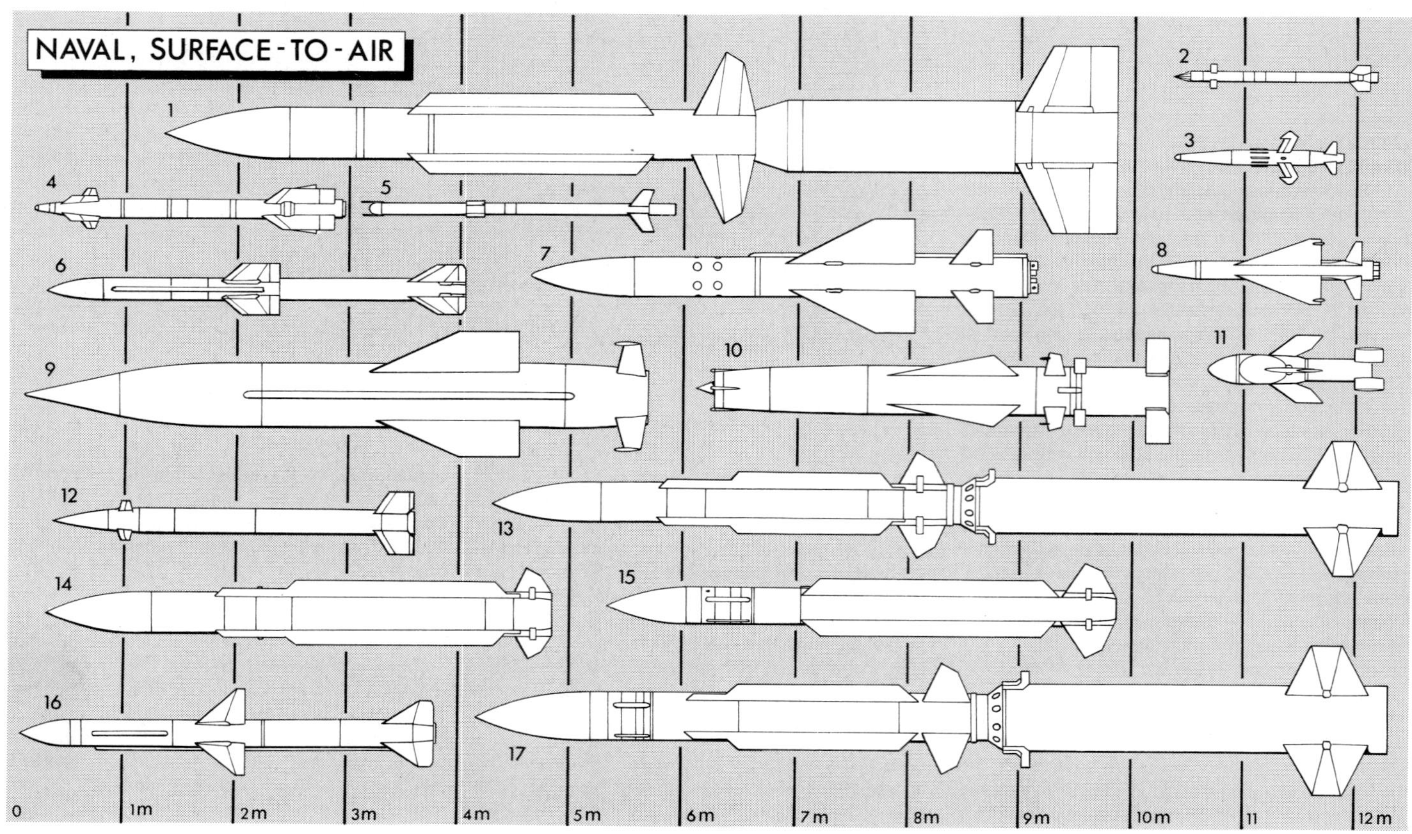

1 Masurca, **2** SADRAL, **3** RBS 70, **4** Crotale, **5** ASMD RAM, **6** Aspide (Albatros), **7** Barak, **8** Seawolf, **9** SA-N-3 Goblet, **10** Sea Dart, **11** Seacat, **12** SA-N-4, **13** Standard SM-2 (ER) **14** Standard SM-1, **15** Tartar, **16** Sea Sparrow, **17** Terrier

CHINA

6188.231

CS-AX-? POINT DEFENCE MISSILE

Photographs released in 1984 by the PLA Navy reveal that the two new 'Jiangdong' class frigates are armed with the first Chinese SAM. The accompanying picture shows two missiles mounted on a launching ramp. It is apparently a naval version of the HQ-61 (**6418.131**) and has been designed to engage targets flying at medium and low altitude. More information can be found in the HQ-61 entry.

STATUS

Probably in late development and trials. Two twin-arm launchers are mounted in 'B' and 'X' positions.

CS-AX-? point defence missiles

FRANCE

4860.231

SAAM NAVAL AIR DEFENCE SYSTEM

In 1983 the French authorities initiated a feasibility study into a family of new air defence weapons suitable for use aboard naval vessels and by land forces to give protection from attacks by aircraft, and missiles in the 1990s and beyond. It is hoped to run the two programmes, known as SAAM (surface-to-air anti-missile) and SAMP (medium range, ground-to-air system) concurrently in the initial stages, and to maintain a considerable degree of commonality between the hardware of the land and naval versions of the system. The land mobile SAMP system is described in the Mobile Surface-to-Air Guided Missile Systems section of this book (**4771.131**) where the common aspects that have been defined to date are summarised.

The threat to be addressed embraces saturation air attacks with aircraft and missiles, in all weathers at all altitudes and from multiple azimuth angles. Targets are expected to be such that the defending missile must be able to defeat missiles capable of manoeuvres up to 15 *g* and speeds of Mach 2.5 or more. Dense and varied ECM and other countermeasures are expected, and high resistance to such challenges is essential, as is maximum autonomy of operation against multiple threats.

The ability to intercept several targets simultaneously and provide a short reaction time will be facilitated by use of a new X-band 3-D radar, Arabel, developed by Thomson-CSF. This will simultaneously carry out surveillance, detection, identification, threat evaluation, designation and tracking of targets, in addition to up-dating the missile's inertial navigation system. Rotating at 60 rpm, this radar will incorporate frequency agility, and will be compatible with a missile launching rate of up to one per second. It is possible that electro-optical sensors may be added as a radar back-up and to provide for loss of target data as a result of jamming.

The Aster missile developed by Aérospatiale has been designed for exceptional manoeuvrability; three times that of potential targets is the estimated requirement, ie about 50 *g*. It will have a hybrid control system incorporating aerodynamic and direct-force control elements. Missile guidance will consist of two steps: the initial guidance phase during which the missile heads toward its target under the control of its on-board inertial system, and the terminal phase during which it will use a K-band active radar seeker.

SAAM

This version of the system is intended principally for the self-defence of ships, and particularly to destroy diving and sea-skimming anti-ship missiles. For this the Aster 15 will have an interception range of 10 km against supersonic and manoeuvring missiles, and 15-17 km against fast aircraft and subsonic missiles with a minimum requirement for destruction at ranges of 2 km. It will be fired from a vertical launcher and will have rocket motor propulsion. The guidance system will be optimised to give minimum miss-distance performance to achieve maximum destructive effect from the warhead. The multi-purpose radar antenna will have a stabilised mount and will be located very high on the ship's superstructure to give the best overall performance against both high- and low-flying targets. Discussions have taken place between the French and Italian Navies about merging the Aster 15 and Idra (**6727.131**) programmes into a joint project provisionally dubbed Astidra.

STATUS

Feasibility study started in 1983. SAAM (Aster 15) development began in 1987 while development of the ground-based version (SAMP) is planned to follow in 1988. The Aster 15 is scheduled to enter service in 1994 aboard the aircraft carrier *Richelieu*.

Aster 15 version of the Aster system

CONTRACTORS

Thomson-CSF, Division Systèmes Electroniques, 1 rue de Mathurins, 92222 Bagneux, France.

Aérospatiale, Division Engins Tactiques, 2 rue Beranger, 92322 Chatillon Cedex, France.

1177.231

MASURCA SURFACE-TO-AIR MISSILE SYSTEM

Masurca forms the main anti-aircraft armament of the French frigates *Duquesne* and *Suffren*, whose principal role is that of escort to the aircraft carriers *Foch* and *Clemenceau*. Masurca has also armed the AA cruiser *Colbert* since 1973 when her modernisation was completed. Each of the Masurca ships is equipped with three-dimensional surveillance radar, a weapon direction system, two independent fire control systems, and a twin launcher. Aircraft and missile targets can be intercepted at ranges of 40 km or more.

The main characteristics of the missile are length (with booster) 8.6 m, (without booster) 5.29 m, diameter 41 cm, span of booster fins 150 cm, weight (with booster) 2090 kg, (without booster) 950 kg. A proximity-fuzed high-explosive warhead is carried, and both boosters and sustainer motors are solid-fuel units.

Two types of guidance system have been employed: the Mk 2 Mod 2 uses radio command, and the Mk 2 Mod 3 is equipped with a semi-active radar homing head, but only the latter is currently in service.

In addition to the missiles themselves, the Masurca weapon system contains the following main elements:

(1) a twin launcher
(2) two storage and handling systems
(3) two directors, each with a tracking radar and a target illuminator.

The DRBR 51 tracking radar has two main functions in the Mk 2 Mod 3 role:

(1) it follows the target continuously
(2) it designates the target illuminator radar.

OPERATION

The DRBR 51 has two antennas, the main one of which is used for target tracking and as an I/J-band target illuminator radar. The missile is equipped with two antenna systems, one at the front of the vehicle, and two fixed antennas mounted on the missile body. The latter are used to receive the illuminating radar signals direct and to provide a reference signal which can be compared with the signals reflected from the target that are collected by the forward missile antenna. This enables a doppler component to be extracted, which is used in the on-board computation of a proportional navigation interception course to the target.

DEVELOPMENT

The Masurca missile was developed by ECAN Ruelle in co-operation with Société des Engins Matra.

Fire control radar, missile homing head and proximity fuze were developed by Thomson-CSF. Solid propellants are manufactured by Société Nationale des Poudres et Explosifs (SNPE). Launch, storage, and handling equipments were designed and developed by ECAN Ruelle.

STATUS

The Masurca weapon system is operational in the Mk 2 Mod 3 version aboard the *Colbert, Duquesne* and *Suffren*. The system is being updated on all three ships, including radars, computers and displays.

CONTRACTOR

Direction des Constructions Navales, 2 rue Royale, 75200 Paris Naval, France.

2111.231

CROTALE NAVAL SURFACE-TO-AIR WEAPON SYSTEM (TSE 5500)

Directly derived from the ground version of the Crotale (**2074.131**), the Crotale Naval weapon system is designed for the self-defence of ships against short range air attacks at medium, low and very low altitude, these attacks being conducted by aircraft, helicopters, air launched and sea-skimming missiles. The Crotale Naval may also be used against surface targets and is capable of artillery control.

The Crotale Naval is now operational with the French Navy and has been ordered by the Saudi Arabian Navy.

A new version, the Compact Modular (formerly Crotale Naval), is now available for installation on board ships of 200 tons or more.

In its Compact version, Crotale Naval is composed of:

(1) a turret assembly with two coaxial turrets: the eight-missile turret and the fire director supporting the radar antenna and receiver, the remote control antenna, the IR receiver for localisation of the missile during the first part of its flight (gathering), the differential IR tracker for guidance of the missile at very low altitude and the TV camera
(2) a shelter housing the electronic equipment data processing, radar and IR processing cabinets, launcher power control cabinet and maintenance unit
(3) an operator console in the combat information centre (CIC) for supervision of the system and sending of firing orders. Operation of the system is fully automatic, pushing the 'Fire' button being the only manual action. In the normal mode, the operator sitting in front of the data processing cabinet has just to start the system, to supervise its working and to depress the 'Fire' button.

Acquisition and tracking of the target on the target designation from the ship (surveillance radar, optical sight, radar detector or another tracking radar) are performed by the radar of the Crotale system.

After launching, the missile is gathered by the large field of the IR localiser RIC (Ralliement Infrarouge Crotale) which locks it within the radar beam. Missile guidance is performed by an improved line-of-sight guidance which optimises the manoeuvrability of the missile. A proportional navigation law is elaborated at the end of the missile flight to reduce the miss-distance in the case of a manoeuvring target. In the case of a target at very low altitude (sea-skimmer), tracking of the target and of the missile is performed by the differential IR tracker, a device developed by

Modular Crotale Naval deployed in a French warship

SAT and known as SEID (Système d'Ecartométrie Infrarouge Différentiel). The missile is fired and guided as in the radar mode.

The Crotale missile is boosted by a solid propellant rocket motor. The warhead, weighing about 14 kg, is triggered by an electromagnetic proximity fuze; the delay of explosion is calculated to concentrate the prefragmented splinters in the most vulnerable zone of the target.

The missile is housed inside a sealed container, which is used for transport, storage and launching. A missile inside its container forms a Crotale round. The 18-missile magazine and loading mechanism is manufactured in French Navy yards.

Modular Crotale Naval is composed of the following parts:

(1) the fire director turret, supporting the radar antenna and receiver, the remote control antenna, the IR tracker (for guidance against sea-skimmer targets) and the TV camera
(2) an independent launcher turret, supporting the IR localiser (gathering) and eight rounds
(3) the data processing cabinet, installed in the CIC
(4) the other electronic equipments (radar, IR and launcher power control cabinets) which are installed in any available compartment(s) of the ship.

Operation of the system does not differ from the Crotale Naval. The fire director of the Modular Crotale Naval, as for the Crotale Naval, can also be used for the direction of naval guns. The Modular Crotale Naval can be installed on ships of 200 tons or more. The light eight-missile loading system is also manufactured by French Navy yards (DCAN).

CHARACTERISTICS

Guidance: Improved line-of-sight command and proportional navigation at end of missile flight
Missile max speed: 750 m/s after 2.3 s
Radar cross-section: 0.1 m²
Max interception range: 13 km (helicopter or non-manoeuvring target); 10 km (manoeuvring Mach 1.2 aircraft); 8.5 km (sea-skimmer)
Minimum interception range: 700 m
Vertical coverage: From 4-5000 m
Target max speed: Mach 2 (radial attack aircraft/sea skimmer); Mach 1.2 (crossing target)
Reaction time: 4.5 s (between target designation reception and first missile start)
Weight
Crotale Naval
Turret assembly: 8.2 tons with 8 rounds
Shelter + electronic equipment: Approx 10 tons
18-missile storage and loading system: Approx 15 tons with 18 rounds
Modular Crotale Naval
Fire director turret: 800 kg
8-missile turret: 4.2 tons with 8 rounds
Electronic equipment: 2.8 tons
8-missile loading system: 3.2 tons with 8 rounds

DEVELOPMENT

Development of the Naval Crotale was carried out under the responsiblity of DCN and DTEN with Thomson-CSF as systems manager and supplier.

STATUS

The Crotale Naval, in its basic version, is now operational in 'Georges Leygues' corvettes and 'Tourville' class destroyers of the French Navy. These units are about to be retrofitted with the anti-sea-skimmer capability (Modular version) and five more systems were ordered by the French Navy in 1982, bringing the total number of systems ordered by the French Navy to 14. Two systems were fitted to carrier *Clemenceau* in 1985-86 and *Foch* will follow in 1987-88.

In late 1987 there were reports that China had purchased two Crotale Naval Systems worth the equivalent of $91.5 million for delivery in 1989.

Five systems were ordered by the Saudi Arabian Navy in 1980 and are now operational. Negotiations are progressing with several other naval forces for Modular Crotale Naval.

CONTRACTORS

System: Thomson-CSF, Division Systèmes Electroniques, 1 rue des Mathurins, 92223 Bagneux Cedex, France.

4776.231
SATAN CLOSE-IN WEAPON SYSTEM

SATAN (Système Autonome Tout temps d'Auto-defense Naval) is a new project for the design and development of an entirely autonomous all-weather close-in defence system for ships to provide protection against aircraft and missiles. The design is intended for fitting in ships which may already be equipped with one or more longer-range missile systems such as the Crotale Naval (**2111.231**), or as the main anti-missile defence in other types of ship. It is completely automatic from detection to firing and has a very short reaction time.

A modular design has been chosen to permit easy installation and rapid changes to accommodate developments in the threats anticipated, and advanced technical features include:

(1) a doppler frequency-agile surveillance radar (Sea Tiger, Triton S or Triton G) which is capable of automatic detection of missile targets in sea or rain clutter and in a heavy jamming environment. An automatic extraction system is associated with an automatic plot tracking device
(2) a Tavitac data processing system with a VDU for threat evaluation, co-ordination with other weapon systems, electronic warfare and target designation
(3) a Castor IIJ fully automatic monopulse doppler tracking radar operating in Ku-band. The system can be provided with an X-band illumination capability to launch semi-active surface-to-air missiles or it can remote control Crotale Naval missiles. An associated digital computer generates the firing data. (In many cases an IR tracking facility will be provided by a ship's Crotale system, but where this is not the case, separate E-O sensors are provided for TV or IR target tracking)
(4) one or two 30 mm Sea Vulcan 30 guns by General Electric. Due to its original process of cancelling 'mirror' effects on radar tracking Castor IIJ can track even sea-skimming missiles.

SATAN prototype installed at the French test range at Gavres

By adopting an off-mounted solution for the surveillance radar and the Castor IIJ fire control radar, it is possible to site the various sensors aboard the ships in the optimum fashion for accurate detection and tracking. The surveillance antenna can be placed at a height that offers maximum sea-skimmer detection range, given the problems posed by geographical range and particularly the interference fringes. (Against a 500 m/s target, every kilometre gained in early detection gives two seconds more for use of defensive weapons). The range of the system is 300 to 1500 m.

As the fire control radar is separate from the gun, it is free from problems such as vibration and debris associated with firing, and may be installed above the gun to improve the field of view or to provide autonomous search facilities. The additional errors introduced by platform movements between the mounting and the radar aiming system are corrected in the firing control loop, and the computer system integrates all platform movements, measured by a strapdown unit.

STATUS

During the autumn of 1987 the SATAN prototype carried out live firing trials destroying seven targets at ranges of 600 to 900 m.

CONTRACTOR

Thomson-CSF, Division Systèmes Defense et Contrôle, 40 rue Grange Dame Rose, BP34, 92360 Meudon-la-Floret, France.

4861.231
SAMOS NAVAL AIR DEFENCE SYSTEM

SAMOS, the SAGEM Optronic Anti-Missile System, is intended to provide a defence against sea-skimming anti-ship missiles and short-range defence against aerial targets. It has been designed in response to a similar CIWS requirement as those covering French projects such as SATAN (**4776.231**) and SADRAL (**4509.231**) in this section.

The SAMOS concept consists of a combination on a single integrated naval mount of a SAGEM Volcan (Viseur Optronique Leger pour Conduite d'Arme Navale) electro-optical fire control system (**3064.293**), a General Electric Sea Vulcan 30 mm gun mount and ground-to-air missiles such as the Matra Mistral used in the SADRAL system mentioned above. This combination promises high effectiveness through accurate stabilisation of the fire control system line-of-sight (LOS), associated with a strap-down inertial reference unit and the low dispersion, high rate of fire (4200 rounds/minute) of the Sea Vulcan gun.

The Volcan FCS includes a lightweight electro-optical sight, the lower section of which accommodates TV, IR and laser rangefinder sensors, while the upper, moving section houses a two-axis gyroscope which ensures an accurate stabilised LOS. A control and display console is provided for operation of the system and its associated electronics.

The inclusion of four launch tubes for short-range surface-to-air missiles on the top of the Sea Vulcan mount is optional. Typically these missiles would be expected to provide a range extension out to 6000 metres using IR homing weapons capable of engaging attacking aircraft or anti-ship missiles during a diving attack.

By using target designation from the ship's surveillance radar and/or a passive IR sensor, SAMOS will provide for acquisition and automatic tracking of the target, by either TV or IR, measurement of target range by the laser rangefinder, with automatic computation of target parameters and weapon control.

SAMOS has been designed as a fully optronic system to perform automatic or semi-automatic anti-aircraft or anti-sea-skimming missile defence. It has a very fast response, is effective in a jamming environment and operates at low elevation and in strong sea clutter.

STATUS

Integrated system firing tests were carried out in October 1987 at the French Navy test range at Gavres.

CONTRACTORS

SAGEM – Societé d'Applications Generales d'Electricite et de Mecanique, Department Marine, 6 ave d'lena, 75783 Paris Cedex 16, France.

General Electric, Armament and Electric Systems Dept, Lakeside Avenue, Burlington, Vermont 05402, USA.

SAMOS combines GE's Sea Vulcan 30 mm gun mount with SAGEM's Volcan optronic fire control, and IR homing SAMs provide for compact naval close-in air defence system

4509.231

SADRAL NAVAL AIR DEFENCE SYSTEM

SADRAL (Système d'Autodéfense Rapprochée Anti-aérienne Léger) is an air defence system based on the Matra Mistral land-based SATCP missile system, modified for naval applications and proposed for the French Navy. The Mistral itself is described in entry **4198.131** and the following entry is concerned with its application to 'close-in' defence of ships from aircraft or anti-ship missiles.

The equipment consists of a remotely operated stabilised mount which carries six Mistral rounds and a closed circuit TV camera. The mount can be trained in azimuth and elevation under the control of the operator who mans a simple control desk below deck. The TV camera is used for target acquisition, and this component can be supplemented by an IR thermal imaging camera for night operations.

For the majority of its components, the SADRAL system uses equipment already certified for marine environments and in use for Otomat systems or those derived from the Totem remote-aiming system. The mass of the carriage fitted with its six rounds of ammunition is approximately 900 kg. Total mass of the SADRAL system (carriage, servos and control desk) is about 1500 kg. It is thus possible to mount two SADRAL systems on low-tonnage vessels, with a mass which is still more than 50 per cent less than that, for example, of a single Phalanx system, according to the manufacturer.

Due to missile accuracy and the penetrating power of the fragments of its charge, SADRAL's efficiency is claimed as superior to that of a gun-based system. In addition, the SADRAL system permits interception of targets at distances much greater than those of gun-based systems, thus offering greater safety for ships. Finally, the time required to pass through the firing envelope being much greater, SADRAL permits engaging several missiles against the same target, or against a wave of aircraft, which a gun-based system does not permit.

Although the priority mission assigned to SADRAL by the French Navy is anti-aircraft defence of small vessels (patrol-boats, mine-sweepers, etc) it will also be mounted on logistic backup and transport ships. It may also be mounted in ships already equipped with a surface-to-air system, such as anti-aircraft corvettes, where its firepower would extend the threshold of saturation of the threat. In addition to defence against helicopters or airplanes, the Mistral could be used to engage sea-skimmer-type missiles flying at low altitude above the sea surface.

Target designation may be performed, according to circumstances, by using the ship's surveillance radar or an electro-optical surveillance system (IR or optical), or, if necessary, simpler designation systems (navigation radar or the ELDO (*equipement leger de designation d'objectif*) a lightweight optical sighting system now under study).

CHARACTERISTICS

Mount: 6-round, stabilised
Weight: 1000 kg (with 6 rounds)
Missile: Mistral
Length: 1.8 m
Diameter: 90 mm
Weight: 17 kg
Warhead: 3 kg
Fuze: Impact, laser proximity
Guidance: IR homing (2-colour)
Range: 300-6000 m
Peak speed: Mach 2.5

STATUS

Development began in 1981. The French Navy is reported to have plans to procure several dozen systems for its vessels with entry into service planned for 1987. The first firing trials at sea were carried out in October 1986.

LAMA

A self-defence missile launcher is under development by CSEE. Called LAMA (Lance-Missiles

SADRAL launcher on top of deck house on FAC. Note eight MM40 anti-ship missile launcher tubes on right

SIMBAD platform carrying two Mistral missiles

d'Autodéfense) the new fire director consists of an optical director for surveillance and target designation, to the sides of which are attached up to six missile launching ramps. Designed for very short range surface-to-air naval missiles such as the Matra Mistral, other types (eg. Stinger, RBS 70, etc) can be accommodated and it is also possible to mount different sensor/missile combinations on the LAMA director. Weight is 1000 kg.

SIMBAD

SIMBAD (Système Intégré de Mistral Bitube d'Auto-Defense) is designed for the protection of smaller vessels, logistics vessels and support vessels. Being operated independently it complements the defence of larger warships. This extremely simple system fires two missiles, and can be readily fitted to any 20 mm gun mounting. Aiming is by simple optical line-of-sight.

CONTRACTORS

Prime: Matra SA, avenue Louis Bréguet, 78140 Velizy-Villacoublay, France.
Homing Head: SAT.
Propulsion: SEP.
Thermal batteries: SAFT.
Warhead, weapon canister: Manhurin.
LAMA Director/launcher: CSEE

LAMA self-contained optical director/launcher under development for use with Matra Mistral by CSEE can also be used for other weapon/sensor combinations

INTERNATIONAL

4775.231

NATO SEA SPARROW SURFACE-TO-AIR MISSILE SYSTEM

The NATO Sea Sparrow Surface Missile System (NSSMS), originally known as the Improved Point Defence Missile System (IPDMS), is a shipboard air defence system employing the Sparrow family of missiles. The system is designed to counter the threat of anti-ship cruise missiles and aircraft. The original shipboard use of the Sparrow missile was in the Basic Point Defense Missile System (BPDMS) which became operational aboard US ships in 1969 but by 1984 was being phased out of service. The NATO Sea Sparrow Surface Missile System (NSSMS) became operational in 1973 using RIM-7H missiles. Modifications to the NATO Sea Sparrow missile systems to accommodate the improved Sparrow AIM/RIM-7M monopulse missile began in 1983. In 1981 Sea Sparrows were successfully fired from Vertical Launchers aboard the Canadian destroyer *Huran*. The Sea Sparrow Vertical Launch System will be operational aboard the new Canadian 'City' class frigates. The Vertical Launch System has also been selected by the Royal Netherlands Navy for its new 'Gröningen' class frigates.

Basic Point Defence Missile System

The BPDMS was assembled from existing hardware. It featured the Sparrow AIM-7E missile launched from a modified eight-tube ASROC missile launcher. The launcher is housed on a modified 3-inch 50 calibre automatic gun mounting. Total weight of the system is 17690 kg.

A CW semi-active radar homing system is used. Target data from the combat information centre is supplied to the manually operated Mk 115 fire control system and the target is acquired and illuminated for homing guidance by a Mk 51 director/illuminator radar.

STATUS

BPDMS development began in 1964. In March 1983 USN ships with BPDMSs totalled 46: (31 in 'Knox' class frigates (one system each); two 'Mount Whitney' class LCCs (two systems each); seven 'Iwo Jima' class LPHs (two each); five 'Saipan' class LHAs (two each); and one CV (two systems)).

CONTRACTORS

Sparrow missile: Raytheon Company.
Solid-state power drive for fire control system: General Electric.
Fire control system: Frequency Engineering Laboratories.

NATO Sea Sparrow Surface Missile System (NSSMS), Mk 57

The NSSMS incorporates a lightweight eight-cell launcher (Mk 29), and a fire control system (Mk 91) utilising digital computers, and powered director/illuminator(s). The RIM-7H and AIM/RIM-7M Sparrow missile family is used in a modified form with folding wings that enable them to fit into the smaller, lighter launcher. NSSMS possesses a range of operational modes from manual to fully automatic with operator intervention.

The new digital Sparrow (AIM/RIM-7M) medium range missile features all-aspects, all-weather, crossing target capability. The 7M Sparrow incorporates an advanced inverse monopulse seeker and a digital computer. This provides substantial performance improvements in seeker sensitivity and enhances target discrimination in heavy clutter, ECM conditions, and extreme low altitude intercepts.

The onboard digital computer, besides being reprogrammable, gives greater immunity in the changing ECM threats, while providing for proportional navigation. Proportional navigation directs the missile to fly a lead angle course to the intercept point, which provides NSSMS equipped ships with the ability to protect other ships in the vicinity. The AIM/RIM-7M also includes a larger boost sustain rocket motor and blast fragmentation warhead.

The NATO Sea Sparrow programme was estab-

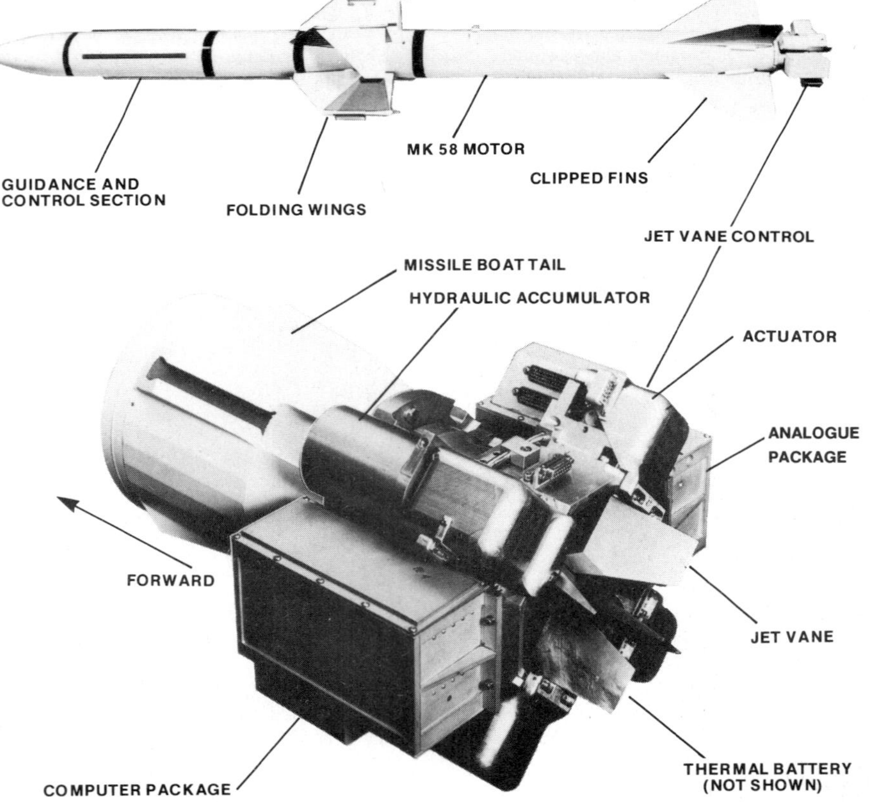

JVC package is attached to tail of AIM/RIM-7M Seasparrow missile to provide for initial tip-over after vertical launch. The 15.8 kg package contains a strapdown inertial reference system, digital microprocessor, independently operated vanes and self-contained actuator system

Vertical launch of Seasparrow from HMCS Huron

lished by a Memorandum of Understanding in 1968 between Denmark, Italy, Norway and the USA. Belgium and The Netherlands joined in 1970. The Federal Republic of Germany in 1977. Canada and Greece acceded in 1982 and Turkey in 1986.

The contract definition phase began in 1968 and in 1969 Raytheon Company was awarded a $23 million prime contract for engineering and development, requiring the production of three engineering and development systems. One system was to be tested by the US Navy, one sent to Norway for operational environment testing, and one remained at Raytheon for systems evaluation.

The first successful shipborne launch took place in March 1972 when a NSSMS on the frigate USS *Downes* (FF-1070) acquired and tracked an incoming target and successfully fired its missile. Another prototype Sea Sparrow System was installed on the Royal Norwegian Navy ship KNM *Bergen* (F-301) in 1973.

NSSMS entered full scale production in 1973 as a co-operative effort by the USA and other NATO countries – Belgium, Denmark, Italy, The Netherlands and Norway. The USA is expected to purchase approximately half of all systems produced. By autumn 1983, 111 systems had been installed with a total of 143 systems being ordered worldwide. Of this total, 87 systems are for the US Navy aboard CV/CVN, DD, AOE and AOR class ships.

Three NATO countries use the full system similar to the US configuration. Denmark has dual director systems in five frigates and a single director system at a land-based site. Italy has single director systems in four frigates. Norway has a single director in five frigates and one land-based site.

Seven other NATO countries use the NSSMS launcher, transmitter and Sea Sparrow missile, but with their own fire control systems: Belgium has four, Greece two. The Netherlands will have 14 plus one land-based site. The Federal Republic of Germany has six plus two land-based sites, Spain has four, Turkey will have four, and Portugal will have three.

Japan (five systems) has also purchased NSSMS.

Sea Sparrow, Guided Missile Vertical Launch System (GMVLS)

Vertical Launch Sea Sparrow was demonstrated aboard the Canadian destroyer HMCS *Huron* with the successful interception of missile targets during a joint at-sea test in 1981 (**2036.231** in the 1983-84 Edition).

The vertically launched RIM-7M Sea Sparrow missile is modified with the addition of a Jet Vane Control (JVC) unit. The JVC provides for initial missile tip-over, orientation, and course control after vertical launch prior to the normal Sparrow guidance control. The GMVLS vertically mounted missile launching canisters are designed to function as the transport and storage units as well as the launch cells.

STATUS

In production.

The AIM/RIM-7M, developed by Raytheon Company, completed a full-scale development test in 1982 and is now in full-scale production, with total production awards of 10 700 missiles. This missile system entered the US Navy inventory in 1983. RIM-7M deliveries to NATO countries also began in 1983.

In 1982 General Dynamics was awarded a second-source contract with total production awards of 6500 missiles. Deliveries began in 1984.

Under a $76 million contract awarded to Raytheon in early 1984 the company is to supply Sea Sparrow VLS missile systems and AN/SPS-49 long range surveillance radars for the Canadian 'Halifax' class frigates. In addition Raytheon has received a $16 million contract for modifications to the RIM-7M Sea Sparrow missile. Both contracts were issued by Paramax Electronics Inc, of Canada, a division of Sperry responsible for integrating the combat systems aboard the Canadian vessels. First ship of the new class is due for delivery in 1988.

The Netherlands also selected the Sea Sparrow VLS system for the new 'Gröningen' class frigates in 1985.

CONTRACTORS

Raytheon Company, Equipment Division, Wayland, Massachusetts, USA.

Raytheon Company, Missile Systems Division, Bedford, Massachusetts, USA. (RIM-7H, AIM/RIM-7M, RIM-7M vertical launch missiles).

General Dynamics Corporation, Pomona Division, Pomona, California, USA. (AIM/RIM-7M missile).

6755.231

NATO ANTI-AIR WEAPON SYSTEMS (NAAWS)

In January 1988 the US Navy Systems Command issued a request for proposals for a NATO Anti-Air Weapon System (NAAWS) with replies scheduled in March. Three concept definition study contracts lasting nine months were expected to be awarded by the middle of the year. Each team will receive $3 million from NATO funds and the members will match this from their own resources.

NAAWS is expected to be an integrated system with electronic and electro-optical sensors, a local area missile system, a point defence missile system and a close-in weapon system, with possible directed energy system. The prime sensor is likely to be an E/F-band phased array radar while the local area missile will be a vertically-launched system capable of high *g* turns and with a range of 20 to 50 km. The system is designed to meet the next generation of supersonic anti-ship missiles capable of low level, manoeuvring and high-angle dive attacks. The system is intended for retrofitting into some 190 US warships and is likely to be the basic system for the NATO Frigate Requirements of the 1990s (NFR-90).

At the completion of the first phase, early in 1989, new proposals will have to be made for the second, demonstration and validation, phase. Two contenders will participate in this phase which will last three years and be followed by a development and production contract.

In January teams led by General Electric's RCA Electronic Systems Department, Martin Marietta Aerospace and Naval Systems and Westinghouse intended responding and it was likely Raytheon would lead a fourth team.

3613.231

GOALKEEPER ANTI-MISSILE CLOSE-IN WEAPON SYSTEM

Goalkeeper is an automatic anti-missile close-in weapon system and consists of the General Electric GAU-8/A 30 mm Gatling gun and Signaal's I-band surveillance radar, X/Ka-band tracking radar and fire control. Operation of the Goalkeeper system is completely automatic from target detection to target destruction, termination of firing and engagement of the next target. In a multiple engagement Goalkeeper automatically designates the higher priority target. During firing the gun aiming is automatically adjusted by means of closed-loop spotting. Automatic kill assessment ensures rapid engagement of successive targets.

For surveillance, the system uses an I-band (3 cm) high-power, pulse-to-pulse coherent search radar to detect small targets in all weather conditions and under severe jamming. A synthesiser-driven travelling wave tube transmitter is used to provide high power for burn-through and to permit flexibility in frequency and PRF to counteract jamming. A 60° vertical beamwidth search antenna is fully stabilised to ensure detection and track information independent of the ship's movements. The antenna has a rotation rate of 60 rpm. Detection probability is further enhanced by the application of processing techniques such as digital pulse-compression and fast Fourier transformation. Digital moving target detection is used to cope with clutter conditions and ECM. ECCM includes amongst others, PRF stagger and frequency diversity. Dual receiver channels are used to speed-up doppler processing, plot extraction and track build-up. This 2-D track data is used for threat evaluation including target priority, followed immediately by target assignment for the tracking radar and weapon. Continuous search while engaging a target allows fast reaction against subsequent targets.

After target designation the target is acquired by the dual frequency (X- and Ka-band) tracking radar. The broad I-band beam provides fast target acquisition. The narrow K-band beam provides accurate and track data down to sea surface level and continuous, automatic comparison is made of the signal to noise ratios of the X- and Ka-band returns to maintain tracking in clutter or degraded conditions. This system has been tested against missile targets flying as low as 5 m. Digital fire control processing including curved path prediction, is used to predict the point of impact and automatic calibration and closed-loop correction of the point of impact are used to compensate for bias type of errors, including inaccurate ballistic data. There is an automatic kill assessment subsystem for use in case of multiple attacks to optimise the system's effectiveness.

The 30 mm GAU-8/A gun has a rate of fire of 4200 rounds per minute. This gun and the prototype EX-83 naval mounting are described in entry **4042.203** in the Naval Guns section. The mounting has an ammunition

Goalkeeper close-in weapon system mounted on Dutch frigate Callenburgh capacity of 1190 rounds carried in a linkless system using a feed and storage drum. To reload the gun a bulk loading system may be employed but manual loading is possible using a manual loading tray. Spent cases are returned to the drum to prevent deck debris after firing. The ammunition capacity is claimed to be sufficient for several target engagements before reloading is necessary. The feed drum and much of the ammunition feed system is protected below deck. The warhead kill capability of Goalkeeper against anti-ship missiles is achieved by use of MPDS rounds, produced by NWM de Kruithoorn, Holland.

This MPDS round has a high density tungsten alloy penetrator. For 'soft' targets HEI and TP type of ammunition can be used.

CHARACTERISTICS

Gun type: 30 mm Gatling-type GAU-8/A, seven barrels
Muzzle velocity: 1020 m/s TP, HEI; 1150 m/s MPDS
Rate of fire: 4200 rounds/minute
Feed system: Linkless, 1190 rounds capacity
Ammunition: MPDS, HEI, TP
Weight: 6372 kg (all-up with 1190 rounds)
Elevation/depression: +80/–20°
Radar (search): I-band
Antenna: Linear array, 2050 × 250 mm
Beamwidth: 1.5° (horizontal); 30 or 60° (vertical)
Stabilisation: 2-axis
Rotation speed: 60 rpm
SLS antenna: Integrated with search antenna
Radar (tracking): Dual frequency, X/Ka-band
Antenna: Cassegrain, 1000 mm diameter
TV camera: Compatible with CCIR, Type B

STATUS

In service with the Royal Netherlands Navy. The Goalkeeper system has been ordered for the Netherlands and British Navies with total orders being for 83 systems. In February 1987 the first system was installed in a Type 22 Batch 3 frigate.

CONTRACTORS

Hollandse Signaalapparaten BV, PO Box 42, 7750 GD Hengelo, Netherlands.

General Electric Company, Armament Systems Department, Lakeside Avenue, Burlington, Vermont 05402, USA.

4016.231

SEAGUARD CIWS

Seaguard CIWS (Close-in Weapon System) is one specific configuration of the modular Seaguard naval combat system (**3509.281**) and consists essentially of a dedicated Search Radar Module (SRM) with its associated Common Support Module (CSM), the fire control tracking module (TMK) and the Sea Zenith 25 mm quadruple-barrelled gun module (25 mm GM). By being modular one or more TMK and 25 mm GM modules may be configured to provide a multiple channel CIWS outfit.

The SRM is a dedicated high data rate, high accuracy, dual beam C-band radar that provides target detection and indication of sea-skimming and high diving missiles, in severe environmental and ECM conditions as well as a near hemispherical general air and surface surveillance.

The CSM is the interface locus for all information to and from the Seaguard search, tracking and weapon modules. It provides the processor capacity for the Seaguard SRM data processing and track-while-scan of air and surface targets. In automatic engagement modes it provides correlation, threat evaluation and weapon assignment of air targets.

The CSM's floppy disk controller also stores the software program related to logistics (diagnostic test) and training programs (training scenario generation).

The TMK is a unique, high dynamic, three-axis multi-sensor (Ku-band or X-band radar, FLIR and laser), all weather, acquisition and tracking system (stabilised with a local strapdown device), impervious to the multi-path effect. The TMK enables tracking down to close-in ranges of high speed manoeuvring missiles throughout the hemisphere including sea-skimming and zenith diving missiles in a dense electronic warfare environment. The multi-sensors (radar and FLIR/laser) are intimately combined with specialised and innovative signal processing and advanced control algorithms into a very modern, high performance, missile/aircraft tracking system.

Detailed monitoring and mode selection of the Seaguard CIWS, is performed by a single operator at the tracker/weapon console (T/WC).

The 25 mm GM is a high dynamic, canted, two-axis gun mount carrying four independent KBB 25 mm cannon having a flexible and combined rate of fire of 3400 rounds per minute, permitting engagement of high speed manoeuvring missiles down to close-in-ranges throughout the hemisphere (–14° to +127°) including sea-skimming and zenith diving missiles, in all conditions of ship motion (stabilisation). The below-deck ammunition feed system, protected from environmental and NBC conditions, has a ready-use capacity of 1660 rounds (18 engagements) and permits topping up during firing.

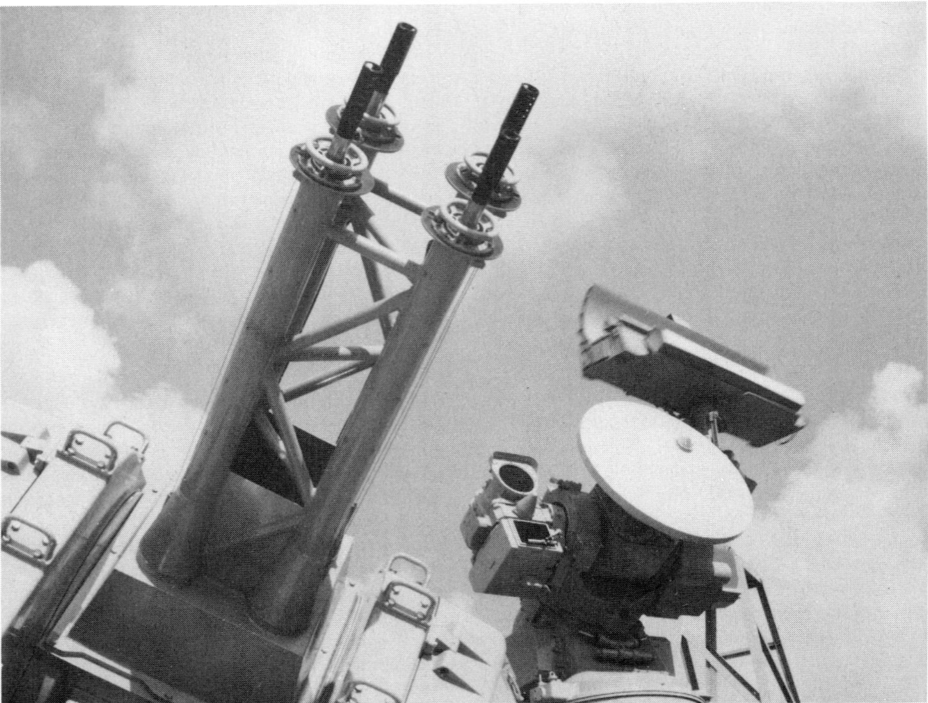

Seaguard CIWS showing quad 25 mm barrels, the TMK tracker and the SRM search radar

The kill mechanism relied upon is based on a single or multiple hit mechanism, detonating the warhead and/or causing catastrophic damage. This is achieved by the very high energy of the 25 mm KBB, AMDS ammunition's subcalibre penetrator. The Sea Zenith gun and its AMDS ammunition are part of the Oerlikon/BMARC development programme where the boundaries of ordnance technology have been considerably advanced. A full range of other ammunition types, eg HEI, may be used according to the target requirement. The overall kill probability of the Seaguard CIWS has been assessed as greater than 90% against a single missile, whatever its terminal trajectory (sea-skimming or zenith diving) with a speed of up to Mach 3. It is also effective against aircraft, helicopters and RPVs.

The modular concept of the Seaguard CIWS provides the integration flexibility necessary for a modern warship to fulfil its dedicated mission with optimum ship survivability.

STATUS

In series production following a contract from a West German shipyard for Seaguard system to be fitted on four 'Yavuz' class frigates under construction for the Turkish Navy.

CONTRACTORS

Contraves AG, Schaffhauserstrasse 580, 8052 Zurich, Switzerland.

Contraves Italiana SpA, Via Affile 102, I – 00131 Rome, Italy.

Machine Tool Works, Oerlikon-Bührle Ltd, Birchstrasse 155, 8050 Zurich, Switzerland.

Plessey Radar Ltd, Oakcroft Road, Chessington, Surrey KT9 1QZ, England.

British Manufacture & Research Company Ltd, Springfield Road, Grantham, Lincolnshire NG31 7JB, England.

ISRAEL

4200.231

BARAK 1 POINT DEFENCE SYSTEM

This is a joint development venture by Israel Aircraft Industries Ltd and Rafael Armament Development Authority for the Israel Defence Forces (IDF) to counter aircraft and missile attacks, including sea skimmers. Described as a compact modular lightweight system, Barak is capable of multiple target engagement under all weather conditions by day or night. It is comprised of a vertical launch unit above deck; search, track and guidance radar on the mast and control units below deck.

Development began in 1979 and the early system comprised three main units: a deck launcher containing eight Rafael point defence missiles (PDMs) on which was mounted an autonomous radar, a fire control console mounted below decks in the ship's combat information centre, and a processing unit. The radar on the launcher unit, in addition to pointing with the launcher, had its own rapid scanning capability. It was a dual-band radar with a single antenna providing both monopulse tracking and target illumination facilities. Eight missiles were carried in the launcher unit, each of these being supersonic and highly manoeuvrable and carrying a blast fragmentation warhead equipped with contact and proximity fuzes.

The missile incorporates cruciform wings and tail control surfaces, the former being positioned in the rear portion of the missile body, close to the tail fin leading edges. Missile propulsion incorporates thrust vector control, for both the jet vanes and tail surfaces, for vertical launch. This unit is ejected after six seconds of flight and the solid-propellant single-chamber triple thrust rocket motor takes over. Separate search and fire control sensors are now used while the vertical-launch installation requires only 2 sq m of deck space.

Barak 1 is the latest version of this system and was publicly revealed for the first time at the 1983 Paris Air Show. It differs in adopting the latest vertical launch technique for the missiles and siting the guidance radars higher up on the ship's superstructure. Apparently, the missile is virtually unchanged, apart from modifications to produce the necessary pitch-over movement when clear of the launching ship toward the target.

Vertical launch units (VLUs) containing up to eight missiles each can be fitted anywhere on the vessel to provide a total of 32 missiles per system, operated from a single below-deck console.

Missile range is about 10 km and both aircraft and anti-ship missile targets can be engaged within 360° in azimuth and between –25 and +85° elevation.

A large warhead (22 kg) is fitted, complemented by a sophisticated adaptive proximity fuze. Flight control of the missile is by a combination of moving tail surfaces which operate aerodynamically and four vanes mounted internally in the rocket motor efflux, the latter probably being necessary to accomplish the initial pitch-over movement as well as assisting in missile manoeuvrability over-all. The aerodynamic tail controls are probably necessary to ensure missile control after rocket motor burn while the missile is 'coasting'. One VLU with eight missiles weighs 1300 kg and the radar and fire control gear accounts for a similar weight.

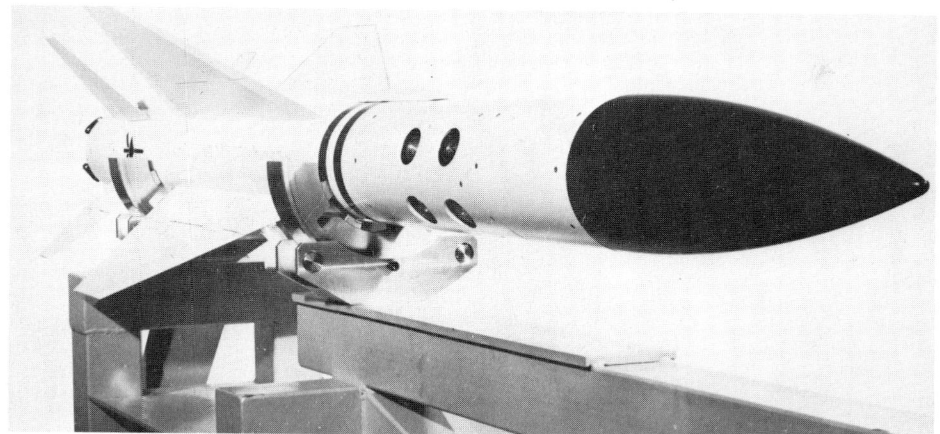
Barak point defence missile

CHARACTERISTICS
Length: 2175 mm
Diameter: 170 mm
Span: 680 mm
Weight: 86 kg
Min/Max range: 0.5-10 km
Time to max range: <30 sec
Speed: 580 m/sec

STATUS
Ground testing began in May 1986 and by the second half of 1987 it successfully intercepted a TOW missile at a distance of 5 to 6 km. The first sea trials were scheduled for 1988-89 and it is to be installed in *Sa'ar* 5' class FAC in 1989. The total development cost is reported to have been the equivalent of $74 million.

CONTRACTOR
Israel Aircraft Industries Ltd, Ben-Gurion International Airport, 70100, Israel.

4563.231
TCM 30 NAVAL AIR DEFENCE SYSTEM

The MBT Weapon System plant, part of the Electronics Division, Israel Aircraft Industries Ltd include among its range of air defence products a version of the Spider AA gun-based air defence system that is intended for use aboard ships of all sizes. The principal weapon employed is the TCM-30G twin 30 mm gun mount (**3729.203**), and up to six of these mounts are combined with a search radar, an optical sight, a tracker-director unit, and control consoles to consititute a flexible autonomous naval version of the Spider II land-based air defence system (**4508.131**). The twin 30 mm gun mount is incorporated in an electrically driven turret, the two automatic cannon and their feed systems being carried on a high angular acceleration slewing and elevating mounting.

It is probable that combined installations have been designed in which both TCM-30G turrets and Barak point defence missiles (**4200.231**) are integrated to form a single ship's air defence system. It is also likely that the recently introduced Elta automatic missile detection radar (AMDR) (**5696.253**) will make an appearance in combination with either or both the gun and missile naval point defence systems. The AMDR is a shipborne radar designed to automatically detect incoming sea-skimming missiles in addition to the normal function of air surveillance with automatic detection and track-while-scan facilities. This will provide rapid and reliable automatic alarm of incoming missiles; automatic detection and tracking of all high speed targets; threat evaluation, designation and data transmission to other defensive systems, such as a ship's ESM and command and control.

Twin 30 mm automatic cannon on TCM-30G mount as used in naval air defence system

CHARACTERISTICS
Main armament: 2 × Oerlikon 30 mm automatic cannon
Traverse: ±160°
Elevation: –20 to +85°
Acceleration (azimuth and elevation): 10 rad/s²
Max speed (azimuth and elevation): 1.5 rad/s
Muzzle velocity: 1080 m/s
Rate of fire: 2 × 650 rounds/minute
Overall length: 3550 mm

STATUS
Some vessels of the *Sa'ar* class and the *Shimrit* class hydrofoils are fitted with the TCM-30G.

CONTRACTOR
MBT Weapon Systems, Israel Aircraft Industries, Industrial Zone, Yahud, Israel.

ITALY

2228.231
ALBATROS NAVAL DEFENCE SYSTEM

Albatros is a medium range SAM system designed to combat threats from aircraft and missiles, either sea-skimming or diving. It is capable of providing either point or limited area defence. The system employs the Aspide missile (**1656.131**).

The concept of the Albatros system is of the 'add-on' SAM system by which the missile system is integrated with the ship's gunnery radar fire control system avoiding the need for duplicating the tracking and illuminating sensors.

In addition to being integrated with the principal current Selenia-Elsag radar fire control systems (NA 21 and NA 30) and Hollandse Signaalapparaten (Mk 2 Mod 9), the Albatros system is compatible with Ferranti, Marconi, PEAB and Thomson-CSF fire control systems. In each configuration Albatros is associated with a single or double tracker system, the latter requiring the use of two CW illuminator transmitters.

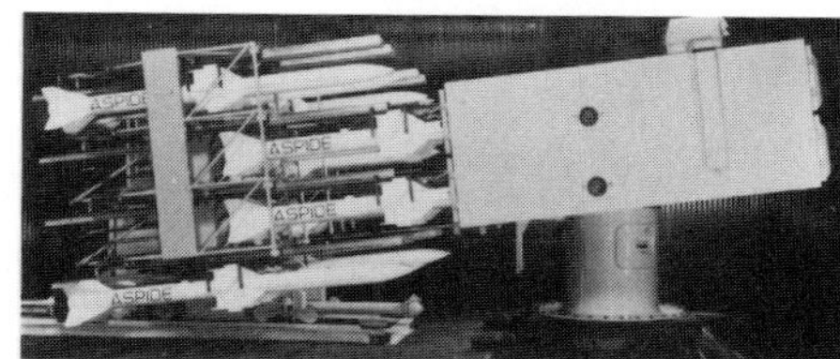

Riva-Calzoni 16-cell missile magazine for Albatros system

The missile basically consists of a guidance section, which interfaces with the ship's radar Fire Control System (FCS) and a launching section. The guidance section includes the CW transmitter and the system control units. The CW transmitter provides the RF power necessary for the missile guidance which is pointed towards the target through the FCS tracker.

The launching section comprises the launcher mounting and relevant control units. Two versions of the launcher can be supplied, the standard 8-cell or lightweight 4-canister.

The storage and reloading section associated with

the 8-cell launcher allows for increased availability of ready use missiles (up to 16) and provides for servo controlled loading and unloading of the launcher.

The Albatros system, in the different available configurations, has been installed in ships varying in size from cruisers to 600 ton corvettes.

STATUS
In current production. More than 100 systems in service with the Italian and 13 other navies.

CONTRACTOR
Selenia-Elsag Naval Systems Consortium, Via di S. Alessandro 28/30, 00131 Rome, Italy.

Albatros 4-cell launcher on Ecuadorean corvette

Launch of Aspide surface-to-air missile from Albatros 8-cell launcher

4497.231
DARDO CLOSE-IN WEAPON SYSTEM

The Dardo system, specifically dedicated to close range anti-missile defence, has been designed and developed by Elettronica San Giorgio – ELSAG, in collaboration with Selenia. Its prime purpose is to counter effectively high speed, late detection attacks requiring very rapid response, and especially attacks by sea skimming missiles, missiles with a diving attack profile, and aircraft-launched rockets and missiles.

Thanks to a direct electronic link with the ship's own main surveillance radar, the Dardo system can operate without human intervention as soon as it is triggered by the detection of a fast low flying target. Manual operation, as a conventional FCS, is also possible. Specifically, the Dardo system carries out the following functions automatically:

(1) correlation of successive target echoes to plot the flight trajectories
(2) assignation of a priority index, continuously updated, based on time remaining to reach own ship, of a significant number of threatening targets
(3) selection of the most threatening target
(4) auto-designation of the target to the system tracking radar
(5) automatic acquisition and tracking initiation
(6) automatic opening of fire at the maximum effective firing range permitted by the tactical situation
(7) auto-spotting on the burst of rounds fired.

The tracking radar is a Selenia Orion RTN 20X (**1935.253**). A TV camera is mounted on the same tracker pedestal.

The Selenia Orion RTN 20X is a modern X-band radar mounted separately from the gun and performs the following functions: automatic acquisition of very low flying and fast targets; autonomous detection of targets at low elevation with a 'guard ring' for target lock-on; automatic target tracking; effective reduction of the 'image effect'; auto-spotting of the burst of rounds fired.

These functions are performed despite clutter and or a hostile ECM environment thanks to: an X-band monopulse antenna system; employment of a coherent MTI for fixed echo cancellation; high PRF; frequency agility and staggered PRF; digital processing of the video inputs.

The data processing and interfacing unit includes a digital computer, specially adapted to fire control problems with special fixed routines; an interfacing electronic module assembly to link the processing section to the peripherals; and a full range of system ON/OFF line test facilities.

This unit, which is directly connected to the ship's surveillance radar, processes the information received, evaluates the threats and performs a decision on which target is to be engaged.

A supervision console, which also allows the operator to carry out a manually assisted firing, is associated with the system for control and supervision of correct operation.

STATUS
In series production and operational with the Italian and several other Navies.

CONTRACTOR
Selenia-Elsag Naval Systems Consortium, Via di S. Alessandro 28/30, 00131 Rome, Italy.

Dardo CIWS on board helicopter carrier Garibaldi

SPAIN

4564.231

MEROKA NAVAL CIWS

Meroka is a novel type of close-in air defence weapon system that has been under development for a number of years by the Spanish authorities. Both land-based and naval versions have been designed; the former is described in entry **3208.131** in the Mobile Guns and Rocket Systems section of this volume. Some of the background to the programme has appeared under that entry number in previous editions.

The weapon used is a twelve-barrelled arrangement of 20 mm/120 calibre Oerlikon automatic cannon, grouped together in two horizontal rows of six. A salvo of 12 rounds can be fired in a time of 0.08 seconds, which amounts to a theoretical rate of fire of 9000 rounds/minute. The main purpose of this high rate is to meet a need for maximum density of fire when engaging certain targets such as low-flying ground attack aircraft and sea-skimming anti-ship missiles. The normal practice is for each salvo to be fired as four groups of three rounds, minimising the total recoil force applied to the mount as well as having other beneficial operating characteristics.

The group of 12 barrels forms a single rigid unit which is traversed and elevated as a conventional gun, mechanical rigidity being aided by a number of steel bands (two in the army version, four in the naval version) that determine the spacing between the individual barrels. This feature is also used to adjust the dispersion of fire and fire pattern. Ammunition is brought to the guns by two belt feeds which are supplied from an hydraulically powered rotating magazine with a capacity of 720 rounds. The gun can be traversed in azimuth through 360 degrees at a rate of 90 degrees/second. The guns fire standard Oerlikon HE, AP and incendiary ammunition although a sub-calibre kinetic energy round is being developed by the Spanish Navy.

The radar selected for target detection was the Selenia RAN-12L L-band search radar, a frequency-agile equipment with a range of about 15 km. Target assessment, threat evaluation and related functions are aided by a 16-bit microprocessor, which can handle data for up to four Meroka turrets. Individual targets are designated to weapons which are automatically laid-on by direct links between the computer and the guns' control systems.

The digital fire control system was designed by Lockheed Electronics who have been associated with the Meroka programme from a very early stage, and the antenna of that company's AN/VPS-2 pulse doppler tracker radar (as used in land-based Phalanx air defence gun system) is mounted on top of the Meroka turret. This X-band radar will automatically track targets from a range of just over five km. A secondary line-of-sight for target tracking is provided by a General Electric low-light TV camera, bore-sighted to the gun axis. The latter can be used by an operator using a joy-stick control for manual target tracking in an emergency.

Meroka CIWS which has been installed in a Santa Maria class frigate

CHARACTERISTICS

Dimensions: 3.2 m × 2.2 m (at deck) 2.9 m (swept radius)
Reloading time: 3 minutes approx
Muzzle velocity: 1215 m/s
Weapon: 12-barrel 20 mm Oerlikon/120 cal
Max rate of fire: 2700-3600 rounds/minute
Reloading time: 3 minutes approx
Barrel length: 2.4 m (120 cal)
Recoiling mass: 300 kg
Magazine capacity: 720 rounds
Elevation: –20 to +85°
Traverse: Unlimited 360°
Effective range: 3000+ m
Turret weight: 4.5 t
Off-mount equipment: 421 kg

STATUS

At least 20 systems have been produced or are on order for the Spanish Navy. These have been, or will be, fitted to the *Principe de Asturias* support carrier, the *Santa Maria* class frigates, and retrofitted to *Descubierta* and *Baleares* class frigates. The first ship with the system was commissioned in September 1987 and a Mk 2 version is under development.

CONTRACTOR

Development by CETME, (Centro de Estudios Tecnicos de Materiales Especiales), Spain.

Production by Empresa Naçional Bazán, San Fernando, Spain.

SWEDEN

4546.231

RBS 70 RAYRIDER MISSILE SYSTEM

The RBS 70 ship launched missile system (SLM) is a recent proposal to adapt the well-established RBS 70 missile system for use on naval vessels. RBS 70 SLM is capable of combating incoming aircraft at long range. The ordinary RBS 70 sight (**2348.131**) is used with its built-in gyro-stabilised optics. The low weight and small dimensions of the system allow it to be fitted on very small vessels such as patrol boats and minesweepers.

During 1983, tests, arranged by the Swedish Navy in collaboration with Bofors, were carried out on a Swedish coastal minesweeper under varying weather conditions and included live firings. The firings took place with the vessel, an 'Arko' class coastal minesweeper, under way at normal speed. A prototype of the RBS 70 SLM installation was mounted on a platform above the large sweeping winches on the after deck. Tests in foreign countries have also been carried out.

The stand for the army version of RBS 70 has been replaced by one specially designed for naval use, and this is fitted directly to the deck. Compared with the army version, RBS 70 SLM has a greatly increased resistance to cross-winds during the initial launching stage.

RBS 70 ship launched missile at sea

The system has been provided with firing limits which prevent the missile being launched on a course which would endanger the vessel or its crew. RBS 70 SLM is operated by one man who sits on the stand seat and operates the sight in exactly the same way as the army version. Reloading can be carried out as in the land version. RBS 70 SLM can receive target data from an on-board radar.

In conjunction with Signaal a quadruple launcher is under development. It allows the operator to fire under cover. Signaal is providing its LIOD tracker.

CHARACTERISTICS
Range: 5 000 m
Height coverage: 3 000 m
Reloading time: Up to 20 s
Time from target detection to launch: From 5 s
IFF: Optional

STATUS
Ready for production. No information on production or orders has been released.

CONTRACTOR
AB Bofors S-691 80 Bofors, Sweden.

UNION OF SOVIET SOCIALIST REPUBLICS

2939.231
GOA SHIPBORNE SURFACE-TO-AIR MISSILE (SA-N-1)

Known by the US alpha-numeric code SA-N-1, the missile, whose NATO code-name is Goa (Soviet designation M1), was for many years the principal surface-to-air missile of the Soviet Navy. It is fitted to the 'Kanin', 'Kashin', 'Kotlin', 'Kresta', and 'Kynda' classes on scales ranging from one to four twin launchers.

The shipborne Goa missile is assumed to be identical with the ground-to-air missile (**2938.131**) but the equipment associated with it is very different. The standard launcher is clearly roll-stabilised as can be seen from the accompanying illustration. Rather different, and not obviously stabilised, launchers have also been observed, but these are most probably not for Goa. The launcher is mounted on top of the missile magazine and is reloaded vertically through small hatches.

Twin launcher for SA-N-1 Goa on 'Kotlin'-class ship (G. Jacobs)

Associated with the missile in all installations is a compound radar system known by the NATO code-name Peel Group (**1323.253**) and described in the Naval Radar section. Guidance is probably by radio command.

CHARACTERISTICS
Range: 31 500 m
Weight: 400 kg
Dimensions: 6.6 m × 701 mm (max)
Wingspan: 1.5 m
Propulsion: Solid fuel rocket with tandem solid fuel boosters
Guidance: Radio command
Warhead: 60 kg conventional

STATUS
Entry into service with the Soviet Fleet was in 1961. There are single systems on each of eight SAM 'Kotlin' class, eight 'Kanin' class, and four 'Kynda' class ships, while 19 'Kashin' class, and four 'Kresta I' class vessels each have two twin launcher installations. One Polish, ex-Soviet 'Kotlin' class, ship is fitted with Goa.

SA-N-1 Goa naval air defence missiles in firing position

2943.231
GUIDELINE SHIPBORNE SURFACE-TO-AIR MISSILE (SA-N-2)

Guideline, also known in its shipborne application by the US alpha-numeric code SA-N-2 (Soviet designation M2), has been installed in only one Soviet ship, so far as is known, the cruiser *Dzerzhinski*.

It has been suggested that one reason for this is the difficulty of providing suitable stable platforms on smaller vessels for the flapping Fan Song (**2866.153**) radars used to guide these missiles. It could also be that the relative difficulty that there appears to be in gathering these missiles to the required flight path presents even more of a problem in shipborne installations than it does on the ground.

So far as is known the general performance of Guideline as a shipborne missile is substantially the same as that described in the entry for the ground-based version.

CHARACTERISTICS
Range: Approx 44 000 m
Weight: 2300 kg
Dimensions: 10.7 m × ?
Wingspan: 2.2 m
Propulsion: Solid fuel rocket with liquid fuel booster
Guidance: Radio command
Warhead: 130 kg conventional
IOC: 1961

STATUS
Obsolescent; presumably retained only for trials.

Close-up of the massive twin launcher for SA-N-2 Guideline on Dzerzhinski (G. Jacobs)

2947.231
GOBLET SHIPBORNE SURFACE-TO-AIR MISSILE (SA-N-3)

The helicopter-carrier *Moskva* and the 'Kresta II' class cruisers were the first ships of the Soviet Navy to be seen equipped with a new surface-to-air missile that is similar in size to Goa (**2939.231**) but possibly larger and certainly launched from a different launcher and directed by a different radar complex. It has since appeared on other classes.

Although there was confusion concerning this missile it now seems to be agreed that it is that classified in the American code as SA-N-3; it is also widely believed – without, of course, any official confirmation – that its NATO code-name is Goblet.

SA-N-3 Goblet air defence missile firing, with another round on the ramp

The launcher does not appear to be roll-stabilised – suggesting that the missile gathering is efficient. The missile is of conventional configuration (see photograph) and the associated radar fire control group has the NATO codename 'Head Light' (**1328.253**). The missile has been credited with a maximum speed of over Mach 3.

The four reload hatches associated with the launcher are generally accepted as indicating that the launcher is a dual-purpose device and that the magazine below deck contains Goblet, anti-aircraft missiles and possibly SS-N-14 anti-submarine weapons also.

CHARACTERISTICS
Range: Approx 55 500 m
Weight: 550 kg
Dimensions: 6.2 m × 335 mm
Wingspan: 1.5 m
Propulsion: Ramjet with solid fuel booster
Guidance: Semi-active homing
Warhead: 80 kg conventional

STATUS
The system entered service in 1967, and it is fitted in the 'Kiev' and 'Moskva' class carriers, and 'Kara' and 'Kresta II' class cruisers.

2954.231

SA-N-4 SHIPBORNE SURFACE-TO-AIR-MISSILE

Since its appearance and entry into operational service in the early 1970s, the SA-N-4 naval surface-to-air system has become the Soviet Navy standard close-in air defence weapon system, fitted to about 120 ships of various classes. The Soviet designation is reported to be P-20 or P-21.

The system quickly became noted for its drum-shaped container magazine/launcher mounting with its 'pop-up' twin ramp launcher, and the associated radar group codenamed Pop Group (**1897.253**). The launcher is about 5 m high, 4.2 m in diameter and can be turned through 360° in azimuth and 90° in elevation. Later analysis of the container credits it with a capacity of 20 reload missiles disposed in four revolving drums in the upper part of the silo. Electro-mechanical and hydraulic support machinery is in the lower part of the silo.

The silo is provided with a flat cover plate in two sections that slide apart to allow the launcher to rise to the firing position. The cover plates close before firing takes place.

The SA-N-4 missile has an estimated maximum speed of around Mach 2.5. The similarities between the SA-N-4 and the SA-8 Gecko land-based SAM have been noted by western observers and it is now generally accepted that the land-based and naval weapons are two versions of the same system. The similarity extends to the associated radar groups of the SA-N-4 and Gecko versions.

CHARACTERISTICS

Range: Approx 14 800 m
Weight: 190 kg
Dimensions: 3.2 m × 210 mm
Wingspan: 0.64 m
Propulsion: Solid fuel rocket
Guidance: Semi-active homing
Warhead: 50 kg conventional

STATUS

Operational since the early 1970s and fitted in a wide variety of ship classes over a relatively long period. Major classes, such as the 'Kiev', 'Kirov', 'Slava' and 'Kara' are each fitted with two SA-N-4 silos equipped with a total of 40 rounds per ship, and the 'Krivak' and 'Koni' class frigates have the same size of SA-N-4 installation. Still more numerous are the 'Grisha' I and II and 'Nanuchka' light missile craft all of which have single silos with 20 rounds. The system is also used to provide air defence for certain important vessels such as the replenishment ship *Berezina*, the two amphibious warfare ships of the 'Ivan Rogov' class and the 'Nanuchka' type corvettes.

SA-N-4 launcher in ready to fire position with missiles on ramp

4201.231

SA-N-5 SURFACE-TO-AIR MISSILE

SA-N-5 is the designation of the naval version of the man-portable anti-aircraft missile SA-7, Grail, (**2941.131**). For use aboard ships the four-round multiple version is employed. It is a relatively simple conversion of the original man-portable battlefield weapon, basically consisting of a framework on which four Grail rocket launcher tubes can be fixed side-by-side and supported on a central pedestal which permits the gunner to slew and train the missiles manually. Performance of the missile (Soviet designation Strela 2M) is likely to be substantially the same as in land-based use, although the task of manually acquiring the target and aiming the weapon cannot have been made any easier for the operator by being mounted on a platform subject to sea motion as well as the ship's own course changes during any engagement.

Nevertheless its potential as a deterrent to close investigation by hostile air forces (especially helicopters), is probably well worth the modest cost of such fittings.

CHARACTERISTICS

Range: Approx 10 000 m
Weight: 9.2 kg
Dimensions: 0.76 m × 69.9 mm
Propulsion: Solid fuel rocket
Guidance: IR homing
Warhead: 2.5 kg

STATUS

Now very extensively fitted in a wide variety of ship classes. *Jane's Fighting Ships* reports at least 169 Soviet vessels fitted with SA-N-5 installations that consist of one, two, three or four quadruple launchers for the system, although fittings of two quadruple launchers per ship are most common.

The most numerous classes of vessel fitted with this system are those built for amphibious operations, such as the 'Ropucha' class (17 ships) some of which have four quadruple mountings, the 'Alligator' class (14) with two or three twin mountings, 'Polnochny' ships (14) with two or four quad mountings, and the two 'Ivan Rogov' class ships which each have two quad SA-N-5 mountings in addition to their SA-N-4 missile silos. Intelligence gathering vessels of many types have been given SA-N-5 installations for their protection; such ships include those of the 'Balzam' type AGIs, 'Primorye', 'Nikolay Zubov', modified 'Pamir', 'Moma', 'Mirny', 'Mayak', and 'Okean' classes. Most of these have two quadruple mountings and between them these classes amount to a total of between 40 and 50 ships.

Other important vessels with SA-N-5 systems are seven 'Ugra' submarine support ships (some with two quad mountings) and 'Lama' class (seven vessels) missile support ships with either four or two quad mountings. Certain light missile craft, such as the 'Osa', 'Tarantul' and 'Pauk' classes are equipped with SA-N-5 mountings also.

4206.231

SA-N-6 (GRUMBLE) SURFACE-TO-AIR MISSILE SYSTEM

SA-N-6 is the designation of the vertically launched naval air defence missile system first reported aboard the Soviet cruiser *Kirov*, commissioned in July 1980, and since noted in her sister *Frunze* and the smaller 'Slava' class cruisers. There are no known photographs of the actual missile but it is generally supposed to be a naval version of the land-based SA-10 air defence system (**3620.131**).

From photographs that have been obtained of the SA-N-6-fitted ships mentioned above, it appears that two kinds of launching arrangements have been devised: the original *Kirov* installation was noted for its array of 12 hatches covering the vertical launchers on the ship's foredeck, immediately forward of the broadly similar (but larger) set of 20 hatches for the SS-N-19 surface-to-surface missiles that arm *Kirov*. The SA-N-6 hatch covers are rectangular in shape and give little indication of the arrangements for firing, loading, replenishing and reloading below deck. However, the 'Slava' class ships which made their appearance somewhat later have a different arrangement for the eight SA-N-6 launcher units that are located abaft the funnels of these ships. In two rows of four running fore-and-aft, these have the appearance of eight sets of eight vertical tubes arranged in a circle, with each set having what could be an external hatch lifting mechanism, pivoted in the centre of the eight tubes and able to be located over each tube. This mechanism might also be connected with replenishing the magazine of each group of eight tubes.

The 'Kirov' class ships are also assumed to have eight rounds per launcher, giving the larger 'Kirov' class 96 missiles per ship, and 'Slava' class 64.

Other external evidence is the Top Dome radar director group (**4088.253**) (two in 'Kirov' class ships, one in 'Slava' class) and various other types of sensor that may play a part in the operational use of the SA-N-6 system. The latter include both passive and active types, some of which undoubtedly provide target detection and designation services for the air defence fire control system. Interpretation of such details is assisted if some assumptions are made concerning the likely operational role of the SA-N-6 system. While by no means necessarily inspired by the American Aegis naval area air defence system (**2507.231**), it is a reasonable guess that the Soviet requirement was for a system to meet a very similar potential threat. This envisages aircraft, aircraft-launched anti-ship missiles, stand-off jamming aircraft, reconnaissance aircraft, and anti-ship missiles launched by both surface vessels and submarines. All-round hemispherical cover is required out to long ranges with a multiple target detection and tracking capability, coupled with high resistance to ECM and jamming. In the Aegis system, the missiles are equipped with mid-course guidance and terminal homing facilities, and these may be expected in the Soviet SA-N-6. At least one respected authority has revised the estimated maximum speed downwards to Mach 3.

Certain aspects of the numerous sensors fitted on the *Kirov*'s superstructure suggest various other possibilities relating to the SA-N-6. Target detection is technically possible by a wide selection of sensors: the Top Pair (**4087.253**) and Top Steer (**4086.253**) air search radar groups are capable of long range air target detection and position fixing and enjoy high placings on the ship, thereby maximising their range potential; also located high on the ship's structure there are dual installations of Round House and Rum Tub passive ESM antenna arrays which obviously provide 360° detection facilities. Lower down on various parts of the upperworks are numerous other shorter-range sensors of different types which could serve for target detection and/or tracking at closer ranges and lower elevation angles. The two Top Dome director radar groups (**4088.253**) are complex systems incorporating a number of different antenna elements, clearly denoting multipurpose operation. The largest of these is that which is concealed behind the large dome from which the group derives its NATO code-name and this probably serves to track both targets and outgoing missiles. Carried on the same trainable pedestal is a group of three semi-cylindrical antennas, the precise function of which is not clear, and above these is a small 'thimble'-shaped unit which might house a command antenna.

CHARACTERISTICS

Range: 55 000 m +

Length: 7 m

Propulsion: Solid fuel rocket

Guidance: Possibly track-via-missile

STATUS

Operational in 1977-78. Fitted in 'Kirov' and 'Slava' class.

Soviet battle cruiser Kirov *shows the varied range of sensors available for use in detecting targets for the SA-N-6 air defence missile system. Top Pair and Top Steer radar groups high on the superstructure give early warning of air targets, while the Top Dome radar director groups (fore and aft) track both target and missile* (Royal Navy)

4473.231

SA-N-7 (GADFLY) SURFACE-TO-AIR MISSILE SYSTEM

SA-N-7 is the number assigned to the naval air defence weapon system associated with the new type of launcher ramp that was first noted in the trials ship *Provorny* (a modified 'Kashin' class destroyer) and subsequently deployed in 'Sovremenny' class vessels. To date, the launcher is all that has been revealed publicly in the West of this new system although official US sources have claimed that the Gadfly missile is probably a naval version of the land-based SA-11 air defence weapon (**4470.131**), with a maximum speed of Mach 3.

Aboard the *Sovremenny* there are fore and aft SA-N-7 launchers, which appear to be single round units, although until photographs of one loaded are obtained it is not possible to be definite on this point. Reportedly associated with the SA-N-7 is what is assumed to be a target tracking/illuminating radar said to bear the NATO reporting name of Front Dome. During Black Sea trials in the *Provorny*, this ship was sighted with no less than eight of these radars, while the *Sovremenny* is fitted with six of them, according to the latest available photographs. Target designation to these individual radars is probably by one or more of the several air surveillance radars fitted in the ship, ie either or both of the Top Steer pair located at the highest point of the superstructure. It has been conjectured that to date the *Sovremenny* has not been seen in its final form, and it is true that there are several possible sites for additional radar antennas to be located.

The 'Sovremenny' class destroyer Otchyannyy, *in addition to starboard group of SS-N-22 anti-ship missile launcher/containers, reveals the new style launcher for the SA-N-7 Gadfly air defence missile. Also visible are Top Steer air surveillance radar group, Band Stand SSM guidance radar, Kite Screech gun FC radar, several of the Front Dome SA-N-7 radars, and various E-O director equipments and EW/ESM antennas* (G. Jacobs)

These provisional details suggest that the SA-N-7 is intended to deal with multiple targets, with a high degree of autonomy after assignment of individual missiles and lock-on to their respective targets. Command guidance is probable and there is also a possibility that the system incorporates provision for a back-up optically-guided CLOS mode to cope with 'saturation' air attacks where there could be more targets than there are tracking radars to deal with them. Such a mode of operation would also be a useful countermeasure against the attacker's ECM, or in the event of loss of own radar for any reason. Official US sources estimate that each ship of the 'Sovremenny' class has a capacity of 44 missiles for its SA-N-7 air defence system.

If the belief that SA-N-7 launcher ramps are single rail units, and that the system is designed to deal with multiple air threats at short to medium ranges, is correct, this implies that each launcher must be capable of rapid firing and reloading. A further reasonable conjecture might be that the overall fire control system has a high level of sophistication, possibly based on digital computers, to ensure optimum co-ordination of such aspects as target designation, threat evaluation, radar assignment, launcher pointing, reloading etc. If, as the above would seem to suggest, a number of simultaneous target engagements are contemplated, by means of a launcher system able to fire several missiles in quick succession, it is therefore likely that tracking radars are assigned to targets individually and the missiles home by semi-active radar, operating virtually autonomously after launch.

CHARACTERISTICS

Range: Approx 28 000 m

Length: 5 m

Propulsion: Solid fuel rocket

Warhead: Conventional

STATUS

Operational in DDG *Provorny* (for trials) and 'Sovremenny' class DDs.

4885.231

SA-N-9 NAVAL SURFACE-TO-AIR MISSILE

This is the short-range SAM fitted in the 'Udaloy' class destroyers, and the new 'battlecruiser' *Frunze*, previously identified as SA-N-8. It is clearly a replacement for the SA-N-4 point defence system but the *Frunze* retains the older system as well. The new system uses vertical launch, with an estimated 128 missiles in the *Frunze* and 64 in the 'Udaloy' class. The missiles are hidden underneath 6 ft diameter cover plates, and the new Cross Sword fire control system is provided for tracking.

STATUS

Apparently operational, but no FCS has been seen so far in the *Frunze*, and the first three of the 'Udaloy' class went to sea with both FCS platforms empty. The two FCS seen in the fourth ship *Admiral Zakorov* then disappeared in 1985, and the fifth ship was seen early in 1986 with only one in position. It was also seen in the sixth ship, *Admiral Tribukh*. All this suggests teething troubles.

The Cross Sword fire control system is reported to have serious technical problems although it is uncertain whether or not this is a matter of software or the missile itself.

Forecastle of the new missile cruiser Frunze *showing the two square hatch covers that are thought to mark the site of the SA-N-9 air defence missile launchers, immediately aft of another hatch cover for an unknown purpose and forward of the ship's SA-N-6 surface-to-air missile launcher tube covers on the foredeck* (via Jane's Defence Weekly)

Frunze's *quarterdeck showing the two covers on either side of the helicopter handling area that are expected to be occupied by vertical launchers for the SA-N-9 air defence missile when fitting out is completed. In the first of class,* Kirov, *these sites are occupied by two 30 mm Gatling guns* (via Jane's Defence Weekly)

2973.231

30 mm CLOSE-IN WEAPON SYSTEM

This 30 mm six-barrel Gatling type weapon has been installed on a large number of ships in the Soviet Navy and those of other countries. The US Navy designation is ADGM-630. It is intended for defence against cruise missiles and 'smart' weapons, and in concept is similar to the US Vulcan-Phalanx system (**2543.231**). The weapon is controlled by a Bass Tilt radar (**4085.253**) and is mounted on at least 21 different classes of ship, ranging from the 'Kirov' class battlecruisers to the 210-ton fast attack craft *Slepen*. This is a modular system without an integral fire control system but controlled by Bass Tilt radar.

The base of the modular mounting appears to be just over two metres across, with a mounting height of about 1.5 metres. The barrel and associated supports are a little under two metres long. The six-barrel mount is fed by a continuous shell feed system. It is probable that this weapon has a cyclic rate of fire of at least 3000 rounds per minute, with an effective range of probably around 500 to 600 metres. The mount is able to elevate to 90° and can swivel between 250 to 330° depending upon location.

The muzzle velocity is probably about 1000 metres per second with a maximum range of about 2500 metres per second but an effective range of 500 to 600 metres.

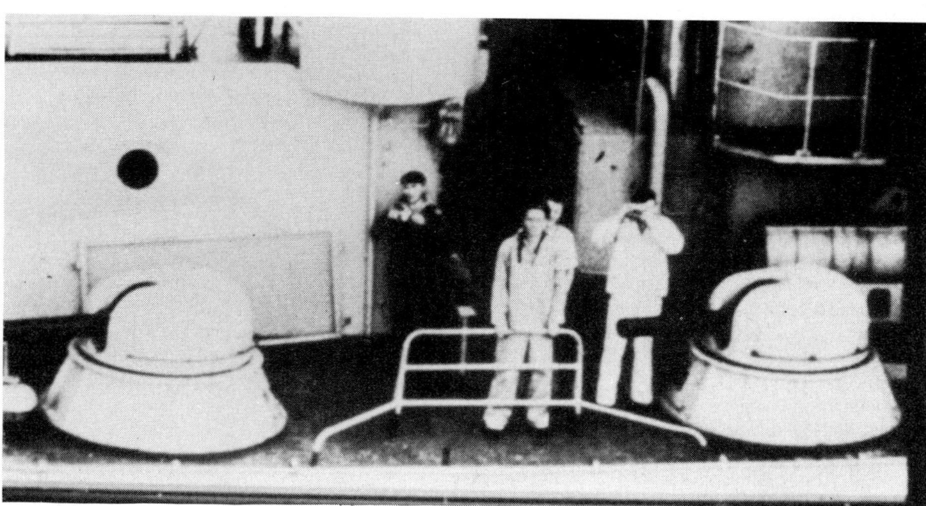

Barbette mountings on 'Kresta II' cruiser

STATUS

In operational service with the Soviet Navy and the 'Tarantul I' class missile corvettes of the East German and Polish navies. It has also been exported to the Indian Navy for its latest 'Mod.Kashin II' class destroyers.

6189.431

SA-NX-? UNDERWATER-TO-AIR MISSILE

The Soviet Union are believed to have a new missile under development, intended to provide submarines with air defence. The round will probably be fired from the sail while the boat is submerged, using data gathered from a periscope TV camera or from a towed sensor. The missile can also be presumed to have autonomous guidance after launch.

What little has been released about this project seems similar to Western projects such as SLAM, Subwinder and the SIAM Self-Initiating AA Missile.

UNITED KINGDOM

1019.231

SEACAT SHIPBORNE SURFACE-TO-AIR MISSILE

Seacat is a close range, shipborne, guided missile system for anti-aircraft defence which may also be used against surface targets within visual range. The guidance system is based on the use of optical tracking and a radio command link, but the system is capable of various degrees of sophistication.

OPERATION

The basic form of the system, introduced into the Royal Navy as Guided Weapons System Mk 20 (GWS 20), consists of separate launcher and operator mounts, the latter carrying sighting binoculars with which the aimer first 'gathers' the missile after launch and subsequently tracks it to the target. A thumb joystick control is provided for missile guidance.

Seacat has since been deployed with the RN GWS 22 (**1242.281** and **1563.253**) fire control system in which a lock and follow tracking radar is linked to the aimer's binocular sight. Subsequent operation is the same as for the GWS 20 system. It has also been combined with Contraves Sea Hunter, Elsag NA9, and Signaal M 40 series fire control systems. A later development is a system that replaces the optical sighting binocular with a closed circuit television (CCTV) system produced by Marconi-Elliot Avionic Systems as the 323 Series (**1021.293**). This enables the aimer to be placed in a much less vulnerable position and also results in important improvements in efficiency. The missile gathering phase has been reduced from about seven seconds (conventional optical technique) to six seconds or less with the aid of CCTV.

Shorts Seacat installed on an RN ship. The system is scheduled for service well into the 1990s

Operational versions of Seacat include:

GWS 20 – Modernised 'Rothesay' class frigates, LPDs *Fearless* and *Intrepid* and older ships since scrapped (simple optical director).

GWS 21 – 'Tribal' class frigates (three now sold to Indonesia), 'County' class DLGs (one sold to Pakistan and three to Chile) and older ships now scrapped (with CRBF director and Type 262 radar).

GWS 22 – 'Leander' class frigates (one, two or three), controlled by Mod MRS3 director with Type 903 radar.

GWS 24 – 'Amazon' class (Type 21) frigates, with WSA-4 and Type 912 radar.

A lightweight version of Seacat was developed for installation in such vessels as 30 m fast patrol boats and inshore minesweepers, giving them guided-weapon defence against low-flying aircraft. It is in service with the Brazilian Navy and the Iranian Navy.

It employs a three-round launcher weighing 2007 kg with all its ancillaries instead of the 4700 kg of the standard four-round Mk 20 launcher. Binoculars mounted on the head of a single manually operated pedestal director form a combined director and missile aiming sight.

The latest Seacat improvement is the introduction of a height control capability which enables the missile to be flown at heights down to 6 m above the sea against surface or sea skimming targets without danger of ditching. Further details will be found in the 1987-88 edition.

CHARACTERISTICS

Max effective range: Approx 5500 m
Warhead: HE
Solid fuel motor
Length: 1.48 m
Max span: 65 cm
Estimated weight: 63 kg

DEVELOPMENT

Development work started in the late 1950s and the first sea trials took place in 1962. The basic system has proved itself adaptable to other roles and a land-based version has been developed as the Tigercat (**2465.131**).

STATUS

In addition to the RN, Seacat was sold to 15 other navies: Argentina, Australia, Brazil, Chile, West Germany, India, Iran, Libya, Malaysia, Netherlands, New Zealand, Nigeria, Sweden, Thailand and Venezuela.

Eight Argentinian aircraft were claimed by Royal Navy Seacat missiles in the Falklands conflict of 1982.

CONTRACTOR

Short Brothers plc, Castlereagh, Belfast BT6 9HN, Northern Ireland.

6004.231

SEA DART SURFACE-TO-AIR MISSILE

Sea Dart (GWS 30) is a third-generation area defence weapon system, capable of intercepting aircraft, from very high to extremely low altitudes, and air and surface-launched missiles. It is also effective against surface vessels thus giving it a triple-role capability. Launch rate is rapid and the weapon system is capable of dealing simultaneously with many targets.

The missile is launched from a twin launcher. Performance data is classified and only the linear dimensions have been released for publication, other data in the accompanying table is taken from published unofficial estimates.

The missile is virtually a thick-walled cylinder built around the Rolls-Royce Odin ramjet engine. The guidance equipment (Marconi Defence Systems), the proximity fuze (THORN EMI), the control system (BAe Bracknell), and the fuel tanks and engine controls are all wrapped around the central air duct and ramjet in the annular space between them and the missile's outer skin. System prime contractor is British Aerospace.

Semi-active homing, proportional navigation guidance is employed. The ramjet engine enables the missile to fly at controlled speeds throughout its flight envelope. The guidance system will detect very small changes in target movement. This capability, combined with very fast control responses, enables very small miss distances to be achieved.

Launch of a Sea Dart missile from the British destroyer Bristol. *Sea Dart is an area defence system against high and low flying aircraft, with an anti-missile and surface-to-surface capability*

CHARACTERISTICS

Designation: GWS 30
Type: Shipborne area defence, surface-to-air, surface-to-surface, and anti-missile guided weapon system
Guidance principle: Radar guided semi-active homing using tracker illuminator radar Type 909
Guidance method: By control of movable tail surfaces
Propulsion: Solid-propellant booster with ramjet sustainer
Warhead: Continuous rod
Length: 4.36 m
Body diameter: 42 cm
Span: 91 cm
Launch weight: 550 kg
Range: 40 km

OPERATION

A fully automatic 22-round magazine is provided below decks. Missiles are stowed vertically and are hoisted through an intermediate position onto the electrically driven twin arm launcher. Targets are designated to the system in three co-ordinates by radar. The system automatically tracks the target and points the launcher. Radar illuminates the target to provide the missile with the RF signal for self-guidance on to the target. The Type 909 target tracking and illuminating radar (**1559.253**) produced by Marconi Radar Systems has been developed from the equipment used with Bloodhound and Thunderbird missile systems. The missile is boosted to speed by a solid fuel tandem boost and speed is sustained by an Odin ramjet burning a liquid fuel.

In 1986 a £100 million programme of warhead improvement was also announced. These include replacing the continuous rod with a blast and fragmentation type. A new Type 996 three-dimensional radar eliminates manual transfer of data from the search radar to the tracker.

Lightweight Sea Dart

In 1978 British Aerospace offered a lightweight version of the Sea Dart system, intended for ships down to 300 tons in size. In place of the GWS 30 missile launcher and missile magazine, the lightweight system employs fixed deck-mounted launchers, each carrying two or four missiles in sealed launch boxes. The Marconi 805 SD (**4302.253**) would be the system fire control radar.

DEVELOPMENT

Development started August 1962. Test firings began in 1965 and the first production order was announced in November 1967.

In mid-1982 BAe revealed a proposed long-range land-based air defence system designed around Sea Dart and called Guardian (**4524.131**).

STATUS

Sea Dart is used in the Royal Navy's 'Invincible' class aircraft carriers and 'Birmingham' (Type 42) class destroyers as well as the Argentinian Navy's 'Hercules' (Type 42) class destroyers.

CONTRACTOR

System contractor: British Aerospace plc, Naval and Electronics Systems Division, PO Box 5, Filton, Bristol BS12 7QW, England.

6751.231
SUPPORT DEFENCE MISSILE SYSTEM

A self-defence missile system for ships with a limited air defence capability is being considered by the Ministry of Defence. The Support Defence Missile System (SDMS) is intended to provide an important layer of naval air defence and its main features are high rate of fire, the capability of multiple engagements, high missile agility and the ability to engage even the most difficult targets.

In June 1987 British Aerospace Naval and Electronics (now Dynamics) Division and GEC Marconi revealed they had been awarded a £652 000 study contract for such a system. The study was to be completed in November. There were reports that Short Brothers had also been awarded a study contract.

2442.231
SEAWOLF SURFACE-TO-AIR MISSILE

The Seawolf close area defence missile was designed to intercept anti-ship missiles travelling at up to Mach 2.

The missile employs command-to-line-of-sight guidance with radar differential tracking and/or TV, both with radio command. Speed and manoeuvrability characteristics are suitable for the engagement of small, Mach 2 missile and aircraft targets under severe weather conditions and sea states.

Seawolf GWS 25 Mod 0

The complete system comprises the following units:
(1) air and low-air surveillance radars, Type 967
(2) radar trackers, Type 910, and TV trackers (GWS 25 Mod 0); radar trackers, Type 911 (GWS 25 Mod 3)
(3) command transmitter
(4) launcher and firing system
(5) missile and handling frame
(6) data handling
(7) guidance shaping unit
(8) operations consoles
(9) magazines.

The Type 910 tracking radar is produced by Marconi and is described more fully in the Equipment Section (**1562.253**). The TV system is produced by GEC Avionics. The Type 911 tracking radar is described more fully in entries **1508.253** and **2439.153** in the Equipment Section. The Type 967 surveillance radar provide both high and low cover, and are also produced by Marconi (**1561.253**). They are of modern design and incorporate features for air target detection up to high elevation angles as well as high performance against low-level and surface targets. Comprehensive precautions against sea and land clutter, as well as natural and man-made interference are incorporated, as is IFF.

The line-of-sight to a target is established by either the tracking radar or the TV system. Error signals proportional to Seawolf missile deviations from this datum are derived from the differential tracking radar or the TV system, and these signals are processed by a guidance shaping unit. Coded correction signals for missile guidance are produced and transmitted by microwave command link to bring the missile to the required flight path. In the GWS 25 system the data processing required to interpret the tracking data and calculate the correction demand signals is based on the use of a Ferranti FM 1600B computer (**1433.063**), which has been adopted as standard in the Royal Navy Type 22 frigates.

Seawolf firing from HMS Brave

A multiple launcher developed by VSEL is common to both standards and consists of six rectangular launch-tubes disposed in two banks of three, one on each side of an azimuth/elevation mounting. Reloading is manual, presumably to avoid the complexity and, particularly, the weight of an automatic system which limits the number of ships which can carry the full Seawolf system. The launcher is separate from the tracking radar. High slewing rate and pointing accuracy are important features of the Seawolf launcher which is equipped for fully automatic firing sequence, with command override.

The missile has four fixed wings and four moving tail fins. A solid booster motor is designed to give minimal launch drop. Successful techniques employed in Rapier (**2424.131**) have been incorporated, and no on-board test or repair facilities for missiles are called for. The HE warhead is provided with both proximity and contact fuzing.

Operation

For the successful interception of an incoming anti-ship missile great accuracy and an extremely short reaction time are required of the system. To achieve this it is arranged that, once a target has been identified as hostile, all subsequent phases of the launch and guidance operation will be carried out automatically and without further manual control.

Other relevant features include the ability to fire salvoes, immediate readiness capability maintained over long periods, and extremely fast data handling facilities in all parts of the system.

Automatic radar guidance is the normal operating mode, with TV tracking by an aimer for low angle of sight and surface target engagement in GWS 25 Mod 0.

The GWS 25 Mod 0 is fitted to the 'Broadsword' (Type 22) and 'Leander' class Batch 3 frigates. Both classes of ship have the six-barrel launcher (one in the Leanders and two in the Broadswords), Type 967 surveillance radar and one or two Type 910 trackers. It entered service in 1978.

The GWS 25 Mod 3 system, fitted to the later 'Broadsword' class, retains the Type 967 surveillance radar and heavyweight launcher but uses two Type 911 Mod 1 (Marconi 805SW) lightweight, dual-frequency (I-band and millimetric) tracking radars which combine major components of the basic Marconi 805 radar and the DN181 Blindfire radar. In addition the FM1600E computers have been upgraded while fast Fourier transform techniques are used in the I-band signal processing.

Variants

The latest version of the Seawolf system is GWS 26 which uses the Type 996 surveillance radar based upon the AWS-5 (see **3172.253** and **4634.253**), Type 911 tracking radars and new launchers.

GWS 26 Mod 2 features a four-barrel, or lightweight, launcher weighing only 2.5 tonnes and the Type 911 Mod 3 tracking radar. The new launcher is a modified version of the Seacat launcher. Development began as a private venture in 1981 and the following year the need to improve the self-defence capabilities of

Seawolf from lightweight 4-barrel launcher

certain major warships led to the Ministry of Defence funding the project. Known commercially as Lightweight Seawolf it features two versions of the launcher; one with an integral barrel in which the missiles are stored separately in magazines and a removable barrel version with the missiles stored in their own launch boxes each weighing 170 kg. The four boxes can be reloaded in two minutes.

In February 1988 it was announced the MoD had placed a £40 million order for GWS 26 Mod 2 which will be installed in the Royal Navy's Type 42 Batch 3 destroyers together with the three 'Invincible' class aircraft carriers. The destroyers will carry two or four launchers while the carriers will have four launchers. Both will have two trackers.

GWS 26 Mod 1 features the Type 911 Mod 2 trackers and the vertical launched Seawolf (Seawolf VL). The system has been developed to eliminate blind engagement arcs, to improve the capability against saturation attacks by improving reaction times as well as extending interception ranges.

The basic Seawolf structure is retained but a strapdown inertial measuring unit and a tandem booster with thrust vector control have been added. The missiles will be stored and transported in sealed canisters with services and control functions provided by a Missile Firing Unit (MFU) which controls up to eight canisters. A module of eight loaded canisters and a MFU weighs 3.4 tonnes. The system has been ordered under a £130 million contract for the 'Duke' (Type 23) class frigates and the new fleet replenishment ships (AOR). Both classes will have a silo of 32 cells and two tracker radars.

Turnover trials have been successfully completed and full system trials began in 1987. The sea acceptance firing trials in HMS *Norfolk*, the first 'Duke' class frigate, are likely to begin in mid-1990.

Early in 1985 British Aerospace received a contract to study a further development of GWS 26 Mod 1, GWS 27, to meet future anti-ship missile threats at the turn of the century but this has now been suspended pending more detailed definition of the NATO Frigate Requirement for the 1990 (NFR-90) local area missile system requirement (see NAAWS).

CHARACTERISTICS

	Seawolf	Seawolf VL
Weight	80 kg	140 kg
Length	2 m	3 m
Diameter (body)	18 cm	18 cm
Speed	Mach 2+	Mach 2+

CONTRACTORS

Missile: British Aerospace plc, Dynamics Division.

Radar: Marconi Radar Systems Ltd.

Television: GEC Sensors Ltd.

Computer: Ferranti Computer Systems Ltd.

Launchers: Vickers Shipbuilding and Engineering Ltd (six-barrel trainable), Rose Forgrove Ltd (vertical launch canister); British Aerospace.

Vertical launch Seawolf

Proximity fuze: THORN EMI Electronics, Defence System Division.

UNITED STATES OF AMERICA

2507.231

AEGIS SURFACE-TO-AIR WEAPON SYSTEM

Aegis is a surface-to-air weapon system primarily designed to defend against anti-ship cruise missiles and missile launcher platforms. It is the USN's primary defensive missile system for the late-1980s and beyond. Aegis is designed to destroy small, fast targets in hostile environments such as severe weather or countermeasure conditions. As an area defence system, Aegis will be capable of defending a task force. The system complements the Tartar (**6006.231**) and Terrier (**6005.131**) missile systems.

Major components of the Aegis system are the missile, its launching system, the fire control, and weapon control systems, the multi-function array radar with its computer control, the system command and control, and the operational readiness test system.

The missile used is the SM-2 version of the Standard (**1122.231**), which uses the same airframe and propulsion system developed for the SM-1 but incorporates significant additional facilities. The SM-2 has a command guidance capability, a self-contained navigation capability provided by an inertial reference unit, and a new terminal homing receiver.

This particular combination of guidance techniques imparts considerable operational benefits. The ability to navigate the SM-2 missile to the target area by means of inertial navigation, plus command corrections, followed by terminal homing, results in a much more efficient trajectory which effectively doubles the range. It also means that the demands on the ship-based fire control/guidance system are appreciably reduced. For example, the target illuminating radar needs to be directed at the target for the homing phase only, leaving it free to engage other targets at other times. Similarly, the command guidance system can be shared between a number of SM-2 targets because the inertial reference unit permits the missile to perform its own mid-course navigation and the telemetry link is used to report missile position back to the ship.

The SM-2 is available in MR (medium-range) and ER (extended-range) models, but the current SM-1 Standard is available in the MR version only.

It is launched from the Mk 26 fully automatic dual-purpose launcher which can be used to launch Asroc (**6001.241**) anti-submarine missiles, but as Aegis fittings increase in the US Fleet more ships will be equipped with the Mk 41 vertical launcher system (**4204.231**).

The MK 26 launcher has a digital interface with the Mk 1 weapons control system. One of the three computerised sub-systems of Aegis, all of which use the AN/UYK-7 digital computer, the Mk 1 system accepts weapon assignment commands and special threat criteria from the Mk 1 command and decision system and tracking data from the multi-function radar. These inputs are processed to determine the possibility of engaging the target and then to generate commands for the Mk 26 launcher and pre-launch orders for the missile, commands for the Mk 99 fire control system for target illumination, commands to the multi-function radar, if mid-course guidance is required, and reports to the Mk 1 command and decision system.

US Navy's second Aegis cruiser, Yorktown, *at sea during builder's trials. Ship was commissioned on 4 July 1984*

The function of the Mk 99 fire control system is to illuminate the target. To do this it uses the AN/SPG-62 (slaved) radar. Inputs to the system come from the Mk 1 weapons control system which, in the case of slaved radar operation, is passing on data from the multi-function radar system.

The multi-function phased-array radar, the AN/SPY-1A (described in **1570.253**) is high-performance electronically-scanned equipment capable of surveillance and the simultaneous detection and tracking of multiple targets; it has been described as the heart of the Aegis system. Associated with it are four AN/UYK-7 computers. The primary function is to search for and acquire targets and track them to whatever extent may be necessary. Beam scheduling for this and other tasks is organised by the computer. The system also handles the two-way link with the missile for mid-course guidance when requested by the Mk 1 weapons control system. Certain special facilities can be provided such as 'burn-through' facility for use in ECM conditions or passive angle tracking. The system accepts general operational commands from the Mk 1 Aegis command and decision system and mid-course guidance commands from the Mk 1 weapons control system; its operational outputs are processed signals giving target detection data to the Mk 1 command system and target and missile track data to the Mk 1 control plus mid-course guidance commands to one or more missiles. Like all other operational sub-systems in the Aegis system the radar control system is also connected to the Mk 1 operational readiness test system.

Together with the AN/SPY-1A radar system and the Mk 1 weapons control, the Mk 1 command and decision system make up the detection and decision loops of the Aegis system. The Mk 1 system is the ship's command and control centre, and targets enter the detection and decision loop from the AN/SPY-1A radar, from other ship's own sensors or from data supplied by other ships or aircraft. What happens next depends on which of the four Aegis operating modes has been selected. The modes are automatic special, automatic, semi-automatic, and casualty. In the automatic special mode targets meeting certain predetermined threat criteria are automatically fired upon unless manual override is invoked; in all other modes positive human action to initiate firing is needed.

In these three modes the Mk 1 weapons control system inserts targets into the engagement queue and schedules equipment for launching and terminal illumination. Trial intercepts are computed and a time to fire predicted. Resulting data are fed back to the Mk 1 command and decision centre which is also receiving target detection data from the radar control system, operational readiness data from the Mk 1 operational readiness test system, and electronic warfare and other data from the ship's command and control centre. Incorporated in the Mk 1 command system is another four-bay AN/UYK-7 computer and a comprehensive display system, the AN/UYA-4. In destroyers for the Aegis Mk 7 the radar is the AN/SPY-1D which is used with the AN/UYK-43 computer. On the basis of all these data, threat evaluation and weapon assignment processes are carried out so that the engagement decision can be taken.

The final major element of Aegis is the Mk 1 operational readiness test system (ORTS). This is linked to all the other major elements of the system and has the functions of monitoring actively and continuously the state of readiness of each, and of reporting the results of this process to the Mk 1 command and decision centre.

Initial planning for Aegis began in 1964 and RCA was awarded the engineering development contract for $252 930 400 in December 1969.

STATUS

In 1983, after completing sea trials and operational tests of Aegis in USS *Ticonderoga*, this ship began operations with the fleet. Present planned construction will result in an eventual total of 27 of these ships. In 1984 Congress authorised construction of the guided missile destroyer *Arleigh Burke* (DDG-51) – lead ship of a new class of 8300 ton Aegis-fitted ships. It is planned to procure 11 DDG-51s over the next five years. In April 1984 RCA announced the receipt of a $233 million contract for the design and development of an Aegis system for the DDG-51 class. The lead ship, *Arleigh Burke*, is due for commissioning in 1989. These ships' Aegis will incorporate the AN/SPY-1D radar and the new AN/UYK-43 computers but performance will be slightly inferior to the full SPY-1A system. Elements of the Aegis system are also to be made available to the Japanese Maritime Self Defense Force for two new missile-armed destroyers.

In July 1984 USS *Yorktown*, the second ship of the 'Ticonderoga' class of Aegis ships was commissioned.

Early in 1988 RCA won a $375.7 million contract for five Aegis Mk 7 systems for three cruisers and two destroyers for delivery by July 1990.

CONTRACTORS

Prime contractor: RCA Government Systems Division, Moorestown, New Jersey, USA. (Three major RCA divisions are engaged in the project.)

FCS Mk 99, radar and SPY-1A high-power transmitter: Raytheon Company.

Mk 26 launchers: Naval Systems Division, FMC Corporation.

Mk 41 VLS: Martin Marietta and FMC, Naval Systems Division.

SM-2 missile: General Dynamics Corporation, Pomona Division; Raytheon Company, Missile Systems Division.

Computer software: Computer Sciences Corporation.

Technical advice to USN: The Johns Hopkins University, Applied Physics Laboratory.

Also more than 600 other suppliers.

2543.231

Mk 15/16 PHALANX CLOSE-IN WEAPON SYSTEM

The purpose of Phalanx is to provide last ditch defence against anti-ship missiles. In its primary mode Phalanx provides continuous surveillance and defence within its engagement envelope, independent of other ship systems. Phalanx is an automatic gun fire control system that uses projectile spotting coupled with closed-loop aiming to provide multiple-target hits and subsequent target destruction. The principal technical feature of the Phalanx system is the closed-loop electronic spotting technique. Aim correction is based on the ability to track both the incoming missile and the outgoing projectile in the vicinity of the missile, and to eliminate the error. The system uses a pulse doppler radar which shares a single transmitter to accomplish the search, acquisition, track and electronic spotting functions.

Phalanx naval close-in weapon system combines on a single mount a six-barrel gatling gun and fire control radars

Incorporating the widely used and service-proven M61A1 Gatling gun used in the Vulcan land mobile system (**2850.131**), Phalanx provides a high fire rate of 3000 rounds per minute. During firing, the ammunition feed system removes rounds from the front of the magazine drum, located under the gun, and feeds them through the chute conveyer system to the gun breech bolt assembly. After each round is fired, the empty case is extracted from the breech and conveyed back into the rear of the magazine. The gun can fire continuously under the control of the system's computer against air targets, or in the operator controlled burst mode against surface targets.

Phalanx is a total weapon system, and as a single modular package, automatically carries out search detection, threat evaluation, tracking, firing, and kill assessment. The system consists of six major assemblies and components: radar-servo assembly, gun assembly, mount and train drive platform, barbette equipment assembly, electronics enclosure and a pair of control panels. Supported by a high-speed digital computer, the fire control assembly performs these functions: target search, detection and declaration, target acquisition, target track and measurement of range, velocity and angle, projectile tracking, and measurement of projectile velocity and angle.

The gun assembly is composed of five major sub-assemblies. The gun, electrically controlled and hydraulically driven, consists of a rotating cluster of six barrels with a breech bolt for each barrel. The ammunition sub-system includes components that store and transport rounds within the system. The gun's effectiveness is due to its firing a high kinetic energy penetrator made of high density metal. Destruction is caused by penetration of the 20 mm projectile into the warhead (of the incoming missile) and the imparting of kinetic energy into it, causing it to blow up before striking the ship.

The dual motor mount drives are torque biased to reduce system backlash, and friction is minimal. The lightweight structure has high stiffness-to-weight ratio in order to minimise impact of vibration on servo performance. The structural design will withstand all shipboard environments, including wave impact and blast forces from adjacent guns.

The Mk 16 system differs in having an improved ammunition supply.

STATUS

In full production with the first system delivered in 1979. The battleships *New Jersey* and *Iowa* have four systems fitted and the 'Aegis' cruiser *Ticonderoga* has two systems. Fittings will be made in 39 classes of ship and the USN is installing it in over 360 ships ranging from aircraft carriers to frigates. In mid-July 1987 General Dynamics delivered the 500th Phalanx system to the US Navy for installation in the USS *Denver* (LPD-9). It is now installed in 200 ships of the US, UK, Australian and Japanese Navies. Also in July 1987 orders for 59 Phalanx worth $154 million were revealed for the United States, the United Kingdom, Australia, Greece, Israel, Japan, Pakistan, Portugal, Saudi Arabia and Taiwan. Eventual orders may reach 1000 mountings.

CONTRACTORS

CIWS system contractor: General Dynamics Corporation, Pomona Division, PO Box 2507, Pomona, California 91769, USA.

Vulcan gun: General Electric.

3983.231

SEA CHAPARRAL SURFACE-TO-AIR MISSILE SYSTEM

The Sea Chaparral MIM-72 surface-to-air missile system is an adaptation of the Chaparral land-based mobile air defence missile system (**2542.131**), which uses a version of the Sidewinder, AIM-9 (**1103.331** etc) air-to-air missile.

The launcher is a modification of that used in Chaparral in the self-propelled (M48) and ground emplaced (M54) land-based systems, but with stabilisation to compensate for sea motion and with provisions for operation either with the gunner in the mount, or with control by a remote gunner in the ship's operations room or other fire control facility.

As with the standard Land Chaparral, the Sea Chaparral system can be enhanced by the addition of forward looking infra-red sensors to provide a night engagement capability, and provisions can be made for target designation to be carried out by other ship's sensors, such as radar or optical target acquisition sights etc.

Details of the missile and its performance will be found in the above-noted entries in this volume.

STATUS

The Sea Chaparral missile system is in service with the Taiwan Navy. The ex-'Fletcher' class destroyers *Kun Yang, Kwei Yang* and *An Yang* and others have been fitted with one launcher aft.

CONTRACTOR

Aeronutronic Division, Ford Aerospace & Communications Corporation, Newport Beach, California 92658-8900, USA.

3017.231
RAM (RIM-116A) SYSTEM

A US Navy requirement for an Anti-Ship Missile Defence (ASMD) system to supplement Phalanx (qv) and Sea Sparrow (**2770.231**) has led to the Rolling Airframe Missile (RAM). The requirement was for high firepower, short reaction time, accurate terminal guidance with proven lethality and minimum dependence upon fire control channels. Originally this was planned to be purely a United States programme and development began in FY 1974 but Congressional scepticism about the project hindered development.

In July 1976 West Germany signed a Memorandum of Understanding to share development costs and later Denmark joined the consortium. A contract for a 50 month full scale engineering development programme was signed in June 1979.

Existing radars, ESM or electro-optical systems designate the target and the missile is launched at a selected elevation in the direction of the target extending its two fins and tail surfaces as it leaves the launcher. As it emerges from the launcher the missile is spun by rifle bands inside the launch canister to simplify guidance. Its rolling airframe autopilot and control system maintain the course until RF homing guidance signals call for course correction. The RF guidance signals also point the IR seeker toward the target. When a sufficient IR signal-to-noise ratio is achieved guidance is switched to the IR seeker for the terminal phase.

The guidance technique has been selected because most missiles emit active radar and IR signals including a significant proportion of Soviet anti-ship missiles. The RAM uses the WDU-17B warhead, the Mk 20 fuze and the same rocket motor as used in the AIM-9L Sidewinder. The IR seeker is the same as the one used in Stinger (**2805.131**).

RAM can be used with the NATO Sea Sparrow (**4775.231**) launcher with insert racks fitting into the Sea Sparrow cells to hold the RAM missile canisters. Five RAM missiles will then fit each cell and two cells in each Sea Sparrow launcher will be assigned to RAM, a configuration known as RAM ORDALT. This launcher system has six Sea Sparrow missiles.

Alternatively, a dedicated launcher, the EX-31 or RAM Standard launcher, has been designed, using the mount, elevation and train drive assemblies of the Phalanx system. It holds 21 missiles and is a bolt-on item which weighs a total of 6464 kg or about 80 per cent of the weight of a Sea Sparrow launcher. For small warships the RAM Alternate Launching System (RALS) has been designed with a capacity for eight missiles and a total weight of some 2950 kg. Both launchers have an elevation of –25 to +80° but the Standard launcher has a train of 360° while the RALS will have a train of ±175°.

CHARACTERISTICS
Length: 2.79 m
Diameter: 12.7 cm
Weight: 70 kg
Speed: Supersonic
Range: 9.6 km
Guidance: Proportional navigation. Dual mode, passive RF, IR homing
Warhead: Focused blast/fragmentation

STATUS
Missile flight tests have been made and in October 1982 a RAM successfully intercepted a drone representing an anti-ship missile in poor visibility over land. In the same month seaborne trials began and by February 1983 there were three consecutive successful interceptions of drones including a supersonic low-altitude drone. In December 1986 11 out of 13 firings were successful.

Dummy RAM EX-31 EDM-2 aboard West German ship Habicht *during testing at sea*

Test firing of ASMD RAM from an EX-31 launcher at a land-based test site

Development was extremely protracted and expensive with the result that Danish interest in the project appears to have waned. However, in July 1987 the United States and Germany signed a Memorandum of Understanding for production of the system. The US Navy will acquire 30 launchers and 4900 missiles while the Germany Navy will acquire 58 launchers and 1923 missiles at a cost of DM1.642 billion. The Germans will fit 38 launchers on three 'Lutjens' class destroyers, eight 'Bremen' class frigates and ten 'Gepard' class FACs while 20 launchers are scheduled to be used by the Type 124 frigates and ten 'Albatros' class FACs.

The German systems will be produced by RAM Systems GmbH, a consortium comprising AEG, BGT, Diehl and MBB. The launchers will be co-produced by General Dynamics as prime and RAM Systems as sub-contractor while the German company will be able to compete against the American company for the manufacture of 70 per cent each of the launchers and missile components needed by both navies. Production will begin in 1989.

CONTRACTOR
Prime: General Dynamics Corporation, Valley Systems Division, PO Box 50-800, Ontario, California 91761, USA.

1122.231
STANDARD (RIM-66A AND RIM-67A) SURFACE-TO-AIR MISSILE

The Standard shipborne surface-to-air missile began life as a two-model range of anti-aircraft weapons, intended gradually to replace Terrier and Tartar with the US fleet. The original two versions are the medium-range RIM-66A, and the extended-range RIM-67A. It was the first to be solid-state and is all electric with no pneumatic or hydraulically powered controls. Power for guidance and control is derived from a 'one-shot' battery which is activated just prior to missile launch.

For both types, fully automatic magazine and loading facilities are provided for twin electrically driven launchers. The RIM-66A had a dual-thrust solid-propellant motor, and the RIM-67A used solid-propellant booster and sustainer rockets. Both versions are equipped with conventional high explosive warheads, with direct action or proximity fuzing.

CHARACTERISTICS

	RIM-66A	RIM-67A
Speed	> Mach 2	> Mach 2.5
Ceiling	> 20 000 m	> 20 000 m
Range	> 18 km	> 55 km
Length	4.47 m	8.23 m
Diameter	34.3 cm	34.3 cm
Weight	Approx 610 kg	Approx 1360 kg

Subsequent developments have somewhat altered the original plan and versions of Standard have been, or are, in the process of development for surface-to-surface, air-to-surface, and new surface-to-air applications. The following list is arranged in logical rather than chronological order.

(1) Medium-range and extended-range

All existing and most contemplated variants of the original missile can be made in either MR or ER versions. This possibility should therefore be borne in

"Why's everyone always picking on me?"

"Cheer up, old bat. A good dose of electronics will soon get you in shape."

New update package to make your Superfledermaus super again.

A joint venture by Philips, Terma and Satt has developed an update package to give the Superfledermaus state-of-the-art performance. News that will prove very welcome on the market, particularly in view of the fact that a new system with equivalent performance can cost three to four times as much as the update package. Deliveries of upgraded Superfledermaus units are already in work in Europe.

The update package involves using a new, decentralized digital computer to replace the electron tubes and electromechanical components of the old system and fitting a new Ku band frequency agility radar with MTI capability.

These improvements will boost hit probability by factors of 5 to 8 and cut reaction time to as little as 5 seconds. And if the poor availability of the system has been a source of irritation in the past, we can quarantee an MTBF of more than 500 hours for our update.

So stop picking on Superfledermaus. Pick up the phone for more information on the update package.

Come and see us at the Farnborough show and we'll tell you more about the new Superfledermaus update package.

SATT Communications
Box 32701, 126 11 Stockholm
Sweden
Telex: 15325 SATTCOM

Philips Elektronikindustrier AB
S-175 88 Järfälla
Sweden
Telex: 11505 Philja S

TERMA Elektronik AS
Hovmarken 4, DK-8520 Lystrup
Denmark
Telex: 68109 terma dk

mind when examining the variants listed below.

(2) Standard 1 Missile (SM-1)

This is the original missile. Its principal advantage over the earlier Terrier or Tartar is that it has a (horizon-limited) surface-to-surface capability. It was expected to remain in production until 1975, but was still being procured in 1978 and is operational in over 70 missile ships and attack carriers.

(3) Standard 2 Missile (SM-2)

This missile completed development in 1977 and is in production. The SM-2 approximately doubles the range of SM-1 in both MR and ER versions. An inertial reference is provided, and to satisfy the requirements of the Aegis system the missile-borne equipment is augmented by a two-way link for the mid-course command guidance. Other improvements include a new semi-active terminal guidance receiver with greater resistance to ECM.

The most important changes embodied in the SM-2 are: the replacement of the SM-1 conical-scan semi-active radar homing receiver by a monopulse receiver; the provision of an inertial reference unit for mid-course guidance; a communications link for mid-course guidance correction/target position data updating and missile position reporting; and a digital guidance computer instead of the SM-1's analogue computer. These improvements make possible a far more efficient trajectory to the target, resulting in an effective 60 per cent increase in range for the MR model and more than 100 per cent increase in that of the ER model.

In the late 1970s, it was decided to expand the number of USN ships capable of employing the added capability of the SM-2 missile, beyond the 'Aegis' class vessels for which it was initially intended. This led to the CG/SM-2 weapon system programme under which it was planned to replace Terrier missiles (**6005.231**) on board 20 ships of the CG-16 and CG-26 classes (33 ships were equipped with Terrier, others being the DDG-37 class of ten vessels, one CG(N)-9 and 2 CV class carriers) with SM-2 Extended-range missiles. This programme entailed modifications to the ships' weapons control systems and radar etc, and computer software and is now largely complete.

Standard 2 missile firing from USS Wainwright *near San Juan, Puerto Rico*

Fiscal year:	**1984**	**1985**	**1986**	**1987**
SM-1 MR	(700) 209.2	(600) 199.3	26.4	(300) 143.5
SM-2 MR	(390) 289.7	(529) 355	(846) 509.7	(900) 527
SM-2 ER	(100) 103	(255) 186.7	(470) 312.2	(600) 365.9
Standard mods	5.1	6.2	17.1	9.3

By means of the inertial reference unit in the SM-2 missile it flies a mid-course phase wherein the missile navigates itself to the vicinity of the target. Semi-active terminal guidance, supported by the target illumination from the original Terrier fire control system, is used to complete the engagement.

The inertial reference unit and guidance computer in the SM-2 are initialised by the new weapon direction system Mk 14. Target data are derived from the modified AN/SPS-48 radar (**1252.253**) and ship attitude and velocity are provided by new shipboard gyro-compass equipment. The Navy Tactical Data System is modified to accommodate the command and control interface facilities required.

This system results in significantly decreased target processing time. Originally, a series of sequential and manual operations was involved between target detection and missile launch, and this can occupy several minutes. With the CG/SM-2 system, all target detections are passed to NTDS and those selected for engagement by the Terrier systems are automatically passed to the weapon direction system. Subsequent to launch, the fire control radar is assigned to provide target illumination for the terminal phase of flight.

The SM-2 monopulse homing receiver is also to be fitted to SM-1 MR missiles intended to be used in upgrading existing Tartar-fitted ships (see **6006.231**) in a broadly similar fashion to the CG/SM-2 programme. Additional information on the Terrier missile system will be found in **6005.231**.

(4) Standard ARM

Essentially an airborne version (see **1123.331**) this has also been adapted for use in a surface-to-surface role as part of the Interim Surface-to-Surface Missile System (**2669.221** and **2808.221**).

(5) Standard Active

The original intention was for this missile to serve as an interim anti-ship weapon until completion of Harpoon development, but delays with Active Standard made both missiles more or less contemporaries. The position at mid-1974 was that Active Standard would be maintained in development only until it was confirmed that it would no longer be required as a back-up to Harpoon. Progress with Harpoon development has led to cancellation of Standard Active.

Brief details of various Standard models and their US Service designations are given in the following list:

YRIM-66A: Standard-1 (MR). Basic development model to replace Improved Tartar. Mk 27 Mod 0 motor

RIM-66B: Standard-1 (MR). Similar to RIM-66A but with improved motor, Mk 56 Mod 0, for better performance

RIM-66C: Standard-2 (MR). Similar to RIM-66B but adapted for Aegis system

RGM-66D: Standard SSM (ARM). Similar to RIM-66B but modified for surface-to-surface anti-radiation roles

RGM-66E: Standard SSM (ARM). Similar to RGM-66D but adapted for use with Asroc launcher

RIM-67A: Standard-1 (ER). Replacement for Terrier. Uses Mk 30 Mod 1 sustainer to give extended range

RIM-67B: Standard-2 (ER). Similar to RIM-67A but used with CG/SM-2 weapon system

In August 1987 Raytheon's Missile Systems Division was awarded a $231 million contract to design the Standard 2 Block 4 or Aegis Extended Range version of the Standard missile. General Dynamics will co-develop the missile, especially on the guidance and control system, then compete against Raytheon for production.

The new version will feature a slip-cast, fused-silica radome, new digital autopilot and signal processor techniques similar to those used in Patriot, AMRAAM and Sparrow. A solid-propellant booster using thrust vector control will extend the missile's range. The missile will also feature larger dorsal fins with insulation to accommodate the heat build up.

The new missile will be used by 'Aegis' class cruisers and 'Arleigh Burke' class destroyers.

STATUS

Development began in 1962 and the Standard missile entered service in the late 1960s more than 10000 have been delivered by General Dynamics to the USN and other navies.

SM-1, SM-2 Block I and SM-2 Block II are operational. SM-2 (MR) and SM-2 (ER) are being upgraded to meet new low-altitude threats. SM-2 Aegis ER is under development to meet high-altitude threats.

CONTRACTOR

General Dynamics Corporation, Pomona Division, PO Box 2507, Pomona, California 91766, USA.

6006.231

TARTAR SURFACE-TO-AIR MISSILE (RIM-24)

Tartar is a medium-range, supersonic surface-to-air shipboard guided missile system. It provides primary air defence for USN destroyers and destroyer escorts and secondary air defence for cruisers. It is frequently forgotten that Tartar, which has a continuous rod warhead, has a surface-to-surface capability. The missile booster and sustainer rockets are combined in a single solid-propellant motor, thus facilitating installation on smaller ships.

OPERATION

Tartar employs a fully automatic magazine handling and loading system. Targets are designated to the system in two or three co-ordinates by radar. The tracker/illuminator radars automatically follow the targets and a computer provides missile launcher pointing orders. Guidance is semi-active homing, probably operating in I/J-band. The missile carries a high explosive warhead with DA and proximity fuzes.

The rocket motor has two levels of thrust to achieve rapid acceleration. The guidance system computes a collision course, then steers to intercept by means of four hydraulically actuated tail fins.

CHARACTERISTICS

Range: > 16 km
Speed: Mach 2
Length: 4.6 m

RIM-24 Tartar surface-to-air missile on Mk 13 GMLS of nuclear-powered cruiser USS California (USN Photo)

Diameter: 34.3 cm
Weight: 680 kg
Motor: Dual-thrust solid-propellant
Guidance: Semi-active homing

DEVELOPMENT

Development began in 1955 shortly after Terrier system and Tartar first entered USN service in 1961.

At mid-1977 there were 49 USN ships deployed with the Tartar weapons system and with Standard SM-1 (medium range) missiles. This includes nuclear-powered cruisers, guided missile destroyers, and guided missile frigates. This fleet could grow to more than 100 vessels in the 1980s with the introduction of the FFG-7 class. This factor and the evolving threat require a continuing programme of ship system and missile improvements. This is applicable in principle to the considerable number of non-US ships fitted with the Tartar/Standard missile system, at present some 29 ships with 11 more planned.

The improvement plan being implemented is similar to that adopted for Terrier-fitted ships and described in the entry for the Standard missile system (**1122.231**), and officially known as the CG/SM-2 (ER) programme. In the case of Tartar ships, such as the DDG-2 class, improvements in the ship's system will entail addition of the AN/SPS-52 three-dimensional radar (**1248.253**), the AN/SPS-65 radar, modifications to the AN/SPS-40 radar (**1746.253**), and integrated automatic detection and tracking of all radar targets by the SYS-1 plot extraction equipment. Control improvements will be provided by the weapons direction system Mk 13 and the DDG tactical data system, which are based on already deployed systems. Engagement from the ship's system aspect is improved by modification to the basic Tartar Mk 74 missile fire control system and addition of the Mk 86 FCS (**1241.281**) with gun and missile capability.

A similar updating and improvement programme will be applied to the SM-1 MR missiles employed. Some of these changes have been implemented and already SM-1 rounds have been deployed incorporating the new, improved Mk 45 fuze and the Mk 90 warhead. Other projected missile improvements include substitution of the SM-2 monopulse seeker receiver for the original conical scan seeker of the SM-1.

Additional details on the Standard Missile programme will be found in **1122.231**, above.

STATUS

Tartar is no longer in production but is still operational in a large number of USN ships and in the Australian, French, Iranian, Italian, Japanese, Netherlands, Spanish and West German navies. Replacement by the Standard missile (**1122.231**) has begun.

CONTRACTOR

Prime contractor: General Dynamics Corporation, Pomona Division, PO Box 2507, Pomona, California 91766, USA.

6005.231

TERRIER SURFACE-TO-AIR MISSILE (RIM-2)

This is a surface-to-air anti-aircraft missile for shipboard use. The Terrier series of missiles has been operational since 1956 and since then it has been the subject of continued development and improvement. The US military designation is RIM-2 and successive models have been denoted by letter suffixes running from RIM-2A to the latest version, the RIM-2F. The latter is also known as the Advanced Terrier.

CHARACTERISTICS

Range: >35 km
Ceiling: >20 000 m
Speed: Mach 2.5
Overall length: 8 m
Length without booster: 4.6 m
Body diameter: 34.3 cm
Weight: 1400 kg (approx)

OPERATION

Beam-rider guidance is employed to direct the missile against a target designated by the ship's tactical data system. A twin launcher is supported by automatic magazine and loading arrangements. Terrier is launched by a solid fuel rocket booster and propelled by a solid fuel sustainer.

The normal warhead is of the HE type with direct action or proximity fuze, but one version of Terrier (RIM-2D) has been produced with a nuclear head.

DEVELOPMENT

Development started about 1951, under the former military designation SAM-N-7, and Terrier was based upon the experimental LARK vehicle. First Terrier became operational in 1956.

STATUS

Advanced Terrier (RIM-2F) became operational in 1963 and is carried by USN cruisers, frigates, and attack carriers. It has also been purchased by the Italian Navy. The system is progressively being replaced by the Standard missile system (**1122.231**). See this entry also for additional details of Terrier weapon system status, in particular in regard to the CG/SM-2 (ER) programme.

CONTRACTORS

Prime contractor: General Dynamics Corporation, Pomona Division, PO Box 2507, Pomona, California 91766, USA.

Launch of Terrier surface-to-air missile from aircraft carrier USS Constellation (USN)

System co-ordination: Vitro Corporation.

4202.231

US MISSILE LAUNCHER SYSTEMS

The Naval Systems Division of the FMC Corporation, as well as providing most of the naval tubed ordnance requirements of the USN and many friendly navies elsewhere, supplies the launcher systems for most of the guided missiles employed by the USN. Guided missile launching systems (GMLS) are fully automatic systems comprising the missile launcher arms complete with slewing and elevation mechanisms, and reloading facilities appropriate to the missile types carried by the ship. FMC believe the GMLS Mk 4, produced about 25 years ago for the USN, to have been the first entirely automatic missile launching system in the world. Since that date the company has designed and produced six different marks of GMLS, each of them in several models, and the main details are outlined in the following paragraphs.

GMLS Mk 4

This system was produced for ships fitted with the Terrier surface-to-air missile system. The GMLS Mk 4 Mods 0 and 1 were fitted in the USS *Boston* and *Canberra* (both now withdrawn) while the Mk 4 Mod 4 was fitted in the Italian cruiser *Giuseppe Garibaldi* (also decommissioned). Twin-ramp launchers were reloaded vertically through two hatches from a magazine below deck, in which the missiles were also stored vertically.

FMC Mk 13 single-ramp launcher in use

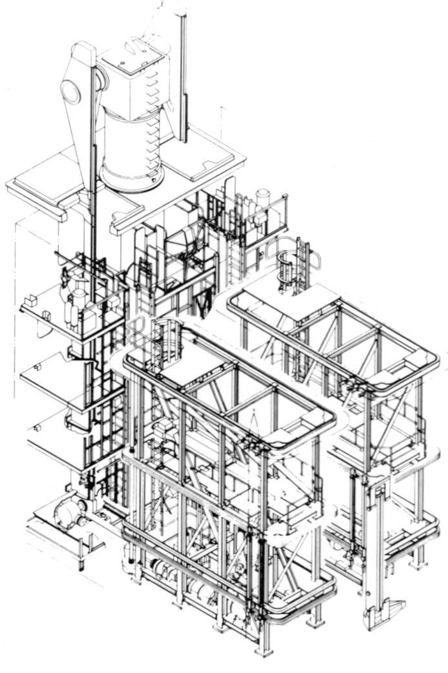

General arrangement drawings of FMC Mk 4 Mod 4 GMLS

No Mk 4 Terrier missile systems remain in active service although a few are used on missile test ranges.

GMLS Mk 7

Also developed for large ship installation, the Mk 7 missile launcher system is now retired from active service, with a few systems in use on missile test ranges. The Mk 7 was designed for use with the heavy long-range Talos surface-to-air missile system. Missiles were stored horizontally in the magazine and the twin launch ramps were reloaded in a horizontal, or near horizontal position. Obsolete.

GMLS Mk 10

There are numerous models of the GMLS Mk 10, a twin-rail launcher system designed for use with Terrier and Standard (ER) surface-to-air missile systems. This design continues the horizontal storage and reloading pattern set by the GMLS Mk 7, although certain installations differ by the amount the loader rail deviates from the true horizontal along the route from the magazine to the missile launch rail. Mods 2, 5, and 7 are the tilt rail system. Various models of the system are found on numerous ships of lesser tonnage. The capacity of the magazines for reload missiles also varies between models, although all

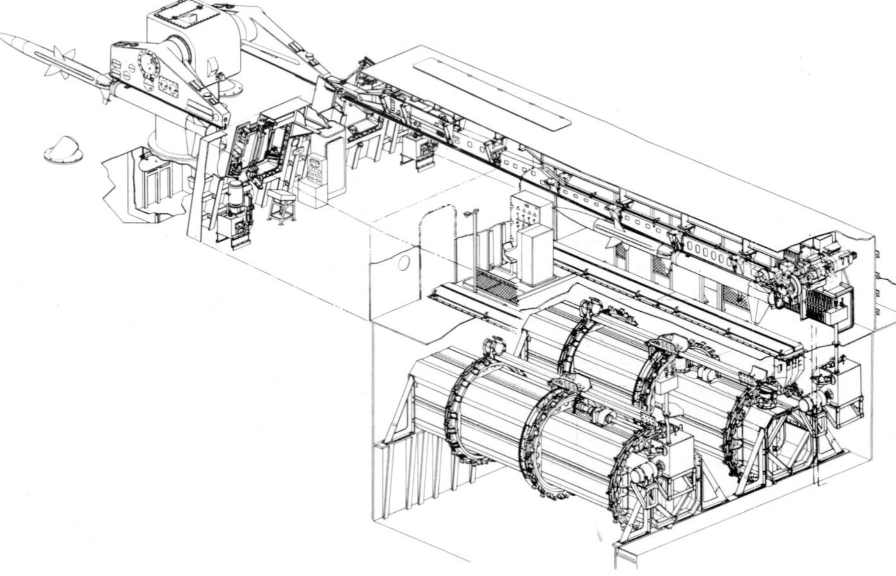

Mk 10 Mod 0 GMLS with two horizontal magazine drums for reload missiles. Other models have more drums

employ a similar form of horizontal drum storage for the rounds to permit mixed salvoes.

Substantial control system and performance improvements are in progress for the Mods 0, 1, 2, 5, 6, 7 and 8. When these are complete the Model designations of these systems will be changed to Mk 10 Mods 10, 11, 12, 13, 14, 15 and 16.

The Mk 10 models 3 and 4 systems originally installed aboard CVAs have been retired from active service.

In addition to the USN fittings of the Mk 10 GMLS the Italian Navy aircraft carrier *Vittorio Veneto* is fitted with a Mk 10 Mod 9 GMLS forward, and the Italian cruisers *Andrea Doria* and *Caio Duilio* were fitted with the Mk 10 Mod 0. The Peruvian ship *Aguirre* was fitted with a Mk 10 Mod 5, but this was removed and returned to the USA.

GMLS Mk 13

The GMLS Mk 13 series marks a return to vertical stowage and reloading of missiles, as well as representing a changeover to single-ramp launcher, embodied in a modular design which combines the launcher, reloading mechanism and magazine in a single unit which can be installed in the ship as an entity. This system has been used by the USN since 1962 in various models (from Mod 0 to 3) and a total of 41 had been supplied to the USN and the navies of six friendly countries by 1979, when the Mk 13 Mod 4 was introduced. This version incorporated improvements in all elements of the electrical control system while retaining the basic Mk 13 mechanical and hydraulic design. Among the benefits from these improvements were reduced reaction time because of remote start and fire capabilities that permit remote operation; reduced manning requirements; ability to fire a wide range of missiles; and ability to withstand inadvertant ignition of the 'hotter' Mk 56 rocket motor by virtue of the addition of ablative material in the magazine plenum chamber. The Mk 13 Mods 0 to 3 were designed to accommodate RIM-24 Tartar and Standard 1 (MR) missiles, but the Mod 4 version handles the later Standard 2 (MR) (with modifications) and Harpoon surface-to-surface missiles also.

By 1981, 41 GMLS Mk 13 Mods 0 to 3 had been built and installed for the USN and the navies of six other countries: Australia, France, Italy, Japan, Netherlands and West Germany. By 1985, 59 GMLS Mk 13 Mod 4 had been built and installed for the USN and the navies of four other countries: Australia, Japan, Netherlands and Spain. Twelve more are planned for delivery by the end of 1987 for the USN, Australia, Japan, Italy, Netherlands and Spain.

The USN Mk 13 Conversion Programme is currently upgrading Mods 0, 1, 2 and 3 systems to include features found on the Mk 13 Mod 4. The refurbishment programme adds ablative materials to the magazine plenum chamber, improves sprinkling characteristics, provides a solid state control system with built-in test equipment, and alters the power system from high to low voltage. When the programme is completed, the Mod 0 GMLS will be designated Mod 5; Mods 1 will become Mods 6 and Mods 3 will become Mods 7. The navies of Australia, Italy and France are also upgrading GMLS Mk 13.

Three main groups of components comprise the Mk 13 Mod 4 GMLS; the launching system control Mk 162 Mod 0, the launcher Mk 126 Mod 4, and the magazine Mk 109 Mod 0, plus certain ancillary units. The first of these electronically directs and monitors operation of the overall GMLS and includes the power panel (EP1), control console (EP2) and electronics panel (EP3). Control of the GMLS normally takes place remotely from the ship's fire control system, but local loading and pointing are possible from EP2, permitting various alternate modes of operation and exercise functions. The Mk 126 single-ramp launcher receives missiles from the ready service ring (RSR), part of the Mk 109 magazine that rotates to present missiles from the magazine via inner and outer rings which provide a total of 40 missiles. A single electric motor and hydraulic pump provide the power to drive either the RSR, or the hoist used to position missiles in the magazine or raise them to the launch ramp. Electric motors and hydraulic power transmission trains are used for actuation of the launcher in train and elevation.

GMLS Mk 22

The GMLS Mk 22 was a lightweight version of the Mk 13, above, having much the same performance and configuration. Only the magazine size was reduced, primarily to facilitate fitting in smaller vessels. Operated from a remote fire control station, the Mk 22 directs and launches 16 Tartar or Standard missiles and can be modified to launch the surface-to-air Standard 2 (MR) and Harpoon missiles for surface-to-surface engagements. No longer in production.

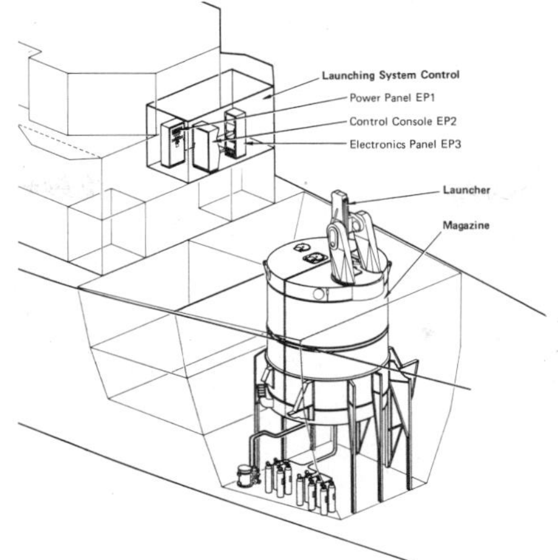

Diagram showing typical location of GMLS Mk 13 Mod 4 components

CHARACTERISTICS

Missile(s): RIM-24 Tartar, Standard (MR), Standard 2 and Harpoon with modification
Ready-use rounds: 16, any mixture
System weight (without missiles and fluids): 4.190 t
System weight (with missiles and fluids): 5.193 t
Training limits: Unlimited
Training velocity/acceleration: 88.5°/s; 132°/s²
Elevation limits: –15 to +95°
Elevation velocity/acceleration: 49°/s; 135°/s²
Tactical crew: 2 (1 panel operator, 1 safety observer)
Ship fittings
6 USN 'Brooke' class missile frigates
5 Spanish 'Baleares' class frigates
System interface: GMFCS Mk 74
Power requirements: 440 V, 60 Hz, 3 phase
Average standby load: 20.6 kW
Average operation load: 39 kW
Peak operation load: 164 kW

GMLS Mk 26

The Mk 26 GMLS is operational with the US fleet and is capable of automatic handling and launching a mixed complement of Standard surface-to-air and ASROC anti-submarine missiles. Modular design and construction enables installations with magazine capacities of 24, 44 or 64 rounds to be fitted. The Mk 26 Mod 5 is the main launching system on the USN's first five 'Ticonderoga' class Aegis cruisers.

CHARACTERISTICS

Missile(s): Standard (MR), RUR-5A ASROC
Ready-use rounds: 24 (Mods 0 & 3); 44 (Mods 1, 4, 5); 64 (Mod 2), any mixture
System weight (without missiles and fluids): 7.303 t (Mod 0); 9.385 t (Mod 1); 12.056 t (Mod 2); 7.210 t (Mod 3); 9.237 t (Mods 4 & 5)

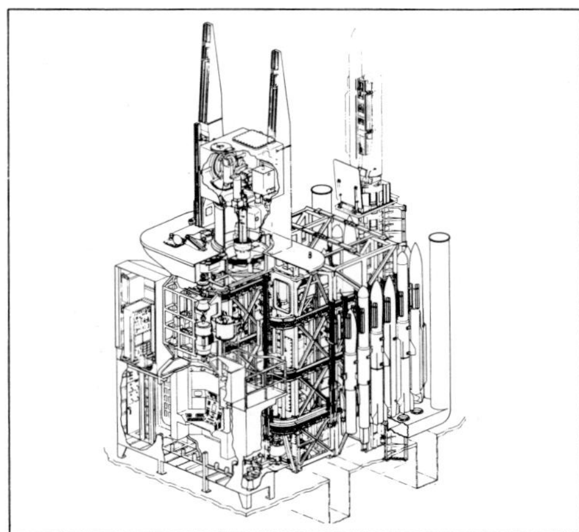

FMC Mk 26 Mod 1 GMLS is a modular system which can be configured with magazine capacities of 24, 44 or 64 missiles

Standard missile launch from GMLS Mk 26 aboard USS Norton Sound

System weight (with missiles and fluids): 8.685 t (Mod 0); 11.920 t (Mod 1); 12.056 t (Mod 2); 8.604 t (Mod 3); 11.792 t (Mods 4 & 5). (Missiles assumed average weight 580 kg each.)
Training limits: Unlimited
Training velocity/acceleration: 90°/s; 180°/s²
Elevation limits: –10 to +90°
Elevation velocity/acceleration: 50°/s; 100°/s²
Tactical crew: 3 (control console operator, 2 fin assembler/folders)
Ship fittings
Mod 0: 4 USN cruisers
Mod 1: 4 USN nuclear cruisers
Mod 2: Land based test unit
Mod 3: 4 USN 'Kidd' class (DD-993) destroyers
Mod 4: 4 USN 'Kidd' class (DD-993) destroyers
Mod 5: USN 'Ticonderoga' class cruisers (CG-47-51)
System interface: GMFCS Mk 74 Mod 5; Mk 116 underwater FCS
Power requirements: 440 V, 60 Hz, 3 phase
Peak standby load: 105 kW
Peak running load: 480 kW

LMOTS (Launcher Maintenance and Operational Training System) provides simulation training for shipboard operators of Guided Missile Launching System Mk 13, Mods 4, 5, 6 and 7 and GMLS Mk 26, Mods 0-5. LMOTS training includes both normal operation, casualty identification and fault isolation procedures.

An LMOTS system consists of the primary tactical GMLS control system elements and a missile simulator plus a Launcher Order Responder Unit (LORU). The use of the standard control system elements is a key feature of the LMOTS concept. They are identical to that which the GMLS operator will be working with aboard ship. When training, the operator receives the same visual indications and simulates the same actions as when operating or troubleshooting a shipboard GMLS system. Hence, control system training is accurate with no compromise for the simulated situation.

The heart of LMOTS is LORU. This is the instructor's console and contains the control system which:

(1) duplicates the electrical, timing and response function of the launcher, magazine and jettison hydraulic/mechanical/assemblies
(2) duplicates weapons control to the extent necessary to generate the proper GMLS and missile orders in the time intervals encountered in the launching system
(3) duplicates the electrical interface characteristcs of missiles to the degree necessary to complete simulated firing responses
(4) provides for insertion of simulated malfunctions of the launching system electrical system and operational conditions.

LMOTS systems are installed in shore-based training facilities that include visual and other instructional aids. Alternative approaches could install LMOTS equipment in moveable trailers along with a mobile instructor team. LMOTS training can be brought directly to pierside for refresher training of ship's personnel or the initial training of operators reporting aboard new-construction ships.

STATUS

The GMLS Mk 4 is now obsolete and few installations remain in service. The situation with the Mk 7 employed for the large Talos air defence missile is also related to the obsolescence of that weapon system. The many existing installations of the various models of the Mk 10 GMLS are likely to remain in service for a considerable period to come, with the systems being updated to accommodate the Standard SM-2 (ER) surface-to-air missile as part of the USN's CGN/New Threat Upgrade programme. FMC was awarded a $36.6 million USN contract in 1983 for continued Mk 13 Mod 4 system production, including two more launchers for FFG-7 class frigates and one launcher for Japan, and with these FMC will have supplied 107 Mk 13 systems since the early 1960s. The Mk 22 is no longer in production, but is being considered for modernisation to accommodate the Harpoon missile. Mk 26 production is complete for the first five 'Ticonderoga' class.

CONTRACTOR

FMC Corporation, Naval Systems Division, PO Box 59043, 4800 East River Road, Minneapolis, Minnesota 55459-0043, USA.

4204.231

MK 41 VERTICAL LAUNCH SYSTEM

The USN's vertical launch systems (VLS) programme achieved operational status at the end of 1986, with the Mk 41 VLS which will provide an improved missile launching system for surface vessels of the US fleet.

The system is based on a modular design, the basic unit being a launcher module containing eight missile cells. Each module, which consists of a structure, independently operating armoured hatch-covers, a gas-management system and requisite electronics, power supplies and support equipment, is assembled ready for installation aboard ship. Each of the eight cells in a module is designed to house a missile canister, and the system design allows any missile adapted for vertical launch to be fired from any cell. The missile canister is of corrugated steel construction and is both protective and reusable, thus serving as the shipping container and as the launcher rail for the missile. The basic eight-cell module is designed as part of a Mk 41 VLS. Each of the two systems for the Aegis cruisers ('Ticonderoga' class) consists of seven eight-cell modules, plus one five-cell 'strikedown' module, thus giving the ship a total capacity of 122 missiles. A complete Mk 41 VLS of seven eight-cell modules and one strikedown module, containing 61 missiles in all, occupies the space required for one Mk 26 Mod 0 conventional GMLS with a capacity of 44 missiles.

The strikedown, or replenishment, module is almost identical to the standard eight-cell module except that a hoist supporting an articulating crane is contained in the space of three cells. The inclusion of this module in each VLS provides a capability for underway replenishment of the system.

The hatch covers of each eight-cell module are individually heated for anti-icing and provided with individual electric motors for opening and closing, these being remotely controlled. A deluge system with individual valves for each canister provides a water deluge to the missile warhead in the event of over-heating or accidental motor ignition. A separate sprinkler system provides fire protection for the other spaces. Each module contains a plenum and uptake to vent the exhaust gases from a fired missile. The plenum and uptake are lined with ablative material to protect the structure against missile exhaust, and each module is capable of a minimum of 64 firings before ablative repairs are needed.

The missile, when ignited, pressurises the canister breaking the lower end cover allowing the gases to fill the plenum and exhaust vertically through the open uptake hatch at deck level. Double elastomeric seals are provided to form gas-tight joints at both ends of the canister to prevent gases entering the launcher spaces.

Two launch control units (LCUs), currently comprising AN/UYK-20 USN digital processors (with upgrade to AN/UYK-44 configuration underway), are located in the ship's weapon control station. The LCUs provide a digital interface with the appropriate ship's weapon control systems and control the launch of missiles. The Mk 41 VLS at present is able to interface simultaneously with up to three weapon control systems: Aegis air defence, Tomahawk Weapon Control System, and Vertical Launch ASROC anti-submarine warfare systems. Beyond its initial capacity for Standard-1 missiles, VLS weapon system capabilities now include the successor of Standard-2 (Block II) and Tomahawk cruise missiles. Integration of the Vertical Launch ASROC (VLA) anti-submarine missile is nearing completion. Integration efforts are also now in process for the Standard-2 (Block IV) Extended Range missile and Tomahawk RGM-109D. The Tomahawk version will accommodate 61 to 122 missiles.

The new Aegis cruiser Mobile Bay *(CG-53) looks underarmed, but her two Mk 41 VLS give her a main armament of 122 Standard SM-2 missiles* (US Navy)

VLS is a strong candidate for inclusion in the USN Short-Range Air-Defense System (SRADS), and the NATO Local Area Missile System (LAMS) as part of the NFR-90 frigate replacement programme.

The modular concept of the Mk 41 system makes it adaptable to many surface combatant platforms, thus extending the mission flexibility of VLS-equipped ships. VLS was introduced into the US Navy in the Aegis cruiser US *Bunker Hill* (CG-52), and has been installed in the USS *Spruance* (DD-963) lead-ship of her class. VLS will also be installed in all 'Arleigh Burke' (DDG-51) guided missile destroyers. The first international sale of VLS was to the Canadian Navy, for its 'Tribal' class destroyer Update and Modernization (TRUMP) programme. The Mk 41 is also being evaluated for adaptation to a variety of other combatant platforms, including the 'Iowa' class battleships, 'Perry' class frigates, carriers, amphibious assault ships and combat support ships.

Martin Marietta have modified VLS so a canister can hold four ship defence missiles. Such a system may be considered for NAAWS.

CHARACTERISTICS

Launcher type: Mk 41 Mod 0 (CG-47 class), Mod 1 (DD-963 class), Mod 2 (DDG-51 class), Mod T (DDH-class) (Canadian)

Missiles: SM-2 (Block II), Tomahawk development – VLA, SM-2 (Block IV) ER, Tomahawk 'D'

Target capability: Air, surface, u/water (development)

System weight: 117 081 kg (empty magazine); 231 971 kg (loaded) (61 cell magazine)

Tactical crew: 0

Anti-icing system: Electric

Power requirements: 60 Hz, 440V AC, 3 phase 207 kW (max); 60 Hz, 115V AC, 3 phase 6 kW (max); 60 Hz, 115V AC, 3 phase 10 kW (max); 60 Hz, 115V AC, 1 phase 5.4 kW (max); 400 Hz, 440V AC, 3 phase 36 kW (max)

STATUS

Developed from 1977 for the USN to specification WS 18968A, the Mk 41 VLS is in full production at Martin Marietta Baltimore (Maryland) Aerospace (lead producer and design agent) and at the FMC Corporation Naval Systems Division (second source producer). Initial full-system firing tests of a prototype launcher module at White Sands Missile Range were completed in January 1981 when six Standard missiles were successfully launched, three as single launches and three in ripple-fire sequence to demonstrate rapid-fire capabilities. Subsequently, SM-2, Tomahawk, VLA and Sparrow-derived SDM (Ship Defense Missile) have been launched from land-based test sites at White Sands or China Lake, or from the test ship USS *Norton Sound*. The first VLS firings from a combatant ship were successfully conducted in May 1986 during sea trials of the USS *Bunker Hill* (CG-52), and followed by *Mobile Bay* (CG-53) firings in September 1986. Five VLS installations have been completed to date for 'Ticonderoga' (CG-47) class ships, with installation planned for an additional 17. Martin Marietta are producing 39 launchers and 1702 canisters for 25 US Navy ships. The first VLS installation in the 'Spruance' class ships took place in December 1986, with all 31 ships in this class to be fitted with this system carrying 61 missiles. The Mk 41 will also be installated in all 29 of the 'Arleigh Burke' (DDG-51) class destroyers. The Canadian Navy contracted for VLS installation in its four 'Tribal' class destroyers in 1986.

CONTRACTORS

Design Agent and Lead Source:

Martin Marietta Baltimore Aerospace, 103 Chesapeake Park Plaza, Baltimore, Maryland 21220, USA.

Second Source: FMC Corporation, Naval Systems Division, PO Box 59043, 4800 East River Road, Minneapolis, Minnesota 55459-0043, USA.

NAVAL GUNS AND BOMBARDMENT ROCKETS

BRAZIL

6193.203

TWIN 20 mm AA TURRET

It is reported that the Brazilian Navy is developing a twin-barrel, radar controlled 20 mm anti-aircraft turret which is being considered as a replacement for the Seacat air-defence missiles in the 'Niteroi' class frigates. The gun type is thought to be a Cockerill revolver design and the electronics are from an unknown European source. Rates of fire during trials are reported to have achieved 6 000 rpm. The status of funding for production is at present unclear and, to date, no Brazilian firm has been selected to manufacture the weapon.

CHINA (PEOPLE'S REPUBLIC)

6055.203

CHINESE NAVAL ORDNANCE

Much of the Chinese PLA Navy is made up of vessels originally designed and constructed in the Soviet Union and subsequently transferred to China. A further section – but one of diminishing importance – consists of ex-Japanese and Second World War ships inherited by the present regime from that of Chiang Kai Shek. Most of these latter vessels are now at least obsolescent if not obsolete.

Since the mid-1950s, however, the Chinese have been building many of their own fighting ships. At first these were mainly built from parts supplied by the Soviet Union, but more recently the Chinese have had to rely increasingly on their own resources.

Because of this overlapping supply it is not possible to say exactly for which vessels the guns have been built in China, or which of those have been entirely manufactured in China. It seems probable, however, that there are Chinese-built guns of at least the following types currently in service: 100 mm dual-purpose (twin and single mounts), 57 mm (twin), 37 mm AA (twin mounts), and 25 mm AA (twin mounts). It is also certain that 130 mm guns (in twin mountings) are being built in China for incorporation in the new 'Luda' class destroyers. 100 mm single mountings too continue to be installed in new Chinese ships. Guns of several other calibres are in service with the fleet, and must therefore be being serviced in Chinese yards, but it cannot be said with certainty that they have been built in China. China is reported to be producing a 30 mm anti-aircraft gun, AK-230, for destroyers and frigates but no further details are available.

French 100 mm Compact guns are being installed in new construction, and we can expect more modern weapons to appear, including CIWS.

14.5 mm calibre AA (twin) mount on P4 Class torpedo boat of Chinese Navy

4513.203

TWIN 130 mm DUAL-PURPOSE GUN MOUNTING

This gun mounting is a copy of the original gun system and mounting used on the Soviet 'Kotlin' class destroyers. Blueprints for the system were probably received prior to 1960, but the Chinese were unable to effect production of the mounting until ten years later when the first of the 'Luda' class destroyers was laid down. The gun house differs slightly in having rounded edges and does not include the fitting of a ranging radar as in the Soviet original.

CHARACTERISTICS

Calibre: 130 mm

Length of barrel: 58 calibres

Elevation: 85° max

Projectile weight: 33.4 kg

Muzzle velocity: 945 m/s

Max range: 29 300 m

Rate of fire: 15-17 rounds/minute

Twin 130 mm gun mount on Chinese 'Luda' class destroyer. Note open rear to turret and basket on back (G. Jacobs)

4514.203

SINGLE 130 mm GUN MOUNTING

This is an old gun mounting originally received with the Soviet 'Gordy' class destroyers transferred to China in 1954-55. Two mountings are also installed in the ex-British 'Castle' class patrol frigate still in Chinese naval use. Although very elderly and relying on the equally elderly Soviet Post Lamp fire control radar for ranging information, the mountings may remain in service for some years since the 'Gordy' class have been fitted with SSMs and this may indicate an extension of their lives beyond the normal length.

CHARACTERISTICS

Calibre: 130 mm

Length of barrel: 50 calibres

Elevation: 45° max

Projectile weight: 33.4 kg

Muzzle velocity: 875 m/s

Max range: 25 000 m

Rate of fire: 12 rounds/minute

Soviet designation: Model B-13-2C

6471.203

SINGLE 100 mm CREUSOT-LOIRE DUAL-PURPOSE GUN MOUNTING

The new 4200-ton destroyers laid down in 1985 are to be armed with the French 100 mm Compact gun supplied by Creusot-Loire. Details will be found under **6194.203**. As only two destroyers are under construction the guns and mountings will be provided from France, but if it is to become the new standard medium-calibre weapon, production in China will almost certainly follow.

6472.203

TWIN 100 mm DUAL-PURPOSE GUN MOUNTING

This mounting appears to have supplanted the venerable single 100 mm mounting (**4515.203**). It is probably the same gun, with a modern Chinese-designed mounting, and may be an interim weapon to bridge the gap until the Creusot-Loire 100 mm is available in quantity.

STATUS

In production, and at sea in 'Jianghu V' and 'Jiangdong' type frigates.

4515.203

SINGLE 100 mm DUAL-PURPOSE GUN MOUNTING

This single mounting was introduced into Chinese naval service in 1956-57 during the assembly of four Soviet 'Riga' class frigates at Hudong shipyard, Shanghai. It was extensively used in the 1960s in rebuilding ex-British and ex-Japanese frigates where all, except for the 'Castle' class, were re-armed with this gun system manufactured in China from blueprints originally sold by the Soviet Union in the late 1950s. It has also been installed in the locally designed 'Jiangnan' class frigates, and also appears to continue to be fitted in the more modern 'Jiangdong' and 'Jianghu' class frigates.

CHARACTERISTICS

Calibre: 100 mm
Length of barrel: 56 calibres
Elevation: 85° max
Projectile weight: 15.9 kg
Muzzle velocity: 875 m/s
Max range: 21 900 m
Rate of fire: 15-18 rounds/minute
Soviet designation: Bu-34

100 mm single dual-purpose mount on stern of a Chinese 'Jianghu' class frigate. Note small size of twin 37 mm mountings where twin 57 mm mounting may have been intended (G Jacobs)

4516.203

SINGLE 85 mm DUAL-PURPOSE GUN MOUNTING

This single 85 mm mounting was first placed in Soviet naval service in 1948 and is controlled by optical directors only. It can be considered obsolescent despite its use in a number of Communist countries. It is fitted to China's 20 'Kronstadt' class patrol vessels.

CHARACTERISTICS

Calibre: 85 mm
Length of barrel: 52 calibres
Elevation: –5 to +85°
Elevation rate: 8°/s
Projectile weight: 9.6 kg
Muzzle velocity: 760 m/s
Max range: 15 000 m
Rate of fire: 15-18 rounds/minute
Soviet designation: Model 90K

4517.203

SINGLE 75 mm DUAL-PURPOSE GUN MOUNTING

The 75 mm dual-purpose gun mounting first became available from captured Nationalist escorts and landing craft after the Civil War. It may subsequently have been manufactured locally from captured blueprints as quantities continue to be seen on refitted American LSTs.

CHARACTERISTICS

Calibre: 75 mm
Length of barrel: 50 calibres
Elevation: –13 to +85°
Projectile weight: 13.05 lb (5.92 kg) (AA); 13.07 lb (5.93 kg) (AP)
Muzzle velocity: 2650 ft/s (808 m/s)
Max range: 14 000 yards (12 800 m)
Rate of fire: 18 rounds/minute
Total mounting weight: 7500 lb (3400 kg) (Mod 17); 8310 lb (3769 kg) (Mod 2)
US designation: Mark 22 Mods 2 and 17

4518.203

TWIN 57 mm AA GUN MOUNTING

This is a water-cooled adaptation of the Soviet 57 mm SIF-31B model. Some 'Luda' class destroyers were initially armed with it as the AA defence system intended for all units of the class. It is, however, now only standard in the 'Hainan' class patrol craft, four of which have been exported to Pakistan. A lead computing optical sight is used for fire control. The magazine on an escort holds about 1600 rounds. The Chinese indicate that they have high explosive, fragmenting, and tracer rounds available.

CHARACTERISTICS

Calibre: 57 mm
Length of barrel: 70 calibres
Elevation: –10 to +85°
Projectile weight: 6.31 kg
Muzzle velocity: 1000 m/s
Max range: 12 000 m
Rate of fire: 2 × 105-120 rounds/minute

Twin 57 mm AA mount on 'Hainan' class. Note the RBU-1200 ASW rocket-launchers on forecastle

6756.203

TYPE 61 TWIN 25 mm AA GUN MOUNTING

This 1753 kg mounting is intended for twin 25 mm guns to intercept air targets at a range of 2.8 km and ground targets at 3 km. It features an hydraulic ammunition reloading device, an hydraulic elevation and bearing system, an automatic halt-fire device together with an automatic firing computer.

CHARACTERISTICS

Calibre: 25 mm
Elevation: –10 to +85°
Max speed in elevation: 40°/s
Max speed in azimuth: 70°/s
Rate of fire: 800-900 rounds per minute
Muzzle velocity: 890 m/s

CONTRACTOR

North China Industries Corporation, 7A Yuetan Nanje, Beijing, People's Republic of China.

4520.203

TWIN 14.5 mm GUN MOUNTING

A light AA mounting fitted in the 'Chengdu' class frigates, the 'Kronstadt' class patrol vessels, the 'Huchuan' torpedo-armed fast attack craft, and the T-43 minesweepers. It is not certain that it is still in production although 'Huchuan' class vessels continue to be built. The ammunition however is available for export. Local fire control only is available.

CHARACTERISTICS

Calibre: 14.5 mm
Length of barrel: 93 calibres
Elevation: –5 to +85°
Projectile weight: 0.06 kg
Muzzle velocity: 1000 m/s
Max range: 7000 m
Rate of fire: 2 × 600 rounds/minute
Soviet designation: 2M-7

4519.203

TWIN 37 mm AA GUN MOUNTING

Both single and twin mountings are standard throughout the Chinese Navy and are currently used by all destroyers, frigates and fleet minesweepers, and some landing ships. Both mounting and ammunition are available for export and equip a number of vessels sold to Bangladesh, Pakistan and Sri Lanka. Only local fire control is available.

CHARACTERISTICS

Calibre: 37 mm
Length of barrel: 63 calibres
Elevation: –10 to +85°
Projectile weight: 1.416 kg
Muzzle velocity: 875 m/s
Max range: 8500 m
Rate of fire: 2 × 160-180 rounds/minute
Soviet designation: B-11-M (twin); Model 70K (single)

Twin 37 mm and twin 25 mm 'over-and-under' mounts on Chinese 'Shanghai' class gunboat (G Jacobs)

FRANCE

6064.203

SINGLE 100 mm DUAL-PURPOSE GUN MOUNTING MODEL 1968-II

The gun mounting Model 1968-II represents the latest version of French 100 mm turret and is a lightened and completely automatic version derived from the Models 1953, 64 and 68-I. This calibre is considered optimum by the French authorities for engaging air targets and surface vessels, as well as shore bombardment. Among the intended targets are missiles, including sea-skimming types.

Features of the Model 1968-II 100 mm gun which suit it for multipurpose roles include: fully automatic remote operation, with the option of autonomous operation with a crew of two; high rate of continuous fire; replenishment of magazine when firing is in progress for long engagements; a container-based munition re-supply system; long bore life due to air purging and water cooling after each round is fired. Both a multipurpose and a prefragmented shell can be used. The multipurpose shell uses either a time or a proximity fuze while the prefragmented shell was primarily designed for proximity fuzing.

CHARACTERISTICS

Calibre: 100 mm
Length of barrel: 55 calibres
Elevation: –15 to +80°
Max speed in elevation: 25°/s
Max acceleration (elevation): 80°/s²
Max slewing speed: 40°/s
Max acceleration (slew): 50°/s²
Projectile weight: 13.5 kg (1.05 kg charge)
Round weight: 23.6 kg
Muzzle velocity: 870 m/s
Max range: 17 000 m (9 nautical miles)
Max practical range against air target: 6000-8000 m
Rate of fire: 60 rounds/minute (80 rounds/minute on recent models)

STATUS

Operational with the French Navy since 1960, this equipment has subsequently undergone several modernisation programmes. It also serves in the navies of the Federal Republic of Germany, Belgium, Portugal and Argentina.

CONTRACTORS

Overall responsibility for the 100 mm Model 1968-II guns and mountings belongs to Direction des Constructions Navales (DCN), 2 Rue Royale, 75200 Paris Navale, France.

Manufacture is shared between Creusot Loire Industrie's Division Méchanique Spécialisée, Immeuble Ile de France, Cedex 33, 92070 Paris la Defense, France (mounting) and ECAN Ruelle, 16600 Ruelle, France (gun). Control gear is made by CSEE, Division Defence, Avenue des Tropiques, ZA de Courtaboeuf, BP 80, 91.943 Les Ulis Cedex, France.

6194.203

100 mm COMPACT GUN MOUNTING

Creusot Loire Industrie has developed a dual-purpose 100 mm Compact Gun Mount, derived from the French Navy's Model 1968, but incorporating important improvements in technology and performances. Main features include:

- Reduction of weight by 7.5 tons, to allow fitting in smaller ships
- A fully automatic, reversible, computer-controlled ammunition feed
- Instant selection of three types of munitions
- Ready-use ammunition increased to 42 or 90 rounds, depending on version
- Replenishing ammunition while firing
- Three rates of fire: 10/40/90 rounds/minute
- Preset bursts of 1/2/3/6 shots
- Increased rates of elevation and slewing

STATUS

In production and in service since 1984. It is used by the Malaysian and Saudi Arabian Navies and one system has been sold to China. May also replace standard 100 mm gun in future French Navy warships.

CONTRACTORS

As with the older 100 mm Model '68 overall design and manufacture of the mounting belongs to Creusot Loire Industrie, Division Mécanique Spécialisée, Immeuble Ile de France, Cedex 33, 92070 Paris la Defense, France.

The gun is made by ECAN Ruelle, 16600 Ruelle, France, and control equipment by CSEE, Avenue des Tropiques, ZA de Courtaboeuf, BP 80, 91.943 Les Ulis Cedex, France.

Creusot 100 mm compact gun on Malaysian frigate

6065.203

TWIN 57 mm AA GUN MOUNTING

This 57 mm gun is now only fitted in the French Navy in the old cruiser *Colbert*. It consists of twin Bofors guns carried on a mount of French design. Each gun is provided with an ammunition reserve of 80 rounds, which can be replenished from a transfer chamber located below the mount. The gun and mount are normally operated under remote control but autonomous operation is also possible.

CHARACTERISTICS

Calibre: 57 mm
Length of barrel: 60 calibres
Elevation: –8 to +93°
Projectile weight: 2.96 kg
Muzzle velocity: 865 m/s
Max range: 13 km (7 nautical miles)
Practical range (air targets): 5000 m (3 nautical miles)
Rate of fire: 2 × 120 rounds/minute
Weight: 16 tons

STATUS

Out of production. Only six mountings in service.

CONTRACTOR

Direction des Constructions Navales (DCN), 2 rue Royale, 75200 Paris Navale, France.

6195.203

STORM ROCKET LAUNCHER SYSTEM

STORM is a naval multiple rocket launcher system, designed to give naval forces a greater fire power capability against shore targets and provide effective mass fire coverage in shore bombardment, thanks to its long-range, low dispersion and powerful rockets.

STORM consists of a computer aimed and stabilised launching mount on which are fitted two expendable multiple launcher pods containing fuzed, sealed-in 160 mm rockets. A standard STORM Launcher Pod Container includes 18 rockets.

All the technical equipment such as power supply box, laying devices, computer, control console, is installed, according to version:

(a) either inside an independent Technical and Control Container, easily mounted on any type of ship – merchant ship, landing ship, support ship, transport, frigate.

(b) or in the ship, as a standard weapon system.

CHARACTERISTICS
Rocket
Calibre: 160 mm
Weight (total): 110 kg
Length: 33 cm
Propulsion: single-stage solid propellant motor
Stabilization: folding wrap-round stabilized fins
Payload: 50 kg
Warheads: High explosive, COFRAM, bomblets, scatterable mines
Launcher Pod Container (standard version)
Length: 3460 mm
Width: 1000 mm
Height: 820 mm
Weight (empty): 540 kg
Launcher and mount
Weight (total): 7400 kg
Weight (empty): 2400 kg
Height (elevation 0°): 31 cm
Overall diameter (loaded): 41.10 cm
Laying: Elevation −15° to +70°, training 350°
System
Maximum range: 35 km
Maximum effective range: 30 km
Rate of fire: 36 rockets per minute
Bursts: from 2 to 36 rockets
Round-to-round fire accuracy: total error 10.8 mils for range 22 km
STATUS
The system is in development. Rocket operational as land based weapon.
CONTRACTOR
Creusot Loire Industrie, Division Mécanique Spécialisée, Immeuble Ile de France, Cedex 33, 92070 Paris la Defense, France.

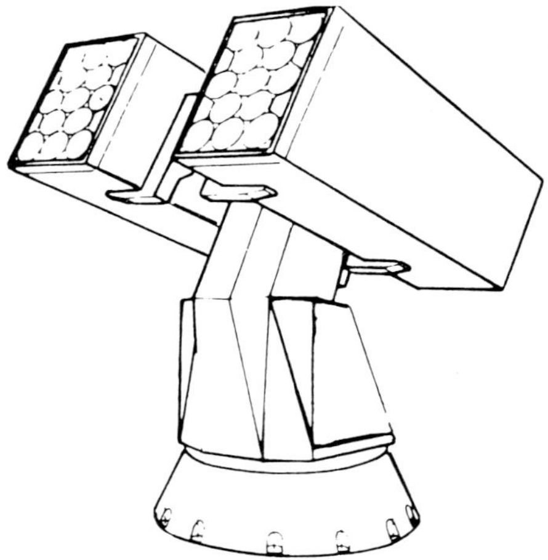

Sketch of Storm launcher mount

3655.203
TYPE A NAVAL MOUNTING FOR 20 mm F2 GUN
This naval mounting is installed in small, lightweight ships, and carries the French F2 20 mm automatic cannon which is also employed in land-based and airborne roles. It is designed for one-man operation, and elevation and slewing motions are manually powered. Two types of ammunition can be employed, carried in a pair of magazines on the mount with a combined capacity of 300 rounds. Continuous, limited burst and single-shot firing can be selected, and the mount can be used for both air or surface target engagement.
CHARACTERISTICS
Calibre: 20 mm
Elevation: −15 to +60°
Muzzle velocity: 1050 or 1250 m/s*
Effective range: 2000 or 1200 m*
Max range: 10 020 or 6770 m*
Rate of fire: 720 rounds/minute
Power requirement: 26 V, 7 A
Weight: 470 kg (360 kg without ammunition)
*Depending upon type of ammunition: explosive or armour-piercing.
STATUS
In series production for 'Tripartite' minehunter vessels of the Belgian, Dutch and French Navies. This gun also exists on a mobile land mounting for army use (**5599.103** in 1982-83 and earlier editions).
CONTRACTOR
Direction des Constructions Navales (DCN), 2 rue Royale, 75200 Paris Navale, France.

Type A naval mounting for 20 mm F2 gun

GERMANY (FEDERAL REPUBLIC)

3177.203
SINGLE 20 mm NAVAL GUN S20
This is the naval mount version of the Rheinmetall Mk 20 Rh 202 20 mm automatic cannon described in **5600.103** (in 1982-83 and earlier editions). S20 naval guns are primarily intended for installation on small and fast ships and as secondary armament on larger vessels as well as for static applications ashore. Low weight and one-man body aimed actuation of the mount assure rapid response for the effective engagement of surface and air targets, whereby the gunner always remains in an upright position.

The weapon itself is the Rheinmetall 20 mm automatic cannon Mk 20 Rh 202 which is particularly suited for these purposes because of its low recoil forces. The Mk 20 Rh 202 is also renowned for its functional safety and reliability in spite of restricted lubrication, sea water contamination, heavy fouling and other extreme environmental conditions. Mount and weapon can be dismantled into their major components without the use of any tools.
CHARACTERISTICS
Type: Mk 20 Rh 202
Calibre: 20 mm
Elevation: −10 to +55°
Traverse: Unlimited
Muzzle velocities: HEI-T, TP-T 1050 m/s; API-T 1100 m/s; APDS-T 1150 m/s
Recoil force: 550 – 750 kgf
Tactical range: 2000 m
Rate of fire: 1000 rounds/minute
Firing: Mechanical
Type of fire: Single or sustained fire
Ammunition supply: 200 rounds belted in exchangeable ammo box
Sight: Mechanical AA sight with parallelogram linkage and shoulder rest. Night vision device and telescope sight are optional
Weight: Mount complete: 334 kg; ammunition: 74 kg
Length: 3070 mm
Width: 1230 mm
Height: 1570 – 1785 mm
STATUS
In production and used by the West German and Hellenic Navies (with Wegmann mounts) and by the navies of the Bahamas, Barbados, Indonesia, Ireland and Thailand (with Rheinmetall mounts).

Rheinmetall S20 single 20 mm gun mount aboard Irish OPV Eithne (AP/Jane's Defence Weekly)

CONTRACTOR
Rheinmetall GmbH, D 4000 Düsseldorf 1, Federal Republic of Germany.

ITALY

5534.203
SINGLE 127/54 GUN MOUNTING
The 127/54 is a dual-purpose gun mounting designed for use as the main armament in frigates and destroyers.

Ready use ammunition is held in three drums just below the turret. A central elevator hoists the ammunition, chosen from one drum, and delivers it to the turret where two oscillating arms perform the final movement to the loading trays. The three drums are automatically reloaded through two hoists manually loaded in the magazine.

This three drum layout makes it possible to stow three different types of ammunition in the ready use magazine from which ammunition can be selected appropriate for the type of action. Remote-controlled fuze setters are provided in the oscillating arms.

The reloading, feeding, loading, and firing sequence is controlled by a control console operated by a single man.

The electrically-operated remote control servosystem is of modular construction using silicon controlled rectifiers. The modules are the same as those used on other modern OTO Melara weapon systems.

The main technical features are the light alloy construction, the compact installation, and the lightweight control mechanism.

The barrel is fitted with a muzzle brake. The shield is of fibreglass and is watertight. Standard USN 5 inch ammunition is used.

127/54 OTO single barrel naval mount

CHARACTERISTICS
Calibre: 127 mm
Elevation: −15 to +83°
Elevation speed: 30°/s
Elevation acceleration: 40°/s²
Traverse: 330°
Traverse speed: 40°/s
Traverse acceleration: 45°/s²
Rate of fire: 45 rounds/minute
Weight: 37.5 tons
Ready ammunition: 69 rounds (3 × 22 + 3)
Trunnion thrust: 24 000 kg
STATUS
In production and in service as follows: Canada – four 'Iroquois' class destroyers; Italy – two 'Audace' class destroyers, eight 'Maestrale' class frigates, four 'Lupo' class frigates; Argentina – four 'Al. Brown' class frigates; Nigeria – frigate 'Aradu; Peru – four 'Lupo' type frigates; Venezuela – six 'Lupo' type frigates; Iraq – four 'Lupo' type frigates.
CONTRACTOR
OTO Melara, Via Valdilocchi 15, La Spezia, Italy.

5532.203
SINGLE 76/62 OTO MMI GUN MOUNTING
The 76/72 OTO MMI single barrel automatic gun was developed to a government contract for ships of the frigate and corvette types as a dual-purpose weapon. It was the forerunner of the successful 76/62 OTO Compact (**5533.203**) described below, which has superseded it. For details see the 1987-88 edition.
CHARACTERISTICS
Calibre: 76 mm
Elevation: −15 to +85°
Elevation speed: 40°/s
Elevation acceleration: 70°/s²
Traverse: 360°
Traversing speed: 70°/s
Traversing acceleration: 100°/s²
Rate of fire: 55 – 65 rounds/minute
Ready to fire rounds: 59
Overall weight: 12 tons
STATUS
No longer in production. There are many in service with the Italian Navy and reconditioned guns are being installed in the new 'Minerva' class corvettes.
CONTRACTOR
OTO Melara, Via Valdilocchi 15, La Spezia, Italy.

4843.203
SINGLE 76/62 SUPER RAPID GUN MOUNTING
Development of the 76/62 super rapid gun mounting began in mid-1980 as an evolution of the 76/62 OTO compact gun mounting (**5533.203**). Its main role is anti-missile defence but it is also effective in either anti-aircraft or anti-ship fire. The new mounting maintains the same architecture and aspect and has identical mechanical and electrical interfaces but it can sustain continuous firing of all 80 ready-use rounds at more than 120 rounds per minute. In addition it incorporates a local stabilisation system for better aiming and firing accuracy and is provided with a secondary feeding capability that allows a rapid change from anti-air/anti-missile automatic fire to anti-surface and semi-automatic fire, and vice versa. The electronic portions of the gun have also been updated with safety firing arcs, pointing accuracy check, status display and BITE functions under mini-computer control.

Two ammunition developments have also been completed. The MOM ammunition is a pre-fragmented projectile designed for use against air targets, with high specific weight tungsten cubes and very high ejection velocity ensuring effective penetration of the target even at large miss-distances. It is fitted with a proximity fuze which is highly effective against sea-skimming missiles and aircraft and can also be used in an anti-personnel role during shore bombardment. The proximity fuze can be paralysed to allow the fuze to operate with impact delay.

SAPOM anti-ship ammunition is semi-armour piercing and optimised against hard surface targets. It is fitted with a base-detonating delayed action fuze and the projectile is made of high quality steel for maximum penetration capability.

Two further types of ammunition are also under development; one based on guided-projectile techniques for anti-missile/anti-aircraft fire and the other an anti-surface extended range round.

OTO 76/62 super rapid gun mounting

CHARACTERISTICS
Calibre: 76 mm
Elevation: −15° to +85°
Elevation speed: 35°/s
Elevation acceleration: 72°/s²
Traverse: Unlimited
Traverse speed: 60°/s
Traverse acceleration: 72°/s²
Rate of fire: greater than 120 rounds/minute
Trunnion thrust: 10 tons (constant)
Overall weight: 7.5 tons
STATUS
Series production for the Abu Dhabi, Canadian, Danish, Italian and Singapore navies.
CONTRACTOR
OTO Melara, Via Valdilocchi 15, La Spezia, Italy.

5533.203
SINGLE 76/62 COMPACT GUN MOUNTING
The 76/62 OTO Compact is a light, small calibre gun mount intended for installation in ships of any type and class down to motor gun boats and hydrofoils as a dual purpose anti-aircraft and anti-ship weapon system.

The gun and mounting were developed from the 76/62 OTO MMI mounting (**5532.203**), the design study being started in 1964, and the first production units came into service in March 1969.

The mounting may be considered in two parts; the shank which is below the weather deck, and the turret which is installed on deck. The shank contains the ammunition feed system, which consists of a rotating platform and a hoist.

It can fire 80 rounds without reloading, and the rate of fire is adjustable from 10 to 85 rounds per minute. A version is available with a rate of fire up to 100 rounds per minute. Existing guns can be modified with a special kit.

Generally, the system is controlled remotely; optionally it can also be fitted with a stabilised line of sight local control system. A reaction time of 2.8 seconds is quoted.

The shield is made of fibreglass, and is watertight. It offers complete protection against nuclear fall-out. The gun barrel has a small multi-hole muzzle brake and a fume extractor.

An Improved Rate Of Fire (IROF) retrofitting kit now exists as an intermediate step to improve fire power and this has now been selected by the Royal Netherlands Navy all of whose Compact gun mountings will be modified.
CHARACTERISTICS
Calibre: 76 mm
Elevation: −15 to +85°
Elevation speed: Max 35°/s
Elevation acceleration: 72°/s²
Traverse: Unlimited

Traverse speed: Max 60°/s
Traverse acceleration: 72°/s²
Rate of fire: 10 – 85 rounds/minute
Overall weight: 7.5 tons

STATUS

In service with, or being delivered to, 37 navies. In the USA this gun is produced under licence by FMC Corporation for the US Navy as the Mk 75 (**3175.203**), and is also manufactured under licence in Australia and Japan, and in Spain for 'Lazaga' and 'Descubierta' classes (see **3230.203**).

CONTRACTOR

OTO Melara, Via Valdilocchi 15, La Spezia, Italy.

OTO 76/62 compact gun mount

6375.203

BREDA TWIN FAST FORTY NAVAL MOUNTING

To take account of the new generation of sea-skimming missiles with supersonic velocities of more than 600 m/s the Breda twin 'fast forty' naval mounting has been developed from the twin 40mm L/70 mounting. It is designed for installation on all types of surface vessels, down to small craft such as fast patrol boats and hydrofoils. The main improvements made to this weapon are a newly-designed recoiling mass capable of developing a rate of fire of 900 rpm, a dual feeding system which allows use of two types of ammunition with intermediate switch from one to the other, proximity fuze or APDS at short ranges.

The recoiling mass has the firing cycle accelerated by 50% with respect to the 40mm L/70 Bofos elevating mass, by reducing the recoil stroke by more than half, and speeding up the breechblock opening and closing cycle and the ramming cycle. These mechanical improvements render the gun capable of firing ready-to-fire ammunition from its large magazine at a rate of 900 rpm.

The dual feed system consists of a lower magazine, fitted below the training platform, and an upper magazine, in two symmetrical sections, supported by the trunnion walls and propelling the rounds in a chain mechanism to the double row symmetrical fanshaped shifter. Both magazines, and the feeding system, are divided into two sections, each feeding one gun. In this way firing can continue in the event of a failure in one gun or gun system.

The new digital servo-control system renders the installation capable of interfacing with the latest fire control systems and gives better tracking accuracy.

CONTRACTOR

Breda Meccanica Bresciana SpA, Via Lunga 2, 25100 Brescia, Italy.

Breda twin 'fast forty' naval mounting

2258.203

BREDA COMPACT TWIN 40 mm L/70 NAVAL MOUNTING TYPE 70

This twin 40 mm L/70 naval mounting has been developed by Breda Meccanica Bresciana in close co-operation with Bofors in Sweden as part of the Breda/Bofors System 75 (see also **2349.131**) which is particularly intended for point defence against aircraft and anti-ship missiles.

The compact mount is fully automatic in operation, employs new remote controlled servo-systems of high performance to improve accuracy – which is also improved by the reductions in weight, inertia, and recoil disturbance resulting from the exceptionally small separation between the two barrels – and features a high rate of fire and a substantial supply of ready-use ammunition.

The weapon is available in two versions, Type A and Type B, the difference residing in the quantity of ready-use ammunition contained in the magazine (736 or 444 rounds) and in some consequential weight and dimension changes which are indicated below. In each type the magazine is divided into two halves each with a hoist serving one barrel. Without manual intervention the two hoists operate in synchronism.

Horizontally, the mounting is also divided into two parts. The upper part contains the cradle, the elevating mass, the automatic feeders, the servo-motors, and associated electronics and the firing, aiming, and limiting devices. The lower structure mainly comprises the magazine and hoists. The magazine is made up of several horizontal layers, there being seven of these in Type A and four in Type B. Before firing and during firing intervals these layers are packed with ammunition in four-round clips.

Breda twin 40 mm L/70 twin mountings

CHARACTERISTICS

Calibre: 40 mm
Elevation: –13 to +85°
Elevation speed: 60°/s
Elevation acceleration: 120°/s²
Traverse: Unlimited
Traverse speed: 90°/s
Traverse acceleration: 120°/s²
Swept radius: 3 m
Muzzle velocity: 1000 m/s
Rate of fire: 2 × 300 rounds/minute
Ammunition type: AP tracer, HE direct action, HE proximity fuze
Ready-to-use ammunition: 736 or 444
Power supplies: 440 V 60 Hz 3-phase; 115 V 400 Hz single-phase; Peak demand 13 kW; Mean demand 4.5 – 8 kW

Weight
Mounting without ammunition: Type A 5500 kg; Type B 5300 kg
Mounting with ammunition: Type A 7300 kg; Type B 6300 kg

STATUS
The mounting is in mass production and in service with more than 20 navies. Widely used as secondary gun system in large ships and as primary armament in small combatants.

CONTRACTOR
Breda Meccanica Bresciana SpA, Via Lunga 2, 25100 Brescia, Italy.

2558.203
TWIN 40 mm NAVAL MOUNTING TYPE 106
This Breda/Bofors 40 mm L/70 mounting comprises two guns and two 32-round magazines. It can be controlled remotely, locally by means of the 'cloche' on the left of the mount, or manually. Ammunition feed may be either automatic by means of the Breda 32-round feeder or manually by inserting a four-round charger into the auto-loader of the elevating mass.

The aiming device is either the NIFE type SRS-5 or the Mirasole line-of-sight type.

CHARACTERISTICS
Generally as for other 40 mm L/70 mountings
Elevation speed: 95°/s
Training speed: 95°/s
Acceleration: 125°/s²
Recoil force: 2 × 2700 kg
Power required: Elevation and training 4 kW each
Gun weight: With feeder and battery, less ammunition 6510 + 100 kg

STATUS
In service with the Federal German Navy but no longer in production.

CONTRACTOR
Breda Meccanica Bresciana SpA, Via Lunga 2, 25100 Brescia, Italy.

2557.203
BREDA/BOFORS NAVAL MOUNTINGS
Breda Meccanica Bresciana make a range of single and twin naval anti-aircraft gun mountings incorporating a variety of different automatic feed systems and Bofors elevating mass. Breda also manufacture a 40 mm field mounting, details of which will be found in entry **2603.103** (in 1982-83 and earlier editions).

Set out in the following entries are brief details of current mountings. All use the 40 mm L/70 Bofors gun, with increased rate of fire (**2360.203**) which has the following general characteristics (which are not repeated in the individual entries below).

CHARACTERISTICS
Calibre: 40 mm
Length of barrel: 70 calibres
Elevation: –10 to +85°
Traverse: Unlimited
Weight of shell: 0.96 kg
Muzzle velocity: 1005 m/s (range 12.5 km max)
Rate of fire: 300 rounds/minute (per barrel)

STATUS
In service.

CONTRACTOR
Breda Meccanica Bresciana SpA, Via Lunga 2, 25100 Brescia, Italy.

2556.203
TWIN 40 mm NAVAL MOUNTING TYPE 64
This is a sophisticated Breda/Bofors mounting comprising two 40 mm L/70 guns and twin 100-round automatic magazines. It can be controlled remotely or locally. A graticule, Mirasole, or NIFE reflection sight can be fitted. Special features of this mounting include barrel cooling arrangements and an automatic fuze-setting device.

CHARACTERISTICS
Generally as other 40 mm L/70 mountings
Elevation speed: 95°/s minimum
Elevation acceleration: 125°/s² minimum
Training speed: 85°/s minimum
Training acceleration: 110°/s² minimum
Rate of fire: 2 × 300 rounds/minute
Power requirement: 6 kW each for training and elevation motors plus 1.6 kW for the feeder motor
Total weight: Excluding ammunition 7900 + 150 kg

STATUS
In service but no longer in production.

CONTRACTOR
Breda Meccanica Bresciana SpA, Via Lunga 2, 25100 Brescia, Italy.

2554.203
SINGLE 40 mm NAVAL MOUNTING TYPE 564
This mounting differs from earlier Breda versions of the Bofors L/70 gun in being equipped with a Model 1971144-round automatic feed of compact design. This permits a more effective use of the gun's high rate of fire and reduces the number of men required on the mount to two, with a third on standby near the mount to reload the automatic feed. A version requiring only one man on the gun is also available. The feed reloading operation can obviously be carried out during lulls in the firing, but with a little practice it is possible to reload while the gun is firing.

A further advantage of this feed system is that it can be fitted retrospectively to gun mountings which incorporate earlier types of feed. Complete modification kits are available to enable a purchaser having access to reasonable engineering facilities to arrange for the modification to be carried out, under Breda supervision, at any convenient location.

It is also known under the designation Breda/Bofors 350P.

CHARACTERISTICS
Generally as for other 40 mm L/70 gun mountings
Elevation speed: Local or remote control 45°/s
Elevation acceleration: 130°/s²
Training speed: Local or remote control 80°/s
Training acceleration: 120°/s²
Power consumption: 6 + 8 kW
Gun weight: With automatic feeder and with battery but excluding ammunition 3300 + 100 kg

STATUS
In service and in production.

Manufacture of the automatic feed device has been licensed by Breda to Bazan for Spanish naval production, incorporated with the Bofors SAK 350.

Breda/Bofors 40 mm L/70 naval mounting with 144-round automatic feed

CONTRACTOR
Breda Meccanica Bresciana SpA, Via Lunga 2, 25100 Brescia, Italy.

5535.203
TWIN 35 mm OE/OTO MOUNTING
The 35 mm OE/OTO mounting has been produced as a private venture and is intended for use in any type of ship down to motor gun boats and hydrofoils, or as local defence for merchant ships in time of war. It is primarily for close anti-aircraft defence, with a secondary anti-ship and anti-shore role.

The mounting utilises a high proportion of light alloys in its construction. The shield is made of fibreglass and is watertight.

The mounting is remotely controlled without local aiming and is fitted with modular servo-systems of very light weight using electric motors with controlled rectifiers developed by OTO Melara.

The Oerlikon KDA 35 mm gun has a high rate of fire and is belt fed. An interesting feature is that two belts are fed to each gun, and either may be selected in about two seconds. Thus the mounting can switch very rapidly from firing, for example, anti-aircraft ammunition to armour-piercing. Each gun is provided with EVA at the muzzle to measure muzzle velocity, and this information is fed back into the computer to permit corrections in laying to be applied.

The Oerlikon guns used are similar to the KDC 35 mm cannons employed in the Oerlikon GDM-A twin 35 mm naval mount described in entry **6036.203** in the

Swiss pages of this section. The GDM-A mount carries KDC cannons while the slightly lighter KDA guns are fitted in the Oerlikon/OTO mount, which has the Swiss designation GDM-C. The two versions of the gun have the same muzzle velocity, rate of fire, barrel length, and gas pressure. The KDA has twin belt feed, selectable left-hand or right-hand; the KDC has automatic clip ammunition feed, independent of the cannon.

Other details will be found in **6036.203** later in this section.

CHARACTERISTICS

Calibre: 35 mm
Length of barrel: 90 calibres (3.15 m)
Elevation: −15 to +85°
Elevation speed: Max 70°/s
Elevation acceleration: 130°/s²
Traverse: Unlimited
Traversing speed: Max 110°/s
Traversing acceleration: 100°/s²
Max recoil: 55 mm
Max range: 6000 m (surface targets); 5000 m (aerial targets)

STATUS

Under series production and in service in Libyan 670 tonne corvettes built in Italy.

CONTRACTOR

OTO Melara, Via Valdilocchi 15, La Spezia, Italy.

OE/OTO 35 mm naval twin mount

2555.203

SINGLE 40 mm NAVAL MOUNTING TYPE 107

This Breda/Bofors mounting comprises a single 40 mm L/70 gun and a 32-round magazine feed. It can be controlled remotely, locally by means of the 'cloche' on the left of the mount, or manually. Ammunition feed may be either automatic by means of the Breda 32-round feeder or manually by inserting a four-round charger into the auto-loader of the elevating mass. The aiming device is either the NIFE type SRS-5 or the Mirasole.

CHARACTERISTICS

Generally as for other 40 mm L/70 mountings
Elevation speed: 95°/s
Acceleration: 125°/s²
Training speed: 95°/s
Recoil force: 2700 kg
Power required: Elevation and training 4 kW each
Gun weight: With feeder and battery, less ammunition 3610 + 100 kg

STATUS

In production and in service with many navies.

CONTRACTOR

Breda Meccanica Bresciana SpA, Via Lunga 2, 25100 Brescia, Italy.

40 mm Breda/Bofors mounting Type 107

3994.203

BREDA TWIN 30 mm COMPACT NAVAL MOUNTING

Breda has designed and developed this mounting, using two Mauser MK 30 mm × 173 guns, set in a Breda twin cradle. Advanced design features of the twin 40 mm L/70 find their place in the new twin 30 mm, features not hitherto evolved in this calibre. Reaction time is cut to the minimum through complete automation, and both training and elevation velocities and accelerations are high (see characteristics). There are large ammunition reserves to meet saturation and successive attacks without human intervention in replenishment. Maximum accuracy is assisted by barrel proximity and mounting rigidity. The mounting is protected against the environment by a fibreglass dome, with one access hatch at the rear, and special attention has been given to sealing the mantlet plate and bearing assembly for watertightness.

The Breda patented magazine structure, containing an exceptionally large number of rounds, trains with the mounting, maintaining a constant relation to the feeders, and is divided into two halves, one half supplying each gun independently. The belted rounds are suspended in vertical folds, each fold being advanced as its predecessor is fired off. Replenishment can take place without firing off all the ammunition.

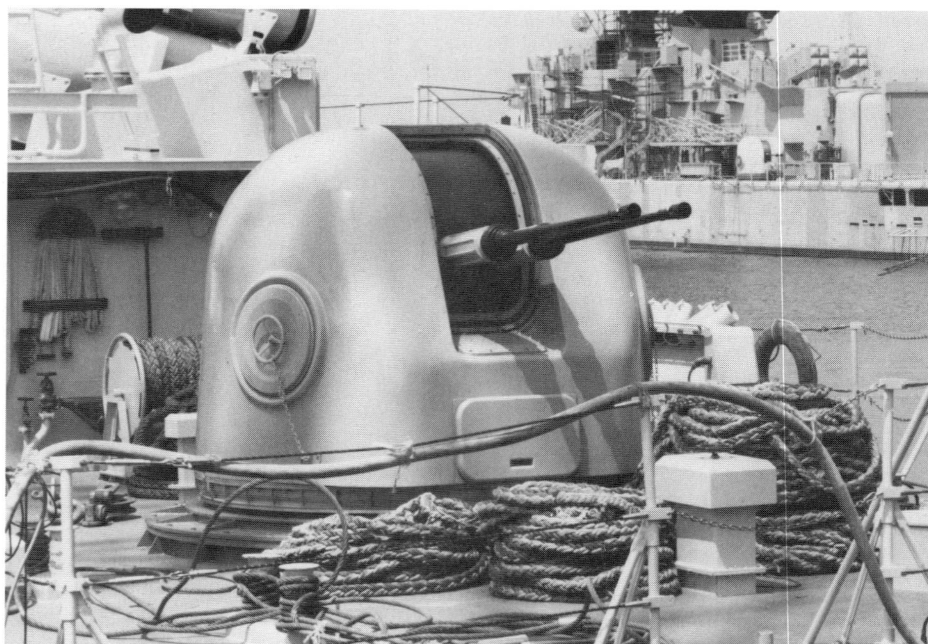

Breda twin 30 mm mounting installed in the Italian Navy patrol boat Saettia

The mounting is supplied in four versions, designed for siting arrangement in various classes of ship. Dimensions are reduced to the minimum possible, aided by barrel proximity, and weight is contained by employing light alloys to the greatest practicable extent. The gun fires GAU-8 ammunition: HE(I), or armour piercing discarding sabot (APDS), with depleted uranium or swaged tungsten penetrator for missile warhead penetration and detonation. Also a frangible round for anti-aircraft fire. Type C is designed without deck-penetration, to permit installation in small craft. Ammunition is housed in two boxes, one on either side of the training platform (500 rounds in all).

CHARACTERISTICS

Elevation limits: −13 to +85°
Elevation speed: 75°/s
Elevation acceleration: 250°/s²
Training limits: Unlimited
Training speed: 130°/s
Training acceleration: 250°/s²
Muzzle velocity: 1040 m/s (APDS M50)
Rate of fire (combined): 1600 rounds/minute
Ready-to-fire rounds: 2000 (1000/gun)
Distance between barrels: 260 mm

STATUS

In service with the Italian Navy.

CONTRACTOR

Breda Meccanica Bresciana SpA, Via Lunga 2, 25100 Brescia, Italy.

4575.203
BREDA SINGLE 30 mm MOUNTING
This uses a single Mauser Model Mk 30 mm × 173 gun, set in a Breda cradle.

The mounting has been designed to provide the lightest vessels with a modern and effective gun, or to give a good auxiliary armament to larger major ships. Its main attributes are: remote control; rate of fire up to 800 rounds per minute; double ammunition feed system; capability for remote control in case of interruption in the power supply; complete protection against environmental conditions; lightweight (1200 kg) due to the extensive use of aluminium alloys; small dimensions and compact design.

The on board installation is simple because it is completely above deck. The ready to fire rounds are contained in two boxes that may be easily reached through two replenishing doors placed on the two sides of the shield. Ammunition of GAU-8 type includes: practice rounds (TP), high explosive incendiary (HEI), armour piercing incendiary (API), high explosive incendiary with self destruction device (HEI-SD), frangible rounds and armour piercing discharging sabot rounds (APDS).

The mounting is also fitted with a Breda 'Hawkeye' optical system.

Breda single 30 mm remote controlled mounting on a Coast Guard patrol boat

CHARACTERISTICS
Elevation arc: –13 to +85°
Elevation acceleration: 300°/s²
Elevation velocity: 80°/s
Training arc: Unlimited
Training acceleration: 300°/s²
Training velocity: 140°/s
Muzzle velocity: 1040 m/s (1150 m/s ROF APDS)
Rate of fire: 800 rounds/minute
Weight (without ammunition): 1200 kg
Weight (fully loaded): 1330 kg
Deck ring (outer diameter): 800 mm
Swept radius: 2720 mm

STATUS
In production for Italian Coastguard.

CONTRACTOR
Breda Meccanica Bresciana SpA, Via Lunga 2, 25100 Brescia, Italy.

6757.203
BREDA LOCAL CONTROL SINGLE 30 mm NAVAL MOUNTING
This is a very lightweight system with certain innovations, for example dual feed and buffer batteries. The former allows automatic selection from two round types out of the GAU-8 range, stowed on the training platform; and the latter allow up to two hundred 360° rotations (or their equivalent in other gun operations) without supply from the ship's electrical system.

The 30 mm single locally controlled naval mounting has been produced in a version which can be operated by a gunner on the mounting by remote control, or as a fully independent installation.

Hawkeye
This is an integrated line-of-sight stabilised optical system on the rear of the mounting. It is controlled by the gunner with the help of the computer. Aim off is automatically calculated from inputs while the target is in sight. Range inputs come from the optical system and when the target image coincides with a circle of predetermined lens magnification the operator opens fire.

The optical system is binocular with magnifications of × 3.2 and × 16. The field-of-view is between 3.8 and 19°.

The Hawkeye computer analyses progressively the target flight path and gives the servos the signals for tracking. The gunner needs only to introduce small corrections.

Breda local control single 30 mm with 'Hawkeye' optical fire control

CHARACTERISTICS
Mounting weight
without ammunition: 1600 kg
with ammunition: 1750 kg
Calibre: 30 mm
Swept radius: 2720 mm
Training arc: unlimited
Max training velocity: 140°/s
Max training acceleration: 250°/s²
Max elevation velocity: 80°/s
Max elevation acceleration: 250°/s²
Rate of fire: 800 rpm
Ammunition: 30 mm × 173 GAU-8/A

STATUS
Ordered by US Navy for trials.

CONTRACTOR
Breda Meccanica Bresciana SpA, Via Lunga 2, 25100 Brescia, Italy.

3995.203
BREDA SCLAR MULTIPURPOSE ROCKET LAUNCHER
The Breda multipurpose naval rocket launcher system normally consists of two Breda rocket launchers for 105 mm countermeasures or assault rockets, an Elsag control unit, and a supply of SNIA rockets. The 20-tube launchers are mounted on each side of the ship and they can be trained and elevated by remote control from the control unit. The latter is connected to the ship's radar and a control panel in the ECM compartment from whence it receives the necessary operating signals. The data needed to stabilise the two launchers against pitch and roll are provided by the ship's gyro-compass.

The SCLAR 105 mm rocket launcher system is intended to provide passive defence against radar and IR homing missiles, confuse enemy radars, and provide illumination for visual fire control at night. To meet these requirements provision is made for loading different types of rockets in mixed arrangements (long range chaff, medium range chaff, illuminating) to permit mixed salvoes. Automatic selection of the rockets to be fired is possible from the remote fire control unit and the time fuzes and launch sequence may be similarly controlled.

A 105 mm rocket, designed and developed by SNIA BDP, is used by the Italian Navy and the various types are listed in the table of characteristics below.

CHARACTERISTICS
Elevation arc: –5 to +60°
Elevation velocity: 30°/s
Elevation acceleration: 50°/s²
Training arc: ±150°
Training velocity: 60°/s
Training acceleration: 60°/s²

Rate of fire: 1 round/s (approx)
Max ready-to-fire rounds: 20/launcher
Ammunition
Chaff distraction: 105LR-C (max range 12000 m)
Chaff seduction: 105MR-C (max range 5000 m)
Illumination: 105LR-I (max range 11500 m)
Launcher weight: 1150 kg (empty); 1750 kg (with 20 rockets)
Trunnion height: 1170 mm
Tubular guide length: 1460 mm
Main power supply: 3 kW

STATUS

Production. About 100 mountings have been installed in ships of the following navies: Argentina, Ecuador, Federal Republic of Germany, Italy, Nigeria, Peru and Venezuela.

CONTRACTORS

Launcher: Breda Meccanica Bresciana SpA, Via Lunga 2, 25100 Brescia, Italy.

Rockets: SNIA BPD, Defence and Space Activities, Via Sicilia, 162-00187 Rome, Italy.

Close-up view of Breda 105 mm rocket launcher

3996.203

BREDA MULTI-CALIBRE ASSAULT LAUNCHER

The Breda assault launcher is a special and improved version of the Breda SCLAR rocket launcher (**3995.203**) of which all the main characteristics are retained. It is specifically designed for conducting coastal bombardment.

This launcher is intended more particularly for installation in landing craft providing these ships with the capability for bombarding longer range targets with 105 mm rockets and those at closer range with 51 mm rockets. The assault launcher differs from the standard multipurpose rocket launcher in the following details: the launching tubes are longer in order to provide guidance necessary for the 51 mm rockets, and there is protective covering of the servo system to improve reliability and maintainability in adverse sea conditions. The ECM and illumination capability remains available with the 105 mm rockets.

STATUS

In production.

CONTRACTORS

Launcher: Breda Meccanica Bresciana SpA, Via Lunga 2, 25100 Brescia, Italy.

Rockets: SNIA BPD, Defence and Space Activities, Via Sicilia, 162-00187 Rome, Italy.

Breda assault launcher showing 105 mm and 51 mm tubes

4574.203

BREDA 81 mm BOMBARDMENT LAUNCHER

This launcher is equipped with a launching pad consisting of 36 stainless steel tubes for 81 mm bombardment rockets of the Oerlikon SNORA type. The launcher is remote controlled and its servo system is installed in a box-shaped light alloy structure for better protection and reliability.

The ammunition (see **1995.303**) is available in different types (PT, PT with mark charge, HE in three different versions and HEAT).

CHARACTERISTICS

Elevation arc: –2 to +55°
Elevation velocity: 30°/s
Elevation acceleration: 50°/s²
Training arc: ±170°
Training velocity: 60°/s
Training acceleration: 60°/s²
Max range: 10 km
Ready-to-fire rounds: 36
Launcher weight: 1270 kg (empty)
Trunnion height: 1170 mm
Swept radius: 1490 mm

STATUS

Ready for production.

CONTRACTOR

Breda Meccanica Bresciana SpA, Via Lunga 2, 25100 Brescia, Italy.

Breda 81 mm bombardment rocket launcher

NORWAY

4041.203

KV-SK/20 DUAL-PURPOSE 20 mm NAVAL MOUNTING

To meet the need for a light anti-aircraft and surface target gun for use aboard ships or shore bases, A/S Kongsberg Vapenfabrikk has developed a naval mounting for the Rh 202 gun (**3177.203**). Development was carried out in co-operation with the Ordnance Section of the Norwegian Naval Logistic Command.

The basic concept was to use the Mk 20 × 139 DM 5 gun (Rh 202) in a slightly modified version of the cradle used in the FK20-2 field gun, and to assemble these on a simple pedestal mount, thereby producing a gun which has approximately the same effect as the present field version, FK20-2. It is very reliable since it extensively uses previously developed and proven components and is quickly available because of its use of standard components. The gun is intended mainly for installation aboard naval vessels such as MTBs and mine-sweepers, fishing boats, and in the defence of coastal forts. The gun can engage fast approaching and slower crossing aircraft and can be used against both land and sea targets.

The pedestal mount consists of the cradle with ammunition feed, the pedestal itself with a fixed base and a rotatable upper part, bowed shoulder-supports and sights. The cradle is basically similar to the existing FK20-2 field gun but where necessary certain components are redesigned, especially those concerned with loading, handling, and servicing. The cradle is supported in bearings at the centre of gravity of the elevating part of the gun and may be locked in four elevation settings (minimum, max, 0° and +30°) for loading and servicing. The elevation locking control is located on the right-hand side of the rotating part of the pedestal. The adjustable shoulder supports are placed at the rear of the cradle. The trigger is located on the left cradle handle and the selector for single shot/automatic fire is located on the right-hand side of the cradle. The complete cradle can be demounted from the pedestal without special tools.

CHARACTERISTICS

Elevation: –15 to +70°
Traverse: 360°
Ammunition capacity (cradle): 2 boxes 100 rounds each
Spent round capacity: approx 200 rounds
Weight: 400 kg (with gun, without ammunition)
Total length: 2870 mm
Max radius: 1730 mm (from end of barrel to axis of rotation)

STATUS
The series produced version is in service in a large number of Norwegian vessels and shore establishments.
CONTRACTOR
Kongsberg Vapenfabrikk AS, 3601 Kongsberg, Norway.

Kongsberg KV-SK/20 20 mm naval mounting

SPAIN

3230.203
OTO 76/62 COMPACT AUTOMATIC GUN
This is a version of the OTO 76/62 compact automatic gun (**5532.203**), built under licence in Spain.
CHARACTERISTICS
As OTO system except:
Elevation: –15 to +85°
Elevation speed: 40°/s
Elevation acceleration: 60°/s²
Traverse: 360°
Traverse speed: 70°/s
Traverse acceleration: 100°/s²
CONTRACTOR
Empresa Nacional Bazán, San Fernando, Spain.

Licence-built OTO-Melara 76 mm Compact gun aboard a Spanish Navy gunboat

2304.203
40 mm AUTOMATIC GUN SAK L/70 BOFORS-BREDA 144 MOD 76
This is a version of the Bofors L/70 gun (**2360.203**) with Breda automatic feed, built under licence in Spain.
CHARACTERISTICS
As Bofors-Breda except:
Elevation: –10 to +85°
Elevation speed: 45°/s
Traverse speed: 85°/s unlimited
Weight: 3.3 tons
STATUS
Operational in ships of the Spanish Navy.
CONTRACTOR
Empresa Nacional Bazán, San Fernando, Spain.

SWEDEN

6073.203
SINGLE 120 mm L/46 GUN MOUNTING
The Bofors L/46 120 mm automatic gun is designed for use against both surface and airborne targets and has a very high rate of fire.

Housed in a 4 mm steel turret mount the gun has two magazines, mounted on the elevating cradle, which are manually filled from a fixed-structure motor-driven hoist. Electro-hydraulic remote control is standard, with the alternative of gyro-stabilised one-man local control. Telescopic sights are also fitted and the hoist and the elevation and traverse mechanisms can be operated by hand. The gun barrel is liquid cooled and has an exchangeable liner.
CHARACTERISTICS
Elevation: –10 to +80°
Elevation speed: 32°/s
Traverse: Unlimited
Traversing speed: 40°/s
Muzzle velocity: 800 m/s
Time of flight to 10 km: 21.7 s
Range: 18.5 km
Rate of fire: 80 rounds/minute
Number of rounds in magazine: 52
Power supplies: 440 V, 3-phase, 60 Hz
Power consumption: (mean when firing) 60 kW
Weight of round: 35 kg
Weight of shell (HE): 21 kg
Weight of charge: 3.15 kg
Weight excluding ammunition: 28.5 tons
STATUS
No longer in production. Installed in Finnish and Indonesian corvettes.
CONTRACTOR
AB Bofors, 0691 80 Bofors, Sweden.

2361.203
SINGLE 57 mm L/70 Mk 1 GUN MOUNTING
This 57 mm Bofors single gun in a plastic cupola is designed for both surface and anti-aircraft fire.

Alternatives of electro-hydraulic remote control or gyro-stabilised one-man local control are available. The gun feed system contains 40 rounds of ready-use ammunition with 128 rounds stowed in racks within the cupola, and there are dual step-by-step fixed supply hoists. The barrel is liquid-cooled.

Two types of ammunition are available: one is a proximity-fuzed prefragmented shell for use against aerial targets; the other is a special surface target shell which penetrates the target and is detonated after a short delay.

The gun can be equipped with rocket-launching rails for two-inch (57 mm) rockets.

CHARACTERISTICS

Elevation: –10 to +75°
Elevation speed: 40°/s
Traverse: Unlimited
Traversing speed: 55°/s
Muzzle velocity: 1025 m/s
Time of flight to 3 km: 3.8 s
Range: 17 km
Rate of fire: 200 rounds/minute
Power supplies: 440 V 3-phase 60 Hz
Power consumption: (mean, when firing) 8 kW
Weight of round: 6.1 kg
Weight of shell: (HE: DA or proximity fuze) 2.4 kg
Weight, excluding ammunition: 6 tons

STATUS

In service in the Swedish Navy's 'Spica' class (12 boats), 'Hugin' class (16 boats) fast attack craft, the 'Nordkapp' class Norwegian coast guard vessels, and with several other navies.

CONTRACTOR

AB Bofors, 0691 80 Bofors, Sweden.

Bofors 57 mm L/70 Mk 1 automatic gun mounting on board 'Spica' class fast attack craft

4031.203

SINGLE 57 mm Mk 2 NAVAL GUN

This unmanned 57 mm all-purpose gun in a compact plastic cupola and with low radar signature is designed for combating both surface and aerial targets. The gun comprises an improved, extremely accurate, electro-hydraulic remote control system with microprocessor-controlled aiming and firing limitations. The new, completely automatic, reloading system allows for 120 rounds available in the cupola, and there are dual step-by-step fixed supply hoists.

Two types of ammunition are available: one is a proximity-fuzed and prefragmented shell for use against aerial targets; the other is a special surface target shell which penetrates the target and is detonated after a short delay.

CHARACTERISTICS

Elevation: –10 to +75°
Elevation speed: 40°/s
Traverse: Unlimited
Traverse speed: 55°/s
Muzzle velocity: 1025 m/s
Time of flight to 3 km: 3.8 s
Range: 17 km
Rate of fire: 220 rounds/minute
Power supply: 440 V 3-phase 60 Hz
Power consumption: 8 kW (mean when firing)
Weight of round: 6.1 kg
Weight of shell: 2.4 kg (HE:DA or proximity fuze)
Weight, excluding ammunition: 6.5 tons

STATUS

In series production and in service with the Swedish Navy. Also selected for the Canadian 'Halifax' class frigates and several other navies.

Bofors 57 mm Mk 2 naval gun on board Swedish 'Stockholm' class fast attack craft

CONTRACTOR

AB Bofors, 0691 80 Bofors, Sweden.

2360.203

SINGLE 40 mm L/70 AUTOMATIC GUN MOUNTING

Bofors make a number of different mountings for their shipborne 40 mm L/70 gun for use against both surface and airborne targets.

Two guns are shown here. Both have electro-hydraulic laying machinery, and can be supplied with both remote and local control. In local control the guns are gyro-stabilised, and reflex sights with speed rings are used for aiming. Both types of gun can be provided with a light plastic cupola for weather protection. A fully automatic gun with a new 100 round magazine is also available.

The proximity-fuzed ammunition can be used in all Bofors 40 mm L/70 guns.

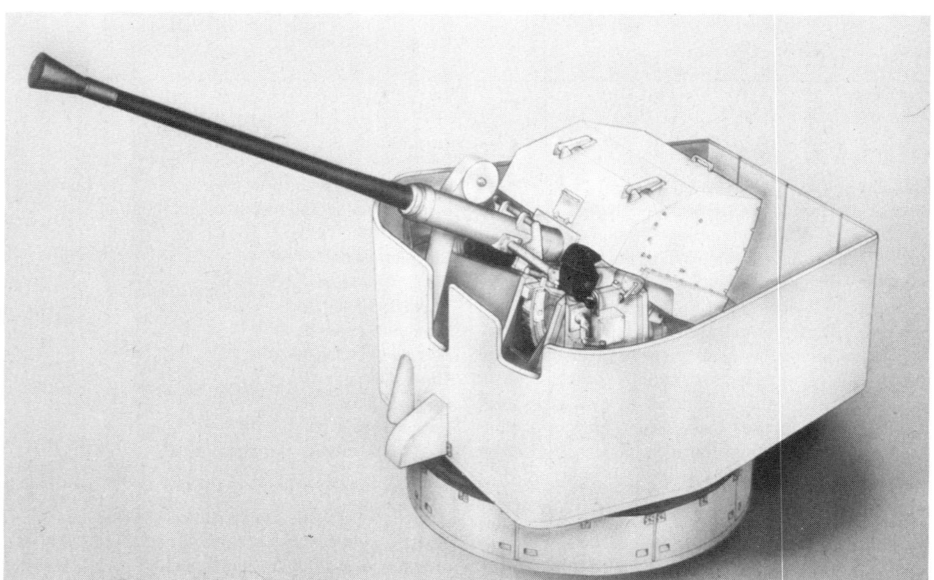

Improved automatic 40 mm L/70-600 Bofors gun

CHARACTERISTICS
Elevation: –10 to +90°
Elevation speed: 45°/s (powered mounts)
Traverse: Unlimited
Traversing speed: 85°/s (powered mounts)
Muzzle velocity: 1005 – 1025 m/s, depending on ammunition type
Range: Tactical range 4000 m (12 km max)
Rate of fire: 300, or 330 rounds/minute
Power supplies: 440 V 3-phase 60 Hz
Power consumption: 5 kW
Weight, excluding ammunition: 2.8, 3, or 3.3 tons
Ammunition: PFHE Mk 2 (principal type), HCHE, APC-T, HE-T, TP-T. Two different types of ammunition can be selected simultaneously.
STATUS
Operational with more than 30 navies, including the Swedish Navy.
CONTRACTOR
AB Bofors, 0691 80 Bofors, Sweden.

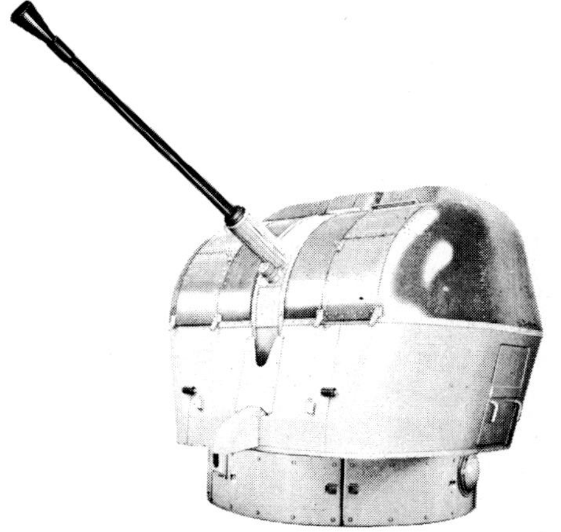

Power-operated 40 mm Bofors gun with plastic cupola

6759.203
TRINITY MOUNTING
A naval mounting for the Bofors Trinity gun system (**4837.131**) is also being offered. It would be an autonomous unit with all fire control sensors on the weapon mount to assure greater accuracy. The upper mounting would comprise a rotating platform with the trunnion mounting and the elevating mass, the gun, the electro-optical sighting system and the tracking radar. In the local control mode the operator would be in an enclosed cab on the left hand side.

The lower mounting would have a heading attitude reference system. No deck penetration is required and the system can be mounted upon a mounting base ring 102 cm in diameter with power and electronics connections through slip rings. If the mounting is remotely controlled the control console, fire control computer, gun computer and interface will be below decks.
CHARACTERISTICS
Calibre: 40 mm
Length of barrel: 70 calibres
Height (without radar): 197 cm
Width: 260 cm
Weight: 3700 kg (without optional search radar)
CONTRACTOR
AB Bofors, S-691 80 Bofors, Sweden.

6088.203
SINGLE 40/60 AA GUN MOUNTING
This 60-calibre version of the well-known Bofors 40 mm AA gun was introduced in 1936 and is still in service in many places. Its performance is, as one would expect, somewhat inferior to that of the modern 70-calibre weapon.

CHARACTERISTICS
Calibre: 40 mm
Length of barrel: 56 calibres
Elevation: –5° to +90°
Projectile weight: 0.9 kg
Muzzle velocity: 800-880 m/s
Max range: 10 km
Tactical range: 3 km
Rate of fire: 120 rounds/minute
STATUS
In service and still fitted in reconditioned form but no longer in production.
CONTRACTOR
AB Bofors, 0691 80 Bofors, Sweden.

SWITZERLAND

6036.203
35 mm TWIN ANTI-AIRCRAFT GUN TYPE GDM-A
This gun is similar in concept and design to the successful Oerlikon field gun. It is primarily intended for convoy or self-protection against air attack, but can also be used to engage sea and land targets.

It is an all-weather weapon system with a bi-axial stabilised mount on deck, and its electronic control-unit below deck. High laying accelerations and speeds are an important feature of the design.

Both 35 mm cannons type KDC are completely interchangeable without the use of additional parts as are the hand cocking devices and the barrels.

Oerlikon 35 mm twin naval AA gun type GDM-A

Three modes of operation are possible:
(1) automatic operation, electrically controlled from fire control equipment or an optical aiming mechanism with auxiliary computer
(2) local operation, electrically controlled from the gun with joystick and the gyro-stabilised gunsight
(3) emergency operation, mechanically controlled from the gun with two hand wheels and the gunsight.

CHARACTERISTICS
Gun type GDM-A
Elevation
Arc: –15 to +85°
Speed: 100°/s
Acceleration: 120°/s²
Traverse
Arc: Unlimited
Speed: 120°/s
Acceleration: 130°/s²
Ready-to-use ammunition: 2 × 56 = 112 rounds
Reserve container: 2 × 112 = 224 rounds
Weight
Gun without ammunition: 5950 kg (approx)
Ammunition (336 rounds): 570 kg (approx)
Servo amplifier cabinet: 272 kg (approx)
Thyristor control unit: 130 kg (approx)
Power supply
Drive voltage: 440 V, 60 Hz, 3-phase
Stabilisation: Bi-axial by gyroscopes
Cannons type KDC
Calibre: 35 mm
Length of barrel: 90 calibres
Muzzle velocity: 1175 m/s
Rate of fire: 2 × 550 = 1100 rounds/minute
Weight
Barrel: 120 kg
Cannon complete: 430 kg (approx)
STATUS
In production. The GDM-A mount is known to be in service with the following countries: Ecuador, Greece, Iran, Libya and Turkey. In collaboration with OTO Melara of Italy, Oerlikon also supplies the OE/OTO 35 mm twin naval mount, the Swiss designation being GDM-C. Additional information can be found in entry **5535.203** in the pages of this section covering Italian naval guns. The gun employed varies slightly and is the slightly heavier type KDA, but the main performance characteristics of both types (KDA and KDC) are the same.
CONTRACTOR
Machine Tool Works Oerlikon-Bührle Ltd, 8050 Zurich, Switzerland.

2395.203
30 mm TWIN NAVAL GUN MOUNTING TYPE GCM-A
Current production gun mountings have been developed from the original version (A01) that was introduced into service in 1968. This was followed by the A02 version of which over 100 are in service with the Royal Navy and 13 other navies. Now the A03 version is being produced in three variants to meet all customer requirements.

All versions feature lightweight, low power consumption and ease of installation requiring no deck penetration. They are, therefore, suitable for installation in a wide variety of ships, as either secondary close range defence armament in larger ships, or as the main armament in smaller ships down to patrol craft size.

All versions are fitted with a pair of 30 mm cannon type KCB, for which a full range of six natures of ammunition is produced, namely TP, TP-T, HEI, HEI-T, SAPHEI and API-T.

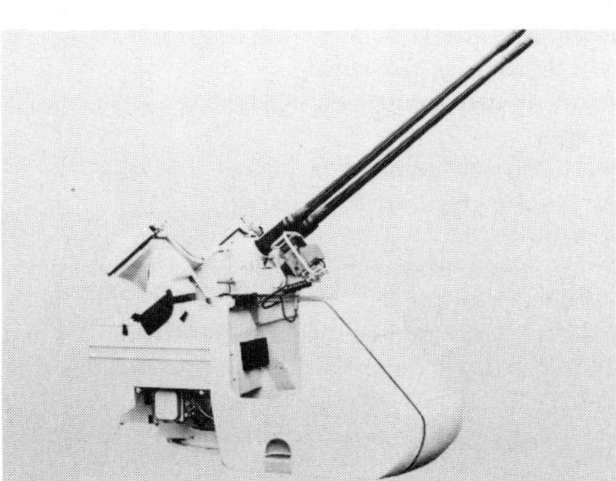

Oerlikon twin 30 mm mounting GCM-A03-3

Manned versions have on-mounting gyro stabilisation and the Ferranti computing gunsight type GSA is fitted as standard. All versions can be remotely controlled from a fire control system.

CHARACTERISTICS

Training arc: Unlimited
Training speed: Max 1.309 rad/s (75°/s)
Elevation arc: –12.5° to +75°
Elevation speed: Max 0.873 rad/s (50°/s)
Control voltage: 115 V, 500 Hz
Drive voltage: 440 V 60 Hz 3 phase
Peak power consumption: 11.5 KVA
Weights (unloaded):
A02: 2000 kg
A03-1: 2150 kg
A03-2: 2050 kg
A03-3: 1900 kg
Swept radius at 0° elevation: 2.52 m
Cannon: 2 × type KCB
Calibre: 30 mm
Rate of fire (2 × 650): 1300 rounds/min
Muzzle velocity: Max 1080 m/s
On-mount ammunition supply
A02 (2 × 160): 320 rounds (350 kg)
A03-1 standard (2 × 160): 320 rounds (350 kg)
A03-1 optional (2 × 250): 500 rounds (510 kg)

Oerlikon twin 30 mm mounting GCM-A03-3

A03-2 (2 × 250): 500 rounds (510 kg)
A03-3 (2 × 320): 640 rounds (660 kg)

STATUS

A01 Out of production. A few still in service.
A02 Out of production. Over 100 in service world-wide.
A03 Current production.

Oerlikon twin 30 mm mounting GCM-A03-1

CONTRACTORS

Machine Tool Works Oerlikon-Bührle Ltd, 8050 Zurich, Switzerland.

British Manufacture and Research Co Ltd, Springfield Road, Grantham, Lincolnshire NG31 7JB.

3657.203

25 mm NAVAL GUN TYPE GBM-A01

The Oerlikon single 25 mm GBM A01 is the most powerful known manually operated gun. The KBA C02 cannon fires NATO standard KBA ammunition at 570 rounds per minute. The twin ammunition feed system gives instantaneous selection from two 70-round boxes, a further 30 rounds can be entered into each flexible feed chute giving a total of 200 rounds of ready-to-fire ammunition of two types.

Due to the KBA cannon's aircraft-type floating recoil system, the deck thrust is only 12 kN.

Optional extras include a choice of day and night sights, including thermal imagers and image intensifiers, plus a protective shield for the aimer.

CHARACTERISTICS

Gun type GBM-A01
Elevation: –10° to +55°
Traverse: Unlimited
Weight: 600 kg (ready to fire)
Ready-to-use ammunition: 200 rounds
Cannon type KBA-CO2
Calibre: 25 mm
Length of barrel: 80 calibres (2173 mm with muzzle brake)
Muzzle velocity: 1100-1360 m/s, depending on ammunition
Rate of fire: 570 rounds/minute

STATUS

No longer in production.

CONTRACTOR

Machine Tool Works Oerlikon-Bührle Ltd, 8050 Zurich, Switzerland. This weapon was also manufactured in the United Kingdom by the British Manufacture and Research Company.

4017.203

SEA ZENITH 25 mm MOUNT TYPE GBM-B1Z

Sea Zenith is the name of the four-barrelled 25 mm gun mounting employed in the Contraves/Oerlikon Seaguard close-in weapon system (**4016.231**). It is named from its ability to bear on targets at the zenith by virtue of a slewing axis which is inclined at 35° to the horizontal. Its function is the close-in destruction of anti-ship missiles using special AMDS ammunition with a forged tungsten core designed to ensure that a single hit will be sufficient to destroy a missile target by exploding its warhead. It also has a performance against aircraft using more conventional ammunition.

The four 25 mm guns have a combined rate of fire of 3400 rounds/minute, each gun being independently fed from ammunition drums which may either be mounted below decks or by the side of the gun. The ready-use supply of 1660 rounds is sufficient for 18 normal engagements of missile targets. Reloading can be achieved during the firing cycle.

The KBB-RO3/LO3 gun is a positively locked gas operated weapon with two breech locking catches and is ruggedly built to ensure operational reliability under extremely adverse environmental conditions.

CHARACTERISTICS

Gun type: Sea Zenith KBB-RO3/LO3
Calibre: 25 mm
Length of barrel: 2300 mm
Elevation
Arc: –14 to +127°
Speed and acceleration: 2.5 rad/s, 10 rad/s²
Traverse
Arc: ±180°
Speed and acceleration: 2.5 rad/s, 10 rads/s²
Muzzle velocity: 1335 m/s
Ready-to-use ammunition: 1660 rounds (415 per gun)
Weight
Mounting with ammunition: approx 4600 kg
Ammunition feed system: 2150 kg

STATUS

In full series production following a contract from a West German shipyard for Seaguard systems to be fitted in four frigates under construction for the Turkish Navy. Two were due for commissioning in 1987.

Seaguard CIWS gun module (Sea Zenith)

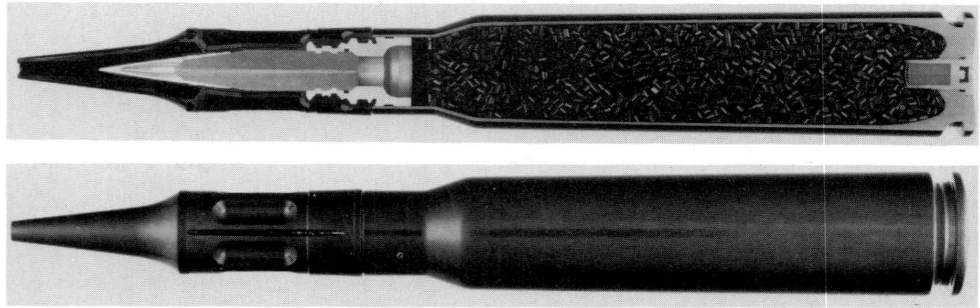

Seaguard CIWS 25 mm KBB AMDS ammunition

CONTRACTOR

Machine Tool Works Oerlikon-Bührle Ltd, 8050 Zurich, Switzerland.

2397.203

20 mm NAVAL GUN TYPE GAM-BO1

The Oerlikon GAM-BO1 is a simple, modern, and efficient 20 mm naval mounting, embodying the well-proven KAA cannon.

All-up weight is low enough to permit installation on any type of naval vessel. The gun can be used for either surface- or AA-fire. No electrical power is needed. The gun is laid by the gunner using a shoulder harness. A choice of day and night sights is available, including thermal imagers and image intensifiers, plus a protective shield for the aimer.

CHARACTERISTICS

Gun type GAM-BO1

Elevation

Arc: –15 to +55°

Training arc: Unlimited

Speed and acceleration: Manual

Ready-use ammunition: 200 rounds

Weight

Gun without ammunition: 410 kg

Ammunition (200 rounds): 90 kg

Cannon type KAA (former design: 204 GK)

Calibre: 20 mm

Length of barrel: 85 calibres (1700 mm)

Muzzle velocity: 1050 m/s (approx)

Rate of fire: 1000 rounds/minute

Weight complete: 92 kg

Barrel: 26 kg

STATUS

Known fittings include the Spanish 'Lazaga' and 'Barcelo' class Fast Attack Craft, most surface ships of the Royal Navy and ships of 17 other navies. In series production.

20 mm Oerlikon naval mounting type GAM-BO1

CONTRACTOR

Machine Tool Works Oerlikon-Bührle Ltd, 8050 Zurich, Switzerland. This weapon is also manufactured in the United Kingdom by the British Manufacture and Research Company.

2396.203

20 mm NAVAL GUN MOUNTING TYPE A41A

The lightest and simplest naval gun mounting in the Oerlikon range, the A41A incorporates the well proven HS804 cannon having a quick exchange drum-feed magazine. Operated by one man, it requires no external power and the very minimum of maintenance.

With a swept radius of 1.7 m, loaded weight of 225 kg and recoil force of only 3.6 kn it can be fitted in smaller assault craft and boats than any other 20 mm gun.

The gun can be supplied with the reflex day sight type RC35 or night sights.

CHARACTERISTICS

Weight including magazine: 225 kg

Swept radius at 0°: 1.7 m

Training arc: Unlimited

Elevation arc: –10° to +50°

Cannon type HS804

Calibre: 20 mm

Rate of fire: 800 rounds/minute

Muzzle velocity: 835 m/s

Magazine capacity: 58 rounds

STATUS

Widely fitted. Available from stock.

CONTRACTORS

Machine Tool Works Oerlikon-Bührle, 8050 Zurich, Switzerland.

British Manufacture and Research Co Ltd, Springfield Road, Grantham, Lincolnshire NG31 7JB.

20 mm Oerlikon naval mounting type A41A

3656.203

20 mm NAVAL GUN TYPE GAM-CO1

The Oerlikon GAM-CO1 is very similar to the type GAM-BO1 (**2397.203**), employing the same kind of sighting with the addition of an optional night sight, and gun laying by means of a shoulder harness. However, in place of the GAM-BO1's KAA 20 mm cannon, the GAM-CO1 uses the 20 mm automatic cannon type KAD-B13-3 (formerly designated HS820).

CHARACTERISTICS

Gun type GAM-CO1

Elevation: –15° to +55°

Traverse: Unlimited

Ready use ammunition: 200 rounds

Weight: 510 kg (ready to fire)

Cannon type KAD-B13-3 (formerly HS820)

Calibre: 20 mm

Length of barrel: 1906 mm

Muzzle velocity: 1050-1150 m/s, depending on ammunition

Rate of fire: 900 rounds/minute

STATUS

No longer in production.

CONTRACTOR

Machine Tool Works Oerlikon-Bührle Ltd, 8050 Zurich, Switzerland. This weapon was also manufactured in the United Kingdom by the British Manufacture and Research Company.

UNION OF SOVIET SOCIALIST REPUBLICS

6038.203

TRIPLE 152 mm GUN MOUNTING

These guns are in service with several of the older cruisers of the Soviet fleet. The 'Sverdlov' class cruisers each carry 12 guns in four triple mountings, except for three ships which have been modified to carry SA-N-2 or SA-N-4 missiles, and have had one or two turrets removed.

Guns in the turrets are mounted in separate sleeves, thus permitting individual elevation to at least 50°. Projectile weight is believed to be about 50 kg and muzzle velocity 915 m/s so that maximum range is probably in the region of 27 km. The gun is 50 calibres long, semi-automatic, and can fire ten rounds/minute. An eight-metre rangefinder is incorporated in each turret.

The triple 152 mm gun mounting is often described as a 150 mm triple mount.

6039.203

TWIN 130 mm GUN MOUNTING

There are two types of twin 130 mm gun in older Soviet destroyers. The oldest, designated B-13 2C, is to be found in the surviving 'Skory' class.

CHARACTERISTICS

Length: 50 cal/6.5 m

Elevation: –5° to +40°

Rate of fire: 6-8 rounds/minute (per barrel)

Range: 24 000 m (effective 14-15 000 m)

Muzzle velocity: 875 m/s

Weight of shell: 33 kg (fixed ammunition)

Weight of mounting: 35 tons

STATUS

No longer in production. Will remain in service until last 'Skory' class disappear from Soviet and Egyptian navies.

6040.203

TWIN 130 mm DUAL-PURPOSE GUN MOUNTING

This mounting first appeared as the main armament of the 'Kotlin' class destroyers in the mid-1950s, and appears to be derived from the 100 mm mountings in the 'Sverdlov' class cruisers. It is fully stabilised and radar-controlled, and is associated with the 'Wasp Head' fire control director and 'Egg Cup' radar on the mounting.

CHARACTERISTICS

Length: 58 cal/7.5 m
Elevation: –5° to +89°
Rate of fire: 15 rounds/minute (per barrel)
Range: 28 000 m (effective 16-18 000 m)
Muzzle velocity: 900 m/s
Weight of shell: 27 kg (HE and SAP)
Weight of mounting: 40 tons

STATUS

No longer in production but still operational in modified 'Kotlin' class destroyers.

4522.203

TWIN 130 mm DUAL-PURPOSE GUN MOUNTING

A new design of automatic, water-cooled, twin mounting was first seen in the early 1980s, in the 'Sovremenny' class destroyers and the 'Kirov' class battlecruiser *Frunze* and the 'Slava' class cruisers.

CHARACTERISTICS

Calibre: 130 mm
Length: 70 cal
Weight: 35 tonnes
Elevation: –15 to +85°
Traverse: 270° (estimated)
Muzzle velocity: 950 m/s (estimated)
Max range: 28 000 m (effective range 14-15 000 m
Rate of fire: 12 to 16 rounds/minute
Ammunition: HE, SAP

6042.203

TWIN 100 mm DUAL-PURPOSE GUN MOUNTING

This twin tri-axially stabilised mounting appeared in the Italian-designed 'Chapaev' and 'Sverdlov' class cruisers but was superseded by the 130 mm version described above. It may derive from German technology handed over in 1938-41 or captured in 1945.

CHARACTERISTICS

Length: 56 cal/6.5 m
Elevation: approx –5° to +80°
Rate of fire: approx 20 rounds/minute (per barrel)
Range: 18 000 m (12 000 m against air targets)
Muzzle velocity: 900 m/s
Weight of shell: 16 kg
Weight of mounting: about 35 tons

STATUS

Out of production for many years, but retained in surviving 'Sverdlov' class cruisers.

6041.203

SINGLE 100 mm DUAL-PURPOSE GUN MOUNTING

This was the 1947 Model of 100 mm gun in an open-backed shield mounting similar to Soviet World War II destroyers.

CHARACTERISTICS

Similar to the 100 mm mounting described above, but with following differences:

Weight of shell: 13.5 kg
Range: 16 000 m (6000 m against air targets)
Muzzle velocity: 850 m/s

STATUS

No longer in production, and obsolescent, but still in service in 'Riga' class in Soviet Navy and elsewhere, and in support ships.

4521.203

SINGLE 100 mm DUAL-PURPOSE GUN MOUNTING (AUTOMATIC)

A new type of single 100 mm dual-purpose mounting, of which little is known, is fitted in the 'Krivak II' frigates, the large destroyer *Udaloy*, and the battle cruiser *Kirov*. It is totally enclosed and is almost certainly automatic unlike the earlier guns of this calibre described in **6041.203**. Control is from Kite Screech radar.

Single automatic 100 mm mountings on the destroyer Udaloy (Federal German Navy)

6043.203

SINGLE 85 mm DUAL-PURPOSE GUN MOUNTING

Fitted on 'Kronstadt' class coastal patrol vessels which are now only in full service in foreign navies and are being converted to torpedo recovery vessels in the Soviet Navy. This gun is presumed to be obsolescent.

CHARACTERISTICS

Soviet designation: 90-K
Calibre: 85 mm
Length of barrel: 4435 mm
Elevation: –5 to +85°
Elevation rate: 8°/s
Traverse rate: 12°/s
Weight of round: 16.17 kg
Muzzle velocity: 800 m/s
Horizontal range: 15.5 km
Rate of fire: 15 to 18 rounds/minute

6044.203

TWIN 76 mm DUAL-PURPOSE GUN MOUNTING

This mounting seems to have first appeared on the 'Kynda' class cruisers in 1962 and is now widely fitted in the Soviet fleet in 'Kiev' class aircraft carriers, 'Kara' and 'Kynda' class cruisers, 'Kashin' and modified 'Kildin' class destroyers, 'Krivak I', 'Koni', 'Mirka' and 'Petya' class frigates and 'Ivan Rogov' class amphibious ships as well as in a number of foreign navies. Doubtless it has been updated during this long period.

The 60-calibre gun can be elevated to about 85°, and this, with a muzzle velocity of about 900 m/s, means that air targets can be engaged at slant ranges up to 14000 metres. Maximum range is 15 km. Rate of fire is said to be 60 rounds per minute. It is normally associated with Owl Screech or Hawk Screech fire control.

Twin 76 mm dual-purpose gun turret on the 'Kashin' class missile destroyer Odarenny (G Jacobs)

2865.203

TWIN 85 mm AA GUN MOUNTING

This is an elderly semi-automatic heavy AA weapon believed to be installed now only in the 'Skory' class destroyers. Barrel length is about 52 calibres, muzzle velocity about 800 m/s, maximum range and height about 10000 and 5000 metres respectively. Maximum barrel elevation is 75°. Rate of fire is about 15 rounds per minute.

4315.203

SINGLE 76 mm DUAL-PURPOSE GUN MOUNTING

This new mounting has appeared recently in the 'Nanuchka III' and 'Tarantul' classes of corvette and the 'Matka' class of missile hydrofoil. It is totally enclosed and therefore most probably automatic, can clearly achieve high elevations, and is suitable for fitting in small craft since 'Matka' displaces no more than 215 tons fully loaded.

6047.203

QUADRUPLE 57 mm AA GUN MOUNTINGS

Quadruple 57 mm mountings are installed in the 'Kanin' and 'Kildin' class destroyers and the 'Lama' class supply ships. The guns are arranged as two pairs mounted vertically one above the other and it is believed that the pairs are essentially similar to the twin 57 mm guns described below (**6046.203**), except that the barrels of the quad-mounted guns do not have muzzle brakes. Fire control is by Hawk Screech.

Two of four quadruple 57 mm AA gun mounts aboard a now obsolete Soviet 'Krupny' class destroyer

2869.203

TWIN 57 mm AUTOMATIC AA GUN MOUNTING

Several of the most modern classes of Soviet ships are equipped with fully-enclosed, fully-automatic 57 mm guns in twin mountings. With 80-calibre barrels and muzzle velocities around 1000 m/s, these guns have maximum horizontal and vertical ranges of 12000 and 5000 metres respectively. Rate of fire is believed to be 120 rounds per minute per barrel. Maximum elevation is around 85°.

Guns of this type are found in at least ten different classes of ship including the 'Moskva', 'Kresta', 'Grisha', 'Nanuchka', 'Poti', 'Ugra', and 'Chilikin' classes. Radar control is normally by Muff Cob (**1611.253**).

Twin 57 mm automatic guns on the submarine tender Ivan Vakrhrameev (G Jacobs)

6045.203

SINGLE 57 mm DUAL-PURPOSE GUN MOUNTING

This 57 mm gun is fitted only to the 'Sasha' class coastal minesweepers and the modified 'Skory' class destroyers. It is believed that the gun performance is the same as that of the twin-mounted 57 mm gun described in **6046.203** but the gun barrel is not fitted with a muzzle brake.

6046.203

TWIN 57 mm AA GUN MOUNTINGS

Twin 57 mm AA guns are fitted to many of the more modern or recently modernised Soviet ships. The guns have 70-calibre barrels with muzzle brakes and a rate of fire of some 120 rounds per minute per barrel is believed to be possible. Muzzle velocity is about 1000 m/s, vertical range is believed to be about 6000 metres, and horizontal range about 9000 metres; the weight of the shell is about 2.8 kg.

Guns of this type are installed in 'T-58', 'Don', 'Oskol', and 'Alligator' classes of ship. Although primarily AA weapons, the 57 mm guns are in some cases the sole armament of a ship and must therefore be intended for use in a dual-purpose role.

CHARACTERISTICS

Soviet designation: SIF-31B
Elevation: –10 to +85°
Elevation rate: 25°/s
Traverse rate: 30°/s
Weight of round: 6.6 kg
Muzzle velocity: 1020 m/s
Range: 8.4 km
Range of fuze: 6950 m
Rate of fire: 100 – 120 rounds/minute

Twin 57 mm AA guns (Novosti)

6048.203

QUADRUPLE 45 mm AA GUN MOUNTS

These mounts are very similar in appearance to the quadruple 57 mm AA gun mounts described above. The only remaining mountings are in the 'Kotlin' class destroyers.

Semi-automatic and with an 85-calibre barrel, each gun can fire about 160 rounds per minute with a muzzle velocity of about 1000 m/s, and vertical and horizontal ranges of some 7000 and 9000 metres.

2870.203

TWIN 37 mm AA GUN MOUNTINGS

Twin 63-calibre AA mountings of various types are found on many of the older cruisers, destroyers, and auxiliaries of the Soviet Navy. The 37 mm AA calibre was first introduced for naval use, adapted from the army weapon, in 1943, but the twin mountings date from some five years later. Fitted in 'Sverdlov', 'Skory' (modified), 'Riga' and 'T43' class.

CHARACTERISTICS

Soviet designation: W-11-M
Elevation: −10 to +85°
Elevation rate: 13°/s
Traverse rate: 17°/s
Weight of round: 1.5 kg
Weight of shell: 732 g
Muzzle velocity: 880 m/s
Range: 4000 m (horizontal), 3000 m (vertical)
Rate of fire: 2 × 160 to 180 rounds/minute

Twin 37 mm AA guns (Tass)

2883.203

SINGLE 37 mm AA GUN MOUNTING 70-K

Derived from an army weapon of the same calibre this gun was introduced for naval use in 1943 and is to be found in open mountings on the older 'Skory' class destroyers, and on 'Kronshtadt' class submarine chasers and in turret mountings on 'T-301' class coastal minesweepers which are now only in service in other navies.

It is also in use in ships of other Warsaw Pact navies.

CHARACTERISTICS

Calibre: 37 mm
Elevation: −10 to +85°
Elevation rate: 15°/s
Traverse: 360°
Traverse rate: 20°/s
Muzzle velocity: 880 m/s
Max range: 4000 m
Rate of fire: 160 – 180 rounds/minute
Weight: 1750 kg
Crew: 6

Single 37 mm AA mounting

2896.203

TWIN 30 mm AA GUN MOUNTING

Widely fitted in many classes of ship including missile boats and other small craft, this is a fully automatic remote-controlled weapon. Barrel length is 65 calibres, muzzle velocity around 1000 m/s, range about 3000 to 4000 metres, and rate of fire around 500 rounds per minute per barrel. Maximum elevation is about 85°.

The gun was first introduced in 1960. It is most commonly associated with the Drum Tilt fire control radar (**1330.253**) or more rarely with Muff Cob (**1611.253**).

Twin fully-automatic 30 mm AA mounting as mounted on missile boats and other small vessels

2897.203

TWIN 25 mm AA GUN MOUNTINGS

LAA guns of this type are mounted in many of the smaller ships of the Soviet Navy and other friendly navies. At least three versions of mounting exist, the simplest being the open type 2-M-3 110PM. This is essentially for local control with visual sighting, although hydraulic power is provided for training and elevation. Later versions, however, may provide for varying levels of remote control.

CHARACTERISTICS

Calibre: 25 mm
Length of barrel: 80 calibres
Elevation: −10 to +85°
Traverse: 360°

Rates:	Hydraulic	Manual
Elevation	2 – 40°/s	15°/s
Traverse	2 – 70°/s	25°/s

Muzzle velocity: 900 m/s
Rate of fire: 270 – 300 rounds/minute (theoretical)
Overall length: 2.845 m
Overall weight: 1500 kg
Crew: 2

This data refers to the open mounting type 2-M-3 110PM; the performance of the alternative versions is likely to be broadly similar.

Later twin 25 mm gun mounting, probably aboard a riverine gunboat

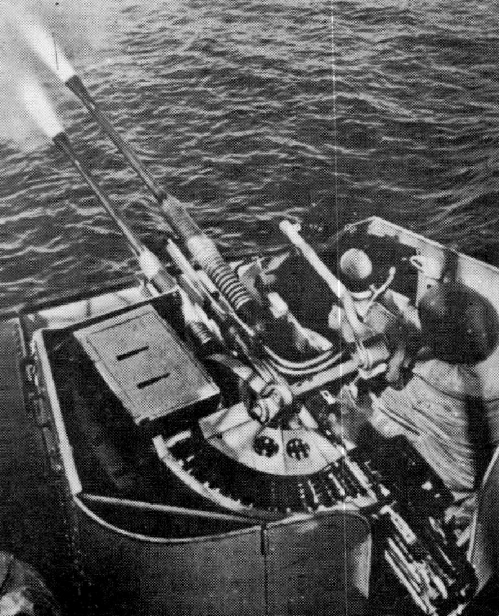

Twin 25 mm AA guns in 2-M-3 110PM open mounting

2899.203

122 mm and 140 mm ROCKET LAUNCHERS

For the past 30 years and more the Soviet Army has been using and arranging for the development of a wide range of unguided barrage rockets (**5555.111**). It is not surprising, therefore, that the Soviet Navy should also use these weapons, indeed it is surprising that they have not made greater use of them.

Systems are fitted in some amphibious ships. The 'Ivan Rogov' class has two 122 mm 20-barrelled launchers, the Type 3 and 4 'Alligator' class have two launchers each and the 'Polnochny' class have two 18-barrelled 140 mm launchers with a range of about 9 km.

UNITED KINGDOM

6012.203

TWIN 4.5 inch GUN MOUNTING Mk 6

This twin-barrelled remotely controlled power operated 4.5-inch (113 mm) naval gun mount was introduced shortly after the Second World War. Electro-hydraulic power control is normally remotely controlled, but can be locally controlled by a joystick. The loading cycle is semi-automatic, shell and cartridge being separately hoisted and then manually loaded into the loading tray. The complete turret weighs about 45 tons.

CHARACTERISTICS

Calibre: 113 mm
Length of barrel: About 45 calibres
Elevation: To 80°
Projectile weight: About 25 kg
Muzzle velocity: About 750 m/s
Max range: About 19 km (10 nautical miles)
Rate of fire: 20 rounds/barrel/minute

STATUS

Operational in British and Commonwealth navies.

CONTRACTOR

Vickers Shipbuilding and Engineering Ltd, Barrow-in-Furness, Cumbria LA14 1AF, England.

6011.203

SINGLE 4.5 inch GUN MOUNTING Mk 8

This 55 calibre long 4.5-inch (114 mm) gun has been designed for the RN by the Royal Armament Research and Development Establishment (RARDE). It is modelled on the British Army's Abbot gun (**5503.103** in 1982-83 and earlier editions) and is fitted with muzzle brake and fume extractor. A range of fixed ammunition has also been developed for use with this mounting.

The 4.5-inch gun mounting, designed by Vickers at Barrow, features a simple ammunition feed system with the minimum number of transfer points. The high pointing accuracy of this weapon is achieved through a remote power control system whose large stability margins are combined with a mounting of stiff structure and low interia. To this end the revolving structure has been kept as light as possible and a sandwich construction glass-reinforced plastic gun-shield fitted. To keep the trunnion height as low as possible the gun is mounted front heavy, balance being restored by an oil spring. To reduce inertia the training gear is fitted to the fixed structure, the training motors are fitted inside the mounting pedestal, and the driving motors are secured vertically to the ring bulkhead which is part of the ship's structure.

The loading system is hydraulically operated from one power plant sited below decks. The system employs only four transfer points between the gunbay (ready use magazine) and the gun, and is designed so that the nature of ammunition may be changed without unloading or firing off a large number of rounds. A stockpile of ammunition is accommodated within the feed system and may be fired remotely from the operations room with no crew closed up at the mounting.

Fail-to-safe interlocks are included in all major operations to safeguard equipment and ammunition, and facilities are arranged for manual control of the power drives for maintenance purposes.

Royal Ordnance are currently developing extended range ammunition for the 4.5-inch gun.

CHARACTERISTICS

Calibre: 4.5 inch (113 mm)
Length of barrel: 55 calibres

Vickers 4.5-inch Mk 8 gun mounting on the Brazilian Mk 10 frigate Liberal

Recoil force (mean): 30 500 kg
Recoil length: 0.38 m
Elevation: −10 to +55°
Training: ±340°
Muzzle velocity: 870 m/s
Chamber pressure (design): 4400 kg/cm²
Max range: 22 km
Rate of fire: 25 rounds/minute continuous
Weight of complete round: 36.5 kg
Weight of HE shell: 21 kg
Weight of HE filling: 2.5 kg
Ammunition: HE, Illuminating, Chaff dispensing, Anti-Aircraft practice, surface practice.

STATUS

The equipment is currently in service with several navies. It is the standard medium calibre gun in the Royal Navy and is fitted to the Type 82 destroyer HMS *Bristol*, the Type 21 and later Type 22 frigates, and the Type 42 destroyers. It will also be fitted in the 'Duke' class (Type 23) frigate and an earlier order for four systems was supplemented, in February 1988, by an order worth £10 million for another four. The Mk 8 is also in service with Argentina on the Type 42 destroyers (1 turret), Brazil on the six 'Niteroi' class frigates (1 or 2 turrets depending on ship), and is now being supplied for a new class of corvette presently under construction in Brazil, with Iran on the four 'Saam' class frigates (1 turret), Libya on the single Vosper Thornycroft Mk 7 frigate (1 turret) and Thailand on the single 'Yarrow' frigate (2 turrets). During the South Atlantic conflict in 1982, the 4.5-inch Mk 8 gun was used to advantage, particularly for naval gunfire support and earned the confidence and commendation of British naval and land forces for its accuracy and effectiveness.

CONTRACTOR

Vickers Shipbuilding and Engineering Ltd, Barrow-in-Furness, Cumbria LA14 1AF, England.

6014.203

TWIN 4 inch GUN MOUNTING Mk 19

Designed by Vickers, this is a twin-barrelled, hand loaded, remotely controlled naval gun mount which is electrically trained and elevated. Normally laid by remote control it can also be locally controlled using joysticks. Fuze setting devices are located alongside each trunnion support. A good gun crew can achieve a rate of fire of up to about 16 rounds per barrel per minute.

CHARACTERISTICS

Calibre: 102 mm
Elevation: To 80°
Projectile weight: 15.9 kg
Muzzle velocity: 760 m/s
Max range: 19.5 km (10.5 nautical miles)
Max altitude: 13 500 m
Rate of fire: Up to 16 rounds/barrel/minute
Date introduced: 1935

STATUS

Operational in Commonwealth and other navies. No longer in production.

6016.203

TWIN 3 inch GUN MOUNTING Mk 6

This is a twin-barrelled 3-inch (70 calibre) automatic dual-purpose, remotely controlled naval gun mount originally intended for the RN Type 12 frigates and RCN 'St Laurent' class, and larger ships. Local control arrangements are provided but the mounting can be operated unmanned. The rate of fire is about 90 rounds per gun per minute and the barrels are water cooled to allow sustained fire. Turret weight is about 38 tons.

CHARACTERISTICS

Calibre: 76 mm
Length of barrel: 70 calibres
Elevation: To 90°
Projectile weight: Probably about 7 kg
Muzzle velocity: 1000 m/s
Max range: Approx 17 km (9 nautical miles)
Rate of fire: 90 rounds/barrel/minute

STATUS

No longer in production. British ships armed with this mounting now scrapped, but it is retained in the Canadian 'Mackenzie' and 'Restigouche' class DDEs.

CONTRACTOR

Vickers Shipbuilding and Engineering Ltd, Barrow-in-Furness, Cumbria LA14 1AF, England.

6010.203

SINGLE 4.5 inch GUN MOUNTING Mk 5

This single-barrelled, remotely controlled, and manually loaded gun mounting was designed in World War II but in the 1950s was modernised for remote power control (RPC). As an alternative to remote control, local control by joystick is possible. Charge and shell are separate and automatically rammed.

CHARACTERISTICS

Calibre: 113 mm
Length of barrel: 45 calibres
Elevation: To about 50°
Projectile weight: About 25 kg
Muzzle velocity: About 750 m/s
Max range: About 19 km (10 nautical miles)
Rate of fire: 14 rounds/minute
Date introduced: 1937

STATUS

No longer operational in the RN but used by Indonesia in three ex-'Tribal' class frigates.

CONTRACTOR

Vickers Shipbuilding and Engineering Ltd, Barrow-in-Furness, Cumbria LA14 1AF, England.

6015.203

TWIN 40 mm (BOFORS) GUN MOUNTING Mk 5

This is a remotely or locally controlled twin Bofors 40 mm 60-calibre close-range naval gun mount. A splinter protection shield to the front of the mounting is installed. It requires one 'on mounting' loader per gun and one operator to lay and train the gun in 'local control' operation. Separate emergency manual laying and training arrangements using two men are provided. Rate of fire is about 120 rounds per minute per barrel.

CHARACTERISTICS

Weight: Approx 3000 kg

Rate of fire: 2 × 120 rounds/minute

Tactical range: 3000 m

STATUS

Operational in many British and Commonwealth naval vessels. Obsolescent.

6013.203

SINGLE 40 mm (BOFORS) Mk 7 AND Mk 9 GUN MOUNTINGS

This is a single-barrelled Bofors 40 mm L/60 power operated ship's gun mounting. It is a lightweight mounting suitable for fitting in FPBs as well as larger vessels. One-man control is by joystick and uses a gyro sight for aiming. One man is necessary 'on mounting' for loading and rate of fire is about 300 rounds per minute. When spares ran out for the World War II Mk 7 the mountings were refurbished for electric drive, becoming Mk 9.

For details of the gun in this mounting see **2360.203**.

Australian Mk 7 mountings are being modernised with a more responsive hydraulic drive, a 440 V AC (three phase) or 240 V AC (single phase) motor, to improve reliability and reduce maintenance needs. The work is being carried out by the Maribyrnong Ordnance Factory.

STATUS

No longer in production but widely fitted in British and Commonwealth naval vessels. Obsolescent, and will be replaced by new 30 mm Mk 1 (see **6196.203**).

6196.203

DS 30 AND 25 SERIES NAVAL GUN MOUNTINGS

The DS 30B Gun Mounting (UK MoD designation Single 30 mm Gun Mounting Mk 1) is a lightweight, power driven, line of sight stabilised gun mounting fitted with the Oerlikon KCB 30 mm cannon. It is intended primarily for local manning by a seated operator, although provision has been made for remote control if required.

The design of the mounting is based on the proven DS 30R (Rarden) mounting which was successfully evaluated by the Royal Navy both at sea in HMS *Londonderry* and on land during the 1970s. It has a low magnetic signature and clean lines which meet NBC washdown requirements. The inclusion of the KCB cannon gives a high rate of fire (650 rpm) with repeat burst (90 rpm) and single shot facilities, thus providing a substantial air defence and surface engagement capability for a wide range of warships and patrol craft.

Compact, and self contained, the system is readily adaptable for a variety of applications, since no deck penetration is required for installation. An integral battery pack provides emergency power in the event of ship's supply failure, providing up to 10 minutes' operation.

Under local control, sighting is by a binocular and disturbed sight line arrangement, with the sight being offset to provide tangent elevation and lead angle correction. On-mount sensor and prediction facilities can be fitted.

The DS 30E naval gun mounting is identical in all major respects to the DS 30B unit but is fitted with the EBO-Mauser 30 mm cannon. This is a joint venture between DES Ltd and Hellenic Arms Industry SA.

The DS 30F naval gun mounting is also identical in all major respects to the DS 30B unit but carries the Mauser 173 Model F 30 mm cannon.

The DS 25M naval gun mounting, also identical in all major respects to the DS 30B unit, carries the McDonnell Douglas 25 mm M242 automatic cannon.

All DS 30 and 25 mountings are designed to be flexible allowing a variety of weapon and sensor packages to be fitted.

CHARACTERISTICS

Operational elevation arc: –20° to +65°

Operational training arc: –165° to +165°

Trunion height: 1.350 m

Slew velocity (training and elevation): 55°/s

Acceleration (training and elevation): 80°/s²

STATUS

The DS 30B has been fully accepted into service by the Royal Navy whose initial order was for 25 to be delivered by 1989. A second order has now been placed.

A DS 30E mounting proposal is currently under evaluation by the Hellenic Navy. The DS 25M mounting is being evaluated for use in Coastguard craft by a NATO navy and is available for customers.

The DS 30F has been ordered for evaluation by the US Navy.

CONTRACTOR

Defence Equipment and Systems Ltd, Salhouse Road, Norwich, Norfolk NR7 9AY, England.

The DS 30B gun mounting, with Oerlikon KCB 30 mm cannon as ordered by the Royal Navy

The DS 25M gun mounting, with the McDonnell Douglas 25 mm M242 automatic cannon

UNITED STATES OF AMERICA

4523.203

TRIPLE 16 inch GUN TURRET Mk 7

With the US Navy decision to re-activate and modernise four 'Iowa' class battleships in 1982, the US fleet re-introduced the powerful 16-inch calibre naval gun to the world's navies. *Iowa*, *New Jersey*, *Missouri* and *Wisconsin* (BB61/2/3/4) all saw action in the Second World War and later during the Korean conflict, and in the case of USS *New Jersey* also in the course of the Vietnam war. The reactivation process involves reducing the 5 inch gun armament, the addition of some new weapons (eg Tomahawk cruise missiles and Phalanx close-in weapon systems), but the nine Mk 7 16-inch guns remain the main armament, arranged in three triple turrets.

These major calibre gun turrets are constructed with armour plate which is 43 cm thick at the sides and front, 30 cm at the rear, and with 18 cm tops. The gun can be removed without dismantling the turret. The 16-inch guns are seen as complementing the Harpoon anti-ship missiles being fitted in these ships with their typical maximum range of about 60 nautical miles.

The two forward triple Mk 7 16-inch guns aboard USS Iowa *which was re-activated in 1985. Each turret weighs over 1700 tons* (L & L van Ginderen)

In mid-1981 it was estimated that there were over 21000 projectiles for Mk 7 16-inch guns in store, of which about 13 000 were then probably still serviceable. Their number includes: 15400 high capacity (HC) projectiles, 3200 armour piercing (AP), 2300 target projectiles, and 138 Mk 19 projectiles. All of the last listed are thought to be unserviceable, while all of the target projectiles and some 90 per cent of the AP rounds are fit for use. Slightly over half the HC projectiles are regarded as serviceable. In the case of charges for these rounds, about 80 per cent of exisiting stocks of 12400 full charges and almost 12600 reduced charges are thought to be fit for use. Each ship has a magazine capacity of 1220 projectiles, and the combined weight of round and full charge is approximately 1525 kg. The HC projectile has a payload of 70 kg of explosive 'D' and the AP round has about 18 kg of the same explosive. US assessments calculate that existing ammunition stocks will be adequate for training and structural test purposes until the end of 1991, after which procurement of target projectiles will be required.

Two new munitions are under development, an anti-personnel round and an anti-armour round, under a 16 inch Naval Gunfire Improvement Project approved in May 1986.

The first is an extended-range anti-personnel round, using a 13 inch discarding-sabot projectile packed with over 600 sub-munitions. Its increased velocity gives nearly double the present range. This programme is in hand.

The second improvement is based on the SADARM Search and Destroy Armor sub-munitions under development for the US Army. It will not be procured until the 1990s, and other payloads under consideration for the 16 inch round are chaff, smoke, and even anti-tank sub-munitions.

The fire control system is also being improved. A new system based on the Mk 160 computing system for the *Arleigh Burke* (DDG-51) is under development. Projectile-tracking radar is also being studied, to do away with the need for forward observers. US Navy statements suggest that the new system will be adapted from counter-battery radar used by the Army and Marine Corps.

CHARACTERISTICS
Calibre: 16 inch (406 mm)
Length of barrel: 50 calibres
Elevation: +45° max at 12°/s
Traverse: 300° at 2°/s
Projectile weights: 862 kg(HC); 1225 kg (AP and drill)
Muzzle velocity: 820 m/s (HC); 739 m/s(AP)
Max range: 38 km (HC); 36.75 km (AP)
Rate of fire: 2 rounds/minute
Weight of gun: 108 510 kg
Assembly weight: 1701 – 1708 tons

STATUS
No longer in production but operational in USS *New Jersey*, *Iowa* and *Missouri*. Recommissioning work is in progress on the fourth ship, USS *Wisconsin* at Ingalls Shipyard, Pascagoula, Mississippi.

4846.203

TRIPLE 6 inch/47 Mk 16 GUN TURRET

These mountings were originally installed on the US Navy's 'Brooklyn' class cruisers, only one of which remains in active service with the Chilean Navy (*O'Higgins*). Five turrets were fitted on the centreline. All turrets were similar with the principal difference being in the ammunition hoist to each.

Each turret carries the three guns in a built up slide. They are elevated, depressed, and trained as a unit but may be fired separately or together, electrically or by percussion. Ammunition is manually served to loading trays from endless conveyor hoists. Each turret is eight feet (2.4 m) high, and has case ejection ports located at the base behind the barbette structure which is 17 feet (5 m) in diameter. Three types of projectile are used – the Mk 34 FCL(VT)/HC; the Mks 35, 36, 37, or 42 AP; or training rounds. Powder case Mk 16 was standard, using either full charge (60 pounds) or reduced charge (51 pounds).

CHARACTERISTICS
Calibre: 152 mm
Length of barrel: 50 calibres
Elevation: –5° to +60°
Traverse: 288° to 300° depending on siting
Muzzle velocity: 762 m/s(AP), 790 m/s(HC)
Max range: 23.9 km(AP), 21 km(HC)
Max altitude: 10.4 km
Assembly weight: 165 to 173 tons
Weight of gun: 4370 kg
Ammunition: AP, HC and drill

STATUS
Originally installed on all light cruisers of the US Navy. The Chilean cruiser *O'Higgins* is the last remaining ship with this gun mounting.

6054.203

SINGLE 5 inch Mk 27 DUAL-PURPOSE GUN MOUNTING

Eight of these guns are fitted to the ex-US 'Brooklyn' class cruiser *O'Higgins*. They are believed to be of US manufacture but 5 inch guns of this type are no longer in service in the US Navy.

4847.203

SINGLE 5 inch/54 CALIBRE GUN MOUNTING Mk 39

The 5-inch/54 calibre Mk 39 is a single enclosed base-ring type mounting and is designed for semi-fixed ammunition. The gun is incorporated in the 5-inch housing Mk 4 with a vertically sliding wedge breech-block mechanism. Ammunition is power-rammed after being placed manually in a loading tray. Gun crews can maintain a rate of fire of 20 rounds per minute.

The single and twin 5-inch/54 calibre Mk 39 mountings were designed during the Second World War as successors to the Mk 32 and Mk 39 dual-purpose guns then in widespread use. Ultimately the single Mod 0 mounting was installed on the 'Midway' class aircraft carriers. Later they were removed from these ships, and transferred to Japan for installation on five destroyers.

CHARACTERISTICS
Calibre: 5 inch (127 mm)
Length of barrel: 54 calibres
Elevation: –10 to +85°
Traverse: 300° plus
Projectile weight: 31.8 kg
Muzzle velocity: 762 m/s
Max range: 23 km
Max altitude: 14.8 km
Rate of fire: 20 rounds/minute max
Gun weight: 2432 kg
Assembly weight: 33 530 to 35 739 kg
Ammunition: AAC/HC, Common, and Illuminating

STATUS
In service in Japanese Navy in 'Akizuki' and 'Murisame' class destroyers. Obsolescent.

6026.203

TWIN 5 inch/38 CALIBRE GUN MOUNTING Mk 28

This is the oldest design of the 5-inch mounting still in service with the USN, but it is also one of the most widely used. The 5-inch/38 calibre gun has indeed been described as 'the prototype of the conventional US naval gun'.

The Mk 28 mounting contains two Mk 12 5-inch/38 calibre guns and an enclosed mounting with ammunition-handling room beneath. It requires a crew of 13 on the mounting and 26 in all. Remotely-controlled, semi-automatic, and dual-purpose, it also has local laying facilities on the mounting. Normal rate of fire is 15 rounds per barrel per minute but the gun's capacity for sustained fire is in excess of this and a good crew can achieve 22 rounds per minute over short periods. Train limits are 300° apart and elevation limits from –15 to +85° but train and elevation velocity and acceleration are significantly lower than those of the later 5-inch mountings.

Semi-fixed ammunition is used, consisting of a projectile weighing about 25 kg (varying according to type) and a case assembly weighing about 13 kg including a full powder charge of 6.8 kg. With a full charge the muzzle velocity is 792.5 m/s giving a maximum horizontal range of 16.5 km (9 nautical miles) and a ceiling of 11400 m. The ammunition is raised to the gun house by hydraulically powered hoists. The complete mounting weighs 153 000 lbs or 169 000 lbs depending on 2 inch or 2.5 inch shield.

CHARACTERISTICS
Calibre: 5 inch (127 mm)
Length of barrel: 38 calibres
Elevation: –10 to +85°
Traverse: +150°
Projectile weight: About 25 kg
Muzzle velocity: 792.5 m/s
Max range: 16.5 km (9 nautical miles)
Max altitude: 11 400 m
Rate of fire: 15 – 22 rounds/minute
Date introduced: 1935

STATUS
Twin mountings were installed in the four 'Iowa' class battleships where they are currently in service. Lighter variants in all World War II cruisers, and later destroyers. Some of these are still in service in ex-US ships of other navies. They are also in service with one Spanish-built ship, and aboard certain Italian-built destroyers and Danish frigates.

4844.203

TWIN 5 inch/38 CALIBRE GUN MOUNTING Mk 38

The 5-inch/38 calibre gun installed in the Mk 38 mounting was designed primarily for destroyer-type ship installation with a base ring and an open or enclosed mounting. It became one of the most widely used gun mountings in the US Navy and was installed on destroyer and auxiliary-type ships throughout the Second World War. It remains in widespread foreign service, mostly on transferred 'Sumner' and 'Gearing' class destroyers.

The Mk 38 mounting contains two 5-inch Mk 12 Mod 1 guns in a semi-automatic dual-purpose mounting with separate ammunition and cartridge hoists for each gun. In automatic control the mounting is operated by a remote fire control director. Four gun shields are provided – the Mks 41, 48, 53 or 62 – and although similar in their fabrication they differ in their degree of thickness for splinter protection with weights from 14500 to 24400 pounds (6577 to 11 068 kg).

Twin 5 inch/38 calibre gun mounting Mk 38 aboard USS Zellers (Stefan Terzibaschitsch)

CHARACTERISTICS

Calibre: 5 inch (127 mm)
Length of barrel: 38 calibres
Elevation: –15 to +85°
Traverse: 328.5°
Projectile weight: 25 kg
Muzzle velocity: 762 m/s
Rate of fire: 15 rounds/barrel/minute practical; 22 theoretical
Assembly weight: 43 466 to 44 600 kg
Gun weight: 1810 kg (Mk 12 Mod 1)
Ammunition: AAC/HC, Illuminating, Drill

STATUS

In service with US Naval Reserve destroyers and in the navies of Argentina, Brazil, Chile, Greece, Denmark, Ecuador, Iran, Italy, South Korea, Mexico, Pakistan, Spain, Taiwan, and Turkey.

6092.203

SINGLE 5 inch/38 CALIBRE GUN MOUNTINGS

In addition to the enclosed twin mounting of the 5-inch/38 calibre gun (**6026.203**) there have been three other general types of mounting for this gun. They are:

(a) **Mk 30** Enclosed single mount with ammunition-handling room below. Originally on destroyers, destroyer escorts, and auxiliaries. Now mainly found on auxiliary vessels of the USN, these single mounts are also the main surface armament of the ex-US 'Fletcher' class destroyers which are still in service in some navies.

(b) Open single mount with ammunition-handling room below. Auxiliary vessels. Probable designation: Mk 30 Mod 24.

(c) **Mk 37** Open single mount without ammunition hoists or handling room. Converted merchant vessels.

In performance there is essentially no difference between the first two of these and one gun of the enclosed twin mounting. The third arrangement is necessarily less efficient in terms of ammunition handling but otherwise the gun characteristics are the same.

Single 5-inch/38 calibre gun aboard the Spanish Navy ship Legazpi (J I Taibo)

The most widespread model is the Mk 30 which carries a single Mk 12 barrel. There are numerous versions, and those of 'Fletcher' class destroyers are referred to as Mods 19, 30 and 31, depending upon the position of the mount aboard ship.

6027.203

SINGLE 5 inch/54 CALIBRE GUN MOUNTING Mk 42

The 5-inch/54 calibre Mk 42 gun mounting was adopted to replace the pre-Second World War 5-inch/38 and wartime 5-inch/54 Mk 39 (**6093.203**), and is the most widely adopted post-war US gun mounting. It is a single-barrel rapid-fire base-ring type mounting used with either local or remote fire control. The Mods 1, 5 and 6 were for command ships (CLC) and the Mod 3 for DL/DLG class ships. Mod 7 was the most common and included CVA, CG, DLG, and DD installations.

5-inch/54 calibre Mk 42 gun on USS Thomas C Hart (Stefan Terzibaschitsch)

All Mods use the 5-inch/54 calibre Mk 18 Mod 0 gun housed in a vertical sliding breech-block. A bayonet type joint attaches the gun to the housing and the entire gun assembly is supported on rollers in the slide. A separate rammer is mounted on the slide. The mounting is fully power driven with an automatically controlled loading system capable of maintaining a sustained rate of fire of 40 rounds per minute. Ammunition is loaded by a dual system with projectiles supplied automatically to the slide alternately from the right and left.

The mounting is enclosed in an armour plate shield assembly with early mountings using the shield Mk 61 Mod 1 and the Mks 7 to 10 mountings using Mod 1 or Mod 4 shields. The weights of the mountings vary from the 145930 pounds (66193 kg) of the Mod 1 to a maximum of 210196 pounds (95343 kg) in the CVA installation. Ammunition includes the Mk 41 AAC/HC, Mk 42 common, and Mk 48 illumination rounds, all of which weigh 70 pounds (32 kg) and have a muzzle velocity of 2650 ft/s (808 m/s).

The Mk 42 Mod 10 is an improved version of the Mod 7 gun mount and over 100 are at sea. A modernisation and overhaul programme substitutes updated control and operating components to incorporate the improved reliability and maintenance features of the Mk 42 Mod 9 gun mount (**6028.203**) into the earlier Mod 7 and 8 gun mounts. The full gun crew complement is reduced to 12 men. Ready-to-use ammunition capacity is 40 rounds and firing rates of 20 or 40 rounds per minute are possible. The associated fire control system is the Mk 68 Mod 11.

CHARACTERISTICS

Calibre: 5 inch (127 mm)
Length of barrel: 54 calibres (269 in)
Elevation: –7 to +85°
Traverse: 360° potential
Muzzle velocity: 2650 ft/s (808 m/s)
Max range: 23 690 m
Max altitude: 14 847 m
Rate of fire: 20 rounds/minute practical; 40 rounds/minute rated max
Assembly weight: 145 930 lbs (66 193 kg) (Mod 1); 146 400 to 183 630 lbs (66 406 to 83 293 kg) (Mod 7)
Weight of gun: 5 615 lbs (2547 kg) (Mod 1); 5 508 lbs (2498 kg) (Mod 7)

STATUS

Operational in ships of the US Navy and elsewhere. Foreign users of the Mk 42 gun include the Royal Australian, Federal German, Japanese and Spanish navies.

6028.203

SINGLE 5 inch/54 CALIBRE GUN MOUNTING

Mk 42 (Lightweight Mod 9)

Functionally similar to the Mk 42 Mod 7 gun mount described above, the Mk 42 Mod 9 is an improved design featuring lower mount weight (58700 kg), nearly ten per cent lower power consumption and a smaller crew requirement. Improvements incorporated include replacement of all electronic components of the earlier mount by solid-state devices. Only two men are needed on the mount as against four for the Mod 7, and the total crew requirement is 12 men instead of 14. The only respect in which the Mod 7 performance is known to be superior is that of elevation acceleration (60°/s² against 40°/s²).

STATUS

Fifty-one units of the above-decks portion of the mount were manufactured by Naval Systems and supplied for use on 'Knox' class frigates. The below-decks portion was produced by the US Navy Naval Ordnance Station, Louisville.

CONTRACTOR

FMC Corporation, Naval Systems Division, Box 55459, Minneapolis, Minnesota 55459-0043, USA.

Lightweight (Mod 9) version of the 5-inch/54 calibre gun mounting Mk 42 in production

6029.203

SINGLE 5 inch/54 CALIBRE LIGHTWEIGHT AUTOMATIC GUN MOUNTING Mk 45

This gun mounting represents a major step forward in medium calibre ordnance for the US Navy. It has been designed primarily for installation in new ships and was required to embody all relevant improvements developed over some 30 years since the 5-inch/38 was first introduced (**6026.203**). It was also required to be light, easily maintained, and exceptionally reliable. The result is a weapon which requires only one-third of the crew of a 5-inch/38 and with which a single man in a control centre can fire a drum load of 20 shells without help. It is also certainly the lightest naval gun mount of its size in the West and probably the lightest in the world.

Specifically, the major characteristics of the Mk 45 mountings are:

Max elevation: 65°
Rate of fire: 20 rounds/minute
Ready service rounds: 20
Mount weight: 25 tons
Crew: Reduced from 14 to 6
Crew on mounting: None

5-inch/54 calibre Mk 45 gun aboard USS Briscoe

The gun is used with the lightweight Mk 86 GFCS (**1241.281**), the Mk 68 digital upgrade, and the Mk 160 GCS of the DDG-51 class. The Mk 45 has also been integrated with the Hollandse Signaalapparaten WM25 weapon control system.

GMOTS (Gun Maintenance and Operational Training System) provides simulation training for shipboard operators of gun system Mk 45 Mods 0 and 1. GMOTS eliminates the need for expensive ordnance hardware for control system training, and provides effective training at reduced cost. This is made possible because GMOTS uses operator control panels that are identical to the control panels that equip the Mk 45 gun system but utilises a simulator in place of the actual gun mount.

A GMOTS training system consists of a power distribution panel (EP1), and operator's control panel (EP2), and a display panel (EP3). Also included are the gun order responder unit (GORU), and electronic fuze setter unit and coil, and an electronic fuze simulator. Operated by a training instructor, the GORU duplicates hydraulic and electrical timing and response functions of the gun mount, ready service loader, lower hoist, and fuze setter. The GORU also simulates ammunition selection, loading, offloading, and strikedown operations. The GORU is also capable of duplicating weapon control orders and interfaces to the trainee.

For maintenance training, the instructor can simulate malfunctions in the gun mount electrical circuits, or in other operational areas. These simulations are fed back to the EP2 panel, where the student is trained to take appropriate action. The GORU duplicates weapons control to the extent necessary to generate proper orders in the approximate time intervals encountered in the gun system. It can also simulate mechanical timing errors.

The GMOTS includes a setter unit, electronic fuze setter, and a dummy round with an electronic fuze to allow trainees to become fully trained on the operation and maintenance procedures for the electronic fuze setter.

The GMOTS provides the trainee with much more thorough training than is possible using an actual gun system because of the ease with which simulated problems and casualties can be introduced into the system. Without GMOTS faults would have to be physically created in the actual ordnance system for the student.

DEVELOPMENT

Development of the Mod 0 weapon started in 1964 on a contract placed with the Naval Systems Division of the FMC Corporation. All current production is by FMC Naval Systems Division.

The Mod 1 gun system entered production in 1983. The new gun has the capability to automatically select, load, and fire all existing types of 5-inch/54 calibre semi-fixed ammunition, including guided projectiles.

The Naval Systems Division were also the developers and the sole manufacturers of the Mk 6 ammunition hoist, which is an integral part of the Mk 45 installations on CGN-41, CGN-36, DD-963, LHA, DD-993 and CG-47 ships. Gun barrels are produced by US Government ordnance factories.

STATUS

Over 125 Mk 45 guns have been delivered to the US Navy. In 1984 FMC was awarded a multi-year contract for 25 Mod 1 guns to be supplied over a three year period. Each of the 'Ticonderoga' class Aegis cruisers are fitted with two Mk 45s, while ships of the 'Arleigh' and 'Burke' class will carry one gun each. In 1984 the Turkish Navy purchased four Mod 1 systems for each of their 'Yavuz' class ships.

A Mk 45 gun system product improvement programme is currently being considered by the US Navy. Advanced controls and mechanical simplification are expected to lead to further increases in reliability, maintainability and mission availability. The improved Mk 45 would enter production in 1989 or 1990.

CONTRACTOR

FMC Corporation, Naval Systems Division, Box 55459, Minneapolis, Minnesota 55459-0043, USA.

6030.203

SINGLE 3 inch/50 CALIBRE GUN MOUNTING

Mk 34

This gun mounting was designed as a dual-purpose weapon with an automatic loader for use with Gun Fire Control Systems Mks 56, 57, or 63 Mods 11 and 22. It can be controlled locally or from a 3-inch Gun Director Mk 51 Mod 3. Installations were intended for destroyer escorts, patrol gunboats, attack transports and other auxiliaries. Many of these mountings were later transferred to US Naval 'Asheville' class fast attack craft and a majority of those still otherwise in use are to be found in former US Naval vessels in service with other navies.

The Mk 34 mounting is commonly found with an environmental shield for protection against icing and bad weather as the Mk 34 Mod 2 version. With a loader Mk 2 it can handle up to 50 rounds per minute in an installation which is identical to US Naval twin 40 mm mountings, designed for easy interchangeability.

CHARACTERISTICS

Calibre: 3 inch (76 mm)
Length of barrel: 50 calibres
Elevation: −15 to +85°
Traverse: 355°
Projectile weight: 5.93 kg
Muzzle velocity: 808 m/s
Max range: 12.8 km
Max altitude: 1492 m
Rate of fire: 50 rounds/minute
Assembly weight: 8677 to 9557 kg
Weight of gun: 564 kg

STATUS

Fitted to US Navy 'Asheville' class and a few other US Navy ships. Also in service with the navies of Indonesia, South Korea, Norway, the Philippines, Thailand, Tunisia, and Turkey.

Single 3-inch/50 calibre Mk 34 Mod 3 mounting on Indonesian frigate Samadikun (G Jacobs)

6031.203

40 mm ANTI-AIRCRAFT GUN MOUNTINGS

There are single, twin, and quadruple barrelled versions of the 40 mm automatic recoil operated gun in service in the US Navy. The twin model is the Mk 1, the quadruple version is the Mk 2, and both types are water-cooled. The single barrel Mk 3 is air-cooled. Most mounts can be either locally or remotely controlled, and are power operated, with emergency hand operation. The gun is copied from the Bofors L/60 design and the accompanying photograph shows a twin-barrel mounting.

CHARACTERISTICS

Calibre: 40 mm
Length of barrel: 56 calibres
Projectile weight: About 0.9 kg
Muzzle velocity: About 820 m/s
Max range: About 11 km but tactical range much less
Rate of fire: 140 to 160 rounds/barrel/minute

US pattern 40 mm Bofors twin-mount

4845.203

TWIN 3 inch/50 CALIBRE GUN MOUNTING Mk 33

This mounting was originally designed as a heavy anti-aircraft gun to replace the 40 mm and 20 mm weapons considered inadequate during the Second World War. Installations did not begin until after the war with the 'Des Moines' and 'Worcester' class cruisers being among the first.

The Mk 33 mounting differs from the earlier Mk 27 in that the slide has been modifed to allow the gun barrel to be removed out of the front of the slide to facilitate barrel changing. In addition it has faster training and elevation speeds but without an alternative hand drive. The guns used are the Mk 22 Mods 4 to 9 and each has an automatic loader Mk 2 capable of handling up to 50 rounds per minute. Either drum or vertical type ammunition magazines are available. The Mods 4 and 7 mountings have environmental shields, which have been made from both grp and aluminium, designed for protection against icing and bad weather. These include provision for a fire control radar to be fitted outside the shield.

CHARACTERISTICS

Calibre: 3 inch (76 mm)
Length of barrel: 50 calibres
Elevation: –15 to +85°
Traverse: 310°
Muzzle velocity: 808 m/s
Max range: 12.8 km
Max altitude: 8951 m
Rate of fire: 50 rounds/minute max
Ammunition: AA, AP, and Illuminating

STATUS

Installed in the US Navy's 'Des Moines' (CA 134), 'Bronstein' (FF-1037), and other classes as well as in various large auxiliaries. Also used by the navies of Argentina, Canada, Chile, Colombia, Greece, Italy, Japan, Norway, Portugal, Spain, Taiwan, and Turkey.

3175.203

76 mm/62 CALIBRE GUN MOUNT Mk 75

FMC Corporation, Naval Systems Division, supplied gun mounts to the United States Government under licence from OTO Melara SpA of La Spezia, Italy. The gun is identical to the single 76/62 compact gun mounting described in entry **5533.203**.

It has been installed in hydrofoils, 'Perry' class frigates and 'Bear' class Coast Guard cutters.

STATUS

Technical evaluation on board the USS *Talbot* was completed in early 1975 with favourable results. 75 guns had been delivered by December 1983. The US Navy continues to buy small numbers of mountings to support Coast Guard and overhaul requirements. The gun mount is not currently planned for any new US Naval construction shipbuilding. Overhaul and refurbishment of the US Navy's Mk 75 guns is being carried out.

CONTRACTORS

FMC Corporation, Naval Systems Division, 4800 East River Road, Box 59043, Minneapolis, Minnesota 55459-0043, USA.

OTO Melara SpA, La Spezia, Italy.

76mm/62 calibre gun mount Mk 75

2584.203

EMERLEC-30 TWIN 30 mm MOUNTING

Emerlec-30 is a twin 30 mm naval ordnance system currently in production at Emerson Electric Co, St Louis, USA. The Emerlec-30 was originally developed as the EX-74 MOD 0 for the coastal patrol and interdiction craft under a USN programme. The system incorporates design improvements that have resulted from land-based and sea testing of the EX-74. Designed for anti-missile, anti-aircraft and surface fire the system comprises two Oerlikon 30 mm cannon, daylight reflex or night image intensifier sight with gyro-stabilised gun line, an integral below-deck magazine with substantial ready-use ammunition and an enclosed environmentally-controlled cabin for the operator. The mounting can be operated by remote control and will interface with any standard fire control system. Ship's power supplies are used in normal operation, but an on-mount battery provides sufficient power to operate the guns and fire a full complement of ammunition in an emergency. Traditional hand crank mechanisms provide further back-up.

CHARACTERISTICS

Calibre: 30 mm
Elevation: –15 to +80°
Traverse: ±360°
Projectile weight: 0.35 kg
Max range: 6 km
Rate of fire: 1200 rounds per minute
Ready rounds: 1970 rounds
Total weight of system (less ammunition): 1905 kg

STATUS

No longer in production and in service with Ecuador, Greece, Malaysia, Nigeria, Philippines and South Korea.

CONTRACTOR

Government & Defense Group, Emerson Electric Co, 8100 W Florissant, St Louis, Missouri 63136, USA.

Emerlec-30 twin mounting aboard 105 ft (32 m) 'Swiftships' fast patrol boat

4042.203

SEA VULCAN 30 LIGHTWEIGHT 30 mm NAVAL GUN MOUNT

The Sea Vulcan 30 lightweight 30 mm naval gun mount designed and developed by the General Electric Company, USA, is a high performance mount embodying the seven-barrel GAU-8/A Gatling gun. It has been designed to provide naval craft with an effective close-in defence against air and anti-ship missile threats.

The production ammunition consists of a new APDS round and a ballistically matched set of 30 mm HEI, API, and target practice rounds. The ammunition feed/storage drum is mounted vertically and rotates in the train axis. This configuration minimises unbalanced loads and inertia changes as the ammunition is expended and the total amount varies. It also reduces the above-deck profile of the mount, and minimises the swept radius required below deck. The below deck location of the feed drum also provides a protected environment for reloading and maintenance of the ammunition replenishment and feed system.

Particular design emphasis was placed on the positioning of the GAU-8/A relative to the mount's train and elevation axes, to ensure reliable ammunition feed to the gun's transfer unit. The elevation axis of the gun passes through the centre of the transfer unit, which minimises chute distortion as the gun is elevated or depressed. Gun firing torque about the elevation axis is eliminated by having the centroid of the firing pressure pulse coincident with the centre-line of the elevation trunnion bearings. The longitudinal axis of the gun assembly is displaced 102 mm from the centre of the train bearing, which positions the firing barrel coincident with the mount's train axis.

CHARACTERISTICS

Gun type: 7-barrel Gatling GAU-8/A
Calibre: 30 mm
Muzzle velocity: 1021 m/s
Rate of fire: 2100/4200 rounds/minute
Drive: Electric
Ammunition types: HEI, API, TP, APDS
Capacity: 1190 rounds
Feed: Linkless
Mount
Weight (loaded): 5295 kg
Height above deck: 1613 mm
Depth below deck: 2200 mm
Max diameter: 2529 mm

STATUS

In quantity production. The prototype system EX-83 was test fired in late 1979 in Den Helder, and used with the Signaal Flycatcher all-weather fire control system (**2301.181**) for a demonstration programme code-named Shortstop. This combination is the basis of the autonomous naval air defence system called Goalkeeper (SGE-30) (**3516.261**). In early 1983 demonstrations were given to representatives of ten countries in the course of trials at Burlington of the Goalkeeper system which is being fitted in ships of the Royal Netherlands Navy and the Royal Navy.

General Electric Sea Vulcan 30 lightweight 30 mm naval gun mount set up for shore firing tests

CONTRACTOR

General Electric Company, Armament Systems Department, Lakeside Avenue, Burlington, Vermont 05402, USA.

4043.203

SEA VULCAN 20 20 mm NAVAL MOUNT

The Sea Vulcan 20 is a lightweight, low profile mount developed to provide firepower for small naval craft. It is designed to accept and fire the 20 mm M197 gun, the XM188E1 gun, the M134 7.62 mm gun, the M19 40 mm grenade machine gun, the Stinger AA missile and other selected weapons. An engineering model, the EX-84, was constructed and tested under a US Navy prototype programme.

The Sea Vulcan 20 mount has many components that are common with the UTS turret as used on the AH-1 Cobra helicopter, and the system is a remotely controlled flexible gun mount with four main elements: a microprocessor-based gun corrector module, mount electronics, ammunition storage and feed system and the mount structure.

The gun corrector is a fire control computer which makes the Sea Vulcan 20 mount self-sufficient, and in addition to the microprocessor it has the necessary components to provide memory and analogue and digital input/output facilities. The corrector compensates for velocity jump, gravity drop, parallax, target motion, and provides self-testing for the whole system.

The ammunition feed and storage system comprises a box assembly and a cover assembly with booster. The box has a capacity of 500 20 mm rounds. This electrically primed ammunition can be handled and stored in a RADHAZ environment. The gun feeder is designed to remove the ammunition from regular links or links with radiation tabs. Flexible chuting provides for transfer of ammunition from the container to the feeder assembly.

The structural components of the mount are the lower support, upper support, and the saddle assembly. The lower support provides the ship interface and mounting. The upper support is mounted to the lower support through a train bearing and drive, and contains an integral gear drive for the elevation axis. The saddle is a frame that supports the weapon which is mounted through the recoil system. The Sea Vulcan 20 is arranged to accept a ring sight, a reflex sight, night vision sight, or any director system.

Components of Sea Vulcan 20 universal naval mount, shown with M197 20 mm gun mounted in turret

CHARACTERISTICS

Mount: Sea Vulcan 20 (with M197 gun)
Gun type: 3-barrel Gatling
Calibre: 20 mm
Elevation: –10 to +55°
Elevation rate: 60°/s
Traverse: ±110°
Traverse rate: 80°/s
Muzzle velocity: 1030 m/s
Recoil force: 545 kg
Rate of fire: 750/1500 rounds/minute
Position accuracy: 3 mils (static in either axis)
Dispersion: 6 mils m (80% rounds fired)
Ammunition: 20 mm M50 series, M 246 HEI-TSD, M 220 TPT, Mk 149 APDS
Capacity: 500 rounds
Weight: 318 kg (empty, without missile launcher); 465 kg (loaded with 500 rounds)
Height above deck: 13.34 cm
Turret diameter: 8.64 cm
Swept radius: 1676 mm

STATUS

In production.

CONTRACTOR

General Electric Company, Armament Systems Department, Lakeside Avenue, Burlington, Vermont 05402, USA.

2836.203

20 mm SEA VULCAN 20P THREE-BARREL PINTLE MOUNT

Designed for installation on a wide range of sea-going and riverine craft, this three-barrel 20 mm gun can provide a small vessel with very substantial fire-power.

The system comprises an M197 three-barrel 20 mm gun and ancillary equipment mounted on a USN Mk 10 stand. The gun and its associated delinking feeder (M89) are currently in use by the USMC in helicopters and in US naval aircraft.

The M197 weapon is electrically operated and power for this is supplied by a battery mounted on the gun stand. Twenty complements of ammunition can be fired without battery recharging; the battery can conveniently be connected to a main power source to provide a 28 V trickle feed, thus maintaining the battery in a charged condition.

The three-barrel gun is derived from the six-barrel M61 (Vulcan) gun (**5547.103** in 1982-83 and earlier editions) and has alternative rates of fire of 600 and 1200 rounds/minute. The M89 delinking feeder includes a declutching arrangement which, when the trigger is released, automatically stops the supply of ammunition to the gun but permits the gun to continue firing long enough to clear the barrels.

The Mk 10 gun mount provides convenient room for the installation of the 300-round ammunition can behind the right-hand shield, the battery and control box being mounted behind the other half.

CHARACTERISTICS

Deck-mounted on Mk 10 mount

Weapon: 3-barrel M197 20 mm
Elevation: –15 to +75°
Traverse: ±180°
Muzzle velocity: 1030 m/s
Rate of fire: 600 or 1200 rounds/minute
Dispersion: 80% in 8 mil circle
Ammunition: M50 series
Feed: XM89
Ready rounds: 300
Battery fire out capability: 6000 rounds
Weight: (Without ammunition) 500 kg approx

CONTRACTOR

General Electric Company, Armament Systems Department, Burlington, Vermont 05402, USA.

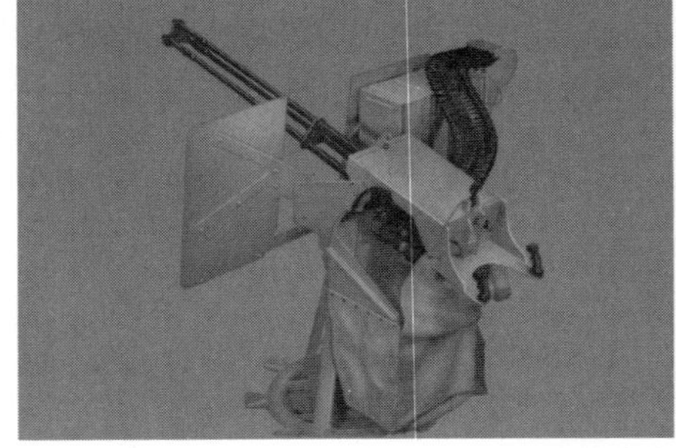

20 mm Sea Vulcan 20 P three-barrel pintle mount

6758.203

SEA VULCAN 25 WEAPON SYSTEM

The Sea Vulcan 25 is a lightweight 25 mm mount developed to protect small surface ships from the threats of close-in air and lightly armoured surface threats. It is designed to be a highly responsive, accurate, reliable and easy to maintain system.

It consists of a 25 mm five-barrel GAU-12/U, with selectable rate of fire, in an autonomous mounting, with environmental protection, which can be either of deck penetration or deck-mounted configuration. The system is slaved to an electro-optic director and features an automatic tracking unit. A dual Stinger missile pack can be attached to the mounting roof or other missiles can be used.

CHARACTERISTICS

Gun: 25 mm GAU-12/U
Weight loaded: 1315 kg
Ready service magazine: More than 500 rounds
Rate of fire: 750-2000 rounds per minute

STATUS

In development. During 1987 the US Navy Sea Systems Command carried out two series of tests with the system mounted in a SES-200 using three electro-optical designators; Mk 24 TDT, CSEE Lynx and Maganvox/Signaal Liod. The tests were carried out against aerial and sea targets and in 213 firings involving 6451 rounds the manufacturers claim they were carried out "without a gun or turret stoppage".

CONTRACTOR

General Electric, Armament Systems Department, Lakeside Avenue, Burlington, VT 05401-4985, USA.

Sea Vulcan 25 naval mount

UNDERWATER WARFARE SYSTEMS

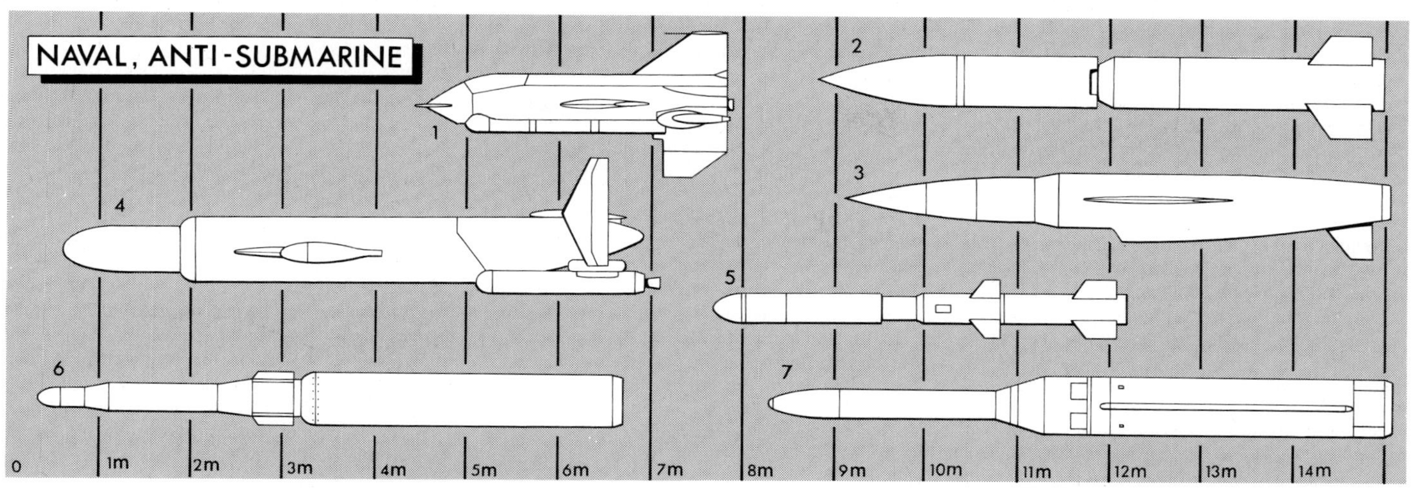

1 Ikara, **2** FRAS-1, **3** SS-N-14 (provisional), **4** Malafon, **5** Asroc, **6** Subroc, **7** Sea Lance

3196.441

UNDERWATER WARFARE SYSTEMS

In this section will be found information on those weapons concerned primarily with underwater targets (submarines) or which operate in the underwater environment for the major portion of their functional life (torpedoes). Under the US heading details will also be found of undersea surveillance projects.

Related entries will be found also under the Naval Fire Control Systems, and Sonar and Underwater Detection sections.

Anti-submarine Torpedoes

The torpedo is a major anti-submarine weapon. It may be launched directly from a torpedo tube on a surface vessel or submarine, from a drone or missile launched from a ship, or from a manned aircraft or helicopter. All available information on torpedoes will be found in this section, and torpedo fire control systems will be found elsewhere in the Systems Section.

Drones and Missiles

Drone torpedo-launchers and torpedo-carrying missiles are also dealt with in the Systems Section. Their names and reference data are listed below.

Name	Entry No	Country
Asroc	**6001.241**	USA
Sea Lance	**4181.441**	USA
FRAS-1	**3968.241**	USSR
Ikara	**6002.241**	Australia
Super Ikara	**6481.241**	Australia/UK
Malafon	**1179.241**	France
SS-N-14	**3969.241**	USSR
SS-N-15	**3970.441**	USSR
SS-N-16	**3970.441**	USSR
Subroc	**1128.441**	USA

Of these it should be noted that Asroc has a dual role – as a carrier of either an A/S torpedo or a depth charge. The submarine-launched missile Subroc (**1128.441**) is not at present a torpedo carrier but implementation of the new US Sea Lance (combined anti-submarine warfare standoff weapon programme) (**4181.441**) will remedy this deficiency as a homing torpedo is under consideration as one of the probable payloads planned for this new weapon.

The details of the Soviet weapons, because of the usual lack of direct evidence, should be regarded as provisional although the information given is considered to be accurate within these constraints.

Depth Charges

Relatively simple cylindrical depth charges that can be rolled or catapulted into the sea are the longest-established anti-submarine weapons: they were first used by the RN in the First World War. They are generally depth-fuzed and have a low sinking rate, thus giving the launching vessel time to get clear. They are still used extensively by many navies: a typical weight of the depth charge is 150 kg and a launcher can project the charge up to about 150 m.

Depth Charge Mortars

Towards the end of the Second World War a more streamlined type of depth charge with a higher rate of sinking was introduced by the RN and subsequently other navies. The intention was to project charges ahead of an attacking ship so that they could be fired more accurately while still in sonar contact, otherwise lost when the attacking ship ran over the submarine to deliver conventional depth charges. Such mortars have been developed in several countries, Squid (**6007.241**) and Limbo (**6008.241**) in the UK being two examples and the Italian Menon being a third. The Menon launcher, which exists in three- and six-barrelled forms, is notable for the length of its tubes and has a range of about 1500 m compared with 350 to 1000 m for the two British devices. A single-barrel mortar is fitted on some Italian frigates and corvettes. This is said to have a range of about 1000 m and a rate of fire of 15 DCs/minute.

Another weapon in this category is the French four-barrelled mortar. This is mounted in a turret and is automatically loaded. It fires a heavier projectile and has a longer range (about 2750 m) than any of the others. All these mortars fire 12 inch (305 mm) depth charges.

It is believed that the Soviet Navy has not adopted the streamlined form of depth charge. They do, however, use depth charge mortars, but it is thought possible that they use compressed air to propel the charge whereas other countries use an explosive cartridge.

Nuclear Depth Charges

Also introduced by the USN is the nuclear depth charge. To take this clear of the launch vessel a rocket is required, and these depth charges have so far been associated only with Asroc (**6001.241**) and Subroc (**1128.441**). In the UK, it is understood that Royal Navy Sea King helicopters can be equipped with 'Bomb 600 lb special ordnance'. Based on the US B-57 depth charge it is intended for the attack of deep-diving submarines which are running too deep or too fast for homing torpedoes. These nuclear devices are of UK manufacture.

Multiple Short-range Rocket Launchers

Another British innovation of the Second World War was the multiple launcher for small (about 25 kg) rockets with impact fuzes. Devices of this nature are widely used by many navies and are commonly known by the name Hedgehog, which was applied to the first US 24-rocket launcher, even though they may not be of identical design. The modern American Hedgehog (which has been considerably elaborated since it was first introduced) has a range of about 350 m; a smaller eight-rocket device known as Mousetrap has a range of about 200 m.

Medium-range Rocket Launchers

One of the most widely adopted developments of the years since the Second World War has been the medium-range (300 to 3600 m) anti-submarine rocket launcher. Brief details of the French system (**2057.241**) and the Swedish system (**6021.241**), on which the French system is based, are given, and although it differs in some important respects from these systems, the Norwegian Terne system (**6022.241**) is in this general weapon category. The USN has also developed systems of this kind, the current one being known as Weapon Alfa, successor to Weapon Able. Many such developments have also taken place in recent years in the Soviet Union.

Other entries give details of the widely used Swedish Bofors four-tube launcher and the more recent two-tube launcher.

AUSTRALIA

2018.241

ANTI-SUBMARINE MORTAR SYSTEM

An improved version of the AS Mk 10 (Limbo) anti-submarine mortar system (**6008.241**) has been developed in Australia. The design of the Limbo system dates from about 1955 and technological advances since then made it possible to overcome certain disadvantages of the system. In particular, the new development eliminates the use of rotating electric machinery in the pitch and roll servo loops – with consequent saving in power, deck-level weight, and noise – to reduce the weight and noise level of the loading mechanisms (pneumatic for Limbo) and to eliminate the mechanical problems that can arise from the use of uniselectors in the fuze-setting system.

In the new design the metadyne servo control system for pitch and roll has been replaced by an electric system using silicon-controlled rectifiers to control the launcher drive motors; the pneumatic rammer has been replaced by a smaller and lighter electrical device, and the fuze setting system has been redesigned to use solid-state logic circuits with which have been incorporated additional supervisory circuits that permit checking of the setting before launching.

One effect of all these changes, apart from reduction in cost, weight, and noise, is a manning reduction from seven to three.

STATUS

All production equipments have been delivered to the Royal Australian Navy.

CONTRACTORS

Sponsoring Organisation: Department of Defence – Advanced Engineering Laboratory, Salisbury, South Australia, Australia.

Contractor: Hawker Siddeley Electronics Ltd, Brookvale, New South Wales, Australia.

6002.241

IKARA ANTI-SUBMARINE WEAPON SYSTEM

The concept of Ikara is the employment of a guided missile to deliver an anti-submarine homing torpedo to the target submarine. The missile is launched from a surface ship which uses a computer to calculate the torpedo dropping position. Target information from the firing ship's own sonar, or via a remote radio linked source, together with information regarding the in-flight position of the missile and other data such as ship's position, wind, etc are processed by the computer.

The guidance system ensures that the missile flies to the continuously updated optimum dropping position to maximise kill potential.

After release from the Ikara vehicle, a lightweight anti-submarine torpedo such as the Mk 44 or Mk 46 descends by parachute. When the torpedo reaches the sea the parachute is discarded and the torpedo carries out a homing attack on the target submarine.

In 1982 a number of additional optional torpedo payloads was announced; these included the British Stingray, Swedish TP42, Italian A244/S and Japanese Type 73.

The Ikara vehicle is 3.42 m long, has a wing span of 1.52 m and an overall height of 1.57 m, and the main structural strength is provided by the two-stage Murawa combined boost and sustainer rocket motor that powers it. Flight speed is high subsonic and maximum range is estimated to be about 20 km. In previous versions of the system Ikara is launched at a fixed elevation of 55°, generally from a trainable launcher, but in 1982 a 'boxed' version of the system was revealed in which Ikara vehicles are carried in individual container/launchers which are bolted to the ship's deck.

OPERATION

Ikara is capable of attacking enemy submarines out to the maximum range of the ship's sonar regardless of the weather conditions.

Target information from the ship's long-range sonar is fed into a computer which calculates the dropping position, taking into account such factors as ship's own course and speed, wind effect and target movement during time of flight. The outputs from the computer are passed to the missile via the guidance system, a ship-mounted radio/radar system enabling the missile to be tracked and guided accurately to the drop zone where command signals initiate the torpedo release sequence.

The Ikara launcher ensures that the missile takes up its correct flight path as quickly as possible, while the automatic handling system ensures rapid reloading from the magazine, where the missiles are stowed with their torpedoes attached.

The layout of the magazine and handling area varies considerably between the classes of ships already fitted with Ikara, and designs are available to cater for the differing requirements of ships ranging from 1500 tons upwards.

Ikara being launched from Royal Australian Navy destroyer. This system is in service with RAN and RN and a variant (BRANIK) has been installed in four Vosper Mk 10 frigates of the Brazilian Navy

DEVELOPMENT

Initial design was undertaken by the Australian Government. The version to meet RN requirements was subsequently developed in a joint Australian/ British programme.

BRANIK

The requirement to fit Ikara to the Brazilian 'Niteroi' class Mk 10 frigates gave rise to the development of a third version of the system. Known as BRANIK, this version again differs from the Australian and RN versions in the way in which the launcher and missile obtain computer service. In the 'Niteroi' weapon control system two fire control computers (Ferranti FM 1600B) are used to control all the ship's weapons. The BRANIK system employs a special-purpose missile tracking and guidance system which is fully integrated with one of these computers. A lightweight semi-automated missile handling outfit is also incorporated in the new version.

BRANIK was the subject of a joint development programme by the Australian Government, Vosper Thornycroft, British Aerospace Dynamics Group (BAeDG), and Ferranti.

In April 1982 the Australian Minister of Defence announced a programme to improve the system and extend its operational life by reducing size and weight and improving reliability and performance. Current Australian plans relate to development of a system in which Ikara may be used in a containerised form (possibly modified to have folding wings) together with SSMs such as Otomat to comprise a ship's weapons fit.

A modified Ikara could carry as its payload any one of the US Mk 44 or 46, Swedish Type 42, Italian A244/S or UK Stingray lightweight torpedoes, and probably the new Italian lightweight torpedo (A-290) believed to be under development, when this is ready.

STATUS

In service with the Brazilian Navy, Royal Australian Navy, the Royal Navy and the Royal New Zealand Navy. The combined value of orders for the Australian, Brazilian, and British navies is in excess of £100 million. A version known as Super Ikara is being developed by British Aerospace and the Office of Defence Production in Australia and is described under entry **6481.241**.

PRODUCTION

Department of Defence Production, Canberra, Australian Capital Territory, Australia.

CONTRACTORS

British Aerospace plc, Dynamics Division, PO Box 5, Bristol BS12 7QW, England.

Office of Defence Production, Canberra, Australian Capital Territory, Australia.

6481.241

SUPER IKARA ANTI-SUBMARINE WEAPON SYSTEM

Super Ikara is an anti-submarine weapon system being developed by British Aerospace in collaboration with the Government of Australia. The new shipborne system will complement the helicopter and significantly contribute to the philosophy of layered defence.

In operation a box-launched air flight vehicle accurately delivers a torpedo to a predicted target position. The vehicle can carry any one of a range of lightweight torpedoes and engage a submarine over 60 miles (96 kilometres) distant from the point of launch. After the torpedo is released the air flight vehicle continues for some distance beyond the seduction range of the torpedo before diving into the sea. Folding wings enable the vehicle to be launched from compact deck-mounted box launchers that can be readily fitted to existing ships down to the size of fast patrol boats.

A unique feature of Super Ikara is the ability to loiter in the vicinity of the target, and mid-course corrections can be provided if the target changes course. It will be possible to hand over control to a helicopter or other forward platform for extended range engagements enabling maximum use to be made of the latest target data.

With emphasis on a compact, versatile, reliable system, which can also be used on oil rigs or shore-based platforms, Super Ikara is designed to be an effective counter to modern submarine threat.

STATUS

In development.

CONTRACTORS

Office of Defence Production, Canberra, Australian Capital Territory, Australia.

British Aerospace plc, Dynamics Division, PO Box 5, Filton, Bristol BS12 7QW, England.

Trials launch of box-launched, folded-wing Ikara missile as part of the Super Ikara development programme

ATILA.
THE MOST ADVANCED BRAINPOWER.
THE MOST ADVANCED FIREPOWER.

ATILA – Artillery Fire Control, Command and Communication System. From Cimsa Sintra. The people who make available to military commanders ten years of success in the supply of automated firepower.

From a one-gun fire mission all the way to the artillery CP at corps level.

ATILA drastically reduces reaction time and considerably improves the efficiency of fire support.

Another big advantage: with ATILA – the first system in service that has automated the command and control functions – you can maneuver your firepower in total liaison with the combat units you're providing with support.

At Cimsa Sintra – world leader in its field – we can supply every link in the chain of command of ground-to-ground artillery.

Our smart terminals incorporate the very latest advances in information and display technology.

And our advanced operational software adapts the system to any artillery organization and to whatever kind of deployment you favor.

ATILA. A lot more power to you. From Cimsa Sintra.

DETECTION, CONTROL
AND COMMUNICATION SYSTEMS GROUP

DIVISION CIMSA SINTRA

10-12, avenue de l'Europe - B.P. 44
78140 Velizy-Villacoublay - FRANCE
Tel.: (1) 30.67.30.67 - Telex: CIMSA 698529 F

THE BRAINPOWER. THE WILLPOWER. THE WINPOWER.

POLARIS EXPANSION ATILA - 44.

CHILE

4423.441

AS-228 DEPTH CHARGE

The AS-228 depth charge is an anti-submarine weapon with a hydrostatic pressure activated fuze that permits its use against targets at depths from 100 to 1600 feet (30 to 490 m) and the detonation depth can be preset to any one of 19 depths between these limits. The detonator also incorporates three safety measures for handling and transportation, inertia, and submarine action. The charge itself can be launched by conventional methods from naval vessels or from aircraft, including helicopters. The fuze, manufactured by Cardoen Explosivos, is also supplied as a separate unit as a replacement for outdated fuzes in depth charges and bombs of other manufacture.

This company also manufactures underwater hand grenades for use as an anti-diver weapon for the protection of ships moored or anchored in insecure waters. These are operated by hydrostatic fuzes and can be set to explode at depths between 4 and 12 m. Thrown overboard at intervals alongside warships they offer a defence against the attentions of frogmen.

Another product is the Bangalore Torpedo demolition charge, consisting of four pipe charges of light aluminium or steel containing 5 kg of Pentolyte and Mexal explosive.

STATUS

Production.

CONTRACTOR

Cardoen Explosivos, Providencia 2237, 6° Piso, Santiago, Chile.

Bangalore Torpedo demolition charge

Cardoen anti-frogman underwater grenades

Cardoen AS-228 anti-submarine depth charge

CHINA

6760.241

CY-1 ANTI-SUBMARINE MISSILE

The CY-1 single-stage anti-submarine ballistic missile was first revealed in 1987. It is designed to be used by surface ships from multi-cell launchers. No further details are available.

CONTRACTOR

China Precision Machinery Import and Export Corporation, 17 Wenchang Hutong Xidan, Beijing, People's Republic of China.

6761.241

CHINESE MCM SYSTEM

China has developed a small, remotely controlled mine countermeasures system believed to be for use in harbours and estuaries. The system uses a 20.94 m long boat with a displacement of 46.95 tonnes which is controlled by radio signals from a shore station or mother ship up to 3 nm (5.55 km) away.

The boat is powered by a 12V150C diesel engine for transit, giving a maximum speed of 11.6 knots, while for operational use an electric motor is used. Acoustic and magnetic sensors are used but it is not clear how the mines which it locates are destroyed.

CONTRACTOR

China Shipbuilding Trading Company.

6762.441

CHINESE NAVAL MINE UPGRADING SYSTEM

The Chinese Navy, which largely relies upon Russian-designed naval mines, has developed an 'intelligent' sea mine actuation system for what are described as 'large' and 'medium' moored mines.

The system features a programmable central processor which can accept inputs from acoustic and magnetic sensors and, optionally, pressure sensors as well. It incorporates a ship-counter system which can permit up to 15 actuations before detonation, a delay mechanism of up to 250 days before arming and a self-destruction timer for up to 500 days. There are eight operating modes which are believed to be mixtures of fuze and logic settings to meet different operational or environmental conditions.

CONTRACTOR

Dalian Warship Institute.

FRANCE

2057.241

ANTI-SUBMARINE WEAPON SYSTEM

This system comprises an anti-submarine rocket launcher associated with a sonar and a computer. The rocket launcher is remotely controlled, aiming and rocket fuzing being determined by the computer, which in turn receives input data from the sonar.

The launcher is made by Creusot-Loire Industrie under licence from Bofors and is a six-tube device with automatic reloading from a magazine. It fires single rockets or salvoes as required, and will accept any of the range of Bofors 375 mm rockets, thereby giving a choice of ranges from about 655 m to about 3625 m. Rate of fire can be up to one round per second.

The most recent version is that equipping the Aviso 69 class of the French Navy and E71 class escorts of the Belgian Navy.

The computer calculates ballistic data for initial velocities of 100, 130, 165, and 205 m/s for the different rockets that may be used with the systems.

Input data from the sonar comprise the location and rate of change of position of the target.

Creusot-Loire Industrie is licensed by Bofors for the manufacture of the twin-tube rocket launcher (**2368.241**) and the corresponding ammunition. Nine equipments have been installed or are in the course of fabrication.

CONTRACTORS

Launcher: Creusot-Loire Industrie, Division Mécanique Spécialisée, Immeuble Ile de France, Cedex 33, 92070 Paris La Defense, France.

Computer and sonar (for French Navy): Thomson-CSF.

Creusot-Loire Industrie six-tube anti-submarine rocket launcher

1179.241

MALAFON SURFACE TO SUB-SURFACE MISSILE

Malafon is a shipborne weapon consisting of a radio command guided winged vehicle carrying a homing acoustic torpedo. It is intended primarily for use from surface vessels against submarines, but may also be used to attack surface targets.

The Malafon missile has the appearance of a small conventional aircraft with short, unswept tapered wings, and a tailplane fitted with endplate fins. Principal dimensions are: length 6.15 m, diameter 0.65 m, wing span 3.3 m, and the launch weight is 1500 kg. Maximum range is about 13 km.

OPERATION

The missile is ramp launched and propelled by two solid-fuel boosters for the first few seconds of flight. Subsequent flight is unpowered. A radio altimeter is fitted to the missile to maintain a flat trajectory at low level. On reaching the target area, approximately 800 m from the target's estimated position, a tail parachute is deployed to decelerate the missile. The homing torpedo is thus ejected from the remainder of the vehicle and enters the water to complete the terminal guidance phase of the attack by acoustic homing.

Target detection and designation in the case of submerged targets is by means of sonar and by radar in the case of surface targets. These sensors, as appropriate, are used during the flight of the missile to provide data on the target for the generation of command guidance signals which are sent via radio command link to guide the missile. Missile tracking is aided by flares attached to the wing tips.

STATUS

Malafon was deployed in an interim form on French Navy vessels, while full development trials were still in progress, these installations being updated as development continued. These fittings have been referred to as Malafon Mk 1 systems. Deployment of the latest versions includes installations in the guided weapons frigate *Suffren*, five modified T47 class destroyers and five corvettes.

CONTRACTOR

Société Industrielle d'Aviation Latecoere, 79 avenue Marceau, Paris (16e), France.

Malafon anti-submarine missile on its mount on the French destroyer Aconit (Stefan Terzibaschitsch)

1163.441

E14 ACOUSTIC TORPEDO

This is a conventionally-shaped, submarine- or surface-ship-launched, anti-ship (or anti-submarine in certain circumstances) torpedo with a strong body in light alloy and a laminated nose cone, with the following five compartments:

(1) acoustic passive self-guidance and electro-magnetic firing device
(2) explosive charge and impact fuze
(3) secondary battery
(4) air tank and automatic pilot
(5) electric motor for propulsion.

CHARACTERISTICS

Length: 4279 mm
Diameter: 550 mm
Weight: 927 kg
Speed: 25 knots
Range: 5500 m
Explosive charge: 200 kg Tolite
Submersion: Can be set between 6 – 18 m (continuously variable)
Guidance: Acoustic, passive, average range 500 m
Firing: Contact (inertial) and influence (magnetic)
Target: Surface vessel from 0 – 20 knots and submarine at shallow depth

STATUS

Quantity production. The equipment is in service with the French forces and supplied for export. About 100 are believed to have been manufactured.

CONTRACTORS

Programme direction: Direction des Constructions Navales (DCN), 8 boulevard Victor, 75732 Paris Cedex 15, France.

Manufacturer: Etablissement des Constructions et Armes Navales (ECAN) de St Tropez, 83990 Saint Tropez, France.

1164.441

E15 ACOUSTIC TORPEDO

A conventionally-shaped, submarine- or surface-ship-launched, anti-ship (or anti-submarine in certain circumstances) torpedo with a strong body in light alloy and a laminated nose cone, the E15 is a lengthened version of the E14 model 1 torpedo. It has the following five compartments:

(1) acoustic passive self-guidance and electro-magnetic firing device
(2) explosive charge and impact fuze
(3) secondary battery
(4) air tank and automatic pilot
(5) electric motor for propulsion.

CHARACTERISTICS

Length: 5900 mm
Diameter: 550 mm
Weight: 1387 kg
Speed: 25 knots
Range: 12 000 m
Submersion: Can be set between 6 – 18 m (continuously variable)
Guidance: Acoustic, passive, medium range
Explosive charge: 300 kg
Firing: Contact (inertial) and influence (magnetic)
Target: Surface vessel from 0 – 20 knots and submarine at shallow depth

STATUS

Quantity production. The equipment is in service with the French forces and supplied for export.

CONTRACTORS

Programme direction: Direction des Constructions Navales (DCN), 8 boulevard Victor, 75732 Paris Cedex 15, France.

Manufacturer: Etablissement des Constructions et Armes Navales (ECAN) de St Tropez, 83990 Saint Tropez, France.

3623.441

F-17 TORPEDO

The F-17 Mod.2 is the lastest version of the F-17 torpedo family which is in current operational service (F-17 Mod.1 and F-17P). The Mod.2 has been upgraded as a result of experience gained, and in the light of recent technological advances in all technical spheres. The F-17 Mod.2 is a multi-mode wire-guided torpedo, capable of submarine or surface-ship launch, and is intended for attack against both underwater and surface vessels. The torpedo is electrically powered by silver-zinc batteries which are activated automatically at launch.

The F-17 Mod.2 carries an extensive warhead charge which enables it to operate successfully against all naval vessels. The sonar homing system can operate in either passive or active mode to give high detection performance. An internal counter-countermeasures complex makes the F-17 Mod.2 able to attain its target in the most adverse conditions. These performance features, associated with a long range, high speed and deep operating depth, plus the ability to operate with considerable stealth, places the F-17 Mod.2 in the premier rank of heavyweight torpedoes.

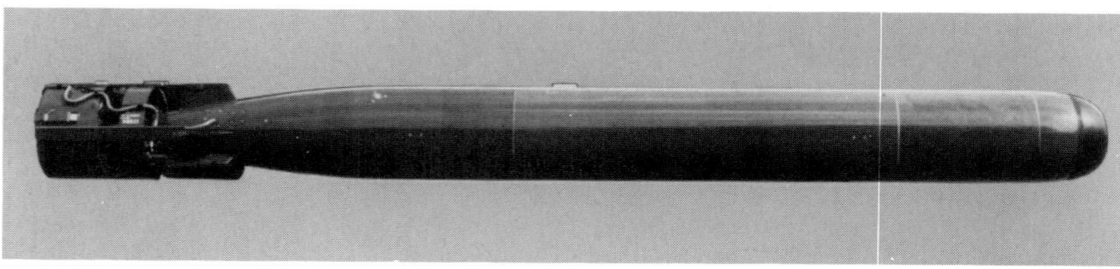

F-17 Mod. 2 torpedo

The F-17 Mod.1 and F-17P can be updated to the Mod.2 standard.

CHARACTERISTICS

Length: 5406 mm
Diameter: 533.4 mm (can be increased to 550 mm if required)
Weight: 1360 kg
Speed: 40 knots
Range: 20 km
Maximum operating depth: 600 m
Target: Surface vessels and submarines
Guidance: Wire and automatic homing

STATUS

F-17 Mod.1 and F-17 Mod.0 are operational with the French Navy and other navies. The F-17 Mod.2 is in production. Spain has signed an agreement with SOFMA to convert 16 E18 torpedoes to F-17 Mod 2 standard.

CONTRACTOR

Direction des Constructions Navales (DCN), 8 boulevard Victor, 75732 Paris Cedex 15, France.

Manufacturer: Etablissement des Constructions et Armes Navales (ECAN) de St Tropez, 83990 Saint Tropez, France.

1165.441

L3 ACOUSTIC TORPEDO

This is a conventionally-shaped, ship-launched or submarine-launched, anti-submarine (to 300 m) torpedo with a strong body in light alloy and a laminated nose cone, with the following five compartments:

(1) acoustic active self-guidance and electro-magnetic firing device
(2) explosive charge and impact fuze
(3) secondary battery
(4) air tank and automatic pilot
(5) electric motor for propulsion.

CHARACTERISTICS

Length: 4300 mm
Diameter: 550 mm
Weight: 910 kg
Speed: 25 knots
Range: 5500 m
Max submersion: 300 m
Explosive charge: 200 kg Tolite A1
Guidance: Acoustic, active, range approx 600 m with favourable inclination of the target submarine. Type AS3T
Firing: Contact (inertial) and proximity (acoustic)
Target: Submarine from 0 – 20 knots and up to 300 m depth

STATUS

Quantity production. The 550 mm diameter version of the weapon is in service with the French forces. A 21 inch (533 mm) version has also been designed and is available for manufacture but not in production. Its performance is the same as that of the 550 mm version, but its length is 4318 mm (170 inches), its weight 900 kg, and its diameter, of course, 21 inches. Approximately 600 have been manufactured.

L3 acoustic torpedo launch

CONTRACTORS

Programme direction: Direction des Constructions Navales (DCN), 8 boulevard Victor, 75732 Paris Cedex 15, France.

Manufacturer: Etablissement des Constructions et Armes Navales (ECAN) de St Tropez, 83990 Saint Tropez, France.

2096.441

L4 AIRBORNE ACOUSTIC TORPEDO

This torpedo is designed to be launched from aircraft or from the ASW missile Malafon (**1179.241**). It is suitable for attacking submarines navigating under water at speeds below 20 knots.

It is a conventionally-shaped torpedo having a body made of removable sections of moulded magnesium alloy and comprising the following main compartments:

(1) head section containing the guidance system, the acoustic firing circuits, and the warhead (which is contained in a removable canister) with its inertial percussive firing mechanism
(2) centre section containing the battery and priming elements
(3) tail section containing the air reservoir, the propulsion unit (electric motor driving two contra-rotating propellers through a reduction gear), and the steering mechanisms.

In addition there are the launching devices designed to insert the torpedo smoothly into the water. These comprise a parachute stabiliser and release mechanism aft and an ejection cap forward.

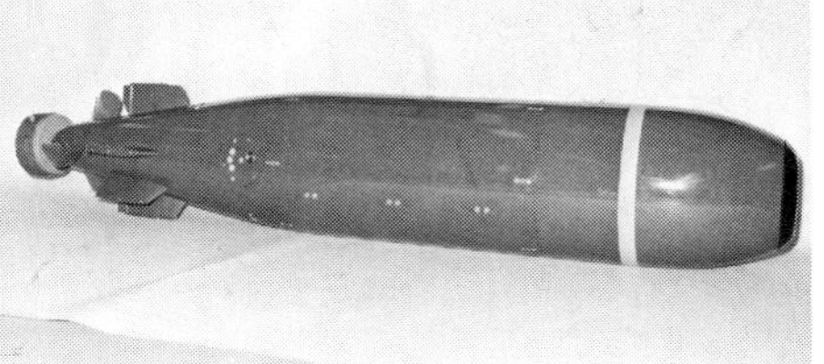

French L4 air-launched torpedo

OPERATION

Once in the water the torpedo describes a circular path until its detection mechanisms locate the target, after which it changes course and homes on the target. On reaching its target the warhead is detonated either by the acoustic proximity mechanism or by an impact fuze.

CHARACTERISTICS

Length: 3.13 m including parachute stabiliser
Diameter: 533 mm
Weight: 525 kg
Warhead: 100 kg HBX
Speed: 30 knots
Guidance: Active acoustic homing
Range: 6000 m
Target: Submerged submarine at up to 20 knots

STATUS

In service with the French Navy. The L4 now exists in a modernised version, capable of operating in shallow water and attacking submarines at depths from periscope depth to very deep cruising. A version suitable for launching from torpedo tubes of surface vessels has also been developed. This differs from its predecessor principally only in length (3.30 m) and mass (570 kg).

CONTRACTOR

Direction des Constructions Navales (DCN), 8 boulevard Victor, 75732 Paris Cedex 15, France.

Manufacturer: Etablissement des Constructions et Armes Navales (ECAN) de St Tropez, 83990 Saint Tropez, France.

2128.441

L5 MULTIPURPOSE TORPEDO

Most recent of the 'L' series of torpedoes, the L5 is powered by silver-zinc batteries which are activated at launch, and provide the power for the motor which drives twin contra-rotating propellors.

There are three models:

L5 torpedo about to enter the water (DCN photo)

L5 Mod.1
L5 Mod.3
L5 Mod.4

The Mod.1 and Mod.3 have been in operational service with the French Navy for a number of years. The Mod.4, derived from the Mod.1, is carried by surface ships of the French Navy and several other navies. All models are designed to attack both underwater or surface vessels. The Mod.3 is submarine-launched and the Mod.1 and Mod.4 are launched by surface ships and submarines. All models are fitted with a passive/active homing head which has various operating modes, including direct attack or programmed search. The warhead consists of 150 kg HE charge which is detonated either on contact or by a electro-magnetic proximity fuze.

CHARACTERISTICS (**L5 Mod.4**)

Length: 4400 mm
Diameter: 533.4 mm (can be increased to 550 mm if required)
Weight: 935 kg
Speed: 35 knots
Range: 7 km
Maximum operating depth: 555 m

STATUS

Operational with the French Navy and other navies. The Spanish Ministry of Defence purchased 9 L5 Mod.4 torpedoes during 1987.

CONTRACTOR

Programme direction: Direction des Constructions Navales (DCN), 8 boulevard Victor, 75732 Paris Cedex 15, France.

Manufacturer: Etablissement des Constructions et Armes Navales (ECAN) de St Tropez, 83990 Saint Tropez, France.

6480.441

MURENE LIGHTWEIGHT TORPEDO

The Murene torpedo is a new generation, lightweight, multipurpose torpedo. It may be launched from aircraft, helicopter, missile or surface ship. It is NATO inter-operable and can be adapted easily to a variety of different platforms. It has been designed with one principal objective, the destruction of the latest generation of nuclear submarines. It is equally effective against other types of submarines, and will eventually replace all current anti-submarine torpedoes in service with the French Navy.

The Murene torpedo is claimed to offer the largest data processing capacity ever installed in a torpedo with a signal processing capacity of 50 mips. This results in a very high degree of analysing power, 12 tracks can be followed simultaneously. The torpedo also offers a very high signal processing capability, and rapid detection at considerable ranges. The combination of these features provides the Murene torpedo with a very high degree of resistance to countermeasures.

Murene can select between one of two speeds; 38 knots during the search and detection phase, and 53 knots for tracking and attack. Its programme controlled CPU provides it with complete autonomy, similar to a 'fire and forget' type missile, throughout the entire firing sequence, and the target search and speed change phases. Endurance is six minutes at maximum speed, or about 12 minutes for a scenario requirement between the two speeds. It has sufficient manoeuvrability to operate at all attitudes, including the vertical, and can attack submarines at all depths from shallow waters of less than 40 m down to 1000 m. The maximum range is said to be 10 km.

The torpedo is electrically driven, with the SAFT motor powered by a SAFT silver-oxide/aluminium battery and is capable of world-wide operation in all waters. Maximum range is understood to be 10 km. The terminal guidance system is such that the target will be attacked at its most vulnerable centre section, in the control room, at an angle of attack which is perpendicular to the hull. It seems most likely that the charge will fire a 'slug' of hot metal through both the outer hull and the inner pressure hull to enable water pressure to complete the destruction.

The SFIM developed central command and control unit, known as Capitole, is located amidships in the torpedo and is designed to guide and control the torpedo towards the target. It incorporates four Motorola 68000 16-bit microprocessors and integrates the functions of attitude determination, automatic flight control and stabilisation, optimal autonomous guidance, target tracking, and automatic programming of the operating modes. The seeker head (acoustic head) has been developed by Thomson Sintra and contains another three 16-bit microprocessors and a 50 mips Mangouste processor. Total capacity is 732 Kwords of 16 bits. The homing system incorporates three separate sonars which use four transducer arrays, two lateral, and one downward pointing each of four transducers, and one forward array with 30 transducers.

Murene lightweight torpedo

DRET are developing a 100 kW AC torpedo motor with infinitely variable speed as well as potentially greater endurance and reliability than DC motors together with lithium thionyl batteries with improved energy storage. It is believed this work relates to further development of Murene.

CHARACTERISTICS

Length: 2.96 m
Diameter: 324 mm
Weight: Approximately 295 kg (major component weights are: seeker 46 kg, charge 59 kg, guidance 13 kg, battery 43 kg, motor 34 kg)
Speed: 38 and 53 knots
Operating range: 6 minutes at 53 knots
Maximum depth: More than 1000 metres
Warhead: 59 kg (shaped)
Propulsion: AgOAl battery with KOH, activated by sea water

STATUS

Approximately six prototypes are under test, with sea trials due to be completed by the end of 1987. A production order will then follow with an in-service date of 1991. The Murene will eventually replace all anti-submarine torpedoes in French service.

CONTRACTORS

Direction des Constructions Navales (DCN), 8 boulevard Victor, 75732 Paris Cedex 15, France.

Manufacturer: Etablissement des Constructions et Armes Navales (ECAN) de St Tropez, 83990 Saint-Tropez, France.

2146.441

Z16 SUBMARINE-LAUNCHED TORPEDO

This is a large submarine-launched anti-ship torpedo. Electrically propelled and free-running, with gyro angle and running depth preset, for target interception, before discharge, it maintains course for a predetermined distance. If no target is encountered the torpedo switches to a zig-zag pattern.

CHARACTERISTICS

Length: 7.2 m
Diameter: 550 mm
Weight: 1700 kg
Speed: 30 knots
Guidance: Preset plus pattern
Range: 10 km
Submersion: To 18 m
Explosive charge: 300 kg
Firing: Magnetic proximity or contact

STATUS

Obsolescent. Only known user is the Pakistan Navy.

3626.441

THOMSON SINTRA SEA MINES

Thomson Sintra is responsible for the production of a variety of sea mines, both war types and for training purposes. The current operational models are as follows:

TSM 3510 (MCC 23)

This is an operational seabed mine designed to be deployed from submarines, and its shape and dimensions are appropriate to its discharge from submarine torpedo tubes (see illustration). Full details of the type of operation employed have not been revealed, but the TSM 3510 probably is of the multiple sensor variety relying upon magnetic, pressure, and acoustic sensors for detection of the target vessel. The sensitivity can be adjusted to suit the depth of the mine and the type of ship(s) intended as targets. The mine is maintained in a passive condition during storage by a locking bar, and is armed prior to insertion into the torpedo tube by the withdrawal of two safety pins.

CHARACTERISTICS

Length: 2368 mm
Diameter: 530 mm
Weight: 850 kg (loaded)

TSM 3530 (MCT 15)

This mine is a seabed defensive mine for deployment from surface ships. It is launched from rails and settles on the bottom after a parachute-retarded descent. The TSM 3530 is armed by clockwork activated time delay, the elapsed time between sowing and arming being preset.

CHARACTERISTICS

Length: 1100 mm
Diameter: 1200 mm
Weight: 1500 kg (loaded)

TSM 3510 (MCC 23) submarine-launched sea mine

STATUS

In service with the French and other navies.

CONTRACTOR

Thomson Sintra Activitiés Sous-Marins, Route de Sainte Anne de Portzic, 29601 Brest Cedex, France.

6763.441

STERNE 1 MINESWEEPING SYSTEM

The Sterne 1 is a towed minesweeping system which simulates both the acoustic and/or magnetic signatures of various types of ship. It has four operating modes; acoustic and magnetic cycle sweeps, multi-influence cyclic sweep and magnetic simulation.

The system comprises the TSM 3800 acoustic sensor and the TSM 3850 magnetic sensor but the TSM 3900 mechanical sweeping device can be integrated with it. On-board equipment includes the Colibri tactical display, the TSM 5730 sonar doppler log and navigation systems such as the Trident III radio location system and Decca 1226 navigation radar.

The system is towed at speeds of 6 to 10 knots at a maximum distance of 100 metres. A mine-avoidance sonar, such as the TSM 5421 or TSM 2022, is integrated into the system to prevent the ship striking moored mines.

STATUS

Under development.

CONTRACTOR

Thomson Sintra, Activitiés Sous-Marins, Route de Sainte Anne de Portzic, 29601 Brest Cedex, France.

4415.441

ECA 38 (PAP 104) MINE DISPOSAL SYSTEM

The ECA 38 (PAP 104) system has been developed as a means of remotely placing a mine disposal charge alongside a mine which has previously been located by a minehunting sonar and identified by the PAP 104 vehicle. The ship equipment consists basically of four main parts: a wire-guided submersible vehicle which carries the destruction charge and is equipped with a TV camera (two vehicles per ship), a control console, a TV monitor, and a remote control box.

Overall dimensions of the submersible vehicle are: length 3 metres, width 1.3 metres, height 1.3 metres, weight 700 kg (including the 100 kg explosive charge). Operating range is 600 metres at depths of 300 metres or less. The maximum speed is 6 knots. Different configurations can be selected according to specific requirements.

The PAP 104 system has been continuously upgraded to meet evolving operational requirements. The latest version, known as the PAP Mk 5, includes the implementation of very powerful electronics combined with the most recent teletransmission devices, and the use of a fibre optic link to reduce further drag of the tether cable, and to increase the range. Vertical thrusters allow mid-water navigation in case of tethered mines. The vehicle can be fitted with two cameras, an identification or relocation sonar, a tracking unit, and a manipulating arm. It also has a very low acoustic and magnetic signature. Cutters can be carried with the charge.

DEVELOPMENT

The system was designed and developed by Société ECA from an original concept of the Direction des Constructions Navales—Groupe d'Études Sous-Marin de l'Atlantique.

STATUS

More than 300 vehicles have been built or ordered for the navies of Australia, Belgium, France, West Germany, Indonesia, Malaysia, Netherlands, Norway, United Kingdom and for the European Tripartite Minehunter.

CONTRACTOR

Société ECA, 17 avenue du Château, BP 16, 92190 Meudon, France.

Preparing to place PAP 104 mine disposal vehicle into the water

6259.421

SM 39 SUBMARINE-LAUNCHED EXOCET

The French Navy employs a submarine-launched version of Exocet, known as the SM 39. The missile is essentially the same as the air-launched AM 39 model (**1770.321**), but for launching from submarines a powered and guided launcher/container, known as the VSM (vehicule sous-marin) has been developed. This is 5.8 m long and can be discharged from a submarine's 533 mm torpedo-tube. In addition to the SM 39 Exocet, with wings folded, it contains a gas generator which powers the VSM in the water and into the atmosphere. The container then breaks up as the missile motor ignites, taking the SM 39 through its in-air phase, towards impact with the target. The weight of the SM 39 in the VSM is approximately 1345 kg.

STATUS

Development of the SM 39 was completed in 1984. Successful launches have been carried out by the diesel-electric submarine *La Praya*, the nuclear-powered submarine *Saphir* and the nuclear-powered ballistic missile submarine *Inflexible* from periscope to maximum launching depth at the Centres d'Essais de la Mediterranée (CEM) or des Langes (CEL). SM 39 is now operational in 15 French submarines and may be sold to Spain for its Agosta type boats.

CONTRACTOR

Aérospatiale, Division Engins Tactiques, 2 rue Béranger, 92322 Châtillon Cedex, France.

GERMANY (FEDERAL REPUBLIC)

2570.441

SURFACE AND UNDERWATER TARGET (SUT) TORPEDO

The SUT torpedo is the latest and most versatile member of the Seal, Seechlange and SST 4 family of torpedoes (**2178.441** and **2000.441**). It is a dual-purpose wire-guided torpedo for engaging both surface and submarine targets. The SUT can be launched from submarines and surface vessels, from fixed locations or mobile shore stations. Its electrical propulsion permits variable speed in accordance with tactical requirements, silent running and wakelessness. The wire guidance gives immunity to interference with a two-way data link between vessel and torpedo. The acoustic homing head has long acquisition ranges and a wide search sector for active and passive operation. After termination of guidance wire, SUT continues operation as a highly intelligent homing torpedo, ie with internal guidance programs for target search, target loss etc. The large payload with combined fuze systems ensures the optimum effect of explosive power. The SUT operates at great depths as well as in very shallow waters. Consort operation permits exploitation of the full over-the-horizon range of the SUT.

CHARACTERISTICS

Length: 615 cm (662 cm with guidance wire casket)
Diameter: 533 mm

STATUS

In service and in production. In early 1986, Indonesia signed a contract for indigenous manufacture of the SUT.

CONTRACTOR

AEG Aktiengesellschaft Marinetechnik, Hafenstrasse 32, 2000 Wedel/Holst, Federal Republic of Germany.

AEG Surface and Underwater Target (SUT) torpedo

2178.441

SEAL/SEESCHLANGE TORPEDOES

These two wire-guided, heavyweight 21 inch (533 mm) torpedoes were developed especially for use in the Federal German Navy; Seal for use against surface ships, and the smaller Seeschlange against submarines. There is a high degree of equipment commonality between the two weapons. Major differences are that the anti-submarine model has half the propulsion battery capacity of the Seal, but is fitted with three-dimensional sonar. The following main features apply to both types.

It is possible to transmit all essential data to the torpedo throughout its run via a dual-core guidance wire. Similarly, actual torpedo running data are simultaneously transmitted to the ship. An active/passive homing head is fitted with a steerable transducer array. Attack options following acquisition are either by manual or computer control from the launch ship, or by self-homing by the torpedo.

Provision is made for a programmed run after guidance-wire pay out or loss of signals from the on-board fire control system. Different programs adapted to various tactical situations are available and can be selected via the guidance wire.

Launch arrangements include compressed air firing from surface ships and swim-out from submarines. There are no limitations on ship movements during the launch and guidance phase. The electric propulsion system employed provides long running distances, permitting launch from beyond target defence area. Torpedo speed is selectable.

Other features are: combined impact and proximity fuze; full performance in shallow or deep water; three-dimensional internal stabilisation; identification of different targets and high hit probability; automatic system check before firing.

CHARACTERISTICS

Seal

Length: 608 cm (655 cm with guidance wire casket)
Diameter: 533 mm
Weight: 1370 kg
Warhead: 260 kg
Firing: Magnetic proximity or impact

Seeschlange

Length: 415 cm (462 cm with guidance wire casket)
Diameter: 533 mm
Warhead: 100 kg
Firing: Magnetic proximity or impact

STATUS

Both torpedoes are in service with the West German Navy aboard Class 206 submarines. Seal is also deployed on surface ships, in particular on the Type 142 and 143 fast patrol boats. Seal has been modified to form the special-surface target torpedo SST 4 (**2000.441**) and the dual-purpose surface and underwater target torpedo SUT (**2570.441**).

CONTRACTOR

AEG Aktiengesellschaft Marinetechnik, Hafenstrasse 32, 2000 Wedel/Holst, Federal Republic of Germany.

4421.441

PINGUIN MINE COUNTERMEASURES VEHICLE

Under a long-term development programme for the Federal German Navy a series of remotely controlled underwater vehicles for mine countermeasures operations has been developed by MBB. Two of these craft have played a significant part in this programme: the first, known as Pinguin A1 was designed by MBB under a pre-development contract awarded by the Government and led to a second type in the shape of the present Pinguin B3. The Pinguin B3 is a remotely controlled, unmanned underwater craft which is a very important element in the complete minehunting system. With the input of target information (derived from minehunting sonars) it is able to travel at any selected distance from the seabed and to be guided almost automatically towards a mine-like object. The Pinguin B3 can then identify the mine, and if necessary, release a mine disposal charge to destroy it. A second charge is carried and the Pinguin craft is intended to continue its mission and proceed to destroy a second mine.

The Pinguin vehicle can be operated freely within the whole operating volume and ranges in the mine-hunting scenario. Equipped with a cutter, it can also counter moored mines. The complete system offers a considerable growth potential for the adoption of alternative sensors, future technology and weapon systems.

CHARACTERISTICS

Vehicle: Pinguin B3
Length: 3.5 m
Hull diameter: 70 cm
Wing span: 1.5 m
Height: 1.4 m
Mass: 1350 kg
Propulsion: 2 variable propulsion electric motors, 1 controllable lift motor
Power supply: Internal battery systems
Diving depth: > 100 m
Speed: (Standard) 6 knots, (max) 8 knots
Endurance: 60/120 minutes
Payload: 240 kg, 2 mine disposal charges, cutter
Sensors: TV camera; sonar

STATUS

In series production.

CONTRACTORS

MBB, Marine and Special Products Division, Hünefeldstrasse 1-5, 2800 Bremen, Federal Republic of Germany.

Messerschmitt-Bölkow-Blohm GmbH (MBB), PO Box 801149, 8000 Munich 80, Federal Republic of Germany.

Pinguin B3 remotely-controlled mine disposal vehicle carries two mine destruction charges and sensors for navigation and identification

---

2000.441

SST 4 SPECIAL SURFACE TARGET TORPEDO

The SST 4 torpedo is comparable in its dimensions, construction, and capabilities to the Seal weapon (**2178.441**) except for those features which can only be applied within the operational area of the West German Navy. In this respect the SST 4 has been adapted to a standard international version.

CHARACTERISTICS

Length: 608 cm (655 cm with guidance wire casket)
Diameter: 533 mm
Guidance: Wire-guided, dual core with two dispensing systems; active/passive homing sonar
Propulsion: Electric
Submersion: To 100 m
Warhead: 260 kg
Fuze: Impact and proximity

STATUS

In service on Class 209 submarines, fast patrol boats of the 'Combattante I and II' and 'Jaguar II' classes, introduced into various NATO and South American navies.

CONTRACTOR

AEG Aktiengesellschaft Marinetechnik, Hafenstrasse 32, 2000 Wedel/Holst, Federal Republic of Germany.

West German class 143 fast patrol boat launches AEG SST 4 torpedo from stern tube

---

3628.241

TROIKA MCM SYSTEM

The problem of sweeping mines without endangering men and material has been solved by the new Troika system. The concept is based on using three unmanned minesweepers which are radar tracked and remotely controlled from a ship-based operations centre, in order to clear specified laps from acoustic and magnetic influence mines.

The self-propelled minesweepers, designated HFG-F1, are equipped with minesweeping facilities which are accommodated in hollow steel cylinders encased by a ship-like hull. The overall design of the minesweepers is characterised by extremely high shock resistance. Its propulsion system consists of a diesel-driven hydraulic power transmision and a combined rudder propeller. For optimum operation, even in rough seas, an autopilot is provided.

The magnetic sweeping field is generated by two coils, mounted forward and aft on the cylinder. For acoustic minesweeping two medium frequency acoustic hammers and one towed low frequency acoustic displacer are provided.

The Troika system as used in the Federal Republic of Germany under the navy designation HL 351 uses a converted coastal minesweeper as carrier for the minesweeping operations centre. The system was defined by AEG-Telefunken which also developed the remote control equipment. The latter comprises a horizon-stabilised X/C-band precision navigation radar, a digital target extractor, a master console and three control displays. Each of the latter is associated with one unmanned minesweeper, on the one hand, and with a remote action system operator console, on the other. The control displays allow the radar image of the swept channel to be greatly magnified so that a high degree of precision is achieved in guiding the unmanned minesweepers along the specified tracks electronically inserted into the radar image.

Control commands are sent to the minesweepers in the form of multiplexed data messages, using a UHF link, which provides high reliability. Their responses and functions are monitored automatically by an integrated monitoring system. The last main feature is the reference buoy which provides a geographically stabilised radar image and constitutes a vital component in this Troika configuration.

STATUS
Delivery since mid-1979.

CONTRACTOR
AEG Aktiengesellschaft, Postfach 1730, 7900 Ulm, Federal Republic of Germany.

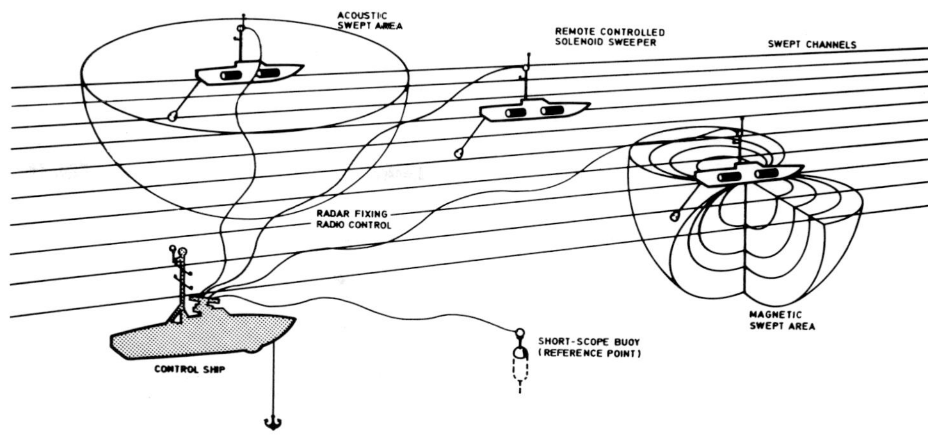

Operating principles of Troika remote minesweeping system

4766.441
FG 1 SEA MINE

The FG 1 is a so-called ground mine, designed to damage or destroy either surface shipping or enemy submarines and warships.

The mine casing is cylindrical configuration, constructed mostly of non-magnetic material, and the mine itself consists of two sections which are combined by means of a clamping ring. The first (smaller) section is the control section which contains the firing mechanism and the safety and arming unit with its ancillary items. The other section contains the explosive charge.

An interface section, located at the bottom of the control section, houses the firing system, which is an electro-pyrotechnic link between the fuze system and the charge. The FG 1 mine is equipped with a proximity fuze. Sensors in the fuze system detect physical fields generated by the hull of a vessel which are then evaluated by the fuze system. If their vessel is within the effective range of the mine, (ie if the signals reach a predetermined level) the charge is detonated.

Depending upon the version, the mine is equipped with various sensors and fuzes. These differences can be established from the external modifications to the mine and the different equipment and sensors attached to it.

Laying can be performed by submarines or surface ships. Prior to laying, the power circuit is connected by inserting a plug into the carrier system. In a live launch, arming of the safety and arming section and water pressure switch is by means of a rope attachment. For surface vessel laying, this function can be performed manually.

CHARACTERISTICS
Length: 2310 mm; 710 mm (control section); 1600 mm (charge section)
Diameter: 534 mm
Total weight: 770.5 kg; 143.5 kg (control); 627 kg (charge)
Charge: 535 kg HE
Service depth: 60 m (max)
Negative buoyancy in water: 291.5 kg

STATUS
Production. For training a recoverable mine, FG 1-Ex, is available.

CONTRACTOR
Faun-Werke, PO Box 8, D-8560 Lauuf a.d. Pegnitz, Federal Republic of Germany.

INTERNATIONAL

6482.441
ADVANCED SEA MINE

A sophisticated new naval mine is to be developed jointly by the United States and the United Kingdom. It is designed to meet the need for a potent and cost effective weapon for use against surface ships and modern submarines in medium depth water. The origins of the project for the Advanced Sea Mine (ASM) lie in a requirement by the Royal Navy for a defensive system to be laid along the Continental Shelf. The US Navy reviewed its requirements and agreed on a joint programme for both development and production.

The first phase of collaborative work included parallel studies of a joint performance requirement by two transatlantic consortia, one led by British Aerospace Dynamics Division with Honeywell and Plessey Marine, and the other led by Marconi Underwater Systems teamed with Loral. This first phase study will be funded by the US Navy. The cost of common elements of subsequent phases of the programme will be shared equally, and will be subject to national approvals at each stage.

The ASM will feature advanced sensors and processors which will identify the target. The target will then be destroyed by a propelled homing device.

STATUS
The two consortia are undertaking competitive project definition, with the winner becoming responsible for complete engineering development leading to acceptance of production standard mines. In-service date is scheduled for the mid-1990s. The programme will be managed by a joint project office under a UK project manager based at Portland, Dorset, UK.

Tenders for project definition were submitted at the end of April 1987. Contracts worth $3.4 million were awarded in September 1987 and full-scale development will begin in 1990 at a probable cost of $80 million. The British Aerospace/Honeywell/ Plessey contender has been named Crusader while the Marconi/Loral contender is Hammerhead.

6484.441
TRAIL BLAZER ROV

Trail Blazer is a remotely operated vehicle (ROV) which has been designed specifically for mine countermeasures. The ROV is linked to its mother ship by an umbilical cable and takes all power and navigation information from this source. The craft is made by Fairey Hydraulics, in conjunction with International Submarine Engineering (ISE) of Canada, which has already sold over 100 of previous model ROVs.

Trail Blazer is capable of six knots on the surface and would normally go out in front of a minesweeper. It is fitted with a forward scan sonar system with a range of about 100 metres, used to locate the mine, and a low-light TV camera for close-in work to place a disposal charge against the mine.

Trail Blazer has five electro-hydraulic propellors, two at the stern for forward and reverse motion, two amidships for sideways movement, and a fifth for up and down movement. It is capable of operating in strong tidal currents down to a depth of 500 metres.

Control of the craft is exercised by an operator on the mother craft, by means of a joystick control which activates and controls the power supplied to the five hydraulic motors.

A derivation of the basic vehicle, known as Mini-Trail Blazer has been produced by ISE, with participation, and eventual production, by Underwater Systems Australia Ltd (USAL).

6764.341
MILAS ANTI-SUBMARINE WEAPON SYSTEM

Milas was first revealed at the Le Bouget Naval exhibition in 1986 and is being developed to meet a joint French and Italian Naval requirement for an air flight vehicle which will be carried by frigates and be capable of carrying a lightweight anti-submarine torpedo.

The air vehicle is derived from Otomat (see **1336.221**) and will have a range of more than 40 km. Milas will be a quick reaction system and target data will be updated in flight. The Milas vehicle will be 6 m long and will weigh 800 kg. French versions will carry the Murene torpedo (**6480.441**) while Italian versions will carry the Whitehead A.244/S (**2004.441**).

STATUS
Under development.

CONTRACTORS
OTO Melara, La Spezia, Italy.
Engins Matra, 78-Vélizy, France.

6765.441
SGM-80 SEA MINE

This ground mine system has been developed under a Memorandum of Understanding of 1986 between Denmark and West Germany concerning the acquisition of mines for use in the Baltic.

SGM-80 (Seegrundmine) is a ground mine designed to defend ports and was developed by Dornier in association with Danish companies. However, the DM560 million production contract was won by Krupp Atlas Elektronik of Bremen who will act as programme managers as well as producing some electronic and electro-mechanical components. The programme costs will be shared with West Germany paying 75 per cent while Denmark pays the remainder. Final assembly will take place in German and Danish naval arsenals.

STATUS

In production with the two governments ordering 2100 mines and 200 practice mines for delivery by 1991. A further contract for 3200 mines and 200 practice mines is likely to follow.

CONTRACTOR

Prime: Krupp Atlas Elektronik, Sebaldsbruker Heerstrasse 235, PO Box 448545, D2800 Bremen 44, Federal Republic of Germany.

ISRAEL

6766.441
ATC-1 TORPEDO DECOY

This Acoustic Torpedo Countermeasures (ATC) system has been developed by Rafael for the Israeli Navy. The ATC-1 has been designed to meet severe operational and environmental requirements while still meeting requirements for a simple effective system which can be installed in a wide range of warships and merchantmen.

The ATC-1 features a towed body 120 cm long and 30 cm in diameter weighing 25 kg together with launch and recovery units weighing a total of 800 kg. Also on board the ship are an electronics cabinet weighing 190 kg and a remote control unit weighing 10 kg, neither of which require much space. The manufacturers claim only a single operator is required for the system.

STATUS

In production.

CONTRACTOR

Rafael, PO Box 2082, Haifa 31021, Israel.

Rafael ATC-1 towed body

ITALY

3624.441
MR-80 SEA MINE

The MR-80 seabed mine is actuated by a combination of influence effects from the target and is able to damage or sink surface vessels and submarines of all types, either conventional or nuclear-powered. Its shape is cylindrical, with the body made of epoxy resin and fibreglass; this renders the mine very resistant to sea corrosion and makes it lighter than previous generation mines. The fore-part contains the explosive charge, while the tail section contains all the actuation, priming and operating devices and it is closed by means of a cover which houses the acoustic magnetic and pressure sensors together with the safety and arming device.

All influence devices are connected to a central unit (the 'Final Logic'), where all delay functions, safety intervals, sterilisation, influence combinations, ship counting, anti-removal and firing control are located.

The sole plate supporting all the electronic modules and the magnetic sensor is linked to the cover by slides that run inside the tail section. By unscrewing the bolts of the crown, the tail cover is easily removed with all the devices linked to it; in this way every part of the mine is accessible for assembly and adjustment operations.

MISAR MR-80 mines at depot

MR-80 seabed mine

CHARACTERISTICS

Type: Seabed mine, actuated by a combination of influence sensors: magnetic, pressure, low-frequency acoustic, audio frequency acoustic
Diameter: 533 mm
Length: Mod A 2750 mm; Mod B 2096 mm; Mod C 1646 mm
Weight: 565 – 1035 kg according to model and type of explosive
Explosive charge: TNT, HBX-3 or similar HE; 380 – 865 kg according to model
Laying: Surface vessels; submarines through torpedo launching tubes; by aircraft
Depth of use: 6 – 300 m for surface targets and for submerged submarines
Power supply: Dry lithium or alkaline-manganese batteries
Selectivity and range: Obtained by adjustments of the discrimination of the signals coming from the sensors
Firing system: Pressure, magnetic, acoustic signals in any combination; modular
Arming and sterilisation delay: Variable up to 999 days. Electronic (quartz) clock
Targets: All surface vessels and submarines
Working temperature: –2.5 to +35°C
Storage life: 30 years with a maintenance mean time of 5 years
Operational life: 500 – 1000 days
Versions: Warhead; inert with data transmission device or smoke signal for exercises

STATUS

In production and in service.

CONTRACTOR

MISAR SpA, ss Goitese, Loc. Fascia d'Oro, 25018 Montichiari (Brescia), Italy.

4863.441
MRP SEA MINE

The MRP advanced general purpose sea mine is a weapon which uses a microprocessor firing device that increases the capability of computation of the system, so achieving the maximum effectiveness and discrimination of the target-countermeasures. Its mechanical, structural and explosive characteristics are the same as used in the MR-80 general purpose ground mine (**3624.441**).

The central programming unit CPU is used to compile the operator settings, mine programming and testing.

MRP advanced general purpose sea mine

CHARACTERISTICS
Type: Advanced general purpose sea mine
Diameter: 533 mm
Length: 2096 mm
Weight: 680-780 kg
Explosive type: HBX-3 or TNT
Charge weight: 520 kg TNT, 620 kg HBX-3
Depth of use: 6-300 m
Firing system: Microprocessor controlled by pressure, ascoustic and magnetic influences. Wide range of selectivity on the threshold, times, combination logistics and localisation logic.

STATUS
In production and in service. MRP customers include India.
CONTRACTOR
MISAR SpA, ss Goitese, Loc. Fascia d'Oro, 25018 Montichiari (Brescia), Italy.

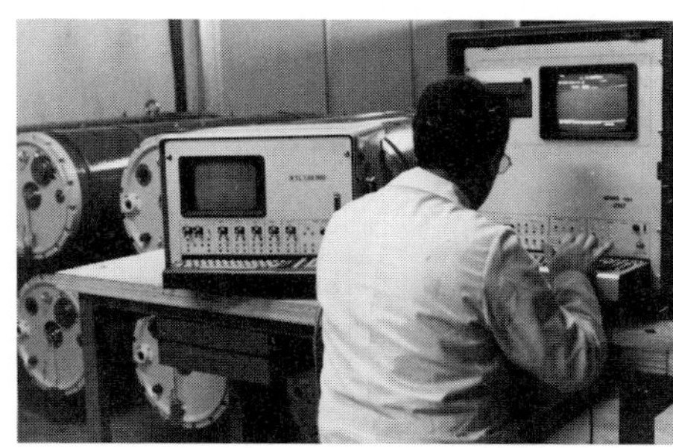

Operator with MRP and UCP central programming unit

4005.441
MANTA MINE

The Manta mine can damage, sink or destroy amphibious and landing craft, small and medium-sized surface vessels, and submarines. It operates at depths between 2.5 and 100 m, and is a bottom dual influence type (acoustic and magnetic). The mine remains effective underwater for more than one year. Its weight and shape are such that it rests firmly on the sea bottom even in running or tidal waters. The mine is 470 mm high and has a diameter of 980 mm. Total weight is about 220 or 240 kg depending on the explosive. The mine consists of two units: the body containing the explosive, and the igniter which comprises all the safety, target detection and firing devices. The mine is equipped with all the safety devices needed for handling and transport. The priming device keeps the detonator away from the explosive until the operating depth is reached so as to prevent any explosion due to a casual ignition.

Electronic circuits include:
(1) Delay circuit: started by the priming device, this enables an actuation delay from 0 to 127 days adjustable by steps of half a day to be set
(2) Pre-alarm device: this detects the noises caused by the passage of the targets and energises the sensing circuit
(3) Sensing circuit: activated by the pre-alarm circuit, this enables the firing circuit
(4) Sterilising circuit: this sterilises the mine after an adjustable preset time.

CHARACTERISTICS
Targets: Landing craft, small and medium-sized surface vessels, submarines
Explosive: 140 kg TNT or 170 kg HBX-3
Laying depth: 2.5 – 100 m
Laying craft: Surface vessels, underwater operators
Diameter: 980 mm
Height: 470 mm
Weight: About 220 kg
Firing system: Pre-alarm and influence firing circuit

MISAR Manta mine

STATUS
In production and in service.
CONTRACTOR
MISAR SpA, ss Goitese, Loc. Fascia d'Oro, 25018 Montichiari (Brescia), Italy.

4006.441
CAM COUNTER-MINE CHARGE

MISAR manufactures this charge for use with automatic mine-clearing systems during minehunting missions. The charge comprises:
(1) explosive charge
(2) igniting device
(3) priming device.

These parts are easily and quickly put together; the charge, with its safety devices, can be positioned fully-assembled aboard the minehunter.

CHARACTERISTICS
Max diameter: 360 mm
Length: 815 mm approx
Total weight: 110 kg
Main charge weight: 80 kg
Max working depth: 150 m

CAM-T Counter-mine Charge
This charge, also produced by MISAR, is similar to the CAM charge, but with some changes to make it more suitable for mine clearance operations carried out by a diving team.

CHARACTERISTICS
Max diameter: 360 mm
Length: 815 mm
Total weight: 110 kg
Main charge weight: 80 kg
Max working depth: 150 m

STATUS
In production.
CONTRACTOR
MISAR SpA, ss Goitese, Loc. Fascia d'Oro, 25018 Montichiari (Brescia), Italy.

MISAR CAM-T counter-mine charge

6272.441
CAP UNDERWATER EXPLOSIVE CHARGE

The Misar CAP charge is particularly suitable for light underwater vehicles for the purpose of detonating mines. The charge is transported by the vehicle by means of a special attachment. When released from the vehicle, hydrostatic pressure activates the system and two detonators automatically advance into a priming position. At the same time a contact is closed completing the sequence for the command to explode. The fire command is activated by dropping an explosive charge into the water at a safe distance.

CHARACTERISTICS
Diameter: 250 mm
Length: 615 mm
Weight: 44 kg (in air)
Main charge weight: 28 kg
Maximum operational depth: 150 m

MDC mine disposal charge
The Misar MDC charge has the same mode of use as the CAP charge

CHARACTERISTICS
Diameter: 360 mm approx
Length: 920 mm approx
Weight: < 120 kg approx
Main charge weight: Not less than 80 kg
Maximum operational depth: 150 m

STATUS
In service.
CONTRACTOR
MISAR SpA, ss Goitese, Loc. Fascia d'Oro, 25018 Montichiari (Brescia), Italy.

CAP underwater explosive charge

2003.441
A 184 SUBMARINE- OR SURFACE-LAUNCHED TORPEDO

The A 184 is a compact dual-purpose, wire-guided, electrically propelled torpedo equipped with an AG 67 panoramic homing head, controlling both course and depth. It is suitable for use against both submarines and surface vessels. It is a dual-speed weapon, carrying a high explosive charge and capable of operating to considerable depths. Dedicated fire control systems exist in different versions for submarines and surface vessels. On-board a number of non-Italian built submarines, the A 184 is controlled by existing fire control systems. The launchers are Whitehead B 512 and B 516 for submarines and surface ships, respectively, although the A 184 may also be used with other launchers for 533 mm torpedoes.

Commands carried by the guidance wires include: course, depth, acoustic mode (active/passive/combined), enabling point, stratum, speed, impact and influence fuze setting, torpedo stop. Replies from the weapon include: course, distance, depth, acoustic mode, speed, other data on interrogation. The fire control system displays the targets, processes their parameters, and computes the torpedo data during its

Type A 184 dual-purpose torpedo

run. This enables the weapon to be wire-guided up to the stage where the target can be spotted acoustically and intercepted.

CHARACTERISTICS

Diameter: 533 mm
Length: 6 m
Weight: 1300 kg
Propulsion: Electric, silver-zinc battery
Warhead: HBX HE
Range: 20 km under wire guidance

STATUS

In production. The A 184 is in service with the Italian Navy and a number of these torpedoes are being sold to a South American country.

Whitehead was one of four companies bidding for a US contract for 2000 torpedoes. Only the Whitehead contender reached the initial evaluation sea trials phase which has been successfully completed. However, the US Navy has not requested funding for the Anti-Surface Warfare (ASUW) torpedo in FY 1988 and FY 1989.

CONTRACTOR

Whitehead, Division of the Gilardini SpA, Via di Levante 48/50, 57100 Livorno, Italy.

2004.441

A 244/S (MOD.0 AND MOD.1) LIGHTWEIGHT TORPEDO

The A 244/S is an electrically propelled lightweight torpedo, suitable for use in normal depth or shallow waters. It is intended for anti-submarine operations and can be launched from surface vessels, helicopters or fixed-wing aircraft.

The torpedo is fitted with an advanced guidance system known as the CIACIO-S. This incorporates a torpedo homing head which is capable of active, passive and mixed modes to close on its target. It is able to discriminate between decoys and real targets in the presence of heavy reverberation by special emitted pulses and signal processing. The head has a large search volume which is covered by multiple preformed beams following a number of self-adaptive search patterns. Other features of the CIACIO-S system are provision of presettable combinations of signal processing, spatial filtering and tactical manoeuvres to match automatically the torpedo performance to the current tactical situation. The latest model (Mod. 1) is characterised by an improved capability against submarines fitted with anechoic coating and computer-based electronics, which makes the overall logics easily reprogrammable whenever so dictated by evolution of threat parameters and ASW tactics.

It has been stated that the A 244/S can be used as the payload of the Ikara anti-submarine weapon system (**6002.241**). Although this has been demonstrated by means of a feasibility study, the programme features, including participants, have not yet been agreed.

CHARACTERISTICS

Diameter: 324 mm
Length: 2.7 m
Propulsion: Electric
Guidance: Active/passive sonar, self-adaptive programmed patterns

Type A 244 lightweight torpedo on helicopter

STATUS

Both the Mod.0 and Mod.1 are in production for various foreign navies.

CONTRACTOR

Whitehead, Division of the Gilardini SpA, Via di Levante 48/50, 57100, Livorno, Italy.

3872.281

B515/M/3 TORPEDO LAUNCHING SYSTEM

The B515 torpedo launching system is designed for the operation of lightweight anti-submarine torpedoes such as the A 244/S, Mk 44, and Mk 46. The system employs two triple-barrel launchers which are pedestal mounted and equipped with thermostatically controlled heaters, and compressed air is used for discharge of the torpedoes from the launch tubes. Launch interlock switches are provided. The system is remotely controlled from a shipborne torpedo control system, also made by the same company, which allows the launcher operator to:

(1) select both the launcher and the individual tube designated for discharge
(2) select the preset data for the torpedo based on information from the combat information centre
(3) initiate the launch command for the designated torpedo.

STATUS

In production for the Italian Navy.

CONTRACTOR

Whitehead, Division of the Gilardini SpA, Via di Levante 48/50, 57100, Livorno, Italy.

Triple-tube launcher of B 515 anti-submarine torpedo system

4412.341

APS SERIES AIRBORNE TORPEDO CONTROL SYSTEMS

These miniaturised systems are designed to allow remote selection, presetting and launch of the A 244/S torpedo, and other existing types of lightweight torpedo, from any type of helicopter or fixed-wing aircraft. The systems are normally supplied in one of two configurations.

Configuration 1: For helicopters fitted with dunking sonar. The torpedo control can be carried out either in the hover or in forward flight. In the former case, data from the dunking sonar are processed by the control panel computer to determine the target course and speed, and the computer then determines the optimum torpedo presetting values which are sent automatically to the selected torpedo. The hit probability is also calculated and displayed on the control panel to assist in decision taking.

Configuration 2: For fixed-wing aircraft and helicopters not fitted with dunking sonars. This configuration is intended for torpedo launch in forward flight rather than both the hover and forward flight conditions.

STATUS

In production.

CONTRACTOR

Whitehead, Division of the Gilardini SpA, Via di Levante 48/50, 57100, Livorno, Italy.

Panel of APS 102 airborne torpedo control system for A 244/S torpedoes

---

4413.441

SPS SERIES SHIPBORNE TORPEDO CONTROL SYSTEMS

This covers a range of modular integrated torpedo control systems produced by Whitehead. The SPS systems allow remote selection, presetting and launch of the A 244/S and other existing types of lightweight anti-submarine torpedoes and can be supplied in different configurations as follows:

Basic configuration: The torpedo remote control is performed by using a control panel installed in the ship's action information room. The control panel computer determines in real time, and continuously updates, the optimum torpedo presetting values which vary with the tactical and displayed situation. The determined optimum presetting values are displayed together with the foreseen hitting probability pertaining to the situation in progress. At the instant of fire, preset data are sent to the torpedo automatically.

Option 1: The control panel is interfaced with the

AIO in this case. When Option 1 is adopted the presetting panel sends data to the AIO which gives it the capability of displaying the ship's recommended launching areas, corresponding respectively to given values of the hit probability.

Option 2: This option is intended for light vessels not fitted with AIO. The remote control panel is directly interfaced with the sonar. It determines and updates the target's course and speed and computes the optimum presetting data which are sent to the selected torpedo. A display is incorporated in the control panel to show the tactical situation, the ship's recommended attack manoeuvre and computed hitting probability.

STATUS

Production.

CONTRACTOR

Whitehead, Division of the Gilardini SpA, Via di Levante 48/50, 57100, Livorno, Italy.

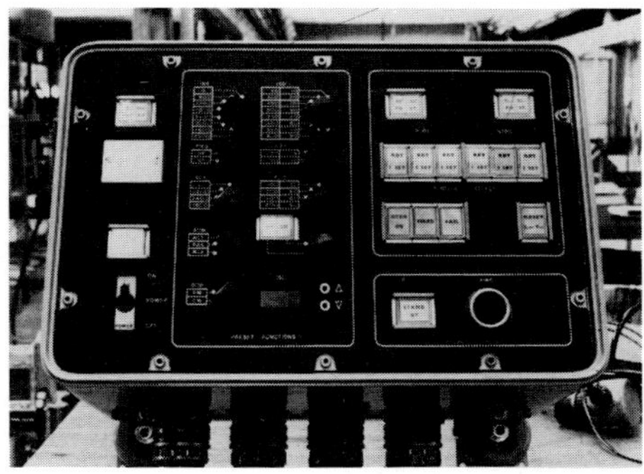

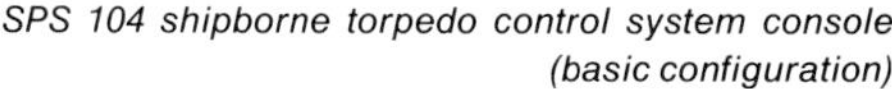

SPS 104 shipborne torpedo control system console (basic configuration)

3873.491

MIN MINE COUNTERMEASURES SYSTEM

The MIN (Mine Identification and Neutralisation) system developed and produced for the Italian Navy by Elettronica San Giorgio (ELSAG) and Riva Calzoni, has the capability of identifying and neutralising both bottom and moored mines. The system consists of:

(1) one self-contained, hydraulically powered, wire-guided submarine vehicle mounting a TV camera and a high definition sonar transceiver
(2) a main console in the minehunter operations room for vehicle guidance and control
(3) one portable auxiliary console for guidance and visual control of the vehicle during launching and recovery operations
(4) operational accessories, such as auxiliary battery charging station, oleo-pneumatic power pack recharging, support system for the vehicle etc.

The vehicle is guided from the ship, via a coaxial cable stored on a reel, through two pairs of thrusters, each consisting of a propeller driven by a hydraulic motor, and a main propeller, fixed to a shaft orientable in the two planes. In the centre of the vehicle there is an oleo-pneumatic power pack to energise all the hydraulic equipment. Below the power pack there is a stowage for the bottom mine destruction charge.

When used against moored mines the vehicle must be equipped with a cutting device carrying an explosive cutter. The vehicle is also provided with a flashing light and an acoustic marker for localisation and recovery; an acoustic transponder to aid manoeuvring of the vehicle from the ship's sonar display; a compass, echo sounder, depth gauge and attitude indication device for supplying information to the main console. This latter is fitted with control levers and a joystick for speed and direction control of the vehicle during the search and approach to the mine.

MIN system underwater vehicle

The TV and the sonar monitors, with their associated control devices on the console, provide the operator with information regarding the underwater scenario.

STATUS

Series production. In service with the Italian Navy.

CONTRACTORS

Consorzio SMIN, Via Panama 52, 00198 Rome, Italy.

Elsag-Elettronica San Giorgio SpA, Via Puccini 2, 16154 Genoa, Italy.

Riva Calyoni SpA.

6262.461

ECP EXPLOSIVE CUTTER DEVICE

The ECP is a specially designed explosive cutter device to be installed on the frontal section of remotely controlled vehicles, and is intended for use in cutting the wire or chain of conventional moored mines. It is capable of cutting wires or chains up to a diameter of 30 mm. Two small guidance conveyors allow the device to be fitted on the wire by the pressure of the sensor against the wire or chain to be cut. At the same time the cutter device is disconnected from the vehicle which can then move back freely leaving the cutter locked in position on the wire. A switch is then activated to run an electral circuit designed to start a time delay from the moment of the vehicle disconnection before the actual cutting operation.

CHARACTERISTICS

Weight: 1.5 kg (in air)
Dimensions: 360 × 220 × 70 mm
Weight of explosive charge: Maximum 150 g
Hydrostatic pressure: More than 150 m

STATUS

In production and in service.

CONTRACTOR

MISAR SpA, ss Goitese, Loc. Fascia d'Oro, 25018 Montichiari (Brescia), Italy.

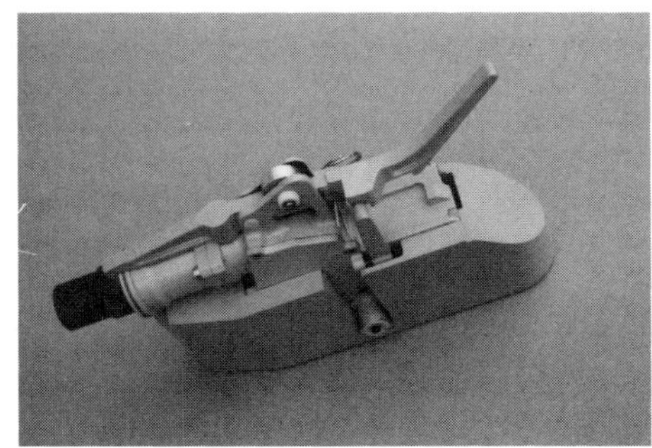

Misar explosive cutter device

NORWAY

6022.241

TERNE III SURFACE TO SUB-SURFACE MISSILE

Terne III is a short-to-medium range anti-submarine weapon system firing rocket-propelled depth charges, in salvoes of six, against submarine targets. It is designed to be the principal anti-submarine weapon for ships up to frigate size, and a secondary weapon for ships carrying long-range anti-submarine armament. The salvo pattern will cover ranges between 400 and 5000 metres, and salvo patterns can be formed to cope with the tactical situation up to the moment of firing. Recoverable practice missiles are available.

The main features of Terne III are:

(a) A unified and balanced design
(b) Low weapons weight and small space requirements. A complete installation weighs less than 10 tonnes.
(c) The weapon part can be interfaced with existing sonars and other target data sources. Norwegian frigates have a new modular fire control unit to which the TERNE III system is interfaced via databus Stanag 4156
(d) Freedom of choice in the position on board
(e) Remote control firing at elevation angles between 47 and 77°, giving variable ranges throughout 360° of bearing
(f) Depth charge is fitted with proximity/time/percussion fuze
(g) High rate of fire.

CHARACTERISTICS
Range: 400 to 5000 m
Explosives: 70 kg TNT equivalent
Length of missile: 2 m
Diameter: 20 cm
Weight: 120 kg
Launcher: Sextuple remote-power-controlled. Automatic loading from manually loaded magazine hopper. Special launcher protection against ice and arctic conditions
Missiles: Length 2 m; diameter 20 cm; Fuzing: proximity, time, and impact

STATUS
In production by Kongsberg since 1960. Fitted in Royal Norwegian Navy ships.

CONTRACTOR
Kongsberg Vapenfabrikk, Postbox 25, 3601 Kongsberg, Norway.

SWEDEN

4862.241
ELMA ASW GRENADE SYSTEM

To meet new threats from intruding small and medium-size submarines into coastal waters the ELMA anti-submarine grenade system has been developed.

The ELMA lightweight ASW system is operational on board all fast attack craft and MCM vessels of the Royal Swedish Navy. The system is the outcome of collaboration between the Navy and an industrial team consisting of FFV for the ammunition, and Saab Missiles for the launcher and electronics.

A ship system comprises four Saab LLS-920 launchers, each with nine tubes, supplementary control electronics and power supply. The tubes and support structure are of non-magnetic stainless steel, and the launchers are mounted on deck with a tube elevation of 30 degrees. Tubes are loaded from the muzzle with the grenades, and from the rear with the propulsive charges. Grenade firing is initiated from the Combat Information Centre in a programmable pattern when a submarine target has been identified and located. A carpet of 36 grenades can be laid over a pinpointed submarine within a range of 350 metres.

The ELMA grenade carries a shaped charge specially designed for good submarine hull penetration qualities. It is stated that the blast jet will make a hole in a boat's pressure hull after passing through the waterfilled superstructure. The grenade has a patented feature that prevents it bouncing or tumbling when striking a hard surface moving through the water.

ELMA ASW grenade system launchers aboard a Swedish FPB

CHARACTERISTICS
System
Number of grenades: 9, 18, 27 or 36 per salvo
Range: 250 – 300 m (approx)
Dispersal: Adjustable
Grenade
Calibre: 100 mm
Length: 278 mm
Weight: 4.35 kg
Equipment weights and dimensions
Launcher: 103 kg, 460 × 920 × 620 mm
Relay unit: 7 kg, 330 × 230 × 110 mm
Firing unit: 6 kg, 330 × 230 × 148 mm
Power supply: 34 kg, 610 × 315 × 210 mm

STATUS
Operational in Sweden since 1984.

CONTRACTORS
Saab Missiles AB, S-58188 Linköping, Sweden – LLS-920 launchers.

FFV Ordnance, S-63187 Eskilstuna, Sweden – Grenades.

6021.241
TYPE 375 ANTI-SUBMARINE WEAPON SYSTEM

This is a shipborne surface to sub-surface short and medium-range anti-submarine warfare rocket system.

OPERATION
Ship's sonar provides submarine position for prediction of Bofors 375 launcher elevation and bearing data. The launcher has either two or four tubes and can fire single- or multiple-shot salvoes. A special design ensures a predictable and accurate underwater trajectory. The launcher is reloaded by automatic means from the magazine which is disposed directly below the launcher.

Rockets have two types of motor giving differing range brackets. The rocket trajectory is flat thus giving a short time of flight to minimise target evasive action. Fuzes are fitted with proximity, time, and DA devices.

DEVELOPMENT
The initial version of the system, comprising the four-tube launcher and the former M/50 rocket, was developed in the early 1950s, and became operational about 1956. The two-tube launcher was developed in 1969-72 and the Nelli rocket in 1969-73. A version made in France by Creusot-Loire under licence from Bofors is described in **2057.241** above.

The 80 mm MI type is launched from a special holder that is handled as a live rocket in the operating room, hoist and launcher.

STATUS
The twin-tube launcher (**2368.241**) and anti-submarine rockets are in production for a number of navies.

CONTRACTORS
Launcher and missile: AB Bofors, 691 80 Bofors, Sweden.

Predictor system: Various: (eg Philips Elektronik AB, Sweden; Plessey and Ferranti, England; and NV Hollandse Signaalapparaten, Netherlands).

CHARACTERISTICS

375mm ASW	Weight (kg) with fuze	w/out fuze	Length (mm) time fuze	proximity fuze	Charge	Max velocity (m/s) 1 motor	2 motors	Final sinking speed (m/s)	Range to 30 m depth (m)
Erika	250	236	2000	2050	107 kg hexotonal	100	130	10.7	655 – 1635
Nelli	230	216	2000	2050	80 kg hexotonal	165	205	9.2	1580 – 3625
Practice									
Type H	100	86.5			—	135	—		980 – 1570
80 mm MI 15.5 (see note under development)	—	—	1120 (indication fuze)	—	—	130	—	—	980 – 1600

2369.241
SHIPBORNE FOUR-TUBE ANTI-SUBMARINE ROCKET LAUNCHER

The Bofors 375 mm four-tube anti-submarine rocket launcher is electro-hydraulically driven, has an integral fixed-structure loading hoist with a rotatable rocket table, and is remotely controlled. For further details see previous editions of *Jane's Weapon Systems*.

STATUS
Design was started in 1948 and the launcher first went into service with the Royal Netherlands Navy, which had inititated the work, in 1954. Since then it has been supplied to Argentina, Colombia, France, Federal Republic of Germany, Japan, Portugal and Sweden.

No longer in production but believed to be still in service with a number of navies. Superseded by the two-tube launcher.

CONTRACTOR
AB Bofors, 691 80 Bofors, Sweden.

2368.241

SHIPBORNE TWO-TUBE ANTI-SUBMARINE ROCKET LAUNCHER

The Bofors ASW rocket launcher is used for firing anti-submarine rockets at close and medium ranges. The launcher has an integral motor-driven twin hoist for loading both tubes at once, and a rotating loading table. The laying mechanism is electro-hydraulic with choice of local or remote operation. Fuze setting is with the rocket in the tube. The rocket table holds four rounds and the total number of rockets in the operating room is 24. The shortest time between successive firings is one second; the time for firing six ready rockets (four on the table and two in the tubes) is one minute and the firing rate for continuous fire is two rockets every 45 seconds. The manning complement during continuous firing is three men.

CHARACTERISTICS

Weight, excluding rockets but including flame guard and deck plate: 3.8 t
Traversing speed: 30°/s
Elevating speed: 27°/s
Traverse limits: Unlimited
Elevation limits: (Mechanical) 0 to +90°; (for firing) 0 to +60°
Power supplies: 440 V, 60 Hz, 3-phase
Power consumption: Tracking 6 kW (mean)

STATUS

In production.

CONTRACTOR

AB Bofors, 691 80 Bofors, Sweden.

Bofors twin-tube ASW rocket launcher

2474.441

TORPEDO TYPE TP 43XO

The FFV Tp 43XO is a homing, wire-guided multi-purpose torpedo for launching from submarines, surface ships and helicopters against surface and underwater targets. It is based on the well-proven FFV Tp 42 and retains the electrical propulsion system with a silver-zinc oxide primary or secondary battery.

FFV Tp 43XO lightweight torpedo

The Tp 43XO is designed around a multiprocessor computer system using Pascal language which gives flexibility to meet changing requirements. The homing system is optimised for use against very silent submarines in shallow water because the acoustic detector distinguishes between target and disruptive noise. The combination of homing and wire-guidance is claimed to result in very high hit probability even at long firing ranges, the torpedo having three speeds. The war torpedo is equipped with both an impact and a computerised proximity fuze. Both fuzes can be controlled from the launcher fire control system. The proximity fuze is tested on every practice run.

The guidance wire communication link permits the launching craft to transmit orders to the torpedo controlling speed, depth, course, target data, etc. The torpedo reports its position, speed, course and depth, homing systems parameters, and target noise. In the event of communications being interrupted, the torpedo computer calculates the expected target position, guides the torpedo to the predicted point of impact and initiates one of the several possible search patterns.

Modular design makes it easy to configure the Tp 43XO for special purposes, eg to accommodate special warheads or two battery sizes. Built-in test equipment is incorporated and computerised test equipment is used in maintenance and preparation. The test equipment is also used for evaluating practice runs. Recorded torpedo data can be presented together with recorded data from the ship fire control system.

CHARACTERISTICS

Length: 2645 mm
Diameter: 400 mm
Weight: 280/310 kg (min/max) including battery and guidance wire
Propulsion: Secondary Ag/ZnO battery 4.2/5.6 kWh (primary battery optional); 3-speed thyristor battery switching unit; electric DC motor with gearbox

STATUS

In production for the Royal Swedish Navy. An order worth SKr 120 million for the Tp 431 torpedo was given by the Swedish Navy in February 1986.

CONTRACTOR

FFV, Ordnance, 631 87 Eskilstuna, Sweden.

2367.441

TORPEDO TYPE TP 617

The FFV Tp 617 is a homing, wire-guided, long-range torpedo for launching from submarines and surface ships against surface targets. It is based on the well-proven FFV Tp 61 which entered service 20 years ago and retains the thermal propulsion system with hydrogen peroxide/alcohol/water as propellants.

The Tp 617 is designed around a programmable digital computer which controls the homing system, communication with the launching fire control system and the torpedo navigation system. The homing system is designed for use against surface targets and has demonstrated excellent acquisition capabilities against a broad range of targets. The combination of homing and wire-guidance results in very high hit probability, even at long firing ranges.

The war torpedo is equipped with both an impact and a computerised proximity fuze. Both fuzes can be controlled from the launching fire control. The proximity fuze is tested on every practice run.

The guidance wire communication link allows the ship to transmit orders to the torpedo controlling speed, depth, course, target data, etc. The torpedo reports its position, speed, course and depth, homing system parameters, and target noise. In the event of communications being interrupted, the torpedo computer calculates the expected target position, guides the torpedo to the predicted point and initiates one of several possible search patterns.

Built-in test facilities are incorporated and computerised test equipment facilitates maintenance and preparation. This test equipment is also used for evaluation practice runs. Recorded torpedo data can be presented together with recorded data from the launching fire control system.

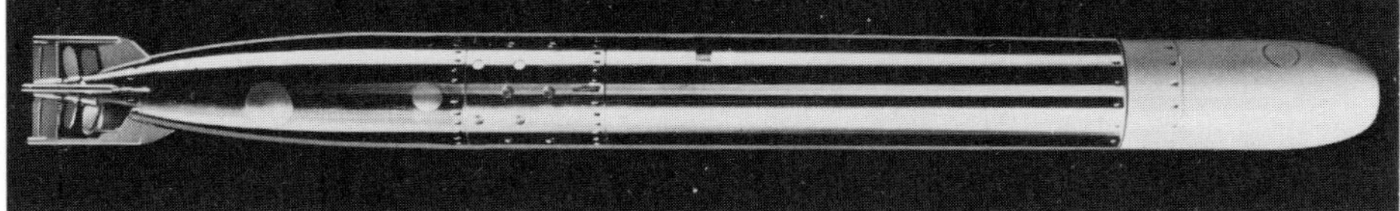

TP 617, latest version of Swedish Type 61 long-range torpedo

CHARACTERISTICS

Length: 6980 mm
Diameter: 533 mm
Weight ready to launch: 1860 kg
Weight of explosive payload: 240 kg approx
Propulsion system: Energy produced by combusting alcohol with hydrogen peroxide as an oxidiser. Two speed valve steam engine with gearbox

STATUS

The Tp 617 has been in operational service with the Royal Swedish Navy since early 1984 and is in service with some other Scandanavian navies.

CONTRACTOR

FFV, Ordnance, 631 87 Eskilstuna, Sweden.

4883.441

BOFORS SEA MINES

Current sea mine production by the Swedish Bofors organisation is concentrated on two types of influence mine introduced in the latter part of 1984: the Rockan GMI 100 ground mine, and the MMI 80 moored mine.

GMI 100 Rockan: This 190 kg weapon is an influence ground mine which is mainly intended for use as an anti-invasion deterrent against small and medium-tonnage vessels in coastal waters, inlets, and confined waters such as harbours. It may also be deployed in deeper waters against submarines to protect friendly routes and bases. A feature of the design is the shape which enables mine laying over a wide area while the minelayer covers the minimum distance. This is achieved by shaping the mine in a casing that enables it to plane, or glide, in the water for a distance of up to twice the depth of the water it is

After resting on the seabed for a time the Bofors Rockan ground mine becomes very difficult to detect both visually and by mine-hunting sonar

being laid in. This also enables mines to be sown in pairs from each side of the ship, or for mines to be spread out directly from the quay.

Another advantage of this shape is its low profile which makes it difficult to detect by mine-hunting sonar and other anti-mine techniques. The mine is constructed of an outer shell of reinforced plastic containing an explosive charge, a fuze with arming unit, and a sensor/electronic unit. The outer shell has sliding runners to enable the mine to be moved on a mine rail, orientated either along or across the rails. The shell can be opened easily for maintenance or for adjustments etc.

CHARACTERISTICS
Length: 1015 mm
Width: 800 mm
Height: 385 mm
Weight: 190 kg
Laying depth: 5 – 100 m
Glide coeffcient: 2 (approx)
Glide speed: 2 m/s (approx)
Weight of explosive: 105 kg
Minimum distance between mines: 25 m

MMI 80 Moored Mine: This is an influence mine for use against small and medium-tonnage surface ships and submarines. It is suitable for laying in water 20 to 200 metres deep. The mine itself consists of a buoyancy section of cellular plastic, which also contains the explosive charge, the fuze and the sensor/electronics package (which is similar to that of the Rockan, but adapted to the moored mine role).

The MMI 80 is programmed to anchor itself automatically at the desired depth which is set on a pressure sensitive device before sinking. A Manostat 80 sinker unit is employed with the MMI 80, and the mine can be laid from rails.

CHARACTERISTICS
Length: 1125 mm
Width: 660 mm
Height: 1125 mm
Total weight: 450 kg
Sinker weight: 240 kg
Weight of explosive: 80 kg
Bottom depth range: 20 – 200 m
Mine depth range: 8 – 75 m
Minimum distance between mines: 25 m

STATUS
Production.

CONTRACTOR
AB Bofors 691 80 Bofors, Sweden.

UNION OF SOVIET SOCIALIST REPUBLICS

3968.241

FRAS-1 ANTI-SUBMARINE WEAPON

FRAS-1 (free-rocket anti-submarine) is the designation that has been assigned to the anti-submarine missile associated with the forward twin launcher (SUW-N-1) in 'Moskva' class helicopter carriers and 'Kiev' class aircraft carriers. FRAS-1 is reported to have a maximum range of up to 30 km and to consist of an Asroc-type weapon (**6001.241**) in which a rocket is used to project a nuclear depth charge to the neighbourhood of the suspected target, as determined by sonar and other means, after which the rocket separates from the charge which sinks to a suitable depth before being detonated by either a straightforward pressure/depth fuze or some type of proximity fuze. It has been conjectured that the probable yield of the nuclear payload is in the region of five kilotons.

Provisional dimensions are: length 6.2 m and diameter 55 cm.

Armament of 'Kiev' class carriers includes RBU-6000 ASW rocket launchers, SUW-N-1 twin launch ramp for long-range ASW weapons, eight SS-N-12 SSMs, twin launch ramp for SA-N-3 air-to-surface missiles, and twin 76 mm DP guns (G. Jacobs)

3969.241

SS-N-14 ANTI-SUBMARINE WEAPON (SILEX)

The SS-N-14 (Silex) is thought to be somewhat similar in concept to the French Malafon (**1179.241**) or the Australian Ikara (**6002.241**) anti-submarine weapons in which a subsonic winged vehicle carries a homing torpedo to the position of the target submarine and can have its course corrected during flight so as to allow for submarine movements and make an accurate drop. In the Soviet case, however, it is believed that a nuclear warhead can be carried instead of the torpedo and has a weight of about 100 kg and a yield in the low kiloton range. No confirmed photographs of the SS-N-14 have been released for general publication, and indeed, for many years, it was thought to be an anti-ship missile system rather than an anti-submarine system. The evidence which has caused this change of assessment has not been released but is believed to be conclusive.

It is also believed possible that the SS-N-14 has an anti-ship capability and may be able to drop a homing torpedo outside a ship's normal defensive cover. No confirmation of this possibility has ever been seen.

CHARACTERISTICS
Length: 7.6 m
Max range: 55 km
Payload: Nuclear warhead or homing torpedo

'Krivak I'-class frigates have a quad SS-N-14 launcher and two forward RBU 6000 rocket launchers for ASW engagements (G. Jacobs)

Trajectory: Programmed/radio command flight to target area under autopilot control with command override capability. Speed is about Mach 0.95 at 750 m above sea

STATUS
Operational since 1968 and fitted in 'Kirov', 'Krivak I', 'Kresta II' and 'Kara' class ships. Later Soviet classes noted with this weapon system include the 'Udaloy' class of anti-submarine destroyers which are fitted with two quadruple launchers.

3970.441

SS-N-15 (STARFISH)/16 (STALLION) ANTI-SUBMARINE WEAPON

The SS-N-15 is reported to be an anti-submarine weapon of the same general type as the American Subroc (**1128.441**), a system in which a submarine-launched missile follows a short underwater path before broaching the surface to follow an airborne trajectory for the major part of the distance to the target area. On reaching the latter, a depth charge bomb is released to continue on a ballistic trajectory until it enters the water near the target, sinking to the optimum depth before detonation. This type of weapon relies upon accurate localisation of the target in the first instance, followed by rapid launch and flight to the target area before the target has time to travel far from its last known position. Maximum range was originally estimated as about 35 km but later official US figures suggest between 45 and 50 km. The same source also confirms the torpedo-tipped version of this weapon which carries a homing torpedo payload instead of the depth charge. In the homing torpedo carrying version the weapon is designated SS-N-16.

STATUS
The SS-N-15 is probably in 'Papa', 'Alfa', 'Victor III', and 'Tango' class Soviet submarines. Other classes of submarine which may be equipped to operate this weapon include some 'Charlie II' class and 'Victor I or II' boats.

4422.241

SOVIET ANTI-SUBMARINE ROCKET LAUNCHERS

There is a wide variety of anti-submarine rocket launchers in the Soviet fleet although all appear to operate on much the same principle of firing charges in a pattern ahead of an attacking ship from a mounting which can be trained and elevated under remote control. Two different calibres of 250 mm and 300 mm have been reported and the weight of the bombs in the 250 mm versions has been estimated as 180 to 200 kg. Few details are available of the many different systems but such as are known are given below. The systems may be designated as either RBU or MBU and the associated number is understood to mean the maximum range possible in metres.

Six-barrell, rocket launcher (USN photo)

MBU 1800: A five-barrelled 250 mm calibre system fitted in some of the smaller and older ships and now probably obsolescent. The launching tubes are in two horizontal rows with three in the top row and two in the bottom row.

MBU 2500: A sixteen-barrelled 250 mm calibre system of much the same vintage and design as the MBU 1800 with two horizontal rows each of eight launching tubes about 1.6 m long. Depth bombs are loaded by hand. The system is fitted in the older cruisers and destroyers although also in the more modern but smaller 'Petya I'.

MBU 4500A: A six-barrelled version hitherto seen only in the one major fleet support ship 'Berezina'. It is thought to be of 300 mm calibre with a launching tube 1.5 m long and to use automatic reloading.

MBU 6000: The most modern system, probably of 300 mm calibre, and fitted very widely. Launch tubes are arranged in circular fashion and it is believed that they are reloaded automatically by bringing them to the vertical and then indexing one by one while depth bombs are loaded from below. Range is 6000 m and the launch barrel length has been extended to about 1.8 m.

MBU 1200: A six-barrelled system which is fitted normally on the quarters of larger ships such as the 'Kirov', 'Kara' and 'Kresta'. Almost certainly it has a range of no more than 1200 m and is probably intended to deal with torpedoes if they can be detected in the closing stages of an attacking run.

STATUS

The MBU 6000 and 1200 are in widespread use throughout the Soviet fleet with the MBU 1200 being reserved for the larger ships.

2995.441

533 mm TORPEDO

Certainly until recently, and possibly on a continuing basis, the standard calibre of torpedo in service in the Soviet Navy has been 533 mm. A torpedo of smaller calibre was in use on torpedo boats and escort vessels before the Second World War, but this is now to be found only on P-4 patrol boats. The standard 533 mm torpedo of the Second World War, which is still likely to be in service, was the Type 3 which was 7.27 m long, weighed 1610 kg (dry) and had a maximum range of 12 km. Maximum speed was 46 knots and the torpedo had a 300 kg TNT/tetryl warhead.

It may confidently be assumed that torpedo development in the USSR has generally proceeded along lines similar to those of developments in Britain, France, and the USA and has probably taken the form of development from wartime German pattern-running and homing torpedoes.

Stern view of 'Kara' class cruiser Petropavlovsk *shows hangar and alighting area for 'Hormone' ASW helicopter, SA-N-3 missile launcher and associated Headlight radar group, and port and starboard 5-tube 533 mm torpedo launchers* (G. Jacobs)

It is, however, worth noting that somewhere around the late 1950s a version of this torpedo with a nuclear warhead was deployed in some submarines, and this was confirmed in 1981 when a Soviet 'Whiskey' class submarine ran aground in Swedish waters. The Soviets would regard it as normal practice to fit alternative nuclear warheads to their torpedoes. Swedish sources estimate a warhead yield of about 15 kilotons.

Two distinct stages have been distinguished in the development of surface vessel launchers for torpedoes of this calibre, and the fact that the launchers are of slightly different lengths at these two stages may indicate that the corresponding torpedoes are also different – but it does not necessarily do so. One 533 mm torpedo believed to be the 1957 pattern, is said to be 825 cm long – which is as large as torpedoes go. It is carried by submarines and torpedo boats. Both surface target and ASW models exist in the Soviet range of 533 mm diameter torpedoes.

The dates of introduction of these two launcher groups appear to have been about 1948/49 and 1957/58. Apart from the difference in length just mentioned the principal change appears to be from manual local training to remote power training. One-, three-, and five-tube launchers are found in the earlier series; two-, three-, four- and five-tube launchers in the later.

STATUS

All the above types of launcher seem still to be in service, as also, presumably are the torpedoes. 533 mm torpedoes are also a standard weapon on Soviet submarines, and the 'Foxtrot' and 'Whiskey' class boats are equipped with four and two 533 mm stern torpedo tubes, respectively.

Other countries operating Soviet 533 mm torpedoes include: Albania, Bulgaria, Chinese People's Republic, Egypt, Finland, German Democratic Republic, India, Indonesia, Iraq, North Korea, Libya, Poland, Romania, Somalia, Syria, Vietnam and Yugoslavia.

6274.441

TYPE 65 660 mm SOVIET TORPEDO

A number of reports have recently indicated that a large diameter torpedo is currently entering service with Soviet submarines. The main data is understood to be a length of some 9144 mm with a diameter 660 mm. The propulsion is given as a closed cycle thermal system or MHD giving a range of 27 nm at 50 knots, and 54 nm at 30 knots. No information is available on the warhead, or on the guidance system, although the latter is likely to be wire guidance with an autonomous final stage.

It is stated that an extensive modification programme is under way to provide the larger torpedo tubes needed for the larger diameter weapon.

STATUS

It has been reported that the torpedo is in the early stages of production. No other information is available.

2997.441

LIGHT TORPEDO

Of comparatively recent development, a 406 mm torpedo is now in service with some submarine chasers and light destroyers of the Soviet Navy. This is launched from trainable tubes that are only about five metres long, so the torpedo must be quite a small one. These torpedoes are believed to be similar in concept to the American Mk 44 and Mk 46 torpedoes (see later in this section) and are principally anti-submarine weapons. They are deployed aboard 'Hotel II and III' class ballistic missile submarines, 'Echo I' class nuclear attack submarines, 'Echo II' class missile submarines, and 'November' class boats, and are fired from special stern countermeasures tubes, similar to those found in British 'Oberon' and 'Porpoise' class submarines. Quintuple mounts are fitted in 'Petya 1' class frigates, and 'Mirka' class frigates. Fixed mounts exist on a number of small patrol classes. Bulgaria is known to operate Soviet 406 mm torpedoes.

3627.441

SOVIET SEA MINES

Soviet sea mines in current use fall into four major categories: moored, seabed, floating, and nuclear, and brief notes on each category appear in the following paragraphs. It is estimated that some 350,000 sea mines, mostly of the latest types, have been stockpiled by the USSR.

Defensive moored mines: The main type is the basic contact moored mine that uses an inertia-type firing mechanism which can either be galvano-contact, contact-mechanical or contact-electrical. The mines come in three sizes: small, medium and large according to the explosive charge carried. The two smallest mines are designated YaRM and YaM. The former has an explosive charge of 3 kg and is used in rivers and lakes; the latter has an explosive charge of 20 kg and is used in lakes and shallow coastal areas. Both have the conventional spherical shape with horns and small sinker units. The medium-sized moored contact mine is believed to be confined to the large stocks of the elderly M08/39 series used for coastal defence barriers. This also is spherical with horns and sinker unit but has an explosive charge of about 120 kg. The largest moored contact mine types are cylindrical with explosive charges in excess of 200 kg. These mines are matched by large stocks of an antenna mine variant for defensive ASW barriers and an acoustic influence activated variant for use in areas of strong tidal or current action. The antenna mine is believed to have a 30 m antenna above and below the mine casing and the acoustic mine has a 30 m target location distance. All three types of moored mine have apparently been exported to Soviet client states and the Warsaw Pact nations.

Offensive bottom mines: There are two basic types of conventional ground influence mines used by the USSR: the AMD-500 and AMD-1000 series. The number refers to their weight in kilograms. The AMD-1000 is reported to be the only one capable of laying by submarine, whereas both the AMD-1000 and AMD-500 can be laid by ship and aircraft. In the last case the designation is reported by Middle Eastern sources to change to KMD-1000 and KMD-500, respectively. The two series are produced in four variants:

(1) magnetic influence that relies on either the intensity of the horizontal or vertical component of the target's magnetic field or the rate of change of the target's field
(2) acoustic influence using either or both low frequency and high frequency noise generated by the target
(3) pressure influence in which the passage of the target over or near the mine causes a reduction in pressure within the water column adjacent to the mine
(4) combination influence in which two or all three of the above influences are combined in a single sensor unit.

For greatest selectivity in target and to maximise sweeping difficulties the combination systems most likely to be used are magnetic-pressure and acoustic-pressure; although the presence of mines with all three influences would be logical for specialist targets such as NATO MCMV forces.

Normal laying depth would be between 4 and 70 m, although the larger AMD-1000 would be effective against submarines down to 200 m. It is assumed that both types can be fitted with ship counter units, anti-handling devices and self-timed neutralisation devices. Both types have been exported widely to Soviet client states and the Warsaw Pact countries. However the pressure influence variants are likely to be restricted to the Soviet Navy and Pact states that are considered trustworthy.

Offensive moored ASW mines: There are three basic types, two of which for delivery by submarine, ship or aircraft are classed as rising mines and the third assessed for delivery by submarine or ship is known as the underwater electrical potential (UEP) mine. All three have their origins as strategic ASW barrier mine types for use in areas adjacent to NATO submarine bases. The rising mines can also be expected to be encountered on transit and choke point zones frequented by NATO submarine and surface units, as they have a secondary anti-ship capability. Of the two rising mine types one is designed for use on the continental shelf (believed to have the NATO code-name Cluster Bay) and the other is an improved version for use on the deeper continental ledge region (believed to be code-named Cluster Gulf). Both are thought to be tethered torpedo-shaped devices fitted with a rocket propulsion unit and an active/passive acoustic sensor device. The target is initially detected by the passive component of the detection system and located by transmissions from the active part. If the target is confirmed as being within the vertical attack zone, the tether is cut and the rocket ignited. The very fast upward speed will allow very little time for the target to evade the device if its launch has been detected. Nothing much is known about the UEP mine other than that it relies on the target's electrical field for detection. It is the presence of these three mines in the Soviet Navy inventory that has prompted the RN to introduce the deep sweeping capability into its MCMV forces to protect the UK's submarine bases.

Nuclear mines: The Soviet Union is believed to have a small stockpile of nuclear mines with yields varying between 5 and 20 kilotons for use against high value surface units, and base targets. Laying of these mines is almost certainly assigned to specially selected SSK/SSN units.

The primary Soviet offensive minelaying platform is the 'Foxtrot', 'Whiskey', 'Tango' and 'Kilo' SSK force because of their covert laying capability. Offensive ASW and ground influence mines would be laid by these boats round European NATO bases and at choke points, while overseas bases, SLOCs and deep water choke points would be assigned to the SSN force. Reactive minelaying and renewal of the fields once laid could be accomplished by the Soviet Naval Air Force with Badger A/G, Blinder A and Backfire B aircraft. Defensive minelaying and selective offensive laying in areas contiguous to the Soviet homeland would be undertaken by the surface fleet and the maritime patrol aircraft.

Designation	Type	Firing Mechanism	Case Depth (m)	Bottom Depth (m)	Explosive (kg)	Layer	Minimum spacing (m)
M 08	Moored	Contact	6.1	110	115	Ship	36
M 12	Moored	Contact	6.1	147	115	Ship	30.5
M 16	Moored	Contact	6.1	366	115	Ship	36
M 26	Moored	Contact	6.1	139	240	Ship	55
M 31	Moored	Contact	6.1	—	200	Ship	—
MKB	Moored	Contact	9.1	272	230	Ship	41
MKB3	Moored	Contact	9.1	272	200	Ship	35
MAG	Moored	Contact	79.3	457	230	Ship	35
AMAG	Moored	Contact	8.5	100	259	Aircraft	44
PLT	Moored	Contact	9.1	137	230	Submarine	80
PLT 3	Moored	Contact	9.1	128	100	Submarine	30.5
R	Moored	Contact	1.8	35	10	Ship	15
R1	Moored	Contact	1.6	35	40	Ship	20
MYaM	Moored	Contact	2.7	51	20	Ship	20
MIRAB	Bottom	Influence	9.1	—	65	Ship	61
KRAB	Moored	Influence	18.3	272	230	Ship	41
MKD	Bottom	Influence	54.9	—	784	Ship/sub	137
AMD 500	Bottom	Influence	24.4	—	300	Aircraft	69
AMD 1000	Bottom	Influence	54.9	—	699	Ship/sub	137
Rising Mine	Moored	Acoustic	609.6	226	227	Submarine	—
Sectionalised	Moored	Influence	488.7	—	227	Submarine	—
BPM 2	Limpet	—	—	—	3.1	Swimmer	—
MZ 26	Moored	Contact	34	46	1.3	Ship	—

UNITED KINGDOM

6008.241

LIMBO ANTI-SUBMARINE MORTAR SYSTEM (AS Mk 10)

This is a shipborne surface-to-sub-surface medium-range anti-submarine mortar system. Mortars are stabilised in pitch and roll by a metadyne system referenced to the ship's stable platform. For details of this operation see the 1987-88 edition.

Weight of the projectile is about 175 kg with a 92 kg HE warhead and range is about 900 m.

DEVELOPMENT

Developed by Admiralty Underwater Weapons Establishment during the 1950s. Ship fitted in the early 1960s.

STATUS

Operational in British and Commonwealth destroyers and frigates and in many other vessels. See also **2018.241** for modified Australian version.

CONTRACTORS

Manufactured to MoD (Navy) designs by several contractors.

2552.441

TORPEDO Mk 8

Quite possibly the longest-lived of all torpedo designs and certainly a remarkable survival in an otherwise reasonably sophisticated navy, the Mk 8 torpedo, which was designed in the early to mid-1930s, was certainly in service in the RN, including nuclear-powered submarines, until 1973, and may possibly still be in service with some minor vessels and is almost certain to be found in some other navies.

Although now replaced by Tigerfish as the primary armament of the principal attack submarines of the RN it should perhaps in justice be said that, within its limitations, the Mk 8 performed well and gave service which was honourable as well as long. The latest known versions are the Mods 2, 3 and 4, all for surface ship targets.

CHARACTERISTICS

Length: 670 cm
Diameter: 21 inches (533 mm)
Weight: 1521 kg
Propulsion: Compressed air
Speed: 40 – 45.5 knots
Range: 6400 – 4570 m
Warhead: HE 340 kg
Submersion: To 18 m
Guidance: Preset course angle and depth

STATUS

Obsolescent. Being replaced in RN service by the Tigerfish torpedo (**2440.441**).

2440.441
TIGERFISH TORPEDO (Mk 24)

Tigerfish is the name given to the production weapon derived from a redevelopment of the Mk 24 torpedo. The weapon is a 21 inch wire-guided/acoustic homing torpedo fitted with impact and proximity fuzes and is designed for submarine launch against submarine and surface targets. In line with the Royal Navy policy of covert submarine operation the torpedo is used in the passive mode whenever possible.

Wire guidance is used in the initial stages of an engagement up to the point where the torpedo's automatic three-dimensional active/passive acoustic homing system can control the run into the target. Wire is dispensed from both torpedo and submarine so as to avoid any wire stress due to their relative motion. The torpedo is roll-stabilised by controlling ailerons on retractable mid-body stub wings and is steered by hydraulically powered cruciform control surfaces mounted at the tail.

The torpedo carries its own computer which is connected through the guidance wire to the computer of the submarine's torpedo fire control system. During the wire guidance phase the on-board computer responds to the demands of the submarine computer, and during the homing run it interprets the data from the homing system sensors and calculates and commands the appropriate course, subject to a priority overriding steer-off azimuth control from the submarine.

The sonar can operate in either passive or active mode. During the attack the interrogation rate is progressively increased, as the torpedo nears the target, so as to improve system accuracy. The interrogation rate is controlled by the on-board computer, which is thus performing several functions during this phase: interrogation control, sonar data computation, torpedo steering control, and data transmission to the submarine to update its own computer memory.

Tigerfish is propelled by a powerful two-speed electric motor driving a pair of high efficiency, low noise, contra-rotating propellers.

CHARACTERISTICS

Length: 6.464 m
Diameter: 21 inch (533 mm)
Weight: 1550 kg (in air)
Speed: Dual high/low selectable at all times, max 50 knots (estimated)
Range: 21 km (estimated)
Fuze: Impact and proximity
Propulsion: Electrically driven contra-rotating propellers; 2 speeds

DEVELOPMENT

Development of Tigerfish Mod 2 was completed in 1985. It incorporates significant homing and wire guidance improvements which enable targets to be attacked successfully in the most difficult operational scenarios. Performance of the Mod 2 was demonstrated successfully in a series of trials which culminated in the sinking of a decommissioned Type 12 frigate. A modification kit has been developed which enables Mod 1 torpedoes to be upgraded to Mod 2 standard.

STATUS

In production for the RN. The Mod 1 version has been in service for some years and the Mod 2, higher performance model is now also in service. Data given above relate to the warshot torpedo; there are also exercise and dummy (handling) versions. The exercise version is similar to the warshot but has rechargeable batteries, becomes buoyant at the end of the run, and has an instrumentation pack for data analysis in place of the warhead. Total cost of developing and producing Tigerfish is reported to be £1 billion at 1984 prices.

Tigerfish has been released for export to some foreign navies. The first of these to be identified being Brazil, which in 1982 ordered the Mk 24 Tigerfish for a new generation of Brazilian submarines designed and built in West Germany.

CONTRACTOR

Marconi Underwater Systems Ltd, Elettra Avenue, Waterlooville, Hampshire PO7 7XS, England.

4559.441
SPEARFISH TORPEDO

Spearfish is a submarine-launched, wire-guided heavyweight torpedo designed to operate from all Royal Navy submarines and interfacing with a minimum of modification to all weapon handling and fire control systems. The torpedo is about 6 m long with a standard diameter of 533 mm and has an in-air weight of approximately 1850 kg. Spearfish meets the Royal Navy staff requirement to counter the forthcoming generation of faster, deeper diving, quieter and stronger-hulled submarines, as well as dealing with surface targets. It will be phased in gradually to replace the present Tigerfish Mod 2 in all nuclear-powered attack submarines (SSNs), the 'Trident' class SSBNs and the new diesel 'Upholder' class SSKs.

The variable speed of Spearfish is achieved by a turbine thermal engine powered by an advanced liquid fuel and oxidiser contained in separate tanks. The top speed (widely reported at 70 knots), endurance and diving depth are understood to be nearly double that of the Tigerfish torpedo. The propulsor itself is a pump jet consisting of a rotor and stator using design techniques from the successful Sting Ray development. The propulsion has been designed to be very quiet in operation to provide the covert operation needed to protect the firing submarine from detection on torpedo launch and subsequent counter-attack.

The torpedo's advanced sonar transducers and electronic system enable it to operate in a primary passive mode. However, when required to operate against a very quiet target, or in the final stages of attack, the active mode is used. In this mode, the powerful transmitters give Spearfish a long detection range and enable it to 'burn through' enemy countermeasures. The detection capabilities are further enchanced by an array offering many times the search volume of Tigerfish, with frequency agile transducers which enable salvo firing to be carried out.

Communication between submarine and torpedo is via a guidewire link. The torpedo normally operates autonomously, relaying data back to the submarine, but the command team can assume control at any time. The wire is dispensed from two reels, one in the torpedo and one in the submarine launch tube through a new design of dispenser which allows discharge at high submarine speeds as well as complete freedom of manoeuvre for the submarine after launch.

Spearfish contains a number of homing and tactics computers to control the weapon, enabling it autonomously to select search, detection and attack modes, to classify signal returns, to decide on appropriate tactics including re-attacks on the target if necessary, and to classify, track and overcome countermeasures and decoys. The torpedo's capability can be enhanced by software updates as target and countermeasure characteristics evolve in the future.

The warhead is a combined blast and shaped charge (directed energy) type, designed to pierce the double hulls of submarines entering service. The advanced homing and computer system enables the torpedo to be guided to the optimum point on the target's hull. Spearfish has an equal capability against surface targets where the fuzing system detonates the warhead under the hull of the target. The blast creates a whipping effect which breaks the ship's back.

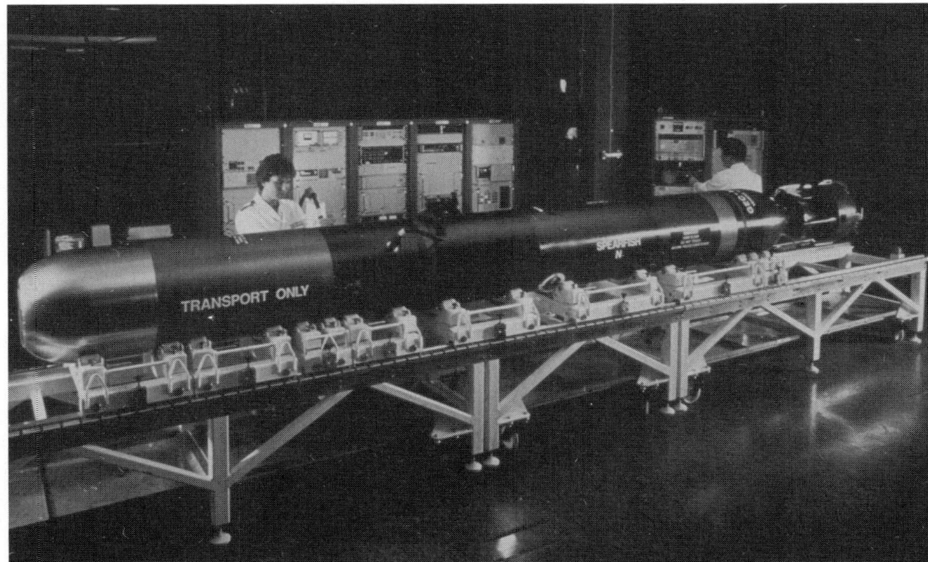

Spearfish production warshot torpedo and associated test equipment

DEVELOPMENT

In 1979-80 there was a competition between the projected Spearfish and the US Mk 48. The UK MoD decided that Spearfish met the design requirements, particularly for covert operation and, taking into account progress made in the Sting Ray programme being managed by Marconi, awarded a fixed price contract for development and procurement understood to be worth about £350 million.

STATUS

Marconi states that development has been to programme with contract acceptance trials taking place in 1987-88. Spearfish is now in production, using additions to the company's manufacturing facility initially purpose-built for Sting Ray, and the first production torpedoes have been delivered and are on contracts acceptance trials. The in-service date is scheduled for early 1989.

CONTRACTOR

Marconi Underwater Systems Ltd, Elettra Avenue, Waterlooville, Hampshire PO7 7XS, England.

2448.441
STING RAY TORPEDO

Sting Ray is an autonomous acoustic homing lightweight torpedo in service with the RN and RAF, and claimed to be the most advanced anti-submarine torpedo in the world with a capability for countering submarine threats effectively until well into the 1990s. It is designed for launching from helicopters, fixed-wing aircraft, and surface ships over a wide range of speeds and sea states. It is claimed to be unique in that its multi-mode multi-beam sonar, quiet high-speed propulsion system, and fully programmable on-board digital computer enable high performance both in shallow water where sonar conditions are notoriously difficult, and also in deep water. Sting Ray's computer enables it to make its own tactical decisions during an engagement to optimise the various homing modes to suit the underwater environment and the behaviour of the target. Both trials and simulation show that the single shot kill probability is very high indeed. The on-board computer enables the torpedo to be tuned to meet a nation's particular requirements by varying the software programmes. A prime feature of Sting Ray is the ability quickly to upgrade the software in the on-board computer to meet changing threats or to meet a national particular requirement.

Sting Ray has very high underwater agility being able to turn at rates as great as that of many in-air guided weapons. This is achieved by an electro-hydraulic proportional control system driving four control surfaces mounted in the efflux of a pump jet propulsor of advanced design.

No official performance details of Sting Ray have been cleared for publication, but informed comments which have been published enable a few deductions to be made. One such comment, in an indirect comparison with the US Mk 46 torpedo, claimed that it and Sting Ray had very similar speed, diving depth and endurance characteristics; from known performance of the Mk 46, this implies a maximum depth capability of as much as 750 m, a maximum speed of at least 45 knots and an endurance of about eight minutes at that speed. It was also hinted that anticipated propulsion improvements might enhance Sting Ray's speed appreciably, and also its range. Elsewhere, it was claimed that the conventional blast type warhead of the Mk 46 is limited in effect against the deeper-diving submarines with extensive outer hulls that are now entering service. Sting Ray is fitted with a directed energy warhead effective against such tougher targets. For maximum effectiveness, precise terminal homing enables the shaped charge to strike accurately at the target's most vulnerable point at the right angle.

STATUS

Marconi was appointed main contractor in October 1977. A £215 million contract from the UK MoD was signed in November 1979. The first production acceptance vehicle produced at the Neston plant was delivered early in August 1981. Contract acceptance trials began April 1982 and the design certificate was issued in December 1982. The torpedo was deployed during the Falkland Islands conflict in 1982.

In full production and in service under a £400 million order from the UK MoD for more than 2000 torpedoes. Sting Ray has also been sold to the Royal Thai Navy and the Egyptian Navy.

CONTRACTOR

Marconi Underwater Systems Ltd, Elettra Avenue, Waterlooville, Hampshire PO7 7XS, England.

Sting Ray lightweight torpedo launch

Sting Ray trials from F27 maritime patrol aircraft prior to entering Royal Thai Navy service

6767.441

BANDFISH EXPENDABLE TORPEDO COUNTERMEASURE

An expendable countermeasures system for use against acoustic torpedoes, Bandfish has been developed by Dowty Maritime Systems. It is designed to be launched from a submarine's submerged signal ejector or from a surface ship launcher and once it leaves the ship its operation is autonomous.

Bandfish, which weighs less than 12 kg and is 99.5 cm long and 10.2 cm in diameter, hovers in mid-water and transmits a high intensity acoustic signal. The operating depth and acoustic pattern can be designed to meet the customers' requirements and it has a minimum shelf life of five years.

STATUS

Available for production.

CONTRACTOR

Dowty Maritime Systems Ltd, Communications Division, 419 Bridport Road, The Metropolitan Centre, Greenford, Middlesex UB6 8UA, England.

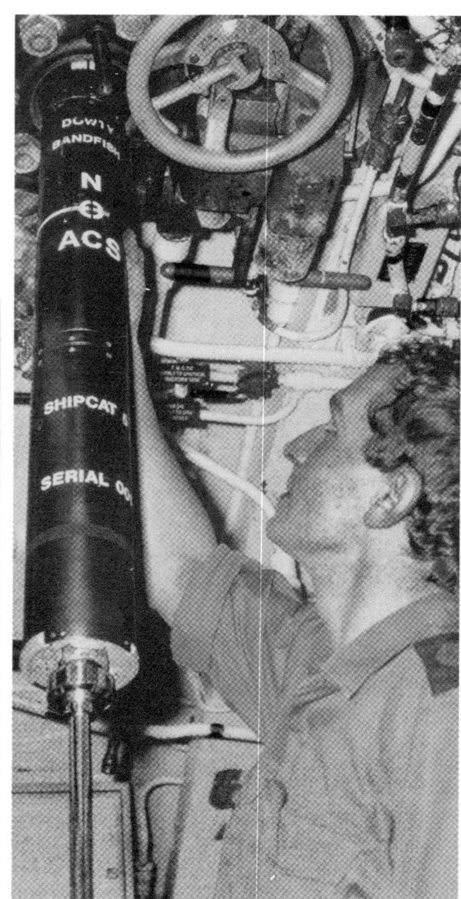

Bandfish expendable torpedo countermeasure

4482.441

SEA URCHIN MINE

Sea Urchin is a family of intelligent mines, which can be programmed to detonate on a range of influence characteristics, including the acoustic signature of a ship, a ship's magnetic influence, or a change in water pressure from a ship's displacement. Its advanced microprocessor control ensures that detonation occurs at the closest approach point of the target within the damage radius of the mine.

Flexibility of warhead charge weight (sizes from 350 to 1200 kg can easily be assembled) enables Sea Urchin to be configured to suit a range of operational requirements in terms of laying depths, target size, and degree of damage to be inflicted. Mine setting and self-testing procedures are simple and secure, being achieved by means of a small plug-in setting box weighing less than 5 kg.

In its ground mine version, Sea Urchin can be deployed in water depths from 5 to 200 metres,

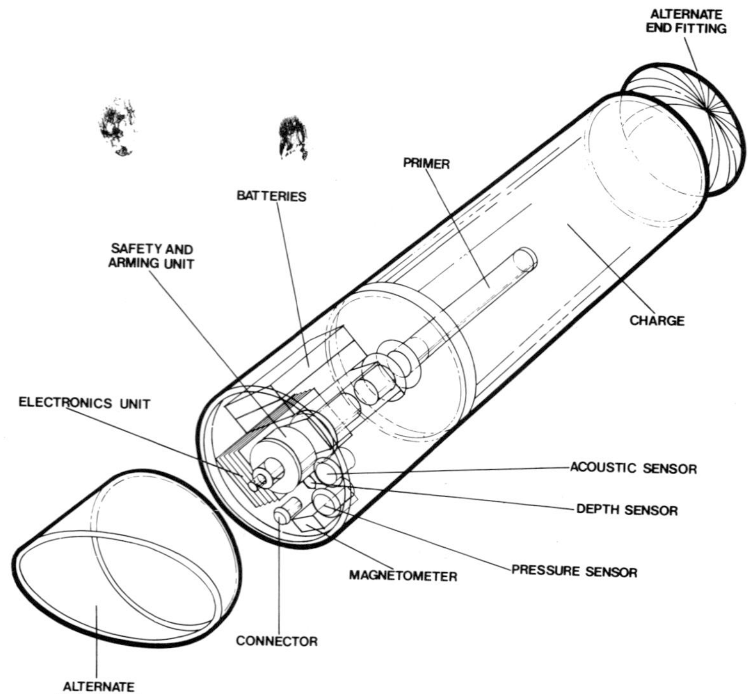

Diagram of internal arrangement of Sea Urchin mine's component parts

various launch attachments are available for the nose and tail of the mine to permit deployment by surface vessels, submarines and aircraft.

The sensing and processing elements, and safety and arming system are very flexible and can be repackaged to fit older generation ground and moored mines to convert them into modern weapons with significantly enhanced performance.

STATUS
Development has been completed but no information is available on production or orders.

CONTRACTOR
British Aerospace plc, Dynamics Division, PO Box 5, Filton, Bristol BS12 7QW, England.

Sea Urchin naval ground mine two 350 kg warheads and SAP module

---

4558.441
STONEFISH MINES

Marconi Underwater Systems Ltd, in conjunction with the Royal Ordnance plc has developed a family of advanced mines comprising warshot, exercise and training versions. Stonefish, as this family is known, is designed on the modular concept enabling mines to be made up to meet any tactical, exercise or training requirement. All variants may be deployed from surface ships, submarines and aircraft.

Stonefish exercise mine externally resembles the live system

The warshot variants are provided with a warhead size of 600 kg. The warhead, together with the appropriate launch kit and a standard tail unit containing the sensors, signal conditioning and processor package, can be configured to cover the range of launch vehicles, depths and targets. The basic sensor pack comprises acoustic, magnetic and pressure sensors. The signal processing includes signal conditioning and information processing, the setting of thresholds, mine logic and sterilisation delay times. The mine programme can be modified by means of a portable pre-setter changing threat and operational conditions before the mine is deployed.

The Stonefish warshot versions have a storage life of over 20 years and an in-water or laid life of over 700 days. They are intended to be laid at depths of between 5 and 90 m for surface targets and down to depths of 200 m for submarine targets. To provide training in mine countermeasures Stonefish is available as an exercise variant.

The exercise variant is capable of assessing minehunting and minesweeping effectiveness in its true environment. When laid it is available for data gathering relevant for target selection, together with a facility to emulate known mine characteristics and will provide proof of minefield effectiveness. The main features are:

(1) command
(2) acoustic command system – enabling mine states to be integrated or changed
(3) digital signal processing
(4) available as a seabed test platform for sensor/processing development.

The assessment variant is used to gather data for use by the tactical planner so that the optimum minefield can be planned.

The training variant is inert but can accept all the available equipment options to facilitate classroom instruction, handling, programming and test trials. The refurbishment life for stored warshot weapons is up to six years.

Mines of the Stonefish family are up to 2.5 m in length, with a diameter of 530 mm and all up weight varying up to 990 kg depending on configuration.

The Stonefish pre-setter, designed as a self-contained unit, employs a microprocessor and incorporates built-in data checking during input, and automatic test facilities. This is a Husky M-280 hand-held computer.

STATUS
In production. Marconi has received a £1.6 million order for Stonefish from the Royal Australian Navy and has sold warshot units to an overseas navy.

CONTRACTOR
Marconi Underwater Systems Ltd, Elettra Avenue, Waterlooville, Hampshire PO7 7XS, England.

6260.441
DRAGONFISH MINE

Dragonfish is a versatile high-technology anti-invasion mine for use in coastal waters and inlets. It is designed to be lethal to all forms of landing craft and small underwater vessels, and can be deployed from surface vessels and aircraft. The mine contains a sophisticated electronics package which enables it to distinguish between a number of signal inputs and to select the nearest target when presented with more than one.

Dragonfish has a presetting facility to initiate the mine's intelligence and to provide an update as necessary for changing conditions. It can self-adjust for environmental background. Acoustic features and magnetic sensors provide data for the microprocessor. Detonation can only take place when both sensors provide a confirmation of each other.

Dragonfish is profiled to present an extremely low acoustic signature and, together with the on-board microprocessor, it is claimed to be virtually immune to mine countermeasures. Shelf life is in excess of 20 years and minimum laid life is 200 days.

STATUS
In late development.

CONTRACTOR
Marconi Underwater Systems Ltd, Elettra Avenue, Waterlooville, Hampshire PO7 7XS, England.

6768.441
BRITISH AEROSPACE MINE MODERNISATION

British Aerospace have designed a new sensor and processing unit using microprocessor technology which can be used to update older mines and extend their service into the next century.

The system is extremely flexible and can be altered to meet existing internal arrangements. At its heart is a target information module housing acoustic and magnetic influence sensors together with a microprocessor-controlled digital electronics unit. A pressure influence sensor is also available. While the original safety and arming mechanism can be retained the company also offers a new device in its system. The package also includes high energy batteries.

Mine preparation is achieved through a hand-held, plug-in setting box which can be used to programme the appropriate mission parameters immediately before the mines are laid. The setting box also tests all the mine electronics automatically before transferring the mission parameters and only if these are fully functional will the data be down-loaded. Testing and setting takes approximately 30 seconds.

STATUS
A £10 million contract has been awarded by the British Ministry of Defence to upgrade Mk 12 and Mk 17 mines.

CONTRACTOR
British Aerospace, Dynamics Division, PO Box 5, Filton, Bristol BS12 7QW, England.

Modified battery chassis is fitted to Mark 12 seabed mine

6589.463

OSBORN MINESWEEPING SYSTEM

Osborn is an advanced acoustic minesweeping system which simulates a wide range of ship's acoustic signatures and can be supplied as either an open loop or closed loop system. It provides a versatile sweeping system capable of very high power output. It is used to activate all acoustic mines, including those with frequency selective triggering characteristics which are designed to be actuated only by certain types or classes of vessels.

The system has been specially designed to withstand the repeated levels of explosive shock likely to be experienced in operation, and is normally deployed in association with a magnetic sweep, to provide a very effective method of sweeping combined influence mines.

STATUS

Osborn is in service with the Royal Navy on its latest MCMVs ('Hunt' class) and is undergoing evaluation trials with the US Navy.

CONTRACTOR

British Aerospace, Dynamics Division, PO Box 5, Filton, Bristol BS12 7QW, England.

Osborn Towed Acoustic Monitor (centre) and Towed Acoustic Generator (left)

6483.241

SEA SERPENT MAGNETIC MINESWEEPING SYSTEM

The Sea Serpent system is a new method of magnetic minesweeping. It consists of a string of towed bodies containing controllable permament magnets, such that the state of each element can be controlled from the towing vessel. Sea Serpent is therefore able to simulate the magnetic characteristics of particular vessels to deceive most intelligent devices. The system capability is further enhanced when it is used in conjunction with an acoustic sweep. Other modes include the ability to produce cyclic waveforms as a conventional sweep, or the possibility of producing screening magnetic characteristics which cause an intelligent underwater threat to ignore a genuine target.

The towed bodies can be either recovered on board the towing vessel or can be moored when not in use, programmed and remain magnetically energised ready for the next mission. Because of the low drag of the system, small vessels can be used to tow Sea Serpent. Depending on the tow vessel and mission, the system requires little or no electric power.

CONTRACTOR

Vosper Thornycroft (UK) Ltd, Southampton Shipbuilding Division, Victoria Road, Woolston, Southampton SO9 5GR, England.

6771.441

LIMPET MINE DISPOSAL EQUIPMENT (LMDE)

The LMDE (Limpet Mine Disposal Equipment) has been designed to destroy limpet mines which are difficult to remove from ships because of their powerful magnets and their anti-handling devices.

The system features a stainless steel barrel bomb disruption device similar to those used against terrorist bombs on land. The barrel is mounted in a moveable sleeve and trunnion on a stand with magnets to attach it to the ship's hull and the whole device, which weighs less than 15 kg, has buoyancy blocks for easier handling.

When a limpet mine is discovered the diver assembles the LMDE out of the water then takes it underwater and places it as close as possible to the mine. The barrel is carefully positioned so it faces the heart of the mine then the diver returns to the surface from where he remotely fires a bridgewire cap cartridge. This propels a powerful slug of water at the mine, smashing open the thin casing and dispersing the main explosive charge.

STATUS

The system has been purchased by five customers.

CONTRACTOR

A B Precision (Poole) Ltd, 134 Stanley Green Road, Poole, Dorset BH15 3AL, England.

Limpet Mine Disposal Equipment

UNITED STATES OF AMERICA

3966.241

LAMPS MK III

The USN LAMPS (Light Airborne Multipurpose System) has been designed to extend and enhance the capabilities of surface ships, particularly in anti-submarine warfare (ASW) and increase effective operational range for the weapon systems fitted in surface vessels of the US fleet. This system extends the electronic and acoustic sensors and provides a reactive weapon delivery capability for destroyers and frigate class ships by operating manned helicopters from them. Sensors, processors, and display capabilities aboard the helicopter enable the three-man crew to extend the ship's tactical, decision-making, and weapons delivery capabilities. Classic line-of-sight limitations for surface ships and limitations to underwater acoustic detection, classification, and localisation are thus mitigated through the use of the manned LAMPS aircraft.

LAMPS is being acquired in two phases. The first, Mk I, involved installation of shipboard equipment and conversion of H-2 helicopters already in the inventory to an SH-2 configuration. Mk I became operational in 1972. The Mk III includes the integration of improved avionics and shipboard systems to be used with the SH-60B Seahawk, a modified version of the Army UH-60A Blackhawk helicopter.

In 1974 the USN selected IBM as the system prime contractor to work with the Naval Air Systems Command in developing LAMPS Mk III. By late 1976, after extensive developmental testing, system capabilities were demonstrated in deep water using two modified helicopters with representative avionic and shipboard electronics. Two open-ocean tests concluded the validation of the LAMPS Mk III mission performance. The first was conducted jointly by the Naval Air Systems Command and IBM. After the data reduction was completed and the technical performance was assessed, the test was judged highly successful. The second test was conducted under the direction of the commander of the operational test and evaluation force to verify the operational suitability and performance of the LAMPS Mk III concept. The Navy was given permission to continue with full-scale development by the Department of Defense in 1978 with IBM as system prime contractor, Sikorsky Aircraft as the air vehicle contractor, and General Electric as the engine manufacturer for the new Mk III air vehicle.

LAMPS helicopter flying over USS McInerney

The LAMPS Mk III weapon system embodies the integration of the parent ship (frigate, destroyer, cruiser) and the manned aircraft (SH-60B Seahawk) operating from that ship for both ASW and anti-ship surveillance and targeting (ASST) missions. As an adjunct of the sensor and attack systems of the parent or similarly equipped ships, the aircraft extends the detection, classification, localisation, and attack capabilities where needed to increase line-of-sight and range capabilities of the parent ship. Radar and ESM sensors extend the ship's line-of-sight against surface threats, whereas sensors that operate relative to the ocean medium (eg sonobuoys and magnetic anomaly detectors) enable the ship to engage the underwater target.

The primary control of aircraft tactics and selection of sensor and weapon modes remain with the parent ship for both the ASW and ASST missions, with an equally important, independent mode of LAMPS aircraft control provided as a backup.

In an ASW mission, the LAMPS aircraft is deployed from the parent ship when a suspected threat has been detected by the ship's towed array sonar, hull-mounted sonar, or by other forces. It proceeds to the estimated target area, where sonobuoys are dropped into the water in a pattern designed to entrap the target. The sounds (acoustic signatures) detected by the buoys are transmitted over a radio frequency link to the aircraft where they are analysed, codified, and retransmitted to the ship for interpretation and analysis. When the location of the threat has been determined with adequate precision, the aircraft descends near the ocean's surface for final confirmation. This may be accomplished using active sonobuoys, passive directional sonobuoys or by trailing a magnetic anomaly detector behind the aircraft. On final confirmation by any of these methods, a torpedo attack can be initiated. The extension of the ship's sensor, tactical control, and attack capabilities is achieved by using a duplex data link that transfers acoustic, electromagnetic, command data, and voice from the Seahawk, and command data and voice to the aircraft. Data is processed and evaluated aboard the parent ship where tactical and command decisions are made, and tactical instructions, weapon delivery information, and transmission of processed data are linked back to the SH-60B Seahawk.

Data transmitted to the parent ship (or other LAMPS-equipped ship) is processed by digital computers and specialised processors, distributed, and evaluated by the parent ship's operators: the LAMPS air tactical control officer, acoustic sensor operator, remote radar operator, and ESM operator. The combat information centre evaluator uses the data in context of the overall tactical situation to determine what actions to take in attacking. Operation of the LAMPS Mk III system was described in detail in previous editions of *Jane's Weapon Systems*.

STATUS

Introduction to the US Fleet of the SH-60B LAMPS Mk III began in 1984, and such helicopters are being deployed aboard about 100 US Navy surface vessels, including DD-963 and DDG-993 destroyers, CG-47 cruisers, and FFG-7 frigates. The earlier LAMPS Mk 1 Seasprite will continue to serve in some older FFG-7s. A total of 204 SH-60Bs is the US Navy planned requirement, and a substantial number are already in service.

CONTRACTORS

System prime contractor: International Business Machines Corporation, Federal Systems Division, 10215 Fernwood Road, Bethesda, Maryland 20034, USA.

SH-60B Seahawk helicopter: Sikorsky Aircraft.

General Electric Company: Engine contractor.

6001.241

ASROC ANTI-SUBMARINE SYSTEM

ASROC (RUR-5A) is an all-weather, day or night, ship-launched ballistic missile carried as the primary anti-submarine warfare (ASW) weapon aboard USN destroyers as well as some cruisers and frigates. The weapon consists of a Honeywell Mk 46 acoustic homing torpedo (**2822.441**) or a nuclear depth charge (estimated yield one kiloton) attached to a solid-propellant rocket motor. It can be fired from an eight-cell launcher Mk 112, the Mk 26 launching system, or from the Mk 10 Terrier missile launcher (ASROC/Terrier system).

Other major components include a fire control computer and an underwater sonar detector.

OPERATION

After launch the weapon follows a ballistic trajectory. The rocket motor is jettisoned at a predetermined point and the payload continues toward the target. If the payload is a torpedo, it is lowered to the water by a parachute, where its homing mechanism is activated upon submersion. Depth charge payloads sink to a predetermined depth before detonating.

Range of the weapon is classified but has been estimated to be from 2 to 10 km. The missile is 4.6 m long, has a diameter of 32.5 cm, and a span of 84.5 cm. Launch weight is about 435 kg.

DEVELOPMENT

ASROC entered service with the USN in 1961, becoming widely fitted in anti-submarine vessels of various classes. Plans to develop the ASROC system to utilise a vertical launch system (VLS), initially for some surface ships but later in attack submarines, were at first suspended on grounds of cost and/or a need to merge the project with the ASW-SOW stand-off weapon programme (**4181.441**). However, changing circumstances in the USN's overall ASW requirement and its inventory of weapons in this category led to a reversal of the original cancellation of vertical launch ASROC (VLA), and it is intended that the new version will ultimately replace the original ASROC deployed on surface ships.

STATUS

Users of the ASROC system include: USA (27 cruisers, 87 destroyers, 65 frigates), Japan (15 destroyers, 11 frigates), Spain (five each destroyers and frigates), Greece and Taiwan (four destroyers each), Canada (four frigates), Federal Republic of Germany (three destroyers), Turkey (three destroyers), Brazil, South Korea and Pakistan (two destroyers each), and Italy (one missile cruiser). In the 1985 fiscal year report by the US Secretary of Defense funding for ASROC (VLA) was included for development with $30.8m assigned for 1984, $26.7m in 1985, and $37.4m in 1986 and $38.1 m in 1987. In FY 1987 the US Navy was granted $39.2 million for continued VLA development and a further $39.8 million for production, the latter figure including the introduction of a second source manufacturer. Having won a competition to become the VLA prime contractor in November 1983, Loral Systems Group (formerly Goodyear Aerospace Corporation) conducted a source selection and selected Martin Marietta as the VLA second source supplier.

8-round ASROC launcher aboard USS David R Ray (W Donko)

The conventional launch ASROC will continue to be supported by Honeywell as prime contractor to the US Naval Ocean Systems Centre, San Diego.

CONTRACTORS

Honeywell Inc, Training and Control Systems Operations, 1200 East San Bernadino Road, West Covina, California 91790, USA (Prime Contractor).

Loral Systems Group, 1210 Massillon Road, Akron, Ohio 44311, USA (prime contracter for VLA development).

1128.441

SUBROC (UUM-44A) ANTI-SUBMARINE MISSILE

This is a submarine launched, rocket-propelled missile which follows a short underwater path before transferring to an air trajectory for the major portion of its journey to the target area. At a pre-determined point during boost the nuclear depth bomb is separated from the remainder of the missile and then follows a semi-ballistic trajectory to the point where it re-enters the water. The W55 nuclear charge then sinks to a predetermined depth before detonation. Estimated yield of the warhead is about one kiloton.

Principal characteristics of SUBROC are: length 625 cm, diameter 53.3 cm, weight 1853 kg (approximately), range 56 km (approximately), and speed is supersonic. It has a solid-fuel motor.

OPERATION

The UUM-44A missile forms part of an advanced anti-submarine system designed for deployment in nuclear-powered attack submarines operating against submerged vessels armed with strategic missiles. This system includes the AN/BQQ-2 Raytheon integrated sonar system and the Mk 113 SUBROC fire control system produced by the Librascope division of the Singer Company.

After detection and location of a target submarine, co-ordinate data are fed into the attack submarine weapon system which programmes an optimum mission profile for the SUBROC missile. It is assumed that this can be accomplished in virtually real-time.

The SUBROC missile is launched horizontally from a standard 21 inch (53.3 cm) torpedo tube, by

SUBROC anti-submarine missile after leaving launch submarine and taking to the air, but before separation of its nuclear depth charge and rocket motor

conventional means. At a safe distance from the launch vessel (which need not be directed towards the target area for firing) the solid-fuel missile motor is ignited and the missile follows a short level path before being directed upward and clear of the water. Missile stability and steering is effected by four jet deflectors, which function in both water- and airborne sectors of the trajectory. Guidance is by means of an inertial system (SD-510) produced by Kearfott.

When free of the water, SUBROC is accelerated to a supersonic speed and guided toward the target area. At a predetermined point separation of the nuclear depth bomb is initiated by explosive bolts and a thrust-reversal deceleration system which enables the warhead to continue to the re-entry point on a trajectory controlled by vanes on the depth bomb.

Impact with the water is cushioned to protect the arming and detonation devices. A preset depth sensor detonates the nuclear charge when the bomb is in the vicinity of the target. It is probable that the target position will need to be established with an accuracy that will permit the warhead to detonate within its estimated lethal radius of 5 to 8 km from the target.

SUBROC missiles can be carried in torpedo tubes without attention for long periods, and launched with minimal preparation time.

DEVELOPMENT

Development of the UUM-44 began in June 1958 at the US Naval Ordnance Laboratory, White Oak, Maryland, under the management direction of the Bureau of Naval Weapons, with Loral Systems Group (formerly Goodyear Aerospace Corporation) as the prime contractor. Technical evaluation was completed in 1964, and production and operational deployment began in 1965.

STATUS

SUBROC is operational on nuclear-powered attack submarines. Each ship carries four to six SUBROC missiles. An improvement programme to sustain the effectiveness of this weapon against new threats was initiated in 1976-77.

This would have entailed replacement of the SUBROC analogue guidance sub-system by a digital one and regraining of the weapon's rocket motors, and it was estimated that to extend the operational life of the system by five years would have cost about $41 million. A Congressional decision was taken in 1979 not to update the system along these lines, and as a consequence the USN turned to the ASW-SOW (Sea Lance) programme (**4181.441**) to meet existing and anticipated operational requirements. The SUBROC system is now considered obsolescent and it is proposed to withdraw the system from the USN inventory during the 1990s. It will be replaced by the Sea Lance weapon.

CONTRACTORS

Prime contractor: Loral Systems Group, 1210 Massillon Road, Akron, Ohio 44315, USA.

Mk 113 fire control system: The Singer Company, Librascope Division, 808 Western Avenue, Glendale, California 91210, USA.

SD-510 inertial guidance system: The Singer Company, Kearfott Division, 1225 McBride Avenue, Little Falls, New Jersey 07424, USA.

Nuclear warhead: Sandia Corporation, Livermore Radiation Labs, Livermore, California, USA.

Solid-fuel motor: Morton Thiokol Inc, Elkton Division, Elkton, Maryland 21921, USA.

4181.441

SEA LANCE ANTI-SUBMARINE WARFARE STAND-OFF WEAPON

The USN ASW-SOW programme was approved by the US Secretary of Defense in 1980 to develop a long-range anti-submarine weapon by the late 1980s. The weapon has now been given the name of Sea Lance, and is required to replace the technologically obsolete SUBROC (**1128.441**) which it is planned to withdraw from the USN inventory during the 1990s. The new weapon is also intended to meet the challenge of new threats such as the latest Soviet 'Alfa' class of submarine, and it will be designed for launching from US nuclear attack submarines, and surface fighting ship vertical launch systems. Specific platform applications of Sea Lance have not been defined more precisely to date.

In USN attack submarines the weapon will be used with existing torpedo tubes and the Mk 117 digital fire control system. (The latter is planned to be the first type of existing shipboard hardware to incorporate Outlaw Shark capabilities, now in research and development to provide a correlated, computer-formated, all-source data handling facility for forces at sea. A major objective of this project is over-the-horizon targeting.) US attack submarines cannot at present launch a weapon against a target unless it closes to within reach of the Mk 48 torpedo. Sea Lance will extend this range appreciably, as well as permitting further attacks on submarines which successfully evade a torpedo attack.

Drawing of the Sea Lance anti-submarine warfare stand-off weapon under development for the US Navy by Boeing

Four separate designs for Sea Lance were proposed but they have in common a combined underwater/air-flight/underwater path from submarine launch platform to target, similar to that of the SUBROC predecessor system, but variations include the method(s) of propulsion employed. Originally alternative payloads consisting of either a nuclear depth bomb or a Mk 50 homing torpedo were considered but only the latter is now under development.

As already stated, the proposals made by the various contractors selected to submit concept formulation plans in response to the operational requirement differed in a number of respects and all four were described in *Jane's Weapon Systems 1981-2*. The team selected to continue development of the project was that linking Boeing, Gould and Hercules Aerospace. Boeing as the leader with responsibility for the vehicle; Gould for the payload, acoustics and signal processing, and guidance; and Hercules for the single solid-propellant rocket motor. The latter is apparently designed to propel the weapon during its initial underwater trajectory (after clearing the launcher torpedo tube) and during the air flight. On clearing the surface of the water, four small wrap-around fins at the rear of the rocket motor casing deploy automatically to aid stability. After booster burnout, the motor is probably jettisoned, leaving the payload section to follow a ballistic path to the target area, where after deceleration it re-enters the water for the payload/warhead section (nuclear or torpedo) to come into action. The missile, including the payload, is 20 ft long.

It will be possible to fire the Sea Lance from surface ships or submarines and rounds will be stored in canisters ready to fire from the vertical launch systems of destroyers (DD 963 class), guided missile cruisers (CG 47 class), and guided missile destroyers (DDG 51 class). From SSN 637/688 class attack submarines, the missile will be carried by capsule to the surface before the solid-propellant rocket motor ignites to propel it to the target area.

Both the surface-launched and submarine Sea Lance missiles will be able to carry either the US Department of Energy-developed nuclear depth bomb or the advanced lightweight torpedo. Surface ships will have a third payload option in the existing Mk 46 Mod 5 torpedo.

STATUS

The Sea Lance programme was initiated in 1980 and it was then estimated that the research, development, test and engineering cost would amount to a total of about $550 million. Funding is under Programme Element 63367N of the US DoD budget and the actual funding for fiscal year 1982 was $35.4 million, with $20.2 million in 1983, $27.4 million in 1984, $51.3 million in 1985 and $75.3 million in 1986. In July 1986 Boeing was awarded a contract valued at $380 million to carry out full-scale development. The 1986 fiscal year report by the US Secretary of Defense stated that Sea Lance will be developed to replace SUBROC aboard USN attack submarines, with provisional entry into service during the early 1990s. The first production buy of 67 units is scheduled to take place in FY 90 and the total requirement is for 1052 units at a cost of $1.9 billion. One of the first classes to be fitted will be the 688 class of nuclear attack submarines, although the weapon will be compatible with surface vessels equipped with vertical launch facilities.

CONTRACTORS

The Boeing Company, Box 3707, Seattle, Washington 98124, USA; with Gould Inc, Government Systems Division, and Hercules Aerospace.

3629.441

TORPEDO Mk 32

The Mk 32 is an acoustic anti-submarine torpedo which is still in use aboard a number of former USN ships now transferred to other navies in various parts of the world, mostly aboard escort types. They are launched by the Mk 4 rack-type launcher unit in most cases, in which a single torpedo is carried in the ready-to-launch position with two more torpedoes stored nearby on deck. The current version is the Mk 32 Mod 2, which is equipped with an active acoustic homing head. It is also capable of air drop launch.

Propulsion is by electric motor, powered by a silver/zinc oxide battery which is activated by sea water. After launch, the torpedo descends to a depth of 6 m and the active sonar begins to operate, transmitting 50 kHz pulses while the torpedo follows a spiral search pattern, circling to left or right as selected. The search pattern is maintained until the sonar obtains a target response and the homing sequence commences. In the event of failure to obtain a target contact, the circular search pattern continues until the battery is exhausted, after which the torpedo dives deeper prior to self-destruction.

CHARACTERISTICS

Length: 2080 mm
Diameter: 483 mm
Weight: 350 kg
Warhead: HE 49 kg
Propulsion: Electric motor
Range: Approx 8600 yards (7864 m) at 12 knots
Max acoustic range: 560 m

STATUS

Operational but obsolescent.

2817.441

TORPEDO Mk 37 Mods 0 and 3

Designed primarily as a submarine-launched anti-submarine torpedo, but suitable for deck launching by the Mk 23 and Mk 25 torpedo launchers, the Mk 37 torpedo is a 19 inch (482.6 mm) weapon that has been described as the first successful high-performance anti-submarine torpedo. The 19 inch diameter of all versions of this torpedo was chosen to enable the torpedo to swim out from a standard 21 inch (533 mm) launch tube, an arrangement with obvious operational advantages.

Mods 0 and 3 are free-running torpedoes. After one has been launched on a target interception course it maintains it until, at a preset range, it 'enables' a process involving arming the warhead and switching in the attack logic circuits. The latter include various preselectable options such as depth limits, search pattern, and type of homing. The final attack is by sonar auto-homing which can be active, passive, or both combined.

Mod 3 is an updated version of Mod 0, the updating consisting of the incorporation of a large number of minor modifications based on operational experience.

DEVELOPMENT

Developed by Westinghouse the Mk 37 Mod 0 torpedo first went into service in 1957. Mk 37 torpedoes have been produced in large quantities by the Naval Ordnance Plant, Forest Park, Illinois, and are still in widespread use in the US fleet.

CHARACTERISTICS

Length: 352 cm
Diameter: 483 mm with guides to fit 533 mm tubes
Weight: 645 kg (warshot); 540 kg (practice)
Warhead: 150 kg HE
Speed: 24 knots (Mod 3)
Submersion: To 270 m (Mod 3)

STATUS

The Mk 37 has been largely replaced by the Mk 37E (**4762.441**) and Mk 48 (**2823.441**), although quantities of the Mk 37 are probably still in service.

2818.441

TORPEDO Mk 37 Mods 1 and 2

These versions of the Mk 37 torpedo differ from Mods 0 and 3 (**2817.441**) in their size and method of guidance; in other respects they are substantially similar.

Both versions are wire-guided, whereas the Mk 37 Mods 0 and 3 are free-running torpedoes. Both are 4.09 m long and weigh 766 kg (warshot) or 657 kg (practice).

Just as Mod 3 is an updated version of Mod 0 so is Mod 2 a version of Mod 1 updated by the incorporation of a number of minor modifications.

CHARACTERISTICS

Length: 409 cm
Diameter: 483 mm with guides to fit 533 mm tubes
Weight: 766 kg (warshot); 657 kg (practice)
Warhead: 150 kg HE
Speed: 24 knots (Mod 2)
Submersion: To 270 m (Mod 2)

STATUS

Operational.

4762.441

TORPEDO NT 37E MODERNISATION

The NT 37E torpedo is a modernised, improved version of the standard, battery-electric propulsion Mk 37 heavyweight torpedo (**2817.441**, **2818.441**). This dual purpose ASW and anti-ship torpedo can be fired from both submarines and surface ships. Incorporating a standard RS-232, RS-422 or Harpoon digital fire control interface, the torpedo is compatible with the following platforms and fire control systems: West German 205/206/MSI-70U; West German 209/HSA-SINBADS (later version), RNLN Swordfish/HSA-M8; British Oberon/TCSS-5, TIOS; US Guppy/US Mk 106; US Mk 32 surface launch tubes; or any class equipped to fire Mk 37 torpedoes.

The NT 37E performance improvement kit and hardware, largely installed in the field, completely upgrades existing Mk 37 torpedo inventories. It consists of three major subsystems:

(1) An Otto fuelled thermo-chemical propulsion system. When compared to the performance of an unmodified Mk 37 silver/zinc oxide battery and electric motor, the new propulsion system demonstrates a 40 per cent increase in speed, a 150 per cent increase in range, and an 80 per cent increase in endurance. The propulsion system utilises existing Mk 37 motor mounting provisions and is compatible with existing Mk 46 fuelling and engine maintenance facilities, engine tools, and engine refurbishment kits.

(2) A solid-state acoustic system and a noise-reduction laminar-flow nose assembly replaces the Mk 37's vacuum-tube acoustic panel and hemispherical nose. The new sonar substantially improves the passive detection range against high-speed surface targets and active detection range against small silhouette submarine targets; in most cases, target acquisition range has been doubled. The new SUBTIP self-noise reduction nose assembly increases transducer isolation while reducing flow noise effects, reducing the likelihood of self-decoying at all depths.

(3) A solid-state computer-based guidance and control system with inertial navigation capability replaces the Mk 37's vacuum-tube analogue system, resulting in greater system stability, reliability and accuracy. The new system utilises inertial sensor assembly and a MIL SPEC 8002 microprocessor, allowing fully programmmable tactics and inertial navigation. Should future naval tactics require modified torpedo command and control, changes can be easily implemented by revision of software and replacement of programmable read-only memory units (PROMs).

Sea trials of NT 37E heavyweight torpedo

CHARACTERISTICS

Type: NT 37E, wire-guided, Mod 2; non wire-guided, Mod 3
Length: 4505 mm (Mod 2); 3846 mm (Mod 3)
Diameter: 485 mm
Weight: 750 kg (Mod 2); 642 kg (Mod 3)
Run modes: (1) Straight run/salvo-anti-ship; (2) straight run with acoustic miss indicator to initiate acoustic re-attack; (3) active snake and circle – ASW; passive snake and circle – anti-ship
Warshot explosive weight: 150 kg HE
Propulsion: Thermo-chemical rotary piston cam engine with Otto fuel

STATUS

The original Mk 37 upgraded version was called the NT 37C and included the thermo-chemical propulsion system and anti-ship kit.

Honeywell, at the request of principal NATO NT 37C user navies, instituted a development programme to improve the logistic support capability and the acoustic and guidance performance of the torpedo. This effort resulted in the new, supportable, solid state acoustic and guidance system and nose assembly.

Advanced development of the solid-state guidance and control system was completed in 1982. Performance of the NT 37E model was demonstrated on a three-dimensional instrumented range and during sea trials with user navies during the period 1981-1985. The tests demonstrated long-range active and passive homing attack on acoustic and live targets. The tests also demonstrated the following improvements: increased dynamic control and accuracy, complete wire guidance capability, ease of changing tactical software, low sonar background noise levels, low acoustic false alarm rate, long range active and passive acoustic attacks on acoustic targets, increased torpedo reliability and reduced maintainance. A semi-automatic test set is available to replace manual test equipment. The Mk 37 family of torpedoes is used by at least 16 navies.

CONTRACTOR

Honeywell Marine Systems Division, 6500 Harbour Heights Parkway, Washington 98204, USA.

2820.441

TORPEDO Mk 44

This is a lightweight torpedo designed for launching from aircraft or helicopters, from surface vessels (using Mk 32 tubes), or by the ASROC rocket system (**6001.241**).

Two models have so far been produced but the differences in dimensions are trivial. Both torpedoes are electrically propelled and their calibre is 12.75 inches (324 mm). Approximate length is 2.56 m and the torpedoes weigh about 233 kg with a 34 kg warhead. Active acoustic homing is used. Depth and course settings are entered by umbilical cable. Estimated maximum submersion depth is approximately 300 m. A range of about 5000 m at a speed of 30 knots has been reported. Arming is by seawater scoop.

STATUS

Obsolescent. Licence production was initiated in a number of foreign countries. Replaced by Mk 46 (**2822.441**) in USA and UK service. Some other states are adopting the Mk 46 but the navies equipped with the Mk 44 are those of Argentina, Australia, Brazil, Canada, Chile, Columbia, Federal Republic of Germany, Greece, Indonesia, Iran, Italy, Japan, Netherlands, New Zealand, Norway, Pakistan, Philippines, Portugal, South Africa, South Korea, Spain, Thailand, Tunisia, Turkey, Uruguay and Venezuela.

2822.441

TORPEDO Mk 46

Torpedo Mk 46 is a lightweight, 324 mm calibre weapon, designed as a successor to the Mk 44 (**2820.441**) for use with several ASW systems. It can be launched from surface vessels, fixed-wing aircraft, or helicopters and can be carried by the ASROC rocket (**6001.241**). Mk 32 tubes are used for surface launching. The latest version is the Mk 46 Mod 5.

The Mk 46 is a deep-diving, high-speed torpedo fitted with an active/passive acoustic homing system and intended for use against submarines. After water entry it searches for, acquires, and attacks its target; if it misses the target it is capable of multiple re-attacks.

Two models of the Mk 46 have been produced. Mod 0 was the first US torpedo to be powered by a solid-fuel motor: it is 2.67 m long and weighs about 258 kg. Immediately it enters the water it starts a helical search pattern. Mod 1 is slightly lighter and is powered by a five-cylinder liquid mono-propellant (Otto) motor. The Mk 46 torpedo also forms the active element of the Captor mine system (**2541.441**, below) in which role it has the designation Mk 46 Mod 4.

CHARACTERISTICS

Type: Mk 46 Mod 1
Length: 2.59 m
Diameter: 32.4 cm

Weight: 230 kg
Max speed: 40 knots
Range: 11 000 m (max)
Acquisition range: 460 m (estimated)
Max depth: Classified
Search patterns: Multiple
Warshot explosive weight: 44 kg
Propulsion: Mono-propellant

STATUS

Operational: the first delivery to the USN was made in October 1965. Following the cancellation of the Mk 31 torpedo a quantity of Mk 46 torpedoes was purchased as an interim measure by the UK MoD for RN use.

The Mod 0 was produced in limited quantities because the Otto fuel was found to be a more practical propellant. It is used in the US fleet only, for aircraft launch applications.

The Mk 46 Mod 1 was introduced in the US fleet in April 1967 and was followed by the Mod 2 in 1972. Production of the former was completed in February 1971 and conversion of 600 Mod 2 units was completed by early 1975.

Recent efforts have been directed to the NEARTIP (Near-Term Improvement Programme) to improve the acoustic performance and countermeasure resistance of the Mk 46 torpedo (Mk 46 Mod 5), and this version is in full production.

In particular, this programme was designed to restore the acquisition range of the Mk 46 torpedo which was reduced by an estimated 33 per cent by the adoption of an anechoic coating for some of their submarines by the USSR. Code-named Clusterguard, this paint reduces the effectiveness of the standard Mk 46's acoustic homing head and NEARTIP restores this aspect of performance to at least the same as the pre-Clusterguard status. Other improvements are to the guidance and control and associated fire control system to achieve improved performance against the more sophisticated threats envisaged, two-speed capability and increased shallow water performance. NEARTIP implementation is in the form of modification kits applied on a retrofit basis to existing Mods 1 and 2 Mk 46 torpedoes. Captor Mk 46 torpedoes (Mod 4) will not be fitted with NEARTIP kits; instead there is a different modification kit for this programme (see **2541.441**, below).

As prime contractor for the Mk 46 since 1965, Honeywell has produced more than 9000 torpedoes for the USN and other navies. The fiscal year 1984 report by the US Secretary of Defense included funding for both new Mk 46 (NEARTIP) torpedoes and conversion kits to upgrade older Mk 46s, an estimated 500 of which will be used in the Captor role. For 1986 and 1987 procurement of 500 torpedoes each year was proposed with annual costs of $132.8 million and $106.8 million, respectively. A contract valued at $295 million was awarded to Honeywell in February 1987 for Mk 46 Mod 5 torpedoes for US Navy, Canada and Turkey. The US Navy is not requesting NEARTIP modifications in FY 1988 and FY 1989 following the FY 1986 order for 500.

A successor to the Mk 46 Mod 5 NEARTIP torpedo is being developed under the Advanced Lightweight Torpedo (ALWT) programme (**4182.441**).

Users of the Mk 46 torpedo include: Australia, Brazil, Canada, France, Greece, Indonesia, Iran, Israel, Italy, Japan, Morocco, Netherlands, New Zealand, Norway, Pakistan, Portugal, Saudi Arabia, South Korea, Spain, Taiwan, Thailand, Turkey, UK and USA.

CONTRACTORS

Aerojet Electro Systems Company, Azusa, California, USA, was responsible for the development and early production of both Mod 0 and Mod 1.

Honeywell Inc and Gould Ocean Systems Division have also been involved in manufacture and the former is now prime contractor for the Mk 46 NEARTIP.

Mk 46 Mod 5 (NEARTIP) torpedo produced by Honeywell which succeeded the Mk 46 Mods 1 and 2

4182.441

MK 50 ADVANCED LIGHTWEIGHT TORPEDO (ALWT)

The Advanced Lightweight Torpedo (ALWT), (once designated EX-50, but now the Mk 50) is being developed for use from ships, submarines and aircraft in order to meet the advancing and sophisticated Soviet submarine threat as a successor to the Mk 46 Mod 5 (NEARTIP) lightweight torpedo (**2822.441**). The design objectives include an enlarged target detection envelope embracing faster submarines operating at greater depths, and other desired features are increased speed and endurance, terminal homing, signal processing, and greater destructive effect. The last objective will be attained by a new design of directed energy warhead which it will be necessary to place on its target with high precision to ensure that the directed energy blast penetrates the hull and does not glance off. It will also be advantageous to hit the target submarine amidships. The ALWT is expected to have similar dimensions and weight to those of the Mk 46 and have a speed of about 40 knots with a maximum submersion of some 600 m.

In addition to the ability to operate against fast, deep diving submarines, there is a requirement for capabilities against shallow, slow-moving submarines and surface ships. The most difficult objective to be met, according to the US DoD, is the realisation of reliable terminal homing which is essential if a directed energy warhead is to be used.

Initially two sets of contractors competed under the A-109 procedure whereby each contractor had broad latitude to develop their own unique torpedo so long as it conformed with the Navy's operational requirements. The combination of Honeywell as prime contractor, with Garrett Pneumatic Systems for the propulsion, was selected as sole contractor during 1981.

A propulsion test vehicle version of the Honeywell/Garrett torpedo was test launched in the latter half of 1981. It used a closed cycle steam turbine engine in a stored chemical energy propulsion system (SCEPS) with the energy being supplied by a chemical reaction. The command/control system was computer based and software controlled and given reference information by an autopilot on which to base its decisions for steering and engine control.

Seven modified Mk 45 torpedoes are being built as high speed mobile torpedo targets (ADMATT) for ALWT testing by Rockwell International. These run at up to 41 knots and tow a hydrophone/echo repeater array that simulates the spatial extent of a real submarine, and this method was chosen as the only feasible way of developing and testing the ALWT terminal homing system. A three-dimensional instrumented shallow water range has also been developed for ALWT evaluation.

Test launch in course of Mk 50 ALWT programme

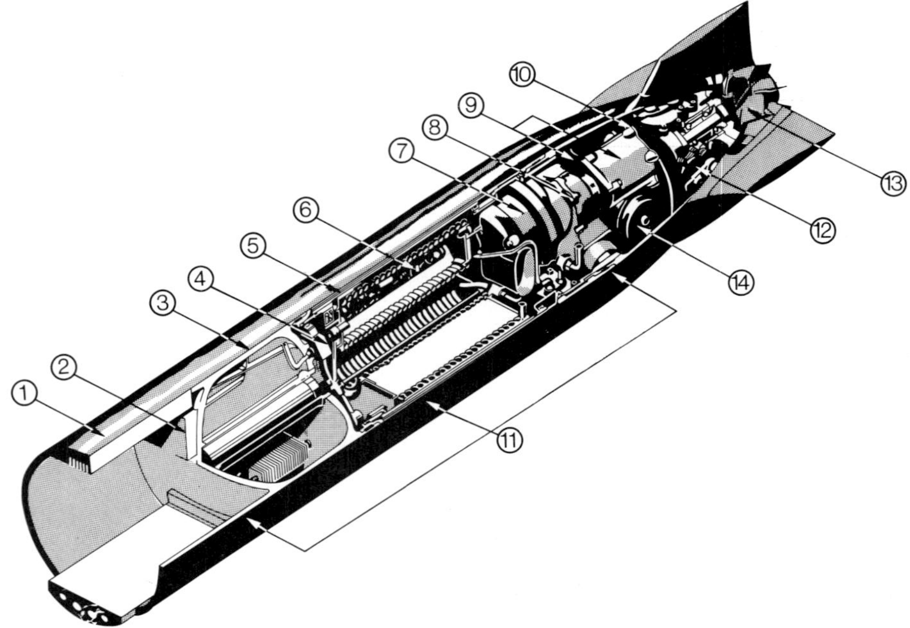

Cut-away diagram of ALWT advanced development model showing SCEPS afterbody. (1) guidance and command section hull (2) full-authority electronic powerplant controller (3) pressure-storage oxidant tank (4) electric-powered oxidant control system (5) high efficiency, hull-integrated condenser (6) high energy, lithium-fuelled boiler/reactor (7) high efficiency, supersonic impulse turbine (8) quiet, lightweight reduction and accessory drive gearbox (9) high speed, rare-earth pm alternators (10) fast start feedwater injector (11) high-strength), lightweight hull structures (12) high-response, electric steering fin actuators (13) pump-jet propulsor (14) vibration isolation engine mounts

STATUS

Now in full-scale development, the Mk 50 is due to join the US fleet early in 1990. Continued development funding budgeting in the four years, 1984-87 was $143.3 million, $146.3 million, $168.4 million and $169.9 million, respectively. The FY88 budget request for the programme was $262 million and 90 test torpedoes are likely to be purchased. Low rate initial production commenced in 1987, with projected procurement of 84 torpedoes at a cost of $124.1 million, and development of a second source for full-scale production. In February 1987 air drop tests were carried out followed, in March, by operational tests. Delivery of warshot torpedoes is anticipated early in 1990.

In June 1987 a team led by Westinghouse Defense and Electronics Center won a $7.5 million initial contract to act as a second source production qualification of the Mk 50. Other members of the team include Martin Marietta (command and control systems) and Argo Tech Corporation, formerly TRW (propulsion system).

CONTRACTOR

Honeywell Inc, Defense Systems Division, Hopkins, Minnesota, USA.

2541.441

CAPTOR—ENCAPSULATED TORPEDO

Captor, a contraction of 'encapsulated torpedo', is the name given to an anti-submarine system comprising a Mk 46 torpedo (**2822.441**) inserted into a mine casing.

Deployment is in deep water, generally in the vicinity of strategic routes travelled by enemy submarines, and submarines are the intended targets. US officials have stated that Captor has the ability to detect and classify submarine targets while surface ships are able to pass over a Captor field without triggering the Mk 46 Mod 4 torpedo which carries the warhead (43.5 kg PBXN103 explosive). This capability is reported to have been tested.

The detection and control unit (DCU) that performs these functions is the most costly sub-assembly of the complete Captor weapon and accounts for about 45 per cent of the total unit cost – approximately $130 000. DoD statements imply that the DCU incorporates facilities for turning itself on and off, in addition to its principal operational functions of detecting possible targets, classifying them by their sound signatures, and initiating release of the Mk 46 torpedo when the target is within range of its homing head. It is probable that the turn-on/turn-off system is quite sophisticated in the interests of power conservation to ensure the maximum operational life for deployed Captor mines. Factors which are likely to be taken into account include the levels of traffic (surface and submarine), ambient conditions, sea-state, etc.

Both active and passive sensor modes are employed and the system first operates in a listening (passive) mode which continues for a certain length of time sufficient to identify the target as a submarine and not a surface vessel. The system then switches to an active mode during which it is assumed target ranging is carried out to determine the optimum release time for the homing torpedo.

The detection and control unit is gated to ignore surface traffic and has an estimated range of about 1000 m on submarine targets. There is no IFF capability and friendly units must be warned of Captor minelaying and positions of deployed mines.

There are presumably provisions for some method of self-deactivation or self-destruct for those Captors which are life-expired and which are not capable of retrieval, and other measures to prevent unauthorised salvage or interference may be expected.

The current deployed life of a Captor mine is thought to be in the neighbourhood of six months, but it is not known if this is the USN's target, although it is a fact that considerable effort has been placed on ensuring maximum shelf and operational life.

Captor mines can be sown by surface ships, submarines, and by aircraft. In the former delivery mode, mine rails or other delivery systems are not required, the chosen technique being by means of an over-the-side boom (or yard and stay) with a capacity of 1045 kg. The Captor is brought to a point about 10 m above the surface before release, which has to be at an angle to ensure proper entry into the water. Any submarine equipped with standard 21 inch (533 mm) torpedo tubes can lay Captor mines, these having the advantage of being capable of covert minelaying. Aircraft employ a parachute technique for delivery of Captor.

According to USN statements, there are two main options for Captor delivery: one utilises P-3 maritime aircraft, nuclear strike submarines and surface ships; the other envisages a combination of USAF B-52 bombers and P-3 aircraft, plus surface ships and nuclear strike submarines. Captor has been tested for delivery on P-3C, A-6, A-7 and B-52 aircraft, LKA and other cargo ships, and aboard ten different classes of submarine, including conventionally powered boats.

Most Captor minefields will be barriers located at some distance from possible enemy defences, and in the more highly defended areas Captor mines would be delivered by aircraft or submarines. The unique capability of submarines to plant mines covertly and under ice would be employed selectively.

CHARACTERISTICS

Length: 3.7 m (overall)
Diameter: 533 mm
Weight: 908 kg (with torpedo and mooring)

STATUS

Initial production was started in March 1976, although the US Secretary of Defense's report of January 1981 said the procurement had been at a low level while development and testing were conducted to correct performance deficiencies. In December 1978 reduced performance against shallow water targets had been indicated. Nevertheless initial operational capability was achieved in September 1979. Further procurement was cancelled after fiscal year 1980 by a decision taken in December 1980, and in January 1980 a Captor improvement programme was approved. The weapon was granted approval for service use in February 1980, but no procurement funds were sought in the fiscal year 1981 budget "because Captor failed to provide the high level of effectiveness we had sought. Subsequent testing has demonstrated that recent modifications have corrected its performance deficiencies.

Despite this somewhat chequered history, in Spring 1980 Admiral Rowden was able to disclose that about 630 Captors had been procured.

CONTRACTOR

Loral Systems Group, 1210 Massillon Road, Akron, Ohio 44315, USA.

2823.441

TORPEDO Mk 48

The Mk 48 torpedo is carried by all US attack and fleet ballistic missile submarines. The weapon can be adapted to launch from surface ships. Currently it is operational only on submarines. The torpedo is wire-guided through a two-way communications link in the current Mod 3 version.

The Mk 48 torpedo is propelled with an axial flow pump-jet propulsor driven by an external combustion gas piston engine. This engine, like that in the Mk 46 torpedo (**2822.441**), is a Gould design. The fuel for the engine is a monopropellant, Otto fuel II.

Data processor cards which are part of digital guidance and control system being developed by Hughes for Mk 48 torpedo ADCAP programme

The Mk 48 torpedo is capable of operation in wire-guided active or passive acoustic and non-acoustic modes. The acoustic modes of operation allow active or passive target detection capabilities. A complete description of the torpedo will be found in the 1987-88 edition.

DEVELOPMENT

Anticipation by the US authorities of impending advances in submarine technology by the USSR noted in the 1960s and 1970s led to studies of the Mk 48's capabilities against likely threats, and in 1975 resulted in an Operational Requirement issued by the Chief of Naval Operations for a programme to develop appropriate modifications to the Mk 48 torpedo to keep pace with anticipated submarine threat developments. The origins of the Mk 48 ADCAP (advanced capabilities) programme lay in this requirement, but the extent and rate of Soviet submarine technology advance hastened both ADCAP progress and another Mk 48 improvement programme.

Recognition (by the USA) of the impressive operational characteristics of the Soviet 'Alfa' class submarine in late spring 1979 resulted in a decision taken in September 1979 to accelerate the ADCAP programme. It was also responsible for an intensive test and analysis programme to determine the true limits of the then current Mk 48 in terms of depth, speed, and acoustic capabilities. This was known as the expanded operating envelope programme, and showed that the Mk 48 was structurally reliable at the depth needed to engage 'Alfa' class submarines; the target speed recognition capability required could be achieved; the vertical coverage was adequate, as was the self-noise at higher speeds wth the existing nose and array; and additional speed could be achieved. Laboratory modifications were made to a few torpedoes for tests, and these changes were implemented in the form of a programme to update fleet Mk 48 torpedoes to what is now the Mk 48 Mod 4 standard.

Of the performance requirements demanded by the ADCAP programme, the most important are:

(1) sustained long acquisition range
(2) minimised adverse environment and countermeasure effects
(3) minimised shipboard tactical constraints
(4) enhanced surface target engagement capabilities.

Hardware changes involved in ADCAP entail replacing the entire nose of the weapon housing the acoustics and beam forming circuits and replacement of the signal processing by the latest electronics. The latter will also incorporate the current command and control electronics. Warhead sensor electronics will be improved.

Application of the expanded operating envelope programme findings (see above) to ADCAP has resulted in the upgraded ADCAP, which incorporates: upgraded acoustics and electronics; an expanded operating envelope (depth, target speed, weapon speed options); increased fuel delivery rate and capacity for optimum speed and endurance; improved surface target capabilities.

CHARACTERISTICS

Length: 5.8 m
Diameter: 21 inch (533 mm)
Weight: About 1600 kg
Max speed: 55 knots
Max range: 38 km
Max depth: 500 fathoms (914 m)

STATUS

The Mk 48 has been in production since 1972 and is used aboard USN attack submarines, and strategic submarines for self defence. By early 1980 more than

1900 torpedoes of this type had been delivered to the USN and an estimated 800 plus were in the production and procurement line. It was then estimated that another 1050 might be required to meet inventory objectives and to allow for peacetime training and testing.

In August 1979, Hughes Aircraft Company received a contract for development of digital guidance and control electronics for the Mk 48 ADCAP programme.

The first test run was carried out by the USN at Nonoose Bay in early 1982, using the inertial guidance system developed for the ADCAP programme. About 240 more runs were programmed before completion of this phase of the programme after which it was expected that entry into service will take place in 1983/84. However, the 1985 report by the US Secretary of Defense amended this to indicate anticipated deployment of the system in the mid- to late-1980s.

In the US Secretary of Defense's Annual Report to Congress in February 1985, it was stated that following completion of a successful test programme in 1984, it had been decided to accelerate production of the ADCAP torpedo. The five-year programme called for production of 1890 ADCAP units, and included 123 in 1986 at a cost of $433 million and 280 units in 1987 costing $671 million. Hughes began production of the ADCAP version in 1985. In 1986 Gould was named the second ADCAP source.

The only known foreign users are Australia and the Netherlands.

Gould is one of four companies bidding for a 2000 torpedo cost-effective programme in the USA. A decision on this programme has been postponed.

CONTRACTOR

Gould Inc, Ocean Systems Division, 18901 Euclid Avenue, Cleveland, Ohio 44117, USA. (Mk 48)

Hughes Aircraft Company, Ground Systems Group, PO Box 3310, Fullerton, California 92634, USA. (Mk 48 ADCAP version)

6769.441

ADC MK 2 EXPENDABLE TORPEDO COUNTERMEASURE

The ADC (Acoustic Device Countermeasure) Mk 2 is a submarine-launched decoy designed to counter acoustic torpedoes. It is an 8 cm diameter device which hovers vertically at a pre-selected depth emitting an acoustic signal.

The uppermost section consists of ceramic transducers and impedance-matching networks for the acoustic transmitter. Below this is the electronics section in which the signals are generated and amplified. At the bottom of this section is a sea-water activated battery which powers a pressure-controlled motor driving a small, shrouded, propeller which keeps the decoy hovering.

STATUS

In production since 1978 and more than 2800 units have been produced.

CONTRACTOR

Emerson Electric Co, Government & Defense Group, 8100 W Florissant, St Louis, Missouri 63136, USA.

3342.441

AMERICAN MINE WARFARE PROGRAMMES

The American mine warfare programme comprises two main elements, mines and mine-countermeasures. The application of mines in weapons against submarines and surface ships has the objective of denying access to and use of ports and to provide offensive and defensive barriers. The current stock of sea mines consists of the Mks 52, 55, 56 and 57 types.

CHARACTERISTICS

Type: Mk 52, aircraft laid bottom mine
Length: 2.25 m
Diameter: 844 mm
Weight: Mod 1, 542 kg; Mod 2, 567 kg; Mod 3, 572 kg; Mod 5, 570 kg; Mod 6, 563 kg
Charge: 270 kg HBX-1
Max depth: 45.7 m (Mod 2, 183 m)
Actuation: Mod 1 acoustic; Mod 2 magnetic; Mod 3 pressure/magnetic; Mod 5 acoustic/magnetic; Mod 6 pressure/acoustic/magnetic

Type: Mk 55, aircraft laid bottom mine
Length: 2.89 m
Diameter: 1.03 m
Weight: Mod 2, 580 kg; Mod 3, 992 kg; Mod 5, 992 kg; Mod 6, 996 kg; Mod 7, 995 kg
Charge: 576 kg HBX-1
Max depth: 45.7 m (Mods 2/7, 183 m)
Actuation: Mod 2 magnetic; Mod 3 pressure/magnetic; Mod 5 acoustic/magnetic; Mod 6 pressure/acoustic/magnetic; Mod 7 dual channel magnetic

Type: Mk 56 Mod 0, aircraft laid moored mine
Length: 3.5 m
Diameter: 1.06 m
Weight: 1010 kg
Charge: 159 kg HBX-3
Max depth: 366 m
Actuation: Total field, magnetic dual channel

Type: Mk 57 Mod 0, submarine or ship laid moored mine
Length: 3 m
Diameter: 510 mm
Weight: 934 kg
Charge: 154 kg HBX-3
Max depth: 350 m
Actuation: total field, magnetic dual channel

Also employed by the USN is a range of air deployed munitions based on modified general purpose low drag bombs and which can be released without requiring a parachute. The modification involves the use of a Mk 75 Mod 0 Destructor Modification Kit which can be added to 500 lb, 1000 lb and 2000 lb Mk 80 series bombs to form the Service Destructors (DST) Mks 36, 40 and 41, respectively. These are mostly intended for use in shallow waters such as estuaries etc, against typical coastal targets. There is also the DST 115A, which can be employed with either aircraft or surface craft for use against surface targets.

CHARACTERISTICS

Type: Mk 36, aircraft laid bottom mine
Length: 2.25 m
Diameter: 400 mm
Weight: 240 kg (with fixed conical fin) 261 kg (with tail retarding device)
Charge: 87 kg H-6
Max depth: 91.4 m
Actuation: Mods 0/3 magnetometer; Mods 4/5 magnetic/seismic

Type: Mk 40, aircraft laid bottom mine
Length: 2.86 m
Diameter: 570 mm
Weight: 447 kg (with fixed conical fin) 481 kg (with tail retarding device)
Charge: 204 kg H-6
Max depth: 91.4 m
Actuation: Mods 0/3 magnetometer; Mod 4/5 magnetic/seismic

Type: Mk 41, aircraft laid bottom mine
Length: 3.83 m
Diameter: 630 mm
Weight: Mods 0/3, 926 kg; Mods 4/5, 921 kg
Charge: H-6
Actuation: Mods 0/3, magnetometer; Mods 4/5, magnetic/seismic

Type: 115A, aircraft laid surface mine
Length: 0.45 m
Diameter: 620 mm
Weight: 61 kg
Charge: 24 kg HBX-3
Actuation: Magnetic/seismic

Quickstrike

The Quickstrike bottom mine development programme embraced a family of mines using different size cases but with common target detection and classification mechanisms. The four members of the Quickstrike family are the Mks 62, 63, 64 and 65. The last of these (Mk 65 Mod 0) is in the 2000 lb (900 kg) class and is in full production by Aerojet Tech Systems in Sacramento, California. The Mk 64 will probably be the next to enter production and this also is in the 2000 lb (900 kg) class, based on a Mk 84 2000 lb bomb and measuring 3.8 m long and 633 mm diameter.

Quickstrike mines are for shallow water deployment (to approximately 100 m) and targets will have to approach to within a few hundred feet for it to act. It will use existing Mk 80 series GP bomb cases as well as a new mine case. Quickstrike mines are deployed by aircraft, surface ships, or submarines, but principally from the former.

This family of mines is based primarily on conversion of existing ordnance (bombs and torpedoes). An exception is the Mk 65 mine which is not a bomb conversion. It has a thinner case than the equivalent bomb and contains the effective underwater PBX explosive. It is 3.25 m long and 533 mm in diameter. The Mk 65 has now been deployed with the US Navy and is being evaluated by the Italian Navy.

Captor

Captor is a deep anti-submarine mine (Mk 60) intended for use in barriers against enemy strategic submarines. It can be emplaced by submarines, surface ships and aircraft. The kill mechanism consists of a Mk 46 Mod 4 homing torpedo (**2822.441**) which is released by a fire control system contained within the detection and control unit of the main Captor unit. Further details of Captor will be found in **2541.441**, above.

SLMM

The submarine-launched mobile mine (SLMM) Mk 67 is intended to provide the US fleet with a capability for planting mines in shallow water (to approximately 100 m) by submarine, using a self-propelled mine to reach water inaccessible to other vehicles. It is also meant for use in locations where covert mining would be particularly desirable from a tactical standpoint. It measures 4.09 m long × 485 mm diameter and weighs 754 kg.

The Mk 67 SLMM consists essentially of a modified Mk 37 torpedo (**2818.441**); alterations involved include some reworking of the Mk 37 torpedo bodies and replacement of the torpedo warhead with the applicable mine components. Tooling and other plant facilties were installed in fiscal year 1978 for production of Mk 67 sub-launched mobile mines in 1979.

Procurement plans for the Mk 67 SLMM for 1987 was 273 but it appears that this has now been cancelled.

Mine Countermeasures

In 1983 construction began of the first in a new class of mine countermeasures ships, the MCM-1 Avenger. Ships of this class are intended to match increasing Soviet mine warfare capabilities and to enhance US minesweeping facilities. Five of these ships have been authorised to 1988 and two have been commissioned. It is planned to authorise the building of nine more ships of this class. A second new class of mine countermeasure ship, the MSH-1 will augment the MCM class ships. The MSH-1 minehunters will be equipped with advanced combat systems similar to those in the MCM-1 ships, but will be smaller and less costly. Congress authorised the lead ship of the class in 1984. To complete the 17-ship programme, the five-year building plan contains another 16 vessels. More detailed information is given in *Jane's Fighting Ships*.

6770.441

US NAVY TORPEDO DECOYS

The current US Navy decoy system for use against acoustic torpedoes is the AN/SLQ-25 Nixie but this is now regarded as inadequate against the present generation of threats. Two programmes, for Anti-Ship Torpedo Defense and for Surface-Ship Torpedo Defense were both cancelled and the US Navy is reported to be interested in a joint programme with the United Kingdom.

However, in June 1987 Sippican Inc and the General Instrument Corporation received contracts worth $4.13 and $4.35 million respectively from the US Naval Weapons Support Center to design, develop and provide data for what is officially described as 'an off-board ship decoy' and this may reflect renewed US Navy interest in a domestically produced torpedo decoy.

4425.441

HONEYWELL MNS

The Honeywell mine neutralisation system (MNS) developed by the company's Marine Systems Division and the US Naval Sea Systems Command, is now in production for the US Navy's new MCM ships. It is the result of more than ten years work to provide a system that surpasses the conventional towed mechanical, magnetic and acoustic sweep techniques. It is designed to detect, locate, classify and neutralise moored and bottom mines, using high-resolution sonar, low light-level TV, cable cutters and mine destruction charges. A special underwater vehicle carries these sensors and countermeasures and is remotely controlled from the parent vessel.

Initial target detection and vehicle guidance information is provided by the ship's sonar and vehicle tracking is aided by a Honeywell RS/900 acoustic position indicator. Vehicle sonar is used during the mid-course search and final homing phases, and high resolution enables operations in poor visibility by sonar guidance alone. Low light TV is used in conjunction with sonar during the precision guidance phase near the target. Underwater launch and recovery of the MNS vehicle assists operations in high sea states.

Vehicle power is provided via a neutrally buoyant umbilical cable which also carries signal and control links between the vessel and the MNS vehicle. Aboard the parent vessel a sonar screen on the control console displays sensor and vehicle status information in digital form. The monitor and control consoles are the focal point for operation and management of the complete system. Display facilities include: sonar, TV, secondary sonar and TV, vehicle control, cutter and dropped-charge actuation.

CHARACTERISTICS

Vehicle endurance: Unlimited
Speed: 6 knots
Weight (in air): 1135 kg
Length: 3.8 m
Propulsion: Hydraulic, twin 15 hp
Thrusters: Horizontal/vertical
Cable length: 1000 m
Total system weight: 11.7 t
Power requirements: 60 kW (peak)

STATUS

In production.

CONTRACTOR

Honeywell Marine Systems Division, 6500 Harbour Heights Parkway, Everett, Washington 98204, USA.

SONAR AND UNDERWATER DETECTION EQUIPMENT

AUSTRALIA

1790.453
MULLOKA SONAR

Mulloka is a lightweight active hull-mounted scanning sonar for installation in Royal Australian Navy anti-submarine escorts. It is a solid-state forced air-cooled equipment of high reliability which has been designed using a standardised packaging approach to aid maintainability and keep production simple.

The equipment includes a dedicated on-line computer which is used to control transmission parameters, perform automatic signal processing functions, and conduct checkout procedures. All signal interface equipment with other ship's systems has been grouped at a single location so that modifications necessary to interface Mulloka with other ship equipment configurations are confined to one unit.

The sonar beam-forming process involves the development of a large cylindrical transducer array consisting of 96 staves with 25 transducers per stave, driven in differing combinations and phases under computer control to form the required sonar beams electronically. The hydrophone elements of the array are driven by 480 power amplifiers during the transmit period, the length of which is a function of the pulse-length of the transmitted signal. The transmit period is selectable over the range of approximately one to four seconds. The transmitter beams are formed sequentially during this period and cover the ocean for 360° in azimuth. The formation of the beam transmit pattern is random in nature from one transmission to the next in order to prevent the submarine commander having any indication of detection. Immediately following the transmission, the sonar reverts to the listening mode for a period determined by the range scale selected by the operator.

STATUS

In 1979 a prototype Mulloka sonar was formally handed over to the RAN after undergoing trials in HMAS Yarra since 1975. A production programme to build systems to equip six RAN 'River' class destroyer escorts has been completed.

In April 1980 it was announced by the Australian Ministry of Defence that an initial contract worth US$1.1 million had been awarded to Honeywell Inc in America to provide a transducer array for the Mulloka sonar. Two Australian companies, Plessey Australia Pty Ltd and Dunlop Industrial and Aviation, were also participating. A contract for the electronics of the system was let to THORN EMI (Australia) Ltd in 1979.

In early 1982, two production contracts were awarded for the 'River' class ships to be fitted with Mulloka sonars. One contract for A$16.1 million was awarded to THORN EMI (Australia) Ltd and the other to Honeywell, USA for A$9.4 million. The THORN EMI contract was for six systems delivered in the period 1983-85 and the Honeywell contract for five transducer arrays delivered in 1983/84.

In 1985 contracts were awarded for two systems to be fitted to the FFG-7 class frigates being built in Australia. As a result of the transferred technology, Plessey Australia was contracted to establish a production facility and to manufacture two transducer arrays and spares to be delivered over the period 1987 to 1989. In mid-1987 THORN EMI Electronics (Australia) was awarded a $A6 million contract to upgrade the system electronics, including a new display system.

1791.453
PROJECT BARRA

Project Barra concerns the development and production of an advanced submarine detection system for use by long range maritime patrol aircraft.

The project evolved from successful development work under Project Nangana, a research and development programme, undertaken by the Weapons Research Establishment in close collaboration with the RAN and RAAF for the development of improved methods of submarine detection.

Following successful completion of an engineering study, initial development and the establishment of production facilities, together with the first production contract, Amalgamated Wireless (Australasia) Limited was appointed prime contractor, and completed the production of this initial Barra sonobuoy contract by the end of 1981. A further A$55 million contract was let in early 1982 for full production against RAF and RAAF requirements. This contract has been fulfilled and further orders have been received.

The project is a collaborative programme with the UK; Australia developed the SSQ-801 Barra sonobuoy and the airborne processing equipment was developed in the UK under the direction of the UK Ministry of Defence. (See **1981.363**, AQS 901 acoustic data processing and display system, in the UK sub-section of this section.) The platform equipment is suitable for installation in a range of aircraft, and has been fitted in RAF Nimrod MR2 and RAAF P-3C Orion maritime aircraft.

Development of the Barra system generated the need for a ground support facility for the P-3C Orion aircraft. AWA designed and built for the RAAF a Compilation, Mission Support, Integration and training facility (CMI). This self-contained modular cabin complex can be reconfigured to suit various RAAF requirements. The CMI is a computer-based system providing training, mission briefing/debriefing and analysis facilities.

CONTRACTOR

Sonobuoys Australia (a consortium of Amalgamated Wireless, Australasia and Plessey Australia), Australia.

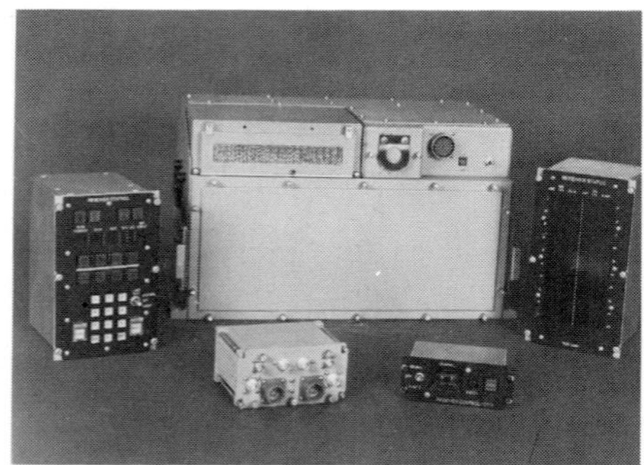

Advanced sonobuoy communication link receiver for Barra sonobuoy data exchange

RAAF P-3C Orion maritime patrol aircraft equipped with Barra ASW system

---

3998.453
SSQ-801 BARRA SONOBUOY

The SSQ-801 sonobuoy was developed specifically for the Australian Project Barra anti-submarine programme (**1791.453**, above) and is a passive directional sensor which may be used either singly or deployed in patterns to detect and locate submarine targets. It provides both range and directional information. Within a standard A size buoy, two parts separate on reaching the water, the upper portion containing radio transmission equipment rises to the surface while the other part housing the acoustic sensor array sinks to a predetermined depth. A cable links the two parts.

The lower portion contains a compass and booms fitted with specially developed hydrophones. Arrays are sequentially switched to give the required directional beaming. Multiplexed signals from the lower section are fed to the floating portion of the buoy for transmission by radio to patrol aircraft, where data are analysed by the Barra system AQS-901 sonics processor (**1981.363**). Each sonobuoy transmits information on a pre-selected VHF radio channel so that the aircraft crew can select a particular buoy's data for analysis. Digital techniques are used in the SSQ-801 to convert the combined sonar and compass information into signals to modulate the buoy's VHF transmissions.

The SSQ-801 is suitable for release from fixed-wing aircraft such as the P-3C Orion or Nimrod, and from helicopters of various types.

CHARACTERISTICS

Type: Passive directional array
Size: A
Length: 914 mm
Diameter: 124 mm
Weight: 13 kg
Operating life: Selectable up to 9 h
RF transmitter frequencies: 136 – 173.5 MHz
RF modulation: Narrow-band FSK
Channel spacing: 0.75 MHz
Power output: 1 W minimum
Operating life: 0.5, 1, 2, 4, 5, 6 or 9 hours
Array depth: 25 or 135 m
Activation time: 60 s at 25 m; 180 s at 135 m
Acoustic bandwidth: 10 to 2000 Hz
Compass accuracy: 1° rms for dip angles less than 65°

DEVELOPMENT

Development of the SSQ-801 sonobuoy was undertaken as part of the Australian Barra project (see above). A derivative, the SSQ-981 (**6861.453**), has been developed by Plessey, UK.

STATUS

Adopted by both the Royal Air Force and the Royal Australian Air Force. A number of other countries are reported to be interested in the Barra system, including France, Japan, the Netherlands and the USA. Amalgamated Wireless (Australasia) Limited is the prime contractor with Plessey Australia, Commonwealth Aircraft Corporation, Tube Makers and Cable Makers Australia, as major subcontractors.

Since the original contract in 1982, additional contracts have been placed for the continuing production of the Barra sonobuoy. A new company, Sonobuoys Australia, has been formed to manufacture the Barra sonobuoy.

CONTRACTOR

Sonobuoys Australia (a consortium of Amalgamated Wireless, Australasia and Plessey Australia).

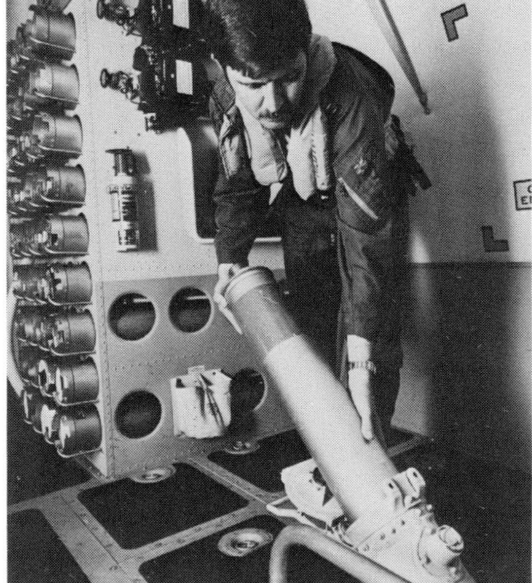

SSQ-801 Barra sonobuoy about to be launched from an RAAF Orion ASW aircraft

5864.463

KARRAWARRA TOWED ARRAY SONAR

The Australian Defence Research Centre has refined a towed array sonar for surface ship or submarine use to what is claimed to be the most compact system available. Code-named Karrawarra, the array carrier is only about 19 mm thick, with sensors the size of a normal cigarette spaced along it. It is stated to be suitable for ships down to small patrol boat size. The system has a thick rubberised coating to make it less vulnerable to predators. No other details are available. The sonar is understood to be in the early development phase.

STATUS

As at the beginning of 1988, the Karrawarra project was still in the development phase. It is expected to be sufficiently developed for installation in the two FFG7-class frigates being built in Australia. It will also be in the new surface combatants currently being studied to replace the RAN's 'River' class destroyer escorts and in the new Type 471 submarines.

CANADA

1636.453

HS-1000 SERIES SONAR

The HS-1000 series of sonars is a family of lightweight omni-directional ASW sonars which have been developed to give small naval vessels a modern sonar capability in search, detection, and attack. The numbering of each sonar designates transducer mounting and/or sonar power:

(1) HS-1001: lightweight variable depth sonar (VDS)
(2) HS-1002: lightweight hull-mounted (HM) sonar
(3) HS-1001/2: lightweight combined VDS/HM sonar with single electronics and a transfer switch
(4) HS-1007: medium-sized sonar available in either the HM or VDS configuration featuring higher power
(5) HS-1007/7: medium-sized combined VDS/HM sonar with single electronics and a transfer switch.

The HS-1001 and HS-1002 sonars are intended for use on patrol craft and corvettes. The HS-1007s are intended for use on corvettes, frigates, and destroyers.

STATUS

The HS-1000 series sonar is available on a current production status and has been delivered in the VDS configuration and in the combined VDS/HM configuration. More detailed information is given under this entry number in previous editions of *Jane's Weapon Systems*.

CONTRACTOR

Westinghouse Canada Inc, Po Box 5009, Burlington, Ontario L7R 4B3, Canada.

3999.353

AN/ASQ-502 SUBMARINE DETECTING SET (MAD)

The AN/ASQ-502 magnetic anomaly detector (MAD) is a sensitive magnetometer system developed for airborne submarine detection roles. A basic sensitivity of 0.01 gamma is claimed for this equipment, which has a total system weight of less than 15.6 kg (34.5 lb).

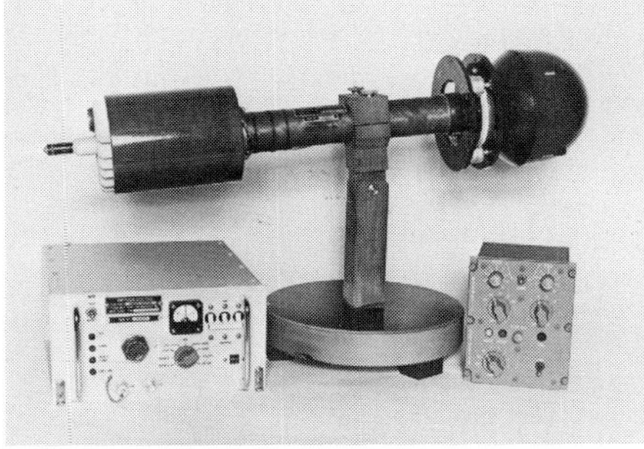

AN/ASQ-502 submarine detecting set (MAD)

In place of the multiple cell detection method, the AN/ASQ-502 employs a single orientated caesium cell as the principal element. It is an optically pumped system, but the unwanted dead zones and heading error effects which are normally minimised by the multi-cell technique, are counteracted in the AN/ASQ-502 by orientating the single caesium cell on a mechanical gimbal system. This, combined with an inherently stable self-oscillator caesium sensor loop and a low-noise, infinite dynamic range, frequency-to-voltage converter, is claimed to yield the most sensitive airborne magnetometer available.

CHARACTERISTICS

Sensitivity: 0.01 γ (10^{-7} oersteds)

Detecting head figure of merit (FOM)

Uncompensated: 1.50 γ or less

Compensated: 0.30 γ or less

Detector: Single orientated caesium cell, optically-pumped self-oscillator Larmor frequency 3.5 Hz/γ

Operating range: 0.20 – 0.75 oersteds

Larmor output (total field): 14 Hz/γ

Analogue output No 1 (minimally filtered): 105 mV/γ uncompensated and 30 mV/γ compensated

Analogue output No 2: 15.0 V p-p fsd

FSD sensitivities: 20, 10, 4, 2, 1, 0.4, 0.2 γ

HP filter: 0.04, 0.06, 0.1, 0.15, 0.3 Hz

LP filter: 0.4, 0.8, 1.2, 1.6 Hz

Military specification: MIL-E-5400

Operating temperature range: –54 to +71°C

Storage temperature range: –62 to +85°C

Power requirements: 115 V AC 400 Hz, 3-phase, 120 VA and 5 V AC/DC 10 VA for panel illumination

Specified reliability: MTBF 1000 h

STATUS

Service test models have been delivered and successfully evaluated by the Royal Navy for use on the Lynx helicopter. Successful evaluation has also been completed by the RAF on the Nimrod, and by the Canadian Forces on the National Aeronautical Establishment MAD aircraft. Production quantities have been delivered for the Canadian Forces' CP-140 Aurora. Several other countries have expressed interest in the equipment for use in their ASW aircraft.

CONTRACTOR

CAE Electronics Ltd, PO Box 1800, St Laurent, Montreal, Quebec H4L 4X4, Canada.

4621.353

AN/ASQ-504(V) MAD SYSTEM

The AN/ASQ-504(V) Advanced Integrated MAD System (AIMS) is designed to provide optimum performance when used on helicopters, fixed-wing aircraft and lighter-than-air platforms. AIMS is a fully automatic magnetic anomaly detection (MAD) system which improves detection efficiency while significantly reducing operator work load. For helicopter installation, the detecting head is fixed to the aircraft body. This provides true 'on top' contact when over a target by eliminating the delay inherent in a towed detecting system.

The AN/ASQ-504(V) system combines sensitivity and accuracy with ease of operation, eliminates aircraft generated interference fields and delivers continuous, automated feature recognition. Contact alert, both visual and audible, is provided. Slant range via an LED display allows the operator to determine if he is within target acquisition range.

AIMS eliminates hazards associated with towed systems and permits high speed surveillance and manoeuvrability, thereby increasing patrol range, improving detectability and substantially reducing false alarms. When used with dipping sonar, transition from one system to the other can be performed quickly and effectively.

The AIMS package weighs 49 lb (21.7 kg) and consists of a 0.005 nano tesla optically pumped magnetometer, a vector magnetometer, amplifier computer and a control indicator. The system operates from a single phase, 115 V, 400 Hz power source and requires less than 200 VA power. AIMS has the flexibility either to perform independently or to accept and execute commands from common control/display units via a 1553 data bus.

CHARACTERISTICS

Detecting head: Caesium, optically pumped, oriented

Sensitivity: 0.01 gamma (in-flight)

Features: Automatic target detection. Operator alert, visual and audible. Slant range estimate to 2800 feet (854 m)

Outputs

Visual: Contact alert and estimated slant range

Analogue: Adaptive MAD signal

Digital: Display interface to MIL-STD-1553

Audio: ICS alert

Slant range: Up to 2800 ft (854 m)

Dimensions: 178 mm diameter × 813 mm long (detecting head); 193 × 257 × 559 mm (amplifier/computer); 152 × 152 × 152 mm (vector magnetometer); 190 × 145 × 145 mm (control indicator)

Weights: 5.6 kg (detecting head); 13.6 kg (amplifier/computer); 0.5 kg (vector magnetometer); 2 kg (control indicator)

STATUS

Production quantities are currently being delivered for requirements on the RAN Sea Hawk, Grumman Turbo-Tracker, RAF Nimrod and RN Lynx and Sea King. The RN contract was awarded in mid-1987 and calls for 235 systems.

CONTRACTOR

CAE Electronics Ltd, PO Box 1800, Saint Laurent, Montreal, Quebec H4L 4X4, Canada.

Units of AN/ASQ-504(V) MAD system

1792.453

AN/SQS-505 SONAR

The AN/SQS-505 is an evolving family of 7 kHz medium-size search and attack sonars intended for use on vessels of 1000 tons and above. It may be used in a hull-mounted and/or variable depth application.

In the Canadian ships, each vessel has two separate but identical SQS-505s on board. One system is variable depth, the other is hull-mounted retractable. Other navies have used the SQS-505 in hull-mounted only applications, with and without hull outfit retracting mechanisms. It is possible to use the 505 in direct communication with a digital computer either for the transfer of range and bearing information or for assistance to the operator in detection and tracking (see **2047.261** underwater combat system).

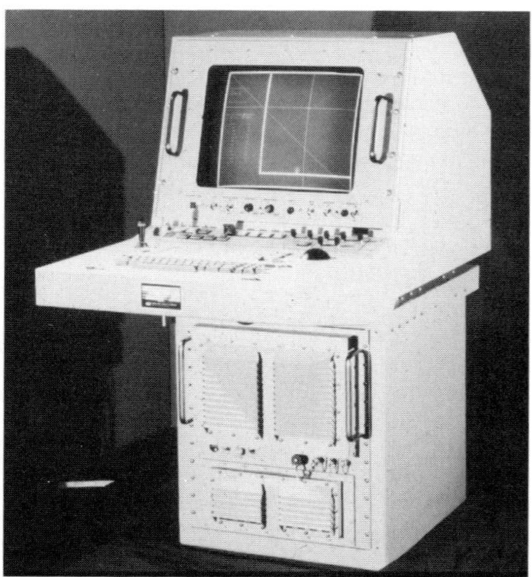

AN/SQS-505(V)6 (AN/UYS-501(V) SHINPADS) control and indicator console

Mounted on major platforms, such as frigates, corvettes and destroyers, the AN/SQS-505 is a deep water sonar surveillance system, proven in operational service and tested to military specifications. It is an omni-directional sonar, also able to operate in a directional mode that can maintain search ability while tracking defined targets. Three selectable frequencies are available so that several ships can survey an area in close proximity without crosstalk. In high sea states which result in excess roll and pitch of the ship, the AN/SQS-505 utilises transmit beam stabilisation. This circuitry delays transmission until near optimum roll and submergence conditions prevail thereby minimising chances of the target being missed altogether.

Hull-mounted Hull Outfits

The SQS-505 hull-mounted transducer can be used in any one of the following applications:

(1) Fixed transducer and dome
(2) Fixed dome/retractable transducer (E-5 hull outfit)
(3) Bow dome.

The requirement for these various alternatives would depend upon the configuration of the ship and the location available for the mounting of the transducer.

CHARACTERISTICS

Coverage: 360° azimuth
Range: 200 to 3200 yards
Doppler range: ±40 knots (7.2 kHz)
Displays: Range (A Scan); Bearing (B Scan); Doppler
Frequenices: 6.4, 7.2 and 8.0 kHz
Source levels: From 224.233 db/puPA/1 m (transmission mode dependent)
Initial detection: 12 ping history of returns
Bearing accuracy: 1 rms
Range accuracy: ±1% range reading
Transmission modes: OMNI, TRDT, DT, ASPECT
Power level: High and low (10 db reduction)

STATUS

The SQS-505 is currently in production. Substantial advancements in the sonar's capability have been incorporated since the system was first installed in ships in the early 1970s.

In addition to the seven Canadian ships outfitted with dual SQS-505 sonars (VDS/HM), four hull-mounted systems have been installed on the latest Belgian 'Weilingen' class (E-71) frigates, the Royal Netherlands Navy has fitted bow dome systems to six of its 'Kortenaer' (F-807) class frigates and has fitted the similar 5 kHz AN/SQS-509 (**4000.453**) to an additional six 'Kortenaer' frigates. The Hellenic Navy has two 'Kortenaer' class frigates fitted with AN/SQS-505 systems.

The latest AN/SQS-505(V)6 sonar system configuration is to be fitted to the six new Canadian patrol frigates of the 'Halifax' class (DDH-330), with fixed, faired dome and retractable transducer.

CONTRACTOR

Westinghouse Canada Inc, PO Box 5009, Burlington, Ontario L7R 4B3, Canada.

4000.453

AN/SQS-509 SONAR

The AN/SQS-509 is a lower frequency version of the SQS-505 (**1792.453**) intended for use on vessels of 2500 tons and above. It has been derived from the SQS-505 by changing the frequency-dependent items, such as the transducer, and consequently there is considerable logistic commonality with the SQS-505. Apart from a lower operating frequency (5.4 kHz), the SQS-509 has the same operational features as the SQS-505 described above.

The transducer is approximately 1.6 m in diameter and 1.6 m high. It consists of 360 individual elements, each of which can be removed while the array is under water. The transmitter group consists of two cabinets, the power supply cabinet and the transmitter cabinet. The power supply converts the ship's prime AC power to DC, required for the operation of the transmitter. The transmitter contains 36 power output modules, each driving one stave of the transducer. It is completely solid-state and contains its own integrated test equipment. The principal transmission modes are OMNI and TRDT.

The receiver is contained in three cabinets of the standard SQS-505 size. It is completely solid-state with its own automatic built-in test equipment which enables the operator to determine the location of a fault down to the replaceable module level.

The control and indicator console consists of a display and controls which enables one operator to have complete control over the operation of the sonar and the ability to detect, classify and track targets. In addition, via separate displays in the console, he is able to maintain continuous independent surveillance of the passive mode. Because of the larger transducer size, the preferred mounting for the transducer is in a bow-dome.

STATUS

The SQS-509 is fitted in six Royal Netherlands Navy 'Kortenaer' class frigates.

CONTRACTOR

Westinghouse Canada Inc, PO Box 5009, Burlington, Ontario L7R 4B3, Canada.

6845.453

SCAN 522/SCAN 545 SHALLOW WATER SONAR

The Scan 522 and Scan 545 enhanced series is intended as a cost-effective solution to the growing need for inshore ASW systems. Available in 22 kHz and 45 kHz models it is designed to operate as a stand-alone search/attack sonar, or in support of other search methods such as dipping sonar. Easily adaptable to 40 to 80 m vessels of 100 to 500 ton displacement, scan technology is a scaled down solution to the ever increasing inshore submarine threat. Search and attack modes are provided and readouts of range, bearing, depth and target strength are displayed to the operator.

The series also features 3 ping history, 3 colour display, true north orientation, vertical beam steering, audio output, horizontal map view, vertical depth view, search and attack modes. Options include true motion, reduced power operation, transmit delay, weapon systems interface, RS 232 output of cursor range and bearing, and slave displays.

STATUS

In production.

CONTRACTOR

Westinghouse Canada Inc, PO Box 5009, Burlington, Ontario L7R 4B3, Canada.

6846.453

AN/SQS-510 DIGITAL SONAR SYSTEM

The AN/SQS-510 is a new digital sonar system that takes advantage of the DND AN/UYS-501 digital signal processor and the AN/UYQ-501 colour display system, both produced by Computing Devices, to replace the receiving and display elements of the current in service AN/SQS-505 sonar (**1792.453**). It is intended to provide surface units with a hull-mounted active detection and localisation system, either as prime sensor, or as a complementary sensor to the AN/SQR-510 CANTASS (**5863.253**).

Under a teaming arrangement, Westinghouse Canada will continue to supply the hull-mounted transducer equipment and the sensor transmitter unit for integration by Computing Devices with the processor and display.

STATUS

A contract to provide an engineering development model was awarded to Computing Devices in mid-1987. This award is part of a Canadian government plan to equip the fleet with a modern sonar system and to meet a NATO committment to provide this sonar to the new Portuguese frigate programme.

CONTRACTOR

Computing Devices Company (a division of Control Data Canada Ltd), PO Box 8508, Ottawa K1G 3M9, Canada.

5863.253

AN/SQR-501 CANADIAN TOWED ARRAY SONAR SYSTEM (CANTASS)

The AN/SQR-501 CANTASS equipment is a critical angle surface ship towed array which uses two Computing Devices standard Canadian Navy products: the AN/UYS-501 array processor and the AN/UYQ-501 SHINPADS display (see illustration in entry **1792.453**). CANTASS employs the Gould AN/SQR-19 (**4463.253**) towing winch and hydrophone subsystem, and modular interfacing and data management electronics. This provides a capable, cost-effective, long-range detection for surface ships.

The system provides frequency and bearing analysis of acoustic emissions from long ranges, in particular for nuclear submarines. Data is presented to operators on programmable CRT displays in a number of alternative formats. A variety of processing algorithms have been developed by the Canadian Navy Defence Research Establishment and Computing Devices Company.

The implementation of the system processing and data manipulation algorithms is entirely in software, providing the system with a growth capability and the adaptability to meet changing needs. The modular physical configuration and the use of standard components allows the system to be adapted to other platforms, such as submarines.

STATUS

In development under a contract from the Canadian Navy for the new Canadian Patrol Frigate. An advanced development model was delivered in January 1988 for trials purposes.

CONTRACTOR

Computing Devices Company (a division of Control Data Canada Ltd), PO Box 8508, Ottawa, Ontario K1G 3M9, Canada.

6233.453

MODEL 971/977 COLOUR IMAGING SONARS

The Mesotech Model 971 colour imaging sonar was designed for use in a variety of ROV applications, including search and recovery, obstacle avoidance and minehunting. The standard 971 system consists of a sonar head, processor and display screen, and is lightweight, compact and versatile. A large selection of head and transducer options are available which can be adapted for specific requirements.

The sonar head has two scanning techniques: imaging and profiling. Imaging uses a fan-shaped beam, usually scanning horizontally to provide an overall picture of the underwater situation. The head can be deployed on a long cable with a compass option that shows orientation of the head with respect to the bottom. The profiling technique uses a cone shaped sonar beam, usually scanning vertically to generate digitised profiles on the bottom.

The display provides a clear detailed image in colour of objects at ranges from 0.5 to hundreds of metres. Display can be in any one of five modes:

(1) Sector – the system scans and displays in a preselected arc
(2) Polar – 360° continuous scan and a PPI type display
(3) Side Scan – the sonar head is locked in one position and images are displayed as scanned by vehicle movement
(4) Linear – gives an accentuated view of close objects
(5) Perspective – gives an impression of depth.

A ×2 or ×4 magnification allows detailed inspection of a target.

The Model 977 fast scanning sonar combines the Model 971 high resolution image technology with significantly increased scanning speed. The 330 kHz version was designed for obstacle avoidance or minehunting applications on faster moving vehicles or submersibles, while the 100 kHz version is used for stationary deployment in harbour surveillance applications.

STATUS

In full scale production and in service.

CONTRACTOR

Mesotech Systems Ltd, 2830 Huntington Place, Port Coquitlam, British Columbia VC3 4T3, Canada.

4222.353

AN/ASA-64 SUBMARINE MAGNETIC ANOMALY DETECTOR

The AN/ASA-64 is a primary signal processor designed to identify and mark submarine-produced local distortions in the earth's magnetic field. When such an event occurs, the operator is alerted by visual and audio alarms, thereby eliminating the requirement for constant signal monitoring.

CHARACTERISTICS

Dimensions	Length (mm)	Width (mm)	Height (mm)	Weight (kg)
Control unit	102	146	90	0.68
ID-1559	229	150	153	2.95

Operating temperature range: –54 to +71°C
Military specification: MIL-E-5400
Power requirement: 115 V AC, 400 Hz, 20 VA

STATUS

Production quantities have been manufactured and delivered for all P-3C users worldwide. CAE has completed a product improvement programme with more processing power. This design improvement has been evaluated by the Royal Navy, Royal Air Force, and the United States Navy. A variant for helicopters is in development.

CONTRACTOR

CAE Electronics Ltd, PO Box 1800, St Laurent, Montreal, Quebec H4L 4X4, Canada.

AN/ASA-64 submarine magnetic anomaly detector equipment

3231.393

AN/ASA-65 COMPENSATION GROUP ADAPTOR

The compensation group adaptor (CGA) has been developed to increase operational efficiency in existing installations of the AN/ASA-65 magnetic compensator group (**1388.393**). By the incorporation of *total system* built-in test features and automatic data gathering, preflight, troubleshooting and compensation update requirements are significantly reduced with a corresponding improvement in performance.

The CGA consists of a small indicator, magnetic field and a computer, magnetic field containing signal conditioning circuits and a special-purpose micro-computer. Installation does not require any modifications to the existing AN/ASA-65 equipment. Input data from the AN/ASA-65 and the detector magnetometer are used by the micro-computer to calculate the numerical values which indicate changes required for each of the nine terms to produce optimum compensation.

No stringent manoeuvre requirements are necessary. Data to update compensation terms require a simple flight programme that involves low amplitude manoeuvres on four headings. The total time required for this programme is no more than six minutes after which the operator is required to adjust the AN/ASA-65 terms as indicated on the CGA control.

CHARACTERISTICS

Figure of merit: Equal or better than best performance obtainable from standard AN/ASA-65 (generally 1 γ or less)

Standard update compensation
Terms compensated: 9
Time: 8 minutes
Manoeuvres required: Small random type on 4 headings

Weapon drop compensation
Terms compensated: 4
Time: 2 minutes
Manoeuvres required: Small manoeuvres on four headings

Dimensions (CGA only)

	Height (cm)	Width (cm)	Depth (cm)	Weight (kg)
Control unit	9.5	14.6	12.7	1.35
Amplifier converter	22.1	22.9	41.8	10.0

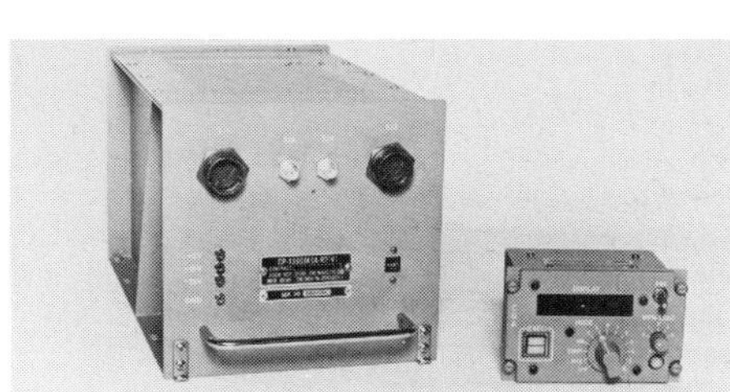

AN/ASA-65 compensator group adaptor

STATUS

Fitted to RAF Nimrod, all P-3C users worldwide, USN-S3A Viking maritime aircraft and JDA P-2J.

CONTRACTOR

CAE Electronics Ltd, PO Box 1800, St Laurent, Montreal, Quebec H4L 4X4, Canada.

1388.393

AN/ASA-65(V) NINE-TERM COMPENSATOR

The CAE nine-term compensator (AN/ASA-65(V)) provides semi-automatic compensation for aircraft-generated manoeuvre interference signals. By minimising unwanted interference signals, it substantially improves MAD detection capability, especially during tactical and turbulent flight conditions. This equipment compensates all significant permanent, induced and eddy-current interference fields in approximately 30 to 45 minutes. Permanent terms can be trimmed up on each flight in about five minutes compared to close to one hour for manual compensation systems.

The AN/ASA-65(V) is compatible with all existing magnetometer systems in present use. Accommodation for the varying compensation requirements of

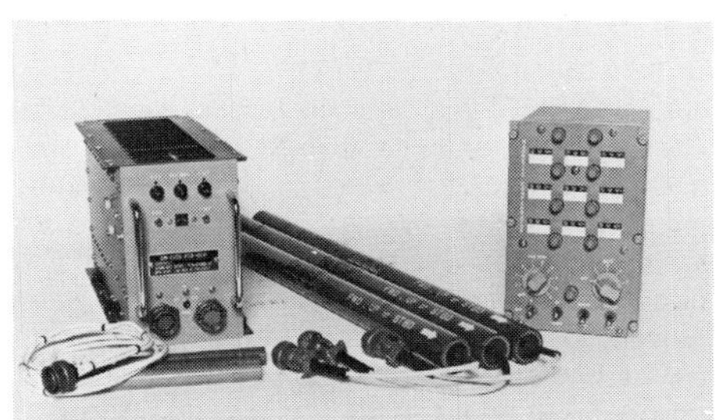

AN/ASA-65(V)2 nine-term compensator

different aircraft types is conveniently accomplished by means of internal patch connectors. The equipment satisfies the requirements of applicable military specifications, and using state-of-the-art techniques has a demonstrated MTBF of more than 1800 hours. The use of completely solid-state circuitry has resulted in a total system weight of less than 13.6 kg.

The compensation figure of merit is stated to be consistently less than 1 gamma, and the equipment is capable of accommodating maximum compensation fields of 50 gamma on each aircraft axis.

CHARACTERISTICS

Dimensions	Height (mm)	Width (mm)	Depth (mm)	Weight (kg)
Control indicator	229	146	165	3.85
Electronic control amplifier	197	149	346	7.94
Magnetometer assembly	152	152	152	0.68
Coil assembly (V)1	89	89	89	0.90

STATUS

Production quantities have been manufactured and delivered to the RAF for the Nimrod maritime aircraft, and the AN/ASA-65(V) has also been delivered for all P-3C users worldwide, USN S-3A and JDA P-2J.

CONTRACTOR

CAE Electronics Ltd, PO Box 1800, St Laurent, Montreal, Quebec H4L 4X4, Canada.

1980.393

0A-5154/ASQ AUTOMATIC MAD COMPENSATION SYSTEM

The 0A-5154/ASQ is another system in the series of CAE compensator systems. This system was developed for use with boom-mounted magnetometers to minimise manoeuvre-related noise caused by the effect of magnetic fields inherent in the aircraft structure from interfering with underwater submarine detection or geological surveying. Excellent compensation combined with the ability to update minor variations rapidly and efficiently increases detection sensitivity to the limits of the magnetometer equipment.

The system is completely self-contained providing signal compensation by means of a micro-computer. The uncompensated 'raw' magnetometer input signal is conditioned and produced in a compensated state. This approach eliminates the requirement for output coils which were necessary to generate opposing fields in previous systems. Other than the 'raw' detector magnetometer signal, input requirements are limited to orthogonal vector magnetometer signals that provide heading, manoeuvre and total Earth's field information. The micro-computer is programmed to process this input data and both generate compensation requirements and update any changes on operator command.

No stringent manoeuvre requirements such as purity or accuracy of aircraft heading are necessary. Data to update all 16 independent compensation terms are simultaneously evaluated during a simple flight programme which involves performing small aircraft manoeuvres on four headings. The total time required for this is no more than eight minutes. The equipment requires no operator involvement other than to initiate the recompensation procedure and update request. Incorporation of built-in test functions and fault indicators allows the operator or technician to perform a systematic checkout of the equipment to the WRA and SRA level.

CHARACTERISTICS

Figure of merit: $<0.4\gamma$

Residual interference signals: $<0.03\gamma$ average

Terms compensated: 16

Compensation time: 8 minutes

Aircraft manoeuvre requirements: 1 minute of manoeuvres on each of 4 approximate orthogonal headings. If desired, other manoeuvres, such as trapping circles, or clover-leafs can be used

Military specification: MIL-E-5400

External input requirements: Minimally filtered analogue output from detector magnetometer

Power requirements: 115 V AC, 400 Hz, and 28 V AC/DC, 10 VA for panel illumination

Specified reliability: 1000 h MTBF

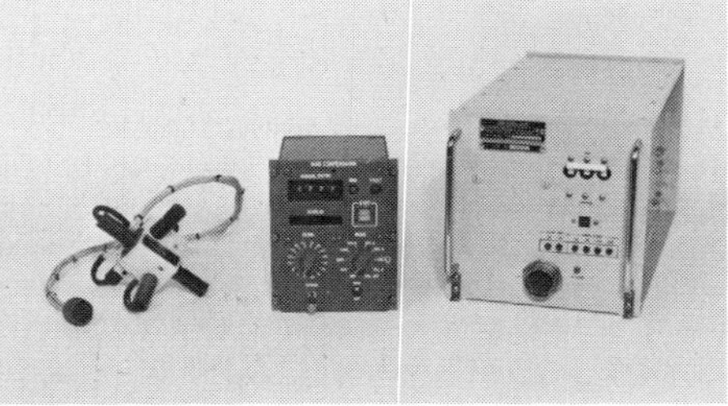

0A-5154/ASQ automatic MAD compensation system

Dimensions	Height (cm)	Width (cm)	Depth (cm)	Weight (kg)
Control indicator	18.2	14.7	16.6	2.2
Electronic amplifier	22.9	22.9	54.6	11.3
Vector magnetometer	15.2	15.2	15.2	0.7

STATUS

Production quantities have been delivered for the Canadian CP-140 Aurora and the Breguet Atlantic for the Netherlands and West German navies.

CONTRACTOR

CAE Electronics Ltd, PO Box 1800, St Laurent, Montreal, Quebec H4L 4X4, Canada.

3086.453

AN/SSQ-522 SONOBUOY

The AN/SSQ-522 is an active sonobuoy developed specially from the AN/SSQ-47B to meet the requirements of the Canadian Forces. The SSQ-47B is produced by both Sparton of Canada and the parent Sparton company in the USA and is described in **1988.453**.

CHARACTERISTICS

Type: Active

Sonar modes: Automatic keyed CW

Transducer depth: 60 or 800 ft (18 or 244 m)

Weight: 24 lb (10.9 kg)

Sonar channels: 6 (HF)

RF channels: 12

Operation of the SSQ-522 is identical to that of the SSQ-47B, except that up to 12 buoys can be launched and operated simultaneously. Two different CW sonar modes are selectable by choice of VHF channel; even channel buoys have sonar modes the same as the SSQ-47B, and odd channel buoys have longer pulse-lengths and intervals.

CONTRACTOR

Sparton of Canada Ltd, London, Ontario, Canada.

3780.453

AN/SSQ-527B SONOBUOY

The AN/SSQ-527B is an omni-directional, passive sonobuoy which enables fixed-wing maritime patrol aircraft, helicopters and surface ships to conduct underwater search and surveillance using low frequency analysis and recording (LOFAR) techniques.

The AN/SSQ-527B is identical in function to the US Navy AN/SSQ-41B but incorporates features which improve detectability of low level, low frequency signals in a noisy sea environment. These features include a sub-surface assembly suspension system which maintains hydrophone depth within specified limits in shear currents up to 2.5 knots and an acceleration noise cancelling hydrophone.

The sonobuoy can be launched at airspeeds between 30 and 300 knots from altitudes of 30 to 7600 metres. Descent is controlled and stabilised by a parachute. Immediately after entry the sea water activated battery energises a mechanism which deploys the sonobuoy. A flotation bag is inflated and carries the upper electronics assembly, containing a VHF transmitter, to the surface while the hydrophone descends to a pre-selected operating depth. Three depths are available: 30, 100 or 300 metres. The sonobuoy becomes operative within 60 seconds of impact at the minimum depth and within 100 seconds at the maximum depth. It remains in operation for a minimum of 1, 3 or 8 hours, according to the pre-selected operating life.

The AN/SSQ-527B can be launched by hand from a ship moving at up to 30 knots without it being necessary to activate mechanical or electrical devices before launch. The electrically activated deployment mechanism functions regardless of attitude on entry since it is not dependent on impact forces acting on the bottom plate.

Rapid Ice Penetration Sonobuoy Systems

The RIPSS consists of a standard SSQ-527B sonobuoy supplemented with a thermochemical charge capable of penetrating arctic ice at better than 1 ft (0.3 m) per minute. The current development is designed to produce a sonobuoy to the experimental development stage capable of penetrating up to 8 ft (2.4 m) in less than 20 minutes. The ice drilling buoys are the joint design of Sparton of Canada and Consolidated Technologies Ltd. A family of 20 thermochemicals has been developed in liquid and solid forms. Design requirements are to effect both downward and upward drilling in self-deployed packages. The current development is being conducted in support of Canada's defence scientists responsible for Arctic operations.

CHARACTERISTICS

Type: Passive omni-directional

Audio frequencies: 10 – 5000 Hz

Depth: 30, 100 or 300 m

Size: A

Weight: 7.5 kg

Operating life: 1, 3 or 8 h

Launch altitude: 30 – 7600 m

Launch speed: 30 – 300 knots

RF channels: 31 (162.25 – 173.5 MHz band)

RF power: 1 W

STATUS

In production.

CONTRACTOR

Sparton of Canada Ltd, London, Ontario, Canada.

4620.453

AN/SSQ-530 DIFAR SONOBUOY

The AN/SSQ-530 is a directional, passive sonobuoy which provides ships, and fixed and rotary wing maritime aircraft with the dual capability of detecting and determining the source direction of underwater acoustic energy radiated by submarines. VHF transmissions containing this information are received by the launch vehicle and are analysed and processed to find the true bearing of the source using directional low frequency analysis and recording (DIFAR) techniques.

The AN/SSQ-530 is an improved version of the AN/SSQ-53A DIFAR sonobuoy (**1990.453**) and is able to select a third intermediate hydrophone operating depth of 100 metres. Depths below the minimum depth setting of 30 metres can be adjusted to suit operating environments. Storage life has been increased from 5 to 7 years.

Launch is possible from aircraft at airspeeds between 30 and 300 knots from altitudes of 30 to 7600 metres. Descent is controlled by parachute. On entry, a sea water activated battery starts deployment of the buoy which becomes operative for one or four hours according to the pre-selected operating life. Ship launch is helped by the electrical activation which ensures reliable deployment regardless of attitude on entry since it is not dependent on water forces acting on the bottom plate.

The sonobuoy is compatible with the AQA-7,

OL-82(AYS) and AQS-901 processors. The Sparton TD-1135A demultiplexer/processor/display can provide a self-contained stand-alone DIFAR capability for ships, helicopters and aircraft.

CHARACTERISTICS

Type: Passive, directional LOFAR (DIFAR)
Function: Search and localisation
Hydrophone depth: 30, 100 or 300 m
Weight: 10.2 kg
RF channels: 31
RF power: 1 W
Launch altitude: 30 – 7600 m
Operating life: 1 or 4 h

STATUS

Developed under Canadian government contract. In production.

CONTRACTOR

Sparton of Canada Ltd, London, Ontario, Canada.

4649.363

AN/UYS-503 SONOBUOY PROCESSOR (SBP)

The SBP sonobuoy processor is a small, lightweight, low cost multi-sonobuoy equipment designed for a variety of ASW platforms. It employs a new concept in processor architecture and uses a Computing Devices proprietary universal demultiplexer to assure future growth capability.

The SBP is made from four identical, independent processor modules, also known as 'slices'. The processing capability of each slice when used with passive sonobuoys is 7 bands from DIFAR (1 surveillance and 6 vernier) and 14 bands from omni (2 surveillance and 12 vernier). The complete system, therefore, has a total of 28 bands when processing four DIFAR buoys and 56 for eight omni-buoys. All bands are processed and stored continuously for real-time viewing. In the active mode, each slice is able to process one range-only active or one DICASS buoy. The latter is processed in both CW and FM modes. The range-only buoy is processed in the same way as the DICASS CW pulse. For bathythermal buoys the data is spectrum analysed and the output tone detected and converted to temperature.

A single SBP system is capable of handling eight omni-passive, four DIFAR, four DICASS, four omni-active, bathythermal buoys and any combinations of these. The processing design of the equipment is such that if one module fails the remaining three are unaffected and continue to function normally. Display formatting is designed to enable the operator to view several of the processing streams simultaneously. The system display monitor is a standard 875-line TV screen with the basic display format dividing the screen into four areas. An operator controlled cursor is provided for alphanumeric readout of any display parameter (frequency, time, range, bearing, etc).

CHARACTERISTICS

Frequency range: 0 – 2560 Hz
Input channels: 8 standard sonobuoy receivers
Size: 1 ATR short – 22.9 × 19.0 × 30.5 cm (9 × 7.5 × 12 in)
Weight: 44 lb (20 kg) – electronics unit; 132 lb (60 kg) – complete system including display and control

STATUS

In production. Three systems delivered to the Royal Swedish Navy in 1985. Selected for the new Australian Navy Sikorsky S-70B helicopter. Eleven systems on order for the Canadian Patrol Frigate and Trump.

CONTRACTOR

Computing Devices Company (a division of Control Data Canada Ltd), PO Box 8508, Ottawa, Ontario K1G 3M9, Canada.

6551.353

AN/AQS-503 HELICOPTER ACOUSTIC PROCESSING SYSTEM

The AN/AQS-503 is a advanced, lightweight sonobuoy/sonar helicopter acoustic processing system (HAPS) which has been designed for the Canadian Department of National Defence. It consists basically of the Plessey Cormorant lightweight dipping sonar (**5637.353**) as the wet-end, a high-speed winch and handling gear developed by Fathom Oceanology (**6229.263**), and the AN/UYS-503 sonobuoy processor (**4649.363**) which will provide the signal processing for sonobuoys and for the Cormorant.

The HAPS system is lightweight, provides comprehensive sonobuoy processing of DIFAR, OMNI, BATHY and VLAD in any mix of buoys, with simultaneous sonobuoy and sonar processing and display. The sonobuoy and sonar outputs can be displayed simultaneously or individually on a TV display. Multi-sensor control is achieved by means of the CDC programmable electroluminescent control panel.

CONTRACTORS

Computing Devices Company (a division of Control Data Canada Ltd), PO Box 8508, Ottawa, Ontario K1G 3M9, Canada.

Plessey Naval Systems, Templecombe, Somerset BA8 0DH, England.

Fathom Oceanology Ltd, 6760 Campobello Road, Mississauga, Ontario L5N 2L8, Canada.

6552.363

BARRA SIDE PROCESSOR (AQA-801)

The Barra side processor is a private venture development of Computing Devices and is designed to provide ASW operators with the capability to process the Barra sonobuoy without modifying existing sonobuoy processors. The processor, which is in production for the Royal Australian Navy S-70B ASW helicopters as the AQA-801, connects to any existing sonobuoy processor and provides a four-Barra buoy processing capability.

The Barra buoy (**3998.453**) has been in operation with the RAAF and RAF for some years and was the first and only passive directional sonobuoy that provides 360° of broadband acoustic data. Using the AQA-801, and two Barra buoys, ASW crews are provided with broadband data that is sufficiently accurate for an attack without alerting the target.

STATUS

In production for the Royal Australian Navy. The US Navy has placed an order for Barra Side Processors in order to evaluate the Barra system, with the processor connected to either an AQA-7 or Proteus (UYS-1).

CONTRACTOR

Computing Devices Company (a division of Control Data Canada Ltd), PO Box 8508, Ottawa, Ontario K1G 3M9, Canada.

6229.263

FATHOM SONAR HANDLING EQUIPMENT

Fathom Oceanology specialise in the provision of sonar towed bodies, handling equipment and sonar domes for underwater defence and commercial applications.

For cable handling and towing systems applicable to variable depth sonar (VDS) systems, Fathom has a range of equipment which is normally mounted at the aft end of the ship and is self-contained, apart from electrical power requirements. The range includes the 6-200, 9-330, 9-600, 15-750 and 10.5-600 systems.

The Towfish and Rayfish are general purpose towed bodies designed to suit a wide range of applications. They are used to house water sampling, seismic side-scan, sub-bottom profiling, video, magnetometer and fish-finding sonar systems. Both bodies have low flow noise characteristics and access to instrumentation through easily removable shells.

Fathom also designs, develops and manufactures a range of sonar domes. These are configured by optimising acoustic, hydrodynamic and structural parameters for specific operational frequency and bandwidth, and incorporate a stainless acoustic window of double skin design.

STATUS

In production.

CONTRACTOR

Fathom Oceanology Limited, 6760 Campobello Road, Mississauga, Ontario L5N 2L8, Canada.

6232.263

DEPTH COMPENSATED RING SHELL PROJECTOR

The Depth Compensated Ring Shell Projector is an underwater sound source that can efficiently produce a high acoustic output near its fundamental resonance, or a moderate output over a broad band. With its passive pressure compensation system, a ring shell projector has an operating depth exceeding 250 m and a safe immersion depth exceeding 500 m.

Depth compensated ring shell projectors are being used in a variety of applications where high power and efficiency are required at relatively low frequencies. As a research tool the projectors have been employed as single units or as part of a large array. Further, passive ASW training is available without the use of a target submarine by generating signals of desired frequencies and levels. The projector can also be adapted for use as a separate active element for current passive arrays, or as part of a low frequency deep submergence sonar.

The resonance frequency of a ring shell projector of a fixed size can be selected at the time of construction within a three octave band by selecting the volume of ceramic in the ring. Furthermore, the bandwidth at a given resonance frequency can be selected by changing the projector diameter. Design information for the construction of the ring is generated by finite element analyses using the DREA finite element program known as MVART. MVART data has been used to design two current productions whose characteristics are listed below.

CHARACTERISTICS

Overall outside diameter: 445 mm (standard); 815 mm (Mk 2)
Thickness: 92 mm (standard); 305 mm (Mk 2)
Weight: 41 kg (standard); 182 kg (Mk 2)
Resonant frequency: 290 Hz at 10 m depth; 325 Hz at 500 m depth (standard); 600 Hz at 30 m (Mk 2)
Power factor: 37% (standard); 47.7% (Mk 2)
Acoustic output: 1580 W

DEVELOPMENT

The ring shell underwater sound projector was developed by the Defence Research Establishment Atlantic, Dartmouth, Nova Scotia, and transfer of technology to Sparton of Canada was initiated in 1981. Since that time models of various sizes and weights have been produced incorporating industrial design improvements to increase the overall operational efficiency.

STATUS

Contracts with USN NUSC and FRG BWB have been negotiated for a total of 31 units.

CONTRACTOR

Sparton of Canada Ltd, PO Box 5125, London, Ontario N6A 4N2, Canada.

6234.263

AN/SQQ-504 TOWING CONDITION MONITOR

The AN/SQQ-504 towing condition monitor is an aid to ship's personnel for successful launch, tow and recovery of the variable depth sonar (VDS) body. With the introduction of this type of equipment, the command and sonar operations are provided with a badly needed means to optimise operations and minimise risk to VDS equipment due to the variability of sea conditions, ship's speed and motion, etc.

The monitor provides VDS tow cable strain information (actual strain when the body is under tow or predicted strain when the body has not been launched or is to be towed at a different cable length) to the operator and ship's command. The system also provides information by which the VDS operator can judge the performance of the VDS hoist system boombobber, and provides information to enable personnel to monitor the degrading effect of ship's motion on the VDS sonar detection.

The system consists of a group of data sensors, a processor unit, two recording units, three identical remote display units, and a VDS operator's control and display unit. These items perform the functions of data acquisition, signal conditioning, data processing, display and recording. Sensors include boom and deck accelerometers, a cable strain guage, a boom switch, a cable length synchro and ship's speed data.

STATUS

The AN/SQQ-504 is fitted to Canadian Navy destroyers.

CONTRACTOR

Westinghouse Canada Inc, Box 5009, Burlington, Ontario L7R 4B3, Canada.

AN/SQQ-504 towing condition monitor, showing primary indicator (left) and remote indicator (right)

FINLAND

6553.453

FISKARS HYDROACOUSTIC SURVEILLANCE SYSTEM

The Fiskars Hydroacoustic Surveillance System (FHS) is a fixed passive surveillance system for the automatic detection and tracking of underwater and surface objects. The system has been optimised for the hydroacoustic conditions encountered in coastal waters and at the entrances to harbours.

The system is based on an array of omni-directional hydrophones which are positioned on the sea bed and connected to the surveillance centre. The cables are specially designed for the system in order to minimise the required investment for the cable network, and the length between a hydrophone and the centre can be up to 30 km long. The FHS can be used as a local surveillance system, or the detection and tracking data can be sent automatically to a higher level surveillance centre. As a modular system the FHS can be customised according to the user's needs. It consists basically of the following main components:

(a) up to 96 expendable hydrophones covering the surveillance area, which can be divided into several surveillance fields
(b) Central unit
(c) Control unit.

The Central unit is a multiprocessor system using parallel processing and distributed functions. The main tasks of the processors are auto alarm at detection; calculation of the position, speed and course of the target; fast Fourier analysis; and communication with other systems and a higher level surveillance centre.

The Control unit features a high resolution CRT display and keyboard, headsets, tape recorders and analysers. Data displayed on the screen includes a map for the surveillance area or field with location of sensors; the sea state around individual sensors; location, course and speed of target; a plot of the route of the target; and the frequency distribution and level of the target's acoustic signature.

CONTRACTOR

Fiskars Oy Ab Elesco, PO Box 19, SF-02361 Espoo, Finland.

FRANCE

3791.363

TSM 8200-8202 AIRBORNE SADANG ACOUSTIC SYSTEM

Installed aboard fixed- or rotary-wing aircraft, this equipment processes and displays signals received from various types of sonobuoy. Modular design permits optimum system configuration to suit all sonobuoy types and the specific number to be processed simultaneously. All signals received are permanently recorded before being processed to allow more extensive utilisation of data at shore-based processing stations. Categories of sonobuoy whose data can be processed include: passive buoys (directional and omni-directional); active buoys (directional and omni-directional); environmental buoys such as bathythermographic, sea noise measuring, etc; and numerous buoys of the same or several different types can be handled simultaneously.

TSM 8200 airborne acoustic system installed on French Navy 'Atlantic' maritime patrol aircraft

Three main elements comprise the TSM 8200 equipment:

(1) Communications unit, in which will be found a 99-channel VHF receiver, a UHF transmitter/receiver for active sonobuoy remote control, and ASSG overall test systems. Each set is associated with a control box for the system operators
(2) Signal processing and display unit. This carries out analogue pre-processing and full processing functions in the active and passive modes using digital processors and appropriate software. Information is presented by a high density graphic recorder and a CRT display. The latter is associated with a high-capacity bulk storage unit and numerous image formats are available by means of elaborate software appropriate to the specific types of buoy and signal processing. The dual processor configuration is normally operated by two operators using an alpha-numeric keyboard and tracker ball. A digital data bus permits dialogue between the operators and the tactical co-ordinator, via the tactical computer. Several signal processing and display units can be associated aboard the same aircraft so that system performance can be enhanced by increasing the number and types of buoys which can be processed simultaneously. An on-top-position indicator is used in conjunction with a radio compass for homing towards the sonobuoy
(3) Magnetic recording unit. This multi-track recorder records all the sonobuoy data at the analogue processing level for subsequent analysis ashore.

Other optional equipment includes an acoustic signal generator simulator for test purposes, and remote control of active sonobuoys.

STATUS

In production.

CONTRACTOR

Thomson Sintra, Activités Sous-Marines, 1 avenue Aristide Briand, 94117 Arcueil Cedex, France.

4347.363

TSM 8220/8221/8222 (LAMPARO) LIGHTWEIGHT AIRBORNE SONOBUOY PROCESSING EQUIPMENT

Derived from the TSM 8200 (**3791.363**), these airborne sonar signal processing and display systems are designed as modern digital equipments for ASW fixed or rotary wing aircraft.

The processing capabilities of these systems are based on a single processing unit in a lighter and more compact packaging. It is manned by a single ASW operator. The same wide panel of facilities and processing modes is available.

The TSM 8220 is orientated towards the processing of omni-directional passive and active buoys.

The signals are displayed on a CRT in TV format and a hard copier can be connected for permanent recording. An on-top-position indicator is used in conjunction with a radio compass for homing on the sonobuoy. Weight of the complete system is 140 kg for the TSM 8222 and 70 kg for the TSM 8220.

STATUS

The TSM 8220 is operational.

CONTRACTOR

Thomson Sintra, Activités Sous-Marines, 1 avenue Aristide Briand, 94117 Arcueil Cedex, France.

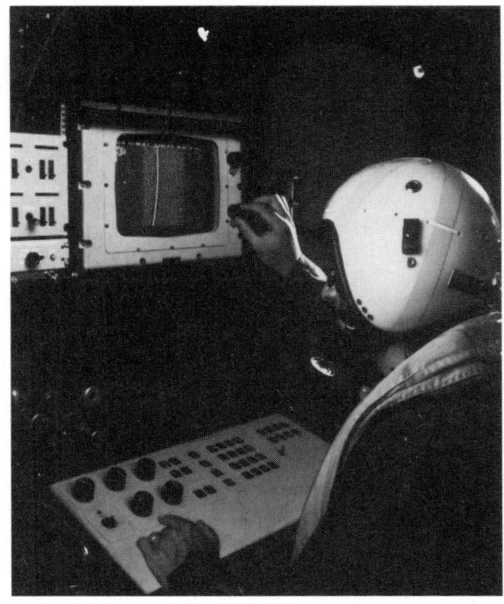

Lamparo sonobuoy processing system

1159.253

DUBV-23D SURFACE VESSEL SONAR

The DUBV-23D is a bow-mounted, low-frequency, panoramic sonar for anti-submarine operations. The 48-column transducer array is housed in a bulb at the forepart of the ship, the bulb being of streamlined design to reduce parasitic noises to permit listening at high speeds. The panoramic sonar is intended for both search and attack roles.

In addition to the transducer array, the equipment includes the transmitter/receiver unit, a computer section for the processing of data being fed to weapons, and control and display consoles at the anti-submarine attack station.

The DUBV-23 is of identical design to the towed sonar DUBV-43 **(1334.453)** and in French vessels the two sonars are used together for anti-submarine warfare.

DUBV-43 sonar control console

OPERATION

Operating modes provide for: panoramic surveillance, sector surveillance, step surveillance, passive surveillance at sonar frequency, panoramic attack transmission, or 'searchlight' attack transmission. In addition to the system's own display devices, the DUBV-23D provides for target data outputs to other ship's systems and repeater PPIs.

CHARACTERISTICS

Transmitter

Frequencies: 4 operating frequencies in the neighbourhood of 5 kHz, of which 2 are operational

Power: 96 kW (2 × 48 kW)

Type: FF (fixed frequency), FM (linear frequency modulation with non-coherent data processing at reception)

Duration: 4, 30, 150 or 700 ms

Scatter echo: With or without rejection

Doppler effect correction: On all 48 channels

Cadence: Adjustable step by step from 1500 to 48 000 yards

Receiver: Panoramic, directional, passive listening in sonar band

STATUS

The DUBV-23D is in service in converted anti-submarine escorts Types T47 and T56, frigates of the 'Suffren' class and Type C67, and corvettes Type C70 of the French Navy.

CONTRACTOR

Thomson Sintra, Activités Sous-Marines, 1 avenue Aristide Briand, 94117 Arcueil Cedex, France.

1334.453

DUBV-43B/C VARIABLE DEPTH SONAR

The DUBV-43B/C variable depth sonar (VDS) consists of a streamlined towed body called the 'fish', containing the sonar transducer array. This is towed from the rear of the parent vessel at distances of up to 250 metres and can be set to run at depths between 10 and 200 metres in the 'B' version and up to 600 metres in the 'C' version. It is equipped with stabilisers providing for control in roll, pitch, and depth. Dimensions of the 'fish' are: length 550 cm, width 170 cm, and submerged weight 7.75 tonnes. The towing cable also incorporates 48 pairs of signal conductors. The range of towing speeds is 4 to 24 knots, and detection ranges of up to 25 km are quoted.

STATUS

The DUBV-43B is in service in converted anti-submarine escorts Types T47 and T56, frigates of the 'Suffren' class and Type F67. The DUBV-43C is in service with corvettes type C70 of the French Navy.

Two updates of the system are in progress:

(a) UTCS 1B which includes improvement of the signal processor

(b) a torpedo alarm receiver is soon to be added to sonars in service.

CONTRACTORS

Direction des Constructions Navales, 2 rue Royale, Paris 8, France.

Thomson Sintra, Activités Sous-Marines, 1 avenue Aristide Briand, 94117 Arcueil Cedex, France.

DUBV-43 VDS towed transducer and towing gear aboard the French corvette Georges Leygues

1356.453

ELEDONE SUBMARINE SONAR

ELEDONE is a family of submarine sonars, the most comprehensive of which combine the main submarine sonar functions within a single integrated system. The system may be fitted to any size of submarine for operational requirement by the simple addition of standard cabinets and the appropriate transducer arrays. Simultaneous use of these various functions is achieved by the use of a colour display which integrates all relevant data. Built-in redundancy and built-in test facilities maintain availability. The system provides:

Passive listening

This facility gives high sensitivity, fine bearing accuracy and great flexibility of installation in all types of submarines. The number of hydrophone staves can be selected, either 32 staves for small diameters down to 1.2 metres, 64 staves for large diameters up to 3.5 metres or 96 staves for the largest. Specific processing channels can provide from four to twelve automatic tracks. Accurate target data are sent continuously to the tactical system to allow effective target motion analysis.

Passive adaptive processing

The adaptive processing function is based upon optimal array processing theory which minimises the effect of jamming by strong signals, when listening to low level signals. It is a very useful facility in the discrimination between two targets when both are within a common limited sector.

Interception

Interception of all active sonar pulses from low frequency surface ship sonars to high frequency torpedo acoustic heads. Very early warning against enemy sonars, with a very low false alarm rate, is provided. Interception warnings are integrated on the passive listening scope with specific colour (visual correlation). Accurate parameters of pulses are presented on digital read-outs and systematically transmitted to the tactical system.

Active capability

Although initial bearings are provided by the passive detection, the operator can also initiate a low frequency wide sector transmission on the same bearing to determine accurate range. In this mode, reception is performed by the passive listening array and associated with high detection sensitivity.

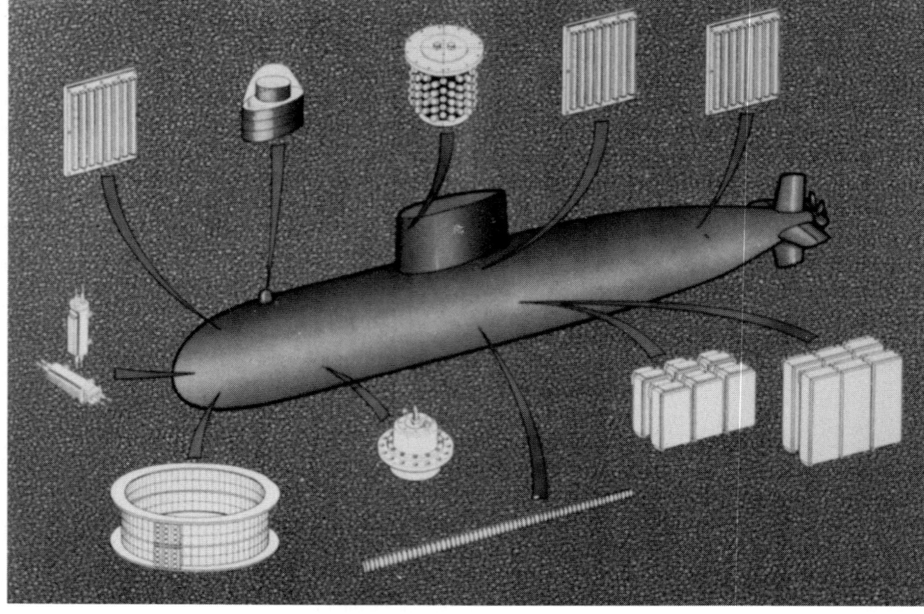

Diagram of ELEDONE submarine sonar system

Spectral analysis

The spectrum analysis facility complements the audio function by providing the operator with specific data in target identification. This analysis can be performed in real-time or played back from memory store. Paper recorders, tape recorders, cavitation warnings, additional mass memory data and other facilities can be provided.

Tactical display

The tactical display system is the operational heart of the submarine and provides a continuous tactical situation review. It provides navigation information, storage and display of sensor data, target parameters analysis, geographical and tactical presentation, target motion analysis and weapons control.

STATUS

In operational service and in serial production for several navies. The sonars are known as Octopus for the version intended for submarines of the 'Walrus' class and Argonaute for the RN '2400' class submarines (**5650.453**).

In mid-1987 the Royal Australian Navy ordered a powerful version of the Eledone sonar for the new Type 471 submarines due to commence fabrication in 1989.

CONTRACTOR

Thomson Sintra, Activités Sous-Marines, Chemin des Travails, BP 53, 06801 Cagnes-sur-Mer, France.

5650.453

ARGONAUTE SONAR SYSTEM

Argonaute is the name of a new sonar system ordered by the Royal Navy (under the designation of Type 2940) for use in its forthcoming Type 2400 class submarines.

Argonaute consists of elements of the ELEDONE sonar which have been modified to interface with Royal Navy equipment and operational requirements.

The ELEDONE system operates in both the active and passive modes, full details are given in the entry **1356.453**.

STATUS

In serial production.

CONTRACTOR

Thomson Sintra, Activités Sous-Marines, Chemin des Travails, BP 53, 06801 Cagnes-sur-Mer, France.

1634.453

DUUA-2A ACTIVE/PASSIVE SONAR

The DUUA-2A sonar is fitted to modernised 'Daphne' class submarines and 1200 tonne submarines, and provides for simultaneous surveillance and attack. It can be used:

(1) For active detection in single (FP), or frequency modulation (FM) modes

(2) For passive detection

(3) As an interceptor, ie location of a sonar source.

The DUUA-2A may also be used for ultrasonic communications purposes and for depth sounding in very deep water.

CHARACTERISTICS

Transmitter: Frequency 8.4 kHz

Power: 20 kW in nominal operation, and 1 kW in reduced operation, SFG and IC. Each emission is manually triggered. Emission duration 30 ms in SFG; 300 ms in FP; 500 ms in FM

Receiver

Pass band: $\pm$170 Hz in FP; $\pm$350 Hz in FM in the neighbourhood of 8.4 kHz; $\pm$500 Hz in EBM in the neighbourhood of a selected frequency between 2.5 and 15 kHz

Range scale in active mode: 3, 6, 12, and 24 km

Transducer: Directional, Type B 88T 8.5, driven by servo mechanism MSS 5 directivity on site and bearing $2\ \theta_3 = 10°$ at 8.5 kHz. Manual orientation by crank on azimuth or automatic orientation at 4 or 8°/s from −175 to +175° on bearing. Reset between stops at 60°/s. Site positioning from +15 to −30°

STATUS

In service in the French 'Daphne', 'Agosta' and 'Narval' classes of submarines. A number of other navies equipped with Type 209 submarines are also fitted with DUUX-2 sonar. These include Ecuador, Columbia, Greece, Peru and Venezuela.

CONTRACTOR

Thomson Sintra, Activités Sous-Marines, 1 avenue Aristide Briand, 94117 Arcueil Cedex, France.

3232.453

DUUA-2B ACTIVE/PASSIVE SONAR

The DUUA-2B sonar can be fitted in all types of submarine and it provides for simultaneous search, attack, sonar interception, and depth sounding in deep water.

Two versions exist:

Model I: Active sonar

Model II: Passive sonar.

Either one or two sonars can be fitted in each submarine, but if two active sonars are fitted only one transmitter need be employed.

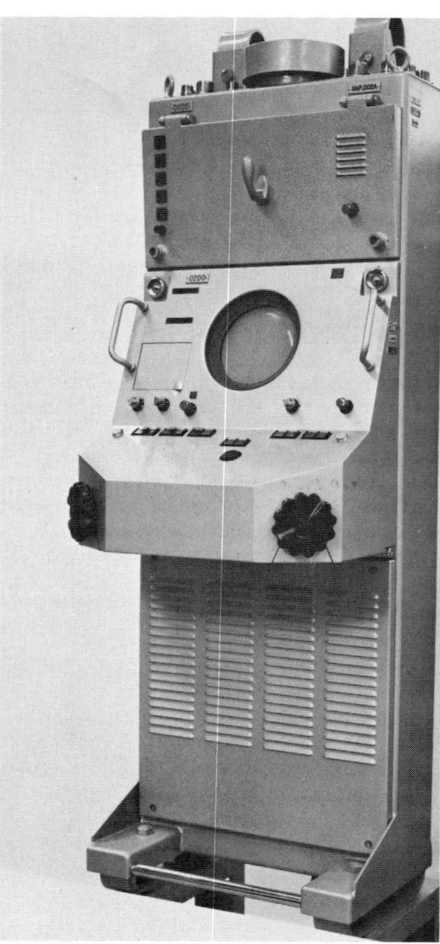

Control console of DUUA-2B active/passive sonar

CHARACTERISTICS
Transducer: Directional. Manual or automatic search in continuous rotation
Transmitter
Frequency: 8 kHz
Power: 3 kW
Pulse: FP or FM mode
Receiver: In passive mode, from 2.5 to 15 kHz

STATUS
In series production and ordered for the French Navy Type SNA72 fleet submarines.

CONTRACTOR
Thomson Sintra, Activités Sous-Marines, 1 avenue Aristide Briand, 94117 Arcueil Cedex, France.

3498.453
FENELON—DUUX-5 PASSIVE SONAR AND ACOUSTIC RANGEFINDER

Fenelon is the name given to a panoramic sonar capable of automatic and simultaneous tracking of three targets. The equipment incorporates a passive acoustic rangefinder for measuring the range of three targets within 120° sectors, and a panoramic sonar interceptor measuring the true bearing of all sonar transmissions received within the 2 to 15 kHz band.

The DUUX-5 equipment was developed by Alcatel as a successor to the DUUX-2 series (**1162.453** *1979-80* and earlier editions), of which at least 120 equipments were built and supplied to 14 navies for fitting in eight types of submarine. The Fenelon equipment enables range and bearing information on targets to be obtained by the submarine without the need for any transmissions and with minimum delay. Speed of operation permits target course and speed to be computed rapidly, also allowing any changes in either speed or course to be detected without delay. Four targets can be tracked simultaneously (three on radiated self-noise, one on sonar pulses), and there is a continuous panoramic bearing display over 360°. Range information is provided over arcs of 120° on each sides of the submarine. There are facilities for transmission of target data automatically to the ship's weapon control system and plotting table.

Performance characteristics include high accuracy and discrimination, immunity against sonar pulse interference, simplified calibration, and integrated test facilities. Under normal conditions, results in the middle sector are 0.3° for bearing accuracy, 2° discrimination accuracy between two targets and 5% of range on radiated noise for a target at 10 km distance. Two types of hydrophones are available. The hydrophonic unit is composed of two bases with three hydrophones each on the starboard and port sides of the submarine. All other technical details remain classified.

STATUS
In series production. Replaces DUUX-2 passive acoustic rangefinder.

CONTRACTOR
Thomson Sintra, Activités Sous-Marines, 1 avenue Aristide Briand, 94117 Arcueil Cedex, France.

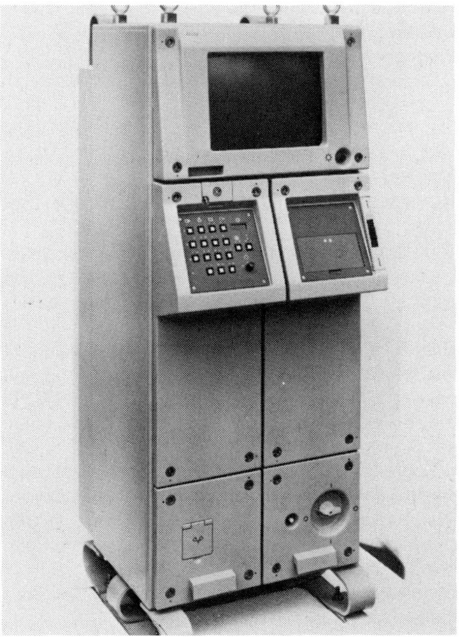

Fenelon – DUUX-5 control and display console

1631.353
DUAV-4 HELICOPTER SONAR (TSM 8251)

The DUAV-4 is an active/passive directive sonar designed for submarine surveillance and location (azimuth, distance, and radial speed). It is specially designed for use on board light, versatile, ship-based helicopters of which the WG13 Lynx is a typical example. It may also be fitted aboard small surface vessels as either a hull sonar or as a VDS system.

The DUAV-4 differs from conventional sonar in its signal processing system which is designed to give improved detection, especially in severe reverberation conditions such as shallow waters. The sonar can be operated in either the active or passive mode, true bearing, range, and radial speed are measured in the former mode, true bearing only in the passive mode.

A combined display unit permits surveillance display (initial detection) or plotting display (precise azimuth determination). Total weight (including electronic rack, cable, and dome) is 250 kg.

STATUS
The DUAV-4 sonar is in active service with the French Navy, the Royal Netherlands Navy and several other navies. Over 90 have been produced. The French Navy has opted to replace it in some of the Lynx helicopters with the HS-12 (**4095.453**).

CONTRACTOR
Thomson Sintra, Activités Sous-Marines, 1 avenue Aristide Briand, 94117 Arcueil Cedex, France.

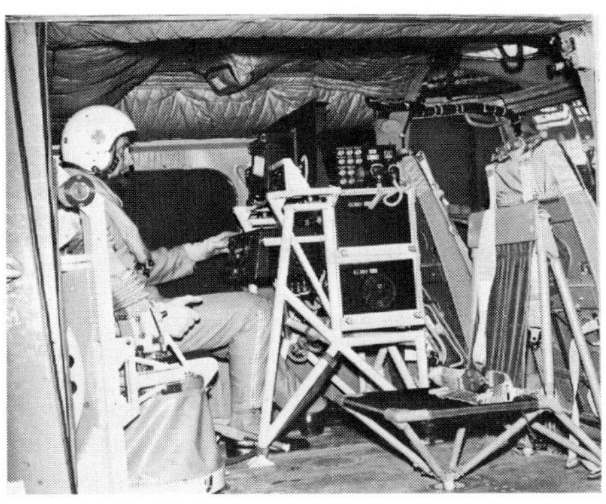

DUAV-4 helicopter sonar installation in WG 13 Lynx

4094.453
SS 12 SMALL SHIP SONAR (HULL AND VDS)

The SS 12 is a panoramic sonar operating at medium frequencies and with preformed beams for fitting in small ships and medium tonnage vessels. It has also been developed in a version suitable for helicopter use (**4095.453**). The use of the same electronics on helicopters and aboard ships yields common training procedures and maintenance facilities.

For surface ships the SS 12 can be supplied as a hull-mounted sonar in either fixed dome or retractable dome configuration, and/or as a variable depth sonar (VDS). In the latter configuration the Mini-Ulis handling system developed and produced by La Société Nouvelle des Ateliers et Chantiers du Havre (ACH) is usually employed, and this has a deck weight of less than four tons.

Thomson Sintra's considerable experience of

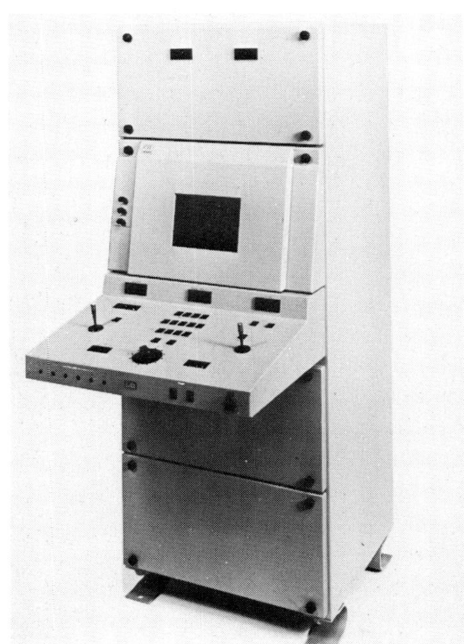

Operating console of SS 12 sonar

SS12 VDS on a 400 ton ship

shallow water sonars has been applied to enable optimisation of the SS 12 for small, noisy ships. Special display provisions for the coherent signal processing of CW and FM data enables the operator to carry out panoramic search without the strain of conventional PPI presentation, and by preserving all modes the sonar realises the maximum figure of merit. Availability of all data permits rapid and reliable classification of targets.

CHARACTERISTICS

Frequencies: 3 in 13 kHz band

Operating modes: CW, rectangular or $\cos^2$ pulse; FM, hyperbolic modulation

Transmission level: 212 dB (ref 1 pascal/yard)

Scales: 2, 4, 8, 16 km

Passive listening: Noise or pinger. Automatic tracking of two echoes and data transmission to the control system (DLA) of the weapon system.

Mini-Ulis

Mechanical System

Fish stabilisation: Servo-control system (fish amplitude about ±2 m)

Towing speed: 16 knots

Max speed on brakes: 30 knots

Hoisting and lowering time: 6 minutes

Max cable speed: 3 m/s

Cable length: approx 200 m

Diameter: 14 mm

Breaking load: 7 tons

STATUS

In production. Two systems were delivered to China in 1987 for use in 'Haigan' class patrol boats.

CONTRACTOR

Thomson Sintra, Activités Sous-Marines, Chemin des Travails, BP 53, 06801 Cagnes-Sur-Mer, France.

4095.453

HS 12 ACTIVE/PASSIVE HELICOPTER SONAR (TSM 8252)

The HS 12 is an active/passive panoramic helicopter version of the SS 12 small ship sonar (**4094.453**) and uses the same electronics as the surface vessels version. It has similar capabilities for operation in shallow/noisy waters, and has a system weight of less than 240 kg, thus suiting it for installation on light naval helicopters such as the Lynx. The HS 12 transducer is raised and lowered by a hydraulic winch at a high speed.

Operation in CW and FM modes is possible and digital signal processing is employed by the system's microprocessor. Automatic tracking of two targets and transmission of elements to an external equipment, such as a plotting table, are provided.

CHARACTERISTICS

Frequencies: Medium frequency

Transmission level: 212 dB (ref 1 pascal/yard)

Passive search: Noise or pinger. Display of 12 preformed beams, each subject to adaptive signal processing. Other characteristics as for the SS 12

Total weight: 250 kg

STATUS

In series production for the French Navy on the Lynx helicopter, and for four other navies.

CONTRACTOR

Thomson Sintra, Activités Sous-Marines, 1 avenue Aristide Briand, 94117 Arcueil Cedex, France.

Typical helicopter installation of the HS 12 sonar on Dauphin naval helicopter

6235.353

HS 312S ASW HELICOPTER SONAR

The HS 312S (TSM 8240/8241/8242/8243) is a medium frequency ASW sonar for helicopters incorporating the facilities of the HS 12 system (**4095.453**) and the Lamparo sonobuoy processing equipment (**4347.363**).

The equipment functions in both passive and active modes (CW and MF) and has a longer range than the HS 12. The acoustic subsystem will process four or eight sonobuoys simultaneously with the dipping sonar.

Only a single operator is required and the light weight and compactness of the system means that the HS 312S can be fitted to any type of light or medium size helicopter. The performances of the basic components have been improved by integration of the processing units, integration of a standardised keyboard and the use of only one display screen.

STATUS

In series production. Three systems were delivered to China in 1987.

CONTRACTOR

Thomson Sintra, Activities Sous-Marines, 1 avenue Aristide Briand, 94117 Arcueil Cedex, France.

HS312S helicopter sonar control/display

4461.453

DSBV 61A TOWED ARRAY SONAR

The DSBV 61A is a passive towed array sonar system consisting of a linear array, a towing cable which is also used for data transmission, a winch for handling and storage of array and cable, and the necessary signal processing and data extraction electronics. The latter is made up of five cabinets, plus two display consoles associated with two TOTES. The DSBV

DSBV 61A linear towed array sonar (serial equipment) on board the Primauguet *C-70 type ASW destroyer*

design allows two selectable towing modes, critical angle towing or towing behind the DUBV-43C VDS body.

The display consoles feature three-colour display of target information and can display, on request, the various sonar memory pages. A built-in fault display allows a quick reconfiguration and/or rapid maintenance of the system.

STATUS

In series production. Operational on the C-70 type ASW corvettes. A derivative, known as Anaconda, is described briefly under entry **6848.453**.

CONTRACTOR

Thomson Sintra, Activities Sous-Marines, Chemin des Travails, BP 53, 06801 Cagnes-sur-Mer, France.

3095.263

SOUND RAY TRACER (TSM 9310)

Derived from the earlier TSM 9300, this equipment uses both a DDA special-purpose processor for sound field computation and a mini-computer for improved data input facilities. Design is to French Navy specifications. The purpose of the equipment is to compute and display the sound rays transmitted or received by an acoustic source, according to a given sound velocity profile. The latter depends upon bathythermic conditions, and can be obtained directly with a sound velocimeter or derived from a bathythermograph profile. On a ship or aircraft, the equipment's ray diagram display shows the actual limitations of sonar equipment and enables ASW personnel to decide upon the best tactics in a given situation.

STATUS

In operational service but no longer in production. More detailed information is given in previous editions of *Jane's Weapon Systems*.

CONTRACTOR

Thomson Sintra, Activités Sous-Marines, Chemin des Travails, BP 53, 06801 Cagnes-sur-Mer, France.

1724.253

DUBA 25 (TSM 2400) SONAR

The Tarpon TSM 2400 (service designation DUBA 25) is a powerful attack sonar fitted in the French 'Aviso' class ships and similar anti-submarine vessels. It is an active sonar with a panoramic transducer array and assembly with sonar dome supplied by the French Navy. Although the sonar is suitable for use as towed VDS equipment, in the DUBA 25 configuration it is hull-mounted, and provides 'Aviso' class ships with all-round surveillance, acquisition and attack facilities for ASW operations. The power employed is sufficient for operating ranges of several kilometres. The transducer array consists of 36 staves arranged to form a cylinder 110 cm in diameter and provided with roll stabilisation.

STATUS

Operational since 1975. No longer in production. More detailed information is given in previous editions of *Jane's Weapon Systems*.

CONTRACTOR

Thomson Sintra, Activités Sous-Marines, Chemin des Travails, BP 53, 06801 Cagnes-sur-Mer, France.

1725.253

TSM 2314 DIODON SONAR

The TSM 2314 Diodon sonar is an advanced equipment for fitting in ASW ships of small or medium tonnage. It is an active, all round system employing digital techniques and advanced signal processing facilities. It performs all round submarine detection, target tracking, and attack operations. It is supplied with automatic integrated test facilities and on-board training equipment. The acoustic transducer array for fitting on board smaller ships is provided with an active roll stabilisation system. It can be fitted within a fixed dome, retractable dome, or supplied as a variable depth (towed) sonar (VDS). Among the optional additional facilities that can be provided with the Diodon system are underwater telephonic equipment, sound ray path indicator, shore training equipment, and maintenance equipment.

The transducer array is cylindrical in shape and consists of 24 identical and interchangeable vertical staves of transducer elements, each stave being capable of radiating 600 watts of acoustic power. The array is pivoted in roll, and active stabilisation is effected by means of an hydraulic actuating ram. The sensor is a pendulum-type device. Three (selectable) transmission frequencies are provided, 11, 12 and 13 kHz, and there are several modes of operation.

Transmission facilities include:

(1) Normal panoramic transmission
(2) Sector or panoramic transmission by the electronic scanning of directional beams (sector beamwidth and sector axis can be controlled throughout 360° in 15° increments)
(3) Normal or reduced power
(4) Pure or modulated frequency
(5) 20, 80 or 200 ms pulses.

The display unit provides the operator with all the necessary controls and indicators, such as:

(1) Panoramic CRT display for active or passive mode acquisition
(2) Target-axis error scope for fine aiming and tracking
(3) Doppler CRT indicator for measurement of target speed
(4) Digital read-out of target data
(5) Audio channel
(6) Controls for the active tracking circuits
(7) Transmission mode controls
(8) Remote control of transducer array assembly
(9) Integrated test system indicator
(10) Control of operational test.

Reception is achieved simultaneously on 24 preformed beams in either of two alternative modes:

(1) Standard filtering (with or without reception), detection and integration
(2) Coherent signal processing.

The sonar can be linked by either synchro or in digital form with other ship's systems, such as ship's log, gyro compass, plotting table, ASW fire control system, tactical displays, etc.

CHARACTERISTICS

Frequency: 11, 12 or 13 kHz, selectable
Beamwidth transmission (3 dB): 15° elevation, 15° azimuth
Panoramic reception: 24 preformed beams
Passive: In band 10 – 14 kHz
Active: Pure frequency with or without rejection (30 Hz). Modulated frequency with coherent signal processing. BT products = 50 on each beam
Range: Fine bearing and range on one target. Semi-automatic or automatic tracking on one target
Dimensions
Display: 180 × 60 × 50 cm, weight 450 kg
Transmitter: 180 × 60 × 50 cm, weight 600 kg
Power supply: 110 × 60 × 50 cm, weight 500 kg
Transducer: Weight 580 kg
Hydraulic unit: 160 × 55 × 121 cm, weight 470 kg

STATUS

In production for at least six navies and operational aboard a number of ships.

CONTRACTOR

Thomson Sintra, Activités Sous-Marines, Chemin des Travails, BP 53, 06801 Cagnes-sur-Mer, France.

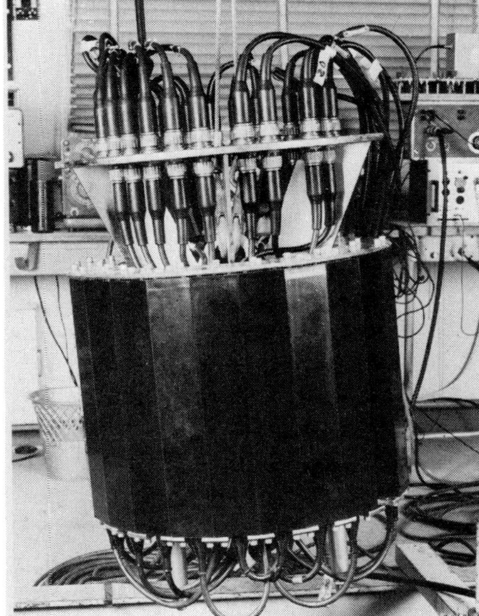

Diodon sonar transducer array

Diodon sonar system, showing (left to right) display console, receiver, and power supply

5646.253

TSM 2630/2633 DIODON/SPHERION SONAR

The TSM 2630 Diodon sonar is a compact panoramic, active sonar for surface ships. The acoustic array may be either hull-mounted or towed (VDS). The towed array is normally the Diodon type, with the spherical shaped Spherion array being hull-mounted. A number of ships are fitted with both arrays and share the same electronics, which are common to both types.

The main features of the TSM 2630 are sophisticated signal processing, both in CW (optional FFT) and FM modes (pulse compression), for detection of slow and fast moving targets; comprehensive data processing with computer aided detection and tracking facilities; advanced multi-colour displays; integrated sound-ray tracing and prediction of detection range; and digital interfaces to international standards.

The basic version of the sonar includes:

(1) Hull-mounted acoustic assembly (dome and array)
(2) Matching unit
(3) Power supply cabinet
(4) Cabinet housing the transmitting, receiving and video processing circuits
(5) Display console incorporating two identical high-resolution multi-colour TV screens.

Spherion hull-mounted array

Optional equipment includes a panoramic passive capability, a video cassette recorder for play-back training and debriefing, and a torpedo or rocket launching control.

CHARACTERISTICS

Acoustic array: 24 staves; source level 230 dB/μ Pa/1 yard; weight 340 kg

Transmitter: Omni-directional transmission in CW or FM

Receiver: 48 preformed beams with true bearing stabilisation; conventional and coherent processing (BxT up to 300); directive and steerable audio

Video processing: Data processing in FM and CW (normalisation, interpolation, correlation, automatic tracking of up to 5 targets); conversational procedure for sonar operation; target analysis (apparent length and aspect angle, doppler measurement)

Display console: 2 identical flicker free, high resolution colour TV monitors

Weight: 1750 kg

STATUS

In production. Over 23 units of the Diodon family are in operation with various navies. Seven TSM 2633 Spherion systems are being delivered to the Royal Norwegian Navy and have been developed in collaboration with Simrad Subsea of Norway.

CONTRACTOR

Thomson Sintra, Activités Sous-Marines, Chemin des Travails, BP 53, 06801 Cagnes-sur-Mer, France.

5648.253

TSM 2640 SALMON SONAR

The TSM 2640 is a series of compact, medium range, variable depth sonar systems for anti-submarine warfare on board light surface patrol craft operating in shallow waters. It provides a high detection capability even against severe background noise and reverberation. The system provides sophisticated signal processing with pulse compression in FM and spectral analysis (FFT) in CW, comprehensive data processing and computer aided facilities, and advanced multi-colour display with high resolution TV monitor.

The basic version consists of an acoustic array of 24 staves in a streamlined fish, a towing winch, an operator console equipped with a multi-colour screen, and a receiver unit and a transmitter unit contained in small cabinets. Optional facilities include automatic extraction, passive listening, sound ray tracer, an additional colour TV screen, video tape recording system and an integrated on-board simulator.

An improved version, the TSM 2643, has been produced and includes a higher level of automation for detection (natural frequency and frequency modulation) for optimum performance on slow and fast targets. Pursuit capabilities have been enhanced and the system can direct beams on specific azimuth bearings. Possibilities for extension have been increased and a towed linear array can be easily configured with the system.

Royal Swedish Navy SPICA corvette fitted with a TSM 2642 SALMON sonar

CHARACTERISTICS

Acoustic array: 24 identical staves; source level 217 dB/μPa/1 yard; weight 120 kg

Transmitter: Omni-directional transmission in CW or FM

Receiver: 48 preformed beams in FM (24 in CW); conventional processing; filtering, detection, integration in FM, FFT in CW; coherent signal processing (pulse compression) with BxT up to 200; noise listening in active band

Video processing: Data processing in FM and CW (normalisation, interpolation, correlation, automatic target tracking of up to 3 tracks simultaneously); conversational procedure for sonar operation; target analysis (apparent length and aspect angle, doppler measurement)

Display: High resolution, flicker-free colour TV monitor

Total weight: 750 kg (additional 5 tonnes for winch)

STATUS

In production. Two TSM 2642 systems have been installed on SPICA III 'Stockholm' class surface attack ships of the Royal Swedish Navy and are in operational service. A further four systems, of the TSM 2643 variant, have been ordered by Sweden for its new generation of 'Goteborg' class corvettes.

CONTRACTOR

Thomson Sintra, Activités Sous-Marines, Chemin des Travails, BP53, 06801 Cagnes-sur-Mer, France.

5649.253

TSM 2820 TARPON SONAR

The TSM 2820 is a long-detection range sonar for surface ships. It is a single operator, low-frequency hull-mounted system (HMS) which can be combined with a medium or a low frequency variable depth sonar (VDS). The HMS and VDS can operate simultaneously in alternate modes.

The main facilities of the TSM 2820 are:

(1) Combined HMS/VDS, medium/low frequency capabilities
(2) Sophisticated signal processing, with pulse compression in FM and spectral analysis (FFT) in CW. The TSM 2820 can operate alternatively in FM and CW for optional detection of slow and fast targets
(3) Comprehensive data processing and computer aided facilities with automatic track extraction
(4) Multi-colour display with high resolution TV monitors
(5) Integrated sound ray trace and prediction of detection range
(6) Digital interfaces to international standards.

The basic version of the hull-mounted Tarpon sonar consists of:

(1) Hull-mounted acoustic assembly (dome and acoustic array)
(2) Matching unit and power transformer unit
(3) Energy storage unit
(4) Transmitter unit, active receiver unit and video processing unit, all working in either HMS or VDS
(5) Display console with two identical multi-colour screens. Optional facilities include panoramic passive receiver for torpedo warning and short-range self protection; 1000 × 1000 colour TV display; and video cassette recorder for training or de-briefing.

CHARACTERISTICS

Acoustic array: 32 identical staves of transducers, high efficiency piezo-electric ceramics, source level 222 dB/μPa/1 yard

Transmitter: Omni-directional transmission in CW or FM

Receiver: 32 preformed beams; simultaneous signal processing on all beams; spectral analysis (CW); 96 (FFT) doppler channels; pulse compression (FM) with BxT 80 – 240 according to range scale; directive and steerable audio
Video processing: Data processing in FM and CW (normalisation, interpolation, correlation, track extraction of up to 25 tracks simultaneously); conversational procedure for sonar operation; target analysis (apparent length and aspect angle, doppler measurement)
Display: 2 identical high resolution colour TV monitors
Weight: 6000 kg

CONTRACTOR

Thomson Sintra, Activités Sous-Marines, Chemin des Travails, BP 53, 06801 Cagnes-sur-Mer, France.

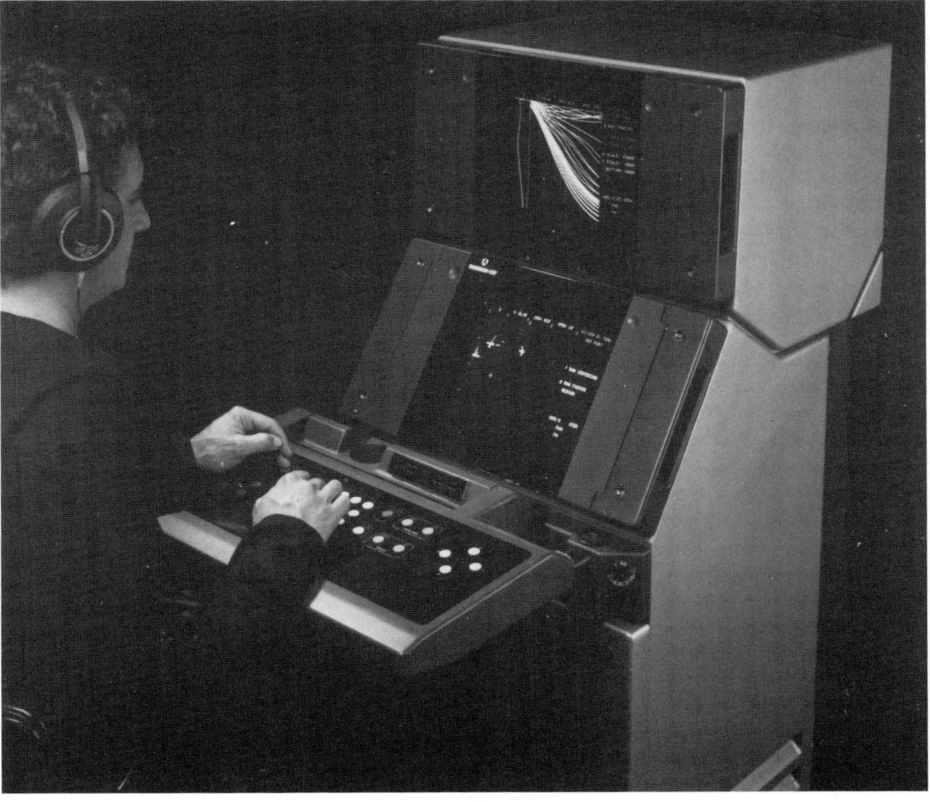

TSM 2820 Tarpon sonar console

6554.463

TSM 2930 LAMPROIE LINEAR TOWED ARRAY

The TSM 2930 Lamproie linear towed array is the latest development in the range of linear towed arrays designed and produced by Thomson Sintra. The array is used for passive detection, classification and position fixing of underwater targets. Panoramic broadband and narrowband analysis of the detected signals are performed continuously and displayed on two high resolution colour TV screens.

The array is designed to be towed by either a submarine or surface ship. When used by a surface ship the array can be towed either behind a long, heavy cable (critical angle tow), or behind a composite system. In the latter case the array is towed by a long, neutrally buoyant cable connected to a depressor which itself is then connected to the towing vessel through a short heavy cable. For submarine use the array is 'clipped-on'.

STATUS

In preliminary production.

CONTRACTOR

Thomson Sintra, Activitiés Sous Marines, Chemin des Travials, BP 53, 06801 Cagnes-sur-Mer, France.

Towed array of the TSM 2930 sonar

1726.453

DUBM 41B SIDE-LOOKING SONAR

The Thomson Sintra TSM 2050 (service designation DUBM 41B) system is a high-resolution side-looking sonar system designed for the location and classification of objects, such as mines, lying on the sea bed. The system consists of three towed sonar vehicles, two consoles, and the necessary cables and shipboard hoists and handling gear. In normal operation two of the three sonar bodies are used, with the third kept as a ready spare. A permanent record of the sonar data gathered is made on a facsimile type recorder and a tape recorder. The information also is presented simultaneously on two CRT storage tubes, these console-mounted displays providing an image of the sea bed. The system is designed for use by low tonnage vessels, and in waters 100 m or deeper. The towed sonars operate at about 5.5 to 7.5 m above the sea bed at speeds of two to six knots. The tow cables are provided with deflector vanes which ensure that the sonar bodies are towed at a distance to left and right of the ship's track, and marker floats are provided to indicate the line of travel for each sonar.

The sonar body is a streamlined vehicle fitted with the side-looking sonars, plus additional sonar transducers for determining height above sea bed and for obstacle detection. Servo-controlled fins are provided for depth control and roll stabilisation is incorporated by means of a separate set of four fins. Dimensions of each sonar vehicle are: length 3.725 m, diameter 36 cm, in-air weight 340 kg. An acoustic pinger is installed as an aid to recovery should the sonar body break its tow. Manual controls are provided at the operating console to permit manual piloting, either to override automatic control or as a standby mode.

Each of the two sonar bodies operates on its own frequency, the two being separated by 50 kHz, with the area of sea bed between the two bodies being scanned by both sonars. A typical scanned area covered by the two bodies amounts to a total width of about 200 m, with maximum sonar range selected. There are two other range settings, 50 and 25 m. At the latter setting, resolution is stated to be better than 5 × 10 cm in the lateral (scan) direction and in the line of travel.

DEVELOPMENT

Development of the DUBM 41B was conducted in collaboration by the French Navy Mine Warfare DCN (GESMA) and Thomson Sintra.

STATUS

In service.

CONTRACTOR

Thomson Sintra, Activités Sous-Marines, Route de Sainte-Anne du Portzic, 29601 Brest Cedex, France.

Towed sonar DUBM 41B on manoeuvres

6847.453

DUBM 42 SIDE-LOOKING SONAR

The DUBM 42 is a side-looking sonar able to detect and classify mines at ranges up to 300 m, with an area coverage of 2000 m²/s. The system is intended for fitment to the future French Navy MCMV *Bamo* for survey of the Atlantic coasts at an operational speed of 10 knots and a height above the sea bed of 30 m.

The basic system consists of one vehicle with three sonars (two lateral and one front-head) on the sonar body, one Colibri sonar display, one Colibri tactical display and one high density magnetic recorder. The data collected by the three sonars is recorded for analysis in a land-based processing centre. In addition a visual display of the sea bed is provided on the operator's console. A magnifier provides a more accurate classification. All navigation functions such as piloting and automatic control are integrated on the same console.

The two lateral and single front-head sonar operate on their own individual frequency, the swept channel being 400 m wide.

STATUS

Development of the DUBM 42 has been carried out jointly by the French Navy Mine Warfare DCN (GESMA) and Thomson Sintra, and the system is in preliminary production.

CONTRACTOR

Thomson Sintra, Activitiés Sous-Marines, Route de Sainte-Anne du Portzic, 29601 Brest Cedex, France.

4348.453

IBIS MINEHUNTING SONAR SYSTEMS

IBIS 3 and 5 are fully integrated minehunting systems using a high performance sonar and a tactical display system which integrates and displays all critical information.

The system consists of:

(1) A minehunting sonar, TSM 2021 (**1977.253**) or TSM 2022 (**4349.453**)
(2) A tactical display console, TSM 2060 NAVIPLOT
(3) A doppler sonar log, TSM 5730 or TSM 5722
(4) Navigation systems, autopilot, windspeed sensors, etc.

Mine detection and classification is the function of the sonar (TSM 2021 or TSM 2022). Detection is performed at distances up to 600 m (2000 m in the case of submarines). Classification is possible in two modes, echo mode and shadow mode, allowing identification of moored and bottom mines.

The action/information data display and storage is performed by the tactical display system which provides ship's track monitoring, location of contacts and records of operational data. Positions of classified mines, possible contacts and non-mine objects are recorded and displayed simultaneously.

STATUS

See under separate TSM 2021 and TSM 2022 entries.

CONTRACTOR

Thomson Sintra, Activités Sous-Marines, Route de Sainte-Anne du Portzic, 29601 Brest Cedex, France.

1977.253

TSM 2021B MINEHUNTING SONAR

The Thomson Sintra TSM 2021B (service designation, DUBM 21B) is a modern minehunting sonar, the design of which applies experience with the DUBM 20A, also developed by Thomson Sintra (**1357.253** *1974-75* and earlier editions). It is a hull-mounted equipment intended for use in conjunction with precision navigation equipment and a mine countermeasure (destruction) system such as the PAP 104 (**1786.453**). A former version is the 2021A with different array housings and minor display variations. TSM 2021 functions are:

(1) The detection of mines at distances up to 600 metres
(2) The classification of such targets by the study of the shape of their echo and acoustic shadow, at distances up to 250 metres.

Each detection and classification sonar chain consists of an acoustic transmitter/receiver transducer; the associated transmission and reception electronics; and a display console. The STH subassembly performs the stabilisation, steering, and retraction of the detector and classifier arrays. This part of the system is the result of studies and development by ECAN at Ruelle.

TSM 2021B minehunting sonar, showing the main electronics cabinet flanked by the classification and detection display consoles (left and right), with the transducer arrays in the foreground

The electro-acoustic assembly (EAA) includes:

(1) The classification sonar transmitter/receiver chain cabinet
(2) A display console which controls the transmission and reception functions and has CRT presentation of mine detection and location by the detector sonar chain
(3) A display console with CRT displays for presentation of mine locations within the classification chain
(4) Bridge repeater display, showing range and bearing of designated targets.

CHARACTERISTICS

Coverage: $\pm 175°$ in sectors of 30, 60 and 90° for the detector chain; 3, 5 or 10° for the classification chain
Elevation: Variable, between –5 and –40°
Stabilisation (roll): $\pm 15°$
Stabilisation (pitch): $\pm 5°$
Detection sonar
Frequency: 100 kHz (modulated $\pm$ 10 kHz)
Pulse duration: 0.2 or 0.5 ms
Range scales: 400, 600 or 900 m
Max sound level: 120 dB
Channels: 20
Beam aperture: 1.5°
Display: PPI
Classification sonar
Range scales: 200 or 300 m
Max sound level: 122 dB
Channels: 80
Beam aperture: 0.17°
Display: CRT and storage magnifier tube

STATUS

The TSM 2021 is in service with the French and other navies. Altogether some 60 systems have been ordered.

CONTRACTOR

Thomson Sintra, Activités Sous-Marines, Route de Sainte-Anne du Portzic, 29601 Brest Cedex, France.

4349.453

TSM 2022 MINEHUNTING SONAR

The TSM 2022 is a lightweight minehunting sonar designed specifically for small and medium sized mine countermeasures vessels. The sonar can also be used for mine avoidance on minesweepers, or in civil applications such as bottom profiling and side-scan surveys.

The TSM 2022 is based largely on experience gained with the DUBM 21B in the field of high resolution beam forming techniques. It is a hull mounted equipment intended also for use in conjunction with precision display and navigation equipment and a mine countermeasures system.

The main feature of the TSM 2022 is its small-size retractable single array assembly, used for both detection and classification, which enables easy installation and maintenance. The high resolution features of the DUBM 21 are maintained and improvements are mainly in the field of digital processing technology.

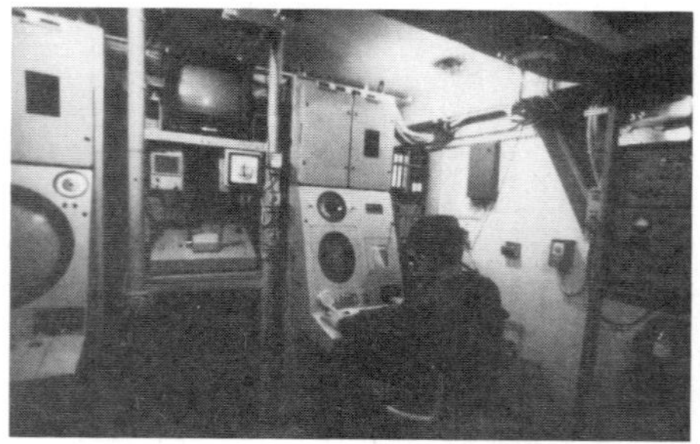

Thomson Sintra minehunting sonar operation room

The main functions of the sonar are: detection and classification of moored and bottom mines, detection at distances up to 600 metres (2000 metres for submarines), classification of targets by analysis of the shape of their echo or shadow at distances up to 250 metres. In the classification mode, horizontal beamwidth is 7°, in detection mode either 14° or 28° beamwidth can be selected. In both modes the vertical beamwidth is 15°.

The main assemblies are:

(1) The hoisting and stabilisation system together with the retractable array, which is installed in the sonar trunk (0.75 metre diameter). The total weight is only 900 kg
(2) Electronics cabinet
(3) Operator's console.

The operator console provides display of sonar images in the various modes, memory and display of former sonar contacts, and includes interfaces with current mine disposal and navigation equipments.

STATUS

Twelve equipments have been ordered by three foreign navies, including the Swedish Navy which has six systems on order for new minehunting vessels. The first systems have been delivered.

CONTRACTOR

Thomson Sintra, Activités Sous-Marines, Route de Sainte-Anne du Portzic, 29601 Brest Cedex, France.

6848.453

ANACONDA TACTAS SYSTEM

Thomson Sintra has recently been contracted to supply a system of hull-mounted and tactical towed arrays, known as Anaconda. This system configures a digital wet-end, currently in service with the French Navy (DSBV-61A), with an advanced technology sonar receiver built around the powerful Mangouste Mk 2 processor. This latter equipment was developed by Thomson Sintra to increase the performance of all sonars, particularly in automatic detection, narrow band analysis, automatic tracking, adaptive beam-forming and self-noise cancellation.

STATUS

Ordered for the Royal Netherlands Navy 'M' class frigates.

CONTRACTOR

Thomson Sintra, 1 avenue Aristide Briand, 94117 Arcueil Cedex, France.

5850.353

MAD MK III MAGNETIC ANOMALY DETECTOR

The MAD Mk III is derived from several years experience with the Mk I and II systems and is specifically designed for inboard use on fixed-wing aircraft and helicopters to detect the presence of a submersible by measuring the disturbance to the earth's magnetic field. It is an integral, digital, solid-state and highly reliable airborne system which becomes operational at switch-on without any warm up time. Target parameters are automatically delivered in real-time on a CRT control and display unit.

The basic equipment can be broken down into three separate units; a detection element, a computer and a control/display unit. A 1553B data bus interface card is included in the system. A graphic recorder is available as an option.

The sensor operates on the nuclear magnetic resonance principle and uses the precession of protons in a liquid, the frequency of which, measured by the pick-up coils, is proportional to the magnetic field to be measured. Compensation is employed to eliminate from the received signal all disturbances created by the magnetic element in the aircraft and their movement in the earth's field. The compensator uses a 16-terms model, representing the magnetic components of the aircraft. The MAD Mk III includes rapid aircraft identification modes.

The target is detected and located automatically. This is a fundamental role, allowing the MAD system to give a high detection probability with a very low false alarm rate, ie to extract the target signal from the background noise. The Crouzet MAD system is based on a mathematical comparision between the current MAD signal and an analytical model of the target signals, as opposed to the more conventional using threshold detection in several frequency bands. All computing tasks are performed by the ALPHA 732 proprietary 1 MOPS digital computer operating on the Pascal language.

CHARACTERISTICS

Dimensions: Detection element 1250 mm long × 125 mm diameter; computer 1/2 ATR long; control/display unit 190 mm long × 146 mm wide × 162 mm high

Weights: Detection element 5.5 kg; computer 10 kg; control/display unit 3.5 kg

STATUS

Crouzet has sold MAD systems in many parts of the world since 1972. This includes the French Navy Alouette III and more recently the Saudi Arabian Dauphin helicopters. The MAD Mk III is now operational with the French Navy Atlantique 2 maritime patrol aircraft.

CONTRACTOR

Crouzet SA, 25 rue Jules Vedrines, 26027 Valence Cedex, France.

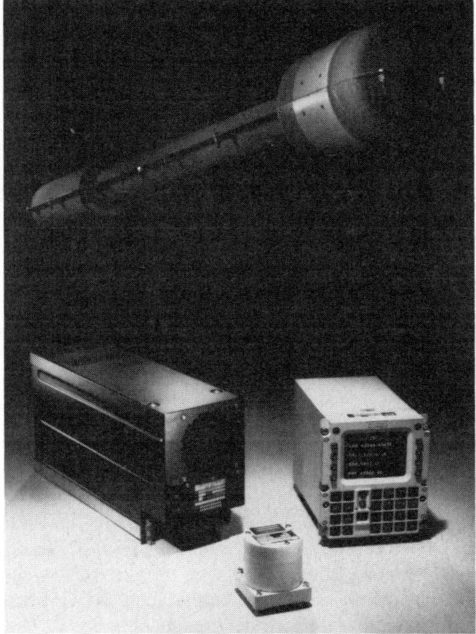

'In-board' MAD Mk III

5851.453

TSM 8020 SONOBUOYS

The A/3-size TSM 8020 (DSTV-7) is a mini omni-directional passive buoy providing a number of additional facilities. These include the on-board selection of the VHF frequency channel from the 31 available before the buoy is dropped, choice of 31 channels from the 99 available between 136 and 173.5 MHz at 375 kHz spacing, a three position selector to give a choice of hydrophone depth immersion (20, 100 or 300 metres), and an increased maximum depth of 300 metres. Other improvements are reduction in overall weight and volume, and the use of a small balloon to act as a brake during the trajectory of the buoy in the air and then as a floater on the water surface. The TSM 8020 replaces the TSM 8010 buoy.

STATUS

In production. Development of other types of sonobuoys is continuing, including a size A DICASS active buoy.

CONTRACTOR

Thomson Sintra, Activités Sous-Marines, 1 avenue Aristide Briand, 94117 Arcueil Cedex, France.

4802.393

ALKAN SONOBUOY LAUNCHERS

Alkan has designed a number of sonobuoy launchers for use in various fixed-wing aircraft and helicopters. They meet a range of operational requirements, such as in-flight reloading, and release or ejection from a pressurised or unpressurised cabin. Various launcher sizes and capacities are available to allow for easy installation in the cabin or equipment bay.

Type 8030

This launcher has been designed specifically for the Breguet Atlantic Mk 1 and Atlantique aircraft which can house up to four launchers. The capacity of each launcher is 18 size A and 18 size F buoys.

Container-launcher

Each buoy is packed in a container-launcher, which provides for its storage and ejection and is equipped with a pressurised vessel at the top and a releasable cap at the bottom. The power for ejection is provided by compressed gas stored in the vessel at the top of the buoy.

STATUS

In series production. A new type of launcher, designed for the launch of F type sonobuoys from the French Navy Lynx helicopters is currently in development and trials.

CONTRACTOR

R Alkan & Cie, BP 23, Rue du 8 Mai 1945, 94460 Valenton, France.

6556.393

SPDT-1A PASSIVE SONAR

The SPDT-1A is a preformed beam passive sonar designed specifically for the detection of the noise level radiated by torpedoes. The level and frequency spectrum is determined by an operator who has a colour video screen for panoramic surveillance, a black/white classification video screen and an audio system for received-noise listening. The system normally uses the array of the on-board active sonar, although a dedicated array can be provided, either in the sonar dome or separately.

CONTRACTOR

Safare Crouzet SA, 98 avenue St. Lambert, BP171, 06003 Nice Cedex, France.

GERMANY (FEDERAL REPUBLIC)

4228.453

DSQS 21 SERIES SONARS

The DSQS 21 series sonars are designed for operation in surface ships as part of an anti-submarine weapon system. For operations against submarines below the layer, the sonar can be supplemented with a towing system providing a variable depth sonar (VDS) capability.

Computer aided detection techniques are used for classification and tracking and the information is presented on colour CRT displays to permit doppler coding and the discrimination of data on the integrated displays. The 'Z' indicates variants equipped with electronic stabilisation to minimise the effects of ship's motion.

The VDS is designed to alternate with the hull mounted sonar, and the displays of the two sonars are integrated. Fitted with automatic depth control and heave compensation for stability of the 'fish', the VDS can be handled automatically or manually.

CHARACTERISTICS

Sonar		Diameter of Transducer Arrays		No of Pre-formed Beams		Class of Ship
		VDS	hull-mounted	VDS	hull-mounted	
DSQS-21	**B**	1.8 m		64		Corvettes frigates
	BZ	1.8 m		64		Corvettes frigates
	B/VDS	1.8 m	1 m	64	32	Corvettes frigates
	BZ/VDS	1.8 m	1 m	64	32	Corvettes frigates
DSQS-21	**C**	1 m		32		Corvettes frigates
	CZ	1 m		32		Corvettes frigates
	C/VDS	1 m	1 m	32	32	Corvettes frigates
	CZ/VDS	1 m	1 m	32	32	Corvettes frigates
DSQS-21	**D**	1 m		32		Up to corvettes
	DZ	1 m		32		Up to corvettes
	D/VDS	1 m	1 m	32	32	Up to corvettes
	DZ/VDS	1 m	1 m	32	32	Up to corvettes

CONTRACTOR

Fried Krupp GmbH, Krupp Atlas-Elektronik Bremen, Sebaldsbrücker Heerstr 235, POB 448545, D 2800 Bremen 44, Federal Republic of Germany.

4229.453

KRUPP ATLAS-ELEKTRONIK SUBMARINE SONARS

Krupp Atlas-Elektronik has developed a range of submarine sonar equipment. Little information has been disclosed but it is possible to give brief summaries of some of the systems.

CSU 3 Sonar

The CSU 3 is a conventional submarine sonar designed for a single operator. It is a fast-scanning medium range active sonar which can be used as a long range passive device, an intercept sonar and as an underwater telephone. In its active mode, the sonar operates using a single beam. The equipment has a cylindrical transducer array with electronic beam steering. The display is a magnified PPI presenting a 20° sector with continuous memory-refreshed display of target echoes. Target information is transferred digitally to the fire control equipment. In the passive mode, the sonar has 360° azimuth coverage with automatic tracking of up to four targets simultaneously. The target information is presented on a CRT display in true or ship's relative bearing and target elevation angle can be determined for estimation of distance and optimum sound raypath. The sonar is equipped with automatic interference suppression against acoustic countermeasures.

PSU 1-2 Sonar

The PSU 1-2 is a passive sonar for use in small conventional submarines. The equipment has the capability for detection, classification, and bearing determination of noise radiating targets with simultaneous azimuth coverage over 360°. It can be used also to receive underwater telephone signals. The target noise information received by the cylindrical hydrophone array (CHA) is processed into 32 pre-formed beams and presented on a CRT display in true or ship's relative bearing. Four targets can be automatically tracked using the ATT equipment and the respective target bearing is transferred to the weapon control systems.

PRS-3 Passive Ranging Sonar

The PRS-3 is a passive submarine sonar designed to be operated in conjunction with the CSU-3. It has two principal operating modes. In the normal mode, the sonar can detect target noise and determine bearing and distance as well as computation of course and speed for a maximum of four targets. A level of confidence in the range measurement is indicated by the equipment. In the automatic mode, the target information is tracked automatically using the ATT channels of the CSU-3 sonar.

OSID Sonar System

OSID is an integrated sonar system developed for use in small submarines. It comprises a passive sonar, data distribution with a tactical display and a torpedo guidance system. The sonar has a broad-band channel for bearing determination and classification, and utilises minimum correlation techniques for the location of weak targets. It features a 360° panoramic presentation of targets on a CRT display, according to azimuth. The processing computer calculates the position and vectors of the target, own submarine, and the torpedo and the information is presented on a tactical display. The torpedo control is a replaceable module designed for each type of torpedo.

SIP 3 Sonar Signal Processor

The SIP 3 is a sonar information processor developed for use with fast-scanning multi-target tracking sonars of the CSU series. The prime function of the equipment is to provide passive classification of targets using low frequency detection and analysis, spectrum survey and comparison, high target resolution and demon analysis. Doppler analysis of target echoes received by the active sonar of the own submarine is used to determine the target's speed. Sound raypath analysis produces an estimate of the conditions for sound propagation. The system has the capability to determine and present specific sonar signal characteristics, comparing them to sound signatures of selected targets held in a data bank.

CSU 83 Submarine Sonar System

The CSU 83 is an active/passive submarine sonar system tailored for use in vessels from 400 tons upwards. The system consists of a fully equipped panoramic sonar, a trainable active sector limited to about 30°, an intercept sonar and an underwater telephone. It is equipped with a CHA and is normally used in conjunction with the SIP 3 sonar signal processor for tracking, analysis and classification. In its active mode the CSU 83 is able to detect targets by measurement of bearing, range, target doppler and echo amplitude. After classification, the information is transferred to the vessel's command system. In the passive mode, the equipment tracks the targets automatically and measures bearing and strength of target noise. Determination of target parameters such as range, course and speed is based on passive target bearing analysis.

In the intercept mode detection, measurement and analysis of other ships' sonar provides bearing, amplitude, frequency, pulse spectra and modulation type, pulse length and PRF. When used as an underwater telephone, the CSU 83 provides intercommunication with other ships and monitoring of foreign underwater transmissions.

CONTRACTOR

Fried Krupp GmbH, Krupp Atlas-Elektronik Bremen, Sebaldsbrücker Heerstr 235, POB 448545, D 2800 Bremen 44, Federal Republic of Germany.

4230.453

KRUPP ATLAS-ELEKTRONIK/MCM SYSTEMS

Krupp Atlas-Elektronik has developed equipment for use in mine countermeasures (MCM). Few details are available but the following are brief summaries of the equipment.

DSQS-IIA Sonar

The DSQS-IIA is a mine avoidance sonar designed for the location of underwater objects, in particular moored mines. It can be used as a navigational aid in mined areas, for surface ships and submarines. When an object is located, the distance, bearing and depth are presented to the operator on a PPI display, with the bearing referenced to the ship, North, or shown in a true motion mode.

DSQS-IIH Sonar

The DSQS-IIH is a mine countermeasures sonar suitable for the detection and classification of surface, moored and ground mines. The equipment features independent operation of detection and classification modes through 360° of azimuth, with simultaneous azimuth coverage of 90° in the detection mode. Target images are stored in the system to facilitate recognition. Electronic beam stabilisation is fitted for both modes and the sonar can be operated at speeds up to ten knots with minimal degradation of performance.

DSQS-21 Sonar

This is a family of hull-mounted and VDS sonars, and includes a D series suitable for installation on vessels up to corvette size.

The DSQS-21 is available in four versions (A to D), with A being the most comprehensive and having an array 3 m in diameter while the D variant has a 0.95 m array. All are available with electronically stabilised beam forming against roll and pitch, target depth evaluation, variable depth and towed array systems as options.

More than 50 systems have been delivered to 15 countries.

MWS 80 Mine Countermeasures

MWS 80 is a fully integrated mine countermeasures weapons system developed for the operations rooms of MCM vessels. It comprises the DSQS-IIH sonar; the command and navigation equipment (CNE) for navigation, co-ordination of the MCM operation, and recording and plotting of the surveyed area; and a remote-controlled mine disposal vehicle which is used for the identification and destruction of mines. The MWS 80 is manned by two operators.

AIS 11 Identification Sonar

The AIS 11 is an active identification sonar intended for use in submersibles. Although the equipment can be used for a variety of underwater tasks, its primary role is that of mine countermeasures, for which purpose it is normally fitted to the PAP 104 mine disposal system (**1786.453**).

The underwater part of the AIS 11 consists of an electronics unit and a transducer unit, the latter being fitted in the hull of the submersible. The parent surface ship is equipped with an evaluation unit, a control panel and a display. Weight of the overall system is 160 kg, that of the underwater part being about 20 kg. Data transfer between the underwater part and the surface vessel takes place along a cable. During operation, a 60° conical beam is transmitted ahead of the vehicle and the acoustic signals reflected by an object are received by a linear antenna. These signals are then processed in real-time, by both analogue and digital methods, and transmitted to the surface vessel where the information is shown on a PPI display in a sector-shaped format with correct angular representation. Fast real-time signal processing enables the underwater scene to be displayed at the frame repetition rate of a normal cine film.

CONTRACTOR

Fried Krupp GmbH, Krupp Atlas-Elektronik Bremen, Sebaldsbrücker Heerstr 235, POB 448545, D 2800 Bremen 44, Federal Republic of Germany.

INTERNATIONAL

5865.353

HELRAS AIRBORNE SONAR

Bendix, British Aerospace and FIAR of Italy are collaborating on the completion of engineering for a helicopter long range dunking sonar known as HELRAS. The new sonar uses a large low frequency volumetric array (when deployed), based on Bendix experience with dunking sonars of which large numbers are in operational service with the US defence forces and 12 allied navies. The 'wet end' consists of a receiver array of eight or 16 hydraulically driven arms which expand to a diameter of 2.8 m when deployed and a transmitter array of eight or ten projector elements. These projectors form a vertical array 6.25 m long that descends from within the 'wet end' central body during deployment. The projectors radiate at high power levels over 360° in azimuth and 10° in elevation. A range of some 200 km has been demonstrated in favourable water conditions.

STATUS

Advanced development is now complete. The system is scheduled to be in production by 1991 and is being proposed for the EH101 helicopters of the Royal Navy and the Italian Navy, as well as the USN advanced low frequency sonar requirement for the V9-22. A flyable evaluation system has been purchased by the Italian Navy and trials were completed successfully in December 1987, using an SH-3 helicopter. Under the terms of an agreement, British Aerospace will undertake a substantial portion of production, will be responsible as lead contractor for marketing, manufacture and support in the UK, and will collaborate with Bendix in a number of other countries. FIAR will oversee the Italian EH101 effort and will also participate in the HELRAS programme on a broad basis. BAe's Dynamics Division has undertaken considerable research in the past in this type of sonar development. Experimental models of HELRAS have already undergone extensive sea trials.

CONTRACTORS

Bendix Oceanics Division (a unit of Allied-Signal Inc), 15825 Roxford Street, Sylmar, California 91342, USA.

British Aerospace plc, Dynamics Division, Filton, Bristol BS12 7QW, England.

FIAR, via G.B.Grassi 93, 20157, Milan, Italy.

Artist's impression of HELRAS system in operation

6887.353

OSPREY DIPPING SONAR

Osprey is a new dipping sonar in development by GEC Avionics of UK and Sanders Associates, USA. The new system is intended for use by helicopters to detect submarines at greater range and depth than is currently possible. It will combine a Sanders high power, lightweight dipping sonar design with system control and data processing provided by one of the GEC Avionics AQS-900 series processors.

Osprey will operate at lower signal frequencies than any current dipping sonar. It will make extensive use of VLSI technology and custom chips to offer a low cost dipping sonar for the 1990s.

STATUS

A Memorandum of Understanding has been signed by Sanders and GEC Avionics to develop the Osprey system.

CONTRACTORS

GEC Avionics Ltd, Maritime Aircraft Division, Airport Works, Rochester, Kent ME1 2XX, England.

Sanders Associates Inc (a Lockheed company), Nashua, New Hampshire 03061-2035, USA.

ITALY

3233.453

IPD70/S INTEGRATED SONAR

The IPD70/S sonar is an improved version of the IPD70 currently installed on the 'Sauro' class submarines and, like the former equipment, it is the result of co-operation between USEA and ELSAG.

The main components of the system are a passive sonar and an active sonar, with three optional equipments: a high precision intercepting equipment, a long-range directional and omni-directional underwater telephone and a spectral analysis sub-system. An additional option, the MD100 passive rangefinder, is also available. This latter option can, however, stand alone and is therefore described in entry **3773.453**.

The IPD70/S uses three transducer arrays to serve the various functions:

(1) a passive low-frequency conformal array disposed along the hull and working from a few hundred Hz up to several kHz
(2) a cylindrical array, common to the active sensor, the high-frequency passive component, the interceptor and the underwater telephone. This transducer is housed in a suitable hull cavity and protected by a high-resistance low flow-noise GRP dome
(3) an optional linear array for the passive rangefinder set.

The equipment can track up to four targets. In the passive operation mode, very long detection ranges can be achieved with the high-directivity characteristics of the conformal array and the signal processing techniques employed. In the active mode bearing, range and doppler data are obtained from a single sonar pulse to minimise the probability of interception by the target. The intercept facilities provide measurement of bearing, pulse length and frequency by enemy sonars within several tens of kilometres.

Long range communication capabilities are given by the underwater telephone which can be operated on either an omni-directional or directional basis. The narrow band spectral analysis sub-system enables the operator to classify a target based on its radiated noise distribution.

The most important change from the earlier IPD70 equipment is the display and control console which has been completely redesigned. The operator's panel now features two high-resolution raster screens on which all relevant data collected and processed by the system can be displayed in an easy-to-read, graphic and alpha-numeric format. The operator controls have been simplified.

STATUS

In operational service with the Italian Navy.

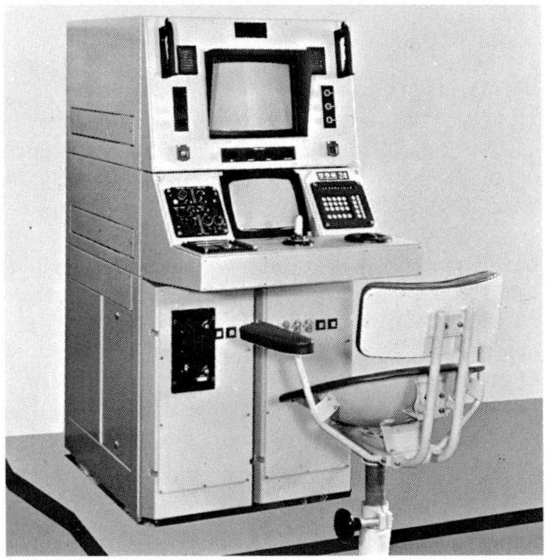

Control console for IPD 70 integrated sonar

CONTRACTOR

Selenia-Elsag Consortium for Naval Systems, Via Panama 52, 00198 Rome, Italy.

3773.453

MD100/S PASSIVE RANGEFINDER SET

The MD100/S is an upgraded version of the MD100 and is the result of co-operation between the companies USEA and ELSAG. The system has been designed for installation on small or medium size submarines and can be supplied either as a stand-alone device or as an optional component of the integrated sonar set IPD70/S (**3233.453**). In the latter case, all necessary space allocation, as well as command and display facilities, are shared on a common basis with the host system.

Though the system has been redesigned, many of the technical solutions employed in the earlier MD100 have been retained. Like its predecessor, the MD100/S is equipped with an independent set of acoustic sensors arranged as two linear arrays, each composed of three equally spaced transducers and steered towards the port and starboard sectors.

The signals collected by the transducers, after initial processing, are fed to the main unit for all the relevant stages of signal processing. Handling of signals entails the evaluation of correlation and coherence relationships existing among the various couples of waveforms involved, so as to obtain bearing and distance estimates of a noise source with high accuracy within the sector coverage of $\pm 60°$ athwartships.

The MD100/S features simultaneous tracking in bearing and range of up to four targets; bearing initiation data can be either collected by the equipment itself or from a surveillance sonar.

The display unit, in both the stand-alone and in the integrated versions, is based on a high-resolution raster-scan type screen, which allows constant monitoring of trajectories and all relevant data (both in graphic and alpha-numeric form) pertaining to the targets being tracked. In addition, continuous indication is given of special parameters, eg reliability coefficients of processed data, tracking losses, etc, which could be valuable in deciding the operator's course of action.

STATUS

The integrated version has been installed in the 'Sauro III' submarines.

CONTRACTOR

SELENIA-ELSAG Consortium for Naval Systems, Via Panama 52, 00198 Rome, Italy.

5754.293

MULTI-APPLICATION SONAR SYSTEM (MAST)

A trainer for sonar operators, known as MAST, has been developed by ELSAG, in co-operation with Raytheon, USA, and delivered to the Italian Navy's training centre. The MAST system consists of the following units:

(1) Trainee equipment consisting of a set of units identical to those used on board
(2) An instructor control station
(3) A computer system
(4) Simulation hardware producing signals relevant to 'own ship's signals' and those of simulated craft. These include all possible variations due to motion, mutual interference, propagation and environmental conditions simulated.

The MAST system is used for the training of operators and maintenance personnel on the DE 1160 and DE 1164 sonar systems. Further improvements will allow training facilities for other passive sonar systems produced by ELSAG and also for the DE 1167 being manufactured under licence from Raytheon.

STATUS

In production.

CONTRACTOR

Selenia-Elsag Consortium for Naval Systems, Via Panama 52, 00198 Rome, Italy.

1723.453

BI SERIES SONOBUOYS

The Servomeccanismi organisation produces three types of passive sonobuoy:

BIT-3: This is an A-size omni-directional passive sonobuoy for use at depths between 20 and 100 m, with a selectable life of one to three hours, and transmitting data over a 31-channel, IW link.

BIT-8: This is similar to the BIT-3 but has different frequency/sensitivity characteristics.

BIR: This is a miniature (500 mm long × 100 mm diameter, three kg) sonobuoy for use with helicopters. It is omni-directional and can be deployed at depths to 20 m. Life is one hour.

Receivers for use with these sonobuoys are the REA-16 and REA-31, having 16 and 31 reception channels, respectively. The BI series buoys have been designed primarily for use in the Mediterranean, under conditions normally found in that sea area.

STATUS

In production for the Italian Navy.

CONTRACTOR

Servomeccanismi, Via Mediana Km 29.3, 00040 Pomezia, Italy.

3234.453

P2072 (AN/SQQ-14IT) MINEHUNTING SONAR

The P2072 is an improved version of the original General Electric AN/SQQ-14 variable depth minehunting sonar and has been selected by the Italian Navy under the designation AN/SQQ-14IT.

The equipment is a dual frequency system having two modes of operation: search mode to detect probable targets up to 900 yards and a classify mode so that a trained operator, who can discriminate between mines and mine-like objects, can classify these after detection. The equipment provides range and bearing information which, related to the wide scanned field of view and narrow resolution in both range and bearing, are used to observe and direct a manned or unmanned neutralisation vehicle to mark or destroy the target.

The FIAR AN/SQQ-14IT is operational either as a hull-mounted or variable depth sonar. The variable depth feature provides for lowering the towed body, including the transducers, to a maximum depth of 150 feet (46 metres). This gives increased depth for minehunting, optimum positioning of transducers in adverse water conditions and favourable acquisition angles. In the variable depth mode the system can function effectively in sea state 5 when the towed body is lowered beneath the ocean surface effects.

In search mode the equipment uses a PPI display and electronically scans a 90° azimuth sector with a narrow beam (1.5 to 2°). In addition, the field may be automatically slewed through 180°, centred as selected on the bow, port bow or starboard bow, at a continuous rate of 15°/second at manually selected ranges of 300, 600 and 900 yards. An azimuth resolution of 1.5° and range resolution of 2.4 feet (0.7 metres) is obtained.

In classify mode the equipment uses a B-scope display associated with a memory display and electronically scans an azimuth sector of 18° with a 0.3° beam. An azimuth resolution of 0.3° and range resolution of approximately seven centimetres is obtained.

Additional details of the AN/SQQ-14 appear in entry **1789.453** in the USA pages of this section.

STATUS

The AN/SQQ-14 is used by the US, Belgian and Spanish navies aboard 'MSO' class minesweepers; the Italian and Saudi Arabian navies have installed it on 'MSC' class ships. In mid-1986 a contract was received from the Italian Navy for supply of the AN/SQQ-14IT for the new MCMVs of the 'Gaeta' (improved 'Lerici' class), and for update of the sonars of the four existing 'Lerici' class MCMVs.

CONTRACTOR

FIAR-Fabbrica Italiana Apparecchiature Radioelettriche SpA, Via Montefeltro 8, 20156 Milan, Italy.

1720.453

IP64-MD64 SONARS

The USEA IP64 sonar set is designed for installation aboard small or medium-size submarines, and it combines passive sonar equipment operating in the band 175 to 5000 Hz, and an active echo-ranging sonar operating at a frequency of 4000 Hz. It is also designed for integration with the MD64 sonar set for range measurement of noise sources by passive means. The receiving transducer elements, which are common to the passive and active mode, are arranged in the bow of the vessel in a long conformal array to enhance the directional capabilities. The transmitting transducer elements are arranged in a circular array located on deck.

Two linear arrays are provided for range measurement by passive means. Each array has a total length of 30 m and is formed by three transducers. The arrays are located on deck, port and starboard.

Search and detection are carried out primarily by means of the passive listening set, which incorporates advanced correlation techniques and records the received signals on a graph plotter. After target detection, a single transmission pulse in the direction of the target by the echo-ranging sonar is sufficient to provide range and relative speed target data.

The MD64 equipment measures the range of a detected noise source and direction is determined by the measurement of the relative time delay of sound waves detected by three hydrophones. Two groups of three hydrophones each are provided, for port and starboard measurements. Each hydrophone uses a number of elements arranged in five vertical staves and phased in the horizontal plane to produce a reception beam, directed toward the source. Main lobe beamwidth is ±15°.

The audio frequency band employed is 1000 Hz wide, centred on any frequency between 4 and 20 kHz. Operation is automatic and after acquisition of the noise source, the equipment locks on and continuous bearing data is produced. Range is measured continuously and is recorded graphically.

CONTRACTOR

USEA, Via G Matteotti 63, 19030 Pugliola di Lerici, La Spezia, Italy.

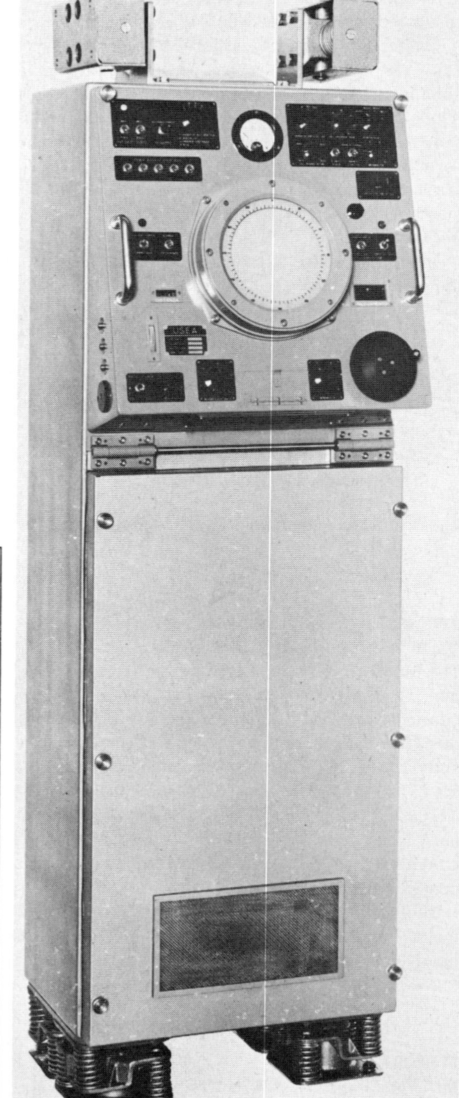

IP64-MD64 sonar display console

1722.353

FALCO SUBMARINE LOCATING SYSTEM

The FALCO system is airborne equipment designed to increase the capability for helicopters and fixed-wing aircraft to detect, classify, and locate submarines. The system operates on the low-frequency noise radiated by targets and gathered by directional low frequency passive sonobuoys.

Noise spectral analysis, target data processing and display are performed by a real-time digital computer.

It is claimed that the FALCO system is able to determine target position with an accuracy sufficient to carry out an attack with automatic homing torpedoes, and with errors due to sonobuoys drifting automatically cancelled out. The equipment is produced in several versions.

CONTRACTOR

Selenia Industrie Elettroniche Associate SpA, Via Tiburtina Km 12400, 00131 Rome, Italy.

4007.453

MISAR MSR-810 SONOBUOY

The MSR-810 passive sonobuoy for detecting and locating submarines from aircraft is manufactured by MISAR under Thomson Sintra licence for the Italian market. It is very similar to the Thomson Sintra DSTV-4L (**3501.453**). The sonobuoy transmits, by VHF/FM radio, the underwater acoustic AF signals received from the hydrophone. The MSR-810 complies with NATO and Italian Navy specifications.

CHARACTERISTICS

Type: Passive

Frequency: Low

Weight: 8 kg

Operating life: 1.3 or 8 h

RF channels: 31

RF output: 500 mW (at antenna)

Antenna: ¼ wave

STATUS

Production and in service.

CONTRACTOR

MISAR SpA, Via Brescia 39, 25014 Castenedolo (Brescia), Italy.

JAPAN

6249.253

QQS-4 SURFACE SHIP SONAR

Nearly all Japanese ASW ships have been fitted with various US types of sonar systems. However the 'Hatsuyuki' class destroyers have been fitted with a new Japanese designed sonar system, the QQS-4. No published data is available but the QQS-4 is a solid-state, low-frequency design which has been developed from the QQS-3 system. This latter sonar is the Japanese designation for the United States AN/SQS-23, built under licence since the late 1960s.

STATUS

In operational service with the 'Hatsuyuki' class destroyers of the Japanese Maritime Self Defence Force. Will eventually be replaced by the AN/SQR-19 TACTAS towed array on these ships.

NETHERLANDS

4464.453

HSS-15 COMPACT SURVEILLANCE SONAR

The HSS-15 is a compact panoramic sonar system for use on fairly small ships in the range of 100 to 800 tons. It fills the gap between commercial fishing sonars and more advanced professional sonar systems and is most useful in ships which need a limited anti-submarine capability, such as small patrol boats on coastal defence work and corvettes or coastguard ships which hitherto could not be fitted with sonar because of size, weight, and cost.

The HSS-15 operates in the 9.3 to 11.7 kHz range with a detection range of four to six km. The design is based on the assumption that the ship in which it is installed will produce fairly high noise levels. The system's main features are: multi-beam panoramic surveillance; relatively low operating frequencies to avoid high transmission loss; three different pulse lengths (25, 50 and 100 ms); three operating frequencies; active or passive operation; ease of operation with light pen control; and small transducer and cabinet dimensions. The overall system weight is about 600 kg.

CONTRACTOR

Hollandse Signaalapparaten BV, PO Box 42, 7550 GD Hengelo, Netherlands.

3142.253

PHS-32 SONAR

The PHS-32 is a medium-range high performance search and attack sonar in which the newest technological developments for signal processing and operation are employed, and helped by the use of a general purpose computer. The computer yields a compact, lightweight sonar for corvettes from 200 tons up to frigate-size ships.

The signal processing facilities provided include Fast Fourier Transformation processing of all pre-formed beam receiving channels. All data is presented on a single TV-type display, while operation has been much simplified by the use of light pen control. Automatic tracking of up to four targets is provided. These features combined with a high accuracy make this sonar very useful in an attack.

A circular transducer permits all round coverage in various modes of transmission such as: OMNI, TRDT, MCC (wide vertical beam), LISTEN (passive with time/bearing recorder presentation). An audio beam is also available. The system can be delivered in a fixed, retractable or VDS-dome outfit.

CHARACTERISTICS

Features: single operator control; display data processing for continuous presentation and memory mode (ping history); four-target automatic tracking; built-in energy storage for lower peak power demands on ships mains; integrated on-line and off-line test systems.

Frequencies: 3

Pulse lengths: 12.5, 25, 50, 100 ms and 400 ms long pulse (CW or FM)

Detection range: > 10 000 yards

Vertical beam width: 12 or 20°

Bearing accuracy: 1° RMS

Range accuracy: ±0.5 to 2%

Notch filtering: Selectable rejection bandwidth

Own doppler correction: On all 60 channels

Roll and pitch performance: Automatic co-ordination of transmitting pulse and vertical beamwidth in rough weather

Weights

Electronic cabinets: (Sonar console, duplexer and amplifier cabinet, transmitter cabinet) 1041 kg

Retractable hull-outfit including transducers: 7900 kg

Fixed hull-outfit including transducer: 2500 kg

Variable depth system: 5000 – 8000 kg

STATUS

In series production and operational service on-board ships from fast patrol boats to frigates.

CONTRACTOR

Hollandse Signaalapparaten BV, PO Box 42, 7550 GD Hengelo, Netherlands.

Lightweight modular transducer of PHS-32 sonar

5636.453

SUBMARINE INTEGRATED ATTACK AND SURVEILLANCE SONAR SYSTEM (SIASS)

The submarine integrated attack and surveillance sonar, developed by Signaal, is available in two versions: SIASS-1 and SIASS-2. The former is a stand-alone system with its own sonar consoles, and interfaces with a submarine data handling and fire control system. SIASS-2 is a sonar system as an integrated part of the Signaal Submarine Sensor Weapon and Command System (SEWACO).

Both SIASS-1 and SIASS-2 perform and integrate the following tasks:

(1) Low (long range) and medium (medium range) surveillance of noise radiating targets through the use of broadband and multi-channel narrowband signal processing for surveillance, analysis and tracking (LOFARGRAMS, ZOOM-FFT, Demon, ALI displays)
(2) Passive range detection
(3) Automatic target tracking for a multitude of targets, including automatic line tracking
(4) Contact motion analysis (position determination on bearings-only information)
(5) Active sonar
(6) LF, MF and HF intercept
(7) Classification
(8) Noise level monitoring and cavitation indication.

The SIASS signal processing is entirely digital and high resolution FFTs are used throughout the system for detection and analysis. Distributed computer architecture and extensive database facilities allow high flexibility, reconfigurability and availability. Analysis and classification aids are provided with electronic libraries and ample history of spectral analysis results (LOFAR) is retained from all pre-formed beams simultaneously for immediate review. Integrated support functions are sound velocity measuring, ray path indication and performance prediction.

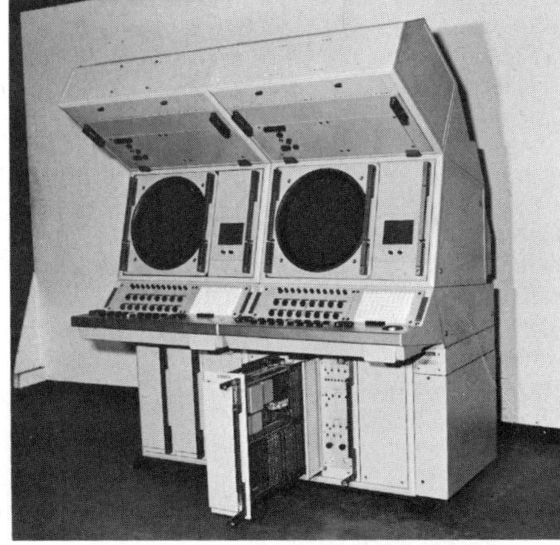

SIASS display and computer consoles

Latest-design acceleration-cancelling hydrophones form the basis of the multi-array configuration which may consist of:

(1) Two low frequency flank arrays
(2) A medium frequency circular array
(3) A high frequency intercept array
(4) Two sets of medium frequency passive ranging arrays
(5) An active transducer array
(6) Noise and cavitation monitoring hydrophones.

Control is exercised from multi-purpose consoles which may be integrated parts of a total SEWACO system (SIASS-2). These display and computer consoles are equipped with a multi-function control panel, a high resolution raster-scan display and an alpha-numeric display. Because of the built-in computer each console is virtually autonomous. As an optional extension, signal processing for a towed array can be offered.

STATUS

In production.

CONTRACTOR

Hollandse Signaalapparaten BV, PO Box 42, 7550 GD Hengelo, Netherlands.

4001.253

PHS-36 SERIES SONARS

The PHS-36 series of sonar systems is modular in design and intended for world-wide operation on any type of surface ship. The series consists of both hull-mounted (HMS) and variable depth (VDS) sonars, with combined HMS/VDS also available. The series provides various active modes (OMNI, OT, SDT, MCC) with simultaneous FM and CW transmission. A passive mode operates simultaneously with the active mode. Combined FM/CW pulses are transmitted for detection of low and high doppler targets under both reverberation and noise limited conditions. The series also includes frequency diversity and provides computer-aided detection and tracking of 12 targets.

A colour-coded presentation on a high-resolution display gives a graphic picture of the sonar information. Integrated support functions such as range prediction and sound ray tracing are included. A number of options are also available, including training facilities, underwater telephone, HF-intercept and shore-based training centres.

The PHS-36 series consists of the following systems:

Type	Frequency	Platform
PHS-24	High	Small ships
PHS-33	Medium	FPB/Corvettes
PHS-34	Low	Corvettes
PHS-35	Low	Frigates
PHS-36	Low	Destroyers

Note: PHS-24, PHS-33 and PHS-34 are HMS and VDS; PHS-35 and PHS-36 are HMS only.

STATUS

The series, which was a result of a joint development programme with the Royal Netherlands Navy, is in production. Nine systems were ordered for the Royal Netherlands Navy in December 1987 for fitment to the new 'M' class frigates.

CONTRACTOR

Hollandse Signaalapparaten BV, PO Box 42, 7550 GD Hengelo, Netherlands.

1525.263

SPI-04 SOUND RAY PATH ANALYSER

The SPI-04 is a small special-purpose computer-based equipment providing instantaneous prediction of sound ray paths calculated directly from sound velocity over long, medium, and short ranges. All electronics, including power supply, are housed in a single cabinet weighing 70 kg. An eight inch (20.3 cm) diameter CRT display is provided, and sound ray paths are displayed with a spacing of 0.5°, the number depending upon beamwidth selection. The SPI-04 is suitable for both anti-submarine operations and oceanography applications. Sound velocity/depth information is provided by equipment such as the Van der Heem/Sippican expendable sound velocimeter (XSV) or similar equipment.

Sound ray path analyser SPI-04

Main features include: use of microprocessors for high speed and reliability; stable presentation due to high display repetition rate; seven adjustable layers with sound velocity settings; variable bottom depth with reflections ON/OFF switch; profile setting display on LED numerical indicators.

Operational applications include:

(1) On surface ships to provide a clear picture of the sonar conditions to enable the best depth for VDS to be established, and to evaluate the possible submarine tactics
(2) In submarines for the determination of optimum listening and escape depths, and defensive tactics against helicopter sonars and sonobuoys
(3) In ASW aircraft and helicopters to establish the best depth for deploying sonobuoy hydrophones and transducers
(4) Minehunting in shallow waters
(5) Instant presentation of sonar coverage.

CHARACTERISTICS

Angular ray spacing: 0.5°
Function switch: Fixed, 4.5 m depth; variable, 0 – 600 m depth
Range/depth scales: 5; 9 km/375 m – 72 km/6000 m
Bottom depth setting accuracy: Better than 2 m

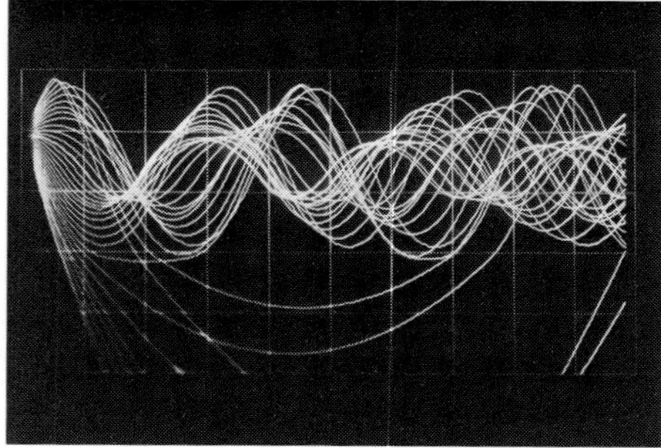

SPI-04 ray path analyser presentation

DEVELOPMENT

Development was undertaken by Van der Heem BV in the Netherlands.

STATUS

The SPI-04 is in operational service with the Royal Netherlands Navy and various other navies.

CONTRACTOR

Van der Heem Electronics, Regulusweg 15, PO Box 16060, 2500 AB, The Hague, Netherlands.

NORWAY

4128.453

SS105 SCANNING SONAR

This sonar is intended to fulfil the sonar requirements of modern ocean going coast guard vessels. It is a 360° scanning sonar with 48 preformed receiving beams with 11° beamwidth and 'split beam' processing in each beam. The working frequency is 14 kHz. The sonar system consists of an operator's console, transmitter unit, receiver unit, hydraulic power unit and hull unit.

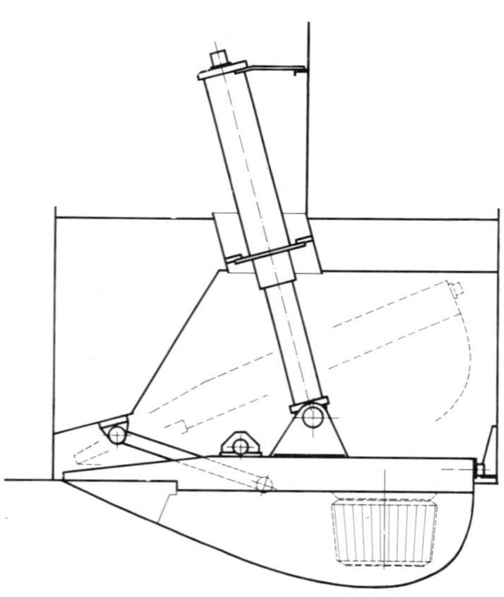

SS105 hull unit

The transmitter consists of 48 switching type amplifiers, 600 W each. Total output power is approximately 15 kW. Both OMNI, Single RDT and Triple RDT are available transmission modes. The main display has a CRT PPI, 280 mm in diameter, with scale ranges from 2 to 16 km. Markers provided are target cursor, stern cursor, transmitting and receiving sectors. An LED display shows target range, relative/true target bearing and ship's speed. Target data is transmitted digitally to other systems on board as required.

The 48 stave transducer is installed in a streamlined retractable sonar dome which will take ship's speed up to 25 knots. A fixed dome-arrangement is also possible.

CHARACTERISTICS

Transmitter
Type: Class S amplifier
Number of channels: 48
Max power output per channel: 600 W
Pulse lengths: 10, 30 and 60 ms
Frequency: 14 kHz
Transducer
Number of staves: 48, circular mounted
Active face per stave: 225 cm²
Beamwidth vertical plane: 12 ± 1°
Resonance frequency: 14 kHz
Tilt: 6° mechanical
Transmitting modes and performance
OMNI
Beamwidth vertical plane: 12 ± 1°
Beamwidth horizontal plane: 360°
Max output level: 219 dB rel. 1 µPa at 1 m
Directivity index: 9.5 dB
SRDT
Beamwidth vertical plane: 12 ± 1°, one beam
Beamwidth horizontal plane: 8.5 ± 1°, scanning a sector variable from 10 – 115°
Directivity index: 25 dB
Max output level: 230 dB rel. 1 µPa at 1 m
TRDT
Beamwidth vertical plane: 12 ± 1°
Beamwidth horizontal plane: 8.5 × 1°, 3 beams each scanning a sector of 120°
Max output level: 230 dB rel. 1 µPa at 1 m
Directivity index: 25 dB
Receiving performance
Number of simultaneous beams: 48
Bandwidths: 400 and 800 Hz
Beamwidth vertical plane: 12 ± 1°
Beamwidth horizontal plane: 11 ± 1°
Directivity index: 26 dB

STATUS

The SS105 is fitted to coast guard vessels of the Royal Norwegian Navy and the Finnish Coastguard service.

CONTRACTOR

Simrad Subsea A/S, PO Box 111, 3191 Horten, Norway.

6555.353

SS240 OMNI SONARS

The SS240 is an active multibeam sonar operating at 24 kHz, and is designed primarily for smaller ships. The transmitting modes are omni-directional, selectoral or directive. By using the Omni mode, targets are easily detected and tracked since the system covers the complete 360° around the ship. In all modes the beams are tiltable from +10° to −20° and, in addition, there is a MCC (Maintenance of Close Contact) mode with a broad vertical beam.

The SS240 consists of four units for easy installation; a hull unit with GRP-dome, a preamplifier unit, a transceiver unit and an operators console. The transducer unit is fixed hull-mounted with a cylindrical active area. It is built of 192 high efficiency ceramic elements with individual transmitters and receivers. The transducer face is covered by an acoustically optimal rubber compound.

The transmitted power is 7 kW with pulse lengths from 2 to 160 ms, both FM and CW. The echoes are presented on a CRT in eight colours with either PPI presentation or combined 180° fore PPI and one beam range-time (echogram) presentation. The presentation is selectable between relative or true motion.

CONTRACTOR

Simrad Subsea A/S, PO Box 111, 3191 Horten, Norway

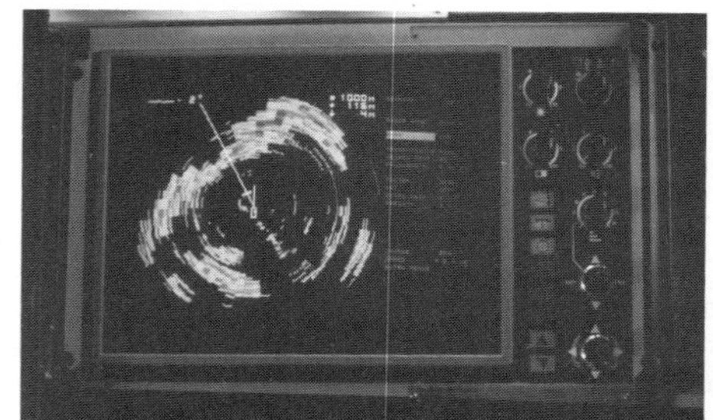

SS240 display (in colour)

5645.253

SS304 SCANNING SONAR

The SS304 is a high resolution scanning sonar suitable for installation in small ships down to 150 tons. It is designed for detection and tracking of submarines and submersibles, as well as for minehunting.

The system consists of a hull-mounted sensor, a transmitter unit, a receiver unit, and a control/display which provides the operator with a six-colour, memory refresh, true motion 14-inch picture of the underwater situation. The fixed streamlined fibreglass sonar dome is capable of withstanding ship speed up to 40 knots.

The sonar system consists of 17 simultaneous receiving beams which in a single ping cover 85°. The 85° fan-shaped beam can be trained and tilted in any direction, for quick searches 360° around the vessel, tiltable down to 90°. The vertical beam width is only 7° which eliminates nearly all interfering noises from the sea bottom.

CHARACTERISTICS

Frequency: 34 kHz

Range: 3000 m

Source level: 232 dB rel 1 Pa, ref 1 m

Transducer beamwidths: 9° horizontal; 7° vertical

Dome: Fibreglass (GRP) max speed 40 knots

Tilt: Upper limit 15° up; lower limit 105° down; speed 10°/s

Transmitter: Output power to transducer – 4 kW

Display: 14-inch colour monitor; 3 colours for symbols, 3 colours for echoes, 50 Hz refresh, 512 lines raster scan

STATUS

In production.

CONTRACTOR

Simrad Subsea AS, PO Box 111, 3191 Horten, Norway.

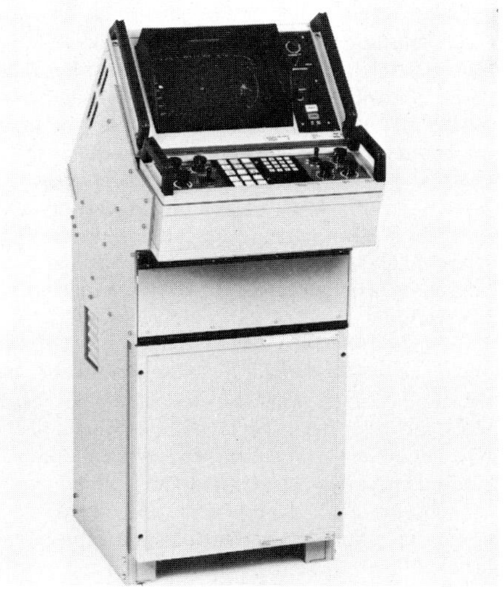

Control console of the SS304 scanning sonar

UNION OF SOVIET SOCIALIST REPUBLICS

3236.453

SOVIET SONAR

Because of Soviet security constraints and the generally inaccesible location of sonar equipment it is not possible to treat this aspect of Soviet underwater equipment in the same manner as has been adopted for that of other nations. Instead, the following notes summarise what little has been gathered, and which is free of Western security restrictions. While this inevitably falls well short of the ideal treatment, it is hoped that readers will find the ensuing paragraphs of some help.

Underwater acoustic experiments, initially in relation to communications applications, by the Soviet Union are widely agreed to date back to the years immediately prior to the First World War, but the earliest references to anything resembling what is now known as sonar occur in the 1930s when research into hydrophones for submarines is mentioned. At this general time, the possibilities of thermal detection devices for both aircraft and surface shipping were pursued. In the same period, the Soviet Union is credited with the production of passive sea bed acoustic detector equipment. Soviet claims record that the Soviet Union's naval forces at the start of the Second World War had available a variety of sonar equipments for shore and ship installations.

Most, if not all, of those reported were apparently passive devices, those for submarine fitting consisting of an elliptical array made of 8, 12 or 16 hydrophone elements. There were also passive sonar sets for surface ships, one of these being named Tamir and which began sea trials in 1940. This equipment is stated by the Soviet Union to have become the standard Soviet sonar employed in the anti-submarine campaign of that time. Neither the Soviet Union's then allies, nor her enemies, appeared to have been unduly impressed by the results achieved, according to historians, official and unofficial. By the

Large sonar arrays feature prominently on bows of Soviet submarines

'Juliett' class submarine bow sonars

'Sierra' class SSN bow with missing plates behind bow sonar belt

'Golf' class submarine bow sonars

'Foxtrot' class submarine bow sonars seen from abeam

post Second World War period, the USSR had gained access to sonar technology originating in America, Britain, and Germany, either by gift or as booty.

Since then, there has been steady and impressive growth of Soviet interest in all aspects of submarine warfare and it is reasonable to suppose that sonar equipment is accorded a priority within these activities at least as great as that given it by the Western navies. The advent of the Soviet ASW helicopter and aircraft carriers, such as the *Moskva*, *Leningrad* and *Kiev*, in advance of comparable ships being commissioned in Western navies, might imply a higher priority.

The appearance of these ships was preceded by smaller vessels designed for ASW operations and special-purpose aircraft and helicopters for naval duties, the latter being deployed in both land-based and embarked formations. These developments occurred in the mid-1950s to mid-1960s period.

Among the ASW ships, 'Petya' and 'Mirka' class vessels have hull-mounted sonars and some of the earlier 'Petya' class have been retro-fitted with a variable depth sonar which replaces the after torpedo tubes. The 'Kresta I' ASW cruisers have hull-mounted sonar and can carry Hormone helicopters which have been equipped with dipping sonar and a magnetic anomaly detector, and can deploy sonobuoys. A bow-mounted sonar is fitted in the later 'Kresta II' class, and the still more recent 'Kara' class ships have a variable depth towed sonar. The 'Krivak' class ships, which appeared at sea in 1971, have both bow sonar and a variable depth sonar.

More recently some information became available on a sonar system fitted to the 'Kashin' class destroyers. This is listed as a keel-mounted HF sonar, which includes Tamir, Pegas and Hercules types. The first two of these systems are very old and it is likely that the system fitted to the 'Kashin' class is a Hercules, probably operating in the 20 to 30 kHz range. The sonar is a unit of four pull-out chassis, with a small (77 mm) CRT behind an orange filter screen. The lower three racks feature simple on/off toggle switches and push buttons, only two multiple selector switches being visible. The only outputs seem to be to the small CRT and a meter which implies a simplistic sonar technology from the 1960s.

Bow sonar on 'Kilo' class submarine

In addition to the Hormone helicopters, the earlier Hound is believed to be capable of operating a dipping sonar, or possibly a towed magnetic anomaly detector (MAD). The MI-14 Haze shore-based helicopter also carries MAD. Other fixed-wing maritime aircraft, such as the Be-12 Mail and the Il-38 May are equipped with MAD, and the latter type also can deploy sonobuoys and process data obtained by this means. Mail has been followed into service by an ASW version of the Tu-20 Bear F which is understood to employ sonobuoys for submarine target detection.

'Foxtrot' class submarine bow sonars

Many submarines of the Soviet fleet prominently feature large sonar arrays and it has been estimated that between 30 and 40 nuclear attack submarines have been commissioned, supported by perhaps as many as 150 diesel attack submarines. These are in addition to the large classes of nuclear and diesel ballistic and cruise missile submarines, all of which have sonar installations.

UNITED KINGDOM

1981.363

AQS-901 SONOBUOY PROCESSING AND DISPLAY SYSTEM

The AQS-901 processes data from all NATO inventory sonobuoys and exploits the considerable advantages gained from advanced buoy such as BARRA, CAMBS and VLAD. It is claimed to be the most powerful and comprehensive system in service.

The aircraft contains two systems, each with two CRT displays. A fifth CRT, which is able to show data from either system is provided for system management. Information is displayed via both hard copy and CRT display media, and is controlled and interrogated by the operator using a keyboard, roll ball and cursor.

STATUS

In production and service. The AQS-901 is fitted to the RAF Nimrod MR Mk 2 and the RAAF P3-C Orion. Deliveries were completed in May 1986.

CONTRACTOR

GEC Avionics Ltd, Maritime Aircraft Systems Division, Airport Works, Rochester, Kent ME1 2XX, England.

AQS 901 sonobuoy processing and display system

3497.363

AQS-902/AQS-920 SERIES ACOUSTIC PROCESSORS

The AQS-902 and its export variants the AQS-920 series systems have been designed and developed for installation in ASW helicopters and maritime patrol aircraft. This range of lightweight and flexible systems will handle all current and projected NATO inventory sonobuoys and will interface with any approved sonobuoy receiver having a 31 or 99 RF channel capability. A maximum of eight sonobuoys may be processed, while for ASW helicopters these systems offer the additional advantage of simultaneous dipping sonar data processing and display.

The modular structure of the AQS-902/920 series acoustic processors provides the capacity for every configuration to be tailored to the aircraft, operational role and specific requirements of each customer. As a result the equipment is suitable for both mid-life updates of existing ASW aircraft and equipment fits in aircraft having no current sonobuoy capability.

The installation comprises four basic units, each of which contains the dedicated modules which allow construction of the selected system configuration at minimum weight and cost. These allow subsequent expansion and modification of the system to suit changing requirements.

The processor unit converts the received sonobuoy signals into digital form, carries out the necessary filtering and analysis, and processes the data into a form suitable for display. It is packaged into a short 1 ATR case.

The display unit may be in the form of a CRT display or a chart recorder, with configurations using either or both of these as required. All CRT and hard copy displays are fully annotated. The operator uses a simple control panel to specify processing modes and parameters, and a roll ball or stiff stick control and cursor to interrogate data on the CRT display. AQS-902 provides numerous standard processing and display facilities, including simultaneous passive and active processing; wide and narrow band spectral analyses; broadband correlation; CW and FM active modes; passive and active auto alerts; auto calculation and tracking of line frequency, target bearing and range, signal to noise ratio, doppler fixing, CPA analysis, bathythermal processing and ambient noise measurement.

A unique feature of the systems is the acoustic localisation plot which presents the operator with a geographical plot of sonobuoy and target positions.

STATUS

Configurations of this range of equipment are in service with the Royal Navy and the Royal Swedish Navy. In production for the Indian Navy and on order for installation in the Italian Atlantic Mk 1 maritime patrol aircraft. AQS-920 series systems will also equip the Fokker Maritime Enforcer. For the Royal Navy GEC Avionics is working on a contract to update the existing equipment on the Sea King Mk 5. 112 systems will be upgraded with 31 additional equipments being supplied.

CONTRACTOR

GEC Avionics, Maritime Aircraft Systems Division, Airport Works, Rochester, Kent ME1 2XX, England.

RN Sea King Mk 5, fitted with AQS 902

5655.363

AQS-903/AQS-930 SERIES ACOUSTIC PROCESSOR SYSTEMS

The AQS-903 and its export variants the AQS-930 series are powerful lightweight acoustic processors currently under development for future helicopter and fixed-wing applications. These systems will process data from a basic four to 64 sonobuoys in any combination. This flexibility is derived from the design based around a number of processing 'pipelines' each able to handle a certain quantity of passive and/or active sonobuoys. Four of these pipelines will provide a processing power eight times that of the AQS-901 at a fraction of the size and weight.

AQS-903/930 series systems make extensive use of distributed processing with the systems incorporating the latest component technology to provide the operator with a wealth of information. Data from up to eight sonobuoys can be processed simultaneously, proving full 360° surveillance. Sector coverage offers additonal classification and/or tracking analysis to allow smooth transition from one mission phase to another, irrespective of the type of buoy in use. Analysis and display cues are in plain language, requiring a minimum of operator training. Control of the twin CRT displays is achieved by the operator using a simple control panel and multi-function keys. AQS-930 systems will interface to other systems responsible for sonobuoy launching, navigation, tactical tracking and weapon release.

STATUS

A configuration from this range is currently under development for the Royal Navy EH101 ASW helicopter, due to enter service in the early 1990s. The basic configuration will process data from the complete range of NATO inventory sonobuoys.

CONTRACTOR

GEC Avionics Limited, Airport Works, Rochester, Kent ME1 2XX, England.

5891.353

ASN-902 TACTICAL PROCESSING SYSTEM (TATTIX)

The ASN-902 Tactical Processing System provides a means of correlating and processing data for display from the wide range of sensors and navigation sub-systems in the modern maritime patrol aircraft and helicopter. Based on proven AQS-902 technology, TATTIX will automate much of the interfacing, processing and display correlation, helping the aircraft tactical commander to solve complex navigation intercept and attack problems.

The flexible design of TATTIX makes it readily adaptable to various levels of system integration, extending to a totally integrated mission management system. It replaces the variety of individual controls and displays, and can provide an integrated control and display, even where the sub-systems are of varying design standards. Standard interfaces of ARINC 419, 429 and 1553 data bus allow installation as original equipment or as a retrofit.

Integrated Systems

GEC Avionics can offer various levels of systems integration. This improves the efficiency and flexibility of the mission avionics suite, minimises the weight of combinations of multiple sensors, and simplifies logistic and training problems with common control units and multi-purpose displays. The company's Maritime Aircraft Systems Division is carrying out systems integration for the Fokker Enforcer maritime patrol aircraft. This programme incorporates an AQS-900 series acoustic processor and the ASN-902 into the total avionics suite.

Integrated mission avionics suite in Fokker Enforcer maritime patrol aircraft incorporating AQS 902 (foreground) and TATTIX (centre)

CONTRACTOR

GEC Avionics Ltd, Maritime Aircraft Systems Division, Airport Works, Rochester, Kent ME1 2XX, England.

5852.363

AD 130 SONAR HOMING AND DF RECEIVER

The AD 130 provides sonobuoy homing and DF facilities on 99 channels over the frequency range 136 to 173.5 MHz at a channel spacing of 375 kHz. It is intended for use on both fixed and rotary wing aircraft. The system consists of a receiver with computer interface, a remote control unit and antenna switching controls. Additional units provided include a DF antenna, a left/right and fore/aft antenna for 'sonobuoy-on-top' indications, and cross-pointer and radio magnetic indicators. Dual receivers, with independent channel selection, may be operated from a single controller.

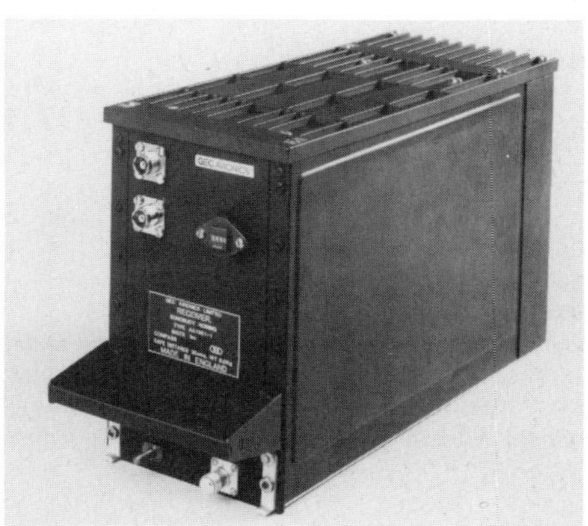

AD 130 sonar homing and DF receiver

The system can be operated in either automatic or manual channel selection modes. In automatic mode the computer carries out selection by decoding the serial bit stream, while manual selection is accomplished by means of pairs of thumbwheel switches. An LED read-out indicates the selected channel. The homing system provides left/right and fore/aft indications on the meter and the DF section detects and processes AM signals from the DF antenna for azimuth display. Built-in test circuitry continuously monitors all modes of operation. The system is stated to be compatible with all known and planned NATO and other allied services sonobuoys.

CHARACTERISTICS

Frequency range: 136 – 173.5 MHz
Channel spacing: 375 kHz
No of channels: 99
Frequency accuracy: 5 parts per million
Selectivity: Minimum 3 dB bandwidth 250 kHz, max 60 dB bandwidth 750 kHz
Dimensions
Receiver: 197 × 125 × 324 mm (½ ATR short)
Controller: 152 × 102 × 302 mm
Switched antennas: Each 50 × 80 × 193 mm
Weights: 8 kg (receiver); 2.3 kg (controller); 0.8 kg (each switched antenna)

STATUS

In production and in service.

CONTRACTOR

GEC Avionics, Airadio Division, Christopher Martin Road, Basildon, Essex SS14 3EL, England.

3268.493

G 738 TOWED DECOY

The G 738 is an improved and re-engineered version of the Royal Navy Type 182 towed torpedo decoy, now being manufactured for several other navies. Known features of the equipment are modern solid-state signal generation equipment and compact, lightweight, deck machinery.

The equipment is designed to decoy both active and passive homing torpedoes from a ship fitted with the G 738/Type 182. Decoy signals are electronically generated within the ship and fed via the towing cable to electro-acoustic transducers within the towed body; the signals thus produced divert passive homing torpedoes astern and confuse or jam the steering mechanisms of active homing torpedoes.

A variety of deck-handling arrangements is produced, and dual systems may be fitted. Remote control of selection of signal characteristics provides for instantaneous changes from operation's room or bridge. Parameters include noise, frequency and amplitude, modulated as well as unmodulated CW random and pre-programmed output beam formations. The equipment can incorporate handling systems for open deck on extended flight deck ships.

The operating characteristics remain confidential between the manufacturer and the purchaser.

STATUS

In production. 14 systems sold overseas.

CONTRACTOR

Graseby Marine (Graseby Dynamics Ltd), Park Avenue, Bushey, Watford, Hertfordshire WD2 2BW, England.

3266.453

G 750 SONAR

The G 750 is a search and attack all-round sonar for ships of frigate size and above. It is a modernised and improved version of the Type 184M (**3265.453**), providing accurate fire control data from two independent automatic tracking systems whilst simultaneously maintaining all-round surveillance, independent doppler search, and classification and continuous torpedo warning.

Type G 750 sonar control console

Three separate processing and display systems are incorporated for:
(1) all-round search and tracking (PPI)
(2) doppler
(3) passive search.

Digital readout displays are used extensively and visual displays are supplemented by audio in all three systems. Nine cabinets are required for the complete electronics. Three are display consoles, three contain transmitter and control equipment, two house receivers, and one is for monitoring purposes. For installation in smaller ships or submarines, single passive or PPI display systems, with smaller transducers, are produced. In the standard G 750 system, an improved version of the Type 184M 32-stave transducer is used to form a large number of beams. Range and bearing measurement is by in-beam scanning, obviating the need for separate expanded displays.

Target range and bearing are measured by either of two methods:

(1) by means of a manually positioned marker technique in which the circular markers are placed over the target echo and it is the marker position that is measured; or
(2) by echo gating where the echo range and bearing are measured directly. This form of tracking is normally on acquisition of a possible target and is available on either computer-assisted or non-computer-assisted basis. Marker boxes representing the gated area replace the circles. In both methods the targets are individually identified by dotted or full markers.

Provisions are made for interfacing the G 750 sonar with computer-based ASW and tactical information systems to provide a source of range, bearing, and target doppler data. Operational capabilities of the equipment include active detection of submarines throughout 360° and the ability to track two targets simultaneously. Similar detection coverage and tracking capabilities are available in respect of torpedo targets, with suitable outputs to torpedo warning and avoidance systems. There is also an active search capability over a 45° arc with facilities for determining target doppler. The system is designed for operation in adverse reverberation conditions, with ripple or omni-direction dual--frequency modulated and doppler CW transmission modes.

STATUS

In production. Three systems delivered to Royal New Zealand Navy and six systems to the Indian Navy. The system is likely to be updated with new displays and improved signal processing.

CONTRACTOR

Graseby Marine (Graseby Dynamics Ltd), Park Avenue, Bushey, Watford, Hertfordshire WD2 2BW, England.

3269.453

G 733 SLUTT AND G 740 SACACS

The G 733 SLUTT (Ship Launched Underwater Transponder Target) is in service with the Royal Navy and other navies throughout the world. SLUTT simulates an underwater target to allow calibration and alignment tests as well as functional training for sonar operators. This rugged system can be tailor-made to any sonar requirements. Some 24 variants are now available.

G 740 SACACS

A transportable long range Sonar Accuracy Check and Calibration System (SACACS) is in service with the Royal Navy. Known as the G 740, about 12 of these systems are now in use for sonar and anti-submarine weapon system acceptance trials and periodic calibration.

Secure Transponder System

This is a hand-held diver's interrogator unit and a seabed transponder unit which provides three microprocessor controlled modes of operation: sonar, bearing and range. It provides a range of applications such as relocation of pre-marked underwater equipment, location of underwater objects, safe path marking and ROV relocation.

CONTRACTOR

Graseby Marine (Graseby Dynamics Ltd), Park Avenue, Bushey, Watford, Hertfordshire WD2 2BW, England.

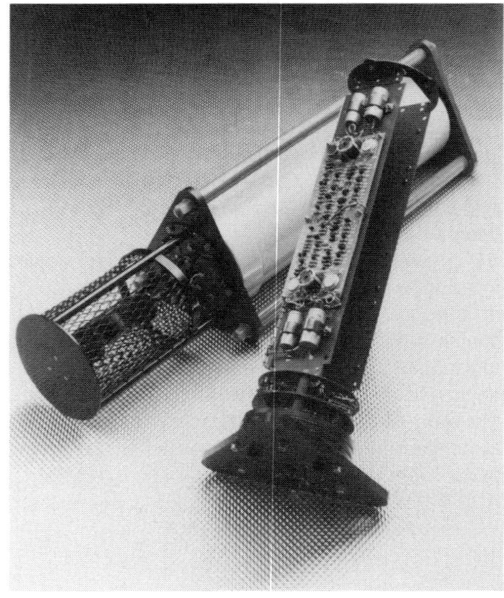

G 733 SLUTT and G 740 SACACS

7002.253

PMS26 and PMS27 SURFACE SHIP SONARS

Hull-mounted ship sonars designed for single operator control in both the surveillance and attack roles, the PMS26 and PMS27 are also used as 'dunking' sonars in certain helicopters, under the designation Type 195.

The PMS26 is a self-contained system for ships and patrol craft down to 150 tons. It provides full 360° coverage in four steps of 90° and may be manually controlled to cover a particular sector, or set to carry out automatic search procedure. It incorporates a 'maintenance of close contact' facility for tracking close or deep targets. The single operator controls the sonar through a special console. He is provided with three sources of sonar information – audio, visual doppler, and visual sector.

The doppler facility provides increased initial detection range and improved classification capabilities compared with conventional small ship sonars. The PMS26 transducer array is mounted within a hull outfit with a glass-reinforced plastic dome.

The PMS27 differs from the PMS26 only in its associated hull outfit. The PMS27 transducer array is mounted in a Royal Navy hull outfit 19 or similar, which makes the equipment suitable for installation in small escorts down to about 650 tons displacement. The PMS27 can be used as a surveillance sonar in association with a separate fire control sonar within the same hull outfit. This enables the ship to continue surveillance for new threats whilst engaging a target already detected.

STATUS

In operational service but no longer in production. More detailed information is given in previous editions of *Jane's Weapon Systems*.

CONTRACTOR

Plessey Naval Systems, Templecombe, Somerset BA8 0DH, England.

6558.253

PMS 56 PURSUIT SONAR

The PMS 56 is the direct successor to the PMS 26 series of sonars (**7002.253**). The system provides all the facilities required for effective ASW operations by small ships and patrol craft. The fixed dome is suitable for installation in vessels up to 3000 tons displacement with maximum speeds up to 30 knots. An optional retractable dome is suitable for installation in vessels up to 1000 tons displacement with maximum speeds up to 18 knots with the dome extended and up to 30 knots with the dome retracted.

A third variant can be supplied without a hull outfit for vessels which are already fitted with a suitable one.

CHARACTERISTICS

Modes: FM, CW, passive and aural

Range scales: 3000, 6000, 12 000 and 18 000 yards

Operating frequencies: 9.5 kHz, 10 kHz, 10.5 kHz

Pulse lengths: 250 ms; 1 s

Weights: Console 220 kg; transmitter rack 50 kg; hull outfit and array 385 kg

CONTRACTOR

Plessey Naval Systems. Templecombe, Somerset BA8 0DH, England.

PMS56 pursuit sonar console

1024.353

TYPE 195/TYPE 2069 HELICOPTER SONAR

The Type 195 sonar is a helicopter dunking system designed principally for Sea King helicopters. It provides full 360° coverage and is understood to be effective at ranges up to 8000 yards (7300 m). Full azimuth coverage is provided in stepped fashion over 90° arcs progressively, but manual control allows the operator to concentrate on any particular sector. Alternatively, the system may be programmed to undertake an automatic search. The operator is provided with audio, visual doppler and visual sector sonar information, and close contact maintenance is provided for the tracking of nearby targets and those at greater depths.

The Type 195 may be employed in either surveillance or attack control, or both simultaneously. Pulse length and detection range settings are operator selectable to optimise the system according to sea conditions and the tactical situation.

The system uses transistorised circuitry, and modular design techniques allow rapid replacement of faulty sections. Weight of the complete system is 269.2 kg.

The current model is the Sonar Type 195M, which incorporates many improvements over the original equipment. Future developments to enhance further the overall performance, and to meet naval requirements, include a longer cable to increase the operating depth of the submersible unit, and integration of the sonar with the latest high performance processing systems.

The Type 2069 sonar is the latest upgrade of the 195M, designed for the Royal Navy Sea King helicopters. The sonar transducer has been re-engineered to provide a greater operating depth and is now integrated with the AQS-902 E/DS acoustic system.

STATUS

Over 250 systems are in service with the Royal Navy and several other navies.

CONTRACTOR

Plessey Naval Systems, Templecombe, Somerset BA8 0DH, England.

Type 195 sonar under a Sea King helicopter

1755.453

TYPE 162M SIDEWAYS-LOOKING SONAR

Sonar Type 162M, which is fitted as standard in ships of the Royal Navy, detects and classifies both mid-water and sea bed targets. It displays port and starboard recordings simultaneously on a single straight-line recorder, which has a maximum range scale of 1200 yards. Operation is simplified by entirely automatic gain control. Reliabiity is enhanced with solid-state technology and maintenance is facilitated by ease of access and comprehensive built-in testing and monitoring features.

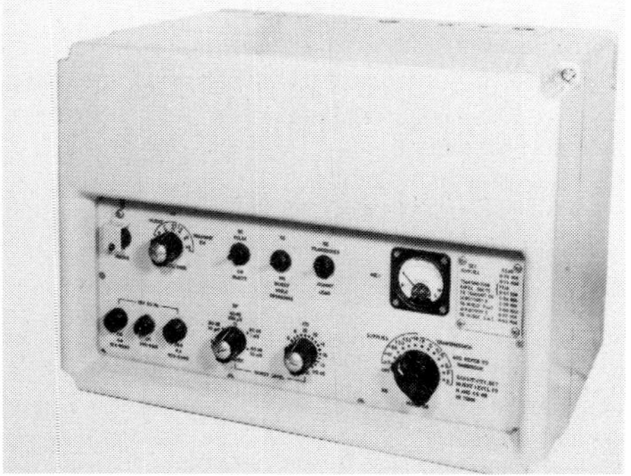

Type 162M sonar transmitter/receiver unit

The three transducers are all similar and employ 49.8 kHz barium titanate elements. Their beam pattern is fan-shaped, about 3° wide and 40° vertical angle; the side-looking elements have their axes 25° below the horizontal.

The recorder design uses a double helix (left and right hand) so that there are two points of contact with the moist electro-sensitive paper. As the helix rotates, the two points of contact move outwards from the centre. Port and starboard signals are fed respectively to the left- and right-hand points of contact so that they are recorded simultaneously. There are two zero lines near the middle of the paper, and the 18 mm gap between them is used for time marks every five minutes. The paper width is 286 mm, and each trace occupies 133 mm.

An electronic oscillator provides a controlled frequency supply for the driving motor and interval marks. Motor speed changes, for the three range scales, are made by frequency division so as to avoid the use of change speed gearbox, and a stroboscopic arrangement is included so that the helix speed may be easily checked.

A loudspeaker and a socket for headphones enable signals to be monitored aurally if required.

The three range scales are 0 to 300, 0 to 600 and 0 to 1200 yards, and accuracy is better than 2 per cent assuming a sound velocity of 4920 feet/second. The paper speed changes automatically when the range scale is selected, and the speeds are 6, 3 and 1.5 inches/minute (152, 76 and 38 mm/minute) respectively. At a ship speed of 10.8 knots the display scales are the same across the paper and vertically. A take-up spool is fitted but its use is optional. A fix marker draws a line across the width of the paper when a button is pressed, and it can be operated in conjunction with a Type 778 echo sounder or other equipment.

A new transducer array has now been developed as a direct replacement unit. The new array incorporates redesigned elements and techniques, using a low voltage device which virtually eliminates insulation problems. Tuning boxes for the three transducers are no longer required since the transducer tuning coils are incorporated into the individual housing gland.

STATUS

In production.

CONTRACTOR

Kelvin Hughes, New North Road, Hainault, Ilford, Essex IG6 2UR, England.

3265.453

SONAR TYPE 184M

The Type 184M is one of the Royal Navy's primary surface ship anti-submarine search and attack sonars.

It is a 360° scanning sonar incorporating both active and passive modes of operation; it provides range, bearing and target doppler data to the fire control computer. The Type 184M was designed in association with the Admiralty Underwater Warfare Establishment (AUWE), and is fitted in the majority of RN surface ships. The equipment has also been supplied to a number of other navies.

Few technical details have been cleared for publication, but it can be stated that the equipment provides a dual frequency transmission and three receiver systems. The latter comprise:

(1) an all-round search and tracking PPI system
(2) a doppler system with B-scan display
(3) a continuous torpedo warning system with its own display.

A circular 32-stave transducer array is employed.

Solid-state modernisation kits to improve the performance and reliability of this sonar are now in production for the Royal Navy.

STATUS

Operational with RN and other navies. RN fittings include Type 21 frigates, Type 42 destroyers and 'Leander' class ships. The G 750 medium-range sonar (**3266.453**) is a solid-state, improved version.

Type 184 sonar control console

CONTRACTOR

Graseby Marine, (Graseby Dynamics Ltd), Park Avenue, Bushey, Watford, Hertfordshire WD2 2BW, England.

1374.253

MINEHUNTING SONAR 193M/193M MOD 1

The 193M is a short-range, high definition sector scanning sonar, operating at two frequencies, providing detection and classification of both moored and ground mines. It has been developed from the Type 193 minehunting sonar; the adoption of solid-state electronics and other advances in technology have resulted in a reduction in installed weight to about 860 kg, which compares with a figure of some 2100 kg for the older Type 193. Among the operational improvements are extensive use of digital displays and facilities for interfacing with computer systems.

The basic concept is to employ high-definition sonar to locate and classify mines which are then destroyed by explosive charges.

The sonar provides both bearing and range data; the fine range and bearing resolution enables the operator to assess accurately the shape of a target and hence its nature. Since the resolution depends both on operating frequency and pulse length, two selectable frequencies are provided, each with a choice of pulse lengths in the classification mode.

Range and bearing data appear on two displays: in the search mode one display shows the total range covered; in the classification mode, a 27 m (30 yards) section of the search display is expanded to fill the second display screen and permit close examination of the target.

Two frequencies are employed by the 193M sonar, 100 kHz for long range search and 300 kHz for short-range search and classification. The transducers for these signals are carried beneath the ship on a stabilised, steerable mounting, the whole assembly contained within a dome. A choice of either inflated fabric or GRP material is provided for the dome. The receiver uses a modulation scanning technique with 15 beams, 1° wide in LF and 0.33° wide at HF giving azimuth coverage of 15° and 6° respectively. The searching of wider areas of the sea bottom is achieved by means of an automatic search sequence selected in accordance with the type of sea bottom. The returned echoes are presented to the operators on separate CRTs at the control console. One of these displays range and bearing of targets within the sector being scanned by the search transducer, while the other is used for the presentation of the classification channel data. The controls for adjusting the transducer position, signal parameters such as frequency and pulse length and for co-operating with the rest of the minehunting and destruction team, are also provided at the console. Type 193M data can also be fed into other ships' systems.

Type 193M sonar operators' console

The Plessey Speedscan system can be fitted to the Type 193M to allow it to be operated in the side-scan mode and to generate a hard copy print-out of the seabed.

Range and bearing data outputs are provided in synchro and digital forms, allowing the sonar to be interfaced with various plotting tables, Action Information Systems and remote indicators. The system is fully integrated with the Plessey Nautis-M system as chosen for the Royal Navy single role minehunter.

DEVELOPMENT

Development of the Type 193M was undertaken by Plessey in 1968. Since then a continuous programme of development and operational trials in conjunction with the UK MoD has resulted in further system improvements defined as the Type 193M Mod 1.

The system offers greatly improved performance in detection and classification by the use of digitally-processed video and display systems, together with computer-aided target classification.

STATUS

In service with the Royal Navy and many other navies, the Type 193M/193M Mod 1 is suitable for installation in all current minehunters.

Type 193M is fitted in the Royal Navy's 'Hunt' class MCMVs. Twelve navies have now purchased some 50 Plessey minehunting sonar equipments, including both the 193M and the earlier version 193. A continuing programme of enhancement will ensure that the efficiency of the equipment is kept continually under review.

The Type 193M Mod 1 enhancements have been retrofitted as a modification to RN ships fitted with 193M, while complete Type 193M Mod 1 sets are available for export.

CONTRACTOR

Plessey Naval Systems, Templecombe, Somerset BA8 0DH, England.

3787.453

TYPE 2016 FLEET ESCORT SONAR

The Royal Navy's Type 2016 fleet escort sonar is a replacement for the existing RN sonars Types 177 and 184. It employs computer-aided techniques which are operator interactive, have automatic detection and tracking, enhanced signal processing and displays complemented by an information storage facility.

The Type 2016 is a hull-mounted panoramic surveillance and attack equipment with facilities for classification and multiple target tracking. Interference between nearby vessels using the Type 2016 sonar is largely eliminated by the use of a new type of broad band transducer. The sonar display console is designed for manning by a single operator under normal cruise conditions, this crew member being able to initiate the preparatory actions necessary for urgent action.

Digital data processing facilities are based on use of Ferranti computer equipment, with other sub-contractors including Marconi Radar.

The Type 2016 system comprises the following main elements: four active receiver cabinets, one passive and control cabinet and one T/R switch and beamformer cabinet, all housed in Plessey MC70 type cabinets. The Ferranti computer suite consists of three D.811A cabinets and a separate cabinet for computer spares. The two transmitters are housed in two standard RN/AUWE cabinets. The solid-state electronics are based on a modular system approach using medium scale integrated devices, printed circuit back planes with wire wrap connections and standard line-drive receivers for all inter-cabinet signal connections. Extensive use is made of hybrid circuit techniques to reduce volume.

The Type 2016 array system, which is roll stabilised, is fitted in a ribless monocoque glass reinforced plastic sonar dome within a fixed hull outfit. The current Royal Navy practice employs a keel dome but the array could equally be mounted in a suitable bow dome.

The display console contains three main displays and a top-mounted versatile console system (VCS) unit. The surveillance display incorporates a special simple keyboard for transmission of information to the command/AIO and for the selection of the appropriate display. The surveillance display also displays monitoring information about the performance of the equipment. When a fault is indicated, the maintainer on stand-by is alerted and proceeds direct to the sonar instrument space where further sub-unit monitoring systems enable him to pinpoint the fault down to board level. The classification display provides expanded presentation of part of the surveillance display in both B-scan and A-scan modes to permit the operator to examine the target structure more closely.

The passive and auxiliary data display with two different integration times enables both torpedoes and more distant targets to be tracked. The auxiliary data display is used for data logging and system information which includes the display of such items as course, speed, weather, sea bottom conditions, and bathymetric information.

DEVELOPMENT

Design was initiated in the late 1960s and an experimental equipment underwent tests in HMS *Matapan*. A development contract was awarded in 1973 and the first prototype was completed in 1978.

In April 1980 Plessey Naval Systems announced that it had received UK Ministry of Defence contracts amounting to more than £50 million, including initial research and development contracts which added up to £10 million. Type 2016 fleet escort sonars have been supplied for the Royal Navy Type 22 frigates. A bow-mounted version is in development for retrofit into RN Type 42 destroyers.

STATUS

In service with the Royal Navy in Type 22 frigates, Batch 3 Leander frigates, 'Invincible' class aircraft carriers and some Type 42 destroyers. Production was completed in mid-1987.

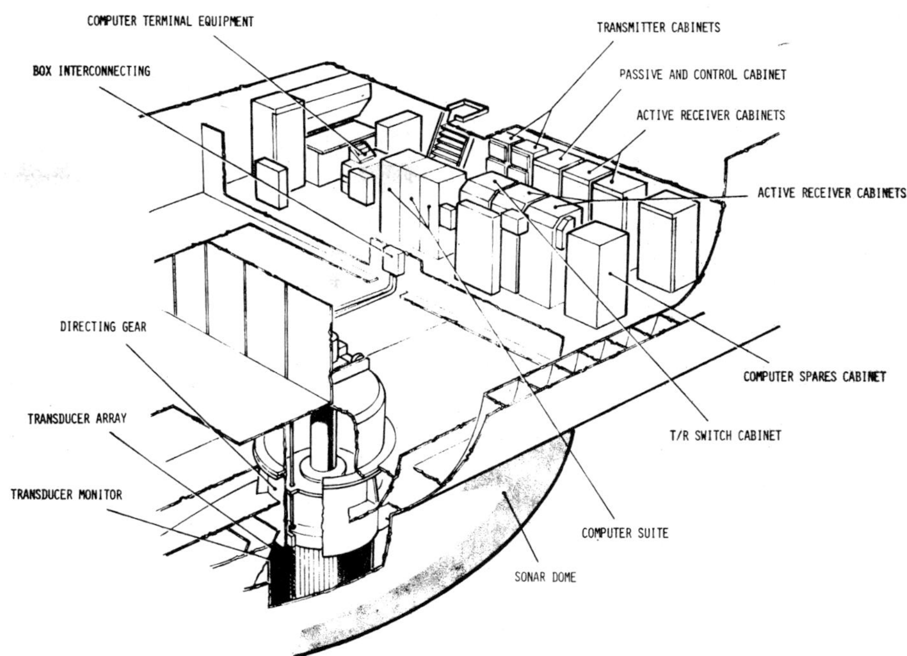

Arrangement of Type 2016 fleet escort sonar equipment and hull outfit compartment

Console of Type 2016 sonar aboard an RN frigate. Three displays (left to right) are for classification, surveillance, and passive presentation

CONTRACTORS

Plessey Naval Systems, Templecombe, Somerset BA8 0DH, England – prime contractor, with Ferranti Computer Systems, Marconi Radar Systems and other sub-contractors.

5651.453

TYPE 2020 SUBMARINE SONAR

The Type 2020 is one of the latest systems from Plessey's range of sonar equipments. It is being fitted in the Royal Navy's hunter-killer nuclear submarines and provides both passive and active detection capabilities from its bow-mounted conformal acoustic array. Although considerable emphasis has been placed on passive detection the equipment incorporates a highly effective computer-aided active capability.

Passive detection is sub-divided into low frequency and high frequency bands with information presented on dual-speed multi-pen recorders. The system is intended primarily for operation in a computer-aided mode but can be automatically reconfigured to a fall-back mode of operation, with target information being passed to the action information organiser in both modes. Facilities for expansion are included in the design to allow interfacing with future sonar systems. Solid-state switching is used for the

Display/console of Type 2020 submarine sonar

transmit/receive function and electronic beam steering techniques are employed. The design provides high detection sonar performance with a full 360° coverage.

A computer-controlled automatic monitoring subsystem is incorporated in the equipment, with a central monitoring unit driving local monitoring units within each equipment cabinet. Diagnosis of computer faults and programme loading is accelerated by the use of a floppy disc.

STATUS

The Type 2020 is intended as the replacement for the currently installed Type 2001 and is being fitted to 'Trafalgar' and improved 'Swiftsure' classes of nuclear-powered submarines. Each system costs £6 million.

The first production model was accepted by the Royal Navy in October 1982 and production deliveries are continuing.

CONTRACTOR

Plessey Naval Systems, Templecombe, Somerset BA8 0DH, England.

5637.353

CORMORANT LIGHTWEIGHT DIPPING SONAR

The Cormorant lightweight dipping sonar is a major advance in helicopter sonar, developed by Plessey Naval Systems to provide a unique, low-frequency active and passive acoustic sensor capable of high performance levels typically twice that of existing systems. When combined with a suitable acoustic processor, Cormorant provides an integrated sonics system which complements the area surveillance of modern sonobuoys to provide the maximum tactical freedom for the ASW helicopter.

The advanced design of Cormorant overcomes the size and weight constraints that have limited the operational performance of previous generations of dipping sonar. Its all-up weight of 220 kg has extended the sonar option to smaller helicopters down to 4000 kg in overall weight.

The Cormorant design is based on an expanding array, provided by a simple folding construction, which achieves a large acoustic receive aperture with a volumetric configuration. The hydrophone array is housed within the protection of the submersible unit for stowage in the aircraft, and during deployment and retrieval operations. Once the submersible unit has reached its operating depth, the hydrophone array is powered out to its operating position.

The projector array, hydrophone array, interface electronics and auxiliary sensors are carried in the submersible unit body in a modular configuration.

The compact transducer body design, constructed in high-strength, corrosion-resistant titanium alloys, achieves a hydrodynamic profile ensuring low drag and stable deployment and recovery. Aerodynamic stability enables 'flight at the trail' at maximum cruising speed, and a minimum winding load gives short recovery and deployment times and long cable life.

STATUS

Cormorant is currently undergoing a full technology demonstrator programme for the Royal Navy and has also been contracted for a 'Foreign Weapons Evaluation' programme for the US Navy. It has been selected by Canada as the 'wet-end' of the AN/AQS-503 HAPS system (**6551.353**).

CONTRACTOR

Plessey Naval Systems, Templecombe, Somerset BA8 0DH, England.

Cormorant sensor array

5866.353

GUILLEMOT DIPPING SONAR

Using and applying the same concepts as Cormorant (**5637.353**) to meet the specifications of larger helicopters, Plessey Naval Systems has also derived a larger family of expanding sonar arrays operating at a lower frequency and providing maximum active range under all types of oceanographic conditions. Of similar size to the current Type 195M dipping sonar (**1024.353**), the Guillemot variants are intended specifically for the EH101. Although still only at the early development stage, the sonar has successfully completed its first sea trials. Guillemot is conceived as a mainly long range surveillance system, with the added capability of providing a dipping sonar of exceptional performance against the anechoically coated third and fourth generation nuclear submarines of the next century.

STATUS

In development.

CONTRACTOR

Plessey Naval Systems, Templecombe, Somerset BA8 0DH, England.

6852.453

TYPE 2026 SONAR

The Type 2026 is a submarine passive towed array sonar and consists of a multi-octave towed array, a tow cable and an inboard processor and display system. The system provides surveillance and classification on data from both towed and flank arrays. It contains tracking, AIO system interface, data recording, demon, aural and hard copy facilities. The signal processor may also be used to provide a signal analysis and display function for data from other sonars.

The sonar may be operated by one or two operators and is controlled using a touch screen joystick and switches. It includes BITE and may be reconfigured in the event of a fault to minimise its effects.

Data is displayed on three acoustic data displays, with comprehensive cursor and annotation facilities. Other facilities for driving remote displays (eg in the submarine's control room) or external video recorders are provided. The signal processor is cooled by on-board chilled water supply, with an optional heat exchanger. The operator's console may be fitted in the sound room, with the signal processor in a remote cabinet space.

The Type 2026 was developed in conjunction with the Admiralty Research Establishment by GEC Avionics, STC and Plessey Naval Systems. GEC Avionics manufactures the signal processor and display system, STC the tow cable and Plessey the array and terminal unit.

STATUS

Systems are operational on a number of Royal Navy SSNs and a variant has been supplied to the Royal Netherlands Navy.

CONTRACTORS

GEC Avionics Ltd, Sonar Systems Division, 26-28 Hydeway, Welwyn Garden City, Hertfordshire, England.

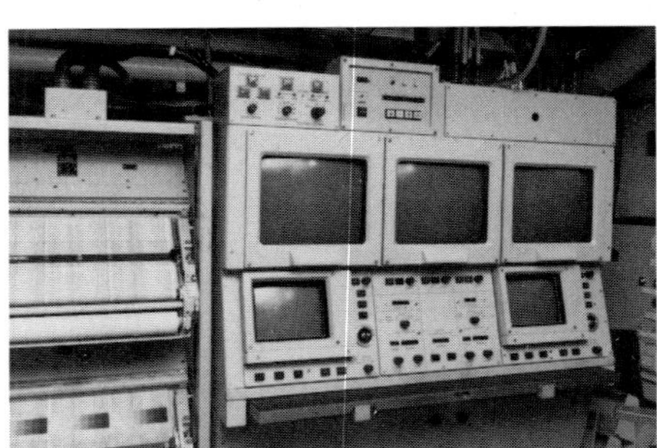

Type 2026 submarine sonar

Plessey Naval Systems, Templecombe, Somerset BA8 0DH, England.

STC Defence Systems, Greenwich, London, England.

6853.453

TYPE 2031I SONAR

The Type 2031I is a passive towed array sonar fitted to Royal Navy ASW frigates. The sonar was developed from earlier 2031X trials and from the Type 2026 (**6852.453**). It consists of a multi-octave towed array, a tow cable, a winch and a self-contained cabin containing signal processing, display and control equipment. The cabin may be installed in any convenient location and is cooled by the ship's chilled water supply.

The sonar provides all-round surveillance coverage and classification facilities, with good low frequency capability. It provides data recording and audio facilities and a display of array parameters including depth, heading and cable tension.

The sonar may be operated by up to three operators. It has three operator stations each including an acoustic display, touch screen and audio output.

STATUS

A number of systems are operational on Royal Navy frigates. The 2031I was developed in conjunction with the Admiralty Research Establishment by GEC Avionics (signal processor and display), Plessey Naval Systems (array and terminal unit), STC (tow cable) and Donkins (winch).

CONTRACTORS

GEC Avionics Ltd, Sonar Systems Division, 26-28 Hydeway, Welwyn Garden City, Hertfordshire, England.

Plessey Naval Systems Ltd, Templecombe, Somerset, England.

6236.253

TYPE 2031Z TOWED ARRAY SONAR

The Type 2031Z is a towed array sonar system currently being fitted to all new ASW frigates of the Royal Navy. Using the innovative 'Curtis' architecture developed at the UK Admiralty Research Establishment in Portland, Dorset, the 2031Z electronic signal processing suite has, because of its compactness, a considerably lower installed cost than other systems, and yet provides a multi-octave broad and narrow band analysis facility capable of tracking the most elusive threats.

The array itself is towed behind the ship and has a very low self noise characteristic and is considerably longer than that used by any other self-contained sonar. It provides good bearing accuracy, even down to the very low frequency end of the acoustic spectrum. The combination of an advanced signal processing architecture and a high performance towed array ensures that the system is a dual capability sensor, able to fulfil both the tactical and surveillance role. A 2031Z equipped frigate can, therefore, act as an ASW picket or as a close escort.

The modular nature of the 2031Z, both as regards the digital signal processing electronics and the towed array itself, makes it possible to configure other types of sonar systems from the same range of modules. As an example, a variant of the 2031Z signal processing suitable for submarine applications is also available and has been delivered to the Royal Navy.

The Waverley Division is able to provide, in collaboration with other suppliers, complete towed array sonar systems for export applications.

DEVELOPMENT

Waverley has worked very closely with the Admiralty Research Establishment throughout the development of the Type 2031Z sonar system.

STATUS

The Type 2031Z is now in series production for the Royal Navy for installation into its new ASW frigates. 22 systems had been ordered as at November 1987.

CONTRACTORS

Prime contractor and manufacturer of the inboard electronics suite – Waverley Division, Dowty Maritime Systems Ltd, Waverley Road, Weymouth, Dorset, England.

Towed Array – Plessey Naval Systems, Templecombe, Somerset BA8 0DH, England.

Winching equipment – NEI Clarke Chapman Limited, Victoria Works, Gateshead, Tyne and Wear, England.

Sonar 2031Z units, including dual display consoles, non-acoustic data displays and hard copy recorders

4127.453

TYPE 2034 SONAR

The Waverley sonar Type 2034 is a short range, high definition, dual side-scan sonar, suitable for establishing the position of ships, aircraft or other objects lost at sea; for charting obstructions in shipping channels or at sea floor engineering sites; and for evaluating underwater topography. Designed for the utmost reliability and to meet exacting military standards, Type 2034 sonars have been in use with the Royal Navy since 1976. The system is now standard equipment on all ocean going and coastal vessels of the survey fleet.

The sideways-looking, left and right, transducers are housed in a towed body, called the 'towfish', the current version of which is the Mk 3. Some of the electronics are contained in the towfish but the majority of this part of the system is housed in the recorder unit which provides for operation and control of the equipment, and display by hard copy, print-out chart recordings.

The sonar operating frequency is 110 kHz with a nominal transmitted pulse length of 100 microseconds. A choice of three range scales, 75, 150 or 300 m and one of three recording paper speeds, 30, 60 or 150 lines/cm can be selected to suit the prevailing operating conditions. The transmission repetition rate depends on the selected range scale, and is 10 pps for 75 m range, 5 pps for 150 m range or 2.5 pps for 300 m range. For all range settings, scale lines are recorded at 15 m range intervals, to aid interpretation of the recordings. Facilities are provided for putting event markers on the recording.

STATUS

Over 50 systems are operational with the Royal Navy Hydrographic Service. Older systems are currently being retrofitted with a microprocessor controlled signal processing system similar to that used in the Waverley 3000 sonar.

CONTRACTOR

Waverley Division, Dowty Maritime Systems Ltd, Waverley Road, Weymouth, Dorset, England.

6237.253

SONAR 3000 PLUS SERIES

The Sonar 3000 PLUS family of side-scan sonars has been developed from the Type 2034 high definition sonar (**4127.453**). The system uses modular construction which allows the assembly of variants to suit a range of applications. The system, originally aimed at the commercial market, is also in use for military applications. One variant is known as the Type 2053 and is in use by the Royal Navy for inshore hydrographic survey work.

The Sonar 3000 PLUS family comprises models ranging from the basic Sonar 3000 to the powerful Sonar 3010. The latter has been developed to provide an affordable solution to the mining threat, it is particularly suited to craft of opportunity and to use by non-specialised personnel. Key features of the Sonar 3010 for their application are:

(a) Automatic adaptive gain so that the operator has no sensitivity settings to adjust and the sonar record is consistent on every pass or survey
(b) Ranges can be varied from 37.5 to 300 m
(c) Video display for classification so that the operator can freeze, zoom, contrast, enhance or colour scale the picture
(d) Target marking facility
(e) Dual frequency based on the Sonar 2034 design but fitted with four transducer arrays for optimum performance at 100 kHz (long range) and 300 kHz (high resolution)
(f) High definition data recorder
(g) High resolution thermal linescan hard copy recorder
(h) Interface compatability with integrated navigation processors such as the Qubit TRAC IV or Racal 900 to provide precise geographical co-ordinate date on marked targets.

STATUS

The Sonar 3000 has been ordered by the Royal Navy Hydrographic Service, under the nomenclature of Type 2053

CONTRACTOR

Waverley Division, Dowty Maritime Systems Ltd, Waverley Road, Weymouth, Dorset, England.

Sonar 3010 displayed in heavy duty, shock mounted cases

6857.363

RASAT SONAR TARGET

RASAT (Radar and Sonar Alignment Target) is a programmable, universal sonar target which is compatible with a wide range of modern, high technology sonars, including hull-mounted and dipping types. It can be used for the alignment of surface ship or helicopter systems, can act as both a transponder for active sonar, and as a triggered noise source for passive sonar and it can be deployed from ship or helicopter. Operating depth is selectable from 15 to 50 m and bandwidth is 3 to 12.5 kHz for the basic model; other operating depths and bandwidths are available.

CONTRACTOR

Waverley Division, Dowty Maritime Systems Ltd, Waverley Road, Weymouth, Dorset DT3 5HL, England.

6238.453

TYPE 2046 SUBMARINE SONAR

The Type 2046 sonar is an integrated, passive system designed for the Royal Navy nuclear powered attack submarines, and is also being fitted into the Type 2400 'Upholder' class of conventional submarines.

Developed from advanced 'Curtis' digital signal processsing technology, and including FMS 12 display processing with Ferranti M700 Argus processors, the Type 2046 provides long range target detection, classification and tracking information.

The system provides 360° bearing cover, multiple beam resolution and broadband surveillance with high resolution narrowband frequency expansion facility. The operator interface consists of a four-screen, two operator console suite with touch sensitive plasma panels.

Like all the Ferranti sonar systems the Type 2046 is

built to a modular design to make maximum use of the latest technology, to save space and cost, and to assist through-life enhancement of the system. It is also equipped with fully integrated built-in-test equipment for ease of maintenance.

STATUS

The first system completed successful sea trials in March 1988 and five systems had been delivered at that time.

CONTRACTOR

Ferranti Computer Systems Ltd, Cheadle Heath Division, Bird Hall Lane, Stockport, Cheshire SK3 0XQ, England.

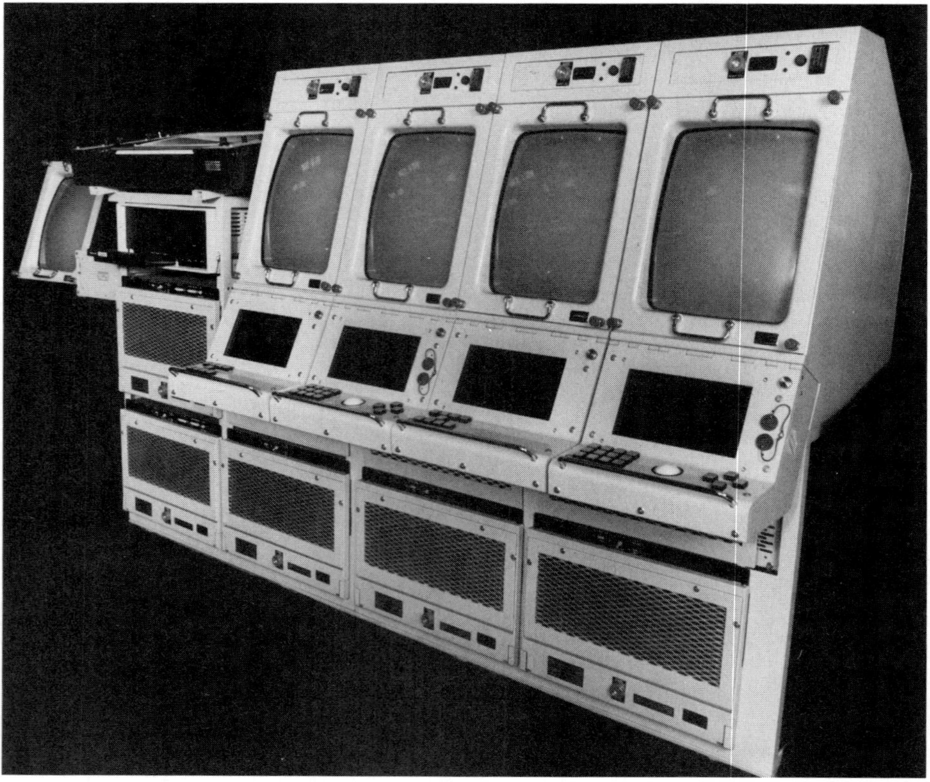

Operators' interface console of Type 2046 sonar

5853.253

TYPE 2050 SONAR

The Type 2050 is the new hull-mounted active sonar for the Royal Navy and will be fitted to the Type 23 and retrofitted to Type 42 and Type 22 frigates. It is a successor to the Type 2016 (**3787.453**) and is compatible with both bow and keel variants of the Type 2016 array. The equipment has been developed from the Ferranti FMS series (**4625.253**), and includes digital signal processing and distributed data processing using Ferranti Digital Signal Processing (DSP) cards and multiple Ferranti Argus M700 processors.

The provision of extensive data processing facilities and improved man-machine interfaces allows one operator to control the complete system. The equipment will interface to the combat system data highway feeding data to the Ferranti CACS action-information system. The suite consists of five processing cabinets and the system uses the same display consoles as will be used in the Type 2054 equipped Trident submarines. These are air-cooled and in monochrome but colour is available in the system for possible export applications.

STATUS

Seventeen Type 2050 sonar systems have been ordered by UK MoD for the Royal Navy, to be fitted in Type 23 frigates and retrofitted in Type 22 and 42 ASW frigates. Delivery commenced in the spring of 1987.

The first production equipment underwent successful sea trials in mid-1987 on board HMS *Scylla*. An export version, the FMS 21, is described in entry **4625.253**.

CONTRACTORS

Ferranti Computer Systems Ltd, Cheadle Heath Division, Bird Hall Lane, Stockport, Cheshire SK3 0XQ, England – prime system contractor.

Marconi Command and Control Systems Ltd – sub-contractor for the transmit/receive function.

Plessey Naval Systems – 'wet ends'.

Operator console of Type 2050 sonar

5858.453

TYPE 2052 SUBMARINE SONAR

The Type 2052 is a passive broadband towed array sonar system designed specifically for submarine use. It is understood that this sonar will be fitted to some of the earlier nuclear-powered fleet submarines of the Royal Navy, such as the 'Valiant' and 'Churchill' classes which were laid down in the 1960s. The system is also understood to be the basis of an export towed array system being developed by Marconi Underwater Systems Ltd. No technical details have been made available.

STATUS

In operational service with the Royal Navy.

CONTRACTOR

Marconi Underwater Systems Ltd, Elettra Avenue, Waterlooville, Hampshire PO7 7XS, England.

5855.453

TYPE 2051 SUBMARINE SONAR

The Type 2051 sonar system is being provided for the update programme of the Royal Navy 'Oberon' class submarines. The prime contractor for the complete sonar programme, known as Triton, is Plessey Naval Systems with Ferranti Computer Systems and Marconi Underwater Systems as co-contractors. The Triton fit is designed to bring the 'Oberon' class submarines up to date with the latest advances in sonar technology, including computer-assisted detection and classification.

The main elements of the system are a compact narrow band sonar which is linked to the submarine's existing flank array or towed array to provide all-round surveillance and classification, broadband passive sonars for both bow and flank arrays, an active sonar with variable power and sector capabilities and an intercept sonar.

STATUS

Nine systems have now been ordered under a contract worth about £25 million. Deliveries began in 1986 when systems were fitted to HMS *Otter* and HMS *Ocelot*. The remaining seven submarines will be fitted by the end of 1988.

CONTRACTORS

Plessey Naval Systems, Templecombe, Somerset BA8 0DH, England – prime contractor.

Ferranti Computer Systems Ltd, Cheadle Heath Division.

Marconi Underwater Systems Ltd, Waterlooville.

5857.453

TYPE 2054 SUBMARINE SONAR

Plessey Naval Systems' contract for the design, development, prototype construction and pre-production of the sonar, Type 2054, which will equip the new Royal Navy Trident ballistic missile-carrying submarines is well advanced. The complete equipment will consist of a suite incorporating passive, active, intercept and towed-array sonars.

Plessey, as the prime contractor and system design authority will supply the bow-mounted hydrophone array together with the necessary beam-forming and signal processing equipment (both narrowband and broadband) and the towed array. Major sub-contractors are Ferranti Computers for the consoles, data processing and displays; Marconi Underwater Systems Ltd for the intercept sonar equipment and STC for the handling system for the towed array.

The signal processing in the Type 2054 is almost entirely digital, the arrays being sampled by the preprocessor and immediately digitised before being passed to the beam former. Signals are then stabilised for both ship's position and geographical location before passing down for further processing in FM, CW and passive modes. These provide bearing and range, target motion and torpedo detection capabilities.

Subsequently the control and monitoring processing adds the ship's data to the detector signals and allows the operator to select particular modes of operation. This part of the system consists of two cabinets, one for target classification and the other for data display handling, which are interlinked using a Eurobus system, and connected to the outside world by a 1553B standard data highway.

STATUS

In development. In May 1986 Plessey announced orders worth over £40 million for enhancements to the Type 2054, plus the first production contract. The total of orders received by Plessey from the UK MoD for the Type 2054 project had totalled £150 million in the 12 months ending May 1986. It is expected that further contracts will be announced in line with the Trident submarine production schedule.

CONTRACTORS

Plessey Naval Systems, Templecombe, Somerset BA8 0DH, England – prime contractor.

Ferranti Computer Systems Ltd.

Marconi Underwater Systems Ltd.

STC.

6870.453

TYPE 2057 TOWED ARRAY SONAR

The Type 2057 is the new passive towed array sonar for the Royal Navy. It is intended for use in a variety of both surface ships and submarines. Engineering features are likely to include arrays, vibration isolation modules and non-acoustic sensors, winch cables, inboard signal and data processing and displays, and recording and systems control.

STATUS

Two competing consortia were due to be chosen in 1988 for a 15 month project definition phase, after which a single consortium would be chosen for full development and initial production.

6868.453

TYPE 2074 SONAR SYSTEM

The Type 2074 submarine sonar system is designed for fitment to three of the Royal Navy's nuclear-powered hunter-killer submarines. It will replace the Type 2001 and provide both improved sonar array performance and will be considerably smaller than the existing equipment. Although no further details are available it is believed that the size reduction in on-board electronics is of the order of 60%.

STATUS

In development. Plessey received a £10 million contract in February 1988.

CONTRACTOR

Plessey Naval Systems, Templecombe, Somerset BA8 0DH, England.

6869.453

TYPE 2075 SONAR

The Type 2075 is the hull-mounted suite for Batch 2 of the Royal Navy 'Upholder' class SSK submarines. The stated requirement is for five integrated active/passive sonar systems for delivery from 1991 onwards.

STATUS

A number of companies have bid for this system but a decision had not been made at the time of writing.

5856.253

SONAR 2093

The Sonar 2093 is claimed to be the world's first variable depth, multi-mode minehunting sonar. Designed by Plessey in partnership with General Electric of USA, Sonar 2093 is currently being fitted to the single role minehunter ships of the Royal Navy. The system is designed to operate in either hull-mount or variable depth mode, deployed through the ship's centre well and will detect, localise and classify all current and future mine threats. With a total system weight of 13 000 kg, Sonar 2093 is the lightest variable depth sonar available.

As the culmination of a six year development programme the system is designed to carry out its mission at high speeds without compromising the ship's safety. A variable depth, dual-frequency search and classification capability provides a sonar able to operate under all bottom and sea conditions at a range claimed to be at least that of hull-mounted sonar and with an increase of three times detection depth. The Sonar 2093 is integrated with the ship's command system to allow simultaneous operation in search, detection and classification modes, while the area coverage rate capability will allow a significant improvement on current route survey operations.

Sonar 2093 employs the unique approach of multiple operating modes, which offer the MCM commander the choice of a number of operational configurations to match the prevailing threat and environment. These multiple operating modes comprise VLF and LF search and moored mine classification, VLF and LF search, search and ground mine classification and search and route survey. Sonar 2093 can combine these modes to enable concurrent operation.

STATUS

Prototype systems available. Plessey has been awarded a contract for ten production systems.

CONTRACTOR

Plessey Naval Systems Ltd, Templecombe, Somerset BA8 0DH, England.

6244.253

COMPACT TOWED ARRAY SONAR SYSTEM (COMTASS)

COMTASS is a cost-effective, compact towed array system which can be installed in a ship or a submarine with the minimum of structural changes, and the 129 m long array has been designed to withstand high speed towing. The processing electronics are contained in two to four 19-inch units, each with integral power supplies and cooling fans. The total modularity of the display and the processor gives scope for the addition of extra processing facilities.

Three versions of COMTASS are available: COMTASS 1 for broadband surveillance in three frequencies, COMTASS 2 which has additional facilities including narrowband surveillance and classification, and COMTASS 3 which also employs automatic detection.

A turnkey system design leads to simple on-deck and below-deck installation, and the critical angle tow design lends itself to ease of handling, deployment and retrieval. The system provides passive detection and tracking of targets with a multi-function operator's console. Detection range is given as 30 nautical miles and the system is claimed to be able to detect surface ships over-the-horizon.

STATUS

COMTASS has been evaluated by the Royal Swedish Navy with extensive sea trials in the Baltic during 1986. An order for at least eight systems is likely to materialise.

CONTRACTOR

Plessey Naval Systems, Templecombe, Somerset BA8 0DH, England.

COMTASS lightweight array and handling system

6854.453

TYPE SP 2110 SONAR

The Type SP 2110 is an extension of the GEC Avionics Type 2031I and 2026 sonars which are in operational service with the Royal Navy. The SP 2110 is smaller with increased processing power and is stated to be considerably cheaper. It is modular and consists of four basic units, the array and handling system, the array terminal unit, processing cabinet and display console. Additional equipment can be supplied and interfaced to provide a complete towed array system providing narrow band surveillance and tracking, broadband, vernier, demon and aural processing. The system can be configured to meet customer requirements.

CONTRACTOR

GEC Avionics, Sonar Systems Division, 26-29 Hydeway, Welwyn Garden City, Hertfordshire, England.

4625.253

FMS SERIES SONAR SYSTEMS

The Ferranti Modular Sonar (FMS) series is an integrated range of low-cost, compact, high performance systems, developed by the company as a private venture to meet the needs of a wide variety of platforms and operational roles; and is based on the new generation of sonar-designs selected by the Royal Navy. The series has been designed to optimise performance, simplify installation and in-service support, and reduce through-life costs.

The systems within the series consist of:

FMS 12 – passive narrowband (2 octave)
FMS 15 – passive narrowband (5 octave)
FMS 20 – active (4 ft hull array with VDS option)
FMS 21 – active (6 ft hull array with VDS option)
FMS 30 – passive broadband (single shelf – 3 ft array)
FMS 31 – passive broadband.

All FMS systems employ the latest advanced distributed digital signal and data processing techniques to provide significant cost, space and performance improvements over the majority of current equipments. These techniques are subject to ongoing development to further improve these factors.

Signal processing, based on advanced 'Curtis' techniques, is achieved using programmable Ferranti digital signal processing modules which can be easily configured to meet the optimum performance requirements of specific array outfits and for easy through-life enhancement.

The data processing equipment is based on multiple M700 processors, arranged in distributed arrays. The use of large numbers of these interconnected in the data processing system gives very high levels of computer assistance and automatic operation. Features such as automatic detection, tracking and classification of multiple targets are provided. Standard programmable modules for both signal processing and data processing functions enable significant reduction in spares holding, simplifies maintenance and reduces overall cost.

The ergonomically-designed operator consoles allow single operator control of the complete system. Each contains two 1000 line high-resolution raster displays and an operator's desk. The man-machine interface includes a trackerball, nudge keys and a touch-sensitive plasma panel. From the plasma panel the operator can select various displays, focus on selected beams, scroll around beams, set verniers, annotate features on the display and select a variety of cursors. The trackerball and nudge keys are used for fast and accurate positioning of the cursor to provide specific displays of frequency, bearing, time and harmonic ratio information.

Each sonar in the FMS series is configured by selecting from a range of standard modules. This allows active or passive sonars for ships and submarines, large or small, to be proposed to suit customer's specific requirements. The modules are the same as, or closely related to equipment in, or ordered for, RN service. This capability enables Ferranti to offer low cost and risk solutions to most sonar requirements.

The processing equipment can be housed in air or water cooled equipment practice cabinets to meet customer's individual requirements.

FMS 12/FMS 15

The FMS 12 and 15 are high resolution narrowband surveillance sonars designed to derive signal inputs from either towed or hull-mounted line arrays. Each system consists of a single electronics cabinet which weighs 300 kg and a single operator's console weighing 350 kg.

FMS 12 provides frequency cover over two octaves, FMS 15 over five. Features common to both systems include: 360 degree bearing cover, 32 beam resolution and multiple beam surveillance display. They also provide high resolution vernier analysis classification, with sonar aural and audio communications interface. Both systems have built-in test equipment, an air-cooled operator console and electronics cabinet.

STATUS

An enhanced version of FMS 12 is included in Project TRITON – the Royal Navy's Type 2051 sonar update programme for the 'Oberon' class submarines. FMS 12 also provides the basis for the RN Type 2046 passive sonar system, 26 of which have been ordered for updating the current and next generation of nuclear powered attack submarines, and for inclusion in the new '2400' class of conventional submarines.

FMS 20

FMS 20 is an active sonar system which combines the range resolution of an FM sweep with the doppler discrimination of a CW pulse. It also has a passive capability for a built-in torpedo alarm facility. In its most comprehensive form, FMS 20 consists of four electronic cabinets, each with a weight of 765 kg and a two-display operator console weighing 340 kg. The associated transducer arrays may be hull mounted (4 ft diameter) with or without an associated variable depth array.

The equipment provides 32 beam resolution, electronic array motion compensation, 360 degree bearing cover, variable range scales and a variable centre frequency of operation. The display formats available to the operator as standard features include active FM and CW information, classification of targets, track and history totes, environmental and ray tracing, and monitoring of the equipment's status.

With Ferranti's concept of using the most modern shelf-based modular technology, the number of electronic cabinets required to drive the FMS 20 system can be reduced to two, albeit with some reduction in overall capability.

STATUS

FMS 20 is currently being considered by several NATO and other navies, for inclusion in both new ship and update programmes.

FMS 21

This system is derived from, and the export variant of, the Royal Navy's new Type 2050 sonar. It is an active sonar system which, like FMS 20, combines the range resolution of an FM sweep with the doppler discrimination of a CW pulse. Associated transducer arrays may be hull-mounted (6 ft diameter), with or without an associated variable depth array, and provide a 64 beam resolution.

The four cabinets which contain the electronics for the FMS 21 system and its two-display operator console (providing single operator control) weigh 765 kg and 340 kg respectively. Equipped with a variety of display formats as standard features, the operator has active FM and CW information, classification of targets, track and history totes at his finger-tips, as well as environmental, ray tracing and system status information. FMS 21 also has a passive capability with an automatic torpedo warning alarm.

STATUS

FMS 21 is currently being considered by the Pakistan and Hellenic Navies for inclusion in their new frigate programmes, whilst 17 systems of Type 2050 have been ordered by the RN.

CONTRACTOR

Ferranti Computer Systems Ltd, Cheadle Heath Division, Bird Hall Lane, Stockport, Cheshire SK3 0XQ, England.

FMS 21 console

5854.253

VIMOS VIBRATION AND NOISE MONITORING SYSTEM

The Ferranti VIMOS is a computer-controlled vibration and noise monitoring system designed to enable the continuous measurement of radiated noise levels at selected positions throughout the ship. In warships, for which the system was originally designed, this allows the increase in these levels to be detected before they adversely affect the ship's sonar capability, or increase the risk of detection. Applications include use by the ship's sonar operators, and monitoring of machinery wear and of detection.

A combination of accelerometers and hydrophones, placed at requisite internal and external points is employed, from which signals are transmitted through amplifiers and switch equipment to one or more central control units. These give data acquisition and signal switch control, Fast Fourier Transform (FFT) signal processing providing both narrow and broadband analysis and back-up facilities.

The control unit incorporates a touch sensitive plasma panel which allows the operator to interact with the software database and control the system. Aural listening and communications facilities are provided at the operator position and at various remote points.

STATUS

Several systems have been ordered by the UK MoD for the Royal Navy Type 23 frigates and for the new Fleet Auxiliary vessels. Three earlier versions of this system, Hull Vibration Monitoring Equipment (HVME) Mk1 were delivered to the Royal Navy in 1985.

CONTRACTOR

Ferranti Computer Systems Ltd, Cheadle Heath Division, Bird Hall Lane, Stockport, Cheshire SK3 0XQ, England.

6240.353

HIGH DEFINITION SCANNING SONAR

This high definition sonar has been developed by Smiths Industries for surveillance, navigation and object location purposes. It can be used both in ships and remotely operated vehicles and is based on a high speed scanning technique used in several sonar systems. Within the duration of a single transmission period it can gather information from a 30° sector and display a colour image of the whole area. In the horizontal plane the combination of high speed scanning with a beam angle of less than 2° gives accurate bearing discrimination, with a range resolution as high as 5 cm. In the vertical plane good discrimination between different targets is achieved by a beamwidth of 10° in conjunction with a 90° tilt facility.

The sonar image is presented in a high resolution TV format on a large screen colour CRT. A microcomputer within the display unit generates operational data such as range markers together with a graphic representation of the transducer pan and tile. The scanning electronics are contained within, or in close proximity to, the transducer housing to avoid long cable lengths.

CHARACTERISTICS

Transducer

Type: Combined 16 element array and curved projector

Pan and tilt: Electric drive, approx 15°/sec

Pan range: ±200°

Tilt range: 0 – 90°

Transmitter

Frequency: 600 kHz

Scanned sector: 30°

Beam angles: 1.6° (horizontal); 10° (vertical)

Pulse length: 100 μs

PRF: 4, 8, 16, 32, 64 Hz

Scanning rate: 15 kHz

Resolution: Range 5 cm; bearing better than 1°

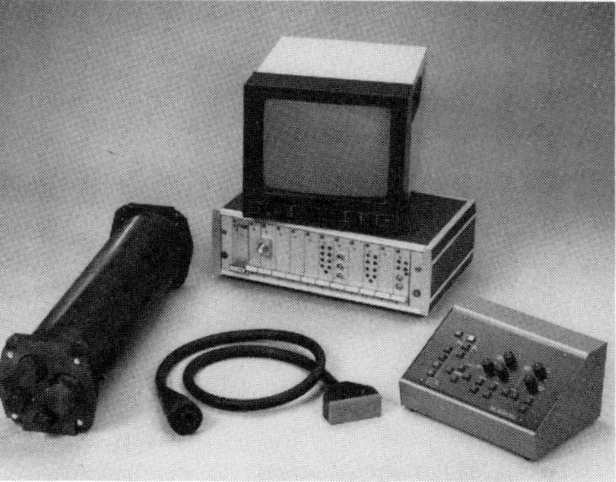

Units of Smiths high definition sonar

CONTRACTOR

Smiths Industries, Bishops Cleeve, Cheltenham, Gloucestershire GL52 4SF, England.

6250.453

ATAS ACTIVE TOWED ARRAY SONAR

ATAS is an active towed array sonar being developed by British Aerospace for ASW operations. It is a hybrid of the variable depth sonar and the passive towed array. The system provides a receiver array, remote from its towing platform, which together with a low frequency, acoustic transmitter may be towed at depths suited to oceanographic conditions. The port/starboard ambiguity normally associated with a towed array is resolved within the systems without the need for a ship manoeuvre.

The compact nature of ATAS permits its deployment from non-specialised ships of 250 tonnes and above, including ships of opportunity and ships taken up from trade, with a minimum of modification to the platform. Hull penetration is eliminated and only minimal space on the stern of the ship is required for winch and handling equipment. The system can also be supplied in containerised form.

The high power low frequency transmitter, operating in the 2-4 kHz band, can be deployed up to 900 m behind the towing vessel, with the in-line receiver array towed a further 300 m behind the transmitter. The system can be operated down to depths of 200 m.

The transmitter has a high-power acoustic sound source consisting of a vertical stave with low frequency flextensional transducers. Transmissions are omni-directional in the vertical plane with a 25° vertical beamwidth. The receiver array is a 20 m long flexible tube housing a line of hydrophones and associated electronics. It gives full azimuth cover and bearing resolution of 0.5° in the broadside beam. Both transmitter and receiver cables have a diameter of 23 mm, the former weighing 1.5 tonnes and the latter 130 kg in air. The weights of the transmitter and receiver respectively (in air) are 200 kg and 125 kg. The shipboard handling gear weighs a further 7 tonnes.

A single operations room console houses the signal and data processing electronics and the sonar display. The latter is in full colour with high resolution and signals are presented in range/bearing or PPI format together with target track information and equipment status.

Artist's impression of ATAS in operation. Inset photograph shows receiver array of hydrophones and electronics

STATUS

In development. In late 1987, BAe received a £600 000 contract to supply flextensional transducers to the US Naval Underwater Systems Center.

CONTRACTOR

British Aerospace, Dynamics Division, Filton, Bristol BS12 7QW, England.

6849.453

DOLPHIN 1000 SERIES SUBMARINE SONARS

The Dolphin sonar series has been designed as a cost-effective solution to meet the requirement to upgrade existing submarine sonar suites by interfacing to existing arrays where possible, or replacing subsystems within the suite. The series is also suitable for fitment to new construction submarines as subsystems or as an integrated sonar suite.

The Dolphin 1000 series comprises broadband bow processing and display, flank array broadband processing and display, narrow band, Demon and Intercept. The subsystems are modular units which can be fitted as stand-alone subsystems or integrated to form a sonar suite. A number of these modules are based on systems in service with the Royal Navy. The GETAS towed array sonar (**6850.453**) has been designed to interface into the Dolphin sonar suite system.

CONTRACTOR

Marconi Underwater Systems Ltd, Elettra Avenue, Waterlooville, Hampshire PO7 7XS, England.

6850.453

GETAS PASSIVE SURVEILLANCE SONARS

The GETAS series of passive sonars has been developed and proven against quiet submarine and surface targets in both a towed array and bottom-laid version. The systems provide a fully integrated comprehensive solution to passive sonar detection, tracking and classification. They are designed for single operator manning.

Marconi Underwater Systems provides the 'dry end' processing and display equipment and Geophysical Company of Norway (GECO A/S) provides the 'wet end' components.

The on-board or shore-based equipment comprises a signal conditioning unit, broad and narrow band data processors, touch sensitive control panels and waterfall displays housed in a single cabinet. A number of features have been incorporated, including BITE, automatic track followers, remote hard copy and tape recording outputs. In addition, real-time clock and date, array heading, depth and tension and seawater temperature and salinity are available on the display.

There are two variants of the shipborne 'wet end' each consisting of a winch and cable with acoustic, vibrator isolator and instrumentation array sections. The all-up weight of the standard system is 8550 kg and the small ship version 2155 kg.

A submarine version is also available. The bottom-laid version is similar to the standard version but modified for abrasion resistance and negative buoyancy. All systems are easily deployed and recovered.

CONTRACTORS

Marconi Underwater Systems Ltd, Croxley Mill, Blackmoor Lane, Watford, Hertfordshire WD1 8YR, England.

GECO A/S, Kjorbokollen, N-1301 Sandvika, Norway.

5861.363

AIRBORNE SONOBUOY TELEMETRY SYSTEMS TYPES 864, 865, 866

Marconi Underwater Systems have designed and manufactured a range of sonobuoy telemetry systems for use in a number of ASW fixed-wing aircraft and helicopters. The range offers multi-channel operation and is fully compatible with the majority of sonobuoys and processing systems in service. An antenna pre-amplifier unit ensures maximum performance with low noise even for large separations between antenna and receiver units.

The Type 864 is capable of handling simultaneously four or eight channels of FM telemetry from deployed sonobuoys transmitting on the 99 VHF channels, including Jezebel, Ranger and Bathythermograph types. The Type 865 extends this range to the DICASS active type buoys. The Type 866 can handle up to eight sonobuoys of all current types, and is a high performance version of the Type 864. Either manual or remote channel selection is available on all versions, via a computer bus, and receiver outputs are fully compatible with a wide range of existing sonics processors.

CHARACTERISTICS

Frequency band: 134.5 – 174.625 MHz

No of channels: 8 (4 per receiver)

Coverage: 99 RF channels plus test channel

Channel spacing: 375 kHz

Modulation: FM (FM/FSK for Type 866)

Dimensions

Pre-amplifier: 177 × 156 × 285 mm

Control unit: 146 × 190 × 142 mm

Receiver: 123 × 193 × 385 mm

Weights: 2.5 kg (pre-amplifier); 1.5 kg (control unit); 10 kg (receiver)

CONTRACTOR

Marconi Underwater Systems Ltd, Elettra Avenue, Waterlooville, Hampshire PO7 7XS, England.

4453.363

ARR 901 SONAR RECEIVER

The ARR 901 is a multi-channel sonar receiver of modular construction for use in fixed-wing maritime patrol aircraft. It simultaneously receives eight VHF channels which are selected by serialised digital commands from a processor or control unit. Advanced filter technology, extensive use of hybrid circuits and wideband techniques ensure high availability. An antenna preamplifier unit gives a low noise figure and the system's high dynamic range and wide bandwidth enables high speed telemetry data to be received.

The receiver provides its full RF performance when installed at up to 50 m from the preamplifier. All interfaces are to ARINC specification 404A. Outputs are compatible with existing aircraft data processing systems and with more advanced systems under development. A built-in test signal generator provides extensive check facilities for the receiver and associated processing equipment and is designed to facilitate servicing by automatic test equipment.

CHARACTERISTICS

Frequency band: 134.5 – 174.625 MHz

Frequency stability: ±13 ppm (at –26°C to +45°C-ambient temperature)

No of channels: 8

Channel coverage: 99 RF channels plus 2 test channels

Channel spacing: 375 kHz

Input impedance: 50Ω

Modulation: FM/FSK

CONTRACTOR

Marconi Underwater Systems Ltd, Elettra Way, Waterlooville, Hampshire PO7 7XS, England.

6851.253

GS 7110 MULTI-CHANNEL VLF/LF RECEIVER

The GS 7110 multi-channel FSK receiver has been designed and developed as a replacement for the earlier generation of receivers fitted to NATO submarines, surface vessels and shore stations. The modular microprocessor-controlled receiver offers full coverage of the VLF and LF bands and interfaces with a wide range of antenna systems and message processing equipment. Its standard features include 100 channel non-volatile memory capacity, local/remote frequency control, modulation mode and receiver status. The receiver has built-in BITE facilities.

A software demodulator gives two or four channel MSK at 50475 band, CW and MCW, while other software options are available to suit other modulation and signal processing requirements. The receiver complies with STANAG 5030.

CHARACTERISTICS

Frequency band: 10 to 200 kHz

Input impedance: 100Ω balance; 10 kHzΩ unbalanced; 50 or 75Ω unbalanced demodulation

Channels: AM, FSK, MSK 2 channel; MSK 4 channel

CONTRACTOR

Marconi Underwater Systems Ltd, Elettra Avenue, Waterlooville, Hampshire PO7 7XS, England.

6856.363

TYPE 605 SONOBUOY RECEIVER

The Type 605 sonobuoy receiver is compatible with all sonobuoys known to be in production or in development. It is a much improved version of the AN/ARR-75 system (**3455.363**) and uses many of its electrical and mechanical design features to simply retrofit applications. The Type 605 receives, demodulates and amplifies FM sonobuoy transmissions in the extended VHF band and provides outputs for analysis and display. It features 99-channel operation with four simultaneous channel operation.

STATUS

The Type 605 is understood to have been selected for the Royal Navy's EH101 helicopter.

CONTRACTOR

Dowty Maritime Systems Ltd, Communications Division, 419 Bridport Road, Greenford, Middlesex UB6 8UA, England.

3502.453

JEZEBEL SONOBUOY

Jezebel is the code name for the standard A-size omni-directional passive sonobuoy system used by the Royal Air Force and the Royal Navy for the detection of submerged submarines. Most details remain classified but it has been stated that these sonobuoys have improved frequency response, a choice of three life settings and a deep hydrophone system which allows various operating depths to be selected.

STATUS

Many thousands of DEL Jezebel sonobuoys have been produced for the British forces over a period of years, and production is continuing, although contracts have been awarded for the newer miniature passive sonobuoy which is about one-third the length of the standard item (**3503.453**).

CONTRACTOR

Dowty Electronics Ltd, Communications Division, 419 Bridport Road, Greenford Industrial Estate, Greenford, Middlesex UB6 8UA, England.

3769.453

SSQ-947B DEL RANGER SONOBUOY

The DEL Ranger is an omni-directional active sonobuoy, based on an American design, the AN/SSQ-47B, and intended for use in the detection, classification, tracking, and location of submarines. It is slightly lighter than the US model, but in most other respects the performance and operational characteristics are believed to be very similar. For comparison, see **1988.453**, later in this section.

CHARACTERISTICS

Type: Active
Length: 914 mm
Diameter: 124 mm
Weight: 13.2 kg
Transducer depth: 60 or 800 ft (18 or 244 m)
Sonar channels: 6 in 13-19 kHz band
Sonar power output: 200 W approx
Duration: 0.5 h
RF power output: 0.25 W
RF channels: 12

STATUS

In production.

CONTRACTOR

Dowty Electronics Ltd, Communications Division, 419 Bridport Road, Greenford Industrial Estate, Greenford, Middlesex UB6 8UA, England.

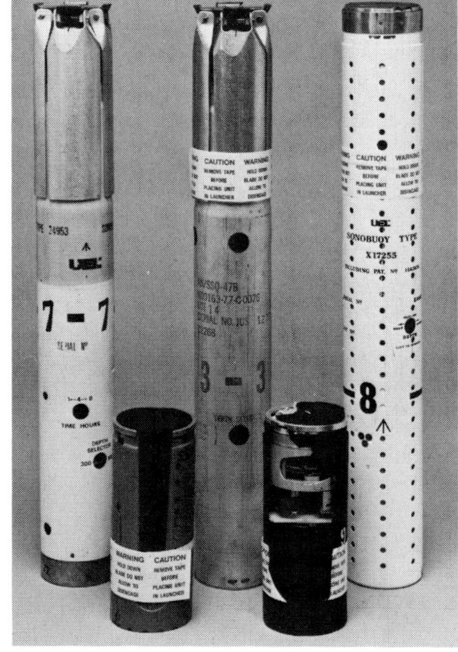

Current DEL sonobuoy production range showing (rear from left to right) Jezebel, Ranger (SSQ-947B), and CAMBS, with new miniature bathythermal sonobuoy (left) and miniature Jezebel sonobuoy (right) in front

3503.453

SSQ-904A/905A/906/907 LOFAR SONOBUOYS

This F-size sonobuoy is designed to be launched from aircraft or helicopters at airspeeds between 60 and 325 knots and at altitudes between about 50 and 10 000 m. Once deployed, it detects and amplifies underwater sounds to modulate the self-contained FM transmitter. The FM signals are received by the aircraft processor, which is currently the GEC Avionics AQS 901, AQS 902 and AQS 903.

A significant feature is the fact that this buoy is only about one-third the length of a standard A-size sonobuoy, and half the weight, yet is claimed to have precisely the same performance characteristics. The 905A and 907 are calibrated versions of the 904A and 906 respectively.

CHARACTERISTICS

Type: Miniature passive omni-directional
Frequency coverage: 10 Hz – 6 kHz
Hydrophone depth: Selectable from either 60 or 300 ft (18 or 91 m); or 60 or 450 ft (18 or 137 m)
Operating life: 1, 4 or 6 h
RF channels: 99 or 31
RF output: 1 W minimum
Dimensions: 30.5 × 12.5 cm (diameter)
Weight: 4.5 kg

STATUS

In production for the Royal Navy. A contract for 10 372 SSQ-904A buoys was awarded to Plessey in November 1986, while a contract for 31 117 SSQ-904A and 1995 SSQ-905A buoys was awarded to Dowty Electronics in 1986. In November 1987 a further contract was awarded to Plessey to supply 48 893 SSQ-906 and 5941 SSQ-907 buoys for delivery in 1988-91. This was part of a total order for SSQ-906, 907, 954B and 937A awarded to Dowty and Plessey.

CONTRACTORS

Plessey Naval Systems, Templecombe, Somerset BA8 0DH, England.

Dowty Electronics Ltd, Communications Division, 419 Bridport Road, Greenford Industrial Estate, Greenford, Middlesex UB6 8UA, England.

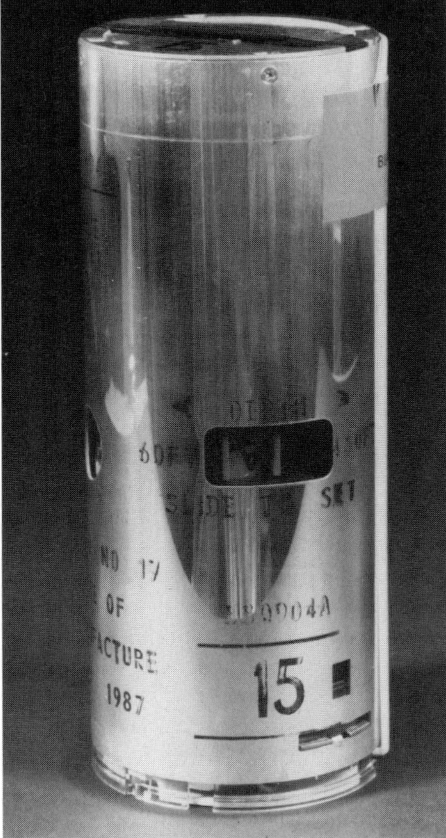

SSQ 904A/905A LOFAR sonobuoy

6243.453

SSQ-954B DIFAR SONOBUOY

The SSQ-954B is a G-size miniature directional passive sonobuoy. It is of unique design and offers the DIFAR capability of target bearing information added to the facility of the SSQ-904 (**3503.453**) in a miniature G-size format (123.825 × 419.1 mm). A further optional facility enables pre-launch selection of any one of 99 RF channels, together with depth and life selection.

STATUS

In service.

CONTRACTORS

Plessey Naval Systems, Templecombe, Somerset BA8 0DH, England.

Dowty Electronics Ltd, Communications Division, 419 Bridport Road, Greenford Industrial Estate, Greenford, Middlesex UB6 8UA, England.

3237.453

SSQ-963 COMMAND ACTIVE MULTI-BEAM SONOBUOY (X17255)

The SSQ-963 command active multi-beam sonobuoy (CAMBS) is a new A-size sonobuoy developed by Dowty Electronics under contract to the British Ministry of Defence.

CAMBS is an advanced active directional sonobuoy, which features a high data rate, and directional information is obtained by the use of a transmitting acoustic projector combined with a directional receiving hydrophone array. Hydrophone depth is adjustable by radio command and there is a 31-channel radio telemetry link operating on standard NATO frequencies for communication with the parent aircraft. The antenna is a monopole with ground plane, erected by and integral with the buoy flotation bag. The design incorporates a flux-gate compass with very rapid stabilisation.

CHARACTERISTICS

Length: 914 mm
Diameter: 124 mm
Weight: 15.9 kg
Transducer depth: Adjustable
RF channels: 31

STATUS

The SSQ-963 sonobuoy is part of the equipment developed for the Mk II version of the Nimrod ASW aircraft. In production. Dowty received a £10 million order from the UK MoD during 1987 for SSQ-963 CAMBS.

CONTRACTOR

Dowty Electronics Ltd, Communications Division, 419 Bridport Road, Greenford Industrial Estate, Greenford, Middlesex UB6 8UA, England.

3770.453

SSQ-937 DEL MINIATURE BATHYTHERMAL BUOY

This air-launched miniature bathythermal buoy is the second A-size buoy to have been reduced to the new F size (A/3). It provides all the facilities of the SSQ-36 but is only one third of its size and half the weight.

CHARACTERISTICS

Length: 305 mm
Diameter: 124 mm
Weight: 5 kg
Operating depth: To 427 m
Descent rate: 1.5 m/s (to 305 m)
Temperature range: –2 to +35°C
Gradient accuracy: ±20%
Modulation frequencies: 1360 – 2700 Hz
Probe time constant: >1 s
RF power output: 0.25 W

STATUS

In full series production against UK MoD contracts.

CONTRACTORS

Dowty Electronics Ltd, Communications Division, 419 Bridport Road, Greenford Industrial Estate, Greenford, Middlesex UB6 8UA, England.

Graseby Dynamics Ltd, Park Avenue, Bushey, Watford, Hertfordshire WD2 2BW, England.

6863.453

SSQ-937A BATHYTHERMAL BUOY

The SSQ-937A air-launched miniature bathythermal buoy is being developed for the UK MoD. The buoy is packaged in an F-size (A/3) configuration and is a further development of the SSQ-937 for deep water operation. It employs a drag fin retarder design to replace the conventional parachute system of current buoys, a concept which reduces both weight and cost.

CHARACTERISTICS

Operating depth: 457 m

Descent rate: 1.5 m/s

Modulation frequency: 1368 – 2700 Hz

Dimensions: 305 mm × 124 mm diameter

Weight: 2.6 kg

STATUS

In production and ordered in large quantity by the UK MoD.

CONTRACTOR

Plessey Naval Systems Ltd, Templecombe, Somerset BA8 0DH, England.

6861.453

SSQ-981 BARRA SONOBUOY

The SSQ-981 is the second generation BARRA sonobuoy to be developed by Plessey for the UK MoD. Derived from the Australian SSQ-801, the SSQ-981 has been designed to reduce manufacturing costs while maintaining the complete performance capability of the BARRA concept.

SSQ-981 is an advanced passive broadband sonobuoy sensor; 25 hydrophones arranged over a large acoustic planar array achieve an enhanced detection gain over coventional buoys. Spectral and broadband energy is digitally telemetred to the ASW aircraft. A BARRA compatible acoustic processor forms 60 beams providing 360° cover to give excellent signal-to-noise discrimination resulting in good azimuth resolution and extremely accurate contact bearings.

SSQ-981 is configured as a conventional A-size sonobuoy which may be launched from helicopter or fixed-wing aircraft using existing stowage and launching systems. It operates with preset VHF channels and allows operator selection of life and depth prior to launch.

STATUS

In production.

CONTRACTOR

Plessey Naval Systems Ltd, Templecombe, Somerset BA8 0DH, England.

SSQ-981 BARRA sonobuoy

6862.453

SSQ-991 SONAR TRANSPONDER

The SSQ-991 air-launched sonar transponder is a first generation helicopter-launched, expendable sonar equipment. It has been developed as a combined radar/dipping sonar alignment aid for Royal Navy ASW helicopters. Based on established sonobuoy techniques, it is housed in an A-size configuration. It may be launched from the helicopter using existing sonobuoy stowage and launch systems.

Although programmed to operate with the Type 195M dipping sonar, the SSQ-991 can be adapted to operate with other sonar equipments in the medium ASW acoustic frequency band. The transponder is capable of many pulse transmission modes, including CW, shaped CW, doppler simulation frequency shifted pulses and FM pulse forms. The octohedral geometry of the radar reflector is optimised around I-band radar frequencies to provide a radar cross-section varying between 3 and 12 m. With an operating range of around 8000 m, the SSQ-991 provides an effective test capability for the ASW system.

CONTRACTOR

Plessey Naval Systems Ltd, Templecombe, Somerset BA8 0DH, England.

3789.453

PLESSEY-SIPPICAN XBT AND XSV

The Plessey-Sippican expendable bathythermograph (XBT) is a surface ship and submarine bathythermograph equipment which uses an expendable probe to obtain accurate and continuous ocean temperature against ocean depth measurements. These can be made in any sea state at ship speeds of up to 56 km/h (30 knots).

XBT/XSV Mk 8 recorder

The Plessey-Sippican expendable sound velocimeter (XSV) is a new development which reduces the difficulties of direct collection of sound velocity data at sea. This system measures sound velocity directly with the ease and accuracy associated with the well known XBT system.

STATUS

XBT is in service with the Royal Navy and has been sold to over 20 countries. The XSV is now being adopted for service in a number of navies. See also information on Sippican sonar products under entry **6246.463**. (Note: Plesssey plc acquired Sippican Corporation in 1987).

CONTRACTOR

Plessey Naval Systems, Templecombe, Somerset BA8 0DH, England.

5638.263

PMS 75 SPEEDSCAN/SONAR 2048

The PMS 75 Speedscan is a self-contained equipment for ships already fitted with a compatible sonar system. It allows the sonar system to be operated in a sidescan mode and generates a hard copy printout of the sea bed. It can be fitted to all types of mine-hunting sonar.

When the sonar is operating with Speedscan, its transducer is trained 90 degrees to port or starboard of the ship's track. As the ship proceeds along a predetermined track, an area of sea bed, parallel to the track, is interrogated by the sonar. The received sonar data and reference timing signals are fed to the Speedscan processor which forms a sidescan beam and presents the processed data as hard copy on continuous recording paper.

The Speedscan equipment consists of two main assemblies, each contained in a portable case. In use the two cases are stacked and locked together with the end covers removed. The upper case houses the processor and the operator's control panel, the lower case the recorder unit.

Speedscan presents information on light-sensitive paper exposed in the recorder. On the record, the sea bed is represented as a series of parallel narrow strips of selected length, perpendicular to the ship's track. The sonar data from the whole azimuth sector scanned by the sonar is extracted by Speedscan to represent a narrow strip of sea bed along the sonar beam axis. This data is extracted by the beamformer to generate a parallel beam of sufficient width to overlap the strips covered by the two previous transmissions, so providing the 3-ping sample of each point of the sea bed along the range axis. The beamforming system is programmed by ship's speed and sonar range to provide a controlled interrogation of the sea bed. This information is used to intensify and modulate a light source in the recorder which draws a line across the record for each sonar transmission, echoes appearing as dark spots on the recording paper.

STATUS

The PMS 75 forms part of the Royal Navy sonar Type 2048 in mine countermeasures vessels.

CONTRACTOR

Plessey Naval Systems, Templecombe, Somerset BA8 0DH, England.

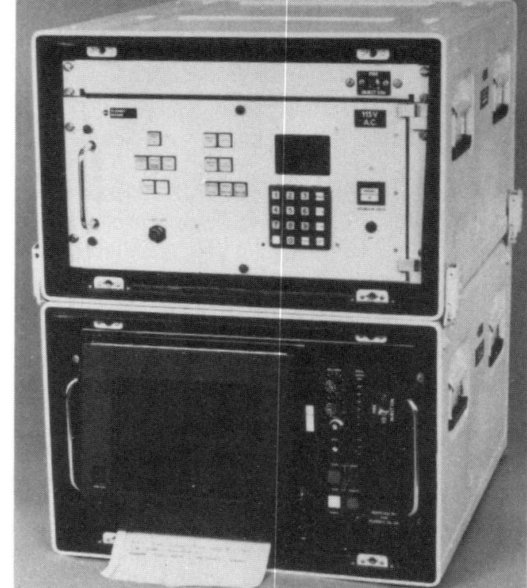

PMS 75 Speedscan

6242.263

SP 2104 FAST TIME ACOUSTIC PROCESSOR

The SP 2104 has been designed to reprocess at high speed the acoustic tape recordings made during airborne anti-submarine operations. The equipment is able to process recorded signals at up to eight times normal time.

The system is used to find targets missed during operational analysis, to confirm the presence and extract more information from specific targets for classification purposes, to provide operator assessment and highlight training requirements and to extract data for library purposes.

The SP 2104 system is modular to allow easier configuration of equipments to meet individual user requirements. Standard interfaces have been used throughout.

CONTRACTOR

GEC Avionics, Sonar Systems Division, 26-28 Hydeway, Welwyn Garden City, Hertfordshire, England.

4624.453

ACOUSTIC INTRUDER DETECTION SYSTEM

The Plessey Acoustic Intruder Detection and Harbour Surveillance system is intended for a variety of safety and security purposes. It provides coverage of jetty and harbour areas against swimmers, surface and subsurface craft by an automatic detection and alarm system. It is a totally intergrated system employing acoustic elements, radar and communications linked to a central command console.

Swimmer detection and craft location form the primary acoustic elements, the frequencies and coverage of these being matched to the environment and mission requirements. The secondary acoustic elements comprise an acoustic fence and VHF system. The fence provides within-harbour security and preliminary warning of possible violation.

The system uses a mixture of active and passive sensors optimised to give protection against underwater intruders from over 30 miles away right up to the inner-harbour zone.

In the early warning zone fixed passive arrays are used. Closer to the harbour entrance the engagement sonar aims at detecting the mini-submarine or free-swimmer. A series of sonar heads is located in a 'daisy-chain' with an overlap between each. Designed for periodic maintenance, the sonar heads can be removed by divers for on-shore maintenance. These sonars detect the target by locating movement.

In the harbour zone the greatest threat is from the swimmer. The pulsed sonar used here has been specifically designed to offer a flexible high definition detector. The mirror assembly or 'head' is a transducer array placed at the foot of a curved metal surface which acts as a beamformer.

Information from all the sensors is correlated in a central processing equipment. Appropriate displays are provided to enable the operator to gain an overall picture of any developing threat. The system is fully automatic.

STATUS

Prototype systems are available.

CONTRACTOR

Plessey Naval Systems, Templecombe, Somerset BA8 0DH, England.

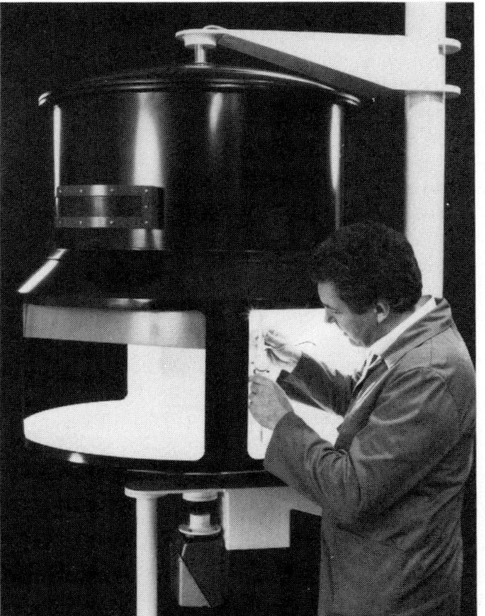

Model of mirror array for Plessey swimmer detection HF sonar

6559.353

TG-1 HF SONAR SYSTEM

Ferranti has designed, developed and produced a new high frequency sonar system which offers electronic scanning at a cost which, the company states, makes it competitive with mechanically scanned equivalents. It has been given a provisional nomenclature of TG-1. The equipment which is understood to have a range of some 700 metres is built to withstand continuous operation down to depths of 500 metres. Targets are identified in both azimuth and depth.

The modularity of the system makes it suitable for a diversity of applications. These include harbour surveillance, offshore platform defence, mine identification and classification, ROV navigation, and oil/gas pipeline profiling.

Typically a single unit comprises two 64-element transducer arrays each connected to its own underwater pod, and using a digital beam-forming technique. An umbilical cable provides both power and data link to the pods from the surface control unit. A standard VDU presents high resolution data which includes the bearing, range and depth of sonar contacts. An auto-alarm system causes an audible alarm to sound when a possible sonar return is received.

In more general surveillance applications up to 255 pods can be linked to form a system. The control system then acts as a master and accesses each of the slave pods in turn.

STATUS

The equipment has already been tested successfully. The basic system was developed by Universal Sonar Limited, Bridlington, which was acquired by Ferranti plc in December 1986.

CONTRACTOR

Ferranti Computer Systems Ltd, Bird Hall Lane, Cheadle Heath, Stockport, Cheshire SK3 0XQ, England.

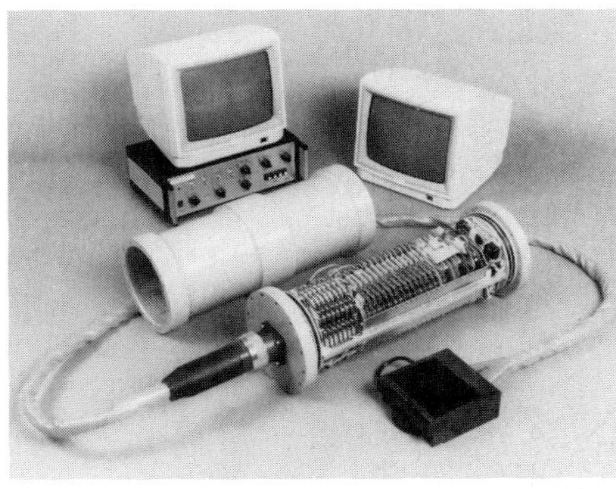

Units of Ferranti TG-1 sonar system

5867.453

SWS 35 MINIATURE TOWED ARRAY

The SWS 35 is a mini towed array developed by Ameeco (now part of the Plessey organisation) and is designed to be operated by as few as two men in a small boat. The 'wet-end' of the system is composed of a streamer approximately one metre long, equipped with hydrophones designed to detect acoustics in the frequency range 25 to 3200 Hz, and is towed by a winch-operated 150 metre long cable. Signals are fed into an on-board processor which produces broadband beams. The outputs are monitored by an operator, both aurally and visually using an FFT dual-channel spectrum analyser with 'waterfall' display. Two high frequency hydrophones, using a separate display, are also incorporated into the array to detect signals from targets using high definition sonar navigation systems. A variant of the system, known as SWS 40, includes a vertical array in the streamer. It is understood that a third variant, SWS 45, is also available and uses a static seabed horizontal array for protection of installations such as oil rigs.

STATUS

In production for an unnamed customer, understood to be the Royal Swedish Navy.

CONTRACTOR

Plessey Naval Systems, 2 North Way, Andover, Hampshire SP10 5AZ, England.

4627.463

TOWED ARRAYS

Ameeco Hydrospace Ltd (now part of the Plessey organisation) has been the main supplier of towed arrays to the defence industry for some years and has supplied all the arrays for the Royal Navy surface ships and submarines. No details are available on technical construction but the company is now concentrating on arrays for the next RN requirement in this field. Plessey is also carrying out research into increased performance and improvements that can be brought about by multiplexing, fibre optics, etc.

Plessey is engaged in a number of other products in the underwater field such as seismic arrays for survey, torpedo firing systems, sound velocity meters and the design and supply of underwater cable assemblies. A subsidiary, Geophysical Systems Ltd, has produced an underwater orientation sensor for use in remote equipment to measure and transmit the heading, pitch and roll angles of the body in which it is positioned.

STATUS

As part of a programme investigating the use of optical fibres as a transduction medium, together with fibre hydrophones, an experimental array is under evaluation by the Royal Navy.

CONTRACTOR

Plessey Naval Systems, 2 North Way, Andover, Hampshire SP10 5AZ, England.

6871.453

PLESSEY ARMS SONAR

A new Advanced Remote Minehunting System (ARMS) has been developed by Plessey for use in warships and vessels of opportunity to locate and evade/destroy mines. The system uses a Remotely Operated Vehicle (ROV) developed by the Straza Division of Ametek, USA for offshore work to carry a sonar and a standard design of mine destructor charge. The ROV swims at a distance of up to 1000 m ahead of the parent ship at depths down to 300 m and at speeds up to 6 knots.

The sonar is a development of the Plessey intruder detection sonar (**4624.453**). In addition the vehicle is fitted with two cameras, one looking forward and one in the belly to give the downward-looking capability for mine classification. Sonar and TV data is passed back to the ship by means of a fibre-optic cable.

The complete ARMS system is containerised in three units for rapid deployment and can be transported to any area in a DC-9 size aircraft.

CONTRACTOR

Plessey Naval Systems, Templecombe, Somerset BA8 0DH, England.

4679.453

THORN EMI UNDERWATER ACOUSTIC RANGES AND SYSTEMS

Systems to monitor accurately the track and performance of underwater weapons are essential to develop fully the effectiveness of torpedoes and other underwater missiles and vehicles.

THORN EMI Electronics has been involved over a long period in the design, supply and operation of such systems for the Royal Navy and has available three types of equipment to meet specific evaluation requirements. All three systems achieve the objective of accurately tracking a torpedo, missile or other underwater weapon system while under the surface of the sea during its final approach to the target, or during the discharge/deployment phase.

The systems available are:
Portable Acoustic Tracking System (PATS)
Fuze Activating Static Target (FAST)

Portable Acoustic Tracking System (PATS)

PATS is a high performance system capable of tracking up to four underwater objects simultaneously over a large area. The objects may include the weapon, the target, the launch vessel or any other suitably instrumented vessel in the firing area. The system is fully transportable and may be deployed as a surface range suspended from anchored buoys, or in shallow water laid on the sea bed.

In the buoy arrangement the mooring buoys are anchored in a triangular form approximately 200 m apart. This then becomes the centre of the tracking range and practice firings can take place in an area 4.5 km around the range centre. The three hydrophones are suspended at the selected depth from small flotation buoys which are in turn attached to each of the mooring buoys when the range is in use. The hydrophones are connected by floating cables to a range support ship normally moored to one of the main buoys. Data is processed on-board the range vessel and will normally be transmitted via a radio link to a completely self-contained base station on land where the tracks are computed and displayed in real time. Alternatively the base station may be taken on-board the range vessel allowing an entirely sea-borne operation to be carried out.

In shallow water an alternative form of PATS may be used whereby the hydrophones are placed on the sea bed, again in a triangular form. Cables from each hydrophone connect to the range vessel which is moored close to the range centre, and data is again relayed to the base station via a radio link.

Submarine-Mounted Underwater Tracking System (SMUTS)

SMUTS is an in-service Royal Navy system developed for the evaluation of underwater weapon systems. It is a dual mode acoustic tracking system capable of short range (300 m) high resolution, high accuracy, three-dimensional tracking coupled with a long range (4.8 km) distance measuring mode. The system may also be used to measure the instant of fuze activation of a suitably instrumented weapon.

SMUTS may be used during the development and proving stages of a new weapon system and for routine in-service weapon effectiveness firings. The 'inboard' system is fully portable and may be rapidly installed on any vessel or installation fitted with an appropriate hydrophone array.

The prime uses are:

(1) to measure and display the approach and three-dimensional trajectory of the terminal phases of an underwater attack on static or moving target submarines, unmanned mobile targets, static targets or fixed installations
(2) to measure and display the discharge trajectory of any underwater weapon (sub-launched missile, stand-off weapon, torpedo, mine, decoy, etc) from a submarine or fixed underwater tubes or dischargers. This may include measurement of the deployment/fall back of torpedo cable dispensers or missile shrouds after water exit separation, etc
(3) to measure and display the deployment and trajectory performance of bottom launched weapon systems.

SMUTS may be supplied as a 'stand-alone' system incorporating on-board computing and display together with digital data storage and recording for subsequent off-line processing. Alternatively it may be supplied as part of a total attack/discharge system interfacing with the ship's weapon and navigation systems either via discrete interfaces or via a highway.

It has been used extensively by the Royal Navy both for torpedo attack and discharge measurement and for sub-launched (Sub-Harpoon) discharge both in the R & D and fleet exercise role. The system has been used in an open ocean environment and on instrumented weapon ranges where compatibility with the weapon range system is required. It is in operation with, or ordered for, the Royal Navy and other NATO navies.

Fuze Activating Static Target (FAST)

An alternative to using a target submarine for underwater weapon evaluation is to use the FAST frame. The frame is covered with wire mesh panels to stimulate the magnetic field of a submarine and is also fitted with an acoustic transmitter to provide a signal for the torpedo's homing system. The system's four hydrophones are mounted on the frame which is lowered to a maximum of 100 m below the surface. All the processing equipment for tracking the torpedo is fitted inside the ship supporting the FAST frame. The frame is designed to fold for easy storage and deployment.

Distance Measuring Equipment

Distance Measuring Equipment forms the basis of many of the available tracking systems and enables the slant range between a mobile or static acoustic transmitter and a receiving hydrophone to be measured with great precision. The equipment is automatic in operation requiring minimal operator attention. Modular design enables tracking systems to be configured to meet the user's immediate needs while allowing expansion later if a more complex tracking system is required.

Acoustic Transmitters (Pingers)

The weapon to be used must be fitted with an appropriate acoustic pinger. This may take the form of a small hull section for a torpedo or missile, or individual sub-assemblies for fitting within the weapon system envelope, discharge capsule or shrouds. The pinger may transmit either HF or LF signals, as appropriate, or in certain applications both transmissions will be required. Each transmission will normally comprise two pulses, one to determine range and one to provide a direct measurement of the weapon operating depth. When operating with a suitable instrumented weapon, the instant of fuze activation may be transmitted by the pinger and registered by the tracking system. The pingers are normally battery powered giving a completely self-contained package incorporating 'power-up' functions.

Noise ranging

The measurement and analysis of underwater noise radiated by ships and submarines is a specialised field involving the latest technologies in a most hostile environment. Ranges for the analysis of underwater radiated noise designed and developed by THORN EMI have included deep water facilities where submarines may be operated at full speeds and static facilities where detailed investigations may be carried out in the absence of noise due to the vessel's movement through the water.

Magnetic ranging

The latest THORN EMI range, the MS90, represents a major advance in degaussing measurement and offers considerable cost savings compared with conventional ranges. It uses extensive mathematical modelling and software techniques and can be packaged into a standard ISO container for transport.

Degaussing ranges are supplied for open sea measurement of MCMVs and steel ships for the eddy current effects of rolling, for land-based measurement, for model testing and for direct treatment ranging. Over 30 degaussing ranges have been supplied to a number of countries including Australia, Norway, UK and NATO.

CONTRACTOR

THORN EMI Electronics Ltd, Defence Systems Division, Albert Drive, Sheerwater, Woking, Surrey GU21 5RU, England.

6855.453

TOWED SONAR SOURCE TYPE 2058

The Sonar Type 2058 broadband towed sound source provides a realistic acoustic target source for operators and command teams, simplified performance assessment of noise ranges and all types of passive sonar, improves training in ASW techniques for ships and aircraft, acoustic and mine countermeasures capabilities and a low frequency sound source for scientific and experimental purposes. The system can simultaneously transmit a mixture of narrow band tones and broadband noise signals over a frequency range covering those met in operation.

The Type 2058 is designed to be deployed from RFAs and warships and towed using the Sonar 182 winch. With suitable winch and handling facilities it can also be deployed from other vessels.

The system consists of a Digital Signal Generator (DSG), power amplifier, towing cable, towed body and test unit. The DSG provides the various low level noise and tone signals required. A VDU and small keyboard mounted on the DSG allows setting and display of the required signal types, levels and frequencies and also displays the towed depth and attitude. The towed body is based on the Sonar 182 Mk 3 body, being free flooding, mainly constructed of glass fibre and housing a hydrosounder and pressure vessel.

STATUS

In service with the Royal Navy.

CONTRACTOR

THORN EMI Electronics Ltd, Naval System Division, Albert Drive, Sheerwater, Woking, Surrey GU21 5RU, England.

Type 2058 towed sonar source

6257.493

SEPADS SONAR PREDICTION SYSTEM

The Type 2068 Sonar Environmental-Prediction and Display System (SEPADS) has been designed by Gresham-CAP to fulfil the requirements for accurate information about the underwater environment. This is a particular requirement for towed array sonars which need accurate information so that they can be used to the best effect.

SEPADS contains a vast quantity of oceanographic, sonar and meteorological information which is stored on a computer disc. This information is added to by more data obtained by expendable bathythermographs and, in conjunction with sound-propagation-loss models, is used to forecast parameters such as the sonar ranges and the optimum depth at which to deploy the towed array. Processing is carried out in a PDP 11/44 computer.

STATUS

In service with the Royal Navy.

CONTRACTOR

Gresham-CAP Limited, Twickenham Road, Feltham, Middlesex TW13 6HA, England.

6858.453

AS360 SONAR SYSTEMS

The AS360 family of sonar systems has been developed primarily for use on unmanned Remotely Operated Vehicles (ROVs). The AS360 is a mechanically scanned sonar of high resolution capability with a range of up to 100 m. A number of versions are available for various applications in the ROV sphere.

CHARACTERISTICS

Frequency: 500 kHz (AS360M and MS1A); 700 kHz (AS360 MS5)

Beamwidth: 1.4 × 27° (AS360M and MS1A); 1.5 × 50° (AS360 MS5)

Depth rating: 300 m (AS360M); 1500 m (AS360 MS1A); 750 m (AS360 MS5)

STATUS

The Type AS360 MS1/1 has been selected for installation on the ECA PAP Mk 5 ROV to be used by the Royal Navy's single-role minehunters.

CONTRACTOR

UDI Group Ltd (a Trafalgar House company), Denmore Road, Bridge of Don, Aberdeen AB2 8JW, Scotland.

6859.453

AS370 SEABED SURVEILLANCE SONAR

The AS370 is a seabed surveillance sonar designed with long term immersion capabilities for coastal waters and harbour surveillance installations. It is designed to detect underwater swimmers out to 500 m in sea-state 3 and small submersibles out to its maximum range of 1 km. It is intended for permanent installation on the seabed and can be deployed with single or with multiple underwater sensors combined to provide a total coverage of an area.

Alarm facilities enable the detection and tracking of sonar targets which satisfy selectable predetermined parameters such as amplitude, target size and doppler content. A surface control unit offers a variety of display and processing features.

CHARACTERISTICS

Range: Up to 1000 m

Beam pattern: 4 × 6°

Scan: Sector or continuous

Depth rating: 60 m

STATUS

The AS370 is currently undergoing long term evaluation trials with the Swedish Defence Material Administration.

CONTRACTOR

UDI Group Ltd (a Trafalgar House company), Denmore Road, Bridge of Don, Aberdeen AB2 8JW, Scotland.

6860.353

AS380 DUNKING SONAR

The AS380 is a combined active and passive dunking sonar providing comprehensive detection and classification capabilities in a single underwater package. It can be deployed from both ships and helicopters. Multiple operating modes allow the system to be used for either ASW or MCM scenarios.

In the detection mode the AS380 provides a rapid search facility to a range of 3 km whereby a 60° sector coverage, using electronic beamforming, can be steered in any direction. In the classification mode a detected target can be examined further for a representation on its size and shape. A single operator console consists of two high resolution monitor displays, a keyboard and a trackerball with push-button controls.

CHARACTERISTICS

Nominal frequency: 50 kHz

Beamwidth: 60° (horizontal); 10° (vertical)

Displayed resolution: 5°

Bearing accuracy: 1°

Modulation: FM (chirp); CW (doppler)

Ranges: 1/2/3 km

Maximum depth: 60 m

CONTRACTOR

UDI Group Ltd (a Trafalgar House company), Denmore Road, Bridge of Don, Aberdeen AB2 8JW, Scotland.

UNITED STATES OF AMERICA

3195.441

AMERICAN UNDERWATER SURVEILLANCE

The current US philosophy regarding defence against the principal ASW threat (Soviet submarines) can be summarised as relying on engaging these submarines in forward areas and at barriers before they can get within attacking range of US forces. To do this reliance rests primarily on American attack submarines and long range P-3 patrol aircraft supported by underwater systems.

The ability to locate enemy submarines within broad ocean areas is essential to the task of containing the large Soviet submarine force, and fixed undersea surveillance systems have played a major role in this for some years. Consolidation and expansion of such systems continues, and funding of the Sound Surveillance System (SOSUS, see below) is sustained at a consistently significant level. To improve US submarine surveillance capabilities, the 1986-90 programme continues funding for mobile SURTASS system. Development of the rapidly deployable air-dropped system (RDSS) has been halted.

Research in the undersea surveillance systems programme, which funds future upgrades, is concentrating on the development of integrated acoustic display and wideband acoustic recall software packages. The US Navy is also conducting experiments with an acoustic survey system in shallow and moderate depth waters.

SOSUS

SOSUS (Sound Surveillance System) consists of fixed undersea acoustic networks of passive hydrophone detector arrays deployed in the Atlantic and Pacific Oceans, and this is stated to provide significant detection capabilities. However, over the years (SOSUS was conceived in the 1950s) it was inevitable that the network gradually expanded as increments were added to enlarge the coverage. Most of these additions were the subject of various classified programmes, of which most details remain secret apart from isolated programme code-names. Among the latter are 'Caesar' and 'Collosus' referring to the US Atlantic and Pacific segments of the SOSUS network. Hydrophones designated AN/FQQ-10(V) are located at intervals of 5-15/miles along a linking cable connected to the shore station(s). Similar shore stations are said to exist beyond the USA (eg the Aleutian Islands, Canada, Denmark, Iceland, Italy, Japan, Korea, Philippines, Spain, Turkey, UK) these being included in programmes code-named 'Barrier' and 'Bronco'. Constantly changing threat characteristics call for significant system improvements and the USN embarked on a two-phase SOSUS improvement programme. The specific aims of this programme remain classified in detail but are known to include improvements to the communications elements of the system and complementary improvements to the signal and data processing segments, thereby facilitating greater operational utilisation of the data obtained by the undersea sensor arrays. In June 1985 a contract awarded to AT & T referred to 'FY 85 SOSUS Backfit Program'. No details of this have been disclosed.

Funding of the SOSUS programme remain fairly constant, but security classification prevents further information. Considerable funding for both development and procurement is likely for at least 10 years. It is understood that the Canadian government is considering the installation of a similar system to SOSUS to detect submarines beneath the ice cap and in Canadian territorial waters.

The ultimate objective for the SOSUS system is the complete integration of all undersea surveillance sensor systems into a fully co-ordinated and centrally controlled network, IUSS.

SURTASS

SURTASS (Surveillance Towed Array Sensor System) is a successor system to the AN/BQR-15 Towed Array Surveillance System (TASS), already operational with the US fleet (see **1984.453**). This mobile surveillance system complements fixed networks by providing the essential flexibility to respond to changes in Soviet submarine deployment patterns and by extending coverage to remote ocean areas not monitored by fixed systems. They can also serve as emergency reserve facilities if fixed networks are disabled. There are two shore-based data processing centres (one on the East Coast of America, and one on the West Coast) to which acoustic target information gathered by the towed arrays is relayed. After processing, target information is transmitted back to operational ships at sea. Five or more ships are planned for each of the two processing centres, operating in waters to the east and west of the American continent, respectively and funds for a total of 18 of these ships had been appropriated by fiscal year 1985.

Whereas the BQR-15 is towed by submarines, slow, small surface ships will tow the long SURTASS arrays back and forth over designated patrol lines. Funds have been set aside for a special type of platform to operate SURTASS, designated T-AGOS. The first seven ships are now operational, with a further 12 building or proposed. More detailed information on these ships is given in *Jane's Fighting Ships*.

SURTASS commenced full-scale development in fiscal year 1976. Testing was carried out at sea during 1977. Technical evaluation of SURTASS was completed in March 1978 and operational evaluation took place from May to June 1978. The first full tests of the SURTASS system, including the satellite link and data processing, took place in July 1979, and the system was declared ready for operational evaluation, which began in March 1980. Introduction to US Fleet service began in late 1984. AT & Technologies was awarded a $33 million contract in September 1987 for an enhanced modular signal processor to update SURTASS and SUBACS.

RDSS

RDSS (Rapidly Deployable Surveillance System) is

designed to use air-deployable, bottom moored, long-life passive acoustic buoys as the main sensor system in a new mobile monitoring system. It will be emplaced by aircraft to provide undersea surveillance coverage on an urgent basis if required in areas of special interest. It could be used to replace or augment existing systems for long periods. RDSS buoys are capable of use in shallow water and areas unsuitable for either SOSUS or SURTASS techniques. The basic design allows for the transmission of data from the RDSS buoy to existing maritime patrol aircraft for either real-time readout or recording for later replay and analysis. Two versions of RDSS buoys have been identified, as Mod 0 and Mod 1, respectively. The former transmits its data to P-3 or S-3 maritime aircraft for analysis and presentation on the aircraft (it is also recorded for later processing), while the Mod 1 design is intended to send information direct to an ASW processing centre which will utilise the data without delay. The Mod 0 buoy is regarded as a 'near-term' item, and the Mod 1 is thought likely to take longer to reach operation. The approval for full-scale development of the system was given in August 1981, after a lengthy study and research programme. Development funding budgeted for fiscal years 1984/85/86/87 $16.2 million, $10.6 million, $15.4 million and $20.8 million, respectively. In December 1984 the US Navy cancelled the programme for a variety of reasons, but 16 days later it was resurrected and re-examined, with a view to restructuring the whole concept. However, development funds for 1987 were deleted by the US Congress and the programme was shelved. It is probable that it will be re-activated at a later date.

TACTAS

TACTAS (Tactical Towed Array Sonar) is intended to remedy the deficiency of current ASW surface ships in passive detection and classification capabilities against submarines. A long-range (direct path, first and second convergence zone) is required which can be used with other systems (LAMPS Mk III). The TACTAS programme will provide a number of different towed array systems for surface ship use. The AN/SQR-18A is a successor development of a towed array tested under a previous (classified) programme, and operational evaluation was completed in February 1976. The array is designed to be towed at high speed and has been installed on all 46 FF 1052 class ships and Naval Reserve FFG-7 frigates. The AN/SQR-18A has been followed by the AN/SQR-19 system which gives active FFG-7, DD-963, DDG-993, DDG-51 and CG-47 classes of ship a passive detection capability. The SQR-19 is in full-scale production and the first two sets were installed aboard two DD-963 'Spruance' class destroyers in early 1987. The US Navy plans to install the AN/SQR-19 aboard all the major battle group ships built since 1970.

3777.363

AN/AKT-22(V)4 TELEMETRY DATA TRANSMITTING SET

The AN/AKT-22(V)4 telemetry data transmitting set forms an essential element of an airborne ASW installation for the operation of sonobuoys and the relaying of sonobuoy data to co-operating ASW surface vessels.

Two AN/ARR-75 radio receiving sets are included in the airborne element, providing eight VHF receiving channels. This configuration permits relay of up to eight passive sonobuoy transmissions, up to six channels of $R\text{-}\theta_3$, and passive data of which up to four channels may be $R\text{-}\theta_3$, or two channels of DIFAR sonobuoy data. An air-to-ship voice channel is available except when the DIFAR mode is used. The AN/AKT-22(V)4 produced by EDMAC consists of the following major units:

(1) transmitter-multiplexer T-1220B
(2) control indicator C-8988A
(3) antenna AS-3033
(4) actuator TG-229.

The multiplexer section of transmitter-multiplexer T-1220B/AKT-22(V)4 uses low-pass filters and VCOs to provide the eight-channel capability. The trigger filter assembly for this set is located in control indicator C-8988A/AKT-22(V)4. Setting any of the trigger-select switches on the control-indicator to the ON position disables data channels 3C and 4C. Composite trigger tones are brought into the multiplexer on a separate line and combined with the sonic and voice channels at the mixer-amplifier to form the composite FM modulation signal. This composite FM signal modulates the transmitter.

In the DIFAR operating mode, the composite FM modulation signal is disconnected from the transmitter input. Two DIFAR sonobuoy transmissions, received on dedicated receiver channels are demodulated in the receivers and are provided to the transmitter-multiplexer. The receiver output is conditioned in an amplifier-adaptor and the DIFAR-A sonic signal is conditioned in a low-pass filter. The DIFAR-B sonic signal is used to drive a VCO centred at 70 kHz and then passes through a bandpass filter. The two filter outputs are linearly combined, and the composite modulation signal is coupled to the transmitter input when the control-indicator mode select switch is in DIFAR position. Composite sonic or composite DIFAR input to the transmitter is selected by a relay in the DIFAR filter controlled by a switch in the control-indicator.

The shipboard receiving equipment required for the eight-channel data link is telemetric data receiving set AN/SKR-( ). The receiver demodulates and demultiplexes the composite sonic or DIFAR transmission, and routes the separate outputs to appropriate processing equipment.

STATUS

In production.

CONTRACTOR

EDMAC Corporation, 7500 Main Street, PO Box 750, Fishers, New York 14453-0750, USA.

Two main elements of the EDMAC AN/AKT-22(V)4 telemetry data transmitting set: transmitter-multiplexer (T-1220B) left, and control-indicator (C-8988A) right

4013.373

AN/ASA-66 TACTICAL DATA DISPLAY

The AN/ASA-66 is a multipurpose cockpit display for the P-3 Orion maritime patrol aircraft. A 9 inch (23 cm) diameter CRT provides pilot and co-pilot with real-time presentation of tactical situations. Graphic and alpha-numeric data are presented on the high brightness and contrast screen for viewing in either high ambient or controlled lighting conditions. The display is designed to be driven from a remote display generator to provide analogue deflection and video drive from computer data or direct from aircraft sensors.

STATUS

In operational service.

CONTRACTOR

Loral Electronic Systems, 999 Central Park Avenue, Yonkers, New York 10704, USA.

AN/ASA-66 tactical data display group

4012.373

AN/ASA-82 TACTICAL DATA DISPLAY SYSTEM

The AN/ASA-82 tactical data display system (TDS) is the primary data display for the US Navy S-3A Viking maritime aircraft. It serves as a real-time link between the four-man crew and the various electronic sensors which the aircraft carries for its task of maritime reconnaissance and anti-submarine warfare. High speed, high density data in the form of alpha-numeric symbols, vectors, conic projections and other appropriate display formats from both acoustic and non-acoustic sensors are presented to the crew. Information is stored, updated and refreshed by the on-board general-purpose digital computer and selectively displayed. Tactical and tabular data,

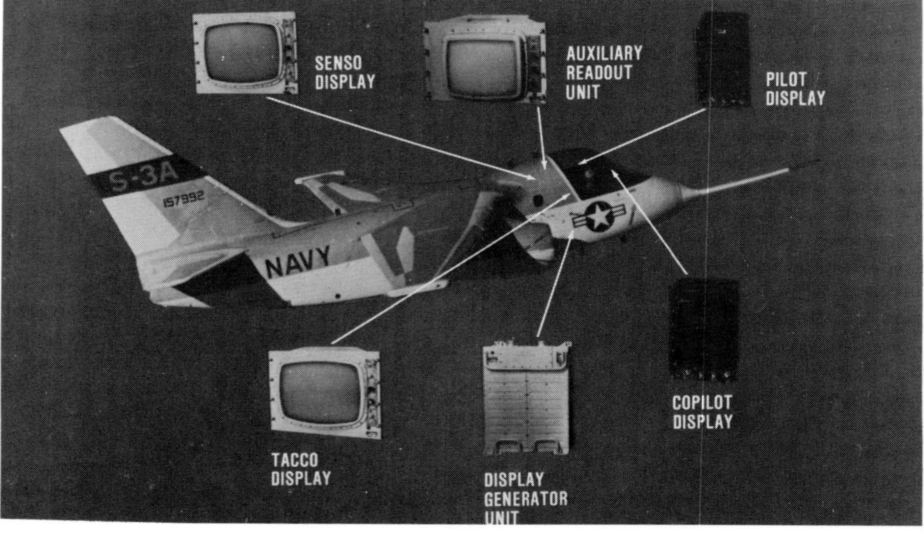

US Navy S-3A Viking ASW/maritime patrol aircraft, showing location of various items of equipment comprising AN/ASA-82 tactical data display system

controlled by the computer, are routed to the display via the display generator unit (DGU), which also provides the computer with display fault status information on a priority basis, controls the routing of display information, and generates the system built-in test functions.

The equipment consists of five CRT displays in addition to the DGU. The TACCO and SENSO (tactical co-ordinator and sensor operator) are each provided with identical multipurpose display units in their respective consoles. In addition, the SENSO has an auxiliary readout unit (ARU), the co-pilot has a multipurpose display, and the pilot has a display which presents a summary tactical plot. The DGU provides all of the displays with digital computer data except the ARU which receives acoustic data direct from the acoustic data processor.

STATUS

In operational service.

CONTRACTOR

Loral Electronic Systems, 999 Central Park Avenue, Yonkers, New York 10704, USA.

1982.363

AN/AYA-8B ASW DATA PROCESSING SYSTEM

Since 1968, GE has been producing the P-3C data processing system (DPS). The DPS equipment constitutes a major portion of the P-3C anti-submarine aircraft avionics.

The purpose of the AN/AYA-8B DPS is to provide an interface between the central computer and other aircraft systems. The main aircraft systems consist of the tactical co-ordinator (TACCO) station, non-acoustical sensor station, navigation communications (NAV/COM) station, acoustical sensor stations, pilot station, radar interface unit, armament/ordnance system, navigation systems, ARR-72 receivers (**3454.363**), submarine anomaly detector, and OMEGA. The DPS consists of four major logic units, as well as a selection of manual keysets and panels.

LOGIC UNIT 1 provides an interface to four types of peripheral information systems:

(1) manual entry
(2) system status
(3) sonobuoy receiver
(4) auxiliary readout display.

The manual entry sub-system provides the communication between the operator stations and the central computer of the man/machine interface. Each operator has a complex of illuminated switches and indicators by which he communicates with the central computer.

System status is received and stored by the status logic sub-unit of logic unit 1 and transmitted to the central computer. The status words are transmitted whenever any status bit changes or upon interrogation by the computer.

AN/AYA-8B pilot keyset (front) with (left to right) ordnance panel, universal keyset, and armament/ordnance panel

AN/AYA-8B modernised logic units for P-3C aircraft

The sonobuoy receiver logic provides for the digital tuning of the sonobuoy receiver sub-system by the central computer. Each of the 20 receiver processor channels can be tuned to one of 31 RF cells by the central computer or manually by the operator.

The auxiliary readout display logic provides the digital interface between the computer and the ARO displays located at the TACCO station and the NAV/COM station.

LOGIC UNIT 2 provides digital communications to three major aircraft sub-systems:

(1) navigation
(2) armament/ordnance
(3) magnetic tape transports.

LOGIC UNIT 3 provides the interface between the computer and three display sub-systems:

(1) TACCO multipurpose display
(2) sensor multipurpose display
(3) pilot display.

LOGIC UNIT 4 provides for the expansion of the computer input/output capability by means of the data multiplexer sub-unit (DMS), furnishes an increased memory capacity by means of the drum auxiliary memory sub-unit (DAMS) and provides an interface between the computer and two aircraft sub-systems:

(1) OMEGA
(2) auxiliary display.

Keysets and Panels

Three universal keysets provide the navigation/communication operator and the two acoustic sensor operators with the capability of entering information into and receiving information from the computer program. The pilot keyset is used by the aircraft pilot to control information presented on his CRT display, to enter navigation stabilisation information into the computer, to drop or cause the dropping of smoke floats and weapons, and to enter information on visual contacts.

The ordnance panel is used to display commands from the computer to the ordnance operator concerning search stores, ie bin and chute number, and status information.

The armament/ordnance test panel provides the capability of monitoring each output to the aircraft armament and ordnance systems from the armament output logic unit and ordnance output logic sub-units of logic unit 2.

STATUS

In production.

CONTRACTOR

GE Aerospace, Utica, New York 13503, USA.

3454.363

AN/ARR-72 SONOBUOY RECEIVING SET

The AN/ARR-72 sonobuoy receiver system is currently used on the P-3C patrol aircraft in conjunction with acoustic signal processors and a digital computer. The receiver system receives, amplifies, and demodulates frequency-modulated radio signals in the 162.25 to 173.5 MHz band transmitted by deployed sonobuoys. These radio signals contain information related to the frequency and directional pattern of the sound spectrum received by each sonobuoy hydrophone array. The baseband information from the receiver is demultiplexed and submitted to a special analysis and/or time correlation and doppler analysis by the acoustic signal processing equipment. Bathythermograph buoys and equipment enable determination of the seawater temperature for acoustic range measurement.

Monitoring of the RF level signals from the 31 on-line receivers enables the digital computer to perform many tactical functions such as the determination of the channel usage or presence of jamming prior to sonobuoy launching, connections of the proper channel RF receiver to each processor, and a determination of remaining sonobuoy life. These same RF level monitoring and channel selection functions may also be accomplished manually through the use of the receiver control-indicator boxes located at the individual sensor stations.

The AN/ARR-72 system is compatible with LOFAR, CODAR, BT, RO, CASS, and DICASS equipment currently being deployed. Specific sonobuoy types include: AN/SSQ-36, AN/SSQ-41, AN/SSQ-47, AN/SSQ-50, AN/SSQ-53, and AN/SSQ-62.

The AN/ARR-72 sonobuoy receiver system is a dual conversion superheterodyne VHF receiver system designed for receiving FM signals from deployed sonobuoys. Radio frequency signals are received at the dual aircraft VHF blade antennas, amplified by the system's AM-4966 dual RF preamplifiers, and provided to the CH-619 31-channel receiver assembly. Within this assembly a multi-coupler distributes the preamplified RF to the 31 fixed tuned receivers where it is further amplified and demodulated to provide baseband audio and RF level signals.

The SA-1605 audio assembly accepts the baseband and RF level outputs of the 31 receivers and outputs them to the computer and processing equipment. Each of the 19 audio channels contains a 31 × 1 switching matrix to select the output of a given receiver. This receiver signal is then amplified and provided through two individually buffered outputs to the processing equipment. Selection of a particular receiver channel may be accomplished by the digital computer or the C-7617 dual channel control indicator. An individual 31 × 1 switching matrix is provided within the audio assembly to service the RF level requirements of the digital computer. This switching assembly is controlled by the computer and provides an integrated RF level for the selected receiver in a digital format upon computer interrogation.

STATUS

Operational. An advanced version sonobuoy receiver, probably the AN/ARR-75( ) is being supplied under the P-3C 'Update III' programme. See **3455.363** and **1982.363**.

CONTRACTOR

EDMAC Corporation, 7500 Main Street, PO Box 750, Fishers, New York 14453-0750, USA.

3455.363

AN/ARR-75 SONOBUOY RECEIVER

The AN/ARR-75 is compatible with all FM sonobuoys in US Navy inventory or in development. Among the improvements incorporated in this receiver are:

(1) single-conversion, frequency-synthesised local oscillator
(2) extremely high dynamic range
(3) linear phase response over broad spectrum
(4) threshold extension of sensitivity through the use of phase-locked demodulation
(5) true, balanced, floating audio output
(6) extremely high rejection of interfering signals
(7) built-in antenna filter utilising helical resonator techniques.

The AN/ARR-75 receiving set was developed as a replacement for the AN/ARR-52A, but may be used to replace the AN/ARR-72. Four AN/ARR-75 equipments can provide the 16 simultaneous audio output channels required in some patrol aircraft configurations with significant size, weight and cost savings. System reliability is improved since any receiver can be tuned to any sonobuoy channel, thus overcoming the limitations of fixed-tuned receivers for each channel.

As a replacement for the AN/ARR-52A (Y configuration), the AN/ARR-75 provides a reduction in volume of approximately 3:1 and a reduction in weight of approximately 2:1. The solid-state design, the use of high reliability components and the use of fixed-tuned circuits to eliminate tuning adjustments has resulted in a tenfold improvement in reliability according to the manufacturer. Operational adequacy of this receiver has been extensively proven in the S2G aircraft. It is currently installed in SH-3H, SH-2F, and LAMPS Mk III.

The AN/ARR-75 receives, demodulates and amplifies FM sonobuoy transmissions in the VHF band between 162.25 and 173.5 MHz, and provides the demodulated output for analysis and display equipment. The receiver is a general-purpose multi-channel FM equipment suitable for use in fixed- or rotary-wing aircraft.

The equipment consists of two units: the receiver group assembly OR-69/ARR-75, in which are the majority of the necessary electronic modules, and the radio set control C-8658/ARR-75. Their respective weights are 9.75 kg and 1.13 kg.

A 99-channel version, the AN/ARR-75 ( ), is also available and probably forms part of 'Update III' to the P-3C aircraft sonar system (see **1982.363**). Export versions include the model R-75 (31 channels) and the model R-75 ( ) (99 channels).

The ARR-75 forms the basis of a new receiver, the R 605, being designed for the Royal Navy's EH 101 helicopter. The new receiver will be designed and developed by Rospatch, USA, and manufactured in UK under licence by Dowty Electronics.

STATUS

Production. Fitted to US Navy SH-2, SH-3 and SH-60 helicopters.

CONTRACTOR

EDMAC Corporation, 7500 Main Street, PO Box 750, Fishers, New York 14453-0750, USA.

4455.453

AN/WQC-2A SONAR COMMUNICATION SET

The AN/WQC-2A is a sonar underwater communication set for surface ships, submarines and shore installations. The system provides an SSB general voice and CW communication set consisting of a control station, remote control station, receiver/transmitter, and LF and HF transducers.

The system employs voice, audio and low speed telegraphy in two frequency bands: high (8.3 to 11.1 kHz) for close range and low (1.45 to 3.1 kHz) for long range. In addition the system facilitates the use, through separate input and output connectors, of a final transmitter amplifier in a frequency range of 100 Hz to 13 kHz.

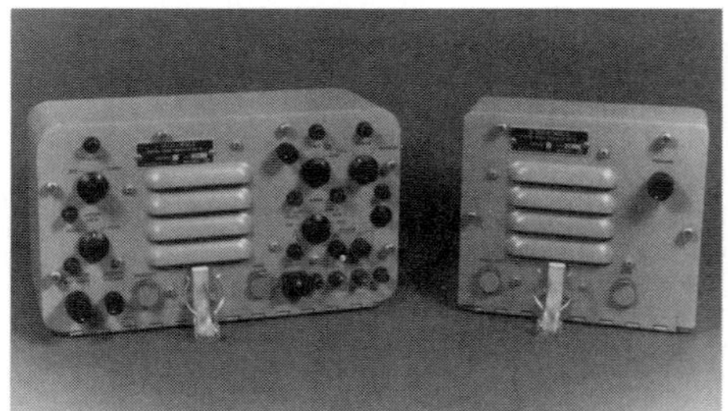

Control and remote control stations for AN/WQC-2A system

The receiver/transmitter contains the main electronic assemblies of the sonar communication system. It consists of a rack-type cabinet with removable drawer assemblies containing the power supply and test panel assembly, the receiver/transmitter assembly and the final amplifier assembly. The primary functions of the receiver/transmitter are to develop the high-powered single sideband transmission from voice, audio and CW signals to drive the LF or HF transducers and to receive and demodulate signals received from the transducers. The receiver and transmitter outputs are made available to external tape recordings and monitoring equipment. The circuits are also capable of muting, or being muted, by other external equipment.

The control station contains the required controls, indicators, microphones, etc. The remote control station is a secondary operating position, if required.

The two transducers (LF and HF) have a horizontal (radial) omni-directional beam pattern. The electromechanical energy conversion is accomplished by piezoelectric ceramic elements which are totally encapsulated and covered by an acoustically transparent neoprene boot.

CHARACTERISTICS

Frequency bands: 1.45 – 3.10 kHz (LF); 8.3 – 11.1 kHz (HF); 100 Hz – 13.0 kHz (auxiliary mode)

Output power: 600 W (LF); 450 W (HF); 1000 VA (auxiliary mode)

Weights

Receiver/transmitter: 223 kg

Control station: 6.8 kg

Remote control: 4.1 kg

Transducers: (with 175 feet of cable) 200 kg (LF) 45.5 kg (HF)

STATUS

In production for US Navy and overseas navies.

CONTRACTOR

General Instrument Corporation, Government Systems Division, 600 W John Street, Hicksville, New York 11802, USA

5640.453

AN/PQS-2A HAND-HELD SONAR

The AN/PQS-2A is a continuous transmission, frequency modulated hand-held non-magnetic sonar set intended for use by divers to locate and close on submerged objects such as mines, lost equipment, downed aircraft and sunken vessels in depths to 300 feet (91 m). Because of its non-magnetic construction, the set can be used safely when working near magnetically-influenced underwater explosive devices.

The equipment has an active (continuous transmission) mode and a passive (listening) mode. In the active mode, one of three range scales is selected (20, 60 or 120 yards (18, 55 or 110 m)) and an acoustic signal is transmitted over a 30 kHz bandwidth, swept from 114 to 145 kHz. When a submerged object is detected a ringing tone, proportional in frequency to the distance of the object, is produced in the earphones. The sonar has a 6° beamwidth and can detect a 12-inch (30.5 cm) diameter air-filled sphere at 120 yards (110 m). In the passive mode, the unit detects active sound sources (pingers) in the 24 to 45 kHz range. The sonar can detect a 39 kHz acoustic beacon at a range of at least 2000 yards (1800 m).

CHARACTERISTICS

Transmission: Continuous transmission frequency modulated (CTFM)

Read-out: Variable audio tone according to range

Active sonar range: Three ranges 18, 55 and 110 m

Passive range: Up to 1800 m

Weight in water: 0.23 kg (positive buoyance collar included in accessories)

Size: 31.75 cm long by 11.43 cm diameter

STATUS

In series production for the US Navy.

CONTRACTOR

General Instrument Corporation, Government Systems Division, 600 W John Street, Hicksville, New York 11802,USA.

AN/PQS-2A hand-held sonar

3788.363

SPARTON AIRBORNE SONAR PROCESSING EQUIPMENT

Sparton's modular airborne sonar processing systems and components are a flexible, lightweight, low-cost means to process and display data from sonobuoys and similar submarine tracking devices. The modular approach allows individual units to be used separately or combined to form a compact system capable of processing and displaying signals from most of the sonobuoy types currently in use or being developed. Typical hardware is described briefly in the following paragraphs although more detailed information is given in earlier editions of *Jane's Weapon Systems*, under this entry number.

Active Processing System

The active processor system consists of five modules: command signal transmitter; command signal generator; active processor; sonar display; power supply. The active processor system is capable of simultaneously processing and displaying two AN/SSQ-47 or AN/SSQ-522 active sonobuoys. The modifications provide the matched filter processing required for the additional CW pulse lengths and an FM processor utilising replica correlation. The system can be further expanded to process and display the AN/SSQ-

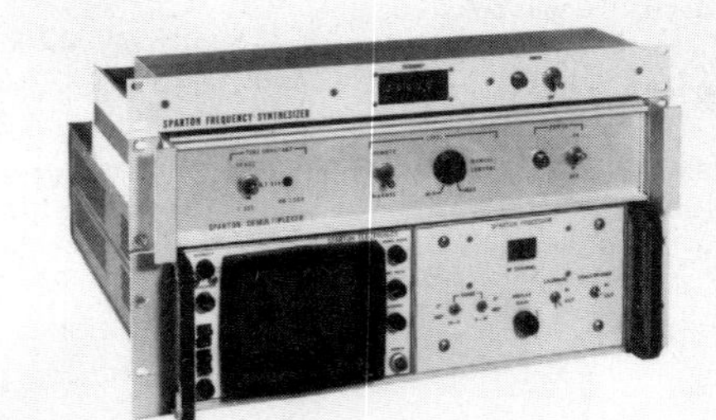

TD-1135/A demultiplexer processor/display

62 or proposed DICASS directional command-active sonobuoys by the addition of a sixth module. This module contains a digital arc tangent computer that calculates and displays bearing with a direct readout in degrees.

Command Signal Generator

The command generator is the active sonobuoy's information source. It is a dual-channel control unit that generates the sonar pulse information for two sonobuoys. Selected channels from the sonobuoy receiver are routed to the active processor. The command signal generator also provides the active processor with trigger pulses and other digital control information.

Active Processor

The active processor is a lightweight, dual-channel unit capable of simultaneously processing two command-active sonobuoys. It automatically analyses information for the detection and localisation of submarines. It is completely solid-state, with integrated circuits used extensively in the analyser portion of the unit. The processed signals from two sonobuoys can be displayed in a split-screen mode on the sonar display.

Sonar Display

The sonar display can be used with the active processor to provide target range and doppler information in the CW mode and target detection and range information in the FM mode. The Sparton sonar display utilises a variable-persistence cathode-ray tube that provides bright, high-quality readouts. In use, the screen is divided vertically into two displays where a combination of A scans or B scans can be displayed for either one of two sonobuoys simultaneously. Persistence time is adjustable from zero to full storage. Stored spot resolution is equivalent to 20 lines/cm.

CHARACTERISTICS

Display: 2 channels, A and/or B scan

CW: Target range and doppler

FM: Target detection and range

Display integration: Variable-persistence CRT for time integration

Controls

Threshold: Provides capability to optimise target detection

TD-1135/A Demultiplexer Processor/Display

The TD-1135/A is a high-quality instrument for testing directional frequency analysis and ranging (DIFAR) and directional command active (DICASS) sonobuoys. The TD-1135/A accepts a composite DIFAR/DICASS signal from any standard sonobuoy receiver, demultiplexes it, and displays the output as a radial vector on a variable-persistence CRT calibrated in degrees from magnetic north. North-south, east-west and omni-directional outputs are provided for computer analysis if greater accuracy is required.

Designed for use with standard sonobuoy monitoring equipment, the TD-1135/A is compatible with receiver types AR/ARR-52, AR/ARR-72, and AR/ARR-75, and with analyser display group types AN/AQA-3, AN/AQA-4 and AN/AQA-5. Included with the TD-1135/A is a Sparton DFS-101A digital frequency synthesiser, with crystal oscillator reference, which is tunable in discrete steps over the entire DIFAR/DICASS frequency band. If an AN/AQA -3, -4 or -5 is not available, the TD-1135/A and frequency synthesiser can be modified by adding a spectrum analysis capability.

The TD-1135/A demultiplexer is available as a separate unit for operation with DIFAR or DICASS sonobuoys.

CHARACTERISTICS

Frequency range: Broadband DIFAR audio

Input impedance: 10 kΩ minimum

Output impedance: 10 kΩ max

Variable gain: 48 dB manual or remote

Time constants: 0.1, 1 and 10 seconds

Phase pilot bandwidth: 1 Hz max

Processor analysis bandwidth: 1 Hz max

MTBF: 1500 h minimum

Outputs available

Demultiplexer: N-S, E-W and omni

Processor: N-S, E-W and omni

Spectrum Analyser

The Sparton spectrum analyser is a low-cost, special-purpose digital processor that performs real-time spectral analysis. The signature of the target being analysed is displayed simultaneously, using both channels of the Sparton sonar display. Discrete Fourier transform technology is used with circuits operating at optimally low bit rates. Features include linear integration of spectral estimates, manual or automatic modes, internal cursor for accurate frequency identification, and total spectrum coverage.

CHARACTERISTICS

Frequency range: 14 – 912 Hz

Frequency bands: 14 – 114 Hz (0.25 Hz resolution); 112 – 912 Hz (2 Hz resolution)

Number of spectral estimates per band: 400

Linear integration of spectral estimates

Manual: 1 – 100 samples

Automatic: 100 samples

Input amplitude dynamic range: 80 dB

DICASS Command Signal Monitor

The Sparton DICASS (directional command-active sonobuoy system) command signal monitor is used to monitor valid command signals during testing of the AN/SSQ-62. The unit is self-contained and requires a UHF antenna and 115 V AC for operation. It provides a permanent printed record of the VHF channel command, identification of command function, RF signal level during command, time of signal transmission, length of command, and RF signal level prior to command.

Sea Noise Directionality and Level Indicator (ID-1872/A MOD)

The ID-1872/A MOD adds the ambient sea noise directional estimator (ANODE) function to a standard ambient sea noise indicator. It maintains the capability of the ASNI ID-1872/A, which measures and displays the omni-directional sea noise level sensed and transmitted by an AN/SSQ-57 sonobuoy and adds the ANODE function.

CHARACTERISTICS

Analysis centre frequency: 50, 100, 200, 440, 1000, 1700 Hz

Analysis bandwidth: 25 Hz

Weight: 3.18 kg

CONTRACTOR

Sparton Corporation, Electronics Division, 2400 E Ganson Street, Jackson, Michigan 49202, USA.

3778.363

AN/ARR-78(V) ADVANCED SONOBUOY COMMUNICATION LINK (ASCL)

The AN/ARR-78(V) radio receiving set, also known as ASCL (Advanced Sonobuoy Communication Link), is for the operation and management of anti-submarine sonobuoys by the crew of ASW aircraft such as the P-3C Orion (Update III) maritime patrol aircraft and S-3B Viking. The equipment comprises five main items (described below) and is employed with an associated on-board ASW signal processor. The main units of the AN/ARR-78(V) are:

(1) the AM-6875 RF preamplifier is an optimised high performance, low noise VHF preamplifier which provides amplification and pre-filtering of received RF signals
(2) the R-2033 radio receiver contains 20 fully synthesised high performance receiver modules (16 acoustic and 4 auxiliary), and one each of the following modules: RF/ADF amplifier multi-coupler, reference oscillator, PROTEUS digital channel, I/O processor, clock generator, BITE, and DC power supply module. Each single conversion receiver module includes mixer conversion, frequency synthesised LO, demodulator, and output interface circuits. Each of the acoustic receiver modules processes FM/analogue or FSK digital signals at any of the 99 channels in the extended VHF band. Each of the four auxiliary receiver modules processes signals at any of the 99 channels and provides one channel for selection and processing of the on top position indicator (OTPI) signals; two channels for the operator to monitor acoustic information; and one channel to monitor the RF signal level in any of the RF channels. Common receiver modules are interchangeable.

   BITE circuits provide comprehensive end-to-end evaluation of each receiver from the VHF preamplifier to the receiver output interface circuits. BITE is initiated automatically by the computer (such as PROTEUS) and/or by the operator through the indicator control unit (ICU). Performance status is displayed on the ICU and routed to the computer
(3) the C-10126 indicator control unit (ICU) provides the operator with a means for manual control of each receiver channel frequency assignment, receiving mode, self test. It also displays status of the receiving set and operator entry information
(4) the ID-2086 receiver status indicator continuously displays the control mode setting, the RF channel number, and the received signal level for each receiver
(5) the C-10127 receiver control unit provides the operator with control over the OTPI channel frequency.

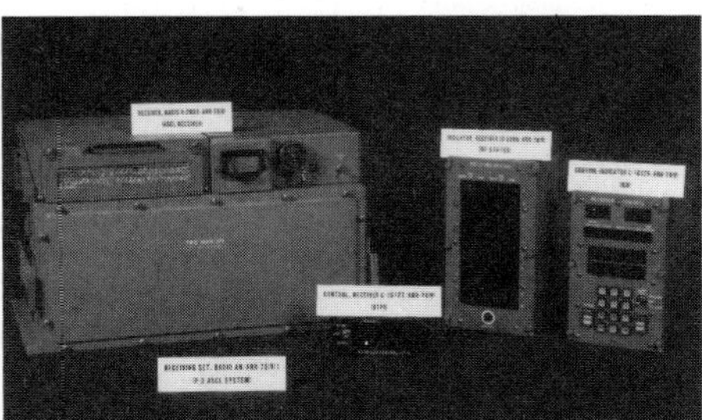

AN/ARR-78(V) advanced sonobuoy communications link equipment (P-3 configuration)

Surface acoustic wave (SAW) IF filters are incorporated to provide high selectivity, coupled with good linear phase characteristics which yield low levels of distortion. Microprocessors are used in the ICU and I/O module to process the commands to, and display status data from the receivers. Versatile programs accommodate changes without requiring aircraft modification.

The auxiliary receivers permit the operator to aurally monitor signals in any of the 99 RF channels without interrupting data processing in that channel.

CHARACTERISTICS

Frequency: Extended VHF

Receivers: 20 (16 acoustic/4 auxiliary)

Channels: 99 per receiver

Audio output

Analogue: 2 V or 4%V rms balanced

Power input: 115 V AC ±10%, 380 – 440 Hz, 3 phase 450 W; 18 – 32 V DC, 7 W; 26.5 V AC, ±10%, 400 Hz, 50 W

Dimensions

AM-6875: 7.6 × 14.6 × 10.8 cm; Wt: 0.95 kg

R-2033: 30.9 × 54.1 × 38.9 cm; Wt: 45.9 kg

C-10126: 22.9 × 14.6 × 16.5 cm; Wt: 3 kg

ID-2086: 26.7 × 14.6 × 12.7 cm; Wt: 1.9 kg

C-10127: 5.7 × 14.6 × 8.3 cm; Wt: 0.6 kg

STATUS

Full-scale development was completed in 1980 under US Navy contracts. The first production contract for P-3C systems and S-3B systems was awarded in December 1982 and further contracts have been awarded up to the end of 1987. Other applications include derivative configurations for the P-3C (Update IV) aircraft and the LAMPS III helicopter. A new version, the ASCL-IV, is in development. This is an upgraded communications link which provides a capability of simultaneous reception of signals from four sonobuoys for helicopter applications.

CONTRACTOR

Hazeltine Corporation (part of Emerson Electric Co.), Greenlawn, New York 11740, USA.

3776.253

RADIO RECEIVER R-1651/ARA SONOBUOY OTPI

The R-1651/ARA is a VHF radio receiver of solid-state construction used in locating sonobuoys. It also acts as a receiving converter when employed with compatible automatic direction finding (ADF) systems; when a command is initiated in the control unit of the ADF, the R-1651/ARA switches the direction finder from a UHF/ADF function to a VHF sonobuoy on-top position indicator/locator (OTPI) mode. The equipment is a replacement for the obsolete R-1047 A/A.

The R-1651/ARA is fitted in ASA patrol aircraft or helicopters and used with the AN/ARA-25 direction finder group. It receives, amplifies, and demodulates RF signals from deployed sonobuoys on any one of 31 RF frequencies within the range 162.25 to 173.50 MHz.

STATUS

Production.

CONTRACTOR

EDMAC Corporation, 7500 Main Street, PO Box 750, Fishers, New York 14453-0750, USA.

EDMAC R-1651/ARA sonobuoy OTPI radio receiver

6562.453

AN/BSY-1(V) INTEGRATED COMBAT SYSTEM

The AN/BSY-1(V) is a submarine integrated sonar/fire control system which will include an active sonar, passive flank array sonar, passive towed array and a combat control system. It will enable later units of the SSN-688 'Los Angeles' class submarines to detect, classify, track and destroy hostile surface and submarine contacts. It will be the US Navy's first submarine system that combines the functions of sonar systems and weapon fire control systems.

The principal subsystem will be the wide aperture array, a passive hull-mounted flank array mounted on both sides of the submarine. Other systems will include the Submarine Active Detection System (SADS) which will probably be a low-frequency bow-mounted sonar used for detection and fire control solutions on targets. A long thin line towed array will probably be used, as well as a mine and ice detection system for close range detection of mines and ice packs.

A more advanced version, the AN/BSY-2, is likely to be installed in later SSN submarines in the early 1990s. This will incorporate upgraded sonars and fire control/display systems, such as the wide aperture array, vertical and horizontal large screen displays, improved long thin line arrays and weapons control systems.

STATUS

In production. The first system, for installation on SSN-751, was delivered in mid-1987.

CONTRACTOR

IBM, Federal Systems Division, 10 215 Fernwood Drive, Bethseda, Maryland 20034, USA.

6248.253

AN/SQR-15 TOWED ARRAY SONAR

The AN/SQR-15 is a surface ship towed array sonar system which has been in operational service for several years. It provides a mobile passive surveillance capability for escort vessels against enemy submarines under most environmental conditions. The system is an improved and redesigned version of the earlier AN/SQR-14 equipment.

STATUS

In operational service.

CONTRACTOR

Gould Inc, Chesapeake Instruments Division, Glen Burnie, Maryland, USA.

6245.263

AN/SQR-17 SONAR SIGNAL PROCESSOR

THE AN/SQR-17 is a four channel sonar signal processor and recorder for the shipboard part of the LAMPS Mk 1 system. The processor receives, analyses, displays and records the acoustic signals detected by sonobuoys, or by sonar devices mounted on the hull of a ship or towed behind the vessel. The sonobuoys may be either placed by the LAMPS helicopter or over the side of the ship. Once the raw acoustic data is received, the AN/SQR-17 displays the signals on the main CRT display in either A-scan or B-scan format. Two auxiliary displays are employed to select the correct system operating mode and for monitoring the status of multiple sensors. The basic equipment can process LOFAR, ACTIVE, DEMON and BT sonobuoys.

The AN/SQR-17 has undergone a series of upgrades and improvements through the (V)1, (V)2 and (V)3 versions. The sixth generation includes the AN/SQR-17A(V)1 and 17A(V)2 models. These add an AN/SQR-18A Tactical Towed Array Sonar (TACTAS) interface kit, plus DIFAR/DICAS/VLAD sonobuoy processing, display and recording capability to the system. This addition improves the redetection, localisation, classification and attack capability both in time and accuracy. The number of acoustic processing channels is increased to eight channels.

STATUS

Over 100 systems are in service with the US Navy and US Coast Guard.

CONTRACTOR

Diagnostic/Retrieval Systems Inc, 16 Thornton Road, Oakland, New Jersey 07436, USA.

AN/SQR-17A sonar signal processor

4622.253

AN/SQR-18A TACTICAL SONAR TOWED ARRAY (TACTAS)

The AN/SQR-18A sonar has been designed to enhance the air/sea warfare capabilities of the US Navy surface fleet. Its long-range, passive detection capability complements the hull-mounted sonar of the US Navy FF 1052 (Knox) class frigates. The 800-foot long array consists of an acoustic section, a vibration isolation module and a rope drogue. It is towed behind the variable depth sonar (VDS) body of the AN/SQS-35(V) (**4179.453**). Preamplifiers are in the VDS body, while signal processing and other electronics are on board the ship.

The AN/SQR-18A is being upgraded to the AN/SQR-18A(V)1, which will include a "noise canceller" that will remove own-ship radiated noise from the operator's display. Another version, the AN/SQR-18A(V)2, will have its own towing and handling capability, permitting it to be used on ships without a VDS.

In both systems bearing data is provided in the narrow and broadband spectrums. The US Navy SQR-18A(V)1 and (V)2 both feature a degree of integration with the SQS-26 hull-mounted sonar and SQR-17 sonobuoy processor. The two variants can use the Advanced Modular Signal Processor (AMSP) – a programmable system using interactive software design.

The Signal Processor Unit 2, manufactured by Diagnostic Retrieval Systems, is a passive system employing digital signal processing techniques to present narrow band and broadband low frequency information. The system may be operated in narrow band and broadband simultaneously. Narrow band outputs are presented on the CRT screens in LOFAR and/or other formats. Up to four channels can be shown on one display. Controls are used to measure the frequency and amplitudes of tonals, harmonic ratio and tag frequencies. Results of measurements, bearings of beams displayed, processor frequency bands and tag numbers are displayed in alphanumeric format.

STATUS

In use with the US Navy, and with the Royal Netherlands and Japanese navies. Approximately 40 AN/SQR-18A systems have been delivered to the US Navy and are being progressively updated to AN/SQR-18A(V)1 standard. One system is being used by the Italian Navy on sea-trials, and another is fitted to the frigate *Isaac Sweers* of the Royal Netherlands Navy.

EDO received multi-million dollar contracts from the US Naval Sea Systems Command for improved AN/SQR-18A(V)1 and (V)2 systems in 1986 and 1987. Contracting is expected to continue in 1988 and beyond with improvements, upgrades and new installations.

CONTRACTORS

EDO Corporation, Government Systems Division, 14-04 111th Street, College Point, New York 11356, USA.

Diagnostic Retrieval Systems Inc, 16 Thornton Road, Oakland, New Jersey 07436, USA (Unit 2 signal processor).

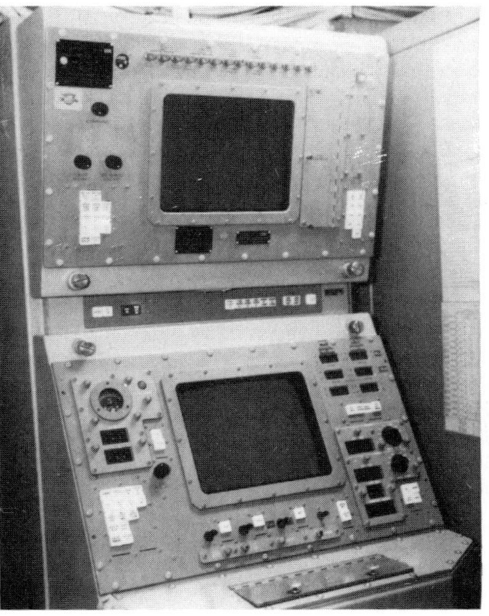

AN/SQR-18A console/display

4463.253

AN/SQR-19 TACTICAL SONAR TOWED ARRAY

The AN/SQR-19 is a passive towed sensor array designed to give surface vessels long range detection and tracking of both submarines and surface ships. It includes a towed sensor array with improved ranging ability and reduced self-noise; a handling system which improves system performance by maintaining array position and depth; and a sophisticated signal processing system providing more sonar information with less operator workload. The 'wet' end consists of a linear array and tow cable, plus the associated handling and stowage equipment. Coaxial conductors within the tow cable carry the flow of telemetred acoustic signals, plus heading depth and temperature data from the array to the ship, and also handle the commands to the array. Located below decks are the signal and data processing equipments, plus the display and control consoles. The displays are shared with the AN/SQS-53 hull-mounted sonar and the LAMPS III ASW helicopter, so that the ship is provided with a single integrated ASW system.

The AN/SQR-19 provides omni-directional long range passive detection and classification of submarine threats at 'tactically significant' own-ship speeds in seas up to state 4, using a 82 mm diameter array towed on a 1700 m cable to provide tow depths down to 335 m.

STATUS

The first operational system was installed on board the USS *Moosbrugger* (DD 980) early in 1982 and has completed both technical and operational evaluation. Installation aboard six US Navy DD-963 'Spruance' class destroyers is underway. It is intended to install the AN/SQR-19 aboard all US Navy DDG-93 and DDG-51 destroyers, CG-47 class cruisers, and the FFG-7 class frigates. Gould has received a number of contracts for the system, including one of $123 million for the Canadian Navy and another in December 1987 valued at $100 million for the US Navy and Spanish Navy.

CONTRACTORS

Gould Inc, Defense Electronics Division, 6711 Baymeadow Drive, Glen Burnie, Maryland 21061, USA.

GE Underseas Systems Dept, Syracuse, New York, USA.

2536.253

AN/SQS-53/26 SONARS

Claimed to be the most advanced surface ship ASW sonar in the US Navy inventory, the AN/SQS-53 is a high-power, long-range system evolved from the AN/SQS-26CX. Functions of the system are the detection, tracking, and classification of underwater targets, and also underwater communications, countermeasures against acoustic underwater weapons, and certain oceanographic recording uses. Target data obtained by the sonar is transmitted to the ship's Honeywell Mk 116 digital underwater fire control system. The latter translates target range, bearing, and depth data sent from the sonar to the ship's central computers into signals controlling the launch of ASW weapons (ASROC or torpedoes).

The AN/SQS-53 can detect, identify, and track multiple targets and is the first USN surface ship sonar designed specifically to interface directly with a vessel's digital computers. The system has a cylindrical array of 576 transducer elements housed in a large bulb dome below the water line of the ship's bow. There are 37 cabinets of signal processing, transmitting, and display equipment. Passive and active operating modes are possible.

There are three active modes:

(1) surface duct
(2) bottom bounce
(3) convergence zone.

The surface duct mode depends upon sound energy being transmitted essentially in the horizontal plane. Because of the high level of noise introduced into the return signals near the surface, this mode is useful only for relatively short distances. Nevertheless, this method is conventional for many surface ship sonars, and the high transmission power of the AN/SQS-53 is stated to provide longer range capability in the surface duct mode than previous sonars.

In the bottom bounce mode the sound energy is directed obliquely toward the sea bed. The energy is reflected upward from the ocean floor toward the surface at considerable distances from the ship. Submarine echoes are received via a similar return path. This method is useful in waters of more than certain minimum depths and where the sea bed has the requisite favourable characteristics.

Convergence zone mode operation takes advantage of the characteristics of very deep water. The sound energy is refracted downward due to the

Bow dome cutaway showing partial view of AN/SQS-53 cylindrical transducer (to right of engineers)

Bow housing of AN/SQS-26 sonar

temperature and pressure conditions near the surface, but, as depth increases, these physical effects change and the sound path alters direction to cause the energy to return to the surface in a coarsely focused convergence zone. This zone can form at great distances from the ship, and provides the longest range of coverage for the sonar when the water conditions are favourable for this mode of operation.

Passive detection gives the bearings of targets based on their own noise generation, rather than by echo location. It has proved a valuable method on the currently deployed AN/SQS-26 equipped ships, especially at low speed. The *Spruance* and her successors are expected to have improved passive detection capabilities at higher speeds due to improved noise suppression measures. The passive mode can be operated simultaneously with the active modes.

STATUS

The AN/SQS-26 is fitted in ten FF1040 'Garcia' class frigates; six FFG-1 'Brooke' class missile frigates; and 46 FF1052 'Knox' class frigates. The AN/SQS-53 is being fitted in DD963 'Spruance' class destroyers as they are built and a total of 30 ships is planned. The five CGN-38 'Virginia' class nuclear cruisers are fitted with AN/SQS-26CX or AN/SQS-53A systems. An updated version, known as the AN/SQS-53B, is in service and $104 million contracts have been awarded to Hughes Aircraft Company by the US Navy for 30 AN/SQS-53B digital display subsystems for DD-963 and CG-47 class ships. In September 1986 a contract valued at $39 million was awarded to General Electric, and a $34 million contract in May 1987.

A new version of the system, the AN/SQS-53C Battle Group Sonar, has been developed by General Electric. The AN/SQS-53C is a much more advanced system which uses 22 fewer cabinets with a consequent reduction in cabling, etc. The new version uses seven Sperry AN/UYK-44(V) computers linked together in a multiple embedded configuration. The ANS/SQS-53C is scheduled to be retrofitted into a number of ships currently equipped with earlier versions.

The SQS-53C incorporates a number of improvements, including a better performance in the active mode, simultaneous operation in active and passive mode, multiple target automatic tracking and improved displays. The first fitment was in USS *Stump* (DD 978) in August 1986.

CONTRACTORS

General Electric Company, Electronic Systems Division, Farrell Road, Syracuse, New York 13201, USA. – Types 53A and 53C

Hughes Aircraft Company, Systems Group, Fullerton, California 92634, USA. – Type 53B.

4179.453

AN/SQS-35 VARIABLE DEPTH SONAR AND AN/SQS-36, AN/SQS-38 HULL-MOUNTED SONARS

These systems have been designed to detect submarines at medium-ranges in both deep and shallow waters. The AN/SQS-35 and the AN/SQS-38 sonars have been in service with the US Navy since the mid 1960s.

The US systems are improved miniaturised solid-state versions of previously developed vacuum-tube equipments manufactured for the US, Italian, Norwegian and Japanese Navies.

Some versions combine both variable depth capability and hull sonar capability, selectable by the sonar operator at the control console.

Weapons associated with the system are the Mk 44 torpedo and ASROC (USA), Terne (Norway) and Lanciabas (Italy). Over 75 systems have been delivered.

CONTRACTOR

EDO Corporation, Government Systems Division, 14-04 111th Street, College Point, New York 11356, USA.

AN/SQS-35(V) variable depth sonar seen aboard escort ship USS Francis Hammond

---

3494.253

RAYTHEON STANDARD SONAR TRANSMITTER (SST)

The Raytheon SST provides sonar systems with the benefits of modern, solid-state technology, for example greatly reduced maintenance costs through increased reliability, plus very significant space savings relative to vacuum tube transmitters. This transmitter sub-system is made available for system modernisation or for new system applications.

Configured to meet a specific system's requirements, the SST consists of multiple one kW modules with associated power supplies. As configured for the AN/SQQ-23 and AN/SQS-23 sonars upgrade, the sub-system consists of two cabinets of 24 modules each and a performance monitor/system interface cabinet. It replaces more than 30 units of the previous transmitter, including all of the energy storage motor generators.,

CHARACTERISTICS

Standard 1 kW module (1 per channel)
Power output: 1 kW
Operating frequency: $\pm$1.5 kHz (nominal)
Distortion: 6% max
Load: Nominal $\pm$100%
Duty cycle: 15%
Linearity: $\pm$1 dB from 0 to −12 dB
Gain adjustment: 3 dB
Protection: Short circuit and over-temperature
Mean time to repair: 3 minutes
Weight and size: 1651 kg in 1.2 × 1.8 m of deck space
SST as manufactured for AN/SQQ/SQS-23
Power requirements: 115V AC, 60 Hz 1-ph, 0.37 kVA, 0.92 PF; 440V AC, 60 Hz, 3-ph, 128 kW (Max 0.9 PF)
Modes of operation

(1) Transmit: same as any associated system such as omni, sector, FM or CW
(2) Performance monitoring: automatic and manual
(3) Self-test: receiving system not required.

STATUS

Current production includes transmitters for seven navies. Similar transmitters are used on the AN/BQS-13, AN/SQS-56, DE1160 series and the AN/SQQ/SQS-23 sonar systems. An air-cooled version of the SST has been delivered in a programme to update the AN/BQS-4 submarine sonar. 72 transmitter subsystems had been sold as at November 1987.

CONTRACTOR

Raytheon Company, Submarine Signal Division, Portsmouth, Rhode Island 02871, USA.

Raytheon standard sonar transmitter

1807.453

AN/SQS-56/DE1160

The AN/SQS-56 is a modern hull-mounted sonar developed as a company funded product by Raytheon's Submarine Signal Division for the US Navy's 'FFG-7' class frigates. The US Navy has provided the AN/SQS-56, via the Foreign Military Sales (FMS) programme, to Saudi Arabia for its PCG ships, to Australia for its 'FFG-7' class frigates and to Turkey for its MEKO 200 programme. Versions commercially exported with the designation DE1160 are operational on the Italian Navy's 'Lupo' class frigates and the Spanish, Moroccan and Egyptian navies 'Descubierta' class frigates, and are being considered for new construction ASW ships of several other navies. The DE1160, when configured with 36 kW transmitters, is identical to the AN/SQS-56. Outfitted with a VDS array and handling subsystem, it becomes Raytheon's DE1164 sonar (**3115.453**) and is installed in the Italian Navy's 'Maestrale' class ships.

The DE1160, when equipped with a larger, low-frequency transducer array and three additional transmitter cabinets, is designated DE1160LF and is capable of convergence zone performance. The DE1160LF was delivered to the Italian Navy for the helicopter carrier *Giuseppe Garibaldi* early in 1984. This system has also been ordered by the Spanish Navy and will upgrade its 'Baleares' class frigates.

A VDS version of the DE1160LF will combine the convergence zone performance of the 'Garibaldi' sonar with the environmental adaptability of the DE1164 under the denomination DE1164LF, a sonar system for major ASW combatants.

The AN/SQS-56 sonar features digital implementation, system control by a built-in mini-computer, and an advanced display system. Digital implementation allows packaging of the complete multi-function active and passive sonar in five medium-size electronic cabinets and one operator's console. Computer controlled functions provide a system that is extremely flexible and easy to operate. The computer is also used to provide automated fault detection and localisation and a built-in training capability. The human-engineered display ensures proper interpretation by operators, even by those with relatively low levels of training.

The sonar is an active/passive, preformed beam, digital sonar providing panoramic echo ranging and panoramic (DIMUS) passive surveillance. All signal processing, except transducer received signal amplification and linear transducer transmit drive, is accomplished in digital hardware, most of which is implemented using US Navy SEMP (standard electronic module programme) components in compact water-cooled cabinets small enough to allow installation through standard size hatches. All visual data are presented on flicker-free, digitally refreshed television type raster scan CRT displays. Complete symbol and alpha-numeric facilities are included. System timing, control, and interface communication are accomplished by a general purpose mini-computer which is a component of the basic system. Both 400 Hz synchro and MIL-STD-1397 Type A or C, category II digital interfaces are available in the basic sonar system. Except for the 400 Hz synchro reference power, the entire sonar operates from 440 to 480 V, three-phase 60 Hz ship's power.

The basic system includes the transducer array, transducer junction box, five electronic cabinets (array interface, transmitter(s), receiver and controller), operator console, sonar dome, and control unit. Options in production include: a loudspeaker/intercom, a water cooling unit, a remote display, and a performance prediction subsystem.

A single operator can search, track, classify, and designate multiple targets from the active system while simultaneously maintaining anti-torpedo surveillance on the passive display. Computer assisted system control permits the operator to concentrate on the sonar data being displayed rather than on the system controls.

STATUS

Approval for service use was issued early in 1980 by the US Navy following successful completion of final operational test and evaluation. Systems ordered as at November 1987 for the AN/SQS-56/DE1160 total 102 and include 53 systems and ten trainers for the US Navy; four systems for Australia (AN/SQS-56); seven systems for Saudi Arabia (AN/SQS-56); four systems for Turkey (AN/SQ-56); six systems for Italy (four DE1160, one DE1160LF and one trainer); 17 systems for Spain (eight DE1160, five DE1160LF, three SQS-56SP and one trainer, for 'FFG-7' class ships); one DE1160 for Morocco.

AN/SQS-56 operator's console

CONTRACTOR

Raytheon Company, Submarine Signal Division, Portsmouth, Rhode Island 02871, USA.

1789.453

AN/SQQ-14 MINEHUNTING SONAR

The AN/SQQ-14 system is essentially a variable-depth, dual-frequency sonar for detecting and classifying bottom mines in shallow water. It utilises a towed body in the shape of an elongated sphere towed through a centre well on a US Navy minesweeper.

A unique aspect of the AN/SQQ-14 design is the towing cable. This consists of discrete, 18 inch (457 mm) sections of articulated struts with universal joints at each section, permitting the cable to flex in any vertical plane, but restraining it from torsional motion. This configuration imparts a constant heading to the towed body, thereby eliminating the need for a gyroscopic heading reference system.

A rubber-jacketed electric cable containing 35 shielded, coaxial, and individual conductors passes through the centre of the strut sections, terminating in a slip ring in the towed body and in a winch with compound winding drum in lieu of a slip ring installed on the 01 deck of a mine countermeasures ship.

The articulated strut type of tow cable posed unique problems related to winching, storage on the winch drum, and ship's tow point construction. Struts, of which there are 117 per system, are stored in a single layer on a cylindrical drum equipped with specially designed pads serving as contact points for the articulated strut knuckles. Since a conventional level wind mechanism could not be used with the struts, the winch drum itself is designed to traverse axially by means of an Acme lead screw and rotating nut, thereby accomplishing the level wind function and maintaining a fixed fleeting point. All equipment for this application must be non-magnetic. Struts, bearings, and knuckles are fabricated of high strength Inconel alloy. The winch drum is cast aluminium and supporting structure is aluminium and stainless steel.

STATUS

In operational service.

AN/SQQ-14 projectors and hydrophone arrays

The SQQ-14 sonar is produced under licence in Italy by FIAR SpA, which has supplied equipment to the Italian Navy. Other users include the Belgian and Spanish Navies. FIAR is offering an improved version, the SQQ-14A, which uses about 40% of the improvements carried out by GE to upgrade the SQQ-14 to the SQQ-30 (**4623.253**).

CONTRACTOR

General Electric Company, Electronic Systems Division, Farrell Road, Syracuse, New York 13201, USA.

4623.253

AN/SQQ-30 MINEHUNTING SONAR

The AN/SQQ-30 is the successor to the AN/SQQ-14 system (**1789.453**) and is a latest technology minehunting sonar. The system consists of two sonars, a search sonar for initial detection and a high frequency, high resolution sonar for classification of targets. The two sonars are separated to give the area coverage needed. The two sonars are housed in a hydrodynamically, egg-shaped vehicle and towed at various speeds by the minehunter. The towed body is streamed from a well in the foward part of the ship, using a winch driving a 3-metre diameter cable drum on the foredeck. Two display consoles are provided for the search and classification sonars.

STATUS

In production. Fitted to the first few US Navy MSM-1 'Avenger' class minehunters.

CONTRACTOR

General Electric Company, Electronic Systems Division, Farrell Road, Syracuse, New York 13201, USA.

5652.253

AN/SQQ-32 ADVANCED MINEHUNTING SONAR SYSTEM

A full scale engineering development programme of an advanced minehunting sonar system (AN/SQQ-32) is nearing completion for the US Navy, by Raytheon teamed with Thomson Sintra, France. The new system is scheduled to be installed in the US Navy's newly constructed minesweepers designated MCM-1 and MSH-1.

The proposed system consists of two separate sonars, a search sonar for initial detection and a high frequency, high resolution sonar for classification of the targets. The two sonars are partially housed in a hydrodynamically shaped vehicle and towed at various speeds by the minehunting ship. The latest technologies in beamforming, signal processing, modular packaging and displays are used throughout the system.

The detection sonar is being designed and manufactured by Raytheon and will be able to detect mines over a wide range of distances and bottom conditions. This system incorporates a computer-aided detection facility which is designed to help the sonar operator to discard non-mine objects detected by the sonar. Raytheon is also providing the computer facilities and display consoles and is responsible for overall system integration.

The classification sonar is being designed and fabricated by Thomson Sintra and will be based on its experience with the DUBM-21 and TSM-2022 minehunting sonars. The classification sonar will provide very high resolution transmission and reception of underwater signals to enable targets to be identified with near-picture quality.

The 'wet end', consisting of a hydrodynamically shaped vehicle, a deployment/retrieval system, a tow cable and a winch is being designed by Charles Stark Draper Laboratory of Cambridge, Massachusetts.

STATUS

The first operational systems are expected to be installed in 1989-90.

CONTRACTORS

Raytheon Company, Submarine Signal Division, Portsmouth, Rhode Island 02871, USA.

Thomson Sintra, Activités Sous-Marines, Route de Sainte-Anne du Portzic, 29601 Brest Cedex, France.

AN/SQQ-32 sonar is to be installed on US Navy minesweepers

6563.353

AN/SQQ-89 INTEGRATED COMBAT SYSTEM

The AN/SQQ-89 provides an integrated detection, location, tracking and fire control system for surface vessels. The system is designed to detect, classify and correlate inputs from all the ship's ASW sensors. There are four variants of the AN/SQQ-89; the (V)1, (V)2, V(3) and V(4) which will be installed aboard various classes of surface vessels in the US Navy.

The AN/SQQ-89 will include the following systems: the AN/SQS-53B or C hull-mounted low-frequency sonar (**2536.253**), the AN/SQR-19 Tactical Towed Array Sonar (**4463.253**), the AN/SQQ-28 LAMPS Mk III acoustic processing system, the Mk 116 Mod 5, 6 or 7 ASW Control System, and the Sonar In Situ Mode Assessment System (SIMAS). The AN/SQQ-28 processes information from all the active or passive sonobuoys carried by the LAMPS Mk III. The Mk 116 correlates information from the ship's different sensors and develops fire control solutions for its ASW weapon systems. SIMAS is an acoustical prediction system which predicts environmental noise conditions. The AN/SQQ-89(V)2 will not carry either the AN/SQS-53 or the Mk 116.

STATUS

The US Navy sees the AN/SQQ-89 as one of its major ASW systems for the late 1980s and early 1990s. The first two production systems were installed in late 1986 and it is planned to install one of the four variants on up to 130 ships by 1995. 23 systems are being procured in 1988 and 19 in 1989.

CONTRACTOR

General Electric Company, Electronic Systems Division, Farrell Road, Syracuse, New York 13201, USA – development/production.

3115.453

DE1164 SONAR

The DE1164 sonar consists of the Raytheon DE1160 hull-mounted sonar (**1807.453**) augmented by a fully integrated variable-depth sonar sub-system. All sonar functions of the DE1164 are identical to those of the DE1160. However, the DE1164 provides transmission and reception via various combinations of the hull-mounted and/or the towed variable depth sonar (VDS) transducer arrays. Addition of the VDS sub-system improves overall sonar flexibility, and allows the VDS transducer array to operate at acoustically favourable depths and in a much quieter environment.

In addition to the components of the full DE1160 hull-mounted system the DE1164 includes one extra cabinet of electronics, a VDS towed body with associated cable, and the electro-hydraulic mechanism associated with launching, towing, and retrieving the VDS body. The VDS handling equipment provides for one-man operation for launching and retrieving and unattended no-power towing. For reliability, two independent hydraulic power supplies are provided, either of which may support the entire operation; an emergency retrieval system is available as an option. VDS body weight, cable size and length and careful attention to drag provide a VDS depth capability greater than 200 metres at 20 knots of ship's speed.

Both the hull-mounted and VDS arrays use the common set of DE1160 transmitting, receiving, and display electronics. Selection of the particular combination of transmit/receive array functions is ordered by the operator via the sonar console input keyboard; the system computer then sets up the required sequence. Alpha-numeric symbols on the display inform the operator about which particular array is in use during any specific ping-cycle.

During the normal operation the power requirements of the DE1164 are identical to those of the DE1160C. A maximum of 50 kW additional power is required from the 440 V, three-phase, 60 Hz power mains during high speed retrieval or launching. The hull-mounted sonar may be operated as a DE1160 when the VDS is stowed or being launched/recovered.

VDS configuration of DE1164 sonar on the Libeccio *of the Italian Navy*

STATUS

Ten DE1164 have been delivered to the Italian Navy and are in operational service on 'Maestrale' and 'Alpino' class ships. On order is one VDS trainer and two DE1164LF systems for use with the Italian Navy new construction 'Animoso' class guided missile destroyer. Under a licensing agreement parts of the DE1164LF system will be co-produced by ELSAG.

CONTRACTOR

Raytheon Company, Submarine Signal Division, Portsmouth, Rhode Island 02871, USA.

3116.453

DE1167 MODERNISATION SONAR

Raytheon's DE1167 family of sonars implements the proven features of the AN/SQS-56/DE1160 systems (**1807.453** and **3115.453**) using advanced microprocessor architecture and state-of-the-art display and transmitter technology. The DE1167 series is based on large modules and air-cooled cabinets, which permit the production of smaller, simpler, cheaper sonar systems, and facilitate in-country manufacturing/repair participation where required. The DE1167 family is designed to satisfy the requirements of most ASW platforms and missions. Configurations include hull-mounted, VDS and integrated HM/VDS systems featuring a 12 kHz VDS and either 12 kHz or 7.5 kHz hull-mounted 36-stave transducer arrays. Configurations using 48-stave arrays, or a frequency lower than 7.5 kHz are possible.

Like the Raytheon DE1160B/C and DE1164 sonars, the DE1167 features primarily digital electronics and an advanced control and display system. The standard inboard electronics consist of three cabinets and a single operator console. Outboard units consist of a transducer array and 2.74 m long dome for the hull-mounted 12 kHz installation (4.17 m long dome for 7.5 kHz) and/or the VDS winch, overboarding assembly, control station, hydraulic power supply and towed body for the 12 kHz variable depth sub-system.

The DE1167 HM is an active/passive, preformed beam, omni and directional transmission sonar which uses three non-interfering 600 Hz wide FM transmission bands centred at 12 or 7.5 kHz and a spatial polarity coincidence correlation (PCC) receiver. The passive mode, which is selected automatically when transmissions are stopped, is primarily useful for torpedo detection. Optional items in production include a performance prediction sub-system, an auxiliary half-frequency passive receiver, an auxiliary display, a remote display, and a training/test target remote control unit. Signal reception and beamforming are accomplished by broadband analogue circuitry followed by clipper amplifiers for perfect data normalisation. Detection processing, display processing, system control/timing and waveform generation are done digitally. Two microprocessors perform display ping history, cursor ground stabilisation, and target motion estimation functions for torpedo direction. The modular air-cooled 12 kW transmitter uses highly efficient class A/D power transistor techniques.

Several operational features are unique to a sonar of this size/range. First, the display processing incorporates a ping history mode through which the sonar data obtained in three of the previous ping cycles may be retained on the viewing surface, allowing the operator to readily differentiate between randomly spaced noise events and geographically consistent, valid acoustic reflectors. Second, the clipped PCC processing permits accurate thresholding of all signals such that the false alarm rate, or number of random noise indications on the screen, remains relatively low and constant over all variations in background noise and reverberation levels, further facilitating contact detection. Third, ground-stabilised cursors and target motion analysis permit rapid determination of contact motion over the bottom, an excellent clue as to the nature of the contact. These three operational features, combined with extremely accurate tracking displays, a built-in fault detection/localisation sub-system, performance verification software and test/training, result in a high performance system which is easy to operate, maintain and support.

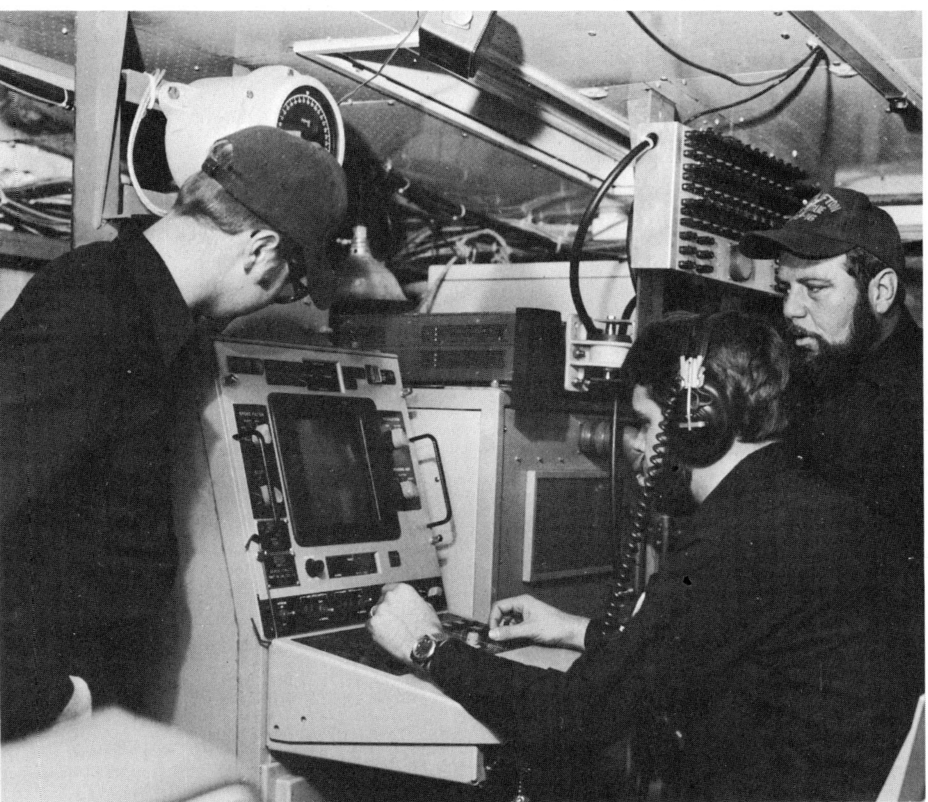

Operator's display/control console of Raytheon DE1167 sonar

CHARACTERISTICS

Centre frequency: 12 kHz (HM) and (VDS), 7.5 kHz (LF) (HM)

Source level: Triple 227 dB (HM), omni 217 dB

Pulse type: 600, 2000 Hz FM sweep 100, 200, 50, 6 ms pulse lengths

Receiver type: Spatial polarity coincidence correlator (PCC) between 36 pairs of half beams

Beam characteristics: 36 sets of right and left half beams for active and passive detection. Selectable 10°H × 13°V sum beam for audio listening. 1.25° bearing interpolation for display

Active display: 300 range cells, 288 bearing cells; single and multiple echo history; 4 intensity levels

Passive display: Electronic bearing time recorder (EBTR) with medium time averaging. DIMUS-type LTA/STA with optional passive receiver

Track displays: Sector scan indicator (1000 yards × 10°) and target doppler indicator (1000 yards × ±60 knots)

Target data: Range: 8 yd (6.1 m) resolution. Bearing 1.25° resolution plus active search display – 0.1° and 3.3 yards on SSI display. ±60 knots of doppler at 1 knot steps on the TDI

Data format: Standard: NTDS ANEW (digital). Optional: NTDS slow, Fast Serial D/S synchro converters

Power requirements: Passive 800 W. Active 20 kVA (pulse) at 10% (max) duty cycle 440 V, 60 Hz, 3-phase, VDS launch/retrieve 75 kVA (max) 440 V, 60 Hz, 3-phase

Weights: Hull-mounted 12 kHz; 1500 kg (nominal); VDS 10 000 kg (nominal) with 200 m cable length

STATUS

Thirty-seven systems were under contract as of November 1987 and included 26 DE1167 for the Korean Navy, two DE1167LF/VDS systems for Spanish corvettes for the Egyptian Navy and five DE1167LF systems for the Italian Navy. ELSAG has implemented a licence agreement with Raytheon for manufacturing DE1167 sonars for the Italian Navy corvette programme. ELSAG has also completed delivery of a multi-application sonar trainer (MAST) built in co-operation with Raytheon (**5754.293**).

CONTRACTOR

Raytheon Company, Submarine Signal Division, Portsmouth, Rhode Island 02871, USA.

6864.453

DE1191 SONAR MODERNISATION

The DE1191 sonar upgrades and improves the performance of older AN/SSQ-23 and AAN/SQS-23 sonars which are being used by navies worldwide aboard numerous FRAM II destroyer class ships.

The DE1191 features advanced inboard electronics which, when coupled with the existing AN/SQS-23 dome and array, comprise a newly configured, lightweight system with the capability to outperform most current modern surface ship sonars. Replacement of the AN/SQS dated inboard electronics with DE1191 hardware, a solid-state transmitter and a modern receiver/display subsystem, eliminates weight totalling 12 tons and reduces on-board space requirements by 450 cu ft. Further, the DE1191 computer assisted receiver display, a slight modification of the proven DE1167 equipment, has notably superior detection range and operability features.

Modernisation can be achieved incrementally in two major stages; the solid-state transmitter, which is always installed first and the receiver/display later. Alternatively the two can be installed at the same time.

CONTRACTOR

Raytheon Company, Submarine Signal Division, Portsmouth, Rhode Island 02871, USA.

2580.253

EDO MODEL 610 LONG RANGE HULL-MOUNTED SCANNING RADAR

Designed for the long-range detection of submarines in deep and shallow water, the EDO Model 610 scanning sonar has two active consoles, enabling it to perform a search-while-track function. Facilities offered include a search capability in three 120° sectors, passive correlation, and reverberation processing. The transmitter and receiver beams are preformed. Output is available for a fire control system. All mode changes and range scale changes are controlled by console push-buttons, and displays include a doppler display on each of the active consoles and a passive sonar bearing-time recorder display.

DEVELOPMENT

Model 610 was developed by the EDO Corporation as a private venture starting in 1965. The first prototype was completed in 1966 and the first production model completed its sea trials in 1969. The Model 610 has been continuously improved, and the current 610E model is of all solid-state construction.

STATUS

Model 610 systems are in operational service with the Royal Netherlands Navy, the Italian Navy and other navies. Associated weapon systems include the Mk 44 torpedo (US), 375 rocket launcher (Sweden), Lanciabas (Italy), and Ikara (Australia).

CONTRACTOR

EDO Corporation, Government Systems Division, 14-04 111th Street, College Point, New York 11356, USA.

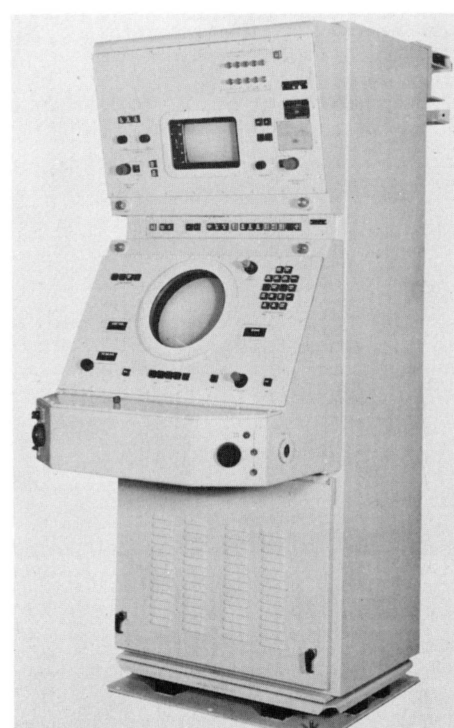

Model 610E sonar control console No 1

1635.453

BATHYTHERMOGRAPH DATA RECORDER

RO-308/SSQ-36

The BT recorder is an integral part of US Navy/ Lockheed P-3C Orion aircraft ASW system. The equipment converts seawater temperature information provided by the AN/SSQ-36 bathythermograph buoy-transmitter set (**3781.453** 1982-83 and earlier editions) and AN/AAR-72 radio receiving set (**3454.363**) to two output forms:

(1) a permanent record of the vertical temperature profile (temperature v depth) on a paper strip chart

(2) a parallel mode, eight-bit binary coded data word for delivery to the AN/AYA-8B data processing system (**1982.363**).

The BT buoy is dropped from an aircraft in the target area. Seawater is utilised as the activating agent and after an initial, predictable delay, the buoy releases a temperature sensing probe (TS probe). The TS probe is the variable element in a frequency generation circuit. A radio frequency signal, transmitted by the BT buoy is modulated at a frequency correlated to the temperature of the water. On board the aircraft the ARR-72 radio is tuned to the BT buoy carrier frequency. Water temperature information is converted to an audio frequency signal and delivered to the BT recorder.

CONTRACTOR

Fairchild Weston Schlumberger, Weston Controls, Archbald, Pennsylvania 18403, USA.

2581.253

EDO MODEL 700 SERIES SONARS

Latest equipment of this range of single operator sonars is the model 700E medium-range hull-mounted sonar. The Model 700/702 hull-mounted VDS uses common 700E electronics with a hull-mounted transducer and a lightweight VDS hoist. Selection of hull-mounted or VDS operation is by push-button on the operator's console. The Model 700/701 is a VDS which provides a capability to detect deep targets when bathythermal conditions are unfavourable for hull-mounted sonars.

The basic equipment has a 254 mm panoramic CRT display and a doppler display. All mode and range-scale changes are made by push-button controls on the operator's console.

DEVELOPMENT

Like the Model 610 (**2580.253**) the Model 700 series has been developed by the EDO Corporation as a private venture for sale on the international market. Model 700 sonars are in operational service with the Brazilian, Royal Netherlands and Japanese Navies. Associated weapon systems include the Mk 44 torpedo (US), the Bofors 375 launcher (Sweden), Lanciabas (Italy), and Ikara (Australia).

CONTRACTOR

EDO Corporation, Government Systems Division, 14-04 111th Street, College Point, New York 11356, USA.

EDO 700E VDS in stowed position aboard Brazilian Mk 10 frigate Niteroi

3495.253

EDO MODEL 780 SERIES SONAR

The EDO Model 780 Series sonar is a family of high performance computer-based sonars designed to maximise the ASW capability of ships ranging from high-speed patrol craft to ASW frigates. Configured as a modular system, the Model 780 Series can be assembled to match ship-size constraints and required ASW capability for hull-mounted or variable depth sonar operation. The following models are available:

780: 13 kHz variable depth sonar (VDS)
786: 13 kHz hull-mounted
795: 5 kHz hull-mounted
796: 7 kHz hull-mounted
7860: Combined 13 kHz hull-mounted with 13 kHz VDS
7867: Combined 13 kHz hull-mounted with 7 kHz VDS
7950: Combined 5 kHz hull-mounted with 13 kHz VDS
7960: Combined 7 kHz hull-mounted with 13 kHz VDS
7967: Combined 7kHz hull-mounted with 7 kHz VDS

Simultaneous active and passive detection capabilities are provided as well as an integral acoustic-intercept receiver. In the VDS configuration its lightweight, compact hoist permits installation on very small ships without sacrifice in performance. The completely solid-state design and integral automatic fault detection and localisation equipment provide for high reliability and ease of maintenance.

In all configurations the Model 780 Series employs the latest state-of-the-art computer-aided control and display equipment. Designed for a single operator, a digital computer, with high speed random access memory, provides direct operator control over every facet of the sonar, and selective display of all target data, system status, and related ship operational information. Centralised communication between the operator and computer is provided by an action entry panel and a track ball. Accuracy figures of the series are $\pm 0.5\%$ of the range scale for active range, $\pm 1°$ for active bearing and $\pm 0.5°$ for passive bearing.

The Model 780 Series electronics is comprised of four major units; the control indicator, a transmitter, a receiver, and a data storage and control unit. A preamplifier unit of small dimensions is separately housed for remote location. All units are air cooled.

The VDS hoist consists of two major units, the winch unit and the towed vehicle. A small operator's console is easily located in close proximity to the handling unit. The simplicity of the system allows towed body depth to be controlled from the bridge as well as the sonar control room. It is light enough for fitting in ships as small as 250 tons.

DEVELOPMENT

The Model 780 Series has been developed by the EDO Corporation for sale on the international market.

STATUS

Systems in the VDS configuration are currently deployed at sea, and the 780 has passed formal technical and operational testing and is fully operational. Two Far Eastern navies are using the hull-mounted configuration.

CONTRACTOR

EDO Corporation, Government Systems Division, 14-04 111th Street, College Point, New York 11356, USA.

Model 780 VDS hoist under test

Model 780 sonar system equipment comprising sonar control console, data storage and control, transmitter/power supply and sonar receiver

5644.453

EDO MODEL 900/910 UNDERWATER DETECTION SYSTEMS

The Model 900 underwater detection set is designed for use on submarines in mine avoidance and underwater obstacle location. It is composed of only three small units; a scanning assembly, a transmitter/receiver and a control indicator. The control indicator contains a 7-inch display which presents target data to the operator in ship-centred plan-position format. The scanning assembly contains the transmitting and receiving transducers, and an electrical drive system to train the transducers over 340° of azimuth. The transmitter/receiver unit incorporates a frequency analyser for high resolution of target data.

The model 910 is a small lightweight system for use on board small surface vessels engaged in minesweeping operations. In all other respects it is basically the same as the type 900.

CONTRACTOR

EDO Corporation, Government Systems Division, 14-04 111th Street, College Point, New York 11356, USA.

6561.493
MODEL 4235 ROV DOPPLER SONAR

Model 4235 is a lightweight Doppler sonar designed for use on remotely operated vehicles (ROVs) and other submersibles where weight and size are critical. It incorporates a significant feature, an operating frequency of 435 kHz which increases the basic resolution capabilities and helps in meeting minimum operating range, depth and speed requirements.

The transducer assembly consists of four sonar elements aligned at 90° increments around a high pressure-resistant rugged housing. The elements are 60° relative to the horizontal. Four preamplifier boards are within the housing, which is designed to withstand pressures of 9000 psi.

The electronic housing is 165 mm diameter (without pressure bottle), 660 mm long and weighs 9 kg. The transducer array with connector is 254 mm diameter, 254 mm deep and weighs 11.3 kg.

CONTRACTOR

EDO Corporation, Government Systems Division, 14-04 111th Street, College Point, New York 11356, USA.

3496.453
EDO MODEL 1102/1105 SUBMARINE SONAR

The EDO Model 1102/1105 is a passive and active sonar system designed for use on medium size submarines. The Model 1102 passive system is the primary detection configuration. In the passive mode, the sound horizon is scanned. Also available to the operator is a manual mode for selection of a specified bearing. The data received from the automatic channel is presented to the operator on a bearing/time recorder which provides a permanent record of all targets detected. An automatic tracking (ATF) channel may be assigned to any target detected for accurate tracking and bearing determination simultaneous with continuation of search. Target bearing outputs are provided for external use.

The Model 1105 active sub-system is designed as an integral supplement to the passive Model 1102 and is intended for use after detection in the passive mode. The Model 1105 uses the same array, preamplifiers, and beam formers of the passive system. Active data received by the tracking channel is fed to a FFT (Fast Fourier Transform) processor. Data is presented to the operator in doppler versus range format on a CRT display. Outputs of target range and range data are available for external use. This equipment is available either in the passive Model 1102 or passive/active Model 1102/1105 configuration.

DEVELOPMENT

The Model 1102/1105 was developed by the EDO Corporation as a private venture for sale on the international market. No information is available on any orders received.

CONTRACTOR

EDO Corporation, Government Products Division, 14-04 111th Street, College Point, New York 11356, USA

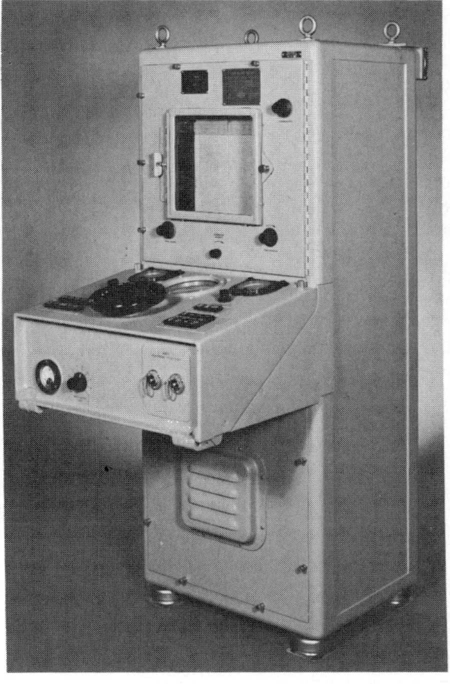

Model 1102 (passive) sonar control console

6865.453
MODEL 1110 SONAR SYSTEM

The Model 1110 uses a passive flank array to enhance the overall sensor capability of submarines. It is designed to be an adjunct to existing sonar, combat and fire control systems already planned or installed. The Model 1110 provides for long range detection, tracking and classification of surface and submarine targets while the existing hull sonar can be optimised for close-in detection.

The Model 1110 is a single operator system built on engineering principles well proven in the naval sphere. It consists of an operator console, processor cabinet, signal conditioning electronics and acoustic sensor arrays.

The system provides capabilities for expanded target detection and analysis through use of low frequency beamforming, split-beam correlation, bearing determination techniques and frequency analysis. It has the capability to track more than 15 targets simultaneously. Target tracking is accomplished by broadband and narrow band signal processing interpolation. Track processing allows the Model 1110 to automatically track many targets with minimal supervision. Selected targets can be assigned to 'high quality' track processors for improved position solution accuracy. The trackers can maintain track even in a crossing target environment without operator intervention. Tracking can be initiated by the operator using the cursor and soft-key controls.

The Model 1110 system single operator console has two high resolution colour displays. In existing submarines the control unit can be located adjacent to the host submarine sonar console and together will create an integrated sonar system. The microprocessor-based host submarine interface allows the Model 1110 to be integrated easily into the combat system of any new or existing submarine.

STATUS

Currently in pre-production evaluation.

CONTRACTOR

EDO Corporation, Government Systems Division, 14-04 111th Street, College Point, New York 11356-1434, USA.

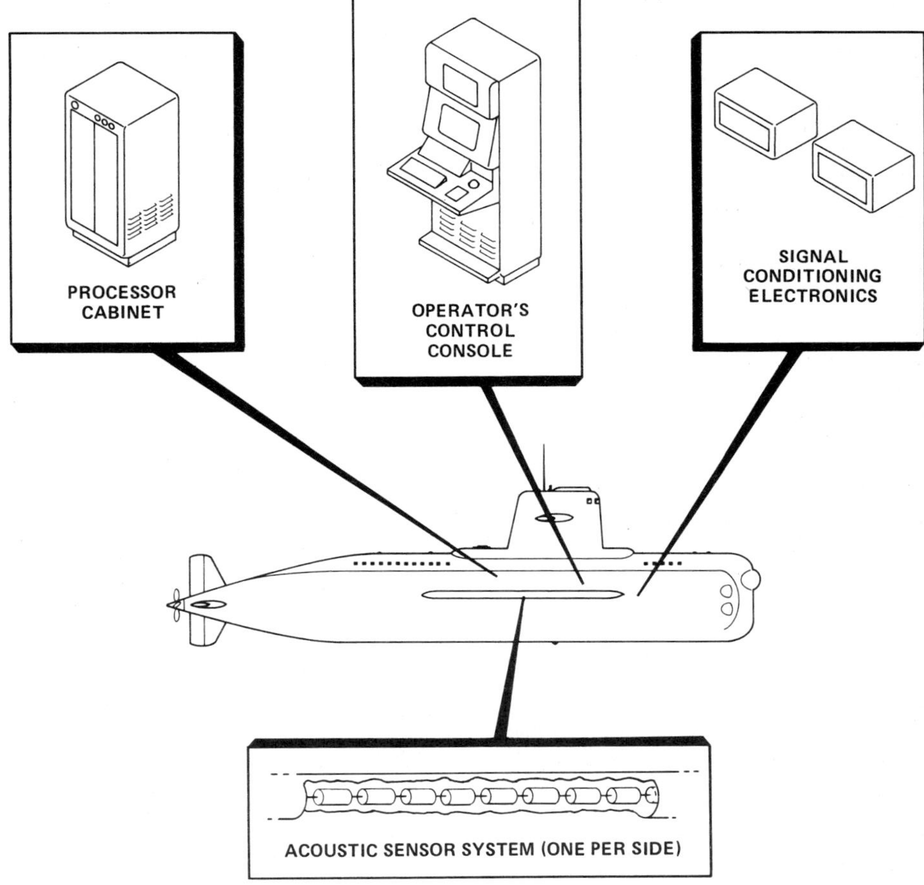

Model 1110 sonar layout

1637.353

AN/AQS-13 HELICOPTER SONAR

The AN/AQS-13 is a helicopter dunking system, the AN/AQS-13B and 13F models being in production. It is one of a series of such equipment which began with the Bendix AN/AQS-10 in 1955, the latest variant being the AN/AQS-13F.

The B system is a long-range, active scanning sonar which detects and maintains contact with underwater targets through a transducer lowered into the water from a hovering helicopter. Opening or closing rates of moving targets can be accurately determined and the system also provides target classification clues.

The AN/AQS-13B has significant advantages in operation and maintenance over earlier systems. To aid the operator, some electronic functions were automated to eliminate several controls. Maintenance was simplified by eliminating all internal adjustments and adding built-in-test (BITE) circuits.

These advantages were brought about by the use of the latest electronic circuits and packaging techniques which also reduced system size and weight.

To enhance detection capability in shallow water and reverberation-limited conditions while essentially eliminating false alarms from the video display, Bendix developed an adaptive processsor sonar (APS) for the system.

The APS is a completely digital processor employing Fast Fourier Transform (FFT) techniques to provide narrow-band analysis of the uniquely shaped CW pulse transmitted in the APS mode. The display retains the familiar PPI readout of target range and bearing but APS adds precise digital readout of the radial component of target doppler.

With APS, processing gains of greater than 20 dB with zero false alarm rates have been measured for target dopplers under 0.5 knot.

The higher energy transmitted with the longer pulse APS mode, combined with the narrow-band analysis, also substantially improves the figure of merit (FOM) in the non-reverberant conditions more typical of deep water operation. Measured processing gains for APS under ambient, wideband noise limited conditions exceeded 7 dB.

The AN/AQS-13F has been designed to provide rapid tactical response against the most advanced submarine threats. It is a sister equipment to the AQS-18 and is identical in many respects. A new state-of-the-art transducer, when lowered to depths of up to 1450 feet, permits instantaneous range improvements of over 100% compared to previous systems. Very high speed reeling allows a dip cycled to maximum depth to be completed in approximately three minutes. The powerful omni-directional transducer providing 216 $\pm$ 1db source level is integrated with a sensitive directional receiver array providing azimuthal resolution in a small rugged unit. The sonar data computer offers digital matched filter processing for 200 m/sec and 700 m/sec sonar pulses, as well as sonobuoy control and processing. The azimuth and range indicator and receiver provides a video display for the operator.

CHARACTERISTICS

Frequencies: 9.25, 10, 10.75 kHz

Sound pressure level: 113 dB ref 1 microbar (13B); 216 db (13F)

Range scales: 1, 3, 5, 8, 12, 20 k yds (0.9, 2.7, 4.6, 7.3, 11, 18.3 km)

Operational modes: Active 3.5 or 35 ms, MTI, APS, passive, voice communications, key communications (13A only). The 13F has the following operational modes: Active 3.5 ms and 35 ms rectangular pulse, 200 ms and 700 ms shaped pulse; MTI; Passive 500 Hz bandwidth 9 to 11 kHz; Communicate SSB at 8 kHz (voice)

Visual outputs: (13A) range, bearing; (13B) range, range rate, bearing, operator verification

Audio output: (13A) single channel with gain control; (13B) dual channel with gain control plus constant level to aircraft intercom

Recorder operation: Bathythermograph, range, aspect, MAD self-test

System weight: (13A) 373 kg; (13B) 282 kg; (13F) 280 kg

Operating depth: 1450 ft at 50 ft hover (13F)

STATUS

The AQS-13 is widely used by American forces and nearly 1000 sets have been supplied or ordered for naval helicopters of 15 foreign navies in Europe, Asia, Middle East, and South America. The AQS-13F has been selected by the US Navy for use in new SH-60F carrier based ASW helicopters. The AN/AQS-18 (**4456.353**) is a sister system to the AN/AQS-13F sonar featuring smaller size, lighter weight and increased performance. It has been developed and is in production, primarily for the Westland Lynx helicopter.

CONTRACTOR

Bendix Oceanics Division (a unit of Allied-Signal Inc), 15825 Roxford Street, Sylmar, California 91342, USA.

4462.353

AN/AQS-14 AIRBORNE MINEHUNTING SONAR

This is a helicopter towed side-looking multibeam sonar, with electronic beam forming, all-range focusing and an adaptive processor, intended for use with the RH-53D Sea Stallion and MH-53E Sea Dragon helicopters used by the US Navy for mine clearance. The underwater vehicle is three metres long and has an active control system which enables it to be run at a fixed height above the seabed, or at a fixed depth beneath the water surface, as chosen by the operator in the helicopter. The tow cable is armoured and is non-magnetic.

Controls in the helicopter include: a TV-type moving window sonar display; underwater vehicle controls and status displays; system status indicators; and a magnetic tape recorder for sonar data recording. The system functions include all that is needed to locate, classify, mark, permanently record, and review records of mines, mine-like objects, and underwater features in the search area. Because the sonar employs multibeam techniques, a rapid search speed is possible. At lower tow speeds the towed sonar automatically reduces the number of transmitted beams. The system is adaptable to surface platforms, especially hovercraft and remotely operated drone vessels for route survey missions.

STATUS

In production for the US Navy MH-53E and RH53 helicopters, with over 30 ordered.

CONTRACTOR

Westinghouse Electric Corporation, PO Box 1488, Annapolis, Maryland 21404, USA.

AN/AQS-14 sonar deployed from US Navy helicopter

4456.353

AN/AQS-18 AIRBORNE SONAR

The AN/AQS-18 is a helicopter-borne, long range active scanning sonar. The system detects and maintains contact with underwater targets through a transducer lowered into the water from a hovering helicopter. Active echo-ranging determines a target's range and bearing, and opening or closing rate relative to the aircraft. Target identification clues are also provided.

The AN/AQS-18 is an advanced version of earlier dunking sonars made by Bendix and includes digital technology, improved signal processing and improved operator displays. The system consists of a small high density transducer with a high sink and retrieval rate, a built-in multiplex system to permit use of a single conductor cable, a 330 m cable and compatible reeling machine and a lightweight transmitter built into the transducer package. The Adaptive Processor Sonar (APS), which provides enhanced performance in shallow water areas, is an integral part of the system.

The AN/AQS-18 offers a number of improvements over earlier dipped sonars. These include increased transmitter power output giving longer range, high speed dip cycle time and reductions in weight of all units.

The APS increases detection capability in shallow water and reverberation-limited conditions while essentially eliminating false alarms from the video display. The APS is a digital processor which uses Fast Fourier Transform (FFT) techniques to provide narrow-band analysis of the uniquely-shaped CW pulse transmitted in the APS mode. The PPI display retains the normal read-out of target range and bearing.

The APS processing gain improvement over the normal AN/AQS-18 analogue processing is 20 dB for a two knot target and 15 dB for a five knot doppler target. The higher energy transmitted with the longer

AN/AQS-18 sonar deployed from West German Navy Westland Lynx Mk 88 helicopter

pulse APS mode, combined with the narrow-band analysis also improves operation in the non-reverberant conditions more typical of deep water. The gain improvement outside the high reverberation zone (ten knots or greater) under wideband noise limited conditions exceeds 7 dB.

CHARACTERISTICS

Operating depth: 330 m
Operating frequencies: 9.23, 10, 10.77 kHz
Sound pressure level: 215 dB ref 1 μPa at 1 yard (0.9 m)
Range scales: 1, 3, 5, 8, 12, 20 kyd
Operational modes: 3.5 or 35 ms MTI, passive, communicate (voice)
Visual outputs: Range, range rate, bearing, operator verification
Audio output: Dual channel with gain control, plus constant level to aircraft intercom system
Recorder operation: Bathythermograph, range, ASPECT, MAD, BITE
System weight: 252 kg plus 13.3 for APS

STATUS

In production and in operational use with the Federal Republic of Germany Navy, and selected by the US Navy in a sister version, the the AN/AQS-13F (**1637.353**) for use in the new SH-60F carrier based ASW helicopter. The South Korean and Taiwanese Navies are also believed to have specified the system for their helicopters.

CONTRACTOR

Bendix Oceanics Division (a unit of Allied-Signal Inc), 15825 Roxford Street, Sylmar, California 91342, USA.

3772.453

MICROPUFFS PASSIVE RANGING SONAR

The MicroPUFFS system is a high accuracy, long-ranging passive sonar designed for installation aboard submarines. The name suggests that this equipment is a derivative of the original US Navy PUFFS (passive underwater fire control feasibility study) hardware; the PUFFS designation was also applied to USN AN/BQG-1/2/3/4 anti-submarine warfare system. It was developed for the detection and tracking of submarine and surface targets, and provides automatic target tracking of up to four targets simultaneously, instantaneous and continuous ranging solution, fire control range and bearing output and secure operation. The basic technique is signal enhancement by cross correlation of the signals received at the arrays with conversion of two measured time differentials into bearing and range data.

The system consists of three hydrophonic baffled arrays mounted on each side of the submarine. The arrays are connected to a display control console, which utilises an analogue processor, digital processor and AN/UYK-26 computer. The console also includes built-in fault localiser equipment. The system is built to US military specifications, and uses standard electronic modules throughout.

STATUS

MicroPUFFS systems are currently installed aboard six 'Oberon' class submarines of the Royal Australian Navy. A subsequent model has been built for the US Navy for test installation on the USS *Barb*.

CONTRACTOR

Sperry Corporation, Electronic Systems, Great Neck, New York 11020, USA.

1983.453

AN/BQQ-5 SONAR

The BQQ-5 sonar system is the principal sensor system of the US Navy's nuclear attack submarines, the SSN-688 class, all of which have been fitted in addition to retrofitting all 'Permit' and 'Sturgeon' class SSNs during regular refits. The BQQ-5 is a digital, multi-beam system employing both hull-mounted and towed acoustic hydrophone arrays. The latter of the two arrays was produced in two types; initially a manually detached and attached receiver array connecting to a fixed towing point was employed, replaced later by a towed array that is retractable and housed along the submarine's hull. Both fixed tow and retractable array versions have provision for mechanically cutting the array adrift in the event of an emergency, except for the first seven equipped with the fixed tow point. The polyethylene-covered tow cable, which has a maximum length of about 800 m, is 9.5 mm in diameter, and the array at the end of the cable where the hydrophones and electronics are located is 82.5 mm in diameter. The array is tapered fore and aft to minimise flow noise. Drag is stated to account for a maximum reduction in speed of 0.5 knot, with no serious inhibition on submerged manoeuvres and little adverse effect on surface manoeuvres with the one exception of those entailing going about.

During the late 1970s the US Navy became concerned that the BQQ-5 would not be able to handle the developing threats and embarked on a programme to update the capabilities of the system. An improved control display console has been developed and sonars that have been upgraded with the new consoles are known as the AN/BQQ-5B. The other major piece of equipment being developed under this programme is the AN/BQQ-5C(V) Expanded Directional Frequency Analysis and Recording System which replaces the standard signal processors. This equipment is expected to become operational in 1988/89 and will upgrade the system from BQQ-5A or BQQ-5B.

The US Navy has also been developing a thin-line tow array and handling system and other modifications for the BQQ-5 system. Although most of this programme is classified it is understood that the development is complete and will enter service in 1988. Systems upgraded will then be known as the AN/BQQ-5D/TBX.

The AN/WLR-9A acoustic intercept receiver, produced by Norden, has been successfully evaluated and forms an integral sub-system of the BQQ-5 and BQQ-6 on new attack and Trident missile submarines. The sub-system features a CRT display, a digital readout and a remote unit for the submarine commander. A 'sensitivity improvement' kit for the AN/WLR-9A has been developed by Norden and deliveries to the US Navy are in progress.

STATUS

The AN/BQQ-5 underwent extensive developmental testing during 1972 and 1973, and was approved for production later in 1973.

Over the period 1973-1986 a number of companies including IBM, Raytheon, Gould, Bendix, Norden and Tracor have been awarded contracts for systems, sub-assemblies and technical support. In December 1986, IBM was awarded a $54 million contract for upgrade kits to bring the AN/BQQ-5 to its latest configuration.

CONTRACTORS

IBM, Federal Systems Division, 10215 Fernwood Road, Bethesda, Maryland 20034, USA.

Raytheon Company, Portsmouth, Rhode Island, USA.

4629.453

AN/BQQ-6 SONAR

The AN/BQQ-6 is an advanced active/passive sonar set developed for the Trident submarine. Forming part of the Trident command and control system it is an integrated complex of command, control, communications and ship defence equipment.

The AN/BQQ-6 is based on the AN/BQQ-5 (**1983.453**) and has many of the same parts. The primary detection group is a digital integrated system employing spherical array, hull-mounted line array and towed array sensors, with an active emission acoustic intercept receiver and high frequency active short range sonar. In the passive mode the system used is identical to that of the AN/BQQ-5. Support equipment has been added to provide for underwater communications, environmental sensing, magnetic recording and acoustic emergency devices. All other technical details are classified.

STATUS

Installed on 'Ohio' class Trident submarines. In December 1986, IBM was awarded a $42 million contract for production of AN/BQQ-6 systems. Various modification and upgrading programmes are on-going in the fields of improved signal processing and towed array technology.

CONTRACTORS

IBM, Federal Systems Division, 10215 Fernwood Drive, Bethesda, Maryland 20034, USA.

Gould Inc, Defense Electronics Division, Glen Burnie, Maryland, USA.

1984.453

AN/BQR-15 SONAR

The BQR-15 is a towed array sonar used by US Navy fleet ballistic missile submarines (SSBNs). Because the array is towed behind the submarine, these sonars provide the first long-range passive detection capability astern. In addition, since the sensors are removed from the submarine hull and internal ship's noise, detection of other submarines is possible at greatly increased ranges. As installed, the BQR-15 includes specially designed hydraulically operated handling equipment that permits the cable array to be partially or completely streamed, retrieved, and adjusted automatically while the SSBN is submerged. The equipment forms part of the SSBN Unique sonar system which also includes the AN/BQR-19 receiving set and the AN/BQR-21 (**1985.453**).

DEVELOPMENT

In May 1972 contracts were awarded for the design and development of SPAD (signal processing and display) systems to exploit more fully the signal gathering potential of the BQR-15 towed array sonar, but difficulties and delays led to the cancellation of this effort in March 1974. Since then funds have been allocated for procurement of a multiple channel signal processing and display equipment for use on FBM submarines and at FBM training facilities with the BQR-15 sonar system.

STATUS

In operational service. A modified array entered production during 1984.

CONTRACTOR

Western Electric Company, Winston-Salem, North Carolina, USA.

1985.453

AN/BQR-21 DIMUS SONAR

The BQR-21 DIMUS sonar is a hull-mounted preformed beam passive sonar with improved digital signal processing for the detection and tracking of quiet nuclear submarines. It is intended for fitting on all pre-Trident SSBNs and selected SSNs, and will be a direct replacement for the AN/BQR-2. The BQR-21 will incorporate the same array configuration as the earlier equipment but will embody significantly enhanced signal processing capabilities. In addition to improvements in the electronics, performance will be improved through platform noise reductions which are being made in the bow dome of the submarine. These include the addition of an array baffle.

STATUS

A total of 56 of this equipment for fitting in fleet ballistic missile submarines and certain attack submarines has now been procured. System delivery began in August 1977 and was completed in December 1979. Two engineering models have been refurbished and delivered for use as trainers. In November 1982 Honeywell was awarded a $5 million contract for engineering support. The system is no longer in series production.

CONTRACTOR

Honeywell Inc, Training and Control Systems Operations, 1200 East San Bernardino Road, West Covina, California 91790, USA.

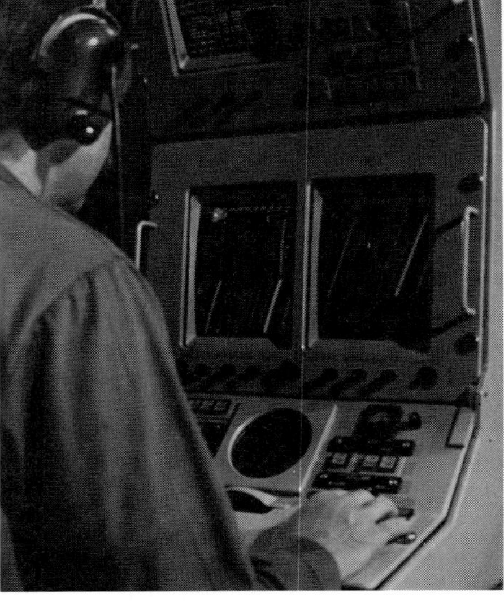

AN/BQR-21 DIMUS sonar display console

1986.453

AN/BQS-13 SONAR

The BQS-13 sonar system is the primary sonar in USN 'Permit Sturgeon' SSN class nuclear attack submarines. The system was service approved in December 1971 and is in operation. A system addition providing new functional capability was service approved in late 1974. This addition, the BQS-13 multipurpose sub-system (BQR-24), is being procured for a limited number of SSNs.

DEVELOPMENT

Acceptance testing of the multipurpose sub-system began in late 1973 and technical and operational testing were conducted aboard the *Archerfish* during 1974. The BQS-13 multipurpose sub-system (BQR-24) was service approved in late 1974. The extended operational test programme delayed procurement to fiscal year 1975/76 when 11 systems plus support were procured for $25.3 million. These sub-systems provide interim improvement in sonar and fire control capability for those submarines not scheduled to receive the BQQ-5 sonar (**1983.453**) until late in that programme. The BQR-24 has been installed only in submarines where it will have a useful life of at least three years prior to receiving a BQQ-5 sonar system.

Commencing in October 1979, Raytheon was awarded a series of US Naval Sea Systems Command contracts calling for refurbishment and conversion of AN/BQS-11, 12, and 13 sonar transmission sub-systems to AN/BQQ-5 transmission sub-system configuration. These sub-systems are part of the AN/BQQ-5 system.

CONTRACTOR

Raytheon Company, Submarine Signal Division, Portsmouth, Rhode Island 02871, USA.

6872.453

AN/BQS-14A SUBMARINE SONAR

The AN/BQS-14A is an upgraded development of the AN/BQS-14 ice detection sonar developed during the late 1960s. In 1979 the US Navy initiated a programme to upgrade the system to provide an improved mine detection and avoidance capability to the 'Sturgeon' and 'Los Angeles' class of nuclear-powered attack submarines. The improved system has been introduced into the fleet as an upgrade and integrated with the AN/BQQ-5.

CONTRACTORS

Hazeltine Corporation, Greenlawn, New York, USA.

IBM, Federal Systems Division, Manassas, Virginia, USA.

6247.453

AN/BQS-15 SUBMARINE SONAR

The AN/BQS-15 is an under-ice, close contact avoidance, submarine sonar system designed to locate and track a target in the active or passive mode. It is intended primarily for operation in heavily mined areas to provide improved mine detection and avoidance capability to the SSN 637/688 class submarines of the US Navy. Using hardware elements in common with the submarine active detective sonar and the AN/BQQ-5, the AN/BQS-15 is being introduced into the submarine fleet as an upgrade, integrated with the AN/BQQ-5, to become part of a submarine advanced combat system.

CONTRACTOR

Ametek Inc, Straza Division, 790 Greenfield Drive, El Cajon, California 92022, USA.

6246.463

SIPPICAN EXPENDABLE ASW DEVICES

Sippican Ocean Systems designs and manufactures a wide variety of expendable devices for use in anti-submarine warfare, which includes sonobuoys, oceanographic instrumentation, communication devices, and a mobile ASW training target.

XBT/XSV Expendable Bathythermograph/Sound Velocity Profiling Systems

The Sippican XBT/XSV system uses inexpensive, expendable probes to obtain vertical profiles of temperature and sound velocity. The temperature data obtained by XBT probes may be used to compute sound velocity, although the XSV probes measure sound velocity directly. Whether computed or measured directly, the sound velocity data obtained assists in the prediction of sound propagation and is crucial to the effective use of sonars. An XBT/XSV system includes XBT or XSV probes, a deck-mounted or through-hull launcher, and a Mk 8 microprocessor-based data processor-recorder.

AN/SSQ-36 aircraft launched expendable bathythermograph (AXBT)

The temperature data obtained by the XBT may be used to compute sound velocity wherever salinity is constant. In areas where salinity may be highly variable, eg the Mediterranean and the Arctic, temperature data may not be sufficient to calculate velocity and to predict adequately acoustic propogation. In these areas sound velocity must be measured directly with the XSV probe. The XSV uses a unique sing-around transducer to obtain this direct measurement.

Both XBT and XSV probes obtain data in real time. Following launch, data collection begins at the surface and as the probe descends wire unreels from a spool located within the probe, as well as unreeling from another spool in the launch cannister. This enables the ship to use the system without restrictions on speed or manoeuvrability.

SUS Mk 84 underwater sound signal

The Mk 8 data processing system, US Navy nomenclature AN/BQH-7, is linked to the descending probe by the hard wire link. The processor provides profiles of temperature or sound velocity to three outputs; digital data stored on magnetic tape, analogue chart paper trace, RS-232 digital interface to external data systems.

Submarine-Launched Expendable Bathythermograph/Sound Velocity (SSXBT/SSXSV) Systems

Profiles of temperature and sound velocity versus depth may also be obtained by a submerged submarine using the SSXBT/SSXSV system. This is similar to the surface system except that the probes are launched through the signal ejector of the submarine and are carried to the surface by a buoy. The buoy releases the probe just before reaching the surface, and the probe then descends measuring temperature or sound velocity on the way. The dual spool technique is also used as in the surface-deployed system. A Mk 8 recorder processes the data in real-time.

AN/SSQ-36 Aircraft-Launched Expendable Bathythermogrpaph (AXBT)

This is the standard sonobuoy used by the US Navy to obtain temperature and computed sound velocity from fixed-wing aircraft and helicopters. The AXBT is packaged in a standard sonobuoy configuration and may be launched from all A-size sonobuoy launch tubes. The AXBT uses an RF link to transmit data in real time over frequencies 170.5, 172 and 173.5 MHz (standard sonobuoy channels 12, 14 and 16). The equipment interfaces directly with standard sonobuoy receivers and data processing equipment, and may be modified to provide a direct measurement of sound velocity.

SUS MK 84 underwater sound signal

The SUS Mk 84 Underwater Sound Signal is an expendable, electro-acoustic device which provides one-way air-to-submarine communications. It may be dropped or deployed from fixed-wing aircraft, helicopters and surface ships, and may also be used to simulate the drop of an ASW weapon during a tactical exercise. It may be launched from altitudes up to 10,000 feet and at speeds up to 380 knots.

The SUS Mk 84 is able to transmit five different acoustic signals, each of which may convey a predetermined message to the submarine. The appropriate signal is selected immediately prior to launch. The submerged submarine may use either passive sonar or underwater telephone to receive the SUS output.

Expendable Current Profiler (XCP)

The XCP is an expendable instrument capable of obtaining profiles of horizontal current direction and speed. The system obtains ocean current data to depths of up to 1500 m, with an accuracy of ±1 cm/sec RMS. It may be deployed from either a surface or airborne platform since it uses an RF link to transmit data. Three channels are available for RF transmission: 170.5, 172 and 173.5 MHz. A complete XCP system includes expendable probes, a Sippican Mk 10 Digital Data Interface and a Hewlett Packard 9816S computer. This latter may be substituted by any computer which accepts the IEEE-488 output and which has a minimum of 384 kbytes of memory available for program software.

Expendable Mobile ASW Training Target (EMATT)

This system is under development for the US Navy and will supplement the actual use of submarines during training. When completed, the development will provide a sophisticated, lightweight, expendable training target easily deployed from surface ships or aircraft.

ASW Sonobuoys

Sippican is a qualified US Navy manufacturer of the AN/SSQ-77A VLAD sonobuoy (**3084.453**), which is the most sophisticated passive acoustic sonobuoy currently in production. The VLAD buoy features a 99-channel selectable RF transmitter, directional and omni-directional hydrophones, and array beamforming electronics.

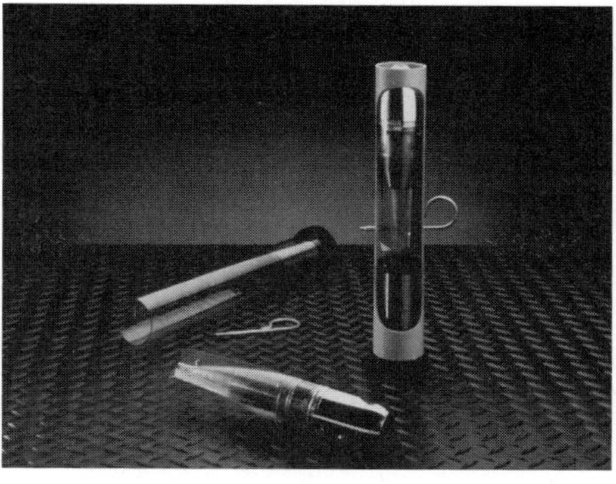

XBT/XSV expendable profiling system

AN/SSQ-103 Low-cost Sonobuoy

Sippican has been developing the AN/SSQ-103 low-cost sonobuoy since 1984. This is a passive device and is expected to be manufactured in very large quantities at a fraction of the unit cost of conventional buoys.

In April 1987, Sippican received a contract from the US Navy for 8250 AN/SSQ-103 sonobuoy launching packaging assemblies. Each assembly contains six buoys in the same volume occupied by one conventional sonobuoy.

CONTRACTOR

Sippican Inc, Seven Barnabas Road, Marion, Massachusetts 02738, USA.

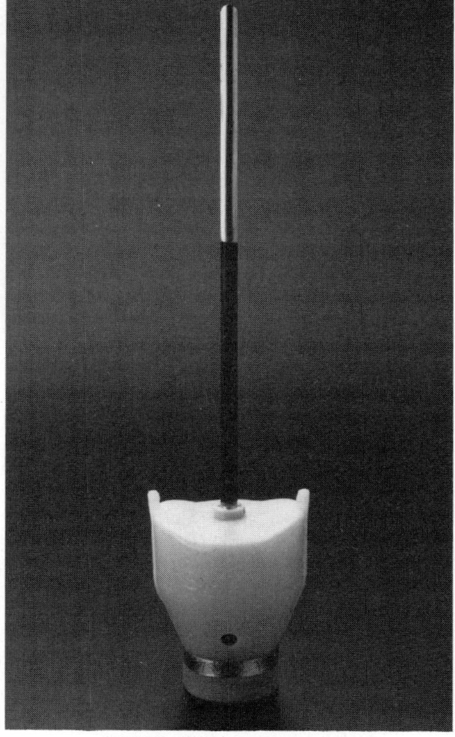

AN/SSQ-103 low cost sonobuoy

1987.453

AN/SSQ-41B SONOBUOY

The Sparton AN/SSQ-41B is an omni-directional, passive sonobuoy with a multiple life and depth capability. In volume production since 1963, the series has been the mainstay of US Navy sonobuoy operations but is gradually yielding this role to the newer, directional AN/SSQ-53B.

The AN/SSQ-41B, extensively redesigned from the earlier AN/SSQ-41, provides an audio bandwidth expanded to ten kHz and improved dynamic range capabilities. The depth selection is 60 or 1000 feet and life selection is one, three, or eight hours.

The AN/SSQ-41B may be launched from an aircraft at airspeeds between 30 and 425 knots and from altitudes of 100 to 40 000 feet (30.5 to 12 200 metres).

After a controlled and stabilised descent and upon impact with the water, the termination mass and the hydrophone are released and descend to a preselected operating depth of either 60 or 1000 feet (18.3 or 305 metres). On contact with seawater, the battery is activated and the sonobuoy becomes fully operative within 60 seconds.

Modulation is accomplished in the crystal oscillator stage by a variable-capacity diode. Two frequency-doubling amplifiers multiply the RF oscillator-doubler frequency to the desired operating frequency of the sonobuoy, which is preset to one of thirty-one channel frequencies within the 162.25 to 173.50 MHz band. Available in both standard and low noise versions.

CHARACTERISTICS

Type: Passive
Audio frequencies: 10 – 10 000 Hz
Depth: 18.3 or 305 m
Size: A
Weight: 9 kg
Operating life: 1, 3, or 8 h
Launch altitude: 30.5 – 12 200 m
Launch speed: 30 – 425 knots
RF channels: 31
RF power: 1 W

STATUS

Although large scale production was contracted through 1982, the US Navy has discontinued further development. This unit will continue to be available.

CONTRACTORS

Sparton Corporation, Electronics Division, 2400 E Ganson Street, Jackson, Michigan 49202, USA.

Magnavox, 1313 Production Road, Fort Wayne, Indiana 46808, USA.

1988.453

AN/SSQ-47B SONOBUOY

The AN/SSQ-47B is an active, non-directional, A-size sonobuoy. The operational function is that of detection, classification, tracking, and location of submarines.

Sonobuoy AN/SSQ-47B provides an active sonar capability for fixed-wing aircraft. It can be operated from a minimum range of zero to ten nautical miles at an altitude of between 500 and 10 000 ft and in sea conditions up to sea state 5. The AN/SSQ-47B permits launching and operation of up to six sonobuoys, either individually or simultaneously, without encountering RF or sonar interference.

CHARACTERISTICS

Type: Active
Sonar modes: Automatic keyed CW
Depth: 18.3 or 244 m
Size: A
Sonar channels: 6 (HF)
RF channels: 12
RF power: 0.25 W
Weight: 14.5 kg

STATUS

In operational service.

CONTRACTOR

Sparton Corporation, Electronics Division, 2400 E Ganson Street, Jackson, Michigan 49202, USA.

1990.453

AN/SSQ-53B DIFAR SONOBUOY

The AN/SSQ-53B DIFAR (Directional Frequency and Recording) sonobuoy is the US Navy's primary sonobuoy sensor. As compared to earlier passive buoy types, the SSQ-53B provides target bearing information as well as improved acoustical sensitivity, particularly in the low frequency ranges.

The SSQ-53B is the result of a US Navy-sponsored development programme to improve the operational capabilities of the earlier SSQ-53. Changes include a triple capability (100, 400 or 1000 ft (30, 120 or 300 m)), an extended launch envelope and improved low frequency performance. The sonobuoy's electrical and mechanical design was extensively revised to improve reliability and to lower production costs.

The AN/SSQ-53B may be dropped from an aircraft at indicated airspeeds of 30 to 425 knots and from altitudes of 100 to 40 000 ft (30 to 12 200 m). Descent of the sonobuoy is stabilised and slowed by a parachute assembly.

Immediately after water entry, the seawater activated battery system is energised. Buoy separation occurs, jettisoning the parachute assembly and erecting the VHF transmitting antenna. This allows the surface assembly to rise and separate from the sonobuoy housing. The housing serves as a descent vehicle and separates from the sub-surface assembly at the operating depth. Prior to launch the buoy is programmed via push-buttons to one of 99 channels, life and depth. Readout is on a LED display contained within the buoy.

CHARACTERISTICS

Type: Directional LOFAR
Size: A
Weight: 8.4 kg
Operating life: 1, 4 or 8 hours
Launch altitude: 100 to 40 000 ft
Launch speed: 30 to 425 knots
RF channels: 99
RF power: 1 W

STATUS

In production. Deliveries began in early 1984 and to date Magnavox has delivered, or has contracts for, a total of 899 027 SSQ-53B sonobuoys.

CONTRACTORS

Sparton Corporation, Electronics Division, 2400 E Ganson Street, Jackson, Michigan 49202, USA.

Magnavox, Fort Wayne, Indiana 46808, USA.

Nippon Electronic, Japan (licence).

Dowty Maritime Systems, UK (licence).

Plessey Electronics Ltd, UK (licence).

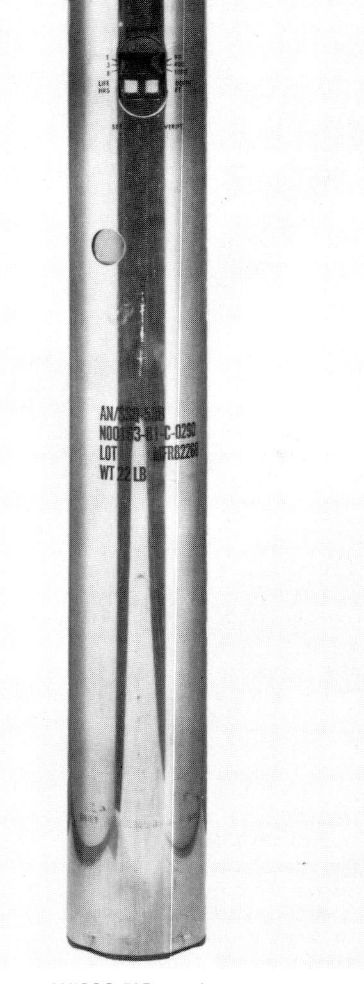

AN/SSQ-53B sonobuoy

3083.453

AN/SSQ-57A SONOBUOY

The AN/SSQ-57A sound reference sonobuoy features a hydrophone that affords exceptional performance stability under temperature and pressure extremes. The hydrophone is a piezoelectric ceramic segmented cylinder with a smooth frequency response to 20 kHz. The sonobuoy is calibrated to allow determination of underwater acoustic sound pressure levels over this wide frequency range.

The AN/SSQ-57A can be air-launched from altitudes between 150 and 10 000 feet (46 and 3048 metres) into sea conditions of up to sea state 5 and is effective at ranges up to ten nautical miles.

Immediately after entering the water the seawater-activated battery system is energised. Gas-filled cylinders inflate the sonobuoy float, jettisoning the parachute assembly and erecting the VHF transmitting antenna. This permits the surface assembly to rise and separate from the sonobuoy housing. The sonobuoy housing serves as a descent vehicle and separates from the sub-surface assembly at the operating depth.

Operating life is one, three or eight hours, selected before launch. At the end of the operating life, the transmitter is shut-off. The watertight sonobuoy housing is equipped with a seawater-soluble plug that dissolves to effect scuttling. The plug's dissolution rate varies as a function of water temperature but is never less than eight hours (maximum selectable operating life) or more than 20 hours.

The AN/SSQ-57A (XN-5) is a version of the AN/SSQ-57A that has a fixed operating depth of 1000 ft (305 m). Available in both standard and low noise versions.

CHARACTERISTICS

Type: Passive, calibrated audio
Audio frequencies: 10 – 20 000 Hz
Depth: 18 – 91 m
Size: A
Weight: 8.6 kg
Operating life: 1, 3, or 8 h
Launch altitude: 30 – 12 200 m
Launch speed: 30 – 425 knots
RF channels: 31
RF power: 1 W

STATUS

The AN/SSQ-57 has been in production since 1968. It is in widespread operational use in the US Navy and a number of other navies.

CONTRACTOR

Sparton Corporation, Electronics Division, 2400 E Ganson Street, Jackson, Michigan 49202, USA.

1991.453

AN/SSQ-62B DICASS SONOBUOY

The AN/SSQ-62B is the sonobuoy component of the Directional Command-Activated Sonobuoy System (DICASS). A high-performance sonobuoy, it detects the presence of submarines using sonar techniques under direct command from ASW aircraft. The AN/SSQ-62B can also determine the range and bearing of the target relative to the sonobuoy's position.

The DICASS sonobuoy is composed of three main sections, an air descent retarder, a surface unit and a sub-surface unit. The air descent retarder consists of a spring-loaded parachute release assembly and a parachute to retard descent immediately after launch. The surface unit receives commands from the controlling aircraft, via a UHF receiver, and sends target information to the aircraft, via a VHF transmitter. The sub-surface unit transmits sonar pulses in the ocean upon command from the aircraft and receives sonar target echoes for transmission to the aircraft.

Command signals are received by the sonobuoy and are accepted if the correct address code is identified by a decoder. The command capability includes depth selection, scuttle, and selection of transducer (sonar) transmission signals. The echoes from the selected activating signal are multiplexed in the sub-surface unit before being transmitted to the receiving station or aircraft.

The AN/SSQ-62B may be launched from fixed- or rotary-wing aircraft within the parameters specified in the launch envelope.

Upon impact with the water, the transducer is released for descent to its shallow operating depth. Immediately after entry, the float is inflated and the VHF-transmitter/UHF-receiver antenna is erected. When the float surfaces, the VHF transmitter begins emitting a continuous FM carrier signal. The sonobuoy is now operating in the passive mode and is ready to receive commands.

The main power source for the sonobuoy comes from a lithium battery pack instead of the more costly silver chloride batteries commonly used in sonobuoys. The sonobuoy is designed for economical volume production without compromising performance or reliability. Electronic design is exclusively solid state with maximum use of multifunction integrated circuits.

CHARACTERISTICS

Type: Commandable, omni-directional active, directional receive
Sonar modes: Pulse CW or linear FM
Depth: Commandable, 27, 119 and 457 m (89, 396, 1518 ft)
Size: A
Life: 30 minutes
Weight: 15 kg
Sonar channels: 4
RF channels: 31
RF power: 0.25 W

CONTRACTORS

Sparton Corporation, Electronics Division, 2400 E Ganson Street, Jackson, Michigan 49202, USA.

Magnavox, Fort Wayne, Indiana 46804, USA.

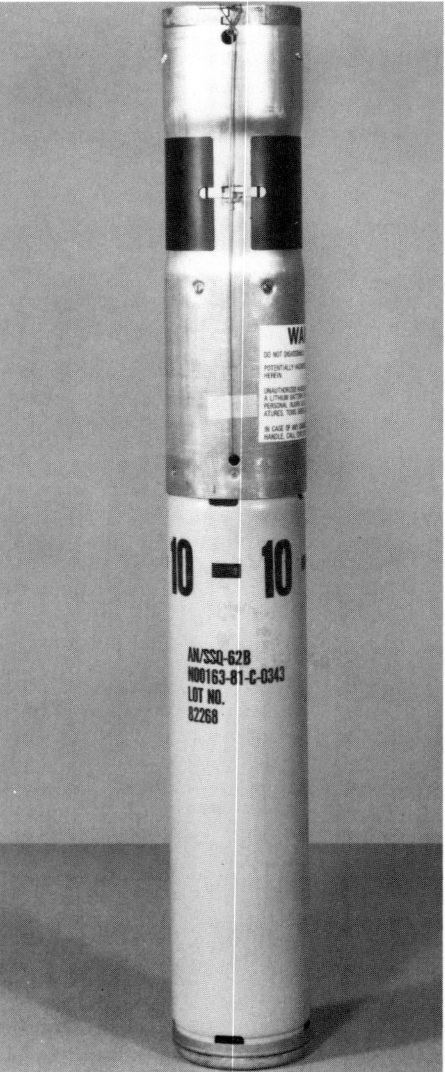

AN/SSQ-62B DICASS sonobuoy

3786.453

AN/SSQ-71 ATAC BUOY

The AN/SSQ-71 air transportable acoustic communication (ATAC) buoy is an expendable two-way sonic RF receive-transmit communications device. It is launched from surface vessels or either fixed-wing or rotary-wing aircraft. Incoming UHF signals are encoded into a multiple-tone format for sonic transmission; incoming sonic signals are likewise processed for VHF transmission. The AN/SSQ-71 borrows heavily from designs developed and refined in other Sparton sonobuoys programmes. The antenna system, for instance, which receives UHF and transmits VHF signals, is derived from the Sparton AN/SSQ-62 sonobuoy. The entire mechanical layout is based upon the AN/SSQ-53A, designed and produced in volume by Sparton.

STATUS

No longer in production.

CONTRACTOR

Sparton Corporation, Electronics Division, 2400 E Ganson Street, Jackson, Michigan 49202, USA.

6564.493

AN/SSQ-75 ERAP SONOBUOY

The AN/SSQ-75 Expendable Reliable Acoustic Path Sonobuoy (ERAPS) is a long-range active search sensor designed to exploit the long-range direct propagation mode known as the reliable acoustic path. It provides a capability to search actively for a submarine that is virtually undetectable by passive sonar methods. The long-range detection is gained by using low-frequency, high-power transmitted pulse and a volumetric receiving array of hydrophones. Range, bearing and Doppler are provided.

STATUS

In development. Preliminary trials have been carried out and production is expected in 1991-92.

CONTRACTOR

Bunker Ramo (a division of Allied Corporation), West Lake Village, California, USA.

3084.453

AN/SSQ-77 (VLAD) SONOBUOY

The AN/SSQ-77 VLAD (vertical line array DIFAR) is a passive, tactical search and surveillance sonobuoy designed to improve detection and tracking capability for the DIFAR system in a noisy, high-traffic environment.

The AN/SSQ-77 concept utilises a vertical line array of omni-directional hydrophones in place of the single omni-directional hydrophone used in the standard DIFAR unit. A directional DIFAR hydrophone mounted at the array phase centre provides target bearing data.

The signal format for the AN/SSQ-77 sonobuoy is identical to that for the AN/SSQ-53 and is compatible with the AN/AQA-7 processor and the Sparton TD-1135/A demultiplexer/processor/display. All beam-forming functions are accomplished within the sonobuoy, with provisions for sea-noise equalisation and omni-directional phase tracking.

The AN/SSQ-77 may be dropped from an aircraft at indicated airspeeds of 45 to 380 knots and from altitudes of 100 to 30 000 feet (30 to 9144 metres). Descent is stabilised and slowed by a parachute assembly.

An internal microprocessor electronically selects the desired RF channel and sonobuoy life. Verification of these settings is provided by an LED display readable through the SLC.

Lithium batteries are activated upon entry to the water. These detonate a squib, causing gas-filled cylinders to inflate the sonobuoy flotation bag and initiate deployment. Inflation of the flotation bag provides a pressure force, releasing the release plate in the top of the sonobuoy and causing the parachute to be jettisoned. This permits the surface assembly to rise and separate from the sonobuoy housing. The sonobuoy housing serves as a descent vehicle and separates from the sub-surface assembly at the operating depth.

CHARACTERISTICS

Type: DIFAR with VLA omni
Size: A
Weight: 13.1 kg
Operating life: 4 or 8 h
Launch altitude: 30 – 9144 m
Launch speed: 45 – 380 knots
RF channels: 99, electronically selected
RF power: 1 W

STATUS

In quantity production.

CONTRACTORS

Sparton Corporation, Electronics Division, 2400 E Ganson Street, Jackson, Michigan 49202, USA.

Magnavox, Fort Wayne, Indiana 46808, USA.

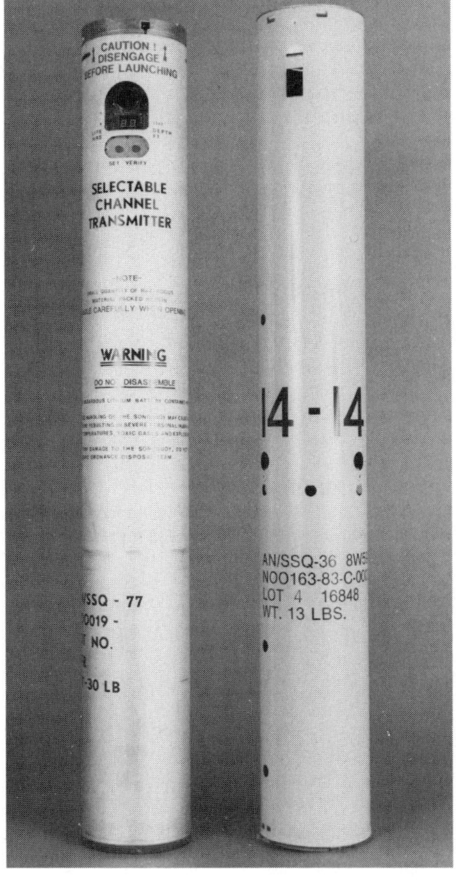

AN/SSQ-77 sonobuoy (left); AN/SSQ-36 sonobuoy (right)

3782.453

AN/SSQ-53C SPARTON DWARF DIFAR SONOBUOYS

The Sparton Dwarf DIFAR sonobuoy is the electrical and acoustical equivalent of a standard AN/SSQ-53B sonobuoy, but occupies only a third of the volume. The Dwarf DIFAR has the 4.87 inch (124 mm) diameter of a conventional A-sized sonobuoy but an overall length of only 12 inches (305 mm). The small size of the Dwarf DIFAR sonobuoy enables a threefold increase in the number of units an ASW aircraft can carry and is particularly valuable with small fixed-wing ASW aircraft and helicopters. Dwarf DIFAR sonobuoys are compatible with existing shipping containers and handling equipment, and the cost saving in shipping and handling is substantial. They can be incorporated into existing fleet operations with minimal changes to associated equipment and deployment philosophy.

The Dwarf DIFAR sonobuoy can be deployed from fixed-wing or rotary-wing aircraft at indicated airspeeds of 0 to 370 knots and from altitudes of 40 to 30 000 feet (12 to 9144 metres). Air descent is stabilised and slowed by a parachute assembly.

The major components of the Dwarf DIFAR sonobuoy are the external sonobuoy housing, the antenna/VHF assembly, the suspension assembly (which consists of the cable pack, the stabilising kite, the damper disc, and the compliant cable), and the integrated hydrophone and electronics assembly.

Upon entry to the water, the Dwarf DIFAR lithium-sulphur dioxide battery pack is activated by a seawater switch that detonates a squib, causing gas-filled cylinders to inflate the sonobuoy flotation bag, jettisoning the parachute assembly and erecting the VHF transmitting antenna. Inflation of the flotation bag provides the pressure force that activates the release plate in the top of the sonobuoy. This causes the parachute to be jettisoned and permits the surface assembly to separate from the sonobuoy housing and the surface housing.The sonobuoy housing serves as a descent vehicle and falls away from the sub-surface assembly at the operating depth, either 30 or 300 metres.

CHARACTERISTICS

Type: Passive
Depth: 30 or 300 m
Size: A diameter × 305 mm
Weight: 4.5 kg
Operating life: 1 or 4 h
Launch altitude: 12 – 9144 m
RF channels: Multi-channel

STATUS

In production.

CONTRACTOR

Sparton Corporation, Electronics Division, 2400 E Ganson Street, Jackson, Michigan 49202, USA.

6560.493

AN/SSQ-58A MOORED SENSOR BUOY

Sparton Electronics has recently announced the existence of its Moored Sensor Buoy, a multi-purpose, building-block system for 'at-sea' measurements. Two separate versions of the buoy are available, a passive system for harbour surveillance and an active system for use in test ranges, etc. Other applications of the buoy include a variety of oceanographic and environmental requirements. The buoy is retrievable and reusable and can also be configured in a free-floating version. Powered by commercially available batteries, or extended life alternative power sources, it has a long operational life. It is available with 99 VHF selectable frequency channels and is compatible with all sonobuoy receivers and processors.

STATUS

The buoy is being procured by the US Navy.

CONTRACTOR

Sparton Corporation, Electronics Division, 2400 East Ganson Street, Jackson, Michigan 49202, USA.

4459.453

AN/SSQ-86(XN-1) DOWN-LINK COMMUNICATION SONOBUOY

The AN/SSQ-86(XN-1) down-link communication (DLC) sonobuoy is a one-way communication device designed to transmit a pre-programmed message to a submerged submarine. Developed to meet the need for a reliable means of transmitting a message without revealing the receiving submarine's position, the DLC is compatible with existing sonobuoy launch platforms. The unit is packaged in a standard A-sized sonobuoy envelope and can be launched from any properly equipped fixed-wing aircraft or helicopter or from a surface ship.

The desired message is programmed into the DLC using a single push-button switch and four 7-segment LED displays. Four groups of three digits make up the message. The message can be verified and corrected if necessary with a second push-button provided to ensure proper data insertion. Once programmed, all remaining operational functions are automatic.

When air launched, the DLC is slowed and stabilised by a small parachute. For launches from surface vessels, it is merely thrown overboard. Upon water entry, the DLC immediately deploys. As soon as the sub-surface unit reaches the shallow operating depth, the message, coded into appropriate tones, is acoustically transmitted. The first transmission is followed by a five-minute pause while the sub-surface unit deploys to the deep depth. The message is repeated at the deep depth; after a second five-minute pause, the message is transmitted a third time. At the end of the final transmission, the DLC automatically scuttles. The nominal life of the DLC from water entry to scuttle is 17 minutes.

For maximum reliability and minimum development time, extensive use has been made of existing technology and hardware from other Sparton sonobuoy designs. Examples include the lithium battery power supply, the air descent and float inflation hardware, and much of the circuit design. The omni-directional projector is a simple resonant bender element designed to operate efficiently at the desired bandwidth.

CHARACTERISTICS

Depth: Shallow and deep
Size: A (123 × 910 mm)
Weight: 11.4 kg
Operating life: 17 minutes nominal

STATUS

In production.

CONTRACTOR

Sparton Corporation, Electronics Division, 2400 E Ganson Street, Jackson, Michigan 49202, USA.

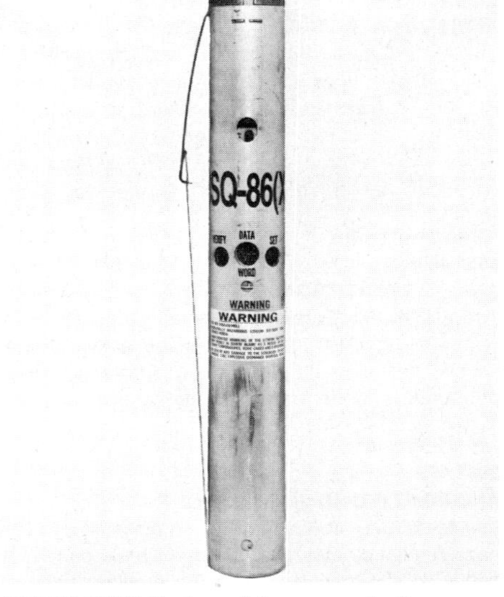

AN/SSQ-86(XN-1) down-link communication sonobuoy

6866.453

AN/SSQ-101 HLA SONOBUOY

The AN/SSQ-101 is an expendable Horizontal Line Array (HLA) sonobuoy which the US Navy intends should be employed on ASW aircraft equipped with the AN/UYS-1, to achieve long range submarine detection. It consists of a self-tensioned horizontal line of passive hydrophones and is designed to transmit multiplexed digital acoustic data to the aircraft for processing. The array is self-deployed after air-drop and is supported until its useful life is exhausted.

STATUS

In early development.

6867.453

AN/SSQ-102 SONOBUOY

The AN/SSQ-102 is a tactical surveillance sonobuoy intended for use by the US Navy P-3 and S-3 aircraft.

The buoy is in early development by Magnavox and a Hazeltine/Sippican joint venture. No further details are available.

STATUS

In engineering development by Magnavox and Sippican/Hazeltine. Development work is expected to continue until the end of 1993.

3599.253

AN/ASQ-81(V) AIRBORNE ASW MAGNETOMETER

The AN/ASQ-81(V) magnetic anomaly detector (MAD) system was developed for the USN for use in the detection of submarines from an airborne platform. The system operates on the atomic properties of optically-pumped metastable helium atoms to detect variations in total magnetic field intensity. The Larmor frequency of the sensing elements is converted to an analogue voltage which is processed by bandpass filters before it is displayed to the operator.

Two configurations of the AN/ASQ-81(V) are available; one for installation within an airframe, and one for towing behind an aircraft. The USN uses the AN/ASQ-81(V)-1 inboard installation with carrier-based S-3A aircraft as well as with the land-based P-3C ASW aircraft. For towing, the configuration is the AN/ASQ-81(V)-2, which is employed by USN SH-3H and SH-2D helicopters. The towed version is also fitted in the USN's latest helicopter the SH-60B Seahawk.

The equipment has been upgraded with a digital implementation version which incorporates microprocessor technology to achieve aircraft compensation and perform all signal processing. An inboard helicopter configuration is also available.

STATUS

The AN/ASQ-81(V) is in production for the USN and a variety of international customers on helicopters such as the WG-13, HSS-2 and 500D.

CONTRACTOR

Texas Instruments Inc, PO Box 660246, Dallas, Texas 75266, USA.

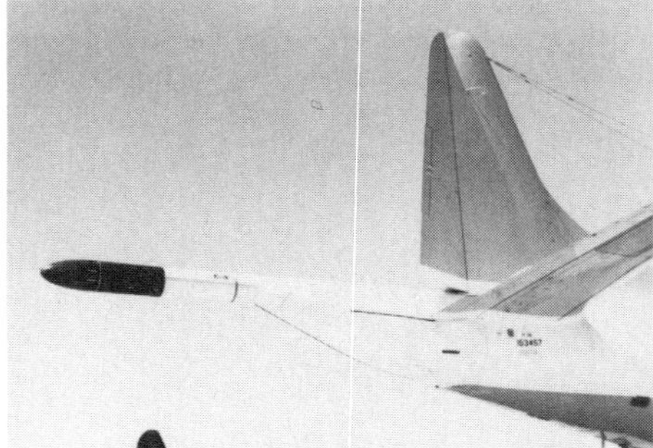

MAD sensor mounted on a P-3C aircraft (W Donko)

4460.453

KLEIN MCM SONAR SYSTEMS

Klein Lightweight Side Scan Sonar Systems are utilized for route survey operations in support of mine counter measures tasks, MCM operations in situations where MCM assets are not available or can not be operated, and for specialized ASW operations to classify bottomed or hovering submarines. The systems utilize a variable depth, towed transducer to provide the capability to be operated from true vessels of opportunity with minimal impact. The towed transducer weighs less than 22 kg (50 lbs) and can be handled on deck by one man without special handling equipment.

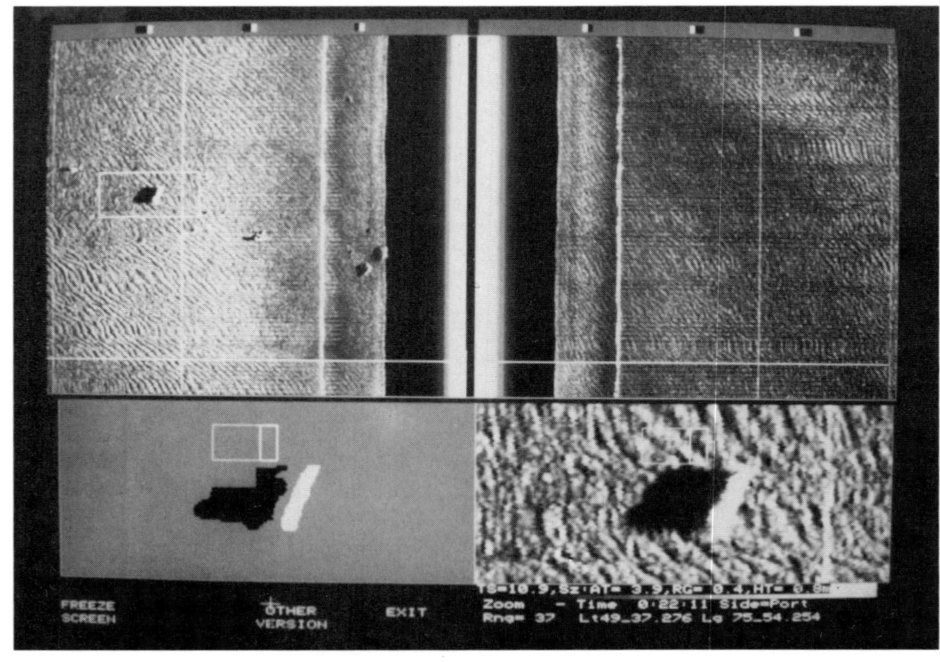

500 kHz Side Scan Sonar display on Model 622 Target Signal Processor showing 50 cm × 1.5 m bottom mine displayed on the port channel within white box in upper left display, same mine target expanded 4× in lower right display, and lower left display, target expanded 4× and computer extracted for ease in operator analysis

The shipboard unit is a lightweight, portable combined sonar transceiver and graphic display unit which contains all processing and display electronics necessary for operation of the sonar. Additionally, the unit has been designed to contain power conditioning equipment to permit operations from 105 to 125 VAC, 210 to 250 VAC 47-420 Hz as well as 23 to 30 VDC. The towed transducer contains transmitting circuitry to energise the dual transducers which project high frequency bursts of acoustic energy in fan-shaped beams, narrow in the horizontal plane and wide in the vertical plane. Echoes from bottom targets are received by the transducers, amplified and transmitted up the electro-mechanical tow-cable to the transceiver/recorder for further processing and display.

Four towed transducers are available for either MCM or ASW operations: 50 kHz for long range operations on larger targets or for rapid selection of potential 'Q' routes; 100 kHz for medium resolution and less range for use on route survey operations or for more detailed investigation of large targets; 500 kHz for the higher resolution investigation of potential MCM targets or detailed inspection of ASW targets; and simultaneous 100 kHz and 500 kHz for the highest probability of detection for the detection of small or unusual MCM targets.

A 3.5 kHz sub-bottom profiler is available for use in conjunction with any of the three frequencies to provide the MCM operator with information on the softness of the bottom and consequently, the potential for mine burial.

A series of Target Signal Processors are available to provide the operator with computer assistance in the detection of underwater targets. The 620 series can process and display either 100 kHz or 500 kHz side scan sonar data, while the 640 series can process and display simultaneous 100 kHz and 500 kHz side scan sonar data for maximum detectability and highest Probability of Detection.

The processors provide the capability to zoom the target image to four times normal size and permit operator interaction in selection of targets through the use of a touch screen on the CRT. Models are available which provide higher levels of signal processing for enhancement or extraction of the target from bottom clutter, as well as computer aided detection of mine targets.

STATUS

All versions of the side scan sonar system are currently in production and are utilised by many navies for MCM applications, ASW applications and hydrographic survey.

CONTRACTOR

Klein Associates Inc, Klein Drive, Salem, New Hampshire 03079, USA.

5643.453

UNDERWATER ORDNANCE LOCATOR MK.24 MOD 0

The Mk.24 Mod 0 ordnance locator is a dual-beam side-scan sonar for use in locating mines and other types of underwater ordnance. It is available with modular accessories which adapt the equipment for use in channel conditioning/route survey mine countermeasures operations as well as for use in explosive ordnance disposal operations.

The Mk.24 system consists of a side-scan sonar towfish, weighing approximately 22 kg (48 lb), a lightweight Kevlar towcable, and a combined transceiver and graphic recorder weighing 44 kg (95 lb). The system and its accessories are lightweight and portable, and can be deployed from any available vessel down to 5 m in length. It can also be towed from any available and unmodified helicopter.

The side-scan towfish contains the circuitry to energise the transducers which project high frequency bursts of acoustic energy in fan-shaped beams, narrow in the horizontal plane and broad in the vertical plane. Echoes from both transponders are processed by the combined transceiver and printed by the graphic recorder. Four types of transducers are available; a 50 kHz long range unit for covering up to 1.2 km swaths of the bottom; a 100 kHz high resolution version for detection operations covering up to 400 m swaths; a 500 kHz very high resolution unit for target classification applications; and a simultaneous 100 kHz and 500 kHz unit for detection and classification of small targets by comparative techniques through the use of frequency diversity. A 3.5 kHz sub-bottom profiler is available for assessing bottom hardness to determine potential buried mine hazards.

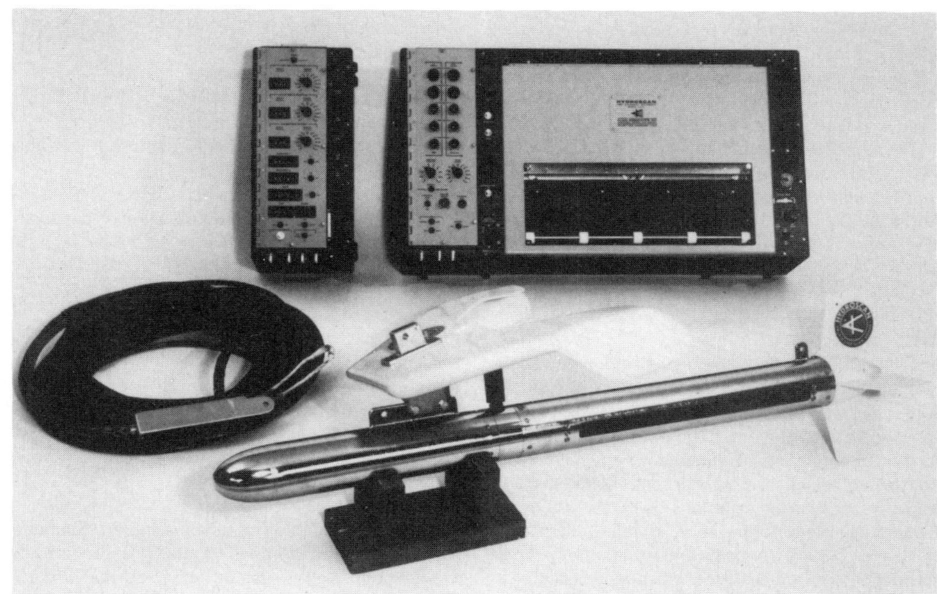

Components of Mk. 24 mine-hunting sonar

The towfish can be towed at speeds up to 16 knots, but for optimum operation against small targets the recommended speed is 5 to 6 knots. Standard towfish can operate down to depths of 1000 m (3280 feet) with options permitting operation down to 2270 m (7445 feet). Special units are available for operations down to full ocean depth.

STATUS

All versions of the system are in full production, and are in use by the US Navy Explosive Ordnance Disposal units.

CONTRACTOR

Klein Associates Inc, Undersea Search and Survey, Klein Drive, Salem, New Hampshire 03079, USA.

6258.493

STRAP PROGRAMME

STRAP (Sonobuoy Thinned Random Array Project) is a programme initiated by the US Navy to emulate the collective processing beamforming techniques of normal sonar systems, but using sonobuoys. The technique involves the use of large numbers of sonobuoys, probably around 15 to 20, which are deployed randomly over a given area. Since it is essential to know the exact position of each sonobuoy, four of the buoys would be fitted with a low power transmitter and deployed in different areas of the pattern. Each sonobuoy would then receive two sets of signals, one from the water and the other from the four emitting sonobuoys. The acoustic processor in the aircraft would then correlate these positions and use the information for beam-forming purposes.

STATUS

The programme is being managed by the US Naval Ocean Systems Command with Lockheed using the AN/UYS-1 Proteus processor for adaptive processing techniques, and Magnavox modifying the sonobuoys. STRAP forms part of an on-going acoustic search sensors programme. Sea trials are expected in the 1987/88 period.

NAVAL RADAR

CANADA

5677.253
AN/SPS-503 SURVEILLANCE RADAR
The AN/SPS-503 is a high performance, lightweight surveillance and target indicating radar operating in the E/F-band. In its standard configuration it is suitable for naval ships with displacement of 500 tons or more. It is of rugged construction and has been designed to operate in very hostile electronic environments.

The main features of this radar are a coherent E/F-band transmitter with broadband frequency agility; pulse compression; a narrow beam, low sidelobe cosecant² antenna; and digital MTI with velocity compensation. The transmitter employs a TWT amplifier fed by a frequency synthesiser with high stability giving very good MTI performance. The low noise, high dynamic range receiver compresses target returns to provide high jamming resistance and range resolution. This pulse compression system has been optimised to ensure very low range sidelobes, even in the presence of large doppler shifts.

The signal processor is an all-digital unit providing MTI, CFAR and PRF discrimination facilities. Two independent signal channels are provided, one for MTI data and one for non-MTI data. The MTI system embodies velocity compensation which shifts the notch of the MTI filter to compensate for the land clutter velocity. In addition, CMC offers additional compensation for weather clutter as an option with the AN/SPS-503. An optional plot extractor will output digital records of the range and azimuth of target centroids. The thresholding system employed gives maximum detection probability for targets and minimal probability of clutter-induced false alarms.

The SPS-503 uses a double curvature reflector antenna with horn feed. The reflector is configured to give a cosec² pattern in the elevated plane. Polarisation is variable from linear to circular and dual rotation rates of 10 and 20 rpm are provided. An IFF antenna is integral with the main antenna. A smaller, lightweight antenna is also available for applications where masthead weight is critical.

CHARACTERISTICS
Peak power: 20 kW
Mean power output: 800 W
PRF: 739 Hz ± 12.5% (stagger)
Antenna rotation speed: 10 or 20 rpm
Range: 130 km (80% probability against a 4m² target)
System weight: 1300 kg

STATUS
This system has been ordered by the Canadian Armed Forces for fitment under the Destroyer Life Extension Programme (DELEX). The AN/SPS-503 has successfully completed qualification testing to MIL-E-16400. Delivery of the qualified system to the Canadian Navy took place in December 1987.

AN/SPS-503 transmitter/receiver/signal processor assembly

CONTRACTOR
Canadian Marconi Company, 415 Legget Drive, PO Box 13330, Kanata, Ontario, Canada K2K 2B2.

6873.253
LN66 RADAR SERIES
The LN66 series is a family of modular radars available in a variety of power levels from 6 to 90 kW and in a number of system configurations tailored to specific requirements. The system applications include general surface surveillance and navigation, coastal surveillance, gun fire control systems and ASW roles. Current missions with the US Navy include use as the prime radar in the LAMPS Mk 1 helicopter (75 kW version), and as the short range navigation radar in a number of surface vessels (10 kW version).

LN66/HP
A rugged five-element radar comprising an antenna, transmitter/receiver, display, north stabilising unit and power supply. The 3 ft wide antenna is a radome end-fed slotted waveguide, horizontally polarised unit and is also available in 14 ft, 8 ft and 5 ft lengths.

CHARACTERISTICS
Frequency: 9375 ±30 MHz
Peak power: 75 kW
Rotation speed: 24 ±2.4 rpm
Pulse width: 0.1 µs or 1 µs
PRF: 500; 2000 Hz
Weights: Antenna 9.7 kg; transmitter/receiver 34 kg; display 22.9 kg

LN66/SP
Designed for operation in a submarine environment, the LN66/SP can be used in any role requiring a lightweight/portable system, either as the primary radar in a small craft or as a secondary system on larger vessels. The LN66/SP is a 6 kW radar comprising only two units; a scanner unit incorporating the transmitter/receiver and the antenna (either a 4 ft slotted waveguide or 3 ft radome), and a display unit. The antenna can be removed from the scanner unit and transferred from the inboard stowage position to the operational position on the navigation bridge.

CHARACTERISTICS
Frequency: 9410 ±30 MHz
Peak power: 6 kW
Rotation speed: 30 rpm
Topside weight: 13.6 kg (including antenna)

STATUS
In operational service in the surface ship, submarine and airborne roles.

CONTRACTOR
Canadian Marconi Company, Radar Division, 415 Legget Drive, PO Box 13330, Kanata, Ontario, Canada K2K 2B2.

CHINA

5842.253
CHINESE NAVAL RADAR
Information on operational naval radar in Chinese ships is very sketchy. A small amount of detail has been released recently, although with the exception of entry **6415.253** which follows, the information provided is insufficient to produce meaningful entries. The Chinese authorities state that a series of navigation radars have been developed for installation on several classes of ships, including torpedo boats, missile equipped destroyers, escort ships, high speed gunboats and submarines, and are known as Types 752a, 756 and 757. Little information has been made available on any surveillance, missile tracking or fire control radar systems which may be in use. Much of the radar currently fitted to Chinese naval vessels is Soviet in origin, some of it being of early design.

The main fire control radar in current use is the Sun Visor system (**1326.253**), known to the Chinese as Wok Won. The Jianghu class frigates are fitted with an air/surface warning equipment, called MX-902, which appears to be a modified version of the Slim Net system (**1322.253**). Other Soviet designed equipments observed on Chinese ships include the Square Tie short-range search radar (**1329.253**), the Knife Rest A long-range equipment, the High Pole and Square Head IFF interrogators (**1595.253**), and the Skin Head system (**1614.253**). Submarines employ either the Snoop Tray or Snoop Plate Systems (**5849.453**). A phased array 3-D radar, known as Rice Screen, has been observed on one of the Luda class destroyers. This appears to be a recent development and it has become apparent that an indigenous programme covering naval, ground and airborne systems is under way, with priority on the shipborne equipment being given to air surveillance and fire-control systems. No doubt these systems will be observed on new classes of warships, although it seems very unlikely that information will be generally released.

6415.253
MODEL 354 SHIPBORNE SURFACE AND LOW ALTITUDE WARNING RADAR
Model 354 is a shipborne C-band search and target acquisition radar normally employed on missile destroyers. It measures the range and bearing of targets with high accuracy and provides information for the ship's information centre, the fire control radar, and the missile guidance radar. Coherent techniques are employed to combat interference and clutter, and the antenna is pitch and roll-stabilised.

CHARACTERISTICS
Frequency: C-band
Peak power: 500 kW
Pulse width: 2µs
PRF: 400 Hz; 800 Hz
Polarisation: Linear
Scan rate: 4 to 10 rpm
Beamwidth: 1.2° horizontal; 5° vertical
Accuracy: 75 m in range; 5 m in azimuth
Resolution: 300 m in range; 1.3° in azimuth
Range: More than 50 nm on an aircraft with effective target area of 10m²

STATUS
Operational.

CONTRACTOR
China National Electronics Import and Export Corporation, 49 Fuxing Road, Beijing, People's Republic of China.

Model 354 shipborne radar

DENMARK

1575.253

TERMA SCANTER MIL NAVIGATION AND SURVEILLANCE RADARS

The Terma navigational radar systems are designed to form an integral part of the total radar system on board naval vessels. They operate in S- and X-bands and are suitable for both navigation and surveillance. A multi-display system capable of operating in various master/slave combinations is available. The operational mode is selected on either display by illuminated push-buttons.

All units are autonomous, ie containing their own power conditioning circuits, and are built up as a modular system using accessible plug-in modules for ease of servicing. The system can be single X-band, single S-band, combined S- and X-bands, double X- and S-bands, or synchronised twin X-band. The modular technique also facilitates system extension. The following features can be supplied as options: electronic bearing line, sector transmission, reflection plotter, and true motion.

Terma Scanter Mil X-band antenna

Scanner Unit

The antenna is a horizontally-polarised slotted waveguide antenna with narrow beam and low sidelobe level. The antenna feed system (including a waveguide rotary joint) offers a low standing wave ratio at the input flange. Two scan rates may be selected by electronic switching of the drive system. The scanner position information is given through synchro transmitters and heading marker signal. The unit is driving-rain proof and a de-icing system is included.

Receiver/Transmitter Unit

The transmitter tube is a pulsed magnetron controlled from a solid-state magnetic modulator. Programmable pulse repetition rates and two pulsewidths may be selected individually. The modulator is divided into two separate assemblies with common input and output circuits, one for long and one for short pulse. The trigger assembly synchronises the total system and selects automatically internal or external synchronising mode. A trigger-pulse shaper and delay assembly perform compensation for transit times in the waveguide and receiver system. The RF system is integrated in one assembly and includes the waveguide duplexer, the T/R switch, the IF mixer, the AFC mixer and the solid-state local oscillator.

Display Unit

A whole range of displays is available all of which have the same physical size and common basic circuitry. The presentation is raster scan daylight and the information consists of a series of displays ranging from navigational to complex tactical with automatic target tracking and target assignment to weapon systems of medium complexity is available.

All displays are fitted for operation from three independent radar sources of virtually any kind.

CHARACTERISTICS

Antenna: 2.1 m slotted waveguide (X-band), 4 m (S-band)

Scanner rotation: Switchable between 24 and 48 rpm $\pm$ 10% at 60 Hz or 20 and 40 rpm $\pm$ 10% at 50 Hz input frequency

Horizontal beamwidth: $<1°$ within 3 dB points (X-band), 2° (S-band)

Vertical beamwidth: 20° within 3 dB points

Polarisation: Horizontal

Gain: >30 dB (X-band), >28.5 dB (S-band)

Receiver/Transmitter unit

Transmitter frequency: 9 GHz (X-band), 3 GHz (S-band)

PRF

Short pulse: Max 4400 Hz

Long pulse: Max 2200 Hz

Pulse length: 0.06 $\pm$ 0.01/µs, or 0.6 $\pm$ 0.1/µs

Peak power: 20 kW $\pm$ 1 dB measured at the output flange

STATUS

In service with the Danish and Swedish Navies.

CONTRACTOR

Terma Elektronik AS, Hovmarken 4, 8520 Lystrup, Denmark.

FRANCE

1894.253

DRBC 32 NAVAL RADAR

The DRBC 32 designation embraces a family of I/J-band fire control radars, of which there are at least five variants (denoted by suffix letters running from A to E) in use in French Navy ships, and which are associated with a variety of director units. The latter may or may not include optical direction facilities. Weapons associated with these directors normally are either 100 mm or 57 mm guns.

A stabilised mount is usual for DRBC 32 installations and, in all but the DRBC 32 A the radar dish is protected by a radome attached to the antenna assembly itself. The 'A' model, as seen on the helicopter-carrier, *Jeanne d'Arc*, has exposed circular reflector and feed assemblies. However, the DRBC 32 A as seen on anti-submarine 'T 47' class escorts do have a protective radome for the antenna. The *Aconit* anti-submarine corvette is fitted with the DRBC 32 B. The 'C 70' class corvettes, the *Colbert* and the *Duperré* have the DRBC 32 C, and the DRBC 32 D installation is aboard the 'F 67' class frigates. The 'A 69' class of Avisos are fitted with the DRBC 32 E.

DRBC 32 D fire control radar

5843.253

DRBC 33 NAVAL RADAR

The DRBC 33 is a fully coherent monopulse tracking and fire-control I-band radar which is in current operational service with the French Navy. The weapon associated with this equipment is the 100 mm gun, although the system can also be used with both 76 mm and 57 mm weapons.

The DRBC 33 will be fitted to the new French Navy 'C-70' class corvettes, and the last anti-submarine corvette. It is also intended to be refitted to the *Suffren* and *Duquesne* missile frigates.

STATUS

In operational service with the French Navy.

CONTRACTOR

Thomson-CSF, Division Systèmes, Defénse et Contrôle, 40 rue Grange Dame Rose, 92363 Meudon-la-Foret Cedex, France.

1593.253

DRBI 10 3-D NAVAL RADAR

The DRBI 10 is a three-dimensional, E/F-band naval air surveillance radar for search and interceptor control functions. It is believed to be a naval version of the land-mobile radar Picador (TH.D 2200) (**2119.153**), which in turn evolved from the TH.D 1940. More detailed information is given under this entry number in *Jane's Weapon Systems 1982-83*.

STATUS

The only known fittings are aboard French vessels. These include the aircraft carriers *Clémenceau* and *Foch*, which each have two; the helicopter carrier *Jeanne d'Arc*; the corvette *Colbert* and the trials ship *Ile d'Oléron*.

CONTRACTOR

Thomson-CSF, Division Systèmes, Défense et Contrôle, 40 rue Grange Dame Rose, 92363 Meudon-la-Foret Cedex, France.

1889.253

DRBI 23 3-D NAVAL RADAR

The DRBI 23 naval radar is a D-band (23 cm) three-dimensional air search and target designation radar forming the main detection element of the French Navy Masurca surface-to-air missile system (**1177.231**). Transmitter peak power is quoted as several MW, and it incorporates a carcinotron oscillator. Six amplification stages are provided, these being capable of virtually identical gain at any frequency within the range of the carcinotron oscillator.

The antenna, which is protected by a large radome, is of the inverse cassegrain type. An array of feed horns directs radiation to a semi-reflective parabolic mirror which in turn returns the energy to a flat plate reflector. This imparts a 45° polarisation change which permits the formed beams to pass through the semi-reflector. Vertical angular information is obtained by the use of monopulse techniques, these data being processed to provide corrected height information on aircraft targets. This information, together with plan position data, is available for use in the ship's action data system and weapon systems.

DRBI 23 radome on the Suffren (Stefan Terzibaschitsch)

Stabilisation against ship's motion is incorporated, and the antenna assembly includes an IFF antenna unit. The system provides for auto plot extraction and the DRBI 23 is interfaced with a digital computer. Associated with it are the Masurca missile tracking and guidance radar groups DRBR 51.

STATUS

In service with the French Navy missile frigates *Suffren* and *Duquesne*.

CONTRACTOR

Thomson-CSF, Division Systèmes, Défense et Contrôle, 40 rue Grange Dame Rose, 92363 Meudon-la-Foret Cedex, France.

3741.253

DRBJ 11 NAVAL RADAR

The DRBJ 11 is a new shipboard electronic phase scanned radar developed by Thomson-CSF for the French Navy. Few technical details have been released for publication, but some indications can be gathered from the company's previous extensive work in this area of radar technology, and from a picture of the experimental antenna unit of the new radar. From the latter it will be seen that the array is composed of numerous separate radiating elements arranged concentrically on a disc surface. The active elements are connected in lines to be fed; in-situ computers determine their phases.

Thomson-CSF has revealed a preference for the use of diode phase-shifters for beam steering in electronically scanned arrays, and in this technique each source element (or reflector) has an associated module which offers the microwaves several alternative paths corresponding to different propagation times. Diodes controlled by DC voltages perform the switching needed to select required phase shifts. In addition, each source contains its own control logic which receives instructions from the beam steering computer circuitry. From the photograph, the individual elements appear to consist of spirals, supported by ceramic bases, but in the absence of anything from which to obtain an indication of scale it is not possible to deduce the likely frequency employed with any certainty.

STATUS

The system is being operationally tested and will be fitted to the 'C70' anti-aircraft destroyer class and the nuclear aircraft carrier *Charles de Gaulle*. The equipment has already been installed on the *Cassard* – the first C70 class to be commissioned.

CONTRACTOR

Thomson-CSF, Division Systèmes, Défense et Contrôle, 40 rue Grange Dame Rose, 92363 Meudon-la-Foret Cedex, France.

Close-up of Thomson-CSF electronic phase scanning naval radar

---

1890.253

DRBR 51 NAVAL RADAR

The DRBR 51 radar group provides for the tracking and missile guidance functions of the French Navy's Masurca surface-to-air missile system (**1177.231**). Search and target designation facilities are provided by the DRBI 23 three-dimensional air search radar (**1889.253**). The DRBR 51 is a multi-mode equipment and associated with the antenna assembly is a TV camera for optical tracking of low-flying targets. The complete system provides for the control of the Masurca missile.

Command signals are transmitted at frequencies in the 7 cm-band, and two missiles may be simultaneously controlled. There are two 5 cm radar tracking channels, 'blue' and 'yellow', one for each missile, and each radio command Masurca round is equipped with a 5 cm transponder to facilitate missile tracking.

When firing the Mk 2 Mod 3 radar homing version of Masurca, the main dish of the DRBR 51 operates as target tracker and I-band illuminator.

STATUS

The missile cruiser, *Colbert*, and the two missile frigates, *Suffren* and *Duquesne*, each have two DRBR 51 systems.

DRBR 51 radar group on the Suffren (Stefan Terzibaschitsch)

CONTRACTOR

Thomson-CSF, Division Systèmes, Défense et Contrôle, 40 rue Grange Dame Rose, 92363 Meudon-la-Foret Cedex, France.

1893.253

DRBV 13 NAVAL RADAR

The DRBV 13 naval radar is pulse-doppler equipment, believed to operate in the E/F-band. It is stated to be a multi-mode radar providing both search and tracking facilities for the SENIT action data automation system installed in the *Aconit* anti-submarine corvette, which is the only vessel known to be fitted with the DRBV 13. The antenna array and turning gear are completely enclosed by a large plastic radome, itself mounted atop a prominent structure which is presumed to house the associated electronics, in an almost amidships location. No technical or performance data have been obtained.

1892.253

DRBV 20 NAVAL RADAR

The DRBV 20 is a long-range search radar operating in the metric waveband. Two antenna outfits are employed. The smaller, DRBV 20 A, version is installed in a variety of French Navy vessels and can be identified by its large open lattice antenna, of rectangular outline, slightly curved, and with a horizontal array of dipole feed elements. The larger, DRBV 20 C, is fitted on French aircraft carriers and has a flat, rectangular antenna array comprising three horizontal rows of dipole feed elements. No performance or other details of technical characteristics have been obtained.

STATUS

The DRBV 20 A is installed in the Tartar-armed 'T 47' escorts. The DRBV 20 C is fitted in the aircraft carriers *Clémenceau* and *Foch* and on the missile cruiser *Colbert*.

CONTRACTOR

Thomson-CSF, Division Systèmes, Défense et Contrôle, 40 rue Grange Dame Rose, 92363 Meudon-la-Foret Cedex, France.

1891.253

DRBV 22 NAVAL RADAR

The DRBV 22 is a conventional naval search radar, having a L-band operating frequency and possibly incorporating dual-beam (high- and low-cover) facilities in some cases. It is widely fitted, both in French Navy vessels and in the ships of other countries. Different versions that have been identified are the DRBV 22 'A', 'C' and 'D' models.

The antenna assembly is very similar in appearance to the American SPS-6 and SPS-12 radars (**1744.253** and **1566.253**, respectively) but the French

DRBV 22 A radar antenna on Aconit (X I Taibo)

equipment can be readily distinguished from both US radars by the two supporting stays for the feed horn that are attached to the upper edge of the reflector of the DRBV radar.

STATUS

Known fittings in French Navy vessels include: DRBV 22 A: five 'T 47' anti-submarine escorts; nine 'A 69' Aviso escorts; and the 'C 65' corvette *Aconit*.

DRBV 22 C: the trials ship *Ile d'Oléron*

DRBV 22 D: the helicopter-carrier *Jeanne d'Arc*; the missile tracker ship *Henri Poincaré*.

Ships of several other nations have been fitted with the DRBV 22, in most, if not all, cases the 'A' version.

CONTRACTOR

Thomson-CSF, Division Systèmes, Défense et Contrôle, 40 rue Grange Dame Rose, 92363 Meudon-la-Foret Cedex, France.

1594.253

DRBV 23 NAVAL SURVEILLANCE RADAR

The DRBV 23 is a L-band (23 cm) long-range naval air search and surveillance radar, similar in appearance to the Jupiter naval radar (**1236.253**). The most readily discerned external difference is the prominent tubular horizontal supports for the DRBV 23 scanner, compared with the lighter structure of the Jupiter.

Two versions have been identified, the DRBV 23 B and DRBV 23 C, the former being that fitted in aircraft carriers of the French Navy, and the latter being a solid-state version fitted in the missile cruiser *Colbert*.

STATUS

The only known fittings are on the following French Navy ships: the aircraft carriers *Clémenceau* and *Foch* and the missile cruiser *Colbert*.

CONTRACTOR

Thomson-CSF, Division Systèmes, Défense et Contrôle, 40 rue Grange Dame Rose, 92363 Meudon-la-Foret Cedex, France.

DRBV 23 L-band air surveillance radar

1062.253

TRITON II SURVEILLANCE RADAR

TRS 3030/TRS 3035

This is a C-band, 5 cm, air and surface surveillance radar for shipboard mounting, and is used for surveillance and target designation functions in the Thomson-CSF Thomsea naval combat systems (**4572.281**). In these systems, Triton can be used for target designation to the fire control units (Castor or Pollux) and with a TWS for surface-to-surface missiles and/or torpedo launching. It may also be used alone as a surveillance and navigation radar. Triton is produced in an MTI version (TRS 3035).

More detailed information is given under this entry number in *Jane's Weapon Systems 1982-83*.

STATUS

No longer in production.

CONTRACTOR

Thomson-CSF, Division Systèmes, Défense et Contrôle, 40 rue Grange Dame Rose, 92363 Meudon-la-Foret Cedex, France.

3685.253

TRITON G DOPPLER SURVEILLANCE RADAR

TRS 3050

TRS 3050 is designed for ships of 150 tons upwards to perform air and surface surveillance, low altitude missile detection, target designation to weapon systems, and emergency navigation under adverse conditions. It is suitable for either original fitting in new ships, or for use during refits.

It is a G-band radar, fully coherent synthesiser driven, with two receiver channels:

(1) air: with digital doppler processing, burst-to-burst frequency agility, automatic tracking and initialisation of targets

(2) surface: with pulse compression, pulse-to-pulse frequency agility, and advanced anti-clutter and anti-jamming processing.

As the adjacent photograph shows, Triton G employs the normal radar reflector of the Triton series of radars but can be fitted with a new antenna allowing detection of modern threats such as diving and sea-skimming missiles. Most of the operational advantages are apparently the result of advances in transmitter/receiver and processor design and/or the antenna feed arrangements.

CHARACTERISTICS

Frequency: G-band

Max detection range: 30 km (Pd $\leqslant$ 0.5, 2 m^2 aircraft target, doppler operation) or 60 km

SCV: At least 45 dB

First blind speed: Over 800 m/s

Antenna gain: At least 27.5 dB

Elevation pattern: 22°

STATUS

In series production. The prototype was installed on board an S148 FAC in 1986 and underwent successful qualification trials. Twenty three Triton G radars were ordered by the West German Navy in 1987 to refit its Type 148 FAC.

CONTRACTOR

Thomson-CSF, Division Systèmes, Défense et Contrôle, 40 rue Grange Dame Rose, 92363 Meudon-la-Foret Cedex, France.

Triton G surveillance radar and Castor IIC fire control system

3254.253

TRITON S SURVEILLANCE RADAR TRS 3033

Triton S is an S-band air and surface surveillance radar for shipboard mounting, used for surveillance and target designation functions in the Thomson-CSF Vega II series of ships' fire control systems.

Specially developed for detection of small air targets (aircraft and missiles) in a heavy clutter environment, Triton S is a fully coherent, digital processing, pulse doppler radar which includes an independent pulse compression, frequency-agile receiver for detection of surface targets.

The antenna is of the non-stabilised type with a broad elevation lobe to counteract the effects of ship motion. The transmitter comprises a TWT and the average power is quoted as 250 W. The radar provides two video signals; an air surveillance video and a surface video, which can be used for display, target designation and track while scan processing.

CHARACTERISTICS

Frequency: S-band – frequency agile

Average power: 250 W

Two receivers: Pulse doppler (processed range 1.5 – 26.5 km); pulse compression

Antenna rotation: 24 rpm

Detection range: Air target (fighter), 33 km; surface target, radar horizon

SCV: At least 50 dB

STATUS

In production. Ordered by Nigeria, Tunisia and other navies.

CONTRACTOR

Thomson-CSF, Division Systèmes, Défense et Contrôle, 40 rue Grange Dame Rose, 92363 Meudon-la-Foret Cedex, France.

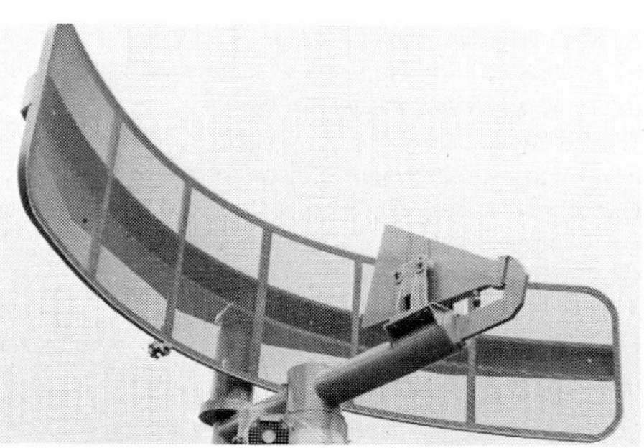

Triton S radar antenna

1063.253

CASTOR II FIRE CONTROL RADARS

Three versions of Castor II are available, Castor IIB, Castor IIC and Castor IIJ. All versions are compact, lightweight and suitable for installation on both small and large vessels.

Castor IIB is an X-band monopulse and MTI tracking radar for weapons control. The radar is managed by a microprocessor which ensures high performance, even in severe environmental conditions against low altitude targets and jamming, by the use of monopulse techniques, broadband operation, high efficiency anti-clutter filters and an integrated TV tracker.

Operationally Castor IIB performs the following functions:

(1) fully automatic acquisition enabling very short reaction times
(2) accurate tracking of targets in the radar mode, TV mode and a combined radar/TV mode
(3) passive tracking of a jammer with range information sent from a tracking device linked to a surveillance radar
(4) display of splashes in the case of anti-surface firing
(5) autonomous surveillance, continuous and by sector, with set absolute elevation.

Castor IIC is an X-band doppler filtering tracking radar for weapons control which is able to acquire and track targets in conditions of active jamming, chaff and heavy clutter. The high performance of the system is achieved by the use of fully coherent transmission, frequency agility, analysis of received jamming signals within all the useful bandwidth, auto-adaptive doppler filtering and central management by computer which matches the radar to the environmental conditions at all times.

Operationally Castor IIC provides:

(1) fully automatic acquisition enabling very short reaction times
(2) short bursts operation (1.5 ms) with simultaneous frequency agility and doppler processing
(3) pulse-to-pulse operation giving true frequency agility
(4) TV tracking or combined TV/radar tracking
(5) passive tracking of a jammer with range information sent from a tracking device linked to a surveillance radar
(6) display of the angular error between the shells and the target
(7) display of the angular tracking error on a jammer
(8) autonomous surveillance, continuous and by sector.

Castor IIJ is a Ku-band radar TV/IR director which forms the basis of an all-weather anti-missile system (see entry **4572.281**). It has been specifically designed for accurate tracking in a high clutter and jamming environment, even against very low altitude targets.

Operationally Castor IIJ provides

(1) fully automatic acquisition enabling very short reaction times
(2) frequency agility and doppler processing
(3) complementary TV or IR tracking
(4) passive tracking of a jammer with range information provided by a tracking device linked to a surveillance radar
(5) autonomous surveillance, continuous and by sector
(6) very low altitude radar tracking due to a patented process which renders the radar insensitive to image effects.

CHARACTERISTICS

Castor IIB and IIC
Frequency: X-band
Peak power: More than 30 kW
Range: 0 – 30 km
Acquisition mode: Fully automatic
Dimensions
Antenna: 1 m diameter
Transmitter: 1 × 0.6 × 0.42 m
Receiver and servo: 1.92 × 0.6 × 0.75 m
Weight: 620 kg
Castor IIJ
Frequency: Ku band for tracking; X-band for illumination channel
Range: More than 25 km
Acquisition mode: Fully automatic
Tracking accuracy: 0.6 mrad (20 m)
Dimensions
Antenna: 2 × 1 m
Transmitter: 1.2 × 1.4 × 0.75 m
Receiver and servo: 1.2 × 11.8 × 0.75 m
System weight: 850 kg

STATUS

In production. Supplied to a number of foreign navies. Twenty Castor 148 systems are to be supplied to the West German Navy for fitment to fast attack craft. Castor IIJ underwent successful trials under operational conditions.

CONTRACTOR

Thomson-CSF, Division Systèmes, Défense et Contrôle, 40 rue Grange Dame Rose, 92363 Meudon-la-Foret Cedex, France.

Castor IIC tracker radar head with TRAKIR (TV + IR) camera

Castor II J tracking radar

1064.253

POLLUX TRACKER RADAR TRS 3220

This is a shipborne tracker radar used in the Thomson-CSF Vega series of naval fire control systems and usually operated in conjunction with a Triton 5 cm surveillance and target designation radar. It is mainly for gun control operations.

The Pollux radar operates on a frequency in the X-band, 3 cm region, and is a fast conical scan tracker radar. Circular polarisation and solid-state circuitry, using silicon semi-conductors, are employed, and peak transmitter power is quoted as 200 kW. Range finding capability extends to 30 km, and target acquisition is stated to remain reliable to 20 km under the most adverse weather conditions.

CHARACTERISTICS

Frequency: X-band
Circular polarisation
Peak power: 200 kW
Rangefinder: 0 – 30 km
Precision
Range: 20 m
Azimuth: 0.5 mrd
Display: Type A/R, A/B or A/E
TV tracking: Optional
Dimensions and weight
Antenna: 450 kg
Transmitter and servo cabinet: 1850 × 600 × 600 mm
Receiver cabinet: 1850 × 600 × 600 mm
Fire control console: 1850 × 600 × 600 mm (computer included)

STATUS

Details of users have not been revealed but contracts for Vega fire control systems have been placed by both French and foreign navies. Known fittings include four Portuguese corvettes of the 'João Roby' class and West German, Greek, and Malaysian fast missile patrol boats. No longer in production.

Pollux tracking radar, with opto-electronic sensor

CONTRACTOR

Thomson-CSF, Division Systèmes, Défense et Contrôle, 40 rue Grange Dame Rose, 92363 Meudon-la-Foret Cedex, France.

1687.253

SEA TIGER II SURVEILLANCE RADAR TRS 3001

Sea Tiger II is a shipborne S-band combined surveillance radar, which can be used in the Thomson-CSF Thomsea naval combat systems. It has been developed to provide detection and tracking of missiles, including sea-skimming anti-ship weapons. By virtue of modular design and operating flexibility, Sea Tiger II can be supplied with either of two antenna types, thus allowing fitting aboard medium or large tonnage ships, or small fast patrol boats. It is designed to perform the following functions in a very severe clutter and jamming environment:

(1) air surveillance
(2) surface surveillance
(3) anti-missile surveillance
(4) target designation for weapon systems (guns; missiles).

Visibility in clutter is achieved through the use of complementary techniques such as circular polarisation, pulse compression, doppler filtering, and an anti-clutter reception chain.

A typical Sea Tiger II installation comprises:

(1) a transmitter with coherent amplifier chain emitting frequency-modulated pulses (pulse compression) over a wide frequency range
(2) a receiver with multiple reception chains: MTI, lin/log, CFAR. The MTI is a digital linear device whose characteristics, associated with the stability of the amplifier chain, make it possible to detect a missile in the most severe clutter environment
(3) an antenna appropriate to the class of ship, either:
(A) stabilised double-curvature antenna with switchable circular/linear polariser and integrated IFF for medium tonnage ships and above
(B) non-stabilised lightweight antenna with a broad elevation lobe to counteract the effects of ship's motion (FPBs).

CHARACTERISTICS

Operational performance

Detection range: 60 nm on 2 m² (fluctuation) target with Pd ⩽ 50%

Altitude detection: 50 000 ft (15 240 m)

Elevation pattern: Cosecant² up to 50°

Sub-clutter visibility: >42 dB

First blind speed: Mach 2

Antenna

Gain: 30 dB

Polarisation: Circular/linear, switchable

Rotation speed: 15 and 30 rpm

Stabilisation: Better than ± 1.5%

Transmitter: Frequency agility, pulse compression, average power 1 kW, 66 kW peak

Sea Tiger stabilised radar antenna head

Receiver: Noise figure better than 5.5 dB, 4 different reception chains – MTI, lin, log, CFAR anti-clutter

STATUS

Derived from the Tiger radar (**2156.153**). In production. Ordered by the French Navy (DRBV 15) and in service with several other navies.

CONTRACTOR

Thomson-CSF, Division Systèmes, Défense et Contrôle, 40 rue Grange Dame Rose, 92363 Meudon-la-Foret Cedex, France.

1236.253

JUPITER NAVAL AIR SURVEILLANCE RADAR TRS 3010

The Jupiter is a L-band (23 cm), long-range air surveillance radar with a peak transmitter output power of 2 MW with a 2.5 microsecond pulse and a PRF of 450 pulses/second. The antenna measures 7.5 × 3 m and has two rotation speeds: 7.5 and 15 rpm. Weight is approximately 1000 kg. Both primary and secondary (IFF) radar functions are provided by the one antenna system. Beamwidth at half power points of the primary radar pattern is 2.5° or less. The elevation pattern is cosecant² up to 50°.

Range performance figures quoted give detection out to at least 200 km on a fluctuating target of two m² cross-section with a detection probability of 50 per cent, and under the same conditions a target with ten metres cross-section can be observed out to 275 km.

Solid-state circuitry is used throughout, with the exception of the microwave stages, and a fixed-frequency, water-cooled magnetron is used in the transmitter. Comprehensive signal processing facilities are incorporated to provide maximum protection from natural interference (such as sea clutter) and ECM.

The receiver section provides multiple reception chains, wide dynamic range receiver and anti-clutter circuits for surface surveillance and detection of low-speed targets. Jupiter is produced in both non-MTI and MTI versions. The latter is equipped with a linear digital MTI receiver for operation in clutter environment.

STATUS

The Jupiter is in use with medium and large vessels of the French Navy, as the DBRV 26A with different antenna, and on South African frigates. A solid-state transmitter version is in late development for the French Navy (DRBV 26C).

CONTRACTOR

Thomson-CSF, Division Systèmes, Défense et Contrôle, 40 rue Grange Dame Rose, 92363 Meudon-la-Foret Cedex, France.

Jupiter long-range surveillance radar antenna

6416.253

RODEO NAVAL AIR SURVEILLANCE RADAR

Rodeo Naval is a shipborne radar derived from the land-based Rodeo 2 system (**5602.153**), itself a variant of the RA20S. Rodeo Naval is a lightweight, pulse doppler, air surveillance equipment and has been ordered by the French Navy.

It is able to track two targets simultaneously, and is reported to be capable of differentiating not only between fixed and rotary wing aircraft but also between various classes of helicopters according to size. The antenna is fully stabilised in pitch and roll and the system is compatible with all types of IFF. Other characteristics are as for Rodeo 2.

STATUS

On order for the French Navy.

CONTRACTOR

Electronique Serge Dassault, 55 quai Marcel Dassault, 92214 St.Cloud, France.

Rodeo Naval surveillance radar

1933.453

CALYPSO III SUBMARINE RADAR TRS 3100

Calypso III is a higher performance radar than Calypso II (**1240.453** in 1979-80 and earlier editions), but with the same general characteristics and same functions (navigation, surveillance, target designation) with a greater detection range and the additional capability of accurate range measurement. It is also of compact design and well fitted to the severe environment characteristics of a submarine.

Calypso III results from the joint studies of a submarine mast specialist (IKL – Ingenieur Kontor Lubeck) and a radar specialist (Thomson-CSF).

Its main features are as follows:

The periscope mast does not rotate, the antenna rotation being achieved by a pressure-tight drive mechanism fitted at the top of the mast. The upper part of the mast is streamlined. The transmitter/receiver is integral with the mast, which allows:

(1) simpler connection and reduced RF losses between T/R and antenna

(2) simpler connection with the fixed part of equipment (flexible cables).

Calypso III comprises four main parts: an antenna (the Calypso II antenna); a non-rotating periscopic mast (hoisting is achieved using the ship's hydraulic system pressure); a transmitter/receiver cabinet (secured at the foot of the mast); and an operational console.

The transmitter is a conventional magnetron transmitter with a frequency adjustable magnetron and the same I/J-band transmitter as Calypso II is employed. The receiver is fitted with RF elements of long service life, a modern mixer and a logarithmic anti-clutter chain. The operational console comprises: the control panel for radar and antenna; the display (16 inch (406 mm) CRT) with an accurate digital rangefinding device; and an operational panel. The optical-periscope direction and the ESM and sonar data can be displayed on the PPI. The accurate range measurement of a detected target can be made during short-time transmission.

The detection range capability of Calypso III for a typical ASW aircraft at 2500 metres altitude is 18 nautical miles, and for a surface vessel depends on the radar horizon.

STATUS

Supplied to three foreign navies. The earlier Calypso II radar was supplied to at least 13 foreign navies. No longer in production.

CONTRACTOR

Thomson-CSF, Division Systèmes, Défense et Contrôle, 40 rue Grange Dame Rose, 92363 Meudon-la-Foret Cedex, France.

3684.453

CALYPSO IV SUBMARINE RADAR TRS 3110

The TRS 3110 Calypso IV radar is an I/J-band equipment designed to provide surveillance and navigation facilities for submarines. It is comprised of three units: an antenna, a transmitter/receiver cabinet and an operating console. Antenna and operating console are similar to Calypso III. To fulfil its functions, the Calypso IV is built around a simple navigation radar equipment, and modular design enables it to carry out additional functions merely by the addition of appropriate functional circuit cards, without changing the basic equipment.

Using a 1 ms pulse, it has a good detection range, limited only by the radar horizon (at least 10 nautical miles in free space), and using the narrow (50 ns) pulse, it has very high resolution over shorter detection ranges (down to about 15 metres). The receiver design has been developed to give good capabilities and performance in the presence of clutter.

The transmitter is a 25 kW (peak) klystron-driven unit providing several transmission options (continuous, sectoral, burst, or receive only – radar silence) giving maximum operational flexibility to meet specific requirements. The overall radar transmission parameters have been chosen to enhance the submarine's security by arranging for them to be similar to those of merchant ship radars, frequency and pulse length, in particular.

The operating console and its data processor permit the presentation, in either raw radar or synthetic form, of information such as target echoes, range and bearing labels, transmission sectors etc, as well as providing for operational control of the radar equipment itself.

STATUS

No longer in production.

CONTRACTOR

Thomson-CSF, Division Systèmes, Défense et Contrôle, 40 rue Grange Dame Rose, 92363 Meudon-la-Foret Cedex, France.

INTERNATIONAL

3312.253

Mk95 MOD 0 RADAR SET

This radar set provides search, acquisition, tracking and illumination for the NATO Sea Sparrow point defence missile system Mk57 developed by Denmark, Italy, Norway and the USA (**2770.231** in 1983-84 and earlier editions). As such, the radar forms part of the Mk91 guided missile fire control system; Mk91 Mod 0 has a single director group, and the Mk91 Mod 1 has dual directors. Each radar has separate antennas for transmission and reception, the dishes of which are protected by radome covers. The radar receiver is carried on the antenna unit. The antennas share a common mounting on the Mk78 Mod 0 missile director. An on-mount TV camera is also provided.

Operation is probably in the I/J-band part of the spectrum.

The receiver is the responsibility of the Danish Radartronics Company, and this unit's functions include clutter filtering, ECM channel, signal conversion and amplification. The Mk83 Mod 0 radar set console has search and tracking displays, status and fault indicators, radar operating controls and built-in test equipment. Operating controls provide for such functions as scan pattern selection, ECM facility selection, check-out etc. The transmitter group Mk73 Mod 0 provides RF power for search acquisition, tracking in air and surface modes, target illumination and modulation for the Sea Sparrow missile; a reference signal for missile tuning; local oscillator for the receiver; range reference for the radar target data processor which provides target detection, doppler tracking, range tracking and angle track signals; visual and aural indications to the radar set console; ECCM; and test functions.

DEVELOPMENT

Development was initially undertaken in the late 1960s by Raytheon, and subsequently on a joint basis by companies in four of the seven nations which form the NATO Sea Sparrow group. The four were Denmark, Italy, Norway and the USA.

STATUS

Operational.

CONTRACTORS

Raytheon, USA – Main contractor.

MBLE SA, France – Printed circuit boards.

Radartronics, Denmark – Radar receiver.

NEA Lindberg A/S, Denmark – Static frequency converter.

Selenia, Italy – Firing officers console.

Bronswerk-Amersfoort, Netherlands – Liquid cooler group.

Fokker VFW BV, Netherlands – Director controller and launcher rails.

Norsk Forsvarsteknologi A/S, Norway – Radar pedestal and fire control computer.

Mk91 FCS for NATO Sea Sparrow system aboard Spruance, *showing the Mk95 radar. Antenna on left is for satellite communications*
(Stefan Terzibaschitsch)

6420.253

EMPAR-EUROPEAN MULTIFUNCTION PHASED ARRAY RADAR

EMPAR is a European G-band phased-array radar in development by Marconi Radar Systems (UK) and Selenia (Italy). The programme provides a new generation multifunction system to meet the requirements of the respective navies, and for particular application in the NATO Frigate Replacement Programme for the 1990s (NFR 90).

The radar system produces and provides both surveillance and tracking facilities. The full phase scan technique employed in the system affords electronic stabilisation of the scanning patterns. Concurrent accomplishment of a variety of functions, ie search, correlation, acquisition and tracking, which are usually carried out by individually tasked systems, will now be superseded by the EMPAR system.

EMPAR is designed to be capable of handling hundreds of targets distributed hemispherically throughout 360°, while simultaneously tracking many targets for missile engagement. Design options include a single rotating antenna array, back-to-back dual array, and multi-array configurations. The system is designed on a modular concept to provide ship fitting versatility and cost-effective solutions to a wide variety of weapon systems applications, including the NFR 90 local area missile systems (LAMS).

CHARACTERISTICS

Frequency: G-band

Antenna: Monopulse phase/phase scan in both planes

Scan angles: Azimuth ±45°; elevation ±60°

Beamwidth: Azimuth 2.5°; elevation 2.5°

Rotation speed: Single array 60 rpm; dual 45 rpm

Masthead weight: Single array 1875 kg; dual 3200 kg

Below deck weight: Single array 3500 kg; dual 6600 kg

STATUS
In development. Scheduled to be in production by 1992.

CONTRACTORS
Marconi Radar Systems Ltd, Writtle Road, Chelmsford, Essex CM1 3BN, England

Selenia-Elsag Consortium for Naval Systems, Via Panama 52, 00198 Rome, Italy.

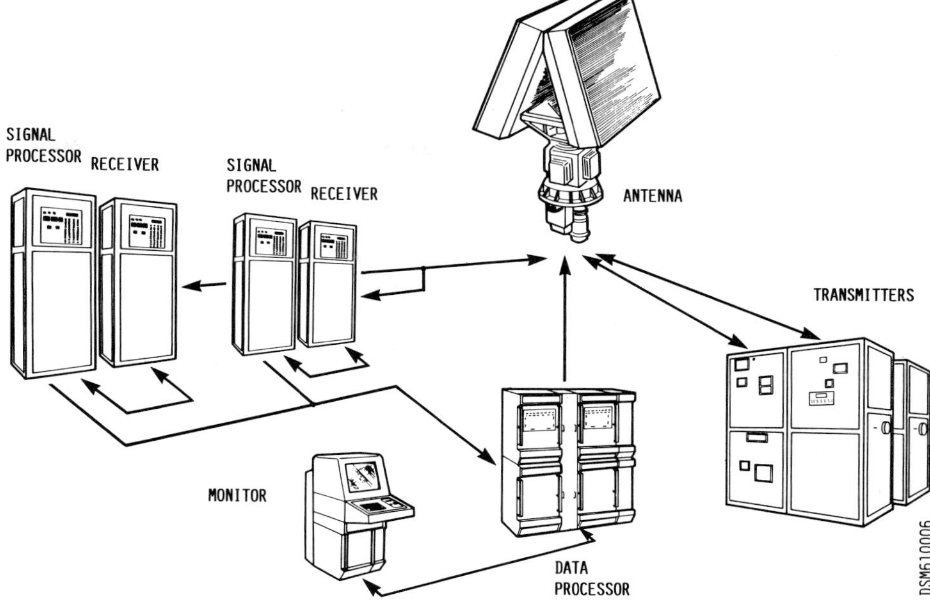

Basic configuration of the EMPAR radar system

ISRAEL

3709.253

EL/M-2221 NAVAL FIRE CONTROL RADAR

The Elta EL/M-2221 is an I/J-band monopulse tracking radar for naval fire control applications. Its operational roles include: surface and air search and acquisition; surface-to-surface missile fire control; surface-to-air missile fire control; gun fire control against surface targets; and gun fire control against air targets. Low weight and compact dimensions make it suitable for installation aboard ships of all sizes, including small patrol craft and missile boats. In addition to its ability to control the Israeli Gabriel anti-ship missile (**6019.221**), both active and semi-active homing missiles can be controlled.

Target acquisition can be initiated by designation from a search radar, other sensors (eg optical, ESM etc), or can occur automatically within a given sector. In the tracking mode there is automatic lock-on, and the radar is TV/optical tracker compatible. In the search mode, the radar performs track-while-scan, and miss-evaluation and ranging facilities are available.

CHARACTERISTICS
Frequency: I/J-band

Dvora fast missile boat uses Elta EL/M-2221 fire control radar to control Gabriel SSMs and guns

Antenna: Monopulse
Diameter: 1 m
Height: 1.5 m
Weight: 230 kg
Radar cabinet
Height: 1.2 m
Width: 70 cm
Depth: 50 cm
Weight: 120 kg

STATUS
Production and in service.

CONTRACTOR
Elta Electronics Industries Ltd, (a subsidiary of Israel Aircraft Industries Ltd), Ashdod, Israel.

5696.253

AUTOMATIC MISSILE DETECTION RADAR

The Automatic Missile Detection Radar (AMDR) has been developed by Elta as a shipborne radar designed to automatically detect incoming sea skimming missiles in addition to normal air surveillance with automatic detection and TWS capability. It is designed for installation on ships from 450-tons upwards.

AMDR is a coherent pulse doppler radar operating in S-band to give fast alarm warning at ranges of over 30 km with a very high probability of detection. The system will detect and warn of an approaching 0.1 m² missile at 15 km with 90 per cent probability of detection. Velocity coverage is 0.3 to 3 Mach. The antenna is a lightweight, stabilised cosecant² unit with a total weight of 220 kg at the masthead. Coherent pulse doppler technique, combined with discrete Fourier transform, doppler filtering processing and programmable data processing enable *a priori* filtering out of slow moving targets (below Mach 0.2) and target radial velocity measurement during a single antenna scan. Display options include a 16-inch PPI, digital scan converter colour TV display and an alpha-numeric tabulated data display.

The AMDR may be installed as a stand-alone system or added to the EL/M-2207, EL/M-2208, or other naval search radars.

CONTRACTOR
Elta Electronics Industries Ltd (a subsidiary of Israel Aircraft Industries Ltd), Ashdod, Israel.

3264.253

ELTA EL/M-2207 NAVAL RADAR

Elta Electronics Industries Ltd produces a number of military radars, details of most of which are classified. However some particulars have been released concerning the EL/M-2207 F-band general purpose naval radar, which is one of the EL/M-2200 range of shore and shipboard radars. The same basic transmitter/receiver unit (either singly or duplicated for diversity operation) is assumed to be used throughout the range, with differences in the antenna employed and the extent of additional facilities incorporated accounting for the main variations between models.

The EL/M-2207 is a medium power equipment offering high performance in both air and surface search roles. A lightweight stabilised antenna is used, the AT-101, which is claimed to require a reflector area only 66 per cent of that required by a conventionally designed antenna to achieve the same performance characteristics.

AT-101 antenna used in EL/M-2207 naval radar system

Operational roles include air early warning surface surveillance, target indication, and navigation. Other features are operator selection of transmitter pulse width, single knob frequency tuning, true motion display, and integrated IFF.

One or two IP-33 41 cm diameter PPI display consoles are normally fitted. Antenna polarisation is switchable linear/circular and there is remote control of antenna elevation angle.

CHARACTERISTICS
Transmitter
Peak power: 425 kW typical
Average power: 425 W typical
Frequency: 3.1 – 3.3 GHz tunable
PRF: 500 – 700 pps, long pulse; 1000 – 1500 pps, short pulse
Pulse length: 1.4 or 0.4μs
Receiver
Type: Linear, with switchable STC, FTC, IAGC, Log-FTC-A Log, CFAR
Noise: 2 dB

IF frequency: 60 MHz
IF bandwidth: 1.4 MHz, long pulse; 3 MHz, short pulse
Antenna
Polarisation: Vertical/circular
Gain: 28 dB (F-band); 17 dB (IFF)
Beamwidth (3 dB): 3.3° azimuth; 10° elevation, cosec² to 30°
Rotation rate: 12 or 24 rpm, selectable
Stabilisation: ± 15°
Dimensions: 2.4 × 1 m (reflector)
Performance
Detection range: 55 km, against fighter aircraft size target
Resolution: < 75 m, short pulse; < 230 m, long pulse

STATUS
No details of ship fittings have been released for publication.
CONTRACTOR
Elta Electronics Industries Ltd (a subsidiary of Israel Aircraft Industries Ltd), Ashdod, Israel.

4610.253
EL/M-2208 SHIPBORNE SEARCH RADAR
The EL/M-2208 is another member of the EL/M-2200 family and is a dual band shipborne radar, with integral IFF, for simultaneous air and surface surveillance. Two transceivers (S- and X-band) use a common lightweight, small size antenna mounted at masthead. Full automatic stabilisation provides complete coverage even in adverse sea states.

Long pulse and DMTI are provided for long range and air surveillance functions. A navigation facility is included by selection of a short pulse. Either separate or combined S- and X-band presentation, plus IFF, can be selected on a 16 inch display.

CHARACTERISTICS
Antenna
Stabilisation: ± 15°
Polarisation: Linear/circular
Transmitter: Magnetron

Technical specification	**S-band**	**X-band**
Frequency	3.1 – 3.3 GHz (tunable)	9.345 – 9.405 GHz (fixed)
Peak power	425 kW	25 kW
Pulse length	1.4 or 0.4 μs	0.25 μs
PRF	500 – 750 pps (long pulse) 1000 – 1500 pps (short pulse)	1000 – 1500 pps

CONTRACTOR
Elta Electronics Industries Ltd (a subsidiary of Israel Aircraft Industries Ltd), Ashdod, Israel.

Detection range: 32 nm (59 km)
Resolution: 1° (azimuth); 50 m (range)
Dimensions: 2.4 × 1 m
Rotation rate: 12 or 24 rpm
Weight: 270 kg (S-band); 200 kg (X-band)

5846.293
AT-101A TRIPLE BAND SURVEILLANCE ANTENNA
The AT-101A surveillance antenna is a triple band unit used in a number of Elta shipborne and land mobile applications. It consists of an S-band cosec antenna with a reflector area only ⅔ the size of a conventional unit, plus X- and L-band feeds which use the same reflector as the S-band. Switchable linear/circular polarisation is incorporated with remote control of antenna elevation angle. The shipborne version is fully stabilised.
STATUS
In service.
CONTRACTOR
Elta Electronics Industries Ltd (a subsidiary of Israel Aircraft Industries Ltd), Ashdod, Israel.

CHARACTERISTICS	**S-Band**	**X-band**	**L-band**
Frequency (MHz)	3100-3300	9345-9405	1030-1090
Gain	28 dB	34 dB	17 dB
Azimuth beamwidth	3°	1°	7°
Elevation beamwidth	10°	5°	20°
Polarisation	Vertical/ Circular	Horizontal	Vertical
Rotation rate	12/24 rpm, remotely selectable		
Stabilisation	± 15°		
Reflector dimensions	2.4 × 1 m		
Overall height	1.75 m (including pedestal)		
Weight	245 kg		

ITALY

1364.253
RAN-3L NAVAL RADAR (MM/SPS-768)
The RAN-3L (Italian Navy MM/SPS-768) is a high power early warning search radar operating in D-band and suitable for installation on large and medium tonnage vessels. The radar combines long range with high precision, coupled with ECCM and anti-clutter facilities.

The system consists of a fully coherent transmitter/receiver chain and a digital processor incorporating a composite coded waveform. It is fully solid-state except for the final RF power amplifier. An IFF antenna can be mounted on the reflector.
CHARACTERISTICS
Frequency: D-band
Antenna: Double curvature parabolic reflector with hog horn feed
Rotation rate: 6 rpm
Polarisation: Linear
Transmitter: Frequency agile, frequency diversity and fixed frequency; multipulse coded radiation
Range: Greater than 110 nm on a fighter aircraft
Range accuracy: 70 m
Angular accuracy: 0.4°
STATUS
In production. In service with the Italian Navy.
CONTRACTOR
Selenia-Elsag Consortium for Naval Systems, Via di S. Alessandro 28/30, 00131 Rome, Italy.

Antenna complex (large antenna on lattice structure) of the early warning search radar MM/SPS-768 (Selenia type RAN-3L) installed on board the Italian Navy cruiser Doria. *The same equipment is installed on* Veneto *and* Audace/Ardito, Garibaldi, *and will be fitted on the new DDGs*

1699.253
RAN-10S NAVAL RADAR (MM/SPS-774)
The RAN-10S (Italian Navy MM/SPS-774) is a high power combined air/surface surveillance radar operating in the F-band and suitable for installation on medium tonnage ships, such as destroyers, frigates and corvettes. The design philosophy is the same as that of the RAN-3L and a number of items are common.

The main features are high elevation coverage, elevated data rate and high precision/resolution. The use of coded waveform in conjunction with digital processing of the received signal gives excellent capabilities of clutter and jamming rejection, enabling it to operate in full frequency agility while maintaining anti-clutter capabilities.

The antenna is roll and pitch stabilised and can accommodate an IFF antenna with ISLS mounted on top of the reflector.

Typical operating roles include air warning against aircraft and missiles, aircraft direction, surface surveillance, navigation, and direction of SSM. The radar is optimised for target designation to guns and/or SAM weapon control systems.
CHARACTERISTICS
Frequency band: F-band
Antenna rotation: 15 and 30 rpm
Polarisation: Linear and circular
Transmitter: Frequency agile, fixed frequency, multi-pulse coded radiation
Range: Greater than 40 nm on a fighter aircraft
Range accuracy: 20 m
Angular accuracy: 0.35°
STATUS
In production. In service with the Italian Navy and several other navies.
CONTRACTOR
Selenia-Elsag Consortium for Naval Systems, Via di S. Alessandro 28/30, 00131 Rome, Italy.

RAN-10S antenna installation aboard a 'Maestrale' class frigate

1365.253

RAN-11 L/X AND RAN-12 L/X NAVAL RADAR

The RAN-11/12 L/X are compact integrated radars which include an I-band (X-band) transmitter and receiver for detection of surface and low flying targets, together with a D-band (L-band) transmitter and receiver for air surveillance. The RAN-12 L/X is the high power version of the RAN-11 L/X, having a high power transmitter in L-band to obtain better air cover. Either version is suitable for installation as the main radar on corvettes, hydrofoils and fast patrol boats, or as a secondary radar on larger vessels. The individual I-band and D-band radars can also be supplied as single equipments.

The D-band transceiver is a completely solid-state radar which results in compactness with no life-limited electronics. It incorporates fully coherent processing with high PRF, multi-pole MTI filtering and pulse compression, implemented by digitised hardware. An associated IFF signal uses the D-band channel.

CHARACTERISTICS

Frequency band: D and I-band
Antenna: Roll and pitch stabilised
Rotation rate: 15 and 30 rpm
Polarisation: Linear and circular
Transmitter: I-band frequency tunable; D-band frequency diversity, waveform coding
Range (on a fighter aircraft): Low altitude (I-band) >20 nm; medium altitude >12 nm (RAN-11 L), >14 nm (RAN-12 L)
Range accuracy: D-band 160 m; I-band 45 m
Angular accuracy: D-band 1.5°; I-band 0.2°

STATUS

In production. In service with several navies.

CONTRACTOR

Selenia-Elsag Consortium for Naval Systems, Via di S. Alessandro 28/30, 00131 Rome, Italy.

Roll and pitch stabilised antenna complex of the RAN-11 L/X search radar produced by Selenia-Elsag. The upper portion of the reflector operates only for the D-band

---

1368.253

ORION RTN-10X NAVAL TRACKING RADAR

This is an I-band (X-band) radar used with different fire control systems to control gun and missile firings. It is based on a conical-scan narrow beam antenna, with a tri-port feed and uses a magnetron with fast tuning capabilities.

The system employs different receiver signal and data processing techniques to ensure a high level protection from jamming and interference and also to maintain accurate tracking of both high and low flying targets.

CHARACTERISTICS

Frequency band: I-band
Antenna type: High gain, tri-port feed
Transmitter: Frequency changeable, high power magnetron
Range: >20 km on missiles

STATUS

In production and in service with the Italian Navy and other navies.

CONTRACTOR

Selenia-Elsag Consortium for Naval Systems, Via di S. Alessandro 28/30, 00131 Rome, Italy.

Orion RTN-10X radar antenna director

1935.253

ORION RTN-20X NAVAL RADAR

This is a J-band (X-band) radar specifically developed for CIWS tasks and currently associated with the DARDO system. The radar uses a monopulse, cassegrain antenna with a polarisation twist and employs a fully coherent chain transmitter using a TWT as the final amplifier. Two different pulse lengths can be selected, both at elevated repetition rates.

The transmission frequency can be changed randomly, on a pulse-to-pulse or burst-to-burst basis, over a wide frequency spectrum. The fully digital receiver uses a multi-canceller MTI and is equipped to minimise the effects of multi-path on low flying targets.

CHARACTERISTICS

Frequency band: J-band
Antenna type: Cassegrain
Tracking method: Monopulse
Transmitter: Coherent chain with TWT amplifier, frequency diversity
Receiver: MTI comptaible with frequency change, anti-jamming features
Range: >10 km on missiles

STATUS

In production and in service with the Italian Navy and other navies.

CONTRACTOR

Selenia-Elsag Consortium for Naval Systems, Via di S. Alessandro 28/30, 00131 Rome, Italy.

Orion RTN-20X tracking radar

6875.253

ORION RTN-25X NAVAL TRACKING RADAR

The RTN-25X is a J-band (X-band) radar that uses the same antenna and transmission chain as the Orion RTN-20X (**1935.253**), but with greater radiating power. In addition two operating modes can be selected; a long pulse and coded waveform and a short pulse using a single pulse.

In both modes the receiving processor comprises a multi-canceller MTI and a data processor enabling the radar to perform accurate tracking on any kind of threat and in various environments. The radar can be applied to different fire control systems to control guns firing at short and medium range.

CHARACTERISTICS

Frequency band: J-band (X-band)
Antenna type: Horn cassegrain
Tracking method: Monopulse
Transmitter: Coherent chain with TWT, dual mode, frequency diversity, pulse compression
Receiver: MTI compatible with frequency diversity, anti-jamming features
Range: 15 km on missiles

STATUS

In development.

CONTRACTOR

Selenia-Elsag Consortium for Naval Systems, Via di S. Alessandro 28/30, 00131 Rome, Italy.

1936.253

ORION RTN-30X NAVAL RADAR

The RTN-30X is a powerful radar operating in I-band (X-band) with a monopulse cassegrain antenna, using polarisation twist, that can also radiate the CW power for surface-to-air missile homing. The transmission chain is fully coherent and uses a high average power transmitter. Two operational modes can be selected; a long pulse mode and coded waveform and a short pulse mode using a single pulse. In both modes the receiving processor comprises a multi-canceller MTI and a data processor that enables the radar to perform accurate tracking of any kind of threat in various environments.

The extended range capability enables this radar to operate with a wide range of weapons (CIWS, guns and missiles) to ensure an in-depth engagement of a single threat and a redundant reaction on multi-threats.

The Orion RTN-30X has many components in common with RAN-30X surveillance radar.

CHARACTERISTICS

Frequency band: I-band
Antenna type: Horn cassegrain
Tracking method: Monopulse
Transmitter: Coherent chain with TWT, dual mode, frequency diversity, pulse compression
Range: >25 km on missiles

STATUS

In production and in service with the Italian Navy.

CONTRACTOR

Selenia-Elsag Consortium for Naval Systems, Via di S. Alessandro 28/30, 00131 Rome, Italy.

6424.253

RAN-30X SURVEILLANCE RADAR

This is a combined surface and air surveillance radar operating in I-band (X-band) and specifically developed for small and medium tonnage ships. The equipment particularly enhances detection of low altitude threats and steep diving missiles in clutter and jamming environments and provides high resolution and data rate.

The RAN-30X makes extensive use of solid-state techniques to achieve a very compact, reliable and easily maintainable system that incorporates the following features:

(a) single or multi-beam antenna for high coverage
(b) fully coherent receiver/transmitter architecture with high internal stability to ensure high clutter cancellation and fast frequency jump capability
(c) pulse compression and adaptive MTI
(d) transmission of high duty ratio coded waveform for improved ECCM performance and low intercept probability.

An IFF antenna unit including ISLS can be integrated with the antenna.

CHARACTERISTICS

Frequency band: I-band
Antenna: Stabilised cylindrical, parabolic reflector
Rotation rate: 15 and 30 rpm
Polarisation: Linear (circular optional)
Transmitter: Frequency agility and diversity, fixed frequency, coded waveform
Range: >20 nm on a fighter aircraft
Range accuracy: 25 m
Angular accuracy: 0.4°

STATUS

In development.

CONTRACTOR

Selenia-Elsag Consortium for Naval Systems, Via di S. Alessandro 28/30, 00131 Rome, Italy.

6876.253

RAN-30S NAVAL RADAR

The RAN-30S is a fully solid-state 3-D radar operating in F-band (S-band) and suitable for installation on large or medium tonnage vessels. It provides a clear and accurate picture of the environment (target and clutter) to ensure a complete threat classification, while at the same time providing accurate designation in range, bearing and elevation to the weapon systems. It is particularly suited for interceptor control and provides accurate TWS of multiple simultaneous targets.

The system can be operated in two search modes; a medium/short range mode with high elevation coverage and high data rate, and a long range mode with extended range capability and low data rate. It incorporates the following features:

(a) single transmitting beam and multiple receiving beams with very low sidelobe levels
(b) pulse compression of the transmitted waveform
(c) advanced anti-clutter with MTI processing
(d) advanced ECCM
(e) automatic target detection and tracking
(f) high performance against targets of small radar cross-section flying at low altitude.

The reflector and stabilised platform are fitted for installation of an IFF ISLS antenna.

CHARACTERISTICS

Operating frequency: F-band
Antenna: Stabilised rectangular planar array
Rotation rate: 5, 15 and 30 rpm
Polarisation: Horizontal
Transmitter: Coherent chain, long pulse frequency modulated, short pulse unmodulated. Operating frequency random in agility mode or constant frequency mode. Staggered PRF
Receiver: Dual frequency conversion, pulse compression, frequency multiplier chains, SAW filtering techniques
Range (on a fighter aircraft): Long range mode 100 nm, medium/short range mode 50 nm
Range accuracy: 50 m
Angular accuracy: 0.1°

STATUS

In development.

CONTRACTOR

Selenia-Elsag Consortium for Naval Systems, Via di S. Alessandro 28/30, 00131 Rome, Italy.

3483.453

MM/BPS-704 SUBMARINE RADAR

The MM/BPS-704 is a naval search and navigation radar for use aboard submarines. The general characteristics are similar to the MM/SPN-703 surface ship version (**3482.253**), using the same 20 kW transmitter and with the same pulse length/PRF combinations. A specially designed high pressure resistant antenna (up to 60 kgm/cm²) is part of the equipment. The principal differences are indicated in the following table.

CHARACTERISTICS

Antenna span: 1 m
Beamwidth: 2.2° horizontal; 11° vertical
Gain: At least 27 dB
Noise figure: Better than 11 dB

STATUS

Italian Navy 'Sauro' class submarines are fitted with the BPS-704.

CONTRACTOR

SMA – Segnalamento Marittimo ed Aereo, Via del Ferrone Soffiano, PO Box 200, 50100 Florence, Italy.

MM/BPS-704 antenna

1702.253

3RM SERIES NAVAL RADARS

The 3RM Series are I-band navigation and surface warning radars, with some air target capability. The equipment has been tested and qualified in full accordance with MIL-E-16400F specifications. The antennas used are of the slotted waveguide type and provide horizontal beamwidths from 0.8 to 2°, depending upon the antenna length, and a vertical beam pattern of 26°, shaped to 40°. Scanner rotation rate is 25 rpm. Two transmitters are available, with outputs of 7 or 20 kW. The display unit has a nine inch (23 cm) diameter PPI.

CHARACTERISTICS

Frequency: 9375 MHz
Peak power: 7 or 20 kW
Antenna: Slotted waveguide
Polarisation: Horizontal
Rotation rate: 25 rpm
Range: 40 nm
Pulse length/PRF: 0.05μs/6000 Hz; 0.15μs/3000 Hz; 0.5μs/1500 Hz; 1.5μs/750 Hz
Receiver: Linear, lin-log
Display: 9 inch (23 cm) PPI; 8 ranges 0.25 – 40 nm
True motion unit: On request

STATUS

In production and fitted to ships of a number of navies, and notably those of the Italian Navy, and on the West German fast patrol boats of the '143' and '148' classes. The US Navy has officially approved introduction of this equipment into its inventory under the military nomenclature AN/SPS-63. A submarine version (3RM20 SMG) is a standard production item.

CONTRACTOR

SMA – Segnalamento Marittimo ed Aereo, Via del Ferrone Soffiano, PO Box 200, 50100 Florence, Italy.

3482.253

MM/SPN-703 NAVIGATION RADAR

The Italian Navy MM/SPN-703 is the latest version of the 3RM navigation and surface search radar (**1702.253**), incorporating the latest state-of-the-art developments in electronic design. Most characteristics are similar or identical to those of the 3RM, above; known differences are listed in the following table.

CHARACTERISTICS

Frequency: 9.375 MHz
Peak power: 20 kW (7 kW optional)
Antenna: Low profile slotted waveguide with cylindrical-paraboloid depolarising reflector
Beamwidth: 1.2° azimuth; 25° vertical
Pulse length/PRF: 0.05μs/5200 Hz; 0.15μs/2600 Hz; 0.5μs/1300 Hz; 1.5μs/650 Hz
Noise figure: 9 dB

STATUS

Adopted by Italian, West German, US and Venezuelan navies.

CONTRACTOR

SMA – Segnalamento Marittimo ed Aereo, Via del Ferrone Soffiano, PO Box 200, 50100 Florence, Italy.

1703.253

SPQ-2D/F NAVAL RADAR

The SPQ-2D/F is a medium power I/J-band radar the principle operation of which is based on time shared transmission of long and short pulses to achieve simultaneous video signals with high sensitivity for surface search and short-range air search, or with high resolution for navigation and close-in control. The last feature is particularly appropriate for rescue operations and the direction of ASW helicopters.

The SPQ-2D equipment is currently integrated as the target designation radar in various electronic combat systems. A later model is the SPQ-2F.

Two control units which include all the operational controls of the radar permit use of remote or local controlling facilities by either CIC or EW operators.

SPQ-2D/F antenna group with stabilised mount

The SPQ-2D/F is designed for IFF compatibility and there is provision for 'Beacon' operation in conjunction with an I/J-band transponder. Among the optional additional facilities which are available are: TWS operation at an antenna rotation rate of 40 rpm; frequency agility; MTI D-band transceiving section integrated with the I/J-band (and with a common SPQ-2 type antenna group) for low-flying target detection; and varying modes of operation.

The antenna assembly is produced in both roll and pitch stabilised and non-stabilised forms. The reflector is a double-curvature lattice and mesh structure of about three metres span, and is illuminated by a feed horn carried by an underslung support boom.

The sensor's output signals can be processed for different types of plot extraction and displayed on SMA-designed and manufactured weapons management consoles.

STATUS

The SPQ-2D/F has been in production for several years and is fitted on ships of several navies including the Italian Navy ('Alpino' class frigates), Canadian Navy (DDH 280 frigates), and the Venezuelan Navy (Vosper FPB).

CONTRACTOR

SMA – Segnalamento Marittimo ed Aereo, Via del Ferrone Soffiano, PO Box 200, 50100 Florence, Italy.

3262.253

MM/SPQ-701 NAVAL RADAR

The SMA MM/SPQ-701 naval radar is an I-band equipment designed for surface and air target detection and primarily intended for fitting in small vessels such as hydrofoils and fast patrol boats. It is essentially a combined system that includes navigational and search components, and the roll-stabilised mount carries both antennas.

Reportedly of advanced design, the MM/SPQ-701 employs frequency agility and is provided with digital MTI. Good performance is claimed, including long and short range target detection in ECM. Provisions include antennas and appropriate rotary joints etc.

STATUS

In production. In service on hydrofoils of the Italian Navy.

CONTRACTOR

SMA – Segnalamento Marittimo ed Aereo, Via del Ferrone Soffiano, PO Box 200, 50100 Florence, Italy.

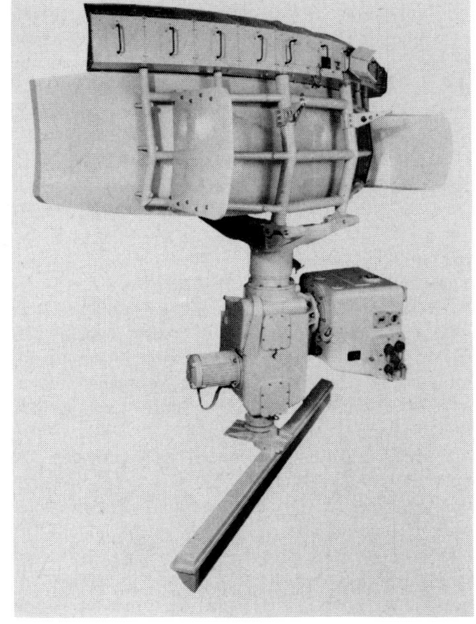

MM/SPQ-701 antenna group

3263.253

MM/SPS-702 NAVAL RADAR

The SMA MM/SPS-702 naval radar is derived from the MM/SPQ-701 described above (**3262.253**), and is essentially comprised of the search components of that equipment. An I-band system, it provides assistance in missile guidance, supplying data for TWS processing and also providing a mount for the command guidance antenna. IFF facilities are incorporated.

STATUS

In service with the Italian Navy.

CONTRACTOR

SMA – Segnalamento Marittimo ed Aereo, Via del Ferrone Soffiano, PO Box 200, 50100 Florence, Italy.

MM/SPS-702 antenna group and stabilised mount, with integrated IFF antenna

6874.253

MM/SPN-728 SEARCH RADAR

The MM/SPN-728 is a simple X-band radar designed to meet the requirements for shipborne surface search. It has been developed by integrating the transceiver of a high performance navigation surface search radar with a very high performance antenna group, including a low noise amplifier. The antenna is a double curvature unit, vertical coverage cosecant squared with a special pattern and an integrated IFF antenna. In addition this antenna system allows operation of the 'Intra' positioning system for helicopters.

The video signal processor and raster scan daylight display is positioned in the ship's operations room and TV presentation also enables display in various parts of the ship by monitor repeaters.

A number of options are available, including different antenna sizes, a roll and pitch stabilised mount and a frequency agile high power transmitter.

CONTRACTOR

SMA – Segnalamento Maritimo er Aereo, Via del Ferrone Soffiano, PO Box 200, 50100 Florence, Italy.

JAPAN

6136.253

OPS SERIES SHIPBORNE SEARCH RADARS

The Japanese Maritime Self-Defence Force employs a number of air and surface surveillance radar systems, some of which have been developed from US systems and others indigenously designed. The modern systems in current operational use include the air search radars OPS-11, OPS-12 and OPS-14, and surface search radars OPS-15, OPS-16, OPS-17, OPS-18 and OPS-28.

The OPS-14 is a medium range air surveillance radar of Japanese design and manufacture which first entered service in 1971 aboard the minelayers *Souya* and *Hayase*. There are a number of variants of this equipment, OPS-14A, B and C. The B version is in service in the Hatsuyuki class of destroyers and the C version is to be installed in the Improved Hatsuyuki class, the first of which was launched in August 1986.

The OPS-18 surface surveillance radar has been developed from the US AN/SPS-10 (**1564.253**) and has been installed in a number of ships, including the Hatsuyuki class. The OPS-28 is an advanced model of Japanese design which first entered service in 1980 in the large helicopter destroyers of the Shirane class. It will replace the OPS-18 in the Improved Hatsuyuki destroyers.

STATUS

All models are in operational service with the Japanese Maritime Self-Defence Force.

NETHERLANDS

1973.253

LW08 NAVAL SURVEILLANCE AND WEAPON DIRECTION RADAR

The LW08 is a high-power, coherent, D-band radar for long-range surveillance and target indication to weapon systems. The lightweight hydraulically stabilised antenna has a low sidelobe level that helps the system to continue operation in an ECM environment, and a synthesiser-driven TWT transmitter results in great frequency flexibility over a wide band. Excellent performance in various clutter conditions is ensured by a high dynamic range receiver with the application of digital video processing (*eg* MTI), supported by circular polarisation. A carbon-fibre or stainless steel antenna is used to counteract the effects of salt atmosphere and corrosive exhaust gases.

CHARACTERISTICS

Antenna: Horn-fed parabolic reflector

Polarisation: Linear and circular

Beamwidth: Vertical cosec up to 40; horizontal 2.2

Transmitter: D-band coherent system

PRF: 1000 and 500 Hz

STATUS

LW08 systems have been delivered for fitting in the Royal Netherlands Navy 'S' type frigates, the Royal Navy anti-submarine cruisers and Type 42 frigates, as well as the Argentinian aircraft-carrier *Veinticinco de Mayo*. In the Royal Navy installations a Marconi antenna is employed and the radar is designated Type 1022 (**3485.253**). An extended range version of the LW08 is applied on the RNIN's air defence 'L' type frigates.

CONTRACTOR

Hollandse Signaalapparaten BV, PO Box 42, 7550 GD Hengelo, Netherlands.

LW08/2S early warning and weapon direction radar

3484.253

SIGNAAL STIR TRACKING RADAR

STIR (Signaal Track and Illuminating Radar) performs automatic simultaneous control of both missiles and guns against high speed sea skimmers and divers for medium to very long range performance. It also includes surface target tracking and full anti-surface weapon control capabilities.

Besides accurate gun control, STIR provides CW illumination guidance for almost all semi-active homing missiles. For enhancement of the low level performance and elimination of the mirror effect, STIR incorporates integrated dual frequency (X/Ka-band) target tracking. As well as target tracking STIR performs sector search, height measurement, splash spotting and correction (automatic) for surface target fire, and curved path prediction in addition to linear path prediction.

All STIR systems have a number of features, including integrated I-band CW illumination, a variety of MTI and ESM/ECM facilities, pilot tone injection for on line boresight calibration, and sector search capability (to search the jammed sector of a surveillance radar). Growth potential and an over-capacity in range are incorporated in the system to cater for future target developments such as higher speed and lower radar cross-section. Operation can be automatic, semi-automatic or manual, either autonomous or in combination with other sensor and control systems.

STIR is available with magnetron or TWT transmitter and with three director types; STIR 1.2, STIR 1.8 and STIR 2.4. This flexibility matches with the various weapons to be controlled and with the area and point defence tasks of the ships.

Surface-to-air missiles currently controlled with STIR are Seasparrow (RIM-7H, RIM-7M and RIM-7M vertical launch); Aspide; Standard missile (SM-1 Block V and VI, SM-2 Block II).

Guns controlled include FMC 5 in/54-cal Mk 45; OTO Melara 127 mm/54-cal and 76/62-cal; Bofors 57 mm Mk 2, 40 mm and 150 mm twin-barrelled; Breda 40 mm twin mount; Oerlikon 25 mm 4-barrelled anti-missile gun; Creusot Loire 100 mm gun mount Mod 68/13.

CHARACTERISTICS

STIR 1.8 (Magnetron)

Antenna: Single dish cassegrain 1.8 m

Polarisation: Vertical

Beams: I, J and K-band pencil beams, widths 1.4, 1.4 and 0.3° respectively

Transmitter: I-band monopulse/magnetron transmitter/receiver chart with FFT processing; K-band monopulse/magnetron transmitter/receiver chain with FFT processing; J-band CW for illumination and rear reference

Peak power: 200 kW I-band; 20 kW K-band, >5 kW average J-band

Instrumental range: 60 km

Mass: 1700 kg

STIR 1.8 (TWT)

Antenna: Single dish cassegrain 1.8 m diameter

Polarisation: Vertical

Beams: Concentric I and J-band beams, width 1.4°

Transmitter: I-band monopulse TWT transmitter, receiver with FFT processing; J-band CW for illumination and rear reference

Average power: 4 kW I-band; >5 kW J-band

Instrumental range: 180 km

Mass: 1700 kg

STIR 2.4 (TWT)

Antenna: Single dish cassegrain 2.4 m diameter

Polarisation: Vertical

Beams: Concentric I and J-band beams, width 1°

Transmitter: I-band monopulse/TWT transmitter/receiver with FFT processing; J-band CW for illumination and rear reference

Average power: 4 kW I-band; >5 kW J-band (depending on CW transmitter)

Target acquisition range: >140 km (for 1 m² target, PD ⩽ 90%)

STATUS

STIR has been widely selected by various navies for corvette and frigate programmes. Up to 100 STIRs have been ordered by nine navies including six NATO countries. As at December 1987 56 STIRs were in operation.

CONTRACTOR

Hollandse Signaalapparaten BV, PO Box 42, 7550 GD Hengelo, Netherlands.

6877.253

STING RADAR FIRE CONTROL SYSTEM

STING is a fully automatic naval fire control derived from the STIR family (**3484.253**). Specifically developed to interface with the British Aerospace Vertical Launch Seawolf missile, STING can also be used to control dual purpose guns and has enough growth potential to make it suitable for course-correctable ammunition when it becomes available.

STING is well suited to small craft such as FPBs and FACs since, depending on the version selected, the director weighs only between 625 and 750 kg.

Image-free tracking down to wave-top level is achieved by Signaal's patented dual frequency (X and Ka) band. The processing unit uses FFT filters and selects signals appropriate for optimal target tracking.

STING is available in four different versions. The basic version incorporates the dual frequency band only, controls a dual purpose gun and offers as a growth potential future application of course-correctable ammunition. The more extended versions include either optronic sensor equipment for passive target tracking, CW target illumination for semi-active homing missiles, or command link equipment for Command-Line-Of-Sight (CLOS) missile systems.

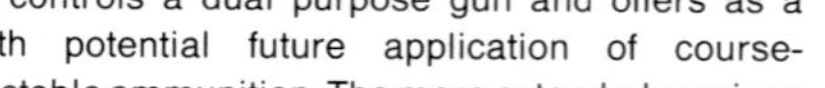

The version for the vertical launch Seawolf integrates the three main radar functions needed into one compact system. These functions are target acquisition and tracking, missile gathering and tracking and guidance of the missile to its target. The missile gathering unit couples information from a wide aperture receiving antenna to a narrow band (FFT) filtering process. This ensures fast and accurate gathering of the vertically launched missile. Moreover the gathering is passive and unaffected by weather and electronic interference.

STATUS

The entire STING system is scheduled to be ready for production in 1988. It has already been offered to three navies in conjunction with Seawolf missiles and to a fourth for use with a dual role naval gun.

CONTRACTOR

Hollandse Signaalapparaten BV, PO Box 42, 7550 GD Hengelo, Netherlands.

1554.253

DA.05 NAVAL SURVEILLANCE RADAR

The DA.05 is a high-power E/F-band radar for medium-range surveillance and target indication to weapon control systems. Capable of accurately detecting a jet fighter at approximately 135 km, the DA.05 is a stable and reliable system, distinguished by its high-gain antenna, powerful transmitter and sensitive receiver. A narrow beamwidth and short pulse length result in clear resolution. This helps ensure accurate target indication. Further enhanced by a choice of frequencies and a video processor, the system is capable of performing effectively in a cluttered environment.

The DA.05 has a magnetron-powered transmitter, a solid-state receiver with digital MTI, and interference

Biaxially stabilised lightweight platform with DA.05 antenna

suppression circuits ensuring high reliability in rain, sea and land clutter. A low sidelobe level helps the system continue operation in ECM environments. The carbon-fibre antenna version, with its low weight, enables installation in positions high on a mast and on smaller ships. A lightweight electro-hydraulic antenna stabiliser is available.

CHARACTERISTICS

Antenna: Horn-fed parabolic reflector
Beamwidth: Vertical cosec up to 40; horizontal 1.7
Polarisation: Linear and circular
Transmitter: E/F band tunable coaxial magnetron
PRF: 1000 and 500 Hz

STATUS

Over 50 DA.05 systems are in use with, or ordered for, the navies of Argentina, Belgium, Malaysia, Netherlands, Finland, Indonesia, South Korea, Ireland, Morocco, Spain and Thailand.

CONTRACTOR

Hollandse Signaalapparaten BV, PO Box 42, 7550 GD Hengelo, Netherlands.

3479.253

DA.08 NAVAL SURVEILLANCE RADAR

The DA.08 is a high-power, coherent F-band radar for medium- to long-range surveillance and target indication. The antenna has a low sidelobe level for high performance in ECM environments, and a synthesiser-driven TWT transmitter results in great frequency flexibility over a wide band. Excellent performance in clutter conditions is ensured by a high dynamic range receiver with the application of digital video processing (*eg* MTI or FFT), supported by circular polarisation and a dual beam antenna. The light weight of a carbon-fibre antenna and a hydraulically controlled stabilisation platform enables installation in positions high on the mast. The DA.08 is available in three versions.

CHARACTERISTICS

Antenna: Horn-fed parabolic reflector
Beamwidth: Vertical cosec up to 40; horizontal 1.55
Polarisation: Linear and circular
Transmitter: F-band coherent system
PRF: 1000 and 500 Hz (version 1); 2300 and 1700 Hz (version 2); 2300 Hz (version 3)

STATUS

The DA.08 system has been fitted to the Type F-122 class frigates of the Federal Republic of Germany, the Argentinian Navy 'Meko 360' frigates, the aircraft carrier *Veinticino de Mayo*, and Royal Malaysian Navy frigates.

CONTRACTOR

Hollandse Signaalapparaten BV, PO Box 42, 7550 GD Hengelo, Netherlands.

DA.08 surveillance radar

1256.253

LW.02 NAVAL AIR SURVEILLANCE RADAR

The LW.02 is a D-band (23 cm), high-power long-range air surveillance radar for use on vessels of frigate size or above. Peak transmitter power is 500 kW and maximum detection range about 100 nautical miles. Vertical cover extends to 18 000 metres.

A double curvature reflector measuring 7 m by 2.9 m is used and a distinguishing feature of this radar is the large 'trunk' at the rear of the centre section of the scanner. There is also an LW.03 version of this radar, the only known difference between the two models being in the scanner rotation rates. The LW.02 rate is between one and ten rpm, and the LW.03 five or ten rpm.

STATUS

The LW.02 has been fitted to a large number of vessels in several countries. These include the navies of the Netherlands, Peru, West Germany, Sweden, Colombia, Malaysia, Australia and Argentina.

Production has ceased, the replacement model being the LW.08 (**1973.253**).

CONTRACTOR

Hollandse Signaalapparaten BV, PO Box 42, 7550 GD Hengelo, Netherlands.

LW.02 antenna

4051.253

ZW SERIES NAVAL RADARS

Signaal offers an entire family of general purpose search radars as follows:

ZW06: I-band naval surface search and navigation radar providing surveillance, navigation, helicopter control, and limited air surveillance facilities. Of compact and lightweight construction, it is suitable for installation on very small vessels. The transmitter/receiver is of solid-state design, except for the magnetron, and the scanner is of stainless steel construction to permit siting in adverse environments.

Operational features include: surface coverage to radar horizon; air coverage sufficient for helicopter guidance; high resolution for navigation, provision for integration of helicopter transponder systems; digital video processor. Anti-clutter measures and ECCM provisions include: circular polarisation; logarithmic receiver with pulse length discrimination; sensitivity time control; suppression of non-correlated pulses; tunable transmitter.

Signaal ZW06 surface search and navigation radar antenna

ZW07: submarine radar for navigation, distance measuring, surface search and limited air warning. The ZW07 is specially suited for use on board submarines. A single shot mode has been adapted for ranging of surface targets. Other operational features are: surface coverage up to radar horizon, limited air warning and high resolution for navigation. Anti-clutter measures and ECCM provisions include: sector scan facilities, logarithmic receiver with pulse length discriminator, suppression of non-correlated pulses and a tunable transmitter.

ZW08: low-level air warning and sea skimmer detection radar. The ZW08 is a system with the special capability of detecting accurately small, fast, low and very low-flying targets at such ranges that successful engagement of these targets is possible. The basic configuration of the radar system consists of a stabilised antenna fed by a high power TWT transmitter and a full coherent chain receiver. The system operates in the I-band.

ZW09: surface search and low-level air warning radar. The configuration comprises a stabilised antenna fed by a magnetron transmitter with matched receiver channels. Its application is on board small ships, giving the following operational features: surface search to the radar horizon, limited air warning and detection capability against low-flying targets. Anti-clutter measures and ECCM provisions are: circular polarisation, DMTI, pseudo jumping frequency, video correlator, linear, logarithmic, Dicke Fix and MTI receivers, pulse length discrimination and suppression of non-correlated pulses.

CONTRACTOR

Hollandse Signaalapparaten BV, PO Box 42, 7550 GD Hengelo, Netherlands.

1590.253

WM20 SERIES FIRE CONTROL RADAR

This radar is the principal sub-system of the Signaal WM20 series of naval fire control systems, and in its full form consists of two radars sharing a common stabilised mount and housed in a near-spherical radome. In the double installation, the tracker/illuminator is carried above the search radar, with the gimbal and stabilising assembly in between. The general arrangement can be seen from the accompanying illustration.

Both antennas are fed from a common I/J-band transmitter or from separated transmitters for surveillance and air target tracking.

The search radar carries out the surface and air search functions, and automatic tracking of surface targets for weapon control purposes. The tracking radar performs tracking functions for weapon control against air targets.

One version of the WM20 series, the WM26, employs only the search antenna, this being mounted on top of the stabilised platform and enclosed in a hemispherical radome. In all other versions the tracking antenna is fitted in the upper position with the search antenna below.

STATUS

The WM20 series of fire control radars is widely fitted and details of known installations are given in the WM20 series fire control systems entry (**1259.281**) in the Naval Section of this edition of *Jane's Weapon Systems*.

Signaal WM20 radars aboard Singapore patrol boats

CONTRACTOR

Hollandse Signaalapparaten BV, PO Box 42, 7550 GD Hengelo, Netherlands.

1589.253

SIGNAAL 3D MULTI-TARGET TRACKING RADAR

The Signaal 3D multi-target tracking radar (MTTR) is an advanced air search and target tracking radar capable of handling more than 100 aircraft tracks. A very high data rate is achieved by the use of an antenna array consisting of back-to-back pairs of parabolic reflectors and planar electronically scanned antennas, all four of which have common turning gear. IFF/SIF secondary radar are integrated with the MTTR. The parabolic dishes are used for search and the electronic-scan antennas for target tracking. High and low cover is provided. Beam steering and data extraction are performed by an SMR series digital computer.

Operational functions include: long range search; search information with high data rate for low-flying aircraft; search information with high resolution of close-in air targets; automatic position and height information; simultaneous tracking of over 100 aircraft targets; target designation facilities for other systems. No performance or other technical details have been obtained. Estimated diameter of the parabolic dishes is in the region of six metres. Both land and naval uses are likely, and the latter is understood to incorporate full stabilisation.

STATUS

The 3D MTTR forms one of the major sensors of the SEWACO action information and fire control system for the 'Tromp' class of Royal Netherlands Navy guided weapon frigates. There are no other known fittings at this stage.

Royal Netherlands Navy ship Tromp, *showing the large radome housing the Signaal 3D MTTR and the smaller one for the WM25 fire control system radar*

CONTRACTOR

Hollandse Signaalapparaten BV, PO Box 42, 7550 GD Hengelo, Netherlands.

5626.253

SMART NAVAL SURVEILLANCE RADAR

SMART (Signaal Multibeam Acquisition Radar for Targeting) is an all-weather three-dimensional naval surveillance radar system intended for all types of naval vessel from fast patrol boats upwards. Its prime application is as the main sensor for data handling and weapon system control and it has a very high performance in the presence of heavy clutter and electronic countermeasures. The equipment has been designed to cope with small high-speed anti-ship missiles, with radar cross sections down to 0.1 m^2 and approach speeds of Mach 3+, which can be either sea skimmers or arriving from high angles of 70° or more.

The SMART system operates in F-band (formerly S-band) where it offers an optimum balance between range, clutter rejection and antenna dimensions. An alternative version operating in G-band is also available. It provides an uninterrupted search envelope from 0 to 90° in elevation with automatic detection followed by automatic initiation and target tracking. It incorporates anti-clutter and ECCM features such as multiple reception beams with ultra low side lobes in elevation and azimuth, a clutter analysis sensor, broadband transmission, PRF and RF agility per burst and a jamming analysis sensor which automatically changes PRF and transmit frequency.

The system comprises an antenna and three below-decks units. The hydraulically-stabilised antenna consists of a single-element squint-free wideband transmitting array, and a multi-element stripline receiving array. The ultra-low sidelobe level single-plane phased array allows accurate 3-D search and tracking. To ensure high sensitivity, pre-processing of the received signals takes place in the antenna unit itself. The output of the 16 antennas is fed to a digital beam forming network in which 12 independent elevation beams are produced, and doppler FFT processing and automatic tracking is carried out. The transmitter is based on a high power, pulse-to-pulse coherent TWT. For the complete integration of any type of IFF, an IFF antenna, rotary joint and an IFF extractor are incorporated.

CHARACTERISTICS

Antenna unit

Rotation speed: 27 rpm

Stabilisation: Hydraulic

Polarisation: Horizontal

Transmitting antenna

Type: Single horn antenna array

Beamwidth: 2° horizontal; up to 90° vertical

Receiving antenna

Type: Stripline array

Beamwidth: 2° horizontal; 100° vertical (processing beamwidth is 9° vertical)

Transmitter

Type: TWT with pulse compression

Frequency: F-band

PRF: 3700 Hz average

Weight

Antenna: 800 kg

Below-deck units: 2800 kg

Antenna dimensions: 360 × 250 cm

STATUS

SMART is developed under contract from the Royal Netherlands Navy. Eleven systems are being produced and will be fitted to 'M' class frigates of the RNIN.

CONTRACTOR

Hollandse Signaalapparaten BV, PO Box 42, 7550 GD Hengelo, Netherlands.

Signaal SMART naval surveillance antenna

6130.253

MW08 3-D SURVEILLANCE RADAR

The MW08 is a G-band 3-D short- to medium-range surveillance and target acquisition derived from SMART (**5626.253**). It enables very rapid target acquisition and tracking by weapon control systems providing extra target information such as elevation, speed and priority. All system functions, including target detection, track initiation, target tracking, and built-in test equipment are automatic. Multi-stripline antennas, a digital FFT beamformer, and doppler FFT processing minimise the effects of clutter and jamming.

The MW08 is capable of tracking up to 30 targets simultaneously. Track data is transferred to the command and control system and, if required, to the weapon deployment console for direct surface-to-target engagement via the databus system or via inter-computer interfaces. With its lightweight hydraulically stabilised platform, the MW08 can be mounted in a high mast position.

CHARACTERISTICS

Antenna: Stripline array

Polarisation: Horizontal

Beamwidth: Vertical up to 70; horizontal 2

Transmitter: Coherent amplifier with TWT output stage

PRF: 3700 Hz

STATUS

MW08 systems have been ordered for the new Portugese 'Meko' type frigates.

CONTRACTOR

Hollandse Signaalapparaten BV, PO Box 42, 7550 GD Hengelo, Netherlands.

SWEDEN

1546.253

9GR 600 NAVAL RADAR

The 9GR 600 is an I-band radar transceiver extensively used in naval radar systems both in the Swedish and other navies.

With the 9GA 205 antenna it is used in the Mareld coastal radar system (**1548.181**), the 9LV 200 fire control system (**1542.281**), the 9LV 200 Mk 2 weapon control system (**3127.281**), the Seafire system (**1543.281** in 1979-80 and earlier editions), the Torci torpedo fire control system (**1541.281**), and with the 9GA 300 antenna in the Subfar submarine radar system (**1545.453**).

The radar can be used either in its broadband frequency agility mode, or in a fixed frequency mode for MTI operation.

CHARACTERISTICS

Transmitter

Operating modes: Frequency agility/fixed frequency MTI

Transmitter tubes (2 off): Magnetron YJ 1180/YJ 1181

Frequency agility range: Minimum 400 MHz, typically 450 MHz within 8700 – 9500 MHz

PEAB 9GA 209 I-band stabilised surveillance antenna

Frequency shift: Randomly within the frequency agility range from pulse to pulse

Pulse peak power: 200 kW

Pulsewidth/PRF: Typically 1μs/1000 Hz and 0.25μs/3000 Hz

Receiver

Frequency range: 8500 – 9600 MHz

IF: 30 dB

Noise figure: Typically 5.5 dB; 7 dB with low-noise preamplifier

Receiver characteristics: Lin and log, log/lin, STC

Operational features: Image rejection mixer, Dicke-Fix receiver, PLD, IAGC, narrow-band jamming suppression, passive ECM mode

Transceiver

Dimensions: 1200 × 635 × 510 mm

Weight: Approx 225 kg

Antenna GA 209

Type: Parabolic, hornfed, stabilised

Reflector material: GRP and metal wires

Frequency range: 8.5 – 9.6 GHz

Rotation speed: Up to 60 rpm

Beamwidths

Horizontal: 1.1°

Vertical: 7°

Gain: 35.5 dB

Sidelobes: –21 dB at ±1.5°; –33 dB beyond ±15°

Helical scan coverage: Up to 40°

Stabilisation: Direct driving hydraulic motors

Stabilisation limits: Roll ±32°, pitch ±12°

Aperture: 2100 × 300 mm

Weight (including turntable): 260 kg

Antenna 9GA 218

Type: Parabolic, dual frequency, stabilised

Reflector type: GRP and metal wires

Frequency range

I-band: 8.5 – 9.6 GHz

Dual-frequency, I- and D-band, stabilised surveillance antenna 9GA 218, for the 9GR 600 naval radar

D-band: 1.01 – 1.11 and 1.25 – 1.35 GHz

Beamwidths

Horizontal: 1° at I-band, 6/8° at D-band

Vertical: 3° at I-band, 20/25° at D-band

Gain

8.5 – 9.6 GHz: 38 dB

1.01 – 1.11 GHz: >17 dB

1.25 – 1.35 GHz: >20 dB

Stabilisation: Direct driving hydraulic motors

Stabilisation limits: Roll ±30°, pitch ±10°

Aperture: 2700 × 700 mm

Weight (including turntable): 260 kg

STATUS

In service.

CONTRACTOR

Philips Elektronikindustrier AB, 17588 Järfälla, Sweden.

1547.253

9LV 200 Mk 2 and Mk 3 TRACKING RADARS

9LV 200 Mk 2

This is a J-band frequency-agile naval fire control radar forming part of the 9LV 200 Mk2 weapon control system (**3127.281**) in which it is used with an I-band search radar. The main role is that of target tracking for gun fire control, and the radar is monopulse equipment with stabilised antenna and a cassegrain scanner system. The radar mounting can also carry TV, IR camera and laser rangefinder for autonomous electro-optical (optronic) tracking. A feature of the mounting is a new type of direct drive hydraulic motor capable of a slewing speed of 85°/s, and an initial acceleration of 500°/s².

9LV 200 Mk 3

The Mk 3 also operates in J-band and is associated with the 9LV 200 Mk 3 weapon control system

(**6449.281**). It has a helix TWT transmitter enabling an ultra-wide frequency agility bandwidth. The antenna has very low sidelobes in both sum and difference pattern by means of a multi-mode feed. The system can operate in pulse-to-pulse frequency agility, MTI or pulse doppler mode. It has a high degree of self adaption to target and environment by automatic selection of pulse length, PRF and radar modes.

CHARACTERISTICS

Type: J-band frequency-agile MTI monopulse
Frequency range: Random pulse-to-pulse in J-band
Transmitter: Philips YJ1321 magnetron (Mk 2); helix TWT (Mk 3)
Peak power: 65 kW (Mk 2); 1.25 kW (Mk 3)
Receiver noise: 11 dB (Mk 2); 10 dB (Mk 3)
Antenna
Diameter: 1 m
Beamwidth (at 3 dB): 1.3°
Gain: 40 dB (Mk 2); 41 dB (Mk 3)

STATUS

The Mk 2 is in service with the Royal Swedish Navy. The Mk 3 is on order for the navies of Sweden, Denmark and Finland with deliveries commencing in 1988.

CONTRACTOR

Philips Elektronikindustrier AB, 17588 Järfälla, Sweden.

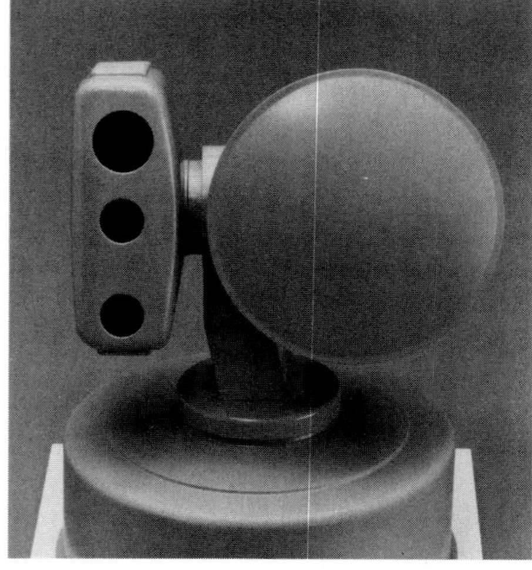

9LV 200 Mk 3 radar/optronic director

4299.253

SEA GIRAFFE NAVAL SEARCH RADAR

The Giraffe family of air defence search radars is based on the fully coherent TWT transmitter concept featuring an extremely good sub-clutter visibility of moving targets. Various versions of the Giraffe radar are in operation in different system configurations in ten countries. The latest version, Sea Giraffe, is a multi-purpose naval search radar designed to meet the operational requirements of a variety of warships from small fast patrol boats to frigate sized ships. There are two basic versions (HC denotes extra height coverage versions):

(1) Sea Giraffe 50 for small and medium-sized patrol boats and fast attack craft
(2) Sea Giraffe 150HC for large fast attack craft, corvettes and frigates

Modified Spica II class fast attack boat of the Swedish Navy fitted with Sea Giraffe

Both versions of the Sea Giraffe are essentially repackaged versions of the Giraffe family to satisfy a variety of requirements in the area defence and point defence roles. They provide all-weather anti-ship missile detection to enable successful defence, long range aircraft detection in severe environmental conditions and ECM, and surface target detection up to the radar horizon. For area defence, Sea Giraffe supplies air and surface data to the ship's fire control system, target information to own surface-to-surface missiles and surface gun fire control with accurate fall-of-shot observations. Ancillary uses for the system include long range navigation and helicopter guidance.

Low altitude coverage is obtained at a considerable range with obvious benefits in the detection of low-flying aircraft and sea skimming missiles.

The radar uses a broadband fully coherent TWT transmitter giving 15 kW output power for Sea Giraffe 50 and 60 kW for Sea Giraffe 150HC. The antenna for the Sea Giraffe 50 is the standard Giraffe family version while a special antenna has been developed for the 150HC. The 150HC has an optional antenna for extra high cover to detect steep diving missiles (60 to 80° dive angle). A high MTI improvement factor ensures excellent low-level detection capability, and frequency agility in both MTI and non-MTI operation is used for high ECCM performance. Agility is on a pulse-to-pulse basis in non-MTI mode and on a pulse burst basis in MTI mode. The adaptive computer based coherent MTI is used to give optimised suppression of mixed clutter from sea, land, rain and chaff. Digital pulse compression is used for high range resolution and improved clutter performance. G/H-band operation enables optimum detection of both air and surface targets:

(1) Sea Giraffe 50 to 15 km for missiles and 35 km for aircraft
(2) Sea Giraffe 150HC to 25 and 55 km respectively.

All versions are normally supplied with a stabilised platform and for operation in extreme winds the antenna should be enclosed in a radome.

CHARACTERISTICS

Frequency: 5.4 – 5.9 GHz
MTI improvement factor: 45 – 50 dB
Antenna rotation rate: 30/60 rpm
Output power (peak): 15 kW (50 version); 60 kW (150 version)
Output power (mean): 300 W (50 version); 1200 W (150 version)
Antenna size: 2.3 × 0.7 m (50 version); 2.4 × 1.8 m (150HC version)
Masthead weight: 250 kg (50 version); 700 kg (150HC version, radome included)

STATUS

Sea Giraffe was first ordered in 1980 by the Swedish Navy for the mid-life refit of the Spica II attack craft, and has been subsequently ordered for the new Spica III 'Stockholm' class fast missile vessels, as well as for the minelayer/training ship *HMS Carlskrona*. Sea Giraffe has been in full operational use with the Swedish Navy since early 1983 in the modified Spica IIs.

At the end of 1983, Sea Giraffe 150HC was ordered for the new Canadian patrol frigate programme. On this vessel, Sea Giraffe will be used as the medium range surveillance radar and as target indication radar to the fire control system.

Negotiations are also being conducted with a number of other customers for various applications of Sea Giraffe.

CONTRACTOR

Ericsson Radar Electronics AB, Defence Systems Division, 431 26 Mölndal, Sweden.

1545.453

SUBFAR 100 SUBMARINE RADAR

The Subfar 100 equipment is an air and surface search radar for submarines and comprises the 9GA 300 antenna, a 9GR 600 I-band frequency-agile transceiver, and a display system.

The 9GA 300 antenna is designed for mounting with its hydraulic turntable on top of the submarine radar mast. This way of mounting eliminates bearing error sources from torsion effects in the radar mast. The hydraulic turntable gives a very low noise level as well as flexible control of antenna rotation speed. The antenna direction can be manually set or slaved to external sources such as the periscope.

Subfar 100 submarine antenna 9GA 300

Sector transmission as well as short time transmission down to emitting a single pulse can be selected. Automatic target detection can be incorporated, which facilitates automatic target tracking by an external fire control computer.

CHARACTERISTICS

9GA 300 antenna
Frequency range: 8.5 – 9.6 GHz
Rotation speed: 0.5 – 24 rpm, and pos, control
Beamwidths
Horizontal: 2.4°
Vertical: 16°
Gain: 26 dB
Sidelobes: 6 to 18 dB
Antenna aperture: 1000 × 140 mm
Height above mast tube: 600 mm
Rotating circle diameter: 1040 mm
Antenna system weight: 95 kg
9GR 600 transceiver
See **1546.253**, above, for full characteristics.
Display system
CRT diameter: 300 mm (12 inch)
Deflection: Magnetic, fixed coils
Scale factors: 1:20 000
1:50 000
1:100 000
1:200 000
True motion: Selectable
Range readout: Digital
Bearing readout: Digital
Dimensions
Height: 1100 mm
Width: 665 mm
Depth: 600 mm plus desk 250 mm
Weight: 135 kg

STATUS

In service.

CONTRACTOR

Philips Elektronikindustrier AB, 17588 Järfälla, Sweden.

SWITZERLAND

6119.253

SEAGUARD TRACKING MODULE (TMX)

Part of the Seaguard range of modular combat systems (**3509.281**), the general purpose tracking module (TMX) is used for accurate long range tracking of air and surface targets for engagement by guns ranging from small (eg Sea Zenith) to large calibre, and/or ship-to-air missiles.

The TMX uses a combination of X-band radar, FLIR (or optionally TV in angles) tracker, and laser rangefinder sensors. The X-band radar is a fully coherent pulse doppler system using pulse compression and a TWT transmitter. Real time FFT doppler processing is used in both acquisition and tracking modes. The radar incorporates a number of advanced features, including random frequency agility over the entire band, MTI, variable pulse compression ratio, CFAR, frequency coded transmission, staggered PRF, and track on jam.

When equipped with electro-optical sensors the combined system processing techniques provide redundancy, and make the TMX impervious to ECM and multipath effects. Optional second radar receiver and CW injection are available for ship-to-air missile guidance purposes.

CHARACTERISTICS

Frequency: X-band

Antenna: Cassegrain

E-O sensors: FLIR spectral bandwidth 8 – 12 μs, Centroid FTT tracking principle; Laser spectral wavelength 1.06 μs with a 25 Hz data rate

Tracker: Unique three-axis, high dynamic (3 rads/s; 15 rad/s^2); hemispherical tracking (elevation: –35 + 120°)

STATUS

In series production

CONTRACTORS

Contraves AG, Schaffhauserstrasse 580, 8052 Zurich, Switzerland.

Contraves Italiana SpA, Via Affile 102, 00131 Rome, Italy.

Seaguard tracking module (TMX)

UNION OF SOVIET SOCIALIST REPUBLICS

Note:

There are a large number of Soviet radar types and in many cases it has not been possible to obtain sufficient information to warrant an entry. A comprehensive list of all known Soviet naval radars is given in the analysis pages later in this book, together with very brief details.

4090.253

BAND STAND NAVAL RADAR

Band Stand is the name given to the radar identified by its large dome shaped radome, and considered to be associated with the ship's overall air surveillance requirements, although there may be some connection with the vessel's main armament.

The latter is thought to be less likely in the light of the continued fitting of the Band Stand radar (or its radome, at least) in the 'Nanuchka' class missile corvettes supplied by the USSR to India, although the latter ships are armed with SS-N-2B anti-ship missiles instead of the six SS-N-9 missiles which arm the Soviet vessels of this class. However, it is probable that in the Soviet ships the Band Stand radar is employed in a complementary fashion in the SS-N-9 system in addition to having air surveillance as its main role. This would include tracking of the out-going SS-N-9 missiles, with guidance signals being transmitted via the two Fish Bowl equipments (which smaller radomes are missing from Indian examples of 'Nanuchka' ships). Band Stand is also fitted to the 'Sovremenny' class in association with the SS-N-22 anti-ship missile.

Both Indian and Soviet ships of the class are fitted with SA-N-4 anti-aircraft missile systems and an air surveillance function of the Band Stand radar, with a fairly high data rate, would be a valuable complement to this system.

The Soviet ship Sovremenny *includes a Band Stand radar in its radar fit. Other items seen here include a Kite Screech fire control radar and Top Steer group*

4085.253

BASS TILT FIRE CONTROL RADAR

Bass Tilt is an H-band Soviet naval fire control radar which resembles Drum Tilt (**1330.253**) and Muff Cob (**1611.253**) in both general appearance and broad functional purpose. Of these three radars, Bass Tilt is the most modern, having entered service in the mid-1970s; it is also decidedly larger than the oldest of the three types, Drum Tilt. They all share a common main operational function of radar direction of small or medium calibre anti-aircraft guns, with Muff Cob generally being associated with Soviet twin 57 mm gun mounts. Both Drum Tilt and Bass Tilt are usually used with smaller calibre weapons, particularly multiple turret twin or single Gatling-type installations. Drum Tilt fittings are mostly confined to smaller ships with a single gun turret such as the common-place 'Osa' class missile boats.

Bass Tilt fittings have been noted on one ship of the 'Kresta I' class (which underwent modernisation in 1975), 'Kresta II' class and 'Kara' class cruisers, some 'Kashin' class destroyers, 'Nanuchka II' and 'Grisha

The Sovremenny *presents a radar identification challenge with multiple fittings of both Muff Cob and Bass Tilt fire control radars (plus one other radome of similar size which is neither of these types but more probably a communications satellite terminal). Other radars are Top Steer air search radar pair, Kite Screech fire control radar, and Band Stand missile director radar*

III' class corvettes, and various miscellaneous types of vessel such as the 'Ivan Rogov' class of amphibious assault ship and the *Berezina* fleet replenishment ship. 'Kresta II' class ships present one of the best opportunities to make direct visual comparisons between the most difficult pair of these three types of fire control radar to identify accurately, Muff Cob and Bass Tilt, as two of each type are fitted in these ships; two Bass Tilt forward for control of four Gatling-gun AA turrets, and two Muff Cob athwart the funnel for the aft twin 57 mm gun turrets. More recent examples of Bass Tilt fittings are the new *Kirov* cruiser and the 'Sovremenny' class destroyers. The latter ship has both Muff Cob and Bass Tilt radars fitted in some profusion, offering something of a challenge to radar 'spotters'. The 'Kiev' class of aircraft carriers each have four Bass Tilt radars for their eight Gatling-gun AA mounts.

1608.253
BIG NET SEARCH RADAR

Big Net is a very large, long-range air surveillance radar operating in D/E frequency band. The open lattice reflector is of elliptical parabolic form, illuminated by an underslung horn feed, carried on a solid boom. Behind the reflector are two balance vanes. Range performance is estimated as more than 100 nautical miles (185 km) against an aircraft at medium altitude, and up to 200 nautical miles (370 km) at higher altitudes.

According to the latest information, all four of the 'Kresta I' class of guided missile cruisers carry Big Net, on which ships it is mounted over the funnel. Some of the 'Kashin' class of guided missile destroyers are fitted with this radar, in which cases it is mounted on a pylon type mast aft of the forward funnel. The cruiser *Dzerzhinsky* has also been seen with Big Net and some of the 'T43/AGR' class of radar picket ships (converted from minesweepers of the 'T43' class) have a Big Net search radar. Other vessels of this class have either a Squat Eye or Knife Rest radar instead of the Big Net. Big Net is also installed on the Kiron class battle-cruiser class ships, as well as other vessels, in conjunction with Top Sail to form Top Pair (**4087.253**).

Big Net long-range search radar

6879.253
CROSS SWORD FIRE CONTROL RADAR

Cross Sword is a fire control radar associated with the SA-N-9 surface-to-air missiles in some of the later Soviet destroyers, including the 'Udaloy' class. No other details are available.

1330.253
DRUM TILT FIRE CONTROL RADAR

This is a pedestal-mounted fire control radar which probably provides acquisition and tracking functions only for 25 and 30 mm AA guns. It is housed in a weather-proof container which is attached to the scanner assembly tilted at about 25°. Maximum acquisition range of aerial targets is approximately 22 nautical miles. Operating frequency is in the I-band. It is principally fitted aboard 'Osa' class missile boats, on a platform between the two aft missile launchers, for the direction of four 30 mm AA guns carried by these craft.

Drum Tilt fittings have also been noted on some 'Sverdlov' cruisers, and ships of the 'Kanin', 'Kotlin', 'Shershen', 'Stenka', 'Turya', 'Natya', 'Yurka', 'Sen-yarin', and 'Polnochniy' classes. The type is gradually being supplanted by later radars such as Muff Cob and Bass Tilt which are fitted in more recent classes of Soviet ship.

This East German 'Koni' class escort vessel at anchor off the Shetland Islands is equipped with a Drum Tilt and a Hawk Screech fire control radar for gun direction, and a Pop Group for use with the SA-N-4 missile system. The air search radar is a Slim Net or Strut Curve (RN photo)

4089.253
EYE BOWL NAVAL RADAR

The Eye Bowl radar is associated with the Soviet SS-N-14 naval missile system fitted aboard 'Krivak' class frigates and the Soviet cruiser *Kirov*, although other classes thought to be fitted with the SS-N-14 system have not yet been noted with the Eye Bowl radar. The precise function of the Eye Bowl radar tends to remain as problematical as that of the SS-N-14 missile with which it is associated. The latter is generally thought to be somewhat similar to the Australian/British Ikara (**6002.241**) in that the payload/warhead is believed to comprise a torpedo, thus making the SS-N-14 principally an anti--submarine weapon although there will also be a certain surface ship target capability. On these assumptions, the main role of the Eye Bowl radar is probably that of tracking and guidance of the outgoing SS-N-14 missile to ensure that it reaches a previously designated target position determined by another of the ship's sensors and the combat information system. This seems to be confirmed by what is discernible from photographs of the Eye Bowl radar. The reflector dish is comparatively small and there appear to be several auxiliary (possibly electro-optical) sensors mounted immediately behind the reflector dish. Command signals to the SS-N-14 missile are probably transmitted as coded information within the radar transmissions, as no separate command antennas can be seen.

The operating frequency is in F-band. This also would suffice for the estimated range of the SS-N-14 system, which is generally thought to be in the neighbourhood of 30 nautical miles.

Beam view of the two Eye Bowl guidance radars on a Soviet 'Krivak' class destroyer, abaft and above the Pop Group radar for the ship's SA-N-4 anti-aircraft missile system (HMS Londonderry)

1605.253
FAN SONG E NAVAL RADAR

This is a shipboard version of the Guideline (SA-N-2) surface-to-air missile fire control and guidance radar (**2868.153**) which is widely used by Soviet land forces. In place of the mobile housing used on land, the shipboard installation is mounted on a stabilised gun mount.

Fan Song E operates in the G/H-band range of frequencies with a maximum acquisition range of 80 nautical miles. The antenna array consists of horizontal and vertical scan Lewis antennas, and three circular parabolic dishes. Two of the latter are for lobe on receive only (LORO) ECCM, and the third dish is for command guidance purposes. Operation of Fan Song E is described more fully in **2868.153**.

The only known fitting of this radar is on the Soviet Cruiser *Dzerzhinsky*, on which ship the Fan Song E was mounted on a platform aft of the after funnel and forward of the twin Guideline missile launcher. It was removed in 1976.

6421.253
FRONT SERIES RADARS

The Front series are F-band missile guidance radars for shipborne weapons. Front Door and Front Piece are primarily for submarine use and provide the mid-course guidance for the SS-N-3 Shaddock and SS-N-12 Sandbox anti-ship missiles. The *Slava* class cruisers are also known to have an installation, for use with the SS-N-12.

Front Dome is a multiple antenna system, operating in H/I-band, which provides target illumination for the SA-N-7 surface-to-air missile. The *Sovremenny* class ships are fitted with six units. See also Trap Door (**6878.253**) which has recently been designated Front Door C by NATO.

STATUS

Operational in a number of ships.

1325.253
HAWK/OWL SCREECH FIRE CONTROL RADARS

Hawk Screech and Owl Screech are fire control radars of conventional design, widely used in many classes of Soviet vessels. They are distinguished by a circular dish scanner, behind which are mounted fairly bulky housings which are assumed to contain transmitter and receiver, and probably turning and stabilisation gear. The complete radar is pedestal mounted. The 2.3 metre diameter scanner has a central feed supported by a four-leg structure. Feed arrangements could incorporate a rotating dipole to provide a conical-scan search pattern. Dish diameter is approximately two metres, with Owl Screech somewhat larger than Hawk Screech, and the operating frequency is in the H/I-band for both types.

Fire control radar for main guns, believed to carry NATO designation Hawk Screech. Compare feed arrangements with those of Owl Screech

The NATO designations, Hawk Screech and Owl Screech, are apparently derived from signal characteristics noted in the course of Elint monitoring. They appear to refer to two versions of the same basic radar, and close examination of photographs of this equipment does reveal external differences to support this. These visible variations do not as yet permit positive identification of either with one of the NATO names although information has been obtained suggesting that Owl Screech is an improved version of Hawk Screech.

The two most obvious physical differences between the two types are in the arrangement of equipment behind the dish and in the mounting of the feed. In Owl Screech, the hardware behind the dish is contained in more numerous housings, giving a less tidy appearance than in the other version. Also, the four legs of the feed support are fixed to the dish at points nearer its centre than in the other radar. The pedestals also differ, Owl Screech being tubular, and that of Hawk Screech being slightly tapered. In the photographs studied by *Jane's Weapon Systems* it has been noticed that the feed housing in the Owl Screech version is at an angle of about 45°, whereas in the other version it is orthogonally mounted. This may be a clue to the signal characteristics of the two types. There is one other feature, which is less obvious, that distinguishes the two types. That believed to be Hawk Screech has a small aperture in the dish, on the centre line and near the upper edge. This could be for a closed-circuit TV camera, and there is a small housing on the rear of the reflector. This could be used for optical target acquisition and/or tracking.

Gun fire control radar believed to carry the NATO designation Owl Screech. Note the angled feed housing, numerous separate equipment housings behind the reflector, tubular pedestal, and feed mounting, which distinguish this radar from that known as Hawk Screech

The principal function is that of gun fire director against aircraft targets, although this type probably also has a certain capability against surface targets, and possibly splash detection.

Hawk Screech is associated with quadruple 45 mm, twin and quadruple 57 mm, and twin 76 mm guns, while Owl Screech is employed with 76 mm twin gun installations. Classes of Soviet Fleet vessels fitted with Hawk Screech include: 'Kanin' (one), 'Kotlin' SAM (one), modified 'Kildin' (two), 'Kotlin' (two), modified 'Skory' (two), 'Lama' (two), 'Koni' (one), 'Don' (one), 'Petya' (one), 'Mirka' (one). Owl Screech fittings include: 'Kiev' (two), 'Kara' (two), 'Kynda' (two), 'Kashin' (two), some 'Kildin' (two), 'Krivak' (two), 'Ivan Rogov' (one).

1328.253
HEAD LIGHT GROUP

This group of radars was first seen on the cruiser helicopter carrier, *Moskva*, but has subsequently made its appearance on its sister ship *Leningrad* and on the 'Kresta II' and 'Kara' classes of guided missile cruisers. Two of these groups of radars are fitted, fore and aft on the 'Kresta II' and 'Kara' vessels, and two forward on the *Moskva*. In very general configuration, and possibly function to some degree, this group resembles the Peel Group missile control radar group for Goa surface-to-air missiles on other Soviet vessels. The Head Light group consists of a group of four radars, apparently comprising two identical pairs of equipment combined on a common mounting. There is a fifth, smaller dish, possibly fulfilling a command link or IFF function. Two versions have been identified, designated Head Light A and Head Light B, respectively. The former is associated with Goblet surface-to-air missiles and the latter with the SS-N-14 missile system.

The four main reflectors of the group are circular dishes, of open mesh construction, one small and one large to each pair, with the smaller in the upper position, and all being disposed symmetrically about the central mounting pillar. Estimated dish sizes are 1.8 and 3.8 m. The electronics for each radar, transmitter/receiver etc, are carried behind the individual reflectors and the whole assembly of four radars is supported from the top of the pedestal turning gear on a yoke of triangulated tube construction.

This photograph of the 'Kara' class cruiser Nikolaev *shows Head Light radar groups for SAM guidance and control, fore and aft. Air surveillance radars are the big Top Sail on the mainmast and a Head Net C back-to-back system* (RAF)

The whole assembly rotates in azimuth, and can also move in elevation. The two upper radars also appear to have provision for individual movement in both axes, possibly limited, and the dynamic balancing vanes fitted to them could support this suggestion.

The upper dishes have unusually large feed arrangements which may be of the cassegrain type, and the larger dishes appear to be front fed. Operating frequencies are reported to be in the G-band for acquisition and H/I-band for tracking.

On the *Moskva* and *Leningrad* there is one Head Light group to each of two surface-to-air missile launchers, each for two missiles. On 'Kresta II' class ships there is one Head Light group associated with similar launchers fore and aft. The arrangement on 'Kara' class ships is similar.

1318.253

HEAD NET A AIR SURVEILLANCE RADAR

This appears to be the most widely fitted of the numerous radars employed by the Soviet Navy, and appears to be of fairly conventional construction. It has a large (about 6 × 1.5 m) elliptical paraboloid reflector of open lattice construction. This is illuminated by a horn feed carried by a boom projecting from below the lower edge of the scanner. An unusual visual feature, compared with practice outside the USSR, is a pair of prominent 'butterfly' vanes extending behind the scanner which are for dynamic balancing.

The scanner mounting is also quite distinctive and has a rather 'lighthouse' like appearance, being quite regular and with a domed top. It is possible, though considered unlikely, that this accommodates some form of scanner stabilisation against ship's motion. It is possible that a relatively high rotation rate is employed.

Operating frequency is E/F-band, with a detection range, on aircraft at medium altitude, of 60-70 nautical miles. No details of the radiated power are available.

The principal function of the Head Net A is assumed to be air search and surveillance and it can be expected to provide target designation facilities for other fire control radars for guns and missiles carried by the vessel.

It is normally mounted on a mast-top location, or other high positions, and both single and double installations have been seen on various classes of Soviet vessel. 'Kynda' and 'Kashin' class guided missile destroyers have one on each of the two masts; the *Bravy* ('Sam Kotlin' class destroyer) has a single installation on the mainmast.

Head Net mountings and scanners are also employed in two types of dual installation. One is a back-to-back configuration (**1319.253**) with one of the scanners at a larger elevation angle (about 15° more than the companion scanner) to provide separate high and lower cover antennas. The second dual configuration (**1320.253**) is also a back-to-back arrangement, but in this case one scanner is tilted laterally from the horizontal with the probable objective of providing a three-dimensional capability.

Head Net A surveillance radar, with Sun Visor radar director below and to right. Note also High Pole IFF and assorted direction-finding and ESM antennas

1319.253

HEAD NET B AIR SURVEILLANCE RADAR

This is a dual installation of the Head Net A radar (**1318.253**) in which two scanners are mounted in a back-to-back configuration. One is tilted at an angle of about 15° elevation in relation to the other scanner to provide separate high and low cover. Operating frequency probably lies within the E/F-band of the spectrum, although D-band is a possibility also.

Functions of the radar are search and surveillance, and target designations for the fire control radars carried for the direction of surface-to-air missiles and guns.

This radar has been seen on a 'Krupny' class destroyer (No 229 – since renumbered), but does not appear to be in widespread use. The general mechanical arrangement is very similar to the Head Net C (**1320.253**), which is also a back-to-back system using dual Head Net A arrays, and it is possible that individual vessels have been converted from one version to the other. Latest information is that of the 'Krupny' and 'Kanin' class ships; *Gnevny* now has Head Net C and the others have a Head Net A. The only known remaining Head Net B installations are on two Soviet range instrumentation ships, *Chazma* and *Chumikan*, which entered service in the early 1960s.

1320.253

HEAD NET C AIR SURVEILLANCE RADAR

This radar is apparently a back-to-back combination of two Head Net A scanners (**1318.253**), similar to Head Net B (**1319.253**) but with the significant difference that one of the scanners has its aperture (ie. its 'span' dimension) tilted from the horizontal by approximately 30°. This has the effect of displacing the resulting fan-shaped elevation beam by the same amount from the vertical. This beam, in combination with the vertical beam produced by the companion scanner, thus provides the means for heightfinding by the so-called 'V-beam' technique, such as is used in the American AN/TPS-34.

There are important differences, however, in the way in which these two radars employ this principle, apart from other variations between them. The most significant is that the US equipment has both beams (slant and vertical) radiating along a common azimuth, rotating in synchronisation, whereas in the Head Net C the beams are separated in azimuth by 180° – as a result of the back-to-back configuration employed.

In practice, the vertical beam fulfils the search function and the operator selects a target for which height data is required, and by placing a marker on this target (or similar technique) this places a range gate in the second (inclined) beam and excludes other targets. Computation to give a height readout can be performed by analogue or digital methods.

This radar has been seen on both 'Kresta I' and 'II' classes of guided missile cruisers, the *Moskva* and *Leningrad* cruiser helicopter carriers, some 'Kashin', 'Kanin' and 'Krupny' GMDs, and 'Kotlin SAM II' GMDs. The newest class of vessel to carry Head Net C is the 'Krivak' general-purpose leader type.

Head Net back-to-back three-dimensional air-surveillance radar on 'Kresta II'

4084.253

KITE SCREECH FIRE CONTROL RADAR

Kite Screech is a gun fire control radar, probably an evolutionary development of the Hawk Screech and Owl Screech family of fire control radars (**1325.253**). A distinguishing feature that enables the Kite Screech to be identified from the other two types is the antenna illuminating feed support assembly which has a noticeably more 'pointed' appearance compared with the Hawk and Owl Screech radars. The latter types also have a comparatively large dielectric housing protecting the actual radiator element, whereas the

Kite Screech fire control radar on the Soviet ship Sovremenny *showing the very pointed appearance of the antenna feed assembly (see text)*

equivalent item on the Kite Screech is much less prominent. G- or I-band operation is probable, and possibly this radar incorporates monopulse tracking instead of conical scan tracking.

The only fittings noted to date are aboard the new 'Kirov' class cruisers, where the Kite Screech radar is used for the direction of two 100 mm gun turrets mounted aft, the 'Sovremenny' class destroyers and the guided missile cruiser *Slava*.

Aboard the Soviet cruiser Kirov *a single Kite Screech radar is used for direction of two single 100 mm dual-purpose gun turrets*

1611.253

MUFF COB FIRE CONTROL RADAR

The Muff Cob fire control radar is similar in general appearance to the Drum Tilt fire control radar (**1330.253**), but the former has provision for elevation motion, being supported by a trunnion yoke carried on the traversing mounting. The operating frequency is in H-band. Some drawings of Muff Cob show what appears to be an on-mounted TV or optical tracking sub-system, but this has not been confirmed.

In all the installations of Muff Cob studied, this radar is employed for gun fire control purposes only. On the helicopter carriers *Moskva* and *Leningrad* there are two Muff Cobs mounted high on each side of the main superstructure for direction of the twin 57 mm guns on either side of these ships. Both 'Kresta I' and 'Kresta II' classes of guided missile cruiser have two Muff Cobs each, carried on platforms on the sides of the funnel. These are for control of twin 57 mm guns on each side. Some of the 'Ugra' and 'Lama' classes of support ships have two Muff Cob installations, and there are single fittings on ships of the 'Poti' class of coastal escorts. The 'T58' class of fleet minesweepers carry a single Muff Cob on top of the bridge house. Some 'Polnochniy' amphibious landing craft (Type IX) carry a Muff Cob for direction of anti-aircraft guns.

Muff Cob fire control radar

1323.253

PEEL GROUP FIRE CONTROL RADAR

This group comprises four distinct radars on a common mounting and is associated with the Goa surface-to-air armament of Soviet vessels. The radars entered operational service in the early 1960s. Within the group are an H/I-band tracking radar for high altitude targets and an E-band guidance radar for lower altitudes. The maximum range is reported as 30 to 40 miles (55 to 75 km). All four scanners are of solid-reflector construction and of elliptical paraboloid shape. There are two large and two small scanners in each group, one of each size being disposed with its major axis horizontal and the other vertical.

The central mounting has noticeably rounded proportions and appears to provide a common axis about which the group rotates in azimuth, and the forward side of this mounting is another large housing providing for rotation of the group in elevation. It cannot positively be ascertained if this provides for separate elevation rotation of the individual elements of the group. Viewed from the front, the large vertically disposed scanner is to the right of the central mounting and the other three scanners are to the left of it.

Although unrestricted rotation of the individual vertical and horizontal elements of the Peel Group is unlikely, sector scanning is probable for the performance of air target tracking. Target designation is probably provided by one of the Head Net series of air surveillance radars, and the Peel Group does not seem to have been designed to give a search capability.

The feed arrangements are a matter of some conjecture. Those of the two smaller scanners are distinguished by large offset boom-mounted illuminator housing boxes the shape of which seems to suggest that they contain multiple horns. One feasible explanation for this is that a monopulse tracking technique is used, with separate radars for each co-ordinate, azimuth and elevation. Operating frequency of the two smaller radars of the group is probably in the I-band.

The feeds of the two larger radars are smaller than those of the other two radars of the group and are not offset. The feed housings are of cylindrical form and may contain devices such as a quarter-wave plate to provide circular polarisation as an aid to higher weather penetration.

Information is too limited to draw firm conclusions concerning the functions of the individual elements of the Peel Group in tracking the target and directing the missile. A likely assessment is that the two larger scanners are employed for the longer range target acquisition and 'coarse' tracking, while the two smaller scanners provide precision tracking.

Between the large vertically oriented scanner and the central mounting a further unit of uncertain function is located. This could be a radiator for guidance signals to the missile until its own homing system comes into operation. It is also a reasonable assumption that some provisions for IFF are made within the system.

The normal deployment arrangements associate one Peel Group unit with one dual Goa surface-to-air missile launcher. 'Kresta I' class guided missile cruisers have two installations, one forward above the bridge and the other aft, atop a short tower, and 'Kashin' class GMDs are similarly fitted. The 'Kynda' class GMDs have a Peel Group installation above the bridge, and on 'Kotlin' class GMDs there is a Peel Group on a square, tapered tower amidships.

The two Peel Group missile control radars stand out, fore and aft, on this modified 'Kashin' class destroyer, at anchor off the Libyan coast. Also prominent are the air search radars Head Net C and Big Net on the two lattice masts, Bass Tilt fire control radars for the Gatling-gun turrets, fore and aft Owl Screech fire control radars for twin dual-purpose gun turrets, and the ship's impressive armament (HMS Intrepid)

1609.253
PLINTH NET SEARCH RADAR

Plinth Net is a medium-range E-band general-purpose search radar with an elliptical lattice reflector illuminated by an underslung boom-mounted feed horn. Functions include air search and target designation. Range against air targets at medium altitudes is about 80 nautical miles (150 km), and 20 nautical miles (37 km) against surface ship targets. There are no balance vanes behind the scanner. The turning gear is housed in a tapered cylindrical mounting, from which the NATO name for this radar is derived. Plinth Net is fitted in certain ships of the 'Kynda' and 'Kresta I' class, 'Talin' class, and 'Sverdlov' class cruisers.

1897.253
POP GROUP FIRE CONTROL RADAR

Pop Group is the NATO code-name for the fire control radar group associated with the Soviet Navy's SA-N-4 surface-to-air missile system (**2954.231**). The name has its origin in the fact that the launcher for this missile system pops up from a silo-type housing preparatory to launching.

Few details have been obtained regarding the operation of the Pop Group radar but at least two antennas are used for radar functions, with a third for the transmission of command signals to the missile. The arrangement consists of a cube-shaped container with sides of about 2.2 m in which is probably housed the bulk of the electronics such as transmitter/receiver units, power supplies, and turning gear, and on the top of this is a trainable radar head assembly. The latter unit, so far as can be ascertained from the limited evidence available, has a parabolic antenna of about two metres aperture which is probably for target search and which may rotate independently of the rest of the radar head (which, as noted above, can be trained in azimuth). On the front face of the head are two circular arrays that are assumed to be for target tracking and missile guidance. Operating frequencies are likely to lie in that part of the spectrum contained within the G- to J-band region. It appears possible that the Pop Group radar may be a monopole, frequency hopping system.

Pop Group radar on a 'Krivak' class destroyer. The Owl Screech gun fire control radar below the Pop Group is facing the opposite way showing the feed arrangements more clearly

The public display by the Soviet Union in November 1975 of the land-mobile SA-8 missile system, complete with its associated radar group, rapidly led observers to comment upon the similarities between this radar and Pop Group. The theory that the two recent sea and land air-defence missiles are related is an attractive one, but it should be noted that whereas the SA-8's radar group has two antennas which are assumed to be for command link signals, the Pop Group set has only one. The fact that the main tracking array of the land radar has cropped sides is not significant, and this measure was probably adopted to enable the overall dimensions of the radar to be reduced to measurements compatible with those of the carrier vehicle.

Operating frequencies for SA-8 are conjectured to be G- or H-band for search and target acquisition with I-band for tracking.

STATUS

Since Pop Group was first detected on Soviet 'Nanuchka' class missile frigates in the late 1960s, it has been fitted to an increasing number of types of vessel of widely differing sizes.

'Nanuchka' class ships have a single Pop Group installation on top of the bridge housing, and a similar arrangement is employed in the 'Grisha' escorts, the SA-N-4 launcher 'bin' being located in the forecastle on both types of vessel. The 'Krivak' missile destroyers have SA-N-4 installations fore and aft, with a Pop Group radar for each system, one on top of the bridgehouse and the other on a raised mounting

Pop Group fire control radar

amidships. 'Kara' class missile cruisers have two SA-N-4 systems, located one on each side of the mast and with their associated Pop Group radars are two automatic anti-aircraft gun turrets on each side, and these would appear to be directed by the Pop Group radars to form (with the SA-N-4 missile launchers) port and starboard integrated close-in autonomous air defence weapon systems.

In the case of the converted 'Sverdlov' cruisers, the largest vessels equipped with SA-N-4, various arrangements exist. *Zhdanov* had its SA-N-4 silo fitted in place of the original 'X' 152 mm gun turret, and the associated Pop Group radar was installed on a platform attached to the aft mast. The *Admiral Senyavin* has its SA-N-4 installation incorporated in the helicopter hangar, with the Pop Group in the same position as on the *Zhdanov*. In the former ship, however, there are four automatic AA gun turrets mounted adjacent to the SA-N-4 system.

The 'Nanuchka' class fittings were the first to appear in quantity, but since those early days many classes and individual ships of the Soviet Fleet have appeared armed with the SA-N-4 missile system and equipped with Pop Group radar. A single Pop Group is fitted to ships of the 'Koni', 'Grisha I' and 'III', and 'Sarancha' classes, as well as the *Ivan Rogov* landing ship and the fleet replenishment ship *Berezina*. Two Pop Group installations are found on ships of the 'Kara', 'Kiev', 'Kirov', and 'Krivak I' and 'II' classes.

1612.253
POT DRUM NAVAL RADAR

Pot Drum is a small surface search radar, probably operating in H/I-band and protected by a flat, slightly domed, radome from which its NATO code name is derived. The diameter of the radome is about 1.5 m and the unit is typically installed at the masthead of small patrol boats. This radar is probably of conventional design with the possible exception of a rather higher than usual rate of scanner rotation and the inclusion of ECCM and perhaps a passive direction finding mode of operation. The transmitter/receiver is housed below decks. The display unit has a small CRT display (about 15 cm), so that short range operation of the radar is likely. In addition to surface target detection, Pot Drum probably is used for pilotage purposes also and may have a limited air warning capability. The principal operational role is that of surface search and fire control for torpedoes carried by fast MTBs. Range is about 20 nautical miles (37 km).

Coastal escorts of the 'Stenka' and 'Kronshtadt' classes are fitted with Pot Drum, as are torpedo boats of the 'Shershen' class, some 'P6' class boats, and the 'Pchela' class of hydrofoil craft.

Soviet 'Shershen' class torpedo boat with Pot Drum surface search radar and Drum Tilt fire control radars for guns

1613.253
POT HEAD NAVAL RADAR

Pot Head is similar in appearance to the Pot Drum surface search radar (**1612.253**) but is slightly smaller with an estimated diameter of 1.2 m. Another distinguishing feature is that the top of Pot Head is flatter than that of Pot Drum. The operating frequency is probably in the I-band but the possibility of E-band operation has been quoted by certain sources. The main function will be that of surface target detection, with pilotage and limited air warning facilities as supplementary capabilities.

Fittings appear to be confined to torpedo boats of the 'P6', 'P8' and 'P10' classes operated by the Soviet and East German Navies.

Pot Head short-range surface radar with IFF antenna

1324.253
SCOOP PAIR SURFACE TARGET RADARS

This radar system is a feature of Soviet vessels equipped with Shaddock surface-to-surface missile launcher tubes. It consists of two apparently identical radars mounted above and below a spherical housing supported at the end of a sponson. Photographs indicate that these sponsons, which are aligned with the ship's fore-and-aft axis, may include provision for rotating the radar group to provide stabilisation against roll motion, and there may also be pitch stabilisation but this is not known for certain.

The scanners are of open lattice construction and of essentially elliptical paraboloid form, although the contours are rather angular. They have quite a high span/depth ratio and the former dimensions are estimated at approximately 4.2 m. They each have the typical Soviet balance vanes behind the scanners, in this case of almost square outline. Each scanner is illuminated by a double horn assembly carried on a boom, coming from beneath the scanner in the case of the upper radar of the group, and above in the case of the lower radar. The disposition of the horns suggests that they are arranged to provide maximum elevation coverage from the combined pair of radars. A further possibility for the use of double horns is that they are associated with special signal processing equipment to provide optimum target detection in the presence of sea clutter returns.

It is not known if the two radars comprising the Scoop Pair group are rotated in azimuth independently, although there is no mechanical reason why they should not be.

Operating frequencies are in the E-band. Probable functions include surface search, and target detection to provide range and bearing data for the Shaddock missile fire control system, and tracking of the missile after launch.

All Shaddock-carrying Soviet vessels appear to be equipped with one or two Scoop Pairs, depending upon the Shaddock launching arrangements. 'Kresta I' class guided missile cruisers have a single installation, high on the main 'tower', and this serves two twin launchers below, and on each side of the bridge. 'Kynda' class ships have two groups, fore and aft, each serving a four-tube launcher.

Scoop Pair radars for surface-to-surface Shaddock missiles. The projection above the upper-scanner could be for IFF. The scanners can be seen to employ a solid reflector in front of which is a parallel grid, presumably to improve clutter rejection

1614.253
SKIN HEAD NAVAL RADAR

The Skin Head radar is principally associated with Soviet torpedo boats and its function is assumed to be surface target detection for use in association with torpedo fire control systems. Its appearance is characterised by a thimble shaped dome from which the NATO code-name is derived, and generally a masthead mounting. Diameter of the radome is estimated to be about 0.8 m. The nature of the scanner assembly inside is not known but a high rotation rate is assumed. Operating frequency is probably in the I-band.

Although Skin Head is mostly associated with torpedo boats of the 'P4' and 'P6' classes, used by the Soviet and friendly navies, the 152 mm gun turrets of Soviet 'Sverdlov' cruisers carry a radome of similar proportions (NATO code-name Egg Cup) which suggests that the same radar may be used for fall-of-shot measurement or other purposes on these ships.

Skin Head radar on masthead of 'P'-type Soviet torpedo boat. The array on top may be an alternative IFF antenna

Skin Head radar used for surface target detection on Soviet torpedo boats. Small antenna is for IFF

1322.253
SLIM NET SURFACE WARNING RADAR

Slim Net is a surface warning, high definition, radar of generally conventional naval design, plus the typical Soviet characteristic of balancing vanes behind the scanner. It also has an air search capability. The scanner is of open lattice construction and approximately 5.5 m span and 1.8 m maximum depth. The general shape is of a tapered rectangle, and the balance vanes also are rectangular. The horn feed is mounted on a boom overhangng the scanner. Operating frequency probably lies within the E/F-band of the spectrum.

Vessels seen with this radar include some ships of the 'Kildin' class of guided missile destroyers (in which it may be associated with the Strela surface-to-surface missile launcher); 'Kotlin' class destroyers; and 'Tallin' class destroyers. In all three cases it is carried on top of the rear tripod mast. It is believed that some 'Riga' class and 'Petya' class escorts have been equipped.

Slim Net scanner assembly, with Yard Rake rotating

Slim Net scanner assembly, with Yard Rake rotating Yagi array (right)

5849.453

SNOOP CLASS SUBMARINE SURVEILLANCE RADARS

A number of surveillance radars have been fitted to Soviet submarines since World War II, some of which have been given a NATO designation starting with Snoop. The early Soviet submarines were equipped with a small antenna designated Snoop Plate. This was apparently fitted to all boats up to the 'Golf/Foxtrot/Romeo' classes and was reputed to have a range of 25 nautical miles against aircraft and 12 nautical miles against surface vessels. Later 'Foxtrot' class vessels and the 'Hotel' generation were fitted with a larger and more powerful radar known as Snoop Tray which entered service in the early 1960s.

Since the early 1960s, all new Soviet submarines have been fitted with Snoop Tray, with the exception of the 'Juliett' and 'Echo' classes of cruise missile submarines which are equipped with a much larger antenna known as Snoop Slab. This antenna is twice the size of the Snoop Tray unit and must be capable of greater definition at longer ranges. Both the 'Juliett' and 'Echo' classes are armed with the surface-launched SS-N-3 Shaddock missile and it is assumed that the Snoop Slab radar will also provide back-up cover for the Front Door/Front Piece missile guidance equipment.

STATUS

In operational service.

1595.253

SQUARE HEAD NAVAL RADAR

The antenna of this system consists of a broadside array of dipoles with a rectangular reflector/support frame measuring about 1.4 m high by 2 m wide. It was originally thought to be a long wavelength search radar of modest power, but later evidence confirms that its function is that of an IFF interrogator antenna, or a directional array for the transmission of guidance signals to surface-to-surface missiles. It is most frequently seen on Soviet missile carrying boats of the 'Osa' class (but apparently not on craft of this class of the navies of countries friendly to the USSR), and these generally have two such arrays carried on sponsons fore and aft of the tubular mainmast. Some 'Kotlin' class destroyers of the Soviet fleet carry four, and Square Head has been seen on vessels of the 'Skory' class.

Recent information suggests that the Square Head antenna is associated with an IFF system operating in the G-band.

1329.253

SQUARE TIE NAVAL RADAR

Square Tie is a small lightweight search radar, probably operating in the I-band to provide short range air and surface search facilities. It is extensively fitted on Soviet vessels of the 'Osa' class armed with anti-ship missiles. The scanner is an elliptical paraboloid reflector illuminated by an overhung horn feed. Functions will include target detection and tracking for anti-ship missile direction and possible target designation for the Drum Tilt gun fire control radar carried by 'Osa' class craft. Maximum range is believed to be about 70 nautical miles (130 km).

1331.253

STRUT CURVE SEARCH RADAR

Strut Curve is a medium-range general-purpose air search radar, probably operating in the E/F-band to provide both air and surface search facilities. An elliptical lattice reflector is used, illuminated by a horn feed carried by a boom projecting from the lower edge of the scanner. The boom is supported by two struts which attach to stays at the upper edge of the reflector. Somewhat unusual for Soviet naval radars, the Strut Curve has no balance vanes behind the reflector.

Range performance against an aircraft target at medium altitude is about 60 nautical miles (110 km) with a likely maximum range of 150 nautical miles (280 km).

Strut Curve search radar

Strut Curve antenna profile

Soviet vessels fitted with Strut Curve include the 'Poti' class of escorts and certain support ships.

1326.253

SUN VISOR FIRE CONTROL RADAR

Sun Visor is identified as a solid paraboloid scanner fixed to the front of a spherical shaped gun fire director station. Two types of director have been identified carrying this type of radar. These have the NATO code names Round Top and Wasp Head, and in the former the radar is carried near the top of the unit compared with a lower location on Wasp Head, which is stabilised. Span of the dish is estimated at approximately 2.2 m, and the depth at 1.3 m. Operating frequency is probably I- or G/H-band. The reflector is illuminated by a dielectric-enclosed feed carried on a solid boom structure projecting from over the upper edge of the reflectors. The disposition of these components suggests that surface target detection is the principal role of this radar.

The fire director units on which Sun Visor is mounted are capable of 360° rotation, but since they are manned units incorporating optical rangefinding equipment also, it is unlikely that it is normally used for target search and surveillance on a continuous basis. A more likely mode of operation is for target co-ordinates to be passed to the system by another of the ship's radars and for the former to operate in a target acquisition and lock-on role to provide either range and bearing data for blind firing or to assist the operator in the training of optical sighting systems.

Sun Visor has been seen on certain vessels of the 'Krupny' class of GMDs, the 'Kanin' class, 'Kotlin' class, 'Talin' class, and 'Sverdlov' class cruisers.

Sun Visor gun fire control radar

4088.253

TOP DOME RADAR GROUP

Top Dome is the name of the radar group associated with the SA-N-6 naval area air defence weapon system fitted in the Soviet cruisers *Kirov* and *Frunze*. This system is thought likely to be an approximate Soviet equivalent of the American Aegis surface-to-air naval air defence missile system, in which case the Top Dome radar group will embody long-range tracking and guidance functions, probably incorporating multi-mode operating facilities of one sort or another. At this early stage it is possible only to embark on preliminary conjecture concerning the functioning of the Top Dome radar group. The 'Kirov' class has two such director groups located fore and aft, this being more to achieve an all-round engagement capability rather than by any requirement to associate the individual directors with missile launcher installations in specific parts of the ship. The SA-N-6 is believed to employ a vertical launch technique so that once missiles are launched from the magazine on the foredeck of the ship and are clear of the ship's superstructure, they can be commanded to turn onto any course required to engage the specified target.

Targets are probably detected by either passive means (of which *Kirov* has an ample selection) or the Top Pair radar group, and targets are designated to either of the two Top Dome director groups which are then laid on to the necessary threat direction. When the Top Dome radar (or passive detection sub-system) acquires the target and locks on and when the target has reached an appropriate engagement range one or more SA-N-6 missiles are launched and commanded to the proper direction, after which automatic guidance to interception takes place. The Top Dome radar group is probably capable of tracking both the SA-N-6 missile and the designated target simultaneously, and there is also the possibility of a

Top Dome radar director group on the Soviet cruiser Kirov *with Kite Screech gun fire control radar to the left*

combined mode of guidance involving target tracking through the SA-N-6 seeker with the command link relaying data back to the ship. In any event, the new system is likely to incorporate several guidance options to cater for changing circumstances and to deal with opposing countermeasures.

The Top Dome group mounting is obviously designed to rotate in azimuth and probably has some horizontal stabilisation against ship's motion also, but continuous rotation of the mount is thought unlikely. Stabilisation gear is possibly housed within the hexagonal portion of the mounting. The 'front' face of the latter carries three antenna housings of hemi-cylindrical shape; these are thought to contain passive sensors or command link elements. A similar purpose is also possible for the small 'thimble' shaped housing carried above the hemi-cylindrical units, at an angle. The contents of the main dome remains uncertain at this stage.

4087.253

TOP PAIR RADAR GROUP

Top Pair is the name of a back-to-back configuration of two very different naval radar types employed for long-range air surveillance. One is the well-known Top Sail 3-D radar which is associated in this configuration with what appears to be the Big Net long-range search radar (or a derivative of it). All known fittings are aboard recent ships known to have a major air role, such as the aircraft carriers of the 'Kiev' class. The Top Pair group has also been noted more recently fitted in the Soviet guided missile cruiser *Slava*, as well as the cruisers *Kirov* and *Frunze*, which in addition to carrying their own helicopters are fitted with the new SA-N-6 surface-to-air missile system and are thought to have an important anti-air escort role in support of the Soviet carrier force. Operating frequencies in the D-band, and possibly E/F-bands, are thought likely for the Top Pair radars. Detection ranges of several hundred kilometres can be expected.

Soviet cruiser Kirov, *showing the Top Pair air surveillance radar group at the very highest position on the ship's superstructure, with another new radar group, Top Steer, to be seen further aft on top of the pyramidal mast. The ship also displays two Top Dome fire control radar groups (fore and aft) for the ship's SA-N-6 vertical launch area defence anti-aircraft missile system. Numerous ESM and EW antennas are clearly visible*

Enlargement of Top Pair radar group on the Soviet cruiser Kirov

6880.253

TOP PLATE/TOP MESH RADARS

Top Plate and Top Mesh are air and surface surveillance radars fitted to some of the later Soviet warships. They are 3-D systems employing back-to-back antennas and operating in D/E-bands. Fitted to the later 'Udaloy' class destroyers.

1327.253

TOP SAIL THREE-DIMENSIONAL RADAR

Top Sail is a very large air surveillance radar, first seen on the Soviet cruiser helicopter carrier, *Moskva*, and her sister ship *Leningrad*, but later on the 'Kresta II' and '*Kara*' classes of guided missile cruisers, the *Kiev* class carriers, and the *Kirov* class battle-cruisers. In all installations the radar is mounted at the highest point of the vessel; in the case of the 'Kresta II' class on a large diameter tubular pedestal atop the principal tower, and on the helicopter cruisers supported by a tubular tripod. In the 'Kara' installation the scanner surmounts a typically Soviet pyramid shaped structure amidships.

The scanner is of cylindrical cross-section, the axis of which is tilted from the vertical at an angle of about 20°. It is of lattice construction and has a frontal aspect of rhombic shape (estimated 5.5 m sides), with 'cropped' corners. Illumination of the reflector is by a linear radiating element located parallel to the reflector's cylindrical axis, and supported top and

The profile of the Soviet aircraft carrier Kiev *clearly illustrates the ship's comprehensive radar fit. Ahead and abaft of the Top Globe on its mast are Top Sail and Top Steer air surveillance radars, respectively, emphasising their size difference; below these radars are fore and aft Head Light missile control radar groups, Owl Screech gun fire control radars, with a navigation radar between the forward Head Light and Owl Screech*

bottom. The feed appears to be from the upper end. It may consist of multiple horns to provide three-dimensional air target data by the 'stacked-beam' method, or some form of electronic scanning in elevation (such as the ridged waveguide) may be used to achieve the same result. Operating frequency is probably in the C/D frequency band, with high transmitter power. The theoretical maximum range is more than 300 nautical miles (555 km).

The 'Kresta II' Top Sail antenna shows what appears to be a second system of waveguide, of smaller dimensions than the main system, apparently feeding a horn at the top of the main illuminator. This might indicate dual band operation facilities for some purposes. There is a subsidiary antenna fixed to the top of the Top Sail scanner which could be for IFF. In addition to the two plates which are the most prominent feature of this secondary antenna there are also three dipoles. Near the bottom of the illuminator of the main Top Sail are what could be a similar group of three dipoles. The function of these elements has not been ascertained.

In all three classes of vessel equipped with Top Sail it is associated with similar radars, a Head Net C back-to-back 'V-beam' set, and Head Light group missile fire control director radars. The Top Sail is probably used for long-range air surveillance and target designation.

4086.253

TOP STEER RADAR GROUP

Top Steer is a 3-D air search radar group comprising a pair of scanners mounted back-to-back, usually high up on the superstructure of the ship. The two scanners are not the same and one is somewhat similar in appearance to the well-known Top Sail 3-D air surveillance radar fitted to several classes of recent major Soviet ships. However the Top Steer antenna is noticeably smaller, as can be seen on ships such as the *Kirov* cruiser which has both Top Sail and Top Steer fitted. The other scanner appears to be either a standard Head Net A search radar or a modified derivative of it. It is not known if the two radars are operated together as part of a single complex system of a new type, or if the two scanners are merely co-mounted as a matter of structural convenience with either one of the two radars being capable of independent operation on its own (apart, of course, from having common turning gear and/or stabilisation).

Operating frequencies of both radars are likely to be in the D/E/F-bands of the radar spectrum and it is possible that the larger antenna may incorporate some form of frequency scanning of the beam in elevation to provide a heightfinding capability. The probable function of this radar group is that of air control of the ship's helicopters, or other aircraft operating with the ship.

Top Steer installations have been noted on the cruisers, *Kirov* and *Frunze*, on the 'Sovremenny' class destroyers, on the recent 'Kiev' class aircraft carriers, and the guided missile cruiser *Slava*.

Top Steer radar group on the Sovremenny

This view of the Soviet aircraft carrier Kiev *clearly shows the considerable difference in size between the Top Sail 3-D radar and the Top Steer radar group mounted aft of the Top Sail and the flaired-in mast that carries the* Top Dome (RAF)

1610.253

TOP TROUGH SURVEILLANCE RADAR

Top Trough is a C-band long-range air search radar with a probable maximum range of up to 300 nautical miles (555 km). The only known installations are on some ships of the Soviet 'Sverdlov' class of cruisers, in particular *Mikhail Kutuzov, Senyavin* and *Zhdanov*, where it is mounted on top of the aft mast.

Soviet 'Sverdlov' class cruiser, Mikhail Kutuzov, *showing Top Trough radar on aft mast. Knife Rest (Uda-Yagi) antenna for air warning can just be discerned forward of aft gun turrets*

6878.253

TRAP DOOR NAVAL RADAR

Trap Door is a missile guidance radar which has been noted on the 'Kiev' class missile cruisers and has recently been redesignated by NATO as Front Door C. It is a development of the Front Door A which is fitted to 'Juliett' and 'Echo II' class guided missile submarines. In the 'Kiev' class Trap Door is located in the foremost part of the ship and is housed in a casing approximately 5 m long and 3 m wide. The upper cover of this casing is designed to hinge up and back. The antenna grid is mounted on the inside of the cover and is thus released as soon as the cover hinges upward. The antenna is concealed when not in use. A more recent version of Front Door C is fitted to the 'Slava' class cruisers.

1900.253

OTHER SOVIET NAVAL RADARS

Photographs have been obtained of a number of other radars fitted to ships of the Soviet Fleet additional to those described in the preceding entries. Insufficient information is available to make possible individual entries, but the photographs are reproduced to assist readers with identification.

Top Bow gun fire control radar

Top Bow is an H/I-band fire control radar for 152 mm guns

High Sieve

Don K navigation radar

Knife Rest

Ball Gun

UNITED KINGDOM

3699.253
CEL 850/4 NAVAL INTERROGATOR

The Cossor CEL 850/4 naval IFF interrogator is the shipboard version of the CEL 850/4 general-purpose interrogator, and additional details will be found in **3692.153** in the Ground Radar section of this volume. In the naval application it is suitable for use in all classes of ship, down to fast patrol craft in size. The transmitter section has an output of 2 kW (peak), and the equipment can operate on Modes 1, 2, 3/A, C and 4. It is designed to accommodate many different interfaces so that integration with numerous types of primary radar presents no problem.

STATUS

In production with deliveries to navies all over the world. The latest systems on order for Abu Dhabi incorporate full IFF Mk XII compatible cryptographic capability.

CONTRACTOR

Cossor Electronics Ltd, The Pinnacles, Harlow, Essex CM19 5BB, England.

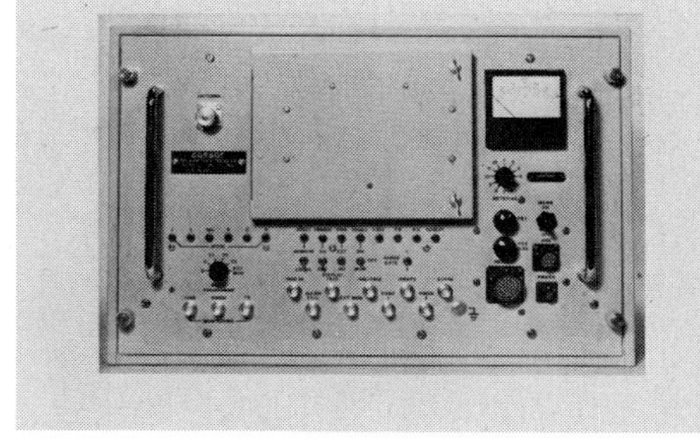
Cossor CEL 850/4 general-purpose interrogator equipment

3701.253
IFF 2750 (TYPE 1013) NAVAL TRANSPONDER

Under contract to the Royal Navy, Cossor modified their IFF 2720 airborne IFF transponder (**3695.353**) for naval applications. The resulting equipment, IFF 2750, has a low magnetic profile, low weight and low volume, thus making it suitable for fitting in all classes of vessel, including minehunters. In the Royal Navy the IFF 2750 is designated the Type 1013. The key features of the IFF 2720, high reliability, ease of access for maintenance, and continuous on-line monitoring, have been retained in the IFF 2750.

Cossor IFF 2750 naval transponder (Type 1013) and control unit Type IFF 2753

CHARACTERISTICS

Modes: 1, 2, 3

Receiver frequency: 1030 MHz

Receiver sensitivity: –76 dBm

Transmitter frequency: 1090 ±3 MHz under all service conditions

Transmitter power output: 750 W (nominal); 500 W (minimum)

Reply coding: 4096 on all modes

Dimensions: 91 × 193 × 381 mm

Weight: 4.6 kg

STATUS

In production and supplied to the Royal Navy and other customers.

CONTRACTOR

Cossor Electronics Ltd, The Pinnacles, Harlow, Essex CM19 5BB, England.

1941.253
RACAL-DECCA SURVEILLANCE AND NAVIGATION RADARS

Racal-Decca radar systems are in world-wide operational use as prime sensors for surveillance, weapon control and navigation. They operate in the F and I frequency bands and are interfaced with a wide variety of fire control, electronic warfare, computer systems and other radar equipments. The commercial frequencies employed provide ESM protection.

The 2459 F/I radar, with its widely spaced dual frequencies, provides a combined air and surface surveillance capacity using a single turning unit, thereby saving top weight and making optimum use of masthead space. The Command Tactical Consoles, either monochrome or colour, are stand-alone auto-tracking consoles which accept video outputs from the majority of surveillance radars and can auto-track up to 20 air or surface targets. Other

Racal-Decca 2459 F/I antenna

facilities include target indication, video mapping, ground stabilisation and digital inputs/outputs.

Racal-Decca radar systems are modular and can be configured from a range of antennas and transceivers with a choice of colour or monochrome displays. They are available in a variety of sizes with relative motion, true motion/anti-collision and ARPA tracking facilities and include well proven advanced video processing techniques to minimise sea and rain clutter and improve the detection of small echoes.

STATUS

In operational service with the navies of more than 90 countries.

CONTRACTOR

Racal Marine Systems Ltd, Burlington House, 118 Burlington Road, New Malden, Surrey KT3 4NR, England.

1563.253

MRS3/GWS22 FIRE CONTROL RADARS TYPES 903/904

The MRS3/GWS22 is standard Royal Navy equipment for the control of guns (MRS3) and Seacat missiles (GWS22). The initial application was gun fire control for quick-firing medium guns of 3, 4.5, and 6 inch calibres. Following application of the system to the Seacat system, the MRS3/GWS22 has been the subject of steady modification and can be seen in several forms.

The radars carried by this director are known as the Types 903 and 904, and both are conventional I/J-band conical scan tracking radars.

More detailed information is given under this entry number in previous editions of *Jane's Weapon Systems*.

STATUS

The MRS3/GWS22 has been extensively fitted in many ships of the Royal Navy. A number of Commonwealth and other navies are equipped with the system, including ships of the Australian, Chilean, Indian, and New Zealand navies. Production is continuing. The systems have now been updated under a contract awarded to British Aerospace (Bracknell).

CONTRACTOR

British Aerospace plc, Bracknell, Berkshire RG12 1QL, England.

3171.253

AWS-4 NAVAL RADAR

The Plessey AWS-4 naval search radar still meets the operational needs for air and surface search, target indication and weapon control of modern navies. It can perform effectively in conditions of jamming, sea clutter and precipitation. With a lightweight, compact antenna system the AWS-4 is readily installed in small ships, such as fast strike craft, and is equally suitable as the search radar for corvettes and light frigates, or as the tactical radar for larger vessels in association with a separate long-range air warning radar. The AWS-4 uses a tunable S-band magnetron transmitter with a low noise amplifier and ECCM receiver to give an effective anti-jamming performance.

AWS-4 masthead equipment fitted in Norwegian coastguard patrol ship K/V Nordkapp

A two-speed antenna, operating with dual pulse lengths and PRFs, optimises the performance of the AWS-4 in the roles of air and surface search and target indication. Variable antenna polarisation enables the operator to minimise the effect of bad weather. IFF facilities are provided by the incorporation of the D-band IFF radiator in the main feed horn. This irradiates the search radar reflector which allows an improved coverage pattern from a lower-power interrogator and reduces masthead weight and windage.

The optional digital moving target indicator (DMTI) provides further protection against clutter caused by sea, land or rain. The DMTI system is self-compensating for clutter velocity, hence also providing good discrimination against chaff. Data from ship's log and gyro is not required.

The radar is able to detect an aircraft target at ranges in excess of 100 km. Against the same target a diversity system will provide increased coverage operationally, enhance anti-jamming performance, increase system reliability and improve availability as one transmitter/receiver may be closed down for maintenance without loss of radar service.

In addition to the optional features of DMTI and diversity operation mentioned above, a two-axis stabiliser is available to improve overall system performance in cases where AWS-4 is used with a computerised AIO/CIC. A modular data distribution rack is also available to provide optimised interfaces between the radar and the rest of the ship's weapon system.

DEVELOPMENT

The AWS-4 has been derived from extended experience with the preceding AWS-1 and AWS-2 radars in service with a number of navies. Programmes for the progressive up-dating of earlier radars to AWS-4 standards are available.

STATUS

Plessey Radar has supplied AWS-4 radars to a number of overseas customers including the Royal Norwegian Navy and the Sultan of Oman's Navy. Over 55 AWS-4 installations had been ordered or supplied as at the end of 1987.

The AWS-4 below-deck equipment is also extensively employed for modernising obsolescent radars such as the Type 293 and the Type 993 (see **3260.253** and **3261.253**) as used by a number of navies. Conversely, AWS-4 antennas are being used to improve the performance of existing radar systems. Eight antennas are being fitted to Royal Canadian Navy ships as part of the DELEX (Destroyer Life Extension) programme.

CONTRACTOR

Plessey Radar Ltd, Oakcroft Road, Chessington, Surrey KT9 1QZ, England.

3172.253

AWS-5 SERIES NAVAL RADAR

The Plessey AWS-5 is an advanced technology S-band tactical radar designed to provide comprehensive surveillance facilities for corvettes and frigates in the range 1000 to 4000 ton. It is also suitable as the search and target indication radar in larger ships, in conjunction with a long-range air warning radar. Advanced features of the AWS-5 enable it to combine the following functions:

(1) air and surface warning, an aircraft will be detected at a range in excess of 180 km
(2) point-defence search, with enhanced missile detection capability against low level attack and detection beyond 30 km against small high elevation diving targets
(3) MTI.

The AWS-5 provides excellent anti-jamming protection and clutter rejection principally through:

(1) high-low beam operation to counter stand-off jamming
(2) circular or linear polarisation by selection to improve rejection against rain
(3) pulse-to-pulse programmable/random frequency agility for ECCM protection and decorrelation of clutter
(4) frequency coded transmissions against deception jamming
(5) high ratio pulse compression for finer resolution and better clutter performance

Nigerian frigate, NNS Aruda, *fitted with AWS-5 radar*

(6) adaptive coherent DMTI to remove echo responses due to sea and land clutter, bad weather and chaff.

The lightweight masthead equipment of the AWS-5 comprises a compact two-axis stabilised mount on which is fitted a single or dual-beam antenna system incorporating variable polarisation and integral IFF facilities.

The main below-decks equipment comprises a medium or high power compact transmitter/receiver cabinet in which are housed the wide-band frequency-modulated transmitter with its drive circuits, power supplies, and a programmable frequency agile source. The receiver chain incorporates pulse compression, CFAR processing, and adaptive MTI. Additional equipment in the AWS-5 outfit includes the stabiliser control system and a data distribution unit with a high-resolution video extractor and comprehensive interface facilities for supply of video and other data to the ship's operations room and AIO/CIC computer.

Time multiplexing and frequency coding techniques are used in both transmitter and receiver chains in conjunction with the dual-beam antenna. This enables the performance characteristics of two radars to be achieved simultaneously from a single AWS-5 outfit. Radar facilities which could previously be accommodated only in large ships are thus now available for ships of the corvette and light frigate categories.

The various modes of video processing provide good protection against clutter and jamming, enhanced by the programmable frequency agility, the clutter-free high beam, accurately controlled antenna characteristics with good sidelobe performance, and high-ratio equalisers giving an exceptionally short compressed pulse. Integrity of performance under hostile conditions was given priority in the development programme.

The AWS-5 is designed for maximum flexibility in use. A modular set of control panels gives the operator a wide choice of facilities including:

(1) independent control of main beam and high beam
(2) active or passive mode

(3) DMTI on either or both beams
(4) separate or combined videos
(5) fixed frequency, limited or full agility
(6) variable main beam polarisation
(7) two data rates.

Certain performance characteristics are programmable during manufacture, enabling the AWS-5 to be closely matched to the role of a particular class of ship.

STATUS
Equipments are in service with the Royal Danish Navy and a containerised system is in use with the Nigerian Navy.

CONTRACTOR
Plessey Radar Ltd, Oakcroft Road, Chessington, Surrey KT9 1QZ, England.

5620.253
AWS-6 SERIES NAVAL RADAR
The AWS-6 is a range of C-band radars designed for naval surveillance combined with maximum missile detection capability. With this combined function it is ideally suited as the primary radar in fast patrol boats, corvettes and other small ships. In larger vessels where multiple radar installations are provided, the AWS-6 will be used in a point defence role only. The equipment is offered in a wide range of configurations and signal processing options and can be profiled to specialised defensive requirements.

The antenna is small and lightweight which allows it to be mounted at masthead with dual-axis stabilisation. IFF and navigational functions are integral. Single- or dual-beam configurations are available; the single beam is used for general surveillance operation and target designation while the dual beam mode enables near hemispherical coverage for threat detection at all angles of elevation. Incoming targets as small as 0.1 m² radar cross-section can be detected, typical of anti-ship missiles. The radar is then used to indicate to the ship's defences – either guns or missiles. One version, the AWS-6/Dolphin missile detection radar, is the search radar module for Seaguard (**3509.281**), the Contraves close-in weapon system, and employs the dual beam configuration for the detection of high diving and sea-skimming missiles.

The AWS-6 has been designed to operate under severe environmental conditions; and frequency agility, pulse compression and within-pulse coding have minimised the effects of ECM. This has been combined with adaptive MTI to give high performance against clutter, chaff and target glint.

The equipment is also available in containerised form for integration with other weapon systems on ships such as armed merchantmen.

STATUS
The equipment is in full production. Seven AWS-6 systems are fitted to Danish fishery protection vessels, and a further seven systems have been ordered for the Danish STANFLEX 300 ships. Other systems are in service, or have been delivered to, the Turkish Navy and other navies.

Antenna unit of the Plessey AWS-6/Dolphin surveillance radar

CONTRACTOR
Plessey Radar Ltd, Oakcroft Road, Chessington, Surrey KT9 1QZ, England.

6101.253
AWS-7 NAVAL RADAR
The AWS-7 is a nomenclature covering a family of Plessey radar systems which, for the first time, incorporate a solid-state transmitter. The systems operate in the S-band frequency range and are designed for medium range surveillance and target indication. The solid-state transmitter concept has a number of features, the two primary ones being graceful degradation and modularity. The latter feature allows the output power to be tailored to meet each individual customer's requirements. Digital electronics are employed throughout to improve reliability and maintainability.

The AWS-7 can use any of the Plessey S-band receivers, processors, single or dual beam antennas and power output options to provide a comprehensive family.

STATUS
A prototype system is currently in operation.

CONTRACTOR
Plessey Radar Ltd, Oakcroft Road, Chessington, Surrey KT9 1QZ, England.

6102.253
AWS-9 NAVAL SURVEILLANCE RADAR
The AWS-9 is the designation for the export version of the Royal Navy Type 996 3-D surveillance and target indication radar (**4634.253**). It is designed so that it can be tailored to meet the performance requirements of particular naval point or area defence weapon systems. It features a stabilised planar array antenna rotating at high speed with an IFF antenna co-mounted. Below decks a high power transmitter, and comprehensive receiver and signal processing equipment provide the inputs to the autonomous tracker required to enable automatic track indication, including height data, to the weapon systems. This output, using the same data highway, can also be employed to provide information to the command display system.

STATUS
In full production for the Royal Navy and under consideration by a number of other navies.

CONTRACTOR
Plessey Radar Ltd, Oakcroft Road, Chessington, Surrey KT9 1QZ, England.

1559.253
RN TYPE 909 RADAR
The Royal Navy Type 909 radar provides target tracking and illuminating facilities for the Sea Dart GWS-30 air defence missile (**6004.231**). This weapon has an anti-ship capability, and the radar is stated to be suitable for gun laying also, so that a surface target capability must be assumed. The antenna is of the cassegrain type and has a diameter of 2.44 m. A small dome mounted near the upper edge probably houses an associated IFF antenna. On board ship, the complete antenna assembly will be protected by a cupola radome. The radar head and the office cabin, containing transmitter/receiver unit and associated electronics, are constructed as a single prefabricated assembly to reduce installation and replacement time and to enable functional testing before fitting.

Few technical particulars have been revealed. A G/H-band operating frequency range is likely and a high transmitter power can be expected. It has been stated that elaborate ECCM are incorporated to counter both active and passive ECM. A photograph appears with the Sea Dart missile entry.

STATUS
The Type 909 is fitted to Type 42 destroyers for the Argentinian and British Navies, and for the RN Type 82 destroyer. Each ship has two radars of this type, mounted in fore and aft radomes. In May 1987 the UK MoD awarded a £35 million contract to Marconi for update of the Type 909 as part of a programme to increase the capability of Sea Dart.

CONTRACTOR
Marconi Command and Control Systems, New Parks, Leicester LE3 1UF, England.

Type 42 destroyer, showing the fore and aft Type 909 Sea Dart missile guidance radars in large radomes. Note missiles on forward launcher (Royal Navy)

1752.253
RN TYPE 901 RADAR
The Royal Navy Type 901 radar equipment provides target tracking and missile guidance for the Seaslug shipborne surface-to-air missile system (**6003.231**). Each of the seven RN 'County' class guided missile destroyers is fitted with a single Type 901 radar, forward of the helicopter deck, for use with the twin Seaslug launcher aft. A stabilised mount is provided and the right-hand side of the director head carries equipment for initial 'gathering' of the beam-riding Seaslug missile. No performance details of the Type 901 have been released. Operation at G/H-band frequencies is probable.

CONTRACTOR
Marconi Command and Control Systems Ltd, New Parks, Leicester LE3 1VF, England.

1558.253
RN TYPE 912 RADAR
Type 912 is the Royal Navy designation for the Selenia Orion RTN-10X fire control radar fitted to RN ships. It is an I/J-band conical-scan pulse radar, and the mounting arrangements include facilities for a closed-circuit TV system for acquisition and missile gathering functions. More details will be found in **1368.253** among the Italian entries of this section.

1562.253

RN TYPE 910 RADAR

The Type 910 is the radar tracking element of the GWS 25 Mod 0 Seawolf point defence missile system. It is used with, and receives target designation data from, the Types 967 and 968 air and surface surveillance radars (**1561.253**). It is a monopulse differential tracker radar, incorporating electronic angle tracking in the tracking loop as a means of correcting the missile guidance line-of-sight for small angles without calling for physical movement of the mount. This benefits both accuracy and the smoothness of tracking. After launch, the Seawolf missile is rapidly acquired by the Type 910 wide-angle antenna and automatically gathered onto the target line-of-sight. Subsequently, target and missile are tracked together, using the same antenna and receiver by means of time division multiplex techniques. Few technical details have been released but operation in the I/J-band has been confirmed. The Seawolf system uses command to line-of-sight guidance with radar differential tracking and a microwave radio command link, and some of the related antennas can be seen flanking the main Type 910 antenna unit. This is a self-adaptive system and is capable of controlling and directing a number of Seawolf missiles to interception. The system also includes a differential tracking optical TV system which monitors the firing and is capable of taking over missile control during flight should circumstances demand.

Information from this tracker could be used for gun control also, and the TV tracker on the same pedestal could also be employed for gunnery purposes.

STATUS

In production. Fitted to all Type 22 and 'Leander' class frigates of the Royal Navy.

CONTRACTOR

Marconi Radar Systems Ltd, Writtle Road, Chelmsford, Essex CM1 3BN, England.

Type 910 tracking radar with type 967/968 antenna in the background

1560.253

RN TYPE 965 RADAR

The Type 965 long-range air search radar of the Royal Navy is widely used on ships of sizes ranging upwards from frigates. The operating frequency range lies in the metric wavelength band and output powers in the megawatt range are assumed. In addition to air surveillance, the Type 965 fulfils target designation functions for guided weapons systems, and on-mounted IFF Mk 10 facilities are provided.

The antenna array consists of a row of eight dipoles, each at the rear of a 'horn' formed of metal mesh, the whole having a characteristic 'bedstead' appearance. The complete antenna scanner assembly may consist of either one or two such groups, and both versions are extensively employed. An IFF antenna is usually mounted on the upper edge of the primary array.

STATUS

The Type 965 has been in production for many years and is widely deployed in either single or double array form on ships of the Royal Navy from frigates upwards, including Type 42 and 82 destroyers. It is obsolescent and is being replaced in the Type 42 ships by the Type 1022 (**3486.253**).

CONTRACTOR

Marconi Radar Systems Ltd, Writtle Road, Chelmsford, Essex CM1 3BN, England.

Close-up view of Type 965 AKE-1 antenna array, with Cossor IFF antenna on upper edge

1561.253

RN TYPES 967 AND 968 RADARS

The Royal Navy Type 967 is an air surveillance radar which is integrated with the Type 968 surface surveillance radar to form a very compact medium-to-short range defence radar. They provide high performance facilities to fulfil the requirements of air and surface warning, search and target designation, from sea level to high elevation angles. These radars with the Type 910 target tracker and missile guidance radar form the radar group for the Seawolf anti-aircraft and anti-ship missile weapon, GWS 25 (**2442.231**).

Scanner and stabilisation arrangements of RN Type 967/968 for Seawolf point defence system. Separate antennas for surface and air search contained in a common housing

The waveguide antennas for both radars are mounted back-to-back in a common housing carried by a single fully stabilised mount protected by a radome. The Type 967 provides air surveillance with cover up to high elevation angles. Operating frequency is in the D-band of the radar spectrum and pulse doppler mode is employed. An enhanced version, known as the Type 967M, is now in production. The Type 968 provides low air cover and surface search, and operates in the E-band. The antenna rotation rate for both is 30 rpm. The Seawolf surveillance radar system is a high-power self-adaptive system, with sophisticated data handling, which resolves both velocity and range ambiguities, initiates track, carries out threat evaluation, takes the engagement decision, and performs attack allocation by assigning a tracker and feeding the track co-ordinates to it. IFF and system ECCM are incorporated. The system has its own digital processor, a Ferranti FM1600 series computer.

STATUS

In production. Fitted to all Type 22 and 'Leander' class frigates of the Royal Navy.

CONTRACTOR

Marconi Radar Systems Ltd, Writtle Road, Chelmsford, Essex CM1 3BN, England.

1253.253

RN TYPE 975 RADAR

The Type 975 is a lightweight, I/J-band, high definition surface warning radar designed for fitting on frigates, coastal and inshore minesweepers, and other naval craft of similar size. Its primary function is navigation, but in some installations it is specially adapted for minehunting in conjunction with a suitable plot and underwater detection equipment. This version is designated Type 975ZW, and incorporates true motion displays and provision for marking sonar targets on the display.

Transmitter power is 50 kW (nominal), and either a 182 or 304 cm slotted waveguide scanner is used. Two models of display unit are available, JUC/1 and JUC/2. The former is a watertight unit with magslip range and bearing, and the latter has synchro range and bearing transmission. Seven range scales provide displays of 0.75 to 48 nautical miles. Minimum detection of the radar is less than 32 metres, and bearing accuracy is within 1°.

STATUS

This radar has been extensively fitted to ships of the RN and other navies.

CONTRACTOR

Kelvin Hughes Ltd, New North Road, Hainault, Ilford, Essex IG6 2UR, England.

1753.253

RN TYPE 992Q RADAR

The Royal Navy Type 992Q radar is a fully stabilised 10 cm (E/F-band), high-power pulse radar equipment, fitted in destroyers and frigates. It is a prime source of surface/air target information for the action information organisation system, and provides space-stabilised output data which can be synchronised in pulse and rotation with that of other radars fitted in the ship. It is of solid-state electronic construction apart from certain circuit elements which retain vacuum tube devices.

STATUS

Fitted in RN 'County' class and Type 82 destroyers.

CONTRACTOR

Marconi Radar Systems Ltd, Writtle Road, Chelmsford, Essex CM1 3BN, England.

'County' class destroyer, HMS Glamorgan, *with Type 992Q scanner on foremast. Below is Type 978 navigation radar, and MRS3 radar director over bridge. Mainmast has Type 965 search radar with AKE (2) antenna outfit. Note ECM arrays*

3260.253

RN TYPE 994 NAVAL RADAR

This new radar incorporates a transmitter/receiver based on the Plessey AWS-4 radar described in **3171.253**. It uses the existing antenna of its predecessor Type 993. The below-decks equipment is the same as for the Plessey AWS-4 radar system, for which Plessey supplies a new high performance lightweight antenna system. The transmitter/receiver is designed to occupy the same space as the original equipment. An interface unit is supplied to simplify the installation work. Widespread use of solid-state technology ensures better reliability and simplifies maintenance procedures.

The radar's performance can be further enhanced by the addition of a digital moving target indicator (DMTI), providing increased protection against clutter caused by sea, land or rain. The DMTI is self-compensating for clutter velocity, hence also providing good discrimination against chaff. Data from ship's log and gyro is not required.

Similar E/F-band tactical radars of various Royal Navy types, both search and heightfinders, are in service with other navies around the world. The introduction of Type 994 means that cost-effective and proven facilities are available for modernising tactical radars on board many different classes of ship, using either the antenna already fitted or the new AWS-4 antenna according to circumstances.

Trials carried out on a typical aircraft target at several different altitudes established clearly that the new equipment not merely matches the improved requirements but exceeds it by a significant margin.

HMS Arethusa *fitted with one of the Type 993/994 conversions showing the antenna at the top of the foremast*

STATUS

In quantity production for fitting aboard Royal Navy frigates and ships of other navies.

CONTRACTOR

Plessey Radar Ltd, Oakcroft Road, Chessington, Surrey KT9 1QZ, England.

4634.253

TYPE 996 NAVAL SURVEILLANCE RADAR

Type 996 is the Royal Navy nomenclature for a new S-band, 3-D surveillance and target indication radar. It is to be fitted to the Type 23 frigates and the 'Fort Victoria' class AORs and retrofitted into the Type 42 destroyers and the 'Invincible' class carriers.

No technical details of the Type 996 are available currently.

STATUS

Plessey Radar received a contract, valued at approximately £70 million, from the UK MoD in September 1983. Additional quantities of the radar have been ordered. Delivery of the first system was made in mid-1986.

CONTRACTOR

Plessey Radar Ltd, Oakcroft Road, Chessington, Surrey KT9 1QZ, England.

3486.253

RN TYPE 1022 NAVAL RADAR

The Type 1022 is a long-range D-band frequency synthesiser-driven travelling wave tube (TWT) radar with pulse-to-pulse coherence. It was developed as a replacement for the Royal Navy's Type 965 surveillance radar. Long-range performance is aided by the use of a high mean-power TWT in the transmitter. A pulse-compression receiver with MTI provides for good performance in rain and in the presence of surface clutter.

The antenna uses a squintless feed to provide tapered illumination of the main reflector. This results in good sidelobe performance across the band. Other technical features include solid-state electronics, frequency agility over a wide band, horizontal and circular polarisation, frequency synthesiser-controlled coherent transmitter, MTI with digital quadrature canceller and digital video processing.

It is understood that the Type 1022 is a joint Anglo/Dutch project in which Signaal is providing the transmitter/receiver and Marconi the antenna. This suggests that the finished equipment will appear on the vessels of both British and Netherlands fleets.

Type 1022 antenna on HMS Nottingham *Type 42 destroyer*

STATUS

In production and fitted to a number of Royal Navy ships.

CONTRACTOR

Marconi Radar Systems Ltd, Writtle Road, Chelmsford, Essex CM1 3BN, England.

1394.253

TYPE 1006 RADAR

The Type 1006 is the standard I-band navigational radar of the Royal Navy.

Antenna Outfit AZJ

The surface role antenna outfit consists of a 2.4 m slotted waveguide linear array rotated at 24 rpm by a turning mechanism. The array has a horizontal beamwidth of 1° and low sidelobe levels to give good bearing discrimination. The vertical beamwidth of 18° gives good performance in conditions of roll and pitch.

Antenna Outfit AZK

This antenna outfit consists of a 3.1 m slotted waveguide linear array rotated by the same type of turning mechanism as outfit AZJ. It is used when the improved bearing discrimination given by the 0.75° beamwidth is required.

Transmitter/Receiver (Surface)

The equipment is non-thermionic with the exception of the magnetron, and operates at a frequency of approximately 9445 MHz.

Transmitter/Receiver (Submarine)

In order to be compatible with the existing submarine antenna system a variant of the transmitter/receiver operating at 9650 MHz is available.

Display Unit JUD

This unit is non-thermionic with the exception of the cathode ray tube. The display unit uses a rotating scanning coil system to provide range scales from ½ to 96 data miles.

STATUS

The Type 1006 radar has been in series production for the Royal Navy and other navies since 1971. It has now been superseded by the Type 1007 (**5847.253**). More detailed information on the Type 1006 is given in previous editions of *Jane's Weapon Systems*.

CONTRACTOR

Kelvin Hughes Ltd, New North Road, Hainault, Ilford, Essex IG6 2UR, England.

5847.253

TYPE 1007 RADAR

The Type 1007 is an I-band navigation radar for naval use, consisting of an antenna unit, transmitter-receiver, and a display.

Three versions of the antenna outfit are available for surface vessels; a 2.4m single array, a 3.1m single array, and a 2.4m dual array for use with helicopter transponders and outfit RRB. Surface ship antenna outfits can operate in winds up to 100 knots and withstand funnel gas temperatures up to 120°C. De-icing systems are available. A fully pressure-tested antenna is also available for fitting to a variety of submarine masts.

The transmitter-receiver is solid-state, with the exception of the magnetron, and operates at a frequency of 9410 MHz with a transmitter power output of 25 kW. A wide dynamic range logarithmic receiver is provided. A built-in monitoring system is included to check that the equipment is operating at peak performance. A low leakage dummy load allows for system testing during periods of radar silence. Centralised emission control circuitry enables command to inhibit transmission immediately. Sector transmission is also incorporated with direct control from the main display. Blanking pulses are incorporated to safeguard sensitive ESM equipment.

The 16-inch main display has nine range scales from 3/8 to 96 nm and displays relative motion, true motion pictures in either head-up, north-up or course-up orientation. Crystal controlled range rings and range marker, fixed or floating electronic bearing line, with digital readout of range, bearing and ship's head form the basic controls and operational features. Additional features include target track history, electronic plotting, target labelling, guard zones, electronic mapping, sector zones, helo track, and sector transmission. Auxiliary video inputs are also included to allow inputs from simulators and helicopter transponders.

Two types of weather-proof auxiliary display are available as options, a polar scan giving a high definition picture with standard operational facilities, and a raster scan high resolution display which provides flicker-free daylight viewing. Both units are portable and are designed for use on an open bridge or submarine fin.

STATUS

Series production started in 1986 for the Royal Navy and other navies.

CONTRACTOR

Kelvin Hughes Ltd, New North Road, Hainault, Ilford, Essex IG6 2UR, England.

1508.253

ST 802 NAVAL TRACKING RADAR

The ST 802 is a lightweight I-band monopulse search and tracking radar for the direction of small and medium calibre guns, and for target tracking in guided missile systems. It is intended for use against air or surface targets, and the weight and dimensions of the system are appropriate to installation on all classes of vessel from patrol boats upwards. The ST 802 is designed with facilities for independent operation and control in weapon systems not equipped with a central computer complex, but can also interface with, and be controlled by, the weapon system's own computer.

The radar director head is a two-axis mount carrying a one metre diameter antenna, and optionally, a TV camera. The mount is controlled by electric servos in azimuth and elevation, with stabilisation of the sight line during tracking by means of rate gyros. The antenna is of the twist cassegrain type. A four-horn monopulse feed and comparator provide three separate RF outputs (sum, elevation difference, and training difference signals) for processing in the IF receivers to produce acquisition and auto-follow data. Accurate angular data result from the combination of a 2.4° pencil beam and monopulse signal processing. With TV camera, the director head weighs 532 kg.

The I-band transmitter incorporates a tunable magnetron, and a short pulse length is used in the interests of range resolution and tracking accuracy. The receiver employs monopulse tracking and embodies features such as the tunable magnetron, monopulse signal processing, MTI, passive tracking of targets using ECM, and CFAR, to reduce the effects of countermeasures. MTI (coherent on receive) provides discrimination against sea clutter when tracking small, low-flying targets.

ST 802 naval tracking radar with IR camera

A frequency agile version of this radar, known as the ST 1802 is also available. It includes a fully coherent transmitter to produce true random frequency agility across a wide bandwidth, offering significant performance advantages, particularly in jamming environments.

OPERATION

Weapons which can be associated with the ST 802 include Seacat, Exocet, and medium and small calibre guns of 20 mm upwards. Acquisition of air or surface targets is automatic, following target designation by search radar or other means. After acquisition, the target is automatically tracked and appropriate data supplied to weapon systems. The antenna head is able to rotate continuously in azimuth at 20 rpm, and this feature in conjunction with MTI is of value in providing additional surveillance cover at low elevation.

If the TV sub-system is fitted, it may fulfil either of two distinct roles, depending upon the weapon system employed. With command line-of-sight missiles, the TV system will permit automatic missile guidance after launch, from a below-decks position. Alternatively it may be used as a back-up to radar tracking in either daylight or low light-level mode. Other electro-optical sensors may also be fitted, such as infra-red or laser rangefinder.

STATUS

The ST 802 is a private development in collaboration with Ericsson Radar Electronics and is in service with a number of navies with various gun and missile fire control systems. In production, it forms part of the Marconi/British Aerospace Sapphire gun control system of the Egyptian '*October*' and '*Ramadan*' fast missile craft.

The S 802 has now been superseded by the S 1802 series (**3687.253**).

CONTRACTOR

Marconi Radar Systems Ltd, Writtle Road, Chelmsford, Essex CM1 3BN, England.

1754.253

S810/820 NAVAL SURVEILLANCE RADARS

Lightweight surveillance radars in the S800 series are comprised of various combinations of two sizes of scanner, and two transmitter/receiver equipments. These are summarised in the table of characteristics.

The Marconi Type S810 is a lightweight stabilised I-band surveillance radar suitable for fitting in small warships down to the size of fast patrol boats where there are limitations of space and weight. I-band gives good performance against surface targets, a major requirement for fast patrol craft, while satisfactory coverage is maintained against air targets by the use of a stabilised antenna in a radome, with vertical fan beam shape. A narrow horizontal beamwidth ensures accurate target indication data for pointing weapons and for putting on a tracking radar.

Detection of surface targets such as fast patrol boats is horizon limited and strike aircraft of 5 m² can be detected at ranges in excess of 25 km with the smaller antenna and up to 40 km with the larger one. A DMTI system is provided to enhance the system performance against small high-speed targets in clutter and the magnetron transmitter is tunable to ECCM or other interference.

S810 series radars use a pulse transmitter of the

CHARACTERISTICS	**S810 (MTI)**	**S820**
Antenna size	1.2 × 0.42 m	2.4 × 0.9 m
Beamwidth		
Vertical	Shaped to 25°	Shaped to 30°
Horizontal	2.2°	3°
Rotation rate	24 rpm	24 rpm
Platform	2 axis stabilised	2 axis stabilised
Radome	1.7 m diameter	3.3 m diameter
Base	1.3 m diameter	3 m diameter
Transmitter/receiver	I-band tunable	E/F-band tunable
Frequency	8.6 – 9.5 GHz	2.7 – 2.9 GHz or 2.9 – 3.1 GHz
PRF	1500/4400 Hz	750/1500 Hz
Pulse length	0.67/0.33μs	1.2/0.6μs
Signal processing	Logarithmic/linear Integrated MTI	Logarithmic/linear and MTI

same design as in the lightweight search/tracker radar Type ST 802 (**1508.253**), resulting in a reduction in the on-board spares if both radars are fitted in a ship. In all versions, the receiver provides either logarithmic/linear or integrated video for the operational displays.

The antenna head, stabilised in roll and pitch, rotates at 20 or 40 rpm and is enclosed in a radome for protection and minimum weight. An elliptical reflector is used, which measures 1.2 by 0.45 m, or 1.8 by 0.6 m. Both types are constructed of glass-reinforced plastic. A single horn feed is located at the focus of the reflector which is contoured to provide a narrow horizontal beam and a fan-shaped vertical beam. The turntable consists of an aluminium alloy centre tube and outer casing, azimuth bearing and azimuth drive and data units. The complete antenna/turntable assembly is supported on a platform giving stabilisation for roll angles up to 25° and pitch angles up to 10°, with acceleration up to 30° per second. The 1.8 or 2.7 m diameter radome which encloses the whole antenna assembly is constructed of a 12 mm thick sandwich of glass-reinforced plastic and foam.

The transmitter employs a tunable magnetron operating within the frequency band 8.6 to 9.5 GHz and delivers a typical peak power of 200 kW. All components are solid-state except the magnetron and the hydrogen thyratron in the pulse modulator.

Two receiver channels, one logarithmic, the other linear, are provided. The required channel is selected by a switch located near the PPI display. MTI processing is available, the DMTI filter being a double canceller employing shift-register storage and feedback. The radar operates with a staggered PRF in the MTI mode in order to overcome the disadvantages of 'blind speeds'. Compensation is provided for ship and wind velocity. When clutter cancellation is not required, the MTI can be switched out and the radar then operates with a jittered PRF of 1.5 kHz allowing a maximum displayed range of 80 km.

For the larger FPB, there is the radar Type S820 operating in the E/F-band. This employs a 2.4 m stabilised antenna enclosed in a radome giving a total masthead weight of 550 kg. The transmitter is tunable within one of two frequency bands. The receiver employs a parametric amplifier and has linear/logarithmic processing and can be supplied with or without MTI.

A recent development is a low cover version of the S810 which has a concentrated low-level beam and a high antenna rotation rate (40 rpm) which is intended for detection of sea-skimming targets.

STATUS

Radars of the S800 series are in production for the UK MoD and other unspecified customers. The S810 is fitted to the *'October'* class and the S820 to the *'Ramadan'* class fast missile craft of the Egyptian Navy.

The S810/820 radars have now been superseded by the S1810 and S1820 systems (**3687.253**).

CONTRACTOR

Marconi Radar Systems Ltd, Writtle Road, Chelmsford, Essex CM1 3BN, England.

4302.253

805 SERIES NAVAL TRACKING RADARS

The 805 series has been developed as lightweight naval tracking radars for the control of missiles and guns. They are intended for use primarily against air targets and are suitable for virtually any type of naval vessel from patrol boats upwards. There are three different types: 805SW, 805SD and 805SS.

The 805SW has been designed for the lightweight Seawolf system and is currently being fitted to ships of the Royal Navy. Details are given under entry **5630.253**, RN Type 911 tracking radar.

The 805SD is designed for use with the lightweight Sea Dart system for smaller warships. The system uses standard modules from the 800 series radars (**1754.253**), and the RN Type 909 (**1559.253**) which is currently in operational use in Royal Navy Type 42 destroyers. The radar director carries a 2.1 metre antenna, which fulfils the functions of target tracking and illumination, plus an optronics package tailored to meet the particular requirement. The system is suitable for ships of 300 tons and above; a containerised fit is available. The 805SD is a private venture equipment intended for the export market.

The 805SS has been designed in co-operation with Raytheon for use with the Sea Sparrow missile. It uses the Raytheon illuminating transmitter, is extremely lightweight and can also be used for gun control.

STATUS

The 805SW has been selected as the tracking radar for the Royal Navy's lightweight Seawolf system and is being fitted aboard Type 22 and 23 frigates. The system is also the subject of considerable interest from a number of foreign countries for smaller types of warship.

The 805SD has been designed for the export market and is currently being studied by a number of defence forces as part of lightweight Sea Dart.

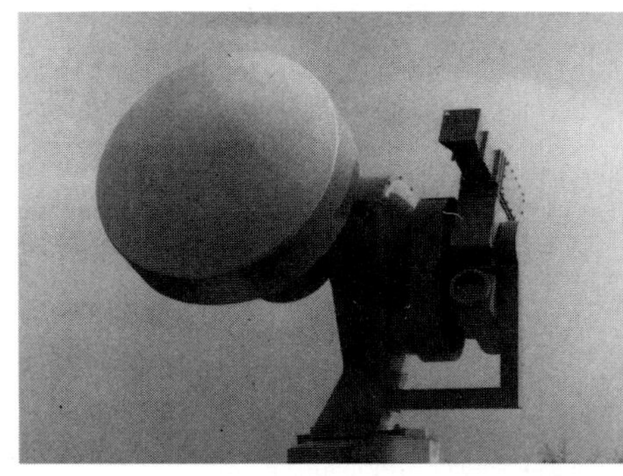

805SD tracking radar for lightweight Sea Dart

CONTRACTOR

Marconi Radar Systems Ltd, Writtle Road, Chelmsford, Essex CM1 3BN, England.

5630.253

RN TYPE 911 TRACKING RADAR

The Marconi 805SW is a lightweight tracking radar for use with the Seawolf short-range point defence system. It is designated Type 911 by the Royal Navy and will be fitted on Type 22 and 23 frigates, Type 42 Batch III destroyers and the 'Invincible' class carriers.

The radar is a dual frequency (I-band and millimetric wave) differential tracking system which includes a command link to control the Seawolf missile in flight. The I-band antenna and transmitter form part of the Marconi ST 800 range and the signal processing uses fast Fourier transform techniques to ensure effective operation in severe clutter environments. The millimetric part of the system is a version of the DN 181 Blindfire radar (**2439.153**) used with the BAe Rapier missile. It provides accurate tracking at low sight angles against targets close to the sea surface, such as sea-skimming missiles and low-flying aircraft. A system of independently illuminating the upper and lower parts of the millimetric antenna ensures that the missile receives good guidance data while flying at low level.

The radar is fully automatic to provide fast reaction time against small targets and is autonomous, requiring only the allocation of the fire control channel to the selected target and information on ship's motion. Excellent clutter rejection is provided on both bands for operation in both open and enclosed waters and ECCM facilities are incorporated to enable use in hostile electronic environments.

STATUS

Over 50 Type 911 trackers, worth more than £200 million, are in production for the Royal Navy. Starting with HMS *Brave* all new build Type 22 frigates will be equipped with a pair of Type 911 trackers. The radar has been designated for all the Type 23 frigates, and the Auxiliary Oil Replenishment (AOR) vessel where it will control vertical launch Seawolf. Other fitments include the three 'Invincible' class carriers and Type 42 Batch III destroyers.

CONTRACTOR

Marconi Radar Systems Ltd, Writtle Road, Chelmsford, Essex CM1 3BN, England.

Marconi lightweight tracking radar 805SW (RN Type 911) for lightweight Seawolf naval SAM system

---

6106.253

ST1802 SERIES FIRE CONTROL RADARS

Two members of the Marconi ST1802 series of fire control radars have now been designated; the 1802SW and the 1802AS. The ST1802 family has been designed to provide lightweight cost-effective fire control for a variety of guns and missiles, in particular those mounted in fast strike craft. A modular format is employed which enables systems to be configured easily to meet the requirements of the particular weapon system.

The ST1802 is a monopulse tracking radar operating in I-band with a switchable moving target indication (MTI) facility. It is autonomous except for the input of designated target range, bearing and ship's reference data such as roll, pitch, heading and speed. The TWT transmitter is fully coherent and produces true frequency agility across a wide bandwidth. The MTI facility is retained in a pulse burst agile mode.

The director is mounted high on the ship's superstructure and carries the radar antenna and an electro-optical package (television or infra-red/laser). It is designed for continuous 360° rotation in azimuth. Elevation coverage for both radar and electro-optics extends from −30 to +85°.

The between decks equipment is contained in two cabinets. A double bay cabinet houses the transmitter, an amplitude comparison monopulse receiver, a fully adaptive digital signal processor and the servo drive equipment. A single bay cabinet houses the autonomous radar control unit which controls the radar during its search and acquisition phase, and provides the space stabilised search patterns.

ST1802 lightweight tracking radar

The radar control console incorporates a colour raster graphics display controlled by soft keys and providing easy operator interface. The radar state, mode and operating conditions are displayed together with the status of the associated weapon system.

The ultra lightweight 1802SW has been designed for the control of the Seawolf missile, in addition to naval guns, on ships from some 450 tons upwards. The 1802SW is based on a number of proven elements, the ST1802 tracking radar, the Type 3800 thermal imager, the TV missile tracking system used with the Javelin air defence missile, and the command link from the 805SW system. In operation, the radar tracks the target and the Javelin TV system tracks Seawolf by means of flares mounted on the missile. The thermal imager is used to support the radar when tracking targets at low angles of sight. The 1802SW is normally used with the Marconi S1820 radar, but any surveillance radar capable of providing track data on small targets is suitable.

The 1802AS system, also known as Albatross, is intended as a radar fire control system for the Selenia Aspide missile. A continuous wave (CW) illumination transmitter is incorporated in the 1802 which then provides both tracking and illumination of the target. The CW signal is fed through the same antenna dish to provide a compact tracker illuminator system.

CHARACTERISTICS (ST1802)

Antenna

Type: Cassegrain
Aperture: 1 metre
Polarisation: Vertical
Beamwidth: 2.4°
Transmitter: Driven TWT, pulse-to-pulse or burst agile
Frequency: I-band
Output power: 50 kW peak
Instrumented range: 40 km (non-MTI); 24 km (MTI)
Azimuth motion: Continuous rotation rate; 20 rpm
Elevation limits: –30 to +85°

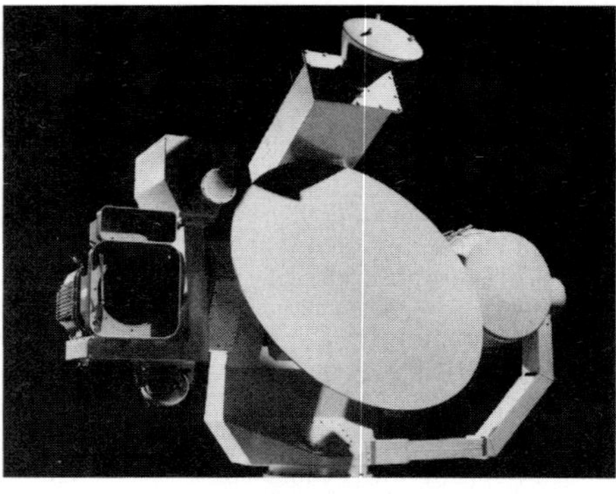

1802SW anti missile system

CONTRACTOR

Marconi Radar Systems Ltd, Writtle Road, Chelmsford, Essex CM1 3BN, England.

3687.253

S1820 NAVAL SURVEILLANCE RADAR

This modular surveillance radar has been developed by Marconi Radar Systems Ltd as a private venture, in association with Canadian Marconi, although some of the technology was developed as part of the RN Type 1030 STIR radar (**3485.253** in 1982-83 and earlier editions).

This model has been described as the minimum configuration version within the S1800 series of equipments, the series itself comprising a range of modular elements, such as antennas, transmitter/receivers, signal processors. The S1820 antenna is a stabilised unit (pitch and roll) of 2.4 m span, very similar to that of the S820 (**1754.253**), housed in a protective radome and associated with a frequency agile transmitter. The latter unit incorporates pulse compression and operates at E/F-band frequencies, and alternative modes of frequency agile working are available. The use of long pulse techniques yields a high mean power which increases range performance while subsequent pulse compression restores range accuracy. Pulse compression on reception also gives additional noiseless gain to genuine echoes.

The MTI processor, particularly when fitted with adaptive velocity compensation, discriminates against land, sea and rain clutter. Combined frequency agility and MTI processing are provided, and CFAR circuits precede the automatic target tracking circuitry. The optional IFF can be integrated with the main antenna.

The S1820 is suitable for fitting aboard ships of 300 tons or above, and provides for air and surface tactical surveillance in addition to target indication data for ship's weapons.

CHARACTERISTICS

Frequency: E/F-band, frequency agile
Transmitter power: 20 kW (peak); 800 W (mean)
Antenna: 2.4 m span stabilised with integral IFF
Antenna weight: 550 kg
Range: Approx 85 km (max)

STATUS

No details of ship fittings are available for publication. Parts of the S1820 system are in production for the Canadian Navy, by Canadian Marconi, under the nomenclature AN/SPS-503 (**5677.253**).

CONTRACTOR

Marconi Radar Systems Ltd, Writtle Road, Chelmsford, Essex, CM1 3BN, England.

6131.293

400 SERIES NAVAL FIRE CONTROL RADARS

The 400 Series is one of the latest in the Marconi family of fire control radars. It is considerably lighter and smaller than many contemporary equipments and, when integrated with a suitable gun armament, provides a cost effective all-weather defence against missiles (sea skimmers in particular), strike aircraft and helicopters. The system features a self-surveillance mode, frequency agility, excellent ECCM characteristics, gyro-stabilised precision tracking and a fast reaction time. It is particularly suited for use in conjunction with rapid-fire and medium-calibre guns.

The fire control system is autonomous against low level targets. The characteristics of this mode of operation are self-surveillance with automatic target acquisition through to track and output gun orders, fast reaction time in bringing guns to bear and automatic sequence. The radar employs clutter suppression and a novel method of reducing angular errors caused by multi-path effects.

CHARACTERISTICS

Antenna

Type: Offset, front fed parabola
Size: 0.6 m diameter
Polarisation: Vertical
Antenna angle control: 60 rpm for surveillance; two axis, gyro stabilised for precision tracking
Transmitter: Frequency agile TWT
Peak power: 2.5 kW
Range: For rapid fire guns 12 km self-surveillance, 10 km for auto-tracking; increased to 24 km for operation with medium calibre guns. Detection range on a strike aircraft – 10 km.
Tracking accuracy: Better than 1 mrad rms in angle, 5 m rms in range
Elevation coverage: –30 to +85°
System weight: 512 kg

CONTRACTOR

Marconi Command and Control Systems Ltd, Weapon Systems Division, New Parks, Leicester LE3 1UF, England.

400 Series fire control radar system

6107.253

S1810 SHIPBORNE AND COASTAL DEFENCE RADAR

The S1810 is a 3 cm (I-band) frequency agile radar which has been optimised for both shipborne and coastal defence applications. The system gives enhanced detection ranges and precise target indication data to weapon systems, and is highly resistant to ECM environments. It has been designed to detect sea skimming missiles well beyond the engagement range.

The S1810 for shipborne applications is a lightweight fully coherent system which provides detection of small aircraft at over 45 km, and surface craft out to the curvature of the earth. The system operates between 8.6 and 9.5 GHz and offers pulse-to-pulse broadband frequency agility. Digital MTI provides improved detection and tracking of low level air targets in clutter and operates with pulse burst frequency agility or fixed frequency.

The S1810 antenna is mounted on a stabilised platform comprising three mutually perpendicular waveguide rotating joints and the azimuth drive. There are three versions:

(a) the standard version with a 2.44 metre reflector and radome, and operating at 24 rpm
(b) lightweight with a 2.44 metre reflector, no radome and operating at 20 or 60 rpm
(c) lightweight compact with a 1.2 metre reflector and radome, operating at 24 rpm.

The S1800CD is the coastal defence variant and incorporates specific features to adapt it for that role. It is designed to detect surface and low level air targets in hostile ECM conditions, and provide accurate target indication data to associated missile and artillery batteries. Digital MTI is available as an option to provide improved detection and tracking of low level air targets in clutter. The antenna rotation rate is 20 or 60 rpm, depending on the role of the radar at that time. The higher rotation rate allows shell splash spotting. No radome is required.

All versions of the S1810 provide excellent low level coverage and angular discrimination. Pulse-to-pulse or short burst frequency agility is achieved using a coherent travelling wave tube transmitter. One of three transmission modes can be selected, including pulse compression to give high effective peak power. A particular transmission mode can be selected within operator defined bearings, with a different mode outside those bearings to achieve, for example, splash spotting within the narrow required sector with full air detection over the rest of the area.

CHARACTERISTICS

Frequency: I-band (3 cm)
Transmission modes: Fixed frequency; pulse burst agility; pulse-to-pulse agility; radar silence; sector transmission
Pulse length: Mode 1 (normal) – 2 μsec compressed to 0.1 μsec (chirped) low PRF; Mode 2 (air) – 1 μsec (high PRF); Mode 3 (surface) 0.2 μsec (high PRF)
Peak power: 50 kW
PRF: 2 kHz (low); 4 kHz (high)

Antenna

Type: Double curvature GRP metallised coating
Beamshape: Cosec² 5 – 30°
Polarisation: Linear horizontal
Beamwidth: 1.1 ± 0.1°
Aperture: 2.44 × 0.84 m or 1.2 × 0.42 m
Rotation rate: 24 or 20/60 (shipborne); 20/60 (coastal)
Weight: 640 or 250 kg (shipborne); 160 kg (coastal)

CONTRACTOR

Marconi Radar Systems Ltd, Writtle Road, Chelmsford, Essex CM1 3BN, England.

6128.253

MULTI-FUNCTION ELECTRONICALLY SCANNED ADAPTIVE RADAR (MESAR)

The Multi-function Electronically Scanned Adaptive Radar (MESAR) is an experimental project being developed jointly by Plessey Radar Limited and the UK Admiralty Research Establishment, Portsdown.

Multifunction Electronically Scanned Adaptive Radar (MESAR)

The project is the result of an investigation into the use of phased array radars for naval use, with an emphasis on a much lower masthead weight than in conventional systems. The phased array radar system, much used in land-based and large ship radars, has considerable advantages in its ability to operate effectively in very high target density and heavy jamming situations. The MESAR system, using gallium-arsenide (GaAs) circuit techniques, is intended to provide all the benefits of a conventional phased array antenna radar, allied with a relatively low masthead weight which would enable such a system to be fitted to smaller ships, as well as providing obvious advantages for larger ships.

The equipment would enable weapon systems to cope with saturation attacks which use missiles of small radar cross-section in a concentrated jamming environment. This would necessitate a large number of fire channels, simultaneous tracking of both hostile and defending missiles, and mid-course guidance commands.

The phased array antenna will be formed of flat sensor panels consisting of arrays of GaAs radiating elements, or modules, sending out variable pulses and building up an in-depth picture of the area in front. Different panels can face different directions, and are linked by computer so that a complete hemispherical radar picture is available at any time. An operational MESAR system would probably consist of about 2000 elements per panel, with four panels mounted at masthead and electronically stabilised. Each array panel can cover 90° in both elevation and azimuth and can be mounted on either fixed or rotating mountings.

Although designed primarily for naval applications, the MESAR system could also be fitted to ground vehicles, or could form part of the next generation of airborne early warning aircraft.

STATUS

Early development. A testbed system is currently undergoing trials at the Plessey Radar facility at Cowes on the Isle of Wight. It is possible that, subject to successful trials, the MESAR system would be bid for a share of the multi-function radar planned for installation on board the future NATO frigates (NFR90).

CONTRACTORS

Admiralty Research Establishment, Portsdown, Hampshire, England.

Plessey Radar Limited, Oakcroft Road, Chessington, Surrey KT9 1QZ, England.

3487.253

TRANSPONDER REPLY RECEIVER RRB

Reply Receiver 'B' (RRB) is a ship and shore fit equipment intended for integration with a wide range of I-band radar systems. Requiring minimal modification to existing radar installations, RRB provides edge of band reception and display of responses from co-operating aircraft fitted with airborne transponder type ARI 5983 or 5984.

The equipment provides aircraft identification and location, typically for naval aircraft, enabling the tactical commander to keep track of and facilitate recovery of his aircraft, particularly in adverse weather conditions, heavy clutter environments and at low altitudes.

The transponder reply is at a separate frequency to the primary radar and this is received by the RRB via a suitable coupler in the radar waveguide. In addition to the increased detection in range provided by the system, the RRB receives individually coded pulse trains from suitably equipped aircraft, allowing simple means of identification and information transfer.

The associated video code suppression unit may be used on any one of the six outputs from the RRB to provide a single video pulse instead of the coded output.

STATUS

In production for the Royal Navy, Algeria, Brazil, India, Greece, Pakistan, Australia, New Zealand and the Middle East.

CONTRACTOR

M/A-Com Ltd, Humphrys Road, Woodside Estate, Dunstable, Bedfordshire LU5 4SX, England.

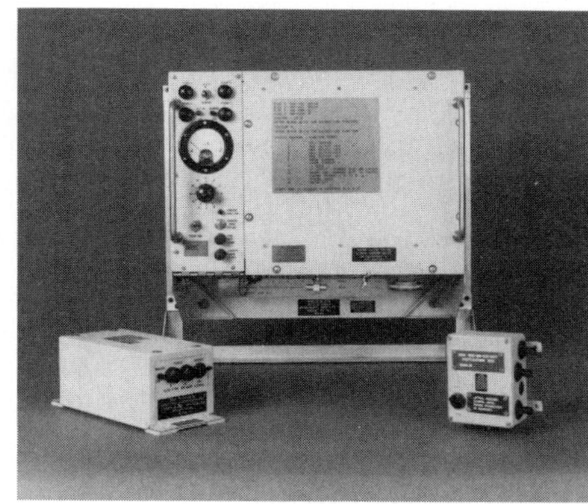

Reply Receiver 'B'

5676.293

RADAR ANTENNA PEDESTALS

Radamec designs and manufactures a wide range of rugged precision director and pedestal platforms to accommodate differing antenna payload configurations, weights and dynamic performance characteristics for radar surveillance, radar tracking, line-of-sight communications, and EW tracker/jammers.

The Radamec 400 and 500 series modular directors and pedestals provide single, dual and three-axis rotation with or without stabilisation. For tracking applications the 400 series directors provide slewing rates up to 360°/s, and reporting accuracies of better than 5 arc seconds are available. For surveillance radar applications, the 500 series pedestals can accommodate payload weights in excess of 600 kg.

The 500 series offers high performance combined with compactness, lightweight and high reliability, with pedestal masthead weights of typically 350 kg making the series suitable for smaller naval vessels. The microprocessor based control electronics system is housed in a single cabinet and incorporates comprehensive BITE and BIST.

All directors meet defence equipment exposed environmental specifications and shock/vibration specifications for naval ships.

STATUS

Over 120 Series 400 systems have been supplied to the UK MoD and other defence agencies. Current contracts include six ship sets of Series 500 radar pedestals.

CONTRACTOR

Radamec Defence Systems Ltd, Bridge Road, Chertsey, Surrey KT16 8LJ, England.

Radamec Type 503 pedestal used with the Ericsson Sea Giraffe

UNITED STATES OF AMERICA

1247.253

AN/SPG-51 RADAR

The AN/SPG-51 pulse doppler radar forms a major element of the missile and gun fire control system Mk 74. It is intended mainly for use with the Tartar (**6006.231**) and Standard (**1122.231**) surface-to-air missile systems, for which it fulfils target tracking and missile guidance functions. Operating frequency is probably in the I-band. A circular parabolic dish reflector with offset horn feed is used, mounted on the Mk 73 gun and missile director (**2632.281**) produced by General Electric.

The radar line-of-sight can be rotated continuously in train, and can be elevated from –30 to +83° relative to the deck plane. Initial pointing information is supplied to the SPG-51 by a search and target designator radar. This information is received via the Raytheon Mk 72 signal data converter and digital fire control computer which generates a search pattern for the SPG-51 until target acquisition is accomplished. Automatic tracking is then initiated and the SPG-51 feeds angle and range information to the computer. The computer generates fire control data for transmission to gun mounts and missile launchers. It also provides missile seeker and angle data.

STATUS

The SPG-51 is widely used on US Navy vessels, and development proceeded through A, B, C and D versions. The last of these began deployment in 1972. Among the other navies using this radar are those of Australia, France, Federal Republic of Germany, Italy, Japan, the Netherlands and Spain.

CONTRACTOR

Raytheon Company, Equipment Divison, Wayland, Massachusetts 01778, USA.

AN/SPG-51C fitted to the Spanish Navy frigate Asturcas (Stefan Terzibaschitsch)

1748.253

AN/SPG-55 RADAR

The AN/SPG-55 radar is the G/H-band guidance radar for the Terrier, Standard SM-1 and Standard SM-2 (ER) surface-to-air missile systems fitted in ships of the US and a number of allied navies. Transmitter power is thought to be of the order of 50 kW and the typical operating range, 50 km. Three versions of this radar have been identified: SPG-55, A and B models. Some references have been found to the use of an associated CW target illuminator system, AN/SPA-44, but no other details have been obtained. Early American Terrier installations employed, instead of the SPG-55, a radar designated AN/SPQ-5, which in appearance is similar to the SPG-49. It is not known how many of these installations remain operational but the SPG-55 remains on three ships of the Italian Navy.

The AN/SPG-55 has been modified several times since its entry into service in 1961. The latest modification is Mod 10 and Unisys Corporation received a $14 400 000 contract in October 1987 for modifications.

STATUS

In operational service.

CONTRACTOR

Unisys Corporation, Gyroscope Division, Great Neck, New York, USA.

SPG-55 Terrier missile guidance radars aboard USS Dahlgren

1932.253

AN/SPG-60 RADAR

The AN/SPG-60 is a monopulse, X-band pulse doppler search and tracking radar forming part of the Mk 86 fire control system (**1241.281**). It is used for acquisition and tracking of air targets to ranges of about 60 nautical miles (110 km), tracking being either automatic or manually aided. The radar's search-to-acquire capability is computer directed, and it accepts two- or three-dimensional target designation co-ordinates in digital or synchro format from either the AN/SPQ-9 radar (**1931.253**), which is also part of the Mk 86 FCS, or other shipboard search radars. The antenna is stabilised.

Acquisition for SPG-60 is automatic, the radar performing its programmed scan search pattern about the designated point under computer control. The computer resolves the problems of blind range and range rates inherent in doppler radars, as well as range ambiguity.

The computer also automatically controls and calibrates receiver gain and monopulse channel balance.

Other features include four-horn monopulse antenna arrangement; continuous azimuth rotation; bore-sighted TV camera; passive angle tracking; adaptive scan patterns for automatic acquisition; adaptive computer control of PRF and pulse-length. The Mk 86 FCS uses the SPG-60 to provide tracking data for the SM-1 and SM-2 SAM systems, and it also provides target illumination by means of CW injection at the SPG-60 antenna.

DEVELOPMENT

Prototypes of the complete Mk 86 FCS were delivered to the US Navy in March 1970, and by December 1986, 77 production units had been delivered for fitting aboard various US Navy ships, or for training and test sites. These include the 'Ticonderoga', 'Virginia' and 'California' class guided missile cruisers, 'Spruance' class destroyers, 'Tarawa' class amphibious assault ships, 'Kidd' class guided missile destroyers and three ships of the 'Charles F. Adams' class guided missile destroyers.

STATUS

The system is now at sea with the US Fleet. Production is continuing.

SPG-60 (STIR) radar aboard USS South Carolina (Stefan Terzibaschitsch)

CONTRACTOR

Lockheed Electronics Company Inc, US Highway 22, Plainfield, New Jersey 07061, USA.

1931.253

AN/SPQ-9 RADAR

The AN/SPQ-9 is a high-resolution, track-while-scan (TWS), pulse compression, X-band naval radar. It forms part of the Lockheed Mk 86 fire control system (**1241.281**) and provides detection and tracking facilities within a range of 20 nautical miles (37 km). The minimum range is 150 yards (137 m). The radar antenna provides surface search and limited air cover (up to 2000 ft, or 610 m). A separate transmitter/receiver interrogates, receives, and decodes replies from transponders such as the AN/TPN-7, AN/TPN-16, or AN/UPN-32. The surface search antenna is mounted on a stabilised platform and protected by a radome.

A high scanning rotation rate is a feature of the SPQ-9 giving a one-second data rate. Real-time signal and data processing permit detection, acquisition, and simultaneous tracking of multiple targets. The basic capacity of the complete Mk 86 SPQ-9 system is four targets, and TWS is expandible to more than 100 targets. The radar operates in any one of five selectable frequency ranges, with pulse-to-pulse frequency agility as an additional countermeasure to jamming and natural clutter. This feature also permits a number of radars to operate without mutual interference. An MTI facility is being developed to further enhance sub-clutter visibility.

Other elements of the Mk 86 system are the AN/SPG-60 radar (**1932.253**) and an electro-optical sensor system, connected to a computer complex and consoles for command and weapon control. Both guns and missiles can be controlled by the system.

DEVELOPMENT

Prototypes of the complete Mk 86 FCS were delivered to the US Navy in March 1970. By December 1986, 77 sets had been delivered for installation aboard various US Navy ships. These include the 'Ticonderoga', 'Virginia' and 'California' class guided missile cruisers, 'Spruance' class destroyers, 'Tarawa' class amphibious assault ships, 'Kidd' class guided missile destroyers and three ships of the 'Charles F. Adams' class guided missile destroyers. The Mk 86 is also fitted in the West German DDG modernisations.

STATUS

The system is now at sea with the US Fleet. Production is continuing.

CONTRACTOR

Lockheed Electronics Company Inc, US Highway 22, Plainfield, New Jersey 07061, USA.

Radars aboard the USS South Carolina *include (from top) Tacan beacon, SPS-40 search radar, and SPQ-9 in radome* (Stefan Terzibaschitsch)

4301.253

AN/SPQ-11 COBRA JUDY RADAR

The Cobra Judy shipborne phased array radar is a detection and tracking system enabling data to be collected on foreign strategic ballistic missiles tests. The system is similar to the land-based Cobra Dane radar (**1949.153**) although considerably smaller.

The large octagonal array of this radar measures 22.5 feet (approximately seven metres) in diameter and is composed of 12 288 antenna elements. The elements are co-ordinated to form transmission patterns by a large scale Control Data CYBER 175-112 computer which allows the radar to detect and track objects at very high rates. After detection and tracking the information is fed to a computer for storage and display as well as post-mission reduction and analysis.

Cobra Judy is installed on the stern of a former merchant vessel, the USNS *Observation Island*, which has been extensively modified. The system weighs 250 tons (approximately 250 000 kg), stands four stories high and is integrated into a steel turret which can be rotated mechanically. The octagonal array makes up one wall of the pyramid-like structure. The complete shipborne sensor is operating in the Pacific Ocean, based at Pearl Harbor, Hawaii.

The Cobra Judy system underwent a major upgrade, managed by the US Air Force Electronic Systems Division. This added an X-band radar, using a parabolic dish antenna, to complement the existing S-band phased-array system. The addition of this new equipment was designed to improve the ability to observe and collect intelligence data on the terminal phase of ballistic missile tests, from the time the re-entry vehicles enter the atmosphere through their descent to the Soviet test range on the Kamchatka Peninsula. Since operation in X-band offers a better degree of resolution and target separation, the additional system may well improve intelligence considerably by the ability to differentiate between multiple warheads and ECM chaff and decoys.

Cobra Judy radar installed on the USNS Observation Island

The X-band dish antenna is installed forward of the phased array, aft of the ship's funnel, and stands five stories high. The dish and pedestal have been supplied by the USAF, and were previously used as a shipborne telemetry radar.

STATUS

The radar has been acquired and operated by the US Air Force Systems Command, Electronic Systems Division. Test and evaluation trials took place during 1981 and the system is fully operational.

CONTRACTOR

Raytheon Company, Equipment Division, Wayland, Massachusetts 01778, USA.

1744.253

AN/SPS-6 RADAR

The AN/SPS-6 is a naval air and surface search radar, operating in the D-band of the radar spectrum and having an output of the order of 500 kW. Approximate range of the equipment is 100 to 200 km. The SPS-6 is widely used on ships of the US Navy and numerous other nations and has been in service for many years. During this period various versions, denoted by suffix letters to the designation, have been evolved. These differ in transmitter/receiver and antenna details.

The SPS-6, SPS-6A, SPS-6B and SPS-6C all employ the same below-decks equipment but have different antennas. The SPS-6C has the same antenna as the SPS-6B (on a different pedestal) and with certain differences in the below-decks equipment. The principal effect of the different antennas used is in the vertical beamwidth for each: SPS-6, 10°; SPS-6A, 20°; SPS-6B/C, 30°. The SPS-6C is believed to incorporate more comprehensive receiver video processing.

STATUS

In addition to the various classes of US Navy vessels fitted with the SPS-6, it is used by the navies of Italy, Japan, Spain, Taiwan, Turkey, Venezuela, and West Germany, and all those navies supplied with 'Fletcher' class destroyers are users.

CONTRACTOR

Westinghouse Electric Corporation, Surveillance Radar Division, PO Box 1897, Baltimore, Maryland 21203, USA.

The larger of the two search antennas is the SPS-6, the higher antenna being the SPS-10. Note also the extensive communications arrays aboard the communication relay ship USS Arlington (Stefan Terzibaschitsch)

1564.253

AN/SPS-10 RADAR

The AN/SPS-10 is a surface search radar operating at G-band frequencies. Limited air search and tracking capabilities are available also. The scanner is of open lattice construction and is illuminated by a horn feed supported by a boom projecting from beneath the lower edge of the scanner. The transmitter/receiver is mounted below decks, so that radar head weight is minimised, thus permitting mounting as high as possible on the ship's upper-works. The span is estimated to be about 2.4 m.

The SPS-10 is widely used and now exists in a variety of forms, denoted by suffix letters to the main designation, and five or six manufacturers have produced one or more models.

STATUS

The SPS-10 has been in service since the early 1960s and is extensively fitted on US Navy ships, and vessels which have been passed on to friendly navies by the US Navy.

The SPS-10 antenna has been the subject of a modification in which a new dual feed is substituted for the original unit to permit integrated use of the SPS-10 scanner for both that radar and the SPS-58 (**1359.253**).

Although still in operational use in a number of older vessels, the AN/SPS-10 can now be considered as obsolescent. It is being replaced in most ships by the AN/SPS-67(V) (**3722.253**).

SPS-10 surface search radar antenna

1565.253

AN/SPS-37 RADAR

The AN/SPS-37 is a metric wavelength, long-range air surveillance ship's radar of conventional design. The antenna consists of lattice structure with an open mesh reflector, and supporting an array of 28 dipoles arranged in four horizontal rows of seven. The upper edge of the scanner usually carries an associated IFF antenna. Approximate dimensions of the SPS-37 array are: span 6 m, and height 3.3 m. This is the same item as employed by the earlier SPS-29, thus complicating the identification problems. No performance details have been disclosed.

STATUS

The SPS-37 entered service in 1958 and production ceased in 1962. It is widely fitted on ships of the US Navy and of some friendly navies. In recent years the driver has been replaced by an all-solid-state version.

CONTRACTOR

Westinghouse Electric Corporation, Surveillance Radar Division, PO Box 1897, Baltimore, Maryland 21203, USA.

1745.253

AN/SPS-30 RADAR

The AN/SPS-30 is a long-range three-dimensional radar fitted in certain aircraft carriers and some guided missile ships of the US Navy. Among its features are a high-gain antenna of characteristic configuration (see illustration), broadband frequency operation, and high power output. The antenna is mechanically stabilised against ship's motion, with compensation for roll and pitch. It is stated to provide range, bearing, and height information on multiple airborne targets with 'extreme' accuracy. Although primarily a high-grade heightfinding radar, the SPS-30 is capable of being used for additional tasks such as long-range air search and target designation.

CHARACTERISTICS

Peak power: Classified
Average power: 9 kW
Frequency: E/F-band
PRF: Classified
Pulse width: Classified
Receiver sensitivity: –117 dBm
Beam characteristics: Pencil beam
Beamwidth: 1.5° azimuth; 1.2° elevation
Antenna azimuth rate: 1, 2, 3, 5, 10 rpm plus sector scan
Elevation rate: 240, 480, 720, 1200, 2400 scans/s
Elevation scan sector: Any 12° sector from 0 – 36°
Antenna gain: 41 dB

STATUS

The SPS-30 was introduced into US Navy service in the early 1960s and is carried on many aircraft carriers and some guided missile ships. No longer in production.

CONTRACTOR

General Electric Company, Electronic Systems Division, Syracuse, New York 13201, USA.

SPS-30 3-D radar head

1249.253

AN/SPS-39A (FRESCAN) RADAR

The AN/SPS-39A is a shipborne 3-D radar for air surveillance. It uses mechanical rotation for azimuth coverage and electronic scanning for beam steering in the vertical plane. Vertical scanning is achieved by selection of the transmitted frequency input to the sinuous feed on the side of the AN/SPA-72B antenna. Although still fitted in US Navy guided missile ships and also to ships of France, Italy and Japan, this radar is now out of production. More information is given in *Jane's Weapon Systems 1982-83* and earlier editions.

CONTRACTOR

Hughes Aircraft Company, Fullerton, California, 92634, USA.

1746.253

AN/SPS-40 RADAR

The AN/SPS-40 is a two-dimensional naval air search and surveillance radar for detection of targets at long and medium ranges. A nominal range of 320 km has been quoted. The AN/SPS-40B baseline (which includes the B, C and D radars) is designed to provide optimum performance capabilities with minimum operator interface. Special features of the AN/SPS-40B include excellent long range resolution and accuracy, lightweight and flexible packaging for easy shipboard installation, field proven high reliability, maintainability and availability. The UHF(B) band operating frequency provides freedom from weather clutter and low vulnerability to anti-radiation missiles. The system's DMTI provides excellent sub-clutter visibility and has solid-state receiver, power supplies and controls. The antenna reflector is of open lattice work construction with an integral IFF antenna.

STATUS

The AN/SPS-40 was first introduced into US Navy service in the early 1960s and is now on numerous classes of ships. Additionally it has been fitted in ships of other navies such as Argentina, Australia, Brazil, Chile, Federal Republic of Germany, Greece and Turkey. The system has been approved for use with the AN/SYS-1 Integrated Automatic Detection and Tracking System. This system correlates inputs from multiple radar sensors (both 2 and 3-D) and combines them into a single non-duplicated track.

Since its introduction into service the AN/SPS-40 has been upgraded on a number of occasions. The latest version, the AN/SPS-40E, will include a solid-state transmitter which is in development by Westinghouse.

CONTRACTOR

Norden Systems, United Technologies Marine/ Ground Systems, 75 Maxess Road, Melville, New York 11747, USA.

SPS-40 search antenna atop the mainmast. Note also Tacan and other antennas (above), intercept radomes, and Mk 68 gunfire control system (below)

1747.253

AN/SPS-43 RADAR

The AN/SPS-43 is a high-power, very long-range search radar, probably operating at metric wavelength frequencies and using a transmitter output of 1 to 2 MW. The large (about 13 metres span) antenna assembly is of open lattice construction and rectangular aspect, but with a characteristic 'sideways-W' profile in side elevation. The upper edge of the array generally carries an on-mounted IFF antenna. Its main function is to provide landing guidance to aircraft on carriers and amphibious assault ships.

STATUS

The SPS-43 has been in operational service with the US Navy since the early 1960s and is fitted to many of the larger ships of the US Navy. In recent years the driver has been replaced by an all-solid-state driver.

CONTRACTOR

Westinghouse Electric Corporation, Surveillance Radar Division, PO Box 1897, Baltimore, Maryland 21203, USA.

SPS-43 antenna aboard USS Springfield (Stefan Terzibaschitsch)

1252.253

AN/SPS-48 AIR SURVEILLANCE RADAR

The AN/SPS-48 is a three-dimensional, long-range air surveillance radar for ship-board applications. It uses a combination of mechanical scanning in azimuth and electronic beam steering in elevation to provide plan position and height information on aircraft targets. Multiple beams can be formed, and elevation scanning is accomplished by changing the frequency under computer control of the E/F-band (10 cm) RF energy to the planar array, frequency-sensitivity antenna, thereby radiating a series of pencil beams. Stabilisation of radiation against ship's motion is achieved electronically. Eight operating modes are available under computer program control, and a range of over 400 km at 30 500 m for target designation and aircraft direction is reported.

In general appearance and in operating principles the SPS-48 is similar to the SPS-52 (which see). Several features enable the two types to be distinguished, however. The SPS-48 is somewhat larger, and there is only one 'end-plate' to the flat antenna array, on the left side when facing the antenna. The scanner is of square proportion, compared with the oblong configuration of the SPS-52. In most installations, the SPS-48 has an on-mounted IFF antenna at its upper edge, whereas with the SPS-52 the IFF is normally mounted at the lower edge.

STATUS

A later development in the SPS-48 radar's evolution is the introduction of the SPS-48(C) model which includes MTI and automatic detection and tracking.

The latter significantly increases the overall effectiveness of the ships' combat systems by providing the capability to detect and track multiple air targets automatically.

The contractor has completed the development of modifications to the radar for the New Threat Upgrade (NTU) programme. Designated the AN/SPS-48E, the radar provides early detection of anti-ship missiles launched from a bomber, such as the TU-26 (Backfire), in heavy ECM conditions; it also provides positional data for mid-course guidance to the extended range SM-2 weapons system. The modifications include a lower sidelobe antenna, increased available energy from the transmitter, adaptive energy management, and computer software interface with the ship's automatic detection and tracking modifications. The US Navy plans to provide NTU modifications to 30 SPS-48 systems serving on Terrier and Tartar missile equipped combatants. Delivery of AN/SPS-48E systems commenced in mid-1986, and the first system was installed on the *USS Biddle* guided missile cruiser in early 1987.

The US Naval Sea Systems Command is projecting a requirement of a further 15 to 20 systems over the next five years, in addition to the 17 AN/SPS-48E systems already under contract to ITT Gilfillan.

CONTRACTOR

ITT Gilfillan, 7821 Orion Avenue, Van Nuys, California 91409, USA.

ITT Gilfillan AN/SPS-48 antenna array

1939.253

AN/SPS-49 RADAR

AN/SPS-49 is a very long range two-dimensional air search radar designed for use as the primary detection radar aboard various combatant ships of several countries. The radar antenna is line-of-sight horizon-stabilised to provide acquisition of low altitude targets in all sea states. The radar is all solid-state with a klystron final amplifier. A new variant (V)5 includes automatic target detection with pulse doppler processing and clutter maps assuring reliable detection in all types of clutter. It has state-of-the-art ECCM capability and high reliability. Each production system is subjected to a minimum of 150 hours of continuous test in a shipboard simulated environmental chamber.

AN/SPS-49 air search radar

STATUS

The AN/SPS-49 is fitted, or due to be fitted to the US, Australian and Spanish FFG 7 class, the new Canadian patrol frigate, Terrier missile firing cruisers and destroyers, Aegis CG 47 class cruisers, battleships, attack aircraft carriers, and LSD and LHD class amphibious ships. By mid-1987, 168 systems had been procured with 128 delivered and 97 operational.

CONTRACTOR

Raytheon Company, Equipment Division, Wayland, Massachusetts, 01778, USA.

5606.253

AN/SPS-52C RADAR

The AN/SPS-52C is a shipborne air surveillance, three-dimensional radar system providing target position in range, bearing and elevation. Three-dimensional coverage from a single antenna is achieved by electronic scanning in elevation and mechanical rotation of the antenna in azimuth. The AN/SPS-52C uses the same AN/SPA-72B antenna as the earlier AN/SPS-52 systems but employs completely different below-decks electronics and incorporates the latest in advanced design and signal processing concepts. As a result, significant improvements have been achieved in the detection performance and in reliability, maintainability and availability.

The antenna assembly is a planar array, tilted back at an angle of 25° from the vertical. This tiltback makes possible coverage to high elevation angles. The array consists of rows of slotted waveguide radiators, fed from a sinuous feed system running the length of one side of the array. Scanning in the vertical plane is achieved by selection of the transmitted frequency. Frequency selection is controlled by a digital computer.

AN/SPA-72 antenna for the SPS-52 3-D radar. This antenna replaced the original units fitted in SPS-39 radars aboard US and other ships

DEVELOPMENT

Development started in 1973 with a study of the system requirements. One of the key requirements resulting from this study was the use of automatic target detection and in 1974 the design of this was commenced. The resulting design was implemented into the pre-production model which underwent extensive testing at a land-based test site and at sea aboard the USS *Towers*. Approval for service use was granted in 1978 and production started that year.

STATUS

The AN/SPS-52C is operational in the US Navy on LHA-1 ships and a number of guided-missile destroyers. It is scheduled for deployment on the first of the LHD-1 class ships. The system is also operational in the navies of West Germany, Japan and Italy, including the Italian VTOL carrier/cruiser *Garibaldi*. The AN/SPS-52C has been fitted to ships of the Australian and Spanish navies.

CONTRACTOR

Hughes Aircraft Company, Fullerton, California 92634, USA.

1248.253

AN/SPS-52/52A/52B RADAR

This series of shipborne radars evolved from the AN/SPS-39A radar. With the addition of a digital stabilisation computer, and transmitter/receiver modifications, the AN/SPS-39A was upgraded to the AN/SPS-52. Minor modifications subsequently upgraded the latter to the AN/SPS-52A. Further modifications produced theAN/SPS-52B which has a clutter rejection capability to enable moving targets to be detected in the presence of natural clutter. More detailed information of the AN/SPS-52/52A/52B is given in previous issues of *Jane's Weapon Systems*. The latest version, AN/SPS-52C, is covered under entry **5606.253**.

STATUS

The AN/SPS-52 and 52A are now out of service. The AN/SPS-52B is scheduled to be used aboard the US FFG-1 class ships. Several 52B systems will also remain in use in ships of the Spanish and Japanese navies. The 52, 52A and 52B are now out of production.

CONTRACTOR

Hughes Aircraft Company, Fullerton, California 92634, USA.

1697.253

AN/SPS-55 NAVAL RADAR

The SPS-55 is a modern solid-state X-band surface search and navigation radar developed as a replacement for the SPS-10. It is designed for service on ships of destroyer size or above. Operational uses are: the detection of small surface targets from ranges of less than 50 m to the radar horizon; navigation and pilotage; tracking of low-flying aircraft and helicopters, detection of submarines at schnorkel and periscope depth.

A lightweight antenna (less than 90 kg) has a low profile configuration to minimise installation space requirements and consists of two (selectable) back-to-back, end fed, slotted arrays, one with circular polarisation and the other linear-horizontal polarised. The horizontal beamwidth is 1.5°, and beam squint compensation is used to optimise bearing accuracy over the operating frequency range. Vertical beamwidth is 20°.

The transmitter/receiver sub-system is housed below decks in a single cabinet, and is capable of operating at any selected frequency in the band from 9.05 to 10 GHz. Two pulse widths (1 and 0.12μs) are provided. The minimum peak transmitter output is 130 kW. Variable sector radiation is also provided.

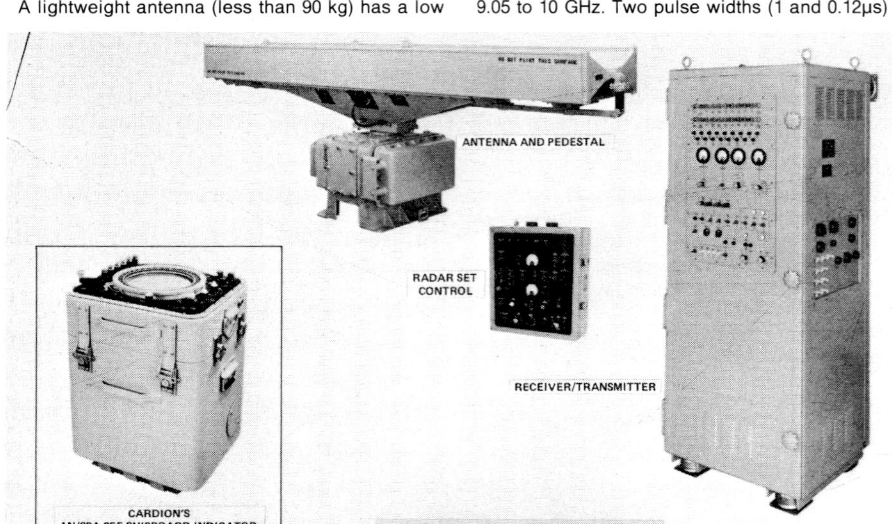

AN/SPS-55 RADAR SET

The SPS-55 set does not normally include its own display, and a separate control unit is provided to permit remote operation of the transmitter/receiver and scanner sub-systems.

The same basic modules can be used with a G-band antenna to form the AN/SPS-502 radar, which has more than 95% parts commonality with the AN/SPS-55.

CHARACTERISTICS

Antenna

Rotation rate: 16 rpm

Polarisation: Circular or linear

Horizontal beamwidth (3 dB): 1.5°

Gain: 31 dB

Transmitter

Frequency band: 9.05 – 10 GHz (a version operating in the band 5.45 – 5.825 GHz is also available)

Peak power: 130 kW

PRF/pulsewidth: 750 pps/1μs; 2250 pps/0.12μs

Receiver

Type: Low-noise, image-suppression mixer

IF: 60 MHz

Bandwidth: 1.2 MHz (long pulse); 10 MHz (short pulse)

Receiver processors: Linear logarithmic, FTC, variable sensitivity time control

STATUS

In production and installed in US Navy ships, including the DD-963, FFG-7, CGN-41, and CGN-47 classes, as well as ships in other navies.

CONTRACTOR

ISC Cardion Electronics, Woodbury, New York 11797, USA.

1359.253

AN/SPS-58/SPS-65 NAVAL RADARS

The basic AN/SPS-58 (latest US Navy version, AN/SPS-65(V)) is a D-band, pulse doppler air search and target acquisition radar, designed to operate with the US Navy point defence surface missile system, but capable of use with a variety of other weapons and produced in several versions. A major design consideration has been the achievement of good low elevation angle performance, high data rate, and clutter rejection characteristics to meet the threat of low-flying aircraft or missile attacks.

This is achieved by means of a modular design, AN/SPS-65 (V1) (V2) models being assembled from various combinations of standard units. The transmitter (common to all four) is a self-contained, air cooled, coherent amplifier unit of solid-state design with klystron power amplifier stages. The receiver/processor also is solid-state and includes a low noise superheterodyne RF receiver. A DMTI filter provides a good anti-clutter performance, and built-in test facilities are incorporated. This unit is common to the 58 and 58A while the 65 V1/V2 configurations utilise a high performance variant of the basic receiver/processor. The 58 and 58A also include a console group for display and weapons interface. The display console provides multiple video display and control functions for manual threat assessment and target designation to weapon systems.

SPS-58 scanner on third-level platform, beneath SPS-10 scanner and above and to left of SPS-30 dish. Large antenna on left is for SPS-43

The AN/SPS-65 (V1) configuration can be installed on ships with existing tactical display systems. Target data is provided on every detected target (ATD) in digital format by a MIL-STD-1397 ANEWS interface. Automatic detection and tracking (ADT) can be performed using this output and an external computer or the V2 hardware which includes its own computer.

The AN/SPS-65 (V2) includes an advanced radar/weapons interface group and is intended for use on non-NTDS ships. Both the V1 and V2 include a radar set control for remote operation of the radar.

A variety of antenna arrangements are available. One is a medium size double-curvature open lattice array of approximately elliptical outline, illuminated by a stacked feed arrangement on a stabilised pedestal. This antenna provides extended vertical coverage, and allows dual operation of radar and IFF system. For smaller ships, a dual-feed unit enables the SPS-58 radar to share an existing SPS-10 lightweight antenna, and permits simultaneous SPS-58, SPS-10, and IFF radar operation.

For installations where there is no SPS-10 antenna, and it is not possible or desirable to mount the larger stabilised antenna unit, a passive planar array is available, this being suitable for back-to-back mounting with almost any existing antenna, or on its own turning gear.

The various configurations of these modules are as follows:

AN/SPS-58: This comprises a stabilised antenna, transmitter, receiver/processor, antenna control unit, signal data converter, and display console.

AN/SPS-58A: This version has the lightweight antenna (SPS-10), transmitter, receiver/processor, signal data converter, and display console.

AN/SPS-65 (V1): This consists of the lightweight, integral antenna, transmitter, high performance receiver/processor, and radar set control unit. Radar data are fed directly to ship's display/action data system.

AN/SPS-65 (V2): This is the same as the V1 version but includes an advanced, radar/weapons interface unit.

The SPS-58/SPS-65 family is fully compatible with Mk 10 and Mk 12 IFF systems, and it provides correlation of IFF and target returns for improved threat reaction time. This latter characteristic is further aided by a high data rate and a low false alarm rate for rapid reaction/threat evaluation and weapon assignment.

Provision has been made in the design for the addition of further facilities such as automatic target designation to weapon systems, or the application of adaptive radar techniques. Modification kits to convert and upgrade AN/SPS-58 units to AN/SPS-65 units are available.

STATUS

The SPS-58 and SPS-65 radars are in service with the US Navy and allied navies. A total of 39 SPS-58 and 15 SPS-65 systems had been produced up to the end of 1986. Systems with which various versions of SPS-58 have been integrated successfully include:

- Mk 37 gun fire control system 5 in/38
- Mk 68 gun fire control system 5 in/54
- SPA-50
- Mk 5 WDS (Tartar, Standard)
- Basic point defence surface missile system
- NATO Sea Sparrow surface missile system
- Naval tactical data system

CONTRACTOR

Westinghouse Electric Corporation, Surveillance Radar Division, PO Box 1897, Baltimore, Maryland 21203, USA.

1940.253

AN/SPS-63 RADAR

AN/SPS-63 is the American designation given to the 3TM20-H version of the 3RM series of radars produced by the Italian company Segnalamento Marittimo ed Aereo (SMA) and described in **1702.253**.

The 3TM20-H model is the selected standard navigation radar for the PHM NATO hydrofoil.

The 3RM series of naval radars are I-band equipments, for which different sizes of scanner and high and lower power transmitters are available.

4300.253

AN/SPS-65(V)ER NAVAL SURVEILLANCE RADAR

The AN/SPS-65(V)ER is a long-range version of the AN/SPS-58/65 family of radars (**1359.253**) currently in the US Navy inventory. This radar uses a 25 kW advanced design solid-state power amplifier, with a duty cycle of 0.005, to achieve very high reliability, maintainability and performance. The radar retains features of the SPS-58/65 family such as the detection of low flying missiles and high performance DMTI, while at the same time providing long-range surveillance capability.

The SPS-58LR/65ER has an instrumented range of 100 nautical miles. It uses a phase-coded, compressed pulse to increase the transmitter average power and to provide good target resolution. Frequency agility on a burst-to-burst basis is provided as an ECCM feature. Advanced DMTI permits operation in severe clutter environments. The radar outputs digital target reports suitable for direct entry into a wide variety of fire control systems.

The enhanced AN/SPS-65(V)ER receiver/processor adds pulse compression techniques to allow maximum range performance with improved resolution, and provides protection from pulse-type interference and jamming. With the addition of a new three-burst transmission programme, the AN/SP-65(V)ER maximises MTI performance and range detection, and provides full doppler coverage.

Westinghouse provides the system with a dedicated antenna assembly consisting of a planar doppler array, pedestal, rotary joint and the necessary filters to accommodate primary and IFF energies. In addition, the system can operate with any antenna which has matching frequency and power handling characteristics. It can use the AN/SPS-6, AN/SPS-12 and AN/SPS-58 antennas which are available on many ships throughout the world. The AN/SPS-65(V)ER can also use the AN/SPS-10 antenna modified to incorporate the Westinghouse designed 1215-1400/5255-5925 MHz band feed horn so that both radars can use the same antenna simultaneously.

AN/SPS-65(V)ER dipole array antenna

CONTRACTOR

Westinghouse Electric Corporation, Surveillance Radar Division, PO Box 1897, Baltimore, Maryland 21203, USA.

3722.253

AN/SPS-67 RADAR

The AN/SPS-67(V) is a solid-state C-band equipment designed to replace the 30-year old SPS-10 (**1564.253**) surface search radar. The antenna of the older radar is retained as an interim measure but will be replaced eventually. The AN/SPS-67 consists of five below-deck cabinets designed for easy installation; transmitter/receiver, video processor, radar set control, antenna controller and antenna safety switch.

Performance of the SPS-67 is improved through the addition of a very narrow pulse mode (0.1 μs) for better navigation and improved resolution of small targets at short ranges. Long and medium pulse (1 and 0.25 μs) modes are used in open sea for detection of long and medium range targets. Performance is further improved by a digital video clutter suppressor and an interference suppressor. Reliability, maintainability and availablity are enhanced by the use of SEM technology and an extensive BIT system.

The basic variant described above is the AN/SPS-67(V)1 radar. A survivable antenna system is available which replaces the AN/SPS-10 unit used with the AN/SPS-67(V)1 in selected ship classes. With this change the radar is designated AN/SPS-67(V)2. An add-on capability is under development to augment the naval gunfire support mission on the US Navy DDG-51. With this capability added the radar is designated as the AN/SPS-67(V)3 and provides DMTI, automatic target detection and track-while-scan for surface targets. Data from the (V)3 configuration can be integrated into the AN/SYS-1 Integrated Automatic Detection and Tracking System for automatic correlation with information from several ships' radars to present a combined single tactical track.

STATUS

AN/SPS-67(V)1 radars ordered from Norden Systems to the end of 1986 total 105. As of September 1987 over 70 systems had been delivered. Of this number four have been supplied to the Federal Republic of Germany and three to the Royal Australian Navy. The US Navy plans to procure a total of about 220 systems.

CONTRACTOR

Norden Systems, United Technologies Marine/Ground Systems, 75 Maxess Road, Melville, New York 11747, USA.

1570.253

AN/SPY-1 MULTI-FUNCTION ARRAY RADAR

The AN/SPY-1 is the electronically scanned fixed array radar for the US Navy Aegis fleet air defence missile system (**2507.231**). It operates in the E/F-band, and the output is several megawatts. The Raytheon transmitter serves several parallel channels simultaneously. The phase scanned arrays are mounted in pairs on each ship, two on the forward deck-house and two on the after deck-house, to provide all-round radar cover. Each array has 4100 discrete elements, and measures 3.65 × 3.65 m. These elements are controlled by AN/UYK-7 digital computers to produce and steer multiple radar beams for target search, detection, and tracking. The SPY-1 also tracks own ship's missiles fired against hostile targets. It also has the function of providing target designation data for the Raytheon target illuminating radars (up to four on a ship) which direct the semi-active homing missiles employed in the Aegis systems.

DEVELOPMENT

The SPY-1 radar was developed by RCA, the prime contractor for the overall Aegis system, under a contract awarded in late 1969. The antenna array began tests in 1972, and by 1973 the transmitter and array were integrated for further tests before embarking on full-scale testing with other elements of the Aegis system.

By early 1974 land-based tests of the radar had been completed and the radar moved from RCA's Moorestown, New Jersey, test site to Long Beach Naval Shipyard, California, for installation in the USS *Norton Sound* in preparation for sea trials. In May 1974 it was announced that while operating in the USS *Norton Sound*, the SPY-1 had detected and automatically gone into the tracking mode to track multiple aircraft flying over the Pacific.

STATUS

The Aegis system (**1834.281**) is now at sea and RCA have announced advanced versions of the SPY-1A radar system. SPY-1B is to be fitted to Aegis cruisers starting with the Princeton (CG-59). It uses a new antenna design to give much lower sidelobes, plus an improved signal processor and a new transmitter tube having double the duty cycle with the same peak power. SPY-1C is a private RCA proposal for use on board aircraft carriers and as yet has no official backing or funding. A fourth version, SPY-1D, is being fitted in the Arleigh Burke (DDG-51) class destroyers and two new Japanese destroyers. It uses only one transmitter with the antennas mounted on a single deckhouse, as opposed to two for the SPY-1A. Considerable space savings have been made in the SPY-1D by using VLSI technology. ECCM capabilities have also been improved.

It is likely that the AN/SPY-1D will be fitted to a number of 6000 ton class destroyers being built by Japan.

CONTRACTORS

RCA Electronic Systems Department, Moorestown, New Jersey 08057, USA – SPY-1 antenna array.

The Raytheon Company, Wayland, Massachusetts, USA – SPY-1 high power transmitter.

Computer Sciences Corporation, Moorestown, New Jersey, USA – computer programs.

AN/SPY-1A phase-scanned array on the US Navy CG-49 (W Donko)

3256.253

W-120 SHIPBOARD FIRE CONTROL RADAR

The W-120 is a lightweight I/J-band acquisition and tracking radar especially adapted from the AN/APQ-120 airborne radar design for small ship gun fire control systems and consists of four major elements:

(1) a lightweight compact equipment enclosure containing both the transmitter and receiver for the radar
(2) a lightweight antenna/pedestal assembly enclosed by a radome
(3) a displays/control assembly providing both 'A' and 'C' scope displays and the operating controls for the radar
(4) an antenna control unit which provides manual control of the antenna.

This radar provides both automatic and manual acquisition and tracking. Automatic acquisition is based on a designation received from the fire control system computer. The automatic acquisition is a result of a search using a selected search pattern around the designation. If automatic track is lost, the W-120 automatically seeks to re-acquire the target. In the automatic mode, when the fire control system computer is not designating or the radar is not processing a current track, the W-120 reverts to ready status wherein the system is fully operational except that the full antenna motion and radiation is inhibited.

In the absence of a fire control system designation, a manually designated bearing, elevation position and range gate can be given and the antenna scans a selected search pattern in azimuth and elevation about the designated position. A 4000 yard acquisition window is positioned to the designated range.

The W-120 provides a low flyer detection mode for acquisition and tracking operations. The off-boresight mode may be selected for tracks lying within 0.8 beamwidths of the sea surface and provides a significant improvement in rejection of multi-path interference as compared with boresight tracking of the same target.

Detection range of the W-120 is 30 nautical miles (56 km) against a 1 m² target. Range accuracy is ± 0.01 per cent to 2.5 nautical miles, then ± 2 per cent over that distance. Tracking accuracy is better than two milliradians in both bearing and elevation.

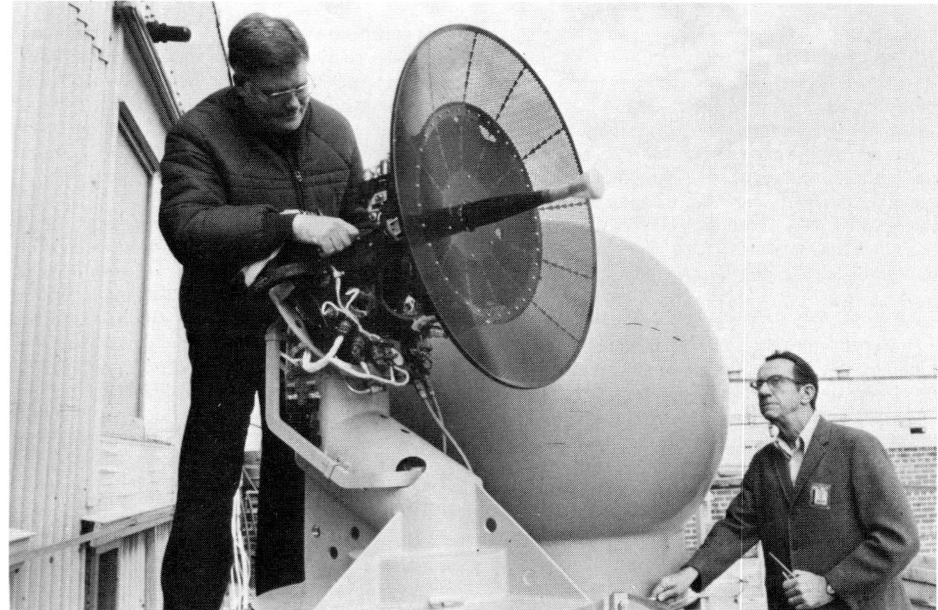

W-120 antenna/pedestal assembly and radome

STATUS

No information on ship fittings is available.

CONTRACTOR

Westinghouse Electric Corporation, Surveillance Radar Division, PO Box 1897, Baltimore, Maryland 21203, USA.

4048.253

W-160 SHIPBOARD FIRE CONTROL RADAR

The W-160 shipboard tracking radar is based on technology used in the AN/APG-66 radar (**3303.353**), built by Westinghouse for the US Air Force F-16 aircraft. The W-160 combines the lightweight and low cost of an airborne system with the capability to track, at all altitudes, small, high speed targets on and over the surface of the sea. A medium-PRF, X-band, multimode, pulse doppler system, the W-160 has two-axis monopulse tracking which provides accurate position data at ranges necessary to achieve a fire control solution. It can acquire and track targets in rain, heavy clutter, and in smooth or rough seas. A collocated and boresighted TV camera is optional.

The W-160 utilises coherent pulse doppler MTI for superior sub-clutter visibility and operation in high clutter environments. In low elevation tracking, the boresight of the antenna is maintained above the target. This off-boresight tracking (or low-flyer) mode reduces tracking inaccuracies caused by reflections from the surface of the sea, an interference problem known as multipath reflection. Interface between the radar and the shipboard fire control system is made possible through the use of modular-digibus techniques.

The W-160 features 15 modes of operation for optimum performance against a wide range of targets and hostile environments. These include five surface search modes, four air search modes, two surface acquisition and track modes, three air acquisition and track modes, and one passive acquisition and track mode. In addition, the W-160 features six special sub-modes for kill evaluation, detection of missile launch from a tracked target, shell spotting, low evaluation track, jammer avoidance and slave to electro-optical systems.

The W-160 consists of three main elements: the antenna/pedestal/radome unit; the microwave assembly which contains the transmitter and receiver elements; and an equipment enclosure which houses the radar computer, digital signal processor and antenna control electronics.

The antenna is a 48-inch (122 cm) diameter paraboloid with a six hour monopulse feed. It is capable of continuous rotation through 360° and has elevation coverage from –30° to +85°. Full gyrocompass stabilisation is provided during all search and track operations. A 72-inch (183 cm) antenna is also available. Use of this antenna extends the radar range to 90 km and permits use of missile illumination.

CHARACTERISTICS

Frequency: X-band (9.7 to 9.9 GHz)
Peak power: 16 kW
Pulse widths: 1, 2, 3, 4 and 8 μs
PRF: 1000 – 14 200 pps

Detection range: 45 km against a 1 m^2 target with 122 cm antenna; 90 km with 183 cm antenna
Tracking accuracy: 1 mrad in bearing and elevation, 15 m in range
Angle tracking capability: Targets with angular rates up to 60°/s and velocities up to 2500 knots
Tracker type: Kalman estimator
Receiver noise: 3.5 dB
Antenna type: 2-axis monopulse
Antenna gain: 37 dB
Antenna beamwidth: 1.8° (1.2° with 183 cm antenna)
Antenna polarisation: Linear (vertical)
Rotation speed: 15, 20 or 25 rpm
Clutter processing: Coherent pulse doppler; 52 dB MTI improvement
Platform speed: Motion compensation for complete range of ship's velocities
Reliability: 600 h MTBF
Weights
Antenna/pedestal assembly: 340 kg
Radar electronics: 955 kg

STATUS
The W-160 is used in the W-1200 series of naval fire control systems (**3514.261**) but no details of ship fittings are available.

W-160 shipboard fire control radar

CONTRACTOR
Westinghouse Electric Corporation, Surveillance Radar Division, PO Box 1897, Baltimore, Maryland 21203, USA.

3258.253
FLEXAR WEAPON CONTROL SYSTEM

FLEXAR (flexible adaptive radar), is proposed as a next generation shore or shipboard weapon control system for short- and medium-range missiles or guns against ships, aircraft or missiles. The system combines fire control functions with surveillance modes that complement existing air search radars. Very fast reaction time to short-range, high-speed targets makes FLEXAR an effective defence against anti-ship missiles. Target acquisition and track initiation are completely automatic. The adaptive radar accepts designation from other active or passive sensors and provides automatic threat evaluation, automatic weapon assignment and rapid kill assessment. It employs multiple waveforms adaptively selected for high performance in clutter, rapid resolution of complex targets and optimum tracking.

FLEXAR display unit

The system uses a 40-inch electronically agile dish antenna that rotates mechanically at 60 rpm to provide hemispheric coverage, but also operates with the antenna stationary in sector fire control modes. The transmitter is capable of handling multiple waveforms and a programmable signal processor provides an optimum match for each waveform. The entire system is under real-time control by a general purpose digital computer with clutter-free data being displayed on a symbolic plan position indicator. The many modes of the system are interleaved by the general purpose computer to provide effective simultaneous search, acquisition, multi-target track and weapon control. The mode of most interest to the US Navy is that in which FLEXAR would serve as a positionable multiple target tracker/illuminator for the latest US Navy ship-launched missiles.

A land-based fire control system, known as Multi-function Radar (MFR) (**5607.153**) is a new application that uses technology derived from the FLEXAR programme. This system, intended for use with existing I-HAWK and other short-to-medium range missiles, can provide a significant reduction in manpower and equipment, greater mobility and transportability, improved operation in ECM and reduced life cycle costs.

STATUS
FLEXAR completed first phase proof of principal search and multi-target track tests in 1981 and intermediate weapon control testing in 1982. Evaluation of the system's multi-function roles with captive I-Hawk and navy standard missiles was completed successfully during 1983. Additional testing and evaluation is continuing. No information is available on probable applications.

CONTRACTOR
Hughes Aircraft Company, Fullerton, California 92634, USA.

3768.253
R-76 NAVAL FIRE CONTROL RADAR

The RCA R-76 naval radar is an I-band monopulse fire control radar with combined search, detection and tracking capabilities. It is employed, with the designation HR-76, as the principal sensor in the Honeywell H-930 series of naval fire control systems (**3749.281** in 1982-83 and earlier editions), in conjunction with other equipments, and also in the RCA R-76 naval fire control system (**3748.281**). Lightweight construction permits mounting high on a ship's superstructure thereby aiding early detection, and the gyro-stabilised antenna provides for continuous 360° search. Dual monopulse tracking enables simultaneous missile and target tracking, with the data displayed on the system's dual A-scope presentation unit, with facilities for instantaneous entry for weapon system control.

The main elements of the HR-76 equipment are the director, the transmitter, receiver/signal processor, computer and a servo controller. A complete ship's installation of the HR-76 typically weighs approximately 1200 kg, of which the above-deck equipment accounts for about 205 kg. The antenna is a four-horn cassegrain unit, housed in a protective radome and mounted on a stabilised elevation-over-azimuth director. The R-76 version uses a director that can accommodate a TV camera and lens. A radome is not required. The weight of this director with TV camera is 360 kg. The transmitter employs a vacuum tube modulator allowing multiple waveforms; other transmitter features include multiple pulse widths and multiple jittered PRFs. Crystal control allows for a narrow-band missile link. A microcomputer performs such functions as: angle and range servo controls; CFAR; AGC; beacon AFC; control; and data correction.

CHARACTERISTICS
Transmitter
Frequency: I-band
Tuning: 12 pre-selected frequencies
Peak power: 250 kW
Pulse width: 0.25, 0.5, 1μs
PRF: 1000 – 3000 Hz
Receiver: 2-channel monopulse
Antenna
Diameter: 1 m
Beamwidth: 2.7°
Signal processor: Digital
Acquisition range: 40 km (on 1 m^2 target)
Tracking range: 40 km (on 1 m^2 target)
Accuracy at 20 km: 10 m range; 1 mrad angular
Search scan period: 4 s
Reaction time: 3.5 s average from designation to tracking

RCA R-76 naval fire control radar

STATUS
Production.

CONTRACTOR
RCA Electronic Systems Department, Moorestown, New Jersey 08057, USA.

4223.253
MBAR FIRE CONTROL RADAR

Multiple Beam Acquisition Radar (MBAR) is a new C-band defensive weapon control system designed to provide smaller ships with self defence capability against small high-speed threats approaching at very low angles, as well as targets diving simultaneously from very high angles.

The MBAR employs a fan beam transmit antenna and a 17-beam receive antenna, which are stabilised electronically in pitch and mechanically in roll. The system can acquire targets in range, azimuth, elevation and range rate during a single scan (1.5 s). The equipment analyses the threat and designates the targets in order of priority to available weapon systems. Up to 200 targets can be accommodated by the system. Target threat assessment is performed on every antenna scan and weapon designation on the following scan.

The receive antenna is equipped with ultra low sidelobes of –50 dB rms for operation in a severe ECM environment. The transmitter employs a high reliability TWT and uses high average power coupled with frequency agility to minimise ECM. The digital processor employs a coherent pulse doppler waveform with fast Fourier transform processing. It is capable of automatic target detection in severe weather and clutter, with a sub-clutter visibility of 60 dB. The equipment can also be operated in a long range search mode or in a navigation surface gun mode.

CHARACTERISTICS

Transmitter: TWT
Frequency: C-band
Peak power: 70 kW
PRF: 17.2 kHz nominal
Receiver: 16 beams time-multiplexed into 8 processor channels
Noise figure: 3.5 dB
Antenna
Azimuth: 1.75° beamwidth
Elevation
Transmit – fan beam, peak gain 26 dB
Receive – beam forming lens, peak gain 32 dB
17 beams varying from 5 – 8° in vertical beamwidth; 16 active, 1 reserved for electronic stabilisation
Dimensions: 2.6 × 0.9 m
Scan rate: 40 rpm (less in long range mode)
Detection range: 55 km (1 m^2 target); 36 km (0.2 m^2 target) (110 nm instrumented range in long range search)
Target speed: Up to 1189 m/s incoming; up to 366 m/s outgoing
Accuracies: ⩾0.4° azimuth; ⩾1° elevation; ⩾61 m in range; ⩾7 m/s in range rate; ⩾0.15° azimuth in surface gun mode

MBAR fire control radar

Angular resolution: 5° elevation beamwidth (low angle); 8° elevation beamwidth (high angle)
Target tracking: 100 targets simultaneously
Weight: 1500 lb (682 kg) at masthead; 4500 lb (2045 kg) below decks

STATUS

In development as a private venture.

CONTRACTOR

Raytheon Company, Equipment Division, Wayland, Massachusetts, 01778, USA.

4608.253

FLANKING-BEAM ARRAY SWITCHING TECHNIQUE RADAR (FAST)

FAST is a shipborne search radar system developed by General Electric, one prototype having been completed in 1982 as a private venture. It employs two phase-scanned antenna faces, mounted back-to-back, which are rotated at either 30 rpm in C-band or 15 rpm in S-band. Each face consists of 5736 elements, scanned only in elevation, and multiple pencil beams (1.4 × 2°) are generated in four different directions at 15 and 45° either side. By the use of eight simultaneous beams, FAST achieves a high data rate to provide precise tracking, mid-course guidance and target illuminator guidance. Peak power is claimed as 1 MW in C-band and 2 MW in S-band, with detection ranges as 143 nautical miles (C-band) and 246 nautical miles (S-band). Pulse compression techniques are used in the signal processor.

STATUS

In development as a private venture.

CONTRACTOR

General Electric Company, PO Box 4840, Syracuse, New York 13221, USA.

4609.253

Mk 23 TARGET ACQUISITION SYSTEM RADAR

The Mk 23 target acquisition system (TAS) is a two-dimensional L-band rotating fan-beam pulse doppler radar intended for use with NATO Sea Sparrow missiles for protection against sea-skimming, high-diving and pop-up missiles. As a secondary role, the system can also be used for air defence surveillance and aircraft control. The 14-foot antenna incorporates 26 flared horns which provide 360 degrees azimuth coverage and up to 75 degrees in elevation while rotating at 15 or 30 rpm. It is attached to a 30 degree stabilised pedestal and is mounted back-to-back with a Mk 12 IFF equipment. The system has a detection range of more than 20 nautical miles on a 1 metre target and a range of nearly 100 nautical miles when used in the secondary role.

Mk 23 TAS radar

The Mk 23 system has four operational roles:

(1) a normal point defence mode where the equipment is used to detect and track missiles at ranges beyond 20 miles and then engage them with anti-missile missiles
(2) A medium range mode for surveillance and aircraft control out to nearly 100 nautical miles
(3) a mixed mode combining (1) and (2)
(4) an emission control mode which allows the operator to scan selected sectors and switch the equipment on and off automatically to avoid detection.

A number of upgrades are likely to be made to the system over the next few years to improve response times, expand radar coverage and add to the system flexibility. These include a new computer to increase the system speed and expand its memory. A new phased array antenna using phase shifters instead of horn antennas began testing in the summer of 1985. According to Hughes this will give the system a four-dimensional capability in bearing, range, elevation angle and range rate to reduce the weapon response time and allow the TAS to be used as a designation radar for 'fire-and-forget' missiles.

See also under entry **1928.211** in the Naval Fire Control Section.

STATUS

In production and fitted to various US Navy warships, including destroyers and aircraft carriers.

CONTRACTOR

Hughes Aircraft Company, Fullerton, California 92364, USA.

5848.293

AIMS ANTENNA SYSTEM

An advanced shipborne antenna system, the AIMS OE-120/UPX group has been developed by Lockheed. The system is also suitable for land-based operations. It consists of an electronically scanned circular antenna which offers multi-function performance and provides military IFF in all interrogation modes, as well as operation with military and commercial air traffic control equipment. It is particularly suited for operation in systems where rapid multi-target identification is necessary, such as track-while-scan phased-array or multiple search radars.

The antenna group consists of an antenna, antenna position programmer and a control unit. The antenna itself measures 3.75 m diameter × 40.6 cms high and has an array of 64 RF radiators on its perimeter. For shipborne applications it is mounted concentrically around the mast to avoid any blockage inherent with mechanical rotating systems. Interrogate Side Lobe Suppression (ISLS) is provided by selectively energising radiating elements to produce the required omni-directional and directional beam patterns. The system can be operated in a continuous, sector or jump-scan mode. In the latter mode, switching from one beam to another can be accomplished within 50 µsec.

The antenna position programmer is located within the perimeter of the antenna. On command from the control unit it distributes RF power to the radiators which, in turn, form and steer the beam. The remotely positioned control unit receives either digital or synchro beam position data and translates this into digital antenna beam steering commands. In the shipborne application, the control unit also provides compensation for roll, pitch and ship's heading. In addition, internally generated beam steering commands may be selected by means of front panel controls.

System maintenance is enhanced by a built-in diagnostic test system which automatically monitors more than 200 operational test points. A front panel digital display provides fast indication of the position of any malfunction within the system.

STATUS

Installed on a number of the US Navy's latest warships.

CONTRACTOR

Lockheed Electronics Company Inc, US Highway 22, Plainfield, New Jersey 07061, USA.

6126.293

AN/UPX-24(V) SHIPBORNE IFF SYSTEM

The AN/UPX-24(V) is a central IFF equipment which operates as part of a shipborne weapon system. Its purpose is to accept central commands from, and provide target reports to, the AEGIS weapon system computer. In performing this purpose, it interfaces with up to four unrelated radars and up to 22 display consoles for normal IFF function.

DEVELOPMENT

In 1968 Litton Data Systems was awarded a contract to design and build a system for operation evaluation. After completion of tests by the US Navy, the company received a contract to build two additional systems incorporating Navy recommendations. These systems were delivered in 1977.

STATUS

Nineteen production systems had been delivered as of December 1986, and they are being installed on the US Navy CG-47, DDG-51 and LHD-1 classes. Ten more systems are under contract and additional systems are being proposed.

CONTRACTOR

Litton Data Systems, 8000 Woodley Avenue, PO Box 7601, Van Nuys, California 91409-7601, USA.

AN/UPX-24(V) shipborne IFF system

ELECTRONIC WARFARE EQUIPMENT (SEA)

AUSTRALIA

5954.293

WINNIN DECOY SYSTEM

The Winnin decoy countermeasures system is an active expendable rocket vehicle with an ECM payload. The version in development for the Royal Australian Navy has a unique active/passive payload, and features a hovering rocket which enables a variety of ECM payloads to be deployed. This is the key to the system for while it is hovering the vehicle can adjust its speed away from the ship by adjusting its degree of tilt from the vertical. This allows the decoy to be deployed rapidly and positioned in the correct sector, while travelling at a speed typical of the warship it is protecting.

The flight profile of the decoy is controlled after launch by an auto-pilot, which is part of a microprocessor-based flight control system. The latter allows data to be transferred to the decoy on launch, which can be supplemented by additional data obtained during flight. This system of autonomous vehicle control dispenses with the need for trainable launchers.

The Winnin system has three components, the electronic payload, the hovering vehicle and a launching system. The rocket can be programmed to hover, move up or down, sideways or rotate on its longitudinal axis. The rocket is a long cylinder and is launched almost vertically. After launch and moving to a preselected position, it manoeuvres away from the ship comparatively slowly along a pre-calculated trajectory. This is chosen to make sea skimming missiles pass well under the decoy to prevent them from re-acquiring a further target.

The decoy system would normally be interfaced with the ship's ESM, radar and combat data system. Two alternative launchers are being developed; a modification of the Super-RBOC chaff grenade launcher (**4656.293**), or a box-launcher which also serves as a storage.

STATUS

The Winnin system has been in development for some years and full-scale engineering development is expected to last until 1988, and will include technical and operational evaluations. Production was scheduled to commence in mid-1988 with operational deployment by the Royal Australian Navy in 1989.

CONTRACTOR

Development is being shared among various Australian Government research establishments of the Defence Science and Technology Organisation, and the Government Aircraft and Ordnance Factories.

CANADA

1714.253

MODEL 100 NAVAL DF SET

The Model 100 is a shipboard direction finding, intercept, and surveillance system covering the frequency range 10 kHz to 180 MHz.

The Model 100 DF set operates on the principle of resolving the electromagnetic field at the antenna location into a set of orthogonal vectors which are then amplified and applied to the plates of a cathode ray tube (CRT), so as to trace a bearing line indicating the azimuth of arrival of the signal.

Two antennas are used: one for the LF to HF ranges and one for VHF. The LF/HF antenna derives the DF vectors from a pair of multi-turn, switched loops while the sense is obtained from a set of balanced dipoles. The VHF antenna comprises 16 separate monopole elements which can be arranged as crossed-Adcock and parallel dipole arrays in both the horizontal and vertical plane of polarisation. The VHF antenna is constructed around a hollow column which permits the passage of cable to the LF/HF antenna and to an aircraft warning light, located uppermost.

The DF signal vectors are amplified by two identical receiver channels which are accurately matched in phase and gain under all conditions of tuning and signal strength. A third channel, matched in phase only and equipped with a separate AGC, serves to amplify the omni-directional sense signal which is applied to the grid of the CRT to blank out the unwanted half of the bearing trace.

In addition to the instantaneous single-signal DF mode, the receiver can be operated in a panoramic mode (amplitude/frequency display), or in a wideband sweep mode, allowing the operator to see simultaneously the bearings and frequencies of all the signals present within the search band. This '3-D' display is made possible by using colour modulation as the frequency parameter.

A phantom omni-directional audio output is available at all times for intelligence interception purposes. This monitor channel can be tuned to any of the signals within the sweep band without disrupting the panoramic bearing display.

Digital, automatic tuning (local or remote) is derived from a phase-locked frequency synthesiser fitted with an oven stabilised frequency standard. The synthesiser is designed for fast access manual tuning, using the 3-DEK decade reed switches which are gear-interlocked to count up or down automatically. The tuning frequency is indicated directly by the decade knobs and in addition there is a projection-type luminous readout which can be duplicated in a remote location.

The equipment is also produced in a land-mobile configuration, under the designation Model 109, to form a complete two-man vehicle-mounted DF and surveillance unit.

STATUS

In operational service with the Canadian armed forces.

CONTRACTOR

General Precision Industries Ltd, PO Box 88, Place Bonaventure, 32 Brome Street, Montreal 114 PQ, Canada.

4161.293

AN/SLQ-501 CANEWS EW SYSTEMS

Specifically designed to perform real time detection, analysis, identification, classification and warning of both hostile and friendly platforms and missiles, CANEWS architecture comprises four major sub-systems:

(a) An IFM based antenna and receiver sub-system which covers the desired frequency bands and provides a very high probability of intercept

(b) A data transfer unit

(c) Twin processors which carry out the real-time analysis and supervisory function, and which provide an interface with the ship command and control system

(d) A display console which provides the operator interface

Built-in test equipment is also provided.

In operation, a main library holds data on numerous radar modes and signatures. Analysis of the detected RF signals identifies the radar type and platform, and provides a confidence level indication. In addition, the operator can build a tactical library of threat radar signatures to meet immediate mission requirements. While processing many automatic functions to reduce operator workload and fatigue, the AN/SLQ-501 provides facilities for manual control if required.

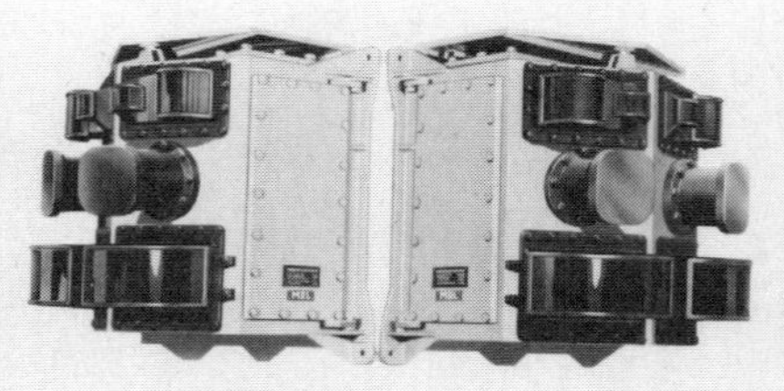

Antenna unit for CANEWS

STATUS

The AN/SLQ-501 shipborne ESM is currently in production for delivery to modernisation and new shipbuilding programmes of the Canadian Navy. The AN/SLQ-501 architecture is adaptable to a variety of other maritime applications, and can be configured for the specific requirements of submarines and aircraft.

CONTRACTOR

MEL Defense Systems Ltd, 1 Iber Road, PO Box 90, Stittsville, Ontario, Canada K0A 3G0.

4054.293

AN/SLQ-503 (RAMSES) SHIPBORNE ECM SYSTEM

The AN/SLQ-503 Reprogrammable Advanced Multi-mode Shipborne ECM System (RAMSES) is a modular naval electronic countermeasures system designed principally to provide protection for ships against I/J-band radar threats. It is an integrated responsive noise and deception repeater jammer which jams search radars at long range to disable or confuse missile launchers (with emphasis on noise jamming) and deception of any missiles launched at shorter ranges, after they have gone into the active radar homing mode where jamming relies more heavily on deception.

In its normal mode of operation, RAMSES is interconnected to and controlled from a system such as CANEWS ESM (**4161.293**), Sphinx ESM (**4055.293**), RAPIDS ESM (**4056.293**) or from a tactical data system. A dedicated ECM operator is not therefore required. As a back-up mode RAMSES can be manually controlled (target designation and jamming) from a remote control panel located conveniently near the EW operator.

RAMSES operates in I/J-band and is capable of high pulse and CW Effective Radiated Power (ERP). Multi-mode jamming operation is possible with multi-target handling capability. The system is reprogrammable and power management is pro-

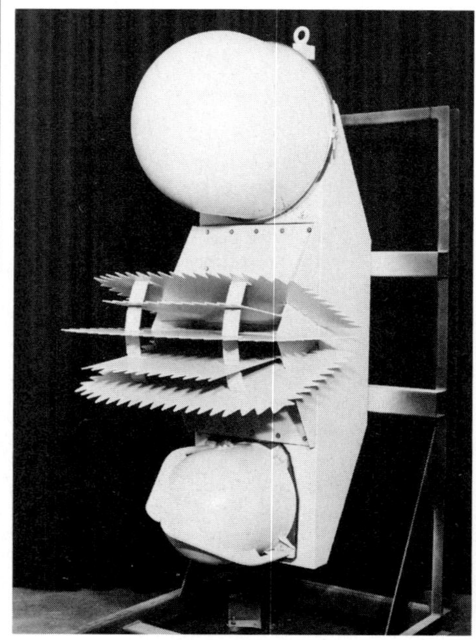

RAMSES antenna assembly

vided. The antennas are fully trainable (port and starboard each through 180°) and are stabilised for roll and pitch.

STATUS

RAMSES is in operational use by the Royal Netherlands Navy and has been sold to Canada for the Canadian patrol frigate programme where the first systems have been delivered. A variant of the system, SCIMITAR, is in operational use with Argentina, Colombia, Malaysia and Peru.

RAMSES and SCIMITAR were designed and developed by Signaal, The Netherlands, in cooperation with MEL (UK). The systems are now manufactured and supported in Canada by MEL Defence Systems Ltd.

CONTRACTOR

MEL Defence Systems Ltd, 1 Iber Road, PO Box 90, Stittsville, Ontario, Canada K0A 3G0.

6162.293

AN/SAR-8 INFRA-RED WARNING SYSTEM

The AN/SAR-8 is an Infra-red Search and Target Designation System (IRSTD) and has been developed to supplement the shipborne radar warning systems to ensure the detection of sea-skimming missiles using infra-red homing techniques. The system is passive, can detect ships, aircraft and missiles and differentiate between them by their IR signatures. The system scans a full 360° and provides azimuth and elevation angles, and associated information to the on-board countermeasures equipment.

The system consists of scanning optics with cryogenic-cooled detector arrays and the necessary electronic processing equipment. Objects are viewed against an ambient temperature background and sensed by thermal contrast. The processing equipment includes low level amplifiers to amplify the IR detector outputs, signal processing, clutter suppression circuitry and track filters to enhance and identify thermal characteristics.

STATUS

The AN/SAR-8 is a joint US Navy/Canadian Navy development and has undergone successful sea trials with both navies.

CONTRACTOR

Spar Aerospace Ltd, Defence Systems Division, Lismer Building, Kanata, Ontario, Canada K2K 1X3.

6866.293

COHERENT COUNTERMEASURES PROCESSOR

The Coherent Countermeasures Processor (CCP) is a deception jammer covering the frequency range 2 to 18 GHz and is effective against both coherent and non-coherent radars, with a library of 500 threats. Telemus claims that the processor can overcome sophisticated ECCM techniques such as pulse coding and compression of extended pulses. The integrated programmable deception techniques generator provides a wide choice of programmes, including the ability to combine range-gate pull-off with constant or swept scan rate modulation.

The DSK-604 digital RF memory incorporated within the CCP has a wide instantaneous bandwidth (10 to 550 MHz). Telemus is also offering an ESM system to provide signal de-interleaving for existing broadband equipments.

STATUS

In production and being retrofitted into 17 Canadian ships to upgrade their AN/ULQ-6 systems. Telemus has also received a study contract for a low cost electronic off-board missile decoy.

CONTRACTOR

Telemus Electronic Systems Inc, 3100 Moodie Drive, Nepean, Ontario K2H 8G3, Canada.

CHINA (PEOPLE'S REPUBLIC)

6485.453

921-A ESM EQUIPMENT

The 921-A is a wideband pulse radar direction finding receiver, and is installed on submarines to detect emitters of airborne, shipborne and shore-based radars. It provides coarse measurement of azimuth, frequency band and operational state of the hostile emitters.

STATUS

Frequency: 2 to 18 GHz in four bands

Sensitivity: 1.5×10^{-3} to 10^{-4} W/M²

Bearing accuracy: Better than ±30°

Dimensions and weights

Antenna unit: 560 mm diameter × 515 mm high; 80 kg

Receiver and display: 450 × 468 × 124 mm; 40 kg

Distribution unit: 145 × 214 × 291 mm; 6kg

CONTRACTOR

China National Electronics Import and Export Corporation, 49 Fuxing Road, PO Box 140, Beijing, People's Republic of China.

921-A ESM Equipment

6498.293

RW-23-1 RADAR WARNING EQUIPMENT

The RW-23-1 is a radar warning system intended for use on small and medium sized ships. It consists of an antenna array and a control, display console and covers the frequency range from 2 to 18 GHz, which is divided into four bands. The antenna is built in two sections for mounting one on either side of the ship. The parameters of up to 15 radars can be stored in the library memory of the systems and are used for comparison with any detected threats. System weight is given as 238 kg. The RW-23-1 is normally used in conjunction with a passive countermeasures system.

CONTRACTOR

China National Electronics Import and Export Corporation, 49 Fuxing Road, PO Box 140, Beijing, People's Republic of China.

FRANCE

3377.293

DAGAIE NAVAL COUNTERMEASURES EQUIPMENT

The Dagaie countermeasure equipment is an automatic dispenser for IR decoy flares and chaff grenades to provide naval vessels with passive self-protection against heat-seeking, radar homing and dual-mode anti-ship missiles. The system will accept threat data from a variety of sources, such as radar, ESM systems, optical, etc, and this information is used with data such as ship's own speed and heading to compute the optimum countermeasures deployment.

The Dagaie launcher comprises ten replaceable containers, each of which is loaded with either IR or chaff rockets. The containers weigh 40 or 45 kg, according to type.

The firing sequence runs automatically and is triggered on a missile alarm from a variety of sources, such as radar, ESM systems, optical sights, thus providing a very short reaction time. The firing direction is optimised in accordance with the threat bearing, wind speed and direction, ship heading and speed data.

The launching of radar and IR decoys is so arranged that the most advanced mixed-guidance missile cannot discriminate them and thus collocation of both decoys is achieved.

The installation is either a double mounting for ships over 800/1000 tons, or a single mounting for smaller vessels, and comprises:

(1) one trainable mounting (or two) which carries ten replaceable containers (suitcases), each of which is loaded with either IR or chaff projectiles. The range of existing ammunition enables decoys to be adapted to the various types of ships
(2) one or two servo-units
(3) one data-processing unit whose purpose is to compute the firing data
(4) one power-supply unit
(5) one supervision unit used to operate the system and display the loading state of the mounting
(6) one manoeuvre indicator which displays the proposed ship's manoeuvre in order to extend the protection time as long as possible without renewing the decoy.

For either the single or double installation a reduced size mounting carrying six suitcases can be supplied.

Dagaie Mk 2 countermeasures equipment

The Dagaie Mk 2 is an improved version of the current system. The Mk 2 has been designed for the following main reasons:

(a) easier integration of the equipment with a central processing system (NTDS), and centralised operation in conjunction with other EW equipments (radar detectors and jammers)
(b) use of a new generation of ammunition LEM and LIR requiring new firing rules and, consequently, more sophisticated software
(c) increase in operational characteristics of Dagaie by use of a new medium range electromagnetic ammunition, the REM, which allows the Dagaie to cover, alone or associated with a jammer, the whole operation of the EW actions possible during an attack sequence (dilution, centroid seduction, dissimulation seduction and range gate stealing seduction).

The organisation of the Dagaie Mk 2 is similar to the present system, the main modification concerning the computer and supervision units. The current system can be easily modified to a Mk 2 version.

CHARACTERISTICS

Resetting time for 90°: <2 s

Medium aiming speed: 1.5 rad/s

Average reaction time: <4 s (reaction time is the time lapse between reception of 'missile alarm' information and the end of IR firing)

STATUS

Ordered or operational with 15 navies including those of France, Argentina, Bahrain, Kuwait, Malaysia, Colombia, Peru, Indonesia, Thailand, UAE, the SAWARI programme of the Royal Saudi Navy, the Combattante II in Qatar, the Combattante III in Tunisia and the Moroccan *Descubierta* Frigate.

CONTRACTOR

Compagnie de Signaux et d'Entreprises Electriques, Division Défense, 17 place Etienne Pernet, 75738 Paris Cedex 15, France.

Dagaie countermeasure launcher in operation during tests of a production equipment off Toulon

4350.293

MAGAIE LIGHTWEIGHT ANTIMISSILE DECOY SYSTEM

Specifically designed for small and medium sized ships, MAGAIE provides an overall protection against missiles fitted with electromagnetic (EM), infra-red (IR) or combined EM/IR seekers. Directly derived from the DAGAIE system (**3377.293**), it also uses the same ammunition suitcases.

With a reduced overall weight, a lighter launcher, only four suitcases and a single compact processing/servo unit, MAGAIE can be installed in single or double launcher configuration. The operating principle is by centroid effect. Ammunition used is standard DAGAIE EM or IR suitcases manufactured by Societe E Lacroix.

The operation is fully automatic from the reception of a missile threat alarm by computation of firing data and selection of ammunition suitcase during the resetting of the launcher to firing of the suitcase.

The MAGAIE processing and launching installation includes, according to the ship's capabilities:

(1) one or two mobile launchers, fully trainable, capable of receiving three suitcases
(2) a processing and servo control unit where all firing data is computed and the mounting reset order is generated
(3) a supervision unit controlling the system operation
(4) a manoeuvre indicator showing the recommended manoeuvre for the vessel.

The system with its simple interfaces is easily adapted to the various sensors (radar, ESM, optical).

CHARACTERISTICS

Average reaction delay: <5 s

Launcher

Max training arc: 330°

Reslewing time for 90°: ≤2 s

Reslewing speed: 1.2 rad/s

Sweeping radius: 400 mm

Height: 1200 mm

Weight: 600 kg

STATUS

Fully developed.

CONTRACTOR

Compagnie de Signaux et d'Entreprises Electriques, Division Défense, 17 place Etienne Pernet, 75738 Paris Cedex 15, France.

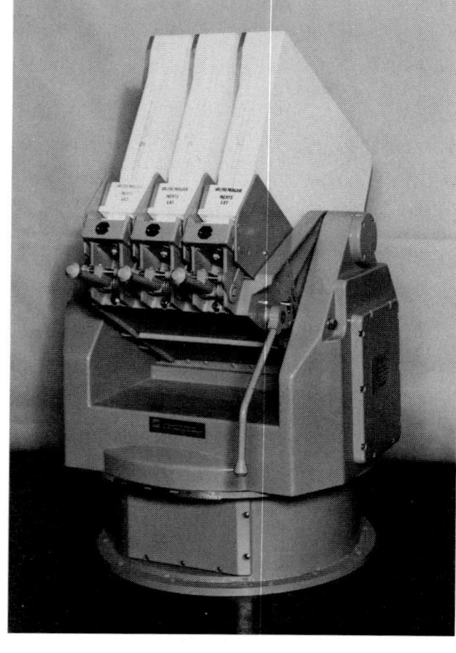

MAGAIE decoy launcher

5656.293

SAGAIE NAVAL COUNTERMEASURES DECOY SYSTEM

SAGAIE is a fully automatic, passive countermeasures system designed to protect medium and large ships from surface-to-surface and air-to-surface missiles. It will provide protection from missiles guided by electromagnetic (EM) or infra-red (IR) seekers, or any combination of these two, even when the missiles are attacking simultaneously over the entire horizon.

The system operates in conjunction with DAGAIE (**3377.293**) and provides long-range defence by confusion and dilution effects against enemy target designating radars, and close-in defence by distraction and seduction against acquisition and tracking systems in missile seekers, in order that DAGAIE can complete its centroid deception effect.

The SAGAIE system is fully automatic from the reception of a missile threat alarm originating from any of the surveillance systems (radar, IR, optical or ESM) and will optimise the use of decoys in a very short reaction time. The equipment can be used alone (confusion and distraction) or jointly with a jammer (dilution, substitution after concealment, substitution after deception) by firing of substitution decoys.

The installation comprises:

(1) one or two fully stabilised launchers loaded with ten rockets in containers and trainable in azimuth and elevation
(2) one or two servo units
(3) one or two aiming and maintenance units
(4) a processing unit which computes the rocket launching sequence and provides information for the supervision unit
(5) a supervision unit with controls and status displays
(6) an interface unit which provides for operation with any tactical data or target designation system.

SAGAIE can fire either EM or IR decoy rocket type ammunition or a combined EM/IR decoy rocket. The rocket is packaged in a waterproof launcher container. This container is used for handling, transport and storage, and when secured to the mounting ensures guidance of the rocket at the moment of launch.

SAGAIE countermeasures system

CHARACTERISTICS
Height: 2000 mm
Sweeping radius: 1600 mm
Weight (empty): 1900 kg
Number of rockets: 10
Container
Length: 1800 mm
Diameter: 330 mm
Weight: 30 kg

STATUS
12 SAGAIE systems were ordered by the French Navy in early 1986 for installation on new anti-aircraft corvettes, the *Foch* and *Clemenceau* aircraft carriers, and the *Suffren* and *Duquesne* destroyers. Systems have also been ordered by another unspecified customer.

CONTRACTOR
Compagnie de Signaux et d'Entreprises Electriques, Division Défense, 17 place Etienne Pernet, 75738 Paris Cedex 15, France.

5784.253
NEWSY NAVAL ELECTRONIC WARFARE SYSTEM
NEWSY is an integrated electronic warfare equipment which is designed for the protection of major warships against missile attack. The heart of the NEWSY system is a dedicated computer and an associated one-man console. The complete system is designed for threat detection, evaluation and classification, selection of appropriate responses and activation of countermeasures against electromagnetic and/or infra-red guided weapons.

NEWSY links together a radar detector for the ESM, a multimodal jammer for the ECM, SAGAIE and DAGAIE launchers for decoying at short, medium and long range, and a control console and display unit. The system is normally completely automatic in operation but manual control can also be exercised if required. Information regarding threats is passed to the operator and displayed on a video screen, which forms part of the ECM control console, in both PPI and alpha-numeric format.

STATUS
In development for the French Navy C70AA destroyers. See also SAPIENS EW system (**5657.293**).

CONTRACTOR
Compagnie de Signaux et d'Entreprises Electriques, Division Défense, 17 place Etienne Pernet, 75738 Paris Cedex 15, France.

5783.253
NADS NAVAL ANTIMISSILE DECOY SYSTEM
The NADS system is an electronic countermeasures equipment using electromagnetic and infra-red decoys and is normally integrated into the complete warship combat system. The equipment is a combination of two decoy launching sub-systems, the SAGAIE for long and medium range and the DAGAIE for short range, plus an action co-ordination computer and a tactical display. Information on the SAGAIE and DAGAIE systems is given under entries **5656.293** and **3377.293** respectively.

NADS is understood to be an export version of NEWSY (**5784.253**) and can be integrated with any type of ESM detection system and jammer. Depending on the size and type of ship requiring protection, NADS is available in two versions with either a single or a double SAGAIE launcher. Without a jammer the system will operate in confusion, dilution, distraction and centroid seduction modes. In association with a jammer other modes such as range gate stealing, seduction and dissimulation seduction are available.

CONTRACTOR
Compagnie de Signaux d'Entreprises Electriques, Division Défense, 17 place Etienne Pernet, 75738 Paris Cedex 15, France.

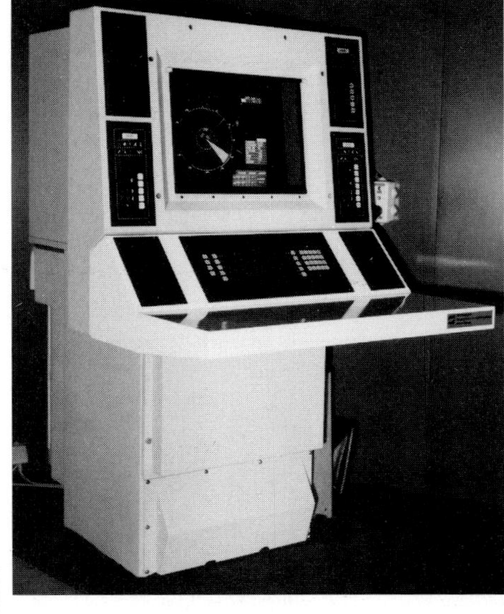
NADS control console

5657.293
SAPIENS EW SYSTEM
SAPIENS (Surveillance Alarm and Protection, Intelligence EW Naval System) is an EW system designed for small and medium size vessels. It is a joint system by Thomson-CSF and CSEE and comprises elements of both companies' EW equipment. SAPIENS 1, for small patrol boats and mine countermeasures craft, consists of a DALIA analyser/identifier with a radar library of at least 1000 possible threats, a DR2000 crystal video ESM receiver and a DAGAIE decoy launcher.

SAPIENS 2, for corvette and frigate sized ships, is basically the same as SAPIENS 1 but uses the DR4000S IFM receiver, instead of the DR2000, to give high sensitivity for longer range surveillance, low frequency radar detection and synthetic data display facilities. A third variant, SAPIENS 3, incorporates a jammer which additionally provides noise and deception jamming or a combination of both.

The system is another variant of the NEWSY equipment (**5784.253**).

CONTRACTORS
Thomson-CSF – Aerospace Group, 122 avenue du Général Leclerc, 92105 Boulogne-Billancourt, France.

Compagnie de Signaux et d'Entreprises Electriques, 17 place Etienne Pernet, 75738 Paris Cedex 15, France.

5813.253
ARBR-17 ESM SYSTEM
The ARBR-17 ESM suite is a high performance radar detection, analysis and threat warning system. The actual frequency coverage has not been disclosed although it is understood to operate over the conventional radar threat frequencies. The system analyses the measured radar parameters and uses these characteristics to determine other parameters such as the radar's type and identity.

The equipment uses two antenna assemblies, an omni-directional system installed on the ship's mainmast and two eight-port directional antennas mounted on the same mast but slightly below. The omni-directional antenna provides initial detection and threat frequency, and carries out primary analysis, while the directional units provide more accurate bearing and frequency information. The antenna assemblies contain preamplifiers for improved performance.

STATUS
The ARBR-17 has been selected by the French Navy for installation on its C70AA air defence frigates which are under construction. The first ARBR-17 detector was delivered to the French Navy in January 1988.

CONTRACTOR
Thomson-CSF – Aerospace Group, 122 avenue du Général Leclerc, 92105 Boulogne-Billancourt, France.

5814.293
ARBB 33 NAVAL JAMMER
The ARBB 33 is a high performance jammer conceived and manufactured for the French Navy, which follows the ARBB 32 family. It is designed for installation on board a variety of surface ships. It has a multiple threat capability and is able to counter up to four threats simultaneously.

The ARBB 33 is capable of jamming all modern threats, target designation radars as well as pulsed fire control radars and seeker heads. It offers a wide range of jamming modes; continuous noise, pulsed noise, cover pulse jamming, synchronous and asynchronous false echoes, and range gate pull-off.

The system can be alerted initially by a threat detection equipment, but also possesses an autonomous detection sub-assembly. It provides a 360° coverage in azimuth by means of electronically switched antennas for both detection and jamming, instead of the conventional directional assembly, in order to allow faster reaction time in any direction.

The system is divided into four elements:
(a) two canisters, port and starboard, each one comprising a transmitter, a receiver, and their electronically switched antennas
(b) a technical cabinet
(c) a control and display console.

Each cannister is divided into two sections, one containing the transmitter units and one housing the receivers. The system operates in H, I and J bands and covers two 180° sectors, 36 ten-degree lobes on each side give complete coverage and direction within 5°. Output power of the jammer is more than 100 kW and the system can handle two different threats simultaneously. Optionally this can be increased to four. A dish antenna is employed to give better coverage than the normal planar type.

CHARACTERISTICS
Frequency: H, I and J bands
Azimuth coverage: 2 sectors of 180°
Reaction time: less than 0.5 sec
Detection sensitivity: better than –50 dBm
Bearing accuracy: better than 5° RMS
Dimensions: Canisters – 1.2 × 2.2 m; cabinet – 1.8 × 0.6 × 0.78 m
Weights: Canisters – 500 kg; cabinet – 400 kg

STATUS
In production. The ARBB 33 has been ordered by the French Navy for its C700A air defence frigates and for several other ships. It has also been selected for the future nuclear aircraft carrier. The first system was delivered to the French Navy in March 1987 and has been fitted to the *Cassard* A/A frigate.

CONTRACTOR
Electronique Serge Dassault, 55 quai Marcel Dassault, 92214 Saint-Cloud, France.

6487.293

ESD LAD OFF-BOARD DECOY SYSTEM

ESD is developing an off-board expendable decoy system for use against anti-ship missiles fitted with electro-magnetic seekers. It is deployed from a rocket fired from the ship's chaff launcher and, after deployment, drifts down on a parachute while jamming the missile seeker. Diameter of the decoy is 127 mm to ensure compatability with all existing launchers and has a height of 900 mm. Because of its wide band design it is able to process all known or foreseeable types of electro-magnetic sensor. The equipment has been described under the title LAD (Leurre Actif Decale), although this is a common term for off-board decoy systems.

STATUS

Development against a requirement by the French Navy.

CONTRACTOR

Electronique Serge Dassault, 55 quai Marcel Dassault, 92214 Saint Cloud, France.

6486.293

DR2000 IFM ESM RECEIVER

The DR 2000 (Mk1 or Mk2)-DALIA is an EW system for use in seaborne, ground-based or airborne applications. Three versions are available, the DR 2000S for surface ship/ground station use, the DR 2000U for submarines and the DR 2000A for airborne fitment. In all cases the basic system consists of a radar intercept receiver (DR 2000) and an analyser (DALIA 1000 Mk1 or Mk2), plus the necessary antenna units.

The DR 2000 intercept receiver is built up of six DF antennas and one omni antenna, and a processing control/display console. It provides a virtual 100% probability of detection over the complete 360° of azimuth. The equipment carries out passive search, detection of all pulse and CW signals, and gives an instantaneous audible and visual alert. The DR 2000 Mk2 is an improved version with better sensitivity and coverage of the frequency bands from D to J and a DF accuracy of 5° RMS.

The DALIA analyser provides alarm analysis and identification facilities, through a library of 1000 radar modes and parameters, which is easily reprogrammable.

STATUS

In production

CONTRACTOR

Thomson-CSF – Aerospace Group, 122 avenue du Général Leclerc, 92105 Boulogne-Billancourt, France.

4351.053

DR4000 IFM ESM RECEIVER (TMV 201)

The DR4000 is a highly sensitive crystal IFM ESM receiver with frequency coverage from C- to J-band, display of graphic and alpha-numeric signals being provided on a three-colour CRT, together with associated data. Symbols, alpha-numeric data and interface comply with NTDS standards. The display shows the tactical situation in the upper part and intercept parameters below.

Very little information is available on this system other than that there are four variants – DR4000S, DR4000A and DR4000U for surface vessels, maritime patrol aircraft or helicopters, and submarines, and a DR4000H version for helicopters without specific display. The system can be interfaced with decoy dispensers or jammers and automatically controls their operation.

The antenna system of the DR4000S includes two DF arrays (2 × 6 DF channels), one omni antenna (frequency channel) and modules housing RF amplifiers. The installation can be at the masthead, around the mast or on both sides of the bridge or superstructure. The processing unit, either frequency band designation or instantaneous frequency measurement, has reprogramming capabilities with deinterleaving, analysis and identification.

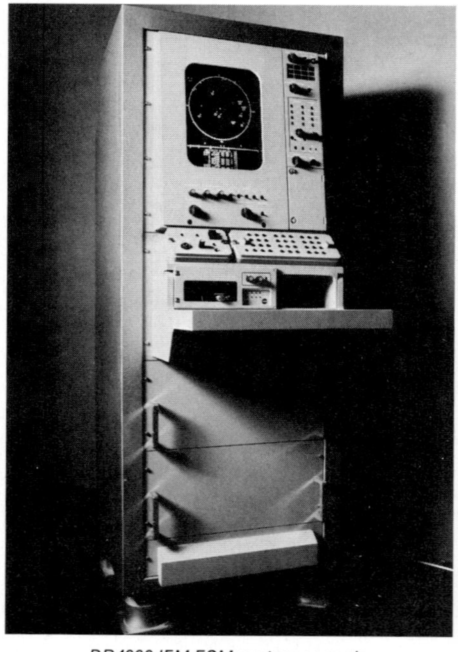

DR4000 IFM ESM system console

CHARACTERISTICS

Dimensions and weight

Omni antenna: 415 mm high × 300 mm diameter; 9 kg

DF antenna: 425 mm high × 480 mm diameter; 2 × 22 kg

IFM processor: 316 × 360 × 520 mm; 25 kg

Console: 590 × 1100 × 1543 mm; 310 kg

STATUS

In production. Fitted to the new FL 2000 frigates of the Saudi Arabian Navy. The first equipment has been installed and is on sea trials.

CONTRACTOR

Thomson-CSF – Aerospace Group, 122 avenue du Général Leclerc, 92105 Boulogne-Billancourt, France.

5785.293

JANET NAVAL JAMMER

Janet is a naval ECM system designed to jam locked-on missile radar seekers as well as surveillance radars operating in the H, I and J frequency bands. It operates against pulsed radars which may use fixed or staggered frequency, frequency agility, or pulse compression transmission modes, and can function as a noise jammer, a deception jammer or a two-mode jammer. Janet is of modular design and comprises up to four modules, each of which is a complete and autonomous jammer with its own two-horn receive antenna for tracking, radar detector, computer/signal generator, and two TWTs and directional transmit antennas. One TWT/directional antenna combination is used for noise jamming and the other for deception jamming. This enables the system to jam simultaneously two radar threats arriving from completely different directions. Each jammer module can also jam two signals if these are located close to each other.

The receiver is used for accurate target bearing and frequency measurement following an alert from an associated radar warning receiver. The computer

Janet naval jammer

controls the signal generator and selects the type of jamming used. This can be a variety of noise signals, including continuous or pulsed noise, as well as dual mode (noise/deception), complex deception and repeater signals. The system has been designed to store received target signals in order to re-create identical ones for deception purposes.

Since Janet is a processor controlled combined noise and deception jammer it provides 'smart' jamming to combat modern ECCM radars. With its modular concept of from one to four modules, it can be installed on many types of ships. Each module has an overall diameter of 1000 mm, a height 270 mm and weighs 250 kg.

STATUS

In production. Fitted to the new FL 2000 frigate of the Saudi Arabian Navy. The first equipment was delivered in 1985.

CONTRACTOR

Thomson-CSF – Aerospace Group, 122 avenue du Général Leclerc, 92105 Boulogne-Billancourt, France.

6488.293

LACROIX NAVAL COUNTERMEASURES

Lacroix manufactures a vast range of expendable munitions for off-board launchers and decoys as follows. These are used in launchers such as DAGAIE, MAGAIE and SAGAIE and consist of electro-magnetic and infra-red decoys both separately and combined. They are rocket propelled for close-in defence by distraction, seduction and concealment seduction against target acquisition and tracking systems used by missile seekers. The electromagnetic decoys are normally made of aluminised fibreglass chaff with a rapid bloom time, and cover the frequency bands I and J. The infra-red decoys consist of an infra-red composition covering the 3 to 5 and 8 to 14 μm bands.

Lacroix also manufactures pyro-acoustic ammunition for anti-submarine warfare. This consists of both bubbles cloud material and acoustic noise generators to provide an increase of ambient noise and mask the ship by false echoes, and active decoys to provide delayed false echoes with a controlled Doppler effect.

STATUS

In production.

CONTRACTOR

Etienne Lacroix, BP 213, 31601 Muret Cedex, France.

Lacroix naval countermeasures

GERMANY (FEDERAL REPUBLIC)

3643.293

BUCK-WEGMANN DECOY SYSTEM

The Buck-Wegmann Hot Dog/Silver Dog equipment is an IR/chaff dispenser decoy system for use on boats and ships as a countermeasure against anti-ship missiles equipped with active radar and/or IR seeker heads. The rapid deployment of IR and radar decoy materials from the vessel, and the good radar returns provided by the chaff cloud cause an attacking missile to lock on to this instead of the vessel itself, allowing the target to take evasive escape action.

The system essentially consists of the reloadable 12- or 24-tube launcher, the main and auxiliary control units, the emergency release, the IR decoy grenades (Hot Dog), and the chaff grenades (Silver Dog). Two or four carrier mountings with three launcher tubes each are deck-mounted on the port and starboard sides of the ships. The launcher tubes may be loaded with either IR decoys or chaff grenades only, or with a combination of both depending on the threat encountered. The grenades can be fired in single shots or in a series of shots. When using the latter mode whereby continuous firing is carried out automatically, the decoy cloud becomes effective after only two seconds, and is sustained over a considerable period. Small dimensions and low weight make the equipment suitable for use aboard vessels of all sizes down to small craft.

CHARACTERISTICS

Type: 76 mm decoy grenade launcher
Number of tubes: 12 or 24 per system
Weight: 115 or 185 kg (approx), complete system excluding grenades
Dimensions: 185 × 710 × 420 mm (w × l × h), 1 group of 3 launch tubes
IR decoy
Grenade type: Hot Dog 76 mm
Weight: 0.74 kg (approx)
Payload weight: 0.41 kg (approx)
Cloud development time: 2 s
Chaff decoy
Grenade type: Silver Dog 76 mm
Weight: 0.95 kg (approx)
Payload weight: 0.45 kg (approx)
Cloud development time: 2 s

Grenade launcher installation for Buck-Wegmann Hot Dog/Silver Dog naval decoy system

STATUS

Current production. In service with West German navy and other NATO navies.

CONTRACTORS

Wegmann & Co GmbH, 3500 Kassel, Federal Republic of Germany – Launch system.

Buck Chemisch-Technische Werke GmbH & Co, 8230 Bad Reichenhall, Federal Republic of Germany – Grenades.

5659.293

SCHALMEI COUNTERMEASURES DECOY SYSTEM

SCHALMEI is an IR/RF countermeasures system designed to provide self-protection for light and medium size naval units against shipborne, aircraft and missile radar and IR sensors by the use of rocket-borne decoy targets. The size and configuration of the targets can be selected either by manually initiated single firing or by an automatic self defence program. The program sequence is controlled by an electronic 'cubicle' which is interfaced with the fire control system. Typical decoy targets are obtainable in the RF range between G and K wavebands and in the IR range of the medium μm bands with maximum radiated power.

The system comprises:

(1) launcher containers and rockets
(2) a CIC control panel
(3) emergency control panel
(4) an electronic 'cubicle'
(5) test and simulation unit.

The rocket is a 70 mm calibre solid-fuel decoy device, supplied in a standard size of 560 mm long and a weight of approximately 2.5 kg, with an IR flare payload (high radiation power and long position hold by means of a parachute), and high performance RF chaff payload (aluminised fibreglass dipoles). An RF decoy rocket of 800 mm length with double chaff is also available. An alternative possibility is arming of the rocket with IR high performance emitter units in the appropriate frequency bands.

Up to a maximum of four launching containers can be installed on each side of the ship, with ten rockets in each container. In order to reduce the flight time the rockets are fired from individual closed launching tubes. Fully loaded weight of each container is 40 kg.

Operation and control is effected by means of the CIC control panel which can be installed in an existing control panel or in a separate console. The panel provides for a maximum of 80 rockets, and launching for all operational types can be manual, or automatic with the fire control system. Configuration of the selected program is controlled and displayed, together with the number of IR and RF decoy rockets ready for launching. The emergency control panel allows the automatic self-defence program to be initiated from the bridge.

STATUS

In production.

CONTRACTOR

AEG Aktiengesellschaft, Marinetechnik, Hafenstrasse 32, 2000 Wedel (Holstein), Federal Republic of Germany.

5820.253

FL1800S NAVAL ESM/ECM SYSTEM

The FL1800S has been designed as a shipborne ESM/ECM suite and is understood to cover the frequency range 7.5 to 17.5 GHz. It offers 360 degree cover against all threats operating over the range and it can be assumed that the system incorporates full threat detection, analysis and jamming capabilities, and can handle several threats simultaneously, even when these differ considerably in either frequency or direction of arrival. The antenna system consists of a DF antenna which is mounted on top of the mast, and five omni-directional antennas, each covering a different frequency band, mounted on a separate pole. Although it is likely that the system operation is primarily automatic a below-decks operator's console provides system control and display of information on a CRT screen. The console is intended to be operated by one person but can accommodate two operators if necessary. No other technical details have been released.

STATUS

Fitted to the Bremen class frigates of the Federal Germany Navy.

CONTRACTOR

AEG Aktiengesellschaft, Postfach 1730, 7900 Ulm, Federal Republic of Germany.

5821.253

FL400 SERIES EW SUITES

The FL400 series is a family of EW suites originally developed for naval applications but now being offered for both ground-based and airborne roles. The series is available in three different basic versions; FL400A ELINT system, FL400U ESM equipment and the FL400G jammer. All versions are capable of operating and handling complex transmission patterns in high density threat conditions.

The FL400A ELINT system covers the frequency range 0.5 to 18 GHz with an optional extension to 40 GHz. It will automatically monitor all radar signals within this band, analyse them to identify and determine their main parameters, and store the information for subsequent use. The equipment uses a superheterodyne receiver coupled to a high-gain reflector antenna and incorporates an omni-directional antenna linked to an IFM receiver and panoramic display to achieve a high probability of intercept. The FL400A can be used as an unmanned ELINT station, forming part of an intelligence gathering network, and controlled from a central station.

The FL400U ESM version also covers the frequency range 0.5 to 18 GHz, with an optional extension to 40 GHz, and is normally used in conjunction with the FL400G jammer. It carries out the duties of signal detection, analysis and identification. It incorporates a large threat library which enables the system to compare incoming signals with those in the computer memory to determine the identity. Any unknown signals are classified and the library is updated for later use. Identified and unknown threats are listed in a continuously updated order of priority and a threat warning is given automatically. When an ECM unit is associated, various ECM options are suggested on the display. Full 360 degree coverage is provided and a rotating antenna is used for accurate DF measurements. As with the FL400A, both a superheterodyne receiver, and an IFM receiver and panoramic display are incorporated.

The FL400G is the ECM jamming version and is intended to be used in association with the FL400U ESM equipment. Its normal frequency coverage is from 7.5 to 17.5 GHz. The FL400G provides full 360 degree coverge, and consists of an antenna assembly, a high power TWT amplifier, a signal modulator and a central processor. An effective radiated power of between 20 and 60 kW can be transmitted and a variety of jamming modes are available for selection by the operator. Both noise and deception jamming is available, either in single or combined modes, and false targets can be generated using analogue or digital signals. The signal modulator is a highly sophisticated unit which can provide frequency agility, scan fast modulation, sweep scan rate and sweep RF techniques. A 'look-through' capability is incorporated to enable update of threat frequency and bearing during jamming.

The FL400 series equipment can be housed in shelters for mobile ground-based, shipborne or airborne purposes. It can also be directly installed in ships, aircraft or fixed ground stations.

STATUS

Fully developed.

CONTRACTOR

AEG Aktiengeselleschaft, Postfach 1730, 7900 Ulm, Federal Republic of Germany.

INTERNATIONAL

6500.293

NATO SEA GNAT

The NATO Sea Gnat programme is intended to develop a standardised interoperable decoy system to provide savings in development, cost, procurement and logistics support. It is a collaborative project involving the USA, UK, West Germany, Norway and Denmark and is managed by the US Naval Electronics Systems Command. The programme is designed to provide electro-magnetic and infra-red decoys to defend NATO ships against guided missiles. The decoys will work in both the distraction and confusion modes and are intended for ships of frigate size and above. The decoys will be compatible with a modified Super Rapid Blooming Off-Board Chaff (SRBOC) launching system (**4656.293**) and should be usable by any ships carrying the SRBOC launcher. There are three variants of the decoy, all of which are specifically designed for compatability with existing launchers.

DEVELOPMENT

The NATO Sea Gnat programme was initiated in the mid-1970s and has now finished engineering development. Limited production began in 1986.

STATUS

A full-scale production decision was expected in 1987.

ISRAEL

4354.293

RAFAEL SHORT RANGE CHAFF ROCKET

This is a short range chaff rocket (SRCR) developed as a ship perimeter defence system to create decoys against missile attacks. Three rockets are encased together in a triple-barrel launcher serving as a loading unit and two sets are mounted on a standard 12.7 mm machine gun stand. This can be rotated to any desired bearing and is normally set to 45 degrees elevation. A remote firing box will operate two such launchers through a rotary selection switch. A solid, single stage rocket motor boosts the rocket into a preselected trajectory and the chaff dispersion takes place at a predetermined distance from the ship.

CHARACTERISTICS

Rocket

Weight: 3.2 kg

Length: 640 mm

Diameter: 90 mm

Chaff content: 1.75 kg

Launcher (three rockets)

Weight: 22 kg loaded

Length: 800 mm

STATUS

In operational service with the Israeli armed forces.

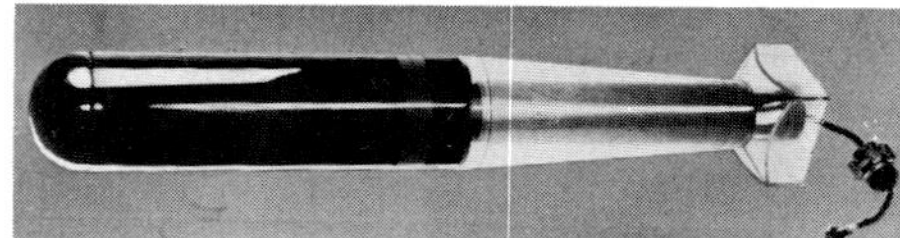

Rafael SRCR

CONTRACTOR

Rafael, Ministry of Defence, PO Box 2082, Haifa 31021, Israel.

4355.293

RAFAEL LONG RANGE CHAFF ROCKET

This is a long range chaff rocket (LRCR) system which serves as a naval search radar decoy. It is normally fired from launcher tubes mounted on the ship's superstructure. Loading is quick and simple and may be carried out by one man so that four tubes are considered sufficient for any one vessel. Rockets are fired from a remote control box with a launch tube selector switch. A two stage motor gives an initial boost thrust and a subsequent sustainer stage. Chaff dispersion is actuated by an altitude fuze which is energised from the launcher so that no internal power supplies are needed. A compact water-tight design ensures a long life and minimum maintenance.

CHARACTERISTICS
Rocket
Weight: 9.4 kg
Length: 922 mm
Diameter: 89 mm
Chaff content: 1.3 kg
Launcher (double barrelled)
Weight: 12.5 kg
Length: 1250 mm
Diameter: 216 mm

STATUS
In operational service with the Israeli armed forces.
CONTRACTOR
Rafael, Ministry of Defence, PO Box 2082, Haifa 31021, Israel.

Rafael LRCR

6489.293
BEAMTRAP CLOSE DEFENCE ROCKET

Beamtrap is a close defence naval chaff decoy system, combat proven by the Israeli Navy. It deceives radar homing sea-to-sea missiles by placing a decoy target at close range. Each launching unit consists of six lightweight launching tubes. In the standard configuration units are installed on each side of the vessel's superstructure. The launching tubes are installed at a fixed elevation, so optimal distance and elevation are obtained.

CHARACTERISTICS
Rocket
Dimensions: 804 mm long × 115 mm diameter
Weight: 6.6 kg
Chaff content: 3.75 kg
Launcher (6 tubes)
Dimensions: 1020 mm long × 270 mm deep × 365 mm wide
Weight: 13 kg
STATUS
In operational service with the Israeli Navy.

Beamtrap close defence rocket

CONTRACTOR
Rafael, Ministry of Defence, PO Box 2082, Haifa 31021, Israel.

5660.293
AUTOMATIC COUNTERMEASURES DISPENSING SYSTEM

The Automatic Countermeasures Dispensing System (ACDS) has been designed for shipboard anti-missile application to control the launch of dispensable electronic countermeasures such as chaff and flares. The system operates automatically thereby reducing reaction time to a minimum, optimising response and providing effective anti-missile protection. It consists of a main unit, distribution units, a remote control unit, and fire control units each of which can control 24 rockets.

Operating in full automatic mode, with manual back-up, the ACDS selects the countermeasures rockets appropriate to the threat definition, range and direction and launches them with the correct priority and timing. The system will adapt to existing rocket decoy launch systems with suitable interfacing, and will adapt to any tactical arena by software adjustment.

STATUS
In production.
CONTRACTOR
Elbit Computers Ltd, Military Systems and Products Division, Advanced Technology Center, PO Box 5390, Haifa 31053, Israel.

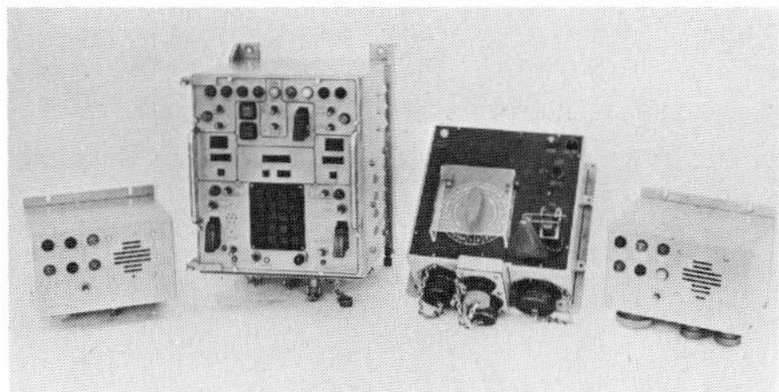

Automatic countermeasures dispensing system

3461.093
ELBIT MN-53 ESM SYSTEM

The Elbit MN-53 is an ESM system for interception and identification of radar emitters. Modular in design, it can be configured to meet a variety of operational requirements. In one configuration, it uses three different functional receivers. The system provides high sensitivity and detection for high intercept probability.

The MN-53 system detects, displays and records radar emitters. These data, when cross-referenced with Elint information, facilitate accurate and quick tactical decision making. The system is readily adaptable to various military environments. Main operational features are:

(1) manual/automatic DF
(2) manual/automatic frequency search
(3) manual/automatic analysis on PRF, PW, PRF wobbulation
(4) automatic printing and recording of RF, PW, PRF, PRF wobbulation, TOA and direction
(5) connection to external computer.

The system incorporates in its design the extensive experience and lessons gained in sea combat during the Yom Kippur war. Human engineering principles applied to the design of consoles, displays and controls allow for simple and comfortable system operation, even under stress conditions.

System components are slide mounted for ease of maintenance and accessibility. The system includes built-in test facilities for fault-detection.

CONTRACTOR
Elbit Computers Ltd, Military Systems and Products Division, Advanced Technology Center, PO Box 5390, Haifa 31053, Israel.

5841.253
ELTA NAVAL EW SYSTEMS

For naval applications Elta Electronics has developed equipment for ESM and ECM, with chaff dispensing. Typical suites cover a wide spectrum of frequencies and are designed to counter modern naval weapon systems. The ESM is performed by the EL/L-8300 line of modular IFM receivers and DF antennas. ECM is carried out by the EL/L-8200 family of noise jammers, the EB series of deception systems and the naval version of the EL/L-7010 series systems (see entry **4225.093**).

EB41 Naval ECM System

The EB41 is designed for protection of ships of all sizes against radar controlled weapon systems. The system features a wide frequency range, automatic or manual operating modes, a high gain antenna and selectable ECM techniques.

EL/L-8303 ESM System

This is a computerised ESM system for shipborne, ground-based and airborne applications. The system is designed to intercept, digitise, identify, display and record radar signals over the frequency range 2 to 18 GHz, with an optional extension down to 0.5 GHz. The processed data of the system can activate chaff or ECM systems, either automatically or manually.

EL/L-8312 ELINT/ESM system

The system provides computer control and processing, instantaneous direction finding over 360 degrees with high accuracy and high intercept probability. A combined alpha-numeric and graphic display, and a built-in emitter library for storage of data, are provided. A range of antennas is available, including single or split DF and omni arrays. Frequency resolution of ± 1 MHz is available with a DF accuracy of 5° (3° optional).

EL/L-8310 ELINT/ESM System

The EL/L-8310 is an advanced microwave superheterodyne receiving system designed for all ELINT/ESM applications. The normal frequency range is from 0.5 to 18 GHz, although this range can be extended at both ends to cover from 60 MHz to 40 GHz. It is one-man operated and capable of shipborne, airborne, fixed or transportable deployment.

EL/L-8312 ELINT/ESM System

This is a fully computerised microwave receiving system based on Elta units such as the EL/L-8310, EL/L-8320 pulse digitiser, EL/S-8610 computer and EL/S-8570 display. It provides signal intercept, analysis, display, emitter classification and identification with a high probability of intercept. Frequency coverage is the same as the EL/L-8310.

STATUS

In production.

CONTRACTOR

Elta Electronic Industries (a subsidiary of Israel Aircraft Industries Ltd), Ashdod, Israel.

6490.253

NS-9001 ELINT SYSTEM

The NS-9001 is a naval ELINT system which provides automatic analysis and identification of radar emitters (including frequency hopping, and staggered or wobbulated PRI) in the 0.5 to 18 GHz frequency band. The system automatically and instantaneously receives all signals in the band and measures and analyses their parameters. Bearing information is provided instantaneously with high accuracy. The analysed information, reflecting the environment situation is presented on a frequency versus azimuth display. ELINT features include magnetic data recording and playback of the entire environment or of any specific emitter, RF intrapulse analyse, time domain video display, automatic emitter triangulation, and several display modes.

The system consists of a mast assembly, receiving and processing equipment installed in a standard 19-inch rack, and an operator console. The mast assembly comprises the DF antenna array with its associated equipment, and an omni-directional antenna. Optional equipment such as a superhet swept receiver, a magnetic instrumentation recorder and a colour display may be integrated into the system.

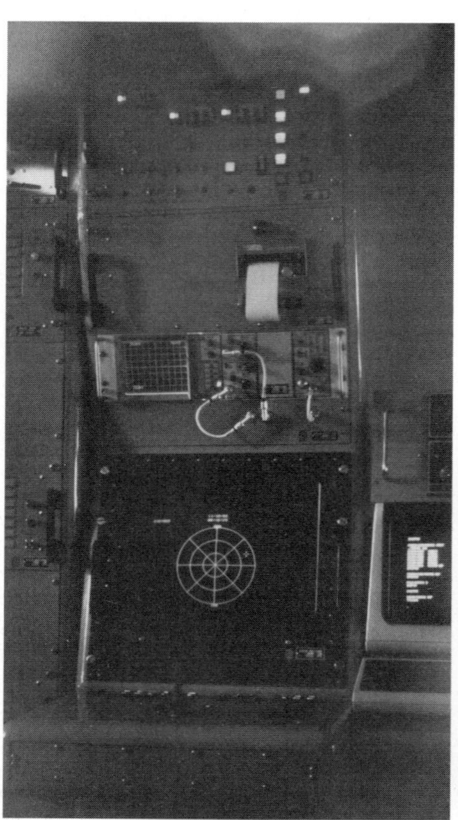

NS-9001 ELINT system

CHARACTERISTICS

Frequency coverage: 0.5 to 18 GHz

DF accuracy: 3° S and C bands; 10° X and Ku bands

Frequency measurement: 2 MHz resolution

Operation modes: IDF, MDF, SDF

Sensitivity: –70 dBm (omni); –80 dBm (SDF)

STATUS

In operational service.

CONTRACTOR

Elisra Electronic Systems (a subsidiary of Tadiran), 48 Mivtza Kadesh Street, Bene Beraq, 51203 Israel.

6491.253

NS-9003 NAVAL ESM SYSTEM

The NS-9003 naval ESM system operates in a multi-radar environment to carry out electronic intelligence, and is able to cope with frequency agility, frequency hopping, and staggered or wobbulated PRI emitters. The system automatically receives, analyses and identifies over-the-horizon radar signals with a 100% probability of intercept over the frequency range from 2 to 18 GHz. The direction of arrival is instantaneously measured per pulse, using an accurate multibeam antenna array. This instantaneous direction finding enables the system to locate any 'exotic' type radar signal in the reception range within a fraction of a second.

The analysed information is presented on a 17-inch graphic situation display in several operater-selected modes for various EW tasks. The system has built-in power management provisions for ECM and chaff/flare activation.

The NS-9003 antenna assembly includes a multi-beam static antenna array with associated RF video and digital DF processing hardware, and an omnidirectional antenna array. Other units in the system are receiving and processing equipment, and an operator console. Interface options include up to six radar blankers, gyro equipment, a weapon system, an ECM jamming and deception system, chaff/flare equipment and a satellite navigation system. An integrated EW suite consisting of the NS-9003 and the NS-9005 jamming and deception system (**6492.253**) is available.

NS-9003 Naval ESM System

CHARACTERISTICS

Frequency range: 2 to 18 GHz

DF accuracy: 1° X and Ku bands; 3° S and C bands

Sensitivity: –70 dBm

Frequency measurement: Instantaneous 2 MHz resolution

STATUS

In service.

CONTRACTOR

Elisra Electronic Systems (a subsidiary of Tadiran), 48 Mintza Kadesh Street, Bene Beraq, 51203 Israel.

6492.253

NS-9005 ECM JAMMER

The NS-9005 is a compact, wideband naval jamming system covering the frequency range 7.5 to 18 GHz, designed to provide self-protection for various types of vessels. The system includes both noise and deception techniques, and is controlled by an advanced digital technique generator. The NS-9005 is intended to operate in conjunction with an automatic, computerised ESM system, and can cope with four simultaneous threats, each with its own optimised ECM programme.

The system can generate spot, barrage or swept noise along with RGPO, RGPI, AGPO deception signals, and can recognise false targets. It uses a 100% duty cycle power amplifier, and a steerable antenna for transmission of the jamming signals. If the foreseen threat environment requires greater probability of intercept, higher detection speeds and a high average ERP transmit capability, the NS-9005 can be integrated with the Elisra innovative Multi-Beam Array Transmitter (MBAT). An integrated EW suite consisting of the NS-9005 and the NS-9003 ESM system (**6491.253**) is available.

CHARACTERISTICS

In service.

CONTRACTOR

Elisra Electronic Systems (a subsidiary of Tadiran), 48 Mivtza Kadesh Street, Bene Beraq, 51203 Israel.

6493.253

NS-9009 NAVAL EW SYSTEM

The NS-9009 is a compact shipborne EW self-defence system designed for small strike craft with limited crew. It has a broadband, sensitive receiver with instantaneous 360° azimuth coverage for the detection of emitters in the frequency range 2 to 18 GHz. Its built-in digital signal processor is capable of analysing and identifying radar signals in a very dense environment with 100% probability of intercept. The operator is presented with a polar display and an alphanumeric parameter table. Other signal parameters can be displayed on request.

The NS-9009 system consists of only two units; an antenna assembly and receivers on the mast, and a processing/display unit below decks. Features of the system include immediate activation of visual and audio alarms upon threat identification, X-Y situation display for order of battle presentation, diagnostic and built-in self test, and simple installation. Chaff dispensing systems can be automatically deployed, enabling the ship to take evasive measures against fire control radars and homing missiles. Interface with other on-board systems includes gyro equipment and radar blankers. The system can be upgraded to include higher sensitivity and accurate instantaneous frequency measurement.

CHARACTERISTICS

Frequency: 2 to 18 GHz

DF accuracy: 8° RMS

Sensitivity: –35 dBm

STATUS

In service.

CONTRACTOR

Elisra Electronic Systems (a subsidiary of Tadiran), 48 Mivtza Kadesh Street, Bene Beraq, 51203 Israel.

6494.253

ND-9010 NAVAL ESM SYSTEM

The NS-9010 is a compact version of the NS-9003 (**6491.253**). It incorporates the sophisticated features of that system, and can be easily installed in a variety of naval vessels, such as fast patrol boats, frigates, and destroyers. The system automatically receives, analyses and identifies over-the-horizon radar signals with a high probability of intercept. Bearing information is gathered by a multibeam static antenna array, using monopulse switching techniques. The analysed information is presented on a 17-inch graphic situation display.

The NS-9010 consists of two main functional and installation groups:

(a) A multibeam static antenna array with associated RF, video and digital processing hardware

(b) An operator console giving control and display functions.

The system has a built-in provision for integration with almost any ECM and chaff/flare equipment which may be operated through automatic, semi-automatic or manual power management.

CHARACTERISTICS

Frequency: 1 to 18 GHz (0.5 to 18 GHz optional)

DF accuracy: 3° X and Ku bands; 5° S and C bands

Operating modes: SDF, MDF, IDF (optional)

Sensitivity: –70 dBm

CHARACTERISTICS

In service.

CONTRACTOR

Elisra Electronic Systems (a subsidiary of Tadiran), 48 Mivtza Kadesh Street, Bene Beraq, 51203 Israel.

6881.463

NS-9034 SUBMARINE ESM/ELINT SYSTEM

The NS-9034 is a compact and sophisticated ESM/ELINT radar warning system for submarines which features instantaneous high sensitivity frequency measurement combined with accurate direction finding. It will carry out accurate and rapid detection and analysis in dense electro-magnetic environments. Targeting is also enhanced by this equipment, using an advanced multi-beam static lens antenna array which enables high bearing accuracy of 1°.

The NS-9034 can be installed in new submarines or retrofitted to existing vessels.

CONTRACTOR

Elisra Electronic Systems (a subsidiary of Tadiran), 48 Mivtza Kadesh Street, Bene Beraq, 51203 Israel.

6208.253

SHIPBOARD ELECTRONIC WARFARE SUITE (SEWS)

SEWS is a fully integrated, modular and computerised shipborne electronic warfare system which combines ESM techniques for the detection, analysis and identification of threat emissions with ECM measures to decoy and/or jam. The ESM subsystem is based on high sensitivity, wide dynamic range and accurate IFM and IDF receivers covering the frequency band of 1 to 18 GHz and a real-time processing system that analyses the information and identifies the threats by comparing the parameters to a stored emitter library. The receiver data includes frequency, bearings, amplitude, time of arrival, pulse width, CW, scan rate and pulse coding. Analysed threat data is displayed superimposed on a map of the area covered, together with true and relative bearing.

Following identification, the ECM subsystem will respond automatically according to the threat or as commanded by the operator. For jamming purposes, a fast set-on frequency transponder and a pulse repeater are used to drive Multi-Beam Array Transmitters (MBAT) with high ERP and fast beam steering. Optionally the Rattler jammer (**6207.153**) can be used. For deception, short and long range chaff rocket systems described under entries **4354.293** and **4355.293** can be activated.

STATUS

In operational service with the Israeli Navy.

CONTRACTOR

Rafael, Ministry of Defence, PO Box 2082, Haifa 31021, Israel.

ITALY

4169.293

ELT/211 NAVAL ESM EQUIPMENT

The ELT/211 equipment was designed to fulfil the operational ESM requirements of naval vessels ranging in size from fast patrol boat to destroyer. It is a highly integrated computer-based digital system employing sophisticated software-controlled de-interleaving and data handling techniques. The processor is of Elettronica design and is capable of effective operation in dense electromagnetic (EM) environments. Digital processing also permits maximum automation of the analysis, identification and tracking processes, enabling single-operator facilities with adequate surveillance and supervision capabilities combined with override options. Designed as a self-contained system, the ELT/211 can also be integrated with active ECM elements such as noise or deception jammers and it forms the ESM element of the Elettronica Newton EW system (**4172.293**).

The above-deck equipment consists of a DF antenna unit, service module, and four omni-directional antennas and RF unit. Masthead units weigh about 60 kg. Below deck is the operator console which incorporates the digital processor and display unit.

Surveillance facilities include:

(1) instantaneous automatic warning of pre-programmed threats

(2) continuous azimuth and frequency surveillance of the EM environment with 100% intercept probability even in very dense environments

(3) automatic tracking of selected emitters

(4) tactical evaluation of the radar scenario by means of threat identification symbols.

The associated analysis functions provide the following additional facilities:

(5) automatic measurement and alpha-numeric presentation of a detected emitter's characteristic parameters

(6) recognition of emission modes

(7) automatic emitter identification

(8) automatic indication of appropriate ECM reaction.

The ELT/211 can be interfaced with the ship's command and control system, FCS, or with a larger EW system, as mentioned above.

STATUS

In production for several navies.

CONTRACTOR

Elettronica SpA, Via Tiburtina Valeria Km 13,700, 00131 Rome, Italy.

4170.293

ELT/318 NAVAL JAMMER

The ELT/318 is a naval noise jamming equipment, generally associated with the Elettronica ELT/211 ESM equipment (**4169.293**) and with that equipment forming the major part of the Newton naval EW system (**4172.293**). The ELT/318 is a dual-band continuous wave noise jammer which employs advanced power management techniques to provide protection within the recognised radar threat bands. Among its facilities are: the ability to inhibit and/or delay weapon system acquisition; jamming of all types of search radar; variable modulation techniques; fast reaction and flexibility to counter new types of threat; full threat frequency band coverage.

The transmitter final output stage incorporates a travelling wave tube amplifier of Elettronica design and manufacture. It is normally employed with one or two ELT/828 stabilised transmit/receive antenna assemblies, although the ELT/318 equipment is suitable for use with medium or high gain antenna sets, depending on the type of ship. The ELT/828 antenna equipment is steerable and this model is generally employed in Newton systems.

STATUS

In production for several navies.

CONTRACTOR

Elettronica SpA, Via Tiburtina Valeria Km 13,700, 00131 Rome, Italy.

4171.293

ELT/521 NAVAL DECEPTION JAMMER

The Elettronica ELT/521 is a naval deception jammer using advanced techniques to ensure the self-protection of naval platforms against all types of radar-supported threat. The transmitter's high peak power, together with the possibility of using omni-directional and/or directional antennas, makes ELT/521 suitable for the protection of ships of any size and tactical role. When associated with omni-directional antennas, the ELT/521 provides completely autonomous operation, while bearing data is required from a DF receiver when operating with the higher gain directional antennas (eg ELT/828). The ELT/521 deception jammer can be associated with virtually any ESM receiving equipment; when integrated with the Elettronica computerised ELT/211 ESM receiving system (**4169.293**) it constitutes a complete electronic warfare system combining the radar detection, threat warning and tracking facilities of the latter with the fast reaction capabilities of the ELT/521. It can also be incorporated in the Newton naval EW system (**4172.293**).

The ELT/521 equipment comprises the following modules: an air- or water-cooled transceiver, and a combined circulator and self-test unit. Various deception techniques can be employed and there are programmable deception routines. Multiple target engagement is possible and there is full threat frequency band coverage, providing protection from all types of fire control radars and active and semi-active homing missiles.

STATUS

In production for several navies.

CONTRACTOR

Elettronica SpA, Via Tiburtina Valeria Km 13,700, 00131 Rome, Italy.

4172.293

NEWTON NAVAL EW EQUIPMENT

Newton is a family of naval ESM/ECM systems designed by Elettronica to meet the operational requirements of naval forces operating in a dense electromagnetic environment. The systems are configured to suit vessels of various sizes from 100 tons upwards. They provide facilities for complete passive surveillance, DF, analysis and identification of the radar emitters in the operational area, and electronic means of protection against modern anti-ship weapons.

The Newton system consists of an ESM section and an ECM section, with the related antennas, controlled from an operator console. The former section is designed to provide 100 per cent intercept probability, passive surveillance, automatic instantaneous warning of pre-programmed threat, automatic analysis of threat parameters, automatic identification and tracking of threats, a synthetic data display, instantaneous DF, and broad frequency coverage. The ECM section provides self-protection by noise and/or deception jamming.

The modular concept is centred on the ELT/211 ESM receiver equipment (**4169.293**) and on several ECM peripheral equipments, in particular the ELT/318 noise jammer (**4170.293**) and/or the ELT/521 deception jammer (**4171.293**) while interface channels are provided to integrate with the ship's command and control as well as other active reaction elements (chaff rocket launchers, FCS etc). The detailed configuration of individual Newton systems is adapted to suit specific ships. Advanced signal processing and handling techniques and software controlled functions, ensure the flexibility necessary to counter the anticipated threat evolution over the next decade. Because of a high level of automation, single-operator control of all ESM and ECM functions from one console is possible, and integration with command system is easily available via computer channels.

The range of Newton configuration possibilities is shown in tabular form. The principal modular elements of Newton systems are shown in the diagrams while additional details of these will be found under the entry numbers mentioned above.

Typical Newton system configurations

	Size of craft						
	100–500 tons		250–1500 tons	1000–3000 tons		2500+ tons	
ESM(ELT/211)*	1	1	1	1	1	1	1
ECM (ELT/318) noise		1	1	1		1	2
ECM (ELT/521) deception	1		1	2	2	3	2
ECM Antennas (ELT/814)	1		1				
ECM Antennas (ELT/828)		1	1	3	2	4	4

*ELT/211 interfaces up to 4 ECMs

STATUS

In production and in service with several navies.

CONTRACTOR

Elettronica SpA, Via Tiburtina Valeria Km 13,700, 00131 Rome, Italy.

Newton integrated naval ESM/ECM system antenna

Newton operator display/control console

5661.253

FARAD NAVAL EW SYSTEM

Designed for corvettes, fast patrol boats, fast attack craft and other small naval ships, FARAD is a complete EW system which can provide passive detection and identification of targets coupled with active self-protection against radar guided weapons.

FARAD is based on four principal units which can be integrated to provide a flexible system adaptable to operational requirements. The principal units are the ELT/123 basic ESM equipment, the ELT/261 analysis module, the ELT/521 deception jammer and the ELT/361 noise jammer with its associated ELT/828 antenna system. The ELT/123 is the fundamental unit. Other units are added as required according to the type of vessel and its operational role.

The ELT/123 supplies radar warning and DF functions to give instantaneous DF of pulse and CW emissions with automatic warning of pre-programmed threat emissions on a CRT display. Information can then be passed to the ELT/261 for automatic analysis of intercept signals and automatic tracking, with continuous surveillance of the radar environment on a panoramic display. The ELT/521 is a high-power deception jammer and is described under entry number **4171.293**. The ELT/361 is a dual-band CW noise jammer with spot and barrage jamming capability. It is driven automatically through the FARAD ESM section (ELT/123 plus ELT/261).

The complete FARAD system is able to operate in a very dense electromagnetic environment and can cope with a large number of simultaneous threats whether by search and acquisition radar or radar homing missiles. In an operational installation, several combinations of the various units described above can be fitted with the ELT/123 either as a stand-alone unit or as the basis upon which the other installations are built.

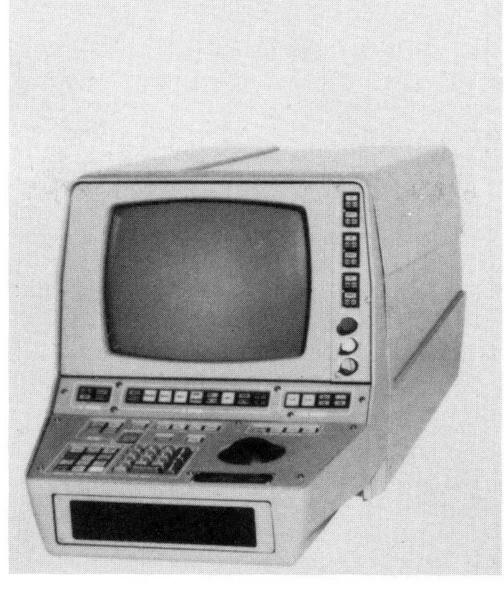

Units of FARAD ESM/ECM system

STATUS

In production and operational service.

CONTRACTOR

Elettronica SpA, Via Tiburtina Valeria Km 13,700, 00131 Rome, Italy.

5662.453

THETIS SUBMARINE ESM SYSTEM

THETIS is a modular ESM system designed to fulfil the EW requirements of a submarine. The system is produced in two configurations. The first and simpler of these, the ELT/124-S, performs high sensitivity instantaneous threat warning and DF functions, while the second, the ELT/224-S, extends these functions to those of a full ESM system or tactical ELINT. Both variants use the same DF antenna as the sensor and both can be fitted with an additional and separate RF pre-amplified omni-directional antenna as an option. This latter antenna is very small and can be installed on top of the submarine's attack or search periscope for primary warning.

The ELT/124-S variant consists of a threat warning and DF receiver system. The hardware comprises a DF antenna and receiver, a processor unit and a warning display. It provides instantaneous, high sensitivity, automatic warning of threats programmed into the processor and continuous surveillance of the radar environment with 100 per cent intercept probability. A dedicated operator is not required.

The ELT/224-S consists of the units of the ELT/124-S plus an IFM receiver and an ESM display. It provides instantaneous DF with 100 per cent intercept probability of pulse and CW emissions, display of all intercepted emissions, IFM, automatic technical analysis and identification of emitters and their platforms. It will also provide warning that the vessel is entering an area under hostile radar surveillance. A single operator is required for this system.

The antenna assembly consists of one conical spiral omni-directional antenna plus eight plane spiral DF antennas assembled in single casing. It has a very low radar cross-section and is also covered with microwave absorbing material. It is light and strong enough to be mounted on the search periscope. The small, lightweight optional antenna can be mounted on the attack periscope and will provide the submarine commander with a first alarm facility but without DF facilities.

THETIS submarine ESM system (display/control console and two antennas)

STATUS

In production and operational service.

CONTRACTOR

Elettronica SpA, Via Tiburtina Valeria Km 13,700, 00131 Rome, Italy.

4173.093

ELETTRONICA EW EQUIPMENT

In addition to those items of electronic warfare equipment described in separate entries in this section, Elettronica produces a comprehensive range of hardware.

A great deal of the detailed information about these equipments and others which have not been identified remains classified, but the following items have been partially analysed and the information concerning them that can be published appears in the following paragraphs.

A typical ship's EW suite might comprise:

(1) ELT/116 radar intercept receiver
(2) ELT/711 radar identification unit
(3) ELT/712 programming unit
(4) ELT/716 data transmission module
(5) ELT/311/511 jammer.

The complete system covers the frequency range 2 to 18 GHz and is based on a four-band instantaneous frequency measuring (IFM) receiver which performs all ESM functions. The four channel outputs are presented to the EW operator on two surveillance CRTs, and it is possible to discriminate between pulse and CW signals. Separate displays are provided for DF and signal analysis and measurements, these two CRTs being positioned below the surveillance CRTs on the EW operator's console; an additional CRT for dealing with the D-band (former L-band) signals is mounted above, if fitted. The former displays are part of the ELT/711 radar identification unit, and the latter is the ELT/214 display unit. There are also threat warning indicator lamps for drawing the operator's attention to about 30 intercepted emissions. These

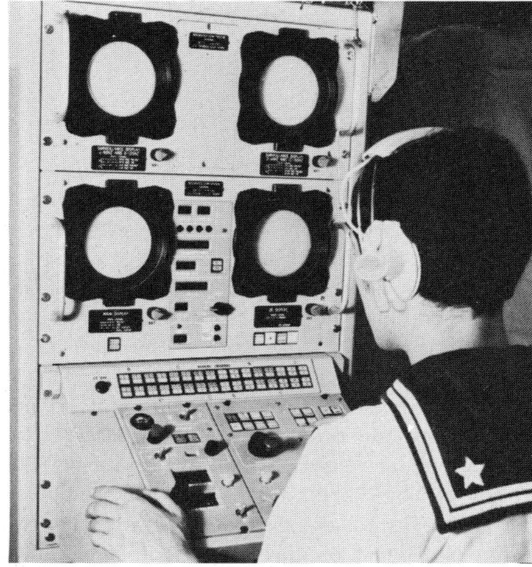

Operator's console of Elettronica 'Gamma' naval EW suite. Two upper CRTs are for surveillance, left lower CRT is main display and lower right CRT is for DF

Elettronica EW antennas aboard an Italian Navy ship; note stabilisation on 2 larger units. Planar array is AN/SPA-72 for SPS-52 3-D search radar

may be selected for individual analysis etc by operation of push-buttons associated with the warning indicators.

The ELT/116 radar intercept receiver system includes omni-directional and DF antennas, an auxiliary DF unit, RF unit and the display and console unit at the EW operator's position. The IFM receiver operates with a crystal video DF receiver in all bands. Frequency measurement accuracy is 0.2%, with a dynamic range of about –70 dBm to zero. Complete digital messages are transmitted to the ship's CIC aboard larger vessels, data transmitted including such items as platform identification, emitter characteristics, lock-on/search warnings and confidence levels etc.

The ELT/311/511 jammers can be used in conjunction with the above passive EW elements to provide a complete EW facility, and jammers can be under operator control or by CIC direction. A wide variety of jamming techniques is employed, including: noise and deception, both automatic and semi-automatic. Range blanking with spot noise or amplitude modulation spot jamming is normally used against search radars, tracking systems and either active or lobe-on-receive-only (LORO) conical scan radars. Angle jamming with spot noise is used against sequential lobing and track-while-scan (TWS) radars. Look-through and time-gated circuits are provided to control jammer operation in conjunction with the system receivers. These circuits are generally employed when the system is required to operate against multiple threats.

STATUS

In production and in operational use.

CONTRACTOR

Elettronica SpA, Via Tiburtina Valeria Km 13,700, 00131 Rome, Italy.

4644.293

CO-NEWS NAVAL ESM SYSTEM

CO-NEWS (Communications Naval Electronic Warfare System) is a naval ESM system which provides full monitoring and analysis of communications in the VHF/UHF frequency band. The system is completely automated to achieve very high intercept probability and to handle high density traffic. It comprises a lightweight antenna assembly, normally mounted at masthead, consisting of three co-planar dipole arrays and a switching matrix, and an operator console located below decks. The equipment is modular in design to cater for expansion and for additional units.

CO-NEWS automatically carries out continuous surveillance of the communications band, technical analysis and bearing of the intercepted emissions and real-time data processing and presentation. The intercepted signals can also be demodulated for monitoring purposes.

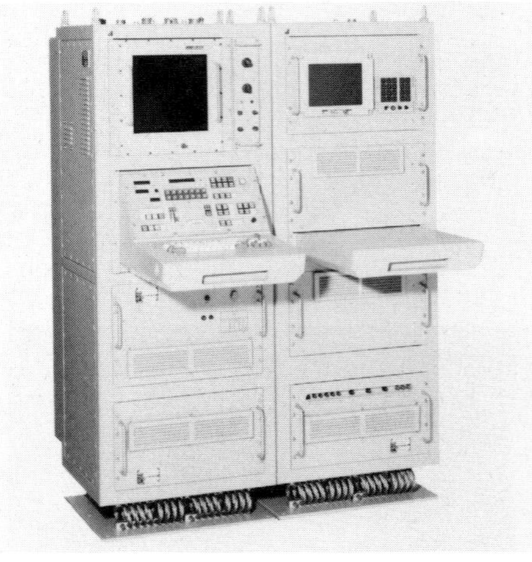

CO-NEWS below-decks equipment console

CO-NEWS antenna assembly

The operator console houses the system modules including:

(1) the surveillance receiver which scans automatically through the frequency bands at a velocity sufficiently high to ensure 100 per cent intercept probability
(2) a direction finder that takes the bearing and performs the technical analysis of intercepted emissions and also serves to monitor these signals
(3) a computer which manages the sensors, and processes and presents the data
(4) displays for the presentation of the measured data, and the operator's command keyboard. Two different displays are provided; a panoramic display dedicated to a synthetic presentation of the search results, and a plasma display for data and situation evaluation.

The system operates automatically under the supervision of an operator who can also over-ride the system's automatic operation to solve ambiguous situations or to monitor the emissions.

STATUS

In operational service.

CONTRACTOR

Elettronica SpA, Via Tiburtina Valeria, Km 13,700, 00131 Rome, Italy.

6882.253

NETTUNO NAVAL EW SYSTEM

Nettuno is a new generation EW system designed for major warships (aircraft carriers, destroyers and large frigates), although scaled-down derivatives for smaller vessels are foreseen.

Basic Nettuno provides integrated radar ESM/ECM, while an independent ESM section with a dedicated operator console is available as an option. The ELINT function does not interfere with the main system tasks.

The system provides 100 per cent intercept of emitters with passive surveillance based on environmental synthesis. A very high level of automation ensures immediate and automatic identification based on a powerful threat library, and accurate tracking of emitters. Real-time pre-processing is achieved by a multi-micro pre-processor using well proven special algorithms that have been extensively tested through computer simulations and sea trials. Processing is computer controlled and employs structured modular software using a high level language. Full interactive integration with the command and control, and flare-chaff launcher systems is provided.

All functions can be controlled by a single operator from a multi-mode synthetic display console with enhanced man-machine interface. Roll-stabilised multi-beam DF antennas provide exceptionally high bearing accuracy and feature very high sensitivity, wide dynamic range logarithmic amplifiers and completely digital processing.

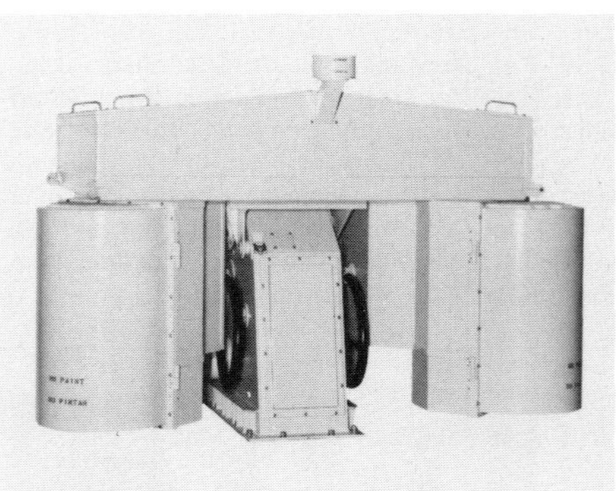

Nettuno transmitter antennas

The ECM section provides self-protection by environment adaptive instant-by-instant automatic jamming, even against simultaneous multiple threats. However, command and control, or the operator, can override the system if necessary. The high power jammer antennas are roll-stabilised and consist of linear arrays that electronically steer a precise and narrow high gain beam. Jamming implementation follows automatic instructions provided by the ESM section and by the power management computer. High ERP is achieved by using a transmitter design employing paralleled matched high power Elettronica-designed TWTs.

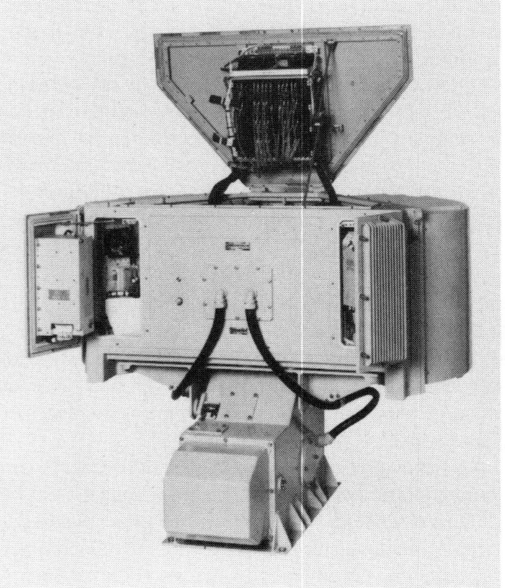

Nettuno multi-beam DF antennas

STATUS

In production and in service.

CONTRACTOR

Elettronica SpA, Via Tiburtina Valeria Km 13,700, 00131 Rome, Italy.

3343.253

RQN-1 RADAR INTERCEPT EQUIPMENT

The RQN-1 radar intercept equipment is a wide-band radar receiver which provides DF, frequency band measurement, video analysis of pulse width and pulse repetition frequency. Being wide open in frequency, it provides high probability of detection of every radar signal within its sensitivity range, which extends well beyond the radar horizon. Its DF accuracy, together with a number of additional features, like pulse length and PRF measurement and frequency band determination, lend themselves to use in both limited ELINT (electronic intelligence) missions and as a quick reaction device for self and mutual protection.

STATUS

In service but no longer in production. More detailed information is given in previous editions of *Jane's Weapon Systems*.

CONTRACTOR

Selenia Industrie Elettroniche Associate SpA, Special Equipment and Systems Division, Via dei Castelli Romani 2, 00040 Pomezia, Rome, Italy.

3528.253

RQN-5 NAVAL ESM/ELINT SYSTEM

The ESM/ELINT integrated naval system, RQN-5, has been developed by Selenia with almost complete system automation which allows a single operator to perform easily the ESM and ELINT functions. The system is characterised by a number of advanced features:

(a) high probability of detection, complete real-time automatic analysis and tracking of ship, aircraft, helicopter navigation and fire control radars, without the need of any pre-loaded information, as well as identification in a completely automatic mode of operation over a wide frequency band and in a very dense electro-magnetic environment
(b) automatic warning capability for threat emitters whose parameters may be stored easily in a dedicated memory, and for any locked-on signals, even of an unknown type
(c) processing and analysis of the radar signals, and storing the analysis results for post-mission data collection
(d) consistent range advance factor versus all the radars present in the scenario
(e) ability to drive automatically active ECM, eg the Selenia TQN family designed to operate against search, tracking and homing radars through sophisticated noise jamming and deception techniques
(f) ability to interface a passive ECM system to initiate chaff and/or infra-red decoys launch
(g) availability of several alternative standard interfaces for integration with a very large variety of peripherals, sensors, weapon systems and command and control systems
(h) the high system modularity enables it to be changed or upgraded at any time by addition or replacement of modules.

The RQN-5 has a complete commonality with the Selenia RQH-5 ESM/ELINT airborne system.

STATUS

In service with several navies.

CONTRACTOR

Selenia Industrie Elettroniche Associate SpA, Defence Systems Division, Via Tiburtina Km 12,400, 00131 Rome, Italy.

5663.253

INS-5 NAVAL ESM/ECM SYSTEM

INS-5 is a total system which allows a single operator to perform both ESM and ECM functions, and is composed of a passive section and an active section. The passive section, RQN-5, provides automatic detection and analysis, and tracking of emissions in a dense electromagnetic environment. It provides warning of emissions in its coverage area and automatic identification of threats based on parameters stored in a dedicated computer library. Information on threat direction, identification parameters, etc is displayed in alpha-numeric and graphic symbology on the display of the INS-5 control console.

The active section, TQN-2, is integrated with the RQN-5 and receives instantaneous data for fast and automatic reaction. It will provide noise and pulse jamming of threat emissions, as well as computer control of passive EW launch systems, eg chaff and flares. It also performs passive tracking in azimuth and elevation of high-speed radar emitters, such as hostile aircraft and anti-ship missiles, and passes this data to other self-defence weapons.

CHARACTERISTICS

ESM operating frequency: 1 – 18 GHz
Receiver: Wide open with IFM and monopulse DF
Horizontal coverage: Omni-directional
Real-time tracking capacity: Over 50 emitters
Automatic identification library: Up to 3000 emitters
ECM operating frequency: I/J-band
Jamming modes: FM-AM-Pulse

STATUS

In production and in operational service.

CONTRACTOR

Selenia Industrie Elettroniche Associate SpA, Defence Systems Division, Via Tiburtina Km 12,400, 00131 Rome, Italy.

ECM antenna of INS-5 system

4641.283

SCLAR Mk 2 ROCKET LAUNCHING SYSTEM

The SCLAR Mk 2 system is a development of the UCLAR System (**1823.281** in 1982-83 and earlier editions) and has been designed and manufactured for installation in the Italian Navy's 'Maestrale' class ASW frigates and the helicopter carrier *Garibaldi*. The SCLAR Mk 2 employs the same rocket launcher as the previous system but data is processed by a digital multiprocessing computer based on the Elsag designed ESA 24 microprocessor.

The system is defined as multipurpose since it provides for the execution of various operations, such as decoying incoming missiles, confusion of hostile radars, illumination or HE bombardment, by simply selecting the type of rocket.

A very large set of decoying patterns can be stored in the digital computer and may be recalled and automatically executed as required. Each stored pattern may be substituted easily and rapidly at any time.

The SCLAR Mk 2 consists of one control console which requires a single operator; one data processing unit; one or two rocket launchers; one or two power supply cabinets; and one or two local control panels. In addition the system requires a set of peripheral panels comprising one alarm loading panel and one acoustic warning device for each launcher. An optional illuminating launching remote control panel is also available.

Two remotely controlled rocket launchers can be driven by the SCLAR Mk 2, which will position them in azimuth and elevation and perform all the necessary logic and computation. This includes rocket pattern programming, azimuth and elevation continuous computation for the accurate launch of each rocket to the programmed burst point and rocket automatic selection and firing.

STATUS

In production.

CONTRACTORS

Selenia-Elsag Consortium for Naval Systems, Via Panama 52, 00198 Rome, Italy.

Breda Meccanica Bresciana SpA, Via Lunga 2, 25100 Brescia, Italy.

SCLAR Mk 2 rocket launcher

3997.293

BREDA LIGHT CHAFF ROCKET LAUNCHING SYSTEM

This system consists of two launchers, installed port and starboard, and a control panel. Each launcher holds six 105 mm short rockets, set in tubes at fixed elevation and bearing (45° and 135° relative to ship's head) so as to launch on each quarter and form a chaff 'cross' pattern over the ship. The control panel monitors the state of the launcher, and commands launching from each launcher independently. Two operational cycles, each consisting of one pattern of four followed by one pair ahead, are performed without reloading. The resulting radar homing missile decoy is accompanied by evasive ship manoeuvres. This launcher is particularly intended for the defence of very light craft. The rockets were designed and developed by SNIA BPD.

CHARACTERISTICS

Launcher
Height: 1700 mm
Length: 480 mm
Width: 300 mm
Weight: 85 kg (without rockets); 125 kg (with rockets)

Control panel
Height: 260 mm
Length: 400 mm
Width: 220 mm
Weight: 8 kg

Rocket
Calibre: 105 mm
Weight: 5 kg (without container); 6.5 kg (with container)
Power consumption: 0.2 kW

STATUS

In production.

CONTRACTORS

Breda Meccanica Bresciana SpA, Via Lunga 2, 25100 Brescia, Italy – Launchers.

SNIA BPD, Defence and Space Activities, Via Sicilia 162, 00187 Roma, Italy – Rockets.

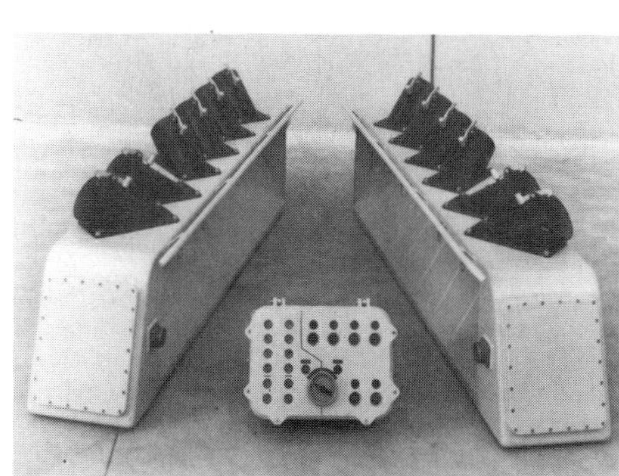

Breda light chaff rocket launcher unit (aft end nearest camera)

JAPAN

6495.253

NOL SERIES NAVAL EW SYSTEMS

The first japanese EW systems closely paralleled those of the United States and it was not until the 1960s that indigenous equipment was developed.

The first Japanese developed shipborne ESM system was the NOLR-1A and NOLR-1B which were fitted to destroyers and used the same antennas and mechanically scanned receivers as comparable US equipment. The next development, the NOLR-5, was fitted to a number of frigates and incorporated some improvements such as new antennas enclosed in radomes. This series did not, however, cope with anti-ship missiles and by the early 1970s a much improved system, the NOLR-6 was introduced. This employed electronic scanning with a frequency coverage up to about 18 GHz. It uses two similar-sized rotating DF antennas to cover the main radar frequencies, with a circular dipole array mounted on the foretop to carry out direction finding in the VHF/UHF band. One unusual system, the OLR-9, was also developed but only one installation was made.

The latest system is the NOLQ-1 which was introduced in 1980 and is fitted in the DDH *Shirane*, and subsequently appeared in the DDGs *Sawakaze* and *Hatakaze*, as well as the DDH *Hatsuyuki*. It consists of an advanced electronic scanning ESM suite integrated with two OLT-2 ECM jammers. The OLT-2, one covering each side of the ship, is believed to be a deception repeater jammer employing advanced automatic DF and 'range-gate pull-off' techniques. The integrated ESM and ECM system, when used in conjunction with the American Mk 36 SRBOC chaff launchers gives excellent protection against anti-ship missiles.

NORWAY

4646.293

SR-1A ESM RECEIVER

The SR-1A is a shipborne ESM radar warning receiver, suitable for installation in larger warships, and designed to intercept, analyse and measure incoming radar signals. It consists of two units; an antenna assembly and a display/control console. The antenna assembly comprises a rotating antenna in a protective radome and a fixed omni-directional octagonal horn array. The fixed antenna gives a coarse bearing to ± 22.5 degrees while the rotating antenna accuracy is approximately ± 3.5 degrees.

Frequency coverage of the system is from 2.5 to 18 GHz with a claimed frequency measurement of better than ± 1 per cent. The display systems give visual indication of DF, pulse analysis and frequency measurement functions. A separate indicator giving radar warning and coarse directional information is also provided and can be installed remotely from the main units.

CONTRACTOR

Nera A/S, Bergen, Norway.

SWEDEN

3538.293

EWS-900 CHAFF LAUNCHING SYSTEM

The Saab EWS-900 shipborne chaff system has been developed principally for use aboard smaller vessels. The main purpose of the system is to establish a chaff radar target, large enough and properly located in space and time, to make the ship and the chaff cloud appear as a single target in the seeker of an approaching missile. Ship manoeuvres, based on wind data, ship's course and line-of-sight to the missile, will cause separation of the false target from the actual ship target. The chaff cloud, having a larger radar target area than the ship, will thus cause the missile to miss the ship.

To minimise reaction time, the system can be operated from any of several locations on board. In a fully automatic version of the system, introduced in September 1979. On-board radar warning facilities can be used to feed a minicomputer with all relevant data for automatic launch of chaff and calculation of information for ship's evasive action.

The system consists of an optional number of launchers, control boxes, and one adaptor unit. The number of modules fitted depends on how many frequency bands are to be covered and on the endurance required of the ship.

The control boxes are located in the CIC near the search radar operator, adjacent to the radar warning operator, and accessible to the lookouts eg on the bridge.

STATUS

A similar system has been developed and delivered to the Swedish Navy. The operational systems have pyrotechnical chaff.

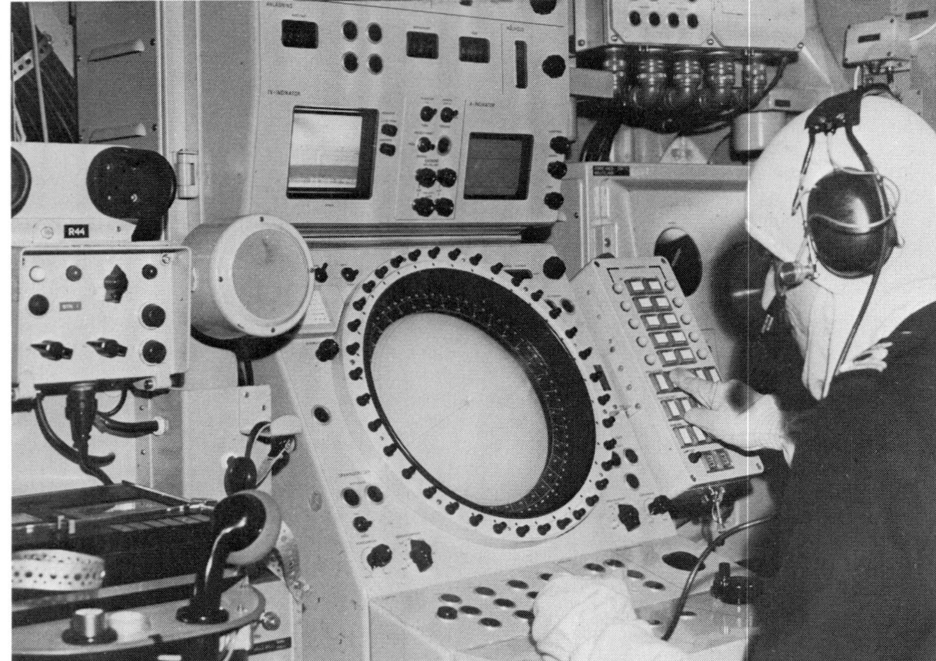

PPI-collar and control/display unit of EWS-900 CA automatic chaff launcher system

CONTRACTOR

Saab Missiles AB, 58188 Linköping, Sweden.

3133.293

9EW COUNTERMEASURES LAUNCHER

9EW is a chaff and IR decoy grenade launching system for the protection of small warships against attacking anti-ship missiles. Of very lightweight construction, it requires only a simple installation.

A ship normally carries two launchers, one on each side, but larger ships might carry four launchers. The launchers can be of the fixed or trainable type as required, but their elevation is fixed.

Each launcher consists of four or eight easily exchangeable chaff magazines, two of which may be replaced by three IR containers, requiring only a removable interface plate (the electrical interface is unchanged). Each chaff magazine contains four groups of grenades, each of which is fired by a single ejection cartridge. The nine grenades in each group are distributed to provide an effective chaff-cloud in space in a very short time, typically less than four seconds.

Each IR container is fired with a single ejection cartridge of the same type as for chaff. The IR decoy consists of sequentially-ignited flares, distributed for optimum deflection of IR or multi-sensor seekers. The IR decoy is effective in both the 3 to 5 and the 8 to 14 micron bands.

A range of associated fire control systems are available from simple manual firing panels, to fully automatic systems, which give information on optimum manoeuvring, launcher selection, etc.

CHARACTERISTICS

Launcher (chaff)
Number of magazines: 6-8
Groups per magazine: 4
Grenades per group: 9
Ejection cartridge: 1/group
Launcher weight: 180 kg
Grenade
Grenade manufacturer: Société E Lacroix, France
Calibre: 40 mm
Grenade weight: 0.37 kg
Payload weight: 0.2 kg
Flight time: 1 s
System data
Grenade flight time: 1 s
Cloud development time: <4 s

9EW chaff and IR decoy launcher

IR decoys
Containers per launcher: 3 or 6
Container dimensions: 131 × 468 × 357 mm
Container weight: 28.5 kg
Payload: 4.5 kg (chemicals and flares)

STATUS

In service.

CONTRACTOR

Philips Elektronikindustrier AB, 17588 Järfälla, Sweden.

5731.253

9EW 300 and 9EW 400 ECM SYSTEMS

The systems are modularly designed for installation on ships ranging in size from patrol vessels (9EW 300) to small frigates (9EW 400). For this reason the system concept features a number of different technical levels and separate modular components to suit the particular vessel and the status of the potential threat.

The 9EW 300 and 9EW 400 systems are based on the philosophy of automating the EW air defence as much as possible. To achieve this both systems are built up around an EW computer for the control of information from the radar warning receivers, ESM system, the laser warning receiver and the ship's own C3 system. This in turn gives a correct counteraction order to the new trainable chaff and IR grenade launcher. The launcher promotes even faster reaction times and optimises the radar cross-section of the chaff in the direction of the threat.

As compared to older systems the chaff round also incorporates improvements to give the latest radar cross-section and blooming time. This has lead to the creation of a system capable of seducing incoming missile threats away from the vessel automatically or after operator decision. Completely new software programmes have been developed to make it possible to install the systems on a large variety of ships with individual characteristics.

STATUS

The basic elements of the system have been in service with the Swedish Navy, as well as navies of other countries, for many years. A contract for four 9EW 300 systems was received from the Finnish Navy in mid-1987.

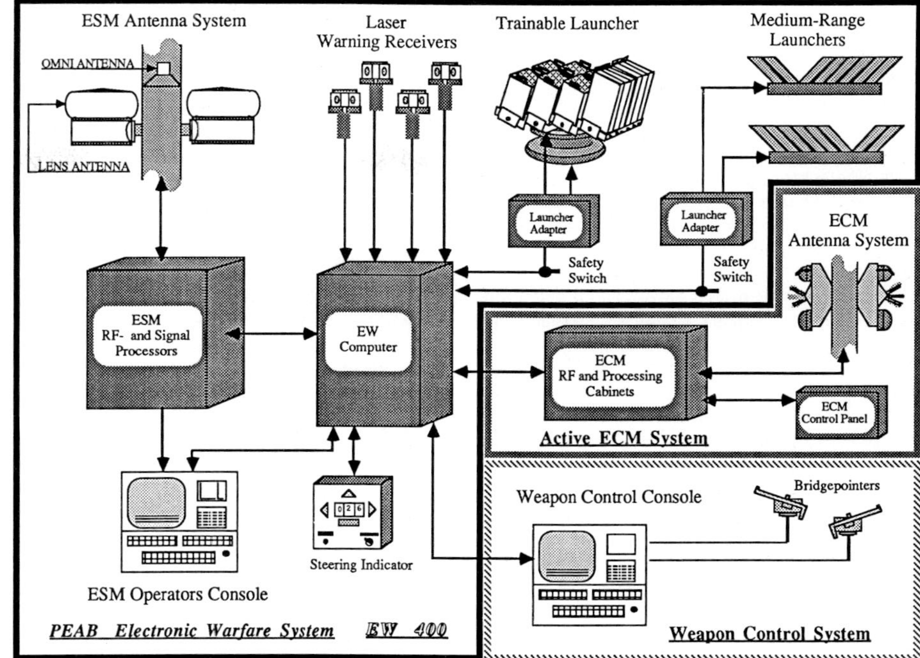

CONTRACTOR

Philips Elektronikindustrier AB, 17588 Järfälla, Sweden.

5836.253

EWS-905 ESM WARNING SYSTEM

This shipborne system is designed to identify targets detected by the vessel's surveillance radar. The antenna turns automatically towards the direction required, using information passed from the ship's tactical display, acquires and analyses threat signals. These are displayed, together with ESM data, and the system will automatically command the chaff launcher and/or other self-protection systems to counter the threat.

Signals received from a limited sector around the designated bearing are analysed and parameters representing bearing, frequency band, pulse interval, pulsewidth and scan time are displayed.

Three modes of operation can be selected:

(1) alert – all bands are covered and the system will detect radars in a selected bearing sector. Frequency and bearing sector are displayed
(2) decision – a selected sector is scanned rapidly and bearing, frequency, pulse interval and width are displayed for every target
(3) surveillance – the selected sector is scanned slowly and the target radar antenna rotation speed is displayed in addition to the parameters in (1).

STATUS

In service with the Royal Swedish Navy.

CONTRACTOR

Saab Missiles AB, 58188 Linköping, Sweden.

UNION OF SOVIET SOCIALIST REPUBLICS

6206.253

SOVIET NAVAL ELECTRONIC WARFARE SYSTEMS

Unclassified information on electronic warfare systems employed on board ships of the Soviet Navy is extremely sparse. A wide range of antenna systems has been observed on various types of ships and there appears little doubt that the Soviet Navy operates a comprehensive suite of ESM and ECM equipment, covering surveillance equipments for ELINT and SIGINT, threat warning systems, jammers and chaff/infra-red flare decoy systems. The following gives brief details of those systems which have been identified, together with their NATO code-names.

Bell Series

This is a series of threat warning and jamming equipments which have been observed on Soviet warships from frigate size upwards. Most of the series are housed in bell-shaped radomes, hence the code-name. Those definitely are Bell Bash, Bell Clout, Bell Shroud, Bell Slam, Bell Squat, Bell Tap and Bell Thump.

Brick Group

This is another series of ESM systems installed in submarines, presumably used for surveillance and threat warning. They are normally used in conjunction with Snoop Group radars (**5849.453**), and include Brick Pulp and Brick Spit.

Cage Pot

A shipborne EW system.

Farm Loaf

An EW antenna housing.

Fig Jar

Another EW antenna housing.

Foot Ball

EW antenna array in small radomes.

Golf Ball

This is a submarine mast-mounted sensor, probably used for surveillance and threat warning of enemy radar systems.

High Ring

An ELINT/SIGINT antenna installed on surface ships.

Rum Tub

ESM antennas, normally in groups of four, disposed so that each covers a 90° sector. Normally found on the larger warships and mounted around a tower.

Ship Globe

These are broadband jammers contained in radomes and mounted in groups of four on either side of the major mast structure in larger warships.

Stop Light

This is a broadband passive ESM system, probably with an electronic intelligence (ELINT) function.

Tilt Pot

An EW system.

Top Hat A and B

Another shipborne EW system.

Watch Dog

This is a rather elderly electronic intercept system first noticed in the 1950s and has been fitted in smaller warships. Although a predecessor of the Bell Series it has been observed on some late construction smaller frigates.

Chaff dispensers

These are in widespread use with Soviet naval ships, mainly as a defence against the growing number of Western anti-ship missiles. Some of the older systems observed seem to consist of only eight tubes, manually loaded, and based on a simple rocket system. However, a new installation noted recently was on-board the frigate *Revnostnyy* and appears to consist of four separate banks of launchers, similar to the Royal Navy EPO naval rocket launcher which is capable of launching chaff and infra-red flares.

UNITED KINGDOM

4226.293

BARRICADE NAVAL DECOY EQUIPMENT

Barricade is designed to provide ships with an effective and lightweight means of deploying chaff and infra-red decoys as countermeasures to missile attack. The system gives a layered defence against the various phases and types of anti-ship missile attack, ranging from the long-range confusion mode, through medium-range distraction and dump modes to the close-in centroid mode.

The complete system consists of port and starboard launchers, safety switches, a control unit, ready-use ammunition lockers and rockets. The launcher consists of six sets of triple barrels, the lower three sets being angled 60° from the next so that a pair of launchers provide all-round coverage. The upper three sets are preset to point to the same bearing, as required. These upper three sets fire short-range Palisade rockets in centroid mode, while the lower barrels fire medium- and long-range Stockade rockets in dump, distraction and confusion modes.

Barricade decoy modes are:

(1) confusion: long-range rockets fired to confuse enemy search radar and make target selection difficult

(2) distraction: medium-range chaff and/or infra-red rockets fired in a pattern of six around the ship to present alternative targets to missiles during their search phase. This mode affords defence against any number of missiles from different directions

(3) dump: if a missile is locked on to a ship and its range gate is removed from the ship by the use of a jammer, the gate may be 'dumped' on a suitably placed chaff cloud, allowing the jammer to be re-deployed. Single medium-range rockets are used

(4) centroid: if a missile is locked on to a ship, a pattern of short-range chaff and/or infra-red rockets is fired quickly to create a large alternative target close to the ship. The ship then manoeuvres away and the missile will home on to the decoy. This is a close-range mode for final defence against single missiles.

The Barricade system is based on a spin-stabilised 57 mm rocket plus infra-red and chaff sources. The rockets are fired by induction ignition and the high-speed spin produces very rapid chaff blooming. Launchers may be used in pairs on fast patrol boats and similar vessels and in larger numbers on bigger ships. Even smaller Stockade and Palisade launcher variants are available for larger ships/auxiliaries and small fast patrol boats/hydrofoils respectively.

A small microprocessor-controlled tactical computer for accurate centroid mode control and manoeuvring is available for siting near the control unit. The complete Barricade system is designed for easy integration with EW/AIO/C² systems.

STATUS

In production with 70 systems sold or in service with eight navies, including the Royal Navy and NATO. Additional orders were received during 1987. A Mk II version is now available with microprocessor control and the ability to fire stored decoy sequences.

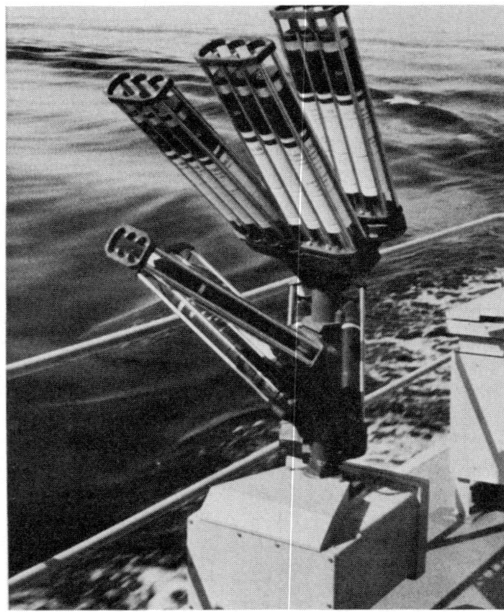

Barricade lightweight 18-barrel launcher

CONTRACTOR

Wallop Systems Ltd, Arkay House, Weyhill Road, Andover, Hampshire SP10 3NR, England.

4651.293

STOCKADE-SEAFLASH DECOY SYSTEM

The Stockade-Seaflash decoy system is intended to place confusion chaff decoys at distances of up to 20 km from its own forces. This is designed to create ship-like echoes on the enemy's radar screen at a considerable distance from the real force without the need for rocket or gun-propelled systems. The system consists of a Flight Refuelling 'Seaflash' remotely-controlled boat fitted with two Stockade decoy launching systems. Both Seaflash and Stockade are under positive secure radio control from a single unit on the controlling ship. Confusion chaff can be dispersed as required from 18 barrels disposed in groups of three at 60 degree bearing intervals.

STATUS

In late development.

CONTRACTORS

Wallop Systems Ltd, Arkay House, Weyhill Road, Andover, Hampshire SP10 3NR, England.

Flight Refuelling Ltd, Brook Road, Wimborne, Dorset BH21 2BJ, England.

3145.293

3-IN WINDOW DISPENSING ROCKET SYSTEM (CORVUS)

This is a simple, lightweight, quick reaction system for the self-defence of surface vessels against surface-to-surface and air-to-surface missiles. Window (chaff)- dispensing rockets are fired to form a radar decoy screen around the vessel. The system can be installed in any surface vessel other than patrol boat types and its simplicity of construction makes it particularly suitable for installation in destroyers and frigates down to the smallest size and retrospective fitting offers no problems.

The latest 8-barrel launcher for Corvus naval radar decoy system

The complete system comprises: two multi-barrelled rocket launchers, launcher control and firing panel, and the 'window' dispensing rockets themselves.

Two modes of operation are available for the system. The modes differ in the azimuth angle at firing and the range at which window is released. The modes of operation are:

(1) Distraction decoy

(2) Centroid decoy.

These are illustrated in the nearby diagrams.

A firing and control panel can be ideally situated in the combat information centre. The maximum loading of 16 'window' rockets provides protection against three missile attacks, before reloading. Any member of the combat information centre crew can control and operate the system.

A cylindrical rotating structure carries eight launching tubes mounted in two sets of three, one above the other, and crossed at 90° in azimuth with two further tubes set above, and aligned midway between the other tubes, all at a fixed elevation of 30°. A deck mounted pedestal supports the rotating structure on its training bearing and also houses a self-contained electrical power conversion unit for the control circuits and the associated electrical equipment. The training drive consists of a gearbox driven by a reversible motor, while braking is achieved by an electromagnetic brake. A slip-ring positioning unit, located within the pedestal, controls the training of the launcher to preselected bearings. The normal limits of the training arc for the launcher are between 60 and 120° but as the lower sets of rocket barrels are angled about the centre line of the rotating structure, the arc covered by the rockets is considerably greater. Within these normal limits, fixed firing bearings for the launcher, at 15° intervals, can be selected by completing the appropriate circuits at the control panel. Training stops and a buffer assembly are incorporated to prevent over-stressing and twisting of the cable bight in the event of an accidental over-run. The launchers can be fitted in any position on No 1 deck or above, providing adequate space is available for slewing and loading, and there is a clear firing arc of approximately 160° in azimuth and 30° elevation.

CHARACTERISTICS

Training

Bearing arc (normal) ie launcher axis: 60 – 120°

Bearing arc (max) ie tube axis: 15 – 165°

Launcher weight: 585 kg

Off-launcher equipment weight: 42 kg

Tube diameter: 195 mm

Tube length: 1.6 m

Weight (empty): 25 kg

No of guides: 4, RH spiral

STATUS

A broadband chaff rocket for use with Corvus launchers has been developed by Plessey (**3650.293**) and is in service with the Royal Navy.

CONTRACTOR

Vickers Shipbuilding and Engineering Ltd, Barrow-in-Furness, Cumbria LA14 1AB, England.

5737.293

RADAMEC LIGHTWEIGHT MISSILE LAUNCHER

The Radamec lightweight stabilised missile launcher uses a Radamec heavy-duty two-axis, stabilised director platform to provide a lightweight, close-range surface-to-air missile system.

The missile system may be configured either as a manually trained or remote controlled weapon system, including associated guidance.

For remote control the system includes Radamec Series 2000 electro-optical fire control equipments to provide an integrated cost-effective defence weapon system for both large and small ships. The system is specifically designed for simple ship installation, both as an initial fit or as a retrofit.

The launcher is mechanically modular to facilitate the use of different payloads. It can be used for launching a variety of small missiles, rockets, infra-red decoys and chaff dispensers. It is able to stabilise payloads in excess of 300 kg in a 30 g shock environment, and can be continuously rotated in azimuth. As an unstabilised unit, the launcher can be deployed as an area weapon for air defence.

CONTRACTOR

Radamec Defence Systems Ltd, Bridge Road, Chertsey, Surrey KT16 8LJ, England.

Radamec manned missile launcher with Shorts Javelin missiles

6143.293

SWORD DECOY LAUNCHER

SWORD (Small Warship Operational Reconnaissance and Decoy Launcher) has been developed jointly by Pains-Wessex and Attewell Engineering. The launcher was intended originally to provide a multi-shot illumination capability but further research and development has produced chaff and infra-red rounds for use in the medium range EW confusion/dilution role, and the addition of a hybrid round to operate in the seduction role.

The basic unit consists of a three barrelled launcher which can be operated manually or commanded automatically from an on-board threat warning system. The barrels can be activated individually or in salvo, and are adjustable in elevation and azimuth. They can be built-up in multiples on the same plinth mounting to provide an all-round launch capability, for use in the decoy or self-protection role. The 51 mm rocket employed can carry an illuminating, chaff or infra-red payload and is designed to be fired at an elevation of 32°. This gives a flight time of 9 seconds, a range of 900 m and a deployment height of 200 m.

STATUS

In production.

CONTRACTORS

Pains-Wessex Ltd (a subsidiary of Chemring Group plc), High Post, Salisbury, Wiltshire SP4 6AS, England – Payloads.

Attewell Engineering Company Ltd, Bath Road, Peasedown St. John, Bath, Avon BA2 8EN, England – Launcher.

3650.293

CHAFF ROCKET

Plessey Aerospace manufactures chaff dispensing rockets for the UK MoD (Navy). The Royal Navy uses the Corvus 3-inch rocket launching system (**3145.293**) for which a Plessey rocket was developed. One rocket, using the same launcher and without modification to the control system, produces a multi-frequency cloud equivalent to the combined clouds of the three previous service round types it replaces.

Each rocket is in three sections:

(1) an electrically ignited solid fuel motor
(2) the payload head which contains chaff packs and dispensing mechanisms
(3) the nose which carries the fuze and head charge.

At a predetermined time the fuze ignites the head charge which separates the nose, chaff packs and dispensing mechanism from the remainder of the rocket. The dispensing of pairs of chaff packs is then evenly and quickly effected to create an efficiently distributed chaff cloud with a rapid bloom characteristic.

The system will provide Distraction or Seduction decoy clouds simply by fuze time adjustment.

STATUS

In production.

CONTRACTOR

Plessey Aerospace Ltd, Abbey Works, Titchfield, Fareham, Hampshire PO14 4QA, England.

5682.293

SHIELD ANTI-SHIP MISSILE DECOY SYSTEM

Shield is a chaff and infra-red decoy system which is designed for both medium/large warships and fleet support ships. A range of modules enables the system to be built up to suit the differing installation requirements and they can be configured to suit ships with, or without, control to command. Launcher barrel modules can be mounted on simple bases in either parallel or crossed formation to give any combination of barrels from 3 to 12 per launcher. Four parallel barrel launchers are normally specified for large ships to give all round protection. The flexibility offered by the Shield module range is such that a complete system can be containerised for use on merchant ships without the need for external services.

The Shield system is suitable for an inventory of different types of decoy ammunition and studies and development programmes are in progress. The current decoy most appropriate for defence against anti-ship missiles is a chaff rocket equipped with variable fuze for medium- or short-range cloud deployment. Its large multiband payload is particularly suited to both distraction and seduction modes.

Distraction

This provides for a number of separate decoy chaff clouds deployed in a pattern around the ship to present several alternative targets to hostile missiles while they are seeking targets.

Seduction-Dump

This mode is intended for use with active ECM radar equipment, which processes the missile's radar signal and transfers range gate to a chaff cloud. The range of deployment depends upon the ECM radar characteristics.

Shield decoy system launcher

Seduction-Centroid

The Seduction-Centroid mode is for use against threats that have a short time interval between detection and arrival at the target and which are locked-on at the time of detection. The decoy is deployed at a range and relative bearing such that it and the ship will combine to become a single target within the missile radar cell. As the combined target elongates, due to continuing chaff cloud bloom and ship movement, the missile radar cell remains centred on the mean cloud/ship signature causing sufficient deviation of the missile flight path to miss the ship.

Two versions are available: a manually fuzed version for use in simple independent systems or a fully-automatic computer-controlled system based on extensive computer tactical modelling of threat scenarios. The command module can be situated ideally in the combat information centre and can be interfaced with the ship's EW suite.

DEVELOPMENT

An infra-red rocket for use with the Shield system is now in production. It is designed to give protection against heat seeking missiles in the seduction mode.

STATUS

In production for the Royal Navy, the Canadian Navy and the new Brazilian frigates.

CONTRACTOR

Plessey Aerospace Ltd, Abbey Works, Titchfield, Fareham, Hampshire PO14 4QA, England.

4056.293

RAPIDS NAVAL ESM

RAPIDS (radar passive identification system) is a shipborne electronic support measures (ESM) system for the interception, analysis and identification of all forms of radar emissions. It is entirely passive and operates autonomously. The system features:

(1) high probability of interception by an instantaneous field of view of 360° and a large instantaneous frequency coverage
(2) high sensitivity
(3) high pulse density capability
(4) unattended operation as far as threats are concerned when integrated with a data handling system
(5) high bearing accuracy.

Operational functions include: interception, threat processing, analysis, identification, tracking, data exchange with ECM and data handling systems, data logging.

The basic configuration of RAPIDS is illustrated.

Signals of linear as well as circular polarisation can be intercepted by the high sensitivity receiver. Signals are processed rapidly by a central management computer which is surrounded by microprocessor-based sub-systems. Intercepted signals are automatically analysed and identified by reference to the emitter library. Processed data is then transferred immediately for presentation and action on the ship's data handling system. The Electronic Data Display provides an overview of the situation by presenting: bearing frequency plots of intercepted signals; alarm cursors for tracking radars locked on to own ship; alarm cursors for radars that trigger the threat warner; symbols to indicate identified interceptions; tell-back data from an ECM when integrated with such a system. Other information displayed includes alpha-numeric data on lock-ons; alpha-numeric data on triggered threat warner channels; alpha-numeric data on identified targets; software keys for the selection of particular system functions.

RAPIDS has provisions for interconnection with a data handling system and an ECM system such as RAMSES (**4054.293**) or Scimitar (**3598.293**). The basic version is illustrated. The frequency range covers 2 to 18 GHz.

STATUS
In production.

CHARACTERISTICS

Dimensions	**Width (mm)**	**Height (mm)**	**Depth (mm)**	**Weight (kg)**
Antenna, radome and base	536	1554	620	94
Receiver unit	597	835	502	145
Processor unit	650	1570	660	250
Display unit	530	813	440	40
Air cooler	597	1045	505	60

CONTRACTOR
MEL, Manor Royal, Crawley, West Sussex RH10 2PZ, England.

3598.293
SCIMITAR JAMMER

The MEL Scimitar jammer derives its name from 'system for countering interdiction missiles and target acquisition radars', and is designed for use with all types of naval on-board ESM system, although specifically intended as a complementary system to the range of major naval ESM equipments produced by MEL, including SCEPTRE, RAPIDS and CANEWS. Scimitar is designed to engage a number of targets simultaneously at short and long range and, when success has been achieved, automatically switches to the next priority threat.

Full jamming capability is provided over the frequency range 8 to 16 GHz to give protection against all types of radar homing missiles and full area defence capability against target indicating radars. Power output is rated at 1.5 kW pulse and 150 W CW. Jamming modes include: CW, programmed band and wide band noise, range-gate pull-off, scan rate and swept scan rate modulation, inverse scan gain and false target generation.

The system consists of two identical antennas, an ECM RF cabinet and a processing cabinet. The antennas provide 360° coverage in azimuth and 0 to 70° in elevation. Each unit comprises a receive and transmit antenna provided with a servo system for beam steering and stabilisation against pitch and roll.

CHARACTERISTICS
Frequency range: 8 – 16 GHz
Output: 1.5 kW pulse; 150 W CW
Frequency channel sensitivity: –60 dBm
Antenna gain: 23 dB
Beamwidth: 7° azimuth; 7° elevation

STATUS
In series production for several navies.

CONTRACTOR
MEL, Manor Royal, Crawley, Sussex RH10 2PZ, England.

6151.253
SCEPTRE ESM SYSTEM

Sceptre is a shipborne family of modular ESM systems designed for specific operational requirements within any prevailing space and weight limitations. The system provides automatic intercept, analysis, classification and identification of all radar emissions.

A wide range of antenna and receiver options is offered to increase frequency coverage and sensitivity, and to enhance bearing and frequency resolution and passive targeting capabilities. Optional digital processing modules are available to extend the pulse density and library handling capabilities. Output data can be stored either on magnetic tape or in hard copy format, and the system can be interfaced with any central data handling system.

Using 'wide-open' ESM antenna technology, Sceptre provides 360° coverage and 100% probability of intercept. Intercepted signals are automatically analysed and identified by reference to a comprehensive library of known radar types. Processed data is immediately displayed to the ESM operator in both cartesian and alphanumeric formats on two interchangeable colour displays.

Sceptre works on the principle of a central management computer surrounded by subsystems, each containing advanced microprocessors which carry out local processing. This modular approach ensures that sub-units and units can be built into a cohesive system to meet operational requirements while matching the available space and power limitations. In addition to the operational flexibility, the modular approach offers lower procurement and through-life costs, and simplifies the fitting and cost of any future options or upgrades.

Three variants of Sceptre are available:
(a) Sceptre O which is a radar threat warner for medium size fast patrol boats and MCM vessels
(b) Sceptre X which provides full ESM for corvettes and frigates
(c) Sceptre XL for large warships.

CHARACTERISTICS
Frequency range: 2 – 18 GHz (optional extensions above and below this range are available)
Dynamic range: 60 dB
Azimuth coverage: 360° instantaneous
Elevation coverage: –10 – +30°
Signal polarisation: Horizontal, vertical, and either LH or RH circular
Bearing accuracy: 6° rms (2 – 6 GHz); 4.5° rms (6 – 18 GHz)
PRF range: 100 Hz – 300 kHz
Pulse width range: 50 ns – 100 µs
Scan types: Circular, sector, steady (simple and complex)
Radar types: Simple pulse, frequency agile, PRF agile, CW high duty cycle, FMOP and PMOP flags
Mission library: 2000 emitter modes – expandable in multiples of 1000 modes
Threat warner: 24 emitters with 6 modes each (144 modes)
Operator library: 100 emitter modes

CONTRACTOR
MEL, Manor Royal, Crawley, West Sussex RH10 2PZ, England.

3539.253
MATILDA RADAR WARNING RECEIVER SYSTEM

Matilda (an acronym for Microwave Analysis Threat Indication and Launch Direction Apparatus) has been designed by MEL to meet the requirement for an effective, low-cost radar threat warning equipment for smaller ships. It operates on frequencies from 7.5 to 18 GHz, and is intended to act primarily as a trigger for the MEL Protean chaff launching system (**3193.293**). Matilda can also be successfully integrated with most types of short-range chaff systems to provide automatic defensive measures against missile attack. Matilda can detect and recognise the threat, indicate the bearing in which it lies and, if required, can activate the chaff launcher. Compact modular design permits location in the most convenient positions on board. The system comprises three basic elements: a small, lightweight antenna covering 360 degrees in azimuth, a processing unit and a small display. It can be fitted to a vessel, such as a merchant ship, in approximately two hours.

The antenna unit, weighing 7 kg, consists of four antenna horns with their associated microwave components, circularly polarised for the detection of vertically, horizontally and circularly polarised signals. It is supplied in two matching halves, each comprising a tube of approximately 115 mm diameter; simple brackets are provided to clamp to either a pole mast or a plated mast structure. A multi-pair screened cable is routed down the mast to feed the processing unit.

The processing unit, which contains the printed circuit cards for signal processing, parameter measurement and the warning unit counter memory, is approximately 320 × 250 × 195 mm in size and weighs 9 kg. The unit is fitted with integral power supplies and may be located in any convenient position. It is able to achieve a processing time of less than 10 µsec and a total warning time of under 1 second.

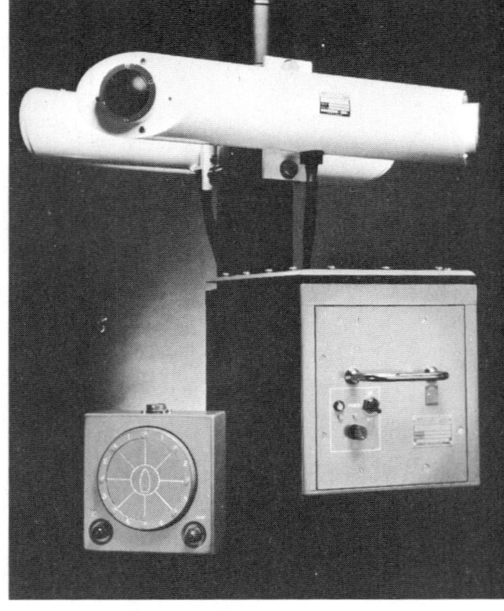
Matilda radar warning receiver system

The warning unit display consists of a small indicator panel designed for location at any convenient position in the operations room, remote from the processing unit if so desired. On detection of a threat, the unit activates an audio warning with the option of a programmable voice module and provides an indication of the threat and bearing sector channel. It may be directly interfaced with the Protean chaff launcher fire control panel. The unit's dimensions are 155 × 155 × 55 mm and it has a weight of 0.45 kg.

CHARACTERISTICS
Frequency range: 7.5 – 18 GHz
Frequency resolution: 1.5 GHz
Pulse width range: 100 ns – 6.4µs
Pulse width resolution: 200 ns
Bearing coverage: 360°
Bearing resolution: 45°
Sensitivity (referred to isotropic): –30 dBm
Dynamic range: 40 dB
Antenna gain 7.5 – 18 GHz: 6 – 8 dB
Voice module: Programmable to any warning required in any language
Dimensions

	Weight kg	**Width mm**	**Height mm**	**Depth mm**
Antenna	7	680	114	114
Processor	9	193	257	318
Display unit	0.45	155	155	55

Total weight: 17 kg approx.

STATUS
In production for the UK MoD as Outfit UAR. Matilda has also been ordered by other customers for installations on-board merchant ships. Four sets were sold to the Finnish Navy in mid-1987 for installation on fast attack craft.

CONTRACTOR
MEL, Manor Royal, Crawley, West Sussex RH10 2PZ, England.

3193.293

PROTEAN CHAFF LAUNCHER

The Protean chaff grenade launcher system is designed for naval use as a countermeasure against active radar homing anti-ship missiles. Two or four Protean launchers may be fitted to give forward or fore and aft protection on the port and starboard sides of the ship.

Each launcher is equipped with four reloadable magazines, each containing 36 chaff grenades to give a total ready-use load per launcher of 144 grenades. The launcher is deck mounted and fixed in elevation and trainable in azimuth. Firing is in groups of nine grenades and up to four groups can be fired simultaneously. Provision can be made to replace one or more of the chaff grenade magazines by IR decoy grenade magazines.

Threat detection and direction can be derived from radar or other sensors, or from an associated ESM warning system.

The Protean launcher appears to be identical to that of the Swedish Philax system (**3133.293**) produced by another member company of the Philips group.

OPERATION

Chaff is one of the primary electronic countermeasures available for use against missiles using active radar seekers for terminal homing, particularly those of the sea-skimming type. With or without the use of active ECM, the deployment of chaff greatly improves the probability of successfully countering the threat, provided that the chaff launching system is capable of reacting quickly, since the flight time from detection to impact of an in-flight missile is normally very short.

For maximum effectiveness, the chaff and own ship's echo must appear simultaneously in the seeker's range resolution cell. Ideally, the chaff cloud should appear at a closer range than the ship and have a larger radar target echoing area. Separation of ship and chaff cloud causes the seeker to lock on to and track the chaff cloud.

Protean chaff grenade launcher

The design of the ejection system enables any number of barrels (1 to 9) per group to be loaded without deterioration in performance, thus providing a useful economy feature for training exercises.

Each launcher is provided with thermostatically controlled heaters in order to avoid icing in the pyrotechnical components. Each also has an isolating switch.

The firing panel provides two basic facilities:

(1) selection of groups of grenades (with clear visual indication of selection) including either IR or chaff selection. Indication is also given of salvoes available of each type in each launcher
(2) firing of groups of grenades.

Firing is manually controlled or automatically initiated by remote sensors, eg ESM or radar.

CHARACTERISTICS

Launcher
Type: MEL AA 1067
Magazines: 4
Groups per magazine: 4
Barrels per group: 9
Pressure source: 1 cartridge per group
Firing interval: 0.1 s between groups
Weight (excl grenades): 180 kg
Dimensions: Width 835 mm; length 750 mm; max height from deck 550 mm
Grenade
Calibre: 40 mm
Weight (total): 0.35 kg
Chaff type: Chemring Ltd
Frequency coverage: Manufactured to order
Weight: 0.175 kg
Flight time: 1 s
Cloud echoing area: 1000 m^2
Cloud development time: <5 s
Cloud height dispersion: 40 – 60 m

STATUS

In production and ordered by several navies, including the Egyptian Navy.

CONTRACTOR

MEL, Manor Royal, Crawley, Sussex RH10 2PZ, England.

4652.493

MANTA SUBMARINE ESM SYSTEM

Manta is a range of advanced ESM systems optimised for submarine applications. Each of the Manta systems intercepts, analyses, classifies and identifies enemy radars.

Instantaneous detection of radar threats at long range provides submarine commanders with the maximum warning of the presence of enemy surface ship or airborne ASW systems. In addition, Manta provides information for tactical battle management, targeting and information gathering. Instantaneous warning enables the submarine commander to maintain a covert posture.

Manta is designed as a modular equipment that enables the most cost effective system to be built up to match the operational requirement for a particular class of submarine.

Different antenna modules provide options for extended frequency coverage and high resolution targeting, while the digital processing modules can extend the pulse density and library capacities. The ESM can be interfaced with the central data handling and the output data may be recorded on either magnetic tape and/or hard copied.

Using wide open ESM techniques, the systems provide 360° coverage and 100 per cent probability of intercept. Intercepted signals are automatically analysed and identified by reference to a comprehensive library of known radar types. Processed data is immediately displayed to the operator in both cartesian and alphanumeric formats on two interchangeable colour displays. Manta uses a central management computer controlling sub-systems, each containing advanced micro-processors carrying out local processing.

CHARACTERISTICS

Frequency: 2 – 18 GHz (optional extensions either end are available)
Azimuth coverage: 360°
Elevation coverage: –10 to +30°
Signal polarisation: Horizontal, vertical and either LH or RH circular
Bearing accuracy: 6° rms
PRF range: 50 ns – 100μs
Scan types: Circular, sector, steady (simple and complex)
Scan periods: 0.02 – 40 secs
Radar types: Simple pulse, frequency agile, PRF agile, CW high duty cycle, FMOP and PMOP flags
Mission library: 2000 emitter modes
Library expansion: In multiples of 1000 modes
Threat warner: 24 emitters with 6 modes each

STATUS

In production.

CONTRACTOR

MEL, Manor Royal, Crawley, West Sussex RH10 2PZ, England.

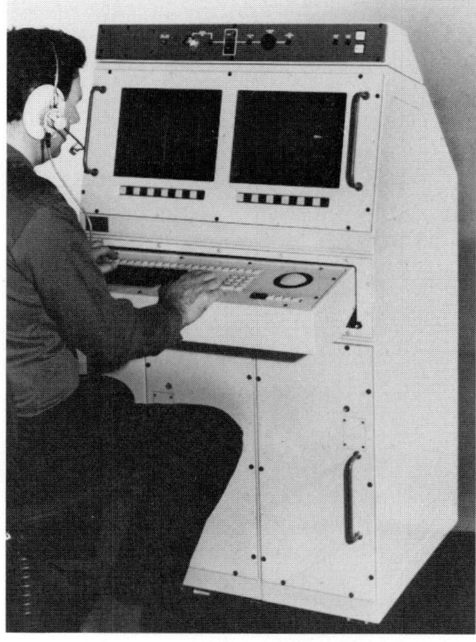

Display console for Manta ESM system

4055.293

SPHINX NAVAL ESM

Sphinx (system for passive handling of intercepted transmissions) is a shipborne electronic support measures (ESM) system for the interception, analysis and identification of all forms of pulsed radar emissions. It is entirely passive and operates autonomously. The system features: a high probability of interception by an instantaneous field of view of 360° and a continuous coverage of the entire frequency range, high sensitivity, high pulse density capability. It is capable of the following functions:

(1) omni-directional interception of signals in the range 1 to 18 GHz and determination of their frequency, and bearing
(2) lock-on detection, ie automatic detection and warning of any radar locked on to own ship
(3) automatic analysis of selected signals in terms of: level, pulse, duration, frequency (max, minimum), PRF (max, minimum), scan period
(4) automatic tracking in bearing of selected signals
(5) blanking of selected signals tracked in bearing
(6) area blanking
(7) identification of signals by comparison with library
(8) data exchange with ECM and data handling system
(9) data logging on cassette of information acquired.

The basic configuration of Sphinx is illustrated and consists of a frequency receiver and an eight-sector bearing receiver. Both are untuned and wide open in all directions so providing simultaneous coverage of 360° in azimuth and 40° in elevation (nominal). Interception probability approaches 100 per cent. Signals of linear as well as circular polarisation can be intercepted.

The operator's console has three displays:

(1) a situation display for presentation of bearing/frequency plots of detected signals and tell-back data from the ECM system
(2) an alpha-numeric display for analyser results
(3) an alpha-numeric display for identification results.

Sphinx has provisions for interconnection with a data handling system and an ECM system such as RAMSES (**4054.293**). One operator can then take care of both ESM and ECM functions on board the ship. Sphinx has built-in test facilities for checking both before and during operation. The bearing array is designed to form part of the load bearing structure of a mast. A WM20 series combined antenna system can be easily mounted on top.

Optional enhancements include: a K-band reception package (33 to 40 GHz) with alarm; a threat alarm, sets of parameters; and a CW detection package. Detailed characteristics of this equipment appear under the above entry number in *Jane's Weapon Systems 1981-82*. Although developed by Signaal, production and marketing has been transferred to MEL, UK.

STATUS

Production.

CONTRACTOR

MEL, Manor Royal, Crawley, West Sussex RH10 2PZ, England.

1743.263

SARIE-SELECTIVE AUTOMATIC RADAR IDENTIFICATION EQUIPMENT

Speed and capacity are all-important in the modern identification system, because in the present environment the rate of radar intercepts can be high and the time that can be permitted to elapse between intercept and threat evaluation may be only a matter of seconds. Dependence upon operator reference to known radar parameters is thus unacceptably time-consuming and automatic identification is essential. The SARIE system produces a solution to the problem by providing the results of automatic comparison of intercepted signal characteristics with a flexible dictionary of radar parameters in approximately 350 msecs. The equipment is also able to extend the existing library of data by analysing and indicating previously unrecorded transmissions, which can be stored by the operator using a built-in scratch pad memory. The new data is then treated as a high priority extension of the library.

The equipment consists of a display unit, control unit, and a data processing unit. The latter houses the dictionary, logic circuitry, magnetic tape reader, and power supply. It can be remotely sited, as access is needed only for maintenance or to modify the dictionary contents. Dimensions are 483 × 355 × 564 mm and the weight 43 kg. The display and control unit carries all the essential controls for operation of the system and the electronic display panel. Separate location of the control and display sections, to suit CIC or EW room requirements, can be accommodated if desired.

SARIE – (Selective automatic radar identification equipment)

SARIE can be used with almost any EW receiver, accepting the video signal derived from the intercepted transmission. It automatically measures the PRF, pulse width and scan period with high accuracy, and detects the presence of PRF jitter. The equipment also accepts data on frequency, frequency agility modulation type, and scan type for use in identification. This additional data, where available from the EW receiver, can be accepted automatically, or may be entered manually by the operator from the control panel. The operator is able to select intercepts for identification. The library is capable of storing parameter information on 1000 radar types; these will be selected by the user to suit the anticipated operational requirements, and are loaded by means of a magnetic tape cassette.

The display presents data in alpha-numeric form, and typical information displayed includes the parameters of the intercepted signal, the results of the automatic comparison with the stored data to give possible identification(s) of the signal, and the associated platform or hostility index. Each comparative identification is given an automatic confidence level figure.

STATUS

In service with the Royal Navy and NATO. Export orders in 1987 exceeded £10 million.

CONTRACTOR

THORN EMI Electronics Ltd, Radar Division, Dawley Road, Hayes, Middlesex UB3 1HN, England.

4653.293

GUARDIAN NAVAL EW SYSTEM

Guardian is an advanced shipborne active and passive EW system in development for the Royal Navy. The major provisions meet a wide range of operational requirements against threats, particularly those of radar-guided missiles. The system is able to counter and confuse surveillance and missile homing radars by the use of jamming techniques for both area and point defence. Azimuth coverage is 360 degrees with an elevation cover of 50 degrees and a detection range of up to 500 km.

The Guardian system has two antenna mounts to ensure uninterrupted all-round view. The mounts can be installed either port and starboard, or at forward and aft positions. Each antenna mount carries a direction-finding receiving antenna together with transmit antennas, using mechanical steering to give full azimuth and elevation coverage.

Received signals are passed from the antenna to a wide-band receiver. They are then analysed to indicate those which may represent a threat and the selected signals are processed to allow generation of the jamming signal.

High power transmitter sub-systems, matched to the antenna, provide jamming capability against several types of target simultaneously. A comprehensive ECM library stores the appropriate countermeasures to a range of potential threats and on receipt of a warning from the ESM system, the ECM processor selects the appropriate jamming mode to give fast response.

System hardware is controlled at high speed by distributed microprocessors, and overall control is carried out by a micro-computer central processor which also controls the console and interfaces with the ship's systems.

The main equipment cabinets can be situated below decks or installed in containers for mounting on the upper deck or superstructure. The container can be customised to the ship's design and provides for easy removal and refit to other ships.

STATUS

Initial units have been delivered to the Royal Navy. A further order received from the UK MoD brings the total order to nearly £60 million.

CONTRACTOR

THORN EMI Electronics Ltd, Radar Division, Dawley Road, Hayes, Middlesex UB3 1HN, England.

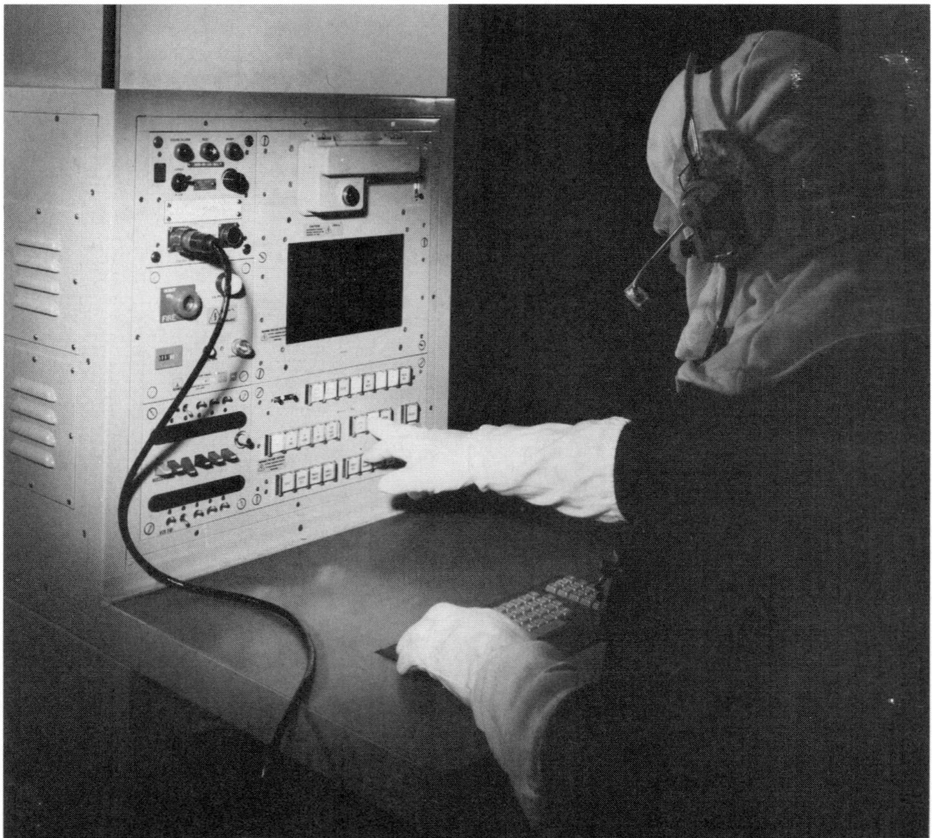

Guardian control console and display

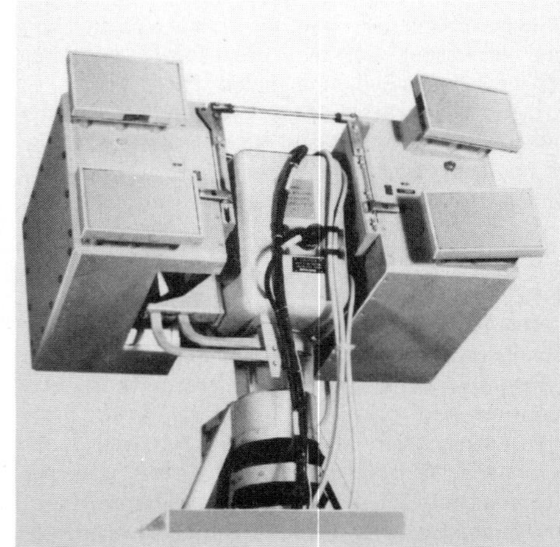

Guardian antenna array

1942.293

RCM SERIES ECM EQUIPMENT

The Racal RCM series of countermeasures equipment covers a range of radar jammers, either ship or land based, and operating in I or J bands to provide both responsive noise and deception jamming. The series is designed to jam early warning and target acquisition radars at long range and to jam and unlock missile guidance radars on launch platforms and missiles.

RCM requires only bearing information to counter a threat but, in addition, the ESM system normally provides warning that action should be taken against a particular radar plus an identification and allocation of priority. Single or dual antenna systems are available to provide 360° cover for small and large platforms. RCM units have their own receivers to provide RF for re-transmission after modulation.

In the responsive noise mode, a frequency measuring unit is used to fast tune a local oscillator to the frequency of the incoming pulse. For deception jamming the same unit can also be used but a frequency memory loop is offered as an option to provide a jamming capability against frequency agile radars. The incoming pulse frequencies are remembered and re-transmitted after modulation to provide a 'range gate pull-off' deception.

RCM-2 is as described above. Types of modulation available include Conscan and 'range gate pull off'.

RCM-3B jammers are improved versions of RCM-2 with a processor and an instantaneous frequency unit giving very efficient processor controlled power management and modulation sequences.

RCM-4B jammers have additional jamming modes such as noise cover pulse and combination jamming and a much higher effective radiated power.

STATUS

In production and in service with various defence forces world-wide.

CONTRACTOR

Racal Radar Defence Systems Ltd, Davis Road, Chessington, Surrey KT9 1TB, England.

1341.253

RDL SERIES ESM EQUIPMENT

The purpose of this equipment is to detect radar transmissions and so provide warning of the approach of surface ships, submarines, or other weapon platforms employing radar controlled weapon systems.

The main RDL systems are: RDL-1, a basic tactical small ship ESM system covering the 2 to 11.25 GHz range, and including an APA-1C pulse analyser; RDL-2 which also provides the frequency of a particular threat; RDL-257 extends the frequency coverage to 1 to 18 GHz; and RDL-268 provides for actual frequency measurement over this extended range. All RDL systems can be enhanced by addition of the SRU-1 signal recognition unit which can store the parameters of up to 500 emitters to permit threat identification. Versions of this equipment are now available for surface ships, submarines, aircraft, or vehicle use.

Surface Ships

A typical system for use on a fast patrol boat is the RDL-1BC, which provides instantaneous bearing, automatic pulse analysis and alarm, together with measurement of frequency band. Frequency coverage is from E to I band. The equipment comprises an antenna, a bearing display, a control unit, a power unit, and a pre-pulse unit, all of which, apart from the display tubes, are solid state. The antenna is light and compact, and has been designed in two halves to wrap around a mast or other mounting and thus avoid the coveted mast head position.

The automatic pulse analyser APA-1C is a self-contained unit which may be used with any suitable EW receiving equipment. Used with the RDL-1BC the unit provides an automatic digital readout of the pulse parameters of all signals or those selected from the switched lower antenna array. The unit is completely automatic and requires minimal operator skill or attention. For Elint purposes a line printer can be used to give a fast permanent record of the measured parameters of the intercepted radar transmissions.

There are also facilities available for the pulse parameters of up to five sets of independent radar transmissions to be set into the unit by plugging in printed circuit cards. When a signal is received which falls within these parameters, the operator is warned by the illumination of a solid state lamp or other remote triggered alarm, that a hostile transmission is present. The unit is completely solid-state and employs LED numerical indicators.

Add-on units are available which convert the RDL-1BC equipment to meet the more exacting requirements of larger ships. The RDL-2ABC equipment has all the facilities of the RDL-1BC and in addition provides frequency measurement, visual pulse analysis, and RF amplification in the analysis channel. The latter increases the range and permits better use of analysis facilities. The equipment consists of the units described previously and in addition contains an RF amplifier unit, a frequency and pulse analyser, and additional power supply units.

RDL console

The antenna is identical to the RDL-1BC unit but contains additional components to upgrade the performance. The initial simple system can therefore be readily converted to the more comprehensive system.

In the RDL system, frequency is displayed panoramically with signals appearing as vertical deflections on the horizontal frequency scale. The operator must first select the azimuth sector containing the required signal and the band he requires to sweep. Then he may select one of two modes, auto-stop or auto-pause. In the auto-stop mode, the receiver stops sweeping automatically on signal reception and the radio frequency is indicated digitally.

Other facilities permit further expansion of the frequency bands, fine tuning of the received signal to obtain better frequency accuracy, and the suppression, if required, of up to three specific frequencies. This enables the operator to prevent the receiver from stopping automatically on the friendly signals or those already analysed.

The RDL-2ABC system contains, in addition, a visual pulse analyser with a dual trace display. This enables two ranges of either pulse width or pulse repetition frequency to be measured visually against an illuminated graticule. This display also enables pulse peculiarities to be identified, including jittered PRF.

Elements of the RDL Series, RDL-5 and RDL-6, provide for the addition of a D-band capability. All facilities available for the RDL-2 series are either usable or available as additions for signals in this frequency band. Similarly RDL-7 and RDL-8 equipment extend the system capability up to band J8.

The system RDL-2ABC 6AC 8AC covers the full 1 to 18 GHz band with frequency measurement and automatic pulse analysis over the whole range. This equipment, with the addition of standard peripheral units such as computer injection panels, printers, and recorders, represents a system comprehensive enough to fulfil the most stringent requirements, but is nevertheless only a few steps away from the simplest tactical system in the series.

Submarines

The system RDL-4BC is a compact, instantaneous bearing measurement and pulse analysis equipment covering the frequency range 2.5 to 20 GHz. It is in full production in a submarine version, RDL-4BCS, where it uses the standard RDL Series units inboard. The bearing display uses coded vectors automatically to identify the frequency band of intercepted signals.

RDL-1BCS is a submarine version of the RDL-1BC and is identical operationally. It uses a novel antenna in which each of 16 antenna elements are individually encapsulated rather than using an overall submersible radome. Other submarine systems are produced ranging from a simple warner, RWR-1, for very light submarines to complex systems for NATO.

STATUS

Production. RDL Series in service or on order for 18 navies.

CONTRACTOR

Racal Radar Defence Systems Ltd, Davis Road, Chessington, Surrey KT9 1TB, England.

3553.293

CUTLASS ESM/ECM EQUIPMENT

The Racal Cutlass series of equipment consists of a range of advanced computer-controlled ESM and ECM systems, primarily intended for use aboard ships, but capable of deployment on other land-, sea- or air-based platforms. Designed for operation in very dense signal environments the equipment receives signals in the 1 to 18 GHz frequency range, measures their parameters, compares these with those in a pre-programmed radar library and displays the information within 1 second. The EW operator is presented with a tabular display for threat identity and threat evaluation and a tactical display giving a pictorial representation of the RF environment. Selected digital outputs can be sent to other local systems, and hard copy printout of the intercepted radar is also available. The tabular display can indicate 150 intercepts, in the order of priority. The Cutlass central processor is very advanced with a library containing the parameters of up to 2000 radars.

Cutlass is wide open in both bearing and frequency, ie it does not employ sweep techniques, giving a very high intercept probability (nearly 100 per cent).

Cutlass is a modular equipment and can be configured to suit fittings in various classes of ships.

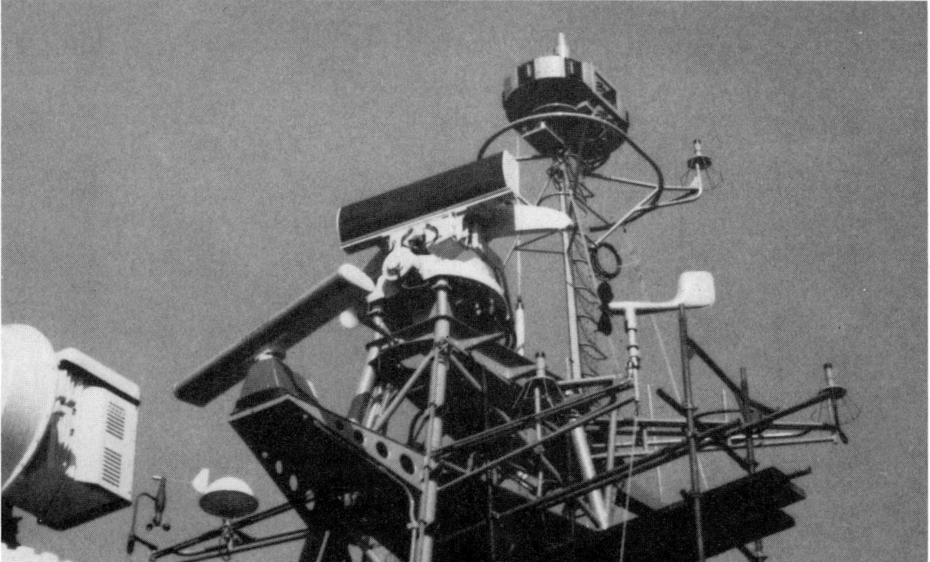

Cutlass antenna system mounted at masthead

The two main variants are Cutlass and Cutlass B1; both use the advanced Cutlass processor which can be integrated with a variety of man-machine interfaces and both use an instantaneous frequency measurement (IFM) receiver. Cutlass has a six-port antenna array for bearing measurement using amplitude comparison techniques, and a separate omni-directional antenna to provide RF for the IFM. Cutlass B1 has a 32-element array antenna to provide bearing measurement by phase analysis techniques. This antenna also provides RF for the IFM. In both systems the processor is provided with fast and accurate information on incoming pulses.

For an integrated EW system, Cutlass can be employed either in conjunction with one of the Racal RCM series of jammers (**1942.293**) or with the Racal Cygnus or Scorpion jammers (**5670.293** and **6884.293**). Cutlass will readily integrate with other ships' systems, such as a tactical data system.

CHARACTERISTICS

Cutlass B1

Accuracy: 2° rms in bearing

Coverage: 360° azimuth; +40 to -10° elevation

Pulse density capability: 500 000 per sec

Radar store capability: 2000 modes

STATUS

In production since 1979 and in service with several overseas navies. In mid-1983, Racal was awarded a £30 million contract for the development and reconfiguration of the Cutlass system for installation on board the Royal Navy Type 23 frigates. A version of Cutlass, known as Sabre, has been developed and is in production for a NATO navy.

CONTRACTOR

Racal Radar Defence Systems Ltd, Davis Road, Chessington, Surrey KT9 1TB, England.

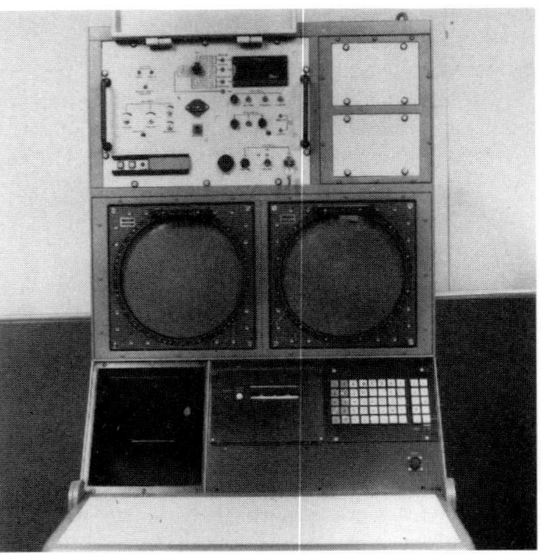

Cutlass console

1876.053

HWR-2 RADAR WARNING RECEIVER

This is a miniature warning receiver, designed specifically for use in environments such as helicopters and small patrol craft where space and weight are invariably at a premium and the installation of sophisticated electronic warfare equipment is uneconomical. In its basic form the receiver is hand-held and operates completely independently of available supplies and mounting arrangements (although it may be hard-mounted in rotary wing aircraft and surface ships).

Held by a pistol-grip handle, it can be used to locate pulsed radar transmissions of any frequency in the range 2 to 11 GHz. The presence of a radar transmission is indicated by listening to the audio note of the transmitted PRF on a headset. The electronic circuits are housed in a cast metal case designed to withstand rugged service conditions.

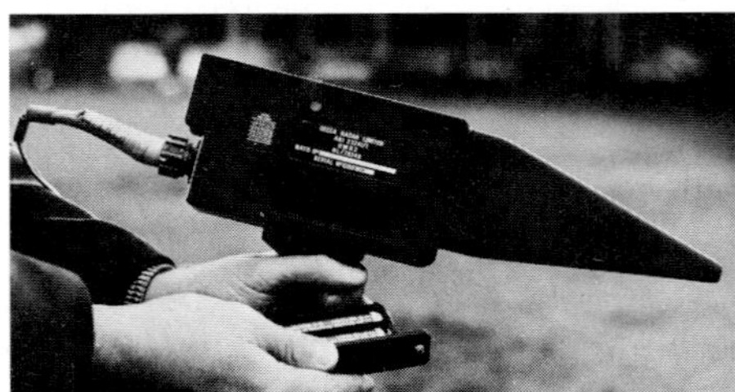

Racal hand-held warning receiver

All controls are grouped on the rear face of the receiver with the single multiway connector for electrical interfaces.

DEVELOPMENT

Newly developed models are available which extend the frequency coverage to 16 GHz and provide for switching over four bands, together with five position sensitivity selection and CW and pulse reception modes. Development of the system up to 18 GHz is continuing. In addition models can be obtained with RF amplification giving greatly increased sensitivity over all bands, or dedicated to a particular 100 MHz bandwidth at customer specified frequencies.

STATUS

In production and over 200 delivered to various users.

CONTRACTOR

Racal Radar Defence Systems Ltd, Davis Road, Chessington, Surrey KT9 1TB, England.

5670.293

CYGNUS ECM SYSTEM

The Cygnus radar jammer is a ship- or land-based equipment operating in the I- or J-bands. Originally designed for integration with Cutlass ESM equipment (**3553.293**), Cygnus uses both responsive noise and deception jamming to provide an effective jamming capability against all types of radar including early warning, target acquisition and missile guidance.

Cygnus has a narrow beamwidth giving very high effective radiated power (300 kW) and is kept on target by a built-in, interferometer type passive tracking system, in both azimuth and elevation. The tracking system may also be used to relay accurate position data to systems of other ships. RF received by the tracking antennas is also used as the basis of transmitted RF and modulation and power management is under the control of a processor which also controls the tracking procedure. Types of modulation sequence available include range gate pull off and false target generation.

STATUS

In production and in service worldwide. The Cygnus system is also being developed to work with Weasel (**5669.153**) as a ground jammer.

CONTRACTOR

Racal Radar Defence Systems Ltd, Davis Road, Chessington, Surrey KT9 1TB, England.

6884.293

SCORPION ECM SYSTEM

Scorpion is an advanced ECM jammer system designed to integrate with the Cutlass range of ESM equipment. It is the next generation of the Cygnus system (**5670.293**) and has techniques for wide and narrow band jamming (ie for area and point defence respectively). Operating over the frequency range 7.5 to 18 GHz, it offers a range of countermeasures designed to confuse long range search radars, target acquisition radars and missile radars operating in both their search and locked-on modes, whether fixed frequency or frequency agile. The ESM library within the Cutlass system contains a customer defined ECM response code for each known threat. This response code is passed to the Scorpion's ECM processor and jamming is initiated immediately.

STATUS

In production for an unspecified customer.

CONTRACTOR

Racal Radar Defence Systems Ltd, Davis Road, Chessington, Surrey KT9 1TB, England.

5671.453

PORPOISE SUBMARINE ESM EQUIPMENT

A submarine version of the Cutlass ESM equipment (**3553.293**), Porpoise is a fully automatic ESM system operating throughout 360° of azimuth. The equipment receives signals in the 2 to 18 GHz frequency range, measures their parameters and compares these with those contained in a pre-programmed radar threat library. Processing of the radar emitter signals is carried out against the library to give the operator an alpha-numeric or graphic display of identification in threat significance order.

Intercepted signals are preamplified in the mast unit before being passed to the processing and analysis equipment inside the hull. Bearing data is extracted using amplitude comparison. Amplitude, pulse width, frequency and time are combined in a single digital word before being passed to the processor unit in the operator's console. There the pulse trains of the different radars are de-interleaved and identified from the library information.

The system is capable of integration with the vessel's fire control and communications systems and may also be integrated with periscope mounted radar warning equipments, eg Sprat (**5672.493**). Porpoise also has the ability to give an alert warning when prime threats, such as helicopter or maritime surveillance radars, reach a pre-programmed danger level.

The Porpoise antenna is a compact 6-port system giving adequate bearing accuracy and may be mounted on either hull penetrating or non-hull penetrating masts. It is built of titanium to reduce weight and overcome corrosion, and is pressure resistant to 60 bar. With the exception of the antenna system, the primary sub-assemblies of Porpoise are fully compatible with those of Cutlass and so enable common logistic facilities.

STATUS

In production and in service. A reduced version has been ordered by the Danish Navy.

Porpoise console unit

CONTRACTOR

Racal Radar Defence Systems Ltd, Davis Road, Chessington, Surrey KT9 1TB, England.

5672.493

SPRAT SUBMARINE ESM SYSTEM

Sprat is a compact submarine ESM receiver designed to give radar warning of possible hostile emissions as soon as the periscope is raised. The system consists of an omni-directional antenna and receiver mounted on top of the search periscope, and an indicator unit in the control room. The receiver operates over the frequency range 8 to 18 GHz. Detected radar signals are passed to the indicator unit and are displayed as relative signal levels with a corresponding audio output denoting radar activity.

The Sprat antenna/receiver may be integrated with the Porpoise ESM system (**5671.453**), enabling the operator to carry out radar interception prior to raising the main ESM mast. The antenna is pressure resistant to 60 bar. Full built-in-test equipment (BITE) is provided, interlocked to the periscope to prevent BITE signal radiation when the periscope is raised.

STATUS

In development.

CONTRACTOR

Racal Radar Defence Systems Ltd, Davis Road, Chessington, Surrey KT9 1TB, England.

5675.293

SEA SAVIOUR EW SYSTEM

Sea Saviour is designed as a radar threat warning system for small warships with a chaff/infra-red decoy trigger capability. It covers the frequency range from 6 to 18 GHz using a 4-port amplitude comparison bearing measurement system to provide bearing over the full 360° in azimuth. The sensor package gives an immediate audible warning when emissions are received indicating that the vessel is being illuminated by a threat radar. An octantal display of the direction of the emission is also provided. Immediate countermeasures to missile or fire control radars can be made by an automatic decoy dispensing system. The Sea Saviour package provides an integrated system which can include sensors, receiver, processor, display and decoy dispensers.

CHARACTERISTICS

Azimuth coverage: 360°

Elevation coverage: –5° to +40° (approximately)

Frequency bands: H, I and J

DF accuracy: to within a 45° sector

Dectection time: within 1s

STATUS

In production.

CONTRACTOR

Racal Radar Defence Systems Ltd, Davis Road, Chessington, Surrey KT9 1TB, England.

Sea Saviour display unit

6883.253

UAF-1 ESM SYSTEM

UAF-1 is an advanced and automated ESM system which is developed under contract from the UK MoD. It measures the parameters of received radar signals and compares them with a pre-programmed library of radar parameters. Signal identity, bearing and threat significance are displayed in tabular format. Raw data on bearing and frequency is given on a display which can be overlaid with symbols and data derived from the Sadie ESM processor. Specific intercepts can be analysed further using a manual pulse analyser.

STATUS

Production systems are being delivered to the Royal Navy for the Type 23 frigate programme.

CONTRACTOR

Racal Radar Defence Systems Ltd, Davis Road, Chessington, Surrey KT9 1TB, England.

6885.293

SADIE ESM PROCESSOR

The Sadie ESM processor has been developed by Racal as a new generation equipment and is an integral part of nearly all the most recent Racal ESM systems. The processor splits into four main processing functions; pulse segregation, known emitter monitoring, new emitter analysis and ESM processor control.

The segregator recognises pulses from emitters which have been previously received and stored. It then segregates them from those new emitters which need to be analysed and classified. The key component is a large scale integrated circuit, known as the extended window addressable memory which performs the segregator matching function.

The known emitter monitor receives sets of data from the segregator, including the emitter number and pulse record. It accumulates the pulse parameters and performs significant change detection and peak detection for scan analysis. The new emitter analyser uses sophisticated algorithms to deinterleave and characterise emitters. Since the segregator has removed the large majority of the pulses, this allows the analyser to concentrate on processing new emitters.

The ESM control processor exercises overall control of the signal processor analyser processor. It uses a Pascal software system, supported by assembler modules.

STATUS

The Sadie processor is in full production and forms an integral part of several new Racal ESM systems.

CONTRACTOR

Racal Radar Defence Systems Ltd, Davis Road, Chessington, Surrey KT9 1TB, England.

5939.253

MENTOR NAVAL ESM SYSTEM

Mentor A and B form part of a family of low cost threat warning and ESM systems, which is ideal for minor warships, fleet auxiliaries and merchant ships. Despite their small size they give an important warner and sensor capability.

Mentor A is a computer controlled ESM system and radar warning receiver. It has a four port wideband crystal video receiver for detection of pulse and CW radar signals. The received signal parameters are measured by a dedicated processor which continuously samples the environment. The data samples are passed to a second processor for signal sorting and preliminary analysis. A third computer then arranges the sorted signals into individual emitter tracks and compares them with a signal library for identification. Signals identified as threats trigger visual and audio alarms. They also enable countermeasures to be initiated.

The identification tag provided by the library is passed to the display processor which arranges the display according to the setting of the controls. Either polar or tabular displays can be selected. Unidentified signals are indicated by displaying their measured parameters so that signals not included in the library can be analysed. The analysis computer can interact with the operator to provide a complete picture of the radar environment and the signal library can be selected to suit a particular mission and loaded onboard.

Mentor A comes in easy-to-fit components that are sited as convenient in the ship. The two antenna detector units are mounted one on each side of the mast or superstructure. The detected RF is then passed to the signal processing unit. The power supply and the Mentor B digital IFM are sited close by. The control and display units are located together, either in the operations room or on the bridge.

CHARACTERISTICS

Frequency coverage: E to J-band

Bearing accuracy: 10° rms

Processing time: 1 s

Library: 1000 radar modes

Azimuth coverage: 360°

STATUS

In production. Marconi has received a contract to supply an ESM system for the first of the Royal Navy's auxiliary oiler replenishment (AOR) vessels. The equipment is designated UAG and is presumably a variant of Mentor.

CONTRACTOR

Marconi Defence Systems Ltd, The Grove, Warren Lane, Stanmore, Middlesex HA7 4LY, England.

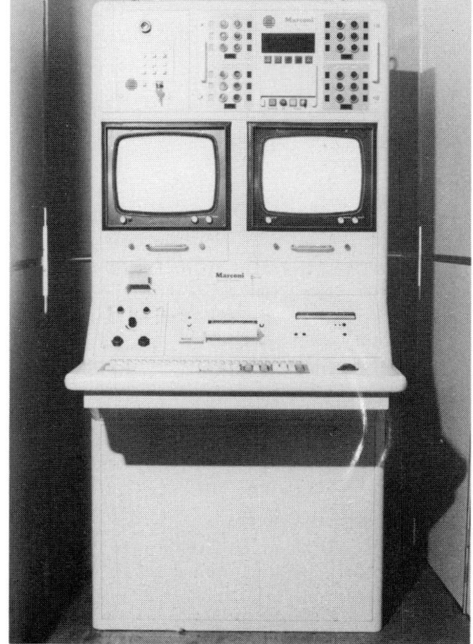

*Mentor console which also incorporates the control panel for the Siren decoy system (**5940.293**)*

5940.293

SIREN NAVAL DECOY SYSTEM

Siren is an offboard countermeasures decoy system designed to counter anti-ship missiles. It consists of an intelligent all-electronic decoy round, a simple multi-barrelled launcher, and a microprocessor-based fire control unit which can control up to four launchers. The system interfaces with the ship sensors to give fully automatic, semi-automatic or manual activation of the Siren decoy. Once launched, and away from the ship under attack, the decoy switches over to an internal power source, and radiates a powerful jamming signal while descending by parachute to seduce the missile away from its target. The total system can be adapted to suit almost any warship design from a patrol boat to a capital ship, and can also be fitted to naval auxiliaries and merchant vessels.

The decoy round consists of a low *g* rocket, a parachute, transmitter, receiver and antennas, control electronics, and a battery source. It is completely independent of the on-board ECM equipment. Dimensions are 1800 mm long × 130 mm diameter, and the weight is 28 kg.

The launcher is a 130 mm calibre fixed equipment in banks of three or six. Standard barrel elevation is 45 degrees. Two launcher positions are proposed for smaller craft and four for larger warships. Chaff and IR rounds can be accommodated.

The controller provides the necessary interface to other ship defence systems. It uses its own microprocessor to determine suitable countermeasures, launcher and deployment positions, and initiates firing. Manual operation can be carried out via a keyboard. Slave controllers are available for local control of individual launchers.

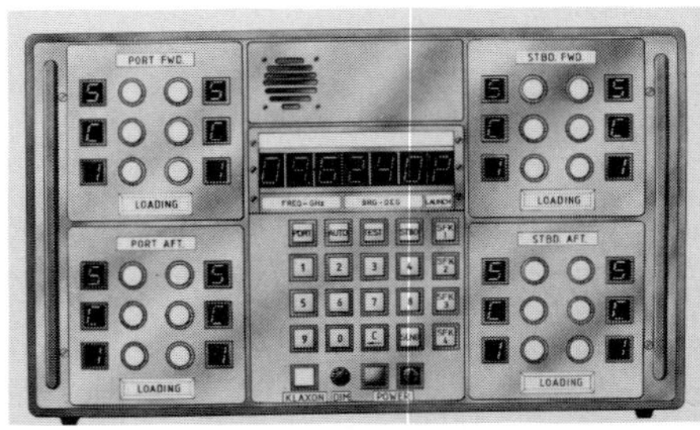

Siren naval decoy controller

STATUS

Preproduction.

CONTRACTOR

Marconi Defence Systems Ltd, The Grove, Warren Lane, Stanmore, Middlesex HA7 4LY.

6161.293

REPLICA NAVAL DECOY

Replica is a rapid broad-band response, RF passive naval decoy intended to provide a ship-like target to decoy anti-ship missiles. It is suitable for the distraction/false target role and its rapid response provides significant capability in the seduction role. It gives full azimuth and elevation coverage. The equipment consists of a launcher and a pair of life-raft type containers, each housing one octahedral shaped deployable radar reflector unit. Deployment in linked pairs ensures excellent azimuth response and a considerable horizontal extent. Operation is fully automatic and immediate with full radar cross-section achieved within a few seconds of hitting the sea. A remote facility can be incorporated allowing units to be pre-positioned and then activated on command. A self-destruct facility could also be incorporated if required. Container size is 1.14 m long by 0.55 m diameter, with a nominal weight of 67 kg including the decoy.

Deployment from a ship is by a simple launching cradle which can be fitted to all classes of ships. In its simplest form the launcher can be operated manually and requires no services. Remote, powered operation can also be provided.

A training system, known as Cygnet, is also available. In addition it is understood that an air-launched version of Replica is under consideration.

STATUS

In operational service with the Royal Navy, under the designation DLF-2.

CONTRACTOR

Irvin Great Britain Ltd, Research and Development Division, Works Road, Letchworth, Hertfordshire SG6 IND, England.

Replica deployed

6150.293

RADAMEC EW ANTENNA DIRECTOR MOUNTS

Radamec designs and manufactures a wide range of rugged precision director mounts to accommodate differing payload configurations, weights and stabilisation requirements. Single, dual and three axis stabilisation are available with slip ring options for continuous rotation. The Types 409, 429, 449 and 459 family of dual axis mounts provide fast high precision directional capabilities with slewing rates of up to 360° per second and positional readout accuracies better than five arc seconds. The Type 459 offers a compact, direct drive, dual axis yoke mount with the payload centred on both axes of rotation.

STATUS

Directors have been supplied for the THORN EMI Guardian jammer supplied to the Royal Navy, the Racal Cygnus jammer systems supplied to Egypt and to other navies, and classified EW systems currently in Royal Navy service.

CONTRACTOR

Radamec Defence Systems Ltd, Bridge Road, Chertsey, Surrey KT16 8LJ, England.

Radamec Type 429 incorporated in Cygnus radar jammer

UNITED STATES OF AMERICA

6223.293

AN/WLR-1H RADAR DETECTION EQUIPMENT

The WLR-1H is a passive radar detection set which has been modified to tie in with existing jammers and missile fire control system to provide an active defence against missiles, in particular the Soviet Styx shipborne surface-to-surface missile (**1155.221**). The system, in various configurations, has been in use for nearly 25 years, although all the earlier versions are being phased out in favour of the latest variant, the AN/WLR-1H. This version features a microprocessor, with a central processor to manage the entire system, a threat library, and an automatic search and signal identification mode. The frequency range covered is from 7 to 18 GHz.

STATUS

In production. ARGO Systems is carrying out modifications to the AN/WLR-H(V)1 and (V)3 systems.

CONTRACTORS

ST Research, Newington, Virginia, USA.

ARGO Systems Inc, 884 Hermosa Court, Sunnyvale, California 94086, USA.

6497.493

AN/WLQ-4 SIGINT SYSTEM

The AN/WLQ-4(V) performs SIGINT missions aboard SSN-637 class submarines of the US Navy. It is an automated, modular signal collection system which allows for the identification of the nature and sources of unknown radar emitter and communications signals. It incorporates a network of minicomputers and microprocessors, and data from these computers is correlated with information received from satellite sensors. The system is part of Sea Nymph, a highly classified US Navy NAVELEX programme.

The AN/WLQ-4 system has a number of key features, including:

(a) Automatic search, acquisition and signal processing
(b) Automatic logging book-keeping and reporting
(c) Semi-automatic correlation of real-time measured data with input from an external system
(d) 400 000 lines of AN/UYK-20 source code
(e) 50 000 lines of executable code in 40 microprocessors
(f) A significant growth capability to handle new threats.

STATUS
In full scale production for the US Navy submarines. In November 1987 GTE received a $10 million contract for repackaging the AN/WLQ-4 for the SSN-21 class submarines.

CONTRACTOR
GTE Government Systems Division, 100 Ferguson Drive, Mountain View, California 94039, USA.

AN/WLQ-4(V) skeletal system in a US Navy SSN-637 submarine

3082.253

AN/WLR-8 TACTICAL EW RECEIVER SYSTEM

The AN/WLR-8 is a tactical electronic warfare and surveillance receiver designed for fitting in both surface ships and submarines of the US Navy. The system is of modular construction and provisions are made for operation in conjunction with numerous types of direction finding or omni antennas and a wide range of optional peripheral equipment to provide comprehensive ESM (electronic support mission) facilities. The WLR-8 is compatible with NTDS (navy tactical data system) and similar action information automation systems. The system can be expanded in frequency or signal handling capability by means of simple additions and/or software changes. Three versions are available; V(1) for submarines, V(2) for SSN-688 class submarines and V(5) for Trident submarines.

Two digital computers are incorporated: a Sylvania PSP-300 for system control, automatic signal acquisition and analysis, and file processing; and a GTE PSP-200 microcomputer for hardware level control functions. Digital techniques are employed throughout the WLR-8 system, which is all solid-state.

Operational facilities provided include:

(1) automatic measurement of signal direction of arrival
(2) signal classification and recognition
(3) sequential or simultaneous scanning over a wide frequency range
(4) signal activity detection for threat warning
(5) analysis of signal parameters such as frequency, PRF, modulation, pulsewidth, amplitude, scan rate etc
(6) logging of signal parameters for display to operator(s), and printout of hard copy to teletype or printer
(7) extensive built-in test equipment
(8) directed priority searches of specific frequency segments.

Direct reporting to on-board computers, such as NTDS, permits response times in the millisecond range with minimal operator involvement. A two-trace CRT is provided for display purposes and this can be supplemented by an optional five-trace panoramic display for presentation of signal activity data. Another CRT display is incorporated if the WLR-8 is used with automatic or manual DF antenna systems.

STATUS
Operational with USN and possibly other navies. More than 2700 sets have been produced. The system is in use on board SSN-688 class submarines, and the V(5) version is fitted to Trident fleet ballistic missile submarines.

CONTRACTOR
GTE Government Systems, 100 Ferguson Drive, Mountain View, California 94039, USA.

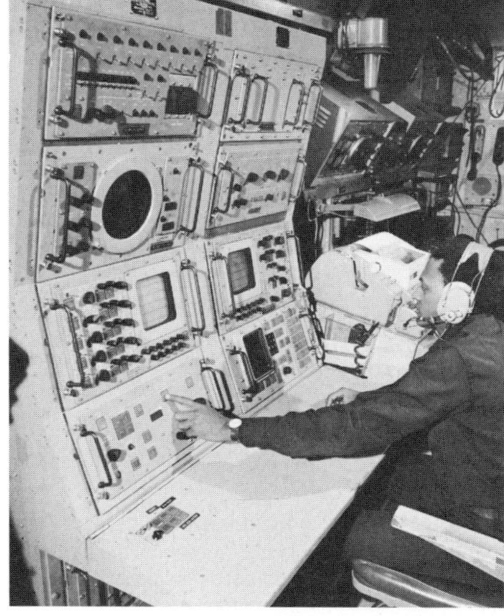

Operating console of AN/WLR-8 tactical EW receiver

6887.453

FERRET SUBMARINE EW SYSTEM

Ferret is a new submarine ESM system which is intended to meet the needs of the latest NATO attack submarines. It features a flexible architecture adaptable to changing operational requirements, in a small configuration with a built-in growth capability for the 1990s and beyond. It provides broadband signal collection and processing, with fully automatic detection, collection, analysis and recording functions over the frequency range from 1 to 40 GHz.

The wideband acquisition subsystem uses a channelised receiver to cover the frequency range and a high performance signal processor to sort and classify intercepted signals. The narrowband acquisition subsystem uses up to four superheterodyne receivers for fine grain operator analysis. The DF processor interfaces to the ESM mast and provides monopulse DF capability.

CHARACTERISTICS
Frequency range: 1 to 40 GHz (wideband); 1 to 18 GHz (narrowband)
Frequency resolution: 5 MHz rms (wideband); 125 kHz (narrowband)
DF accuracy: 5 to 10° over the frequency range 1 to 18 GHz
Weight: 681 kg

STATUS
GTE states that the Ferret equipment has been developed for the export market. It is understood that, as at the beginning of April 1988, negotiations were in progress with a number of countries, including Canada and Israel.

CONTRACTOR
GTE Government Systems, 100 Ferguson Drive, PO Box 7188, Mountain View, California 94039, USA.

5835.293

AN/WLR-11 IFM RECEIVER

The AN/WLR-11 is an IFM receiver intended for shipborne applications to detect and provide a rapid analysis of anti-ship missile threats. It operates over the frequency range 7 to 18 GHz with a 100 per cent probability of intercept. The equipment is employed with an omni-directional and a DF antenna system which is capable of scanning at 300 rpm. It is used with an automatic processor which can call-up emitter parameters from the computer memory and use these to improve the overall reaction time. As an interim measure, to outfit the first 10 US Navy DD-963 'Spruance' class destroyers, the AN/WLR-11 is being integrated with the AN/WLR-1G.

CONTRACTOR
ARGO Systems Inc, 884 Hermosa Court, Sunnyvale, California 94086, USA.

3551.293

AN/SLQ-17 NAVAL ELECTRONIC COUNTERMEASURES SUITE

The AN/SLQ-17 naval countermeasures set is a deception jammer for shipboard use. It provides the appropriate threat detection and jamming techniques to counter threats and is equipped with the requisite microwave sources and antennas. When coupled with the AN/WLR-8 ECM receiver the SLQ-17 forms the AN/SLQ-29 suite.

The system is designed to protect carriers and other major ships against cruise missiles and other intruders by developing an electronic "image" of the target ship as it is sensed by the enemy's guidance radar. It then offsets the image so that the weapon guides itself to a false "ghost" target some distance from the real one.

The automated SLQ-17 system is designed to operate in dense electromagnetic environments. It detects and tracks signals of several score missiles of

An AN/SLQ-17 antenna mounted on the superstructure of a US Navy ship

many types, and will track up to several hundred signals from other platforms. The system keeps track of signals used for navigation, search, and tracking in the frequency bands that are in the emission range of present and future hostile platforms. It can intercept and track potential missile launch platforms beyond the ranges at which the attacking missiles lock on to the ship and automatically identify both threat and friendly platforms.

As a signal is received on either of the two antennas, the SLQ-17 automatically measures the radar parameters emitted and the computer searches its memory files to separate friend from foe. If it identifies the signal as a missile radar, it generates the most effective deception for the missile, causing it to strike the ocean safely away from the carrier. Meanwhile, the system automatically keeps track of signals in its area, displaying their platforms for the operator to review on his tactical situation display. Information on all the platforms is displayed for the operator's action in order of their decreasing threat level.

STATUS

In operational service. Eight systems had been delivered by the latter part of 1985 and a further five systems were delivered in 1986. No further orders are envisaged.

CONTRACTOR

Hughes Aircraft Company, Fullerton, California 92634, USA.

4388.293

AN/SLQ-30 TRUMP

The AN/SLQ-30 is an integrated shipborne defensive electronic countermeasures system designed for the US Navy Electronic Systems Command. The name TRUMP is an acronym of Threat Reactive Update Modernisation Programme and the system is designed to improve effective ECM action and response times while still using existing hardware. It comprises extensively modified ULQ-6B transmitters and SLA-12 antennas, a broadband interception receiver and a processor. The digital processor carries out threat analysis, assigns priority of action, selects the appropriate ECM response and operates overall system command.

STATUS

No longer in production but still in service operationally.

CONTRACTOR

SEDCO systems Inc (a Raytheon company), EW Technology Center, 1578 Rt. 23, Wayne, New Jersey 07470, USA.

4389.293

SR-200 ESM SYSTEM

The SR-200 is a modular electronic warfare support system for signal and threat identification and location. The system is designed for land-based, naval and airborne applications. It performs automatic threat detection and identification, and provides alarm signals on a priority basis, using emitter profiles stored in the memory of its high-speed digital processor. The system covers the frequency range 500 MHz to 18 GHz with an optional extension up to 40 GHz. Bearing accuracy in the I/J band is better than $\pm 2°$ rms.

An omni-directional antenna is used to pass on signals of interest to a rotating direction finding antenna which provides outputs into the equipment. A display unit provides information on the hostile transmissions, as well as giving an order of priority on those considered most hostile.

STATUS

The SR-200 is fitted to a number of naval forces and is also in production for a land mobile application.

CONTRACTOR

Sanders Associates Inc, Countermeasures Division, Nashua, New Hampshire 03061-2004, USA.

SR-200 ESM system

6156.253

GUARDIAN STAR SHIPBORNE EW SYSTEM

Guardian Star is a family of EW systems designed to cover requirements from threat detection to electronic intelligence (ELINT). Various configurations allow the system to be configured to the specific requirements of surface ships and submarines.

The Mk 1 system offers early warning to small surface craft or patrol boats, with an average bearing accuracy of $\pm 10°$ provided by octave frequency measurements. The Mk 2 system is designed to meet the basic ESM needs of surface ships and submarines. It provides YIG tuned frequency measurement and emitter average bearing accuracy of $\pm 5°$. The Mk 3 is an ELINT system for surface ships and submarines. It carries out instant threat warning and accurate frequency measurement by IFM devices in the receiver front end. Average bearing accuracy of less than $\pm 5°$ is achieved.

The systems consist basically of an antenna assembly, an RF/digital interface unit (RFDIU) and a display/controller unit (DCU). The antenna assembly comprises an omni-directional and six spiral DF antennas covering the frequency range from 2 – 18 GHz. All the necessary preamplifiers and preprocessing components are included in the assembly. The RFDIU processes all signals before transfer to the main digital processor in the DCU. All signal conversion electronics, auxiliary outputs and power supplies are contained in this RFDIU. The DCU includes the input/output section, main processor, magnetic tape unit, keypad and all required operator interfaces. The operational program and library file data are loaded into the processor memory via the magnetic tape unit.

In the surveillance mode the system operation is broadband from 2 – 18 GHz. This mode is entirely automatic and the operator has only to view the situation summary display. This display will provide data on up to 50 emitters (10 per page). The displayed emitter characteristics are frequency, PRF/PRI, pulse width, pulse amplitude, true bearing (instantaneous and averaged), scan rate, emitter name (if matched in the library), threat level (if matched in the library). The library file handles up to 2000 sets of emitter parameters. The operator can store previously known emitter data together with preassigned threat levels in the file. The system matches incoming emitter data with the file and issues an immediate alert on high interest threats. Library data can be changed or entered via the display/controller unit.

CHARACTERISTICS

Frequency range: 2 – 18 GHz

Library: 2000 emitter parameters

Dimensions: Antenna assembly 360 mm high × 15 cm diameter (including radome); RFDIU 430 × 360 × 350 mm; DCU 250 × 430 × 530 mm

Weights: Antenna 5 kg; RFDIU (Mk 3) 33.5 kg; DCU 27 kg

CONTRACTOR

Sperry Corporation, 1725 Jefferson Davis Highway, Suite 401, Arlington, Virginia 22202, USA.

4671.253

S-2150 ESM SYSTEMS

The S-2150 ESM systems are shipborne threat warning equipments normally fitted to smaller ships and submarines. They are automatic in operation and provide warning and direction finding facilities.

The S-2150-01 provides instantaneous direction finding cover over the frequency range 2 to 18 GHz, with an accuracy of 10 degrees rms. The system consists of a masthead unit containing the antenna array and receiver, and a processor/display and control panel. An emitter library is referenced with each signal intercepted and an audio alarm is activated if the signal is matched with a possible threat. Information is also displayed in alpha-numeric and graphic form on the display panel.

The S-2150-03 is also an automatic surveillance system that detects and analyses signals in the 2 to 18 GHz band and provides direction finding with a five degree accuracy. The equipment operates independently or as part of an overall EW suite. An operator display provides threat symbology, bearing and approximate range as well as other data.

CONTRACTOR

EM Systems Inc, 45757 W. Northport Loop, Fremont, California 94538, USA.

3349.293

AN/SLQ-32 SHIPBOARD EW EQUIPMENT

The AN/SLQ-32 is a naval electronic warfare equipment developed to replace AN/WLR-1 receivers and AN/ULQ-6 deception jammers aboard many USN ships. The equipment is the result of a USN/industry programme lasting three or four years, in the course of which Hughes and Raytheon designed and built competitive models for test and evaluation. The Hughes system was designated AN/SLQ-31, and the successful Raytheon version, the AN/SLQ-32.

The AN/SLQ-32 consists of three modular variants of increasing levels of complexity and cost. The basic version, AN/SLQ-32(V)1, provides for warning, identification and direction finding of incoming radar-guided anti-ship missiles; the (V)2 version provides the facilities of the (V)1 plus early warning, identification and direction finding of those radars associated with the targeting and launch of the missiles; (V)3 includes the (V)2 capability to which is added a jamming capability for the prevention or delay of targeting and launch of missiles, and also for the deflection of any missiles that are launched away from their target. The (V)1 was designed for 'Knox' class frigates and smaller auxiliary ships; the (V)2 with its expanded ESM capability for DDGs, FFGs, and 'Spruance' class destroyers; and the (V)3 version with the active jamming system will be fitted in cruisers and certain other larger vessels.

All of the AN/SLQ-32 systems employ multiple beam antennas for reception in all bands except the lowest frequency band, and the lens-fed multiple beam array is a major feature of the system. The antenna consists of an array of elements fed through coaxial cables by a multiple beam parallel-plate lens which is constructed in stripline form using printed circuit techniques. This array provides a set of individual, contiguous high-gain beams, all existing simultaneously, and with each beam having the full gain of the array aperture. For a single array, more than an octave of frequency coverage is provided.

AN/SLQ-32(V)1 Suite

This version is designed to detect all RF signals in Band 3 at all azimuths, covering all potential enemy radar-guided anti-ship missiles and their associated supporting radars. By this ESM the (V)1 suite provides alert facilities to an attack on the ship and also allows for the control of the Mk 36 chaff rocket system (**3363.293**) carried by the ship. The (V)1 receivers are wide-open in both angle and frequency, and the complete suite consists of two antenna assemblies each having two Band 3 direction finding receiver arrays and lenses covering 90° each. One Band 3 semi-omni antenna covering 180° is located on each antenna assembly. In parallel with the multi-beam antenna and direction finding receivers the system employs a semi-omni antenna to sense the dominant pulse and feed it to the IFM receiver.

A special purpose digital processor, known as the presorter, consists of a direction/frequency correlator and digital tracking unit sections; coarse frequency data from the IFM receiver and amplitude data from the direction finding receiver are correlated in the direction/frequency correlator. Time of arrival data is added to form a pulse descriptor word, which is then stored by frequency and angle cell in the emitter file memory section of the digital tracking unit. If three or more pulses with this angle and frequency signature are received within a programmable time interval (up to 32 milliseconds), the digital tracking unit informs the computer that a new emitter is present. The computer then commands the digital tracking unit to provide enough additional pulses to permit further extensive analysis. The computer calculates from this data the PRF, type of scan, scan period and frequency; these parameters are usually sufficient to characterise an emitter.

Identification is completed by comparing the observed signal characteristics with parameters stored in a library of friendly and threat emitter characteristics within the computer memory. On completion of the identification processs, the computer initiates appropriate alert signals and other actions.

AN/SLQ-32(V)2 Suite

The increased capability of this version is achieved by the addition of two receiving sub-systems. Two Band 2 direction finding receivers and arrays and lenses each covering 90° are added to each of the port and starboard antenna assemblies, and a Band 2 semi-omni antenna covering 180° is added to each

One of the antenna arrays of the AN/SLQ-32 naval EW system (Stefan Terzibaschitsch)

assembly. Four small Band 1 spiral antennas are located on a yardarm (port and starboard) to cover 360°. As well as the necessary extra electronics needed to handle the increased input data, the AN/UYK-19 computer remains the same as in the (V)1 version apart from having an additional 16 K of storage capacity.

AN/SLQ-32(V)3 Suite

The ESM capability of the (V)3 suite is the same as that of the (V)2 suite. The active electronic countermeasures (AECM) function of this suite requires two more racks of equipment in the EW room. The main additions are: eight high-voltage power supplies for the TWTs, a transponder, a digital switching unit and a techniques generator. The computer remains the same except for a further increase in memory capacity from 64 K to 80 K. The outboard antenna assemblies are enlarged to accommodate Band 3 transmitter antennas and electronics; two are located in the lower portion of each assembly.

The (V)3 antenna assemblies are hydraulically roll-stabilised.

The AN/SLQ-32(V)3 AECM mode can be operated semi-automatically, where the operator initiates jamming against a given emitter; or automatically, under computer control, where the system initiates countermeasure action as soon as a threat is identified. In each case the actions of the system are essentially the same. The library identifies the jamming technique most appropriate to the particular threat encountered. The parameters of the technique designated in this way are passed to the technique generator unit (TGU), which synthesises the waveforms and applies them as modulation to voltage controlled oscillators in the transponder and the associated drive unit. The output of the drive unit is switched to the appropriate input of the transmission lens to ensure that the energy of the output tubes is directed in the required beam direction. The computer schedules the transmission of energy associated with a given target so that several hostile emitters can be engaged at one time.

STATUS

In full production for the US Navy with over 300 systems delivered. Raytheon was awarded two additional contracts in September 1987 and January 1988 totalling more than $80 million.

CONTRACTOR

Raytheon Company, Goleta, California, USA.

6220.293

AN/SSQ-72 OUTBOARD SYSTEM

The AN/SSQ-72 is designed to provide technical and tactical intelligence data such that the force commander can analyse enemy dispositions and intentions for over-the-horizon (OTH) detection and identification of surface ships for targetting purposes. It is composed of the AN/SRD-19 shipborne direction finding set, the AN/SLR-16 countermeasures receiving set and the OK-324/SYQ System Supervisor Stations. The Systems Supervisor Station integrates and controls Outboard sensors as well as external data links and communication to provide information processing and technical dissemination. The system uses hull and mast-mounted antenna arrays.

The system is being fitted in pairs to 24 guided missile frigates, destroyers and cruisers to provide signals intelligence, early warning and OTH capabilities to two surface ships that accompany each US Navy carrier force.

STATUS

In production.

CONTRACTOR

Sanders Associates Inc, Surveillance Systems Division, Hudson, New Hampshire 03051-907, USA.

3910.253

SLR-600 SERIES NAVAL ESM EQUIPMENT

This range of naval ESM equipment at present comprises three models, of which the base model is the SLR-600. This is a fully automatic and compact equipment primarily intended for fitting in fast patrol boats and similar classes of naval craft, where limited crew numbers are usual and fully automatic operation is of particular value. The SLR-600 combines a broadband E- to J-band microwave receiving system with a specially designed EW processor to ensure both maximum signal intercept probability and rapid system response. The display facilities present signal activity in range and bearing format, and the operator may also select a display which gives a readout at the bottom of the CRT of the operating parameters of selected emitters.

The normal installation aboard a ship includes four combined antenna/receiver units which can be mounted either as a group of four around a mast to give 360° azimuth coverage, or at the ends of yardarms etc to achieve the same result. The display is available as a single console for fitting in the bridge or wheelhouse, or alternatively the display can be configured for use as a separate unit for shelf or overhead mounting in the bridge or as a stand-up bridge control. Display modes include:

(1) radar identification, which gives specific radar identification based on programmed radar data
(2) unknown characterisation, providing automatic characterisation (pulse width, PRF, frequency, and scan) for unprogrammed radars
(3) radar parameter readout, which provides means for the operator to select any displayed radar for detailed parameter readout on the CRT.

Alarm audio and visual warning indications are given also.

CHARACTERISTICS

Frequency coverage: Continuous E- to J-band in 4 bands, C and D band detection available
Receiver type: Crystal video
Azimuth coverage: 360°
DF accuracy: Better than 15°
Shadow time: 2μs
Processor memory: 18 000+ words E² PROM/RAM/ROM
Emitter library storage: 100+ emitters; 1000 operational modes
Emitters displayed: Up to 15 simultaneously, with identity, status, range and bearing

SLR-610 naval ESM system

Display: CRT with computer-controlled symbology
Weight: 56.7 kg
Power consumption: 465 W

SLR-610

The SLR-610 provides the same performance as the SLR-600 but with extended frequency coverage and an alternative method of data display. This display is similar to the SLR-600 except that the radar parameter readout is provided by a column of multiple digit LED panels which show the parameters of the selected radar. There are also facilities for operator monitoring of intercepted radar audio signals.

SLR-640

The SLR-640 provides the highest level of ESM capability in the SLR-600 series and is designed to provide a complete ESM facility for fast patrol boats, frigates and destroyers. It provides broadband, high sensitivity signal detection and frequency measurements together with precision direction of arrival information. A dual-mode CRT display combines a polar plot of total signal activity with a detailed radar parameter readout on a selected list of active emitters over C to J band.

STATUS

Production. Selected by an overseas navy.

CONTRACTOR

General Instrument Corporation, Government Systems Division, 600 West John Street, Hicksville, New York 11802, USA.

3363.293

RBOC II

The Rapid Bloom Off-board Countermeasures (RBOC) launching system comprises a family of mortar launchers, countermeasures cartridges, and appropriate control and support arrangements for the protection of ships of all sizes. The number of launchers per ship, and the number of tubes per launcher, can be tailored to the needs of each particular vessel, as can the cartridge payload. An enhanced version, RBOC II Decoy Launching System, is now in service. Loral Hycor has added a proprietary management processor, CAMLAC, for fully automatic operation of the launchers. All RBOC installations make use of standardised launch tubes, firing circuits, and cartridges. Mk 33 is the designation for a four-launcher system; there is also a two-launcher system designated Mk 34.

A typical arrangement is shown in the block diagram.

The cartridges employed are the Mk 181 Mod 0 chaff load, the HIRAM II (Hycor Infra-red Anti-Missile) decoy, the GEMINI which combines chaff and IR payloads in a single cartridge, LOROC (Long Range Offboard Chaff) a rocket propelled chaff decoy with user selectable ranges 1 to 4 kilometres from the ship, and CHAFFSTAR, a chaff payload with a four times greater radar cross-section than earlier cartridges. These cartridges have been upgraded over the past few years to provide a greater countermeasures capability.

When launched from a shipboard mortar (Mk 135 or equivalent) the Mk 171 cartridge rapidly forms a large chaff cloud which confuses radar seekers and fire control systems. Small vessels, such as fast patrol boats, are protected by firing a single Mk 171 cartridge. Ships of larger size are protected by firing an appropriate number of cartridges. The radar cross-section of the Mk 171 chaff cloud is large enough to be effective without any special manoeuvring by the launching ship in most cases.

The Mk 171 Mod 0 chaff cartridge is a cylinder of 112 mm diameter with an overall length of 412 mm. The cylinder is closed at one end by an aluminium cap and at the other by a base plate. A plastic obturating skirt at the base acts as a gas seal to contain the propelling gases. Threaded to the base plate is an assembly containing the firing coil, propelling charge, and pyrotechnic time delay. A burst charge is located along the centreline of the cylinder, with the chaff payload packed around the charge. Inductive initiation is employed to eliminate electric connectors and provide complete HERO and RADHAZ safety.

The HIRAM decoy cartridge is designed to attract anti-ship missiles that employ IR seekers. When launched from a shipboard mortar (Mk 135 or equivalent) the HIRAM decoy cartridge quickly deploys an IR flare which is equipped with a float. The flare presents an attractive alternative target to the missile's IR seeker. The output of a single flare is sufficient to simulate the radiant intensity of even a large ship. The HIRAM II decoy cartridge upgrades the earlier HIRAM I by combining longer coverage time with greater radiant output.

The HIRAM decoy cartridge is a cylinder of 112 mm diameter and overall length of 417 mm. The flare is located within a tube along the centreline of the cylinder. Within the flare is a deployment charge, ignition mix, and the flare fuel mixture. Attached to the flare are a CO_2 cartridge, a rubber float (deflated), a folded parachute, and a folded self-erecting staff.

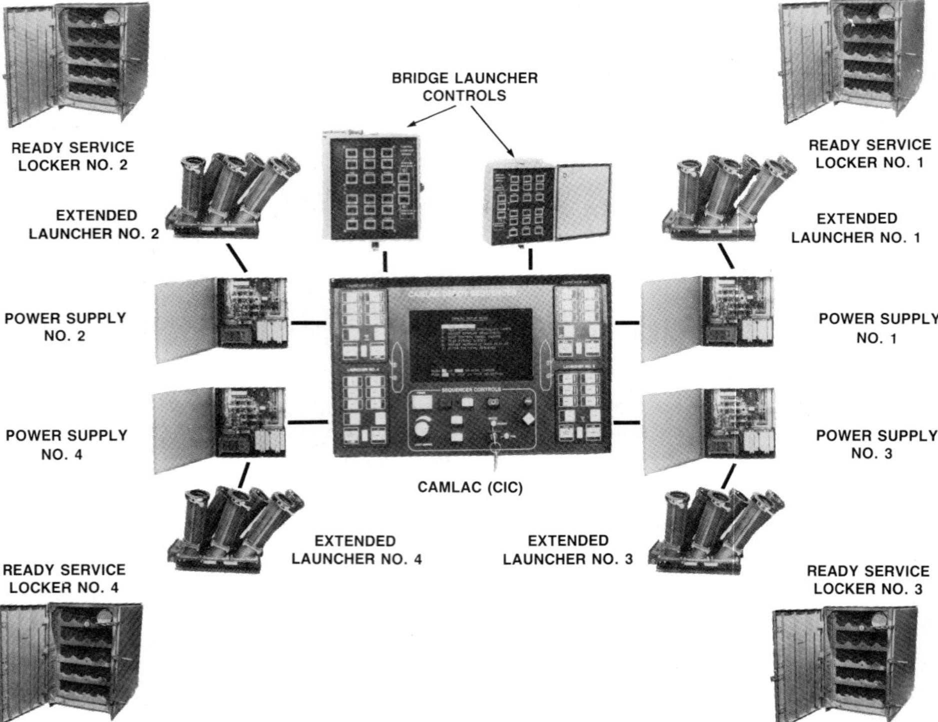

RBOC II launching system

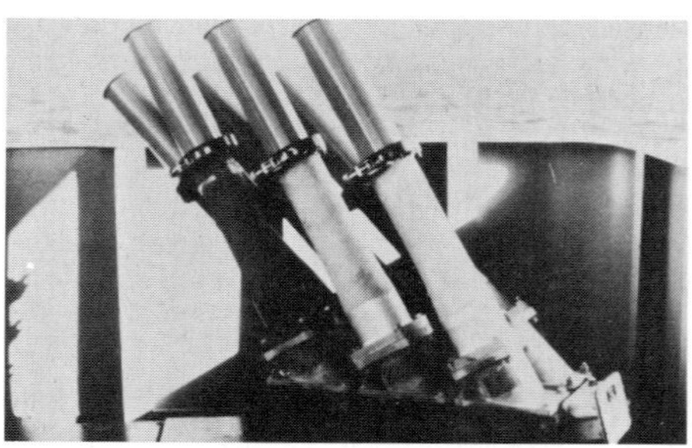

Loaded Super-RBOC launcher

Within a few seconds after launch, the GEMINI decoy cartridge develops a large radar echo and simultaneously generates a highly effective IR decoy signal. A small vessel (eg a fast patrol boat) may be protected from radar guided anti-ship missile by firing a single GEMINI cartridge. Larger ships are protected by firing additional cartridges.

The IR flare, similar to the HIRAM unit, presents an attractive alternative target to a missile employing an IR seeker. The output of a single GEMINI flare is sufficient to simulate the radiant intensity of even a large ship.

The GEMINI cartridge is a cylinder of 112 mm diameter with an overall length of 434 mm. The chaff payload is primarily packaged within the bottom portion of the cartridge, identical to the Mk 171 chaff cartridge. The IR flare, its ignition mechanism and folded parachute occupy most of the upper section of the cartridge.

The LOROC cartridge is a rocket propelled chaff decoy providing a large radar cross section at user selectable ranges of 1 to 4 kilometres from the ship. When combined with traditional close-in chaff decoys, LOROC provides optimal protection against missiles or radar guided artillery attack.

The CHAFFSTAR chaff decoy cartridge incorporates recent breakthroughs in chaff dipole dispersion techniques. CHAFFSTAR produces an exceptional rapid deploying chaff cloud with a radar cross-section approximately four times greater than earlier chaff cartridges. An upgraded version, CHAFFSTAR II is now available for use with RBOC II.

STATUS

Supplied to the US Navy and the navies of other countries. Loral Hycor Inc produces the Mk 33/Mk 34 and RBOC II launch system. HIRAM, GEMINI, LOROC, CHAFFSTAR, CAMLAC and RBOC II are proprietary developments by Loral Hycor Inc.

CONTRACTORS

Loral Hycor Inc, 10 Gill Street, Woburn, Massachusetts 01801, USA.

Lundy Electronics & Systems Inc, 3901 NE 12th Avenue, Pompano Beach, Florida 33064, USA (Mark 171 cartridges).

Tracor Aerospace Inc, Advanced Systems Operation, Bollinger Canyon Road, PO Box 196, San Ramon, California 94503, USA.

4656.293

MK 36 SUPER RAPID BLOOM OFF-BOARD COUNTERMEASURES

The Mk 36 Super Rapid Bloom Off-board Countermeasures system (SRBOC) provides a deck-launched self-protection against radar or IR guided anti-ship missiles. SRBOC can be activated automatically by the shipborne countermeasures system or by a dedicated microprocessor programmer with interface to the on-board radar warning receiver, IR warning receiver, ship computer, fire control system or other emitter threat data sensors.

The system has large-size expendable decoys and is designed for ships more than 550 feet long. SRBOC is available in either Mod 1 (two launchers) or Mod 2 (four launchers) configuration. This includes six tube launchers, internal battery power for each launcher, bridge remote controls, master launcher control and ready storage locker for each launcher. The decoys are identical in size and weight and use an induction firing system.

The Mk 36 SRBOC is a development from the SRBOC described above and uses similar cartridges. These include the SUPER CHAFFSTAR decoy chaff cartridge, the SUPER GEMINI which generates a large radar cross-section and simultaneously deploys a co-located infra-red flare, the SUPER HIRAM III which ignites on water impact and generates large flame area to decoy the infra-red seeker in a missile, the SUPER HIRAM IV which deploys a parachute-suspended infra-red flare, and the SUPER LOROC chaff cartridge. SUPER CHAFFSTAR, SUPER GEMINI, SUPER HIRAM III and IV, and SUPER LOROC are proprietary developments of Loral Hycor.

STATUS
In operational service.
CONTRACTORS
Loral Hycor Inc, 10 Gill Street, Woburn, Massachusetts 01801, USA.

Tracor Aerospace Inc, Advanced Systems Operation, Bollinger Canyon Road, PO Box 196, San Ramon, California 94583, USA.

6888.053
DIRECT COUPLED JAMMING/SURVEILLANCE SYSTEM

The Direct Coupled Jamming/Surveillance System (DCJSS) is a command, control and communications countermeasures system intended to disrupt enemy voice and data communications, as well as providing intercept capability, with simultaneous jamming and surveillance over the frequency range 100 to 500 MHz. This is achieved by directly coupling two monitor receivers and two 400 W jammer combinations (receiver/exciter/transmitter) in the basic system. The system can be used in a shipborne, airborne and ground-based role, and is modularised for easy expansion to up to 14 monitor receivers (2 to 1100 MHz), up to 10 receiver/exciter combinations, up to 10 high power amplifiers with power combiner options and a direction finding system. The DCJSS equipment has been developed from a similar system which was originally designed for training purposes.

STATUS
In evaluation trials by the US Navy. A $10 million order from another NATO country is being negotiated as at April 1988.
CONTRACTOR
Tracor Flight Systems Inc, 1241 E.Dyer Road, Santa Ana, California 92705, USA.

4739.293
TRITON THREAT WARNING SYSTEM

Triton is a digital threat warning system for small combat ships, such as fast patrol boats, to provide warning of attack by air-to-surface and surface-to-surface radar-guided weapons, as well as the presence of unknown threats. It is virtually identical to the AN/APR-39A (**5692.353**), but repackaged for naval use. It features user-programmable synthetic voice alerts and visual presentation of multi-sensor threats, including the type of threat, its direction of arrival and whether it is tracking or still searching for the target. An indication is also given when the hostile radar lock has been broken.

The system contains a threat library for comparison and analysis which can be reprogrammed on board within minutes to provide for new and updated threats. The display symbols and audio alerts can be adapted quickly to suit user requirements. The system is designed as a direct replacement for older equipments, using the same electrical connections and interfaces.

STATUS
In production.
CONTRACTOR
The Singer Company, Dalmo-Victor Division, 1515 Industrial Way, Belmont, California 94002, USA.

5838.293
S-3000 SURVEILLANCE SYSTEM

The S-3000 operates over the frequency range 0.5 to 18 GHz and is designed as a radar surveillance system for rapid, high-sensitivity signal interception, and to provide an audio warning of threats. It uses a narrow beam, high-gain scanning antenna in conjunction with an omni-directional antenna and an IFM receiver. The system contains a comprehensive emitter library which is used to compare parameters of the intercepted signal. These parameters are also displayed on an operator's CRT. The operator controls the DF antenna, choosing between scan, sector and point modes. The system is designed to operate in dense signal environments and will identify priority signals as defined by the user.

STATUS
EM Systems is currently working on a $20 million contract to supply systems to MEL (UK) for retrofitting in the RN Oberon class submarines and in nuclear-powered SSNs.
CONTRACTOR
EM Systems Inc, 45757 W. Northport Loop, Fremont, California 94538, USA.

6125.293
D-ESM SHIPBORNE RECEIVER

The Anaren digital ESM receiver is designed for use in the protection of ships and fast patrol boats. It covers the range 2 – 18 GHz and provides frequency, angle of arrival, time of arrival, pulse width and pulse amplitude information in standard digital format. The receiver, when coupled with an effective processor and display, is an integral part of a tactical electronic warfare system, and is suited for passive launch of missiles. A masthead unit, consisting of a high band antenna and a low band antenna, is protected by a radome.

CHARACTERISTICS
Frequency coverage: 2 – 7.5 GHz (Band 1); 7.5 – 18 GHz (Band 2)
Azimuth coverage: 360°
Elevation coverage: –10 to +40°
Polarisation: Linear, Slant 45°
Bearing resolution: 8 bits (1.4°)
RMS bearing accuracy: 2° (Band 1); 2.25° (Band 2)
RMS frequency accuracy: 3.5 MHz (Band 1); 5 MHz (Band 2)
System senstivity: –60 dBmi (2-7.5 GHz); –55 dBmi (7.5-16 GHz); –50 dBmi (16-18 GHz)
Dimensions: 840 × 610 × 510 mm (receiver); 0.6 m diameter × 1.1 m height (masthead unit)
Weight: 130 kg (receiver); 60 kg (masthead unit)

STATUS
In operational service with the navies of five countries.
CONTRACTOR
Anaren Microwave Inc, 6635 Kirkville Road, E. Syracuse, New York 13057, USA.

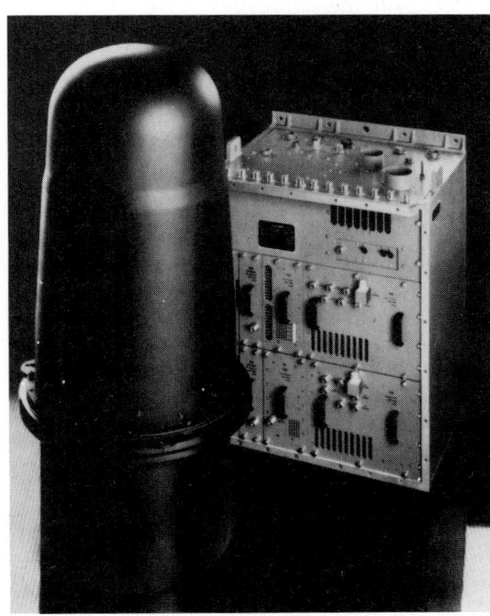

D-ESM shipborne receiver

6496.293
ANAREN SHIPBORNE TRACKING RECEIVER

As a major subsystem designed for self protection of fast patrol boats, the Anaren Tracking Receiver operates in co-operation with an ESM system and jammer to counter fire control and guidance radars, and assist in passive tracking of emitting platforms.

The receiver has a number of design features which enable correct performance of its function. Highly accurate DF information (less than 0.2 rms) is provided and enhances precise target discrimination and tracking. A unique antenna 'step-zoom' approach to the acquisition and high gain tracking process allows the receiver to operate with an ESM receiver having relatively crude DF capability. The use of a unique antenna main lobe/side lobe identification process obviates the need for a separate omni antenna. The high gain nature of the narrow beam high gain tracking array provides significant immunity to multi-path and jamming. A high receiver sensitivity allows long range detection and tracking for maximum reaction capability.

CHARACTERISTICS
Frequency coverage: 8 to 16 GHz
Polarisation: Linear; Slant 45°
Spatial acquisition range: Elevation/azimuth ±20°
DF boresight accuracy: 1°
Frequency accuracy: 2 MHz

STATUS
In production.
CONTRACTOR
Anaren Microwave Inc, 6635 Kirkville Road, PO Box 178, E. Syracuse, New York 13057, USA.

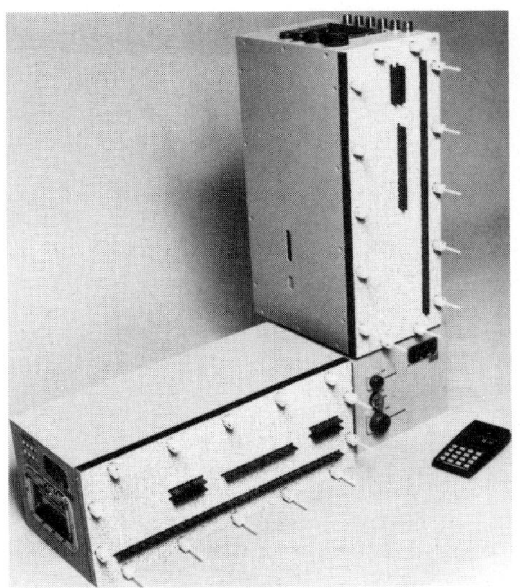

Anaren shipborne tracking receiver

6499.253

APECS II ESM/ECM SYSTEM

APECS II is an advanced modular ESM/ECM system operating over the frequency range 0.5 to 18 GHz. It is designed for use on a variety of vessels from fast patrol boats up to destroyers and is intended to counter most known radar threats. It is a fully automatic system which will identify and jam threat emitters using a mixture of IFM receivers and micro-processor controlled jammers. The antenna is a phased array assembly and high speed switching enables up to 16 different threats to be jammed simultaneously.

STATUS

Eight systems have been ordered by the Royal Netherlands Navy for its 'De Zevon Provincia' (M class) frigates under a $30 million contract.

CONTRACTOR

ARGO Systems Inc, 884 Hermosa Court, Sunnyvale, California 94086, USA.

6163.493

PHOENIX ESM SYSTEM

The Phoenix IIU ESM system has been designed for use in submarines to provide automatic identification and bearing of intercepted radar emissions in the frequency range 2 – 18 GHz. The system is fully automatic, and comprises an antenna array and a highly sensitive IFM receiver which provides precise measurement of the threat radar operating frequency. A detailed display is provided to the operator giving all significant parameters of the incoming signals. The system is modular in concept and can be configured to meet individual customer requirements.

STATUS

In production.

CONTRACTOR

ARGO Systems Inc, 884 Hermosa Court, Sunnyvale, California 94088, USA.

4407.293

AN/KAS-1 CHEMICAL WARFARE DIRECTIONAL DETECTOR

The AN/KAS-1 is a passive infra-red imaging sensor capable of detecting certain chemical agents. It consists of a sensor mounted on a two-axis pivot and a power conversion unit, and can operate over extended periods by the use of a miniature Stirling-cycle cooler.

The detector sensor unit is based on the man-portable thermal imager (**3823.093**) developed for the US Army. This system is well-proven in the field, is lightweight and rugged, and is in volume production.

The AN/KAS-1 has been designed for use on ships, and to complete the system for shipborne chemical warfare roles, the original man-portable imager has been modified to include a spectral filter wheel, a power conversion unit and a two-axis mount.

In addition to its primary role in chemical warfare detection, the AN/KAS-1 can also be utilised for other operational roles, such as reconnaissance, base defence, surveillance, threat detection, pilotage, and search and rescue. The AN/KAS-1 is available in either a direct-view configuration, a remote-view configuration which uses a charge-coupled device (CCD) camera to relay the image to a remote monitor on or below deck, or a gun-mount configuration.

STATUS

The AN/KAS-1 is in service with the US Navy.

CONTRACTOR

Texas Instruments Inc, PO Box 660246, Dallas, Texas 75266, USA.

6154.093

ACTIVE ELECTRONIC BUOYS

The Active Electronic Buoy (AEB) is a ship or air launched expendable decoy buoy similar in packaging to the A-size sonobuoy. It consists of a receiver, TWT transmitter, deployed antenna, associated power supplies and a programmable digital processor. The buoy, which also contains an electronic warfare device, is released into the ocean where it looks like an entire fleet of ships on enemy radar.

STATUS

Dalmo Victor, with Magnavox and Varian as subcontractors, is producing 45 engineering models under a $6.8 million development contract from the US government awarded in June 1984. It is understood that the US Space and Naval Warfare Systems Rewson project office is seeking sources to manufacture expendables similar to those being developed. Main production was scheduled to commence in the latter part of 1987 after evaluation.

CONTRACTOR

The Singer Company, Dalmo Victor Division, 1515 Industrial Way, Belmont, California 94002, USA.

ELECTRO-OPTICAL EQUIPMENT (SEA)

DENMARK

3857.293

TYPE 771 LOW LIGHT TV CAMERA

The Type 771 low light-level television camera was developed under a Royal Danish Navy contract. It is designed to conform to DEF 133N2 and to be used in conjunction with a television tracking system for automatic fire control applications. A filter wheel permits use over a wide range of ambient lighting conditions, the extremes of attenuation provided being from zero to 10^6. The sensor employed is an ISIT (intensified silicon intensifier target) tube which enables useful pictures to be obtained down to face-plate illumination levels of 5×10^{-4} lux. Lenses are selected to meet user requirements, as are the housings. A window with conductive coating and a washer/wiper are incorporated for protection.

CHARACTERISTICS

Sensor: RCA 4849H, EEV P8065, or Westinghouse WX 32890

Limiting resolution: 550 lines

Scanning: 625 lines 50 Hz 2 :1 interlaced CCIR or 525 lines 60 Hz 2 :1 interlaced EIA

Black level control: Automatic

AGC amplifier: Normal range 10 dB, max 20 dB

Weight: 11 kg camera with filter wheel; 22 kg lens with housing

STATUS

'Willesmoes' class fast patrol boats, and other ships, of the Royal Danish Navy are equipped with fire control systems that incorporate the Type 771 camera in a target tracking facility.

CONTRACTOR

JAI (Jorgen Andersen Ingeniorfirma AS), Produktionsvej 1, 2600 Glostrup, Copenhagen, Denmark.

FRANCE

3064.293

FRENCH NAVAL ELECTRO-OPTICAL SYSTEMS

The DTCN (the Technical Department for Naval Construction) is energetically continuing research and development into a number of aspects of electro-optical systems. These include the use of lasers for telemetry, guidance, aiming, illumination, and communications; passive infra-red for guidance, tracking, detection, thermal imaging, and surveillance. Light intensification and low light-level television techniques are being developed and manufactured, as are the uses of thermal imaging for surveillance, target designation, and guidance.

These programmes include:

LASER RANGEFINDER

The laser rangefinder is designed as a compact and high data rate unit for installation on all old and new rangefinding mountings to provide accurate range information which can be used directly by the gunnery control computer. A fleet-compatible prototype of this laser rangefinder has been built by Compagnie Générale d'Electricité to STCAN specifications. TAON (télémétrie d'artillerie optronique pour navire) equipment is designed for TTAC remote aiming turrets. Its main functions are ranging of air targets and shore targets. Rangefinding is possible between 1.5 and 16 km to an accuracy of ± 5 metres. Beamwidth is one milliradian and pulse rate 30 pulses/second. The equipment weighs about 15 kg.

IR RADAR (IRDAR)

DTCN is considering using a 10.6 micron laser source and a heterodyne receiver. This would be a pulse doppler laser radar operating in the infra-red. Although the source and receiver are still at the design stage, the experimental prototype IRDAR equipment was delivered to DTCN by SAT for tests in late 1980 as an exploratory development.

VOLCAN

VOLCAN (viseur optronique leger pour conduite d'arme navale) is an experimental infra-red bispectral sensor for naval surveillance and/or fire control applications. The experimental prototype was built for DTCN by SAGEM, aided by SAT, CILAS, CSEE and SOFRETEC who provided the optronic sensors. The equipment is suitable for ships of any size, and with sensors weighs about 80 kg. It takes the form of a small turret, within which are housed eight and twelve micron wavelength infra-red cameras, a one-inch vidicon television camera, and a laser rangefinder. The last of these is capable of pulse rates of 2 to 20 Hz using a YAG laser. The equipment also includes an automatic tracking system for night and day optical surveillance, tracking of air targets, aircraft and missile control, particularly low-altitude weapons. The VOLCAN sight is gyro-stabilised.

CONDOR

CONDOR is the French acronym for radar target designation optronic night camera, and its function is to replace manual mountings, to enable a target to be designated and identified, and enable a laser to be laid on. In addition, and in the context of comprehensive integration with radar, it can provide sighting facilities or automatic tracking in the event of radar failure, radar silence, loss of track or loss of accuracy at low elevation.

DTCN has designed the operational version of CONDOR. Now in production by SINTRA, it is a TV camera with two different optical paths: one for daylight, a vidicon, and for night use (ISIT). These cameras are to equip all the new French Navy frigates, corvettes and Avisos.

In addition to these lines of research, DTCN is studying potential applications for light amplification and image intensification techniques such as night surveillance from ships and commando operations.

IR PERISCOPE

To increase the capabilities of conventional submarine periscopes, STCAN has designed and is producing a new periscope that permits night vision. PIVAIR (périscope intégré de veille à infra-rouge) (**6546.493**) has been built by SAGEM in collaboration with SOPELEM to STCAN specifications and is fitted to all new type French Navy submarines and will be retrofitted to others.

A submersible optronic system used in mine disposal operations

Thermal imaging by infra-red has appreciable potential, especially where hot-spots and temperature contrasts can be expected, and there is considerable interest in this area of opto-electronics also. A development contract awarded to SAGEM includes a SAT thermal camera operating in the eight to twelve microns band with parallel scanning for target detection over 360° in azimuth in the surveillance mode and target identification in the thermal imaging mode.

3427.293

PIRANA INFRA-RED TRACKER

The PIRANA infra-red tracker has been developed for use on combined electro-optical naval fire control equipments such as the CSEE Totem (**3424.293**), in which it is employed with television and a laser rangefinder for the direction of anti-aircraft weapons, or for use on its own as an ancillary to autonomous weapons. The equipment provides a means of passive search, acquisition and target tracking, especially useful in low-altitude low elevation engagements. It has two IR channels for simultaneous tracking at two different wavelengths.

The configuration of the PIRANA equipment can be seen from the nearby illustration and it will be carried on one side of the Totem E-O director unit. The main modules of the infra-red unit are the receiver optical assembly, analysis section, IR detector, cooling, and tracking electronics module. Power supply unit and control panel are located below deck.

PIRANA infra-red tracker and VAMPIR surveillance equipment mounted on the frigate Cassard

CHARACTERISTICS
Optical aperture: 170 mm
Field of view: ±8 mrad
Wavelength: 4 – 5 and 8 – 12 micron bands
Range: 10 – 20 km, depending on type of target
Cooling: Joule Thomson cycle to about 80 K
Dimensions: 75 × 41 × 41 cm
Weight: 50 kg (tracking telescope)
Power supply: 115 V, 400 Hz, 1 kVA

DEVELOPMENT
Development was initiated in collaboration with the French Naval authorities in the late 1960s and early 1970s. (See entry **3064.293** at the start of the French entry for further historical details on this and related naval opto-electronic projects.)

STATUS
Experience with the PIRANA I and II resulted in the PIRANA III version which is now operational with the French Navy in the frigate *Cassard*. This version is a bi-spectral equipment with two channels working simultaneously together with appropriate signal processing to achieve minimal sensitivity to jamming and environmental noise.

CONTRACTOR
SAT (Société Anonyme de Télécommunications), 41 rue Cantagrel, 75624 Paris Cedex 13, France.

6548.293
SEID INFRA-RED TRACKER

SEID is an infra-red differential tracker which has been designed to equip the Thomson-CSF Crotale Naval turret. It has been developed through various study programmes performed by SAT for the French Navy and basically features common modules used in the PIRANA infra-red tracking system (**3427.293**). It is intended to increase the anti-aircraft and anti-missile capability of the Crotale Naval weapon system at low and medium elevation ranges.

Following target designation by the search system, SEID enables the target to be tracked by supplying the required data to the turret servo-mechanism. Once the Crotale missile is fired and picked up within the tracker field, the SEID system provides enemy target differential offset data continuously to the missile. This data allows guidance commands to be formulated for the automatic interception of the enemy target through the cancellation of the Crotale enemy target offset.

STATUS
This equipment was originally developed for Saudi Arabia and is now also operational aboard French Navy warships.

CONTRACTOR
SAT (Société Anonyme de Télécommunications), 41 rue Cantagrel, 75624 Paris Cedex 13, France.

SEID infra-red tracker

4106.293
VAMPIR SURVEILLANCE EQUIPMENT

VAMPIR (veille air mer panoramique infra-rouge) is a passive infra-red naval surveillance equipment which provides for surveillance and search of both air and surface targets. It is also capable of target designation to either radar or optronic fire control/director systems or equipment.

The configuration of the equipment consists of a cylindrical housing carried on a vertically stabilised mount which is designed to be fixed on the ship's superstructure in a high position to ensure maximum range and freedom from obstruction. Stabilisation of the housing automatically compensates against ship motion, while the optical head rotates at a rate of about 60 rpm to provide all-round scanning of the horizon. The stabilisation horizontal datum can be adjusted up to a maximum elevation of +20°. The equipment is a bi-spectral device operating in two distinct IR bands, thus achieving a very low false alarm rate in the presence of environmental noise. The two separate channels also enable target detection in a variety of meteorological conditions. Signal processing is performed by a sub-system provided by CSEE.

CHARACTERISTICS
Wavelength: 4 – 5 and 8 – 13 micron (simultaneous channels)
Field of view: 360° × 0 to +24° elevation
Detection range: At least 10 km for frontal attack by missile or aircraft
Target designation accuracy: 1 mrad in elevation and bearing
Sensor cooling: Joule Thomson cycle to about 80 K
Dimensions of stabilised unit: 125 × 60 cm diameter
Rotation rate: 60 rpm
Weight: 450 kg
Power: 115 V, 400 Hz, 300 VA

STATUS
Vampir version has been qualified for operation on-board large warships and has been selected for the French Navy anti-aircraft corvettes and is operational in the frigate *Cassard*.

CONTRACTOR
SAT (Société Anonyme de Télécommunications), 41 rue Cantagrel, 75624 Paris Cedex 13, France.

6549.293
MURENE THERMAL CAMERA

The Murene thermal camera is a naval version of the Thermal Modular System (SMT) (**4107.093**) which has been developed jointly by SAT and TRT. Murene has been designed especially to respond to specific imaging applications in naval forces and is compatible with all current turret designs. It is incorporated in Sadral mounts (**4509.231**).

CHARACTERISTICS
Waveband: 8 to 11 microns
Field of view: 1.9° × 2.9° narrow; 5.7° × 8.7° wide
Dimensions: 955 × 210 × 310 mm
Weight: 35 kg

CONTRACTOR
SAT (Société Anonyme de Télécommunications), 41 rue Cantagrel, 75624 Paris Cedex 13, France.

3566.293
CANOPUS (TRS906) LIGHTWEIGHT ELECTRO-OPTICAL DIRECTOR

This equipment is a line-of-sight stabilised E-O director designed for use aboard warships for surveillance, target indication, fire control or missile guidance purposes. (See **3102.281**).

It comprises a two-axis director stabilised by rate gyros, and a servo control box. The director consists of a fixed pedestal in which is centred a bearing rotating unit which supports the elevation axle actuator. The elevation axle is fitted at both ends with flanges to provide the mechanical interface with the E-O sensor units.

The director is fitted with a joystick, and with a collimated grid sight for local control. The director can be controlled by:

(1) a deck operator, using the joystick
(2) a remote control unit
(3) the elevation and bearing errors signals generated by the E-O equipment.

Slewing speeds and accelerations are 3 rad/s and 7 rad/s² in bearing and 1 rad/s and 3 rad/s² in elevation. Slewing time for 180° is 1.8 s in bearing. Pointing stability is better than 0.2 mrad rms. Data take-off accuracy is 0.5 mrad peak.

A variety of options exist for using various sensor combinations; in addition to television trackers, infra-red automatic tracking with laser rangefinding or a K-band (millimetric waveband) radar can be employed (see Canopus fire control system **3102.281**).

CHARACTERISTICS
Director (fitted with a TV camera)
Height: 0.77 m
Rotation diameter: 0.8 m
Weight: 120 kg
Servo control box: 0.86 × 0.6 × 0.3 m
Weight: 80 kg

STATUS
In production.

CONTRACTOR
Thomson-CSF, Division Systèmes, Défense et Contrôle, 40 rue Grange Dame Rose, BP34, 92360 Meudon-la-Foret, France.

3426.293
LYNX OPTICAL SIGHT

The CSEE Lynx is a self-contained lightweight optical sight designed for use aboard surface ships of all sizes. It provides for air and surface surveillance and may be used in conjunction with other optical equipments or with a ship's command information centre, and can also be employed for weapon control. In the RADOP 10 naval fire control system (**4049.261**), Lynx is the main sighting element. For further details see *Jane's Weapon Systems 1987–88*.

The Lynx director may also be equipped with a television camera and a monitor.

CSEE Lynx optical sight

CHARACTERISTICS
Magnification: × 2.5 or × 8
Field of view: ±9.5° or ±3°
Scanning field: Unlimited in azimuth, –15 to +75° in elevation
Max speed in elevation: 1 rad/s
Max acceleration: 2 rad/s² (elevation)
Height of eyepieces: Adjustable 1.38 – 1.63 m
Weight: 195 kg

STATUS
In production for Bahrain, Kuwait, Oman, Saudi Arabia, Finland and Morocco.

CONTRACTOR
CSEE (Compagnie de Signaux et d'Entreprises Electriques), 17 place Etienne Pernet, 75738 Paris Cedex 15, France.

3425.293
PANDA MK2 OPTICAL DIRECTOR

This is a naval optical director equipment capable of use for weapon control, air and surface target search and designation, and for practice firing observation. It is the main optical sight and director of the RADOP 20 naval fire control system (**4049.261**). The general arrangement of the unit can be seen from the accompanying illustration. The deck-mounted sight is provided with power driven azimuth movement and a stabilised line-of-sight by means of integral rate gyros. There is also provision for remote control of the sight to enable it to be laid on from an external reference.

When operating in the weapon control mode the sight is used to track the target. The co-ordinates transmitted by the sight convey all the necessary firing corrections based on target velocity, weapon characteristics and parallax between weapon and sight. In the search mode of operation the sight scans the surveillance zone in bearing and elevation.

CHARACTERISTICS
Magnification: × 8 or × 2.66
Fields of view: ±3° and ±9°
Azimuth aiming arc: Unlimited
Elevation arc: –25 to +65° (relative to deck)
Max velocity: 1 rad/s (bearing); 0.5 rad/s (elevation)
Weight: <330 kg

The above description applies to the basic optical version of this director, which is referred to as the Panda. A more sophisticated version, with additional facilities such as laser rangefinder, automatic tracking, local/remote control transfer, multiple weapon capability etc, is also produced as the Naja electro-optical director (**3561.293**).

STATUS
In production for 16 navies.

CONTRACTOR
CSEE (Compagnie de Signaux et d'Entreprises Electriques), 17 place Etienne Pernet, 75738 Paris Cedex 15, France.

CSEE Panda Mk2 optical director

3561.293
NAJA ELECTRO-OPTICAL DIRECTOR

This is a naval electro-optical director equipment usable for air and surface surveillance, automatic TV tracking at high and low altitude, and accurate fire control by laser telemetry. The Naja equipment contributes to the self-defence of surface ships against aircraft or missile attacks.

In addition to being employed on its own, the Naja director is the principal fire control element of the RADOP 30 naval weapon control system (**4049.261**). For further information see *Jane's Weapon Systems* 1987-88.

CHARACTERISTICS
Optical equipment
Magnification: × 2.5 (field ± 9.5°); × 8 (field ±3°)
Daylight camera or LLTV camera
CILAS laser
Azimuth aiming arc: Unlimited
Elevation arc: –25 to +70°
Max velocity: 60°/s (bearing); 60°/s (elevation)
Weight: 500 kg approx

STATUS
In production. The Naja director has also been offered as part of a land-based air defence system based on the Breda twin 40 mm naval system.

CONTRACTOR
CSEE (Compagnie de Signaux et d'Entreprises Electriques), 17 place Etienne Pernet, 75738 Paris Cedex 15, France.

CSEE Naja electro-optical director

4255.293
NAJIR ELECTRO-OPTICAL DIRECTOR

The Najir director contributes to the self-defence of surface ships of any displacement against aircraft or missile attacks. It can be used against naval or land targets by night or day. It is unaffected by enemy jamming.

The Najir director operates as an autonomous unit with integrated stabilisation and a digital computer. It is manned by a single operator in two principal modes: surveillance and target tracking, or fire control of several weapons of different calibres.

In the former mode the operator can: receive and re-copy a target designation transmitted by other surveillance equipment; by means of the auxiliary sight, immediately acquire a target designated by a watch keeper; quickly identify the target (optical or IR display); decide to select automatic tracking either TV by day or IR by night (or by day in disturbed atmosphere); obtain the target range by laser rangefinder.

In the fire control mode the operator can select the firing mode (against aircraft or ships), order the simultaneous coupling of two guns, decide to open fire on receipt of 'gun ready' and 'target within range' information and observe the results of firing.

The IR camera is the CT 15 produced by SAT. Weight of the camera is 10 kg and display is on a micro-monitor or on a remotely located standard CCIR TV monitor.

CHARACTERISTICS
Aiming arcs: unlimited (bearing); –20 to +70° (elevation)
Max velocity: 60°/s (bearing); 60°/s (elevation)
Max acceleration: 60°/s² (bearing); 60°/s² (elevation)
Optical equipment
Binocular telescope: × 2.5 (field ±9.5°); or × 8 (field ±3°)
Daylight TV camera
IR camera
Laser rangefinder
Weight: 550 kg approx

STATUS
Development.

CONTRACTOR
CSEE (Compagnie de Signaux et d'Entreprises Electriques), 17 place Etienne Pernet, 75738 Paris Cedex 15, France.

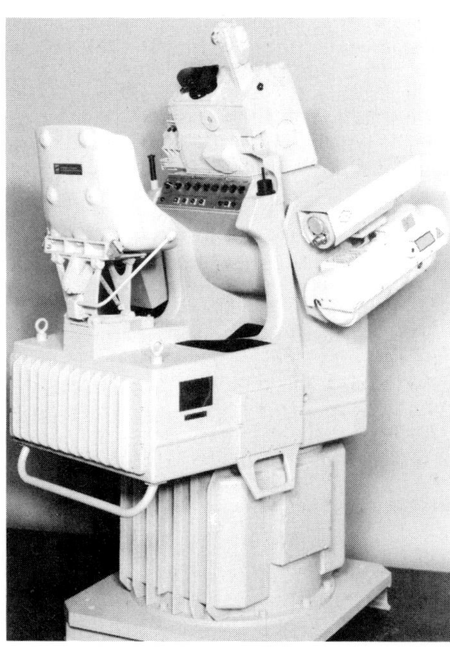

CSEE Najir electro-optical director

3424.293

TOTEM NAVAL FIRE CONTROL EQUIPMENT

The Totem electro-optical fire control director has been developed by CSEE to provide warships with passive air defence surveillance, target designation and tracking for use with air defence weapons. A passive infra-red tracker, a laser rangefinder, and either a normal or low light-level television camera with automatic tracker are mounted on a rotating mounting with an elevating head. Below decks are the servo rack and control and monitoring panels. The control and monitor panels may be incorporated in the weapon consoles of the related gun or missile systems.

The Totem sensor head is stabilised by integral gyros. The infra-red tracker is the SAT PIRANA equipment (**3064.293** and **3427.293**) which is also suitable for mounting on guns or separately. The television sub-system is by SINTRA and a CILAS laser rangefinder is employed. Alternative sensors may be used and CSEE has a combined electro-optical/radar version of the Totem FCS under development; the radar is a Ka-band equipment manufactured by Thomson-CSF, similar to that used in the TRS 906 (**3566.293**).

CHARACTERISTICS

Turret weight: <200 kg (plus weight of sensors)
Max slewing speed: 2 rad/s (elevation and azimuth)
Max slewing acceleration: 2 rad/s² (elevation and azimuth)
Angular tracking accuracy: 1′
Power supply: 115 V, 3 phase, 3 kVA

STATUS

In development.

CSEE Totem electro-optical fire control system with SAT infra-red tracker (left) and LLTV and laser rangefinder (right)

CONTRACTOR

CSEE (Compagnie de Signaux et d'Entreprises Electriques), 17 place Etienne Pernet, 75738 Paris Cedex 15, France.

4716.293

VISON OPTICAL DIRECTOR

As a periscopic optical director, Vison carries out surveillance, target designation and acquisition functions, and automatic tracking of targets. With the incorporation of an optional fire computer it can also be used as a fire control system.

The basic system consists of a periscope providing an optical channel from an exterior-mounted stabilised mobile head. The orientation in bearing and elevation of the mobile head line of sight is controlled from a data processing console at the foot of the periscope. The operator uses a control stick to control the head and an eyepiece which provides optical observation of the surveyed sector (30° or 7.5° according to magnification). In addition to direct optical observation, a TV camera is installed in the optical path and provides a video screen display to the operator. Automatic TV tracking of the target can be used when available.

With the reduced dimensions of Vison, installation of the complete system in the ship structure is relatively simple.

CHARACTERISTICS

Magnification: ×2 or ×8
Field of view: 30° or 7.5°
Bearing scan: Unlimited
Elevation scan: –25 to +70°
Weight: Approx 120 kg (sub-assembly head and periscope); 300 kg (complete system)

STATUS

In development.

CONTRACTOR

CSEE (Compagnie de Signaux et d'Entreprises Electriques), 17 place Etienne Pernet, 75738 Paris, Cedex 15, France.

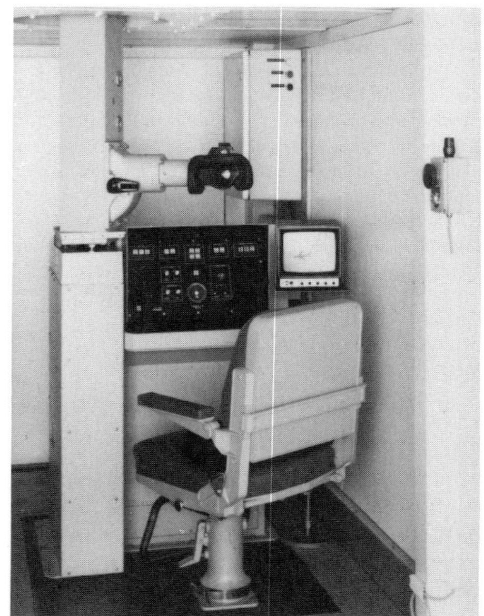

Vison optronic director (interior view)

4717.293

COTOL OPTRONIC FIRE CONTROL SYSTEM

Cotol is a lightweight optronic fire control equipment derived from the Totem system (**3424.293**), and is used for the control of guns of all calibres against both air and surface targets. The system is normally installed as part of the ship's overall anti-aircraft defence system, particularly against low flying aircraft, but it can also be used as a control unit against surface targets. The equipment is fully self-contained, with TV and IR cameras, a laser rangefinder and a remote control console. Facilities provided include target acquisition, automatic IR or TV tracking, memory tracking, computation of firing data for two weapons, and control and firing of the guns.

CHARACTERISTICS

Bearing scan: Unlimited
Elevation scan: –30 to +85°
Weight: 370 kg

STATUS

In development.

CONTRACTOR

CSEE (Compagnie de Signaux et d'Entreprises Electriques), Division Défense, 17 place Etienne Pernet, 75738 Paris Cedex 15, France.

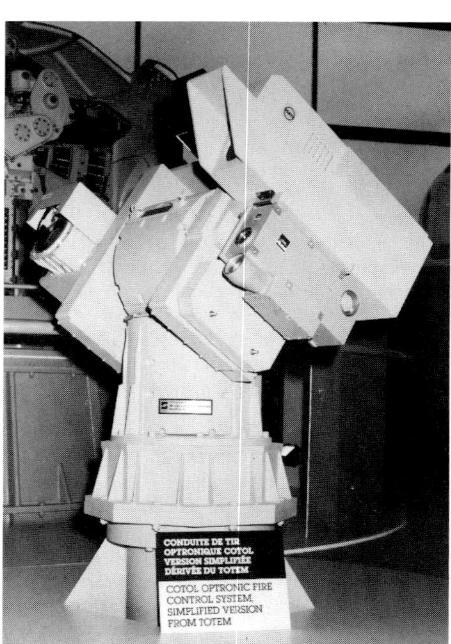

Cotol optronic fire control system

5917.293
COTROL OPTRONIC FIRE CONTROL SYSTEM

Cotrol is a lightweight radar/optronic system which can be used as the main or secondary fire control to provide day and night surveillance, automatic tracking of targets, and computation of firing data and fire control for two weapons. It comprises a Ka band radar, IR and TV cameras and trackers, a remote control console and a laser rangefinder.

CHARACTERISTICS
Bearing: Unlimited
Elevation: –30 to +85°
Weight: 460 kg

STATUS
In development.

CONTRACTOR
CSEE (Compagnie de Signaux et d'Entreprises Electriques), 17 place Etienne Pernet, 75738 Paris Cedex 15, France.

Cotrol optronic fire control system

3147.293
TAON TMY83 LASER RANGEFINDER/ILLUMINATOR

The TMY83 TAON (télémètre d'artillerie optronique pour navire) is a Nd-YAG laser rangefinder developed for naval applications such as integration into electro-optical fire control systems for air defence weapons. The rangefinder and its power supply are housed within a single casing, which is designed for ease of fitting to director mounts or directly on to weapons such as guns. The laser repetition rate is adjustable between ten and twenty pulses per second.

CHARACTERISTICS
Laser type: Nd-YAG, electro-optical Q-switching
Wavelength: 1.06 micron
Pulse duration: $\leqslant 20 \times 10^{-9}$ s
Output: >5 MW
Beam divergence: 0.5 mrad
Range: 500 – 20 000 m
Range resolution: ±5 m
Dimensions: 460 × 265 × 160 mm
Weight: 15 kg

DEVELOPMENT
Design and development has been carried out under contract from the French Navy.

STATUS
The TMY83 replaces the TMV26 naval laser rangefinder (**1799.293** in 1976 and earlier editions).

CONTRACTOR
CILAS (Compagnie Industrielle des Lasers), route de Nozay, 91460 Marcoussis, France.

TAON TMY83 laser rangefinder

3662.293
TMY113 LASER RANGEFINDER/ILLUMINATOR

The TMY113 is a high repetition rate laser rangefinder designed for integration into the Naja naval electro-optical fire director equipment (**3561.293**). The rangefinder, electronic timing clock, and power supply are all housed within the same casing.

CHARACTERISTICS
Laser material: Nd-YAG
Wavelength: 1.06 micron
Pulse duration: 20×10^{-9} s
Peak power: >5 MW
Accuracy: ±5 m
PRF: 20 pps
Detector: Avalanche photodiode
Detector field: 2 mrad
Dimensions: 470 × 230 × 170 mm
Weight: <14 kg
Power supply: 115, V/400 Hz, 3-phase

STATUS
Series production.

CONTRACTOR
CILAS (Compagnie Industrielle des Lasers), route de Nozay, 91460 Marcoussis, France.

5918.293
TMY 156 LASER RANGEFINDER

The TMY 156 is a high repetition rate laser rangefinder/target illuminator designed for integration into target designation optronic systems. It can be used for acquisition, designation and tracking of air, surface and land targets. The equipment consists of a single unit containing the rangefinder and power supply.

CHARACTERISTICS
Laser type: Nd:YAG; Q-switched
Wavelength: 1.06 μm
Pulse duration: 20×10^{-9} s
Detector: Avalanche diode
Field angle: 1 mrad
Dimensions: 600 × 225 × 195 mm
Weight: 25 kg

STATUS
Prototype.

CONTRACTOR
CILAS (Compagnie Industrielle des Lasers), route de Nozay, 91460 Marcoussis, France.

4108.493
SOPELEM SUBMARINE PERISCOPES

SOPELEM designs and manufactures all types of submarine periscope, and while details of specific equipments have not been obtained, current production includes the following basic categories:

Surveillance Periscopes
Periscopes of this type provide for observation at fixed navigation periscope depth and can be fitted with a rangefinding radar antenna and a radar search scanner, and several electro-optical sensors. They are designed for day and night observation and may be fitted with a gyro-sextant for star sighting navigational purposes. These periscopes are generally equipped with a recording photographic camera.

Attack Periscopes
With such instruments the crew is able to make an observation of the surface area in all directions from fixed position at all submarine navigation periscope depths. The periscope is completely controlled by hydraulic means giving maximum flexibility and speed of operation. The periscope upper section is of very small cross-section and thus creates minimum wake and also makes detection by radar much more difficult.

All these periscopes are designed and manufactured by SOPELEM which also produces the hydraulic controls and the associated electronic equipment. They equip all submarines of the 'Daphne', '1200 T Agosta' classes, and French nuclear powered submarines.

Pivair periscope
This is a joint development by SAGEM and SOPELEM using an SAT thermal imager. It is described under entry **6546.493**.

STATUS
In production.

CONTRACTOR
SOPELEM, 19 boulevard Ney, 75018 Paris, France.

6546.493
PIVAIR SUBMARINE PERISCOPE

The PIVAIR periscope is a new generation equipment designed by SAGEM and SOPELEM which incorporates a CTX thermal imager manufactured by SAT. It offers substantial increase in the utilisation time of the equipment due to the thermal imager system which allows day and night vision in all-weather conditions. It also provides an improved image quality through the two-axis stabilisation of both the optical and the infra-red channels, as well as possible remote control and monitoring of images. The periscope can be adapted to any type of submarine.

PIVAIR consists of five main assemblies:

(a) The masthead comprising the ESM antenna, the gyro-stabilised unit for both optical and infra-red channels, the magnification control, an infra-red camera, and the optical channel with filters.
(b) The mast housing the optical transfer system, wave guides and cables.
(c) The mast base which houses a high accuracy sextant with its associated vertical sensor, a 35 mm camera, the infra-red monitor, binocular viewing, and a microphone/speaker. Provisions

are made for an optional TV camera (normal or low-light level) and for an optional rangefinder.
(d) The hydraulic raising and rotating control system. An optional shroud can be provided.
(e) An optional remote control and monitoring console which includes all the controls to remotely operate the periscope, and monitor screens.

CHARACTERISTICS
Optical channel
Magnification: × 1.5 to × 6 (× 12 optional)
Field of view: 26°, 9° and 4.5°
Sighting angle: –10° to + 80°
IR channel
Wavelength: 8 to 12 micron
Field of vision: 3° × 6°
Sighting angle: –10° to + 80°
CONTRACTORS
SAGEM, 6 avenue d'Iéna, 75783 Paris Cedex 16, France.
SOPELEM, 19 boulevard Ney, 75018 Paris, France.

ISRAEL

4718.293
GYRO STABILISED SIGHTING SYSTEM

The Gyro Stabilised Sighting System (GSSS) is used for surveillance, identification and target tracking. The system consists of a dual magnification periscope with a gyro-stabilised upper mirror. The mirror coating is effective over a wide spectral range from the visual to near infra-red, and permits optional use of laser or infra-red trackers in parallel with the visual path. High magnification by the stabilised optics enables the observer to identify and lock-on to the target at considerable distances.

The sight is an optical periscope, roof or floor mounted with its rotating viewing head outside. The head includes the upper periscope mirror which is free to rotate in azimuth and elevation. The system can be integrated with a laser rangefinder and/or a low light TV camera.

CHARACTERISTICS
Magnification: × 10 or × 2.5
Field of view: 5° or 20°
Scan: 360° azimuth; –30 to + 60° elevation
Tracking rate: 30°/s (azimuth and elevation)
Weight: 65 kg
CONTRACTOR
Israel Aircraft Industries Ltd, Tamam Precision Instruments, Ben Gurion Airport, Israel.

5921.293
DS-35 (SPIRTAS) TARGET ACQUISITION SYSTEM

The SPIRTAS passive IR target acquisition system (designated DS-35) is a thermal detection equipment designed for ship protection against sea-skimming missiles. It operates in the three to five micron wavelength band using a multi-element array detector head, cooled by a closed-circuit cryogenic cooler.

The system consists of a sensor mounted at masthead which contains the stabilised scanning head, a below-decks control and display console and an auxiliary electronics unit. The console contains signal processing units, a TV video screen, and a control panel. The monitor screen can present three types of information: a low resolution situation display, an enlarged FLIR type thermal display and an auxiliary display. The latter will provide a variety of data including, the IR intensity behaviour of selected targets, information transmitted to other on-board systems, thermal image position and cursor position.

CHARACTERISTICS
Wavelength: 3 – 5 microns
Scanner coverage: Azimuth 360°; elevation –2 to + 20° in full search mode
Scan period: Full search mode 8s; anti-missile mode 1.25s
Accuracy: 7 mrads in both azimuth and elevation
Dimensions: Sensor unit: 1.2 m high; 0.8 m wide
Weight: Sensor unit 120 kg; control/display console 170 kg; electronics unit 90 kg
STATUS
Believed to be in production for the Israeli Navy for a number of warships.
CONTRACTOR
ELOP (Electro-Optics Industries Ltd), PO Box 1165, Rehovot 76110, Israel.

Sensor for SPIRTAS DS-35 system

ITALY

3585.293
OGR7/2 OPTICAL DIRECTOR

The optical director Type OGR7 is a shipborne equipment of small size and weight for the control of up to 3 inch calibre guns against air and surface targets. It consists of a tracking system, provided with a high luminosity binocular telescope and open sight, and a computer through which the train and elevation lead angles and the super-elevation are computed with high accuracy. A single gunlayer/operator controls the speed in train and elevation by means of a joystick. The line-of-sight is stabilised so that the operator has only to account for the motion of the target.

CHARACTERISTICS
Telescope
Magnification: × 2.3 and × 7
Optical field: 30° and 10°
Max train speed: 100°/s
Max elevation speed: 60°/s
Max train lead angle: ± 25°
Max elevation lead angle: ± 20°
Crossing range limits: 100 – 2000 m
Target speed limits: 60 – 350 m/s
STATUS
More than 250 sets of the OGR7 series have been supplied to main navies, among them those of West Germany, Greece and Peru. A special version of the OGR7, known as the OGR7 Seacat, has been developed which provides an additional telescope for Seacat missile guidance as well as gunfire direction. This model also bears the designation OG20. A later development, the OGR7/2-A, is described under entry **5919.293**.
CONTRACTOR
Officine Galileo SpA, Via Einstein 35, 50013 Campi Bisenzio, Florence, Italy.

5919.293
OGR7/2-A OPTICAL DIRECTOR

The optical director OGR7/2-A is the latest version of the OGR7/2 equipment (**3585.293**), and is a shipborne lightweight fire control system (FCS) mounted on a revolving platform and manned by a single operator. It consists basically of a pedestal on which an optical head with an external image intensifier, an autonomous integrated two-axis stabilisation system and a digital computer for lead angle computation and FCS management are mounted. The azimuth movement is provided by an electro-hydraulic unit while elevation is controlled by a servo.

CHARACTERISTICS
Telescope
Magnification: × 2.3 and × 7
Optical field: 30° and 10°
Maximum train speed: 100°/s
Maximum elevation speed: 60°/s
Target speed limits: 60 – 350 m/s
STATUS
In production for the Italian Navy.
CONTRACTOR
Officine Galileo SpA, Via Einstein 35, 50013 Campi Bisenzio, Florence, Italy.

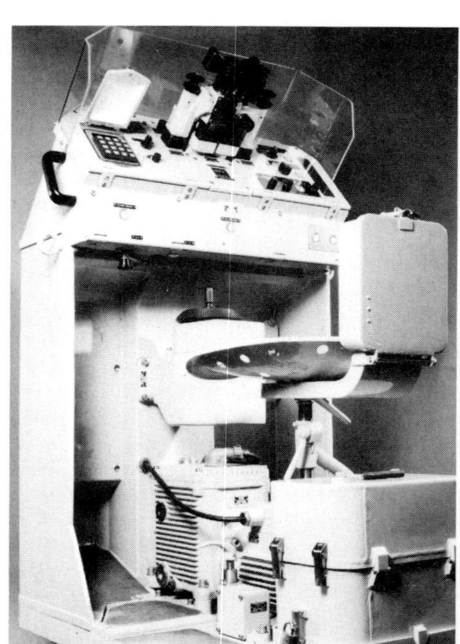
OGR7/2-A optical director

6547.253

MEDUSA OPTICAL DESIGNATION AND FIRE CONTROL SYSTEM

The Medusa optical fire control system can be considered either an extremely simple and easy-to-use fire control system for remote control of small calibre guns, or a sophisticated optical aid to more complex fire control systems.

Medusa consists of two units, a manned director and a control console. The director mounts a twilight or low-light level TV camera and a 7-inch black-and-white CRT monitor. Facilities are provided at the director for designating the target to another fire control system and for activating a firing action. The director operator is thus able to carry out the following functions; surveillance, target search and detection, non-stabilised tracking of a target, direct weapon control (weapon aiming over line-of-sight), and as a navigation aid.

The control console, which can be located in a suitable operational area, consists of another 7-inch black-and-white CRT monitor and has operating controls to carry out all the system functions. The control console thus enables the command to supervise and direct the functions of the director with the addition of a calculator the system is capable of simple ballistic calculations.

STATUS

In series production.

CONTRACTOR

Consorzio Sistemi Navali Selenia-Elsag, Via Panama 52, 00198 Rome, Italy.

4262.293

OG30B SHIPBORNE THERMAL IMAGER

The OG30B thermal imaging equipment has been developed for integration with the Selenia-Elsag 30B radar tracker pedestal (**3861.281**). It can also be fitted to any fire control system to perform panoramic surveillance, observation and identification, and acquisition and automatic tracking of both surface and airborne targets. These functions are achieved by transmitting the thermal image of the target(s) on to an appropriate display on the control console. By converting the imager video into a conventional TV signal format, automatic TV tracking techniques can be employed.

The system consists of an IR camera, compressor, digital video converter, TV contrast tracker and a panoramic surveillance signal processor. The camera comprises a hard carbon coated optical head, a scanner of the parallel scan type with two servoed mirrors providing 25 frames per second image scanning, a detector-dewar-cryoengine assembly, pre-amplifiers and control electronics. The panoramic surveillance function is performed by circular scanning of the IR scene. The cooling system is of the closed circuit type with one fluid compressor and one cooling unit.

CHARACTERISTICS

Wavelength: 4 – 14 microns
Image format: 1:2
Instantaneous field of view: 0.25 mrad
Frame rate: 25 Hz
Lines per frame: 200
IR detector: CMT linear array
System weight: 65 kg

STATUS

In service with the Italian Navy

CONTRACTOR

Officine Galileo SpA, Via Einstein 35, 50013 Campi Bisenzio, Florence, Italy.

6326.293

P0701 LASER RANGEFINDER

Two new Nd:YAG laser rangefinders have been developed by FIAR, the P0701 for naval applications and the P0702 for airborne use. The P0701 has been designed primarily for naval fire control systems, in particular for use on small vessels, and can also be used in land-based and airborne fire control. The P0702 is described in the Electro-Optical (Air) section.

CHARACTERISTICS

Output power: 12 MW
Pulse duration: 12 $\pm$5 ns
PRF: 1 to 20 Hz, selectable by remote control
Range: 400 to 20 000 m
Accuracy: 3 m
Dimensions: Length 590 mm, diameter 240 mm
Weight: Approx 20 kg

STATUS

In production.

CONTRACTOR

FIAR (Fabbrica Italiana Apparecchiature Radioelettriche) Spa, Via Montefeltro 8, 20156 Milan, Italy.

P0701 laser rangefinder

6327.293

P4674 THERMAL IMAGER

The FIAR P4674 Thermal Imager has been designed for naval applications and can be used in fire control systems and for general surveillance in the detection and tracking of targets at long range. The output from the camera is a CCIR TV video and, when presented on the TV monitor, provides the operator with an image of the scene together with an adaptive reticle. The reticle is generated electronically and preprogrammed, thus ensuring that it can adapt to almost any pattern.

CHARACTERISTICS

Spectral range: 8 – 12 μm
Field of view: Horizontal 4°; vertical 2.6°
Magnification: Horizontal × 2.6; vertical × 1.3
Range: At least 18 km for fixed-wing aircraft and 13 km for helicopters
Dimensions: 330 × 200 × 475 mm
Weight: 30 kg

STATUS

In production.

CONTRACTOR

FIAR (Fabbrica Italiana Apparecchiature Radioelettriche) SpA, Via Montefeltro 8, 20156 Milan, Italy.

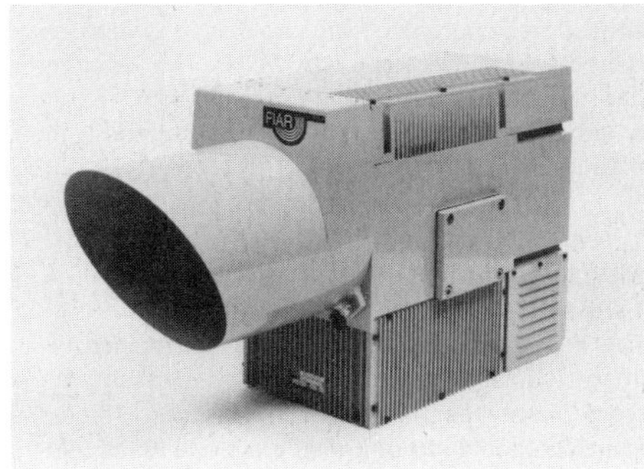

P4674 thermal imager

NETHERLANDS

3143.293

LIOD LIGHTWEIGHT OPTRONIC FIRE CONTROL SYSTEM

The LIOD (Lightweight Optronic Director) has a tracking unit which includes TV and IR cameras together with a Nd YAG laser rangefinder and is an unmanned remotely-controlled director for the automatic tracking of air and surface targets.

Tracking is by the optical contrast technique whereby the target is tracked against its background as viewed by a TV or infra-red camera. The TV camera has automatic filters to adjust for light conditions. Alternatively a correlation tracker can be implemented which is capable of automatic target acquisition. Range information is provided by a laser rangefinder. Alternatively a 3-D laser-tracker can be fitted. Both the camera and rangefinder are co-mounted on the director head unit, which is provided with appropriate servos and anti-vibration mountings.

The director head is also used as a mounting for combined radar and E-O sensors in the LIROD (**3767.281**), to provide all-weather capability.

Because of its modular design, LIOD can be used for autonomous or integrated weapon control systems, as an auxiliary sensor for various weapon control or data systems. The system includes a console with 15 in TV display.

The daylight version is fitted with a TV camera and laser, while the high-performance model has a TV camera, infra-red camera and laser rangefinder.

With its independent computer capacity using a SMR-MU computer and operational control facilities, it could well be used as an autonomous short-range gun control system.

The LIOD sensor, in its basic version, employed as an autonomous weapon control system, is capable of performing the following functions:

(1) optical surveillance for air and surface targets

LIOD lightweight optronic director

(2) target designation for anti-air and anti-surface engagements
(3) air or surface fire control
(4) automatic or rate-aided tracking of one air or surface target
(5) air or surface target gun prediction calculation
(6) independent control of two or more dual-purpose guns of the same or different calibre
(7) engagement monitoring, kill assessment, and gun fire correction
(8) system status monitoring.

CHARACTERISTICS
Field of view – TV camera: 3 – 30°
Field of view – IR camera: 3°
Elevation scan: -15 to +65°

STATUS
In production and delivered to various naval forces.

CONTRACTOR
Hollandse Signaalapparaten BV, PO Box 42, 7550 GD Hengelo, Netherlands.

5722.293
IRSCAN INFRA-RED SURVEILLANCE SYSTEM

IRSCAN is a fast-reaction infra-red surveillance system for air and surface targets. It has been developed for naval applications but can also be used in the land-based role. The scanner is of modular construction with three sensors covering 3 to 5 and 8 to 12 μm wavelengths, and includes elevation drive, self-supporting closed-cycle stirling-type coolers and is mounted on an azimuth drive. In the naval role a stabilisation system is normally provided with shock and vibration isolating mounting.

The system has a dual wavelength capability to cover a variety of infra-red target signatures and environmental conditions, and special techniques for spatial filtering and signal correlation are used to give the system a highly reliable and accurate target detection and tracking performance. A large number of air targets can be auto-tracked while scanning. A display console is equipped with a 15-inch (38 cm) TV monitor, memory supported with alpha-numeric data presentation and light-pen control keys. The console also contains the system computer and all signal processing.

IRSCAN can be used with systems such as LIOD (**3143.293**), LIROD (light-weight radar optronic director) and Goalkeeper (**3616.231**), as well as other anti-missile defence systems.

CHARACTERISTICS
Horizontal coverage: 360°
Elevation scan patterns: Low, medium, high (up to 40°)
Stabilised platform weight: 350 kg
Stabilisation: 2-axis gimbal for roll and pitch

STATUS
Under development.

CONTRACTOR
Hollandse Signaalapparaten BV, PO Box 42, 7550 GD Hengelo, Netherlands.

IRSCAN scanner in naval configuration

3431.293
TYPE UA 9053 THERMAL IMAGING CAMERA

The UA 9053 camera has been designed to enable passive target tracking by both day and night. Its application is either as an auxiliary E-O sensor readily integrated in a radar weapon control system, or as a sensor for an autonomous short-range weapon control system.

The camera produces a standard CCIR video output. Resolution and sensitivity are such that small low-flying targets can be observed and tracked. Tracking of surface targets is also possible. The sensor is equipped with a closed cycle cryo-generator, Philips type UA 7011. The equipment has been designed for application in maritime environments and operates for long periods without maintenance.

CHARACTERISTICS
Field of view: 42 × 23 mrad (azimuth × elevation)
Instantaneous field of view: 0.13 × 0.13 mrad
Detector cooling: Closed cycle type UA 7011
Cool-down time: Approx 10 minutes
Output signal: CCIR compatible
Input power: +24 V DC, –24 V DC
Power consumption: 250 W
Dimensions: 470 × 200 × 250 mm
Weight: ≤36 kg

STATUS
In production.

CONTRACTOR
Philips Usfa BV, Meerenakkerweg 1, PO Box 218, 5600 MD Eindhoven, Netherlands.

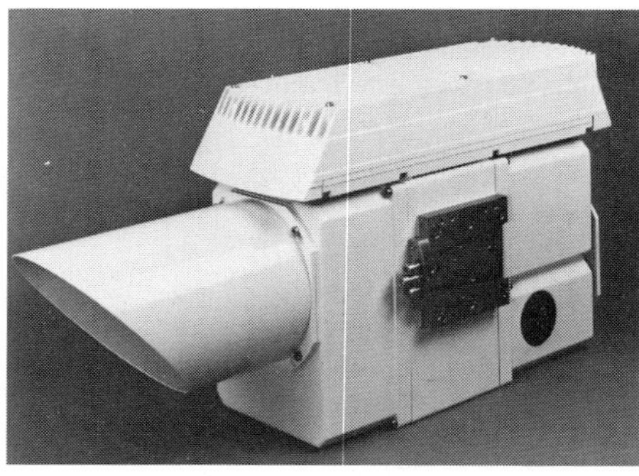

Type UA 9053 thermal imaging camera

3559.293
TYPE MS7TS MARINE NIGHT SIGHT

The MS7TS marine night observation device is an image intensifier system for medium- and long-range night observation at sea. Unlike radar it gives a visual presentation of a ship's silhouette facilitating discrimination between classes of vessels. It makes unilluminated objects immediately recognisable, for instance a blind fishing boat, a floating trunk, rocks, buoys.

The double-hinged bracket permits fast scanning of the horizon from a pedestal on the bridge or on deck. The night observation device is suitable for use on board all kinds of seagoing ship. Its versatile and sturdy design is well adapted to internal and outside use. The system comprises a high speed mirror objective, a three-stage fibre-coupled image intensifier with power supply, a bi-ocular system for comfortable viewing, or if desired a monocular eyepiece. The nickel cadmium battery can be recharged by a special battery charging unit. It is also possible to feed the observation device from the ship's main supply (optional).

A distortion-compensated image over the entire field of view is offered. The device is provided with an automatic brightness control. Detection of unilluminated ships ranges from 8 to 12 km, depending on ambient conditions. The weight is less than 20 kg.

CHARACTERISTICS
Magnification: ×7
Field of view: 7°
Focus: adjustable, 50 m – infinity
Detection range: 8 – 12 km for unilluminated ships

STATUS
In production for various defence forces.

CONTRACTOR
Oldelft (BV Optische Industrie De Oude Delft), PO Box 72, 2600 MD Delft, Netherlands.

NORWAY

5728.293
TC10 THERMAL IMAGER

The TC10 is a thermal imager designed specifically for naval applications to provide high penetration of smoke, haze and camouflage. It is a passive system giving high resolution with a minimum resolvable temperature difference of 0.03°C which offers long-range detection and identification. Detection is independent of ambient light and the system will, therefore, operate as efficiently by night as by day. The thermal picture is displayed on a TV monitor.

The TC10 consists of a thermal camera, signal processing unit and a TV monitor. In the shipboard application it is normally mounted on a stabilised platform.

CHARACTERISTICS
Field of view: 1° (vertical); 3° (horizontal)
Resolution: 0.25 mrad
Focusing: 25 m – infinity
Detector: Cadmium mercury telluride
Spectral band: 8 – 12 μm
Mission time: 20 hours with 5 litre cooling bottle at 300 atmospheres
Detector cooling: Compressed gas
Weight: 8 kg (thermal camera); 7.5 kg (signal processing unit)

CONTRACTOR
Simrad Optronics, PO Box 6414, Etterstad, Oslo 6, Norway.

SWEDEN

1231.283

SAAB TVT-300 FIRE DIRECTOR SYSTEM

The TVT-300 is a system for automatic tracking of objects, in which the optical contrasts of the object against its background as seen by a TV camera are utilised. The basic system consists of a TV camera on a gyro-stabilised platform, a monitor, and an electronics unit containing video processing and servo control circuits. The equipment can be engineered and assembled to serve either fire control or missile guidance applications.

OPERATION

With the target within the view of the TV camera the operator places an electronic 'window' on the target by means of a joystick. The operator then initiates automatic tracking whereby the servo control circuits keep the 'window' centred on the target contrast and also align the target to the centre of the camera field of view.

Signals corresponding to the deviation of the 'window' from the optical axis of the camera are used to control the platform servo motors which, triggered by the deviation signals, train the platform to keep the camera continuously directed towards the selected target. Tracking can be performed on dark or bright contrasts, or on very slight contrast levels. The platform angular positions in bearing and elevation are used for automatic tracking of the guns.

In the Saab TVT-300 the TV camera bearing and elevation values are used to control directly the aiming of AA guns which are slaved to the system. The system replaces the conventional fire direction and correction devices. Target acquisition and corrections are quickly and accurately made from a close-up view of the target seen on the TV monitor.

STATUS

The Saab TVT-300 is operational in several navies.

CONTRACTOR

Saab Missiles AB, 58188 Linköping, Sweden.

Saab TVT-300 automatic TV tracker on board Swedish fast patrol boat. Other versions are used in the Kalle system and those of other users. Various laser rangefinders are employed

UNITED KINGDOM

1207.293

TARGET DESIGNATION SIGHT (TDS)

The target designation sight is used for visual acquisition and tracking of surface and air targets. The sight consists of a binocular mounted on a sight head fitted to a pedestal which is bolted to the deck. The binocular platform is fitted with handgrips with which the binocular may be rotated and the sight head raised or lowered on an elevating shaft to provide height adjustment. Defence Equipment and Systems Ltd acquired sole manufacturing rights from Short Brothers in 1987 with a 20-year licence agreement.

DEVELOPMENT

This equipment, which weighs 77 kg, has been evolved from an element of the Seacat missile system (**1019.231**), and is a modification of the pedestal-mounted visual director used with that system.

STATUS

In production and in service with nine navies.

CONTRACTOR

Defence Equipment and Systems Ltd, Salhouse Road, Norwich NR7 9AY, England.

TDS

3055.293

DES OPTICAL FIRE DIRECTOR

The DES Optical Fire Detector (OFD) is a lightweight, line-of-sight stabilised director fitted with binoculars for operation by a standing operator. The design concept of the director is to produce a low weight equipment of modular construction, easy to install and maintain in surface vessels up to corvette size. It may be used as a main gunfire/missile control system or as a back-up to more sophisticated weapon systems.

The director has duplicate sets of coarse/fine synchro transmissions for both azimuth and elevation. It can therefore be used to control two separate weapon systems or alternatively provide independent transmissions to one weapon for damage control purposes. It may be controlled remotely in either training and elevation, or in both by coarse/fine synchro transmissions. Stabilisation about the line of sight is controlled by two rate integrating gyros.

For local operation, control of the director is by two pistol type handgrips. The righthand grip contains a thumb operated joystick which governs training and elevation. Controls for rate of movement, target acquisition and firing are housed in the lefthand grip. Other OFD controls, and those for linked weapons, are mounted on an easily accessible panel between the handgrips.

Where remote operation is required, the OFD can be controlled by two separate methods:

(a) by slave drive from the ship's surveillance radar (target indication and training only), or tracking radar

(b) by remotely controlled console where sensors suitable for visual monitoring are mounted on the OFD.

In addition to the standard binocular fit, sensor packages up to a 150 kg balanced load may be added. These could include TV, laser rangefinders and thermal imagers. A close range fire control predictor can be fitted on the mount if required.

DES optical fire director

The OFD is constructed from cast aluminium, non-magnetic stainless steel, and GRP to minimise weight and magnetic signature. The director is self-contained, the switch/fuze and distribution box are normally remotely mounted.

CHARACTERISTICS

Electrical requirements:

(a) 380/415/440 V, 60 Hz, 1 kVA

(b) 115V/400 Hz single phase 50 VA

(c) 24 V DC 10 A (including heated suit)

(d) 115V/50 or 60 Hz conditioning heater supply

Training system performance

Arc: Unlimited

Slew rate: 96°/s

Max acceleration: 120°/s²

Elevation system performance:

Arc: –20° to 70°

Max acceleration: 120°/sec²

Weight: 485 kg (including platform)

Below decks assembly weight: 182 kg

Dimensions: Overall height 2.057 m; swept radius 0.762 m

STATUS

Design and development was a private venture project. Adopted by the Royal Navy as a director for the GSA7 fire control system. It has also been selected by ten other navies.

CONTRACTOR

Defence Equipment and Systems Ltd, Salhouse Road, Norwich, Norfolk NR7 9AY, England.

3829.493

BARR & STROUD SUBMARINE ATTACK PERISCOPE

These periscopes have a full torque drive facility enabling complete remote control. The optical system includes line-of-sight stabilisation and anti-vibration features designed to give maximum resolution in high and low powers with quasi-binocular viewing to ease operation. Top window heating is controlled by a solid-state unit in the periscope and the eyepiece heating system prevents condensation.

A television camera is fitted to the rear of the ocular box to relay the optical system to the remote control console and the thermal scene is viewed in the eyepiece direct and relayed to the remote console for viewing or recording, when alphanumeric data can be inserted. A motorised Olympus OM2 Camera is fitted with a Polaroid option available.

Bearing transmission is from an optical encoder in the crosshead and bearing or elevation is displayed in the right eyepiece. Range from the range estimator is calculated by microprocessor, target height entry is on the right periscope control panel.

The periscope has its own continuous BITE system which displays command level faults and the system can be used for diagnostic purposes by selecting an extended test mode.

ESM and communications antennae can be incorporated to suit operational requirements.

Barr & Stroud also offer 190 mm diameter attack and search periscopes for submarines with a displacement of more than 1000 tonnes and 127 mm diameter periscopes for submarines in the 100-400 tonne displacement range.

CHARACTERISTICS

Main tube diameter: 254 mm
Top tube diameter: 70 mm nominal
Overall length: 14 m approx
Well radius: 356 mm
Magnification: × 1.5, × 6
FOV: 32° × 8°
Laser protection: Filters
TV: Visual spectrum plus lowlight capability
Image intensifier: Mullard 50/40 mm Type XX 1332
Thermal imager: 8–13 micron band IR 28 system

STATUS

The CH 84 is in production for the Royal Navy's 'Upholder' class diesel electric and 'Trafalgar' class nuclear submarines. Also available are the CH 85 and CH 86. The system has been ordered for six Australian submarines under a £30 million contract. Deliveries will take place between 1991 and 1996.

CONTRACTOR

Barr & Stroud Ltd, Caxton Street, Anniesland, Glasgow G13 1HZ, Scotland.

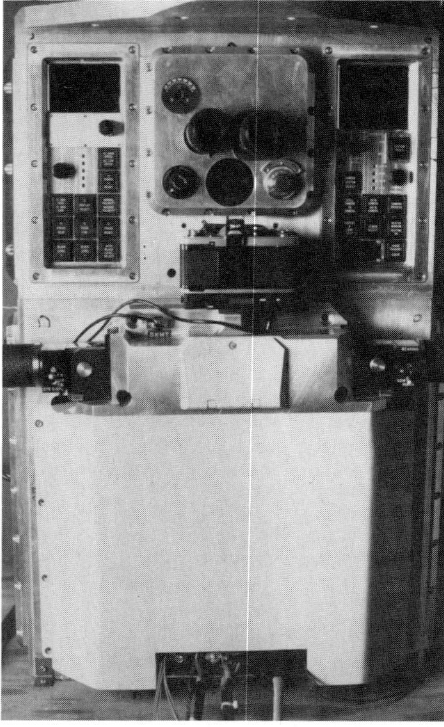

Barr & Stroud attack periscope

3830.493

BARR & STROUD SUBMARINE SEARCH PERISCOPE

Designed to provide the necessary optical performance for visual observation in poor light conditions, high resolution photography and operation of the new Artificial Horizon Periscope Sextant (AHPS). This periscope shares the following features with its attack partner; line-of-sight stabilisation, anti-vibration optics, top window heating, eyepiece demisting, range estimation and transmission, full torque drive and still camera facility.

Previously mechanically operated systems are now operated by electro-servo mechanisms including top elevating prism, change of magnification power, sextant filters, graticule in/out. As with the attack periscope all these features have a 'fail to safe' design for operation during total loss of power supplies. The periscope is quasi-binocular with a choice of four magnification powers. The AHPS has been simplified in operation such that a sight can be taken within a few seconds limiting exposure to a minimum and print-out provides hard copy of the observed altitude, bearing and time of sight.

ESM and communications antennae can be incorporated to suit operational requirements.

CHARACTERISTICS

Main tube diameter: 254 mm
Overall length: 14 m approx

Barr & Stroud search periscope

Well radius: 356 mm
Magnfication: × 1.5, × 3, × 6, × 12
FOV: 24°, 12°, 6°, 3°
Laser protection: Filters
TV: 16 mm SIT vidicon

STATUS

The CK 34 is in production for the Royal Navy's 'Upholder' and 'Trafalgar' class submarines. Also available are the CK 35 and CK 36 versions. The system has been ordered for six Australian submarines, together with attack periscopes. Initial units will be made in Scotland while the remainder will be produced in Australia.

CONTRACTOR

Barr & Stroud Ltd, Caxton Street, Anniesland, Glasgow G13 1HZ Scotland.

3831.493

SUBMARINE PERISCOPE IMAGE INTENSIFIER

Barr & Stroud has developed a submarine periscope image intensifier equipment which can be retrofitted to existing periscopes such as the CH74, 75, 77, and 81, and others with a tube diameter of 180 mm or more, which can accept a large aperture window, to provide an enhanced night watchkeeping capability. This equipment is offered as a refit package with a complementary laser rangefinder (**3832.493**) for periscope modernisation programmes. A low light television system that can be used with the Attack normal viewing optics or with the image intensifier, is also understood to be available.

CHARACTERISTICS

Tube type: Mullard 50/40 mm diameter second-generation channel tube with encapsulated power supply
Magnification: × 6 nominal
Angular field: 7° nominal
Line of sight referred to horizon: Elevation ⩾ 15°; depression ⩾ 7½°
Graticule: LED illuminated crossline

CONTRACTOR

Barr & Stroud Ltd, Caxton Street, Anniesland, Glasgow G13 1HZ, Scotland.

4721.493

BARR & STROUD SUBMARINE OPTRONIC MAST

The Barr & Stroud electro-optic mast concept brings together four sensors; thermal imaging, visual (TV), ESM and communications navigational facilities, when a submarine is at periscope depth or snorting.

The system can directly replace a periscope or dedicated ESM mast, be fitted on a non-hull penetrating mast saving critical space in the strict confines of the control room, or fitted on top of an existing periscope to extend the submarines operational capabilities as a mid-life improvement. It includes a remote control and display console.

Based on modules currently in production and proven in the service environment, the electro-optic mast is in pre-production testing. The configuration of individual systems depends on the operational requirements, eg thermal imaging coverage to the zenith dictates overall diameter of approximately 340 mm but with restricted elevation of line-of-sight to some 50°, means the diameter reduces to 220 mm. The ESM antenna configuration selected by the customer may also affect the overall diameter.

All surfaces of the electro-optic mast liable to radar reflection can be coated with Radar Absorbant Material (RAM).

STATUS

Pre-production testing. The system is understood to have been selected for the 'Vanguard' class ballistic missile nuclear submarines.

CONTRACTOR

Barr & Stroud Ltd, Caxton Street, Anniesland, Glasgow G13 1HZ, Scotland.

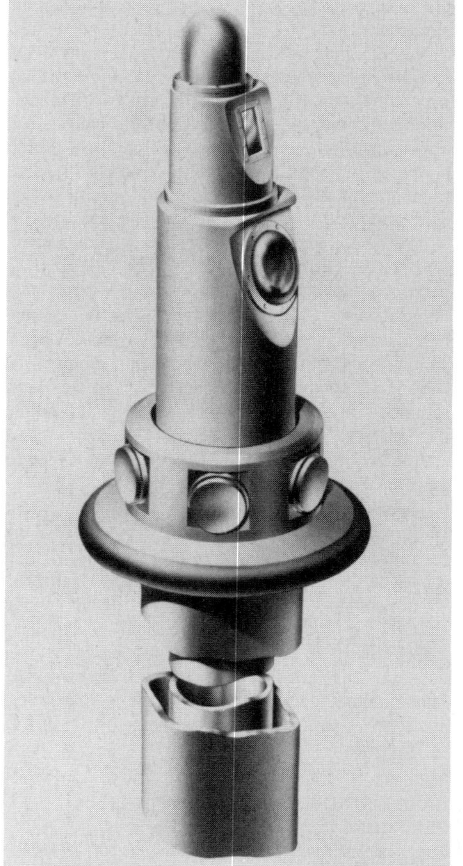

Barr & Stroud optronic mast

6772.293

DIRECTOR-MOUNTED THERMAL IMAGING SYSTEM (DiMTIS)

The DiMTIS system has been designed by Barr & Stroud as a dedicated naval thermal imaging system for use in surveillance, the detection and identification of threats as well as weapon control. It comprises the Radamec Type 409 dual axis stabilised director mount (**6773.293**) together with the IR 26 thermal imager (**6592.093**) with the addition of lowlight television camera, laser rangefinder and an automatic tracking capability as optional extras.

CONTRACTOR

Barr & Stroud Ltd, Caxton Street, Anniesland, Glasgow G13 1HZ, Scotland.

3405.293

IRCOM

The IRCOM system was developed by PPE to meet the requirement for directional ship to ship covert signalling. A PPE night vision device is used to sight the receiving or transmitting ship and to aim the ship's own Francis signalling lamp. The signal beam is viewed by the night sight which is sensitive to the infra-red light and is visible at up to 6 n miles in daylight and on the horizon at night.

The night sight can be used for general surveillance at night and will enable recognition of a frigate at 3 to 4 n miles under starlight conditions. Targets can be readily detected by their phosphorescent wash or shape.

The signalling light can be used to provide additional illumination for more positive identification of intruders, etc, where the sight is used in the surveillance role.

CHARACTERISTICS

Night vision system

Field of view: 22°

Magnification: × **1.4**

Range (moonlight): Frigate at 4 nm

Resolution (starlight): 0.7 mrad

Tube: 3-stage 25 mm extended S20; XX1060/01 Mullard or equivalent; automatic brightness control

Weight: 4.7 kg

Size: 720 × 100 mm

Mounting: Fitted to searchlight

Signalling searchlight

Diameter: 300 mm

Power: 150, 250, or 400 W

Beam divergence: 2 – 3° depending on light source

Mounting: Pintle

Electrical

Sight: 6.7 V battery or 24 V ship supply

Searchlight: 24 V ship supply

CONTRACTOR

Pilkington PE Ltd, Glascoed Road, St Asaph, Clwyd LL17 0LL, Wales.

3794.093

GEC SENSORS V300 SERIES CCTV EQUIPMENT

GEC Sensors Ltd has designed and produces a wide range of closed-circuit television (CCTV) equipment which is used for a variety of military applications, including missile and weapon control and guidance, target damage assessment, manual or automatic target tracking, surveillance, and security uses. Development of the range began in the early 1960s and variants have been in service for some years. The V323, for example, was supplied to the Royal Navy and other navies for use in Seacat missile guidance and control and this version was described in *Jane's Weapon Systems 1974-75* and earlier editions under entry number **1021.293**.

V334

The V334 is the latest daylight/low light TV camera channel equipment in this series. It is a single unit camera of modular design as described below.

The camera is assembled from a range of modules to form either a daylight or low light camera, using a selection of lenses to provide a wide angle or narrow angle field-of-view with a suitable aperture. The camera head incorporates the camera and optical assembly within a totally enclosed housing for full environmental protection and electromagnetic shielding.

The front section of the housing is sealed with a special window module having integral heating and wiping facilities. The optical assembly combines the objective lens and elements of the light control system, including neutral density filters, spectral filters, or focal plane apertures. The camera module provides the necessary scanning, processing circuits and supplies for a daylight TV camera. The light control module contains the additional facilities required to operate an intensifier Ebsicon low light sensor tube and allows automatic exposure control over the full range of naturally occurring illumination. Additional features include spectral and polarising filters, each of which offers conditional enhancement of video signal contrast while focal plane apertures of different sizes aid in the rejection of interfering light sources from within the field-of-view. A simple mechanical interface is provided for attachment of the camera to most director platforms.

Four cameras have been identified:

V3341: Low light; narrow angle field-of-view

V3342: Daylight; narrow angle field-of-view

V3343: Daylight; wide angle field-of-view

V3344: Low light; wide angle field-of-view

STATUS

The V334 daylight camera is being supplied for the RN lightweight Seawolf system, and for other RN applications. It has been selected for various export requirements and also forms part of the Ferranti combat weapon system for four new Brazilian corvettes.

CONTRACTOR

GEC Sensors Ltd, Electro-Optical Products Division, Christopher Martin Road, Basildon, Essex SS14 3EL, England.

Several generations of GEC V300 Series military television equipment are illustrated in these three typical applications. (Right) V323 used on Orion tracker radar for Seacat missile direction; (centre) V0084 low light TV system, on GEC ST801 tracker radar; (left) similar radar but this time with latest V334 TV camera

---

5729.293

V3800 THERMAL IMAGING SENSOR

The V3800 has been designed specifically for long-range detection and automatic target tracking either in support of naval tracking radars or as the prime sensor in an electro-optical fire control system. It is designed as a 'stand-alone' unit capable of incorporation into various defence systems to give indirect-viewing capability in all weathers.

The sensor is based on the UK Thermal Imaging Common Module programme class II (**4280.193**) and uses a zoom telescope giving 2 × 3° to 8 × 12° fields of view. The high sensitivity and resolution of the V3800, together with full environmental protection, makes the equipment equally suitable for coastal defence or border surveillance applications.

STATUS

In production. The V3800 has been ordered as part of a fire control system being supplied to Ferranti to equip four new corvettes for Brazil, and by Contraves, Switzerland for its Seaguard close-in-weapon system. The UK MoD establishments have ordered special versions for precision target tracking requirements and as a safety aid to aircraft landing systems. Similarly USA is evaluating various V3800 imagers for incorporation into several programmes. In the Far East imagers will be integrated into weapon control systems for the Royal Thai Navy, and will be installed into the Hong Kong patrol craft operated by the Royal Navy.

CONTRACTOR

GEC Sensors Ltd, Electro-Optical Products Division, Christopher Martin Road, Basildon, Essex SS14 3EL, England.

4720.293

DIRECTOR AIMING SIGHT (DAS)

The Director Aiming Sight is a power driven, line of sight stabilised director. It comprises a deck-mounted pedestal with rotating head, and a platform which carries a seated operator. A binocular is mounted at eye level on an elevating arm. Electro-optical sensors may be added together with auto-track and on-mount prediction facilities up to a balanced load of 150 kg.

Two direct drive servo systems are employed. One trains the rotating head on to the target bearing, while the other positions the elevation arm and sensors to acquire the target in elevation.

The sight is capable of being driven in the following modes:

(a) **Local.** The director is controlled manually by a seated operator who tracks the target through the binocular. Director movement is controlled by a thumb operated joystick. Outputs from the system can be fed to the ship's fire control system, or to gun or missile mounts.

(b) **Remote.** With the addition of a suitable 'on-mount' sensor package, control can be exercised from a remote console weighing 182 kg.

(c) **Remote Follow.** The director can be slaved to external fire control radar. The operator can act as an observer, verbally recording fall-of-shot and other information.

(d) **Remote Target Indication.** The director may be automatically rotated on to a bearing by a surveillance radar, while the target is acquired by operator joystick movement.

CHARACTERISTICS

Overall height: 1.820 m

Weight: 650 kg (without sensors)

STATUS

In production.

CONTRACTOR

Defence Equipment and Systems Ltd, Salhouse Road, Norwich, Norfolk NR7 9AY, England.

Director aiming sight

4129.093

RADAMEC SERIES 2000 FIRE CONTROL AND SURVEILLANCE SYSTEMS

For naval use the Radamec 2000 series (see **6300.193**) includes multiple electro-optic sensors, video automatic tracker, gun lead predictor computer and interface unit, a tracking and weapon control console together with optional Ka-band radar tracker and target designator sights.

For target acquisition the director slews automatically towards the target bearing and elevation using data from the radar or target designator systems. The target automatic tracker is then used with data from the radar tracker or electro-optical sensors which also provide target prediction data for the computer. This data is then processed by the ballistic computer to calculate super elevation and azimuth lead angle.

For smaller warships a Type 409.035 manned fire control director may be used.

To meet the various mounting requirements the Series 2000 will use Radamec's Type 409, 429, 449 and 459 family of dual axis mounts and the Type 469 mirror mount, details of which will be found either in **6300.193** or **6772.293**.

STATUS

The Series 2000 is in operational use in a large number of fire control and surveillance applications. These include the Royal Navy, US Navy and Finnish Frontier Guard. New tracking systems and fire control systems are currently being supplied to South Korea.

CONTRACTOR

RADAMEC Defence Systems Limited, Bridge Road, Chertsey, Surrey KT16 8LJ, England.

Radamec Series 2000 fire director Type 409-029

6773.293

RADAMEC NAVAL DIRECTORS

A wide range of directors is manufactured by the Radamec company, brief details of which are given below.

Type 409

This mounting, which weighs 60 kg, can accept payloads of 70-100 kg and has a maximum slewing speed of 30-150°/sec in azimuth and 30-100°/sec in elevation. The drive system incorporates low inertia DC motors and fail-safe electromagnetic brakes. The velocity and payload configurations can be adapted to meet customer requirements.

Type 409.035

This version of the Type 409 is a fully autonomous self-stabilised manual director with television or lowlight television camera and a gun control. Other sensor options include thermal imager, rangefinder and minocular or binocular optical devices.

Type 429

This is a heavy duty version of the Type 409 with a range of drive ratios to meet customer requirements. The basic mounting weighs 100 kg and it can accept payloads of up to 200 kg. The maximum slewing speed is 120°/sec in both azimuth and elevation.

STATUS

In production. More than 100 Type 409s have been sold while the Type 409.035 is used by the Danish Navy and Jordanian forces.

CONTRACTOR

RADAMEC Defence Systems, Bridge Road, Chertsey, Surrey KT16 8LJ, England.

4719.293

FERRANTI MULTI-ROLE NAVAL LASERS

Ferranti has developed a number of lasers which can be used in naval applications. The Type 908A rangefinder is modular in concept and is one of a family of naval lasers using common modules. The constituent parts come from the extensive range of successful systems used worldwide in various operational roles. The basic version uses a Type 629 general purpose laser transmitter (**4787.193**) operating at medium power in the wavelength of 1.064 µm and delivering high stability coded pulses.

The Type 908A is a twin aperture equipment, with transmit optics which have a variable beamwidth to allow for rangefinding or illumination. A high sensitivity wide aperture receiver enables rangefinding at extended ranges, the system using the Type 629 transmitter being able to range to 20 km and designate to 9 km. An alternative, the Type 905 high power laser module, is directly interchangeable with the Type 629 and gives the Type 908A a capability of rangefinding out to 35 km and target designation to 16 km. The Type 905 has an output energy of up to 300 mJ, a PRF of up to 20 pps and a beam divergence of 0.25 mrad. Accuracy of range measurement is 2.5 metres.

A simple external attachment to the Type 908A can convert it into a homing beacon for those aircraft which are fitted with either a Ferranti Laser Ranger and Marked Target Seeker (**1196.393**), or a simple infra-red detector with small field-of-view.

Ferranti is also developing an interchangeable receiver module which will provide angle error information to the director servo system, and thus an auto tracking capability, by the use of a quadrant photodiode in place of the existing ranging diode.

STATUS

In production. The Type 908A has been designed for use with the Contraves Seaguard CIWS.

Type 908A laser rangefinder

CONTRACTOR

Ferranti plc, St. Andrews Works, Robertson Avenue, Edinburgh EH11 1PX, Scotland.

6324.293

FERRANTI LIGHTWEIGHT ELECTRO-OPTIC MULTISENSOR

The Ferranti lightweight electro-optic multisensor is fully stabilised and is mounted in a ball-shaped head. The equipment has been designed to provide passive surveillance using both thermal imager and TV as required. Target designation and range capability is provided by a variant of the Ferranti Type 118 laser (**5930.393**). The equipment can be joystick or auto-track controlled with below decks equipment as necessary to suit the operational requirements.

An important advance in the application of the sensors is the optional integration of integration with the Ferranti Helmet Acquisition and Tracking Indication system (HATIS) (**4696.093**). By being linked with HATIS, an operator, from the bridge for example, is able to point the sensor by his head movement only onto a potential target, giving him an immediate TV/thermal picture or laser range and/or designation.

CHARACTERISTICS

Passive surveillance

Thermal imager: 8 – 14 µm to customer requirement

TV: Newvicon or CCD camera with zoom lens to suit

Field of view: Typically 5° narrow; 30° wide

Stabilisation: Four axis

Azimuth freedom: 220°

Elevation: –60° to +90°

Angle output accuracy: 3 mrad

Slew rate: 1 rad/s

Acceleration: 1 rad/s²

Laser rangefinder/designator

Type: Nd:YAG

Beam divergence: 0.4 mrad

Output power: 100 mJ

Range accuracy: 5m

Tracker: Autotracking under specified conditions 2 pixel, ie 150 µrad in narrow field of view
Configuration: Sensor head diameter 425 mm, height 520 mm
Mass: 32 kg
Power requirements: approx 400 W at 24V

CONTRACTOR
Ferranti plc, St Andrews Works, Robertson Avenue, Edinburgh EH11 1PX, Scotland.

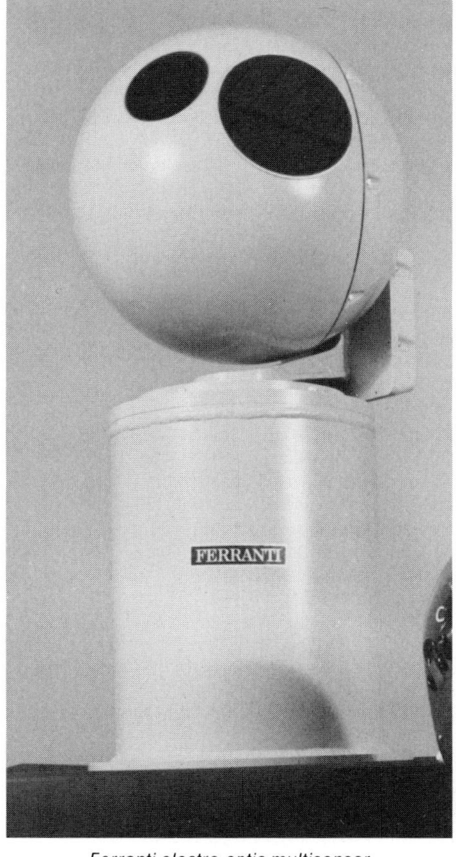

Ferranti electro-optic multisensor

6774.293
FERRANTI FD 5000 PERISCOPE VIDEO CAMERA
This lightweight (135 g) monochrome video camera has been designed to be mounted within a periscope head to assist in periscope observation. It measures 37 × 52 × 55 mm and features a 24 mm lens and offers a resolution of 400 lines.

Submarines equipped with the system would come to periscope depth long enough for a scan of the horizon. The submarine can then dive while the images are examined on a TV monitor linked to the camera by a screened cable.

CONTRACTOR
Ferranti Defence Systems Ltd, Display Systems Department, Ferry Road, Silverknowes, Edinburgh EH4 4AD, Scotland.

6323.292
THORN EMI NAVAL THERMAL IMAGERS
THORN EMI has developed two naval thermal imaging systems basd on the MRTI equipment (**4260.193**). The first of these is a multi-role system (NTIS) used for a variety of purposes in surveillance, target tracking, and weapon fire control. The second is a thermal imaging gun sight (TIGS) for use with 20, 25 and 30 mm gun mountings.

THORN EMI thermal imaging gun sight

The NTIS consists of an environmentally protected housing which contains the MRTI, an operations console containing the remote controls and displays, a deck-mounted air compressor to provide pure air to the Joule Thomson mini-cooler in the imager, and a servo controlled, dual axis director stabilised by gyros.

The TIGS equipment is a compact high performance thermal imaging gun sight employing the THORN EMI handheld variant of the MRTI. It provides day and night aiming capability, dual field of view and a choice of graticule.

CHARACTERISTICS
Naval Thermal Imaging Systems (NTIS)
Field of view: NTIS/1 – 4.3° × 2.8°; 11.1° × 7.4°
NTIS/2 – 3.3° × 2.2°; 8.1° × 5.4°
Director travel: Elevation –35° to +65°; azimuth 190°
Minimum resolvable temperature difference: 0.3 K
Weight: 100 kg (director and imager); 115 kg (compressor); 130 kg (console)
Thermal Imaging Gun Sight (TIGS)
Application: 20, 25, and 30 mm gun mountings
Field of View: Wide 20° × 8.6° (×2 magnification); Narrow 8° × 3.4° (×5 magnification)
Display: Red LED raster with illuminated graticule and warning symbols for low battery and low bottle
Dimensions: 470 mm long (including eye cup) × 210 mm wide × 160 mm high
Weight: 5 kg complete with 0.55 Ah battery and 0.6 litre air bottle

STATUS
Fully developed.

CONTRACTOR
THORN EMI Electronics Ltd, Electro-Optics Division, Forest Road, Feltham, Middlesex TW13 7HE, England.

UNITED STATES OF AMERICA

4024.293
MODEL 975 OPTRONIC DIRECTOR
The Model 975 optronic director has the function of acquiring and tracking surface or air targets for surveillance or directing the ship's defensive weapons. The platform's line of sight is internally stabilised in two axes to the deck plane. Under control of an operator, automatic tracking and range determination of targets are available using the television system or infra-red target pointer and the laser rangefinder. The ship's weapons can be slaved to the optronic director through the fire control system. The director optical scanning capability can be combined with inputs from an existing ship's fire control system or it can be integrated with the Kollmorgen fire control system (**4019.281**).

The director can be utilised in three modes of operation:
(1) search (surveillance mode)
(2) weapon slave (manual or automatic target tracking mode)
(3) director slaved (external designation).

Operational features include: stabilised line of sight; low light-level television with automatic tracking; infra-red target pointer or thermal imaging; laser rangefinder; optional visual capability.

CHARACTERISTICS
Electrical
Angular coverage: –25 to +70° elevation; n × 360° azimuth
Slew rate: 120°/s² (acceleration); 80°/s (velocity)
Tracking rate: 80°/s² (acceleration); 80°/s (velocity)
Tracking accuracy: ±20 mrad
Television
Field of view (FOV): 30° or 7.5°
Resolution (angular): 1 mrad
Scan rates: 525/60 (other scan rates available)
Video output: 1 V P-P, 75 Ω termination
Thermal imaging sensor
Wavelength: 8 – 12 µm
Field of view: 12° wide; 4° narrow
Detector: MCT (Sprite)
Cooler: Closed cycle (Split Sterling)
Laser
Type: Nd-YAG
Power: 4 mW
Pulse energy (radiated): 60 mJ
Repetition rate: 10 pps/continuous
Range: 12 km (with good visibility)

STATUS
Production.

CONTRACTOR
Kollmorgen Corporation, Northampton, Massachusetts 01060, USA.

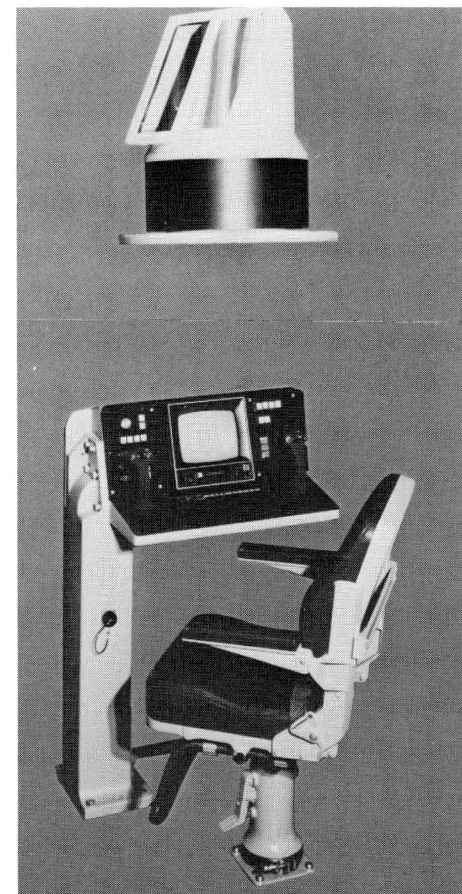
Kollmorgen Model 975 optronic director showing above-deck optical head and operator's console

4023.293

MODEL 985 OPTICAL DIRECTOR

The Model 985 optical director has the function of acquiring and tracking surface and air targets. It is an internally stabilised optical aiming instrument which provides target bearing and elevation signals to the ship's weapon. It consists of:

(1) the director, including all required components for self-stabilisation, power change, and focusing
(2) control console, including all controls, indicators, and read-outs
(3) junction box through which all inputs and outputs are interfaced. Miscellaneous electrical and electronic components are housed in the junction box
(4) cabling to interface the junction box to the console and the director.

CHARACTERISTICS

Angular coverage: –25 to +70° (elevation); n × 360° (azimuth)
Slew rate: 120°/s² (acceleration); 80°/s (velocity)
Tracking rate: 80°/s² (acceleration); 80°/s (velocity)
Tracking accuracy: ±2 mrad
Magnification: ×2, ×8
Field of view: 30°; 7.5°
Focus: 10 m – infinity

STATUS

In production.

CONTRACTOR

Kollmorgen Corporation, Northampton, Massachusetts 01060, USA.

3097.293

Mk 5 LOW LIGHT-LEVEL TV SYSTEM

The Mk 5 LLLTV system can be used for day and night target detection and recognition, extremely long-range detection of faint light sources at night, damage assessment, multiple target discrimination, plus general surveillance and navigation purposes.

The system normally operates at standard line rates, thus permitting easy use with additional remote monitors, video tape recorders, and video trackers, but simple modifications are available for operation at other than standard line rates.

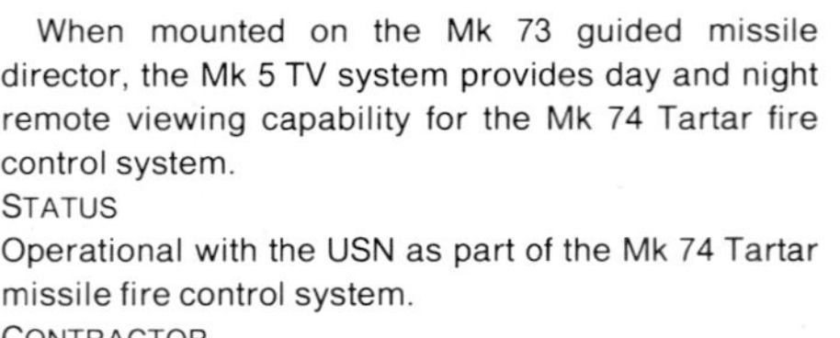

When mounted on the Mk 73 guided missile director, the Mk 5 TV system provides day and night remote viewing capability for the Mk 74 Tartar fire control system.

STATUS

Operational with the USN as part of the Mk 74 Tartar missile fire control system.

CONTRACTOR

GE Aerospace, French Road, Utica, New York 13503, USA.

Mk 5 low light-level television system mounted on USN Mk 73 guided missile director

6775.293

KOLLMORGEN MODEL 76 PERISCOPE

The Kollmorgen Model 76 is a modular periscope system with common components for the attack and search versions. The basic difference is that the attack periscopes have smaller heads while the search periscopes have larger heads to act as multi-purpose reconnaissance platforms.

The system consists of a mast unit (whose weight depends upon customer requirements) with optical train, a display and control unit including a split beam binocular eyepiece, a 35 mm camera, training handles. In addition to the mast unit there is a hoisting yoke weighing 90 kg, a control unit, and a junction box unit weighing 86 kg.

The control unit includes a control panel, system focus, mode select, stadimeter control and microphone. The attack periscope includes a broadband antenna and crystal video receiver ESM system together with a display and control panel on the control unit. The search periscope also includes a system focus control.

The periscopes can be fitted with a Galilean telescope to provide ×12 magnification, line-of-sight stabilisation, a power rotation drive, a remote control unit with 9 in television monitor, an image intensifier, a television camera, thermal imager, video, laser or radar rangefinders.

CHARACTERISTICS

Diameter: 190.42 mm
Elevation: -10 to +74° (attack); -10 to +60° (search)
Magnification: ×1.5, ×6
Field of view: 8°, 32°

STATUS

In production for the US Navy and export customers. In June 1987 the company, which has sold three periscopes to the Dutch Navy, won a $6.6 million contract to provide another two. The company revealed in July 1987 it had won a $19 million contract to supply nine Type 18 periscopes to the US Navy for delivery by 1989.

CONTRACTOR

Kollmorgen Corporation, Electro-Optical Division, Northampton, Massachusetts 01060, USA.

Kollmorgen Model 76 attack periscope

AIRBORNE WEAPON SYSTEMS AND ASSOCIATED EQUIPMENT

This main section contains airborne weapons and their associated equipments. It covers air-to-air weapons and air-to-surface missiles, aircraft armament, weapon aiming systems, surveillance and fire control radar systems, IFF, electronic warfare equipment (including jammers, radar and laser warning systems, electronic intelligence gathering systems and passive countermeasures), electro-optical equipments (including FLIR systems, thermal imagers, laser rangefinders and target designators, head-up displays and other optical systems), and remotely-piloted vehicles. This alteration in the running order of *Jane's Weapon Systems* has been made to enable systems and equipments of each of the four main operational areas, ie strategic, land, sea and air, to be brought together. The numerical system remains identical, thus making it simple to cross-refer to previous editions.

AIR-TO-AIR MISSILES

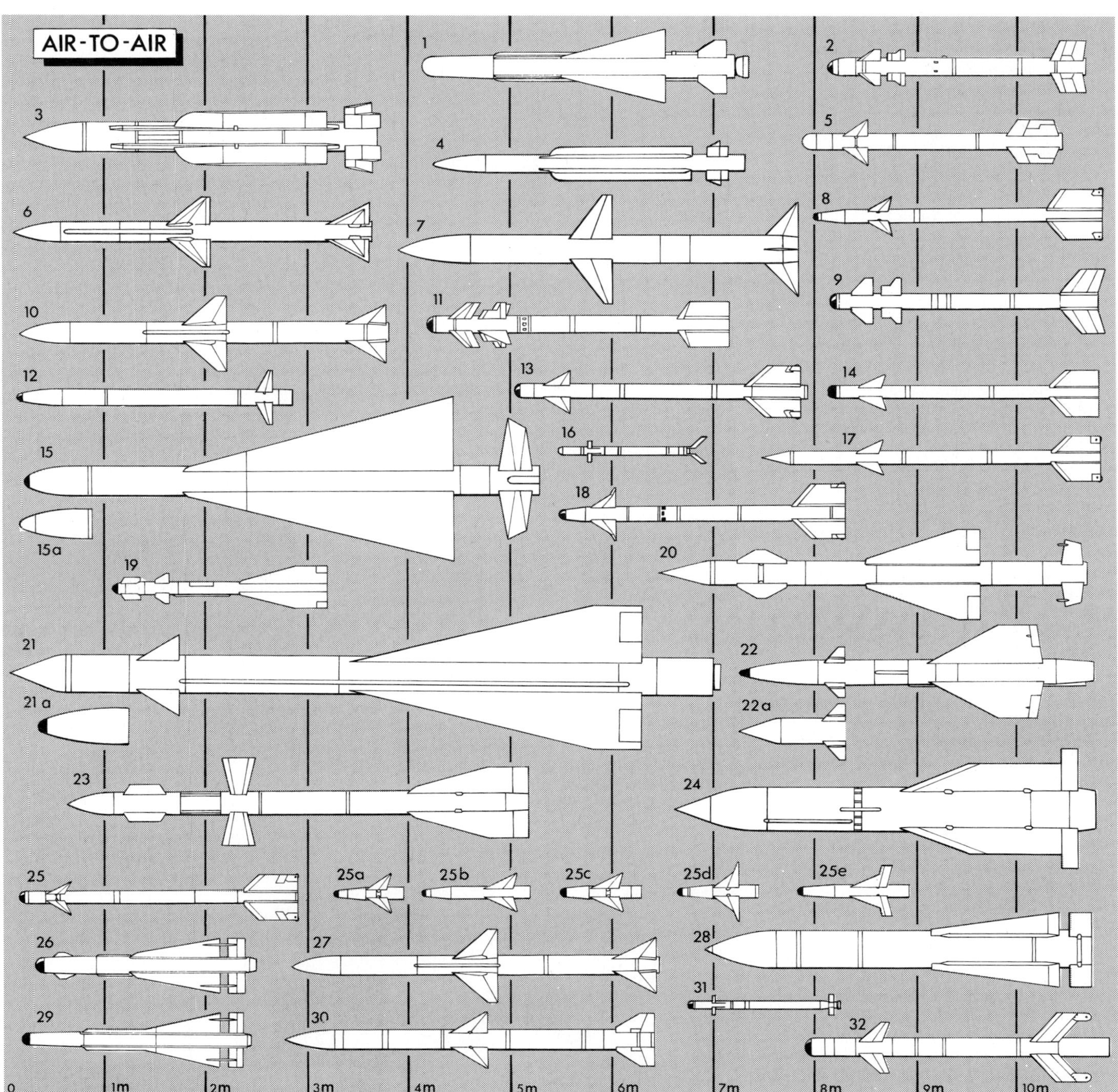

(1) R.530, **(2)** R.550 Magic, **(3)** Super 530, **(4)** Matra Mica, **(5)** Shafrir, **(6)** Aspide, **(7)** PL-9, **(8)** PL-2, **(9)** PL-7, **(10)** Sky Flash, **(11)** Kukri, **(12)** ASRAAM, **(13)** MAA-1, **(14)** AA-2 'Atoll', **(15)** AA-5 'Ash' (IR), **(15a)** AA-5 'Ash' (Radar), **(16)** SA-7 'Grail', **(17)** AA-2 'Advanced Atoll', **(18)** PL-5B, **(19)** AA-8 'Aphid', **(20)** AA-7 'Apex', **(21)** AA-6 'Acrid' (Radar), **(21a)** AA-6 'Acrid' (IR), **(22)** AA-3 'Anab' (IR), **(22a)** AA-3 'Anab' (Radar), **(23)** AA-10 'Alamo', **(24)** AA-9 'Amos', **(25)** AIM-9B Sidewinder, **(25a)** AIM-9D, **(25b)** AIM-9E, **(25c)** AIM-9H, **(25d)** AIM-9L/M, **(25e)** AIM-9N/P, **(26)** AIM-4D Falcon, **(27)** AIM-7F Sparrow, **(28)** AIM-54C Phoenix, **(29)** AIM-26 Falcon, **(30)** AIM-120 AMRAAM, **(31)** AIM-92 Stinger, **(32)** Python 3

BRAZIL

4192.331

MAA-1 AIR-TO-AIR MISSILE

The MAA-1 is an air-to-air infra-red guided missile which has been designed under contract to the Brazilian Air Force for use by the AMX and other fighter aircraft. The programme has been active for several years and the missile is essentially the same as the Piranha which has been in development since 1976.

As an infra-red homing weapon the missile has a range of about 5 km carrying a fragmentation warhead of approximately 12 kg. As can be seen from the illustration, the missile configuration is very similar to that of Sidewinder, and uses a solid propellant motor.

In combat use, the target is acquired either autonomously (visual target acquisition mode), or by slaving the missile's seeker to the aircraft radar (expanded acquisition mode). MAA-1 has a very wide acquisition field and advanced target acquisition modes which enhance combat efficiency. It can be launched at large boresight angles, and the sensitive seeker and active proximity fuze makes it effective against head-on targets.

CHARACTERISTICS

Length: 2824 mm
Fuselage diameter: 152 mm
Wing span: 652 mm
Guidance: IR homing
Launch weight: 90 kg
Propellant weight: 27 kg
Range: 5 km
Mission duration: 40 seconds
Proximity fuze: Active laser
Warhead: 12 kg HE fragmentation

STATUS

Development began in 1976, but was halted at least three times because there was, at the time, no firm Brazilian Air Force requirement. This requirement was established about two years ago and the missile is now in an advanced development and trials stage before being brought into operational service on the Brazilian Air Force's Mirage III and Northrop F-5 aircraft initially, and later on the new AMX fighter. Although there has been no official confirmation, it is understood that a number of live firing trials were carried out during 1986/87. A new company has been formed to produce this weapon as well as the MSS.1.1 anti-tank missile and other rockets and missiles.

CONTRACTOR

Orbita Sistemas Aeroespaciais SA, Engexco Exportadora SA, Avenue Tucunare, 125/211, PO box 152, 06400 Barueri-SP-Brazil.

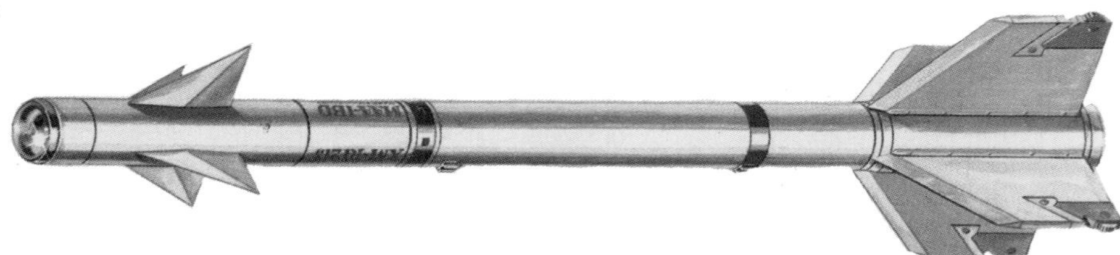

MAA-1 air-to-air missile

CHINA (PEOPLE'S REPUBLIC)

6601.331

PL-2 AIR-TO-AIR MISSILE

Little is known about the Chinese early air-to-air missiles, except that there was a series of missiles PL-2A, 2B and possibly more that were developed from the Soviet Atoll AA-2 missile, itself a copy of the earlier Sidewinder AIM-9. These missiles would have been developed in the 1960s and have been fitted to the Chinese J-5 (MiG-17), J-6 (MiG-19) and J-7 (MiG-21) aircraft.

The PL-2 is similar in appearance to the Sidewinder AIM-9B version, with a length of 2.83 m, a body diameter of 127 mm and a wing span of 0.64 m. The missile weight is around 76 kg. The performance will be limited to tail aspect engagements only, with a range of between 1 and 3 km.

STATUS

PL-2 probably entered service with the Chinese Air Force in the late 1960s and there have been several improvements since then. Unlikely to be still in production.

CONTRACTOR

CATIC, Beijing, People's Republic of China.

6565.331

PL-5B AIR-TO-AIR MISSILE

The PL-5B air-to-air missile is infra-red guided and is designed to fit on virtually any fighter aircraft. The missile has the capability of off-boresight firing and, at its operational altitude, can be snapped-up and snapped-down. A newly designed HE fragmentation warhead is paired with an infra-red integrated circuit proximity fuze which has high counter-countermeasures capability and detonation reliability. Alternatively a continuous rod HE warhead can be fitted, with a radio proximity fuze.

CHARACTERISTICS

Guidance: Infra-red seeker
Length: 2892 mm
Diameter: 127 mm

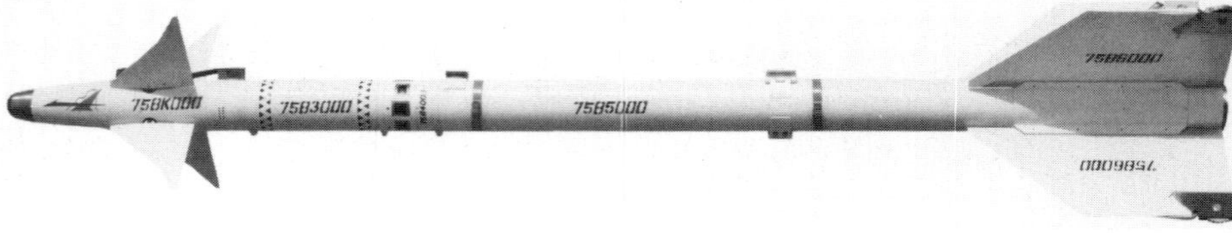

PL-5B air-to-air missile

Wing span: 657 mm
Launch weight: 85 kg
Range: 5 km
Effective lethal radius: 10 m
Warhead: Either fragmentation or continuous rod HE
Fuze: IR or RF

STATUS

In service.

CONTRACTOR

China National Aero-technology Import and Export Corporation, PO Box 1671, Beijing, People's Republic of China.

6602.331

PL-7 AIR-TO-AIR MISSILE

The PL-7 missile was probably developed in the early 1980s and it is believed to be based upon the technology of the R550 Magic 1 missile (entry **1348.331**). This missile would probably be carried by the J-7 (MiG-21) and the new fighter development the J-811.

The missile is similar in appearance to the Magic 1, with a length of 2.72 m, a body diameter of 157 mm and a wing span of 0.66 m. The missile weighs 89 kg and has the distinctive double nose fins of the Magic. This missile will be capable of tail aspect engagements only.

STATUS

Development started in the early 1980s and there is insufficient evidence to know whether or not it is yet in service.

CONTRACTOR

CATIC, Beijing, People's Republic of China.

Chinese PL-7 air-to-air missile

6603.331

PL-9 AIR-TO-AIR MISSILE

There have been reports for several years concerning the development of a medium range air-to-air missile for the Chinese Air Force, to arm the J-811 fighter which first flew in 1984. An air intercept radar is being developed by US contractors for the J-811 and this has a CW illuminator capability which implies a monopulse semi-active homing missile must also be in development. Discussions have been reported concerning the use of Sparrow, Sky Flash or Aspide components by the Chinese and the recently displayed HQ-61 surface-to-air missile (**6418.131**) confirms that the Chinese have obtained the necessary capabilities.

PL-9 is the air-to-air variant of the HQ-61 missile, which is similar in appearance to Sparrow, but larger. The missile is 3.99 m long, with a body diameter of 286 mm and a wing span of 1.17 m. The missile weighs 300 kg. Guidance uses a CW doppler semi-active radar, which together with ECCM gives a good low level performance.

Chinese PL-9 air-to-air variant of HQ-61 missile

STATUS

In development, with initial deliveries expected in about 1989.

CONTRACTORS

CATIC, Beijing, People's Republic of China.

CPMIEC, Beijing, People's Republic of China.

FRANCE

4882.331

MICA AIR-TO-AIR MISSILE

MICA (Missile d'Interception et de Combat Aerien) is a Matra project to develop a new air-to-air missile that is expected to enter service in the mid-1990s. The missile is expected to be smaller than AMRAAM (**3340.331**) but larger than ASRAAM (**3622.331**), and to have a performance somewhere between these two.

To fulfil these functions the development programme envisages an operating range capability covering from a few hundreds of metres to 50 km depending on the homing head in use, exceptional manoeuvrability and high speed, within a 3.1 metre, 110 kg airframe. Other features include a multiple target capability and multi-mode operation. The latter will offer:

(1) long range interception, with inertial guidance plus inflight updates (data link from the launching aircraft) and homing head terminal guidance

(2) medium range interception, using inertial guidance and the homing head for terminal guidance; the missile thereby being fire and forget and fully self-contained after launch

(3) dogfight combat and close range interception, in which homing head 'lock-on' is achieved before firing and release of the missile from the aircraft. The missile is fire and forget from this point and no other co-operation from the launch aircraft is needed.

The multi-mode operation requirement will probably be met by providing interchangeable homing heads for the MICA so that it may be equipped with either an active radar head or a passive IR seeker. Both types will be comprehensively protected against enemy countermeasures, and only switch on for a few seconds after the inertial guidance phase (which is impervious to jamming).

The projected configuration of the MICA can be seen from the photograph. The 3.1 metre long missile will be controlled by a jet-deflector system operating in the efflux of the solid propulsion rocket motor, and is expected to achieve a manoeuvring capacity under load factors greater than those of Matra's Magic air-to-air missile (**1348.331**).

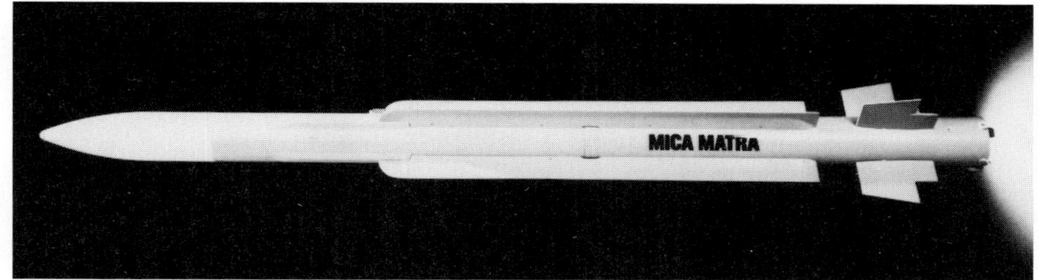

Matra's MICA multi-purpose air-to-air missile concept for combat aircraft of 1990s

STATUS

Preliminary development started early in 1985, and full scale development was authorised by DGA in 1987. MICA is expected in service around 1995.

CONTRACTOR

Matra SA, 37 avenue Louis-Bréguet, 78140 Velizy-Villacoublay, France.

4510.331

AATCP MISTRAL HELICOPTER MISSILE SYSTEM

This is a project to employ the short-range anti-aircraft SATCP Mistral missile (**4198.131**) for helicopter applications such as air-to-air combat between battlefield helicopters, or for the self-protection of helicopters employed in an anti-armour role. In French service, Mistral is known by the initials AATCP from the French for 'very short-range air-to-air'.

The concept is similar to the Soviet use of SA-7 Grail missiles with the Hind helicopter. This heliborne version of the Mistral missile can be mounted on a support protection helicopter as main armament with a gun as the complementary weapon. It can also be mounted on anti-tank helicopters as self-protection armarment. It is likely to be used initially with an SFIM roof-mounted sight of the type used with HOT missiles. With later helicopters it will use a helmet-mounted sight.

The possibility of homing head pre-pointing makes the missile firing easier, and further improvement derives from the ability to choose the scanning pattern of the homing head. Thanks to the fire and forget guidance employed by the Mistral, the launching helicopter is free to manoeuvre immediately after firing the missile. Other significant operational characteristics are the 3 kg warhead and proximity fuze, short system reaction time, high missile velocity, and these limit the target's ability to take cover before missile arrival.

The missiles are carried in a twin-round launcher and the number of these launchers fitted depends on the type of helicopter and other weapon/equipment

French Army Gazelle helicopter with Mistral missiles loaded

loading factors. Typically, four launchers would be fitted on a support protection helicopter.

CHARACTERISTICS
Missile: AATCP (Mistral)
Length: 1.8 m
Diameter: 90 mm
Weight: 18 kg
Warhead: 3 kg
Fuze: Laser
Guidance: IR homing
Range: 3 km

STATUS
In development and early production phase. Firing trials have been taking place during 1986-87 and the missile was due to enter service with the French services in 1988.

CONTRACTOR
Matra SA, 37 avenue Louis-Bréguet, 78140 Vélizy-Villacoublay, France.

1176.331
R.530 AIR-TO-AIR MISSILE

This is an all-weather all-aspect air-to-air missile system produced with alternative semi-active radar homing or IR guidance heads. Principal characteristics are: length 328 cm, diameter 26 cm, wing span 110 cm, weight 195 kg. The range is around 18 km for the radar version and 3 km for the IR. Apart from the homing heads the two versions of the R.530 are identical. Propulsion is by a two-stage solid-propellant Hotchkiss Brandt motor of 8500 kg static thrust. Either of two types of high explosive warhead produced by the same concern, and weighing 27 kg, can be fitted.

The IR homing version is limited to tail aspect engagements only.

The semi-active radar homing head is produced by Electronique Serge Dassault under the designation AD26 and this is for use with an I-band target illuminating radar.

STATUS
Production ceased in 1980, when deliveries ammounted to at least 2400 missiles, of which 45 per cent are stated to have been exported to 12 countries. Many of these weapons are likely to remain in service.

CONTRACTORS
Matra SA, 37 avenue Louis-Bréguet, 78140 Velizy-Villacoublay, France.

AD26 homing head: Electronique Serge Dassault, 55 quai Marcel Dassault, 92 Saint-Cloud, France.

IR homing head: Société Anonyme de Télécommunications, 41 rue Cantagrel, 75013 Paris, France.

1348.331
R550 MAGIC AIR-TO-AIR MISSILE

The Matra R550 Magic is designed for 'close-combat' operations (from less than 500 m to 5 km), with consequent emphasis upon the ability to withstand high load factors imposed by the severe manoeuvre demands required. The general configuration of the weapon is shown in the accompanying photograph. Propulsion is by a Hotchkiss Brandt solid rocket motor. Twist and steer control of this missile is effected by the canard arrangement of fins at the forward end of the weapon. A SAT IR homing head provides guidance, and the missile launcher houses a liquid nitrogen container to provide cooling for the IR sensor.

CHARACTERISTICS
Type: R550 Magic
Length: 275 cm
Wing span: 66 cm
Diameter: 157 mm
Weight: 90 kg
Guidance: IR homing

STATUS
Both the French Air Force and Navy have the R550 in their inventories and Magic has been operational since 1975. By January 1987 approximately 8300 missiles had been ordered, with some 75 per cent being for export to 16 countries, including: Ecuador, Egypt, Greece, India, Iraq, Kuwait, Libya, Oman, Pakistan, Saudi Arabia, Spain, South Africa, Syria, and United Arab Emirates. Among the aircraft types that have been fitted are: Mirage III, V, F1 and 2000, Crusader, Jaguar, Sea Harrier, Super Etendard, Macchi 326K, Hawk and Alpha Jet.

Magic 2 series production started in 1984. A batch of missile front sections containing electronics and controls await testing

Magic 2
In mid-1983 it was revealed that an improved version known as Magic 2 was under development to go into production in 1984. This model employs a new solid-propellant rocket motor, a new IR homing head using the latest technology for high sensitivity, and a new RF proximity fuze.

The main operating characteristics remain very similar to those of the R550 Magic but with the important addition of a front quadrant attack capability to give an all-aspect engagement facility.

Series production of Magic 2 began in 1984 and this model has been ordered by several countries. Adaptation of Magic 2 to the Mirage 2000 aircraft has also been completed, and it is now in service.

CONTRACTOR
Matra SA, 37 avenue Louis-Bréguet, 78140 Velizy-Villacoublay, France.

Matra's two newest air-to-air missiles in current service: Super 530D left, and Magic 2 right

1349.331
SUPER 530 AIR-TO-AIR MISSILE

This is a successor to the Matra R.530 weapon, intended to meet the higher speed and altitude performance requirements of the latest generation of interceptor aircraft.

Two versions have been produced: the Super 530F, which has been operated with the Mirage F1 since 1979; and the Super 530D. The latter is actually a new missile, though with some common items of equipment with the Super F and the same apparent configuration. However, digital processing, a new monopulse doppler homing head, and increased motor thrust lead to increased performance.

With a design concept for reaching targets flying at very high altitude and at high Mach numbers, the Super 530F comprises: a 'long-wing' of low width, which gives it exceptional aerodynamic qualities, in particular in the area of load factors at altitudes greater than 70 000 feet (21 330 metres); a motor with a very high acceleration rate, which allows the interceptor aircraft to fire with vertical separations of more than 25 000 feet (7620 metres).

The Super 530 is equipped with a semi-active CW radar homing head developed by Electronique Serge Dassault (ESD) and a Matra autopilot, and the missile employs proportional navigation. Steel and steel honeycomb is used for the structures, and this with a ceramic radome allows the Super 530 to withstand higher speeds and manoeuvres of up to 25 *g*.

CHARACTERISTICS (Super 530F)
Length: 354 cm
Tail-fin span: 88 cm
Body diameter: 263 mm
Weight: 250 kg
Homing head: ESD, semi-active radar

Proximity fuze: Thomson-CSF radar
Propulsion motor: Thomson-Brandt, solid composite
Warhead: Thomson-Brandt, 30 kg, fragmenting
Range: 25 km
Performance: Shoot-up, shoot-down capability permits engagement of targets more than 7000 m above or below launch aircraft, ECCM incorporated.

STATUS

The French Air Force awarded contracts for the start of series production of the Super 530F at the end of 1977, and French squadrons were equipped from January 1980 onward. The Super 530D is intended for use with the Mirage 2000 with increased range and snap-up engagement capability, advanced technology, and enhanced ECCM. This version entered service with the French Air Force in 1987.

In test firings of the Super 530D with the Mirage 2000 aircraft and the RDI radar, effective ranges of more than 40 km have been achieved. The first firing with a real warhead took place at the Landres test centre in early 1986. Test firings continued during 1986 and 1987.

CONTRACTOR

Matra SA, 37 avenue Louis-Bréguet, 78140 Velizy-Villacoublay, France.

Mirage F1 carrying Matra Super 530 (underwing) and Matra 550 Magic (wingtip) air-to-air missiles

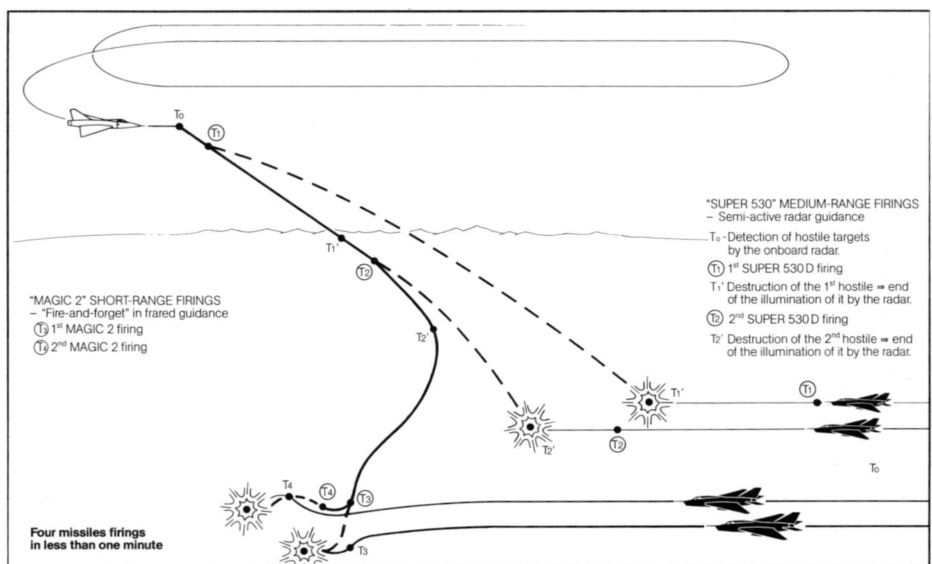

Matra interpretation of how a Mirage 2000 with two Super 530D missiles and two Magic AAMs detecting low-level intruders (Mach 0.9 at 300 ft) could deal with them

INTERNATIONAL

3622.331

ASRAAM (AIM-132)

The AIM-132 Advanced Short-Range Air-to-Air Missile (ASRAAM) is a European development programme started in 1982, as the second part of the four power air-to-air family of weapons. The MoU was signed by the Federal Republic of Germany, France, UK and USA (AMRAAM AIM-120 **3340.331** is the first part). The ASRAAM programme is aimed at developing a successor to the very successful Sidewinder AIM-9 series of air-to-air missiles, with improved manoeuvrability, off-boresight capability, multi-target discrimination, higher flight speed and a smaller miss-distance. The development is now being shared between FRG, UK and Norway, with initial production in Europe planned for the early 1990s followed by possible production in the USA. The improved performance will be coupled with a more advanced IR 'fire and forget' guidance system, and decoy discrimination. Considerable development effort is being made to produce a low cost and reliable missile system that will be completely interchangeable with existing Sidewinder aircraft interfaces. The ASRAAM system will be able to engage multiple targets simultaneously from all aspects.

ASRAAM is a small wingless missile with aerodynamic control by means of tail fins. The missile is expected to be 2.50 m long, with a body diameter of 150 mm and a fin span of 0.45 m. The missile will weigh around 65 kg.

Advanced short-range air-to-air (ASRAAM) test vehicle

STATUS
Feasibility studies completed in 1983 led to project definition which was to have been completed in 1987 but has been extended for a further year. Engineering development should be authorised in 1988, with production starting around 1993. A new company, Bodenseewerk British Aerospace Gerätetechnik (BBG), was formed in 1984 to carry out the development programme for ASRAAM, and is jointly owned by Bodenseewerk Gerätetechnik and British Aerospace. The contracting authority for the programme is the Bundesamt für Wehrtechnik und Beschaffung (BWB) in the FRG. BBG has placed contracts with BAe, Bodenseewerk, Raufoss in Norway, MBB and Junghans in Germany, THORN EMI in UK and Garrett in Canada.

CONTRACTOR
BBG (Bodenseewerk British Aerospace Gerätetechnik), PO Box 1120, 7770 Überlingen, Federal Republic of Germany.

ISRAEL

4185.331

PYTHON 3 AIR-TO-AIR MISSILE

Python 3 is the third-generation successor developed to follow the Israeli Shafrir 2 air-to-air missile. It bears a close resemblance to the Shafrir, which in turn reveals much of the influence exerted by the US Sidewinder on its design. Python 3 has a passive IR homing head, solid propellant motor, and similar gyro stabilised rolleron assemblies on the rear wings that are found on Sidewinders and the Shafrirs. Improvements have been made to the IR homing head, including a longer cooling time and more sensitive detector giving an all-aspect engagement capability.

There are three modes of target acquisition: boresight, uncaged and radar slaved. Maximum and minimum engagement ranges are around 5 km and 500 m, respectively. The major difference between the Python 3 and the Sidewinder family lies in the former's considerably larger warhead, designed to cause catastrophic damage to a target.

CHARACTERISTICS
Length: 3 m
Diameter: 160 mm
Wing span: 0.85 m
Weight: 120 kg
Range: 5 km
Fuze: Active laser

STATUS
In operational service with the Israeli Air Force, Python 3 was used in the Bekaa Valley conflict in 1982 and credited with around 50 kills at a kill ratio of over 80%. Python has been cleared for carriage on the Kfir, F-4 Phantom, F-15 and F-16 aircraft.

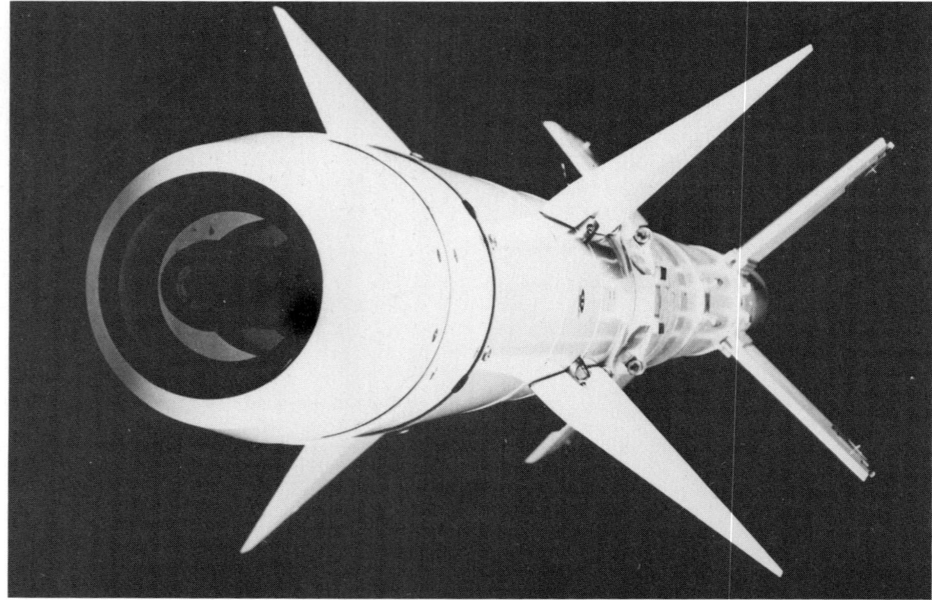

Python 3 air-to-air missile

CONTRACTOR
Rafael Armament Development Authority, Ministry of Defence, PO Box 2082, 31021 Haifa, Israel.

1659.331

SHAFRIR AIR-TO-AIR MISSILE

Shafrir is an air-to-air, IR homing missile for use against high performance aircraft at heights up to 18 000 metres (60 000 ft). It is a relatively small missile, 16 cm diameter and 260 cm long, with pneumatically-operated control fins at the fore-end and stabilising surfaces at the rear. The latter are fitted with the gyroscopically-actuated tabs found on the Sidewinder and Soviet Atoll missiles, for roll stabilisation. Launch weight is 95 kg, of which the warhead accounts for 11 kg. Contact and proximity fuzing of the warhead are provided.

Simplicity was a major design objective, and both missile and aircraft-fitted elements of the Shafrir weapon system have been kept to the minimum. Electronics are solid-state. Apart from the firing circuit, no aircraft-installed equipment is needed.

The current version is the Mk 2, the Mk 1 model having been replaced in service by Shafrir Mk 2 and Python 3.

OPERATION
Firing of Shafrir is 'fire and forget' for when a target is detected within the firing range, the audio signal is heard and an indicator light in the cockpit is automatically operated. This indicates that the missile launch button can be used. From this point, the guidance system is completely independent of the launching aircraft.

Shafrir air-to-air missile on Israeli aircraft

STATUS
Shafrir has been launched in air combat with a high kill ratio (75 per cent) on numerous occasions, according to official Israeli sources and more than 200 kills were credited to Shafrir during the October 1973 conflict. Details of export sales have not been released officially but among countries believed to have been supplied are: Chile, South Africa and Taiwan.

The successor is the Python 3 (**4185.331** above).

CONTRACTOR
Rafael Armament Development Authority, Ministry of Defence, PO Box 2082, 31021 Haifa, Israel.

ITALY

1656.331

ASPIDE MULTIROLE MISSILE

Aspide is a development of the Sparrow 3 missile, first used in the surface-to-air role and more recently developed for the air-to-air role. It uses an Italian semi-active CW radar guidance head developed by Selenia, employing monopulse techniques to improve the performance against low level targets.

Principal characteristics are: length 3.7 m, body diameter 20.3 cm, wing span 102 cm (air-to-air) and launch weight 220 kg. Propulsion is by means of a single-stage solid-propellant rocket motor.

Aspide will be effective both at very high altitudes and for very low level engagements (few metres) as well as in snap-up and snap-down geometries. It has been included in the weapon inventory of the Italian Air Force for the F104S-ASA Starfighter interceptor (improved version of the original F104S). In the surface-to-air role, Aspide is operational in the Albatros naval point-defence weapon system (**1551.281**), and in the Spada ground-based point-defence weapons system (**2250.131**). The two above mentioned systems are in service with the Italian Navy and Air Force respectively.

Technical areas in which major objectives have been achieved are seeker range, sub-clutter visibility, fuzing system, multiple target effectiveness, ECCM capabilities, functional modularity of construction, reliability, and ease of maintenance. Thin-film microwave circuitry developed by Selenia is incorporated in the design.

DEVELOPMENT

Development was started in late 1969 and flight tests began in July 1974. The complete test programme was regarded as highly successful and proved the full realisation of design objectives.

The contracting agency is the Italian MoD with direction of the programme under the General Direction of Aeronautic Constructions (Costarmaereo).

STATUS

Full operational firings from vessels of the Italian and foreign navies equipped with the Albatros system were successfully carried out beginning in 1979. Some firings were also made from ships equipped with the NATO Sea Sparrow weapon system or its Dutch variant (**4775.231**). This was made possible by an adaptation kit developed by Selenia. The system adaptation is readily reversible in either direction and can be carried out even at sea.

Several operational firings were also carried out from the Spada system during 1983 and 1984. A number of systems are now operational for the defence of permanent Italian Air Force installations.

Demonstration firings were made in October 1981 with a Skyguard system (**2377.181**) modified for the guidance of the Aspide missile fired from a four cell launcher developed for this purpose. This system, named Skyguard/Aspide, has been selected by the Italian Army for air-defence on the battlefield.

Italian F-104S launching Aspide air-to-air missile

For the airborne deployment, carriage flights were carried out to demonstrate the transportability of the Aspide missile from the F104S-ASA, alone or in different armament configurations. Subsequently a programme of autopilot firings were successfully made to verify the correct execution of the missile separation and the various interfaces between missile and aircraft. In 1985/86 a number of operational guided firings were carried out against live targets. However, the Aspide air-to-air variant has not yet entered service. An active radar guided variant, called IDRA, is under development with a notional in-service date of the mid-1990s.

In mid-1986 the People's Republic of China awarded Selenia a first-stage contract for integration of Aspide into the Chinese inventory. If successful, the contract would be followed by orders for Aspide, including local production and participation in future development.

CONTRACTORS

Selenia, Industrie Elettroniche Associate SpA, Via Tiburtina Km. 12 400, 00131 Rome, Italy. (Prime)
SNIA-BPD SpA Rome, Italy; Microtecnica SpA Turin, Italy. (Main sub-contractors)

JAPAN

6605.331

XAAM-3 AIR-TO-AIR MISSILE

Little is known about the development of air-to-air missiles in Japan. While Japanese industry has built several models of Sidewinder AIM-9 and Sparrow AIM-7 missiles under licence and sometimes these have been described as Japanese missiles, there has been little development work on national designs. Reports were published in 1985 concerning the development of a new missile, XAAM-3, as a possible national replacement for their Sidewinder missiles. The Technical Research and Development Institute has been awarded a contract and it is expected that Mitsubishi Heavy Industries would act as the prime contractor.

It is expected that the XAAM-3 missile will be about 2.60 m long and weigh around 70 kg. Reports state that alternative IR and active radar seekers are being examined for the missile.

STATUS

Feasibility studies started in 1985 and if successful would lead to full development contracts by 1988-89.

CONTRACTOR

Mitsubishi Heavy Industries, Tokyo, Japan.

SOUTH AFRICA

4526.331
KUKRI AIR-TO-AIR MISSILE

Kukri is the export version of the South African Air Force's (SAAF) V3B IR-guided dogfight missile and like V3B, it is used with a helmet-mounted sight. Compatibility with existing aircraft and weapon systems was maintained by conforming to the Sidewinder and Magic launcher mechanical and electrical interfaces. Development of the earlier V3A system started in the late 1960s and it went into production in 1975. It was the first air-to-air missile to be coupled to a helmet sight and is still operational. The SAAF operational requirements evolved over the years and work on V3B, an upgraded version of V3A, was started in 1975. System improvements included an IR-head with increased look angle and sensitivity, an increased helmet sight designation angle and an upgraded motor. V3B went into production in 1979. The export version, Kukri, is now available to selected customers, and the system is cleared for use on the Dassault-Breguet Mirage III and Mirage F1 types. Trial installations have been made on other aircraft types. Both V3A and V3B are currently operational with the SAAF.

Kukri is a highly manoeuvrable dogfight missile with an asymmetric double-canard aerodynamic configuration. In one plane there are two fixed canards in front with two double-delta movable canards aft and in the other plane there are two simple delta movable canards. The canards are actuated by a hot gas servo and twist and steer control is employed. The triangular canards give roll control and the double-delta canards control pitch acceleration. The tail fins rotate freely around the rear end of the rocket motor. A double base solid-propellant rocket motor and a conventional fragmentation warhead are used.

The missile is delivered in two sections, the first being the motor and warhead assembly and the second the guidance, control and fuzing assembly (GCF). Assembly of the two sections can be effected in a few minutes using standard tools and a simple alignment jig.

A helmet sight is coupled to the missile, thereby giving the pilot an off-boresight target acquisition capability. The helmet sight consists of an IR diode array attitude measurement system and interface electronics to the missile.

The launcher contains the clean air cooling system for the missile IR detector as well as the necessary power supplies and helmet sight interface electronics. It is mounted on the wingtip in the case of the Mirage F1 and on an underwing strong-point in the case of the Mirage III.

OPERATION

Besides the special flying helmet and consequent modifications to the personnel connector on the ejection seat, installation of the Kukri missile system requires some minor aircraft wiring modifications and the addition of three black boxes – the reference unit, the pre-amplifier and the data processing unit. The first two of these must be mounted in the cockpit while the data processing unit can be installed in any suitable location.

This system was designed to be as simple as possible in operational usage and all the controls related to mode switching, target designation and missile firing are situated on the joystick and the throttle. The pilot simply has to look at the enemy aircraft and when target acquisition is signalled by an audio tone in his headset, he designates the target, thereby putting his missile in the tracking mode. Once the missile is tracking the target the pilot can move his head at will. The missile can be fired while tracking or even in the aiming mode, provided that the audio tone is present.

Indigenously developed V3 Kukri missile on SAAF aircraft

South African V3 Kukri air-to-air missile

The pilot's decision making is further simplified by the aiming reticle extinguishing automatically when the target is outside the helmet sight designation limits.

The missile can be fired over the entire flight envelope of the clean aircraft (any speed, altitude and load factor). Minimum operational range is 300 m and the maximum varies between 2 km at low altitudes to in excess of 4 km at high altitudes.

CHARACTERISTICS

Missile
Length: 2944 mm
Diameter: 127 mm
Tail fin span: 530 mm
Fixed canard span: 420 mm
Weight: 73.4 kg
Minimum range: 300 m
Max range: 2 km at sea level; 4 km + at high altitudes
Manoeuvrability: 25 *g* sustained; 35 *g* peak
Max speed: Launch velocity + 500 m/s
Launch limitations
Max load factor: 7 *g*
Max altitude: 50 000 ft (15 240 m)
Max speed: Mach 1.8

STATUS

The V3A went into production in 1975 and both it and the later V3B are operational with the SAAF. Production of the V3B began in 1979. The export version of the V3B, Kukri, has been available since 1982.

CONTRACTOR

Armscor Corporation of South Africa Ltd, Pretoria 0001, South Africa.

TAIWAN

6305.331
SKY SWORD AIR-TO-AIR MISSILE

Revealed in May 1986, Sky Sword is the first indigenous air-to-air missile to be produced in Taiwan. Developed at the Chung Shan Institute of Science and Technology, the Sky Sword is an IR homing weapon bearing a close external resemblance to the US built Sidewinder. Published photographs show it mounted on the wing tip of an F-5E fighter and successfully engaging a target drone. The missile is 2.87 m long, has a body diameter of 127 mm and a wing span of 0.64 m. The missile weighs around 90 kg.

STATUS

Sky Sword was successfully test fired for the first time in mid-April 1986. Assuming that this is a development of Sidewinder AIM-9L, production could be expected to start around 1988-89.

Taiwan developed Sky Sword air-to-air missile mounted (top left) on F-5E, and being launched against target drone

UNION OF SOVIET SOCIALIST REPUBLICS

1146.331

AA-2 (ATOLL) AIR-TO-AIR-MISSILE

Atoll is the NATO code-name assigned to a Soviet air-to-air missile, believed to bear the USSR designation SB06 and/or K13A.

This missile closely resembles the American AIM-9B, IR homing Sidewinder and is of similar dimensions and (estimated) weight. Atoll dimensions are: length 2.90 m, diameter 130 mm, forward control surfaces span 45 cm, tail plane span 0.65 m. The so-called Advanced Atoll, ie the semi-active radar homing model, is an estimated 60 cm longer than the IR guided versions, the extra length consisting of an extended nose section which is presumed to house the radar receiver and antenna package. Solid-propellant and conventional high explosive warhead are assumed. The rear wings have the familiar Sidewinder rollerons fitted.

STATUS

Atoll has been widely deployed with MiG-21 (Fishbed) interceptors of the Soviet home forces and on export versions of this aircraft. Like Sidewinder, Atoll has been developed through many generations. Known foreign users include Egypt and India. The latter country, which has over 50 MiG-21s, has facilities for Atoll production under licence.

In addition to the Warsaw Pact countries, there is an impressive list of foreign users which includes: Afghanistan, Algeria, China, Cuba, Egypt, Finland, India, Iraq, North Korea, Syria, and Vietnam, and the full list of non-Soviet operators of Atoll is thought to number 20 or more countries. In addition to the Indian licence-built version, there is also an indigenous Chinese model called PL-2 (entry **6601.331**).

It is still in service with Warsaw Pact air forces, but is being replaced by the AA-8 Aphid (**3339.331**).

MiG-21 Fishbed interceptor is armed with both standard IR homing AA-2 Atoll air-to-air missile and radar guided Advanced Atoll (on outer wing weapon stations)

Libyan MiG-23 Flogger E armed with four standard AA-2 Atoll IR homing air-to-air missiles (USN)

1144.331

AA-3 (ANAB) AIR-TO-AIR MISSILE

Anab is the NATO code-name assigned to the Soviet air-to-air missile first seen carried by the Yak-28 (Firebar) in 1961. It was subsequently seen on the Su-9 (Fishpot) interceptor, and is known to have been adopted as a standard weapon by the Soviet forces.

The existence of both I-band radar and IR homing versions has been reported. A radar system with which the AA-3 Anab has been employed is known within NATO by the code-name Skip Scan. Length of the radar version is estimated at approximately 360 cm, diameter 28 cm, wing span 130 cm. The IR version is 330 cm long, but otherwise similar. Anab has a cylindrical body with a large cruciform wing assembly at the rear of the missile, and a set of four in-line fins ahead of the wings, and about one-quarter of the missile length from the nose end.

Solid fuel propulsion is assumed, and the range of the radar version is thought to be 10 km, with the IR version range at around 3 km.

This weapon has been in operational use for a considerable time and it is reasonable to assume that it has been the subject of periodic improvements and updates in the same way as some long-life western counterparts have been modified.

STATUS

Believed to be in the process of withdrawal from Soviet forces, in favour of the AA-7, but this could take a considerable time to complete. In addition to the USSR, other nations supplied include Bulgaria, Czechoslovakia, East Germany, Hungary, Poland and Romania. Aircraft types which are, or have been associated with either or both versions of the AA-3 Anab include: Su-9 Fishpot, Su-15 Flagon, and Yak-28 Firebar.

Yak-28P Firebar armed with AA-3 Anab air-to-air missile

1145.331

AA-5 (ASH) AIR-TO-AIR MISSILE

Ash is the NATO code-name assigned to large missiles carried beneath the wings of Soviet Tupolev Fiddler long-range interceptor aircraft. Four such missiles, which are assumed by Western observers to be air-to-air weapons, can be carried by each aircraft. Both I-band radar-guided and IR homing

Radar and IR guided Ash air-to-air missiles on Soviet Fiddler aircraft

versions are in use, two radar homing missiles generally being carried on the wing outer pylons and two IR homing missiles on the inner pylons of the Fiddler interceptor.

Estimated dimensions are: length 530 cm, wing span 130 cm, diameter 30 cm. Cruciform wing and tail surfaces, in-line, are fitted, the former having a sharply swept delta planform. The tail fins have little sweep on either leading or trailing edges and are mounted close behing the wing trailing edges.

The size of the missiles, parent aircraft, and the large nose radome of the latter, suggests that they are intended for long-range interception. It is expected that the radar version has a range of 20 km and the IR version 5 km.

STATUS

Ash has been reported in service with the forces of Bulgaria, Czechoslovakia, East Germany, Poland and Romania, in addition to the Soviet Union. It is reported that several thousand were built.

3337.331

AA-6 (ACRID) AIR-TO-AIR MISSILE

The AA-6, NATO code-name Acrid, was the first of a new family of Soviet air-to-air missiles to be seen by the West, when the first picture of it on a MiG-25 Foxbat aircraft was published in Western magazines in the closing months of 1975.

The resemblance to the earlier AA-3 Anab (**1144.331**) is quite clear, but AA-6 is appreciably larger, having an estimated length of 6.0 m. This dimension refers to both the radar-guided version of the missile and the IR homing model. Two of each type are carried on the Foxbat, the latter type of missile carried on the inner pylons. The launch weight is likely to be in the region of 460 kg.

The AA-6 is thought to be the Soviet first approximation to the American Phoenix long-range air-to-air missile, although it is doubted if it has that weapon's multiple target capability. However it is possible that the IR homing version is launched very shortly after the initial radar-guided Acrid missile to a similar point in space where it is anticipated that the IR seeker will detect and lock-on to its own target. Such a mode of operation would be feasible and useful against multiple targets approaching together. For maximum efficacy, the AA-6 would require an inertial- or autopilot-controlled mid-course phase, and it would enable the radar-guided version to fly to the target area without requiring target illumination by the aircraft's radar if an active homing head is incorporated. The size and weight of the AA-6 tend to support this suggestion.

Estimates of range for both versions have varied between sources but 30 km is likely.

MiG-25 Foxbat E interceptor armed with AA-6 Acrid missiles

The wings of the Acrid missile plainly have control surfaces and it is conjectured that the canards are also used for control purposes. Solid propulsion is assumed.

STATUS

Acrid is the standard armament of the MiG-25 Foxbat, which can carry four such missiles. In addition to the Soviet air forces, other users include the Libyan Air Force and perhaps Algeria and Iraq. The AA-6 has been reported more recently on Su-15 Flagon D and E aircraft.

3338.331

AA-7 (APEX) AIR-TO-AIR MISSILE

Apex, AA-7, is one of three later generation Soviet air-to-air missiles which became known in the West in 1976. Preceded slightly by the AA-6, Acrid, (**3337.331**) which was revealed in late 1975, Apex is comparable to the American Sparrow and is thought to be a successor to AA-3, Anab, (**1144.331**) to which there is some resemblance. However, Apex is clearly of superior performance and somewhat larger. A 'look-down', 'shoot down' capability against low-level targets is also claimed for this missile.

Both radar-guided and IR homing models have been reported, and the US Defense Intelligence Agency refers to AA-7a and AA-7b versions. However, it is thought that the IR version was not produced and the Aphid AA-8 was used instead.

Estimated length of the radar version is about 4.3 m. Weight is probably around 275 kg, of which the warhead could account for up to 30 kg. Solid propulsion is assumed, and range for the radar version is probably in the region of 20 km.

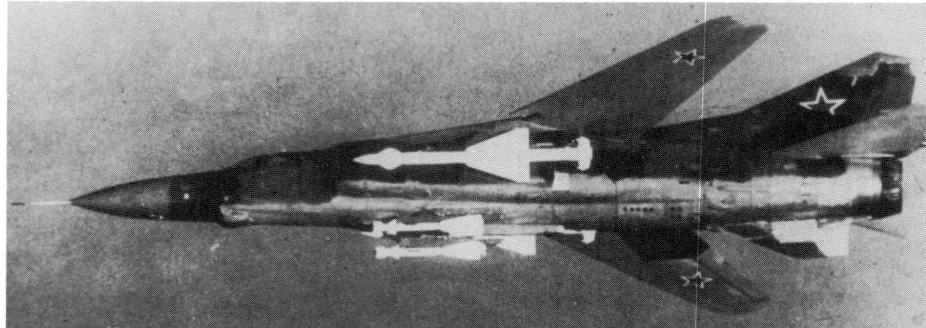

Soviet MiG-23 Flogger B all-weather interceptor carries varied armament of air-to-air missiles with AA-7 Apex radar homing weapons on outer weapon pylons and smaller, IR guided AA-8 Aphid missiles on inboard stations

STATUS

Apex is used on MiG-23 Flogger, MiG-25 Foxbat E and MiG-29 Fulcrum.

3339.331

AA-8 (APHID) AIR-TO-AIR MISSILE

Aphid, AA-8, is thought to be a small close combat air-to-air missile and a replacement for Atoll AA-2.

Radar and IR homing models are reported to exist, but to date the only photographs seen are of what is assumed to be an IR guided version. Estimated length is about 2.1 m. Launch weight is about 65 kg, with a warhead of some 6 kg. Probable range is about 5 km.

STATUS

In use with MiG-23 Flogger, MiG-21 Fishbed, MiG-25 Foxbat, MiG-29 Fulcrum, MiG-31 Foxhound, Su-17 Fitter, Su-22 Fitter J and Yak-38 Forger aircraft.

MiG-23 Flogger G armed with AA-7 and AA-8 air-to-air missiles

4768.331

AA-9 (AMOS) AIR-TO-AIR MISSILE

AA-9 is the numerical designation assigned by the US authorities to the air-to-air missile deployed with the MiG-31 Foxhound interceptor aircraft now in Soviet service. According to US official sources, the AA-9 was designed for use specifically with the MiG-31. This aircraft has a capacity of four AA-9s. The weapon bears a marked resemblance to the American Phoenix and is intended for long-range interception with a 'snap-down' capability. Guidance could be inertial with active radar terminal homing or semi-active CW radar with monopulse for improved performance against low-level targets. The missile is 4.0 m long, has a body diameter of 400 mm and a wing span of 1.0 m. It is estimated that the missile weighs 450 kg. Range is estimated at 75 to 150 km.

MiG-31 Foxhound carrying four AA-9 air-to-air missiles under the belly

4820.331

AA-10 (ALAMO) AIR-TO-AIR MISSILE

The new missile designed for use with the recently introduced Soviet MiG-29 Fulcrum and Su-27 Flanker combat aircraft, has been assigned the number AA-10 in the US sequence. There are two versions of Alamo AA-10, one IR and the other semi-active radar guided. The semi-active radar version is longer at 4.0 m, has a body diameter of 185 mm and a wing span of 0.7 m. The IR version is shorter at 3.20 m, but is otherwise very similar. It is estimated that the radar version weighs 200 kg and the IR weighs 155 kg. Ranges for these missiles are expected to be in the order of 30 km for the radar version and 8 km for the IR.

Photographs have shown the Su-27 Flanker carrying four radar missiles under the fuselage and two IR missiles on outer wing pylons. The MiG-29 Fulcrum has only been seen with two radar missiles on the inboard wing pylons.

AA-10 missiles carried by MiG-29 Fulcrum aircraft with AA-8 missiles outboard (Jane's Defence Weekly)

6171.331

AA-11 (ARCHER) AIR-TO-AIR MISSILE

The only reference to AA-11 has been made in a US Department of Defense report and no details have been given. It is thought to be a short range IR missile to replace Apid AA-8.

6606.331

GRAIL SA-7 AIR-TO-AIR MISSILE

Grail SA-7 is the NATO code and designation for this first generation man-portable Soviet surface-to-air missile (entry **2941.131**). Developed in the 1960s, this missile entered service in the late 1960s and has been progressively improved since. Grail SA-7 has been reported as being used on Soviet helicopters since the late 1970s. The missile has been seen carried on Mi-24 Hind-E helicopters. Gremlin SA-14 is the replacement system for Grail and although this is now in service there have not as yet been any reports of it being helicopter carried as well.

Grail SA-7 is a tube-launched missile, 1.29 m long, with a body diameter of 100 mm and a wing span of 0.3 m. The missile weighs 9 kg. Guidance is by an IR seeker, with later versions having a two colour flare discrimination system. The missile is limited to tail aspect engagements only.

STATUS

Grail is in service with many countries and may still be in licenced production. The Soviet armed forces are the only known users of this missile in the air-to-air role.

UNITED KINGDOM

1774.331

SKY FLASH AIR-TO-AIR MISSILE

Sky Flash is a medium range all-weather air-to-air missile based on the Sparrow missile but with a completely new electronics suite and of improved performance. It is a semi-active radar guided missile capable of attacking both subsonic and supersonic targets from low to high altitudes and has an all-round attack capability. It employs a new advanced guidance system developed by Marconi Defence Systems and a new advanced fuze system developed by THORN EMI Electronics Ltd. The MDS seeker employs the monopulse technique which provides resistance to ECM. The missile has a snap-up capability which allows launches from low levels (100 metres or less) to intercept high altitude targets; a complementary snap-down facility enables targets to be engaged from above against a background of ground clutter. The autopilot and power systems have been updated by British Aerospace Dynamics to include solid-state electronics and thermal batteries. BAe also manufactures the missile structure and carried out final assembly and testing of the complete weapon.

Sky Flash firing from Tornado ADV aircraft

CHARACTERISTICS
Length: 3.66 m
Wing span: 1.02 m
Diameter: 203 mm
Weight: 195 kg
Range: 40 km
Guidance: Semi-active radar, monopulse
Warhead: 30 kg
Fuze: Radar

DEVELOPMENT
In mid-1985 details emerged of a potential Swedish/UK collaborative project to develop a new version known provisionally as Sky Flash 90. A major difference would be the adoption of an active radar homing head, to be developed by MDS, and the current Aerojet propulsion motor would be replaced by a higher performance unit called Hoopoe produced by Royal Ordnance. Swedish interest in the project is related to the RSAF JAS 39 Gripen combat aircraft.

STATUS
Project definition was completed in 1973 with the development programme following immediately.

R & D and RAF evaluation trials were completed in 1977. These trials entailed a variety of attack situations and were completed successfully.

In December 1978 a £60 million contract with the Swedish Forsvarets Materielverk for the supply of Sky Flash missiles to arm RSAF JA37 Viggen all-weather fighter aircraft was confirmed. This followed a two-year integration programme and deliveries of the Sky Flash (Swedish designation, RB71) began in the third quarter of 1980. In the course of the JA37 integration programme, and during routine RAF squadron practice firing, the weapon's high performance obtained in evaluation trials was confirmed.

In December 1983 BAe announced the successful conclusion of a reliability verification and firing programme for the Swedish authorities.

Sky Flash is in service with RAF Phantoms and with the Tornado F2/F3. It has also been ordered by both Saudi Arabia and Oman.

CONTRACTORS
Prime contractor: British Aerospace Dynamics, Manor Road, Hatfield, Hertfordshire AL10 9LL, England.

Homing head: Marconi Defence Systems Ltd, The Grove, Warren Lane, Stanmore, Middlesex HA7 4LY, England.

Fuze: THORN EMI Electronics Ltd, Defence Systems Division, Victoria Road, Feltham, Middlesex TW13 7DZ, England.

UNITED STATES OF AMERICA

3340.331
AMRAAM (AIM-120A)

The AIM-120A advanced medium range air-to-air missile (AMRAAM) programme originated in the United States as a joint US Air Force/US Navy study and development project for a new beyond-visual-range air-to-air weapon to replace the AIM-7 Sparrow missile. The AMRAAM programme together with the Advanced Short-Range Air-to-Air Missile (ASRAAM) project (**3622.331**), forms an important NATO 'family of weapons' programme. Under this arrangement covering future air-to-air missile requirements, the US is responsible for development of AMRAAM while European NATO members are free to undertake development of a new generation of advanced short-range air-to-air missiles.

The Memorandum of Understanding (MoU) covering the AMRAAM development was initiated in April 1978; progress of the project up to that date, when it was pursued as a joint US Air Force/US Navy programme, is described under the above entry in previous editions of *Jane's Weapon Systems*. The air-to-air family of weapons MoU negotiations were completed in February 1980.

AMRAAM is being cleared for carriage with F-14, F-15, F-16 and F-18 aircraft and design studies have been made to consider the modifications required for carriage by Tornado F3, Sea Harrier, Phantom F4F, Mirage 2000, Eurofighter and Rafale. It satisfies the NATO staff requirement for medium-range air-to-air missiles, and the AMRAAM staff target was written by NATO Air Forces Armaments Group SG/13 and is approved by all participating states (with the exception of France). The design calls for an all-weather, all-aspect, radar-guided missile capable of engaging numerically superior aircraft forces before they come within visual range, and there will be a capability for multiple launches at ranges greater than this and for these launches to become autonomous soon after launch to enable the launching aircraft to manoeuvre and/or engage more targets quickly. It is being developed as a successor to the AIM-7F/M Sparrow. Experimental models were built competitively by Hughes and Raytheon and after a fly-off competition, the Hughes design was chosen by the US Air Force in December 1981.

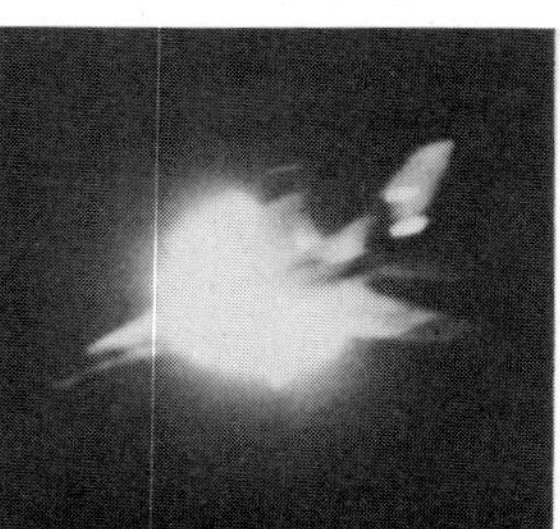

An AMRAAM round destroys a QF-100 drone over the White Sands Missile Range during a test in late 1985. Launched from a USAF F-16 (top picture) the instrumented AIM-120A missile switched to its on-board active radar to guide it to the target (lower left and centre). The AMRAAM, which was fired at medium range, made a direct hit on the drone (lower right)

CHARACTERISTICS
Length: 3.57 m
Diameter: 178 mm
Wing span: 526 mm
Tail span: 627 mm
Weight: 150 kg
Guidance: Command, inertial and active radar

The guidance system will employ digital techniques based on a microprocessor and other engineering details include a TWT transmitter, an integrated RF processor on the seeker gimbal, and a steel fuselage structure requiring no thermal protection. The multi-function microprocessor combines navigation, autopilot, data link, radar, fuzing, missile sequencing and self-test in one data processor.

US Air Force F-16 has Hughes AIM-120A AMRAAM on mid-wing pylon, prior to test flight at Holloman AFB

Information given to the US Congress contains details of how AMRAAM is likely to be used with the F-16 aircraft, and this is summarised here. The aircraft radar will be modified to provide a track-while-scan (TWS) facility and includes a programmable signal processor. The latter will allow multiple target discrimination and tracking, provide software control of radar signal processing to improve responses to ECM, and provide improved clutter rejection features. The TWS mode will be capable of simultaneously tracking multiple targets while continuing to search a large airspace volume for other potential targets. Simultaneous launches of up to eight missiles at multiple targets will also be possible with AMRAAM, and its active radar seeker permits the aircraft to follow a launch-and-manoeuvre pattern, which in the case of a single target engagement follows the following sequence:

(1) in the pre-launch phase the aircraft radar acquires and tracks the target, and an on-board computer calculates the launch acceptability zones and displays the information to the pilot
(2) during the launch phase the pilot first makes his decision to launch the AMRAAM, and inertial reference data on the target and the launch aircraft are fed into the missile computer to provide the necessary initial information
(3) the AMRAAM leaves the aircraft
(4) the inertial mid-course phase is carried out using

inertial data provided by the aircraft's avionics system prior to launch and guidance laws stored in the missile computer to navigate the missile to a calculated point in space where the missile seeker will go into the active mode. The aircraft avionics system will display to the pilot the time-to-go to initiation of the active mode, and he continues to track the target on his radar until that point is reached

(5) the terminal phase begins when, at the calculated point, the seeker transmitter is turned on, and the target is detected. The seeker then tracks the target and the missile completes the intercept.

DEVELOPMENT

In 1983 Hughes embarked on a six-month feasibility study into the use of the AIM-120 missile as the basis of a new point defence weapon for ships. Sea AMRAAM, as the project was dubbed, was the subject of a contract awarded by US Navy Sea Systems Command, and announced in August 1983.

STATUS

A four-power air-to-air family of weapons MoU was signed by the Federal Republic of Germany, France, the UK and USA in August 1980, under which the US will develop AMRAAM and Britain and Germany will develop the next short-range missile. France adopted 'special observer' status and is not committed to full participation. In June 1984 British Aerospace was chosen as lead contractor in a European consortium of four companies (British Aerospace, Marconi Defence Systems, MBB and AEG-Telefunken) to study European manufacture of AMRAAM.

The AMRAAM programme completed the initial concept phase in February 1979 when the US Air Force selected Hughes and Raytheon from five bids for validation phase contracts. During this 33-month phase both contractors built and demonstrated their respective models of the AMRAAM, and in December 1981 the Hughes design was selected for full-scale development.

A $421 million full-scale development contract was awarded to Hughes in 1981, to cover the period 1981-86, with 87 missile firings planned. In 1985 the programme was delayed and the in-service date put back three years from 1986 to 1989, as well as an extensive cost reduction exercise carried out during 1985-86. The development contract provided for the introduction of a US second source for production, and Raytheon was selected. Raytheon was awarded a qualification lot contract for 15 missiles in November 1985. In July 1986 the US Air Force contracted with Raytheon for a major portion of the producibility projects designed to reduce costs of the missile. In December 1986, Raytheon was awarded a contract for the initial procurement of long lead items for a Lot 1 production of 75 missiles.

The MoU provides for a European production line, a third source, to be set-up later if required.

The missile has been in full flight testing and has apparently achieved a high success rate. Low rate initial production of AMRAAM was authorised in 1987, with deliveries starting in 1989 at around 180 missiles per year, increasing over five years to the full planned production rate of 3000 missiles a year. There is a long term requirement by the USAF and US Navy of about 24 000 missiles.

AMRAAM is scheduled for deployment on the USAF/USN F-14, F-15, F-16 and F-18 aircraft, UK and West German Tornados, West German F4-Fs, and Royal Navy Sea Harriers.

CONTRACTORS

Hughes Aircraft Company, Missile Systems Group, Canoga Park, California 91304, USA.

Raytheon Co., Bedford, MA, USA.

First fully guided prototype AIM-120A AMRAAM at Hughes' Canoga Park plant

1085.331

FALCON AIM-4/AIM-26 AIR-TO-AIR MISSILE

Development of the original Falcon AIM-4 missiles started in 1947, with the first production missiles delivered to the US Air Force in 1954. Falcon missiles have been produced with both IR and semi-active radar seekers and there was a 1.5 KT nuclear warhead version in service during the 1960s. There were three designations given to Falcon missiles, AIM-4, AIM-26 and AIM-47. Three versions remain in service with the Air National Guard in the US, the AIM-4D, 4F and 4G. Falcons are also used by the Swedish, Finnish and Swiss Air Forces and these exported missiles are the IR homing HM-58 based on the AIM-4D with the semi-active radar homing HM-55 missiles based on the AIM-26 version. Falcons have been cleared for carriage on many aircraft, including the F-101 Voodoo, F-102 Delta Dagger, F-4 Phantom, F-106 Delta Dart, J-35 Draken and Mirage 3.

California Air National Guard F-106 fires radar-guided AIM-4F Falcon missile during firing practice

Chronologically the AIM-4F was developed before the 4G and the 4D was developed later. The AIM-4F has a semi-active radar seeker and a range of about 8 km. The AIM-4D and 4G both have the same IR seekers, but the 4D was given better manoeuvrability to improve its performance against fighter targets. The range of these two IR missiles is in the region of 3 km and limited to tail aspect engagemetns only. The AIM-26 version has an improved semi-active radar seeker in a larger missile, with a long tapering nose and blunt radome. The AIM-4 missiles are all in the 60 to 70 kg weight bracket, whilst the larger AIM-26 weighs 115 kg.

STATUS

The AIM-4F and 4G entered service around 1960 and the AIM-4D and AIM-26 about three years later. The AIM-4D, 4F and 4G are used by the US Air National Guard. All Falcon missile production ceased in the late 1960s. Exports were made to Canada, Sweden, Finland, Switzerland, Greece, Taiwan, Turkey and Japan.

CONTRACTOR

Hughes, Canoga Park, California, USA.

1099.331

PHOENIX (AIM-54) AIR-TO-AIR MISSILE

This is a long-range, high performance air-to-air weapon, now in use with US Navy F-14 Tomcat interceptor aircraft. Officially issued characteristics of the AIM-54A include: length 396 cm, diameter 38 cm, wing span 92.0 cm, weight 443 kg, semi-active mid-course and active terminal guidance. Estimated range is 150 km.

The complete F-14 weapon system consists of the AWG-9 fire control and armament system (referred to as WCS – weapon control system), the AIM-54, AIM-7 and AIM-9 missiles, and the M61 Vulcan 20 mm cannon. The following paragraphs are concerned with the AIM-54 Phoenix missile, details of the other missiles (Sparrow and Sidewinder) will be found later in this section of *Jane's Weapon Systems*.

OPERATION

During the cruise phase of missile flight, Phoenix guidance is by sample data, semi-active radar homing, the AWG-9 system's radar serving as target illuminator.

The Phoenix missile incorporates an active pulse doppler radar which is switched on to provide target illumination during the terminal guidance phase and in short-range engagements.

The AWG-9 can launch up to six missiles simultaneously against six separate targets, and the AIM-54's principal operational role is to attack enemy bombers at long range before they are able to launch cruise missiles against ship targets.

DEVELOPMENT

The Phoenix concept was initiated in 1960 and Hughes was selected as prime contractor by the US Navy in 1962. Flight testing began in 1965, and the first successful intercept was in September 1966.

Phoenix procurement was initiated during fiscal year 1971. F-14 flight trials started in April 1972, and in December of that year four jet drone targets were successfully engaged by four Phoenix missiles launched and directed by the AWG-9 system of an F-14 Tomcat.

The current and projected threat for the 1980-90

period includes platforms and weapons not originally considered, all of which may be encountered in a severe ECM environment. To meet these threats the US Navy initiated development of an improved Phoenix (AIM-54C) in 1977, with flight trials spread from 1979 to 1985.

The Phoenix improvement programme involved a new programmable digital signal processor, a digital autopilot with a strap-down inertial reference for the missile, and a solid-state transmitter/receiver for the active radar terminal guidance mode. The digital signal processor will perform many autopilot functions, offering expanded high-altitude performance capacity for increased ECCM logic, flexibility for the future, and an increase in reliability by using fewer parts. The mechanical flipper actuators and servo amplifier functions are not changed. The Naval Weapons Centre at China Lake, California, also developed a new active radar fuze for AIM-54C.

CHARACTERISTICS AIM-54C

Length: 3.96 m

Diameter: 380 mm

Weight: 454 kg

STATUS

The first Phoenix AIM-54A missiles were operational by spring 1974.

In the course of fiscal year 1977, the Phoenix Improvement Programme was started to improve the reliability and performance capability of the missile to cope with the projected threats of the 1990s. This became the AIM-54C version, which entered service in 1985.

In early 1985 flight tests began in connection with a further enhancement of Phoenix. This programme will integrate the 'sealed' AIM-54C Phoenix missile with US Navy F-14A and F-14D fighter aircraft and ensure compatibility. A major feature of the sealed Phoenix is the addition of internal heaters that eliminate the need for the F-14D to provide the missile with temperature conditioning liquid prior to launch.

In 1986, Raytheon was awarded a US Navy contract for second source production qualification. Qualification will be via directed buys in US fiscal years 1986-88. Full competition will commence in fiscal year 1989.

CONTRACTORS

Prime contractor, missile system and launcher: Hughes Aircraft Company, Missile Systems Group, Canoga Park, California 91304, USA.

Missile, second source: Raytheon, Missile Systems Division, Bedford, Massachusetts.

Third engineering model of improved AIM-54C Phoenix air-to-air missile after launch from US Navy F-14 Tomcat over Pacific Missile Range

4767.331

SIDEWINDER (AIM-9) AIR-TO-AIR MISSILE FAMILY

The family of air-to-air missiles that evolved from the original AIM-9B weapon developed by the US Navy in the late 1940s and early 1950s is now an extensive one comprising numerous versions, with the latest production model having the designation AIM-9M. Most of the earlier variants have been described and/or illustrated individually in previous editions of this book under various entry numbers from **1103.331** to **3812.331**. (All previous entries for Sidewinder are included in the Numerical List of Entries at the back of this volume).

The Sidewinder family originated from a US Navy design from the Naval Weapons Centre at China Lake, California. Production contracts have been placed with many US companies over the 30 years span, but the guidance and control sections have been manufactured by General Electric, Ford Aerospace and Raytheon in the USA. Several versions have been produced under licence in other countries, and notably AIM-9L versions were built in Japan by Mitsubishi Industries and in Europe by Bodenseewerk Gerätetechnik (BGT) leading a consortium of companies from the Federal Republic of Germany, UK, Italy and Norway. The first model to go into major production was the AIM-9B, and this was used by both the US Navy and USAF. The US Navy and USAF then developed separate versions, with the US Navy having AIM-9D, 9G and 9H, and the USAF having AIM-9E, 9J, 9N and 9P versions. From the early 1970s the US Navy and USAF programmes merged again, and both services have received the latest versions, the AIM-9L and 9M. In Europe, BGT developed an improved version of the AIM-9B, with CO_2 cooled PbS detector, for European weather and low altitude operation. Of this version, known as the AIM-9B/FGW Mod 2, 7700 were built and shipped to various NATO nations.

DEVELOPMENT

The AIM-9A was the prototype of the Sidewinder and was first fired successfully in September 1953. The first production version (AIM-9B) joined the US Air Force in 1956, but had some limitations in use (inability to engage targets close to the ground or head on, for example). Subsequent versions eliminated these shortcomings, and most 'B' models were later upgraded to the 'E' or AIM-9P export version status.

The AIM-9J modification of the AIM-9B/E, with greater manoeuvring capability, higher speed and increased range, was first issued in 1977 to equip the F-15 and other Sidewinder compatible aircraft.

Production of the AIM-9L began in 1976. This has a more powerful motor, improved tracking and manoeuvring ability. A more advanced active optical fuze increases missile lethality and resistance to ECM. An FM-AM conical scan pattern increases seeker sensitivity and gives better tracking stability. This is the first model Sidewinder with the ability to attack from all angles, including head on. The AIM-9M was also in production in 1982, and this model has the all round capability of the AIM-9L, plus higher performance and increased resistance to IR countermeasures.

The AIM-9P is an upgraded version of the AIM-9J which enable targets to be engaged at longer ranges, and deliveries of this model began in 1978. The AIM-9P-1 has an active optical target detector instead of the IR fuze; the AIM-9P-2 has a reduced smoke rocket motor; the AIM-9P-3 has both the above modifications and the latest version, AIM-9P-4, has an all-aspect engagement capability.

AIM-9B Sidewinder air-to-air missile mounted on wing of F-104A Starfighter

AIM-9R

Due to become operational in the early 1990s the AIM-9R will show a significant improvement in target acquisition range and improved ECCM performance. Originally known as the AIM-9M product improvement programme, the AIM-9R had its functional data package prepared by the Naval Weapons Centre (NWC), China Lake in 1986, and the decision to proceed with full scale development was made in early 1987 by the US Navy. Ford Aerospace has been chosen as the primary development support contractor for the AIM-9R.

CHARACTERISTICS (AIM-9L and 9M)

Length: 2.87 m

Diameter: 12.7 cm

Fin span: 0.64 m

Launch weight: 87 kg

Guidance: IR

Warhead: Blast fragmentation HE

STATUS

The AIM-9 Sidewinder, in its variants, continues in production in the USA. The missile has also been sold to other countries, the latest being Egypt and Oman. The AIM-9R is likely to replace other variants with the US defence forces and procurement of these will probably be cut back until the AIM-9R comes into service.

CONTRACTORS

Ford Aerospace & Communications Corporation, Ford Road, Newport Beach, California 92660, USA.

Raytheon Company, Missile Systems Division, Bedford, Massachusetts 01730, USA.

Test firing of Sidewinder AIM-9P from Northrop F-20 Tigershark tactical fighter aircraft at the Naval Weapons Centre, China Lake, California

1106.331
SPARROW (AIM-7) AIR-TO-AIR MISSILE

The Sparrow missile is one of the most widely-used US weapons. Its versatility has been demonstrated in surface-to-air applications in addition to its original role as an air-to-air missile. The present production version, the AIM/RIM-7M is the fifth model of the Sparrow III family which includes the AIM-7C, D, E and F models.

The AIM-7E and F are still in the operational inventory of the US Air Force, US Navy and several allied nations. A variant of the AIM-7E designated the RIM-7E-5 is used in the surface-to-air role in the US Navy's Basic Point Defense System. Another AIM-7E variant, the RIM-7H-5, is used in the NATO Sea Sparrow Surface Missile System. The RIM-7H-5 has folding wings and clipped fins to reduce the size and weight of the deck launcher/containers.

The Sparrow AIM-7F is a medium-range, all-weather, all-aspect, semi-active guided missile that is compatible with both CW and pulse doppler illumination. The AIM-7F is similar in external appearance to earlier Sparrow missiles. It is 3.66 m long and 203 mm in diameter. The AIM-7F has a wing span of 1.02 m, a tailspan of 0.8 m and a total weight of 227 kg.

It features a more powerful rocket motor that provides both extended range and faster speed to intercept, an improved 39 kg continuous rod warhead that is triggered by either contact or proximity fuzes, and greatly increased lethality. An all solid-state seeker provides greater sensitivity and reliability and permits snap starting the missile in less than two seconds.

After a competitive flyoff with General Dynamics, Raytheon was awarded a contract in 1978 for the full-scale development of the AIM/RIM-7M Sparrow missile. That development provided an inverse monopulse seeker that includes a digital signal processor, a new autopilot and a new fuze. The results are performance improvements in seeker sensitivity that permit picking out targets from the heavy clutter in look-down, shoot-down environments. It also provides greater immunity to the growing and changing countermeasures threat. The design features isolation mounting, reduced parts count and built-in test, which contribute to improvements in reliability and maintainability. The AIM/RIM-7M is a common missile with all-weather, all-aspect capability in both air-to-air and surface-to-air applications.

The AIM-7M has the same outside dimensions as the AIM-7F. The RIM-7M version, by adding the folding wings and clipped tail fins will be compatible with the NATO Sea Sparrow launcher.

The digital processor which controls the AIM/RIM-7M is reprogrammable as the ECM threat changes and offers potential for further growth in missile capability.

The latest version, AIM-7P/RIM-7P, full development programme will run from 1987 to 1991 and introduces further improvements to the fuze and electronics aiming to provide an improved performance against cruise missiles and sea-skimming anti-ship missiles.

F-20 Tigershark tactical fighter fires AIM-7F Sparrow missile, MQM-74C target drone almost 13 miles away was destroyed by direct hit

STATUS

The AIM/RIM-7M is in full rate production and began entry into active US Air Force and US Navy service inventories in 1982. Current US procurement plans call for the US Air Force and US Navy to be equipped with more than 17 000 AIM/RIM-7M missiles. AIM-7P development will continue until 1991 and is seen as a prudent hedge against further problems or delays with AMRAAM.

CONTRACTORS

Raytheon Company, Missile Systems Division, Bedford, Massachusetts, 01730, USA.

General Dynamics Corporation, Pomona Division, Pomona, California 91766, USA.

6566.331
AAAM LONG RANGE AIR-TO-AIR MISSILE

The US Navy is about to start the development of a new advanced long-range air-to-air missile (AAAM) which is intended to replace the AIM-54C Phoenix in the mid- to late-1990s. A technology validation phase is planned to run until 1991, with two competing industrial design teams, with a selection for full scale development being made in 1992. The two competing teams are Hughes, Raytheon and McDonnell Douglas, with the other team comprising General Dynamics, Westinghouse and Morton-Thiokol.

The new missile is designed to combat increased threats and would enable the US Navy's F-14 and F/A-18 aircraft to carry more missiles wth longer ranges. It is understood that a missile with a diameter of some 23 cm, a weight of less than 300 kg, and a speed of Mach 3 is the requirement. The missile range would need to be 'significantly greater than the Phoenix'.

STATUS

The programme is in the technology validation phase, which is planned to continue until 1991. Full scale development should then start in 1992.

6604.331
STINGER (AIM-92) AIR-TO-AIR MISSILE

Although the portable surface-to-air Stinger FIM-92A (entry **2805.131**) has been in production since 1980, full scale engineering development for the air-to-air missile system did not start until 1984. The Air-To-Air Stinger (ATAS) programme provides a sensitive lightweight missile for use at short range against low flying aircraft and helicopters. Stinger improvements include Stinger Post and Stinger RMP, the former incorporating a dual colour IR and UV seeker for decoy discrimination and the latter a reprogrammable microprocessor for subsequent counter--countermeasures modifications. These improved versions entered production in 1986. The Stinger AIM-92 has been carried on the OH-58C Kiowa, OH-58D AHIP, UH-60 Black Hawk, AH-64A Apache, AH-1 Cobra and the Lynx 3 helicopters.

The Stinger missile is tube-mounted and the missile is 1.52 m long, has a body diameter of 70 mm and a wing span of 0.14 m. The missile weighs 16 kg and a twin launcher with two missiles weighs 47 kg. The later Stinger Post has a sensitive two colour IR and UV seeker and microprocessor capable of acquiring most helicopters from all aspects as well as discriminating between decoys and ground clutter.

OH-58C Kiowa helicopter firing air-to-air Stinger missile

The ATAS has a simple boresight acquistion system, with the pilot uncaging the missile seeker and firing when he has a clear audio tone.

STATUS

The air-to-air Stinger AIM-92 programme entered full scale engineering development in 1984 and the missile is due to enter service with the US Army in 1988. Stinger FIM-92B surface-to-air missiles to Stinger Post/RPM standard began delivery in 1987 and may be used in the air-to-air role. Although the surface-to-air missiles have been widely exported, there are no known exports of the air-to-air variant.

CONTRACTOR

General Dynamics Valley Systems, Ontario, California, USA.

AIR-TO-SURFACE MISSILES

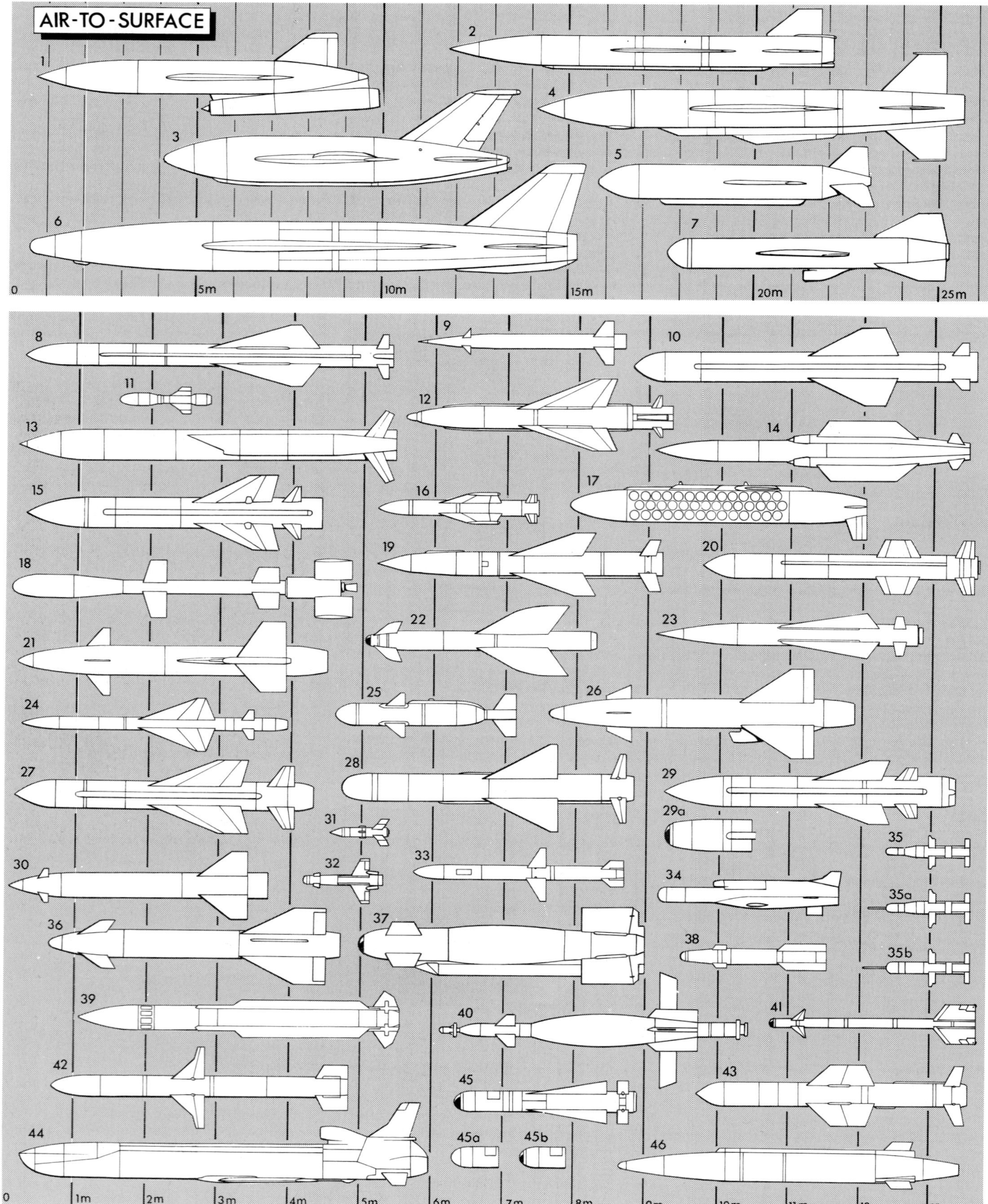

1 AS-2 Kipper, **2** AS-6 Kingfisher, **3** AS-5 Kelt, **4** AS-4 Kitchen, **5** C-601, **6** AS-3 Kangaroo, **7** HY-4, **8** Exocet AM39, **9** Martin Pescador, **10** C-801, **11** HOT, **12** AS-30L, **13** ASMP, **14** ANS, **15** Armat, **16** AS-15 TT, **17** Apache/CWS, **18** Marte 2, **19** Kormoran 1, **20** Gabriel 3, **21** Rb 04, **22** Penguin 3, **23** Rb 05, **24** ALARM, **25** Sea Skua, **26** RBS-15, **27** Sea Eagle, **28** ASM-1, **29** Martel AS-37, **29a** Martel TV AJ-168, **30** AS-7 'Kerry', **31** AT-3 'Sagger', **32** AT-2 'Swatter', **33** AGM-45 Shrike, **34** Tacit Rainbow, **35** TOW, **35a** ITOW, **35b** TOW 2, **36** AS-9 'Kyle', **37** AGM-130, **38** AGM-114 Hellfire, **39** AGM-78 Standard, **40** AGM-123 Skipper, **41** AGM-122 Sidearm, **42** AGM-88 Harm, **43** AGM-84 Harpoon, **44** AGM-86 ALCM, **45** AGM-65B Maverick, **45a** AGM-65E, **45b** AGM-65 D/F/G, **46** AGM-69 SRAM I

ARGENTINA

3678.311

MARTIN PESCADOR AIR-TO-SURFACE MISSILE

The Martin Pescador is a supersonic aircraft-launched tactical missile for use over short ranges (2.5 to 8 km). It is powered by a single-stage solid-propellant rocket motor, and carries a conventional high explosive warhead weighing 40 kg.

The configuration of the Martin Pescador consists of the following major components: the fore-body containing radio receiver, power source, control circuits and logic devices; centre section with warhead and fuze; aft body section containing the rocket motor, propellant and flares; four wings attached to the aft body; and four canard control surfaces actuated by the fore-body control sub-system.

The approximate time of flight is 13 s.

OPERATION

Radio command is used to direct and steer the Martin Pescador to its target. Launching is possible from any aircraft capable of a speed of Mach 0.5 or more. Launch from a hovering helicopter is possible also, in which case a range of 4.3 km can be achieved. The rear of the missile carries three flares to assist in tracking. Command signals are sent to the missile from the launching aircraft which carries the command radio transmitter. The pilot has to keep the missile on his line of sight to the target during its flight time and a button on the aircraft control column is provided for this purpose. In order that a number of Martin Pescador missiles can be launched simultaneously from several aircraft, co-ordinated changes of radio command frequency have to be taken into account; facilities for this are provided on a missile control panel in the cockpit.

DEVELOPMENT

The Martin Pescador missile has been developed by the Instituto de Investigaciones Cientificas y Tecnicas de las Fuerzas Armadas (CITEFA) on behalf of the Argentinian Forces. It should be noted that the word 'Martin' does not signify any connection with the well-known US defence contractor, or any of the missiles produced by the American concern; *Martin Pescador* is Spanish for 'Kingfisher'.

CHARACTERISTICS

Length: 294 cm
Diameter: 220 mm
Wing span: 73 cm
Total weight: 140 kg
Warhead: HE 40 kg
Range: 8 km

STATUS

In full production. New versions are under development, with increased range and with a heavier warhead. Other developments include helicopter-launched models with longer range.

CONTRACTOR

Instituto de Investigaciones Cientificas y Tecnicas de las Fuerzas Armadas (CITEFA), Zufriategui y Varela, 1603 Villa Martelli, Provincia de Buenos Aires, Argentina.

CHINA (PEOPLE'S REPUBLIC)

6473.311

C-601 ANTI-SHIP MISSILE

The C-601 is an anti-ship guided missile intended primarily to attack surface targets as large as a light cruiser. The missile has a liquid propellant rocket motor. Terminal guidance is by means of a monopulse radar which is equipped with ECCM to combat countermeasures. Two missiles are normally carried by a H-6 Xian aircraft (TU-16 Badger A), one under each wing. The missile is 7.38 m long, has a body diameter of 920 mm and a wing span of 2.80 m. The missile weighs 2440 kg, of which the HE blast warhead contributes some 400 kg. Speed is said to be about Mach 0.8.

STATUS

Operational with the Chinese Air Force.

CONTRACTOR

Chinese Precision Machinery Import and Export Corporation, 17 Wenchang Hutong Xidan, Beijing, People's Republic of China.

C-601 anti-ship missile mounted on wing pylon of Xian H-6 aircraft of Chinese Air Force

6607.311

HY-4 AIR-TO-SURFACE MISSILE

The HY-4 missile is clearly a development from the earlier HY-2 Styx based missile with the Chinese designation C-601. The HY-4 retains the familiar body shape, but the available information suggests that the whole missile has been modernised.

The HY-4 programme has been developed primarily as a coastal defence surface-to-surface weapon and the airborne variant has been developed since about 1983 with no clear indication that it is actually in service.

HY-4 is 7.36 m long, with a body diameter of 760 mm and a wing span of 2.8 m. The missile weighs 1740 kg in the air-to-surface mode, which is without the solid propellant boost motor used for surface launching. Inertial guidance is used for the mid-course phase, followed by an active radar terminal seeker. The missile is subsonic, powered by a turbofan engine and programmed to fly at low level. A range in the order of 150 km is claimed. HY-4 has a shaped charge HE warhead of around 500 kg.

STATUS

The HY-4 missile has been developed through the 1980s as a coastal defence weapon, but is unlikely to be in service yet as an airborne system.

Chinese HY-4 air-to-surface missile

CONTRACTOR

CPMIEC, Beijing, People's Republic of China.

6476.311

C-801 AIR-TO-SURFACE MISSILE

The C-801 is a missile intended primarily for the anti-shipping role. It was first shown publicly in Beijing in 1984 and bears some resemblance to Exocet and Harpoon. The cruising altitude can be as low as 20 to 30 metres and the warhead is semi-armour piercing. It has a solid propellant motor. It appears to be a versatile weapon capable of being launched from ships, submarines and coastal batteries as well as from aircraft. The only known pictures of launch platforms show the C-801 being carried by a Q-5 Nanchang aircraft and the 'Wuhan' class ships. The missile is 4.80 m long, has a body diameter of 550 mm, a wing span of 1.65 m and a weight around 1025 kg.

STATUS

The first ship-launched missiles were seen in 1986 and it is thought that the air-launched version is not yet in operational service.

CONTRACTOR

Chinese Precision Machinery Import and Export Corporation, 17 Wenchang Hutong Xidan, Beijing, People's Republic of China.

FRANCE

4887.311

ARMAT ANTI-RADAR MISSILE

Armat is a successor to the AS37 anti-radar version of the Martel missile jointly developed and produced by France and Britain (**1022.311**), and is stated to be an improved and advanced development of that weapon. Electronique Serge Dassault, which was responsible for the passive radar homing head of Martel, now produces an enhanced seeker for Armat. Other improvements claimed by the missile's manufacturers include a much greater range, depending upon launch altitude, but without significant increase in weight or size. The configuration is very similar to that of the Martel. It is 4.20 m long, 0.4 m in diameter, and has a wingspan of 1.2 m. Launch weight is 550 kg of which 160 kg belongs to the HE warhead.

Armat has been designed for use with Mirage F1, Jaguar and Mirage 2000 combat aircraft.

STATUS

Operational status was reached in 1984 with the French Air Force with export deliveries being made to several countries including Iraq and Kuwait. The Iraqis have already used it in combat against Iranian air defence radar targets using Mirage F1EQ aircraft as the launch platform.

CONTRACTOR
Matra SA, 37 avenue Louis-Bréguet, 78140 Vélizy, France.

Armat anti-radiation missile beneath this Mirage 2000 is the larger weapon inboard of the Magic air-to-air missile nearer the camera

3335.311
AS.30 LASER AIR-TO-SURFACE MISSILE
Successor to the command-guided AS.30 (**1171.311** in 1982-83 and earlier editions) tactical missile which has now ceased production, the AS.30 Laser is a supersonic, homing missile designed for attacking surface point targets such as ships, bridges, concrete bunkers etc.

The missile is in the form of a streamlined body, fitted with cruciform, swept back wings. Propulsion is by two-stage (boost and sustainer) solid-propellant motors.

CHARACTERISTICS
Length: 3.65 m
Diameter: 342 mm
Wing span: 1 m
Launch weight: 520 kg
Warhead weight: 240 kg
Max range: 10 km

Due to high impact velocity the hardened steel coated GP warhead can penetrate up to two metres of concrete before detonation, using a delayed fuze.

Homing by laser illumination gives the weapon system high accuracy and allows the launch aircraft to take evasive action after launch, to remain outside the effective range of enemy defences. The missile flight path comprises two phases: the mid-course phase which is conducted by gyro reference; and homing onto the laser-marked target by means of a Thomson-CSF Ariel homing head. The laser target designation pod is the ATLIS equipment.

STATUS
The French Air Force has ordered a number of AS.30 Laser missiles for use with Jaguar aircraft, target designation being provided by the ATLIS pod equipment (**3136.393** in the Electro-optical (Air) section of this volume). Adaptation to the Mirage F1 is complete with Iraqi Air Force F1EQ variants having already used the missile on combat missions against Iranian targets. Adaptation to the Mirage 2000 was completed in 1986 for export customers. Following a successful feasibility study, adaptation to the F-16 and other US aircraft is now in progress. Orders have been placed by the air forces of France, Iraq and Egypt.

CONTRACTOR
Aérospatiale, Division Engins Tactiques, 2 rue Béranger, 92322 Chatillon Cedex, France.

Mirage F1 launching Aérospatiale AS.30 laser guided air-to-surface missile

3359.311
AS.15 TT AIR-TO-SURFACE MISSILE
The AS.15 TT (TT: tous temps – all weather) is an anti-ship missile for use in maritime helicopters, slow-flying fixed wing aircraft, fixed or mobile coastal batteries, or small surface craft. A coastal defence system based on the AS.15 TT has been designed jointly by Aérospatiale and Thomson-CSF, but further work has been suspended for the time being. This system was described in entry **3974.121** in 1980-81 and earlier editions of *Jane's Weapon Systems*.

The AS.15 TT has a similar warhead arrangement to that of the AS.12 (**1174.311** in 1984-85 and earlier editions), which it is intended to replace, and can accommodate payloads weighing up to 30 kg. Range is 10 km.

The guidance system is command-to-line-of-sight in azimuth with the missile slaved to the aircraft radar. The radar, Thomson-CSF Agrion (**3298.353**), incorporates pulse compression and frequency agility, and is equipped for automatic tracking of the target and differential target/missile range and bearing measurements required for guidance of the missile. The guidance signals from the aircraft are passed in coded form within the radar's normal transmissions. A sea-skimming trajectory is followed in the vertical plane, with guidance data obtained from a radio altimeter.

CHARACTERISTICS
Length: 2.16 m
Diameter: 185 mm
Wing span: 0.53 m
Weight: 96 kg
Warhead: 30 kg HE
Range: 10 km
Guidance: Radar command
Propulsion: Solid

Dauphin 2 SA365F armed with AS.15 TT missiles and equipped with Agrion radar

STATUS
Announced in late 1976, development began in 1978, and flight tests began in 1980. The first flight of the missile took place on 22 June 1981. The AS.15 TT is deployed on Dauphin 2 helicopters and the AS 332 Super Puma carries AS.15 TT and AM 39 Exocet missiles at the same time. The first production AS.15 TT missile launch from an SA365 Dauphin naval helicopter took place in February 1984, with first deliveries being made in 1985. Operational with the French and Saudi Arabian defence forces and an order was announced for India in 1988.
CONTRACTOR
Aérospatiale, Division Engins Tactiques, 2 rue Béranger, 92322 Chatillon Cedex, France.

1770.321
AIR-LAUNCH EXOCET (AM 39)
AM 39 is the air-to-surface version of the all-weather anti-ship Exocet missile. The same AM 39 missile is designed to be launched against naval surface targets from helicopters, maritime patrol and surveillance aircraft, and jet strike/attack aircraft.

The Exocet AM 39 weapon system comprises:
(1) a command panel for insertion of operational and tactical data, and orders
(2) a missile adaptor kit which allows missile selection, initiates the firing sequence, and translates for the missile the information coming from the command panel and aircraft sensors
(3) one or more missile launchers under the aircraft wings, or alongside the fuselage, depending on aircraft type.

The weapon system uses the target range and bearing given by the aircraft's air-to-surface radar, and possibly updated by an inertial platform.

More details of the missile itself will be found in the entries dealing with the ship and land-based applications of Exocet (**2118.121** and **1156.221**).
CHARACTERISTICS
Length: 4.70 m
Diameter: 35 cm
Wing span: 1.10 m
Weight: 652 kg
Propulsion: 2-stage solid rocket motor
Range: 50 km
Flight speed: High subsonic
DEVELOPMENT
By late 1979 the AM 39 had reached operational status and tests with the Sea King and Super Frelon helicopters and Super Etendard naval attack aircraft followed. Details of earlier development history appeared under this entry in previous editions of *Jane's Weapon Systems*.

AM 39 Exocet anti-ship missile on AS 332 Super Puma helicopter

STATUS
The decision to arm aircraft of the French Navy with the AM 39 Exocet was taken by the French Government in May 1974. At present those aircraft are the Super Etendard and the Atlantique maritime patrol aircraft. Eight foreign governments, Argentina, Egypt, Iraq, Kuwait, Oman, Pakistan, Peru and South Africa, have ordered the system for their own helicopters and for Super Etendard and Mirage strike aircraft.

The AM 39 scored its first successes in the Gulf War during 1980 and since then well over 100 have been fired by the Iraqis from Aérospatiale Super Frelon helicopters, Super Etendards and Mirage F1EQ-5 jet fighter-bombers against Iranian warships, oil rig platforms and various merchant ships. In 1982 the Argentinian Navy used AM 39s from its Super Etendards against Royal Navy ships during the Falklands War to cause the loss of HMS *Sheffield*, a Type 42 air defence destroyer and the MV *Atlantic Conveyor*, a merchant ship taken up from trade.
CONTRACTOR
Aérospatiale, Division Engins Tactiques, 2 rue Béranger, 92322 Chatillon Cedex, France.

3635.311
ASMP AIR-TO-SURFACE MISSILE
ASMP is the acronym under which the French Air-Sol-Moyenne-Portée project to develop a new tactical nuclear stand-off missile is known. In general concept it is similar to the American SRAM (**1107.311**), although there are naturally certain differences in performance and in other respects.

The project was given formal approval in April 1978 when a design and development contract for an air-to-ground medium range missile for the French nuclear tactical force was awarded to Aérospatiale. The ASMP has been deployed on a force of 18 Mirage IVP bomber conversions which will gradually be replaced by Mirage 2000 Ns. The French Navy will also use it to replace AN-22 nuclear bombs carried by Super Etendards.

The ASMP is a supersonic ramjet powered missile with pre-programmed inertial guidance and capable of a stand-off range of 80 to 250 km depending upon launch altitude. It is understood that several flight profiles are possible. A nuclear warhead of 300 KT yield was developed by the government agency Commissariat à l'Energie Atomique.

ASMP air-to-surface missile being carried under a Mirage IV aircraft

The propulsion system is a liquid-fuelled ramjet with integral solid propellant booster. The latter, which is used to accelerate the missile to supersonic speed at which ramjet action can be sustained, is ignited in the main combustion chamber of the motor. A smaller exhaust nozzle is employed for the acceleration phase and this is rapidly enlarged for ramjet propulsion. Two inlets are provided to admit air for combustion and to derive the necessary pressurisation. Length of the missile is 5.38 m, span 1.20 m and diameter 0.38 m.

Guidance is by SAGEM inertial platform and digital computer.
STATUS
The ASMP programme was officially initiated in 1976, following studies which began in 1974. Aérospatiale became prime contractor with the award of a contract valued at Fr500 million. Development flight tests began in 1980. Operational deployment was in May 1986.
CONTRACTOR
Aérospatiale, Division Engins Tactiques, 2 rue Béranger, 92322 Chatillon Cedex, France.

GERMANY (FEDERAL REPUBLIC)

1180.321
KORMORAN AIR-TO-SURFACE ANTI-SHIP MISSILE
Kormoran is an air-to-surface missile, designed principally for strikes against surface shipping. The guidance system employed provides a useful stand-off capability for the launch aircraft. Main characteristics of the Kormoran 1 missile are: length 440 cm, diameter 34.4 cm, wing span 1 metre, weight 600 kg and range 30 km.

Propulsion is by solid-fuel motors, and a hybrid guidance system is employed. Cruciform, broad chord swept wings are located just aft of the cylindrical missile body midpoint, and these are followed by a set of four in-line tail fins. A specially developed 165 kg high explosive warhead is carried.

The Kormoran is roll stabilised and is steered by aerodynamic control surfaces at the rear. The missile is constructed in three main sections: the nose, containing the homing head; warhead; and motor, which also includes the rear structure and the main navigation package that is carried ahead of the sustainer motor.

The warhead is designed to penetrate hulls without being damaged, and detonation is initiated by an impact fuze with a pyrotechnic delay set to explode the warhead approximately in the centre of the target. Liners arranged around the warhead form projectiles on detonation and their resultant kinetic energy is stated to be sufficient to penetrate up to seven bulkheads. The basic explosive effect of the warhead is sufficient to cause extensive damage, not least because of a special system of simultaneous ignition at three places, which results in detonation of the entire explosive in a shorter time.

The navigation package consists of a twin-gyro platform, an inertial computer, associated electronics, and a time switch. There is a separate course and position computer in the tail section and also a radar altimeter.

Terminal phase guidance employs an active radar seeker based on the RE 576 developed by Thomson-CSF. This is stated to be capable of passive homing onto jamming radiation.

OPERATION

A typical Kormoran sortie envisages a low-level approach to a known target followed by a brief climb to above the radar horizon when the aircraft radar is turned on briefly to acquire the target. Upon acquisition, the pilot transmits the relative position of the target into his autonomous navigation system and turns off the nose radar. The aircraft then manoeuvres into an appropriate attack position and when this has been reached another brief radar transmission is made to establish target position. This and other necessary initial data are then automatically inserted into the Kormoran's navigation system. On reaching a predetermined launching range the missile is released and the aircraft makes an immediate break away.

After launch, the missile descends to its programmed flight level. In the first part of its flight it is under inertial guidance, aided by the radar altimeter. At a prescribed distance the inertial system of the missile homing head is activated and an active radar search pattern is initiated. After lock-on has been achieved, the inertial guidance is slaved and corrected in azimuth and range by the seeker head and the missile heads for the target. At a short distance from the target, the missile descends to its final flight level in order to hit the target just above the water line. This is the standard mode, but others are possible.

DEVELOPMENT

The Kormoran project was started in 1964 to meet a West German Navy requirement. Development started in 1967-68 with MBB as prime contractor.

Production was completed at the end of 1983, by which time studies and design work on a Kormoran 2 version had been in progress for several years. The new missile is principally intended to arm Bundesmarine Tornado aircraft in the 1990s. Kormoran 2 will have a digitised active radar seeker head, now under development by Thomson-CSF. It will be an all solid-state system. Comparable improvements will be incorporated to other missile sub-systems such as a larger 220 kg warhead, improved propulsion and guidance.

STATUS

The first air launch of Kormoran 1, from an F-104G, took place successfully in March 1970. Kormoran 1 has been used initially to arm West German F-104 G aircraft, followed by Tornados of the Naval Air Arm, until Kormoran 2 becomes available. A total of 350 Kormoran 1 missiles were ordered for West German use. Kormoran has also been ordered for the Italian Air Force, which uses it on Tornado aircraft.

The development contract for Kormoran 2 was announced in June 1983, when the following improvements were listed: increased hit probability and penetration, better ECM protection, optimised trajectory, target selection, multiple launch, increased range, and simpler operation. Kormoran 2 is expected to enter service in 1989/90.

CONTRACTORS

Prime contractor: Messerschmitt-Bölkow-Blohm GmbH, (MBB), Post Box 8011 49, 8000 Munich 80, Federal Republic of Germany.

Aérospatiale, Division Engins Tactiques, 2 rue Béranger, 92322 Chatillon Cedex, France.

Thomson-CSF – Aerospace Group, 122 avenue du Général Leclerc, 92105 Boulogne-Billancourt, France.

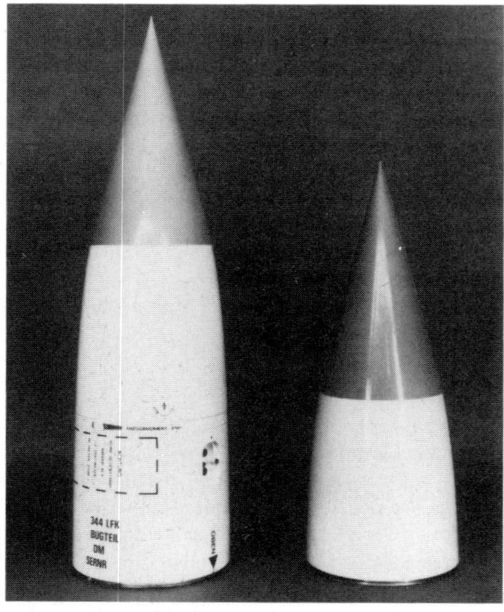

Comparison of the radar seeker heads of Kormoran 1 and 2 anti-ship missiles. Kormoran 2 head on right is all solid-state

West German Naval Air Arm Tornado armed with four Kormoran air-to-ship missiles

INTERNATIONAL

4450.321

AIR-LAUNCHED ANS

This is an air-launched version of the supersonic anti-ship missile, ANS (Anti-Navire Supersonique) being developed jointly by France and the Federal Republic of Germany as a successor to the AM 39 Exocet and Kormoran, respectively. The new weapon is an air-breathing ramjet powered missile which is intended to fly at speeds of Mach 2 at sea level and 2.5 at altitude. Other operational requirements are an ability to manoeuvre at 10 *g*, 180 km range, autonomous operation and programmable trajectories. The missile is expected to be 4.50 m long and to weigh around 650 kg.

Model of ANS supersonic anti-ship missile

The two contractors leading the project for the French and West German governments are Aérospatiale and MBB, respectively.

Additional information will be found in the entry for the ship-launched model (**4446.221**), together with photographs of a test vehicle.

STATUS

Start of development is expected in 1988 with possible entry into service around 1995. Final choice between a ramjet engine derived from the power plant of the ASMP (**3635.311**), or a new ramjet is to be made in 1988.

CONTRACTORS

Messerschmitt-Bölkow-Blohm GmbH (MBB), Post Box 801149, 8000 Munich 80, Federal Republic of Germany.

Aérospatiale, Division Engins Tactiques, 2 rue Beranger, 92322 Chatillon Cedex, France.

1022.311

MARTEL AS.37/AJ.168 AIR-TO-SURFACE MISSILES

Martel is an air-to-surface tactical missile with two alternative terminal guidance systems capable of offering a considerable stand-off capability. Both versions are designed to operate in an ECM environment, to which a high resistance is claimed. The two forms of terminal guidance are: passive homing on to electromagnetic radiation in the AS.37, anti-radar version; and visual guidance to a selected target by means of a nose-mounted TV camera and a data link over which both video and command signals are passed between aircraft and missile, in the AJ.168 variant. The Martel system is the product of a joint Anglo/French development programme with prime responsibility for the AS.37 resting with Matra, and for the AJ.168 with British Aerospace Dynamics.

General configuration of the missiles can be seen from the photograph on the next page. Length of the AJ.168 is 390 cm, and of the AS.37, 420 cm; body diameter and wingspan, respectively, for both versions are 40 and 120 cm. No details of performance have been revealed, but range has been officially stated both as 'several tens of miles' and 'several tens of kilometres'.

Both versions of Martel employ the same basic autopilot which provides guidance until either TV or radar homing systems assume control for the terminal guidance phase. There are some indications that the autopilot can be pre-programmed to provide several options of mission profile. A fundamental difference, in addition to the guidance system, is related to the two-stage propulsion assembly, which is adapted to the mission requirements specific to each version.

Buccaneer armed with one anti-radiation Martel and two TV-guided Martels. Note Martel data-link/control pod on starboard inner wing section

STATUS

In service with the British and French forces. The Martel system is used on the Mirage IIIE, Atlantique, Jaguar and Buccaneer Mk II.

Use of both anti-radiation and TV-homing versions is restricted to British Forces; the TV-homing AJ.168 is not used on French aircraft.

In mid-August 1977 the UK MoD announced an RAF requirement for a new anti-ship missile to arm Buccaneer and Tornado aircraft. This led to the Sea Eagle (P3T) development (**3630.311**). The French in turn developed Armat (**4887.311**) to replace Martel AS-37.

CONTRACTORS

Prime Contractors:

British Aerospace Dynamics, Manor Road, Hatfield, Hertfordshire AL10 9LL, England.

Matra, 37 avenue Louis-Breguet, 78140 Vélizy, France.

TV guidance system: GEC Sensors, Basildon, Essex, England.

AS.37 homing head: Electronique Serge Dassault, 55 quai Marcel Dassault, 92214 Saint-Cloud, Paris, France.

1771.311

HOT AIR-TO-SURFACE MISSILE

The increase in interest regarding the use of helicopters in a specifically anti-tank role has led to the development of a helicopter installation for the HOT anti-tank weapon. The basic HOT missile is described in **2212.111** in an earlier section of this book, and the following refers to the helicopter version.

Depending upon the type of helicopter used, four, six or eight launchers to accept the HOT combined container-launch-tube can be fitted, and these are elevation-slaved to the line-of-sight. For helicopter operation, a stabilised optical sighting system, APX M397, is used, this being derived from the APX M334 sight described in **7032.393** in the Equipment Section of this book.

The guidance system includes a triple-field IR 'localiser', with a 10° acquisition field, a 5° tracking field, and a 1° tracking field. The function of the IR localiser is to establish the position of the HOT missile relative to the line-of-sight (which normally will be maintained on target by the operator using his optical sight). Appropriate guidance signals are computed and transmitted via the command wire to maintain the missile on the line-of-sight to the target. An improved missile, the HOT-2, with an enhanced K150 warhead has also been developed for use against the new generation of Soviet armour.

A prototype night-firing device is currently undergoing trials. This is the roof-mounted VIVIANE system, designed for installation on Gazelle or Dauphin helicopers after the initial trials of VENUS (viseur écarto-métrique de nuit stabilisée) nose-mounted day/night sight fitted in a Dauphin helicopter.

Night view of Dauphin helicopter equipped with two quadruple HOT anti-tank missile launcher pods and VENUS night firing equipment in ball turret

STATUS

By December 1985 more than 59 000 HOT missiles (of all versions) were on order and well over 80% of that number had been produced for 14 countries: Egypt, France, Iraq, Kuwait, Saudi Arabia, Spain, Syria, West Germany, Qatar, Morocco, Gabon, Cameroon, United Arab Emirates and one undisclosed nation. The missile has seen combat in the Gulf War, the 1982 Peace for Galilee War and the Moroccan-Polisario Front confrontation.

CONTRACTOR

Management, sales, and production responsibilities: Euromissile, 12 rue de la Redoute, Fontenay-aux-Roses, France.

4554.311

APACHE/CWS 50

Under an international agreement announced in May 1983 between MBB of the Federal Republic of Germany and Matra in France, the two companies agreed on a collaborative programme leading to a stand-off weapon system for use by tactical aircraft against a variety of targets including airfields, aircraft and vehicle concentrations, armoured formations,

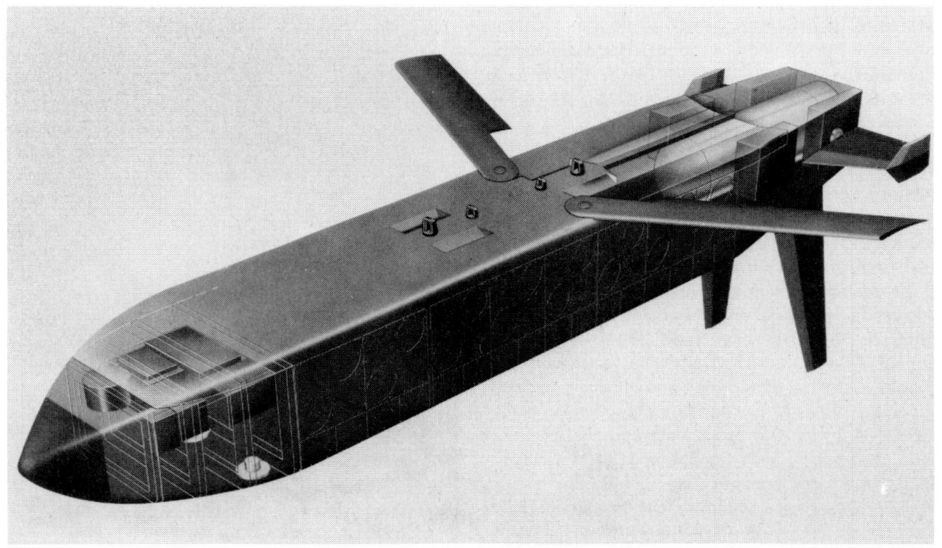

'Ghost' drawing of the Matra/MBB Apache/CWS modular stand-off weapon

etc. The project combines the German container weapon system (CWS) (**4449.311**) work that MBB has been carrying out since 1978 with the Matra company's Apache project to develop a multi-purpose modular weapon. It is envisaged that a family of 1200 kg-class modular weapons will evolve capable of stand-off ranges from ten to several tens of kilometres. The full range of submunitions available with the MW-1 system (**3517.311**) ie KB 44, MIFF, MUSA, MUSPA and STABO, can be deployed with the CWS.

The basic vehicle is largely based on the MBB CWS, with an overall length of 4.5 m, with a 2 m centre section designed to house and dispense the submunitions. Folding wings will deploy after release to give a high-wing monoplane configuration. The guidance package will differ according to the mission, based on an inertial navigation system. Additional terminal guidance systems can be used in a modular way.

CHARACTERISTICS

Version	**10**	**50**
Range (low-level)	10 km	80 km
Propulsion	No	Turbofan
Weight	1200 kg	1200 kg
Guidance	Inertial	Inertial + terminal guidance
Pattern size	Length up to 1000 m × width up to 350 m (optimised for each submunition type)	

STATUS

In development since 1984, with flight tests having been made since 1986. The status of this programme may be affected by MSOW (entry **6608.311**).

CONTRACTORS

Messerschmitt-Bölkow-Blohm GmbH (MBB), Post Box 8011 49, 8000 Munich 80, Federal Republic of Germany.

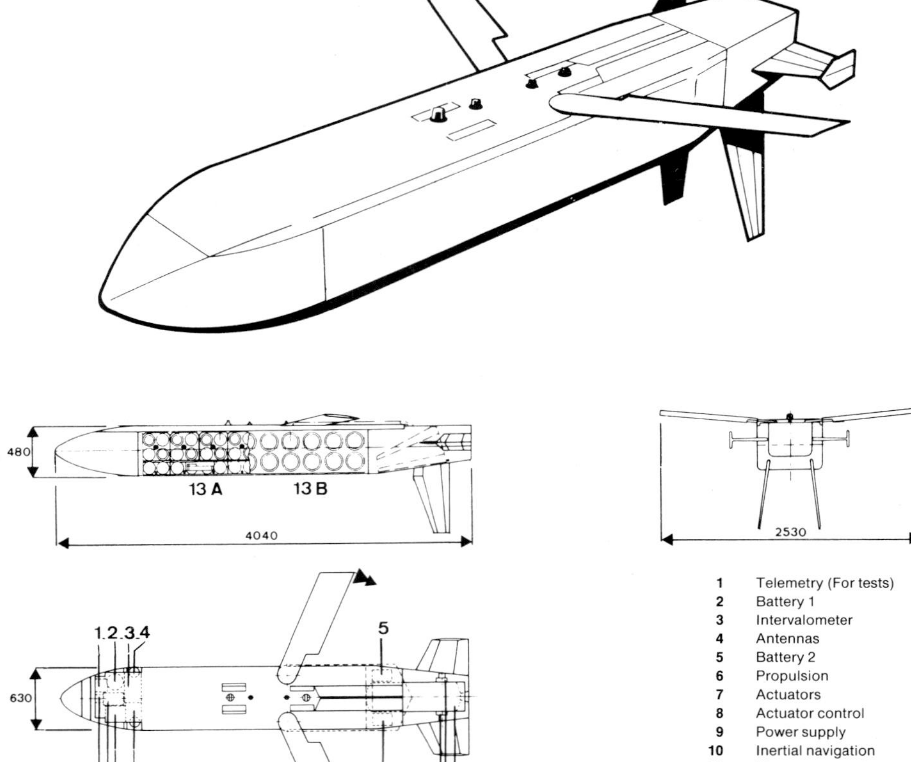

General arrangement drawing of Apache/CWS 50

Matra SA, 37 avenue Louis-Bréguet, 78410 Vélizy, France.

6608.311

MODULAR STAND-OFF WEAPONS

A Memorandum of Understanding was signed by seven nations in August 1987, outlining a basic working agreement for the development of a family of Modular Stand-Off Weapons (MSOW) for conventional sub-munitions. The seven nations involved are the USA, Canada, Federal Republic of Germany, France, Italy, Spain and the UK. RFPs were issued for consortia bids in October 1987 and it is expected that the project definition and demonstration phase contracts will be awarded late in 1988. The project definition phase is planned to last 30 months and then a further MoU will be signed by the nations to proceed into full scale development. FSD is expected to start in 1992.

Three different variants of stand-off weapon are required and it can be assumed that they will be broadly similar in body shape though with differing propulsion systems. There are two medium range systems required, one with a range of 15 to 30 km and the other with a range of 30 to 50 km. The third version is a long range missile, with a range of from 185 to 500 km and capable of carrying either sub-munitions or a single warhead. Early proposals have included a Tomahawk cruise missile shape for the long range version.

STATUS

This MSOW programme is one of the NATO programmes qualifying for US funds under the Senator Nunn collaborative development initiative. It is expected to reach project definition in late 1988, start full scale development in 1992 and a provisional in service date would look to be around 1997.

CONTRACTORS

There are two competing consortia, that will both complete the project definition phase and these consortia are being lead by General Dynamics and Rockwell International. In addition, there are about seven consortia bidding for the propulsion systems contracts.

ISRAEL

4529.321

GABRIEL MK III AIR-LAUNCHED ANTI-SHIP MISSILE

An air-launched version of the successful Gabriel III anti-ship missile (**6019.221**) was revealed by the Israeli manufacturers in the last quarter of 1982 as the Gabriel Mk III A/S, together with photographs of test and development models on an Israeli Phantom aircraft. The weapon can also be carried by A-4 Skyhawk, Kfir C2, Sea Scan and other aircraft types, principally for use in anti-shipping roles.

The missile itself was derived directly from the ship-launched Gabriel mentioned above, but has different aerodynamic surfaces, more suitable for carriage and launching from high performance aircraft. With an extended range propulsion rocket motor, this gives a range of approximately 40 km.

The system provides all-weather, day/night, anti-ship capability against targets ranging from small fast patrol boats to destroyers and larger ships. The Gabriel Mk III A/S is an active homing, sea-skimming missile with 'fire-and-forget' and 'fire-and-update' operating modes. The I-band frequency agile radar seeker performs data processing and control functions, providing good all-weather and ECCM performance. Inertial components, a radar altimeter and a digital processor are employed for missile guidance, and four tail control surfaces steer the missile and control its flight path. A high explosive warhead weighing 150 kg ensures maximum target damage.

Gabriel air-launched anti-ship missile mounted under wing of F-4 Phantom aircraft

The solid-propulsion motor provides over-the-horizon ranges at low level and the terminal attack altitude is set before take-off in accordance with the prevailing sea state to one of three values. Other characteristics of the missile are given below and can be gathered from the adjacent illustrations.

The complete system comprises one or more missiles (depending on the aircraft type, but typically two) and the missile fire control system (MFCS) fitted in the aircraft. In the case of a Gabriel Mk III A/S installation in a fighter aircraft the main MFCS elements include: a pilot fire control panel (PFCP), missile processing unit (MPU), power supply unit (PSU), and armament interface hardware (AIH). Built-in test facilities are incorporated and operate automatically as part of the firing sequence. The AIH units are incorporated in missile loading hardpoints and the other MFCS items are fitted in various fuselage locations.

OPERATION

As mentioned briefly above, there are two operating modes. 'Fire-and-forget' missiles, to accomplish their objective with high probability of target acquisition have to accommodate wide search areas. The actual size of the search area depends on the missile-target range at launch. To minimise the uncertainty of target position at the time of missile seeker activation, a 'fire-and-update' facility is provided in the Gabriel Mk III A/S system in addition to the basic 'fire-and-forget' mode. 'Fire-and-update' is used on larger maritime patrol aircraft as an optional mode of deployment, and relative missile/target locations are monitored, calculated and transmitted by the MFCS to the missile in flight. For this mode, in addition to the track-while-scan facilities of the aircraft's surveillance radar, a command link to the missile is added to the MFCS installation on the aircraft.

By this technique the seeker search zone in bearing and range is reduced considerably, so that the updating facility ensures that each missile is directed toward its designated target and not against 'potential targets' (which may or may not be significant) which may be close to the actual target area. After launch, data transfer between missile and launch aircraft are entirely automatic. Thus, 'fire-and-update' is an enhanced version of 'fire-and-forget' with the advantage of a reduced search zone and improved target lock-on.

In the vertical plane, at high altitudes, the missile pitch angle is controlled by a preset angle program for the descent to the midcourse phase. At low level the missile guidance system switches to altitude command to maintain a midcourse cruise altitude of about 20 metres, before descending to 1.5, 2.5 or 4 metres for the terminal phase under pre-programmed control. In the horizontal plane, inertial guidance is used for the initial and midcourse phases for 'fire-and-forget' operation, and inertial is also used in the 'fire-and-update' mode for the midcourse phase. For both modes, seeker switch-on is under the control of pre-programmed logic in the computer. Target detection in the search area, lock-on and terminal homing are by active radar, operating in the horizontal plane only.

Target data (range/bearing) can be supplied from sensors aboard the launch aircraft, or from external sources. Altitude, speed and attitude data required for missile launch are provided from aircraft systems. The MFCS is operated by the pilot, using the PFCP, the other units (MPU, PSU and AIH) compute and supply the power and pre-launch data required by the missile. Missile release is activated by the pilot, by pressing a missile fire button on the control stick.

Missiles operated from fighter aircraft use the 'fire-and-forget' mode for one of the following procedures:

(1) radar-range and bearing launch (RBL)
(2) manual RBL
(3) bearing only launch (BOL)

Radar RBL For this technique, the aircraft radar detects potential targets and the range and azimuth data are transferred to the MPU by using the appropriate push-button on the PFCP. Target position, relative to the aircraft, is then calculated automatically and fed into the missile up to the moment of launch. The pilot receives an indication that the predetermined launch criteria have been met, by the flashing of a fire indicator light on the PFCP.

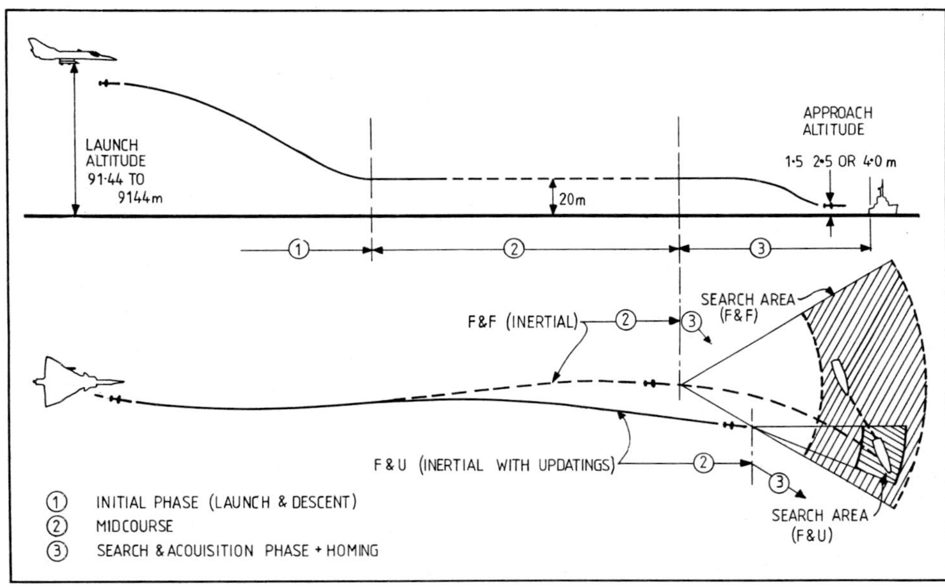

Gabriel launch and attack profile and plan

Gabriel Mk III A/S air-launched anti-ship missile

Manual RBL This method allows target location information to be fed to the missile by manual means, by either reading this information from aircraft navigation displays etc, and entering it manually on the PFCP, or receiving target data from an external source (eg a maritime patrol aircraft, ship, or other aircraft), entering this on the PFCP, while present aircraft position is updated by using an initiation point (entered via the PFCP before take-off). The MPU will continue to calculate the target location and transfer the required target range and bearing data to the missile. When the aircraft reaches the release range (as calculated by the MPU and indicated on the PFCP) the pilot can initiate missile firing.

Bearing Only Launch In this mode the pilot uses an estimated range together with a defined target bearing. Both estimated bearing and measured range are entered via the PFCP and transferred to the missile via the MPU. The estimated range is defined as Low, Medium or High, corresponding to 20, 20 to 40, or 40+ km respectively. This operating mode is automatically selected if no other choice is made, as a fall-back mode, and in this case the aircraft is aligned to the target direction and for range the minimum weapon range is chosen.

CHARACTERISTICS

Length: 3.85 m
Diameter: 34 cm
Max span: 1.1 m
Launch weight: 560 kg
Warhead: HE 150 kg
Fuze: Impact, delayed action
Propulsion: Solid
Range: 40 km
Launch altitude: 90 – 9000 m
Cruise altitude: 20 m
Terminal phase altitude: 1.5/2.5/4 m (pre-set)
Guidance: Inertial (plus optional command update) midcourse; active radar homing

STATUS

In production.

CONTRACTOR

Israel Aircraft Industries Ltd, Ben Gurion International Airport 70100, Israel.

ITALY

1651.321

MARTE ANTI-SHIP MISSILE SYSTEM

The purpose of the Marte system is the destruction or disabling of naval craft in 'all weather' conditions.

Installation of the Marte weapon system has been carried out on the Agusta/Sikorsky SH-3D helicopter, but the system is also suitable for installation on light shipborne helicopters, such as the Agusta AB 212.

In addition to the two missiles, the Marte Mk 1 equipment carried by the helicopter consists of an SMA MM/APQ-706 I-band radar (**3309.353**) for target search, acquisition and missile guidance; auxiliary launch and guidance equipment (ALGE) which includes the missile control console (MCC); command transmitter and antenna; optical tracker and 'joystick' control; and a pilot's display unit. An Elettronica ESM system is also normally fitted to provide passive detection and other electronic warfare facilities.

The APQ-706 radar provides two different operational modes, one for operations not involving the launching mission, and the other for missile guidance. There are back-to-back antennas and two transmitter/receivers with different output powers. The radar includes a data distribution and processing sub-system and a tactical PPI display with true and relative motion presentation and a micro-B display.

The MCC is part of the tactical display PPI console and contains the operational firing panel, synchro clock unit, launching computer and power supply. Using this and other ALGE items the crew are able to either perform or monitor pre-launch sequences, and radar or optical launches. Provisions are also made for elaboration of data. The pilot's display is used to select the optimum course during the launch approach. Radar and optical command guidance modes were available in the original Marte Mk 1 system, but the newer Mk 2 system (see below) employs a different technique incorporating active radar homing.

MARTE Mk 2

Essentially, the Mk 2 version of Marte consists of the Mk 1 missile to which an active radar homing head based on that of the Otomat anti-ship missile (**1336.221**) is added to replace the command guidance system employed in the Marte Mk 1 system. Thus the Mk 2 weapon becomes a 'fire-and-forget' missile using radar homing, leaving the launching aircraft free to take evasive manoeuvres immediately after weapon release. The complete Marte Mk 2 system consists of two or four missiles, a missile firing box, a pilot's display and a data processing unit (DPU), and the system is interfaced with the aircraft search radar, gyrocompass, altimeter and doppler navigator.

There are two versions of Marte Mk 2, one launched

from fixed wing aircraft and a version for helicopter launch with an additional boost motor assembly. The range of both versions is around 20 km. The payload is an HE semi-armour piercing warhead weighing 70 kg, activated by an impact/proximity fuze. In the latter of these two fuzing modes, activation is by means of a proximity sensor associated with the missile radio altimeter, which provides the necessary signal when over-flying the target.

Guidance in the vertical plane is by means of the CW radio altimeter which keeps the missile at a controlled constant height above sea level up to the moment of impact. This height selected between 3, 4 and 5 metres depends on sea state and target dimension. Guidance in the horizontal plane during the initial phase of flight is gyroscopically controlled towards the target, following preset direction data already provided by the system DPU before launch. Inertial guidance maintains the missile on an optimal trajectory for detecting and tracking the target in the final phase of the attack and this occupies approximately the first 14 km of flight. The active radar homing head, after target acquisition, is responsible for control in the horizontal plane for the last 6 km or so until impact.

Other modes of operation can be employed according to tactical circumstances, for example: launch on ESM data (in radar silence conditions), or launch on radar data from another co-operating aircraft or helicopter.

CHARACTERISTICS

Type: Marte Mk 2
Length: 4.80 m (helicopter-launched) or 3.90 m (aircraft-launched)
Wing span: 0.978 m
Diameter: 316 mm (max)
Launch weight: 345 kg (helicopter-launched) or 260 kg (aircraft-launched)
Range: 20 km
Average speed: High subsonic
Warhead: High explosive, semi-armour piercing
Fuze: Impact/proximity
Guidance: Inertial and radio altimeter, active radar homing

STATUS

The Italian Navy received the first operational Marte 1 system, for use on the SH-3D helicopter, at the end of 1977.

The Marte Mk 2 project was revealed in mid-1983 and on March 23, 1984 a 'fire-and-forget' test launch at the Salto di Quirra, Sardinia range, resulted in a successful hit on a ship target at a range of 19 km.

In June 1985 OTO-Melara and Aermacchi revealed plans to deploy the Marte system on the latter company's MB.339C trainer/light attack aircraft. An Italian Navy SH-3D helicopter launched a Marte missile against a target tug in June 1985.

Marte Mk 2 entered service with the Italian Navy in 1987, with the helicopter carried version of the missile.

CONTRACTOR

OTO-Melara SpA, 15 via Valdilocchi 1, 19100 La Spezia, Italy.

Marte Mk 2 anti-ship missile loaded on Italian naval helicopter

---

JAPAN

1653.311

ASM-1 AIR-TO-SURFACE MISSILE (TYPE 80)

Under the designation ASM-1 the first Japanese tactical air-to-surface anti-ship missile has been formally introduced as the Type 80 for use with the Mitsubishi F-1 close-support and Phantom F-4J aircraft. No official details have been obtained but provisional characteristics are: range approximately 50 km, length 4.0 m, body diameter 335 mm, wing span 1.02 m and a weight of 610 kg. An inertial system is used for mid-course guidance and an active radar seeker for terminal homing.

STATUS

In November 1973 Mitsubishi Heavy Industries Ltd was selected as prime contractor for the development of the ASM-1. Development was completed in 1979 and production commenced in 1980 to meet the requirements of the Japan Self Defence Agency only.

It is believed that the missile entered service in 1983.

CONTRACTOR

Mitsubishi Heavy Industries Ltd, 5-1 Marunouchi 2-chome, Chiyoda-ku, Tokyo 100, Japan.

JASDA ASM-1 air-to-surface missile on laboratory support trolley. Stencilled designation on missile suggests US practice, in which case this is probably an early experimental round, denoted by prefix letter 'X'

NORWAY

3070.321

PENGUIN MK 3 AIR-TO-SURFACE MISSILE

The Penguin Mk 3 missile is a growth development of the existing Penguin Mk 2 surface-to-surface missile (**1339.221**). This missile is already in service with several navies as main armament on fast patrol boats. The Penguin Missile System Mk 3 has demonstrated its ability to successfully attack surface targets in both open and confined waters with minimum exposure of the launching platform. It was developed for the Royal Norwegian Air Force.

Penguin Mk 3 missile is a 'fire-and-forget' weapon allowing aircraft break-off immediately after launch. Following the trajectory instructions the missile will find its way to the target position using inertial navigation plus altimeter for reference. The missile can be launched over land and programmed to fly via a waypoint where it may descend to sea-skimming altitude and turn more than 90° on to its final heading. Once in the target area the missile's IR area seeker is activated. When the target is acquired, the seeker locks on and guides the missile to target impact.

The Penguin Mk 3 missile system is designed to interface with existing aircraft fire control systems with minimum aircraft modification requirements. This is achieved by utilising an adaptor between the missile and the aircraft weapon station. The adaptor containing the retention and release mechanism for the missile also provides space for the avionic components and attachment hardware for interface with the parent aircraft. This solution makes interface with various aircraft types possible with little or no impact on either missile or aircraft.

The air-launched version of Penguin is expected to have a useful range of 40 km. The modified Bullpup warhead has been retained. The missile is equipped with one solid-propellant rocket motor only. Using aircraft in the high subsonic speed range allows the booster motor to be omitted.

CHARACTERISTICS

Length: 3.18 m
Diameter: 28 cm
Wing span: 1 m
Weight: 380 kg
Warhead: Modified Bullpup, 120 kg
Range: 40 km

Guidance: Programmed inertial, IR homing in terminal phase

OPERATION

Target designation to the missile can be accomplished using the launch aircraft radar, by placing a marker over the target on the display and pressing a button to insert this data into the missile. The same technique can also be employed to insert a waypoint into the missile guidance system program. A similar technique can be used with the aircraft head-up display, in which case the pilot first acquires the target optically. This method can also be adopted using an aircraft gunsight in lieu of the head-up display. After release the Penguin missile follows an inertially guided flight path to either the designated waypoint or to the target area. In the case of reaching the waypoint, the missile then changes course and possible altitude, to fly on to the target area. On reaching the target area the IR seeker is switched on and the missile commences the terminal homing phase. This employs one of several alternative search patterns which can be selected by the pilot before the missile is launched.

STATUS

Studies for an air-launched version of Penguin were carried out in the mid-1970s, using an F-104 as the weapon platform. Following competitive evaluation, Penguin Mk 3 was chosen by the Norwegian authorities in 1979 to arm F-16 aircraft. The possibility of using it with the P-3 Orion maritime patrol aircraft has also been studied.

Full-scale development was decided on in 1980 and contracts were awarded in that year and 1982. First live air launches took place in 1984. Entry into service took place in 1987.

CONTRACTOR

Norsk Forsvarsteknologi A/S, PO Box 1003, 3601 Kongsberg, Norway.

Penguin Mk 3 air-launched anti-ship missile mounted on F-16

6265.321

PENGUIN MK 2 MOD 7 AIR-TO-SURFACE-MISSILE

Penguin is being adapted for use with the American SH-60B Seahawk naval helicopter, which can carry two of these missiles. Penguin is integrated into the helicopter avionic system, utilising existing equipment for missile control and operation. This version is designated Penguin Mk 2 Mod 7.

As Penguin is a 'fire-and-forget' system, the helicopter exposure time can be reduced to a minimum with a corresponding reduction in helicopter vulnerability. Seahawk aircraft equipped in this way can operate autonomously or in a co-ordinated attack with ship-launched over-the-horizon missiles for cost-effective suppression of enemy defensive systems.

CHARACTERISTICS

Length: 2.96 m
Diameter: 28 cm
Wingspan: 140 cm; 70 cm (folded)
Weight: 385 kg
Speed: High subsonic
Range: 30 km

Other details are the same as for the Penguin Mk 3, above.

Helicopter-launched version of Penguin is being developed for SH-60B and similar helicopters

STATUS

The US Navy and the Royal Norwegian Navy have signed a contract for full scale adaptation of Penguin to the Seahawk helicopters of the LAMPS III programme.

CONTRACTOR

Norsk Forsvarsteknologi A/S, PO Box 1003, 3601 Kongsberg, Norway.

SWEDEN

3975.321

RBS 15 ANTI-SHIP MISSILE

RBS 15 is the designation of a new Swedish anti-ship missile for use by Swedish aircraft, coastal defence units and naval vessels. The ship-launched variant (**3976.221**) has been addressed first, with deployment aboard 'Spica' class fast patrol boats in 1985.

The general configuration of the RBS 15 can be seen from the photograph, and the air intake for the turbofan engine will be noted beneath the missile body. An air-breathing propulsion system was selected to provide long range with a high subsonic speed at sea-skimming heights. In the surface-launched mode, the RBS 15 is assisted by two solid-propellant booster motors, but these have been deleted in the air-launched mode.

The RBS 15 will have a launch and leave capability with inertial mid-course guidance and an active radar terminal seeker developed by Philips Elektronikindustrier. The homing head is digitally programmable with selectable search patterns, target choice logic, frequency agility and variable ECCM. The trajectory can be altered in accordance with operational demands. Prelaunch operations will be minimal in scope and all calculations will be performed automatically by an on-board computer. A radio altimeter will ensure safe missile cruise at low levels above the sea surface. A 'heavy' warhead is produced by Förenade Fabriksverken for the RBS 15.

CHARACTERISTICS

Length: 4.35 m
Diameter: 50 cm
Wing span: 1.4 m
Weight: 595 kg
Guidance: Inertial and active radar terminal
Range: 70 km

STATUS

The ship-launched variant entered service with the Royal Swedish Navy in 1985 and the air-launched variant RBS-15F is in development with an expected in service date of 1990.

CONTRACTOR

Saab Missiles AB, 58188 Linköping, Sweden.

RBS 15 missile beneath Saab AJ 37 Viggen aircraft during test flights

1189.321

SAAB 04E AIR-TO-SURFACE MISSILE

This is an air-to-surface missile developed principally for anti-shipping strikes. The version in current Royal Swedish Air Force service is the Saab 04E which has the RSAF designation RB 04E. Principal characteristics are as follows: length 445 cm, body diameter 50 cm, wing span 2 m, and total weight about 600 kg. These figures are the same as for the earlier models but the RB 04E is technically more advanced. After launch, the missile acts independently of the launch aircraft and is automatically guided to a low-level altitude where target search and acquisition are performed at a high subsonic speed.

Guidance in the approach phase of an anti-shipping attack with the RB 04E is by means of a high-quality autopilot towards the target area. Precision terminal homing is by an active radar homing head, produced by Philips in Sweden.

DEVELOPMENT

The Robot RB 04 series of missiles was initiated in the 1960s and developed by the Missile Bureau of the Swedish Air Force Board (now the Defence Materiel Administration – Missiles Directorate). This organisation was responsible for the development of the 04C and 04D versions, but subsequent R & D has been performed by Saab-Scania.

STATUS

The RB 04E is operational with the AJ 37 Viggen all-weather attack aircraft, which can carry three missiles.

Saab RB 04E anti-shipping missile

CONTRACTOR

Saab Missiles AB, 58188 Linköping, Sweden.

1190.311

SAAB 05A AIR-TO-SURFACE MISSILE

The Saab 05A is a radio-command guided, manually controlled tactical air-to-surface missile. It is for use against both sea and land targets but may also be used in certain air-to-air roles. Principal characteristics are: wing span 80 cm, diameter 30 cm, length 360 cm, and total weight 305 kg.

The airframe of the Saab 05A is made of conventional aircraft materials and consists of a pointed cylindrical body with long-chord cruciform wings and aft-mounted cruciform control surfaces. A liquid rocket motor is centrally located and the motor casing forms part of the outer skin and load-carrying structure of the missile. The armament system is located in the nose and most of the control equipment at the rear of the missile. The VR 35 rocket motor is fitted with a tail-pipe which passes through the rear of the body. A pre-packaged liquid-propellant motor, supplied by Volvo Flygmotor AB, is used. The propellant tanks are placed around the centre of gravity to avoid CG-shift. Electrical power is supplied by a thermal battery. Before launching, some necessary electrical preheating of the missile is supplied from the aircraft. The hot gas for the control surface actuators is provided from a solid-propellant gas generator. An effective proximity-fuzed armament system is used.

The auxiliary airborne missile system equipment comprises a control stick, coding unit, transmitter, and antenna.

After launching, the missile is guided manually by the pilot who visually lines up the missile tracking flares and the target. The control signals are transmitted over the radio link with the transmitter in the aircraft and the receiver in the missile aft section. The guidance function from a control stick is specially designed to provide a high degree of accuracy for both small and large off-set angles. The coded command link is difficult to jam. Guidance signals received by the missile are converted by an autopilot to control surface deflections by means of four hot gas actuators. The autopilot initially guides the missile into the pilot's field of view.

Saab 05A air-to-surface missiles carried by AJ 37 Viggen aircraft

DEVELOPMENT

Developed by Saab-Scania on behalf of the Royal Swedish Air Force, the system entered advanced testing in 1968-69. In June 1969, Saab was awarded a large contract by the Swedish Government for further development and production and the missile entered service in the early 1970s.

STATUS

The RB 05A is in service with the AJ 37 Viggen aircraft, J-35 Draken and the Sk-60 (Saab 105). Quantity production was completed in 1977.

CONTRACTOR

Saab Missiles AB, 58188 Linköping, Sweden.

UNION OF SOVIET SOCIALIST REPUBLICS

1150.311

KIPPER AIR-TO-SURFACE MISSILE (AS-2)

Kipper is the NATO code-name of an anti-shipping stand-off missile, which entered service in 1960. Official US information recently obtained indicates a length of 10 metres. Kipper, with a wing span of about 4.9 metres and a diameter of 0.9 m, also has an aircraft-like configuration with swept wing and tail surfaces. Propulsion is by a turbojet motor suspended beneath the missile fuselage.

Recent analyses have deduced that Kipper flies to its target area under control of a programmed autopilot, to which there are command override options available. Terminal homing relies on active radar homing. A nuclear or 1000 kg high explosive warhead is carried, with the missile having a range of 120 km and maximum speed of Mach 1.2 from a high altitude launch. The only known carrier platforms are the Tu-16 Badger-C and Badger-C Mod which carry the 4200 kg weapon under the aircraft fuselage. Still operational but being phased out and replaced by the AS-4 (**1151.311**).

1147.311

KANGAROO AIR-TO-SURFACE MISSILE (AS-3)

Kangaroo is the NATO code-name assigned to the largest Soviet air-to-surface missile so far revealed. Entering service in 1961, it is associated with the Tu-95 Bear-B bomber aircraft and is generally assumed to have retained operational status though its operational importance has probably declined significantly as newer weapons have been deployed.

As the photograph shows, the Kangaroo has the configuration of a conventional swept-wing fighter aircraft and similar dimensions, approximately 15 metres (US official sources quote 14.95 m) in length and with a span of 9.15 m. The propulsion used is a turbojet engine. Recently obtained information indicates that mid-course guidance is controlled by an autopilot with command override facilities for navigational corrections, but no terminal homing facilities are carried. Maximum range of the 11 000 kg weapon is 650 km using a high altitude launch.

The role of the AS-3 Kangaroo is that of a

long-range stand-off nuclear weapon for use against area targets, such as major cities or ports. For this a 2300 kg 800 KT yield thermonuclear warhead is carried. At present the Bear-B carrying platform is being phased out and converted to Bear-G AS-4 Kitchen carriers.

Kangaroo stand-off missile beneath Tu-95 Bear bomber aircraft

1151.311

KITCHEN AIR-TO-SURFACE MISSILE (AS-4)

The AS-4 Kitchen has a fuselage of about 11.3 m long to which are attached 3 m span delta planform wings and a tail assembly. Propulsion is by a single stage rocket, and inertial guidance is used for the mid-course phase of flight, after which an active radar homing provides for terminal guidance to the target. In 1981 the existence of a passive radar homing version was reported. American DoD sources have stated that a number of AS-4 versions are in use, possibly forming two distinct families. One associated with the original Tu-22 Blinder-B carrying platform and the other with the later Tu-26M Backfire-B and Tu-95 Bear-G. The strategic version logically has no terminal homing system and is equipped with a 350 KT nuclear warhead whilst the anti-shipping variant can use either a 1000 kg HE one or a similar nuclear one with an active radar terminal homing seeker. The defence suppression passive radar homing variant would probably use only the HE warhead. Range of the 5900 kg weapon varies from 300 km using a low altitude launch profile to 400 km at high altitude. The missile is designed for high level release, cruising at around 35 000 ft followed by a steep terminal dive onto a target.

First seen carried by the Tu-22 Blinder-B, the AS-4 has armed the Tu-95 Bear-G, and the Tu-26 Backfire-B variable-geometry aircraft. On the Bear, the AS-4 is carried beneath the aircraft wings, on the Tu-22 Blinder and Tu-26 Backfire-B in a semi-recessed position on the underside of the fuselage and more recently on underwing pylons on the Tu-26 Backfire-B.

AS-4 Kitchen carried in semi-recessed position under fuselage of Tu-26 Backfire-B strategic aircraft

1149.311

KELT AIR-TO-SURFACE MISSILE (AS-5)

Kelt is the NATO code-name of a Soviet air-to-surface weapon, first deployed in 1966 on Tu-16 Badger-G bombers, for anti-shipping and ground attack roles. It has superficial but pronounced similarities with the Kennel (**1148.311**) which also was used with the Badger, and with the Styx shipborne surface-to-surface missile (**1155.221**). A length of 8.59 metres has been officially reported for Kelt in an American government publication, with wing span of 4.8 m. The fuselage centre-body and wings of Kelt appear to be the same or very similar to those of Kennel, while the nose section of Styx is employed. A liquid fuel rocket propulsion motor is used and the tail control surface arrangements of Kelt differ from those of both Styx and Kennel. Guidance is by an autopilot with an active radar terminal homing seeker. The latter was used in combat by the Egyptians against the Israelis in the 1973 war. Used by both the Soviet Air Force and Navy, it has a 1000 kg HE warhead as the payload. Missile weight is 3000 kg whilst the range is about 180 km.

AS-5 Kelt air-to-surface missile loaded on to underwing pylon of Egyptian Air Force Tu-16 Badger aircraft

3633.311

KINGFISH AIR-TO-SURFACE MISSILE (AS-6)

Kingfish, AS-6 in the American numerical sequence, is a high performance air-to-surface missile first deployed in 1970 and carried by Tu-16 Badger-G Mod and C Mod aircraft of the Soviet Navy and Air Forces. The configuration is a cylindrical fuselage supported by short-span, long-chord stub wings of delta planform and is very similar to Kitchen AS-4. Dimensions are: length 11.0 m, wing span 2.5 m and diameter 90 cm.

Propulsion is by single stage solid propellant motor and an estimated range of 400 km has been reported.

An American report also revealed that one role for this missile is anti-shipping missions in which a high-altitude supersonic cruise approach trajectory is followed by a diving attack onto the target. Inertial mid-course guidance is employed with an active radar terminal homing phase.

It is also multi-mission capable with similar variants to the AS-4 Kitchen and the same guidance techniques. Warheads can either be 1000 kg HE or 350 KT nuclear according to the target to be attacked.

AS-6 Kingfish missiles carried beneath wing of Soviet Tu-16 Badger (JASDF)

6266.311

SOVIET TACTICAL AIR-TO-SURFACE MISSILES

The first Soviet tactical air-to-surface missiles were in fact modifications of air-to-air missiles and did not receive US/NATO reporting names or numbers. The first tactical weapon to actually get such identifiers was the AS-7 (NATO codename Kerry). Used by both the Soviet Air Force and Soviet Navy on their tactical fighter-bombers the AS-7 has also been exported to a number of countries including Yugoslavia and India. It entered service in the mid-to-late seventies and is a single stage, radio command guided, solid propellant missile weighing approximately 295 kg and having a range of about 8 km with a 60 kg HE warhead. The 3.5 m long AS-7 was followed into service by the AS-9 (NATO codename Kyle), which is an anti-radar missile designed to attack fire control radars. The Kyle is 6.0 m long, has a body diameter of 500 mm and a wing span of 1.5 m. With a launch weight of about 750 kg, this solid propellant missile probably has a range of around 75 km. The AS-9 is used by Soviet Air Force and Navy Su-24 Fencer, Tu-16 Badger and Tu-26 Backfire bombers for defence suppressions tasks when penetrating hostile airspace or attacking carrier battle groups.

The AS-10 (Karen) is a 3.5 m long missile with a body diameter of 300 mm and a wing span of 0.9 m. The missile weighs about 300 kg and is believed to be the Soviet equivalent to the Maverick AGM-65 family of missiles. Initial versions were TV guided and later versions now have semi-active laser seekers. The AS-11 (NATO codename Kilter) is believed to be an improved version of AS-10 Karen, with an IR seeker. AS-12 (NATO codename Kegler) is believed to be an anti-radar missile, to replace AS-9 Kyle. This missile is reported as weighing 350 kg and to be capable of launch from low level over a range of about 35 km.

The AS-14 (formerly known as the Advanced Tactical ASM, NATO codename Kedge) is thought to be a further improvement upon AS-10 Karen and AS-11 Kilter, with mid course guidance using a data link pod and electro-optical homing for the terminal phase. Maximum range is said to be 30 km.

6267.311

AS-13 AIR-TO-SURFACE MISSILE

The designation AS-13 is believed to refer to the long reported Mach 3.5, 800 km range successor to the AS-4 Kitchen and AS-6 Kingfish, which are carried by Tu-26 Backfire B/C, Tu-22 Blinder-B and Tu-95 Bear-G strategic bombers.

6475.311

KENT AIR-LAUNCHED CRUISE MISSILE (AS-15)

The designation AS-15 is the American number for the Soviet air-breathing, turbofan-powered, low altitude, air-launched cruise missile equivalent to the BGM-109 Tomahawk. The AS-15 entered service in 1984 as the main offensive armament for Tu-95 Bear-H bombers. It is also expected to arm the new variable geometry Tupolev Blackjack supersonic bomber in the early nineties. According to American intelligence sources by mid-1987 over 50 new build Bear-H launch platforms were in service with some examples already having been detected flying simulated cruise missile attack missions against the continental United States. The missile is said to have a range of 3000 km and be armed with a 200 KT nuclear warhead. Some type of terrain matching guidance unit with autopilot is assumed. From data released by the Soviet authorities, with the December 1987 INF Treaty, the missile is expected to be similar to the GLCM RK-55 system with a length of 8.09 m, a body diameter of 510 mm and a launch weight of 1700 kg.

4556.311

SOVIET AIR-LAUNCHED ANTI-TANK MISSILES

The Soviets use at least three types of helicopter launched anti-tank guided weapons.

The AT-2 Swatter (**2985.111**) first entered service in the early 1960s and the helicopter-launched versions have radio command or semi-active laser guidance. Swatter has been reported carried on Mi-8 Hip E and Mi-24 Hind A/D helicopters. Each of these can carry up to four on their outboard weapons racks. The missile is 1.16 m long and weighs about 29 kg for the radio command version. The semi-active laser version, known as Swatter C, weighs 29.8 kg.

Gazelle military helicopters operated by Yugoslavia are equipped to carry Soviet AT-3 Sagger anti-tank missiles. Note optical sight in cockpit roof to aid weapon guidance

For non-Warsaw Pact Mil helicopter users the Soviets have exported both the AT-3 Sagger A/B manual wire guided command-to-line-of-sight variants and the AT-3 Sagger C semi-automatic wire guided command-to-line-of-sight version (**2950.111**). The latter is capable of engaging targets between 250 and 3000 metres, whilst the former have a greater minimum range of 500 metres. The Mil Mi-2 Hoplite can carry up to four missiles on the sides of its cabins whilst export Mil Mi-24/Mil M-25 Hind-D carry a similar number. The largest number, six, is carried by the heavily armed Mil Mi-8 Hip-F assault helicopter. Yugoslavia has also successfully mated Saggers to its license built SA341H Gazelle helicopters, for which a roof sight is mounted over the left hand seat to assist the gunner in target acquisition and missile guidance. Sagger is 0.86 m long, has a wing span of 0.38 m and weighs 11.3 kg.

The latest missile fielded is the AT-6 Spiral which is a tube launched weapon and is at present deployed in pairs on the wing racks of Soviet, Polish and East German Mil Mi-24 Hind-E attack helicopters. It is also expected to arm the Mil Mi-28 Havoc and Kamov Hokum attack helicopters as well. The Spiral is tube-mounted and reported to be radio command or semi-active laser guided. It has a single round kill probability of over 90 per cent with range capabilities of 500 to 5000 metres. The large (8-10 kg) HEAT warhead ensures that any armoured vehicle hit will either be severely damaged or destroyed.

Loading Soviet AT-3 Sagger anti-tank missile on Yugoslav Gazelle helicopter

More information on Soviet anti-tank missiles is given in the Anti-Tank Assault Weapons section of this book.

UNITED KINGDOM

4528.311

ALARM AIR-TO-SURFACE MISSILE

ALARM is the air-launched anti-radar missile being developed by British Aerospace (BAe) and Marconi for the RAF to enable it to meet an essential requirement to provide a deep penetration strike capability. ALARM is designed to counter enemy surface-to-air gun and missile defences by destroying their associated radars. A fixed price bid for development and production was submitted to the UK MoD early in December 1982. Two competitive solutions were considered and ALARM selected in July 1983. The alternative was procurement of the American AGM-88A, HARM (**1769.311**) with some licence production in the UK.

BAe in association with Marconi Defence Systems (MDS) and other sub-contractors are working on the design and development of the system and have produced sub-systems which are undergoing airborne tests and trials.

Other significant contributors are THORN EMI Electronics, who won a contract worth more than £4 million for the development and supply of the proximity fuze, and MBB of the Federal Republic of Germany who are developing the missile's warhead.

The fuze being developed by THORN EMI will be based on the company's laser experience and at least two years private-venture research had already been completed before the contract award in May 1984. The new generation fuze will provide autonomous classification of targets and precise trigger point selection.

ALARM is a microprocessor-based software-controlled missile which does not require complex aircraft systems or interfaces, and it can be updated to meet changing threats and scenarios by software changes alone. The Marconi anti-radar seeker exploits UK investment in seeker technology and EW.

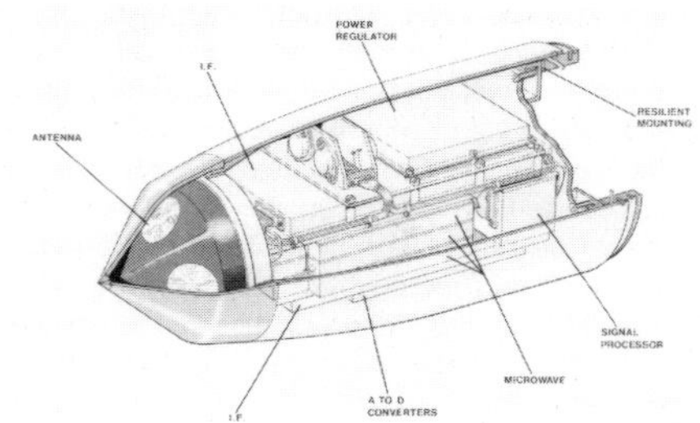

Schematic diagram of Marconi seeker head for ALARM

The complete seeker head will be programmable, enabling operators of the ALARM system to react autonomously and rapidly to threat changes as they arise. This facility derives from the employment at the core of the seeker receiver of a complex digital processor that incorporates modern LSI techniques and new microwave IC materials. All software is produced in-house by MDS. The seeker package can be programmed before a mission with a list of target types and priorities. Re-programming during flight is also possible. The associated broadband microwave receiver is the first in the UK to utilise novel RF processing techniques, similar to those employed in EW radar homing and warning receivers.

The combined weight of ALARM plus its launcher is 280 kg which will allow up to four 200 kg, 4 m long, 0.23 m diameter, 0.73 m wing span ALARM missiles to be carried by the Tornado GRl in addition to its normal attack payload. ALARM does not require dedicated aircraft, and has been designed from the outset with a high degree of autonomy. It can also be fitted to fixed-wing aircraft down to BAe Hawk size as well as large aircraft and helicopters.

ALARM has a number of operational modes, among them direct attack where the missile flies to the selected target, and a loiter mode where the missile climbs to altitude and deploys a parachute from which it remains suspended until a suitable target is identified. The parachute is then jettisoned and the missile descends under gravity to home onto the target from overhead.

STATUS

A detailed joint feasibility study by BAe and MDS was completed in 1982 and a submission was made to the UK MoD after certain company-funded development work had been carried out. The choice between ALARM and HARM procurement to meet ASR 1228 was announced in July 1983 and a contract formally signed the following month. A UK requirement for about 750 missiles has been quoted with entry into RAF service originally scheduled for 1987. The first three firings were carried out late in 1986 at the UK MoD Aberporth range. Two were unguided; the third was guided and homed successfully into its target.

Problems with the development of the rocket motor have caused delays to the programme and a new motor has to be proven before the renegotiated contract is signed. It is expected that ALARM will not now enter service until 1989.

CONTRACTORS

British Aerospace Dynamics, Manor Road, Hatfield, Hertfordshire AL10 9LL, England.

Marconi Defence Systems Ltd, Stanmore, Middlesex, England.

Launch of the first fully-guided ALARM from a Tornado GR1 aircraft

Tornado GR1 aircraft carrying load of seven ALARMs

1530.321

SEA SKUA HELICOPTER AIR-TO-SURFACE MISSILE

Sea Skua is an all-weather, helicopter-launched, sea-skimming anti-ship guided weapon system which is in operational use with the RN and for three other navies. Initially designed for fitment to the naval variant of the Westland Lynx helicopter fitted with the Ferranti Seaspray radar, it is the RN's principal air-to-surface light strike weapon for use against a variety of targets ranging from missile-firing fast patrol boats to coastal escorts, frigates and destroyers. Sea Skua is also suitable for fitting on other types of helicopter and fixed wing aircraft. Studies have also been carried out on a ship-launched version and this is described in entry **4553.221**.

The helicopter/Sea Skua combination provides a rapid-reaction, surface attack capability up to, and well beyond, the radar horizon of the parent ship which retains the option of remaining passive and therefore undetected.

The missile is light enough to permit a four-Sea Skua fit on a small helicopter. Solid-propellant boost and sustainer motors are used to give the missile a range in excess of 15 km, sufficient to provide a good 'stand-off' capability for the helicopter with consequent protection from counter attack. To commence an engagement the helicopter would close the target to enable its radar to lock on and automatically track. On release from the helicopter the missile drops for a short distance under autopilot control maintaining attitude angle stabilisation in roll, pitch and yaw before the rocket motors are ignited. The Sea Skua missile then descends in stages, under control of a radio altimeter, to one of four terminal sea-skimming heights selected by the pilot prior to missile release, depending on the sea state or size of target. The missile, guided by the semi-active radar homing head in azimuth, will fly on a proportional navigational course to hit the target. The warhead is designed to explode within the target to give high lethality. Sea Skua missiles can be fired in rapid succession if required and the helicopter is free to manoeuvre after the last missile release within the limits required to maintain target illumination.

System control equipment associated with armed release of the missiles has been designed for rapid removal and replacement, thereby reducing to a minimum any weight penalty to the helicopter when it is not required to operate in the strike role. The missile can be treated as a round of ammunition as no 'on-board' testing is required. It is delivered in a wheeled 'palletrolley' fitted with a shock absorbing system suitable for shipborne magazine stowage.

CHARACTERISTICS

Length: 2.5 m
Diameter: 0.25 m
Span: 0.72 m
Weight: 145 kg
Range: 18 km
Warhead: 20 kg SAP HE

STATUS

In June 1981, a multi-million pound contract for Sea Skua production for the RN was announced, and it was confirmed that initial deliveries were already being made. In May 1982 the weapon scored its first operational successes in the fighting to regain the Falkland Islands, despite not having completed the RN's formal acceptance procedure.

In addition to the RN, Sea Skua has been selected to arm naval helicopters of the Federal Republic of Germany, Brazilian and Turkish navies.

Royal Navy Lynx helicopter armed with Sea Skua anti-ship missiles

CONTRACTORS

Prime contractor: British Aerospace Dynamics, Manor Road, Hatfield, Hertfordshire AL10 9LL, England.

Homing head: Marconi Defence Systems Ltd, The Grove, Warren Lane, Stanmore, Middlesex HA7 4LY, England.

3630.321

SEA EAGLE AIR-TO-SURFACE MISSILE

The Sea Eagle (formerly P3T) missile is described as a second generation, air-launched, sea-skimming, anti-ship missile. It is intended as a successor to Martel and arms Buccaneer and Sea Harrier aircraft employed by UK forces for anti-shipping operations. It may later arm RAF Tornado GR1 aircraft. A zero speed launch version for use with helicopters or as a ship-launched weapon has also been developed (**4447.221**). A new multi-role active radar homing head has been developed by Marconi Defence Systems Ltd, specially for use against ship targets and incorporating a digital computer that provides for over-the-horizon capability against multiple targets. In addition, the active seeker's computer serves as the central processor for controlling pre-launch information, management of the transmit and receive functions, target detection, acquisition and tracking. This enables the Sea Eagle to engage a single designated target within a convoy of other, but potentially less valuable, targets. Certain ECCM functions are also under the control of the missile computer.

Flight trajectories of the Sea Eagle will also be under computer control, the computer's re--programming facility being of value in this regard, and possible flight paths will include a variety of sea-skimming modes. A digital inertial autopilot/navigation system is employed, with a radar altimeter for height data. The Ferranti F100-L microprocessor is used in the Sea Eagle to carry out the digital data processing and control functions mentioned above.

The turbine used to power the Sea Eagle is the French Microturbo TRI 60 turbofan which has a thrust of 3 kN at 28 500 rpm. The Sea Eagle airframe is very similar to that of the Martel (**1022.311**) from which it was derived, but Sea Eagle incorporates an advanced warhead developed by the Royal Ordnance. The zero speed launch and ship-launched versions employ a slightly modified Sea Eagle missile with two strap-on booster rockets.

CHARACTERISTICS

Length: 4.14 m
Wing span: 120 cm
Diameter: 40 cm
Range: 110 km
Guidance: Inertial navigation plus active radar homing
Propulsion: Microturbo TRI 60 turbofan
Weight 600 kg

STATUS

Development was started in 1979. Flight tests began in 1981, and the first control and propulsion test firing was completed successfully in April 1981. The go-ahead for full production was announced on 10 February 1982.

A salvo firing of two missiles from a Buccaneer took place successfully in October 1982 at a range of more than 50 km. In July 1983 an Indian Navy contract estimated at about £200 million was announced for the supply of 12 Mk 42B Sea King helicopters (with an option on eight more) which will have a version of Sea Eagle missiles as their armament.

Deliveries of evaluation missiles for trials by No 31 Joint Service Trails Unit from Buccaneer and Sea Harriers began in early 1984. First production models left the assembly line in November 1984. Service evaluation of the missile was completed in December 1985 and Sea Eagle entered service in 1986.

CONTRACTOR

British Aerospace Dynamics, Hatfield, Hertfordshire AL10 9LL, England.

Homing head: Marconi Defence Systems Ltd, The Grove, Warren Lane, Stanmore, Middlesex HA7 4LY, England.

RAF Buccaneer carrying four British Aerospace Sea Eagle air-to-surface missiles

Two British Aerospace Sea Eagle air-to-surface missiles with additional boost motors, on Westland Sea King helicopter

UNITED STATES OF AMERICA

1766.311

ALCM (AGM-86) AIR-LAUNCHED CRUISE MISSILE

The ALCM is the result of an initiative by the US Air Force to provide an air-launched strategic weapon for deployment on the B-52 and B-1B bombers.

The AGM-86B ALCM is larger than the original AGM-86A, which was designed to be compatible with the SRAM (**1107.311**) rotary launcher. Dimensions were increased and a larger fuel tank was fitted to produce the AGM-86B, and in addition the wing sweep angle was changed from 35° to 25°. For internal stowage, all flying and control surfaces are folded or retracted, being deployed automatically after release from the aircraft. Dimensions in flying configuration are: length 6.32 m, span 3.65 m, diameter 0.69 m and weight 1458 kg. The engine is a Williams International Corporation F107-WR-101 turbofan developing about 600 lb (272 kg) of thrust for a weight of 66 kg. Operating speed is 'high subsonic'. An internal rotary launch rack fitted to the B-52H enables eight ALCMs to be launched in rapid succession. The design is intended for high-speed cruise flight at very low altitudes for distances of up to 2500 km and to aid penetration of opposing defences a very low radar cross section has been achieved.

A W-80-1 200 kT nuclear warhead is used. The missile has an inertial navigation system which is updated by using a map-matching position updating technique called TERCOM (Terrain Contour Matching). A terrain-following system is also incorporated which allows low-level flight. The complete navigation package is integrated by the Boeing Aerospace Company into the ACLM. Boeing also provides the complete software for the system. The navigation hardware consists of a gimballed inertial platform and digital computer, both provided by Litton Industries, a radar altimeter, provided by Honeywell or Kollsman; a pitotstatic probe, air data computer and temperature probe provided by Rosemount.

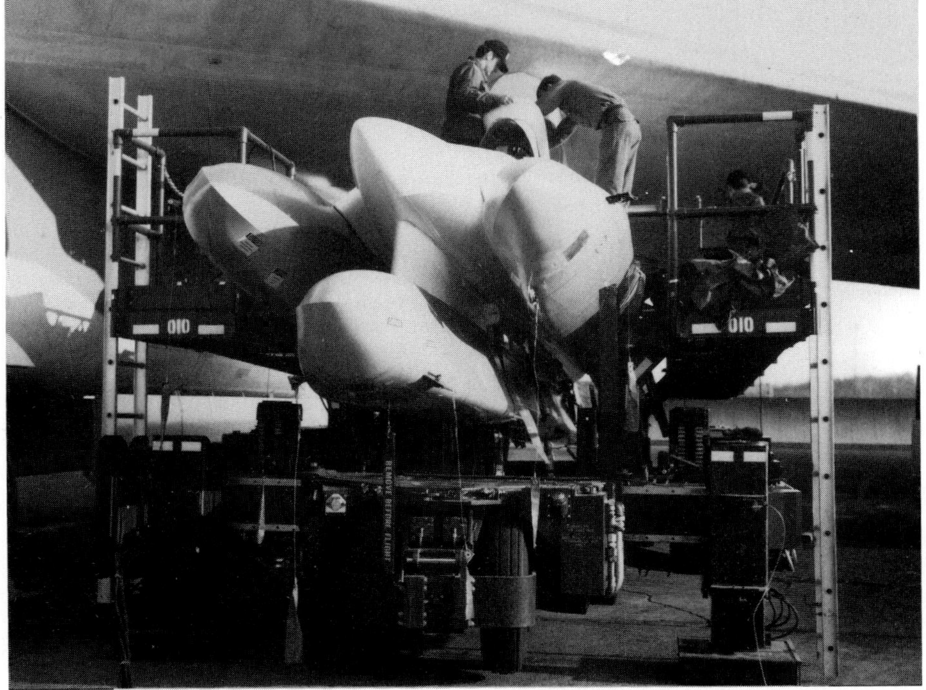

B-52G at Griffiss AFB, New York being loaded with pylon carrying six ALCMs by airmen of SAC's 416 Bombardment Wing

The terrain correlation updates are provided by software which processes the outputs of the air data system, radar altimeter, and the inertial platform and compares the resultant sampled terrain elevation data to a stored terrain elevation map. The correlation of the elevation profile data with the reference map provides the missile with an axis position update. The terrain correlation updates provide a very accurate position update to the missile. The inertial navigation accuracy is further improved by processing the position update through a Kalman filter in the software which refines the platform alignment, its velocity, and its instrument error sources. The initial alignment of the platform is done by an in-air Kalman filter transfer alignment to the B-52 OAS (Offensive Avionics System).

The ALCM terrain-following system is a hybrid analogue/digital system. It uses a rate gyro and body accelerometer package supplied by Northrop and a Boeing-supplied electronics package. The principal input to the terrain-following system is the radar altimeter but the air data system and inertial system also provide inputs.

STATUS

The first powered ALCM flight was made from a B-52G release over the White Sands Missile Range in March 1976. In March 1980 the Boeing ALCM was selected for deployment with the US Air Force.

The first squadron became operational in December 1982, each B-52G aircraft carrying 12 ALCMs on two six round underwing pylons. The more modern B-52H carries the same underwing load but has an additional 8 ALCMs on a rotary launcher in a modified bomb bay which on the B-52G is loaded with nuclear bombs and/or SRAM. The Rockwell International B-1B bomber can carry eight AGM-86B ALCMS internally with a further 14 AGM-86B ALCMS under its fuselage. The final production model AGM-86B, the last of 1715 missiles produced by Boeing, was handed over to the USAF in October 1986.

Following an assessment of the Air Force requirements for a successor to the ALCM in 1982, and an industry contest, General Dynamics was selected to develop an advanced cruise missile (ACM) in 1983.

CONTRACTOR

Boeing Aerospace Company, PO Box 3999, Seattle, Washington 98124, USA.

6610.311

ADVANCED CRUISE MISSILE AGM-129

The Advanced Cruise Missile (ACM) AGM-129A entered development in 1983, with a contract won by General Dynamics. Required as a replacement for ALCM AGM-86 (entry **1766.311**) this missile is expected to have longer range, better accuracy, lower radar cross section and lower IR signature. AGM-129A is planned to be carried on the B-52H, B-1B and the B-2 bomber aircraft.

There are no photographs available and few details have been released on this classified programme. An artist's impression has shown a delta planform, with an integral wing and body. The Williams International F112-WR-100 turbofan engine has been selected to power the missile. It is expected that the nuclear warhead will be in the 200 KT class.

STATUS

ACM entered development in 1983 and was planned to enter service in 1990. However in 1987 it was reported that flight trials had been delayed and that the programme was now one or two years behind schedule. McDonnell Douglas were awarded a contract in 1987 to qualify as a second source for the production of the missile. It is expected that around 1000 missiles will be required by the USAF.

CONTRACTORS

General Dynamics, San Diego, California, USA.

McDonnell Douglas, St Louis, Missouri, USA.

1107.311

SRAM (AGM-69) SHORT RANGE ATTACK MISSILE

The Short Range Attack Missile (SRAM) is a supersonic air-to-surface nuclear weapon carried by B-52 G/H and FB-111A bombers of the US Air Force. The former type of aircraft can carry up to 20 missiles, 12 on wing pylons and eight internally, and the FB-111A is able to load two missiles under each wing with two more in the weapons bay. The Rockwell International B-1B bomber can carry 24 AGM-69 SRAM internally with a further 14 being carried externally under its wings. An advanced version, SRAM 2, is in development (see later).

SRAM is designed to attack and neutralise enemy terminal defences, particularly the Soviet surface-to-air missile (SAM) defences. The weapon is capable of penetrating terminal defences and striking mission targets while the bombers stand off outside the range of enemy defences. It can also attack enemy SAM and anti-aircraft sites so that bombers can strike primary targets with other SRAM or conventional bombs.

The SRAM 1 missile is 426.7 cm in length (for internal carriage), 482.6 cm in length (for external carriage), 44.5 cm in diameter and weighs 1016 kg at launch. The missile is supersonic, with a range of 200 km. The CEP value is said to be around 450 m.

The W-69 nuclear warhead carried is stated to be of the same 170 kT size as that of the Minuteman 3 with the option of either a contact ground burst or preset altitude airburst type of nuclear explosion. The actual type chosen depending on the target to be attacked and the damage level required.

A KT-76 inertial guidance system is used, and the propulsion system is a two-stage solid rocket motor developed by Lockheed. The first stage provides initial acceleration; the second stage is activated as the missile approaches its target. Trajectory can be changed in mid-flight, and the missile can be launched to hit targets in any direction regardless of the bomber's flight path. The missile can be launched from either a high or low attack mode. The high mode is said to yield the greatest range but also increases the missile's vulnerability to enemy defences.

Four basic modes of flight trajectory can be employed: semi-ballistic, altimeter-controlled terrain following, pull-up from behind radar screening terrain followed by inertial, and a combination of inertial and terrain following. Lateral deviations in flight profile can also be programmed.

STATUS

The first production SRAM was delivered to the US Air Force in March 1972. The 750th production missile was delivered in February 1974, and a total of 1500 SRAMs have been deployed with Strategic Air Command units. The last missile was produced in 1975.

SRAM 2

A contract to develop an improved version, SRAM 2 (XAGM-131), for use in the B-1B and B-2 bombers has been won by Boeing. The new weapon is a supersonic air-to-surface missile carrying a nuclear warhead and is intended to replace the current SRAMs. The SRAM 2 incorporates a new rocket motor for improved speed and range, an improved guidance system, and a new warhead. It will be the same length as the SRAM 1, but only two-thirds of the diameter, with a weight of approximately 1800 lbs (818 kg). First flight test of SRAM 2 is scheduled for 1989 with an initial in-service date of early 1992. The US Air Force has a requirement for more than 1500 of the new missiles. Total cost of the programme is expected to be $2.4 billion.

CONTRACTOR

Main contractor: Boeing Aerospace Company, PO Box 3999, Seattle, Washington 98124, USA.

Strategic Air Command B-52 bombers can carry 20 AGM-69 SRAMs, eight internally and another 12 on the wings

1102.311

SHRIKE (AGM-45) AIR-TO-SURFACE MISSILE

This is an anti-radiation air-to-surface missile for the destruction of ground defensive radar installations. Official details include: length 304.8 cm, diameter 20 cm, wing span 91.4 cm, weight 177 kg. A solid propellant rocket motor gives Shrike a range of about 12 km. A 65.8 kg high-explosive fragmentation warhead is carried.

OPERATION

Initial warning of illumination of Shrike-carrying aircraft by ground defence radars is obtained from ESM/ECM receivers installed in the aircraft. When within appropriate range, the Shrike sensor heads are switched on and the missiles fired when target acquisition has been achieved.

Shrike guidance heads have been developed that are effective against enemy early warning, ground control intercept, and SAM guidance radars, each of which covers a different frequency range. The frequency range of a particular Shrike version is denoted by a suffix number to the AGM-45 designation, and there are at least 12 of these. Based on the specific radar to be attacked, the operational commander selects the appropriate seeker head option. The US Air Force use Shrike on modified F-4 aircraft, termed Wild Weasel aircraft, with special radar homing and warning equipment. The US Navy employs the missile on A-4, A-6 and A-7 aircraft.

STATUS

Production complete. In service with the US Armed Forces, Israel and the United Kingdom (limited stocks only). Used widely during the Vietnamese War, and by Israel during the War of Attrition, the Yom Kippur War and the Peace for Galilee operation. Limited use was made by the United Kingdom during the 1982 Falklands War on Vulcan bombers. The latest use has been by the US Navy during the April 1986 attack on Libya.

CONTRACTORS

Joint prime contractors: Texas Instruments and Sperry Rand/Univac.

Shrike AGM-45A air-to-surface anti-radar missile prepared for flight testing at China Lake, US Naval Weapons Centre (USN)

1098.311

MAVERICK (AGM-65) AIR-TO-SURFACE MISSILE SYSTEM

The AGM-65 Maverick is a precision guided tactical missile developed by Hughes Aircraft Company under contract to the US Air Force for use against hard targets such as tanks, armoured vehicles, field fortifications, gun positions, concrete communication centres and aircraft shelters. The system is currently operational on the F-4, A-7, A-10, F-111, F-16, F-5, AV-8B, A-6, A-4, F-18, Hunter and AJ 37 aircraft.

CHARACTERISTICS

Length: 2.49 m

Diameter: 30.5 cm

Wing span: 71.9 cm

Weight: A/B 210 kg, D 220 kg, E 293 kg, F/G 307 kg

Guidance: TV (A/B models); Laser (E model); Infra-red (D/F/G models)

Range: 8 – 25 km

The basic warhead is a 56.7 kg forward-firing shaped charge, designed for penetration through heavy armour and reinforced concrete. An alternative kinetic energy penetrator blast-fragmentation warhead weighs 136 kg, has a design optimised for effectiveness against major combatant ships and buried concrete fortifications and is fitted to the later E/F and G versions. The weapon is powered by a two-stage solid fuel rocket motor and is rail launched from either single-missile or triple launcher units. Variants of the Maverick are described separately.

TV Maverick (AGM-65A/B)

This basic missile is precision guided by a miniature TV homing system in the nose section. On his TV monitor, the pilot sees the picture transmitted from the missile seeker. He slews the seeker to the target, locks the TV tracker on target, and launches the missile. The aircraft can take immediate evasive action (launch-and-leave) or fire successive missiles while the first missile homes automatically on to the target.

Tracker logic steers the missile for impact at the centroid of the target in order to maximise kill probability. The B model has a modified optics system with a 2.5° field of view for greater magnification of the target area and longer range identification of small targets.

AGM-65D IR Maverick launched from an A-10 aircraft

More than 30 000 AGM-65A/B Mavericks have been produced. Of the total, 100 were combat firings, which yielded 87 per cent direct hits. In addition to the USA, TV Maverick has been ordered by: Egypt, Federal Republic of Germany, Greece, Iran, Israel, Jordan, Morocco, Pakistan, Portugal, Saudi Arabia, Singapore, South Korea, Sweden, Switzerland, Taiwan and Turkey.

Laser Maverick (AGM-65C/E)

Laser Maverick provides close air support against laser-designated targets in support of ground forces. Mounted forward of missile centre and aft sections that are common to all Mavericks, the semi-active laser seeker homes on to the energy reflected off the target by the designator.

The C model, containing the basic shaped-charge warhead, completed development flight testing with over 90 per cent direct hits on target. Targets were illuminated with various laser designators. This version did not go into production.

The production E model utilises laser guidance with the alternative 136 kg warhead in order to meet operational requirements of the US Marine Corps. Operational flight testing of that configuration was completed in the summer of 1982 with a score of 15 out of 15 direct hits.

US Air Force analysis indicates that Laser Maverick effectiveness in the close air support role is dependent on the availability of stabilised ground target designators such as the Ground Laser Locator Designator (GLLD) or airborne laser designators, such as the A-6E TRAM, the OV-10D NOS or the US Air Force Pave Tack system. The capability of hand-held designators is adequate for use with Laser Maverick against large or static targets. This model will be integrated with A-4M Skyhawk, A-6E Intruder, AV-8B Harrier and F/A-18 Hornet aircraft.

IR Maverick (AGM-65D/G)

An imaging infra-red (IR) system provides autonomous launch-and-leave guidance both day and night, and under low visibility conditions. IR Maverick retains hardware commonality with the TV and Laser Mavericks. In addition, IR Maverick's pilot operating procedures are similar, allowing the AGM-65D to be used on all aircraft types already equipped to operate TV Maverick. It can be used autonomously to locate and strike targets in the combat zone, or in the traditional night interdiction of roads and railroads; or it can be precisely slaved to other target acquisition aids such as forward looking infra-red (FLIR), radar, INS, LORAN, or radiation direction finders (anti-SAM).

Target recognition ranges with IR, day or night, are about twice the range for corresponding daylight visual recognition of typical tactical targets. The imaging IR seeker has performance superior to TV or laser in low visibility. The long wave spectrum used penetrates battlefield smoke and dust which is opaque to visual systems. Thermal signature also provides unique advantages in detection of hidden/camouflaged targets and the determination of live from dead (cold) targets.

The AGM-65D entered service in 1983 and an improved version, AGM-65G, is now in development.

The AGM-65G uses the alternate 136 kg warhead with the IR seeker and propulsion section of the AGM-65D. Algorithms are optimised for larger land targets.

Navy Maverick (AGM-65F)

In late 1980 Hughes received a contract for the development of anti-ship algorithms to be incorporated in the IR seeker to give the IR Maverick additional capabilities against ship targets. Designated the AGM-65F, this model combines the IR guidance unit of the US Air Force AGM-65D modified with additional tracking logic for ship targets, with the warhead and propulsion sections of the AGM-65E model. IR guidance will allow US Navy pilots to undertake attacks in darkness and poor weather, and the heavy blast penetration warhead will be used with a selectable fuze for optimum effectiveness against various types of target. Integration of the AGM-65F with A-6E and F/A-18 aircraft is complete and production started in 1987.

STATUS

AGM-65D/G: Large quantities of AGM-65D IR Maverick have been delivered by Hughes under a number of contracts over the period from 1982. The fiscal year 1987 budget called for 2755 missiles at a cost of $381 million. For 1988 the USAF infra-red Maverick requirements are for 2100 missiles, with another 1900 requested in 1989.

In May 1983, Raytheon was selected by the US Air Force as the second source for production of AGM-65D IR Maverick. After successful completion of qualification test launches, full funding was provided for an option of 800 missiles.

AGM-65E: This version entered service in 1985. The US Navy funded 1500 Laser Mavericks in the 1986 budget and 1800 in the 1987 budget. A final 1099 have been requested in the 1988 budget.

AGM-65F: The US Navy has completed a three-round DT&E flight test programme of the Navy IR Maverick which resulted in three direct hits against ship targets. Production started in 1987 and the 1988 budget is for 601 missiles, with 731 requested in 1989. This version is due in service in 1989.

AGM-65G: This version is still in development.

CONTRACTORS

Hughes Aircraft Company, Missile Systems Group, Canoga Park, California 91304, USA.

Raytheon Company, Lexington, Massachusetts 02173, USA.

1123.311

STANDARD AGM-78 ANTI-RADAR MISSILE

The Standard AGM-78 missiles were originally developed in the late 1960s, based upon the RIM-66 Standard surface-to-air missile for the US Navy. There were four versions of the AGM-78 delivered into service over the period 1969 to 1975, the AGM-78A/B/C/D. The missile has been carried by F-105 Thunderchief, F-4 Phantom, A-6 and EA-6 Intruder aircraft.

The AGM-78 is 4.57 m long, has a body diameter of 305 mm and a wing span of 0.91 m. The missile weighs 815 kg. Guidance is by passive radar homing, programmed before launch by an aircraft-mounted Target Identification and Acquisition System (TIAS).

STATUS

Standard AGM-78 entered service with the US Navy in 1969 and finished production in 1975. They are still in operational use.

CONTRACTOR

General Dynamics, Pomona, California, USA.

Standard AGM-78 anti-radar missile beneath A-6 Intruder aircraft

6477.311

AGM-130 AIR-TO-SURFACE MISSILE

A powered version of the GBU-15 (**4190.311**), designated AGM-130, is under development and promises to provide about three times the low-altitude stand-off of the GBU-15 for enhanced aircraft survivability. The rocket motor, developed and certified by Hercules Corporation expressly for the AGM-130, is mounted on the bottom of the weapon. The nine-inch (22.86 cm) diameter motor is designed for low altitude operation and weighs approximately 480 lbs (218 kg). In addition to the rocket motor, features unique to the AGM-130 include a radar altimeter and a digital autopilot. These items are designed to pave the way for future system growth by allowing the missile flight profile to be reprogrammed to meet evolving requirements. AGM-130 uses the GBU-15 short-chord airfoil group, and commonality between other GBU-15 systems and the AGM-130 has reduced development costs and time.

In addition to the AGM-130A (Mk-84 warhead), other variants are under consideration, including a SUU-54 dispenser version (AGM-130B) and a version utilising the BLU-109 unitary warhead (AGM-130C). In addition a $42 million full-scale development contract for an improved, jam-resistant data link was awarded in November 1986 to a team consisting of Harris Corporation and Magnavox Corporation. The data link development is scheduled for completion in April 1990.

Aircraft with GBU-15 capability are inherently compatible with AGM-130 due to similarities in the electrical and physical interfaces, and the unchanged data link pod installation. In addition to the US Air Force F-4E and F-111 aircraft, AGM-130A is anticipated on the F-15E and B-52G.

STATUS

The AGM-130 full-scale development award was made in late 1984, and the first flight tests occurred in late 1985. Current US Air Force documents indicate that some 6000 AGM-130s will be ordered, with a mix of IR and TV seekers. Early production deliveries were expected in 1990, but the programme was halted at the time of the US DoD FY 1989 budget proposals and its future is uncertain.

CONTRACTOR

Rockwell International, Missile Systems Division, 1800 Satellite Boulevard, Duluth, Georgia 30136, USA.

6609.311

SIDEARM AGM-122 ANTI-RADAR MISSILE

The Sidearm AGM-122 is a development programme using the near 1000 Sidewinder AIM-9C semi-active radar homing heads, that have been in storage since the mid-1970s. Development of the original semi-active seekers into broadband passive receivers was carried out at the US Naval Weapons Centre, China Lake. Production will be carried out by Motorola, who made the original AIM-9C seekers. The development programme was required as a low cost and lightweight anti-radar missile to attack SAM and gun control radars. It is planned that Sidearm will be carried by F-18 Hornet, AV-8B Harrier, F-4G Phantom, A-4 Skyhawk and the AH-1 SeaCobra helicopter. Further studies are examining the development of an all new passive seeker for the 1990s.

Sidearm will have the same remaining assemblies as the Sidewinder AIM-9L and 9M versions and will weigh 91 kg. Guidance is by a single broadband passive radar seeker, incorporated into the original AIM-9C guidance and control section.

STATUS

Sidearm AGM-122A is in development and early production and is expected to enter service in early 1989. The reworking of the guidance and control sections is being carried out at about 250 assemblies per year.

CONTRACTOR

Motorola Inc, Tempe, Arizona, USA.

6268.311

AGM-123A SKIPPER II GUIDED MISSILE

The Skipper II Guided Missile was developed by the US Naval Weapons Center, China Lake, California, as a modification to the Paveway II GBU-16. The Naval Avionics Center proved the technical data package by building initial pre-production quantities.

Skipper II and Paveway II both automatically seek a laser designating signal sent by the launching aircraft or other source. Primary mission of Skipper II is anti-ship. Compared to alternatives, Skipper II provides a stand-off launch capability through modification of existing rounds at reduced cost.

Skipper II improvements over Paveway II include the addition of a rocket booster and modifications to the guidance electronics. These improvements effectively more than double the range while enabling low altitude launch. Together these improvements greatly enhance launch aircraft survivability. Other improvements in the Skipper II make it compatible with shipboard operations.

The US Navy version of Skipper II adapts to a standard Mk 83 1000-pound bomb. It has been deployed on carrier-based A-6, A-7 and on the F/A-18. The same type of modification could be made to the 500-pound or 2000-pound warhead Paveway II for the US Air Force and those countries that have those versions of the Paveway II.

Skipper II is 4.33 m in length, has a wing span of 1.60 m and weighs 582 kg.

STATUS

In March, 1985, Emerson was awarded the first contract to produce 2500 modification kits to convert Paveway II bombs to the new AGM-123A Skipper II guided missile system configuration. Delivery has now been completed.

In addition to the USA, 32 other countries have Paveway II bombs which could be converted to the new Skipper II design.

CONTRACTOR

Emerson Electric Co, Government & Defense Group, 8100 W. Florissant, St. Louis, Missouri 63136, USA.

AGM-123A Skipper II guided missile being launched from US Navy A-6 Intruder aircraft

6611.311

TACIT RAINBOW AGM-136 ANTI-RADAR MISSILE

Details of this previously classified programme were released in 1987 and indicate that full scale development started around 1984 with completion due in 1988/89. Northrop have received an early production contract, to continue on from their development contract. Tacit Rainbow is planned as a low cost anti-radar missile with the ability to loiter in the vicinity of any target radars, waiting for them to transmit and then homing passively onto these transmissions. It is expected that carriage will be on the B-52, F-111 and A-6 Intruder aircraft initially.

Tactic Rainbow AGM-136 has folding rectangular wings under the body, with an air inlet for the Williams International turbofan engine located forward on the upper side of the body. The missile is 2.54 m long, has a body diameter of 686 mm and a wing span of 1.56 m. The missile weighs in the region of 575 kg. Guidance is by a broadband digital passive radar seeker, developed by Texas Instruments. It is assumed that an HE blast fragmentation warhead is fitted and that the range of the Tacit Rainbow is in the region of 90 km.

STATUS

Full scale development started around 1984 and is due to be completed and a production decision made in 1988-89. The missile is expected to be in service in 1991.

CONTRACTOR

Northrop Ventura, Newbury Park, California, USA.

Tacit Rainbow AGM-136, prepared for test flight from A-6E Intruder aircraft

1301.321

HARPOON (AGM-84) AIR-TO-SURFACE ANTI-SHIP MISSILE

Harpoon is an all-weather anti-ship missile for use against surfaced submarines, patrol craft, destroyers and larger vessels, merchant ships and trawlers.

The missile is capable of being air-launched or, with the addition of a rocket booster, ship or submarine-launched. Further details of Harpoon in these versions will be found in **2641.221** and **2797.421** in 1976 and earlier editions.

Missile dimensions are: length 385 cm, body diameter 34 cm, span 91 cm, and weight 522 kg. Ship and submarine-launched versions have an additional booster motor, with four fins, attached to the rear of the missile. Mid-course guidance is performed by a system utilising a digital processor, and a strap-down sensor/guidance sub-system. This system uses computer techniques to process the data from the strap-down gyros to sense missile velocity in three axes and derive control signals from this information.

The missile employs an active radar terminal guidance seeker. During the March 1986 Gulf of Sidra incidents US Navy A-6 Intruders used air-launched Harpoons to sink two Libyan warships, a Combattante II missile boat and a Nanuchka II missile corvette. A third vessel, another Nanuchka, may have been badly damaged by other air-launched Harpoons.

The original AGM-84A used a terminal 'pop-up' manoeuvre to counter close-in defences. This was eliminated following improvements to low-altitude penetration capability. The current production version has increases in range, waypoints and selectable terminal trajectories. An improved Harpoon, known as SLAM (Stand-off Land Attack Missile), fitted with a Maverick imaging infra-red seeker, is in development.

STATUS

In October 1972 the first powered flight was conducted at the US Navy Pacific Missile Test Centre at Point Mugu, California. Entry into service began in 1977.

It was announced in April 1983 that three anti-ship Harpoon missiles had been launched successfully from US Air Force B-52 bombers against a surface target in the Pacific. Two B-52 squadrons are being equipped to deploy Harpoons. Each aircraft can carry 12 Harpoon missiles on external wing mountings.

Royal Air Force Nimrod maritime aircraft having carried two bomb-bay loaded Harpoon missiles as an emergency measure during the Falklands conflict, are to retain the Harpoon capability and additional missiles valued at $248 million have been ordered. Royal Australian Air Force F-111C fighter bombers became certified to launch Harpoon in mid-1985. The F-111C retrofit programme is in progress.

Harpoon has been cleared for carriage on P-3 Orion, Nimrod MR-2, A-6 Intruder, B-52, F-111C, F-27 and F-18 Hornet aircraft.

F/A-18 aircraft carrying two McDonnell Douglas Astronautics Harpoon air-to-surface anti-ship missiles

CONTRACTOR

McDonnell Douglas Astronautics Company, PO Box 516, St Louis, Missouri 63166, USA.

1769.311

HARM (AGM-88A) HIGH-SPEED ANTI-RADAR MISSILE

The high-speed anti-radar missile (HARM) has entered service as a replacement for the existing Shrike (**1102.311**) and Standard ARM (**1123.311**). It is between these two missiles in terms of overall size but has a number of operational advantages, including higher speed, faster reaction, longer range and a more destructive warhead. In addition, unlike the Shrike which is configured individually to specific frequency bands, HARM has a broad-band capability, thus enabling a single missile to engage any anticipated air defence-associated surface radar (ship- or land-based). HARM also has software-controlled processing which enables new threats to be addressed by reprogramming instead of hardware modifications as enemy weapons develop. This will enable HARM to be used for the suppression or destruction of emitters other than those associated with surface-to-air systems, including early warning radars, ATC radars and weather radars whose elimination might offer tactical dividends.

The missile seeker, control section, wings and fins are produced by Texas Instruments, and the warhead, rocket motor and fuzing are being produced by other contractors. Technical direction is by the US Naval Weapons Centre, China Lake. The single seeker head employs a fixed antenna, thereby eliminating the complication and cost of gimbals and steering controls, and all pneumatic and hydraulic systems have been eliminated. It comprises the antenna array, ten microwave circuit boards and a video processor. This assembly provides direction finding facilities without requiring a moving antenna array, the latter consisting of a planar spiral-helix associated with other elements housed within a radome which itself forms an integral part of the overall RF antenna sub-system.

A 66 kg prefragmented HE warhead is carried which is triggered by a laser fuze that senses missile height above the terrain and detonates the warhead at one of two settings, depending on whether an overshoot or an undershoot is sensed. These two values are established so that maximum effect of the known dispersion pattern of the warhead fragments is utilised. It is likely that in future there will be more than two settings, which may be inserted automatically prior to launch in accordance with the target detected and engaged.

The other major part of the HARM weapon system consists of the aircraft avionics and controls; the US Navy uses the AN/AWG-25 equipment installed in the A-7 aircraft, and the US Air Force utilises the avionics and displays of the Wild Weasel system to interface with HARM.

The command launch computer interfaces with the existing radar warning system and other aircraft avionics, while the cockpit control panel entails simple modifications to the existing panel to adapt it for operation of the anti-radiation missile function. The computer has sufficient memory capacity to include all land- and sea-based emitters that US Navy task forces are expected to encounter. The US Air Force has a similar capability in the Wild Weasel system.

OPERATION

Three basic modes of operation for the HARM system are to be provided for the tactical air crew:

(1) *self-protect mode:* this is the most fundamental mode, and uses the aircraft radar warning receiver to detect threatening emitters. The HARM launch control computer then sorts and assigns priorities to these emissions and provides in digital language a complete set of instructions for the missile. This is achieved in a very short space of time and the pilot can fire the HARM almost immediately

(2) *target of opportunity mode:* in this mode the very high sensitivity of the missile seeker head to detect certain parameters of radar operation and also those transmissions associated with other parts of a radar installation, is utilised. It permits the engagement of targets by homing onto emissions that could not be detected and handled by Shrike or Standard ARMs

(3) *pre-briefed mode:* this employs the long range of HARM to engage targets whose signals can be detected but are not adequate for missile lock-on, or it may be used to fire HARM on a pre-computed trajectory into the vicinity of known targets. If these targets emit while HARM is in flight, they are attacked immediately; if the target does not radiate, the HARM destroys itself.

CHARACTERISTICS

Length: 4.17 m
Diameter: 25.4 cm
Wing span: 1.13 m
Launch weight: 361 kg
Guidance: Broad-band passive radar homing seeker
Warhead: 66 kg prefragmented HE
Fuze: Laser proximity
Range: 25 km

DEVELOPMENT

First contracts in support of HARM research and development were awarded in 1972. In May of 1974 Texas Instruments was awarded a contract to serve as the weapon system integration contractor for HARM.

STATUS

The first production HARM was delivered to the US Navy on December 2, 1983. In the second and third years of production, Texas Instruments delivered 256 and 283 missiles respectively. HARM is in full-scale production at Texas Instruments with production rates contracted to reach 210 missiles per month for the USAF and USN.

HARM has been cleared for carriage on F-4, A-6, EA-6B, A-7, F-18 and Tornado aircraft.

HARM has been purchased by West Germany for use on the Tornado, with an order for more than 350 missiles. Spain proposes to purchase 88 missiles, plus launchers, spares and test equipment.

First combat use was during the March and April 1986 Gulf of Sidra incidents when HARMs were fired from A-7s and F/A 18s at missile acquisition radar targets.

CONTRACTOR

System integration contractor: Texas Instruments Incorporated, PO Box 405, Lewisville, Texas 75067, USA.

F/A-18 of the US Navy launching a AGM-88A HARM missile

2831.311

TOW AIR-TO-SURFACE APPLICATION

As noted in the main description of the TOW BGM-71 anti-tank weapon system (**2830.111**), this weapon, like several other anti-tank weapons, can be used as armament for fighting helicopters.

Both the gunner's sight and the optical sensor are mounted on a specially developed gyrostabilised platform, which gives the tracking and control system a constant reference surface. The sight is pointed at the target by means of a joystick.

The HNVS is being produced as a low-cost, microprocessor-controlled forward-looking infra-red (FLIR) system that can be used for a variety of missions. It is designed to reduce a pilot's work load on night missions, including low-level and nap-of-the-earth flights. It can be used as a navigation aid, for target recognition and for search and rescue.

As the pilot skims over trees and through valleys, a FLIR image is projected on his helmet visor to reveal the world outside his cockpit. He can glance in any direction and his head movements are instantly transmitted through a helmet servo linkage to a FLIR turret mounted under the nose of the helicopter. The turret is automatically aimed along his line-of-sight (LOS). The image contrast and brightness are automatically kept at the ideal viewing level, and the pilot can call for flight symbology to be superimposed, further reducing his need to look down at cockpit displays as he flies.

The mast-mounted sight version of the airborne TOW system enables the helicopter aircrew to use hills and trees for cover while scouting for targets. Once a target is spotted, the gunner can fire a TOW missile, exposing only the sight to enemy detection during the entire operation. The missile climbs into the gunner's LOS following launch, while the helicopter remains hidden.

British Army Lynx helicopter is equipped with a roof-mounted telescopic sight to fire TOW anti-tank missiles

The mast-mounted sight is an adaptation of a roof-mounted sight Hughes Aircraft developed for the British Army Lynx helicopter. The roof-mounted sight is being produced in the United Kingdom by British Aerospace Dynamics under licence from Hughes Aircraft.

The sight is stabilised to compensate for aircraft movement and the vibration of the rotor blades. The design allows for future installation of a laser rangefinder, an automatic target tracker, multi-launch logic, and a FLIR system.

The initial airborne TOW system built in 1975 is the anti-tank system for the AH-1 Cobra series helicopters used by the US Army and Marine Corps as well as other nations. The system also has been installed on a variety of foreign helicopters.

Under a contract from the US Army Aviation Systems Command, disclosed in late 1984, pilots of Cobra attack helicopters will be able to fly combat missions round-the-clock using a night vision system produced by Hughes Aircraft. This will give Cobras a FLIR capability and the programme is known as C-NITE (Cobra-Night).

The contract also calls for modifications to enable the Cobras to fire and guide the new TOW 2 missile, day or night. TOW 2 features a more potent warhead, designed to defeat advanced enemy armour, and an improved guidance system.

The FLIR device, which makes it possible for gunners to see through darkness, smoke or haze, will be installed in the telescopic sight for the Cobra's airborne TOW missile system. In addition to firing the missile, the sight is used to accurately direct cannon and rocket fire.

There are two versions of the TOW sight now in use by the US Army, the standard optical sight and the Laser Augmented Airborne TOW (LAAT) sight. Production on the LAAT sight, which contains a mini-laser rangefinder, began in 1979.

The anti-armour version of the Bell Cobra is equipped with the M65 airborne TOW system, and up to four two-round modular TOW launch pods can be carried, while still leaving the inboard pylons free for rocket or 20 mm minigun pods. The M28 turret with a 7.62 mm minigun and 40 mm grenade launcher is retained. In addition to stabilised optics and a missile tracker, the modified M65 sight has provisions for a laser rangefinder. Helmet sights allow the crew to control the turret directly or provide target acquisition for the stabilised optics.

In the air-launched role, TOW has a maximum direct range of 4 km. It has seen combat use with the US forces in Vietnam, the Israeli Air Force during the 1973 Yom Kippur War and the 1982 Peace for Galilee campaign, the Moroccan Air Force against the Polisario Front and with the Iranian Army against Iraq.

STATUS

Hughes has delivered more than 1500 airborne TOW systems throughout the world and TOW has been deployed with air and ground forces of at least 33 countries. A number of nations operating TOW are using the airborne version, including Italy, Sweden and the UK. The latter producing a special roof-mounted optical sight for British Army Lynx helicopters under licence in the UK and a Further Improved TOW (FITOW) version is being developed in the UK to incorporate a proximity fuze and guidance changes to give it the capability of over-head top attack.

CONTRACTOR

Hughes Aircraft Company, Missile Systems Group, El Segundo, California 90045, USA.

1391.311

HELLFIRE (AGM-114) TACTICAL AIR-TO-SURFACE MISSILE

The Hellfire Modular Missile System (HMMS) is the US Army's latest anti-armour weapon system. Hellfire is an acronym derived from 'heliborne-launched fire and forget'. From 1976, when the US Army selected the Missile Systems Division of Rockwell International to undertake engineering development, numerous Hellfire 'prototype' missiles were launched in army concept feasibility tests. Flight and ground tests were conducted during that period at the US Army Missile Command. Operational tests were conducted in 1974, 1980 and 1981 by the US Army at Fort Hunter Liggett, California.

For its initial application, Hellfire has been designated as the point target weapon system for the AH-64 Advanced Attack helicopter. The AH-64 carries up to 16 missiles. These are distributed equally on four wing-mounted launchers, two inboard and two outboard. The launchers are modular in design so that the two lower rails may be removed.

Maximum length and weight of the missile is 1.63 m and 43 kg respectively, while the diameter is 177.8 mm with wing span of 330 mm. The Hellfire missile is also modular and will accept a variety of mission-tailored guidance modules, but the production version is the laser homing model. The tactical prototype flight test phase was conducted using the Hellfire semi-active laser seeker developed by Martin Marietta, Orlando, Florida.

Other guidance modules studied for Hellfire are the Air Defense Suppression Missile (ADSM) seeker and the imaging infra-red (IIR) seeker.

Hellfire is not limited to direct line-of-sight attack. Numerous indirect flights have been performed. In the indirect mode, the laser Hellfire missile is launched without seeker lock-on. It climbs over obstacles, searches for its target, then locks on automatically and impacts with no degradation in terminal accuracy.

US Army AH-64 Apache attack helicopter with eight Hellfire missiles loaded (US Army)

A factor in reducing helicopter exposure time during direct fire launches and in minimising total engagement time for all missions is laser Hellfire's broad launch window. This means that the launch heading of the helicopter and the azimuth to the target may differ considerably. Thereby, precision attitude alignment criteria, common to many other systems, are essentially eliminated.

In May 1984, laser Hellfire missiles demonstrated the 'ripple fire' technique of target engagement. With this technique, missiles are launched only fractions of a second apart, at different targets, designated by laser designators on separate codes. If only one designator is used, multiple targets can still be engaged; however, time between launches increases to a few seconds.

STATUS

In production and in service with the US Army (on AH-64) and US Marine Corps (modified AH-1) helicopters. The US Army had procured 15 082 Hellfire missiles up to the end of 1985, the 1986 procurement was for 6576, and no requirements were made for 1987. The 1988 request is for 5000 missiles. Total US Army procurement objective is for 48 686 missiles. Hellfire has also been test launched from a UH-60 Black Hawk helicopter.

Combined procurement and funding for Hellfire for Army and Marine requirements for 1988 is 6393 missiles ($212 million).

In 1983 firing trials of Hellfire installed in a British Lynx helicopter and using a Ferranti 306 laser target designator were carried out in Norway. This aircraft can carry up to eight Hellfire missiles.

In March 1985 Martin Marietta received a $126.2 million US Army contract to build Hellfire missiles as a second source supplier to Rockwell International. The US Army and US Marine Corps are expected to buy more than 60 000 missiles over the life of the programme.

Hellfire AGM-114 missiles are being 'product-improved'. A 'minimum smoke' rocket is being fitted and a digital autopilot is being developed to replace the AGM-114A analogue device.

CONTRACTORS

Rockwell International Missile Systems Division, 1800 Satellite Boulevard, Duluth, Georgia 30136, USA.

Martin Marietta, PO Box 5837, Orlando, Florida 32855, USA.

AIRCRAFT ARMAMENT

ARGENTINA

3806.303

CBAS-1 ALBATROS AIRCRAFT ROCKET

The Argentinian CBAS-1 Albatros is an unguided aircraft rocket used for typical air-to-surface tactical roles but primarily in naval applications for which it was originally designed. The rocket is of 70 mm calibre and the solid rocket motor case is constructed of aluminium. A peripherically inhibited double-base extruded propellant weighing 2.8 kg is used in the rocket motor, and the wrap-around forged steel tail fins are provided with a mechanical opening device which operates after launch.

CHARACTERISTICS

Calibre: 2.75 inch (70 mm)

Propulsion: Solid rocket motor

Launch weight (less warhead): 5.5 kg

Warhead

Training (smoke): 2.95 kg

Anti-tank/anti-personnel (ATAP): 3.80 kg

Fragmentation: 4.75 kg

DEVELOPMENT

The CBAS-1 was designed by the Argentinian Scientific and Technical Research Institute of the Armed Forces (CITEFA) and produced by EDESA.

STATUS

Operational with the Argentinian Navy since 1973. More than 78 000 rockets have been exported. The contractor (EDESA) also produces integrated weapon systems which employ this 2.75 inch rocket.

CONTRACTOR

Empresa de Desarrollo Especiales Sociedad Anonima (EDESA), Avenida España 3101 (1107), Buenos Aires, Argentina.

CBAS-1 Albatross aircraft rocket

BRAZIL

4145.303

AIRCRAFT ARMAMENT

A number of aircraft armament systems have been developed in Brazil, and are currently in production and in local service and abroad. Brief details follow:

37 mm SBAT-37 rocket

This spin-stabilised air-to-ground rocket is fired from the seven-round LM-37/7 launcher, and can be fitted to light aircraft and helicopters.

70 mm SBAT-70 rocket

This solid-propellant air-to-ground folding-fin rocket is similar to the US-made 2.75 inch FFAR, and can be used with re-usable multiple launchers of two (LM-70/2), four (LM-70/4), seven (LM-70/7) or nineteen (LM-70/19) rounds. A disposable 19-round launcher is also available.

SBAT-70 rockets are produced in two main versions: the SBAT-70 M1 non-spinning type generally used with high-speed aircraft: and the SBAT-70 M2 spinning model for low-speed aircraft and helicopters, but also providing greater accuracy and longer stand-off ranges for fighter-bombers etc. The current range of warheads for these rockets is as follows:

Warhead	Type
AVC-70/AC	HEAT anti-tank
AVC-70/AP	HE fragmentation anti-personnel
AVC-70/AC/AP	Combined anti-armour/ anti-personnel
AVC-70/F	Flechette anti-personnel/ anti-materiel
AVC-70/FB	Smoke WP (white phosphorus)
AVC-70/EF	Smoke practice
AVC-70/E	Inert practice (dummy)

The double base motor weighs 4.98 kg, has a thrust of 470 kg, a burn time of 1.2 seconds and uses BD-108 propellant: the alternative composite motor weighs 5.15 kg, has a thrust of 525 kg, a burn time of 1.2 seconds and uses CV-7 propellant.

127 mm SBAT-127 rocket

In the same class as the US-built five inch (127 mm) HVAR and Zuni Mk 16, this air-to-ground rocket has a 20 kg warhead and cruciform fins at the tail. It may be used with two or four tube rocket launchers.

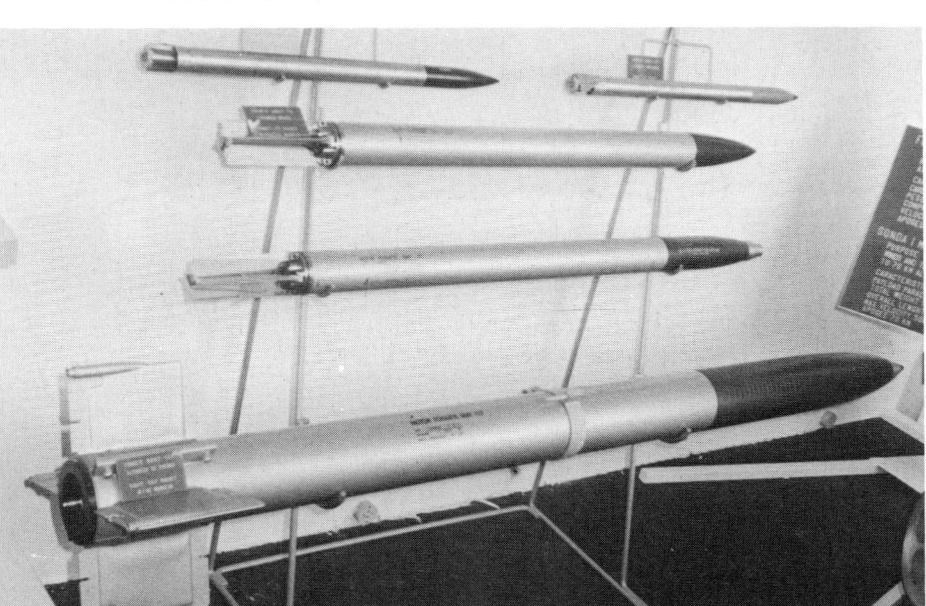

Avibrás aircraft rockets (top to bottom) SBAT-37, SBAT-70, SBAT-127 (Ronaldo S Olive)

Avibrás 500 lb general-purpose bomb (Ronaldo S Olive)

Avibrás LM-70/7 multiple launcher on Brazilian Navy Wasp helicopter (Ronaldo S Olive)

Helicopter armament system

This consists of a flexible mounting for a 7.62 mm FN MAG machine gun and rack for an LM-70/7 launcher for seven SBAT-70 70 mm FFARs. In Brazilian use this system has been fitted to Air Force Bell UH-1D/Hs and to Navy Helibrás Esquilos (locally-built Aérospatiale Écureil), Westland Wasps and Westland Lynx. It has been exported to several countries.

7.62 mm machine gun pods

Pods for two 7.62 mm machine guns for aircraft use are operational and in production for local use and export.

Aircraft bombs

Five types of aircraft bomb are known to be in production in Brazil for its air forces and for export:

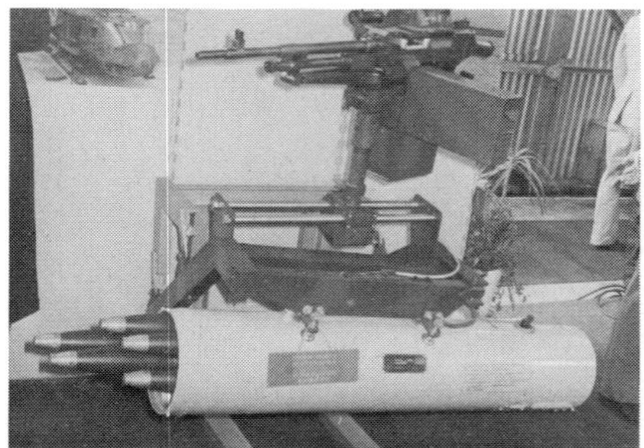

Avibrás helicopter armament system combining 7.62 mm machine gun mounting and 7-round rocket launcher pod (Ronaldo S Olive)

Avibrás LM-70/19 multiple launcher on Brazilian Air Force F-5E (Ronaldo S Olive)

Avibrás twin machine-gun pod for 7.62 mm machine guns (Ronaldo S Olive)

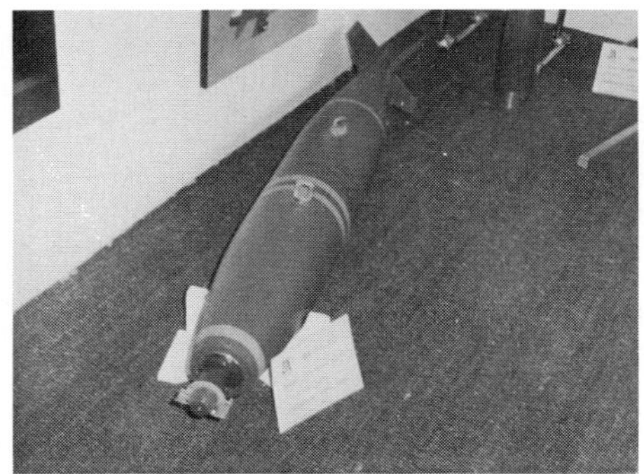

AV-BAFG-120 250 lb low-drag general-purpose bomb (Ronaldo S Olive)

CHARACTERISTICS

AV-BAFG-120
Length: 1.8 m
Body diameter: 230 mm
Total weight: 120 kg (equivalent to US Mk 81)

AV-BAFG-250
Length: 2.2 m
Body diameter: 270 mm
Total weight: 250 kg (equivalent to US Mk 82)

AV-BI-200
Length: 1.81 m
Body diameter: 420 mm
Total weight: 177 kg (equivalent to US 400 lb fire bomb)

AV-BI-250
Length: 3.46 m
Body diameter: 400 mm
Total weight: 271 kg (equivalent to US BLU-32)

AV-BI-375
Length: 3.63 m
Body diameter: 480 mm
Total weight: 362 kg (equivalent to US BLU-27)

CONTRACTOR

Avibrás – Indústria Aeroespacial SA, Caixa Postal 229, 12200 São José dos Campos, SP, Brazil.

CANADA

4539.303

CRV7 AIRCRAFT ROCKET

Designed and developed by the Canadian Department of National Defence, the CRV7 2.75 inch (70 mm) wrap-around fin aircraft rocket operated by Canadian Forces aircraft is claimed to possess a number of advantages compared with similar rockets such as the Mk 4/40 series. These include: higher velocity, giving a flatter trajectory and three times the impact energy; greater stand-off range; more precise control of rocket motor ignition timing, giving reduced dispersion and enhanced accuracy; and improved aerodynamic stability through decreased deployment time for deployment of the wrap-around fins.

The CRV7 rocket weapon system has two motors: the RLU-5001/B (C14) and RLU-5002/B (C15). The C14 consists of a case-bonded high energy composite propellant which uses a hydroxyl-terminated polybutadiene (HTPB) binder with an 88 per cent solids content of ammonium perchlorate, aluminium and ferric oxide. This is primarily for fixed-wing use. The C15 motor is the same as the C14 with the exception that the aluminium powder is omitted. Consequently the production of primary smoke has been virtually eliminated. This is intended primarily for helicopter use. Weight of the assembled motor is 14½ pounds (6.6 kg). The propellant for both motors is an internal burning cylinder 83.8 cm long with a bore diameter of 20.5 mm.

The fibreglass/phenolic moulded nozzle has metallic inserts to provide strength, electrical contacts and a shear pin release system. Integrally moulded vanes induce spin prior to exit from the launcher. Aluminium wrap-around fins are fully deployed within 36 cm of exit from the launcher and further stabilise the rocket in flight.

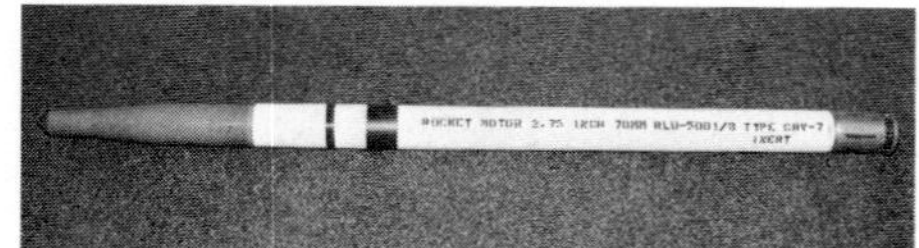

Complete CRV7 rocket fitted with practice warhead

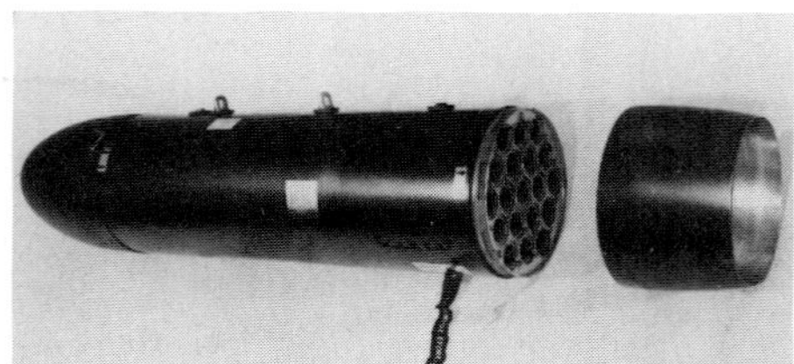

BAL 19 19-tube re-usable training/operational launcher

Operationally the CRV7 is fired from the 19-tube LAU-5003 A/A launcher. In training roles the CRV7 is fired from the six-tube LAU-5002 A/A. The SUU-20 bomb/rocket dispenser has been modified so that it is compatible with the CRV7.

The launchers are also compatible with the standard 14-inch (35.6 cm) suspension systems, use standard intervalometers and the standard five-pin electrical connections all of which conform to STANAG 3763AA.

CHARACTERISTICS

Calibre: 70 mm
Length: 104.2 cm (without warhead)
Rocket weight: 6.6 kg (without warhead)
Action time: 2.2 s
Max thrust: 7 kN
Spin: 40 cycles/s (with 4.5 kg warhead)
Max velocity: 1250 m/s (a/c speed 540 KTAS, 4.5 kg warhead); 1500 m/s (a/c speed 540 KTAS, 3 kg warhead)

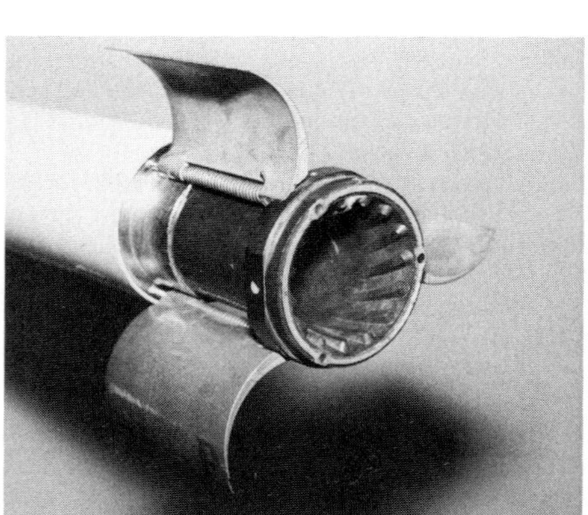

Deployed wrap-around tail fins of CRV7 aircraft rocket and integrally moulded vanes in motor exit cone to induce spin

Loading CRV7 rockets into LAU 5002A/A launcher on BO-105 helicopter

Effective range: Air-to-ground 6500 m; ground-to-ground 18 km

Launchers
LAU 5002A/A 6-tube re-usable training/operational launcher
LAU 5003A/A 19-tube disposable combat launcher
SUU 20 4-tube training dispenser
BAL 19 19-tube re-usable training/operational launcher
Warhead weights: 3, 4.5, 7 kg

A practice warhead, WTU 5001/B, has been developed specially for spinning 2.75 inch rockets. The design eliminates dynamic instability problems. It weighs about 4.25 kg and is also able to penetrate light and medium armour.

DEVELOPMENT
Development and flight proving was carried out by the Canadian Department of National Defence. Primary research was performed by the Valcartier Defence Research Establishment in association with Bristol Aerospace Ltd.

STATUS
The Canadian Forces have chosen the CRV7 as a replacement for earlier versions of the 2.75 inch rocket. In full production.

CONTRACTOR
Bristol Aerospace Limited, PO Box 874, Winnipeg, Manitoba R3C 2S4, Canada.

CHILE

6113.303
FAMAE 3.5 kg ANTI-PERSONNEL BOMB

This anti-personnel bomb relies upon its steel body fragmentation effects for its destructive effects although there are some steel pellets packed along the centre of the bomb body which contains 0.4 kg of TNT. The bomb has two safety devices. The first is a pin which is removed when the bomb is loaded onto an aircraft or helicopter, and the second is another pin which is removed as the bomb is released. A nose-mounted propellor then arms the fuze for operation after 3 or 5 seconds and the bomb detonates on contact.

CHARACTERISTICS
Total weight: 3.5 kg
Weight of TNT charge: 0.4 kg
Length overall: 429 mm
Length of body: 229 mm
Length of tail: 145 mm

STATUS
In production for Chilean Air Force.

CONTRACTOR
FAMAE Fabricaciones Militares, Avenida Pedro Montt 1568/1606, Santiago, Chile.

6173.303
FAMAE GENERAL PURPOSE LOW-DRAG BOMBS

FAMAE produce four sizes of these general purpose low drag bombs. The sizes are 250 lb, 500 lb, 750 lb and 1500 lb. All are intended to be carried externally on high speed aircraft and all use nose and/or tail impact fuzes. For increased target penetration the nose fuzes may be plugged. All four have cruciform fins.

STATUS
In production.

CONTRACTOR
FAMAE Fabricaciones Militares, Avenida Pedro Montt 1568/1606, Santiago, Chile.

4426.303
PJ-1 MANUALLY OPERATED BOMB

The PJ-1 is a manually dropped bomb which can be used from almost any type of aircraft or helicopter, including civil aircraft in an emergency. It measures 500 mm long by 85 mm in diameter and weighs 3 kg including 800 grammes of high explosive. The main explosive charge is surrounded by 110 steel fragments which are projected to an effective operating range of 35 metres.

The bomb is of simple construction and safety is ensured through safety locks which do not allow the fuze to activate until three seconds after leaving the aircraft. The head may be copper covered to allow perforation of armoured vehicles or installations, or alternatively fitted with a hollow charge device to give considerable anti-armour effect.

CONTRACTOR
Industrias Cardoen S.A., Los Conquistadores 1700/Piso 28°, Santiago, Chile.

Cardoen PJ-1 bomb with storage and transportation container

4598.303
CB 130 & CB 500 CLUSTER BOMBS

Promising experience with the earlier Cardoen 100 lb and 500 lb cluster bombs (**4427.303** and **4487.303** in the *1983-84* edition) led the company to embark on development of improved models, the CB 130, CB 500 and CB 500-K which are 130 lb, 500 lb and 1000 lb cluster bombs respectively. All feature bomblets of a new indigenous design and a new programmable fuze designed by Cardoen.

Using these bombs, low or high altitude attacks can be made at high aircraft speeds against targets dispersed over wide areas without the need for extreme delivery accuracy. After release, the bombs are opened in mid-air by an electronically programmed fuze, and a number of armour-piercing bomblets are dispensed; 50 from the CB 130, 240 from the CB 500 and 450 from the CB 500-K. The surface area covered by the bomblets has an elliptical shape of between 15 000 and 90 000 sq m, depending on the size and operational use of the bomb. Release height also has a considerable effect. The altitude at which the bomb opens after release is controlled by the programmable electronic fuze which provides a variety of opening delay times.

The Cardoen bomblet is intended for use against tanks and armoured vehicles, troop concentrations etc. It is stated armour plate of more than 203 mm can be penetrated. Used against troop concentrations, shrapnel spreads over an area of up to 50 000 square metres. The complete cluster bomb is intended to withstand sustained forces of 6 *g* encountered by combat manoeuvres of a high speed attack aircraft. The CB 130 and CB 500 are fitted with lugs for the standard NATO 14 inch, Warsaw Pact 250 mm spaced suspensions and the single point bomb lug.

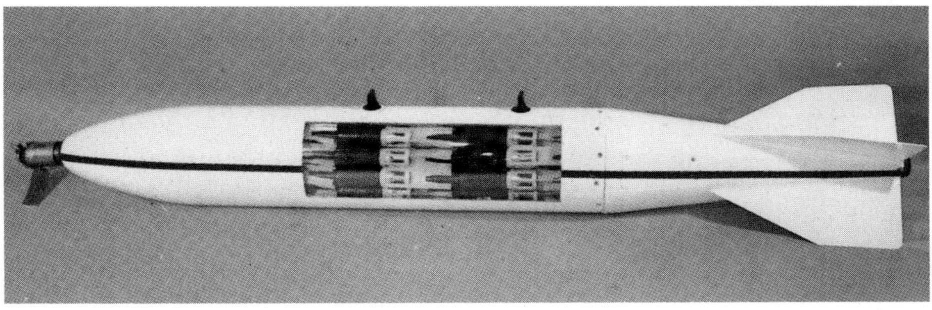

CB 130 cluster bomb with cutaway skin

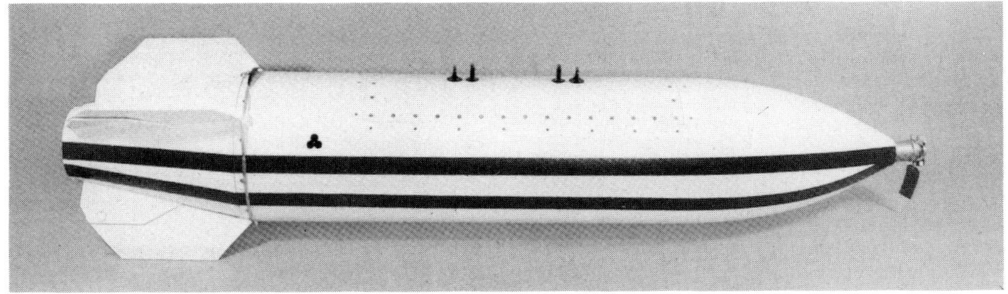

CB 500 cluster bomb

CHARACTERISTICS

Type	CB 500	CB 130	CB 500-K
Length	2.64 m	2.05 m	—
Diameter	446 mm	253 mm	—
Fin span	650 mm (closed) 1140 mm (open)	— 400 mm (open)	—
Weight	245 kg	60 kg	454 kg

Fuze: Nose mounted, adjustable delay 3 – 18 s
Bomblet (for all versions)
Weight: 740 g
Length: 360 mm
Diameter: 48 mm
Tail diameter: 48 mm

STATUS
Production.

CONTRACTOR
Industrias Cardoen S.A., Los Conquistadores 1700/Piso 28°, Santiago, Chile.

4488.303

CARDOEN GENERAL-PURPOSE BOMBS

Cardoen manufactures a standard range of general-purpose aircraft bombs suitable for use with bomber or fighter aircraft. Various sizes are produced, standard items being the 250 lb (113 kg) Mk 81, 500 lb (227 kg) Mk 82, and the 1000 lb (454 kg) Mk 83. All are constructed of forged seamless steel bodies with a tempered steel nose section and cruciform tail assembly for flight stabilisation. Standard explosive fillings are employed and mechanical time fuzes are fitted normally. Bombs are provided with NATO standard suspension lugs at 14-inch (35.6 cm) centres.

Parachute or metallic tail retarders may be fitted in place of the normal tail fins.

STATUS

In production.

CONTRACTOR

Industrias Cardoen S.A., Los Conquistadores 1700/Piso 28°, Santiago, Chile.

Cardoen GP bombs are produced in three standard sizes: 250 lb, 500 lb and 1000 lb

6174.303

FERRIMAR WB-500-F AVISPA CLUSTER BOMB SYSTEM

The FERRIMAR WB-500-F Avispa (Wasp) cluster bomb system is produced in three sizes, 500 lb, 1500 lb and 2250 lb. The bomb bodies are produced using aluminium alloys, fibreglass, carbon and plastic materials and can carry between 100 and 248 Avispa bomblets that rely partially upon the bomb body rotation for lateral dispersion. The bomblets vary in warhead diameter between 48 mm and 53 mm and the developed range includes eight types, all electronically fuzed: HEAT, HEAT-AP.30, HEAT-AP.50, HEAD (high explosive area denial), HEAD-AP.30, HEAD-AP.50, AP-Prox.30 and AP-Prox.50. Some of these bomblets may not enter full production. A system of larger bomblets for use against submarines is under development.

STATUS

Only Avispa bomblets in production.

CONTRACTOR

Ferrimar, c/o L. Thayer Ojeda 113, PO Box 16789, Santiago-9, Chile.

FERRIMAR Avispa bomblets

FERRIMAR WB-500-F Avispa cluster bomb units (T J Gander)

CHINA (PEOPLE'S REPUBLIC)

4437.303

57 mm AIRCRAFT ROCKET

Among the range of aircraft armament indigenously produced in the People's Republic of China, the 57 mm aircraft rocket is employed on Chinese-built F6 fighter/ground attack aircraft for use against either aircraft targets or surface targets such as naval craft or battlefield strong points, tanks, personnel concentrations, fuel dumps etc. On the F6 aircraft the 57 mm rockets are carried in underwing pods, each of which contains eight rounds. These may be fired (electrically) separately or in salvo. The rocket is stabilised in flight by a set of folding tail-fins.

CHARACTERISTICS

Calibre: 57 mm

Length: 915 mm (overall)

Span: 230 mm

Weight: 3.992 kg (complete)

Warhead weight: 1.372 kg (HE)

Operating temperature range: +50 to −60°C

STATUS

Production.

AERO-AMMUNITION 30-1

The F6 fighter aircraft mentioned above has as its normal armament three 30 mm cannon and weapons of this calibre are found on other aircraft of the Republic's air forces. A range of ammunition is manufactured for use in 30 mm cannon against air and surface targets; standard types include: high explosive incendiary tracer (HEIT), high explosive incendiary (HEI), armour piercing explosive (APE), and practice. All types are supplied with cartridge cases of 70/30 brass.

CHARACTERISTICS

Type	HEIT	HEI	APE
Calibre	30 mm	30 mm	30 mm
Weight of round	840 g	840 g	828 g
Total length	266 mm	266 mm	266 mm
Projectile weight	418 g	418 g	407 g
MV	795 m/s	795 m/s	800 m/s
Pmax (kg/cm^2)	3050	3050	3050
Penetration	—	—	10 mm*

*Penetration means the depth of penetration of the shell when encountering a uniform steel target at a distance of 50 to 75 m with a velocity of 360 m/s and an angular incidence of 30°.

STATUS

Production.

CONTRACTOR

China North Industries Corporation, PO Box 30X, 2137 Beijing, People's Republic of China.

6175.303
CHINESE AIRCRAFT CANNON

Currently on offer for export are two Chinese-produced 23 mm automatic cannons, the Type 1 and Type 2. Both are intended for fixed or flexible mountings.

The 23 mm Type 1 has a rate of fire between 800 and 950 rounds per minute. It fires Type 1 HEI or HEI-T rounds with a muzzle velocity of 680 m/s. Weight is 39 kg and the recoil forces are stated to be up to 2600 kg.

The 23 mm Type 2 is intended primarily for fixed mountings and fires Type 2 API or HEI rounds at a muzzle velocity of 705 m/s. Rate of fire is approximately 1200 rpm. The Type 2H is the same weapon fitted with a long perforated muzzle brake. The Type 2 weighs 43 kg and the Type 2H 47 kg.

A 30 mm Type 1 aircraft cannon is known to exist. It fires a 0.418 kg HEI-T projectile at a muzzle velocity of 795 m/s. Its effective range is stated to be 800 m.

STATUS

The above cannon are in production.

CONTRACTOR

NORINCO, China North Industries Corporation, 7A Yue Tan Nan Jie, PO Box 2137 Beijing, Beijing, China.

Type 2 23 mm aircraft cannon

6176.303
100 kg AIRCRAFT BOMBS

The 100 kg high explosive general purpose bomb is the Type 2 which has a payload of 39.8 kg of TNT. It is fitted with the Hang-Yin-1 tail or nose-mounted propellor-armed fuze and uses drum fins. It is fitted with a single suspension lug and has a total weight of 140 kg. Body diameter is 280 mm and total length (without fuze) is 1080 mm.

A generally similar marker bomb is designated the Type 1 and is filled with a red, green or white pyrotechnic mixture that burns for up to 10 minutes. Weight of the Type 1 is 80 kg and it is fitted with a nose-mounted Hang-Yin-4B fuze.

STATUS

In production and offered for export.

CONTRACTOR

NORINCO, China North Industries Corporation, 7A Yue Tan Nan Jie, PO Box 2137 Beijing, Beijing, China.

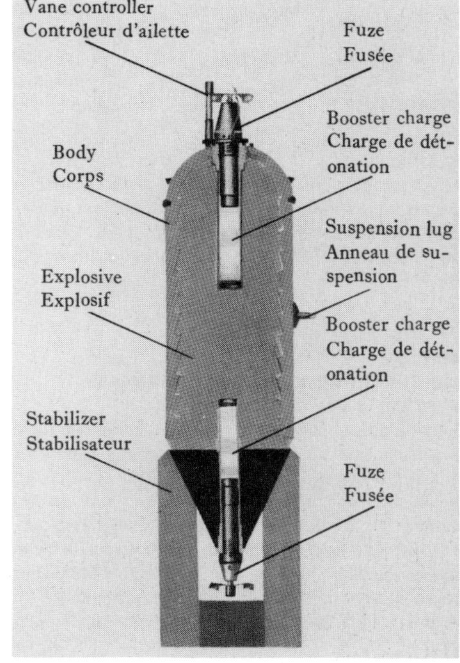

Cutaway drawing of 100 kg Type 2 general purpose bomb

FRANCE

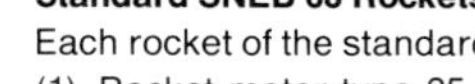

4833.303
SNEB 68 AIRCRAFT ROCKETS

Thomson Brandt Armements (TBA) produces a variety of 68 mm aircraft rockets to suit a number of operational requirements, and suitable for use with different classes of combat aircraft.

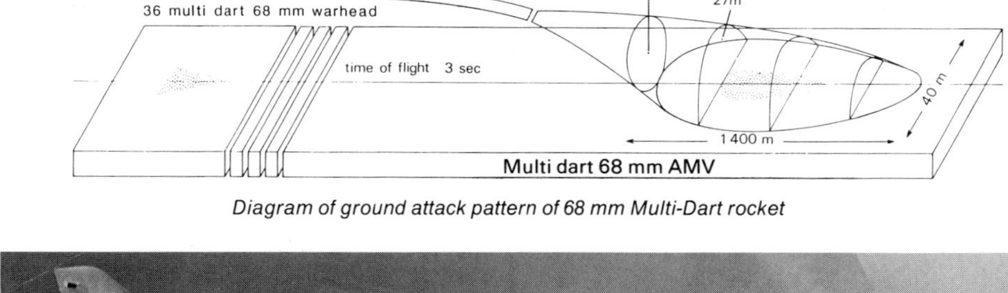

Diagram of ground attack pattern of 68 mm Multi-Dart rocket

Ripple of 68 mm rockets fired from Mirage 2000

Standard SNEB 68 Rockets

Each rocket of the standard SNEB series includes:

(1) Rocket motor type 25 F1B for fixed wing aircraft and type 25 H for helicopters

(2) An explosive warhead designed for a direct hit and whose effectiveness results from impact with target.

Rocket launchers used are:

Fixed wing aircraft	**Helicopters**
Matra F1 – 36 rockets	TBA 68-12 – 12 rockets
Matra F2 – 6 rockets	TBA 68-22 – 22 rockets
Matra F4 – 18 rockets	

Multi-dart 68 Weapon system for Fixed Wing Aircraft

This anti-armour area attack system is designed for low level, high speed, straight attack. Rockets fitted with multi-dart warheads are fired in ripple mode with a stand-off range from 1600 to 4000 m. The effectiveness of the system depends on the ejection of the large number of high velocity sub-projectiles (or darts) organised in a series of tight sheaves, horizontally scattered over an extended ground pattern.

The darts are ejected from the rocket warhead following the operation of a variable setting electronic time fuze.

This weapon system uses the current Matra 155 18-tube rocket launcher.

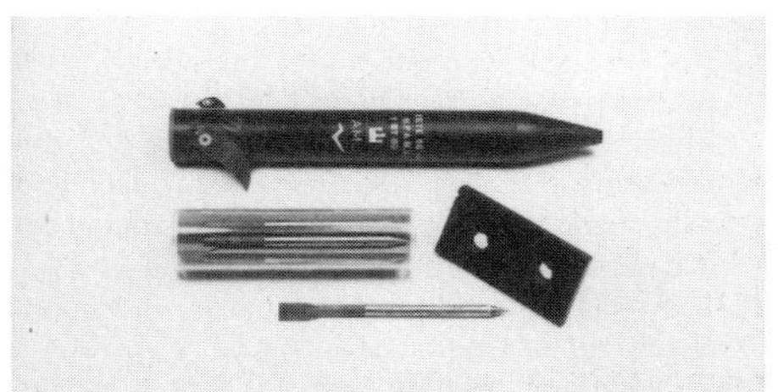

68 mm AMV warhead with cutaway showing anti-armour darts and typical effect on steel plate

CHARACTERISTICS

Warhead	Type	Effect	Rocket motor	Fired from
		Standard SNEB 68 Rockets		
HEAT type 23	Anti-armour hollow charge	Armour-piercing >400 mm	25 F1 B	FW A/C
HE type 26P	General purpose fragmentation	440 main fragments	25 F1 B or 25 H	FW A/C and helicopters
29 AMV	Multi-projectile	Anti-equipment and vehicle areas	25 H	Helicopters
29 E	Illuminating	1 000 000 candelas 55 seconds	25 F1 B or 25 H	FW A/C and helicopters
22-4	Practice target marker	Same ballistics as HE type 26P and 29E	25 F1 B or 25 H	FW A/C and helicopters
22-5	Practice target marker	Same ballistics as HEAT type 23	25 F1 B	FW A/C
29 X FUM	Practice target marker	Same ballistics as 29 AMV	25 H	Helicopters
		Multi-dart 68 Weapon System		
68 ABL	Multi-projectile	Anti-armour area weapon	F 2	FW A/C
68 AMV	Multi-projectile	Anti-equipment and vehicle area weapon	F 2	FW A/C and helicopters
68 MX FUM	Practice target marker	Same ballistics as 68 ABL and 68 AMV	F 2	FW A/C and helicopters

STATUS

Mass production. Since 1955 at least 3 million explosive charges have been made for first generation rockets. Second generation rocket development continued from 1977-1979, and qualification from 1982 to 1984.

CONTRACTOR

Thomson Brandt Armements, 204, Rond-Point du Pont de Sèvres, 92516 Boulogne-Billancourt Cedex, France.

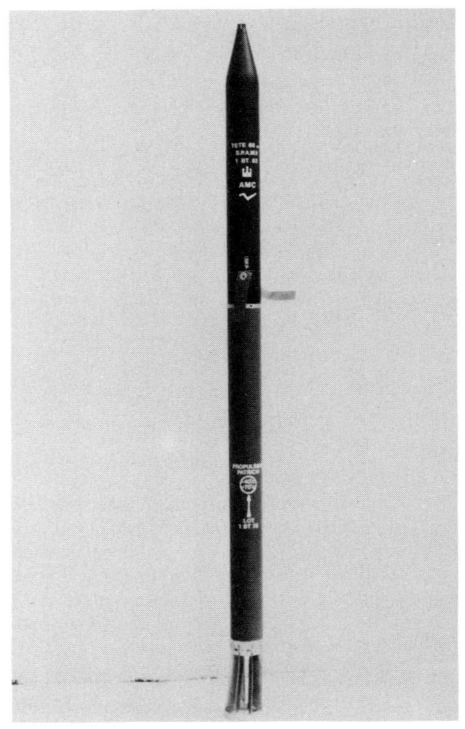

68 mm AMV Multi-Dart rocket with F2 motor

4834.303

MULTI-DART HELICOPTER ANTI-ARMOUR SYSTEM

Recent advances in fire control techniques have made it possible to develop a new weapon system for attack helicopters such as the Eurocopter HAP, AH-64, UH-60, Cobra, Augusta 129, and after modification, the Puma, Panther, Gazelle, BO 105 etc. The system employs 68 mm aircraft rockets with multi-projectile warheads and is called the Multi-Dart rocket system by Thomson Brandt Armements (TBA).

In addition to the fire control unit, the system consists of: a specific rocket launcher (68-12L or 68-22L) and a multi-projectile munition. The munition is made of two parts: (a) an AMV multi-projectile armour-piercing warhead using kinetic energy to acquire its piercing power, and (b) an improved F2 type 68 mm rocket motor.

Each tube of the rocket launcher is fitted with an electronic remote control which sends an electromagnetic impulse to the warhead fuze. The transmitted signal defines the trajectory time of flight before the burst of the military payload: dart sub-projectiles (other versions can use flares and decoys).

OPERATION

The warhead is controlled by a remotely settable time fuze. At a preset time after rocket motor ignition, the warhead ejects a sheaf of armour-piercing projectiles in the vicinity of the target. The programmed time of flight is initiated from the cockpit through a manual operator box or through the fire control system.

Considerable penetration is the result of the dart thinness (L/D ratio 10:1) and of the very high velocity provided by the new composite propellant 68 mm rocket motor. The darts have an impact velocity of 600 to 300 m/s:

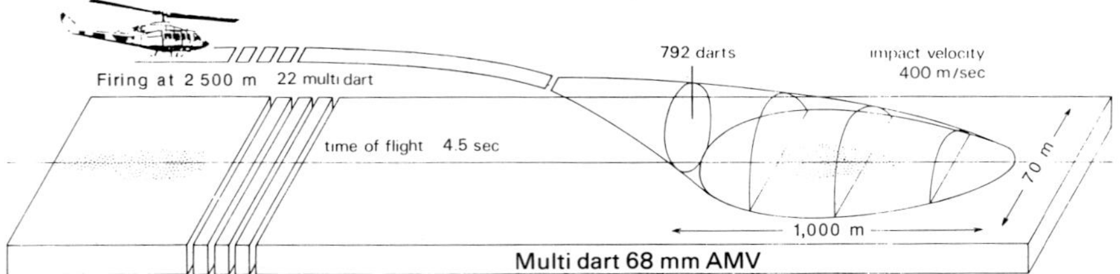

Diagram of Multi-Dart helicopter ground attack pattern

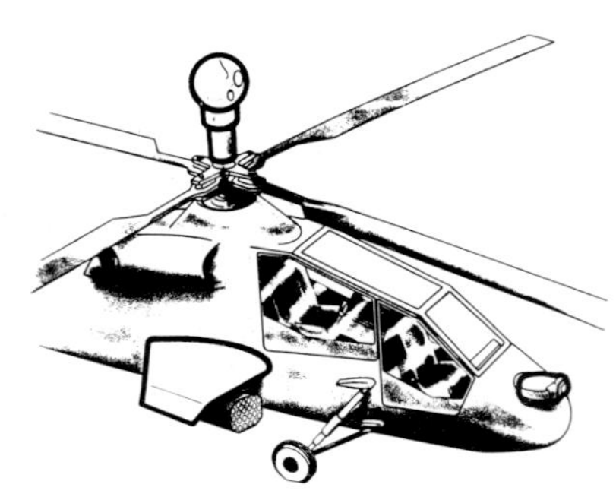

Impression of Multi-Dart helicopter anti-armour weapon system

AMV darts (anti-equipment and vehicle), diameter: 9 mm, weight 35 g, perforate 8 mm of steel plate.

The useful range varies from 1200 to 3000 m. This weapon is a current threat to armoured air defence vehicles and all armoured infantry fighting vehicles.

Munition characteristics are essentially the same as those given in the table in **4833.303**, above, apart from a reduction in maximum range of 2000 m for first generation rockets and 1000 m for second generation rockets when used in the helicopter application, and a slightly lower maximum impact velocity (600 m/s) in the latter role.

The new Multi-Dart 68 mm munition is fired in ripples, 12 to 18 rounds, from new TBA rocket launchers (compatible with the first generation rockets, SNEB 68 mm). Two versions of rocket launcher are available, according to the carrying capacity of helicopters:

Type 68.12 L: Capacity 12 munitions

Type 68.22 L: Capacity 22 munitions

STATUS

Mass production. 25 H powered rockets began development in 1977 and production commenced in 1979. Second generation development followed with mass production starting in 1987.

CONTRACTOR

Thomson Brandt Armements, 204, Rond-Point du Pont de Sèvres, 92516 Boulogne-Billancourt Cedex, France.

4835.303

TBA 100 mm AIRCRAFT ROCKETS

Thomson Brandt Armements (TBA) has a range of 100 mm aircraft rockets, similar to the company's 68 mm rockets (**4833.303**). Two generations of rockets are currently available:

(1) The explosive warhead rocket, the effectiveness of which results from impact on the target and which is designed for a direct hit: (ECC) the anti-armour rocket; (DEM) the demolition and anti-shelter rocket; and (EEG) the anti-equipment and personnel rocket.

(2) The multi-projectile warhead rocket, multi-dart, with an armour-piercing capability.

The latter is an area weapon. It is fired in ripples with a stand off range from 2000 m to 4000 m, at a very low altitude. The effectiveness of the multi-dart warhead results from the ejection of a high number of sub-projectiles, long darts with an L/D ratio of 15:1. These armour-piercing darts, organised in series of tight sheaves, horizontally scatter over large areas and strike all vehicles they encounter with a high velocity.

The 100 mm multi-dart warhead exists in two main versions:

(a) 36 darts ABL (anti-light armour), armour-piercing power: 15 mm of steel plate

(b) 192 darts AMV (anti-equipment and vehicles), armour-piercing power: 8 mm of steel plate

TBA 100-4 rocket launcher pods on Jaguar aircraft

Even when fired on a single pass this area weapon does not require any stabilised aiming.

This universal munition, which can be adapted to every combat aircraft, is used with TBA 100.4 or Matra C6 rocket launchers.

STATUS

First generation rounds were developed between 1968 and 1969, and have been produced since 1974. The second generation development lasted from 1979 to 1984, and production began in 1985.

CONTRACTOR

Thomson Brandt Armements, 204, Rond-Point Du Pont de Sèvres, 92516 Boulogne-Billancourt Cedex, France.

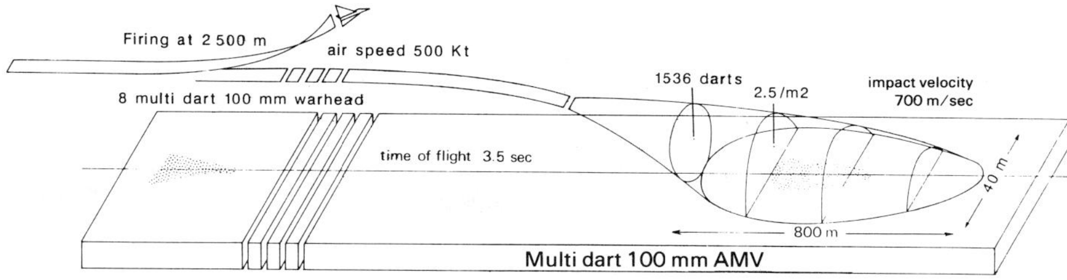

Diagram of 100mm Multi-Dart ground attack

CHARACTERISTICS

Warhead	Type	Effect	Rocket motor
		TBA 100 Rockets	
ECCAP	High explosive anti-tank/ anti-personnel	Combining effect of armour-piercing >600 mm and fragments	F 1A
EEG	High explosive general purpose	Blast and fragments: instant or delayed fuzing	F 1A
DEM	Demolition	Hard target piercing: fuze delay 5 or 20 ms	F 1A
		Multi-dart 100 Weapon system	
100 ABL	Multi-projectile	Anti-light armour area weapon	F 1A
100 AMV	Multi-projectile	Anti-equipment and vehicle area weapon	F 1A
		Practice/Marker Rockets	
X FUM F3	Practice/target marker	Same ballistics as TBA 100 ECCAP, Multi-dart 100 ABL, Multi-dart 100 AMV	F 1A
X FUM F4	Practice/target marker	Same ballistics as TBA 100 EEG and TBA 100 DEM	F 1A

3151.303

TBA 100 mm ROCKET LAUNCHERS

In addition to supplying 100 mm rockets to the producers of a variety of rocket pods and launchers the Thomson Brandt Armements organisation also manufactures its own launcher pods. The Type 100-4 (F3) is intended for use with any of the Thomson Brandt 100 mm diameter aircraft rockets described in **4835.303**, above. This supersonic four-tube launcher is suitable for any 14-inch (356 mm) NATO standard store station. It has been selected by the French Air Force in which service it is known as the F3.

CHARACTERISTICS

Type 100-4 (F3)
Length: 290 cm
Width: 23 cm
Height: 24 cm
Weight: 70 kg (empty); 240 kg (loaded)
Rate of fire: 400 rounds/minute
Number of tubes: 4

STATUS

Production.

CONTRACTOR

Thomson Brandt Armements, 204, Rond-Point du Pont de Sèvres, 92516 Boulogne-Billancourt Cedex, France.

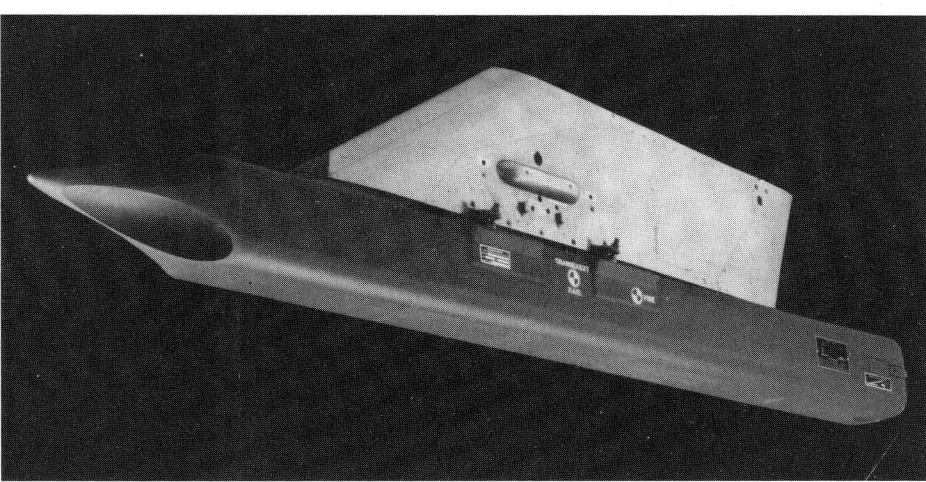

TBA 100-4(F3) rocket launcher

3867.303

HELICOPTER GUN MOUNTING 19A

The 19A mounting is designed for use in SA 330 Puma helicopters to carry a 20 mm automatic gun Type 621 which is located in the right-hand fuselage door of the helicopter, enabling the machine to be used for attacking ground troops, armed reconnaissance, and similar air support actions. The Type 621 automatic gun is equipped with the left-hand flat feed mechanism, and a muzzle brake with two vent ports. The ammunition supply consists of four boxes of 240 rounds and the gun is fitted with an SFOM optical sight. Firing is under the pilot's overriding authority and an ADS CPL8 control box is provided which enables firing in a free burst, or in eight-round limited bursts at a rate of 740 rounds/minute. There is compensation for aerodynamic forces acting on the gun by means of an adjustable device which is electrically and mechanically controlled. There is electro-mechanical locking of the fork after traverse setting.

CHARACTERISTICS

Calibre: 20 mm
Elevation: +5 to −40°, relative to aircraft floor
Traverse: ±45° from axis of aircraft R/H door port
Ammunition: 4 × 240 rounds
Weight of mounting with gun: 165 kg
Overall weight: 632 kg (with 960 rounds)

STATUS

Series production. Operational with the French Army.

Type 621 20 mm automatic gun mounted on helicopter

CONTRACTOR

GIAT (Groupement Industriel des Armements Terrestres), 10 place Georges Clémenceau, 92211 Saint-Cloud, France.

1271.303

30 mm AIRCRAFT GUN DEFA 552A

The 30 mm DEFA gun is the French development of the German Mauser revolver gun principle and is similar to, and virtually interchangeable with, the British ADEN Mk 4. Over 10 000 units have been manufactured and the gun has been proved in combat in a number of theatres. Since 1971 a long-life version has been developed, called Canon Automatique de 30 mm, 550-F2A (DEFA 552A) which replaces the gun type 552 and with which it is directly interchangeable.

The principal characteristics of the DEFA 552 and the DEFA 552A are light weight, a rate of fire of 1300 rounds per minute, and provision for automatic recocking in the event of a stoppage. The gun uses 30 mm ammunition developed for the DEFA/ADEN weapons which is available in the usual variety of types: armour piercing, HE incendiary, etc. Belted ammunition may be fed to either side of the gun. Spent cases are ejected from the same side of the weapon as accepts the feed. Links are ejected on the opposite side. The ammunition is electrically fired.

The DEFA 552 revolver gun incorporates safety devices to ensure that firing is impossible unless the cartridge is in the firing position and the control slide, which carries the round firing contact, is in the forward position. In operation the whole gun recoils about 12 mm. On the ground a special device is used to cock the gun by hand. A single pyrotechnic cartridge is provided for automatic recocking during firing if a stoppage should occur.

CHARACTERISTICS

Calibre: 30 mm

Gun weight: 81 kg

Rate of fire: 1300 rounds/minute

Chrome barrel life: 5000 rounds

STATUS

The DEFA 552 and DEFA 552A cannon are in service with the French Air Force, the Israeli Air Force and others throughout the world. Two 552s are standard equipment on all Mirage III and Mirage Vs, and other aircraft types fitted include the Jaguar, Etendard IV M, Fiat G91 and Super-Etendard.

CONTRACTOR

GIAT (Groupement Industriel des Armements Terrestres), 10 place Georges Clémenceau, 92211 Saint-Cloud, France.

1273.303

30 mm AIRCRAFT GUN DEFA 553

The DEFA 553 30 mm aircraft cannon is a direct development of the DEFA 552. Its performance is similar to the 552 but design modifications have been made to improve service life and facilitate installation. A new barrel in nitro-chrome steel is used with a muzzle device to reduce muzzle pressure. Other changes have been made to simplify installation and the DEFA 553 weapon will now accept ammunition feed from either side without modification.

Other features, including in-flight recocking, correspond to the DEFA 552 model, and the same ammunition is used.

The electric control for the DEFA 553 provides the following facilities:

'OFF' – firing of the gun completely inhibited.
'Continuous fire' – not normally used in action.
'0.5 s' – maximum continuous burst length 0.5 s.
'1 s' – maximum continuous burst length 1 s.

In twin installations a unit is needed for each gun and an additional control and junction box providing:

Position 1 – both guns inhibited.
Position 2 – continuous fire, both guns.
Position 3 – both guns in burst lengths (which may be different) set into each individual control unit.
Position 4 – gun camera switched on.

CHARACTERISTICS

Calibre: 30 mm

Gun weight: 81 kg

Rate of fire: 1300 rounds/minute

Chrome barrel life: 5000 rounds

STATUS

The DEFA 553 cannon is in production and is fitted in Mirage F1, Jaguar, Alpha-Jet, Macchi 339K, Casa 101, and Pucara aircraft.

CONTRACTOR

GIAT (Groupement Industriel des Armements Terrestres), 10 place Georges Clémenceau, 92211 Saint-Cloud, France.

5246.303

30 mm AIRCRAFT GUN DEFA 554

The new DEFA 554 lightweight aircraft gun is a further development of the DEFA 553 (**1273.303**) gun and has been designed for installation on high performance aircraft. It has been selected for the Mirage 2000 fighter and each of these will have two fuselage mounted guns. The gun, which can also be pod mounted, weighs only 85 kg and has two rates of fire, 1800 rounds a minute for use in the air-to-air role and 1100 rounds a minute for the air-to-ground role. Muzzle velocity is about 820 m/s.

CHARACTERISTICS

Calibre: 30 mm

Gun weight: 85 kg (approx)

Rate of fire: 1100/1800 rounds/minute

STATUS

In production for French Air Force.

CONTRACTOR

Groupement Industriel des Armements Terrestres (GIAT), 10 place Georges Clémenceau, 92211 Saint-Cloud, France.

30 mm aircraft gun DEFA 554

1275.303

20 mm M621 CANNON POD

The 20 mm M621 cannon pod has been designed for use on light fixed-wing aircraft and helicopters. The moulded structure of the pod was computer-designed to reduce drag and uses the NATO 14-inch mounting system for attachment to the aircraft by lugs or saddles. There is provision for emergency jettisoning.

On the ground a manually-operated interlock on the trigger mechanism of the weapon causes the breechblock to lock and thus avoid any possibility of accidental firing. In flight a safety device cuts out the main power supply, and hence the firing mechanism, and causes locking of the breechblock extending the recocking cylinder. The electro-hydraulic device makes recocking possible both on the ground and in the air and can perform two recocking operations in less than one second. The number of recocking operations during a mission is unlimited. Reloading takes less than two minutes.

CHARACTERISTICS

Calibre: 20 mm

Weight empty: 93 kg

Weight loaded with 180 rounds: 175 kg

Length: 3.516 m

Diameter: 400 mm

STATUS

In mass production.

20 mm M621 cannon pod

CONTRACTOR

Groupement Industriel des Armements Terrestres (GIAT), 10 place Georges Clémenceau, 92211 Saint-Cloud, France.

3865.303

BM 400 MULTI-MUNITION BOMB

The BM 400 multi-munition bomb is designed as a low altitude air-to-ground area saturation weapon for the destruction of targets such as armoured vehicles, road or rail convoys, ammunition dumps, radars, missile launchers and similar high value equipments. This is achieved by successively ejecting three modules from the bomb body after its release, at preset times. Each module consists of a pre-fragmented warhead. The tail parachute, attached to each bomb module, slows it and ensures that its final descent onto the target is almost vertical.

On impact with the ground, each anti-armour module ejects a horizontal sheet of high velocity calibrated fragments. Two fragment generators are available comprising respectively about:

(1) 800 main fragments, able to perforate 17 mm of steel at 50 metres from the explosion point (12 mm at 100 metres)

(2) 1500 main fragments, able to perforate 12 mm at 50 metres from the explosion point (7 mm at 100 metres).

An area of 600 metres long by 100 metres wide (or more depending on the perforation power and fragment density needed) can be covered using only two 400-kg modular bombs with suitable module spacing. Launching can be effected either by releasing the bomb(s) from an aircraft flying over the target area, or by the "toss-bombing" method. The stand-off range of the bomb varies from 500 metres to more than 5 km. A rocket assisted BM 400, currently under development, will achieve a stand-off range of 10 km.

Two new sub-munitions under development for use with the BM 400 (and other direct attack dispensers) are the SMAB armour-piercing sub-munition derived from the ABL warhead used with the BAT 120 anti-armour weapon system (**3863.303**), and the SMAP cratering sub-munition derived from the warhead used with the BAP 100 area denial and cratering weapon system (**3152.303**). The SMAB weighs 20 kg and has a body diameter of 113 mm. It will produce more than 500 calibrated fragments. The SMAP weighs 23 kg and the body diameter is 95 mm. It will be able to penetrate 300 mm of concrete and produce a 30 square metre area of damage. Both sub-munitions could be in mass production by 1989.

Mirage 2000 wing stations carry six BM 400 multi-munition bombs for ground attack and a pair of Magic air-to-air missiles

CHARACTERISTICS

Length: 3.2 m
Diameter: 0.32 m
Weight: 390 kg
Weight of single module: 100 kg
Number of modules: 3
Attack altitude: 30 m
Firing speed: 350 – 550 knots

STATUS

Production.

CONTRACTOR

Thomson Brandt Armements, 204, Rond-Point du Pont de Sèvres, 92516 Boulogne-Billancourt Cedex, France.

3864.303

RETARDED 125 kg CLOSE SUPPORT BOMB

The 125 kg retarded close support bomb developed by Thomson Brandt Armements has two main elements: an impact fuze, and a retarder package. The company designed an impact fuze/retarder package for use with standard 125 kg fragmentation bombs to enhance their tactical performance and versatility. The 125 kg model derives from the fragmentation bomb type 81A thick wall ammunition, designed to scatter numerous fragments and manufactured by SAMP for the French Air Force.

The complete package comprises two distinct parts: the fuze which is inserted into the bomb, and the retarder tail that is attached to the rear of the bomb body.

Retarder Unit. Specially designed by Thomson Brandt for the 125 kg bomb, the retarder tail unit appreciably improves the aerodynamic efficiency of the bomb. A near-vertical final trajectory is achieved by use of a parachute retarder, which is locked into the bomb body only after appropriate safety measures have been completed. The parachute opens after a preset delay, which is selected on the ground before take-off, according to mission requirements.

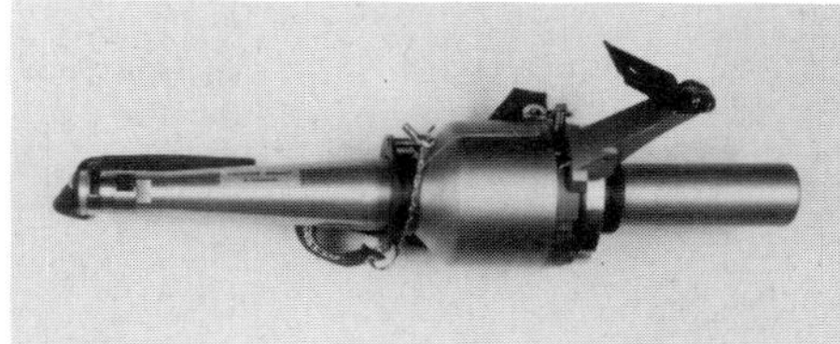

Thomson Brandt Armements FP 78 fuze

CHARACTERISTICS

Length: 770 mm (outside bomb body)
Span: 310 mm (packed); 640 mm (deployed)
Weight: 12 kg
Parachute delay: 13 – 17 s (preset on ground)

FP 78 Fuze. This is a high sensitivity super-rapid impact fuze, which can also be used with standard 250 kg bombs. It ensures that the bomb will explode at optimum height by means of a rod which protrudes from the nose of the weapon. Arming of the fuze is achieved at a safe distance by means of a faired air-driven motor and an accelerometer-based locking device.

Retarded 125 kg close support bomb under pod

CHARACTERISTICS

Length: 450 mm (350 mm clear of bomb body)
Weight: 2.6 kg
Safety arming delay: 7 s for 125 kg bomb at 500 knots

STATUS

The complete fuze/retarder package is in production.

CONTRACTOR

Thomson Brandt Armements, 204, Rond-Point du Pont de Sèvres, 92516 Boulogne-Billancourt Cedex, France.

4836.303

FEU 80 ELECTRIC MULTI-PURPOSE BOMB FUZE

Existing stocks of older free fall aircraft bombs are retained for use as conventional or retarded HE bombs, and to maximise their operational usefulness Thomson Brandt Armements (TBA) offers the FEU 80 fuze which was developed for new generation bombs but can be fitted to earlier munition stocks. Four versions are available:

FEU 80 BO: impact nose fuze
FEU 80 BC: impact tail fuze
FEU 80 PLRF: adjustable proximity fuze
PRL: ballistic bomb proximity fuze.

The FEU 80 fuze can fulfill by itself the missions previously requiring the use of two old generation fuzes (nose and tail). It meets all requirements and works in any climatic conditions.

All the functions are performed through advanced electro-mechanical or electronic circuits ensuring safety, reliability, adaptability and effectiveness.

Safety, both when stored and when mounted, is ensured through the total absence of power in the arming mechanism and firing circuit; the presence of visual-check mechanical locks; and immunity to electro-magnetic radiation.

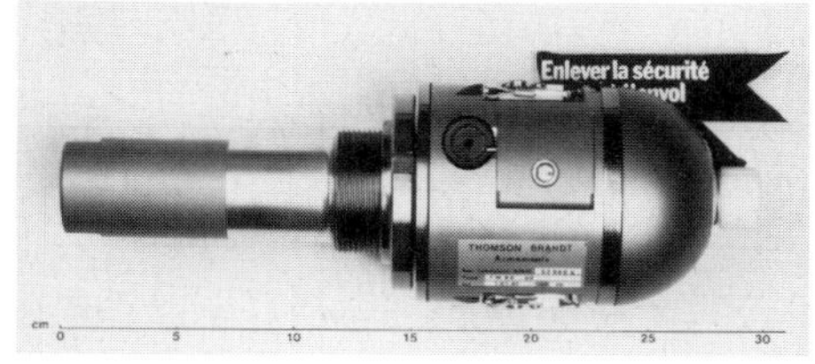

FEU 80 multi-purpose bomb fuze

Reliability is guaranteed through: electronic circuits which can be checked automatically without destructive effects, being air-tight and water-tight when mounted and on the ground (bombs as complete rounds can thus remain under the aircraft or placed on the ground out-of-doors); a calculated service-life of 20 years in a container; and the position of the firing and priming system inside the bomb body. On account of this mechanical and electrical protection, operation is ensured after impact, irrespective of the hardness of the target.

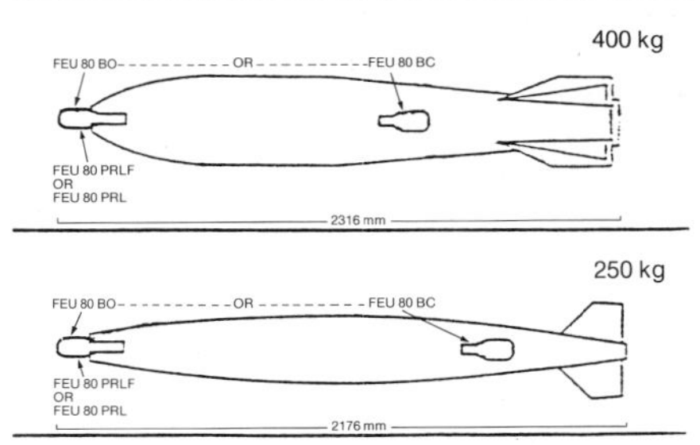

Typical arrangements of FEU 80 fuzes in bombs

Adaptability is ensured through the possibility of displaying the adjustable parameters without removing the fuze and modular design (straightforward adaptation of the fuze functions for use on new type bombs). The control circuits will permit in-flight adjustment of the arming-time, the priming delay and the explosion-height (PRLF) when the associated avionics become available.

Effectiveness, using electrical power affords the possibility of priming the tail fuze with the firing order being generated remotely or on impact by the nose detector for greater effect in the event of an attack from above.

The fuze is countermeasure resistant.

CHARACTERISTICS

Overall length: 122 mm
Diameter: 85 mm
Assembly thread diameter: 2 or 3 inches
Weight: 2.4 kg

STATUS

In series production. Entered service with French armed forces in 1984.

CONTRACTOR

Thomson Brandt Armements, 204, Rond-Point du Pont de Sèvres, 92516 Boulogne-Billancourt Cedex, France.

3863.303

BAT 120 ANTI-ARMOUR WEAPON SYSTEM

The Thomson Brandt Armements BAT 120 is a 120 mm calibre super-retarded weapon system designed to engage surface targets such as convoys of vehicles, parked aircraft and missile sites. This unmatched weight-effective weapon has two high technology, focused, pre-fragmented warheads for scattering horizontal sheets of calibrated fragments:

(1) AMV anti-vehicle equipment. This delivers more than 2600 main fragments regularly distributed and designed to perforate 4 mm of steel at a distance of 20 m from detonation point
(2) ABL anti-lightly armoured vehicle. This version ejects about 800 main fragments, each capable of penetrating 7 mm of steel at 20 m from impact point.

In both cases, the resulting explosions produce numerous secondary fragments which may be effective against many targets. As the weapon is very light, it is suitable for long range missions and can be carried in large numbers (up to 36 on a 5 ton class aircraft) in clusters of nine or 18 rounds, thus saving store stations. A stick of 18 rounds scatters 45 000 main fragments over 35 000 sq m.

BAT 120 tactical retarded bombs loaded on BAe Hawk aircraft

BAT 120 are released in sticks from the same Thomson Brandt Armements carriers as BAP 100 cratering weapon (**3152.303** below). For example, 72 BAT 120s could be released from only four store stations on a single aircraft. An electronic intervalometer integrated in each carrier ensures correct sequencing of bomb release.

CHARACTERISTICS

Bomb length: 1500 mm
Bomb weight: 34 kg
Warhead weight: 24 kg
Weight of HE: 6 kg
Time of fall: 6 sec
Pitch down attitude on impact: 85°
Operating temperatures: -40 to +70°

STATUS

Series production for French Air Force and several other air forces.

CONTRACTOR

Thomson Brandt Armements, 204, Rond-Point du Pont de Sèvres, 92516 Boulogne-Billancourt Cedex, France.

3152.303

BAP 100 mm AIRFIELD DENIAL AND ANTI-RUNWAY WEAPON SYSTEM

The Thomson Brandt Armements BAP 100 area denial and cratering weapon system is specially designed to neutralise airfield runways and taxiways. It can be used against other hard targets such as highroads, concrete buildings and shelters.

A stick of bombs is released to provide a very high hit probability despite low level altitude aiming difficulties. In order to achieve one cut, it has been calculated and demonstrated that 12 to 18 bombs are needed depending on the aircraft sighting system and the size of the target. After a parachute retarding phase, a rocket assisted acceleration phase ensures concrete penetration. The bomb explodes under the concrete slab thanks to an optimised delay. A long delay fuze bomb is also available.

A specific multi-bomb carrier system attached to a single weapon station or an aircraft can accommodate nine or 18 BAP 100 or BAT 120 weapons. A quick hooking system allows the loading of 18 bombs in less than ten minutes without any special training.

F-5 releases BAP 100 bombs

The main adaptors designed for the BAP 100/BAT 120 system are:

	Lug spacing	Maximum load	Total weight (kg)
14-3-M2	14 inches	9 bombs	320
30-6-M2	30 inches	18 bombs	710

CHARACTERISTICS

Length: 1780 mm
Weight: 32.5 kg
Warhead weight: 18 kg
Charge weight: 3.5 kg
Penetration: 300 mm of concrete
Crater coverage: Over 40 m²
Alignment of pyrotechnic train: 2.25 sec after ejection
Rocket motor ignition delay: 3.75 sec after ejection
Burning time: Approx 0.3 sec
Impact velocity: Over 260 m/s
Operating temperatures: –40° to +70°

STATUS

In series production and operated by five countries.

CONTRACTOR

Thomson Brandt Armements, 204, Rond-Point du Pont de Sèvres, 92516 Boulogne-Billancourt Cedex, France.

French Air Force Jaguar aircraft with 18 BAP 100 area denial and cratering bombs mounted on the under-fuselage attachment point

---

4597.303

74 mm ANTI-ARMOUR GRENADE

Thomson Brandt produces a 74 mm shaped charge anti-tank grenade which also has a significant secondary anti-personnel effect due to an explosive charge scattering lethal fragments to a distance of 10 m. The grenades are retarded in order to achieve a near vertical attitude for improving their penetrating power against MBTs top side armour plate.

The cartridge, housing the grenade, can be fired from all 74 mm diameter launchers such as Alkan types 501, 530 (20 or 40 cartridges) and 5120-CEM1 (59 cartridges).

CHARACTERISTICS

Calibre: 74 mm (cartridge); 70 mm (grenade)
Length: 214 mm (cartridge); 190 mm (grenade)
Weight: 2 kg (cartridge); 1.6 kg (grenade)
Release speed: 350 – 600 knots
Release height: 50 m
Main effect: > 200 mm armour steel perforation
Secondary effect: lethal fragments up to more than 10m

STATUS

Production.

CONTRACTOR
Thomson Brandt Armements, 204, Rond-Point du Pont de Sèvres, 92516 Boulogne-Billancourt Cedex, France.

Alpha Jet with anti-armour grenade dispensers on wing stations

Thomson Brandt 74 mm anti-armour cartridge showing parachute deployed form

---

1514.393
MATRA TYPE 200 BOMB RETARDING SYSTEM
The Matra Type 200 bomb retarding system was approved for use by the French Air Force in 1964. It can be used with SAMP-made 250 and 400 kg GP bombs (**5214.303**). The equipment consists of a cruciform nylon parachute packed into a container within the tail of the bomb, together with the appropriate fuzing equipment according to the mission being flown and the characteristics of the retarding system. In typical operations the bomb is released at an altitude of about 30 m (100 ft) at a forward speed of between 400 and 600 knots. The bomb will then hit its target, exploding on impact, at a time when the attacking aircraft is about 480 m (1500 ft) ahead of the explosion. This ensures complete safety for the attacker. Provision can be made before loading the weapon on to the aircraft to adjust the fuze operation for different release conditions.

Comprehensive safety provisions are made in the Matra 200 system. The parachute is locked to the bomb body only after release has taken place, so that should the parachute deploy while the weapon is still being carried by the aircraft it will pull away before dangerous turning moments are built up by the drag of the 'chute'. Similarly the parachute does not open until it is well clear of the aircraft. Provision is also made for the parachute to rotate with respect to the bomb while it is in flight eliminating the possibility of parachute twist.

Nose and tail fuzes are fitted with the retarding mechanism. These normally operate in the 'instantaneous' mode unless safety requirements are not met, ie when parachute functioning, release altitude, and release speed are all correct. If safety requirements are not met then there is a danger that the attacking aircraft will be damaged by the bomb explosion, and the fuze is automatically set to operate after a delay of 15 s. Provision is made within the fuze to prevent inadvertent instantaneous explosion in the case of a ricochet when conditions required for normal operation have not been met and the system is operating within 15 s delay.

STATUS
In production and in service with the French and other Air Forces.

CONTRACTOR
Matra SA, BP No 1, 37 avenue Louis Breguet, Velizy-Villacoublay Cedex 78146, France.

Type 200 retarded bomb

Release of Matra Type 200 retarded bombs by Mirage aircraft

4596.393
MATRA SFA BOMB RETARDING SYSTEM
The Matra 200 bomb retarding system (**1514.393**) has been in use since 1965 with a variety of 250 and 400 kg 'conventional' bombs of French and other manufacture. In order to extend the release envelope of such weapons to fit smaller bombs and also to permit air burst detonations, Matra has developed a new system, SFA, which consists of a new retarding and arming system.

SFA is designed to fit on SAMP 125 or 250 kg general purpose or reinforced splinter effect bombs and allows bombing from heights half those of earlier systems and over a much wider speed range (in particular low speeds). The SFA system permits bombing from a distance by delaying the retarding function.

The system is delivered as a single rear-end unit to be connected onto the bomb frame as any conventional tail unit would be. There is a choice of operating mode; a proximity fuze for initiating detonation above ground level, or an impact fuze for detonation on contact. Performance characteristics are adequate for release heights ranging from 20 to 30 metres, and permit diving bombing approaches. It is also possible to employ 'toss-bombing' stand-off attack techniques using the ability to vary the retardation delay from zero (ie at release) to 24 seconds.

Other adjustable parameters include: release speed, 350 – 650 knots; release height, down to 24 metres; and proximity or impact fuzing. Safety features ensure that detonation is inhibited if any of the following conditions are encountered: insufficient altitude, inadequate speed, parachute failure etc.

DEVELOPMENT
Development was undertaken by Matra as a private venture using company funding, and by early 1984 was virtually complete.

STATUS
The in-flight qualification has been completed and production started.

CONTRACTOR
Matra SA, BP No 1, 37 avenue Louis Bregeut, Velizy-Villacoublay Cedex 78146, France.

SFA 250 retarded bombs on Mirage 2000 aircraft

1999.303
DURANDAL PENETRATION BOMB
The Matra Durandal penetration bomb has been designed particularly for the disablement of airfields by low-level bombing attacks in which the objective is to render the runways unusable for a long period. Naturally it can also be employed against other targets of a similar nature. The technique of Durandal consists of an initial rapid retardation of the bomb after release until it adopts the optimum non-ricochet angle, followed by prompt acceleration to a very high impact velocity to ensure penetration of the runway. It is stated that concrete slabs up to 40 cm thick can be penetrated in this way. The ensuing explosion of the warhead from beneath the runway surface results in extensive damage over a wide area because of the heave effect, according to the makers.

Durandal system qualification has already been carried out for the Mirage III, Mirage 5, Mirage 2000 and Jaguar aircraft, and adaptation and qualification testing under other aircraft are continuing. The

Mirage III or 5 can carry between six and ten Durandal while the Jaguar can carry between six and eleven Durandal.

CHARACTERISTICS

Length: 250 cm
Max diameter: 22.3 cm
Max wingspan: 43 cm
Weight: 219 kg
Max carrying speed: Mach 1.8 at 10 000 m
Release conditions: Minimum launching altitude 75 m, launching speed 450 – 600 knots
Effectiveness: Destruction of 150 – 200 m² of concrete runway

STATUS

Series production began in 1977 and over 29 000 units have so far been ordered for use by 12 countries. The first operational Durandals were delivered in 1978. Operational on Mirage III, 5, Jaguar and other aircraft types. The USAF has now adopted the Durandal system.

CONTRACTOR

Matra SA, BP No 1, 37 avenue Louis Breguet, Velizy-Villacoublay Cedex 78140, France.

Multiple load of Matra Durandal penetration bombs carried by Mirage aircraft

---

3150.303

BELOUGA WEAPONS DISPENSER (BLG 66)

Belouga is an area weapon system. It is suitable for attacking stationary or moving targets and can be released from high speed, low flying aircraft. Three types of grenade have been developed for the following attack missions:

(1) General purpose (fragmentation) for soft-skinned vehicles, equipment, fuel stores and parked aircraft etc
(2) Anti-tank (armour piercing) for tanks, AFVs, APCs etc
(3) Area interdiction for use against airfields, harbour areas, shunting yards, highway junctions etc.

After release the Belouga bomb is retarded to enable the aircraft to clear the target area before the grenades are ejected. The grenades are then ejected by cartridge radially in 16 directions from the bomb axis to ensure a homogeneous ground pattern over either a long carpet (> 10 000 sq m) or a high impact density short carpet (>5000 sq m) for maximum effect. After ejection the individual grenades are braked to allow impact with target in almost the vertical position to achieve best possible results with the horizontal scattering of fragments and vehicle roof attack with the HEAT grenades. The weapon is supplied as a complete round, armed and ready for use, in a special packing, and a standard NATO 14-inch (356 mm) attachment is used. The pilot can select either of the two pattern configurations prior to release.

CHARACTERISTICS

Overall length: 3300 mm
Diameter: 360 mm
Span: 550 mm
Number of grenades: 151
Grenade calibre: 66 mm
Grenade weight: 1.3 kg each

Belouga weapons submunition dispenser on Jaguar

STATUS

In production and in operational service with French and other air forces.

CONTRACTORS

Matra SA, BP No 1, 37 avenue Louis Breguet, Velizy-Villacoublay 78140, France.

Thomson Brandt Armements, 204, Rond-Point du Pont de Sèvres, 92516 Boulogne-Billancourt Cedex, France.

5213.303

SAMP PRACTICE BOMBS

All these bombs can be provided with a spotting charge of either the visual or acoustic type, and can be used with any type of bomb rack. The type 2 is used to simulate anti-personnel and anti-materiel bombs, type 6 is used to simulate retarded bombs and types 7 and 8 are used to simulate low altitude and dive bombing attacks.

CHARACTERISTICS

Type	Weight	Length	Diameter
2	2.2 kg	489 mm	98.5 mm
6	6.1 kg	489 mm	98.5 mm
7	5.3 kg	610 mm	98.5 mm
8	12.2 kg	610 mm	98.5 mm

CONTRACTOR

Société des Ateliers Mécaniques de Pont-sur-Sambre, 37 Grande Rue, 59138 Pont Sur Sambre, France.

5214.303

SAMP GENERAL-PURPOSE BOMBS

50 to 120 kg bombs. These are intended for use on light attack aircraft and are carried on underwing pylons, and in addition they can be carried on multiple ejector racks of fighter-bombers. They all have NATO suspension lugs and a standard two-inch well for the fuze.

250 to 500 kg bombs. These can be carried under any NATO bomb rack and can also be provided with a retarding device for low altitude bombing and an integrated arming control device.

CHARACTERISTICS

Nominal weight	Type	Weight	Explosive content	Length	Body diameter
50 kg	BL 5	50 kg	40%	1.32 m	176 mm
120 kg	BL 6	118 kg	45%	1.778 m	228 mm
120 kg	BL 7*	115 kg	50%	1.66 m	228 mm
250 kg	EU 2	250 kg	40%	2.253 m	273 mm
250 kg	25	247 kg	50%	2.121 m	324 mm
400 kg	T 200	345 kg	50%	2.195 m	403 mm
500 kg	EU 3	452 kg	45%	2.855 m	356 mm
1000 kg	BL 4	1000 kg	55%	3.5 m	457 mm

*Only for use with SF 260 aircraft.

1000 kg bomb. This can be carried by any aircraft with 30-inch (762 mm) suspension lugs and can also be provided with a retarding system or a guidance system, such as that developed by Matra.

CONTRACTOR

Société des Ateliers Mécaniques de Pont-sur-Sambre, 37 Grande Rue, 59138 Pont Sur Sambre, France.

Type 25 (250 kg) general-purpose bomb

5215.303

SAMP FRAGMENTATION BOMBS

These bombs have been designed for delivery against light vehicles, light structural framed buildings and un-sheltered personnel. All have a standard two-inch well for a nose fuze. The BL 9 can be used only with the Italian SM 260 aircraft and T 15 bombs are normally dropped in a cluster of four. The BL 18 125 kg fragmentation bomb can be provided with a French proximity fuze to give an airburst capability.

CHARACTERISTICS

Type	Weight	Explosive content	Length	Body diameter
T 15	15 kg	20%	752 mm	119 mm
BL 8	120 kg	17%	1584 mm	195 mm
BL 9	120 kg	17%	1380 mm	195 mm
BL 18	125 kg	20%	1805 mm	195 mm

CONTRACTOR

Société des Ateliers Mécaniques de Pont-sur-Sambre, 37 Grande Rue, 59138 Pont Sur Sambre, France.

4189.311

MATRA LASER-GUIDED BOMBS (LGB)

Matra has developed a series of laser-guided bombs similar to the American Paveway munitions (**1534.311**) in which laser seeking guidance packages are added to standard bombs to provide increased accuracy when used against laser-marked targets. In the Matra system target illumination may be either by an aircraft-mounted laser marker or by a ground laser transmitter. In the former case a version of the ATLIS II target designator pod (**3820.393** in 1981-82 and earlier editions) is normally used, and the laser guidance sub-system for fitting on the bomb is a version of the Ariel seeker for the AS.30 laser missile (**3335.311**, below) known as the EBLIS and also produced by Thomson-CSF.

In the LGB application, the EBLIS is fitted in an aerodynamically shaped housing equipped with a wind-vane and carried on the nose of the bomb. The laser error detector measures the angle between the bomb velocity vector and the line between the aircraft and the target, and makes corrections by actuating electric motors to move the fins or flying surfaces to adjust the bomb's trajectory to bring it to the target. Other guidance components in the package are a guidance order computer and a roll gyroscope. These are housed in a streamline container mounted on the fore-part of the bomb and with four control vanes disposed on it in a cruciform fashion. Another package of the LGB kit consists of four larger aerofoil surfaces mounted on an extension fitted to the rear of the bomb body; these too are arranged in a cruciform pattern.

Kits have been designed for fitting to several types of bomb, ie the 400 kg and 1000 kg types, and for some types semi-folding wings facilitate carriage on the underwing armament stations of most aircraft types. The complete 1000 kg LGB is also known by the name Arcole.

A feature of the Matra LGB system emphasised by the manufacturer is its ability to be used in a low-level delivery mode, in level flight and at speeds up to Mach 0.9, enabling stand-off ranges of 4 to 10 km. There may also be a lateral offset of several degrees without impairing accuracy. Weapon flight times of up to 30 s after release are provided in the design, power for control actuation and electronics etc being supplied by a battery.

STATUS

The Matra LGB programme was started in 1978 in response to a French Government requirement, and the first phase of development was completed in late 1980 with a number of successful launches of 400 kg LGBs. A special 1000 kg bomb body has been designed, specifically for the destruction of hard targets, as a French Air Force requirement. This LGB was engaged on final flight trials in 1986 and is now entering service. The 400 kg version entered service in 1985.

By mid-1985 500 LGBs had been ordered or were on option, with the 400 kg model in production for several foreign customers. Approximately 100 1000 kg LGBs have been ordered for the French Air Force for use on Mirage 2000s. The 1000 kg version has also been ordered by foreign customers.

CONTRACTORS

Matra SA, 37 avenue Louis-Bréguet, 78140 Vélizy, France.

Thomson-CSF – Aerospace Group, 122 avenue du Général Leclerc, 92105 Boulogne-Billancourt, France.

4864.303

CH-TABO BOMB DISPERSAL POD

The Swiss Federal Aircraft Factory at Emmen and the Federal Ammunition Factory at Houne collaborated in the development of an aircraft armament pod to be used as a dispenser for bombs and/or submunitions for low-level attacks against lightly armoured vehicles, gun emplacements, radio/radar stations, and similar ground targets. Troops may also be attacked with the weapon. This weapon is now marketed by Matra SA.

The system comprises an aerodynamically shaped container which is attached to the carrier aircraft by normal weapons hard points, and bombs which are spherical in shape and are equipped with a rigid cast aluminium tail assembly. The bomb case consists of an aluminium matrix containing 8000 cast steel balls. These are explosively dispersed after detonation by either mechanical detonators that are actuated when a telescopic probe touches the ground, or by electronic fuzes set to operate at any chosen height above the surface. Using a proximity fuze to achieve an optimum height air burst, two CH-TABO container

Hunter aircraft equipped with CH-TABO dispensers releases bomblets at low level

loads of bombs can penetrate 5 mm steel plate armour with a density of four fragments per square metre over an area of 200 × 50 m.

The bombs are carried inside the pod and ejected from the open rear end of the pod by dynamic pressure. After release they are aerodynamically braked. The container payload can be dispensed at any airspeed up to Mach 0.6 – 0.85. The container may also be employed for dropping other stores, such as medical or other supplies, communications equipment, life-saving items etc. The CH-TABO container is not jettisoned after dropping bombs or stores.

STATUS

Early production.

MARKETED BY

Matra SA, BP No 1, 37 avenue Louis Breguet, Velizy-Billacoublay Cedex 78146, France.

CH-TABO bomb dispenser with two ejected bombs

6402.303

BAVARD LXT 462 ADJUSTABLE PRACTICE BOMBS

The BAVARD LXT 462 range of adjustable ballistic practice bombs is designed to train pilots in the effective use of all types of live aircraft stores. Individual bombs can be adjusted by varying weight, centre of gravity and drag to simulate the ballistic performance of the required live store.

The bombs are moulded from pressure-injected high resistance macralon. Ballast is contained in the hollow central section and the nose. The tail houses the impact marker and ignition assembly. Both bomb and ballast disintegrate on impact to eliminate the danger of ricochets and reducing clearing problems.

The BAVARD bombs can be carried on the same carrier points as live stores or suitable rack adaptors may be used.

CHARACTERISTICS

Max length: 900 mm
Min length: 720 mm
Diameter: 98.5 mm
Weight: 2 to 30 kg

STATUS

In production.

CONTRACTOR

Etiennes Lacroix SA, Route de Toulouse, 31600 Muret, France.

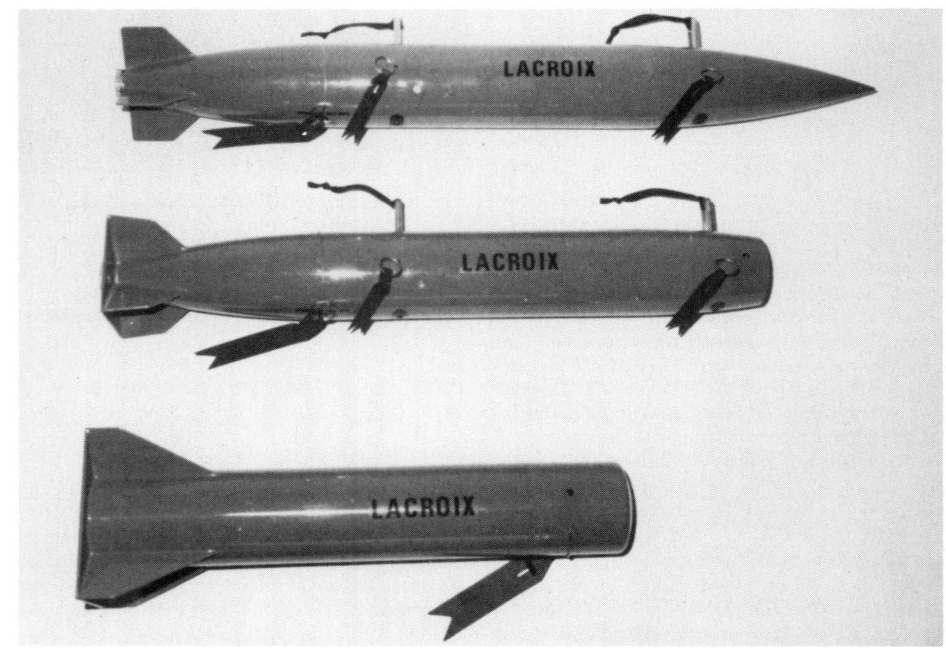

BAVARD LXT 462 adjustable practice bombs

4541.303

LU250 EG-FT BOMB

The LU250 EG-FT bomb is a general-purpose low-drag explosive bomb complying with American Mk 82 bomb specifications (MIL-B-82548). The bomb is designed to meet the requirements of the largest possible majority of bombing missions using both the blast and fragmentation effects along with the concussion effect produced in the ground and buildings.

The fuzes used for 'smooth' bombs with conical fins are generally mechanical. The bomb bodies are equipped to receive either mechanical or electrical fuzes. The first type is assembled using booster adaptors.

A BSU 49/B retarder is also fitted to use the LU250 EG-FT in low-altitude attacks. Two versions are offered:

(1) bomb body with conical fin
(2) bomb body with retarder.

The BSU 49/B inflatable retarder is a cruciform tail stabiliser that can be used in either the 'low-drag' configuration without inflating the retarder, or in the 'high-drag' configuration with the retarder deployed. The inflatable retarder is used in low-altitude bombing with a high drag to allow the bombing aircraft to escape the blast and fragmentation effects of the bomb. This high drag is achieved by opening a balloon-parachute, called ballute, enclosed in the container fin.

When the bomb is dropped, the spring which holds the arming wire stretched, unwinds. When completely unwound, it frees two pin safeties and opens the cover. The ballute assembly is then freed under the effect of the external springs on the cover and of the internal ejection spring, and is immediately inflated by the external airflow, which comes in through four symmetrical inlets. The cover remains integral with the ballute, thus avoiding any risk of projection to the rear which might strike the bombing aircraft.

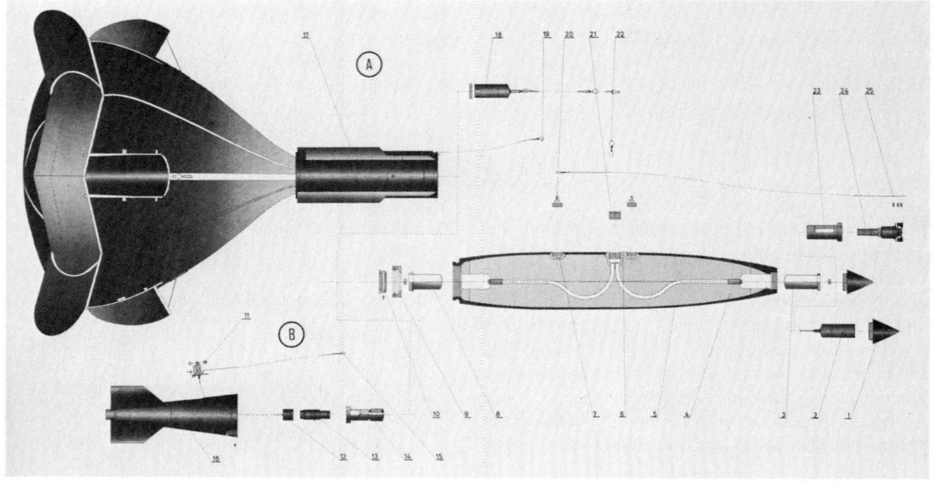

LU250 EG-FT bomb with retarder or conical fin, electrical or mechanical fuzes

(A) *LU250 EG-FT bomb with retarder* **(B)** *LU250 EG-FT bomb with conical fin* **(1)** *nose cap* **(2)** *nose electrical fuze* **(3)** *nose adaptor bushing* **(4)** *explosive charge* **(5)** *bomb body* **(6)** *NATO lug and arming wells for electrical fuzes* **(7)** *NATO suspension lug insert* **(8)** *tail adaptor bushing* **(9)** *rear blanking plate* **(10)** *sleeve coupling* **(11)** *wind vane* **(12)** *coupling regulator* **(13)** *fin mechanical fuze* **(14)** *priming relay* **(15)** *fin mechanical fuze arming wire* **(16)** *conical fin* **(17)** *retarder* **(18)** *electrical fuze* **(19)** *retarder opening control wire* **(20)** *NATO suspension lug* **(21)** *electrical circuit wells closure* **(22)** *fuze arming swivel hook* **(23)** *priming relay* **(24)** *nose mechanical fuze* **(25)** *nose mechanical fuze arming wire*

CHARACTERISTICS

Type LU250 EG-FT	(conical fin)	(with retarder)
Length	2.154 m	2.162 m
Diameter	273 mm	273 mm
Fin span	380 mm	385 mm
Weight	240 kg	254 kg

The US also produce a similar design but larger retarder known as BSU-50/B for use with the 2000 lb Mk 84 bomb. Retarder characteristics are as follows.

CHARACTERISTICS

Retarder type	**BSU-49/B**	**BSU-50/B**
Diameter	222 mm	403 mm
Span	381 mm	940 mm
Length	660 mm	775 mm
Weight	25.4 kg	43.5 kg

STATUS

In service with French Air Force.

CONTRACTORS

Luchaire SA, 180 Boulevard Haussmann, 75008 Paris, France.

GERMANY (FEDERAL REPUBLIC)

3805.303

HBS 202 HELICOPTER ARMAMENT SYSTEM

The helicopter armament system HBS 202 has been designed specifically for installation on the MBB helicopter BO 105. It is a rigid mounted system which is aimed by helicopter manoeuvring. It consists of a frame fixed underneath the fuselage between the skids, a cradle for gun installation, an electro-hydraulic cocking and triggering device and the control panel. The ammunition is stored in an in-board container with a flexible connecting chute to the gun. A reflex sight is employed.

The weapon itself is the Rheinmetall 20 mm automatic cannon Mk 20 Rh202 (**5600.103** in 1982-83 and earlier editions) which is particularly suited to these purposes because of its low recoil forces and its high reliability under extreme service conditions.

CHARACTERISTICS

Weapon

Type: Mk 20 Rh202

Calibre: 20 mm × 139 NATO

Rate of fire: 1000 rounds/minute

Muzzle velocity: HEI-T, TP-T 1050 m/s; API-T 1100 m/s; APDS-T 1150 m/s

Recoil force: 550 – 750 kgf

Tactical range: 2000 m

Type of fire: Single fire; or controlled burst

Cocking: Electro-hydraulic

Triggering: Electro-hydraulic

Ammo storage capacity: 525 rounds

Sight: Reflex

Helicopter armament system HBS 202 with Mk 20 Rh202 mounted below fuselage. Note flexible ammunition feed from inside aircraft

Weight

Gun mount: 96 kg (without gun and ammo)

Mk 20 Rh202: 75 kg (approx)

525 linked cartridges: 175 kg

Combat weight: 346 kg

STATUS

Production.

CONTRACTOR

Rheinmetall GmbH, Ulmenstrasse 125, 4000 Dusseldorf 30, Federal Republic of Germany.

5186.303

27 mm MAUSER CANNON

Mauser in participation with Diehl and Dynamit Nobel (combat ammunition, propellant and primer cap) developed the Mauser 27 mm weapon system under contract with NAMMO (NATO MRCA Development and Production Management Organisation).

The main characteristics of this cannon are high muzzle velocity, high rate of fire from the first round, high degree of accuracy and reliability, favourable installation parameters such as low weight, small volume envelope, low reaction forces and for all types of ammunition identical exterior ballistics (permit mixed belts) and fuze response at extreme angle of impact (HE). The five types of ammunition are: AP, APHE, HE, TP and TP-F.

The MRCA Tornado aircraft is armed with two of these installed cannons and the German version of the Alpha-Jet with a single cannon in a pod. The new Swedish multi-role combat aircraft (JAS-39 Gripen) now under development will also be equipped with one internally installed cannon.

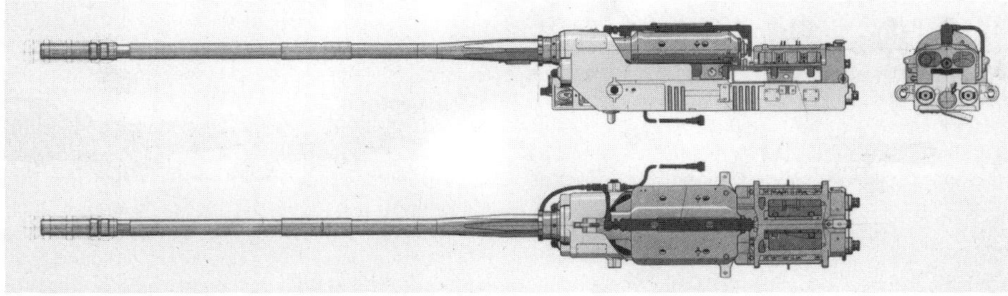

General arrangement drawings of the 27 mm Mauser cannon showing the main components of the weapon

CHARACTERISTICS

Calibre: 27 mm

Length with barrel: 2.3 m

Length of barrel: 1.7 m

Weight of gun (incl barrel): 100 kg

Power supply: 24 – 28 V DC

Rate of fire: 1000 or 1700 rpm

STATUS

In production.

CONTRACTOR

Mauser-Werke Oberndorf GmbH, Postfach 1349/1360, D-7238 Oberndorf, Federal Republic of Germany.

4866.303

MODULAR DISPENSER SYSTEM (MDS)

The Modular Dispenser System is a dispenser weapon for use on tactical fighter aircraft. It fits on all standard bomb racks. The basic ejector modules can be assembled in different configurations to suit aircraft type(s) and operational function(s). Various submunitions may be carried and dispensed with the MDS to suit it for use against runways, armoured fighting vehicles, soft targets, surface-to-air missile sites, and other battlefield objectives.

MDS Modular Dispenser System mounted on F-5 aircraft centre-line pylon

Operational features include low-level submunition delivery and automatic pattern control. The system may be re-used for numerous sorties.

A dropped container version of the Modular Dispenser System is now under development. Several of these drop containers may be carried by an aircraft (depending on the aircraft load capacity) and they fall in free flight after dropping. The submunitions are dispensed over the target area after a pre-planned time.

CHARACTERISTICS

Captive version

Overall length: 2.5 – 4.5 m (variable, depending upon aircraft type)

Width: 630 mm

Height: 340 – 660 mm (variable, depending upon aircraft type)

Total weight: 500 – 2000 kg

Submunition weight: 70 – 80% of total

Aircraft speed at submunition ejection: Mach 0.6 – 0.9

Altitude at ejection: Approx 50 m

Drop version

Overall length: approx 4 m

Width: 630 mm

Height: approx 350 mm

Weight: 600 kg

Aircraft speed at drop point: Mach 0.6–0.9

STATUS

In development. Flight tested. The Swedish government has awarded a contract to MBB for the development of a new dispenser weapon system for the JAS 39 Gripen aircraft. The system will be based on the MDS.

CONTRACTOR

Messerschmitt-Bölkow-Blohm GmbH (MBB), Post Box 80 11 49, 8000 Munich 80, Federal Republic of Germany.

4770.311

VERTICAL BALLISTIC WEAPON/VBW AIRBORNE ANTI-TANK WEAPON

Vertical Ballistic Weapon/VBW is the name of an experimental airborne anti-tank weapon system being developed in West Germany under a Ministry of Defence contract awarded to MBB. The system is intended to provide an aircraft with a capability to carry out top attacks on multiple armour targets on a single pass over the target area.

It differs from cluster weapons and other area weapons in that a relatively small number of unguided anti-tank projectiles is used to combat whatever armour is encountered within a swath of terrain on each side of the aircraft's flight path, rather than dispensing large numbers of specialised submunitions which have varying degrees of sophistication. To achieve this the VBW system consists of integrated pods in which the multi-sensor package and the anti-tank projectiles are located, and which are mounted beneath the aircraft. During the experimental phase two pods were required; one housed a multi-sensor package for detecting, classifying and accurately locating the target(s), while the other carried anti-tank projectiles which are launched downward and rearwards to attack the target's vulnerable top surface.

Sensors incoporated in the detector/locator pod include an IR linescan equipment, laser and radar, and a radiometer. The last of these plays an important part in identification of the target by means of its thermal characteristics. This process and other signal processing and data handling functions are performed by digital microprocessor, which also is responsible for carrying out the calculations needed to ensure the precise instant to fire the projectiles to ensure hitting the target(s) on the first pass.

A typical VBW pod designed for West German Air Force Alpha jets would contain about 18 tubes and would be suitable for carriage on standard NATO mountings on most types of aircraft. Experimental installations had been made on a Luftwaffe F-4F aircraft.

Reports state that operation is fully autonomous after the system is armed by the pilot prior to commencing his run over the target area, at low level and high speed. Any target encountered thereafter is automatically acquired and attacked. At a flight level of 250 feet (76 metres) the sensor pod scans a band of terrain about 100 feet (30 metres) wide.

STATUS

Development was started in 1977 under a contract thought to be worth about DM7 million. By late 1983 more than 500 trial sorties had been flown by the contractor, in the course of which some 100 live firings were made. The VBW is scheduled for entry into service by the end of this decade.

CONTRACTOR

Messerschmitt-Bölkow-Blohm GmbH, (MBB), Post Box 8011 60, 8000 Munich 80, Federal Republic of Germany.

3517.311

MW-1 MULTI-PURPOSE WEAPON

The MW-1 (Mehrzweckwaffe) is a multi-purpose cluster weapon designed and developed to engage battlefield targets such as soft- and hard-skinned vehicles moving in tactical formations over open ground; vehicles and personnel in various battlefield or support structures; vehicles, aircraft and runways on airfields. To compensate for the limitations of the majority of earlier cluster weapons which normally carry only one type of single-purpose submunition, and often have only a limited distribution pattern, the MW-1 is provided with a range of five types of submunition, and the distribution pattern can be readily varied to suit the nature of the target(s) and the operational environment.

As employed with the Tornado aircraft, the system consists of a single dispenser 5.3 metres long and large enough to contain sufficient submunitions to successfully engage a tactical target (in the anti-tank role one dispenser on a single aircraft can deploy 4536 anti-armour projectiles). The dispenser consists of four sections each containing 28 transverse tubes, ie 112 tubes with 224 apertures. To ensure uniform distribution of submunitions laterally, four different ejection charges are used. This results in different ejection velocities and thus a varying spread of the submunitions on reaching the surface. The submunitions are ejected from both sides of the dispenser without recoil effects. The width of cover from a single pass can be preset between 55 and 500 metres and the length of pattern is variable between 200 and 2500 metres. This linear variation can be adjusted in flight by the pilot who can therefore select the required density of cover for a given target.

The MW-1 system includes five types of submunition, which can be dropped in any combination. These are:

(1) KB 44 armour piercing bomblet
(2) MIFF anti-tank mine
(3) MUSA fragmentation munition
(4) MUSPA, an airfield denial munition similar to the MUSA except for a complex sensor system
(5) STABO runway penetrator

The aircraft crew can adapt the coverage area to the target area until immediately prior to weapon release. Coverage width depends on the type of target, the two main types being armoured and mechanised units (Main Target Group 1 – MTG 1) and airfields (Main Target Group 2 – MTG 2). KB 44, MIFF and MUSA submunitions are used against MTG1, and STABO, MUSPA, MUSA and MIFF for engaging MTG 2.

Like the submunitions for the British JP233 system (**5218.303**) fuze details are classified. Other known features of each are summarised in the following paragraphs.

KB 44 This 44 mm calibre, 0.6 kg shaped charge armour piercing bomblet is capable of penetrating the top and side armour of all known tanks and is a sub-calibre projectile. Normally seven of them are bundled together to fill the dispenser tube cross-section, each tube containing three bundles on each side, ie 42 odd bomblets per tube or a total of 4536 per dispenser.

MIFF is a 3.4 kg, 132 mm calibre anti-tank mine with two plate charges, which are fired either by magnetic sensors in response to a tank hull or by ground vibration. Seven or eight mines are housed in each tube making a total of 892 per dispenser. The mine is automatically stabilised, (either end uppermost) on impact. It will detonate after an initial delay of a few seconds whenever the sensor systems are activated. There is also a self-destruct time.

This weapon is not designed to create a tank barrier but to be employed in conjunction with the KB 44 against a tank formation, or with other submunitions against an airfield.

MUSA is a 132 mm calibre 4.2 kg fragmentation mine for destroying soft-skinned and semi-hardened targets and is suitable for use against depots or airfields where aircraft or vehicles might be in the open. It is parachute retarded in flight and rights itself on reaching the ground for maximum fragmentation effect, and then detonates. Each dispenser can carry 668 MUSA submunitions.

MUSPA This is an airfield denial mine similar to the MUSA, but with an acoustic sensor that will detect aircraft landing, taking off or taxying, and if the aircraft comes within its effective range it will detonate. There is also a self-destruct device with a random time delay to deter airfield clearance personnel.

STABO is a 16 kg runway penetrator and cratering bomb. Its operating sequence begins with its ejection, stabilisation in flight and parachute retarded descent. During this phase a small sensor is exposed at the bottom of the charge, which when it touches the ground results in detonation of the charge. This opens up a hole in the concrete through which a follow-through projectile is fired to detonate beneath the surface to generate heave and create a crater. 200 are carried in one dispenser.

In October 1986, the Swedish government awarded a contract for the development of a new dispenser weapon system for the JAS 39 Gripen multi-role combat aircraft to MBB. Design of the system will be based on the MBB MDS (**4866.303**) which itself derived from the MW-1.

STATUS

MW-1 is in production for the West German forces.

Development of the system for use against MTG 2 was completed by the end of 1985, with first deliveries to operational units planned for 1987. Apart from West German procurement, a MoU signed with the Italian Government ensures that Italy will also receive MW-1 dispenser sets. The West German Air Force expects to receive some 1000, the Italian Air Force some 100 MW-1 systems.

CONTRACTORS

RTG Raketen-Technik GmbH, Oberweg 8,D-8025 Unterhaching, Federal Republic of Germany (jointly owned by Messerschmitt-Bölkow-Blohm GmbH (MBB) and Diehl GmbH & Co.)

ISRAEL

6403.303

OPHER TERMINAL GUIDANCE KIT FOR GENERAL PURPOSE BOMBS

The OPHER terminal guidance kit for general purpose bombs can be used with bombs of the Mk 82 and 83 type and uses an infra-red seeker/sensor mounted in the nose to detect infra-red-producing targets and guide the bomb towards them. The kit consists of an infra-red seeker/sensor, an acquisition and guidance computer, control surfaces and a fixed tail. The infra-red sensor assembly is mounted in a stabilised gimballed housing at the end of a probe at the nose of the bomb. During flight the airstream passing through the ring tail and the rear of the detector causes the seeker/sensor to 'weather vane' and align approximately along the velocity vector of the bomb. OPHER centres the target on the detector optical axis by commanding movements of the bomb's control surfaces. These control surfaces are just behind the seeker/sensor head and are under the control of the

OPHER terminal guidance kit fitted to test bomb carried by an F-4 Phantom aircraft

system guidance computer via a servo mechanical flight control unit.

OPHER can be delivered using a number of attack profiles. Once the bomb is dropped it is a 'fire and forget' system as the OPHER seeker/sensor head and system takes control of the bomb to make trajectory alterations. In a release from up to 7600 metres the bomb can travel at speeds up to Mach 0.95 and homing onto a tank-sized target will commence from approximately 1000 metres. OPHER-equipped bombs can be carried by and dropped from any type of strike aircraft.

OPHER is stated to be a low cost system. Flight testing began in April 1985 and the first full homing bomb drop was made in March 1986. The final report on the flight tests was delivered in December 1986 and the prototype was demonstrated in 1987. Series production will commence in mid 1988 with the first deliveries being made at the end of the same year.

STATUS

Early production was due to commence in mid-1988.

CONTRACTOR

Elbit Computers Limited, Advanced Technology Center, PO Box 5390, Haifa 31053, Israel.

4428.303

TAL CLUSTER BOMBS

The TAL-1 cluster bomb is designed to achieve a high spin rate when dropped so that, when the fuze is activated after a preset time delay, sub-munition bomblets will be scattered evenly over a wide area in a circular pattern. The pattern size is governed by the point of fuze activation and can be a maximum of 53 000 m². 279 bomblets are carried, each with a weight of 500 grammes containing 160 grammes of high explosive and with a kill radius of eight metres.

TAL-1 is designed for external carriage by tactical ground support aircraft. It may be carried singly on a standard bomb rack pylon or in multiple mounting on tandem, TER, or MER type bomb rack. Safe carriage is ensured by a single arming wire which locks the time fuze and, after separation from the aircraft, a safety and arming mechanism allows a safe distance between the cluster and the aircraft. Each bomblet also has its own arming wire which is withdrawn by aerodynamic forces produced by the free air-stream when the cluster is activated.

CHARACTERISTICS

Weight: 250 kg

Length: 2159 mm without fuze, 2345 mm with fuze

Diameter: 406 mm

Fin span: 560 mm

Suspension: 356 mm (14-inch) twin lug

Payload: 270 × 500 g bomblets (TAL-1); 315 × 400 g bomblets (TAL-2)

Bomblet explosive: 160 g HE

Number of fragments: 1800

Between 270 and 315 bomblets are released when the bomb splits open, and these are armed by aerodynamic forces produced by the airstream and they explode on impact.

TAL-2 cluster bomb, after pre-set delay, splits open and its spin scatters between 270 and 315 bomblets

STATUS

First revealed in 1981 and believed to be operational. The TAL-2 model was revealed in 1983.

CONTRACTOR

Rafael, Ministry of Defence, PO Box 2082, Haifa, Israel.

ITALY

4867.303

AEREA MULTI TASK POD

Aerea has developed a combination aircraft armament pod for use on fixed-wing aircraft and helicopters, which incorporates one .50 inch calibre machine gun and six tubes for launching 2.75 inch rockets. The new unit is known as a multi task pod (MTP), and by proper selection of both the rocket warheads and round types, a variety of missions can be fulfilled with only a single pod.

The MTP can accept any .50 inch machine gun that is suitable for airborne use, together with 250 rounds of ammunition. Other features include in-flight re-cocking, rocket intervalometer incorporated in pod, suitable for integrated store management system, lightweight, small cross section and standard NATO 14-inch hook spacing.

CHARACTERISTICS

Length: 1.8 m

Height: 410 mm

Width: 410 mm

Weight: 79 kg (empty); 175 kg (loaded, 250 rounds + 6 rockets)

Aerea Multi Task Pod contains .50 inch M3 machine gun and 250 rounds plus six 2.75 inch aircraft rockets

Firing rate: 650 – 1000 rounds/minute

Temperature range: –20 to +50°C

STATUS

Production.

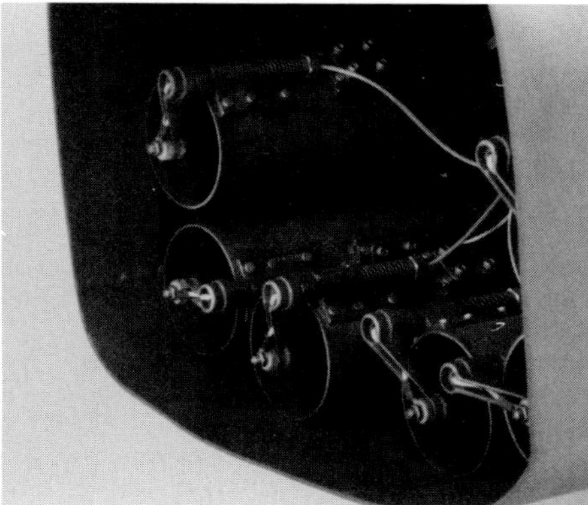

Rear of Multi Task Pod showing ends of rocket launcher tubes and firing mechanism

CONTRACTOR

Aerea SpA, via Cefalonia, 18-20156 Milan, Italy.

4074.303

AEREA TWIN GUN POD 7.62

The Aerea twin gun pod P/N 781 586 is designed to carry two 7.62 mm MAG 58 machine guns with a total ammunition supply of 1000 rounds (500 rounds per gun). The pod may be carried as an external store on any aircraft pylon equipped with two standard NATO 14-inch (356 mm) lugs. Each gun is supported by a damping cradle to minimise recoil effects, and can be easily collimated using a double-axis front support. Cartridge cases and coupling links are normally ejected through special ducts, but an improved pod with retained cases and links can be provided. The normal bomb jettison feature may be used to jettison the pod in an emergency.

CHARACTERISTICS

Length: 1750 mm

Height: 290 mm

Width: 395 mm

Weight: 55 kg (empty, with guns); 84 kg (loaded)

Rate of fire: 650 – 1000 rounds/minute

STATUS

Production.

Aerea twin gun pod on light trainer aircraft

CONTRACTOR

Aerea SpA, via Cefalonia, 18-20156 Milan, Italy.

4073.303

BRD BOMB AND ROCKET DISPENSERS

Aerea SpA currently produces a range of four bomb and rocket dispensers; BRD-4-250, BRD-4-450, BRD-4-268 and BRD-4-270. The BRD-4-250 bomb and rocket dispenser is used for training pilots, and this model can be loaded with four practice bombs mixed with 2 inch, 2.75 inch or 68 mm SNEB rockets. It is constructed of light alloy metal and its weight makes it suitable for use with light aircraft and helicopters as well as other tactical aircraft. A newer version, the BRD-4-450, can carry four practice bombs and four rockets of the same calibre as the BRD-4-250. Suspension is by means of standard NATO 14-inch (356 mm) lugs.

CHARACTERISTICS

BRD-4-250
Length: 2400 mm
Width: 432 mm
Weight: 73 kg (empty)
Payload: 4 × Mk 76 practice bombs, and/or Mk 106 bombs, 2 × 2 or 2.75 inch rockets or 68 mm SNEB rockets

BRD-4-250 bomb and rocket dispenser with typical load

STATUS
Production.

CONTRACTOR
Aerea SpA, via Cefalonia, 18-20156 Milan, Italy.

6178.303

BPD 122 mm FALCO AIR-TO-GROUND ROCKET SYSTEM

The BPD 122 mm Falco air-to-ground rocket system is based upon the use of the 122 mm rockets derived from those used with the BPD FIROS 25 field rocket system (see **5230.111**). The system can be used on fixed wing aircraft (4-tube launcher) and helicopters (3-tube launchers).

The rockets used with the system can be fitted with HE or HEAT warheads or with sub-munition warheads. There are two sub-munition rocket warheads, the APAMB carrying 77 anti-personnel bomblets and the ATM carrying six anti-tank mines.

A system management unit is available for system control under all flight conditions and the same unit can be used to control any other BPD rocket system fitted.

CHARACTERISTICS

Warhead type	HE, HEAT	Sub-munition
Calibre	122 mm	122 mm
Length	2.5 m	3 m
Weight	60 kg	66 kg
Burn time	1.21 s	1.21 s
Burnout velocity	750 m/s	670 m/s
Launcher weight loaded:		
3-tube	239 kg	257 kg
4-tube	309 kg	333 kg

Operating temperature range: –40 to +60° C

BPD 122 mm Falco launchers fitted to an AMX light attack aircraft

STATUS
In production.

CONTRACTOR
SNIA BPD, Settore Difesa e Spazio, via Sicilia 162, 00187 Rome, Italy.

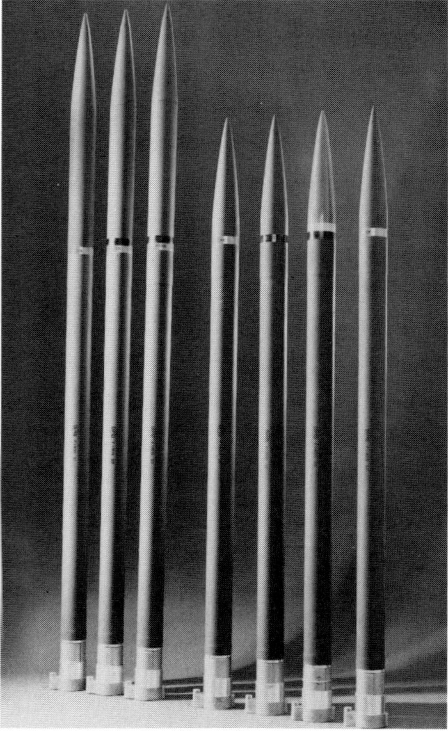

BPD 122 mm air-to-ground rocket system rockets

4869.303

2-inch BPD AIRCRAFT ROCKETS

The 2-inch BPD air-to-ground rockets system has been developed to provide effective attack by fixed-wing aircraft and helicopters against a variety of battlefield targets. Systems consist of 2-inch BPD rockets and launchers of various sizes to suit different operational tasks. Two-inch ARF/8M3 and ACR/M4 rocket models are currently available for use.

Several types of warhead are available: HEI, PFF, AT-AP, SP (spotting), illuminating, practice (smoke, flash and inert) and all weigh 2.2 kg each. At present, 3.0 kg warheads with higher terminal effects are being developed.

The current range of launchers consists of six models: 14, 18 or 28 tubes for helicopters; 29 tubes for helicopters and light aircraft; and 18 and 25 tubes for aircraft.

CHARACTERISTICS

Rocket types: ARF/8M3 and ACR/M4
Calibre: 51 mm
Length: 1048 mm
Weight: 4.8 kg
Burn time: 1.1 s
Total impulse: 2.05 kN.s
Max pressure: 11 771 K Pa
Burnout velocity: 470 m/s

STATUS
In production.

CONTRACTOR
SNIA BPD, Settore Difesa e Spazio, via Sicilia 162, 00187 Rome, Italy.

4429.303

DOOR GUNNER POST

The Door Gunner Post (DGP) is a device used to mount 12.7 mm (0.5 inch) machine guns in helicopters and capable of rapid fitting or removal in either single or double installations. DGP can fire at up to 70° downwards with the gunner always seated. A wide

Aerea Door Gunner Post installation for mounting in ground attack helicopters

variety of helicopters including the very smallest can be fitted and the system includes a rotating support which incorporates a damping cradle to minimise recoil effects; a gas shock absorber to balance the gun for better aiming; a gunner's seat with inertia belts; and a base fitting sliding on rails with a locking device. An ammunition box is mounted on the gun with other boxes being located within easy reach of the gunner. A flexible chute is provided for case ejection and a net box collects the ammunition links.

CHARACTERISTICS

Gun: FN Browning 0.5 inch or equivalent

Ammunition: 100 rounds on mounting with stowage for 400 – 1000 rounds off mounting

Weight (with gun): 70 kg

STATUS

Production.

CONTRACTOR

Aerea SpA, via Cefalonia, 18-20156 Milan, Italy.

4868.303

BPD MEDUSA 81 mm AIR-TO-GROUND ROCKET SYSTEM

The BPD Medusa 81 mm air-to-ground rocket system has been designed to meet a wide range of employments and consists of an electronic management system, rocket launchers and the rockets themselves. A laser rangefinder, an optical sight and a head-up display are available as options.

The electronic management system has been designed to meet three different possible installations requirements; to unarmed aircraft, to already armed aircraft without on-board data processing fitted and to aircraft already provided with on-board data processing. All components are thus modular and are designed to interface with a variety of other processors. The system can manage rockets with different warheads, control up to six fire stations, maintain a continuous fire station inventory, select ripple timing, select warheads for firing, set time and delayed-action fuzes, check for missfires, carry out automatic troubleshooting and interface with laser rangefinders and on-board instruments.

Several types of launcher have been designed and are in production. They can be produced for helicopters, subsonic and supersonic aircraft. Re-usable launchers for 6, 7 and 12 rockets are available.

Standard rockets use an extruded double base propellant rocket motor with high total impulse. Several types of warhead are available, both unitary and sub-munition. Unitary warheads, all weighing 7 kg, include PFF (pre-formed fragmentation) HE, HEAT and SP (spotting). The sub-munition warhead, weighing 10 kg, is loaded with 11 anti-personnel/anti-materiel bomblets. The warhead opens under the control of an electronic time fuze set by the computer unit within the electronic management system.

SNIA BPD has under development a hypervelocity rocket (max velocity 1450 m/s) with a calibre of 81 mm and a high energy/high burn rate composite propellant. The rocket uses a carbon fibre motor case and can have either a flechette warhead with an electronic time fuze or a PFF warhead with a proximity fuze. The rocket is optimised for the air-to-air role against helicopters or aircraft.

The BPD Medusa 81 mm system has been installed and tested on many types of helicopter and aircraft and has carried out firing tests from the Agusta A128 Mangusta attack helicopter.

CHARACTERISTICS

	Unitary warhead	Submunition warhead
Rocket type		
Calibre	81 mm	81 mm
Total length	1.57 m	1.9 m
Total weight	15.9 kg	18.9 kg
Warhead weight	7 kg	10 kg
Max range*	12 000 m	10 300 m
Elevation for 6000 m range	9°	14°

*ground to ground

STATUS

In production.

CONTRACTOR

SNIA BPD, Corso Garibaldi 22, 00034 Colleferro (Rome), Italy.

A BPD Medusa 81 mm 12-tube rocket launcher mounted on a light aircraft

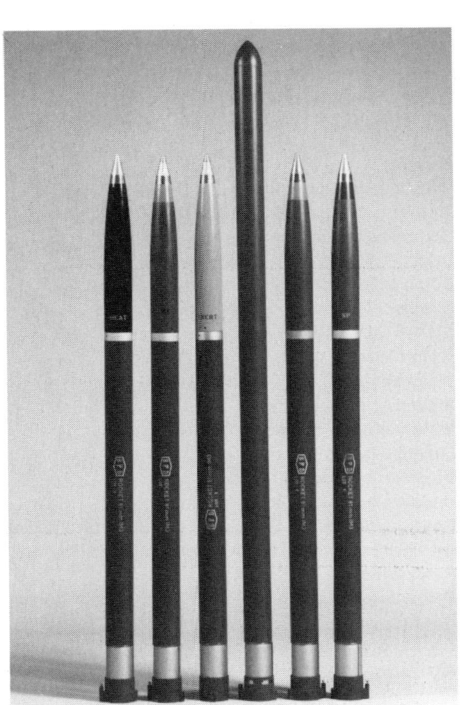

BPD Medusa 81 mm rockets

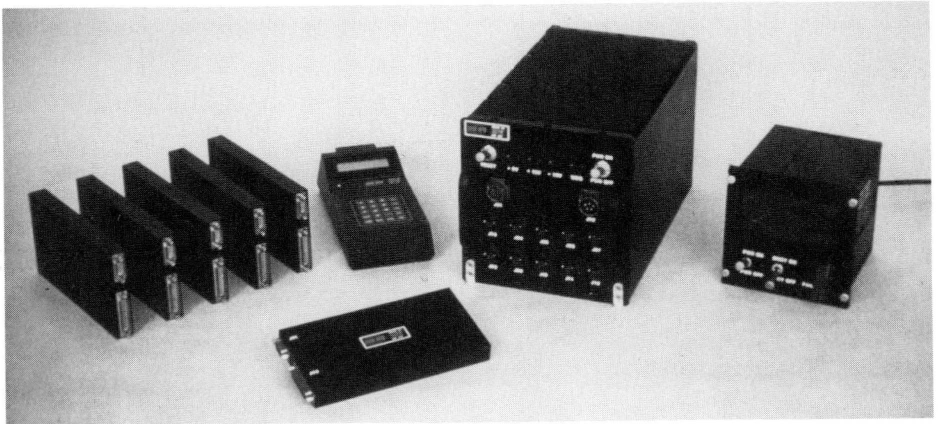

Components of the electronic management system for the BPD Medusa 81 mm air-to-ground rocket system

SOUTH AFRICA

6179.303

GA1 20 mm AUTOMATIC CANNON

The 20 mm GA1 automatic cannon is an improved version of the German WW2 MG151 cannon and is designed to be extremely compact and lightweight in comparison with other cannons of the same calibre. It is intended for installation in aircraft and helicopters as well as light vehicles and infantry fighting vehicles. Ammunition feed can be from left or right and recoil forces are low. The cannon fires from an open breech and firing may be remotely controlled. Stripping and maintenance are simple.

Ammunition available for the GA1 includes HEI and TP. Under development are HEIT, TPT and APHEI.

CHARACTERISTICS

Calibre: 20 mm
Bore groove diameter: 20.7 mm
Number of grooves: 8
Gun length: 1760 mm
Gun width overall: 244 mm
Gun height overall: 190 mm
Muzzle velocity: 720 m/s
Rate of fire (cyclic): 700 rpm
Weight of weapon assembly with barrel: 39 kg
Weight of barrel: 10 kg
Recoil force: 2700 N
Power requirements: 22 – 29 V DC

GA1 20 mm automatic cannon on Rattler servo-assisted weapon system intended for use in helicopters

STATUS

In production.

CONTRACTOR

Armscor, Private Bag X337, Pretoria 0001, South Africa.

6180.303

CB 470 CLUSTER BOMB SYSTEM

The bomblets used with the CB 470 cluster bomb system had their design origins in Rhodesia, during the period of UDI. The Rhodesian armed forces required some form of aircraft-delivered air burst weapon but had no money or expertise to produce proximity fuzes. Instead they developed a form of small spherical 'bouncing bomb' with a time fuze that initiated as the bomblet struck the ground to detonate the bomblet some 3 – 5 m in the air. The spherical steel casing of the bomblet ensured a large number of fragments and a considerable destructive and anti-personnel effect.

When Rhodesia became Zimbabwe the designers moved to South Africa and development continued to the point where the bomblet became known as 'Alpha'. Initially Alpha was dispensed at low altitude from basket-type racks carried in bomb bays but further development has resulted in its use in the CB 470 cluster bomb system.

In the CB 470 each bomblet weighs 6.2 kg, 1.4 kg being the RDX/TNT explosive filling. The body is roughly spherical and is made of pressed steel (cast steel may be used in future) covered with a thin rubber coating. Each CB 470 container carries 40 bomblets arranged in radial rows of four that are ejected by a standard ejection cartridge (albeit with a smaller propellant load than normal) 0.8 seconds after the container has been dropped from an aircraft – other ejection times can be pre-set. The radial rows are ejected at differing velocities of from 2 metres a second to 12 metres a second. After one second the bomblets are armed in flight and upon hitting the ground the bomblets lose about half their forward momentum but by spinning 'bounce' forwards and upwards, even from mud or water. As the bomblet strikes the ground a 0.65 second delay is initiated after which the bomblet explodes, scattering fragments at velocities between 1200 and 1300 metres a second. The actual explosion height can vary from 3 – 5 metres, depending on the nature of the ground. A single container load will provide a shrapnel footprint 70 metres wide and 250 metres long, ie an area of 17 500 square metres. Delivering multiple containers in ripples can cover much larger areas – eight CB 470 units rippled at 0.4 second intervals in pairs can cover 54 000 square metres.

This footprint assumes a release speed of 1000 km/h from a height of 30 metres. This low release altitude is made possible by a combination of safety factors built in the CB 470 system. One is that the drag of the bomblets combined with the loss of velocity after hitting the ground added to the 0.65 second delay gives a high speed aircraft plenty of time to get clear of the resultant explosions and fragments. Slow speed aircraft can deliver their loads from higher altitudes. Another safety feature is that the air driven safety and fuctioning unit (SAFU) only functions when the primary fuzing wire is withdrawn. A secondary fuzing wire further isolates the electrical ejection charge. A bomblet cannot be armed unless it has been correctly ejected by the ejector charge and a further one second fuze arming delay is to eliminate the functioning of the fuze during a possible mid-air collision.

The safe release speed bracket for a CB 470 to be dropped from a height of 30 metres is from 850 to 1100 km/h. It is possible to release the CB 470 from a minimum height of 25 metres.

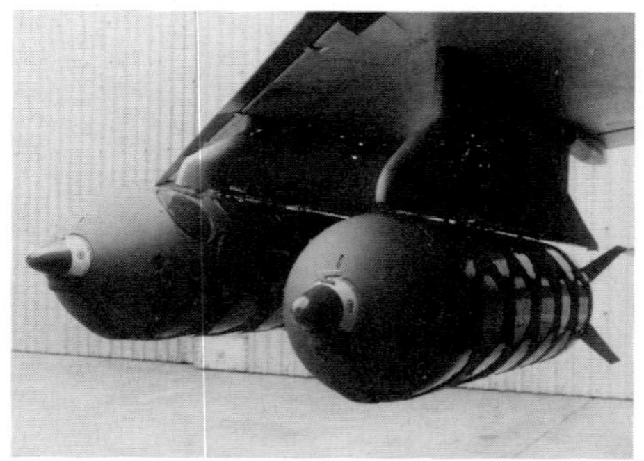

CB 470 cluster bombs in position on an aircraft's wing pylons

A CB 470 container together with its load of 40 bomblets

CHARACTERISTICS

Bomblet
Weight, fully loaded: 6.2 kg
Weight, explosive filling (RDX/TNT): 1.4 kg
Bomblet ejection time: 0.8s minimum
Bomblet arming delay time: 1 sec
Bomblet functioning delay time: 0.65 sec
Number of bomblets per container: 40
CB 470 Container
Weight loaded: 450 kg
Weight empty: 170 kg
Weight of tail unit: 11 kg
Weight of other components: 4 kg
Length overall: 2600 mm
Diameter of body: 419 mm
Distance across tail fins: 640 mm
Suspension points: Twin lug NATO Standard (355.6 mm/14 in) or twin lug Warsaw Pact Standard (250 mm)
Release height: 30 – 300 m
Release speed range: 850 – 1100 km/h at 30 m height

STATUS

CB 470 is in production and is already in service with an unspecified air force.

CONTRACTOR

Armscor, Private Bag X337, Pretoria 0001, South Africa.

4540.303

SA AIRCRAFT BOMBS

Various types of free-fall bombs and special to purpose fuzes of indigenous design and manufacture, meeting NATO standards, are produced for use by aircraft of the South African forces. Current production includes the following:

12.5 kg Practice Smoke and Flash Aircraft Bomb

This 12.5 kg practice bomb is intended for day and night high speed bombing practice, simulating the use of the SA 460 kg aircraft bomb up to a release speed to Mach 0.95. It has a cast iron body containing the impact striker mechanism, safety and arming devices and the suspension lug. The vaned aluminium alloy tail section contains 0.92 kg of pyrotechnic filling and the detonator/burster. Length is 765 mm and diameter 95 mm.

120 kg Aircraft Bomb

This is a conventional free-fall bomb consisting of a cast or forged body filled with RDX/TNT or TNT in the HE model, or filled with HE substitute in the practice model. The body can be fitted with a penetration nose cone and cup support, a charging tube (front and rear) for electronic fuzing, and an arming wire assembly for mechanical fuzing. The filling is either RDX/TNT 60/40 or TNT and the standard fuze is the SA771 mechanical nose fuze. The latter is a selective arming delay time point detonating instantaneous or delay action fuze with back-up from the SA772 tail fuze. This tail fuze is impact firing and also arming delay time selective, with either non-delay or various delayed firing options in the millisecond range.

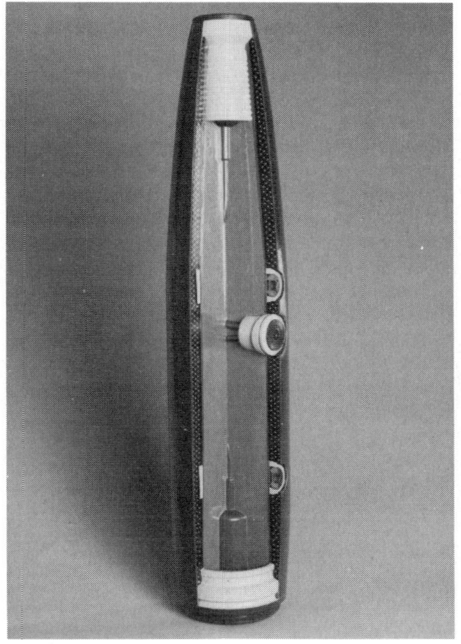

Cutaway model of 120 kg shrapnel aircraft bomb

Production line for 250 kg aircraft bombs

The empty bomb case weighs approximately 60 kg, or 107 kg assembled and filled.

120 kg Shrapnel Aircraft Bomb
This is a low-drag bomb designed to be carried by high performance aircraft. It can be used with existing 120 kg low-drag bomb weapon systems without adjustment, clearing or qualification. The bomb is filled with 27 kg of RDX/TNT but other types of explosives and mixtures are available. Pre-fragmentation is achieved by a layer of steel balls cast in epoxy resin between the outer fibreglass skin and the explosive core. The bomb is produced with charging tubes in order to accommodate fuzes with internal arming wires. The number of shrapnel balls carried by the bomb will vary with ball diameter as follows: 6.7 mm diameter, 42 000; 7.9 mm. 26 000; 8.7 mm, 19 500. Bomb weight filled and plugged is 114.5 kg (plus or minus 4 kg) and the 702 mm long tail (span 325 mm) weighs a nominal 7.5 kg. A quick-acting retard tail for use at altitudes of 60 m or under is under development for this bomb.

250 kg Aircraft Bomb
This is of similar construction and configuration to the 120 kg bomb, but has an empty weight of about 116 kg, 219 kg assembled and filled, with a tail assembly weighing 9 kg.

250 kg Practice Aircraft Bomb
This is a practice low-drag bomb designed to be carried externally by high performance jet aircraft and can be used with existing 250 kg low-drag bomb weapon systems. The body of the bomb is filled with steel shot and constructed from reinforced fibreglass and polyurethane foam, with two embedded steel rings and lug wells. On impact the bomb disintegrates, leaving few parts to clear up or create ricochets on subsequent practice sorties. The bomb is fitted with standard NATO lugs although other fittings can be fitted. Weight plugged is 215 kg (plus or minus 6 kg) and the tail unit is a nominal 9 kg.

460 kg Aircraft Bomb
This bomb is made as either the HE version with a filling of Torpex 4B for the cast steel body, or HE substitute for the practice version. Other components are a penetration nose and cup support, front and rear charging tubes for electronic fuzing, and an arming wire assembly for mechanical fuzing. The same fuzing arrangements as for the 250 kg bomb are available for the heavier weapon. Weights are: empty bomb, approximately 225 kg; bomb body assembled and filled, approximately 438 kg; tail assembly 28 kg.
STATUS
In production and available for export to approved customers.
CONTRACTOR
Armscor, Private Bag X337, Pretoria 0001, South Africa.

SPAIN

1843.303
LANCO RC-06-100
This re-usable unit is designed to carry and fire six heavy 100 mm INTA S12 rockets. It can be fitted to any aircraft with 14- and 30-inch (356 and 762 mm) NATO attachments. The conversion from 14 to 30 inches is carried out by changing the position of the rear attachment.
OPERATION
Rocket firing is controlled by an electronic intervalometer incorporated in the launcher. Single rocket firing or continuous ripple firing can be selected by the pilot or set in on the ground if this unit is not available. The firing sequence can be controlled between 20 and 70 ms by adjusting the intervalometer without removing it from the rocket launcher.
CHARACTERISTICS
Length: 3.4 m approx
Weight (empty): 100 kg
Weight (loaded): 460 kg
CONTRACTOR
Construcciones Aeronauticas SA, Rey Francisco 4, Madrid-8, Spain.

LANCO RC-06-100

6181.303
EXPAL BME 330 CLUSTER BOMB
The BME 330 cluster bomb weighs approximately 330 kg and contains 180 bomblets. As the bomb is released from an aircraft a parachute deploys to cause the bomb casing to fall apart to allow ejection of the bomblets under gas pressure. Each bomblet is armed for activation on contact by the deploying of its own individual parachute. Each cluster bomb has its own inertial safety with a clockwork mechanism keeping the firing train out of line for a safety period after dropping. A typical area covered with bomblets could be 100 m long and 40 m wide, depending on aircraft speed and altitude.

Three types of bomblets can be used, with identical dimensions. There is an anti-personnel bomblet filled with hexolite and using a fragmentation case; an incendiary filled with white phosphorous (WP); and a hollow charge warhead bomblet using RDX.
STATUS
In production.
CONTRACTOR
Expal Explosivos Alaveses SA, Apartado 198, 010 80 Vitoria, Spain.

5207.303
EXPAL BRFA 330 BOMB
The BRFA 330 bomb has been designed specifically for attacking aircraft runways and can be carried by high performance aircraft. Once released from the aircraft a parachute is deployed which retards the bomb and ensures that it is in the vertical position. A rocket motor is then ignited and propels the bomb to the target. The bomb penetrates the runway to a depth of some 600 mm before it explodes. Area of damage is quoted as 180 m²–250 m² depending on the reinforcements in the concrete.
CHARACTERISTICS
Length: 3.2 m
Diameter of bomb: 300 mm
Diameter over fins: 600 mm
Weight of bomb: 330 kg
Weight of explosive: 75 kg
CONTRACTOR
Expal Explosivos Alaveses SA, Apartado 198, 010 80 Vitoria, Spain.

5245.303

EXPAL BRP RETARDED BOMBS

The BRP range of low drag parachute retarded bombs have been designed specifically for fast low-level bombing by aircraft such as the F-4 Phantom. When the bomb is released, the arming wires activate the parachute system, braking the bomb while the aircraft pulls away. The bombs have safety devices which prevent explosion of the bomb in the event of any failure in any of its components, or if launching altitude does not guarantee a minimum safety distance. All Expal bombs (**5244.303**) can be transformed into BRP retarded bomb configuration by replacing the standard tail unit by the BRP tail and fitting a KAPPA nose fuze.

For use at very low altitudes there are 'super' retarded bombs with a rapid-acting parachute tail. The types involved are the BRP.S 500, BRP.S 250 and BRP.S 125.

STATUS

In production.

CONTRACTOR

Expal Explosivos Alaveses SA, Apartado 198, 010 80 Vitoria, Spain.

CHARACTERISTICS

Type of bomb (kg)	**500**	**375**	**250**	**50**
Diameter of body (mm)	360	330	290	180
Length (m)	3	2.855	2.285	1.425
Diameter over fins (mm)	510	470	404	250
Distance between lugs (mm)	355.6	355.6	355.6	355.6
Weight of explosive (kg)	206	170	112	20

5244.303

EXPAL GENERAL-PURPOSE LOW DRAG BOMBS

The BR range of general-purpose low drag bombs have been designed to be carried by high performance aircraft such as the F-4 Phantom used by the Spanish Air Force. The bombs are delivered with a nose and tail fuze of either the instantaneous or delayed action type. The bombs can be fitted with a variety of lugs including a single central lug or two lugs. The tail of the bomb is detachable and can be replaced by a braking kit as fitted to the Expal BRP retarded bomb (**5245.303**).

CHARACTERISTICS

Type of bomb (kg)	**1000**	**500**	**375**	**250**	**125**	**50**
Diameter of bomb (mm)	460	360	330	290	240	180
Length (m)	3.82	2.94	2.805	2.15	1.82	1.395
Diameter over fins (mm)	538	500	500	440	360	275
Distance between lugs (mm)	762	355.6	355.6	355.6	355.6	355.6
Weight of explosive (kg)	475	210	170	115	55	20

STATUS

In production.

CONTRACTOR

Expal Explosivos Alaveses SA, Apartado 198, 010 80 Vitoria, Spain.

SWEDEN

5224.303

SWEDISH AIRCRAFT ARMAMENT

Listed below is a resumé of Swedish aircraft armament.

Bombs

75 kg M62 (Flash)
80 kg M60 LYSB (HE-fragmentation)
100 kg (Flash)
120 kg M51 SB (HE)
120 kg M63 FFV Virgo (para-fragmentation)
250 kg M50 SB (HE)
500 kg M56 MB (HE)
600 kg M50 MB (HE)

Cannon

20 mm – four 20 mm in Saab-32 Lansen
30 mm – two 30 mm in Saab-35 Draken
30 mm – one Swiss 30 mm KCA cannon (**1269.303**) mounted in pod under Saab-JA-37

Rockets

75 mm M57AC anti-tank (single or in pods of 19 rockets)
75 mm M57B fragmentation (single or in pods of 19 rockets)
75 mm M57 7 kg (single or in pods of 19 rockets)
135 mm M56 42 kg HE (single or in pods of 6 rockets)
135 mm M56B fragmentation (single or in pods of 6 rockets)
135 mm M60 42 kg HE (single or in pods of 6 rockets)
135 mm M70AP fragmentation (single or in pods of 6 rockets) (**1513.303**)
145 mm M49 PSRAK anti-tank (single)
150 mm and 180 mm (no details available)

5216.303

FFV UNI-POD 0127

The FFV Uni-Pod 0127 has been designed for installation on jet trainers, light aircraft and helicopters. The pod is fitted with a single 0.5 inch (12.7 mm) Browning AN-M3 machine gun which has a rate of fire of up to 1250 rounds per minute. A total of 220 rounds of belted ammunition are carried and the spent cartridge cases and links are ejected or, as an option, may be collected in the pod. The machine gun can fire a variety of ammunition including ball, AP, API, APIT, and the API (HC); the latter will penetrate 13 to 19 mm armour plates at a range of 800 m at impact angles between 40 and 90°. A special damper on the upper part of the pod attenuates the recoil forces and the load on the suspension lugs is less than 2000 N (204 kg).

The latest version of the pod is equipped with a remotely controlled gun charger which makes it possible to charge the gun, and automatically recharge it in the air in case of a normal jam, make it safe, and count the rounds fired while airborne. The drag factor is as low as 0.13.

CHARACTERISTICS

Length: 1916 mm
Diameter: 360 mm
Suspension: 14-inch (356 mm) MIL-A-8591E
Weight loaded: 118 kg
Weight empty: 90 kg
Power supply: 28 V DC

FFV Uni-Pod 0127

STATUS

Certified for various aircraft and helicopters.

CONTRACTOR

FFV Aerotech, S-581 82 Linköping, Sweden.

4430.303

FFV 30 mm ADEN GUN POD

The FFV 30 mm Aden gun pod has been designed for installation on jet aircraft for air-to-air and air-to-ground use. It is of the so called self-contained type and houses the 30 mm Aden gun, a magazine with 150 rounds and an option for 200 rounds, an electrical power unit, and a relay box. The gun is cocked pneumatically on the ground and a safety device prevents firing until the aircraft is airborne. The suspension system may be either 30- or 14-inch (762 or 356 mm) NATO standard lugs, MIL-A8591E, or Douglas T-shape lugs.

CHARACTERISTICS

Length: 3850 mm
Diameter: 500 mm
Weight loaded: 364 kg
Weight empty: 290 kg
Power supply: 28 V DC and 400 V AC, 400 Hz, 0.5 A 3-phase

STATUS

Produced in large numbers to arm the Saab-Scania AJ 37 Viggen and 105 aircraft.

CONTRACTOR

FFV Aerotech, S-581 82 Linköping, Sweden.

FFV 30 mm Aden gun pod on AJ 37 Viggen

1513.303

135 mm BOFORS AIR-TO-GROUND ROCKET SYSTEM M70

The 135 mm air-to-ground rocket system manufactured by AB Bofors is intended as part of a range of armament developed for the Saab AJ 37 Viggen. It consists of a launcher carrying six rockets and a pulse distributor of electronic type, which can be set for single or ripple firing. The launcher has a low-drag design. The rockets are protected against thermal heating during high and low altitude flight at supersonic speeds. Standard NATO-type suspension lugs can be provided so that the system is compatible with a wide range of tactical aircraft.

The rockets can be provided with either a general-purpose or an armour-piercing fragmentation warhead. A practice head is also available. Two fuzes can be used: a direct-action fuze with post-impact delay, or a highly sensitive direct-action fuze.

The rocket motors give high velocity and low dispersion, which gives a time of flight of approximately 3.5 s at 2000 m slant range in typical launching conditions.

CHARACTERISTICS

Total weight with six rockets: 400 kg
Weight (empty): 125 kg

Length: 3226 mm
Diameter: 486 mm
Pulse distributor for single or ripple firing
Firing interval in ripple firing: 0.1 s
135 mm rocket
Weight of rocket motor: 25 kg
Burning time: 2 s
Velocity at burn-out: 600 m/s
Weight
GP warhead with fuze: 21 kg
AP fragmentation warhead with fuze: 20 kg
GP warhead charge: 3.7 kg
AP fragmentation warhead charge: 5 kg
Practice head: 21 kg

STATUS

In service with the Swedish Air Force on the Viggen aircraft. Successful installations and trials have also been made with the F-4 Phantom, A-7 Corsair, A-4 Skyhawk and Hunter.

CONTRACTOR

AB Bofors, S-691 80 Bofors, Sweden.

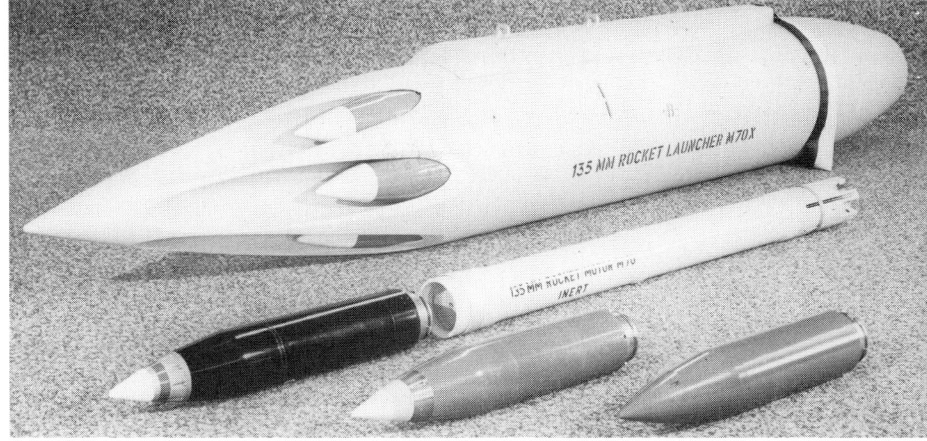

Bofors 135 mm rocket system M70X showing launcher and rockets

6182.303

BOFORS 80 kg AIRCRAFT PARACHUTE FLARE SUPER LEPUS

The 80 kg aircraft parachute flare SUPER LEPUS is an improved version of the LEPUS parachute flare and all types of aircraft for which LEPUS has been approved can carry and release the SUPER LEPUS. The flare is used to illuminate target areas at night and can be carried and released from aircraft flying at supersonic speeds. It can also be dropped from low altitude. It has an extremely high luminous intensity of more than 7 million candela and at a burning height of 1500 m can illuminate an area of approximately 20 square km.

Among aircraft tested and approved to carry the SUPER LEPUS are the Hunter, F-104 Starfighter, F-4 Phantom, Jaguar, Tornado, Harrier, A-4 Skyhawk, F-5, MB-326 and MB-339.

CHARACTERISTICS

Total weight: 77 kg
Length: 1970 mm
Body diameter: 201 mm
Tail span: 408 mm
Weight of flare composition: 26 kg
Luminous intensity: 7-8 Mcd
Burning time: 3 m/s
Mean descent time: 4 m/s
Maximum carrying speed at ground level: Mach 1.3
Maximum release speed: Mach 1.1
Fuze setting, manual: 2.5 to 28 sec continuous
Delay arming: 2 sec
Electrical connector location: 55 mm in front of rear lug (length of electrical cable adjustable)

Bofors 80 kg aircraft parachute flare SUPER LEPUS

Arming circuit resistance: 6.3 ohms
Power supply required: 28 V dc

STATUS

In production. In service with the air forces of France, West Germany, Sweden and the United Kingdom.

CONTRACTOR

AB Bofors, S-691 80 Bofors, Sweden.

SWITZERLAND

4134.303

OERLIKON 81 mm AIRCRAFT ROCKET

The Oerlikon 81 mm SURA-D aircraft rocket is an upgraded version of the SURA-FL. This unguided solid fuel rocket with sliding fins is intended for air-to-ground use by fighter aircraft, and has proved effective when fired from helicopters and light reconnaissance aircraft. It can be used to engage point or area targets.

The fin assembly is a sliding fit on the rocket body and has a dual function; before firing the assembly acts as the front suspension, engaging with the front suspension lugs on the aircraft underwing rail. Lower rockets in the same cluster can also be engaged by means of the fin slots and by an additional integral suspension ring located forward of the main fin ring. The rockets are fired electrically, and on firing the sliding fin ring acts as a guide until it is met by the rear cone of the motor. It is then carried on with the rocket, stabilising it in flight.

Based on a standard propulsion element, the TWK 007, a variety of rocket types is produced by the use of differing fuze arrangements and warheads. These are summarised in tabular form.

Type RAK 052 rocket launchers with NATO 14 inch 1000 lb class suspension for helicopters

Rocket launchers for fighter aircraft with standard suspension

Calibre: 81 mm
Length: 1077 – 1212 mm (depending on type of warhead)
Initial mass: 12.7 – 14.2 kg (depending on type of warhead)
Max velocity at +18°C: Approx 530 – 595 m/s
Ignition voltage: 24 – 28 V
Propulsion unit mass: 8.4 kg
Mean thrust at +18°C: 7300 N
Burning time: 0.9 s

SURA-D CONFIGURATIONS

SURA-D-rocket with:	Rocket type	Warhead type	Propulsion unit
Practice shell	3 kg RAK 051	USE-3	TWK 007
	4.5 kg RAK 048	UGK 033	
Marker shell	4.5 kg RAK 049	USK 035	
Fragmentation explosive shell	3 kg RAK 053	US-3	
	4.5 kg RAK 050	SSK 031	
Incendiary blast shell	3 kg RAK 052	UIS-3	
Hollow charge warhead	3 kg RAK 047	PI-3	

STATUS
In production. SURA rockets are in service with numerous air forces. Installations have been made on many aircraft and helicopter types.

CONTRACTOR
Machine Tool Works Oerlikon-Bührle Ltd, 8050 Zurich, Switzerland.

1995.303
OERLIKON SNORA 81 mm ROCKET

The Oerlikon 81 mm SNORA rocket is an unguided solid fuel rocket with folding fins. It has a pure internal burning propulsive element. This rocket can be fired from multi-tube launchers both in subsonic air-to-ground engagements and ground-to-ground roles. Pods are used for supersonic air-to-ground missions. The design has been carried out with the needs of fighter aircraft, helicopter, tracked and wheeled vehicle installation in mind. The SNORA rocket is provided with the necessary safety devices for storage, transportation, handling, and firing. Correct functioning is ensured over a temperature range of –45 to +65°C. Based on a standard propulsion element, the TWK 006, a variety of types of rocket are produced by the use of differing fuze arrangements and warheads. These are summarised in the following paragraphs.

CHARACTERISTICS
Propulsion element type TWK 006
Weight: 8.7 kg
Mean thrust at +18°C: 11 000 N
Total impulse at +18°C: 910 kg/s
Action time at +18°C: 0.85 s
Length: 1033 mm

Rocket type RAK 022
Length: 1548 mm
Weight: 15.7 kg
V max from ground at +18°C: Approx 670 m/s
Range ground-to-ground at +18°C: Approx 10 km
Flight time over 10 km at +18°C: 50 s
Warhead type UGK 028: Practice
Weight: 7 kg

Rocket type RAK 023
Length: 1416 mm
Weight: 13.2 kg
V max from ground at +18°C: Approx 820 m/s
Range ground-to-ground at +18°C: Approx 10 km
Flight time over 10 km at +18° C: Approx 50 s
Warhead type SSK 031: Fragmentation explosive
Weight
Trotyl explosive: 1 kg
Warhead ready to fire: 4.5 kg
Number of spherical fragments: 702
Fragment weight: 2 g
Nose fuze type RKX332: Mechanical impact
Length of fuze: 46 mm
Muzzle safety: Minimum 30 m
Armed after: Max 100 m
Weight
Explosive: 9.5 g
Fuze ready to fire: 320 g
Response angle: 15°
Response sensitivity on 2 mm aluminium: 100 m
Detonator safety
Rain safety

Rocket type RAK 024
Length: 1548 mm
V max from ground at +18°C: Approx 670 m/s
Range ground to ground at +18°C: Approx 10 km
Flight time over 12 km at +18°: Approx 50 s
Warhead type SSK 029: Fragmentation explosive
Weight
Trotyl explosive: 1.7 kg
Warhead ready to fire: 7 kg
Number of spherical fragments: 1215
Fragment weight: 2 g
Nose fuze type RKX 332: Mechanical impact

Rocket type RAK 025
Length: 1783 mm
Weight: 19.7 kg
V max from ground at +18°C: Approx 520 m/s
Range ground to ground at +18°C: Approx 10 km
Flight time over 12 km at +18°C: Approx 50 s
Warhead type SSK 032: Fragmentation explosive
Weight
Explosive: 2.8 kg
Warhead ready to fire: 11 kg
Number of spherical fragments: 2133
Fragment weight: 2 g
Nose fuze type RKX 332: Mechanical impact

Rocket type RAK 026/RAK 054
Length: 1510 mm
Weight: 13.2 kg
V max from ground at +18°C: Approx 820 m/s
Flight time over 2 km at +18°C: Approx 3.2 s
Warhead type PHK 030: Hollow charge
Weight
Explosive: Approx 1 kg
Warhead ready to fire: 4.5 kg
Warhead with fragmentation effect
Penetration in steel: Minimum 350 mm
Response angle: 15°
Ignition via contact cap
Base fuze type RBK 331: Electro-mechanical, with mechanical and electronic time element
Detonator safety : to 30 m
Armed after: 100 m
Rain safety
Ignition energy from propulsion element

A range of rocket launchers suitable for fitting to fighter aircraft and helicopters is produced by Oerlikon, and these are detailed in entry **4135.303** which follows.

STATUS
No longer in production.

CONTRACTOR
Machine Tools Works Oerlikon-Bührle Ltd, 8050 Zurich, Switzerland.

4135.303
OERLIKON ROCKET PODS

A range of rocket pods for firing Oerlikon SNORA aircraft rockets is produced for fitting in various fighter aircraft and helicopters. The main characteristics of these are listed in the following tables.

Rocket Launcher Type SAL-12-80 (RWK 020)
Suitable for use on all types of sub- and supersonic aircraft, the SAL-12-80 pod is of lightweight metal construction with a replaceable frangible nose fairing to close the 12 firing tubes during flight. It is fitted with a standard NATO and centre lugs suspension system.

CHARACTERISTICS
Calibre: 81 mm
Number of tubes: 12
Length: 2550 mm (approx)
Diameter: 396 mm (approx)
Weight: 69 kg (empty); 227 – 305 kg (with rockets, depending on type of shell)
Firing: Single, or salvoes of 2, 6, or 12 rockets
Rate of fire: 12 rounds/s

Rocket Launcher Type SAL-6-80 (RWK 021)
This is a six-tube launcher for use on supersonic aircraft.

CHARACTERISTICS
Calibre: 81 mm
Number of tubes: 6
Length: 2600 mm
Diameter: 296 mm
Weight: 48 kg (empty); 127 – 166 kg (with rockets, depending on type of shell)

Type SAL-12-80 rocket launcher installed on a fighter aircraft

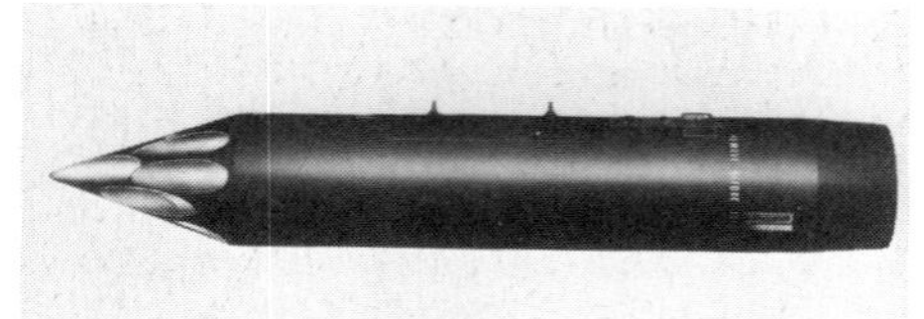

Type SAL-12-80 rocket launcher (RWK 020)

Type AL-6-80 rocket launcher for subsonic and light aircraft

Type SAL-6-80 rocket launcher for supersonic aircraft

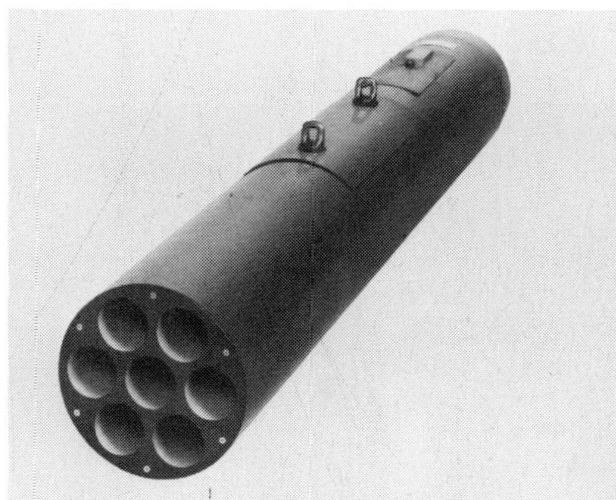

Type HL-7-80 helicopter rocket launcher

Firing: Single, or salvoes of 2, 3, or 6 rockets
Rate of fire: 10 rounds/s

Rocket Launcher Type AL-6-80 (RWK 024)
This is a six-tube launcher for use on subsonic and light aircraft.
CHARACTERISTICS
Calibre: 81 mm
Number of tubes: 6
Length: 2214 mm
Diameter: 296 mm
Weight: 38 kg (empty); 117 – 132 kg (with rockets, depending on type of shell)
Firing: Single, or salvoes of 2, 3, or 6 rockets
Rate of fire: 10 rounds/s

Rocket Launcher Type HL-7-80 (RWK 022)
This launcher is designed for use on helicopters.
CHARACTERISTICS
Calibre: 81 mm
Number of tubes: 7

Type HL-7-80 rocket launcher installed on a helicopter

Length: 1800 mm
Diameter: 296 mm
Weight: 33 kg (empty); 125 – 143 kg (with rockets, depending on type of shell)
Firing: Single, or salvoes of 2, 3, or 7 rockets
Rate of fire: 10 rounds/s

STATUS
No longer in production.
CONTRACTOR
Machine Tools Works Oerlikon-Bührle Ltd, 8050 Zurich, Switzerland.

1269.303
OERLIKON 30 mm GUN TYPE KCA
The aircraft gun type KCA is of 30 mm calibre and has been developed by Oerlikon, Switzerland. The weapon is a further example of the development of the Mauser revolver principle which has been almost universally adopted for single barrel aircraft cannon since the end of the Second World War.

The KCA differs, however, from almost all comparable guns in the present generation in using a new design of ammunition in which the ratio of propellant charge to projectile weight is over 2:1 whereas in most other guns, the ratio is likely to be 3:2. In addition, the KCA ammunition has a higher projectile weight than other 30 mm types.

As a result of these factors and lightweight design of the moving parts of the weapon, a firing rate of 1350 rounds per minute is obtained and a muzzle velocity of over 1030 m/s. The weapon is therefore well suited to ground attack as well as air-to-air engagements, and for the former role a long gun range is required to give the attacker a wide firing bracket.

The Oerlikon KCA has a four chamber revolver mechanism so that before firing the gun mechanism has to be cycled three times to charge all chambers. This can be done in the air by a pneumatic re-loading system.

Ammunition for the KCA is available in a range of types: TP, HEI, SAPHEI, and AP.
CHARACTERISTICS
Calibre: 30 mm
Length of barrel: 1976 mm
Length of gun overall: 2691 mm
Width of gun: 242.5 mm

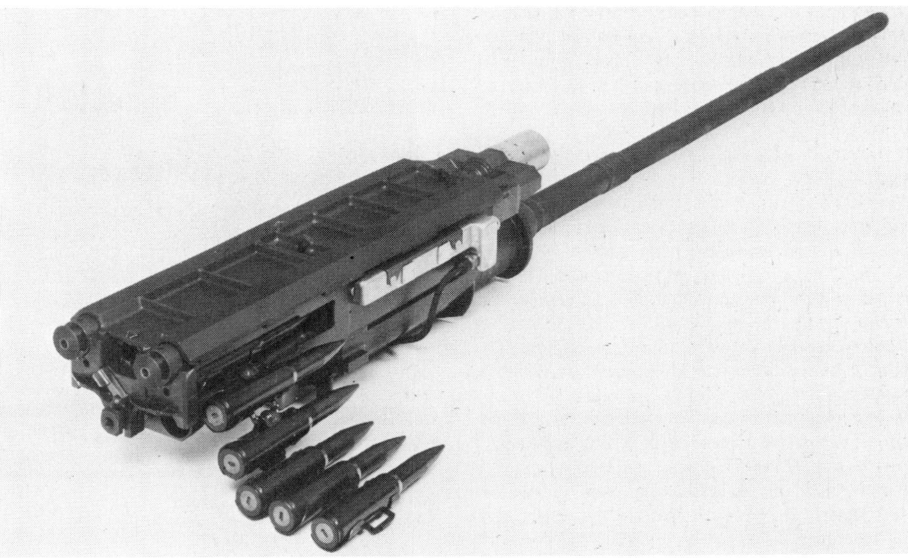

Oerlikon Type 30 mm KCA cannon

Height of gun: 249.5 mm
Gun weight: 136 kg
Muzzle velocity: 1030 m/s
Rate of fire: 1350 rounds/minute
Accuracy: 50% rounds inside 2.5 mils
Round weight: 0.89 kg with steel cartridge
Projectile weight: 0.360 kg
Range: Over 2000 m gun stationary

STATUS
The Oerlikon KCA is in series production, and has been selected by the Swedish Air Force for installation under the Saab Viggen attack aircraft.
CONTRACTOR
Machine Tool Works Oerlikon-Bührle Ltd, 8050 Zurich, Switzerland.

1272.303
HISPANO OERLIKON 20 mm TACTICAL AIR ARMAMENT
Two types of 20 mm calibre aircraft cannon were manufactured by Hispano Suiza in Switzerland and by their UK company, British Manufacture and Research Co Ltd of Grantham, Lincolnshire. Both types, the 20 mm Mk 5 and 20 mm Mk 2* are similar and are derived from the original Hispano type 404 weapon. The Mk 5 model is lighter in weight than the Mk 2* and has a higher rate of fire. Although no longer part of the company's programme, this weapon remains in extensive service in many parts of the world.

A wide range of 20 mm ammunition types is available, including armour piercing, tracer, ball, and HE incendiary. Different types of fuze are also available. The round has a brass case and is percussion fired. Ammunition is normally supplied in 60-round belts or as individually packed rounds.

A wide variety of tools and accessories are available for 20 mm Mk 2* and Mk 5 guns.
CHARACTERISTICS
20 mm Mk 5 gun
Calibre: 20 mm
Gun weight with feed: 42 kg
Muzzle velocity: 850 m/s
Rate of fire: 580 – 640 rounds/minute
Length of gun: 2052 mm
Round weight: 1367 g
Projectile weight: 138 g
Range: 1600 m, to velocity 300 m/s, gun stationary
STATUS
Hispano 20 mm aircraft cannon are in service with over 30 air forces. Over 98 000 20 mm guns of all types have been made in the UK alone. It is now regarded as obsolete though many examples can be expected to remain in service for a considerable period to come.

UNION OF SOVIET SOCIALIST REPUBLICS

3181.303

SOVIET AIRCRAFT ARMAMENT NOTE

Unclassified information on Soviet aircraft armament tends to be limited in both quantity and precision but the increasing emphasis on this aspect which has been detected by observers of Soviet military activities calls for a summary of the known characteristics of Soviet equipment of this nature.

Bombs

The following types of free-fall bomb are known to have entered service with Soviet forces and those of some allied nations:

(1) HE – High explosive bombs are available in 100, 250, 500, 750 and 1000 kg sizes
(2) Incendiary – Incendiary bombs are available in 100, 250 and 1000 kg sizes
(3) AP – Armour piercing bombs are available in 250 and 500 kg sizes
(4) Concrete dibbers are known to be used in both 250 and 500 kg sizes, the latter being designated the M62
(5) Cluster bombs have been in service for a number of years and the following varieties are known to exist: HE, incendiary, HEI, fragmentation, and anti-tank (shaped charge)
(6) Napalm bombs are available.

Categories (4) and (5) probably include one or more types of retarded bomb mechanism.

Unguided Rockets

Unguided rockets are employed in both air-to-air and air-to-ground roles and are carried by fixed-wing aircraft and helicopters. In the case of the heavier rockets, use is restricted to ground attack (or anti-ship) roles. The following types have been identified:

57 mm: The rocket itself is designated S-5 and is of 55 mm diameter for use with 57 mm launcher tubes. A variety of warheads is available including: chaff, anti-tank, HE, and fragmentation. Installations have been noted on such types as the Fitter and Fishbed fighter aircraft and the Hind helicopter. A range of launcher pods has been observed of 8, 16, 19 and 32 round capacities and designated: UV-8-57; UV-16-57; UV-19-57; UV-32-57.

137 mm: This rocket is known as the M-100 and has a launch weight of 15.6 kg.

160 mm: No information is available on this rocket apart from its designation (S-16) and the fact that it can be launched from the Hind-A helicopter.

190 mm: Of 190 mm calibre, and designated TRS 190, this rocket has a launch weight of 46 kg and can be launched by the MiG-19.

212 mm: Little is known apart from the designation, ARS 212, launch weight of 116 kg and use with the MiG-19.

220 mm: No information is available, but some sources report the existence of a 210 mm rocket known as the S-21 which is used with the Hind-A.

240 mm: Known as the S-24 and associated with the MiG-17 and MiG-21 (one rocket per pylon) and the Hind helicopter.

325 mm: No details available.

Cannon and Machine Guns

The following types are known to be in service:

12.7 mm: UBK machine gun has been used in the MiG-15UTI trainer aircraft and more recently in the Mi-24 helicopter. The latter is probably the most important application of this calibre of machine gun in aircraft, beginning with the single chin mounting in the Hind. This was followed by a multi-barrelled, Gatling-type gun with four 12.7 mm barrels mounted in a turret that provided coverage of 70° left and right of the aircraft datum in azimuth, and elevation movement from –60 to +15°. The estimated range of this weapon is about 1500 m. It has been suggested that the four-barrel 12.7 mm gun may be succeeded on the Hind helicopter by the GSh-23 (see below) as used on the MiG-21.

23 mm: Two 23 mm single-barrel cannon have been identified, the NS-23KM (Nudelman-Suranov) and the NR-23 (Nudelman-Rikter). In many installations the two are probably interchangeable. The NS-23 is understood to be of earlier design and to have a lower rate of fire, ie 550 rounds/minute compared with about 850 rounds/minute for the NR-23. Widely used in fighter, bomber, and certain transport aircraft.

Mi-24 Hind ground attack helicopter is armed with four UV-32-57 rocket launcher pods, rails for anti-tank guided missiles, and a four-barrel Gatling-type 12.7 mm machine gun in a remotely aimed chin turret

Reloading UV-32-57 rocket launcher pods on Mi-24 Hind ground attack helicopter with 57 mm unguided rockets

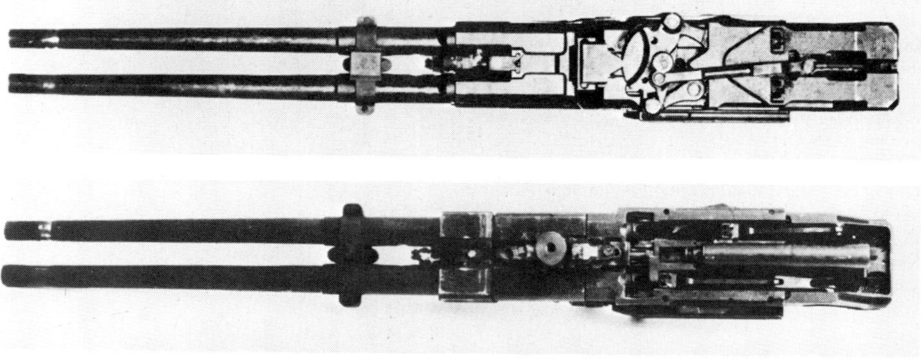

Top and bottom views of the Soviet GSh-23 23 mm twin-barrel gun

Badger bomber's twin NR-23 23 mm guns in remotely controlled ventral turret

Nose-mounted 12.7 mm machine gun on Mi-6 (Hook) helicopter

23 mm GSh-23: This is a twin-barrel 23 mm gun employed both as an internal weapon and in the GP-9 gun pack for belly mounting. In the latter case, as used on the Fishbed 'E', the pack has a capacity of 200 rounds. Cyclic rate of fire is about 3000 rounds/minute and effective range 3000 m. A version of this weapon has been noted that is designed for mounting within a pod for carriage on an aircraft underwing attachment point. This variant has been seen on some Yak-36 Forger carrier aircraft.

23 mm: A 23 mm six-barrelled cannon is believed to be in use on some recent Soviet aircraft but the only known installation of this is on MiG-27 aircraft.

30 mm: The NR-30 cannon is used on the Su-7, Su-20, Yak-28, and MiG-19S, and has a rate of fire of 850 rounds/minute with ball, HE, and HEI ammunition. A capacity of 70 rounds is typical.

30 mm: The 2NR cannon is used in Fitter A and Fitter C aircraft.

37 mm: The N-37 (or NS-37) cannon has a cyclic rate of fire of 400 rounds/minute and is used in such aircraft as the MiG-15, MiG-17, MiG-19, Yak-25, and Yak-27R.

Close-up of Badger tail gun installation with two NR-23 23 mm cannon

UNITED KINGDOM

5185.303

30 mm ADEN CANNON

The 30 mm Aden cannon was developed shortly after the end of the Second World War by the ADE (now the Royal Armament Research and Development Establishment at Fort Halstead), and the Royal Small Arms Factory (RSAF) at Enfield Lock. The first production model was known as the Mk 1 and the current production cannon is the Mk 4. An improved Mk 5 version was also developed, known as the Straden, which had a higher rate of fire (see below).

The design of the weapon is simple with the whole assembly contained within the cradle unit; installation mounting points are at the rear and front barrel entry position of the cradle. Mounting of the cannon is possible at any angle on the longitudinal axis to suit the specific aircraft requirement, and both left- or right-hand feed versions are available. Cocking is pneumatic with provision for manual operation, and the ammunition, which is belt fed (disintegrating link), is electrically fired.

The Aden operates in the following manner. A cylinder (revolving drum) has five cartridge chambers, into which the cartridges are loaded in two stages, indexed by the movement of a gas operated slide. Firing takes place when a round is in line with the barrel, the whole mechanism being safely locked at the moment of firing; a round cannot be fired if the barrel/cylinder is not correctly locked in position, and with a round in line with the barrel. When a round is fired, the gas pressure passing through the barrel gas port acts via a piston onto a slide mechanism, which, travelling rearward within the cradle unit revolves the breech cylinder and feed sprockets, bringing a round into position for feeding into a chamber. The slide mechanism is then moved forward by the action of return springs, further indexing the cylinder and feeding into a chamber. The Aden cannon can be mounted internally in an aircraft, in a pod for installation under a pylon, or in a blister pod which is mounted under the fuselage of the aircraft.

TYPICAL INSTALLATIONS

Harrier with two cannon, each with 130 rounds of ammunition.

Hawk trainer with a blister pod with one 30 mm cannon.

HAL Gnat with two cannon, each with 115 rounds of ammunition.

HF-24 Marut with four Mk 4 guns, each with 120 rounds of ammunition.

Hunter with four cannon, each with 150 rounds of ammunition.

Jaguar (RAF) with two cannon each with 150 rounds of ammunition per cannon.

CHARACTERISTICS

Type: Aden Mk 4
Calibre: 30 mm
Gun length: 1590 mm
Total gun weight: 87 kg
Barrel length: 1080 mm
Barrel weight: 12.25 kg
Muzzle velocity: 790 m/s
Rate of fire: 1200/1400 rounds/minute
Recoil load: 31.4 kN
Cocking: Pneumatic
Firing: Electric (26 V DC)
Ammunition
Practice round (Mk 4*Z)
High explosive round (Mk 6*Z)
Armour piercing (Mk 1*Z)
Armour piercing incendiary (under development)
Typical system weight: 196 kg (gun + 200 rounds)

DEVELOPMENT

Enfield Lock has completed the development programme designed to increase the effectiveness of the Aden Mk 5 cannon by increasing its rate of fire from 1200/1400 rounds/minute to 1500/1700 rounds/minute combined with improved reliability. Guns modified to this standard are termed Aden Mk 5 (Straden) cannon. This work has now been absorbed into the 25 mm Aden programme (see **4870.303**, below).

CONTRACTOR

Royal Ordnance, Small Arms Division, Enfield Lock, Middlesex, England. Enquiries to Royal Ordnance, Griffin House, 5 Strand, London WC2N 5BB, England.

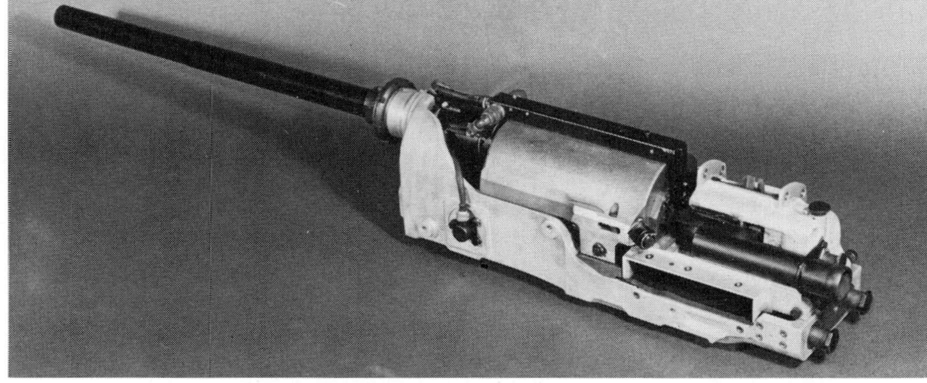

30 mm Aden cannon

4870.303

25 mm ADEN CANNON

The 30 mm Aden cannon (**5185.303**) was developed shortly after the Second World War at the RSAF, Enfield, and subsequently enjoyed a considerable reputation in a wide variety of aircraft and other applications, but it was realised that the scope for improvements to overall system performance was limited. It was therefore decided to modify the 30 mm Aden to fire the more powerful NATO 25 mm ammunition. The result is the 25 mm Aden cannon, which has been selected for RAF Harrier II (GR5).

The design criteria laid down during feasibility studies that led to the Aden 25 were:

(1) The space envelope and mounting points of the gun (excluding the barrel) should be the same as those of the 30 mm Aden.
(2) It should have a rate of fire of 1650 – 1850 rounds/minute, so that in a two-gun configuration its performance would match that of the US GAU 12.
(3) Weight should not be more than ten per cent above that of the 30 mm Aden.
(4) Loads on the aircraft airframe should be similar to those for the 30 mm Aden.

The studies also revealed that the longer length and cartridge case diameter of the 25 mm round could be accommodated within the 30 mm cradle length; higher strength materials would enable the higher pressure of the 25 mm round to be restrained within the same dimensions as the 30 mm Aden gun; and the rate of fire of about 1800 rounds/minute could be achieved using the technology already developed for the Straden (stretched Aden) project of the early 1970s.

In conjunction with Royal Ordnance Small Arms Division, BAe are developing a pod to mount the gun and to accommodate the storage and feed for the larger 25 mm round. Other aspects, such as the suppression of blast and gas ingestion, involve RAE, Farnborough and A&AEE, Boscombe Down.

CHARACTERISTICS

Calibre: 25 mm
Gun length: 2285 mm
Barrel length: 1700 mm
Operation: Gas, revolver
Mass: 92 kg (including barrel)
Max cross section: 246 × 240 mm
Rate of fire: 1650 – 1850 rounds/minute
Time to rate: 20 ms
Muzzle velocity: 1050 m/s
Ammunition: 25 mm NATO Stanag 4173

Electromagnetic hazard: None (percussion fired)
Power requirement for drive: Self-powered
Trunnion force: 26 kN (peak); 21 kN (average)

STATUS

The British MoD has selected the 25 mm Aden for the Harrier GR5 being jointly developed by BAe and McDonnell Douglas. The pilot model has completed a significant number of trial firings with good results and prototype guns have been built at Enfield Lock for air firing (in Harrier GR3) and other trials in 1985. The pre-production and qualification phases are in progress. In RAF service the Harrier GR5 will employ twin Aden 25s carried in pods beneath the fuselage and just to the sides of the centre-line. Total weight of the twin Aden 25 pod with 200 rounds is 430 kg.

CONTRACTOR

Royal Ordnance, Small Arms Division, Enfield Lock, Middlesex, England.
Enquiries to Royal Ordnance, Griffin House, 5 Strand, London WC2N 5BB, England.

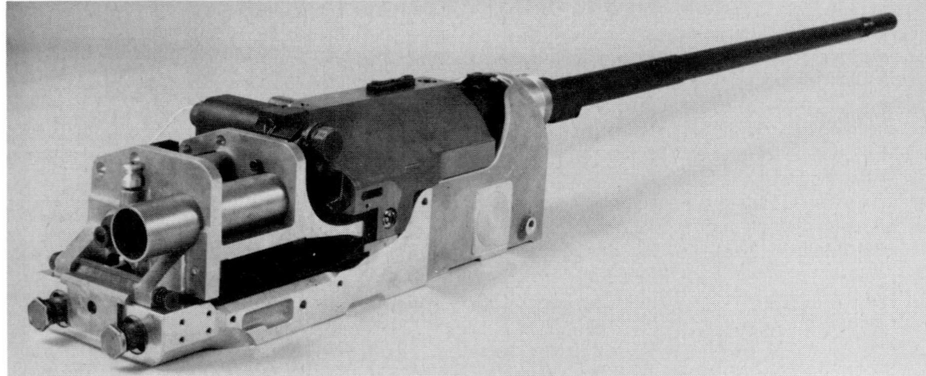

25 mm Aden aircraft cannon without muzzle brake

1277.303

CLUSTER BOMB No 1 Mks 1, 2, 3, and 4 (BL 755)

BL 755 has been developed to meet a requirement for a weapon that will yield a high kill probability against a range of small hard and soft targets encountered in the battlefield and immediate tactical area. Because of the efficiency of modern surface-to-air weapon systems it is operationally necessary for air-to-ground attacks, in support of ground forces, to be carried out at very low level. Any other form of attack necessitates prolonged exposure to the enemy ground defences and negates any element of surprise. This will result in unacceptable losses to the attacking aircraft.

To compensate for the aiming errors inherent in low-level attack, BL 755 covers the target area with a pattern of bomblets the dimensions of which are proportional to the aiming errors. The dual function bomblets are effective against armoured and soft-skinned vehicles, parked aircraft and personnel, and are distributed evenly within the pattern.

Before take-off, one of four time delays is set to ensure a safe weapon/aircraft separation distance. When the bomb is released, its primary striker is armed and, on completion of the selected timer interval, the striker operates and fires the primary cartridge. The gas pressure generated by this cartridge releases the two-part bomb body skins before activating the main gas cartridge to eject the bomblets. 147 bomblets are carried in seven bays with 21 bomblets in each bay of the bomb body. The bomblet ejection velocity is varied between the bays to ensure an even ground pattern. Each bomblet is armed after a short period of flight and, on impact, uses a shaped-charge effect to give a penetration of armour of at least 250 mm. In addition to its armour penetration capability, each bomblet casing explodes into over 2000 fragments which have an effective penetration against non-armoured vehicles and other targets.

BL 755 is compatible with current and future strike aircraft. It can be carried in a bomb bay or externally with twin 14-inch (356 mm) or single suspension. It can be delivered using a simple or sophisticated sighting system; it is immune to electronic countermeasures and incorporates a safety mechanism which provides protection for the aircraft during carriage and release and the deployment of bomblets.

CHARACTERISTICS

Weight: 277 kg
Weapon payload: 147 bomblets
Length: 2451 mm
Diameter: 419 mm
Fin span: 566 mm
Suspension: Twin lug at 14-inch (356 mm) or 250 mm spacing or single lug

Main variations from Mk 1

Mk 2, shorter time delays
Mk 3, addition of 250 mm twin lug suspension
Mk 4, incorporates same time delays as Mk 2

BL 755 No 2

The interim RAF requirement to increase its anti-armour capability against MBTs with improved armour that are entering service is to be met by a BL 755 No 2.

The BL 755 No 2 system retains the shot-gun principle of compensating for aiming errors but takes advantage of the higher weapon aiming accuracy of modern aircraft by modifying the pattern of bomblets.

The submunition is retarded to increase angle of attack thus permitting later target acquisition and better target penetration whilst the warhead incorporates advances made in shaped charge technology and explosives since the introduction into service of the BL 755 No 1.

Other BL 755 design features are unaltered so that full aircraft compatibility and flexibility of operation of the current weapon are retained.

STATUS

BL 755 has been in service with the Royal Air Force since the early 1970s and has been supplied to eight primary NATO air forces and those of other nations. It is cleared for carriage and release from numerous aircraft types. BL 755 No's 1 and 2 are in production and in service with the RAF and are available for export to approved clients.

CONTRACTOR

Hunting Engineering Ltd, Reddings Wood, Ampthill, Bedfordshire MK45 2HD, England.

Improved BL 755 bomblet

Tornado GR1 with BL 755 anti-armour weapons

1267.393

BRITISH BOMB RETARDING SYSTEM

The British bomb retarding system was developed in the UK by Hunting Engineering Ltd. The retarder is incorporated in a special bomb-tail assembly which can be readily applied to a wide range of existing ballistic bombs.

The retarding tail utilises a combination of a ribbon parachute and air-brakes (retarder arms) formed from the structure of the tail. This technique has the advantages of providing very high drag after a closely predictable time delay, essential for accurate control of bomb stick spacing leading to improved probability of target destruction.

The basic structure of the retarder tail consists of a base ring made in forged steel in which are incorporated eight securing pads for attaching the tail to the bomb. When closed the skin sections form the tail cone enclosing the folded ribbon fabric.

The design incorporates various safety measures which ensure that at no time does the bomb constitute an unacceptable hazard to the aircraft during carriage or after release.

A timer mechanism which controls the operation of the unit is housed within the tail. As the bomb is released from the aircraft this timer is initiated by a lanyard. After a short interval to allow the bomb to fall clear of the aircraft the timer triggers the release of the retarder arms. These arms are forced into the airstream by a spring and then aerodynamic forces extend them to the fully open position. The energy released in this operation is absorbed by a tear-webbing shock absorbing device.

At the base of the tail, couplings are provided to connect with the tail fuze of the bomb, so that the arming of the latter is controlled by the action of the retarding tail mechanism.

While being carried aboard the aircraft, retarding tail bombs resemble conventional fixed fin types. The fins are provided to stabilise the bomb aerodynamically immediately after release. In the event of the bomb being jettisoned the retarding tail is not operated and neither is the fuze armed. Retarded bombs can be carried externally or internally in subsonic or supersonic aircraft.

CHARACTERISTICS

Bomb Retarder Tail Type 117

This tail is 1035 mm long and 584 mm span. It was developed initially for use with British 1000 lb HE medium capacity Mk 6 and 9-19 series bombs. Two types of tail, Mk 3 and Mk 4, are available. The Mk 3 with a tail fin span of 584 mm is suitable for external carriage while the Mk 4 with a tail fin span of 419 mm is suitable for both external and internal bomb carriage.

Bomb Retarder Tail Type 118

This tail is 970 mm long and 463 mm span. It was originally designed for use with the British 540 lb HE medium capacity bomb.

Multiple release of retarded bombs from Phantom aircraft

Adaptors are available to couple type 117 and 118 retarding tails to many other types of HE bombs, including the principal US types such as Mk 82 and 83, Mk 64, Mk 65, and Mk 117.

Both types of retarder are designed for a shelf life of at least five years without attention.

STATUS

The retarding system is in service with the Royal Air Force and is being supplied to air forces overseas.

CONTRACTOR

Hunting Engineering Ltd, Reddings Wood, Ampthill, Bedfordshire MK45 2HD, England.

5218.303

JP233 LOW ALTITUDE AIRFIELD ATTACK SYSTEM

The JP233 airfield attack weapon system has been developed for the Royal Air Force for use in counter air operations. The operational concept of the weapon is the suppression of an enemy's ability to mount aircraft sorties by damaging the aircraft operating surface of his airfields and then preventing its repair, to an extent that prevents take-off and landing for prolonged periods.

The weapon is designed for the high speed, low level, simultaneous delivery of two complementary weapons – a cratering weapon to attack and render unusable, runways, taxiways and grass operating strip, and an area denial weapon to pose a continuing threat to vehicles and crew engaged in airfield repair. The JP233 weapon comprises a cratering weapon designated SG357 and an area denial weapon designated HB876. Each of the weapons consists of a dispenser loaded with submunitions, 30 for the SG357 and 215 for HB876. A Tornado carries two JP233 systems mounted in tandem on the port and starboard shoulder pylons with the forward positions occupied by the two HB876 dispensers and with two SG357 dispensers aft.

To ensure effective ground cover both types of submunitions are dispensed simultaneously in a co-ordinated sequence. The SG357 submunitions disrupt the operating surfaces by causing craters, heave and fractures over a large area whilst the HB876 mines overlay the cratered area to delay runway repair operations for some considerable time. The submunitions also cause a considerable amount of consequential damage to secondary targets such as aircraft, vehicles or buildings adjacent to the area being attacked.

The system can be configured for carriage by a range of advanced aircraft including Tornado, F-111 and F-16 where multiple dispensers are employed on either fuselage or wing mountings.

Apart from denying the enemy the use of his airfields, JP233 can equally well be employed against other important targets such as railway marshalling yards, transport and tactical vehicle concentrations, road networks, supply support depots, areas to which the enemy requires freedom of access and use.

It is claimed that typically, JP233 is five times more effective per aircraft load than other systems.

STATUS

In production and in service with the RAF and one overseas customer to date. The HB 876 area denial munition has been designated for use with BKEP in the proposed USAF Direct Airfield Attack Combined Munition (DAACM).

HB876 area denial submunition of JP233 airfield attack weapon

PRIME CONTRACTOR

Hunting Engineering Ltd, Reddings Wood, Ampthill, Bedfordshire MK45 2HD, England.

SG357 runway cratering submunition for JP233 airfield attack weapon

JP233 full load release from Tornado GR1 aircraft

6406.303

HUNTING AREA-DENIAL SYSTEM (HADES)

The Hunting Area-Denial System (HADES) is an air-delivered solution to the attack and closure of high-value fixed targets and choke points such as airfield, road and rail concentrations, river crossings, munitions dumps and HQ sites.

HADES combines the proven BL755 cluster dispenser with the HB876 area denial munition (part of the JP233 airfield attack system – see previous entry). Each dispenser contains 49 separate HB876 munitions and is suitable for high speed and low level delivery by current and future front line aircraft. The use of an existing dispenser together with identical mass characteristics to BL 755 does away with the need for expensive aircraft clearance trials in most cases.

After release, gas pressure strips the dispenser skins and ejects the munitions which are then parachute retarded to reduce the ground impact velocity. Spring steel legs turn the munition upright with its warhead uppermost soon after ground impact. Detonation can occur either when the munition is initiated by sensors, or by a randomly timed self-destruct signal. On detonation the warhead casing has a dual effect: it produces a high-velocity slug capable of penetrating the blade of a clearance vehicle and damaging the engine, and the wall fragments pose a threat over long distances to targets such as vehicles and parked or taxiing aircraft.

STATUS

Both BL 755 and HB 876 components are in full scale production. HADES is available for export to approved overseas customers and will be used by the USAF Armament Division for HB 876 trials purposes in 1988 as part of the DAACM development programme.

CONTRACTOR

Hunting Engineering Ltd, Reddings Wood, Ampthill, Bedfordshire MK45 2HD, England.

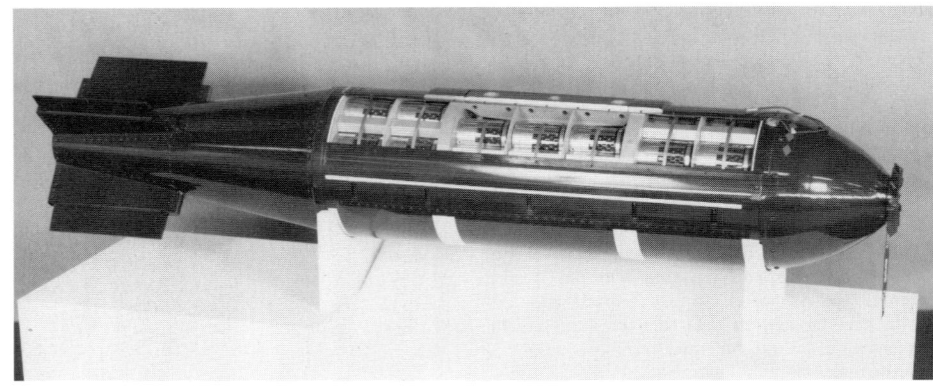

Hunting area-denial system – HADES

UNITED STATES OF AMERICA

5225.303

UNITED STATES AIRCRAFT ARMAMENT

The range of American aircraft armament is without doubt the most extensive of any nation in the world, and probably the most varied. It includes well over 200 types of bomb, among them conventional free-fall bombs of many weights and sizes, as well as 'cluster' bombs containing sub-munitions of all types. In the following paragraphs certain specific types have been described in greater detail, eg Mk 20 cluster bomb (**1998.303**), while in the Analysis section of this volume will be found a comprehensive table (**1850.314**) listing more than 300 types of air-deliverable munitions produced in the USA. Elsewhere in this book there are other entries dealing with related aspects such as the use of add-on guidance packages to convert conventional 'iron' or 'dumb' bombs into guided weapons, or 'smart' bombs (**1597.311**), HOBO (EO) and laser-guided smart bombs and Paveway laser-guided bombs (**1534.311**) in the Systems section.

The American inventory of air-deliverable weapons includes free-fall nuclear bombs which may be capable of both tactical and strategic use, and these also are listed in the table. Additional details of these and other payloads in the strategic context will be found in appropriate entries in the Systems section dealing with US strategic weapon systems.

Another type of aircraft-carried munitions dealt with in other parts of *Jane's Weapon Systems* are those used for anti-submarine warfare, eg mines and depth charges.

US airborne munitions have been widely supplied to foreign nations, and therefore can be found in the inventories of many countries; for this reason, and because of the long period over which such supplies may have been delivered, no attempt has been made to compile extensive lists of users.

General Dynamics F-16 with Mk 82 (500 lb) bombs

General Dynamics F-111 with B-61 nuclear weapons

4865.303

ISCB-1 AREA DENIAL CLUSTER WEAPON

The ISCB-1 is a newly introduced, air-delivered, area denial cluster weapon designed to fulfil such missions as runway repair, interruption of enemy traffic over bridges etc. This capability is achieved by enhancing ISC Technology's Rockeye cluster bomb (**4599.303**) with computer techniques. The result is an air-delivered fragmentation weapon featuring variably programmable detonation of individual area denial mines.

Designed, developed and manufactured within ISC, the ISCB-1 consists of a Mk-7 dispenser, Mk 339 nose fuze, a tail cone assembly, and 160 electronically timed mines plus 65 dummy mines. The Mk-7 dispenser gives the same weapon/aircraft interface, aerodynamic compatibility, and ground dispersion patterns as the Rockeye. This permits immediate certification with at least 14 current aircraft types.

As with Rockeye, after launching from the aircraft the Mk 339 fuze initiates a linear shaped charge, the Mk-7 dispenser splits open and the mines are released. However, differing from the Rockeye, the ISCB-1 tail cone assembly houses a bomb panel which receives information to programme the delay times of individual mines. This programming is done on the ground using ISCT's delay programmer or fuze set key to detonate on impact or in 10-second increments for up to 24 hours to deny enemy access to target areas covering 3000 to 5000 square metres. Programming can be carried out until the time of take-off.

STATUS

In production.

CONTRACTOR

ISC Technologies Inc, Ordnance Systems, 3700 Electronics Way, PO Box 3040, Lancaster, Pennsylvania 17604-3040, USA.

4599.303

ROCKEYE II Mk 20 CLUSTER BOMB

The Rockeye II Mk 20 cluster bomb is an unguided free fall weapon, delivered in the same manner as a conventional bomb, for use with all bomb rack configurations. Rockeye dispenses 247 dual-purpose armour-piercing shaped charge bomblets, over a variable attack area. Since precise aiming is not required, the pilot need maintain delivery attitude for only a short duration prior to release of the bomb, thereby reducing the aircraft's vulnerability to ground defences and improving pilot safety.

The Mk 339 dispenser opening fuze provides pilot options for timing dispenser openings, ensuring safe separation from the aircraft prior to arming, and initiating a linear shaped charge to split open the container. The bomblet fuze discriminates between soft and hard targets, detonates the bomblet, and provides additional safety for pilot and aircraft.

Rockeye is smaller and lighter but contains more net explosives than similar cluster bombs. Its smaller size and weight permit as much as 800 lbs more fuel per mission according to the weapon's manufacturer,

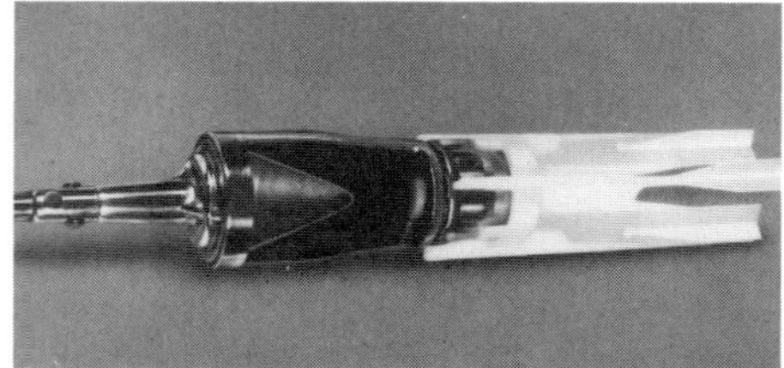

Single dual-purpose Rockeye II bomblet

thus considerably increasing aircraft range and endurance.

No special storage facilities or handling equipment are required, and no periodic maintenance is necessary. The store is delivered ready to use and service life exceeds 20 years it is claimed. Only visual inspection is needed before flight.

OPERATION

Delivery is literally the same as that for a conventional 500 lb bomb. The sizes and shapes of impact areas can be changed according to aircraft speed, altitude, dive angle and the dispenser opening delay time. In heavily defended target areas, Rockeye can be delivered at high speeds and at altitudes as low as 30 metres pitch up or 76 metres in level flight.

CHARACTERISTICS

Length: 2.33 m (without fuze cover); 2.4 m (with fuze cover)
Diameter: 335 mm
Fin span: 437 mm (closed)
Lug spacing: 355 mm
Centre of balance: 165 mm (fwd of aft lug)
Weight: 222 kg (assembled)
Explosive payload of bomb cluster: 45.3 kg
Bomblet weight: 151.5 kg (total)
Release altitude: 76 m (minimum in level pitch); 30 m (minimum pitch up)
Typical bomblet impact pattern: 4800 m^2 (released from 152 m)

STATUS

Operational. Entered service in early 1970s. In use by US forces and more than ten other countries throughout the world. Modern tactical aircraft that have carried Rockeye include: A-4, A-6, A-7, A-10, F-4, F-5, F-8, F-15, F-16, F-111, Mirage, Jaguar, and Hawk.

ISC is the main international contractor for Rockeye, and assumes total responsibility for the supply of all contracted goods and services.

CONTRACTOR

ISC Technologies Inc, Ordnance Systems, 3700 Electronics Way, PO Box 3040, Lancaster, Pennsylvania 17604-3040, USA.

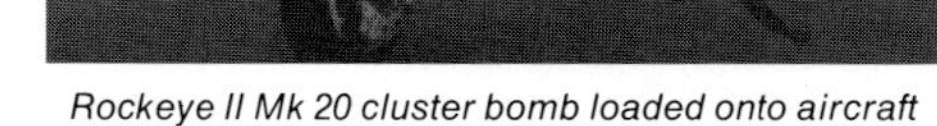

Rockeye II Mk 20 cluster bomb loaded onto aircraft

4190.311

GBU-15 MODULAR GLIDE BOMB

The GBU-15 is a precision modular glide bomb for attacking heavily-defended fixed targets, including bridges, tunnels, airfields, control centres and surface-to-air missile sites. It provides stand-off for increased aircraft survivability while preserving an extremely high degree of accuracy. Like its predecessor, the GBU-8 HOBO, the GBU-15 is designed around the standard 2000-lb Mk-84 general purpose warhead, but includes a data link feature that eliminates the need to lock-on before launch. This allows blind launches at low altitude or extreme stand-off ranges, as well as aim point refinements until shortly before target impact.

As a modular weapon system, the GBU-15 is easily adapted to meet evolving tactical requirements. The weapon currently has several seeker/warhead options to respond to changing weather conditions and targeting needs. The GBU-15(V)1/B features a TV seeker for daylight attacks, while the GBU-15(V)2/B uses the imaging infrared seeker from the AGM-65D Maverick for night attack. These seeker modules are interchangeable and an integral part of the GBU-15 family of weapons.

In addition to the basic Mk-84 warhead, the GBU-15 has been demonstrated successfully with the BLU-109 improved 2000-lb warhead for increased penetration against hardened targets. The GBU-15 is also compatible with the SUU-54 dispenser for a submunition capability for attacking area targets such as surface-to-air missile sites.

The major components of the GBU-15 are:

(1) 2000-lb warhead (Mk-84, BLU-109 or SUU-54), and the appropriate fuze
(2) guidance module (DSU-27A/B TV seeker or WGU-10/B IR seeker)
(3) adaptor to join the guidance section to the warhead (varies according to warhead shape)
(4) control module (WCU-8/B; contains battery, flight control sensors, autopilot, and control surface actuators)
(5) weapon data link receiver/transmitter module
(6) cruciform airfoil group (consisting of wings, strakes, and control surfaces, together with electrical conduit and umbilical).

Two interchangeable airfoil groups have been produced for the GBU-15. The original is called the 'long-chord wing' and comprises most of the existing inventory; however, current production is solely of the smaller 'short-chord wing', which was introduced as a cost-saving measure. Although it is about one-half the wing area of the original airfoil, its low altitude performance is the same. The short-chord wing is also lighter and stronger than the original, and compatible with the AGM-130 weapon (**6477.311**). The new wing also required a change in the strake design. Both airfoils use the same wing control surfaces and have the same wing span. Required equipment for the GBU-15 control aircraft includes a compatible data link pod (such as the USAF AXQ-14), cockpit pod control, suitable E-O display and controls, and a hand controller for the seeker. Video is transmitted continuously from the weapon to the aircraft; commands from the aircraft to the weapon are transmitted only when the weapon operator is slewing the seeker or commanding lock-on.

OPERATION

The GBU-15 is normally employed in the indirect attack mode where a weapon is launched towards the target without a lock-on. After launch the weapon performs a programmed G-bias climb and the operator acquires the target shortly afterwards. The operator then has the option to lock-on to the aim point, or manually track it. If a lock-on is taken, he can subsequently break lock to refine his aim point or transition to a manual track. In addition to the lock-on after launch mode, the GBU-15 has retained the direct attack (or lock-on before launch) feature of the GBU-8 weapon, allowing GBU-15 employment by aircraft having E-O capability but not a data link. Alternatively, the GBU-15 may be used in the 'buddy' control mode, where one aircraft delivers the weapon and another performs the control functions (the delivery aircraft may or may not have a data link capability). The control aircraft can then take full advantage of the data link's transmissions range and can even be outside hostile airspace. Other tactics afforded by the data link include having one aircraft deliver two or more weapons on a single pass, with one weapon controlled by the delivery aircraft and the others by 'buddy' aircraft located elsewhere. Each weapon must operate on a separate frequency during multiple releases.

CHARACTERISTICS

Length: 3.94 m
Diameter: 46 cm (Mk-84)
Wing span: 1.5 m
Weight: 1140 kg (Mk-84 with long-chord wings)
Guidance: TV (GBU-15(V)1/B); IR (GBU-15(V)2/B)
Range: 1.5 to 80 km (highly dependent on launch speed and altitude

STATUS

The system has been certified on the following aircraft: USAF F-4E, F-111F and B-52D; RAAF F-111C; Israeli F-15. Integration is under way with the USAF F-15E and B-52G, and the Israeli F-16C/D aircraft.

GBU-15 has been in full scale production since 1980, with orders received for 3200 weapon kits. Since 1975 over 100 TV and IR weapons have been launched in operational tests and training, achieving an overall success rate of more than 85%. Weapon kits have been procured by the USAF and the Israeli Air Force.

CONTRACTORS

Rockwell International, Missile Systems Division, Defence Electronics Operations, 1800 Satellite Boulevard, Duluth, Georgia 30136, USA – prime contractor.

Hughes Aircraft Company, Missile Systems Group, Canoga Park, California 91304, USA – weapon data link and control pod.

1534.311

PAVEWAY LASER-GUIDED BOMBS

The family of laser-guided bombs (LGBs) uses a common laser guidance and control sub-assembly, with only the aerodynamic surfaces (control fins and aerofoil group) changed to match the particular size warhead. The system allows for stand-off capability while providing precision weapon delivery against a wide spectrum of targets.

The original LGB conversion kits have become known as Paveway I, following the introduction of a new set of components in 1978, the latter being known as Paveway II. One notable feature of the later system is the substitution of extending tail fins for the original fixed fins of Paveway I conversions. See *Jane's Weapon Systems 1987-88* for references and diagrams.

OPERATION

The guided-bomb kit directs the weapon toward a target which has been illuminated by laser energy. The desired target can be selected by the bombing aircraft, a companion aircraft, or a ground observer. The laser energy is reflected from the target and detected by the laser guidance system. The system processes the information, computes appropriate control commands, and applies these commands to movable surfaces to effect changes in trajectory, thereby guiding the weapon to the target.

The system requires no electrical connection to the unmodified delivery aircraft and is delivered in the same manner as a conventional bomb. There is no requirement for tracking or lock-on before launch, which minimises the time required for delivery with concomitant reduction in aircraft exposure to hostile ground fire.

DEVELOPMENT

The US Air Force Armament Development and Test Centre (ADTC), Eglin AFB, initiated a competitive evaluation of two laser guidance concepts for standard bombs in 1965; the first LGB flight occured in April 1965. This evaluation proved the feasibility of the concept and was followed by establishment of the Paveway programme office at Wright-Patterson AFB. Texas Instruments was selected to conduct engineering development to supply additional laser guidance kits for M117 GP bombs.

Utilising the laser guidance kit developed by Texas Instruments, two additional GP bombs (Mk 82 and Mk 84) and the M118 demolition bomb were later successfully adapted to laser guidance and operationally deployed.

The Paveway II version was developed in the mid-1970s and production began in 1977. Similar to Paveway I in appearance and operational concepts, Paveway II has more advanced electronics and a folding wing aerofoil group which improved sensitivity and manoeuvrability. Paveway II also incorporates many cost-reduction features such as plastic lenses instead of glass, a moulded-plastic ringtail instead of metal, and intergrated circuitry to improve reliability and reduce manufacturing time. The folding wings also permit increased load density on the delivery aircraft.

The development of the third generation of LGBs began in 1980-81. Known as the low-level laser-guided bomb (LLLGB) or Paveway III, this weapon is designed to be delivered from very low altitudes and significant stand-off ranges. The LLLGB incorporates state-of-the-art microprocessor technology, high-lift folding wings, and an improved scanning seeker. After a successful development flight test programme, the system has undergone initial and follow-on operational test and evaluation and production approval was granted in mid-1983. There are US Air Force plans to embody Paveway III know-how in an improved 2000 lb bomb.

STATUS

The US Air Force and US Navy utilise the GBU-10E/B (Mk 84), GBU-16B/B (Mk 83), and GBU-12D/B (Mk 82) versions of Paveway II LGBs. These weapons are compatible with the F-4, F-111, F-5, A-6, A-37, A-7, A-4, A-10, F-16, AV-8A (Harrier) and B-52 aircraft. The F-14, F-15, F-18, Mirage Tornado, Buccaneer, and Jaguar aircraft types have also been added to this list.

The LGB has been introduced into the RAF inventory and also other NATO countries. The RAF system for the Mk-13/18 1000lb warhead was developed jointly by the Royal Aircraft Establishment, Texas Instruments and Portsmouth Aviation Ltd.

In addition to US and UK, operators of Paveway II LGBs include Australia, Canada, Greece, Israel, the Netherlands, Saudi Arabia, South Korea, Spain, Taiwan, Thailand, Turkey and 20 other nations.

Successful completion of follow-on test and evaluation of Paveway III was announced in August 1986. During these tests the US Air Force dropped 47 LLLGBs from F-4 and F-111 aircraft and scored 44 successes.

CONTRACTOR

Texas Instruments Incorporated, 2501 South Highway 121, PO Box 405, Lewisville, Texas 75067, USA.

6183.303

DIRECT AIRFIELD ATTACK COMBINED MUNITION (DAACM)

The Direct Airfield Attack Combined Munition (DAACM) is a package of two types of munition contained within a standard Tactical Munitions Dispenser (TMD). The DAACM contents of the TMD are eight BLU-106/B Boosted Kinetic Energy Penetrators (BKEP) and 24 Hunting Engineering Limited HB876 area denial munitions (see **5218.303**).

The BKEP is a boosted kinetic energy penetrator designed specifically for airfield attack as part of a 28-month full scale development programme awarded in March 1983 by the US Air Force Armament Division at Eglin Air Force Base. As the BKEPs are released from their dispenser their fins deploy and a tail drogue is released to pull out the main parachute. The main parachute slows down the BKEP until it is pointing downwards at an angle of 65° when the rocket motor will fire and jettison the parachute. On impact with the runway the rocket motor will separate but the main warhead will not detonate until the BKEP body has penetrated some way under the surface to create a runway crater.

The BKEP is compatible with the SUU-20 test bed dispenser and the GBU-15/SUU-54 (AGM-130).

STATUS

In development. The BKEP is currently under contract.

CONTRACTOR (BKEP)

Avco Systems Textron, 201 Lowell Street, Wilmington, Massachusetts 01887, USA.

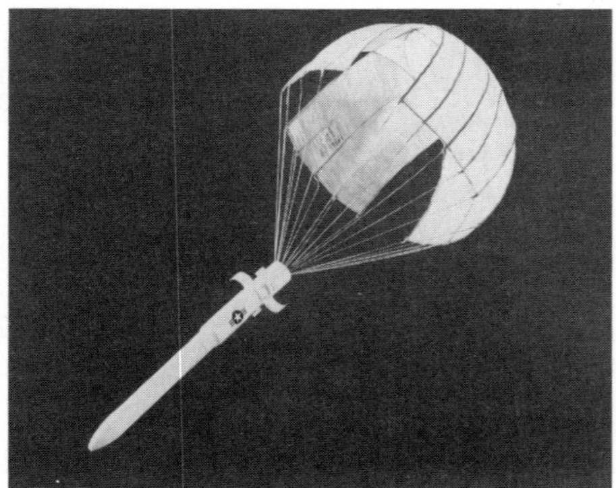

Model of Boosted Kinetic Energy Penetrator (BKEP) with main parachute deployed

Model of Direct Airfield Attack Combined Munition showing HB876 area denial munitions stowed towards the tail and BLU-106/B Boosted Kinetic Energy Penetrator (BKEP) forward

3637.311

WAAM PROGRAMME

The Wide Area Anti-armour Munitions programme (WAAM) has evolved somewhat since its initiation in the late 1970s as a result of changes in emphasis of the overall US Army/US Air force efforts in the area of tactical/theatre anti-armour and changing circumstances over the intervening years. The background to these events has been summarised under this entry number and related entries in previous editions, including the varying relationships with other complementary programmes which are connected with the main purpose of this system. The following paragraphs represent the latest known position on the current WAAM programme.

In the fiscal year 1986 USAF statement on the budget estimates it was stated that development of the sensor fuzed weapon is being continued and will provide a significant increase in effectiveness against enemy armour. The sensor fuzed weapon (**6478.311**) employs Avco Skeet IR fuzed submunitions packaged in a tactical munitions dispenser.

This concept has its origin in the Avco system previously known as ERAM (Extended Range Anti-armour Munition) in which submunitions are delivered by parachute from dispensers released by tactical aircraft flying close support or interdiction missions. Dispersed along an enemy line of advance, or directly over an enemy concentration, the ERAMs engage and destroy armoured and mechanised units. The ERAM is designed to fit into the USAF SUU-65/B dispenser which can be carried on a variety of US strike and attack aircraft, and this combination of dispenser and submunitions has been designated the CBU-92/B by the USAF.

It is capable of delivery over a wide range of altitudes and airspeeds, and once released, the dispenser splits open low over the target and ejects its ERAM submunitions. Each of the latter then deploys a parachute designed to disperse it randomly over a wide area.

DAACM

Also in full-scale development is the Direct Airfield Attack Combined Munition (DAACM) which is basically a combination of Avco's BLU-106/B boosted kinetic energy penetrator (BKEP) submunition, mines such as Skeet and ERAM, and a tactical munitions dispenser, designed for use against enemy airfields. The intention is to crater runways and penetrate aircraft protective shelters.

BKEP uses a rocket motor to accelerate its warhead to the velocity necessary to penetrate the runway surface, after which it explodes causing maximum damage to the paved surface. Though originally designed for compatibility with the now-cancelled MRASM weapon, BKEP is suitable for use with the Tactical Munition Dispenser (TMD) and the GBU-15/SUU-54 (AGM-130) delivery systems. The TMD has been designated as the DAACM carrier by the USAF and this contains eight BKEP submunitions plus a quantity of mines.

Submunition ground impact patterns are controlled by varying the parachute opening times of individual submunitions. On ejection from the dispenser, fins deploy to stabilise the submunition and parachute deployment retards forward motion. The rocket motor ignites at a programmed time causing the parachute to be jettisoned and the warhead to be accelerated to impact. A time fuze detects impact and sets off warhead detonation below the runway surface for optimum heave and cratering effects.

STATUS

The WAAM programme was started in 1975 and work is concentrated at Eglin AFB, at the Armament Development Test Centre there.

6478.311

SENSOR FUZED WEAPON (SFW)

The Sensor Fuzed Weapon (SFW) is a direct application of the Avco Systems anti-armour Skeet smart warhead being developed for the US Air Force Armament Division. The SFW consists of a SUU-64/B Tactical Munitions Dispenser (TMD) which contains ten BLU-108/B submunitions, and is delivered by a tactical aircraft such as the F-16. Each submunition carries four Skeet warheads for a total of 40 warheads per TMD.

The SFW can be delivered at high speed and low altitude, or from high altitude on both column and array armoured targets. When delivered by a tactical aircraft, a parachute system decelerates and orientates the submunitions vertically over the targets. At a preset altitude, the parachute is jettisoned and a rocket motor reverses the fall, while spinning up the submunition. The four Skeet warheads per submunition are thrown outward and upward in search of targets over a wide area. The built-in infra-red sensor in each warhead finds a target and fires an explosively formed penetrator into the vulnerable area on the topside of the AFV.

STATUS

The SFW is in full-scale development for the US Air Force.

CONTRACTOR

Avco Systems (a subsidiary of Textron Inc), 201 Lowell Street, Wilmington, Massachusetts 01887, USA.

6377.303

TROUND INTERNATIONAL 12.7 mm AIRCRAFT MACHINE GUN

During the early 1960s David Dardick developed a radical type of ammunition that was first used with a pistol of Dardick's design. The ammunition cross section is that of a triangle with rounded sides and is called a Tround, for triangular round. This shape allows for the side feeding feature which is a key to the 'open chamber' design, and its symmetry allows for alignment in any of three positions.

One feature of the Tround is its injection-moulded cartridge case, currently made of engineered thermoplastic. The projectile is fully telescoped within the case and acts as its own firing chamber when supported by the gun cylinder and frame.

The Tround and its associated pistol failed to find commercial or military acceptance and the principle faded from public view, even though a Tround-firing eight-barrel Gatling-type gun did achieve a cyclic rate of fire of 30,000 rounds a minute under test. The Tround principle re-emerged during 1986 in a new 12.7 mm machine gun produced by Tround International Inc who produced a weapon for trials with the US Navy at China Lake with a view to further development resulting in a 20 mm cannon for possible installation in the F/A-18B strike fighter.

The Tround International machine gun has a cyclic rate of fire of 1250 rounds per minute and is motor driven, although a self-powered version is possible. A cam driven system is used to index the gun cylinder (which houses the Tround cartridge) with a smooth motion. There is a dwell period when the Tround is in the battery (firing) position, positively aligned with the barrel. Feed from the 100-Tround drum, currently located under the barrel, is carried out using cams and ramps. With no need for the usual reciprocating components the gun has fewer parts than an equivalent or orthodox weapon and has improved fire power per unit weight. It is claimed that the rate of fire of a Tround International gun is limited only by the heat and wear limitations of the barrel rather than by the feed mechanism.

Other advantages claimed for the gun is that it is not prone to cook-offs as the plastic case acts as an effective thermal insulator. The Trounds are safe to store and handle, and plastic cases are cheaper to produce than conventional rounds. Ammunition feed problems are virtually eliminated due to the cross-sectional shape which correctly aligns itself to any prepared feed-way.

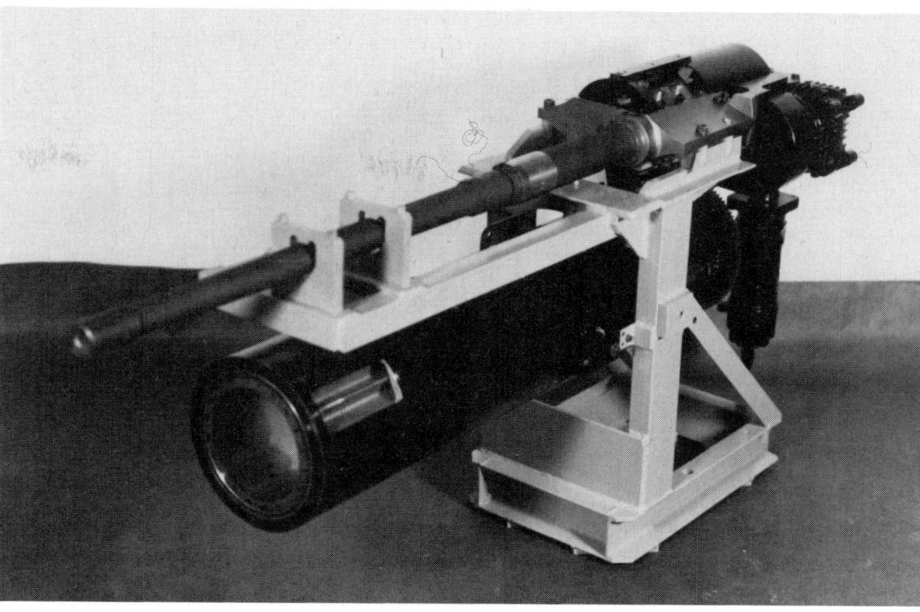

Tround International 12.7 mm machine gun

The Tround principle can be adapted to any calibre and it is possible for a single Tround to contain up to three projectiles with a single charge firing through an equivalent number of barrels simultaneously.

The data provided in the Characteristics table refers to the 12.7 mm gun under test with the US Navy. The gun system weight could be reduced by 15.875 kg if existing off-the-shelf components were replaced by specially-designed equivalents.

CHARACTERISTICS

Tround

Projectile calibre: 12.7 mm
Cartridge length: 127 mm
Weight: 113.5 g
Weight of projectile: 42 g
Propellant: WC-860
Primer: Percussion
Muzzle velocity: 975 m/s

Gun

Weight: Bare gun (frame cylinder and barrel) 25.85 kg; feeder mechanism and guides 1.81 kg; cam index drive 9.07 kg; motor drive unit 17.7 kg
Motor drive: 28 V DC
Rate of fire: 1250 rpm
Time to rate: Less than 100 ms (gun rate by second shot)
Time to first shot: Less than 30 ms
Burst control: 2 – 99 rounds
Peak current load: Motor start 900 A
Average current load: 250 A
Safe/arm: Independent electrical and mechanical

CONTRACTOR

Tround International Inc, Alpha Industrial Park, RD1, Box 13-T, Phillipsburg, New Jersey 08865, USA.

1996.303

GAU-8/A 30 mm GUN SYSTEM

The GAU-8/A 30 mm gun system is designed to provide a cost-effective weapon to engage the full array of ground targets encountered in the close air support role. The system has an ammunition capacity of 1174 rounds and weighs 1723 kg when fully loaded. Aluminium cased ammunition with either armour piercing incendiary, HE incendiary, or target practice projectiles can be fired at either 2100 or 4200 shots per minute. US ammunition for the GAU-8 weapon is manufactured by the Aerojet Ordnance and Manufacturing Company of Downey, California and Honeywell Inc of Hopkins, Minnesota. Ammunition is also produced by Royal Ordnance in the UK and De Kruithoorn in the Netherlands. The family consists of the following rounds: armour piercing incendiary (API), high explosive incendiary (HEI) and training practice (TP). The API round has a lightweight body which contains a sub-calibre high-density penetrator of depleted uranium. In addition to its penetration capability, depleted uranium is a naturally pyrophoric material which enhances incendiary effects. The HEI round employs a standard M505 fuze and explosive mixture with a body of naturally fragmenting material that is effective against lighter vehicle and material targets. The TP projectile simulates the exterior ballistics and provides a ballistic match to the HEI and is used for pilot training and gunnery practice. Development of the aluminium alloy case material instead of the conventional brass or steel resulted in a 272 kg reduction in weight of a fully loaded system. The GAU-8/A gun system is composed of four major subsystems: gun, feed system, drive system and an electrical control system.

The gun is an externally-powered, Gatling-type mechanism. Each of the gun's seven barrels fires only once during each revolution of the barrel cluster. Barrels are attached to the gun rotor by quick release interrupted lugs. The gun rotor is journalled within a stationary housing and contains the seven gun bolts. The bolts slide fore and aft on tracks and provide the ram, lock, fire, unlock, and extracting functions. The stationary outer housing contains a cam which drives the bolts through their respective functions. The GAU-8/A uses a rotary-lock bolt which provides ample locking area and support for the high pressure, aluminium cased 30 mm ammunition. The gun design features a three-shaft transfer unit which transfers rounds coming from the feed system into the gun rotor.

A new reverse clearing technique is used to clear ammunition from the gun after each burst. An electro/hydraulic drive system controls the clearing function and assures consistent positioning of the first live round on the feed side of the gun. Live rounds which have cycled into the weapon but have not yet fired at trigger release are returned to the feed side of the gun in less than one second ready to fire at the next trigger application.

The GAU-8/A gun system is driven by hydraulic drives operating off the aircraft hydraulic system. Dual motors are employed operating from independent hydraulic systems. Only one motor is energised to fire the system at half rate. The hydraulic drive units include controls for the two firing rates and the reverse clearing function.

A linkless ammunition feed system is used to store and feed ammunition. Spent cartridge cases and any unfired rounds are returned to the storage drum after passing through the gun.

General Electric GAU-8/A 30 mm gun system

The major components are the cylindrical drum unit in which the ammunition is stored; the exit unit which removes rounds from the drum unit and inserts them into an endless conveyor belt; the conveyor system comprising the conveyor belt and the chuting through which it moves; the transfer unit which removes rounds from the gun end of the conveyor belt, feeds them into the gun, and re-inserts the spent cartridge cases into the belt for return to the storage drum; and the entrance unit which transfers cases from the conveyor system back into the drum unit. Each round and case is in smooth, continuous motion when the system is feeding and the gun is firing.

CHARACTERISTICS
Firing rate: 2100/4200 rounds/minute
Length: 6400 mm
Weight: 1723 kg
Ammunition capacity: 1174 rounds
Power requirements: 77 hp
Muzzle velocity: 1066 m/s
Barrel life: 20 000 rounds (minimum)
Time to rate: 0.55 s
Dispersion: 5 mil

STATUS
Operational testing at Nellis AFB has verified its ability to defeat medium tanks. Over eight million rounds have been fired. The General Electric Company has delivered more than 700 for the A-10 close air support aircraft.

CONTRACTOR
General Electric Company, Armament Systems Department, Lakeside Avenue, Burlington, Vermont 05402, USA.

4061.303
GAU-12/U EQUALISER 25 mm AUTOMATIC GUN

General Electric's 25 mm five-barrel Gatling gun (GAU-12/U Equaliser) was designed making full use of the technology developed and proved on the GAU-8/A 30 mm gun programme. High rate of fire, combined with high muzzle velocity and excellent ballistics of the 25 mm round, make it a candidate for numerous applications in either air-to-air, air-to-ground, or ground-to-air roles.

The outstanding terminal ballistics are attributable to the standard 25 mm round of ammunition. This round produces a muzzle velocity of 1097 metres per second which, when combined with the rate of fire of up to 4200 rounds/minute, makes the GAU-12/U Equaliser extremely effective for a variety of combat missions.

The lightweight (125 kg) externally powered weapon features reverse clearing, a 20 000-round maintenance cycle, and an MRBF that exceeds 15 000 rounds. Although providing much more effective firepower than the 20 mm Vulcan, the GAU-12/U Equaliser is similar in size and power requirements to the 20 mm.

The standard 25 mm round family includes TP, HEI and APDS rounds with steel cases and percussion primers. In addition, a full-bore API round has been developed for use in high performance aircraft.

CHARACTERISTICS
Type: 5-barrel Gatling, 25 mm
Length: 2134 mm
Max diameter: 279 mm
Weight: 125 kg
Power to drive: 10.4 kW
Ammunition: Standard 25 mm (TP, HEI, APDS)
Muzzle velocity: 1097 m/s HEI
Dispersion: 6 mil (80% of rounds fired)
Rate of fire: Variable up to 4200 rounds/minute
Time to rate: 0.4 s
Drive type: Externally powered
Feed system: Linkless or linked
Clearing: Gun reversal
Recoil force
Peak max: 9000 lb (40 kN)
Peak recurrent: 8000 lb (36 kN)
Average at 3600 rounds/minute: 4000 lb (18 kN)

STATUS
In production and in US Marine Corps service.

General Electric GAU-12/U Equaliser 25 mm automatic gun and ammunition

CONTRACTOR
General Electric Company, Armament Systems Department, Lakeside Avenue, Burlington, Vermont 05402, USA.

3891.303
GAU-13/A 30 mm LIGHTWEIGHT GUN

The GAU-13/A 30 mm lightweight aircraft gun is a four-barrel, Gatling design, which although based on the General Electric GAU-8/A seven-barrel gun (**1996.303**), incorporates a number of innovations to improve serviceability and reduce overall weight while retaining comparable ballistic performance. With a firing rate of up to 3000 rounds/minute, the GAU-13/A employs the same reverse clearing operation as the GAU-8/A. The gun is suitable for internal or external mounting on an aircraft, and GE has developed a special pod to house the GAU-13/A.

GPU-5/A
The GPU-5/A is a compact external store incorporating the GAU-13/A 30 mm four-barrel gun to provide a significant anti-armour and ground attack capability for attack or fighter aircraft. The pod can be used with a wide variety of aircraft types such as the F-4, F-5, A-7, F-15, F-16, F-20, A-4, A-10 and F-18. The main components are the GAU-13/A gun, a helical ammunition feed system which surrounds the gun, a self-contained pneumatic drive, and a supporting structural assembly. The ammunition feed has space for 353 rounds, enough air capacity to fire two full complements of ammunition, and the only power required from the aircraft is 7.5 A of 28 V DC or 115 V AC, 400 Hz, 300 W. The support structure can be configured for use as either a standard 30-inch (762 mm) rack fitting or as an adaptor for direct mounting on specific aircraft. A novel boresighting procedure in effect boresights the receiver aircraft rack thereby eliminating the need to boresight the gun each time it is re-attached to the aircraft.

CHARACTERISTICS
Gun: GAU-13/A (4-barrel)
Calibre: 30 mm
Length: 2794 mm
Max diameter: 304 mm
Ammunition: GAU-8/A TP, HEI, API
Rate of fire: Up to 3000 rounds/minute
MV: 1037 m/s TP/HEI; 987 m/s API
Gun life: 250 000 rounds
Pod: GPU-5/A
Length: 4320 mm
Diameter: 610 mm
Weight: 862 kg (loaded); 621 kg (empty)
Ammunition capacity: 353 rounds

STATUS
In USAF service.

GPU-5/A has been successfully flight-tested against ground targets on the Northrop F-5 and F-20, the Vought A-7, the General Dynamics F-16 and the McDonnell Douglas F-4E and F-15.

CONTRACTOR
General Electric Company, Armament Systems Department, Lakeside Avenue, Burlington, Vermont 05402, USA.

Four-barrel GAU-13/A Gatling 30 mm lightweight gun with GAU-8/A range of ammunition

GPU-5/A lightweight 30 mm gun pod on centre-line station of USAF F-5

1678.303
GPU-2/A 20 mm LIGHTWEIGHT GUN POD

The GPU-2/A 20 mm pod is a lightweight self-contained gun system requiring only a trigger signal from the aircraft. This pod can be mounted on a wide variety of existing helicopters and fixed-wing aircraft by using standard suspension racks. The total system weight when loaded with 300 rounds is less than 600 lb (272 kg).

General Electric Company has designed and produced a quantity of GPU-2/A pods for the USN. These systems use the three-barrel M197, 20 mm gun which has been in military inventory since 1969. The firing rate for the GPU-2/A is selected at either 750 or 1500 rounds per minute. Rounds are fed to the gun through a linkless ammunition storage and feed

GPU-2/A 20 mm lightweight gun pod

system. This system maintains positive round control throughout all storage and feed operations.

The design concept of this system has been proven by combat experience with the SUU-11/A mini-gun pods. Power for the system is provided by a 32 V DC rechargeable nickel-cadmium battery located in the aft section. This battery will fire up to three ammunition complements without a recharge. If the aircraft has sufficient power available, provision is made to trickle charge during operation of the aircraft.

The pod can be loaded with ammunition while suspended from the aircraft. The storage drum is made accessible by simply removing the aft fairing. A belt of ammunition is attached to a loader mechanism which strips the links and feeds rounds into the storage drum in the proper attitude.

The first maintenance required for the GPU-2/A is at 15 000 rounds. At this time the front track bolts on the gun rotor are re-torqued. Barrel life is 15 000 rounds; however, the barrels can be changed within two minutes while the pod is still suspended from the aircraft. No special tools are required for maintenance.

CHARACTERISTICS
Weight: Approx 270 kg
Length: 304 cm
Diameter: 48 cm
Ammunition capacity: 300 rounds
Rate(s) of fire: 750 or 1500 rounds/minute
Gun: 20 mm M197
CONTRACTOR
General Electric Company, Armament Systems Department, Lakeside Avenue, Burlington, Vermont 05402, USA.

4434.303
GPU-5/A LIGHTWEIGHT 30 mm GUN POD

The GPU-5/A is a lightweight 30 mm gun pod with a high rate of fire developed to destroy heavy armour and a wide variety of mobile and fixed targets at extended ranges. It is adaptable to many fighter and attack aircraft, weighs 862 kg when loaded, and is completely self-contained.

The system comprises the GAU-13/A 30 mm four-barrel Gatling gun (**3891.303**), a helical closed loop ammunition feed system, a self-contained pneumatic drive with microprocessor controls, and a strongback and structural assembly. The linkless ammunition feed system holds 353 rounds in two helical layers of ammunition carriers surrounding the gun and extensive use is made of fibre reinforced plastics to save further weight. The pneumatic drive uses a reversible turbine capable of generating 60 hp at 9000 rpm and sufficient air is stored for two loads of ammunition to be fired in one-second bursts.

The primary structural component is the strongback which is a one-piece aluminium housing used to mount the pod to the aircraft with the recoil adaptors mounted inside it and connected by linkage to the gun. The skin of the pod is made of aircraft-type structural aluminium honeycomb and the entire pod is designed to be capable of withstanding supersonic flight speeds.

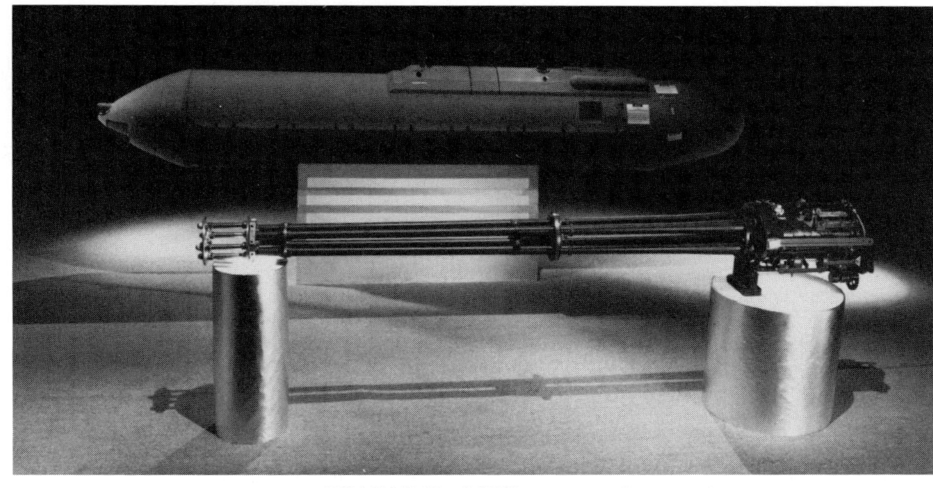
GPU-5/A lightweight 30 mm gun pod

CHARACTERISTICS
Gun: GAU-13/A 4-barrel Gatling
Calibre: 30 mm
Rate of fire: 2400 rounds/minute
Ammunition capacity: 353 rounds
Length: 4.3 m
Height: Approx 610 mm
Weight (loaded): 862 kg
Weight (empty): 621 kg
Power requirements: 115 V, 400 Hz, 300 W, 3-phase or 28 V DC, 210 W
STATUS
In production and in service with USAF. Can be installed on wing or centre body stations of a variety of tactical aircraft including the A-4, A-7, AV-8B, F-4, F-5, F-15, F-16, F-18 and the A-10. Other applications are being investigated.
CONTRACTOR
General Electric Company, Armament Systems Department, Lakeside Avenue, Burlington, Vermont 05402, USA.

4433.303
AV-8 HARRIER 25 mm ARMAMENT SYSTEM

The AV-8 armament system is a lightweight 25 mm gun system with a high rate of fire designed to provide an effective air-to-air and air-to-ground attack capability. It was developed for the USN and uses the GAU-12/U gun.

The GAU-12/U gun is a five-barrel Gatling type which is similar in overall dimensions to the M61A1 Vulcan gun (**5547.103** in 1982-83 and earlier editions) and fires the US standard family of ammunition which includes practice, HE incendiary, armour piercing incendiary and armour piercing discarding sabot ammunition and is also compatible with European-made 25 mm ammunition. A full-bore armour piercing incendiary round has been developed as a part of the AV-8 armament contract.

The system consists of two pod-like structures bolted to the bottom of the left and right of the aircraft. The pod on the left contains the GAU-12/U gun, blast deflector and drive mechanism. That on the right contains a compact 300-round linear linkless feed system. A cross-over fairing connects the two pods and contains the ammunition chuting and drive shaft. The pneumatic drive employs a rotary drive motor turning at 9000 rpm and developing 35 hp. The production version uses bleed-air from the aircraft engine. Gun recoil forces are transferred directly to the left forward aircraft mounting points via recoil adaptors.

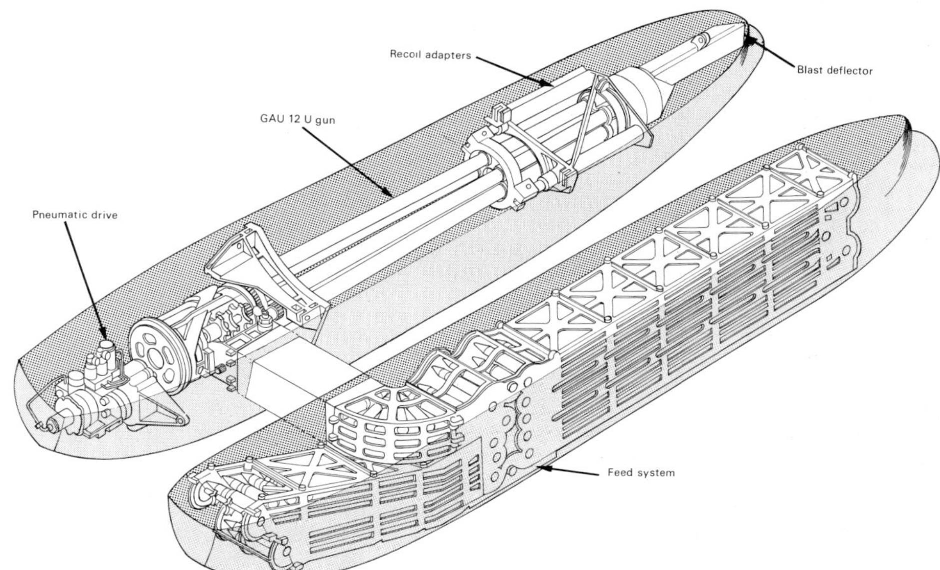

Internal arrangement of AV-8 Harrier armament system

CHARACTERISTICS
Gun: GAU-12/U
Calibre: 25 mm
Muzzle velocity: 1097 m/s (depending on ammunition used)
Rate of fire: Up to 4200 rounds/minute
Ammunition capacity: 300 rounds
Weight
Gun: 125 kg
System (loaded): 558.4 kg
System (empty): 408.6 kg
Power requirements: 28 V DC 15 A max
STATUS
In production for the USN.
CONTRACTOR
General Electric Company, Armament Systems Department, Lakeside Avenue, Burlington, Vermont 05402, USA.

6407.303

GENERAL ELECTRIC 25 mm LIGHTWEIGHT GUN – GE 225

The General Electric 25 mm lightweight gun, also known as the GE 225, is a general purpose weapon based on the Gast principle of operation. The two barrels are aligned side-by-side and use a common gun housing with the critical working mechanisms interconnected by a reciprocating lever which synchronises both firing barrels. One side of the lever permits the chambering of a live round while the other is extracting a spent case. The barrels fire alternately and remain essentially fixed. Each of the two firing mechanisms has a linkage arrangement that provides the reciprocating action. A cam at the rear of the gun provides the necessary motion multiplication and the controlled acceleration of the moving parts.

The action produces low recoil forces, allowing the weapon to be installed in light aircraft and helicopters. The weapon can be produced in both external drive and self-powered drive configurations. All live rounds are automatically cleared from the chambers at the end of each burst. The feeder is simple, requiring only one shaft and sprocket to service both gun bolts from a single, linked ammunition belt. The feed system can be installed to feed either side of the gun and spent cases are ejected downwards.

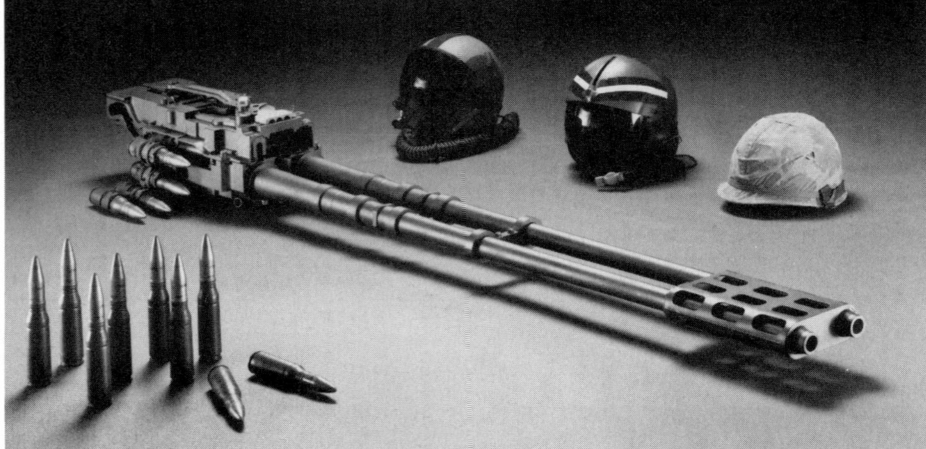

General Electric 25 mm Lightweight Gun – GE 225

CHARACTERISTICS

Calibre: 25 mm
Weight: 81.7 kg
Length: 2210 mm
Width: 216 mm
Height: 203 mm
Power to drive: 25 hp at 2000 rpm, 3 hp at 750 rpm
Muzzle velocity: 1097 m/s
Rate of fire: variable to 2000 rpm
Dispersion: 2 mil

STATUS

Ready for production.

CONTRACTOR

General Electric Company, Armament Systems Department, Lakeside Avenue, Burlington, Vermont 05402, USA.

1287.303

VULCAN GUN PODS SUU 16/A AND SUU 23/A

These gun pods are of identical size and weight and both incorporate six-barrel Vulcan 20 mm cannon and ammunition. Designed for installation on standard external store pylons, these pods provide a ready means of greatly increasing the cannon fire power of subsonic and supersonic fighter and ground attack aircraft. The pod mounted Vulcan gun is particularly well suited to the engagement of ground targets from ranges of 700 m or more. The pods are of aerodynamic shape and attachment to the aircraft is by bomb lugs at standard 760 mm (30-inch) centres.

Both pods are of similar internal layout, a cylindrical ammunition drum containing 1200 rounds occupying the rear portion of the pod. A linkless feed system couples this to the gun.

The pods differ in the type of Vulcan gun fitted but their performance and hitting power are identical. In the earlier SUU 16/A design the cannon is the standard M61A1 mechanically driven by a ram air turbine. This turbine is mounted on a hinged skin panel of the pod and is lowered into the airstream prior to firing. The use of a ram air turbine restricts the rate of fire at aircraft speeds below 350 knots (650 km/h). A further consequence of the turbine is that pod drag is somewhat greater than for the SUU 23/A pod in which a self-powered GAU-4 model Vulcan is used. For initial rotation of the GAU-4 gun in the SUU 23/A pod an electric inertia starter is provided.

In operation Vulcan gun pods are aimed using a gunsight or fire control system. During firing spent cartridge cases are ejected overboard and not retained. The gun is cleared automatically after the firing button has been released so that 'cook-off' is eliminated. During the clearing process a small number of live rounds are ejected.

A number of accessories are available to simplify loading, maintenance, and testing of the pods.

SUU 23/A Vulcan 20 mm gun pod

CHARACTERISTICS

	SUU 16/A	SUU 23/A
Length	5.05 m	
Diameter	560 mm	
Weight with 1200 rounds	780 kg	785 kg
Weight empty	484 kg	489 kg
Boresight adjustment		
elevation	±1°	±1°
azimuth	±0.5°	±1°
Accuracy	80% shots inside 8 mil	
Rate of fire	6000 rounds/minute	
Ammunition	20 mm M 50 series	
Power input from aircraft		
at 208 V 400 Hz 3 phase	7 A	10 A
at 28 V DC	—	3 A

STATUS

Both SUU 16/A and 23/A pods are in service with US forces. They are compatible with a wide range of tactical aircraft including F-100, F-105, F-4, A-4D, and F-111. The SUU 23/A pod is in service on Royal Air Force McDonnell Douglas Phantoms.

CONTRACTOR

General Electric Company, Armament Systems Department, Lakeside Avenue, Burlington, Vermont 05402, USA.

1270.303

GENERAL ELECTRIC VULCAN 20 mm AIRCRAFT GUN AND LINKLESS FEED SYSTEM

The General Electric M61A1 Vulcan aircraft cannon is of 20 mm calibre and employs the multi-barrel principle first used by Dr Gatling. The Vulcan is the standard USAF aircraft gun and is increasingly being used by the USN.

The gun has a rotary action and in the M61A1 is externally powered from the aircraft hydraulic or electric supply. A self-powered version, the GAU 4 is similar in all respects except that barrel rotation and operation of the action is derived from gun gas bled from four of the six barrels.

The outstanding features of the Vulcan are its high rate of fire, normally 6000 rounds per minute, the high muzzle velocity of about 1036 m/s and its exceptional reliability in comparison with more conventional weapons.

Ammunition linked into belts has been used with Vulcan guns but feed reliability was poor at the high rates of fire characteristic of the weapon. There are also problems of link disposal. For these reasons in the majority of applications Vulcan cannon are associated with a linkless feed system. There are considerable variations in the detailed layout of gun and feed systems in service as each is tailored to the particular requirements of the aircraft installation. The essential elements are the M61A1 or GAU 4 gun, the linkless feed ammunition chutes which contain the ammunition conveyor, the cylindrical ammunition canister, and the drive power coupling between the gun and the ammunition drum.

The Vulcan gun has six rifled barrels and is externally powered. All barrels are rigidly clamped together and attached to the forward end of the breech rotor which rotates in a stationary housing. The rotor revolves anti-clockwise looking in the direction of fire. Cam followers operate the bolt for each barrel successively chambering, firing, and extracting the rounds as the gun rotates. The linear actions needed for these operations are imparted to the followers by an elliptical slot machined in the rotor housing.

The multi-barrel design gives long weapon life by reducing heat dissipation and barrel erosion. A variety of barrel muzzle clamps are available to give different shot dispersion patterns. When the minimum dispersion clamp is in use 80 per cent of shots are within an eight mil cone.

Standard M50 series electrically primed ammunition is used which is available in a variety of types: incendiary, armour piercing with and without tracer etc.

In the self-powered variant, the GAU 4, gun gas is bled from four of the six barrels to operate the mechanism by means of a gas piston and drive which is mounted between the barrels ahead of the gun action. The GAU 4 develops sufficient power to operate the feed system. The gun is initiated by an electrically operated inertia starter.

Linkless Feed System

The linkless feed ammunition drum may be tailored to meet specific ammunition capacity requirements or to fit a specific envelope. Linking the drum to the gun is a single or twin ammunition conveyor belt contained in flexible chuting. With a single conveyor belt the spent cases are ejected overboard, but in many installations a return conveyor for spent cartridge cases is provided.

The ammunition drum carries the rounds stored

between radial partitions. A central rotor in the form of a helical archimedean screw then moves the rounds from the drum into the conveyor in a multi-stage operation.

Many accessories are available for loading and maintaining Vulcan cannon and their feed systems.

CHARACTERISTICS

The self-powered GAU 4 variant of the basic M61A1 Vulcan cannon has virtually identical performance and characteristics, except that it is approximately 5 kg heavier.

Calibre: 20 mm
Weight: 120 kg
Length: 1875 mm
Recoil travel: +6 mm
Average recoil force: 1810 kg
Max rate of fire: 6000 rounds/minute
Time to max rate: 0.3 s
Stopping time: 0.5 s
Ammunition: M53, M55AZ, M56A3 etc
Muzzle velocity: 1036 m/s
Round weight: 0.25 kg
Projectile weight: 0.1 kg

20 mm Linkless Feed System

The exact specification for the 20 mm linkless feed depends upon aircraft design considerations. The data below relate to the installation on the LTV A-7D aircraft which is typical:

Ammunition capacity: 1020 rounds
Firing rate: 6000 or 4000 rounds/minute
Ammunition weight: 114 kg
Feed system weight (less ammunition): 190 kg
Drive system: Shaft coupling to gun hydraulic motor
System type: Double ended – cases retained in drum
Total ammunition chuting length: 4.57 m

STATUS

The M61A1 and GAU 4 20 mm Vulcan guns with their associated linkless feed systems are in service with the USAF, the Royal Air Force, and with other air forces operating F-104 and F-4 aircraft. The weapon and its associated feed are in service on A-7, F-104, F-111, F-105, F-14, F-15, F-16, F-18 and other aircraft. Podded installations are also available and are in service.

CONTRACTOR

General Electric Company, Armament Systems Department, Lakeside Avenue, Burlington, Vermont 05402, USA.

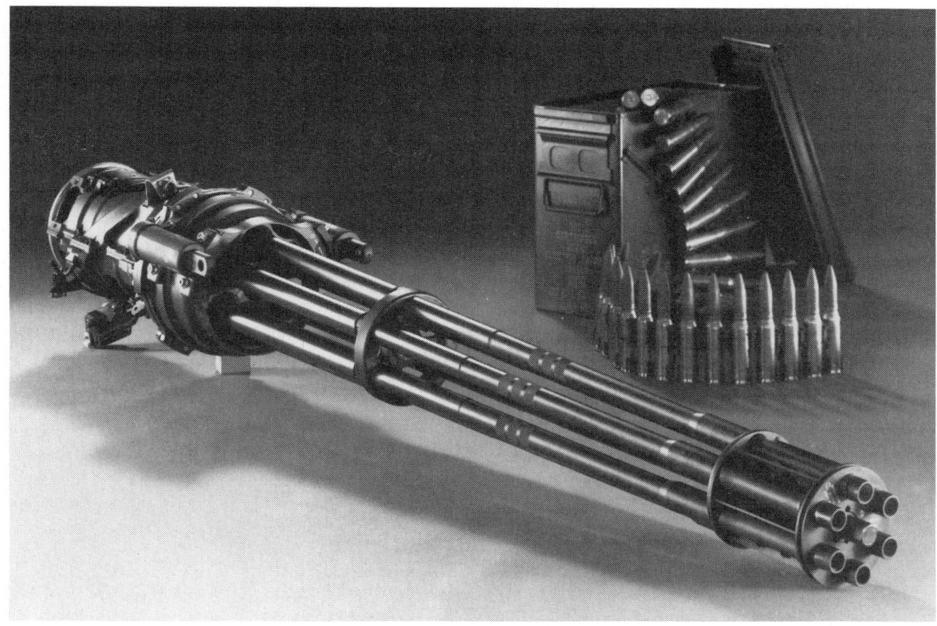

20 mm M61 A1 Vulcan gun

Linkless ammunition feed system and 20 mm M61A1 Vulcan gun for F-4E aircraft

6408.303

GENERAL ELECTRIC 20 mm VULCAN GUN SYSTEM FOR AMX FIGHTER

The AMX fighter aircraft for the Italian Air Force uses a 20 mm General Electric Vulcan Gun system that involves a linked ammunition feed system – the links are removed by the system as they are loaded into the gun. The gun involved is the 20 mm M61A1 which uses 20 mm ammunition carried in a drum magazine holding 403 rounds of M50-series and Raufoss multi-purpose ammunition. The drive for the gun system is hydraulic.

CHARACTERISTICS

Calibre: 20 mm
Weight: total 278.2 kg
Rate of fire: 4000 rpm
Ammunition capacity: 403 rounds
Drive system: hydraulic
Acceleration time at 4000 rpm: 0.35 s max
Deceleration time at 4000 rpm: 0.4 s max
Gun life: 150,000 rounds
System life: 100,000 rounds
Reliability: up to 15,000 MRBF

STATUS

Production.

CONTRACTOR

General Electric Company, Armament Systems Department, Lakeside Avenue, Burlington, Vermont 05402, USA.

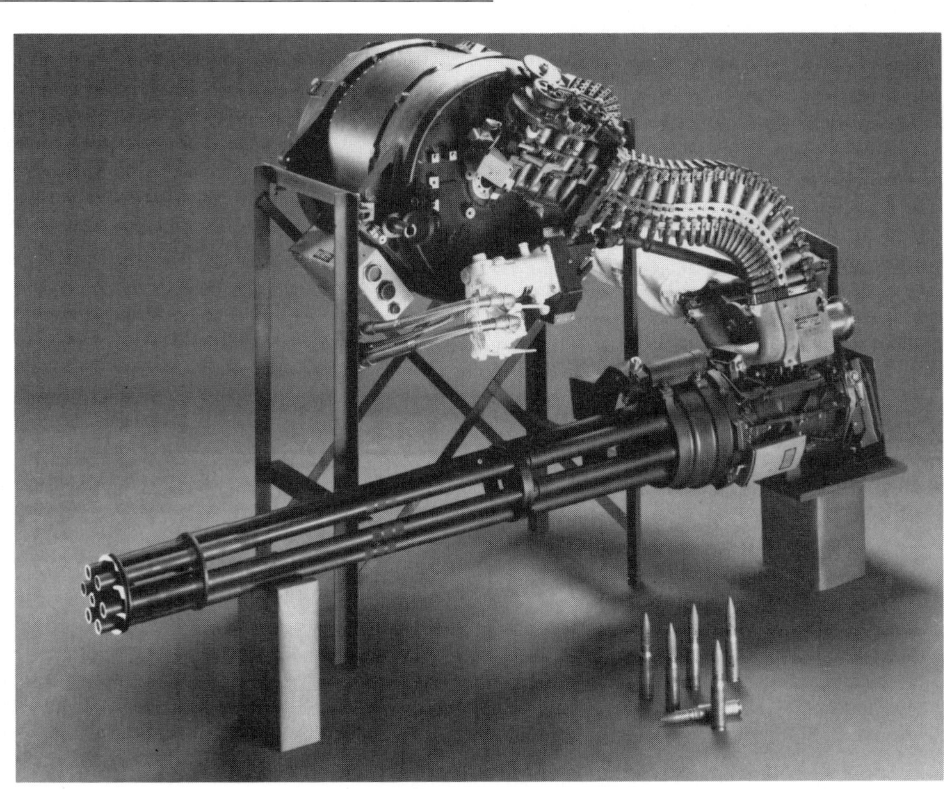

General Electric 20 mm Vulcan gun system for AMX fighter

1118.303

XM35 ARMAMENT SUB-SYSTEM

The XM35 is a fixed mounted armament sub-system located under the left weapon sponson on an AH-1G helicopter. The system carries one 20 mm automatic gun XM195, a Vulcan type with six barrels. This is a modified M61A1 gun with blast deflectors on the end of 102.6 cm barrels. The 20 mm ammunition is link-fed to the right side of the gun from ammunition containers mounted on the left and right sides of the helicopter outside the aircraft. The weight of the system with ammunition is 530 kg; without ammunition 245 kg. The system carries approximately 950 rounds of ammunition with a firing rate of 750 ± 100 rounds/minute. The 20 mm cannon has a maximum effective range of 3500 m. The sight used is an XM73 reflex type.

DEVELOPMENT

The XM35 system was developed by the US Army Weapons Command and the General Electric Company.

STATUS

In service with US forces.

CONTRACTOR

General Electric Company, Armament Systems Department, Lakeside Avenue, Burlington, Vermont 05402, USA.

1182.303

Mk 4 MOD 0 GUN POD

The Mk 4 Mod 0 pod is a self-contained and self-powered 20 mm gun system. The pod contains a twin-barrelled 20 mm Mk 11 Mod 5 gun. The container is a low-drag cylindrical shape with detachable nose and tail cones, designed for supersonic flight in fighter and ground attack aircraft. Belted ammunition is carried forward from the rotary magazine through dual feed chutes. Expended cases and links are ejected downward from the bottom of the pod at a velocity of 23 m. A salient feature of the gun pod is the instantaneous full rate-of-fire. The pod is equipped with mounting provisions for both 30- and 14-inch (762 and 356 mm) suspension systems and is normally attached to the aircraft on standard ejector racks such as the Aero 7A, Aero 27, MAU 9A/A etc.

CHARACTERISTICS

Primary kill mechanism: Blast fragments, or penetration (depending on whether HEI or API ammunition is used)

Weapon type: Anti-materiel and anti-personnel

Payload fuzing: Contact

Range: 950 m

Firing rate: 700 or 4200 rounds/minute

Delivery: Strafe

Weight

Total, loaded: 630 kg

Total, empty: 357 kg

Payload (750 rounds): 269 kg

OPERATION

The Mk 4 gun pod is aimed with either a fixed sight or a fire control system. It is fired by pressing the control stick trigger. Gun reaction time is instantaneous. Delivery is normally made in a shallow dive angle of 5 to 30°. Slant range for a 'point' target should be 900 m or less. Area targets may be profitably attacked from a longer range because accuracy is not of primary importance. Normally delivery speed is 400 knots but may be at any speed within the aircraft flight envelope.

STATUS

The Mk 4 gun pod has been flown and fired from the following aircraft: A-4 (1, 2, and 3 pods), F-4 (1 and 3 pods), A-7 (1 and 2 pods), A-6 (1, 2, 3, and 4 pods), OV-10A (1 pod), F-100 (1 and 2 pods), H-34 (1 pod).

The first production pod was delivered in August 1965, and by that time more than $12 million had been expended on development at MDHC and about one million rounds of 20 mm TP ammunition had been fired. The first USN production contract was for 347 pods and this was closely followed in June 1965 by another for a further 482 pods. The last deliveries were made in October 1967 but MDHC continued active in-service development and support under later USN contracts.

Active service on USN and USMC A-4, F-4, and OV-10 aircraft took place in SE Asia, and a number of pods were transferred to the Israeli Air Force with which service they were used on A-4 aircraft.

CONTRACTOR

McDonnell Douglas Helicopter Company, Mesa, Arizona 85205, USA.

Mk 4 gun pod

1183.303

Mk 11 MOD 5 GUN

This weapon is intended for aircraft air-to-ground, air-to-air and ground vehicle installation. Each installation consists of a 20 mm Mk 11 Mod 5 gun mechanism and a Mk 2 Mod 1 loader. The gun mechanism consists of the following major components: receiver assembly, breech assembly, revolver cylinder assembly, and one Mk 19 Mod 3 and Mk 20 Mod 3 barrel.

The Mk 11 Mod 5 gun is a twin-barrel, air-cooled, belt-fed, combination gun-gas and recoil-operated automatic weapon firing electrically primed ammunition from an eight-chamber revolver cylinder. The gun is fed by two belts of Mk 6 links which enter the mechanism on opposite sides of the loader. Belts are simultaneously advanced by a single sprocket within the loader which is driven by the revolver cylinder shaft. The mechanism simultaneously rams two cartridges, fires two cartridges, and ejects two cases which are relinked and ejected from the gun at approximately 30 m/s.

Mk 11 Mod 5 20 mm gun

CHARACTERISTICS

Gun length: 199 cm (with loader); 179 cm (without loader)

Gun weight

Gun mechanism: 65.3 kg

Barrels: 23.1 kg

Loader: 20.4 kg

Rate of fire: 4200 or 700 rounds/minute (other rates available)

Time to rate: Instantaneous (0.003 s)

Chamber pressure: 4.077 kg/cm²

System of operation: Recoil gun-gas boosted

System of locking: Fixed breech revolver principle

System of feeding: Self-feeding through gun-driven sprockets

Method of loading: Gun-gas powered dual rammers

Location of feed opening: Left and right side of loader, 2 belts (top and bottom feed optional)

Location of ejection opening: Rear of loader, 2 openings

Method of charging: Integral, air operated valve

Method of cooling: Air

Recoil distance: 28.5 mm

Distance between barrels: 104.7 mm

Number of chambers in revolver: 8

Belt velocity, average: 1.5 m/s (each belt)

Ignition system: 120 V, 400 Hz

Barrel length: 143.5 cm

Actual length: 143.5 cm

Effective length: 161.3 cm (including revolver)

Barrel removal: Quick disconnect lever, 60° interrupted threads

Rifling

Barrel: None (smoothbore system)

Revolver: 8 rifled, replaceable inserts

Number of grooves: 9

Twist: 26°, right hand

Recoil

Average: 1134 kg

Peak: Recoil 3628 kg; counter-recoil 2268 kg

Ammunition: 20 mm Mk 100 series; 20 mm Mk 50 series

STATUS

Developed for USN as part of the Mk 4 gun pod programme (see **1182.303**, above). Although demonstrated successfully as an internally fitted aircraft gun in the F-8, the Mk 1 gun has only seen active service in the podded application. Details of these roles with the USN, USMC, and Israeli Air Force are noted in the Mk 4 pod entry.

CONTRACTOR

McDonnell Douglas Helicopter Company, Mesa, Arizona 85205, USA.

1119.303

M156 ARMAMENT MOUNT

The M156 multipurpose armament mount consists of two separate mountings, one fitted to each side of the UH-1 helicopter. Each mount is capable of carrying and firing the M158A1 and M200A1 2.75-inch rocket launchers or carrying external stores suspended from standard 14-inch (356 mm) bomb racks, MA-4A. External stores must not exceed 245 kg per mount. A product improvement proposal to enable the M156 to carry a fully loaded 19-tube rocket launcher using the later heavy warhead rocket has been evaluated.

Both mounts are fixed in elevation and traverse, and sighting is by means of an XM60 reflex fixed sight. Total empty weight is 37 kg.

STATUS

In service with US forces.

4432.303

XM188E1 30 mm THREE-BARREL AIRCRAFT GUN

The XM188E1 is a lightweight 30 mm three-barrel variant of the basic General Electric 20 mm M61A1 Vulcan design (**5547.103** in 1982-83 and earlier editions) and succeeds the earlier XM188 (**1285.303** in 1981-82 and earlier editions). It is designed to fire the XM-788/789/799 ADEN/DEFA ammunition at rates of fire variable up to 2000 rounds/minute. Its low recoil force attenuation system and lightweight design make it suitable for air-to-air and air-to-ground applications on small fixed-wing aircraft and helicopters as well as for surface-to-surface and surface-to-air applications on small naval craft. The gun is completely interchangeable with the 20 mm M197 weapon (**2856.103** in 1982-83 and earlier editions) with no modifications to the vehicle. Its variable rate of fire allows the optimum rate to be selected to avoid resonance effects of the vehicle.

CHARACTERISTICS

Type: Externally powered 3-barrel Gatling
Calibre: 30 mm
Muzzle velocity: 792 m/s
Rate of fire: Variable up to 2000 rounds/minute
Recoil force: From 696 kg max at 730 rounds/minute to 1814 kg max at 2000 rounds/minute
Length: 1440 mm
Max diameter: 234 mm
Weight: 50 kg

STATUS

Prototype weapons were built and tested by the US Army, but no production plans have been revealed to date.

XM188E1 30 mm three-barrel aircraft gun

CONTRACTOR

General Electric Company, Armament Systems Department, Lakeside Avenue, Burlington, Vermont 05402, USA.

4435.303

UNIVERSAL TURRET FOR HELICOPTER GUNSHIPS

This is essentially the same turret as that now in production for the US Army's AH-1J and AH-1T helicopters but in production for the AH-1S Cobra helicopter. It is a fourth-generation system incorporating numerous improvements over the initial production version.

The basic components consist of the turret, linked feed system, and three electronics boxes containing the turret, gun, and logic controls. The turret aim is controlled by the pilot or co-pilot through helmet sights or by the co-pilot through a dedicated pantograph sight or a companion missile sub-system sighting station such as the TOW sight unit. The turret is electrically driven by two servo motors which receive position commands from the sighting system and give quick response and safe reliable operation.

The turret can operate the 7.62 mm Minigun, the 20 mm M197 Vulcan, the 30 mm XM188E1, 25 mm GE225, and the .50 calibre GECAL 50 and all can be interchanged without using special tools in less than 30 minutes. 750 rounds of 20 mm ammunition or 500 rounds of 30 mm ammunition can be carried. The prime armament to date has been the M197 Vulcan 20 mm gun and the following characteristics apply to this installation.

General Electric universal turret for helicopters, with selection of various guns which can be fitted. (Top to bottom: M197, GE225, XM188E1, GECAL 50 and 7.62mm Minigun)

CHARACTERISTICS

Gun: M197
Calibre: 20 mm
Rate of fire: 750 or 1500 rounds/minute
Ammunition capacity: 750 rounds
Elevation
Limits: +21 to –50°
Slewing rate: 60°/s
Training
Limits: ±110°
Slewing rate: Over 80°/s
Pointing accuracy: <3 mil in each axis

STATUS

In production for the US Army AH-1S Cobra helicopter and the US Marine Corps AH-1W.

CONTRACTOR

General Electric Company, Armament Systems Department, Lakeside Avenue, Burlington, Vermont 05402, USA.

1286.303

M197 20 mm THREE-BARREL AIRCRAFT GUN

The M197 aircraft gun is a three-barrel, fully automatic gun. It is a lightweight derivation of the 20 mm Vulcan M61A1 six-barrelled weapon. The M197 utilises many of the M61A1 components and has a rate of fire variable to 3000 rounds per minute.

Like other guns in the Vulcan series the M197 needs external power – about 3 hp – for its operation.

The M197 fires standard US 20 mm ammunition of the M.50 series. Ammunition can be fed to the gun as a standard linked belt or by a linkless feed system.

CHARACTERISTICS

Calibre: 20 mm
Rate of fire: 400 – 3000 rounds/minute
Muzzle velocity: 1036 m/s
Number of barrels: 3
Weight: 66 kg
Length: 1892 mm
Average recoil force: 68 kg (at 1500 rounds/minute)

STATUS

In production since October 1969 and in service on the USMC, South Korean, Taiwan and Italian AH1 J/S/T/W Cobra attack helicopters and the OV-10A aircraft. The M197 forms part of the GE flexible weapon system and is used in the GE lightweight gun pod.

CONTRACTOR

General Electric Company, Armament Systems Department, Lakeside Avenue, Burlington, Vermont 05402, USA.

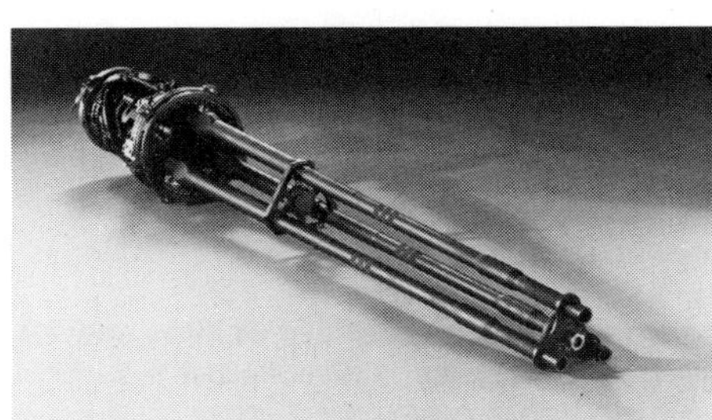

20 mm M197 aircraft gun

1112.303

M21 MAMEE HELICOPTER ARMAMENT SYSTEM

The M21 aircraft armament system is made up of two 7.62 mm M134 machine guns (miniguns) and two M158 pods with seven 2.75 inch folding fin rockets each. MAMEE comprises a pair of novel ammunition storage and feed containers which provide a capacity of 3000 rounds of 7.62 mm ammunition in each of the two 11-inch (30 cm) diameter cylindrical containers. The system can fire six pairs of rockets in one second or 4000 rounds of 7.62 mm ammunition a minute. The rocket pods are aimed by manoeuvring the aircraft to which they are fixed. The machine guns, however, traverse 12° inboard to 70° outboard and +15 to –90° in elevation and depression. When the guns traverse inboard to 12° they operate a limit switch that stops the firing on the inboard pointing gun. The gun on the opposite side then increases its rate of fire from 2400 to 4000 rounds/minute, compensating for the non-firing gun. This allows a heavy volume of fire to be placed on the target.

A complete M21/MAMEE installation consists of the following major components:

(1) two 2.75 inch rocket launchers (M158A1)
(2) two 7.62 mm six-barrelled machine guns (M134)
(3) two MAMEE ammunition storage and feed systems
(4) one pilot's rocket launcher reflex sight and intervalometer
(5) one gunner's sighting station for machine guns
(6) one gunner's control panel for guns
(7) control box assembly for 7.62 mm guns
(8) two gun mount assemblies
(9) two rocket rack assemblies.

STATUS

American and Italian Forces. Over 600 M21 systems and about 1000 MAMEE installations have been produced.

CONTRACTOR

Emerson Electric Co., Government & Defense Group, 8100 West Florissant Ave, St Louis, Missouri 63136, USA.

1116.303

M28A1 ARMAMENT SYSTEM

The M28A1 armament sub-system is an electrically controlled, hydraulically operated, dual weapon that provides wide angle coverage and rapid fire for the AH-1G Hueycobra. The M28A1 consists of a power operated chin turret, mounting the 7.62 mm machine gun M134 on the right and the 40 mm grenade launcher M129 on the left. Ammunition is stored in the ammunition bay and moved to the turret through fixed and flexible chuting. Fire control equipment is located at the gunner/co-pilot station and allows the gunner to train the weapons in azimuth and elevation. Provisions are made for the pilot to fire the weapons in the stow position. The system provides the AH-1G Hueycobra with area and point fire for attack on personnel and lightly armoured and unarmoured materiel targets during escort and armed reconnaissance missions.

CHARACTERISTICS

Max effective range: 1500 m (7.62 mm gun); 1500 m (40 mm grenade launcher)

Elevation/depression: +12 to +17.5/–50°

Muzzle velocity: 838 m/s (MG); 240 m/s (grenade launcher)

Rate of fire: 2000 or 4000 SPM (MG); 420 SPM (grenades)

Traverse: 114° l or r of aircraft datum

Weight: 203 kg (empty); 407 kg (loaded)

Ammunition

Capacity: 4000 (M134), 300 (M129) rounds

Type: Linked 7.62 and 40 mm

Sighting: M73 pilot's reflex with lighted reticle and manual range adjustment; gunner's flexible reflex sight with automatic airspeed compensation

STATUS

In service with US forces.

1117.303

XM28E1 ARMAMENT SUB-SYSTEM

The armament sub-system XM28E1 is similar to the M28A1 (**1116.303**) except for minor changes in the drive motor, which is a motor with two distinct speeds as opposed to a motor which produces two speeds through use of a control resistor.

STATUS

In service with US forces.

1836.303

XM30 AIRCRAFT ARMAMENT SUB-SYSTEM

The XM30 sub-system is a two-gun installation of XM/140 30 mm guns, one on each side of the UH-1B helicopter. Ammunition is stored inside the aircraft. The sub-system weighs approximately 1800 lb (816 kg) when fully loaded. It is capable of firing at a combined rate of approximately 850 rounds per minute. The guns are remotely controlled and may be fired by the co-pilot. The pilot can also fire the guns when they are in the fixed forward (stowed) position.

6409.303

GECAL 50 AIRCRAFT MACHINE GUNS

Two versions of the GECAL 50 (General Electrical Calibre .50) aircraft machine gun are produced, one with six barrels and the other with three barrels. Both can fire NATO standard 12.7 mm ammunition and can be fitted with barrels from 914 mm to 1295 mm long to allow ballistic performance to be optimised for a mission. The cyclic rate of fire for the six-barrelled model is 8000 rpm while that for the three-barrel model is 4000 rpm.

Gun drive power can be derived from electrical, pneumatic or hydraulic sources depending on the application. Both versions have the add-on capability of a self-starting gas drive for applications where external power is not available. A 'controlled burst' firing rate control allows the firing of a salvo of approximately ten rounds at the maximum rate of fire.

Using a delinking feeder the gun can handle M9 links. The gun can also be fitted with a linkless feed system. Gun clearing is accomplished by a declutching feeder in linked systems and through gun reversal in linkless systems.

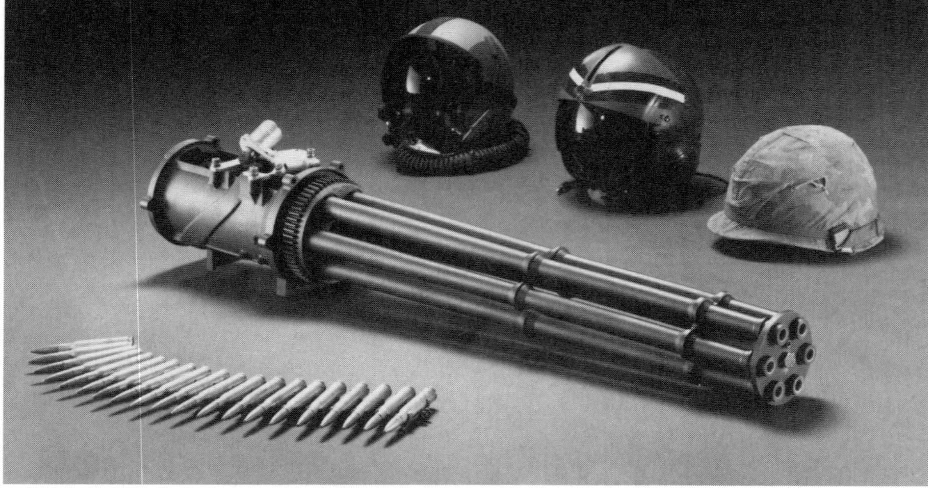

GECAL 50 aircraft machine gun, six-barrel version

CHARACTERISTICS

Type	**6-barrel**	**3-barrel**
Calibre	12.7 mm	12.7 mm
Weight	43.6 kg	30 kg
Length	1181 mm	1181 mm
Length of body	295 mm	295 mm
Height	119 mm	119 mm
Width	204 mm	204 mm
Rate of fire	8000 rpm	4000 rpm
Time to rate	0.4 s	0.4 s
Muzzle velocity	884 m/s	884 m/s
Dispersion	3 mil	3 mil

STATUS

Ready for production.

CONTRACTOR

General Electric Company, Armament Systems Department, Lakeside Avenue, Burlington, Vermont 05402, USA.

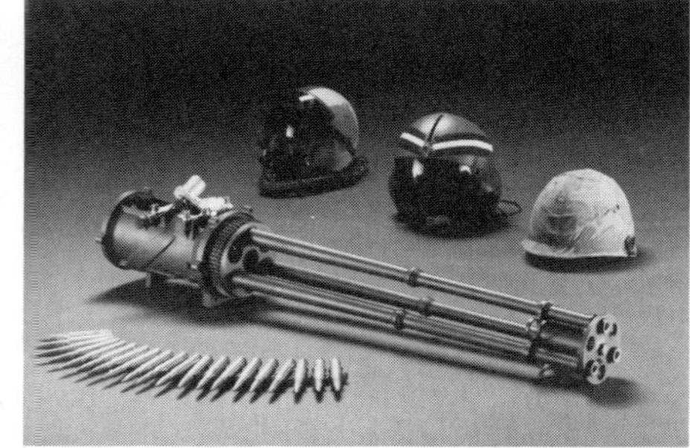

GECAL 50 aircraft machine gun, three-barrel version

6413.303

GECAL 50 COMPOSITE TURRET WEAPON SYSTEM

The GECAL 50 Composite Turret Weapon System has been developed by General Electric for the next generation of lightweight aircraft. The aerodynamically engineered turret is constructed from lightweight composite materials to give maximum strength with minimum weight. It can be used for forward and side field of fire coverage.

The turret incorporates the GECAL 50 three- or six-barrel machine gun with a selectable rate of fire for air-to-air and air-to-ground engagements. The turret can be equipped with various capacity linkless feed systems for maximum round capacity. Manual in-flight loading of the linkless feed system with linked M9 ammunition can be accomplished using a speed wrench and a General Electric-designed delinking loader.

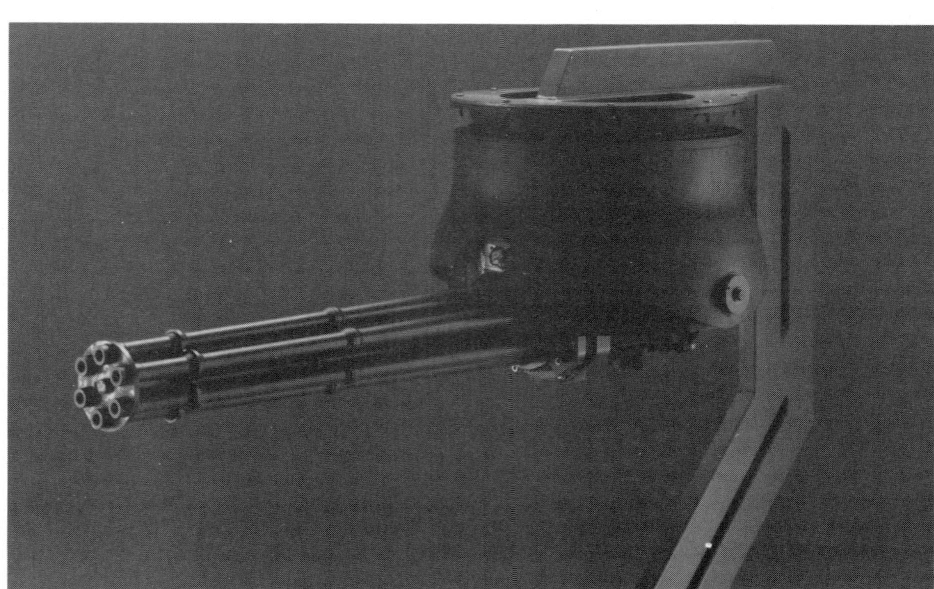

GECAL 50 Composite Turret Weapon System

CHARACTERISTICS
Weight: complete with 1500 rounds 141.9 kg
Turret diameter: 508 mm
Length: centre of turret axis to muzzles 1041 mm
Slew rates: 60°/s minimum azimuth and elevation
Azimuth: 200°
Elevation: –45° to + 15°
Recoil load: 158.9 kg peak at 1500 rpm
Rate of fire: variable to 8000 rpm
Muzzle velocity: 884 m/s

STATUS
Ready for production.

CONTRACTOR
General Electric Company, Armament Systems Department, Lakeside Avenue, Burlington, Vermont 05402, USA.

1268.303
MINIGUN POD TYPE SUU 11B/A

The General Electric M134/GAU 2B/A Minigun is a 7.62 mm calibre development of the 20 mm M61 series of 'Vulcan' cannons. It has outstanding firepower and the SUU 11B/A pod provides a convenient means of giving its power to any tactical aircraft. The pod is light in weight and small in size and is equipped with standard bomb lugs for attachment to the weapon pylon. The pod can be carried on a wide range of fixed-and-rotary wing aircraft up to a maximum speed of Mach 1.2.

The gun is mounted in the forward half of the pod and fires through the stainless steel nosecap. The centre section of the pod contains the linkless ammunition feed system. A drum magazine is employed containing 1500 rounds of standard NATO 7.62 mm ammunition Type M59, M80, or M60. In the rear of the pod is the electrical control package. The linkless feed system is essentially the same as the larger scale one already developed for the Vulcan. In the Minigun pods the feed is single-ended and spent cartridge cases are ejected overboard through a port on the bottom of the pod. The ammunition is conveniently loaded from standard belted ammunition using a hand operated de-linking loader.

Provision for gun harmonisation is built into the gun mounting. A screw adjustment provides movement of the gun line through approximately ±¾°. Shot dispersion can be varied to suit operating conditions and target characteristics. At minimum dispersion 80 per cent of rounds are within a 6.5 mil included angle.

SUU 11B/A Minigun pod

CHARACTERISTICS
Designation: USAF Minigun pod SUU 11B/A; US Army Minigun pod M18A1
Gun: 7.62 Minigun M134/GAU 2B/A
Feed: Linkless type MAU 7
Drive: SUU 11B/A electric. AC or DC: 28 V at 15 A or 208 V AC 400 Hz 3-phase 2 A per phase.
XM 18 E1 electric, 28 V DC 15 A only
Length: 2159 mm
Diameter: 305 mm
Weight: 147 kg (loaded); 111 kg (empty)
Ammunition capacity: 1500 rounds
Rate of fire: SUU 11B/A 3000 or 6000 rounds/minute; M18A1 2000 or 4000 rounds/minute
Average recoil force: 136 kg at 6000 rounds/minute

STATUS
In service with US and other forces.

CONTRACTOR
General Electric Company, Armament Systems Department, Lakeside Avenue, Burlington, Vermont 05402, USA.

3890.303
HUGHES LIGHTWEIGHT ROCKET LAUNCHER M260

Hughes Aircraft Company, under US Army contract, has developed a lightweight launcher for the 2.75 inch rocket system to be used on Army attack helicopters. It will be fitted to both the AH-1 Cobra, and the AH-64 Apache attack helicopter. The lightweight rocket launcher (LRL) is produced in two models, with capacities of seven and 19 rounds respectively and both are lighter than their predecessors. With a combat ordnance load of four 19-tube launchers, the LRLs provide a gross weight saving of about 120 kg per aircraft. This reduction is effected by the inclusion of the stores management, fuzing and fire control system, thus allowing an increase in fuel load. Fuze-timing selection is possible from the cockpit and the LRL is capable of launching 2.75 inch rockets powered by either the standard Mk 40 or its replacement higher thrust, Navy-developed, Mk 66 motor.

US Army AH-64 Apache attack helicopter fires a pair of 2.75 in rockets from launchers on outboard wing pylons

CHARACTERISTICS

	M260	**M261**
	7-tube	19-tube
Length	1651 mm	1651 mm
Diameter	254 mm	406 mm
Weight (empty)	15.5 kg	35.9 kg
Rocket capacity	7	19
Suspension	MIL-A-8591	MIL-A-8591
Firing interval	0.06 s	0.06 s

DEVELOPMENT
Development was carried out by Hughes Aircraft's Missile Systems Group for the US Army Missile Command, Redstone Arsenal, Alabama.

STATUS
The initial production contract awarded in late 1979 called for delivery of 1500 M-260 (7-tube) and 1200 M-261 (19-tube) models. Production was completed six months ahead of schedule in October 1981. A second production contract for nearly 4000 M-261 and 600 M-260 models was placed in May 1983, with deliveries to be completed in 1986. A third contract was given in 1987.

CONTRACTOR
Hughes Aircraft Company, Missile Systems Group, Canoga Park, California 91304, USA.

6414.303
GENERAL ELECTRIC 7.62 mm MINIGUN

The General Electric 7.62 mm Minigun (USAF GAU-2B/A, US Army M134) is a six-barrel externally powered weapon evolved from the M61A1 Vulcan gun. It has numerous light aircraft and helicopter

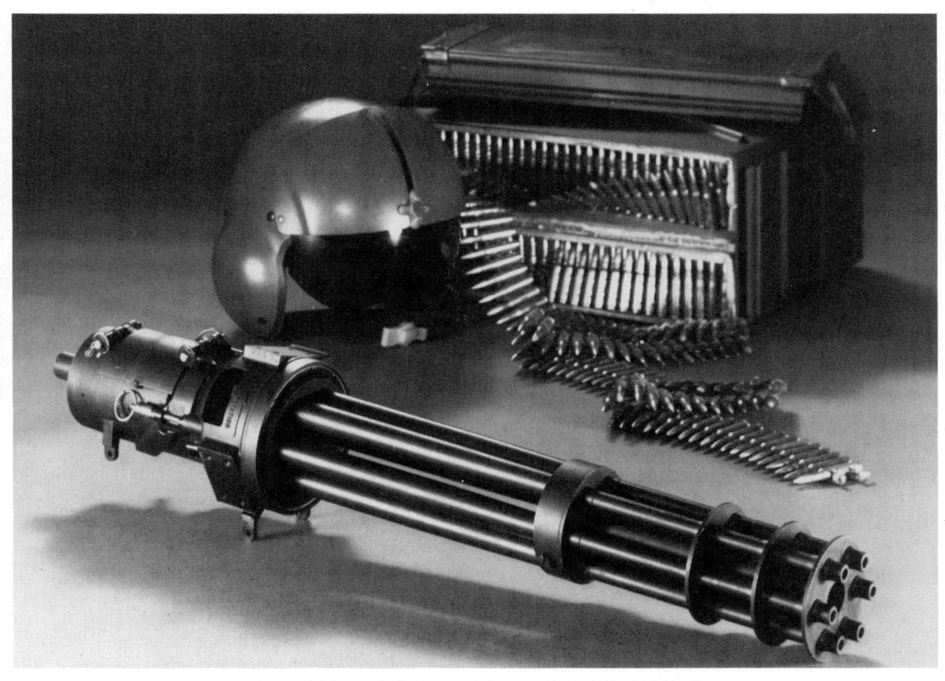
General Electric 7.62 mm Minigun (M134, GAU-2B/A)

applications and can fire up to 6000 rounds per minute. The Minigun can be adapted for mounting in various turrets, gun pods, pintle mounts, internal installations and surface vehicles. It is part of the US Army Black Hawk Weapon System.

The Minigun is used for the following American armed forces applications and is in service with some other nations:

US Army
AH-1G, turret M28
Helicopter aircraft pod XM18E1
UH-1B/1D, turret M21
LOH-58, turret M27
OH-6, turret M27E1

US Air Force
HH-53, helicopter pintle system
UH-60, pintle system
UH-1N, helicopter pintle M93
UH-1H, helicopter pintle M93E1
AT-38, pod SUU-11B/A
AT-37, internal system MAU-58A/A
AC-119, module MXU-470

US Marine Corps
AH-1G, turret M28
UH-1N, pintle A/A49E3

US Navy
SH-3A helicopter pintle system
River boat pintles and turrets

CHARACTERISTICS
Calibre: 7.62 mm NATO
Number of barrels: 6
Weight: 18.8 kg
Length: 801.62 mm
Length of barrels: 559 mm
Muzzle velocity: 838 m/s
Rate of fire: selectable to 6000 rpm
Reliability: 75 000 mrbf

STATUS
In production for US armed forces and for other nations.

CONTRACTOR
General Electric Company, Armament Systems Department, Lakeside Avenue, Burlington, Vermont 05402, USA.

1115.303
XM27E1 7.62 mm HELICOPTER ARMAMENT SUB-SYSTEM

The XM27E1 was developed by McDonnell Douglas Helicopter Company (MDHC) under contract from the US Army Weapons Command for use with the OH-6A/OH-58A Scout helicopters. The basic concept was an outgrowth of the MDHC XM7 system which was proposed as a part of the original US Army Scout helicopter competition. The XM7, with similar characteristics, mounted two 7.62 mm M60 machine guns. The XM27E1 system, on the other hand, mounted a GFE General Electric M134 7.62 mm Minigun which was modified by MDHC to incorporate a dual speed drive assembly consisting of an electric motor, speed control unit, and a gear reduction assembly.

The XM27E1 is a compact, self-contained unit which is positioned on the left floor of the rear compartment. The unit can be mounted or demounted in less than ten minutes by two men and attachment is accomplished through use of three quick-release pins. The gun is flexible in elevation only (+10°, –24°) and fired at a 'two-step trigger' selectable rate of 2000 or 4000 rounds/minute. System capacity is 2000 rounds of belted 7.62 mm NATO ammunition. The MDHC-developed XM70E1 reflex sight was provided with the system for fire control purposes. This sight is an optical beam splitter type which is mechanically linked to the system.

CHARACTERISTICS
Armament sub-system
Weight (without ammunition): 48.5 kg (107 lb)
Weight (with ammunition): 106.6 kg (235 lb)
Effective range: 1100 m
Ammunition capacity: 2000 rounds
Elevation limits: +10°, –20°
Elevation rate: 30°/s
Operating voltage: 22 – 30 V DC

Automatic gun M134 (Minigun)
Calibre: 7.62 mm
Rate of fire: 2000/4000 rounds/minute
Weight: 30.4 kg (67 lb)
Muzzle velocity: 869 m/s
Cooling: Air
Feed: Gun-driven de-linking feeder MAU 56/A
Links: M-13
Power: Electric, dual-speed drive motor

Helicopter reflex sight XM70E1
Type: Collimated, illuminated reticle
Weight: 2.2 kg (4.8 lb)
Width (extended/stowed): 254/356 mm (10/14 in)
Height: 228 mm (9 in)
Range: Adjustable, 250 – 1500 m
Drive: Mechanically coupled to gun
Lamp power requirements: 0.68 A (each filament)
Range compensating for 7.62 or 40 mm ammunition

STATUS
Over 10 000 000 rounds were fired from XM27E1 kits mounted on OH-6A helicopters during the Vietnam conflict. To our knowledge, the XM27E1 was never utilised on the OH-58A in combat, although it was checked out and cleared for use in US testing. The XM27E1 earned a reputation as one of the Army's most reliable helicopter weapon systems. This same weapon system, under MDHC's designation of HGS-5, is now in limited production for sales world-wide for use on the MDHC 500 series light attack helicopter.

MDHC XM27E1 7.62 mm armament sub-system

CONTRACTOR
McDonnell Douglas Helicopter Company, Mesa, Arizona 85205, USA.

4436.303
HGS-55 7.62 mm ARMAMENT SUB-SYSTEM

Capitalising on the XM27E1 technology (**1115.303**), the McDonnell Douglas Helicopter Company has developed and built a light combat helicopter armament sub-system designated the HGS-55 which mounts the MDHC 7.62 mm EX-34 Chain Gun weapon. As adapted to MDHC 500 series helicopters, the reliable EX-34 provides accurate point target suppressive firepower at ranges out to 900 metres. In highly manoeuvrable helicopters, the EX-34 is extremely effective against non-armoured soft targets such as general materiel and transportation systems.

In the HGS-55 system, the EX-34 fires 550 rounds a minute of NATO standard 7.62 mm ammunition. Over 500 000 rounds have been fired through a series of EX-34 prototypes and all firings confirm the exceptional reliability of this weapon. The contractor claims the potential for combat availability several orders of magnitude better than any comparable weapon of this calibre.

CHARACTERISTICS
Configuration: HGS-55
Gun: EX-34
Calibre: 7.62 mm
Rate of fire: 550 rounds/minute
Muzzle velocity: 856 m/s
Ammunition: 2000 rounds NATO standard
Weight (with ammunition): 98.9 kg
Weight (without ammunition): 40.8 kg
Power supplies: 26.5 V DC at 0.22 kW

7.62 mm EX-34 Chain Gun in HGS-55 configuration

HGS-55 helicopter armament system equipment

500MD helicopter equipped with HGS-55 armament sub-system

STATUS
Following extensive tests in 1978-79 by the British Royal Small Arms Factory, a number of these guns were purchased by the British Army. Subsequently the EX-34 was adopted by the UK MoD for use on the MCV-80 vehicle and future AFVs. The Royal Small Arms Factory has begun series production of the EX-34 under licence, with deliveries commencing in 1985. The Canadian Forces are also evaluating the EX-34 for various applications. An experimental version capable of 1000 rounds/minute is under development, and in addition to the HGS-55 model featured in this entry there are other versions for fixed wing aircraft and land-based applications.
CONTRACTOR
McDonnell Douglas Helicopter Company, Culver City, California 90230, USA.

1108.303
M5 AIRCRAFT ARMAMENT SYSTEM
The M5 is used on the UH-1B/M helicopters. It consists of a flexible remote controlled, servo-power driven gun turret mounted in the nose of the UH-1B/M. The turret incorporates one M75 40 mm grenade launcher. Linked ammunition is stored in a rotary ammunition drum in the cargo hook hole, pulled through a flexible chute by an ammunition booster, and fed to the grenade launcher. The system also has a box feed system in addition to the rotary ammunition drum. A master armament control panel is located in the instrument console and is accessible to both the pilot and co-pilot gunner. A flexible hand control sight assembly mounts above the co-pilot's seat from which the co-pilot can sight and fire the system. The sub-system can also be fired in the stow position by either the pilot or co-pilot by means of a trigger switch on both cyclic stick grips. In this mode, the turret is flexible in elevation. An MWO has been applied to the sub-system to give the sight lead angle compensation. A dual-range reticle has been applied for more accuracy at long range.
CHARACTERISTICS
Effective range: 1500 m
Elevation/depression: + 15°, –35°
Muzzle velocity: 240 m/s
Rate of fire: 230 rounds/minute
Traverse: 60° right and left
Ammunition
Type: Linked 40 mm
Capacity: 150 rounds; 300 rounds
Sighting: Reflex type: gunner/co-pilot operated in elevation and deflection

Weights	**Box fed**	**Drum fed**
Empty	106 kg	101 kg
Loaded	152 kg	208 kg

STATUS
In service with US forces.

1284.303
XM214 5.56 mm AUTOMATIC GUN
The General Electric Company has designed and developed a 5.56 mm automatic gun. The gun design is based on the proven principles of the operational M61A1, 20 mm Vulcan gun, and the GAU-2B/A, 7.62 mm Minigun. Selectable firing rates are available up to 10 000 rounds per minute, permitting optimum engagement of a variety of targets. As few as three rounds to as many as 1500 rounds can be fired in single bursts. The weapon is a multi-barrelled machine gun externally powered and suitable for aircraft, helicopter, or ground vehicle installation. The 5.56 mm M193 (M196 Tracer) round weight is approximately half that of the 7.62 mm round and is the same as is used on the M16 rifle.

A single, rugged main spring is used for actuation of the firing pins. Elimination of the springs from the individual bolts allows the bolts and entire gun to be simpler, smaller, and lighter. A de-linking feeder provides a declutching action which, when the trigger is released, immediately stops the feeding of rounds. The gun continues to fire for a fraction of a second, firing out remaining rounds. This operation automatically clears and 'safes' the gun at the end of each burst. A combination access handle and safing lever separates the main spring from the six gun bolts and provides a positive visual verification of the safe position.

A side-stripping feeder strips and disposes the ammunition links and feeds rounds to the gun. This feeder permits rapid installation of flexible ammunition belts or ammunition cans. The gun's access cover, bolts, and feeder can be removed and installed without tools or timing procedures. Reliability and maintainability of the 5.56 mm gun are expected to be equal to or better than that of the 7.62 mm Minigun.

The power input needed by the gun varies between 0.75 and 3.2 hp according to rate of fire selected. Electric or hydraulic motor drives are available for this purpose. A self-powered propellant gas-driven variant of the Minigun has also been developed. For the electric drive a range of nickel-cadmium batteries is available giving a duration of fire up to 80 000 rounds at 1000 rounds per minute.
CHARACTERISTICS
Calibre: 5.56 mm
Muzzle velocity: 990 m/s
Rate of fire: Up to 10 000 rounds/minute
Dispersion: 80% shots within 4 mils
Length: 686 mm
Weight (gun and drive): 15 kg
Recoil force, average: 110 kg (at 10 000 rounds/minute)
STATUS
Weapons have been assembled and delivered.
CONTRACTOR
General Electric, Lakeside Avenue, Burlington, Vermont 05402, USA.

XM214 5.56 mm automatic gun

1846.303
M230 30 mm CHAIN GUN WEAPON
Developed by McDonnell Douglas Helicopter Company, the 30 mm M230 chain gun is a single-barrel, externally powered weapon which incorporates a rotating bolt mechanism driven by a simple and reliable chain drive. The entire gun weighs 127 lb (57.5 kg) and will achieve firing rates from single round to 625 rounds per minute. The chain drive principle permits a simplified gun cycle which operates safely from an open bolt without requirement for chargers, declutching feeders, or other special devices. Since the M230 is motor driven, it provides inherent high belt pull, thereby eliminating the need for a powered booster. Long bolt-lock time insures an essentially gasless action and the resultant long dwell following firing provides for hangfire safe weapons.

AH-64 Apache attack helicopter armed with 30 mm M230 chain gun/cannon and 16 Hellfire missiles

As all the moving parts are totally keyed together, each and every motion within the chain gun is precisely timed and fully controlled. Further, 100 per cent positive round control within the M230's ballistically independent system eliminates those malfunctions due to conventional ammunition problems. These factors, coupled with the smooth power flow from its drive motor, ensure the highest reliability at all rates of fire.
DEVELOPMENT
McDonnell Douglas Helicopter Company initiated development of a 30 mm chain gun weapon in December 1972 and after an accelerated engineering and fabrication effort, the weapon was first fired in April 1973 and burst fired in May of that year. A 2500-round feasibility trial sponsored by the US Army was successfully completed in September 1973 and by December 1973, 4000 rounds had been fired. Subsequent models of XM230s were built and tested and a quarter of a million rounds of XM552/XM639 ammunition were fired through these several chain guns. Early in 1976, the DoD directed that the XM230 be rechambered to fire ADEN/DEFA type ammunition for purposes of NATO interoperability. To ensure interoperability, McDonnell Douglas Helicopter Company was given the responsibility for development of a family of US 30 mm rounds: M788 (TP), M789 (HEDP) and M799 (HE). The modified XM230 – designated the M230 – was first fired in March 1978. While the basic M230 gun can fire either linked or linkless ammunition through the use of a lightweight interchangeable transfer unit, the AH-64 will employ a linear linkless conveyor system.

Another member of the Chain Gun family, the M242, is in production for the US Army as the primary weapon for installation in the M2 and M3 Bradley fighting vehicles and the Light Armoured Vehicle (LAV-25) for the US Marine Corps. Additionally, the US Navy has incorporated the M242 in the Mk 38 naval system. Initial use is for fast patrol boats, however, the M242 is scheduled to replace all the older 20 mm Mk

16 guns in the Navy inventory over the next five years. This weapon (**5237.103** in 1982-83 and earlier editions) fires both European and US 25 mm ammunition.

CHARACTERISTICS

Calibre: 30 mm
Combat ammunition: ADEN/DEFA/M789/M799
Practice ammunition: M788
Weight of receiver: 37.1 kg
Weight of barrel: 15.0 kg
Weight of recoil adaptor: 5.4 kg
Weight of linked or linkless transfer unit: 5.9 kg
Total gun system weight: 57.5 kg
Length: 1675 mm
Width: 267 mm
Height: 290 mm
Barrel life: 10 000 rounds
Rate of fire: 625 ± 25 rounds/minute
Time to fire: 0.2 s
Time to stop: 0.1 s
Clearing method: Open bolt
Effective impulse (with muzzle brake): 18.12 kg
Power required: 3.0 hp
Reliability predicted: 15 000 MRBF

STATUS

In production.

CONTRACTOR

McDonnell Douglas Helicopter Company, Mesa, Arizona 85205, USA.

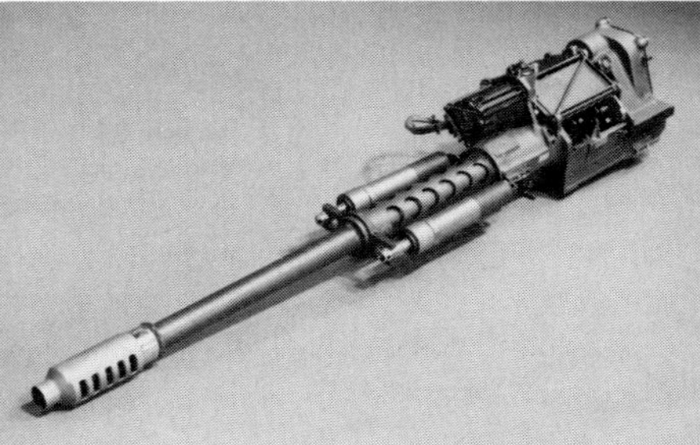

M230 cannon

1110.303

M16 ARMAMENT SUB-SYSTEM

Armament sub-system M16 contains two 7.62 mm M60CAL machine guns and a M158 7-tube rocket launcher mounted on each side of the UH-1B/C helicopter. Each set of guns is flexible in both elevation and azimuth. Capacity for the machine guns is 6000 rounds which are link-fed to the left side of each gun. The machine guns fire 2000 to 2600 rounds/minute with a maximum effective range of 1000 m. Elevation of the guns is from +11 to –63° while traverse ranges from 12° inboard to 70° outboard. Machine gun firing cuts off electrically or mechanically when guns on a given side of helicopter are in danger of firing into the aircraft.

The rockets from the M158 launchers have an effective range of 3000 m. The pilot aims through a manual reflex sight for both machine guns and rockets. The co-pilot uses a flexible reflex sight for machine guns only. Machine guns cut off while rockets fire.

STATUS

US forces.

1113.303

M23 ARMAMENT SUB-SYSTEM

The M23 armament sub-system consists of two M60D 7.62 mm machine guns and two mounts, one on each side of the aircraft in the doorways. The pintle post assembly is attached to a base tube assembly which is attached to hard points on the helicopter fuselage. The machine gun, which is a belt fed, gas operated, air-cooled automatic weapon, is attached to a pivot cradle on top of the pintle post. The base can be employed at either the right or left side, and the pedestal is designed for right or left installation. The machine gun installs on either right or left pintle mounts without adaptation, and is a modified machine gun, M60, with aircraft ring type sights, spade grips and an improved feed system. A rapid reloading capability is provided by a 550-round ammunition box attached to the weapon by a flexible chute. A canvas bag attaches to the right side of the receiver to catch links and ejected cartridges. Free traverse and elevation is allowed within fixed stop limits to prevent self-inflicted damage to the helicopter. The sub-system utilises all standard 7.62 mm ammunition.

CHARACTERISTICS

Max effective range: 1100 m
Elevation: +3.5° forward, +6.5° aft, –80° depression
Traverse: 2 – 178° in azimuth
Muzzle velocity: 838 m/s
Rate of fire: 550 rounds/minute
Ammunition
Capacity: 550 rounds per gun
Type: Linked 7.62 mm
Weight: 78 kg (empty); 95 kg (loaded)

STATUS

In service with US forces.

1114.303

M24 ARMAMENT SUB-SYSTEM

The armament sub-system M24 is similar to armament sub-system M23 except that it is mounted on the CH-47A helicopter and carries 200-round ammunition boxes. One machine gun mounts in the left forward escape hatch, the other across the right forward door.

STATUS

In service with US forces.

5220.303

M41 ARMAMENT SYSTEM

The M41 armament system consists of a pintle mount, a 7.62 mm M60D machine gun, link and brass retainer, ammunition box and a gunner's safety harness, and the complete system is mounted on the rear ramp of the Boeing CH-47 helicopter used by the United States Army. The M60D is provided with positive mechanical stops to limit the elevation and traverse of the weapon.

CHARACTERISTICS

Ammunition capacity: 200 rounds
Weight without ammunition: 19 kg
Maximum effective range: 1100 m
Muzzle velocity: 838 m/s
Rate of fire: 550 rounds/minute
Elevation: +12.5 to –69°
Traverse: 47° left and right
Sight: Aircraft ring and post type

STATUS

In service with the United States Army.

5221.303

XM94 ARMAMENT SUB-SYSTEM

Each sub-system provides two identical pintle installations which can be interchangeably mounted on either the right or left doorway compartment of a helicopter. These weapons can be utilised as a flexible pintle installation, or fixed in a forward attitude and remotely fired by the pilot. The grenade launcher used in the XM94 sub-system is the M75.

CHARACTERISTICS

Ammunition capacity: 800 rounds
Weight without ammunition: 540 kg
Effective range: 1500 m
Muzzle velocity: 240 m/s
Rate of fire: 400 rounds/minute
Elevation: +3.5 to –60°
Traverse: From fixed forwards to 135° aft

5222.303

XM59 ARMAMENT SUB-SYSTEM

The XM59 sub-system is an exterior mount sub-system for mounting one 7.62 mm M60D and one 0.50 XM213 (modified AN-M2) machine gun, in each side door of the UH-1D or UH-1H helicopter. The XM59 is basically an M23 system adapted for use with the 0.50 machine gun. The 0.50 kit pintle post assembly is attached to the M23 sub-system base tube assembly. A link receptacle and an expended brass deflector are provided to eliminate any hazard to the aircraft as expended cases and links are ejected from the weapon. Mechanical stops are provided to prevent any self-inflicted damage to the aircraft.

CHARACTERISTICS

Ammunition capacity: 100 rounds of 0.50 ammunition; 550 rounds of 7.62 mm ammunition
Weight: 126 kg with ammunition; 98 kg without ammunition
Range: 0.50 weapons 3000 m; 7.62 mm weapons 1100 m
Muzzle velocity: 0.50 – 857 m/s; 7.62 mm – 838 m/s
Elevation: +6.5 to –80°
Traverse: 2 – 178° in azimuth

6410.303

LORAL AIR INFLATABLE BOMB RETARDERS – BALLUTE

The Loral air inflatable bomb retarders, or Ballutes, are cruciform tail stabilisers that can be used in either the 'low drag' configuration without inflating the retarder, or in the 'high drag' configuration with the retarder deployed. The retarders are used in low-altitude bombing with a high drag to allow the bombing aircraft to escape the blast and fragmentation effects of the bombs. This high drag is achieved by opening a balloon-parachute called a Ballute, enclosed in the container fin. The retarder is certified for use up to 700 knots air speed and at an altitude of 18.3 metres.

When the bomb is dropped, the spring which holds the arming wire is stretched and unwinds. When completely unwound, it frees two safety pins and opens the cover. The Ballute assembly is then freed under the combined effects of the external springs on the cover and the internal ejection spring, and is immediately inflated by the external airflow which enters through four symmetrical inlets. The cover remains integral with the Ballute, thus avoiding any risk of projection to the rear which might strike the bombing aircraft.

The BSU-49/B retarder is a Loral Systems Group design which is employed with the American Mk 82 500 lb general purpose bomb. The company also produces the BSU-50/B for use with the 2000 lb Mk 84 bomb. Both retarders are certified and used by the US Air Force and are also used by other air forces. The BSU-85/B is under development for US Navy use with the Mk 83 1000 lb bomb.

CHARACTERISTICS

Type	BSU-49/B	BSU-50/B	BSU-85/B
Diameter	222 mm	403 mm	326 mm
Span	381 mm	940 mm	516 mm
Length	660 mm	775 mm	995 mm
Weight	25.4 kg	43.5 kg	40.8 kg

STATUS

The BSU-49/B and BSU-50/B are in production and in service with the US Air Force and some other air forces. The BSU-85/B is presently in low rate initial production for the US Navy.

CONTRACTOR

Loral Systems Group, Engineered Fabrics Division, Goodyear Street, Rockmart, Georgia 30153, USA.

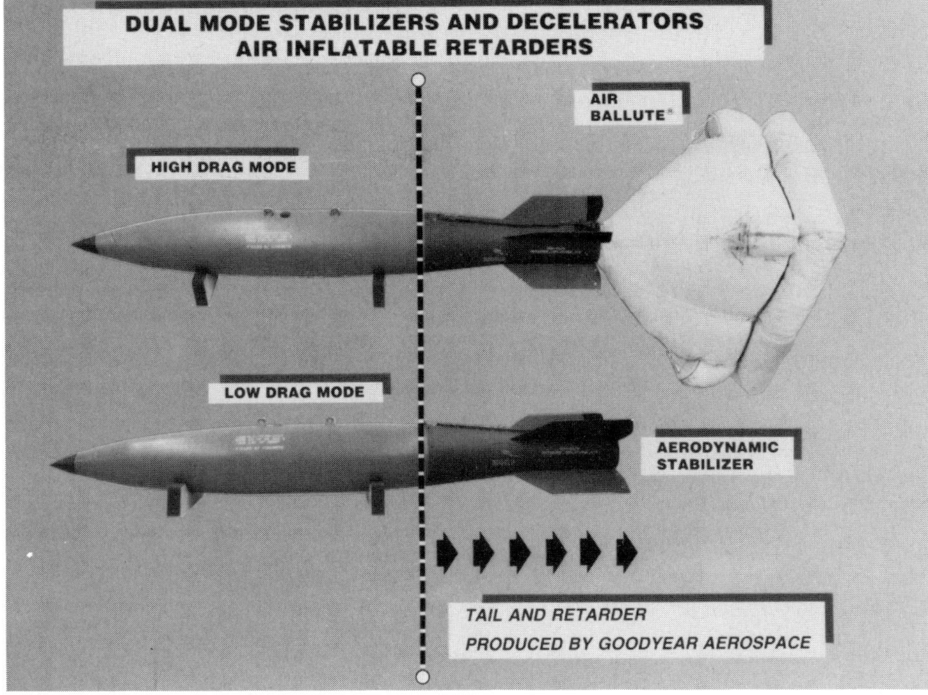

Basic principles of the Loral air inflatable bomb retarder – Ballute

YUGOSLAVIA

5223.303

YUGOSLAV AIRCRAFT ARMAMENT

Listed below is a resumé of aircraft armament used by the Yugoslav Air Force. It should be noted that the Soviet designations for bombs are very similar to those of Yugoslavia, for example Soviet HE general-purpose bombs are the FAB-100, FAB-250 and FAB-500 etc, Soviet incendiary bombs are the ZAB-100 etc, Soviet HC anti-tank bomblets are the PTAB-5 etc and the Soviet chemical bombs are the VAP and ZAP series.

BOMBS:

1.5 kg PTAB-1.5 HC anti-tank bomblets
2.5 kg PTAB-2.5 HC anti-tank bomblets
2.5 kg RAB-2.5 fragmentation bombs
3.5 kg RAP-3.5 fragmentation bombs
16 kg RAP-16 fragmentation bombs (normally used in clusters of eight)
45 kg ZAB-45 incendiary bomb
50 kg SAB-50 parachute flare
50 kg FOTAB-50 photo-flash
120 kg RAB fragmentation cluster bombs
250 kg FAB-250 HE general-purpose bombs
500 kg FAB-500 HE general-purpose bombs
DPT-150 cluster bomb unit with 54 PTAB 1.5 kg or 44 RAB 2.5 kg or 34 RAP 3.5 kg
150 litre PLAB-150L napalm

ROCKETS:

57 mm VRZ-57 high explosive (or anti-tank) in pods of 12 (pod is Lanser L57)
127 mm VRZ-127 high explosive (or anti-tank)

AIRBORNE RADAR

CANADA

5874.353

IRIS INTEGRATED RADAR IMAGING SYSTEM

The MacDonald Dettwiler IRIS (Integrated Radar Imaging System) is a radar reconnaissance system which, due to a unique form of on-board synthetic aperture processing, combines the tactical advantage of a standard sideways-looking airborne radar (SLAR) with the high resolution of a synthetic aperture radar (SAR) into a single system.

Used as a tactical radar, it provides front-line field commanders with high resolution, real-time imagery of both fixed and moving targets. The SAR principles provide range independent resolution, allowing the IRIS airborne segment to operate at safe stand-off distances. The acquired imagery is stored and displayed in real-time on board the aircraft. At the same time the imagery is downlinked to simple cost-effective transportable ground stations.

Used in a strategic reconnaissance role, where fine resolution change detection imagery is required, the airborne segment can either store, or simultaneously downlink, the imagery to a precision analysis facility. This analysis facility provides the interpreter with the ability to analyse the radar imagery in detail.

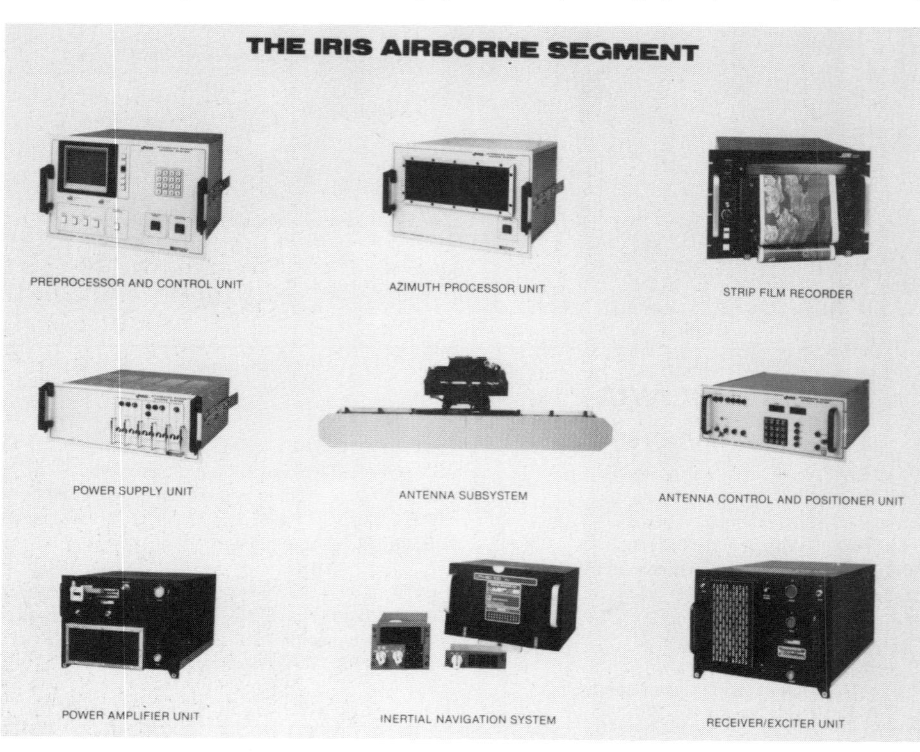

The self-contained airborne segment consists of an airborne imaging synthetic aperture radar with on-board digital processing, downlink transmission, high density magnetic tape data recording, and image production capabilities. Images are processed in real-time and displayed on a video monitor and hard copy paper strip. Two operational modes provide the capability of switching in-flight from a wide swath surveillance mode, with 18 × 18 metre resolution, to a high resolution reconnaissance mode with a 3 × 3 metre resolution. The airborne segment has been packaged in compact, rugged modules for installation in a wide variety of aircraft, although optimised in design and construction for the high performance executive-size turboprop and turbojet aircraft.

The transportable ground station consists of a downlink receiver, digital image processor, and tactical work station, and is typically mounted on a two-to-five ton truck. The tactical work station features high density magnetic tape image recording, continuous image and frame image video display capabilities.

The precision analysis facility provides radar interpreters with the ability to analyse reconnaissance images received downlink or recorded on board the aircraft and the transportable ground segment. The facility can archive and retrieve multiple digital data sets, including maps, multi-sensor data and other interpretation aids entered at the facility. The precision analysis facility allows for the in-depth interpretation of imagery, change detection, map mosaicing, and preparation of precision hard copy film products.

STATUS

In production and service. Six IRIS (and IRIS prototypes) are currently operational, with three additional systems under construction.

CONTRACTOR

MacDonald Dettwiler Technologies Ltd, 3751 Shell Road, Richmond, British Columbia V6X 2Z9, Canada.

6423.353

LN66 RADAR SYSTEM

The LN66 is a modular radar system which can be used for general and specific radar requirements in the airborne and shipborne roles. The system has a high performance record in a variety of applications including general surface surveillance and navigation, coastal surveillance, gun fire control systems and ASW roles. It is in use with the US Navy as the prime radar in the LAMPS Mk.1 helicopter, and as the surface surveillance and short range navigation radar in a number of surface vessels. More detailed information is given under entry **6873.253** in the Naval Radar section.

STATUS

In operational service in both the airborne and surface ship roles.

CONTRACTOR

Canadian Marconi Company, Commercial Communication Division, 2442 Trenton Avenue, Montreal, Quebec H3P 1Y9, Canada

---

FRANCE

3297.353

MIRAGE 2000 RDI AND RDM RADARS

New radars have been developed for the multi-role and all-altitude interceptor versions of the Mirage 2000 aircraft. The RDM radar is a low PRF doppler system which is fitted to the multi-role aircraft, while the RDI is a high PRF pulse doppler radar designed for the interceptor version. The Mirage 2000s of the French Air Force are equipped with the RDM or RDI system. Both radars are illustrated over the page, enabling both the external differences and similarities to be seen. Few technical details have been cleared for publication but the following paragraphs contain summaries of the salient features of each.

RDI

This version is equipping the all-altitude air superiority interceptor variant of the Mirage 2000 for the French Air Force. It uses a TWT X-band coherent transmitter, and features a flat slotted plate antenna with a diameter of 67.4 cm. The antenna has an integrated IFF antenna and the overall system incorporates an integrated IFF transmitter/receiver. The equipment is optimised for air superiority in all weather conditions at any height, and range is stated as being in the region of 60 nm against a head-on target. TWS facilities are available and the RDI radar incorporates an internal digital data bus to facilitate sophisticated radar signal processing. Other facilities include digital doppler filtering circuits and higher resistance to ECM. Total weight of the 12 line-replaceable units is 255 kg. The available air-to-air modes will allow the following functions to be performed:

(1) air-to-air search at all altitudes
(2) long-range continuous tracking and missile guidance
(3) automatic short-range tracking for missiles or guns

Air-to-ground modes are for:

(1) ground mapping
(2) contour mapping
(3) air-to-ground ranging.

RDM doppler radar for the multi-role version of the Mirage 2000

RDI pulse doppler radar for the air superiority versions of the Mirage 2000

STATUS

The first production systems were delivered in 1986. Thomson-CSF is the prime contractor for this programme and co-operates with Electronique Serge Dassault.

RDM

This low PRF doppler radar is fitted in multi-role variants of the Mirage 2000. It employs a TWT coherent transmitter in conjunction with an inverse cassegrain antenna. Features emphasised in the design are operational versatility, extensive options and growth potential. The radar is capable of accommodating a CW illuminator for the guidance of missiles that use a doppler homing head, an IFF interrogator, doppler beam sharpening circuits for high resolution mapping and ground target identification, as well as new high reliability sub-systems for low level penetration and sea surface search and tracking. The basic operational functions provided are:

(1) air-to-air search (all-altitude, all sectors) and interception. Detection range is given as 60 nautical miles in clear air and 25 nautical miles looking down against ground clutter (5 m² cross-section target)
(2) low level strike (mapping, terrain avoidance, blind let-down)
(3) air-to-ground attack
(4) sea surface search and attack.

For air-to-air combat, the RDM offers a 60 degree cone of coverage, the antenna scanning at either 50°/s or 100°/s, with ±60°, ±30° or ±15° of scan. A 5 m² target can be detected at up to 60 nautical miles range. For air-to-air gun attacks, the 3.5° beam-width radar can be locked on to the target at 10 nautical miles range, with automatic tracking within the head-up display's field of view or in a 'super-search' area, or in vertical search. In a look-down air-to-air situation, a 5 m² target can be detected at 25 nautical miles range.

TWS facilities are available, and the design incorporates an internal digital data bus, and built-in test equipment (BITE).

For both RDI and RDM, radar data are displayed on a HUD and on a trichromatic multi-mode CRT head-down display, linked to an interception and firing computer.

STATUS

The first flight tests of the RDM model began in 1979 and first production deliveries were made in January 1983. Apart from France, 124 Mirage 2000 aircraft have been ordered by Abu Dhabi, Egypt, Greece, India and Peru and are, or will be, equipped with RDM radars.

CONTRACTOR

Thomson-CSF, Aerospace Group, 122 Avenue du Général Leclerc, 92105 Boulogne-Billancourt, France.

6137.353

RDX AIRBORNE RADAR

Thomson-CSF is developing a new range of airborne radar systems intended for use in combat aircraft of the 1990s, including the French ACT tactical combat aircraft. The RDX is a fire control radar with an electronic scanning antenna which is designed to search and track several airborne and naval targets simultaneously. It is capable of operating in an interlaced mode which enables the radar, for example, to carry out terrain-following for low level flight, while at the same time searching for and tracking airborne targets.

Two other radars are also in early development, a multi-mode system with an active module antenna and electronic scan intended for surveillance and tracking during air-to-ground missions (see entry **6428.353**), and a model with mechanical antenna scan in the horizontal plane and electronic scan in the vertical.

STATUS

In early development.

CONTRACTOR

Thomson-CSF, Aerospace Group, 122 Avenue du Général Leclerc, 92105 Boulogne-Billancourt, France.

RDX two-axis electronic scanning radar

6890.353

RDY MULTI-FUNCTION RADAR

The RDY is a multi-function radar designed for the Mirage 2000 aircraft. In the air-to-air mode the system offers long range detection of very low and very high altitude penetrating targets, irrespective of their angle of approach. The RDY's track-while-scan function enables target tracking, presents the pilot with tactical situation analysis data and identifies targets through interactive dialogue with IFF interrogators.

The RDY offers several facilities in the air-to-ground mode, including:

(a) all-weather low altitude penetration with ground and contour mapping functions
(b) air support and direct-on-sight assault against stationary or moving targets in virtually all weather conditions
(c) resetting of navigation system.

In the air-to-sea mode the RDY will detect targets, even in high seas, and will be used for firing long range missiles, such as AM39 or Kormoran.

In the air-to-air mode the RDY covers all waveforms currently in use, including multiple range-gate high, medium and low repetition frequency.

The RDY is equipped with a flat-plate low-inertia aperture antenna of an entirely new design, which offers excellent agility and very low levels of secondary lobing, using a conical scan of 60° in all directions. This antenna may later be replaced by an electronic scanning antenna.

A compact dual peak power transmitter optimises power output across a broad spectrum of frequencies, irrespective of the operating mode of the radar, and at the heart of the radar is a programmable signal processor with a very large computational capacity.

STATUS

In development with several prototype systems already test flown.

RDY multi-function radar

CONTRACTOR

Thomson-CSF, Aerospace Group, 122 Avenue du Général Leclerc, 92105 Boulogne-Billancourt, France.

6428.353

RACAAS AIRBORNE RADAR

Thomson-CSF is developing a new programmable multi-mode airborne radar, designed for the next generation of combat aircraft. The equipment is part of the RACAAS (air-combat and ground-attack radar) programme and is being developed under contract from the Service Technique Télécommunications et des Équipments Aéronautiques (STTE).

The RACAAS radar uses virtually all the techniques needed for the combat aircraft of the future. It has a TWT transmitter with two peak-power values, allowing for selection of appropriate waveforms with each operational configuration, eg a wide selection of PRF, pulse compression, etc. Multiple RF receivers with wide dynamic range are associated with large capacity, programmable signal processing systems. High speed information processing is designed to handle all the different operating modes (air-to-air, air-to-ground and air-to-surface).

DEVELOPMENT

Flight tests are in progress using a Mystere 20 from the Brétigny test centre. The RACAAS radar is initially fitted with a low-inertia planar-array antenna for high scanning agility. After the current phase of testing, a new electronic-scan antenna will be fitted.

STATUS

In development.

CONTRACTOR

Thomson-CSF, Aerospace Group, 122 avenue du Général Leclerc, Boulogne-Billancourt, France.

3298.353

IGUANE/AGRION/VARAN AIRBORNE RADARS

Under French Government contract, a family of airborne radars has been developed for sea surface surveillance and maritime warfare applications. Three versions have been named:

Iguane: This model is in production to replace the DRAA 2A ASV radar (**1961.353** in 1981-82 and earlier editions) fitted in the Breguet Alize carrier-borne maritime and ASW aircraft, and is also being fitted to the Breguet Atlantic ATL2 long-range maritime patrol aircraft.

Varan: This is essentially an Iguane radar with an antenna which makes it suitable for virtually all the medium size maritime patrol aircraft (fixed-wing and helicopter). The Varan is the radar on the French Navy Gardian maritime surveillance aircraft. An option with this version is a software package giving synthetic aperture capability.

Agrion: This model exists in several versions but is primarily designed for use aboard helicopters or light aircraft forming a part of task forces employed for support at sea or for coastal protection. An important function of the Agrion radar is AS 15TT missile (**3359.311**) guidance in strike missions against small surface vessels and patrol boats, where it is fully integrated via data link, to the ship's weapon system. Agrion is being supplied for several foreign navies including that of Saudi Arabia for Dauphin shipborne helicopters.

Antenna assembly of the Varan radar

Each member of the Iguane family can perform the following operational missions:

(1) surface and anti-submarine warfare
(2) over-the-horizon targeting for shipborne SSMs
(3) search and rescue
(4) marine environmental protection
(5) maritime law enforcement
(6) navigation
(7) weather avoidance.

All three models operate in the X-band, using pulse compression and frequency agility to ensure high performance on maritime targets in all combinations

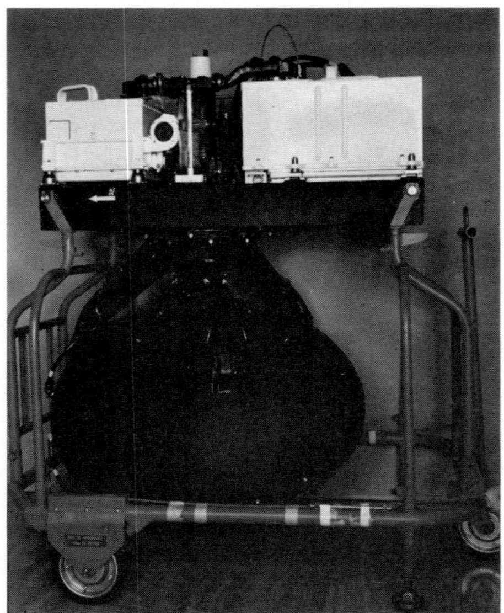

Iguane airborne radar

of weather, sea state, and operating altitude. These same techniques also provide the Iguane family of radars with maximum protection against ESM and ECM. The Iguane and Varan radars are also capable of providing sideways-looking radar imagery in real-time using the basic antenna as an optional extra facility.

Last design of the family is the ARCANA radar for the French Air Force.

STATUS

The Iguane is in production for the Alize update programme and the Breguet Atlantic ATL2, as noted above. The Varan version is fitted to the sea surveillance model of the Gardian aircraft. The Agrion radar is in service on Dauphin helicopters of the Saudi Arabian defence force. The Agrion 15 version is also designed for use with OTHT ship-launched surface-to-surface missile.

Agrion radar antenna on a Dauphin helicopter

CONTRACTOR

Thomson-CSF, Aerospace Group, 122 Avenue du Général Leclerc, 92105 Boulogne-Billancourt, France.

4797.353

RAPHAEL TH AIRBORNE SURVEILLANCE RADAR

Raphael TH is a sideways-looking airborne radar employing techniques based on synthetic aperture techniques. It was originally designed for land area surveillance but is also capable for use in the maritime surveillance role. The system is pod-mounted and employs pulse compression techniques to provide accurate and sharp mapping. Radar information is transmitted by a high-speed link to a ground or ship-based command centre, which can be up to several hundreds of kilometres away, where the information is processed and displayed in virtual real-time.

STATUS

In production for the Mirage F1 of the French Air Force and other countries. The system is also being fitted to the Mirage 2000 (SLAR 2000). First production units were delivered in late 1986.

CONTRACTOR

Thomson-CSF, Aerospace Group, 122 Avenue du Général Leclerc, 92105 Boulogne-Billancourt, France.

Raphael TH SLAR 2000 radar

1960.353

ORB-31 RADAR SYSTEMS

The ORB-31 is a range of airborne radar systems designed to carry out the following missions:

(1) detection of surface ships
(2) target designation
(3) weather mapping
(4) radar navigation
(5) search and rescue at sea.

All the ORB-31 system radars operate in the I-band and are especially suited to helicopter-borne uses.

The configurations, based on modular units, are calculated according to the mission and the carrier. The different sub-systems available are described in the ORB 31 radar (**1793.353**), which exists in the following configurations:

ORB 31 S1 Alouette III (SNIAS Helicopter Division)
ORB 31 W French Navy WG 13 helicopter
ORB 31 AS anti-submarine
ORB 31 D Super Frelon (Exocet system).

ORB 31 S1

The ORB 31 S1 system is an airborne surveillance radar fitted to the Alouette III helicopter. The choice of the transmitter (0.5 or 4μs pulsewidth) enables the system to be adapted to the scheduled mission.

ORB 31 W

The ORB 31 W system is an airborne surveillance system fitted to the French Navy's WG 13 helicopter (design undertaken under STTA contract).

The ORB 31 W system uses two transmitter/receivers. The first is intended for long-distance detection, and the second, with its higher resolution, is less sensitive to clutter.

ORB 31 AS

Anti-submarine warfare model for the detection, identification, location and station-keeping of other helicopters of a tactical ASW group. A continuous and updated display of sonar contact and the tactical situation is provided, and features of the large-display console include time, motion, offset, scales up to 100 nautical miles, and all controls positioned on the front panel.

ORB 31 D

The ORB 31 D system is a target designation airborne radar fitted to the Super Frelon helicopter carrying the Exocet system. Its function is to detect, track, and then locate accurately a target on the sea in order to supply the data necessary to the weapon system with which it is integrated.

The system consists of two transmitter/receivers, three indicators (two PPI and one B-scope), and a telemetry box.

STATUS

In operational service with a number of defence forces but no longer in production.

CONTRACTOR

Omera-Segid (part of Thomson-CSF), 55 rue Maurice-Gunsbourg, 94204 Ivry-s/Seine Cedex, France.

6425.353

ORB 32 RADAR SYSTEMS

ORB 32 is a range of I-band airborne radars, based on the considerable experience acquired by Omera during the development, manufacture and operation of ORB 31, of which it is a logical extension. It is built from modular sub-assemblies compatible with one another, each performing one of the following functions:

(1) maritime patrol
(2) anti-submarine warfare (ASW)
(3) active missile fire control (TD)
(4) search and rescue (SAR)
(5) radar navigation
(6) weather radar.

Reduced weight and dimensions facilitate installation on all types of aircraft.

OPERATIONAL CHARACTERISTICS

Pitch and roll antenna stabilisation
360° azimuth and 30° elevation scanning
Sectoral scans of 60, 120, 180 or 240°
Azimuth, bearing or true motion stabilisation
Peak power: 80 kW, fixed, tunable, frequency agile
Range: 90 km on 500 m target
Special-to-type characteristics appear below.

ORB 32 AS

This is one of the elements of the anti-submarine warfare weapon system in the SA 321 Super Frelon helicopter. It performs the following functions: helicopter station holding in ASW, tactical situation information, guidance and helicopter attack on a designated target, navigation, weather mapping. The ORB 32 AS consists essentially of a system performing both primary radar and secondary radar functions. The use of transponders allows the localisation and identification of helicopters flying at low altitudes by eliminating sea-clutter.

The ORB 32 has been ordered by the People's Republic of China for the upgrading of its SA321J helicopter avionics fit.

ORB 32 airborne maritime surveillance radar

ORB 32 WD

The ORB 32 WD is an airborne reconnaissance and target designation radar. When integrated in a weapon system, its purpose is to detect, designate and accurately track two sea-targets. Target co-ordinates may be automatically transmitted to the active missile carried by aircraft, or helicopter or to a launch vessel (over the horizon surface-to-surface attack).

ORB 32 WL

This is a simple, compact and lightweight system, suitable for small aircraft and helicopters, specially designed for surface reconnaissance, search and rescue (SAR), radar navigation, weather radar.

ORB 32 JN

This system, which uses one of the larger display consoles, performs all the functions required for maritime patrol missions.

ORB 32 ASD

Combination of the AS and WD types.

One version of the ORB 32 has been selected by the Swedish Navy for integration into a new anti-surface, anti-submarine and search rescue system for helicopters known as the 9HCI 100.

It has been developed in co-operation with Philips Elektronikindustrier, with the latter supplying a new generation radar display known as RADDS, and MEL of UK supplying transponders.

CONTRACTOR

Omera-Segid (part of Thomson-CSF), 55 rue Maurice-Gunsbourg, 94204 Ivry-s/Seine Cedex, France.

6426.353

ORB 37-HL AIRBORNE RADAR

The ORB 37-HL is a high performance navigation and weather radar, designed specifically for helicopters, and has been developed from the previous generations of ORB-37 radars. It is equipped with a digital indicator in order to provide a stable, high brightness display. Main applications are mapping, navigation with beacons, terrain following assistance and weather avoidance. The system consists of an antenna, a transmitter-receiver, and a digital indicator with processing unit.

CHARACTERISTICS

Frequency: 9375 MHz
Power output: 10 kW
Antenna: 400 x 300 mm, stabilisation limits of $\pm 25°$

STATUS

Fitted to the Puma and Super Puma of the French ALAT.

CONTRACTOR

Omera-Segid (part of Thomson-CSF), 55 rue Maurice-Gunsbourg, 94204 Ivry-sur-Seine Cedex, France.

ORB 37-HL airborne navigation/weather radar system

6124.353

ANTILOPE V TC AIRBORNE RADAR

The Antilope V TC radar has been designed to equip the penetration version of the Mirage 2000N. It consists of line easily accessible replaceable units assembled in the nose cone of the aircraft. The basic functions of the Antilope V are the generation of flight-control orders for very low-altitude and high-speed flight (terrain following); manual or automatic navigation system updating by surface correlation; air-to-air search, tracking and combat; air-to-surface search and tracking; and air-to-ground ranging. The terrain following function, for which safety is paramount, has continuous automatic test and redundant architecture to counter possible malfunction. The system uses a TWT J-band coherent transmitter and features a flat slotted plate antenna. The radar data is displayed on a HUD and on a trichromatic multi-mode CRT head-down display.

STATUS

Deliveries for the French Air Force Mirage have now commenced. Electronique Serge Dassault is the main contractor, with Thomson-CSF as a co-contractor.

CONTRACTOR

Electronique Serge Dassault, 55 quai Marcel Dassault, 92214 St. Cloud, France.

Antelope V terrain-following and navigation radar

6427.353

ANEMONE AIRBORNE RADAR

Intended for the modernised Super Etendard, the Anemone radar has been developed under the prime management of Electronique Serge Dassault, with the participation of Thomson-CSF. It has the following functions:

(1) air-to-surface (main function)
(2) air-to-ground
(3) air-to-air.

It consists of a nose cone (antenna and circuitry), an aircraft/radar interface unit and radar controls.

The Anemone radar is equipped for operation in a severe EW environment and allows linear scan search, continuous tracking, single target track and track-while-scan, and ground mapping and ranging.

STATUS

The Anemone is to be fitted to the modernised Super Etendard of the French Navy.

CONTRACTORS

Electronique Serge Dassault, 55 quai Marcel Dassault, 92214 St.Cloud, France – prime contractor

Thomson-CSF, Aerospace Group, 122 avenue du Général Leclerc, 92105 Boulogne-Billancourt, France – sub-contractor.

Anemone airborne radar

3131.353

CYRANO RADAR FAMILY

The Avionics Division of Thomson-CSF designed and developed the Cyrano family of airborne fire control radars for the Mirage fighter aircraft. The first one of the family, Cyrano I, was fitted on the Mirage III C. The subsequent versions were:

Cyrano II

Multi-function radar for the Mirage III E.

CHARACTERISTICS

I/J-band; monopulse; cassegrain antenna; peak power 200 kW; liquid cooling; mounted in a pressurised radome. Data are displayed on a CRT and a head-up display. Weight of equipment: 188 kg in nosecone; 29.5 kg in equipment bay and cockpit.

DEVELOPMENT

The first prototype was completed in 1962 and the radar put into service in 1963.

STATUS

Cyrano II is used on the Mirage III E aircraft of the French Air Force and a number of other countries.

Cyrano III

Cyrano III radar has a design which is not specific to a given aircraft. It was developed as a prototype and should be considered merely as a development phase of the Cyrano IV.

Cyrano IV

Information on the Cyrano IV series is given in the next entry.

CONTRACTOR

Thomson-CSF, Aerospace Group, 122 Avenue du Général Leclerc, 92105 Boulogne-Billancourt, France.

1396.353

CYRANO IV/CYRANO IV M ATTACK RADAR

The Cyrano IV is a development of the Cyrano series of airborne radars and is a multi-role system. The Cyrano IV M is the newest version of this series.

Basic functions performed by the radar are:

(1) air-to-air search
(2) automatic tracking
(3) interception and fire domain computations
(4) dog-fight engagements
(5) home-on-jam mode
(6) ground mapping

Additional options are:

(7) contour mapping
(8) terrain avoidance
(9) blind let-down
(10) air-to-ground ranging.

The Cyrano IV M model incorporates TWS facilities and is also suitable for air-to-sea search and tracking roles in addition to those listed for the Cyrano IV.

Data can be presented by means of a Type 196 gunsight or CRT HUD, and inputs to weapon systems are available also. Other aircraft systems providing inputs to the radar system include an inertial or gyro platform for altitude reference information, and an air data computer for aircraft performance and ambient parameters.

Operational missions include interception, air superiority or interdiction using guns or missiles; all-weather penetration; air-to-ground attack with guns, bombs and rockets.

The Cyrano IV M, as well as being a multi-function radar, differs from its predecessor in the embodiment of new technology which confers improved reliability and maintainability on the later model.

STATUS

The Cyrano IV was produced in quantity for the Mirage F1 aircraft of the French Air Force and those of other countries. All French aircraft and many of the others have been fitted with the Cyrano IV M version. Production of the IV M began in 1980. A version of this radar has been developed for adaptation to Mirage III, Mirage V and Mirage 50. The Cyrano IV MR is for the reconnaissance version of the Mirage F1.

CONTRACTOR

Thomson-CSF, Aerospace Group, 122 Avenue du Général Leclerc, 92105 Boulogne-Billancourt, France.

Cyrano IV and Cyrano IV M attack multi-function radars pictured side-by-side for comparison. The later Cyrano IV M (right) incorporates TWS facilities and improved reliability and maintainability due to new technology

1672.353

AGAVE AIRBORNE RADAR

The Agave is a lightweight, multi-role radar, fitted to the Super-Etendard carrier-based aircraft and the Jaguar International strike aircraft. It has been optimised for naval applications, but its air-to-air and air-to-ground functions give it appreciable interception and ground attack capabilities.

Basic functions fulfilled are:

(1) search: air-to-surface
air-to-air
(2) target designation for the homing head of a long-range active missile or to a HUD
(3) automatic tracking: air-to-surface
air-to-air
(4) ranging: air-to-air
air-to-ground
(5) map display.

CHARACTERISTICS

X-band; monopulse; inverse cassegrain antenna stabilised in roll and pitch; PRF and pulsewidths adapted to operating conditions; beam shaping; solid-state construction; integrated circuits; built-in test equipment. Weight of equipment: 48 kg in nosecone, 17 kg in cockpit and fuselage.

OPERATION

In the search mode, the scanned sector is 140° in azimuth and 60° in elevation. After target identification, the pilot switches to automatic tracking. Target data are supplied either to the homing head of an active missile, or to a HUD. In the air-to-air mode, the detection range for a fighter aircraft is within 10 and 15 nautical miles (18 to 28 km). In the air-to-surface mode, the detection range for a patrol boat is within 22 and 30 nautical miles (40 to 55 km).

The angle of search in azimuth as noted above is ±70° forward from the line-of-flight, thus giving 140° sweep coverage ahead from left to right. In elevation, the angle of search can be either 6 or 12° depending on whether a one- or two-bar scan pattern is used. In practice, a one-bar pattern is selected automatically for low level altitudes and, at higher altitudes, a two-bar scan cuts in to improve near coverage. The pilot, however, has the option of moving this scan pattern up or down through 60°.

Agave airborne radar

All essential radar controls are positioned on a control stick, the handgrip controlling not only the antenna elevation, designation marker bearing and range, but also the commencement of lock-on for the automatic tracking sequence which, once the target has been pinpointed, will provide the pilot with continually updated range and bearing information.

In the model for the Jaguar, all radar information will be displayed either on a raster head-up display by means of a scan converter on a combined map/radar head-down display, thus retaining the projected map display in its normal place. The radome is fitted in the nose of the aircraft in place of the laser ranger and marked target facilities of the 'S' version of the aircraft.

STATUS

Agave is fitted to the French Navy Super Etendard aircraft, the international version of Jaguar and the export versions of the Mirage F1 and Mirage 50. It is also fitted to the Dassault-Breguet Lancier, the combat version of the Alpha Jet trainer. Over 170 Agave systems have been ordered. An updated version, called Anemone, is described under entry **6427.353**.

CONTRACTORS

Thomson-CSF, Aerospace Group, RCM Division, 178 Boulevard Gabriel Peri, 92242 Malákoff Cedex, France.

Electronique Serge Dassault, 55 quai Marcel Dassault, 92214 St Cloud, France.

6889.353

DAV SEARCH AND ALERT RADAR

DAV is a search and alert radar which is installed on, and rotating with, the rotor of a helicopter. It is intended for use by ground support helicopters to detect air threats within a radius of 7 km and also to provide omnidirectional search and target designation with an extremely short reaction period, particularly against other helicopters flying at low altitude.

STATUS

In development.

CONTRACTOR

Electronique Serge Dassault, 55 quai Marcel Dassault, 92214 St Cloud, France.

DAV search and alert radar

6122.393

NRAI-4 IFF TRANSPONDER

The NRAI-4 transponder is the airborne element of the Mk X IFF unit, and is compatible with the Mk XII IFF system. It includes Mode C altitude coding and an interrogator sidelobe suppression (ISLS) function that inhibits replies to sidelobe signals from the interrogator. Friendly aircraft can be identified without manoeuvring, and an identification of position capacity allows specific friendly aircraft to be identified on the radar screen.

CHARACTERISTICS

Frequency: 1090 MHz (transmit); 1030 MHz (receive)
Power output: 500 W peak
Modes available: 1, 2, 3A/C and Mode 4 compatibility
Codes: 32 on Mode 1, 4096 on Modes 2 and 3/A, 2048 on Mode C
Dimensions: 127 × 129 × 145 mm
Weight: 2.6 kg

STATUS

In service with the Jaguar, Mirage F1 and Super Etendard fixed-wing aircraft, and the SA 330 helicopter.

CONTRACTOR

LMT Radio Professionelle, 46 quai Alphonse Le Gallo, 92103 Boulogne-Billancourt Cedex, France.

5871.353

NRAI-7 AIRBORNE IFF TRANSPONDER

The NRAI-7 is a solid-state diversity transponder which inhibits replies to interrogator side-lobe transmissions, and automatically codes special replies to provide assistance in the position identification of particular aircraft and in emergencies. The diversity function is provided by a dual receiver with inputs connected to upper and lower antennas, a system for comparison of received signals and an antenna switch that directs the response to the antenna that has received the strongest interrogation signal. This allows more accurate identification, particularly during aircraft manoeuvres which can blanket or interrupt signals. The pilot is also able to insert codes such as radio failure alert and warning of hijackers aboard. A single or two-unit format is offered and a naval version is also available.

CHARACTERISTICS

Peak power: 500 W
Frequency: 1030 MHz (receive); 1090 MHz (transmit)
Modes available: 1, 2, 3A/C and Mode 4
No. of codes: 32 in Mode 1, 4096 in Modes 2 and 3/A, 2048 in Mode C
Dimensions: 130 × 127 × 145 mm
Weight: 3 kg

STATUS

In production and in service with Mirage 2000, Mirage F1, Mirage III, Mirage IV, AS 332 and AS 335.

CONTRACTOR

LMT Radio Professionelle, 46 quai Alphonse LeGallo, 92103 Boulogne-Billancourt Cedex, France.

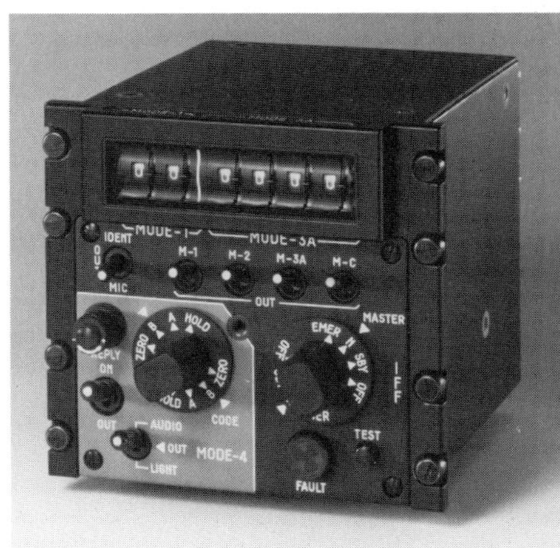

NRAI-7 airborne IFF transponder

6123.393

NRAI-9A IFF TRANSPONDER

The NRAI-9A is essentially a two-box version of the NRAI-7 (**5871.353**) which incorporates a number of improvements. These include the elimination of sidelobe response, and automatic special-code referral, with positive identification permitting a ground operator to locate a particular aircraft. Special emergency codes may also be employed, chosen by the pilot, such as radio failure. Dual receiver channels connected to upper and lower antenna and comparison circuits provide a diversity function.

CHARACTERISTICS

Frequency: 1030 MHz (receive); 1090 MHz (transmit)
Power output: 500 W peak
Modes available: 1, 2, 3A/C and Mode 4 compatibility
Codes: 32 on Mode 1, 4096 on Modes 2 and 3/A, 2048 on Mode C
Dimensions: Transmitter/receiver 58 × 193 × 361 mm; control unit 127 × 130 × 80 mm
Weight: Transmitter/receiver 2.5 kg; control unit 1.4 kg

STATUS

In production for the ATL2 maritime patrol aircraft.

CONTRACTOR

LMT Radio Professionelle, 46 quai Alphonse Le Gallo, 92103 Boulogne-Billancourt Cedex, France.

NRAI-9A IFF transponder

5872.353

NRAI-10A AIRBORNE IFF INTERROGATOR

The NRAI-10A is an airborne IFF interrogator designed primarily for maritime surveillance purposes, to enable the radar operator to identify aircraft and ships passing through his control area. Coded interrogation signals are transmitted synchronously with pulses from the aircraft's radar, the coded reply supplying the identity, or its altitude for aircraft, according to the interrogator mode selected. The interrogator is a ⅜ ATR single box unit weighing 5.9 kg, which incorporates a coder/decoder, transmitter and dual receiver. The equipment is normally used with a monopulse antenna having two different radiation patterns, one with a maximum and one with a minimum on the antenna axis. This enables unwanted replies due to sidelobe interrogation to be suppressed and valid replies to be sharpened on the display.

CHARACTERISTICS

Frequency: 1030 MHz (receive); 1090 MHz (transmit)
Power output: 1 kW
Modes available: 1, 2, 3/A and C. Mode 4 is available with suitable coder/decoder
Weight: 4.2 kg

STATUS

In production and in service for the Dassault-Breguet Mystere-Falcon 20 F, and Gardian. In production for Dassault-Breguet ATL2.

CONTRACTOR

LMT Radio Professionelle, 46 quai Alphonse LeGallo, 92103 Boulogne-Billancourt Cedex, France.

5873.363

NRAI-11A IFF INTERROGATOR-DECODER

The NRAI-11A is an interrogator-decoder for identifying and determining the range and bearing of any friendly mobile unit within the surveillance area of the IFF system with which it is operating. The equipment consists of an encoder, a transmitter, an antenna switch, two receivers, an analogue processing unit, a defruiter, a passive decoder, an evaluator, an extractor, a vobulator and an automatic self test. Peak output power of the system is 1 kW with a duty cycle of 1%.

The dual receiver system determines bearing of the target by monopulse techniques and an antenna switch is incorporated for suppression of sidelobes during ISLS interrogation. Interrogation mode is 1, 2, 3 without interlacing, with passive decoding on mode 1, 2 or 3 and 4096 combinations. The extractor has a memory capacity of 255 plots.

CHARACTERISTICS

Peak power: 1 kW
Frequency: 1030 MHz (transmit); 1090 MHz (receive)
Dimensions: 124 × 80 × 194 mm
Weight: 12 kg

STATUS

In production and in service with the Mirage 2000.

CONTRACTOR

LMT Radio Professionnelle, 46 quai Alphonse LeGallo, 92103 Boulogne-Billancourt Cedex, France.

1962.353

ESD 3300/3400 IFF/SIF TRANSPONDERS

The range of type 3000 IFF airborne transponders designed by ESD are used for the identification of any type of fixed- or rotary-wing aircraft by interrogation from a secondary radar or air traffic control system. These micro-miniaturised equipments represent a new generation of airborne transponders and can be installed easily in the cockpit of any aircraft or helicopter. Their modular design facilitates maintenance and a built-in automatic test system allows permanent monitoring of operation.

The type 3000 range has been designed for association, if required, without any internal modification, with an automatic code switching system, increasing airspace protection. In addition, the ESD 3400 possesses a space diversity system maintaining contact with interrogating stations independently of aircraft manoeuvres, thereby reinforcing mission safety.

The range of transponders meets military specifications: Stanag 5017 (Mk XA/IFF 3300; Mk XII/IFF 3400), Air 7304 and civilian specifications ICAO (Appendix 10) and ARINC 572. They are used as standard equipment by various air forces.

ESD 3300 IFF/SIF transponder

Modular design has permitted the two functionally identical versions to be packaged in either the ESD 3300 single-box model or the ESD 3400 two-box configuration.

CHARACTERISTICS

Reception
Frequency: 1030 MHz
Sensitivity: –77 dBm
IF: 60 MHz
SLS: 3-pulse
Transmission
Frequency: 1090 MHz
Power output: 500 W peak
Power consumption: 28 V DC 50 W
Dimensions
ESD 3300: 133 × 146 × 145 mm
ESD 3400: 133 × 146 × 75 mm (controller)
194 × 90 × 319 mm (transmitter/receiver) ⅜ short ATR
Weight
ESD 3300: 4 kg
ESD 3400: 7.8 kg
MTBF: Given as 1500 h

CONTRACTOR

Electronique Serge Dassault, 55 quai Marcel Dassault, 92214 St Cloud, France.

5875.353

ORCHIDEE BATTLEFIELD SURVEILLANCE RADAR

The Orchidee system is a battlefield surveillance and target designation radar designed to provide tactical information to the battlefield commander. The name is an acronym for Observatoire Radar Coherent Heliporté d'Investigation Des Elements Ennemis and is intended to be helicopter-mounted on the AS 332 version of the Super Puma to provide a deep penetration battlefield surveillance system for the French Army.

The system consists of two elements, airborne and ground, the former comprising the helicopter itself, a pulse doppler X-band radar transceiver and antenna, and a data link for two-way communication with the ground element. The radar is designed to detect moving targets only, between speeds of 6 to 400 km/h, with the information transmitted over the data link to a mobile ground station and shown on a map display. LCTAR is responsible for the radar development, ESD for the communication links and the ground station, and Aérospatiale as the prime contractor responsible for total system co-ordination.

The airborne radar system is a pulse doppler, multi-mode type using a large span flat antenna and a coherent TWT equipped transmitter. Range is understood to be in the region of 150 km. The antenna system folds under the tail of the helicopter during take-off and landing.

STATUS

Development began in 1986, with test flying of an early prototype still in progress. The complete Orchidee system is due to be operational in the mid-1990s.

CONTRACTOR

Responsibility for the overall programme is with the Delegation Générale pour l'Armement (DGA) – part of the French Ministry of Defence.

INTERNATIONAL

1784.353

AN/APS-503 and LASR-2 RADARS

The AN/APS-503 airborne search radar is installed in Sea King helicopters of the Canadian Forces (Sea Element) in service with DDH280 class destroyers.

The AN/APS-503 radar is lightweight (45 kg) with high power (50 kW peak) for its size. It comprises five units: a transmitter/receiver, an antenna complete with mounting, an azimuth range indicator (ARI), a radar control unit, and a range bearing unit. The ARI is seven inches (18 cm) P-7 phosphor, designed for use by a radar operator, and possesses three scales each with range markers spaced appropriately.

The range and bearing unit positions a range strobe and an azimuth marker on the PPI, at the same time transmitting this information to ancillary equipment such as navigation systems.

The transmitter operates in the I-band, with a pulse-width and pulse recurrence frequency optimised for maritime reconnaissance, and a 50 kW peak power. The receiver has a solid-state local oscillator with logarithmic response and sensitivity time control and fast time constant.

The antenna is a 24 × 18 inch (61 × 46 cm) parabola, providing a 4° beam in azimuth and a 5° beam in elevation. The antenna is stabilised for pitch and roll and has a manual tilt control. Top-mounted, belly-mounted, and nose-mounted antenna versions are also available. A pilot's bright display is available as an option.

STATUS

The system is no longer in production but is still in service. More detailed information is available in earlier editions of *Jane's Weapon Systems*.

CONTRACTORS

Littor Systems Canada Ltd, 25 Cityview Drive, Rexdale, Ontario M9W 5A7, Canada.

Eaton Corp, Command Systems Division, Broad Hollow Road, Farmingdale, New York 11735, USA.

3473.353

AN/APS-504(V) SERIES AIRBORNE RADAR

The APS-504(V) series of airborne search radars have been designed for tactical transport and maritime surveillance applications. They can be employed in either a fixed- or rotary-wing type of aircraft. The systems are operationally suitable for use in a sea search, ground mapping, navigation, station keeping or weather avoidance role.

The APS-504(V)3 provides digital signal processing and scan conversion allowing the use of standard RS343A high-resolution 875 line TV displays. This system also allows recording and playback of mission radar data and alphanumerics on compatible TV systems.

The system consists of six basic units:

(1) receiver/transmitter
(2) antenna/pedestal assembly

(3) digital TV displays
(4) radar control unit with joystick cursor control
(5) radar processor and converter
(6) key pad entry unit.

The antenna is stabilised for pitch and roll and is designed to make use of available radome volume. Features include variable width sector scan, range delay, offset sweep, sweep expansion, true motion display using inputs from the navigation system and a 1553B interface to the tactical mission computer.

The APS-504(V)5 is the latest system in the series. It enhances the detection capability of the original (V)2 system and couples it with the digital processing, high definition and raster display of the (V)3. The V(5) version incorporates frequency agility, pulse compression techniques and a TWT amplifier. An SAR derivative of this version is used in the IRIS reconnaissance radar (**5874.353**).

CHARACTERISTICS (AN/APS-504(V)5)

Frequency: 8.9 – 9.4 GHz
Peak power: 8 kW
Antenna: Parabolic, selectable polarisation
Beamwidth: Horizontal 2.5°; vertical 7°
Beacon mode/search mode/simultaneous beacon/search
Linear and logarithmic receivers
Digital X-Y displays viewable in high ambient light
Multi-function display

STATUS

In production.

CONTRACTORS

Litton Systems Canada Ltd, 25 Cityview Drive, Rexdale, Ontario M9W 5A7, Canada.

Eaton Corporation, Command Systems Division, 815 Broad Hollow Road, Farmingdale, Long Island, New York 11735, USA.

5621.053

NATO IDENTIFICATION SYSTEM (NIS)

NATO's future IFF system, the NATO Identification system (NIS), is in the design phase with studies and development being carried out in USA, UK, West Germany and France. Present Mk X and Mk XII IFF systems in use with NATO are becoming outdated. Both use a fixed frequency system which can easily be jammed by hostile action. Another problem is that the current systems can only identify friendlies and the lack of a response leaves the operator uncertain whether he is interrogating a hostile aircraft or a friendly one whose IFF is faulty, turned off or jammed.

The new system being developed, under NATO Standardisation Agreement (Stanag) 4162, and known as Mk XV, will use spread spectrum and additional ECCM techniques to counter jamming, and will have a high level of message security. It is intended to combine the basic information with other data sources to provide a high accuracy IFF. After considerable discussion the four countries concerned have agreed on the development of a system in D-band (1 to 2 GHz). The UK has insisted on the incorporation of a modified D-band which involves the addition of a radar mode whereby a ground radar transmission is modulated with IFF information and a transponder in the aircraft responds when scanned by the radar beam.

The main companies involved in development are two teams in the US – Bendix/Raytheon and Texas Instruments/E-Systems/Teledyne; Cossor and Plessey in UK, Siemens in West Germany and Thomson-CSF in France. Plessey and Cossor have formed a partnership to undertake a contract from the UK MoD for the project definition phase in the United Kingdom. Time scales are difficult to assess, the two teams in the US are being funded by the US Air Force to carry out development work in two phases. Phase 1, which consists of building 'breadboard' models, has been completed and the two US teams have moved into the second phase with both working on parallel contracts to produce advanced development models for flight testing. This development and flight test phase is expected to last until the end 1988. The US Air Force was expected to award contracts in 1988 with production deliveries probably starting around 1990. It is understood that the European companies, which are also being funded by their individual governments, are on a similar time scale.

STATUS

In the USA, Texas Instruments has delivered a development model of the Mk XV IFF to the USAF for flight testing. NIS hardware has also been submitted by Plessey and Cossor for testing for interoperability with the Texas Instruments equipment. The Plessey equipment was delivered in September 1987 and the Cossor system in November 1987. In both cases the tests are reported to have been satisfactory. The Texas Instruments system has also been trialled with Bendix equipment and with Siemens units. Compatibility has been established with existing Mk XII IFF systems in use on US Navy ships and US Army I-Hawk missile systems.

Progress towards the establishment of the NIS system is gathering momentum. In the USA requests for proposals for full scale development have been issued and nearly 20 companies are expected to respond. A selection was due to be made in mid-1988. Europe is about six months behind the USA and should award its development contracts by the end of 1988.

6429.353 EFA RADAR

The airborne radar for the European Fighter Aircraft (EFA), designed for operations in the 1990s, has been the subject of bids by a number of companies. Although RFPs had been issued to some 15 companies, the ultimate selection will be between two consortia. The first is that known as ECR 90 and comprising Ferranti (UK), FIAR (Italy) and Inisel (Spain), the second is called Emerald and consists of AEG (West Germany), Marconi Defence Systems (UK), FIAR and Inisel.

The ECR 90 system is an X-band radar featuring wide-band spread-spectrum transmission and very powerful processing. New generation programmable signal processors will provide fast and flexible changes of mode and function. The Emerald system is a derivative of the Hughes AN/APG-65 radar which is described in detail under entry **3302.353**.

STATUS

As at April 1988, no decision had been made as to which system would be finally developed.

ISRAEL

3305.353

EL/M-2001B RADAR

The Elta EL/M-2001B dual-mode I/J-band pulse doppler radar for combat aircraft is a range-only equipment for air-to-air and air-to-ground operations. Of compact dimensions and weighing less than 50 kg, it is suitable for installation in the nose position of small new combat aircraft or for retrofitting to earlier types.

It is constructed of six line replaceable units: transmitter, receiver, power supply, RF exciter, RTU, and servo unit. These are attached to the radar chassis/mounting unit which also carries the beam directing unit, gear box, RF head and TWT. The last of these items is the only component that is not solid-state.

OPERATION

Modes of operation are air-to-air, all altitudes and all aspects; air-to-ground, manual or computer controlled. Target detection is by visual means and the radar will automatically acquire and track any target viewed in the HUD by the pilot. It operates in heavy ground clutter and remains clutter-free even in very low level air combat, according to Elta.

Target acquisition is also automatic in the air-to-ground mode, although in both modes there is provision for pilot intervention for selection of targets of interest. Information provided by the EL/M-2001B can either be displayed on the aircraft HUD or fed into the weapons' control computer where it may be used for CCIP and CCRP weapon delivery computation.

EL/M-2001B installed in an Israeli Air Force Kfir aircraft

CHARACTERISTICS

Frequency: I/J-band
Form factor: Frustum of a cone
Length: 49 cm (including connectors)
Base diameter: 45 cm
Antenna aperture: 195 mm
Weight: <50 kg
Power requirements: 115 V, 400 Hz, 3-phase, 1 kVA, DC 30 W

STATUS

In production and in operational service with the Israeli Air Force.

CONTRACTOR

Elta Electronics Industries Ltd, (a subsidiary of Israel Aircraft Industries Ltd), Ashdod, Israel.

3306.353

EL/M-2021 RADAR

The Elta EL/M-2021 multi-role combat radar operates in the I/J-band of the radar spectrum and provides facilities for a variety of air-to-air and air-to-ground missions. These include air interception, close-in combat, air-to-ground bombing and gunnery, terrain following, avoidance and mapping.

In the air-to-air mode it performs search and automatic tracking, and in the air-to-ground mode it supplies ranging information to the bombing computer (both CCIP and CCRP weapon delivery modes are available). Head-up and head-down displays provide the pilot with data for interception of targets and provide fire control for guns. A stabilised radar picture is produced which is of value in the navigation mode using radar mapping.

The equipment is constructed of line replaceable units which include transmitter, processor, modulator, missile illumination unit, RF head, weapon delivery unit, receiver, RF exciter, servo amplifier unit, power supply, interface computer unit, and missile delivery computer. An inverse cassegrain antenna is employed and solid-state circuitry is used.

Coherent processing using digital techniques is used, and both look-up and look-down waveforms are understood to be provided.

CHARACTERISTICS

Transmitter

Frequency: I/J-band

Bandwidth: 500 MHz

Peak power: 3 kW

Average power: 200 W

Receiver noise: –5 dB

Antenna: Inverse cassegrain

Scan	Tracking	Search/AI	Dogfight
Azimuth	±70°	±45°	±10°
Elevation	+80 to –40°	±5°	–0 to +60°

Range tracking accuracy: 10 m ± 10%

Processor: CCD processor for signal extraction; digital data processor for avionics interface

Dimensions (excluding antenna)

Max diameter: 74 cm

Minimum diameter: 44 cm

Height: 1.295 m

Weight: 120 kg (including antenna)

STATUS

Development. The EL/M-2021B has been flight tested in an F-4 Phantom. Although designed originally for the F-16 aircraft of the Israeli Air Force, a version was developed for the cancelled Israeli Lavi fighter (**5876.353**).

CONTRACTOR

Elta Electronics Industries Ltd, (a subsidiary of Israel Aircraft Industries Ltd), PO Box 330, Ashdod 77102, Israel.

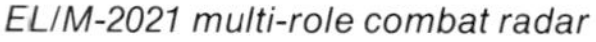

EL/M-2021 multi-role combat radar

6891.393

EL/M-2011 AND EL/M 2032 FIRE CONTROL RADAR

The EL/M-2011 is a lightweight multi-mode fire control radar designed for upgrading light attack, close air support and advanced trainer aircraft. The EL/M-2032 serves a similar purpose for fighter attack aircraft. Both are derivatives of the EL/M-2035 system (**5876.353**) and can be customised to meet specific user requirements.

The main features of the equipments include a low sidelobes planar array antenna, pulse doppler operation, a TWT coherent transmitter and programmable signal processing. They provide all-aspect, all altitude, look-up and look-down detection capability, with air-to-air detection and combat facilities, and air-to-ground ranging. Weight of the complete system is 67 kg in the case of the EL/M-2011 and 98 kg for the EL/M-2032. Typical detection ranges are given as 8 nm and 18 nm respectively.

STATUS

In development.

CONTRACTOR

Elta Electronic Industries Ltd (a subsidiary of Israel Aircraft Industries Ltd), PO Box 330, Ashdod 77102, Israel.

EL/M-2011 multi-mode fire control radar

5876.353

EL/M-2035 MULTI-MODE FIRE CONTROL RADAR

Elta Industries has developed an airborne multi-mode radar which was intended for the Israeli Lavi fighter aircraft. It is a version of the EL/M-2021B radar (**3306.353**) and is adaptable to a wide range of combat and surveillance missions. As a sensor for the avionics system, the equipment determines both airborne and ground target data such as position, velocity and acceleration. A coherent transmitter and stable multi-channel receiver are incorporated to provide reliable look-down performance over a broad band of frequencies. In the air-to-air mode it includes track-while-scan facilities and performs search and automatic target acquisition. For air-to-ground use, beam-sharpened ground mapping provides accurate resolution, and terrain avoidance and sea search modes are incorporated. The signal processing unit is software programmable and is backed up by a distributed computer network to allow optimal allocation of computing power for the various modes.

STATUS

Flight testing of prototypes has been carried out, but with the cancellation of the Lavi aircraft no further information has been made available.

CONTRACTOR

Elta Industries Ltd (a subsidiary of Israel Aircraft Industries Ltd), PO Box 330, Ashdod 77102, Israel.

EL/M-2035 multi-mode radar

6892.353

PHALCON AEW RADAR

Phalcon is a solid-state AEW radar system, developed by Elta, for use on the Boeing 707 SIGINT aircraft. It is also adaptable for other airframes. The system uses six fixed antenna arrays, two on each side of the fuselage, one in an enlarged nose cone and one under the tail. Each array consists of several hundred antenna elements. One transmitter/receiver module is provided for each antenna element, thereby achieving a total of several hundred solid-state transmitters and receivers in the system.

A six-way switch allows transmit/receive modules to be switched to the six antenna arrays. Each array scans a given azimuth sector, giving a total coverage of 360°. Scanning is carried out electronically in both azimuth and elevation. Scan rate is variable, for each operator selected sector. Typically 2-4 seconds scan rates are used in high priority sectors and 10-12 seconds in low priority sectors. A special full track mode in addition to track-while-scan, allows the radar to lock on specific targets and track them with very high probability of detection and high accuracy.

CONTRACTOR

Elta Electronic Industries Ltd (a subsidiary of Israel Aircraft Industries Ltd), PO Box 330, Ashdod 77102, Israel.

ITALY

3307.353

MM/APS-705 RADAR

The MM/APS-705 is a multi-purpose radar which can be installed on both fixed- and rotary-wing aircraft. The main functions are navigation, search and detection, target localisation, mapping, and target tracking and designation to weapon systems.

The system is provided with two transmitter/receivers for frequency diversity operation. Interfaces are included to ensure integration with sonar, doppler navigator, beacon, IFF and ESM. Extractors and track-while-scan units are available for weapon fire control and data transmission. Various sizes of antennas are available with 360° of view, stabilised in roll and pitch.

The MM/APS-705 is intended primarily for use in embarked helicopters, land-based helicopters and maritime patrol aircraft. In addition a pod-based version is available and can be fitted onto the aircraft pylons as an 'add-on' radar for anti-ship missions.

STATUS

Standard equipment in Italian Navy AB.212 and SH-3D helicopters, and procured by other navies.

MM/APS-705 airborne radar installation

CONTRACTOR

SMA (Segnalamento Marittimo ed Aereo), PO Box 200, Via del Ferrone Soffiano, 50100 Florence, Italy.

3308.353

MM/APS-707 RADAR

The MM/APS-707 is a search radar suitable for fixed- or rotary-wing aircraft where reduced weight and size, low power consumption and low cost are required. Several modes of operation are available in order to satisfy the requirements of different operational roles, and features are also provided for integration of the radar into complex avionics mission systems.

The system consists of four units which can be installed in the aircraft. It is provided with a sophisticated video signal processor which includes the presentation management on a TV monitor. A cockpit TV repeater can be installed. The radar presentation can be integrated with graphic or alphanumeric data from other sensors.

Main operational features include: long range 360° coverage, detection of small targets, target designation for weapon systems, radar mapping, and various ECCM and graphic facilities for tactical data presentation.

Technical characteristics incorporated are: a rapid rotation rate stabilised antenna, X-band operation, pulse-to-pulse broad-band frequency agility, high transmission power, scan-to-scan integration, and 'freeze' display capability with TV format presentation.

STATUS

In production for an unspecified country.

CONTRACTOR

SMA (Segnalamento Marittimo ed Aereo), PO Box 200, Via del Ferrone Soffiano, 50100 Florence, Italy.

3309.353

MM/APQ-706 RADAR

The MM/APQ-706 search and attack radar is used in the Italian Navy Marte helicopter-launched anti-ship missile system (**1651.321**). It consists of two I-band transmitter/receiver channels operating in frequency diversity, with frequency agility facilities on one channel, and is provided with a data processing and extraction system and tactical display console. The MM/APQ-706 can fulfil typical naval helicopter roles such as surface search, navigation, ASW etc in addition to providing target detection, acquisition and missile guidance facilities for the Marte system.

STATUS

Installed on Italian Navy SH-3D helicopters for Marte anti-ship missile system but no longer in production.

CONTRACTOR

SMA (Segnalamento Marittimo ed Aereo), PO Box 200, Via del Ferrone Soffiano, 50100 Florence, Italy.

6893.353

AM/APS-717 SEARCH AND WEATHER RADAR

The AM/APS-717 is an X-band radar suitable for both fixed-and rotary-wing aircraft. It provides:

(a) search, detection and localisation of different targets in adverse weather conditions and an aid to identification for search and rescue operations

(b) navigation and weather data, and ground mapping.

The radar is easy to install in the aircraft and can be integrated with the FLIR sensor as an aid to target identification. The antenna is a flat plate, planar array which can provide sectorial or 360° scanning, with nose or ventral/dorsal installation.

Information is presented by means of colour monitors suitable for cockpit use.

STATUS

In production for the Italian Air Force.

CONTRACTOR

SMA (Segnalamento Maritimo ed Aereo), PO Box 200, Via del Ferrone Soffiano, 50100 Florence, Italy.

6201.353

GRIFO AIRBORNE RADAR

Grifo is a multimode radar designed for air-to-air and air-to-ground combat missions. In the air-to-air mode the radar combines search and tracking functions and is configured for operation in conjunction with cannon and AIM-9L Sidewinder missiles. In the look-down air-to-ground mode, Grifo is capable of ground mapping, ground ranging, and search and track of moving ground targets. It is a coherent pulse-doppler radar operating in X-band and has modular architecture enabling adjustments in configuration for various platform applications. The antenna is of a flat plate type with a diameter of up to 60 cm. The transmitter is frequency agile and uses pulse compression techniques. The processor uses low and medium PRF waveforms and the look-down function is provided by a doppler filter with FFT filters.

STATUS

The Grifo radar is in an advanced development and is aimed primarily as a small, relatively cheap radar for the retrofit market. No orders have been announced. See also SC-01 radar (**6894.353**).

CONTRACTOR

Fabbrica Italiana Apparecchiature Radioelettriche SpA (FIAR), Via Montefeltro 8, 20156 Milan, Italy.

6202.353

ALTAIR AIRBORNE RADAR

In conjunction with Westinghouse of USA, FIAR is developing a lightweight multi-mode fire-control radar under the name of Altair. FIAR is responsible for the receiver and the radar digital assembly unit which are being based on the Grifo radar (**6201.353**). The system will also use elements of the Westinghouse AN/APG-66 radar (**3303.353**), the principal combat sensor of the F-16 aircraft.

STATUS

In development. Potential application includes the AV-8B/GR.5 Harrier 2.

CONTRACTOR

Fabbrica Italiana Apparecchiature Radioelettriche SpA (FIAR), Via Montefeltro 8, 20156 Milan, Italy.

6203.353

SETTER FIRE CONTROL RADAR

Setter is a frequency-agile airborne fire control radar with a look-down capability to detect and track low-flying aircraft. It is an updated version of the FIAR R21G radar installed in the F-104S aircraft of the Italian Air Force, using a new magnetron transmitter with a non-coherent output, but which will provide look-down air-to-air target detection. The target information in the background clutter returns is derived by using the clutter itself as a reference to enable the doppler frequency shift produced by moving targets to be extracted.

The original R21G antenna has been retained. The MTI feature operates within $\pm 45°$ of scan, although the mechanical scan limits are still $\pm 60°$. A new processor is incorporated which uses advanced integrated circuits to carry out signal analysis.

STATUS

A R21G prototype equipment has been flight tested by the Italian Air Force. Fiar has a contract to supply 200 kits to modify R21G radars. The first production deliveries were made in the autumn of 1986.

CONTRACTOR

Fabbrica Italiana Apparecchiature Radioelettriche SpA (FIAR), Via Montefeltro 8, 20156 Milan, Italy.

6894.353

SC-01 (SCIPIO) MULTI-MODE RADAR

The SC-01 has been developed by a combination of the Brazilian Air Force, SMA and Tecnasa and although tailored for the AM-X aircraft of the Brazilian Air Force can be installed in many existing aircraft. It is a multi-mode, low weight radar system and is believed to be a variant of the Grifo system (**6201.353**). It provides the following modes of operation:

(a) air-to-air mode with look-down capabilities
(b) air combat mode with automatic detection, designation and tracking
(c) sea mode for target detection and tracking in ASV operations
(d) automatic multiple target track-while-scan
(e) ground ranging and mapping.

STATUS

In development.

CONTRACTORS

SMA (Segnalamento Marittimo ed Aereo), PO Box 200, Via del Ferrone Soffiano, 50100 Florence, Italy.

Tecnasa, Brazil.

Brazilian Air Force.

6895.353

MM/APS-784 SEARCH RADAR

The MM/APS-784 has been developed jointly by FIAR and SMA for the Italian version of the EH-101 helicopter. It is a X-band system, using pulse compression and automatic track-while-scan techniques. It consists of four line replaceable units with an integrated IFF interrogator. It employs a coherent TWT transmitter with pulse-to-pulse frequency agility, with linear and circular polarised radiation. Its main functions are 360° surface search and tracking, missile launch assistance and weather detection.

STATUS

In development.

CONTRACTOR

Consorzio Eliradar, Via V.E. Orlando, 00185 Rome, Italy (Eliradar is a consortium established by FIAR SpA and SMA SpA to develop the radar for the Italian Navy EH-101).

3488.353

SIT 422 SHORAD INTERROGATOR

The Italtel short range air defence (SHORAD) SIT 422 interrogator is a compact rugged equipment designed to operate and survive in tactical battlefield environments.

It provides weapon systems with an all-weather fire capability enabling rapid engagement of hostile aircraft at maximum weapon system range. It is particularly suitable for installation in land mobile systems, small ships, and for all those applications where a need for short range identification exists, and where small size, light weight and low power consumption are especially important. The SIT 422 interrogator can be controlled directly from an associated fire control system computer, or by means of a control unit. Associated equipment is a control unit (where applicable), and a Model AS-131 75 cm antenna.

CHARACTERISTICS

Frequency: 1090 ±0.2 MHz (receiver); 1030 ±0.2 MHz (transmitter)
Sensitivity: –76 dBm
Output power: 200 W
Duty cycle: 1% max
Weight: 12.71 kg

STATUS

In production.

CONTRACTOR

ITALTEL, Defense Telecommunications Division, 66 via due Macelli, 00187 Rome, Italy.

3489.353

SIT 421 (MM/UPX-709) TRANSPONDER

The SIT 421 (MM/UPX-709) is a 'one-box' airborne IFF transponder, suitable for fitting in fixed-wing aircraft or helicopters. It operates in Modes 1, 2, 3/A, 4 and C. The receiver/transmitter includes a 500 W solid-state transmitter, dual-channel receiver and RF interface module. The first of these comprises a delay line oscillator, modulator, driver and power amplifier.

The controls for operation of the transponder, code and mode selection, etc, are mounted on the front of the equipment, which is designed for cockpit mounting.

Two remote-mounted versions are also in production:

(a) SIT 421T controlled via multi-wire cable from the SIT 901 control unit
(b) SIT 421T/1553 controlled via data bus MIL 1553B.

CHARACTERISTICS

Frequency: 1030 MHz (receiver); 1090 MHz (transmitter)
Sensitivity: –77 dBm (adjustable 69 – 77)
Dynamic range: 55 dB
Output power: 27 ±3 dBW at 1% duty cycle
Weight: SIT 421 3.5 kg; SIT 421T 4.1 kg (transponder) 0.8 kg (controller)

STATUS

In production.

CONTRACTOR

ITALTEL Defense Telecommunications Division, 66 Via due Macelli, 00187 Rome, Italy.

SIT 421 (MM/UPX-709) IFF transponder

6896.393

SIT 434 IFF INTERROGATOR

The SIT 434 is a modular architecture IFF interrogator, intended for both fixed- and rotary-wing aircraft and has considerable growth potential. In its basic configuration, the equipment operates under control of the SIT 905 control box.

It is capable of operating in each of the SIF Modes 1, 2, 3/A and in Mode 4, used either separately or interlaced (1 SIF mode + Mode 4). For Mode 4 operation an associated crypto unit, with its mounting and key-loading devices, must be used.

Challenge control is possible by means of an enabling signal coming from the control box or from the radar system. Video output signals are generated as the result of the processing relevant to the last interrogation cycle. Different symbols are provided for target, SIF friend or Mode 4 friend.

CHARACTERISTICS

Transmitter power (peak): 28 dbW minimum
Transmit frequency: 1030 + 0.2 MHz
Duty cycle: 1% max
Decoding: Passive decoding (4096 codes for SIF modes; dual code capacity for M3/A overlap period)
Size: 1 ATR short
Weight: 15 kg including controller and mounting

CONTRACTOR

ITALTEL, Defense Telecommunications Division, 66 Via due Macelli, 00187 Rome, Italy.

SIT 434 interrogator and SIT 905 controller

JAPAN

6899.353

MITSUBISHI AIRBORNE RADAR

A new airborne radar for future fighter aircraft has been developed by Mitsubishi Electronics Corporation under the management of the Japanese Defence Agency's Technical Research and Development Institute. The radar has an active phased array antenna, 66 cm in diameter, made up of 750 modules. It is reported to have track-while-scan facilities with each module generating computer-controlled radar beams.

STATUS

Fl ght tests of the new radar began in 1986 aboard a C-1 transport aircraft and are scheduled to continue ur til early 1989.

SWEDEN

1210.353

PS-37/A AIRBORNE ATTACK RADAR

This is a multi-mode airborne attack radar, operating in the I/J-band, and using the monopulse technique. High output power is employed to provide long range performance. Cassegrain or parabolic dish scanners with fixed feed have been developed for this radar, but the latter is the preferred antenna system.

The PS-37/A is an important element in the attack system of the Saab Viggen aircraft, and its design provides for considerable integration with the navigation, display, and the digital computer-based data processing sub-systems.

The radar is comprised of two main units: the electronics package and the scanner assembly. The former is made up of 13 replaceable units housed in a main assembly which carries the interconnecting cable looms. The division of electronic functions between the individual replaceable units has been arranged so that, as far as possible, related functions are housed within one unit, enabling the replacement of units to be effected without the need for subsequent trimming or adjustment. Each of the replaceable units contains between three and eight sub-units, which provide further facility for servicing.

With the exception of certain high-frequency devices, the PS-37/A is a completely solid-state equipment. The mechanical design incorporates an hydraulic drive for the antenna scanning and stabilisation system, providing accurate and rapid movement over wide angles.

In the interest of high performance and optimum operational flexibility, including conditions of high interference (natural or ECM), elaborate signal processing facilities at both the radio-frequency and video levels are provided in the PS-37/A. Another technique employed to increase accuracy and resolution is lobe-shaping, which synthetically produces the effect of increased antenna aperture and reduced sidelobes.

In the Viggen installation both head-up and head-down display of radar information is provided, the data presented varying in accordance with the operational mode employed.

STATUS

Operational but no longer in production. A version for the Swedish fighter version of the Viggen (JA 37) is described under **1964.353**.

CONTRACTOR

Ericsson Radar Electronics AB, Airborne Electronics Division, 431 26 Mölndal, Sweden.

1209.353

UAP 13 SERIES AIRBORNE ATTACK RADARS

This is a range of nose-mounted, forward-looking radars for search and fire control applications in interceptor aircraft. The radars in this series are pulse radars operating in the I/J-band of the spectrum.

The basic model of the range is the UAP 13102, which is also designated PS-01/A. It is fitted to the Saab Draken J35F. The UAP 13103 (PS-011/A) is similar to the UAP 13102 but is designed to be used with the S71N infra-red search and track set. Both these models are intended for use in aircraft having a fire control computer.

STATUS

UAP 13102 and 13103 models are fitted to Swedish Air Force Draken aircraft, the latter variant of radar being the one used with the S71N infra-red search and track set. The UAP 13104 model equips Drakens of the Finnish Air Force. No longer in production.

CONTRACTOR

Ericsson Radar Electronics AB, Airborne Electronics Division, 431 26 Mölndal, Sweden.

1964.353

PS-46/A MULTI-MODE AI RADAR

PS-46/A is the multi-mode AI radar fitted to the JA 37 fighter version of the Saab-Scania Viggen. The pre-production models of this radar were described in **1741.353** (*Jane's Weapon Systems,* 1974-75) under their developmental designation, DAX-100/200.

To the general operational requirements of all-weather capability, ability to operate effectively in an ECM environment, and high availability, the PS-46/A specification covers the following particular requirements for air-to-air roles:

(1) capability against high-performance aircraft, transports, and helicopters
(2) all hemisphere coverage
(3) look-down capability
(4) air-to-ground ranging with (optional) mapping.

Ericsson PS-46/A multi-mode AI radar installed in Viggen

In considering the transmitter waveform options to meet the above multi-role requirements, a number of techniques are available.

The air-to-ground requirements can be met by the conventional non-coherent pulsed waveform whereas the air-to-air requirement calls for a more sophisticated waveform. Waveforms that provide look-down capability are:

HPD: high PRF pulse doppler
MPD: medium PRF pulse doppler
LPD: low PRF pulse doppler

These waveforms are mainly characterised by their pulse repetition frequencies since this parameter is the determining factor on overall performance.

HPD and MPD require an internal radar frequency reference and subsequently coherent transmission, whereas LPD in addition can use a pseudo-coherent transmission (coherent-on-receiver) or a non-coherent transmission resulting in auto-coherent or non-coherent airborne moving target indicator. A careful review of the pros and cons of these waveforms led to the selection of the two waveforms:

(1) low PRF waveform for look-up modes
(2) medium PRF waveform with doppler filters and signal processing to resolve range ambiguities for look-down modes.

The radar is designed as a 'cartridge' fixed to the forward aircraft bulkhead with four bolts. The aircraft systems supply the necessary electric power, cooling air, and hydraulic power for the hydraulic vane motors that drive the two-axis antenna. The radar consists of ten line replaceable units (LRUs) and these are supported or housed in a lightweight frame which is in itself an LRU. The construction provides for simple removal and replacement of units.

The main design characteristics of the PS-46/A radar are summarised as:

I/J-band
medium PRF pulse doppler operation
coherent transmitter (TWT)
large antenna with low sidelobes
digital signal and data processing
integrated CW target illumination
built-in test.

Target data extraction and smoothing are carried out by Kalman filtering in stabilised Cartesian co-ordinates.

OPERATION

The PS-46/A radar can be configured to provide the following modes of operation:

(1) target search
(2) target acquisition
  (a) automatic using HUD
  (b) semi-automatic using HUD
(3) target tracking
  (a) TWS mode
  (b) continuous track
(4) target illumination
(5) air-to-ground ranging.

Single- and multi-bar scans, wide or narrow, are provided as well as spiral or raster scans for HUD lock-on in dogfights.

STATUS

In production for the JA37 Viggen.

CONTRACTOR

Ericsson Radar Electronics AB, Airborne Electronics Division, 431 26 Mölndal, Sweden.

4083.353

ERICSSON SLAR

Ericsson SLAR (side-looking airborne radar) is a lightweight system for maritime surveillance and other reconnaissance purposes. It is an I-band equipment and consists of a three metre long antenna pod which is usually mounted underneath the aircraft, with the other electronics carried in standard 19-inch (483 mm) racks in the aircraft cabin. The total weight of the equipment, including the antenna housed in the pod, is less than 70 kg.

The radar uses a standard TV display, which gives a true real-time presentation of flicker-free images with 64 grey-tone levels. Digital signal processing is used, and digital recording of radar video is possible on the aircraft for further processing on the ground later. The TV images may also be recorded on a standard tape recorder for replay of sequences of interest on the ground. All images are annotated with alphanumeric information, including time, position and heading etc. The TV image display gives a rolling

Ericsson SLAR in use aboard a Swedish Coast Guard Cessna 337 engaged on maritime surveillance

map presentation covering either one or both sides of the aircraft, and medium or large ships can readily be detected at ranges of up to 100 km. An MTI filter can be provided.

DEVELOPMENT

Development was carried out by Ericsson in co-operation with the Swedish Coast Guard.

STATUS

In production and in service with the Swedish Coast Guard and several other countries. The SLAR has been installed in a Metro III aircraft for trials in the coastal surveillance role.

CONTRACTOR

Ericsson Radar Electronics AB, Defence Systems Division, 431 26 Mölndal, Sweden.

5684.353

JAS-39 AIRBORNE RADAR

The JAS-39 Gripen is the replacement aircraft for the Viggen. Prototypes are due to fly in 1988 and the aircraft is scheduled to enter operational service in the early 1990s. Ericsson is currently engaged in the development of an advanced multi-mode, pulse doppler radar for this aircraft, which has been given the nomenclature PS-05. The system uses the planar array antenna and employs FM pulse compression for long-range detection. It is understood that the transmitter employs a liquid-cooled TWT. Ferranti of UK is a nominated sub-contractor and is providing the antenna pedestal and is participating in design and development of the signal processor, which can be programmed for a considerable number of functions to allow a true multi-mode operation. In the air-to-air role the new radar will perform long-range search and multi-target TWS, short range wide-angle quick scan, and missile and weapon control. The air-to-ground functions are sea and ground target search, ground mapping and missile fire control.

STATUS

System evaluations of a complete radar have been carried out in a test rig, and flight tests in a Viggen aircraft are in progress. Production of series-like prototypes has commenced. A standardised computer system, Ericsson SDS8D, has been developed.

CONTRACTOR

Ericsson Radar Electronics AB, Airborne Electronics Division, 431 26 Mölndal, Sweden.

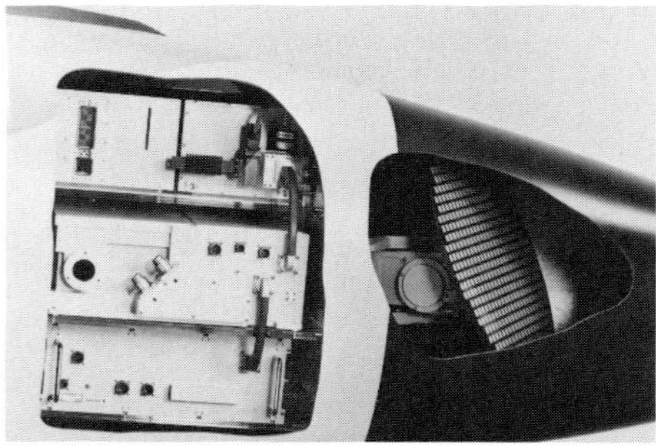

Mock-up of the JAS-39 Gripen radar

6196.353

ERICSSON AEW RADAR

Ericsson has commenced development of a new type of airborne early warning radar for the Swedish Air Force. The radar can be carried by small commuter aircraft, eg SAAB SF340 or Fairchild Metro III. Flight tests were made successfully in late 1985 with a dummy antenna mounted on top of the fuselage of a Metro III.

The radar is all solid-state with electronic scanning and is designed to operate in an intense EW environment. Operation in S-band, with a wide aperture antenna, provides good ECCM performance. The range performance against fighter-size targets is stated to be approximately 300 km. New techniques have been employed in the transmitter and signal processor. In the version for the Swedish Air Force, the radar information will be down-linked to a combat control centre.

STATUS

In development. Ground testing has been completed and initial flight tests in a Metro III are in progress.

CONTRACTOR

Ericsson Radar Electronics AB, Airborne Electronics Division, 431 26 Mölndal, Sweden.

AEW antenna unit mounted on a Fairchild Metro III

5887.353

HERA AIRBORNE RADAR

Philips of Sweden has developed a new surveillance radar system designed for use by the Royal Swedish Navy in helicopters and potentially as an airborne targeting aid for anti-ship missile systems. The radar uses a transceiver supplied by Omera of France and is allied to a display system developed by Philips. The transmitter has a power output of 60 kW and employs frequency agility techniques to counter ECM. A frequency-agile transponder system is incorporated. The Hera system will be used in the anti-submarine role, as well as anti-ship missile targeting.

STATUS

Sixteen systems have been ordered and deliveries of the first systems have been made.

CONTRACTOR

Philips Elektronikindustrier AB, 17588 Jarfalla, Sweden.

UNION OF SOVIET SOCIALIST REPUBLICS

6430.353

BIG BULGE AIRBORNE RADAR

Big Bulge is a surface surveillance and targeting radar, the name arising from the large under-fuselage radome on the TU-20 Bear-D aircraft. It is also fitted to the Ka-25 Hormone-B helicopters in a chin-mounting. The system operates in I/J-band and has a range of approximately 350 km. It is used, via a data link, to transmit information of target vessels to missile-equipped aircraft and/or submarines.

6431.353

FLAT JACK AWACS RADAR

Flat Jack is the NATO nomenclature for the airborne warning and control system (AWACS) mounted in the Tu-126 Moss aircraft. The antenna is fitted in a rotating radome mounted above the after fuselage, and is very similar to the Boeing E-3A AWACS aircraft. The system is reported to have a limited operational capability, particularly in the over land role. A later and improved AWACS aircraft, based on a version of the IL-76, is understood to have an upgraded variant of Flat Jack, and has recently entered service.

6432.353

HIGH LARK AI RADAR

High Lark is an I/J-band air intercept (AI) radar which is fitted to a number of Soviet aircraft and is associated with several air-to-air missiles. It is fitted to the MiG-23 Flogger, and an improved version may be fitted to the MiG-29 Fulcrum and the Su-27 Flanker. Range is in the region of 80 km for search and 50 km for tracking.

3311.353

JAY BIRD AIRBORNE RADAR

Jay Bird is the NATO reporting name of the air-to-air ranging radar fitted in the intake nosecone of the MiG-21 Fishbed-J and MiG-23 Flogger-E multi-role fighter aircraft. It is not clear if the NATO designation derives from the 'J' of the aircraft model name, or from the fact that the radar operates in the J-band, on frequencies between 12.88 and 13.2 GHz according to one source.

Antenna gain has been calculated at about 33 dB for an approximate diameter of 40 cm. Three PRF ranges have been quoted: 2042 to 2048 pps; 1592 to 1792 pps; 2716 to 2724 pps; and the provisional search and tracking ranges are 30 and 20 km, respectively.

Raster scanning is used for the search mode and lobe switching for tracking. A transmitter peak power in the region of 100 kW is assumed, in a 3.5° square beamwidth with pulse widths from 0.3 to 1 µsec. The antenna is vertically polarised. The system does not appear to employ pulse compression techniques and no significant look-down facility is apparent.

The Fishbed-J armament is said to include a radar-homing version of the Atoll air-to-air missile, in which case a target illuminating radar mode will be required of the Jay Bird.

The NATO designation, Jay Bird B, applies to an alternative version which may be the export version for MiG-19M Fishbed-J and/or MiG-19SMT Fishbed-K used by India and Warsaw Pact states, respectively. Alternatively, this designation may refer to a version of Jay Bird said to be fitted to the MiG-25 Foxbat in one of its versions.

1475.353

LOOK TWO AIRBORNE RADAR

Look Two is the NATO designation for an I-band Soviet bombing and navigation radar system. No mechanical or technical details are available beyond the signal characteristics which have been obtained by 'Western' Elint (electronic intelligence) operations. These include operating frequencies in the band 9245 to 9508 MHz, with frequency agility capability and four PRFs: 320 to 336 pps, 619 to 632 pps, 1247 to 1253 pps, and 1871 to 1879 pps. Presumably, pulse length is also varied, changes being made according to operating mode and height. Fittings are apparently limited to certain versions of the Yak-28 Brewer twin-engined aircraft.

1476.353

PUFF BALL AIRBORNE RADAR

Puff Ball is the NATO designation given to an I-band Soviet search radar which has been stated to equip the Bison long-range, four-jet reconnaissance bomber. It is also fitted in certain versions of the Badger and is used in the Ka-25 Hormone-A ASW helicopter. Its principal function is long-range surface mapping, and among its uses may be the direction of air-launched missiles such as Kangaroo, Kennel, Kelt, and Kipper.

American reports state that Puff Ball has facilities for providing friendly surface-to-surface missile batteries with target designation services, and that this is accomplished by transmitting via a data link the relative positions of target and missile battery as presented on the radar PPI display. Range is approximately 250 km.

Puff Ball PRFs are 414 to 418 and 621 to 628 pps.

1477.353

SCAN FIX AIRBORNE RADAR

Scan Fix is the NATO designation given to the AI (airborne intercept) radar fitted to MiG-17 and MiG-19 interceptors. Like those aircraft, the Scan Fix must be considered obsolescent. Both I-band and E/F-band versions are believed to have been produced. The latter equipped the Frescoe series of MiG-17s, while the MiG-19 Farmer A, C, and D versions were given I-band models, these being known by the NATO designation High Fix. American reports confusingly state that both versions employ fixed scans. This may mean that there is no antenna stabilisation, or that the radars may provide only target ranging and illumination facilities. The system is believed to be obsolete.

1479.353

SCAN ODD AIRBORNE RADAR

Scan Odd is the NATO designation for the AI (airborne intercept) radar fitted in MiG-19, Farmer B, Soviet interceptors. It is an I-band set, operating at frequencies between 9300 and 9400 MHz. This is apparently a later system than the Scan Fix AI radar, and is reputed to have an unusually complex scan pattern. There are differing PRFs for search and tracking modes of operation. Search/track range is given as 15/5 km.

1478.353

SHORT HORN AIRBORNE RADAR

Short Horn is the NATO designation given to an airborne bombing and navigation radar, which, from the Elint (electronic intelligence) data available, appears to be an example of some of the more recent Soviet radar technology. The operating frequency is in the J-band, between 14 and 15 GHz, and there are frequency agility and frequency diversity facilities. There are four PRF/pulse-width combinations for various operational modes: 313 to 316/1 to 1.8, 496 to 504/0.5 to 1.4, 624 to 626/0.4 to 1.3, and 1249 to 1253/0.01 to 0.9 pps/µs. Both circular and sector scans have been recorded.

No information is available on the physical configuration but the parameters above suggest that its uses include air-to-surface vessel and maritime applications. Aircraft types reported to be equipped with Short Horn include the Brewer Yak-28 in its B, C, D and E variants, the Tu-22 Blinder A, C and G models, and the Tu-16 Badger H. Range is approximately 200 km.

1481.353

SKIP SPIN AIRBORNE RADAR

Skip Spin is the NATO designation for the AI (airborne intercept) radar installed in Su-15, Flagon-A, and Yak-28P, Firebar interceptors. The Soviet designation is understood to be RP11. Introduction took place in the mid-1960s. The operating frequency is in the I-band, and estimated transmitter power is in the region of 100 kW, giving a reported range of 40 km. The frequency has been quoted as being between 8690 and 8995 MHz, pulse width about 0.5 µs, and PRF 2700 to 3000 pps.

The Flagon-A and Firebar interceptors generally are armed with Anab air-to-air missiles (**1144.331**), (though the IR homing Atoll air-to-air missile (**1146.331**) is also carried) and therefore Skip Spin can be assumed to provide search and tracking modes for both these weapons, and target illumination for the radar homing version of Anab. The Yak-28P has also been observed with a significantly longer nose radome but the reason for this is not known.

1482.353

SPIN SCAN AIRBORNE RADAR

Spin Scan is the NATO designation given to a series of AI (airborne intercept) radars. The Soviet designation is believed to be R1L, and in the export version (eg, India) the R2L. They operate at I-band frequencies, and have a transmitter power in the 100 kW region.

Soviet aircraft fitted with Spin Scan radars are the D and F models of the MiG-21, Fishbed interceptor, and the Su-9, Fishpot B. In all cases, the radar installation is of the type where it is housed in an aerodynamic pod in the nose air intake of the aircraft. These differ in size, that of the Fishbed D being larger than the Fishbed F version. Spin Scan A/B fitted in Fishbed D and Fishpot B, has two PRFs, 825 to 950 pps for search, and 1750 to 1850 pps for the intercept and tracking modes. Search/track range is 40/30 km. A rotating scan pattern is employed.

Another version known to NATO as Spin Scan B is designated by Soviet Forces RP9.

4616.353

SUAWACS AIRBORNE AEW RADAR

A second generation SUAWACS (Soviet Union Airborne Warning and Control System) has recently been developed to follow-on to the Tu-126 system illustrated below. Very little is known about the radar other than it is carried in an AWACS version of the Il-76 aircraft (known to NATO as 'Mainstay'), and is coupled with the MiG-25M Foxhound aircraft which has a nose radar able to display and track 20 targets simultaneously. The radar is probably an upgraded version of the Flat Jack system (**6431.353**) used in the

Il-76 Mainstay second generation Soviet AEW aircraft

Tu-126 aircraft, and uses a similar antenna in a 'saucer' type rotating radome. The latest information available is that Il-76 Mainstay aircraft are deployed at three airfields in the west of the Soviet Union. Production rate of the aircraft is not definitely known but is believed to be at least five per year.

A new Soviet AEW aircraft also appeared recently, based on the An-74 light STOL transport, and highlights the intense activity being undertaken by the Soviet Union in airborne early warning and control. The new aircraft has a code name of Madcap and features a large radome mounted on top of the heavily modified fin. The association of a large AEW antenna with a STOL airframe suggests that the aircraft could be designed for areas where long runways are not available. It could also be assumed that Madcap has the same relationship with the Il-76 Mainstay as the Grumman E-2C Hawkeye does to the E-3B AWACS.

It is believed that the Soviet Union is devoting a considerable amount of effort into the development of highly capable AEW aircraft and it has been suggested that the Il-76 Mainstay is only an interim solution. Certainly the latter is reported to have been three years late entering operational service, having probably run into development problems, possibly in the data processing area.

NOTE: Other Soviet airborne radars will be found listed in the Airborne Radar Table (**1194.354**) in the Analysis Section of this volume. Separate entries have not been made for this equipment as it has not been possible to establish reliable data in sufficient quantity to justify such action. However the tabular references contain the salient information that has been obtained.

View of Bee Hind installation in Tu-22 Blinder aircraft

Kamov Ka-25 Hormone ASW/maritime helicopter with prominent chin radar for surface search. This is probably the Puff Ball system

Close-up of Bee Hind radar in tail of Badger bomber

Hormone helicopter has relatively large radar for surface vessel and submarine detection

Wet Eye J-band maritime surveillance radar fitted to the Il-38 May aircraft

Spin Scan AI radar installation in MiG-21 interceptor

Another view of Badger-D showing Puff Ball nose radar. Bee Hind tail radar. Foremost of three ventral radomes may house Short Horn bombing and navigation radar; the other two have not been identified

Badger-D maritime/electronic reconnaissance aircraft. Large nose radar installation is probably Puff Ball

UNITED KINGDOM

1785.353

FOXHUNTER AIRBORNE INTERCEPTION RADAR

The Foxhunter airborne interception radar was first publicised in early 1976 when GEC Avionics was awarded a major development contract for the AI (airborne intercept) radar of the RAF air defence version of the Tornado. Ferranti is the major sub-contractor and supplies the transmitter, illuminator and antenna platform. The Foxhunter system made its first flight in a Tornado in June 1981 after extensive proving trials in other types of aircraft. Responsibility for the Foxhunter programme has now been transferred from GEC Avionics to Marconi Defence Systems.

The design of this AI radar provides a multi-mode system compatible with the size and weight limits of the RAF air defence variants of Tornado and the operational requirements of the next two decades. A substantial part of the signal processing is performed digitally, in addition to digital radar data handling. The equipment anticipates trends in offensive tactics such as low-level penetration, use of ECM, etc, and has the flexibility to operate as part of ground or AEW based control environments while retaining the ability to operate autonomously.

Foxhunter is designed to detect and track subsonic and supersonic targets at ranges in excess of 100 nautical miles at both low and high level. Although it is intended to form part of an overall air defence system, ie in a ground or AEW control environment, the equipment has been designed to operate independently since either or both of the former may be nullified or degraded by enemy action. In normal circumstances, however, Foxhunter will be integrated within the overall UK air defence system, unified by the JTIDS – ECM resistant data link fitted in the Tornado ADV aircraft.

Foxhunter operates in I-band (3 cm) using a pulse-doppler technique known as frequency modulated interrupted continuous wave (FMICW). At the heart of the radar is a master timing and synchronising unit, employing phase-locked loop techniques, which generates the complex transmission waveforms for amplification by the transmitter, and the accurately related reference signals and sets of precise timing pulses employed within the receiver and signal processing circuits.

High performance microwave integrated circuits are used in the receivers giving benefits in size and weight. Compact and lightweight surface acoustic wave devices, developed by Marconi, provide signal waveforms for the pulse compression modes of the radar. The antenna uses the 'Elliott' twist-reflecting cassegrain principle which combines rigidity and light weight with extremely low levels of spurious radiation lobes, a key feature in the rejection of ground clutter and jamming signals. This type of antenna has been developed to a very advanced level of performance. The scanner, built by Ferranti, employs a hydraulic drive mechanism and very high grade servos to achieve the speed and precision of beam pointing and stabilisation demanded while the aircraft is manoeuvring under conditions of high *g* and rapid rates of roll.

Foxhunter radar mounted in the nose of an RAF Tornado (courtesy of RAF Coningsby)

The transmitter, which is also the responsibility of Ferranti, has been designed with the flexibility to provide the optimum form of transmission for the wide variety of operational tasks. The heart of the signal processing system is a signal processor embodying fast Fourier transform techniques of frequency analysis to filter the signal returns into narrow frequency channels. Target echoes are segregated from clutter returns and a particularly uniform detection threshold is achieved against all target velocities.

Foxhunter also provides target illumination for the Sky Flash medium range air-to-air missiles which have semi-active homing heads and, since heavy jamming is virtually certain, the system incorporates strong ECCM facilities.

STATUS

Foxhunter is in service with the RAF Tornado F2 and F3 aircraft and will be fitted to export versions of the aircraft.

As at March 1988 nearly 100 production units had been delivered. The UK MoD and Marconi have reached agreement on the total forward programme of Foxhunter for the Tornado F3 aircraft. The final standard of radar will have considerably more capability than earlier versions, including more automation and reserve computing capacity.

CONTRACTOR

Marconi Defence Systems Ltd, The Grove, Warren Lane, Stanmore, Middlesex HA7 4LY, England.

5612.353

SKYRANGER AIRBORNE RADAR

Skyranger is a new, lightweight airborne weapon control radar developed for light fighter and light attack aircraft and with the retrofit market very much in mind. It consists of three main units: antenna, transmitter/receiver and signal processor/power supply, and since the amount of space available for retrofit programmes can often be limited and irregular in shape, the modularity of Skyranger has been established at printed circuit card level. The individual cards can, therefore, be packaged into housings designed for the space available.

Skyranger accepts discrete digital commands from a cockpit-mounted control panel and provides output data in the form of a digital serial link (ARINC 429) to a HUD and other weapon aiming systems. It has two main modes, guns and missiles, the former having a shorter range, wide-angle beam. In the missile mode the radar energy is fed from the feed horn and reflected back from the parabolic antenna in a 6 degree beam with a maximum range of 15 km. For gun attacks, the radar energy is fed directly out from the antenna through a polarised window. This results in an 18 degree beam width and a range of 5 km. Minimum range is 300 m for guns and 150 m for missiles, with ranging accuracy in the order of ± 15 m below 3 km and ± 30 m above. Target relative velocities of from –500 to + 1000 metres/second may be handled.

The equipment operates in I-band and has five per cent pulse-to-pulse agility. MTBF is given as 200 hours and the equipment contains built-in test systems. The current version has a fixed antenna but a scanning antenna is planned as an option.

Skyranger radar in format for the Chinese Type 7 aircraft

CHARACTERISTICS

Frequency: I-band

Range: 15 km for missiles; 5 km for gun operation

Range resolution: 150 m

Pulse-to-pulse agility: 5%

Power requirements: 27 V DC, $<$50 W; 115 V 400 Hz single phase, $<$400 VA

Weight

Antenna: 4 kg

Transmitter/receiver: 25 kg

Signal processor/power supply: 8 kg

Total installed weight: 40 kg

STATUS

In production for the Type 7 fighter aircraft of the People's Republic of China, as part of a Marconi integrated avionics system.

CONTRACTOR

Marconi Defence Systems Ltd, Elstree Way, Borehamwood, Hertfordshire WD6 1RX, England.

1740.353

SEARCHWATER SURVEILLANCE RADAR (ARI 5980)

Searchwater is the commercial name for the long range maritime surveillance radar which, under the designation of ARI 5980, entered service in 1979 as standard equipment on the Nimrod MR Mk 2 aircraft of the Royal Air Force. Subsequently, in 1982, during the Falklands campaign, two Sea King helicopters were fitted with modified Searchwater radars in response to an urgent RN requirement for carrier-borne fleet AEW. In this configuration, the operator can detect, track and plot low level and surface targets at long ranges, and control fighter aircraft. The unique features of the maritime reconnaissance system (long range surface surveillance, target classification by profile and over-the-horizon targeting) are retained. The Sea King AEW helicopter was highly successful and has now been fully integrated for carrier and fleet defence, particularly in joint operation with Sea Harriers. Each carrier is provided with a dedicated AEW flight of three aircraft.

In the maritime role, Searchwater as fitted to the Nimrod has a very high performance in terms of detection, target classification, information extraction and ergonomics, enhanced by several features:

Operator's console of the AEW version of the Searchwater radar on a Sea King helicopter of the Royal Navy

(a) frequency agility, pulse compression, integrated IFF system, ground stabilised tactical display and pitch and roll stabilised antenna
(b) a signal processing system which enhances detection of targets, particularly periscopes in high sea states at long ranges. Presentation and classification of targets and IFF transponder returns are highlighted by the use of an integrating scan converter. The radar operator is presented with a selection of bright, flicker-free, TV type PPI, B-scope and A-scope displays and a variety of interactive operating modes. Weather radar and navigation facilities are also provided within the system
(c) a real-time digital computer which continuously and automatically tracks, stores and analyses data on all targets
(d) built-in-test equipment providing automatic detection and diagnosis of faults
(e) a compact heat extractor to enable the system to operate efficiently over the range of temperatures encountered.

In the surface surveillance role these facilities enable patrol aircraft to operate in a stand-off mode and to ranges beyond the surface horizon, thereby avoiding the need to overfly the target for visual confirmation. They also provide over-the-horizon targeting to cue the guidance systems of anti-ship missiles.

All units are designed on a modular basis, with the interface and mechanical construction optimised for ease of fault-location and replacement. Major units are designed to be functionally self-contained as far as possible with a minimum of inter-connections to other parts of the system. Extensive use is made of hybrid and integrated circuit micro-electronic techniques.

The radar system and its services consist of separate line-replaceable units, most of which contain built-in test equipment which is connected to a central processing and display unit at the operator's position. The radar is controlled by a single operator using separate control panels arranged around the display screen.

The transmitter uses solid-state frequency generators and mixers, followed by two cascaded TWTs. A fluorocarbon liquid cooling system is used.

Sikorsky SH3D helicopter of the Spanish Navy fitted with Searchwater AEW radar

The scanner, which transmits and receives both radar and IFF signals, uses a reflector of lightweight construction based upon resin-bonded carbon-fibre techniques. Control of the pitch, roll and azimuth axes is effected by an hydraulic servo mechanism. Surface acoustic wave technology is employed in the pulse compression system.

DEVELOPMENT

Improvements to the Searchwater radar are being delivered to both Nimrod and Sea King fleets to meet the increasing complex threats being encountered in maritime scenarios.

STATUS

In service contracts totalling over £100 million have been placed for Searchwater radars. These cover both the standard radar and the modified AEW system. Interest from other countries includes an order from the Spanish Navy for the AEW variant which is installed on Sikorsky SH3D helicopters. Deliveries to the Royal Navy began in mid-1984 and to the Spanish Navy in October 1986.

CONTRACTOR

THORN EMI Electronics Ltd, Radar Division, Dawley Road, Hayes, Middlesex UB3 1HN, England.

5886.353

SKYMASTER MULTI-ROLE AIRBORNE RADAR

With the success of the Searchwater in an AEW configuration (**1740.353**), THORN EMI re-assessed the requirement for a system effective in the detection of airborne targets overland, while retaining the maritime performance. To achieve this in one system is of significant advantage to the surveillance and support role of assault forces in sea/land operations, air defence against the overland threat and the patrolling of national borders.

It was recognised that a small, lightweight system could be developed at a fraction of the price of larger AEW systems and the Skymaster system has been developed to give an effective performance, adequate to counter the overland threat and complement the Searchwater system.

The Skymaster radar uses techniques and software-controlled system management to provide multi-role capability. It is compact in size and is suitable for installation in helicopters and low-cost aircraft, providing an economic solution to many airborne surveillance requirements.

In the AEW role the radar provides detection and automatic tracking of small airborne targets at all altitudes over both land and sea, and incorporates facilities enabling control of airborne interceptions. Small surface targets can also be detected and tracked in the maritime surveillance mode at long range and in high sea states. In-built facilities enable over-the-horizon targeting, maintenance of a tactical plot, and control of engagements.

Operator's console of the Skymaster radar

The radar system employs an I-band travelling wave tube transmitter capable of operation in the pulse doppler mode for look-down detection of airborne targets against ground clutter, and in a frequency agile conventional pulse mode for the detection of maritime surface targets and look-up detection of airborne targets. A fully stabilised antenna provides a narrow beam 360° search, with continuous and variable scan capabilities. The polarisation can be adjusted by the operator to reduce clutter while optimising detection probability. Track-while-scan facilities enable specific targets to be tracked while the radar continues to scan for other targets.

Where circumstances permit, up to five operator consoles can be used, each equipped with a multi-colour TV display and touch sensitive control panel. Ground stabilised or rolling map displays are provided, with accurate latitude/longitude or grid referencing of targets. Control of other on-board systems is possible, from the radar console, through the MIL-STD 1553B serial interface.

STATUS

Installed in a Defender aircraft, the radar is undergoing flight trials prior to customer demonstrations.

CONTRACTOR

THORN EMI Electronics Ltd, Radar Division, Dawley Road, Hayes, Middlesex UB3 1HN, England.

4613.353

ASTOR – AIRBORNE STAND-OFF RADAR

ASTOR is a UK MoD sponsored project to assess the effectiveness of long range stand-off radars in over-the-border and battlefield surveillance roles.

The prime objective of the radars is to deny the element of surprise by monitoring force build-up and defensive force multiplication through accurate and timely information on ground movements and deployment. Such a battlefield capability is sought by both the army and air force in the UK.

Following evaluation of several development contracts, competitive proposals were sought to produce two separate technology demonstrators. The Radar Division of THORN EMI has been successful in winning the contracts for both radars.

The demonstrator for the army role is based on the privately funded THORN EMI Skymaster radar (**5886.353**) which is installed in a Pilatus Britten-Norman Defender. This multi-role surveillance radar is already flying in the AEW version of this aircraft. The software driven I-band system uses advanced signal processing and a stabilised narrow beam antenna providing a selectable 360° or designated sector scan. The battlefield surveillance mode complements the existing airborne early warning and

Defender aircraft equipped with THORN EMI multi-mode radar

maritime surveillance modes. The radar can be installed in most twin-engined fixed-wing aircraft and medium lift helicopters.

The demonstrator for the RAF requirement is a dual mode synthetic aperture and moving target indication radar mounted in a converted Canberra aircraft to provide long range reconnaissance capability. High resolution static imagery and the capability to detect moving targets allow a 24 hour observation and targeting intelligence of enemy first and second echelon forces.

Related equipment to the ASTOR programme includes a secure air/ground data link to a mobile ground station which incorporates computer assisted facilities for the interpretation of the airborne radar data.

STATUS
The two airborne technology demonstrator systems are in development. Evaluation of the systems is scheduled for completion in early 1990.

CONTRACTOR
THORN EMI Electronics, Radar Division, Dawley Road, Hayes, Middlesex UB3 1HN, England.

3310.353
ARI 5983 AIRCRAFT TRANSPONDER

The ARI 5983 transponder is used in RN helicopters operating beyond visual range of their parent or co-operating ships to provide a means of enabling interrogation by primary radars on shore and aboard ships. This provides navigational assistance, enhances ship's radar range performance and gives an identification facility in the course of patrol, ASW, and sonar operations involving ships and helicopters.

The equipment operates at frequencies in the I-band and consists of two units, transponder and control unit. These operate in conjunction with two antennas mounted on the exterior of the helicopter.

The transponder receives interrogating signals, via the antenna, from pulse radars at any frequency in two bands, 100 MHz wide. On the receipt of such interrogation the transponder will transmit a reply at a closely controlled fixed frequency in the same region. This will be either a single RF pulse, which provides enhancement of the radar return, or a coded group of up to six pulses which provide identification. Sixteen different reply codes are available. These facilities may be selected by the operator at the transponder control unit.

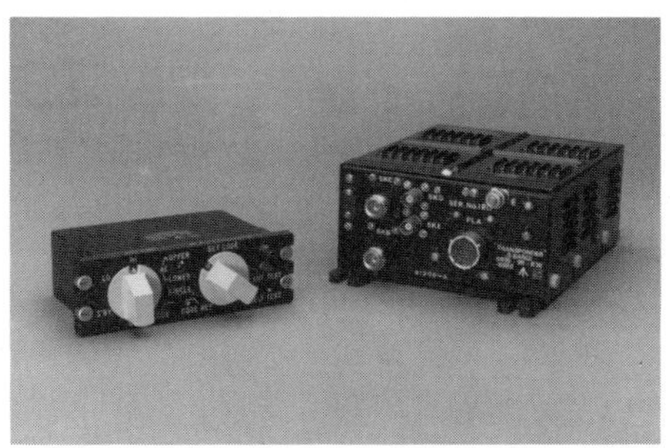

ARI 5983 I-band transponder and control unit

Reception and transmission can take place via either of two antennas, the required antenna being selected by means of an RF switch in the transponder which in turn is controlled from the control unit. Another switch causes a reduction of the transmitted power of approximately 11 dB. The equipment is arranged so that it will be suppressed during the operation of other I-band equipment in the aircraft. Similarly, a suppression pulse is fed to other I-band equipment when the transponder is replying.

Each transponder can transmit a choice of differently coded replies, selected by the operator simply setting a switch on the operator's control unit. A self-test generates an interrogate signal which is fed into the system input and if the transponder is functioning correctly a green acceptance lamp is illuminated.

The same basic transponder design can have several forms. For example, the system can be designed so that it requires double or multiple pulse interrogation to provide a reply selectively where there are many interrogating radars (such as in a harbour or other busy area). Where there are few radar-bearing vessels, a single pulse is usually sufficient.

CHARACTERISTICS
Transponder: I-band
Receiver frequency: 2 bands, centre $F_1 \pm$ 85 MHz
Bandwidth: $\pm$50 MHz
Sensitivity: –93 dBW sufficient to trigger reply
Transmitter frequency: $F_2 \pm$ 7 MHz
Output power: 135 W minimum to 300 W max peak
Pulse duration: 0.5 μs $\pm$ 0.1 μs
Reply code: 6-pulse code, 16 settings, single-pulse reply capability
Pulse spacing: 2.9 μs nominal
Duty cycle: 0.005% max
Weight: 2.5 kg

STATUS
In production for the RN. It is fitted to the Sea Harrier, Sea King and Lynx and is proposed for the Royal Navy EH-101. The equipment is currently in service with the Royal Marine Helicopter Squadron and the navies of India, Australia, Brazil, China, Denmark and Pakistan, and also in the Middle East.

See also transponder reply receiver RRB (**3487.253**) in the Naval Radar section of this volume.

CONTRACTOR
M/A-Com Ltd, Dunstable, Bedfordshire LU5 4SX, England.

3695.353
IFF 2720 AIRBORNE TRANSPONDER

The Cossor IFF 2720 is a general-purpose secondary surveillance radar transponder which is now in service with all types of aircraft from helicopters to interceptors. The equipment is housed within a $^3/_8$ ATR case size and provides 4096 reply code capability on all modes 1, 2, 3/A, B, and with 2048 codes on mode C. The basic equipment operates, without the need for cooling air or pressurisation, up to an altitude of over 15 000 m. For operation above this altitude a pressurised version is available.

The complete equipment consists of the type 2720 transponder with an IFF 2743 control unit. Single or dual antenna installations can be provided, in the latter case an automatic antenna switch is available. Electronic warfare provisions include resistance to CW, MCW, and pulse jamming, sidelobe rate limiting, short pulse and spurious interference protection, single pulse rejection and long pulse discrimination.

CHARACTERISTICS
Transmitter frequency: 1090 MHz
Transmitter power: 27 dBW (500 W)
Receiver frequency: 1030 MHz
Sensitivity: –76 dBm
Dynamic range: ⩾50 dB
Sidelobe suppression: 3 pulse
Low sensitivity reduction: 14 dBm
Dimensions: 90 × 194 × 314 mm
Weight: 4.6 kg
IFF 2743 control unit: 146 × 95 × 76 mm; 0.91 kg
IFF 2733 control unit: 146 × 57 × 114 mm; 0.56 kg

STATUS
In quantity production and in service with British and other air forces.

CONTRACTOR
Cossor Electronics Ltd, The Pinnacles, Harlow, Essex CM19 5BB, England.

IFF 2270 airborne transponder showing internal accessibility

5603.353
IFF 2743 ELECTRONIC CONTROL UNIT

The Cossor IFF 2743 is a microprocessor-based digital controller used with IFF 2720 and other transponders. In a panel height of 95 mm, it provides full 4096 code selection on modes 1, 2, 3/A and B, together with height encoding. Emergency and Identification of Position (IP) selection facilities are provided. Code readout is by means of large easy-to-read LED displays. LED dimming is microprocessor controlled and is provided by a flick switch. Code selection is by flick switches, which offer many advantages over older thumbwheel switches.

STATUS
In production for many customers.

CONTRACTOR
Cossor Electronics Ltd, The Pinnacles, Harlow, Essex CM19 5BB, England.

Cossor IFF 2743 control unit

3696.353
IFF 3100 AIRBORNE TRANSPONDER

The Cossor IFF 3100 transponder was designed initially as a compact single unit for cockpit mounting in the Tornado multi-role combat aircraft. Full 4096 code capability on modes 1, 2, 3/A, and B and 2048 codes on mode C are provided together with space for future expansion. Other features of this equipment include ECM/interference protection measures such as long pulse elimination, short pulse rejection, single pulse rejection, and echo suppression.

CHARACTERISTICS
Transmitter frequency: 1090 MHz
Transmitter power output: 27 dBW
Receiver frequency: 1030 MHz
Sensitivity: ⩾–76 dBm
Dynamic range: ⩾50 dB
Sidelobe suppression: 3 pulse; rate limiting provided
Dimensions: 146 × 132 × 165 mm
Weight: 5.3 kg (max)

STATUS
In production for the Tornado IDS and F2 aircraft of the RAF, and the 72 Tornados and 36 Hawks ordered for the Royal Saudi Air Force.

CONTRACTOR
Cossor Electronics Ltd, The Pinnacles, Harlow, Essex CM19 5BB, England.

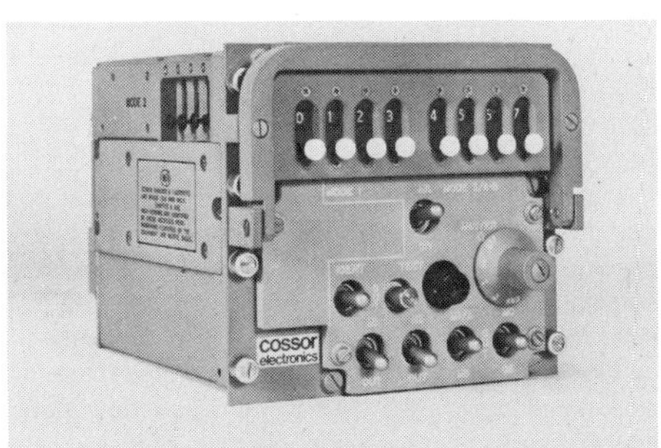

Cossor IFF 3100 transponder for RAF Tornado

3697.353

IFF 3500 AIRBORNE INTERROGATOR

The Cossor IFF 3500 airborne interrogator has been designed for use in any type of aircraft, including high-performance interceptors and helicopters. It is regarded as having previously been impossible to achieve a narrow beamwidth for airborne IFF systems, but in this equipment Cossor claims to have solved the problem. The company says that techniques have been developed that yield extremely high resolution, even when operating within small radomes. Collaboration with radar manufacturers and aircraft constructors ensures that the system achieves the required performance.

Within a 1 ATR case size, the IFF 3500 contains a transmitter, logarithmic receivers, encoder, decoder, defruiter and improved video processing. Automatic code changing is provided to reduce pilot workload, to eliminate incorrect code selection and to improve the security of the IFF Mk 10A system. The automatic code change units are suitable for use, not only with interrogators but also with transponders. Coding information in either serial or parallel format ensures commonality with all in-service transponders.

Special video processing circuits in IFF 3500 provide an indication if incoming signals are garbled, and echo suppression and multipath suppression circuits reduce the effects of reflections from ground and sea surfaces.

After extensive studies, Cossor rejected such previously used methods known as receiver sidelobe suppression, centre beam marking, sliding window techniques, and others as unsuitable to achieve high bearing resolution. The transmitter employs P2 emphasis to provide antenna beam sharpening. P1 and P3 are transmitted on the antenna sum channel and P2 on the difference channel, and 6 dB of emphasis is available, variable in steps of 3 dB.

CHARACTERISTICS

Transmitter frequency: 1030 ± 0.75 MHz

Power output: P1, P3: 30.5 dBW
P2: 36 +3–0 dBW selectable –3 and –6 dBW

Receiver frequency: 1090 ± 1.5 MHz

Sensitivity: –80.5 dBm

Dynamic range: 60 dB

Weight: No more than 15.9 kg

STATUS

In production for the Phantom and Tornado ADV aircraft. Deliveries to the RAF are in progress. A variant, known as IFF 3570 has been fitted to Sea King AEW helicopters and integrated with Searchwater radar.

CONTRACTOR

Cossor Electronics Ltd, The Pinnacles, Harlow, Essex CM19 5BB, England.

Cossor IFF 3500 airborne interrogator designed for use in any class of aircraft and being produced for Tornado ADV, and Phantom for the RAF

6434.353

IFF 4700 SERIES AIRBORNE TRANSPONDERS

With the new NATO Standardisation Agreement STANAG 4193 coming into effect, Cossor Electronics has developed a family of new technology equipments, the 4700 series, to meet this demanding IFF Mk XII specification. Three variants are available to meet any installation requirement. Each variant uses a set of modules which is common to the family, hence all types of transponder can be supported by a common inventory of spares and test equipment. An interface card to suit the control function is fitted. Compatability with 1553B data bus installations is available. The three variants are:

(a) IFF 4720 – form, fit and function for the Cossor IFF 2720 type of transponder and suitable for a wide variety of installations

(b) IFF 4740 – form, fit and function for the Cossor IFF 3100 type of transponder or any single box installation requirement

(c) IFF 4760 – remote mounted, data bus controlled equipment for the new style avionics installations.

The 4700 series is an all solid-state equipment with true diversity operation, large-scale integration digital processing and a full IFF/ATC capability. The equipment has a built-in ACC option and is normally supplied with a crypto facility.

STATUS

In production. 103 IFF 4760 systems have been ordered for fitment to the RAF GR5 version of the Harrier.

CONTRACTOR

Cossor Electronics Ltd, The Pinnacles, Harlow, Essex CM19 5BB, England.

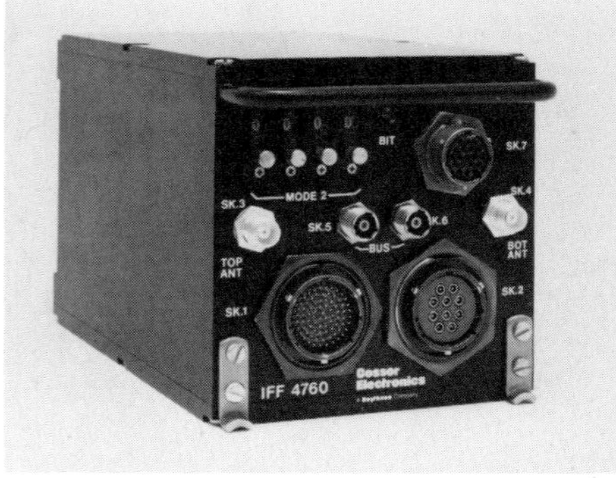

IFF 4760 remote mounted transponder

4341.353

MAREC II MARITIME RECONNAISSANCE RADAR

The MAREC II (MAritime REConnaissance radar), based on the MEL helicopter radar system, was developed to meet the requirements for a small, efficient, low-cost means of conducting sea patrol duties, including coastguard surveillance, search and rescue, fishery and oil rig protection, and pollution control. Suitable for fixed-wing aircraft and helicopters, it has a true-motion plotting table display covering 360° in azimuth and a range of 250 nm, together with a pilot's display.

The system has flown in the BAe Coastguarder demonstrator (a version of the HS 748), the prototype Airship Industries Skyship 500 aircraft and for Dornier 128 Skyservants for maritime patrol operations. A version of Marec II, the Super Marec system (**6433.353**) is in service on the Do 228.

CHARACTERISTICS

Frequency: I-band (9345 MHz)

Power output: 80 kW

Pulse widths: 0.4 and 2.5 μs

PRF: 200 and 400 Hz

Plotting table ranges: 17 to 219 nm

Plotting table size: 430 × 430 mm

Pilot's indicator ranges: 10 to 250 nm

STATUS

In production.

CONTRACTOR

MEL, Manor Royal, Crawley, Sussex RH10 2PZ, England.

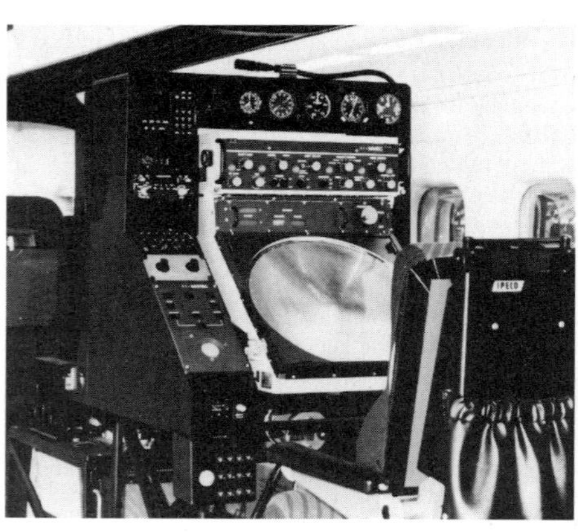

Marec II installed in an Indian Coastguard Do228 aircraft

6433.353

SUPER MAREC AIRBORNE RADAR

The Super Marec system combines the proven radar sensor components of Marec II (**4341.353**) with a scan converted colour TV raster display similar to that developed for the Super Searcher equipment (**5625.353**). This saves weight and space, and increases the tactical navigation facilities available to the operator, so reducing his workload on long patrols. Additional facilities include automatic multiple target tracking, and navigation overlay data.

STATUS

In production. The Indian Coast Guard is fitting Super Marec to its complete fleet of HAL-built Dornier 228 aircraft, and will also retrofit its existing aircraft.

CONTRACTOR

MEL, Manor Royal, Crawley, Sussex RH10 2PZ, England.

5614.353

ASR 360 AIRBORNE SURVEILLANCE RADAR

The ASR 360 is an X-band radar designed for maritime surveillance in small to medium size low performance fixed-wing aircraft and medium size helicopters. It is based on the Racal standard range of solid-state marine radars which are currently in service with many naval forces and merchant ships throughout the world. The system has now been adapted for airborne use and consists of three main units: antenna and gear box, transmitter/receiver and true motion display.

The antenna unit consists of the antenna itself, which is a centre fed slotted waveguide radiator with an aperture of 30 inches (76 cm), and the necessary gearing and drive. The antenna rotates at 23 rpm around a coaxial feed which runs concentrically through the main gear wheel, and is coupled into the waveguide between the magnetron and the transmitter/receiver. The radiation pattern at the 3 dB (half-power) points has a beamwidth of 3° in the horizontal plane, and 27° in the vertical plane for a stable picture under pitch and roll conditions.

The transmitter/receiver unit contains the 25 kW transmitter, duplexer and receiver, together with the power supply and the processing circuits. The transmitter consists of a trigger-controlled thyristor modulator driving a magnetron which produces the short bursts of RF energy. The PRF and the pulse lengths are controlled from the display unit.

The true motion display provides a bright, clear, high-definition picture based on the Clearscan video processing system to give high resolution for small targets to be detected against rain and sea clutter. Eight range scales are provided with range rings at fixed intervals; all of these, with the exception of the longest range scale, can be off-centred.

CHARACTERISTICS

Antenna

Type: Centre-fed slotted waveguide contained in fibreglass radome

Aperture: 760 mm

Rotation speed: 23 rpm

Polarisation: Horizontal

Beamwidth: 3° horizontal; 27° vertical

Transmitter

Peak power: 25 kW nominal

Pulse lengths: 0.05, 0.25, 1 µs

Display

Minimum range: 23 m on 10 m² targets

Discrimination: 9 m on the 0.5 nm scale with 10 m² targets

Range scales: 8 ranges covering 0.5 to 95 nm

STATUS

ASR 360 has been sold to the Royal Navy for use in Jetstream 31 aircraft, the Royal Norwegian Navy's Navajo, the Royal Swedish Air Force, the Finnish Air Force and the Sultanate of Oman for installation in Short Skyvan aircraft.

CONTRACTOR

Racal Avionics Ltd, Burlington House, Burlington Road, New Malden, Surrey KT3 4NW, England.

3710.353

SEA SEARCHER AIRBORNE RADAR ARI 5991

The Sea Searcher radar is a lightweight, high performance, 360° system for ASW and ASV roles, with over-the-horizon target tracking. It is suitable for both helicopter and fixed-wing aircraft and is in production for the Royal Navy Sea King helicopters. The system employs digital processing techniques permitting easy connection with other avionic equipment and was specifically designed to interface with sonar equipment and the Racal TANS navigation computers.

The north-orientated ground-stabilised (true motion) display is formed on a 430 mm flat plotting table, which allows transparent navigation charts to be overlaid. This provides the operator with extra tactical information and allows navigation data to be plotted without the need to transfer radar data to a separate chart or vice-versa. This facility is enhanced by the inclusion of an I-band (3 cm) transponder channel for identification of airborne and surface support craft; coded transponder replies being displayed on the clutter-free plotting-table surface. The dual pulse width transceiver includes built-in test circuits which, together with the radar display, enables in-flight or ground testing of the complete system.

In order to eliminate unintentional jamming by other radars operating in the vicinity, an in-flight manual adjustment is available to subject the transmitter PRF to random jitter. This facility also provides a first level of ECCM protection.

A dedicated computer allows the automatic tracking of two targets simultaneously, permitting over-the-horizon targeting to be performed. The high accuracy of this digital tracking system enables it to be used for pre-launch guidance of missile systems such as Otomat and Exocet, or via a telemetry link to a surface weapon system.

To meet the installation needs of different aircraft, a family of antenna sizes is available, ranging from 610 to 107 mm horizontal aperture, the largest offering optimum performance. The antenna normally operates in a continuous 360° scan mode, but sector scanning is available whereby the antenna scans about any bearing selected by the operator. This has a two-fold effect of maintaining radar silence in the non-sectored area and increasing the data rate over the illuminated sector.

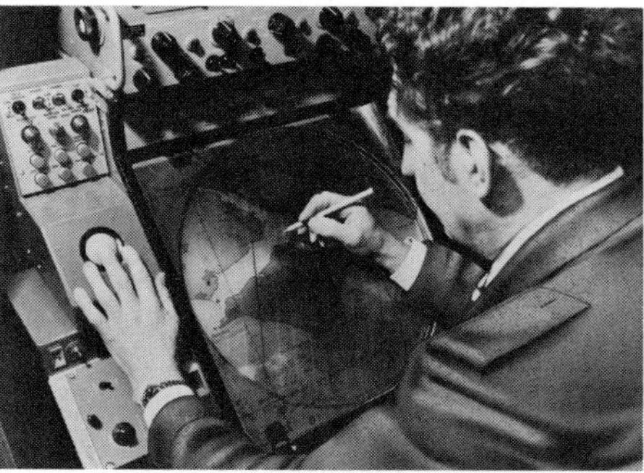

Display and control section of the MEL Sea Searcher ARI 5991 fitted to RN Sea King helicopters

STATUS

In production and in service.

CONTRACTOR

MEL, Manor Royal, Crawley, Sussex RH10 2PZ, England.

5625.353

SUPER SEARCHER AIRBORNE RADAR

Super Searcher is a development of Sea Searcher, and is a member of the MEL range of airborne radars. Designed for multi-threat maritime operations, it is a lightweight I-band command and control radar with a unique 42-inch (1067 mm) horizontal aperture antenna providing a high definition display on 14-inch (356 mm) colour, raster scan, cathode ray tube that can show true motion, or centre PPI, with variable sector scan.

By comparison with Sea Searcher, this latest version has a greater detection, target-tracking and guidance performance. The colour display replaces the earlier manned PPI plot. The system has an in-built guidance capability for the Sea Skua anti-surface-vessel sea-skimming missile.

The system incorporates three selectable pulse-widths, including an ultra-short width to give high definition of small targets in bad weather. Contact recognition is also improved by the use of microprocessor techniques and signal processing algorithms. The colour CRT includes freeze-frame and memory storage facilities, as well as a moving target indicator. It can be set to give either true motion/offset modes or relative motion latitude/longitude lines, grids, previous-mission intelligence, or other reference frames, and provides accurate and permanent plotting with numerous suite wave-point markers. The system also has a multiple track-while-scan capability.

The Super Searcher radar operates in primary and secondary modes, either separately or in combination. A comprehensive library of symbols that can be vectored as required eases the operator's task, particularly in intelligence storage and extraction. The system is compatible with IFF/SSR interrogators and can be supplied with an optional electronic surveillance measures antenna covering the C/D and E/J frequency bands.

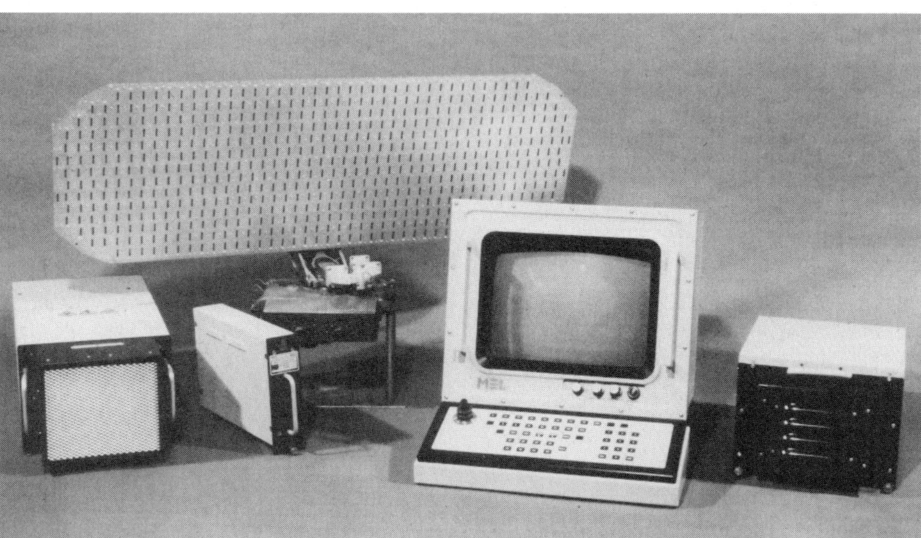

Super Searcher airborne radar

An advanced derivative of Super Searcher, known as 'Triple S' was intended for a radar update programme on the Royal Navy Lynx helicopters. This programme has now been cancelled.

STATUS

In production. Super Searcher is the principal sensor on board the 20 Sea King Mk 42B ASW helicopters ordered by the Indian Navy. It has also been selected for the eight Sikorsky S-70B helicopters for the Royal Australian Navy, with delivery due to be completed by 1988.

20 Super Searchers have also been ordered by a South American country for maritime surveillance use. This is believed to be Brazil for fitment to Embraer EMB-111 maritime patrol aircraft.

CONTRACTOR

MEL, Manor Royal, Crawley, Sussex RH10 2PZ, England.

6382.353

MEL AEW RADAR

Shortly before the 1986 Farnborough Air Show MEL announced the development of a new type of airborne early warning radar based on work carried out by a number of Philips companies. The system is based on MEL EW equipment and airborne radar experience, and a Signaal E/F band 3-D radar with an antenna/receiver optimised for long-range detection of small supersonic targets in high clutter. It would perform automatic detection, followed by automatic track initiation and automatic target tracking to provide a high data rate and accurate information on every track in terms of range, bearing, height and velocity.

The radar antenna consists of a single-element, squint-free wide-band transmitting array, and a multi-element stripline receiving array. The single-plane phased array has been designed with ultra-low sidelobe levels and allows accurate 3-D search and tracking of numerous targets. A rotodome type housing is proposed for the antenna system, mounted on top of the fuselage of an aircraft such as the A 320 Airbus or the Transall. Advanced processing techniques, including digital beam forming and fast Fourier Transform processing ensures accuracy and target discrimination in dense threat environments.

STATUS

In development. MEL is also developing a low-cost early warning system for installation in a Sabreliner, based on the Super Searcher equipment (**5625.353**).

CONTRACTOR

MEL, Manor Royal, Crawley, West Sussex RH10 2PZ, England.

1342.353

SEASPRAY HELICOPTER ASV RADAR

Seaspray was specifically developed for, and is fitted to, the RN's Westland Lynx helicopter, in which application it carries the designation ARI 5979. The equipment has been designed to detect and track small high-speed surface craft of the type increasingly used as platforms for anti-shipping missiles. Other possible applications include general maritime surveillance and over-the-horizon targeting.

The system consists of five line replaceable units (LRUs): scanner assembly, transmitter, receiver, processor, and a display and control unit. The system operates at I-band frequencies with a high transmitter power, and employs frequency agility operation. This is employed for reasons of improved detection of targets in sea and weather clutter and ECM conditions, and also improves bearing resolution by reducing the effect of target 'glint'. Built-in test facilities are incorporated.

The scanner LRU is located in a nose radome on the Westland Lynx naval helicopter, and is capable of providing a wide angle search (±90° relative to aircraft heading), or narrow angle search (±30° on selected bearing). The observer is provided with a scan-converted, bright TV-type display unit, on which target range and bearing information are displayed in alphanumeric form in addition to PPI of radar returns. In addition to the display controls, this unit also carries push-buttons for radar operational mode selection and other functions. The display can also be used for other sensors, such as FLIR and low-light television.

Two advanced models, designated Mks 2 and 3 are available. The Mk 2 incorporates TWS of multiple targets to permit tracking of these on the observer's display. The processor LRU is replaced to allow TWS processing and a control panel is added. The Mk 3 version incorporates 360° scanning (as well as TWS) for use in the control of the Sea Skua missile (**1530.321**), and can be fitted to a variety of helicopters and fixed-wing aircraft, or to small naval vessels for ship-launched missile control.

Seaspray naval helicopter radar system

OPERATION

Eight operational modes are provided as follows:

(1) Search: used for the detection of low-flying aircraft, submarine periscopes, and large and small vessels in bad weather and/or high sea conditions
(2) Weapon systems integration: this employs monopulse tracking techniques and provides guidance for the Sea Skua air-to-surface missile, and accurate target range and bearing inputs for surface-to-surface missiles via a data link
(3) Identification: used in an interrogate mode for target indentification
(4) Navigation: normal operation and by integration with a Racal TANS system
(5) Search and rescue
(6) Station keeping
(7) Weather warning
(8) Let-down and recovery to base in bad visibility.

CHARACTERISTICS

Frequency: I-band
Weight: 140 lb (64 kg)
Antenna dimensions: 686 mm wide × 254 mm deep
Antenna scan: 90° about fore and aft axis

STATUS

Seaspray is in production and in service with Sea Lynx helicopters of the Royal Navy and eight other countries. Seaspray Mk 3 has been ordered by the West German Navy for Sea King Mk 41 helicopters and another NATO country for Agusta Bell 212 helicopters. A land-based, coastal-sited version of Seaspray has been developed as a checksite device to monitor the effectiveness of shipborne ECM equipment.

CONTRACTOR

Ferranti Defence Systems Ltd, Radar Systems Department, Ferry Road, Edinburgh EH5 2XS, Scotland.

4615.353

BLUE VIXEN RADAR

Ferranti has been awarded a contract from the UK MoD to develop a new radar system, known as Blue Vixen, for the update programme for Sea Harrier, in conjunction with the fitment of the AMRAAM air-to-air missile, to provide a look-down/shoot-down capability. Like several other Ferranti airborne radars, this system was developed from a private venture programme known as Blue Falcon. The design of Blue Vixen is a X-band multi-mode coherent pulse doppler radar for all weather operation in air-to-air and air-to-surface modes, providing look-up and look-down capability over land and sea against airborne targets. It incorporates high resolution land and sea search modes for navigation and surface target detection, with air-to-surface ranging for weapon delivery. Track-while-scan and single target track operational modes are available. Ericsson Radio Systems of Sweden is working on the antenna and parts of the electronics as a sub-contractor to Ferranti.

STATUS

In development. The first development model has been delivered to the Royal Aircraft Establishment, Bedford for flight trials in a BAC One-Eleven. The Sea Harrier FRS 2 aircraft is due to enter service in 1989 and the production programme for Blue Vixen will be phased into the schedule.

CONTRACTOR

Ferranti Defence Systems Ltd, Radar Systems Department, Ferry Road, Edinburgh EH5 2XS, Scotland.

Blue Vixen radar system

1972.353

BLUE FOX RADAR

Blue Fox is a lightweight, high-performance, radar designed to fulfil the dual role of airborne interception and air-to-surface search and strike. It has been developed primarily for the RN Sea Harrier FRS Mk 1 V/STOL aircraft in which it forms part of a fully-integrated weapon system.

Blue Fox operates in I-band and uses frequency agility to enhance the radar's immunity to ECM, and improve its ability to detect small targets in bad weather or rough sea states.

For air-to-air interception, Blue Fox can be used for lead-pursuit or chase attacks and incorporates a transponder mode for identifying friendly aircraft or ships.

Using a flat aperture, slotted array antenna stabilised in pitch and roll, Blue Fox fits neatly into the nose of the Sea Harrier. Built on the line replacement unit principle, each component part of the radar (transmitter, receiver, processor, amplifier) can be easily checked or removed for servicing independently.

The radar display provides a bright, digital

scan-converted picture, which can be used to display other sensors requiring a TV-type display. Superimposed on the display are the flight symbols showing aircraft attitude, speed, heading etc, so that the pilot can monitor and control the aircraft's manoeuvres while using the radar.

STATUS
Deliveries to the Royal Navy are complete. Production is continuing for deliveries to the Indian Navy for its Sea Harriers.
CONTRACTOR
Ferranti Defence Systems Ltd, Radar Systems Department, Ferry Road, Edinburgh EH5 2XS, Scotland.

Blue Fox radar showing the distinctive flat aperture antenna

5877.353
RED FOX AIRBORNE RADAR
Red Fox is a lightweight, high performance I-band pulse modulated radar designed to fulfil the roles of airborne interception, air-to-sea surface search and attack plus an overland capability. It is basically a repackaged version of Blue Fox (**1972.353**) and is suitable for other aircraft types with a different nose profile to the Sea Harrier FRS Mk 1. It employs a two plane monopulse flat aperture antenna which is roll stabilised for air-to-surface ranging accuracy. Frequency agility is incorporated for ECM packaged in modular line replacement units for ease of maintenance and handling.
STATUS
Fully developed. No orders have been announced.
CONTRACTOR
Ferranti Defence Systems Ltd, Radar Systems Department, Ferry Road, Edinburgh EH5 2XS, Scotland.

Red Fox radar system

5685.353
BLUE KESTREL HELICOPTER RADAR
Ferranti Radar Systems is developing the I-band Blue Kestrel radar for the UK MoD version of the EH 101 helicopter. The system is designed for a variety of operational tasks, including surface surveillance, anti-submarine warfare, anti-surface vessel, over-the-horizon targeting, search and rescue, and navigation. This radar has a flat aperture antenna of advanced design which, together with a TWT transmitter, frequency agility, multiple target track-while-scan pulse compression, sophisticated digital data and signal processing, will provide a high performance radar for the EH 101 in its anti-submarine and anti-surface vessel roles.
STATUS
Blue Kestrel is another development based on the Blue Falcon programme. The first model for flight trials on a modified Sea King helicopter has been delivered and trials are in progress. The EH 101 is scheduled to enter service with the Royal Navy in 1990/91.
CONTRACTOR
Ferranti Defence Systems Ltd, Radar Systems Department, Ferry Road, Edinburgh EH5 2XS, Scotland.

Key elements of the Blue Kestrel radar

UNITED STATES OF AMERICA

1765.353
AN/APG-63 AIRBORNE RADAR
The APG-63 is an advanced multi-mode nose radar installed in the US Air Force F-15 supersonic fighter aircraft. It is a pulse doppler system that operates in I-band. A gridded TWT transmitter is employed, with digital doppler signal processing, and digital mode/data management. These features permit operation over a wide range of PRFs, pulse widths, and processing modes. The I-band antenna is a planar array type, carried on a three-axis gimbal system.

Radar information is digitally processed and there are two types of display employed: one is a small CRT located at the upper left-hand corner of the instrument panel and called the vertical situation display (VSD); the other is the HUD. In general, the VSD is employed for the longer-range, initial stages of an interception, while the HUD is for use during actual engagements or close-in encounters.

The VSD presents the pilot with a 'cleaned' synthetic display of computer-processed radar video data, together with alphanumerics and symbols.

Controls for the APG-63 are located at three positions in the cockpit. The main control panel is on the console on the left side of the pilot. This console also carries the two throttles and key radar operating controls are located on them. The third location for radar controls is the aircraft control stick.

Typical alphanumeric information that can be presented on the VSD are target altitude, ground-speed, heading, range, aspect angle, closure rate, and *g*-force. Ground-speed and *g*-force data give valuable indications of the kind of target being tracked. A satisfactory response to an IFF interrogation is indicated by a symbol displayed on the VSD.

An automatic acquisition switch on the aircraft control stick enables the pilot to lock the radar on to targets within close-in ranges. There are three modes: in the boresight mode, the radar locks on to the first target that enters the aircraft boresight, as designated by the gun reticle on the HUD; the second, called super-search, locks the radar on to the first target that comes within the HUD field of view; the third, vertical scan, locks the radar on to the first target that enters an elevation scan pattern normal to the aircraft's lateral axis.
STATUS
New software-controlled programmable signal processors were incorporated in new F-15 APG-63 radars in mid-1980 and retrofitting programmes have been implemented for earlier F-15 aircraft. Features include doppler beam-sharpened ground mapping. An upgrade of the AN/APG-63, designated the AN/APG-70 (**5879.353**), is being produced for use in the F-15E and in late model F-15C/Ds. Delivery of the AN/APG-70 commenced in 1986.

An AN/APG-63 radar, with modifications to incorporate synthetic aperture radar for real-time all-weather high resolution ground mapping and air-to-ground weapons delivery, was flown in a Hughes/McDonnell Douglas sponsored F-15 advanced fighter capability demonstrator for several years. The modified radar utilises doppler processing to distinguish closely spaced targets within an antenna beamwidth. Aircraft motion provides the doppler spectrum to be processed and the resultant radar picture is synthesised electronically.

Trials of ground mapping have produced results claimed to have a resolution ten times better than that of any previous airborne tactical radar maps. The system produced high resolution ground maps, as if seen from overhead.

The AN/APG-63 radar has been installed in US Custom Service P-3A Orion aircraft for use against drug smuggling. The first of four P-3A aircraft so equipped was put into service in 1984. The last two were delivered in 1986.

CONTRACTOR

Hughes Aircraft Company, Radar Systems Group, El Segundo, California 90245, USA.

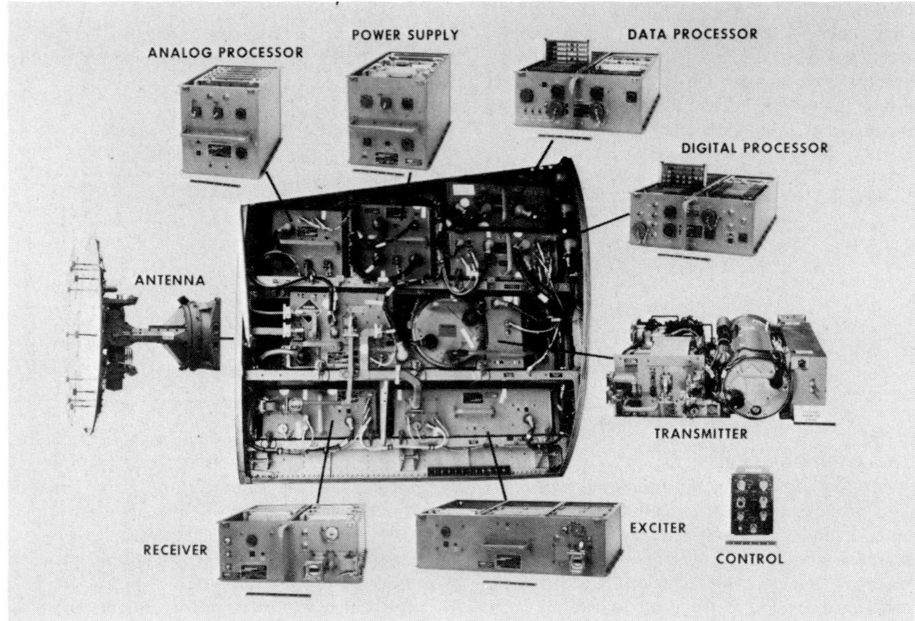

AN/APG-63 radar showing individual sub-systems and their location in the complete equipment

3302.353

AN/APG-65 AIRBORNE RADAR

The APG-65 radar was developed by Hughes for use in the US Navy's F/A-18 Hornet strike fighter aircraft. It is an I-band equipment designed as an all-digital, multi-mode system suitable for both air-to-air combat and air-to-ground weapon delivery missions.

It provides radar information for the control of the F/A-18's 20 mm gun, Sparrow and Sidewinder missiles in aerial combat, and a full range of conventional and precision guided weapons in the ground attack role.

In its air-to-air role, the APG-65 presents a 'clean-scope', synthetic scan-converted display against airborne targets in all aspects, all altitudes and through all target manoeuvres. It incorporates complete search, track and air combat mode variations.

The radar includes a velocity search mode to provide maximum detection range capability against nose aspect targets; a range-while-search mode to detect all-aspect targets; a TWS mode which, combined with a future autonomous missile such as AMRAAM, will give the aircraft a launch-and-leave capability; a single target track mode; a gun director mode; and a rapid assessment mode which enables the operator to expand the region centred on a single tracked target, permitting radar separation of closely spaced targets.

Three air combat manoeuvring modes provide automatic target acquisition in various search volumes; the HUD acquisition mode to scan the entire HUD volume and to lock onto the first target located within a specified range; the vertical acquisition mode in which the radar scans vertically in a narrow width volume and automatically acquires the first target found within a specific range; and a boresight acquisition mode to allow the pilot to point the aircraft at the desired target and acquire it automatically. The pilot can step through successive targets until he acquires the one he wants.

Surface attack modes include long range, high resolution surface mapping which, combined with other modes, gives the pilot the ability to detect and track fixed or moving targets on land or sea. The APG-65 radar includes a precision velocity update feature to improve navigational accuracy; a terrain avoidance mode for low-level penetration missions in limited visibility conditions; ground moving target indication/track or fixed target modes which the pilot may select depending on the target tactical situation; air-to-surface ranging; and a sea surface mode which enables the radar to detect ship targets regardless of sea condition.

AN/APG-65 radar, extended from the nose of an F/A-18 aircraft for inspection at Patuxent River, Maryland, US Naval Air Test Center

STATUS

In operational service with the US Navy F/A-18 aircraft. Canada ordered 138 CF-18 aircraft, and initial deliveries of the AN/APG-65 radars were made in 1982 and are in service. Additional units will be delivered to Canada until 1988. Australia and Spain also ordered the F/A-18 to serve as their new combat aircraft. The first of 75 Hornets ordered by Australia was delivered in 1984, and Spain has ordered 72, with an option on 12 more. Delivery began in 1986. The first EF-18 for Spain was rolled out in 1985. Selection of the AN/APG-65 to upgrade the F-4F Phantoms of the Federal Republic of Germany was announced by the German government in early 1985. Seventy-five F-4s of West Germany's two tactical fighter wings will be equipped with the APG-65, with an option to equip two Phantom fighter-bomber wings.

CONTRACTOR

Hughes Aircraft Company, Radar Systems Group, El Segundo, California 90245, USA.

3303.353

AN/APG-66 F-16 RADAR

The Westinghouse AN/APG-66 radar is an X-band coherent pulse doppler system for the F-16 aircraft. The system is a direct development of an in-house design called the WX-200 (**1673.353**), which is claimed by the company to be the first airborne radar with a programmable signal processor. Designed to operate in conjunction with Sparrow medium-range and Sidewinder short-range missiles, the AN/APG-66 has a look-down range in ground clutter of 20 to 30 nm, and a look-up range of 25 to 40 nm. The production system has no associated hydraulics, rate gyros or roll gyro, and there are only 9500 components. This is reflected in a formally demonstrated MTBF of 141 hours and a mean-time-to-repair of five minutes.

There are ten operating modes, several being associated with frequency agility to resist jamming. All functions, including self-test, are controlled by a computer via a digibus serial digital data-bus. Westinghouse claims that the WX-200 was the first airborne radar to incorporate a programmable signal

F-16 radar mounted in ground rig

processor, and that this technology was absorbed into the AN/APG-66.

The AN/APG-66 is of modular design with six functional line-replaceable units, each with its own power supply, so that unforeseen improvements can be easily incorporated. One such improvement has been chosen by the Japanese Self-Defence Force to replace the AN/APQ-120 fire control system on its F-4EJ aircraft. Another variant, the AN/APG-66Z has been selected by the RNZAF for its A-4 aircraft, while yet another version incorporates the APG-66, a new fire control computer and a CW illuminator in the F-8 aircraft of the People's Republic of China.

Line replaceable units are co-produced under licence by the European F-16 operators; The Netherlands, Belgium, Norway and Denmark. The system entered service in 1978.

The radar has been installed in six Cessna Citation and eight modified Piper Cheyenne III aircraft operated by the US Customs for anti-smuggling and illegal immigration surveillance duties.

CHARACTERISTICS

Frequency: X-band

Transmitter: TWT

Antenna: Planar array 740 × 480 mm

Search angle: 120° in azimuth and elevation

Range: 148 km

Operating modes: (air-to-air) look-up, look-down, search and track, air combat search, automatic tracking; (air-to-ground) real beam mapping, 8:1 doppler beam sharpening, scan freeze, ranging, beacon homing, sea search

Weight: 134.5 kg

Volume: 0.102 m^3

STATUS

In production for US Air Force and the air forces of Belgium, Netherlands, Denmark, Norway, Venezuela, Israel, Pakistan, Egypt, Indonesia, Thailand and Singapore. By 1987 more than 2000 systems had been built. Deliveries of the APG-66 variant for the Chinese F-8 aircraft is scheduled to begin in March 1990.

The APG-66 has also been selected by British Aerospace for the Hawk 200 single seat fighter. It is understood that 18 sets are to be delivered initially.

F-16 aircraft with radar receiver and transmitter in lower section for easy accessibility

CONTRACTOR

Westinghouse Electric Corporation, Defense and Electronics Center, Aerospace Divisions, Friendship Site, PO Box 746, Baltimore, Maryland 21203, USA.

5610.353

AN/APG-67 MULTIMODE AIRBORNE RADAR

The AN/APG-67 multimode airborne radar is a coherent X-band pulse doppler system which has been designed for both air-to-air and air-to-surface modes for the F-20 Tigershark aircraft. The system comprises four line replaceable units (LRUs): antenna, transmitter, radar target data processor and a radar data computer containing two central processing units (CPUs) with an overall weight of 270 lb (123 kg).

The antenna is a flat plate slotted array which features low sidelobes and full azimuth and elevation monopulse operation. Scan capability covers ±60° in both azimuth and elevation. Variable sector width scans are selectable and elevation search can be accomplished in either 1, 2 or 4 bar modes. The antenna can be either line-of-sight or rate gyro stabilised. The radar receiver is mounted on the antenna assembly and features a low noise front end which gives additional detection range. The receiver consists of two channels which provide the monopulse sum and multiplexed difference channels (azimuth and elevation).

The transmitter is coherent, containing a TWT operating in X-band, and is programmable providing a high degree of mode flexibility. It operates at low, medium and high PRFs and other features include variable output power and pulse widths.

The radar data computer is the heart of the system and contains two CPUs. One CPU handles radar mode control via the internal radar data bus and interfaces with the F-20 mission computer. This CPU also conducts the built-in-test function to determine operational status without the need for special flight line test equipment. The second CPU performs the radar data processing requirements associated with motion compensation and target tracking, and provides raster scan output of the radar data and symbology to the cockpit display. The radar target data processor receives raw data from the receiver and integrates this to establish firm target reports. To achieve its various capabilities, this unit performs a variety of functions including fast Fourier transformation, MTI, pulse compression, motion compensation, variable waveforms and variable bandwidth operation.

AN/APG-67 multimode airborne radar installed in an F-20 Tigershark

OPERATION

The AN/APG-67 is designed for five main missions: supersonic intercept, combat air patrol and air superiority in the air-to-air mode with interdiction, close air support and sea detection in the air-to-surface mode.

Air-to-air

In the air-to-air role the AN/APG-67 provides all-aspect, all-altitude coverage in three main modes of operation:

(1) search and track in the look-up (up look) (above the horizon)
(2) search and track in the look-down (down look) (below the horizon)
(3) automatic acquisition and tracking for air combat.

The pilot selects the appropriate operating mode and mode parameters by controls on the radar control panel or the fire control computer automatically selects a mode appropriate to the tactical situation. In either look-up or look-down modes the pilot can select antenna scan limits of ±10, ±30 or ±60° and elevation can be set at 1, 2 or 4 bars. Range scales are set at 5, 10, 20, 40 and 80 nautical miles although ranges against a typical fighter are 30 nautical miles in the look-up mode and 22 nautical miles for look-down.

In the air combat mode, the radar scans a 20 × 20° field of view and automatically acquires the target. The pilot can also select an alternative vertical scan of 7 × 60°, or boresight. In the track-while-scan mode the radar tracks and displays all aircraft within the search area and gives priority to ten of these on the basis of time-to-go.

Air-to-surface

Six air-to-surface modes are available:

(1) ground mapping using a real beam mapping mode with azimuth scans of ±10, ±30 or ±60° and range scales of 5, 10, 20, 40 and 80 nautical miles
(2) an expanded map mode using doppler beam sharpening techniques to improve resolution for navigation and/or target location

(3) an air-to-ground ranging mode providing real-time measurement to a designated ground point

(4) a freeze mode whereby the ground map is frozen on the display in a quasi-silent operation, with the transmitter on stand-by. Aircraft movement relative to the map is shown from outside information and the map can be updated periodically by operating a radar scan

(5) (6) two sea surface search modes for surface vessel detection. Sea 1 employs frequency agility techniques to detect stationary and moving targets with a detection range of 35 nautical miles for a 50 m² target in sea state 1. Sea 2 uses a coherent MTI mode with a detection range of 30 nautical miles for the same target in sea state 4.

STATUS

In pilot production to support the F-20 Tigershark. It was announced in February 1988 that the Taiwanese indigenous defence fighter radar will be based on the APG-67.

CONTRACTOR

GE Aerospace, French Road, Utica, New York 13503, USA.

5881.353

AN/APG-68 AIRBORNE RADAR

The AN/APG-68 is a development based on the AN/APG-66 (**3303.353**). The system was developed as a substantially improved version for the F-16 as part of a multi-national staged improvement programme (known as MSIP). It is intended to improve the aircraft capability to deliver air-to-air and air-to-ground weapons in all-weather conditions. In particular it will allow F-16 pilots to launch the AIM-120 AMRAAM advanced medium range air-to-air missile beyond visual range, and will improve the ground mapping performance. The major change over the AN/APG-66 is the incorporation into one unit of all digital processing activities. The new processor weighs 45 kg, occupies 0.03 m³ and dissipates 1 kW. The new radar has more than 22 air-to-air and air-to-ground modes. All programme instructions are stored in a 416 k word, non volatile, block-orientated random access memory. The radar computer is programmed in Jovial J73, the USAF approved high-order language. Initial operating capability for the F-16C and -D aircraft, the first to have the AN/APG-68, was in December 1984.

The new capabilities are gained by the addition of a programmable signal processor and a dual-mode transmitter. The former incorporates advanced modular processing architecture and a reliable high-density, solid-state memory. The latter selects the best waveform for each mode of operation, ranging from low PRF in air-to-surface engagements, to medium and high PRF for long-range air interception.

In the air-to-air mode, the range at which on-coming targets are detected is increased by using a high PRF velocity searching mode, ie the radar is searching for targets with high relative velocities. Once detected a medium PRF range-while-search mode can be employed against targets with any aspect to gain additional range and angle information. In the track-while-scan mode the radar can track 10 targets simultaneously, assess the degree of threat from each, and launch missiles as appropriate. By using high-resolution doppler techniques, closely spaced targets can be distinguished and tracked separately.

In the air combat mode the radar scans selected airspace and automatically acquires the nearest target. In look-down situations land-based targets, such as moving vehicles or vessels are ignored because the radar rejects returns with less than a specified threshold. This ground-mapping terrain radar technique ensures that only airborne targets are portrayed on the radar scope.

The air-to-ground performance of the radar has been improved by an eightfold increase in the ground mapping resolution and substantial spare computing power is available for future growth, which could include, for example, automatic terrain-following, terrain avoidance or very high resolution synthetic aperture mapping.

CHARACTERISTICS

Frequency: X-band

Transmitter: Gridded, multiple peak power TWT

Antenna: Planar array 480 × 740 mm

Search range: 160 nm

Operating modes: (air-to-air) range-while-search, track-while-scan (10 targets), velocity search, look-up search, raid clutter resolution, multiple auto-acquisition air combat; (air-to-ground) real beam mapping with doppler beam-sharpening and scan freeze, sea surface search, beacon homing, ground moving target indication and tracking, fixed-target tracking.

Weight: 153 kg

STATUS

In production with orders for the US Air Force, Israel, Korea, Turkey, Greece and Egypt. More than 1500 US Air Force F-16s are due to have the AN/APG-68. A version, designated AN/APQ-164 (**4647.353**) forms part of the avionics suite in the B-1B aircraft.

CONTRACTOR

Westinghouse Electric Corporation, Defense and Electronics Center, Aerospace Divisions, Friendship Site, PO Box 746, Baltimore, Maryland 21203, USA.

5882.353

AN/APG-69(V) MULTIMODE AIRBORNE RADAR

The AN/APG-69(V) multimode radar is a pulse-doppler, digital, coherent air-to-air and air-to-surface fire control system. It has been under company-funded development since 1976.

The radar weighs less than 80kg and has a mean-time-between failure of 200 hours. The technical approach has been to develop a fire control system that exploits the significant capabilities of a coherent system while physically occupying the same space as the lower performance, non-coherent AN/APQ-159 system used in the F-5 aircraft.

The six line replacement units (LRUs) consists of an antenna, receiver-transmitter, radar processor, signal data converter, video indicator and radar set control. The modular AN/APG-69 is designed for LRU-for-LRU replacement of currently installed F-5 radars with no structural, wiring or cooling changes necessary. This modular approach makes the radar adaptable for other types of aircraft, eg AV-8B, Mirage, Alpha Jet, Hawk 200, AMX, A-4, MiG-21 and F-7.

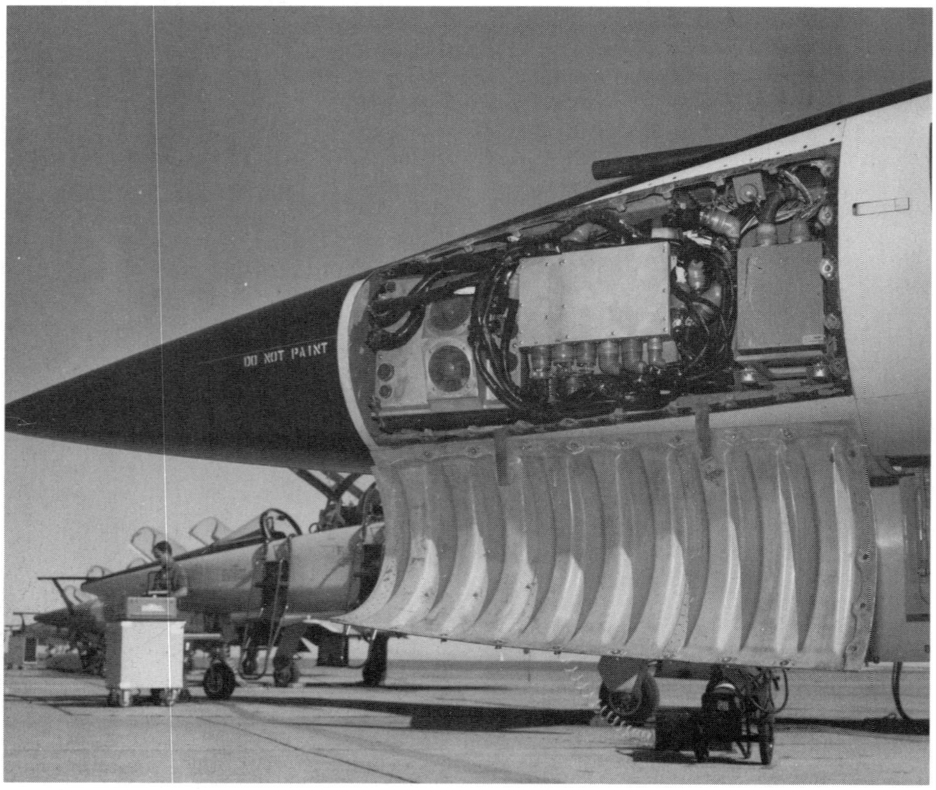

AN/APG-69 fire control radar

The RF section of the system features a low-noise, crystal-derived exciter amplified by a TWT amplifier and received through low noise GaAS FET RF amplifiers. The high voltage for the TWT is derived from a low noise power supply/modulator. It uses advanced techniques in pulse-width modulation to provide voltage regulations for the high voltage for the transmitted pulse. Pulse compression is used to allow the use of longer pulsewidths for increased average power at moderate levels of transmitted peak power (8 kW) while retaining (upon processing of the return) the finer resolution normally associated with short pulse systems. The compressed waveform is used in the low PRF (pulse) mode to achieve unambiguous range while maintaining the high average power required for detection purposes.

The RF power is radiated through the sum channel of a slotted array, flat-plate antenna. The return power is received through the antenna, where it is separated into sum and difference channels in track mode, and through the sum and guard horn in search. The sum and difference/guard energy then passes through the two-channel receiver. The receiver has a wide-range automatic gain control (AGC) for improved dynamic range, and a sensitivity time control for reduction of second-time-round echoes. The signal processing includes digital processing of the video signals, including a 32-point fast Fourier transform doppler processor, a digital AGC, a sliding-window constant false alarm rate threshold, and angle-track processing of the two-channel video. Other functions of the signal processor include range, velocity and voltage-controlled crystal oscillator clutter loop control, as well as system mode control.

The radar system has a balanced gimbal antenna driven by direct-current drives for the azimuth and elevation gimbals. Use of this design yields low power consumption, resulting in low heat dissipation in the amplifier and high system reliability. The video display processing is carried out in the signal data converter. The video output is in RS-170 format.

The AN/APG-69 provides air-to-air search, acquisition and track; air-to-surface detection and mapping; and self-test. In the air-to-air search the radar detects targets in both clutter-free and ground clutter environments. It will acquire and track a target which has been manually designated by the operator using the radar display. Acquisition is automatic in the 'dogfight' modes. Detection of targets in ground clutter uses coherent processing, while a pulse-compressed waveform with non-coherent processing provides extended range performance in lower-clutter conditions.

The radar has a real-beam ground map capability that supplies video to the radar/EO display. This capability is enhanced by frequency diversity which allows sea target detection performance in lower sea state environments. Air-to-ground ranging at head-up-display commanded angles is provided.

CHARACTERISTICS

Operating frequency: X-band
Power: 1380 W
Transmit power: 160 W
Weight: 175 lb (80 kg)

STATUS

Extensive flight testing has been in progress since 1983 with over 700 test missions, of which more than 200 were carried out in an F-5 aircraft.

CONTRACTOR

Emerson Electric Company, Government and Defense Group, 8100 West Florissant Avenue, St. Louis, Missouri 63136, USA.

5879.353
AN/APG-70 AIRBORNE RADAR

The AN/APG-70 is an upgrade of the AN/APG-63 (**1765.353**) and is intended for the F-15E aircraft and late models of the F-15C/D. Four advanced design, higher performance line replacement units (LRUs) succeed the previous five units, and two other LRUs are modified. Features in the modified transmitter include high average power gridded TWT, multiple PRFs and increased stability giving increased dynamic range. The newly designed receiver/exciter has increased bandwidth, improved tracking in ECCM, greater sensitivity and longer range detection. Other newly designed units are the radar data processor, programmable signal processor and the analogue signal converter. The antenna and power supply are the same as with the AN/APG-63.

STATUS

In production for the F-15E aircraft and operational in late model F-15C/Ds. Successful trials with an AN/APG-63 demonstrated a synthetic aperture radar for real-time all-weather high resolution ground mapping and display (see **1765.353** Status) which is included in the AN/APG-70 fit for the F-15E. The first engineering development models and the first production model were delivered in 1986.

In December 1987 Hughes received a $58 million contract to develop a new radar for the proposed US aerial gunship, the AC-130U. This will be a modified version of the AN/APG-70.

CONTRACTOR

Hughes Aircraft Company, Radar Systems Group, El Segundo, California 90245, USA.

AN/APG-70 radar

5880.353
AN/APG-71 AIRBORNE RADAR

The AN/APG-71 is an upgraded version of the radar portion of the AN/AWG-9 airborne weapon control system (**1100.353**) which is in development to satisfy US Navy air superiority requirements for the new F-14D Super Tomcat. Key elements in the upgrade are a fully programmable processor and a companion radar data processor which replace the AN/AWG-9's four analogue processors and provide digital capability. Digital radar displays are incorporated and the system includes sophisticated ECCM provisions. Detection and tracking modes have also been improved over those of the AN/AWG-9. Operational details and use are the same as for the AN/AWG-9.

STATUS

In development under a contract from Grumman Aerospace Corporation. The first engineering development models were delivered in late 1986 and are now in test in the Hughes test facility in El Segundo. They are also in flight test in a TA-3B at Point Mugu Naval Air Test Center. The TA-3B was modified for installation of the radar and associated electronics. The first radar tests in an F-14D were scheduled for the spring of 1988.

CONTRACTOR

Hughes Aircraft Company, Radar Systems Group, El Segundo, California 90245, USA.

6435.353
MULTI-MODE RADAR FOR F-5E

GE has developed a lightweight, advanced multi-mode radar suitable for retrofitting into the F-5E aircraft. The equipment operates in X-band, uses digital pulse compression and coherent signal integration techniques, and has a high antenna gain as well as low sidelobes.

The equipment has two main modes of operation, air-to-ground and air-to-air, with the antenna scanning over $\pm 40°$, $\pm 30°$, or $\pm 10°$ in azimuth. In the air-to-ground mode facilities include Doppler beam sharpening to 20 nm range, ground track moving target indication (MTI) up to 40 nm range, air-to-ground ranging for target acquisition, ground mapping to 80 nm range, freeze frame (which is radar silent), and scale expansion. Similar capability is possible over the sea, with detection of a moving or stationary patrol boat at 31 nm range in seas of less than one metre. For high sea states, a coherent MTI mode can be selected to detect moving targets over eight knots.

In the air-to-air mode facilities include look-up and look-down search and track, air combat automatic acquisition in boresight, super search, and vertical scan. In super-search the radar scans the $20° \times 20°$ field of view of the head-up display, vertical scan is $10° \times 40°$. In these sub-modes, and in boresight, the radar automatically tracks onto a target and switches to single target track. Average detection range against a typical fighter aircraft is 30 nm in the look-up mode, and 19 nm in look-down.

In air-to-search angle accuracy is given at 3 milliradians, range accuracy at 15 metres or 1% of range, and velocity accuracy at 15 metres/second or 0.4% of range. In air-to-ground the range resolution is 45 metres and accuracy is 15 metres.

CHARACTERISTICS

Frequency: X-band
Power output: 160W average
Antenna dimensions: 279 x 400 mm
Weight: 104 kg
Volume: 0.06 cubic metres
MTBF: 150 hr

STATUS

Production.

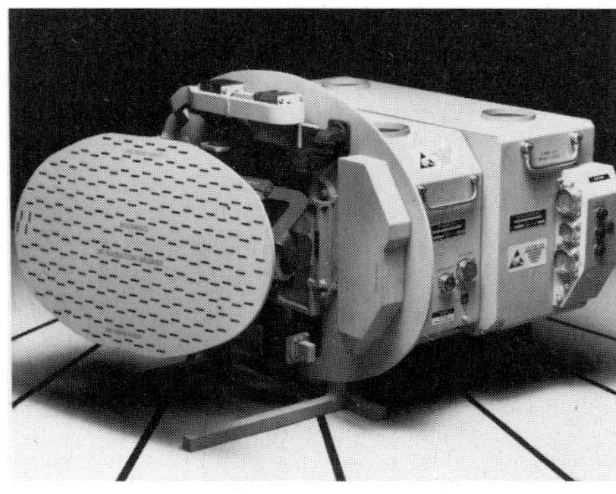

Multi-mode radar for F-5E

CONTRACTOR

GE Aerospace, French Road, Utica, New York 13503, USA.

4333.353
TORNADO NOSE RADAR

The nose-mounted radar for the interdictor/strike version of the Tornado aircraft has been developed and manufactured by Texas Instruments under contract to Panavia. The radar is essentially two separate systems which share a common mount, power supply and processor/computer. The two systems are the terrain following radar (TFR) and the ground map radar (GMR). Both systems are designed to operate in severe ECM conditions.

The TFR is an advanced terrain following radar enabling the aircraft to make blind, low-level, high speed terrain following attacks using natural geography masking to avoid radar detection.

The GMR is the primary attack sensor for the Tornado and operates in the air-to-ground and air-to-air modes to provide high resolution mapping for navigation update and target identification, air-to-air fire control and air-to-ground fire control. The system enables the pilot to have accurate fixes that update the navigation system, perform air-to-ground ranging and target tracking for weapon delivery, and air-to-air ranging and tracking for self-defence.

The radar also provides obstacle warning above a

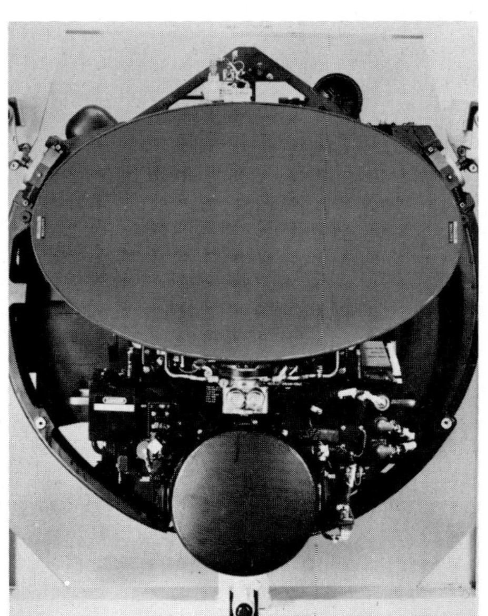

Head-on view of Tornado nose radar installation

zero height clearance plane which is horizontally stabilised. Video signals of all obstacles penetrating this clearance are displayed simultaneously with ground map video. Terrain following can be either automatic or manual.

For air-to-ground attack the GMR automatically acquires, locks-on and tracks designated ground targets. In the air-to-air mode, the radar locks on to designated airborne targets in range and angle for weapon delivery.

Both radars incorporate a high degree of advanced technology and feature broadband frequency agility, monopulse (four-lobe for GMR), wide and narrow antenna scan, fast and slow antenna scan and ECCM features. The system operates in the Ku-band.

STATUS

Currently in production for the Tornado interdictor/strike aircraft in Europe, under licence from Texas Instruments.

CONTRACTOR

Texas Instruments Inc, PO Box 600246, Dallas, Texas 75266, USA.

1585.353

E-3 (AWACS) RADAR AN/APY-2

The primary radar of the E-3 airborne warning and control system (AWACS) is the Westinghouse AN/APY-2. This radar was specially developed for airborne early warning and control applications and for mounting on a modified Boeing 707-320B airframe. The radar antenna is mounted back-to-back with a complementary IFF/SSR antenna, both of which are contained in a saucer-shaped radome carried above the fuselage of the aircraft. Within the fuselage are radar transmitters, computers, displays and other associated electronics. The radome has a normal rotation rate of six rpm, to give a scan rate of one every ten seconds. Operating frequency of the AN/APY-2 is in the 10 cm wavelength band and there are seven operating modes. On any azimuth scan, the surveillance volume can be divided into as many as 32 sub-sectors, each with its own set of operating modes and conditions. These modes can be accommodated on subsequent scans or re-arranged to vary the type of coverage for any given area of interest, or to accommodate changes in operating conditions. The seven modes are summarised briefly as follows:

(1) pulse doppler non-elevation scan (PDNES): this provides surveillance of aircraft down to the surface using pulse doppler, with narrow doppler filters and a sharp beam, to eliminate ground clutter. Target elevation is not measured
(2) pulse doppler elevation scan (PDES): radar operation in this mode is similar to PDNES, but target elevation is derived by electronic scanning of the beam in the vertical plane
(3) beyond-the-horizon (BTH): the BTH mode uses pulse radar – without doppler – for extended range surveillance where ground clutter is in the horizon shadow
(4) passive: the radar transmitter can be shut down in selected sub-sectors while the receivers continue to process ECM data. A single strobe line passing through the position of each jamming source is generated on the display console
(5) maritime: this involves use of a very short pulse to decrease the size of the sea clutter patch to enhance the detection of moving or stationary surface ships. An adaptive digital processor adjusts automatically to variations in sea clutter, and blanks land returns by means of computer-stored maps of land areas
(6) test/maintenance: control is delegated to the radar technician for maintenance purposes
(7) standby: radar kept in a warmed operational condition, ready for immediate use. Receivers are shut down.

Two of the multipurpose consoles aboard the US Air Force AWACS aircraft

E-3 Sentry (AWACS) aircraft take-off showing rotating radome of AN/APY-2 early warning radar

PDES and BTH can be used simultaneously or alone; either or both can be active or passive, as required. PDNES may be used simultaneously with the maritime mode. Blanking commands in BTH and pulse doppler modes can be used in each of the 32 sub-sectors, thus enabling the maximum potential of the system to be concentrated in those sub-sectors of greatest interest.

New colour displays are to be retrofitted to the US Air Force's AWACS fleet under a $16 million contract awarded to Hazeltine in late 1984. These high resolution displays will equip the aircraft's 14 command and control consoles.

The technical functioning of the AN/APY-2 radar and its associated signal processing was described at some length under this entry number in the 1977 edition of *Jane's Weapon Systems* and other details appeared in earlier editions. The system was initially known as the AN/APY-1 but the addition of a maritime surveillance facility resulted in a change of nomenclature to -2.

There is also a programme called Salty Net designed to enable AWACS to operate with the 412L, NADGE, and other systems. The NATO AWACS will also interface with 407L in the American Tactical Air Control System.

STATUS

AWACS requirements for CONUS air defence depend on the air defence alert condition. In peace time, a total of five AWACS aircraft would be deployed on a rotational basis, one at each of the US Region operations control centres of the Joint Surveillance System (JSS). Four are in the continental USA, and the fifth in Alaska.

The first AWACS aircraft were delivered to the US Air Force in March 1977. The 552nd AWACS Wing (now the 28th Air Division) at Tinker Air Force Base became operational in the spring of 1978. All 34 aircraft, now designated the E-3A Sentry, have been delivered to the US Air Force, and are being progressively updated to E-3B and E-3C standard of electronic fit.

In December 1978, the NATO Defence Planning Council, made up of the Defence Ministers of the member nations, gave a go-ahead to acquire a fleet of 18 E-3A aircraft for the NATO AEW requirement. This was an $1800 million undertaking. Boeing was the major contractor for the programme.

In addition to the sub-contract team assisting with the E-3A Sentry, a team of West German and Canadian sub-contractors was engaged on the NATO AEW aircraft. These included AEG-Telefunken, Ulm, which co-operated with Westinghouse in the manufacture of the E-3A radar system. Dornier was assembly and checkout contractor. First delivery of a NATO system was made on 22 January 1982 and the final E-3A was delivered to NATO in April 1985.

Five E-3 aircraft, also fitted with the AN/APY-2 system, were ordered by the Royal Saudi Air Force. The first aircraft was handed over in June 1986 and all five aircraft, plus eight E-3 derivative tanker aircraft (KE-3), were delivered by September 1987.

As a result of a lengthy competitive evaluation, the E-3 AWACS aircraft, complete with the AN/APY-2 radar, has been selected for the Royal Air Force AEW fleet. Seven aircraft have been ordered with an option for one more. The first aircraft will be delivered in January 1991. France has also ordered five E-3s, with an option to buy two more at a later date.

CONTRACTORS

Westinghouse Defense and Electronics Center, Baltimore, Maryland 21203, USA.

Boeing Aerospace Company, PO Box 3999, Seattle, Washington 98124, USA.

4340.353

AN/APN-215 AIRBORNE RADAR

The AN/APN-215 is a lightweight, digital, colour-display, multi-mode radar, derived from the commercial RDR-1300 system. Designed for weather avoidance and terrain mapping, the AN/APN-215 is suitable for a range of aircraft from light general aviation to heavy twins.

Weather is shown on the display in three discrete colours. When used with other on-board equipment, pictorial navigation data can be superimposed over the weather or terrain display. When used with a Checklist Control Unit, the system also provides a pilot-programmable alphanumeric display presentation or operational checklists and other pertinent data.

STATUS

Operational in the US Army and Coast Guard. As the AN/APN-234, the system adds a sea-search function and is operational with the US Navy C-2A COD aircraft.

CONTRACTOR

Allied-Signal Inc, Bendix Avionics Division, PO Box 9327, Fort Lauderdale, Florida 33310, USA.

6105.353

AN/APN-234 MULTI-MODE RADAR

The AN/APN-234 is a lightweight digital multi-mode, colour display radar derived from the commercial RDR-1300 system. It is identical to the AN/APN-215 radar (**4340.353**) with the addition of a sea search function.

STATUS

Operational in the US Navy C-2A carrier on-board (COD) aircraft, which are designed specifically to deliver cargo to air groups deployed on board US Navy carriers.

CONTRACTOR

Allied-Signal Inc, Bendix Avionics Division, PO Box 9327, Fort Lauderdale, Florida 33310, USA.

6104.353
RDR-1500 MULTI-MODE SEARCH AND SURVEILLANCE RADAR
The RDR-1500 is a lightweight 360° multi-mission radar designed specifically for both helicopters and fixed-wing aircraft. It can be used for a variety of maritime missions, including search and rescue, surveillance and patrol, fisheries protection and anti-surface ship operations.

Standard radar modes include weather detection, ground mapping, search, beacon detection and identification. A selectable ground stabilised reference display mode is available, in addition to aircraft heading or north orientation. The RDR-1500 also has the facility to offset the sweep centre to any location on the display and provide target marker capability. Information from a variety of navigation sensors, such as Omega, VOR, DME and ARINC 565 IAS, can be displayed as overlays. The system modular design concept provides additional growth capability.

CHARACTERISTICS
Frequency: 9375 ± 5 MHz – transmitter/receiver (weather/map); 9310 ± 5 MHz – receiver (beacon)
Power output: 10 kW nominal
System weight: 32.4 kg
CONTRACTOR
Allied-Signal Aerospace Company, Bendix/King Air Transport Avionics Division, PO Box 9327, Fort Lauderdale, Florida 33310, USA.

1970.353
AN/APQ-99 RADAR
The AN/APQ-99 is a forward-looking radar developed by Texas Instruments for the RF-4 version of the Phantom aircraft. It is an elevation (two-lobe) monopulse system with ground-mapping and terrain-following capabilities. Operation is at J-band frequencies.
STATUS
The AN/APQ-99 has been supplied to the air arms of the United States, Federal Republic of Germany, Iran, and Japan. In March 1985 Texas Instruments received $30 million as part of a $96 million multi-year contract to increase the reliability and maintainability of the AN/APQ-99 for the US Air Force's RF4C aircraft. Modifications to the system are currently in progress and following these the AN/APQ-99 will be redesignated AN/APQ-172 **(6117.353)**.
CONTRACTOR
Texas Instruments Inc, PO Box 660246, Dallas, Texas 75266, USA.

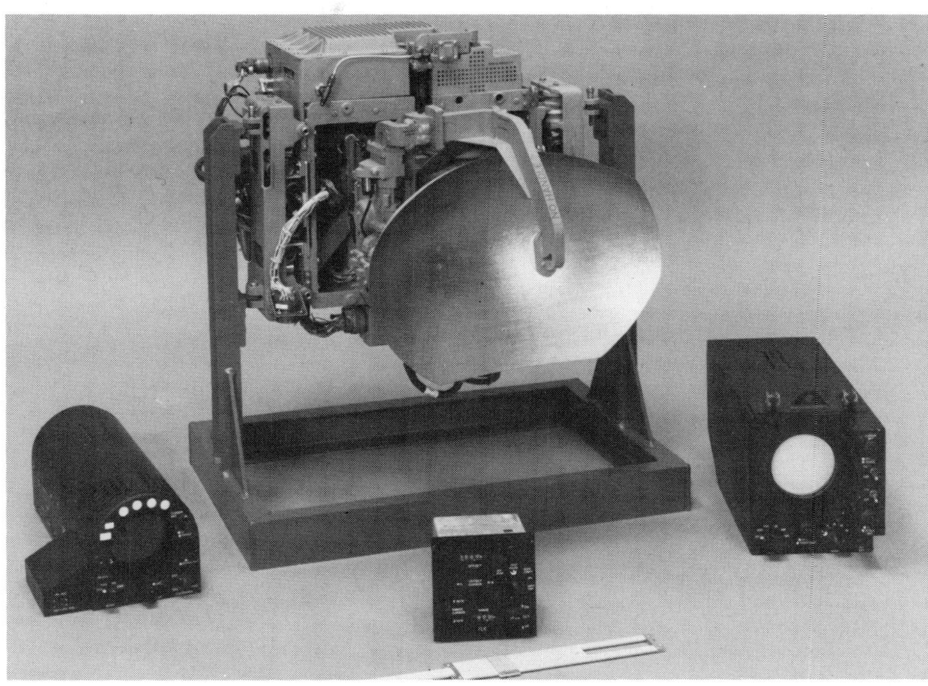
Texas instruments AN/APQ-99 radar

1966.353
AN/APQ-113 RADAR
The APQ-113 radar is a multi-mode nose radar system developed in the early 1960s for the F-111A, C, and E aircraft. The radar operates in the J-band frequency range and provides both air-to-ground and air-to-air facilities. In the air-to-ground mode the system is used for navigational position fixing, all-weather weapon delivery, and air-to-ground ranging. The air-to-air mode is compatible with the AIM-9 Sidewinder missile systems and provides automatic range search, target acquisition, and track, while angle tracking in a TWS mode.

The APQ-113 radar is the first of a family of attack radars which includes the AN/APQ-114, the AN/APQ-144, AN/APQ-161, AN/APQ-165 and AN/APQ-169.
CHARACTERISTICS
Frequency: 16 – 16.4 GHz
Tuning: Frequency agility/manual tuning
PRF: 337/674/2022 pps
Pulse width: 2.4/1.2/0.4 μs
Noise figure: 10 dB
IF: 40 MHz
Scan width: ±10 or ±45°
Scan speeds: 40 or 74°/s
Range scales: 5, 10, 30, 80, 160 nm
Dynamic range: 80 dB

AN/APQ-113/114 radar installation on the General Dynamics F-111 aircraft

STATUS
In service with the F-111A and F-111E of the US Air Force and the F-111C of the Royal Australian Air Force.
CONTRACTOR
GE Aerospace, French Road, Utica, New York 13503, USA.

1967.353
AN/APQ-114 RADAR
The AN/APQ-114 is a member of the APQ-113 family (**1966.353**) and was developed in the mid-1960s specifically for use in the FB-111A strategic bomber. In addition, some minor changes to the basic characteristics, such as pulse width and range scale, were incorporated to add the following facilities:
(1) a north-oriented display
(2) beacon mode of operation
(3) automatic photo recording.

This radar is being upgraded by USAF F/FB-111 AMP to the AN/APQ-169(V1/V2) radar.
CONTRACTOR
GE Aerospace, French Road, Utica, New York 13503, USA.

1310.353
AN/APQ-120 FIRE CONTROL RADAR
This is the latest in the APQ series of aircraft fire control radars produced by Westinghouse for the various versions of the F-4 Phantom. In 1959/60 the first of this series, the APQ-72, was delivered to the US Navy in the F-4B which remains a fundamental item in the US Fleet's interceptor and tactical aircraft inventory. Participation by the US Air Force in the Phantom aircraft programme from 1962 onwards resulted in three versions of the initial weapons' control radar – the APQ-100 in the F-4C, APQ-109 in the F-4D, and finally the APQ-120. The various versions which followed the APQ-72 were all the result of advancing performance and air-to-ground operational mode requirements additional to the original air interception functions.

The APQ-120 offers considerable improvements in weight (290 kg), volume, performance, and reliability through the use of solid-state circuitry. It is also designed to operate reliably in close proximity to the nose-gun installation of the F-4E.
STATUS
In production since 1967, nearly 2000 examples of the APQ-120 had been built when production was completed in 1980. It is in the inventory of nine nations. The APQ-120 has now been modified with a new digital computer produced by Westinghouse.
CONTRACTOR
Westinghouse Defense and Electronic Systems Center, Baltimore, Maryland 21203, USA.

1487.353

AN/APQ-122(V) AIRBORNE RADAR

The AN/APQ-122(V) is a dual frequency nose radar developed for use in the US Air Force AWADS (adverse weather aerial delivery system) programme for installation in C-130E transport aircraft. This long range navigation sensor is used for weather avoidance and navigation in supply dropping missions. The equipment provides ground mapping out to more than 200 nautical miles, weather information up to 150 nautical miles and beacon interrogation up to 240 nautical miles when using the I-band frequency radar. K-band frequencies are used when short range high resolution performance and target location are required. In the K-band role the radar provides a high-resolution ground map display to permit target identification and location for position fixing and aerial delivery missions. In this situation, the radar detects and displays a target with a radar cross section of 50 m² while operating in a rainfall environment of 4 mm per hour.

In addition to the dual frequency system designed for AWADS (designated APQ-122(V1)), three other configurations have been developed. A single frequency (I-band) radar has been designed under the designation AN/APQ-122(V5) as a direct replacement for the AN/APN-59 radar for use in C-130 and E-4B aircraft. Facilities include long range mapping, weather evaluation and avoidance, and beacon rendezvous. A navigation trainer version of the V5, the AN/APQ-122(V7), has been designed, for use in T-43A aircraft.

Another dual frequency radar, the AN/APQ-122(V8) incorporates a terrain following capability in the basic radar for use in C-130 aircraft. In addition to the normal ground mapping, weather information and beacon interrogation modes, this configuration provides terrain following, terrain avoidance and cross-scan. In the terrain following mode the radar supplies commands to fly the aircraft at a fixed distance above the terrain while in the terrain avoidance mode the radar displays all terrain at and above the altitude of the aircraft. The cross-scan mode combines the terrain following and avoidance functions on a time-shared basis. The K-band radar supplies the high resolution map display for target identification and location.

In the terrain following, terrain avoidance and cross-scan modes a smaller circular antenna is connected to the I-band receiver/transmitter. Both antennas are mounted on the forward assembly.

STATUS

The APQ-122(V) has been supplied to the US Air Force as well as Argentina, Australia, Bolivia, Cameroon, Congo, Denmark, Ecuador, Eqypt, Gabon, Greece, Indonesia, Iran, Israel, Italy, Jordan, Libya, Malaysia, Morocco, Nassau, New Zealand, Niger, Nigeria, Oman, the Philippines, Portugal, Singapore, Spain, Sudan, Thailand, Venezuela, Saudi Arabia and Zaire. Installations have been carried out in C-130H, RC-135C, RC-135A, KC-135A, RC-130A, and B747 (AABNCP) aircraft.

CONTRACTOR

Texas Instruments Inc, PO Box 660246, Dallas, Texas 75266, USA.

4334.353

AN/APQ-126(V) AIRBORNE NAVIGATION AND ATTACK RADAR

The AN/APQ-126(V) is a forward looking, variable configuration, airborne navigation and attack radar in production for the US Navy A-7E and the US Air Force A-7D aircraft. Operating frequency of the APQ-126(V) is in J-band and its primary functions are ground mapping, air-to-ground ranging and safety of flight (TF/TA). The units of this radar provide the following modes of operation:

(1) ground mapping (GMP/GMS) – pencil-shaped beam
(2) air-to-ground ranging (AGR)
(3) air-to-air boresight ranging
(4) terrain avoidance (TA)
(5) terrain following (TF)
(6) cross-scan terrain following/terrain avoidance (CSTA)
(7) cross-scan terrain following/ground map pencil (CSGMP)
(8) beacon (BCN)
(9) television (TV)
(10) radar homing and warning (RHAW).

The AN/APQ-126(V) also features adverse weather 'look-through' using selectable circular polarisation, slaved antenna pointing in air-to-air ground ranging and variable antenna tilt control which allows the system operator to optimise ground map displays and highlight targets of interest.

AN/APQ-126 airborne radar

STATUS

The AN/APQ-126(V) has been supplied to the US Navy and Air Force, as well as Greece and Portugal. A variant, known as the AN/APQ-158, has been developed for the HH-53 helicopter (**4336.353**).

CONTRACTOR

Texas Instruments Inc, PO Box 660246, Dallas, Texas 75266, USA.

1968.353

AN/APQ-144 RADAR

The APQ-144 is another in the APQ-113 family of forward radars (**1966.353**) and was developed in the late 1960s for use in the F-111F tactical bomber. The basic features of the APQ-113 and APQ-114 (**1967.353**) radars were retained with a number of performance enhancement features added. The principal among these were the addition of a 0.2 μs pulse width and a 2.5 mile (4 km) display range. It has subsequently been upgraded to AN/APQ-161 (Pave Tack).

DEVELOPMENT

Programmes for adding DMTI and K-band transmitter facilities were carried out and successfully flight tested, but these features were not incorporated in production units. A modified APQ-144 radar was produced in small numbers for the B-1 bomber prototype programme. This was given the designation AN/APQ-163 (**3682.253**).

CONTRACTOR

GE Aerospace, French Road, Utica, New York 13503, USA.

4335.353

AN/APQ-146 AIRBORNE RADAR

The AN/APQ-146 is a dual channel, forward looking, multi-mode radar developed for the F-111F. The radar's primary function is to provide proper commands to automatically control the F-111 aircraft at a pre-selected set clearance (altitude) above the terrain. This allows the F-111 to follow a low altitude contour to the earth's surface, thereby avoiding detection. Each channel of the terrain following radar (TFR) system is identical, providing system flexibility and improving overall system reliability and mission success. The units of this radar provide the following modes of operation:

(1) terrain following
(2) situation (terrain avoidance)
(3) ground map.

AN/APQ-146 airborne radar

GET THE ULTIMATE SUPPORT FROM JANE'S MILITARY LOGISTICS
If you need to compare and evaluate items of logistical equipment, you need an accurate, up-to-date reference source.
Jane's Military Logistics—formerly called Jane's Military Vehicles and Ground Support Equipment—contains the latest comparative data on over 2,200 items of equipment from around the world—including Soviet Bloc countries.
You'll find comprehensive information on specifications, capabilities, dimensions and operating methods for every non-combat vehicle in production, plus mines and mine warfare equipment, bridging systems, field fortifications, fuel storage systems, portable runways and water supplies.
Whether you are involved in procurement, manufacturing or supply, Jane's Military Logistics allows you to make detailed comparisons of all types of equipment.
You can't get better support than Jane's Military Logistics.
■ Compare and evaluate competing equipment
■ Keep up with worldwide developments
■ Identify unknown equipment
■ Monitor your competitors
The only comprehensive analysis of military vehicles and ground support equipment—anywhere.
JANE'S MILITARY LOGISTICS 1988

There are two F-111 attack radar backup modes which may be selected by equipment external to the APQ-146. One is an air-to-ground ranging mode which uses the elevation monopulse resolution improvement capability of the TFR, and is slaved in elevation to an externally generated position taken from the lead-computing optical sight, the azimuth axis being aligned with the drift angle. The other is a ground map backup mode where the TFR transmitter/receiver is routed through the attack antenna.

DEVELOPMENT

Earlier versions of the APQ-146 were the APQ-110 designed for use on the F-111A, F-111C and F-111E aircraft; the APQ-128 designed for use on the F-111D aircraft and the AQP-134 designed for use on the FB-111. Texas Instruments is currently under contract to develop a modification to increase reliability and maintainability of the F-111 airborne radars. Following modification of the AN/APQ-146, AN/APQ-110, AN/APQ-128 and AN/APQ-134 the systems will be redesignated AN/APQ-171 **(6116.353)**.

CONTRACTOR

Texas Instruments Inc, PO Box 660246, Dallas, Texas 75266, USA.

1675.353

AN/APQ-153 FIRE CONTROL RADAR

The AN/APQ-153 ASTAR (airborne search target attack radar) is a lightweight search and range tracking radar used aboard F-5E aircraft. This system operates in the I-band frequency range and provides stabilised search, automatic acquisition and target illumination, and automatic target ranging with boresight steering in the missile mode for heads-down launch of the Sidewinder AIM-9 series missiles. In the gunnery modes, the radar automatically provides range and range-rate outputs for targets within the sight lead angle computation envelope.

The system comprises five main units: antenna assembly, receiver/transmitter, radar processor, indicator, and set control. Combined weight is 110 lb (50 kg). The antenna is a parabolic 12 × 16 inch (30 × 40 cm) dish with horizontal polarisation. The indicator uses a 5-inch (127 mm) direct view storage tube and provides a 'B' type search display and a lock-on missile mode display. Built-in test is provided for radar performance verification.

OPERATION

The APQ-153 has the following principal modes of operation:

Search Mode

In search operation, a 7° elevation beamwidth is stepped up 3° at the right azimuth limit and down 3° at the left azimuth limit, providing a two-bar scan coverage of 10° in elevation. This two-bar 10° coverage can be adjusted ±45° in elevation by the pilot. Azimuth search coverage is ±45°. The search pattern is space stabilised for aircraft pitch and roll motion to allow searching a given volume of space and to prevent loss of the target and/or smearing of the display. Range of search coverage extends to 20 nautical miles (37 km).

Boresight Missile Mode

The missile mode enables the pilot to lock-on to targets out to 10 nautical miles (19 km) and provides aircraft steering information to align the acquisition envelope of the AIM-9 missile with the target. Once a target has been acquired a lock-on occurs, and azimuth and elevation steering data are provided on the indicator display in the form of a target steering bar. The aircraft is flown towards the bar to align it within the allowable AIM error circle scribed on the overlay. Missile acquisition occurs when the steering bar is within the allowable AIM error circle and the IN-RANGE indicator is illuminated.

Air-to-Air Gunnery Modes

There are two air-to-air gunnery modes in the APQ-153: dogfight, and AA1/AA2. Both are external commands to the radar. The gunnery modes are heads-up and the radar automatically provides range and range rate information to the sight. Activation of a gunnery mode causes the antenna to align to a boresight in azimuth and a 2.2° down in elevation. The range gate will automatically slew from 500 to 6000 feet at 22 000 feet per second (152 to 1830 m at 6700 m/second). Acquisition is automatic for the first target encountered.

STATUS

Produced and in operational use aboard Northrop F-5E aircraft of 14 countries.

CONTRACTOR

Emerson Electronic Company, Government and Defense Group, 8100 West Florissant Avenue, St Louis, Missouri 63136, USA.

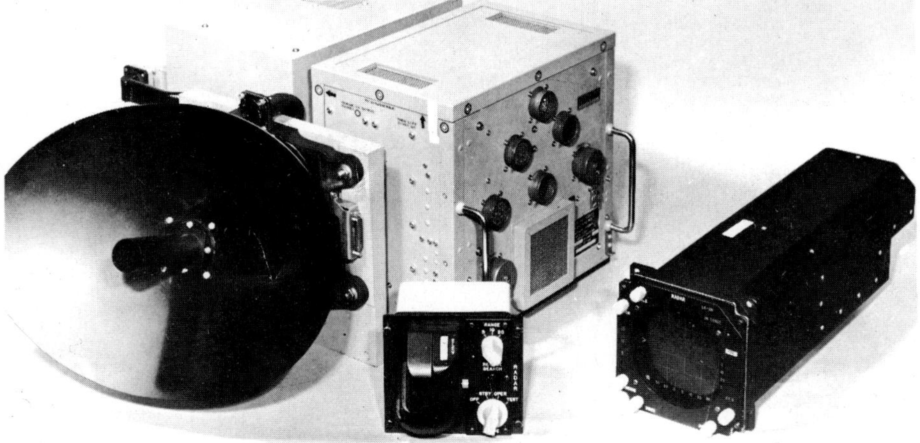

APQ-153 fire control radar

1567.353

AN/APQ-156/148 MULTI-MODE RADAR

The AN/APQ-156/148 is a J-band airborne multi-mode radar specially developed to combine the functions of two radars previously required by US Navy A-6A all-weather attack aircraft in a single radar. These two radars are the APQ-92 (search, and terrain avoidance/following) and the APQ-112 (target tracking and ranging). The APQ-156 system is a modification of the APQ-148 to accommodate the addition of a FLIR/laser target recognition attack multi-sensor system to the A-6E aircraft. Functions performed by the APQ-156/148 include:

(1) search
(2) ground mapping
(3) tracking and ranging of fixed or moving targets
(4) terrain avoidance or terrain following
(5) beacon detection and tracking.

Norden AN/APQ-156/148 multi-mode radar for A-6E aircraft

The TWS capability provides simultaneous range, azimuth, and evaluation data for weapon delivery. As in other systems, range and azimuth markers must be placed on the target, but elevation data are available on a continuous basis and are derived from a separate phase interferometer array carried below the main scanner dish. The latter has a width of about one metre and is illuminated by a conventional horn feed, to produce a very narrow beam in azimuth.

The beam has a cosecant^2 profile in elevation and this, with the interferometer elevation data provided, eliminates the need for mechanical scanning in the elevation plane. The interferometer array consists of two adjacent rows of 32 horns and moves with the main dish. Energy reflected from ground targets arrives at the upper and lower rows with a time difference which is measured by phase comparison techniques and translated into angular information.

There are two cockpit displays in the A-6E APQ-148 installation: a 5-inch (127 mm) storage tube unit for the pilot, and a 7-inch (178 mm) direct view radar indicator (DVRI) for the bombardier/navigator. In the case of the APQ-156, only the DVRI is provided. Terrain data from the radar system are also presented on a vertical display for the pilot. The system incorporates comprehensive built-in test facilities. System weight is about 500 lb (230 kg).

STATUS

In 1970 Norden received a $5.946 million contract for the delivery of 13 systems, and the first production equipment was handed over in 1971. Approximately 200 systems had been developed by the end of 1975, some of which are being used to retrofit equipment in A-6A aircraft.

In 1978 the manufacturer was awarded a $394 000 contract for the first 18 production MX-9736/APQ video processor-airborne moving target indicator (AMTI) units for A-6E aircraft. Six pre-production examples were developed in 1976. The AMTI enables the radar to detect moving surface targets in the presence of clutter.

In June 1980 Norden received a $5.2 million US Navy contract for 295 AMTI units, and in July there was a further award of $3.8 million for six complete radars for new aircraft and 56 units for updating AN/APQ-148 radars of older A-6E aircraft to AN/APQ-156 standard. In April 1981 a $15 million contract was received which included updating 39 AN/APQ-148 radars, and a number of new AN/APQ-156 systems.

In early 1985, Norden received a contract from Grumman to supply a new radar for the A-6E upgrade programme. The time scale envisaged deliveries in 1987 for flight testing, and up to 300 systems may be required for new and retrofitted US Navy aircraft. The new radar is designed to have improved resolution detection, tracking; air-to-air, ship classification, and ECCM facilities and easier maintainability.

A variant of the AN/APQ-148, known as the AN/APS-130 (**3520.353**), is in operational service with the US Navy EA-6B electronic warfare aircraft. However, in 1987, Norden delivered six AN/APQ-156C radars to the US Navy for the EA-6B fleet.

CONTRACTOR

Norden Systems, United Technologies Corporation, Norden Place, Box 5300, Norwalk 06856, Connecticut, USA.

3113.353

AN/APQ-157 FIRE CONTROL RADAR

The AN/APQ-157 is a dual control lightweight search and range tracking radar used aboard F-5F aircraft. This system operates in the I/J-band frequency range and provides stabilised search, automatic acquisition and target illumination, and automatic target ranging with boresight steering in the missile mode for heads-down launch of the Sidewinder AIM-9 series missile. In the gunnery modes, the radar automatically provides range and range-rate outputs for targets within the sight lead angle computation envelope.

The system comprises eight main units: radar processor, antenna assembly, receiver/transmitter, front indicator, rear indicator, coupler power supply, and two set controls. Combined weight is 140 lb (64 kg). The indicators use a 5-inch (127 mm) direct view storage tube and provide a 'B' type search display and a lock-on missile mode display. Video trim is provided for harmonisation of video displays. Each display can be adjusted to suit the individual pilot operational preference and will correlate information within 1°. The set controls are transferable for individual cockpit control. The antenna is a parabolic 12 × 16 inch (30 × 40 cm) dish with horizontal polarisation. Built-in test is provided for radar performance verification.

STATUS

Produced and in operational use aboard Northrop F-5F aircraft.

CONTRACTOR

Emerson Electric Company, Government and Defense Group, 8100 West Florissant Avenue, St Louis, Missouri 63136, USA.

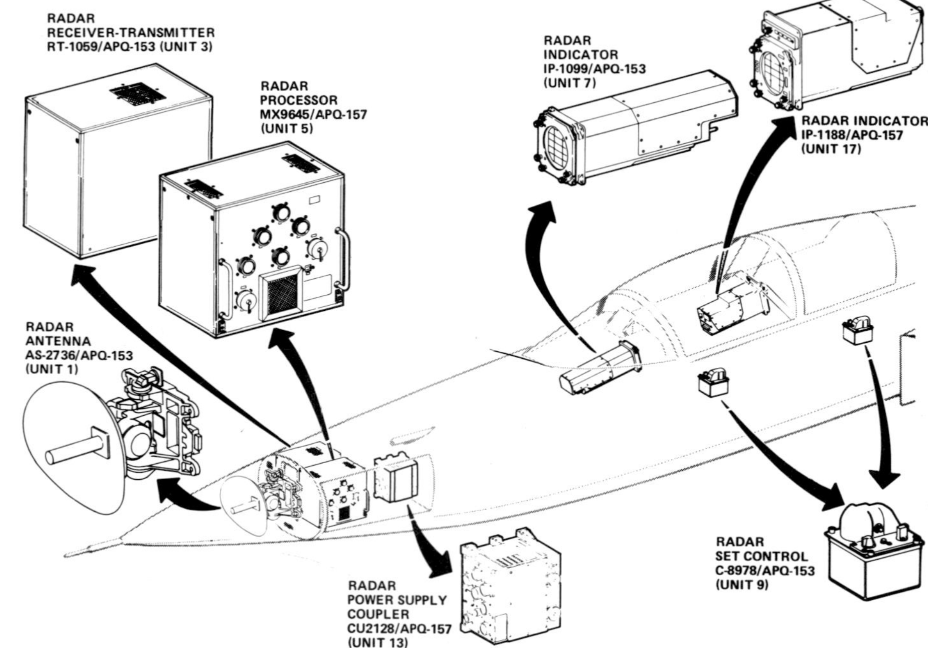

Emerson AN/APQ-157 radar and location of units in F-5F aircraft

4336.353

AN/APQ-158 AIRBORNE RADAR

The AN/APQ-158 is a multi-mode forward-looking radar used primarly for terrain following/terrain avoidance at low altitudes, in the Pavelow III Night/Adverse Weather Search and Rescue helicopter, HH-53. The equipment is similar to the AN/APQ-126 but is modified for compatibility with unique helicopter characteristics and Pavelow III mission requirements. The radar contains 15 line-replaceable units which provide the same basic ten modes of operation as detailed for the AN/APQ-126 (**4334.353**).

System upgrades provide this radar with the capability to supply updates in all modes except TF and to perform TF missions over very high clutter areas such as cities.

CONTRACTOR

Texas Instruments Inc, PO Box 660246, Dallas, Texas 75266, USA.

AN/APQ-158 airborne multi-mode radar

3114.353

AN/APQ-159 AIRBORNE RADAR

The AN/APQ-159(V)-1 is a lightweight multi-mode forward-looking I/J-band pulse radar used on the F-5E aircraft. This system incorporates a dual PRF frequency agility transmitter and provides stabilised search, automatic and manual acquisition and tracking on and off boresight, and target spotlight with automatic target ranging and tracking for air targets. In the missile mode the radar provides range, range rate and angular position from boresight, for heads-down launch of the Sidewinder AIM-9 series missiles. In the gunnery modes the radar automatically provides range and range-rate outputs for targets within the sight lead angle computation envelope. In the electro-optical (EO) mode the radar displays a normal or expanded image from the AGM-65A Maverick missile, and control is provided for the EO sensor.

The AN/APQ-159(V)-2 is a dual control radar system with characteristics the same as the (V)-1, and is used on the F-5F aircraft. In addition it provides independently selectable radar/EO presentations with transferable functions on the set controls and video indicators.

The (V)-1 system comprises six main units: antenna assembly, receiver/transmitter, radar processor, signal data converter, set control, and video indicator. Combined weight is 140 lb (64 kg). The antenna is an 11 × 19 inch (28 × 48 cm) planar array with vertical polarisation. The video indicator uses a 5-inch (127 mm) high brightness CRT and provides a radar 'B' type search display, a lock-on missile mode display, and an electro-optic TV image.

The (V)-2 system consists of the (V)-1 units plus the addition of another set control and video indicator. Combined weight is 155 lb (70 kg). Built-in test is provided in both systems for radar performance evaluation and signal data converter and video indicator fault isolation.

There are also (V)-3 and (V)-4 models which have identical performance to the (V)-1 and (V)-2 sets, and

AN/APQ-159 fire control radar

which are also configured for single-seat and twin cockpit aircraft types, respectively. The essential difference is in the display arrangements; in place of the 12 × 12 cm rectangular CRTs used for the (V)-1 and (V)-2, the (V)-3 and (V)-4 versions have one or two 8.6 cm circular direct view storage-tube (DVST) displays; and in the absence of EO data presentation facilities, the signal data converter unit is not required in the later two versions of the APQ-159. System weights for the (V)-3 and (V)-4 are 56 kg and 71 kg, respectively.

In September 1986, Emerson was awarded a contract to upgrade the F-5 Aggressor avionics with the AN/APQ-159(V)-5 fire control radar. The (V)-5 is configured the same as the (V)-3 but has two minor modifications, one in the receiver to increase its detection and acquisition capabilities by approximately 7 to 8%, and the other in the processor to increase the maximum search range from 20 to 40 nm and the maximum acquisition from 10 to 20 nm. In addition, the contract requires a minimum system MTBF of 100 hours to reduce the logistics support cost.

STATUS
Currently in production for use aboard the F-5E and F-5F aircraft.

CONTRACTOR
Emerson Electric Company, Government and Defense Group, 8100 West Florissant Avenue, St Louis, Missouri 63136, USA.

3301.353
AN/APQ-161 RADAR

The APQ-161 is an update and replacement of the APQ-144 radar (**1968.353**) for use in the Pave Tack F-111F tactical bomber. Modifications to the radar include a digital scan converter and a dual-tube display which enables the operator to display multi-sensor data simultaneously and record flight data on a VCR.

CONTRACTOR
GE Aerospace, French Road, Utica, New York 13503, USA.

4647.353
AN/APQ-164 AIRBORNE RADAR

The AN/APQ-164 is the radar developed for the US Air Force B-1B aircraft. This radar combines technology from the F-16 AN/APG-68 radar (**5881.353**) and the Electronically Agile Radar (EAR) programme of the US Air Force.

The B-1B radar generates data for navigation, penetration, weapon delivery, and for certain other functions such as air refuelling. There are really four modes in the AN/APQ-164 system that provide the navigation capability. The primary mode is a high-resolution synthetic aperture radar mapping mode, backed-up by a real beam ground mapping mode. The system also detects weather ahead and can display ground beacon returns over any radar image. The penetration functions of the radar include automatic terrain-following and terrain-avoidance. For weapon delivery the radar provides three different functions. The first is a velocity update mode, similar to a doppler navigator, which generates velocity information for the inertial navigation system. Secondly, there is a ground moving-target detection and tracking capability. There is also a high-altitude altimeter function that provides a very accurate measure of height above the ground.

The synthetic aperture mode provides the operator a high-resolution image of an area of ground that can be chosen by the inertial navigation system or the operator. Long-range maps can be made and five different map scales displayed. The synthetic aperture mapping mode accepts the coordinates of a waypoint from the inertial navigation system and makes a map centred on that point. To make an image the antenna is electronically scanned out in the direction the waypoint should be located. The radar transmits a burst of energy, gathers data for the image, and then switches itself off. At the same time, the image is stored in the radar and presented on the display in a rectangular, ground coordinate display.

The radar provides the basic data required for automatic terrain-following. It scans the ground in front of the aircraft and measures the terrain in a range versus altitude profile out to 10 nm and stores that data in the computer. The profile data is sent across the multiplex bus to the terrain-following control unit where the data is used to generate climb-dive commands. This flight profile is then automatically fed into the pilot's flight control system. Since the radar is not continuously scanning in terrain-following, a very low update is used helping to reduce the risk of detection. This rate is variable and depends on aircraft altitude, terrain roughness and aircraft groundspeed. Under normal conditions updates are made at 3 to 4 seconds intervals. However if the terrain demands it, data can be gathered continuously.

The AN/APQ-164 in the B-1B is a dual redundant system, with two complete and independent sets of line-replaceable units, except for the phased array antenna. This was the first airborne application of this technology for combat aircraft. Only one channel is used at a time, the other being maintained on standby.

The phased array is an outgrowth of the antenna developed on the EAR programme. It contains 1526 phase control modules and allows virtually instantaneous beam movement to any point in the antenna field of regard. When the radar computer commands the desired pointing angle, the antenna is physically movable to three different positions on a roll detent mount. The radar can, therefore, look off to either side of the aircraft or forward by rolling the antenna about an axis. The normal antenna position is looking forward. However when the antenna is rolled to one side, the field of view extends from the aircraft nose back to about 105°, permitting a look-off to the side of interest without having to change aircraft heading. Once physically moved to one of the three available positions, the antenna is locked into a detent. From the fixed spot, it can be scanned electronically ±60° in azimuth and elevation by means of a unit on the antenna called the beam steering control, which controls all 1526 phase modules.

CHARACTERISTICS
Frequency band: X-band
Transmitter: Gridded, multiple, peak-power TWT (identical to the AN/APG-68 transmitter)
Antenna: Phased-array electronically scanned, 1118 × 559 mm
Operating modes: (air-to-ground) high resolution mapping, real beam mapping, automatic terrain-following, manual terrain-avoidance, velocity update, ground moving target detection and track, high altitude calibrate, ground beacon; (air-to-air) weather mapping, air-to-air beacon, rendezvous mode. Growth for full conventional stand-off capability and a full air-to-air mode complement is provided.
Weight: 570 kg, including both channels

STATUS
In production for US Air Force B-1B bomber. 100 of these aircraft are due to be acquired, the first of which was delivered in June 1985. Flight trials using a BAe. One-Eleven airliner were carried out from July 1984 to December 1985. This aircraft is still being used to evaluate operational improvements to the radar system.

CONTRACTOR
Westinghouse Electric Corporation, Defense and Electronics Center, Aerospace Divisions, Friendship Site, Box 746, Baltimore, Maryland 21203, USA.

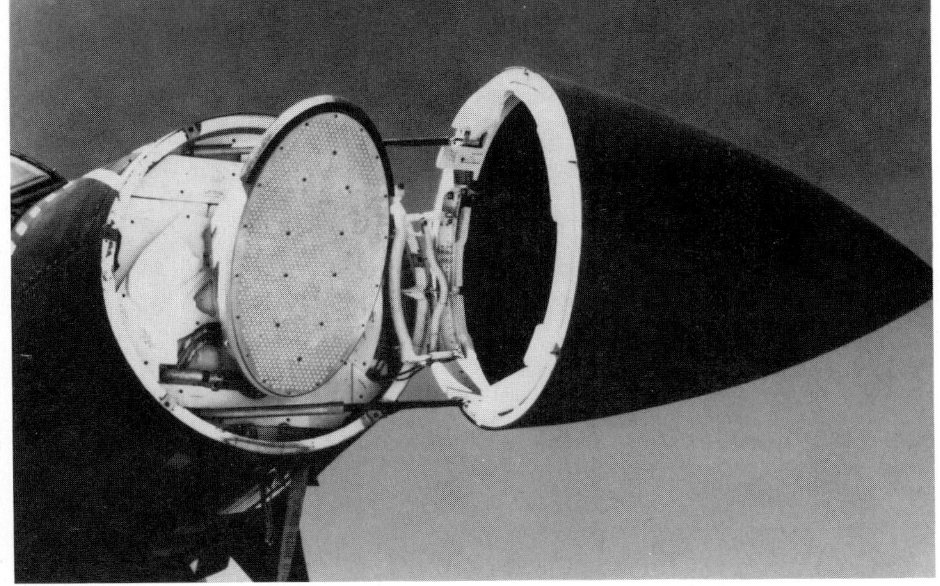

AN/APQ-164 airborne radar

6897.353
AN/APQ-165 AIRBORNE RADAR

The AN/APQ-165 is an update of the APQ-113 radar used by the RAAF in the Pave Tack F-111C tactical bomber and is a variant of the APQ-161. Modifications to the radar include a digital scan converter and a dual-tube display which enables the operator to display multi-sensor data simultaneously and record flight data in real-time on a VCR.

CONTRACTOR
GE Aerospace, French Road, Utica, New York 13503, USA.

6441.353

AN/APQ-168 AIRBORNE RADAR

The AN/APQ-168 is a multi-mode radar under development for the US Joint Services Advanced Vertical Lift Aircraft (JVX), or Bell/Boeing V-22 Osprey. Original known as the MMR, the radar has now been designated AN/APQ-168 and is considered to be the key avionics system component. The system is a derivative of the Texas Instruments LANTIRN terrain following radar, and will provide both terrain following and terrain avoidance facilities during low-level, adverse weather and hostile threat operations.

STATUS

The system is in development. Flight trials in a HH-60 helicopter were scheduled to commence in early 1988 and in a V-22 Osprey in 1989.

CONTRACTOR

Texas Instruments Inc, PO Box 660246, Dallas, Texas 75266, USA.

6116.353

AN/APQ-171 AIRBORNE RADAR

The AN/APQ-171 is the designation of the modified AN/APQ-146 series of airborne radar systems (**4335.353**). Following modifications to the AN/ALQ-146, AN/ALQ-110, AN/ALQ-128 and AN/ALQ-134 systems for the F-111 aircraft the equipments will be redesignated AN/APQ-171. These modifications have been incorporated to improve reliability and maintainability on the F-111.

CONTRACTOR

Texas Instruments Inc, PO Box 660246, Dallas, Texas 75266, USA.

6117.353

AN/APQ-172 AIRBORNE RADAR

The AN/APQ-172 is the latest designation for the airborne radar fitted on-board the RF-4 version of the Phantom aircraft. Following modifications to the AN/ALQ-99 system to improve reliability and maintainability, the radar will be redesignated to AN/APQ-172.

STATUS

In operational service.

CONTRACTOR

Texas Instruments Inc, PO Box 660246, Dallas, Texas 75266, USA.

6898.353

AN/APQ-174 MULTI-FUNCTION RADAR

The AN/APQ-174 is another derivative of the LANTIRN TFR and maintains commonality with five of the six LANTIRN line replaceable units. It is intended for combat rescue and special operations in aircraft such as the HH/MH-60, CH/MH-47, HH-53 and the V-22. It provides terrain following, terrain avoidance, ground mapping and air-to-ground ranging.

CONTRACTOR

Texas Instruments Inc, PO Box 660246, Dallas, Texas 75266, USA.

1484.353

AN/APS-94F SIDE-LOOKING RADAR

The AN/APS-94F is a side-looking airborne radar carried by US Army OV-1 Mohawk aircraft and used for battlefield and forward area reconnaissance and surveillance. The radar also has survey and mapping applications. The main difference between the AN/APS-94F and earlier versions is the inclusion of an ECCM capability.

The aircraft installation comprises five main sub-assemblies:

(1) antenna
(2) receiver/transmitter
(3) power supply/mount
(4) signal processor
(5) interconnecting box
(6) cockpit complex.

AN/APS-94F side-looking radar mounted on a US Army OV-1 Mohawk

The antenna unit consists of a pod containing two slotted waveguide arrays mounted back-to-back on a gyro-stabilised assembly that pivots at the centre. In flight, the antenna arrays are yaw stabilised to preserve the quality of the radar picture.

The receiver/transmitter is tunable within a portion of the radar band. It is located in the Mohawk equipment bay, with the signal processor and the interconnecting box. The antenna and other units can be simply removed and replaced on the aircraft to permit its use for alternative duties at short notice. Apart from high-power microwave devices, and certain display components, the APS-94F is of solid-state design.

The cockpit complex consists of the units necessary for operator control of the radar, recording and presentation of the data gathered. Two radar area maps are available to the operator. One depicts the entire sensed area, including all detected ground targets as though they are stationary; this is called the fixed target map. The other radar map shows detected moving targets and displays them against a suppressed background map of the area resulting in a moving target map. The radar imagery is developed in the air using a dry silver film process.

An area up to 100 km on each side of the aircraft can be mapped when both of the antenna arrays are in use. Either one can be selected if it is desired to map only one side. A range control determines the width of the target area to be mapped and presented on the photo-radar map. This control has three settings corresponding to 25, 50, and 100 km wide scans by each antenna. When used in conjunction with the antenna switch to select either left, right, or both arrays, maps corresponding to four standard scales can be presented on the display: 1:250 000, 1:500 000, 1:1 million and 1:2 million. Provisions are made for the insertion of fiducial markers and other data.

DEVELOPMENT

The AN/APS-94F was developed under contract to the US Army Electronics Research and Development Command.

STATUS

The AN/APS-94F is in service on US Army OV-1D Mohawk reconnaissance aircraft. The AN/APS-94D has also been fitted in an Argus of the Canadian Forces and has been installed in B-26 and P-3 aircraft, and the UH-1 helicopter.

CONTRACTOR

Motorola Inc, Government Electronics Group, Radar Systems Division, 2100 East Elliott Road, Tempe, Arizona 85252, USA.

1969.353

AN/APS-115 AIRBORNE RADAR

The AN/APS-115 is one of the Texas Instruments family of airborne search radars, and is an I-band frequency-agile system of modular design. It is employed principally for ASW and maritime roles. The AN/APS-115 is a dual system to provide 360° coverage for the P-3C 'Orion' land-based ASW aircraft. One antenna is mounted in the nose and the other in the rear. In addition to the underslung, stabilised antenna assemblies, the equipment includes dual receiver/transmitters, an antenna position programmer, dual radar set controls and a common antenna control unit.

STATUS

In production and in operational service.

CONTRACTOR

Texas Instruments Inc, PO Box 660246, Dallas, Texas 75266, USA.

1485.353

AN/APS-116 AIRBORNE RADAR

The AN/APS-116 is the ASW search radar currently in use in the US Navy's fleet of S-3A 'Viking' carrier borne ASW aircraft. It was specifically designed for detecting periscope-type targets of limited exposure time in high sea states. Its suitability in meeting the design objectives and requirements has been demonstrated in sea trials and fleet operations. It operates in the I-band with the scanner housed in a nose radome. Principal features include high transmitted energy (obtained from a high-power TWT driving a cross-field amplifier), low-noise receiver, linear FM pulse compression, rapid-scan antenna, scan-to-scan integration for sea clutter decorrelation, and multiple display formats (PPI, B scan, scan converter, raw) available to the sensor stations.

The system is integrated with the aircraft's ASW data processing and display system, and in addition to periscope and other small target detection, it provides mapping, navigation, and general surface vessel detection.

STATUS

The AN/APS-116 is currently being upgraded to the AN/APS-137(V)1 as part of the weapon system improvement programme to upgrade the 'Viking' to the S-3B configuration. See entry **6436.353**.

CONTRACTOR

Texas Instruments Inc, PO Box 660246, Dallas, Texas 75266, USA.

3518.353

AN/APS-124 AIRBORNE RADAR

The AN/APS-124 search radar was designed and developed specifically for use in the US Navy SH-60B Seahawk helicopter (also known as LAMPS Mk III). It is specifically designed for helicopter installation featuring a low-profile antenna and radome. Optimum detection of surface targets in high sea states is accomplished by several unique features including fast scan antenna and interface with the auxiliary OU/103 signal data converter (digital scan converter) to achieve scan-to-scan integration.

The system is integrated with a multipurpose display and with the LAMPS data link for display of airborne radar video on board LAMPS equipped ships.

STATUS

The equipment is currently in main production and operational service to support the LAMPS Mk III system production schedule. It is likely to be supplied to a number of other countries as part of an SH-60B export programme.

CONTRACTOR

Texas Instruments Inc, PO Box 660246, Dallas, Texas 75266, USA.

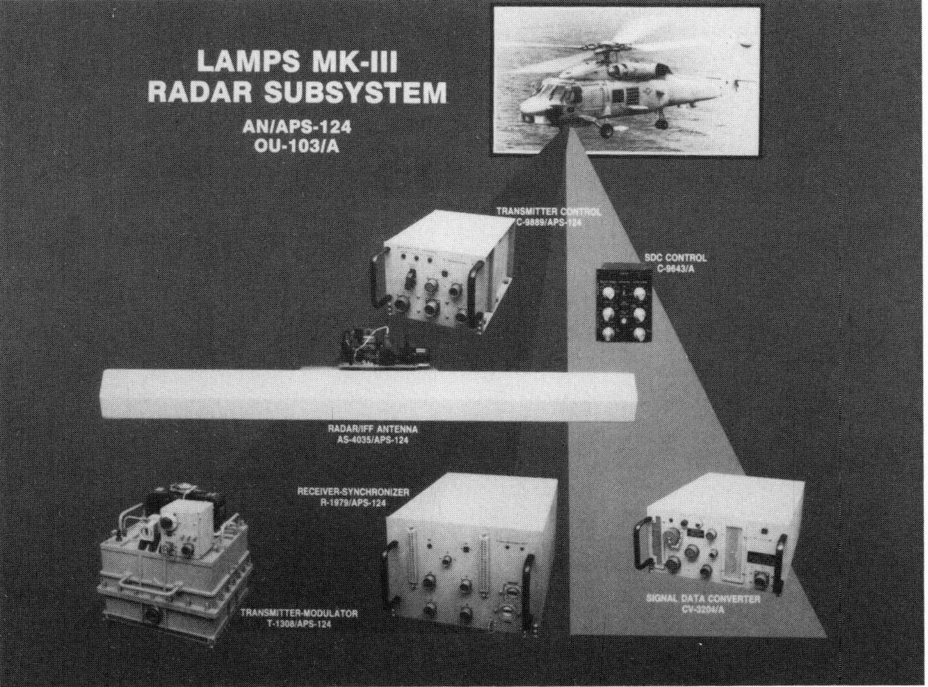

AN/APS-124 search radar equipment for US Navy LAMPS helicopter

3519.353

AN/APS-127 AIRBORNE RADAR

The AN/APS-127 search radar is produced for the US Coast Guard for use in the Dassault-Breguet HU25A Gardian aircraft. The system was developed to support search and rescue missions as well as for enforcement of laws and treaties. A unique feature that characterises the AN/APS-127 is the stabilised, lightweight, high efficiency fast scanning antenna. Scan-to-scan integration is accomplished by interface with separate direct-view storage-tube displays for both the pilot and the surveillance system operator.

STATUS

The equipment is in operational service.

CONTRACTOR

Texas Instruments Inc, PO Box 660246, Dallas, Texas 75266, USA.

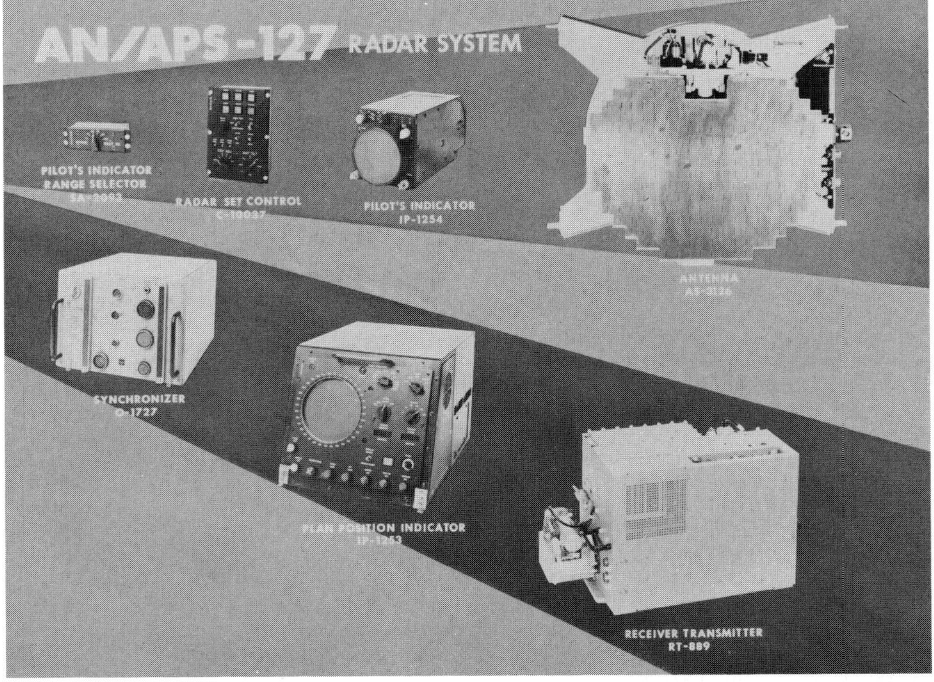

AN/APS-127 radar equipment

3132.353

AN/APS-128 MODEL D SURVEILLANCE RADAR

The AN/APS-128 Model D surveillance radar is an all-digital equipment using a scan converter to present data in TV raster format. It contains target enhancement and clutter reduction circuitry consisting of frequency agility, sensitivity time control, CFAR and scan-to-scan integration.

The system consists of a rectangular flat-plate antenna and pedestal, a transmitter/receiver, a radar control unit, a digital scan converter, a trackball or joystick cursor and a bright display. Dual display for cockpit weather presentation is available as well as add-on cabin displays. The system uses fully programmable microprocessor operational features with alphanumeric and graphic options to meet varied mission requirements. A number of antenna array sizes are available to suit aircraft and radome configurations. An alternative parabolic antenna features a dual-polarisation capability in one moulded dish to provide pencil beam for sea search and shaped beam for mapping.

Target detection of the system is given as:

(1) 30 nautical miles on a snorkel or fishing vessel (assumed six metres long), 10 m² cross-section, in sea state 3
(2) 60 nautical miles on a trawler (160 metres long), 150 m² cross-section, in sea state 5
(3) 100 nautical miles on a freighter (360 metres long), 500 m² cross-section, in sea state 5
(4) 120 nautical miles on a tanker (600 metres long), 1000 m² cross-section, in sea state 5.

The system also functions as a weather radar with a range of approximately 200 nautical miles.

CHARACTERISTICS

Frequency: 9375 MHz
Frequency agility: 85 MHz peak/peak
Power output: 100 kW peak
Pulse width: 2.4 and 0.5 μs (0.1 μs pulse compression system optional)
PRF: 400, 1200, 1600 Hz
Noise figure: GAS-FET 3.5 dB low noise RF amplifier
Pulse compression: Retrofit transmitter/receiver to provide linear chirp capability for pulse width of 0.1μs. Provides detection of 1 – 5 m² targets in heavy sea states

AN/APS-128 Model D airborne surveillance radar

Antenna rotation rate: 15 or 60 rpm continuous. 15 rpm sector scan from 30 – 360° selectable
Bright display: High ambient light, 875 line high resolution
System weight: 91.8 kg (202 lbs) with 8 inch display

STATUS

Over 90 AN/APS-128 systems have been sold worldwide and the radar is operational with the Brazilian Air Force, Chilean Navy, Gabon Air Force, Indonesian Air Force, Japanese Maritime Safety Agency, NASA, Royal Malaysian Air Force, Spanish Air Force, Uruguayan Air Force, Uruguayan Navy and the US Customs Service.

CONTRACTOR

Eaton Corporation, Command Systems Division, 815 Broad Hollow Road, Farmingdale, Long Island, New York 11735, USA.

4612.353

DIGITAL TACTICAL SYSTEM (DITACS) RADAR

The Digital Tactical System (DITACS) radar is a full AN/APS-128 Model D surveillance radar with added tactical sensor capability for enhanced multi-mission applications.

Tactical navigation and sensor plot data are displayed alternately or in combination on a bright display, together with radar target data. Control and management data are stored, recalled and displayed in operator-oriented formats. Both display and radar are under microprocessor control providing bright presentation with alphanumeric designations, tailored graphic formats and multi-sensor overlays.

Keyset functional firmware is based on specialised mission operational sequences. DITACS requires only the addition of a keyset control panel in hardware, the remaining features being added in software ROM and RAM programming capability. This provides a stand-alone tactical control and display system.

STATUS

DITACS is installed on the Beech multi-mission 200T aircraft and is under development for the Grumman S-2E Tracker aircraft as well as several helicopter installations.

Digital Tactical System (DITACS) radar

CONTRACTOR

Eaton Corporation, Command Systems Division, 815 Broad Hollow Road, Farmingdale, Long Island, New York 11735, USA.

3520.353

AN/APS-130 AIRBORNE RADAR

The AN/APS-130 radar is in service on the US Navy EA-6B electronic warfare aircraft. It is a variant of the Norden Systems AN/APQ-148 equipment (**1567.353**), which is used by the US Navy A-6E Intruder all-weather attack aircraft. However, since the EA-6B carries no offensive weapons, the APS-130 uses only the navigational features of the APQ-148. It replaces an older radar system, also supplied by Norden Systems, which was used on the EA-6B.

STATUS

In operational service with the US Navy.

CONTRACTOR

Norden Systems, United Technologies Corporation, Norden Place, Box 5300, Norwalk 06856, Connecticut, USA.

5883.353

AN/APS-131 SIDE-LOOKING RADAR

The AN/APS-131 is a side-looking airborne radar (SLAR) used for the detection of ships and boats, search and rescue, and the detection of oil pollution. The aircraft installation consists of six main sub-assemblies; antenna, receiver/transmitter, synchroniser, amplifier, recorder-processor-viewer and a control unit.

The antenna unit consists of a pod containing two eight-foot (2.44 metre) yaw stabilised, vertically polarised, slotted waveguide arrays mounted back-to-back on a gyro-stabilised assembly that pivots at the centre. The remaining equipment is mounted inside the cabin. The synchroniser provides radar timing and control functions and, based on inputs from the aircraft inertial navigation system, creates latitude and longitude lines for display on the radar imagery. The recorder-processor-viewer contains the components necessary to create the film imagery. Video data is impressed on the film by applying it to two CRTs as the film is pulled across the CRT faces. Typically, each CRT generates an image of the earth's surface on either side of the aircraft. The dry silver film is developed in near real-time.

An area of up to 108 nautical miles on either side of the aircraft can be mapped when both arrays are in use and either can be selected if mapping of only one side is required. A range control determines the width of the target area to be mapped and presented on the photo-radar map. This control has four settings corresponding to 13.5, 27, 54 and 108 nautical mile wide scans by each antenna. When used in conjunction with the antenna switch to select either left, right or both arrays, maps corresponding to four standard scales can be presented on the display: 1:250 000, 1:500 000, 1:1 000 000 and 1:2 000 000. Radar and aircraft operational data is annotated on the film. This data, together with the latitude and longitude printed on the film, helps the measurement of map co-ordinates for any feature observed on the radar image.

AN/APS-131 installed in a US Coast Guard HU-25A aircraft

STATUS

The AN/APS-131 was developed under contract to the US Coast Guard and is in service on HU-25A aircraft. A similar radar, the AN/APS-135 (**5884.353**) is used on the US Coast Guard HC-130s.

CONTRACTOR

Motorola Inc, Government Electronics Group, Radar Systems Division, 2100 East Elliot Road, Tempe, Arizona 85252, USA.

4337.353

AN/APS-133 AIRBORNE RADAR

The AN/APS-133(V) multi-mode airborne radar is designed to provide high performance weather detection, precision terrain mapping and beacon homing for a variety of military aircraft, including tankers and transports.

The complete system consists of four units: receiver/transmitter, display unit, control panel and a fully stabilised antenna. The colour display presents levels of precipitation in red, yellow and green (weather mode) and red, yellow and blue (map mode). The display may also be used in conjunction with an auxiliary interface unit to display various types of information from other on-board sensors, eg: INS, Omega, IKK, SKE and FSAS. These can be superimposed on the weather or terrain map, or can be displayed in various combinations.

CHARACTERISTICS

Frequencies: 9376 $\pm$ 5 MHz – transmitter/receiver (weather/map); 9310 $\pm$ 5 MHz – receiver (beacon).
Power output: 65 kW
Range: 300 nm
Antenna: 30-inch (76 cm) parabola with 2.9° beam and CSC² Cos fan beam.
Weight: 118 lb (53.6 kg).

DEVELOPMENT

The AN/APS-133(V) was derived from the commercial RDR-1F radar which is in widespread use in wide-body aircraft. In addition to many other applications, it has been fitted to both the C-141 and KC-10 aircraft, and an improved land mapping version is being fitted to the US Marine Corps KC-130T tankers and the US Air Force E-4B National Emergency Airborne Command Post (NEACP).

STATUS

In operational service.

CONTRACTOR

Allied-Signal Inc, Bendix Avionics Division, PO Box 9327, Fort Lauderdale, Florida 33310, USA.

4338.353

AN/APS-134(V) MARITIME SURVEILLANCE RADAR

The AN/APS-134(V) ASW and maritime surveillance radar is the successor to the AN/APS-116 system (**1485.353**) used in the US Navy for airborne surveillance. It is designed specifically to detect periscopes under high sea conditions and incorporates all the features of the previous generation with improved performance and a unique maritime surveillance mode. The principal features include a fast scan antenna and associated digital signal processing which form an effective means of eliminating sea clutter. Two other techniques, pulse compression and scan-to-scan processing, are used to combat the inherent problems of detecting small targets in the sea clutter environment.

Three modes of operation are available:

(1) periscope target detection in sea clutter using fast scan (150 rpm), high PRF (2000 pps) and high resolution with display ranges selectable up to 32 nautical miles

(2) maritime surveillance out to 150 nautical miles with high resolution, low PRF (500 pps) and medium scan speed (40 rpm)

(3) long range search and navigation out to 150 nautical miles with medium resolution, low PRF and slow scan (6 rpm).

Long range performance is provided by a high power transmitter (500 kW peak power) coupled with a high antenna gain. A radar control/display gives video presentation on a ten inch square CRT, with the necessary controls and indicators.

CHARACTERISTICS

Transmitter frequency: 9.5 – 10 GHz (modes A & B) 9.6 – 9.9 GHz (mode C)
Power output: 500 kW, peak, average 500 W
Antenna beamwidth: 2.4° azimuth, 4° elevation
Polarisation: Vertical

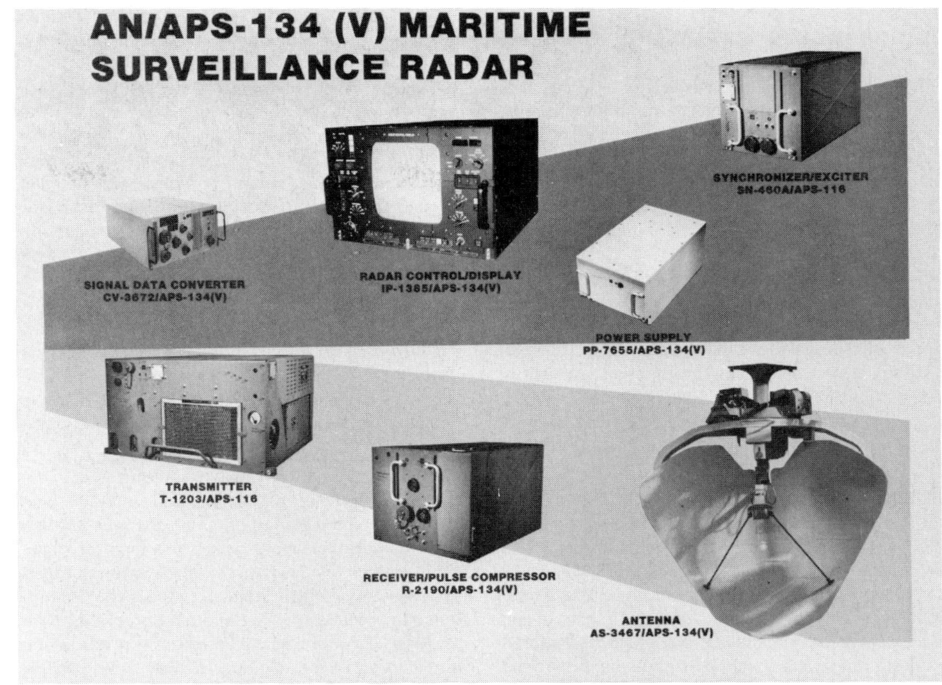

Elements of the AN/APS-134(V) radar installed in the Federal German Atlantic maritime aircraft

STATUS

The AN/APS-134(V) is in production to meet several national requirements, including the Federal German Atlantic maritime patrol aircraft and the Royal New Zealand Air Force P-3Bs. The system is in full operational service.

CONTRACTOR

Texas Instruments Inc, PO Box 660246, Dallas, Texas 75266, USA.

5884.353

AN/APS-135 SIDE-LOOKING RADAR

The AN/APS-135 is a side-looking airborne radar (SLAR) used for the detection of ships and boats, search and rescue, and ice patrol in the North Atlantic. Apart from the position of the antennas, the system is virtually identical to the AN/APS-131 (**5883.353**). The aircraft installation consists of seven main sub-assemblies: antennas, antenna switching unit, receiver/transmitter, synchroniser, amplifier, recorder-processor-viewer and a control unit.

The antenna unit consists of two pods, mounted either side of the aircraft to provide an unobstructed view, each containing a sixteen foot (4.88 metre) horizontally polarised slotted waveguide array. In flight the arrays are yaw stabilised to preserve the radar picture quality. The remaining equipment is mounted on a pallet inside the aircraft for easy removal. The synchroniser provides radar timing and control functions and, based on inputs from the aircraft inertial navigation system, creates latitude and longitude lines for display on the radar imagery. The receiver/transmitter contains the magnetron and low noise receiver. The antenna switching unit directs the radiated power to the left or right antenna.

AN/APS-135 installed in a US Coast Guard HC-130 aircraft showing the left side antenna

The process for creation of the film imagery and the mapping areas covered is identical to the AN/APS-131 system.

STATUS

The AN/APS-135 has been developed under contract to the US Coast Guard and is currently in service on HC-130 aircraft. The AN/APS-131 is used on the US Coast Guard HU-25A aircraft.

CONTRACTOR

Motorola Inc, Government Electronics Group, Radar Operations, 2100 East Elliot Road, Tempe, Arizona 85252, USA.

6436.353

AN/APS-137(V) AIRBORNE RADAR

The AN/APS-137(V) is an upgraded version of the AN/APS-116 (**1485.353**) as part of the weapon system improvement programme to upgrade the S-3A Viking carrier-based aircraft to S-3B configuration. The most significant of the radar improvements is the addition of inverse synthetic aperture radar imaging to provide long-range ship classification capabilities. The AN/APS-137(V) is also to be fitted to the US Navy P-3C aircraft.

STATUS

In early 1987, Texas Instruments received contracts totalling over $80 million for production of the new radar for fitment to the US Navy S-3B and P-3C aircraft, and US Coast Guard C-130s. A further contract for $28 million was awarded in November 1987 and another for $21 million in April 1988.

CONTRACTOR

Texas Instruments Inc, PO Box 660246, Dallas, Texas 75266, USA.

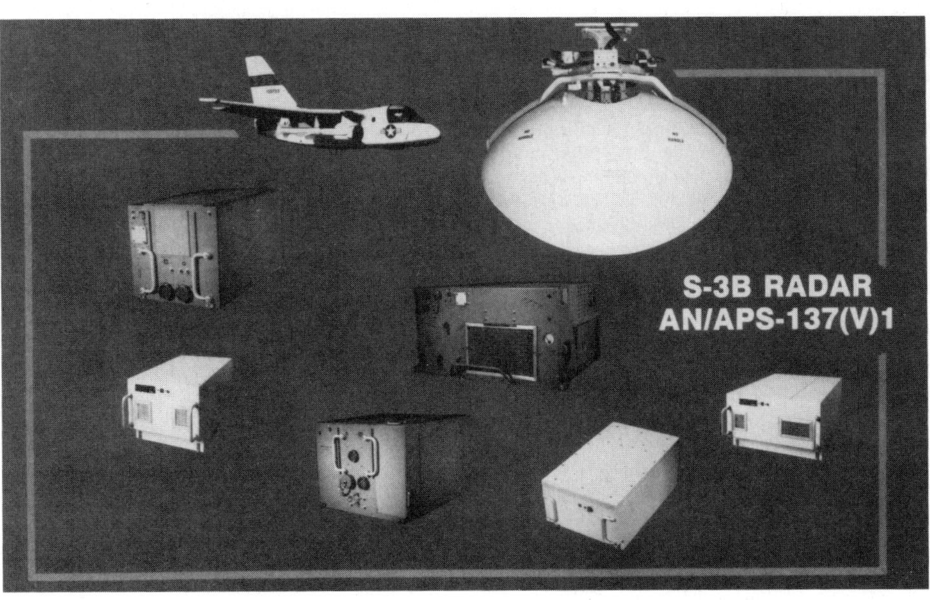

AN/APS-137(V)1 radar system

5885.353

SLAMMR SIDE-LOOKING RADAR

SLAMMR (Side-Looking Airborne Multi-Mission Radar) is a versatile long-range surveillance radar intended for use in maritime patrol, border surveillance and mapping. It can also be used for ice patrol missions, oil pollution detection, search and rescue and fishing area patrol. It is a member of the same SLAR family as the AN/APS-131 (**5883.353**) and the AN/APS-135 (**5884.353**) and is identical in many respects.

The aircraft installation consists of eight main sub-assemblies; antennas, antenna switching unit, receiver/transmitter, synchroniser, amplifier, recorder-processor-viewer, control unit and an optional moving target processor. The antenna consists of two yaw stabilised, horizontally polarised, slotted waveguide arrays. These can be mounted back-to-back within a single pod, or individually mounted on either side of the aircraft to provide an unobstructed view.

The remaining units are either pallet or rack mounted inside the aircraft for easy removal. The synchroniser provides radar timing and control functions and, based on inputs from the aircraft inertial navigation system, creates latitude and longitude lines for display on the radar imagery. The receiver/transmitter contains the magnetron transmitter and the low noise receiver. The antenna switching unit directs the radiated power and received signal to either the left or right antenna.

The process for the creation of the film imagery, and the mapping areas covered, are identical to those of the AN/APS-131 and 135 systems. A real-time CRT colour display is available which stores and displays the data. The data can also be recorded for future use as well as transmitted to ground stations, in real-time, via a radio link. The moving target processor is an MTI option that is available for the detection of moving targets at long-range for border surveillance. By the use of this processor, surveillance can be carried out at ranges up to 80 nautical miles. A high quality fixed target map is also available for radar mapping and geological exploration. Current upgrades include a colour CRT display and high resolution digital recorder.

STATUS

SLAMMR is in service on C-130 and Boeing 737 aircraft.

CONTRACTOR

Motorola Inc, Government Electronics Group, Radar Systems Division, 2100 East Elliot Road, Tempe, Arizona 85252, USA.

1965.353

AN/APS-138/125 AIRBORNE RADARS

ARPS (Advanced Radar Processing Sub-system), is an improvement to the AN/APS-120 (**1783.353**) which combines increased sensitivity in noise and clutter conditions with sophisticated false alarm control and significant ECCM advances to provide the E-2 aircraft early warning system with a truly automatic overland capability. It also provides a digital airborne MTI in place of the analogue MTI of the AN/APS-120. Another capability is the suppression of sidelobe jamming. Implementation of ARPS into the APS-120 occasioned the US Navy nomenclature of APS-125 to denote the improved model. Two sets of ARPS modified APS-120 radars existed in 1974 and, after laboratory testing, underwent flight test at Grumman. Testing continued until early 1978 when US Navy requirements were met completely.

STATUS

A retrofit programme for existing APS-120 radars to bring them up to APS-125 standards started in 1977 and was completed in 1984. Production of the APS-125 itself was completed in 1982 and APS-138 was initiated in 1983. Radar circuitry changes for the APS-138 were relatively minor, the main changes being addition of a new, low sidelobe antenna called the TRAC-A (Total Radiation Aperture Control-Antenna) produced by the Randtron Corporation. The last APS-138 radar was delivered early in 1987 and a retrofit programme to upgrade all APS-125s to APS-138s is continuing. The APS-139 (**6438.353**), another relatively minor update, has finished US Navy flight evaluation and production deliveries started in the latter half of 1987 and will continue until 1989 when APS-145 production is due to commence. APS-139 modifications will not be retrofitted into the APS-138 equipped E-2Cs but total upgrades to the APS-145 configuration are planned. The APS-145, currently in flight test, will be produced until the mid-1990s when another major modification, with resultant new nomenclature, is predicted. It will address low observable target detection and further ECCM protection. Basic E-2C radar production is expected to continue for the US Navy until the year 2000 and the E-2C is accordingly expected to fly well beyond that time before a replacement aircraft takes over.

E-2C Hawkeye aircraft equipped with AN/APS-125 airborne early warning radar

Four aircraft fitted with APS-125 are being operated by Israel, eight have been bought by Japan and up to five more are being considered. Egypt has procured five and Singapore four. Pakistan may also decide to buy E-2Cs with the APS-138 and there is interest by other countries. A US Navy furnished APS-125 is also being fitted to a P-3 aircraft by Lockheed under contract to the US Customs Service for a drug enforcement mission. The Customs Service plans to procure several more and the P-3 AEW configuration involving various E-2C radars is being proposed by Lockheed to the international market.

The US Navy recently awarded Westinghouse-Airship Industries company a five year Airship Operational Development Model (ADM) programme which will use initially a modified E-2C avionics suite with an APS-138 or APS-139 radar with an enhanced aperture (10 × 40 ft) antenna for operational testing with the new fleet. A modified version of the E-2C radar, or an entirely new system could be used in a final airship configuration.

GE has embarked on an Advanced Phase Array Radar (APAR) technology programme (**5609.353**) with Grumman. It is anticipated that technology from this basic programme will variously support new radar projects such as the next major radar update to the E-2C system, the USAF Atmospheric Surveillance Technology/Airborne Radar (AST/AR) programme for which the GE/Grumman team is under contract and further into the future a new radar for the AEW version of the US Navy Advanced Tactical Support Aircraft (ATSA), the projected replacement for the E-2C.

CONTRACTOR

GE Aerospace, French Road, Utica, New York 13503, USA.

6438.353

AN/APS-139 AIRBORNE RADAR

The AN/APS-139 is a development of the AN/APS-138 (**1965.353**) E-2C radar, and is being evaluated by the US Navy. It is intended to improve on the ECCM capability and surface surveillance and detection of the AN/APS-138.

STATUS

US Navy evaluation is completed and production deliveries have commenced. Expected to enter service in 1988. See AN/APS-145 under entry **6437.353**.

CONTRACTOR

GE Aerospace, French Road, Utica, New York 13503, USA.

6437.353

AN/APS-145 AEW RADAR

The AN/APS-145 is an advanced version of the AN/APS-138 system (**1965.353**) which is the prime sensor on board the Grumman E-2C AEW aircraft. The equipment incorporates all the modifications developed for the AN/APS-138 and AN/APS-139 radars and will be produced and retrofitted to all US Navy E-2Cs from 1990. The development specifically addresses the problem of overland clutter since the early system is reported to perform well over sea and desert but performance rapidly degrades as the terrain becomes more rugged. To reduce the false alarm rate a feature, known as environmental processing, is being developed which adjusts sensitivity of the radar cell by cell, according to the clutter and traffic in each cell.

To enable large aircraft to be detected at ranges out to 350 nm a new lower PRF is being introduced, and to match this development the E-2Cs rotodome will be slowed from 6 to 5 rpm. A third PRF modification is also being introduced allowing the radar to operate with different PRFs during scanning, eliminating problems with target 'blind speed' caused by single PRF operation.

STATUS

The AN/APS-145 is in the flight test stage and is scheduled to enter operational service in 1990. See also AN/APS-138/125 under entry **1965.353**.

CONTRACTOR

GE Aerospace, French Road, Utica, New York 13503, USA.

5609.353

ADVANCED PHASED ARRAY RADAR (APAR)

GE is engaged in an Advanced Phased Array Radar (APAR) programme (formerly known as Conformal Radar System) with Grumman, originally to the US Navy only. The concept involves a prospective radar embedded in the wings and side of an aircraft using solid-state transmit/receive modules. This would eliminate the need for a rotodome antenna for early warning radar in the US Navy and USAF aircraft to follow the E-2C.

DEVELOPMENT

The first phase of the development programme, the passive receive only portion of the concept, was pursued in an aircraft wing structure and was ground tested in 1984. Flight tests were to follow but it was decided to proceed with provision of an active system for roof top testing, followed by technology demonstration flight tests in 1990-91. Reference should also be made to the STATUS section of the AN/APS-138/125 radar entry (**1965.353**).

STATUS

Development in conjunction with Grumman Aerospace.

CONTRACTOR

GE Aerospace, French Road, Utica, New York 13503, USA.

3688.353

AN/APX-72 AIRBORNE IFF TRANSPONDER

The AN/APX-72 is an airborne IFF transponder which is widely used on US military and certain civil aircraft. It provides automatic radar identification, emergency signals, altitude reporting (when used in conjunction with a suitable altitude digitiser) and responds to interrogations in modes 1, 2, 3/A, C and 4. The last mode requires use of a KIT-1A/TSEC computer.

CHARACTERISTICS

Frequency: 1090 MHz
Transmitter power output: 27 dBW nominal
Receiver frequency: 1030 MHz
Dynamic range: >50 dB
Sensitivity: –90 dBV nominal
Dimensions: 146 × 162 × 340 mm
Weight: 6.8 kg

STATUS

Operational in US fixed-wing aircraft, helicopters and shipborne installations, and in the aircraft of a number of other countries. Well over 45 000 transponders have now been manufactured.

CONTRACTOR

Hazeltine Corporation (part of Emerson Electric Company), Greenlawn, New York 11740, USA.

AN/APX-72 IFF transponder

3690.353

AN/APX-76 AIRBORNE IFF INTERROGATOR

The AN/APX-76 airborne interrogator exists in two main versions, the original electron tube model and a solid-state transmitter version. For both versions, a further product improvement programme began in 1983 for component modernisation and technical improvements to the system. In 1986 production was commenced to implement these programmes. The earlier model, AN/APX-76A is described first.

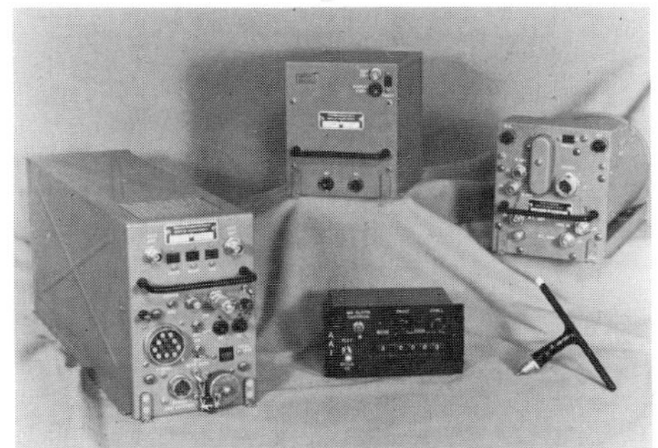

AN/APX-76 IFF interrogator system

AN/APX-76A

The AN/APX-76A is an air-to-air interrogator for all-weather interceptor and other tactical aircraft, and has full AIMS capability in modes 1, 2, 3/A and 4. It achieves a narrow antenna beamwidth, and a reduction in 'fruit' through ISLS and RSLS circuits (interrogation and receiver sidelobe suppression, respectively), in conjunction with special antennas having sum (main-lobe) and difference (sidelobe) suppression patterns. Bracket-decoded video and discrete code-decoded video are displayed on the radar screen to provide unambiguous correlation between IFF and primary radar targets. The equipment is fully IFF Mk XII compatible.

AN/APX-76B

Hazeltine has developed an all solid-state transmitter and power supply unit for the APX-76, under a US Navy contract, thus yielding a number of advantages, both operational and in respect of maintenance and reliability; MTBF increased to 400 hours, compared with 225 hours for the AN/APX-76A.

The new version of the AN/APX-76 is directly interchangeable with its predecessor, and no re-alignment is necessary, the complete operation requiring only 15 minutes.

AN/APX-76C/D

The nomenclature AN/APX-76C designates a tube model incorporating the improvement programme noted in the first paragraph, while the AN/APX-76D is the solid-state transmitter version with these improvements.

STATUS

The AN/APX-76 entered production in 1967, and it is estimated that more than 6000 of these sets have since been supplied to the air forces of the USA and friendly nations. Orders have been placed for the APX-76B version and the US Navy is retrofitting all APX-76As with solid-state transmitter/power supply assemblies. The APX-76A/B is installed in the F-4B, C, E, J, EC-121, P-3C Orion, CP-140, E-2C, F-111D, F-14, F-15, SH-60B and others. In addition to American forces, the equipment has been supplied to Australia, Canada, Greece and South Korea.

CONTRACTOR

Hazeltine Corporation (part of Emerson Electric Company), Greenlawn, New York 11740, USA.

1569.353

AN/APX-83 AIRBORNE IFF INTERROGATOR

The AN/APX-83 is an airborne IFF interrogator installed in EC-121 airborne early warning aircraft to upgrade their surveillance capability. Each aircraft is equipped with an interrogator system and control/indicator equipment groups for each operator position in the aircraft. Each group semi-automatically processes and displays IFF responses from aircraft, ships, and ground transponders within the EC-121 interrogation range. There are provisions for expansion to a fully automated processing and display system.

A complete equipment set consists of either a synchroniser/coder or coder, decoder, decoder controls, decoder displays, interrogator control, receiver/transmitter, and RF switch.

The coder interfaces with the associated radar, R/T unit, RF switch, decoder, decoder control, and the aircraft's APX-72 transponder. It accepts an external pre-trigger from the associated radar set and generates synchronous pulse-pairs and SLS pulses. These pulses are sent to the R/T unit for modulation of the RF signals. A sync-trigger is sent to the decoder for range correlation of replies. Mode coding on a per sweep basis is done on two data lines which are also connected to the decoder. Video is steered through the coder to the decoder.

The decoder accepts raw video and mode triggers from the coder and generates bracket decodes, delayed raw video and identity, or altitude data for processing by the decoder control.

STATUS

In production and fitted to EC-121 early warning aircraft.

CONTRACTOR

Eaton Corporation, Command Systems Division, 815 Broad Hollow Road, Farmingdale, Long Island, New York 11735, USA.

4339.353

AN/APX-100(V) AIRBORNE IFF TRANSPONDER

The AN/APX-100(V) transponder is a panel mounted IFF transponder using micro-miniature technology in both digital and RF circuitry, and is currently in production for a number of US military aircraft to enable them to operate in the military ATC surveillance system.

The system is a completely solid-state, modular constructed equipment with a complete dual channel diversity system, comprehensive built-in test, digital coding and encoding and high anti-jamming capability. Two antennas form part of the equipment and the diversity system receives signals from each and switches the transmitter output to the antenna which received the strongest interrogation signal. This system is designed to cure the problems of poor coverage with a single antenna which suffers on occasions from aircraft manoeuvre and antenna 'shadowing'.

AN/APX-100(V) IFF transponder

The transmitter is all solid-state with its 500 watt peak power output obtained from four parallel microwave transistors. Two additional transistors complete the transmitter oscillator and driver stages. The diversity system provides improved antenna coverage and allows improvement in performance in overloaded, jamming and multipath environments. Automatic overload control and anti-jamming features are also incorporated.

For aircraft that have no cockpit space available, a remote configuration, RT-1157/APX-100(V), is available. In addition, the RT-1471/APX-100(V), a data bus version, operating in accordance with MIL-STD 1553B, is available.

STATUS

Currently in production for all new US tri-service aircraft including the F-18, AV-8B, C-12, EC-130, SH-60B, UH-60, AH-1S, HH-65A, AH-64A, T-46A, SH-65A, C-5B, HH-60A and US Navy 'Sea Fox' surface craft. It is also being retrofitted to several other aircraft, including the OH-58 and CH-47.

CONTRACTOR

Allied-Signal Inc, Bendix Communications Division, 1300 E Joppa Road, Baltimore, Maryland 21204, USA.

3300.353

AN/APX-103 AIRBORNE INTERROGATOR

The AN/APX-103 interrogator set is the first airborne IFF interrogator set to offer complete AIMS Mark X SIF air traffic control (ATC) and Mark XII military IFF processing capabilities in one integrated system. Simultaneous Mark X and Mark XII multi-target and multi-mode operations permit an operator to obtain instantaneously the position (range, azimuth and elevation), code identification, and friend/foe status for all targets within the radar surveillance volume. This information can be used in conjunction with the radar to perform ATC, AEW, intercept tracking and escort surveillance missions.

OPERATION

Command words received from the E-3A display processor functional group initiate processor action and direct the mode of operation. The processor generates interrogations to the active receiver/transmitter (R/T), which modulates interrogations and sends them via an RF switch to sum and difference antenna channels. Target aircraft within the surveillance volume respond with coded RF replies via the antenna and RF switch. Replies are demodulated and detected by the active R/T and are sent as video pulses to the processor.

The processor detects, degarbles, and decodes replies and forms target records for replies which meet certain criteria. Completed target records are sent to the E-3A DPFG as target reports.

STATUS

In operational use in the E-3A AWACS aircraft.

CONTRACTOR

Eaton Corporation, Command Systems Division, 815 Broad Hollow Road, Farmingdale, Long Island, New York 11735, USA.

6900.393

TELEDYNE IFF INTERROGATORS/TRANSPONDERS

Teledyne designs and manufactures a wide range of IFF interrogators and transponders for airborne applications. Brief details of some of the latest equipments is given below:

AN/APX-101 airborne transponder

This is an all solid-state ATC transponder widely adopted by the US forces. It consists entirely of replaceable modules and consists of a 500 W transmitter and a dual channel TRF receiver in a single unit of standard dimensions. The transponder operates in Modes 1, 2, 3/A, 4 and C. It receives RF interrogations from two antenna systems, decodes the interrogations into the proper mode, encodes the selected reply and transmits this through the correct antenna. It operates with an air data computer for Mode C, with a transponder computer for Mode 4.

AN/APX-108 Mk XII IFF diversity transponder

The single unit configuration and reduced size of the AN/APX-108 and the complementary reductions in weight and power result from applications of the latest techniques. Using advanced microwave packaging, high speed A/D converters, advanced digital signal processing and CMOS LSI gate array circuitry, the component count, volume, weight and power dissipation of the AN/APX-108 has been minimised. Features include a COMSEC applique that uses electronic key fill to eliminate the requirement for a KIT-1A computer. For interface flexibility, the equipment can be interfaced with either a 1553B MUX bus or a C-6280 control box.

TEC-60i combined interrogator/transponder

The TEC-60i is an all solid-state microprocessor-controlled Eastern IFF interrogator and transponder housed in a single unit. The unit is designed to be placed in the electronics bay without need of cooling air. The control box is available in various configurations to meet user requirements. Frequencies and coding are programmed from the control box. Coding techniques can be user-controlled by both software and hardware development.

CIT combined interrogator/transponder

CIT is an efficient solid-state designed equipment that uses advanced microwave packaging, surface mount devices and CMOS LSI gate array circuits to reduce significantly the component count, volume and weight of the IFF system. It offers a prioritised four channel AOC, a first in IFF system design, adaptive thresholding of the received video to improve performance in high noise and jamming environments and monopulse processing of received target video for enhanced azimuth accuracy. The CIT also uses a KIV-3 module that incorporates the functions of KIT-1A as well as the KIR-1A cryptographic computers. This module directly interfaces and attaches to the CIT. The unit can be interfaced to either a mechanically or electronically scanned antenna.

Teledyne CIT equipment

STATUS

All units are in production.

CONTRACTOR

Teledyne Electronics, 649 Lawrence Drive, Newbury Park, California 91320, USA.

3691.353

MM/UPX-709 AIRBORNE IFF TRANSPONDER

The MM/UPX-709 is an airborne multiplex diversity IFF transponder providing civil and military air controllers with full aircraft identity information and automatic altitude reporting. The equipment can be fitted in aircraft, helicopters and ships, and coverage is all-round. The diversity features give full spherical coverage, and continuous track data is supplied to automatic processing systems regardless of aircraft attitude: antenna blind spots are virtually eliminated because of the antennas located above and below the aircraft fuselage.

CHARACTERISTICS

Frequency: 1030 MHz receive; 1090 MHz transmit

Sensitivity: Adjustable (–90 dBV matched load); normal –77 dBm

Modes: 1, 2, Test, 3/A, C, AIMS mode 4

Dimensions: 136 × 136 × 213 mm (remote version); 133 × 146 × 160 mm (panel mounted)

Weight: 4.1 kg (remote); 3.2 kg (panel)

STATUS

Production. Licence manufacture is undertaken in Italy by Italtel (**3489.353**).

CONTRACTOR

Hazeltine Corporation (part of Emerson Electric Company), Greenlawn, New York 11740, USA.

MM/UPX-709 IFF transponder — panel mounted version

1100.353

AN/AWG-9 AIRBORNE WEAPON CONTROL SYSTEM

This is the aircraft-fitted element of the Phoenix (AIM-54) weapon control system (which see) developed for the US Navy F-14 aircraft. It comprises a fire control and target illuminating radar, digital computer and displays. Provision is also included for the automatic exchange of data link information between the AWG-9 and the naval tactical data system (NTDS) and airborne tactical data system (ATDS) for target designation and other functions.

For on-board target acquisition, the AWCS includes a long range, high-power pulse doppler radar. This system has a look-down capability that enables it to pick moving targets out of the ground clutter that normally obscures targets in a conventional radar.

In addition to its long range, the AWCS introduced into the fleet for the first time in an aircraft, the ability to track many targets at once, with computer-aided selection of target priority. The system is designed for

AN/AWG-9 weapon control system, with Phoenix missile in foreground

consecutive launch and for simultaneous guidance of up to six AIM-54 missiles against separate targets.

The radar antenna is a planar, slotted-plate array, and represents a significant step forward in design, with a high aperture efficiency for increased radar range.

The principal weapon is the long-range Phoenix, AIM-54 (**1099.331**) air-to-air missile, but the AWG-9 system can be used also for launching Sparrow (**1106.331**), and Sidewinder (**1308.331** in 1983-84 and earlier editions) missiles, and control of an M-61 Vulcan 20 mm cannon.

OPERATION

The AWCS uses a pulse doppler radar as its primary target sensor, a multipurpose digital computer, and associated control and display sub-systems. The pulse doppler radar system represents several years of development in transmitter tubes, crystal filters, and planar array antenna techniques to enable the long range requirements of Phoenix to be met. A low noise parametric amplifier in the receiver section contributes to this long range capability. Advanced doppler techniques make look-down target acquisition possible. A flexible, high capacity computer permits simultaneous track of a large number of targets and aids the radar intercept officer (RIO) in the assignment of kill priorities and in missile firing.

Target detection, tracking, and ranging functions for all F-14 air-to-air weapon configurations (6 × Phoenix; 6 × Sparrow; a mixture of these; plus 4 × Sidewinder; and the 20 mm Vulcan cannon) are handled by the AWG-9 radar. It can operate in either pulse doppler or conventional pulse modes. A separate TWT provides CW illuminating energy for the semi-active homing Sparrow (AIM-7E or -7F CW). Time-sharing techniques allow up to six Phoenix missiles to be given pulse doppler mid-course guidance simultaneously.

Data processing in the AWG-9 AWCS is performed by a general-purpose digital computer that features high speed operation and a large memory capacity in an extremely compact package. The central computer keeps track of targets detected by the radar while the radar continues to search. Based on pre-programmed logic the computer evaluates threats, generates steering information for the pilot, and paints a complete tactical situation for the RIO in standard NTDS symbology, all based on data generated either internally or obtained through external data links.

For control of the Phoenix weapon system, the aircraft's RIO is provided with two CRT display units. One, a 12.7 cm diameter unit, is used as a multi-mode display for the presentation of raw radar derived target information and IFF returns, while a larger (25.4 cm diameter) unit is used for the display of processed data.

The latter includes target track information, alphanumeric and symbolic data obtained via data link from other units of a naval force. The display is also used as a computer read-out device for the presentation of computer-generated missile/target assignments.

STATUS

US Navy squadrons now operational with the F-14A are embarked on four aircraft carriers in the Pacific and four in the Atlantic. More than 700 F-14s are planned by the US Navy to support its 15 carrier oriented force structure. The US Navy has completed a retrofit programme to double the core memory in the central processor.

In 1984, Hughes was awarded a development contract from Grumman Aerospace Corporation to design, develop and test an upgraded version designated the AN/APG-71 radar subsystem (**5880.353**). Key elements in the upgrade are a fully programmable signal processor and a companion radar processor to replace the original four analogue processors and provide digital capability, digital displays, ECCM and improvements to expand detection and tracking performance.

CONTRACTOR

Hughes Aircraft Company, Radar Systems Group, El Segundo, California 90245, USA.

3018.353

AN/AWG-10, 11 and 12 MISSILE CONTROL SYSTEMS

The AN/AWG-10 was designed as the airborne, multi-mode missile control system to provide fire control for guns, and Sparrow and Sidewinder air-to-air missiles on the F-4 aircraft. The first radars were delivered in 1966 and became standard equipment. The AN/AWG-10 was claimed as the first interceptor radar to feature transistorised circuitry.

AN/AWG-10

During 1966, the US Navy announced a look-down requirement for its F-4 aircraft and the AN/AWG-10 became the first multi-mode radar with pulse-doppler look-down capabilities and a comprehensive built-in test system. The United Kingdom bought the AN/AWG-11 and -12 modified versions of the -10 for its F-4K and F-4M Phantoms.

AN/AWG-10A

During 1973 Westinghouse was contracted by the US Navy to improve the performance and reliability of the AN/AWG-10. The most significant change affecting reliability was the substitution of a solid-state transmitter using a klystron power amplifier. A digital computer was added for the more effective solution of launch equations, for example target manoeuvring information could now be used. A servoed optical sight display was also added and these modifications also allowed the full range of air-to-ground ordnance to be exploited.

Only three of the AN/AWG-10 line replacement units (LRU) remain unmodified, there are six new LRUs, and seven units have been deleted.

The success of the AN/AWG-10A computer led to its procurement by the US Air Force for the AN/APQ-120 radar in the F-4E to provide computer-aided target acquisition. The West German Air Force, in a programme known as Peace Rhine, also procured the computer to upgrade its F-4F aircraft.

AN/AWG-11/12

The AN/AWG-11/12 are the designations given to the AN/AWG-10 radars fitted to the F-4K Phantoms of the Royal Air Force and the F-4Ms of the Royal Navy in the 1970s. Ferranti is the UK design authority. The updates which gave rise to the AN/AWG-10A were incorporated into the -11 and -12.

STATUS

The AN/AWG-10 family has long been out of production but is still in widespread service. In early 1984, the US DoD announced the possibility of flight demonstrating F-4s with new engines and avionics to create a new market for F-4s currently in the US inventory. Fundamental changes in the avionics would be substitution of the AN/APG-66 radar as used in the F-16.

AWG-10 missile control system radar

CONTRACTOR

Westinghouse Electric Corporation, Defense and Electronics Centre, Aerospace Divisions, Friendship Site, PO Box 746, Baltimore, Maryland 21203, USA.

5611.353

MULTIBEAM MULTIMODE SURVEILLANCE RADAR

The Multibeam Multimode Surveillance Radar (MSR) is an air-to-ground system for battlefield surveillance applications.

The MSR system consists of a multibeam electronically scanned phased array antenna and a real-time all digital LSI radar signal processor. It is a computer controlled multi-mode radar that operates in the MTI search, MTI area track, synthetic aperture radar spotlight, and doppler beam-sharpened modes, and includes extensive ECCM capabilities.

The MSR uses multiple waveforms and includes a reconfigurable real-time radar signal processor. The system consists of nine core line replaceable units:

(1) pod-mounted multibeam electronically scanned phased array antenna
(2) coherent transmitter
(3) exciter/synchroniser
(4) multi-channel receiver/converter
(5) preprocessor
(6) digital pulse compressor
(7) spectral analyser
(8) target detect and data report
(9) radar controller.

The preprocessor, spectral analyser, digital pulse compressor, and target detect and data report form the radar signal processor.

The system operates at Ku-band, using linear and non-linear frequency modulation plus binary coded waveform with bandwidths of 5 and 20 MHz. The pod-mounted, side-looking multibeam antenna has a 2.5 metre aperture length and provides ±60° azimuth scan coverage in both the single and multibeam modes. Real-time radar signal processing, especially pulse compression and spectral analysis, is accomplished by using fast Fourier transform hardware.

The radar controller contains two computers; one controls the operation of all units over a data bus, the other implementing the required motion compensation of the data for the imaging modes.

Except for the antenna, standardised packaging has been used throughout the units with microwave, analogue, digital, LSI and power supply circuitry all packaged in standard ¾ ATR modules, making them applicable in other programmes.

STATUS

The MSR has been under development by General Electric since 1978. The antenna was developed under sponsorship of the Defense Advanced Research Projects Agency (DARPA), with the balance of the radar being designed on a private venture basis.

CONTRACTOR

GE Aerospace, French Road, Utica, New York 13503, USA.

3304.353

HOWLS EXPERIMENTAL AIRBORNE RADAR

An experimental airborne radar system for detecting and locating hostile weapons has been developed by GE as part of the hostile weapons location systems (HOWLS) programme under the sponsorship of the US Defense Advanced Research Projects Agency (DARPA) and management of MIT's Lincoln Laboratory. The HOWLS radar is an extremely versatile coherent J-band phased-array radar, which is designed to investigate techniques from which solutions may be defined to detect and locate fixed tactical targets in a clutter environment. Particular emphasis is being given to the detection of artillery.

Key to the new radar system is a lightweight, phased-array antenna designed and developed by GE. The antenna system utilises a low cost, highly reproducible phased array element that combines the radiating element and PIN diode phase shifter in a single integrated structure to minimise weight and cost.

HOWLS has six basic modes of operation. These are: contextual ground map, fixed target detection, ground moving target detection, low level doppler spectrum signature analysis, projectile detection and tracking, and doppler beam sharpening. Through microprocessor control, the radar has mode parameter flexibility and is capable of interleaved multi-mode operation. Selectable parameters include antenna scanning pattern and rate, PRF, data sampling window position, and data sampling pattern. The radar is capable of switching among parameters in an orderly manner so that up to four operating modes can be interleaved. Characteristics of one of the experimental HOWLS radars appeared in *Jane's Weapon Systems 1978*.

Results of the experimental HOWLS programme will provide the technology base from which radars for RPVs, helicopters, and fixed-wing aircraft can be configured to meet a variety of operational missions.

CONTRACTOR

GE Aerospace, French Road, Utica, New York 13503, USA.

1673.353

WESTINGHOUSE WX SERIES FIRE CONTROL RADARS

The Westinghouse WX Series is a private venture project for the development of a range of forward radars for fighter and attack aircraft, based on the use of modular design. The latter feature is intended to enable radars appropriate to varying roles and with differing degrees of sophistication to be constructed from mostly standard units, with consequent cost benefits.

The 'base' model of the range is the WX-200 and this is a pulse doppler system intended for air-to-air combat purposes. Beyond a provisional system weight of about 350 lb (160 kg), without cockpit display, few specific details have been revealed, but the following general design characteristics are known. A TWT will be used in the transmitter to provide wide bandwidth, a high peak-to-average power ratio, and to permit the adoption of frequency agility techniques of ECM.

Wherever possible digital circuitry is used, with incoming returns being converted to digital format as near the receiver front as convenient. The receiver employs a parametric amplifier. A cassegrain scanner is employed to eliminate the cost, weight, and reliability considerations of a rotary joint. The scanner itself is not stabilised, this function being performed electronically to provide a stabilised cockpit display, the data presumably being derived from the aircraft's own attitude sensors.

The Westinghouse development plan is to evolve other radars from the WX-200 which are both simpler and more sophisticated than the base model. The WX-50, a lightweight air-to-ground system for close-air-support applications, has also been fabricated. The WX-50 is a non-coherent Ka-band radar with a reflector antenna for ground mapping and a receive-only elevation phase monopulse antenna for letdown terrain clearance. Additional growth modes include ground MTI, terrain avoidance, air-to-ground ranging, and EO cueing. The production WX-50 will weigh 64 kg.

The other models of WX radar would be achieved either by the addition of modules to provide extra desired operational features, or by a simplification of the basic WX-200 to satisfy less elaborate requirements.

STATUS

Two models of the WX-200 have been fabricated and have undergone extensive engineering testing at the Westinghouse facility. A pod-mounted version of the WX-50 has been flight tested by the US Navy on a TA-4J, on OV-10 and a UH-1N helicopter. The 208 kg prototype system demonstrated excellent performance and achieved a field MTBF of 160 hours without any contractor support.

Westinghouse is currently producing the F-16 radar which is based on the WX series. Information on the F-16 system is given under entries **3303.353** (AN/APG-66) and **5881.353** (AN/APG-68).

CONTRACTOR

Westinghouse Defense and Electronics Center, Baltimore, Maryland 21203, USA.

1676.353

AN/UPD-4 SIDE-LOOKING RADAR SYSTEM

The AN/UPD-4 side-looking radar system is an advanced design, all-weather, high-resolution reconnaissance sensor system for airborne collection of tactical and strategic intelligence information. Utilising I/J-band radar energy to illuminate selected terrain swathes, it operates equally as well in darkness as in daylight and in adverse weather.

The AN/UPD-4 consists of an AN/APD-10 synthetic aperture radar set mounted in an RF-4 aircraft, a ground-based correlator/processor set and test consoles for the maintenance of airborne and ground equipment. Airborne data link transmitters and ground receiving and processing systems are optional additions to provide radar-collected information to ground commanders more rapidly than the basic system.

The airborne system includes multi-element, phased, linear waveguide antenna arrays individually gimballed for left or right side operation. Look-angle of the antennas is maintained in flight by a control system that receives error signals from the aircraft inertial navigation system and gimbal-mounted gyroscopes and accelerometers.

The frequency converter/transmitter generates the stable radio frequency used for phase locking the radar signals and contains the transmit/receive and antenna switching components. It also supplies the swept frequency for modulating the transmitted pulses, amplifies the reflected signals and converts the received RF to IF. The amplifier/modulator is a high pulse RF pulse amplifier.

The signal data generator receives information relative to the aircraft velocity and mode, and uses this data to establish the radar PRF and generate the basic timing reference signals required throughout the system.

The radar mapping recorder contains the optical, electronic and electro-mechanical assemblies required to record the radar video data and associated coded data on 241 mm wide photographic film. These data are displayed on two 5-inch CRTs as four intensity modulated traces that are transferred by two mirror assemblies and four recording lenses to focus the images on the film. Mode strips that convey essential operational information and multi-element blocks containing information pertinent to the mission are recorded on the film. For data link operation, an identical recorder is used at the correlator/processor on the ground to establish a continuous recording/correlation path. Film speed is carefully controlled in either application to maintain a fixed relationship to aircraft ground speed.

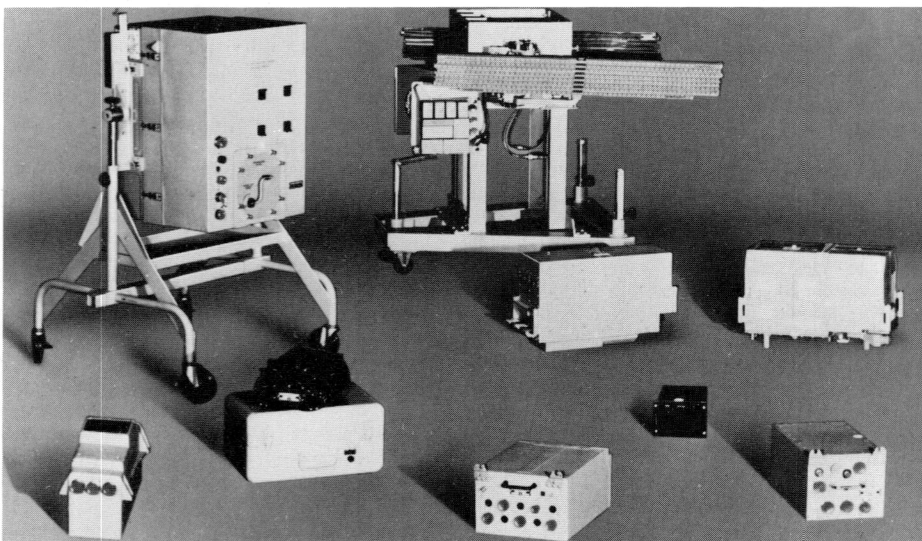

Basic components of the airborne equipment of the AN/UPD-4 radar reconnaissance system with data link equipment

OPERATION

The equipment has several modes of operation, providing a variety of stand-off distances and altitudes, and the option of recording only fixed target imagery (FTI), or both FTI and MTI. Imaging of the terrain at either side can be obtained at the discretion of the operator. With data link equipped systems, the information may be transmitted to a ground station in real-time for recording and processing. The final imagery is recorded on 241 mm wide film in four channels at a scale of 1:100 000 for all modes. In the azimuth (along track) direction, targets and terrain features are imaged in terms of distance travelled and represent true ground separation. In the range (across) track direction, the imagery is recorded in slant range or the distance from the aircraft to the target.

STATUS

The AN/UPD-4 is in service in several versions with Phantom aircraft of the US Marine Corps and the Japanese Air Force. The system was developed for the US Air Force as a successor to the AN/APD-102A; but has since been enhanced and designated AN/UPD-8 (**5686.353**).

CONTRACTOR

Loral Systems Group, Defense Systems Division, Litchfield Park, Arizona 85340, USA.

5686.353

AN/UPD-8 AND AN/UPD-9 SIDE-LOOKING RECONNAISSANCE RADARS

The AN/UPD-8 is a side-looking, synthetic aperture airborne reconnaissance radar system for the US Air Force. It is, in fact, an updated and improved version of the AN/UPD-4 equipment (**1676.353**), and gives a much greater range. The system provides an all-weather, day/night, stand-off, tactical and strategic reconnaissance capability, and with its extended range antenna pods and airborne data link electronics is the latest in a sequence of Goodyear airborne reconnaissance radars to have been operational since 1964.

Operating in the I/J band, the system records with equal effectiveness during daylight and darkness, and through periods of adverse weather that would render other types of reconnaissance systems ineffective. The airborne radar system's stand-off capability, which is increased by 60 per cent in the new equipment, allows coverage of border regions, harbours and coastal areas without violating another nation's airspace or national waters. Maximum range, limited by power, is about 130 km. At such distances resolution is 6.1 m in azimuth and 4.6 m in range, but resolution increases at shorter ranges. At 50 to 90 km the radar can resolve to 3 m, allowing large aircraft to be identified and showing clear images of vehicle formations.

Loral has also developed a pod-mounted version of

the UPD-8. Similar in size to a 1250 litre fuel tank and weighing less than 682 kg, the pod includes the main antenna, support electronics and a data link. A version of this pod is to be supplied for the US Navy RF-18 aircraft. The podded system can also be installed on an executive type jet aircraft.

Another upgrade of the UPD-4 is the UPD-9 developed for the US Marine Corps. It is similar to the UPD-8, with a data link, but does not have the long range features. Basic stand-off range is about 80 km.

OPERATION

The airborne sensor part of the overall system, the AN/APD-12, operates in eight modes selectable by the operator according to aircraft altitude and the distance to the target area. The system can record to left or right of the aircraft with the normal antennas or with the extended-range antennas which are mounted in a centre-line pod. Target information can be transmitted to the ground in real-time through a data link, or can be recorded in flight for later processing on the ground. The former facility enables a commander on the ground, perhaps several hundred kilometres away, to evaluate detected targets while the aircraft continues its mission.

STATUS

In operational service with the US Air Force and US Marine Corps. Loral is also carrying out trials on mini-SLARs for use by unmanned aircraft.

CONTRACTOR

Loral Systems Group, Defense Systems Division, Litchfield Park, Arizona 85340, USA.

6440.353

AIRSHIP SURVEILLANCE RADAR

The United States Navy is studying proposals for an airship surveillance system. Three companies Goodyear, Boeing, and Airship USA Inc. (Westinghouse and Airship Industries) have been bidding for the complete contract which covers the airship, the radar system and associated ground equipment. The primary function is detection and tracking of low-altitude cruise missiles, with intelligence gathering and anti-submarine warfare as secondary roles.

Although competitive radar bids from Hughes, Westinghouse and RCA were evaluated, the US Navy has specified the combat suite from the E-2C AEW aircraft as the baseline system. This is presumably a variant of the AN/APS-138 family of radars described under entry **1965.353**. The antenna array will use a modified E-2C rotodome contained within the airship envelope but is likely to be superseded by a conformal array at a later date.

STATUS

In early development.

4618.353

RCA BALLOON-BORNE RADAR

Project Seek Skyhook is a tethered balloon-borne radar system in operation in Florida to provide surveillance of the state's coastal regions. The radar system used is an RCA AN/DPS-5 equipment which has a range of 150 nautical miles at an altitude of 12 000 feet. It is designed to detect low-flying unidentified aircraft for the US Air Force and other users. It is the latest in a series of aerostat developments that began in 1969, and four systems are in operation in Florida.

The AN/DPS-5 radar is an S-band equipment using a rotating parabolic dish measuring 22 × 12.5 feet which returns an inverted cosecant signal with two selectable polarisations that can be directed circular and horizontal or vertical and horizontal.

The radar ground sub-system has an RCA digital processor with the capability of simultaneous operation with two aerostats. Normal and MTI radar video are selectable with or without CFAR. Display can be either PPI or digitised video. The radar beacon digitiser includes capabilities for clutter mapping, target sorting and use of target track history.

CHARACTERISTICS

Frequency: 3100 MHz
Range: 150 nm (278 km) at 12 000 feet altitude
Resolution: Range 500 feet, azimuth 1°
Antenna: Rotating parabolic dish
Scan speed: 5 rpm
Weight: 1000 lb (455 kg) including mounting, gimbal, antenna, transmitter, receiver and power supply
Endurance aloft: 120 h normal, 143 h peak

STATUS

Operational. In September 1986, RCA was awarded a contract worth $42 million to supply three sea-based aerostat systems for the US Coast Guard. They will be equipped with Eaton Corporation radars, customised by RCA for sea-based surveillance.

CONTRACTOR

RCA, Aerospace and Defense, Missile and Surface Radar Division, Moorestown, New Jersey 08057, USA.

RCA sea-based surveillance aerostat

3162.353

WESTINGHOUSE BALLOON-BORNE RADAR

Under contract to the US Defense Advanced Research Projects Agency (DARPA), Westinghouse has provided radars for use with advanced aerostat or balloon platforms. In the mid-1960s DARPA initiated an R and D programme to develop a stable balloon platform and to exploit the use of such platforms for sensor and communications applications. Early work was carried out using modified British barrage balloons and these were later superseded by purpose-designed aerostats, of various sizes. Typical dimensions range between 1415 and 16 990 cm; 35 and 8.5 metres in length; and 9 and 30 metres in diameter.

This family of tethered balloons is stated to be capable of operations at altitudes of 3000 and 4500 metres. A powered tether system and low helium leakage allow the aerostat to operate at altitude for extended periods, providing greatly increased radar surveillance or communications coverage due to the increased line-of-sight ranges obtainable. Typical radar applications proposed include surveillance and AEW, location, and tracking of targets, fire control and weapons direction. Communications, EW, and target location/designation by other means, eg electro-optical sensors, are other possibilities.

Radars employed at first were all of the non-coherent variety but Westinghouse later received a contract from the US Air Force Rome Air Development Center to build a coherent system to detect airborne moving targets.

STARS (small tethered aerostat relocatable system)

Airship Industries has signed an agreement with Westinghouse to co-operate in plans for a 'battle surveillance airship system' for the US Navy. However, US DoD defence budget cuts in early 1988 have placed this project in jeopardy.

DEVELOPMENT

The following brief details of this programme refer to the Westinghouse balloon-borne radar projects which have been identified.

Low Altitude Surveillance Radar (LASR)

This is a static tethered balloon-borne system which is able to carry a radar such as a slightly modified version of the AN/TPS-63. It can carry 1800 kg to a height of several thousand metres for periods of up to 20-30 days. Westinghouse has received a $12.6 million contract from Saudi Arabia for a demonstration of a LASR system to supplement the E-3A airborne early warning aircraft. The system consists of a modified AN/TPS-63 suspended from a 365 000 cubic feet aerostat, having a nominal altitude of approximately 10 000 feet (3048 metres). The demonstration and trials took place in Saudi Arabia in 1986.

Westinghouse is under contract to the US Customs Service (USCS) to provide LASS for Project SOWRBALL. This is a project to deploy aerostats along the southern USA border and in the Caribbean. The first system became operational at the end of 1987. Another LASS has been operational in the Bahamas for over three years and has completed 10 000 hours service. Information from these aerostats is shared by the USCS and Coast Guard for drug interdiction and by the USAF for strategic and tactical surveillance.

Small Tethered Aerostat Relocatable Systems (STARS)

A tactical, transportable system built by Westinghouse. Normal use is for battlefield and coastal surveillance and at 750 metres altitude a small balloon-borne radar can detect a target with a radar cross-section of 100 m² at ranges of up to 110 km. The sensor selected for the system is the Westinghouse AN/APG-66 fire control radar, with a small modification to the antenna system. Westinghouse has linked with LMT Radio Professionelle (a Thomson-CSF subsidiary) to offer the RASIT battlefield surveillance radar for tactical use (see entry **2141.153**).

STATUS
Westinghouse aerostat surveillance systems are in use around the world for air defence functions. These systems are able to detect cruise missile size target and low flying aircraft at ranges of 260 km.

CONTRACTOR
Westinghouse Defense and Electronics Company, 7301 Parkway Drive, South Hanover, Maryland 21076, USA.

5888.353
B-52 AIRBORNE RADAR

The radar system currently carried by the US Air Force B-52 bomber force, which is nearly 30 years old, is being upgraded to provide greater reliability and capabilities. In particular the system will be enhanced in terms of terrain avoidance and ground mapping capabilities, also analogue units will be replaced by digital. The new system will consist of eight units in place of the 50 currently installed in the aircraft, a reduction which highlights the advances made in radar technology over the past three decades. Boeing is the prime contractor for the updated radar and will be designing the antenna, interfaces and displays, but the transmitter, radar and processor will be provided by Norden Systems.

STATUS
Boeing has received a £109 million contract which covers 11 systems, training, spares and logistics support. As many as 264 B-52s may eventually be scheduled for upgrading.

CONTRACTORS
Boeing Military Airplane Company, Wichita, Kansas, USA.

Norden Systems Inc, Norden Place, Norwalk, Connecticut 06856, USA.

6439.353
ASARS-2 SYNTHETIC APERTURE RADAR

The ASARS-2 advanced synthetic aperture radar is a high altitude, side-looking reconnaissance system that provides real-time, high-resolution radar ground maps in all weathers and at ranges well in excess of electro-optical devices. Hughes developed the system under contract from the US Air Force Aeronautical Systems Division. Performance, cost and production schedules are highly classified.

STATUS
The ASARS-2, flown on the USAF TR-1 aircraft is in production.

CONTRACTOR
Hughes Aircraft Company, Radar Systems Group, PO Box 92426, Los Angeles, California 90009, USA.

5624.353
JOINT SURVEILLANCE AND TARGET ATTACK RADAR SYSTEM

The Joint Surveillance and Target Attack Radar (J-STARS) project is a joint US Air Force and US Army development which combines the former's Pave Mover programme with the latter's Battlefield Data Systems projects (Assault Breaker programme). J-STARS was scheduled for development commencing in 1983, under the control of a joint US Air Force/US Army team, as an airborne anti-tank radar system for the two services. Various designs were proposed and four of these were flight tested during 1982.

The radar is primarily intended to detect enemy armour on and behind the battlefield and also to direct attacks against it. The system is designed to direct real-time attacks against moving ground targets by low-flying aircraft and missiles, and also to provide guidance of missiles carrying sub-munitions to the target locale. Tactical information on targets would either be processed on board the aircraft or relayed to the ground via a data link.

It was originally intended that requests for proposals (RFP) be issued in 1983 but due to disagreements between the two services this date slipped and an RFP was finally issued in August 1984. Originally the US Army wanted a wide area surveillance and an MTI capability, and the US Air Force wanted a fixed target indicator and a data link to guide weapons. This issue was settled and the Boeing E-8A has been agreed as the platform to carry the synthetic aperture and MTI radar. In a typical US Army mission data would be sent directly to a ground station via the secure data link for processing and dissemination to the field commanders. On the other hand, the US Air Force would process the information on board so that the operators could make target decisions and relay these to existing $C^3$1 networks.

The J-STARS requirement calls for the development of four systems, two to be tested initially in the USA before being sent to Europe for demonstrations in an operational environment. The other two will be kept back for full-scale development testing.

STATUS
Four industry teams bid for the J-STARS contract and in October 1985 the team of Grumman Corporation/Norden Systems was awarded a contract valued at $657 million, with Grumman as prime contractor. Under the programme, Norden will develop and produce an advanced multi-mode radar designed to classify and track fixed and moving targets deep behind the forward edge of the battle area. Once a target has been identified, the system will alert either airborne or ground-based stations and provide guidance to the aircraft or surface-to-air missiles assigned to engage the target. The development of J-STARS is expected to take about five years, with an engineering flight test scheduled for mid-1988. Production of the airborne segment is expected to begin in early 1990, with ground station production beginning in 1987. The system will initially be based aboard eight E-8A aircraft which are being modified by Boeing from Boeing 707-323Cs purchased from American Airlines to accept the Norden radar. Grumman will integrate and install the various sub-systems into the aircraft, develop the necessary software and conduct the test programme. The first two E-8A aircraft were delivered to Grumman in 1987. The overall cost of the complete development and production J-STARS programme is expected to reach $1.5 billion. Motorola has received two contracts to provide the ground station equipment and Cubic Defense Systems is developing the surveillance and control data link.

The first flight of the E-8A with its J-STARS radar was originally scheduled for March 1988. However, a six to nine months slippage in the radar programme has delayed airborne radar tests until the last quarter of 1988.

CONTRACTORS
Grumman Aerospace Corporation, Bethpage, New York 11714, USA (prime contractor).

Norden Systems, Inc (subsidiary of United Technologies), Norden Place, Norwalk, Connecticut 06856, USA.

Boeing Military Airplane Company, Wichita, Kansas, USA.

4648.353
COVERT STRIKE RADAR

Covert Strike is a bistatic radar system project under development by the USAF Avionics Laboratory at Wright Patterson Air Force Base. The system provides for illumination of a target by a stand-off radar installed on either a satellite or an aircraft, with only a receiving system required on board the strike aircraft. This allows the latter to approach the target in a passive, or covert, manner, and is also designed to seriously hamper the enemy in his jamming techniques. No technical details have been disclosed other than that the radar is probably a synthetic aperture system.

STATUS
A long-term project in the design phase with development and trials continuing until about 1990.

ELECTRONIC WARFARE EQUIPMENT (AIR)

CHILE

5824.393
CAIQUEN AIRBORNE RADAR WARNING RECEIVER
The Caiquen II radar warning receiver has been developed by ENAER to meet the requirements of the Chilean Air Force. The receiver operates over the 2 to 18 GHz frequency range and provides 360° coverage in azimuth, and ±40° in elevation. The equipment is fully automatic and uses a high-speed digital processor to evaluate and classify the threat, determine the radar type and also the direction of arrival. Information is presented to the pilot on a CRT display divided into 45° sectors and also indicates whether the threat is from a search, fire control or acquisition type and whether the transmissions are continuous wave. The equipment can also be preprogrammed to determine relative threat priorities.

The Caiquen system consists of eight line-replaceable units with a total weight of 8 kg. Wideband, high sensitivity crystal-video receivers, which can cope with PRFs within the range 200 to 12 000 Hz, are incorporated.

STATUS
Installed on Hunter aircraft of the Chilean Air Force and is being offered for export.

CONTRACTOR
ENAER (Empresa Nacional de Aeronautica), Gran Avda J M, Carrera No 11087, Santiago, Chile.

5825.393
ECLIPSE CHAFF/IR DECOY LAUNCHER
Eclipse is an airborne passive self-protection chaff/IR flare dispenser system consisting of up to four dispensing units mounted externally and a control unit in the aircraft cockpit. Each dispenser can carry either 40 RR-170 chaff cartridges, 20 MJU-7/B IR flares or a combination of the two types. Either semi-automatic or manual mode is possible, and in the former mode the pilot can select launch intervals of between 0.25 and 3.5 seconds. A digital processor in the cockpit control unit maintains an updated display of the number and type of remaining loads and carries out the required operating mode and launch sequence selection. The processor also allows the Eclipse system to be interfaced with the Caiquen II (**5824.393**) RWR for automatic reaction to a detected threat. The complete Eclipse system of five line-replaceable units weighs 60 kg.

STATUS
The Eclipse system is fitted to Hunter aircraft of the Chilean Air Force and is being offered in the export market.

CONTRACTOR
ENAER (Empresa Nacional de Aeronautica), Gran Avda J M, Carrera No 11087, Santiago, Chile.

Control unit for Eclipse decoy launcher

5826.353
ITATA AIRBORNE ELINT SYSTEM
The Itata Elint system is another development by ENAER and is a high-sensitivity electronic intelligence gathering system that can detect, locate and measure the parameters of emissions from search, acquisition and fire control radars. Itata consists of a fully programmable superheterodyne receiver, a digital pulse analyser, and a high-gain, wideband, rotating dish antenna which provides 360° coverage and bearing information to within a few degrees accuracy. Although intended primarily for light transport type aircraft, the equipment can also be installed in shipborne or ground vehicle configurations.

The receiver operates over the wide frequency range of 3 MHz to 18 GHz in six bands. It can be used in either a wide-open mode over the complete frequency range or in a selective mode over a single band. After detection of a transmission of interest, the receiver locks onto it automatically and measures its frequency and other parameters. Digitised data of each intercepted signal can be recorded automatically for subsequent analysis.

STATUS
The Itata system is operational with the Chilean Air Force and is being offered for export.

CONTRACTOR
ENAER (Empresa Nacional de Aeronautica), Gran Avda J M, Carrera No 11087, Santiago, Chile.

Itata receiver and analyser units

6505.393
MEDUSA JAMMER
A prototype jamming system was shown at the 1986 Feria Internacional del Aires (FIDA) exhibition under the code name of Medusa. It is a VHF jammer which produces up to 1 kW of 'white noise' and was produced, states the manufacturer, to show a capability to the Chilean armed forces. No other information has been made available and it is not known whether development of this system will be continued.

STATUS
Prototype. ENAER also manufactures a range of intelligence gathering receivers and jammers, but no further information has been made available.

CONTRACTOR
ENAER (Empressa Nacional de Aeronautica), Gran Avda J M, Carrera No 11087, Santiago, Chile.

CHINA (PEOPLE'S REPUBLIC)

6504.393
GT-1 CHAFF/IR FLARE DISPENSER
The GT-1 is a chaff and IR flare dispensing set which has been developed to provide self protection to fixed and rotary-wing aircraft. It provides chaff dispensing countermeasures in the 2 to 18 GHz band and IR countermeasures in the 1.5 to 5 μm wavelength. The complete system consists of a programme controller, an operations control, dispensers and cartridges, and is intended to be interconnected with an appropriate radar warning receiver. The system is able to dispense 68 chaff cartridges or 32 IR cartridges. Chaff and IR cartridges cannot be mixed in the same dispenser.

STATUS
In production

CONTRACTOR
China National Import and Export Corporation, 40 Fuxing Road, PO Box 34, Beijing, People's Republic of China.

GT-1 dispenser set

6901.393
KJ 8602 RADAR WARNING RECEIVER

The KJ 8602 is a radar warning receiver designed for tactical and other combat aircraft. It consists of a digital signal analyser, a CRT display unit, control box, several receivers and a number of antenna units. It is stated to be a wide band unit, capable of dealing with multiple threats, with automatic sorting and identification, and automatic audio alarm and recording. Although no frequency range is quoted it is assumed that the receiver covers the normal band from 2 to 18 GHz. Flexible reprogramming is claimed, with a low weight and built-in test facilities.

CONTRACTOR
Southwest China Research Institute of Electronic Equipment, PO Box 853, Guanxian, Sichuan, People's Republic of China.

FRANCE

3374.353
TYPE BF RADAR WARNING RECEIVER

The Type BF radar warning receiver is for airborne applications in aircraft of the Mirage and similar types. It provides the pilot with warning of most categories of airborne or surface threat radars and an indication of their direction.

The system consists of four wide-band antennas, a video receiver, and optionally, a synchronisation unit. In aircraft other than the Mirage, in which case the system controls and indicator unit are integrated into the cockpit by the aircraft manufacturer, these items are supplied as separate units.

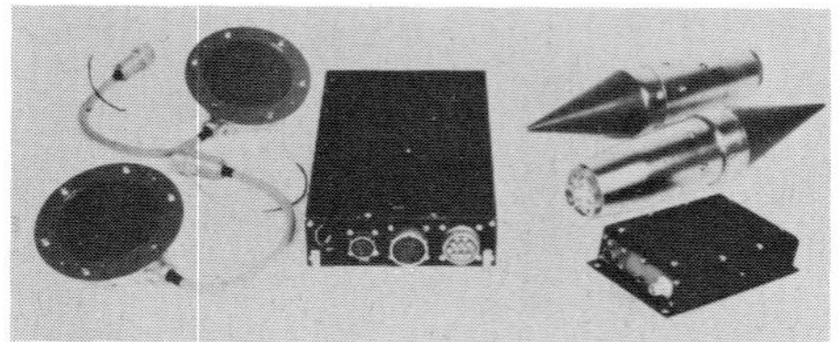
Type BF radar warning receiver equipment

Each antenna unit contains a photographically etched spiral antenna and microwave circuits such as: RF test oscillator; limiter modulator diodes; high-pass filter; detector circuits; video modulation unit; video preamplifier. Packing of the antenna units is of two types, one having a flush-mounting radome for the left- and right-hand units, and the other having a conical radome and housing suitable for mounting on the aircraft fin to provide fore and aft coverage. Each antenna has a 90° cover, to provide all-round search capability with the full system.

The BF equipment will detect signals of pulse, CW, ICW, or TWS radars, either fixed frequency or frequency-agile. PRFs between 600 and 12 000 Hz can be accommodated.

An audio alarm is given for each radar threat detected, and the general direction of the threat is given by the illumination of one or more of four signal lamps around the aircraft outline on the indicator. The nature of the threat is shown by illumination of one of three indicator lamps denoting:

(1) conventional pulse radar
(2) CW or ICW radar
(3) TWS ground threat radar.

CHARACTERISTICS
Antenna (L and R) dimensions: 53 × 148 mm diameter
Antenna (F and R) dimensions: 360 × 82 mm diameter
Video receiver dimensions: 322 × 193 × 57 mm
Synchroniser dimensions: 209 × 110 × 45 mm
Indicator dimensions: 68 × 61 × 61 mm
Control box dimensions: 146 × 95 × 40 mm
Combined weight: 9.2 kg

STATUS
Production. Operational in Mirage F1-C and Super Etendard, also Mirage F1-A, Netherlands F-5, and Mirage IIIZ aircraft. A helicopter version is produced as the TMV 008H.

CONTRACTOR
Thomson-CSF – Aerospace Group, 122 avenue du Général Leclerc, 92105 Boulogne-Billancourt, France.

3379.393
DB 3163 JAMMING POD (HIGH-BAND REMORA)

The DB 3163 jammer pod, known also as Remora, is designed to provide self-protection for tactical aircraft against both air and ground radar threats. Both pulse and CW emitters can be detected, identified and countered automatically by means of the power management system incorporated in the pod.

A superheterodyne receiver performs a frequency scan search on emissions received by antennas located at both ends of the pod, this being carried out simultaneously. Frequency bands can be selected by the pilot from six bands pre-programmed before flight. Threat data are stored digitally and up to three can be jammed at the same time by means of electronic switching and power management of the TWT transmitter system. A 'boot-strap' system of air cooling is employed.

CHARACTERISTICS
Frequency: I/J-band. Pre-programmed
Pod length: 3.52 m
Diameter: 254 mm
Weight: 175 kg
Power consumption: <1.7 kVA (transmit); <0.7 kVA (receive)

STATUS
In production.

DB 3163 jammer pod

CONTRACTOR
Thomson-CSF – Aerospace Group, 122 avenue du Général Leclerc, 92105 Boulogne-Billancourt, France.

5786.393
DB-3141 JAMMING POD (LOW-BAND REMORA)

The DB-3141 is an airborne noise jamming pod operating in the H to I frequency band, and is a lower band version of the DB-3163 (Remora) pod (**3379.393**). It contains a single receiver, a travelling-wave-tube jammer and fore-and-aft transmitter antennas. The DB-3141 provides a simple active EW capability for Dassault Mirage fighter aircraft and is also fitted to the Super Etendard deck-landing strike aircraft. The system has a 'look-through' capability which enables it to discontinue jamming as soon as threat reception ceases.

CHARACTERISTICS
Dimensions: 3.5 m long × 0.25 m diameter
Weight: 175 kg
Power required: 1.7 kVa at 200 V 400 Hz

STATUS
In production.

CONTRACTOR
Thomson-CSF – Aerospace Group, 122 avenue du Général Leclerc, 92105 Boulogne-Billancourt, France.

DB-3141 jamming pod

5787.353
CAIMAN NOISE/DECEPTION JAMMING POD

Caiman is an airborne noise jamming pod which is probably developed from the earlier Alligator system. It is intended for dedicated EW aircraft which are operating to protect groups of similar aircraft in ground attack roles by jamming surveillance and target designation radars. Within the pod are two jammers, each with its own receiver, and fore-and-aft receiver antennas. Power output from each jammer is 250 watts in I-band (10 GHz) or 1 kW in D-band (1 GHz). The jamming signal bandwidth is programmable before the mission. Three modes of operation are available; in manual mode the pilot can select any one of three predetermined frequencies at will; in semi-automatic mode a signal indicates to the pilot when a transmission is being received so that he can take the appropriate action; and in fully automatic mode the jammers are automatically activated when such a signal is received.

Caiman jamming pod on a Mirage F1

The pod is self-contained with ram-air entering the unit through an annular intake to drive a power turbine and provide cooling. It can be installed either underwing or on a fuselage pylon.

CHARACTERISTICS
Dimensions: 5.95 m long × 0.41 m diameter
Weight: 550 kg

STATUS
In production. Fitted to the Mirage F1 aircraft and supplied to other countries for several aircraft types, including the Mirage 2000 and the Jaguar.

CONTRACTOR
Thomson-CSF – Aerospace Group, 122 avenue du Général Leclerc, 92105 Boulogne-Billancourt, France.

5788.393

SYREL ELINT POD (TMV 018)

Syrel is a fully-automatic electronic reconnaissance system pod which is pylon-mounted under the fuselage of the fighter. It is designed primarily for use during tactical penetration missions at medium or low altitudes to acquire and record automatically data relating to the identification and location of ground-based electronic systems. It is intended to provide reliable information of early warning, search and acquisition, GCI and fire control radars associated with anti-aircraft missiles or artillery.

In normal operations, the aircraft will carry out direction finding and position fixing of transmitters by flying along a path and taking a series of bearings to establish the position of the emitter.

The pod is 3.57 m long × 0.42 m diameter and weighs 265 kg. It has antennas at both front and rear, with receiver units, an amplifier and recorders in the centre section. The pylon houses a cooling system which has a ram-air intake in the leading edge. High speed of the system is assisted by thick and microwave circuit assemblies on ceramic substrates.

Syrel ELINT pod mounted on Mirage F1

STATUS

In production and in operational service in the Mirage F1 and Mirage 2000.

CONTRACTOR

Thomson-CSF – Aerospace Group, 122 avenue du Général Leclerc, 92105 Boulogne-Billancourt, France.

5946.393

BAREM SELF-PROTECTION JAMMER

Barem is a pod-mounted self-protection jamming system designed for tactical/strike aircraft. It is equipped with travelling wave tubes and provides instantaneous coverage over a wide frequency band. It can detect, identify and jam simultaneously a number of threats from a variety of ground-to-air and air-to-air pulse, Doppler and CW radars. The equipment has an extensive memory capacity and a modular software design which is reprogrammable to cater for future threat developments. Operation is fully automatic. Barem has been developed using much of the experience gained with Remora (**3379.393**). It is, however, considerably lighter. It can be packaged in two units for internal installation, if required.

STATUS

In operational service with the French Air Force.

CONTRACTOR

Thomson-CSF – Aerospace Group, 122 avenue du Général Leclerc, 92105 Boulogne-Billancourt, France.

Barem self-protection jammer

5947.393

BASILISK JAMMING POD

Basilisk is another pod-mounted Thomson-CSF jamming system which has been designed especially for support jamming, ie against surveillance radars, including battlefield and AWACS type systems. It instantaneously covers three full octaves of hostile frequencies, and can jam up to 20 threats simultaneously. The pod which contains the transmitters, power supplies and antennas can be carried either under the wing or the fuselage of the aircraft. Pod dimensions are 5.98 m long × 0.41 m diameter, with a weight of 550 kg. For jamming tracking radars, Basilisk operates over a wider frequency range than Caiman (**5787.353**), and has an output jamming power in excess of 500 W. Basilisk incorporates an extensive memory capacity, and modular design software for reprogramming to meet additional threats. No other details are currently available, although it can be assumed that the system operation can be automatic, semi-automatic or manual.

STATUS

In development. In addition to the various types of jamming equipment in service or in development, Thomson-CSF is known to be developing a solid-state jamming array which will be ready for fitment to the Rafale aircraft.

CONTRACTOR

Thomson-CSF – Aerospace Group, 122 avenue du Général Leclerc, 92105 Boulogne-Billancourt, France.

5789.353

SERVAL RADAR WARNING RECEIVER

The Serval radar warning receiver is a new generation detector which provides the pilot with warning when his aircraft is illuminated by surface or airborne threat radars. Frequency coverage is from 2.5 to 18 GHz. Displayed information is threat direction, level and identification. The latter is obtained by comparison of the received signal with a threat library contained in the system. The display unit is a panel-mounted CRT where several threats can be displayed simultaneously. At the same time, an audio alarm is generated in the pilot's headset.

Four antennas are used, located at the aircraft wingtips and vertical fin. They are connected to an analogue and digital processing unit.

STATUS

Fitted to both French Air Force and export versions of the Mirage 2000 and a number of other aircraft types. A later version, known as Sherloc, is described below (**6251.393**).

CONTRACTOR

Thomson-CSF – Aerospace Group, 122 avenue du Général Leclerc, 92105 Boulogne-Billancourt, France.

6251.393

SHERLOC RADAR WARNING SYSTEM

Sherloc is an airborne crystal video radar warning receiver intended for fighter/attack aircraft and helicopters. It is designed for instantaneous detection and identification of radar emitters in complex signal environments. Sherloc is also designed to integrate data from the Thomson-CSF laser warning receiver (**6252.393**). This provides the crew with comprehensive information regarding the overall threat environment of current and future missions.

The basic system covers the frequency range 2–18 GHz, with extensions available at both ends of the range to cover from 1 to 40 GHz. A high speed analogue and digital processor is incorporated with a threat library of 100 modes which can be readily reprogrammed on the flight line. The library can be extended to 200 modes.

The system has very high sensitivity, is of small size and weighs only 10 kg. It is capable of detecting all types of radar, whether rotating, scanning or tracking. Sherloc can be fitted as a new installation or can be retrofitted to nearly any type of aircraft. Radar data is presented to the pilot in the form of alphanumeric symbology on a CRT display. These symbols correspond to the various signal identities, and their actual position on the display indicates signal bearing and approximate range. An alternative LED display shows threat bearing, level and identification in tabular format. In both cases a total of eight threat emitters can be presented simultaneously to the pilot. An aural warning is also incorporated, and an optional synthetic voice is available. This latter function is particularly useful on helicopters where the pilot is not able to look at the display for long periods.

Units of Sherloc radar warning receiver

STATUS

At least 50 systems have been sold and the equipment is being fitted to the Mirage F1.

CONTRACTOR

Thomson-CSF – Aerospace Group, 122 avenue du Général Leclerc, 92105 Boulogne-Billancourt, France.

6904.353

MSPS EW SYSTEM

MSPS is a very lightweight, easy to install EW suite mainly designed for retrofit of existing aircraft. It combines the Sherloc RWR with the Barem jamming pod, providing a combined crystal video/superheterodyne receiving system for accurate threat description and fast data transmission between the two equipments for fast and accurate reaction.

The MSPS system can be installed on a wide variety of fixed-wing aircraft and helicopters. The jamming part can be installed internally if enough room is available. The system is fully automatic and reprogrammable and can accept additions such as chaff and flare countermeasures, laser detection equipment and support jamming systems. The weight of the system is 100 kg when the jammer is pod-mounted and 80 kg when internally installed.

STATUS

In production.

CONTRACTOR

Thomson-CSF – Aerospace Group, 122 avenue du Général Leclerc, 92105 Boulogne-Billancourt, France.

6905.393

MIRIADE RADAR WARNING SYSTEM

Miriade is a warning receiver in the millimetre wave region. It is designed for the protection of helicopters, light aircraft or armoured vehicles from 'smart' weapons using millimetre wave seekers. It provides immediate warning over a 360° coverage and weighs approximately 8 kg.

STATUS

In development.

CONTRACTOR

Thomson-CSF – Aerospace Group, 122 avenue du Général Leclerc, 92105 Boulogne-Billancourt, France.

6252.393

THOMSON-CSF LASER WARNING RECEIVER

Thomson-CSF is currently developing a laser warning receiver which is suitable for fitment to a number of different platforms, including fixed-wing aircraft, helicopters and armoured vehicles. It covers the band from 0.69 to 1.06 microns and will detect any laser emissions from target markers, rangefinders and beamriders, with a 360° coverage in azimuth by means of either four or eight sensors. Laser threats are displayed on the Sherloc RWR screen (**6251.393**). Detection range is of the order of 10 km. A further development programme is in process to cover the infra-red frequency band for protection against the latest CO_2 laser systems which are likely to come into general service in the near future.

STATUS

In late development.

CONTRACTOR

Thomson-CSF – Aerospace Group, 122 avenue du Général Leclerc, 92105 Boulogne-Billancourt, France.

6253.393

THOMSON-CSF ACTIVE EXPENDABLE JAMMERS

A range of active expendable jammers has been designed by Thomson-CSF. Three main types are being developed, two airborne and one for naval applications.

The first of these is a self-protection jammer which can be fitted in a normal flare container and would cover the C to Ku frequency bands. The jammer has delta-wing configuration with an ERP of about 10 W, and is designed to 'fly' for a reasonable period of time. This type of jammer is much more effective against doppler radars than the conventional chaff.

The second airborne active jammer is for offensive purposes, covers L to S bands, and creates false targets to protect air corridors, etc. It is of a delta-wing configuration and can stay airborne for up to 20 minutes.

The naval jammer is designed to be fired from a ship at slow speed to make an angular capture of a missile radar seeker. It is fired by a rocket of the same size as that employed in the Dagaie cartridge.

CONTRACTOR

Thomson-CSF – Aerospace Group, 122 avenue du Général Leclerc, 92105 Boulogne-Billancourt, France.

5831.353

MIRAGE 2000 INTEGRATED EW SUITE (ICMS)

Thomson-CSF, Electronique Serge Dassault, SAT and Matra are developing in co-operation, a new integrated, internally mounted EW suite for the Mirage 2000 aircraft. This is a highly sophisticated system, known as ICMS (Integrated Counter Measures Suite), where all parts are linked to a central interface and management unit.

The system incorporates three warning receivers, plus an IR threat detector which is currently being designed by Matra. These receivers consist of a version of the Serval equipment (**5789.353**), a superheterodyne receiver to detect CW radar, pulse-compression signals and low-power pulse Doppler signals, and a receiver/processor mounted in the aircraft nose to detect missile command links. The latter equipment is designed to carry out initial countermeasures when a locked-on missile is detected.

Two detector-jammers are included, each with its own receiver which allows them to operate should the radar warning receiver be out of action. These jammers consist of a high band sub-unit (designed by Electronique Serge Dassault – see entry **5941.393**) to counteract airborne and surface-to-air threats, and a low band sub-unit (designed by Thomson-CSF) to operate against surface-to-air missiles. The Matra Spirale chaff-IR flare dispenser is also included in the overall system to provide passive countermeasures. Spirale is an internally-mounted equipment which dispenses stores through openings in the aircraft structure. The SAT IR detector is described in entry **6903.393**.

STATUS

In development.

CONTRACTORS

Thomson-CSF – Aerospace Group, 122 avenue du Général Leclerc, 92105 Boulogne-Billancourt, France.

Electronique Serge Dassault, 55 quai Marcel Dassault, 92214 St Cloud, France.

S A Matra, Centre de Montigny, 3 avenue du Centre, 78182 Saint-Quentin Yvelines Cedex, France.

SAT (Société Anonyme de Télécommunications), 41 rue Cantagrel, 75631 Paris Cedex 13, France.

6254.353

ASTAC TACTICAL ELINT SYSTEM

ASTAC (previously known as STAR) is a tactical airborne electronic reconnaissance system consisting of a pod-mounted airborne sensor and a ground processing station. It is intended to perform detection, identification and localisation of any radar type in a very dense environment. A data link between the pod and the ground station enables a very rapid build-up of the electronic order of battle of the observed area.

The main characteristics of the system are a very wide frequency coverage, wide instantaneous bandwidth, high sensitivity, high discriminating power and high direction measurement accuracy by interferometer. The system is fully automatic, fully reprogrammable and possesses very high speed processing capability of up to 20 radars per second. It can process pulse modulated radar with pulse repetition internal diversity or agility, or radio frequency diversity or agility, as well as pulse compression, CW and interrupted CW systems. Two SAW receivers are used, one wide band and one narrow band. The control unit consists of two displays, one CRT and the other tabular, and a keyboard.

The pod, which is 4.5 m long × 0.53 m diameter weighing 420 kg, is intended for fitment to the Mirage F1 and Mirage 2000. The building blocks of the system can be installed internally if enough room is available.

STATUS

In late development and preproduction.

CONTRACTOR

Thomson-CSF – Aerospace Group, 122 avenue du Général Leclerc, 92105 Boulogne-Billancourt, France.

3376.353

TMV 026 AIRBORNE ESM SYSTEM

The TMV 026 airborne ESM system consists of a DR2000 radar intercept receiver, an alarm, analysis, and identification equipment (DALIA 1000, DALIA 500 or ARIAL 15) and a set of six DF antennas and one omni-directional antenna. The equipment is suitable for use in both helicopters or fixed-wing maritime patrol aircraft.

The DR2000 provides all-round passive search, detection of all pulse and CW signals, and gives an instantaneous audible and visual alert.

Additional alarm and analysis capabilities are given by the DALIA (or ARIAL – *analyse radar pour identification et alerte*) equipment. This provides specific threat warning signals; automatic analysis of

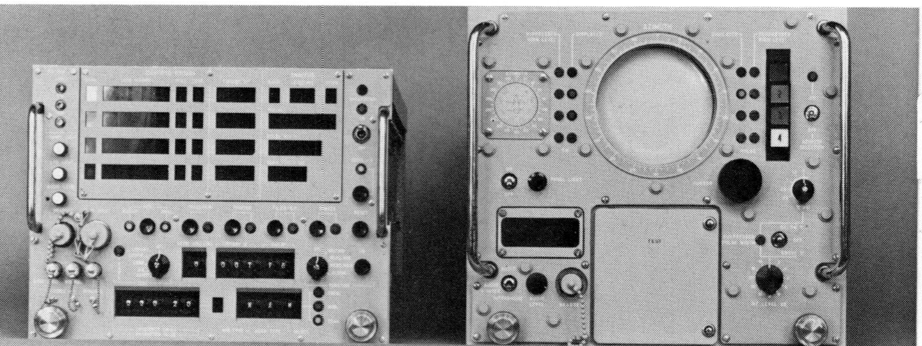

DALIA processor (left) and DR 2000 display/control unit form part of TMV 026 system

received signals to yield frequency, PRF, pulse width, RF level, antenna rotation rate, jitter, etc; and automatic identification of emitters by comparison with a library including 1000, 500 or 15 radars (DALIA 500 or ARIAL 15)

STATUS

Series production. Limited numbers only were delivered with the ARIAL analyser, most employing the higher capacity DALIA equipment. An improved version of DR 2000, the Mark 2, is now in production. This is characterised by a wider frequency coverage and higher sensitivity (see entry **6486.293**).

CONTRACTOR

Thomson-CSF – Aerospace Group, 122 avenue du Général Leclerc, 92105 Boulogne-Billancourt, France.

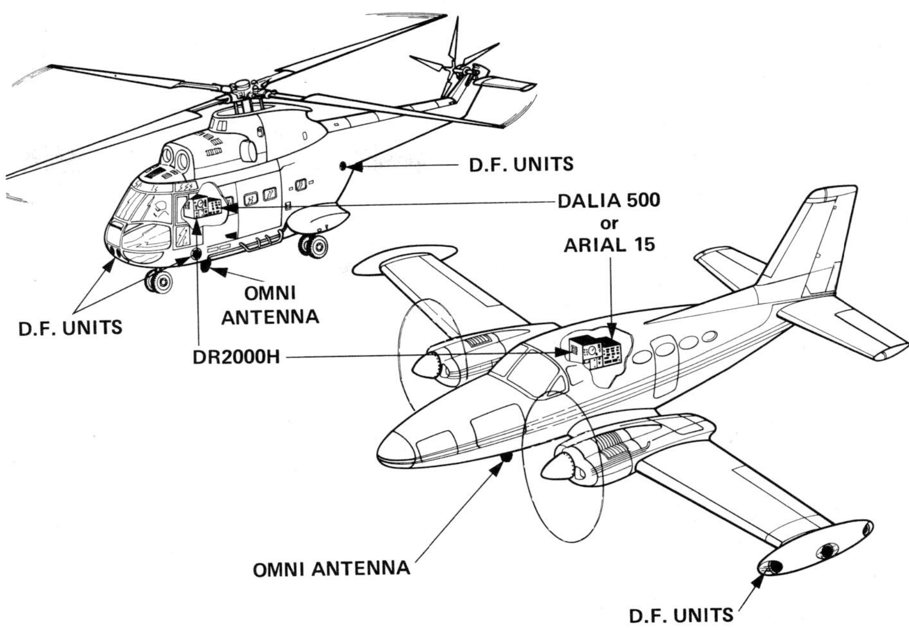

Location of equipment in typical helicopter and fixed-wing aircraft installations of TMV 026 airborne ESM system

6506.353

TMV 202 (DR 4000A) ESM SUITE

The TMV ESM 202 suite is a new generation designed for high sensitivity, instantaneous intercept probability and automatic processing against electromagnetic threats. It is intended for use in fixed and rotary-wing aircraft and provides threat detection, surveillance, identification, ELINT and data handling system integration. It is a version of the DR 4000 series of equipments, described under entry **4351.053**. The basic system consists of antennas (2 × 6 DF antennas, one omni-directional antenna, MIC modules housing RF amplifiers), an IFM processing unit, and an operator console containing controls and three-colour graphic and alphanumeric displays.

CONTRACTOR

Thomson-CSF – Aerospace Group, 122 avenue du Général Leclerc, 91205 Boulogne-Billancourt, France.

DR 4000 ESM suite

6906.393

THOMSON-CSF SIGINT AIRCRAFT RECEIVERS

Thomson-CSF has developed a range of receivers for large aircraft ranging from the Falcon type to the Boeing 707 or DC-8. They are based on the Elisa family (**3534.053**) and feature wide band superheterodyne receivers and interferometry for high probability of intercept, high sensitivity and very high localisation accuracy.

STATUS

In production.

CONTRACTOR

Thomson-CSF – Aerospace Group, 122 avenue du Général Leclerc, 92105 Boulogne-Billancourt, France.

6907.353

MARITIME PATROL ESM

ESM is one of the major systems on board maritime patrol aircraft and Thomson-CSF has developed systems for this specific purpose. The DR 2000 system is one such and a new system, the DR 3000A, is being developed. Technical details are highly classified but it is known that it provides 360° coverage over the D to K frequency bands with a DF accuracy of approximately 6°. Weight of the new system is given as 50 kg.

The accompanying picture shows a typical ESM system, as installed in the Atlantique ATL-2 aircraft.

CONTRACTOR

Thomson-CSF – Aerospace Group, 122 avenue du Général Leclerc, 92105 Boulogne-Billancourt, France.

ESM system for Atlantique ATL2 aircraft

5941.393

ABD 2000 DETECTOR-JAMMER SYSTEM

The ABD (Airborne Broadband Defence) detector-jammer system is an internally-mounted system designed primarily for the Mirage 2000 aircraft. It includes a signal receiver for piloting a jammer capable of countering present and foreseeable future air-to-air and surface-to-air threats. The system consists of four line replaceable units; a main unit including rear coverage antennas, two auxiliary units, and a fin fitted antenna module providing forward coverage. The ABD system is programmable as a function of the threats present in a particular operational theatre and is capable of operating in a fully automatic mode. A built-in-test function is provided.

STATUS

In production for the air-defence and strike versions of the Mirage 2000 aircraft. It is in service in several air forces.

CONTRACTOR

Electronique Serge Dassault, 55 quai Marcel Dassault, 92214 St Cloud, France.

ABD internal detector-jammer for Mirage 2000

6902.393

BARAX DETECTOR-JAMMER

Barax is a pod-mounted airborne detector-jammer intended for the self-protection of combat aircraft. It has a supersonic capability, with a diameter of 157 mm and a weight of 85 kg. Possessing a wide frequency band, and multiple and independent jamming techniques, it is capable of processing all types of threat. It is functionally independent, able to perform watch, alert and jamming functions automatically over the complete frequency band. A threat library is preprogrammed and loaded before take-off. Reaction time is very short and, since it is entirely controlled by microprocessors, it can be adapted to future threat developments.

STATUS

In full production.

CONTRACTOR

Electronique Serge Dassault, 55 quai Marcel Dassault, 92214 St Cloud, France.

6903.393

DDM IR DETECTOR

The DDM equipment is intended for the Mirage 2000 and will detect the launch of any IR guided missile directed against the aircraft. As soon as the DDM detects the threat it supplies information on the threat position to ensure that the countermeasures system computes the most efficient moment for releasing decoys. A warning of the attack is also given to the pilot.

DDM is a bispectral system with panoramic or 180° sectorial azimuth coverage and 60° coverage in elevation. It has a very fast reaction time, a low false alarm rate and weighs less than 17 kg. All other technical details are classified.

STATUS

In full development with prototypes undergoing airborne testing by the French Air Force. Main production is scheduled for the early 1990s.

CONTRACTOR

SAT (Société Anonyme de Télécommunications), 41 rue Cantagrel, 75631 Paris Cedex 13, France.

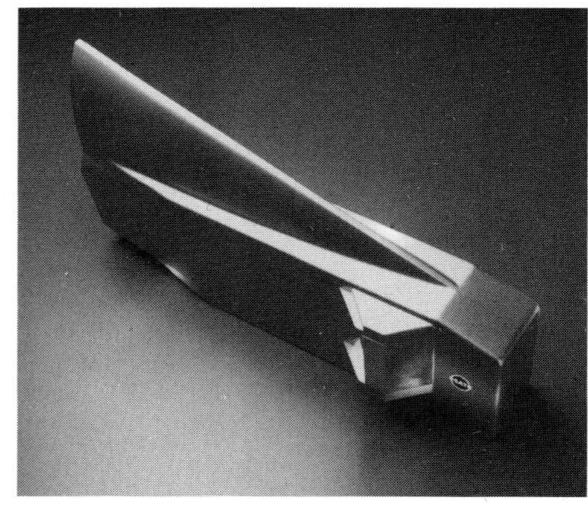

DDM IR detector for Mirage 2000

5665.393

ALKAN CHAFF/INFRA-RED COUNTERMEASURES DISPENSERS

ALKAN has developed a generation of countermeasures dispensers that can accommodate either chaff or infra-red flare cartridges, or a combination of both. The cartridges are arranged in interchangeable, easily handled magazines loaded in the modular dispenser. Typically, a single dispenser will accommodate five to seven modules. Each module houses a magazine containing, for instance, either eight 60 mm diameter infra-red or eighteen 40 mm diameter chaff cartridges.

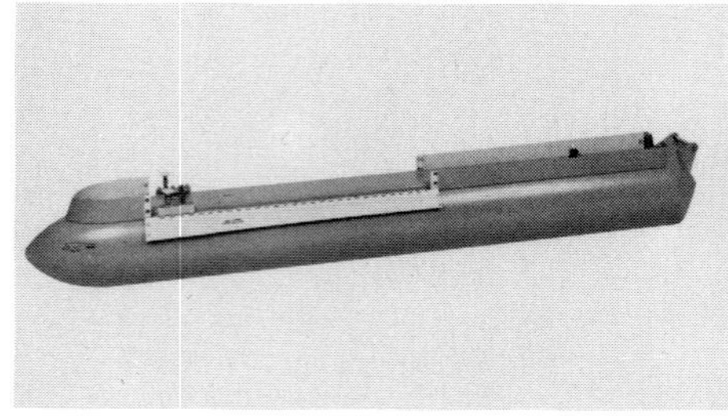

Decoy dispenser Type 5080

The electronic management system of the dispenser, whether connected or not to a radar warning receiver, performs the firing sequence created by software, permanently manages the inventory of available cartridges, and provides the necessary information to a cockpit control unit which displays the status of the complete equipment.

STATUS

The system designed for the Jaguar aircraft, which is now in series production, includes two dispensers under each wing in a conformal installation near the aircraft fuselage. Each dispenser contains seven modules (type 5020). The same system is used for the Mirage F1 and Mirage 2000 aircraft as the heart of the complete Matra self protection system, called respectively Sycomor and Spirale. Another example of integrated installation is the Type 5013 which is under development for retrofit of the Mirage III/5 aircraft. The unit compries four modules and is installed in the rear part of the aircraft fuselage.

The Alkan pod type 5080 is designed to fit either to the JATO point of the MiG-21 or to any 14-inch standard armament hard point. It is in series production.

Various other applications of the same concept are in development for a wide range of aircraft, including military lift, maritime patrol or light ground attack aircraft.

CONTRACTORS

R. Alkan & Cie, rue du 8 Mai 1945, 94460 Valenton, France.

4635.393

LACROIX AIRBORNE COUNTERMEASURES AMMUNITION

Société E Lacroix manufactures a range of electromagnetic (EM) and infra-red (IR) passive countermeasures ammunition for combat aircraft. These munitions are used in aircraft such as the F-16, Tornado, Viggen, Jaguar, Mirage, Alpha-Jet, Phantom and a number of helicopters. The company also manufactures 'bright flash' cartridges designed to blind temporarily or to disorientate weapon systems using optical seekers.

The EM cartridges, Types 622, 651, 657, 677, 728 and 888L, are made up of aluminium coated fibre-glass, with the length cut according to the individual customer requirements, and are electrically ejected. The IR munition, Types 407, 587, 623, 658 and 659, consists of one or two IR flares fitted with a pyrotechnic part containing ignition and safety devices and is ejected electrically from the cartridge. A new chaff cartridge, known as Type 700 (or ZCH), is being developed in collaboration with Philips of Sweden. The cartridge consists of three groups of three subloads each which are ejected in sequence by pyrotechnic squibs. New IR flare cartridges are the Types 698 and 750.

STATUS

Mass production.

CONTRACTOR

Société E Lacroix, Departement Contre-mesures, Route de Toulouse, 31600 Muret, France.

5790.393

MATRA PASSIVE COUNTERMEASURES

MATRA is concerned with the development of passive countermeasures equipment, based primarily on the company's experience with PHIMAT, the electromagnetic decoy chaff system in use with a number of air forces. The company is now engaged in the design and production of an intelligent system family for analysing and dealing with threats, and for control and programming.

Sycomor

This has been developed for the Mirage F1 and other aircraft in service to jam radars, radar seekers and infra-red guided missiles by ejecting clusters of chaff and/or cartridges. The equipment is available in the form of containers and/or gondolas fixed under the wing. It is in operational service with the Mirage F1.

Spirale

This fulfils the same requirements as Sycomor and is in development as an internal system for the Mirage 2000. Dispensers and IR warning receivers are integrated into the fuselage, with chaff and cartridges ejected through openings in the aircraft structure. It has been ordered by the French Air Force and a number of other air forces.

Saphir

A self-protection system in development for helicopters.

CONTRACTOR

S A Matra, Centre de Montigny, 3 avenue du Centre, 78182 Saint-Quentin Yvelines Cedex, France.

Sycomor countermeasures pod under Mirage F1

GERMANY (FEDERAL REPUBLIC)

6908.353

ROHDE & SCHWARZ COMINT SYSTEM

A powerful, computer-controlled COMINT system, covering the VHF/UHF frequency band, has been developed by Rohde & Schwarz, and consists of several airborne stations and a semi-mobile ground station for evaluation of the collected data. The airborne system is automatic and only one operator is required to handle the airborne equipment. Detection and subsequent recording of the signals is computer-controlled and very fast, the data being fully available to the ground station immediately the mission is completed.

The computer-controlled automatic receiver ESP (see entry **4139.093**) used for the detection of the radio signals is the heart of the airborne monitoring system. Ten VHF/UHF ESM 500 receivers (**4138.093**) tuned to active frequency channels are provided to monitor the radio signals over the frequency range 20 to 1000 MHz.

The semi-mobile ground station is accommodated in a double shelter and carries out processing of the data recorded on audio cassettes in the airborne station. Audio records of particular interest can be transmitted in flight to the ground station via the aircraft communications system.

CONTRACTOR

Rohde & Schwarz, Muhldorfstrasse 15, 8000 Munich 80, Federal Republic of Germany.

INTERNATIONAL

6157.393

AWARE-3 RADAR WARNING RECEIVER

AWARE-3 is an improved version of the DATAR system developed jointly by Ferranti Instrumentation of UK and E-Systems of USA.

The equipment is a radar warning receiver with facilities for integration with electronic countermeasures subsystems. It has been specifically designed to operate in a dense radar environment, particularly in helicopters where weight, size and ease of installation are critical. It can also be used in other applications such as light strike and transport aircraft, land vehicles and ships.

The AWARE-3 system uses aircraft-mounted antennas, coupled with crystal video receivers to detect radar pulses with the high probability of intercept necessary for immediate warning. The equipment operates in both pulse and CW environments and covers the E to J frequency bands with provision for extension. It analyses received signals to determine frequency and PRI and compares this information with data held in its threat library. One of the features of the system is the ability to update the threat library in the field and to determine priorities on a mission to mission basis. The threats are displayed to the crew on a small 3-inch CRT indicator in plan position format. The display is heading compensated showing the relative range and bearing of the threat with symbology depicting the type. An imminent threat triggers an audio alarm. The system can also be used to trigger ECM subsystems.

AWARE-3 is packaged into four line replaceable units, a receiver, central processor, indicator and control unit. A typical installation employs five antennas. Interfaces are provided for blanking and bus control through MIL-STD-1553B, and also for laser warning systems and passive ECM (chaff and flares).

CONTRACTORS

Ferranti Instrumentation Ltd, Aircraft Equipment Department, Lily Hill House, Lily Hill Road, Bracknell, Berkshire RG12 2SJ, England.

E-Systems Inc, ECI Division, 1502 72nd Street N, Po Box 12248, St Petersburg, Florida 33733-2248, USA.

ISRAEL

5948.393

EL/K 7032 COMINT SYSTEM

The EL/K 7032 is an airborne communications intelligence (COMINT) system designed for electronic surveillance and interception of signals within the 20 to 500 MHz frequency range. It is performed by local or remote digitally controlled communications receivers operating in the HF, VHF and UHF bands and featuring high speed switching between bands to allow very fast and reliable coverage of the complete spectrum. Computer-controlled automatic operation enables a minimum number of individual operators to carry out multi-task COMINT missions.

STATUS

In production.

CONTRACTOR

Elta Electronics Industries Ltd (a subsidiary of Israel Aircraft Industries Ltd), Ashdod, Israel.

EL/K-7032 airborne COMINT system

4357.393

EL/L-8202 ECM POD

The EL/L-8202 is a compact ECM pod designed to provide self-protection for combat aircraft against both surface and air radar threats. It has a wide frequency coverage and a high broadband output with selectable ECM techniques. A number of optional versions are available which include specified frequency ranges, an enhanced receiver/processor unit and interfaces which are adaptable to a particular aircraft's avionics.

The EL/L-8202 pod comprises the following major units: antennas and radomes, RF unit, RF power amplifier, logic unit, pod shell and cooling unit. The RF unit generates electronic countermeasures, jamming signals according to the required ECM programme. These signals are further amplified by the power amplifier, employing broadband high-power TWTs and transmission is through the antennas which are located at both ends of the pod and features special beam-shape patterns. The logic unit generates the logic and video signals required for the selected ECM programme and provides interface functions to the aircraft radar warning system and to the cockpit control unit.

The pod itself is carried on the normal ordnance station of the aircraft and is designed to have a minor impact on aircraft performance. Cooling is performed in the normal ECM pod method, namely an integral ram/air liquid unit.

CHARACTERISTICS

Frequency range: F – J-bands
Power consumption: 1.7 – 2.3 kVA (0.5 kVA standby)
Dimensions: 26 × 39 × 290 cm
Weight: 200 kg

STATUS

In operational service with the Israeli Air Force.

CONTRACTOR

Elta Electronics Industries Ltd (a subsidiary of Israel Aircraft Industries Ltd), Ashdod, Israel.

E/L-8202 ECM pod on Kfir C-2 aircraft

6153.393

EL/L-8231 ECM SYSTEM

The EL/L-8231 is an internally mounted miniaturised ECM system covering the H to J frequency range and designed to protect combat aircraft or helicopters against missiles using CW radar guidance. The equipment is small and lightweight and consists of an antenna assembly and RF transmission line, an equipment unit containing RF, logic and control circuits, an RFI unit for interface with aircraft power lines, and a cockpit-mounted control unit. The antenna system can be single or separate transmit and receive arrays, depending on aircraft installation constraints. The complete system weighs 18 kg.

Radar signals are received and analysed, and jamming starts automatically if the signal is identified as a threat. The incoming RF is amplified by the power amplifier, employing a broadcast TWT transmitter. The logic unit generates the logic and video signals required for ECM signal generation.

STATUS

In service.

CONTRACTOR

Elta Electronic Industries Ltd (a subsidiary of Israel Aircraft Industries Ltd), PO Box 330, Ashdod 77102, Israel.

Equipment unit of EL/L-8231 ECM system being installed

5730.393

EL/L-8230 AIRBORNE ECM SYSTEM

The EL/L-8230 is a combined noise and repeater jammer designed to protect combat aircraft against both surface and air radar threats. It is internally mounted and operates over the G to J frequency band with a power output of nearly 1 kW. The system consists of antennas, an RF unit, an RF power amplifier and a logic unit. The latter unit provides logic and video signals for the application of the required ECM programme, and also provides interface functions to the aircraft radar warning receiver. The system contains its own receiver and is capable of automatic or manual operation modes. It can be packaged in different configurations for a variety of aircraft installations.

STATUS

In operational service with the Israeli Air Force.

CONTRACTOR

Elta Electronics Industries Ltd (a subsidiary of Israel Aircraft Industries Ltd), Ashdod, Israel.

6909.353

EL/L 8240 ECM SYSTEM

The EL/L 8240 is an internally-mounted self-protection ECM system designed principally for advanced combat aircraft. The system consists of a fully integrated radar warning and radar jamming/deception equipment and can be adapted for installation on any combat aircraft. It provides wide frequency and angular coverage, with a short response time and a high transmitted power in the jamming mode. A substantial number of jamming and deception techniques are included. Total weight of the system is 136 kg.

CONTRACTOR

Elta Electronic Industries Ltd (a subsidiary of Israel Aircraft Industries Ltd), PO Box 330, Ashdod 77102, Israel.

5949.393

EL/L 8312A ELINT SYSTEM

The EL/L 8312A system is an airborne computerised electronics intelligence (Elint) equipment. It comprises units such as the EL/L 8312R modular microwave receiver, the EL/L 8321 pulse digitiser, EL/S 8610 computer and the EL/S 8570 alpha-numeric display. These can be configured according to the operational requirement, and as part of the overall system Elbit Computers has developed a multipurpose tactical display that can serve various intelligence users. The overall equipment forms part of the EL/L-8300 SIGINT system.

The system is an automatic microwave receiving system which detects, measures parameters, classifies and identifies targets at long distance. The target bearing is measured instantaneously and the range/location is established after a series of bearings have been taken along the aircraft's track.

STATUS

In production.

CONTRACTOR

Elta Electronics Industries Ltd (a subsidiary of Israel Aircraft Industries Ltd), Ashdod, Israel.

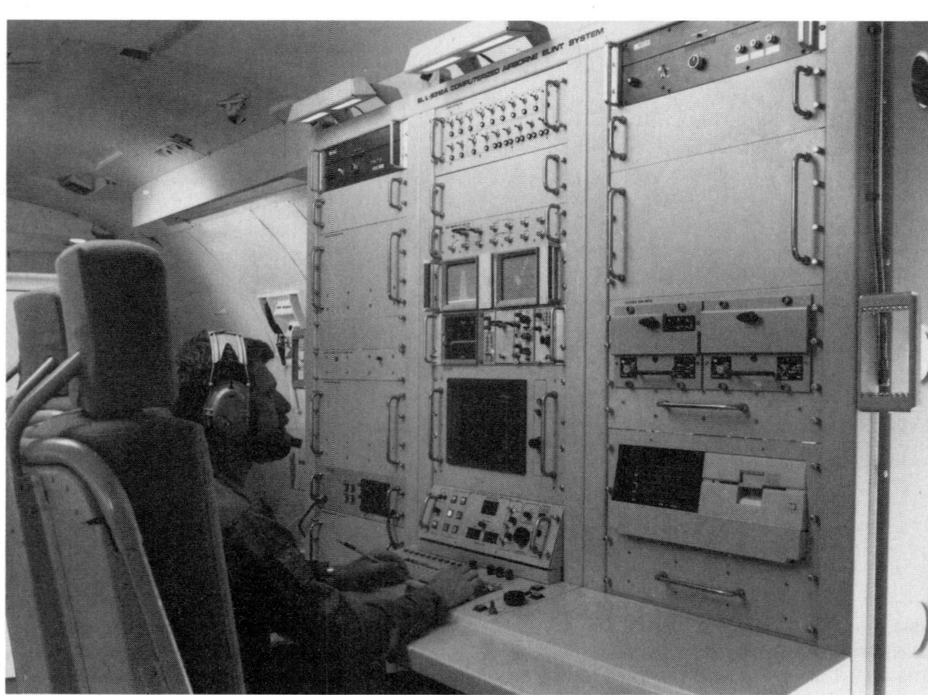

EL/L-8312A airborne ELINT system

6152.353

EL/L-8300 AIRBORNE SIGINT SYSTEM

The EL/L-8300 is an airborne signals intelligence (SIGINT) system that provides strategic, operative and tactical information on opposing forces by means of electronic intelligence (ELINT) and communications intelligence (COMINT). The airborne part of the system is the EL/L-8312A (described under entry **5949.393**), the EL/K-7032 (**5938.393**), and the EL/L-8350 Command and Analysis Station. This latter station provides mission supervision and guidance, integration of ELINT and COMINT intelligence, and real-time reporting to the ground stations. The airborne component of the EL/L-8300 features high sensitivity, 360° coverage, detection of signals up to ranges of 450 km, fast reaction time, accurate measurement of signal parameters and graphic/alphanumeric man-machine interfaces for the operators.

The ground components of the overall system are the EL/L-8353 Tactical Ground Station, the EL/L-8352 ELINT Post-Mission Analysis System and the EL/L-8351 ELINT Trainer Simulator. The ground station provides data processing, evaluation and dissemination of intelligence, while the analysis system also updates the intelligence database and produces post-mission reports.

CONTRACTOR

Elta Electronic Industries Ltd (a subsidiary of Israel Aircraft Industries Ltd), PO Box 330, Ashdod 77102, Israel.

El/L-8300 airborne SIGINT system

4640.353

SPS-20 AIRBORNE SELF PROTECTION SYSTEM

The SPS-20 is a wideband warning system, designed to fit existing airborne installations while providing improved characteristics and additional capabilities. The system detects and displays pulsed radar threats operating within the 0.7 to 18 GHz frequency range. CW radar threats are detected by a dedicated high sensitivity receiver. A high sensitivity, 3-inch display unit provides an alpha-numeric presentation of the type, coarse angle-of-arrival, relative lethality and status of the analysed radar threats.

The SPS-20 features an advanced digital signal analyser which executes data processing and interfacing tasks. It also provides an extensive emitter library file which can be easily updated for new and changing environments and/or systems. The system was originally designed for helicopters but has also been used in fighter aircraft.

A more sophisticated version, the SPS-20/SRS-25/LWS-20, has been developed as an integrated helicopter self-protection system. The equipment has an improved capability to intercept, detect, analyse and display battlefield threats, with the incorporation of the laser warning system LWS-20 to provide detection of laser emissions. The SRS-25 is described under entry **6507.393** and the LWS-20 under **6508.393**.

STATUS

In service.

CONTRACTOR

Elisra Electronic Systems (a subsidiary of Tadiran), 48 Mivtza Kadesh Street, Bene Beraq 51203, Israel.

SPS-20 self protection system

6507.393

SRS-25 AIRBORNE RECEIVER

The SRS-25 is an airborne superhet receiver forming part of the overall SPS-20 integrated helicopter self-protection system. It features detection of CW, high PRF and medium ERP radars, frequency measurement and direction finding of received signals, and sophisticated signal processing. It detects and processes those CW radar and high PRF signals which are not detected by conventional radar warning systems. The SRS-25 scans the frequency range, measures frequency, calculates the angle of arrival, analyses emitter parameters and provides data for display.

The SPS-1000 integrated system combines the SRS-25 receiver with the SPS-200 self-protection system (**5793.353**) and is intended primarily for the retrofit and upgrading market.

CHARACTERISTICS

Frequency: 6.5 to 18 GHz

Dimensions: 101 × 122 × 250 mm

Weight: 4 kg

CONTRACTOR

Elisra Electronics Systems (a subsidiary of Tadiran), 48 Mivtza Kadesh Street, Bene Beraq 51203, Israel.

SPS-1000 system combines SRS-25 and SPS-200

6508.393

LWS-20 LASER WARNING SYSTEM

In the LWS-20 airborne laser warning system, four dual laser sensors receive and detect laser pulsed signals and send detected pulses to the warning analyser. The analyser characterises each pulse for its angle of arrival, relative time of arrival and amplitude. The analyser processes these pulses, single pulses or pulse trains, and identifies the emitter threat based on preprogrammed library information. A display of angle of arrival in alphanumeric symbology is provided, as well as a visual and audible alarm. Weight of the system is 3.4 kg, including four sensors and the analyser. The LWS-20 forms part of the SPS-20 integrated self-protection system.

CONTRACTOR

Elisra Electronic Systems (a subsidiary of Tadiran), 48 Mivtza Kadesh Street, Bene Beraq 51203, Israel.

6510.353

SPS-65 HELICOPTER SELF-PROTECTION SYSTEM

The SPS-65 is a high sensitivity, lightweight system capable of intercepting and analysing the widest range of known or anticipated emissions in the battlefield, including laser radiation sources. It comprises a low volume, lightweight radar warning system which detects, processes and displays radar threats operating in the 0.7 to 18 GHz frequency range.

A high sensitivity superheterodyne receiver subsystem detects and processes CW radar signals in the 6.5 to 18 GHz range which are not recognised by conventional radar warning systems. A laser warning receiver detects, identifies and locates hostile laser-guided weapons.

A 3-inch display unit provides an alphanumeric readout of the type, relative bearing, relative lethality, and status of up to 16 analysed threats. In addition an audio warning is provided through the aircraft intercom system. A flare/chaff dispensing system, automatic or manual, is available for countermeasures.

Software and emitter tables can be field loaded into the emitter library in seconds, thus providing a complete and immediate adaptability to changing threat situations.

CONTRACTOR

Elisra Electronic Systems (a subsidiary of Tadiran), 48 Mivtza Kadesh Street, Bene Beraq 51203, Israel.

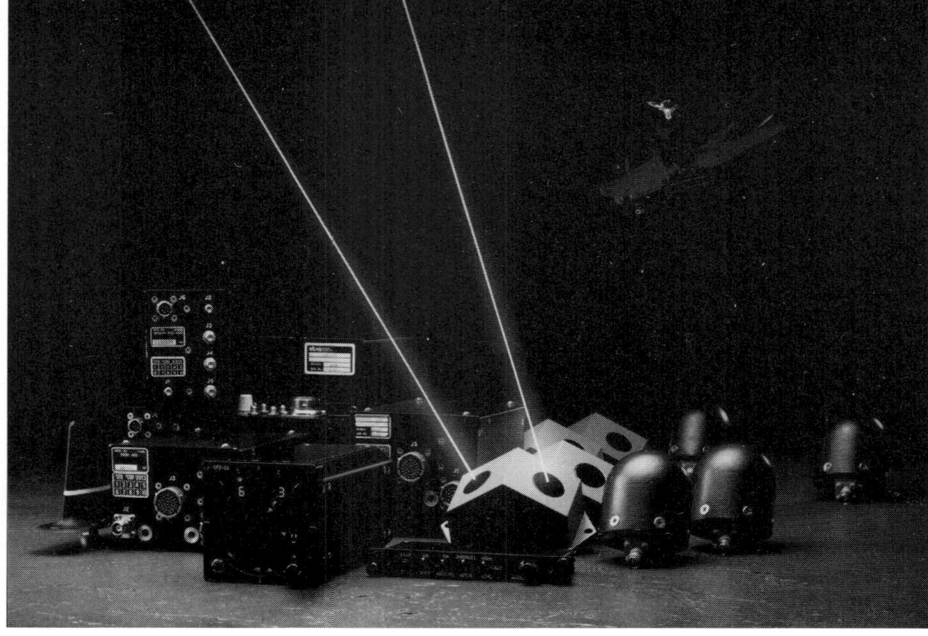

SPS-65 integrated helicopter self-protection system

6264.353

CR-2800 ESM/ELINT SYSTEM

The CR-2800 is an airborne computer-controlled distributed processor architecture system incorporating the features of both ESM and ELINT systems. The equipment has been designed specifically to meet both overland and maritime intelligence requirements, and detects, analyses and identifies radar emissions up to 330 km range, providing instantaneous 360° azimuth coverage over the frequency band 0.5 to 18 GHz. The CR-2800 performs bearing measurements, computes location, builds up and provides EOB reports and warnings against radar threats. It features high probability of intercept, and is capable of tracking up to 64 radar emissions.

Installation of the antennas is adaptable to meet the individual aircraft requirements, and consists normally of one cylindrical shaped omni antenna, four cavity-backed 10-inch diameter spiral antennas and eight cavity-backed 3-inch diameter spiral antennas. They are flush-mounted to the aircraft in a pod or blister type installation. The operational display modes include polar display plus tabulated data. A geographical display can be provided as an option. The CR-2800 is fully automatic but the operator has full control over system operation and system display. Weight of the complete system is approximately 300 kg, including rack-mounted equipment, two DF receivers and the antennas.

CONTRACTOR

Elisra Electronic Systems (a subsidiary of Tadiran), 48 Mivtza Kadesh Street, Bene Beraq 51203, Israel.

CR-2800 ESM/ELINT equipment rack and operator's console

5793.353

SPS-200 AIRBORNE SELF PROTECTION SYSTEM

The SPS-200 is an airborne self protection equipment to detect and display pulsed and CW radar threats in the frequency range 0.7 to 18 GHz. As with the SPS-20 it has been designed to fit existing installations, to provide improved characteristics and additional capabilities, as well as being suitable for new installations in the latest aircraft.

The system incorporates an advanced digital signal analyser that provides data processing and interfacing tasks. It operates in conjunction with an optional MPMN-36(V) flare/chaff dispensing subsystem and can interface with on-board radars and jammers. A 3-inch CRT azimuth indicator provides an alpha-numeric display of the type, angle-of-arrival, relative lethality and status of the threats. An extensive threat emitter library, which can be easily updated, provides the necessary comparison information with the incoming signals.

The SPS-1000 integrated system combines the SPS-200 with the SRS-25 (**6507.393**).

STATUS

In production and in operational service with the Israeli Air Force and a number of other countries.

CONTRACTOR

Elisra Electronic Systems (a subsidiary of Tadiran), 48 Mivtza Kadesh Street, Bene Beraq 51203, Israel.

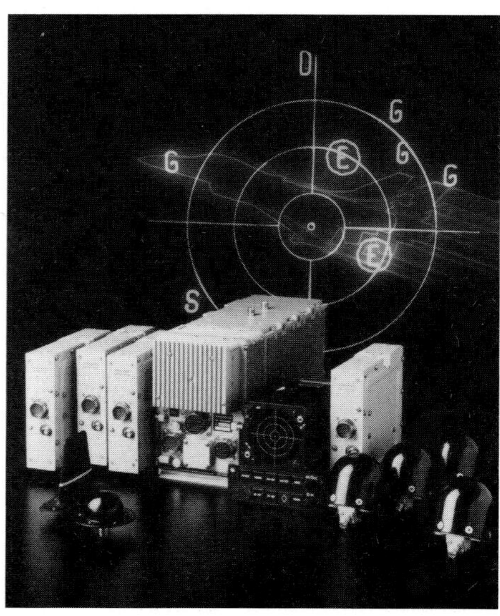

SPS-200 self protection system

6509.353

SPS-2000 AIRBORNE SELF-PROTECTION SYSTEM

The SPS-2000 is an advanced fully integrated self-protection system which provides unambiguous emitter identification of surface-to-air missiles, anti-aircraft artillery and airborne interceptors in dense hostile environments. The system is available for use in a wide variety of tactical aircraft.

The digital signal analyser unit accepts and processes video pulse output received from the detector-amplifiers in the 2 to 18 GHz frequency range. Threat recognition is carried out based on comparison with pre-programmed library. CD pulse signals in the 0.7 to 1.3 GHz range (optionally up to 2 GHz) are detected and characterised by the CD-DF subsystem. The receiver subsystem detects CW radars, high PRF and low PRF signals in the 2 to 18 GHz band and, in addition, provides for integration with power-managed active jammers.

Up to 16 threats can be displayed on the alphanumeric display unit screen. An audio warning is also given over the aircraft intercom system. The display unit shows relative bearing, relative lethality and emitter type, and up to 50 different symbols can be programmed to represent specific threat types.

Software and emitter tables can be field-loaded into the emitter library in a matter of seconds, thus providing complete and immediate adaptability to changing threat environments. A chaff/flare dispensing subsystem, either automatic or manually activated, is integrated with the SPS-2000 for countermeasures capability.

CONTRACTOR

Elisra Electronic Systems (a subsidiary of Tadiran), 48 Mivtza Kadesh Street, Bene Beraq 51203, Israel.

SPS-2000 self-protection system

ISRAELI SIGINT AIRCRAFT

Israel is known to have outfitted at least one Boeing 707 for SIGINT purposes for the Argentine Air Force. Although no information has been received regarding other conversions, for this aircraft, the accompanying diagram gives some idea of the installation layout proposed for the Boeing 737.

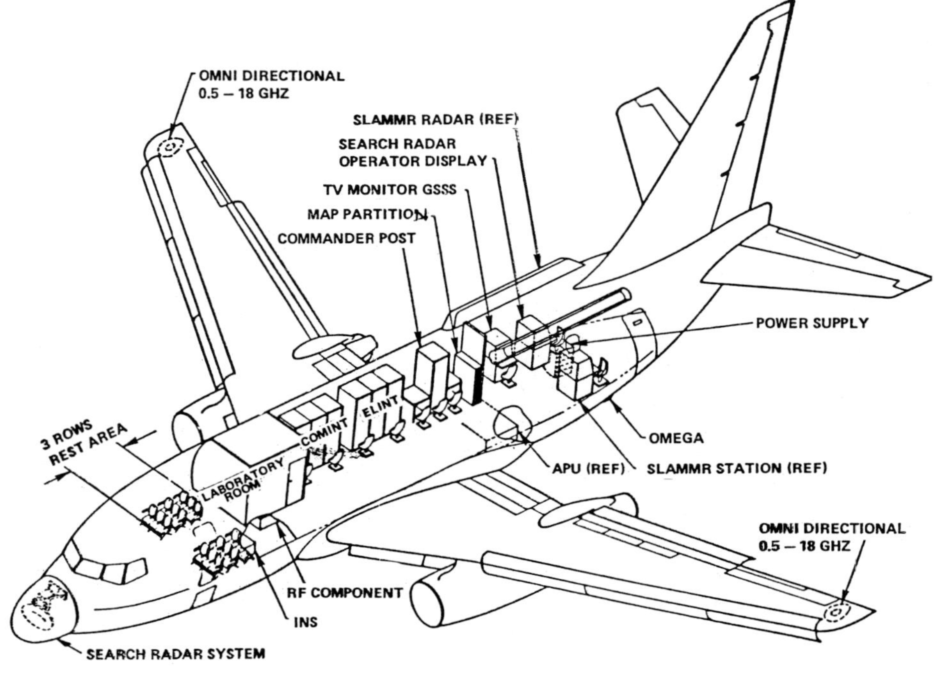

ITALY

4358.393

ELT/156 RADAR WARNING RECEIVER

The ELT/156 airborne passive warning receiver is designed to provide rapid non-ambiguous warning of illumination by an enemy search or tracking radar transmission. It is intended for use on light fighter/ground attack aircraft and helicopters employed in the low altitude ground support activities.

The ELT/156 provides warning against radar threats through the frequency coverage of the appropriate radar bands, full 360° azimuth and elevation coverage and bearing/range presentation on a cockpit mounted dual mode bright display. An audio warning facility is also provided to alert the pilot of possible danger. Simultaneous warning is given of multiple threats.

The system is computer-controlled and provides a very high measure of protection and warning for aircraft forced to enter areas protected by radar controlled surface-to-air and air-to-air missiles or anti-aircraft artillery. The software is easily reprogrammable to combat new types of threats.

CHARACTERISTICS

Azimuth coverage: 360°

Weights

Antenna (4 off): 0.15 kg each

RF head (2 off): 1 kg each

Signal processor: 5.2 kg

Display: 1.9 kg

Control panel: 0.6 kg

Total system weight: 10.3 kg

STATUS

The system is currently in production for a number of national air forces.

CONTRACTOR

Elettronica SpA, Via Tiburtina Valeria Km 13,700, 00131 Rome, Italy.

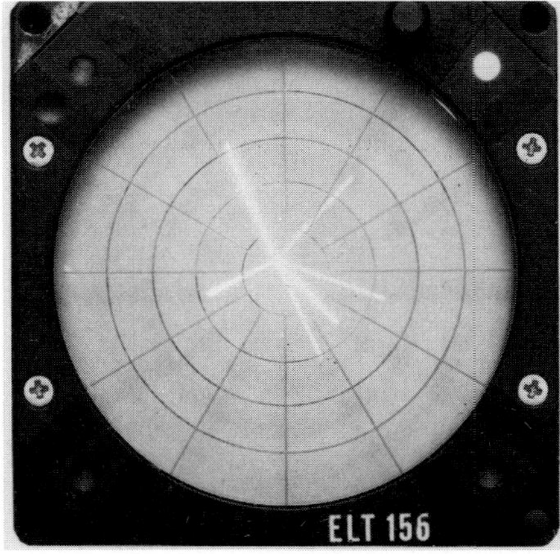

ELT/156 radar warning receiver display (synthetic mode)

4166.393

ELT/263 AIRBORNE ESM EQUIPMENT

The ELT/263 airborne ESM system is designed principally for use in maritime aircraft such as the Beech 200T but has been configured for other types such as the Guardian, Casa C212, Learjet 36A, F27 Maritime, Bandeirante, Piaggio P166, and Britten Norman Maritime Defender. The surveillance of coastal water is the main function, and detection, analysis and identification of emission in the E- to J-bands is performed in addition to bearing measurement, which permits emitter location by using a series of successive bearing measurements.

The ELT/263 equipment, in addition to its surveillance tasks, provides for emitter location by the triangulation method. The information gathered (emission analysis, identification, bearing, etc) can be transferred to external users such as ground stations, naval units and the patrol aircraft's radar operator/tactical navigator, and can also be recorded on an optional printer. The system's coverage is depicted in the diagram and the associated tables.

The main items of equipment comprise: a set of four DF antennas; an omni-directional antenna set; DF receiver; IFM receiver; ESM display and control console; a radar warning processor; and a radar warning display unit.

STATUS
Production and in operational service.

CONTRACTOR
Elettronica SpA, Via Tiburtina Valeria Km 13,700, 00131 Rome, Italy.

Radar type	Flight altitude (feet)	Speed (knots)	Instantaneous surveillance coverage Radius nm/km	Area nm²/km²	Area covered in 1 hour's linear flight nm²/km²	Area covered in 4 hour patrol mission nm²/km²
I-band navigation	2000	180	42/77	5500/18 600	20 600/70 000	66 000/224 000
	6000	200	53/98	8800/30 200	30 000/102 800	93 600/320 300
F-band search	2000	180	65/120	13 300/45 200	36 700/125 300	106 900/365 400
	6000	200	100/186	31 400/108 600	71 400/246 500	191 400/659 900

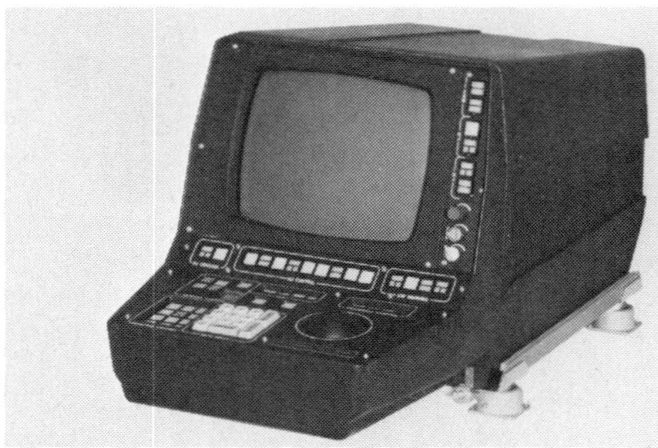

Display/control console for ELT/263 airborne ESM equipment

ELT/263 receiver antennas mounted in wing tip pods and tail cone of Learjet 35A

4168.393
ELT/555 JAMMER POD

The ELT/555 is a pod-housed airborne deception jammer and warning system designed for the self-protection of modern attack and fighter aircraft from enemy anti-aircraft systems (surface-to-air and air-to-air, and anti-aircraft artillery). The system is completely self-sufficient as regards cooling and primary power and makes no demands whatsoever on the carrier aircraft's power supply. The philosophy adopted for the ELT/555 system ensures that the jammer will automatically tailor its reaction to the type of detected enemy radiation. The system operates against both pulsed and CW radar emissions and presents the following salient operational features: simultaneous multi-threat reaction; very high resistance to ECCM, including any type of frequency agility; full automation; optimised threat deception programs; multi-band frequency coverage (H- to J-band).

The equipment is effective against all types of fire control radars, say the manufacturers, and active or semi-active homing missiles. Various deception techniques are employed, and multiple target engagement is possible.

Air speeds of up to Mach 1.1 at sea level and Mach 1.5 at 40 000 feet were tested during the qualification of the pod. The central main body section and tail module is completely clean, except for the two bottom radomes and stabilising fins. The latter, located toward the rear of the main body section, serve to increase the safety envelope during jettison. The pod can be jettisoned by the pilot and is provided with a self-destruction device. Self-destruction occurs at a presettable interval after the jettison command. The device can be installed or removed at flight-line level.

The pod is provided with a cooling system and self-sufficient power generating system which supplies the electric power necessary to the pod's equipment, independent of the carrier aircraft's supply. The prime mover of the power generating system consists of a ram air turbine (RAT) which drives a four-pole permanent magnet generator. The RAT also drives the compressor of the pod's cooling system, which operates on a closed circuit fluid recycling principle.

There are forward and rear facing CW transmission antennas on the under-surface of the pod body, and nose and tail radomes house antennas for transmission and reception of pulse signals, and reception of CW signals.

The ELT/555 pod measures 300 cm in length, with a body diameter of 27 cm (34 cm RAT maximum) and a weight of 140 kg. Types of aircraft nominated by Elettronica as capable of using the ELT/555 include: F-5, Mirage III, Alpha Jet, MB 326, MB 339, Jaguar, MiG-21, Phantom, Mirage 5, Hawk, Saab 105, etc.

STATUS
In production and operational service. No contractual information is available.

CONTRACTOR
Elettronica SpA, Via Tiburtina Valeria Km 13,700, 00131 Rome, Italy.

ELT/555 jammer mounted on ventral pylon of BAe Hawk

4359.393
ELT NOISE JAMMER PODS

A range of pod-contained automatic airborne noise jammers, ELT 457, 458, 459 and 460 are designed by Elettronica for installation in high performance ground attack and fighter aircraft. They are intended for operation against ground based air defence radars to mask target infomation on attacking and thereby neutralise anti-aircraft systems. Instantaneous automatic warning is given against pulsed radar emissions, with a very high resistance to ECM, including any type of frequency agility. Multi-threat reaction capability is provided and each jammer contains a large 'library' which can be programmed before flight with a selection of probable radar threat characteristics and relative priority levels.

The systems incorporate power management techniques, by which jamming power and optimised modulation are allocated on a threat priority basis, ie the signals for jamming are radiated in the correct sector, at the correct time and at the appropriate frequency. The power management philosophy therefore ensures that the entire electromagnetic resources of the system are directed against the most dangerous programmed threat or threats.

Each of the four system variants operates on a different radar frequency band and pod deployment can be optimised in view of the expected threat; simultaneously multi-band frequency coverage can be achieved by deploying more than one pod on one or more aircraft.

The pods can be jettisoned by the pilot and are provided with a self-destruct device. Self-destruction occurs after a presettable interval after the jettison command.

The main unit of the ELT range is the external pod designed for supersonic speeds up to Mach 1.1 at sea level and Mach 1.5 at 40 000 feet. The pod houses the receivers, transmitters and relevant antennas, the warning processor and the cooling and power generators. The antennas, which are protected by radomes, differ for each type of noise jammer pod. A control panel is installed in the cockpit containing the commands for two pods.

The jammer electronics have been designed for easy maintenance with modular construction and easy access. The pods are provided with a self-

ELT/460 supersonic ECM pod

sufficient power system in order to render them independent of the aircraft supply. The prime mover of the power generating system is a ram air turbine (RAT).

CHARACTERISTICS
Length: 312 cm
Diameter: 27 cm (34 cm RAT)
Weight: 145 kg

STATUS
In production and operational service.
CONTRACTOR
Elettronica SpA, Via Tiburtina Valeria Km 13,700, 00131 Rome, Italy.

5795.393
ELT/562 AND ELT/566 DECEPTION JAMMERS

These are deception jammers designed for internal installation in both fixed and rotary wing aircraft. The ELT/562 is intended to counter pulse radar threats while the ELT/566 operates against CW systems. One or both jammers may be associated with an airborne ESM system such as the Colibri (**4360.353**). The jammers are effective against threats operating in the H to J bands (6 to 20 GHz), and are intended to protect aircraft in battlefield environments.

STATUS
In production.
CONTRACTOR
Elettronica SpA, Via Tiburtina Valeria, Km 13,700, 00131 Rome, Italy.

5796.353
ELT/999 AIRBORNE COMINT SYSTEM

The ELT/999 is designed specifically for airborne applications as a strategic communications intelligence system able to perform surveillance over very large areas at the border of a country, or in difficult access areas, and to classify and recognise a variety of emitters. It provides fast surveillance to ensure interception of communications emissions in the VHF and UHF bands, even those of short duration, and direction finding and location of the detected emitters. Technical analysis and demodulation of the intercepted signals is performed for the purpose of monitoring and/or recording on the intelligence information. The collected data is computer-processed and presented to the operators on several displays, including alpha-numeric plasma displays, panoramic and graphic presentations and CRT displays for monitoring.

The ELT/999 is designed for medium-sized aircraft, is of a modular concept, and can be easily configured to fulfil all the different operational requirements for COMINT purposes. The instantaneous surveillance coverage of the system against a typical emitter is approximately 20 per cent greater than the geometrical horizon of the airborne platform, eg at an altitude of 4000 feet (1200 metres) the coverage radius is about 160 km.

The equipment makes use of advanced signal processing and handling techniques, as well as software controlled functions, which provide the high degree of flexibility to render it suitable for different types of missions. It incorporates a high level of automation and advanced human engineering design for easy operation. A single supervisor can control the complete system while up to three operators perform monitoring tasks. The weight of the system depends on its configuration, the maximum being approximately 900 kg.

STATUS
In production and operational use.
CONTRACTOR
Elettronica SpA, Via Tiburtina Valeria, km 13,700, 00131 Rome, Italy.

4360.353
ELT COLIBRI HELICOPTER ESM/ECM SYSTEM

The Colibri system is an integrated ESM/ECM equipment designed for installation on any type of medium sized helicopter operating in the naval surveillance, anti-submarine and anti-ship roles, thereby extending the ESM/ECM range of the parent ship. The system provides a helicopter with the facilities for passive surveillance, DF analysis, identification and location of radar emitters, even when the system is operating in a dense electromagnetic environment. It also provides for effective self or mutual protection against radar guided missiles.

Colibri can be configured to meet particular operational roles by combining a variety of basic modules. The basic configuration is represented by the ELT/161 ESM and radar warning system. A helicopter's radar warning function can then be supplemented by the ELT/562 deception jammer to obtain effective self-protection and the configuration can be further expanded by the addition of the ELT/261 for ESM capability and ELINT roles. An additional unit, the ELT/361, can be added to achieve even greater protection capability.

In the basic configuration, Colibri comprises the ELT/161, consisting of a receiver, the antenna system, a warning processor and a radar warning display. The ELT/161 operating in conjunction with the ELT/261 system, gives overall capability in that field and is designed to fulfil the maritime surveillance and EW roles to extend the electronic warfare range of the parent ship. In this role, the system supplies instantaneous DF and display of intercepted emissions, frequency measurement of pulse and CW emissions, automatic analysis, identification of source and automatic tracking of selected emissions. The addition of the ELT/562 deception jammer provides the helicopter with self-protection against sea-to-air and air-to-air missiles. The ELT/361 noise jammer is added to suppress enemy surveillance radars.

ELT naval Colibri antenna installation on AB-212 helicopter

STATUS
In production and in service with several navies.
CONTRACTOR
Elettronica SpA, Via Tiburtina Valeria Km 13,700, 00131 Rome, Italy.

4642.353
SMART GUARD HELICOPTER ESM SYSTEM

Developed for helicopter operation, Smart Guard provides continuous and automatic surveillance of VHF/UHF communications bands. In particular, the system can intercept, analyse and DF any signal in these bands, even those of extremely short duration. Smart Guard will automatically correlate successive intercepts of the same emitter for location purposes.

Basic elements of the system are:

(1) a high scan velocity surveillance receiver
(2) an analysis and monitoring receiver with associated IFF display
(3) a very high precision direction finder
(4) a computer and data processing system
(5) a plasma display
(6) a dual-track audio recorder
(7) a digital recorder for mission data.

Smart Guard performs automatic surveillance of one or more frequency ranges, automatic control of predetermined channels, demodulation of intercepted emissions for operator audio or recording, and display of the data acquired.

STATUS
In production.
CONTRACTOR
Elettronica SpA, Via Tiburtina Valeria, Km 13,700, 00131 Rome, Italy.

4643.353
FAST JAM HELICOPTER ECM SYSTEM

Fast Jam is a communications ECM system developed specifically for helicopter installation. The system provides surveillance and automatic jamming of hostile emissions with provisions to safeguard friendly emissions.

System elements consists of:

(1) an analysis and monitoring receiver with associated IFF display. The receiver is computer-controlled and operates in look-through mode during jamming to confirm the continuing presence of the hostile emission
(2) a computer for system management which controls the status of the channels under surveillance, processes and displays the collected data and activates jammers following stored instruction
(3) one or more solid-state jammers
(4) a plasma display.

The system normally operates in an automatic

mode under the supervision of the operator who assigns the initial instructions, updates of which can be made even during a mission. In the automatic mode the system has an extremely short reception and transmission time. The system may also be wholly or partially manually controlled. Sophisticated jamming techniques allow surveillance of the hostile emission (look-through) and the simultaneous jamming of more emissions without loss of effectiveness.

STATUS

In production.

CONTRACTOR

Elettronica SpA, Via Tiburtina Valeria, Km 13,700, 00131 Rome, Italy.

3382.393

RQH-5(V) ESM/ELINT AIRBORNE SYSTEMS

The RQH-5(V) family of systems has been developed from the RQH-5 ESM system designed and developed in the late 1970s time frame and currently in operation within the ESM/ECM system IHS-6. The system is capable of fully automatic ESM performance, and in a suitable form for user interpretation and/or digital recording. The extensive modularity adopted for the front-end (two interchangeable IFD modules), the automatic processing levels and interfacing peripherals (displays, controls, recording devices, ECM interfaces, high DF accuracy antennas and receivers) have generated three RQH-5 families:

(a) Series 719 RWR for fighter aircraft
(b) Series 730 for maritime patrol, ASW or other multi-role aircraft
(c) Series 780 for ESM/ECM dedicated operational missions.

This modular architecture also provides for growth capability or updating of the system.

The basic configuration consists of an antenna group (omni and DF), IFM receiver in two versions, video receiver unit, automatic extractor processor and a multi-function control unit. This enables the RQH-5(V) to provide wide frequency coverage, high sensitivity with real-time automatic extraction, analysis and tracking of all incoming radar signals. The equipment has full capability on frequency and PRF agile signals and automatic CW measurement facilities. Computer assisted identification of low priority radars within a very wide identification library is provided and the ELINT capability allows the system to characterise and analyse the radar signals and keep records of them for post-flight data collection.

The RQH-5(V) is able to drive an active ECM system automatically and will also interface with a passive countermeasures system in order to initiate chaff and/or flare decoys launch.

Units of the RQH-5(V) system

CONTRACTOR

Selenia Industrie Elettroniche Associate SpA, Defence Systems Division, Via Tiburtina Km 12400, 00131 Rome, Italy.

4174.393

SL/ALQ-234 ECM POD

The SL/ALQ-234 is designed to provide protection against AA artillery radar directors and surface-to-air missile systems, and it also provides a radar warning facility for the pilot, to indicate any threat which has locked onto the aircraft. It is self-contained and incorporates its own broad-band ESM section, using a wide-open receiver, and intercepted data are fed to the pod's on-board digital computer for use in threat list compilation and comparison against an existing threat library. Intercepted threats are arranged in order of priority automatically. The computer also performs jamming power management against several threats simultaneously (where necessary) in order of priority. Travelling wave tube (TWT) transmitters are used for jamming, covering a band about 6 GHz wide in the fire control region of the radar spectrum. Closed-loop liquid cooling combined with air cooling allows aircraft speeds and altitudes ranging from low level/speed to supersonic/high altitude to be accommodated.

Each pod contains three jammers, a receiver/processor and antennas at each end. One jammer, at the front, is used to counter doppler and CW threats. Two centrally located jammers are 'smart-noise' transmitters. The rear section houses the superheterodyne radar warning receiver (RWR) with an instantaneous frequency measurement (IFM) system.

CHARACTERISTICS

Pod

Length: 3.825 m
Diameter: 414 mm
Weight: 270 kg
Power generation: RAT, 7.5 kVA
Cooling: Dual circuit with heat exchanger

Pulse countermeasures

Frequency band: I-J
Sector coverage: Fore and aft
Detection: Built-in RWR with automatic designation and reaction management
Jamming: Noise and deception. Multiple threat capability by computer controlled power management

CW countermeasures

Frequency band: H-J
Sector coverage: Fore
Detection: Built-in RWR with automatic designation and reaction management
Jamming: Deception. Multiple threat capability

STATUS

Earlier ALQ-234 pods have been supplied to air forces of other countries and it is still in production. A new version has completed trials and is now in production. Development is being carried out on versions with a lower frequency capability (eg C, D, E, F-bands) for use against air defence search radars. Later developments include the upgrading of the computer, based on a multiprocessor system which draws information from a single database.

Selenia SL/ALQ-234 self-defence ECM pod for carriage by high performance aircraft

CONTRACTOR

Selenia Industrie Elettroniche Associate SpA, Defence Systems Division, Via Tiburtina Km 12400, 00131 Rome, Italy.

JAPAN

4176.093

JAPANESE EW EQUIPMENT

In the past many of the Japanese requirements for airborne EW equipment has been met by US developed systems. An increasing range of these systems are now being manufactured indigenously. This commenced with the J/APR-2 developed for the RF-4E aircraft of the Japanese Self Defense Force, and the J/APR-3 for the F1 support fighter. Both these equipments are still in service in some aircraft. Later versions of these systems are described below.

J/APR-4

This equipment was designed for fitment on the F-15J Eagle aircraft. It is able to process multiple inputs simultaneously in a dense electromagnetic environment, and has a digital computer with a reprogrammable software package to allow reconfiguration for future requirements. The indicator provides for daylight viewing and multi-threat data presentation in alphanumeric and graphic format. The system is also designed to interface with other EW equipment such as the J/ALQ-8.

J/APR-4A, -5 and -6

These are a family of advanced J/APR-4 based types, with greater capability to cope with current threats. All F-15J/DJ aircraft which are due to be built will be equipped with the J/APR-4A. All RF-4Es currently equipped with J/APR-2 are scheduled to be refitted with J/APR-5. The decision has also been made to install the J/APR-6 on all modified F-4EJ aircraft. These equipments differ in their size and configuration due to the restriction on space provision of the aircraft concerned.

Other airborne EW equipment

Various other items of airborne equipment in the J/ALQ series have been identified although details remain classified. The latest systems include the J/ALQ-6 fitted to the F-4 and F1, and the J/ALQ-8 for the F-4 and F-15J. The J/ALQ-8 countermeasures set is understood to operate in three bands; 1-4, 4-8 and 7.5-18 GHz.

It has been reported recently that Mitsubishi has

developed a tail-mounted surveillance radar for combat aircraft. It is described as a 2-D system capable of detecting targets with a radar cross-section as small as 10 cm, and is presumably designed to detect and warn of approaching missile threats. It has a miniature antenna which will not require a significant radome. The system provides both a visual indication and an audio alert, and can initiate passive countermeasures such as chaff and flares.

STATUS

In production and service.

CONTRACTORS

Tokyo Keiki Company Ltd, 16,2-chome, Minami-Kamata, Ohta-ku, Tokyo, Japan.

Mitsubishi Electric Corporation, Mitsubishi Denki Building, 2,2-chome, Marunouchi, Chiyoda-ku, Tokyo 100, Japan.

J/APR-5 radar warning receiver for RF-4E

SPAIN

5815.053

SPANISH EW PROGRAMME

Until recently the Spanish electronics industry has designed and manufactured little or nothing in the electronic warfare field. The ground-based, naval and airborne EW systems in use by the armed forces are of foreign design, primarily US or French. However the Spanish Ministry of Defence is now engaged on the design of a comprehensive EW programme. It is understood that this programme is of the highest priority and combines the efforts of private and state-owned companies and research laboratories throughout the country. The objectives are to obtain the widest range of electronic measures and countermeasures, support systems, intercept/ELINT systems and threat warning equipment.

The Ministry of Defence will co-ordinate the programme which will involve 42 armed forces research centres, 51 university research centres and 130 private and state-owned companies. The programme is understood to be worth some 678 million pesetas (approximately $6 million as at April 1988), two-thirds of that sum being in the form of government subsidies. The research and development investment will be devoted to work on ELINT equipment, radar threat warning systems and frequency hopping transceivers. One of the aims of the programme is reported to be the provision of monitoring systems for the Spanish Air Force capable of observing military operations in North Africa.

The principal company likely to be involved in the programme is probably EISA (Experiencias Industriales SA) of Aranjuez, Madrid.

STATUS

The design and development programme was scheduled to stretch over the period 1985 to 1987. No further information has been released.

CONTRACTOR

Spanish Ministry of Defence.

SWEDEN

3025.393

AR 753 JAMMER SET-ON RECEIVER

The AR 753 set-on receiver has been designed for use with the AQ 31 range of pod mounted jammers (**3024.393**). As all components employed in the AR 753 have good wide-band characteristics, the operating ranges can readily be varied. Other features claimed include high sensitivity and good selectivity, with digital indications. The cockpit unit, which is common to all pod configurations, contains the displays, the antenna selector switch, and the gain control. Pod dimensions are 1600 mm long, by 140 mm diameter, and the weight is 25 kg. All other information on the AR 753 remains classified.

AR 753 jammer set-on receiver pod

STATUS

In production and operational.

CONTRACTOR

SATT Communications AB, PO Box 32701, 126 11 Stockholm, Sweden.

3537.393

AR 765 RADAR WARNING RECEIVER

The AR765 is a lightweight radar warning receiver (RWR) specially designed for fitting on the Agusta Bell 206A Jet Ranger helicopter, but also suitable for use with other types of fixed-wing aircraft and helicopters. The AR 765 provides visual and aural threat warning and indication against surveillance as well as fire control radars, and specific features include: DF and homing facilities; multi-octave band frequency coverage; built-in test equipment (BITE). In the case of the Jet Ranger, installation is very simple as the main unit is designed as a direct replacement for the battery access panel in the front of the helicopter fuselage, and the RWR is bolted on in place of the normal battery hatch cover plate.

CHARACTERISTICS

Frequency coverage: Multi-octave or narrow-band
Warning sectors: 2 in front $\pm 45°$ relative to line of flight
Azimuth coverage: $\pm 90°$
Elevation coverage: $\pm 45°$
Antennas: Broad-band circularly polarised spirals
Receiver: Crystal video, 2 channels
Sensitivity: –40 dBm (TSS)
Dynamic range: >40 dB
Minimum pulsewidth: 0.1μs
Threshold adjustment range: 35 dB (from cockpit)
Threat indication: Lamps on panel, and headset
Supply voltage: 28 V DC, 40 VA
Options: CW detection, rejection on non-hostile emitters, laser warning

STATUS

In service but no longer in production.

CONTRACTOR

SATT Communications AB, PO Box 32701, 126 11 Stockholm, Sweden.

3904.393

AR 777 AIRBORNE MICROWAVE RECEIVER

The AR 777 is a rugged, computer-controlled airborne microwave receiver intended for signal acquisition and jammer set-on. The system performs rapid and unambiguous detection and identification of emitters in dense signal environments.

The AR 777 is a superheterodyne receiver with a high performance YIG-tuned RF head. Due to linear YIG oscillators, a special IF processor and a microcomputer-controlled calibration routine, an excellent frequency accuracy and resolution has been obtained over the entire frequency range.

It covers the frequency range 1 to 8 GHz with three RF-heads, each covering one octave. An RF-head includes YIG-filters, YIG-tuned oscillators and drive circuits. By adding or replacing RF-heads the frequency range can be extended or changed. The receiver uses several modes of tuning and sweeping which are controlled by a microprocessor.

The receiver uses a microprocessor to perform control functions and to process, store and display data. All RF circuits are housed in a standard ATR package called the receiver unit. The microprocessor and associated circuitry are housed in a tailor-made unit called the display unit.

The two units communicate through a digital 32-bit serial word data highway which contains information for the tuning (12 bits), control functions and parity check. A D/A converter converts the digital data to tuning voltages for the YIG components. A keyboard on the display unit is used to communicate with the microprocessor. This is also used to select the centre frequency. The received signals are displayed on a spectrum-type display consisting of two rows of lamp indicators (see figure). The upper row has a memory function and stores the displayed data for a preset time while the lower row gives a momentary indication. Different search patterns can be initiated as well as a fixed tuned mode. In this (identification) mode the PRF is evaluated and the result is displayed on two digital displays. The displays can simultaneoulsy display two PRFs. If more than two PRFs are present at one RF-frequency they are displayed sequentially on one of the displays. A separate display is used to indicate if garbled or jittered PRFs are present. The spectrum analysis mode is used for recording the output spectrum from a co-operating jammer system.

A digital output is provided from the microcomputer. All essential panel settings and all received emitter information can be transmitted in serial form to an external digital recorder. Stored data can be further analysed in a ground based evaluation equipment.

CHARACTERISTICS
Frequency range: 1 – 8 GHz (std)
Sensitivity: –65 dBm
Dynamic range: 65 dB
Sweep width: 100/300 MHz
Sweep rate: 10 sweep/s
Frequency accuracy: 5 MHz
Frequency resolution: 5 MHz
PRF measurement: 200 – 2000 Hz

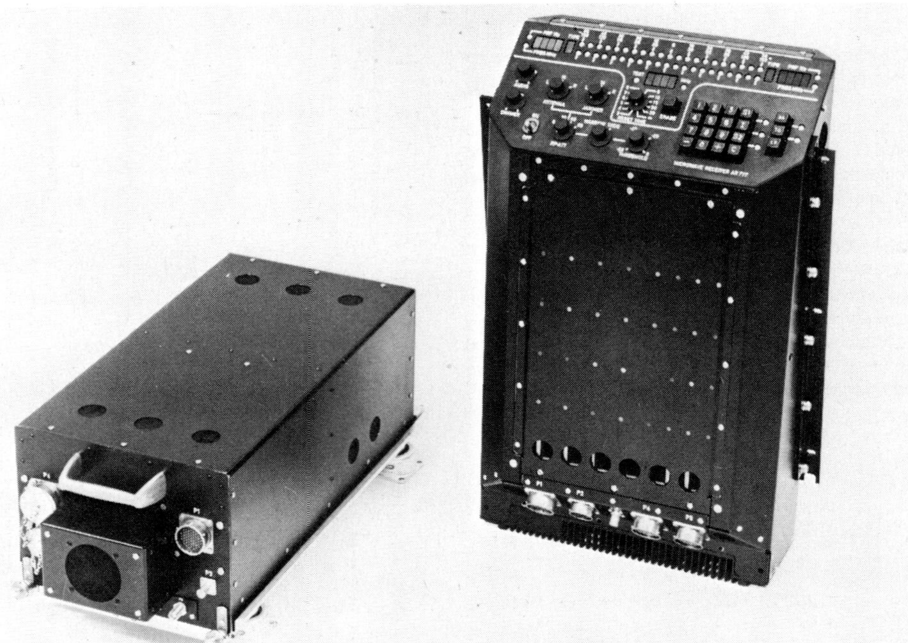

SATT AR 777 airborne microwave receiver with display unit on right

PRF accuracy: 1 Hz
Power supply: 200 V 3-phase 400 Hz 300 VA
Dimensions
Receiver unit: 497 × 257 × 194 mm
Display unit: 580 × 355 × 173 mm

STATUS
In production.

CONTRACTOR
SATT Communications AB, PO Box 32701, 126 11 Stockholm, Sweden.

4733.393

AR 830 AIRBORNE WARNING SYSTEM

The AR 830 is a fully programmable threat warning system for the detection of radar and laser transmissions. It incorporates a special purpose, high speed signal processor, a general-purpose computer and broadband receivers for both radar and laser signals. It is modular in design to allow updating of the system to handle new threats and can be interfaced to jammers, chaff dispensers, flare dispensers, recorders and other systems.

The signal processing provides rapid and unambiguous threat identification in dense electromagnetic environments and an extensive emitter library allows comparison and fast analysis of threats. The library can be easily reprogrammed and updated with new threat information. The incoming threat data is either displayed on a separate CRT or integrated with the aircraft display system.

STATUS
In production but no details of operational use have been released.

CONTRACTOR
SATT Communications AB, PO Box 32701, 126 11 Stockholm, Sweden.

3024.393

AQ 31 JAMMER

The AQ 31 is CW radar jamming equipment designed for aircraft pod mounting and up to full band coverage in the E- to J-bands can be provided. Few specific details have been released but AM, FM noise/sawtooth jamming modes are known to be available and high power output is claimed. Ram-air cooling is employed. Digital control of all major functions is provided and interface facilities are available for use of the AQ 31 with digital signal processing equipment. A single control panel in the cockpit enables operation of one or two pods, and pre-programmable routines can be employed. The manufacturers also offer a bolt-on companion set-on receiver for use with AQ 31 jamming pods, designated AR 753 (**3025.393**). No other details are available for publication.

AQ 31 jammer pod

STATUS
In production and operational.

CONTRACTOR
SATT Communications AB, PO Box 32701, 126 11 Stockholm, Sweden.

4361.393

AQ 800 JAMMER

The AQ 800 jammer is intended for use in the training of radar operators and for evaluation of radar station performance. It is pod-mounted and hung from a pylon under the wing or fuselage of an aircraft, the pod is ram-air cooled and has its own built-in cooling system. The method of mounting is flexible to allow mounting on many different types of aircraft.

The system consists of a pod and a cockpit unit. The pod contains a TWT type transmitter and a sweeping superheterodyne set-on receiver. Each pod covers one frequency band, ie S-, C-, X- or Ku-band. The cockpit unit can handle two jammer pods of any frequency combination. The transmitter can generate up to three simultaneous jamming signals. The modulation, centre frequency and bandwidth of each signal are independent of each other. Available modulation types are AM-noise, FM-noise and FM-sawtooth plus noise.

AQ 800 airborne jammer

Both the receiver and the transmitter have two antennas, one pointing forward and the other pointing aft. Either direction of operation is selected from the control panel. The system can be operated by a special instructor operator or by the pilot himself.

CHARACTERISTICS

Frequency range: S-band (2-4 GHz), C-band (4-8 GHz), X-band (8-12 GHz), Ku-band (12-18 GHz)

Signal generators: Each pod contains 3 signal generators, each generator covering the following part of the band: S-band 700 MHz, C-band 1200 MHz, X-band 2000 MHz, Ku-band 2000 MHz

Antennas: 2; beamwidth 70° horizontal, 30° vertical. Polarisation is slant linear

Weight: 220 kg

Dimensions: 3.9 m long × 0.43 m wide × 0.52 m high

CONTRACTOR

SATT Communications AB, PO Box 32701, 126 11 Stockholm, Sweden.

6511.393

AQ-861 NOISE AND DECEPTION JAMMING SYSTEM

As with the AQ-800 (**4361.393**), the AQ-861 is intended for training purposes but can also be used in various platforms such as small fixed-wing aircraft, helicopters or even ships. It covers the I/J frequency bands.

The complete system consists of a receiver, a signal generator unit, a power amplifier and two antennas (one for the receiver and one for the transmitter). Weight of the system is 60 kg. Incoming signals of interest are received and displayed to the operator.

Two basic signal modes are available, white noise and range-gate pull-off. In the latter mode the unit makes use of a frequency memory loop to store the frequency of an incoming pulse. This means that every incoming pulse will be delayed and retransmitted at its original RF frequency. All incoming signals within the bandwidth of the frequency memory loop will be pulled off simultaneously.

STATUS

In production

CONTRACTOR

SATT Communications AB, PO Box 32701, 126 11 Stockholm, Sweden.

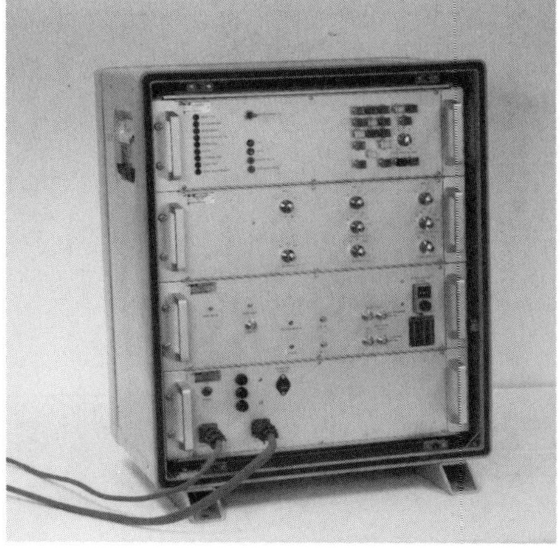

AQ-861 jamming system

4734.393

AQ 900 AIRBORNE SELF-PROTECTION SYSTEM

The AQ 900 is a pod-mounted automatic jamming system which is available in different wideband versions covering the frequency range from S-band to Ku-band. The complete system consists of two pods, operated from one cockpit control unit, one pod for jamming search radars in S-band and one for jamming fire control radars in X- or Ku-band. The pods are systemwise interchangeable and are designed for pylon mounting on the aircraft.

The S-band and X/Ku-band versions are essentially identical in operation and consist of a receiver, a threat analysis unit, a decision and control unit and a transmitter. Each pod has its own ram air turbine generator and cooling system.

The receiver contains all the RF and signal processing circuitry for reception and threat identification. The signal is then passed to the analysis unit where the signals are analysed and compared with data stored in the system library. The decision and control unit includes a microcomputer that gives the system tactical flexibility and allows threat priority, technique selection, power management, signal generation and control, and cockpit control and indicator interface.

The transmitter contains all the signal generating circuitry and the high power TWT amplifier. The X- and Ku-band version also contains a unit for various deception techniques to counter fire control and tracking radars.

CONTRACTOR

SATT Communications AB, PO Box 32701, 126 11 Stockholm, Sweden.

6512.393

ERIJAMMER A100

The ERIJAMMER A100 is a manually or automatically controlled jammer pod for active ECM and training of air defence fighters and AAA operators. It can also be used operationally as an escort jammer. The pod has a radar warning receiver with single or multi-threat capability. Noise jamming and several advanced range and angle deception programmes are available. The pod can be used for radar signature generation.

The ERIJAMMER A100 consists of two units, the pod with built-in antennae and electronics, and the control panel. The pod is entirely self-contained and requires only power from the aircraft. The system is controlled by the ECM operator who choses one of the three 120° antenna lobes. Other than this the analysis and subsequent jamming of incoming signals is initiated either manually or automatically. A frontal antenna with a higher gain and a more narrow beam can be chosen for specific missions such as escort jamming and jamming of AA artillery fire control systems.

CHARACTERISTICS

Frequency coverage: H, I and part of J band

Output power: 220 W

Noise bandwidth: 5 to 500 MHz

Antenna coverage: 360° horizontal (three antennas with selectable direction); ±20° vertical

Polarisation: Circular (or frontal antenna linear)

Cooling: Ram-air

Dimensions: 3.4 m long × 0.42 m diameter

STATUS

In production.

Erijammer A100 jammer pod

CONTRACTOR

Ericsson Radio Systems AB, 163 80 Stockholm, Sweden.

6513.393

ERIJAMMER 200

The ERIJAMMER 200 is a pod-mounted fully automatic system effective against pulse and CW radar transmissions. Velocity, range and angular deception modes are available, and the equipment can communicate with the aircraft avionics via a standard data-bus. The system consists of two units; the pod-mounted jammer with built-in electronics and antennas, and a cockpit-mounted control panel.

CHARACTERISTICS

Frequency coverage: Extended H, I and J bands

Output power: 300 W

Antennas: Elliptically polarised horn type

Speed: Mach 1+

Pod dimensions: 4.25 m long × 0.425 m diameter

Weight: 350 kg

STATUS

In production.

CONTRACTOR

Ericsson Radio Systems AB, 163 80, Stockholm, Sweden.

5798.353

ERIJAMMER 39

Ericsson has been selected for the integration of the total EW suite for the JAS 39 Gripen multi-role combat aircraft. The system will be an adaptive, power managed multimode system tailored to meet the requirements to combat end-of-the-century threats.

In addition to its role as systems integrator, Ericsson is currently designing the active ECM part of the system.

STATUS

In development.

CONTRACTOR

Ericsson Radio Systems AB, 163 80 Stockholm, Sweden.

5666.393

BO SERIES COUNTERMEASURES DISPENSERS

A range of aircraft countermeasures dispensers is currently in production or development by Philips Elektronikindustrier.

BOZ 100

This is a pod-mounted advanced ECM dispenser originally developed for the Swedish Air Force but also supplied to export customers. It is capable of sophisticated break-lock and corridor operation at subsonic and supersonic speeds and provides self-protection against electromagnetic and heat-seeking missiles. The unit has a comprehensive EW system interface and is suitable for use in deep-penetration strike, reconnaissance and EW roles. It is microprocessor controlled and has a reprogrammable memory.

STATUS

In service with the Swedish Air Force (under the nomenclature BOX 9), and also fitted to the Tornado and the French Air Force Jaguar.

BOZ 3

BOZ 3 is a high capacity chaff pod dispenser originally designed to a Swedish Air Force requirement and since adapted for training use. It can be manually or automatically initiated and has both break-lock and corridor chaff modes.

STATUS

In production.

BOP 300

The BOP 300 is designed for the protection of light- to medium-weight fighter aircraft. It is modular in design and can be adapted to various requirements by the addition of one to eight dispensing blocks to dispense both chaff and IR flares. The dispensing blocks are designed to give maximum mounting flexibility to avoid interference with the aircraft stores capability. The unit can be adapted to different types of chaff and flare ammunition.

STATUS

In development.

BOH 300

A chaff/flare dispenser for helicopters and designed to provide protection in hovering and forward flight. It is processor controlled and has reprogrammable operation features. The BOH 300 is suitable for radar, optical and IR deployment. The dispensers can be mounted on the landing gear, fuselage or weapon hard points.

STATUS

In development.

BOL

This is a chaff dispensing system specially designed by Philips to integrate with the standard Sidewinder missile LAU7A launcher. No other information is available other than a contract offered by the UK MoD for development. It is understood that five launchers will be modified for flight trials and qualification, with a probable production run of between 130 and 450 systems.

CONTRACTOR

Philips Elektronikindustrier AB, 17588 Järfälla, Sweden.

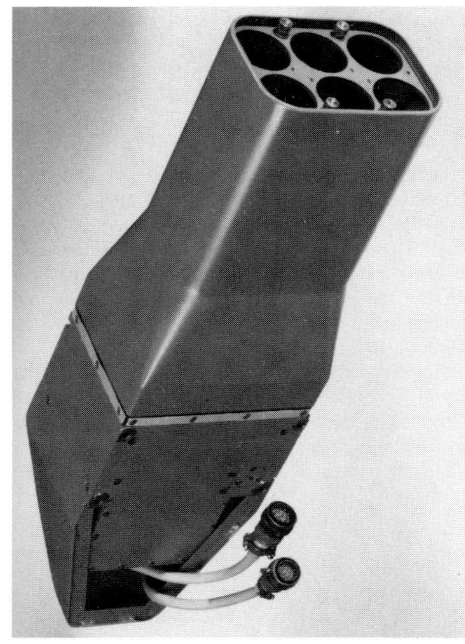

BOP 300 dispenser

BOZ 100 pod dispenser

BOH 300 helicopter dispenser

UNION OF SOVIET SOCIALIST REPUBLICS

The Soviet Union employs large numbers of electronic warfare systems in the ground, airborne and shipborne applications. This covers ESM surveillance systems, radar warning receivers, ECM systems and jammers. In addition, both ships and aircraft have chaff and decoy launchers to deal with radar and heat seeking missiles. Although no definite information has emerged from USSR, there seems little doubt that the Soviets are also developing warning receivers to counter the threats from laser target markers and rangefinders. Very few details of the electronic warfare systems have become available on an unclassified basis, and information on individual equipments is insufficient to warrant separate entries in this section. A list of those systems definitely identified is given in the Electronic Warfare Analysis section later.

UNITED KINGDOM

3383.393

ARI 23246/1 ECM POD (SKYSHADOW)

The ARI 23246/1 (Skyshadow) ECM pod was first shown publicly at the 1976 Farnborough Air Show when it was stated that it was being developed as a modern technology ECM pod for RAF strike aircraft. A multi-million pound contract was awarded to a consortium of British companies led by Marconi Defence Systems, which included Racal Radar Defence Systems, Hawker Siddeley Aviation, GEC Avionics and Plessey. Marconi Defence Systems is the prime contractor with responsibility for overall design and development and subsequent testing.

A high-power TWT amplifier is used with a dual-mode capability for deceptive jamming and CW operation. A voltage-controlled oscillator uses a varactor-tuned Gunn diode covering the full frequency band. The set-on receiver is of Racal design and signal processing uses GEC Avionics hardware and MDS software.

The pod is a fully autonomous jammer which can jam multiple and simultaneous threats. Its dual receivers and jamming transmitters are connected to a dedicated central processor. The processor and central computer identify the threat, establish priorities on lethality and also on optimum jamming strategy. Once a threat is identified, classified and given a priority, the Skyshadow system automatically sets the jamming frequency and optimum modulation technique to counter the enemy radar.

The system is software programmable to allow modification of the threat library, and will easily accommodate changes in operating environment or mission requirements. The equipment digitises receiver outputs which are then passed on to the processor in digital form. The processor maintains a file of all active emitters detected by the system, identifies emissions not previously detected, and then displays the identity and bearing of those categories requiring display. Additional inputs to the processor come from other EW sensors on board the aircraft. The on-board displays provide pilot and navigator with both tabular and graphic presentations of threats, as well as the overall radar environment.

ARI 23246 Skyshadow pod mounted beneath wing of RAF Tornado

Skyshadow countermeasures pod for UK strike aircraft

The pod has radomes at both ends and is capable of countering multiple ground and air threats, including surveillance, missile and airborne radar types. Automatic power management can be assumed, and modular construction probably allows for variations in the pod's configuration for differing operational missions. Approximate dimensions are: length 3.35 metres and diameter 38 cm.

STATUS

The system is in full production for the UK MoD and fitted to the IDS version of the RAF Tornado. An upgrading programme is in progress to improve the operational capability to deal with the threats of the 1990s. Deliveries of new modules and modification kits are scheduled to begin in 1992 to phase in with the Tornado mid-life update programme.

CONTRACTOR

Marconi Defence Systems Ltd, The Grove, Warren Lane, Stanmore, Middlesex HA7 4LY, England.

ARTILLERY FIRE CONTROL SYSTEM

Hit at the first round!

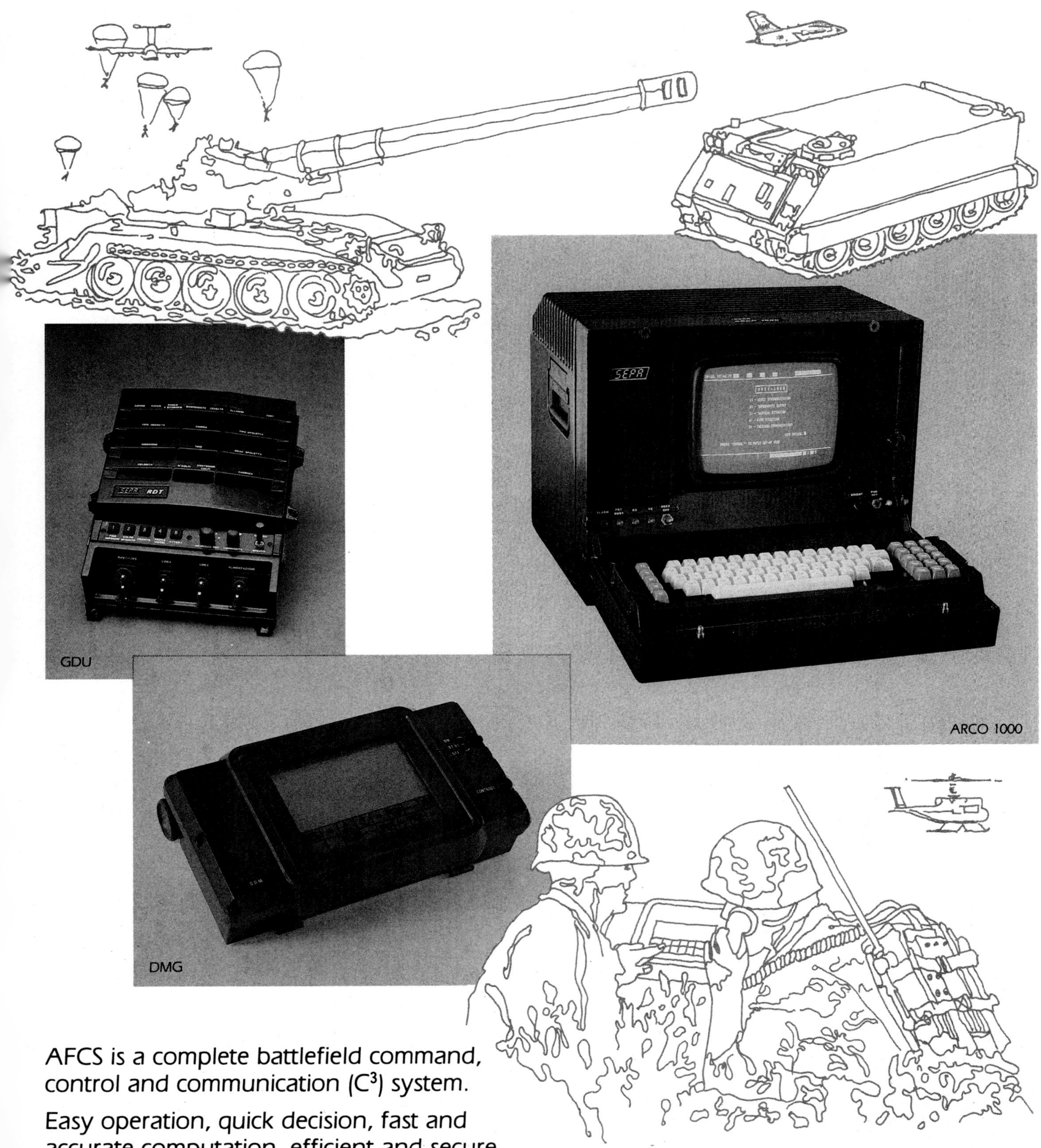

GDU

ARCO 1000

DMG

AFCS is a complete battlefield command, control and communication (C^3) system.

Easy operation, quick decision, fast and accurate computation, efficient and secure communication.

AFCS components:

- ARCO 1000 artillery computer
- GDU gun display unit
- DMG digital message generator

SEPA

Società di Elettronica per l'Automazione S.p.A.

Corso Giulio Cesare 294 - 10154 Torino (Italy)
Tel. (011) 205.3371 - Telex 221527 Sepa I

A Spectrum of Capabilities.
for tank fire control systems.
MSZ-2-TANK FIRE CONTROL SYSTEM
When you're seeking a source for computerized Tank Fire Control systems you demand a spectrum of capabilities. Definition, specification, research, development, production, testing must all be linked together - and forged by hard facts - the facts of combat.
El-Op TFC Systems meet these standards and are available for most tanks now in use. Developed in response to active combat feedback, they significantly increase first round hit probability, feature very short response time and simple operation and maintenance. Add El Op's thermal imaging option and aiming can be carried out 24 hours a day and under all weather conditions.
Thousands of El Op TFCs are fully operational today. They are the heart of new MBT's and have upgraded older MBT's into powerful fighting systems.
Proven repeatedly in combat, they provide the armored corps of tomorrow with the spectrum of capabilities vital to mission mastery.
elop
elop Image of Quality
ELECTROOPTICS INDUSTRIES LTD.
Advanced Technology Park, KIRYAT WEIZMANN • P.O.B. 1165
REHOVOT 76110, ISRAEL • Phone 08-486211 • Telex 361944
Facsimile No. 08-486748
350 Fifth Avenue New York N.Y. 10118 U.S.A
Phone 212-947 4600 Telex 237048 B TADUS UR

3038.353

MARCONI RADAR WARNING RECEIVERS

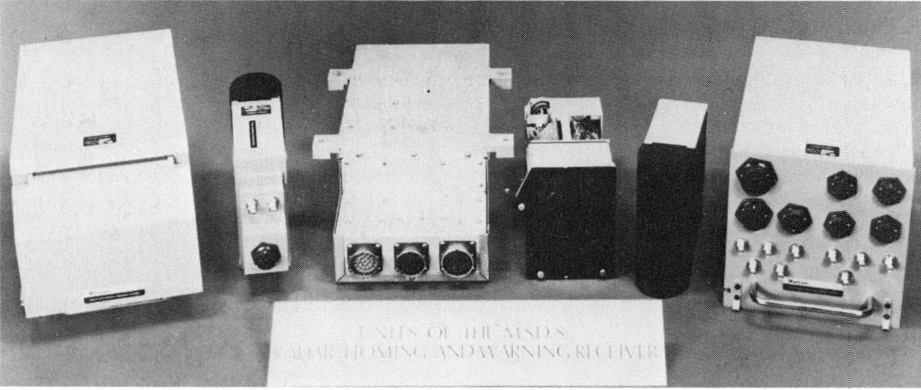

MDS radar warning receiver equipment units

Marconi Defence Systems Ltd has developed and produced in quantity the two types of radar warning receivers (RWR) fitted to the Harrier, Jaguar, Buccaneer and Phantom strike aircraft of the Royal Air Force.

Two versions of the RWR were needed to meet the separate requirements of the multi-seat and single-seat aircraft. Whilst both types use the same antenna/RF sub-system, the Jaguars and Harriers are fitted with a lamp type display which shows the quadrant in which the priority threat is located, whilst the Buccaneer and Phantom have the multi-threat CRT displays. Both systems indicate frequency band and type of signal being detected. Frequency coverage is E- to J-band. Audio alarm is also available.

For the Tornado F2 a multi-threat warning system, has been developed to provide threat analysed readouts in alpha-numeric form. The RHWR detects, identifies and provides accurate bearing of emissions out to the aircraft radar horizon, so that the crew can correlate this information with the aircraft AI radar tracks in order to classify and allocate priority to enemy airborne threats. The RHWR is understood to operate in either a passive mode, or can 'see through' enemy jamming attempts to degrade the AI radar. The system provides for long range radar surveillance and accurate direction finding. A processor unit allows parametric measurement of intercepted signals with automatic comparison of these with information contained in a comprehensive emitter library. Direction finding accuracy is enhanced by using a time-shared vernier interferometer to augment the more conventional amplitude comparison techniques. Among the parameters measured are RF frequency, RF agility, PRI, pulse width and scan characteristics.

STATUS

In operational service. The RAF Buccaneer radar warning/ESM suite is being up dated (see **5804.393**).

CONTRACTOR

Marconi Defence Systems Ltd, The Grove, Warren Lane, Stanmore, Middlesex HA7 4LY, England.

5689.353

ZEUS AIRBORNE ECM SYSTEM

Zeus is a compact multi-purpose EW countermeasures system for modern combat aircraft, designed by Marconi Defence Systems and using major components from Northrop Defense Systems Division, USA. It provides a complete defensive system consisting of a radar warning receiver (RWR), and a jammer. These may be fitted as a 'stand-alone' RWR, or as a full ECM system with the RWR also providing the jammer control inputs.

The Zeus RWR is the latest in a series of MDS passive intercept systems. It is based on the best features of IFM and fast superheterodyne techniques and is able to intercept and measure the characteristics of all radar controlled systems likely to be a threat to the aircraft. These include radar-controlled anti-aircraft guns, surface-to-air missile radars, air-to-air missile radars and airborne intercept radars. Having measured the threat parameters, the receiver passes the signal to the Zeus digital processor which identifies the radar type and causes it to be shown on the display. This digital processor can also be used to control the Zeus jammer and other means of countermeasures (eg chaff, flares and decoys).

Zeus airborne ECM system

The Zeus jamming system is controlled directly by the RWR digital processor and the transmitter chains have alternative configurations to meet specific tasks of each type of aircraft to which it is fitted. Signals from the processor set all the necessary parameters to allow the transmitter chains to jam both pulse and CW radars, while control features ensure that home-on jam weapons are not allowed enough time to acquire the aircraft. A range of different jamming modes is available with each installation tailored to the operational requirements of the individual customer.

Typically, a full Zeus system (RWR and jammer) would weigh approximately 110 kg although the actual weight would depend on the number of transmitter chains and the exact configuration.

STATUS

The system was fitted to the UK Harrier GR-3s and GR-5s under a £100 million contract received in April 1984, and the two companies are proposing the equipment for the AV-8B aircraft. Installations have also been designed for the Hawk, F-16, F-18, F-5 export series, A-4, Jaguar and Mirage.

CONTRACTORS

Marconi Defence Systems Ltd, The Grove, Warren Lane, Stanmore, Middlesex HA7 4LY, England.

Northrop Corporation, Defense Systems Division, 600 Hicks Road, Rolling Meadows, Illinois 60008, USA.

5732.053

HERMES ESM SYSTEM

Hermes is a range of ESM surveillance systems which provide the information necessary for planning operations and defence by the interception and analysis of enemy signals. These systems can be part of air defence and coastal defence land-based nets, can be used in the airborne role in many types of aircraft and also have naval applications.

In the airborne role, Hermes can be configured into a range of equipments from a basic radar warning receiver in light strike aircraft and helicopters to a full ESM system for reconnaissance or AEW aircraft. A version can also be fitted to remote piloted vehicles for battlefield surveillance, coupled with a data link to pass vital information to the ground.

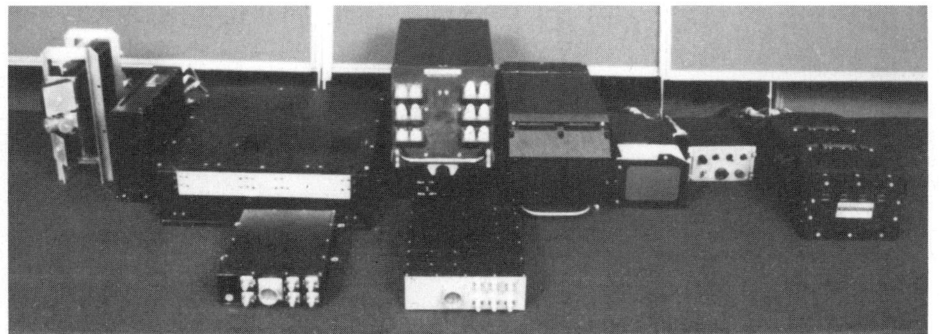

Units of Hermes ESM system

In the land-based role, Hermes is known as Sentry (**5803.153**) and can be configured to meet a variety of battlefield surveillance roles to provide signal intercept and analysis capability. For naval applications the system would normally be employed in smaller ships, such as fast patrol boats, to provide interception, analysis and display, and also to activate chaff/flare dispensers if necessary.

CHARACTERISTICS

Frequency coverage: C to J bands

Angular coverage: 360°

STATUS

The radar warning and homing receiver is based on that used in the Tornado F2 version for the RAF. Hermes ESM systems have been supplied to an export customer for fitment to operational helicopters.

CONTRACTOR

Marconi Defence Systems Ltd, The Grove, Warren Lane, Stanmore, Middlesex HA7 4LY, England.

5804.393

SKY GUARDIAN AIRBORNE RADAR WARNING RECEIVER

Sky Guardian is a family of radar warning receivers (RWR)/ESM systems based largely on experience gained with the ARI 18223/228 systems now in service with a variety of aircraft. The system is designed to combine accurate analysis and threat warning with the high probability of signal interception provided by a crystal video receiver.

There are three standard systems for different aircraft types: Series 100 for helicopter and light aircraft; Series 200 which provides a radar warning receiver with many features of an ESM system; and Series 300 for larger platforms which is an ESM system with an RWR capability.

Parameters of the intercepted signal are measured and passed to a software-controlled digital processor which analyses and identifies the threat. Information is displayed to the aircraft crew by one of several types of display, alpha-numeric presentation on a CRT display, LED display, central warning display, or on the head-up display. The processor also has sufficient capacity to manage a range of counter-measures such as jammers, chaff dispensers, IR flares, etc, and provides the passive intercept and power management element for the Apollo system (**5805.353**).

STATUS

The Sky Guardian system is in production for update programmes on the RAF Buccaneer S2B low-level strike aircraft, and the Sea Harrier.

CONTRACTOR

Marconi Defence Systems Ltd, The Grove, Warren Lane, Stanmore, Middlesex HA7 4LY, England.

5805.353

APOLLO AIRBORNE ECM SYSTEM

Apollo is a version of the Zeus system (**5689.353**) and is a comprehensive ECM equipment for both passive and active countermeasures on strike aircraft. The radar warning capability is provided by the Marconi Sky Guardian receiver (**5804.393**) and, using the radar warning receiver processor for management, a transmitter is added to provide a jamming system capable of countering SAM, AAM and radar controlled gun systems. Apollo can be fitted conformally, as a complete on-board system, or can be pod-mounted, depending upon the aircraft type. The system also provides facilities for the control of additional countermeasures such as IR flares and chaff dispensers.

STATUS

Fully developed.

CONTRACTOR

Marconi Defence Systems Ltd, The Grove, Warren Lane, Stanmore, Middlesex HA7 4LY, England.

5807.353

MONARCH RPV EW SYSTEM

Marconi Defence Systems is engaged in the development of both passive and active EW systems for RPVs. Equipped with a passive ESM system, the RPV can be pre-programmed to carry out surveillance tasks in areas of interest and pass the processed data direct to the ground by means of data links. Control of the on-board system can also be exercised by the same method. A range of expendable jammers for use against hostile communications and radar systems is also being developed. These jammers are also suitable for use in the ground environment.

Although very few details are available on these systems, they are known to be based on the latest technology, and are lightweight and miniaturised to be suitably packaged for fitment into the very restricted payload areas of an RPV.

STATUS

In development. The expendable jammers are undergoing operational trials in the RPV and ground environments.

CONTRACTOR

Marconi Defence Systems Ltd, The Grove, Warren Lane, Stanmore, Middlesex HA7 4LY, England.

RPV engaged in ESM trials

6514.393

MARCONI LASER WARNING SYSTEM

Marconi has developed a laser warning system suitable for use on a variety of aircraft. It combines the use of silicon detectors with wide-bandwidth logarithmic amplifiers to give high dynamic range, low noise and a low false-alarm rate. The system is designed to give visual and aural warning of a hostile laser emission at a range of several kilometres. The display unit gives a visual indication of threat sector of a cockpit-mounted LED display and, operationally, would probably use the same display as the radar warning receiver. The company states that spectral resolution is typically 100 nanometres, giving non-ambiguous identification of most military laser types.

STATUS

Successful airborne trials were carried out in mid-1986 using a Sea King helicopter.

CONTRACTOR

Marconi Defence Systems Ltd, The Grove, Warren Lane, Stanmore, Middlesex HA7 4LY, England.

5667.393

MASQUERADE AIRBORNE DECOY SYSTEM

Masquerade is a chaff and infra-red decoy defence pod for combat aircraft, and provides self-defence against air-to-air and ground-to-air radar, infra-red guided missiles and anti-aircraft shells. It carries 48 Wallop 'Radashield' chaff decoy cartridges and 42 Wallop 'Infrashield' infra-red decoy cartridges and can dispense any combination at varying time intervals as required. The pod is an adaption of ML Aviation's light stores bomb carrier CBLS 200 already in service with the RAF and other air forces. Masquerade will fit on the stores station of virtually any aircraft and requires little or no aircraft modification.

The pod has a microprocessor-controlled decoy dispensing system which can be initiated manually or by a radar warning receiver. This latter may be mounted in the pod if required. The cartridges are contained in three separate modules, each containing 30 decoys. These are loaded before being inserted into the pod to give an extremely low operational turn-round time.

STATUS

In production.

CONTRACTOR

Wallop Systems Ltd, Arkay House, Weyhill Road, Andover, Hampshire SP10 3NR, England.

Masquerade airborne decoy system

5668.393

CASCADE HELICOPTER DECOY SYSTEM

Cascade is a chaff and infra-red decoy system for helicopters. It fires chaff and/or infra-red decoys to protect the aircraft from surface-to-air and air-to-air missile attacks. The decoys are designed specifically for helicopter protection and burst clear of the aircraft to afford protection without the need to manoeuvre, and are also designed to eliminate the danger of matter being ingested by the engines.

The system consists of a small cockpit unit, a lightweight launcher with a rapid reload facility and an automatic sequencer which can disperse any number or variation of decoys as required. The disposable magazine contains completely encapsulated firing circuits and charges for operation under all environmental conditions. Cascade can be fitted to most types of military helicopter and is also easily integrated with a radar warning receiver to provide automatic or semi-automatic firing.

STATUS

Fully developed.

CONTRACTOR

Wallop Systems Ltd, Arkay House, Weyhill Road, Andover, Hampshire SP10 3NR, England.

4650.393

EVADE AIRBORNE DECOY SYSTEM

Evade is an airborne decoy system which provides a highly effective means of firing chaff and infra-red decoys as false targets against radar-guided and heat-seeking missiles. Although intended primarily for fighter aircraft, the system is suited for any fixed-wing type. It comprises four or more lightweight 'scab-on' dispensers and a cockpit control unit. The system is very easy to install and operate and the dispensers can be fitted on any suitable external surface. Pylon mounting is suitable and does not affect the store carrying capacity of the aircraft. On fighter aircraft better chaff dispersion is achieved by mounting on the fuselage underside as far aft as possible.

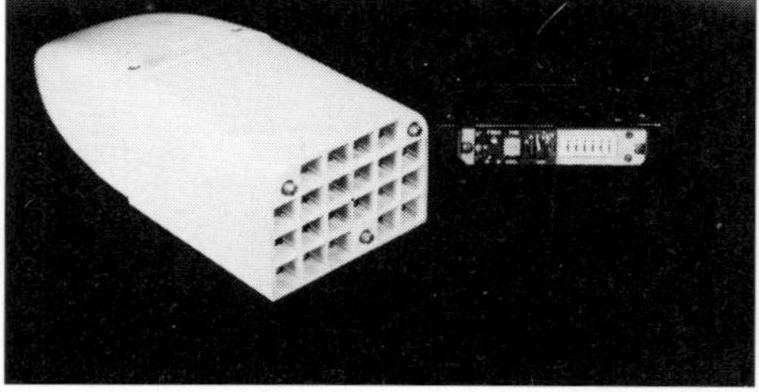

Evade dispenser unit and cockpit control

The aerodynamically-designed dispenser unit is a two-piece container which comprises a firing unit containing the electronic circuitry, and either a chaff or flare pack (the standard Evade suite being two of each). The firing unit is aircraft mounted and connected to the cockpit control unit. The chaff or flare pack is fixed to the firing unit with quick release bolts for rapid rearming. The chaff pack contains 21 cartridges which fire to the rear of the aircraft where the chaff is dispersed in the surrounding turbulence. Dipole lengths are supplied to individual requirements. The chaff is slip coated to maximise rapid blooming and minimise 'birdsnesting'. The flare pack contains 10 infra-red cartridges which ignite when just clear of the aircraft and burn from 3 to 5 seconds. The cartridges in both packs are ejected by electronically detonated squibs.

A compact lightweight microprocessor in the cockpit gives the pilot complete control of the system. Firing sequences can be preset before flight to give single, burst, salvo or continuous firing modes. Display lamps indicate the system state. Built-in-test is incorporated and can be interfaced with other on-board systems.

STATUS

In full production.

CONTRACTOR

Wallop Systems Ltd, Arkay House, Weyhill Road, Andover, Hampshire SP10 3NR, England.

5734.353

KATIE AIRBORNE RADAR WARNING SYSTEM

KATIE (Killer Alert Threat Identification and Evasion) is an airborne radar warning receiver to provide warning of illumination by pulsed or continuous wave radar-directed weapon systems. Four antennas and associated receivers provide all-round azimuth coverage using amplitude comparison techniques to obtain accurate threat-bearing information. The system has a digital processor in which radar parametric data is stored and used to identify radars by type, bearing and threat mode. Threat radars are identified on a heading-oriented cathode-ray tube display which uses alphanumeric symbols, and is readable in bright sunlight or using night vision goggles. A voice synthesiser output is also available to give speech identification and warnings, or warning tones.

Units of KATIE airborne RWR

Radar detection is available over a frequency bandwidth from 2 to 18 GHz, with optional extension down to 0.5 GHz. KATIE is suitable for helicopters or light strike aircraft and the system comprises a processor, two receiver units, a control unit, display and four separate antennas. Weight of the system is approximately 6 kg. The threat-identification library is held in EPROM storage for rapid field reprogramming. The display can also accept data from other sensors such as laser warning receivers, while the processor can provide data for automatic operation of chaff, smoke, flare and expendable decoy stores, or an integrated jammer.

STATUS

In joint development by MEL and Dalmo Victor. MEL is designing a new receiver for a more advanced version, while Dalmo Victor will further develop the processor and display system of the present Mark 3 system.

CONTRACTOR

MEL, Manor Royal, Crawley, West Sussex RH10 2PZ, England.

4410.393

KESTREL AIRBORNE ESM

The Kestrel airborne ESM system is intended to provide electronic intelligence (ELINT) during all phases from peace to active wartime operations. The system enables staff to gather information on possible hostile threats, their deployment and movements, at long range without the hazards associated with other methods or reconnaissance.

Kestrel receives and processes radar emissions over C to J bands. A multi-port amplitude comparison bearing measurement system is used providing instantaneous digital bearing over 360 degrees in azimuth. At the same time a frequency measurement receiver provides instantaneous digital frequency. The pulse-by-pulse digitised information is then passed to the processor which de-interleaves the overlapping pulse trains from the different radars, deriving their pulse repetition, frequency agility, scan type and identification. The most likely identification of the radars is made by comparison of the measured and derived parameters with those stored in the library of known emitters.

Full information about the radar signal environment is presented to the operator on an ordered tabular or tactical display within one second of intercept. Alternatively the information can be displayed on remote displays via a standard data highway to allow the use of common avionic displays. The measured information includes bearing, frequency, frequency agility, PRI, PRI agility, pulse width, scan type and scan period.

CHARACTERISTICS

Frequency: 0.6 to 18 GHz
Bearing accuracy: 7° rms
Frequency accuracy: 5 MHz rms
Radar store capacity: 500 modes
Processing time: 1 s max
Weight: 70 kg (depending on system configuration)

STATUS

In production. In late 1987, Racal received an order from the Royal Danish Navy for its fleet of Lynx helicopters.

CONTRACTOR

Racal Radar Defence Systems Ltd, Davis Road, Chessington, Surrey KT9 1TB, England.

3554.393

MIR-2 ESM EQUIPMENT

The MIR-2 ESM equipment was originally designed for use in helicopters but is also suitable for use on fixed wing aircraft and small naval craft where lightness and compact dimensions are important. The MIR-2 uses an advanced digital receiver, covering a wide range of frequencies from C- to J-band, around which a fully solid-state wide-band system has been built. It features lightweight antennas and a cockpit display which incorporates a very compact solid-state LED (light emitting diode) presentation of signal intercept data.

Six antenna packages, which can be flush mounted or externally mounted, are fitted on the aircraft with their boresights at 60° intervals in azimuth. Each antenna package consists of two antennas, both cavity-backed spirals. The modules contain all RF components and received signals are transmitted to the main processing unit. The illustration shows the forward antennas fitted to the nose of a Sea King, the rear antennas mounted just in front and below the roundel with the sideways antennas above the rear wheel.

The pulse receiver processes the signals received from the antennas and presents the information on the control indicator unit. Signals are displayed to the operator by frequency band, bearing and amplitude on the solid-state activity display. The total frequency coverage is divided into four frequency ranges with three amplitude levels indicated in each band at each bearing.

MIR-2 antennas on RN Sea King

STATUS

Developed for the Royal Navy and in operational service with Lynx and Sea King helicopters, under the code name Orange Crop. Total sales exceed 300 sets and the equipment is in full production. The Lynx system is to be upgraded.

CONTRACTOR

Racal Radar Defence Systems Ltd, Davis Road, Chessington, Surrey KT9 1TB, England

5673.353

PROPHET RADAR WARNING RECEIVER

The Prophet system is a radar warning receiver for helicopters and light fixed-wing aircraft designed to reduce vulnerability to radar associated threats on the battlefield by providing the pilot/observer with a real-time and unambiguous threat warning. It is a modular, easily interpreted equipment and system design has taken into account the possible requirement to integrate laser warning, hostile fire indicator and frequency extensions.

The receiver detects radiation in C to J bands and, by means of a processor using programmable threat data, warns of the presence of imminent threats. The system can operate in a dense RF environment with a low false alarm rate. A 4-port antenna system is used, bearing being derived by amplitude comparison and a digital IFM is used to provide frequency data. The processor extracts the data necessary to recognise any threat emission when compared with its field reprogrammable threat library. Audible warning of danger is provided and fed to the aircraft internal communication system.

The display uses light emitting diodes to form an array so that a number of threats can be displayed simultaneously. The information displayed consists of an arrow indicating the direction of the threat relative to the aircraft heading with octantal resolution, together with a two or four character alpha-numeric identifier. The colour and brightness of the display is such that it can be viewed in bright sunlight and is compatible with night vision goggles. Display controls are mounted on the display panel, chaff dispensing controls may also be collocated if required. The total equipment weight is 7 kg and power consumption is 70 W.

Prophet radar warning receiver

STATUS

In production. The Prophet system has been chosen for the RN Sea King, and is in service with a fixed-wing combat aircraft.

CONTRACTOR

Racal Radar Defence Systems Ltd, Davis Road, Chessington, Surrey KT9 1TB, England.

6515.393

RACJAM AIR

RACJAM AIR is a helicopter mounted VHF responsive jammer system which provides an airborne ERCM capability without the need for aircraft to be permanently assigned to the jamming role. The equipment consists of an ESM subsystem used to find targets for the RJS3100 (AN/ULQ-19) 100 W VHF responsive jammer, and a blade antenna fitted to the aircraft.

In the ESM subsystem target signals within the jammer operating band can be identified using the RA1796 receiver and the MA1119 signal display unit. In the jammer subsystem, 16 selected target frequencies in the 20 to 80 MHz band may be entered into the jammer which can then automatically detect and jam the signal on a predetermined priority basis. Because jamming occurs only when targets are active it is possible to provide effective disruption of more than one net simultaneously. Output power of the jammer is 100 W.

The fully ruggedised system consists of two shock-resistant cases each weighing less than 91 kg secured to the aircraft floor. It can be installed in an appropriate helicopter, to which an appropriate blade antenna has been fitted, within 15 minutes. (See also entry **5809.393**.)

CONTRACTOR

Racal Communications Ltd, Bracknell, Berkshire RG12 1RG, England.

RACJAM AIR installed in helicopter

3369.393

HOFIN HOSTILE FIRE INDICATOR

The MS Instruments Ltd hostile fire indicator (HOFIN) is a passive warning system designed to alert a helicopter pilot that his aircraft is under fire. The first indication is an audible warning which draws the pilot's attention to a visual display which shows the general direction of the threat. This enables evasive action to be taken according to the current operational situation.

Three small units comprise the complete system, which weighs less than 6 kg: sensor array, computer unit, and the indicator unit.

The sensor array is mounted beneath the helicopter, and is aligned to the longitudinal airframe axis. It detects the shock wave front generated by the projectile and converts the impulsive pressure change to electrical signals. The electrical signals are fed to the computer unit for further processing.

The computer unit is housed in a 3/8 ATR short case, and is normally mounted in the equipment bay of the aircraft. This unit processes the signals provided by the sensor array, and computes the general direction of the hostile fire. The computer unit generates the one second audible warning signal, which is fed into the intercom system of the helicopter and also drives the indicator unit.

The indicator unit is housed in a 4 ATI instrument case which is normally located on the instrument panel, and consists of a circular red display which is divided into eight 45° octants. When displaying information to the pilot, four adjacent octants are illuminated for five seconds. The illuminated arc rotates in 45° steps and shows the source of the hostile fire relative to the longitudinal airframe axis. The indicator unit houses the operating controls of the system and consists of the display dimmer control, power on/off switch, system test switch, and the reset switch.

STATUS

Supplied to the British and Canadian Forces and evaluated by a number of other nations.

CONTRACTOR

MS Instruments Ltd, Rowden Road, Beckenham, Kent BR3 4NA, England.

5735.393

BAe INFRA-RED JAMMER

British Aerospace has developed an infra-red jamming system designed to protect helicopters operating in forward battle areas from attack by surface-to-air heat-seeking missiles. The jammer consists of an electrically powered infra-red source, an optical enhancement system, and a mechanical modulation assembly. The infra-red source is a graphite radiating element contained in a sapphire envelope. The optical assembly surrounding the lamp is rotated at high speed to provide a shutter action as the assembly passes between the lamp and sixteen slots in the outer drum. This modulates the infra-red radiation at a rate which degrades the tracking performance of the missile. Mechanical and electrical complexity has been avoided by the design of the optical enhancement system to provide maximum modulation for increased jammer efficiency.

CHARACTERISTICS

Azimuth cover: 360°

Elevation cover: ± 25°

Dimensions: 40 cm × 20 cm diameter

Weight: 15 kg

STATUS

Successful airborne trials carried out by UK MoD. A production order has been received from an unspecified customer, believed to be the UK MoD for helicopter operations, and the equipment is now in service.

CONTRACTOR

British Aerospace, Naval and Electronic Systems Division, PO Box 5, Filton, Bristol BS12 7QW, England.

BAe infra-red jammer

6160.393

FERRANTI LASER WARNING RECEIVER

A laser warning receiver is under development by Ferranti, designed for use in fixed-wing aircraft, helicopters and armoured fighting vehicles. It is intended as a countermeasure to laser assisted weapon systems, and provides an audible alarm and visual display showing the direction from which the threat has originated. Output from the receiver can be combined with radar warning equipment to give an integrated battlefield threat warning system.

The system uses a number of dispersed sensors to detect incident laser radiation. The sensor head configuration can be tailored to provide spherical or hemispherical cover, and presents the laser threat bearing to the crew in 45° sectors. The number of sensors required would be tailored to the aircraft type. The sensor protrudes through the aircraft skin as a 25 mm hemisphere and occupies a space of 50 mm depth behind the skin.

The sensors are passive. Laser radiation is routed to the central processing unit by fibre optic cableforms. This feature, which Ferranti believes is unique, eliminates risks of false alarms being generated by radio frequency interference.

STATUS

In development.

CONTRACTOR

Ferranti Defence Systems Ltd, Electro-optics Department, St. Andrew's Works, Robertson Avenue, Edinburgh EH11 1PX, Scotland.

5739.393

VICON 77 ESM POD

Vicon 77 is a 406 mm (16-inch) diameter chaff and flare dispensing pod for use in the self-protect and confusion roles. Two versions are available, the Series 101 and the Series 102.

The Series 101 is configured with eight common dispenser modules and an integral radar warning receiver to provide a self-contained system for retrofit to multi-role aircraft. The Series 102 has 12 dispensers with a maximum chaff payload of 384 cartridges for use in the confusion and corridor laying role. This latter version may also be used for aircraft self-protection, in association with an existing radar warning receiver. Manual or automatic threat adaptive options are available.

CONTRACTOR

W Vinten Ltd, Western Way, Bury St Edmunds, Suffolk IP33 3TB, England.

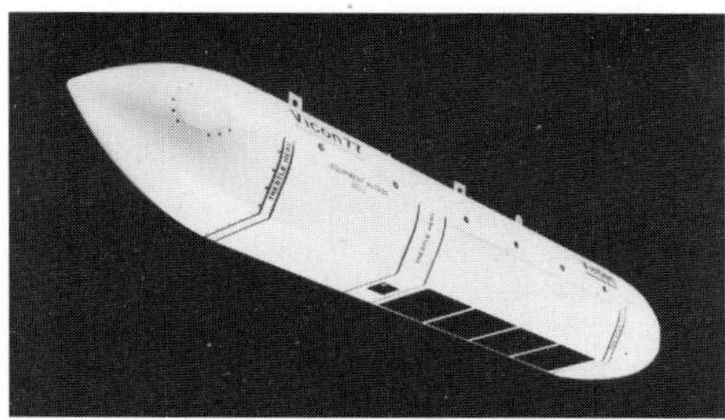

Vicon 77 ESM pod

5822.393

VICON 78 AIRBORNE DECOY DISPENSER

Vicon 78 is a family of lightweight chaff and flare dispensing systems for use in the self-protect mode on fixed or rotary-wing aircraft. These systems may be mounted internally or externally on the aircraft. A common dispenser module accepts interchangeable chaff or flare magazines, each loaded with 32 chaff cartridges or 16 infra-red decoy flares respectively. Externally mounted dispensers are enclosed in streamlined fairings individually tailored to each aircraft type. A typical system comprises two or four dispensers mounted in or on the airframe. These dispensers are independent of the aircraft external store-stations. System control may be manual, or fully automatic when interfaced via a serial data link with advanced radar threat warning to provide threat adaptive dispensing.

A typical two dispenser system, consisting of six line-replaceable units having a total mass of 17 kg, is easy to fit and requires 28 V DC power supply.

STATUS

In production.

CONTRACTOR

W Vinten Ltd, Western Way, Bury St Edmunds, Suffolk IP33 3TB, England.

Typical Vicon 78 line replaceable units

5799.393

CHEMRING COUNTERMEASURES

Chemring specialises in the research, development and manufacture of a wide range of radar reflective materials and their applications. The company has developed and manufactured various types of radar reflective chaff. The chaff available includes aluminium foil and aluminised glass. These and other types of ECM chaff are produced and marketed under the trade name Chem-Chaff, and can be packaged to customer specific requirements for use by land, air and sea forces in any of the current dispensing systems. Chemring also provides design expertise on chaff payloads to systems manufacturers from the early stages of system development.

The company also supplies Radasound chaff-packed rockets for distress signalling, range marking and radar operator training.

In other related spheres, Chemring produces radar corner reflectors for meteorological purposes, radar reflectors for ships, expanding reflective balloon covers, metallic mesh for radio frequency screening and a range of other specialised products. The company has also expanded its research and development in passive ECM materials, using trials and test ranges to gain a better understanding of Chem-Chaff deployment, operation and response characteristics.

The company has been assessed by the UK MoD to NATO quality assurance AQAP-1, and organises and manages trials for both government departments and systems manufacturers in a number of countries.

STATUS

In volume production.

CONTRACTOR

Chemring Ltd, Alchem Works, Fratton Trading Estate, Portsmouth, Hampshire P04 8SX, England.

6155.393

WALLOP INFRA-RED DECOY FLARES

Wallop Industries manufactures a wide range of infra-red decoy flares for use in a variety of fixed-wing aircraft and helicopters. These include Hot-Spot square and round flares used in a range of dispensers such as the AN/ALE-29/39, AN/ALE-40, Wallop Evade, Wallop Masquerade, Philips BOZ 107, BO 300, etc.

Round types include the Infrashield 57 mm cartridge, the Infrastar CART CM 40 mm, Hot-Spot 5 and 6, and the CART CM55. Square types consist of the HS1, 2 and 4. All types use the latest pyrotechnic compositions giving a rapid rise time and high power output at all altitudes, and are fitted with a dual safety and initiating mechanism.

STATUS

In production and in operational service.

CONTRACTOR

Wallop Systems Ltd, Arkay House, Weyhill Road, Andover, Hampshire SP10 3NR, England.

UNITED STATES OF AMERICA

6147.393

QRC 83-05 INFRA-RED COUNTERMEASURES SYSTEM

The QRC 83-05 was first introduced as an internal configuration on US Air Force EB-66 aircraft for deployment in South-East Asia. Subsequently the system was reconfigured for helicopter applications and updated with multi-threat capabilities for re-deployment on US Air Force H-53 helicopters. In the helicopter configuration, the dual transmitter provides protection on bands I and II without engine suppressors. The system electronically modulates a vistal caesium infra-red source to produce a highly effective jamming system.

STATUS

In operational service.

CONTRACTOR

Northrop Corporation, Defense Systems Division, Rolling Meadows, Illinois 60008, USA.

6148.393

QRC 84-02 INFRA-RED COUNTERMEASURES POD

The QRC 84-02 is a multi-threat infra-red countermeasures system capable of operating in a supersonic environment. This pod is a second generation system to meet new and continuing threats. The system is mounted in an aerodynamically faired pod and can be configured with a ram-air turbine to allow protection independent of the aircraft prime power and cooling resources.

STATUS
In operational service.
CONTRACTOR
Northrop Corporation, Defense Systems Division, Rolling Meadows, Illinois 60008, USA.

6149.393

MODULAR INFRA-RED TRANSMITTER SYSTEM (MIRTS)

The Modular Infra-red Transmitter System (MIRTS) can be pod or internal mounted depending on the aircraft requirements. It utilises advanced jammer technologies, including a variable optics/reflector design to provide optimum aircraft infra-red signature coverage combined with the latest digital electronics and mode switching power supplies to enhance reliability, maintainability, flexibility and growth potential. The system is modular and threat programmable with built-in test.

STATUS
The MIRTS is under consideration for a variety of foreign aircraft, including special VIP types.
CONTRACTOR
Northrop Corporation, Defense Systems Division, Rolling Meadows, Illinois 60008, USA.

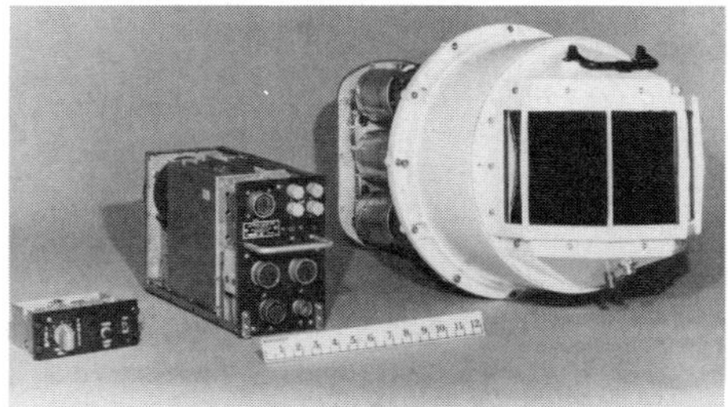
Modular infra-red transmitter system (MIRTS)

4662.393

AN/AAR-34 INFRA-RED RECEIVING SET

The AN/AAR-34 is an infra-red receiving set intended for aircraft self-protection against threats from heat seeking missiles. It consists of four units: a scanner and cryogenic converter normally mounted on the tail unit, a processor and a cockpit controller. The system gives automatic pilot warning on a CRT type display, and automatic countermeasures command.

Multiple threat detection and multiple discrimination modes are provided and the equipment is insensitive to clouds, sun and ground clutter.

STATUS
In production.
CONTRACTOR
Cincinnati Electronics Corporation, 2630 Glendale-Milford Road, Cincinnati, Ohio 45241, USA.

AN/AAR-34 infra-red receiving set

3901.393

AN/AAR-44 INFRA-RED WARNING RECEIVER

The AN/AAR-44 airborne infra-red warning receiver was developed under sponsorship of USAF/ASD to provide warning of launched missile attacks. The set continually searches the lower hemisphere while tracking and verifying missile launches. It warns the aircrew of the position of missiles, and automatically controls countermeasures to neutralise threats and enhance survivability.

Design features of the equipment include: automatic pilot warning and countermeasure command; continuous track-while-search processing; multiple missile threat capability; discrimination against solar radiation, terrain and water backgrounds and countermeasures; multiple discrimination modes; lower hemispherical coverage.

CHARACTERISTICS
Control and display unit: 104 × 145 × 79 mm
Weight: 0.64 kg
Processor: 191 × 168 × 259 mm
Weight: 4.22 kg
Sensor: 366 × 376 mm (dia)
Weight: 15.56 kg
Displays: Sector threat indicators; external countermeasures command and audio tone

STATUS
Production.
CONTRACTOR
Cincinnati Electronics Corporation, 2630 Glendale-Milford Road, Cincinnati, Ohio 45241, USA.

AN/AAR-44 infra-red warning system

6911.393

AN/AAR-47 MISSILE WARNING SET

The AN/AAR-47 missile warning set is a small, lightweight, passive, electro-optical threat warning device to protect aircraft against surface-to-air missile threats. The system detects potential incoming threats, discriminates against false targets and commands an associated countermeasure dispenser for automatic decoy deployment. A very low false alarm rate is achieved through the use of algorithms and signal processing techniques.

The total system consists of a central processor and four sensors and weighs approximately 14 kg. It is ideally suited for a variety of tactical aircraft platforms.

CONTRACTOR
Honeywell Electro-Optics Division, 2 Forbes Road, Lexington, Massachusetts 02713, USA.

3109.393

LUNDY COUNTERMEASURES

Lundy Electronics and Systems Inc has been a major supplier of penetration aids and expendable countermeasure systems and equipment for over 30 years, and claims to be the world's largest producer of metallised glass chaff. So far as is known, Lundy concentrates on the passive aspects of EW, and in addition to the Minuteman III, Mk 12 re-entry vehicle penetration aids sub-system for which the company was awarded a contract in 1962, a full range of dispenser systems has been produced for aircraft and drone applications. The full list includes the following, some of which are described in more detail in individual entries, as indicated:

High capacity chaff/flare dispenser system on EA-6A

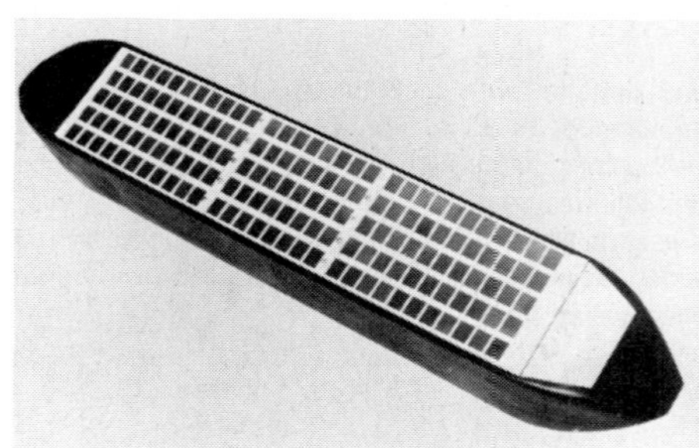
Universal pod for expendable countermeasures RPV

Mk 12 penetration aids system for Minuteman ICBM

AN/ALE-1 and -2: Dispensing sets, pioneer pod system
AN/ALE-24: Chaff dispensing system for B-52G and H strategic bombers
AN/ALE-27: Chaff dispensing system for B-52C to F model bombers
AN/ALE-29A: Countermeasure dispensing set; deployed on various tactical aircraft (**3360.393**)
AN/ALE-32: Chaff dispensing system for EA-6A aircraft
AN/ALE-33: Chaff dispenser for BQM-34 target drone
AN/ALE-43: Chaff cutter dispenser pod (**3110.393**)
AN/ALE-44: Chaff and flare dispensing system, housed in supersonic pod (**3111.393**)
AN/ALE-XX: Universal pod for expendable countermeasures for RPV applications.

CONTRACTOR
Lundy Electronics & Systems Inc, 3901 NE 12th Avenue, Pompano Beach, Florida 33064, USA.

AN/ALE-27 chaff dispenser for B-52C and F aircraft

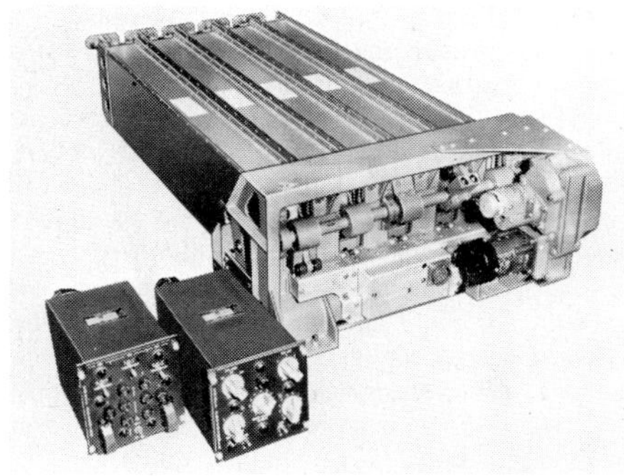

AN/ALE-24 chaff dispenser for B-52G and H aircraft

3360.393
AN/ALE-29A/29B COUNTERMEASURES DISPENSER SYSTEM

The AN/ALE-29A countermeasures dispensing equipment is used on tactical aircraft for the controlled pyrotechnic ejection of chaff and IR decoy flares as a means of self-defence against radar directed and IR homing missiles. The equipment is comprised of one programmer; two sequencer switches; two dispensers (each consisting of a block and printed circuit board); two dispenser housings; and interconnecting cable. The ALE-29A system carries 30 chaff or flare payloads per dispenser, and RR-129 and RR-144 chaff cartridges, and Mk 46 and Mk 47 IR flares are used.

The programmer is normally located in the aircraft cockpit and carries the necessary controls for selection of dispensing patterns, timing sequences, etc. An optional feature is a dual-channel programmer, which allows the simultaneous dispensing of chaff and flares via separate programmes. The AN/ALE-29B is an improved version allowing greater payload flexibility.

The AN/ALE-29A/B system can be interfaced with threat warning devices and forms part of a tactical jamming system on the US Navy EA-6B Prowler aircraft.

STATUS
Operational on US Navy tactical aircraft.

CONTRACTOR
Lundy Electronics & Systems Inc, 3901 NE 12th Avenue, Pompano Beach, Florida 33064, USA. (Lundy is the latest of several manufacturers to build this system.)

AN/ALE-29A dispenser and dispenser housing, showing sequencer mounted on back of housing

4657.393
AN/ALE-36 COUNTERMEASURES DISPENSING POD

The AN/ALE-36 (QRC-490) is a modified AN/ALE-38/41 tactical fighter countermeasures dispenser pod adapted to carry RR-136 and RR-137 chaff units. The QRC/TBC-600 version is a modified AN/ALE-38 pod carrying 600 RR-170 chaff units for saturation/corridor chaff applications. Payloads are carried in 10 dispenser blocks of 60 payloads each and include an integral programmer. Each dispenser can be loaded with chaff units or flares. The fully loaded weight is 680 lbs (309 kg).

STATUS
In production.

CONTRACTOR
Tracor Aerospace Inc, 6500 Tracor Lane, Austin, Texas 78725-2070, USA.

1904.393
AN/ALE-37 CHAFF DISPENSER

The AN/ALE-37 is a high-capacity chaff dispenser for large area coverage, and although not officially confirmed, it is assumed to be carried internally by the seeding aircraft. It is capable of dispensing chaff or expendable jammers for anti-radar purposes, or flares. The probable use for this equipment is not so much self defence for the seeding aircraft as an element of full-scale EW support for a large-scale bomber attack or other large formation operation. Although it cannot be confirmed, it is probable that there are provisions in the equipment for programmed shot dispensing of the chaff or other stores, and that a range of radar frequency bands can be handled.

4654.393
AN/ALE-38/41 BULK CHAFF DISPENSER

The AN/ALE-38 (US Air Force) and AN/ALE-41 (US Navy) equipments are similar high capacity bulk chaff dispensers. They employ unique dispensing techniques and provide continuous dipole dispersal and instantaneous bloom for laying chaff corridors. They can also be used for aircraft self-protection and can be turned on automatically by the radar warning receiver. Pre-cut dipoles are sandwiched between two wraps of mylar film; six 50 lb (22.7 kg) rolls are carried in each pod. The AN/ALE-38/41 can be fitted to aircraft such as the F-105F, F-4, A-3, EA-4, A-7 and AQM-34H.

STATUS
In operational service but probably no longer in full-scale production.

CONTRACTOR
Tracor Aerospace Inc, Bollinger Canyon Road, PO Box 196, San Ramon, California 94583, USA

1905.393
AN/ALE-39 ECM DISPENSER

The AN/ALE-39 ECM dispenser is a third generation expendable countermeasures dispenser system for protection of tactical aircraft from ground and air launched missile threats and radar directed anti-aircraft guns. It represents the culmination of over 10 years' evolution of dispenser system refinement by the US Navy, and is designed with flexibility of response to meet present and future needs in a changing threat environment.

The system is capable of accommodating up to three types of expendable payloads (chaff, IR flares and expendable jammers), loaded in any combination of multiples of 10. All three types of payload can be dispensed manually (single payload) or automatically in accordance with preset programs. The dispensing function can be initiated by the pilot (or EWO officer in the case of the F-14). The system is also capable of accepting dispense commands from aircraft warning receivers.

The programming flexibility, payload loading versatility, operational modes and wide selection of payload dispensing options and sequences (which includes simultaneous multiple flare ejection, random chaff ejections and rapid dump of flares) combine to make the AN/ALE-39 one of the better first generation analogue programmable countermeasures dispensing systems in the US inventory.

The system is operational on A-4, A-6, A-7, and F-14 aircraft and also on US Marine Corps Harriers and the F-18. The system now also dispenses POET and GENE-X expendable countermeasures.

STATUS
More than 900 systems have been delivered since the initial production release in October 1973 and the equipment has been retrofitted to nearly all the US Navy 117 SH-2F LAMPS 1 helicopters.

CONTRACTOR
Tracor Aerospace Inc, Bollinger Canyon Road, PO Box 196, San Ramon, California 94583, USA. (Tracor is the latest of several manufacturers to build this system.)

1906.393
AN/ALE-40 COUNTERMEASURES DISPENSER SYSTEM

The AN/ALE-40 countermeasures dispenser system, RR-170 chaff cartridge and MJU-7/B flare cartridge were developed for the F-4 family of aircraft. Each AN/ALE-40 consists of four dispensers, a chaff/flare programmer and a cockpit control unit (CCU). One dispenser is mounted on each side of the inboard armament pylons. A dispenser consists of a mounting plate to provide attachment to pylon structure, an aerodynamic nose fairing for drag reduction, and a detachable magazine for carrying the chaff cartridge. Each dispenser accommodates 30 RR-170 chaff cartridges for a total of 60 cartridges per pylon and 120 per aircraft. The outboard dispensers can carry 15

MJU-7/B IR flares. System drag is comparable to a Sidewinder missile and launcher. No weapons are displaced. Loaded weight is under 60 kg.

The RR-170 cartridge is square in cross section, 23 mm on each side and 184 mm long. The MJU-7/B flare is rectangular in cross-section, 23 × 50 × 184 mm.

Utilising the technology base of the AN/ALE-40 system, various configurations have been developed for adaptation to other aircraft types. With each new configuration, every attempt is made to install the system in such a way as to avoid interference with weapons stores/capabilities. Installation configurations vary from internal, flush mounted, skin mounted or scab-on, to semi-internal mounting.

An example of an internal flush mounted system is the AN/ALE-40(V) 4, 5, 6 installed on F-16 aircraft which consist of CCU, chaff/flare programmer, sequencer switch, EMI filter and two dispensers. The dispensers are mounted internally in the aft fuselage. Each dispenser carriers 30 RR-170 chaff units or 15 MJU-7/B IR flares.

Examples of the semi-internally mounted systems are the AN/ALE-40(V) 7, 8, 9 installed on F-5E/F aircraft. The system consists of CCU, chaff/flare programmer, sequencer switch, EMI filter and one dispenser housing containing two dispenser magazines. Each dispenser magazine carries 30 RR-170 chaff units or 15 MJU-7/B IR flares. The dispenser is mounted in the belly of the aircraft near the left wing root. The AN/ALE-40(V)10 is an internally mounted installation for A-10 aircraft, and the AN/ALE-40(V)11 is an internal, flush mounted system for the A-7D.

The AN/ALE-40(N) is the adaption of AN/ALE-40 for skin mounting on the NF-5 (Royal Netherlands AF) aircraft. A system consists of cockpit controls, chaff/flare programmer and two dispensers. Each dispenser carries 30 RR-170 chaff units or 30 flares.

AN/ALE-40 countermeasures dispensers mounted on aircraft stores pylon

The dispensers are skin mounted onto aft fuselage. Drag is negligible. Loaded weight is less than 30 kg.

A similar AN/ALE-40(N) installation has also been adapted and flight tested on F-104 aircraft. Skin mounting is in the same aft fuselage area with similar controls and dispenser components providing 30 RR-170 chaff units or flares.

An upgrade to the system, known as the AN/ALE Threat Adaptive Dispenser System (TADS) has been developed. This latest upgrade is software controlled and provides for automatic, semi-automatic or manual dispensing.

Other variants of the AN/ALE-40 are operationally deployed on Hunter and Mirage aircraft.

STATUS

The AN/ALE-40 system is in production and operationally deployed world-wide.

CONTRACTOR

Tracor Aerospace Inc, 6500 Tracor Lane, Austin, Texas 78725-2070, USA.

5688.393

THREAT ADAPTIVE COUNTERMEASURES DISPENSER SYSTEM (TADS)

TADS is an exact mechanical replacement for AN/ALE-40(V) systems (**1906.393**), and uses solid-state technology to incorporate a threat adaptive programmer. The microprocessor controlled programmer accepts information from the aircraft radar warning receiver and tail warning system if available, air data computer and throttle quadrant power setting, and enables the optimum deployment routine for available expendables.

TADS is capable of the following operating modes:

(1) off/standby

(2) manual – in this mode the pilot initiates a pre-selected dispensing programme

(3) automatic – this is a fully automatic mode and causes automatic selection of the appropriate expendables as well as automatically selecting the optimum dispensing programme for the threat engagement at hand

(4) semi-automatic – this requires pilot initiation of the automatically selected optimum dispensing routine.

Mode selection is made by a single switch on the cockpit-mounted control and display unit. An emergency jettison mode is available to dispense all expendables on command.

TADS is a next generation countermeasures dispensing system fully suited to the latest aircraft. It includes growth capabilities that allow quick updates to programmer decision logic, by modifying microprocessor software for updating threat characteristics or future mission scenarios. The additional mechanical interchangeability feature with conventional AN/ALE-40(V) systems enables inexpensive and rapid retrofit installation in aircraft equipped with the latter system, as well as providing a retrofit for unequipped aircraft requiring sophisticated self-protection.

STATUS

In development for the USAF.

TADS countermeasures dispenser system

CONTRACTOR

Tracor Aerospace Inc, 6500 Tracor Lane, Austin, Texas 78725-2070, USA.

3362.393

FAC COUNTERMEASURES DISPENSER

The Lundy FAC countermeasures dispenser is a compact, lightweight system which fires both chaff and flares. Designed for the US Air Force to dispense the percussion initiated Mk 50 flare, the FAC dispenser has been successfully flight tested on the 0-2 and UH-1 aircraft.

To augment its operational usefulness, Lundy developed and flight tested a percussion initiated chaff unit which is compatible with the FAC dispensing mechanism. The chaff unit provides effective protection against known radar controlled threats operating in 3 to 16 GHz frequency range.

The basic FAC system contains a cockpit control/display unit, two payload modules, and two firing mechanisms. A cascade feature in the control circuitry allows any number of dispenser assemblies to be added to the basic system. The payload module holds 20 chaff units or flares interchangeably and is removed from the aircraft during ground operations for loading. The FAC dispenser is configured for both internal and external mounting.

Although primarily developed for use with forward air control aircraft types, this dispenser can be used to fire any 40 mm munitions from air or land vehicles.

CONTRACTOR

Lundy Electronics & Systems Inc, 3901 NE 12th Avenue, Pompano Beach, Florida 33064, USA.

Lundy FAC countermeasures dispenser equipment

3110.393

AN/ALE-43 CHAFF CUTTER DISPENSER POD

The AN/ALE-43 chaff cutter dispensing system is believed to be the first cutter-type dispenser to receive an American 'AN' nomenclature. It provides the EW operator with the option of selecting, while in flight, various tuned dipole lengths to meet changing threats in frequency ranges from A- to K-band.

The concept of cutting chaff from a continuous supply to various dipole lengths, while in flight, has numerous advantages. The limitations in utilising the 'cutter' concept for chaff countermeasures in the past was the lack of a suitable chaff supply material. Lundy has developed a chaff cutter roving which is now available in a variety of filament counts and package configurations.

The AN/ALE-43 is designed for two types of installation: an external pod or an internal mount. In the external pod installation, the chaff cutter assembly is mounted on a structural bulkhead at the rear of the pod centre section. This assembly consists of a drive motor, clutch/brake unit, cutting mechanism and supporting framework. The cutting mechanism contains a rubber platen roller and three cutter rollers. The cutter rollers are designed with blades that yield three specific combinations of dipole lengths as commanded. In the internal installation the cutter assembly is mounted directly to the aircraft structure.

For either installation, the system consists of a chaff roving supply, chaff cutter assembly and a cockpit programmer. The chaff supply is made of metallised glass roving packages contained within a chaff hopper. The chaff is dispensed when rovings are drawn simultaneously from each of the roving supply packages. Each roving passes through a guide tube and is drawn between the platen roller and the cutter rollers for dipole cutting. The cut dipole lengths of chaff are then discharged at the exit which opens into the airstream.

A control unit allows the pilot to select dispensing programs and to start and stop dispensing sequences. The pod has mounting provisions for standard NATO 14 inch and 30 inch stores suspension hardware and can be mounted conveniently on existing stores stations.

CHARACTERISTICS

Pod length: 337 cm

Pod diameter: 48 cm

Weight: pod 138.5 kg (empty); Internal unit 36 kg

Chaff payload: pod 145 kg (8 roving packages); internal 164 kg (9 packages)

Dispensing rate: pod 7.2 × 10^6 dipole inches/s; internal 8.1 × 10^6 dipole inches/s

STATUS

In production.

CONTRACTOR

Lundy Electronics & Systems Inc, 3901 NE 12th Avenue, Pompano Beach, Florida 33064, USA.

AN/ALE-43 chaff cutter dispensing pod

3111.393

AN/ALE-44 DISPENSER POD

The AN/ALE-44 is a dispensing pod for use on tactical, support, drones and strike aircraft capable of supersonic flight. It is a lightweight unit for use with both chaff and flare countermeasure stores, and meets 'HERO' requirements. A typical system consists of a control unit and two dispenser pods. Each pod houses two dispenser modules and a sequencer.

The pod features single- or dual-channel dispensing and quick turn-around time. A low frontal profile, small size, and central mounting pod provide for pod mounting on a wingtip, stores pylon, or fuselage. Fully laden weight is only 20 kg.

Payloads for the system, the RR-129 chaff package and Mk 46 IR flare, are both US Navy inventory items. The payload capacity for each pod is 32 chaff and/or flare expendables which are ejected by impulse cartridges. The dispensers (loaded) can be installed in the pod without removal from the aircraft.

A cockpit control unit permits selection of various burst rates, burst intervals, and units per burst, with simultaneous dispensing of flares and chaff.

CHARACTERISTICS

Pod dimensions: 206 × 10.7 × 18 cm

Weights: 13.6 kg (empty); 19.9 kg (chaff loaded); 22.6 kg (flare loaded); 1.1 kg (control unit)

Mode: 1 or 2 units per burst

Program: 1, 2, 4, 8 or continuous bursts per program

Rate: 4, 2, 1, or ½ bursts per s

CONTRACTOR

Lundy Electronics & Systems Inc, 3901 NE 12th Avenue, Pompano Beach, Florida 33064, USA.

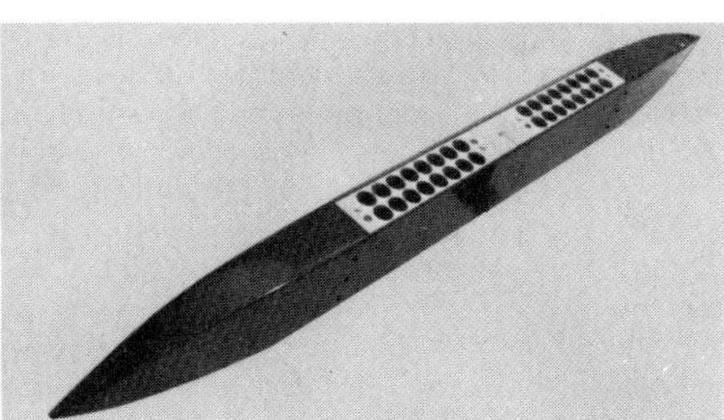

AN/ALE-44 dispenser pod

5828.393

AN/ALE-45 COUNTERMEASURES DISPENSER SYSTEM

The AN/ALE-45 is a solid-state microprocessor controlled dispenser for high performance aircraft. It is automatic in operation in response to threat information from the radar warning receiver, and can also be initiated by the pilot. A dispensing programme is incorporated which can be easily changed by replacing a plug-in threat matrix memory. The system is designed for the US Air Force F-15 aircraft where it is linked to the aircraft's tactical electronic warfare system (TEWS).

STATUS

In production for the US Air Force F-15 aircraft.

CONTRACTOR

Tracor Aerospace Inc, 6500 Tracor Lane, Austin, Texas 78725-2070, USA.

5829.393

AN/ALE-47 COUNTERMEASURES DISPENSING SYSTEM

The AN/ALE-47 is a new concept in countermeasures dispensing which is intended to replace the AN/ALE-40 on the F-16, A-10 and other aircraft. The line replaceable units are electrically and mechanically interchangeable with the existing units. The ALE-47 is intended to provide protection from both airborne and ground-based threats and will require significantly greater processing power to provide the necessary threat adaptive and automatic capabilities. The system will be completely automatic in operation based on information supplied by an associated threat warning receiver.

STATUS

Under a Phase 1 contract both Tracor and Goodyear produced competing prototype sets for demonstration. The US DoD is planning to acquire 26 preproduction sets.

3345.393

M-130 COUNTERMEASURES DISPENSER

The M-130 aircraft general purpose dispenser is a lightweight countermeasures dispenser system developed using AN/ALE-40 (**1906.393**) technology for the US Army. The system is designed for employment on tactical rotary or fixed wing aircraft, such as the AH-1, CH-47, OH-58, UH-1, RU 21 and OV-1D, to provide self-protection from radar directed weapons by dispensing of chaff, and/or to provide protection from infra-red homing weapons by dispensing flares.

The M-130 system is of a modular design to allow flexibility of operational configuration. The modules consist of a cockpit control unit, and electronics module, one or more dispensers, cabling, and aircraft adaptors. The cabling and adaptors are unique to the aircraft type, whereas the other modules are common among all aircraft types.

The M-130 uses the M-1 countermeasures chaff and the M-206 aircraft countermeasures flare as payloads. Each payload module accommodates 30 payloads of a single type. The system configuration may be either a single dispenser of 30 payload capacity, or a double dispenser of 60 payload capacity. The system weight, with 60 payloads, is 21.8 kg.

The M-130 will function with input from the infra-red threat warning device directly to the electronic module, obviating the need for pilot action.

The payloads, the M-1 countermeasures chaff and the M-206 countermeasures flare, have a nominal 25 by 25 mm square configuration, and are 210 mm in length. The payload form factor is identical to the RR-170A/AL chaff used in the AN/ALE-40 and is functionally interchangeable.

CONTRACTOR

Tracor Aerospace Inc, 6500 Tracor Lane, Austin, Texas 78725-2070, USA.

6214.393

POET EXPENDABLE JAMMER

The Primed Oscillator Expendable Transponder (POET), nomenclature AM-6988A, is an expendable jammer which will fit into a standard 1.4 inch (36 mm) diameter and 5.8 inch (147 mm) long cartridge, and can be carried in many existing countermeasures dispensers. It is specifically compatible with the AN/ALE-29 and AN/ALE-39 dispensers, which allow for the control and dispensing of three expendables: chaff, flares and POET.

STATUS

In service but no longer in production. An advanced system, known as GENE-X, is in production by Texas Instruments (**5618.393**).

CONTRACTOR

Sanders Associates Inc, CS 2004, Nashua, New Hampshire 03061, USA.

5618.393

GENE-X EXPENDABLE CARTRIDGE

The Generic Expendable (GENEX) cartridge is a new, small active-type expendable countermeasures cartridge that is ejected from aircraft and emits radar-like signals to lure a radar-guided missile from its intended target. It is a follow-on from the POET system with a wider frequency range, and will be dispensed from aircraft using equipment such as the AN/ALE-39.

STATUS

Texas Instruments was awarded a $117 million contract in August 1987 to develop and produce the GENE-X.

CONTRACTOR

Texas Instruments Inc, PO Box 226015, Dallas, Texas 75266, USA.

4655.393

MODULAR CHAFF/FLARE DISPENSER

This is a pyrotechnic cartridge type chaff and IR flare dispensing system which has been developed for service with both United States and NATO forces for passive aircraft protection. Because of its modular design, the system can be adapted to a great variety of aircraft and a short cartridge length allows flush mounting to F-16 and other types.

Chaff cartridges developed for use in the MCFD incorporate several advanced features including transverse chaff dipole orientation for rapid bloom and frequency coverage versatility. The flare cartridge which was designed in conjunction with the equipment is twice the size of the chaff cartridge and is replaceable on a two-to-one exchange basis. The flare incorporates improved composition and a rapid ignition system for fast time rise.

STATUS

In production for the J-35 Draken aircraft of the Royal Danish Air Force and under evaluation for the Harrier aircraft of the Royal Air Force.

CONTRACTOR

Tracor Aerospace Inc, Advanced Systems Operation, Bollinger Canyon Road, PO Box 196, San Ramon, California 94583, USA

3112.393

LUNDY CHAFF AND FLARES COUNTERMEASURES SYSTEM

Lundy produces a form of unitary design for expendable countermeasures systems. The basic element of the system is the dispenser unit, which contains 14 flares or chaff cartridges.

Positioned in a back-to-back pair, the two dispenser units form an externally mounted dispensing assembly, having a low aerodynamic profile. Dispenser units are attached to the aircraft with a contoured mounting plate. The plate serves as both a mechanical and electrical interface. This allows mounting options at numerous locations on the surface of the aircraft. Only the contour of the mounting plate is altered to accommodate the skin line curvature at any location.

The dispenser unit contains the payload module and sequencer. These components are handled as a single unit during ground operations. As a result the electrical initiators (squibs) are shorted at all times. This design feature provides immunity to HERO environment. In addition, the signal routeing circuits of the dispenser unit conform to the highest level of EMI specifications.

A particular advantage of this system is that it offers a mission-oriented flexibility, wherein any aircraft can be provided with a flare and chaff capability by the simple installation of mounting plates. These aircraft can be flown with or without dispenser units, reducing the quantity of dispenser units required to support a fleet of aircraft.

A cockpit mounted control/display unit allows the pilot to preselect chaff and flare dispensing programs, initiate and terminate dispensing activity, and monitor the status of the system operation.

The control display unit responds to input commands from warning receivers (automatic mode) and to input commands from the air crew (manual mode).

The performance of the flare (intensity, burn-time) is sufficient for the self-protection of military jet aircraft (F-4, F-16), during maximum cruise conditions. The chaff cartridge responds at threat frequencies in the E, F, G, H, I and J bands. The radar cross-section (RCS) created by the chaff cartridge immediately after ejection, 'less than 2 seconds', will exceed the average RCS of an F-4 aircraft.

CHARACTERISTICS

Dispenser assembly
Dimensions: 71 × 13 × 15 cm
Weights: 5.5 kg (empty)
14.2 kg (flare loaded)

IR flare
Dimensions: 127 × 40 mm (dia)
Weight: 0.32 kg
Output, average: 15 kW, 3 – 5 microns
Burn time: 3 s minimum

Chaff unit
Dimensions: 127 × 40 mm (dia)
Weight: 0.30 kg
RCS, average: 20 m²; 1 – 2 s
Frequency bands: E, F, G, H, I and J

CONTRACTOR

Lundy Electronics & Systems Inc, 3901 NE 12th Avenue, Pompano Beach, Florida 33064, USA.

3541.393

AN/ALQ-78 COUNTERMEASURES SET

Countermeasures set AN/ALQ-78 is used by the US Navy aboard the P-3C maritime aircraft as the ESM sensor. It automatically detects and measures the characteristics and bearings of intercepted radar signals of ASW and EW interest.

The measured parameters and bearing of the intercepted signals are supplied to the aircraft central data processing system for evaluation, recording and presentation on the aircraft displays.

The system uses a high-speed rotating antenna and a scanning, superheterodyne receiver for acquisition of signals in specific frequency bands of particular interest to the P-3C. Operation is to a great extent automatic, based on parametric data. The countermeasures set normally operates in an omnidirectional search mode. When a radar signal of interest is acquired and analysed, the AN/ALQ-78 automatically initiates a direction finding routine. The signal data is processed by the central data computer and formatted for readout on a multipurpose display.

STATUS

Operationally deployed with the US Navy since 1969. The AN/ALQ-78 is currently being upgraded to the 'A' configuration. The system will be replaced eventually in the P-3C Update programme by the AN/ALR-66(V)5 (**3642.353**).

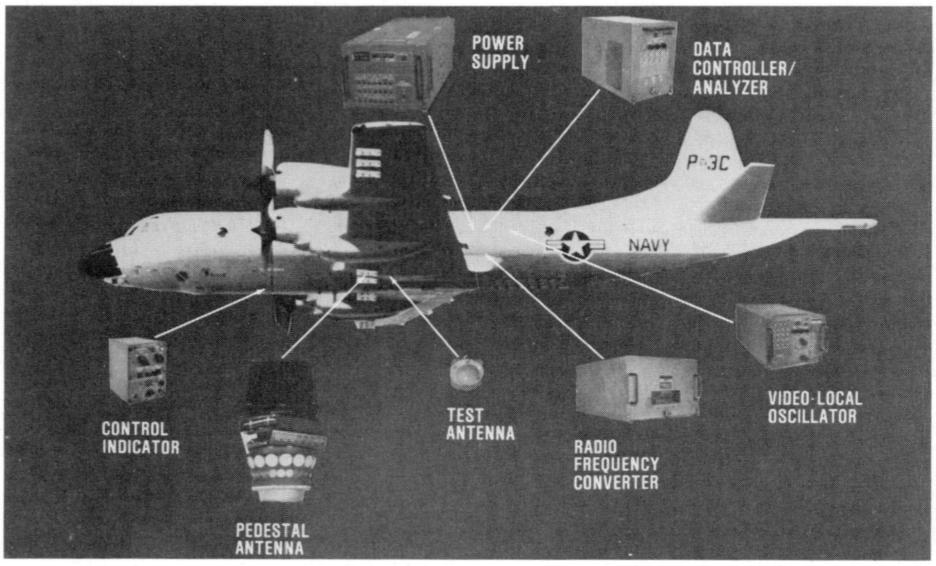

AN/ALQ-78 countermeasures set showing location of units on P-3C Orion maritime aircraft

CONTRACTOR

Loral Electronic Systems, Ridge Hill, Yonkers, New York 10710, USA.

1908.393

AN/ALQ-99 ECM EQUIPMENT

The AN/ALQ-99 is a large and sophisticated tactical noise jamming system forming the major portion of the US Navy EA-6B ECM aircraft's operational payload. An internally-mounted version, AN/ALQ-99E, is also installed in the USAF EF-111A aircraft, and is described later.

The EA-6B version comprises five external pods, each capable of housing two very high powered jamming transmitters, a tracking receiver, the requisite antennas and a ram-air turbine to provide electrical power. Other associated equipment is housed in the large fin-top blister of the EA-6B, and in the aircraft itself there is further ALQ-99 equipment which includes a digital computer and display, and control facilities for two operating crew members.

The pods are detachable from the aircraft and are capable of housing various combinations of transmitters to cover any desired frequency bands. These can be arranged to provide an aircraft with any combination of duplicated and different operating bands, within the total capacity of the equipment and as required by operational necessities.

An IBM 4 Pi general purpose digital computer inside the aircraft carries out processing for the tactical jamming system as well as performing a navigational function. The latter facility enables the computer to perform a transmission beam steering (or transmission blanking) function so that jamming radiation can be restricted to a given 30° azimuth sector. The computer also interfaces with the display sub-systems.

In the automatic mode, one of three possible operating modes for the ALQ-99, the computer performs the sorting of detected signals and directs the selection and activation of jamming against threats. In this mode, the two operators monitor system operation. In the semi-automatic mode, the computer identifies and indicates threats and the operators select and indicate ECM actions. In the manual mode the operators each search their apportioned parts of the spectrum, identify threats, and assign jammers to them.

The basic EA-6B operates with the original AN/ALQ-99 system and is limited to four jamming frequency bands. Since then a number of modifications have been made to the system which double the number of frequency bands and also incorporate other improvements in operational use and reliability. These later upgrades, known as AN/ALQ-99B and AN/ALQ-99C provide advanced software techniques and improved systems response for the expanded capability EA-6B. Major improvements have been made in travelling wave tubes and transmitters, new software allows multiple signal jamming via a frequency selector mode, and a faster system encoder and clock improves the system response. The computer memory has been expanded by 8000 words. Some of the frequency bands have been combined giving an overall coverage from 64 MHz to 10.5 GHz.

The latest version of the AN/ALQ-99, known as ADVCAP, is designed to provide an advanced capability and involves upgrading of the processing equipment, jammers and receivers. Litton Amecom Division has been selected by Grumman Aerospace to develop the ADVCAP version for the EA-6B aircraft.

The version used in the USAF EF-111A, AN/ALQ-99E, is an internally-mounted version which uses up to ten transmitters mounted on a pallet in the aircraft as opposed to the five of the EA-6B version. It employs five exciters, each powering two of the transmitters. It is also understood to have reduced time for receivers to search the radar bands, and an increase in the number of radars that can be jammed by each transmitter by increasing the jamming bandwidth and generating modulated instead of noise jamming signals. It has increased operational flexibility by allowing several jammers with different frequency coverage to employ the same exciter or signal source and has further extra operational flexibility by adding omni-directional antennas to permit selection of either directional or omni-directional transmissions.

STATUS

The AN/ALQ-99 is the heart of both the US Navy EA-6B Prowler and the USAF EF-111A electronic warfare aircraft, and will continue in production for several years. It has been upgraded on a number of occasions over the past 20 years, the latest upgrade version being known as ADVCAP. The EA-6B version of the ADVCAP, developed by Litton, was due to be flight tested in 1988 with full scale production in 1991. Grumman has a $225 million contract for full scale development of airframe modifications to accommodate the extra electronic equipment.

A competition for the EF-111A version upgrade was cancelled by the USAF in 1987 on the grounds of cost. However, the US Congress supported the upgrade in December 1987 and it would appear that this will go ahead. The time scale on this has slipped and initial testing will probably not start until 1989, with production commencing in 1991/92.

CONTRACTORS

A large number of contractors and sub-contractors are involved in the programme. Foremost amongst these are:

Eaton Corporation, Deer Park, Long Island, New York 11729, USA – prime contractor of AN/ALQ-99 and ALQ-99E

Litton Systems Inc, Amemcom Division, College Park, Maryland, USA -prime contractor of ADVCAP developments

Raytheon Company, Electromagnetic Systems Division, 6380 Hollister Avenue, Goleta, California 93117-3197, USA – transmitters/exciters.

AN/ALQ-99 ECM system used in US Navy EA-6B aircraft undergoing final test at Eaton Long Island plant

1910.393

AN/ALQ-101 ECM PODS

The AN/ALQ-101 designation embraces a large family of jamming pods with varying capabilities, which have evolved over a period of about 20 years.

All earlier versions of the ALQ-101 were of circular cross-section and were about 25 cm in diameter, the first model having been approximately 230 cm long and the three succeeding versions about 390 cm in length. The final model of the series, the ALQ-101(V)-10, was the first of the so-called 'gondola' type in which the underside of the pod has throughout its length a trough-shaped compartment which considerably increases the available volume of the pod without serious increase in cross-sectional area. This model enlarged the frequency coverage once again and also incorporated other improvements to provide 'full capability', or the ability to counter all known anti-aircraft threats. Earlier ALQ-101 pods were updated to (V)-10 standard and a total of more than 400 was procured. This final model of ALQ-101 is approximately equivalent to the ALQ-119 which followed after the USAF awarded a contract in 1970.

The ALQ-101 pod is usually employed on F-4 aircraft and has been supplied to US forces and other operators of this type. One Buccaneer squadron of the Royal Air Force is equipped with this pod (Mk 10 version) and it is believed that the RAF has 22 further sets.

One other derivative of the original pod should be mentioned. This is the ALQ-105 which consisted of a repackaged version that amounted to a longitudinal splitting of the pod, the two halves being grafted on to the sides of the F-105 fighter-bomber aircraft. About 90 were produced.

STATUS

In widespread operational use on several types of aircraft.

AN/ALQ-101(V)-10 countermeasures pod on F-4

CONTRACTOR

Electronic Warfare Division, Westinghouse Electric Corporation, Baltimore, Maryland 21203, USA.

6221.393

AN/ALQ-108 JAMMING POD

The AN/ALQ-108 is a jamming pod used primarily on ASW and ELINT aircraft to jam IFF transmissions. Major components are the R-1672 receiver/processor, the T-1164 transmitter, the C-8490 control/indicator, plus an equipment rack and shockmount. No other technical details are available.

STATUS

More than 300 units have been produced, of which over 250 were for use on C-2, E-2C, EP-3A and S-3A aircraft. West Germany has purchased a number of units for its F-4 aircraft.

CONTRACTOR

Magnavox Government and Industrial Electronics Company, 1313 Production Road, Fort Wayne, Indiana 46808, USA.

6222.393

AN/ALQ-117 JAMMER

The AN/ALQ-117 is an active X-band jammer developed to add protection to the B-52 aircraft against radar-directed missiles and guns. The system, previously designated QRC-510, is a dual-mode noise/deception tail-mounted jamming system designed for both the B-52G and B-52H. The installation is in the form of pairs in a 40-inch extension of the aircraft tail. The system is now being modified to the AN/ALQ-117 Pave Mint configuration with improved line-replaceable units and threat capability. A new embedded digital processor, capable of handling threat files, is also incorporated. Pave Mint is an interim modification between the AN/ALQ-117 and the AN/ALQ-172 (**5693.393**) designed to provide near-term improvements.

STATUS

The AN/ALQ-117 has been in production since the mid-1970s. The Pave Mint improvement programme began in 1981 and reached initial operational capability in September 1986. Although the B-52G aircraft will use the modified AN/ALQ-117, the B-52H are being fitted with the AN/ALQ-172.

CONTRACTOR

ITT Corporation, Avionics Division, Nutley, New Jersey, USA.

1911.393

AN/ALQ-119 ECM POD

The AN/ALQ-119 is a dual-mode jammer pod which was developed for the USAF to provide 'full-capability' protection for strike aircraft. By full-capability is meant a complement of countermeasures against known radar threats. Various types of aircraft can be equipped with the ALQ-119 but the principal mode is the F-4 Phantom. The system is also fitted to, or specified for, A-7, A-10, F-15 F-16 and F-111 aircraft. It is reported to be a three-band system, and qualifies for dual-mode status by virtue of facilities for both noise jamming and repeater deception jamming.

When initiated as QRC-522 in 1970, the equipment was given extended frequency capabilities and enhanced modulation performance to become the USAF production version designated ALQ-119. The three frequency bands covered by the equipment have not been disclosed but are probably E/F-, G/H-, and I-bands.

DEVELOPMENT

Development was initiated by QRC-522 and R & D was carried out by Westinghouse. In 1970 a contract for 20 prototype pods was awarded. It entered service in Vietnam in 1972.

It is claimed to be the first dual-mode jammer to enter volume production, a claim which is also made for the dual-mode TWT employed.

STATUS

More than 1600 ALQ-119 pods have been produced, at rates from 10 to 30 per month.

In addition to US Forces, the ALQ-119 is in service on aircraft operated by Israel, West Germany, Egypt and Turkey.

The pod has been continually updated to combat new threats. The current model (ALQ-119(V)-15) is physically different from earlier versions in that it has a short radome mounted below the forward end of the gondola. All previous models are being upgraded to (V)-15 configuration. A physically shorter version, designated ALQ-119(V)-17, is also employed by the USAF.

A development contract was awarded to Raytheon for update of the AN/ALQ-119. The resulting system is re-nomenclatured the AN/ALQ-184 (**6142.393**).

CONTRACTOR

Electronic Warfare Division, Westinghouse Electric Corporation, Baltimore, Maryland 21203, USA.

AN/ALQ-119 ECM pod on USAF A-10 aircraft

5950.393

AN/ALQ-122 ECM SYSTEM

The AN/ALQ-122 is the power-managed radar false-target generator system in use on the B-52 bomber. The equipment searches for, acquires and tracks radar threat signals, and generates narrow-band, low duty-cycle ECM signals to deny range and azimuth to enemy radars. Signals are linear amplified by solid-state amplifiers in the aircraft.

STATUS

In operational use on B-52 aircraft. The last contract for 38 receivers and 28 processors was due to be completed in May 1988.

CONTRACTOR

Motorola Inc, Government Electronics Group, Scottsdale, Arizona, USA.

1912.393

AN/ALQ-123 INFRA-RED COUNTERMEASURES

The AN/ALQ-123 is airborne IR jamming equipment, developed to provide aircraft with protection from IR-homing missiles. It is pod-mounted and powered by a ram-air turbine generator, and was designed for carriage on A-6 and A-7 type attack aircraft. Modification of the system for internal installation in the tail of fighter aircraft has been given some consideration by the US Navy. The system is completely self-contained and modular in construction. All components within the pod extend on slide rails for easy access and module replacement.

The ALQ-123 is not an IRCM of the flare dispensing type (which is another technique of decoying IR-homing missiles), and it is understood to generate a train of light pulses appropriately coded to confuse or break the lock-on of an IR seeking missile. The system was updated in the mid-1980s with new modulation rates to enable it to cope with new types of IR guided missiles.

CHARACTERISTICS

Dimensions: 102 inches long × 10 inches diameter (2591 × 254 mm)

Weight: 378 lbs (172 kg)

MTBF: 150 hours

MTTR: 30 minutes

Lamp life: 150 hours

STATUS

In operational service.

CONTRACTOR

Loral Electro-Optical Systems, 300 N Halstead Street, Pasadena, California 91107, USA.

AN/ALQ-123 installed on US Navy fighter

4366.353

AN/ALQ-125 TACTICAL ELECTRONIC RECONNAISSANCE SENSOR

The AN/ALQ-125 Tactical Electronic Reconnaissance (TEREC) sensor detects, identifies, locates and reports on threats, including surface-to-air missiles and anti-aircraft artillery batteries. The sensor can be installed in any manned or unmanned aircraft. It is fully automatic; provisions are available for an operator to determine the geographical area and frequency band of coverage and monitor sensor operation, including threat display selection. The TEREC sensor provides the tactical commander the capability to keep pace with the dynamics of the modern battlefield by providing rapid threat recognition and location information. In addition to real time cockpit display of threat information, the data can be transmitted via data link to small mobile terminals at various levels of tactical command.

TEREC uses two antennas to provide coverage for both sides of the aircraft. A computer directed frequency search is provided over a broad bandwidth and under computer control the receiving sub-system scans through the pre-programmed RF ranges of interest. Once the data of a threat emission is in the TEREC computer, a direction of arrival is calculated on the basis of quantitised values as measured by the interferometer array. The system general purpose computer enables automatic data collection, data processing and reformatting of the data for display, data linking and magnetic tape recording. Operator controls include the tape recorder, data link, system power and coverage side. In addition a built-in-test function can be performed at the operator's discretion.

STATUS

22 USAF RF-4Cs are equipped with TEREC. Production is believed to have been completed.

CONTRACTOR

Litton Systems Inc, Amecom Division, College Park, Maryland, USA.

AN/ALQ-125 TEREC system

1913.393

AN/ALQ-126A/B ECM EQUIPMENTS

The AN/ALQ-126 is a deceptive electronic countermeasures (DECM) system developed by the US Naval Air Systems Command for the Tactical Air Electronic Warfare programme. It provides a wider coverage of radar frequency bands than the preceding AN/ALQ-100 system, and was initiated in response to new threats from AA, SAM and AI systems.

AN/ALQ-126 equipment

The AN/ALQ-126A is internally fitted in US Navy tactical aircraft, such as the A-4, F-4, A-7 and A-6, and has replaced the AN/ALQ-100. More than 1100 AN/ALQ-126A have been delivered to the US Navy, plus a small number for the Royal Netherlands Air Force F-104 aircraft.

An improved version, the AN/ALQ-126B, is now in full production. In addition to increased frequency coverage, the AN/ALQ-126B incorporates improved deception techniques, more modern construction, packaging and cooling arrangements. The AN/ALQ-126B includes a distributed microprocessor control system to enable the system to be reprogrammed to meet changing threats, and also provides considerable improvements in signal processing. The AN/ALQ-126B operates either autonomously or when fully integrated with the AN/ALQ-45F/APR-43 or the AN/ALR-67 radar warning equipments, the AN/ALQ-162 CW system and the HARM missile. The AN/ALQ-126B is compatible with provisions for the A-4, F-6, A-6, A-7, F-14 and F/A-18.

CHARACTERISTICS

(AN/ALQ-126B)

Dimensions: 411 × 270 × 680 mm

Volume: 67 589 cc

Weight: 86.3 kg

STATUS

The US Navy has awarded contracts totalling nearly $500 million for production of the AN/ALQ-126B. Systems have been delivered and production is to continue until the end of 1988 under continuing contract awards to reach a total of more than 1100 sets of this version. The system is also in production for other air forces for the protection of CF-18s of the Canadian forces, F/A-18s of the Royal Australian Air Force, and the EF-18s of the Spanish Air Force. The 250th system was delivered in February 1987. In July 1987 Sanders received a $37 million contract to supply 71 systems for Spain and Australia.

CONTRACTOR

Sanders Associates Inc, Countermeasures Division, Nashua, New Hampshire 03061-2004, USA.

4668.393

AN/ALQ-128 WARNING RECEIVER

The AN/ALQ-128 is a multi-mode threat warning receiver forming part of the Tactical EW System (TEWS) used on the US Air Force F-15 aircraft. It is combined with the AN/ALQ-135 (**3346.393**) and the AN/ALR-56 (**3347.353**). No details have been released regarding this equipment but it is known that recent updating changes have been included to increase the number and type of threats that the equipment can recognise. The equipment probably covers the area beyond the H/I/J range of the AN/ALR-56.

STATUS

Most units required for the US Air Force F-15 aircraft have now been procured. A contract for 48 systems for the USAF was awarded in January 1987. The system is being upgraded as part of the F-15 TEWS improvement.

CONTRACTOR

Magnavox Electronic Systems, Fort Wayne, Indiana 46808, USA.

1914.393

AN/ALQ-130 ECM EQUIPMENT

The AN/ALQ-130 is tactical communications jamming equipment for use by US Navy fighter and attack aircraft, including A-4, A-6, A-7 and F-4 types. The principal function is probably the disruption of the communications links associated with enemy defence networks and surface-to-air missile formations. The jamming may include broad-band electrical and/or acoustic noise or spot frequency jamming, but the former is probably more likely.

Development of the ALQ-130 was initiated as an updating programme for the ALQ-92 tactical communications jammer, which on the USN EA-6B is internally mounted, suggesting that the ALQ-130 will also be carried this way.

CONTRACTOR

Eaton Corporation, Deer Park, Long Island, New York 11729, USA.

1915.393

AN/ALQ-131 ECM POD

The AN/ALQ-131 is an advanced ECM, jamming pod developed under Programme Element 64739F which covers protective systems for various aircraft types. Dimensions of the ALQ-131 shown here are 285 cm × 63 cm deep × 30 cm wide, and weight is 255 kg. It is modular equipment, described in US defence circles as a full-capability-plus system, and is based on the use of a digital computer contained within the pod. Adoption of the modular technique will allow dual-mode (noise/deception) jamming pods covering from one to three frequency bands to be assembled to meet the specific requirements of varying missions and theatre of operations. Particular targets are the radars and guidance systems associated with anti-aircraft defensive systems. The power-management facility of the ALQ-131 is a software re-programmable processor. There are a number of different modules for one version of the pod, providing for jamming of all known anti-aircraft missile system frequency bands. The modular design allows assembly of up to 16 structural configurations to meet various requirements. Smaller or larger pod configurations can be used ranging from 217 to 392 cm long.

The pod is a software re-programmable model which allows a tactical commander to tailor his available pod equipment for his mission requirement(s). Another important feature of the ALQ-131 is its ability to be re-programmed, on the flight-line, in response to threat changes by making changes in the pod's digital computer. This will obviate long and expensive hardware modifications when threat techniques or parameters change and can even be made on the flight-line. The final capability of this pod is the self-contained power management feature. This is included in a single receiver/processor module which not only detects radar threats but also directs the optimum jamming programme.

AN/ALQ-131 ECM pod mounted on F-16 aircraft

DEVELOPMENT

Development was undertaken in response to a USAF RFP, (QRC-559) in the 1971-72 period, when proposals were put up by a number of US industry teams. That headed by Westinghouse was chosen to develop the ALQ-131. Other members of this team were Motorola and Dalmo Victor whose modules did not go into production. Present power management module is being produced by Loral Electronic Systems.

In August 1980 the manufacturer was awarded a USAF contract worth $15.6 million to carry out a classified modification to the ALQ-131 under the Have Exit programme. This is understood to embrace the addition of facilities giving protection against new threats.

An update of the system, known as the AN/ALQ-131 Block II, incorporates new countermeasures techniques and hardware, and is the updated system referred to in the previous paragraph. Field qualification tests and evaluation of the Block II system began in November 1985 in USAF F-16 and A-10 aircraft, although 178 systems had been delivered to the USAF by January 1987.

STATUS

As at the end of 1987, over 1000 AN/ALQ-131 Block I and II systems had been ordered by the USAF, plus orders from the Netherlands, Israel, Japan, Egypt and Pakistan.

In mid-summer 1977, the ALQ-131 was officially designated as being compatible with the F-16 fighter aircraft by the USAF. Other types of aircraft associated with this pod include the F-4, F-111, A-7, A-10, C-130, and F-15.

CONTRACTOR

Westinghouse Electric Corporation, Electronic Warfare Division, Baltimore, Maryland 21203, USA.

3877.363

AN/ALQ-131 RECEIVER/PROCESSOR

The receiver/processor is a self-contained, single modular package that fits within the AN/ALQ-131 jamming pod (**1915.393**). It enhances the operation of this ECM system by maximising its jamming capability and effectiveness against a multiple radar threat environment. This is accomplished through the concept of power management.

The receiver conducts a signal search within a prescribed frequency range, under control of the processor. When a signal representing a threat is acquired the signal is analysed for parameter values, formatted, and used to control the operation of the jammers so that jamming energy is applied with optimum timing and is better concentrated within the emitters RF, IF and servo bandwidths. The concept of power management thereby assures that energy waste, normally inevitable in conventional jamming, is minimised. A look-through feature of the receiver/processor permits continuous surveillance of the radar environment while jamming is in process.

CONTRACTOR

Loral Electronic Systems, Ridge Hill, Yonkers, New York 10710, USA.

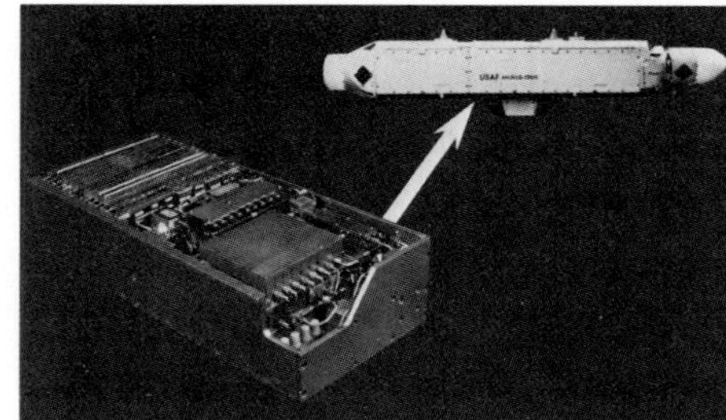

Loral AN/ALQ-131 receiver/processor showing its location in ALQ-131 jamming pod

---

1916.393

AN/ALQ-133 ELINT EQUIPMENT

The AN/ALQ-133 (Quick Look II) is a tactical Elint system deployed in OV-1D Mohawk aircraft for emitter detection and passive surveillance by the US Army. The ALQ-133 has also been proposed for use in a number of A-10 tactical aircraft.

The configuration of the ALQ-133 consists of two pods: one contains Elint receivers needed to cover a frequency range extending from 400 MHz to 18 GHz, and equipped with the appropriate broad-band antennas; the other contains data processing equipment housed in the aircraft fuselage. The pod antennas will be phased interferometer elements, capable of providing direction-of-arrival measurements over a 90° sector abeam of the aircraft to a typical accuracy of 0.5°. If some degradation of bearing accuracy is acceptable, the coverage sector can be enlarged to 120°.

The data processor is responsible for control of the search receivers, analysis of intercepted signals, and comparison of these against a file of known hostile radar characteristics. Other data processing options include housing this part of the equipment in a third pod (this is proposed for the A-10), and locating it on the ground, the raw Elint being passed from the aircraft by data link.

STATUS

In operational service with the US Army.

CONTRACTOR

United Technology Laboratories, 410 Kirby, Garland, Texas 75042, USA.

3346.393

AN/ALQ-135 ADVANCED JAMMING SYSTEM

The internal countermeasures set (ICS) is a component of the tactical electronic warfare system for the US Air Force F-15 Eagle fighter. It is installed in varying configurations for the F-15A, F-15C and F-15E. The AN/ALQ-135 operates with the AN/ALR-56 radar warning system and the AN/ALQ-45 countermeasures dispenser.

The AN/ALQ-135 is an advanced jamming system employing high powered transmitters. All equipment is mounted internally and jamming management is self-contained for the F-15C and F-15E.

The basic AN/ALQ-135 consists of seven line replaceable units (LRUs) plus associated waveguides and antennas. Three LRUs are control oscillators and the other four are amplifiers. Over the years, the system has continued to evolve with the capability of the aircraft and changes in the threat. While maintaining commonality with the original system and support electronics, the ALQ-135 has been updated to include full band coverage and highly effective technique flexibility.

STATUS

In production. Initial development funding commenced in August 1974 and led to a $25 million production in September 1975. This has been followed by further production contracts totalling over $2 billion for more than 1000 sets up to the end of 1986. In February 1987, the US Air Force awarded a $333 million contract for Lot III. Work on this contract should be completed by April 1990. Additional countermeasures improvements are being developed for this system which will extend the survivability of the F-15 through the 1990s.

CONTRACTOR

Northrop Corporation, Defense Systems Division, 600 Hicks Road, Rolling Meadows, Illinois 60008, USA.

4369.393

AN/ALQ-136 JAMMER

The AN/ALQ-136(V) is an airborne jamming system designed for helicopters operating in the tactical role. Its primary role is to protect helicopters against anti-aircraft missile radar systems and operates in the I/J band. It is intended only as a short term jammer to enable the aircraft to carry out a mission from a low-level approach and to 'pop-up' and complete its operational role before the hostile radar can initiate countermeasures.

The system consists of three units: a control unit, two spiral antennas (one each for transmit and receive) and a transmitter/receiver. The latter analyses received signals and responds with the appropriate deception jamming. When the helicopter is illuminated by a hostile radar, the jammer automatically operates, analyses the received pulses and provides the necessary electronic countermeasures.

The entire system weighs approximately 18 kg and is a microprocessor-based system, software programmable and designed for future expansion requirements.

STATUS

Approximately 600 systems have been procured by the US Army, and the system has been evaluated by the US Marine Corps for its AH-1T helicopters. Delivery of equipment for the US Army AH-1S Cobra helicopters is in progress and the system is also likely to be fitted to the AH-64A.

A contract for the latest version, the AN/ALQ-136(V)1/5 was awarded to ITT in early 1988. This covers 114 systems for US Army combat helicopters.

AN/ALQ-136 receiver/transmitter

CONTRACTOR

ITT Avionics Division, Nutley, New Jersey, USA.

4669.393

AN/ALQ-137 AIRBORNE JAMMER

The AN/ALQ-137 is an updated version of the AN/ALQ-94 and fills the self-protection jammer role for the F/FB-111 and EF-111A aircraft of the US Air Force. It is a power-managed deception jammer using a wide-band crystal-video set-on receiver and can be operated in three modes: repeater, CW noise and transponder. The system is broken down into three sub-systems covering E/F, G/H and I/J bands. A pulse power of more than 1 kW is provided at 4 to 5 per cent duty cycle, with a power output on CW of 100 watts. Jamming techniques available include RGPO, VGPO, and selective range jamming. The jammer response time is approximately 100 nanoseconds with a memory duration of about 2.5 microseconds. The AN/ALQ-137 works in conjunction with the AN/ALR-62 radar warning receiver (**4377.393**).

STATUS

In production.

CONTRACTOR

Sanders Associates Inc, Countermeasures Division, Nashua, New Hampshire 03061-2004, USA.

AN/ALQ-137 airborne jammer

4367.353

AN/ALQ-142 ESM SYSTEM

The AN/ALQ-142 is a lightweight, high performance ESM system which provides surveillance in the S, C, X and K-bands in the US Navy LAMPS Mk III ASW helicopter. It acts as the airborne front end of a total ship/air EW system and is data-linked to the ship's AN/SLQ-32 EW system (**3349.293**), and correlated with the on-board database to enhance system response. It is also suitable for other helicopters, marine patrol aircraft and small vessels (fast patrol boats, etc). The system consists of three variants; ASW operations, surface ship surveillance and missile targeting, and airborne area surveillance and missile targetting. Each system has been designed to fulfil a specific mission role. Intercept and DF are performed over 360° in azimuth, with elevation coverage tailored to LAMPS requirements. Four Rotman-lens array antenna units, each providing 90° azimuth coverage, are located fore and aft on the helicopter with the receiver processor centrally located inboard.

The ASW operations variant provides real-time identification and bearing of surveillance radars and their associated platforms. It is the basic helicopter configuration to which modules can be added for increased performance. Sorting techniques are used to detect, identify and analyse all electromagnetic radiation from aircraft and surface vessels in the tactical area.

The surface targetting variant is formed by the addition of receiver modules. This allows surveillance and missile targetting from a small surface ship at ranges beyond the radar horizon, including the detection of long range cruise missiles.

The maritime patrol variant provides additional frequency coverage for assessment of enemy forces from the air. Airborne area surveillance and missile targetting use the receivers in the ALQ-142 to identify emissions at extended ranges, to measure its bearing accurately and provide targetting information for air/sea and surface/surface missile systems.

STATUS
In production.

CONTRACTOR
Raytheon Company, Electromagnetic Systems Division, 6380 Hollister Avenue, Goleta, California 93117-3197, USA.

3546.393
AN/ALQ-144 IR COUNTERMEASURES SET

The AN/ALQ-144 is an electrically powered infra-red countermeasures set to provide helicopters with protection against heat-seeking missiles. Three variants are available; ALQ-144(V)1, ALQ-144(V)3 and an export model ALQ-144(XE). It is an omni-directional system consisting of a cylindrical source surrounded by a highly efficient modulation system to confuse the seeker of the incoming missile. It may be fitted to either the top or bottom of the aircraft. The electrically heated source is carbon, and the complete equipment weighs less than 14 kg. It is one of two systems, together with the ALQ-147, to comprise the 'Hot Brick' IR countermeasures system.

CHARACTERISTICS
Weight: 12.7 kg (transmitter)
0.5 kg (control unit)
Size: 24.1 × 24.1 × 33.6 cm (transmitter)
Field of view: 360° azimuth

STATUS
Two production contracts have been awarded: 822 sets for the US Army for use on AH-1S and UH-60 helicopters and 182 sets for the US Marine Corps for use on AH-1J, AH-1T and UH-1N. Deliveries commenced in March 1981 and over 1200 systems have been delivered to US and allied forces. An improved version, known as the AN/ALQ-144A was cancelled by the US Army in 1986. Two versions, the ALQ-144(V)1 and ALQ-144(V)3 are now being produced by AUL and Numax Electronics. In addition the export model, ALQ-144(XE) has been sold to Jordan and Pakistan.

CONTRACTORS
Sanders Associates Inc, Countermeasures Division, Nashua, New Hampshire 03061-2004, USA.

AUL Instruments Inc, 1055 Stewart Avenue, Garden City, New York 11530, USA.

Numax Electronics, Hauppage, New York, USA.

AL/ALQ-144 IR countermeasures set

3545.393
AN/ALQ-147A IR COUNTERMEASURES SET

The AN/ALQ-147A(V)2 is one of two IR countermeasures equipments referred to under the code name 'Hot Brick'. It is a directable system that fits the aft end of the 150-gallon external fuel tank of the OV/RV-1D aircraft. Because of limited electrical power resources on these aircraft, the ALQ-147A employs a fuel-fired source, and only 200 W of electrical power is required. The heart of the system is a ceramic IR source which is heated by burning a small amount of JP fuel with ram air. The IR output from the source is then modulated mechanically to provide the jamming signal. The system contains a covert filter to minimise the visible emissions, and it cannot be seen at night.

The system is modular in that the fuel tank or modulator can be removed by means of a single clamp. It is this feature which facilitated the later development of a 'stand-alone' version designated AN/ALQ-147(V)1. This is a podded equipment specifically developed for use on the OV-1D aircraft, which is thereby enabled to carry IR protection without the need for heavy external fuel tanks.

STATUS
In service but no longer in production.

CONTRACTOR
Sanders Associates Inc, Countermeasures Division, Nashua, New Hampshire 03061-2004, USA.

AN/ALQ-147A(V)1 pod mounted version of Hot Brick IR countermeasures set

4678.393
AN/ALQ-149 JAMMER

The AN/ALQ-149 is designed to jam hostile communications and long-range early warning radars, and has been scheduled for the US Navy EA-6B Prowler aircraft ADVCAP programme. Sanders is in the engineering development model phase of the on-board portion of the system. The on-board elements include an acquisition subsystem which has separate communications and radar intercept, and processing elements, that co-ordinate their activities with the AN/ALQ-99 radar jammer. The analysis subsystem acquires communication and radar signals and defines appropriate responses. Results are transferred to the central processing subsystem which interfaces to other aircraft systems, including the on-board mission computer, thus enhancing the effectiveness of the EA-6B.

The AN/ALQ-149 will use the AN/ALQ-99 display and a common base of other hardware and software. The transmitters for the AN/ALQ-149 will be housed in a pod. The pod portion of the system has not yet been awarded by the US Navy, but a decision was due in early 1988.

STATUS
Engineering development model phase for seven sets. The original contract was awarded in 1983, and renegotiated during 1986. The first engineering model was delivered in December 1987, with full scale production likely in 1991.

CONTRACTOR
Sanders Associates Inc, Surveillance Systems Division, Hudson, New Hampshire 03051-907, USA.

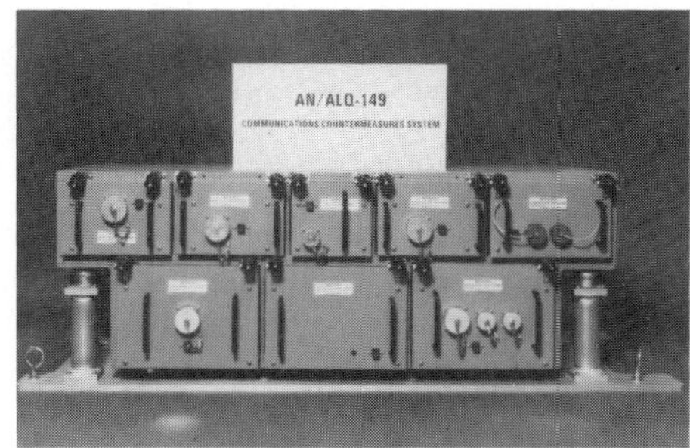

AN/ALQ-149 jammer

4370.353

AN/ALQ-150 ESM RECEIVER

The AN/ALQ-150 is a tactical airborne ESM system used as part of the Cefire Tiger programme of electronic warfare. It is employed against multi-channel communication transmitters and has an electronic support measures capability for integration with jamming systems. The complete system is deployed in four modular units in three US Army RU-21 aircraft and can be used against frequency division multiplexed (FDM), frequency modulated (FM) and time division multiplexed (TDM) emitters.

The Cefire Tiger airborne ESM system interfaces with the Le Fox Grey control processing centre and the forward control and analysis centre to provide overall control of all ECM missions on a real-time basis. The control processing centre has wide-band data links which communicate with up to three Cefire Tiger systems simultaneously. The effective radiated power is from 3 to 10 kW, depending on frequency. Frequency ranges per system are:

(1) Aircraft 1: 60 to 115 and 1500 to 9000 MHz
(2) Aircraft 2: 115 to 480 MHz
(3) Aircraft 3: 450 to 1500 MHz.

STATUS

The first units were deployed during mid-1984.

CONTRACTOR

GTE Government Systems, PO Box 7188, Mountain View, California 94039, USA.

4371.393

AN/ALQ-151 AIRBORNE EW SYSTEM

The AN/ALQ-151, otherwise known as Quick Fix, is a direction finding, intercept and electronic countermeasures system fitted on standard US Army helicopters. It operates in the frequency range from 2 to 76 MHz and has a jamming output capability of 40 to 150 W. The system can interface with all other tactical army aircraft using a secure communications link, and also interfaces with the operations centre.

Quick Fix is normally deployed in the US Army Blackhawk utility helicopter EH-60A/UH-60A and is operational up to 10 000 feet under nearly all weather conditions. Three units are normally deployed with division support companies and two units with brigade support companies. As a communications jammer, the AN/ALQ-151 has been designed to be interoperable with the US Army Aircraft Survivability Equipment (ASE) and will be mounted on the same platform. See also entry **5811.393**.

STATUS

In production. ESL carried out all the design, development and initial manufacture of the AN/ALQ-151 system and was selected as prime contractor by the US Army in early 1984. However, in September 1984, Tracor was awarded a $51 million contract to produce the system and to integrate it on the UH-60A aircraft.

Tracor is currently under contract to provide EH-60A aircraft, modified to a Quick Fix II system, to the US Army. This is under a fixed-price contract of over $87 million, and the first was delivered in July 1987.

CONTRACTORS

ESL Incorporated, Sunnyvale, California, USA.

Tracor Aerospace Inc, 6500 Tracor Lane, Austin, Texas 78725-2070, USA.

3384.393

AN/ALQ-153 TAIL WARNING SET (TWS)

The ALQ-153 tail warning set (TWS) is a rear-looking pulse doppler radar that detects and discriminates approaching missiles and aircraft. It is a range-gated doppler system which automatically ejects chaff and flares against hostile threats.

Development tests were completed successfully in 1976. Competitive 'fly-off' testing against the AN/ALQ-154 on a B-52 aircraft began in 1976 and was completed in 1977, after which Westinghouse was selected for full-scale production.

The first two systems were delivered in April 1980 and were installed on B-52G/H aircraft. By the end of 1985, 321 systems had been delivered and aircraft installations are now completed.

STATUS

System production is complete with the delivery of 321 systems. Spare line replacement unit production is ongoing.

CONTRACTOR

Westinghouse Defense and Electronics Center, Electronic Warfare Division, Baltimore, Maryland 21203, USA

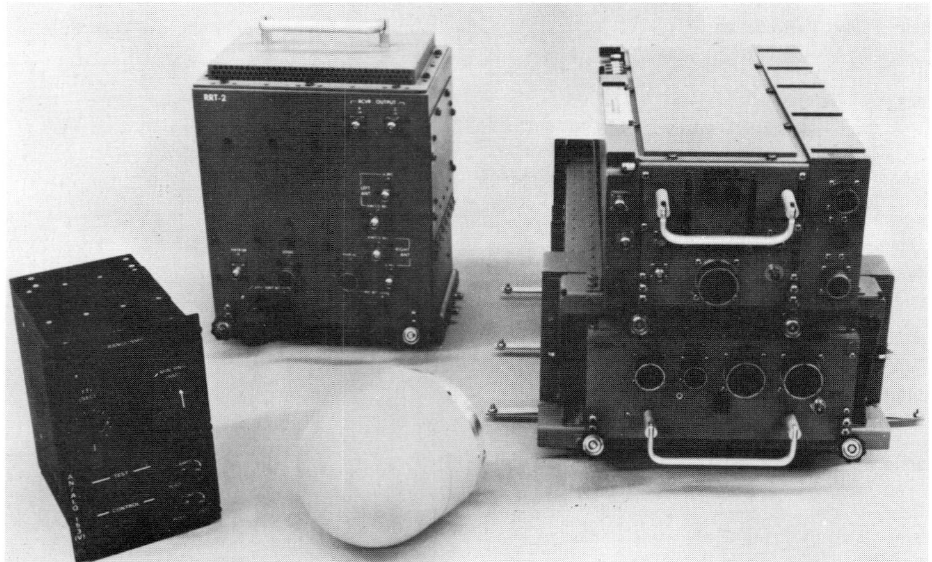

AN/ALQ-153 tail warning set in B-52 aircraft configuration

3552.393

AN/ALQ-155(V) EW POWER MANAGEMENT SYSTEM

The AN/ALQ-155(V) countermeasures power management system forms part of the upgrading programme to improve the defence avionics of the US Air Force B-52 bomber. It provides integral set-on receivers for each jamming transmitter, plus increased effective radiated power density through accurate frequency set-on. The system is a power-managed evolution for the ALT-28(V) set providing automated hands-off from the AN/ALR-46 radar warning receiver with near instantaneous jammer response. It is computer managed and field programmable. The system contains automatic frequency control in all modes and a wide variety of ECM techniques that are automated, semi-automated or manual. A 12-transmitter upload capability is provided.

A variety of improvements are incorporated in the ALQ-155(V), including frequency agility against multiple threats, pulse repetition interval trackers, cover pulse jamming techniques, false target generation through pseudorandom noise, coherent and incoherent jamming, and down-link jamming. It has a hybrid IFM receiver and central receiver capability, programmable noise optimisation, increased pulse-up power for CW to pulse operations, electronically steerable antenna system compatability, and compatability with the AN/ALQ-117 deceptive I/J band jammer.

STATUS

Total contracts since 1975 have exceeded $160 million for design, development and production of the AN/ALQ-155(V). About 300 systems have been delivered but production is probably now complete.

CONTRACTOR

Northrop Corporation, Defense Systems Division, 600 Hicks Road, Rolling Meadows, Illinois 60008, USA.

Dual heat exchanger element of new AN/ALQ-155(V) computer-managed ECM set for USAF B-52 bomber

3547.393

AN/ALQ-156 MISSILE DETECTION SYSTEM

The AN/ALQ-156 missile detection system was developed to provide a means of automatically triggering the ejection of IR decoys from an aircraft (specifically the CH-47C helicopter and RU-21 aircraft) in response to an attack by heat-seeking missile(s). The system is also compatible with the AN/ALE-39 and AN/ALE-40 dispensers.

The ALQ-156 is a pulse doppler radar which effectively creates a continuous protection ring around the aircraft and senses an incoming missile. This results in automatic triggering of an ECM dispenser to eject IR decoy flares. The AN/ALQ-156 is also in use on US Navy P-3Cs and other tactical aircraft. The system evaluates the threat by comparison of the closing rates and other parameters, and is stated to be capable of operating in close proximity to the ground without detriment to the performance. The equipment consists of four units: a receiver/transmitter, a control indicator, and two identical antennas. The complete installation weighs about 20 kg.

The use of a microprocessor and a programmable digital signal processor permits rapid adaptation of the ALQ-156 from a helicopter to a fixed-wing aircraft configuration and it can be used in conjunction with other systems such as laser or radar warning receivers and RF decoys.

CHARACTERISTICS

Weight: 22.5 kg
Size: 51.8 × 25.9 × 19.3 cm (receiver/transmitter)
MTBF: 300 h
Power: 425 W

STATUS

Operational on US Army CH-47 helicopters and fixed-wing special mission aircraft, USAF C-130 transports, US Navy P-3Cs, and is currently in continuous production. Development programmes were initiated in 1986 to adapt the system for supersonic applications for possible use on USAF F-16s.

CONTRACTOR

Sanders Associates Inc, Countermeasures Division, Nashua, New Hampshire 03061-2004, USA.

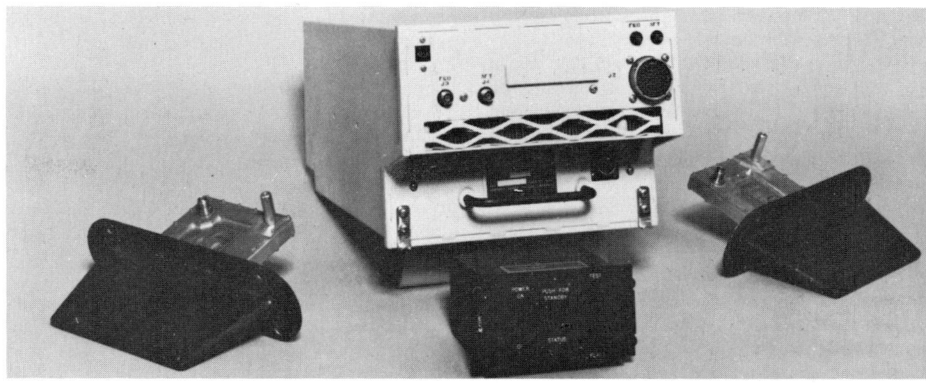

AN/ALQ-156 missile detection set hardware

4375.393

AN/ALQ-157 INFRA-RED COUNTERMEASURES SYSTEM

The AN/ALQ-157 is designed to provide protection against enemy infra-red heat-seeking missiles, and is being manufactured for the US Naval Air System Command for use on large troop-carrying helicopters. Installation is simple and straightforward and the system is adaptable to other types of aircraft. The accompanying illustration shows installation on the sail of a helicopter. Two transmitters are installed, one on each side of the sail for unobstructed protection in azimuth.

The AN/ALQ-157 installation consists of four basic sub-systems: two transmitters, a control power supply, EMI filter assembly and a pilot's control indicator. A switch on the control power supply allows in-flight selection of any one of five pre-programmed jamming codes. Additional codes can be pre-programmed as new threats are defined. The microprocessor also directs all operational sequences of the system.

The system ensures full-time protection against threats, easy access to all components and employs built-in test circuits to perform operational readiness tests automatically. Weight of the complete system is 160 lbs (72.7 kg).

STATUS

Production commenced in December 1983 under a contract to fit the system to US Navy CH-46 helicopters. It is also likely to be ordered for the US Navy CH-53A/D helicopters.

CONTRACTOR

Loral Electro-Optical Systems, 300 North Halstead Street, Pasadena, California 91107, USA.

AN/ALQ-157 system installed on sail of heavy-lift helicopter

3034.393

AN/ALQ-161 B-1 RFS/ECM SUB-SYSTEM

The US Air Force announced the selection of Eaton Corporation in late 1973 to develop the defensive avionics sub-system for the B-1 strategic bomber.

The $70 million cost-plus-incentive-free contract award is for the development, fabrication, and testing of the B-1 radio frequency surveillance/electronic countermeasures sub-system. This sub-system is designed to monitor radio frequencies and employ electronic countermeasures. Eaton is one of three other Air Force contractors presently working on the B1 development programme. These include Rockwell International Corporation of Los Angeles, system contractor; General Electric Company of Evendale, Ohio, engine contractor, and The Boeing Company of Seattle, avionics sub-systems interface contractor.

The Air Force Systems Command's Aeronautical Systems Division at Wright-Patterson Air Force Base, Ohio, is responsible for overall B-1 system development.

The B-1's defensive avionics system is intended to counter surface-to-air missile, anti-aircraft, and air-to-air missile fire control radars and to degrade by noise jamming early warning and GCI radars.

A single AN/ALQ-161 system contains and controls a large quantity of jamming transmitters and antennas. In addition to the jamming hardware, a sophisticated control system managed by a network of digital computers is employed. This can jam signals from many radars simultaneously. The numerous jamming chains are deployed around the periphery of the B1-B to jam signals in frequencies from 200 MHz up to K-band and coming from any direction.

Integrated with the jamming control system is a network of separate receiving antennas, receivers and processors which act as the 'ears' of the system. By means of this receiving sub-system, new signals can be picked up, identified and then jammed, with optimised jamming techniques, in a fraction of a second. One of the advantages of having the receiving function completely integrated with the jamming function is that it allows the receiving system to detect new signals and continue to monitor old signals while jamming in the same frequency band. A special sub-system allows this to be accomplished by monitoring the output of the jamming transmitters and continuously adjusting the receivers. The key to the system's near real-time response to threats is the complete integration of the receiving and jamming sub-systems.

The AN/ALQ-161 system is made up of over 100 units which, exclusive of cabling, displays and controls, weighs approximately 5200 lbs (2363 kg) and consumes about 120 kW of power in the 'all-out' jamming mode.

An operator will have displays and controls by which he can monitor and direct the system. The computer will perform the power management role based on radar threats sensed by the RFS system and pre-stored information about the threats.

Terminal threat jammers, phased-array antennas and other hardware are being provided by sub-contractors.

USAF and Eaton are relying on two features to extend the life and effectiveness of the ALQ-161 system against changing threats in the future in a way that was technologically impossible with the original B-52 ECM hardware. The equipment uses versatile components capable of being adapted. Because the system is digital, it can be software-controlled, permitting changes to be made more easily. Modulation techniques can be added or modified, for example, by changes in the software or by the addition of plug-in modules. There is also growth space to accommodate hardware growth if this becomes necessary. A tail warning function is being integrated.

STATUS

Two contracts totalling more than $657 million for full-scale development and initial production were awarded to Eaton Corporation in mid-1982. The original development contract for $198.8 million included modifying one of the original B-1A bombers and conducting a flight test programme. The production contract totals $458.9 million for building the initial systems. Delivery of the first line replaceable units began in February 1984 and three years flight testing has been carried out. The system is now in full production.

CONTRACTOR

Eaton Corporation, Deer Park, Long Island, New York 11729, USA.

4373.393

AN/ALQ-162 RADAR JAMMER

Northrop Corporation is under contract with the US Navy for production of the AN/ALQ-162.

Development of the system was initiated in 1979 by the US Navy and was subsequently co-sponsored by the US Army. The AN/ALQ-162 reprogrammable jammer is designed to fit a range of US Navy aircraft,

AN/ALQ-162(V) CW radar jamming equipment

the US Army's special electronic mission aircraft, and other US and foreign tactical fighters.

Packaged in a single 40 lb (18.2 kg) box, with a volume of less than one-half cubic foot, this Northrop self-protection system automatically scans the threat signal environment, identifies the threat, and generates specific countermeasures against the highest priorities. The internally mounted receiver/transmitter system is fully software programmable and, with additional growth hardware, can meet future advanced threats or interface with future on-board equipment.

STATUS

In production. The AN/ALQ-162 has been thoroughly evaluated and proven in both threat simulation and flight testing by the US Navy and Army. The system's market potential is valued at more than $360 million. Delivery of the system was due to commence in 1988.

Initial production units will be installed on AV-8B aircraft, and US Army RC-12D, RV-1D and RU-21 fixed wing, and EH-1 and E-60 rotary-wing special electronic mission aircraft. In addition the AN/ALQ-162 is to be installed on NATO aircraft such as Denmark's F-16s and J-35s, Spain's F/A-18 Hornets and Canada's CF-18s.

CONTRACTOR

Northrop Corporation, Defense Systems Division, Rolling Meadows, Illinois 60008, USA.

4740.393

AN/ALQ-164 ECM POD

The AN/ALQ-164 is an airborne pod-mounted jamming system for both pulse and CW modes. It is reprogrammable to provide for threat changes and features mid and high band pulse capabilities developed from the AN/ALQ-126 system (**1913.393**). It is also designed to accommodate the AN/ALQ-162 jammer (**4373.393**) in the pod. It is intended for fitment on the centre-line pylon of the AV-8B VTOL aircraft and has applications for a wide variety of tactical and transport aircraft.

STATUS

Operational evaluation has been carried out on a US Marine Corps AV-8B.

CONTRACTOR

Sanders Associates Inc, Countermeasures Division, Nashua, New Hampshire 03061-2004, USA.

AN/ALQ-164 ECM pod

4374.353

AN/ALQ-171(V) TACTICAL ECM SYSTEM

The AN/ALQ-171(V) is a fully automatic electronic countermeasures system. It consists of a self-contained receiver/processor jammer with an advanced power management capability to ensure the most effective use of available jamming power. The system is available in two configurations, a pod version and a conformal countermeasures system (CCS).

The conformal configuration was developed to provide protection for Northrop's family of F-5 aircraft. Designed to fit on the underside of the aircraft fuselage, the CCS version allows the full use of all weapons and stores stations without affecting the performance and manoeuvrability of the aircraft. The pod version consists of identical electronics and is available for a number of other countries' tactical aircraft. Each system contains multiple transmitters and receivers that share common support electronics, such as a system controller, signal generator and processors, antenna subsystem and cooling. Pod and conformal versions of the AN/ALQ-171 have common logistic support.

The jamming modes and techniques are software reprogrammable, each mode using an extensive range of jamming techniques and complex waveforms. Jamming waveforms and frequencies are programmed from threat data provided by the user. Jamming modes may be combined in response to specific threats, eg. deception to produce false target data combined with noise jamming to suppress real radar returns.

STATUS

In production for the export market. The system was developed as a joint programme with the Swiss Air Force who is installing it in the F-5E aircraft. Portugal has selected the podded system for its A-7s and Taiwan may also order the CCS version.

CONTRACTOR

Northrop Corporation, Defense Systems Division, Rolling Meadows, Illinois 60008, USA.

AN/ALQ-171(V) under aircraft fuselage in conformal configuration

3894.393

AN/ALQ-165 AIRBORNE SELF PROTECTION JAMMER

The AN/ALQ-165 Airborne Self Protection Jammer (ASPJ) is the next generation electronic countermeasures system designed for the US Air Force F-16 and US Navy F-18 tactical aircraft, and also for other fleet aircraft such as the F-14, A-6E and AV-8B. The equipment is being developed jointly by the US Navy and Air Force and will incorporate the latest technology in travelling wave tubes, microwave components and packaging. It can be electrically reprogrammed on the flight line and the built-in-test and modular plug-in design permits rapid replacement of assemblies at the operational level without external test equipment. Weight of the system is between 91 and 150 kg, depending on configuration, and it occupies a space of 2.3 ft³.

The AN/ALQ-165 has the ability to select automatically the best jamming techniques to use against any given threat, based on the system's own computer data and real-time data of the threat signal from the receiver/processor. The computer software can be modified to accommodate new threats as they arise. The equipment covers the frequency range in only two bands, and is technically expansible to cover a greater frequency range, if required.

The transmitters with the system can jam a large number of threats simultaneously over various ranges and in different modes. The computer selects the power and duty-cycle criteria. An augmented version, with an additional transmitter power-booster is also available.

Two configurations are being designed, one will be inside the aircraft, while the other will be installed in an external pod. The internal system for the F-16C will use an augmentation set, consisting of two LRUs, installed in the aircraft tail root that will increase the jammer signal power for the aft-end antenna and will also reduce the time in which the system can respond to hostile radars. The remainder of the system, consisting of five LRUs, will be mounted in the forward fuselage compartment. The F/A-18 aircraft is destined to receive an internally mounted ASPJ having five LRUs as shown in the illustration.

The F-14 models will house the system internally,

probably with an augmentation set similar to the F-16. The US Navy AV-8B Harrier II will also use an externally mounted pod, which will house the two ASPJ five LRUs. The pod was developed as part of the ASPJ development programme. Its length is 99 inches (251 cms) in order to be accommodated in the AV-8B centreline station.

STATUS

Consolidated Electronics Countermeasures, a joint venture between ITT Avionics and Westinghouse, was awarded a Product Verification Contract valued at over $376 million in August 1987. The award is based on the success of the full scale development programme for the ASPJ system. Twelve engineering development models have already been delivered to the USAF and USN and have been undergoing extensive tests and evaluation. The Product Verification programme requires the delivery of six systems in Phase 1 and 14 additional systems in Phase 2. The Phase 2 part is to be awarded during 1988 after the successful completion of 800 hours reliability testing and initial flight tests. Following Product Verification, the US DoD programme calls for independent manufacture by ITT and Westinghouse of 96 systems each beginning in 1989. Thereafter, each company will bid competitively for each fiscal year's production. The USAF and USN plan to acquire over 2500 systems by 1997.

CONTRACTORS

Westinghouse Electronic Warfare Division, Baltimore, Maryland 21203, USA.

ITT Avionics Division, Nutley, New Jersey 07110, USA.

5693.393

AN/ALQ-172 NOISE DECEPTION JAMMER

This is an advanced ECM system intended for installation in the US Air Force B-52 strategic bomber fleet and is a modified version of the AN/ALQ-117 noise deception jammer currently operational in the B-52G/H fleet.

The improvements incorporated in the AN/ALQ-172 include specific threat identification, extended frequency coverage, dense environment operation, advanced CW and pulse jamming techniques, a reprogrammable automatic operation and an improved power management system. There are two versions of the system, (V)1 and (V)2, the latter being upgraded with a phased array antenna.

STATUS

In production. More than $800 million had been awarded to the end of 1986, for development and preproduction. As at the end of 1987, more than two squadrons of B-52Gs had been fitted with the AN/ALQ-172(V)1. The (V)2 version is being fitted to the B-52Hs and was expected to go into service in late summer 1988.

CONTRACTOR

ITT Corporation, Avionics Division, Nutley, New Jersey, USA.

4100.393

AN/ALQ-176(V) ECM POD

The AN/ALQ-176(V) is a pod-mounted jamming system for ECM support, stand-off jamming, and combat evaluation and training. The pod system is composed of transmitter modules and complementing antennas, housed in cannisters to form a slim pod with a 10-inch (25 cm) diameter. The AN/ALQ-176(V) mounts onto standard aircraft store/munitions stations, either wing or fuselage. The pod offers the option of using aircraft internal power or a ram air turbine. It can also be used on RPVs.

The inherent modular design of the AN/ALQ-176(V) pod allows flexibility in ECM configurations. A two or three cannister arrangement is available. The two cannister configuration (AN/ALQ-176(V)1) houses up to three voltage tuned magnetron (VTM) transmitters. The three cannister version (AN/ALQ-176(V)2) provides housing for five transmitters. Dependent on frequency band and tube selection, each transmitter produces 150 to 400 watts CW (up to 30 per cent bandwidth) at efficiencies of greater than 50 per cent. The transmitters provide ECM across the specified threat frequencies. Transmitter protection circuits safeguard the system and provide fault status to the cockpit control panel.

The system has been engineered specifically to allow rapid turnround in meeting the demands of new threat frequencies. The AN/ALQ-176(V) can accommodate new demands through selection of different stored parameters or through easy replacement of the transmitter tube or antenna. Transmitter spares, maintenance, costs, time and training skill are greatly reduced by the use of standard transmitter design. In addition, the design incorporates built-in test and provides for parameter/mode selection at the cockpit control panel. The flexibility in ECM pod configurations and standard aircraft stores stations installation enables the pod to be used on a wide variety of tactical and support aircraft.

AN/ALQ-176(V)1 in two canister configuration with ram air turbine power supply and internal electronics tray

CHARACTERISTICS

Frequency coverage: VTM transmitters from 0.8 – 15.5 GHz

Transmitter power output: 150 – 400 W per tube CW (dependent on frequency and tube selection

Modulation: Noise (various)

Weight of transmitter module: 15 kg

System weights: (V)1 pod – 100 kg; (V)2 pod – 145 kg

Diameter: 25 cm

Length: (V)1 pod – 199 cm; (V)2 pod – 259 cm

STATUS

Thirty systems have been ordered by the US Air Force and delivery is in progress. Additional orders have been placed by the Royal Norwegian Air Force. The pod has also been successfully evaluated by the US Navy and the Canadian Air Force.

The AN/ALQ-176(V) is currently being adapted for use on the Boeing Robotic Air Vehicle (BRAVE), and tests of the system, known as Pave Cricket have been successful. Use on other RPVs is considered likely.

CONTRACTOR

Sperry Corporation, Electronic Systems, Great Neck, New York 10020, USA.

4009.393

AN/ALQ-178 ECM SYSTEM

The AN/ALQ-178 ECM system is an integrated RWR/ECM suite for the F-16 aircraft. Based upon tested components and techniques of the Rapport II system (**3543.393**), the ALQ-178 provides the F-16 with the automatic countermeasure capability to survive in the increasingly deadly, and sophisticated, hostile air defence environments. The system utilises a central programmable computer for data analysis and system control, with independent microprocessors to direct RWR, display and jamming functions. The wide-band RWR continuously scans the threat radar environment. Detected signals are de-interleaved and identified as to radar type and displayed to the pilot on a daylight viewable CRT, in legible, unambiguous format. The power management algorithm optionally matches the countermeasures to the RWR's constantly changing threat picture. Separate forward and aft jammers are used for maximum spatial coverage. The jammers are accurately set for maximum power output at each victim radar's frequency to maximise ERP and jammer effectiveness. The jammer frequency range provides coverage of the entire threat band while the jammer power is sufficient to counter these radars throughout their lethal area.

Although the AN/ALQ-178 consists of both an RWR and an active jammer, the former can be installed as an independent threat warning system, with the jammer added later.

AN/ALQ-178 ECM system units and their location in F-16 aircraft

CHARACTERISTICS

Receiver: Superheterodyne and crystal video

Memory: 128 k, 2-page

Dimensions: 158 cm³

Weight: 188 kg

STATUS

In production.

CONTRACTOR

Loral Electronic Systems, Ridge Hill, Yonkers, New York 10710, USA.

6142.393

AN/ALQ-184 JAMMING POD

The AN/ALQ-184 ECM pod is designed to provide effective countermeasures against surface-to-air missiles, radar directed gun systems and airborne interceptors. An upgrade of the AN/ALQ-119(V) pod (**1911.393**), the AN/ALQ-184 functions as a repeater, transponder or noise jammer. Multi-beam architecture based on proven Rotman-lens antenna technology provides substantially reduced countermeasures response time with a tenfold increase in ERP at 100 per cent duty cycle. High reliability is inherent in the architecture due to redundant mini-TWTs, reduced operating voltages and lower operating temperatures. Maintainability has been improved by the use of digital circuitry, automatic gain optimisation, elimination of manual adjustments and extensive computer driven built-in test.

STATUS

In full production and intended for installation of USAF F-4, F-15, F-16, A-7, A-10 and F-111 aircraft. The pod has flown over 80 successful flight tests on combat aircraft in multi-threat environments.

CONTRACTOR

Raytheon Company, 6380 Hollister Avenue, Goleta, California 93117-3197, USA.

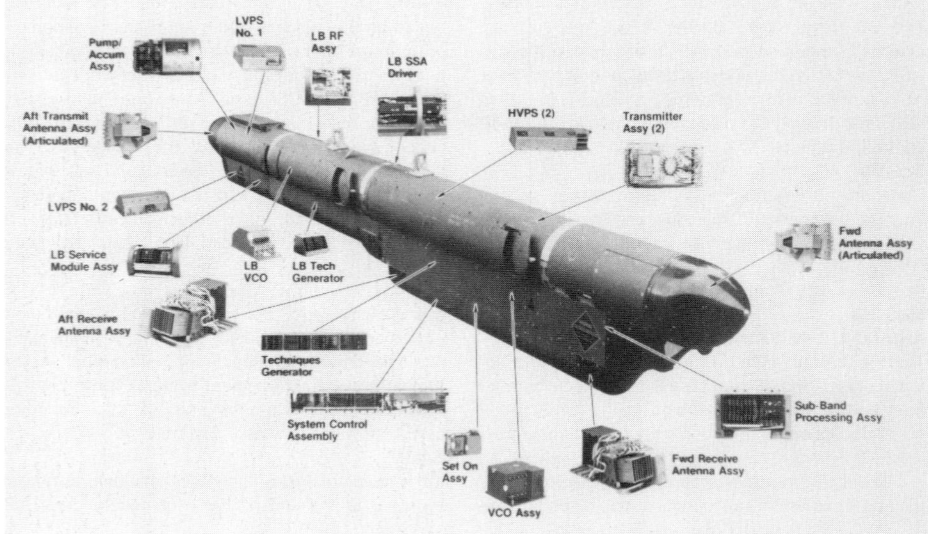

AN/ALQ-184 jamming pod

6516.393

AN/ALQ-186(V) JAMMER

The AN/ALQ-186(V) is an ECM pod system designed for testing air-defence radar systems. The pod design permits mission flexibility and test platform carriage variation during test and evaluation of advanced radar air defence designs. A unique feature of the jammer is the selection of different transmit antenna arrays which use multibeam technology. This technology employs a bank of mini-TWTs which feed high-gain antenna arrays resulting in very high effective radiated power without increasing demands on aircraft prime power and cooling capability. Simple and complex ECM techniques are programmed into the pod via a ground control unit. Selection of various techniques during flight is achieved via an aircraft control unit mounted in the cockpit, or by telemetry if the system is mounted on drone aircraft.

STATUS

The AN/ALQ-186(V) is in production for the US Army.

CONTRACTOR

Raytheon Company, Electromagnetic Systems Division, 6380 Hollister Avenue, Goleta, California 93117-3197, USA.

AN/ALQ-186(V) ECM pod jammer

6517.353

AN/ALQ-187 COUNTERMEASURES SYSTEM

The AN/ALQ-187 is an internally housed, fully automatic countermeasures system integrated with radar warning and flare/chaff systems for tactical aircraft self-protection. Primarily intended for F-4, F-16, A-7 and Mirage 2000 type aircraft, it detects and defends against surface-to-air missiles, anti-aircraft artillery and air-to-air interceptor weapon systems. It can interface with the AN/ALE-39 or AN/ALE-40 chaff/flare dispensers and the AN/ALR-69, AN/ALR-66 or AN/ALR-74 radar warning receivers.

The fully software programmable system automatically detects single or multiple threat radars of same or different technologies, and selects ECM programmes to counter against pulse, pulse doppler or CW ground-based, shipborne or airborne emitters. The system is flight line reprogrammable, allowing immediate incorporation of the latest intelligence threat data and EM techniques. Advanced power management techniques ensure maximum jamming effectiveness. The AN/ALQ-187 is an advanced version of the Raytheon ASPECS systems.

STATUS

In production and operational with a NATO nation.

CONTRACTOR

Raytheon Company, Electromagnetic Systems Division, 6380 Hollister Avenue, Goleta, California 93117-3197, USA.

AN/ALQ-187 countermeasures system with aft transmitter

1917.353

AN/ALR-45 AND AN/ALR-45F RADAR WARNING AND CONTROL SYSTEMS

The AN/ALR-45 is the standard US Navy tactical radar warning system and has been in use since the early 1970s. The system was a replacement for the AN/APR-25, the first standard tri-service radar warning receiver in service. The AN/ALR-45 is a crystal-video receiver using a hardwired programmable signal processor, four cavity-backed planar spiral antennas with preamplifiers and processor. The AN/ALR-45F replaces the AN/ALR-45 processor and azimuth indicator with a dual ATAC-16M computer-based threat processor and an alphanumeric microprocessor-based, MIL-STD 1553 compatible, display. The processor features threat software reprogrammability, on-board avionics and EW suite communications, and interface control of such systems as HARM, AN/APR-43, AN/ALQ-126A/B, AN/ALE-29/39, AN/ALQ-162, AN/ALQ-164 and AN/ALQ-165. The processor and display terminal are standard AN/ALR-67 weapons-replaceable assemblies and can be used interchangeably in the AN/ALR-45F configuration of the F-4J, A-4M, A-7E and the AV-8C aircraft without kit wiring changes.

STATUS

The AN/ALR-45 is in widespread use with the US Navy and NATO. Several thousand equipments have been produced. It is being replaced in the US Navy by the AN/ALR-67(V) and the AN/ALR-45F. The first three AN/ALR-45F systems were delivered in April 1983 and another 105 by September 1984. Two follow-on contracts were awarded in January 1984 and July

1985. The 55 systems on the first contract were delivered in mid-1985, the remaining systems delivery was completed in 1987. In March 1987, Litton received a $164 million contract for 194 AN/ALR-45F systems and 220 ALR 67s.

CONTRACTOR

Litton Applied Technology, 645 Almanor Avenue, Sunnyvale, California 94088-3478, USA.

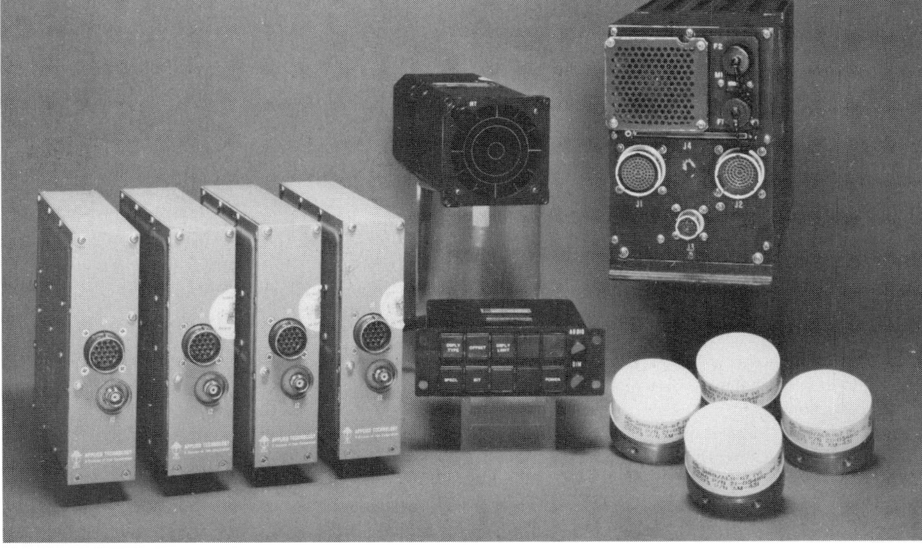

AN/ALR-45F radar warning and control system

3035.353

AN/ALR-46 RADAR WARNING SYSTEM

The AN/ALR-46 is a digital threat warning receiver used in most front line aircraft to replace the AN/APR-25/26 and AN/APR-36/37 systems. The AN/ALR-46(V)3 version is virtually a digital version of the AN/APR-36/37, the main factor in the change from analogue to digital being the addition of a DSA-20 processor. The AN/ALR-46(V)6 is available with an analogue processor.

The AN/ALR-46 uses a wide open, front-end, crystal-video type receiver, operating over the frequency band 2 to 18 GHz, with a digital processor which can be reprogrammed in the field to accept new threat data. In addition to providing cockpit visual and audio warning cues, it can directly control jamming systems.

STATUS

In service in the United States and the aircraft of several other countries.

AN/ALR-46 radar warning system

CONTRACTOR

Litton Applied Technology, 645 Almanor Avenue, Sunnyvale, California 94088-3478, USA.

1862.353

AN/ALR-47 RADAR HOMING AND WARNING SYSTEM

The AN/ALR-47 countermeasures receiving set has been developed for the US Navy's ASW aircraft, the S-3A Viking (**1403.302** in 1974-75 and earlier editions). In the Viking the ALR-47 obtains emitter signature and directional data which is fed into the aircraft's general-purpose digital computer for display on a multi-function display console. Among new features of the ALR-47 are automatic operation with variable control for frequency band limits, speed of tuning, and signal selection. Fixed antennas are used (four in each wing-tip), as are superhet receivers, and a digital processor.

Twin narrow-band receivers are incorporated in each ALR-47, and these incorporate an IF discriminator, out-of-band rejection logic, and voltage controlled oscillators which are tuned directly by computer control. The receiver channels are designed for high sensitivity and there is logarithmic compression of video outputs.

The characteristics of video outputs from four monopulse receiver channels are derived in a signal comparator which converts the data to digital format. All subsequent signal processing is performed digitally under computer control.

The two antenna groups are each comprised of two pairs of cavity-backed planar spiral broad-band antennas, arranged in a near orthogonal disposition for monopulse direction finding.

Control of the ALR-47 system is by means of computer programs which provide for receiver tuning, signal processing, and initiating warnings of significant contacts or known threats identified by stored signal characteristics. Provision is made for human intervention and manipulation of the software in the course of a mission to meet specific operational needs. Examples of this are facilities for filtering out unwanted signals of various types so that the overall processing and display load is reduced. Similarly parameters such as scan speed and frequency scanning limits, PRF, pulse-length, or bearing limits can all be varied at the discretion of the operator.

STATUS

In service with the US Navy S-3A Viking ASW aircraft, and in the Canadian CP-140 P-3C as the ALR-502. The system is no longer in production and all equipments are likely to be upgraded into the AN/ALR-76 (**5833.353**).

CONTRACTOR

IBM, Federal Systems Division, 10215 Fernwood Road, Bethesda, Maryland 20034, USA.

1918.353

AN/ALR-50 ECM EQUIPMENT

The AN/ALR-50 is a surface-to-air missile alert and launch warning receiver set. Official details of its frequency coverage range have not been released but the ALR-50's coverage is believed to provide for warning of the radars associated with such Soviet surface-to-air missiles as the SA-6 Gainful as well as the earlier SA-2 and SA-3 weapons. It is possible, however, that whereas systems of this type (ALR-50) normally provide the aircrew with some indication of the status of the threat (eg, search, tracking, lock-on, etc), in the case of the SA-6 Gainful the use of as many as five bands for this missile's associated radars may prevent the ALR-50 from providing complete threat status information.

STATUS

In service with various US Navy aircraft types, including A-4, RA-5, A-6, A-7, F-4 and F-8. No longer in production and being replaced in most aircraft types by the AN/APR-43 and AN/ALQ-126.

CONTRACTOR

Magnavox Government and Industrial Electronics Company, 1313 Production Road, Fort Wayne, Indiana 46808, USA.

1919.353

AN/ALR-52 ECM RECEIVER

The AN/ALR-52 is a multi-band IFM (instantaneous frequency measuring) receiver for airborne or land mobile applications. A naval variant is also produced under the nomenclature AN/WLR-11. Typical systems cover the microwave frequency band from 0.5 to 18 GHz, using receiver modules that cover octave bandwidths. Additional capabilities include provision for selection of a particular signal or frequency band, blanking of signals which are of no further interest, separation of interleaved pulse trains, measurement of radar parameters, analysis of CW as well as pulse signals, and direction-of-arrival (DF) measurement. The complete equipment provides for two operator positions, each equipped with a control unit and a display. The two IFM control units are capable of complete control over any band selected. If the same band is selected by both units, the video processing unit gives priority control to one position only. All intercepted signals, however, are presented on the displays at both positions.

A modified version of the ALR-52 has been developed and deployed, in which one of the operator positions has been replaced by an interface unit to link the IFM receiver with a digital computer. This system also employs a 1 to 18 GHz DF antenna system. The computer uses the data to 'tag' every received pulse with its frequency, pulse width, time of arrival, and azimuth. This information is stored and used to develop an emitter parameter.

STATUS

In operational service. Fitted to the EP-3E Elint aircraft operated by the US Navy.

CONTRACTOR

ARGO Systems, 1069 East Meadow Circle, Palo Alto, California 94303.

3347.353

AN/ALR-56 RADAR WARNING RECEIVER

The AN/ALR-56 radar warning receiver (RWR) is used with the AN/ALQ-135 internal countermeasures set jamming equipment (**3346.393**) to form the tactical electronic warfare system (TEWS) of the F-15 fighter aircraft. There are now three versions of the ALR-56; -56A, 56C and 56M (see later).

The main items of equipment comprising the AN/ALR-56 are: processor and low-band receiver (R-1867); high-band receiver (R-1866); countermeasures display (IP-1164); receiver control (C-9428); receiver control (C-9429); power supply (PP-6968); and the antenna system.

The processor and low-band receiver unit contains three major sections; a single-channel low-band superheterodyne receiver, a dual-channel IF section, and a processor.

The low-band receiver is electronically tuned under control of the processor. The dual channel IF section operates with either the high-band dual-channel receiver (R-1866/ALR-56) or the low-band single-channel receiver. Selection of the receiver to operate with the dual-channel IF section is under the control of the processor.

The processor contains a pre-processor and a general purpose digital computer. The pre-processor contains all the video circuits for analysing intercepted signals. It also provides digital outputs to the computer which represent the measured signal parameters. On the basis of these measurements, a digital output is provided to the display and an audio signal generated. The computer also controls all AN/ALR-56 system functions and is software programmable for the expected threat environment.

The solid-state, digitally controlled, dual-channel high-band receiver is capable of scanning an extremely large portion of the electromagnetic spectrum. Its single conversion superheterodyne design provides high sensitivity; the use of dual YIG RF preselection affords excellent selectivity and spurious rejection.

High frequency accuracy is obtained through the use of a frequency synthesised local oscillator.

The RF input ports are configured to accept two main antenna inputs to each channel and two additional RF inputs which may be designated to either channel. All switching functions are provided internally.

The receiver is capable of receiving signals over a large dynamic range for analysis and DF measurements. A self-contained precision signal source is provided for calibration and/or BIT over the entire tuning range. Provision is also made for multiplexing out important receiver functions.

All receiver functions are controlled by serially generated NRZ Manchester coded data. The precise data rate also serves as the reference clock for the receiver frequency synthesiser.

On the countermeasures display, rapid threat evaluation for aircraft defensive manoeuvres, at night or in direct sunlight, is accomplished through automatic intensity control utilising ambient light sensing techniques. Sharp, high contrast, unambiguous alpha-numerics and special symbology allow immediate assessment of the overall tactical threat, without requiring pilot display adjustments as external ambient light fluctuates. Phosphor screen-optical filter matching provides contrast enhancement to overcome direct sunlight contrast degradation.

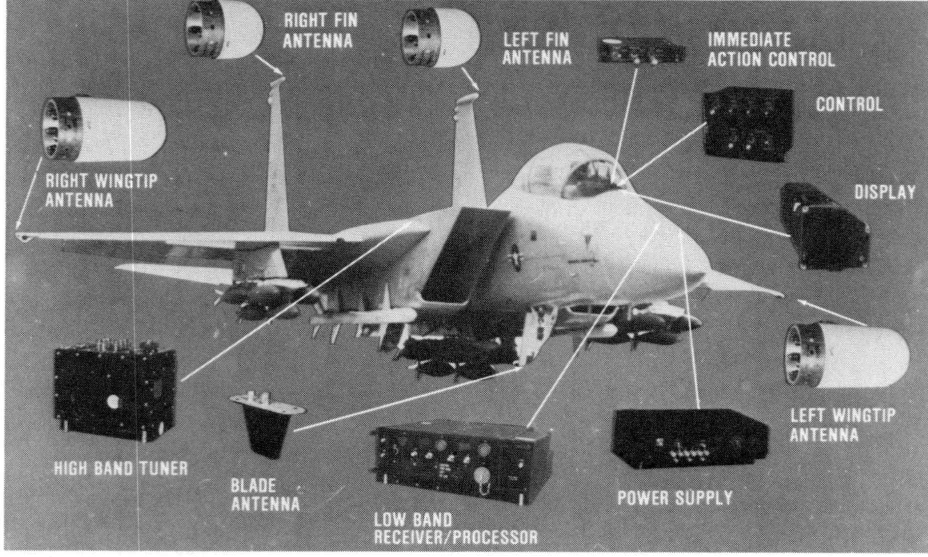

AN/ALR-56 internal countermeasure set equipment, showing positions of main units in F-15 Eagle

A range-bearing data presentation is provided with special clutter elimination programs as well as special threat status. Built-in test circuits determine display system malfunction and automatically indicate 'GO' conditions. An integral lighting panel provides control illumination for night time viewing.

The antenna system consists of four circularly polarised spiral antenna assemblies, each within its own radome, and a blade antenna. Collectively this antenna system provides omni-directional acquisition and direction finding over the operating frequency range of the RWR. The four spiral antennas, which cover the high band frequency range, are mounted in a manner that provides 360° azimuth coverage. DF operation is accomplished under computer control. The blade antenna provides omni-directional coverage of the low-band frequency range.

Of the two receiver control units, one provides for control of the RWR as well as other countermeasure sub-systems. A single switch controls the entire RWR on/off function, and separate switches and relays are provided for the sub-systems under RWR control. Audio tone cues are controlled from this unit also.

The other receiver control panel is for activating in-flight functions that require instant pilot reaction. In addition to countermeasures sub-system mode control, interrogated data readout on this unit can be called up from the sub-systems by means of push-button operation.

A major update known as the ALR-56C has now been completed with improvements in the processor to handle new threats, greater signal densities and other changes. The ALR-56C is a digitally controlled, dual-channel superheterodyne receiver covering the E to J frequency bands, and is capable of sorting and identifying all current and projected threats to the aircraft. RF input ports are configured to accept two main antenna inputs to each channel and two additional RF inputs that may be designated to either channel, with all switching functions provided internally. The system's major functions are broadly similar to those of the ALR-56A and include programmed frequency search, signal acquisition and analysis, establishment of priorities, jammer management, passive countermeasures management, and visual and audible displays/warnings.

A miniaturised version of the ALR-56C system, known as the ALR-56M, is in development and a competitive flight evaluation against the ALR-74 began in July 1987 for an installation contract on the F-16C/D. This will require a reduction of about 40 per cent in volume from the ALR-56C.

STATUS

In production and operational on F-15 Eagle aircraft. The ALR-56C system is to be fitted to all new F-15 aircraft, and retrofitted to most existing types replacing the ALR-56A.

CONTRACTOR

Loral Electronic Systems, Ridge Hill, Yonkers, New York 10710, USA.

4376.353

AN/ALR-60 COMMUNICATIONS INTERCEPT AND ANALYSIS SYSTEM

The AN/ALR-60 communications intercept and analysis system is carried on the US Navy EP-3E electronic reconnaissance aircraft and is used primarily for tracking and identifying Soviet warships. It is a huge complex and has multiple operator positions, equipped with video terminals that allow the operator to randomly access raw audio data and processed text data. Digitally controlled receivers are operated on a priority basis.

Only seven systems have been built, as far as is known. They use the latest advances in computer technology to provide the maximum information to the operators. Computer formatted CRT pages, and overall control by the computer of the receiving, recording, priority and processing functions, are used to ensure more complete coverage of all signals intercepted.

STATUS

In operational service.

CONTRACTOR

GTE Government Systems, 100 Ferguson Drive, Mountain View, California 94039, USA.

4377.393

AN/ALR-62(V) RADAR WARNING SYSTEM

The AN/ALR-62(V) is the standard radar warning set in the US Air Force F/FB-111 and EF-111D aircraft. It is a third-generation development that includes a complete antenna set, forward and aft receivers, a digital signal processor and cockpit-mounted threat indicator/countermeasures control units. The AN/ALR-62(V)3 is used in the F-111D aircraft while the AN/ALR-62(V)4 is fitted to the EF-111A electronic warfare aircraft. The AN/ALR-62(V)4 constitutes a major part of the EF-111A terminal threat warning system. It monitors and analyses radar threats and has the capacity to 'look through' high power jamming from the AN/ALQ-99 transmitters carried on the aircraft. Dalmo Victor is updating the system into the AN/ALR-62I for deployment in the 1990s.

STATUS

In production and in operational service. In November 1987, Singer was awarded a $19 million contract for AN/ALR-62 systems for USAF F-111 aircraft. The first ALR-62I prototypes were delivered in 1987 and full scale production is scheduled for 1988/89.

CONTRACTOR

The Singer Company, Dalmo Victor Division, 1515 Industrial Way, Belmont, California 94002-4095, USA.

3642.353

AN/ALR-66 SERIES RADAR WARNING RECEIVERS

The AN/ALR-66(V) is a series of radar warning receivers based on the original AN/ALR-66 system.

AN/ALR-66A(V)1

This is a fully programmable radar warning receiver that incorporates a digital computer specially designed to provide rapid, unambiguous identification of emitters in complex signal environments. The system computer/processor accepts up to half-a-million pulses per second from a mixed scenario of sea, air and land based radar emitters. Programmable alphanumeric symbology for up to 15 emitters is presented on a polar high brightness CRT display. The system has an E²PROM memory and the user can reprogramme the complete library on the flight-line or in-flight. This can be done in less than 90 seconds using a General Instruments Computer Memory Loader.

The AN/ALR-66A(V)1 is a compact and lightweight system and can be used on a wide variety of maritime patrol aircraft. It is the standard ESM system of the US Navy SH-2F Lamps Mk I ASW ASMD helicopters, and is used by other navies on HSS-2 and Sea King helicopters.

CHARACTERISTICS

Receiver: Crystal video, high sensitivity receivers available

Frequency: Contiguous over E to J bands

Warning/ID: All pulsed radars including pulse doppler, CW, ICW, LPI, 3-D, Jitter/stagger pulse compression, and frequency agile radars

Azimuth cover: 360°

Emitter storage: In excess of 1800 emitter modes

Reprogramming time: 90 seconds per platform on the flight line and in flight

Symbology: 1, 2 or 3 symbols per emitter as desired; programmable

Weight: 27 kg

MTBF: Documented over 630 hours

AN/ALR-66(VE) System

The AN/ALR-66(VE) is a lightweight radar warning receiver designed for use in fighter aircraft and helicopters. It provides enhanced performance and extended versatility due to the use of E²PROM in the computer memory. The system is designed mechanically as a 'drop-in' replacement/upgrade of older analogue and digital receivers in combat aircraft. The use of E²PROM in the memory section of the unit allows complete reprogramming in ninety seconds on the flight line by customer personnel. The system is capable of automatic recording of all major parameters of detected radars for post-flight analysis.

CHARACTERISTICS

Receiver: As for AN/ALR-66A(V)1

Frequency: Contiguous over E to J bands, with C/D band, omni or directional capability as options

Warning/ID: As for AN/ALR-66A(V)1

Radar Storage: More than 1000 emitters

Programming time: As AN/ALR-66A(V)1

Symbology: As AN/ALR-66A(V)1

Mission recording: Solid-state, automatic in-flight radar parameter recording optional

Weight: 29.3 kg

MTBF: Over 725 hours

AN/ALR-66(V)2 Surveillance and Targeting System

The AN/ALR-66(V)2 is a recent development for the US Navy and is an important enhancement in capability for its multimission aircraft. The equipment forms an integral part of its ocean surveillance system and can be used on offensive as well as defensive missions. Derived from the AN/ALR-66A(V)1, the equipment provides extremely advanced capabilities in such areas as precision DF accuracy, high sensitivity for over-the-horizon detection, precise frequency measurement, advanced signal processing concept coupled with expanded data memory, multimode operator-interactive display and controls, precision emitter parameter measurement, capability to integrate with other aircraft primary sensors, and E²PROM in-flight reprogramming capability.

AN/ALR-66(VE) radar warning receiver

AN/ALR-66(V)3 Surveillance and Targeting System

The AN/ALR-66(V)3 is a successor to the US Navy's AN/ALR-66(V)2 systems, and has been designed specifically for installation on-board the P-3C Orion anti-submarine patrol aircraft. The system detects, identifies and locates radars in the C-J band frequency range, and provides over-the-horizon targeting through features such as an ultra-high sensitivity receiver, precision DF accuracy, positive emitter identification in high-density environments, an interactive plasma display and control, self-protection to the aircraft, completely automatic operation, and a built-in-test capability.

The basic modes of operation are:

(a) Surveillance mode – the highest priority emitters are presented on the plasma display in positions corresponding to their range and bearing, using unique symbology which is either emitter or platform-related

(b) Targeting mode – precise targeting data on any emitter is presented on the plasma display. This operating mode provides the over-the-horizon targeting data required for weapon activation

(c) Emitter waveform analysis – raw video is accepted from the computer/converter and data is presented directly on the plasma display, allowing the operator to analyse emitter waveforms.

AN/ALR-66(V)5 Surveillance and Targeting System

This is the latest and most technically advanced version in the AN/ALR-66(V) family. It is in the development phase to meet the US Navy's requirements for future anti-submarine aircraft applications.

STATUS

The AN/ALR-66A(V)1 is in operational service on the US Navy SH-2F Lamps Mk I helicopters, and on HSS-2 and Sea King aircraft of other navies.

The AN/ALR-66(VE) is in operational use in a number of fighter type aircraft. It has also been selected for the A-4 upgrade by the RNZAF, as well as for tankers and transport aircraft of the Royal Air Force.

The AN/ALR-66(V)2 is in worldwide operational use with the US Navy and a contract valued at $20 million was received in December 1987 from Grumman to supply systems for the S-2 maritime patrol aircraft.

The AN/ALR-66(V)3 has been procured by the US Navy and the Japanese Maritime Self Defense Force for their P-3C aircraft, and the Portuguese Air Force for its P-3Bs.

The AN/ALR-66(V)5 is in development as part of the US Navy P-3 Update IV programme.

CONTRACTOR

General Instrument Corporation, Government Systems Division, 600 West John Street, Hicksville, New York 11802, USA.

AN/ALR-66(V)3 surveillance and targeting system

4372.393

AN/ALR-67(V) RADAR WARNING SYSTEM

The AN/ALR-67(V) is a countermeasures warning control system which is the US Navy's front line threat warning system in tactical aircraft. The programme for development of this system began in March 1975 and is currently in the operational evaluation phase by the US Navy. The system consists of broadband crystal-video receivers, a superheterodyne receiver, an integrated low-band receiver antenna array and azimuth indicator; all under the control of the ATC-16M threat processor. The AN/ALR-67(V) is designed for integration with the ASPJ (AN/ALQ-165) in various US Navy aircraft. A sub-set of the AN/ALR-67 is the AN/ALR-45F which consists of the central digital processor and the azimuth indicator. The AN/ALR-45F system is replacing the AN/ALR-45 equipment currently deployed.

STATUS

In production. Designed for installation in the US Navy F/A-18 Hornet aircraft and for the A-6E, F-14 and AV-8B. Litton has received contracts totalling nearly $300 million for the AN/ALR-67(V) from Australia, Canada and Spain as well as the US Navy. Dalmo Victor was selected in 1986 as second source for production.

CONTRACTOR

Litton Applied Technology, 645 Almanor Avenue, Sunnyvale, California 94088-3478, USA.

The Singer Company, Dalmo Victor Division, 1515 Industrial Way, Belmont, California 94002-4095, USA.

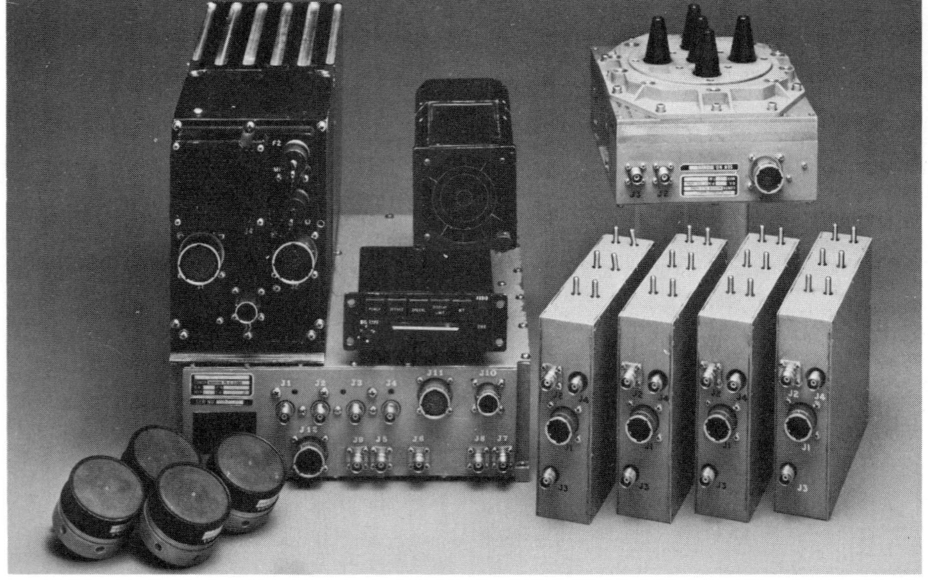

AN/ALR-67(V) developed by Itek for US Navy

3525.353

AN/ALR-68 ADVANCED RADAR WARNING SYSTEM

The AN/ALR-68 is an advanced digital threat warning receiver designed specifically for the West German Air Force. It is based on improvements incorporated in the AN/ALR-46, a major factor in the development being the conversion to digital operation by the addition of the Applied Technology Advanced Computer (ATAC).

The equipment is a wide-open, crystal-video, field programmable system. It provides in-cockpit threat parameter programming and hand-off to tactical ECM systems. The AN/ALR-68 uses a digital threat processor containing software provisions unique to the West German threat scenario. It is the first wide-open digital, general purpose computer radar warning receiver to become operational in Europe.

STATUS

In operational service with the F-4F and RF-4E aircraft of the West German Air Force.

CONTRACTOR

Litton Applied Technology, 645 Almanor Avenue, Sunnyvale, California 94088-3478, USA.

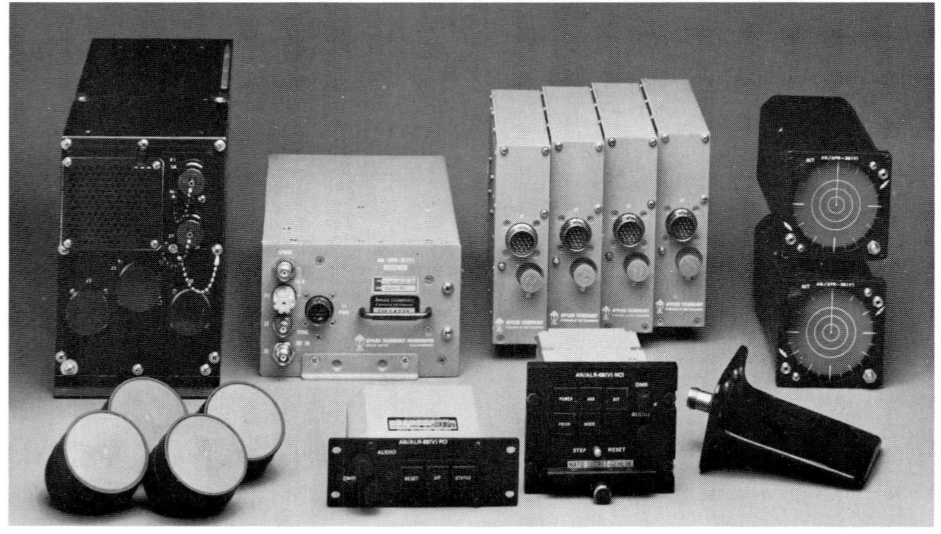

AN/ALR-68 digital threat warning receiver system

4379.353

AN/ALR-69 RADAR WARNING RECEIVER

The AN/ALR-69 radar warning receiver is a development of the earlier versions of the AN/ALR-46 equipment. It is essentially an ALR-46 with a frequency selective receiver system (FSRS) and a low-band missile alert receiver (AN/ALR-64 Compass Sail) (**4378.353**) added to the basic system. The Compass Sail equipment detects and analyses surface-to-air missile guidance emissions and provides a warning to the pilot of the direction and nature of the threat.

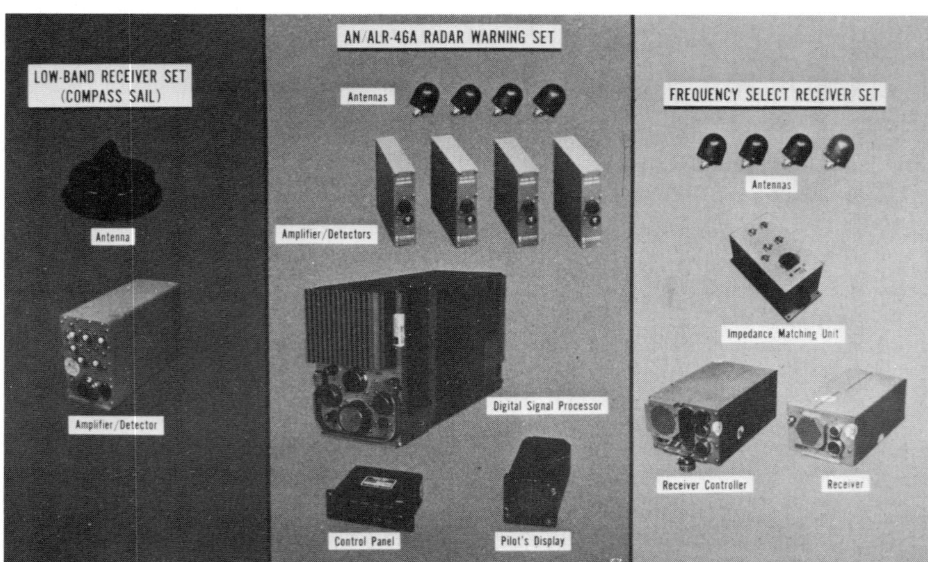

AN/ALR-69 radar warning receiver

The FSRS detects and analyses higher frequency missile guidance radiations that are not associated with the tracking radar and gives relative bearing of the missile approach. The system is designed to activate ECM resources automatically. The signal processor provides executive control for the FSRS. It accepts video inputs from five receivers and processed information from the FSRS, sorts and analyses the data, identifies and labels the radar signals found, tracks the status of these signals, and provides threat warning to the operator. Functions performed by the FSRS include warning and direction finding on CW signals, accurate frequency measurements on pulse signals for ambiguity resolution, threat antenna scan and rate analysis, and jammer frequency set-on for jammer power management and blanking functions.

STATUS

In production since 1978. More than 3500 systems have been delivered to the US Air Force and other NATO countries. The AN/ALR-69 is used on the F-16, A-10, F-4 and C-130 aircraft. The US Air Force announced in late 1986 that it was planning a major upgrade to make the system more effective against the latest threats. Update kits will be developed and procured competitively.

CONTRACTOR

Litton Applied Technology, 645 Almanor Avenue, Sunnyvale, California 94088-3478, USA.

4164.393

AN/ALR-73 PASSIVE DETECTION SYSTEM

The AN/ALR-73 passive detection system (PDS) is an airborne ESM equipment developed for the US Navy's E-2C airborne early warning aircraft. It is the improved version and successor to the AN/ALR-59 (**5932.353** in previous editions of *Jane's Weapon Systems*) and because of the extensive update of that system, the ALR-73 was given its own AN designation. The primary operational roles assigned to the E-2C consist of surveillance of both airborne and surface, hostile and friendly forces; early warning of hostile aircraft in order to protect the fleet; and the exercise of real-time control of the carrier tactical aircraft. The ALR-73 is intended to augment the AEW, and surface, sub-surface, command and control role of the E-2C by enhancing the threat detection and identification performed by the aircraft. It is a completely automatic, computer-controlled, superheterodyne receiver/processing system that communicates directly with the E-2C aircraft command and control central processor. The design of the system was motivated by four major considerations:

(1) very high probability of intercept in dense environments
(2) automatic system operation
(3) high reliability
(4) ease of maintenance.

Features of the ALR-73 which are related to its intercept probability performance include: four quadrant 360° antenna coverage; four independently controlled receivers; dual processor channels; digital closed loop rapid tuned local oscillator. Others concerned with automated system operation are: low false alarm report rate; automatic overload logic; AYK-14 computer adaptively controls hardware; degraded mode operation.

The PDS system can measure DOA, frequency, pulse width and amplitude, and PRI simultaneously. Scan rate information is also available if called for by the central processor. Special emitter tags can also be provided. A functional block diagram of the PDS is shown in Figure 1. The PDS detects and analyses electromagnetic radiations within the microwave spectrum. It sends emitter data reports (pulse width, PRI, DOA, frequency, pulse amplitude and special tags) to the E-2C's central processor via the PDS data processor. The central processor performs the identification function. The PDS immediately reports new emitters to the central processor. It eliminates redundant data on emitters for a programmable period of time, thus significantly reducing the data rate to the central processor. The PDS – a multiband, parallel scan, mission programmable, superheterodyne receiving system – covers the frequency range in four bands through step sweeping. Programmable frequency bands and dwell time permit very rapid surveillance of priority threat bands. Non-priority bands are also monitored, but at a reduced rate. Probability of intercept is increased without sacrificing sensitivity through the detection of both real and image sidebands.

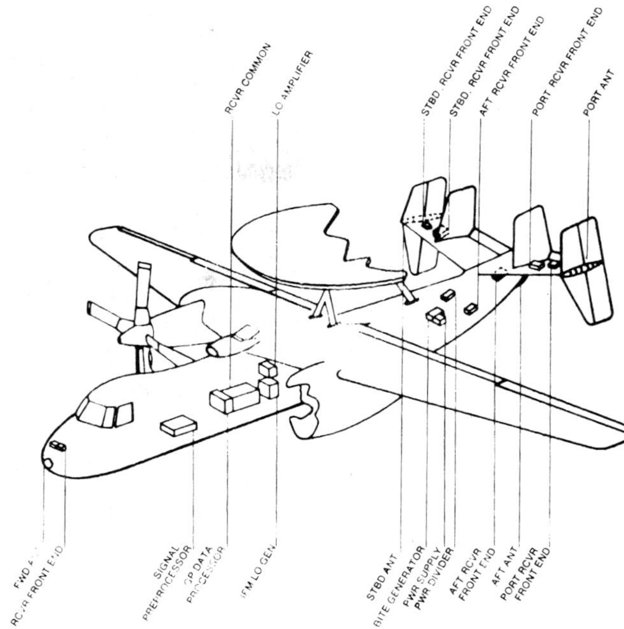

Fig 2: installation of AN/ALR-73 equipment on E-2C aircraft

Installation of the ALR-73 on an existing aircraft required a functional and mechanical modularity in the system partitioning to facilitate installation on the aircraft (Figure 2). Antenna packages located in the four quadrants of the aircraft provide 360° azimuth coverage. The large phase tolerance of the binary beam minimises the sensitivity of the DOA system to radome distortion and aircraft reflection. All four bands in the normal mode of operation scan through their respective frequency limits independently and simultaneously.

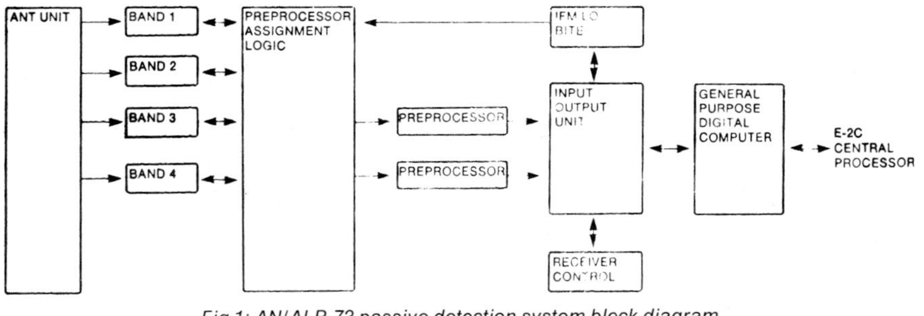

Fig 1: AN/ALR-73 passive detection system block diagram

Activity indications may be obtained on any of the four bands. Following an indication of activity, permitted dwell time is increased, and processing of the intercepted signal is started. Dual-signal processing circuits allow intercepted signals in any two of the four bands to be processed simultaneously.

The signal processor is a special-purpose logic processor which performs pulse train separation, DF correlation, band tuning and timing, and built-in test equipment (BITE) logic functions. The signal processor also contains the I/O circuitry necessary for the computer to communicate with both the signal processor and the aircraft central processor. The AYK-14 provides the system control function, data storage, and formatting. Variable frequency coverage, along with variable dwell times and processing times, provide a means for optimising probability to intercept for any given theatre.

The local oscillator system consists of three units: the IFM/LO generator, LO amplifier and LO power divider. The system is unique in that it is extremely fast and accurate. The key unit in the LO system is the instantaneous frequency measurement (IFM) receiver, which samples the LO frequency and converts it to a digital tuning command. This closed loop system permits frequency measurement accuracy, while being largely insensitive to environmental variations. The local oscillator signals from the IFM/LO are amplified by the LO amplifiers to overcome the long LO cable losses, and are then routed, via the power divider, to the various receiver front ends.

STATUS

128 systems have been delivered.

CONTRACTOR

Litton Systems Inc, Amecom Division, 5115 Calvert Road, College Park, Maryland 20740, USA.

5694.353

AN/ALR-74 RADAR WARNING SYSTEM

Providing self-protection for the F-16 and other tactical aircraft, the ANR/ALR-74 was formerly known as the AN/ALR-67/69 update, and is designed with advanced technologies to achieve commonality between US Air Force and US Navy radar warning systems. The system consists of an amplifier/detector, a processor, a low band receiver, a frequency measurement receiver system, controller, control indicator and azimuth indicator.

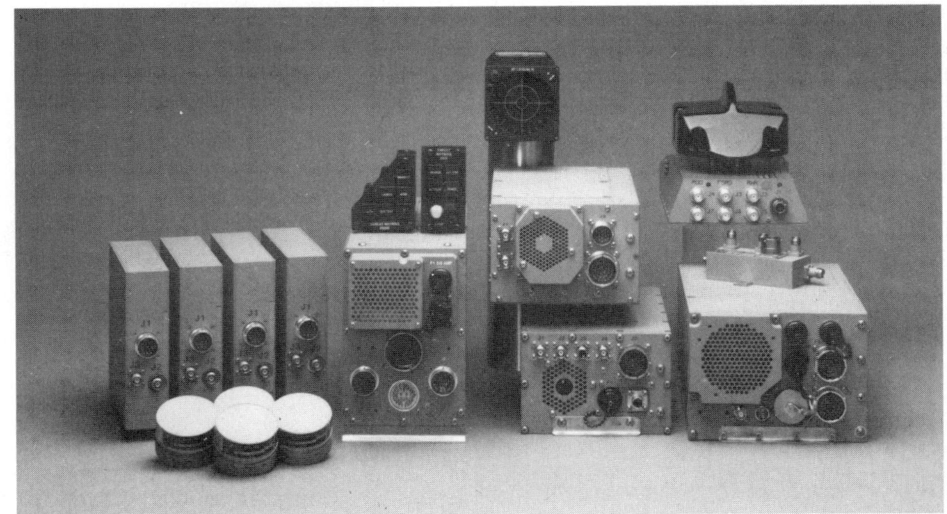

AN/ALR-74 radar warning system

STATUS
Litton has been awarded more than $1100 million for this system. Delivery of prototype models was completed in late 1983 and flight testing on F-4 and F-16 aircraft by November 1984. The first preproduction contract was awarded in December 1984 for the initial RF-4, F-4 and F-16 requirements. Under this contract a number of hardware and software improvements are being incorporated as required to meet projected threats of the 1990s. 15 preproduction models are being upgraded for testing purposes under a contract awarded in December 1986. A competitive evaluation of the AN/ALR-74 commenced in July 1987, against the AN/ALR-56M. A decision is due to be made later in 1988.
CONTRACTOR
Litton Applied Technology, 645 Almanor Avenue, Sunnyvale, California 94088-3478, USA.

6256.393
AN/ALR-75 (V) SURVEILLANCE RECEIVER
The AN/ALR-75(V) is a surveillance receiver providing spectrum surveillance over the frequency range 0.1 to 18 GHz. It consists of an aircraft-style set of control panels and a number of remote control units in ATR configurations. The system includes complete digital control, computer compatability, scratch-pad memory and digital refreshed displays. The AN/ALR-75(V), also known as the SCR-2100, is one of a number of equipments in the SCR series of receiving equipments and is described under entry **6159.153.**
STATUS
In operational service in the ERA-3B, NKC-135A and EC-24A aircraft, and also as a ground-based equipment in the AN/ULQ 13(V)1 vans.
CONTRACTOR
Scientific Communications Inc, 2908 National Drive, Garland, Texas 75041, USA.

5833.353
AN/ALR-76 ESM SYSTEM
This is a successor to the AN/ALR-47 (**1862.353**) and is intended for fitment to the US Navy S-3B aircraft. It is automatic in operation and provides techniques for ambiguity resolution. With considerably greater processing power than the ALR-47 the system carries out continuous direction finding to provide the exact location of all emitters automatically. The equipment is modular in construction and weighs approximately 59 kg.
STATUS
IBM was awarded a contract for 32 AN/ALR-76 systems for the S-3B Viking in December 1987, under a $30 million contract. Deliveries are due to commence in January 1989.
CONTRACTOR
IBM Federal Systems Division, Owego, New York 13827, USA.

5690.393
AN/ALR-77 ESM SYSTEM
The AN/ALR-77 ESM system was intended for the P-3C Orion anti-submarine warfare aircraft update. The antennas are housed in the wing tips of the aircraft. When in operation, the system detects shipborne and airborne radar signals and feeds this information to an on-board computer. The information is then processed and a readout is provided for the crew with such parameters as the signal location and threat potential. The system also provides targetting data for the radar-guided Harpoon missiles.

The ALR-77 consists of interferometer antennas, multiband converters, processor units, an IFM encoder, IF synthesiser and controller. The antennas are wingtip-mounted with a total of 36 antennas, 18 on each wingtip. Each 90-degree quadrant has nine antennas and these are futher broken down into three sets, each covering one of three bands for bearing determination. A multiband converter is also housed in each wingtip assembly to control the antenna selection, and perform fine direction-finding quadrant selection. The system weight is about 365 lb (166 kg).
STATUS
Eight systems were produced during a preproduction period. A P-3C was modified by Lockheed to accept the new ESM system during 1985 and was delivered to the US Navy for flight tests in November 1985. However, after Boeing won the contract for the P-3C update, the General Instruments AN/ALR-66(V)5 was selected for the programme.
CONTRACTOR
Eaton Corporation, Comac Road, Deer Park, New York 11729, USA.

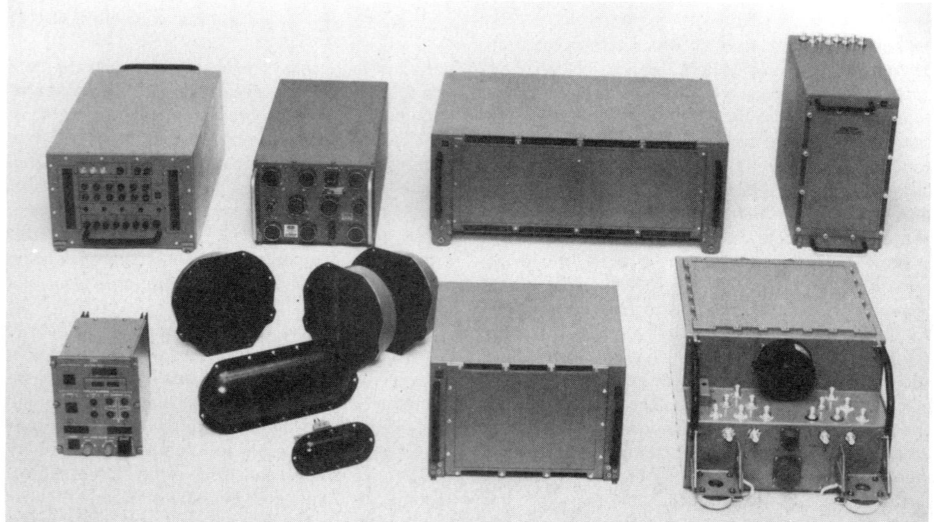

Units of the AN/ALR-77 ESM system

5942.393
AN/ALR-80(V) RADAR WARNING RECEIVER
The AN/ALR-80(V) is the latest US government nomenclatured lightweight digital radar warning receiver designed for use in fighter aircraft and attack helicopters. The system is an outgrowth of the AN/ALR-66(VE) system (**3642.353**). The AN/ALR-80(V) provides enhanced performance and flexible, modular versatility due to the use of E^2PROM in the computer memory, and new powerful software. The system is designed mechanically as a drop-in replacement/upgrade of older analogue and digital radar warning receivers used in combat aircraft.

With a library storage capacity of more than 1800 radar modes, the AN/ALR-80(V) is claimed to be the most powerful and economical RWR system available. The use of E^2PROM in the memory section of the unit allows complete memory reprogramming in 90 seconds on the flight line. The AN/ALR-80(V) is capable of automatically recording all major parameters of detected radars for post-flight analysis. An option is available to fit the receiver with a special DF antenna for enhanced performance in the C/D band. It is available for fitment to a wide range of aircraft such as the F-16, F-4, F-5, A-4, Mirage, Jaguar, etc.

Units of the AN/ALR-80(V) radar warning receiver

CHARACTERISTICS
Receiver: Crystal video
Frequency: Contiguous over C to J bands (C/D band): Omni or directional
Warning: All pulse radars; CW, ICW and pulse doppler, LPI, 3-D, jitter/stagger, pulse compression and frequency agile radars
Radar storage: More than 1800 emitters
Programming time: 90 seconds per aircraft, on flight line by the customer
Symbology: 1, 2 or 3 symbols per emitter as desired, programmable by customer
Interface: RS-232 or MIL-STD-1553 as required, for use with jammers, chaff/flare or data links
Weight: 29.3 kg
MTBF: 725 hours (MIL-STD-781)

STATUS
In production. The AN/ALR-80(V) was announced in 1985 and is fully approved. Several enquiries have been received from NATO and other countries.
CONTRACTOR
General Instrument Corporation, Government Systems Division, 600 West John Street, Hicksville, New York 11802, USA.

5687.353
ENHANCED RADAR WARNING SYSTEM
The Enhanced Radar Warning Equipment (ERWE) is a fully programmable digital threat processor and three separate receiver elements which provide C-D band coverage, and both wideband and narrowband coverage for E-J band. A crystal-video receiver is employed to provide wideband coverage for rapid threat response, and a superheterodyne receiver provides high sensitivity and selectivity. Both receiver sections employ the Applied Technology ATAC computer. The equipment provides a prioritised alphanumeric display and a high-intensity azimuth

indicator, and is capable of rapid change of threat parameters and priorities from mission to mission. The ERWE was developed for the European threat scenario and is now operational with the West German Tornado aircraft.
STATUS
In operational service with the West German Air Force. A co-production agreement exists with AEG Aktiengesellschaft, Ulm.
CONTRACTOR
Litton Applied Technology, 645 Almanor Avenue, Sunnyvale, California 94088-3478, USA.

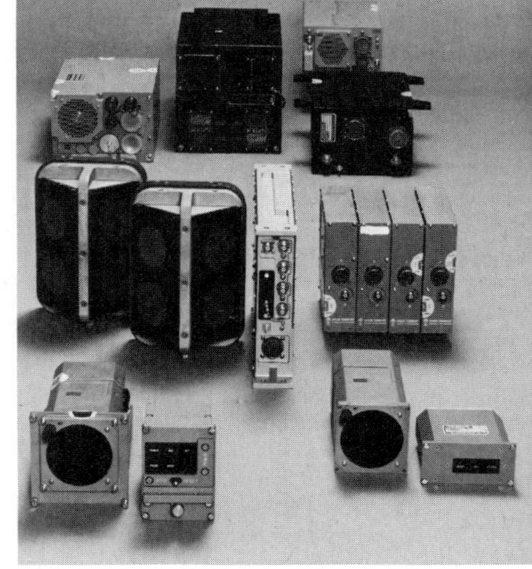

Enhanced radar warning system

3908.353
AN/ALR-606(V)1 AND (VE) RADAR WARNING RECEIVERS
The ALR-606(V)1 and (VE) radar warning receivers are the latest models to be introduced by General Instrument Corporation. They share the same hardware configuration as the AN/ALR-66A(V)1 and (VE) respectively (**3642.353**) but software enhancements provide additional facilities. The main items of equipment are a digital signal processor; a high intensity CRT cockpit display; four cavity-backed planar spiral antennas and their associated receivers covering the E to J-bands; a power supply unit; and a control indicator unit. The processor separates and categorises the characteristics of received signals, and unique symbols are presented on the display to indicate the range and bearing of each detected target. Symbol blinking and an audio alarm denote high priority radar types. The operator is able to select standard and special equipment operating modes. Radar data is presented in the form of alpha-numeric symbols which indicate specific parameters such as pulse width, PRF, and frequency band.
CHARACTERISTICS
Frequency: E to J-band continuous; C/D band optional
Radar types: All pulse types, CW detection
Receiver type: Crystal video
Azimuth coverage: 360°
DF accuracy: Better than 15°
Shadow time: 2μs
Processor memory: 18 000 words E²PROM/RAM/ROM 68 K total capacity

ALR-606(VE) radar warning receiver system

Radar capacity: 15 symbols can be displayed simultaneously. System signal tracking capacity is 35 radars simultaneously
Weight: 29.22 kg ALR-606(VE); 27 kg ALR-606(V)1
STATUS
The ALR-606(V)1 and -606(VE) were designed specifically for the export market and have been selected by a number of air forces for fighter aircraft fitment.
CONTRACTOR
General Instrument Corporation, Government Systems Division, 600 West John Street, Hicksville, New York 11802, USA.

5952.393
AN/ALT-28 NOISE JAMMER
The AN/ALT-28 is a carcinotron noise jammer carried in pairs aboard the B-52 aircraft and operates over the frequency range from UHF to 100 GHz. It was developed as part of a later series of modifications to the EW suite of the B-52. The equipment depends on the ALQ-155 power management system (**3552.393**) to improve efficiency in multiple threat environments. The system has been continuously updated, the latest being the addition of the ALQ-155 which ties in the AN/ALT-28 with the ALR-46 radar warning receiver.
STATUS
The system is in service on the B-52 aircraft but is no longer in production.
CONTRACTOR
Northrop Corporation, Defense Systems Division, 600 Hicks Road, Rolling Meadows, Illinois 60008, USA.

3544.393
AN/APR-38 WILD WEASEL CIS
The AN/APR-38 control indicator set (CIS), designed to be the "eyes and ears" for the Wild Weasel II aircraft, is composed of seven line replaceable units (LRUs) that receive, display and transmit real-time data, in association with other on-board equipment. It provides the aircraft commander with computer-supplied data that permits him to complete his mission successfully. The system can be operated in either automatic or semi-automatic mode, at the pilot's option. The antenna/receiver systems are installed on the nose and the tail fin. The frequency coverage is classified but is likely to extend from UHF band up to J-band.

The CIS receives computer-generated information on detected radar emitters. The digital information is transmitted serially. The highest priority emitters are classified as to threat type, priority of threat, range, magnetic bearing and other pertinent information. A threat library is contained within the central computer.

The CIS can select the types of emitters to be displayed on a PPI display and can be controlled by the computer semi-automatically or manually at the option of the operator.
STATUS
The CIS is currently operational in the F-4G aircraft. A performance upgrade programme, including frequency coverage and processing capability, is in progress. This programme is in two phases; Phase 1 concentrates on expanding on-board computer capability to interface with the HARM missile. This has now been tested and is in production. Phase 2 performs the other hardware/software upgrading. As a result of the performance upgrade programme a successor equipment, the AN/ALR-47, is in development and preproduction (see **6519.393**).

CONTRACTOR
Loral Electronic Systems, Ridge Hill, Yonkers, New York 10710, USA.

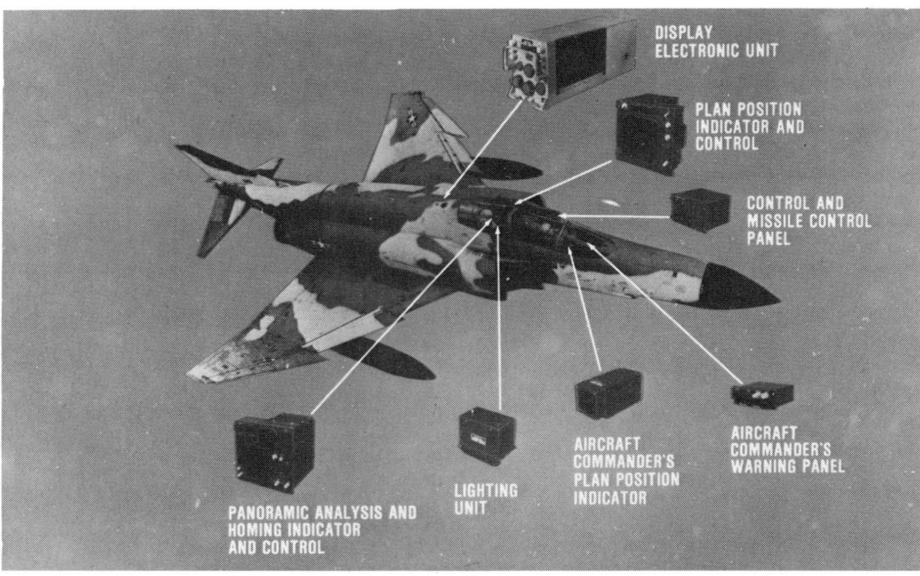

Location of units of APR-38 control indicator set (CIS) in F-4G Wild Weasel aircraft

3542.393
AN/APR-39(V) RADAR WARNING SYSTEM
The AN/APR-39(V)1 radar warning receiver equipment provides automatic warning of emitters in the E, F, G, H, I and most of J radar bands, as well as the appropriate portions of C and D-band radar associated signals. It is intended for use on either fixed-wing or helicopter type aircraft. The equipment provides indications of bearing, identity, and the mode of operation of detected signals, this being displayed on a cockpit indicator. Proportional PRF of displayed signals, and alarm tones, are presented to the aircraft crew via the audio system also.

The equipment of the basic APR-39(V)1 system comprises: two dual video receivers, four spiral cavity-backed antennas, one blade antenna, an indicator unit, a comparator, and a control unit. Without cables and brackets, this weighs about 3.63 kg, and either internal or pod mounting is possible.

An updated version, APR-39(V)2, is in production. In the APR-39(V)2 the comparator is replaced by a Loral CM-480/APR-39(V) digital processor. This performs signal sorting, identification of emitters, bearing computation, and character generation for the presentation of threat details in alpha-numeric form on the cockpit display. This unit weighs only 6.5 kg, and incorporates an adaptive noise threshold and angle-gate, programmable pulse repetition interval filters, and coded emitter outputs. A 19 K word programmable read-only memory/random access memory is provided.

DEVELOPMENT
The AN/APR-39(V)1 was designed and developed by E-Systems. The APR-39(V)2 processor was designed and developed by Loral under contract to the US Army Electronic Command. A third version, the AN/APR-39A(V)1 has now completed development and is in production by Dalmo Victor. This version incorporates a number of updates, including the provision of warning of modern threat radars operating in the millimetric wave spectrum (see **5692.393**).

STATUS
Both the APR-39(V)1 and APR-39(V)2 are in production. Deliveries of the APR-39(V)1 totalled 6000 systems by January 1988. It has been fitted to 15 aircraft types, as well as fast patrol boats and gun boats. The West German MoD is installing 230 systems on PAH-1 anti-tank helicopters.

Units of AN/APR-39(V)1 system

CONTRACTORS
E-Systems, Melpar Division, Falls Church, Virginia 22046, USA.
Loral Electronic Systems, Ridge Hill, Yonkers, New York 10710, USA.

5692.353
AN/APR-39A(V) THREAT WARNING SYSTEM
The AN/APR-39A(V)1 is a lightweight, digital threat warning system (TWS), developed for use with helicopters and light fixed-wing aircraft, in particular those flying on nap-of-the-earth missions. It provides aural and visual warning of any radar energy incident on the host aircraft over a very wide frequency range. Detected radiation is compared with emitter characteristics contained within the system's threat library, and each emitter is assigned a priority in accordance with the magnitude of the threat. The system displays multiple threats, although the highest priority threats are highlighted while keeping track of other emitters in the threat buffer. The system displays threat type, bearing from the aircraft, and degree of danger, on azimuth indicator displays.

The AN/APR-39A is an upgrade of the earlier analogue AN/APR-39, and is designated for potential application on US Air Force, US Army and US Marine Corps aircraft. The system has completed US Army testing and has demonstrated MTBF of greater than 900 hours.

STATUS
Production quantities began to the US Army, US Marine Corps and US Air Force in 1987, under a $63 million contract awarded in November 1986. In March 1987, Singer was awarded a $8 million contract for development of a special version of the APR-39A for US Army and Marine Corps use. It is intended for use in the V-22 tilt rotor aircraft.

CONTRACTOR
The Singer Company, Dalmo Victor Division, 1515 Industrial Way, Belmont, California 94002-4095, USA.

AN/APR-39A radar warning system

4010.393
AN/APR-43 TACTICAL RADAR WARNING RECEIVER
The AN/APR-43 is the warning receiver developed under the Compass Sail Clockwise programme to provide US Navy operational tactical aircraft with threat warning and identification, direction finding and approximate range information against pulse and CW missile weapon systems. The system consists of two units (weighing approximately 11 kg); a multi-element antenna WRA and multiband receiver WRA containing a crystal video receiver for pulse emitter detection and a superheterodyne receiver for the CW emitters. The APR-43 directly interfaces to the AN/ALR-45 and AN/ALR-45F (ALR-67 processor) warning receivers and defensive jammers AN/ALQ-

126 and AN/ALQ-162 to form the USN tactical aircraft ECM protection system. The APR-43 also contains those ALQ-162 antenna output elements required for CW jamming.

STATUS

In production. In February 1987, American Electronics Laboratories was awarded a $21.5 million contract for R-2087A/APR-43 receiver interface units/radar warning receivers and antenna interface units.

CONTRACTORS

Loral Electronic Systems, Ridge Hill, Yonkers, New York 10710, USA.

American Electronic Laboratories Inc, 305 Richardson Road, Lansdale, Pennsylvania 19446, USA.

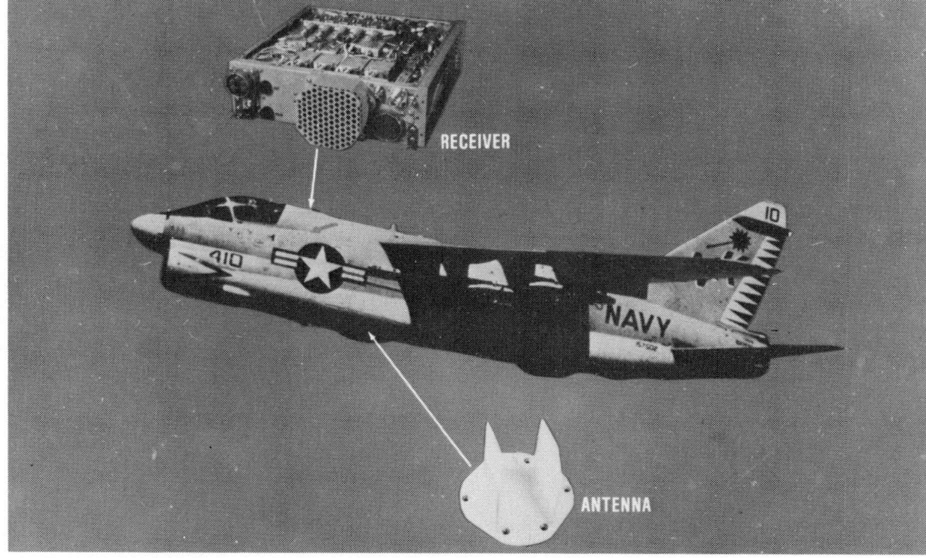

US Navy A-7 Corsair aircraft showing location of AN/APR-43 Compass Sail Clockwise tactical warning receiver and its antenna

4381.353

AN/APR-44 RADAR WARNING RECEIVER

The AN-APR-44(V) is a small power-frugal, lightweight radar warning receiver designed to detect specific radar signals with omnidirectional coverage. Both visual and aural warnings of threat radar signals are provided. The system is available in three configurations, (V)1, (V)2 and (V)3, for different frequency band coverage, with the basic system consisting of monopole antennas, a receiver and a control unit. The RF circuitry, including switches, is a microwave integrated circuit module. False alarms are eliminated by special pulse rejection circuitry.

The (V)1 system employs a R-2097 low-band receiver covering the H/I band, the (V)2 version an R-2098 high-band receiver covering J-band, while the (V)3 system employs both the R-2097 and R-2098. Other coverages are available.

STATUS

In operational service with US forces. In September 1987, AEL was awarded a contract valued at over $9 million to supply 229 systems.

CONTRACTOR

American Electronic Laboratories Inc, 305 Richardson Road, Lansdale, Pennsylvania 19446, USA.

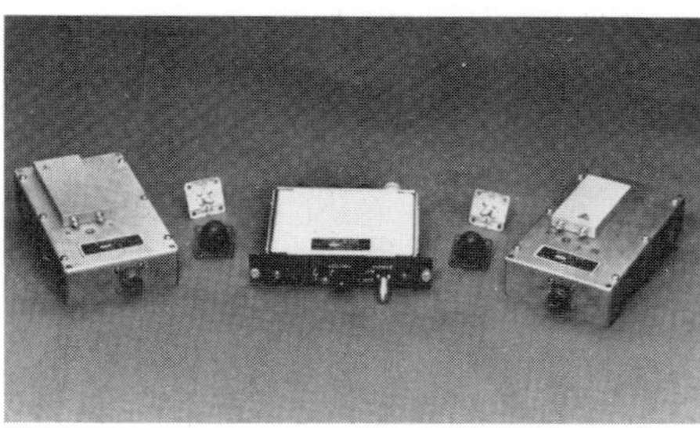

AN/APR-44(V)3 radar warning system

6519.393

AN/APR-47 WILD WEASEL CIS

McDonnell Douglas announced in October 1986 that it had been awarded a contract to produce the AN/APR-47, a new sensor system designed to replace the AN/APR-38 in F-4G Wild Weasel aircraft. The APR-47 is as a result of the APR-38 update programme and will use some elements of the latter equipment, allied with the Sperry CP-1674 signal processor and an E-Systems directional receiver subsystem. The new processor reportedly features five times faster signal processing, eight times the memory, 20 times the display capability and 10 times faster data processing than the APR-38.

STATUS

In development and pre-production.

CONTRACTOR

McDonnell Douglas Corporation, St Louis, Missouri 63166, USA – prime contractor.

3365.353

CMR-500B WARNING RECEIVER

The CMR-500B is a small, low-cost warning receiver, designed to detect CW signals. The warning receiver provides a visual and aural output of detected CW RF signals within the frequency range determined by an input filter. The CMR-500B is a modification of the AN/APR-42 (CMR-400) warning receiver (**3364.353** in 1979-80 and earlier editions) and is electrically interchangeable and compatible with the AN/APR-42 for full threat detection capability. The system consists of the following three parts: monopole antenna, providing omni-directional vertically polarised coverage; receiver, which contains an RF filter, detector limiter switch, video amplifier, processing and output circuitry; control panel containing alert indicator lamp, ON/OFF volume control, and phone jack.

The AEL Model AOS-1543 antenna is an omni-directional vertically polarised antenna, designed to mount on any flat, horizontal surface on the underside of an aircraft fuselage. When mounted on the conductive ground plane the antenna provides coverage in the solid angle defined by an elevation sector 50° wide and omni-directional in azimuth. The gain is 3.5 dB at the beam peak which occurs 15 to 25° below the horizon as a function of the ground plane shape.

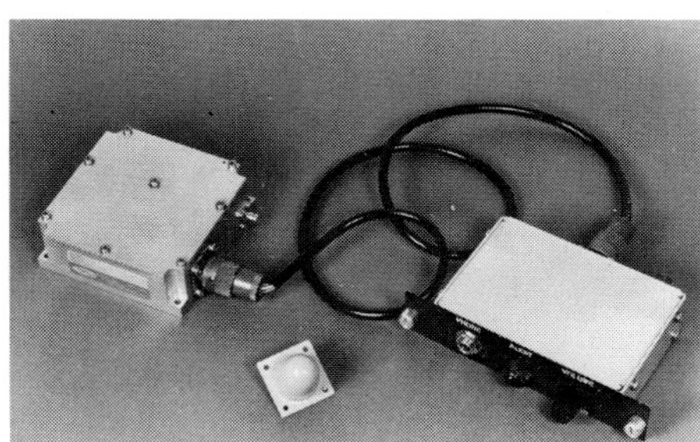

CMR-500B CW radar warning equipment

The receiver is an RF-chopped, crystal video type with a CW sensitivity of –45 dBm. The receiver consists of a bandpass filter followed by an RF switch/detector module (MIC format), a linear video amplifier and processing circuitry. The output of the video amplifier is routed to a comparator, and then to the processing logic where it is compared to the RF chopping signal to detect the presence of a CW signal. The processing circuitry provides a 3 kHz audio output, a 1.5 kHz logic output, an alert lamp driver, and a logic blink signal (2 Hz), upon reception of a CW signal. The receiver RF frequency is restricted by the bandpass filter to the frequencies of interest. Special pulse suppression circuitry is incorporated in the receiver to prevent false alarms. A power supply regulator is integral with the receiver for operation from 28 V DC aircraft power.

The incorporation of a control panel is optional, determined by the mission requirement. The control panel provides an alert light indicator and a phone jack for a headset for alert aural indication. The control panel also provides an ON/OFF switch and two interface cables.

CHARACTERISTICS

Receiver frequency and bandwidth: Selectable

Sensitivity: –45 dBm (minimum)

Max RF input level: 1 W CW (integral limiter)

Receiver dimensions: 91 × 91 × 33 mm

Receiver weight: 354 g

CONTRACTOR

American Electronic Laboratories Inc, 305 Richardson Road, Lansdale, Pennsylvania 19446, USA.

6217.393

AN/AVR-2 LASER WARNING RECEIVER

The AN/AVR-2 is a laser warning receiver, normally integrated with the AN/APR-39 radar warning receiver to provide combat helicopter pilots with warning of hostile laser-aided weapons. The system detects, locates and identifies laser threats for warning and target cueing, and provides visual and audible alerts through the radar warning receiver cockpit display.

The system consists of two sensor heads giving 360° azimuth protection, and a comparator which processes the data and passes information to the RWR for display.

STATUS

In operational service with US Army and US Marine aircraft.

CONTRACTOR

Perkin-Elmer, Electro-Optics Division, 100 Wooster Heights Road, Danbury, Connecticut 06810, USA.

6166.095
AN/ASQ-191
COMMUNICATIONS/COUNTERMEASURES SET

The AN/ASQ-191 has been developed for use in an airborne, fixed ground or ground mobile application and covers the frequency range 225 – 400 MHz. It has two basic modes of operation: a communications mode for normal voice communications or imitative communications deception, and a C³CM mode where it provides monitoring or jamming capability. It can operate against fixed frequency or slow frequency-hopping radios. The AN/ASQ-191 incorporates a number of automatic scanning, monitoring and jamming functions which cover the complete frequency range, parts of it or selected frequencies. The jammer has an output of about 100 W and the jamming sequence includes a 6 m/s look-through mode. A 400 W jamming variant is available.

STATUS
Fully developed.

CONTRACTOR
Rockwell International, Collins Division, 400 Collins Road NE, Cedar Rapids, Iowa 52498, USA.

3543.393
RAPPORT ECM SYSTEMS

The Rapport II ECM system was developed for the Belgian Air Force Mirage V aircraft. It provides pilot warning and automatic, optimal countermeasures required to survive the present day and future hostile air defences. The system weighs less than 100 kg, and consists of separate RWR and jammer sub-systems. The RWR, which includes a programmable core memory computer, provides power management capability required for optimal use of the jammer sub-system. Functionally, the RWR provides the means to sample and sort the environment. It uses a programmable digital processor for all signal analysis, and provides the pilot with an easily readable, unambiguous warning display. The power management capability, incorporated in the RWR processor, matches the countermeasures response to the constantly changing threat picture, for an optimal jamming response. The jammer sub-system, under RWR control, provides the actual response to the hostile radar environment. To conserve jamming power and maximise ERP, the jammers are accurately set to the exact frequencies of the victim radars. The jammer frequency range provides coverage of all threat radars, while its power output is sufficient to counter these radars at lethal range.

A development of the system, known as Rapport III, is being produced for the F-16 aircraft ordered by the Belgian Air Force. This latest version is an internal integrated EW system having both radar warning and jamming functions controlled by a central processor, and is described under entry **4009.393** (AN/ALQ-178).

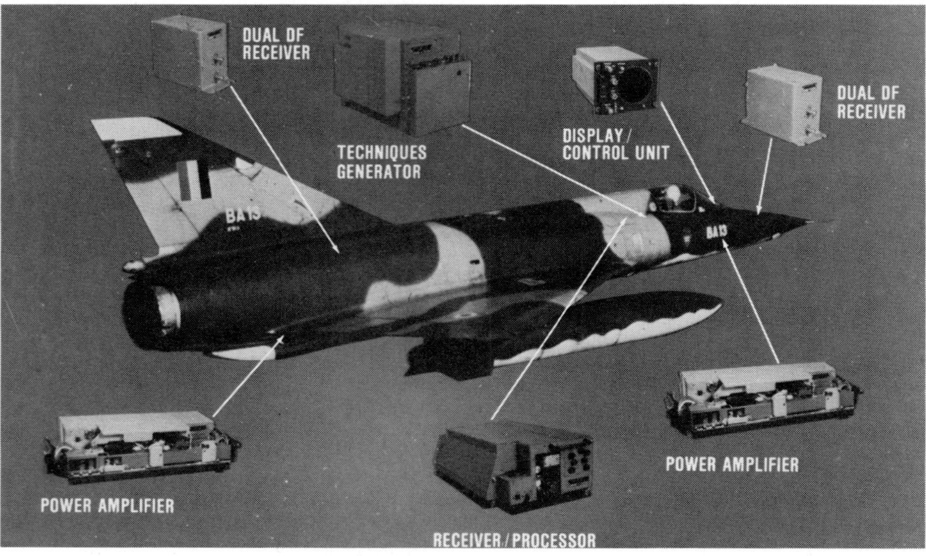

Rapport II ECM equipment and its location in Mirage V aircraft

STATUS
The Rapport II system is operational with Mirage V aircraft of the Belgian Air Force. Loral is developing Rapport III for the Belgian F-16 aircraft under contract, and also received in September 1983 an $88 million contract from Israel, to fit modified Rapport III systems in its F-16s. Flight testing of the Rapport III prototype commenced late in 1985. A decision by Belgium is still awaited.

CONTRACTOR
Loral Electronic Systems, Ridge Hill, Yonkers, New York 10710, USA.

4160.393
ARI 18240/1 ESM

A version of the EW-1017 ESM (electronic support measures) equipment (**4382.353**) has been developed for the Nimrod Mk 2 MR under the designation ARI 18240/1. Few details have been cleared for publication.

Special wingtip pods were built by British Aerospace to house the 16 cavity-backed planar spiral antennas and the two front-end receivers (LRU-1). Two other line-replaceable units (LRU-2 and LRU-3) are housed in the fuselage. These probably include the signal processors and comparator circuitry, as well as providing the requisite interfaces with the sensor elements and receivers and the associate control and display facilities. The last of these facilities almost certainly includes provision for integrating ESM data with that from other sources on the aircraft's main tactical display sub-system.

The 16 antenna elements comprise two sets of eight high-band and low-band signal intercept antennas, these being arranged to produce separate fore and aft coverage quadrants on each side of the aircraft in each of the broad frequency bands covered by the equipment. No information has been given of the frequency limits but the upper limit is practically certain to be 18 GHz and may well be higher. Superhet receivers are believed to be used.

STATUS
In production.

CONTRACTOR
Loral Electronics Systems, Ridge Hill, Yonkers, New York 10710, USA.

4382.353
EW-1017 ELECTRONIC SURVEILLANCE SYSTEM

As an airborne electronic surveillance the EW-1017 automatically acquires and identifies those emissions over the known frequency band that are liable to provide a threat to the aircraft safety, eg surveillance and tracking radars, both ground-based and airborne. The system is designed to receive and identify all those emissions illuminating the aircraft, including short bursts, in particular when the aircraft is operating in very dense signal environments. The warning of possible danger is given both visually and aurally on a display unit; preferential scan is used to ensure immediate recognition of possible lethal threats.

The system consists of antenna arrays, receiver, a processor system and a cockpit display/control unit. Broadband spiral antennas are used to provide omni-directional coverage and, together with their separate multi-band receivers, are mounted in pods on each wingtip. This location drastically limits aircraft 'shadowing' and the proximity of the receiver cuts signal losses to a minimum. Angular bearing of the emissions is determined by using selected pairs of antennas.

The hybrid superheterodyne receiver combines acquisition probability, high sensitivity, frequency accuracy and a high degree of frequency selection and selectivity to measure emissions. A broad bandwidth is used in the acquisition mode to obtain the initial intercept, and a narrow bandwidth for accurate bearing measurement and analysis. To ensure processing capability in highly dense signal conditions a high speed digital computer performs the data processing functions, supplemented by microprocessors. This enables the receivers to scan the frequency band continuously, on a reprogrammable basis, so that the conventional CW and agile signals are processed for identification.

A control/display unit allows the operator to monitor and control the automatic surveillance function to resolve possible ambiguities and evaluate and use to the best advantage, the data displayed.

STATUS
Fitted to the Mk 2 maritime reconnaissance Nimrods of the RAF, and the BR1150 Atlantic aircraft.

CONTRACTOR
Loral Electronic Systems, Ridge Hill, Yonkers, New York 10710, USA.

6218.393
ARR-81 COMINT RECEIVER

The ARR-81 is an airborne system composed of digitally controlled receivers and a microprocessor-based control/display unit, and is designed for communications intelligence applications. The receivers cover the frequency range 1 kHz to 500 MHz, with an optional extension receiver covering up to 2 GHz. Each receiver in the system covers 1 kHz to 32 MHz, 20 to 500 MHz or 500 MHz to 2 GHz. Maximum tuning time is approximately 5 ms.

STATUS
In operational service in US Navy SIGINT aircraft.

CONTRACTOR
Magnavox Government and Industrial Electronics Company, 1313 Production Road, Fort Wayne, Indiana 46808, USA.

5812.353

INTEGRATED ELECTRONIC WARFARE SYSTEM

The Integrated Electronic Warfare System (INEWS) programme is directed by the US Air Force Aeronautical Division as a joint US Air Force/Navy project to provide an overall EW system for combat aircraft in the 1990s. INEWS is investigating multi-spectral warning and response. The system will analyse threats, activate countermeasures automatically, display information on the threats to the crew, and recommend evasive action. INEWS is a totally new concept in that it uses interactive architecture, ie all the aircraft electronics such as navigation, EW, communications and fire control will interact.

INEWS entered the demonstration/validation phase in June 1986, and no details are likely to emerge until later in development, although various sources have suggested a blend of VHSIC, fibre optics and Ada software. Two teams were awarded contracts of about $50 million each for risk reduction demonstrations, trade studies, design refinement, and full scale development planning. Originally, five teams competed for the INEWS contract, the two remaining are:

(a) TRW/Westinghouse with Honeywell, Perkin-Elmer, GTE, Northrop and Tracor

(b) Sanders/General Electric with Motorola and HRB Singer.

One of the joint venture teams above is likely to be picked for full-scale development (FSD), and its members will compete for the production awards. Although the procurement strategy is currently undergoing review, it is likely that FSD and production will be conducted under sub-contract(s) to the airframe prime contractor(s). Total development costs are expected to be about $1 billion.

STATUS

The two teams received 31-month $48 million contracts in June 1986 to advance into Phase 1B demonstration/validation. One team will be chosen in 1989 for Phase 2 full scale development. Full production is planned for 1993 with contracts worth several billion dollars. Lockheed has chosen the Sanders/GE team to provide the INEWS for its YF-22A prototype in the USAF ATF fly-off.

CONTRACTORS

See above contracting teams. Direction is by US Air Force Aeronautical Division, Wright-Patterson Air Force Base, Ohio, USA.

5817.353

SCARECROW RPV EW SYSTEM

Scarecrow is an ECM system intended for use in RPVs over the battlefield area. The system includes a directional high gain antenn and pattern steerability in both the transmit and receive modes. Full data link capability is interfaced for control of the system from the ground and transmission of data from the RPV. A relatively low powered set-on jammer of less than 1 kW is employed for radar neutralisation purposes at close range. Scarecrow incorporates a broadband high dynamic range receiver which is used to detect threat signals intercepted by a high-gain antenna array. This latter is directional and can discriminate between emitters on the same frequency but on different bearings. The signals are checked against a pre-selected priority list of possible threats and the required jammer modulation techniques are selected. Equipment weight is under 18 kg.

STATUS

In flight trials phase. A number of countries have expressed interest, including UK, Canada and various Middle East defence forces.

CONTRACTOR

Sanders Associates Inc, Countermeasures Division, Nashua, New Hampshire 03061-2004, USA.

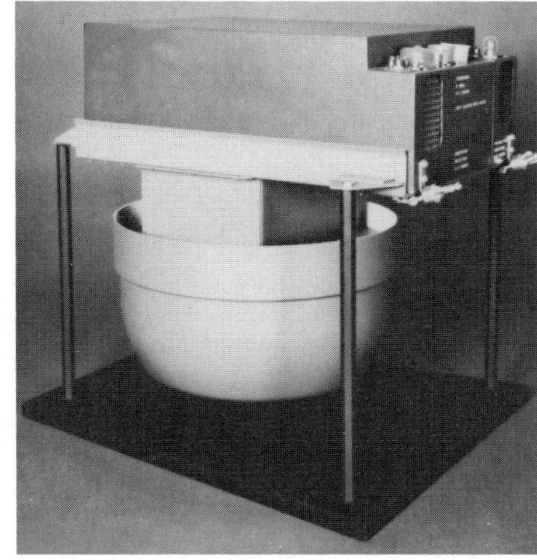

Scarecrow RPV EW system

6141.393

FAIRCHILD WESTON ECM PAYLOAD PACKAGES

This low cost suite of three ECM payload packages for manned and unmanned platforms consists of a homer, communications jammer and a radar jammer.

Homer

The homer is an amplitude-based direction finder and a video pulse train sorting signal processor. Once in a target area the homer detects radar emissions and using its DF capabilities, guides a pilot or RPV to set up a loiter pattern above the site. It operates over the frequency range from 1 to 10 GHz, with an extension to 18 GHz in development.

VTM radar jammer

The VTM transmitter can be operated over any 500 MHz band in the range 1 to 10 GHz, with a typical CW output power of 200 W RF. It is effective in jamming conventional and frequency agile multi-pulse or multi-beam radars. The VTM is modulated as a high rate swept carrier device and induces high PRF streams instantly into any radar. The result is a continuous stream of high density false targets which jams the radar completely. An extension of the frequency range to 18 GHz is in development.

Communications jammer

The communications jammer is really two jammers; a repeater equipment for FM voice communications and a barrage type for voice and data communications. The repeater is a multi-signal equipment capable of jamming conventional and frequency agile radios. Even multiple frequency hop modulations are jammed simultaneously. The two data link barrage jammers and repeaters operate in selected bands between 20 and 500 MHz.

CONTRACTOR

Fairchild Weston Systems Inc, 300 Robbins Lane, Syosset, New York 11791, USA.

4099.393

JAM PAC(V) ECM JAMMING SYSTEM

This is a compact jamming system developed for light strike/attack aircraft and helicopter applications, and especially for export versions of aircraft such as the F-5. Jam Pac units can be attached to aircraft singly, or in sets of up to three (typically) or four (maximum).

Jam Pac provides self-protection as well as combat systems evaluation or ECCM training. Within each unit is a transmitter operating in the range 0.8 to 15.5 GHz, using 30 per cent bandwidth voltage tuned magnetrons (VTMs). Two high efficiency VTM transmitters can be accommodated in one Jam Pac. Power output per jammer is 150 to 400 watts CW and power-managed receivers control jammer frequency tuning. Where a radar warning receiver is already installed in an aircraft, the power management receivers can be left out.

Jam Pac units are carried under wing or fuselage and should not interfere with stores carriage. Aircraft performance deterioration with Jam Pac fitted is stated to be negligible.

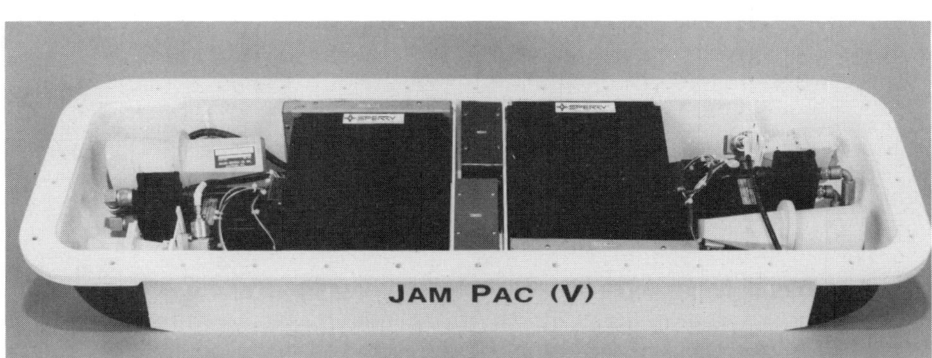

Internal view of Sperry Jam Pac(V), showing two voltage tuned magnetron transmitters, each with pair of forward- and rear-facing horn antennas, designed for under-fuselage or under-wing mounting on fighter aircraft or helicopters

CHARACTERISTICS

Frequency range: 0.8 – 15.5 GHz with 30% bandwidth VTMs

Power output: 150 – 400 W CW

Weight: 60 lb (27 kg)

Unit volume: 0.84 ft^3 (0.027 m^3)

STATUS

Fully developed.

CONTRACTOR

Sperry Corporation, Electronic Systems, Great Neck, New York 10020, USA.

5945.393

LORAL INFRA-RED COUNTERMEASURES

In addition to the AN/ALQ-123 (**1919.393**) and AN/ALQ-157 (**4365.393**) infra-red countermeasures (IRCM) equipments, Loral produces a number of other IRCM systems. Brief details of these are given below:

Defender

This is a family of pod-mounted systems which can be used for applications where an internal-mounted system is not practical. They can be supplied in various configurations, eg an aft-mounted jammer for aft protection only, an aft mounted jammer with side windows for aft and side protection; a nose-mounted jammer for forward protection only; and aft and nose jammers for all-round protection.

Challenger

A small omni-directional IR jammer, especially suitable for helicopters and light aircraft, which can be installed in fixed aircraft apertures or deployed on retractable platforms. The jammer head weighs about 9 kg and power consumption is within the range 1 to 2 kW. The jammer assembly is driven by a compact controller in the aircraft.

Matador

Matador is described under entry **5823.393**.

STATUS

In production.

CONTRACTOR

Loral Electro-Optical Systems, 300 North Halstead Street, PO Box 7101, Pasadena, California 91109, USA.

5811.393

QUICK FIX IIB COUNTERMEASURES SYSTEM

Tracor is in production of the EW system, nomenclatured the EH-60A (Quick Fix IIB). The EH-60A is an ECM/ESM system intended to provide tactical SIGINT/EW and DF capabilities, and interfaces with similar airborne and ground units via secure voice and data links. The mission electronics are configured in a standard UH-60 US Army utility helicopter and include intercept/DF and intercept/jamming capabilities.

STATUS

Tracor Aerospace is currently under contract to provide EH-60A aircraft to the US Army under a firm fixed-price contract valued at more than $87 million. The first system was delivered to the US Army in July 1987.

CONTRACTOR

Tracor Aerospace Inc, 6500 Tracor Lane, Austin, Texas 78725-2070, USA.

5743.053

ESCORT ESM SYSTEM

This is a family of electronic support measures equipment for maritime patrol aircraft, small ships and ground-based vehicles. The systems provide signal detection, sorting, analysis, identification and reporting of radar frequency emitters. Wide instantaneous spatial and frequency coverage combined with sophisticated processing techniques provide high probability of intercept. All mission data can be recorded on tape cartridge for post-mission analysis and reduction. The ELINT configuration provides precision direction finding by using phase interferometer antennas in conjunction with the basic receiver processor and display system.

The basic system operates over the 2 to 18 GHz frequency band with growth options available. It uses two antenna units, each consisting of a four element antenna array, a high speed RF switch and various control and amplification circuitry. The receiver unit accepts RF from port and starboard antenna units and performs frequency measurement, TOA, amplitude and AOA on each received pulse. The processor consists of a 16-bit general purpose computer with 128 000 16-bit words of memory and a variety of interface circuits. The search and dwell process used by the receiver as it steps between sub-bands to intercept and collect emitter data, is controlled by the processor software. Because of the wide bandwidth of the receiver, the search cycle over the entire frequency/azimuth domain is typically accomplished within one main beam illumination of a radar, ensuring high probability of intercept and a very low intercept time for new emitters.

The ESM operational software programme maintains an extensive emitter data file which includes the measured parameters as well as time of initial and most recent intercepts. Formats of CRT displayed outputs vary according to user requirements.

CHARACTERISTICS

Frequency coverage: 2 to 18 GHz

Polarisation: Any linear

Measurement accuracy: 8° (DF); 5 Mz (frequency); 1 μs (PRI)

Processing capability: 100 emitters

Volume: 4 cu ft

Weight: 91 kg (complete system)

CONTRACTOR

Raytheon Company, Electromagnetic Systems Division, 6380 Hollister Avenue, Goleta, California 93117-3197, USA.

4384.393

AN/ULQ-11 AIRBORNE JAMMING SYSTEM

This airborne jamming system, known as the AN/ULQ-11 Cefirm Leader, is an ECM equipment system operated by the US Army and comprises nine RU-21 aircraft fitted with various types of electronic countermeasures (ECM) equipment. It is designed to conduct ECM missions, ESM missions, jamming, initiate communications deception and direction finding. Of the nine aircraft, four are designated as 'A' aircraft which carry out airborne direction finding, three are 'B' aircraft conducting intercept and acting as system control, and two are 'C' aircraft which carry out jamming.

The types of communication against which the AN/ULQ-11 operates include amplitude modulation, frequency modulation, frequency shift key, CW, ICW, MCW and single side-band. The system is operated by RU-21 US Army utility aircraft.

CHARACTERISTICS

System frequency: 2 – 80 MHz

Jamming power: 500 W at 2 MHz and 70 W at 80 MHz

STATUS

In operational service.

5809.393

RACAL HELICOPTER-BORNE JAMMING SYSTEM

This system is a helicopter-borne transportable 'carry-on/carry-off' intercept and jamming equipment designed to provide signal intercept and surveillance over the frequency range 2 MHz to 1 GHz, and communications jamming from 20 to 79.975 MHz. Signals are intercepted and analysed using a 2 MHz to 1 GHz receiver and an associated signal/display unit. The frequencies, thresholds and priorities of signals to be jammed are placed in memory for later recall. When the jamming part of the system is activated, the stored frequencies appear on the display and the signals are then automatically jammed in accordance with the priority assigned.

The system is shock-mounted in two cargo cases which allows it to be carried on board and installed quickly. This is easily accomplished since the feet of the cargo cases are designed to fit into the standard cargo hold-downs of the helicopter.

CONTRACTOR

Racal Communications Inc, 5 Research Place, Rockville, Maryland 20850, USA.

5823.393

MATADOR

Matador is an engine-mounted family of infra-red countermeasures systems which has eleven different configurations. The system is suitable for all types of large transport aircraft with unsuppressed engines, one transmitter per engine being suggested for maximum protection. The basic system consists of a transmitter, an electronics control unit (ECU), and an operator's controller. Transmitters are electronically synchronised by the ECU which controls and monitors one or two transmitters. The operator's controller is common to all configurations, controls from one to four transmitters, and incorporates a system status display.

Each transmitter contains an IR source which emits pulsed radiation to combat a heat-seeking missile. Pre-programmed multi-threat jamming codes are provided, selectable on the operator's control unit, and all new codes can be entered as required to cope with new threats. Each transmitter weighs 22 kg nominal and is normally pod-mounted for aicraft attachment.

STATUS

In production.

CONTRACTOR

Loral Electro-Optical Systems, 300 North Halstead Street, Pasadena, California 91107, USA.

6211.393

IRCM SELF-DEFENSE SYSTEM

The IRCM self-defense system (ISDS) is a fuselage or pylon-mounted equipment designed to provide protection from anti-aircraft missiles employing infra-red homing devices. The system units consist of a transmitter, electronic control unit and operator control unit. The system configuration varies according to the type of aircraft but for most installations there would be one transmitter per engine, with an associated electronic control unit. A single operator control unit located in the cockpit controls one to four transmitters. The system has also been flight-tested in a ram-air turbine-powered pod configuration for fighter aircraft. Weight of the transmitter is 65 lb (29.5 kg), and of the electronic control unit 5 lb (2.3 kg). The operator control unit is a standard panel-mounted box measuring 5.7 × 4.94 × 2.25 inches (146 × 125 × 57 mm) and weighing 3 lb (1.56 kg).

STATUS

Operational on fixed-wing turbojet and jet aircraft, and in production for US and allied forces.

CONTRACTOR

Sanders Associates Inc, Countermeasures Division, Nashua, New Hampshire 03061-2004, USA.

IRCM self-defence system

6219.393

DRAGONFLY AIRBORNE VHF/UHF ECM SYSTEM

GTE Government Systems has developed an airborne VHF/UHF communications surveillance and jamming system against a requirement by the UK MoD. The equipment is code-named Dragonfly, and is designed to fulfil both a training role and possible use in tactical operations. It is intended to be operated automatically under the control of a computer, and will monitor selected frequencies and jam up to six of these on a priority basis. Manual control can be exercised at any time by the operator.

The system consists of two antennas, two receivers, an exciter/power amplifier, digital processor, and a control/display unit. The latter provides all operator control functions and displays the signal activity as required.

STATUS

In production under a $16 million contract from the UK MoD.

CONTRACTOR

GTE Government Systems, 100 Ferguson Drive, Mountain View, California 94039, USA.

4670.353

GUARDRAIL AIRBORNE SIGINT SYSTEM (AN/USD-9(V)2)

Guardrail is an airborne signal intelligence (SIGINT) system that intercepts, locates and classifies target systems and transmits data to ground processors to provide real-time intelligence information. The Guardrail suite consists of a transportable ground-based system and RU-21H and RC-12D aircraft carrying remotely controlled mission equipment.

The Guardrail system operates in three frequency bands, 20 to 75 MHz, 100 to 150 MHz and 350 to 450 MHz.

STATUS

Guardrail has been in operational service with the US Army for a number of years. The original Guardrail II was updated to IIA, and later was updated to Guardrail IV. Further updating has provided the latest version, Guardrail V, with better facilities, under the nomenclature AN/USD-9(V)2. Development is proceeding on a remote relay system for Guardrail V using satellite communications. The US Army has begun work on a modification to Guardrail V that includes Advanced Quick Look (AN/ALQ-133) (**1916.393**) and the communications high accuracy airborne location system (CHAALS).

CONTRACTOR

ESL Inc, Sunnyvale, California, USA.

5837.353

LR-5200 TACTICAL ESM SYSTEM

This system is intended for reconnaissance aircraft to provide real-time intelligence information over the frequency range 0.5 to 18 GHz. Intelligence is gathered by on-board equipment, which may be either internal or pod-mounted, and transmitted to an operations centre by data link. Information is also recorded on magnetic tape for post-flight analysis. The airborne system is normally installed in an unmanned aircraft.

The complete system includes all airborne, ground processing and maintenance support equipment. Ground processing and analysis facilities include a completely self-contained data analysis centre, which can be either fixed or transportable, and a remote data terminal for installation on command aircraft, land vehicles, naval vessels or other forward area platforms. The airborne part of the system weighs 102 kg.

CONTRACTOR

Litton Systems Inc, Amecom Division, College Park, Maryland, USA.

5839.353

COMPASS CALL ECM SYSTEM

Compass Call is a specialised ECM system installed in US Air Force EC-130H aircraft, and works with ground mobile C³CM systems to confuse enemy defences and to jam their C³ systems. A number of large blade antennas have been added to the basic C-130 aircraft configuration, including under each outer wing and above the dorsal fin, with a smaller horizontal blade antenna on each side of the rear fuselage. Bullet-shaped cannisters outboard of each underwing antenna and the extreme tail of the aircraft house trailing wire antennas that will extend to several hundred feet behind the aircraft. No technical details of the on-board ECM systems, or the contractors, have been released.

STATUS

In service.

3354.053

WJ-1740 RECEIVING SYSTEM

The WJ-1740 is a parallel-scanned, digitally controlled superheterodyne receiver system which can include as many as eight tuners (12 for special applications) in any combination of frequency bands. The tuners cover the 0.03 to 40 GHz range in nine frequency bands. Tuners are scanned synchronously, in parallel, although the operator can call any one tuner from the scan for signal analysis.

Selection of a tuner for manual analysis switches the IF output of the tuner to an analysis demodulator channel that provides IF attenuation control, IF bandwidth control, and linear with or without AGC, log, or FM video outputs. The tuner selected can also be phase-locked to a stable crystal-controlled reference in a local oscillator synchroniser to provide frequency accuracy and stability of one part in 10^7 per day.

The analysis video is presented on the analysis display, and the tuner selected for analysis can be operated in manual, sector scan, or band scan modes. In the first of these modes, the analysis display provides an IF pan signal presentation; when the tuner in the analysis channel is in either of the other two modes, the display indicates signal activity versus frequency, with the horizontal frequency scale automatically expanded to cover the full width of the CRT.

The system provides auxiliary IF outputs from each tuner, and auxiliary scan demodulator log video outputs for each tuner. It also provides IF and video outputs from the analysis channel.

Individual tuners can be placed in the Hold mode to provide continuous monitoring of signals, and up to 16 frequencies can be stored in memory for instantaneous push-button controlled recall. In the Hold mode, a tuner is tuned to one frequency and its IF and video outputs can be applied to either a tape recorder or special analysis equipment.

An optional memory scan unit provides 64 memory cells and a memory scan mode permitting a programmed search strategy.

The complete system was designed to permit full manual or computer-assisted operation and control by as many as four different operators through individual receiver control panels.

In a computer-assisted WJ-1740 system, the computer monitors the receiver scan, monitors signal activity, makes tentative signal identifications, and records data, leaving the operator free to identify signals and perform detailed analysis tasks.

DEVELOPMENT

The WJ-1740 receiver system was developed for a USAF airborne application.

CONTRACTOR

Watkins-Johnson Company, 2525 North First Street, San Jose, California 95131, USA.

4666.393

SMALL PLATFORM EW SYSTEM

A tactical jamming system to be carried by a remotely piloted vehicle is being developed by Motorola under contract from the US Air Force. This system is known as the Small Platform EW (SPEW) and is intended for use against ground-based radars in support of the US Air Force EF-111A aircraft. The proposed vehicle is the Pave Tiger RPV. Although no details have been released the system appears to consist of a combined transmitter/receiver/processor and a four quadrant antenna.

STATUS

In development.

CONTRACTOR

Motorola Inc, Government Electronics Group, Scottsdale, Arizona, USA.

4667.393

AIRBORNE RADAR JAMMING SYSTEM

The Airborne Radar Jamming System (ARJS) is a new high-power standoff jammer intended to be used against air defence, surveillance and target acquisition radars. It is designed to be helicopter mounted, probably in a UH-60 Apache attack aircraft, and would be deployed behind the forward troop positions.

The technology employed would be based on the AN/ALQ-99 jamming system used by the US Navy in its EA-6B aircraft and the US Air Force in its EF-111As. This would provide detection of a threat signal, location of the emitter, comparison against a threat library held in the system's computer and transmission of a jamming signal.

STATUS

In development.

CONTRACTOR

Grumman Aerospace, South Oyster Bay Road, Bethpage, New York 11714, USA.

6910.393

FIBRE-OPTIC LASER WARNING SYSTEM

The fibre-optic laser warning system developed by Varo has applications for airborne, shipborne and land-based systems. The basic system covers wavelengths from 400 to 100 nanometres, with modular expansion available for other bands. The tiny sensor system can be flush-mounted anywhere on the platform and gives an angular coverage of approximately 90 degrees. Four to six sensors would be required on most platforms to provide all-round warning of laser illumination.

Information of wavelength and direction of the threat is provided, with the direction normally displayed on the radar warning receiver display unit.

STATUS

In late development. Trials by the US Navy are currently under way and are primarily aimed at wavelength detection and measurement.

CONTRACTOR

Varo Inc, Systems Division, Electro-Optical Warfare Systems, 2800 W Kingsley Road, Garland, Texas 75046-9015, USA.

6210.393

AIRBORNE REMOTE CONTROLLED ESM SYSTEM

This is an airborne remotely controlled system for ESM surveillance. It consists basically of a number of radio receivers and antennas operated by remote radio control from a ground control facility which is equipped with the necessary processing and display equipment. It is intended to be carried in small, relatively inexpensive aircraft and the system configuration can be tailored to the particular aircraft type. On-board equipment would normally consist of radar detection units covering the frequency band 0.5 – 18 GHz, radio monitoring receivers covering 1.5 MHz – 2 GHz and radio direction finding systems covering 20 – 1200 MHz. The data link is a full-duplex, highly directional system.

STATUS

Fully developed as a private venture.

CONTRACTOR

E-Systems Inc, 7700 Arlington Boulevard, Falls Church, Virginia 22046, USA.

6520.393

AIRBORNE EXPENDABLE DECOYS

The United States Naval Research Laboratory has initiated a programme to develop a series of advanced airborne expendable decoys (AAED). Like the POET and GENE-X expendable cartridges, the AAED system is intended to provide a radar target decoy to an incoming missile. Two basic systems are being studied, a rocket propelled system fired forward of the aircraft and a version towed behind the aircraft.

The forward-fired version uses a rocket filled with electronic hardware and is designed to fly ahead of the aircraft using an autopilot programmed to keep it flying as long as possible. The towed version uses a decoy towed far enough behind the aircraft such that an explosion, should the missile hit the decoy, will not damage the aircraft.

STATUS

In early development. It is understood that the first contract for design and development of the on-board electronics has been awarded to Teledyne-CME.

YUGOSLAVIA

5943.393

SO-1 RADAR WARNING RECEIVER

The SO-1 is a conventional radar warning receiver designed to warn the pilot that his aircraft is being illuminated by a threat tracking radar. The equipment covers pulse type radar systems in the I and J bands with an azimuth coverage of 360 degrees and an elevation coverage of ± 45 degrees. DF accuracy is 17 degrees. The visual alarm consists of a panel-mounted instrument containing four lamps which give quadrantal direction indication. An aural alarm is also incorporated in the aircraft audio system. Weight of the system is approximately 60 kg, and it operates from a 27.5 DC power supply with a consumption of 10 W.

STATUS

In production.

CONTRACTOR

Federal Directorate of Supply and Procurement, PO Box 308, 9 Nemanjina Street, 1105 Beograd, Yugoslavia.

ELECTRO-OPTICAL EQUIPMENT (AIR)

Note: This section covers a variety of airborne electro-optical systems, including weapon control systems, lasers, FLIR systems, sights, thermal images and Head-Up Displays (HUDs). Except where they are integrated with weapon aiming systems, HUD equipments are dealt with only briefly. More detailed information on HUD systems is provided in our companion yearbook *Jane's Avionics*.

CANADA

6776.393

OPTECH LIDAR

Optech Systems of Ontario, a research and development organisation, are developing a range of laser radar or Lidar (Light detection and ranging) devices for military use.

The laser radar involves transmitting a pulsed laser beam across a strip of sea with the wavelength selected for minimum propagation losses in water. As the transmitted laser pulse strikes the surface there is a strong back-scatter and by comparing this with the subsequent back-scatter signature, data on sub-surface targets can be acquired through analysis of the signature.

In 1982 Optech produced a neodymium-YAG scanning operational prototype, the Larsen 500 airborne scanning Lidar bathymeter, for the Canadian Hydrographic Service. This was used to investigate a defence application detecting potential targets including mines, torpedoes, submarine wakes, nets and fish schools. Submarines were detected and located in coastal waters down to 30 m and submarine wakes at greater depths.

Under a US$6 million DARPA contract awarded in August 1987 Optech has developed, in association with Northrop Electronics Division's Electronics Systems Group, a copper vapour laser version for mine detection. This laser has a 510.6 mm wavelength, a PRF of 10 KHz and transmits a pulse of 0.3-0.5 m with a pulse length of 8 ns. The system, known as ALARMS (Airborne Laser Radar Mine Sensor), also features a Litton LTN-90 ring laser gyro strapdown inertial navigation system and a Raydist-T navigation system sensor. This is the first stage of development which involves an installation in a Sikorsky MH-53E Sea Dragon mine countermeasure helicopter.

This pallet-mounted system, which weighs less than 454 kg (1000 lb) can operate day or night with the aircraft flying at 55 knots at an altitude of 615 m. The beam covers a 400 m wide strip and is capable of detecting mines at depths of between 20 and 40 m. The system can operate in Continental Shelf and open ocean waters. The company claim it can operate in tropical waters above and below the thermocline, in sub-tropical waters above the thermocline and in temperate sub-Arctic and Arctic waters. The wavelength of the system can be modified to meet water absorption problems and for tropical and sub-tropical waters the minimum absorption occurs in the 460 to 480 nanometres wavelength range while for the other waters the transmission window widens into the 460 to 535 nanometres, or blue-green spectrum, range.

Negotiations are now underway for Phase II, a pre-full scale development of an engineering model which is likely to cost US$8 million. This would feature an Excimer laser to probe deeper, a redesigned sensor head, an agile scanner and an intelligent console. The company hope this will be followed by Phase III, a US$20 million programme for a full military specification system, with production beginning in 1993. They have already proposed to DARPA an ASW version, which would be fitted into a SH-60, although a 'commercial' standard system could enter production in 1992. The company also believe the technology could be applied to underwater communications either from aircraft or from satellites.

STATUS

The US Army will use the Larsen 500 system for river survey work and a unit for use in a Bell 204 helicopter would be delivered in the autumn of 1989. The system has also been used in experimental ASW work for an unidentified European country and an operational prototype is under development.

Trials with the ALARMS Phase I are being carried out at Panama and San Diego with the objective of developing algorithms to detect various types of mine. These will be completed in September.

CONTRACTOR

Optech Systems, 701 Petrolia Road, Downsview, Ontario M3J 2N6, Canada.

FRANCE

7032.393

APX M334 SERIES GYRO-STABILISED SIGHTS

The APX M334 sight has been designed for the surveillance, detection and acquisition of targets, and guidance of missiles in manual remote control.

OPERATION

The equipment consists of a monocular, angularly positioned single lens sight, a body made of a prism correcting the image slanting and a panoramic head including a gyroscope for stabilisation. Two magnifications, × 3.2 and × 10.8, are provided and the mirror scanning speed is adjustable from 1.5′/s to 13.5°/s at the lower magnification, and from 1.5′/s to 4.5°/s at × 10.8 magnification. The field of view is 17.1° at magnification × 3.2 and 5.2° at magnification × 10.8; the total scanning field is 262° in azimuth and 72° in elevation.

In the wide field of view, a luminous circle represents the narrow field, and angular differences between the line of sight and the axis of the helicopter can be indicated. For safety reasons the eyepiece is retractable when the equipment is not in use to enable the operator to have normal unobstructed vision outside the aircraft.

The designation APX M334 covers a variety of roof-mounted sight systems for helicopters. Variants of the basic system have specific capabilities, e.g. M334-04 ATHOS, M334-25, M397-HOT, M334-05 VISIGOTH.

STATUS

The sight has been fitted in its various versions on different helicopters, such as Aerospatiale Gazelle 341 and 342, Super Frelon SA 321, Westland Wasp, Lynx, Bell 204/205/206, 212, Sikorsky SH-3D and MBB BO105.

CONTRACTOR

SFIM (Société de Fabrication d'Instruments de Mesure), 13 avenue Ramolfo-Garnier, 91344 Massy Cedex, France (under DTAT-APX licence).

APX M334-04 gyro-stabilised sight and control box

3661.393

TCV 115 AIRBORNE LASER RANGEFINDER

The TCV 115 laser rangefinder was designed for integration into the gyro-stabilised APX M334-04 helicopter sight (**3763.393**), and consists of a TCV 80H laser rangefinder and an optical assembly which adapts the rangefinder to the sight and also ensures the projection of the rangefinder cross-hairs in the eyepiece of the sight. The TCV 80H laser rangefinder is also employed in the APX M550 rangefinder/sight (**3414.193**) for land-based use.

CHARACTERISTICS

Wavelength: 1.06 micron
Pulse duration: 25 × 10^{-9} s
Energy: 100 mJ
Peak power: 4 MW
Divergence of emitted beam: ⩽0.7 mrad
Detector: Avalanche photodiode
Minimum detectable power: ⩽2 × 10^{-8} W
Receiver field: ⩽0.5 mrad
Operating voltage: Rangefinder 19 – 28 V DC; graticule projector 115 V/400 Hz
Dimensions: 390 × 190 × 180 mm
Weight: 8kg (with optical assembly)
Measurement range: 450 – 19 900 m
Accuracy: ±10 m
Repetition rate: 1 measurement every 2 s for 3 successive shots; 12 measurements/minute

An optical attenuator in the eyepiece ensures protection for the operator's eye during laser firing and provides attenuation of 70 dB at 1.06 microns.

STATUS

In series production.

CONTRACTOR

CILAS (Compagnie Industrielle des Lasers), route de Nozay, 91460 Marcoussis, France.

CILAS TCV 115 airborne laser rangefinder

3763.393

APX M334-04 GYRO-STABILISED SIGHT

The APX M334-04 gyro-stabilised sight is the basis of the M334 series of sights.The main items of the equipment are: the sight itself incorporating a gyro-stabilised mirror allowing the aiming axis to be kept stable irrespective of pitch and yaw movements; an aiming control stick which controls the sight in azimuth and elevation; an optional direction indicator for use by the pilot; and an electronics unit.

The APX M334-04 sight can be fitted if needed with a luminous reticle generator, a fire simulator device, a laser rangefinder, a special eyepiece bearing arm, a TV or cine camera and electronics for sighting assistance.

The APX M334-04 sight is of the monocular type with two magnifications, ×3.2 and ×10.8, the respective fields of view being 300 and 90 mrad. The total scan field is 262° in azimuth and 72° in elevation. The azimuth scan raster can be varied in a stepless fashion over the range 0.2 to 230 mrad with ×3.2 magnification, and 0.2 to 80 mrad with ×10.8 magnification. Stabilisation and aiming accuracy is better than 0.1 mrad.

STATUS

Production.

CONTRACTOR

SFIM (Société de Fabrication d'Instruments de Mesure), 13 avenue Ramolfo-Garnier, 91344 Massy Cedex, France (under DTAT-APX licence).

SFIM APX M334 Series stabilised sight in French military helicopter

6330.393

APX M334-25 GYRO-STABILISED SIGHT

The APX M334-25 is a specialised sight designed for range-finding and localisation of targets up to 10 km or more range, for adjusting artillery fire. When used in connection with a navigation system, it constitutes the OSLOH III system giving absolute position of the target.

OSLOH III consists of the APX M334-25 sighting system which measures the relative position of the target with respect to the helicopter together with the navigation system providing the helicopter's current position. The advantages of this system are primarily in the operation of light observation helicopters (LOH), particularly in artillery operations to give target absolute position in UTM coordinates with a very short reaction time and high accuracy.

STATUS

In production and in operational service in five countries.

CONTRACTOR

SFIM (Société de Fabrication d'Instruments de Mesure), 13 avenue Ramolfo-Garnier, 91344 Massy Cedex, France.

APX M334-25 OSLOH sighting system

6331.393

V3G HELICOPTER DAY/NIGHT SIGHTS

Based on previous developments of electro-optical stabilised platforms, SFIM has designed a lightweight series of sights to meet the requirements of modern combat helicopters.

The V3G is a third-generation sight which can be roof, chin or mast-mounted and can incorporate a variety of sensors. According to the type of sensors fitted the sight can perform day and night observation, rangefinding, provide anti-tank fire control by day and night, and provide air-to-air headseeker missile sighting.

Among the series is the Viviane sight (**3856.393**) selected for the addition of night capability to the Euromissile HOT anti-tank missile system.

STATUS

In development. A SFIM spherical mast-mounted sight has been flight tested in a Dauphin helicopter. The sight is equipped with a TV camera and FLIR sensor and a laser rangefinder. The sight can accommodate a sensor package with a useful weight of about 50 kg.

CONTRACTOR

SFIM (Société de Fabrication d'Instruments de Mesure), 13 avenue Ramolfo-Garnier, 91344 Massy Cedex, France.

SFIM V3G helicopter day/night sight

3856.393

VIVIANE/HOT HELICOPTER DAY/NIGHT SIGHT

SFIM has developed and manufactured a wide range of gyrostabilised sights for observation and anti-tank missile firing by day. One of these, the APX M397-02, a variant of the APX M334 sights family, is fitted on the Gazelle AS 342M and on the BO 105/PAH-1 in the HOT missile system.

By night the first HOT missile was fired from a Dauphin helicopter SA 361. The APX M397 was completed by a SFIM gyrostabilised platform Venus located in the nose of the helicopter and carrying a TRT Hector thermal camera.

The sighting system was too heavy for light helicopters like Gazelle. The new Viviane/HOT helicopter night sight is now under development by SFIM. The normal version is a roof-mounted sight but it can be either mast of chin-mounted for any type of light helicopter. It includes in the same platform a TRT Victor thermal camera, a CILAS rangefinder, a MBB/SFIM missile localiser and a direct optical new block.

The firing installation needs no modification in the standard equipment. The operator uses the same eyepiece switching quickly from the visible image to the infra-red or inverse.

Viviane will replace the APX M397 on the Gazelle and the BO 105 and is also suitable for intermediate/medium helicopters like Dauphin.

SFIM Viviane night sight for helicopters

CHARACTERISTICS
Platform assembly movement: Azimuth ±120°, elevation –25° to +30°
Target detection range: 5.5 km
Target engagement range: 4000 m
Diameter: 650 mm (on roof)
Weight: 58 kg

STATUS
The original HOT sight has been in operational service for a number of years. The Viviane sight is in development and has been chosen by the French Army Aviation to equip its new Gazelle helicopters. In service date is planned for 1989.

CONTRACTORS
Main contractor: Aerospatiale-Euromissile
Principal subcontractors: SFIM, SAT, TRT

5923.393
TANGO THERMAL IMAGER

Tango is an airborne IR system intended for use in maritime applications. It can be used in the search mode to detect surface vessels, schnorkels, etc and for search and rescue missions. It has been designed for installation in a ball-shaped gyro-stabilised platform mounted under the nose of the French Navy Atlantique 2 aircraft. The complete Tango system incorporates a turret and a camera, the latter using modules developed under the SMT common module programme (**4107.093**). These modules operate in the 8 to 13 micron waveband and are cooled by a split-Sterling system. The turret is a 60 cm diameter gimballed platform which rotates in azimuth and elevation.

The system converts the thermal radiation of a target into an image which, when incorporated with the necessary symbology, is displayed on TV monitors in the aircraft. Normally the system is slaved to the line-of-sight but it is possible to automatically designate a target using co-ordinates generated by the aircraft weapon system.

CHARACTERISTICS
Wavelength: 8 – 13 microns
Magnification: Partial × 2 magnification is provided in a window in the centre of the image
Camera lens: Bifocal type with 6.45° × 4.3° and 2.15° × 1.43° fields-of-view
Image scanning: Serial-parallel
Platform assembly movement: Azimuth ±110°; elevation +15° to –45°
Stabilisation accuracy: 1 mrad
Weight: 150 kg

STATUS
In operational service in the Atlantique 2 aircraft of the French Navy.

CONTRACTOR
TRT (Télécommunications Radioélectriques et Téléphoniques), Defence and Avionics Commercial Division, 88 rue Brillat-Savarin, 75460 Paris Cédex 13, France.

Tango thermal imager installed under the nose of an Atlantique 2 aircraft

---

3136.393
THOMSON-CSF ATLIS LASER TARGET DESIGNATION POD

ATLIS (automatic-tracking laser illumination system) is a high precision air-to-ground designation system mounted in a pod. It can be mounted on single-seater combat aircraft. The system consists of:

(1) TV/laser fire control pod with TV camera operating with wide or narrow field of view, laser illuminator/rangefinder, automatic two-mode TV tracking, stabilising system, and air-conditioning system
(2) monitoring and control panel and a hand controller in the cockpit, associated with a TV monitor if no TV-type display equipment is already available. The system can be connected to the head-up display, or the radar navigation system or helmet sight for target predesignation.

The stabilisation system located in the front of the pod gives a stabilised optical path for live scene data processing, laser designation/reception, and provides mechanical steering for the optical line-of-sight. The live scene data is reflected from a stabilised mirror into a fixed optical assembly which folds and focuses the image into the TV camera. The output of the camera provides two field-of-view video to the display in the cockpit and to the automatic tracker. Laser designation energy is reflected to the dichroic portion of a combining glass located within the live scene data path. The combining glass passes shorter wavelength scene data but reflects laser energy so that the laser energy exits co-linear with the incoming scene data.

The stabilised mirror provides a stable optical line-of-sight path when high frequency vibration and structure induced motion are present, and enables external steering of the line-of-sight. A pitch/yaw rate stabilised inertial platform affords rejection of high frequency dynamic motions, and is combined with a roll turret drive to provide line-of-sight steering.

The dual mode tracker provides both area co-relation and point tracking. The former mode is used to stabilise the scene and provide designation for area or low contrast targets. Point tracking is used for stationary or moving point targets. The TV camera operates in both the infra-red and visual spectra. A digital processor manages all the pod functions and provides automatic test, rate-aided tracking signals, and fire control computation capabilities.

OPERATION
ATLIS is used for low altitude, high speed, autonomous attack of surface targets with laser-guided weapons; it reduces risks for the attacker owing to the immediate breakaway and long stand-off ranges allowed by the system.

ATLIS laser designation pod mounted on a Mirage 2000

ATLIS laser target designation pod mounted on a USAF F-16 aircraft

CHARACTERISTICS
Weight: 178 kg
Diameter: 322 mm
Length: 2.50 m

DEVELOPMENT
In 1975 Thomson-CSF awarded a $3.7 million contract to Martin Marietta Aerospace for the development and test of the ATLIS pod. Thomson are responsible for manufacturing and marketing the pod. The French CILAS company provides the ITAY 139 laser illuminator-rangefinder.

The first prototype system, known as ATLIS 1, was flight tested during the late 1970s. These flight tests were highly successful with some 700 passes against a wide variety of targets which confirmed the system's ability to locate and mark targets for attack by a laser guided missile such as the AS.30 (**3335.311**), from a range of 10 km. This programme's results were such that the French Air Force decided to equip a number of Jaguar aircraft with ATLIS pods, and cockpit viewing and control equipment. Development and production of suitable laser guided bombs and rockets by Matra/Thomson-CSF and Thomson Brandt has continued in parallel with AS.30 development by Aerospatiale using the Thomson-CSF Ariel laser seeker.

Following completion of the ATLIS pod development, Thomson-CSF and Martin Marietta developed and flight tested a new system, ATLIS 2, for use with Jaguar and F-16 aircraft. ATLIS 2 weighs 170 kg, compared with the 240 kg of the original ATLIS, and includes a computer to interface directly with a digital bus-bar.

STATUS
ATLIS 2 is in full production by Thomson-CSF for the French Air Force, and has been delivered to several other countries' air forces for Mirage 2000 and F-16, the customers including Pakistan.

CONTRACTOR
Thomson-CSF, Aerospace Group, General Avionics Division, 122 Avenue du Général Leclerc, 92105 Boulogne-Billancourt, France.

3423.393
THOMSON-CSF VE 110 HEAD-UP DISPLAY

This display comprises two units: a pilot display unit and an electronic unit operating in conjunction with the aircraft nav-attack system. These are described separately in the following paragraphs.

Pilot Display Unit: Collimated images are generated by a pilot display unit which includes: high brilliance and long-life CRT with its deflection control coils, providing stroke written or raster images; the EHT generation circuits; a collimation optical system with a large-diameter lens (110 mm for the VE 110 model which is a simplified version adapted to light combat or training aircraft); the power supply; and push-button test. A sight recorder mount and pilot controls are provided at the rear part of the pilot display unit. A special mount permits instantaneous replacement of the pilot display unit without any adjustment or re-harmonisation.

Electronic Unit: The symbol generation, ballistic and flight circuits, the interface with the digital data transmission system, and an integrated test system are contained in the electronic unit. The electronic circuitry is of solid-state, integrated circuit type. All circuits are distributed on plug-in cards. The technology used in this equipment gives maximum design flexibility.

STATUS
The VE 110 is in production for the modernised Mirage III/IV, the Alpha Jet NG and other light tactical support/trainer aircraft. It is also fitted to the Mirage III G.

CONTRACTOR
Thomson-CSF, Aerospace Group, 122 Avenue du Général Leclerc, 92105 Boulogne-Billancourt, France.

Thomson-CSF VE 110 head-up display

1317.393

THOMSON-CSF SERIES 121-196 ATTACK SIGHTS

These displays comprise three units: a collimating head projecting a coloured display; an electronic unit, for signal processing; a pilot's control unit (not required if the head-up display (HUD) is operated through the control unit of the nav/attack system). Further details will be found in the 1987-88 edition.

STATUS
The 121 sight has been fitted in the Jaguar mainly for air-to-ground missions; the 196 sight was employed in the Mirage F1 principally for air-to-air roles. Both attack sights are being replaced on these aircraft by electronic head-up displays of Series VE110, VE120 and VE130.

CONTRACTOR
Thomson-CSF, Aerospace Group, 122 Avenue du Général Leclerc, 92105 Boulogne-Billancourt, France.

3135.393

THOMSON-CSF SERIES VE 120/130/VEM 130 HEAD-UP DISPLAYS

These displays comprise two units: a pilot display unit and an electronic unit operating in connection with the aircraft nav/attack system.

Pilot Display Unit: Collimated images are generated by a pilot display unit which includes: high brilliance and long-life CRT with its deflection control coils, providing stroke written or raster images; the EHT generation circuits; a collimation optical system with a large-diameter lens and combiner giving a large IFOV and a push-button built-in test. A sight recorder mount and pilot controls are provided at the rear part of the pilot display unit. A special mount permits instantaneous replacement of the pilot display unit without any adjustment or pre-harmonisation.

Electronic Unit: The symbol generation, fire computation and navigation computation circuits, the interface with the digital data transmission system, and an integrated test system are contained in the electronic unit. The electronic circuitry is of solid-state, integrated circuit type. All circuits are distributed on plug-in cards. The technology used in this equipment gives maximum design flexibility and allows modifications in symbology.

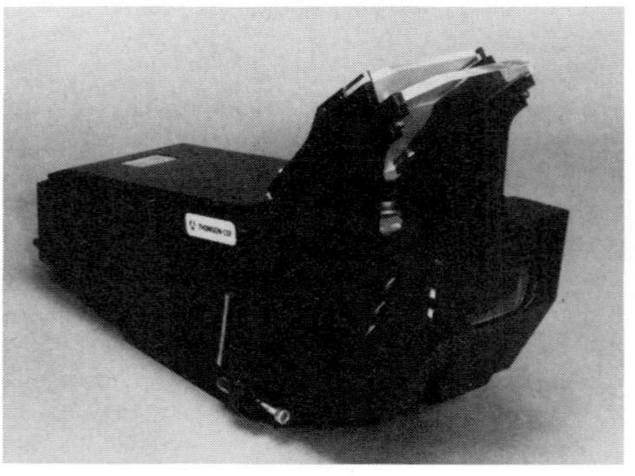

VEM 130 head-up display combined with an eye-level display

The VEM 130 is the latest model HUD and is a raster/stroke system with conventional lens. It is associated with the TMM 1410 head-down display to provide the pilot with two collimated images.

Thomson-CSF VE 120 head-up display unit

STATUS
The systems are in production for the Super-Etendard of the French Navy, for Mirage F1, Alpha Jet and, integrated in an all-electronic multi-display system, on the Mirage 2000 (**4723.393**).

CONTRACTOR
Thomson-CSF, Aerospace Group, 122 Avenue du Général Leclerc, 92105 Boulogne-Billancourt, France.

4723.393

TMV 980A ELECTRONIC MULTIDISPLAY SYSTEM

The TMV 980A is a highly flexible multimode, multicolour digital display system for the Mirage 2000 aircraft. It consists of an integrated head-up display (HUD), a head-down display (HDD), a complementary HDD in monochrome and an electronics unit/digital computer. The system is intended for use by the pilot in most stages of the mission, using either the HUD or HDD as appropriate. Information on the displays is provided by the various sensors on board and is designed to give the pilot all the information required for a particular phase of his flight.

The VE 130 HUD (**3135.393**) has a large field of view through the windscreen and provides information symbols collimated on the combining glass at infinity. It uses a 3 inch high-resolution, high-brightness, monocolour (green) cathode ray tube associated with the collimating optical system. The 5.1 inch diameter collimating lens gives wide instantaneous field of view which is enlarged in elevation by the use of a twin glass combiner with an optical transmit factor of 80 per cent. An automatic brightness control and manual contrast allows the symbols to be read in ambient light up to 100 000 lux.

The Type VMC 180 HDD is a multifunction unit whereby information is presented in TV raster and/or calligraphic writing with coding and colour for rapid assimilation of the visual data. Two video inputs are provided (radar and TV) plus a link to the electronics unit for generation of coloured alpha-numerics for superimposition on either the radar or TV frame. Information presented includes a radar map, synthetically derived tactical situations, TV raster images (from TV sensors and flight instruments), and tactical data from the system itself or from external sources. The unit uses a 5 × 5 inch multimode cathode ray tube screen with red, amber and green colours. For brightness control, automatic photo cells and manual override are incorporated on the control panel.

Units of the TMV 980A multidisplay system

The complementary Type VCM 65 HDD provides the same information on a 2.5 inch circular monocolour screen.

The electronics unit provides the interface with the aircraft systems, carries out flight and weapon aiming computations, and processes and generates the display symbology.

Weights of the various units are 13 kg for the HUD, 14 kg for the VMC 180 HDD, 4 kg for the VCM 65 HDD and 9 kg for the electronics unit. The latter is of ½ ATR format.

STATUS
In production for the Mirage 2000 multi-role fighter aircraft.

CONTRACTOR
Thomson-CSF, Aerospace Group, 122 Avenue du Général Leclerc, 92105 Boulogne-Billancourt, France.

1736.393

TAV-38 LASER RANGEFINDER

The TAV-38 laser rangefinder is designed to provide air-to-ground ranging facilities up to ranges of 10 km in clear-sky conditions. The equipment comprises two units: laser head and power unit. The latter unit weighs 12 kg and has no moving parts. It provides the high energy DC voltage to the flash exciter tube and the auxiliary supplies to the system. The laser head weighs 8.5 kg, and houses the laser cavity, with its neodymium rod, flash exciter, and triggering device. It also contains the receiver circuitry which incorporates an avalanche photodiode, the digital range counter, and transmission/reception optics.

Among the design features contributing to light weight and modest dimensions is an original type of exciter yielding more power to the neodymium rod, thus reducing the heat that needs to be carried away. The laser/internal cooling arrangements make the unit suitable for use under a wide range of ambient temperature conditions.

OPERATION

Ranging is effected by measuring the time interval between the emission of an energy pulse and the return of reflected energy. Measurement is by means of a digital counter, with a clock frequency of 29.98 MHz, which corresponds to a distance increment of five metres. The digital output is converted into serial binary format for output. The wavelength of the emissions is 1.06 microns.

The deflector system developed by Thomson-CSF enables the direction of the beam emitted by the laser rangefinder to be referenced to the boresight provided for the pilot by a collimator. The angle of deviation thus obtained is $\pm 10°$ in both elevation and azimuth.

DEVELOPMENT

The type TAV-38 is the result of collaboration between Thomson-CSF and Marcoussis Laboratories CGE Research Centre, the former being responsible for the beam deflector system and the latter for the TAV-34 laser rangefinder.

STATUS

In operational use with a number of exported Mirage F-1 aircraft and in the French Air Force Jaguar.

CONTRACTORS

Thomson-CSF, Aerospace Group, 122 Avenue du Général Leclerc, 92105 Boulogne-Billancourt, France.

CILAS (Compagnie Industrielle des Lasers), route de Nozay, 91460 Marcoussis, France.

5744.393

TMV 630 LASER RANGEFINDER

This equipment has been designed to meet single-unit, small installation requirements and can be easily fitted to a wide range of aircraft.

The rangefinder provides high precision aircraft-to-target range measurement and is claimed to increase considerably the performance of conventional weapon-aiming systems. The large field-of-view provided is compatible with all head-up displays and the electrical interfaces are compatible with nearly all aircraft types. The high speed and accurate laser beam steering is specifically adapted for continuously computed impact point attacks irrespective of terrain or the nature of the weapons, or aircraft altitude.

STATUS

Operating wavelength: 1.06 microns
Maximum range: 19 km
Accuracy: > 1 milliradian
Dimensions: 190 × 190 × 520 mm
Weight: 15 kg

STATUS

In production as part of nav/attack system on Dassault Mirage and Dassault-Breguet/Dornier Alpha Jet aircraft.

CONTRACTOR

Thomson-CSF, Aerospace Group, 122 Avenue du Général Leclerc, 92105 Boulogne-Billancourt, France.

3562.393

THOMSON-CSF TYPE 902 GUNSIGHT

The 902 gunsight is one of the series R gunsights (**1518.393** in 1979-80 and earlier editions) produced by Thomson-CSF and sold in many countries as well as in France. The 902 assures maximum performance efficiency in air-to-ground attacks with guns, rockets and clean or retarded bombs and also in the air-to-air gun fire mode with visual stabilised tracking. It is specially adapted to the sighting problem of light strike aircraft. Further details will be found in the 1987-88 edition.

STATUS

No longer in production.

CONTRACTOR

Thomson-CSF, Aerospace Group, 122 Avenue du Général Leclerc, 92105 Boulogne-Billancourt, France.

4724.393

HUD SIGHT REPEATER

The Head-up Display (HUD) Sight Repeater is designed for use by the rear seat occupant in two-seat aircraft and provides a televised, collimated image of the display seen through the combiner glass of the pilot's HUD. The rear occupant, normally an instructor pilot, can then follow the symbolic HUD data superimposed on the landscape, control the actions of the front pilot as necessary and check target identification or obstacle avoidance.

The sight repeater consists of two elements:

(1) an electronic scanning unit installed in the front cockpit and consisting of a vidicon pick-up tube, an optical lens, an automatic shutter and a 4-filter filtering unit

(2) the display sub-assembly mounted in the rear cockpit and comprising the display unit itself and a control panel for remote control of the scanning unit operation. The display unit is available in two sizes.

STATUS

In production.

CONTRACTOR

Thomson-CSF, Aerospace Group, 122 Avenue du Général Leclerc, 92105 Boulogne-Billancourt, France.

HUD sight repeater

4794.393

THOMSON-CSF VEH 3020 HOLOGRAPHIC HEAD-UP DISPLAYS

Thomson-CSF is developing a holographic head-up display (HUD) with a field-of-view of 30 × 20 degrees, and an increased combiner transmissibility and reflectivity. It consists of the VEH 3020 wide field-of-view holographic HUD and the TMM 1410 head-down display. Prototypes have been flight tested with successful results, and one of these is to be installed on board the Dassault Rafale experimental combat aircraft.

Thomson-CSF is also developing monochrome wide field-of-view collimated displays which the company intends to be associated with either the holographic HUD or the VEM 130 (**3135.393**).

STATUS

In development.

CONTRACTOR

Thomson-CSF, Aerospace Group, 122 Avenue du Général Leclerc, 92105 Boulogne-Billancourt, France.

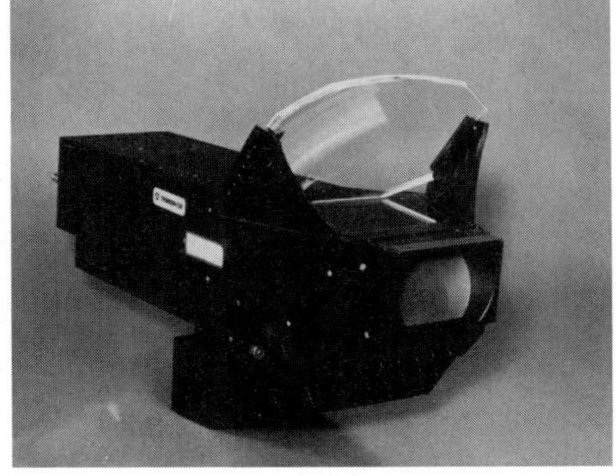

The VEH 3020/TMM 1410 combined displays

VH 100 head-up display for helicopters

6328.393

VH 100 SERIES HELICOPTER HEAD-UP DISPLAYS

Under the designation of VH 100, Thomson-CSF is developing a new series of head-up displays for helicopter use. These head-up displays will form part of the aircraft weapon control system for air-to-ground rocket firing, or for air-to-air missile fire control with such weapons as the Stinger or the Matra Mistral.

STATUS

The series has been undergoing flight trials and certification programmes on a Gazelle helicopter in France, and a Bell 406 helicopter in the United States. Eight successful firings were carried out on the Bell 406 with the Stinger AAM, and the VH100 has been certified for use in the Bell OH 58 of the US Army.

CONTRACTOR

Thomson-CSF, Aerospace Group, 122 Avenue du Général Leclerc, 92105 Boulogne-Billancourt, France.

6329.393

THOMSON-CSF HELMET MOUNTED SIGHTS

A series of helmet mounted sights and display systems is being developed by Thomson-CSF, using a variety of technical and physical configurations, which can be adapted to a wide choice of operational requirements. The series is suitable for combat aircraft and helicopters and has been flight tested.

For advanced helicopter programmes, Thomson-CSF is developing a very wide field-of-view helmet visor display.

STATUS

In development

CONTRACTOR

Thomson-CSF, Aerospace Group, 122 Avenue du Général Leclerc, 92105 Boulogne-Billancourt, France.

Thomson-CSF helmet-mounted sight

6777.393

CONVERTIBLE LASER DESIGNATION POD (CLDP)

Thomson-CSF's Optronic Department are developing an electro-optical Convertible Laser Designation Pod (CLDP) for the export market in which a laser designator is supplemented by either a TV or a thermal imager sensor.

The options are being offered because the company recognises that there are environmental limitations to thermal imagers operating in the 8 to 12 micron range in certain parts of the world, especially in the Middle East. The CLDP will feature a common body with laser transceiver, electronic assembly and environmental control system and two separate nose sections which can be changed in two hours.

The TV head will feature a TV camera and gimballed mirror and a roll-stabilisation device. It will also have a field-of-view (FOV) selector and a visible or near infra-red (0.7 to 0.9 microns) spectrum selector. This will have four magnifications (×2.5, ×5, ×10, ×20) and a 1.5° FOV and can be used in daylight irrespective of humidity conditions.

The TC head features a gimballed optical head with laser and thermal imager optics and roll-stabilisation device. There are two FOVs; 12° for navigation and 4° for target acquisition and tracking. This system is for use at night and in a very humid environment.

STATUS

Flight tests of the CLDP with TV head began in mid-1986 and the system is available for production. Flight tests of the TC head version began in January 1988 using a Jaguar aircraft. It is anticipated development should be completed by the middle of 1988.

CONTRACTOR

Thomson-CSF, General Avionics Division, Optronic Department, 52 rue Guynemer, 92130 Issy-les-Moulineaux, France.

5925.393

CROUZET TYPE 200 HELICOPTER SIGHT

The Type 200 lightweight, easy to install, user friendly Crouzet helicopter system has been developed to provide air-to-air defence and air-to-ground attack support for all types of helicopters.

It is an electro-mechanical head-up weapons sight showing a collimated reticle which can be moved between −10° and +7° in elevation and +6° in azimuth, angles compatible with the requirements for air-to-air and air-to-ground weapons launch.

The sighting system is based on a modular concept; the basic component being the sight head (single or dual-axis monocular system) which can be used for both day and night weapon firing in conjunction with second or third generation micro-channel night vision goggles.

The sight head is suspended from the canopy frame of the helicopter. With this system the pilot can fire the various weapons, including rockets and missiles, without any manipulation other than manual selection of functions in the control unit. The sight head weighs 2 kg and the control unit weighs 1.5 kg.

Sight recording by CCD camera has been validated for both training and operational firing and is available as an option.

The second component of the system is a control unit which combines command and calculation functions. It includes the necessary controls for moving the sight head, symbology animation controls, firing tables stored in an internal memory, weapon firing computation systems (guns and air-to-air missiles) and system interfaces. The system is equipped with a built-in automatic test facility, with status information being displayed on the sight head.

STATUS

The Type 200 helicopter sighting system has been tested on MBB BO 105, Sikorsky S76 Eagle, Hughes 500/530, Bell 412. It has been selected by French Army Light Aviation (ALAT) to equip the Gazelle armed with guns and very short range air-to-air missiles. It has also been chosen to equip the Ecureuil of the French Air Force.

CONTRACTOR

Crouzet SA, 25 rue Jules Vedrines, 26027 Valence Cedex, France.

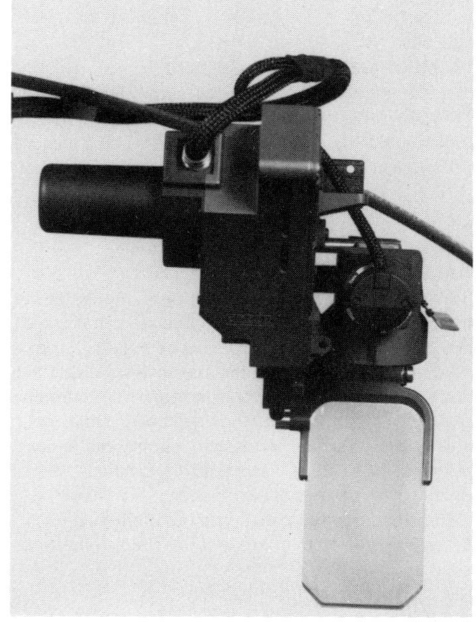

Crouzet Type 200 helicopter sighting system

6567.393

SUPER CYCLOPE AND CORSAIRE INFRA-RED SCANNERS

The Super Cyclope is a high performance line scanner used for airborne reconnaissance. The equipment can be either pod-mounted or integrated into the general configuration of the aircraft. The payload consists basically of an infra-red line-scan head operating in the 8 to 12 μm spectral bandwidth, a signal processing and addressing reporting unit for overall information management, and a magnetic recording unit. It equips the Mirage IV, Mirage F1, Mirage F1CR and Mirage 2000 aircraft.

Corsaire is a wide field-of-view infra-red reconnaissance system which is especially adapted for the operational requirements of the Canadair/Dornier CL289 RPV programme.

Both systems can be associated with a real-time transmission system which consists of a transmitter/encoder unit with its antenna, and a ground receiver station covering the surveillance, reception, demodulation and restitution facilities. Technical details on both systems are classified.

STATUS

Super Cyclope SCM 2450 is in service with Mirage F1CR aircraft and the Corsaire will enter production in 1988.

CONTRACTOR

SAT (Société Anonyme de Télécommunications), 41 rue Cantagrel, 75624 Paris Cedex 13, France.

6568.393

ASPIC INFRA-RED DETECTION SYSTEM

Aspic is an airborne infra-red equipment designed for the detection and acquisition of airborne targets at long range, and for tracking such targets at medium and short ranges as an aid to combat and fire control. The system employs a version of the SAT CTX family of thermal imaging cameras which are based on a parallel scanning technique whereby energy received by the camera's optical system is passed to a vertical strip of N detectors. Two consecutive horizontal sweeps enable a complete image to be formed which consists of two interwoven half-images, each with an N number of lines. A complementary electronics unit enables automatic tracking. No other details are available.

CONTRACTOR

SAT (Société Anonyme de Télécommunications), 41 rue Cantagrel, 75624 Paris Cedex 13, France.

Aspic infra-red detection system

GERMANY (FEDERAL REPUBLIC)

3853.393

CE626 LASER TRANSMITTER

This unit has been developed for the Panavia Tornado multi-role combat aircraft system.

The transmitter is a nitrogen-cooled neodymium-YAG laser operating with a variable pulse repetition frequency between 9 and 11 Hz and ascertaining together with the receiver the exact range to mobile and stationary targets. The display/operating and control unit and the corresponding power supply units form part of the laser transmitter.

Since its electronics and technology are tailored to system integration, it can be used in many fields, eg ships, aircraft, helicopters, tracked and wheeled vehicles as well as spacecraft. It has been developed for a PRF of 40 Hz.

STATUS

In production.

CONTRACTOR

Eltro GmbH, Postfach 102120, 69 Heidelberg 1, Federal Republic of Germany.

ISRAEL

3801.393

WDNS-391 WEAPON DELIVERY SYSTEMS

WDNS-391 is the name of a family of weapon delivery systems produced for fighter/bomber aircraft by Israel Aircraft Industries Ltd. They are designed to assist the pilot by presenting essential data on the head-up display (HUD) for use in navigation, air-to-air and air-to-surface combat missions.

Typically, the system comprises sensors, processors, display and control units. Among the sensors are radar, inertial measuring units, the angle-of-attack systems and a flux-valve compass. In addition to the central processor, computers will include an auxiliary computer and an air data computer. Display equipment may comprise the HUD, horizontal situation indicator, attitude direction indicator and a terminal unit.

The WDNS-391s used by the Israeli Air Force are capable of the following operating modes:

(1) Air-to-air: gunnery, including Hotline (snap shooting), and/or Hotpoint; missile fire; override of air-to-air mode.

(2) Air-to-ground: continuously computed impact point immediate or delayed release; continuously computed release point designated toss bombing (Des-toss), Dive-toss, Toss; manual altitude or range release.

(3) Navigation with possible storage of up to 40 waypoints.

(4) Integration with stores management and release subsystem.

It is possible to use the system to 'mark and store' targets of opportunity.

STATUS

In production and operational.

CONTRACTOR

Israel Aircraft Industries Ltd, Military Aircraft Marketing, Ben Gurion Airport, Israel.

6579.393

SLOS – STABILISED LONG-RANGE OBSERVATION SYSTEM

SLOS is a TV-based surveillance system, designed for long range optronic observation. It consists of a triple-magnification telescope mounted on the floor of the aircraft, and uses a gyro-stabilised mirror housed inside an external rotating viewing turret. The mirror can rotate in azimuth and elevation for landscape scanning. The coating of the mirror is effective over a very wide spectral range from the visual to infrared. This allows the optional use of lasers and infrared sensors in parallel with the visual optical path. The system is designed principally for use by turboprop aircraft, where it is installed under the belly, but can also be used on other types of fixed-wing aircraft and helicopters. A data link is provided for transmission of data to a ground station.

STATUS

In production for the Israeli Air Force.

CONTRACTOR

Israeli Aircraft Industries Ltd, Electronics Division, Ben Gurion Airport, Israel.

4241.383

LOW COST WEAPON DELIVERY SYSTEM (LCWDS)

Based on the company's experience in developing and producing systems for A-4, F-4 and Kfir aircraft operated by the Israeli Forces, Elbit have introduced a low cost weapon delivery system (LCWDS) equipment which represents a compromise between the expensive, full performance weapon delivery and navigation system and the very low cost, low performance gunsight. The LCWDS is designed to enhance aircraft operational performance capabilities, while the use of digital computation with a head-up display (HUD) improves pilot performance. It is particularly suitable for use on light attack aircraft, advanced trainers and interceptors.

Modes of operation provided include: air-to-ground, air-to-air, and navigation. The navigation mode requires the aircraft to be fitted with a doppler velocity sensor. Facilities provided in the air-to-ground mode are continuously computed impact point, CCRP and manual. There are two uses of the air-to-air mode, Hotline (snap shooting) and LCOS. Computation and display of steering commands for pre-selected waypoints is the main function of the navigation mode.

Elbit LCWDS units, left to right: central mission computer, head-up display, control terminal unit

The LCWDS is based on the installation of three items of equipment in the aircraft and interfacing them with existing aircraft sensors and indicators. The added equipment consists of: the central mission computer (CMC), containing a digital processor, symbol generator and aircraft interface unit; HUD; and the control terminal unit (CTU). Existing aircraft systems interfaced to the LCWDS comprise flux valve, angle of attack, air data computer and gyro reference unit sensors; control facilities such as an armament programmer, and stick and throttle switches; and displays (attitude direction indicator and the horizontal situation indicator).

CHARACTERISTICS

Accuracies: 10 mrad (A/G); 3 mrad (A/A) dependent on accuracy of aircraft gyro reference unit

MTBF: 800 h minimum

Dimensions: CMC 249 × 194 × 140 mm; CTU 167 × 175 × 127 mm; HUD 320 × 120 × 150 mm

Weight: CMC 9 kg; CTU 2.5 kg; HUD 5 kg

Environmental Conditions: CMC MIL-E-5400 Class II; CTU MIL-E-5400 Class IA; HUD MIL-E-5400 Class IA

STATUS

In production.

CONTRACTOR

Elbit Computers Ltd, PO Box 5390, Haifa 31053, Israel.

4242.383

SYSTEM 81 WDNS

The Elbit weapon delivery and navigation system (WDNS) 81 is a modern IMU-based digital computerised weapon delivery and navigation system for use in advanced combat aircraft for both air-to-air and air-to-ground operations. There are three principal modes of operation provided by the System 81: air-to-air, ground attack, and navigation. In the first of these, the missile launching envelope is computed and displayed to the pilot by a head-up display (HUD) for engaging missiles. When using guns in this mode, Hotline (snap shooting) is the normal method but there is also a lead computing optical sight (LCOS) back-up facility available. For delivering ground attack weapons there are facilities for continuously computed impact point (CCIP), CCRP, toss bombing (toss), release of 'smart bombs', guided munitions etc (special ordnance), and a direct back-up visual delivery capability. The navigation mode computes and displays steering commands for pre-selected destinations, with up to 40 waypoints.

These facilities enable a pilot to navigate to his target accurately and on time, to release a wide range of airborne weapons in 'uncanned' paths, obtain air-to-air missile firing envelope information presented on the HUD, employ 'snap shooting' and LCOS air-to-air firing modes, and to enter before-flight data and control commands. Other features include in-flight position and velocity update, rapid warm-up/alignment, and the design is based on the HOTAS (hands on throttle and stick) principle.

System 81 equipment comprises four main items: fire control and navigation computer (FCNC), inertial measurement unit, control terminal unit (CTU) and the HUD. The first of these comprises the following five units: central processing unit and program store, converter unit, symbol generator, core store and the power supply unit. The CTU is used for entering data to the central computer and the presentation of certain data from the FCNC to the pilot. The CTU controls system alignment, operation, test, and selection of sub-modes. Its external features can be seen from the nearby illustration. The HUD provides the pilot with all relevant operational information and is designed to employ the minimum number of signals and to have the least possible interference with the pilot's normal viewing. It provides warning information symbology.

Other aircraft systems, indicators and sensors which are linked with the System 81 central computer as peripherals include search radar (range, range rate, antenna position, target data, altitude, Mach number, aspect angle), radar altimeter, inertial measurement unit, flux valve heading reference, air data computer, ADF, attitude direction indicator, horizontal situation display, armament programmer, 'smart' weapons interface, stick and throttle switches.

CHARACTERISTICS

Central computer: 16-bit with 17 32-bit registers and 7 16-bit registers

Elbit WDNS 81 equipment with head-up display (top), control terminal unit (left) and computer (right)

Converter: A/D: 28 channels; D/A: 6 channels; S/D: 3 channels; Synchro buffers: 2 channels; Input discrete: 48 channels; Output discrete: 24 channels
Symbol generators: 3000 mrad/20 ms, straight lines, scales, digits, letters, circles
Core store: 16 K
Memory: EPROM 48 K × 16 bits; RAM 6 K × 17 bits; Non-volatile RAM 2 K × 17 bits
Dimensions: 273 × 382 × 195 mm
Weight: 20 kg
Control terminal unit
Serial interface to main computer: 1 Mbit/s (16-bit words)
Dimensions: 175 × 167 × 127 mm
Weight: 2.5 kg

STATUS

Production.

CONTRACTOR

Elbit Computers Ltd, PO Box 5390, Haifa 31053, Israel.

4002.383

SYSTEM 82 WDNS

The Elbit System 82 weapon delivery and navigation system is an enhanced version of the company's System 81 (**4242.383**, above), providing automated weapon aiming and aircraft navigation facilities for advanced combat aircraft employed on air-to-air and ground attack missions using missiles, guns, conventional bombs, and 'smart' weapons. Compared with the System 81, the WDNS 82 differs in having a stores management sub-system, consisting of an image (TV) display capability for all 'smart' weapons, a separate computer for stores management, a microprocessor-based back-up computer for the fire control and navigation computer (FCNC) and a CRT-based armament control and display panel.

The stores management computer (SMC) manages weapon release with maximum efficiency and safety, taking into account the conditioning and limiting factors. It includes the following units: dual redundant symbol generator, common memory, fault monitor and power supply unit. The back-up computer (BUC) calculates and generates the display of LCOS air-to-air and air-to-ground back-up mode functions to the FCNC. The armament control and display panel (ACDP) is controlled by both the pilot and the SMC. It displays on a TV-type screen continuously updated weapons inventory, video from a variety of sensors (such as forward looking infra-red imaging equipment fitted to the aircraft, or E-O sensors in the homing heads of 'smart' weapons carried by the aircraft), selected program parameters, and warnings in case of erroneous programming.

The extra capacity permits much greater exploitation of the weapons and other related systems and equipment fitted in the aircraft as well as allowing greater flexibility in the choice of weapons carried and sophistication in their employment. The system is also understood to incorporate a considerable growth potential. Where the System 82 characteristics differ from those of the System 81, they are listed in the table that follows.

CHARACTERISTICS

SMC
Memory: EPROM 58 K; RAM 3 K; Non-volatile RAM 1 K
Inputs: 160 discrete
Outputs: 120 discrete
Real-time clock: 1 kHz
Dimensions: 321 × 194 × 172 mm
Weight: 13.6 kg
ACDP
Scanning: 525 or 875 lines
Bandwidth: 20 MHz
Contrast ratio: 7:1
Dimensions: 172 × 147 × 267 mm
Weight: 7 kg
Back-up computer: Based on 8086 microprocessor
Memory: EPROM 16 K; RAM 1 K
Inputs: 8 analogue, 4 synchro, 10 discrete
Symbol generator: 2300 mrad/20 ms
Dimensions: 241 × 194 × 140 mm
Weight: 9 kg

STATUS

Production.

CONTRACTOR

Elbit Computers Ltd, Advanced Technology Center, PO Box 5390, Haifa 31053, Israel.

Elbit WDNS System 82 in KFIR C-7 cockpit

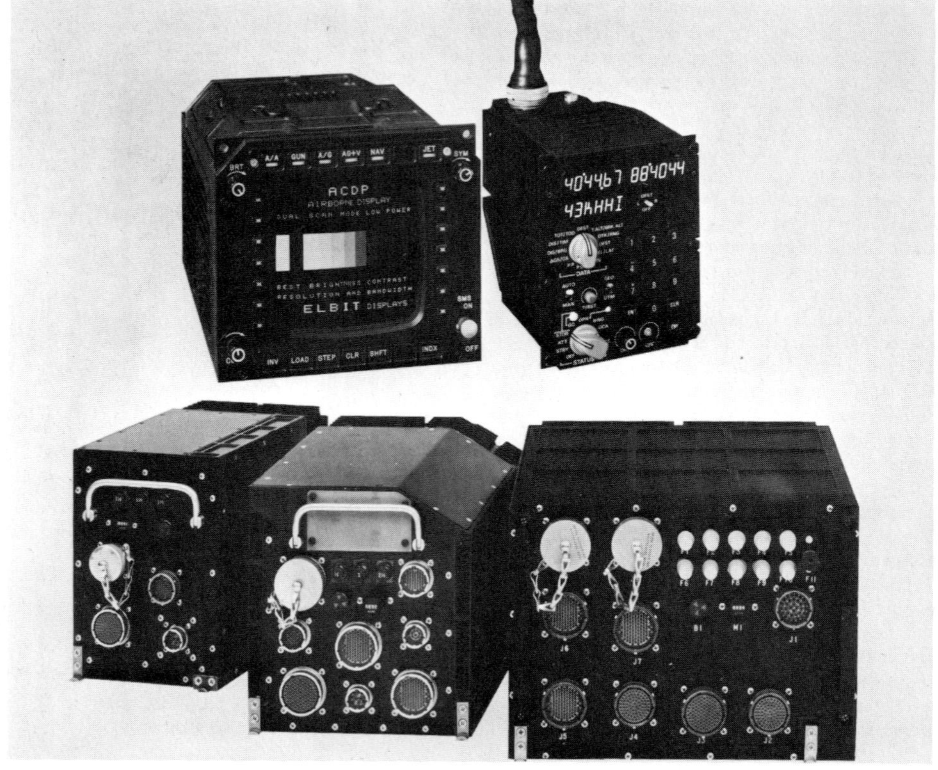

Elbit System 82 WDNS showing (left to right) armament control and display panel and control terminal units (top); back-up computer, stores management computer and fire control and navigation computer (below)

6332.393

ELBIT NIGHT VISION HEAD-UP DISPLAY

The Elbit night vision head-up display (HUD) is an advanced electro-optical system that combines the vision seen through the night vision goggles with computer-generated graphic and digital symbology. Vital information is presented in front of the pilot, such as heading, attitude and navigational data as well as other pertinent flight details collimated with the external view. The HUD is intended primarily for helicopter operations and forms part of a computerised cockpit system to offer improved performance in flight management, navigation guidance, communication control and reduction in pilot workload.

The information gathered by the computer from the helicopter's various systems is processed and displayed on a CRT. From there the symbology is combined with the image seen through the night vision goggles. The system is lightweight with a total weight on the helmet of 150 grammes, and features symbology in colour contrast to the image.

STATUS

In development.

CONTRACTOR

Elbit Computers Ltd, Advanced Technology Center, PO Box 5390, Haifa 31053, Israel.

6569.393

DASH (DISPLAY AND SIGHT HELMET) SYSTEM

DASH is a helmet display and sighting system which has been designed to enhance overall performance and achieve superiority in air-to-air combat. The basic design feature is to exploit the pilot's head direction line-of-sight to direct aircraft systems, ie radar, missiles, etc, to the target and lock them on to it.

The system measures the pilot's line-of-sight in relation to the aircraft and transfers this information to other aircraft systems. Aircraft sensors, avionics and weapons are thereby enslaved to the target. With DASH, the lock-on envelope is expanded and aerial engagement is shortened. Crucial information such as line-of-sight, status, flight information and warnings is displayed directly in front of the pilot on his helmet visor, eliminating the need to glance inward at the instrument panel or head-up display.

DASH consists of four major sub-assemblies; a helmet assembly including standard helmet, sensor, display extension cable and visor protective cover; a radiator with cockpit compensation unit; a control panel; and a mission computer.

CHARACTERISTICS

Instantaneous field-of-view: 12 to 15°

Angular coverage: ±160° azimuth; ±70° elevation; ±60° roll

Resolution: 500 TV lines

Contrast: 1.2 at 10 000 ft/lux

Added weight on helmet: 150 grammes

STATUS

The first series production system has been in operation on board an F-15 aircraft since October 1986.

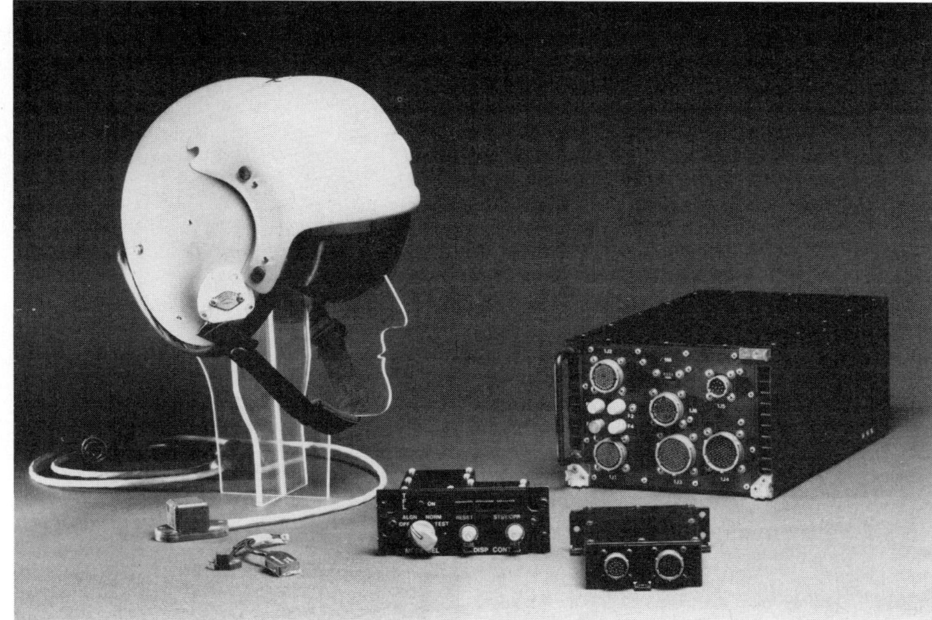

Units of the DASH system

CONTRACTOR

Elbit Computers Ltd, Advanced Technology Center, PO Box 5390, Haifa 31053, Israel.

6570.393

HALO HELICOPTER AVIONICS

HALO is an advanced avionics suite enabling a helicopter pilot to fly at night, at low level and high speed, with increased safety. The system ensures accurate navigation, as well as threat and obstacle avoidance.

The HALO system integrates three basic systems: night vision goggles/head-up display which provides flight symbology collimated with the external view; a Doppler radar; and a Global Positioning System (GPS). The main features of the equipment include digital computers, multi-function cockpit displays for both the pilot and co-pilot, a head-up display projected on to night vision goggles, digital discrete communication and an in-flight data recorder. A control/display unit provides overall management of the system.

STATUS

Deliveries to the Israeli Air Force were scheduled for mid-1988.

CONTRACTOR

Elbit Computers Ltd, Advanced Technology Center, PO Box 5390, Haifa 31053, Israel.

6333.393

ELOP HEAD-UP DISPLAYS

ELOP provides a variety of head-up displays (HUD) which can be installed in many types of fixed and rotary wing aircraft. The combined electro-optical equipment in front of the pilot's head provides him with the necessary flight information, and a computerised weapon delivery and/or navigation data is available for his selection while searching or tracking targets. Optionally, as a background, a video picture can be projected in rastermode, based on FLIR or TV-type sources to provide enhanced adverse weather/night capability.

To reduce pilot workload, while in combat conditions, the Up Front Control Panel (UFCP) can be adapted as an additional attachment to the HUD, giving the pilot the ability to communicate with the mission's computers and other avionics systems, without having to look down into the cockpit.

CHARACTERISTICS

Total field of view: Up to 30° horizontal and 22° vertical

Aperture of exit lens: 2.5 – 7 inches diameter

STATUS

Several models in production.

CONTRACTOR

ELOP (Electro-Optics Industries Ltd), Advanced Technology Park, Kiryat Weizmann, PO Box 1165, Rehovot 76110, Israel.

ELOP head-up display with UFCP, for the Israeli Kfir aircraft

6334.393

BLUE THUNDER HADAS (HELMET AIRBORNE DISPLAY AND SIGHT)

The ELOP Blue Thunder HADAS is a combined electronic/optical system providing holographic visual display of navigation and delivery flight information (raster/stroke) on the pilot's helmet visor. In order to ensure a lightweight helmet, information gathered by the system computer is first processed and displayed on a CRT. The CRT picture is then transmitted to a normal-sized helmet visor by fibre-optic techniques.

The system features, and operational capabilities include:

(a) Information on point sensors, turreted weapons, missile seekers and crew members cueing signals to the operator's line-of-sight by simply looking at the target.

(b) Display of sensor, weapon, missile and crew line-of-sight information for target acquisition by the pilot.

(c) Operation with global head-up display capability, allowing the display of collimated symbols throughout the entire field-of-view, representing critical flight, navigation and fire control information.

(d) Display of normal sized video imagery from night vision sensors (FLIR) directly in front of the operator's eye.

(e) Reduction in weapon launch time and improvement in weapon delivery accuracy.

(f) Reduction in workload of the crew members.

CHARACTERISTICS

Instantaneous field of view: At least 30° horizontal by 22° vertical

Field of regard: As wide as the ability to move the head in the cockpit

Contrast ratio: 1.3 against 10 000 fL ambient

Shades of gray: 6 against 50 fL ambient

STATUS

In development.

CONTRACTOR

ELOP (Electro-Optics Industries Ltd), Advanced Technology Park, Kiryat Weizmann, PO Box 1165, Rehovot 76110, Israel.

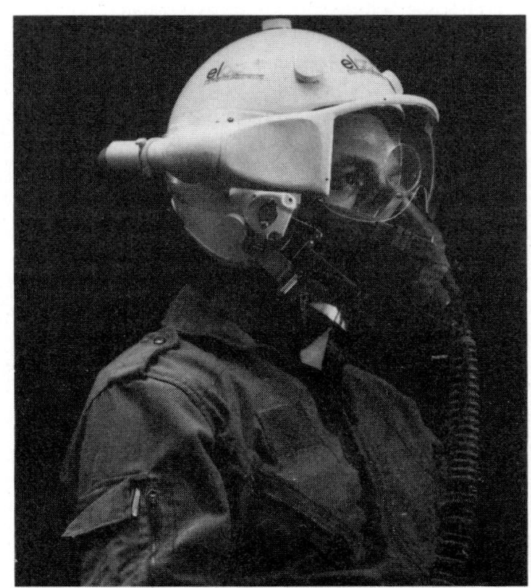

HADAS helmet display and sight

6778.393

LITENING INFRA-RED TARGETING AND NAVIGATION POD

Litening advanced airborne infra-red pod for targeting and navigation purposes is a low cost modular system, based on a miniature FLIR. Different versions are available for various applications, such as night vision, navigation and targeting, from a variety of platforms, including light aircraft, helicopters and high performance aircraft. The system is fully operational 24 hours a day, and in adverse weather conditons. Litening incorporates the MTIS high performance FLIR with three fields-of-view. The system has modular electronics design, featuring line replaceable units (LRUs) and interfaces with Mil-Bus 1553 B.

The FLIR is mounted on a four axis stabilised platform, with a 10°/s LOS slewing rate. Miniaturisation of the payload allows for smaller dimensions, lighter weight, modularity and low cost, but with performance equivalent to, or higher than competitive systems.

Rafael has past experience with electro-optical equipment for day and night applications. Litening makes extensive use of existing subsystems and know-how, along with manufacturing and logistic support facilities.

Because of its compact design, lightweight, stabilised image and wide field-of-view, Litening is a powerful instrument for both day and night navigation.

STATUS

Under development.

CONTRACTOR

Rafael Armament Development Authority, PO Box 2082, Haifa 31021, Israel.

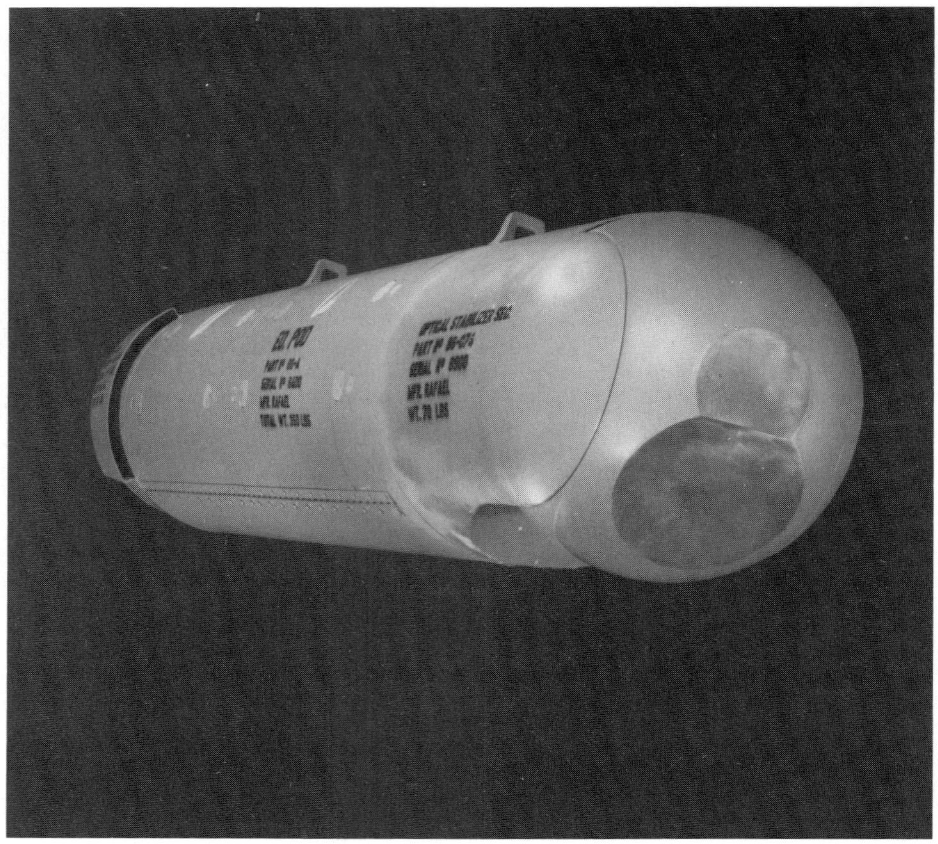

Mock-up of the Rafael Litening pod

ITALY

6335.393

P0702 AIRBORNE LASER RANGEFINDER

The P0702 is a variant of the P0701 laser rangefinder (**6326.293**) developed by FIAR for airborne applications. It is made up of a single unit which includes the laser transmitter, receiver, cooling system, processing electronics, lenses and power supply. Communication between the rangefinder and the fire control system is carried out by a single cable.

CHARACTERISTICS

Type: Nd:YAG

Wavelength: 1.064 μm

Output power: >2 MW

Pulse duration: 12 ± 5 ns

Beam divergence: 1 mrad nominal

Repetition frequency: 10 Hz

STATUS

In production.

CONTRACTOR

FIAR (Fabbrica Italiana Apparecchiature Radioelettriche) SpA, Via Montefeltro 8, 20156 Milan, Italy.

SWEDEN

3819.393

HELIOS HELICOPTER OBSERVATION SYSTEM

The Saab Helios is a roof-mounted observation system for scout/rescue helicopters. The system has × 3 and × 12 magnification, high resolution optics (by Pilkington P.E., UK) and high stabilisation accuracy. The eyepiece arm protruding into the cabin can be stowed close to the roof when not in use and the eyepiece height is adjustable. The system itself comprises five units; roof-mounted sight head, electronics unit, control panel, control unit and line-of-sight indicator.

Helios has provision for a laser rangefinder/designator and a thermal imaging system for night vision. It can accept commands from a laser-warning receiver, helmet sight or avionic systems for accurate target positioning, and generate outputs for weapon aiming. A camera port for a film- or CCD-camera is included to allow recording of the screen.

HELITOW Weapon Sight

The HELITOW sighting system has been developed from Helios and, together with the HELITOW missile launching and guidance sub-system (by Emerson Electric, USA), it forms a helicopter system for firing any TOW missiles. The system is fully prepared for introduction of IR vision and a laser rangefinder/designator. A TOW missile simulation system BT49H, by Saab Training Systems, can be attached for cost effective gunnery training.

STATUS

In production for the Swedish, Danish and Italian armies.

CONTRACTOR

Saab Instruments AB, Box 1017, S-551 11 Jönköping, Sweden.

Helios sight

1730.393

RGS 2 OPTICAL SIGHTING SYSTEM

The RGS 2 is a three-unit optical sighting system comprising a sight head, a computer and gyro unit and a control unit.

Following a modular design concept and using servo techniques, all but the most important elements of the optical system have been eliminated from the sight head and put in the computer and gyro unit. Likewise most of the controls are found on a separate module (normally installed adjacent to the weapon fire control system in the aircraft).

The results of this design concept are a very compact sight head base enabling it to be installed into the smallest cockpit and the most crowded instrument panel. Despite the small dimensions, the sight head is designed to provide a full two-eyed viewing capability through the combining glass.

STATUS
The Saab RGS 2 is certified for 15 different types of aircraft and has been in operational service for nine years.
CONTRACTOR
Saab Instruments AB, Box 1017, S-551 11 Jönköping, Sweden.

Twin installation of Saab RGS 2 sights in Italian Aermacchi 339 jet trainer aircraft

5745.393
ERICSSON LASER RANGEFINDER
A compact airborne high repetition rate laser rangefinder has been developed by Ericsson for easy up-grading of existing systems. It can be integrated into any sighting and weapon delivery system to give higher accuracy and enhanced safety in dive attacks. The rangefinder uses a modular design consisting of transmitter, receiver, range counter and deflection unit. The laser is aimed at the target by slewing the deflection unit to the aircraft sighting system.

The transmitter is a Nd:YAG type with a PRF of 5 Hz continuously and 10 Hz in bursts. Typical range is 7 km or greater at an optical visibility better than 20 km. The range counter resolution is 5 m.
CONTRACTOR
Ericsson Radio Systems AB, 16380 Stockholm, Sweden.

3387.353
GREEN BARON POD
The Green Baron pod is a reconnaissance system for all-weather missions. It is equipped with a Texas Instruments infra-red linescanner and Omera 70 mm panoramic camera. For medium range reconnaissance missions a 5-inch Omera camera with a focal length of 300 mm is mounted in the rear part of the pod. This aft part is turnable in roll during flight to any oblique or vertical position for maximum tactical flexibility.

In the front of the pod a forward looking camera, FLIR or LLTV can be mounted. The pod has a flexible control system with the possibility to use information regarding positions in roll, angle of attack, altitude, etc, either from the aircraft or from sensors mounted in the pod.
CHARACTERISTICS
Dimensions: Length 2840 mm, diameter 500 mm
Weight: 230 kg
Power: 200 V 400 Hz and 28 V DC
STATUS
Flight tested on Saab J35 Dragon aircraft. Discussions are underway concerning its use with the JAS-39 combat aircraft.
CONTRACTOR
FFV Aerotech, 581 82 Linköping, Sweden.

FFV Green Baron pod

3001.353
RED BARON POD
The Red Baron pod is a reconnaissance system for tactical operations at high speed and low level during day and night. It contains a Texas Instruments infra-red linescanner, roll-stabilised and provided with oblique slew capability. The pod is self-contained to a large extent, eg it is equipped with a vertical reference gyro and radar altimeter to supply signals for roll stabililsation and film velocity control.
CHARACTERISTICS
Dimensions: Length 2955 mm, diameter 400 mm
Weight: 140 kg
Spatial resolution: 0.5 mrad
Power: 115 V 400 Hz 3-phase and 28V DC
STATUS
Sold to a number of countries including Denmark and Switzerland.
CONTRACTOR
FFV Aerotech, 581 82 Linköping, Sweden.

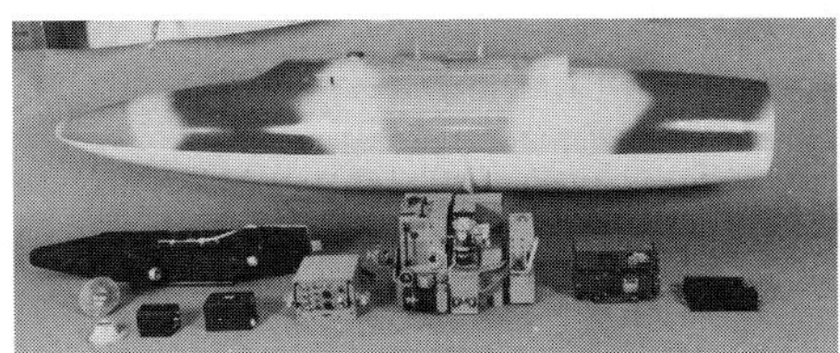

Red Baron pod

6374.393
HELICOPTER SELF SCREENING LAUNCHER
The FFV Helicopter Self Screening Launcher (HSSL) is an optical countermeasures system, designed to cope with urgent protection requirements in both forward flight and hovering modes. It lays down a covering smokescreen 100 m ahead of the helicopter within 2 seconds of actuation. To minimise the recoil effect, the screen is deployed by rockets and is efficient throughout the spectral range 0.4 to 15 μm. The standard version HSSL unit contains ten smoke rounds in two launcher sections. In addition to, or in place of, the visual/IR smoke type rounds, the ammunition payload may contain IR or radar decoys.
CHARACTERISTICS
Rocket calibre: 84 mm
Weight: 7 kg (empty launcher section); 2.8 kg (rocket including launch tube)
Dimensions: 830 mm wide × 125 mm high
CONTRACTOR
FFV Aerotech, 732-81 Arboga, Sweden.

Helicopter Self Screening Launcher

UNITED KINGDOM

3057.393

F-16 A/B HUD SIGHT

The head-up display (HUD) system in the General Dynamics F-16 A/B model aircraft consists of a pilot's display (PDU) and the electronics unit (EU). The PDU is ruggedised to withstand the airloads generated in the canopy-off situation. The EU provides the aircraft interface, symbol generation and air-to-air weapon aiming computation.

Variants of the system have been adapted for a number of other aircraft types with minor modifications. The majority of variants have exploited the computing power of the EU to provide full air-to-air and air-to-ground weapon aiming solutions using algorithms suitable for current weapon inventories.

STATUS

In full production at GEC Avionics, Atlanta, Georgia facility. In addition, full scale production of electronics units by Kongsberg Vapenfabrikk, Norway and pilot display units at NV Optische Industry, The Netherlands has been instituted for NATO F-16 A/B models.

CONTRACTOR

GEC Avionics Ltd, Airborne Display Division, Airport Works, Rochester, Kent, England.

GEC Avionics F-16 HUD sight

6336.393

F-16 C/D HUD SIGHT

The F-16 C/D model head-up display sight is a dual mode system which in addition to supplying all the facilities of the A/B HUD sight system (**3057.393**) extends the aircraft attack capability in the night mode. The pilot's display unit provides a raster picture of the scene from an on-board sensor such as an FLIR. The flight symbology is overwritten cursively on this outside world scene image such that the independent control of symbology brightness relative to the raster image is possible, in addition controls are provided to control the brightness and contrast of the raster scene. This facility allows existing weapon aiming solutions to be used for night ground attack, and the system is capable of modification to fit most fighter aircraft. The pilot's display unit has a total field of view of 25 degrees and the electronics unit has 1750A processor architecture, 48 K of memory and a 1553B databus.

STATUS

In full production at GEC Avionics, Rochester and Atlanta. A variant of the F-16 C/D sight has been retrofitted to the A-7D aircraft for use by the US Air National Guard. Another variant, the F-16 LANTIRN sight is described under entry **4314.393**.

CONTRACTORS

GEC Avionics Ltd, Airborne Display Division, Airport Works, Rochester, Kent, England.

GEC Avionics Inc, 2975 Northwoods Parkway, Norcross, PO Box 81999, Atlanta, Georgia, USA.

F-16 C/D HUD system

4314.393

F-16 LANTIRN HOLOGRAPHIC HUD SIGHT

The LANTIRN head-up display sight was developed in response to a USAF requirement for a dual mode HUD system which provided a large field of view. The pilot's display unit has a field of view of 30 degrees in azimuth by 17 degrees in the vertical. It is constructed with a holographic combiner and features a generous head motion box which has proved completely operationally acceptable during low level night missions.

This HUD is considered essential in the LANTIRN (Low Altitude Navigation and Targeting Infra-Red for Night) system and is claimed to be the only fully mission proved wide angle holographic HUD in existence at this time.

STATUS

In full production. A contract valued at $72 million was awarded to GEC Avionics in November 1986 to supply more than 450 sights for the USAF F-16C aircraft. Deliveries were due to begin in spring 1988. Prototypes of the HUD have flown over 1000 sorties with the F-16 LANTIRN Combined Test Force.

CONTRACTOR

GEC Avionics Ltd, Airborne Display Division, Airport Works, Rochester, Kent, England.

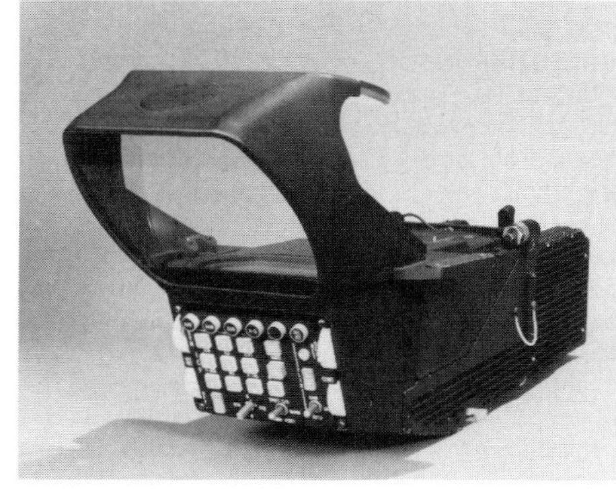

GEC Avionics HUD sight for F-16 LANTIRN aircraft

6337.393

F-5 E/F HEAD-UP DISPLAY WEAPON AIMING COMPUTER (HUDWAC)

GEC Avionics has developed a head-up display and weapon aiming computer (HUDWAC) for the Northrop F-5E and F-5F that, as an avionics upgrade, allows the performance and manoeuvrability of the F-5 to be combined with the advantages of 'head-up, eyes-out' flight.

The F-5 HUD uses technology and hardware from other GEC Avionics HUD programmes, and performs air-to-air and air-to-surface weapon aiming calculations, as well as delivering symbology for flight and navigation modes. The HUD comprises a pilot display unit (PDU), an electronics unit (EU), and a weapon data input panel (WDIP).

The PDU fits into an existing mounting tray. The PDU's 25° total and 15.75° vertical by 16.73° azimuth instantaneous fields of view are achieved without impinging on the F-5's ejection plane. A 16 mm cine camera is included in the PDU, but a TV camera is optionally available. The PDU control panel includes switches for controlling symbol brightness, standby sight selection, and symbol declutter. Data entry is via a key pad on the same control panel.

The EU will fit in all existing configurations of the F-5 E/F in the same location as the LCOSS gyro lead computer, which is removed for HUDWAC installation. It contains power supply integral cooling fans, and the circuit cards for processing, symbol generation, and interfacing with the F-5 avionics systems. The HUDWAC in its baseline configuration interfaces with the F-5 power supply, CADC, AHRS, fire control system, AN/APQ-153/-157 or -159 radar, and LN-33 INS. Optionally the HUDWAC will interface with the AN/APG-69(V) radar and any inertial navigation system. The system software provides weapon aiming calculations for weapons certified for F-5 carriage and is written to display symbology which complies with MIL-STD-1787. The system has the capability to handle AIM-9-4 line-of-sight and off-boresight aiming.

F-5 E/F head-up display weapon aiming computer (HUDWAC)

The system operates in three modes; navigation, surface attack and air combat. In navigation (and a landing sub-mode), the display symbology shows speed, altitude, heading, velocity vector, pitch, bank, Mach no, and *g*. INS information, when available, is also displayed. Surface attack modes include Continuously Computed Impact Point (CCIP), Continuously Computed Release Point (CCRP) displays. There are three air-to-air modes; Missiles, Lead Computing Optical Sight (LCOS), for tracking 20 mm cannon attacks, and Dogfight, which combines Missiles, LCOS, and the snapshoot or Continuously Computed Impact Line (CCIL) on the same display.

The system has built-in test facilities to isolate faults to a particular line replaceable unit. Predicted mean-time-between-failures for the HUDWAC is in excess of 750 hours. The support equipment includes a semi-automatic test system set which uses a 16-bit microprocessor.

STATUS

In production. The system is available through the United States FMS organisation, by direct sales, or through the UK/MS organisation.

CONTRACTORS

GEC Avionics Ltd, Airborne Display Division, Airport Works, Rochester, Kent, England.

GEC Avionics Inc, 2975 Northwoods Parkway, Norcross, PO Box 81999, Atlanta, Georgia, USA.

6571.393

A-6E/F HUD SIGHT

This sight has been designed to retrofit into the A-6E/F aircraft. A dual mode sight with a unique optical design, it provides a total field of view of 30 degrees in azimuth by 23.5 degrees in elevation. The sight mounts a dual combiner with graded coatings which guarantee continuity of the outside world scene as viewed across the complete field of view. The pilot's display unit is driven by an electronics unit which has been developed from an early successful design, and is capable of providing full mission weapon aiming computation. The small size and low profile of this system makes it suitable for retrofit into most fighter aircraft in order to provide a night attack capability.

STATUS

In flight evaluation at the United States NATC, Patuxent River.

CONTRACTOR

GEC Avionics Ltd, Airborne Display Division, Airport Works, Rochester, Kent, England.

Head-up display sight for the A-6E/F aircraft

6572.353

MIG-21 HUDWAC SYSTEM

The Type 006 head-up display/weapon aiming computer (HUDWAC) system has been designed as a direct replacement for the existing gunsight in the MiG-21 aircraft. Two line replaceable units constitute the HUD set, a pilot's display unit (PDU), and an electronics unit.

The PDU mounts a full up-front control panel with a keyboard, plus housing for a cockpit video camera. The pilot's display unit fits into a mounting tray which is pre-boresighted to allow PDU change without the necessity for re-harmonisation, a feature of all GEC Avionics HUDs. The electronics unit acts as a symbol generator, weapon aiming and navigation computer, and interface unit to other aircraft systems.

The system operates in three modes; navigation, surface attack and air combat. Weapon delivery accuracies are increased by a factor of four by the retrofit of this equipment.

CONTRACTOR

GEC Avionics Ltd, Airborne Display Division, Airport Works, Rochester, Kent, England.

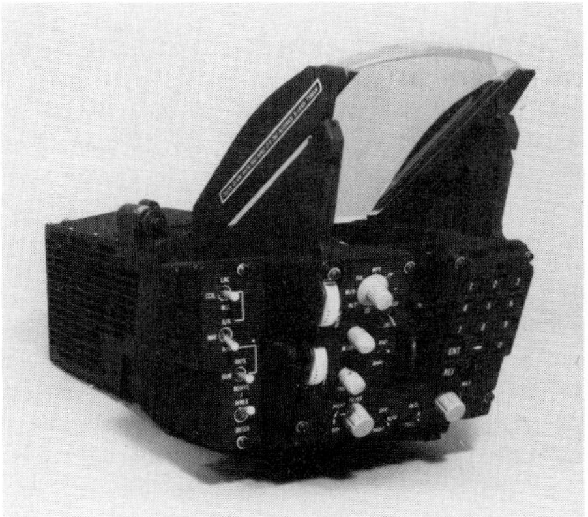

HUDWAC sight for the MiG-21 aircraft

4726.393

THERMAL CUEING UNIT

The GEC Sensors Thermal Cueing Unit (TCU) is intended to assist the pilot to detect a potential target at the earliest moment, particularly when flying low and fast over unfamiliar terrain by day or night.

Used in conjunction with a forward looking infra-red (FLIR) imaging sensor and a head-up display (HUD), the TCU alerts the pilot to the presence of up to 16 tracked targets simultaneously which he might otherwise not detect. It extracts the location of up to eight potential targets from the FLIR video signal and generates cue symbols for display at the appropriate position on the HUD, overlaying the real scene by day or the synthetic FLIR image by night.

In sophisticated systems, target position co-ordinates generated by the TCU and designated by the pilot can be further processed to allow automatic target acquisition and weapon handover. Position data can be used to direct a narrow field-of-view imager at a point of interest for target recognition on a head-down display (HDD).

The TCU, which weighs 12 kg, is fully compatible with the UK Thermal Imager Common Module (TICM) system (**4280.193**) and the GEC Avionics LANTIRN HUD (**4314.393**). Alternative input and output interfaces can be made available to allow the TCU to operate with other types of FLIR sensor and HUD.

In systems where a thermal image is not required on any display, the TCU can operate directly from the FLIR scanner and pre-amplifiers, thus avoiding the need for scan conversion to raster display format.

STATUS

The Thermal Cueing Unit is now in development following trials and evaluation, under UK MoD contract. It is included in a number of FLIR systems being developed by GEC Sensors – see entry **5748.393**.

CONTRACTOR

GEC Sensors Ltd, Electro-Optical Systems Group, Christopher Martin Road, Basildon, Essex SS14 3EL, England.

Thermal Cueing Unit display

5748.393

GEC SENSORS FLIR SYSTEMS

The Electro-Optical Surveillance Division of GEC Sensors has developed a number of forward looking infra-red (FLIR) systems for a variety of applications. These have been developed using the Class II thermal imager modules described under entry **4280.193**.

Internally-mounted FLIR system

This system is being developed for the UK Tornado IDS and Harrier GR5 aircraft under a contract from the UK MoD. The same system will also be installed in the US Marine Corps AV-8B aircraft by McDonnell Douglas. A number of successful flights have taken place in the United States and the United Kingdom with prototype systems.

The FLIR system, based on the Class II thermal imager, is installed inside the fuselage with the optics looking forward through a small blister near the front of the aircraft. It provides an automatically optimised high resolution image of the terrain ahead, with automatic cueing of potential targets, for viewing on the pilot's HUD. Regarded as a major 'force multiplier' this passive system will enable close support missions to be undertaken at night and in poor weather, and will significantly increase front line aircraft utilisation.

Atlantic FLIR system

The Atlantic FLIR is a pod-mounted system designed to give night and poor weather capability to ground

GEC Sensors Atlantic FLIR System on an F-16

attack aircraft on high speed low level missions.

A new carbon-fibre composite pod designed to house the modular FLIR is being developed.

Atlantic has been involved in on-going flight trials and demonstrations in the USA and Europe, fitted to a General Dynamics F-16 aircraft to assess suitability for export F-16s.

Navigation FLIR pod

The Navigation FLIR pod is a flexible solution to the pilot night vision and target detection requirements in a variety of aircraft. The high sensitivity image can be viewed on either a head-down or a head-up display with a 1:1 overlay on the outside world. The equipment is designed for role flexibility and easy retrofit on the aircraft fuselage or wing stores pylon.

STATUS

The Navigation FLIR pod is in production. The other two systems are in development. A £48 million contract was received from the UK MoD in December 1985 for the FLIR system to be installed in the Tornado IDS and Harrier GR5 aircraft. This order also covers the US Marine Corps AV-8B aircraft.

CONTRACTOR

GEC Sensors Ltd, Electro-Optics Systems Group, Christopher Martin Road, Basildon, Essex SS14 3EL, England.

Navigation FLIR pod with front cover removed

6338.393

SEA OWL PASSIVE IDENTIFICATION DEVICE SENSOR TURRET

The Passive Identification Device (PID) is under development for installation on Lynx helicopters of the Royal Navy to provide a day or night long range target detection and identification capability. The system, named Sea Owl, uses a GEC Sensors thermal imaging sensor mounted in a highly stabilised turret platform sited on the nose of the aircraft.

The high resolution image is automatically optimised for maximum picture quality and is then displayed in the cockpit. The employment of advanced signal processing provides potential automatic target cueing to reduce cockpit workload. The system will be completely integrated with the aircraft avionics, including auto operation and manual control via a joystick, if required.

STATUS

In development. A £34 million contract for the system was received in October 1985.

CONTRACTOR

GEC Sensors Ltd, Electro-Optical Systems Group, Christopher Martin Road, Basildon, Essex SS14 3EL, England.

Sea Owl System mounted in a Lynx helicopter

6573.393

MULTIROLE FLIR TURRET

The Multirole FLIR turret is a compact and lightweight (35 kg) system for installation on a variety of fixed wing aircraft and helicopters. It consists of a gyro-stabilised and steerable sensor platform into which is mounted a TICM II based thermal imager, a continuous zoom telescope from ×2.5 to ×10 and a mini compressor cooling unit.

The system is especially designed for airborne roles and will enable operations such as surveillance, target detection, and search and rescue to be carried out at night or in poor visibility conditions.

The turret employs production standard UK thermal imaging modules (TICM II) (**4280.193**). The field-of-view ranges from 9 × 3.3° to 19.4 × 13° with 525 line systems and from 6 × 4° to 24 × 16° with 625 line systems.

STATUS

In development. The turret has been selected as the sensor package for the Phoenix RPV and has also been supplied to a European country for a similar project.

CONTRACTOR

GEC Sensors Ltd, Electro-Optical Systems Group, Christopher Martin Road, Basildon, Essex SS14 3EL, England.

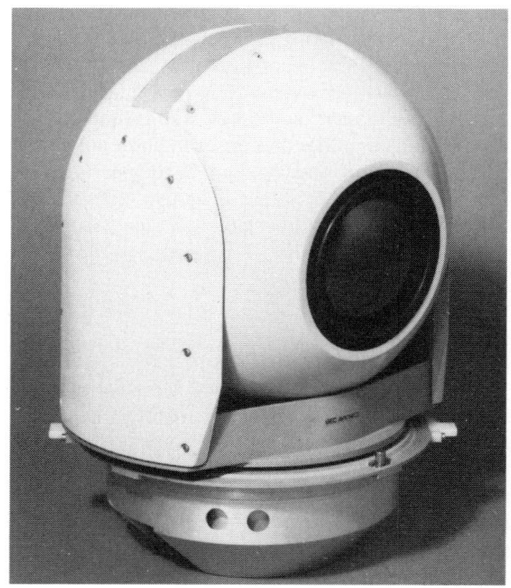

Multirole FLIR turret

6340.393

FRAMING THERMAL IMAGER

The GEC Sensors Framing Thermal Imager (FTI) is a small, lightweight, high resolution, low-cost thermal imager designed primarily for use in unmanned vehicle applications. The unit has its own framing element which allows generation of a frame scene and makes the operation of the equipment independent of the airborne vehicle's motor.

The FTI contains the sensor itself as well as the processing electronics. The sensor is of conventional design, the low cost and high resolution (1024 × 640 pixels) being achieved by the use of a low frame rate. The effects of visual distortion resulting from vehicle motion are minimised automatically by electronic compensation. Because of the low frame rate, the video information is stored in a digital frame store and then read out and transmitted to the ground station by means of a narrow bandwidth data link. The processing electronics unit contains all the circuitry required to sample and process the video image data prior to transmission. The high resolution of the FTI, with suitable electronic processing on the ground offers a high probability of target detection and recognition.

STATUS

The FTI has been developed under a UK MoD contract and has been undergoing evaluation flight trials.

CONTRACTOR

GEC Sensors Ltd, Electro-Optical Systems Group, Christopher Martin Road, Basildon, Essex SS14 3EL, England.

3795.093

GEC SENSORS CCD CAMERAS

GEC Sensors manufactures a range of TV camera equipment based on the use of solid-state area imaging devices known as charge-coupled devices (CCD). A practical feature of this equipment is the combination of very small dimensions and extreme ruggedness. Among other valuable attributes are low power requirements and long operating life.

CCDs are said to yield an order of magnitude improvement over vidicon camera tubes in dynamic range; they have the broad spectral range of silicon, are free from microphonic effects, have zero lag and inherent geometric accuracy. Cameras based on CCDs can be used with standard TV monitors and video tape recorders and can be employed as discrete items or as part of a larger system.

A typical application is the AN/AXQ-16V cockpit TV system which provides a real-time recording of gunsight head-up display image data, as well as cockpit audio data, and can provide a real-time presentation of visual data on a rear cockpit display. The system is based on an all-solid-state TV camera which uses a high resolution CCD photosensor array to achieve outstanding sensitivity, resolution, and dynamic range. A split camera design minimises obstruction of the pilot's view, the small camera head being installed on the head-up display.

The use of an on-board video casette recorder (V-1000 AB-R) ensures that a complete video and audio record of engagements is available for immediate replay as soon as the aircraft lands.

Monochrome and colour variants of these systems can be supplied for installation in many aircraft types and fitments.

STATUS

The system is flight proven, qualified and now in production for military aircraft worldwide.

CONTRACTOR

GEC Sensors Ltd, Electro-Optical Systems Group, Christopher Martin Road, Basildon, Essex SS14 3EL, England.

3056.393

AV-8A HEAD-UP DISPLAY AND WEAPON AIMING COMPUTING SYSTEM

The Smiths Industries HUDWAC system which is currently fitted to the AV-8A aircraft of the USMC and the Matador aircraft of the Spanish Navy provides a comprehensive display of weapon aiming, flight and navigation symbology, as well as performing all the weapon aiming calculations in a separate dedicated computer. The basic system comprises a pilot's display unit (PDU), waveform generator (WFG), interface and weapon aiming computer (IWAC), pilot's control panel (PCP) and extra high tension unit (EHTU). Further details will be found in the 1987-88 edition.

The original WFG and IWAC are now being replaced by a signal data converter (SDC) and digital interface and weapon aiming computer (DIWAC), both of which have been designed by the company's Clearwater Division in Florida, USA. The new units are 'form, fit and function' replacements of the original units but employ later technology.

Extra High Tension Unit: The EHTU provides a high voltage supply to the CRT in the PDU. To maintain the high level of accuracy required for weapon aiming, the 15 kV output from the EHTU has a very high degree of stabilisation over the full operating temperature range.

STATUS

In production.

CONTRACTORS

Smiths Industries, Aerospace & Defence Systems Company, Cheltenham Division, Bishops Cleeve, Cheltenham, Gloucestershire GL52 4SF, England.

Smiths Industries Aerospace & Defence Systems Inc, PO Box 5389, Clearwater, Florida 33518, USA.

4130.393

AV-8B HEAD-UP DISPLAY

Smiths Industries (UK) and Smiths Industries Inc (USA) were selected to supply the pilot's display unit (PDU) and the display processor respectively, for the AV-8B version of the Harrier.

The PDU, designed to US MIL specifications, has a $4\frac{1}{2}$ inch lens and is a line replaceable unit. A feature is the precision dual combining glass assembly. There is also an electronically depressible stand-by sight. The high grade optical system and the HUD electronics are housed in separate castings mounted one above the other. This eliminates many of the normal fasteners and mating surfaces and allows easy access to all components for servicing. This construction technique and the use of lighter castings, have made it possible to design a HUD with the same mechanical strength as conventional equipment but weighing 20 per cent less.

Periodic and initiated BIT routines provide up to 98 per cent detection of failures.

STATUS

In production. Smiths Industries was awarded a contract in November 1985 by McDonnell Douglas to provide modified versions of the Head Up Display (HUD) and Display Procesor (DP) which the company already supplies to the AV-8B Harrier II. In addition to providing the necessary information to the HUD the display processor also drives the new Smiths Industries multi-purpose colour CRT displays which have been ordered for the AV-8B Harrier II.

AV-8B HUD pilot's display unit

The modified HUD retains its normal symbology displays for daylight operation, but will also be able during night operations to present symbology superimposed on a FLIR (forward looking infra-red) image of the outside world. The modified HUD is also fully compatible with a pilot's night vision goggles because it uses refractive optics, albeit of a novel design.

CONTRACTORS

Smiths Industries, Aerospace & Defence Systems Company, Cheltenham Division, Bishops Cleeve, Cheltenham, Gloucestershire GL52 4SF, England.

Smiths Industries Aerospace & Defence Systems Inc, PO Box 5389, Clearwater, Florida 33518, USA.

1805.393

TORNADO HEAD-UP DISPLAY

Smiths Industries Ltd, in collaboration with Teldix GmbH and Ottica Mechanica Italiana (OMI), currently supplies the head-up display and weapon aiming computing (HUDWAC) system for the multi-national Tornado programme. The system consists basically of a pilot's display unit (PDU) and an electronic unit (EU). In addition, Smiths Industries is sub-contracted by Litef-Litton Technische Werke for manufacture of the Spirit 3 airborne computer for the Tornado.

STATUS

Production.

CONTRACTOR

Smiths Industries, Aerospace & Defence Systems Company, Cheltenham Division, Bishops Cleeve, Cheltenham, Gloucestershire GL52 4SF, England.

Tornado pilot's head-up display unit

3090.393

HEAD-UP DISPLAY FOR HARRIER AND JAGUAR

From inception the Harrier and Jaguar weapon systems have featured a Smiths Industries electronic HUD. The system, which comprises a pilot's display unit (PDU), waveform generator (WFG), pilot's control panel (PCP) and extra high tension unit (EHTU), is designed to provide the pilot with accurate analogue and alpha-numeric symbol displays of primary flight data and navigation and weapon aiming information.

Pilot's Display Unit: This consists essentially of two modules: the optical system and the CRT assembly. The optical system consists of a collimating lens assembly and a combining glass. The 100 mm, f0.97 lens assembly provides a 25° total field of view. The installation provides an instantaneous binocular field of view in the region of 18° in azimuth and 16° in elevation. All glass-to-air surfaces are treated with an anti-reflective coating to reduce spurious reflections.

STATUS

Production, with over 600 systems supplied for the Harrier and Jaguar programmes to date.

CONTRACTOR

Smiths Industries, Aerospace & Defence Systems Company, Cheltenham Division, Bishops Cleeve, Cheltenham, Gloucestershire GL52 4SF, England.

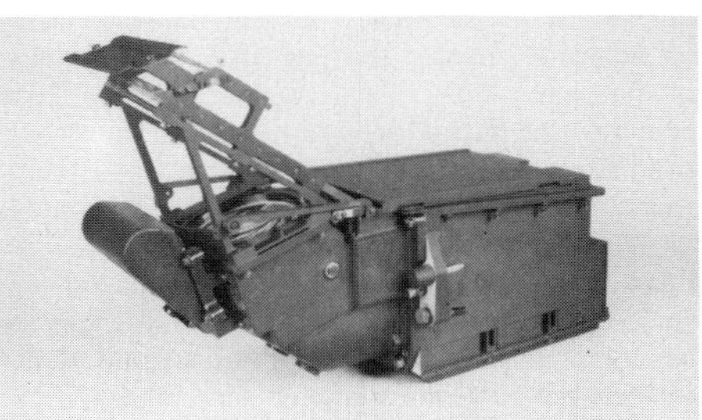

Head-up display for Indian Jaguar aircraft

3117.393
HEAD-UP DISPLAY FOR SEA HARRIER

The Smiths Industries' head-up display and weapon aiming computing (HUDWAC) system which is currently being supplied for the Sea Harrier, provides a comprehensive display of weapon-aiming, flight and navigation symbology as well as performing all the weapon aiming computations. The system comprises a pilot's display unit (PDU), an electronic unit (EU) and a pilot's control panel (PCP).

Pilot's Display Unit: This is a fully line-replaceable PDU featuring integral deflection amplifiers, bright-up and power supply modules (including EHT) and standby sight. A special feature of this PDU is that it can display symbology and outside world pictures in raster form in addition to traditional cursive symbology. It is of modular construction for ease of maintenance and consists essentially of two main sub-assemblies: the electronic (front) module and the optical module.

The optical module features two optical channels: 'Main' CRT and 'Standby'. In addition it also houses the electronics needed to convert the 28 V aircraft supply to the six volts required to illuminate the quartz halogen lamp used for the standby sight. The combiner glass mounting forms an integral part of the module and is so arranged that the combiner can hinge downwards to allow both it and the aircraft windscreen to be cleaned. Provision is also made for the mounting of a pilot's display recording camera. The high grade optical system has a 114 mm exit lens providing a total field of view of 22°. Careful design throughout has ensured that optical performance is of a very high standard.

STATUS
Production.

CONTRACTOR
Smiths Industries, Aerospace & Defence Systems Company, Cheltenham Division, Bishops Cleeve, Cheltenham, Gloucestershire GL52 4SF, England.

Head-up display for Sea Harrier

3568.393
SAAB JA37 VIGGEN HEAD-UP DISPLAY

Smiths Industries currently supplies the pilot's display unit (PDU) for the JA37 version of the Viggen. This is a fully line-replaceable PDU consisting essentially of two modules; an electronic or front module and an optical module (with integral control panel).

The optical module consists of a combining glass assembly and a high accuracy 100 mm lens system. The lens system features precision ground input and exit lens groups and a prism, the whole providing a total field of view of 28°. In addition the optical module is provided with a standby sight in order to permit reversionary weapon aiming in the event of a HUD failure.

The electronic module forms the main structure of the PDU and houses the deflection amplifiers, power supplies (including EHT), BITE, CRT assembly, brightness control, combiner servo-amplifier and all the peripheral electronic circuitry. For ease of maintenance the electronics are contained within the PDU in modular form.

STATUS
Production.

CONTRACTOR
Smiths Industries, Aerospace & Defence Systems Company, Cheltenham Division, Bishops Cleeve, Cheltenham, Gloucestershire GL52 4SF, England.

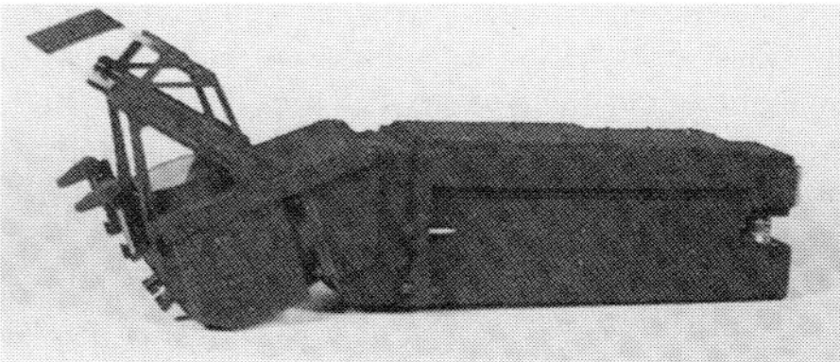
Pilot's display unit on the Swedish Viggen aircraft

6576.393
ZHUD SIGHT

The ZHUD is an innovative wide field of view raster head-up display in development by Smiths Industries. A fundamental feature of the sight is that the reflector, or collimator, is quasi-axial with the incident and reflected axis of the optical system subtending an angle of less than 10 degrees. This results in a minimum of optical aberrations, generated by the collimator, that need to be corrected in the remainder of the optical system. The results, Smiths states, is a fundamentally simpler and more economic optical system.

ZHUD has a large exit pupil allowing good freedom of head movement. This is the result of the low aberrations of the quasi-axial collimator which also allow the wide field of view (WFOV) to be maintained up to the edges of the exit pupil. Generally speaking, the current range of WFOV HUDs has provided a total field of view of some 30 degrees in azimuth by 22 degrees in elevation. With the ZHUD at least 40 degrees is achievable without a reduction in head freedom.

STATUS
Smiths Industries has carried out a collaborative project with the Royal Aircraft Establishment, Farnborough and Pilkington Defence Systems to design and manufacture a ZHUD for evaluation. Flight trials have been carried out in a Jaguar aircraft.

CONTRACTOR
Smiths Industries, Aerospace & Defence Systems Limited, Cheltenham Division, Bishops Cleeve, Cheltenham, Gloucestershire GL52 4SF, England.

4178.383
FIN 1064

The Ferranti FIN 1064 digital inertial navigation and attack system (DINAS) was developed for the RAF Jaguar refit programme to replace the original NAVWASS equipment (**1312.383**).

In the Jaguar, five separate units were replaced by one unit, with a reduction in volume of two-thirds and a weight-saving of some 50 kg compared with the previous system. The FIN 1064 system incorporates the computer, platform, power supply, electronics and interface modules into the same line replaceable unit (LRU).

FIN 1064 interfaces with all the other weapon-aiming and navigation facilities in the RAF's Jaguars, including the Ferranti LRMTS (laser ranger and marked target seeker) (**1196.393**) which is standard equipment in the single-seat version. Built-in system test software enables the FIN 1064 to self-check automatically.

For the pilot, the Ferranti FIN 1064 offers significant advantages over older systems, in terms of both ergonomics and performance. Pre-flight alignment times have been minimised and the advanced design of the inertial platform and computer maximises weapon-aiming and navigation accuracy. Navigation and steering commands are shown on the HUD, providing the pilot with vital data super-imposed on his view of the outside world. With the single-pass, low-level ground attack very much in mind, FIN 1064 has been developed to provide a comprehensive range of 'hands-on-stick and throttle' weapon delivery profiles.

The digital computer is addressable up to 64 K of memory – a large increase on previous systems.

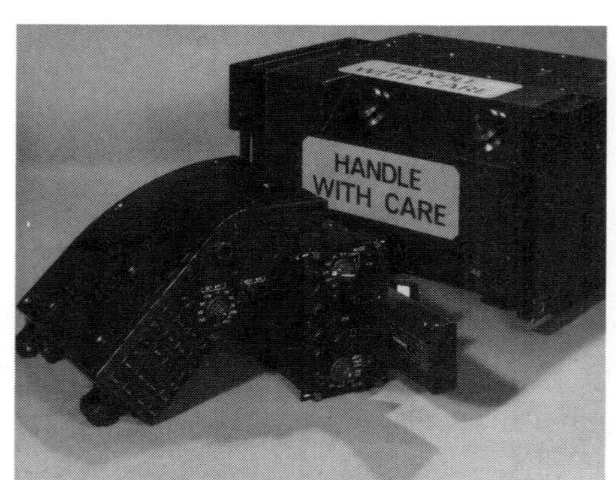

Three units comprising FIN 1064 for RAF Jaguar (left to right) pilot's display, control unit, and one-box LRU incorporating computer, inertial platform interface, electronics and power supply modules

System reliability is significantly improved, based on the experience gained by similar Ferranti systems and the early FIN 1064 models.

DEVELOPMENT
FIN 1064 is one of the FIN 1060 family of inertial nav/attack systems derived from the FIN 1000 series. These are in production at Ferranti's Silver Knowes, Edinburgh factory for the Tornado, Sea Harrier, Nimrod MR2 and Mitsubishi F1 aircraft, as well as Jaguar. FIN 1064 can accept PODS (portable data store) for in-cockpit transfer of mission data pre-loaded into PODS by computer-assisted mission planning equipment such as Ferranti Autoplan or Tacplan.

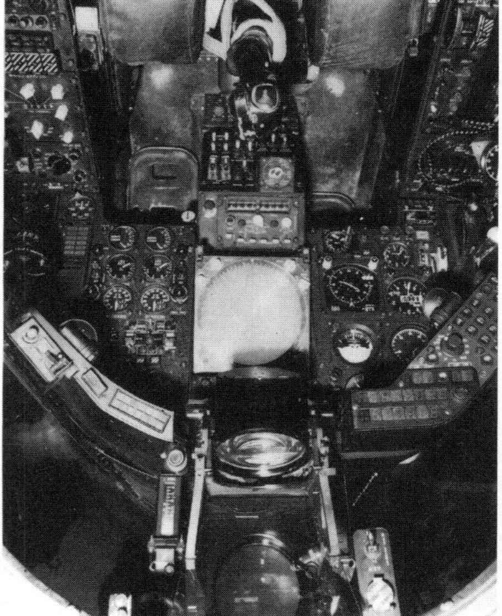
FIN 1064 in an aircraft cockpit

STATUS
The FIN 1064 is now in operational service.

CONTRACTOR
Ferranti plc, Navigation Systems Department, Silver Knowes, Edinburgh EH4 4AD, Scotland.

4730.383

FERRANTI FINAS

FINAS (Ferranti Nav/Attack System) is an integrated system which includes the Type 4510 cursive/raster head-up display and weapon aiming computer (HUDWAC), the FIN 3020 ring laser gyro strapdown inertial navigator and the Type 105D laser rangefinder. It is intended both for new aircraft and as a retrofit package for current types.

The Type 4500 HUDWAC was first displayed in public at the Farnborough Air Show in 1982. It consists of two units, the HUD display itself and a ½ ATR weapon aiming computer. The HUD has an unusually large field of view, 24° × 18°, and incorporates its own controls for mode selection. Dual combiners are provided and can be replaced without reharmonisation being necessary. The HUD projects a comprehensive variety of information onto the pilot's line of vision and a number of air-to-air and air-to-ground modes provide data for the automatic or manual release of weapons. It is compatible with night vision goggles.

The Type 105D laser rangefinder has already been produced for the Royal Danish Air Force Draken aircraft and is described under entry **4003.393**.

STATUS

The complete system is on offer for both new aircraft and retrofit programmes. The Type 4500 is undergoing a series of flight trials and has been proposed for a number of aircraft. The FIN 3020 has been on flight trials in fixed wing aircraft of the RAF. The Type 105D laser rangefinder is in operational service with the Danish Draken aircraft. The FD 4510 has been specified for F-5 and A-4 avionics upgrade programmes.

CONTRACTOR

Ferranti plc, Navigation Systems Department, Silver Knowes, Edinburgh EH4 4AD, Scotland.

1012.393

ISIS WEAPON AIMING SYSTEM

The ISIS (integrated strike and interception systems) series of equipment comprises a range of aircraft sighting systems based on the use of gyro-controlled, lead computing optical techniques. Different versions are available to fulfil air-to-air and air-to-ground missions with varying degrees of complexity, according to operational requirements. Initially intended for the aiming of guns, rockets, missiles and bombs, the ISIS sight can use range information from a laser rangefinder. It can also serve as a low-cost alternative to a CRT type HUD by displaying pointing data from a laser seeker such as the Martin Marietta Pave Penny or the Ferranti LRMTS.

Common to all versions of the ISIS family of weapon aiming sights is the use of a two-axis rate gyro, the movement of which is controlled by variations in the magnetic restraint imposed upon it in accordance with the requirements of the aiming process. The gyro is located in the sighting head, and has two principal parts, concerned with display and control, respectively. The display portion consists of a circular mirror centred on the rotational axis of the gyro. The aiming pattern is reflected from this into the pilot's forward view so that any deflection of the gyro relative to its housing results in a corresponding deflection of the aiming pattern relative to the centre of the field of view. The control portion of the gyro rotates in a controllable magnetic field, the strength of which determines the amount of deflection produced by a given aircraft rate of turn. The control field may also be intentionally unbalanced to produce required deflections of the gyro, and thus the aiming pattern.

The complete range of equipments consists of:

ISIS D-101, F-107, D-126R, F-126, M126WL, HD-158, D-195R-6, F-195R, D-195R Mk 3/5/7, D-209RM and D-282 Mk2 systems.

ISIS D-195R sight for Hawk aircraft

ISIS D-209RM sight for Hunter aircraft

The M126WL and HD-158 systems are designed specifically for helicopters.

Details of operation will be found in the 1987-88 edition.

DEVELOPMENT

The basic concept of the ISIS has been developed with the object of enabling the requirements of different users to be met from a range of standard systems, several of which are capable of subsequent extension.

STATUS

The ISIS system first entered operational service in 1968 and current users include the air forces of Argentina, Australia, Austria, Canada, India, Indonesia, Italy, Kenya, Norway, Singapore, UK and Venezuela, and the USMC.

ISIS equipment has been fitted to 25 different aircraft types, including the SAAB 105, (ISIS F-126), Fiat G91Y (ISIS B), Aeromacchi MB 326 (ISIS F-126), McDonnell Douglas A-4S (ISIS D-101), Northrop CF-5A (ISIS N), Northrop NF-5A (ISIS F-195R), BAe Hawk (ISIS D-195R), HAL Ajeet (ISIS F-195R), HAL HF24 (ISIS F-124) MiG 21FL, Mostar Galeb, and LAS A-4C (D-126R).

The ISIS D-101 has been delivered to the USN for use by the USMC, in conjunction with a USN laser target designator system.

The BAe Hunter aircraft is in service throughout the world, using the GGS MK8S weapon aiming system. As this system was designed in 1953, most of its components are now obsolescent. Ferranti has designed the ISIS D-209RM (three units) weapon aiming system to replace the GGS MK8S (13 units) system in the Hunter so that cockpit modification work is kept to a minimum.

In April 1987, Ferranti announced that an improved version of ISIS had been selected for use on the SIAI-Marchetti S211 basic jet trainer.

The ISIS F-195R/3 is in licence production at the Hindustan Aeronautics Ltd, Lucknow plant.

CONTRACTOR

Ferranti plc, Navigation Systems Department, Silver Knowes, Edinburgh EH4 4AD, Scotland.

4261.393

AF532/AF533/AF580 HELICOPTER ROOF SIGHTS

These Ferranti helicopter sights are designed for use in surveillance, target identification and missile guidance. A built-in interface allows a laser ranger or designator module to be added. Facilities are also provided for the attachment of a recording TV camera. The sight can be adapted for use with night vision equipment, helmet sights and weapons. The AF532 version is in production for the UK Army Air Corps and the AF580 has been evaluated by the US Army as part of an automatic target hand-off system in a Bell OH-58C Kiowa helicopter (see under STATUS below).

The sight provides for two magnifications. For initial surveillance ×2.5 is provided, with a field of view of 22°, and for target recognition and identification, × 10, with a field of view of 5.5°. The choice of these magnifications are based on optical considerations such as specified identification and recognition criteria as well as the need to achieve a compact and light design.

The sight is installed with its head and most of the optics carried above the normal roof line. Only the down-tube and eyepiece extend down in front of the operator when the sight is in use. The down-tube is adjustable in length which enables the installation to be readily tailored to fit most helicopters. When not required the down-tube is swung up and locked in position, out of the way, against the roof.

The down-tube may be adjusted as necessary for use then pushed aside until required. Also, if the pilot were to be incapacitated, it would be easy for the sight operator to push aside the down-tube and take over control immediately. The control handle is adjustable, both in length and tilt, so that it may always be used with the right forearm resting on the knee for comfortable control. A thumbstick controller on the control handle is used to steer the sightline through ±120° in azimuth and ±30° in elevation, movement of which is achieved by precession of the gyro which directly stabilises the mirror mounted within the rotating head. At × 10 magnification the slew rate is variable up to 5°/s; at ×2.5 magnification the slew rate is progressively variable up to 5°/s but 75°/s fast-slew can be obtained with full deflection of the thumbstick. The two-degree-of-freedom gyro is prevented from toppling under all operational conditions by limit and logic switches associated with a caging mechanism.

CHARACTERISTICS

Weight of installed system: 28 kg (33 kg with laser rangefinder/designator)

Resolution: 20 line pairs/mrad

Magnification: ×2.5, ×10

Field of view: 22° (×2.5 magnification); 5.5° (× 10 magnification)

Total sightline exploration: Azimuth ±120°; elevation ±30°

Exit pupil size: 5 mm diameter

Eye relief: 35 mm

Eyepiece focus: Fixed at −1 dioptre

Vertical adjustment of eyepiece: Variable over 110 mm

DEVELOPMENT

The AF533 sight was developed from the AF530 on which work was commenced in 1970 for use with the BAe Hawkswing missile. Successful trials were carried out before termination of that system's development by the UK MoD in 1975. It is a refined

AF533 gyro-stabilised helicopter observation sight installed in Gazelle helicopter

version of the AF530 updated and incorporating improved electronics. Sights with other AF500 designations are being prepared for various helicopters.

The AF120 first-generation stabilised sight, which uses similar stabilisation techniques, is still in service with the British Army.

STATUS

The AF532 is now in production, under a UK MoD contract for fitment to Gazelle helicopters of the Army Air Corps.

AF580 systems have been installed in two US Army Kiowa helicopters and integrated with Rockwell Collins automatic target hand-off equipment. This allows information on target range and bearing to be transmitted by a data burst to an atttack helicopter. All versions of the sight are designed to accept a Ferranti Type 118 laser rangefinder and designator, which is a repackaged variant of the Type 629 (**4787.193**) plus a telescope.

CONTRACTORS

Ferranti plc, is responsible for the electronics and gyro subsystems and is the main contractor for the sight. The optical subsystem of this equipment is manufactured by Avimo Ltd, Taunton, Somerset, England.

1196.393

FERRANTI LASER RANGER AND MARKED TARGET SEEKER

This is an aircraft ranging and detection system designed primarily for air-to-surface attacks, when it is used either as a self-contained laser ranger or, when operating in conjunction with a laser target marker (designator), as a simultaneous rangefinder and target seeker. This latter feature, coupled with high accuracy, gives this laser system a distinct advantage over radar ranging equipment. The laser ranger and marked target seeker is specifically intended for use with weapon systems optimised for low level, first pass attack when targets are difficult to detect, tracking time is necessarily short and air-to-surface grazing angles are very low. Details of the operation were given in the 1987-88 edition.

The laser beam is fully stabilised against aircraft movements by mounting the transmit and receive optics within a three-axis gimbal head. The head may be used to direct the beam +3 to –20° in elevation, ±12° in azimuth and is roll stabilised to ±90°. The target seeker coverage is considerably more than the limits of head movement. Range is quoted as being greater than pilot's visual range up to 9 km.

The main unit weighs 21.5 kg and is 607 mm long by 269 mm in diameter. The total height, including the heat exchanger, is 300 mm. The electronics unit, weighing 14.5 kg and measuring 432 mm long by 330 mm wide by 127 mm high, contains ranging, seeking, and power supply circuits.

STATUS

The equipment is now in service with the RAF and several other air forces. It is operational in Jaguar, Harrier and Tornado aircraft. Over 500 systems have been produced.

CONTRACTOR

Ferranti Defence Systems Ltd, Electro-optics Department, St Andrew's Works, Robertson Avenue, Edinburgh EH11 1PX, Scotland.

3046.393

TYPE 105 AIR-TO-SURFACE LASER RANGER

The Type 105 equipment is a series of high repetition rate, Nd:YAG, steerable laser rangefinders, developed by Ferranti as a private venture, to provide compact, low-cost, accurate target ranging sensor for ground attack aircraft. A prototype flew in the Fairchild Republic night attack A-10 aircraft in 1979. The D version is in production for the Draken aircraft of the Royal Danish Air Force retrofit and has also contributed to the more recent Fairchild single seat night attack programme.

The 105 series includes the D, E and S variants. The D and S versions are single packages combining the Ferranti Type 629 transceiver, a beam steering unit and beam steering electronics. The E version, virtually identical, has the beam steering unit packaged separately.

Target range is measured to 3.5 m standard deviation to a range of 10 km, effectively removing the largest source of error in air-to-surface weapon delivery. Other features include low power consumption, ease of integration with aircraft avionic systems, flexible configuration, and small size and frontal area.

CHARACTERISTICS

Wavelength: 1.064 μm
PRF: 10 pps
Angular coverage: Within 10° semi-apex angle cone
Range: < 10 km
Accuracy: 3.5 m standard deviation
Power: 200 W 28 V DC
MTBF: 1000 hours
Dimensions: 198 × 261 × 357 mm
Weight: 1.3 kg

STATUS

In operational service with the Draken aircraft of the Royal Danish Air Force.

CONTRACTOR

Ferranti Defence Systems Ltd, Electro-optics Department, St Andrew's Works, Robertson Avenue, Edinburgh EH11 1PX, Scotland.

3119.393

TYPE 106 LASER TARGET SEEKER

The Type 106 laser target seeker was conceived in answer to a requirement for a simple laser seeking device for installation in light, ground attack aircraft. It is designed to integrate with the Ferranti ISIS series of gunsights head-up display systems.

The concept of use is described under the item dealing with the Ferranti laser ranger and marked target seeker (**1196.393**). The incorporation of this system significantly increases the effectiveness of close air support against difficult targets, while reducing the required briefing time between forward air controller and pilot.

The main unit comprises the stabilised optical head of the Ferranti laser ranger and marked target seeker equipment contained in a short, pressurised canister. An electronics unit contains power supplies and target seeking circuits and is also a variant of the RAF's equipment.

The optical head is roll stabilised ±90° by the aircraft reference and is capable of being directed and stabilised +20 to –25° in elevation, relative to equipment roll axis. The azimuth coverage is ±18° relative to aircraft heading with automatic scanning. This coverage may be further increased by simply manoeuvring the aircraft.

The main and electronics unit weigh approximately 10.5 kg each. The main unit is 312 mm long with a diameter of 247 mm. The total height is 275 mm. The electronics unit is 432 mm long by 330 mm wide by 127 mm high, though it may be repackaged to approximately half this volume if so desired.

In September 1978 a podded version of the Type 106 was announced by Ferranti and Portsmouth Aviation. The pod weighs 125 kg, inclusive of air conditioning, and is about 2.3 m long and 37 cm in diameter.

The Type 106 is also used in combination with the ISIS D-209RM gyro gunsight to form a low-cost head-up air-to-ground target acquisition system for ground attack aircraft.

DEVELOPMENT

The equipment is a private venture variant of similar devices for Jaguar and Harrier.

STATUS

Several other air forces are showing interest in the Type 106 but it is not yet in production other than in its RAF form.

CONTRACTOR

Ferranti Defence Systems Ltd, Electro-optics Department, St Andrew's Works, Robertson Avenue, Edinburgh EH11 1PX, Scotland.

5929.393

TYPE 116 LASER RANGEFINDER

The Type 116 is a high repetition rate CO_2 laser rangefinder which has been designed initially for airborne applications. The equipment is 'eye-safe', is compatible with forward looking infra-red (FLIR) systems and provides high performance in smoke and battlefield conditions. It contains an integral power supply and receiver electronics.

CHARACTERISTICS

Type: Sealed CO_2 TEA
Wavelength: 10.59 microns
Peak power: 250 kW
Pulse energy: 30 mJ nominal
PRF: 20 – 50 Hz
Beam divergence: 0.4 mrad
Laser Endurance: $> 10^6$ shots
Receiver: Photoconductive cadmium mercury telluride
Range: 0.5 – 10 km
Accuracy: 2.5 m
Dimensions: 275 × 230 × 230 mm
Weight: 17.5 kg

STATUS

In development for the Royal Aircraft Establishment, Farnborough.

CONTRACTOR

Ferranti Defence Systems Ltd, Electro-optics Department, St Andrew's Works, Robertson Avenue, Edinburgh EH11 1PX, Scotland.

6779.393

FERRANTI HELMET POINTING SYSTEM

Ferranti, who have developed a helmet pointing system for use with Tracked Rapier (**4696.093**), have now developed a dual head helmet-mounted sighting system for use with the Lucas Aerospace helicopter turret.

The system uses an aircrew helmet which has been adapted to incorporate an aiming reticle and sensor. It provides steering commands by measuring the orientation of the sensor to the helicopter within a reference electro-magnetic field. The turret-sighting system is for air-to-surface roles with guns but can also be used in the air-to-air role when the gun in the turret is replaced by Stinger missiles. The turret has been fitted to Bell 406, Bell 412 and BK 117 helicopters.

STATUS

Under development.

CONTRACTOR

Ferranti Defence Systems Ltd, Electro-optics Department, St Andrew's Works, Robertson Avenue, Edinburgh EH11 1PX, Scotland.

6575.393

TIALD DESIGNATOR POD

Ferranti Defence Systems is developing, under UK MoD contract, a Thermal Imaging Laser Designator (TIALD) pod. The system is intended for use in demonstrating the accurate delivery of laser guided bombs, and for further research into advanced weapon delivery under various conditions of day and night visibility.

TIALD will have an integrated sensor package housed in a roll gimbal containing sightline stabilisation in the front of the pod. An important feature of the system is the ability to engineer a combined optical path for the different wavelengths of the laser and thermal imager, in order to save space and weight. The main elements of the package are a UK common modules (TICM II) thermal imager, a video auto tracker, and a Ferranti laser target designator/ranger. The sensors are integrated with a common optical path, with sightline stabilisation provided by a Ferranti gyro-stabilised mirror.

In operation, TIALD will be directed on to the target area by other aircraft sensors, or by the navigator, as the aircraft makes its final approach. Once a target has been selected the system locks on to the video signal provided by the thermal imager and the aircraft is free to manoeuvre without losing sight of the target. At this time the laser beam can also be fired to acquire target range. By updating the aircraft's weapon aiming computer with range and bearing information, an attack could proceed using conventional weapons. Alternatively the laser can be used to designate the target for laser-guided weapons from either the attacking or accompanying aircraft.

The pod is 30.5 cm diameter and 2.6 m long, and will be fitted with standard NATO attachments. It weighs 150 kg. Apart from electrical connections for 1553B bus interfaces and AC power supplies, the design is self-contained making the pod easily transferable between aircraft.

TIALD designator pod

STATUS

In development under a UK MoD £2 million contract. The demonstration pod was scheduled for delivery in April 1988 for evaluation and trials by the Royal Aircraft Establishment, Farnborough. The system is intended to meet Staff Requirement (Air) 1015.

CONTRACTORS

Prime contractor: Ferranti Defence Systems Ltd, Electro-optics Department, St Andrew's Works, Robertson Avenue, Edinburgh EH11 1PX, Scotland.

Subcontractors: GEC Sensors Ltd, Christopher Martin Road, Basildon, Essex SS14 3EL, England.

British Aerospace, Dynamics Division, Downshire Way, Bracknell, Berkshire, England.

4727.393

TYPE 117 LASER DESIGNATOR

The Type 117 laser designator/ranger has been designed by Ferranti as an integral part of the forward looking infra-red (FLIR) pod for the US Navy F/A-18 Hornet aircraft. No technical details have been released other than that it is a high power, high-repetition rate Nd:YAG device with an output power in excess of 100 mJoules.

STATUS

In development under contract to Ford Aerospace and Communications Corporation. The initial contract, worth over $2 million, covers engineering developments and provision of units for qualification and flight testing. The first development model has been delivered to Ford Aerospace.

CONTRACTOR

Ferranti Defence Systems Ltd, Electro-optics Department, St Andrew's Works, Robertson Avenue, Edinburgh EH11 1PX, Scotland.

5930.393

TYPE 118 LASER DESIGNATOR/RANGER

The Type 118 laser designator/ranger is a lightweight system and is intended primarily for use in helicopter applications. It is of modular design, has common transmit/receive optics and can be easily integrated into a variety of sights or combined with other electro-optical sensors. The Type 118 can operate with all existing laser guided weapons and laser spot trackers.

CHARACTERISTICS

Type: Nd:YAG
PRF: 20 pps max
Beam divergence: 2 mrad (raw beam)
Output energy: 130 mJ
Range: 0.3 – 10 km
Resolution: 5 m
Dimensions: 270 × 215 × 160 mm
Weight: 5.9 kg

STATUS

In production for the mast-mounted sight for the US Army OH58 helicopters. First production equipment was delivered in mid-1985.

CONTRACTOR

Ferranti Defence Systems Ltd, Electro-optics Department, St Andrew's Works, Robertson Avenue, Edinburgh EH11 1PX, Scotland.

5955.393

TYPE 122 HELICOPTER LASER RANGEFINDER

The Type 122 is a low-cost Nd:YAG laser rangefinder designed for helicopter use and suitable for integration with the visual optics of both existing and future sights. It is designed to provide accurate range measurement from an aircraft in a variety of tactical situations. A Type 520 laser transceiver (**5757.193**) is used.

The Type 122 rangefinder is similar in many respects to the Type 121 (**5931.393**) and can also be installed in the Ferranti AF500 series of roof mounted sights (**4261.393**).

STATUS

In production.

CONTRACTOR

Ferranti Defence Systems Ltd, Electro-optics Department, St Andrew's Works, Robertson Avenue, Edinburgh EH11 1PX, Scotland.

5747.393

HELICOPTER LASER DESIGNATOR (HELD)

As with its two ground-based variants, HOLD and FOLD, the Ferranti helicopter laser designator (HELD) is based on the Type 306 designator which has been in service with the British Army since 1980 under the name Laser Target Marker and Ranger (LTMR) (**1742.193**). In the HELD version the Type 306 is fitted to a simple helicopter mount which requires virtually no modification to the helicopter. A gyro-stabilised mirror is added to the mount to remove the worst of the vibration and motion effects on the laser spot. The system has been successfully tested with laser guided bombs and spot trackers.

CONTRACTOR

Ferranti Defence Systems Ltd, Electro-optics Department, St Andrew's Works, Robertson Avenue, Edinburgh EH11 1PX, Scotland.

5928.393

TYPE 234 SENSOR PAYLOAD SYSTEM

The Ferranti type 234 sensor ball is a fully stabilised optronic sensor package mounted in a ball-shaped head. The system consists of a laser rangefinder/designator, a thermal imager such as the Barr & Stroud IR18 and a TV camera in a combination of one or two from these three sensors. It is suitable for fixed wing aircraft, helicopters or RPVs and a version has also been developed for shipborne applications.

The Nd:YAG laser rangefinder/designator is a variant of Type 118 (**5930.393**) and operates at 1.06 μm. The thermal imager can be selected to suit the customer's requirements. See also entry **4785.393**.

CHARACTERISTICS

Sensor head
Angular freedom: Full lower hemisphere coverage; pan ±220°; tilt +20° to –120°
Pointing accuracy: 3 mrad
Slew rate: 1 rad/s
Slew acceleration: 1 rad/s²
Stabilisation: Better than 50 μrad
Dimensions: 350 mm diameter × 470 mm high
Weight: 24 kg

Laser rangefinder/designator
Type: Nd:YAG
Waveband: 1.06 μm
Beam divergence: 0.4 mrad
Range accuracy: ±5 m
Output energy: 100 mJ
PRF: 20 pps

Thermal imager
Wavelength: 8 – 14 microns
Field-of-view: Dual switched or zoom
TV system: New vicon or CCD with zoom lens

CONTRACTOR

Ferranti Defence Systems Ltd, Electro-optics Department, St Andrew's Works, Robertson Avenue, Edinburgh EH11 1PX, Scotland.

Ferranti stabilised electro-optic sensor package

5749.393

TYPE 221 THERMAL SURVEILLANCE SYSTEM

The Type 221 thermal imaging surveillance system is designed for use with military helicopters but can also be installed in fixed-wing aircraft and surface vessels. Developed by Ferranti in conjunction with Barr & Stroud, the system incorporates an IR18 thermal imager and telescope by Barr & Stroud (**3834.093**), with sightline stabilisation steering provided by a Ferranti stabilised mirror. The assembly is contained in a pod which can be mounted either beneath the nose of a helicopter or may project through an aperture in the aircraft floor.

The Barr & Stroud IR18 imager unit provides a normal field-of-view (FOV) of 38° in azimuth and 25.5° in elevation. In the Type 221 application, users can choose a telescope magnification of either ×2.5 or ×9, with corresponding wide or narrow FOV. These are 15.2° azimuth and 10.2° elevation for the wide FOV and 4.2° × 2.6° for the narrow. With the Ferranti sighting mirror in the pod installation the system has fields of regard of +15 to –30° in elevation and ±178° azimuth. The entire sensor system is vertically mounted above the mirror which is angled, periscope fashion, at 45° to the horizontal to provide views in the horizontal plane.

The mirror, produced by Ferranti, has an aluminium reflective element which is diamond-machined to a flatness of two fringes at 550 nanometres. The mirror has a reflectivity of greater than 97.5 per cent at 45° incidence, averaged over 8 to 12 μm. Its stabilisation system comprises a two-axis device with integrating-rate gyros as rate sensors. The mechanism is driven in each axis by a direct-drive torque motor while steering is obtained by torquing the integrating-rate gyro. Angular information is derived from a resolver fitted to each axis. The mirror sightline is controlled by means of signals from an electronics unit which also provides power and signals to the main turret azimuth drive on the pod. System control may be exercised either through a digital computer or by a hand control unit.

The display may be presented on either 525 or 625 line TV monitors. The output is either in CCIR or EIA composite video formats as required. This TV-compatible output may be displayed on one or more monitors.

STATUS

In production and operational service with the RAF.

CONTRACTORS

Barr & Stroud Ltd, Caxton Street, Anniesland, Glasgow G13 1HZ, Scotland.

Ferranti Defence Systems Ltd, Electro-Optics Department, St Andrew's Works, Robertson Avenue, Edinburgh EH11 1PX, Scotland.

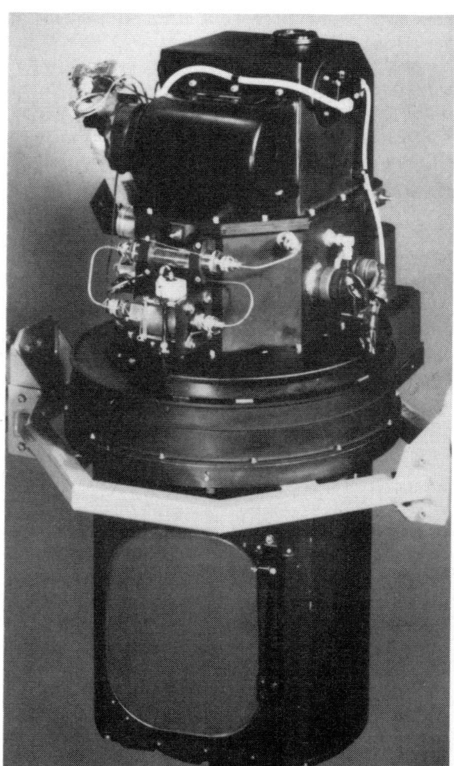

Type 221 thermal surveillance system

6574.393

TYPE 311/318 CO_2 LASER DEVICES

The Type 311 equipment is an advanced self-contained and miniature carbon dioxide (CO_2) laser transceiver. It is eye safe for training and has excellent smoke penetration for use in battlefield conditions.

The techniques developed for the Type 311 have also been used for Type 318, a compact, self-contained scanning CO_2 laser source capable of covertly attracting the attention of aircraft equipped with 8 to 13 micron FLIRs, over a defined arc. The scanning source appears on an airborne FLIR display as a flashing image of sufficient intensity to be easily discernable against terrain background under day, night and adverse weather conditions.

STATUS

In development.

CONTRACTOR

Ferranti Defence Systems Ltd, Electro-optics Department, St Andrew's Works, Robertson Avenue, Edinburgh EK11 1PX, Scotland.

6344.393

BAe IMPROVED TOW SIGHT

British Aerospace is carrying out a mid-life improvement programme to the optical TOW roof sight mounted on British Army Lynx helicopters. A night vision capability is being developed, operating in the infra-red waveband, to enable armoured targets to be engaged at night or in bad visibility. The improvement consists of a thermal imaging system which will be incorporated in all new TOW sights and retrofitted into existing roof sights.

British Aerospace has co-operated with Hughes Aircraft Company, manufacturer of the TOW missile and the basic sighting system, in the development of the improved sight. The thermal imaging subsystem has been developed by Rank Pullin Controls. Because of the limited space available in the sight, it has been necessary to develop a special system which is significantly more compact.

See also entries **3815.393** and **6341.393**.

STATUS

In production.

CONTRACTORS

British Aerospace plc, Dynamics Division, Six Hills Way, Stevenage, Hertfordshire SG1 2DA, England.

Rank Pullin Controls Ltd, Langston Road, Loughton, Essex IG10 3TW, England – thermal imager.

BAe improved TOW sight

3404.393

NIMTAN AIRBORNE NIGHT SIGHT

The PPE Nimtan sight is a modified Spartan battlefield sight (**3403.193**) incorporating a straight eyepiece and pintle mounting. The sight was designed for the MoD for use by the RAF in maritime patrol aircraft. It can also be used in other patrol, supply, and spotter aircraft, helicopters, land vehicles, ships, and hovercraft with or without the pintle.

The sight interfaces with aircraft window shelves provided with bearing potentiometers. The mounting is provided with azimuth locking, horizontal scanning, and a safety lanyard. Sight controls consist of on-off switch, focus, shutter, and light control setting. Power is obtained from a self-contained battery. The tube is fitted with automatic brightness control.

The sight is of modular construction for ease of maintenance. Both the equipment and storage case meet the environmental requirements of DEF 133.

The range and field of view provide excellent surveillance and reconnaissance capability and the magnification (× 1.7) is sufficiently low not to cause vibration of the picture.

CHARACTERISTICS

Magnification: × 1.7

Field of view: 20°

Image tube: 25 mm 3-stage cascade

Weight: 4.5 kg (7.4 kg with support bracket)

STATUS

In service with the RAF.

CONTRACTOR

Pilkington PE Ltd, Glascoed Road, St Asaph, Clwyd LL17 0LL, Wales.

5926.393

VINTEN VICON 90 SERIES 101 FLIR POD

The Vinten Vicon 90 is a high resolution miniature forward looking infra-red (FLIR) pod designed for use in an airborne reconnaissance role. The FLIR sensor is a lightweight thermal imager manufactured by Rank Pullin Controls and based on the Sprite detector which, with the incorporation of advanced signal processing, provides high levels of thermal sensitivity and spatial resolution. The imager comprises a scanner and processing electronics unit incorporating a cryogenic cooling system, and is fully automatic in operation.

The pod is self-contained with the aircraft interface requirement being limited to the installation of a control panel. The imagery is recorded on a video cassette recorder fitted in the pod, and can also be displayed in real-time on a video monitor in the aircraft.

CHARACTERISTICS

FLIR head depression angle: 0 or 10° fixed

FLIR window: Zinc sulphide

Pod dimensions: 1345 mm long × 203 mm diameter

Weight: 37 kg approx

CONTRACTOR

W Vinten Ltd, Western Way, Bury St. Edmunds, Suffolk IP33 3TB, England.

Vicon 90 mini surveillance FLIR pod

UNITED STATES OF AMERICA

4243.383

TOMAHAWK CCD TV SENSOR

A miniature, solid-state charge-coupled device (CCD) TV camera has been designed and built by Fairchild Imaging Systems Division under contract to McDonnell Douglas Astronautics Corporation for use as the sensor in the Tomahawk cruise missile (**3993.011** and **4197.221**) digital scene matching area correlator (DSMAC) guidance system.

In conjunction with suitable optics and an image intensifier, the CCD TV camera presents real-time video imagery to the processor designed by McDonnell Douglas Astronautics. The latter, using this video information in a correlation mode, supplies final guidance corrections in the terminal phase of the cruise missile flight to its target. The camera uses a CCD sensor developed by Fairchild.

STATUS

In volume production

CONTRACTOR

Fairchild Weston Systems Inc, 300 Robbins Lane, Syosset, New York 11791, USA.

1008.393

XM76 ANTI-OSCILLATION SIGHTING SYSTEM

The XM76 is a telescopic target acquisition and identification equipment, providing stabilisation against vibrating environments such as helicopters, vehicles, boats, and ships.

OPERATION

The sighting system senses motion in two axes by means of rate gyros, and the resulting signals are used to control torque motors acting upon a fluid prism optic system to compensate for the sensed motion, thus stabilising the image of the desired field of view. There are separate, and virtually identical, servo-loops for each axis control. The motion signals measured at the optical head are applied to a control module as error signals, and are then demodulated, amplified, and stabilised, power amplified and fed to the torque motors as correction signals. The polarity and magnitude of the corrective signals determines the direction and speed of the torque motors to compensate for the motion of the optical head.

The optical device responsible for effecting motion correction is a fluid prism.

The anti-oscillation sighting system has also been adapted to armed helicopter fire control. There have been two applications. The first was to improve the accuracy of the fire by providing the gunner with a stabilised image and × 1.5 to × 12 magnification optics on the gun sight. The second was also to provide a further increase in accuracy by incorporating resolvers into the optical servo loop, and thus slaving the turret to the optical line-of-sight of the XM76 mounted on the sighting station in the cockpit.

The sight also incorporates a monocular zoom lens, controlled by two push-buttons on the unit. These control the direction of the zoom; the × 1.5 for zoom-out (decreased magnification) and the × 12 for zoom-in (increased magnification). Speed of zoom is determined by the detent to which the × 1.5 or × 12

Stabilised sight head for the M97 20/30 mm armament system on the Bell AH-1J Twin Cobra

button is pressed: first detent for slow zoom, second detent for fast zoom.

STATUS

Users include the US Army, USAF, USMC, USN, and the Government of Israel. Vehicles fitted with the XM76 sighting system include the OV-2, AHIG, OH6A, OH-58A, and the UH-1 series helicopters for observation, target acquisition, attack and damage assessment.

Fixed-wing aircraft include the A-4 and F-4, and other types of vehicle fitted range over AFVs and various types of naval craft. Recent helicopter applications are the stabilisation of laser beams for both target designation and weapon simulation. The Dynalens optical element which corrects for motion has also been incorporated in the pantograph sight of the M97 20/30 mm flexible armament system produced by the General Electric Company (**1288.303** in 1983-84 and earlier editions).

CONTRACTOR

Sigma Dynamics Corporation, 2300 Computer Avenue J-54, Willow Grove, Pennsylvania 19090-1732, USA.

1630.393

HONEYWELL VISUALLY COUPLED SYSTEMS

The Honeywell electro-optical and magnetic helmet sights have evolved with the related objectives of simplifying target acquisition for airborne weapons, and also improving this function by utilising the wide search-angle and flexibility of the human eye. This is accomplished by arranging for the pilot's line-of-sight to be determined in relation to aircraft axes. This data can be used as pointing information for a wide range of aircraft systems (AI radar, cameras, infra-red seekers, etc) and for weapon aiming.

The Honeywell helmet sight system consists of the following major elements: helmet-mounted unit (HMU), sensor surveying units (SSU), sensor electronics unit (SEU), and controls mounted on the instrument panel and control stick. Details of the operation will be found in the 1987-88 edition.

In versions being supplied to the USN, a Honeywell HDC-202 digital computer is included.

AMHMS

The Advanced Magnetic Helmet Mounted Sight System (AMHMS) has been qualified for USAF applications through a contract with the USAF Aeronautical Systems Division. The system uses a visor projected reticle and a line-of-sight computation, based on magnetic fields surrounding the pilot's helmet, to allow off-boresight target engagement by merely turning the head to place the reticle on the target.

VTAS

A total of 500 electro-optical helmet sight systems were supplied to the USN for fitting in F4J aircraft. The visual target acquisition system (VTAS) is used for off-boresight targeting of the Sidewinder air-to-air missile.

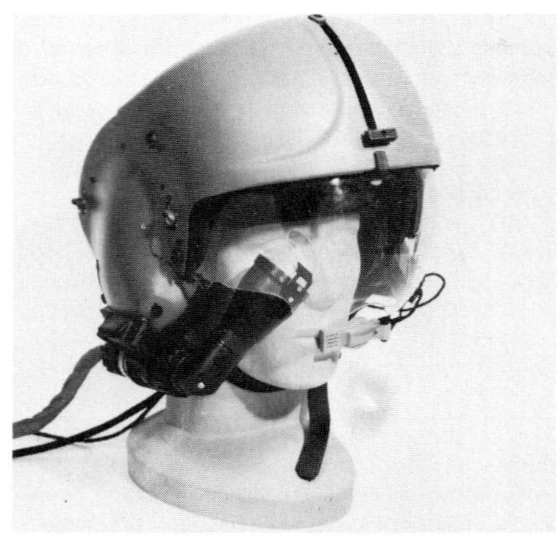

Honeywell IHADSS helmet sight/display

IHADSS

The latest version of the Honeywell electro-optical system is the Integrated Helmet and Display Sighting System (IHADSS) which is now in production for the AH-64A advanced attack helicopter. Both the pilot and co-pilot/gunner are provided with helmet units and controls to allow independent and co-operative use of the system. The IHADSS sight component provides off-boresight line-of-sight information to the fire control computer for slaving weapon and sensor to the pilot's head movements. Real world sized video imagery from the slaved and gimballed infra-red sensor is overlaid with targeting as well as flight information symbology, and projected on a combiner glass immediately in front of the pilot's eye. The IHADSS allows night nap-of-the-earth flight at below treetop altitudes without reference to cockpit instruments, and rapid target engagement.

Advanced Concepts

Honeywell is developing advanced helmet-mounted sight and display concepts which include wide FOV binocular displays, NVG compatible/integrated displays, NBC mask compatible/integrated displays, eye tracking, and improved head tracking methods for both aircraft and ground vehicle applications.

STATUS

The company has delivered 500 VTAS systems for the F4J aircraft.

The IHADSS is currently in production for the Hughes AH-64A and A129 Mangusta advanced attack helicopters. The system has also been integrated into the CH-53E, HH-60D, AH-1S and PAH 2 aircraft.

The AMHMS has been military qualified and is undergoing flight evaluation in the AFTI-16 aircraft. It is also in use in the Helmet Mounted Oculometer System for the tactical combat trainer at the US Wright Patterson AFB.

CONTRACTOR

Honeywell Military Avionics Division, 2600 Ridgway Parkway, Minneapolis, Minnesota 55413, USA.

5753.393

SPASYN VISUAL TARGET ACQUISITION SYSTEM

The SPASYN visual target acquisition system is an advanced, ultra-light helmet-mounted sighting system. SPASYN (space synchro) measures the location and orientation of an object in space by means of radiating directional magnetic fields in three orthogonal co-ordinates. Applied to a helmet-mounted sight, this closed-loop transducing system provides a full 6° of freedom in measuring a pilot's line-of-sight to an accuracy of better than 0.5° within a head motion envelope of 5 ft³ (0.14m³). The system enables a pilot to achieve visual acquisition of a given target and lock on automatic tracking systems while remaining in the head-up position and free to perform evasive manoeuvres or other in-flight tasks.

A small transmitter (radiator) mounted on the cockpit frame emits magnetic field vectors which induce voltages (signals) in a sensor mounted in the helmet. These signals are processed by a computer in the system's electronics unit to define the sensor's translation and orientation, ie the operator's line-of-sight angles. The sensor mounted on the helmet weighs approximately ½ ounce (14 grams) and does not encumber the wearer. The SPASYN-based system can be used in both fixed- and rotary-wing aircraft to command weapons and sensors. It can also be used on land or at sea as a target acquisition system, can be vehicle-mounted or can act as an alternative fire-control system in naval vessels.

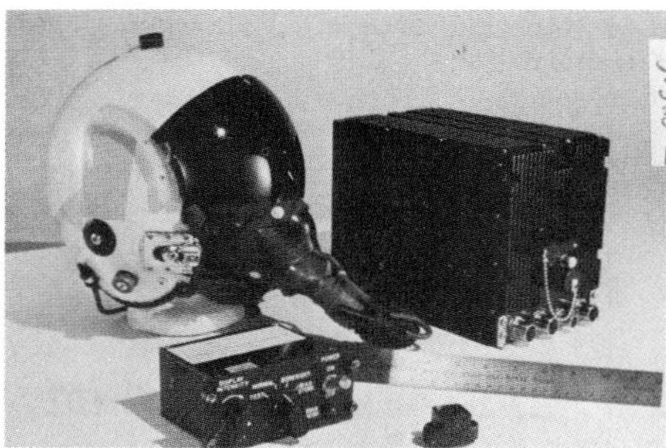

SPASYN visual target acquisition sight

In January 1988 McDonnell Douglas won a contract to develop a standardised Magnetic Helmet-Mounted Sight for all future US Navy aircraft. The system will measure the pilot's line-of-sight and transmit the data to devices controlling weapons, sensors and graphic display generators. The system will interface with advanced helmet display devices.

CHARACTERISTICS

Weight: Sensor ½ oz (14 g); Radiator 1 oz (28 g); Electronics unit 22 lb (10 kg)

Dimensions: Sensor ½ in (1.26 cm) cube; Radiator 1 in (2.54 cm) cube; Electronics unit 13$\frac{1}{10}$ × 7$\frac{3}{5}$ × 4$\frac{4}{5}$ in (33.3 × 19.3 × 12.2 cm)

Accuracy: Nominal head position 0.1°; max error condition 0.5°

STATUS

Although originally developed by Austin Electronics, the SPASYN product line was sold to McDonnell Douglas in 1985.

CONTRACTOR

McDonnell Douglas Corporation, Electro-Mechanical Division, Box 1704, Grand Rapids, Michigan 49501, USA.

4136.393

AN/ASG-26A LEAD COMPUTING OPTICAL SIGHT

The General Electric Company has developed an improved lead computing optical sight (LCOS) which enables air-to-air attacks to be made without the need for continuous target tracking. This capability is incorporated in the LCOS system for the McDonnell Douglas F-4E aircraft. The complete equipment consists of a pilot's head-up display, a two-axis lead computing gyroscope, gyro mount, and a lead computing amplifier.

Against airborne targets, the equipment displays gun and missile fire control information by means of a servoed aiming reference. Against ground targets, the pilot adjusts the aiming reference manually to control gunnery, rocket firing and bombing displays. An optical path in the centre of the sight-head presents radar aiming data. The head-up display projects the aiming reference by means of a collimated reticle display reflected from the combining glass into the pilot's field of view. The image is focused at infinity.

The aircraft's own manoeuvres generate rate and acceleration signals in the gyro lead computer, range to target is measured by radar, and angle of attack, air density and airspeed data are obtained from the air data computer. Using these parameters, the pilot's aiming reference is displaced in the proper direction to introduce lead angle and gravity corrections. Analogues of the aircraft's roll angle and range information are also projected onto the combining glass. In ground attack modes, other aircraft systems are used to generate corrections for drift, and offset bombing is also possible.

STATUS

In production.

CONTRACTOR

General Electric Company, Aircraft Equipment Division, Binghampton, New York 13902, USA.

4137.393

AN/ASG-29 LEAD COMPUTING OPTICAL SIGHT

General Electric produces the AN/ASG-29 lead computing optical sight (LCOS) for the F-5E aircraft. Compared with the GE AN/ASG-26A, which has a family relationship with the AN/ASG-29, the latter comprises two main units only; a pilot's optical display and a lead computer with its associated mounting base. The equipment provides all weapon delivery modes applicable to the aircraft type, including air-to-air and ground attack engagements with both guns and missiles. Provision is also made for the inclusion of a CCIP bombing mode. In most other respects the AN/ASG-29 is broadly similar to the AN/ASG-26A described in **4136.393**.

CHARACTERISTICS

Total field of view: ±6° azimuth; +2 to −12° elevation
Instantaneous field of view: 7.5°
Collimating lens: 10.4 inches (264 mm) focal length, 4 inches (102 mm) clear aperture
Weights: 6.9 kg (sight head); 7.9 kg (gyro lead computer); 0.8 kg (mounting base)
MTBF: Over 300 h

STATUS

In production for F-5E aircraft.

CONTRACTOR

General Electric Company, Aircraft Equipment Division, Binghampton, New York 13902, USA.

3815.393

LAAT SIGHT

An improvement to the M-65 optical sight is known as the Laser Augmented Airborne TOW (LAAT) sight is produced by Hughes for use on US Army AH-1S Cobra attack helicopters. The M-65L contains a thermal imaging system comprised of standard US Army common modules, and Hughes has designed and developed a laser transmitter capable of being fitted in the small space between the gimbal assembly and the housing plate of the existing night turret for the TOW anti-tank missile system. This unit is approximately 13 × 13 × 4 cm in size and has a repetition rate of four pulses per second for five seconds on and five seconds off. A multi-pulse laser rangefinder is necessary due to the helicopter's forward motion during target sighting, requiring several range measurements to update the fire control computer.

A novel cooling technique is embodied to reduce size and weight; the laser's flash tube is embedded in a heat conductive and highly reflective layer to avoid the need for more conventional cooling techniques.

STATUS

In operational service. More than 1500 TOW sights have been delivered by Hughes since 1979. An improved version of the LAAT sight, known as C-NITE, has been developed by Hughes and is described under entry **6341.393**.

CONTRACTOR

Hughes Aircraft Company, Electro-Optical and Data Systems Group, PO Box 902, El Segundo, California 90245, USA.

The thousandth telescopic sight for the airborne TOW system, at the Hughes plant prior to delivery to the US Army

6341.393

C-NITE TOW MISSILE SIGHT

Hughes has developed an improved version of the LAAT sight (**3815.393**) with a FLIR capability which is used to fire the airborne TOW 2 missile. The sight, known as C-NITE (COBRA-NITE), will be used by pilots of the AH-1S Cobra attack helicopter to flight combat missions by day and night using a new night vision system which will be used to upgrade both M-65 and M-65L.

The C-NITE equipment incorporates modifications to the basic LAAT that give the aircraft a FLIR capability. It is also equipped with a laser rangefinder and laser training device. The sight will also enable the pilot to fire the new TOW 2 missile which features a more potent warhead and an improved guidance system. C-NITE is designed to upgrade the two versions of the TOW fire control system now in use, the M-65 with the standard optical sight and the M-65L with the laser augmented airborne TOW (LAAT) sight.

STATUS

A $67 million contract to produce 52 C-NITE systems was awarded to Hughes in August 1987 for completion by June 1990. It is anticipated the production potential will be for 500 systems.

CONTRACTOR

Hughes Aircraft Company, Electro-Optical and Data Systems Group, PO Box 902, El Segundo, California 90245, USA.

C-NITE missile sight fitted to an AH-IS

1806.363

AN/ASX-1 TARGET IDENTIFICATION SYSTEM, ELECTRO-OPTICAL (TISEO)

The target identification system, electro-optical (TISEO) is a high resolution closed-circuit television sensor used for target acquisition, recognition and tracking. The sensor contains a single telescope with two fields of view mounted on a stabilised platform.

TISEO can be slaved to the aircraft radar or operate in an independent manual mode. The system is used for air-to-air and air-to-ground target spotting, lock-on and continuous tracking at ranges far beyond the capability of the pilot's unaided eye.

Four line-replaceable-units (LRUs) comprise the total system. The sensor head/stabilised platform is mounted in the left leading edge wing of the F-4E aircraft. Separate video processor and power supply units are stored within the aircraft. A control panel is integrated in the F-4E rear cockpit. TISEO operator viewing is shared with the F-4E radar monitor.

STATUS

TISEO was developed by Northrop under contract to the USAF and has been in service for over a decade. It is operational as standard equipment on USAF F-4E aircraft and is also deployed on F-4E aircraft of five other countries. Over 500 systems were delivered throughout the 1970s. No longer in full production but Northrop continues to actively support the programme and produce spares.

CONTRACTOR

Northrop Corporation, Electro-Mechanical Division, 500 East Orangethorpe Avenue, Anaheim, California 92801, USA.

6783.363

IRVAT (INFRA-RED VIDEO AUTOMATIC TRACKER)

IRVAT is the computerised tracking component of the A-6 Target Recognition Attack Multi-sensor (TRAM) weapon system. It is being designed to monitor and track ground targets automatically.

STATUS

Under development following the award, by Grumman, of a $8 million contract. The system has reached the flight test stage and pre-production systems were delivered late in 1987. Production is scheduled to begin in mid-1988.

CONTRACTOR

Northrop Electronics Division, Electronic Systems Group, 2301 West 120th Street, Hawthorne, California 90250, USA.

Prototype IRVAT system

6784.353
NORTHROP MAST-MOUNTED SIGHT (MMS) SENSOR SUITE

Northrop is producing the sensor suite for the Mast-Mounted Sight (MMS), a passive multi-sensor system mounted above the rotor of the Bell Textron OH-58D Advanced Scout Helicopter. The MMS gives the crew improved survivability and effective stand-off ranges while acquiring and designating targets.

The sensor suite consists of a FLIR, a day TV sensor, a laser rangefinder/designator (**5758.393**), an advanced digital tracker and an automatic boresight device. The sensors provide for day/night operation through adverse weather and battlefield smoke.

STATUS
In production. The MMS is operational with the US Army. Total production will be 578 units.

CONTRACTOR
Northrop Electronics Division, Electronics Systems Group, 2301 West 120th Street, Hawthorne, California 90250, USA.

Bell OH-58D helicopter with the Northrop MMS sensor suite

3098.353
AN/ASQ-145 LOW LIGHT-LEVEL TV SYSTEM

A passive TV system, the AN/ASQ-145 is a USAF system utilising dual cameras to provide wide and narrow fields of views simultaneously. In addition to production, GE was responsible for system assembly, alignment, and test. The ASQ-145 is a primary sensor of the airborne fire control system used on board the USAF AC-130 aircraft.

STATUS
Operational on board AC-130 aircraft.

CONTRACTOR
GE Company, Aerospace, French Road, Utica, New York 13503, USA.

AN/ASQ-145 wide and narrow field of view cameras for the passive LLLTV system

AN/ASQ-145 low light-level television system (without camera heads)

6580.363
EAGLE EYE III ELECTRO-OPTIC SENSOR

The Eagle Eye III is an electro-optic sensor used to visually identify aircraft beyond normal eye range. It has been designed to provide the F-15 aircraft with the same type of daylight electro-optic long range capability as the TISEO system on the F-4E aircraft (**1806.363**). It consists of a lens assembly and a video camera mounted in the wing-root of the aircraft which provides the pilot with a video image on a cockpit screen.

The lens assembly and camera package is about 60 cm long and has a 25 cm diameter. The lens has both a narrow and wide field of view, selected through the use of a mirror assembly, with the complete unit boresighted on the flight line of the aircraft.

STATUS
In development.

CONTRACTOR
Perkin-Elmer, Applied Optics Division, 7421 Orangewood Avenue, Garden Grove, California 92641, USA.

3569.393
VIRTUAL IMAGE DISPLAY

The Virtual Image Display (VID) system is a cockpit TV-display developed to enable fighter aircraft to use the precision-guided Maverick missile (**1098.311**) or other air-to-surface missiles with TV or imaging infra-red-type guidance.

The display, which uses only a quarter of the space required by a conventional unit, consists of a miniature CRT 2.54 cm in diameter with optics to provide a pilot with a high resolution image more than twice the size of the display unit's lens. This enables the pilot to quickly identify and lock-on to targets at long stand-off ranges.

The optics are designed so that the pilot may look outside the cockpit, then view the display without re-focusing his eyes. A built-in contrast enhancement filter provides a clear display in bright sunlight without using a hood. Remote electronics are used to drive the CRT so that only the optics and the CRT need to be mounted on the cockpit panel. The system is flexible enough to be modified to fit other applications where a compact remote TV display is needed. The electronics unit does not require special cooling systems and can be located remotely in the aircraft up to a 6.1-metre cable length from the display unit.

The display was designed by Hughes under contract from SRA Communications AB, Stockholm and the Swedish Air Material Department to enable the Swedish Air Force Viggen fighter to carry the Maverick missile without costly and extensive modification to the aircraft's cockpit.

STATUS
Hughes manufactured and delivered 110 sets of the VID system, which is currently operational with the Viggen. The system is also operational with the air forces of Singapore and Switzerland.

CONTRACTOR
Hughes Aircraft Company, Radar Systems Group, El Segundo, California 90245, USA.

4215.393
MODEL 200 TISS

The Model 200 thermal imaging sensor set (TISS) provides a passive target imaging capability for naval fire control. The equipment is designed to replace or complement conventional television target sensors on naval fire directors or gimbal sets. Existing video displays, auto-trackers and symbology generators can be used without modification.

The Model 200 incorporates an IR scanner and two-fields-of-view telescope optics. The IR scanner has 20 mercury cadmium telluride detectors arranged in two arrays which are cooled to between 77 and 90 K. Detector cooling can be provided by either a split-cycle refrigeration machine or a high pressure bottle of nitrogen and a Joule-Thompson cryostat. The equipment is designed to be mounted on a two-axis gimbal without loss of imaging performance. Boresight stability is maintained by an internal raster control circuit. For target recognition and surveillance, a built-in grey scale display is provided, to aid the operator in making control settings for optimum imaging performance.

CHARACTERISTICS
Spectral response: 8 – 11.5 µm
Output: Direct video 625 line 50 Hz
Fields of view: 2.4 × 1.8° (narrow); 9.7 × 7.3° (wide)
Resolution: 0.12 mrad horizontal (narrow field-of-view)
Noise equivalent temperature: 0.3° C
Minimum resolvable temperature: 0.5° C
Power: 115 V AC, 60 Hz
Weight: 28 kg (not including cryo-cooler)
Operating temperature range: –25 to +55° C

CONTRACTOR
Ford Aerospace & Communications Corporation, Ford Road, Newport Beach, California 92658-8900, USA.

3809.363

NORTHROP TCS (TELEVISION CAMERA SET)

The Northrop television camera set (TCS), formerly TVSU television sight unit, is an electro-optical tracking system with functional characteristics similar to those of the basic Northrop TISEO (**1806.353**). It features a stabilised closed-circuit television system and precision optics that provide passive target acquisition, identification and tracking at ranges far beyond the capability of the unaided eye. It is packaged in two weapon replaceable assemblies which interface with the avionics and electrical power supply system of the aircraft. TCS interfaces with an on-board video recorder to retain target tracking data. It can operate in an independent manual mode or be slaved to the aircraft radar. The TCS also contains the capability of automatic search and automatic acquisition of airborne targets.

STATUS

TCS was developed by Northrop for installation on the USN F-14 Tomcat fighter to satisfy both air-to-air and air-to-ground missions and an initial production contract was awarded in 1981 to start a multi-year installation programme on new and existing F-14 aircraft. By the end of 1988. TCS will be installed in all operational F-14A aircraft of the US Navy.

CONTRACTOR

Northrop Corporation, Electro-Mechanical Division, 500 East Orangethorpe Avenue, Anaheim, California 92801, USA.

TCS on a F-14A Tomcat

4728.393

NORTHROP FOCAL PLANE ARRAY

The Northrop focal plane array system is an experimental infra-red sensor that enables optical target acquisition systems to find distant targets at night and in adverse weather. The system consists of 16 384 sensors, each 0.001 inch in diameter, arranged in a small cluster and linked to a microprocessor. Infra-red radiation detected by the sensors is processed for conversion into a picture display on the host optical unit's television monitor.

STATUS

Flight testing has been carried out successfully in a US Army helicopter. The system is privately funded by the contractor and no details have yet been disclosed, although possibilities include the Northrop TCS and TAS.

CONTRACTOR

Northrop Corporation, Electro-Mechanical Division, 500 East Orangethorpe Avenue, Anaheim, California 92801, USA.

4011.383

AQUILA MISSION PAYLOAD SUB-SYSTEM

Under contract to Lockheed Missile and Space Company, Westinghouse has developed the sensor package for the US Army's Aquila remotely piloted vehicle (**3168.351**). The system will be used for aerial target acquisition, designation and reconnaissance missions.

The Mission payload sub-system weighs only 60 lb (27.3 kg) and consists of 14 modules that make up the TV camera, target tracker, and laser ranger/designator (**4729.393**), as well as an optical stabilisation system and microprocessor electronics for control and processing. Modular construction of the MPS facilitates adaptation to day/night (FLIR) operation and helps minimise life cycle cost. The Aquila mission payload sub-system embodies previous work carried out by the company in developing the Blue Spot target designator equipment (**3050.383**) described in previous editions of *Jane's Weapon Systems*.

The FLIR system is in development by Honeywell, teamed with Ford Aerospace, using a version of the mini-FLIR system described under **5932.393**.

CONTRACTOR

Westinghouse Electric Corporation, Development and Operation Division, PO Box 1693, MS112, Baltimore, Maryland 21203, USA.

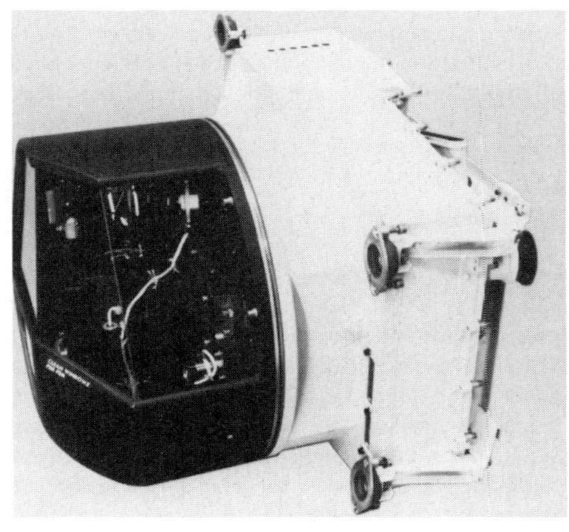

Assembled Aquila payload

4113.393

ROCKWELL MAST-MOUNTED SIGHT

Rockwell International's Missile Systems Division, working in co-operation with ARRADCOM (US Army Armament Research and Development Command), Dover, New Jersey, has developed and tested a mast-mounted sight and target designator for use on ground attack helicopters. The device is mounted at the rotor hub, above the blades, to permit the helicopter crew maximum observation height while enabling the aircraft to remain behind cover as far as possible.

The mast-mounted sight sensor head contains stabilised optics for target detection and recognition at ranges of 3 to 4 km and laser target designation for weapons such as the Hellfire and Copperhead guided weapons. The equipment has been installed and flown on UH-1M and OH-58C helicopters.

STATUS

In performance tests on a UH-1M the equipment has demonstrated its ability to detect targets at a range of 4 km while remaining undetected by 7 × 10 power optical devices on the ground.

CONTRACTOR

Rockwell International, Missile Systems Division, 4300 East Fifth Avenue, Columbus, Ohio 43216, USA.

5758.393

LITTON MAST-MOUNTED SIGHT LRF/D

Litton Laser Systems is producing the laser rangefinder/designator (LRF/D) for the Northrop/US Army Mast Mounted Sight (MMS) (**6784.353**) which enhances the capability of US Army Aviation hunter/killer helicopter teams. Mounted above the rotor, the MMS allows the Aeroscout helicopter to acquire and illuminate targets at greater standoff ranges while concealed from the battle area. Hellfire, Copperhead and other laser guided munitions can now be guided to their targets using the MMS-LRF/D.

By efficient packaging design, the MMS-LRF/D incorporates all the electronics, high voltage power supply, cooling system, range receiver and laser transmitter in one compact lightweight line replaceable unit. The micro-processor controlled LRF/D features a serial asynchronous digital interface for precise communication with the MMS computer. Internally, all sub-assemblies interface via a common bus under control of the main processor, which maintains all system timing through software control. This maximises system flexibility for expansion and alternate functions at minimum cost and complexity.

CHARACTERISTICS

Type: Nd:YAG
Wavelength: 1.064 μm
Pulse rate: All standard codes
Weight: 5.45 kg (12 lb)

CONTRACTOR

Litton Laser Systems, PO Box 7300, Orlando, Florida 32854, USA.

4729.393

AQUILA LASER TRANSMITTER

Litton Laser Systems has developed the laser for the Westinghouse payload on the Lockheed Aquila remotely piloted vehicle for the US Army. The laser operates in conjunction with a television camera to provide intelligence gathering, navigation up-date, range to target information and target illumination.

Lightweight construction and compact packaging are featured in the unit which weighs only just over 3 kg, making it suitable for all types of RPVs or applications where weight is critical. The laser consists of a heat exchanger, cooling system, power supply, pulse forming network, optics and associated electronics.

CONTRACTOR

Litton Laser Systems, PO Box 7300, Orlando, Florida 32854, USA.

3053.383

ATAFCS EQUIPMENT

The airborne target acquisition and fire control system (ATAFCS) is an advanced electro-optical system containing a television camera for daylight operations, a forward looking infra-red (FLIR) equipment for day/night operations, a laser spot detector for target acquisition, a laser transmitter/receiver for target designation and ranging, and an autotracker with both day and night mission capability. It is designed for installation on both attack and scout helicopters, and is intended to increase the combat effectiveness of US Army helicopters.

Development was undertaken by Ford Aerospace & Communications Corporation, and the project was originally known as ALLD (airborne laser locator designator). In August 1974 the US Army Material Command Precision Designator Project Office, Redstone Arsenal issued a follow-on contract valued at about $3.5 million for the building of another four sets of ATAFCS equipment. This made a total of six sets ordered, two having been called for in the initial award.

STATUS

The first equipment was delivered during 1974 and installed on a US Army Bell Cobra helicopter for flight test and evaluation by the US Army Missile Command. Further deliveries were made in 1974 and in 1975.

The system has been tested successfully in both a pod-mounted version and in a configuration integral with the helicopter nose (as shown).

A helicopter mounted ATAFCS has been used successfully in a night test at the White Sands Missile Range, to acquire and mark a moving tank target for a US Army cannon launched guided projectile (CLGP) fired from a 155 mm howitzer at a range of 5 miles (8 km).

The ATAFCS equipment now forms an integral part of the projected Seafire naval fire control sytem (**3727.281**).

US Army Cobra AH-1G ground attack helicopter showing nose mounting of ATAFCS

CONTRACTOR

Ford Aerospace & Communications Corporation, Ford Road, Newport Beach, California 92658-8900, USA.

3981.393

ALT AIRBORNE LASER TRACKER

The Rockwell airborne laser tracker (ALT) is an automatic laser seeker equipment used to automatically search and track laser marked targets. The ALT is in production for use on the AH-1S Cobra helicopter. In addition ALT components are being produced for the AAH-64A helicopter's target acquisition designation system/pilot night vision system (TADS/PNVS) competition (**3609.393**) and units have also been delivered for other helicopter programmes.

The equipment consists of three units: an electronics unit, a control panel, and a receiver unit. The gimballed optics in the last of the three have a five-inch (127 mm) aperture, and the equipment field of regard is 180° in azimuth and +30° to –60° in elevation.

CHARACTERISTICS

Detector: 4 quadrant silicon

Dynamic range: 120 dB (10^6)

Instantaneous field of view: 20° azimuth × 10° elevation

Logic: Last pulse

Optics: 5 inches (127 mm) f 0.3

Primary power: 28 V DC and 400 Hz AC

STATUS

In production.

CONTRACTOR

Rockwell International Corporation, Missile Systems Division, 1800 Satellite Boulevard, Duluth, Georgia 30136, USA.

3609.363

TADS/PNVS

The target acquisition designation sight and the pilot night vision sensor (TADS/PNVS) are two independently functioning systems designed and under development for use with US Army advanced attack helicopters (AAH) to enable these aircraft to operate at extended stand-off ranges, day or night, under adverse weather conditions. TADS provides the gunner with a search, detection, and recognition capability with direct view optics, television, or forward looking infra-red (FLIR) sighting systems; these may be used singly, or in combination, depending on tactical, weather, or visibility conditions. Once acquired, targets can be tracked manually or automatically for autonomous attack with 30 mm guns, rockets, or Hellfire missiles. The laser may also be used to designate targets for remote attack by other helicopters or by artillery units firing laser-guided Copperhead munitions. The TADS equipment consists of a rotating turret on the nose of the helicopter. It houses the sensor sub-systems, an optical relay tube (ORT) at the co-pilot/gunner's position, four electronic units in the avionics bay, and cockpit-mounted controls and displays.

PNVS or AN/AAQ-11 is used by the pilot for night navigation, and consists of an FLIR sensor system packaged in a rotating turret mounted above the TADS; an electronic unit located in the avionics bay; and the pilot's display and controls. The PNVS is slaved to the pilot's helmet display line-of-sight and provides imagery that allows the helicopter to be flown in a 'nap-of-the-earth' fashion at altitudes likely to avoid, or at least minimise, enemy detection.

The laser rangefinder/target designator, provided by Litton Laser Systems, is integrated with the system and provides navigation and attack capability in low visibility and night conditions. It consists of two line replaceable units, a laser transceiver and an electronics unit.

TADS/PNVS system on AH-64A Apache advanced attack helicopter

CHARACTERISTICS

TADS

System coverage: ±120° azimuth; +30 to –60° elevation

Fields-of-view (day): TV 0.9° narrow, 4° wide; Direct view 3.5° narrow (× 18), 18° wide (× 3.5)

Fields-of-view (night): 3.1° narrow; 10.1° medium; 50° wide

PNVS

System coverage: ±90° azimuth; +20 to –45° elevation

Field of view: 30 × 40°

STATUS

In production and in service. Martin Marietta received a contract worth $365 million for 572 systems, the first of which became operational in 1986. This made a total of $1.4 billion for the development/production of TADS/PNVS. In September 1987 the 500th production system was delivered and the contract will be completed in March 1989. In November 1987 Martin Marietta received a $53.77 million contract for another 46 systems with deliveries to be complete by October 1989.

CONTRACTOR

Martin Marietta, Electronics and Missiles Group, PO Box 5837, Orlando, Florida 32855, USA

5932.393

HONEYWELL FLIR SYSTEMS

Honeywell produces a range of forward looking infra-red (FLIR) systems for both manned and unmanned fixed wing aircraft and helicopters. Honeywell FLIRs are serial scan thermal imaging sensors. Energy from the scene is scanned in the horizontal direction with a continuously rotated eight-faceted mirror, and a flat mirror in the vertical direction. The output signal from the detectors is amplified and processed, and then displayed in the standard US and European TV format.

The use of standard TV format allows ready integration with helmet and remote displays, automatic trackers and target cues, and video recorders and processors. Optical modules, monolithic electronics and mechanical cooling allow greater system production and installation flexibility.

HNVS

The Helicopter Night Vision System (HNVS) is a high performance miniature FLIR with a single or multiple field-of-view optics mounted on a two-axis gimbal. It is a compact, spherical shaped sensor package with a diameter of 203 mm and a weight of 11 kg.

FLIR Mini-Pod

The TV compatible FLIR Mini-Pod was designed to aid navigation both day and night. The system's small diameter and low weight preserve vehicle aerodynamics and payload capability. The mini-pod is a simple system and therefore has a lower life cycle cost and higher reliability than more complex gimballed systems, yet it extends a helicopter mission envelope to perform transport ambulance sorties, and to engage in assault missions around the clock. The video output signal can be specified for either the US standard EIA RS170 (525 lines) or the European CCIR-B (625 lines). Dimensions of the system are 292 × 152 mm with a weight of 8.6 kg.

Honeywell Airborne FLIR System

The Honeywell Airborne FLIR System is an infra-red system for day/night navigation, search and reconnaissance. The system is a high performance Honeywell miniature thermal imager mounted on two-axis stabilised gimbal. With a spherical diameter of 305 mm, and a total weight less than 19 kg, aircraft performance and critical payload capacities are virtually unaffected.

The system may include a number of additional features such as automatic correlation and feature match modes, several types of alpha-numeric script, electronic zoom fields-of-view, and various reticles which can be either standard NATO or customer programmed. Operating modes can be controlled by autotracker or autosearch, or helmet sight coupled. Display can be TV compatible, head-up, helmet-mounted or panel-mounted.

The mini-FLIR, plus a Honeywell helmet-mounted display (**1630.393**) and a Ferranti (UK) steerable platform are integrated to form the pilot's night vision system on the Agusta A129 anti-tank/scout helicopter.

Aquila FLIR mission payload

Honeywell is teamed with Ford Aerospace to field the FLIR mission payload system for the US Army Aquila RPV programme. This ultra-lightweight day/night targeting system provides the RPV with a 24-hour capability for target detection, recognition and identification, artillery fire adjustment, laser target detection, reconnaissance and damage assessment.

The Honeywell mini-FLIR is a proven sub-system for the Aquila FLIR mission payload system. It presents the ground-based operator with real-time passive thermal imaging in a television formatted display.

CONTRACTOR

Honeywell Aerospace and Defense, Electro-Optics Division, 2 Forbes Road, Lexington, Massachussetts 02173, USA.

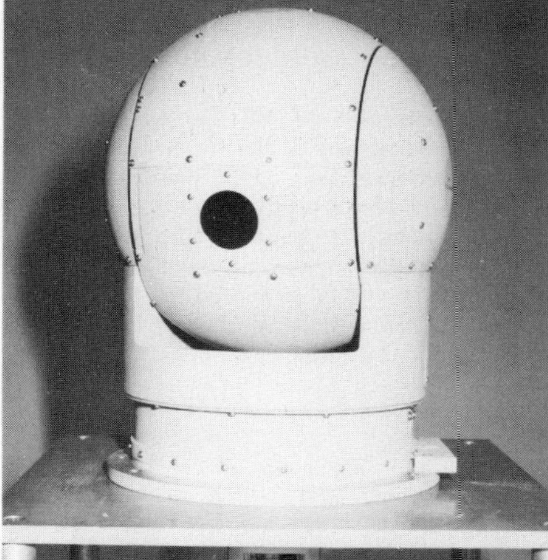

Honeywell HNVS

6781.353

STABILISED THERMAL IMAGING SYSTEM (STIS)

The Honeywell Stabilised Thermal Imaging System (STIS) is an infra-red system for both day and night navigation, reconnaissance, and search and rescue operations. The system uses a Honeywell mini-FLIR mounted on a four axis stabilised platform. Three fields-of-view and a 12 power telescope are integrated into this package to accommodate navigation, as well as target detection and recognition applications. System size (a spherical diameter of 356 mm) and weight (less than 34 kg), make the stabilised STIS applicable to RPVs as well as fixed and rotary wing aircraft.

In addition, the system may include a number of additional features such as multi-mode tracking, several types of alphanumeric script, a laser rangefinder, along with various reticles that can either be standard NATO or customer programmed. Operating modes can be controlled by the optional autotracker, autosearch, or helmet sight coupled. Displays can either be TV compatible, head-up, helmet or panel-mounted.

STATUS

Under development.

CONTRACTOR

Honeywell Military Avionics Division, 2600 Ridgway Parkway, PO Box 312, Minneapolis, Minnesota 55440, USA.

Honeywell Stabilised Thermal Imaging System

6782.353

HELICOPTER INFRA-RED NAVIGATION SYSTEM (HIRNS)

The mini-FLIR and associated electronics are integrated into a 216 mm diameter dome to form the Helicopter Infra-Red Navigation System (HIRNS). This pointing system is IHADSS compatible, weighs 6.35 kg and is readily adaptable to a Ferranti steerable two axis unstabilised platform. Applications for this system include navigation and observation missions, with compatibility to both fixed and rotary wing aircraft. Both HIRNS and the Ferranti gimbal are integrated to form the pilot's night vision system on the Agusta A129 Mangusta anti-tank/scout helicopter.

STATUS

Under development.

CONTRACTOR

Honeywell Military Avionics Division, 2600 Ridgway Parkway, PO Box 312, Minneapolis, Minnesota 55440, USA.

Honeywell Helicopter Infra-Red Navigation System

3807.353

AN/AAS-40 SEEHAWK FLIR

Seehawk (AN/AAS-40) is a thermal imaging system using US DoD Common-Module FLIR components to provide high resolution imagery. Designed for aircraft or surface vessels, the system's current principal application is aboard a US Coast Guard Sikorsky HH-52A helicopter serving the primary role of search and rescue, law enforcement, maritime environmental control, reconnaissance, fisheries control, disaster control, marine and border control and navigational assistance.

The Seehawk system has also successfully completed two years service aboard a Beech 200T aircraft. The system was fully operational in conjunction with other on-board avionics systems, demonstrating maritime patrol, surveillance and other reconnaissance applications. The latest generation

AN/AAS-40 Seehawk is currently installed aboard a modified Sikorsky S-76 helicopter being used as a demonstrator in support of the US Army's LHX programme.

The Northrop AN/AAS-40 Seehawk lightweight FLIR system consists of the turret assembly, the control electronics unit and the power supply unit. FLIR imagery is fed to a cockpit-mounted display.

An automatic scan capability provides constant search coverage in elevation and azimuth. The autosearch mode is enhanced by inclusion of automatic lock-on which reacts to either large or small targets, as selected by the operator.

CHARACTERISTICS

System coverage: Azimuth ±190°; elevation +30 to −105°

Field of view: Wide 30° × 40°; medium 15° × 20°, narrow 5° × 6.7°

STATUS

In development for future service on HH-65A Dolfin helicopters of the US Coast Guard. The AN/AAS-40 Seehawk has also been selected by the China Air Force for installation on the S-2G(T) ASW aircraft. The system is scheduled for delivery in late 1988/early 1989.

Northrop Corporation, Electro-Mechanical Division, 500 East Orangethorpe Avenue, Anaheim, California 92801, USA.

Seehawk system on an HH-52 helicopter

3808.363

AN/AVQ-27 AIRBORNE LASER TARGET DESIGNATOR SET (LTDS)

The AN/AVQ-27 LTDS is rail mounted in the rear cockpit of the Northrop F-5F. The system features small size, light weight, low power, simple operation, and ease of installation and removal. It is packaged and installed so that it does not interfere with other cockpit operations. The LTDS may be installed in most two-seat fighter or fighter-bomber aircraft with only minor wiring modification and without affecting the aircraft's suitability for other missions. The LTDS concept allows the aircraft to be equipped with target designation capability as required by the mission and have it removed when not in use.

The system's principal elements are a stabilised operator sight unit with a retractable periscopic telescope, a laser transmitter, and a 16 mm mission recording camera, all of which are boresighted through common optics in a single self-contained package. The stabilised sighting assembly includes a rate-stabilised mirror that is controlled by a two-axis hand control and provides two tracking modes (rate and rate-aided), a retractable periscopic telescope, and associated optics. The hand control grip contains all the switches necessary to allow for one-hand operation. The laser transmitter consists of a single package that houses a hermetically sealed laser and its associated optics and electronics. The 16 mm mission camera integrated into the LTDS is a version of the same KB-25A/KB-26A standard gun camera now used in the Northrop F-5E/F and other aircraft.

STATUS

LTDS was developed by Northrop to expand the capabilities of F-5B and F-5F aircraft for foreign military sales. Various foreign operators of F-5 aircraft are using, or have awarded contracts for, LTDS.

CONTRACTOR

Northrop Corporation, Electro-Mechanical Division, 500 East Orangethorpe Avenue, Anaheim, California 92801, USA.

LTDS mounted in rear cockpit of F-5F provides a single-package laser designator for laser guided weapons

3065.393

AN/AVQ-26 PAVE TACK

The Pave Tack system is a 24-hour, adverse weather, electro-optical acquisition, laser designator and weapon delivery system produced for the USAF.

The Pave Tack system provides 24-hour capability in clear and adverse weather from all altitudes, including extremely low altitude penetration and interdiction. It provides precise target acquisition and laser target designation for a variety of existing and projected laser, electro-optical and IR weapons, against fixed and moving targets. The system additionally provides target location data to the aircraft weapon delivery digital computer to permit more accurate delivery of free-fall conventional ordnance.

The laser rangefinder/designator (LRF/D), is designated the AN/AVQ-25, designed and produced by Litton Laser Systems. It provides target designation capability for the delivery of laser-guided weapons and range to target information for navigation and conventional munition delivery. The LRF/D system consists of flow line replaceable units, a laser transmitter, a power supply, a laser receiver and a range counter.

The forward-looking infra-red (FLIR) sensor is produced by Texas Instruments.

General Electric produces the virtual image display which is used in the rear cockpit to magnify the image presented on the display. Initially engineered for USAF F-111F aircraft, it is now being applied to expand the weapons delivery capabilities of the RF-4C and F-4E night attack systems.

The Pave Tack pod is divided into two basic sections, a base section assembly and a head section assembly. The latter contains an optical bench on which are mounted the FLIR, laser, range receiver and the stabilised sight assembly. The optical bench is mounted on the turret structure, which rotates in pitch and is driven by turret servo drive motors. The turret is mounted to the head structure, which rotates in roll with respect to the base section. Roll motion of the head and pitch motion of the turret provide complete unvignetted lower hemisphere coverage and a highly stabilised line-of-sight which is fully controllable over the entire lower hemisphere of the aircraft.

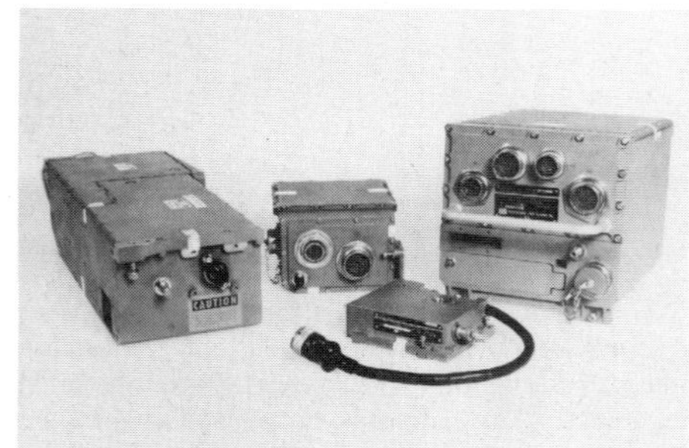

AN/AVQ-25 Pave Tack laser rangefinder/designator

The FLIR is boresighted to (and stabilised with) the laser, and the entire assembly is controlled by the Pave Tack operator aided by a pod-mounted digital aircraft interface unit. The stabilised image and rate-aided tracking mechanisation provide accurate target tracking, laser designation, and weapon delivery throughout a wide spectrum of evasive aircraft manoeuvres. The FLIR has a wide field-of-view display for target acquisition. A narrow field-of-view provides high magnification to enhance target identification and allow precise target tracking.

The electronic assemblies to operate the head section are located in the base section. In addition to the power supply and control electronics, the base section contains a digital aircraft interface unit and a CRT display interface unit. The digital aircraft interface unit is integrated with the ARN-101 digital avionics on board the F-4E and RF-4C aircraft as well as with the digital avionics on board the F-111 aircraft. It provides precise target location data to the aircraft weapon delivery digital computer. The computer additionally provides complete built-in testing for all components on the pod.

A video tape recorder records the video displayed in the cockpit for use in bomb damage assessment and location of targets for future missions.

STATUS

The USAF has taken delivery of 149 pods, the last of these being accepted into service in July 1982. The number of aircraft converted to accept Pave Tack pods is believed to be 60 RF-4Cs, 180 F-4Es and 100 F-111s. Total cost of the 149 systems, plus modifications to the various aircraft is nearly $240 million. In addition, Ford Aerospace has produced pods worth $26.8 million for the RAAF, the last of these were delivered in September 1983.

CONTRACTORS

Ford Aerospace & Communications Corporation, Ford Road, PO Box A, Newport Beach, California 92658-8900, USA.

Texas Instruments Inc, PO Box 660246, Dallas, Texas 75266, USA – subcontractor.

Litton Laser Systems, 2707 South Orange Blossom Trail, Apopka, Florida 32703, USA – subcontractor.

3154.393

AN/ASQ-153 PAVE SPIKE LASER TARGET DESIGNATOR

Pave Spike is a day tracking/laser target designator system that provides the F-4D and F-4E aircraft with a self-contained laser-guided ordnance delivery capability. It is being produced by Westinghouse under production contracts monitored by the Aeronautical Systems Division of the USAF. Designated AN/ASQ153(V), the Pave Spike system is able to acquire, track, and designate tactical surface targets from a manoeuvring aircraft at stand-off ranges usual with laser-guided munitions.

The Pave Spike pod contains an optical sub-system, stabilisation and beam pointing sub-system, TV tracking sensor, laser designator/ranger sub-system, environmental control system, and associated electronics. The aircraft-mounted parts of the system comprise line-of-sight indicator, control panel, and range indicator, and a modified radar control handle and weapons release computer.

Dimensions of the pod are: length 366 cm, diameter 25.4 cm, and weight 193 kg. It is divided into three major sections:

(1) Forward nose/roll section
(2) Umbilical section
(3) Aft electronics section.

The forward nose/roll section rotates about the pod's axis, and the rotating assembly contains the electro-optical sensors and the stabilised gimbal. It is sealed and pressurised with nitrogen and temperature controlled for optimum sensor performance. The umbilical section provides physical and electrical interconnections between the rotating nose section, the aircraft, and the aft electronics section. The last of these contains a thermal cold plate that runs the length of the section. Six major electronic LRUs are mounted on it.

Pave Spike laser target designator pod AN/ASQ-153(V) mounted in one of the Phantom's missile wells

The AN/ASQ-153(V) designates the Pave Spike system as configured for the F-4D and F-4E aircraft, the pod being mounted in the left front Sparrow missile well by means of a standard 30 inch (76.2 cm) centre ECM pod adaptor. It has also been integrated with other NATO aircraft. The Pave Spike pod designation is AN/AVQ-23.

STATUS

The last Pave Spike pod was delivered in August 1977. The designator has also been supplied to other countries including the United Kingdom and Turkey. This system continues in production for FMS (Foreign Military Sales) requirements.

CONTRACTOR

Westinghouse Electric Corporation, Development and Operation Division, PO Box 1693, MS112, Baltimore, Maryland 21203, USA.

4305.393

LANTIRN

LANTIRN (Low Altitude Navigation and Targeting Infra-red system for Night) is a pod-mounted system that will permit attack aircraft to acquire, track, and destroy ground targets with both guided and unguided weapons at speeds exceeding 500 mph. It is planned for the single-seat F-16C/D and F-15E aircraft, but is applicable to other one- or two-seat aircraft in the air-to-ground attack role. With the option to use either or both pods, depending on the particular mission requirement, the system provides operational flexibility and minimises system support impact.

The navigation pod (AN/AAQ-13) contains a wide field-of-view forward-looking infra-red (FLIR) and a terrain-following radar. The FLIR uses a single mercury-cadium-telluride array and the picture is displayed to the pilot on a new wide field-of-view head-up display, providing him a night window for low-level flight. The terrain-following radar enables the pilot to operate at very low altitudes.

The targeting pod contains a stabilised wide and narrow field-of-view targeting FLIR and a laser designator/rangerfinder, in the nose section, Maverick hand-off unit boresight correlator and computers in the electronics centre section and an environmental control unit in the tail section. The targeting pod interfaces with the aircraft controls and displays and with the aircraft fire control system to permit low-level, day/night target acquisition and precision weapon delivery.

The disposition of individual items of equipment can be seen from the illustrations.

In operational use the LANTIRN target acquisition FLIR's two fields-of-view are optimised for target acquisition and recognition. The line-of-sight is cued by the aircraft inertial navigation system. The sensor will be properly oriented for the planned attack and selected weapon delivery profile. This ensures maximum stand-off and first-pass attack for guided and unguided weapon delivery. The system is particularly effective when used for the delivery of such weapons as the infra-red Maverick missile or laser-guided bombs.

The laser rangefinder/designator (LRF/D) has been developed by Litton Laser Systems. It provides the laser designation/ranging essential to the effective delivery and guidance of laser terminally guided weapons.

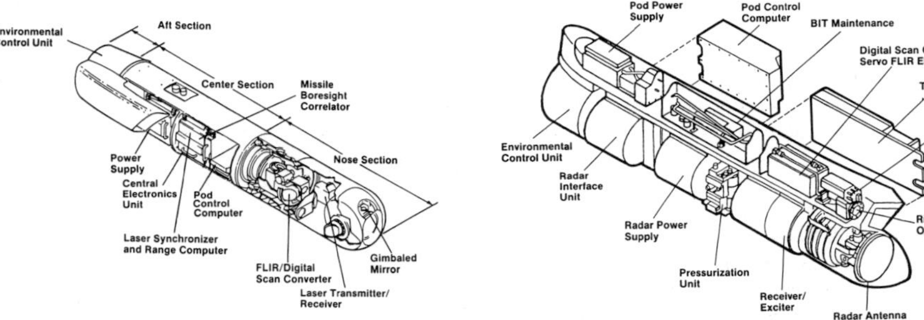

The internal arrangements of the LANTIRN targetting (left) and navigation (right) pods

F-16 fighter with the two-pod LANTIRN system mounted beneath the engine inlet

Modularised, microprocessor-controlled, lightweight and compact, the LANTIRN LRF/D is one of the most advanced systems of its kind. Having a digital interface, the LRF/D provides serial or parallel commands directly to the Pod control computer through the system data busses. The LRF/D is equipped with self-contained, automatic, and initiated BIT functions. All major signals are monitored during operation. The system is equipped with connectors for automatic-test-equipment fault isolation.

The LANTIRN LRF/D consists of three Line Replaceable Units (LRUs): a laser transmitter/receiver, an electronics unit and a high-voltage power supply.

STATUS

The LANTIRN system is in initial production by Martin Marietta. Initial flight testing began in September 1982 using an F-16B outfitted with a wide field-of-view head-up-display manufactured by GEC Avionics (UK) (**4314.393**). Flight testing of operational pods began in late 1983 and continued for three years. In February 1985 Martin Marietta was awarded an $87 million contract to begin production of navigation pods. The first navigation pod was delivered in March 1987 and the first targeting pod in June 1987. The pods will be produced at a rate of 20 per month for each system until November 1990 for the navigation system and July 1991 for the targeting system. The USAF plans to buy 700 pods and 29 field support systems by 1993.

In November 1987 Martin Marietta Overseas Corporation and THORN-EMI Electronics received a project definition contract from the British MoD for a LANTIRN-based targeting system to be fitted to Tornado GR.1.

CONTRACTORS

Prime contractor: Martin Marietta Aerospace, PO Box 5837, Orlando, Florida 32855, USA.

Other suppliers include Texas Instruments (terrain-following radar), Delco Division of General Motors (computers), Sundstrand Aerospace (environmental control systems) and Litton Laser Systems (laser designator).

The LANTIRN programme is directed by the USAF Aeronautical Systems Division, Deputy for Reconnaissance/Strike and Electronic Warfare Systems, Wright Patterson AFB, Ohio, USA.

6780.363

PATHFINDER NAVIGATION POD

Martin Marietta have applied the technology of the LANTIRN system (**4305.393**) to produce Pathfinder (Passive Thermal Forward-Looking Infra-red for Navigation, Detection and Enhanced Resolution), a navigation system for a variety of fixed wing aircraft permitting them daytime tactics during the night.

The pod consists of a forward-looking infra-red (FLIR) system, the power supply and environmental control unit and uses mature line replaceable units from production LANTIRN hardware. The pod is 198 cm long, 24.8 cm in diameter and weighs 84 kg. It has two fields-of-view; 9 × 7° (with an option of 7 × 5°) and 28 × 21° (20 × 15° optional). The wider field-of-view provides the pilot with a general picture of the terrain while the narrower one provides additional resolution and greater detection range.

STATUS

Two systems have been produced for flight demonstrations. One was fitted, in October 1987, to a YA-7E while the second was fitted to a F-16 in November 1987.

CONTRACTOR

Martin Marietta, Electronics and Missile Group, PO Box 5837-MP 455, Orlando, Florida 32855, USA.

Pathfinder system carried under the wing of a YA-7E Corsair

3203.383

AN/ASB-19 ARBS ANGLE-RATE BOMBING SET

The AN/ASB-19 ARBS is basically a simple angle-rate system. The tracker, after locking on a target, provides the aircraft-to-target line-of-sight angle and angle rate to the weapon delivery computer (WDC). This information, combined with the true airspeed and altitude (from the air data computer), is processed by the WDC, yielding the weapon delivery solution. Target position, weapon release, and azimuth steering information are generated for display to the pilot via a head-up display (HUD). Unlike most current operational bombing systems neither measurement of range to the target nor inertial quality platform inputs are required for ARBS.

Once the tracker is locked on to the target, in either the laser or TV mode of operation, the pilot is free to manoeuvre within the gimbal limits of 70° nose down, 35° either side of the nose, 10° nose up, and 450° of roll. This freedom allows him to fly an erratic, unpredictable flight path to confuse enemy AAA/SAM tracking, while simultaneously providing target tracking information to the computer. The accuracy of ARBS is equal to the ballistic dispersion CEP of all free-fall weapons in modern inventories. A stand-off, accurate delivery and identification of targets is made possible by the × 6 magnification of the TV tracker.

ARBS is in production for the US Marine Corps A-4M Skyhawk and AV-8B Harrier, the UK GR.Mk 5 Harrier and the Spanish Harrier II.

In the A-4M, ARBS consists of a dual-mode laser/TV tracker (DMT) designed by Hughes, a System/4 Pi Model Sp-1 general purpose digital computer built by IBM Federal Systems Division under US Navy contract, and a control sub-system. Its weight is about 58 kg. The system works in conjunction with the existing cockpit controls and aircraft sensors. An AN/AVQ-24 HUD by Marconi Avionics is used to display steering information to the pilot.

US Marine Corps AV-8B Harrier equipped with ARBS

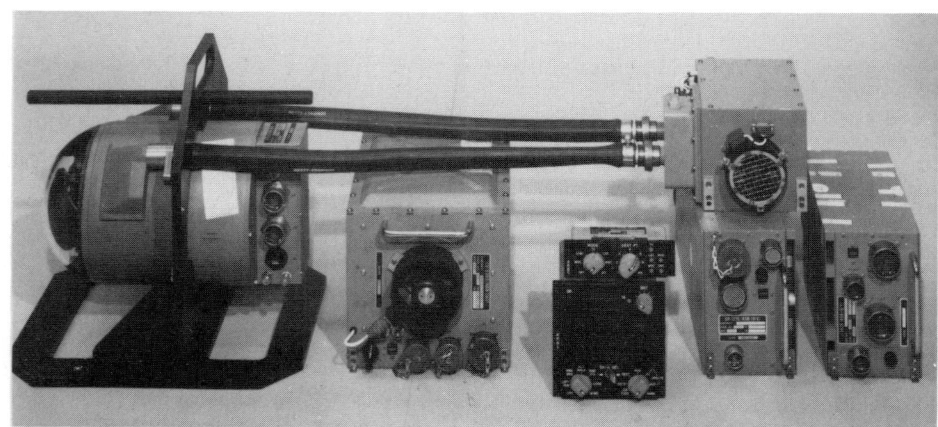

Components of the ARBS showing (from left), receiver/processor, signal data converter, bombing set control (top), control indicator, heat exchanger (top), computer and computer interface

With the AV-8B/GR.Mk 5, Hughes supplies the DMT to McDonnell-Douglas Aircraft Corporation in St Louis where it is integrated into the Harrier's on-board mission computer. ARBS controls, data imagery, video imagery and steering information have been integrated into the HUD, the multi-purpose display (video) and the up-front control set.

The DMT is the main element of the system. A modification of the Hughes STALOC II TV tracker with TV camera shares a common optical element with a laser spot tracker detector on a three-gimbal stabilised platform. Tracking filter electronics, an output signal converter, power supplies, and control logic complete the DMT. A dichroic filter behind the optics separates the laser energy from the visible light for sensing by a four-quadrant laser detector. The visible light is directed to form an image on the TV vidicon.

Operating in the laser mode, the sensor automatically acquires the target and, via the WDC, presents steering signals for the pilot on the HUD. After head-up TV acquisition, the pilot is shown a magnified image of the target on the cockpit TV monitor. Weapon release in both laser and optical tracking modes is either automatic or manual. A weapon data insert panel provides entry into the weapon delivery computer of weapon characteristics and rack type for ordnance being carried on a particular mission.

In addition to the fuselage installation in the A-4 and the AV-8B/GR.Mk 5, a podded configuration is under consideration to give aircraft with radars or other equipment in the nose an accurate, precise air-to-ground capability. Aircraft such as the F-5E and the F-4 are candidates for the pod.

ARBS can be employed with the complete inventory of guided and unguided weapons. Only software update is required to adapt it to new weapons such as the laser and infra-red guided Maverick air-to-surface missiles. It can be given a further night attack capability by replacing the TV vidicon with a Hughes infra-red seeker.

STATUS
The US Marine Corps is accepting deliveries of ARBS for retrofit into the A-4M. Deliveries from the production line are also being made for the AV-8B/GR.Mk 5 Harrier for the US, UK and Spain. Production is expected to last until the early 1990s with a potential of over 500 units.

CONTRACTOR
Hughes Aircraft Company, Missile Systems Group, Canoga Park, California 91304, USA.

5935.393
AN/AAQ-16 NIGHT VISION SYSTEM

The Hughes AN/AAQ-16 night vision system is a microprocessor-controlled, two-axis gimballed forward looking infra-red (FLIR) equipment suitable for a variety of surveillance applications. It consists of a turret/FLIR unit (TFU), a systems electronics unit (SEU), a control unit, and a display unit(s) which can be either panel or helmet mounted. The TFU is the sensor which is used to collect and convert IR radiation to the electronic signals employed by the SEU. It has a 30° × 40° wide field-of-view (FOV) at × 1 magnification, and a narrow FOV at × 6 magnification. A triple FOV model giving magnifications from × 1 to × 9 is also available.

The SEU converts the signals received from the TFU into standard TV format video. Avionic data processing, symbol generation, video processing, turret pointing and built-in-test are microprocessor controlled.

The FLIR turret mounted under the nose of the helicopter or fixed-wing aircraft is controlled via a servo linkage to the pilot's helmet such that his head movements are instantly transmitted to keep the turret aimed along his line-of-sight. A FLIR image is projected on to a holographic combiner on his helmet visor, and can be superimposed with flight symbology if required. A panel-mounted CRT display can also be fitted to the aircraft, although the system can be interfaced with display systems already installed. A pod-mounted version for use on high performance aircraft has also been developed.

Turret/FLIR and electronics unit of the AN/AAQ-16 night vision system

CHARACTERISTICS
Field-of-view: 30° × 40° (wide-angle)
Telescopes: Dual FOV × 1 to × 6 magnification; triple FOV × 1 to × 9 magnification (optional)
Turret angular movement: Azimuth ±210°; elevation +85° to −180°
Slew speed: 2.5 rad/s
Weight: Turret/FLIR unit 19.1 kg; electronics unit 17.3 kg; panel display unit 5 kg

STATUS
In production. In service with the US Army and Customs Service. A derivative has been produced for the V-22 Osprey aircraft and four full scale engineering systems have been delivered. A production contract is anticipated in 1989 with deliveries in 1991.

CONTRACTOR
Hughes Aircraft Company, Electro-Optical and Data Systems Group, PO Box 902, El Segundo, California 90245, USA.

4114.393
OR-89/AA FLIR

The OR-89/AA forward-looking infra-red (FLIR) equipment is a direct derivative of the AN/AAD-4 and AN/AAD-7 FLIR sensors extensively used for the USAF 'gunship' programmes in South-east Asia. The OR-89/AA version is adapted primarily to meet the needs of the S-3A Viking maritime aircraft.

The equipment comprises three main items:

(1) the infra-red sensor package which is designed for gimbal mounting in the aircraft fuselage, with the optical sensor elements projecting below the fuselage underside, and with self-contained stabilisation and servo units
(2) power supply/video converter
(3) infra-red control converter.

Control of the system is via a general purpose digital computer (GPDC) to the FLIR system control converter. The former transmits digital words to the FLIR converter which translates the data into set control commands or analogue servo commands. Position data is supplied by the FLIR system control converter, which converts the data into digital format for transmission to the GPDC for status purposes.

The IR sensors employ HgCdTe detectors and there is automatic optical temperature compensation. The output from the system is in television format.

CHARACTERISTICS
Azimuth coverage: ±200°
Elevation coverage: 0 to −84°
Video output: Composite TV, 875-line RS-343
Weight: 120 kg

STATUS
USN fleet introduction of the OR-89/AA began in February 1973 with deployment continuing up to January 1978. At least 12 other types of aircraft have been fitted with OR-89/AA derivative equipments.

CONTRACTOR
Texas Instruments Incorporated, PO Box 660246, Dallas, Texas 75266, USA.

4120.393
AN/AAR-42 FLIR

The AN/AAR-42 is a pod-mounted forward-looking infra-red (FLIR) equipment, initially designed for use on the A-7E but later modified and used on the A-10 aircraft class. It provides two fields of view displayed on the pilot's head-up display (HUD): the wide FOV gives a 1:1 image of the scene permitting pilot orientation, navigation updating, and target acquisition while the × 4 narrow FOV is employed for target identification and weapon delivery. This two-axis stabilised FLIR sensor enables the A-7 pilot to carry out close air support, interdiction, and surveillance/reconnaisance missions by day or night and in adverse weather conditions.

Other features of the equipment are automatic thermal focus compensation and a slewable line of sight.

Components of the AN/AAR-42 FLIR produced for pod-mounting on A-7E aircraft

CHARACTERISTICS
Azimuth coverage: ±20°
Elevation coverage: +5 to −35°
Video output: Composite TV, 875-line RS-343
Weight: 114 kg

STATUS
Limited production of the AN/AAR-42 for the A-7E began in 1977, with entry into USN service later that year. In an evaluation programme for the USAF in 1979, the equipment was adapted for use on the two seat N/AW A-10 aircraft. This system was pod-mounted and boresighted with the GAU-8 gun and interfaced via a 1553A data bus. Many evaluation programmes have been carried out at Edwards AFB and Elgin AFB during 1983 with satisfactory results. Production is now complete.

CONTRACTOR
Texas Instruments Incorporated, PO Box 660246, Dallas, Texas 75266, USA.

6345.393
AN/AAR-45 FLIR POD

The AN/AAR-45 system is a forward looking infra-red (FLIR) system for the A-7 Corsair II aircraft. The pod consists of a gimballed optics package with thermal sensors, and with refrigeration and electronics further back in the pod. In the AN/AAR-45 system the thermal images are displayed on a raster HUD provided by GEC Avionics (**6336.393**) with other symbology superimposed. This differs from the normal arrangement in most twin-seated FLIR-equipped aircraft which have the image displayed on the radar scope or on a separate display. By displaying the image on the HUD, the pilot can easily keep his eye on both the target image and the relevant instrument readings without moving his head. The pod is 490 mm long, 53 mm in diameter, and weighs 327 kg.

STATUS
In production and in operational service.

CONTRACTORS
LTV Aerospace, Dallas, Texas, USA – prime contractor.

Texas Instruments – design and manufacture of pods.

GEC Avionics – HUD.

4115.393

OR-5008/AA FLIR

The OR-5008/AA is a special derivative of the OR-89/AA forward-looking infra-red (FLIR) equipment (**4114.393**) adapted for use on the Lockheed CP-140 maritime aircraft used by the Canadian Forces. It is mounted in a similar position to the AN/AAS-36 FLIR (**4116.393**) fitted to American P-3 aircraft. The Canadian version is used for search and rescue, maritime surveillance, mapping, ice reconnaissance and various defence tasks.

Additional requirements beyond the facilities of the OR-89/AA included the use of a USN P-3 interface casting; provision for upward viewing (to +5°); additional composite video outputs; facilities for retraction and extension of the sensor head; and reduced system maintenance requirements. The equipment comprises three main items, as does the OR-89/AA, and in this and most other respects the version for the CP-140 is essentially the same as the S-3A Viking equipment.

STATUS

In production.

CONTRACTOR

Texas Instruments Incorporated, PO Box 660246, Dallas, Texas 75266, USA.

4118.393

AN/AAQ-9 FLIR

The AN/AAQ-9 FLIR detecting set was developed for use in the USAF AN/AVQ-26 Pave Tack target designator pod (**3065.393**). The AN/AAQ-9 uses a common module infra-red sensor to provide the aircraft weapon system officer with facilities for location, acquisition, and tracking of enemy targets in day, night, or adverse weather conditions, and to direct weapons against these targets.

The equipment has a two-field-of-view optical system, and there is automatic compensation for aircraft roll movement (derotation). The IR sub-system incorporates automatic thermal and range optical compensation, and has high resolution with very high sensitivity.

CHARACTERISTICS

Magnification: × 2 or × 1

Derotation limits: ± 190° to 100°/s

Video output: Composite TV, 525-line RS-170 or 875-line RS-343

Cooling: Forced air

Weight: 48.6 kg (receiver); 12.7 kg (control/electronics)

STATUS

Entered production in 1977. Operationally deployed by USAF on F-4E, RF-4C and F-111 aircraft. The Pave Tack pod with the AN/AAQ-9 has been purchased for use on Australian F-111 aircraft and other countries have disclosed possible applications.

CONTRACTOR

Texas Instruments Incorporated, PO Box 660246, Dallas, Texas 75266, USA.

4117.393

AN/AAQ-10 FLIR

The AN/AAQ-10 is another derivative of the OR-89/AA FLIR (**4114.393**) which has been developed specially for use on helicopters performing search and rescue missions. A single-window turret design was employed to obtain the increased field of regard needed to utilise the helicopter's capability. The vibration characteristics of the vehicle required particular design emphasis on the stabilisation servo elements. The video output is in TV format to EIA standard RS-343A 875-line specification.

The AN/AAQ-10 FLIR configured for use on the HH-53 helicopter consists of an infra-red receiver, power supply, electronic control amplifier, control indicator, and mounting base. All controls are contained in the control indicator, providing the operator with facilities for one-handed control of sensor operation and servo slew commands.

STATUS

Thirty-one sets were ordered for the USAF and it is understood that these have been delivered.

CONTRACTOR

Texas Instruments Incorporated, PO Box 660246, Dallas, Texas 75266, USA.

3202.383

AN/AAS-33 TRAM TARGET RECOGNITION AND ATTACK MULTI-SENSOR

The A-6E aircraft TRAM programme incorporates an electro-optical sensor package, AN/AAS-33 TRAM detecting and ranging set, into the basic A-6E. The sensors are contained within a 51 cm diameter stabilised turret, which extends below the aircraft's nose radome. Inside the turret are a forward-looking infra-red receiver (FLIR), a laser ranger/designator, and a separate laser receiver for detecting energy from ground or airborne forward air controller lasers which is reflected by designated targets. The basic operational requirement for TRAM is to improve the USN capability for night surveillance and attack and to deliver laser-guided weapons. Naval applications were foremost in design but the system is capable of all general attack missions.

To fulfil these requirements, the system must have gimbal coverage adequate to permit manoeuvring after weapon release, and stabilisation and pointing accuracies consistent with sensor resolution and weapon performance.

In operation, the operator can simultaneously observe both radar and FLIR displays. For a radar significant target, he would normally use radar for acquisition and hand over to FLIR for identification and attack. The turret can sweep through the full lower hemisphere beneath the aircraft, allowing for free manoeuvre after weapon release.

The FLIR has a continuous optical zoom from a wide field of view with × 2.6 magnification to a narrow × 13 field.

STATUS

In 1976 a $21 million contract for production of 36 TRAM equipments was awarded to Hughes, and an extensive technical and operational evaluation programme took place in 1977. The first production TRAM was delivered in early 1979 and a total of about 400 systems have been ordered under current contracts. Although US Naval Air Systems Command issued requests for proposals for a replacement system the US Navy intends purchasing TRAM until FY 1989 for the A-6F.

CONTRACTOR

Hughes Aircraft Company, PO Box 902, El Segundo, California 90245, USA.

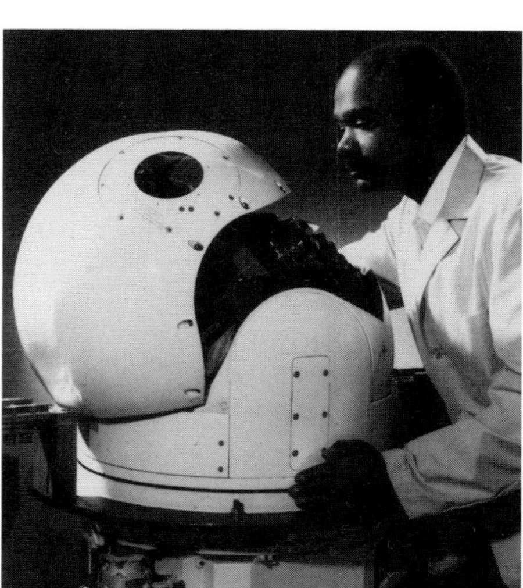

AN/AAS-33 TRAM detecting and ranging set incorporates FLIR, a laser rangefinder/designator and a separate laser receiver

3204.383

AN/AAS-35 PAVE PENNY LASER TRACKER

Pave Penny is the code name of a laser tracking pod initially for use on the A-10 close-support attack aircraft and later the A-7D, F-16 and other types. Target designation is accomplished by forward air controllers on the ground or in another aircraft, and the function of the Pave Penny pod is to acquire and track such designated targets, providing outputs to the pilot's head-up display and/or to weapon delivery system avionics.

The pod is 20 cm in diameter and 83 cm long. Weight is about 14.5 kg. It contains optics gimballed in roll and pitch to focus the incident energy on to a silicon PIN diode detector housed within the pitch (inner) gimbal. In a typical installation, this arrangement permits tracking targets within –90 and + 15° of the aircraft horizontal datum. Azimuth cover is 90° left or right of the aircraft centre line. The pilot is provided with controls that permit selection of seeker scanning patterns and limits, and choice of various modes of operation.

STATUS

Martin Marietta has been awarded at least six contracts worth a total of more than $80 million for production of Pave Penny pods and associated aircraft and support equipment for the A-10 and A-7 aircraft. In March 1977 the first operational Pave Penny/A-10 was delivered, when it was stated that the USAF intends to equip all of the planned 733 of these aircraft with Pave Penny. The first operational unit of the USAF was the 354th Tactical Fighter Wing, in January 1978.

Production of Pave Penny was completed in 1983 with over 700 pods delivered.

CONTRACTOR

Martin Marietta, Electronics and Missiles Group, PO Box 5837, Orlando, Florida 32855, USA.

Pave Penny pod mounted on pylon on USAF A-10 aircraft. Mounts are pre-boresighted so that pods can be installed in a matter of minutes

4116.393

AN/AAS-36 IRDS

The AN/AAS-36 infra-red detecting set (IRDS) is an airborne equipment using IR common modules as the basis of a passive detection system for use on the USN P-3C maritime aircraft. The IR receiver is mounted in an azimuth-over-elevation stabilised gimbal system attached to the underside of the aircraft fuselage to provide extensive lower hemisphere coverage. This permits covert detection of surface ships, surfaced or snorkel-depth submarines, or other targets during darkness or in limited visibility.

In addition to the sensor turret, other items of the equipment include two electronics boxes for control and servo circuitry, and power supplies and video conversion; a video CRT indicator unit; a control panel; and a control stick for target tracking.

CHARACTERISTICS

Azimuth coverage: ±200°

Elevation coverage: +16 to –82°

IR detectors: HgCdTe

Video output: Composite TV, 875-line RS-343

Weight: 136 kg

STATUS

The AN/AAS-36 entered production for USN P-3C aircraft in 1977 and became operational in early 1979. In addition to the USA, numerous other countries equipped with the P-3 aircraft are fitting the AN/AAS-36. Various other FLIRs have evolved from it, such as the AN/AAS-37 (**4119.393**) fitted in the OV-10D aircraft, and the sensor turret has been employed in various other types, including the CH-53 helicopter, Beechcraft E-90, King Air 200 and Cessna Citation II. Total deliveries of AN/AAS-36 equipments to US and foreign services were expected to exceed 400 by the end of 1987.

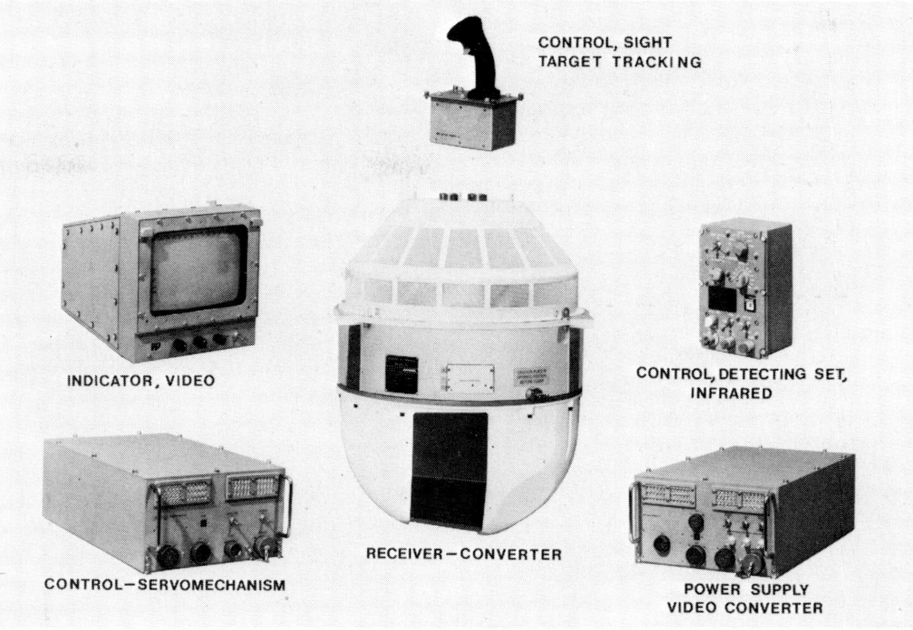

AN/AAS-36 infra-red detecting set (IRDS) system components

CONTRACTOR

Texas Instruments Incorporated, PO Box 660246, Dallas, Texas 75266, USA.

4119.393

AN/AAS-37 DRTS

The AN/AAS-37 detecting, ranging, and tracking system (DRTS) is a special derivative of the AN/AAS-36 infra-red detecting set (**4116.393**) to which a number of additions have been made to meet the requirements of USMC OV-10D multiple mission aircraft. Its capabilities include day or night detection and identification of targets and terrain for surveillance, observation and/or interdiction. A laser provides for ranging and designation of ground targets for laser-guided weapons such as Paveway bombs or the Hellfire missile. The electronics are contained in three boxes; there is a separate pilot's video indicator and three small extra control boxes for the additional facilities of this version.

A direct readout of laser ranges is available, and the designation facility uses PRF/PIM coding. There is rate-aided target tracking (adaptive gate or centroid tracker) and offset tracking from the target or a waypoint is possible.

CHARACTERISTICS

Azimuth coverage: ±200°

Elevation coverage: +15 to –84°

Video output: Composite TV, 875-line RS-343

STATUS

In operational use with the USMC but no longer in production.

CONTRACTOR

Texas Instruments Incorporated, PO Box 660246, Dallas, Texas 75266, USA.

4110.393

(AN/AAS-38) F/A-18 FLIR

The F/A-18 FLIR equipment AN/AAS-38 provides the single-seat, twin-engine F/A-18 Hornet with a 24-hour strike capability in support of USN/USMC requirements for attack aircraft to perform interdiction and close air support missions. The FLIR system provides a day, night and adverse weather attack capability by presenting the pilot with real-time passive thermal imagery, in a TV formated display, for locating and identifying targets. A Laser Target Designator/Rangefinder (LTD/R) has successfully completed flight tests and became operational in 1987.

The 33 cm diameter, 183 cm long, 170 kg system attaches to the lower left side of the aircraft at the Sparrow missile station. The pilot can select a displayed field-of-view of 3 × 3° or 12 × 12° and can control the system's precisely stabilised line-of-sight over a field of regard of 30° × 150°, with ±540° of roll freedom. Accuracies are: stabilisation 35 microradians; pointing 400 microradians; and tracking 50 microradians.

The F/A-18 FLIR system interfaces with the aircraft and other on-board avionics systems through the mission computer, receiving commands directly over a MIL-STD-1553A multiplex data bus. The system provides the aircraft mission computer with accurate target line-of-sight pointing angles, angle rates and a complete initiated and periodic BIT evaluation of system readiness. It is the first Navy FLIR attack sensor designed to perform in a supersonic flight environment. An automatic in-flight boresight compensation for dynamic pod flexure is employed to assure pointing accuracies over the full flight regime.

STATUS

Full scale development began in March 1978 and flight testing began in 1980. Development was completed in 1983 and the first production system was delivered in December 1983.

The F/A-18 FLIR system was operationally deployed by the US Navy in February 1985. Of 201 systems ordered 74 had been delivered by the end of 1987.

AN/AAS-38 FLIR mounted under the fuselage of an F/A-18

A derivative of the AN/AAS-38, Nite Owl (**6577.393**), has been developed by Ford Aerospace and Texas Instruments. It provides increased capability by inclusion of a laser, self-contained cooling and universal mounting provisions.

CONTRACTORS

Ford Aerospace & Communications Corporation, Ford Road, Newport Beach, California 92658, USA.

Texas Instruments Incorporated, PO Box 66046 Dallas, Texas 75266, USA.

Ferranti Defence Systems Ltd, Robertson Avenue, Edinburgh EH11 1PX, Scotland.

6577.393

NITE OWL TARGETING POD

Nite Owl is a derivative of the AN/AAS-38A F/A-18 FLIR LTD/R (Laser Target Designator/Ranger) pod (**4110.393**). The system provides a day, night and adverse weather attack capability by presenting the pilot with real-time passive thermal imagery, in a TV formated display, for locating and identifying targets. Nite Owl consists of 12 line replaceable units (LRUs) which can be readily accessed and replaced. Commonality with the F/A-18 FLIR is more than 85 per cent. Of the 12 LRUs, eight are identical with the AN/AAS-38A, two have minor modifications (Power Supply and Controller Processor) and two are of different design (Environmental Control Unit and Mid-section).

Nite Owl interfaces with the aircraft and other on-board avionics systems through the aircraft fire control computer, receiving commands directly over a MIL-STD-1553 multiplex data bus. The system provides the fire control computer with extremely accurate target line-of-sight pointing angles, angle rates, laser range and a complete initiated and periodic built-in-test evaluation of system readiness. The Nite Owl is designed to perform in a supersonic environment with in-flight boresight compensation for dynamic pod flexure provided by a built in autocollimator system to assure pointing accuracies over the full flight envelope.

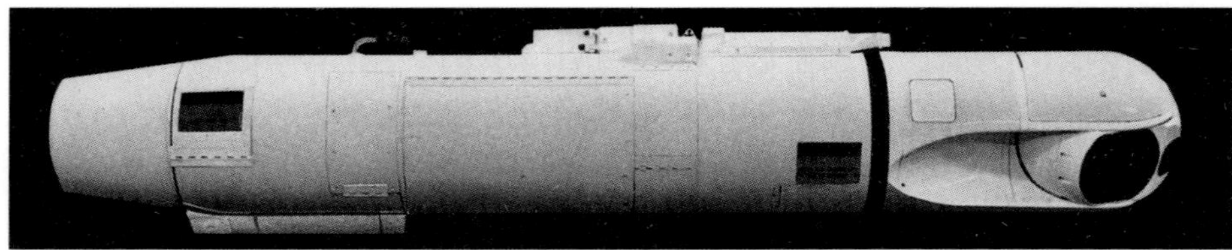

Nite Owl electro-optical targeting pod

CHARACTERISTICS

Field-of-view: Narrow 3° × 3°; wide 12° × 12°

Field of regard: Pitch +30° to −150°; roll ±540°

Dimensions: Length 229 cm; diameter 33 cm

Weight: 178 kg

STATUS

Under a teaming arrangement with Ford Aerospace, Nite Owl has been offered by Ferranti Defence Systems, UK as the targeting pod for the Royal Air Force Tornado aircraft. Extensive flight demonstrations have been carried out on F-15 and F-16 aircraft.

CONTRACTOR

Ford Aerospace & Communications Corporation, Ford Road, Newport Beach, California 92658, USA.

Texas Instruments Inc, PO Box 660246, Dallas, Texas 75266, USA.

6342.393

NIGHT ATAC FIRE CONTROL SYSTEM

The Texas Instruments Night Airborne Target Acquisition Capability (ATAC) fire control system is engineered from US DoD night vision 'common modules' to provide a night targeting system with an autonomous facility for controlling the firing of multiple missiles such as TOW, Hellfire, or Sidewinder from an attack helicopter. This enables the helicopter to be used for anti-armour and close air support purposes in adverse weather and under day and night visibility conditions.

The Night ATAC sensors, comprising a digital FLIR and a three field of view FLIR/day TV, are carried in a nose-mounted turret. Associated electronics/control/interface units are carried separately in the helicopter fuselage. Additional optional facilities include: modular direct viewing; automatic target acquisition; inflight video playback; fire control automation; improved FLIR performance; and enhanced digital processing.

CHARACTERISTICS

Field of regard: ±120° deg (azimuth); +30° to −70° deg (elevation)

Field of view (FLIR): 27.8° × 37° (VWFOV); 6° × 8° (WFOV); 1.8° × 2.4° (NFOV)

Field of view (TV): 6° × 8° (VWFOV); 1.8° × 2.4° (NFOV); 1.3° × 1.75° (VNFOV)

Slew rate: 160°/s

Detection range: Over 6 km

Recognition range: Over 3 km

Turret weight: 150 kg (approx)

CONTRACTOR

Texas Instruments Inc, PO Box 660246, Dallas, Texas 75266, USA.

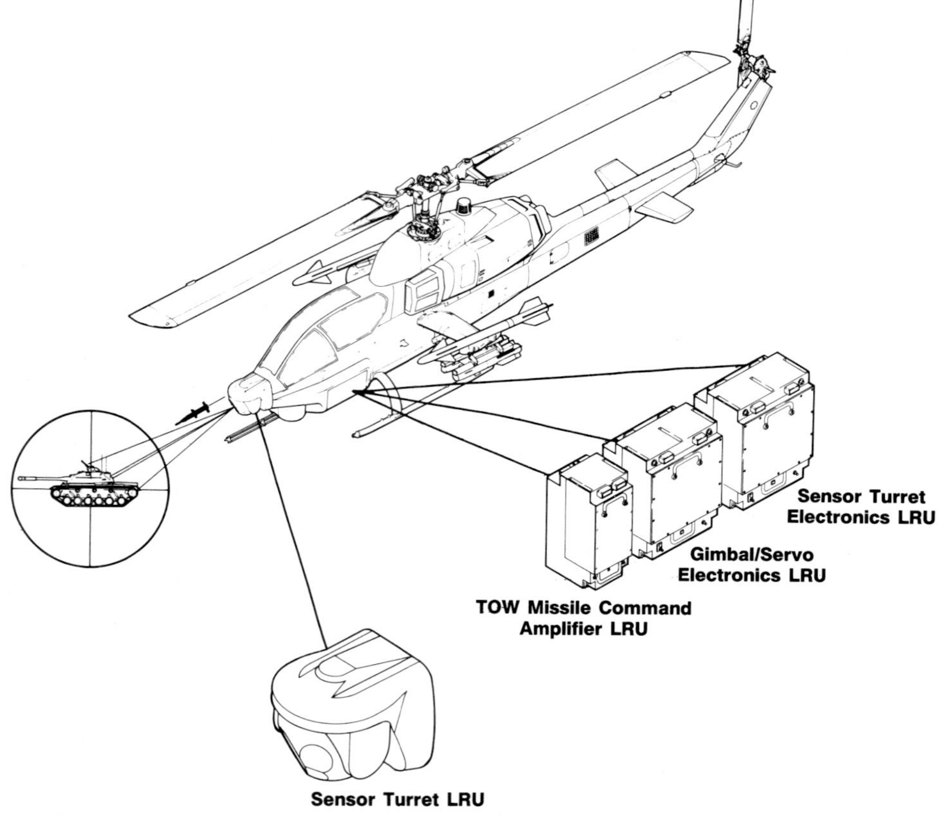

Texas Instruments Night ATAC fire control system for attack helicopters

5933.393

2000 SERIES THERMAL IMAGING SYSTEMS

FLIR Systems is a manufacturer of both standard and specialised infra-red imaging equipment, primarily for airborne platforms. The earlier models 100A and 1000A have been superseded by the 2000 series which had expanded to five models by 1986. In addition to these standard systems, the company has developed special OEM packages and is currently developing a land-based, surveillance-orientated system.

The 2000 series of airborne thermal imaging systems consists of five models varying in resolution and field of view. The simplest system is the basic model 2000A which has two fields of view (28° × 15° and 7° × 3.5°) and excellent resolution at 1.9 mrad (wide) and 0.47 mrad (narrow). The system is mounted in an unstabilised, unaided positioning system and is used primarily for surveillance, law enforcement, and search and rescue. The most sophisticated system is the 2000G, also with two fields of view (28° × 15° and 5° × 2.6°) but with higher resolution, 1.4 mrad in wide and 0.25 mrad in narrow field of view. The 2000G is also equipped with an inertial pointing to aid tracking, and electronic caging to lock the system in a fixed orientation relative to the axis of the aircraft. This allows the 2000G to be readily adapted for low-cost fire control systems.

All models are available in both US and European video standard (EIA RS170, CCIR) and are compatible with conventional video accessories, including recorders, downlinks and digital video processors.

STATUS

In production and operational service.

CONTRACTOR

FLIR Systems Inc, 16505 S.W. 72nd Avenue, Portland, Oregon 97224, USA.

5750.383

IRSTS (AIRBORNE INFRA-RED SEARCH AND TRACK SYSTEM)

An airborne infra-red search and track system (IRSTS) is under development by GE which will conduct the design, fabrication and flight evaluation of the system. The equipment employs modern focal plane array technology and sophisticated signal/clutter processing techniques. Relatively unknown as compared with FLIR technologies, the IRSTS sensors are seen as the key in combating the threats anticipated from the mid-1980s onwards.

IRSTS is able to track targets of thermal energy at extremely long ranges with high accuracy and represents a significant 'force multiplier' effect in that a multiplicity of targets can be tracked simultaneously so that weapon management can be optimised effectively. The IRSTS sensor is also passive in operation and consequently relatively immune to countermeasures.

STATUS

GE received a $10 million contract to design, develop and produce IRSTS with deliveries from March 1988 while Martin Marietta was selected as second source producer in June 1987. Total production options are reportedly worth $100 million.

CONTRACTOR

GE Aerospace, French Road, Utica, New York 13503, USA.

3837.093

CONTRAVES DAY/NIGHT REMOTE VIDEO EQUIPMENT

This is a product of the Contraves Goerz concern which consists of a long range video surveillance system capable of operating under all lighting levels. Its electro-optical system is completely dome-enclosed on an elevation platform that provides a gyro-stabilised line-of-sight together with pan and tilt functions.

The system permits the use of separate electro-optical systems for day and night. A choice of different day or night modes, focal lengths, filtering, and focusing are controlled remotely from the control console.

There is a portable version which is also remotely operated and battery powered. In this model the TV camera is tripod mounted and controlled from a console which can be located up to 50 m away.

No further installation, sensor, or performance details are available for publication.

STATUS

No details have been made available.

CONTRACTOR

Contraves Goerz Corporation, 610 Epsilon Drive, Pittsburgh, Philadelphia 15238, USA.

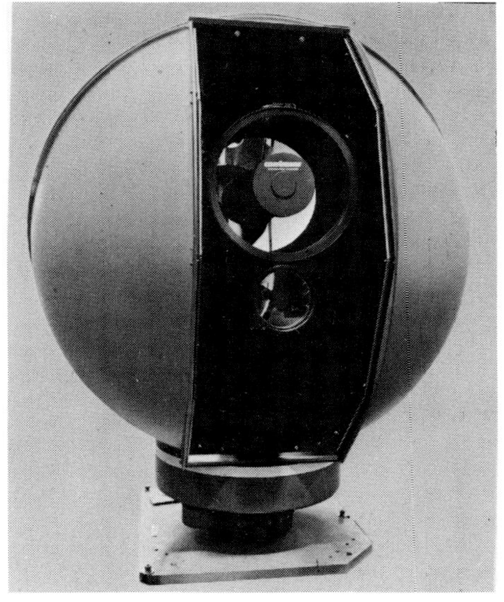

Spherical housing of Contraves Goerz day/night remote video surveillance equipment

3846.393

AN/AXQ-14 WEAPON CONTROL EQUIPMENT

The AN/AXQ-14 is a two-way data link weapon control and guidance system developed by Hughes Aircraft Company for the GBU-15 precision-guided weapon system (**4190.311**). It consists of three major sub-systems: aircraft data link pod, aircraft data link control panel, and weapon data link. The aircraft pod, developed by Hughes and certified by the USAF for supersonic flight, contains four line-replaceable units: an electronics unit containing all RF generating and receiving equipment, two antennae, a de-multiplexer to decode aircraft commands and pod control signals, an encoder, and antenna controls; an electronically phase-scanned antenna which tracks the weapon during normal operation; and a second forward horn antenna to provide additional antenna coverage. Provision has also been made for inclusion of a mission tape recorder to maintain a permanent record of video data provided by the electro-optical seeker in the weapon. The aircraft data link control panel is used with an existing display in the command aircraft to enable the weapon systems operator to interface with the weapon's guidance system during a mission. The weapon data link mounted on the rear of the GBU-15 provides the link between the weapon and the command aircraft's data link. The system has been successfully flight tested on a variety of USAF and USN aircraft and is compatible with modern high performance aircraft including F-4, F-15, F-16, F/A-18, F-111, A-7 and B-52.

STATUS

The AN/AXQ-14 is currently in production for the USAF and other customers.

CONTRACTOR

Hughes Aircraft Company, Radar Systems Group, El Segundo, California 90245, USA.

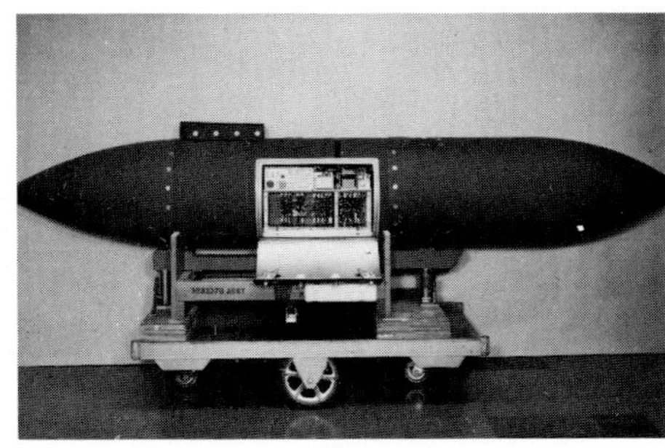

AN/AXQ-14 data link pod contains electronics, coder and decoder units, an electronically-scanned antenna for tracking the weapon in normal operations, a forward horn antenna for extended coverage, and recording equipment

AN/AXQ-14 data link system providing two-way video and command link facilities is used for direction of GBU-15 'launch-and-leave' weapon seen beneath this F-4 Phantom's wing. AN/AXQ-14 data link pod is on aircraft fuselage centre-line

5752.393

HUGHES HEAD-UP DISPLAY

This latest head-up display (HUD) developed by Hughes uses diffraction optics techniques to provide pilots with a clearer and wider field-of-view. The prime advantage of diffraction optics is that the combiner can be produced with a wide field-of-view such that IR or low-light TV imagery can be displayed on the HUD for high speed low altitude night missions.

The combiner acts as a mirror that reflects only the selected wavelength of light used to produce the symbology on the display. This increases the pilot's ability to see through it to the outside world (transmissivity) to 85 per cent as compared with 50 to 70 per cent for convential HUDs.

Pilot orientated features designed into the HUD include a single piece combiner to avoid bulky support structures, symbology bright enough to be read in very high ambient light conditions, no cockpit reflections and no spillage from the HUD into the cockpit.

STATUS

The new diffraction optics HUD has been selected for the JAS-39 Grippen, Sweden's new multi-role combat aircraft which is due to fly in 1988 and enter service in the 1990s. A full-scale development and production contract was awarded to Hughes by SRA Communications AB, Sweden in January 1983. The first flight test model of the HUD was delivered in November 1984.

CONTRACTOR

Hughes Aircraft Company, Radar Systems Group, El Segundo, California 90245, USA.

JAS-39 head-up display undergoing calibration

6343.393

SRL NIGHT VISION/HEADS-UP DISPLAY

The SRL Night Vision/Heads-Up Display (NV/HUD) provides high resolution flight and navigation data superimposed within the field-of-view of night vision goggles. The system collects the display information from the aircraft sensors, in both analogue and digital form and processes this to provide the desired visual presentation.

The equipment consists of a Video Graphics Generator (VGG), a dual output Video Symbology Display (VSD), two coherent fibre-optic cables and combiner assemblies and a panel-mounted control unit. The dual display provides simultaneous HUD capability to both the pilot and co-pilot.

The VGG digitises and formats the received data into composite high resolution video graphics. The data is then passed to the VSD, which is a raster scan video monitor with two 1-inch CRTs and is converted to an optical display. An optical relay lens system focuses the CRT image onto the fibre-optic cable for transmission to the combiner system which presents a focused-at-infinity image to the input lens of the night vision goggles.

In addition to providing the NV/HUD graphic display, a composite video signal is available for head-down presentations on a standard 875 line video monitor.

CONTRACTOR

Systems Research Laboratories (a division of Arvin/Calspan), Defense Electronic Systems, 2800 Indian Ripple Road, Dayton, Ohio 45440-3696, USA.

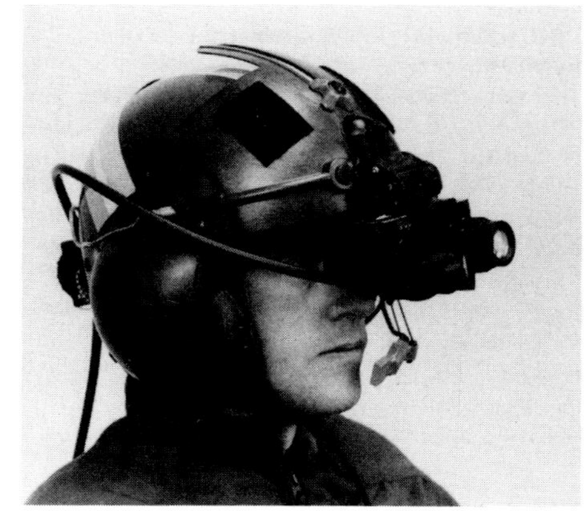

SRL night vision goggle with optical combiner for HUD

6578.393

ES-250 STANDOFF ELECTRO-OPTICAL RECONNAISSANCE SYSTEM

This standoff imaging system incorporates a 10 240 element charge coupled device (CCD) in the focal plane. With a 100-inch (254 cm) focal length, the system is operational at medium to high altitudes, and has the capacity of dawn to dusk operation. Temperature preconditioning or pressure altitude corrections are not necessary. The system has depression angles that can sweep a 170° arc. The imagery can be viewed in real-time in the cockpit and can also be transmitted to a ground station, via a data link. A digital magnetic tape recorder is incorporated for retrospective analysis. The camera and electronics can be mounted either internally or externally in a pod.

CONTRACTOR

Litton Itek Optical Systems, 10 Maguire Road, Lexington, Massachusetts 02173-3199, USA.

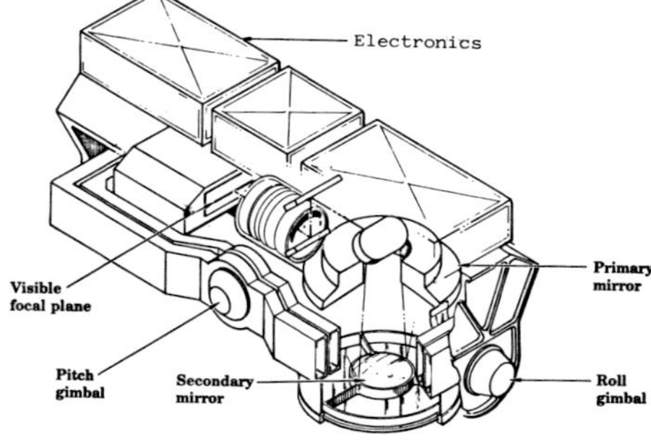

Drawing of the ES-250 reconnaissance pod

5934.393

ITT ANVIS

The ITT ANVIS (Aviator's Night Vision Imaging System) is a helmet-mounted, binocular-sized, high-performance night vision goggle designed for covert work. It incorporates two third-generation image intensifier tubes, although second-generation tubes can be fitted for applications not requiring such a high performance. The goggle provides full peripheral vision and will give full vision even under overcast starlight conditions.

CHARACTERISTICS

Field-of-view: 40°

Scene illumination: 10^{-5} to 1 ft candles

Weight: 0.86 kg

STATUS

In service.

CONTRACTOR

ITT Electro-optical Products Division, 76356 Plantation Road, Roanoke, Virginia 24019, USA.

5936.393

HUGHES ANVIS

The Hughes Aviator's Night Vision Imaging System (ANVIS) is a helmet-mounted, unity power image intensifier binocular which allows night-time, low flying or contour flying in either fixed-wing aircraft or helicopters. The system is compatible with either second or third-generation image intensifier tubes, and can be used in conjunction with FLIR systems. The binocular is lightweight, can be fitted to any aviator helmet and provides full peripheral vision. The complete ANVIS system consists of the binocular assembly, image intensifier tubes, helmet visor interface and a battery power pack.

CHARACTERISTICS

Field-of-view: 40°

Magnification: × 1

Binocular weight: 463 g

STATUS

In service with the US Army.

CONTRACTOR

Hughes Optical Products Inc, 2000 S Wolf Road, Des Plaines, Illinois 60018, USA.

6785.363

LASER DETECTOR-TRACKER/STRIKE CAMERA (LDT/SCAM)

The Laser Detector-Tracker/Strike Camera (LDT/SCAM) is an integrated sensor pod system designed specifically for the McDonnell Douglas F/A-18 Hornet to enhance its operational effectiveness irrespective of the time or weather. The pod, which is 228.6 cm long, 20.3 cm in diameter and weighs 73.5 kg, consists of five weapon-replaceable assemblies; the LDT, an interface unit, an adapter, the SCAM and a rotary mount.

The LDT consists of an inertially stabilised, gimballed optical receiver with a four quadrant photodiode, sensors and a laser pulse decoder. It seeks, acquires and tracks pulse-coded energy reflected from targets designated by ground-based or airborne forward observers. When a target is acquired and tracking begins the data is fed to the aircraft's mission computer for weapon aiming and release. The LDT provides visual target cueing to the pilot through the HUD.

The SCAM is a 35 mm variable frame-rate panoramic camera housed in a rotating unit in the rear of the pod. The SCAM automatically tracks the target and is independently controlled by the mission computer. It provides high resolution photographic coverage of targets before, during and after an attack removing the need for separate damage-assessment sorties.

STATUS

Under development. Martin Marietta have received a $50.2 million contract from McDonnell Douglas to manufacture 84 LDT/SCAM systems.

CONTRACTOR

Martin Marietta Aerospace, PO Box 5837, Orlando, Florida 32855, USA.

6786.393

AIRBORNE ELECTRO-OPTICS AND SDI

A series of airborne experiments and concepts with Strategic Defense Initiative (SDI) implications are underway or being proposed.

Airborne Optical Adjunct (AOA) Experiment

This is to evaluate the feasibility of using Long Wavelength Infra-Red (LWIR) sensors on an airborne platform to support late mid-tier and terminal defences against ballistic missiles. The experiment is to validate the capability of detecting and tracking incoming warheads from an aircraft flying at 40 000 ft (12 192 m), distinguishing them from decoys then passing the data to a ground-based radar system which will then conduct interception operations using non-nuclear interceptors.

The AOA uses a Boeing 767 with a Hughes-made sensor subsystem of cryogenically cooled LWIR detectors mounted on a focal plane assembly. This will be housed in a 2.43 m high cupola. There will be a Honeywell data processor, while the navigation, sensor update and communication equipment together with the control display consoles will be provided by Boeing Aerospace. Boeing will provide the mission data recording equipment.

Airborne Laser Experiment (ALE)

ALE is designed to validate the performance of a long range measurement sensor as it operates with the AOA LWIR. Specifically it will investigate the passive-to-active sensor target hand-over and the functional requirements for operating a laser sensor on an aircraft.

The system includes a laser transmitter-receiver and associated beam control unit together with the hardware and software for processing the signal and data. Also provided are power generation equipment and cooling equipment. The programme is being conducted by Boeing and will be completed by the end of 1988.

Airborne Optical Surveillance (AOS) Concept

This is the subject of a concept definition study for a future SDI tactical subsystem using the result of the AOA. Under consideration are the payloads of the integrated passive and active optical sensors, the processors and communications equipment together with the high altitude aircraft required.

6787.353

ADVANCED TACTICAL AIR RECONNAISSANCE SYSTEM (ATARS)

The Advanced Tactical Air Reconnaissance System (ATARS) is to be an electro-optical sensor suite which will provide battlefield commanders with target data through near real-time tactical reconnaissance. The sensor will include an infra-red sensor for night time operations and two visual-spectrum sensors for operations in the day. The system must be capable of operating at any time and in all weathers and incorporate a wide field-of-view, low altitude sensor and a stand-off sensor for medium altitudes.

The project is led by the US Air Force but will be applicable to the US Navy. It is intended to retrofit ATARS into the RF-4C then into a US Navy mid-range RPV. Subsequently ATARS will be used to upgrade the US Navy's Tactical Air Reconnaissance Pod System (TARPS) on the F/A-18 Hornet and F-14 Tomcat.

6788.363

CRUISE MISSILE ADVANCED GUIDANCE (CMAG)

In June 1987 McDonnell Douglas and General Dynamics were awarded cost plus award fee/cost share contracts worth \$7 million and \$5.5 million respectively to develop a carbon dioxide laser radar Cruise Missile Advanced Guidance (CMAG) system. The systems are designed to permit cruise missiles to be flown in poor weather at low level with mid-course navigation made possible by position-fix and doppler velocity updates. Development is to be completed by August 1989.

ADDENDA

FINLAND

6789.261

NOKIA COASTAL DEFENCE AND SURVEILLANCE SYSTEM

This Finnish system is designed to provide efficient means of surveillance and control of high density maritime traffic but also has coastal defence applications. In these it provides both coastal surveillance and automatic fire control for artillery and missile units. It consists of Local Command Centres (LCC) and District Command Centres (DCC) supported by a range of surveillance and target acquisition sensors, together with a communications system with ECCM features.

The LCC consists of a computer system for data acquisition, data processing, presentation and communications, a main control desk with 16 inch (40.6 cm) PPI displays, an alphanumeric display, a remote radar control panel, joystick and control keys, an auxiliary display and control desk with PPI display, joystick and control keys, alphanumeric terminals for command and measurement data input and optional multi-colour CRTs. Communications include telephone and radio links for speech and digital data communications, encryption equipment and a printer. Communications equipment includes the Nokia DA 8520/8522 digital message terminals and the Nokia MS 8525 modem. The LCC is linked to unmanned radar stations, MA 8530 laser rangefinders and electro-optical directors/night observation devices by cable or radio link.

DCCs have similar equipment, although they are rarely linked to sensors, but they have improved processor capacity. They control a number of LCCs and their role is to obtain the overall picture using synthetically generated maps and symbols on multi-colour graphic displays.

STATUS

In Finnish service.

CONTRACTOR

Nokia Industrial Automation, PO Box 780, SF-00101, Helsinki 10, Finland.

GERMANY (FEDERAL REPUBLIC)

6790.461

KRUPP ATLAS ISUS (INTEGRATED SENSOR AND UNDERWATER SYSTEM)

ISUS, which is used by the German Navy as SLW 83, is a compact, lightweight fire control system developed by Krupp Atlas Elektronik (KAE) for submarine use. It incorporates the CSU 83 sonar (**4229.453**) with modular hardware and software elements for tactical data handling and weapon control, all linked to sensors and data links by a digital highway.

Sonar data is subject to Target Motion Analysis and automatic target classification before being combined with ESM and periscope data in the Tactical Data Handling Section (TDHS). The TDHS then generates the combat situation display together with geographical data. The system, which can analyse data from 24 targets simultaneously, can control up to three weapons simultaneously either from one console or a number of positions.

The system uses Multi-Function Common Consoles (MFCCs) whose numbers vary with the size of boat; two to three in boats under 500 tonnes, three to four in boats of 500 to 1000 tonnes, five in boats of 1000 to 1800 tonnes and six in larger vessels. Each console has two 15 inch (38.1 cm) triple-colour raster-scan displays arranged vertically, the top display is usually used for tabular data and the lower one is normally a PPI display. They are software-run and are usually operated in a dedicated role, eg in a submarine of less than 1000 tonnes two of the MFCCs will be used with the sonar suite and the remainder for tactical data handling and fire control.

The total system has 23 KAE EPR 1300M/2300M processors based upon the Motorola MC68020 microprocessor. There are about 15 Mbytes of memory in the system.

STATUS

In production and in service with the German Navy in Type 206A submarines.

CONTRACTOR

Krupp Atlas Elektronik, Sebaldsbrucker Heerstr 235, PO Box 448545, D 2800, Bremen 44, Federal Republic of Germany.

ITALY

6912.153

RAT-33S/M COASTAL RADAR

A medium range S-band combined air and search radar designed for use in coastal and off-shore surveillance and control centres. It employs an advanced technique for data extraction, suitable for narrowband transmission, with a high probability of detection while maintaining a low false alarm rate even in adverse environments. It can be remotely controlled via a narrowband data link.

The system reacts to the continuously changing environment and operates a wide range of automatically selected techniques to ensure maximum performance. In practice this means that it is able to distinguish between the various types of echo (clear areas, ground clutter, interference etc) and, having recognised the type, automatically selects the best possible configuration for the environment and geographical zone.

CHARACTERISTICS

Frequency: E-band (S-band)
Antenna: Dual beam type
Polarisation: Linear, circular and elliptical
Rotation rate: 15 rpm
Transmitter: High power, PRF staggering
Receiver: STC, MTD processor, doppler filters adaptive selection, azimuth correlator
Range: Greater than 70 nm on a fighter aircraft
Range accuracy: 50 m
Angular accuracy: 0.35°

STATUS

In production.

CONTRACTOR

Selenia-Elsag Naval Systems Consortium, Via di S. Alessandro 28/30, 00131 Rome, Italy.

UNITED KINGDOM

6913.253

HYDRA SONAR SYSTEM

Hydra is a highly versatile and advanced family of sonar systems designed for both surface ship and submarine applications. It is based on the latest acoustic, data and software modules. The modular architecture of the system makes it possible to address the complete spectrum of submarine applications, from self-protection passive systems for the smallest 100 ton boats, through the powerful active/passive systems required by the largest conventional submarines, to the highly sophisticated equipment required by nuclear-powered hunter/killer vessels.

STATUS

A £4 million contract has been received from the Swedish MoD to update the Sjoeormen (Sea Serpent) submarines of the Royal Swedish Navy.

CONTRACTOR

Plessey Naval Systems, Templecombe, Somerset BA8 0DH, England.

UNITED STATES OF AMERICA

6791.293

KOLLMORGEN TYPE 8/15/18 PERISCOPES

In addition to the Model 76 periscope (**6775.293**) Kollmorgen have three other periscopes in production; the Type 8L, Type 15L and Type 18.

The Type 8L and Type 15L are widely used by the US Navy and the Mod (T) version is in production for the 'Ohio' class of SSBNs, which carry the Trident ballistic missile. A Trident periscope shipset consists of two periscope sets for general surveillance. Each consists of a periscope assembly, electrical and electronic adapter assembly, travelling slide connector assembly, junction box indicator and control box.

The periscopes are located in the command and control centre with periscope No 1, which includes the AN/BPS-15 radar antenna on the starboard side and periscope No 2, with ESM antennas, on the port side. The periscopes also include antennas for external communications and navigation satellite communication. A Nikon F 35 mm camera and a television camera are used with these periscopes.

Both periscopes have a Field-Of-View (FOV) of 32° (x1.5 magnification) and 8° (x6 magnification). Their line-of-sight elevation is +60 to −10°.

The Type 18 periscope has been designed for use by the 'Sturgeon' and 'Los Angeles' class SSNs and has also been exported for use in diesel-electric submarines. The Type 18 is a multi-purpose instrument with multiple levels of optical magnification for use as a surveillance or attack aid. It features gyro-stabilised optics, image motion compensated photography, image intensifiers and low light television as well as providing a broad spectrum of RF intercept capability.

Two versions are produced; the Type 18B with an overall length of 36 ft 10 in and the Type 18D with an overall length of 42 ft. Otherwise their characteristics are identical.

The Type 18 magnification rates and FOVs are; x1.5 (32°), x6 (8°), x12 (4°), x24 (2°). The elevation and depression characteristics are identical with the Type 8L/15L.

STATUS

In production.

CONTRACTOR

Kollmorgen Corporation, Electro-Optical Division, Northampton, Massachusetts 01060-2390, USA.

ANALYSIS

Italic type indicates that equipment is described in detail in previous editions of Jane's Weapon Systems.

Missiles

- Strategic 906
- Tactical land-based surface-to-surface 907
- Tactical shipborne surface-to-surface 910
- Naval surface-to-air 912
- Air defence 913
- Air-to-air 915
- Air-to-surface 916

Radar

- Land-based 919
- Naval 930
- Airborne 935

Sonar equipment 940

Miscellaneous underwater detection equipment 944

Torpedoes 945

Electronic warfare equipment 947

NATO designations of Soviet systems and equipment 955

Readers requiring tabular information on land artillery, armoured fighting vehicles, naval vessels, and military aircraft will find illustrations and information in the appropriate companion Jane's year book.

2202.114

Strategic Missiles

Name or NATO code	Entry no	US code	No of stages	Propellant	Type of warhead	Single warhead	Alternative warheads	Range (km)	Status	Deployment	CEP
Strategic/Intercontinental Ballistic Missiles (ICBM)											
CHINA, People's Republic											
DF-4	**4210.111**	CSS-3	2	Liquid	Nuclear	Estimated 2 MT	None known	7000	Operational	8	N.A.
DF-5	**4211.111**	CSS-4	3	Liquid	Nuclear	Estimated 1–5 MT	—	10 000	Operational	5 launchers	N.A.
UNION OF SOVIET SOCIALIST REPUBLICS											
Sego	**4207.111**	SS-11	2	Liquid	Nuclear	950 KT (Mod 1) 1 MT (Mod 2)	3 MRVs at 100-300 KT (Mod 3)	10 000 (Mod 1) 13 000 (Mod 2) 10 600 (Mod 3)	Operational	470	1400 m (Mod 1) 1100 m (Mods 2/3)
Soviet RS-12/ SS-12 (Savage)	**2958.111**	SS-13	3	Solid	Nuclear	750 KT	None known	9400	Operational	60	1800 m
Soviet RS-14 (Sinner)	**6190.111**	SS-16	3	Solid	Nuclear	N.A.	—	9000	Uncertain	50 dismantled at Plesetsk	N.A.
Soviet RS-16 (Spanker)	**2991.111**	SS-17	2	Liquid (cold-launch)	Nuclear	3.6 MT (Mod 2)	MIRV 4 × 200 KT (Mods 1 and 3)	10 000 (Mods 1 and 3) 11 000 (Mod 2)	Operational in 1975	150 in converted SS-11 silos	N.A.
Soviet RS-20 (Satan)	**2992.111**	SS-18	2	Liquid (cold-launch)	Nuclear	25 MT (Mod 1) 20 MT (Mod 3)	MIRV 8-10 × 500 KT (Mods 2 and 4)	12 000 (Mod 1) 11 000 (Mods 2 and 4) 16 000 (Mod 3)	Operational in 1974	308	250 m (Mod 4)
Soviet RS-18 (Stilleto)	**2993.111**	SS-19	2	Liquid	Nuclear	5–10 MT (Mod 2)	MIRV 4-6 × 200 KT (Mod 1) 6 × 500 KT (Mod 3)	10 000	Operational in 1974	360	300 m
Soviet SS-23 (Scalpel)	**4778.111**	SS-24	3	Solid (cold-launch)	Nuclear	—	MIRV 8-10 × 300-500 KT	10 000	Operational in 1987	Rail-mobile	<200 m
Soviet SS-12M (Sickle)	**4779.111**	SS-25	3	Solid	Nuclear	550 KT	—	10 500	Operational	100+	200 m
UNITED STATES OF AMERICA											
Minuteman II	**2716.111**	LGM-30F	3	Solid	Nuclear	W-56 2 MT	None	12 500	Operational	450	370 m
Minuteman III	**2717.111**	LGM-30G	3	Solid	Nuclear	—	MIRV 3 × Mk 12A W-78 335 KT	11 000	Operational	528	200 m
Peacekeeper (formerly MX)	**4561.111**	MGM-118	4	Solid	Nuclear	—	MIRV 10 × Mk 21A W-87 500 KT	8000	Operational	50	100 m
Intermediate-range Ballistic Missiles (IRBM)											
CHINA, People's Republic											
DF-2	**2010.111**	CSS-1	1	Liquid	Nuclear	Possibly 20 KT	None	1200	Possibly operational	50	N.A.
DF-3	**2049.111**	CSS-2	1	Liquid	Nuclear	1-3 MT	3 MRV 50-100 KT	2700	Operational further development	15-20	N.A.
FRANCE											
SSBS S-3	**2163.111**	—	2	Solid	Nuclear	1.2 MT	None known	3000	Operational in 1980	18	N.A.
UNION OF SOVIET SOCIALIST REPUBLICS											
Soviet R-12 (Sandal)	**2952.111**	SS-4	2	Liquid	Nuclear	1 MT	—	1500-2000	Operational	65	2400 m
Soviet: RSD-10 Pioner (Saber)	**3219.111**	SS-20	2	Solid	Nuclear	—	MIRV 3 × 150 KT	5000	Operational in 1977	405	400 m
Submarine-launched Ballistic Missiles (SLBM) & Sea-launched Strategic Weapons											
CHINA, People's Republic											
JL-1	**4893.411**	CSS-N-3	2	Solid	Nuclear	2 MT	—	2700	Operational	24	N.A.

Name or NATO code	Entry no	US code	No of stages	Propellant	Type of warhead	Single warhead	Alternative warheads	Range (km)	Status	Deployment	CEP
FRANCE											
MSBS	**1134.411**	M20	2	Solid	Nuclear	1.2 MT	None	3000	Operational	48	N.A.
		M4	3	Solid	Nuclear	—	MIRV 6 × 150 KT	4500	Operational in 1985	16	N.A.
UNION OF SOVIET SOCIALIST REPUBLICS											
Soviet R-21 (Sawfly)	**1154.411**	SS-N-6	—	Liquid	Nuclear	1 MT (Mods 1 and 2)	MRV 2 × 500 KT	2400 (Mod 1) 3000 (Mods 2 and 3)	Operational in 1968	272 in 'Y' class submarines	1300 m
—	**3078.411**	SS-N-8	—	Liquid	Nuclear	1 MT (Mod 1)	MRV 2 × 800 KT	7800 (Mod 1) 9100 (Mod 2)	Operational in 1973	200 in 'D I' and 'D II' class submarines	400 m
Serb	**1153.411**	SS-N-5	2	Liquid	Nuclear	1 MT	—	1400	Obsolescent entered service 1964	39 in Golf submarines	2800 m
Snipe	**3356.411**	SS-N-17	—	Solid	Nuclear	N.A.	—	3900	Considered operational	12 in one 'Y' class submarine	1400 m
Soviet RSM-50 (Stingray)	**3357.411**	SS-N-18	2	Liquid	Nuclear	450 KT (Mod 2)	MIRV 3 × 200 KT (Mod 1) 7 × 200 KT (Mod 3)	6500 (Mods 1 and 3) 8000 (Mod 2)	Operational in 1979	224 in 'D III' class submarines	900 m (Mods 2/3)
Soviet RSM-52 (Sturgeon)	**4208.411**	SS-N-20	—	Solid	Nuclear	—	MIRV 6-9 × 100 KT	8300	Operational in 1984	80 in 'Typhoon' class submarines	500 m
Skiff	**4795.411**	SS-N-23	—	Liquid	Nuclear	—	MIRV 10 ×	8300	Operational in 1986	48 in 'D IV' submarines	900 m
UNITED KINGDOM											
Polaris A-3 TK	**2390.411**	UGM-27C	2	Solid	Nuclear	—	Chevaline MRV 3 × 200 KT	4600	Operational in 1969	64 in 4 RN submarines	900 m
UNITED STATES OF AMERICA											
Poseidon C-3	**1133.411**	UGM-73A	2	Solid	Nuclear	—	MIRV (10 × Mk3 W-68 40 KT)	4600	Operational in 1971	256 operational	450 m
Trident C-4	**2840.411**	UGM-93A	3	Solid	Nuclear	—	MIRV 8 × W-76 100 KT	7400	Operational	384 in Ohio, Franklin and Madison class submarines	460 m
Tomahawk TLAM-N	**3993.001** **4197.221**	BGM-109A	1	Turbojet (solid booster)	Nuclear	W80 200 KT	—	2500	Operational in 198?	164: In Los Angeles class submarines and several surface ship types	280 m

3229.114 Tactical Land-based Surface-to-surface Missiles

Designation	Entry no	Dimensions l x d – span	Weight	Warhead	Propulsion	Range	Guidance	Deployment
Battlefield support								
CHINA								
Type M	**6470.111**	910 × 100	6200	HE	Solid	600 km	Inertial	Development
EGYPT								
Sakr 80	**6712.111**	—	660	HE	Solid	80 km	—	Development
FRANCE								
Hades	**4777.111**	Pluton successor programme		Nuclear 10–25 KT	Solid	480 km	Inertial	Development
Pluton	**2130.111**	764 × 65–141 cm	2423 kg	Nuclear 25 or 15 KT	Solid	120 km	Inertial	French Army
INDIA								
Prirhvi	**6713.111**	—	—	HE	Liquid	240 km	Inertial	Development
TAIWAN								
Ching Feng	**4547.111**	Similar to US Lance		—	—	—	—	Operational

Designation	Entry no	Dimensions l x d – span	Weight	Warhead	Propulsion	Range	Guidance	Deployment
UNION OF SOVIET SOCIALIST REPUBLICS								
FROG-7	**2926.111**	910 × 55 cm	*c*2300 kg	HE or nuclear	Solid	70 km	Unguided	USSR & client states
SS-1B (Scud A)	**3327.111**	*c*1050 × 85 cm	*c*4500 kg	HE or nuclear	Storable liquid	80-150 km	Radio command	Warsaw Pact. Possibly obsolescent
SS-1C (Scud B)	**3327.111**	*c*1125 × 85 cm	*c*6370 kg	HE or nuclear	Storable liquid	160-280 km	Simplified inertial	Warsaw Pact, Egypt, Iraq, Syria
SS-12A (Scaleboard A)	**2959.111**	1200 × 100 cm	9700 kg	Nuclear	Probably liquid	900 km	Inertial	USSR
SS-12B (Scaleboard B)	**4549.111**	1238 × 101 cm	8800 kg	Nuclear	Solid	900 km	Inertial	USSR. Replacement for SS-12A
SS-21 (Scarab)	**4548.111**	600 × 85 cm	3000 kg	HE or nuclear	Solid	120 km (est)	Inertial	Replacement for FROG-7
SS-23 (Spider)	**4550.111**	—	—	HE or nuclear	Solid	500 km (est)	Inertial	USSR
UNITED STATES OF AMERICA								
GLCM (BGM-109)	**4194.111**	640 × 53–250 cm	1773 kg	Nuclear	Turbofan	2500 km	Inertial + TERCOM	USA
Lance (MGM-52)	**2682.111**	614 × 56 cm	1527 kg	Nuclear Other options	Pre-packed liquid	130 km	Simplified inertial	Belgium, Germany (FR), Italy, Israel, Netherlands, UK
Pershing 1a (MGM-31A)	**2765.111**	*c*1050 × 100 cm	*c*4600 kg	Nuclear 60-400 KT	2-stage solid	740 km	Inertial	Germany (FR),
Pershing II	**2767.111**	Improved version of Pershing 1a with terminal guidance				1800 km		Germany (FR)
Coastal defence missiles								
BRAZIL								
SM-70	**6748.121**	600 × 30 cm	—	HE	—	70 km	—	Development
CHINA, People's Republic								
HY-2	**3984.121**	609 × ?	2500 kg	HE	Turbojet	95 km	Active radar homing	China and possibly some African states
HY-4	**6723.121**	—	—	HE	Turbojet	150 km	—	Development
FRANCE								
Exocet MM 40	**2118.121**	578 × 35–100 cm	850 kg	165 kg HE	2-stage solid	70 km	Inertial + active radar homing	Early production
INTERNATIONAL								
Otomat	**1337.121**	438 × 40–124 cm	700 kg	210 kg HE	Turbojet + boosters	160 km	Autopilot + active radar homing	Egypt
NORWAY								
Penguin	**3074.121**	300 × 28–140 cm	385 kg	120 kg SAP	Solid	30 km	Inertial + IR homing	Development
SWEDEN								
RB 08	**2372.121**	572 × 66–301 cm	900 kg	HE	Turbojet + boosters	150 km	Autopilot + active radar homing	Sweden
RBS 15	**6185.121**	435 × 50 cm	600 kg	HE	Turbojet + boosters	150 km	Autopilot	Development
RBS 17	**4878.121**		49 kg		Coastal defence version of Hellfire anti-tank weapon			Sweden
TAIWAN								
Hsiung Feng	**4442.121**	Based on coastal defence version of Israeli Gabriel ASM						
UNION OF SOVIET SOCIALIST REPUBLICS								
SSC-3	**4890.121**	700 cm	—	Nuclear		*c*300	Active radar, IR option	Germany (DR), Poland, Yugoslavia
Samlet (SSC-2B)	**2951.121**	*c*700	*c*3000 kg	HE	Turbojet + boosters	*c*200 km	Command, probably active radar homing	USSR & some client states

Designation	Entry no	Dimensions l x d – span	Weight	Warhead	Propulsion	Range	Guidance	Deployment
USSR cont								
Sepal (SSC-1B)	**2975.121**	*c*1000 cm	—	HE or nuclear	Air breathing	*c*450 km	Radio command IR homing	USSR
Anti-tank/Assault								
ARGENTINA								
Mathogo	**3677.111**	99.8 × 10.2–47 cm	11.3 kg	HEAT	Solid	350–2100 m	Wire-guided CLOS	Argentina
CHINA								
Red Arrow 8 (HJ-8)		87.5 × 12–32 cm	11.2 kg	HEAT	Solid	100-3000 m	Wire CLOS	China
FRANCE								
SS-11	**2139.111**	120 × 16.4–50 cm	29.9 kg	HEAT	Solid	500-3000 m	Wire-guided CLOS	France & 30 countries
ACL-STRIM	**2055.111**	60 × 9 cm	2.2 kg	HEAT 0.565 kg	Solid	400-500 m	—	France & other countries
ERYX	**4552.111**	88.5 × 13.5 cm	14.4 kg	HEAT	Solid	25-600 m	CLOS	France
GERMANY, Federal Republic								
Cobra 2000	**2181.111**	95 × 10–48 cm	10.3 kg	HEAT 2.7 kg 3 types	Solid	2000 m	Wire CLOS	Many countries
Mamba	**2188.111**	95 × 12–40 cm	11.2 kg	2.7 kg 3 types	Solid	300 - 2000 m	Wire CLOS	Various
INTERNATIONAL								
ADATS	**4159.111**	205 × 15 cm	51.4 kg	12 kg+	Solid	8000 m	Laser beam-rider	Canada
ATGW 3-MR	**4545.111**	—	16 kg	—	Solid	2000 m	Optical beam rider	Early development
ATGW 3-LR	**4545.111**	—	—	—	Solid	5000 m	Infra-red homing	Early development
HOT	**2212.111**	128 × 14–31 cm	23.5 kg	6.48 kg HEAT	2-stage solid	4000 m	Wire CLOS	France, Germany (FR) & others
MILAN	**2215.111**	77 × 9–27 cm	6.70 kg	HEAT (K115 introduced in 1984)	Solid	25 - 2000 m	Wire automatic CLOS	France, Germany (FR), UK & other countries
MSS 1.1/MAF	**6466.111**	138 × 13 cm	14.5 kg	4.1 kg HEAT	2-stage solid	70 - 3000 m	Laser beam rider	Development
ISRAEL								
MAPATS	**4859.111**	145 × 14.8 m	18.5 kg	HEAT 3.6 kg	Solid + booster	4500 m	Laser LOS	Israel
ITALY								
Folgore	**2257.111**	74 × 8 cm	5.2 kg	HEAT	Solid	50 - 1000 m	—	Italy
JAPAN								
KAM-3D (Type 64)	**2262.111**	100 × 12–60 cm	15.7 kg	HEAT	2-stage solid	350 - 1800 m	Wire CLOS	Japan
KAM-9 (Type 79)	**2263.111**	156.5 × 15.2–33 cm	42 kg	HEAT	Solid	4000 m	SACLOS	Japan
SWEDEN								
Bantam (RB 53)	**2363.111**	85 × 11–40 cm	11.5 kg	1.9 kg HEAT	2-stage solid	2000 m	Wire CLOS	Sweden, Switzerland
RBS 56 (Bill)	**4544.111**	90 × 15–41 cm	10.7 kg	HEAT	Solid	150-2000 m	Wire CLOS	Sweden
Strix	**4557.111**	81 × 12 cm	17.6 kg	HEAT	Mortar launched	600–7500 m	IR homing	Development
TAIWAN								
Kun-Wu	**4438.111**	Taiwan produced version of Soviet AT-3 Sagger			—	—	—	Taiwan
UNION OF SOVIET SOCIALIST REPUBLICS								
AT-2 (Swatter)	**2985.111**	116 × 13–66 cm	29.4 kg	HEAT	Solid	500–3000 m	Wire CLOS possibly with IR homing	Warsaw Pact & others
AT-3 (Sagger)	**2950.111**	86 × 11.9–47 cm	11.3 kg	HEAT	Solid	500–3000 m	Wire CLOS	Warsaw Pact and others
AT-4 (Spigot)	**3638.111**	*c*98 × 13 cm	10-12 kg (est)	HEAT	Solid	*c*2000 m	Semi-automatic CLOS	Warsaw Pact & others
AT-5 (Spandrel)	**3608.111**	*c*100 × 16 cm	12-18 kg (est)	HEAT	Solid	4000 m (est)	CLOS	Warsaw Pact & others

Designation	Entry no	Dimensions l x d – span	Weight	Warhead	Propulsion	Range	Guidance	Deployment
USSR cont								
AT-6 (Spiral)	**3639.111**	—	—	HEAT	Solid	—	CLOS	Warsaw Pact & others
AT-8 (Songster)	**6224.111**	—	25 kg	HEAT	Solid	4000 m	Radio command	Soviet Union
UNITED KINGDOM								
LAW 80	**3986.111**	95 × 9.4 cm	4.6 kg	HEAT	Solid	500 m	—	UK & Jordan
Merlin	**4756.111**	90 × 8.1 cm	6 kg	HEAT	Mortar launched	4500 m	IR homing	Development
Swingfire	**2450.111**	107 × 17–37 cm	34 kg	HEAT	Solid	150 - 4000 m	Wire CLOS	Belgium, Egypt, UK & Sudan
Vigilant	**2475.111**	107 × 11–28 cm	14 kg	HEAT 6 kg +	2-stage solid	1375 m	Wire manual or auto CLOS	Finland, Kuwait, Libya, Saudi Arabia, UAE
UNITED STATES OF AMERICA								
Dragon I (M47)	**2573.111**	74 × 13–33 cm	13.8 kg	HEAT	Multiple solid	1000 m	Wire CLOS	USA & others
Hellfire (AGM-114A)	**4877.111**	c1700 × 17.8–32.6 cm	c46 kg	Twin HEAT	Solid	7000 m	Laser/IIR/IRIS	Development in limited numbers
TOW 1 (BGM-71A)	**2830.111**	117 × 15 cm	18.9 kg	HEAT	Solid	3750 m	Wire Auto missile tracking CLOS	USA and 35 countries
ITOW (BGM-71C)	**2830.111**	140 × 15 cm	19.1 kg	HEAT	Solid	3750 m	Wire Auto missile tracking CLOS	USA and 35 countries
TOW 2 (BGM-71D)	**2830.111**	140 × 15 cm	21.5 kg	HEAT	Solid	3750 m	Wire Auto missile tracking CLOS	USA and 35 countries
Copperhead (M712)	**3212.111**	137 × 15 cm	63.5 kg	6.4 kg HEAT	Cannon launched	16 km	Laser homing	USA

1763.224 Tactical Shipborne Surface-to-surface Missiles

Designation	Entry no	Dimensions l x d – span	Weight	Warhead	Propulsion	Range	Guidance	Deployment
CHINA, People's Republic								
FL-1 (CSS-N-2)	**3973.221**	Chinese-built version of Soviet SS-N-2, Styx anti-ship missile						In several classes of Chinese ship, also 2 Egyptian frigates
C801	**4891.221**	581 × 36–118	815 kg	HE	—	50 km	Inertial + active radar	Reported operational in one 'Jianghu' class frigate
FRANCE								
MM 38 Exocet	**1156.221**	521 × 35–100 cm	735 kg	165 kg HE	2-stage solid	42 km	Inertial + active radar	France & 14 others. Still in production
SM 39 Exocet	**1156.221**	580 × 35–101 cm	655 kg	165 kg HE	2-stage solid + booster	50 km +	Inertial + active radars	Sub version. French submarines
MM 40 Exocet	**1156.221**	578 × 35–113 cm	850 kg	165 kg HE	2-stage solid	70 km +	Inertial + active radar	Production. France & 8 countries
INTERNATIONAL								
ANS	**4446.221**	Anti-Navires Supersonique project by Aérospatiale and MBB						Development
Otomat	**1336.221**	446 × 46–135 cm	770 kg	210 kg SAP HE	Solid boosters, turbojet cruise	60-180 km	Autopilot + active radar	Eqypt, Iraq, Italy, Kenya, Libya, Nigeria, Peru, Saudi Arabia and Venezuela. Mk 2 & Téséo versions also

Designation	Entry no	Dimensions l x d – span	Weight	Warhead	Propulsion	Range	Guidance	Deployment
ISRAEL								
Gabriel Mk I Mk II Mk III	**6019.221**	335 × 34–135 cm 341 × 34–135 cm 381 × 34–135 cm	430 kg 520 kg 560 kg	100 kg HE 100 kg HE 150 kg HE	2-stage solid	18 km 36 km 36 km +	Autopilot/ command + IR or radar	Argentina, Israel, S. Africa, Singapore, Taiwan (as Hsiung Feng) Longer-range Mk II & III in production
Gabriel Long Range	**6187.221**				Turbojet	200 km	Inertial + active radar homing	Development
ITALY								
Otoma	**4195.221**	Italian project for successor to Otomat						Study
Sea Killer Mk 2	**2253.221**	470 × 21–100 cm	300 kg	70 kg SAP HE	2-stage solid	25 km	Beam-rider/ radio command CLOS radar	Iran. Used in Marte helicopter anti-ship system
NORWAY								
Penguin	**1339.221**	304 × 28–142 cm	360 kg	120 kg HE	2-stage solid	30 km	Inertial + IR homing	Greece, Norway, Turkey. Coastal-defence & air-launched versions
SOUTH AFRICA								
Skorpioen	**4444.221**	Believed licence-built version of Israeli Gabriel II ASM						S Africa, MoD class FAC
SWEDEN								
RB 08A	**2366.221**	572 × 66–301 cm	1215 kg	*c*250 kg	2 × solid boosters, turbojet cruise	—	Command/ autopilot + active radar homing	No longer operational with Swedish Navy. Also used for coastal defence
RBS 15	**3976.221**	435 × 50–140 cm	600 kg	—	Turbojet solid boosters	150 km (est)	Probably autopilot + active radar homing	Production for Sweden & Finland
TAIWAN								
Hsiung Feng	**4443.221**	Version of Israeli Gabriel ASM						Republic of China (Taiwan) Navy
UNION OF SOVIET SOCIALIST REPUBLICS								
SS-N-2 Styx	**1155.221**	580/655 × 75–275 cm (est)	2300 kg	HE 400 kg (2A/2B); 450 kg (2C)	Solid booster + internal cruise motor	40-80 km	Probably radio command + IR or active radar homing	Algeria, Bulgaria, China, Cuba, Egypt, Germany (DR), India, Indonesia, N Korea, Poland, Romania, Syria, USSR, Yugoslavia. China has own production
SS-N-3 Shaddock	**2976.221**	1020 × 86 cm (est)	4700 kg	HE or nuclear	Solid boosters + internal turbojet	460-735 km	Probably radio command	USSR only
SS-N-7 Starbright	**2987.411**	700 cm	3375 kg	HE or nuclear	Probably solid	55-64 km	Radar homing	Submarine-launched
SS-N-9 Siren	**1760.221**	884 cm	2950 kg	HE or nuclear	Solid	110 km	IR + radio command	Fitted in 'Nanuchka' missile boats and 'Sarancha' hydrofoils
SS-N-12 Sandbox	**4445.221**	1170 cm	—	HE or nuclear	Turbojet	550 km (est)	Radio command + active radar homing	In 'Kiev' carriers and possibly Echo II subs
SS-N-19 Shipwreck	**4196.221**	Vertically-launched anti-ship missile arming 'Kirov' class cruisers and possibly 'Oscar' class submarines						
SS-N-22 Sunburn	**4761.221**	—	—	HE or nuclear	Solid	110 km	—	'Sovremenny' and 'Tarantul II' class ships

Designation	Entry no	Dimensions l x d – span	Weight	Warhead	Propulsion	Range	Guidance	Deployment
UNITED KINGDOM								
Sea Eagle SL (P5T)	**4447.221**	Ship-launched version of Sea Eagle air-launched ASM						Development but not ordered
Sea Skua (SL)	**4553.221**	Ship-launched version of air-launched anti-ship missile						Development but not ordered
UNITED STATES OF AMERICA								
Harpoon (RGM-84A)	**2641.221**	384 × 34–83 cm	519 kg	HE	Solid booster, turbojet cruise	90 km	Programmed inertial + radar	Production for USA & others. Air- & sub-launched versions
Tomahawk (BGM-109)	**4197.221**	640 × 53–261 cm	1773 kg	HE or nuclear	Solid booster, turbojet cruise	450 - 2500 km	TERCOM-aided inertial or active radar homing	Production
Deadeye 5-in Guided Projectile	**4213.221**	154 × 12.7 cm	47.5 kg	13.6 kg	Solid	24 km +	Laser homing	Production

3471.234 Naval Surface-to-air Missiles

Designation	Entry no	Dimensions l x d – span	Weight	Warhead	Propulsion	Range	Guidance	Deployment
FRANCE								
Masurca	**1177.231**	860 × 41–150 cm	1850 kg	HE	Solid	40 km	Semi-active radar	In 2 frigates & 1 cruiser
Crotale	**2111.231**	290 × 15–54 cm	80 kg	*c*15 kg HE	Solid	18 km	Radio command	French Navy, Saudi Arabia
SAAM	**4860.231**	Naval element of Aster air defence missile project			Solid	10-17 km		Project
SADRAL	**4509.231**	180 × 9 cm	17 kg	HE 3 kg	Solid	300-6000 m	IR homing	Development
INTERNATIONAL								
NATO Seasparrow (Mk 57)	**4775.231**	366 × 20–102 cm	200 kg	HE	Solid	25 km	Semi-active radar	Belgium, Canada, Denmark, Germany (FR), Italy, Netherlands Norway, USA
RIM-116A CRAM	**3017.231**	279 × 12.7 cm	708 kg	HE	Solid	9.6 km	Proportional navigation + dual mode homing	USN/Germany(FR) Development
ISRAEL								
Barak	**4200.231**	217 × 17 cm	86 kg	HE	Solid	10 km	Semi-active radar	Development
ITALY								
Albatros	**2228.231**	370 × 20–80 cm	220 kg	HE	Solid	*c*25 km	Semi-active radar	Italy, Greece, Peru, Spain, Thailand, Venezuela and nine others
SWEDEN								
RBS 70	**4546.231**	132 × 11 cm	25 kg	HE	Solid	5 km	Laser beam rider	Development
UNION OF SOVIET SOCIALIST REPUBLICS								
SA-N-1 (Goa)	**2939.231**	660 × 70.1 cm	400 kg	60 kg HE	Solid	31 500 m	Beam rider, semi-active homing	Soviet and Polish fleet
SA-N-2 (Guideline)	**2943.231**	1070 × 50–220 cm	2300 kg	130 kg HE	Solid booster	*c*44 000 m	Radio command	1 ship fitted Obsolete
SA-N-3 (Goblet)	**2947.231**	620 × 33.5–150 cm	550 kg	80 kg HE	Solid	*c*55 km	Semi-active homing	In later Soviet ships
SA-N-4	**2954.231**	320 × 21 cm	190 kg	50 kg HE	Solid	14 800 m	Semi-active homing	Mostly post-1970 Soviet ships
SA-N-5	**4201.231**	Quadruple mount of SA-7 man-portable SAM (**2941.131**)				10 000 m	IR homing	Some AGIs & certain support ships

Designation	Entry no	Dimensions l x d – span	Weight	Warhead	Propulsion	Range	Guidance	Deployment
USSR cont								
SA-N-6 (Grumble)	**4206.231**	700 cm	—	HE	Solid	55 000 m	—	'Kirov' and 'Slava' classes
SA-N-7 (Gadfly)	**4473.231**	500 cm	—	HE	Solid	28 000 m	—	'Sovremenny' class
SA-N-9	**4885.231**	New system						*Frunze* battlecruiser
UNITED KINGDOM								
Seacat	**1019.231**	148 × –65 cm	c63 kg	HE	Solid	c5.5 km	CLOS	Used by UK and 15 other navies; obsolescent
Sea Dart	**6004.231**	436 × 42–91 cm	550 kg	HE	Solid + ramjet	40 km +	Semi-active radar	Argentina & UK. Lightweight version designed but not produced
Seaslug Mk 2	**6003.231**	600 × 41 cm	—	HE	Solid	c45 km	Beam rider	Chile
Sea Wolf	**2442.231**	200 × 18–55 cm	80 kg	HE	Solid	—	Command	RN
Sea Wolf VL	**2442.231**	300 × 18–55 cm	140 kg	HE	Solid	—	Command	Development
UNITED STATES OF AMERICA								
Sea Chaparral	**3983.231**	Naval version of Chaparral land-based air defence system						ROC (Taiwan) Navy
Standard (RIM-66A)	**1122.231**	447 × 34.3 cm	610 kg	HE	Solid	c18 km	Semi-active radar	USN AAW ships
Standard (RIM-67A)	**1122.231**	447 × 34.3 cm	1360 kg	HE	Solid	c55 km	Inertial reference + command	USN AAW ships
Tartar (RIM-24)	**6006.231**	460 × 34.3 cm	680 kg	HE	Solid	16 km	Semi-active radar	Australia, France, Germany (FR), Italy, Japan, Netherlands, Spain, USA (replaced by Standard)
Terrier	**6005.231**	800 × 34.3 cm	1400 kg	HE	Solid	c35 km	Beam rider, semi-active homing	Italy, Netherlands, USA. Obsolescent
ASMD RAM	**3017.321**	279 × 12.7 cm	70 kg	HE	Solid	—	Passive	Development

3472.314 Air Defence Missiles

Designation	Entry no	Dimensions l x d – span	Weight	Warhead	Propulsion	Range	Guidance	Deployment
CHINA, People's Republic								
HQ-2J	**4889.131**	Chinese produced version of Soviet SA-2						Production
HQ-61	**6418.131**	399 × 29 cm	300 kg	HE	Solid	10 km	Semi-active radar	Late development
HN-5A	**6479.131**	Chinese version of SA-7 Grail						
EGYPT								
Aybal-Sakr	**4841.131**	134 × 10 cm	15 kg	HE	Solid	4.4 km	IR homing	Egypt
FRANCE								
Crotale	**2074.131**	294 × 16–54 cm	80 kg	15 kg HE	Solid	11 km	Command	France & others. Cactus version for S Africa, & Shahine for Saudi Arabia
SAM 90	**4771.131**	Version of Aster missile system			—	30 km	—	Feasibility studies
Mistral	**4198.131**	180 × 9 cm	17 kg	3 kg	Solid	500-6000 m	IR homing	Production for France.
Shahine	**3972.131**	—	—	14 kg HE	Solid	11 km	Command	Saudi Arabia

Designation	Entry no	Dimensions l x d – span	Weight	Warhead	Propulsion	Range	Guidance	Deployment
GERMANY, Federal Republic								
MFS-2000	**4772.131**	Design/feasibility study programme for Hawk replacement						Study project
INTERNATIONAL								
ADATS	**4158.131**	205 × 15 cm	51.4 kg	12 kg + HE	Solid	8 km	Laser beam rider	Canada
Roland 1/2	**2218.131**	240 × 16–50 cm	66.5 kg	6.5 kg HE	Solid	6.3 km	Command	France, Germany (FR), & 8 others
Roland 3	**2218.131**	260 × 27–50 cm	85 kg	9 kg HE	Solid	8 km	Command	France, Germany (FR), & 8 others
ISRAEL								
ADAMS	**4888.131**	Mobile VLS system based on Barak point defence missile (PDM)						Project
ITALY								
Aspide	**1656.331**	370 × 20–80 cm	220 kg	HE	Solid	—	Semi-active radar	Italy
JAPAN								
Tan-SAM (Type 81)	**3458.131**	270 × 16 cm	100 kg	HE	Solid	7 km	Command + IR homing	Production
SWEDEN								
RBS 70 (Ray Rider)	**2348.131** **3617.131**	132 × 10.6 cm	*c*20 kg	HE	Solid	5 km	Laser beam rider	Production. Several versions
UNION OF SOVIET SOCIALIST REPUBLICS								
SA-1 (Guild)	**2944.131**	*c*1200 × 71 cm	—	—	Liquid	*c*32 km	—	USSR
SA-2 (Guideline)	**2942.131**	*c*1070 × 70 cm	*c*2300 kg	130 kg HE	Solid with liquid sustainer	40-50 km	Radio command	USSR & many others
SA-3 (Goa)	**2938.131**	670 × 60 cm	636 kg	HE	Solid	25 km	Probably radio command	USSR & others
SA-4 (Ganef)	**2934.131**	*c*880 × 90–260 cm	*c*2500 kg	135 kg HE	Ramjet + solid boosters	*c*70 km	Radio command	USSR, & some Warsaw Pact states
SA-5 (Gammon)	**2940.131**	*c*1060 × 85–290 cm	*c*10 000 kg	HE or nuclear	Solid	*c*300 km	—	USSR, Syria
SA-6 (Gainful)	**2930.131**	580 × 33.5–124 cm	580 kg	*c*40 kg HE	Rocket ramjet	Up to 60 km	Command + radar homing	USSR & others
SA-7 (Grail)	**2941.131**	134 × 10 cm	*c*9 kg	1.1 kg HE	Solid	3.5-5.5 km	IR homing	USSR & others
SA-8 (Gecko)	**3209.131**	*c*320 × 21–60 cm	est 170 kg	HE	Solid	12 km	Command	Warsaw Pact
SA-9 (Gaskin)	**3072.131**	180 × 12 cm	30 kg +	2.6 HE	Solid	8 km	IR homing	Warsaw Pact
SA-10 (Grumble)	**3620.131**	700 × 45 cm–10 (est)	1500 kg	HE or nuclear	Solid	100 km (est)	—	USSR
SA-11 (Gadfly)	**4470.131**	Probably new mobile short-range SAM, Mach 3, 3-28 km range						Development probably early deployment
SA-12 (Gladiator)	**4774.131**	750 × 50–135 cm	2000 kg	150 kg HE	Solid	80 km	Semi-active radar	USSR
SA-13 (Gopher)	**4471.131**	220 × 12–40 cm	55 kg	4 kg HE	Solid	10 km	IR homing	USSR
SA-14 (Gremlin)	**6270.131**	Replacement for SA-7 (Grail)						
SA-X-15	**6227.131**	Medium-range SAM, probably based on the SA-N-9 naval SAM						Development
ABM-1B (Galosh)	**2932.131**	*c*2000 × 275 cm	Anti-ballistic missile					USSR
UNITED KINGDOM								
Bloodhound	**2406.131**	846 × 55–283 cm	—	HE	Ramjet + solid boosters	*c*80 km	Semi-active radar homing	UK, Singapore, Switzerland
Blowpipe	**2409.131**	139 × 7.6–27 cm	20.7 kg	HE	Solid	Short	Radio command	UK, Canada & others
Javelin	**4533.131**	140 × 8 cm	—	HE	Solid	4 km +	SACLOS	Production for UK

Designation	Entry no	Dimensions l x d – span	Weight	Warhead	Propulsion	Range	Guidance	Deployment
UK cont								
Rapier	**2424.131**	224 × 13.3–38.1 cm	42.6 kg	HE 0.55 kg	Solid	6 km +	CLOS	UK & 13 others
Tigercat	**2465.131**	148 × 19 cm	63 kg	HE	Solid	—	Command	Argentina & 6 others
Starstreak	**6271.131**	—	—	—	Solid	4 km +	—	Development
UNITED STATES OF AMERICA								
Chaparral (MIM-72)	**2542.131**	291 × 13–64 cm	c84 kg	HE	Solid	Short	IR homing	USA & others
Hawk (MIM-23B)	**4177.131**	503 × 36–119 cm	627.3 kg	HE	Solid	40 km	Semi-active homing	USA & over 20 others
MSAM	**6419.131**	Conceptual development to replace Hawk and Chaparral in the 1990s						Study/early design
Nike-Hercules (MIM-14B)	**2723.131**	1210 × 80 cm	4858 kg	Nuclear/HE	Solid	140 km	Command	NATO & some other states
Patriot (MIM-104)	**2800.131**	531 × 41 cm	—	Nuclear/HE	Solid	—	Command + TVM	Entering service
Redeye (FIM-43A)	**2784.131**	122 × 7 cm	13 kg	HE	Solid	c3 km	IR homing	Australia, Sweden, USA & others
Saber	**4534.131**	109 × 12 cm	18.1 kg	HE	Solid	—	Command	Demonstrations
Stinger	**2805.131**	152 × 7 cm	15.8 kg	HE	Solid	Short	IR homing	USA

1757.334 Air-to-air Missiles

Designation	Entry no	Dimensions l x d – span	Weight	Warhead	Propulsion	Range	Guidance	Deployment
BRAZIL								
MAA-1 (Piranha)	**4192.331**	282 × 15–65 cm	90 kg	12 kg HE	Solid	5 km	IR	Development
CHINA (PEOPLE'S REPUBLIC)								
PL-2	**6601.331**	283 × 13–64 cm	76 kg	HE	Solid	3 km	IR	Operational
PL-5B	**6565.331**	289 × 13–66 cm	85 kg	HE	Solid	5 km	IR	Operational
PL-7	**6602.331**	275 × 16–66 cm	90 kg	HE	Solid	5 km	IR	Development
PL-9	**6603.331**	399 × 29–117 cm	300 kg	HE	Solid	—	Semi-active radar	Development
FRANCE								
MICA	**4882.331**	310 × ?? cm	110 kg	HE	Solid	50 km	Active radar & IR versions	Development
R.530	**1176.331**	328 × 26–110 cm	195 kg	27 kg HE	Solid	18 km	Semi-active radar & IR versions	Operational
Mistral	**4510.331**	180 × 9 cm	18 kg	3 kg HE	Solid	3 km	IR	Development
Super 530	**1349.331**	354 × 26–88 cm	250 kg	30 kg HE	Solid	25 km	Semi-active radar	Operational
R.550 (Magic)	**1348.331**	275 × 16–66 cm	90 kg	HE	Solid	5 km	IR	Operational
INTERNATIONAL								
ASRAAM (AIM-132)	**3622.331**	250 × 15–45 cm	65 kg	HE	Solid	—	IR	Development
ISRAEL								
Shafrir	**1659.331**	260 × 16 cm	95 kg	11 kg HE	Solid	—	IR	Operational
Python 3	**4185.331**	300 × 16–85 cm	120 kg	HE	Solid	5 km	IR	Operational
ITALY								
Aspide	**1656.331**	370 × 20–102 cm	220 kg	HE	Solid	—	Semi-active radar	Development
JAPAN								
XAAM-3	**6605.331**	260 cm	70 kg	HE	Solid	—	IR and active radar versions	Development

Designation	Entry no	Dimensions l x d – span	Weight	Warhead	Propulsion	Range	Guidance	Deployment
SOUTH AFRICA								
Kukri (V3B)	**4526.331**	294 × 13–53 cm	73.4 kg	HE	Solid	4 km	IR	Operational
UNION OF SOVIET SOCIALIST REPUBLICS								
Anab (AA-3)	**1144.331**	360 × 28–130 cm	—	HE	Solid	10 km	Semi-active radar & IR versions	Operational
Ash (AA-5)	**1145.331**	530 × 30–130 cm	—	HE	Solid	20 km	Semi-active radar & IR versions	Operational
Atoll (AA-2)	**1146.331**	290 × 13–65 cm	—	HE	Solid	—	IR & semi-active radar versions	Operational
Acrid (AA-6)	**3337.331**	600 × ?? cm	460 kg	HE	Solid	30 km	Semi-active radar & IR versions	Operational
Alamo (AA-10)	**4820.331**	400 × 19–70 cm	200 kg	HE	Solid	30 km	Semi-active radar & IR versions	Operational
Amos (AA-9)	**4768.331**	400 × 40–100 cm	450 kg	HE	Solid	150 km	Inertial, semi-active and active radar	Operational
Apex (AA-7)	**3338.331**	430 × ?? cm	275 kg	HE	Solid	20 km	Semi-active radar	Operational
Aphid (AA-8)	**3339.331**	210 × ?? cm	65 kg	6 kg HE	Solid	5 km	IR	Operational
Archer (AA-11)	**6171.331**	—	—	HE	Solid	—	IR	Presumably operational
UNITED KINGDOM								
Sky Flash	**1774.331**	366 × 20–102 cm	195 kg	30 kg HE	Solid	40 km	Semi-active radar	Operational
UNITED STATES OF AMERICA								
AMRAAM	**3340.331**	357 × 18–63 cm	150 kg	HE	Solid	—	Command, inertial & active radar	Development
Falcon (AIM-4D)	**1085.331**	198 × 16–51 cm	60 kg	HE	Solid	3 km	IR	Operational
Falcon (AIM-26)	**1085.331**	213 × 29–63 cm	115 kg	HE	Solid	8 km	Semi-active radar	Operational
Phoenix (AIM-54C)	**1099.331**	396 × 38–92 cm	454 kg	HE	Solid	150 km	Semi-active and active radar	Operational
Sidewinder (AIM-9M)	**4767.331**	287 × 13–64 cm	87 kg	HE	Solid	—	IR	Operational
Sparrow (AIM-7M)	**1106.331**	366 × 20–102 cm	227 kg	39 kg HE	Solid	45 km	Semi-active radar	Operational
Stinger (AIM-92)	**6604.331**	152 × 7–14 cm	18 kg	HE	Solid	—	IR	Development
AAAM	**6566.331**	—	300 kg	HE	—	—	—	Validation phase

1756.314 Air-to-surface Missiles

Designation	Entry no	Dimensions l x d – span	Weight	Warhead	Propulsion	Range	Guidance	Deployment
ARGENTINA								
Martin Pescador	**3678.311**	294 × 22–73 cm	140 kg	40 kg HE	Solid	8 km	Radio command	Operational
CHINA (People's Republic)								
C-601	**6473.311**	738 × 92–280 cm	2440 kg	400 kg HE	Liquid	—	Active radar	Operational
HY-4	**6607.311**	736 × 76–280 cm	1740 kg	500 kg HE	Turbofan	150 km	Inertial & active radar	Development
C-801	**6476.311**	480 × 55–165 cm	1025 kg	HE	Solid	—	Active radar	Development

Designation	Entry no	Dimensions l x d – span	Weight	Warhead	Propulsion	Range	Guidance	Deployment
FRANCE								
Armat	**4887.311**	420 × 40–120 cm	550 kg	160 kg HE	Solid	—	Passive radar	Operational
AS-15 (TT)	**3359.331**	216 × 19–53 cm	96 kg	30 kg HE	Solid	10 km	Radar command	Operational
AS-30L (Laser)	**3335.311**	365 × 34–100 cm	520 kg	240 kg HE	Solid	10 km	Semi-active laser	Operational
ASMP	**3635.311**	538 × 38–120 cm	—	Nuclear 350 KT	Solid boost & Ramjet	250 km	Inertial	Operational
Exocet (AM 39)	**1770.321**	470 × 35–110 cm	652 kg	165 kg HE	Solid	50 km	Inertial & active radar	Operational
GERMANY, Federal Republic								
Kormoran 1	**1180.311**	440 × 34–100 cm	600 kg	165 kg HE	Solid	30 km	Inertial & active radar	Operational
Kormoran 2	**1180.311**	440 × 34–100 cm	630 kg	220 kg HE	Solid	—	As Kormoran 1	Development
INTERNATIONAL								
ANS	**4450.321**	450 cm	650 kg	HE	Ramjet	180 km	—	Development
Apache CWS 50	**4454.311**	450 cm	1200 kg	Submunitions	Turbojet	80 km	Inertial	Development
HOT 2	**1771.311**	130 × 15–31 cm	23.5 kg	HE	Solid	4 km	Wire CLOS	Operational
Martel (AJ 168)	**1022.311**	390 × 40–120 cm	520 kg	HE	Solid	20 km	TV command	Operational
(AS 37)		420 × 40–120 cm	535 kg	HE	Solid	55 km	Passive radar	
MSOW	**6608.311**	—	—	—	Turbofan	15–500 km	—	Project definition
ISRAEL								
Gabriel III	**4529.321**	385 × 34–110 cm	560 kg	150 kg HE	Solid	40 km	Inertial + active radar	Operational
ITALY								
Marte Mk 2	**1651.321**	480 × 32–98 cm	345 kg	70 kg HE	2-stage solid	20 km	Inertial + active radar	Operational
JAPAN								
ASM-1	**1653.311**	400 × 34–102 cm	610 kg	HE	Solid	50 km	Inertial & active radar	Operational
NORWAY								
Penguin Mk 3	**3070.321**	318 × 28–100 cm	380 kg	120 kg HE	Solid	40 km	Inertial + IR	Operational
Penguin Mk 2 Mod 7	**6265.321**	296 × 28–140 cm	385 kg	120 kg HE	Solid	30 km	Inertial + IR	Development
SWEDEN								
RB 05A	**1190.311**	360 × 30–80 cm	305 kg	HE	Liquid	—	Radio command	Operational
RB 04E	**1189.321**	445 × 50–200 cm	600 kg	HE	Solid	—	Inertial + active radar	Operational
RBS 15F	**3975.321**	435 × 50–140 cm	595 kg	HE	Turbofan	70 km	Inertial + active radar	Development
UNION OF SOVIET SOCIALIST REPUBLICS								
Kipper (AS-2)	**1150.311**	1000 × 90–490 cm	4200 kg	HE/Nuclear	Turbojet	120 km	Command, autopilot, active radar	Operational
Kangaroo (AS-3)	**1147.311**	1500 × 185–915 cm	11 000 kg	HE/Nuclear 800 KT	Turbojet	650 km	Programmed autopilot + command option	Operational
Kitchen (AS-4)	**1151.311**	1130 × 95–300 cm	5900 kg	HE 1000 kg Nuclear 350 KT	Solid	400 km	Inertial + active radar	Operational
Kelt (AS-5)	**1149.311**	859 × 100–480 cm	3000 kg	1000 kg HE	Liquid	180 km	Autopilot + active radar	Operational
Kingfish (AS-6)	**3633.311**	1100 × 90–250 cm	—	HE 1000 kg Nuclear 350 KT	Solid	400 km	Inertial + active radar	Operational
Kerry (AS-7)	**6266.311**	350 × 30–? cm	295 kg	HE 60 kg	Solid	8 km	Radio command	Operational
Kyle (AS-9)	**6266.311**	600 × 50–150 cm	750 kg	HE	Solid	75 km	Passive radar	Operational
Karen (AS-10)	**6266.311**	350 × 30–90 cm	300 kg	HE	Solid	—	TV or semi-active laser	Operational

Designation	Entry no	Dimensions l x d – span	Weight	Warhead	Propulsion	Range	Guidance	Deployment
USSR cont								
Kilter (AS-11)	**6266.311**	350 × 30–90 cm	300 kg	HE	Solid	—	Imaging IR	Operational
Kegler (AS-12)	**6266.311**	—	350 kg	HE	Solid	35 km	Passive radar	Operational
AS-13	**6267.311**	Believed to be Mach 3.5 successor to the AS-4			—	—	—	Development
Kedge (AS-14)	**6266.311**	350 × 30–90 cm	300 kg	HE	Solid	30 km	Imaging IR	Operational
Kent (AS-15)	**6475.311**	809 × 51–?? cm	1700 kg	Nuclear 200 KT	Turbofan	3000 km	Inertial and terrain matching	Operational
Swatter (AT-2)	**4556.311**	116 × ?? cm	29 kg	HE	Solid	—	Radio command or semi-active laser	Operational
Sagger (AT-3)	**4556.311**	86 × ?? cm	11.3 kg	HE	Solid	3 km	Wire guided	Operational
Spiral (AT-6)	**4556.311**	—	—	HE	Solid	—	Radio command or semi-active laser	Operational
UNITED KINGDOM								
ALARM	**4528.311**	400 × 23–73 cm	200 kg	HE	Solid	—	Passive radar	Development
Sea Eagle	**3630.321**	414 × 40–120 cm	600 kg	HE	Turbofan	*c*110 km	Inertial + active radar	Operational
Sea Skua	**1530.321**	250 × 25–72 cm	145 kg	HE 20 kg	Solid	18 km	Semi-active radar	Operational
UNITED STATES OF AMERICA								
AGM-130	**6477.311**	394 cm	1323 kg	HE	Solid	—	TV or imaging IR	Development
ALCM (AGM-86B)	**1766.311**	632 × 69–365 cm span	1458 kg	Nuclear 200 KT	Turbofan	2500 km	Inertial + TERCOM	Operational
ACM (AGM-129)	**6610.311**	—	—	Nuclear 200 KT	Turbofan	—	—	Development
Standard (AGM-78)	**1123.311**	457 × 31–91 cm	815 kg	HE	Solid	—	Passive radar	Operational
Sidearm (AGM-122)	**6609.311**	287 × 13–64 cm	91 kg	HE	Solid	—	Passive radar	Development
Shrike (AGM-45A)	**1102.311**	305 × 20–91 cm	177 kg	66 kg HE	Solid	12 km	Passive radar	Operational
Maverick (AGM-65)	**1098.311**	249 × 31–72 cm	A/B 210 kg D 220 kg E 293 kg F/G 307 kg	A/B/D 57 kg HE E/F/G 136 kg HE	Solid	8-25 km	A/B TV E semi-active laser D/F/G Imaging IR	A/B/D/E operational F/G development
HARM (AGM-88A)	**1769.311**	417 × 25–113 cm	361 kg	66 kg HE	Solid	25 km	Passive radar	Operational
Harpoon (AGM-84A)	**1301.311**	385 × 34–91 cm	522 kg	HE	Turbofan	120 km	Inertial + active radar	Operational
Hellfire (AGM-114)	**1391.311**	163 × 18–33 cm	43 kg	HE	Solid	—	Semi-active laser	Operational
Tacit Rainbow (AGM-136)	**6611.311**	254 × 67–156 cm	575 kg	HE	Turbofan	90 km	Passive radar	Development
SKIPPER 2 (AGM-123)	**6268.311**	433 × 36–160 cm	582 kg	HE	Solid	—	Semi-active laser	Operational
SRAM (AGM-69A)	**1107.311**	427 × 45 cm	1016 kg	Nuclear 170 KT	Solid	200 km	Inertial	Operational
TOW	**2831.311**	117 × 15–45 cm	19 kg	HE	Solid	4 km	Wire-guided	Operational

1195.154

Land-based Radar

Designation	Entry no	Description	Contractor(s)
AUSTRALIA			
Jindalee	**6121.153**	Over-the-horizon backscatter radar programme	—
BRAZIL			
ASR-BR1	*4133.153*	S-band airport surveillance radar	IPD/PEA
Meteorological radar	**4132.153**	S-band weather and sonde tracking radar	IPD/PEA
CANADA			
CMR-3	**5678.153**	Coastal surveillance radar	Canadian Marconi
CHILE			
Aqila	**6383.153**	Mobile air surveillance radar	ASMAR
CHINA, People's Republic			
Chop Rest	**2019.153**	Early warning radar similar to Soviet Spoon Rest	—
Cross Legs	**2019.153**	Early warning radar used for GCI	—
Cross Slot	**2019.153**	Coastal defence/air search radar	—
Gin Sling	**2019.153**	SAM guidance radar similar to Soviet Fan Song	—
HN-401R	**6386.153**	Mobile 2-D air defence radar	—
Hunting Dog	**6389.153**	Gun fire control radar	—
Model 17-C	**6385.153**	Early warning radar	—
Moon Cone	**2019.153**	Air early warning radar. Stacked dipole arrays	—
Moon Face	**2019.153**	Similar to Moon Cone but only 1 array	—
Moon Mat	**2019.153**	Similar to Moon Cone but only 1 array	—
Moon Plate	**2019.153**	Stacked dipole array on trailer mounting	—
OTH	**6384.153**	Over-the-horizon early warning radar	—
Slot Rest	**2019.153**	Similar to Soviet Spoon Rest. VHF early warning	—
Team Work	**2019.153**	AA gun fire control radar. E-band	—
JPL-40	**4812.153**	Air defence surveillance radar	—
Type 581	**4811.153**	Air warning radar	—
JLG-43	**4813.153**	Height finding radar	—
JY-9	**6370.153**	Low-altitude search radar	—
Type 404A	**4814.153**	Coastal surveillance radar	—
Coast radar	**5761.153**	Coastal tracking radar	—
Eagle	**4810.153**	Coastal radar for artillery use	—
Cheetah	**4809.153**	Mobile battlefield artillery radar	—
MW-5	**4815.153**	Mobile artillery fire control radar	—
Type 311-A	**4816.153**	AA fire control radar	—
Type 378	**6807.153**	Portable battlefield radar	—
Type 408-C	**6387.153**	Mobile long-range air defence radar	—
Type 571	**6388.153**	Low-altitude L-band air defence radar	—
Type 701	*4804.153*	Wind finding radar	—
Type 711	*5759.153*	Meteorological radar	—
Type 713	*5760.153*	Meteorological radar	—
Type 714	*4803.153*	Meteorological radar	—
Type 796	*4806.153*	Air route surveillance radar	—
Type 793	*4805.153*	Terminal area radar	—
Type 791-A	**4807.153**	Precision approach radar	—
JY-10 RIPP	**6127.153**	Radar information processing post	—
CZECHOSLOVAKIA			
OPRL-4	*3284.153*	Surveillance & PAR	Tesla
RL-41	*3285.153*	Air search & surveillance radar	Tesla
RL-42	*4331.153*	Successor to RL-41	Tesla
RP-4G	*4332.153*	Surveillance & PAR successor to OPRL-4	Tesla
DENMARK			
DR 513	*1576.153*	Artillery radar	Dannebrog Elektronik
DR 810	*2286.153*	Ballistic radar	Dannebrog Elektronik
DR 532	*1954.153*	Ballistics radar for guns or rockets	Dannebrog Elektronik
M/532	*1577.153*	Doppler radar for projectile velocity measurement	Dannebrog Elektronik
Pointer	*1573.153*	Portable independent Terma radar I-band	Dannebrog Elektronik
Terma doppler	**5629.163**	Projectile velocity measuring radar	Terma
TSSR	*1574.153*	Terma splash spotting radar	Dannebrog Elektronik
FRANCE			
Adour	**2056.153**	TH.D 1215 automatic multi-mission missile tracking radar, 200 km range on passive target	Thomson-CSF
Antares	**2060.153**	High precision 3-D E/F-band monopulse radar with electronic scan, in elevation	Thomson-CSF
Aquitaine II	*2062.153*	Precision missile/satellite tracking monopulse radar	Thomson-CSF
Arabelle	*2066.153*	Portable transponder	Thomson-CSF
Artois	**2063.153**	(Under development) Missile tracking radar, using electronic scanning techniques, with capacity to track several targets simutaneously	Thomson-CSF
Atlas	**1943.153**	TH.D 2503 G-band monopulse range tracker	Thomson-CSF
Bearn	**2065.153**	Precision missile/satellite tracking radar	Thomson-CSF
Bretagne	*2067.153*	TH.D 1801 tracking radar	Thomson-CSF
Champagne	*2069.153*	Missile tracking radar, using 300 km range on 1 m² passive target and to measure target speed to nearest 10 cm/s	Thomson-CSF
Domino 20	*2073.153*	D-band surveillance radar. Coherent, pulse-doppler; 17 km range; 10° beamwidth; coherent pattern	Thomson-CSF
Domino 30	*2083.153*	TRS 2101 as Domino 20, but 6.5° beamwidth and 30 km max range	Thomson-CSF
Domino 40N	*2084.153*	As Domino 20 but 3° pencil beam and 80 km max range	Thomson-CSF

Designation	Entry no	Description	Contractor(s)
FRANCE cont			
Eldorado	*2080.151* *2082.153*	I/J-band tracking radar for light anti-aircraft operations against low-flying targets. Associated with MIRADOR surveillance radar as mobile combination	Thomson-CSF
ER-116-A	**2157.153**	IFF interrogator	Thomson-CSF
Gerfaut	**4224.153**	See TRS 2620	Thomson-CSF
LDTR	—	G/H-band scanning radar for special application. Designed to work under high level ECM conditions, has an exceptionally narrow antenna beam	Thomson-CSF
Louxor	**3275.153**	Experimental electronic scan short-range air defence radar	Thomson-CSF
LP23 (TRS 2050)	*2100.153*	Long-range D-band surveillance radar	Thomson-CSF
Matador	*2102.153*	3-D E/F-band radar (Mobile and tridimensional air defence operations radar). Range about 250 km. Electronic elevation scan for accurate height measurement	Thomson-CSF
Mirador	*2080.151* *2116.153*	Pulse-doppler surveillance and target designation radar E/F-band; 18 km range; pencil beam 5.6° wide. Associated with ELDORADO tracking radar for mobile operations for defence against low-flying aircraft	Thomson-CSF
Mirador II	*2080.151* **2074.131** *2117.153*	Mobile E/F-band pulse-doppler surveillance radar for defence against low-flying aircraft. Associated with Crotale surface-to-air guided weapons system. Dual, back-to-back, 4° beamwidth antennas. Range 18 km	Thomson-CSF
NR-AI-5A	**6396.153**	IFF interrogator-decoder	LMT
NRAI-6A	**6397.153**	IFF interrogator-decoder	LMT
NRS 1-3	**5782.153**	IFF interrogator	LMT
Oeil Noir 1 (DR-VC-1A)	*2129.153*	D-band pulse-doppler surveillance radar. 10° cosecant pattern beam. Range 15 km. Mounted on AMX-30 tracked AFV with twin 30 mm cannon to form mobile anti-aircraft system	Thomson-CSF
Oeil Vert	**3197.153**	Later version of Oeil Noir	Thomson-CSF
Olifant	*2131.153*	Lightweight battlefield radar	Thomson-CSF
Olifant II	*2132.153*	Lightweight man-portable battlefield patrol radar	Thomson-CSF
ORP-105 (DLM-10-E)	*1945.153*	E-band meteorological radar	Omera (Thomson-CSF)
ORP-310	**1945.153**	I/J-band meteorological radar	Omera (Thomson-CSF)
ORP 340	**1945.153**	I-band meteorological radar	Omera (Thomson-CSF)
ORV-31	**1945.153**	I/J-band meteorological radar	Omera (Thomson-CSF)
Palmier	—	20MW 3-D E/F-band surveillance radar	Thomson-CSF
Perceval/Spartiate	**2120.153**	Helicopter recovery radar	Thomson-CSF
Picador	**2119.153**	Mobile 3-D radar	Thomson-CSF
RA 20	*3717.153*	Mobile AA fire control radar	ESD
RA 20S	**5601.153**	Surveillance and target designation radar for low-level air defence	ESD
RA 21	*3718.153*	Mobile AA fire control radar	ESD
Racine	*2136.153*	I/J-band non-recurrent pulsed doppler surveillance and target acquisition radar	LCT
RALF	*2152.153*	Lightweight I/J-band low altitudes surveillance radar using Racine techniques	LCT
Ramsa	**3274.153**	TRS 2140 low-level air defence gap-filler	Thomson-CSF
Rapace	*2137.153*	J-band AFV-mounted surveillance radar	ESD
Rappist	**6823.153**	I/J-band search and fire control radar	Thomson-CSF
Rasit	*2142.153*	I/J-band portable medium-range battlefield surveillance radar	LMT
Rasit 72A1 (Rapière)	**2141.153**	I/J-band ground surveillance radar. 20 km range	LCT
Rasit 72B	*2153.153*	Lightweight version of Rasit 72A	LCT
Rasit 72C	*2154.153*	Short-range (10 km) version of Rasit 72A	LCT
Rasura	*2143.153*	I/J-band portable or vehicle-mounted battlefield surveillance radar	ESD
Rasura II	*3716.153*	Battlefield surveillance radar	ESD
Ratac	**1528.153**	Battlefield surveillance. Also produced in USA and West Germany	LMT/SEL/ITT
RB12A (Rassur)	**6132.153**	Ku-band pulse doppler ground surveillance radar	LMT
Rodeo 2	**5602.153**	Fire control radar for low-level air defence	ESD
Rodeo 20	*3719.153*	Anti-aircraft fire control radar	ESD
Siclamen	**3173.153**	Battlefield IFF	LMT
Savoie	**1944.153**	Long-range tracker	Thomson-CSF
SB14	**6809.153**	IFF interrogator for SHORAD weapon systems	LMT
Stentor	**3287.153**	Battlefield surveillance radar, I/J-band	LCT
TH.D 1012	*2133.153*	PAR	Thomson-CSF
TH.D 1013	*2098.153*	Light PAR	Thomson-CSF
TH.D 1021	*2058.153*	Airfield surveillance radar	Thomson-CSF
TH.D 1060	—	D-band medium-range surveillance radar. Transportable and suitable for use as gap-filler or as surveillance element of GCA system. 200 km range on small targets; digital MTI system; 2-beam antenna	Thomson-CSF
TH.D 1940	—	Low cost 3-D surveillance radar with electromechanical elevation sweep. Transportable by cargo aircraft, helicopter, or truck	Thomson-CSF
TH.D 1955	*2155.153*	High-power 3-D E/F-band monopulse stacked-beam radar for surveillance or GCI. 400 km range capability	Thomson-CSF
Tigers (TRS 2100)	**2156.153**	E/F-band lightweight surveillance radar for detection of high-flying targets	Thomson-CSF
Tiger G (TRS 2105/TRS 2106)	**6395.153**	Low altitude surveillance radar	Thomson-CSF
TRS 2050	*2100.153*	See LP 23	Thomson-CSF
TRS 2052	—	Successor to TRS 2050	Thomson-CSF
TRS 2053	**3099.153**	Smaller antenna version of TRS 2052 for use as gap-filler	Thomson-CSF
TRS 2054	**3683.153**	Long-range D-band surveillance radar	Thomson-CSF
TRS 2055	*2104.153*	2 MW Terminal Area surveillance radar. MTI system. Uses antenna type TH.D-284B	Thomson-CSF
TRS 2056	**3099.153**	Centaure. Transportable recovery radar based on TRS 2052	Thomson-CSF
TRS 2060	*2121.153*	Mobile or static E/F-band surveillance radar	Thomson-CSF
TRS 2065	*2059.153*	ATC approach radar	Thomson-CSF

Designation	Entry no	Description	Contractor(s)
FRANCE cont			
TRS 2201	**2155.153**	Long-range E/F 3-D air defence radar	Thomson-CSF
TRS 2215	**3272.153**	Transportable version of TRS 2230 E/F-band 3-D radar. TRS 2215D and TRS 22XX are later versions	Thomson-CSF
TRS 2230	**3272.153**	Fixed site 3-D air defence radar. Derived from Satrape	Thomson-CSF
TRS 22XX	**6808.153**	Long range 3-D surveillance radar	Thomson-CSF
TRS 2310	**3492.153**	PAR	Thomson-CSF
TRS 2620 (Gerfaut)	**4224.153**	E/F-band pulse doppler target designation radars	Thomson-CSF
TRS 2630 (Griffon)	**6394.153**	Short range target acquisition radar	Thomson-CSF
TRS 2730	*3276.153*	G-band weather radar	Thomson-CSF
TRS 3405	**3100.153**	I-band coastal surveillance radar	Thomson-CSF
TRS 3410	**3273.153**	Transportable version of TRS 3405	Thomson-CSF
Type 3737A	—	IFF interrogator antenna	LMT
Volex III (TRS 2205)	*2160.153*	Medium-range 2-beam 3-D radar. Fast electromechanical elevation sweep with narrow (2 × 2°) beam	Thomson-CSF
Volex IV	**3271.153**	TRS 2206. Improved version of Volex III	Thomson-CSF
—	—	D-band panoramic radar with double-beam AT 431 HB (9 × 13 m) antenna; 400 km range; high ECM resistance	Thomson-CSF
—	—	Phased array 3-D radar for air defence and ballistic missile detection. Possible replacement for PALMIER radar	Thomson-CSF
—	—	400-600 MHz surveillance radar for special applications Range better than 1000 km	Thomson-CSF
GERMANY, Federal Republic			
ASR 920	*3707.153*	Mobile radar	AEG-Telefunken
DII 111/DII 511	**3286.153**	Lightweight IFF systems	Siemens
DII 211	**3286.153**	Mobile IFF interrogator	Siemens
DR 151	**3286.153**	Longer range version of MPDR 30	Siemens
DR 161	**3286.153**	Extended range version of DR151	Siemens
DR 172	**3286.153**	Extended range version of DR151	Siemens
DR 621/641	**3286.153**	Search radar for B2 35 mm Flakpanzer	Siemens
MBVR 120	*3478.153*	Balloon/drone tracking radar	Philips
Meteor 300	*5613.153*	Meteorological radar	Gematronik
MPDR 12	**3286.153**	J-band tracker for B2 35 mm Flakpanzer	Siemens-Albis
MPDR 16	**3286.153**	E/F-band search radar for Roland mobile SAM	Siemens
MPDR 30	**3286.153**	Mobile air search radar. Gap-filler	Siemens
MPDR 3002 S	**3286.153**	Medium-range air search radar	Siemens
OREST	**3706.153**	Mobile electronic scan air defence radar	Siemens
PAR-T4	—	PAR	AEG-Telefunken
Ratac-S	**6390.153**	Battlefield surveillance radar	SEL
SRE-A4	—	10 cm E/F-band medium-range surveillance radar. 500 kW	AEG-Telefunken
SRE-A5	—	Medium-range (60 nm) surveillance radar	AEG-Telefunken
SRE-LL1	—	23 cm D-band long-range surveillance radar. Back-to-back 14.5 × 9 m reflectors 2 × 0.24 MW. Range 255 nm	AEG-Telefunken
SR-M2B	—	23 cm D-band long-range surveillance radar. IMW. Range 245 nm	AEG-Telefunken
SRT-2	—	SSR	AEG-Telefunken
—	—	Surveillance radar for anti-aircraft fire control system	Siemens
TRMS	**3016.153**	Transportable electronic scan air defence radar	AEG-Telefunken
1990D	**3286.153**	IFF interrogator	Siemens
Wimera	*4140.153*	I-band met balloon tracking radar	Elektro Spezial
INDIA			
Loward	**6205.153**	Series of air defence radars	Bharat
PSM33	**6120.153**	Transportable E/F band 3-D radar developed in cooperation with Thomson-CSF	Bharat
INTERNATIONAL			
COBRA	**6103.153**	Weapon locating radar—various consortia tendering for contract	
RATAC	**1528.153**	Battlefield surveillance radar	LMT/SEL/ITT-Gilfillan
ISRAEL			
EL/M-2106	**3646.153**	D-band battlefield air warning radar	Elta
EL/M-2108	*3647.153*	Portable infantry radar	Elta
EL/M-2216(V)	**6108.153**	Coastal surveillance radar variant of the EL/M-2215	Elta
EL/M-2121	**4330.153**	Battlefield surveillance radar	Elta
EL/M-2200	**1958.153**	Range of E/F-band search radars	Elta
EL/M-2205	**1958.153**	ATC search radar	Elta
EL/M-2206	**1958.153**	Coastal surveillance radar	Elta
EL/M-2207	**3708.153**	Shipborne model of EL/M-2215	Elta
EL/M-2208	**3708.153**	Shipborne version EL/M-2215	Elta
EL/M-2210	**1958.153**	Mobile search radar	Elta
EL/M-2215	**3708.153**	Surveillance radar	Elta
EL/M-2216	**3708.153**	Coastal surveillance radar	Elta
EL/M-2220	**4606.153**	Mobile air surveillance radar	Elta
ITALY			
Argos RAT-5C	*2239.153*	Coastal defence radar	Selenia
Argos RAT-6L	*2243.153*	Short-range L-band acquisition radar for AA targets Suitable for mountainous terrain	Selenia
Argos RAT-7S	*2282.153*	E/F-band 450 W low-coverage radar	Selenia
Argos RAT-8S	*1952.153*	Mobile coastal surveillance version of RAT-10S	Selenia
Argos RAT-10S	*2237.153*	Coastal defence radar	Selenia
Argos RAT-15	—	10 cm E/F-band search and surveillance radar. 600 kW	Selenia
Argos RAT-20L	*2281.153*	D-band long-range surveillance radar	Selenia
Argos RAT-21C	*2280.153*	C-band high-power low-altitude surveillance radar	Selenia
Argos RAT-31S	**1953.153**	E/F-band mobile 3-D radar	Selenia
Argos 10	**1959.153**	D-band high power early warning radar. GE transmitter	Selenia

Designation	Entry no	Description	Contractor(s)
ITALY cont			
Argos-45	**6170.153**	Tactical radar for forward area low altitude defence	Selenia
Argos 800	—	Coastal defence radar	Selenia
ATCR-22	*2248.153*	High-power D-band civil/military ATC surveillance radar	Selenia
ATCR-33	*2248.153*	Medium-power E/F-band civil/military ATC surveillance radar	Selenia
ATCR-44	*2248.153*	Medium-power D-band civil/military ATC surveillance radar	Selenia
HSS-1	**4884.153**	Harbour surveillance radar	SMA
Indigo FC Centre (CT40-GM)	*2291.153*	Missile control system. Italian version of Superfledermaus	Contraves
LPD-20	**1529.153**	Pulse-doppler and acquisition radar	Contraves
MLV-4	—	Doppler infantry portable battlefield surveillance radar	Selenia
MM/SPQ-3	**3278.153**	I/J-band multipurpose coastal radar	SMA
Pagoda	**6391.153**	Towed version of Shorar	Contraves
Pluto	**3277.153**	Low-cover surveillance radar for air and coastal defence	Selenia
RAT-31SL	**6169.153**	Long range 3-D, E/F band radar	Selenia
RAT-3SS/M	**6912.153**	S-band air and surface search radar	Selenia-Elsag
RIS-3E	*2244.153*	Monopulse skin or beacon tracking radar	Selenia
RIS-4C/A	*2245.153*	I/J-band pulsed tracking radar	Selenia
RIS-5X	*2246.153*	Drone tracking radar	Selenia
Selenia GCA	**5770.153**	Mobile GCA radar	Selenia
Selenia SIR	*2248.153*	Secondary surveillance radar	Selenia
Sentinel RQT-9X	*2249.153*	Portable infantry radar	Selenia
RQT-10X	*1951.153*	Portable infantry radar	Selenia
Shorar	**5618.153**	Search acquisition and target designation radar	Contraves
SIT 431 (AN/TPX-54)	**3279.153**	IFF interrogator	Italtel
TPN-720	**6810.153**	Mobile control and landing system	SMA
JAPAN			
JAN/MPQ-N1	**4081.153**	Type 92 mortar locating radar	—
J/MPQ-P7	**4141.153**	Artillery locating radar	Toshiba
NPG-360	**2297.153**	E/F-band surveillance radar	NEC
NPG-434	**2297.153**	High-power ARSR	NEC
NPG-435	**2297.153**	PAR	NEC
NPG-454	**2297.153**	GCA system	NEC
NPG-460	**2297.153**	E/F-band airport surveillance radar	NEC
NPG-542	**2297.153**	IFF/SSR system	NEC
NPG-554	**2297.153**	GCA radar	NEC
NPG-630	**2297.153**	ARSR	NEC
NPG-864	**2297.153**	Fixed GCA system	NEC
NPG-880	**3702.153**	Static 3-D air defence radar, electronic scan	NEC
NPG-905A	**2297.153**	Secondary surveillance radar	NEC
NPG-1240	**2297.153**	Airport surveillance radar	NEC
NPM-510	**2298.153**	Mobile 3-D radar electronic elevation scan	NEC
NPM-554	**2297.153**	Mobile GCA system	NEC
XJ-FPS-3	**4605.153**	Air surveillance radar project	Mitsubishi
NETHERLANDS			
Flycatcher	**3283.153**	Low-level anti-aircraft fire control system	Signaal
L4/5AA Weapon Control System	—	Weapon control system for use with 40 mm anti-aircraft guns for close air defence. Search and tracking radar antennas on compound mounting which, together with transmitter-receivers computer and displays are mounted on single trailer. Search radar employs FASCAN principle to achieve early detection of very low-flying aircraft	Signaal
LAR, 8GR 200	—	Long-range civil or military air surveillance radar	Signaal
LDTR	—	Long-dwell time radar	Signaal
Signaal Coastal	**6168.153**	Coastal radar system	Signaal
SLAM	*1557.153*	Signaal low air-defence module	Signaal
Star, 8GR 550	*1556.153*	Terminal area radar with single transmitter-receiver	Signaal
Star, 8GR 552	*1556.153*	Terminal area radar with dual diversity transmitter-receiver	Signaal
Stola	—	Target acquisition system for use against very low flying aircraft. Believed to be I/J-band with back-to-back antenna arrangement. Antennas mounted on telescopic mast which can raise them (apparently) about 10 m to give good low-angle cover and clear obstructions	Signaal
Tank Radar	**2302.153**	Integrated radar system for 35 mm AA tank	Signaal
Vanguard	**6167.153**	Mobile 3-D radar operating as the heart of the Signaal Falconer system	Signaal
VI, SGR 104	—	Heightfinding radar	Signaal
VI, 8GR 551	—	Automatic heightfinding and long-range tracking radar	Signaal
—	—	Gap-filling radar	Signaal
POLAND			
AVIA B	**2340.153**	23 cm D-band surveillance radar. 2 1.5 MW transmitter/receivers in frequency diversity. Range 240 km	Industrial Institute of Telecomms, Warsaw
Gap Gate	**6811.153**	D/E-band mobile early warning radar	—
SPAIN			
IRS-10	**5763.153**	Secondary surveillance radar	CECSA
IRS-20MP	**6125.153**	Monopulse SSR radar system	CESELSA
Compact	**5764.153**	IFF/secondary radar	CESELSA
SWEDEN			
9GR 600	*1549.153*	Frequency-agile 200 kW search radar	PEAB
9KR 400	**4029.153**	Coastal surveillance radar	PEAB
Domti	*2354.153*	Mobile coherent G/H-band pulse doppler search radar for use with AA guns or missiles	Ericsson
Eagle	**5771.153**	Fire control radar	Ericsson

Designation	Entry no	Description	Contractor(s)
SWEDEN cont			
Ecstra	*2357.153*	I/J-band coherent search and fire control radar for AA gun or missile systems. Separate search and tracking antennas	Ericsson
GA 250K/X	**5773.153**	Antenna systems	Philips
Giraffe (PS-70/R)	**1957.153**	Mobile G/H-band search radar	Ericsson
Hard	**5772.153**	3-D search radar	Ericsson
Isidor	*1955.153*	Muzzle velocity measuring radar	SATT
Isidor	*2481.153*	J-band portable intrusion radar	Ericsson
Mareld	—	I/J-band coastal defence radar with closed-circuit TV for visual tracking and 30 km range laser rangefinder (by Ericsson)	Philips Teleindustri
PE-48/T	*2325.153*	Mobile I/J-band AA fire control radar	Ericsson
PE-452/T	*2327.153*	I/J-band range-only AA fire control radar	Ericsson
PE-453/T	*2328.153*	Mobile I/J-band AA fire control radar	Ericsson
Peder II	*2355.153*	J-band monopulse fire control tracking radar	PEAB
PS-171/R	*2326.153*	Mobile E/F-band search radar	PEAB
TX 745	*1956.153*	Lightweight portable IFF for AA guns and missiles	SATT
UAR 1021	*2357.153* **2377.181**	I/J-band combined search and track pulse doppler radar	PEAB
UAR 1022	*2356.153*	Tracking sub-system of the UAR 1021	PEAB
SWITZERLAND			
CONAR	*2551.153*	Artillery radar fire control system	Contraves
Fieldguard/Skyguard	**2377.181**	Solid-state successor to Superfledermaus; uses separate search and tracking radars whose antennas are concentrically mounted on single pedestal, as with Sea Hunter Shipborne tactical air defence system. Radar is by Ericsson; computer and control circuits by Contraves	Contraves/Ericcson
Superfledermaus	**2376.181**	Anti-aircraft fire control system incorporating tracking radar by Albiswerk with visual attachment. Equipment uses valve circuits and has analogue computer. Latest version has MTI addition by Ericsson	Contraves/Albiswerk/Ericsson
TAIWAN			
Tien Kung	**6813.153**	Mobile phased array radar for Tien Kung 1 missile	Sun Yat-sen Institute
UNION OF SOVIET SOCIALIST REPUBLICS			
Back Net	**2860.153**	Early warning GCI search radar	—
Barlock	**2861.153**	GCI and air search radar. Dual antennas	—
Beam Track	—	Multiple Yagi tracking array	—
Big Bar	**2862.153**	Similar to Barlock but V-beam antennas for heightfinding	—
Big Fred	**4329.153**	Artillery/mortar locating mobile radar	—
Big Mesh	**2863.153**	Similar to Back Net but V-beam antennas	—
Bill Fold	—	Search radar. Transportable	—
Box Brick	—	Trailer-mounted	—
Cat House	—	Ballistic missile defence tracking radar	—
Bread Bin	—	Meteorological radar for artillery I-band	—
CBR	—	Counter battery radar. E-band 2/300 kW artillery/mortar locating	—
Cheese Brick	—	Truck-mounted	—
Clam Shell	—	I-band 3-D radar for SA-11 missile	—
Dog House	**2864.153**	Large fixed, phased-array radar for ABM system	—
Dome Glass	—	Trailer-mounted tracker or fire control radar	—
Dry Rack	—	Yagi array	—
Dumbo	—	Compound array of multiple UHF & VHF Yagi arrays	—
End Tray (RMS-1)	—	Parabolic dish tracker. Van-mounted	—
Fan Song	**2866.153**	Series of SAM guidance radars	—
Fan Song B	**2867.153**	E/F-band missile guidance radar	—
Fan Song C	—	Similar to Fan Song B	—
Fan Song E	**2868.153**	G-band missile control radar	—
Fire Can	**2871.153**	E-band trailer-mounted radar similar to US SCR-584	—
Fire Dish	**2872.153**	Similar to Fire Can	—
Fire Wheel	**2873.153**	Conical scan gun fire control radar	—
Flap Track	—	Air search and target acquisition radar	—
Flap Wheel	—	Parabolic dish	—
Flat Face	**2874.153**	Air search and target acquisition radar	—
Flat Twin	**6821.153**	Ballistic missile defence system tracking radar	—
Fork Rest A	—	Array of 4 VHF Yagi antennas	—
Four Stack	—	Array of 8 co-mounted Yagi antennas	—
Gage	**2875.153**	Acquisition radar. Stacked feeds	—
GS-11	—	Battlefield surveillance radar. J-band, 10 kW, 1.2 km range. Manpack	—
GS-12	—	Battlefield surveillance radar. J-band, 25 kW, 3.5 km range. Manpack or vehicle	—
GS-13	—	Vehicle-mounted battlefield surveillance radar. J-band, 50 kW, 12 km range	—
Gun Dish	**2876.153**	Fire control radar for ZSU-23-4 SP AA tank	—
Hen Egg	**2877.153**	Early warning radar	—
Hen House	**2879.153**	Early warning radar for ABM system	—
Hen Nest	**2877.153**	Early warning radar	—
Hen Roost	**2877.153**	Early warning radar	—
Knife Rest	**2881.153** **2880.153**	VHF early warning radar. Soviet designation P-10	—
Landfall	—	Coastal warning radar	—
Land Roll	—	Target acquistion radar for SA-8 missile	—
Long Brick	—	Van-mounted array	—
Long Eye	—	Back-to-back dish radars on trailer	—
Long Talk	**3294.153**	Aircraft recovery and GCA search radar	—
Long Track	**2937.153**	Tactical surveillance radar for SAMs	—
Long Trough	—	Trough-shaped air search radar on half-tracked vehicle	—

Designation	Entry no	Description	Contractor(s)
USSR cont			
Low Blow	**2884.153**	I-band radar group for low-level SAMs	—
Mercury Noseband	—	Truck-mounted	—
Mesh Brick	—	Probably mortar locating radar. Offset spiral feeds	—
Mound Brick	—	Trailer-mounted radar with hemispherical dome	—
Nysa B	—	Nodding heightfinder	—
Nysa C	**3296.153**	Long-range air search radar. Probably VHF	—
Odd Lot	—	E-band early warning radar and heightfinder	—
Odd Pair	—	E-band heightfinder	—
One Eye	—	Air search radar with supplementary antenna	—
OR-2	—	Probably airfield surveillance radar	—
OTHR	**3288.153**	Over-the-horizon radar. Pulse type. HF	—
Pat Hand	**2936.153**	H-band missile guidance radar for SA-4 Ganef	—
Patty Cake	**2878.153**	Back-to-back nodding heightfinders	—
Pawn Shop	**6822.153**	Missile guidance radar	—
Pole Dish	—	—	—
Pork Trough	**4328.153**	Similar to Long Trough on fully tracked vehicle	—
PSNR-1	**4602.153**	Battlefield surveillance radar	—
Pushkino	—	Ballistic missile defence battle management radar	—
RL-1	—	Airfield surveillance radar	—
RL-2D	—	Air search radar	—
Rock Cake	—	Transportable nodding heightfinder	—
RP-2E	—	Twin PAR installation	—
Scoreboard	**2974.153**	Broadside array of stacked dipoles. A and B versions	—
Side Net	**3289.153**	Transportable nodding heightfinder	—
Small Yawn	—	Tracker(possibly conical scan) on tracked van	—
Sponge Cake	**2878.153**	Trailer-mounted nodding heightfinder	—
Spoon Rest A	**2889.153**	VHF air warning radar. Vehicle mounted	—
Spoon Rest B	—	VHF air warning radar. Semi-fixed version	—
Square Pair	**3290.153**	Target tracking and missile guidance radar for SA-5	—
Squat Eye	**3291.153**	Air search and target acquisition radar	—
Squint Eye	**2891.153**	Target acquisition radar	—
Stone Cake	**2878.153**	Trailer-mounted nodding heightfinder	—
Straight Flush	**2885.153**	Fire control radar group for SA-6	—
Strike Out	—	Similar to Barlock	—
Tachi I	—	Multiple Yagi arrays associated with searchlight	—
Tall King	**3239.153**	Long-range early warning radar. Soviet designation P-14	—
Thin Skin	**3292.153**	H-band nodding heightfinder	—
Token	**2892.153**	E/F-band early warning radar. V-beam antennas	—
Track Dish	—	Dish tracker on tracked van	—
Try Add	—	Ballistic missile defence guidance and engagement radar	—
Tub Brick	—	Search radar with radome on trailer	—
Twin Eye	—	Back-to-back dish antennas on building	—
Two Spot	**3295.153**	PAR group	—
Whiff	**2893.153**	Van-mounted fire control radar. E/F-band	—
Whiff Brick	—	Dish-type tracker on trailer van	—
Witch Eight	—	Eight-element dipole-plus-ground-plane array	—
Witch Five	—	Array of five Yagis	—
Witch Four	—	Quadruple dipole array	—
Wood Gage	—	Similar to OR-2	—
Yo Yo	**2895.153**	Early missile guidance radar for SA-1 Guild	—
UNITED KINGDOM			
AA No 3 Mk 7	—	Mobile AA fire control radar. Lock and follow	AEI
AA5504	*5767.153*	Muzzle velocity radar	GEC Sensors
ACR-6	—	Medium-range 10 cm search radar. Similar to CR 787 but with additional display and processing facilities	Cossor
ACR-7C	—	Mobile short-range I/J-band surveillance radar	Cossor
ACR 430	*2420.153*	Series of I-band air search and coastal defence radars	Plessey
AR-1	*1140.153*	E/F-band medium-range surveillance radar. 650 kW Single or dual transmitter/receivers with frequency diversity. Range 75 nm	Plessey
AR-3D	**2484.153**	E/F-band 3-D air surveillance radar	Plessey
AR-5	*2416.153*	D-band long-range surveillance radar. 2 MW Single or dual transmitter/receivers. Range 200 nm	Plessey
AR-15/2	*2456.153* and **1141.153**	E/F-band medium-range surveillance radar similar to and superseding AR-1. Available in static (*2456.153*) or mobile (**1141.153**) versions	Plessey
AR-320	**3649.153**	E/F-band surveillance radar. Used in UKADGE update programme	Plessey
AR-325 (Commander)	**6376.153**	Long range S-band air defence radar based on AR-320	Plessey
AWS-7	*3281.153*	10 cm primary air and surface search radar	Plessey
AWS-8	*3480.153*	Coastal surveillance radar	Plessey
CEL 850	**3692.153**	IFF Interrogator	Cossor
Cervantes	*5631.153*	Mortar and rocket locating radar	Thorn-EMI
Coastal artillery radar	—	3 cm target acquisition and fire control for use with several coastal batteries	Racal
Cossor Monopulse	**4604.153**	Secondary surveillance radar	Cossor
CR 21	—	Medium-range 10 cm surveillance radar. Similar to CR 787 but less microwave preamplifier	Cossor
CR-62	**3715.153**	PAR. UK version of FPN-66	Cossor
CR 787	—	Medium-range 10 cm surveillance radar	Cossor
CR 901	—	23 cm long-range high-cover surveillance radar. 2 MW	Cossor
CRI 600	—	Microminiature IFF Mk 10 interrogator	Cossor
CRS 512	**5622.153**	Large aperture IFF/SSR antenna	Cossor
Cymbeline	**1018.153**	Mortar locating radar	EMI
DASR-1	—	10 cm back-to-back long-range/high-cover surveillance radar. 2800 kW transmitter/receivers. Range 110 nm	Plessey

Designation	Entry no	Description	Contractor(s)
UK cont			
DASR-3	—	Improved performance version of DASR-1 2.5 MW transmitter/receivers. Range 150 nm	Plessey
Falconer	**3715.153**	Precision approach radar	Cossor
Fastar	**4600.153**	Forward area surveillance and target acquisition radar	GEC Sensors
FCE.AA No. 7	—	Anti-aircraft fire control radar for use with guns or missiles at close range	EMI
Firelight	*2413.153*	Type 86 (which see) target illuminating radar	Ferranti
Green Archer (FA No 8 Mk 1)	*2412.153*	Mortar locating radar	EMI
GS No 14 Mk 1	**2490.153**	See ZB 298	Marconi
GS No 18 Mk 1	*2483.153*	Infantry combat radar. Successor to ZB 298	Marconi
GS No 20	**3490.153**	Claribel hostile fire locating radar	Racal-MESL
Guardsman	**4607.153**	Point defence radar. Guardsman coastal defence radar (**5775.153**) is a variant	Plessey
HF-200	*1142.153*	E/F-band height finder radar. 2.5 MW. 10.6 × 2.4 m antenna	Plessey
IFF 800	**2410.153**	Solid-state military IFF/SIF secondary surveillance radar developed from SSR. 700	Cossor
LAR 200	**6361.153**	Low-cost local area radar. Operates in S-band and can be used for battlefield aircraft control	Racal
LC 150	—	10 cm low cover coastal surveillance radar 800 kW. (10.6 m antenna)	Racal
Marconi OTH	**5869.153**	Over-the-horizon radar	Marconi
Martello (S713)	**3491.153**	Mobile 3D surveillance radar. Used in UKADGE update programme	Marconi
Martello (S723)	**4798.153**	New version of S713 Martello. Used in UKADGE and NADGE update programme	Marconi
Mk 10 IFF/SIF	*2418.153*	IFF Mk 10(SIF) interrogation and passive and active decoding equipment	Plessey
Messenger	**4603.153**	Monopulse SSR	Marconi
MFU 402	**6817.153**	Fire control radar for guns and SAMs	Marconi
MSR	**5766.153**	Millimetric surveillance radar	Racal
MSTAR	**6398.153**	Man-portable surveillance and target acquisition radar requirement	—
NIS	**5621.153**	Range of IFF interrogators and transponders for NATO indentification system (NIS)	Cossor
Pacer	*2435.153*	Muzzle velocity measuring radar	Ferranti
Pacer 2	**5627.153**	Muzzle velocity measuring radar	Ferranti
Prowler	*1946.153*	Infantry radar	Marconi
Rapier	**2425.153**	Search and target acquisition radar for Rapier AA missile system	Racal
Rapier 2000	**6399.153**	Search and target acquisition radar for Rapier 2000 system	Plessey
Rapier (DN 181)	**2439.153**	Tracking radar (Blindfire) for Rapier AA missile system	Marconi
S 105	—	3 cm fixed or mobile harbour and coastal (and Rapier 2000) surveillance radar. 70 kW.	Marconi
S 244	—	10 cm heightfinder radar. 2.5 MW 12.2 × 2.1 m antenna	Marconi
S247/S 266	**2458.153**	Back-to-back early warning radar. Two 2.25 MW transmitter/ receivers, one 10 and one 23 cm	Marconi
S 259	*2480.153*	Air-transportable D-band 2 MW tactical radar with limited ground mobility	Marconi
S 264	—	Range of 50 cm long-range surveillance radars	Marconi
S 314	—	10 cm coastal defence radar. 800 kW. 8.8 × 3 m antenna 40 nm range	Marconi
S 330	*2463.153*	Land mobile tactical surveillance radar. Uses 2 low-profile back-to-back antennas, E/F-band, and D-band plus IFF	Marconi
S 404	*2463.153*	Land mobile G/H-band height finder radar	Marconi
S 500 Series	**5615.153**	Primary airfield surveillance radars, S511 and S512	Marconi
S 505	—	3 cm harbour and coast watch radar. 20 kW	Marconi
S 513	*5641.153*	Primary surveillance radar	Marconi
S 600 Series	*1168.153*	Modular systems of transmitter/receivers, static and mobile antenna heads, displays and processing equipment capable of deployment to produce wide range of air defence radar systems. Main elements of S 600 Series are as follows:	Marconi
S 1006	—	10 cm static heightfinder antenna. 2.4 × 12.1 m cylindrical parabolic reflector with linear slotted feed	Marconi
S 1010	—	10 cm transportable surveillance antenna, 5.48 × 1.8 m single curvature parabolic reflector, linear feed, frequency diversity capability	Marconi
S 1011	—	10 cm static surveillance antenna. 13.7 × 4.57 m parabolic reflector. Can be used in back-to-back configuration	Marconi
S 1012	—	10 cm static surveillance antenna. 13.7 × 4.57 m cosecant-squared reflector. Can be used in back-to-back configuration	Marconi
S 1013	—	23 cm static surveillance antenna.13.7 × 4.57 m parabolic reflector. Can be used in back-to-back configuration	Marconi
S 1014	—	23 cm static surveillance antenna 13.7 × 4.57 m cosecant-squared reflector. Can be used back-to-back	Marconi
S 1015	—	10 m transportable surveillance antenna. 5.48 × 1.8 m cosecant-squared reflector	Marconi
S 1016	—	23 cm transportable surveillance antenna. 5.48 × 1.8 m cosecant-squared reflector	Marconi
S 1017	—	5.5 cm transportable heightfinder antenna. 4.27 × 1.3 m double curvature, point fed reflector	Marconi
S 1061	—	D-band surveillance. 7.6 × 2.4 m	Marconi
S 2010	—	1 MW 10 cm transmitter/receiver. Tunable within band	Marconi
S 2011	—	2 MW 23 cm transmitter/receiver. Tunable within band	Marconi
S 2012	—	2.5 MW 10 cm transmitter/receiver	Marconi
S 2013	—	1 MW 5.5 cm transmitter/receiver	Marconi
S2018	*2458.153*	E/F-band; 3.3 MW surveillance radar	Marconi

Designation	Entry no	Description	Contractor(s)
UK cont			
S2019	*2458.153*	D-band; 3.3 MW surveillance radar	Marconi
S 613	**2479.153**	G/H-band mobile/transportable heightfinding radar. 1 MW	Marconi
S 631/S 647	*2458.153*	E/F- and D-band back-to-back 2-plus MW surveillance radars	Marconi
S 650	—	50 cm D-band surveillance radar. 16 × 3.7 m reflector 500 kW. 150–160 nm range	Marconi
S 654	*2459.153*	23 cm surveillance radar 11 × 4.6 m reflector 2 MW, 90 nm range	Marconi
S 669	*2462.153*	E/F-band 2.25 MW 150 nm heightfinding radar	Marconi
S 670	—	50 cm surveillance radar 20.7 × 3.7 m reflector 500 kW. 180 nm range	Marconi
S 690	*2458.153*	E/F-band back-to-back surveillance radar	Marconi
S700 series	**4630.153**	Family of mobile surveillance radars in D- and E/F-bands. Comprises S704, S705, S706, S711 and S712	Marconi
S 841/S 842	**2464.153**	'800-series' mobile coastal defence I/J-band surveillance radars. Type S 842 has tunable transmitter	Marconi
ST850	**2466.153**	'800-series' I/J-band tracking radar with TV for use with Tigercat AA missile system (**2465.131**)	Marconi
SLA-3C	—	PAR	STC
Scorpion (Type 87)	**2428.153**	Target tracking and illuminating radar for Bloodhound II surface-to-air missile system	Marconi
Sentinel	*5619.153*	Mobile E/F-band tactical radar	Plessey
Shrimp	*2482.153*	J-band surveillance radar	GEC Avionics/RRE
Stingray (Type 83)	—	Target acquisition, tracking, and illuminating mobile radar for use with Bloodhound I and Thunderbird surface-to-air missiles. Used by UK, Swedish, and Australian forces	Marconi
Transportable Watchman	**4631.153**	Mobile version of Watchman	Plessey
Type 13	—	Heightfinder	—
Type 80	—	Long-range surveillance radar. Range 200 nm plus. Probably D-band	Plessey
Type 82	*2457.153*	'Volumetric', stacked beam 3-D surveillance radar. Used for tactical control of Bloodhound I surface-to-air missiles	Marconi
Type 83	—	Mobile target tracking and illuminating radar for Bloodhound I	Marconi
Type 84	*2457.153*	High-power D-band surveillance radar	Marconi
Type 85	*2457.153*	High-power stacked beam 3-D surveillance radar similar to AEI 4502. Used in Linesman/Mediator UK air defence/ATC system	Marconi
Type 86 (Firelight)	*2413.153*	Mobile CW target illuminating radar for Bloodhound II surface-to-air missiles	Ferranti
Type 87 (Scorpion)	**2428.153**	Target tracking and illuminating radar for Bloodhound II. CW	Marconi
Type 200	**1142.153**	E/F-band heightfinder radar. 2.5 MW. 10.6 × 2.4 m antenna	Racal
Type 430	**2420.153**	Airfield radar	Plessey
Type 431(M)	*3280.153*	Coastal radar. Fixed and mobile (M) versions	Plessey
Watchman	**5608.153**	E/F-band medium-range surveillance radar. Watchman coastal defence radar (**5774.153**) is a variant	Plessey
ZB 298	**2490.153**	Man-portable non-coherent I/J-band pulsed doppler battlefield surveillance infantry radar for location of men and vehicles out to 10 m	GEC Avionics
870 Series	*3693.153*	IFF interrogators	Cossor
880 Series	**3694.153**	IFF interrogators	Cossor
890 Series	**6373.193**	IFF interrogators for weapon systems	Cossor
4502	—	Stacked beam 3-D long range 10 cm surveillance radar 12 lobes. Elliptical parabolic reflector. 17.7 × 6.6 m. 4 MW	Marconi
40 Series	*2478.153*	Range of 3-D radars developed from 4502	Marconi
UNITED STATES OF AMERICA			
ACWAR	**6400.153**	Agile CW acquisition radar – development of Hawk CWAR radar	Hughes
ADAR	*2506.153*	Advanced design array radar. Electronically steered beam for tracking multiple targets	Hughes
AMSR	**6111.153**	Advanced Missile Surveillance radar – variant of the AN/TPS-70(V)1 for missile detection	Westinghouse
AN/-	—	US joint service nomenclature for many systems listed below uses AN/- prefix. For convenience this prefix has been omitted. Thus AN/CPN-6 is listed simply as CPN-6	—
ARBAT	**4079.153**	J-band ballistic testing instrumentation radar	ITT-Gilfillan
ARSR-3/ASR-30	**5623.153**	D-band surveillance radar	Westinghouse
ARSR-3D	*5779.153*	3-D surveillance radar	Westinghouse
ASR-7	—	E/F-band surveillance radar giving 100 km range on 3 m² target. All digital signal processor. Frequency diversity and 1 MW transmitter available	Texas Instruments
Astar-2 & 3	**5868.153**	Surveillance radars for tactical use	Aydin Corp
AWCS	**2533.181**	Air weapons control system. Code number 412L, otherwise known as QUICKDRAW. Mobile control and warning system	General Electric
Baltic Approach	**4082.153**	E/F-band Danish coastal defence radar (DCR)	Cardion
BMEWS	**2525.181**	Ballistic missile early warning system to give warning of ICBM/IRBM attacks on USA/UK. Sites at Thule, Greenland (equipped with FPS-49 and FPS-50); Clear, Alaska (equipped with FPS-50 and FPS-92); and Fylingdales Moor, UK (equipped with FPS-49) System code 474L	General Electric, RCA and Western Electric
BPS-1000	*3149.153*	Long-range D-band surveillance radar	Bendix
Cobra Dane (663 A)	**1949.153**	Phased array space tracking radar	Raytheon
Conus OTH-B	**5617.153**	Over-the-horizon surveillance radar	General Electric
CPN-4	—	Mobile GCA	ITT-Gilfillan
CPN-18	—	ATC surveillance radar	Bendix
CPS-1	—	Long-range, semi-transportable early warning radar	General Electric
CPS-5	—	ATC radar	Bendix
CPS-6B	—	Long-range search and heightfinder for early warning	General Electric

Designation	Entry no	Description	Contractor(s)
USA cont			
CPS-9	—	Weather radar	Raytheon
DEW	**2567.181**	Distant early warning system. Code number 413L	Western Electric
DCR	**4082.153**	See Baltic Approach	Cardion
DR810 Mk V	**5778.153**	Muzzle velocity radar	Lear Siegler
Dual-band	**6401.153**	Dual-band surveillance radar operating simultaneously in S-band and L-band	Westinghouse
EMS	—	Emergency mission support system. Air transportable, system designed to set up ATC, communications and varied facilities to support operations in areas where such facilities are not available. Code number 842L. Also known as Project FOUR WHEELS. Radars used TPN-14, TPS-35	Chu Associates, ITT-Gilfillan
ETA S	**6818.153**	Ground surveillance radar as part of the US Army Airland Battle 2000 concept	Eaton
FAAR	**1526.153**	Forward area alerting radar. Detects very low-flying high-speed aircraft; checks IFF identification; and relays data by VHF link to anti-aircraft sites equipped with Chaparral Redeye or Vulcan. Mobile system with high-mast antenna	Sanders Associates/Sperry
Falcon series (GPS-100)	**4322.153**	Frequency-agile low-coverage netted radar	ITT-Gilfillan
FPN-1A	—	Airfield approach and landing guidance radar system	Bendix
FPN-16	—	GCA-radar. Similar to CPN-4 and MPN-11	ITT-Gilfillan
FPN-28	—	GCA radar	Bendix
FPN-33	—	GCA radar	ITT-Gilfillan
FPN-34	—	Long-range D-band ATC radar	Bendix
FPN-36	**4125.153**	QUAD radar. Tactical GCA system	ITT-Gilfillan
FPN-40	—	GCA radar with IFF capability	ITT-Gilfillan
FPN-47	—	E/F-band medium-range ATC radar (ASR-5)	Texas Instruments
FPN-48	—	Landing Control Central GCA. Similar to MPN-13	ITT-Gilfillan
FPN-50	—	Landing Control Central GCA. Similar to CPN-4A	ITT-Gilfillan
FPN-51	—	E-band airport surveillance radar. Similar to FPN-47 but with ECM capability	Texas Instruments
FPN-52	**4076.153**	GCA PAR system based on FPN-16	ITT-Gilfillan
FPN-55	—	Similar to FPN-47	Texas Instruments
FPN-62	**3714.153**	PAR	Raytheon
FPQ-6	*2508.153* **2846.153**	MIPIR (Missile Precision Instrumentation Radar) G/H-band monopulse missile tracking radar. Transportation version is TPQ-18	RCA
FPQ-13	*2857.153*	G/H-band monopulse missile tracking radar	RCA
FPQ-14	*2508.153*	Modified version of FPQ-6	RCA
FPS-3, 3A	—	D-band long-range radars	Bendix
FPS-5	—	Long-range radar	Hazeltine
FPS-6, 6A, 6B	**2492.153**	E/F-band long-range heightfinding radars. Superseded by FPS-89. Mobile version is MPS-14	General Electric
FPS-7, 7A, 7B, 7C, 7D	**2526.153**	10 MW 3-D search/GCI radar. Cosecant elevation pattern with 7 stacked beams. Search range 500 km. Elevation cover 45 km. MTI, frequency diversity. Part of SAGE system	General Electric
FPS-8	**2493.153**	D-band long-range search radar with MTI. Superseded by FPS-88. Mobile version is MPS-11	General Electric
FPS-10	—	See FPS-27	
FPS-14	*2527.153*	E/F-band 500 kW medium-range (110 km) search radar used in SAGE system	Bendix
FPS-16	**4327.153**	G/H-band monopulse missile-tracking combined primary/ secondary radar. Transportable version is MPS-25	RCA
FPS-17	—	Long range early warning radar for missile detection	General Electric
FPS-18	*2527.153*	1.2 MW E/F-band search radar used in SAGE	—
FPS-19	—	DEW LINE radar	Raytheon
FPS-20, 20A, 20B	**2528.153**	Long-range, dual-channel search and GCI radar with MTI 2 MW, D-band; range 350 km. FPS-20 is used in SAGE, FPS-20A in AWCS and FPS-20B in both. GPS-4 is mobile version	Bendix
FPS-23	—	Aircraft alarm system. VHF multiple station doppler aircraft intruder detection system	Motorola and Western Instruments
FPS-24	**2494.153**	Long-range search radar, used in SAGE	General Electric
FPS-26	—	Long-range heightfinder for SAGE. Possible application as early warning radar for SLBM after modification	Avco
FPS-27	—	Replacement for FPS-10. Search radar for SAGE. E-band	Westinghouse
FPS-28	—	Search radar for SAGE	Raytheon
FPS-30	—	Long-range duplex search radar. D-band	Bendix
FPS-33	—	D-band long-range, high-altitude tracker for supersonic targets Includes 2 × FPS-8 or 2 × MPS-11	General Electric
FPS-35	—	Long-range SAGE or ATC surveillance radar	Sperry
FPS-36	**2723.131**	Nike-Hercules radar. Others are FPS-56, FPS-61, FPS-69-71, FPS-75-76	Raytheon
FPS-37	**2499.153** **2723.131**	High-power acquisition radar for Nike-Hercules	General Electric
FPS-41	—	Long-range meteorological recording radar	Raytheon
FPS-46	—	Long-range electronically steerable array radar (ESAR) See FPS-85	Bendix
FPS-49	**2509.153**	Very long-range tracking for missile early warning (BMEWS) and satellite tracking (SPADATS)	RCA
FPS-50	**2511.153**	BMEWS surveillance radar	General Electric
FPS-56	**2723.131**	See FPS-36	
FPS-59	—	Tracking radar	General Electric
FPS-60	—	SAGE heightfinding radar	—
FPS-61	**2723.131**	See FPS-36	
FPS-63	—	DEW line gap-filler, being modified to FPS-74	Budd
FPS-64, 65, 66, 67, 67A	—	Search radars for SAGE, FPS-67A is also used in AWCS	Bendix
FPS-69, 70, 71	**2723.131**	See FPS-36	

Designation	Entry no	Description	Contractor(s)
USA cont			
FPS-74	—	See FPS-63	
FPS-75, 76	**2723.131**	See FPS-36	
FPS-77	—	Weather radar	Lear Siegler
FPS-78, 79, 80	—	Tracking radars. FPS-79 and FPS-80 are part of SPACETRACK	General Electric
FPS-85	**2546.153**	D-band 3-D phase array radar for satellite detection, identification and tracking. Based on FPS-46 and part of SPADATS/SPACETRACK	Bendix
FPS-87A	—	Long-range D-band surveillance radar. Similar to FPS-91	Bendix
FPS-88, 89	**2512.153** **2513.153**	Modernised versions of FPS-6, FPS-8	General Electric
FPS-90	—	High-power heightfinder, version of FPS-6 ( ). Similar to APS-507 another version of FPS-6B	General Electric and Canadian Marconi
FPS-91	—	Long-range search and tracking radar. Version of FPS-20	Bendix
FPS-92	**2509.153**	BMEWS scanner/tracker radar. Improved version of FPS-49	RCA
FPS-93	—	Modification of FPS-20	Raytheon
FPS-95	—	Tracking radar	RCA
FPS-100	—	Modernised FPS-20	Bendix
FPS-103	—	Weather radar	Bendix
FPS-105(V)	**2847.153**	G/H-band range instrumentation radar	RCA
FPS-115	**3174.153**	Pave Paws missile warning radar	Raytheon
FPS-117	**4326.153**	Long-range 3-D air defence radar for Seek Igloo programme	General Electric
FPS-118	**5617.153**	Continental United States over-the-horizon radar system	General Electric
FPS-507	—	See FPS-90	Canadian Marconi
FPT-5, 7	—	Long-range surveillance and tracking radar transmitters	Ling Temco Vought
FSS-7	*2538.153*	Radar surveillance central for SLBM system	Avco Cincinnati
GE 592	**4321.153**	Export version of TPS-59	General Electric
GPN-2	—	Mobile ATC radar	Bendix
GPN-6	—	E/F-band ATC surveillance radar 150 km range	LFE
GPN-22	**3713.153**	PAR. Improved TPN-25	Raytheon
GPS-3	—	D-band back-up surveillance radar	General Electric
GPS-4	—	See FPS-20	
GPS-57A	—	Surveillance radar	Bendix
GSN-5	—	Landing Control Central automatic landing system	Bell
GSS-1	—	Mobile gap-filler radar	Raytheon
GSS-7	—	Mobile air defence radar	Raytheon/Hazeltine
HR-3000	**3721.153**	Version of HADR for NATO's air defence system	Hughes
HADR	**3721.153**	Electronic scan multi-mode 3-D air defence radar	Hughes
HIPAR	**2499.153**	High-power acquisition radar for Nike Hercules See FPS-37	General Electric
LAADS	**3123.153**	Low-altitude aircraft detection system	Sanders
LASR	**4323.153**	Low-altitude surveillance radar	Hughes
MFR (Multi-function radar)	**5607.153**	Weapon control radar	Hughes
Model 2679	**4632.053**	IFF interrogator	Hazeltine
MPG-1	—	Coastal artillery fire control radar	Bendix
MPN-1, 11, 13, 14, 15	—	Mobile GCA radars	ITT-Gilfillan
MPN-3, 5	—	Mobile GCA radars	Bendix
MPQ-4	**2497.153**	Mortar-locating radar	General Electric
MPQ-10A	**2500.153**	E/F-band mortar locating radar	Sperry
MPQ-12, 18	—	E/F-band 300 km range missile tracking radars	Reeves
MPQ-14	—	Radar Course Directing Central	Advance/Ultrasonic
MPQ-25	—	Trailer-mounted radar for Corporal missile system	ITT-Gilfillan
MPQ-29	—	Drone control radar	Sperry
MPQ-31	—	E/F-band 4000 km range 3-D tracking and guidance radar	Canoga
MPQ-32	—	Field detection and tracking radar	Sylvania
MPQ-33	*2640.131*	CW illuminator for Hawk missile system. Also MPQ-39	Raytheon
MPQ-34	*2640.131*	CW target acquisition radar for Hawk	Raytheon
MPQ-35	*2640.131*	Pulse target acquisition radar for Hawk	Raytheon
MPQ-37	*2640.131*	Range radar for Hawk	Raytheon
MPQ-39	*2640.131*	See MPQ-33	
MPQ-43	**2723.131**	HIPIR mobile high-power acquisition radar for Nike Hercules	General Electric
MPQ-49/54	**1526.153**	See FAAR	
MPQ-53	**4478.153**	Multi-function radar group for Patriot air defence missile system	Raytheon
MPQ-501	—	Mortar locating radar. Used in Canada	Raytheon
MPS-3, 4	—	Heightfinding radars	Raytheon
MPS-4, 4A, 4C	—	Heightfinding radars	Hazeltine
MPS-7	—	Long-range D-band search radar	Bendix
MPS-9	—	E/F-band drone and missile control radar. Range 300 km	Reeves
MPS-11, 11A	**2493.153** **2495.153**	See FPS-8	
MPS-14	**2492.153** **2496.153**	See FPS-6	
MPS-16	—	Nodding heightfinder for AWCS G/H-band; 300 km range	Crosley
MPS-19	—	Mobile tracking radar. E-band	Reeves
MPS-21	—	3-D search radar. See also TPS-34	Sperry
MPS-23	—	FRESCAN-elevation frequency scanning 3-D search radar	Hughes
MPS-24	—	See TPS-28	
MPS-25	**4327.153**	See FPS-16	
MPS-30	—	Mobile surveillance radar; trailer-mounted TPS-25	Hazeltine, FMC
MPS-36	**2514.153**	Instrumentation tracking radar	RCA
MPS-39	**6815.153**	Instrumentation and tracking radar	RCA
MPS-117	**5776.153**	IFF tracking system for FPS-117	Eaton
MRSR	**6112.153**	Multi-role survivable radar for air defence	Westinghouse
MSG-1	—	Transportable AA artillery director radar	Sperry
MSQ-1A	—	Aircraft ground attack director system	Reeves
MSQ-2	—	Mobile ground attack director system	Reeves

Designation	Entry no	Description	Contractor(s)
USA cont			
MSQ-4	—	Mobile ATC system	Adler Communication
MSQ-35	—	300 km range I/J-band tracking radar	Reeves
MSQ-39	—	Radar Bomb Scoring Central	Reeves
MSQ-44	—	Radar Bomb Scoring Central	Reeves
MSQ-51	—	Target drone control system	Reeves
MSQ-77	—	Radar Bomb Directing Control	Reeves
MSR	*2791.153* **2798.131**	Missile site radar for Safeguard ABM system. Phased array type. Prototype at Kwajalein test area	Raytheon
MSW-1	**2723.131**	Acquisition radar for Nike Ajax and Nike Hercules systems	—
MTAS	**6110.153**	Multi-sensor battlefield target acquisition system	Rockwell
MTTS	—	Mobile target tracking system for drone tracking	Epsco
OTH R	*1948.153*	Over-the-horizon radars	—
OX-60/FPS-117	**4317.153**	IFF interrogator for FPS-117 Seek Igloo radar	Hazeltine
Pave Paws	**1949.153**	Missile warning radars	GE, Raytheon, RCA
PAR	*2790.153* *2798.131*	Perimeter Acquisition Radar for Safeguard ABM systems Under development	—
PAR 80	**4077.153**	Fixed site J-band PAR	ITT-Gilfillan
PPS-3	—	Portable surveillance radar	Sperry
PPS-4	—	Silent Sentry portable forward area surveillance radar	Sperry
PPS-5/5B	**2540.153**	Transistorised versions of PPS-4	AIL
PPS-6	*5604.153*	Portable battlefield surveillance radar	General Instrument Corp
PPS-9	—	Portable battlefield surveillance radar	RCA
PPS-11/12	**5616.153**	Hand-held battlefield surveillance radar	General Instrument Corp
PPS-15A	**1947.153**	I/J-band infantry radar	General Dynamics
PPS-15B	**4324.153**	Development of PPS-15B	General Dynamics
PPS-55	—	Portable forward area surveillance radar	—
PSS-10	*5605.153*	Miniature battlefield radar receiver	General Instrument Corp
PPX-3B	**6816.153**	Lightweight IFF interrogator	Teledyne
Quadradar Mk V	*4125.153*	Multi-function ATC radar	ITT-Gilfillan
Quadradar Mk VI	*6118.153*	Development of Quadradar Mk 5 with MTI	ITT-Gilfillan
R-200 Series	*3640.153*	J-band infantry radar	General Instrument Corp
R-2080	*4316.153*	Battlefield surveillance radar	General Instrument Corp
RAIDS	**6814.153**	Detection radar against ultra-light aircraft, parachutists and hang gliders	Sanders
RCA 2019 M2	*2783.153*	I/J-band hand-held tactical radar. System weight 4 kg; T/R weight 1 kg; range 500 m on 0.5m² target. Resolution 25 m. Aural output	RCA
RCA 4019 M2	*2783.153*	I/J-band hand-held tactical radar. System weight 6.35 kg including accessories; T/R weight 3.6 kg; range 1500 m, resolution 25 m. Aural and/or visual output	RCA
Real-time velocimeter	**5628.153**	Transportable artillery tracking radar	Datron
ROTHR	**5781.153**	Relocatable over-the-horizon-radar	Raytheon
SAGE	**2803.181**	Semi-automatic ground environment system for detecting, identifying and tracking aircraft threats to North America and directing weapon interception. Code number 412L	Western Electric, System Development Corporation Burroughs
SAM-D	**2800.131**	Phased array radar for SAM-D guided missile system now under development	Raytheon
Sentry	**5777.153**	Tactical defence radar-licensed version of Elta EL/M-2106	Lear Siegler
Series 52	**4076.153**	Remote controlled PAR	ITT-Gilfillan
Series 320	**4078.153**	Transportable 3-D E/F-band air defence radar	ITT-Gilfillan
Site Defence Radar	*2792.153*	Phased-array radar similar in purpose to MSR but smaller	General Electric
SLBM	—	Sea-launched ballistic missile detection system. Purpose is to detect and identify missile attack on North America and warn NORAD and SAC. Code number 474N	Avco
Spacetrack	**2825.181**	USAF world-wide detection, identification, tracking, and reporting system. Records data on all space objects Spacetrack is US section of NORAD SPADATS system Spacetrack code number 496L	General Electric, Philco-Ford, RCA, Bendix, System Development Corporation
SPADATS	**2825.181**	NORAD space object detection and recording system of which SPACETRACK is large part	—
STARTLE	*3648.153*	AFV surveillance and target acquisition radar Development suspended late 1982 for 12 months	Martin Marietta
TIRS	**6109.153**	Terminal imaging radar system	—
TMQ-19	*3477.153*	Meteorological radar	Servo Corps
TPN-8	—	Helicopter-transportable approach radar	ITT-Gilfillan
TPN-12	—	ATC GCA	LFE, Bendix
TPN-14	—	Transportable GCA for EMS	ITT-Gilfillan, RCA
TPN-17	—	Air-transportable landing control central	LFE
TPN-18A/18B	**3129.153**	Lightweight I-band air-transportable ATC radar	ITT-Gilfillan
TPN-19	**2537.153**	All-weather transportable tactical ATC radar with E/F-band terminal area surveillance and I/J-band,dual-mode (search and track) 6 target. PAR. Cassegrain feed antenna	Raytheon
TPN-22	**4075.153**	Automatic all-weather landing radar system	ITT-Gilfillan
TPN-24	**3711.153**	Airfield surveillance radar element of TPN-19 system	Raytheon
TPN-25	**3712.153**	PAR used in TPN-19	Raytheon
TPQ-10	**2498.153**	Transportable radar course direction central for control of close support bombing operations	General Electric
TPQ-14, 15, 16	*2640.131*	Transportable versions of MPQ-33, 34, 35	Raytheon
TPQ-18	**2846.153**	Transportable version of FPQ-6	RCA
TPQ-19	*2640.131*	Transportable version of MPQ-39	Raytheon
TPQ-27	*3720.153*	Radar Course Directing Central	RCA
TPQ-28	—	Omni-directional mortar-locating radar	ITT-Gilfillan
TPQ-31	*1950.153*	Mortar-locating radar	Raytheon
TPQ-32	**1526.153**	See FAAR	
TPQ-36	**2848.153**	Electronically-scanning mortar locating radar. Used by US and Australia	Hughes

Designation	Entry no	Description	Contractor(s)
USA cont			
TPQ-37	**1976.153**	Artillery locating radar	Hughes
TPQ-39	**2849.153**	G/H-band digital instrumentation radar	RCA
TPS-1	*3128.153*	Transportable D-band air search radar. Version of FPS-36	
TPS-10D	—	I/J-band heightfinder. 220 km range	RCA
TPS-21	**2529.153**	Portable I/J-band battlefield surveillance and reconnaissance radar. Range 9 km. Also TPS-33	Admiral
TPS-22	—	Tactical early warning radar for AWCS system Complemented by TPS-27	Westinghouse
TPS-25, 25A	*3130.153*	Battlefield surveillance radars	Hazeltine
TPS-27	*2515.153*	3-D tactical radar for AWCS. Complemented by TPS-22	Westinghouse
TPS-31	—	Harbour surveillance radar	Raytheon
TPS-32	**2516.153**	Automatic 3-D tactical long-range surveillance radar	ITT
TPS-33	**2529.153**	See TPS-21	Admiral
TPS-34	**2530.153**	3-D (V-beam) tactical long-range early warning radar	Sperry
TPS-35	—	D-band 500 km surveillance radar for UMS. Modernised version of UPS-1	RCA
TPS-37	—	Transportable version of MPS-16	Avco
TPS-40	—	Long-range heightfinder. Transportable. Similar to MPS-16 and TPS-37	General Electric
TPS-43	**2517.153**	Transportable 3-D tactical tracking radar measuring range, altitude, and azimuth plus IFF identification ECM-resistant for forward area operation	Westinghouse
TPS-44	**2518.153**	Transportable tactical surveillance radar (ALERT)	Cardion
TPS-48	*2519.153*	Transportable 3-D tactical radar	Westinghouse
TPS-58	**1528.153**	US designation for RATAC battlefield radar	ITT-Gilfillan
TPS-59	**3282.153**	3-D tactical radar	General Electric
TPS-61	*2520.153*	Tactical surveillance radar	Westinghouse
TPS-63	**3475.153**	Tactical surveillance radar. Operated by USMC	Westinghouse
TPS-65	**3476.153**	Tactical surveillance radar	Westinghouse
TPS-70 (V)-1, -2, -3, -4	**4319.153**	Mobile tactical radar. Air surveillance. Developed from TPS-43	Westinghouse
TPX-19, 20, 22	—	Mk X IFF interrogator-responsors	Raytheon
TPX-42	**2535.153**	IFF interrogator system	AIL
TPX-44	—	Interrogator used with TPN-18	ITT-Gilfillan
TPX-45	—	IFF for Hawk missile system	Hazeltine
TPX-46	**3703.153**	IFF interrogator used with various systems	Hazeltine
TPX-50	*3704.153*	IFF interrogator for FAAR system	Hazeltine
TPX-54	**3705.153**	IFF interrogator	Hazeltine
Trackstar	**6129.153**	Self-contained, armoured integrated radar/command system for low altitude defence	Sanders
TSQ-81	—	Radar Bomb Directing Central	Reeves
TSQ-96	—	Bomb Directing Central	Reeves
TSW-5	—	Tactical approach radar	RCA
TTQ-28	—	Mortar-locating radar	ITT-Gilfillan
TWS-QR	*4320.153*	Track-while-scan-quiet radar development	Hughes
Type 386	**4325.153**	Battlefield ground surveillance radar	General Dynamics
UPS-1	—	Gap-filler radar for AWCS. See also TPS-35	RCA, Bendix
UPW-1	—	Drone tracking and control radar	Ford, Sperry
UPX-6	—	Mk X IFF interrogator-responsor	Raytheon
VPS-2, 5	**2547.153** *2844.153* **2850.131**	Vulcan air defence system radar	Lockheed
Vigilant	**5870.153**	Tactical surveillance radar. Development of TPS-70	Westinghouse
VSTAR	**4080.153**	Transportable E/F-band air defence radar, based on TPQ-37	Hughes
W-2000	**5780.153**	3-D surveillance radar	Westinghouse

1332.254 Naval Radar

Designation	Entry no	Description	Contractor(s)
CANADA			
AN/SPS-503	**5677.253**	Naval surveillance radar (see also S1800-entry **3687.253**)	Canadian Marconi
LN-66	**6873.253**	Family of surveillance radars	Canadian Marconi
CHINA, People's Republic			
Model 354	**6415.253**	Surface and low altitude warning radar	—
Type 756	*4808.253*	Marine-navigation radar	—
DENMARK			
Scanter Mil	**1575.253**	I-band navigation radar	Terma Elektronik
Scanter Mil Data	*4633.253*	Tactical data system	Terma Elektronik
FRANCE			
ELI 4	*6133.293*	Naval IFF interrogator	LMT
ELR-3	*6134.293*	IFF transponder	LMT
Triton II	**1062.253**	TRS 3030 5 cm air and surface surveillance radar. Used with Castor or Pollux in Vega Series. 205 kW. TRS 3035 is MTI version	Thomson-CSF
Triton C	**3685.253**	TRS 3050 doppler surveillance radar	Thomson-CSF
Triton G	**5844.253**	C-band air and surface surveillance radar	Thomson-CSF
Triton S	**3254.253**	TRS 3033 pulse doppler version of Triton	Thomson-CSF
Triton X	*3481.253*	TRS 3040 used in Vega-Canopus fire control	Thomson-CSF
Castor II	**1063.253**	TRS 3201-4 I/J-band target tracking radar used in some versions of Vega Series fire control system 20 kW	Thomson-CSF

Designation	Entry no	Description	Contractor(s)
FRANCE cont			
Pollux	**1064.253**	TRS 3220 I/J-band target tracking radar used in some versions of Vega Series fire control systems. 200 kW	Thomson-CSF
Pollux II	*1686.253*	Improved version of Pollux	Thomson-CSF
Calypso II	*1240.453*	TH.D 1030. I/J-band submarine radar for surveillance and navigation. 70 kW	Thomson-CSF
Calypso III	**1933.253**	TRS 3100 I/J-band submarine radar. Improved version of Calypso II	Thomson-CSF
Calypso IV	**3684.253**	TRS 3110 submarine navigation radar	Thomson-CSF
Jupiter	**1236.253**	TRS 3010 Long-range surveillance radar. D-band. 2MW	Thomson-CSF
Ramses	*1239.253*	TH.D 1022. Short-range navigational and surveillance radar. I/J-band. 35 kW. 60 nm range	Thomson-CSF
Rodeo Naval	**6416.253**	Air surveillance radar developed from Rodeo 2	ESD
Lynx	*1238.153*	Th.D 1051. Dual radar coastal mine-watching system	Thomson-CSF
Saturne	*1237.253*	TRS 3043 E/F-band medium-range air and surface surveillance radar. 1 MW peak	Thomson-CSF
Sea Tiger II	**1687.253**	TRS 3001 surveillance radar. Ordered for new Colombian frigates	Thomson-CSF
Triskel	**5845.253**	Multi-function C-band radar	Thomson-CSF
DRBC 32	**1894.253**	Series of gun fire control radars. A, B, C, D, and E versions fitted in various classes of French ships	—
DRBC 33	**5843.253**	Fire control I-band radar	Thomson-CSF
DRBI 10	**1593.253**	E/F-band 3-D air surveillance radar. Robinson scanner	Thomson-CSF
DRBI 23	**1177.231**	3-D surveillance and target designator	Thomson-CSF
DRBJ 11	**3741.253**	Electronic scan naval radar	Thomson-CSF
DRBV 13	**1893.253**	E/F-band pulse-doppler air search radar. Multi-mode operation	—
DRBV 20	**1892.253**	Long range metric air surveillance radar A and C versions in service	—
DRBV 22	**1891.253**	Search radar. A, C, and D versions in service on French and foreign vessels	—
DRBV 23	**1594.253**	D-band long-range air search radar. B and C versions in service	Thomson-CSF
DRBV 26	**1236.253**	See Jupiter	Thomson-CSF
DRBV 27	**1236.253**	Solid-state version of Jupiter. In development	Thomson-CSF
DRBR 51	**1177.231**	I/J-band tracker and target illuminator. 3 scanners	Thomson-CSF
	1890.253	Part of Masurca surface-to-air missile system	
	1889.253	radar. D-band 23 cm. Stacked-beam system	
3973A	*3255.253*	IFF interrogator antenna system	Thomson-CSF
INTERNATIONAL			
EMPAR	**6420.253**	New generation multifunction radar in development	Marconi/Thomson-CSF/Selenia
Mk 95 Mod O	**3312.253**	Director radar group for NATO Sea Sparrow	NATO Consortium
ISRAEL			
AMDR	**5696.253**	Automatic missile detection radar	Elta
AT-101A	**5846.253**	Triple-band (L, S & X) surveillance antenna	Elta
EL/M-2207	**3264.253**	F-band general purpose naval search radar	Elta
EL/M 2208	**4610.253**	Air & surface surveillance radar	Elta
EL/M-2221	**3709.253**	J-band naval fire control radar	Elta
ITALY			
Argus 5000	—	High-power ship's early warning radar. 5 MW peak power	Selenia
MM/BPS-704	**3483.253**	Italian Navy version of 3RM I-band radar for submarine use	SMA
MM/SPN-703	**3482.253**	Italian Navy version of 3RM for surface roles	SMA
MM/SPN-728	**6874.253**	X-band surface search radar	SMA
MM/SPQ-701	**3262.253**	I-band search and target designation radar	SMA
MM/SPS-702	**3263.253**	Simplified version of MM/SPQ-701	SMA
MM/SPS-768	**1364.253**	Italian Navy version of RAN-3L	Selenia
MM/SPS-774	**1699.253**	Italian Navy version of RAN-10S	Selenia
Orion 250	—	I/J-band fire control radar. Used in NA9 system. Conical scan. 200 kW	Selenia
Orion RTN-10X	**1368.253**	Fire control radar. Used by RN and Brazil	Selenia
Orion RTN-16X	*1367.253*	I/J-band monopulse fire control radar	Selenia
Orion RTN-20X	**1935.253**	I/J-band fire control radar. Used in DARDO system	Selenia
Orion RTN-30X	**1936.253**	I/J-band fire control radar. Used in Albatros system. Production for Italy 1982	Selenia
Orion RTX-25X	**6875.253**	X-band naval tracking radar	Selenia
RA-30S	**6876.253**	S-band search and interceptor control system	Selenia
RAN-2C	*1700.253*	G/H-band dual purpose air and surface surveillance radar	Selenia
RAN-3L	**1364.253**	D-band early warning radar. Digital signal processing	Selenia
RAN-7S	*1527.253*	10 cm air and surface search radar	Selenia
RAN-10S	**1699.253**	E/F-band dual purpose air/sea search radar	Selenia
RAN-11L/X and 12L/X	**1365.253**	D- and I/J-band system for air warning and weapons control on small ships. Operational Denmark, ordered by Peru and Venezuela	Selenia
RAN-13X	*1701.253*	I/J-band search radar	Selenia
RAN-14X	*1366.253*	I/J-band low altitude and surface search radar	Selenia
Sea Hunter	*1704.253*	Search radar	Contraves
Sea Hunter	*1705.253*	Tracker radar	Contraves
SPQ-2D	**1703.253**	I/J-band search radar	SMA
3RM	**1702.253**	Series of I-band navigation and surface warning radars	SMA
OPS Series	**6136.253**	Series of air and surface surveillance systems in use by Japanese Self-Defense Force	—
NETHERLANDS			
DA.05/2	**1554.253**	E/F-band high cover air warning and surface search radar. Medium-range	Signaal
DA.05/3	**1554.253**	STAR, Signaal terminal area radar	Signaal
DA.05/4	**1554.253**	Medium-range air warning and surface search radar. E/F-band	Signaal
DA.08/1	**3479.253**	Medium-range air warning and surface search radar. E/F-band	Signaal
DA.08/2	**3479.253**	High cover version of DA.08/1	Signaal

Designation	Entry no	Description	Contractor(s)
NETHERLANDS cont			
DA.08/3	**3479.253**	Air warning radar	Signaal
LW.02	**1256.253**	Long-range D-band air surveillance radar	Signaal
LW.04	*1553.253*	Long-range D-band air search radar	Signaal
LW.08/1	—	Prob D-band	Signaal
LW.08/2	**1973.253**	Long-range air warning and surface search radar. D-band	Signaal
LW.08/3	**1973.253**	Long-range air route surveillance radar (LAR)	Signaal
M40	*1591.253*	Fire control radars for guns/Seacat missiles	Signaal
MW08	**6130.253**	G-band surveillance radar for weapon control	Signaal
SGR-109	*1257.253*	E/F-band heightfinder	Signaal
SMART	**5626.253**	3-D surveillance radar for data handling and weapon control	Signaal
STING	**6877.253**	Fire control radar	Signaal
STIR	**3484.253**	Tracking and illuminating radar. Used in NATO Sea Sparrow and Seawolf VM40 system	Signaal
VM40	**3484.253**	See STIR	
WM20	**1590.253**	Series of small vessel fire control radars	Signaal
ZW.03	*1258.253*	I-band surface search and navigation radar	Signaal
ZW.06/1	—	Surface search, navigation and helicopter control radar. E/F-band. Operational India, Morocco and Spain; ordered South Korea and Thailand	Signaal
ZW.06/2S & 3/S	**4051.253**	Stabilised surface search, navigation and helicopter control radar. I-band	Signaal
ZW.06/4S	—	Stabilised I-band surface search and navigation radar	Signaal
ZW07/5	**4051.253**	Submarine surface search and navigation radar. I-band	Signaal
ZW08/2S & /3S	**4051.253**	Stabilised I-band surface search, navigation and helicopter control radar	Signaal Signaal
ZW08/4S	**4051.253**	Version of ZW.08 with sea-skimmer detection	Signaal
ZW09/1	**4051.253**	Surface search, navigation and helicopter control radar. I-band	Signaal
ZW09/2S & /3S	**4051.253**	Stabilised versions of ZW.09/1	Signaal
ZW09/4S	**4051.253**	Sea skimmer version of ZW.09/3S	Signaal
3D MTTR	**1589.253**	3-D multi-target search and tracking radar	Signaal
SWEDEN			
Sea Giraffe	**4299.253**	Air defence radar. Retrofitting in Spica IIs, to be procured for Spica IIIs	L M Ericsson
Subfar 100	**1545.453**	I/J-band submarine radar	PEAB
9GR 600	**1546.253**	I-band. Frequency-agile 200 kW	PEAB
9LV 200	**1547.253**	J-band monopulse frequence-agile tracking radar	PEAB
SWITZERLAND			
Seaguard TMX	**6119.253**	General purpose tracking module for Seaguard	Contraves
UNION OF SOVIET SOCIALIST REPUBLICS			
Ball End	—	High definition I-band navigation and surface search radar	—
Ball Gun	**1900.253**	Similar to Ball End	—
Band Stand	**4090.253**	Search radar group on Nanuchka missile corvettes Housed in large radome. Used with SSN-9 surface-to-surface missile	—
Bass Tilt	**4085.253**	Gun fire control radar. Similar to Drum Tilt, Muff Cob	—
Big Net	**1608.253**	Long-range air surveillance	—
Boat Sail	*1507.253*	Air search radar for submarine pickets	—
Bow Series	—	Target designation radars	—
Cheese Cake	—	Missile targeting radar	—
Cylinder Head	*1596.253*	Gun fire control group on optical director	—
Cross Bird	—	Obsolescent early warning radar	—
Cross Sword	**6879.253**	Fire control radar for SA-N-9 SAM	—
Dead Duck	—	IFF system	—
Don Series	**1900.253**	I/J-band navigation and target designation radar	—
Drum Tilt	**1330.253**	I/J-band fire control. Enclosed scanner	—
Egg Cup	—	Splash spotting radar mounted on gun turrets	—
Eye Bowl	**4089.253**	Missile guidance radars on 'Krivak' class destroyers	—
Fan Song E	**1605.253**	Missile control and tracker group for Guideline SAM on *Dzerzhinski*	—
Flat Spin	*1321.353*	Long-range air search radar	—
Front Dome	**6421.253**	Target illumination and tracking radar	—
Front Door	**6421.253**	Submarine missile guidance radar for SS-N-3 and SS-N-12	—
Front Piece	**6421.253**	Parabolic cylinder reflector. Missile guidance radar for submarines	—
Hair Net	*1607.253*	General purpose E-band search radar. Obsolescent	—
Half Bow	—	I/J-band torpedo target designation radar	—
Hawk Screech	**1325.253**	Gun fire control radar	—
Headlight	**1328.253**	Missile fire control group. G/H/I-band operation	—
Head Net A	**1318.253**	Long-range air surveillance radar	—
Head Net B	**1319.253**	Back-to-back installation of 2 Head Net A for low and high cover	—
Head Net C	**1320.253**	Dual V-beam 3-D installation of Head Net A	—
High Lune	*1606.253*	Nodding heightfinder	—
High Pole	—	IFF antenna	—
High Scoop	—	Elliptical paraboloid reflector. Surveillance radar	—
High Sieve	**1900.253**	Surface search radar	—
Horn Spoon	—	Navigation radar	—
Kite Screech	**4084.253**	Gun fire control radar	—
Kivach	—	Navigation radar for smaller vessels	—
Knife Rest B	**1900.253**	Long wavelength early warning radar	—
Long Bow	—	I/J-band target designation radar for torpedoes or guns	—
Low Sieve	—	Surface search radar	—
Muff Cob	**1611.253**	Gun fire director radar	—
Neptune	—	Navigation radar	—
Okean	—	Search and navigation radar for auxiliary vessels	—
Owl Screech	**1325.253**	Gun fire control radar	—
Palm Frond	—	Search and navigation radar	—
Peel Group	**1323.253**	Missile fire control group for Goa	—

Designation	Entry no	Description	Contractor(s)
USSR cont			
Peel Pair	—	Surface search and navigation radar	—
Plank Sheve	—	Air and surface search radar	—
Plinth Net	**1609.253**	Medium-range general purpose E-band search radar	—
Pop Group	**1897.253**	Fire control radar group for SAN-4 missile system	—
Post Lamp	—	I/J-band target designation radar	—
Pot Drum	**1612.253**	Short-range surface search radar. H/I-band	—
Pot Head	**1613.253**	Short-range surface search radar	—
Punch Bowl	—	Mast mounted submarine radar	—
Round Top	—	Similar to Wasp Head	—
Salt Pot	—	IFF radar	—
Scoop Pair	**1324.253**	Twin E-band radar group for SSM	—
Sea Gull	—	P-band long-range air search radar	—
Sheet Bend	—	General purpose navigation and pilotage radar	—
Sheet Curve	—	General purpose navigation and pilotage radar	—
Ship Globe	—	Large tracking radar for use in missile tests and satellite activities	—
Skin Head	**1614.253**	Surface target detection radar for torpedo boats	—
Ski Pole	—	IFF antenna. On 'Osa' missile boats	—
Slim Net	**1322.253**	High definition surface target radar	—
Snoop Plate	**5849.253**	Submarine surveillance radar	—
Snoop Slab	**5849.253**	Submarine surveillance radar	—
Snoop Tray	**5849.253**	Submarine surveillance radar	—
Spin Trough	—	I-band search and navigation radar	—
Square Head	**1595.253**	IFF interrogator array	—
Square Slot	—	Surface detection radar	—
Square Tie	**1329.353**	General purpose search radar	—
Strut Curve	**1331.253**	Lightweight search radar	—
Strut Pair	—	Version of Strut Curve using back-to-back antennas	—
Sun Visor	**1326.253**	Optical director-mounted gun fire control radar	—
Top Bow	**1900.253**	Target acquisition radar for naval guns	—
Top Dome	**4088.253**	SAN-6 missile guidance group	—
Top Mesh	**6880.253**	Long-range air surveillance radar	—
Top Pair	**4087.253**	Long-range air surveillance radar	—
Top Plate	**6880.253**	Long-range air surveillance radar	—
Top Sail	**1327.253**	Long-range 3-D air surveillance radar	—
Top Steer	**4086.253**	Air surveillance radar	—
Top Trough	**1610.253**	C-band long-range air search radar	—
Trap Door	**6878.253**	SS-N-12 mid-course guidance radar	—
Wasp Head	—	Gun fire control radar	—
Witch Five	—	IFF system	—
UNITED KINGDOM			
400 Series	**6131.293**	Family of fire control radars	Marconi
805 Series	**4302.253**	Naval tracking radar for fire control	Marconi
AWS-1	*1139.253*	Naval version of land-based AR-1	Plessey
AWS-2	*1751.253*	Improved version of AWS-1	Plessey
AWS-3	*3170.253*	Advanced version of AWS-2	Plessey
AWS-4	**3171.253**	Lightweight antenna version of AWS-2. Integrated IFF	Plessey
AWS-5	**3172.253**	Advanced version of ASW-4. Integrated IFF	Plessey
AWS-6	**5620.253**	Search and surveillance radar for naval point defence	Plessey
AWS-7	**6101.253**	Family of S-band radars incorporating a solid-state transmitter for surveillance & target indication	Plessey
AWS-9	**6102.253**	Export version of the RN Type 996 radar	Plessey
CEL 850	**3699.253**	Naval IFF interrogator	Cossor
Cossor IFF	*1571.253*	Naval IFF Mk 10 (SIF) 800 Series, 825, 825M	Cossor
Decca	**1941.253**	I/J-band navigation/tactical radars	Racal
Dolphin	*4004.253*	Target detection radar for Seaguard weapon system	Plessey
IFF 877	*3700.253*	Naval IFF interrogator	Cossor
MESAR	**6128.253**	A multi-function electronically scanned adaptive radar being developed with Admiralty Research Establishment. Operates in S-band	Plessey
MRS3/GWS 22	**1563.253**	Seacat and gun fire control radar	Sperry
PTR 461	*1896.253*	Shipborne IFF transponder	Plessey
PTR 826	*3259.253*	Shipborne IFF interrogator	Plessey
RRB	**3487.253**	Transponder reply receiver for use with ARI 5983/5954	Microwave Assoc
S604 HN	*1895.253*	Search radar. Fitted Egyptian Ramadan class ships	Marconi
S810	**1754.253**	Surveillance radar. Fitted Egyptian Ramadan class ships	Marconi
S820	**1754.253**	E/F-band surveillance radar	Marconi
S1100	**5632.253**	Radar series for navigation and short-range surveillance	Marconi
S1820	**3687.253**	Series of naval surveillance radars (S1820, S1821 and S1840) based on S810/820	Marconi
SNG 20	—	Fire control system	Marconi
SNW 10	—	Metric wavelength early warning radar	Marconi
SNW-12	—	Early warning radar. 150 cm. 450 kW. 4.28 × 2.14 m scanner	Marconi
SNW-20	—	Early warning radar	—
ST802	**1508.253**	Lightweight fire control radar	Marconi
ST1802	**6106.253**	Series of fire control radars	Marconi
S1810	**6107.253**	I-band frequency agile radar for shipborne and coastal defence	Marconi
ST 804	*3686.253*	Naval tracking radar	Marconi
TM626	—	Small navigational radar	Racal
Type 262	—	Short-range anti-aircraft fire control radar	EMI
Type 277	—	Nodding heightfinder radar	—
Type 293	—	General purpose E/F-band search radar	—
Type 294	*3261.253*	Plessey adaptation of RN Type 293	Plessey
Type 901	**1752.253**	Seaslug missile target tracking and guidance radar	Marconi
Type 903	**1563.253**	J-band conical scan tracker used with MRS3 director	Sperry/Plessey
Type 904	**1563.253**	J-band conical scan tracker used with MRS3 director	Sperry/Plessey
Type 909	**1559.253**	Target tracker and illuminating radar for Sea Dart missile system (GWS 30). 8ft cassegrain antenna	Marconi

Designation	Entry no	Description	Contractor(s)
UK cont			
Type 910	**1562.253**	Tracking radar for Seawolf missile	Marconi
Type 911	**5630.253**	Tracking radar for lightweight Seawolf (previously 805SW)	Marconi
Type 912	**1558.253**	RN designation of Selenia RTN-10X	—
Type 944	—	IFF	—
Type 965	**1560.253**	Primary search radar	Marconi
Type 967	**1561.253**	Compact air surveillance radar. Combined in Sea Wolf system	Marconi
Type 968	**1561.253**	Surface surveillance radar	Marconi
Type 975	**1253.253**	Lightweight I/J-band high definition surface warning radar. 50 kW	Kelvin Hughes
Type 975ZW	**1253.253**	Minehunting version of Type 975. True motion display plus sonar contact working	Kelvin Hughes
Type 978	*1254.253*	3 cm navigational radar. Double-cheese scanner. Successor to Type 974	Racal
Type 979	—	Similar to 978 but with additional B-scope display	—
Type 982	—	Aircraft direction	—
Type 983	—	Nodding heightfinder radar	—
Type 984	*1041.253*	3-D air surveillance long-range radar	Marconi
Type 992	*1255.253*	Long-range surveillance radar. Obsolescent	EMI
Type 992Q	**1753.253**	Solid-state general purpose radar	Marconi
Type 993	—	E/F-band search radar	—
Type 994	**3260.253**	Combination of Type 993 antenna with AWS-2 transmitter/receiver	Plessey
Type 996	**4634.253**	3-D surveillance and target indication radar for Type 23 frigates, Type 42 destroyers and Invincible class carriers	Plessey
Type 1006	**1394.253**	Successor to Type 975. Solid-state. 305mm display Max range 64 nm	Kelvin Hughes
Type 1007	**5847.253**	I-band surveillance radar	Kelvin Hughes
Type 1010	*3698.253*	RN version of Cossor IFF 800	Cossor
Type 1013	**3701.253**	Naval version of Cossor IFF 2750	Cossor
Type 1022	**3486.253**	Replacement for Type 965 long-range radar. Fitted in various classes	Marconi
Type 1030	*3485.253*	Surveillance and target indication radar. Type 992Q replacement	Marconi
Type 1800	*3485.253*	Export version of Type 1030	Marconi
UNITED STATES OF AMERICA			
BPS-1	—	Submarine I-band search and fire control radar	Western Electric
BPS-2	—	Submarine D-band air search radar	Raytheon
BPS-3	—	H-band submarine air search radar	Western Electric
BPS-4	—	Short-range submarine air and surface search radar	Westinghouse
BPS-5	—	Short-range surface search radar for SSN 593 subs	Lockheed Electric
BPS-9	—	Submarine I-band air search radar	Western Electric
BPS-11	—	Submarine surface-search radar for SSB 616 subs	Western Electric
BPS-12	—	Medium range submarine surface-search radar	Fairchild
BPS-13	—	Similar to BPS-12	Fairchild
BPS-14	—	Similar to BPS-12. I/J-band	Fairchild
BPS-15	—	Submarine search and navigation radar. I-band	Sperry
FAST	**4608.253**	C and S-band search radar	General Electric
FLEXAR	**3258.253**	Weapon control system radar	Hughes
HR-76	**3768.253**	I-band combined search and track FCS radar for H-930 system	RCA
MBAR	**4233.253**	Multiple beam acquisition radar	Raytheon
RTN-10	*1245.253*	I/J-band fire control radar for Sea Sparrow III	Raytheon
Ship OTHR	*1937.253*	Shipboard over-the-horizon radar project	—
SPA-72	—	3-D fixed planar array surveillance radar	Hughes
SPG-49	*1749.253*	Guidance radar for Talos and Terrier surface-to-air missiles Used with SPW-2	Sperry
SPG-51	**1247.253**	Tartar missile guidance radar. I/J-band. Part of Mk 74 FCS	Raytheon
SPG-55	**1748.253**	Terrier missile fire control radar. G/H-band	Sperry
SPG-56	*1750.253*	Talos fire control radar. G/H-band	Sperry
SPG-59	—	Typhon missile radar. Performs all functions. Luneberg lens	Westinghouse
SPG-60	**1932.253**	Mk 86 FCS search and track radar	Lockheed
SPQ-5	*1938.253*	Early guidance radar	Sperry
SPQ-9	**1931.253**	Track-while-scan surface fire control radar used in Mk 86 FCS	Lockheed
		Housed in 300 cm radome	—
SPQ-11	**4301.253**	Missile space tracking radar. Cobra Judy. Operational	Raytheon/RCA/CSC
SPQ-55	—	Terrier tracking and guidance radar	Sperry
SPQ-60	—	Pulse doppler air target tracking radar used in Mk 86 FCS	—
SPS-3	—	Surveillance radar	Sylvania
SPS-4	—	Air and surface surveillance radar	Raytheon
SPS-5	—	Surface surveillance radar	Raytheon
SPS-6	**1744.253**	Search radar	Westinghouse
SPS-10	**1564.253**	G/H-band surface search radar	Sylvania/Raytheon
SPS-12	*1566.253*	D-band medium/long-range surveillance radar	RCA
SPS-17	—	Long-range surveillance radar	General Electric
SPS-23	—	Heightfinder	General Electric
SPS-29	—	Air surveillance radar for guided missile destroyers	Westinghouse
SPS-30	**1745.253**	Long-range 3-D radar	General Electric
SPS-32	*1250.253*	3-D fixed planar array electronic scanning air and surface search radar	Hughes
SPS-33	*1251.253*	3-D fixed planar array electronic scanning target tracking radar	Hughes
SPS-34	—	Shipborne version of FPS-7 stacked beam 3-D radar	General Electric
SPS-37	**1565.153**	Air search radar	Westinghouse
SPS-39	**1249.253**	'Frescan' hybrid 3-D air target search radar	Hughes
SPS-40	**1746.253**	Air search radar. 'A' version has IFF	Norden
SPS-42	—	Hybrid electronic scanning 3-D radar	Hughes
SPS-43	**1747.253**	Air search radar	Westinghouse
SPS-48	**1252.253**	E/F-band long-range frequency scanned search and target acquisition radar. C and E versions now in production	ITT-Gilfillan
SPS-49	**1939.253**	Long-range air search radar for fitting in 5 Spanish FFGs	—

Designation	Entry no	Description	Contractor(s)
USA cont			
SPS-52/52A/52B	**1248.253**	Rotating planar (hybrid) 3-D surveillance radar	Hughes
SPS-52C	**5606.253**	Development of SPS-52	Hughes
SPS-55	**1697.253**	I/J-band surface radar	Cardion
SPS-58	**1359.253**	D-band air search radar used in H-930 FCS	Westinghouse
SPS-65(V)ER	**4300.253**	Long-range version of SPS-58	Westinghouse
SPS-59	**3749.281**	Surface search radar used in H-930 FCS	—
SPS-63	**1940.253**	US designation for Italian 3RM series radars (SMA) Used on NATO hydrofoil	—
SPS-65	**1359.253**	Variant of SPS-58	Westinghouse
SPS-67	**3722.253**	Solid-state transmitter/receiver to update SPS-10. New antenna being designed	Norden
SPW-2	—	Talos guidance radar. Used with SPG-49	Sperry
SPY-1	**1570.253**	Multi-function array radar for Aegis system. Electronic scan	RCA/Raytheon
Type 23 TAS	**4609.253**	2-D acquisition radar for NATO Sea Sparrow	Hughes
UPX-24(V)	**6126.293**	Control IFF system operating as part of a weapon system	Litton
W-120	**3256.253**	Shipboard version of APQ-120 fire control	Westinghouse
W-160	**4048.253**	J-band fire control radar	Westinghouse
W-611	*3257.253*	Shipboard version of TPS-61 D-band search radar	Westinghouse

1194.354 Airborne Radar

Designation	Entry no	Description	Contractor(s)
CANADA			
APS-504	**3473.353**	ASV search radar	Litton
IRIS	**5874.353**	Integrating synthetic aperture imaging system	McDonald-Dettwiller
LH66/HP	**6423.353**	Airborne and shipborne surveillance radar	CMC
FRANCE			
Agave	**1672.353**	I/J-band helicopter radar. Fitted French Navy Super Etendards, Jaguar International, all export Mirage FL and 50	Thomson-CSF/ESD
Aida I	—	Boresight ranging radar. Fitted French Navy Etendard IVM. I/J-band	Electronique Serge Dassault
Aida II	**1211.353**	Improved version of Aida I	Electronique Serge Dassault
Anemone	**6427.353**	Updated version of Agave for the new generation Super Etendard	ESD
Antilope	—	All weather strike radar, J-band. 4-lobe cassegrain antenna. Terrain following, mapping, ground attack	Electronique Serge Dassault
Antilope II	—	Simplified version of Antilope for light aircraft	Electronique Serge Dassault
Antilope IV	*1963.353*	Nose radar development for penetrator version of Super-Mirage	ESD/Thomson-CSF
Antilope V TC	**6124.353**	Strike radar for Mirage 2000N	Thomson-CSF
Cobra	—	Low-level strike and terrain following. Experimental system. Some features used in Cyrano III	Thomson-CSF
Cyrano I bis (RA 536)	*1050.353*	Monopulse fire control radar. Used in Mirage IIIC. I/J-band	Thomson-CSF
Cyrano II (RA 537)	*1051.353*	Monopulse fire control radar. Cassegrain antenna. Used in Mirage IIIE. I/J-band	Thomson-CSF
Cyrano III (RA 538)	**3131.353** *1052.353*	9 models in range, providing number of options up to full multi-role capability including air and ground attack and terrain following	Thomson-CSF
Cyrano IV	**1396.353**	Attack radar, air-air/air-ground. French F1s retrofitted with Mk IVM	Thomson-CSF
DAV	**6889.353**	Search and alert radar mounted on helicopter rotor	ESD
DRAA-2B	*1961.353*	ASV and navigation radar for Atlantic ASW aircraft	Thomson-CSF
DRAC-25	—	I/J-band fire control radar used in French Vatour fighters	Thomson-CSF
ESD 3300/3400	**1962.353**	IFF transponders	ESD
Heracles	**1960.353**	System of modular tactical radars based on ORB-31	Omera-Segid (Thomson-CSF)
Iguane/Agrion/Varan	**3298.253**	ASV radars in production for fixed and rotary wing aircraft	Thomson-CSF
Madrague	—	Pulse doppler early warning radar	Electronique Serge Dassault
Mirage 2000	**3297.353**	Pulse-doppler I/J-band radar for Mirage 2000. Multi-mode RDM version delivered to French AF 1983	Thomson-CSF/ESD
NRA1-3A	**6122.393**	Airborne IFF interrogator	LMT
NRA1-7A	**5871.353**	Airborne IFF transponder	LMT
NRA1-9A	**6123.393**	IFF transponder	LMT
NRA1-10A	**5872.353**	Airborne IFF interrogator	LMT
NRA1-11A	**5873.353**	IFF interrogator/decoder	LMT
ORB-31	**1960.353**	Helicopter and light aircraft I-band ASV, navigation, and met radar	Omera-Segid (Thomson-CSF)
ORB-32	**6425.353**	Development of ORB-31	Omera-Segid (Thomson-CSF)
ORB 37-HL	**6426.353**	Navigation and weather radar	Omera-Segid (Thomson-CSF)
Orchidee	**5875.353**	Battlefield surveillance radar	LCT
Oryx	*1351.353*	J-band terrain following and attack radar	Electronique Serge Dassault
RACAAS	**6428.353**	Multi-mode radar in development	Thomson-CSF
Rafal	—	I/J-band coherent sideways looking radar	Thomson-CSF
RDX	**6137.353**	Range of airborne systems for combat aircraft of the 1990s	Thomson-CSF
RDY	**6890.353**	Multi-function radar for Mirage 2000	Thomson-CSF
RH 370	*1395.353*	I/J-band helicopter ASV radar	Thomson-CSF
Saiga	*1352.353*	Helicopter terrain avoidance radar. K-band	Electronique Serge Dassault
Saiga II	—	Simplified, lightweight version of Saiga	Electronique Serge Dassault
INTERNATIONAL			
APS-503	**1784.353**	Helicopter ASV, search, and navigation radar. I-band	AIL/Litton Canada
APS-504(V)3	**3473.353**	Airborne search radar	Litton Canada/Eaton
EFA	**6429.353**	Airborne fire control radar for European Fighter Aircraft	—
NIS	**5621.053**	NATO identification system (IFF). In development	—

Designation	Entry no	Description	Contractor(s)
ISRAEL			
EL/M-2001B	**3305.353**	I- or J-band ranging radar	Elta
EL/M-2011	**6891.353**	Lightweight fire control radar	Elta
EL/M-2021	**3306.353**	Multi-mode, pulse-doppler, I/J-band attack radar. A version is under development for the Israeli Lavi fighter (**5876.353**)	Elta
EL/M-2032	**6891.353**	Lightweight fire control radar	Elta
EL/M-2035	**5876.353**	Multi-mode fire control radar for the Lavi fighter	Elta
Phalcon	**6892.353**	Solid-state AEW radar for use on SIGINT aircraft	Elta
ITALY			
Altair	**6202.353**	Multi-mode fire control radar in development with Westinghouse	FIAR
AM/APS-717	**6893.353**	X-band search and navigation radar	SMA
Grifo	**6201.353**	Multi-purpose doppler radar using a coherent TWT transmitter	FIAR
MM/APQ-706	**3309.353**	I-band helicopter radar for Marte anti-ship missile system	SMA
MM/APS-705	**3307.353**	I-band helicopter radar. Dual transmitter/receivers	SMA
MM/APS-707	**3308.353**	I-band helicopter radar. Single channel version of APS-705	SMA
MM/APS-784	**6895.353**	X-band radar for Italian EH-101	Fiar/SMA
MM/UPX-709/SIT 421	**3489.353**	IFF transponder	Italtel
SC-01	**6894.353**	Multi-mode radar intended primarily for the Brazilian Air Force AM-X aircraft	SMA/Tecnasa/Brazilian Air Force
Setter	**6203.353**	Frequency agile fire control radar-update of R21G radar	FIAR
SIT 422	**3488.353**	Short range air defence interrogator	Italtel
SIT 434	**6896.393**	IFF interrogator	Italtel
JAPAN			
Mitsubishi	**6899.353**	Airborne radar for future fighters	Mitsubishi
SWEDEN			
DAX-100/200 Series	*1741.353*	Fire control radars	Ericsson
Ericsson AEW	**6196.353**	Airborne early warning radar for Swedish Air Force	Ericsson
Ericsson SLAR	**4083.353**	J-band maritime surveillance radar	Ericsson
Hera	**5887.353**	Helicopter surveillance radar	Philips
PS-03/A (UAP 12)	*1208.353*	I/J-band AI radar, conical scan. Cassegrain antenna. Used in J35B and J35D Swedish AF Drakens	Ericsson
PS-01/A (UAP 13)	**1209.353**	Improved version of PS-03/A. Used in J35F Draken	
PS-05	**5684.353**	Nose radar for JAS-39	Ericsson
PS-011/A (UAP 13)	**1209.353**	Version of PS-01/A used with S71N IR search and track set in J35F	Ericsson
PS-37/A (UAP 1011)	**1210.353**	I/J-band. Monopulse. Modes include search, ground ranging, mapping, and obstacle avoidance	Ericsson
PS-46A	**1964.353**	Multi-mode radar for Viggen interceptor	Ericsson
JAS-39	**5684.353**	Nose radar and AEW radar for JAS-39 Grippen aircraft	Ericsson
UNION OF SOVIET SOCIALIST REPUBLICS			
Bee Hind	—	Tail warning and gun fire control	—
Big Bulge	**6430.353**	Large ventral radar on Tu-95 Bear bomber	—
Big Nose	—	AI radar used in Tu-28 Fiddler strike aircraft	—
Crown Drum	—	Surface search radar associated with AS-3 missile	—
Down Beat	—	Bombing and navigation radar	—
Fan Tail	—	Tail radar for Tu-26 Backfire bomber	—
Flat Jack	**6431.353**	AWACS radar on TU-126 aircraft	—
Fox Fire	—	AI radar. I-band	—
High Lark	**6432.353**	J-band AI radar. Used in Flogger B/D/F	—
High Fix	—	Airborne ranging radar for Fitter-A and Fitter-C	—
Jay Bird	**3311.353**	AI radar for MiG-19	—
Look Two	**1475.353**	I/J-band bombing and navigation radar	—
Mushroom	—	Helicopter ASV radar	—
Puff Ball	**1476.343**	I/J-band search radar used in Bison aircraft	—
Scan Can	—	AI radar	—
Scan Fix	**1477.343**	AI radars for MiG-17 and 19 interceptors	—
Scan Odd	**1479.353**	I-band AI radar for MiG-19	—
Scan Three	*1480.353*	I-band AI radar for Yak-25	—
Shore Walk	—	Bombing radar for Yak-28 Brewer-B	—
Short Horn	**1478.353**	J-band bombing and navigation radar	—
Skip Spin	**1481.353**	I-band AI radars used in Su-11 and Yak 28	—
SLAR	—	Side-looking radar 1MW. J-band	—
Spin Scan	**1482.353**	Series of AI radars used in Su-9 and MiG-21	—
SRO-1/2	—	IFF transponder. SRO-1 obsolescent	—
SUAWACS	**4616.393**	Airborne AEW system for IL-76	—
Toadstool	—	Used on Coke transport aircraft	—
Twin Scan	—	AI radar	—
Wet Eye	—	Surface search radar	—
UNITED KINGDOM			
UK AEW	*1971.353*	Airborne early warning radar using Nimrod aircraft – now cancelled	GEC Avionics
Airpass I	*1031.353*	I/J-band AI radar. Search, tracking, and ranging. Used in RAF Lightning interceptors. Produced in 4 versions RAF designation AI23	Ferranti
Airpass II	*1033.353*	Development of Airpass I with added air-to-ground modes	Ferranti
Airpass III	*1032.353*	Surface strike radar for Buccaneer aircraft. Multi-mode I/J-band cassegrain antenna	Ferranti
APY 920	*5878.353*	Version of UK AEW radar for C-130 now cancelled	GEC Avionics
ARI 5955/5954	*1027.353*	I-band system for use in ASW helicopters. Enables helicopter to be used as tactical control centre for ASW operation. Secondary radar capability. Fitted RN Sea Kings	MEL
ARI 5983	**3310.353**	I-band transponder for naval helicopters	Microwave Assoc
ARI 5991	**3710.353**	Sea Searcher. Replacement for ARI 5955/5954	MEL
ASR 360	**5614.353**	Maritime surveillance radar for medium sized fixed wing aircraft and helicopters. Developed from Decca Marine systems	Racal

Designation	Entry no	Description	Contractor(s)
UK cont			
ASTOR	**4613.353**	Airborne battlefield surveillance radar	Ferranti/Thorn EMI
ASV-21D	—	Air-to-surface vessel search radar used in Nimrod maritime aircraft	EMI Electronics
Blue Fox	**1972.353**	Version of Seaspray. Production for RN and Indian Sea Harriers	Ferranti
Blue Kestrel	**5685.353**	ASV/ASW system in development for EH101 helicopter	Ferranti
Blue Vixen	**4615.353**	Multi-role pulse doppler system for Sea Harrier update programme. In development	Ferranti
E290M	*1028.353*	Combined weather and tactical radar. RCAF	
Foxhunter	**1785.353**	UK airborne interceptor radar. In production for RAF Tornado ADV aircraft	Marconi
IFF 2720	**3695.353**	IFF transponder	Cossor
IFF 2743	**5603.353**	Electronic control unit for use with IFF transponders	Cossor
IFF 3100	**3696.353**	IFF transponder. Selected for P110	Cossor
IFF 3500	**3697.355**	IFF airborne interrogator. Selected for P110	Cossor
IFF 4700	**6434.353**	Mk XII airborne transponder series	Cossor
LRR	*1035.353*	Lightweight ranging radar. Air and ground ranging Terrain warning	Ferranti
MAREC II	**4341.353**	Maritime reconnaissance radar	MEL
PTR 820	*1224.353*	IFF airborne interrogator	Plessey
P111	—	Coherent pulse multi-mode strike. AI and terrain following radar. Developed for classified project	Marconi
P391	*1023.353*	K-band sideways looking recce radar. Used in UK Phantom recce pod. Licensed to Electronique Serge Dassault	EMI Electronics
Red Fox	**5877.353**	Repackaged version of Blue Fox airborne radar	Ferranti
SATF	*1216.353*	Strike and terrain following radar. I/J-band multi-mode	Ferranti
Searchwater	**1740.353**	ASV and maritime radar. Fitted RAF Nimrods and RN AEW Sea Kings.	Thorn EMI
Sea Searcher	**3170.353**	See ARI 5991. Operational	MEL
Seaspray	**1342.353**	I/J-band helicopter ASV radar	Ferranti
Sky Guardian	*4614.353*	Private venture AEW system based on cancelled UK AEW radar	Marconi
Skymaster	**5886.353**	Surveillance and early warning radar	Thorn EMI
Skyranger	**5612.353**	Lightweight weapon control system for small fighter and ground attack aircraft. Aimed at retrofit market	Marconi
Sky Searcher	*4617.353*	AEW radar based on Sea Searcher	MEL
Super Marec	**6433.353**	Development of MAREC II	MEL
Super Searcher	**5625.353**	Lightweight maritime surveillance system developed from Sea Searcher	MEL
UNITED STATES OF AMERICA			
AIXP	*6115.393*	IFF interrogator transponder	Teledyne
Airship radar	**6440.353**	Proposed airship surveillance radar based on APS-138 series	—
APD-7	—	Sideways looking radar. Used RA-5C	Westinghouse
APD-8	—	Sideways looking radar for RF-111A. Not deployed	Westinghouse
APG-22	—	CW ranging radar. USN	Raytheon
APG-31	—	Ranging radar. USAF	Raytheon
APG-43	—	CW AI radar	Raytheon
APG-55	—	AI intercept radar for USAF pulse doppler	Westinghouse
APG-59	—	Fire control radar. USN, USAF	Westinghouse
APG-63	**1765.353**	Multi-mode fire control radar for F-15	Hughes
APG-65	**3302.353**	I/J-band multi-mode radar. Operational in F/A-18 Hornet	Hughes
APG-66	**3303.353**	I/J-band multi-mode F-16 radar	Westinghouse
APG-67	**5610.353**	Multi-mode nose radar for F-5G Tigershark	GE Aerospace
AN/APG-68	**5881.353**	Development of AN/APG-66 for F-16 C/D aircraft	Westinghouse
AN/APG-69(V)	**5882.353**	Multi-mode airborne radar	Emerson
AN/APG-70	**5879.353**	Upgrade of AN/APG-63 for F-15E aircraft	Hughes
AN/APG-71	**5880.353**	Upgrade of AN/AWG-9 for the US Navy	Hughes
APN-59	—	Search and mapping radar	Sperry
APN-170	*1037.353*	Terrain following radar	General Dynamics
APN-171	*3124.353*	Terrain following radar	General Dynamics
APN-215	**4340.353**	Weather and terrain mapping radar	Bendix
APN-234	**6105.353**	Lightweight multi-mode radar for sea search	Bendix
APQ-55	—	Surveillance mapping radar. US Army	
APQ-56	—	Sideways looking radar for RB-47E. USAF	Westinghouse
APQ-67	—	AI radar. USAF	Raytheon
APQ-72	—	I/J-band AI radar. Target illuminator for Sparrow III missile Fitted F-4B	Westinghouse
APQ-81	—	Track-while-scan pulse doppler. USN	Westinghouse
APQ-83	—	Fire control radar for Sidewinder missile	Magnavox
APQ-86	—	Sideways looking battlefield surveillance radar	Texas Instruments
APQ-92	*1489.353*	Search radar used in USN A-6A aircraft	Norden
APQ-94	—	Fitted in F-8D aircraft	Magnavox
APQ-97	*7014.353*	Sideways looking radar. US Army	Westinghouse
APQ-99	**1970.353**	Forward radar for RF-4	Texas Instruments
APQ-100	—	Fire control radar. Used in F-4C. USAF	Westinghouse
APQ-102A	—	Recce radar. Used by USAF, USMC RF-4Cs Modified version fitted to some West German RF-104Gs	Goodyear
APQ-109	—	Fire control radar F-4D. USAF	Westinghouse
APQ-110	—	J-band terrain following and avoidance radar for F-111A Mounted below APQ-113	Texas Instruments
APQ-112	*1490.353*	Tracking radar on A-6A. Used with APQ-92 search radar	Norden
APQ-113	**1966.353**	J-band multi-mode radar for F-111A. Operational USAF and RAAF	GE Aerospace
APQ-114	**1967.353**	Multi-mode nose radar for FB-111A	GE Aerospace
APQ-116	—	Fire control radar. Ground ranging, terrain following AI Used on A-7A	Texas Instruments
APQ-119	—	Modified version of APQ-113. Limited procurement for F-111 programme	GE Aerospace
APQ-120	**1310.353**	AI terrain following radar for F-4E	Westinghouse
APQ-122	**1487.353**	Dual frequency I/J- and K-band radars for AWADS C-130 aircraft	Texas Instruments

Designation	Entry no	Description	Contractor(s)
USA cont			
APQ-124	—	Doppler AI ranging radar. USN	Magnavox
APQ-125	—	Doppler ranging radar for F-8J	Magnavox
APQ-126	**4334.353**	Used in USN A-7 aircraft	Texas Instruments
APQ-127	—	Forward looking radar	Sperry
APQ-128	—	F-111 terrain following radar	Sperry
APQ-130	—	F-111D attack radar for Mk II avionics	Autonetics
APQ-136	—	Nose radar for AC-119 gunship	Texas Instruments
APQ-137	*1401.353*	K-band MTI radar for detection of ground targets	Emerson
APQ-139	*1486.353*	J-band AMTI terrain avoidance/mapping radar Used in Tropic Moon 3 programme	Texas Instruments
APQ-140	*1488.353*	J-band multi-mode electronic scan radar	Raytheon
APQ-144	**1968.353**	Multi-mode nose radar for F-111F	GE Aerospace
APQ-146	**4335.353**	Forward looking multi-mode nose radar	Texas Instruments
APQ-148/156	**1567.353**	J-band multi-mode radar for A-6E aircraft	Norden
APQ-152	—	All-weather topographic mapping radar set	Goodyear
APQ-153	**1675.353**	Fire control radar	Emerson
APQ-156	**1567.353**	See APQ-148	Norden
APQ-157	**3113.353**	Dual control search and range tracking radar for F-5F	Emerson
APQ-158	**4336.353**	Forward looking multi-mode radar used for Pavelow III	Texas Instruments
APQ-159	**3114.353**	Lightweight multi-mode nose radar. (V)-1 model on F-5E. (V)-2 on F-5F	Emerson
APQ-161	**3301.353**	Update of APS-144 for Pave Tack F-111F	GE Aerospace
APQ-163	**3682.353**	Experimental nose radar for B-1	GE Aerospace
APQ-164	**4647.353**	Navigation and weapon delivery radar for B-1 supersonic bomber	Westinghouse
APQ-165	**6897.353**	Update of the APQ-113 for RAAF F-111C aircraft	GE Aerospace
APQ-168	**6441.353**	Multi-mode radar for the JVX project – derived from LANTIRN terrain-following radar	Texas Instruments
APQ-171	**6116.353**	Modified version of ALQ-146	Texas Instruments
APQ-172	**6117.353**	Modified version of ALQ-99	Texas Instruments
APQ-174	**6898.353**	Derivative of the LANTIRN TFR	Texas Instruments
APS-20	—	Airborne early warning radar	
APS-42	—	Navigation and search radar	Bendix
APS-45	—	Airborne heightfinder radar	Texas Instruments
APS-49	—	High scan-rate ASV	Hazeltine
APS-70	—	Airborne early warning radar	GE Aerospace
APS-73	—	Series of SLARs developed for USAF and flown on C-97, B-58, and C-135 aircraft for test and evaluation of coherent, focused, synthetic aperture side-looking radars	Goodyear
APS-81	—	Nose radar for B-52	—
APS-85	—	Sideways looking MTI radar	Motorola
APS-88A	—	ASV search radar for S-2E tracker aircraft	Texas Instruments
APS-91	—	Airborne early warning radar	—
APS-94	**1484.353**	Long-range sideways looking radar. I/J-band	Motorola
APS-96	*1781.353*	Airborne early warning radar for E-2A	GE Aerospace
APS-111 (XN-1)	*1782.353*	Early warning radar for detection of aircraft targets against surface background. Used in E-2A. 2 built	GE Aerospace
APS-113	—	Lightweight helicopter radar	GE Aerospace
APS-115	**1969.353**	ASV search radar for P-3C ASW aircraft. I-band	Texas Instruments
APS-116	**1485.353**	ASV search radar for S-3A ASW aircraft. I-band	Texas Instruments
APS-119 (XN-1)	*1311.353*	Experimental version of APS-96 to provide overland target detection	GE Aerospace
APS-120	*1783.353*	Airborne early warning radar for E-2C aircraft	GE Aerospace
APS-124	**3518.353**	Search radar for LAMPS III helicopter	Texas Instruments
APS-125	**1965.353**	Improved version of APS-120 with ARPS (advanced radar processing sub-system). Fitted E-2C aircraft. Operated US, Israel, Japan etc	GE Aerospace
APS-127	**3519.353**	Search radar for USMC medium-range surveillance aircraft	Texas Instruments
APS-128	**3132.353**	Maritime air search radar. See also DIGITACS (**4612.353**)	AIL
APS-130	**3520.353**	Navigation radar for EA-6B	Norden
APS-131	**5883.353**	Side-looking surveillance radar	Motorola
APS-133	**4337.353**	Weather and terrain mapping radar	Bendix
APS-134(V)	**4338.353**	ASV/maritime reconnaissance radar for BR-1150	Texas Instruments
APS-135	**5884.353**	Side-looking surveillance radar	Motorola
APS-137(V)	**6436.353**	Update of APS-116 for S-3A ASW aircraft	Texas Instruments
APS-138	**1965.353**	Updated APS-120/125 radar for E-2C	GE Aerospace
APS-139	**6438.353**	Development of APS-138	GE Aerospace
APS-145	**6437.353**	Advanced version of APS-138	GE Aerospace
APX-72	**3688.353**	Airborne IFF transponder	Hazeltine
APX-76	**3690.353**	Airborne IFF interrogator	Hazeltine
APX-83	**1569.353**	Airborne IFF interrogator for E-2C aircraft	AIL
APX-100(V)	**4339.353**	Airborne IFF transponder	Bendix
APX-101	**6900.393**	Solid-state ATC transponder	Teledyne
APX-103	**3300.353**	IFF interrogator system for E-3A AWACS	AIL
APX-104	*3689.353*	Airborne IFF interrogator	Hazeltine
APX-108	**6900.393**	Mk XII IFF diversity transponder	Teledyne
APY-1	**1585.353**	Airborne warning and control system (AWACS) radar for E-3A	Westinghouse
ASARS-2	**6439.353**	Side-looking, synthetic aperture reconnaissance radar	Hughes
ASG-14	—	Fire control radar for nuclear armed F-104A	Texas Instruments
ASG-15	—	Rear fire control radar for B-52	Arma
ASG-17	—	Fire control radar. Nuclear weapons	—
ASG-18	—	Fire control radar for AIM-47A and AIM-54 missile Fitted FY-12A	Hughes
ASG-19	—	Fire control radar for F-105D (Thunderstick)	GE Aerospace
ASG-25	—	Fire control radar. Non-nuclear arms	—
ASQ-112	—	MTI radar. USN	Norden
AWG-9	**1100.353**	Missile fire control radar system. F-14	Hughes
AWG-10	**3018.353**	AI-terrain avoidance, mapping radar for F-4J. High prf pulse doppler	Westinghouse

Designation	Entry no	Description	Contractor(s)
USA cont			
AWG-11	**3018.353**	Version of AWG-10 for RN Phantoms	Westinghouse
AWG-12	**3018.353**	Version of AWG-10 for RAF Phantoms	Westinghouse
CIT	**6900.393**	Combined IFF interrogator/transponder	Teledyne
DPS-5	**4618.353**	Balloon-borne radar	RCA
UPD-4	**1676.353**	Side-looking radar. Used on USAF RF-4Cs and Japanese RF-4E	Loral
UPD-5	—	High resolution side-looking radar with MTI for USAF and West Germany. Development	Westinghouse
UPD-6	*3474.353*	Reconnaissance radar for West German AF	Loral
UPD-8	**5686.353**	Side-looking synthetic aperture radar. Update of UPD-4	Loral
UPD-9	**5686.353**	Medium range version of UPD-8	Loral
UPX-709	**3691.353**	Airborne IFF transponder	Hazeltine
AWACS	**1585.353**	Airborne early warning radar for E-3A	Westinghouse
B-52	**5888.353**	Upgrade of B-52 radar	Boeing/Norden
Cobra	*1706.353*	Fire control radar	Hughes
Conformal radar	**5609.353**	Design of surveillance radar embedded in nose and wings with solid-state modules	GE Aerospace
Covert strike radar	**4648.353**	Stand-off radar project	—
DIGITACS	**4612.353**	Improved version of APS-128	Eaton
F-16	**3303.353**	I/J-band pulse-doppler, multi-mode radar for F-16 aircraft	Westinghouse
Egyptian Goose	**3162.353**	Balloon-borne radar project. ARPA	Westinghouse
ESAIRA	*1491.353*	Experimental electronic scanned antenna	Hughes
HOWLS/ULTRA	**3304.353**	Experimental ultra lightweight transmissive array for Hostile Weapons Location Systems programme	GE Aerospace
JSTARS	**5624.353**	Joint USAF/US Army requirement for airborne anti-tank radar. 4 contenders bidding: General Electric, Westinghouse, Hughes, Grumman/Norden	US DoD
MERA	—	Molecular electronics for radar applications Experimental phase scanning system	Texas Instruments
Multibeam Survivable radar	**5611.353**	Air-to-ground weapon control and battlefield surveillance radar. Contender for JSTARS	GE Aerospace
Multi-mode radar	**6435.353**	GE multi-mode X-band radar for retrofit into F-5E	GE Aerospace
Pocket Veto	**3162.353**	Surveillance radar on ARPA Family II aerostat	Westinghouse
Project Seek Skyhook	**4618.353**	Balloon-borne radar	RCA
NASARR	*1568.353*	North American search and ranging radar for F-104	Autonetic/NATO
RARF	—	Radome, antenna, RF. Refractive lens phase scanning J-band radar. Experimental	Emerson
RARF	—	Radome, antenna, RF. Reflecting lens phase scanning I- and J-band versions built	Raytheon
RDR-1500	**6104.353**	Multi-search and acquisition radar	Bendix
Balloon-borne radar	**3162.353**	Improved Pocket Veto radar for surveillance and clutter measurement of USAF	Westinghouse
SLAMMR	**5885.353**	Side-looking surveillance radar	Motorola
TEC-60i	**6900.393**	Combined IFF interrogator/transponder	Teledyne
Tornado nose radar	**4333.353**	Terrain following/mapping radar for strike version of Tornado	Texas Instruments
EC-153	*3299.353*	I/J-band pulse-doppler multi-mode derivative of APQ-153 and APQ-159.	Emerson
WX Series	**1673.353**	Fire control radar series. Developed version selected for F-16	Westinghouse

1521.454

Sonar Equipment

Designation	Entry no	Description	Contractor(s)
AUSTRALIA			
Barra (SSQ-801)	**1791.453**	Project Barra. RAAF/RAN advanced sonobuoy and airborne detection system	Amalgamated Wireless
Karrawarra	**5864.463**	Towed array sonar for surface ship/submarine use. In early development	—
Mulloka	**1790.453**	Hull-mounted sonar for RAN	—
SSQ-801	**3998.453**	Barra system passive directional sonobuoy	Amalgamated Wireless
CANADA			
AN/AQS-503	**6551.353**	Helicopter acoustic processing system	CDC
AN/UYS-503	**4649.363**	Sonobuoy processor	CDC
Barra processor	**6552.353**	Side processor for Barra sonobuoys	CDC
HS-100	**1636.453**	Lightweight search/attack sonar. Hull-mounted and towed versions	Canadian Westinghouse
AN/SQQ-504	**6234.263**	Towing condition monitor	Westinghouse
Model 971/977	**6233.453**	Colour imaging sonar for minehunting	Mesotech
Scan 515	**6230.253**	High resolution omni-sonar for patrol vessel	Scannar
Scan 522	**6845.453**	Shallow water sonar system	Canadian Westinghouse
Scan 525	**6231.253**	Surface vessel search/attack sonar	Scannar
Scan 545	**6845.453**	Enhanced version of Scan 522	Canadian Westinghouse
SQR-501	**5863.253**	CANTASS (Canadian towed array sonar system) for surface ships	CDC
SQS-505	**1792.453**	Medium search/attack sonar	Canadian Westinghouse
SQS-507 (Helen)	*2038.453*	Lightweight variable-depth towed sonar	Canadian Westinghouse
SQS-509	**4000.453**	Lower frequency version of SQS-505	CAE
SQS-510	**6846.453**	Digital sonar for surface ships	CDC
SSQ-517	*3085.453*	Passive sonobuoy	Sparton
SSQ-518	*3085.453*	Passive sonobuoy. Long-life version of SSQ-517	Sparton
SSQ-522	**3086.453**	Active sonobuoy. Canadian version of SSQ-47	Sparton
SSQ-523	*3087.453*	CANCASS, Canadian command-active sonobuoy system	Sparton
SSQ-527	**3780.453**	Improved version of SSQ-517B	Sparton
SSQ-529	**4619.453**	Directional sonobuoy	Sparton
SSQ-530	**4620.453**	Directional passive sonobuoy	Sparton
SSQ-531	*6550.453*	Directional sonobuoy	Sparton
FINLAND			
Fiskars	**6553.253**	Fixed passive surveillance system	Fiskars
FRANCE			
Anaconda	**6848.453**	Hull-mounted and tactical towed array series (TACTAS)	Thomson-Sintra
Argonaut	**5650.453**	Sonar for RN Type 2400 submarine	Thomson-Sintra
Beluga	*3774.453*	Panoramic sonar	Thomson-Sintra
Crouzet Mk 3	**5850.353**	Magnetic anomaly detection system	Crouzet
DHAX-1	*1389.353*	Magnetic anomaly detection system	Crouzet
DHAX-3	*1390.353*	Magnetic anomaly detection system	Crouzet
Diodon (TSM 2314)	**1725.253**	Surface vessel sonar. Active	Thomson-Sintra
DSBV 61A	**4461.453**	Experimental towed linear array sonar	Thomson-Sintra
DST A 3	*1067.453*	Active sonobuoy used with DSAA-4 system	Thomson-Sintra
DSTV-4M/DSTV-7Y	**3501.453**	Passive sonobuoys	Thomson-Sintra
DUBA-3A	*1157.253*	Surface vessel attack sonar	Thomson-Sintra
DUBM-20A	*1357.253*	Active minehunting sonar	Thomson-Sintra
DUBM-21B (TSM 2021)	**1977.253**	Mine countermeasures sonar. (Ibis). TSM 2019/2021/2022 variants also	Thomson-Sintra
DUBM-40	*1358.253*	Active minehunting sonar, towed	Thomson-Sintra
DUBM-41B (TSM 2050)	**1726.453**	High-resolution side-looking sonar	Thomson-Sintra
DUBM-42	**6847.453**	Side-looking sonar for mine hunting	Thomson-Sintra
DUBV-23D	**1159.253**	Active surface vessel search/attack sonar	Thomson-Sintra
DUBV-24C	*1217.253*	Low-frequency panoramic search/attack sonar	Thomson-Sintra
DUBV-43B/C	**1334.453**	Variable depth towed sonar. Active or passive. Used with DUBV-23	Thomson-Sintra
DUBY-24C	*1355.453*	Active submarine sonar. Panoramic, sector or tracking modes	Thomson-Sintra
DUUA-1	*1161.453*	Submarine sonar. A, B and C versions	Thomson-Sintra
DUUA-2A	**1634.453**	Simultaneous search and attack sonar for modernised 'Daphne' class submarines	Thomson-Sintra
DUUA-2B	**3232.453**	Development of DUUA-2A	Thomson-Sintra
DUUX-2	*1162.453*	Passive submarine detection system	Thomson-Sintra
DUUX-5	**3948.453**	Successor to DUUX-2. Sea Fenelon	Thomson-Sintra
DUAV-4	**1631.353**	HS-71 helicopter sonar	Thomson-Sintra
DUAV-18	*1160.353*	Helicopter sonar. Superseded by HS.70	Thomson-Sintra
Eledone	**1356.453**	Passive sonar for submarine	Thomson-Sintra
Fenelon	**3498.453**	Passive rangefinder for submarines	Thomson-Sintra
HS.12	**4095.453**	Helicopter version of SS12 sonar	Thomson-Sintra
HS.70	*1219.353*	Helicopter sonar	Thomson-Sintra
HS 312S	**6235.353**	Integrated ASW equipment for helicopters, including HS12 and Lamparo processor	Thomson-Sintra
IBIS	**4348.453**	Minehunting sonar system	Thomson-Sintra
MAD Mk3	**5850.353**	In-board magnetic anomaly detector	Crouzet
Pascal	*1633.253*	Surveillance and tracking sonar for small and medium ships	Thomson-Sintra
Piranha (TSM 2140)	*1978.253*	Small ship attack sonar	Thomson-Sintra
Premo	*1218.253*	Panoramic search/attack sonar	Thomson-Sintra
SPDT-1A	**6556.393**	Passive sonar for torpedo detection	Crouzet
SS12	**4094.453**	Panoramic sonar for small/medium ships	Thomson-Sintra
SS24	*4096.453*	Panoramic sonar more powerful version of SS12	Thomson-Sintra
SS 24LF	*5633.453*	Passive panoramic sonar	Thomson-Sintra
SS48	*4097.453*	Panoramic surface ship sonar	Thomson-Sintra
SQS-17A	*1158.253*	Panoramic sonar. Superseded by Premo	Thomson-Sintra
TSM 2019	**1977.253**	Version of DUBM-21 minehunting family	Thomson-Sintra
TSM 2021	**1977.253**	DUBM-21B minehunting sonar for Triparatite Minehunter (F, Nl, B)	Thomson-Sintra

Designation	Entry no	Description	Contractor(s)
FRANCE cont			
TSM 2022	**4349.453**	Derivative of TSM 2021 (above)	Thomson-Sintra
TSM 2400/DUBA 25	**1724.253**	Surface vessel sonar (Tarpon)	Thomson-Sintra
TSM 2600	*5647.253*	Lightweight version of TSM 2630	Thomson-Sintra
TSM 2630	**5646.253**	Improved version of Diodon sonar	Thomson-Sintra
TSM 2633	**5646.253**	Spherion version of TSM 2630	Thomson-Sintra
TSM 2640	**5648.253**	Passive sonar for patrol craft	Thomson-Sintra
TSM 2820	**5649.253**	Active panoramic sonar (Tarpon)	Thomson-Sintra
TSM 2930	**6554.463**	Lamproie linear towed array for surface ships and submarines	Thomson-Sintra
GERMANY, Federal Republic			
ASO4-2 Mod	—	Active sonar for small vessels. Hull mounted	Krupp Atlas-Elektronik
CSU 3	**4229.453**	Passive/active/intercept sonar for submarines	Krupp Atlas-Elektronik
DSQS-11A	**4230.453**	Hull-mounted mine avoidance sonar for minesweepers	Krupp Atlas-Elektronik
DSQS-11H	**4230.453**	Hull-mounted minehunting sonar	Krupp Atlas-Elektronik
DSQS-21B/C/D	**4228.453**	Series of active panoramic anti-submarine sonar for surface ships Hull-mounted and towed versions	Krupp Atlas-Elektronik
PSU 1-2	**4229.453**	Passive submarine sonar	Krupp Atlas-Elektronik
PRS 3	**4229.453**	Passive ranging sonar for submarines	Krupp Atlas-Elektronik
SIP 3	**4229.453**	Sonar processor for passive classification	Krupp Atlas-Elektronik
INTERNATIONAL			
HELRAS	**5865.353**	Long-range helicopter dunking sonar	Bendix/BAe
Osprey	**6887.453**	Dipping sonar for helicopter use	Sauders/GEC Avionics
ITALY			
BI68	*1721.453*	Sonobuoy. Double number of RF channels	SELENIA-ELSAG
BIR	**1723.453**	Miniature sonobuoy for helicopter use	Servomeccanismi
BIT-3	**1723.453**	Passive sonobuoy	Servomeccanismi
BIT-8	**1723.453**	Passive sonobuoy	Servomeccanismi
FALCO/GIARDA	*3235.453*	Submarine detection and location system	SELENIA-ELSAG
IP64-MD64	**1720.453**	Submarine sonars	SELENIA-ELSAG
IPD-70/S	**3233.453**	Integrated sonar for Sauro submarines	SELENIA-ELSAG
MD 100/S	**3773.453**	Passive submarine sonar	SELENIA-ELSAG
MLS/1A	*3235.453*	Sonar	SELENIA-ELSAG
MSR-810	**4007.453**	Passive sonobuoy	MISAR
P2072	**3234.453**	Upgraded version of GE AN/SSQ-14	FIAR
SQQ-14	**3234.453**	Mine detection and classification sonar	FIAR
JAPAN			
QQS-4	**6249.253**	Solid-state, low frequency surface ship sonar	—
NETHERLANDS			
HSS-15	**4464.453**	Compact panoramic sonar for small ships	Signaal
LW-30	**3141.253**	Passive sonar/intercept system	Signaal
PHS-32	*3142.253*	Search/attack sonar	Signaal
PHS-34	*5635.453*	Long-range panoramic sonar	Signaal
PHS-36	**4001.453**	Active panoramic sonar	Signaal
SIASS	**5636.453**	Submarine attack and surveillance sonar	Signaal
NORWAY			
SK3D	*3644.453*	Higher frequency version of SQ-D	Simrad
SQ-D	*3644.453*	Hull mounted sonar	Simrad
SQ3D/SF	*3644.453*	Medium range 'searchlight' sonar	Simrad
SS 105	**4128.453**	Coastguard 360° scanning sonar	Simrad
SS 240	**6555.353**	Active omni multibeam sonar for small ships	Simrad
SS 304	**5645.253**	Small ship scanning sonar	Simrad
ST	*3644.453*	Bulkhead mounted sonar	Simrad
SU	*3644.453*	Bulkhead mounted version of SU-R	Simrad
SU-R	*3644.453*	Similar to SU-RS without transducer stabilisation	Simrad
SU-RS	*3644.453*	Hull mounted long range sonar	Simrad
UNITED KINGDOM			
AD 130	**5852.353**	Sonar homing and DF receiver	GEC Avionics
AQS 901	**1981.363**	Acoustic data processing and display system	GEC Avionics
AQS 902	**3497.363**	Lightweight acoustic processing and display system	GEC Avionics
AQS 903	**5655.363**	Acoustic processing and display system	GEC Avionics
ARMS	**6871.453**	Advanced remote minehunting system for ROVs	Plessey
AS360	**6858.453**	Mechanically scanned sonar for use on ROVs	UDI (Marconi)
AS370	**6859.453**	Seabed surveillance sonar to detect swimmers and small submersibles	UDI (Marconi)
AS380	**6860.353**	Active and passive dunking sonar – can be displayed from ships or helicopters	UDI (Marconi)
ATAS	**6250.453**	Active towed array sonar for small surface vessels	Ferranti/BAe
DOLPHIN 100	**6849.453**	Processing and display upgrading equipment for submarines	Marconi
SSQ-937	**3770.453**	F-size version of SSQ-36	Dowty
SSQ-963	**3237.453**	Command Active Multi-Beam Sonobuoy	Dowty
COMTASS	**6244.253**	Compact towed array systems (COMTASS 1, 2 & 3)	Plessey
FMS Series	**4625.253**	Family of active and passive sonar systems for surface vessels	Ferranti
FMSS	*6241.253*	Fast minehunting sonar for small ships	Marconi
G738	**3268.493**	Towed system for decoying active and passive homing torpedoes	Graseby
G750	**3266.453**	Multipurpose all-round active/passive sonar for corvettes and above. Based on RN Type 184. Auto-tracking	Graseby
G768	—	Derived from 750 series for use in smaller ships	Graseby
G777	*3267.453*	Compact sonar for patrol craft down to 100 tons	Graseby
G780	—	Passive sonar for very small submarines	Graseby
GETA S	**6850.453**	Passive surveillance sonar for towed and bottom laid array applications (GECO A/S provides 'wet end')	Marconi/GECO A/S (Norway)

Designation	Entry no	Description	Contractor(s)
UK cont			
GS 7110	**6851.253**	Multi-channel FSK receiver	Marconi
Guillemot	**5866.353**	Dunking sonar	Plessey
Cormorant	**5637.353**	Passive/active helicopter sonar	Plessey
Hydra	**6913.353**	Surface ship and submarine sonar	Plessey
Jezebel	**3502.453**	Passive sonobuoy	Dowty
PMS26/27	**7002.253**	Lightweight search/attack sonar	Plessey
PMS32	*1221.253*	Active panoramic sonar	Plessey
PMS35	*3094.253*	Small frigate digital sonar	Plessey
PMS40	**5639.253**	Series of modular sonar systems	Plessey
PMS 56	**6558.353**	Lightweight search/attack sonar–successor to PMS 26	Plessey
PMS75	**5638.263**	Sidescan sonar equipment	Plessey
PSC 8400	*6557.393*	Passive sonar surveillance system for coastal defence	Marconi
RASAT	**6857.363**	Programmable universal sonar target	Dowty
MOSAIC	*5654.363*	ASW avionics integration system	GEC Avionics
MS70	**1374.253**	Solid-state version of Type 193 minehunting sonar	Plessey
SSQ-947B	**3769.453**	SSQ-947B sonobuoy	Dowty
SADE	*1788.453*	Sensitive Acoustic Detection Equipment. Intruder detection system	Plessey
Sonar 3000	**6237.253**	Side-scan sonar developed from the Type 2034	Dowty
SP2100	**6242.253**	A range of signal processors	GEC Avionics
SSQ-904	**3503.453**	Miniature passive (Jezebel) sonobuoy	Dowty
SSQ-937A	**6863.453**	Air-launched miniature bathythermal buoy	Plessey
SSQ-954	**6243.463**	G-size miniature directional passive sonobuoy	Dowty
SSQ-981	**6861.453**	Second generation BARRA sonobouy derived from SSQ-801	Plessey
SSQ-991	**6862.453**	Air-launched sonar transponder	Plessey
STS 3000	*5859.453*	Submarine integrated sonar system	Marconi
STS 3100	*5860.453*	Surface ship/submarine towed array sonar	Marconi
SWS-35	**5867.453**	Miniature towed array sonar	Plessey
T17164	*1787.453*	Mk IC active sonobuoy	Plessey
TATTIX	**5891.353**	Tactical processing and display system	GEC Avionics
TG-1	**6559.353**	High frequency sonar system	Ferranti
Type 162M	**1755.453**	Side-looking and bottom profile sonar	Kelvin Hughes
Type 170	—	Surface ship attack sonar	Graseby
Type 176	—	Passive protection sonar for surface ships	Graseby
Type 177	—	Surface ship medium-range searchlight sonar	Graseby
Type 182	**3268.453**	(GI738) towed torpedo decoy sonar	Graseby
Type 183	—	(GI732) emergency underwater communications	Graseby
Type 184M	**3265.453**	RN surface vessel ASW search and attack sonar	Graseby
Type 185	—	Underwater communications sonar	Graseby
Type 186	—	Submarine sonar	Thorn EMI
Type 187	—	Submarine sonar	Thorn EMI
Type 193	*1373.253*	Minehunting sonar of RN Mk 1 system	—
Type 193M	**1374.253**	Solid-state improved version of Type 193 minehunting sonar	Plessey
Type 195	**1024.353**	Helicopter sonar based on MS 26/27	Plessey
Type 199	*2447.253*	Variable depth towed sonar	Thorn EMI
Type 605	**6856.363**	Sonobuoy receiver (improved version of the ARR-75)	Dowty
Type 719	—	Submarine sonar	Thorn EMI
800 Series	**3269.453**	Series of new generation sonars. Modular	Graseby
Types 864, 865, 866	**5861.363**	Airborne sonobuoy telemetry systems	Marconi
Type 2001	—	Nuclear submarine sonar	Graseby
Type 2004	—	Conventional/nuclear submarine sonar	Graseby
Type 2007	—	Conventional submarine sonar	Graseby
Type 2008/9	—	Underwater communications sonar	Marconi
Type 2015	**3789.453**	XBT bathythermographic buoy	Plessey
Type 2016	**3787.453**	Advanced fleet escort attack sonar	Plessey
Type 2020	**5651.453**	Passive/active submarine sonar	Plessey
Type 2026	**6852.453**	Submarine passive towed array sonar	GEC Avionics
Type 2028	—	Surface ship sonar	Graseby
Type 2031I	**6853.453**	Passive towed array sonar on RN frigates	GEC Avionics
Type 2031Z	**6236.253**	Towed array sonar for all new Royal Navy ASW frigates	Dowty
Type 2034	**4127.453**	Short-range bottom profile sonar	Dowty
Type 2046	**6238.453**	Integrated passive system for submarines	Ferranti
Type 2050	**5853.253**	Hull-mounted active sonar for Type 23 and 22 frigates	Ferranti
Type 2051	**5855.453**	Sonar system for RN Oberon class submarine update	Plessey
Type 2052	**5858.453**	Passive broadband towed array submarine sonar	Marconi
Type 2054	**5857.453**	RN Trident submarine bow-mounted sonar	Plessey
Type 2057	**6870.453**	New passive towed array sonar for RN	—
Type 2058	**6855.453**	Broadband towed sonar target source	Thorn EMI
Type 2068	**6257.493**	Environmental prediction and display system	Gresham-CAP
Type 2074	**6868.453**	Submarine sonar for RN hunter-killer vessels	Plessey
Type 2075	**6869.453**	Hull-mounted suite for RN Upholder class SSK submarines	—
Type 2093	*5856.253*	Minehunting sonar	Plessey
Type 2110	**6854.453**	Variant of Types 2031I and 2026	GEC Avionics
UNITED STATES OF AMERICA			
AMSS	**5652.253**	Advanced minehunting sonar system development	Raytheon/ Thomson-CSF
AQS-13	**1637.353**	Helicopter sonar	Bendix
AQS-14	**4462.353**	Helicopter towed side-looking sonar for mine clearance	Westinghouse
AQS-18	**4456.353**	Helicopter panoramic sonar	Bendix
BQG-1/4	—	Submarine passive fire control sonars	Sperry/Raytheon
BQQ-1	—	Search and fire control sonars	Raytheon
BQQ-2	—	Sonar for Subroc system	Raytheon
BQQ-5	**1983.453**	Nuclear attack submarine sonar	IBM/Raytheon
BQQ-6	**4629.453**	Active/passive sonar for Trident submarines	IBM
BQR-2	—	Submarine passive sonar	EDO/Raytheon
BQR-3	—	Submarine passive sonar	Raytheon

Designation	Entry no	Description	Contractor(s)
USA cont			
BQR-7	—	Passive sonar. Part of BQQ-2 system. Fitted FBM subs	EDO/Raytheon
BQR-15	**1984.453**	Towed submarine sonar. Fitted FBM subs	Western Electric
BQR-19	—	Submarine sonar	Raytheon
BQR-21	**1985.453**	Submarine passive detection and tracking set (DIMUS). Fitted FBM subs	Honeywell
BQS-4	—	Adds active sonar capability to BQR-2B	EDO
BQS-6	—	Active submarine sonar. Part of BQQ-2 system	Raytheon
BQS-8	—	Under-ice navigation sonar	Hazeltine/EDO
BQS-13	**1986.453**	Submarine search sonar. Passive/active	Raytheon
BQS-14A	**6872.453**	Submarine navigation and mine-detection sonar for under-ice passage	Hazeltine
BQS-15	**6247.453**	Submarine sonar for mine detection in heavily mined areas. Active & passive	Amtek
BRT-1	*3785.453*	Sonar radio transmitting buoy	Sparton
BSY-1(V)	**6562.353**	Sonar/fire control integrated combat system	IBM
DE1164	**3115.453**	SQS-56 plus variable depth sonar	Raytheon
DE1167/1191	**3116.453**	Small ship sonar based on SQS-56	Raytheon
DE 1191	**6864.453**	Upgrade suite for SSQ-23 and SQS-23 sonars	Raytheon
Dwarf	**3782.453**	Reduced length, standard diameter miniature passive sonobuoy	Sparton
Dwarf DIFAR	**3782.453**	DIFAR version of Dwarf	Sparton
Dwarf DIFAR(VLAD)	**3782.453**	Vertical line array DIFAR version of Dwarf	Sparton
Dwarf omni	*4458.453*	Dwarf omnidirectional passive sonobuoy	Sparton
Hydroscan	**4460.453**	Towed side-scan seabed mapping sonar	Klein
Micro PUFFS	**3772.453**	Submarine passive ranging sonar	Sperry
Mk 24	**5643.453**	Underwater ordnance location sonar	Klein
Model 1110	**6865.453**	Passive flank array system	EDO
Model 4235	**6561.493**	Short range doppler sonar for ROVs	EDO
Moored sensor buoy	**6560.493**	Surveillance sonobuoy	Sparton
PQS-2	**5640.453**	Hand-held active/passive diver sonar	General Instrument Corp
SQA-10	—	Variable depth sonar	Litton
SQA-13	—	Variable depth sonar	EDO/Litton
SQA-14	—	'Searchlight' sonar	Raytheon
SQA-16	—	'Searchlight' sonar	Raytheon
SQA-19	—	Variable depth sonar	Litton
SQG-1	—	Anti-submarine attack sonar	Raytheon
SQQ-14	**1789.453**	Minehunting and classification sonar	GE
SQQ-23	—	Sonar for A/S patrol ships	—
SQQ-30	**4623.253**	Minehunting sonar	General Electric
SQQ-32	**5652.253**	Minehunting sonar	Raytheon
SQQ-89	**6563.353**	Sonar/fire control integrated combat system	General Electric
SQR-14	—	Surface sonar	—
SQR-15	**6248.253**	Surface ship towed array sonar	Gould
SQR-17	**6245.263**	Four-channel processor forming part of LAMPS Mk1	Diagnostic
SQR-18A	**4622.253**	TACTAS–Tactical Towed Array Sonar	EDO
SQR-19	**4463.253**	Tactical towed array sonar	Gould
SQS-4	—	Short-range active sonar	Sangamo/GE
SQS-23	—	Long-range active sonar	Sangamo
SQS-26	**2536.253**	Bow-mounted, 'bottom bounce' mode sonar to replace SQS-23	EDO/GE
SQS-29/32	—	Surface vessel active sonars. Numbers relate to differing frequencies. 'B' models are associated with variable depth sonar	Sangamo
SQS-35	**2534.253**	Variable depth sonar	EDO
SQS-36	**2534.253**	Medium-range hull sonar	EDO
SQS-38	**2534-253**	Medium-range hull-mounted version of SQS-35	EDO
SQS-53	**2536.253**	Development of SQS-26	GE
SQS-56	**1807.453**	Lightweight sonar for USN PF (Patrol Frigate) ships and other navies (DE1160B)	Raytheon Sparton
SSQ-36	*3781.453*	Bathythermograph sounding buoy	Sparton
SSQ-41	**1987.453**	Sonobuoy, passive	Magnavox/Sparton
SSQ-47	**1988.453**	Sonobuoy, active	Sparton
SSQ-50	*1989.453*	CASS-Command Active Sonobuoy System	Sparton
SSQ-53A	**1990.453**	Sonobuoy, passive directional (DIFAR)	Sparton/Magnavox/Sanders
SSQ-53B	**4457.453**	Successor to SSQ-53A	Sparton
SSQ-57A	**3083.453**	Sound reference sonobuoy	Sparton
SSQ-62	**1991.453**	Sonobuoy, directional version of SSQ-50 (DICASS)	Sparton/Raytheon/Magnavox
SSQ-71	**3786.453**	ATAC, air transportable acoustic communication buoy	Sparton/Sanders
SSQ-75	**6564.493**	Long range, low frequency active sonobuoy	Bunker Ramo
SSQ-77	**3084.453**	VLAD-vertical line array DIFAR passive sonobuoy	Hazeltine/Sparton
SSQ-79	—	Steered vertical line array sonobuoy. Development	Hazeltine
SSQ-86	**4459.453**	Down-link communications sonobuoy	Sparton
SSQ-101	**6866.453**	Expendable HLA sonobuoy for long range detection	—
SSQ-102	**6867.453**	Tactical surveillance sonobuoy	Magnavox/Sippican/Hazeltine
SST	**3494.253**	Advanced sonar standard transmitter	Raytheon
STRAP	**6258.493**	Experimental programme to use sonobuoys for beam forming	US Navy/Lockheed/Magnavox
UQS-2	—	Minehunting sonar	GE
610	**2580.253**	Long-range hull sonar	EDO
700 Series	**2581.253**	Medium-range. Hull and variable depth versions	EDO
780 Series	**3495.253**	Variable depth sonar. 13 kHz	EDO
786 Series	**3495.253**	Hull-mounted sonar. 13 kHz	EDO
795 Series	**3495.253**	Hull-mounted sonar. 5 kHz	EDO
796 Series	**3495.253**	Hull-mounted sonar. 7 kHz	EDO
900 Series	**5644.453**	Submarine mine avoidance	EDO
910 Series	**5644.453**	Surface ship mine avoidance	EDO
1102/1105 Series	**3496.453**	Submarine active/passive medium range	EDO
7860 Series	**3435.253**	Combined 13 kHz hull sonar with 13 kHz VDS	EDO
7950 Series	**3495.253**	Combined 5 kHz hull sonar with 13 kHz VDS	EDO
7960 Series	**3495.253**	Combined 7 kHz hull sonar with 13 kHz VDS	EDO
TVLAD	**3783.453**	Tuned vertical array omni sonobuoy	Sparton
Widetrac	*3784.453*	Sonar communications buoy	Sparton

4212.454 Miscellaneous Underwater Detection Equipment

Designation	Entry no	Description	Contractor(s)
CANADA			
ASA-64	**4222.353**	Anomaly detection signal processor	CAE
ASA-65	**3231.393**	Compensator group adapter for ASA 65(V)	CAE
ASA-65(V)	**1388.393**	Nine-term compensator for MAD system	CAE
ASQ-502	**3999.353**	Submarine detecting set (MAD)	CAE
ASQ-504(V)	**4621.353**	Submarine detecting set (MAD)	CAE
OA 5154/ASQ	**1980.393**	Automatic MAD compensation system	—
Trail Blazer	**6484.441**	Mine countermeasures ROV	ISE
FRANCE			
Crouzet MAD	*1979.353*	ASW helicopter MAD equipment	Crouzet
DHAX-1	**1389.353**	Airborne MAD equipment	—
DHAX-3	**1390.353**	ASW helicopter MAD equipment	—
Lamparo (TSM 8210/8220)	**4347.363**	Airborne digital sonar data processor	Thomson-CSF
PAP 104	**1786.453**	Mine disposal system	Société ECA
Sonar processor	*1719.263*	Computer for analysis of sonar coverage	SINTRA-ALCATEL
TSM 8200	**3791.363**	Airborne sonar processing equipment	Thomson-CSF
TSM 9310	**3095.263**	Sound ray tracer for analysis of sonar equipment performance	Thomson-CSF
ITALY			
CIACIO	**3235.453**	Torpedo self-homing system	SELENIA-ELSAG
ELT/810	**5653.263**	Sonar prediction system	Elettronica
FALCO	**1722.353**	Airborne submarine locating system	SELENIA-ELSAG
FALCO/GIARDA	**3235.453**	Submarine detection and localisation system	SELENIA-ELSAG
P MICCA	**3235.453**	Remote-control of deep sea mines	SELENIA-ELSAG
SFM-A	**3235.453**	Underwater target identification system	SELENIA-ELSAG
NETHERLANDS			
SP1-04	**1525.263**	Sound ray path analyser	Van der Heem
XSV-01	**3493.453**	Expendable sound velocimeter for sound ray path measurement	Van der Heem
UNITED KINGDOM			
AQS-901	**1981.363**	Airborne acoustic data processing and display system	Marconi/Computing Devices
Area defence	**4626.453**	Passive sonar for area defence	Marconi
ARR-901	**4453.363**	Airborne sonar data receiver	McMichael
AQS-902 LAPADS	**3497.363**	Lightweight acoustic processing and display system for ASW helicopters	Marconi
AQS-903	**5655.363**	Lightweight helicopter processing and display system	Marconi
Deep Mobile target	**4680.493**	Instrumented ship-launched target	Thorn-EMI
MCM System	*3161.253*	Mine countermeasures system	Plessey/Racal-Decca/Sperry/ Vosper Thornycroft
G740	**3269.453**	Transportable long-range calibration system	Graseby
Nantis-M	**4681.273**	Action information system	Plessey
G733	**3269.453**	Underwater transponder target	Graseby
Sea Serpent	**6483.441**	Magnetic minesweeping system	Vosper Thornycroft
Towed Arrays	**4627.463**	Towed arrays	Plessey
UNITED STATES OF AMERICA			
Active processing system	**3788.363**	Airborne system for processing sonobuoy data	Sparton
AKT-22(V)4	**3777.363**	Telemetry data transmitting set	EDMAC
ARR-72	**3454.363**	P-3C Orion sonobuoy receiver system	EDMAC
ARR-75	**3455.363**	Sonobuoy receiver	EDMAC
ARR-78(V)	**3778.363**	Advanced sonobuoy communication link (ASCL)	Hazeltine
ASA-66	**4013.373**	P-3C Orion ASW tactical data cockpit display	Loral
ASA-82	**4012.373**	S3A Viking ASW tactical data display system	Loral
ASQ-81(V)	**3599.253**	Airborne MAD system	Texas Instruments
AYA-8B	**1982.363**	P-3C Orion ASW data processing system	General Electric
DICASS command signal monitor	**3788.363**	SSQ-50 test equipment in the Directional Command Active Sonobuoy System (DICASS)	Sparton
Directional sonar processor	**3788.363**	Airborne system for processing DIFAR data	Sparton
OL-5003(–)/ARR	**3788.363**	Sonobuoy signal processor	Sparton
R-1651/ARA	**3776.253**	VHF radio receiver for sonobuoy location	EDMAC
RO 308/SSQ-36	**1635.453**	P-3C Orion bathythermograph data recorder	Western Components
Spectrum analyser	**3788.363**	Airborne digital processor for real time spectral analysis	Sparton
TD 1135A	**3788.363**	Demultiplexer processor/display	Sparton
WQC-2A	**4455.453**	Sonar underwater communications set	General Instruments Corp

2296.444

Torpedoes

Designation	Entry no	Dimensions (d × l)	Description	Contractor(s)
FRANCE				
E14	**1163.441**	21 in × 4.29 m	Passive acoustic homing. Contact and magnetic fuze	DCN/CIT-ALCATEL
E15	**1164.441**	550 mm × 6 m	As E14 but 300 kg charge instead of 200 kg	DCN/CIT-ALCATEL
F17	**3623.441**	21 in × 5.9 m	Wire-guided or automatic homing. Surface or submarine targets. F17P multimode version	DCN
L3	**1165.441**	21 in × 4.32 m	Active acoustic homing. Contact and magnetic	DCN/CIT-ALCATEL
L3 Mod 1	**1165.441**	550 mm × 4.3 m	fuze. 200 kg charge 21 in version available for manufacture but not in production	DCN/CIT-ALCATEL
L4	**2096.441**	21 in × 3.13 m	Circular search. Active acoustic homing. Contact and acoustic proximity fuze. Air-launched. Used in Malafon	DCN
L5 Mod 1	**2128.441**	21 in	Active/passive acoustic homing. Direct attack programmed search. Surface launch weight 1000 kg	DCN
L5 Mod 3	**2128.441**	21 in	As Mod 1 but submarine launch weight 1300 kg	DCN
L5 Mod 4	**2128.441**	21 in	Anti-submarine model	DCN
L5 Mod 4P	**2128.441**	21 in	Multi-purpose model	DCN
Z16	**2146.441**	550 mm × 7.2 m	Preset course and depth. Runs zig-zag pattern if no target encountered after predetermined distance run. Magnetic and contact fuze. 300 kg charge. Obsolescent	—
Murone	**6480.441**		Lightweight torpedo for air- and surface-launch	DCN/ECAN
GERMANY, Federal Republic				
Seal	**2178.441**	21 in × 6.08 m	Wire-guided heavy surface target torpedo. Surface launch Magnetic and impact fuzes. 260 kg warhead	AEG/Krupp Atlas Elektronik
Seeschlange	**2178.441**	21 in × 4.15 m	Wire-guided, surface launch, submarine targets. 100 kg warhead	AEG/Krupp Atlas Elektronik
SST-4	**2000.441**	21 in × 6.5 m	Wire-guided; active/passive acoustic homing. Torpedo sonar linked to shipborne FCS. 260 kg warhead	AEG
SUT	**2570.441**	21 in × 6.1 m	Wire-guided; active/passive acoustic homing. Similar to SST-4 but dual purpose	AEG
ITALY				
A.184	**2003.441**	21 in × 6 m	Wire-guided, automatic acoustic homing	Whitehead-Moto Fides
A.244	**2004.441**	324 mm × 2.7 m	Acoustic homing. Shallow water and anti-reverberation capabilities	Whitehead-Moto Fides
A.244/S	**2004.441**	324 mm × 2.7 m	A.244 with more sophisticated CIACIO homing head	Whitehead-Moto Fides
SWEDEN				
Type 41	*2323.441*	400 mm × 2.44 m	Passive homing in azimuth and depth. Shallow water capability Impact and proximity fuze. Electric propulsion. All-up weight 250 kg	Förenade Fabriksverken (FFV)
Type 42	**2474.441**	400 mm × 2.44 m	Successor to Type 41 with helicopter-launch capability and optional wire guidance (length then 2.6 m)	FFV
TP 421/422	**2474.441**	—	Swedish Navy versions of Type 42	FFV
TP 427	**2474.441**	—	Export version of Type 42	FFV
TP 61	**2367.441**	21 in × 6.98 m	Long-range heavyweight passive homing	FFV
TP 613	**2367.441**	—	Swedish Navy version of Type 61	FFV
TP 617	**2367.441**	—	Export version of Type 61	FFV
UNION OF SOVIET SOCIALIST REPUBLICS				
21 in Torpedo	**2995.441**	21 in	Standard fit for submarines. Alternatives available for surface vessel launching	State
Airborne Torpedo	**2996.441**	—	—	State
Light Torpedo	**2997.441**	406 mm × 5 m approx	Training deck launchers	State
660 mm Torpedo	**6274.441**	660 mm × 9144	Reported large diameter torpedo	State
UNITED KINGDOM				
Mk 8	**2552.441**	21 in × 6.7 m	1930s design. Compressed air propulsion, free-running. In RN service until 1973.	UK MoD
Mk 24 Mod 0	**2472.441**	21 in	Wire-guided active/passive acoustic homing	UK MoD
Tigerfish (Mk 24)	**2440.441**	21 in × 6.46 m	Long-range wire-guided active/passive acoustic homing. Anti-surface, anti-submarine. Computer allows autonomous homing and re-attack. Entered RN service 1980; Mod 2 operational 1986	MUSL
Spearfish	**4559.441**	21in × 6 m(app)	Advanced heavy-weight sub-launched torpedo. Scheduled to enter Royal Navy service in early 1989	MUSL
Stingray	**2448.441**		Lightweight air and surface launched torpedo. In service with the Royal Navy	MUSL
UNITED STATES OF AMERICA				
Mk 14	*2813.441*	21 in × 5.25 m	1930s design. Free-running with preset depth and course angles. Compressed air propulsion. Weight 1780 kg. Withdrawn from USN service in 1973 but still in service elsewhere	—
Mk 27 Mod 4	*2165.441*	19 in × 3.23 m	Passive acoustic homing. Electric propulsion. Used in USN service as training round prior to Mk 37 introduction but some warshot torpedoes sold elsewhere	—
Mk 32	**3629.441**	483 mm × 2.08 m	Acoustic anti-submarine. Operational but obsolescent	—
Mk 37 Mod 0	**2817.441**	19 in × 3.52 m	Active/passive acoustic homing. Electric dual-speed propulsion Warshot weight 648 kg. Mod 3 is improved version	Westinghouse
Mk 37 Mod 1	**2818.441**	19 in × 4.09 m	Wire-guided. Electric dual-speed propulsion Warshot weight 766 kg. Mod 2 is improved version	Westinghouse
Mk 37 Mod 2	**2818.441**	19 in × 4.09 m	Updated version (minor modifications) of Mod 1	Westinghouse
Mk 37 Mod 3	**2817.441**	19 in × 3.52 m	Updated version (minor modifications) of Mod 0	Westinghouse
NT 37C Mods 2 & 3	*3971.441*	19 in × 3.52 m 19 in × 4.09 m	Improved versions of Mk 37 Mods 2 and 3. Better speed, range, and acoustic performance including shallow water capacity. Sold to Canada and Netherlands	Northrop/Honeywell
NT 37D	*4180.441*	19 in × 4.5 m (Mod 2) 19 in × 3.8m (Mod 3)	Further improvement of NT 37 with Mk 46 OTTO engine and enhanced guidance	Honeywell

Designation	Entry no	Dimensions (d × l)	Description	Contractor(s)
USA cont				
NT 37E	**4762.441**	483 mm × 4.5 m	Latest modernised model of Mk 37	Honeywell
Mk 44	**2820.441**	19 in × 2.56 m	2 versions differing slightly in length and both weighing about 233 kg. Active acoustic homing. Electric propulsion. Replaced by Mk 46 in USN and UK service but still in service elsewhere	Several in USA also licence-built overseas
Mk 46 Mod 0	**2822.441**	324 mm × 2.67 m	Deep-diving, high-speed, active/passive acoustic homing. Weight about 258 kg. First US torpedo with solid-fuel propulsion	Aerojet-General and others
Mk 46 Mod 1	**2822.441**	324 mm × 2.59 m	As Mod 0 but slightly larger and with liquid mono-propellant (OTTO) motor	Aerojet-General and others
CAPTOR	**2541.441**	324 mm × 3.7 m	Mk 46 Mod 4 torpedo inserted in a mine casing and released when target detected	Goodyear
Mk 48 Mod 0	**2823.441**	21 in	AS-only version. Superseded during development by Mod 2	Westinghouse
Mk 48 Mod 1	**2823.441**	21 in × 5.8 m	Deep-diving (914 m) high-speed (93 km/h) active/passive acoustic homing, wire-guided, long-range (46 km) weapon. Weight about 1600 kg. Said to be most complex torpedo ever designed	Gould (formerly Clevite)
Mk 48 Mod 2	**2823.441**	21 in	Competing against Mod 1 which was selected after comparison at pilot production stage	Westinghouse
Mk 50	**4182.441**	—	Formerly EX-50 Advanced Lightweight Torpedo (ALWT). Successor to Mk 46 (NEARTIP)	Honeywell
Freedom Torpedo	*2618.441*	19 in × 4.83 mm	Private-venture development. Wire-guided, pattern-running, electrically propelled. Contact fuze. Warshot weight 1237 kg. Charge 295 kg. Range 11 km Adapts to 21 in tubes	Westinghouse

4234.094

Electronic Warfare Equipment

Designation	Entry no	Description	Contractor(s)
AUSTRALIA			
Winnin	**5954.293**	Rocket based shipborne decoy system	—
CANADA			
AN/SAR-8	**6162.293**	Infra-red warning system	Spar
Model 100	**1714.253**	Naval direction finder	General Precision
CANEWS	**4161.293**	Naval integrated ESM/ECM system	MEL/Westinghouse
CHINA (PEOPLE'S REPUBLIC)			
Model 970	**6451.153**	Radar jammer	—
GT-1	**6504.393**	Chaff/IR flare dispenser	—
KJ 8602	**6901.393**	Radar warning receiver for tactical aircraft	—
921-A	**6485.453**	Submarine ESM system	—
RW-23-1	**6498.293**	Surface ship radar warning receiver	—
CHILE			
Caiquen	**5824.393**	Airborne radar warning receiver	ENAER
Eclipse	**5825.393**	Chaff/IR decoy launcher	ENAER
Itata	**5826.353**	Airborne ELINT system	ENAER
Medusa	**6505.393**	Prototype VHF jammer	ENAER
FRANCE			
ARB-33	**5814.293**	Naval jamming system	ESD
ABD	**5941.393**	Detection/jamming system	ESD
ARBR-17	**5813.253**	Naval jamming system	Thomson-CSF
Barax	**6902.393**	Pod-mounted detector/jammer for Mirage 2000	ESD
Barem	**5946.393**	Airborne self-protection system	Thomson-CSF
Basilisk	**5947.393**	Airborne jamming pod	Thomson-CSF
Binoc	**3913.193**	Jammer	—
Bromure	**3913.193**	Communications jammer	—
Caiman	**5787.393**	Noise/deception jamming pod	Thomson-CSF
Dagaie system	**3377.293**	Naval countermeasures dispensing system	CSEE
DB 3141	**5786.393**	Noise jamming pod	Thomson-CSF
DB 3163	**3379.393**	Airborne I/J-band jamming pod	Thomson-CSF
DDM	**6903.393**	Airborne IR detector for Mirage 2000	SAT
DR 2000	**6486.293**	Broadband receiver	Thomson-CSF
DR 3000A	**6907.353**	ESM system for maritime patrol aircraft	Thomson-CSF
DR 4000	**4351.053**	Video/IFM ESM receiver	Thomson-CSF
Elebore	**3913.193**	Computer-based HF monitoring system	—
Elfa	**3913.193**	Intercept system	—
Elodee	**3913.193**	Computer-based communications intercept system	—
Emeraude	**3913.193**	Computer-based VHF monitoring system	—
Expendables	**6253.393**	Series of active expendable jammers	Thomson-CSF
Janet	**5785.293**	Naval jammer	Thomson-CSF
Laser Warner	**6252.393**	Laser warning receiver	Thomson-CSF
Magaie	**4350.293**	Naval countermeasures decoy system	CSEE
Miriade	**6905.093**	Radar warning receiver operating in the millimetric waveband	Thomson-CSF
MSPS	**6904.353**	EW system for retrofit market-continues Sherloc RWR with Barem jammer	Thomson-CSF
NADS	**5783.253**	Naval anti-missile decoy system	CSEE
NEWSY	**5784.253**	Naval EW system	CSEE
Sagaie	**5656.293**	Naval countermeasures decoy system	CSEE
Saphir	**5790.393**	Helicopter chaff/IR dispenser	Matra
Sapiens	**5657.293**	Naval EW system	Thomson-CSF
Serval	**5789.353**	Airborne radar warning receiver	Thomson-CSF
Sherloc	**6251.393**	Airborne radar warning receiver	Thomson-CSF
Spirale	**5790.393**	Mirage chaff/IR dispenser	Matra
STAR	**6254.393**	Airborne reconnaissance system developed from Syrel	Thomson-CSF
Sycomor	**5790.393**	Mirage 2000 chaff/IR dispenser and IR warning receivers	Matra
Syrel	**5788.393**	ELINT pod (TMV 018)	Thomson-CSF
TMV 026	**3376.353**	Airborne ESM	Thomson-CSF
TMV 200	**3534.053**	Elisa ELINT receiver	Thomson-CSF
TMV 202	**6506.353**	DR4000A airborne Thomson-CSF ESM suite	—
TMV 433	*3533.293*	Naval countermeasures set incorporating DR 2000 receiver, DALIA or ARIAL analysis equipment	Thomson-CSF
TRC 243	**6452.153**	HF radio direction finder	Thomson-CSF
TRC 281	**6452.153**	HF tactical jammer	Thomson-CSF
TRC 284	**6452.153**	V/UHF tactical jammer	Thomson-CSF
TRC 285	**6452.153**	VHF tactical jammer	Thomson-CSF
TRC 296	**6452.153**	V/UHF direction finder	Thomson-CSF
TRC 296D	**6452.153**	Lightweight antenna version of TRC 296	Thomson-CSF
TRC 298A	**6452.153**	V/UHF intercept receiver	Thomson-CSF
TRC 298B	**6452.153**	V/UHF monitoring receiver	Thomson-CSF
TRC 2981	**6452.153**	V/UHF panoramic adaptor	Thomson-CSF
TRC 3941B	**6452.153**	HF panoramic adaptor	Thomson-CSF
Type BF	**3374.353**	Airborne radar warning receiver	Thomson-CSF
XR 100	*3535.053*	VHF/UHF communications intelligence receiver system	SINTRA
XR 100 SA	*3535.053*	Automatic receiving system	SINTRA
XRA 100	*3535.053*	Analysis receiver	SINTRA
GERMANY, Federal Republic			
Alberich	**6843.193**	Laser/radar warning receiver for AFVs	Eichweber/Rheinmetall
Babylon	**6840.193**	Jammer used against HF communication links	Rohde & Schwarz
Blast	**6164.193**	Mobile acquisition and jamming system	AEG
Hot Dog/ Silver Dog	**3643.293**	Naval IR and radar countermeasures dispensing system	Wegmann/Buck

Designation	Entry no	Description	Contractor(s)
GERMANY, Federal Republic cont			
Schalmei	**5659.293**	Naval countermeasures decoy system	AEG
E. Series	**4352.053**	Range monitoring receivers	AEG
EB 100	**1901.053**	Portable, battery operated ESM receiver	Rohde & Schwarz
H. Series	**4636.193**	Range of antenna systems	Rohde & Schwarz
ET 001	*3381.053*	VHF/UHF receiving system	Rohde & Schwarz
HELAS	**6462.153**	Vehicle-based ESM system	Siemens
Hummel	**4353.153**	VHF jamming transmitters	AEG
ESM 500 Series	**4138.093**	Family of receivers for communications and ESM applications	Rohde & Schwarz
ESM 1000	**6838.193**	Family of ESM receivers (20 to 1000 MHz)	Rohde & Schwarz
ESP	**4139.093**	Fast scan automatic intercept receiver	Rohde & Schwarz
FL 400	**5821.253**	Naval EW suite	AEG
FL 1800S	**5820.253**	Naval ESM/ECM system	AEG
GA 082	**5791.193**	FSK analyser	Rohde & Schwarz
HE 101	*3531.093*	Active receiving antenna	Rohde & Schwarz
PA 055	**3532.093**	Broad-band doppler direction finder	Rohde & Schwarz
PA 007	**1901.053**	VHF doppler direction finder	Rohde & Schwarz
PA 008	**1901.053**	VHF doppler direction finder	Rohde & Schwarz
PA 009	**1901.053**	UHF doppler direction finder	Rohde & Schwarz
PA 510	**4637.193**	Mobile HF direction finder	Rohde & Schwarz
PA 555	**4637.193**	Mobile version of PA 055	Rohde & Schwarz
PA 010	**3532.093**	HF doppler direction finder	Rohde & Schwarz
PA 2000	**1901.153**	Direction finder	Rohde & Schwarz
Plath DF	**5792.193**	Plath direction finding systems	Plath
AC002	**5658.153**	Directional antenna system	Rohde & Schwarz
AC008	**6453.193**	Microwave directional antenna	Rohde & Schwarz
EP 1650	**3536.193**	Automatic VHF/UHF radio reconnaissance system	AEG
Pr 017	**6454.193**	Azimuth analyzer	Rohde & Schwarz
INTERNATIONAL			
COLDS	**6464.193**	Laser detector system	MBB/Tracor
Datar	**6157.393**	Radar warning receiver	Ferranti/E-Systems
Sibyl	**5683.293**	EW Naval decoy system	BAe/Brandt
NATO Sea Gnat	**6500.253**	Standardised decoy system for NATO	—
ISRAEL			
Beamtrap	**6489.293**	Medium-range chaff rocket	Rafael
MN-53	**3461.093**	ESM system for radar intercept	Elbit
EB 41	**4225.093**	Naval ECM system	IAI
EL	**4225.093**	Range of EW systems	IAI
Rafael	**4354.293**	Short-range chaff rocket	Rafael
Rafael	**4355.293**	Long-range chaff rocket	Rafael
Timnex-4CH	**4356.093**	Automatic Elint/ESM system	Elbit
ACDS	**5660.293**	Naval dispenser system	Elbit Computers
AT-4910	**5937.153**	Comint/jamming system	Elisra
CR 2700	**6455.153**	Vehicle-mounted mobile ELINT system	Elisra
CR 2740	**4731.153**	Mobile ELINT system	Elisra
CR-2800	**6264.353**	Airborne ESM/ELINT system	Elisra
EL/L-8202	**4357.393**	ECM pod	Elta
EL/L-8231	**6153.393**	Internally mounted ECM system	Elta
EL/L-8240	**6909.353**	Internally mounted airborne ECM system	Elta
EL/L-8300	**6152.353**	Airborne SIGINT system	Elta
EL/L-8303	**4225.093**	ESM system	Elta
EL/L-8310	**4225.093**	ELINT/ESM system	Elta
EL/L-8312	**4225.093**	ELINT/ESM system	Elta
EL/L 8312A	**5949.393**	Elint system	Elta
EL/K 1150/1160/1250	**4639.193**	COMINT receivers	Elta
EL/K-7001	**4225.093**	Communications monitoring & jamming system	Elta
EL/K-7010	**4225.093**	VHF Communications jammer	Elta
EL/K-7012	**4225.093**	Communications jammer exciters	Elta
EL/K-7020	**4225.093**	EW system	Elta
EL/K 7032	**5948.393**	Comint system	Elta
EL/L-8230	**5730.393**	Airborne ECM system	Elta
EL/S-8811	**6824.193**	Tactical intelligence system	Etta
EL/S-8812	**6824.193**	Manpack tactical intelligence system	Etta
Lavi	**5794.393**	Self-protection system for Lavi fighter	Elta
LWS-20	**6508.393**	Airborne laser warning system	Elisra
NN-9034	**6881.463**	Submarine ESM/ELINT system	Elisra
NS9001	**6490.253**	Naval ELINT system	Elisra
NS9003	**6491.253**	Naval ESM system	Elisra
NS9005	**6492.253**	Naval ECM jammer	Elisra
NS9009	**6493.253**	Naval EW system	Elisra
NS9010	**6494.253**	Naval ESM system	Elisra
RAJ 101	**4638.193**	Land-based jammer	Rafael
Rattler	**6207.193**	Jamming system	Rafael
SEWS	**6208.253**	ECM shipborne EW suite	Rafael
SPS20	**4640.353**	Airborne radar warning receivers	Elisra
SPS-65	**6510.353**	Helicopter self-protection system	Elisra
SPS-200	**5793.353**	Airborne self-protection system	Elisra
SRS-25	**6507.393**	Airborne ELINT receiver	Elisra
SPS-2000	**6509.353**	Airborne radar warning system	Elisra
TDF-205	**6165.093**	Communications DF equipment	Tadiran
ITALY			
ALQ-X	*4175.093*	Series of airborne jammer pods	Selenia
ALQ-234	**4174.393**	Airborne jammer pod	Selenia
ELT/116	**4173.093**	Naval ESM/ECM system	Elettronica

Designation	Entry no	Description	Contractor(s)
ITALY cont			
ELT/128	**4645.093**	Communications direction finder	Elettronica
ELT/132	*4167.193*	Lightweight man-portable radar warning receiver	Elettronica
ELT/211	**4169.293**	Naval ESM equipment	Elettronica
ELT/214	**4173.093**	Naval ESM	Elettronica
ELT/263	**4166.393**	Airborne ESM system (Gufo)	Elettronica
ELT/311	**4173.093**	Jammer	Elettronica
ELT/318	**4170.293**	Naval noise jammer. Dual-band CW	Elettronica
ELT/457	**4359.393**	Airborne noise jammer pod	Elettronica
ELT/458	**4359.393**	Airborne noise jammer pod	Elettronica
ELT/459	**4359.393**	Airborne noise jammer pod	Elettronica
ELT/460	**4359.393**	Airborne noise jammer pod	Elettronica
ELT/511	**4173.093**	Jammer	Elettronica
ELT/555	**4168.393**	Airborne deception jammer and warning pod	Elettronica
ELT/526	**5792.393**	Airborne deception jammer	Elettronica
ELT/566	**5792.393**	Airborne deception jammer	Elettronica
ELT/711	**4173.093**	Naval ESM radar identification module	Elettronica
ELT/712	**4173.093**	Naval EW programming unit	Elettronica
ELT/716	**4173.093**	Naval EW data interface unit	Elettronica
ELT/814	**4172.293**	Naval ECM antenna	Elettronica
ELT/999	**5796.353**	Airborne COMINT system	Elettronica
ESS-2	**5664.153**	Surveillance & analysis system	Selenia
IGS-1/IGS-3P	**3526.193**	Land-mobile ECM system	Selenia
IGS-3	**4175.093**	Computerised mobile integrated ESM/ECM system	Selenia
IHS-6	**3382.393**	Airborne integrated ESM/ECM system	Selenia
INS-1	**4175.093**	Integrated naval EW system for fast patrol boats	Selenia
INS-3	**5663.253**	Computer-assisted naval integrated ESM/ECM system	Selenia
Model 700	*5830.353*	Airborne EW suite	Selenia
NETTUNO	**6882.253**	Integrated ESM/ECM system for major warships	Elettronica
PAW-1	**4175.093**	Radar pulse analyser and warning equipment	Selenia
RQN-1	**3343.253**	Radar intercept equipment	Selenia
RQN-3	**3528.253**	Naval radar intercept equipment	Selenia
TQN-1	**4175.093**	Naval jammer	Selenia
TQN-2	**4175.093**	Modular multi-band airborne jammer	Selenia
TQN-3	**4175.093**	Jammer	Selenia
TQN-4	**4175.093**	Jammer	Selenia
Breda LCRS	**3997.293**	Lightweight chaff rocket launching system	Breda/SNIA Viscosa
Colibri	**4360.353**	Helicopter ECM/ESM system	Elettronica
CO-NEWS	**4644.293**	Naval ESM system	Elettronica
FARAD	**5661.253**	Naval EW system	Elettronica
Fast Jam	**4643.353**	Helicopter-borne jammer	Elettronica
ISIS	**5953.093**	SIGINT system for air, ground & naval use	Elettronica
NEWTON	**4172.293**	Naval integrated EW system incorporating ELT/211/318/521	Elettronica
RIDE	*4165.393*	Radio intercept and direction-finding equipment	Elettronica
SCLAR Mk II	**4641.281**	Rocket launching system	Selenia-Elsag
Smart Guard	**4642.353**	VHF/UHF helicopter ESM system	Elettronica
THETIS	**5662.453**	Submarine ESM system	Elettronica
JAPAN			
J/ALQ-2	**4176.093**	Airborne EW used on T-33 aircraft	—
J/ALQ-3	**4176.093**	Airborne EW used on C-46 and YS-11 aircraft	—
J/ALQ-4	**4176.093**	Airborne EW used on F-104 aircraft	—
J/ALQ-5	**4176.093**	Airborne EW used on C-1 aircraft	—
J/ALQ-6	**4176.093**	Airborne EW used on F-4 and ET-2 (F1) aircraft	—
J/ALQ-8	**4176.093**	Airborne countermeasures set used on F-4 aircraft	—
J/APR-1	**4176.093**	Airborne radar warning receiver used on F-104	—
J/APR-2	**4176.093**	Airborne radar warning receiver used on F-4EJ	Tokyo Keiki
J/APR-3	**4176.093**	Radar warning receiver	Tokyo Keiki
J/APR-4	**4176.093**	New airborne RWR for Japanese F-15J aircraft	Tokyo Keiki
NOL Series	**6495.253**	Naval EW system	—
NETHERLANDS			
Ramses	**4054.293**	Multi-mode naval ECM system	Signaal
Sphinx	**4055.293**	Naval ESM system	Signaal
RAPIDS	**4056.293**	Radar passive identification system. Naval ESM system	Signaal
SPECTRA	**4735.193**	Communications ELINT system	Signaal
NORWAY			
RL 1	**3344.153**	Laser warning receiver	Simrad
SR-1A	**4646.293**	Naval ESM radar warning receiver	Nera
SPAIN			
EN/TSD	**6826.193**	Range of DF systems for COMINT applications	Electtronica Ensa
SWEDEN			
BO series	**5666.393**	Pod dispensers for fixed and rotary wing aircraft	Philips
UA 731	**3022.163**	Microwave pulse analyser	SATT
AR 765	**3537.393**	Lightweight RWR for Agusta Bell 206A helicopter	SATT
AR 777	**3904.393**	Airborne computer-controlled ESM receiver	SATT
AR 830	**4733.393**	Airborne warning system	SATT
AQ 800	**4361.393**	Airborne noise jamming pod	SATT
AQ 861	**6511.393**	Noise and deception jammer	Ericsson
AQ 900	**4734.393**	Airborne pod-mounted jammer	SATT
AQ 31	**3024.393**	Airborne CW jamming pod	SATT
Erijammer A100	**6512.393**	Airborne jammer pod	Ericsson
Eirjammer 200	**6513.393**	Airborne jammer pod	Ericsson
POD KA	*4362.393*	Airborne noise jamming pod	SRA

Designation	Entry no	Description	Contractor(s)
SWEDEN cont			
Lake 2000	*4363.393*	Airborne noise jamming pod	SRA
AR753	**3025.393**	Airborne jammer set-on receiver	SATT
Type A	**5798.353**	Integrated EW suite for Gripen aircraft	Ericsson
EWS-900	**3538.293**	Naval chaff countermeasures dispenser system	Saab-Scania
EWS-905	**5836.253**	Naval ESM warning system	SAAB
Philax	**3133.293**	Naval IR and radar countermeasures system	PEAB
9CM 100/200	**5731.253**	Naval ECM systems	Philips
UNION OF SOVIET SOCIALIST REPUBLICS			
Bell Series	**6206.253**	Naval threat warners & jammers	
Brick Group	**6206.253**	Submarine ESM systems for surveillance & threat warning	
Cage Pot	**6206.253**	Shipborne EW system	
Farm Loaf	**6206.253**	EW antenna housing	
Fig Jar	**6206.253**	EW antenna housing	
Fix	**6827.193**	Range of HF/DF systems	
Foot Ball	**6206.253**	Shipborne EW system	
Golf Ball	**6206.253**	Submarine surveillance & threat warning system	
High Ring	**6206.253**	Shipborne ELINT/SIGINT antenna	
Krug	**6827.253**	Ground-based HF/DF system	
Loop Series	—	Ground-based HF/DF system (static or mobile)	
Rum Tub	**6206.253**	ESM antenna group	
Ship Globe	**6206.253**	Broadband jammer	
Sirena 2/3	—	Airborne radar warning receiver	
Stop Light	**6206.253**	Broadband passive ESM system	
Tilt Pot	**6206.253**	Shipborne EW system	
Top Hat	**6206.253**	Shipborne EW system	
Watch Dog	**6206.253**	Seaborne surveillance system	
UNITED KINGDOM			
ACEWS	**6829.153**	Land-based tactical intelligence gathering system	Plessey
Apollo	**5805.353**	Airborne ESM system	Marconi
Barbican	**4163.193**	Mobile tactical microwave ESM system	MEL
Barricade	**4226.293**	Naval decoy equipment	Wallop
Cascade	**5668.393**	Helicopter decoy system	Wallop
CCS 3900	**5681.153**	EW command and control system	Racal
CERES	**3370.053**	Computer enhanced radio emission surveillance system	Redifon
CORVUS	**6828.193**	Transportable ELIN system for radar detection and analysis	Thorn EMI
Corvus	**3145.293**	Naval decoy system	Vickers
Cutlass	**3553.293**	Range of computer-based ESM/ECM systems	Racal
Cygnus	**5670.293**	I/J band radar jammer	Racal
EPO	**5737.293**	Naval rocket launcher	Evershed
Evade	**4650.393**	Airborne decoy system	Wallop
Fullback	**4411.093**	Radar warning receiver	Racal
Guardian	**4653.293**	Shipborne active/passive ECM system	Thorn EMI
HERMES	**5732.053**	ESM system	Marconi
Hofin	**3369.393**	Hostile fire indicator	MS Instruments
Hunting 6	**5802.153**	EW DF system	Racal
HWR-2	**1876.053**	Hand-held radar warning receiver	Racal
Jamcat	**5808.193**	Adaptor to convert a combat net radio into a jamming system	Racal
Katie	**5734.353**	Airborne radar warning system	MEL
Kestrel	**4410.393**	Airborne ESM system	Racal
Laser Warner	**6160.393**	Laser warning receiver	Ferranti
MA1122	**5801.153**	Tactical DF Processing/display	Racal
Manta	**4652.493**	Submarine-borne passive ESM system	MEL
Masquerade	**5667.393**	Airborne decoy system	Wallop
Matilda	**3539.253**	Naval radar threat warning system	MEL
Matlock	**5733.153**	Land-based radar warning system	MEL
Mentor	**5939.253**	Naval ESM system	Marconi
Miranda	**3912.193**	Mobile microwave intercept system	MEL
MIR-2	**3554.393**	Helicopter/light aircraft ESM system	Racal
MS Series	**6216.193**	Microprocessor controlled receivers	Thorn EMI
Peregrine	**4411.093**	Pod-mounted warning receiver	Racal
Pinemartin	**5695.153**	Land-based tactical ESM system	Racal
Porpoise	**5671.453**	Submarine version of Cutlass	Racal
PR2250	**3905.153**	HF communications and monitoring receiver	Plessey
Prophet	**5673.353**	Airborne radar warning receiver	Racal
Protean	**3193.293**	Naval chaff decoy dispensing system	MEL
PVS860/1120	**1712.153**	HF monitoring and DF system	Plessey
R405	**3371.053**	Multi-purpose anti-jamming radar receiver	Plessey
R505	*3373.053*	Anti-jamming radar receiving module	Plessey
RA 1779	**3921.153**	Programmable HF surveillance receiver	Racal
RA 1792	**3922.153**	Programmable synthesised surveillance receiver	Racal
RA 1794	**3919.153**	Tactical receiver used in Racal Series 3000 EW systems	Racal
RA 1795	**5679.153**	ELINT VHF/UHF receiver	Racal
RA 1796	**4737.053**	ELINT HF/VHF/UHF receiver	Racal
RA 3701	**6457.153**	HF surveillance receiver	Racal
RACEWS	**5800.153**	EW and C³I system	Racal
RACJAM-AIR	**6515.393**	VHF responsive jammer for helicopters	Racal
Rampart	**4227.193**	Land-based chaff and decoy dispensing system	Wallop
RDL-1	**1341.253**	Tactical ESM system for small ships (2-11.25 GHz)	Racal
RDL-2	**1341.253**	As RDL-1 plus threat evaluation	Racal
RDL-257	**1341.252**	Frequency extended to 1–18 GHz	Racal
RDL-268	**1341.253**	As RDL-257 but includes frequency measurement	Racal
RCM-1	**1942.293**	Naval and land-based I-band jammers	Racal
RCM-2	**1942.293**	Jammers with or without stabilised antennas. Variable bands	Racal
Replica	**6161.293**	Inflatable decoy system	Irvin

Designation	Entry no	Description	Contractor(s)
UK cont			
RJS 3100	**4736.193**	Mobile automatic VHF jamming system	Racal
RJS 3105	*4738.193*	VHF/UHF jammer	Racal
RJS 3140	**4736.193**	Unattended expendable jammer	Racal
RL1	*4365.193*	Laser warning instrument	Lasergage
RL2	*4365.193*	Laser warning instrument	Lasergage
RPV EW	**5807.353**	EW surveillance for RPVs	Marconi
RS 3153	**3918.393**	Modular EW and radio surveillance range of equipments	Racal
RTA 1470	**3920.153**	Radio DF equipment	Racal
RWR-1	**4411.093**	Radar warning receiver	Racal
S373	**4162.193**	Mobile multi-band ESM and ECM system	Marconi
SADIE	**6885.293**	ESM processor	Racal
SARIE	**1743.263**	Naval selective automatic radar identification equipment	Thorn EMI
SAT3311	**5680.153**	HF signal analysis terminal	Racal
Saviour	**5674.193**	Armoured vehicle radar/laser warning receiver	Racal
Sceptre	**6151.253**	Shipborne ESM system	MEL
Schermuly	**5806.193**	IR countermeasures systems	Pains-Wessex
Scimitar	**3598.293**	Naval jammer	MEL
SCORPION	**6884.293**	ECM jammer designed to integrate with Cutlass range	Racal
Seafan	**3555.293**	Naval rocket decoy system	Vickers/Plessey
Sea Saviour	**5675.293**	Naval radar warning receiver	Racal
Seeker	*5938.153*	Mobile tactical EW sub-system	Racal
Sentry	**5803.153**	Vehicle-based ESM intercept system	Marconi
Series 500	**6830.153**	Mobile ECM system	Plessey
Shield	**5682.293**	Naval missile decoy system	Plessey
Siren	**5940.293**	Naval decoy system	Marconi
Sky Guardian	**5804.393**	Airborne radar warning receiver	Marconi
Skyshadow	**3383.393**	ECM pod for Tornado and other strike aircraft (ARI 23246/1)	Marconi
Sprat	**5672.453**	Submarine ESM system	Racal
Stockade	**3911.293**	Naval chaff and IR decoy dispensing system	Wallop
Stockade-Seaflash	**4651.293**	Remotely controlled decoy system	Wallop/Flight Refuelling
SUSIE	*1707.253*	Series of Naval ESM systems	MEL
Sword	**6143.293**	Small warship decoy system	Pains Wessex
TAC-Weasel	**5669.153**	Version of Weasel	Racal
Type 405J	**3372.193**	Transportable land-based radar jammer	Plessey
UAF-1	**6883.253**	Automated ESM system for the Royal Navy Type 23 frigates	Racal
Vicon 70	*5944.393*	Airborne IR countermeasures pod	Vinten
Vicon 77	**5739.393**	Airborne ESM pod	Vinten
Vicon 78	**5822.393**	Airborne decoy dispenser	Vinten
Wavefinder	*6138.193*	Multi-mode passive ESM system	Thorn EMI
Weasel	**5669.153**	Ground-based combined ESM and ELINT system	Racal
Zeus	**5689.353**	Airborne ECM system	Marconi
UNITED STATES OF AMERICA			
AD/EXJAM	**6139.193**	Artillery delivered expendable countermeasures	Fairchild Weston
AN/AAQ-4	**6147.393**	Airborne infra-red countermeasures (internal)	Northrop
AN/AAQ-8	**6148.393**	Pod-monted infra-red countermeasures systems	Northrop
AN/AAR-34	**4662.393**	IR warning receiver	Cincinnati
AN/AAR-44	**3901.393**	Infra-red warning receiver	Cincinnati Electronics
AN/AAR-47	**6911.393**	Missile warning system against electro-optical guided threats	Honeywell
AN/ALE-1	**3109.393**	Countermeasures dispensing pod	Lundy
AN/ALE-2	**3109.393**	Countermeasures dispensing pod	Lundy
AN/ALE-24	**3109.393**	Chaff dispensing system for B-52G/H aircraft	Lundy
AN/ALE-27	**3109.393**	Chaff dispensing system for B-52C-F aircraft	Lundy
AN/ALE-29A/B	**3360.393**	Chaff and IR dispensing set for USN tactical aircraft	Lundy
AN/ALE-32	**3109.393**	Chaff dispensing set for EA-6A aircraft	Lundy
AN/ALE-33	**3109.393**	Chaff dispenser used on BQM-34 target drone	Lundy
AN/ALE-36	**4657.393**	Countermeasures dispensing pod	Tracor
AN/ALE-37	**1904.393**	High-capacity chaff dispenser development	—
AN/ALE-38	**4654.393**	Bulk chaff dispenser (USAF)	Tracor
AN/ALE-39	**1905.393**	Expendable countermeasures dispenser for USN/USMC tactical aircraft	Loral
AN/ALE-40	**1906.393**	Countermeasures dispenser system for various aircraft	Tracor
AN/ALE-40(V)	**5688.393**	Countermeasures dispenser system	Tracor
AN/ALE-41	**4654.393**	Bulk chaff dispenser (USN)	Tracor
AN/ALE-43	**3110.393**	Chaff cutter/dispenser pod	Lundy
AN/ALE-44	**3111.393**	Countermeasures dispenser pod	Lundy
AN/ALE-45	**5828.393**	Airborne countermeasures dispenser	Tracor
AN/ALE-47	**5829.393**	Airborne countermeasures dispenser	—
AN/ALQ-78	**3451.393**	Airborne ESM set used in P-3C aircraft	Loral
AN/ALQ-99	**1908.393**	Tactical noise jamming system for EA-6B	Various
AN/ALQ-100	*1909.393*	Deception jammer	Sanders
AN/ALQ-101	**1910.393**	Range of modular jamming pods	Westinghouse
AN/ALQ-108	**6221.393**	Jamming pod	Magnavox
AN/ALQ-117	**6222.393**	X-band jammer on B-52 aircraft	ITT
AN/ALQ-119	**1911.393**	Dual-mode jamming pod	Westinghouse
AN/ALQ-122	**5950.393**	B-52 radar jamming system	Motorola
AN/ALQ-123	**1912.393**	Airborne IR deception jammer	Xerox
AN/ALQ-125	**4366.353**	Tactical electronic reconnaissance sensor	Litton
AN/ALQ-126	**1913.393**	Deception jammer developed under USN Charger Blue programme	Sanders
AN/ALQ-128	**4668.393**	Multimode threat warning receiver for USAF F-15	Magnavox
AN/ALQ-130	**1914.393**	Tactical communications jamming equipment for USN aircraft	Eaton Corp
AN/ALQ-131	**1915.393**	Self-protection jamming pod	Westinghouse/Loral
AN/ALQ-133	**1916.393**	Airborne tactical Elint system	United Technology Labs
AN/ALQ-134	*4368.393*	Expendable jammer	Sanders
AN/ALQ-135	**3346.393**	Internal tactical EW system for F-15	Northrop
AN/ALQ-136	**4369.393**	Airborne jammer	ITT
AN/ALQ-137	**4669.393**	Airborne jammer for F/FB-111 and EF-111A aircraft	Sanders

Designation	Entry no	Description	Contractor(s)
USA cont			
AN/ALQ-140	*3548.393*	Internal IR countermeasures set	Sanders
AN/ALQ-142	**4367.253**	Naval ESM system	Raytheon
AN/ALQ-144	**3546.393**	Deception IR jammer for helicopters	Sanders
AN/ALQ-146	*3550.393*	IR countermeasures set for CH-46D helicopter	Sanders
AN/ALQ-147	**3545.393**	Hot Brick IR countermeasures equipment	Sanders
AN/ALQ-149	**4678.393**	Airborne communications jammer for EA-6B aircraft	Sanders
AN/ALQ-150	**4370.353**	ESM receiver	—
AN/ALQ-151	**4371.393**	Airborne EW system	ESL
AN/ALQ-153	**3384.393**	Radar tail warning set for B-52 aircraft	Westinghouse
AN/ALQ-155	**3552.393**	Computer-managed power management system for B-52	Northrop
AN/ALQ-156	**3547.393**	Missile detection system for aircraft self-protection	Sanders
AN/ALQ-157	**4375.393**	Infra-red countermeasures system	Xerox
AN/ALQ-161B	**3034.393**	Defensive elctronics suite for B-1 bomber aircraft	Eaton
AN/ALQ-162	**4373.393**	Airborne radar jammer	Northrop
AN/ALQ-164	**4740.393**	Airborne pod-mounted jammer for AV-8B	Sanders
AN/ALQ-165	**3894.393**	Airborne self-protection jammer for USN aircraft	Northrop/Sanders/Westinghouse/ITT
AN/ALQ-171	**4374.353**	Conformal countermeasures system	Northrop
AN/ALQ-172	**5693.393**	Noise deception jammer	ITT
AN/ALQ-176(V)	**4100.393**	Lightweight jamming pod system	Sperry
AN/ALQ-178	**4009.393**	Integrated RWR/ECM suite for F-16 aircraft	Loral
AN/ALQ-184	**6142.393**	Upgrade of the AN/ALQ-119	Raytheon
AN/ALQ-186	**6516.393**	ECM pod for testing out defence systems	Raytheon
AN/ALQ-187	**6517.353**	Airborne countermeasures system	Raytheon
AN/ALR-45	**1917.353**	Aircraft RWR	Itek
AN/ALR-46	**3035.353**	Airborne digital threat warning receiver	Itek
AN/ALR-47	**1862.353**	Radar homing and warning set for USN S-3A aircraft	IBM
AN/ALR-50	**1918.353**	Aircraft SAM alert and launch warning receiver	Magnavox
AN/ALR-52	**1919.353**	Multi-band IFM receiver for air or land-mobile applications	ARGO Systems
AN/ALR-56	**3347.353**	Airborne RWR	Loral
AL/ALR-59(V)	*5932.353*	Airborne ESM system	Litton
AN/ALR-60	**4376.353**	Communications intercept and analysis system	GTE Sylvania
AN/ALR-62(V)	**4377.393**	Radar warning system	Singer (Dalmo Victor)
AN/ALR-64	*4378.393*	Compass sail warning system	Singer (Dalmo Victor)
AN/ALR-66	**3642.353**	Programmable airborne RWR	General Instruments
AN/ALR-67(V)	**4372.393**	Radar warning system	Itek
AN/ALR-68	**3525.353**	Airborne digital threat warning receiver	Itek
AN/ALR-69	**4379.353**	Radar warning receiver	Applied Technology
AN/ALR-73	**4164.393**	Airborne ESM for USN E-2C early warning aircraft	Litton Systems
AN/ALR-75(V)	**6256.393**	Airborne ESM surveillance receiver	Scientific Comms
AN/ALR-80	**5942.393**	Airborne radar warning receiver	General Instruments
AN/ALR-606	**3908.353**	Aircraft RWR	General Instruments
AN/ALR-646	**3909.353**	Airborne ESM system based on ALR-66	General Instruments
AEL 6040	**4403.093**	ECM universal jammer	American Electronic Labs
AN/ALR-46A	*3892.363*	Airborne digital signal processor	Singer (Dalmo Victor)
AN/ALT-28	**5952.393**	B-52 noise jammer	Northrop
AN/APR-38	**3544.393**	Control indicator set for Wild Weasel II aircraft	Loral
AN/APR-39(V)	**3542.393**	Airborne radar warning system	E-Systems/Loral
AN/APR-43	**4010.393**	RWR developed for USN Compass Sail Clockwise programme	Loral
AN/APR-44	**3365.353**	CMR-500B low-cost RWR	American Electronic Labs
AN/APR-47	**6519.393**	Wild Weasel CIS – replacement for AW/APR-38	McDonnell-Douglas
AN/ASQ-171	*4742.393*	Airborne ELINT system for EP-3E ECM aircraft	IBM
AN/ASQ-191	**6166.095**	Communications jammer for air, ground or naval use	Rockwell
AN/AVR-2	**6217.393**	Laser warning receiver for helicopters	Perkin-Elmer
AN/GLQ-3B	**4401.193**	HF/VHF communications jammer	Fairchild
AN/KAS-1	**4407.293**	Chemical warfare detection system	Texas
AN/MLQ-33	**4741.193**	Search and jamming system against air/ground communications	GTE Sylvania
AN/MLQ-34	**3895.193**	Mobile tactical communications countermeasures system for US Army (TACJAM)	GTE Sylvania
AN/MLQ-T6	**4664.193**	Communications data link jammer	Cincinnati Electronics
AN/MSQ-103	**4405.193**	TEAM PACK ESM monitoring system	Emerson
AN/PRD-11	**6458.093**	Man-portable DF equipment	Watkins-Johnson
AN/SLQ-27	**3349.293**	Shortstop naval defensive EW system. Superseded by SLQ-32	—
AN/SLQ-30	**4388.293**	Naval electronic countermeasures systems (TRUMP)	Kuras-Alterman
AN/SLQ-32	**3349.293**	Naval ESM/ECM system. Replacement for WLR-1. Modular	Raytheon
AN/SSQ-72	**6220.293**	Shipborne ESM system	Sanders
AN/TLQ-15	**3367.193**	Communications and countermeasures equipment	American Electronic Labs
AN/TLQ-17A	**4406.093**	Countermeasures and jamming set	Fairchild
AN/TSQ-109	**4408.153**	Sigint system (AGTEWS)	Bunker Ramo
AN/TSQ-112	**4409.153**	ECM ground warning and jamming system	GTE
AN/TSQ-114A	**4409.153**	Automatic ESM system (Trailblazer)	GTE
AN/ULQ-11	**4384.394**	Airborne jamming system (cefirm header)	—
AN/ULQ-14	**4385.153**	Radar countermeasures system	—
AN/ULQ-19	**5834.193**	Tactical communications jammer	Racal
AN/VLQ-5(V)1	**6836.193**	Part of US ADEWS/Comfy Challenge project (COMFY SWORD)	Fairchild Weston
AN/WLQ-4	**6497.453**	SIGINT system for US Navy SSN-637 submarines	GTE
AN/WLR-1H	**6223.293**	Passive radar detection set	ST Research
AN/WLR-8	**3082.253**	Naval tactical EW/ESM receiver system	GTE
AN/WLR-11	**5835.293**	Shipborne IFM receiver	Argo
ADEWS	**6835.153**	Air defence EW system for the US Army. In development	Raytheon
AOCM	*4383.393*	Advanced optical countermeasures pod	Martin Marietta
APECS II	**6499.253**	Naval ESM/ECM system	Argo
ARI 18240/1	**4160.393**	ESM suite for UK Nimrod aircraft	Loral
ARJS	**4667.393**	High power stand-off jammer	Grumman
ASAS/ENSCE	**6842.153**	All sources analyses system for the US Army. In development	—
CEWS	**6140.153**	Communications EW system	Fairchild
ARR-81	**6218.393**	Airborne COMINT receiver	Magnavox

Designation	Entry no	Description	Contractor(s)
USA cont			
Chief	**5951.193**	Vehicle-mounted EW combat system	HRB-Singer
CMR-610A	**4397.053**	Wideband direction-finding receiver	American Electronic Labs
Compass Call	**5839.353**	Airborne EW system	—
C204/WJ-1140	**4391.053**	Analysis control unit	Watkins-Johnson
D1800	**5810.153**	DF and monitoring system	EM Systems
D5100	**5819.093**	Family of DF systems	Tech-Comm
OE-320/A	**6146.193**	DF system	EM Systems
DF 8000	*4665.093*	Communications direction-finding system	Cincinnati Electronics
E-ESM	**6215.293**	Shipborne ESM receiver	Anaren
Dragonfly	**6219.393**	Communications jamming and surveillance in VHF/UHF band	GTE
ESCORT	**5743.353**	Airborne Elint system (M-204)	Raytheon
EW-1017	**4382.353**	Electronic surveillance system	Loral
FAC	**3362.393**	Lightweight countermeasures dispenser system	Lundy
Ferret	**6887.453**	Submarine ESM system	GTE
Guardian Star	**6156.253**	Family of ESM systems	Sperry
Guardrail	**4670.353**	Airborne SIGINT system	ESL
HAWCS	**4380.393**	Helicopter all-weather cueing system	Texas Instruments
INEWS	**5812.353**	Integrated EW system in development	—
IRCM	**6211.393**	Engine-mounted IR self defence system	Sanders
Jam Pac (V)	**4099.393**	Family of low-cost jammer pods	Sperry
J-3400/3410	**4677.153**	VHF mobile ECM systems	GTE
LQ-102A	**4663.193**	Hand-emplaced expendable jammer	Cincinnati Electronics
LR-5200	**5837.353**	Tactical ESM system	Litton
L6-24	**3875.093**	Direction-finding system	Watkins-Johnson
L-6/A-WJ-1140	**1381.053**	Microwave direction-finder	Watkins-Johnson
MEWSS	**6463.153**	Mobile EW support system	—
MIRTS	**6149.393**	Internal or pod-mounted IR jamming system	Northrop
MPI-5	**3105.093**	Miniature pulse analyser	American Electronic Labs
MRES	**4658.053**	ECM and C³CM threat generators	Tracor
Phoenix	**6163.293**	Submarine ESM system for radar detection	ARG
Piranha	**6158.193**	Applique VHF communications jammer	Fairchild
POET	**6214.393**	Dispenser launched expendable jammer	Sanders
QRC-259	**4396.053**	Superheterodyne receiver system	Watkins-Johnson
Quick Fix IIB	**5811.393**	Airborne countermeasures system	Tracor
R-281	**1875.153**	Computer-controlled search receiver	Cincinnati Electronics
R-1849	*3033.053*	Remote-controlled HF search receiver	Cincinnati Electronics
R-1850	*3033.053*	Remote-controlled VHF search receiver	Cincinnati Electronics
R-1851	*3033.053*	Remote-controlled UHF search receiver	Cincinnati Electronics
R-3030	**5816.053**	Surveillance receiver for ground naval and airborne use	Cubic Communications
R4000	**6145.193**	ESM surveillance and monitoring system	EM Systems
R-5500	**3529.093**	Control and display system for ESM	GTE
Rapport II	**3543.393**	ECM system developed for Belgian Mirage V aircraft	Loral
RA-6778	**4676.093**	Communications monitoring equipment	Racal
RA-6793A	**4675.093**	HF monitoring receiver	Racal
RBOC Mk33	**3363.293**	Naval rapid bloom off-board countermeasures decoy system	Hycor/Lundy
RD-280	**1874.153**	HF/VHF surveillance receiver	Cincinnati Electronics
RE-1320A	**4400.093**	Spectrum display unit	Racal
RG Series	**4672.093**	Family of spectrum monitors	Racal
RG-1340	**4399.093**	Spectrum surveillance controller	Racal
RG-5202A	**4673.093**	VHF monitoring receiver	Racal
RG-5540	**4398.053**	VHF/UHF monitor receiver	Racal
RG-5545A	**4674.093**	VHF/UHF monitoring receiver	Racal
RPV EW	**6141.393**	Suite of ECM packages for RPVs	Fairchild Weston
RWR 1000	**6844.193**	Hand-held radar warning equipment for combat troops	GTE
S-2150	**4671.353**	Threat warning systems for small ships and submarines	EM Systems
S-3000	**5838.293**	Shipborne surveillance receiver	EM Systems
S-500	**6144.193**	ESM monitor receiver (500 MHz to 18 GHz)	EM Systems
Scarecrow	**5817.353**	RPV EW system	Sanders
SCR Series	**6159.193**	Range of surveillance receivers	Scientific Communications
Sea Sentry	*4020.293*	Series of three naval ESM systems	Kollmorgen
Series 2000	**3903.053**	Range of receivers, control and display units for ESM	Norlin
SHIELDS	**5840.253**	Shipborne EW system	Sanders
SLR-600 Series	**3910.253**	Range of naval ESM systems based on SLR-600	General Instruments Corp
SR Series	**3037.053**	Range of surveillance receivers for Comint and Elint	Norlin
SPEWS	**4666.393**	Tactical jamming systems for RPVs	Motorola
SRBOC Mk 36	**4656.293**	Naval super rapid bloom off-board countermeasures decoy system	Tracor
SR 200	**4389.293**	Naval ESM system	Sanders
SR 2020	**4659.093**	Tactical ESM receiver	Applied Communications
SR 2152	**4661.093**	VHF/UHF COMINT receiver	Applied Communications
SR 2500	**4660.193**	VHF/UHF monitoring receiver	Applied Communications
TEREC	**4387.153**	Tactical electronic reconnaissance remote terminal (TEREC) & (TRT)	Texas Instruments
TN-1000/WJ-1840	*4395.053*	Minceiver TM tuner	Watkins-Johnson
Triton	**4739.293**	Naval threat warning system for small vessels	Singer (Dalmo Victor)
WJ-927	**4390.053**	Microwave and direction-finding signal analysis system	Watkins-Johnson
WJ-945	**3664.093**	Microwave surveillance and direction-finding system	Watkins-Johnson
WJ-1205B	**3665.093**	Computerised pulse train analyser	Watkins-Johnson
WJ-1240	**1716.053**	Computer-controlled microwave recce system	Watkins-Johnson
WJ-1424	**3666.193**	Microwave ESM equipment	Watkins-Johnson
WJ-1440	**4394.053**	Microwave receiver system	Watkins-Johnson
WJ-1740	**3354.053**	Parallel-scanned digitally controlled microwave receiver	Watkins-Johnson
WJ-1840	**3355.053**	Broadband superhet microwave receiver	Watkins-Johnson
WJ-1920	**4392.093**	Distributed processing system	Watkins-Johnson
WJ-1988	**4393.053**	Direction-finding system	Watkins-Johnson
WJ-1996	**6459.093**	ESM receiver	Watkins-Johnson
WJ-3600	**6834.193**	EW alert and collection system	Watkins-Johnson

Designation	Entry no	Description	Contractor(s)
USA cont			
WJ-4810	**6833.193**	Communications jamming system (88 to 500 MHz)	Watkins-Johnson
WJ-8976	**6461.093**	Direction finding system	Watkins-Johnson
WJ-8990	**6831.193**	Tactical intelligence system (MANTIS)	Watkins-Johnson
WJ-9045	*6460.093*	Tactical receiving system	Watkins-Johnson
XM-130	**3345.393**	Countermeasurs system based on ALE-40 for US Army	Tracor
YUGOSLAVIA			
LID	**6371.193**	Laser irradiation detector	SPQR
SO-1	**5943.393**	Airborne radar warning receiver	Iskra

1505.094

NATO Designations of Soviet Systems and Equipment

Designation	Description
Acrid	Air-to-air missile (AA-6)
Alamo	Air-to-air missile (AA-10)
Alkali	Air-to-air missile (AA-1)
Amos	Air-to-air missile (AA-9)
Anab	Air-to-air missile (AA-3)
Apex	Air-to-air missile (AA-7)
Aphid	Air-to-air missile (AA-8)
Archer	Air-to-air missile (AA-11)
Ash	Air-to-air missile (AA-5)
Atoll	Air-to-air missile (AA-2)
Awl	Air-to-air missile (AA-4)
Backfire	Tupolev VG bomber
Back Net	Early warning and GCI radar
Badger	Tu-16 bomber/reconnaissance aircraft
Ball End	Naval surface warning radar
Ball Gun	Naval surface warning radar
Band Stand	Naval radar for SSN-9 missile
Bank	B-25 bomber aircraft
Barge	Tu-85 bomber aircraft
Bark	Il-2 bomber aircraft
Barlock	P-50 Twin scanner land-mobile early warning and GCI radar
Bass Tilt	Naval gun fire control radar. Similar to Drum Tilt and Muff Cob
Bat	Tu-2 bomber aircraft
Beacon	Airborne IFF
Beagle	Il-28 twin-jet tactical bomber, recce, EW aircraft
Beam Track	Yagi tracking array. Mobile
Bean Shell	Aircraft discone VHF antenna
Bear	Tu-20 bomber aircraft
Beast	Il-0 bomber aircraft
Beauty	Formerly Tu-22 (see Blinder)
Bee Hind	Tail warning radar on Blinder, Bison
Bell Bash	Naval ECM
Bell Clout	Naval ECM
Bell Jar	Naval ECM housing
Bell Shroud	Naval ECM
Bell Slam	Naval ECM
Bell Squat	Naval ECM
Bell Tap	Naval ECM
Bell Thump	Naval ECM
Big Bar A	Early warning and GCI radar
Big Bird	Associated with SA-10 SAM
Big Bulge	Airborne radar on Bear bomber
Big Mesh	Early warning and GCI radar
Big Net	Naval surveillance radar
Big Nose	Airborne radar on Fiddler
Bike Pump	Broadband end-fed dipole antenna
Bill Fold	Transportable search radar
Bill Spring	Helix radiator with plate reflector
Bird	Shipboard dipole and Yagi antenna array
Bison	Mya-4 bomber-reconnaissance aircraft
Blackjack	Long range bomber aircraft
Blinder	Tu-22 bomber aircraft
Blowlamp	Il-54 bomber aircraft
Boat Sail	Submarine radar
Bob	Il-4 bomber aircraft
Bob Tail	Naval radio sextant
Boot	Tu-91 bomber aircraft
Bosun	Tu-14 bomber aircraft
Bounder	Prototype bomber
Bowl Mesh	Shipboard tracking radar
Bow Mat	Naval Elint/Sigint antenna
Bow Tie	Broadband shipboard antennas
Box	A-20 aircraft
Box Brick	Mobile radar
Brassard	Formerly Yak-28 (see Brewer)
Brawny	Il-40 bomber aircraft
Bread Bin	Trailer-mounted radar
Brewer	Yak-28 attack aircraft
Brick Group	Submarine ECM system incorporating Brick Pulp and Brick Spit
Brick Pulp	Submarine EW antenna
Brick Round	Naval antenna with circular dish reflector
Brick Spit	Submarine EW antenna
Brick Spring	High-gain naval helix antenna with ground plane
Brick Square	Similar to Brick Round but with different feed and reflector
Buck	Pe-2 aircraft
Bull	Tu-4 bomber aircraft
Butcher	Formerly Il-28 (see Beagle)
Buzz Stand	Battlefield surveillance radar
Cab	Li-2 (DC-3) transport aircraft
Cage Bare	Naval Elint/Sigint antenna
Cage Cone A/B	Naval IFF antenna
Cage Pot	Shipborne EW system
Cage Stalk	Naval broadband dipole communications antenna
Camber	Il-86 transport aircraft
Camel	Tu-104 airliner. Civil version of Badger
Camp	An-8 transport aircraft
Candid	Il-76 transport aircraft
Careless	Tu-154 transport aircraft
Cart	Tu-70 transport aircraft
Cash	An-28 transport aircraft
Cat	An-10 transport aircraft
Cat House	Missile defence system radar
Charger	Tu-144 supersonic transport
Chekhov	ABM system radar
Cheese Brick	Truck-mounted version of Box Brick
Chuck Luck	Similar to Cage Stalk but different polarisation
Chuck Tube	Naval communications dipole antenna
Clam	Il-18 transport aircraft
Clamshell	3-D acquisition radar for SA-11
Clank	An-30 survey version of Coke
Classic	Il-62 transport aircraft
Cleat	Tu-114 transport aircraft
Cline	An-32 transport aircraft
Clobber A	Yak-42 jet transport aircraft
Clod	An-14 general purpose aircraft
Cluster Bay	'Rising' sea mine
Cluster Guard	Anti-sonar hull coating for submarines
Cluster Gulf	'Rising' sea mine for deep water
Coach	Il-12 transport aircraft
Coaler	An-72 transport aircraft
Cock	An-22 transport aircraft
Cod Eye A/B	Submarine sensors
Codling	Yak-40 transport aircraft
Coin Grass	Corner reflector land-based VHF array
Coke	An-24 transport aircraft
Colt	An-2 general purpose aircraft
Condor	An-124 heavy transport aircraft
Cooker	Tu-110 transport aircraft
Cookpot	Tu-124 transport aircraft
Coot	Il-18 transport aircraft
Cork	Yak-16 transport aircraft
Crab Pot	Microwave communications link antenna
Crate	Il-14. Improved Coach
Creek	Yak-12 light aircraft
Crib	Yak-8 transport aircraft
Cross Bird	Naval early warning radar
Cross Fork	Truck-mounted Yagi array
Cross Legs	Transportable search radar
Cross Loop A	Naval direction-finder
Cross Out	Trailer-mounted search radar
Cross Slot	Ground radar
Cross Sword	SA-N-9 fire control radar
Cross Up	Airborne IFF
Crow	Yak-12 transport aircraft
Crown Drum	Aircraft antenna system
Crusty	Tu-134 transport aircraft
Cub	An-12 transport aircraft
Cuff	Be-30 transport aircraft
Curl	An-26 transport aircraft
Cylinder Head	Ship's gun fire director radar
Dawn	I-band navigation radar
Dead Duck	Naval IFF
Dog House	Phased-array missile and space tracking radar
Dome Glass	Trailer-mounted optical tracker system
Don A/B/2/ Kay	Naval navigation radars
Donets	Series of ship's navigation radars
Down Beat	Navigation/bombing radar for Backfire
Drum Tilt	AA fire control radar
Dry Rack	Pole-mounted quadruple Yagi array
Dumbo	VHF early warning and search radar
Dustbin	Submarine mast sensor
Egg Ball	Naval antenna housing
Egg Cup A/B	Ship's splash spotting radar
End Tray	Ground radar used with Scud
Eye Bowl	Naval missile guidance radar (SS-N-14)
Faceplate	MiG YE-2A fighter aircraft
Fagot	MiG-15 jet fighter
Faithless	Mikoyan STOL fighter aircraft
Farm Loaf	Shipborne IFF and antenna housing
Fang	La-10 fighter aircraft
Fan Song A	Tracking radar
Fan Song B	Target tracking radar
Fan Song E	Target tracking radar
Fan Song Series	Range of electro-mechanically scanned target tracking radars
Fan Tail	Tail fire control radar for Backfire B
Fargo	MiG-9 fighter aircraft
Farmer	MiG-19 jet fighter
Farm Loaf	Naval EW antenna housing
Feather	Yak-17 fighter aircraft
Fencer	New swing-wing fighter
Feniks	Submarine sonar
Fiddler	Tu-28 strike aircraft
Fig Jar	Naval ECM antenna housing
Fin	La-7 fighter aircraft
Firebar	Yak-28P all-weather fighter aircraft
Fire Can	Land-mobile AA fire control radar
Fire Dish	Trailer-mounted tracking radar
Fire Iron	Naval gun fire control array of 4 Yagi antennas with reflectors
Fire Wheel	AA fire control radar
Fishbed	MiG-21 interceptor aircraft
Fish Bowl	Naval fire control radar
Fish Net	Pole-mounted twin Yagi array possibly IFF
Fishpot	Su-9 interceptor aircraft
Fitter	Su-7 interceptor aircraft
Fix Eight	Land-transportable HF direction-finding system
Fix Four A	Fixed HF direction-finding system
Fix Four B	Fixed HF direction-finding system
Fix Four C	Fixed Adcock HF direction-finding system
Fix Four D	Fixed VHF direction-finding system
Fix Six A	Similar to Fix Eight with 6 elements
Flagon-A	Su-11 all-weather fighter aircraft
Flagon-B	Sukhoi STOL fighter
Flagon-E	Su-15 interceptor aircraft
Flanker	Su-27 combat/fighter aircraft
Flap Lid	Tracking radar for SA-11
Flap Track	Tracking radar mounted on tracked vehicle
Flap Wheel	AA fire control radar
Flashlight	Yak-25 all-weather interceptor
Flat Face	P-15. Land mobile target acquisition radar
Flat Spin	Naval air search radar
Flat Twin	Tracking radar (ABM-X-3)
Flipper	Experimental fighter
Flora	Yak-23 fighter aircraft
Flogger	Mikoyan variable-geometry fighter (MiG-23)
Foil Two	Ground based IFF
Foot Ball	Shipborne EW antenna array in small radomes
Forger	Yak-36 V/STOL carrier aircraft
Fork Rest A	VHF Yagi array
Four Eyes	Naval electro-optical fire control
Four Stack	UHF Yagi array

Designation	Description
Foxbat	MiG-25 interceptor aircraft
Fox Fire	I-band AI radar
Foxhound	'Super-Foxbat' with new look-down
MiG 31	radar and AA-9
Frame Spring	Helical tracking antenna with ground-plane. Shipboard
Frank	Yak-9 fighter aircraft
Fred	P-63 (US) fighter aircraft
Freehand	Experimental VTOL aircraft
Fresco	MiG-17 jet fighter
Fritz	La-9 fighter aircraft
Frog-1	Surface-to-surface missile
Frog-2, 3, 4, and 5	Surface-to-surface missiles
Frog-7	Surface-to-surface missile
Frogfoot	Su-25 ground attack aircraft
Front Dome	Naval FCS radar for SA-N-7
Front Door	Guidance radar for SS-N-3
Front Piece	Guidance radar for SS-N-3
Fulcrum	MiG-29 combat aircraft
Gadfly	SA-11 surface-to-air missile
Gage	Ground target acquisition radar
Gainful	Surface-to-air missile (SA-6)
Galosh	Surface-to-air missile
Gammon	Surface-to-air missile (SA-5)
Ganef	SA-4 surface-to-air missile
Gaskin	SA-9 air defence missile
Gazelle	SAH-08 anti-ballistic-missile missile
Gecko	Mobile AA missile system (SA-8)
Gin Pole	Naval ECM
Gladiator	SA-12 SAM system
Goa	Surface-to-air missile (SA-3)
Goblet	SAN-3 naval missile
Golf Ball	Submarine mast sensor
Gopher	SA-13 mobile SAM
Grail	Man-portable AA missile (SA-7)
Gremlin	SA-14 surface-to-air defence missile
Grid Bow	Directional broadband naval communications antenna
Grid Shield	Naval Elint/Sigint antenna
Griffon	Surface-to-air missile
Grumble	SA-10 surface-to-air missile
Guideline	Surface-to-air missile (SA-N-2)
Guild	Surface-to-air missile
Gun Dish	AA fire control radar
Hair Net	Naval search radar
Half Bow	Naval target designation radar
Halo	Mi-26 helicopter
Hare	Mi-1 light helicopter
Harke	Mi-10 helicopter
Harp	Ka-25 helicopter
Hat Rack	Pole-mounted directional antenna
Hat	Ka-10 helicopter
Havoc	Mi-28 helicopter
Hawk Screech	Naval fire control radar
Hay Pole	Vertically polarised omnidirectional antenna
Hay Rick	Truck-mounted omni array
Hay Ring	Microwave link relay antenna
Hay Wain	Omnidirectional discone antenna
Haze	Mi-14P amphibious helicopter
Headlight	Naval fire control radar
Head Net A/B/C	Naval long-range air surveillance radar series
Helix	Ka-32 ASW helicopter
Hen	Ka-15 helicopter
Hen Egg	Ballistic missile early-warning radar
Hen House	Ballistic missile early-warning radar
Hen Nest	Ballistic missile early-warning radar
Hen Roost	High power early-warning radar
Hercules	Submarine sonar
High Brick	Truck-mounted radar
High Fix	Ranging radar for Fitter A/C
High Lark	J-band AI radar
High Lune	Ship's nodding heightfinder
High Pole	Naval IFF transponder antenna
High Ring	Naval Elint/Sigint antenna
High Scoop	Naval search radar
High Sieve	Naval search radar
Hind	Mi-24 attack helicopter
Hip	Mi-8 helicopter
Hog	Ka-18. Obsolete development of Hen helicopter
Hokum	Ka designed combat helicopter
Home Guide	Yagi array for aircraft homing system
Homer	Mi-12 helicopter
Home Talk	PRL-5M precision approach radar
Hoodlum	Ka-26 helicopter
Hook	Mi-6 helicopter
Hoop	Attack helicopter (poss Mi-29)
Hoplite	Mi-2 helicopter. Enlarged Hare
Hormone	Ka-25 helicopter
Horn Spoon	Ship's navigation radar
Horse	Yak-24. Obsolete military helicopter
Hound	Mi-4 helicopter
Jay Bird	J-band radar used on MiG-25 Foxbat
Kangaroo	Air-to-surface missile
Karen	AS-10 air-to-surface missile
Kedge	AS-14 air-to-surface missile
Kegler	AS-12 air-to-surface missile
Kelt	AS-5 air-to-surface missile
Kennel	AS-1 air-to-surface missile
Kent	AS-15 air-launched cruise missile
Kerry	AS-7 air-to-surface missile
Kilter	AS-11 air-to-surface missile – Imaging IR guided
Kingfish	AS-6 air-to-surface missile
Kipper	AS-2 air-to-surface missile
Kitchen	AS-4 air-to-surface missile
Kite Screech	Naval gun fire control radar
Knife Rest A, B, C	P-10 Series. VHF truck-mounted early warning radar
Krug	Very large fixed HF direction-finding system
Kyle	AS-9 anti-radiation missile
Land Fall	Coastal surveillance radar
Leningrad	Shipboard high definition surface search radar
Light Bulb	Naval search radar
Long Bow	Naval target designation radar
Long Brick	Small truck-mounted radar/comms system
Long Bull	Target designation radar. Back-to-back antennas
Long Ears	Naval optical fire control system
Long Eye	Land-mobile radar. Back-to-back antennas
Long Talk	Land-mobile multi-radar system
Long Track	Search and acquisition radar for SA-4 and SA-6 missiles
Long Trough	Mortar locating radar
Look Two	Bombing and navigation radar
Loop Three	Trailer-mounted direction-finding system
Low Blow	Radar for SA-3 missile system
Low Sieve	Naval surface search radar
Madcap	An-24 STOL AWACS aircraft
Madge	Be-6 flying boat
Maestro	Trainer for Firebar/Brewer aircraft
Magnet	Yak-17 utility aircraft
Magnum	Yak-30 aerobatic competition aircraft
Maiden	Two-seat version of Fitter
Mail	Be-12 maritime reconnaissance amphibian
Mainstay	Il-76 AWACS aircraft
Mallow	Be-8/10 jet flying boat
Mandrake	High flying recce aircraft
Mangrove	Yak-26 tactical reconnaissance aircraft
Mantis	Yak-32 single-seat jet
Mare	Yak-14 assault glider
Mark	Kay-7V aircraft
Mascot	Il-28U trainer version of Beagle
Max	Yak-18 trainer and transport versions
May	Il-38 maritime recce version of Coot
Mercury Grass	Land-mobile system using VHF Yagi arrays
Mercury Noseband	Land-mobile electronic system
Mesh Brick	Land-mobile radar
Midas	Il-76 tanker aircraft
Midget	MiG-15UTI trainer version of Fagot
Mink	UT-2 (YA-20) aircraft
Mist	TS-25 assault glider
Mole	Be-8 utility amphibian aircraft
Mongol	Trainer version of Fishbed
Moon	Fixed HF direction-finding array
Moon Cone	Large planar array of dipoles with reflector
Moon Face	Similar to Moon Cone
Moon Mat	Similar to Moon Face
Moon Plate	Similar to other Moon systems but land-transportable
Moose	Yak-11 trainer
Mop	Amphibian bomber/recce aircraft
Moss	AEW version of Cleat
Mote	MBR-2 utility amphibian
Mowik	Su-7U utility aircraft
Mound Brick	Trailer-mounted tracking system
Muff Cob	Naval fire control radar
Mug	MDR-6 amphibian bomber/recce aircraft
Mule	Trainer
Mushroom	Helicopter radar
Neptun	Ship's navigation radar
Nysa B	Trailer-mounted radar
Nysa C	Land-transportable radar system
Odd Lot	E-band GCI radar system
Odd Pair	Heightfinder radar. Land-based
Odd Rods	Aircraft multi-band EW antennas
Oka	Ship's navigation radar
One Eye	Surveillance radar. Land-based
OR-2	Semi-fixed search radar
Owl Screech	Naval fire control radar
Paddle Wheel	Shipboard monitoring antenna
Palm Frond	Naval navigation radar
Park Lamp	Submarine mast direction-finding loop antenna
Pat Hand	Radar for SA-4 surface-to-air missile
Patty Cake	Heightfinder radar
Pawn Shop	Missile guidance radar (ABM-X-3)
Pea Pod	Aircraft antennas
Peel Group	Naval fire control radars for SAN-1 surface-to-air missile
Pill Box	Phased array ballistic missile tracking radar at Pechora
Plank shave	Naval search radar
Port Spring	Submarine mast conical spiral antenna
Plinth Net	Target acquisition radar
Plum Jar	Shipboard EW antenna system
Pole Dish	Possible microwave link antenna
Pop Group	Naval radar for SAN-4 missile
Pork Trough	Mortar locating radar
Post Bow	Naval target designation radar
Post Lamp	Naval target designation radar
Pot Drum	Naval fire control radar
Pot Head	Naval surface search radar
Prong Rest	VHF Yagi array
Puff Ball	Airborne search radar
Punch Bowl	Submarine mast radome
Quint Spring	Navigational tracking antenna
Ram J/K	New Soviet fighter aircraft designs
Ram L	MiG-29 fighter aircraft
Ram M	High altitude reconnaissance aircraft
Ram P	Aircraft thought to be equivalent of US B-1 bomber
Rib Cone	Shipboard broadband omni-directional VHF antenna
Ring Two	Trailer-mounted direction-finding system
Rock Cake	Heightfinder radar
Rod Mat	Ground-based IFF
Round Top	Naval optical fire control system
Rum Tub	Shipborne ESM antennas

Designation	Description
Saber	Strategic missile (SS-20)
Saddler	SS-7 ICBM
Sagger	AT-3 anti-tank missile
Salish	Coastal defence missile
Salt Pot	Shipborne IFF antenna
Samlet	Surface-to-surface version of Kennel missile (SSC-2B)
Sandal	SS-4 surface-to-surface missile
Sandbox	SS-N-12 anti-ship missile
Sapwood	Rocket booster
Sark	Obsolete SLBM
Sasin	ICBM
Satan	Strategic missile (SS-18)
Savage	ICBM (SS-13)
Sawfly	Submarine-launched ballistic missile (SS-N-6)
Scale	IRBM
Scaleboard	SS-12A/B surface-to-surface missile
Scalpel	Strategic missile (SS-X-24)
Scamp	IRBM (SS-14)
Scan Can	Aircraft fire control radar
Scan Fix	AI radar
Scan Odd	AI radar
Scan Three	I/J-band AI radar
Scapegoat	IRBM
Scarab	SS-21 battlefield support missile
Scarp	ICBM
Scoop Pair	Naval target acquisition radars
Score Board	Ground-based IFF
Score Board A	Naval IFF
Scrag	ICBM
Scrooge	IRBM
Scrubber	Naval anti-shipping missile
Scud A and B	Surface-to-surface missiles
Sea Gull	Long-range naval search radar
Sea Net	Naval search radar
Sego	SS-11 ICBM
Sepal	Coastal defence version of Shaddock (SSC-1B)
Serb	Submarine-launched ballistic missile (SS-N-5)
Shaddock	Naval surface-to-surface missile
Sheet Bend	Naval search radar
Sheet Curve	Naval search radar
Sheet Nest	Shipboard radar reflector/enhancement system
Shelf	Aircraft antenna
Shika	ZSU-23-4 vehicle with quad-23mm AA system
Ship Globe	Shipborne missile and space tracking radar
Ship Wheel	Shipboard space tracking radar circular reflector
Shipwreck	SS-N-19 vertically-launched shipborne anti-ship missile
Shore Walk	Communications relay system antenna
Short Horn	K-Band bombing and navigation radar
Shyster	Obsolete ballistic missile
Sickle	Strategic missile (SS-25)
Side Globe	EW housings on large Soviet warships
Side Net	Heightfinder radar
Silex	SS-N-14 ASW missile
Sinner	Strategic missile (SS-16)
Siren	SS-N-9 anti-ship missile
Sirena	Aircraft radar warning set
Skean	IRBM
Skiff	Submarine-launched ballistic missile (SS-NX-23)
Skin Head	Target acquisition radar for torpedo boats
Ski Pole	Ship's IFF
Skip Spin	AI radar
Slant Blade	Shipboard broadband VHF antenna
Slim Net	Naval search radar
Small Cross	Rotating dipole direction-finding system
Small Yawn	Artillery locating radar
Snapper	Anti-tank missile
SNAR-6	J-band mortar locating radar
Snipe	SS-N-17 SLBM
Snoop Plate	Submarine surveillance radar
Snoop Slab	Submarine surface search radar
Snoop Tray	Submarine surveillance radar
Snow Net	Shipboard ring and dipole VHF omni antenna
Snow Shoe	Similar to Snow Net
Songster	AT-8 anti-tank missile
Spandrel	AT-5 anti-tank missile
Spanker	Strategic missile (SS-17)
Spider	SS-23 battlefield surface-to-surface missile
Spigot	AT-4 anti-tank missile
Spike Two	Truck-mounted twin dipole array
Spin Scan	AI radar
Spin Trough	Naval IR searchlight or gun director
Spiral	AT-6 anti-tank missile
Sponge Cake	Heightfinder radar
Spoon Rest A	P-12 VHF early warning radar
Sprage Star	Shipboard VHF dipole array
Sprat Star	Shipboard communications array
Square Four	Crossed loop direction-finding array. Land-based
Square Head	Naval IFF interrogator antenna
Square Pair	Target and missile tracking radar for SA-5 missile
Square Pick	Array of horizontal and vertical dipoles. Land-based
Square Slot	Naval search radar
Square Tie	Naval GP fire control radar
Square Twin	Sigint/Elint antenna
Squat Eye	Search radar. Land-based
Squint Eye	Surveillance radar
Starbright	SS-N-7 submarine launched anti-ship missile
Stick Tree	Land-based Yagi array
Stiletto	Strategic missile (SS-19)
Stingray	SS-N-18 SLBM
Stone Cake	Heightfinder radar
Stop Light	Broadband passive ESM
Straight Flush	Radar for SA-6 missile
Straight Key	Sigint/Elint antenna
Strella 10	Track-mounted surface-to-air (SA-13)
Strike Out	Land-mobile search radar similar to Barlock
Strut Curve	Naval search radar
Stub Brace	Shipboard broadband crossed dipole array
Sturgeon	Submarine-launched ballistic missile (SS-N-20)
Styx	Naval anti-shipping missile
Sunburn	SS-N-22 shipborne tactical missile
Sun Hat	Pole-mounted antenna
Sun Visor A/B	Naval fire control radar
Swatter	AT-2 anti-tank missile
Tachi 1	Multiple Yagi arrays associated with searchlight
Tall King	Land-based early warning radar
Tamir	Submarine sonar
Tee Plinth	Naval E-O system
Thick Eight A	Fixed HF direction-finding array
Thin Skin	Heightfinder radar
Till Pot	Probably naval ECM
Toadstool	Airborne radar on Coke
Token	Early warning GCI radar
Top Bow	Naval long-range tracking radar
Top Dome	Naval SA-N-6 missile guidance group
Top Hat	Shipboard EW antenna system
Top Knot	Naval tactical air navigation radar
Top Mesh	3-D air and surface surveillance radar
Top Net	Naval search radar
Top Pair	Naval back-to-back long-range search radar group
Top Plate	3-D air and surface surveillance radar
Top Sail A/B	Naval 3-D air search radar
Top Steer	Naval back-to-back air surveillance radar group
Top Trough	Naval air search radar
Track Dish	Ground tracking radar
Trap Door	Naval missile fire control radar
Tread Mill	Shipboard Wullenweber VHF dipole array for direction-finding
Try Add	Long-range early warning radar
Tub Brick	Trailer-mounted radar
Tube Tree	Multi-element antenna array
Twice Up	Dipole antenna
Twin Eye	Back-to-back dish antenna array
Twin Scan	I/J-band AI radar
Two Spot	Truck-mounted PAR radar
Vee Cone A/B/C	Shipboard conical spiral antenna system
Vee Tube	Naval communications antenna
Wasp Head	Naval optical fire control director
Watch Dog	ESM multi-band intercept antenna array
Whiff	AA fire control radar
Whiff Brick	Trailer-mounted dish radar
Witch Eight	8-element dipole array for IFF
Witch Five	Naval IFF
Witch Four	Four-element dipole array for IFF
Wood Gage	Semi-fixed search radar
Yard Rake	Shipboard rotatable Yagi array (IFF)
Yew Loop	Shipboard radar
Yo Yo	Target tracking radar

NUMERICAL LIST OF ENTRIES

Deleted entries appear in italics, the date given shows the edition in which the entry last appeared

Entry number	Subject	Page No/ Last edition*
1001.281	*AWCLS All-Weather Carrier Landing System*	*1970-71*
1002.283	*SPN-41 Carrier Approach & Landing System*	*1970-71*
1003.283	*SPN-42 Carrier Automatic Landing System*	*1970-71*
1004.183	*TRN-28 Land-Based Version of SPN-41*	*1970-71*
1006.181	*TILS Tactical Instrument Landing System*	*1970-71*
1007.161	*TSQ-82 Photo Interpretation System*	*1970-71*
1008.393	XM76 Anti-Oscillation Sighting System	887
1009.393	*Type AF 120 Gyro-Stabilised Sight*	*1982-83*
1010.351	*AN/USQ-28 Geodetic Mapping & Survey System*	*1970-71*
1011.453	*UDACS-1 Underwater Detection & Classification System*	*1970-71*
1012.393	ISIS Weapon Aiming System	883
1013.341	*HS801 Nimrod Mr 1 ASW Aircraft*	*1969-70*
1014.373	*DS 5500 Tactical Display System*	*1970-71*
1015.341	*P-3 Orion ASW Aircraft*	*1969-70*
1016.341	*Breguet 1150 Atlantic ASW Aircraft*	*1969-70*
1018.153	Cymbeline Mortar Locating Radar	262
1019.231	Seacat Shipborne Surface-to-Air Missile	476
1020.281	*NA 9 Ships Fire Control System*	*1980-81*
1021.293	*Marconi 323 Series Military TV*	*1979-80*
1022.311	Martel AS.37/AJ.168 Air-to-Surface Missile	710
1023.353	*EMI Sideways Looking Recce Radar Type P391*	*1979-80*
1024.353	Type 195/Type 2069 Helicopter Sonar	565
1025.273	*Outfit JYA Automatic Surface Plotter*	*1970-71*
1026.341	*SH-3D Sea King ASW Helicopter*	*1969-70*
1027.353	*ARI 5955/ARI 5954 Helicopter Tactical Radar*	*1981-82*
1028.353	*E290M Combined Weather & Tactical Radar*	*1979-80*
1029.383	*Ekco Missile MDI Systems*	*1970-71*
1030.231	*Talos Shipborne Surface-to-Air Missile*	*1981-82*
1031.353	*Airpass I*	*1982-83*
1032.353	Airpass III	559
1033.353	*Airpass II*	*1976*
1035.353	*Ferranti Lightweight Ranging & Terrain Warning Radar*	*1971-72*
1036.383	*FE 541 Inertial Nav/Attack System*	*1970-71*
1037.353	*AN/APN-170 Terrain Following Radar*	*1976*
1038.183	*Chieftain Tank Gun Control Equipment*	*1974-75*
1039.393	*NV-100 Low Light-Level TV Camera*	*1973-74*
1040.261	*ADA Action Data Automation System, HMS Eagle*	*1972-73*
1041.253	*Type 984 Naval Air Surveillance Radar*	*1974-75*
1042.263	*Ferranti Poseidon Computer*	*1970-71*
1043.261	*ADA Mk 1*	*1982-83*
1044.261	ADAWS 2 Action Data Automation Weapons System	436
1045.263	*Ferranti FM 1600 Digital Computer*	*1970-71*
1046.261	*CAAIS-Computer-Assisted Action Information System*	*1980-81*
1047.263	*Ferranti FM 1600B Digital Computer*	*1970-71*
1048.191	*Action Speed Tactical Trainer (ASTT)*	*1970-71*
1049.281	*Gun System Automation (GSA4/GWS24)*	*1971-72*
1050.353	*Cyrano Ibis Fire Control Radar*	*1972-73*
1051.353	*CyranoII Airborne Attack Radar*	*1984-85*
1052.353	*Cyrano III Forward Looking Radar*	*1976*
1053.281	*Vega Series Ships Fire Control System*	*1983-84*
1054.281	*Vega-Castor Fire Control System*	*1973-74*
1055.281	*Vega-Pollux Fire Control System*	*1973-74*
1056.281	*Vega-Pollux PC Fire Control System*	*1973-74*
1057.281	*Vega-Pollux PCE Fire Control System*	*1973-74*
1058.281	*Vega-Pollux PCET Fire Control System*	*1973-74*
1059.281	*Vega-Triton BF Fire Control System*	*1970-71*
1060.281	*Vega-Triton E Fire Control System*	*1973-74*
1061.281	*Vega-Triton T Fire Control System*	*1973-74*
1062.253	Triton II Surveillance Radar TRS 3030/TRS 3035	603
1063.253	Castor II Fire Control Radars	604
1064.253	Pollux Tracker Radar TRS 3220	604
1065.373	*Tactical Display, ASW*	*1970-71*
1066.383	*Alkan Type J Submarine-Detection Launcher*	*1973-74*
1067.453	*Sonobuoys Types DSTA3 & DSTV 2K*	*1978*
1068.353	*Omera-Segid Reconnaissance Cameras*	*1970-71*
1069.363	*Sagem AE 51 Digital Computer*	*1970-71*
1070.353	*Cyclope Airborne IR Recce System*	*1982-83*
1071.363	*LCT 825 Digital Computer*	*1970-71*
1072.363	*LCT CS2 Digital Computer*	*1970-71*
1073.393	*SFENA 12B Miss-Distance Acoustic Detector*	*1970-71*
1074.393	*Type 295 Missile Fire Simulator*	*1972-73*
1075.163	*TIV 1050 Digital Computer*	*1970-71*
1076.383	*Twin-Gyro Platform Type 250*	*1970-71*
1077.191	*Sintra Simulators*	*1970-71*
1078.311	*Blue Steel Stand-off Bomb*	*1974-75*
1079.331	*Firestreak Air-to-Air Missile*	*1980-81*
1080.331	*Red Top Air-to-Air Missile*	*1987-88*
1081.311	*Condor (AGM-53A) Air-to-Surface Tactical Missile*	*1978*
1082.331	*Falcon Air-to-Air Missile Series*	*1973-74*
1083.331	*Falcon (AIM-4A) Air-to-Air Missile*	*1977*
1084.331	*Falcon (AIM-4C) Air-to-Air Missile*	*1976*
1085.331	Falcon AIM-4/AIM-26 Air-to-Air Missile	703
1086.331	*Super Falcon (AIM-4E/F) Air-to-Air Missile*	*1987-88*
1087.331	*Super Falcon (AIM-4G) Air-to-Air Missile*	*1987-88*
1088.331	*Falcon (AIM-4H) Air-to-Air Missile*	*1972-73*
1089.331	*Falcon (XAIM-26A & AIM-26B) Air-to-Air Missile*	*1982-83*
1090.331	*Falcon (AIM-47A) Air-to-Air Missile*	*1977*
1091.331	*Genie (AIR-2A) Air-to-Air Missile*	*1982-83*
1092.311	*HART/ZAP Air-to-Surface Missile*	*1971-72*
1093.311	*Hound Dog (AGM-28A & B) Air-to-Ground Missile*	*1980-81*
1094.383	*KS-120/KS-140 Astro Tracking Sub-System*	*1970-71*
1095.311	*Blue Eye (XAGM-79A) Air-to-Surface Missile*	*1972-73*
1096.311	*Viper (XAGM-80A) Air-to-Surface Missile*	*1971-72*
1097.311	*Hornet (ZAGM-64A) Air-to-Surface Missile*	*1976*
1098.311	Maverick (AGM-65) Air-to-Surface Missile System	722
1099.331	Phoenix (AIM-54) Air-to-Air Missile	703
1100.353	AN/AWG-9 Airborne Weapon Control System	810
1101.391	*Quail (ADM-2C) Decoy Missile*	*1977*
1102.311	Shrike (AGM-45) Air-to-Surface Missile	722
1103.331	*Sidewinder 1A (AIM-9B) Air-to-Air Missile*	*1983-84*
1104.331	*Sidewinder 1C (AIM-9C & D) Air-to-Air Missile*	*1982-83*
1105.331	*Sidewinder (AIM-9E) Air-to-Air Missile*	*1982-83*
1106.331	Sparrow III (AIM-7) Air-to-Air Missile	705
1107.311	SRAM (AGM-69) Short Range Attack Missile	721
1108.303	M5 Aircraft Armament System	769
1109.303	*XM8 40mm Armament Subsystem, Helicopter*	*1985-86*
1110.303	M16 Armament Sub-System	770
1111.303	*XM18E1 Armament Pod*	*1976*
1112.303	M21 MAMEE Helicopter Armament System	765
1113.303	M23 Armament Sub-System	770
1114.303	M24 Armament Sub-System	770
1115.303	XM27E1 7.62mm Helicopter Armament Sub-System	768
1116.303	M28A1 Armament System	766
1117.303	XM28E1 Armament Sub-System	766
1118.303	XM35 Armament Sub-System	764
1119.303	M156 Armament Mount	764
1120.303	*XM140 Automatic Gun*	*1978*
1121.303	*XM93/XM94 Armament Subsystem*	*1976*
1122.231	Standard (RIM-66A & RIM-67A) Surface-to-Air Missile	480
1123.311	Standard ARM AGM-78 Anti-Radar Missile	723
1124.193	*AN/VGS-3 (XM23E2) Laser Range Finder*	*1973-74*
1125.161	*FADC System*	*1980-81*
1126.193	*M43 Periscope*	*1973-74*
1127.373	*Specto Head Up Display*	*1970-71*
1128.441	SUBROC (UUM-44A) Anti-Submarine Missile	533
1129.311	*Walleye (AGM-62A) Air-to-Surface Missile*	*1984-85*
1130.411	*Polaris A-2 (UGM-27B) Fleet Ballistic Missile*	*1976*
1131.411	*Polaris A-3 (UGM-27C) Fleet Ballistic Missile*	*1981-82*
1132.414	*US Navy FBM Submarines*	*1971-72*
1133.411	Poseidon C-3 (UGM-73A) Fleet Ballistic Missile	27
1134.411	MSBS Submarine Launched Ballistic Missile Family	4
1135.173	*Plessey Mk 7 Tabular Viewing Unit*	*1970-71*
1136.173	*Type PSD 100 Synthetic Display Drive*	*1970-71*
1137.173	*Plessey Mk 5 Radar Displays*	*1970-71*

Entry number	Subject	Page No/ Last edition*
1138.173	*Plessey Mk 8 Modular Display System*	1970-71
1139.253	*Plessey AWS-1 Naval Radar*	1972-73
1140.153	*Plessey AR-1 Radar*	1980-81
1141.153	AR-15M Mobile Surveillance Radar	267
1142.153	*Plessey Type HF-200 Height Finder Radar*	1983-84
1143.331	*Alkali (AA-1) Air-to-Air Missile*	1980-81
1144.331	AA-3 (Anab) Air-to-Air Missile	699
1145.331	AA-5 (Ash) Air-to-Air Missile	699
1146.331	AA-2 (Atoll) Air-to-Air Missile	699
1147.311	Kangaroo Air-to-Surface Missile (AS-3)	716
1148.311	*Kennel Air-to-Surface Missile*	1985-86
1149.311	Kelt Air-to-Surface Missile (AS-5)	717
1150.311	Kipper Air-to-Surface Missile (AS-2)	716
1151.311	Kitchen Air-to-Surface Missile (AS-4)	717
1152.411	*Sark Submarine Launched Ballistic Missile*	1980-81
1153.411	SS-N-5 (Serb) Submarine Launched Ballistic Missile	13
1154.411	SS-N-6 (Sawfly) Submarine Launched Ballistic Missile (Soviet designation R-21)	14
1155.221	SS-N-2 (Styx) Shipborne Surface-to-Surface Missile	455
1156.221	MM 38 Exocet Surface-to-Surface Missile	450
1157.253	*DUBA-3A Surface Vessel Sonar*	1973-74
1158.253	*SQS-17A Surface Vessel Sonar*	1973-74
1159.253	DUBV-23D Surface Vessel Sonar	547
1160.353	*DUAV-1B Helicopter Sonar*	1970-71
1161.453	*DUUA-1A/B/C Submarine Sonar*	1973-74
1162.453	*DUUX-2A/B/C Passive Sonar*	1979-80
1163.441	E14 Acoustic Torpedo	514
1164.441	E15 Acoustic Torpedo	514
1165.441	L3 Acoustic Torpedo	515
1166.421	*Z16 Torpedo*	1970-71
1167.453	*MF65 & MF65-P1 Hydrophones*	1970-71
1168.153	*Marconi S600 Series Radars*	1980-81
1169.173	*Marconi Photographic Positive/Negative Viewer*	1970-71
1170.311	*AS.20 Air-to-Surface Missile*	1978
1171.311	*AS.30 Air-to-Surface Missile*	1982-83
1172.311	*AS.30L Air-to-Surface Missile*	1972-73
1173.311	*AS.11 Air-to-Surface Missile*	1984-85
1174.311	*AS.12 Air-to-Surface Missile*	1984-85
1175.221	*SS.12M Surface-to-Surface Missile*	1970-71
1176.331	R.530 Air-to-Air Missile	694
1177.231	Masurca Surface-to-Air Missile System	463
1178.331	*R.511 Air-to-Air Missile*	1978
1179.241	Malafon Surface to Sub-Surface Missile	513
1180.321	Kormoran Air-to-Surface Anti-Ship Missile	709
1181.181	NADGE - NATO Air Defence Ground Environment	46
1182.303	Mk 4 Mod 0 Gun Pod	764
1183.303	Mk 11 Mod 5 Gun	764
1184.231	*Murene/Mureca Surface-to-Air Missile System*	1973-74
1185.311	*Hellcat Air-to-Surface Missile*	1969-70
1186.331	*Taildog Air-to-air Missile*	1970-71
1187.331	*Mitsubishi AAM-1 Air-to-Air Missile*	1985-86
1188.331	*Mitsubishi AAM-2 Air-to-Air Missile*	1984-85
1189.321	Saab 04E Air-to-Surface Missile	716
1190.311	Saab 05A Air-to-Surface Missile	716
1191.004	*US Missile Names and Designations*	1973-74
1192.004	*US Rocket & Missile Designation System*	1971-72
1194.354	Airborne Radar Equipment Table	Analysis (1045)
1195.154	Land-based Radar Equipment Table	Analysis (1029)
1196.393	Ferranti Laser Ranger and Marked Target Seeker	884
1197.373	*Unihedd Multi-Mode Display*	1970-71
1198.153	*MV.21 Series Instantaneous Frequency Measuring Receivers*	1972-73
1199.353	*Hawker Siddeley Dynamics Type 201 Infra-Red Linescan*	1979-80
1200.331	*AIM-82A Short-Range Air-to-Air Missile*	1972-73
1201.183	*Omega Navigation System*	1970-71
1202.283	*AN/SRN-12 Omega Navigation Receiver*	1970-71
1203.193	*Harrier Flight Simulator*	1973-74
1204.351	*UK Airborne Early Warning System*	1970-71
1205.373	*Elliott Head Down Display*	1970-71
1206.373	*Elliott Head Up Displays*	1970-71
1207.293	Target Designation Sight (TDS)	685
1208.353	*UAP 12 Series Airborne Attack Radars*	1977
1209.353	UAP 13 Series Airborne Attack Radars	783
1210.353	PS-37/A Airborne Attack Radar	783
1211.353	Aida II Fire Control Radar	846
1212.373	*AJ37 Viggen Head-Up and Head-Down Display System*	1970-71
1213.253	*UAL 10102 Marine Laser Range Finder*	1979-80
1214.193	*SFENA LPCB II Bomb Scoring Equipment*	1970-71
1215.193	*SFENA Type 14 Miss Distance Indicator*	1970-71
1216.353	*Ferranti Strike & Terrain Following Radar*	1977
1217.253	*Sonar DUBV-24C*	1982-83
1218.253	*Alcatel Premo Search & Attack Sonar*	1973-74
1219.353	*HS-70 Helicopter Sonar*	1973-74
1220.391	*Scad Decoy Missile*	1974-75
1221.253	*PMS 32 Sonar*	1980-81
1222.453	*XBT Expendable Bathythermograph*	1970-71
1223.263	*Plessey Acoustic Ray Trace Indicator*	1977
1224.353	*PTR 820 Airborne IFF*	1987-88
1225.393	*Model 800B-7 Miss Distance Indicator*	1970-71
1226.393	*Digidops Miss Distance Indicator*	1970-71
1227.393	*Model 800B-4 Miss Distance Indicator*	1970-71
1228.393	*Hayes TA-8 Towbee Targets*	1970-71
1229.393	*Hayes TJT-1 Towed Target*	1970-71
1230.393	*Hayes Target AZC-1*	1970-71
1231.283	Saab TVT-300 Fire Director System	685
1232.281	*Alcatel DLT-L-4C Torpedo Launch Control System*	1979-80
1233.481	*TIOS Tactical Information Organisation System*	1980-81
1234.261	ADAWS 4 Action Data Automation Weapon System	436
1235.261	ADAWS 5 Action Data Automation Weapon System	436
1236.253	Jupiter Naval Air Surveillance Radar TRS 3010	605
1237.253	*Saturne (TH.D 1041) Naval Radar*	1977
1238.153	*Lynx (TH.D 1051) Mine Watch Radar*	1972-73
1239.253	*Ramses (TH.D 1022) Naval Radar*	1972-73
1240.453	*Calypso II (TH.D 1030) Submarine Radar*	1979-80
1241.281	Gun Fire Control System Mk 86	442
1242.281	*MRS3/GWS22 Fire Control System*	1982-83
1243.281	*Argo 10 Fire Control System*	1971-72
1244.281	*DLB-1 Anti-Submarine Weapon Control System*	1982-83
1245.253	*Raytheon RTN-10 Fire Control Radar*	1977
1246.331	*Air-to-Air Missile*	1977
1247.253	AN/SPG-51 Radar	635
1248.253	AN/SPS-52/52A/52B Radar	640
1249.253	AN/SPS-39A (Frescan) Radar	638
1250.253	*SPS-32 (Scanfar) Radar*	1982-83
1251.253	*SPS-33 (Scanfar) Tracking Radar*	1982-83
1252.253	AN/SPS-48 Air Surveillance Radar	638
1253.253	RN Type 975 Radar	630
1254.253	*RN Type 978 Radar*	1982-83
1255.253	*RN Type 992 Radar*	1973-74
1256.253	LW.02 Naval Air Surveillance Radar	613
1257.253	*SGR 109 Naval Height-Finding Radar*	1982-83
1258.253	*Signaal Surface Warning & Navigation Radar*	1980-81
1259.281	M20 Series Fire Control Systems	420
1260.281	*Gun Fire Control System Mk 87*	1982-83
1261.281	M40 Series Fire Control Systems	423
1262.283	*Sagem LH-C Electromagnetic Log*	1970-71
1263.303	*SURA 80mm Aircraft Rocket*	1980-81
1264.393	*ML Aviation Lightweight Low-Drag Twin Store Carrier*	1987-88
1266.393	*Hunting Hivolt Weapon Release Equipment EMRU No 20 Mks 1 to 5*	1987-88
1267.393	British Bomb Retarding System	754
1268.303	Minigun Pod Type SUU 11B/A	767
1269.303	Oerlikon 30mm Gun Type KCA	751
1270.303	General Electric Vulcan 20mm Aircraft Gun and Linkless Feed System	762
1271.303	30mm Aircraft Gun DEFA 552A	734
1272.303	Hispano Oerlikon 20mm Tactical Air Armament	751
1273.303	30mm Aircraft Gun DEFA 553	734
1274.303	*68mm SNEB Rocket*	1984-85
1275.303	20 mm M621 Cannon Pod	734
1276.273	*TTF/TTG Tactical Plotting Table*	1970-71
1277.303	Cluster Bomb No 1 Mks 1, 2, 3, and 4 (BL 755)	754
1278.393	*Matra 1869mm Rocket Launchers Types 150 & 155*	1970-71
1279.393	*Matra Expendable Launcher Type 116*	1970-71
1280.311	*Bullpup (AGM-12) Air-to-Surface Tactical Missile*	1984-85
1280.393	*ML Ejector Release Unit No 122 Mk 2*	1981-82
1281.303	*Matra 30mm Gun Pods*	1982-83
1282.393	*ML Bomb Carrier Universal Two Store*	1981-82
1283.393	*Stores Rack AW1 Models 11-005-001 and 11-006-001*	1970-71
1284.303	XM214 5.56mm Automatic Gun	769
1285.303	*XM 188 30mm Three-Barrel Aircraft Gun*	1981-82
1286.303	M197 20mm Three-Barrel Aircraft Gun	765
1287.303	Vulcan Gun Pods SUU 16/A and SUU 23/A	762

Entry number	Subject	Page No/ Last edition*
1288.303	M97 20/30mm Flexible Weapon System	1983-84
1289.283	Sagem CG-J Gyro Compass	1970-71
1290.283	Sagem VG-C Vertical Gyro	1970-71
1291.463	Alcatel Sound Field Plotter	1972-73
1292.261	Alcatel Anti-Submarine Action Data System	1972-73
1293.193	Avimo Weapon Simulators	1980-81
1294.383	SRAM (AGM-69A) Guidance and Control Unit KT76	1970-71
1295.383	AN/ASN-90 Inertial Measuring Set	1970-71
1296.383	AN/APN-190(V) Doppler & Signal Converter	1970-71
1297.383	AN/APN-185 Doppler	1970-71
1298.383	ASN-57 & ASN-58 Gyro-compassing Attitude Reference System	1970-71
1299.383	AN/ASN-100 GAVRS	1970-71
1300.383	Orion Doppler/Inertial Systems APN-187/ASN-84	1970-71
1301.321	Harpoon (AGM-84) Air-to-Surface Anti-Ship Missile	724
1302.331	Agile Air-to-Air Missile	1976
1303.331	Bomber Defence Missile	1973-74
1304.351	AWACS - Airborne Warning & Control System	1970-71
1305.351	Moss Soviet Early Warning Aircraft	1970-71
1306.283	Arma Brown Heading Reference System	1970-71
1307.231	SAMID-Ship Anti-Missile Integrated Defence	1977
1308.331	Sidewinder (AIM-9G/H/J) Missile	1983-84
1309.481	Submarine Torpedo Fire Control System DLT-S-4E	1978
1310.353	AN/APQ-120 Fire Control Radar	799
1311.353	AN/APS-119 (XN-1) Airborne Surface Search Radar	1977
1312.383	Jaguar Nav/Attack System	1970-71
1313.353	AHV/6 Radio Altimeter	1970-71
1314.163	Sperry 1412 Digital Processor	1970-71
1315.363	Sagem MS-A-200 Airborne Disc Memory	1970-71
1316.383	Sagem NS1 Inertial Navigator	1970-71
1317.393	Thomson-CSF Series 121-196 Attack Sights	870
1318.253	Head Net A Air Surveillance Radar	620
1319.253	Head Net B Air Surveillance Radar	620
1320.253	Head Net C Air Surveillance Radar	620
1321.253	Flat Spin Naval Radar	1980-81
1322.253	Slim Net Surface Warning Radar	623
1323.253	Peel Group Fire Control Radar	621
1324.253	Scoop Pair Surface Target Radars	623
1325.253	Hawk/Owl Screech Fire Control Radar	619
1326.253	Sun Visor Fire Control Radar	624
1327.253	Top Sail 3-D Radar	625
1328.253	Head Light Group	619
1329.253	Square Tie Naval Radar	624
1330.253	Drum Tilt Fire Control Radar	618
1331.253	Strut Curve Search Radar	624
1332.254	Naval Radar Table	Analysis (930)
1333.091	Electronic Warfare Systems, Note	1978
1334.453	DUBV-43B/C Variable Depth Sonar	547
1336.221	Otomat Anti-Ship Missile (France/Italy)	451
1337.121	Otomat Coastal Defence System (France/Italy)	161
1338.321	Otomat Anti-Ship Missile (France/Italy)	1987-88
1339.221	Penguin Mk 2 Anti-Ship Missile	453
1340.293	Look-Out Aiming Sight Mk 1	1983-84
1341.253	RDL Series ESM Equipment	667
1342.353	Seaspray Helicopter ASV Radar	792
1343.321	HLM-Helicopter Launched Missile	1972-73
1344.331	SRAAM-100 Air-to-Air Missile	1980-81
1345.391	GUFO Tactical Reconnaissance System	1982-83
1346.353	Type 401 Infra-Red Linescan	1982-83
1347.353	Type 601 Infra-Red Linescan	1971-72
1348.331	R550 Magic Air-to-Air Missile	694
1349.331	Super 530 Air-to-Air Missile	694
1350.321	Albatros Anti-Ship Missile	1973-74
1351.353	Oryx Attack Radar	1977
1352.353	Saiga Helicopter Radar	1980-81
1353.231	Hirondelle Air Defence System	1977
1354.393	LT 102 Laser Range-Finder	1972-73
1355.453	DUBY-24 Sonar	1972-73
1356.453	ELEDONE Submarine Sonar	548
1357.253	DUBM-20A Mine Hunting Sonar	1974-75
1358.253	DUBM-40X Mine Hunting Sonar	1980-81
1359.253	AN/SPS-58/SPS-65 Naval Radars	640
1360.173	Radar Display Console IP-17-A	1972-73
1361.273	Sintra Marine Display Unit	1972-73
1362.273	CT-EE Display Unit	1972-73
1363.261	Commodore AIO System	1974-75
1364.253	RAN-3L Naval Radar (MM/SPS-768)	608
1365.253	RAN-11L/X and RAN-12L/X Naval Radar	609
1366.253	RAN-14X Naval Radar	1972-73
Entry number	**Subject**	**Page No/ Last edition***
1367.253	Orion RTN-16X Naval Radar	1974-75
1368.253	Orion RTN-10X Naval Tracking Radar	609
1369.173	ATAC Display Equipment	1972-73
1370.273	PLAD Automatic Data Plotter	1972-73
1371.393	EMI Miniature Moonlight Television	1977
1372.063	AIL Type 707 Spectrum Analyser	1979-80
1373.253	RN Mine Hunting System (Acoustic) Mk 1	1979-80
1374.253	Minehunting Sonar 193M/193M Mod 1	566
1375.481	M8 Submarine Torpedo Fire Control System	423
1376.393	LME Airborne Laser Range-Finder	1981-82
1377.393	TA 101 & TAY 18 Airborne Laser Range-Finders	1976
1378.091	ELINT Equipment	1976
1379.091	EW Direction Finders	1976
1380.091	EW Antennas	1976
1381.053	L-6/A-WJ-1140 Microwave Direction Finding System	334
1382.063	WJ-1007 Microwave Collection System	1979-80
1384.393	GE Low Light-Level TV Camera	1976
1385.173	Sintra Colour Display Console	1972-73
1386.353	ASQ-501 Magnetic Anomaly Detector	1980-81
1387.393	AMPC for Magnetic Anomaly Detectors	1973-74
1388.393	AN/ASA-65(V) Nine-Term Compensator	543
1389.353	DHAX-1 Magnetic Anomaly Detector	1987-88
1390.353	DHAX-3 Magnetic Anomaly Detector	1987-88
1391.311	Hellfire (AGM-114) Tactical Air-to-Surface Missile	726
1392.351	147 Series Reconnaissance Drones	1982-83
1393.351	154 Firefly Reconnaissance Drones	1976
1394.253	Type 1006 Radar	632
1395.353	RH 370 Helicopter Radar	1972-73
1396.353	Cyrano IV/Cyrano IV M Attack Radar	776
1397.151	Igloo White Surveillance System	1974-75
1398.302	LAMPS Helicopter	1974-75
1399.302	INFANTUH-1M Iroquois Helicopter	1973-74
1400.302	SMASHSE Asia Multiple Sensor Armament System Helicopter	1973-74
1401.353	APQ-137 Airborne Radar	1985-86
1402.302	Orion P-3 Aircraft	1974-75
1403.302	S-3A ASW Aircraft	1974-75
1404.302	Hawkeye E-2 AEW Aircraft	1974-75
1405.302	Mohawk OV-1 Reconnaissance Aircraft	1974-75
1406.302	Mirage Milan Aircraft	1974-75
1407.302	Tomcat F-14A Aircraft	1974-75
1408.302	Phantom II F-4E	1974-75
1409.302	RB-57A Aircraft	1974-75
1410.302	RA-5C Aircraft	1974-75
1411.302	SR-71 Reconnaissance Aircraft	1974-75
1412.302	Kiowa OH-58A Helicopter	1974-75
1413.302	Blackhawk S-67 Helicopter	1973-74
1414.302	Skyhawk A-4M Aircraft	1974-75
1415.063	UYK-501 Computer	1972-73
1416.063	IP 08C Computer	1973-74
1417.063	Camille Computer	1973-74
1418.063	AE-51 Computer	1973-74
1419.063	EMD Type A Computer	1973-74
1420.063	EMD Type B Computer	1973-74
1421.063	CNM-1 Computer	1973-74
1422.063	Iris 55M Computer	1973-74
1423.063	Iris 35M Computer	1973-74
1424.063	Type SMR Computer	1973-74
1425.063	Type SM-3 Computer	1973-74
1426.063	CK-37 Computer	1973-74
1427.063	Censor 908 Computer	1973-74
1428.063	Censor 932 Computer	1973-74
1429.063	Argus 400 Computer	1971-72
1430.063	FM 1200 Computer	1971-72
1431.063	F 1600 Computer	1973-74
1432.063	FM 1600 Computer	1973-74
1433.063	FM 1600B Computer	1973-74
1434.063	FM 1600C Computer	1973-74
1435.063	FM 1600D Computer	1973-74
1436.063	Poseidon Computer	1973-74
1437.063	Myriad II Computer	1973-74
1438.063	Myriad III Computer	1973-74
1439.063	AA6701 Computer	1973-74
1440.063	920 M Computer	1972-73
1441.063	MCS 920M Computer	1973-74
1442.062	920C Computer	1973-74
1443.063	Type 1412 Computer	1973-74
1444.063	D26J Computer	1973-74
1445.063	D37B Computer	1973-74
1446.063	D200 Computer	1973-74
1447.063	Verdan Computer	1973-74
1448.063	BR-1018 Computer	1973-74
1449.063	D84 Computer	1973-74

Entry number	Subject	Page No/ Last edition*
1450.063	Alpha Series Computers	1973-74
1451.063	Alert Computer	1973-74
1452.063	H4400 Computer	1973-74
1453.063	HCM-205 Computer	1973-74
1454.063	L-300 & L-3000 Series Computers	1973-74
1455.063	NDC-1051 Computer	1973-74
1456.063	NDC-1060 Computer	1973-74
1457.063	NDC-1070 Computer	1973-74
1458.063	SKC-2000 Computer	1973-74
1459.063	GPK-10 Computer	1972-73
1460.063	GPK-20 Computer	1973-74
1461.063	GPK-33 Computer	1973-74
1462.063	Micro-Minac Computer	1972-73
1463.063	ASN-24(V) Computer	1973-74
1464.063	ASN-25(G) Computer	1973-74
1465.063	L90 Series Computers	1972-73
1466.063	ASP Computers	1973-74
1467.063	UYK-7 Computer	1973-74
1468.063	Type 1230 Computer	1973-74
1469.063	Type 1832 Computer	1973-74
1470.063	CNM-2 Computer	1973-74
1471.063	LCT 825P Computer	1973-74
1472.063	LCT CS2 Computer	1973-74
1473.063	T/V 1050 Computer	1973-74
1475.353	Look Two Airborne Radar	785
1476.353	Puff Ball Airborne Radar	785
1477.353	Scan Fix Airborne Radar	785
1478.353	Short Horn Airborne Radar	785
1479.353	Scan Odd Airborn Radar	785
1480.353	Scan Three Airborne Radar	1986-87
1481.353	Skip Spin Airborne Radar	785
1482.353	Spin Scan Airborne Radar	785
1483.363	Sensor Data Processing System	1974-75
1484.353	AN/APS-94-F Side Looking Radar	804
1485.353	AN/APS-116 Airborne Radar	804
1486.353	APQ-139 Airborne Radar	1981-82
1487.353	AN/APQ-122(V) Airborne Radar	800
1488.353	APQ-140 Airborne Radar	1987-88
1489.353	APQ-92 Airborne Radar	1981-82
1490.353	APQ-112 Airborne Radar	1981-82
1491.353	Esaira Airborne Radar	1972-73
1493.261	US Naval Tactical Data System	446
1495.281	Vega-Pollux SC Fire Control System	1973-74
1496.281	WSA-4 Fire Control System	1985-86
1497.091	Passive Electronic Counter-Measures	1976
1498.094	US Electronic Intelligence & Monitoring Equipment Table	1983-84
1499.094	US Electronic Warfare Direction Finding Equipment Table	1983-84
1500.094	US Noise Jammers Table	1983-84
1501.094	US Deception Jammers Table	1983-84
1502.094	US Chaff Dispenser Equipment Table	1983-84
1503.351	Joint Services In-Flight Data Transmission System (JIFDATS)	1973-74
1504.351	Compass Reconnaissance Programmes	1976
1505.094	NATO Designations of Soviet Systems & Equipment Table	Analysis (955)
1506.351	Reconnaissance Satellites	1971-72
1507.253	Boat Sail Submarine Radar	1977
1508.253	ST 802 Naval Tracking Radar	632
1509.303	Giboulee Dispenser	1983-84
1510.303	Brandt 100mm Aircraft Rocket	1984-85
1511.393	Matra Unguided Aircraft Rocket Launchers	1987-88
1512.303	Streuwaffen	1979-80
1513.303	135mm Bofors Air-to-Ground Rocket System M70	748
1514.393	Matra Type 200 Bomb Retarding System	737
1515.393	ML Lightweight Twin Store Carrier (High Strength)	1974-75
1516.393	ML ERU No 123 (MACE)	1979-80
1517.393	ML Light Four-Store Carrier	1971-72
1518.393	Thomson-CSF R Series Gunsights	1979-80
1519.393	AGAT Air-to-Surface Weapon Aiming System	1974-75
1520.193	Sintra Simulators	1982-83
1521.454	Sonar Equipment Table	Analysis (940)
1522.261	SENIT Naval Tactical Data Handling Systems	410
1523.391	Turana Target Drone	1979-80
1524.261	WSA-400 Series Naval Weapon Control System	1985-86
1525.263	SPI-04 Sound Ray Path Analyser	560
1526.153	Forward Area Alerting Radar AN/TPQ-32/MPQ-49/MPQ-54	287
1527.253	RAN-7S Naval Radar	1976
1528.153	RATAC Battlefield Radar	239
1529.153	LPD-20 Search Radar	243
1530.321	Sea Skua Helicopter Air-to-Surface Missile	719
1531.311	Strebo Air-to-Surface Area Weapons	1978
1532.311	Jumbo Air-to-Surface Missile	1977
1533.311	Paveway Air-to-Ground Weapon Delivery System	1978
1534.311	Paveway Laser Guided Bombs	757
1535.331	South African Air-to-Air Missile	1976
1536.351	KAD Reconnaissance Drone	1973-74
1537.351	Dornier Aerodyne	1981-82
1538.351	L450F Reconnaissance Drone	1978
1539.351	Beech QU-22B Bonanza Drone	1977
1540.391	SPT-B Selectable Performance Target Ballistic (BATS)	1982-83
1541.281	TORCI Torpedo Fire Control System	1986-87
1542.281	9 LV 200 Fire Control System	427
1543.281	Seafire Weapon Control System	1979-80
1544.481	9 TCI 210 Submarine Torpedo Fire Control System	1981-82
1545.453	Subfar 100 Submarine Radar	616
1546.253	9 GR 600 Naval Radar	615
1547.253	9 LV 200 Mk 2 and Mk 3 Tracking Radar	615
1548.181	9KA 400 Kastell Coast Fire Control System	91
1549.153	9 GR 600 Radar	1980-81
1550.281	Argo NA 10 Fire Control Systems	1982-83
1551.281	Albatros Weapon System	419
1552.281	Sea Hunter-4 Fire Control System	1978
1553.253	LW.04 Naval Air Surviellance Radar	1980-81
1554.253	DA.05 Naval Surveillance Radar	612
1555.253	ZW.06 Surface Search and Navigation Radar	1987-88
1556.153	STAR - Signaal Terminal Area Radar	1977
1557.153	SLAM - Signaal Low Air Defence Module	1976
1558.253	RN Type 912 Radar	629
1559.253	RN Type 909 Radar	629
1560.253	RN Type 965 Radar	630
1561.253	RN Types 967 and 968 Radars	630
1562.253	RN Type 910 Radar	630
1563.253	MRS3/GWS22 Fire Control Radars Types 903/904	628
1564.253	AN/SPS-10 Radar	637
1565.253	AN/SPS-37 Radar	637
1566.253	SPS-12 Radar	1981-82
1567.353	AN/APQ-148/156 Multi-Mode Radar	801
1568.353	NASARR Airborne Radar	1977
1569.353	AN/APX-83 Airborne IFF Interrogator	809
1570.253	AN/SPY-1 Multi-Function Array Radar	641
1571.253	Cossor Naval IFF	1979-80
1572.353	Cossor Airborne IFF Transponders	1980-81
1573.152	Pointer - Portable Independent Terma Radar	1979-80
1574.153	Terma Splash Spotting Radar	1978
1575.253	Terma Scanter Mil Navigation and Surveillance Radars	601
1576.153	DR 513 Ballistic Radar	1976
1577.153	M/532 Doppler Radar	1974-75
1578.253	EX 77 Mod 0 Director Group	1977
1579.302	HS 1182 Strike/Trainer Aircraft	1974-75
1580.302	Alpha Jet	1974-75
1581.302	Panavia 200 MRCA	1974-75
1582.302	B-1 Strategic Bomber	1974-75
1583.302	105 G Aircraft	1974-75
1585.353	E-3 (AWACS) Radar AN/APY-2	798
1586.302	MFI-17 Army Co-op/Trainer Aircraft	1974-75
1587.302	WG-13 Lynx Helicopter	1974-75
1588.302	SA.330 Puma Helicopter	1974-75
1589.253	3D Multi-Target Tracking Radar	614
1590.253	WM20 Series Fire Control Radar	614
1591.253	Signaal M40 Series Fire Control Radar	1981-82
1592.303	Emerson Flexible Turret System (FTS)	1986-87
1593.253	DRBI 10 3D Naval Radar	601
1594.253	DRBV 23 Naval Surveillance Radar	603
1595.253	Square Head Naval Radar	624
1596.253	Cylinder Head Fire Control Radar	1980-81
1597.311	Smart Bombs	1980-81
1598.393	Alkan Type 65 Bomb Rack Adapter	1981-82
1599.393	Alkan Type PM3 Bomb Rack	1981-82
1600.393	Alkan Store Ejectors	1987-88
1601.393	CASA 04.080 Rocket Rack Launcher	1987-88
1602.393	HASA Honeycombe Rocket Launchers	1973-74
1603.393	CASA AH 039220 Universal Pylon	1987-88
1604.393	ML Carrier Bomb Light Stores No. 200	824
1605.253	Fang Song E Naval Radar	619
1606.253	High Lune Height-Finder Radar	1980-81
1607.253	Hair Net Naval Radar	1980-81
1608.253	Big Net Search Radar	618
1609.253	Plinth Net Search Radar	622
1610.253	Top Trough Surveillance Radar	626

Entry number	Subject	Page No/ Last edition*
1611.253	Muff Cob Fire Control Radar	621
1612.253	Pot Drum Naval Radar	622
1613.253	Pot Head Naval Radar	622
1614.253	Skin Head Naval Radar	623
1615.063	*Teledyne 20000 Series Computers*	*1973-74*
1616.063	*Teledyne TDY-300/TDY-310 Computers*	*1973-74*
1617.063	*RCA 195 Military Computer*	*1973-74*
1618.063	*RCA 215 Military Computer*	*1973-74*
1619.161	FLER-H Tank Fire Control Computer	73
1620.281	*SATIN A Fire Control System*	*1973-74*
1621.281	*SATIN B Fire Control System*	*1973-74*
1622.281	*Sea Killer Integration Kit*	*1976*
1623.351	*Military Satellites*	*1972-73*
1624.351	*American Satellite Programmes*	*1977*
1625.351	*Cosmos Series Soviet Satellites*	*1977*
1626.161	Falke Artillery Computer System	71
1627.063	*Uni-Comp Signal Analysis Systems*	*1974-75*
1628.153	*UAL 11201 Tank Laser and UAL 11501 Artillery Laser*	*1981-82*
1629.393	*BT9R Bombing System*	*1978*
1630.393	Honeywell Visually Coupled Systems	888
1631.353	DUAV-4 Helicopter Sonar (TSM 8251)	549
1632.253	*Active Buoy System Tyope DSAA-4*	*1979-80*
1633.253	*Pascal Sonar*	*1978*
1634.453	DUUA-2A Active/Passive Sonar	548
1635.453	Bathythermograph Data Recorder RO-308/SSQ-36	590
1636.453	HS-1000 Series Sonar	541
1637.353	AN/AQS-13 Helicopter Sonar	592
1638.353	*Linescan Type 210 Series*	*1979-80*
1639.353	*RCA Laser-TV System*	*1973-74*
1640.353	*Dornier Reconnaissance Pod*	*1973-74*
1641.393	*RCA ASD 3000 Series TV Cameras*	*1976*
1643.091	*American Electronic Warfare Projects*	*1978*
1644.091	*Soviet Electronic Warfare*	*1978*
1645.193	*LMT Simulators*	*1981-82*
1646.193	*Florett*	*1982-83*
1647.193	*Redifon Simulators*	*1982-83*
1649.193	*TACDEW - Tactical Advanced Combat Direction and Electronic Warfare Simulator*	*1982-83*
1650.321	*Airtos Air-to-Surface Anti-ship Missile*	*1974-75*
1651.321	Marte Anti-Ship Missile System	713
1652.311	*Bantam (Rb 53) Airborne Anti-tank Missile*	*1978*
1653.311	ASM-1 Air-to-Surface Missile (Type 80)	714
1654.311	*Hawkswing Helicopter Anti-tank Missile*	*1976*
1655.311	*Bulldog (AGM-83A) Air-to-Surface Missile*	*1976*
1656.331	Aspide Multirole Missile	697
1657.331	*Brazo Air-to-Air Missile*	*1979-80*
1658.331	*Seek Bat (XAIM-97A) Air-to-Air Missile*	*1977*
1659.331	Shafrir Air-to-Air Missile	696
1660.351	*Short Skyspy Surveillance Drone*	*1979-80*
1661.261	*ASWDS Anti-Submarine Warfare Data System*	*1980-81*
1662.481	SINBADS - Submarine Integrated Battle and Data System	423
1663.261	CIDIS - Combat Information Display System	421
1664.281	SEWACO - Sensor, Weapon, Control and Command System	421
1665.261	*Niteroi Class AIO*	*1980-81*
1666.261	*Gun Director Mk 35 Mod 0*	*1980-81*
1667.302	*SF 37 and SH 37 Viggen Reconnaissance Aircraft*	*1974-75*
1668.302	*SA 341 Gazelle Helicopter*	*1974-75*
1669.302	*F-15A Eagle Aircraft*	*1974-75*
1670.302	*A-10A Close Support Aircraft*	*1974-75*
1671.302	*AX-2 (IA-58) Attack Aircraft*	*1974-75*
1672.353	Agave Airborne Radar	776
1673.353	Westinghouse WX Series Fire Control Radars	812
1674.353	*NR-AI-3A-Airborne IFF Interrogator*	*1984-85*
1675.353	AN/APQ-153 Fire Control Radar	801
1676.353	AN/UPD-4 Side-looking Radar System	812
1677.303	*GAU-7/A Aircraft Gun System*	*1973-74*
1678.303	GPU-2/A 20mm Lightweight Gun Pod	760
1679.393	*ML Triple Store Carrier*	*1987-88*
1680.393	*Miles Free Fall MACE*	*1974-75*
1681.393	*Hunting Hivolt Carrier Bomb Light Stores*	*1987-88*
1682.393	*Alkan ASW Aircraft Armament Equipment*	*1980-81*
1683.393	*Alkan Cartridge Launchers*	*1981-82*
1684.393	*Saab Ejector Release Unit ULM-1*	*1973-74*
1685.393	*Saab Bomb Lock ULM-3*	*1978*
1686.253	*Pollux II Tracker Radar*	*1973-74*
1687.253	Sea Tiger II Surveillance Radar TRS 3001	605
1688.253	*ELI 4 IFF Interrogator*	*1984-85*
1689.363	*12/12 Airborne Computers*	*1973-74*
1690.063	*Selenia CDG-3032 Computer*	*1973-74*
1691.063	*T-VT BCH Computer*	*1973-74*
1692.063	*UTD Computer*	*1973-74*
1693.063	*CP-16A Computer*	*1973-74*
1694.063	*CP-32A Computer*	*1973-74*
1695.063	*SKC-300 General Purpose Computer*	*1973-74*
1696.393	*2in ARF/8 M2 Rocket*	*1984-85*
1697.253	AN/SPS-55 Naval Radar	640
1698.063	*ISAC-77 Computer*	*1973-74*
1699.253	RAN-10S Naval Radar (MM/SPS-774)	608
1700.253	*RAN-2C Naval Radar*	*1976*
1701.253	*RAN-13X Naval Radar*	*1976*
1702.253	3RM Series Naval Radars	610
1703.253	SPQ-2D Naval Radar	610
1704.253	*Sea Hunter Search Radar*	*1978*
1705.253	*Sea Hunter Tracking Radar*	*1978*
1706.353	*Hughes Cobra Radar*	*1976*
1707.253	*SUSIE ECM Receiver System*	*1985-86*
1708.353	*Orpheus Aerial Reconnaissance System*	*1982-83*
1709.353	*IRLS-5 Infra-red Linescan*	*1982-83*
1710.353	*TA-7M2 Aerial Camera*	*1979-80*
1711.353	*TA-8M2 Aerial Camera*	*1982-83*
1712.153	Plessey HF Monitoring Equipment	321
1713.253	*SR-1 Radar Search Receiver*	*1973-74*
1714.253	Model 100 Naval DF Set	646
1715.253	*MM/SPR-A Naval Radar Intercept System*	*1980-81*
1716.053	WJ-1240 Receiving System	334
1717.053	*AN/GLR-9 Receiving System*	*1982-83*
1718.053	*ND210 and ND260 Monitoring Equipment*	*1974-75*
1719.263	*Sonar Coverage Computer*	*1982-83*
1720.453	IP64-MD64 Sonars	558
1721.453	*BI-68 Sonobuoy*	*1978*
1722.353	FALCO Submarine Locating System	558
1723.453	BI Series Sonobuoys	558
1724.253	DUBA 25 (TSM 2400) Sonar	551
1725.253	TSM 2314 Diodon Sonar	551
1726.453	DUBM 41B Side-Looking Sonar	553
1727.193	Driver's Periscope Type BM 8005	361
1728.453	*AN/WQX-1 Underwater Damage Assessment TV system*	*1973-74*
1729.453	*TC-125-SIT Low Light-level TV Camera*	*1973-74*
1730.393	RGS 2 Optical Sighting System	876
1731.393	*Type 664 HUD-WAS*	*1979-80*
1732.153	*Type LAR Laser Rangefinder*	*1979-80*
1733.153	*Type LAT Tank Laser Rangefinder*	*1981-82*
1734.193	*Laser Gunnery Trainer*	*1977*
1735.393	*Submarine Signal Simulator*	*1982-83*
1736.393	TAV-38 Laser Rangefinder	871
1737.193	*APX M.396 Stabilised Sight*	*1985-86*
1738.331	*Viper Air-to-Air Missile*	*1976*
1739.461	DLT-D-3A Submarine Torpedo Fire Control System	410
1740.353	Searchwater Surveillance Radar (ARI 5980)	788
1741.353	*DAX-100/200 Series Airborne Radar*	*1974-75*
1742.193	Type 306 Laser Designator/Ranger	384
1743.263	SARIE - Selective Radar Identification Equipment	666
1744.253	AN/SPS-6 Radar	637
1745.253	AN/SPS-30 Radar	638
1746.253	AN/SPS-40 Radar	638
1747.253	AN/SPS-43 Radar	638
1748.253	AN/SPG-55 Radar	636
1749.253	*SPG-49 Radar*	*1983-84*
1750.253	*SPG-56 Radar*	*1974-75*
1751.253	*AWS-2 Naval Radar*	*1985-86*
1752.253	RN Type 901 Radar	629
1753.253	RN Type 992Q Radar	631
1754.253	S 810/820 Naval Surveillance Radars	632
1755.453	Type 162M Sideways-Looking Sonar	565
1756.314	Air-to-Surface Missiles Table	Analysis (916)
1757.334	Air-to-Air Missiles Table	Analysis (915)
1758.211	*US Navy Cruise Missile*	*1979-80*
1759.411	*SLCM Sea-Launched Cruise Missile*	*1980-81*
1760.221	SS-N-9 (Siren) Shipborne and Submarine Launched Surface-to-Surface Missile	457
1761.211	*SSN-10 Shipborne Surface-to-Surface Missile*	*1978*
1762.221	*SSN-11 Shipborne Surface-to-Surface Missile*	*1978*
1763.224	Tactical Shipborne Surface-to-Surface Missiles Table	Analysis (910)
1764.231	*Naval Blowpipe Missile System*	*1974-75*
1765.353	AN/APG-63 Airborne Radar	793
1766.311	ALCM (AGM-86) Air-Launched Cruise Missile	721
1767.311	*AMX Airborne Missile X*	*1974-75*
1768.311	*Close Air Support Weapon System (Laser Maverick)*	*1974-75*

Entry number	Subject	Page No/ Last edition*
1769.311	HARM (AGM-88A) High Speed Anti-Radar Missile	724
1770.321	Air-Launch Exocet (AM 39)	709
1771.311	HOT Air-to-Surface Missile	711
1772.311	*American Air-to-Surface Improvement Programme*	1977
1773.311	*SAGMI Surface Attack Guided Missile*	1976
1774.331	Sky Flash Air-to-Air Missile	701
1775.193	*SABCA Talafit*	1982-83
1776.193	*Tank Gunnery Trainer Type DX. 150*	1982-83
1777.193	*BT 33 Artillery Fire Control Simulator*	1981-82
1778.193	*Link-Miles Simulators*	1982-83
1779.281	*Vega Tactical Information Unit Type I*	1978
1780.281	*Vega Tactical Information Unit Type II*	1978
1781.353	*APS-96 Airborne Radar*	1980-81
1782.353	*APS-111 (XN-1) Airborne Radar*	1980-81
1783.353	*APS-120 Airborne Radar*	1986-87
1784.353	AN/APS-503 and LASR-2 Radars	778
1785.353	Foxhunter Airborne Interception Radar	786
1786.453	*PAP 104 Mine Disposal Weapon*	1987-88
1787.453	*Mk 1C Active Sonobuoy T17 164*	1976
1788.453	*Plessey Underwater Programme*	1979-80
1789.453	AN/SQQ-14 Minehunting Sonar	587
1790.453	Mulloka Sonar	540
1791.453	Project Barra	540
1792.453	AN/SQS-505 Sonar	542
1793.353	*ORB-31 Airborne Radar*	1985-86
1794.353	*Type AF530 Gyro-Stabilised Sight*	1981-82
1795.393	*TAV 19 Helicopter Laser Rangefinder*	1976
1796.193	*APX M401 Optical Sight and Rangefinder*	1976
1797.193	*APX M409 Optical Sight and Rangefinder*	1986-87
1798.193	*IPY & Portable Laser Target Marker*	1976
1799.293	*TMV 26 Naval Laser Rangefinder*	1976
1800.193	*Type HV7X200AT Night Sight*	1987-88
1801.193	Type TP1MS Night Periscope	373
1802.193	*RZ 502 Thermal Imaging Equipment*	1976
1803.193	AN/VVS-1 Tank Laser System	397
1804.193	LP3 Laser Rangefinder	374
1805.393	Tornado Head-Up Display	881
1806.363	AN/ASX-1 Target Identification System, Electro-Optical (TISEO)	889
1807.453	AN/SQS-56/DE1160	586
1808.391	*Halbran Target Missile*	1982-83
1809.351	*Argus Reconnaissance System*	1984-85
1810.391	*Dornier Aerial Target System (DATS)*	1982-83
1811.391	*AEL Falcon Drone*	1976
1812.391	*LAST Low-Altitude Supersonic Target (XBQM-8)*	1980-81
1813.391	*Beech Model 1092*	1974-75
1814.391	*Beech Model 1089 Army MQM-107A/C VSTT*	1987-88
1815.391	*Beech Model 1070 HAST*	1979-80
1816.351	*Model 235 (YQM-98A) Compass Cope R Drone*	1977
1817.351	*Compass Cope B Drone (YQM-94A)*	1977
1818.311	*Model 234 (BGM-34A/B) Ground Attack Drones*	1979-80
1819.391	*PQM-102 Target Drone*	1987-88
1820.351	*RPAOD Remotely-Piloted Aerial Observation Designation System*	1977
1821.391	*Kaman RPV & Drone Programme*	1977
1822.391	*Northrop NV-123 VSTT*	1976
1823.281	*Italian Navy Multi-Purpose Rocket Control System (ULCAR)*	1982-83
1824.281	AGIS Combat Information System	411
1825.281	SATIR Action Information System	412
1826.481	*CICAS Computer Integrated Command and Attack System*	1979-80
1827.481	*TCS-9 Torpedo Control System*	1983-84
1828.261	Marconi/BAe Lightweight Gunfire Control System (Sapphire)	439
1829.281	*Marconi Weapon Control Systems*	1982-83
1830.281	ADAWS6 and 10 Action Data Automation Weapon Systems	437
1831.281	*Mine Countermeasures CAAIS*	1985-86
1832.261	*Gun Fire Control System Mk 37*	1981-82
1833.261	*Gun Fire Control System Mk 68*	1979-80
1834.281	Aegis Command and Control System	448
1835.281	Fire Control System Mk 92	443
1836.303	XM30 Aircraft Armament Sub-System	766
1837.303	*XM 41 Armament Subsystem*	1976
1838.303	*XM 59 Armament Subsystem*	1976
1839.303	*CASA Type 06.070 Rocket Launcher*	1987-88
1840.303	*CASA Type 18.037 Rocket Launcher*	1987-88
1841.303	*CASA Type 18.070 Rocket Launcher*	1987-88
1842.303	*CASA Type 54.037 Rocket Launcher*	1987-88
1843.303	Lanco RC-06-100	747
1844.393	*EMRU No 22*	1987-88
1845.393	*Westland-Frazer Nash High Strength MACE*	1987-88
1846.303	M230 30mm Chain Gun Weapon	769
1847.303	*Fuel/Air Explosive Weapons*	1984-85
1848.303	*HSM Hard Structure Munitions*	1981-82
1849.303	*Gator Anti-tank Weapon*	1982-83
1850.314	*Air-to-Surface Tactical Munitions Table*	1987-88
1851.302	*MiG-23 Flogger Aircraft*	1974-75
1852.302	*Backfire Variable Geometry Bomber Aircraft*	1974-75
1853.302	*Hind Helicopter*	1974-75
1854.302	*Shackleton AEW Mk 2 Aircraft*	1974-75
1855.302	*Il-38 May Maritime Patrol Aircraft*	1974-75
1856.064	*Table of Military Computers*	1976
1857.253	*SR-1-A Radar Search Receiver*	1982-83
1858.253	*VR-30 Series Radar Search Receivers*	1976
1859.253	*VR-1-B/C Radar Search Receivers*	1982-83
1860.253	*VR-2 Portable Radar Search Receiver*	1980-81
1861.053	GTE EW Antennas	329
1862.353	AN/ALR-47 Radar Homing and Warning System	853
1863.363	*DSA-20 Digital Signal Analyser*	1979-80
1864.353	*Jaguar Reconnaissance Pod*	1982-83
1865.363	*Vinten Reconnaissance Pods*	1982-83
1866.353	*Vinten Airborne Cameras*	1982-83
1867.173	*Vinten Photo-Interpretation Equipment*	1982-83
1868.163	*Patricia Reconnaissance Data Processing System*	1982-83
1869.173	*RCA Laser Beam Image Reproducers*	1976
1870.173	*RCA Laser Radar Recorder*	1974-75
1871.353	*RBV Camera System*	1976
1872.153	*R-200 Instantaneous Frequency Measuring Receiver*	1976
1873.153	*Model R-1437 Receiver*	1979-80
1874.153	*RD-280 Receiver and Display*	1981-82
1875.153	*R-281 Search Receiver*	1981-82
1876.053	HWR-2 Radar Warning Receiver	668
1877.253	*EMI Signal Selection and Pulse Analysis Equipment*	1982-83
1878.353	*F126 Reconnaissance Camera*	1982-83
1879.353	*F135 Reconnaissance Camera*	1982-83
1880.353	*Fairchild Panoramic Cameras*	1982-83
1881.353	*F-639 Medium/High Altitude Mapping Camera*	1982-83
1882.353	*Minipan Camera and Mount*	1981-82
1883.353	*Pave Spike Video Film Recorder*	1982-83
1884.391	*Condor Target Drones*	1976
1885.391	*Petrel Target Drone*	1984-85
1886.153	*FSRMXB Series Broadband Receiving Equipment*	1980-81
1887.153	*FAIRS Sigint System*	1976
1888.153	*Model 3600 Microwave Receiving System*	1976
1889.253	DRBI 23 3-D Naval Radar	601
1890.253	DRBR 51 Naval Radar	602
1891.253	DRBV 22 Naval Radar	602
1892.253	DRBV 20 Naval Radar	602
1893.253	DRBV 13 Naval Radar	602
1894.253	DRBC 32 Naval Radar	601
1895.253	*S604 HN Naval Radar*	1979-80
1896.253	*PTR461 Shipborne IFF Transponder*	1980-81
1897.253	Pop Group Fire Control Radar	622
1898.351	*US Navy Ocean Surveillance Programme*	1978
1899.351	*Pave Strike Programmes*	1978
1900.253	Other Soviet Naval Radars	626
1901.053	Rohde & Schwarz Direction Finding Equipment	310
1902.053	*EZF/EZFU Radio Reconnaissance & Monitoring System*	1981-82
1903.053	*ND210 & ND260 Radio Monitoring Equipment*	1981-82
1904.393	AN/ALE-37 Chaff Dispenser	839
1905.393	AN/ALE-39 ECM Dispenser	839
1906.393	AN/ALE-40 Countermeasures Dispenser System	839
1907.393	*ALQ-87 ECM Pod*	1980-81
1908.393	AN/ALQ-99 ECM Equipment	842
1909.393	*ALQ-100 ECM Equipment*	1986-87
1910.393	AN/ALQ-101 ECM Pods	843
1911.393	AN/ALQ-119 ECM Pod	844
1912.393	AN/ALQ-123 Infra-Red Countermeasures	844
1913.393	AN/ALQ-126A/B ECM Equipments	845
1914.393	AN/ALQ-130 ECM Equipment	845
1915.393	AN/ALQ-131 ECM Pod	845
1916.393	AN/ALQ-133 Elint Equipment	846
1917.353	AN/ALR-45 and AN/ALR-45F Radar Warning and Control Systems	852
1918.353	AN/ALR-50 ECM Equipment	853

Entry number	Subject	Page No/ Last edition*
1919.353	AN/ALR-52 ECM Receiver	853
1920.281	Dardo Fire Control System	415
1921.281	*Lightweight Director Mk 80*	*1980-81*
1922.481	Trident Fire Control System Mk 98	444
1923.281	*Lockheed Sharpshooter*	*1984-85*
1924.281	*Radar Gun Fire Control System Ex 93*	*1982-83*
1925.281	Passive Fire Control System Mk 105 (TAC)	445
1926.281	*Updated Gun FCS Mk 68*	*1982-83*
1927.281	*9AU-300 Fire Control System*	*1983-84*
1928.281	Mk 23 Target Acquisition System (TAS)	441
1929.481	*MSI-7OU Submarine Fire Control System*	*1983-84*
1930.281	*CLIO-BS Otomat Fire Control System*	*1977*
1931.253	AN/SPQ-9 Radar	636
1932.253	AN/SPG-60 Radar	636
1933.453	Calypso III Submarine Radar TRS 3100	606
1934.235	*ELR-3 IFF Transponder*	*1984-85*
1935.253	Orion RTN-20X Naval Radar	609
1936.253	Orion RTN-30X Naval Radar	609
1937.253	*Shipborne OTH Radar*	*1979-80*
1938.253	*SPQ-5 Radar*	*1977*
1939.253	AN/SPS-49 Radar	639
1940.253	AN/SPS-63 Radar	641
1941.253	Racal-Decca Surveillance and Navigation Radars	627
1942.293	RCM Series ECM Equipment	666
1943.153	Atlas Instrumentation Tracking Radar	226
1944.153	Savoie Instrumentation Tracking Radar	227
1945.153	*Omera Meteorological Radars*	*1983-84*
1946.153	*Prowler Infantry Radar*	*1979-80*
1947.153	AN/PPS-15A Infantry Radar	284
1948.153	*Over-the-Horizon Radar*	*1979-80*
1949.153	Cobra Dane Radar	276
1950.153	*TPQ-31 Mortar Locating Radar*	*1979-80*
1951.153	*Sentinel RQT-10X Portable Infantry Radar*	*1978*
1952.153	*RAT-8S Mobile Coastal Surveillance Radar*	*1983-84*
1953.153	RAT-31S Transportable 3-D Radar	245
1954.153	*DR 532 Ballistic Radar System*	*1979-80*
1955.153	*Isidor Muzzle Velocity Radar*	*1978*
1956.153	*TX 745 IFF Interrogator*	*1978*
1957.153	Giraffe Search Radar	250
1958.153	Elta Surface Radars	242
1959.153	Argos-10 Early Warning Radar	245
1960.353	ORB-31 Radar System	774
1961.353	*DRAA 2B Airborne Radar*	*1981-82*
1962.353	ESD 3300/3400 IFF/SIF Transponders	778
1963.353	*Antilope IV Airborne Radar*	*1980-81*
1964.353	PS-46/A Multi-mode AI Radar	783
1965.353	AN/APS-138/125 Airborne Radars	808
1966.353	AN/APQ-113 Radar	799
1967.353	AN/APQ-114 Radar	799
1968.353	AN/APQ-144 Radar	800
1969.353	AN/APS-115 Radar	804
1970.353	AN/APQ-99 Radar	799
1971.353	*UK AEW Radar*	*1986-87*
1972.353	Blue Fox Radar	792
1973.253	LW08 Naval Surveillance & Weapon Direction Radar	612
1974.281	*M9 Torpedo Fire Control System*	*1976*
1975.281	MSI-80S Fire Control System	425
1976.153	AN/TPQ-37 Weapon Locating Radar	290
1977.253	TSM 2021B Minehunting Sonar	554
1978.253	*Piranha (TSM 2140) Attack Sonar*	*1981-82*
1979.353	*Crouzet Magnetic Anomaly Detector*	*1983-84*
1980.393	OA-5154/ASQ Automatic MAD Compensation System	544
1981.363	AQS 901 Sonobuoy Processing and Display System	562
1982.363	AN/AYA-8B ASW Data Processing System	581
1983.453	AN/BQQ-5 Sonar	593
1984.453	AN/BQR-15 Sonar	593
1985.453	AN/BQR-21 DIMUS Sonar	593
1986.453	AN/BQS-13 Sonar	594
1987.453	AN/SSQ-41B Sonobuoy	595
1988.453	AN/SSQ-47B Sonobuoy	595
1989.453	*AN/SSQ-50 CASS Sonobuoy*	*1983-84*
1990.453	AN/SSQ-53A DIFAR Sonobuoy	595
1991.453	AN/SSQ-62 DICASS Sonobuoy	596
1992.393	*Alkan Light Aircraft and Helicopter Suspensions*	*1981-82*
1993.393	*ML Ejector Release Unit No 120*	*1987-88*
1994.393	*ML Angled Twin Store Carrier (YDC)*	*1981-82*
1995.303	Oerlikon SNORA 81mm Rocket	750
1996.303	GAU-8/A 30mm Gun System	759
1997.303	*RBU-1/A Pave Rocket*	*1981-82*
1998.303	*Rockeye Cluster Bomb Mk 20*	*1983-84*
1999.303	Durandal Penetration Bomb	737
2000.441	SST 4 Special Surface Target Torpedo	518
2001.441	*Submarine-launched Torpedo Type G6e*	*1974-75*
2002.441	*Submarine-launched Torpedo Type G62ef*	*1978*
2003.441	A 184 Submarine or Surface-Launched Torpedo	521
2004.441	A 244/S (Mod.0 and Mod.1) Lightweight Torpedo	522
2005.102	*Main Battle Tank Leopard (Italy)*	*1979-80*
2006.102	*Modified M-47 Tank (Italian)*	*1977*
2007.102	*Wheeled Armoured Vehicles (Italian)*	*1973-74*
2008.102	*Chinese Armoured Personnel Carriers*	*1973-74*
2009.102	*Israeli Armoured Fighting Vehicles*	*1977*
2010.111	CSS-1 MRBM (Chinese designation DF-2)	2
2011.102	*Patton Tank (Israel)*	*1979-80*
2012.103	*Albanian Army Ordnance*	*1976*
2013.103	*45mm Anti-tank Gun M-37*	*1982-83*
2015.102	*Austrian Medium Tank*	*1976*
2016.181	*Hubcap*	*1978*
2017.391	*Jindivik*	*1987-88*
2018.241	Anti-Submarine Mortar System	511
2019.153	Chinese Military Radar	219
2020.351	*Epervier Surveillance Drone*	*1982-83*
2022.103	*105mm Field Howitzer 1840*	*1977*
2023.103	*106mm Recoilless Rifle M40A1 (Towed)*	*1977*
2024.103	*RM-130 Artillery Rocket Launcher*	*1982-83*
2025.111	*Anti-Tank Weapon System*	*1976*
2026.103	*Artillery Rockets*	*1980-81*
2027.203	*Shipborne Bombardment Rocket R-115*	*1976*
2028.102	*Brazilian Armoured Vehicles*	*1976*
2029.103	*Twin 30mm Anti-aircraft Cannon M-5359*	*1982-83*
2030.103	*Spanish Ordnance*	*1976*
2031.103	*105mm - 26 Howitzer*	*1982-83*
2032.103	*75mm Light Field Gun*	*1982-83*
2033.111	*Artillery Rockets*	*1984-85*
2034.351	*Canadair AN/USD-501 Reconnaissance Drone*	*1987-88*
2035.221	*Canadian Sea Sparrow Missile System*	*1976*
2036.231	*Canadian Sea Sparrow Missile System*	*1983-84*
2037.103	*122mm Field Gun M-60*	*1982-83*
2038.453	*AN/SQS-507 Helen Sonar*	*1982-83*
2039.103	*105mm Light Field Howitzer M-6137 & M-37/10*	*1982-83*
2040.183	*Land Navigation Systems LNS 101 & 102*	*1970-71*
2041.103	*95mm Recoilless Anti-Tank Gun M-58*	*1978*
2042.103	*Finnish Army Mortars*	*1978*
2043.103	*55mm Recoilless Anti-Tank Grenade Launcher M-55*	*1978*
2044.103	*158mm Artillery Rocket System BR 51 GS*	*1981-82*
2046.261	CCS-280 Command & Control Systems (DDH-280) Class	408
2047.261	Underwater Combat System (UCS) 257/280	408
2048.111	*Chinese ICBM*	*1976*
2049.111	CSS-2 IRBM (Chinese designation DF-3)	2
2050.131	*Note on Chinese Surface-to-Air Weapons*	*1984-85*
2051.411	*Note on Chinese Subsurface-to-Surface Weapons*	*1979-80*
2052.181	Chinese Air Defences	40
2053.153	*Telsa Radars*	*1972-73*
2054.111	*Ballistic Rockets (Egypt)*	*1972-73*
2055.111	ACL-STRIM Anti-Tank Rocket Launcher	138
2056.153	Adour II Instrumentation Tracking Radar	225
2057.241	Anti-Submarine Weapon System	513
2058.153	*Airfield Surveillance Radar TH.D 1021*	*1972-73*
2059.153	*ATC Approach Radar TRS 2056*	*1979-80*
2060.153	Antares Heightfinding Radar	225
2061.183	*Automatic Gyro Compass CDN 10 & 20*	*1970-71*
2062.153	*Aquitaine II Tracking Radar*	*1974-75*
2063.153	Artois Instrumentation Tracking Radar	226
2064.153	*Automatic Tracking Radar TH.D 1213*	*1972-73*
2065.153	Bearn Instrumentation Tracking Radar	226
2066.153	*Arabelle Portable Transponder*	*1978*
2067.153	*Bretagne Tracking Radar TH.D 1801*	*1971-72*
2068.103	*ACLAPX 80 Light Collective Anti-Tank Weapons*	*1976*
2069.153	*Champagne Tracking Radar*	*1971-72*
2070.453	*Acoustic Range*	*1969-70*
2071.111	*ACRA Anti-Tank Weapon System*	*1976*
2072.453	*DG Ranges*	*1970-71*
2073.153	*Domino Surveillance Radar*	*1972-73*
2074.131	Crotale Low-Altitude Ground-to-Air Weapon System (TSE 5000)	169
2075.293	*Degaussing Equipment*	*1969-70*
2076.393	*CTB 2530 Target Towing System*	*1970-71*
2077.393	*ECA 57 Glider Target*	*1970-71*
2078.393	*ECA 58 Towed Target*	*1970-71*
2079.491	*ED 28 Exercise Mine*	*1969-70*
2080.151	*Eldorado-Mirador Fire Control System*	*1973-74*
2081.111	*Entac Anti-Tank Missile*	*1978*

Entry number	Subject	Page No/ Last edition*
2082.153	Eldorado Fire Control Radar TH.D 1229	1973-74
2083.153	Domino 30 Radar	1979-80
2084.153	Domino 40 Radar	1973-74
2085.103	ARPAC 68mm Anti-Tank Rocket Launcher	1977
2086.103	120mm Mortar M-65	1978
2087.103	120mm Heavy Mortar Am-50	1978
2088.103	120mm Light Mortar M-60	1978
2089.103	120mm Rifled Mortar	1978
2090.103	SARPAC 68mm Anti-Tank Rocket Launcher	1978
2091.103	MAS Type A Individual Anti-Tank Weapon	1973-74
2092.453	Harbour Protection Equipment	1977
2093.193	IR Aiming Telescope OB-17A	1979-80
2094.193	IR Binocular Aiming Telescope OB-23-A	1979-80
2095.111	Harpon Battlefield Missile	1981-82
2096.441	L4 Airborne Acoustic Torpedo	515
2097.441	Light Acoustic Torpedo	1972-73
2098.153	Light Precision Approach Radar TH.D 1013	1972-73
2099.193	IR Periscope OB-11-A	1979-80
2100.153	Long-Range Radar LP 23	1976
2101.193	Rangefinder M292	1977
2102.153	Matador 2000 3-D Radar TRS 2215	1978
2103.193	Binocular Telescope M267 and Prism Head M270	1977
2104.153	Medium Range Radar TA 23 (TRS 2055)	1977
2105.193	Telescope-Periscope M112/3	1985-86
2106.181	Midas Air Defence System	1983-84
2107.193	Aiming Telescope M262	1977
2108.193	IR Aiming Telescope (M267 Modified)	1977
2109.193	IR Day/Night Driving Periscope	1979-80
2110.131	Javelot Surface-to-Air Weapon System	1981-82
2111.231	Crotale Naval Surface-to-Air Weapon System (TSE 5500)	463
2112.491	MCC 23C War Mine	1969-70
2113.491	MCT 15 War Mine	1969-70
2114.231	Catulle Surface-to-Air Weapon System	1980-81
2115.181	STRIDA-Compact Air Defence System	1983-84
2116.153	Mirador Radar	1973-74
2117.153	Mirador II	1982-83
2118.121	Exocet Coastal Defence Missile System	161
2119.153	Picador Mobile 3-D Radar	231
2120.153	Spartiate Recovery Radar TRS 2300	231
2121.153	ATC Approach Radar TRS 2060	1984-85
2122.231	Hirondelle Surface-to-Air Weapon System	1977
2123.351	Nord 510 Flying Platform	1974-75
2124.391	Nord C 30 Drone	1973-74
2125.391	Aérospatiale CT.20	1987-88
2126.391	Nord CT 41 Drone	1972-73
2127.351	Nord R20 Drone	1979-80
2128.441	L5 Multi-Purpose Torpedo	515
2129.153	Oeil Noir 1 (DR-VC-1A) Radar	1982-83
2130.111	Pluton Battlefield Support Missile	119
2131.153	Olifant Radar	1979-80
2132.153	Olifant II Radar	1979-80
2133.153	Precision Approach Radar TH.D 1012	1978
2134.103	RAP 14 Artillery Rocket System	1982-83
2135.203	RAP 14 Naval Weapon System	1985-86
2136.153	Racine Non-Recurrent Pulsed Doppler Radar	1973-74
2137.153	Rapace (DR-VT-3) AFV Surveillance Radar	1980-81
2138.131	AMX30 SA 30mm Twin Gun AA Weapon System	1982-83
2139.111	SS 11 Battlefield Missile	137
2140.111	SS 12 Battlefield Missile	1974-75
2141.153	Rasit E Ground Surveillance Radar	232
2142.153	Rasit Battlefield Surveillance Radar	1973-74
2143.153	Rasura (DR-PT-2A) Battlefield Surveillance Radar	1980-81
2144.441	R3 Experimental Lightweight Torpedo	1973-74
2145.111	SSBS Medium-Range Ballistic Missile	1982-83
2146.441	Z16 Submarine-Launched Torpedo	516
2147.103	105mm Tank Gun D1504	1982-83
2148.103	Tank Turret Type FL-12	1982-83
2149.103	Tank Turret Type FL-10	1982-83
2150.103	20mm AA Turret TG-521-F1	1977
2151.103	60mm Vehicle Mortar (60-MC-A1)	1978
2152.103	Attila Mk II Rocket	1982-83
2152.153	RALF Low-Altitude Acquisition Radar	1976
2153.153	Rasit 72-B Battlefield Surveillance Radar	1976
2154.153	Rasit 72-C Battlefield Surveillance Radar	1976
2155.153	TRS 2201 Three-Dimensional Radar	235
2156.153	Tiger S (TRS 2100 Series) Radar	235
2157.153	LMT Ground-Based IFF Interrogators	1984-85
2158.102	Panhard New Generation AFVs	1976
2159.193	Rangefinder M.208 for AMX-30 Tank	1977
2160.153	Volex III Surveillance Radar TH.D 1945	1979-80
2161.181	STRIDA Air Defence System	41

Entry number	Subject	Page No/ Last edition*
2162.283	Optical Fire Control and Target Designation Equipment	1974-75
2163.111	SSBS Intermediate Range Ballistic Missile Type S-3 D/TN-61	4
2164.131	M-3 VDA 20mm AA Gun System	192
2165.441	Torpedo Mk 27	1982-83
2166.181	DACTA ATC & Air Defence System	1984-85
2167.181	Air Defence System Chile	1978
2168.103	130mm Field Gun Type 59	1982-83
2169.103	76mm Light Field Gun Type 54	1978
2170.103	75mm Recoilless Rifles Types 52 & 56	1978
2171.103	57mm Anti-Tank Gun Type 55	1978
2172.103	37mm Light Anti-Aircraft Gun Type 55	1978
2173.103	82mm Mortar Type 53	1978
2174.103	90mm Mortar Type 97	1978
2175.103	120mm Mortar Type 5355	1978
2176.103	160mm Mortar Type 60	1978
2177.181	Introduction Land-Based Air Defence Systems	1976
2178.441	Seal/Seeschlange Torpedoes	517
2179.103	155 & 105mm Field Howitzers	1978
2180.103	90mm SP Anti-Tank Gun	1976
2181.111	Cobra 2000 Anti-Tank Missile	138
2182.131	20mm Twin Gun Anti-Aircraft System	194
2183.351	Dornier DO 32K Kiebitz Drone	1979-80
2184.103	Armbrust-300 Infantry Anti-Tank Weapon	1978
2185.193	Passive Night Viewing Unit NYX 2002	1973-74
2186.153	IR Gunsight Type BM 8001	1980-81
2187.153	IR Gunsight Type BM 8004	1979-80
2188.111	Mamba Portable Anti-Tank Weapon System	139
2189.102	FUG (OT-65) Armoured Personnel Carrier	1979-80
2191.103	New High-performance Guns	1976
2192.103	155mm SP Howitzer M-109G	1982-83
2193.103	155mm SP Howitzer M-109 Modified	1982-83
2194.103	RS 80 Artillery Rocket System	1980-81
2195.103	155mm Towed Howitzer FH 70	1982-83
2196.103	155mm SP Howitzer SP 70	1982-83
2197.103	110mm Tank Gun L 10	1976
2198.153	Zenda Battlefield Radar Project	1973-74
2199.221	Ship Martel Surface-to-Surface Missile	1972-73
2200.111	Introduction Strategic Surface-to-Surface Missiles	1982-83
2201.004	Table of Principal Missiles	1970-71
2202.114	Strategic Missiles Table	Analysis (906)
2203.114	Table of Anti-Tank Missiles	1971-72
2204.114	Table of Tactical Surface-to-Surface Missiles	1971-72
2205.134	Table of Strategic Surface-to-Air Missiles	1971-72
2206.134	Table of Tactical Surface-to-Air Missiles	1971-72
2207.114	Table of Surface-to-Sub surface Missiles	1971-72
2208.314	Table of Strategic Air-to-Surface Missiles	1971-72
2209.314	Table of Tactical Air-to-Surface Missiles	1971-72
2210.334	Table of Air-to-Air Missiles	1971-72
2211.181	ADGE (IPG) Air Defence System	1973-74
2212.111	HOT Anti-Tank Missile	140
2213.131	Helip Hawk Improvement	1976
2214.111	Atlas Anti-Tank Missile	1972-73
2215.111	MILAN Anti-Tank Missile	142
2216.231	Seacat-Bofors Gun Launcher	1972-73
2218.131	Roland Mobile AA Weapon System	171
2219.421	Submarine-Launched Exocet	1976
2221.411	Introduction Sub surface-to-Surface Weapons	1976
2222.444	Table of Subsurface-to-Subsurface Missiles	1971-72
2223.304	Table of Drones	1971-72
2224.304	Table of Towed Targets and Tow Systems	1970-71
2225.304	Table of Miss-Distance Indicators and Scoring Systems	1970-71
2226.111	Israeli Battlefield Support Missiles	1978
2226.181	Israel Air Defence System	44
2227.111	Anti-Tank Weapon (Italy)	1972-73
2228.231	Albatros Naval Defence System	469
2229.281	ARGO Naval Gun and Guided Missile Control Systems	1971-72
2230.281	Autonomous Naval Fire Control System	1979-80
2231.111	Artillery Rockets (Italy)	1972-73
2232.111	Battlefield Support Weapons	1973-74
2234.283	Gun Remote Control Equipment	1974-75
2235.131	Indigo Mobile AA Weapon System	1978
2236.391	Meteor Drones	1981-82
2237.153	RAT 10S Coastal Surveillance Radar	1983-84
2238.111	Mosquito Anti-Tank Missile	1978
2239.153	Argos RAT-5C Coastal Defence Radar	1972-73
2240.221	Nettuno (Sea Killer Mk 1)	1977
2241.221	Sea Killer Missile Systems	1976
2242.221	Sea Killer Mk 3 Shipborne Surface-to-Surface Missile	1977

Entry number	Subject	Page No/ Last edition*
2243.153	*RAT-6L Short-Range Low Altitude Coverage Radar*	*1979-80*
2244.153	*RIS-3E Monopulse Tracking Radar*	*1974-75*
2245.153	*RIS-4CA Tracking Radar*	*1974-75*
2246.153	*RIS-5X Drone Tracking Radar*	*1974-75*
2247.293	*Albatros Missile Launcher*	*1974-75*
2248.153	*Selenia CivilMilitary ATC Radars*	*1987-88*
2249.153	*Sentinel RQT-9X Portable Infantry Radar*	*1987-88*
2250.131	Spada Point Defence Missile System	173
2251.111	*Sparviero Anti-Tank Missile*	*1983-84*
2252.103	*Attila Mk II Rocket*	*1982-83*
2253.221	*Sea Killer Mk 2*	*1985-86*
2254.181	BADGE Air Defence System	46
2255.103	*Type 30 Artillery Rocket (Japan)*	*1982-83*
2256.181	*Second Generation Air Defence System*	*1979-80*
2257.111	Folgore Anti-Tank Weapon	144
2258.203	Breda Compact Twin 40mm L/70 Naval Mounting Type 70	490
2259.203	*Twin 40mm Bofors Mountings*	*1973-74*
2260.203	*Single 40mm Bofors Mountings*	*1973-74*
2261.203	*Twin 35mm Anti-Aircraft Gun Mounting*	*1979-80*
2262.111	KAM-3D Anti-Tank Missile (Type 64)	145
2263.111	KAM-9 (Type 79) Battlefield Missile	145
2264.391	*KAQ-5 Drone*	*1972-73*
2265.103	*Triple 20mm Anti-Aircraft Cannon M-55*	*1982-83*
2266.111	128mm Rocket Launcher M-63	135
2267.103	*105mm Field Howitzer M-56*	*1982-83*
2268.103	*76.2mm Mountain Gun M-48 B1*	*1982-83*
2269.103	*82mm Recoilless Anti-Tank Gun M-60*	*1978*
2270.103	*120mm Heavy Mortar UDM-52*	*1978*
2271.103	*81mm Medium Mortar M-68*	*1978*
2272.103	*81mm Medium Mortar M-31*	*1978*
2273.103	*60mm Mortar M-57*	*1978*
2274.103	*88mm Coastal Defence Gun (Flak 36)*	*1982-83*
2276.103	*Yugoslav Ordnance*	*1976*
2277.103	*44mm Recoilless Anti-Tank Grenade Launcher M-57*	*1978*
2278.203	*Combined Chaff/Illuminating Rocket Launcher*	*1979-80*
2279.203	*SCLAR Multi-Purpose Naval Rocket System*	*1979-80*
2280.153	*RAT-21C Low-Altitude Surveillance Radar*	*1974-75*
2281.153	*RAT-20L Long-Range Surveillance Radar*	*1977*
2282.153	*RAT-7S Low Coverage Radar*	*1977*
2283.111	Introduction Tactical Surface Weapons	113
2284.111	*Introduction Anti-Tank/Assault Weapons*	*1981-82*
2285.111	*Introduction Battlefield Support Weapons*	*1981-82*
2286.153	*DR 810 Ballistic Radar*	*1976*
2287.131	*Indian Surface-to-Air Missile*	*1981-82*
2288.111	*Battlefield Support Missile (1973)*	*1977*
2289.111	*Indian Anti-Tank Missiles*	*1979-80*
2290.131	Introduction Land-Mobile Surface-to-Air Weapons	166
2291.153	*Indigo Fire Control Centre (CT-40-GM)*	*1977*
2292.131	*MIFLA Surface-to-Air Weapon System*	*1976*
2293.241	*Introduction Anti-Submarine Weapons*	*1976*
2295.441	*Introduction Torpedoes*	*1976*
2296.444	Torpedoes Table	Analysis (945)
2297.153	*Air Traffic Control Radars*	*1986-87*
2298.153	NPM-510 Mobile 3-D Radar	247
2299.203	*Norwegian Naval Guns*	*1979-80*
2300.151	*L45 Weapon Control System*	*1980-81*
2301.181	Flycatcher AA Fire Control System	82
2302.153	Integrated Radar System for 35mm AA Tank	84
2303.203	*Single 3in (76mm) 50-calibre Gun Mounting Mk 34*	*1982-83*
2304.203	40mm Automatic Gun SAK L/70 Bofors-Breda 144 MOD 76	495
2305.103	*57mm Recoilless Rifle*	*1978*
2306.103	*152mm Howitzer M-1846*	*1982-83*
2307.103	*105mm Howitzer M-1849*	*1977*
2308.103	*100mm Field Gun M53*	*1982-83*
2309.103	*85mm Field Gun M-52*	*1982-83*
2310.103	*82mm Recoilless Anti-Tank Rifle*	*1978*
2311.103	*57mm Anti-Aircraft Gun*	*1982-83*
2312.103	*155mm M-68 Howitzer*	*1982-83*
2313.103	*81mm Tampella Mortar Type C*	*1978*
2314.103	*60mm Tampella Mortar*	*1978*
2315.103	*81mm Tampella Mortar*	*1978*
2316.103	*120mm Light Tampella Mortar*	*1978*
2317.103	*120mm Tampella Mortar*	*1978*
2318.103	*160mm Tampella Mortar*	*1978*
2319.103	*155mm M-68 Self-Propelled Gun Howitzer (L-33)*	*1982-83*
2321.181	*South African Air Defences*	*1983-84*
2322.181	Combat Grande Air Defence System	49
2323.441	*Torpedo Type 41*	*1982-83*
2324.181	*Anti-Aircraft Fire Control Systems*	*1979-80*
2325.153	*PE-48/T AA Fire Control Radar*	*1981-82*
2326.153	*PS-171R AA Search Radar*	*1979-80*
2327.153	*PE-452T AA Fire Control Radar*	*1978*
2328.153	*PE-453T AA Fire Control Radar*	*1979-80*
2329.153	*C-Band Search Radar Type PS 70R*	*1974-75*
2330.103	*89mm Anti-Tank Rocket Launcher M-65*	*1978*
2331.103	*155mm Medium Howitzer*	*1982-83*
2332.103	*155mm Brigade Gun 77A*	*1982-83*
2333.103	*90mm Recoilless Anti-Tank Rifle*	*1978*
2334.103	*105mm Light Field Howitzer M40*	*1973-74*
2335.103	*105mm Light Field Howitzer 4140*	*1982-83*
2336.103	*57mm Anti-Aircraft Gun M54*	*1982-83*
2337.103	*40mm Anti-Aircraft Gun L60*	*1982-83*
2338.103	*20mm Anti-Aircraft Cannon M40-70*	*1982-83*
2339.103	*75mm Coastal Defence Gun*	*1982-83*
2340.153	Polish Radar	249
2341.131	Cactus	175
2342.103	*34mm Anti-Aircraft Gun M-38*	*1976*
2343.103	*105mm Tank Gun L74*	*1982-83*
2344.161	*Field Artillery Fire Control Equipment GFA 101*	*1981-82*
2345.161	*Field Artillery Computer Equipment SAAB ACE-30*	*1977*
2346.181	*Tank Fire Control Equipment Bofors Types S and C*	*1976*
2347.193	*Tank Laser Rangefinder*	*1978*
2348.131	RBS 70 Anti-Aircraft Missile System	213
2349.131	Bofors 40 mm BOFI Gun System	201
2350.181	*AVISTA Aircraft Tracking and Data Recording System*	*1976*
2351.193	*BT16 & BT17 Field Scorers*	*1979-80*
2352.193	*BT19 & BT19A Gunfire Simulators*	*1979-80*
2353.153	*L-Band Search Radar*	*1979-80*
2354.153	*DOMTI C-Band Search Radar*	*1973-74*
2355.153	*Peder II Coherent-on-Receive Fire Control Radar*	*1977*
2356.153	*UAR 1022 Track Only Radar*	*1976*
2357.153	UAR 1021 Combined Search & Track Pulse Doppler Radar	251
2358.181	STRIL Air Defence System	49
2359.153	*UAL 11105 Laser Rangefinder*	*1974-75*
2360.203	Single 40mm L70 Automatic Gun Mounting	496
2361.203	Single 57mm L70 Mk 1 Gun Mounting	495
2362.203	*Single 3in Automatic Gun Mounting*	*1985-86*
2363.111	Bantam Anti-Tank Missile (BS 53)	147
2364.203	*Shipborne 120mm L46 Automatic Gun Mounting*	*1969-70*
2365.203	*Single 120mm L50 Automatic Gun Mounting*	*1972-73*
2366.221	*RB 08A Anti-Ship Missile*	*1987-88*
2367.441	Torpedo Type TP 617	525
2368.241	Shipborne Two-Tube Anti-Submarine Rocket Launcher	525
2369.241	Shipborne Four-Tube Anti-Submarine Rocket Launcher	524
2370.131	*AA Tank System*	*1977*
2371.23	*Combined Sea Indigo/Sea Killer Mk 2 Weapon System*	*1972-73*
2372.121	RB 08 Coastal Defence Weapon	162
2373.181	Florida Air Defence Command and Control System	50
2374.131	*Anti-Aircraft Field System Type 2 ZLa353 MK*	*1982-83*
2375.131	*MICON Surface-to-Air Missile*	*1972-73*
2376.181	Super-Fledermaus	94
2377.181	Skyguard Anti-Aircraft Fire Control System	93
2378.181	BOFI Anti-Aircraft Fire Control System	88
2379.103	*30mm Automatic Cannon KCB*	*1982-83*
2380.103	*30mm Anti-Aircraft Mounting Type 661*	*1981-82*
2381.103	*20mm Automatic Cannon Type 804*	*1976*
2382.103	*20mm Automatic Cannon Type KAD*	*1982-83*
2383.103	*Single 20mm Mounting Type GAI-C*	*1981-82*
2384.103	*Twin 20mm Mounting Type GAI-DO1*	*1982-83*
2386.103	*DIRA 8mm Rockets*	*1982-83*
2387.103	*105mm Light Howitzer M-46*	*1982-83*
2388.103	*105mm Machine Gun M-35*	*1982-83*
2389.103	*150mm Medium Howitzer M-42*	*1982-83*
2390.411	UK Polaris Missiles	20
2391.103	*81mm Medium Mortars Mw 33*	*1978*
2392.103	*120mm Heavy Mortar Mw 41*	*1974-75*
2393.103	*90mm Anti-Tank Guns M-5057*	*1982-83*
2394.103	*93mm Portable Anti-Tank Rocket Launcher M-58*	*1978*
2395.203	30mm Twin Naval Gun Mounting Type GCM-A	497
2396.203	20mm Naval Gun Mounting Type A41A	499

Entry number	Subject	Page No/ Last edition*
2397.203	20mm Naval Gun Type GAM-BO1	499
2398.203	*140mm Rocket Launcher*	*1978*
2399.181	*Tank Fire Control System*	*1981-82*
2400.161	AMETS - Artillery Meteorological System	99
2401.293	*Frigate ASW Command & Operator Trainer*	*1982-83*
2402.193	*Helicopter ASW Crew Trainer*	*1982-83*
2403.193	*AFV No 52 Gunners Sight*	*1984-85*
2404.193	*AFV Nos 68/71/75 Commanders Sight*	*1984-85*
2405.193	*Passive Night Driving Periscope for Chieftain Tank*	*1984-85*
2406.131	Bloodhound Mk 2 Surface-to-Air Weapon System	183
2407.131	*Falcon SP 30mm AA Weapon System*	*1978*
2408.221	*Blowpipe Shipborne Surface-to-Surface Application*	*1972-73*
2409.131	Blowpipe Portable Anti-Aircraft Missile System	214
2410.153	*Cossor IFF Ground Equipment*	*1980-81*
2411.393	*Equipment Recovery Beacon*	*1969-70*
2412.153	*Green Archer Mortar Locating Radar*	*1979-80*
2413.153	*Firelight Target Illuminating Radar*	*1985-86*
2414.193	*Fighting Vehicle Driving Simulator*	*1970-71*
2415.161	FACE - Field Artillery Computer Equipment	100
2416.153	*Plessey AR-5D Air Defence Radar*	*1983-84*
2417.181	*Hubcap Air Defence System*	*1976*
2418.153	*Plessey IFF Mk 10 (SIF) Interrogation & Decoding Equipment*	*1981-82*
2419.253	*Plessey Naval IFF Mk 10 (SIF)*	*1980-81*
2420.153	*Plessey 430 Airfield Radar*	*1981-82*
2421.181	*Nomad Air Defence System*	*1976*
2422.153	*IRIS Infra-Red Intruder System*	*1977*
2423.183	*Rapier Command Guidance Equipment*	*1970-71*
2424.131	Rapier Mobile Low-Level Anti-Aircraft Missile System	183
2425.153	Rapier Search & Acquisition Radars	270
2426.181	*Marconi Tactical Air Defence Systems*	*1978*
2427.393	*Rushton Towed Target*	*1970-71*
2428.153	Scorpion Target Illuminating Radar	271
2429.181	*Marconi Static Air Defence Systems*	*1978*
2430.193	*Lightweight AFV Cupola No 16*	*1977*
2431.193	*AFV Cupola No 22*	*1977*
2432.193	*LF2 Laser Rangefinder for Chieftain Tank*	*1976*
2433.193	*IF Binocular Sights CU-10 and CU-13*	*1976*
2434.193	*Long-Range Viewer CU-15*	*1976*
2435.153	*Muzzle Velocity Measuring Equipment*	*1977*
2436.193	*MEL IR Sights for Tanks*	*1980-81*
2438.193	*Gun Control & Stabilisation System EC215*	*1973-74*
2439.153	Rapier Blindfire Tracking Radar	272
2440.441	Tigerfish Torpedo (Mk 24)	529
2441.421	*Underwater-to-Surface Guided Weapon System*	*1976*
2442.231	Sea Wolf Surface-to-Air Missile	477
2443.181	*Linesman Air Defence System*	*1980-81*
2444.181	UK Air Defence System	54
2446.231	*SLAM Close-Range Weapon System*	*1983-84*
2447.253	*Sonar Type 199*	*1979-80*
2448.441	Sting Ray Torpedo	529
2449.493	*Submarine Command Team Trainer*	*1982-83*
2450.111	Swingfire Anti-Tank Missile	151
2451.103	*105mm Tank Guns L7A1, L7A2*	*1977*
2452.103	*105mm Tank Guns L7A3*	*1977*
2453.103	*120 Tank Gun L11A3*	*1982-83*
2454.103	*Single 20mm Pedestal Mounting Type A41B for Land Rover*	*1974-75*
2455.103	*40mm Light Anti-Aircraft Gun*	*1982-83*
2456.153	*Plessey AR-15/2C Surveillance Radar*	*1984-85*
2457.153	*Marconi High-Power Static Radars*	*1982-83*
2458.153	*Marconi High-Power Static Surveillance Radars (Linear Feeds)*	*1987-88*
2459.153	*Marconi L-band Surveillance Radar Type S654*	*1979-80*
2460.131	*Thunderbird 2 Surface-to-Air Missile*	*1977*
2461.181	*Thunderbird 22 Air Defence System*	*1976*
2462.153	*Marconi Long-Range Heightfinding Radar Type 669*	*1987-88*
2463.153	*Marconi Mobile & Transportable Tactical Control Radars*	*1987-88*
2464.153	Marconi 800 Series Radars (Land-Based)	264
2465.131	Tigercat Close-Range Air Defence System	186
2466.153	ST850 Tigercat Radar	272
2467.183	*Gun Control System EC5 70*	*1973-74*
2468.183	*Power Traverse System for Light Fighting Vehicles*	*1974-75*
2469.183	*Weapon Stabilisation System*	*1974-75*
2470.153	*Tobias Intruder Alarm System*	*1977*
2471.441	*Torpedo Mk 20 (Improved)*	*1983-84*
2473.441	*Torpedo Mk 31*	*1971-72*
2474.441	Torpedo Type 43XO	525
2475.111	Vigilant Anti-Tank Missile	152
2476.103	*Mobat Recoilless Anti-Tank Gun*	*1978*
2477.103	*Wombat Recoilless Anti-Tank Gun*	*1978*
2478.153	*Marconi 40-Series Volumetric Radars*	*1979-80*
2479.153	S613 Heightfinding Radar	265
2480.153	*Marconi Transportable Tactical Radar Type S-259*	*1977*
2481.153	*Isidor Portable Intrusion Radar*	*1977*
2482.153	*Marconi/RRE Shrimp Battlefield Surveillance Radar*	*1976*
2483.153	*Infantry Combat Radar GS No 18 Mk 1*	*1974-75*
2484.153	AR-3D Air Defence Radar	266
2485.103	*3in Mortar*	*1978*
2486.103	*17-pounder Anti-Tank Gun*	*1982-83*
2487.103	*6-pounder Anti-Tank Gun*	*1982-83*
2489.103	*3.7in Anti-Aircraft Gun*	*1982-83*
2490.153	ZB 298 Short-Range Ground Surveillance Radar	272
2491.103	*90mm Mecar Light Gun*	*1982-83*
2491.441	*Drill and Practice Torpedo MW 30 (Mk 44)*	*1979-80*
2492.153	AN/FPS-6 Heightfinding Radar	278
2493.153	AN/FPS-8 Search Radar	279
2494.153	AN/FPS-24 Search Radar	279
2495.153	AN/MPS-11 and AN/MPS-11A Search Radars	283
2496.153	AN/MPS-14 Heightfinding Radar	283
2497.153	AN/MPQ-4 Mortar Locating Radar	282
2498.153	AN/TPQ-10 Radar Course Directing Control	287
2499.153	HIPAR (High-Power Acquisition Radar)	300
2500.153	*AN/MPQ-10A Mortar Locating Radar*	*1985-86*
2501.393	*AA 37U-15 Tow System*	*1970-71*
2502.393	*Acoustiscore Scoring System*	*1970-71*
2503.393	*Acoustiscore*	*1970-71*
2504.191	*Acoustiscore Type DA2*	*1970-71*
2505.193	*Acoustiscore Type DA3*	*1970-71*
2506.153	*ADAR Advanced Design Array Radar*	*1977*
2507.231	Aegis Surface-to-Air Weapon System	478
2508.153	*AN/FPQ-14 Monopulse Tracking Radar*	*1981-82*
2509.153	AN/FPS-49 Early Warning & Tracking Radar	280
2510.163	*AN/GPA-122 Coder-Decoder Group (IFF)*	*1970-71*
2511.153	AN/FPS-50 Early Warning Radar	280
2512.153	AN/FPS-88 Surveillance Radar	281
2513.153	AN/FPS-89 Heightfinding Radar	281
2514.153	AN/MPS-36 Instrumentation Tracking Radar	283
2515.153	*AN/TPS-27 3-D Tactical Radar*	*1973-74*
2516.153	AN/TPS-32 3-D Long Range Surveillance Radar	291
2517.153	AN/TPS-43E Tactical 3-D Radar	293
2518.153	AN/TPS-44 Tactical Surveillance Radar (ALERT Mk II A/O)	294
2519.153	*AN/TPS-48 3-D Tactical Radar*	*1974-75*
2520.153	*AN/TPS-61 Tactical Surveillance Radar*	*1981-82*
2521.391	*Beech Cardinal Drone*	*1974-75*
2522.391	*Beech Model 1107 (AQM-37A)*	*1987-88*
2523.391	*Beech Sandpiper Drone*	*1972-73*
2524.391	*Beech Stiletto Drone*	*1973-74*
2525.181	Ballistic Missile Early Warning System (BMEWS)	56
2526.153	AN/FPS-7 3-D Search & GCI Radar	278
2527.153	*AN/FPS-14 & AN/FPS-18 Medium-Range Search Radars*	*1982-83*
2528.153	AN/FPS-20 & AN/FPS-100 Family of Long-Range Search & GCI Radars	279
2529.153	AN/TPS-21 & AN/TPS-33 Pulse Doppler Battlefield Radars	291
2530.153	AN/TPS-34 Transportable Air Defence Radar	293
2531.131	*Bomarc B Long-Range Air Defence Weapon*	*1972-73*
2532.391	*Bomarc Target Drone*	*1972-73*
2533.181	*Air Weapons Control System 412L*	*1982-83*
2534.253	*AN/SQS-36 & AN/SQS-38 Medium-Range Hull-Mounted Scanning Sonars*	*1980-81*
2535.153	AN/TPX-42 IFF-ATCRBS Equipment	298
2536.253	AN/SQS-53/26 Sonars	585
2537.153	AN/TPN-19 Landing Control Central	285
2538.153	*AN/FSS-7 SLBM Detection & Warning Radar*	*1980-81*
2539.103	*DICORAP Directional Controlled Rocket-Assisted Projectile*	*1972-73*
2540.153	AN/PPS-5 Combat Surveillance Radar	286
2541.441	Captor Encapsulated Torpedo	537
2542.131	Chaparral Surface-to-Air Weapon System	186
2543.231	Mk 15/16 Phalanx Close-In Weapon System	479
2544.131	*GLADS Low-Altitude Defence System*	*1978*

Entry number	Subject	Page No/ Last edition*
2545.293	*Missile Launcher/Canister*	*1974-75*
2546.153	AN/FPS-85 Long-Range Phased Array Radar	280
2547.153	AN/VPS-2 Ranging Radar	299
2548.183	*All-Electric Stabilised Gun Drive*	*1974-75*
2549.183	*Hydromechanical Transmission*	*1974-75*
2550.341	*DASH (Anti-Submarine Drone System)*	*1973-74*
2551.153	*CONAR Artillery Radar Fire Control System*	*1978*
2552.441	Torpedo Mk 8	528
2553.203	*Shipborne Bombardment Rocket FB-R 127*	*1976*
2554.203	Single 40mm Naval Mounting Type 564	491
2555.203	Single 40mm Naval Mounting Type 107	492
2556.203	Twin 40mm Naval Mounting Type 64	491
2557.203	Breda/Bofors Naval Mountings	491
2558.203	Twin 40mm Naval Mounting Type 106	491
2559.203	*Single 30mm Mounting Type CAS 62*	*1982-83*
2560.393	*D-100A/1 Radop Scorer*	*1970-71*
2561.103	*75mm Recoilless Rifle M-20*	*1978*
2562.393	*DF-4MFC Towed Target*	*1970-71*
2563.393	*DF-6MFC Towed Target*	*1970-71*
2564.393	*DF-14 Towed Target*	*1970-71*
2565.391	*DH2C Target Drone*	*1972-73*
2566.351	*Surveillance Drone System*	*1972-73*
2567.181	North Warning System (DEW Line)	57
2568.393	*DL-45 Tow Target Launcher*	*1970-71*
2569.393	*DL-14 Tow Target Launcher*	*1970-71*
2570.441	Surface and Underwater Target (SUT) Torpedo	517
2571.103	*Rocket-Assisted Projectiles*	*1979*
2572.103	*155mm GCT/Leopard SP Gun*	*1980-81*
2573.111	Dragon Anti-Tank/Assault Weapon (FGM-77A/FTM-77A)	154
2574.131	*Advanced Ballistic Missile Defence*	*1976*
2575.131	*Homing Intercept Target (HIT) Warhead*	*1976*
2576.393	*DX-4A Tow Reel*	*1970-71*
2577.393	*DX-6A Supersonic Tow Reel*	*1970-71*
2578.393	*DXL-6B Tow Target Weapon Training System*	*1970-71*
2579.393	*DXL-6G Towed Target Weapon Training System*	*1970-71*
2580.253	EDO Model 610 Long Range Hull-Mounted Scanning Sonar	589
2581.253	EDO Model 700 Series Sonars	590
2582.103	*105mm Howitzer M-50*	*1977*
2583.103	*20mm Gun M621 for Vehicle or Naval Mounting*	*1976*
2584.203	Emerlec-30 Twin 30mm Mounting	508
2585.103	*106mm SP Recoilless Rifle Type 60*	*1982-83*
2586.103	*South African Ordnance*	*1981-82*
2587.103	*90mm Tank Gun, Bofors KV90 S73*	*1982-83*
2588.103	*106mm Recoilless Rifle M40A2*	*1978*
2589.103	*Introduction Portable Unguided Anti-Tank Weapons*	*1978*
2590.203	*Rafale Naval Rocket System*	*1979-80*
2591.103	*Panzerfaust Leicht Portable Anti-Tank Weapon*	*1974-75*
2592.103	*RL-83 Blindicide Rocket Launcher*	*1978*
2593.103	*Wolf Battlefield Missile*	*1982-83*
2594.103	*Bora Artillery Rocket*	*1982-83*
2595.103	*Mira Artillery Rocket*	*1982-83*
2596.183	*Gun Control & Stabilisation Systems GCE 576 & GCE 581*	*1974-75*
2597.183	*Power Traverse Systems for Light Fighting Vehicles*	*1974-75*
2598.181	SABCA Tank Fire Control Systems	63
2599.103	*105mm Mortar ECIA Model L*	*1978*
2600.103	*120mm Mortars ECIA Models L & SL*	*1978*
2601.103	*81mm Mortars ECIA Models L & LI*	*1978*
2602.103	*60mm Mortar ECIA Model L*	*1978*
2603.103	*40mm Breda/Bofors LAA Field Mountings*	*1982-83*
2604.153	*Isidor Muzzle Velocity Meter*	*1974-75*
2605.103	*40mm Anti-Tank Grenade Launchers Type 56 & 69*	*1978*
2606.103	*90mm Anti-Tank Rocket Launcher Type 51*	*1978*
2607.203	*20mm Naval Mount Type GAM-B*	*1978*
2609.111	*Rafale Artillery Rockets*	*1985-86*
2610.351	*Fairchild Hiller Bikini Surveillance Drone*	*1972-73*
2611.103	*WP-8 Artillery Rocket Launcher*	*1982-83*
2612.103	*M-51 Artillery Rocket Launcher*	*1982-83*
2613.103	*60mm Light Mortar Type 63*	*1978*
2614.103	*82mm Mortars*	*1978*
2615.391	*Firebee I (BQM-34A/BQM-34S/MQM-34D)*	*1987-88*
2616.391	*Firebee II (BQM-34E/F/T)*	*1987-88*
2617.351	*Firebee (Reconnaissance) Drone*	*1970-71*
2618.441	*Freedom Torpedo*	*1982-83*
2619.103	*60mm Commando Mortar EC1A Model C*	*1978*
2620.103	*Rattlebox Lightweight Infantry Rocket Launcher*	*1982-83*
2621.103	*51mm Light Mortar*	*1978*
2622.103	*2in Mortar*	*1978*
2623.103	*Mortars Introduction*	*1978*
2624.103	*50mm Light Mortar M-8*	*1978*
2625.103	*120mm Heavy Mortar Mir 64*	*1978*
2626.103	*160mm Tampella Heavy Mortar M66*	*1977*
2627.103	*120mm Mortar M40*	*1978*
2628.103	*81mm Mortars M38 & M56*	*1978*
2629.103	*120mm Mortar M-41C*	*1978*
2630.103	*81mm Mortar M-29*	*1978*
2631.283	*Gun Director Mk 37*	*1969-70*
2632.281	Gun & Guided Missile Director Mk 73	442
2633.281	Gun Fire Control System Mk 56	441
2634.281	*Gun & Guided Missile Fire Control System Mk 74*	*1976*
2635.103	*81mm Light Mortar M44*	*1978*
2636.353	*Gunfire Warning & Detection Equipment*	*1969-70*
2637.103	*81mm Mortar Mk 72*	*1978*
2638.391	*Gyrodyne QH-50D Drone Helicopter*	*1979-80*
2639.131	*Hardsite ABM System*	*1976*
2640.131	*Hawk Surface-to-Air Weapon System*	*1980-81*
2641.221	Harpoon Shipborne Surface-to-Surface Missile (RGM-84A)	460
2642.181	Army Fire Control Systems - Introduction	63
2643.181	Integrated Tank Fire Control System	105
2644.161	BATES - Battlefield Artillery Target Engagement System	100
2645.161	Odin Artillery Fire Control System	85
2646.161	*AN/UMC-7 Meteorological Data Sounding System*	*1982-83*
2647.181	IFCS - Improved (Tank) Fire Control System	102
2648.181	*Kalle Light Anti-Aircraft Fire Control System*	*1977*
2649.103	*20mm Gun Turret Type GAD-AOA*	*1982-83*
2652.111	Honest John Battlefield Support Missile	132
2668.193	*Integrated Night Obse, vation System*	*1982-83*
2669.221	*Interim Surface-to-Surface Missile*	*1976*
2670.111	*Janus Project Dual Role Strategic Missile*	*1970-71*
2682.111	Lance Battlefield Support Missile	131
2683.111	*Lance-Derived Experimental Battlefield Missile*	*1974-75*
2684.131	*Land Sparrow Mobile AA Weapon System*	*1973-74*
2685.131	*Laser-Aided ABM Missile*	*1972-73*
2690.111	*Little John Surface-to-Surface Missile*	*1970-71*
2694.391	*Locat Target Missile*	*1972-73*
2702.111	*Mace B Surface-to-Surface Missile*	*1970-71*
2707.111	*Triad Strategic Deterrence Combination*	*1976*
2708.111	*Ballistic Missile Re-Entry Vehicles*	*1982-83*
2709.111	*ABRES Advanced Ballistic Re-Entry Systems*	*1980-81*
2710.111	*Minuteman ICBM*	*1982-83*
2711.383	*Minuteman II Guidance & Control Equipment*	*1970-71*
2712.111	*Minuteman Rebasing*	*1972-73*
2713.293	*Missile Launcher R-202*	*1974-75*
2714.293	*Missile Launcher R-204*	*1974-75*
2715.111	*Minuteman I Intercontinental Ballistic Missile*	*1976*
2716.111	Minuteman II Intercontinental Ballistic Missile	26
2717.111	Minuteman III Intercontinental Ballistic Missile	26
2718.111	*Missile-X Intercontinental Ballistic Missile*	*1982-83*
2719.111	*MARV (Manoeuvrable Re-Entry Vehicle) Systems*	*1978*
2720.111	*Minuteman IV Intercontinental Ballistic Missile*	*1974-75*
2721.111	*Nemesis Project Strategic Missile*	*1972-73*
2722.411	*Nemesis Project (Sub-Surface)*	*1970-71*
2723.131	Nike Hercules (MIM-14B) Air Defence Missile	190
2724.153	*Night Observation Device Long-Range NODLR*	*1974-75*
2725.103	*Miniature Nuclear Warheads*	*1978*
2730.391	*Northrop MQM-33 Drone*	*1973-74*
2731.351	*Northrop MQM-57 Drone*	*1973-74*
2732.391	*Chukar II Target Drone*	*1987-88*
2733.391	*Northrop RP-76 Drone*	*1973-74*
2734.391	*Northrop RP-76-4 Drone*	*1973-74*
2740.183	*NWL Weapon Stabilisation System*	*1974-75*
2750.181	Over-the-Horizon (OTH) Radar Systems	57
2763.103	*Ontos 106mm SP Recoilless Rifle*	*1973-74*
2764.111	*Pershing 1 Battlefield Support Missile*	*1973-74*
2765.111	Pershing Ia Battlefield Support Missile	131
2766.153	*PERSID-4A Intruder Alarm System*	*1977*

Entry number	Subject	Page No/ Last edition*
2767.111	Pershing II Battlefield Support Missile	132
2770.231	*Point Defence Missile System (Sea Sparrow)*	*1983-84*
2779.103	*60mm Light Mortar T18 E6*	*1976*
2780.391	*Radop Weapon Training System*	*1970-71*
2781.193	*Raytector Radar Monitor*	*1969-70*
2783.153	*RCA Portable Battlefield Radars*	*1974-75*
2784.131	Redeye Portable Anti-Aircraft Missile	216
2785.103	*81mm Mechim Mortar M1*	*1978*
2786.103	*60mm Mortar XM224*	*1978*
2787.103	*60mm Light Mortar M-19*	*1978*
2788.103	*60mm Light Mortar M-2*	*1978*
2789.391	*Roadrunner Drone*	*1977*
2790.153	*Safeguard Perimeter Acquisition Radar*	*1977*
2791.153	*Safeguard Missile Site Radar*	*1982-83*
2792.153	*Site Defence Radar*	*1976*
2793.441	*New Torpedo Projects*	*1980-81*
2794.131	*SABMIS Anti-Ballistic Missile System*	*1972-73*
2795.103	*SMAWT Anti-Tank Rocket Launcher*	*1974-75*
2796.103	*Stabilised Armament Station*	*1981-82*
2797.421	*Submarine-Launched Harpoon Missile*	*1976*
2798.131	*Safeguard Anti-Ballistic Missile System*	*1977*
2799.131	*Safesam Anti-Ballistic Missile System*	*1972-73*
2800.131	Patriot (MIM-104) Tactical Air Defence System	189
2801.181	*SAFOC Semi-Automated Flight Operations Centre*	*1976*
2802.483	*SD-510 Inertial Guidance System*	*1970-71*
2803.181	*SAGE (Semi-Automatic Ground Environment) Air Defence System*	*1979-80*
2804.111	*Sergeant Battlefield Support Missile*	*1980-81*
2805.131	Stinger Portable Anti-Aircraft Missile (FIM-92A)	217
2806.131	*Site Defence of Minuteman (SDM)*	*1974-75*
2807.131	*Sprint II Anti-Missile Missile*	*1977*
2808.221	*Standard Arm Surface-to-Surface Application*	*1972-73*
2809.111	*Shillelagh Close Support Weapon System*	*1982-83*
2810.181	SLBM Detection System 474N	58
2811.131	*Spartan Anti-Missile Missile*	*1977*
2812.131	*Sprint Anti-Ballistic Missile Missile*	*1977*
2813.441	*Torpedo Mk 14*	*1982-83*
2814.441	*Torpedo Mk 16*	*1977*
2815.441	*Torpedo Mk 18*	*1978*
2816.441	*Torpedo Mk 28*	*1977*
2817.441	Torpedo Mk 37 Mods 0 & 3	534
2818.441	Torpedo Mk 37 Mods 1 & 2	535
2820.441	Torpedo Mk 44	535
2821.153	Flat Twin Radar	255
2821.441	*Torpedo Mk 45 (ASTOR)*	*1979-80*
2822.441	Torpedo Mk 46	535
2823.441	Torpedo Mk 48	537
2824.181	*Tactical Air Control System (TACS)*	*1983-84*
2825.181	Spacetrack	59
2826.111	*Titan II Intercontinental Ballistic Missile*	*1987-88*
2827.161	TACFIRE Tactical Fire Direction System (AN/GSG-10)	107
2828.193	*Stabilised Gun Control System*	*1974-75*
2829.193	*Add-on Weapon Stabilisation System*	*1974-75*
2830.111	TOW Anti-Tank Missile (BGM-71)	155
2831.311	TOW Air-to-Surface Application	725
2832.103	*90mm Tank Gun M-41*	*1982-83*
2833.103	*152mm Tank Gun XM-162*	*1979-80*
2834.103	*Rocket-Assisted Projectiles (RAP)*	*1979-80*
2835.181	AN/TSQ-73 Air Defence Missile Control and Co-ordination System	62
2836.203	20 mm Sea Vulcan 20P Three-Barrel Pintle Mount	509
2837.111	*Rocket-Assisted Guided Projectiles*	*1976*
2838.111	*Terminally-Guided Sub-Missile*	*1982-83*
2839.111	*Pave Pepper Re-Entry Project*	*1974-75*
2840.411	Trident C-4 (UGM-93A) Strategic Weapon System	28
2841.131	*Alternate Stinger*	*1980-81*
2842.131	*SHORADS/LOFADS Air Defence Missile System*	*1976*
2843.131	*AVADS Automatic Vulcan Air Defence System*	*1978*
2844.153	*AVDS Fire Control Radar*	*1979-80*
2845.221	*US Naval Surface-to-Surface Missiles*	*1973-74*
2846.153	AN/FPQ-6 and AN/TPQ-18 Precision Instrumentation Radars	278
2847.153	AN/FPS-105(V) Range Instrumentation Radar (Capri)	281
2848.153	AN/TPQ-36 Weapon Locating Radar	289
2849.153	AN/TPQ-39(V) Instrumentation Radar	290
2850.131	Vulcan Air Defence System	208

Entry number	Subject	Page No/ Last edition*
2851.203	*20mm Vulcan Gun (naval application)*	*1974-75*
2852.111	*Vulcan ICBM Project*	*1972-73*
2853.102	*XM-1 Main Battle Tank*	*1979-80*
2854.103	*105mm Tank Gun M-68*	*1982-83*
2855.103	*90mm Tank Gun M-36*	*1982-83*
2856.103	*20mm Cannon M-197 Pintle Mounted*	*1982-83*
2857.153	*AN/FPQ-13 Tracking Radar*	*1981-82*
2858.203	*20mm Naval Vulcan Air Defence Systems*	*1982-83*
2859.103	*30mm Chain Gun*	*1978*
2860.153	Back Net GCI Radar	253
2861.153	Barlock GCI Radar	253
2862.153	Big Bar A GCI Radar	253
2863.153	Big Mesh GCI Radar	253
2864.153	Dog House ABM Radar	254
2865.203	Twin 85mm AA Gun Mounting	501
2866.153	Fan Song Missile Control Radars	254
2867.153	Fan Song B Missile Control Radar	254
2868.153	Fan Song E and F Missile Control Radars	254
2869.203	Twin 57mm Automatic AA Gun Mounting	501
2870.203	Twin 37mm AA Gun Mountings	502
2871.153	Fire Can Fire Control Radar	254
2872.153	Fire Wheel Fire Control Radar	255
2873.153	Flap Wheel Fire Control Radar	255
2874.153	Flat Face Target Acquisition Radar	255
2875.153	Gage Acquisition Radar	255
2876.153	Gun Dish Fire Control Radar	255
2877.153	Hen Series Early Warning Radars	256
2878.153	Heightfinding Radars	256
2879.153	Hen House ABM Radar	256
2880.153	Knife Rest A Early Warning Radar	256
2881.153	Knife Rest B & C Early Warning Radars	256
2883.203	Single 37mm AA Gun Mounting 70-K	502
2884.153	Low Blow Tracking & Missile Control Radar	257
2885.153	Straight Flush Fire Control Radar	258
2886.103	*152mm Gun Howitzer M-37 (M20)*	*1982-83*
2887.103	*37mm Automatic AA Gun M38/39*	*1982-83*
2888.103	*100mm SP Anti-Tank Gun SU-100*	*1982-83*
2889.153	Spoon Rest A Early Warning Radar	259
2890.111	*FROG Battlefield Missiles Introduction*	*1980-81*
2891.153	Squint Eye Acquisition Radar	258
2892.153	Token Early Warning & GCI Radar	260
2893.153	Whiff Fire Control Radar	261
2894.103	*122mm Gun M-1931/37 (A19)*	*1982-83*
2895.153	Yo-Yo Missile Control Radar	261
2896.203	Twin 30mm AA Gun Mounting	502
2897.203	Twin 25mm AA Gun Mounting	502
2898.203	*Twin MG Mountings*	*1979-80*
2899.181	*Russian Air Defence Systems*	*1976*
2899.203	122mm and 140mm Rocket Launchers	502
2900.111	*Introduction USSR Surface-to-Surface Missiles*	*1979-80*
2901.241	*140mm Rocket Launcher*	*1976*
2902.241	*250mm Anti-Submarine Rocket Launcher*	*1981-82*
2903.241	*300mm Anti-Submarine Rocket Launchers*	*1981-82*
2904.103	*107mm Mortars M38 & M107*	*1978*
2920.103	*FROG 1 Battlefield Support Missile*	*1974-75*
2921.103	*FROG 2 Battlefield Support Missile*	*1974-75*
2922.111	*FROG 3 Battlefield Support Missile*	*1980-81*
2923.111	*FROG 4 Battlefield Support Missile*	*1980-81*
2924.111	*FROG 5 Battlefield Support Missile*	*1980-81*
2925.111	*FROG 7 Guided Battlefield Support Missile*	*1976*
2926.111	FROG-7 Battlefield Support Missile	128
2927.103	*180mm S23 Gun*	*1982-83*
2928.103	*76mm Recoilless Rifle SPG-9*	*1974-75*
2929.103	*160mm Heavy Mortar M-1943*	*1973-74*
2930.131	SA-6 (Gainful) Surface-to-Air Missile	179
2931.103	*1160mm Heavy Mortar M-1953*	*1973-74*
2932.131	ABM-1B (Galosh) Anti-Ballistic Missile Missile (Soviet Designation: UR-96)	17
2933.131	*Improved Galosh Anti-Ballistic Missile*	*1977*
2934.131	SA-4 (Ganef) Surface-to-Air Missile	177
2935.131	*New ABM Missile*	*1977*
2936.153	Pat Hand Fire Control Radar	257
2937.153	Long Track Surveillance Radar	257
2938.131	SA-3 (Goa) Surface-to-Air Missile	177
2939.231	Goa Shipborne Surface-to-Air Missile (SA-N-1)	472
2940.131	SA-5 (Gammon) Long-Range Air Defence Missile	178
2941.131	SA-7 (Grail) Man-Portable Anti-Aircraft Missile	214
2942.131	SA-2 (Guideline) Medium-Range Surface-to-Air Missile	177
2943.231	Guideline Shipborne Surface-to-Air Missile (SA-N-2)	472
2944.131	SA-1 (Guild) Surface-to-Air Missile	176
2945.181	*Moscow Air Defences*	*1976*

Entry number	Subject	Page No/ Last edition*
2946.411	*Soviet Submarine-Launched Ballistic Missiles*	*1979-80*
2947.231	Goblet Shipborne Surface-to-Air Missile (SA-N-3)	472
2948.131	*Russian Anti-Ballistic Missile Systems*	*1973-74*
2949.221	*Russian Naval Surface-to-Surface Missile*	*1976*
2950.111	AT-3 (Sagger) Anti-Tank Missile	149
2951.121	Samlet Cruise Missile (SSC-2B)	163
2952.111	SS-4 (Sandal) MRBM (Soviet designation R-12)	6
2953.121	*Salish Cruise Missile*	*1983-84*
2954.231	SA-N-4 Shipborne Surface-to-Air Missile	473
2956.111	*Sasin ICBM (SS-8)*	*1980-81*
2958.111	SS-13 (Savage) ICBM (Soviet designation RS-12/SS-12)	7
2959.111	SS-12A (Scaleboard A) Battlefield Support Missile	127
2960.111	*Scamp Mobile Strategic Missile (SS-14)*	*1976*
2961.111	*Scapegoat IRBM (SS-14)*	*1984-85*
2962.111	*Scarp ICBM (SS-9)*	*1980-81*
2963.111	*New Strategic Missile*	*1973-74*
2964.131	*SA-9/10 Surface-to-Air Missile Systems*	*1974-75*
2965.111	*Scrag ICBM (SS-10)*	*1980-81*
2966.103	*100mm Tank Gun D10 Series*	*1982-83*
2967.111	*Scrooge Mobile Strategic Weapon System (SS-XZ)*	*1978*
2968.221	*Scrubber Shipborne Surface-to-Surface Missile*	*1978*
2969.111	*Scud A Battlefield Support Missile*	*1977*
2970.111	*Scud B Battlefield Support Missile*	*1977*
2971.111	*Scud C Battlefield Support Missile*	*1974-75*
2972.103	*115mm Type 62 Tank Gun*	*1982-83*
2973.231	30mm Close-in Weapon System	475
2974.153	Score Board IFF Radar	258
2975.121	Sepal Mobile Cruise Missile (SSC-1B)	164
2976.221	SS-N-3 (Shaddock) Shipborne Surface-to-Air Missile	456
2977.111	*Shyster MRBM*	*1976*
2978.111	*SS-9 Mod 3 ICBM*	*1979-80*
2979.111	*SS-9 Mod 4 ICBM*	*1979-80*
2980.111	*Cold-Launched Ballistic Missiles*	*1974-75*
2981.111	*Skean MRBM (SS-5)*	*1984-85*
2982.111	*SS-7 (Saddler) ICBM*	*1980-81*
2983.111	*AT-1 Sapper Anti-Tank Missile*	*1984-85*
2984.111	*SS-11 (Sego) Mod 1/2 ICBM*	*1980-81*
2985.111	AT-2 (Swatter) Anti-Tank Missile	148
2986.221	*Strela Shipborne Surface-to-Surface Missile*	*1973-74*
2987.411	SS-N-7 (Star Bright) Submarine-Launched Cruise Missile	457
2988.111	*SS11 (Sego) Mod 3 ICBM*	*1980-81*
2989.181	*Tallinn Air Defence System*	*1976*
2990.111	*SS-16 ICBM*	*1982-83*
2991.111	SS-17 ICBM (Spanker) (Soviet designation RS-16)	8
2992.111	SS-18 ICBM (Satan) (Soviet designation RS-20)	8
2993.111	SS-19 ISBM (Stilleto) (Soviet designation RS-18)	8
2994.221	*Russian Shipborne Surface-to-Surface Missiles*	*1973-74*
2995.441	533mm Torpedo	527
2996.441	*Airborne Torpedo*	*1980-81*
2997.441	Light Torpedo	527
2998.103	*155mm SP Medium Howitzer M-109U*	*1982-83*
2999.181	*Iran Air Defence System*	*1980-81*
3000.353	*Blue Baron Reconnaissance Pod*	*1982-83*
3001.353	Red Baron Pod	877
3002.353	*Minipan Strike Cameras*	*1977*
3003.353	*AN/UXD-1 Reconnaissance System*	*1981-82*
3004.353	*KA-102A 66in Camera and Pod*	*1982-83*
3005.353	*RS-700 Infra-Red Linescan*	*1982-83*
3006.193	*Weston Simfire Solartron Weapon Simulators and Trainers*	*1982-83*
3007.393	*AF530 Sight Simulator*	*1980-81*
3008.293	*SSS Series Tactical Simulators*	*1982-83*
3009.193	*RIS Series 10 GCI Simulator*	*1982-83*
3010.351	*CL-227 (Sentinel) Reconnaissance and Target Acquisition Drone*	*1987-88*
3011.351	*Dornier UKF*	*1978*
3012.351	*FR 500 Drone*	*1979-80*
3013.391	*Multi-Mission RPV BGM-34C*	*1979-80*
3014.351	*Teledyne Ryan Mini-RPV*	*1979-80*
3015.351	*US RPV Programmes Note*	*1976*
3016.153	TRM/TRS C-Band Radar Series	238
3017.231	RAM (RIM-116A) system	480
3018.353	AN/AWG-10, 11 and 12 Missile Control Systems	811
3019.221	*Saab SKA Surface-to-Surface Missile*	*1978*
3020.204	*Fighting Ships Table*	*1987-88*
3021.453	*NE10A Radar Search Receiver*	*1982-83*
3022.163	*UA731 Pulse Analyser*	*1981-82*
3023.353	*SATT Radar Warning System*	*1983-84*
3024.393	AQ31 Jammer	830
3025.393	AR735 Jammer Set-on Receiver	829
3026.293	*TERSI EW Simulator*	*1979-80*
3027.353	*Kiebitz ESM System*	*1979-80*
3028.153	*FSR-100 Surveillance Receiver*	*1980-81*
3029.053	*Model R-47 Microwave Tuners*	*1978*
3030.053	*R75 Receiving System*	*1978*
3031.053	GTE Computer Controlled Receiving Systems	332
3032.153	Circularly Disposed Antenna Array (CDAA)	331
3033.053	*Cincinnati Surveillance Receiver Equipment*	*1983-84*
3034.393	AN/ALQ-161 B-1 RFS/ECM Sub-System	849
3035.353	AN/ALR-46 Radar Warning System	853
3036.053	*SR-1000 Computer-Controlled Receiving System*	*1976*
3037.053	SR-Series Surveillance Receivers	339
3038.353	Marconi Radar Warning Receivers	833
3039.193	*TCV 29 and TCV 30 Tank Laser Rangefinders*	*1986-87*
3040.193	TM 17 (TM 12) Artillery Sight/Laser Rangefinder	350
3041.153	Ericsson Coastal Artillery Laser Rangefinder	378
3042.193	Ericsson Anti-Aircraft Laser Rangefinder	377
3043.193	Ericsson Artillery Laser Rangefinder	377
3044.193	Integrated Laser for Main Battle Tanks	377
3045.193	*No 18/LV2 Tank Laser Rangefinder*	*1980-81*
3046.393	Type 105 Air-to-Surface Laser Ranger	884
3047.393	*Inertial Navigation and Attack System INAS*	*1976*
3048.193	AN/VVS-2 Driver's Night Vision Viewer	397
3049.353	*AN/ASQ-151 Electro-Optical Viewing System*	*1982-83*
3050.383	*Blue Spot Target Designator*	*1980-81*
3051.193	AN/TVQ-2 Ground/Vehicular Laser Locator Designator (G/VLLD)	401
3052.393	*TOW/Cobra Sight*	*1979-80*
3053.383	ATAFCS Equipment	891
3054.053	*LRR-103 Artillery Laser Rangefinder*	*1985-86*
3055.293	DES Optical Fire Director	685
3056.393	AV-8A Head-Up Display and Weapon Aiming Computing System	881
3057.393	F-16 A/B HUD-Sight	878
3058.393	*MEAS Helmet Sight*	*1979-80*
3059.053	*PPE Night Vision Equipment*	*1977*
3060.193	*SS20 Individual Weapon Sight*	*1979-80*
3061.193	*SS30 Heavy Weapon Sight*	*1979-80*
3062.193	*SS32 Night Viewing Device*	*1979-80*
3063.193	SS100/110 Night Sight	389
3064.293	French Naval Electro-optical Systems	677
3065.393	AN/AVQ Pave Tack	894
3067.331	*Saab 372 (RB 72) Air-to-Air Missile*	*1980-81*
3068.331	*Sidewinder (AIM-9L and AIM-9M) Air-to-Air Missile*	*1983-84*
3069.311	*RB 05B Air-to-Surface Missile*	*1977*
3070.321	Penguin Mk 3 Air-to-Surface Missile	714
3071.311	*LVRJ Test Vehicle*	*1980-81*
3072.131	SA-9 (Gaskin) Surface-to-Air Missile	180
3073.181	Tactical Air Defence System - TADS	59
3074.121	Penguin Coastal Defence System	162
3075.351	*USAF Tactical Reconnaissance*	*1978*
3076.351	*USN Tactical Reconnaissance*	*1978*
3077.111	*Chinese ICBMs*	*1980-81*
3078.411	SS-N-8 Submarine Launched Ballistic Missile	14
3079.411	*SSN(X)-13 Submarine-Launched Missile*	*1976*
3080.404	*Submarines Table*	*1987-88*
3081.304	*Military Aircraft Table*	*1987-88*
3082.253	AN/WLR-8 Tactical EW Receiver System	671
3083.453	AN/SSQ-57A Sonobuoy	596
3084.453	AN/SSQ-77 (VLAD) Sonobuoy	597
3085.453	*AN/SSQ-517 and AN/SSQ-518 Sonobuoys*	*1983-84*
3086.453	AN/SSQ-522 Sonobuoy	544
3087.453	*AN/SSQ-523 (Dicancass) Sonobuoy*	*1985-86*
3088.391	*AEL Snipe Target Drone*	*1981-82*
3089.183	*Tracker Infra-Red Guided Missile Type SU-36/P*	*1979-80*
3090.393	Head-Up Display for Harrier and Jaguar	881
3091.193	LP3 Laser Rangefinder	380
3092.193	LV2 Laser Rangefinder	380

Entry number	Subject	Page No/ Last edition*
3093.163	*Matra Mobile Reconnaissance Laboratory*	*1982-83*
3094.253	*PMS 35 Sonar*	*1977*
3095.263	Sound Ray Tracer (TSM 9310)	551
3096.193	RZ1001 Thermal Imaging Equipment	360
3097.293	Mk 5 Low Light-Level TV System	690
3098.353	AN/ASQ-145 Low Light-Level TV System	890
3099.153	TRS 2052/TRS 2053/TRS 2056 D-band Early Warning Radar	227
3100.153	Coastal Surface Surveillance Radar TRS 3405	231
3101.391	*Super Condor Mk 3 Target Drone*	*1982-83*
3102.281	Canopus Series Weapon Control System	409
3103.481	*STWS-1 Shipborne Torpedo Weapon System Mk 1*	*1985-86*
3104.481	*MCS10M Mortar Mk 10 Control System*	*1985-86*
3105.093	MPI-5 Miniature Pulse Analyser	332
3106.181	*Conar Artillery FCS*	*1977*
3107.131	*MEI Air Defence System*	*1981-82*
3108.351	*Aeronutronic Ford Mini-RPV*	*1977*
3109.393	Lundy Countermeasures	838
3110.393	AN/ALE-43 Chaff Cutter Dispenser Pod	840
3111.393	AN/ALE-44 Dispenser Pod	841
3112.393	Lundy Chaff and Flares Countermeasures System	842
3113.353	AN/APQ-157 Fire Control Radar	802
3114.353	AN/APQ-159 Airborne Radar	802
3115.453	DE1164 Sonar	588
3116.453	DE1167 Modernisation Sonar	588
3117.393	Head-Up Display for Sea Harrier	882
3118.393	*Ferranti Laser Ranger/Marked Target Receiver*	*1979-80*
3119.393	Type 106 Laser Target Seeker	884
3120.193	*Ferranti Type 504 Surface-to-Air Laser Ranger*	*1984-85*
3121.193	*Commando Day-Night Vision System*	*1985-86*
3122.393	*HOT Missile Training Simulator DX.147*	*1982-83*
3123.153	LAADS Low Altitude Aircraft Detection System	288
3124.353	*AN/APN-171 Terrain Following Radar*	*1982-83*
3125.161	AWDATS - Artillery Weapon Data Transmission System	101
3126.161	BODIL Coast Artillery FCS	85
3127.281	9 LV 200 Mk 2 Weapon Control System	428
3128.153	*AN/TPS-1 Search Radar*	*1983-84*
3129.153	AN/TPN-18A/18B Radar	285
3130.153	*AN/TPS-25 Battlefield Surveillance Radar*	*1982-83*
3131.353	Cyrano Radar Family	776
3132.353	AN/APS-128 Model D Surveillance Radar	805
3133.293	9EW Countermeasures Launcher	660
3134.193	Thomson-CSF Canasta Day/Night Aiming Equipment (DIVT 13)	357
3135.393	Thomson-CSF Series VE 120/130/VEM 130 Head-up Displays	870
3136.393	Thomson-CSF Atlis Laser Target Designation Pod	869
3137.193	Barr & Stroud Tank Laser Sight	382
3138.193	*BT 39 AA Gunfire Simulator*	*1982-83*
3139.391	*MATS-B Target Drone*	*1979-80*
3140.391	*QF-86E Drone Aircraft Target*	*1982-83*
3141.253	*LWS-30 Passive Sonar/Intercept System*	*1979-80*
3142.253	PHS-32 Sonar	559
3143.293	LIOD Lightweight Optronic Fire Control System	683
3144.193	RIA Gunsight	376
3145.293	3in Window Dispensing Rocket System (Corvus)	662
3146.193	APX M504 Optical Sight and Rangefinder	351
3147.293	TAON TMY83 Laser Rangefinder/Illuminator	681
3148.393	*MAHAT Weapons Delivery System*	*1983-84*
3149.153	*BPS-1000 Surveillance Radar*	*1985-86*
3150.303	Belouga Weapons Dispenser (BLG 66)	738
3151.303	TBA 100mm Rocket Launchers	733
3152.303	BAP 100mm Airfield Denial and Anti-Runway Weapon System	736
3153.393	*Aircraft Rear Viewer*	*1977*
3154.393	AN/ASQ-153 Pave Spike Laser Target Designator	895
3155.193	*LP7 Laser Rangefinder*	*1987-88*
3156.193	*LV3 Laser Rangefinder*	*1979-80*
3157.193	AN/PAQ-1 Laser Target Designator	402
3158.293	*Type 21 Operations Room Simulator*	*1980-81*
3159.393	*Synthetic Flight Training Systems (SFTS)*	*1982-83*
3160.393	*Air Combat Manoeuvring Range (ACMR)*	*1982-83*
3161.253	*UK Mine Countermeasures*	*1983-84*
3162.353	Westinghouse Balloon-Borne Radar	813
3163.193	*TAST Air Defence Simulator*	*1982-83*
3164.391	*Hayes Target Systems*	*1987-88*
3165.391	*Mitsoubac Decoy and Target Drone*	*1982-83*
3166.351	*VFW-Fokker UKF*	*1983-84*
3167.351	*Aerodyne (Dornier/HSD)*	*1978*
3168.351	*Aquila (YMQM-105) Mini-RPV*	*1987-88*
3169.351	*Aequare Mini-RPV*	*1981-82*
3170.253	*AWS-3 Naval Radar*	*1980-81*
3171.253	AWS-4 Naval Radar	628
3172.253	AWS-5A Series Naval Radar	628
3173.153	Siclamen IFF System	233
3174.153	AN/FPS-115 Pave Paws Radar	277
3175.203	76mm 62 Calibre Gun Mount Mk 75	508
3176.203	*Meroka Naval Air Defence System*	*1987-88*
3177.203	Single 20mm Naval Gun S20	488
3178.304	*Inventory of Aircraft Types Table*	*1987-88*
3179.104	*Land Forces Inventory Table*	*1987-88*
3180.393	*ML Ejector Release Unit No. 125*	*1979-80*
3181.303	Soviet Aircraft Armament Note	752
3182.281	*Plessey Lightweight Weapon Control Systems (WS 80 Series)*	*1979-80*
3183.261	*DECCA Swordfish NAV/TAC System*	*1977*
3184.261	IPN Tactical Data Display System	419
3185.281	*Kalle Ship LAA Fire Control System*	*1977*
3186.261	EPLO Tactical Data System	427
3187.161	Sirocco Meteorological Radar Station (QR-MX-2)	68
3188.161	Atila Artillery Automation System	69
3189.161	*Sycomore Intelligence and Operational Command System*	*1981-82*
3190.381	*M.435 Helicopter Survey System*	*1981-82*
3191.191	*Hadar Digital Message Device*	*1980-81*
3192.161	DAVID Artillery Battery Computer	75
3193.293	Protean Chaff Launcher	665
3194.053	*Series 80 Receiving System*	*1979-80*
3195.441	American Underwater Surveillance	579
3196.441	Underwater Warfare Systems	511
3197.153	Oeil Vert Radar	229
3198.131	*Mobile Spada Air Defence System*	*1979-80*
3199.391	*Andromeda RPV System*	*1987-88*
3200.193	AN/GVS-5 Laser Rangefinder	395
3201.383	*LATAR Laser Augmented Target Acquisition and Recognition equipment*	*1981-82*
3202.383	AN/AAS-33 TRAM Target Recognition and Attack Multi-Sensor	898
3203.383	AN/ASB-19 ARBS Angle-Rate Bombing Set	896
3204.383	AN/AAS-35 Pave Penny Laser Tracker	898
3205.311	*ASALM Advanced Strategic Air Launched Missile*	*1982-83*
3206.311	*New Soviet Air-to-Surface Missiles*	*1985-86*
3207.331	*New Soviet Air-to-Air Missiles*	*1977*
3208.131	Meroka 20 mm Anti-Aircraft System	199
3209.131	SA-8 (Gecko) Surface-to-Air Missile	179
3210.231	*Shield Surface-to-Air Weapon*	*1979-80*
3211.231	*Sea Phoenix Surface-to-Air Missile*	*1977*
3212.111	Copperhead CLGP (M712)	158
3213.181	Joint Surveillance System	58
3214.181	Soviet Air Defence	51
3215.181	Air Defence Systems Note	34
3216.131	US Ballistic Missile Defence Programme	30
3217.111	*Chinese Strategic Missiles*	*1980-81*
3218.161	*Field Artillery Fire Control Computer 9FA301*	*1981-82*
3219.111	SS-20 (Saber) Mobile IRBM (Soviet designation RSD-10)	9
3220.481	*Submarine Fire Control Mod 0*	*1984-85*
3221.481	Fire Control System Mk 113	445
3222.281	ASROC Mk 114 Fire Control System	445
3223.281	Mk 116 Fire Control System	445
3224.224	*Reconnaissance Systems, Introduction*	*1978*
3225.051	*US Defence Intelligence*	*1978*
3226.051	*Defence Mapping Agency*	*1978*
3227.151	*US Army Tactical Reconnaissance*	*1978*
3228.351	*Precision Location Strike System PLSS*	*1978*
3229.114	Tactical Land-based Surface-to-Surface Missiles Table	Analysis (907)
3230.203	OTO 76/62 Compact Automatic Gun	495
3231.393	AN/ASA-65 Compensation Group Adaptor	543
3232.453	DUUA-2B Active/Passive Sonar	548
3233.453	IPD70/S Integrated Sonar	557
3234.453	P2072 (AN/SQQ-14 IT) Minehunting Sonar	558
3235.453	*Selenia-Elsag Sonar Activities*	*1987-88*
3236.453	Soviet Sonar	561
3237.453	SSQ-963 Command Active Multi-Beam Sonobuoy (X17255)	575
3238.351	*AN/USD-502 (CL-289) Reconnaissance Drone*	*1987-88*
3239.351	*Aérospatiale C.22 Target Drone*	*1987-88*

Entry number	Subject	Page No/ Last edition*
3240.391	*Vandal Interim Anti-Ship Missile Target (MQM-8G)*	*1987-88*
3241.351	*Compass Cope (HALE) RPV*	*1979-80*
3242.391	*Northrop VLCEHV*	*1980-81*
3243.351	*Wisp RPV*	*1980-81*
3244.391	*HDH-10 Enmoth Target Drone*	*1982-83*
3245.391	*Soviet Drones and RPVs Note*	*1982-83*
3246.281	Sea Archer 1 FCS	440
3247.281	*WS Torpedo Fire Control System*	*1979-80*
3248.281	9LV 100 Fire Control System	427
3249.281	*OTO Melara Local Control Unit*	*1983-84*
3250.281	*Italian Navy Action Data Systems*	*1980-81*
3251.281	*Hydra Naval Data Automation System*	*1979-80*
3252.481	*A184 Submarine Torpedo FCS*	*1985-86*
3254.253	Triton S Surveillance Radar (TRS 3033)	603
3255.253	*IFF Interrogator Antenna 3973-A*	*1987-88*
3256.253	W-120 Shipboard Fire Control Radar	642
3257.253	*W-611 Naval Radar*	*1986-87*
3258.253	FLEXAR Weapon Control System	643
3259.253	*Plessey IFF Interrogator/Receiver Type PTR826*	*1983-84*
3260.253	RN Type 994 Naval Radar	631
3261.253	*Plessey Type 294 Radar*	*1983-84*
3262.253	MM/SPQ-701 Naval Radar	611
3263.253	MM/SPS-702 Naval Radar	611
3264.253	ELTA EL/M-2207 Naval Radar	607
3265.453	Sonar Type 184M	566
3266.453	G 750 Sonar	564
3267.453	*GI 777 Sonar*	*1983-84*
3268.493	G 738 Towed Decoy	564
3269.453	G 733 SLUTT and G 740 SACACS	564
3270.391	*Firebrand Target Drone (ZBQM-111A)*	*1981-82*
3271.153	Volex IV 3-D Surveillance Radar TRS 2206	235
3272.153	TRS 2230/TRS 2215 3-D Radars	228
3273.153	Mobile Coastal Surface Surveillance Radar TRS 3410	232
3274.153	Ramsa Low-altitude Surveillance Radar TRS 2140	226
3275.153	Louxor Air Defence Radar	226
3276.153	*Weather Radar TRS 2730*	*1987-88*
3277.153	Pluto Low-Coverage Surveillance Radar	244
3278.153	MM/SPQ-3 Coastal Radars	242
3279.153	SIT 431 (AN/TPX-54) IFF Interrogator	246
3280.153	*Plessey Type 431 and Type 431M Coastal Defence Radars*	*1984-85*
3281.153	*AWS-7 Coastal Search Radar*	*1980-81*
3282.153	AN/TPS-59 3-D Tactical Radar	294
3283.153	Flycatcher Radar	248
3284.153	*OPRL-4 Surveillance PAR*	*1982-83*
3285.153	*RL-41 Surveillance Radar*	*1982-83*
3286.153	Siemens Military Radars	236
3287.153	Stentor Battlefield Radar	234
3288.153	*Soviet OTH Radar*	*1979-80*
3289.153	Side Net Heightfinder	258
3290.153	Square Pair Guidance Radar	258
3291.153	Squat Eye Acquisition Radar	258
3292.153	Thin Skin Heightfinder Radar	260
3293.153	Tall King Early Warning Radar	260
3294.153	Long Talk Recovery Radar	256
3295.153	Two Spot PAR	260
3296.153	NYSA C Long Range Radar	257
3297.353	Mirage 2000 RDI and RDM Radars	772
3298.353	Iguane/Agrion/Varan Airborne Radars	773
3299.353	*EC-153 Airborne Hunter Radar*	*1985-86*
3300.353	AN/APX-103 Airborne Interrogator	810
3301.353	AN/APQ-161 Radar	803
3302.353	AN/APG-65 Airborne Radar	794
3303.353	AN/APG-66 F-16 Radar	794
3304.353	HOWLS Experimental Airborne Radar	811
3305.353	EL/M-2001B Radar	779
3306.353	EL/M-2021 Radar	780
3307.353	MM/APS-705 Radar	781
3308.353	MM/APS-707 Radar	781
3309.353	MM/APQ-706 Radar	781
3310.353	ARI 5983 Aircraft Transponder	789
3311.353	Jay Bird Airborne Radar	785
3312.253	Mk95 Mod 0 Radar Set	606
3313.181	9AA 100 Fire Control System	90
3314.181	Fieldguard Artillery Fire Direction System	93
3315.391	*Flight Refuelling Targets*	*1987-88*
3316.161	*Tactical Computer System (TCS) AN/UYQ-19(V)*	*1986-87*
3317.161	Tactical Computer Terminal (TCT) AN/UYQ-30	108
3318.181	*AGA Type E Tank Fire Control System*	*1981-82*
3319.161	*Marine Tactical Command and Control Systems*	*1980-81*

Entry number	Subject	Page No/ Last edition*
3320.161	Wavell ADP System	96
3321.181	*SFCS 600 Tank Fire Control System*	*1984-85*
3322.181	*GCE576 and GCE581 Gun Control Systems*	*1982-83*
3323.181	Soviet Army Fire Control	95
3324.181	Mobile Automatic Reporting Station (Mars-402)	45
3325.181	Oman Air Defence	48
3326.181	Austrian Air Defence (MRCS-403)	38
3327.111	SS-1 (SCUD) Battlefield Support Missile	127
3328.221	*Mariner Anti-Ship Missile System*	*1983-84*
3329.193	*Flakpanzer Training Equipment (33mm AA-Tank)*	*1982-83*
3330.131	Flakpanzer 35 mm AA Tanks	203
3331.221	*ASSM (Anti-Ship Supersonic Missile)*	*1981-82*
3332.131	*AA 20 Weapon System*	*1987-88*
3333.131	*Vadar AA Gun System*	*1980-81*
3334.231	*Shipboard Intermediate Range Combat System (SIRCS)*	*1979-80*
3335.311	AS.30 Laser Air-to-Surface Missile	708
3336.311	*LUZ 1 Air-to-Surface Missile*	*1979-80*
3337.331	AA-6 (Acrid) Air-to-Air Missile	700
3338.331	AA-7 (Apex) Air-to-Air Missile	700
3339.331	AA-8 (Aphid) Air-to-Air Missile	700
3340.331	AMRAAM (AIM-120A)	702
3341.241	*140mm Anti-Submarine Rocket Launcher*	*1979-80*
3341.441	*Torpedo Mk 23*	*1983-84*
3342.441	American Mine Warfare Programmes	538
3343.253	RQN-1 Radar Intercept Equipment	658
3344.153	Simrad RL I Laser Warning Device	318
3345.393	M-130 Countermeasures Dispenser	841
3346.393	AN/ALQ-135 Advanced Jamming System	846
3347.353	AN/ALR-56 Radar Warning Receiver	854
3348.053	*R-5000 Receiving System*	*1979-80*
3349.293	AN/SLQ-32 Shipboard EW Equipment	672
3350.063	*PSP Series Programmable Signal Processors*	*1981-82*
3351.093	*Sylvania ECM Power Amplifiers*	*1981-82*
3352.053	*WJ-940 Receiving System*	*1979-80*
3353.053	*WJ-1440 Receiving System*	*1979-80*
3354.053	WJ-1740 Receiving System	865
3355.053	WJ-1840 Receiving System	335
3356.411	SS-N-17 (Snipe) Submarine Launched Ballistic Missile	15
3357.411	SS-N-18 (Stingray) Submarine Launched Ballistic Missile (Soviet designation: RSM-50)	15
3358.311	*LASSO (AM.10) Air-to-Surface Missile*	*1979-80*
3359.311	AS.15 TT Air-to-Surface Missile	708
3360.393	AN/ALE-29A/29B Countermeasures Dispenser System	839
3361.393	*Lundy Supersonic Countermeasures Pod*	*1983-84*
3362.393	FAC Countermeasures Dispenser	840
3363.293	RBOC II	674
3364.353	*APR-42 (CMR-400B) Warning Receiver*	*1979-80*
3365.353	CMR-500B Warning Receiver	861
3366.153	*CMR-610A Wide-Band DF Receiver*	*1979-80*
3367.193	AN/TLQ-15 Communications and Countermeasures Set	341
3368.193	Combat Vehicle SACS	346
3369.393	HOFIN Hostile Fire Indicator	836
3370.053	Ceres Communications Monitoring System	321
3371.053	*Type R405 ECCM Receiver*	*1983-84*
3372.193	Type R405J Mobile ECM Equipment	320
3373.053	*Type R505 ECCM Receiver*	*1987-88*
3374.353	Type BF Radar Warning Receiver	816
3375.353	*TMV 008H Helicopter Radar Warning Receiver*	*1980-81*
3376.353	TMV 026 Airborne ESM System	818
3377.293	Dagaie Naval Countermeasures Equipment	647
3378.293	*DR 3012 Naval Countermeasures Set*	*1981-82*
3379.393	DB 3163 Jamming Pod (High-Band Remora)	816
3380.053	*EK 049 Receiver*	*1981-82*
3381.053	*ET 001 VHF/UHF Receiving System*	*1983-84*
3382.393	RQH-5(V) ESM/ELINT Airborne Systems	828
3383.393	ARI 23246/I ECM Pod (Sky Shadow)	832
3384.393	AN/ALQ-153 Tail Warning Set (TWS)	848
3385.353	*Super Cyclope Airborne Infra-red Reconnaissance System*	*1982-83*
3386.353	*FFV Uni-Recon Pod*	*1982-83*
3387.353	Green Baron Pod	877
3388.353	*Model 20076 Real-Time Reconnaissance Television Sensor*	*1979-80*
3389.163	*Telepims Photo-Interpretation Equipment*	*1982-83*
3390.163	*PIMS 220 Photo-Interpretation Table*	*1982-83*
3391.353	*Matra Lightweight Photopod*	*1980-81*
3392.353	*Type 155R Reconnaissance Pod*	*1978*

Entry number	Subject	Page No/ Last edition*
3393.163	*REVI 111 Reconnaissance Film Viewer*	*1982-83*
3394.193	*Type RS4TS Individual Weapon Sight*	*1987-88*
3395.193	*Type GS6TS Crew-Served Weapon Sight*	*1987-88*
3396.193	*Type HV5X80AT Mini-Weapon Sight*	*1985-86*
3397.193	*Type PG1MS Night Vision Goggles*	*1987-88*
3398.193	*Type PHIMC Night Vision Goggles*	*1979-80*
3399.193	*Type PMIMS Passive Viewer*	*1979-80*
3400.193	*Type PB4DS Passive Binoculars*	*1987-88*
3401.193	*PPE Trilite Night Sight*	*1981-82*
3402.193	*PPE Lolite Night Surveillance Device*	*1979-80*
3403.193	No 63 Passive Night Sight (L3A1)	388
3404.393	Nimtan Airborne Night Sight	887
3405.293	IRCOM	687
3406.193	*Night Observation Device NOD-A*	*1987-88*
3407.193	*PPE Hawklite Night Observation Device*	*1983-84*
3408.193	*PPE Pocketscope*	*1985-86*
3409.193	*PPE Night Vision Goggles*	*1987-88*
3410.193	PPE Passive Driving Periscope	387
3411.193	GSA Series Gunsights	385
3412.393	*H200 Stabilised Sight*	*1983-84*
3413.393	*APX M397 Stabilised Sight*	*1979-80*
3414.193	APX M550/TCV 80 Laser Rangefinder	351
3415.193	*APX M546 Optical Sight and Rangefinder*	*1986-87*
3416.193	TM 18 (TPV 89) Laser Rangefinder	351
3417.193	*OB-25 Night Sight*	*1980-81*
3418.193	*OB-44 Night Observation Binoculars*	*1987-88*
3419.193	*VN-4 Night Sight*	*1979-80*
3420.193	*CN2.8 Passive Vehicle Periscope*	*1979-80*
3421.193	*TN2-90 Passive Sight*	*1979-80*
3422.193	*OB-46 Night Sight*	*1981-82*
3423.393	Thomson-CSF VE 110 Head-Up Display	869
3424.293	Totem Naval Fire Control Equipment	680
3425.293	Panda Mk 2 Optical Director	679
3426.293	Lynx Optical Sight	678
3427.293	PIRANA Infra-Red Tracker	677
3428.353	*TRT FLIR*	*1982-83*
3429.193	UA 9124/9126 Day/Night Periscope System	372
3430.193	UA 9630 Series Universal Passive Periscope	372
3431.293	Type UA 9053 Thermal Imaging Camera	684
3432.193	*Type UA 1116 Individual Weapon Sight*	*1987-88*
3433.193	*Type UA 1242 Night Observation Sight*	*1987-88*
3434.093	Eltro Electro-Optical Equipment	360
3435.193	Saab IRS-700 IR Surveillance Equipment	379
3436.083	*SAAB IKA-211 IR Tracker*	*1981-82*
3437.193	Soviet Sights - Note	379
3438.193	*TP-1050L Tank Sight*	*1982-83*
3439.193	*Bantam Simulator*	*1982-83*
3440.193	*RBS70 Training Equipment*	*1982-83*
3441.193	*Mosquito Training Equipment*	*1978*
3442.193	*SURO Target Simulator*	*1982-83*
3443.193	*KAM-3TP Missile Simulator*	*1982-83*
3444.193	*KAM/9T Trainer*	*1982-83*
3445.193	*British Aerospace (BAC) Training Systems*	*1982-83*
3446.193	*Rutger ECM Simulator*	*1982-83*
3447.093	*DX.175 Tactical Training Monitor*	*1982-83*
3448.193	*DX.143 Milan Simulator*	*1982-83*
3449.093	*GEC-Marconi Simulation and Training Equipment*	*1982-83*
3450.093	*AVST Simulator*	*1978*
3451.293	*Sperry Mine Clearance Simulation*	*1982-83*
3452.093	*ILS Laser Fire Simulator Equipment*	*1982-83*
3453.193	*USMC Air Operations Simulator*	*1982-83*
3454.363	AN/ARR-72 Sonobuoy Receiving Set	581
3455.363	AN/ARR-75 Sonobuoy Receiver	582
3456.481	*TIOS D Submarine Weapon Control System*	*1981-82*
3457.293	*Seafan Rocket Decoy System*	*1978*
3458.131	*Tan-SAM Surface-to-Air Missile*	*1978*
3459.193	*Israeli Tactical Training Centre*	*1982-83*
3460.281	Elbit Israeli Action Information System	413
3461.093	Elbit MN-53 ESM System	653
3462.393	*System 80 Nav/Attack System*	*1981-82*
3463.393	*Elbit PWRS-1*	*1980-81*
3464.351	*Asmodée Reconnaissance Drone*	*1982-83*
3465.103	*Viper Anti-Tank Weapon*	*1978*
3466.103	*SAMM S365.3 Light Turret*	*1979-80*
3467.103	*SAMM S20 Turret*	*1979-80*
3468.102	*LVTP-7 Improvement Programme*	*1978*
3469.102	*Merkava Mk 1 MBT*	*1979-80*
3470.102	*BDX Armoured Personnel Carrier*	*1979-80*
3471.234	Naval Surface-to-Air Missiles Table	Analysis (912)
3472.314	Air Defence Missiles Table	Analysis (913)
3473.353	AN/APS-504(V) Series Airborne Radar	778
3474.353	*AN/UPD-6 Side Looking Airborne Radar*	*1982-83*
3475.153	AN/TPS-63 Tactical Surveillance Radar	295
3476.153	AN/TPS-65 Tactical Surveillance Radar	296
3477.153	*AN/TMQ-19 Meteorological Radar*	*1982-83*
3478.153	*MBVR 120 Tracking Radar*	*1981-82*
3479.253	DA.08 Naval Surveillance Radar	613
3480.153	*AWS-8 Coastal Radar*	*1979-80*
3481.253	*Triton X Naval Radar (TRS 3040)*	*1987-88*
3482.253	MM/SPN-703 Navigation Radar	610
3483.253	MM/BPS-704 Submarine Radar	656
3484.253	Signaal STIR Tracking Radar	612
3485.253	*RN Type 1030 STIR*	*1982-83*
3486.253	RN Type 1022 Naval Radar	631
3487.253	Transponder Reply Receiver RRB	635
3488.353	SIT 422 Shorad Interrogator	782
3489.353	SIT 421 (MM/UPX-709) Transponder	782
3490.153	Radar GS20 Hostile Fire Locator Claribel	263
3491.153	5713 Martello 3D Surveillance Radar	263
3492.153	*Precision Approach Radar TRS 2310*	*1987-88*
3493.453	*XSV-01 Expendable Sound Velocimeter*	*1987-88*
3494.253	Raytheon Standard Sonar Transmitter (SST)	586
3495.253	EDO Model 780 Series Sonar	590
3496.453	EDO Model 1102/1105 Submarine Sonar	591
3497.363	AQS-902/AQS-920 Series Acoustic Processors	562
3498.453	Fenelon - DUUX-5 Passive Sonar and Acoustic Rangefinder	549
3499.453	*SS12 Sonar*	*1979-80*
3500.453	*SS24 Sonar*	*1979-80*
3501.453	*DSTV-4L Sonobuoy*	*1984-85*
3502.453	Jezebel Sonobuoy	574
3503.453	SSQ 904A/905A/906/907 Sonobuoys	575
3504.281	Class 122 Frigate Combat System	412
3505.281	WSA-420 and CAAIS 450 Series Fire Control and Action Information Systems	431
3506.281	Ferranti Mine Countermeasures Control System (MCCS)	439
3507.481	*Control for Submarine Discharged Torpedoes (CSDT)*	*1982-83*
3508.281	Control for Surface Launched Torpedoes (CSLT)	435
3509.281	Seaguard Naval Combat System	412
3510.161	Battery Computer System	104
3511.481	Ericsson Submarine Action Information and FCS	430
3512.281	Ericsson Command and Control Information System (CCIS)	427
3513.281	*Plessey Autonomous Display (A10) Systems*	*1983-84*
3514.261	W-1200 Series Fire Control Systems	447
3515.481	*Gresham Lion Submarine Weapon Control Systems*	*1982-83*
3516.261	Goalkeeper Naval FCS	422
3517.311	MW-1 Multi-Purpose Weapon	742
3518.353	AN/APS-124 Airborne Radar	805
3519.353	AN/APS-127 Airborne Radar	805
3520.353	AN/APS-130 Airborne Radar	806
3521.181	Australian Air Defence	37
3522.181	*GCE620 Gun Control and Stabilisation System*	*1982-83*
3523.161	Morcos Mortar Computing System	99
3524.161	Gun Display System	76
3525.353	AN/ALR-68 Advanced Radar Warning System	856
3526.193	IGS-1/IGS-3P Mobile Integrated ESM/ECM System	318
3527.153	Selenia Electronic Surveillance System	318
3528.253	RQN-5 Naval ESM/ELINT System	659
3529.093	*R-5500 ESM Control Equipment*	*1982-83*
3530.093	*R-57 Series RF Tuners*	*1981-82*
3531.093	*HE 101 Active Receiving Dipole*	*1983-84*
3532.093	PA 005 and PA 010 Doppler Direction Finders	311
3533.293	*TMV 433 Naval Countermeasures Set*	*1984-85*
3534.053	TMV 200 Elisa Elint Receiver	306
3535.053	*XR100 COMINT Receiving System*	*1987-88*
3536.193	EP 1650 Automatic VHF/UHF Radio Reconnaissance System	309
3537.393	AR765 Radar Warning Receiver	829
3538.293	EWS-900 Chaff Launching Equipment	660
3539.253	Matilda Radar Warning Receiver System	664
3540.093	*Plessey Laser and IR Searchlight Detector*	*1980-81*
3541.393	AN/ALQ-78 Countermeasures Set	842
3542.393	AN/APR-39(V) Radar Warning System	860
3543.393	Rapport ECM Systems	862
3544.393	AN/APR-38 Wild Weasel CIS	859
3545.393	AN/ALQ-147A IR Countermeasures Set	847
3546.393	AN/ALQ-144 IR Countermeasures Set	847
3547.393	AN/ALQ-156 Missile Detection System	848

Entry number	Subject	Page No/ Last edition*
3548.393	*AN/ALQ-140 IR Countermeasures Set*	*1985-86*
3549.393	*AN/ALQ-132 IR Countermeasures Set*	*1982-83*
3550.393	*AN/ALQ-146 IR Countermeasures Set*	*1985-86*
3551.293	AN/SLQ-17 Naval Electronic Countermeasures Suite	671
3552.393	AN/ALQ-155(V) EW Power Management System	848
3553.293	Cutlass ESM/ECM Equipment	667
3554.393	MIR-2 ESM Equipment	835
3555.293	*Seafan 105mm Rocket Decoy System*	*1981-82*
3556.353	*Heli-Tele Airborne Reconnaissance Equipment*	*1982-83*
3557.193	TCV 107 Tank Laser Rangefinder	350
3558.193	VAQ-3/VAQ-33 Tank Laser Rangefinders	366
3559.293	Type MS7TS Marine Night Sight	684
3560.193	VS 580 Gyro-Stabilised Vehicle Sight	355
3561.293	Naja Electro-Optical Director	679
3562.393	Thomson-CSF Type 902 Gunsight	871
3563.193	*Hand-Held Laser Rangefinder*	*1981-82*
3564.393	*RGS3 Aircraft Sight*	*1985-86*
3565.053	*Model NT/RR Laser Rangefinder*	*1982-83*
3566.293	Canopus (TRS906) Lightweight Electro-Optical Director	678
3567.293	*V0063 Stabilised Electro-Optical Sight*	*1979-80*
3568.393	Saab JA37 Viggen Head-Up Display	882
3569.393	Virtual Image Display	890
3570.193	AN/VSG-X (TIS) Thermal Imaging System	401
3571.193	AN/VVS-501 Night Driving/Observation Viewer	397
3572.193	*BER-40-4 Bi-Ocular Eyepiece*	*1987-88*
3573.393	*RGS-4 Weapon Aiming System*	*1981-82*
3574.083	*EOS-500 Video Tracking Equipment*	*1981-82*
3575.083	VTS-800 Video Tracking Equipment	378
3576.193	Minilaser Rangefinder	371
3577.293	*Type MS7TS Night Observation Device*	*1979-80*
3578.193	Type TS7TS Passive Aiming Sight	370
3579.193	LP8 Laser Rangefinder	374
3580.193	*LV22 Laser Rangefinder*	*1979-80*
3581.093	RLD-2 Hand-Held Laser Rangefinder	406
3582.193	TLMD Series Tank Laser Rangefinders	406
3583.193	*ALD-M2 Lightweight Laser Rangefinder*	*1987-88*
3584.193	*MR36 Aiming System*	*1984-85*
3585.293	OGR7/2 Optical Director	682
3586.293	*OGR7/3 Optical Director*	*1983-84*
3587.193	*OG14 Tank Fire Control Equipment*	*1982-83*
3588.193	*VAQ-3/VAQ-33 Tank Laser Rangefinder*	*1984-85*
3589.193	P56 Series Computing Sight and Gun Drive	366
3590.193	*P75 Laser Computing Sight and Gun Drive*	*1984-85*
3591.193	*Weapon Delivery Simulator*	*1982-83*
3592.193	*Air Navigation Procedures Trainer*	*1982-83*
3593.193	*BOFI Gun Training Simulator*	*1982-83*
3594.293	*Operations Room Procedural Trainers*	*1982-83*
3595.293	*Action Speed Tactical Trainer (ASTT)*	*1982-83*
3596.293	*Trials Recording and Analysis Equipment*	*1979-80*
3597.293	*Elettronica Naval EW Equipment*	*1980-81*
3598.293	Scimitar Jammer	664
3599.253	AN/ASQ-81(V) Airborne ASW Magnetometer	598
3600.451	*Dornier LA-RPV*	*1980-81*
3601.351	*Dornier Experimental Spähplattform*	*1987-88*
3602.351	*Dornier Mini-Drone/RPV*	*1987-88*
3603.351	*Tadiran Mastiff Mini-RPV Mk III*	*1987-88*
3604.391	*MBB Locust Mini-Drone*	*1983-84*
3605.351	*MBB RPV Programmes*	*1980-81*
3606.351	*Wideye/Supervisor Surveillance RPH System*	*1980-81*
3607.181	TMLD Low-Level Reporting and Control System	43
3608.111	AT-5 (Spandrel) Anti-Tank Missile	150
3609.363	TADS/PNVS	892
3610.111	*AHAMS Advanced Heavy Anti-Tank Missile System*	*1980-81*
3611.111	*ATEM Anti-Tank Missile*	*1982-83*
3612.231	*Seafox Naval Point Defence System*	*1980-81*
3613.231	*Vanessa Surface-to-Air Missile*	*1981-82*
3613.231	Goalkeeper Anti-Missile Close-In Weapon System	467
3614.231	*Flash Naval Air Defence System*	*1979-80*
3615.231	*NATO 6S Naval Air Defence System*	*1979-80*
3617.131	Vehicle-Mounted RBS 70 Anti-Aircraft Missile	175
3618.131	*DIVAD Sgt York Gun System*	*1985-86*
3619.131	Tan-SAM Surface-to-Air Missile (Type 81)	174
3620.131	SA-10 (Grumble) Surface-to-Air Missile	180
3621.131	*SIAM Air Defence Missile*	*1982-83*
3622.331	ASRAAM (AIM-132)	695
3623.441	F-17 Torpedo	514

Entry number	Subject	Page No/ Last edition*
3624.441	MR-80 Sea Mine	520
3625.441	*VS SM600 Sea Mine*	*1987-88*
3626.441	Thomson-Sintra Sea Mines	516
3627.441	Soviet Sea Mines	527
3628.241	Troika MCM System	518
3629.441	Torpedo Mk 32	534
3630.321	Sea Eagle Air-to-Surface Missile	720
3631.311	*Sabre Anti-Armour Air-to-Surface Missile*	*1979-80*
3632.311	*RB83 Air-to-Surface Missile*	*1979-80*
3633.311	Kingfish Air-to-Surface Missile (AS-6)	717
3634.311	*Kerry Air-to-Surface Missile (AS-7)*	*1985-86*
3635.311	ASMP Air-to-Surface Missile	709
3637.311	WAAM Programme	758
3638.111	AT-4 (Spigot) Anti-Tank Missile	149
3639.111	AT-6 (Spiral) Anti-Tank Missile	150
3640.153	*R-200 Series Battlefield Radars*	*1982-83*
3641.153	*AN/PPS-10 Radar Signal Detector*	*1982-83*
3642.353	AN/ALR-66 Series Radar Warning Receivers	854
3643.293	Buck/Wegmann Decoy System	651
3644.453	*Simrad Sonars*	*1982-83*
3645.351	*Stabileye Mini-RPV*	*1987-88*
3646.153	EL/M-2 106 Point Defence Alert Radar	241
3647.153	*EL/M2108 Battlefield Radar*	*1982-83*
3648.153	*STARTLE Radar*	*1987-88*
3649.153	AR-320 Three-Dimensional Air Defence Radar	269
3650.293	Chaff Rocket	663
3651.131	*SA 10 Surface-to-Air Missile*	*1979-80*
3652.391	*QF-100 Full-Scale Aerial Target (FSAT)*	*1987-88*
3653.353	*Infra-Red Linescan 204*	*1982-83*
3654.353	*Model 24129 Real-Time Reconnaissance Television Sensor*	*1982-83*
3655.203	Type A Naval Mounting for 20mm F2 Gun	488
3656.203	20mm Naval Gun Type GAM-CO1	499
3657.203	25mm Naval Gun Type GBM-AO1	498
3658.203	*LS30R 30mm Naval Gun Mounting*	*1985-86*
3659.203	*105mm Autonomous Naval Patrol Gun*	*1982-83*
3660.193	*TCV 109 Tank Laser Rangefinder Sight*	*1985-96*
3661.393	TCV 115 Airborne Laser Rangefinder	867
3662.293	TMY113 Laser Rangefinder/Illuminator	681
3663.193	TCY 133 Laser Rangefinder/Illuminator	351
3664.093	WJ-945 Receiving/DF System	353
3665.093	WJ-1205B Video Digitiser Unit (VDU)	333
3666.193	WJ-1424 Microwave Receiver (TLR-31)	335
3667.103	*TRIM F2 Turrets*	*1982-83*
3668.103	*SAMM ALTO Light Turrets*	*1982-83*
3669.103	*SAMM SABRE Twin 20mm Turrets*	*1982-83*
3670.103	*155mm Howitzer M-68*	*1982-83*
3671.103	*155mm Howitzer M-74*	*1982-83*
3672.111	*140mm Multiple Rocket Launcher*	*1985-86*
3673.111	Type 75 130 mm Multiple Rocket Launcher	124
3674.111	SLAM-Pampero 105 mm Multiple Rocket System	113
3675.111	SAPBA-1 127 mm Multiple Rocket System	114
3676.103	*Model 77 155mm Towed Howitzer L.33 X.1415*	*1982-83*
3677.111	Mathogo Anti-Tank Missile	136
3678.311	Martin Pescador Air-to-Surface Missile	707
3679.131	*Dragon 30mm Twin Gun AA System*	*1985-86*
3680.131	*Twin 20mm AA Weapon System (TSE6200)*	*1982-83*
3681.131	Wildcat Twin 30 mm AA Weapon System	193
3682.353	*AN/APQ-163 Radar*	*1987-88*
3683.153	TRS 2054 Long Range Surveillance Radar Family	228
3684.453	Calypso IV Submarine Radar TRS 3110	606
3685.253	Triton G Doppler Surveillance Radar TRS 3050	603
3686.253	*ST804 Naval Radar*	*1982-83*
3687.253	S1820 Naval Surveillance Radar	634
3688.353	AN/APX-72 Airborne IFF Transponder	809
3689.353	*AN/APX-104 Airborne IFF Interrogator*	*1986-87*
3690.353	AN/APX-76 Airborne IFF Interrogator	809
3691.353	MM/UPX-709 Airborne IFF Transponder	810
3692.153	CEL 850/4 IFF Interrogator	274
3693.153	*870 Series IFF Interrogators*	*1986-87*
3694.153	880 Series IFF Interrogators	274
3695.353	IFF 2720 Airborne Transponder	789
3696.353	IFF 3100 Airborne Transponder	789
3697.353	IFF 3500 Airborne Interrogator	790
3698.253	*IFF 800 Naval Interrogator (Type 1010)*	*1980-81*
3699.253	CEL 850/4 Naval Interrogator	627
3700.253	*IFF 877 Naval Interrogator*	*1982-83*
3701.253	IFF 2750 (Type 1013) Naval Transponder	627
3702.153	NPG-880 3-D Air Defence Radar	248
3703.153	AN/TPX-46(V) IFF Interrogator	298

Entry number	Subject	Page No/ Last edition*
3704.153	*AN/TPX-50 IFF Interrogator*	*1984-85*
3705.153	AN/TPX-54 IFF/SSR Interrogator	298
3706.153	*OREST Mobile Air Defence Radar*	*1984-85*
3707.153	*ASR 920 Mobile Radar*	*1983-84*
3708.153	EL/M-2215 Surveillance Radar	241
3709.253	EL/M-2221 Naval Fire Control Radar	607
3710.353	Sea Searcher Airborne Radar ARI 5991	791
3711.153	AN/TPN-24 Airport Surveillance Radar	286
3712.153	AN/TPN-25 Precision Approach Radar	287
3713.153	AN/GPN-22 Precision Approach Radar	282
3714.153	AN/FPN-62 Precision Approach Radar	277
3715.153	Falconer Precision Approach Radar	263
3716.153	*RASURA II Battlefield Surveillance Radar*	*1982-83*
3717.153	*RA20 Fire Control Radar*	*1982-83*
3718.153	*RA21 Fire Control Radar*	*1982-83*
3719.153	*Rodeo 20 (DD VC-1) Fire Control Radar*	*1982-83*
3720.153	*AN/TPQ-27 Radar Course Directing Central*	*1981-82*
3721.153	Hughes Air Defence Radar (HADR)	300
3722.253	AN/SPS-67 Radar	641
3723.131	Oerlikon Escorter Twin 35 mm SP Air Defence System	203
3724.131	AW-X Self-Propelled Twin 35 mm AA Gun System	199
3725.103	*TF 20.15 One-Man Turret*	*1982-83*
3726.103	*TS 7 Two-Man Turret*	*1982-83*
3727.281	*Seafire Naval Fire Control System*	*1987-88*
3728.103	*Emerson Cannon Turret*	*1982-83*
3729.203	*TCM 30 T7in 30mm Naval Gun*	*1986-87*
3730.391	*Skeet Target Drone*	*1987-88*
3731.391	*BQM-74C Aerial Target*	*1987-88*
3732.391	*Lockheed Army RPV*	*1982-83*
3733.391	*Dornier Do 34 Kiebitz*	*1984-85*
3734.391	*Dornier Aerial Target System (DATS III)*	*1984-85*
3735.391	*Lockheed RPV programmes*	*1982-83*
3736.391	*Scout Mini-RPV*	*1987-88*
3737.391	*Shrike Target RPV*	*1984-85*
3738.391	*Sky-Eye Mini-RPV*	*1984-85*
3739.391	*E-Systems E-45 Mini-RPV*	*1983-84*
3740.391	*FQM-117A Target (RCMAT)*	*1984-85*
3741.253	DRBJ 11 Naval Radar	602
3742.481	*STWS-2 Shipborne Torpedo Weapon System Mk 2*	*1986-87*
3743.481	DLA Anti-Submarine Fire Control System	410
3744.281	NAVKIS	425
3745.261	*Thomson-CSF Tactical Data Handling System*	*1980-81*
3746.481	Outfit DCC Submarine TDHS	432
3747.281	EOS-400 Fire Control System	429
3748.281	R-76 Fire Control System	441
3749.281	*H-930 Combat Control System*	*1982-83*
3750.281	*OFLA-M Fire Control System*	*1983-84*
3751.193	CN2-500 Series Passive Vehicle Periscope	353
3752.193	OB-47 Night Sight	352
3753.193	*OB-48 Night Sight*	*1980-81*
3754.193	*TJN2-90-A Combined Day/Night Sight*	*1985-86*
3755.193	LV5 Laser Rangefinder	374
3756.193	LV3 Laser Rangefinder	380
3757.193	*LV10 Laser Rangefinder*	*1985-86*
3758.193	*LV14 Laser Rangefinder*	*1980-81*
3759.193	*LV15 Laser Rangefinder*	*1985-86*
3760.193	*LP7 Laser Rangefinder*	*1985-86*
3761.193	*LP8 Laser Rangefinder*	*1985-86*
3762.193	*LV5 Laser Rangefinder*	*1985-86*
3763.393	APX M334-04 Gyro-Stabilised Sight	868
3764.111	FIROS 6 51 mm Field Rocket System	122
3765.093	Vinten Tracking Mount and Tripods	395
3766.481	Gipsy Submarine Data Handling and Weapon Control System	423
3767.281	LIROD 8 Naval Director	420
3768.253	R-76 Naval Fire Control Radar	643
3769.453	SSQ-947B Del Ranger Sonobuoy	575
3770.453	SSQ-937 DEL Miniature Bathythermal Buoy	575
3771.441	*CI-1 Submarine Target*	*1987-88*
3772.453	Micropuffs Passive Ranging Sonar	593
3773.453	MD100/S Passive Rangefinder Set	557
3774.453	*Beluga Panoramic Sonar*	*1981-82*
3775.391	*KT-04 Towed Target*	*1982-83*
3776.253	Radio Receiver R-1651/ARA Sonobuoy OTPI	584
3777.363	AN/AKT-22(V)4 Telemetry Data Transmitting Set	580
3778.363	AN/ARR-78(V) Advanced Sonobuoy Communication Link (ASCL)	583
3779.351	*Tucan Mini-RPV*	*1987-88*
3780.453	AN/SSQ-527B Sonobuoy	544
3781.453	*AN/SSQ-36 Bathythermograph Buoy*	*1982-83*
3782.453	AN/SSQ-53C Sparton Dwarf DIFAR Sonobuoys	597
3783.453	*Tuned Vertical Array Omni Sonobuoy*	*1987-88*
3784.453	*Widetrac Sonobuoy*	*1987-88*
3785.453	*AN/BRT-1 Radio Transmitting Buoy*	*1982-83*
3786.453	AN/SSQ-71 ATAC Buoy	597
3787.453	Type 2016 Fleet Escort Sonar	566
3788.363	Sparton Airborne Sonar Processing Equipment	582
3789.453	Plessey-Sippican XBT and XSV	576
3790.453	*SS 12/24/48 Surface Ship Sonars*	*1980-81*
3791.363	TSM 8200-8202 Airborne Sadang Acoustic System	546
3792.391	*ASAT Target Drone*	*1987-88*
3793.393	*FFV Uni-Flir Pod*	*1987-88*
3794.093	GEC Sensors V300 Series CCTV Equipment	687
3795.093	GEC Avionics CCD Cameras	881
3796.093	Argus Day/Night Sight	389
3797.161	Lemstar Tank Fire Control System	72
3798.353	*Rigel IR Surveillance Equipment*	*1982-83*
3799.353	*IR Linescan 214*	*1982-83*
3800.261	*MBT Tads*	*1983-84*
3801.393	WDNS-391 Weapon Delivery System	873
3802.093	Gyro-Stabilised Sighting System (GSSS)	362
3803.303	*PB-101 Practice Bomb*	*1983-84*
3804.103	*TCM-20 20mm Twin Anti-Aircraft Gun*	*1982-83*
3805.303	HBS 202 Helicopter Armament System	741
3806.303	CBAS-1 Albatros Aircraft Rocket	727
3807.353	AN/AAS-40 Seehawk FLIR	893
3808.363	AN/AVQ-27 Airborne Laser Target Designator Set (LTDS)	894
3809.363	Northrop TCS (Television Camera Set)	891
3810.193	BM 8025 Passive Night Sight	361
3811.353	*Corsaire Airborne Infra-Red Linescan*	*1982-83*
3812.331	*Sidewinder (AIM-9N, AIM-9P) Air-to-Air Missile*	*1983-84*
3813.093	*Type L AA Laser Sight*	*1987-88*
3814.193	M1 Laser Rangefinder	400
3815.393	LAAT Sight	889
3816.193	AN/PAQ-3 Modular Universal Laser Equipment (MULE)	402
3817.193	SS130 Night Driving Periscope	390
3818.393	*RGS1 Optical Sight*	*1985-86*
3819.393	Helios Helicopter Observation System	876
3820.393	*ATLIS II*	*1981-82*
3821.393	*AN/AXQ-15 Helicopter Tow Training Device*	*1982-83*
3822.193	*Model 350 Night Driving Sight*	*1987-88*
3823.093	MCTNS Man-Portable Common Thermal Night Sights	403
3824.193	AN/TAS-4 TOW Night Sight	403
3825.193	*AN/TAS-5 Dragon Night Sight*	*1985-86*
3826.193	AN/TAS-6 (AN/UAS-11) Night Observation Device, Long Range	404
3827.093	*EPO409/010 Optical Fire Director*	*1980-81*
3828.393	*Martin Marietta Mast-Mounted Sight*	*1983-84*
3829.493	Barr & Stroud Submarine Attack Periscope	685
3830.493	Barr & Stroud Submarine Search Periscope	686
3831.493	Submarine Periscope Image Intensifier	686
3832.493	*Submarine Laser Rangefinder*	*1985-86*
3833.093	*Types LF6/LF7 Laser Designator/Rangefinder*	*1981-82*
3834.093	IR18 Mk II Thermal Scanner	381
3835.393	*PGS 400/600 Sensor Platforms*	*1981-82*
3836.093	Ericsson Laser Tracker	378
3837.093	Contraves Day/Night Remote Video Equipment	901
3838.193	APX M363 Fire Control	356
3839.193	*Cotac APX M401*	*1980-81*
3840.193	APX M520 Panoramic Sight	356
3841.193	APX M527 Tank Sight	356
3842.193	*Cotac APX M581*	*1980-81*
3843.181	APX M586 Tank FCS Update	67
3844.181	*Costac Tank Fire Control Equipment*	*1983-84*
3845.193	Mira Thermal Imager	354
3846.393	AN/AXQ-14 Weapon Control Equipment	901
3847.193	*Chamois IR Imaging Equipment*	*1981-82*
3848.193	B171-II IR Aiming Unit	359
3849.193	CE632 Laser Rangefinder	359
3850.193	CE624 Laser Rangefinder/Sight	359
3851.193	CE628 Laser Rangefinder	359
3852.093	CE618 Laser Rangefinder	359
3853.393	CE626 Laser Transmitter	873
3854.393	*Hector Imaging IR Equipment*	*1984-85*
3855.393	*Hades Helicopter Laser Angle Tracker*	*1985-86*
3856.393	Viviane/Hot Helicopter Night Sight	868
3857.293	Type 771 Low Light TV Camera	677

Entry number	Subject	Page No/ Last edition*
3858.353	*Omera Reconnaissance Cameras*	1982-83
3859.163	*TI Advanced Sensor Data Processing Equipment*	1982-83
3860.353	*ESSWACS Camera System*	1982-83
3861.281	NA 30 Series Fire Control System	415
3862.303	*Helicopter Weapon System Z*	1983-84
3863.303	BAT 120 Anti-Armour Weapon System	736
3864.303	Retarded 125 kg Close Support Bomb	735
3865.303	BM 400 Multi-Munition Bomb	735
3866.393	*Alkan Twin-Store Carriers*	1987-88
3867.303	Helicopter Gun Mounting 19A	733
3868.303	*81mm SNORA Rocket (Italy)*	1984-85
3869.303	*Aramis Helicopter Weapons Pod*	1982-83
3870.161	Elbit Fire Control System	76
3871.104	*Armoured Fighting Vehicles Table*	1987-88
3872.281	B515/M/3 Torpedo Launching System	522
3873.491	MIN Mine Countermeasures System	523
3874.261	*NA21 Fire Control System*	1980-81
3875.093	*L6-24 DF System*	1986-87
3876.193	*EW Engagement Simulator*	1982-83
3877.363	AN/ALQ-131 Receiver/Processor	846
3890.303	Hughes Lightweight Rocket Launcher M260	767
3891.303	GAU-13/A 30mm Lightweight Gun	760
3892.363	*ALR-46A Digital Signal Processor*	1985-86
3893.053	*Dalmo Victor Mk III Miniature Digital Radar Warning System*	1986-87
3894.393	AN/ALQ-165 Airborne Self Protection Jammer	850
3895.193	AN/MLQ-34 TACJAM	340
3896.093	*Model 150-GL-5 Wideband Amplifier*	1981-82
3897.053	*GTE Bragg Cell Receiver*	1987-88
3898.193	*FIDS*	1982-83
3899.193	*Radar Warning Trainer*	1981-82
3900.311	*Advanced Airborne Anti-Armour Weapon (AST 1227)*	1982-83
3901.393	AN/AAR-44 Infra-Red Warning Receiver	838
3902.193	*LQ-101 Hand-Emplaced Jammer/Training Unit*	1982-83
3903.053	Series 2000 Receiver System	339
3904.393	AR 777 Airborne Microwave Receiver	829
3905.153	*PR 2250 Surveillance Receiver*	1984-85
3906.193	*LH 100 Laser Rangefinder*	1983-84
3907.093	*Signaal EW Equipment*	1980-81
3908.353	AN/ALR-606(V)1 and (VE) Radar Warning Receivers	859
3909.353	*ALR-646 Airborne ESM Equipment*	1981-82
3910.253	SLR-600 Series Naval ESM Equipment	673
3911.293	*Stockade Naval Decoy Equipment*	1982-83
3912.193	MEL Modular EW Programme	327
3913.193	French Land-Based Communications EW Systems	306
3914.153	*Thomson-CSF Communications Intelligence Equipment*	1986-87
3915.193	*Alligator Jammer*	1981-82
3916.193	*DB 3141 Jammer Pod*	1981-82
3917.193	RACJAM-V Responsive VHF Jammers	323
3918.193	RS 3153 Radio Surveillance Equipment	323
3919.153	RA 1794 HF/VHF/UHF Tactical EW Receiver	321
3920.153	Racal Tactical Direction Finding Systems	322
3921.153	*RA1779 Programmable Receiver*	1982-83
3922.153	RA1792 Programmable Receiver	321
3923.193	*SAAB Portable Target System*	1982-83
3924.193	*SAAB Stationary Target System*	1982-83
3925.193	*BT41 Tank Combat Simulator*	1982-83
3926.193	*RMS/ACMI Multi-Lateration Position Location System*	1982-83
3927.353	*AADS-1220 Infra-Red Linescan*	1982-83
3928.193	*Visor Laser Recorder*	1982-83
3929.163	*Goodyear Modular Processing Systems*	1982-83
3930.163	*ES-85A Strike Camera System*	1982-83
3931.163	*ES-95A Photo/Radar Reconnaissance Processing System*	1982-83
3932.161	*C12 Field Artillery Computer*	1984-85
3933.393	*SFIM Helicopter Training Sight*	1982-83
3934.161	ATIBA Artillery Fire Control System	69
3935.161	CAC 101 Field Artillery Computer	68
3936.151	REMBASS	109
3937.161	AN/GYG-501 MILIPAC Artillery Computer	65
3938.161	*TOS Tactical Operating System*	1980-81
3939.161	BC81 (Re 83) (Fargo) Artillery Computer	74
3940.161	US Tactical Data Handling Projects	110
3941.151	*SOTAS (Stand-Off Target Acquisition System)*	1983-84
3942.161	*Portable Intelligent Field Terminals*	1985-86
3943.161	MIFASS	104
3944.161	PLRS (AN/TSQ-129)	108
3945.181	GEADGE	43

Entry number	Subject	Page No/ Last edition*
3946.193	*TS2 Skyguard Simulator*	1982-83
3947.193	*Weibull AA Gunnery Trainers*	1982-83
3948.193	*Submarine Operations Room Simulator*	1981-82
3949.193	*AN/BQR-T4 Sonar On-Board Trainer*	1982-83
3950.193	*F-18 Weapons Tactics Trainer*	1982-83
3951.193	*Gabriel Weapon System Trainer*	1982-83
3952.193	*ASWT-2 Anti-Submarine Warfare Trainer*	1982-83
3953.193	*TALISSI Shot Simulation System*	1982-83
3954.193	*Vulcan Training System*	1982-83
3955.193	*Tactical Radar Threat Generator*	1982-83
3956.193	*Sperry Air Simulation and Training Equipment*	1982-83
3957.193	*Sperry Naval Training Equipment*	1982-83
3958.193	*Sperry Army Training Equipment*	1982-83
3959.193	*MILES*	1982-83
3960.193	*Tank Crew Trainer*	1982-83
3961.193	*MKT Automatic AA Weapon Trainer*	1982-83
3962.193	*Honeywell Sonar Operator Trainers*	1982-83
3963.193	*Device 14A2*	1982-83
3964.193	*RASIF Simulator*	1982-83
3965.441	*TAR 6 & TAR 16 Mines*	1984-85
3966.241	LAMPS Mk III	532
3967.451	*FORACS*	1982-83
3968.241	FRAS-1 Anti-Submarine Weapon	526
3969.241	SS-N-14 Anti-Submarine Weapon (Silex)	526
3970.441	SS-N-15 (Starfish)/16 (Stallion) Anti-Submarine Weapon	526
3971.441	*NT-37C Torpedo*	1983-84
3972.131	Shahine Air Defence System (TSE 5100)	170
3973.221	FL-1 (CSS-N-2) Naval Surface-to- Surface Missile	449
3974.121	*AS.15TT Coastal Defence System*	1980-81
3975.321	RBS 15 Anti-Ship Missile	715
3976.221	Saab RBS 15 Anti-Ship Missile	454
3977.131	*Mobile Blowpipe AA Missile System*	1981-82
3978.193	*OB-49 Combined Day and Night Aiming Sight*	1985-86
3979.181	SOPTAC 11 Fire Control System	67
3980.093	*SIRETRAC Infra-Red Tracker*	1981-82
3981.393	ALT Airborne Laser Tracker	892
3982.393	*Firefly III*	1987-88
3983.231	Sea Chaparral Surface-to-Air Missile System	479
3984.121	HY-2 Coast Defence Missile System	160
3985.303	*MATRA BGL Laser-Guided Bombs*	1982-83
3986.111	LAW 80 Light Anti-Armour Weapon	151
3987.111	*US Rocket Assisted Projectiles (RAP)*	1986-87
3988.111	SADARM System	158
3989.111	*Assault Breaker*	1983-84
3990.111	*SMAW Shoulder-Launched Multi-Purpose Assault Weapon*	1980-81
3991.111	*Picket Anti-Tank Missile*	1986-87
3992.111	*Viper Anti-Tank Missile*	1983-84
3993.001	Cruise Missile Programme	27
3994.203	Breda Twin 30mm Compact Naval Mounting	492
3995.203	Breda SCLAR Multipurpose Rocket Launcher	493
3996.203	Breda Multi-Calibre Assault Launcher	494
3997.293	Breda Light Chaff Rocket Launching System	659
3998.453	SSQ-801 Barra Sonobuoy	540
3999.353	AN/ASQ-502 Submarine Detecting Set (MAD)	541
4000.453	AN/SQS-509 Sonar	542
4001.253	PHS-36 Series Sonars	559
4002.383	System 82 WDNS	874
4003.393	*Ferranti Type 105D Air-to-Surface Laser Ranger*	1985-86
4004.253	*Dolphin Search Radar Module for Seaguard*	1982-83
4005.441	Manta Mine	521
4006.441	CAM Counter-Mine Charge	521
4007.453	MISAR MSR-810 Sonobuoy	558
4008.193	SNAP-1 Steerable Null Antenna Processor	345
4009.393	AN/ALQ-178 ECM System	851
4010.393	AN/APR-43 Tactical Radar Warning Receiver	860
4011.383	Aquila Mission Payload Sub-System	891
4012.373	AN/ASA-82 Tactical Data Display System	580
4013.373	AN/ASA-66 Tactical Data Display	580
4014.281	Mk 82 (AEGIS) Gun & Missile Director	442
4015.131	PIVADS (Product Improved Vulcan Air Defence System)	209
4016.153	Seaguard CIWS	468
4017.203	Sea Zenith 25mm Mount Type GBM-B1Z	498
4018.153	*Alerter Radar*	1986-87

Entry number	Subject	Page No/ Last edition*
4019.281	*Sea Serpent FCS*	*1983-84*
4020.293	*Sea Sentry Naval ESM*	*1987-88*
4021.293	*Model 910 Stabilised Sight*	*1987-88*
4022.293	*Mk 945 Optical Gun Director*	*1982-83*
4023.293	Model 985 Optical Director	690
4024.293	Model 975 Optronic Director	689
4025.281	9 LV 400 Series C^{3}I System	429
4026.281	9 LV 300 Naval Weapon Control and C^3 System	428
4027.281	9CSI 500 STINA	429
4028.281	9CSI STINA	429
4029.153	9KR 400 Coastal Radar	251
4030.103	*155mm Field Howitzer 77B*	*1982-83*
4031.203	Single 57 mm Mk 2 Naval Gun	496
4032.203	*20mm Naval Gun Type GAI-BM1*	*1981-82*
4033.103	*35mm Gun Turret Types GDD-A & GDD-B*	*1982-83*
4034.103	*25mm Gun Turret Type GBD*	*1982-83*
4035.103	*105mm Self-Propelled Howitzer Type 74*	*1982-83*
4036.391	*Firebolt Model 305 HAHST (AQM-81A)*	*1987-88*
4037.391	*Locust Harassment Drone Programme*	*1983-84*
4038.391	*VFW Locust Harassment Drone*	*1983-84*
4039.391	*E-Systems E-55 Mini-RPV*	*1983-84*
4040.391	*E-Systems Multiple Mission RPVs*	*1987-88*
4041.203	KV-SK/20 Dual-Purpose 20mm Naval Mounting	494
4042.203	Sea Vulcan 30 Lightweight 30mm Naval Gun Mount	509
4043.203	Sea Vulcan 20 20mm Naval Mount	509
4044.481	*TIOS E Tactical Information Organisation System*	*1983-84*
4045.261	CACS 1	431
4046.261	ADLIPS	408
4047.281	*LINCE FCS*	*1987-88*
4048.253	W-160 Shipboard Fire Control Radar	642
4049.261	RADOP Series Naval Fire Control Systems	408
4050.311	*SOM (Stand-Off Missile)*	*1981-82*
4051.253	ZW Series Naval Radars	613
4052.283	*Toucan E-O Mounting*	*1984-85*
4053.261	TAVITAC Tactical Data Handling Systems	409
4054.293	AN/SLQ-503 (RAMSES) Shipborne ECM System	646
4055.293	Sphinx Naval ESM	665
4056.293	RAPIDS Naval ESM	663
4057.261	DG Naval Fire Control System	414
4058.281	NA 18 Fire Control System	414
4059.281	NA 21 Fire Control System	414
4060.111	Multiple Launch Rocket System (MLRS)	133
4061.303	GAU-12/U Equaliser 25mm Automatic Gun	760
4062.131	*Vulcan/Commando Air Defence System*	*1986-87*
4063.493	*Submarine Attack Teacher*	*1982-83*
4064.293	*Mine Countermeasures Operations Room Trainer*	*1982-83*
4065.093	*Part-Task Radar Trainer*	*1982-83*
4066.493	*Sonar Analysis Initial Trainer (SAINT)*	*1982-83*
4067.493	*Ferranti Sonar Trainers*	*1982-83*
4068.103	*Rheinmetall Rifled or Smooth Bore Tank Gun Rh 105-30*	*1982-83*
4069.103	*20mm AA Twin Gun*	*1982-83*
4070.303	*Oerlikon Helicopter Armament*	*1987-88*
4071.303	*Aerea Rocket Pods*	*1987-88*
4072.303	*SUU-21/A Bomb Dispenser*	*1983-84*
4073.303	BRD Bomb and Rocket Dispensers	744
4074.303	Aerea Twin Gun Pod 7.62	743
4075.153	AN/TPN-22 Precision Approach and Landing Radar	285
4076.153	Series 52 Precision Approach Radar	277
4077.153	PAR-80 Precision Approach Radar	286
4078.153	Series 320 3-D Air Defence Radars	291
4079.153	AN/MPQ-63 Instrumentation Ballistic Radar (ARBAT)	292
4080.153	VSTAR/VSTAR-PT Air Defence Radar	301
4081.153	JAN/MPQ-N1 (Type 92) Mortar Locating Radar	247
4082.153	ISC Cardion Coastal Radar System	303
4083.353	Ericsson SLAR	783
4084.253	Kite Screech Fire Control Radar	620
4085.253	Bass Tilt Fire Control Radar	617
4086.253	Top Steer Radar Group	626
4087.253	Top Pair Radar Group	625
4088.253	Top Dome Radar Group	624
4089.253	Eye Bowl Naval Radar	618
4090.253	Band Stand Naval Radar	617
4091.293	*Helicopter Sonar Simulator*	*1982-83*
4092.193	*B-52 Weapon Systems Trainer*	*1982-83*
4093.193	*Naval Multi-Environmental Trainer*	*1982-83*
4094.453	SS 12 Small Ship Sonar (Hull and VDS)	549
4095.453	HS 12 Active/Passive Helicopter Sonar (TSM 8252)	550
4096.453	*SS 24 Sonar*	*1985-86*
4097.453	*SS 48 Sonar*	*1985-86*
4098.111	Israeli 290 mm Medium Range Rocket (MAR 290)	121
4099.393	JAM PAC(V) ECM Jamming System	863
4100.393	AN/ALQ-176(V) ECM Pod	851
4101.193	*AV No. 62 Commanders Sight*	*1984-85*
4102.193	*Type DC1032/DC1033 General Purpose Surveillance Sight*	*1984-85*
4103.193	LP6 Laser Rangefinder	380
4104.193	*Digital Goniometer*	*1985-86*
4105.193	*LH80 Hand-Held Laser Rangefinder*	*1987-88*
4106.293	VAMPIR Surveillance Equipment	678
4107.093	SMT Modular Thermal Imaging System	356
4108.493	SOPELEM Submarine Periscopes	681
4109.293	*ROD Remote Optronics Director*	*1985-86*
4110.393	(AN/AAS-38) F/A-18 FLIR	899
4111.193	*TADPOLE*	*1982-83*
4112.193	AN/VSG-2 Tank Thermal Sight	396
4113.393	Rockwell Mast-Mounted Sight	891
4114.393	OR-89/AA FLIR	897
4115.393	OR-5008/AA FLIR	898
4116.393	AN/AAS-36 IRDS	899
4117.393	AN/AAQ-10 FLIR	898
4118.393	AN/AAQ-9 FLIR	898
4119.393	AN/AAS-37 DRTS	899
4120.393	AN/AAR-42 FLIR	897
4121.193	*BER-25 Series Bi-ocular Eyepieces*	*1987-88*
4122.193	*BER-47-1 Bi-ocular Eyepiece*	*1987-88*
4123.111	Oerlikon 81mm Type RWKO14 Rocket Launcher	125
4124.193	*USAF 485L Stem Simulator*	*1982-83*
4125.153	*Quadradar Mk V ATC Radar*	*1987-88*
4126.193	*IMPACT*	*1982-83*
4127.453	Type 2034 Sonar	569
4128.453	SS105 Scanning Sonar	560
4129.093	Radamec Series 2000 Fire Control and Surveillance Systems	688
4130.393	AV-8B Head-Up Display	881
4131.193	L1A2 Artillery Director	386
4132.153	S-Band Meteorological Radar	218
4133.153	*ASR-BR1 Airport Surveillance Radar*	*1987-88*
4134.303	Oerlikon 81mm Aircraft Rocket	749
4135.303	Oerlikon Rocket Pods	750
4136.393	AN/ASG-26A Lead Computing Optical Sight	888
4137.393	AN/ASG-29 Lead Computing Optical Sight	889
4138.093	ESM 500 Series VHF/UHF Receivers	311
4139.093	ESP Scanning Receiver	311
4140.153	Wimera Tracking Radar	236
4141.153	J/MPQ-P7 Artillery Locating Radar	247
4142.161	*EPCS Electronic Plane Conversion System*	*1984-85*
4143.161	*DF1-5 Fire Control Systems*	*1986-87*
4144.161	HEROS Army Command, Control and Information System	71
4145.303	Aircraft Armament (Brazil)	727
4146.111	AVIBRAS Artillery Rockets	114
4147.181	COTAC FCS	66
4148.161	Ferranti Fire Control System	103
4149.193	Model 9898 Gunner's Periscope	399
4150.193	Model 9894 Gunner's Periscope	398
4151.193	Model 9897 Laser Rangefinder	399
4152.193	Model 9896 Night Vision Elbow	399
4153.193	Model 9892 Driver's Viewer	398
4154.193	Model 9891 Commander's Periscope	398
4155.193	M32E1 Model 9880 Night Vision Elbow	398
4156.181	*VARO Modular Tank FCS*	*1986-87*
4157.181	PADS Position and Azimuth Determining System	100
4158.131	ADATS Missile System	171
4159.111	ADATS (Air Defence Anti-Tank System)	140
4160.393	ARI 18240/1 ESM	862
4161.293	AN/SLQ-501 CANEWS EW System	646
4162.193	Type S373 ECM Equipment	326
4163.193	Barbican EW System	326
4164.393	AN/ALR-73 Passive Detection System	857
4165.393	*RIDE ESM Equipment*	*1983-84*
4166.393	ELT/263 Airborne ESM Equipment	825
4167.193	*ELT/132 Portable Radar Warning Receiver*	*1984-85*
4168.393	ELT/555 Jammer Pod	826
4169.293	ELT/211 Naval ESM Equipment	655
4170.293	ELT/318 Naval Jammer	656
4171.293	ELT/521 Naval Deception Jammer	656
4172.293	Newton Naval EW Equipment	656
4173.093	Elettronica EW Equipment	657
4174.393	SL/ALQ-234 ECM pod	828

Entry number	Subject	Page No/ Last edition*
4175.093	*Selenia EW Equipment*	*1987-88*
4176.093	Japanese EW Equipment	828
4177.131	Hawk Surface-to-Air Missile	187
4178.383	FIN 1064	882
4179.453	AN/SQS-35 Variable Depth Sonar and AN/SQS-36, AN/SQS-38 Hull-Mounted Sonars	586
4180.441	*Torpedo NT37D Modernisation*	*1983-84*
4181.441	Sea Lance Anti-Submarine Warfare Stand-Off Weapon	534
4182.441	Mk 50 Advanced Lightweight Torpedo (ALWT)	536
4183.441	*Other UK Underwater Weapons*	*1983-84*
4184.441	*WP-900 Ground Mine*	*1984-85*
4185.331	Python 3 Air-to-Air Missile	696
4186.181	ASAT Anti-Satellite Weapon	33
4187.331	*AIAAM (Advanced Intercept Air-to-Air Missile)*	*1982-83*
4188.311	*RBS 70 Helicopter Self-Defence Missile*	*1985-86*
4189.311	Matra Laser-Guided Bombs (LGB)	739
4190.311	GBU-15 Modular Glide Bomb	757
4191.311	*MRASM Medium-Range Air-to-Surface Missile (AGM-109)*	*1983-84*
4192.331	MAA-1 Air-to-Air Missile	692
4193.111	*Brazilian Cobra 2000 Anti-Tank Missile*	*1986-87*
4194.111	GLCM Ground Launched Cruise Missile	130
4195.221	*Otomach 2 Anti-Ship Missile*	*1984-85*
4196.221	SS-N-19 (Shipwreck) Anti-Ship Missile	458
4197.221	Tomahawk Sea-Launched Cruise Missile (BGM-109)	459
4198.131	Mistral Air Defence Missile	212
4199.131	Man-Portable Surface-to-Air Missile	213
4200.231	BARAK 1 Point Defence System	468
4201.231	SA-N-5 Surface-to-Air Missile	473
4202.231	US Missile Launcher Systems	482
4203.231	*Sea Sparrow Vertical Launch System*	*1983-84*
4204.231	Mk 41 Vertical Launch System	484
4205.231	*Outer Perimeter Defence Missile*	*1983-84*
4206.231	SA-N-6 Grumble Surface-to-Air Missile System	473
4207.111	SS-11 (SEGO) ICBM	7
4208.411	SS-N-20 (Sturgeon) SLBM (Soviet designation: RSM-52)	15
4209.411	UK Trident SLBM System	21
4210.111	CSS-3 ICBM (Chinese designation DF-4)	2
4211.111	CSS-4 ICBM (Chinese designation DF-5)	3
4212.454	Miscellaneous Underwater Detection Equipment Table	Analysis (1053)
4213.211	Deadeye 5-Inch Guided Projectile	461
4214.181	TAOC Tactical Air Operations Central	60
4215.293	Model 200 TISS	890
4216.193	C215 Articulated Telescopic Sight	365
4217.193	P 186 Telescopic Sight	368
4218.193	P 192 Night Driving Periscope	368
4219.193	P 204 Periscope	369
4220.193	V 200 Vision Block	368
4221.193	T178/278 Goniometer	370
4222.353	AN/ASA-64 Submarine Magnetic Anomaly Detector	543
4223.253	MBAR Fire Control Radar	643
4224.153	TRS 2620 Radar GERFAUT	231
4225.093	Elta Ground-Based EW Systems	314
4226.293	Barricade Naval Decoy Equipment	662
4227.193	Rampart	320
4228.453	DSQS 21 Series Sonars	555
4229.453	Krupp Atlas-Elektronik Submarine Sonars	556
4230.453	Krupp Atlas-Elektronik/MCM Systems	556
4231.181	Combat Monitoring System	71
4232.161	OG 14 L2B Tank Fire Control System	78
4233.161	*CMP 32 Field Artillery Computer*	*1984-85*
4234.094	Electronic Warfare Equipment Table	Analysis (1056)
4235.193	UA 9092 Tank Thermal Imager	372
4236.193	*LP150 Laser Rangefinder*	*1985-86*
4237.193	Type 772 Sighting System	349
4238.293	*Lightweight Remote Optronics Director*	*1987-88*
4239.393	*SFIM Stabilised Optronic Systems (Ophelia/Venus)*	*1985-86*
4240.183	TAS Tracking Adjunct System (Hawk)	405
4241.383	Low Cost Weapon Delivery System (LCWDS)	873
4242.383	System 81 WDNS	873
4243.383	Tomahawk CCD TV Sensor	887
4244.193	TD22 Thermal Imaging System	362
4245.193	OB-31 Night-Driving Periscope	354
4246.193	*OB-41 Night-Driving Binoculars*	*1987-88*
4247.193	*OB-42 Night Observation Binoculars*	*1987-88*
4248.193	*ELDO Target Designator*	*1985-86*
4249.193	CDN 22 North Seeker	356
4250.193	*CDN 13 North Seeker*	*1986-87*
4251.193	*Marilu 80 × 240 Telescope*	*1987-88*
4252.193	*BM 8028 Image-Intensifier Goggles*	*1987-88*
4253.193	VTG 120 Light Thermal Imager	366
4254.293	*Tantale Director Turret*	*1984-85*
4255.293	Najir Electro-Optical Director	679
4256.193	*Mk IIIm Stedi-Eye Stabilised Monocular*	*1987-88*
4257.193	*Mk IIImg Stedi-Eye Stabilised Monocular*	*1987-88*
4258.193	*Mk V Stedi-Eye Stabilised Day/Night Monocular*	*1987-88*
4259.193	M61/M61A1 Fire Control Sight	400
4260.193	THORN EMI Multi-Role Thermal Imager	391
4261.393	AF532/AF533/AF580 Helicopter Roof Sights	883
4262.293	OG30B Shipborne Thermal Imager	683
4263.281	OG 30A Naval Multi-Sensor Director	416
4264.193	*MS4GT Passive Mini-Weapon Sight*	*1987-88*
4265.281	*OG 20 Naval Fire Control System*	*1983-84*
4266.103	*25mm Automatic Cannon Mk 25 Rh 205*	*1982-83*
4267.181	*SOPTAM FCS*	*1987-88*
4268.181	SOPTAA FCS	62
4269.183	SOPTAL 29 Fire Control System	352
4270.281	Challenger SA-2	444
4271.281	Gemini Naval Fire Control System	420
4272.481	KAFS Submarine A10 & FCS	432
4273.281	F100-L Fire Control Predictor	439
4274.181	Bofors AFV Fire Control System	90
4275.181	SILLACS Low-Level Air Defence System	42
4276.181	Siemens Battery Co-ordination Post (BCP)	44
4277.281	*Interceptor Series Naval FCS*	*1982-83*
4278.161	SACU Command and Control	75
4279.083	*EOS-450 Fire Director*	*1982-83*
4280.193	TICM II Thermal Imaging Systems	386
4281.193	*LRR-104 Mk IV Laser Rangefinders*	*1987-88*
4282.193	*Model TK/60 Tank Laser Rangefinder*	*1983-84*
4283.193	Vipere Infra-Red Surveillance Equipment	357
4284.193	Passive Infra-Red Surveillance Equipment	383
4285.193	*KN150 Night Observation Device*	*1987-88*
4286.183	Land Navigation System (LANS)	373
4287.183	Tamam PADS	363
4288.163	*Air Transportable Reconnaissance & Mission Preparation System*	*1982-83*
4289.353	*Agiflite Hand Held Camera*	*1982-83*
4290.353	*Large Format Camera*	*1982-83*
4291.353	*PC-202 72-in Lorop Panoramic Camera*	*1982-83*
4292.353	*Metritek 21 Cartographic Camera*	*1982-83*
4293.353	*24-in Optical Bar Panoramic Cameras*	*1982-83*
4294.351	*Dornier Experimental Mini-Telecopter*	*1987-88*
4295.351	*(Fuji) Mini-RPV Research Test Vehicle*	*1987-88*
4296.391	*AEL Snipe Target Drones*	*1987-88*
4297.391	*AEL 4600 Streek Target Drone*	*1987-88*
4298.391	*Augmented RCMAT*	*1984-85*
4299.253	Sea Giraffe Naval Search Radar	616
4300.253	AN/SPS-65(V)ER Naval Surveillance Radar	641
4301.253	AN/SPQ-11 Cobra Judy Radar	636
4302.253	805 Series Naval Tracking Radars	633
4303.383	*Marconi LLTV*	*1982-83*
4304.363	*BAe Target Data Acquisition System*	*1987-88*
4305.393	LANTIRN	895
4306.261	*CANE*	*1985-86*
4307.161	9FA 302 Field Artillery Fire Control Computer	89
4308.093	*Thompson-CSF Simulators*	*1982-83*
4309.193	*HOT Missile Training Simulators DX 131, DX 132, and DX 133*	*1982-83*
4310.193	*Safari Artillery Observer Trainer*	*1982-83*
4311.193	*ATA Tank Gunnery Sub-Calibre Range Moving Target Mechanism*	*1982-83*
4312.393	*Laser Training Aid*	*1982-83*
4313.393	*AN/AXQ-16(V) Solid-State Cockpit Television Sensor*	*1982-83*
4314.393	F-16 LANTIRN Holographic HUD Sight	878
4315.203	Single 76mm Dual-Purpose Gun Mounting	501
4316.153	*R-2080 Battlefield Surveillance Radar*	*1982-83*
4317.153	OX-60/FPS-117 IFF Interrogator	298
4318.281	9KA 500 Kardinal Coastal Fire Control System	92
4319.153	AN/TPS-70(V)-1 Tactical Radar	296
4320.153	*TWS-QR Tracking Radar*	*1985-86*
4321.153	GE 592 Transportable Air Defence Radar	295
4322.153	Falcon Series Surveillance Radar	292
4323.153	Hughes Low Altitude Surveillance Radar	288
4324.153	AN/PPS-15B Base Security Radar	285
4325.153	Type 386 Ground Surveillance Radar	285
4326.153	AN/FPS-117 Air Defence Radar	282
4327.153	AN/FPS-16 Instrumentation Radar	279

Entry number	Subject	Page No/ Last edition*
4328.153	Pork Trough (SNAR-2) Counter-Battery Radar	258
4329.153	Big Fred Artillery/Mortar Locating Radar	253
4330.153	EL/M-2121 Battlefield Surveillance Radar	240
4331.153	*RL-42 Surveillance Radar*	*1987-88*
4332.153	*RP-4G Precision Approach Radar*	*1987-88*
4333.353	Tornado Nose Radar	797
4334.353	AN/APQ-126(V) Airborne Navigation and Attack Radar	800
4335.353	AN/APQ-146 Airborne Radar	800
4336.353	AN/APQ-158 Airborne Radar	802
4337.353	AN/APS-133 Airborne Radar	806
4338.353	AN/APS-134(V) Maritime Surveillance Radar	807
4339.353	AN/APX-100(V) Airborne IFF Transponder	809
4340.353	AN/APN-215 Airborne Radar	798
4341.353	MAREC II Maritime Reconnaissance Radar	790
4342.103	*GBT 155 SP Howitzer System*	*1982-83*
4343.103	*Breda Twin 40 L70 Field Mounting*	*1982-83*
4344.103	*127mm Artillery Rocket System*	*1982-83*
4345.111	Working BEE-6 117mm Artillery Rocket	126
4346.111	126mm Kung Feng III/IV Multiple Rocket System	126
4347.363	TSM 8220/8221/8222 (Lamparo) Lightweight Airborne Sonobuoy Processing Equipment	547
4348.453	IBIS Minehunting Sonar Systems	554
4349.453	TSM 2022 Minehunting Sonar	554
4350.293	MAGAIE Lightweight Antimissile Decoy System	648
4351.053	DR4000 IFM ESM Receiver (TMV 201)	650
4352.053	AEG Receivers and Ancillary Systems	308
4353.153	Hummel VHF Jammer	309
4354.293	Rafael Short Range Chaff Rocket	652
4355.293	Rafael Long Range Chaff Rocket	652
4356.093	Timnex-4CH Automatic ELINT/ESM System	316
4357.393	EL/L-8202 ECM Pod	821
4358.393	ELT/156 Radar Warning Receiver	825
4359.393	ELT Noise Jammer Pods	826
4360.353	ELT Colibri Helicopter ESM/ECM System	827
4361.393	AQ 800 Jammer	830
4362.393	*Pod KA Noise and Deception Jammer*	*1984-85*
4363.393	*Lake 2000 Airborne Jamming Pod*	*1984-85*
4364.393	*Airborne Matilda Warning System*	*1982-83*
4365.193	*RL1 and RL2 Laser Warning Instruments*	*1987-88*
4366.353	AN/ALQ-125 Tactical Electronic Reconnaissance Sensor	844
4367.353	AN/ALQ-142 ESM System	846
4368.393	AN/ALQ-134 Expendable Jammer	1985-86
4369.393	AN/ALQ-136 Jammer	846
4370.353	AN/ALQ-150 ESM Receiver	848
4371.393	AN/ALQ-151 Airborne EW System	848
4372.393	AN/ALR-67(V) Radar Warning System	855
4373.393	AN/ALQ-162 Radar Jammer	849
4374.353	AN/ALQ-171(V) Tactical ECM System	850
4375.393	AN/ALQ-157 Infra-Red Countermeasures System	849
4376.353	AN/ALR-60 Communications Intercept and Analysis System	854
4377.393	AN/ALR-62(V) Radar Warning System	854
4378.393	*AN/ALR-64 Compass Sail Warning System*	*1986-87*
4379.353	AN/ALR-69 Radar Warning Receiver	856
4380.393	*Helicopter All-Weather Cueing System*	*1982-83*
4381.353	AN/APR-44 Radar Warning Receiver	861
4382.353	EW-1017 Electronic Surveillance System	862
4383.393	*Advanced Optical Countermeasures Pod*	*1983-84*
4384.393	AN/ULQ-11 Airborne Jamming System	864
4385.153	AN/ULQ-14 Radar Countermeasures	329
4386.093	Sperry Antenna Systems Equipment	331
4387.153	Tactical Electronic Reconnaissance (TEREC) Remote Terminal (TRT)	332
4388.293	AN/SLQ-30 Trump	672
4389.293	SR-200 ESM System	672
4390.053	WJ-927 ESM Equipment	333
4391.053	C-204/WJ-1140 Analysis Control Unit	333
4392.093	WJ-1920 Multiparameter Distributed Processing System	333
4393.053	WJ-1988 Amplitude Monopulse DF Processor	334
4394.053	WJ-1440 Microwave Receiver System	335
4395.053	*TN-1000/WJ-1840 MINCEIVER*	*1987-88*
4396.053	QRC-259 Receiver System	335
4397.053	CMR-610A Wide Band DF Receiver	337
4398.053	RG-5540 VHF/UHF Monitoring Receiver	337
4399.093	RG-1340 Spectrum Surveillance Controller	338
4400.093	RG-1320A Spectrum Display Unit	338

Entry number	Subject	Page No/ Last edition*
4401.193	AN/GLQ-3B Communication Countermeasures Set	338
4402.193	*Pirahna Applique ECM System*	*1983-84*
4403.093	AEL Type 6040 Universal Jammer	341
4404.193	*AN/MLQ-33 ECM System*	*1982-83*
4405.193	AN/MSQ-103 Teampack ESM System	341
4406.093	AN/TLQ-17A Countermeasures Set	342
4407.293	AN/KAS-1 Chemical Warfare Directional Detector	676
4408.153	*AN/TSQ-109 Agtelis Sigint System*	*1987-88*
4409.153	AN/TSQ-112 Tacelis ESM System	343
4410.393	Kestrel Airborne ESM	835
4411.093	Racal EW Programmes	328
4412.341	APS Series Airborne Torpedo Control Systems	522
4413.241	SPS Series Shipborne Torpedo Control Systems	557
4413.441	SPS Series Shipborne Torpedo Control Systems	522
4414.441	*MR-80/I Exercise Mine*	*1987-88*
4415.441	ECA 38 (PAP 104) Mine Disposal System	516
4416.441	*TVX Underwater Test Vehicle*	*1987-88*
4417.441	*Deep Mobile Target*	*1987-88*
4418.441	*THORN EMI Underwater Acoustic Ranges, Systems, and Targets*	*1987-88*
4419.441	*Portable Underwater Tracking Range*	*1982-83*
4420.441	*Versatile Exercise Mine System*	*1987-88*
4421.441	Pinguin Mine Countermeasures Vehicle	518
4422.241	Soviet Anti-Submarine Rocket Launchers	527
4423.441	AS-228 Depth Charge	513
4424.441	*SPAT*	*1983-84*
4425.441	Honeywell MNS	539
4426.303	PJ-1 Manually Operated Bomb	729
4427.303	*100 lb Cluster Bomb*	*1983-84*
4428.303	TAL Cluster Bombs	743
4429.303	Door Gunner Post	744
4430.303	FFV 30mm Aden Gun Pod	748
4431.393	*ML Bomb Ejector Rack BRU-36A/A and Ejector Release Unit No 126*	*1987-88*
4432.303	XM188E1 30mm Three-Barrel Aircraft Gun	765
4433.303	AV-8 Harrier 25mm Armament System	761
4434.303	GPU-5/A Lightweight 30mm Gun Pod	761
4435.303	Universal Turret for Helicopter Gunships	765
4436.303	HGS-55 7.62mm Armament Sub-System	768
4437.303	57mm Aircraft Rocket	730
4438.111	Kun Wu Anti-Tank Missile	148
4439.111	*Rattler Anti-Tank Weapon*	*1983-84*
4440.111	*GEMSS*	*1985-86*
4441.111	*Corps Support Weapon System*	*1983-84*
4442.121	Hsiung Feng Coastal Defence Missile	163
4443.221	Hsiung Feng Anti-Ship Missile	455
4444.221	Skorpioen Anti-Ship Missile	454
4445.221	SS-N-12 (Sandbox) Anti-Ship Missile	457
4446.221	ANS Anti-Ship Missile	451
4447.221	Sea Eagle SL Anti-Ship Missile	458
4448.311	*Hypervelocity Missile (HVM) System*	*1987-88*
4449.311	*Container Weapon System (CWS)*	*1985-86*
4450.321	Air-Launched ANS	710
4451.453	*Type 2093 Mine Hunting Sonar*	*1982-83*
4452.453	*Marconi Small Ship's Sonar*	*1987-88*
4453.363	ARR 901 Sonar Receiver	574
4454.461	*Mains 500 Mine Hunting System*	*1983-84*
4455.453	AN/WQC-2A Sonar Communication Set	582
4456.353	AN/AQS-18 Airborne Sonar	592
4457.453	AN/SSQ-53B Sonobuoy	596
4458.453	*Sparton Dwarf Omni Sonobuoy*	*1987-88*
4459.453	AN/SSQ-86(XN-1) Down-Link Communication Sonobuoy	598
4460.453	Klein MCM Sonar Systems	598
4461.453	DSBV 61A Towed Array Sonar	550
4462.353	AN/AQS-14 Airborne Minehunting Sonar	592
4463.253	AN/SQR-19 Tactical Sonar Towed Array	585
4464.453	HSS-15 Compact Surveillance Sonar	559
4465.131	*Kürassier/Sabre Twin 30mm AA System*	*1984-85*
4466.131	GEMAG 25 AA Gun System	209
4467.203	*25/30mm Zenith Naval Mount*	*1982-83*
4468.203	*155mm Vertical Load Gun System (VLGS)*	*1986-87*
4469.181	Reporter Tactical Early Warning System	83
4470.131	SA-11 (Gadfly) Surface-to-Air Missile	181
4471.131	SA-13 (Gopher) Surface-to-Air Missile	182
4472.111	*Tank Breaker Anti-Tank Missile*	*1984-85*
4473.231	SA-N-7 (Gadfly) Surface-to-Air Missile System	474
4474.153	*Quad-Radar with MTI ATC Radar*	*1982-83*
4475.391	*Heat Drone*	*1984-85*
4476.391	*EMI Argus UMA*	*1984-85*
4477.391	*NV-144 RPV/Target System*	*1987-88*

Entry number	Subject	Page No/ Last edition*
4478.153	AN/MPQ-53 Guidance Radar	282
4479.391	*FA-25 Surveillance RPV*	*1983-84*
4480.391	*ML Aviation Sprite RPH*	*1987-88*
4481.391	*Pave Tiger Mini-Drone (UCGN-121A)*	*1984-85*
4482.441	Sea Urchin Mine	530
4483.441	*Towed Unmanned Submersible (TUMS)*	*1987-88*
4484.393	*ML Aviation Carrier Bomb Twin Store 30°*	*1987-88*
4485.393	*ML Aviation Twin Store Integrated Carrier*	*1987-88*
4486.303	Alkan Ground Attack Cartridge Launchers	798
4487.303	*500-Pound Cluster Bomb*	*1983-84*
4488.303	Cardoen General-Purpose Bombs	730
4489.281	Sea Tiger	413
4490.161	SEDA and SEDAB Artillery Automation Systems	77
4491.281	H-930 Series Combat/Weapon Control Systems	444
4492.161	Combat Artillery C³ System	74
4493.161	AS80 Artillery Fire Control System	86
4494.161	S700 Meteorological Station	87
4495.161	EMVA Mk 10B Muzzle Velocity Analyser	87
4496.261	DAISY and Foresee Systems	421
4497.231	Dardo CIWS	470
4498.461	SACTIS Combat Information System	416
4499.261	MM/SSN-714 Minesweeper Data Processing System	418
4500.261	MM/SSN-715 Navigation and Tracking System	418
4501.481	Submarine Integrated Combat System	411
4502.131	Twin 20 mm AA Weapon System TA20/RA20 S	192
4503.114	Self-Propelled Artillery Table	Analysis (1066)
4504.134	Self-Propelled Anti-Aircraft Guns Table	Analysis (1067)
4505.114	Towed Artillery Table	Analysis (1069)
4506.134	Towed Anti-Aircraft Guns Table	Analysis (1068)
4507.114	AFV Armament Table	Analysis (1072)
4508.131	Spider II Air Defence System	196
4509.231	SADRAL Naval Air Defence System	465
4510.331	AATCP Mistral Helicopter Missile System	693
4511.131	Artemis 30 twin 30 mm Light Anti-Aircraft Gun System	195
4512.131	Sabre Twin 30 mm AA Gun Turret System	192
4513.203	Twin 130mm Dual-Purpose Gun Mounting	485
4514.203	Single 130mm Gun Mounting	485
4515.203	Single 100mm Dual-Purpose Gun Mounting	486
4516.203	Single 85mm Dual-Purpose Gun Mounting	486
4517.203	Single 75mm Dual-Purpose Gun Mounting	486
4518.203	Twin 57mm AA Gun Mounting	486
4519.203	Twin 37mm AA Gun Mounting	487
4520.203	Twin 14.5mm AA Gun Mounting	486
4521.203	Single 100mm Dual-Purpose Gun Mounting (Automatic)	500
4522.203	Twin 130mm Dual-Purpose Gun Mounting	500
4523.203	Triple 16in Gun Turret Mk 7	505
4524.131	*Guardian Air Defence System*	*1986-87*
4525.131	Oerlikon 25 mm Twin Field AD Gun (Diana)	204
4526.331	Kukri Air-to-Air Missile	698
4527.111	Valkiri 127mm Artillery Rocket System	124
4528.311	ALARM Air-to-Surface Missile	719
4529.321	Gabriel Mk III Air-Launched Anti-Ship Missile	712
4530.311	*LAD Low Altitude Dispenser*	*1987-88*
4531.181	Classic (RGS 2740)	104
4532.181	CATRIN Communications System	79
4533.131	Javelin Man-Portable Anti-Aircraft Missile	215
4534.131	*Saber Portable Anti-Aircraft Missile*	*1986-87*
4535.131	SIDAM 25 AA Turret	197
4536.131	Breda Lightweight 30 mm AA Field Mounting (Sentinel)	198
4537.131	Oerlikon Twin Field 35 mm Air Defence Gun Type GDF	202
4538.261	AN/USQ-81(V) Command Control System	443
4539.303	CRV7 Aircraft Rocket	728
4540.303	SA Aircraft Bombs	746
4541.303	LU250 EG-FT Bomb	740
4542.303	TBA 68mm Helicopter Rocket Launchers	795
4543.131	*Low Strike Surface-to-Air Missile System*	*1984-85*
4544.111	RBS 56 Bill Anti-Tank Missile	147
4545.111	Trigat Programme	142
4546.231	RBS 70 Rayrider Missile System	471
4547.111	Ching Feng (Green Bee) Surface-to-Surface Missile	126
4548.111	SS-21 (SCARAB) Battlefield Support Missile	128
4549.111	SS-12B (SCALEBOARD B) Battlefield Support Missile	128
4550.111	SS-23 (Spider) Battlefield Support Missile	128
4551.111	*Guided Anti-Armour Mortar Projectile (GAMP)*	*1986-87*
4552.111	Eryx Anti-Tank Missile	138
4553.221	Sea Skua SL Anti-Ship Missile System	458
4554.311	Apache/CWS 50	711
4555.311	*Pegase*	*1985-86*
4556.311	Soviet Air-Launched Anti-Tank Missiles	718
4557.111	Strix Anti-Tank Mortar Bomb	146
4558.441	Stonefish Mines	531
4559.441	Spearfish Torpedo	529
4560.111	US ICBM Modernisation Programme	22
4561.111	Peacekeeper (MGM-118) ICBM	23
4562.311	*US Strategic Bomber Force*	*1987-88*
4563.231	TCM 30 Naval Air Defence System	469
4564.231	Meroka Naval CIWS	471
4565.231	*SA-N-8 Naval Surface-to-Air Missile*	*1986-87*
4566.391	*Machan Research RPV*	*1987-88*
4567.203	*LS30 Series Seated Operator Gun Mounts*	*1985-86*
4568.281	9LV + 50 ASW System	429
4569.391	*Raven 1 RPV*	*1987-88*
4570.391	*MBB-UM Attack Drone KDH and DAR*	*1987-88*
4571.181	T.ACCS	40
4572.281	Thomsea Naval Combat Systems	409
4573.481	MSI-90U Submarine and Weapon Control System	425
4574.203	Breda 81mm Bombardment Launcher	494
4575.203	Breda Single 30mm Mounting	493
4576.481	*Submarine Tactical Command and Weapon Control System*	*1985-86*
4577.481	VSEL Combat System Data Bus	434
4578.281	NAUTIS Command and Control Systems	437
4579.281	Reshet Naval Command and Control System	413
4580.161	ATAC Artillery Mini Computer	68
4581.181	Marconi DFCS	103
4582.181	EFCS 600	101
4583.181	*Marconi AFCS*	*1985-86*
4584.161	ARS Field Artillery Computer System	111
4585.161	SUVOA Fire Control System	111
4586.181	Gun King Sighting System	95
4587.481	Spectrum I and II Submarine SEWACO System	424
4588.131	Marksman Twin 35 mm Anti-Aircraft Turret	205
4589.131	Breda Twin 40L/70 AA Field Mounting	198
4590.131	TCM Mk 3 Light AA System	196
4591.391	*Changcheng B-2 Target Drone*	*1987-88*
4592.351	*Phoenix RPV*	*1987-88*
4593.391	*CBT BQM-1 BR RPV*	*1985-86*
4594.131	Otomatic 76 mm 76/62 SPAAG	197
4595.391	*Eyrie Mini-RPV*	*1987-88*
4596.393	Matra SFA Bomb Retarding System	737
4597.303	74mm Anti-Armour Grenade	736
4598.303	CB 130 & CB 500 Cluster Bombs	729
4599.303	Rockeye II Mk 20 Cluster Bomb	756
4600.153	FASTAR Battlefield Surveillance Radar	273
4601.153	*EMVA Mk 10B Muzzle Velocity Radar*	*1986-87*
4602.153	PSNR-1 Battlefield Radar	261
4603.153	Messenger SSR	276
4604.153	Cossor Monopulse SSR Systems	275
4605.153	XJ/FPS-3 Air Defence Radar	248
4606.153	EL/M-2220 Mobile Surveillance Radar	241
4607.153	Guardsman Defence Radar	268
4608.253	Flanking-Beam Array Switching Technique Radar (FAST)	644
4609.253	Mk 23 Target Acquisition System Radar	644
4610.253	EL/M-2208 Shipborne Search Radar	608
4611.153	*Helicapture Tactical Radar System*	*1984-85*
4612.353	Digital Tactical System (DITACS) Radar	806
4613.353	Astor - Airborne Stand-Off Radar	788
4614.353	*Sky Guardian AEW Radar*	*1986-87*
4615.353	Blue Vixen Radar	792
4616.353	SUAWACS Airborne AEW Radar	785
4617.353	*Sky Searcher Airborne AEW Radar*	*1985-86*
4618.353	RCA Balloon-Borne Radar	813
4619.453	*AN/SSQ-529 (DICANCAS) Sonobuoy*	*1986-87*
4620.453	AN/SSQ-530 DIFAR Sonobuoy	544
4621.353	AN/ASQ-504(V) MAD System	541
4622.253	AN/SQR-18A Tactical Sonar Towed Array (TACTAS)	584
4623.253	AN/SQQ-30 Minehunting Sonar	587
4624.453	Acoustic Intruder Detection System	577
4625.253	FMS Series Sonar Systems	572
4626.453	*Marconi Area Defence Sonar*	*1987-88*
4627.463	Towed Arrays	577
4628.063	*MUSL Sonar Products*	*1984-85*
4629.453	AN/BQQ-6 Sonar	593
4630.153	S700 Series Mobile Radar Systems	265

Entry number	Subject	Page No/ Last edition*
4631.153	Transportable Watchman Defence Radar	267
4632.053	Model 2679 IFF Interrogator	299
4633.253	*Terma Scanter Mil Plot Data System*	*1987-88*
4634.253	Type 996 Naval Surveillance Radar	631
4635.393	Lacroix Airborne Countermeasures Ammunition	820
4636.193	Rohde & Schwarz Antenna Systems	312
4637.193	PA 510 Mobile Direction Finding System	313
4638.193	RAJ 101 Ground Radar Jammer	315
4639.193	*Elta COMINT Receivers*	*1984-85*
4640.353	SPS-20 Airborne Self Protection System	823
4641.283	SCLAR Mk 2 Rocket Launching System	659
4642.353	Smart Guard Helicopter ECM System	827
4643.353	Fast Jam Helicopter ECM System	827
4644.293	CO-NEWS Naval ESM System	658
4645.093	ELT/128 Communications Direction Finder	317
4646.293	SR-1A ESM Receiver	660
4647.353	AN/APQ-164 Airborne Radar	803
4648.353	Covert Strike Radar	814
4649.363	AN/UYS-503 Sonobuoy Processor (SBP)	545
4650.393	Evade Airborne Decoy System	834
4651.293	Stockade-Seaflash Decoy System	662
4652.493	Manta Submarine ESM System	665
4653.293	Guardian Naval EW System	666
4654.393	AN/ALE-38/41 Bulk Chaff Dispenser	839
4655.393	Modular Chaff/Flare Dispenser	842
4656.293	Mk 36 Super Rapid Bloom Off-Board Countermeasures	674
4657.393	AN/ALE-36 Countermeasures Dispensing Pod	839
4658.053	Multiple Radar Emitter System	346
4659.093	SR-2020 Tactical ESM System	340
4660.193	SRS-2500 VHF/UHF Receiving System	340
4661.093	SR-2152 VHF/UHF Receiver	340
4662.393	AN/AAR-34 Infra-Red Receiving Set	838
4663.193	LQ-102A Hand-Emplaced Expendable Jammer	346
4664.193	AN/MLQ-T6 Communications Data Link Jammer	340
4665.093	*DF-8000 Direction Finding System*	*1987-88*
4666.393	Small Platform EW System	865
4667.393	Airborne Radar Jamming System	865
4668.393	AN/ALQ-128 Warning Receiver	845
4669.393	AN/ALQ-137 Airborne Jammer	846
4670.353	Guardrail Airborne SIGINT System (AN/USD-9(V)2)	865
4671.253	S-2150 ESM Systems	672
4672.093	RG Series Spectrum Monitors	339
4673.093	RG-5202A VHF Scanning Receiver	337
4674.093	RG-5545A VHF/UHF Receiver	337
4675.093	RA-6793A HF Receiver	338
4676.093	RA6778-C Receiver/RA6778-Q Converter	337
4677.153	J-3400/J-3410 Mobile ECM Systems	330
4678.393	AN/ALQ-149 Jammer	847
4679.453	Thorn EMI Underwater Acoustic Ranges and Systems	578
4680.493	*Deep Mobile Target*	*1987-88*
4681.273	*NAUTIS-M Command and Control System*	*1987-88*
4682.193	*TN21 Night Aiming System*	*1987-88*
4683.193	*Circe Portable Night Sight*	*1987-88*
4684.193	*BM 8042 Infra-Red Target Marker*	*1987-88*
4685.193	OFRIS Target Tracker	360
4686.193	Rainbow Laser Rangefinder	362
4687.193	*Matador Fire Control System*	*1984-85*
4688.193	OG-P20 Periscope	369
4689.193	P 233 Night Vision Periscope	368
4690.193	*GS6MC Crew-Served Weapon Sight*	*1987-88*
4691.193	TP1MC Night Periscope	373
4692.193	LP100 Laser Rangefinder	375
4693.193	TC10 Thermal Imager	375
4694.093	Ericsson Thermal Imager	378
4695.193	*Wild Heerbrugg Electro-Optical Systems*	*1987-88*
4696.093	Ferranti Helmet Pointing System	386
4697.193	*LT 1065 Hand-Held Thermal Imager*	*1987-88*
4698.193	*PVS5A Night Vision Goggles*	*1987-88*
4699.193	*PVS4 & TVS5 Weapon Sights*	*1987-88*
4700.193	*NOD2 Night Observation Device*	*1987-88*
4701.193	HHI-8 Hand-Held Thermal Imager	387
4702.193	*Eagle Long-Range Night Sight*	*1987-88*
4703.193	Observer Thermal Imaging System (OTIS)	392
4704.193	THORN EMI Hand-Held Thermal Imager	393
4705.193	L20 Laser Rangefinder Sight	393
4706.193	L50 Conversion Unit	394
4707.193	*W201 Individual Weapon Sight*	*1987-88*
4708.193	*Solartron Night Vision Systems*	*1987-88*
4709.193	*LRR-104 Mk V Laser Rangefinder*	*1987-88*
4710.193	*LRR-104/20 Artillery Rangefinder*	*1987-88*
4711.093	TK-640 Laser Rangefinder	396
4712.093	N-90 Series Laser	396
4713.193	*GP/NVG-1 Night Vision Goggles*	*1987-88*
4714.193	*GP/NVB-4 and -6 Night Vision Binoculars*	*1987-88*
4715.193	*SPDR Sighting Devices*	*1987-88*
4716.293	Vision Optical Director	680
4717.293	Cotol Optronic Fire Control System	680
4718.293	Gyro Stabilised Sighting System	682
4719.293	Ferranti Multi-Role Naval Lasers	688
4720.293	Director Aiming Sight (DAS)	688
4721.493	Barr & Stroud Submarine Optronic Mast	686
4722.393	*CN²H Night Observation Binoculars*	*1987-88*
4723.393	TMV 980A Electronic Multidisplay System	870
4724.393	HUD Sight Repeater	871
4725.393	*Scotch Weapon Delivery System*	*1986-87*
4726.393	Thermal Cueing Unit	879
4727.393	Type 117 Laser Designator	885
4728.393	Northrop Focal Plane Array	891
4729.393	Aquila Laser Transmitter	891
4730.383	Ferranti FINAS	883
4731.153	CR-2740 ELINT System	316
4732.253	*NS-9000 Series Naval EW Systems*	*1986-87*
4733.393	AR 830 Airborne Warning System	830
4734.393	AQ 900 Airborne Self-Protection System	831
4735.193	Spectra ELINT System	318
4736.193	RJS3140 Unattended Expendable Jammer	324
4737.053	RA 1796 HF/VHF/UHF Tactical EW Receiver	322
4738.193	*RJS3105 Series Communications Jammers*	*1987-88*
4739.293	Triton Threat Warning System	675
4740.393	AN/ALQ-164 ECM Pod	850
4741.193	AN/MLQ-33 ECM System	348
4742.393	*AN/ASQ-171 Airborne ELINT System*	*1986-87*
4743.193	LRS 4 Mk II Fire Control System	348
4744.193	LRS 5 Fire Control System	348
4745.193	*LRF 104 Hand-Held Laser Rangefinder*	*1987-88*
4746.193	Castor Thermal Imaging Sight	354
4747.193	*RS4MC Small Arms Weapon Sight*	*1987-88*
4748.193	*PC1MC Night Vision Goggles*	*1987-88*
4749.193	*PB4SL Night Binoculars*	*1987-88*
4750.193	Thermal Observation Gunnery Sight (TOGS)	382
4751.181	PYK-80 Field Artillery Computer	80
4752.181	GSQ-250K Remotely Monitored Sensor System	81
4753.161	Quickfire Artillery FCS	97
4754.181	Marcal Gun Calibration System/Mogul	98
4755.181	*Kollmorgen CVFCS*	*1987-88*
4756.111	Merlin Anti-Tank Mortar Bomb	152
4757.111	*Lawmine Anti-Tank Weapon*	*1985-86*
4758.111	*XM815 Heat-MP-T Projectile*	*1985-86*
4759.111	LAR 160 160 mm Light Artillery Rocket System	122
4760.111	*Joint Tactical Missile System*	*1984-85*
4761.221	SS-N-22 (Sunburn) Anti-Ship Missile	458
4762.441	Torpedo NT 37E Modernisation	535
4763.441	*Transportable Acoustic Range*	*1987-88*
4764.441	*Depth Charge Mk 11 (Mod 3)*	*1987-88*
4765.441	*Kingfisher Launch System*	*1987-88*
4766.441	FG 1 Sea Mine	519
4767.331	Sidewinder (AIM-9) Air-to-Air Missile Family	704
4768.331	AA-9 (Amos) Air-to-Air Missile	701
4770.311	Vertical Ballistic Weapon/VBW Airborne Anti-Tank Weapon	742
4771.131	SAAM Surface-to-Air Missile System	168
4772.131	*MFS-2000 Surface-to-Air Missile*	*1985-86*
4773.131	*Landwolf Air Defence System*	*1984-85*
4774.131	SA-12 Gladiator Giant Surface-to-Air Missile System	181
4775.231	NATO Sea Sparrow Surface-to-Air Missile System	466
4776.231	SATAN Close-In Weapon System	464
4777.111	Hades Tactical Surface-to-Surface Missile	119
4778.111	SS-24 (Scalpel) ICBM (Soviet designation SS-23)	10
4779.111	SS-25 (Sickle) ICBM (Soviet designation SS-12M)	11
4780.193	*LT 1085 Thermal Imager*	*1987-88*
4781.193	*LWS 1060 Night Sight*	*1987-88*
4782.193	*MEL Weapon Sight*	*1984-85*
4783.193	VM Series AFV Equipment	393
4784.193	AV 62 Commanders Sight	380
4785.193	Barr & Stroud Lightweight Electro-Optical Sensor	927

Entry number	Subject	Page No/ Last edition*
4786.161	Lancelot Tank FCS	76
4787.193	Type 518 and Type 531 Laser Systems	384
4788.193	SS200 Fire Control System	390
4789.193	Rank Thermal Imager	391
4790.193	SS500 Armoured Vehicle Sight	391
4791.193	SS120 Night Sight	389
4792.193	SS141, SS142, and SS162 Commanders Periscopes	390
4793.393	*MBB Night Sight Attack System*	*1985-86*
4794.393	Thompson-CSF VEH 3020 Holographic Head-Up Displays	871
4795.411	SS-N-23 (Skiff) SLBM	16
4796.411	SS-NX-21 Cruise Missile	16
4797.353	Raphael TH Airborne Surveillance Radar	774
4798.153	S723 Martello 3-D Surveillance Radar	264
4799.393	Apollo Airborne ECM System	977
4800.393	Guardian Airborne Radar Warning Receiver	977
4801.193	Sentry ESM System	977
4802.393	Alkan Sonobuoy Launchers	555
4803.153	*Type 714 Meteorological Radar*	*1987-88*
4804.153	*Type 701 Wind Finding Radar*	*1987-88*
4805.153	*Type 793 Terminal Area Radar*	*1987-88*
4806.153	*Type 796 Air Route Surveillance Radar*	*1987-88*
4807.153	Type 791-A Precision Approach Radar	224
4808.253	*Type 756 Naval Radar*	*1985-86*
4809.153	Cheetah Artillery Radar	223
4810.153	Eagle Coastal Radar	222
4811.153	Type 581 Air Warning Radar	221
4812.153	JLP-40 Surveillance Radar	220
4813.153	JLG-43 Heightfinding Radar	222
4814.153	Type 404A Coast Surveillance Radar	222
4815.153	MW-5 Fire Control Radar	223
4816.153	Type 311-A Fire Control Radar	224
4817.153	JY-8 Mobile Tactical 3-D Radar	219
4818.181	*GM-09 Tank FCS*	*1984-85*
4819.001	Soviet Cruise Missile Programme	12
4820.331	AA-10 (Alamo) Air-to-Air Missile	701
4821.391	*AEL Sparrowhawk*	*1987-88*
4822.481	Submarine Combat Control System Mk 2	446
4823.181	9AA 200 Fire Control System (Trackfire)	90
4824.281	SEN Series Combat Systems	416
4825.281	*SER 100 Naval FCS*	*1987-88*
4826.481	IMICS (Integrated Mine-Hunting Combat System)	418
4827.481	Submarine Integrated Combat System (SICS)	417
4828.281	SEPA Naval Fire Control System	417
4829.181	Army Command and Control System Programmes (FRG)	72
4830.161	FIST Artillery System	82
4831.181	Marconi Gun Control and Stabilisation Equipment	102
4832.181	Computer Sighting System (CSS)	102
4833.303	SNEB 68 Aircraft Rockets	731
4834.303	Multi-Dart Helicopter Anti-Armour System	732
4835.303	TBA 100mm Aircraft Rockets	732
4836.303	FEU 80 Electric Multi-Purpose Bomb Fuze	735
4837.131	Bofors 40 mm Armour Trinity	200
4839.231	*TA25/RA20 S 25 mm Air Defence System*	*1987-88*
4840.131	BPTSG Modular FCS for AA Guns	66
4841.131	AYB Al-Saqr Portable Anti-Aircraft Missile	212
4842.311	*Mobidic*	*1987-88*
4843.203	Single 76/62 Super Rapid Gun Mounting	489
4844.203	Twin 5 in/38 Calibre Gun Mounting Mk 38	506
4845.203	Twin 3 in/50 Calibre Gun Mounting Mk 33	508
4846.203	Triple 6 inch/47 Mk 16 Gun Turret	505
4847.203	Single 5 in/54 Calibre Gun Mounting Mk 39	505
4848.351	*Brevel Reconnaissance Drone*	*1987-88*
4849.351	*NPU Type D-4 RPV*	*1987-88*
4850.391	*BTT-3 Banshee Target*	*1987-88*
4851.391	*RS Systems Aerial Targets*	*1987-88*
4852.391	*Intimidator SLAT (YQAM-127A)*	*1985-86*
4853.351	*Skyeye R4E-40*	*1987-88*
4854.391	*Brave-200 Tactical Unmanned Vehicle*	*1987-88*
4855.161	CMB8 Mortar Computer	77
4856.161	TURMS Laser Tank Fire Control System	78
4857.161	ATREOS Laser Tank Fire Control System	79
4858.161	FLP 10 Tank Fire Control System	73
4859.111	MAPATS Anti-tank Weapon	144
4860.231	SAAM Naval Air Defence System	462
4861.231	SAMOS Naval Air Defence System	464
4862.241	ELMA ASW Grenade System	524
4863.441	MRP Sea Mine	520
4864.303	CH-TABO Bomb Dispersal Pod	739
4865.303	ISCB-1 Area Denial Cluster Weapon	756
4866.303	Modular Dispenser System (MDS)	741
4867.303	Aerea Multi Task Pod	743
4868.303	BPD Medusa 81mm Air-to-Ground Rocket System	745
4869.303	2-inch BPD Aircraft Rockets	744
4870.303	25mm Aden Cannon	753
4871.111	Teruel 140 mm Multiple Rocket System	125
4872.111	130 mm Type 70 Multiple Rocket Launcher	115
4873.111	107 mm Type 81 Multiple Rocket Launcher	117
4874.111	VAP 80 mm Multiple Rocket Launcher	118
4875.111	122 mm Multiple Rocket Launcher	118
4876.001	*Volcano Multiple Delivery Mine System*	*1985-86*
4877.111	Hellfire Modular Missile	153
4878.121	RBS 17 Hellfire Shore Defence System	162
4879.111	Rheinmetall Anti-Armour Ammunition Developments	139
4880.111	US ICM	157
4881.281	THORN EMI Torpedo Fire Control Equipment	435
4882.331	MICA Air-to-Air Missile	693
4883.441	Bofors Sea Mines	525
4884.153	HSS-1 Harbour Surveillance Radar	243
4885.231	SA-N-9 Naval Surface-to-Air Missile System	475
4886.311	*SRSOM and LRSOM Stand-Off Missiles*	*1987-88*
4887.311	Armat Anti-Radar Missile	707
4888.131	ADAMS Air Defence Missile System	173
4889.131	HQ-2J Air Defence Missile	167
4890.121	SCC-3 and SSC-4 Coast Defence Missile Systems	164
4891.221	C801 Surface-to-Surface Anti-ship Missile	449
4892.221	*MM 15 Anti-ship Missile*	*1987-88*
4893.411	CSS-N-3 Submarine-Launched Ballistic Missile (Chinese designation JL-I)	3
4894.111	Soviet ICBM Developments	11
4895.131	Soviet Advanced Defence Projects	18
4896.131	US Strategic Defence Initiative (SDI)	31
4897.111	USAF Small ICBM	25
4898.181	Siscea Integrated Airspace Control System	38
4899.391	*BQM-126A Missile Target*	*1987-88*
5001.102	*Saracen Armoured Personnel Carrier*	*1979-80*
5002.102	*Ferret Scout Car Mks 1 & 2*	*1979-80*
5003.102	*STRV 103 (S Tank) Main Battle Tank*	*1979-80*
5004.102	*Centurion Main Battle Tank*	*1979-80*
5005.102	*Saladin Armoured Car*	*1979-80*
5006.102	*Chieftain Main Battle Tank*	*1979-80*
5007.102	*FV432 Armoured Personnel Carrier*	*1979-80*
5008.102	*SPZ 12-3 (HS-30) Armoured Personnel Carrier*	*1979-80*
5009.102	*AMX 10P Mechanised Infantry Combat Vehicle*	*1979-80*
5010.102	*AMX-VTP M-65 Armoured Personnel Carrier*	*1979-80*
5011.102	*DAF YP408 Armoured Personnel Carrier*	*1979-80*
5012.102	*Hotchkiss-Rive Light Fighting Unit*	*1972-73*
5013.102	*Marder-Schutzenpanzer APC*	*1979-80*
5014.102	*Rakete M-1966 GW Launcher*	*1979-80*
5015.102	*Leopard Main Battle Tank*	*1979-80*
5016.102	*M113 Armoured Personnel Carrier*	*1979-80*
5017.102	*Shorland Mk 3 Armoured Car*	*1979-80*
5018.102	*AMX 30 Main Battle Tank*	*1979-80*
5019.102	*Kanone JPZ 4-5 Self-Propelled Anti-Tank Gun*	*1979-80*
5020.102	*AMX 13 Light Tank*	*1979-80*
5021.102	*AML 245 Armoured Car*	*1979-80*
5022.102	*VTT-AML Armoured Personnel Carrier*	*1979-80*
5023.102	*Pbv 302 Armoured Personnel Carrier*	*1979-80*
5024.102	*M60 Main Battle Tank*	*1979-80*
5025.102	*M60 A1 E2 Main Battle Tank*	*1979-80*
5026.102	*Type 60 Armoured Personnel Carrier*	*1979-80*
5027.102	*Type 61 Main Battle Tank*	*1979-80*
5028.102	*MBT-70 Main Battle Tank (Germany, FR)*	*1971-72*
5029.102	*MBT 70 Main Battle Tank (USA)*	*1972-73*
5030.102	*M551 Sheridan Light Tank*	*1979-80*
5031.102	*Pz 61 Main Battle Tank*	*1979-80*
5032.102	*Tank Destroyer/Light Tank Jagdpanzer Kurassier*	*1979-80*
5033.102	*V-100 Commando Armoured Personnel Carrier*	*1977*
5034.102	*BTR-60 Armoured Personnel Carrier*	*1979-80*
5035.102	*BTR-50P Armoured Personnel Carrier*	*1979-80*
5036.102	*BRDM-2 Reconnaissance Car*	*1979-80*
5037.102	*BRDM-1 Reconnaissance Car*	*1979-80*
5038.102	*PT-76 Light Amphibious Tank*	*1979-80*
5040.102	*Scorpion Light Tank*	*1979-80*
5041.102	*Vijayanta (Victorious) Main Battle Tank*	*1979-80*
5042.102	*T59 Main Battle Tank*	*1979-80*
5043.102	*M113CR Command & Reconnaissance Carrier*	*1979-80*

Entry number	Subject	Page No/ Last edition*
5044.102	*M114 Command & Reconnaissance Carrier*	*1978*
5045.102	*Tornado Mechanised Infantry Combat Vehicle*	*1979-80*
5046.102	*M47 Main Battle Tank*	*1979-80*
5047.102	*T62 Main Battle Tank*	*1979-80*
5048.102	*T.10 Heavy Tank*	*1979-80*
5049.102	*T54/T55 Main Battle Tanks*	*1979-80*
5050.102	*ASU 57 Assault Gun*	*1979-80*
5051.102	*ASU 58 Assault Gun*	*1979-80*
5052.102	*M60 A1 (Italian) Main Battle Tank*	*1976*
5054.102	*BMP-1 Armoured Personnel Carrier*	*1979-80*
5055.102	*XM2 Infantry Fighting Vehicle*	*1979-80*
5056.102	*M48 Main Battle Tank*	*1979-80*
5057.102	*Vickers Main Battle Tank*	*1979-80*
5058.102	*FN4 RM62 Light Armoured Car*	*1979-80*
5059.102	*Berliet VXB Armoured Personnel Carrier*	*1979-80*
5060.102	*Bussing 8-wheeled Armoured Car*	*1972-73*
5061.102	*M4 Super Sherman (T1-67) Medium Tank*	*1979-80*
5062.102	*Sabra (or Tsabar) Main Battle Tank*	*1977*
5063.102	*1 Kv 91 Light Tank*	*1979-80*
5064.102	*Fox Armoured Car*	*1979-80*
5065.102	*Ferret Mk 5 Missile Launch Vehicle*	*1979-80*
5066.102	*Armoured Car V200 Commando*	*1977*
5067.102	*M 132 A1 Flame Thrower*	*1978*
5068.102	*M577 A1 Command Post Vehicle*	*1979-80*
5069.102	*M106 A1, M 125 A1 Mortar Carriers*	*1979-80*
5070.102	*XM 800 Armoured Reconnaissance Scout Vehicle*	*1976*
5071.102	*LVTP7 Tracked Landing Vehicle*	*1979-80*
5072.102	*M 113 and Variants Armoured Personnel Carriers*	*1979-80*
5073.102	*Type 74 Main Battle Tank*	*1979-80*
5074.102	*OT 64 Armoured Personnel Carrier*	*1979-80*
5075.102	*Steyr Schutzenpanzer 4K4FA*	*1979-80*
5076.102	*Main Battle Tank Pz 68*	*1979-80*
5077.102	*Puma Amphibious Personnel Carrier*	*1976*
5078.102	*V-150 Commando Armoured Personnel Carrier*	*1977*
5079.102	*SWAT Armed Personnel Carrier*	*1972-73*
5080.102	*Riverine Utility Craft (RUC)*	*1973-74*
5081.102	*Chinese Light Tanks*	*1973-74*
5082.102	*Armoured Personnel Carrier BTR-152*	*1979-80*
5100.102	*Cutia-Vete TIAI*	*1979-80*
5101.102	*EE-9 Cascavel*	*1979-80*
5102.102	*EE-11 Urutu*	*1979-80*
5103.102	*T-60 Light Tank*	*1979-80*
5104.102	*T-62 Tank*	*1979-80*
5105.102	*T-63 Tank*	*1979-80*
5106.102	*Type 55 APC*	*1979-80*
5107.102	*Type 56 APC*	*1979-80*
5108.102	*M-1967 or M-1970 APC*	*1979-80*
5109.102	*OT-62 APC*	*1979-80*
5110.102	*AMX-10RC*	*1979-80*
5111.102	*Saviem VAB*	*1979-80*
5112.102	*T-64 MBT*	*1978*
5113.102	*M-1970 Light Tank*	*1979-80*
5114.102	*Leopard 2 MBT*	*1979-80*
5115.102	*Spähpanzer 2*	*1979-80*
5116.102	*UR 416 APC*	*1979-80*
5117.102	*AT 104 APC*	*1979-80*
5118.102	*Infantry Armoured Fighting Vehicle*	*1979-80*
5119.102	*Fiat 6616 Armoured Car*	*1979-80*
5120.102	*Armoured Infantry Fighting Vehicle*	*1979-80*
5121.102	*Fiat 6614 APC*	*1979-80*
5122.102	*Israeli Centurion*	*1979-80*
5123.102	*Israeli T-54 and T-55*	*1979-80*
5124.102	*Type 73 APC*	*1979-80*
5125.102	*Lockheed XM-800 ARSV*	*1976*
5126.102	*FMBT-80*	*1979-80*
5127.102	*MOWAG Roland*	*1979-80*
5128.102	*AT-105 IS Vehicle*	*1979-80*
5129.102	*MOWAG Piranha*	*1979-80*
5130.102	*Shorts SB-301 IS Vehicle*	*1979-80*
5131.102	*M60 APC*	*1979-80*
5132.103	*122mm Self Propelled Howitzer M-1974*	*1982-83*
5133.103	*152mm SPG M-1973*	*1982-83*
5134.102	*FUG-70 Scout Car*	*1979-80*
5135.103	*105mm Recoilless Rifle M68*	*1978*
5136.102	*M-1970 (GT-T) Tracked Vehicle*	*1977*
5137.111	122 mm Rocket Launcher M-1972 (RM-70)	118
5138.102	*107mm Rocket Launcher Type 63*	*1979-80*
5139.102	*Lohr VPX110 Lightweight Infantry Tracked Vehicle*	*1979-80*
5140.102	*Ramta RBY Mk 1*	*1979-80*
5141.103	*Indian Artillery*	*1982-83*
5143.103	*Mobile Test Unit/Laser Weapon*	*1979-80*
5144.103	*105mm Recoilless Rifle M-65*	*1978*
5145.103	*130mm Coastal Gun SM-4-1*	*1982-83*
5147.103	*100mm Anti-Tank Gun T-12*	*1982-83*
5148.103	*General Support Rocket System*	*1980-81*
5149.103	*Bushmaster Cannon*	*1980-81*
5150.103	*Tampella 120mm Mortar*	*1978*
5151.103	*Soltam 155mm Gun/Howitzer*	*1982-83*
5152.102	*Timoney APC*	*1979-80*
5153.111	YMRL 32 128mm Multiple Rocket Launcher M77	135
5154.111	107 mm Type 63 Multiple Rocket Launcher	116
5155.103	*New American Lightweight Gun*	*1979-80*
5156.102	*1M-1975 APC*	*1977*
5167.102	*XM966 High Mobility Weapons Carrier*	*1979-80*
5168.103	*100mm FG M-1944*	*1982-83*
5169.102	*MOWAG Gepard Tank Destroyer*	*1979-80*
5170.103	*Rheinmetall Tank Guns*	*1982-83*
5171.103	*CNMP Turrets*	*1982-83*
5172.103	*Trilateral Tank Gun Trials*	*1978*
5173.102	*MOWAG Taifun MICV*	*1979-80*
5174.102	*MOWAG Grenadier*	*1979-80*
5175.102	*Commando Vehicles*	*1979-80*
5176.102	*EE-17 Sucuri Tank-Destroyer*	*1979-80*
5177.102	*Shir Main Battle Tank*	*1979-80*
5178.103	*British 120mm Tank Gun*	*1982-83*
5179.102	*Grief ARV*	*1978*
5179.103	*155mm Self-Propelled Howitzer*	*1982-83*
5180.102	*Canadian AFV*	*1979-80*
5181.102	*Commando Scout*	*1979-80*
5182.102	*Patel Armoured Car*	*1979-80*
5183.102	*TAM Tank*	*1979-80*
5184.102	*German Light Tank*	*1979-80*
5185.303	30mm Aden Cannon	753
5186.303	27mm Mauser Cannon	741
5187.103	*Type 53 T2 Anti-Aircraft Mount (Tarsque)*	*1982-83*
5188.102	*MT-LB Vehicle*	*1979-80*
5189.102	*M980 MICV*	*1979-80*
5190.103	*Cadillac Gage Turrets*	*1982-83*
5191.103	*76mm gun L23*	*1982-83*
5192.103	*60mm Tampella Mortar Type A*	*1978*
5193.103	*60mm Tampella Mortar Type C*	*1978*
5194.103	*TG-120 Turret*	*1982-83*
5195.103	*105mm L7 Tank Gun*	*1982-83*
5196.103	*152mm M-1943(D-1) Howitzer*	*1982-83*
5198.103	*Oerlikon 35mm AAG*	*1982-83*
5199.103	*MCSMAW*	*1978*
5200.103	*Falcon AAG System*	*1979-80*
5201.111	*Frog-9*	*1979-80*
5202.103	*60mm Long Range Mortar*	*1978*
5203.103	*60mm Gun Mortar Model LR*	*1978*
5204.103	*Panhard ERC Armoured Car*	*1979-80*
5205.103	*Panhard VCR APC*	*1979-80*
5206.103	*AMX-32 MBT*	*1979-80*
5207.303	EXPAL BRFA 330 Bomb	747
5208.303	*Lucas Gun Turret*	*1979-80*
5209.303	*Thomson Brandt 68-12 Rocket Launcher*	*1982-83*
5210.303	*Thomson Brandt 68-22 Rocket Launcher*	*1982-83*
5211.303	*KCA Gun Pod*	*1982-83*
5212.303	*Brandt 68mm Rocket Launcher*	*1984-85*
5213.303	SAMP Practice Bombs	738
5214.303	SAMP General-Purpose Bombs	738
5215.303	SAMP Fragmentation Bombs	739
5216.303	FFV Uni-Pod 0127	748
5217.303	*Liquid Propellant Aircraft Guns*	*1982-83*
5218.303	JP233 Low Altitude Airfield Attack System	755
5219.303	*Compact High-Performance Aerial Cannon (CHAG)*	*1981-82*
5220.303	M41 Armament System	770
5221.303	XM94 Armament Sub-System	770
5222.303	XM59 Armament Sub-System	770
5223.303	Yugoslav Aircraft Armament	771
5224.303	Swedish Aircraft Armament	748
5225.303	United States Aircraft Armament	756
5226.102	*Condor APC*	*1979-80*
5227.102	*TM 170 APC*	*1979-80*
5228.102	*TM 125 APC*	*1979-80*
5229.102	*Cadillac Gage Armoured Patrol Car*	*1979-80*
5230.111	FIROS 25/30 122 mm Field Rocket System	123
5231.103	*SYRA MRS*	*1982-83*
5232.103	*TS 90 Turret*	*1982-83*
5233.103	*Cerbere 20mm AA Mount*	*1982-83*
5234.103	*Centaure 20mm AA Mount*	*1982-83*
5235.103	*122mm Field Howitzer D-30*	*1982-83*
5236.103	*155mm Howitzer M65*	*1982-83*
5237.103	*25mm M242 Bushmaster Cannon*	*1982-83*
5238.103	*57mm Aux Prop AT Gun Ch-26*	*1982-83*
5239.103	*M114/39 Modified Gun System*	*1982-83*
5240.103	*155mm GC 45 Gun/Howitzer*	*1982-83*

Entry number	Subject	Page No/ Last edition*
5241.102	*T-64/T-72/T-80 MBTs*	*1979-80*
5242.102	*New British MBT*	*1979-80*
5243.303	*ML Advanced Ejector Release Unit*	*1979-80*
5244.303	EXPAL General-Purpose Low Drag Bombs	748
5245.303	EXPAL BRP Retarded Bombs	748
5246.303	30mm Aircraft Gun DEFA 554	734
5501.103	*105/14 Model 56 Pack Howitzer*	*1982-83*
5502.103	*British 25-Pounder Gun*	*1982-83*
5503.103	*105mm SP Gun Abbot*	*1982-83*
5504.103	*Rarden 30 Automatic Gun*	*1982-83*
5505.103	*105mm Light Gun*	*1982-83*
5506.103	*115mm Gun M2*	*1982-83*
5508.103	*155mm SP Howitzer M109*	*1982-83*
5509.103	*105mm SP Howitzer M108*	*1982-83*
5510.103	*155mm Howitzer M114A1*	*1982-83*
5511.103	*105mm Howitzer M101A1*	*1982-83*
5512.103	*Lightweight 105mm Howitzer M102*	*1982-83*
5513.103	*Soviet Ordnance, Note*	*1977*
5514.103	*57mm AA Gun S-60*	*1982-83*
5515.103	*57mm Anti-Tank Gun M.1943 (ZIS-2)*	*1982-83*
5516.103	*76mm Divisional Gun M.1942 (ZIS-3)*	*1982-83*
5517.103	*85mm Divisional Gun D-44*	*1982-83*
5518.103	*100mm Anti-Tank Gun M.1955*	*1982-83*
5519.103	*122mm Howitzer D-74*	*1982-83*
5520.103	*122mm Howitzer M-1938 (M30)*	*1982-83*
5521.103	*152mm Gun-Howitzer D20*	*1982-83*
5522.203	*Twin 30mm Gun Mounting GCM-A01*	*1978*
5523.103	*8in SP Howitzer M110*	*1982-83*
5524.103	*175mm Gun M107*	*1982-83*
5525.103	*280mm Gun M66*	*1972-73*
5526.103	*Anti-Aircraft SP Gun, Twin 40mm M42*	*1982-83*
5527.103	*155mm SP Automatic Gun L/50 (VK 155 L/50)*	*1982-83*
5528.103	*40mm Automatic Gun L/70*	*1982-83*
5529.103	*Communist China, Ordnance, Note*	*1982-83*
5530.103	*Ex-Japanese 70mm Howitzer Type 22*	*1969-70*
5531.103	*75mm Mountain Gun Type 94 (ex-Japanese)*	*1978*
5532.203	Single 76/62 OTO MMI Gun Mounting	489
5533.203	Single 76/62 Compact Gun Mounting	489
5534.203	Single 127/54 Gun Mounting	488
5535.203	Twin 35mm OE/OTO Mounting	491
5536.103	*155mm SP Howitzer Mk F3*	*1982-83*
5537.103	*105mm SP Howitzer AMX 105A*	*1982-83*
5538.103	*105mm SP Howitzer AMX 105B*	*1976*
5539.103	*SP Twin 30mm AA Guns AMX DCA 30*	*1982-83*
5540.103	*203mm Gun-Howitzer M55*	*1977*
5541.103	*Tracked Vehicle for 155mm Howitzer*	*1982-83*
5542.103	*105mm Tank Gun*	*1982-83*
5543.103	*90mm AFV Gun*	*1982-83*
5544.103	*SAMM S.530 Turret*	*1982-83*
5546.103	*BL 5.5 in Gun*	*1982-83*
5547.103	*M61 A1 Gun on Vulcan XM 163*	*1982-83*
5548.131	23 mm ZSU-23-4 SP AA Vehicle	205
5549.103	*8in Howitzer M.115*	*1982-83*
5550.103	*155mm SP Howitzer M.44*	*1982-83*
5551.103	*105mm SP Howitzer M52 & M52A1*	*1982-83*
5552.103	*105mm Light Gun*	*1976*
5554.103	*130mm Field Gun*	*1982-83*
5555.111	Artillery Rockets	129
5556.103	*105mm Howitzer with Auxiliary Propulsion (Major-Minor Wheels)*	*1976*
5557.111	84mm RCL Carl-Gustav	146
5558.103	*Miniman Recoilless Anti-Tank Weapon*	*1978*
5559.103	*81mm Light Mortar*	*1978*
5560.103	*60mm Light Mortar*	*1978*
5561.103	*60mm Commando Mortar*	*1978*
5562.103	*60mm CS Mortar*	*1978*
5563.103	*81mm Medium Mortar*	*1978*
5564.103	*81mm Mortar M-29*	*1978*
5565.103	*107mm Mortar M-30*	*1978*
5566.103	*82mm Medium Mortars*	*1978*
5567.103	*120mm Heavy Mortars*	*1978*
5568.103	*160mm Heavy Mortar*	*1978*
5569.103	*240mm Heavy Mortar*	*1978*
5570.103	*155mm SP Howitzer M109A1*	*1982-83*
5571.103	*155mm M198 Howitzer*	*1982-83*
5572.103	*105mm Field Artillery Direct Support Weapon System XM204*	*1980-81*
5573.103	*8in SP Howitzer M110E2*	*1982-83*
5574.103	*155mm GCT SP Gun*	*1982-83*
5575.103	*German Artillery General*	*1976*
5576.103	*105mm Light Field Howitzer*	*1982-83*
5577.111	LARS - 110mm Artillery Rocket Launcher	120
5578.103	*130mm AA Gun KS-30*	*1982-83*
5579.103	*100mm AA Gun KS-19*	*1982-83*
5580.103	*85mm AA Gun M-1944*	*1982-83*

Entry number	Subject	Page No/ Last edition*
5581.103	*Twin 23mm Automatic ZU-23 AA Cannon*	*1982-83*
5582.103	*Quad-Mounted 14.5mm AA Machine Gun ZPU-4*	*1982-83*
5583.103	*Recoilless Anti-Tank Guns*	*1978*
5585.103	*Artillery Rockets*	*1982-83*
5586.103	*Oerlikon 25mm AA Gun Turret*	*1980-81*
5587.103	*20mm AA Gun Type GAI-BO1*	*1982-83*
5588.103	*Czechoslovak Ordnance*	*1973-74*
5589.103	*20mm Gun M621*	*1982-83*
5590.103	*Samurai Artillery Rocket*	*1977*
5591.103	*Light Anti-Tank Grenade Launcher PZF-44*	*1978*
5592.103	*Light Anti-Tank Weapon (LAW) M-72*	*1978*
5593.103	*90mm Anti-Tank Rifle M-67*	*1978*
5594.103	*3.5 in Rocket Launcher M20 A1*	*1978*
5595.103	*100mm Anti-Tank Rocket Launcher*	*1978*
5596.103	*Recoilless Anti-Tank Rifle M-18A*	*1978*
5597.103	*Anti-Tank Grenade Launchers RPC-2 and RPG-7*	*1978*
5598.103	*Anti-Tank Grenade Launcher P-27*	*1978*
5599.103	*20mm Gun M693*	*1982-83*
5600.103	*20mm Rapid-Fire Gun Mk 20 Rh202*	*1982-83*
5601.153	RA 20S Fire Control Radar	230
5602.153	RODEO 2 Fire Control Radar	230
5603.353	IFF 2743 Electronic Control Unit	789
5604.153	*AN/PPS-6 Battlefield Surveillance Radar*	*1985-86*
5605.153	*AN/PSS-10 Battlefield Radar Receiver*	*1985-86*
5606.253	AN/SPS-52C Radar	639
5607.153	Multi-Function Radar (MFR)	302
5608.153	Watchman Air Surveillance Radar	267
5609.353	Advanced Phased Array Radar (APAR)	808
5610.353	AN/APG-67 Multimode Airborne Radar	795
5611.353	Multibeam Multimode Surveillance Radar	811
5612.353	Skyranger Airborne Radar	787
5613.153	*Meteor 300 Weather Radar System*	*1987-88*
5614.353	ASR 360 Airborne Surveillance Radar	790
5615.153	S500 Series Primary Airfield Surveillance Radars	265
5616.153	AN/PPS-11/12 Battlefield Surveillance Radars	284
5617.153	AN/FPS-118 Over-the-Horizon Radar	302
5618.153	SHORAR Search Radar	244
5618.393	Gene-X Expendable Cartridge	841
5619.153	*Sentinel Mobile Radar*	*1983-84*
5620.253	AWS-6 Series Naval Radar	629
5621.053	NATO Identification System (NIS)	779
5622.153	CRS 512 SSR Antenna	276
5623.153	ARSR-3/ASR-30 Surveillance Radars	286
5624.353	Joint Surveillance and Target Attack Radar System	814
5625.353	Super Searcher Airborne Radar	791
5626.253	SMART Naval Surveillance Radar	614
5627.153	Pacer Mk 2 Muzzle Velocity Radar	273
5628.153	Real-Time Velocimeter Radar (RTVS)	282
5629.153	Terma Radar and Ballistic Instrumentation	225
5630.253	RN Type 911 Tracking Radar	633
5631.153	*Cervantes Mortar & Rocket Locating Radar*	*1983-84*
5632.253	Type S1100 Navigation and Surveillance Radar	552
5633.453	SS 24A LF Sonar	728
5634.293	*Thomson Sintra Ancillary Sonar Equipment*	*1987-88*
5635.453	*PHS-34 Sonar*	*1984-85*
5636.453	Submarine Integrated Attack and Surveillance Sonar System (SIASS)	559
5637.353	Cormorant Lightweight Dipping Sonar	568
5638.263	PMS 75 Speedscan/Sonar 2048	576
5639.253	PMS 40 Series Sonar Systems	612
5640.453	AN/PQS-2A Hand-Held Sonar	582
5641.153	*S513 Primary Surveillance Radar*	*1983-84*
5642.153	*Marconi Secondary Surveillance Radar*	*1983-84*
5643.453	Underwater Ordnance Locator Mk.24 Mod 0	599
5644.453	EDO Model 900/910 Underwater Detection Systems	590
5645.253	SS304 Scanning Sonar	561
5646.253	TSM 2630/2633 Diodon/Spherion Sonar	552
5647.253	*TSM 2600 Diodon Sonar System*	*1983-84*
5648.253	TSM 2640 Salmon Sonar	552
5649.253	TSM 2820 Tarpon Sonar	552
5650.453	Argonaute Sonar System	548
5651.453	Type 2020 Submarine Sonar	567
5652.253	AN/SQQ-32 Advanced Minehunting Sonar System	587
5653.263	*ELT/810 Sonar Prediction System*	*1985-86*
5654.363	*Mosaic ASW Integration System*	*1985-86*
5655.363	AQS 903/AQS-930 Series Acoustic Processor Systems	563
5656.293	SAGAIE Naval Countermeasures Decoy System	648

Entry number	Subject	Page No/ Last edition*
5657.293	SAPIENS EW System	649
5658.153	*Directional Antenna System AC 002*	*1986-87*
5659.293	SCHALMEI Countermeasures Decoy System	651
5660.293	Automatic Countermeasures Dispensing System	653
5661.253	FARAD Naval EW System	657
5662.453	THETIS Submarine ESM System	657
5663.253	INS-5 Naval ESM/ECM System	659
5664.153	ESS-2 Electronic Surveillance System	318
5665.393	Alkan Chaff/Infra-Red Countermeasures Dispensers	820
5666.393	BO Series Countermeasures Dispensers	831
5667.393	Masquerade Airborne Decoy System	834
5668.393	Cascade Helicopter.Decoy System	834
5669.153	Weasel ESM & ELINT Collection System	328
5670.293	Cygnus ECM System	668
5671.453	Porpoise Submarine ESM Equipment	668
5672.493	Sprat Submarine ESM System	669
5673.353	Prophet Radar Warning Receiver	836
5674.193	Saviour Laser Warning System	328
5675.293	Sea Saviour EW System	669
5676.293	Radar Antenna Pedestals	635
5677.253	AN/SPS-503 Surveillance Radar	600
5678.153	*CMR-3 Coastal Surveillance Radar*	*1986-87*
5679.153	RA1795 VHF/UHF Receiver	322
5680.153	SAT 3311 HF Signal Analysis Terminal	324
5681.153	CCS 3900 Command & Control System	624
5682.293	Shield Anti-Ship Missile Decoy System	663
5683.293	Sibyl EW Naval Decoy System	650
5684.353	JAS-39 Airborne Radar	784
5685.353	Blue Kestrel Helicopter Radar	793
5686.353	AN/UPD-8 and AN/UPD-9 Side-Looking Reconnaissance Radars	812
5687.353	Enhanced Radar Warning System	859
5688.393	Threat Adaptive Countermeasures Dispenser System (TADS)	840
5689.353	Zeus Airborne ECM System	833
5690.393	AN/ALR-77 ESM System	858
5691.193	*Exjam Expendable Communications Jammer*	*1983-84*
5692.355	AN/APR-39A(V) Threat Warning System	860
5693.393	AN/ALQ-172 Noise Deception Jammer	851
5694.353	AN/ALR-74 Radar Warning System	857
5695.153	*Pinemartin Ground-Based ESM System*	*1984-85*
5696.253	Automatic Missile Detection Radar	607
5697.193	*CC-35 Direction Reference System*	*1987-88*
5698.193	*OB-50 Night Aiming Telescope*	*1987-88*
5699.193	Daldo Target Acquisition Aid System	358
5700.193	*Selenia Laser Units*	*1985-86*
5701.193	*GM 79 Goniometer*	*1987-88*
5702.193	RSI Tank Laser Rangefinder	371
5703.193	LP160 Laser Rangefinder	374
5704.193	*E4020 Miniature Combat Night Sight(MNV)*	*1987-88*
5705.193	E5010 Laser Rangefinder	376
5706.193	E2030 Night Driverscope	376
5707.193	ENOSA Electro-Optical Systems	376
5708.193	PKD-8/KOL-2 Sighting System for Artillery	379
5709.193	SRS-5 Reflecting Sight	379
5710.193	Condor Commander's Day/Night Sight	388
5711.193	*W401 Night Observation Device*	*1987-88*
5712.193	TL10-T Articulated Telescope Laser Rangefinder	381
5713.193	Type 312 Laser Beacon and Target Simulator	384
5714.193	NVL53 Day/Night Sight and Rangefinder	381
5715.193	Thermal Imager Vehicle Sight	387
5716.193	*Rank Pullin Sights*	*1987-88*
5717.193	M19A1 Driver's Night Periscope	400
5718.193	*Mini Electro-Optical Imaging System*	*1987-88*
5719.193	Kollmorgen Electro-Optical Systems	399
5720.193	*Litton Night Vision Products*	*1987-88*
5721.193	SECO Electro-Optical Systems	406
5722.293	IRSCAN Infra-Red Surveillance System	684
5723.193	SUVOA Artillery Fire Control System	407
5724.193	*M61 Stereoscopic Rangefinder*	*1987-88*
5725.193	Vickers/GEC CO_2 Tank Laser Rangefinder	393
5726.193	Combat Vehicle Night Sight	757
5727.093	*GSL Gyro Stabilised Laser Rangefinder*	*1985-86*
5728.293	TC10 Thermal Imager	684
5729.293	V3800 Thermal Imaging Sensor	687
5730.393	EL/L 8230 Airborne ECM System	822
5731.253	9EW 300 and 9EW 400 ECM Systems	661
5732.053	Hermes ESM System	833
5733.153	*Matlock Radar Warning System*	*1984-85*
5734.353	KATIE Airborne Radar Warning System	835
5735.393	BAe Infra-Red Jammer	836
5736.053	BAe ESM System	329
5737.293	Radamec Lightweight Missile Launcher	663
5738.053	*THORN EMI EW Systems*	*1987-88*
5739.353	Vicon 77 ESM Pod	837
5740.093	MEL IFM Receivers	327
5741.193	Hand-Emplaced Expendable Jammer	346
5742.393	*Northrop Infra-Red Countermeasures Systems*	*1986-87*
5743.353	Escort ESM System	864
5744.393	TMV 630 Laser Rangefinder	871
5745.393	Ericsson Laser Rangefinder	877
5746.383	*V325 Low Light TV System*	*1983-84*
5747.393	Helicopter Laser Designator (HELD)	886
5748.393	GEC Sensors FLIR Systems	879
5749.393	Type 221 Thermal Surveillance System	886
5750.383	IRSTS (Airborne Infra-Red Search and Track System)	900
5751.393	*Tigershark Head-Up Display*	*1986-87*
5752.393	Hughes Head-Up Display	901
5753.393	SPASYN Visual Target Acquisition System	888
5754.293	Multi-Application Sonar System (MAST)	557
5755.193	Forward Observer Laser Designator (FOLD)	385
5756.193	Battlefield Operations Laser Designator (BOLD)	385
5757.193	Type 520 Laser Rangefinder	384
5758.393	Litton Mast-Mounted Sight LRF/D	891
5759.153	*Type 711 Meteorological Radar*	*1987-88*
5760.153	*Type 713 Meteorological Radar*	*1987-88*
5761.153	Coast Tracking Radar	222
5762.193	Siemens Artillery Surveillance Radar	237
5763.153	IRS-10 Secondary Surveillance Radar	249
5764.153	Compact IFF/Secondary Radar	250
5765.153	Possum Air Defence Surveillance Radar	270
5766.153	Zulu Millimetric Surveillance Radar	271
5767.153	*Type AA5504 MV Radar*	*1987-88*
5768.153	*B & P Mobile Surveillance Radar*	*1987-88*
5769.153	*IFF Interrogator Antenna 3937A*	*1987-88*
5770.153	Selenia Mobile GCA System	246
5771.153	Eagle Fire Control Radar	251
5772.153	HARD 3-D Search Radar	251
5774.153	Watchman Coastal Defence Radar	268
5775.153	Guardsman Coastal Defence Radar	268
5776.153	AN/MSQ-117 IFF Tracking System	299
5777.153	AN/UPS-3 Tactical Defence Radar	301
5778.153	DR-810 Mk V Muzzle Velocimeter	304
5779.153	*ARSR 3-D Surveillance Radar*	*1985-86*
5780.153	W-2000 3-D Surveillance Radar	286
5781.153	Relocatable OTH Radar	303
5782.153	IFF Interrogator NSRI-3	233
5783.253	NADS Naval Anti-Missile Decoy System	649
5784.253	NEWSY Naval Electronic Warfare System	649
5785.293	Janet Naval Jammer	650
5786.393	DB-3141 Jamming Pod (Low-Band Remora)	816
5787.353	Caiman Noise/Deception Jamming Pod	816
5788.393	Syrel Elint Pod (TMV 018)	817
5789.353	Serval Radar Warning Receiver	817
5790.393	Matra Passive Countermeasures	820
5791.193	GA 082 FSK Analyser	313
5792.193	Plath Direction Finding Systems	313
5793.353	SPS-200 Airborne Self Protection System	824
5794.353	Lavi Self-protection EW System	678
5795.393	ELT/562 & ELT/566 Deception Jammers	827
5796.353	ELT/999 Airborne COMINT System	827
5797.393	*Loke Radar Jammer Series*	*1986-87*
5798.353	ERIJAMMER 39	831
5799.393	Chemring Countermeasures	837
5800.153	RACEWS EW System	324
5801.153	MA 1122 Tactical DF Processing & Display Unit	325
5802.153	Hunting Direction Finding System	325
5803.153	Sentry ESM System	320
5804.393	Sky Guardian Airborne Radar Warning Receiver	834
5805.353	Apollo Airborne ECM System	834
5806.193	Schermuly IR Countermeasures	329
5807.353	Monarch RPV EW System	834
5808.193	JAMCAT Jamming Unit	325
5809.393	Racal Helicopter-Borne Jamming System	864
5810.153	D1800 DF & Monitoring Systems	344
5811.393	Quick Fix IIB Countermeasures System	864
5812.353	Integrated EW System	863
5813.253	ARBR-17 ESM System	649
5814.293	ARB 33 Naval Jammer	649
5815.053	Spanish EW Programme	829
5816.053	R-3030/3050/3080 Surveillance Receivers	332

Entry number	Subject	Page No/ Last edition*
5817.353	Scarecrow RPV EW System	863
5818.393	*Lancer Radar Warning Receiver*	*1986-87*
5819.093	TC-5100 Series DF Systems	344
5820.253	FL1800S Naval ESM/ECM System	652
5821.253	FL400 Series EW Suites	652
5822.393	Vicon 78 Airborne Decoy Dispenser	837
5823.393	Matador	864
5824.393	Caiquen Airborne Radar Warning Receiver	815
5825.393	Eclipse Chaff/IR Decoy Launcher	815
5826.353	ITATA Airborne Elint System	815
5827.393	*GTE Airborne VHF/UHF Jammer*	*1987-88*
5828.393	AN/ALE-45 Countermeasures Dispenser System	841
5829.393	AN/ALE-47 Countermeasures Dispensing System	841
5830.353	*Model 700 Series EW Suites*	*1987-88*
5831.353	Mirage 2000 Integrated EW Suite (ICMS)	818
5832.353	*AN/ALR-59(V) ESM System*	*1986-87*
5833.353	AN/ALR-76 ESM System	858
5834.193	AN/ULQ-19 Tactical Communications Jammer	329
5835.293	AN/WLR-11 IFM Receiver	671
5836.253	EWS-905 ESM Warning System	661
5837.353	LR-5200 Tactical ESM System	865
5838.293	S-3000 Surveillance System	675
5839.353	Compass Call ECM System	865
5840.253	*Shields Shipborne EW System*	*1986-87*
5841.253	Elta Naval EW Systems	653
5842.253	Chinese Naval Radar	600
5843.253	DRBC 33 Naval Radar	601
5844.253	*Triton G Naval Radar TRS3050*	*1985-86*
5845.253	*Triskel Multi-function Naval Radar*	*1987-88*
5846.293	AT-101A Triple-band Surveillance Antenna	608
5847.253	Type 1007 Radar	632
5848.293	AIMS Antenna System	645
5849.453	Snoop Class Submarine Surveillance Radars	624
5850.353	MAD Mk III Magnetic Anomaly Detector	555
5851.453	TSM 8020 Sonobuoys	555
5852.363	AD 130 Sonar Homing & DF Receiver	563
5853.253	Type 2050 Sonar	570
5854.253	VIMOS Vibration and Noise Monitoring System	573
5855.453	Type 2051 Submarine Sonar	571
5856.253	Sonar 2093	571
5857.453	Type 2054 Submarine Sonar	571
5858.453	Type 2052 Submarine Sonar	570
5859.453	*STS 3000 Integrated Sonar System*	*1987-88*
5860.253	*STS 3100 Towed Array Sonar*	*1987-88*
5861.363	Airborne Sonobuoy Telemetry Systems Types 864, 865, 866	574
5862.453	*Marconi 360-Degree Surveillance Sonar*	*1987-88*
5863.253	AN/SQR-501 Canadian Towed Array Sonar System (CANTASS)	543
5864.463	Karrawarra Towed Array Sonar	541
5865.353	HELRAS Airborne Sonar	556
5866.353	Guillemot Dipping Sonar	568
5867.453	SWS 35 Miniature Towed Array	577
5868.153	ASTAR-3 & ASTAR-2 Air Defence Radars	297
5869.153	Marconi OTH Radar	265
5870.153	Vigilant Tactical Surveillance Radar	296
5871.353	NRAI-7A Airborne IFF Transponder	777
5872.353	NRAI-10A Airborne IFF Interrogator	778
5873.363	NRAI-11A IFF Interrogator-Decoder	778
5874.353	IRIS Integrated Radar Imaging System	772
5875.353	Orchidee Battlefield Surveillance Radar	778
5876.353	EL/M-2035 Multimode Fire Control Radar	780
5877.353	Red Fox Airborne Radar	793
5878.353	*APY920 Mission Avionics System*	*1986-87*
5879.353	AN/APG-70 Airborne Radar	797
5880.353	AN/APG-71 Airborne Radar	797
5881.353	AN/APG-68 Airborne Radar	796
5882.353	AN/APG-69(V) Multimode Airborne Radar	796
5883.353	AN/APS-131 Side Looking Radar	806
5884.353	AN/APS-153 Side Looking Radar	807
5885.353	SLAMMR Side Looking Radar	808
5886.353	Skymaster Multi-Role Airborne Radar	788
5887.353	HERA Airborne Radar	784
5888.353	B-52 Airborne Radar	814
5889.193	*OIP Miscellaneous Electro-Optics*	*1987-88*
5890.153	Chinese Electro-Optical Systems	349
5891.353	ASN-902 Tactical Processing System (TATTIX)	563
5892.193	M 389 Panoramic Sight	359
5893.193	Vassir 85 Anti-Aircraft Optronic Sight	358
5894.153	LGA 389 Artillery Goniometer Sight	359
5894.353	*LAVI Self Protection EW System*	*1987-88*
5895.193	M 590 Gyrostabilised Sight	359
5896.193	VISAA Anti-Aircraft Sight	355
5897.193	TCV 186 Laser Rangefinder	350
5898.193	TCY 167 Miniature Laser Rangefinder	352
5899.093	*TRM 7401 Thermal Imager*	*1985-86*
5900.193	MT-18 and MT-19 Laser Rangefinders	364
5901.193	ELOP Tank Laser Rangefinder	364
5902.193	Vanth Optronic Fire Control System	367
5903.193	Madis Sighting and Drive System	367
5904.193	SAMTOR Thermal Observation System	371
5905.193	*Type 630 CO_2 Laser*	*1986-87*
5906.193	TOTE Missile Sight	383
5907.193	Trigat Sighting System	361
5908.193	Raven Day/Night Sight	388
5909.193	*M-931 Ranging Nightscope*	*1987-88*
5910.193	AN/VAS-3 Driver's Thermal Viewer	402
5911.193	AN/PVS-6 Melios Hand-Held Laser Rangefinder	403
5912.193	Rokit Tank Sight	401
5913.193	*Pilar Fire Control System*	*1986-87*
5914.193	OMU-2 Artillery Laser Rangefinder	407
5915.193	*LH-83 Laser Rangefinder*	*1987-88*
5916.293	*SAMOS Anti-missile Optronic System*	*1985-86*
5917.293	Cotrol Optronic Fire Control System	681
5918.293	TMY 156 Laser Rangefinder	681
5919.293	OGR7/2-A Optical Director	682
5920.293	*Manned Fire Control Director*	*1987-88*
5921.293	DS-35 (SPIRTAS) Target Acquisition System	682
5922.293	*Sea Lite Beam Director*	*1985-86*
5923.393	Tango Thermal Imager	869
5924.393	*Calipso Thermal Imager*	*1987-88*
5925.393	Crouzet Type 200 Helicopter Sight	872
5926.393	Vinten Vicon 90 Series 101 FLIR Pod	887
5927.393	*PPE Optical Modules*	*1987-88*
5928.393	Type 234 Sensor Payload System	886
5929.393	Type 116 Laser Rangefinder	884
5930.393	Type 118 Laser Designator/Ranger	886
5931.393	*Type 121 Laser Designator/ranger*	*1985-86*
5932.393	Honeywell FLIR Systems	892
5933.393	2000 Series Thermal Imaging Systems	900
5934.393	ITT ANVIS	902
5935.393	AN/AAQ-16 Night Vision System	897
5936.393	Hughes ANVIS	902
5937.153	AT-4910 Comint/Jamming System	317
5938.155	*Seeker EW System*	*1985-86*
5939.253	Mentor Naval ESM System	669
5940.293	Siren Naval Decoy System	670
5941.393	ABD 2000 Detector/Jammer System	820
5942.393	AN/ALR-80(V) Radar Warning Receiver	858
5943.393	SO-1 Radar Warning Receiver	866
5944.393	*Vicon 70 Series 33 IRCM Pod*	*1987-88*
5945.393	Loral IR Countermeasures	863
5946.393	Barem Self-Protection Jammer	817
5947.393	Basilisk Jamming Pod	817
5948.393	EL/K 7032 COMINT System	821
5949.393	EL/L 8312A Elint System	822
5950.393	AN/ALQ-122 ECM System	844
5951.153	Chief EW Combat System	345
5952.393	AN/ALT-28 Noise Jammer	859
5953.093	ISIS SIGINT System	317
5954.293	Winnin Decoy System	646
5955.393	Type 122 Helicopter Laser Rangefinder	886
6001.241	ASROC Anti-Submarine System	533
6002.241	Ikara Anti-Submarine Weapon System	512
6003.231	*Seaslug Surface-to-Air Missile*	*1987-88*
6004.231	Sea Dart Surface-to-Air Missile	476
6005.231	Terrier Surface-to-Air Missile (RIM-2)	482
6006.231	Tartar Surface-to-Air Missile (RIM-24)	481
6007.241	*Squid Anti-Submarine Mortar System*	*1980-81*
6008.241	Limbo Anti-Submarine Mortar System (AS Mk 10)	528
6009.231	*STAAG Mk 2 & 3 Stabilised Tachymetric Anti-Aircraft Gun System*	*1972-73*
6010.203	Single 4.5in Gun Mounting Mk 5	503
6011.203	Single 4.5in Gun Mounting Mk 8	503
6012.203	Twin 4.5in Gun Mounting Mk 6	503
6013.203	Single 40mm (Bofors) Mk 7 and Mk 9 Gun Mountings	504
6014.203	Twin 4in Gun Mounting Mk 19	503
6015.203	Twin 40mm (Bofors) Gun Mounting Mk 5	504
6016.203	Twin 3in Gun Mounting Mk 6	503
6017.203	*Twin 6in Gun Mounting Mk 26*	*1982-83*
6019.221	Gabriel Shipborne Surface-to-Surface Missile	452
6021.241	Type 375 Anti-Submarine Weapon System	524

Entry number	Subject	Page No/ Last edition*
6022.241	Terne III Surface to Sub-Surface Missile	523
6023.203	*Triple 16in Gun Mounting*	*1974-75*
6024.203	*Triple 8in Mk 15 Gun Mounting*	*1974-75*
6025.203	*Triple 6in Mk 16 Gun Mounting*	*1982-83*
6026.203	Twin 5in/38 Calibre Gun Mounting Mk 28	505
6027.203	Single 5in/54 Calibre Gun Mounting Mk 42	506
6028.203	Single 5in/54 Calibre Gun Mounting Mk 42 (Lightweight Mod 9)	507
6029.203	Single 5in/54 Calibre Lightweight Automatic Gun Mounting Mk 45	507
6030.203	Single 3in/50 Calibre Gun Mounting Mk 34	507
6031.203	40mm Anti-Aircraft Gun Mountings	508
6032.203	*20mm Rapid Fire Gun*	*1971-72*
6033.203	*0.50 calibre Machine Gun*	*1971-72*
6034.203	*20mm Gun Mount Mk 56*	*1974-75*
6035.203	*SAMM/HS 30mm Gun Mount Type TM 1-30*	*1974-75*
6036.203	35mm Twin Anti-Aircraft Gun Type GDM-A	497
6037.203	*Triple 130mm Gun Mounting*	*1977*
6038.203	Triple 152mm Gun Mounting	499
6039.203	Twin 130mm Gun Mounting	499
6040.203	Twin 130mm Dual-Purpose Gun Mounting	500
6041.203	Single 100mm Dual-Purpose Gun Mounting	500
6042.203	Twin 100mm Dual-Purpose Gun Mounting	500
6043.203	Single 85mm Dual-Purpose Gun Mounting	500
6044.203	Twin 76mm Dual-Purpose Gun Mounting	500
6045.203	Single 57mm Dual-Purpose Gun Mounting	501
6046.203	Twin 57mm AA Gun Mountings	501
6047.203	Quadruple 57mm AA Gun Mountings	501
6048.203	Quadruple 45mm AA Gun Mounts	501
6049.203	*Light Anti-Aircraft Guns (USSR)*	*1972-73*
6050.203	*Single 5in Dual-Purpose Gun Mounting*	*1981-82*
6051.203	*Twin 4.5in Gun Mounting*	*1974-75*
6052.203	*Single 3in Dual-Purpose Gun Mounting*	*1974-75*
6053.203	*Twin 3 inch AA Mounting*	*1986-87*
6054.203	Single 5 inch Mk 27 Dual-Purpose Gun Mounting	505
6055.203	Chinese Naval Ordnance	485
6056.203	*Twin 6in Gun Mounting*	*1971-72*
6057.203	*Twin 120mm Dual-Purpose Gun Mounting*	*1971-72*
6058.203	*Twin 8in Gun Mounting*	*1978*
6059.203	*Single 120mm AA Gun Mounting*	*1977*
6060.203	*Twin 120mm Gun Mounting*	*1979-80*
6061.203	*Single 105mm Gun Mounting*	*1974-75*
6062.203	*Single 5in Gun Mounting*	*1972-73*
6063.203	*Twin 5in Dual-Purpose Gun Mounting*	*1981-82*
6064.203	Single 100mm Dual-Purpose Gun Mounting Model 1968-II	487
6065.203	Twin 57mm AA Gun Mounting	487
6066.203	*Twin 135mm Dual-Purpose Gun Mounting*	*1972-73*
6067.203	*Twin 76/62 Dual-Purpose Gun Mounting*	*1972-73*
6068.203	*Quadruple 40/56 Bofors Gun Mounting*	*1976*
6069.203	*Single 6in Gun Mounting*	*1979-80*
6070.203	*Triple 6in Gun Mounting*	*1985-86*
6071.203	*Single 120mm Dual-Purpose Gun Mounting*	*1978*
6073.203	Single 120mm L/46 Gun Mounting	495
6074.203	*Twin 120mm Dual-Purpose Semi-Automatic Gun Mounting*	*1979-80*
6075.203	*Single 105mm Dual-Purpose Gun Mounting*	*1972-73*
6076.203	*Modified Single 5in 54 calibre Gun Mounting Mk 42*	*1976*
6077.203	*Single 5in 54 calibre Mk 65 Gun Mounting*	*1972-73*
6078.203	*Twin 3in 70 calibre Mk 23 Gun Mounting*	*1972-73*
6079.203	*Single 120mm Gun Mounting*	*1974-75*
6080.203	*Triple 6in Gun Mounting*	*1978*
6081.203	*Single 4.7in (120mm) Gun Mounting*	*1974-75*
6082.203	*Single 4in Dual-Purpose Gun Mounting*	*1977*
6083.203	*HS 30mm Gun Mountings*	*1974-75*
6084.203	*Single 37mm AA Gun Mounting*	*1974-75*
6085.203	Twin 6in Gun Mounting	372
6086.203	*Twin 120mm Dual-Purpose Automatic Gun Mounting*	*1985-86*
6087.203	*Twin 57mm AA Gun Mounting*	*1985-86*
6088.203	Single 40/60 AA Gun Mounting	497
6089.203	*Triple 6in Gun Mounting*	*1981-82*
6090.203	*Triple 8in Mk 16 Gun Mounting*	*1974-75*
6091.203	*Triple 8in Mk 15 Gun Mounting (Modified)*	*1976*
6092.203	Single 5in/38 Calibre Gun Mountings	506
6093.203	*Single 5in/54 Calibre Gun Mounting Mk 39*	*1986-87*
6094.203	*Single 8in Major Calibre Lightweight Gun Mount Mk 71*	*1986-87*
6095.203	*Shipborne Anti-Submarine Weapons Introduction*	*1976*
6096.203	*Russian Anti-Submarine Rocket Launcher*	*1976*
6097.241	*Soviet Anti-Submarine Weapon Systems*	*1979-80*
6101.253	AWS-7 Naval Radar	629
6102.253	AWS-9 Naval Surveillance Radar	629
6103.153	COBRA Weapon Locating Radar	239
6104.353	RDR-1500 Multi-Mode Search and Surveillance Radar	799
6105.353	AN/APN-234 Multi-Mode Radar	798
6106.253	ST1802 Series Fire Control Radars	633
6107.253	S1810 Shipborne and Coastal Defence Radar	634
6108.153	EL/M-2216(V) Coastal Surveillance System	241
6109.153	Terminal Imaging Radar System	304
6110.153	MTAS Target Acquisition System	305
6111.153	Advanced Missile Surveillance Radar	297
6112.153	Multi-Role Survivable Radar	297
6113.303	FAMAE 3.5 kg Anti-Personnel Bomb	729
6114.393	*IFF TEC-60 Combined Interrogator/ Transponder*	*1987-88*
6115.393	*AIXP Advanced Interrogator/Transponder*	*1987-88*
6116.353	AN/APQ-171 Airborne Radar	804
6117.353	AN/APQ-172 Airborne Radar	804
6118.153	*Quadradar Mk VI ATC Radar*	*1987-88*
6119.253	Seaguard Tracking Module (TMX)	617
6120.153	PSM 33 Mobile Radar	239
6121.153	Jindalee OTH Radar	218
6122.393	NRAI-3A IFF Transponder	777
6123.393	NRAI-9A IFF Transponder	777
6124.353	Antilope V TC Airborne Radar	776
6125.193	IRS-20MP Monopulse Secondary Surveillance Radar	250
6125.293	D-ESM Shipborne Receiver	675
6126.293	AN/UPX-24(V) Shipborne IFF System	645
6127.153	JY-I0RIPP Radar Information Processing Post	220
6128.253	Multi-Function Electronically Scanned Adaptive Radar (MESAR)	635
6129.153	Trackstar Low Altitude Air Defence System	288
6129.193	Trackstar Low Altitude Defence System	291
6130.253	MW08 3-D Surveillance Radar	615
6131.293	400 Series Naval Fire Control Radars	634
6132.153	RB12A (Rassur) Ground Surveillance Radar	233
6133.293	*ELI-4 (NRBI-21) Shipborne IFF Interrogator*	*1987-88*
6134.293	*ELR-3 (NRBI-26/NRUI-26) IFF Transponder*	*1987-88*
6135.153	Soviet OTH Radar	261
6136.253	OPS Series Shipborne Search Radars	611
6137.353	RDX Airborne Radar	773
6138.193	*Wavefinder ESM System*	*1987-88*
6139.193	Artillery Delivered Expendable Jammer (AD/EXJAM)	346
6140.153	Communications EW System (CEWS)	339
6141.393	Fairchild Weston ECM Payload Packages	863
6142.393	AN/ALQ-184 Jamming Pod	852
6143.293	SWORD Decoy System	663
6144.193	S500 ESM System	344
6145.193	R4000 Superheterodyne Receiver	344
6146.193	OE-320/A DF System	344
6147.393	QRC 83-05 Infra-Red Countermeasures System	837
6148.393	QRC 84-02 Infra-Red Countermeasures Pod	838
6149.393	Modular IR Transmitter System (MIRTS)	838
6150.293	Radamec EW Antenna Director Mounts	670
6151.252	Sceptre ESM System	714
6151.253	Sceptre ESM System	664
6152.353	EL/L-8300 Airborne SIGINT System	822
6153.393	EL/L-8231 ECM System	822
6154.093	Active Electronic Buoys	676
6155.393	Wallop IR Decoy Flares	837
6156.253	Guardian Star Shipborne EW System	672
6157.393	AWARE-3 Radar Warning Receiver	821
6158.193	Piranha Add-On Jammer	343
6159.193	SCR Series Receiving & Monitoring Equipment	344
6160.393	Ferranti Laser Warning Receiver	836
6161.293	Replica Naval Decoy	670
6162.293	AN/SAR-8 IR Warning System	647
6163.293	Phoenix ESM System	676
6164.193	Blast Jamming System	309
6165.093	Tadiran COMINT Systems	317
6166.095	AN/ASQ-191 Communications/Countermeasures Set	862
6167.153	Vanguard 3-D Radar	248
6168.153	Signaal Coastal Radar	249
6169.153	RAT-31SL Long Range Radar	245
6170.153	Argos-45 Tactical Radar System	246
6171.331	AA-11 (Archer) Air-to-Air Missile	701
6172.111	130 mm (30-round) Type 82 Multiple Rocket System	116
6173.303	FAMAE General Purpose Low-Drag Bombs	729
6174.303	FERRIMAR WB-500-F Avispa Cluster Bomb System	792

Entry number	Subject	Page No/ Last edition*
6175.303	Chinese Aircraft Cannon	731
6176.303	100 kg Aircraft Bombs	731
6177.303	*Aerea External Store Rack Type CLDERU*	*1987-88*
6178.303	BPD 122mm Falco Air-to-Ground Rocket System	744
6179.303	GA1 20mm Automatic Cannon	746
6180.303	CB 470 Cluster Bomb System	746
6181.303	Expal BME 330 Cluster Bomb	747
6182.303	Bofors 80 kg Aircraft Parachute Flare Super Lepus	749
6183.303	Direct Airfield Attack Combined Munition (DAACM)	758
6184.121	C801 Coastal Defence Misile System	160
6185.121	RDS 15 Coastal Defence Weapon	162
6186.121	Excalibur Coast Defence Missile System	165
6187.221	*Gabriel Long Range Anti-Ship Missile*	*1987-88*
6188.231	CS-AX-? Point Defence Missile	462
6189.431	SA-NX-? Underwater-to-Air Missile	475
6190.111	SS-16 ICBM (Soviet Designation RS-14)	7
6191.411	SS-NX-24 Cruise Missile	16
6192.131	SH-08 (Gazelle) Anti-Ballistic Missile Missile	18
6193.203	Twin 20mm AA Turret	485
6194.303	100mm Compact Gun Mounting	517
6195.203	Storm Rocket Launcher System	487
6196.203	DS 30 and 25 Series Gun Mountings	504
6196.353	Ericsson AEW Radar	784
6197.281	Sea Cobra Fire Control System	413
6198.281	Launching System for Lightweight Torpedoes	436
6199.281	CANE/MAINS	438
6200.261	*AN/SSQ-91 Combat Simulation Test System*	*1987-88*
6201.353	Grifo Airborne Radar	781
6202.353	Altair Airborne Radar	781
6203.353	Setter Fire Control Radar	781
6204.153	Type 282 Ranging Radar	273
6205.153	INDRA Air Defence Radar	239
6206.253	Soviet Naval EW Systems	661
6207.193	Rattler Jammer	316
6208.253	Shipboard EW Suite (SEWS)	655
6209.153	AN/TSQ-114 Intercept and DF System	343
6210.393	Airborne Remote Controlled ESM System	866
6211.393	IRCM Self-Defense System	864
6212.153	Tactical Communications Surveillance System	342
6213.193	Tactical Communications Jammer System	342
6214.393	POET Expendable Jammer	841
6215.293	D-ESM Shipborne Receiver	675
6216.193	MS Series ELINT/ESM Systems	327
6217.393	AN/AVR-2 Laser Warning Receiver	861
6218.393	ARR-81 Comint Receiver	862
6219.393	Dragonfly Airborne VHF-UHF ECM System	865
6220.293	AN/SSQ-72 Outboard System	673
6221.393	AN/ALQ-108 Jamming Pod	843
6222.393	AN/ALQ-117 Jammer	843
6223.293	AN/WLR-1H Radar Detection Equipment	670
6224.111	AT-8 (Songster) Anti-Tank Missile	151
6225.111	Advanced Anti-Tank Weapon System - Medium (AAWS-M)	153
6226.111	Tactical Missile System	156
6227.131	SA-X-15 Low-to-Medium Altitude Surface-to-Air Missile System	182
6228.131	Forward Area Air Defense System	206
6229.263	Fathom Sonar Handling Equipment	545
6230.253	*Scan 515 ASW Sonar*	*1987-88*
6231.253	*Scan 525 Search/Attack Sonar*	*1987-88*
6232.263	Depth Compensated Ring Shell Projector	545
6233.453	Model 971/977 Colour Imaging Sonar	543
6234.263	AN/SSQ-504 Towing Condition Monitor	546
6235.353	HS 312S Helicopter Sonar	550
6236.253	Type 2031Z Towed Array Sonar	568
6237.253	Sonar 3000-Plus Series	569
6238.453	Type 2046 Submarine Sonar	569
6239.353	*Magnetic Anomaly Detector Mark 3*	*1986-87*
6240.353	High Definition Scanning Sonar	573
6241.353	*Fast Minehunting Sonar*	*1987-88*
6242.263	SP 2104 Fast Time Acoustic Processor	577
6243.453	SSQ-954B DIFAR Sonobuoy	575
6244.253	Compact Towed Array Sonar System (COMTASS)	571
6245.263	AN/SQR-17 Sonar Signal Processor	584
6246.463	Sippican Expendable ASW Devices	594
6247.453	AN/BQS-15 Submarine Sonar	594
6248.253	AN/SQR-15 Towed Array Sonar	584
6249.253	QQS-4 Surface Ship Sonar	559
6250.453	ATAS Active Towed Array Sonar	573
6251.393	Sherloc Radar Warning System	817
6252.393	Thomson-CSF Laser Warning Receiver	818
6253.393	Thomson-CSF Active Expendable Jammers	818
6254.353	ASTAC Tactical Elint System	818
6255.393	*AN/ALQ-189 Airborne Jammer*	*1987-88*
6256.393	AN/ALR-75(V) Surveillance Receiver	858
6257.493	SEPADS Sonar Prediction System	579
6258.493	STRAP Programme	599
6259.421	SM 39 Submarine-Launched Exocet	517
6260.441	Dragonfish Mine	531
6261.441	*Minnow ROV Countermeasures Vehicle*	*1987-88*
6262.461	ECP Explosive Cutter Device	523
6263.181	Modular Control Equipment	61
6264.353	CR-2800 ESM/ELINT System	824
6265.321	Penguin Mk 2 Mod 7 Air-to-Surface Missile	715
6266.311	Soviet Tactical Air-to-Surface Missiles	718
6267.311	AS-13 Strategic Air-to-Surface Missile	718
6268.311	AGM-123A Skipper II Guided Missile	724
6269.321	*Helitow Airborne TOW System*	*1987-88*
6270.131	SA-14 (Gremlin) Man-Portable Anti-Aircraft Missile	214
6271.131	Starstreak High Velocity Missile System	216
6272.111	AT-7 (Saxhorn) Anti-Tank Missile	150
6272.441	CAP Underwater Explosive Charge	521
6274.441	Type 65 660 mm Soviet Torpedo	527
6275.481	COMKAFS (Outfit DCH) Submarine Combat System	433
6276.281	Pegaso Fire Control System	416
6277.153	M60A3 Laser Tank Fire Control System	401
6278.193	*Mk X Stedi-Eye Stabilised Binocular*	*1987-88*
6279.193	*KN200/250 Night Sights*	*1987-88*
6280.193	LV350 Series Laser Rangefinders	375
6281.193	VML23 Vehicle Laser Rangefinder	381
6282.193	*UA 1119 Target Designation and Aiming Sight*	*1987-88*
6283.193	UA 9127 Thermal Imaging/Daylight Aiming and Observation System	373
6284.193	*PB6SL Passive Lightweight Binoculars*	*1987-88*
6285.193	SOPTAC 36 Fire Control System	353
6286.193	SOPTAC II Fire Control System	353
6287.193	SOPTAC 18 Fire Control System	353
6288.193	M 391 Episcopic Sight	353
6289.193	TJN2 71 Mixed Day/Night Sight	352
6290.193	ELOP Mini-Laser Tank Rangefinder	364
6291.193	HRLR-04 AAC Laser Rangefinder	365
6292.193	BAT-30 Fire Control System	365
6293.193	Matador Fire Control System	365
6294.193	OG-P101 Periscope Sight	369
6295.193	LWDR Lightweight Designator/Rangefinder	396
6296.193	*VARO Individual Night Vision Equipment*	*1987-88*
6297.193	Tank Thermal Sight	393
6298.193	Surveillance Thermal Imager	393
6299.193	*Marinit Night Vision System*	*1987-88*
6300.193	Radamec Electro-Optical Directors	394
6301.193	SGS-55 Tank Day/Night Sight	406
6302.193	TS-M Thermal Sight	407
6303.193	EFCS-2 Tank Fire Control System	407
6304.193	10 Hz Laser for Anti-Aircraft Systems	407
6305.331	Sky Sword Air-to-Air Missile	698
6306.193	Osprey Day/Night AFV Sight	388
6307.193	Sabre Day/Night Vehicle Sights	389
6308.193	*AN/PVS-7B Night Vision Goggles*	*1987-88*
6309.391	*Boeing Hind-D Target*	*1986-87*
6310.391	*YMF BJ7104 Target Drone*	*1987-88*
6311.391	*NAI Changkong IC Target Drone*	*1987-88*
6312.391	*ADE Experimental Mini-RPV*	*1987-88*
6313.391	*ADE Pilotless Target Aircraft (PTA)*	*1987-88*
6314.351	*Pioneer Mini-RPV*	*1987-88*
6315.391	*Tamnar Target Drones*	*1987-88*
6316.391	*Fuji Model 820 Target Drone (XJ/AWM-1)*	*1987-88*
6317.391	*BTT-2 Demon*	*1987-88*
6318.391	*Martin Marietta Orlando Slat (YAQM-127A)*	*1987-88*
6319.351	*Hynes H-2T and H-5T*	*1987-88*
6320.352	*ASVEC TCIB Merlin*	*1987-88*
6321.391	*Swift Target Drone/RPV*	*1987-88*
6322.351	*AMT-RPV1 Vehicles*	*1987-88*
6323.292	THORN EMI Naval Thermal Imagers	689
6324.293	Ferranti Lightweight Electro-Optic Multisensor	688
6325.253	*IVISS*	*1987-88*
6326.293	P0701 Laser Rangefinder	683
6327.293	P4674 Thermal Imager	683
6328.393	VH 100 Series Helicopter Head-Up Displays	871

Entry number	Subject	Page No/ Last edition*
6329.393	Thomson-CSF Helmet Mounted Sights	871
6330.393	APX M334-25 Gyro-Stabilised Sight	868
6331.393	V3G Helicopter Day/Night Sights	868
6332.393	Elbit Night Vision Head-Up Display	874
6333.393	ELOP Head-Up Displays	875
6334.393	Blue Thunder HADAS (Helmet Airborne Display and Sight)	875
6335.393	P0702 Airborne Laser Rangefinder	876
6336.393	F-16 C/D HUD Sight	878
6337.393	F-5 E/F Head-Up Display Weapon Aiming Computer (HUDWAC)	878
6338.393	Sea Owl Passive Identification Device Sensor Turret	880
6339.393	*Phoenix RPV Sensor Payload*	*1987-88*
6340.393	Framing Thermal Imager	880
6341.393	C-Nite TOW Missile Sight	889
6342.393	Night ATAC Fire Control System	900
6343.393	SRL Night Vision/Head-Up Display	901
6344.393	BAE Improved Tow Sight	887
6345.393	AN/AAR-45 FLIR Pod	897
6346.181	AVIBRAS EDT-FILA Anti-Aircraft Fire Control System	63
6347.181	AMC-11 Mortar Computer	66
6348.181	AVC-101 Artillery Fire Control Computer	66
6349.181	Sopelem SOPTAC 36 Day/Night Fire Control System	67
6350.181	ADIVA Artillery Division Automation System	70
6351.181	Artillery Fire Control System	79
6352.161	Integrated Tracking and Fire Control System For Wildcat	83
6353.161	SATT Update Package for Anti-Aircraft Fire Control System	92
6354.161	AN/PSC-2 Digital Communications Terminal	106
6354.171	Briefcase Terminal	106
6356.161	Intelligent Terminal	106
6357.161	Lightweight LTACFIRE	107
6358.193	Honeywell AFV Thermal Sights	405
6359.193	M6043 Fire Control Director Sight	400
6360.183	Infra-Red Scanning Seeker Head	394
6361.193	OG-PO7B Periscope	379
6362.193	TIMS	404
6363.181	Watchdog Coastal Defence Systems	48
6364.111	70mm RADIRS Multiple Rocket System	132
6364.351	*Raider Tactical Drone*	*1987-88*
6365.351	*PRIAMOS Reconnaissance System*	*1987-88*
6366.161	Seimor Mortar Computing System	88
6367.161	Morzen II Mortar Computing System	99
6368.161	Gunzen Artillery Fire Data Computer	98
6369.161	MAMBA Artillery and Mortar Ballistic Aid	99
6370.153	JY-9 Low-Altitude Search Radar	219
6371.193	Laser Irradiation Detector	347
6372.193	Marconi Unattended Expendable Jammer	329
6373.193	890 Series IFF Interrogators	275
6374.393	Helicopter Self-Screening Launcher	877
6375.203	Breda Twin Fast Forty Naval Mounting	490
6376.153	AR-325 Commander Air Defence Radar	269
6377.303	Tround International 12.7 mm Aircraft Machine Gun	759
6378.303	*Opher Bomb Guidance Kit*	*1986-87*
6379.193	Oldelft AIFV Thermal Sights	373
6380.351	*Raider Tactical Drone*	*1986-87*
6381.153	LAR 200 Local Area Radar	271
6382.353	MEL AEW Radar	791
6383.153	Aquila Mobile Radar	218
6384.153	Chinese Over-The-Horizon Radar	222
6385.153	Model 17-C Surveillance Radar	220
6386.153	HN-401R Mobile Surveillance Radar	221
6387.153	Type 408C Long Range Air Warning Radar	221
6388.153	Type 571 Surveillance Radar	221
6389.153	Hunting Dog Radar	224
6390.153	RATAC-S Battlefield Radar	238
6391.153	Pagoda Search Radar	244
6392.151	Radar Elevator System	249
6394.153	TRS 2630 GRIFFON Shorads Radar	232
6395.153	Tiger G (TRS2105/TRS2106) Low Altitude Radars	236
6396.153	NR-AI-5A IFF Interrogator-Decoder	234
6397.153	NRAI-6A IFF Interrogator-Decoder	234
6398.153	MSTAR Man-Portable Radar	273
6399.153	Rapier 2000 Surveillance and Target Acquisition Radar	270
6400.153	ACWAR - Agile CW Acquisition Radar	304
6401.153	Dual Band Radar	296
6402.303	Bavard LXT 462 Adjustable Practice Bombs	740
6403.303	OPHER Terminal Guidance Kit for General Purpose Bombs	742
6404.303	*Area Air-to-Air Missile Launcher*	*1987-88*
6405.303	*Area Defence Helicopter Weapon Conversion Kits*	*1987-88*
6406.303	Hunting Area-Denial System (Hades)	756
6407.303	General Electric 25 mm Lightweight Gun - GE 225	762
6408.303	General Electric 20 mm Vulcan Gun System for AMX Fighter	763
6409.303	GECAL 50 Aircraft Machine Guns	766
6410.303	Loral Air Inflatable Bomb Retarders - Ballute	770
6411.411	CSS-NX-4 Submarine-Launched Ballistic Missile	4
6412.111	SSBS Intermediate Range Missile Type S-4 (SX)	4
6413.303	GECAL 50 Composite Turret Weapon System	766
6414.303	General Electric 7.62 mm Minigun	767
6415.253	Model 354 Shipborne Surface and Low Altitude Warning Radar	600
6416.253	Rodeo Naval Air Surveillance Radar	605
6417.131	General Electric Air Defence Turrets	207
6418.131	HQ-61 Surface-to-Air Missile System	167
6419.131	Medium Surface-to-Air Missile (MSAM) System	190
6420.253	EMPAR - European Multifunction Phased Array Radar	606
6421.253	Front Series Radars	619
6422.153	Circe National Secure IFF Systems	275
6423.353	LN66 Radar System	772
6424.253	RAN 30X Surveillance Radar	610
6425.353	ORB-32 Radar Systems	775
6426.353	ORB 37HL Airborne Radar	775
6427.353	Anemone Airborne Radar	776
6428.353	RACAAS Airborne Radar	773
6429.353	EFA Radar	779
6430.353	Big Bulge Airborne Radar	784
6431.353	Flat Jack AWACS Radar	784
6432.353	High Lark AI Radar	784
6433.353	Super Marec Airborne Radar	790
6434.353	IFF 4700 Series Airborne Transponder	790
6435.353	Multimode Radar for F-5E	797
6436.353	AN/APS-137(V) Airborne Radar	807
6437.353	AN/APS-145 AEW Radar	808
6438.353	AN/APS-139 Airborne Radar	808
6439.353	ASARS-2 Synthetic Aperture Radar	814
6440.353	Airship Surveillance Radar	813
6441.353	AN/APQ-168 Airborne Radar	804
6442.161	Eagle Eye Anti-Aircraft Fire Control System	74
6443.181	PAC 101 - Portable Artillery Computer	80
6444.181	Linda C3I Artillery System	88
6445.161	MIA Fire Control Computer and Communication Unit	89
6446.161	Mistral Passive Meteorological System	97
6447.221	C.101 Supersonic Anti-Ship Missile	449
6448.221	FL-7 Anti-Ship Missile	450
6449.281	9LV Mk3 Weapon Control System	428
6450.281	Sea Archer 30 Fire Control System	440
6451.153	Model 970 Radar Jamming System	306
6452.153	TRC Series Communications Intelligence Equipment	306
6453.193	AC 008 Microwave Directional Antenna	310
6454.193	PV 017 Azmiuth Analyser	313
6455.153	CR-2700 Mobile ELINT System	317
6456.153	*Rediffusion VHF/UHF Direction Finder*	*1987-88*
6457.153	RA 3700 Series Modular HF Receivers	322
6458.093	AN/PRD-11 Man Portable DF Equipment	337
6459.093	WJ-1996 ESM Receiver	336
6460.093	*WJ-9045 Tactical Receiving System*	*1987-88*
6461.093	WJ-8976 Direction Finding System	336
6462.153	HELAS ESM System	308
6463.143	Mobile EW Support System (MEWSS)	346
6464.193	COLDS Laser Detection System	314
6465.111	Red Arrow 8 (HJ-8) Anti-Tank Missile	137
6466.111	MSS 1.1/MAF Anti-Tank Missile	140
6467.111	Advanced Anti-Tank Weapon System-Heavy (AAWS-H)	157
6468.111	273 mm (4-Round) Type 83 Multiple Rocket System	115
6469.111	122 mm (40-Round) SP MRS	116
6470.111	Chinese Tactical Missiles	117
6471.203	Single 100 mm Creusot-Loire Dual-Purpose Gun Mounting	485
6472.203	Twin 100 mm Dual-Purpose Gun Mounting	485

Entry number	Subject	Page No/ Last edition*
6473.311	C-601 Anti-Ship Missile	707
6474.311	*Star Anti-Radar Missile*	*1987-88*
6475.311	Kent Air-Launched Cruise Missile (AS-15)	718
6476.311	C-801 Air-to-Surface Missile	707
6477.311	AGM-130 Air-to-Surface Missile	723
6478.311	Sensor Fuzed Weapon (SFW)	758
6479.131	HN-5A Portable Anti-Aircraft Missile System	211
6480.441	Murene Lightweight Torpedo	515
6481.241	Super Ikara Anti-Submarine Weapon System	512
6482.441	Advanced Sea Mine	519
6483.241	Sea Serpent Magnetic Minesweeping System	532
6484.441	Trail Blazer ROV	519
6485.453	921-A ESM Equipment	647
6486.293	DR 2000 IFM ESM Receiver	650
6487.293	ESD LAD Off-Board Decoy System	650
6488.293	Lacroix Naval Countermeasures	651
6489.293	Beamtrap Close Defence Rocket	653
6490.253	NS-9001 ELINT System	654
6491.253	NS-9003 Naval ESM System	654
6492.253	NS-9005 ECM Jammer	655
6493.253	NS-9009 Naval EW System	655
6494.253	ND-9010 Naval ESM System	655
6495.253	NOL Series Naval EW Systems	660
6496.293	Anaren Shipborne Tracking Receiver	675
6497.493	AN/WLQ-4 SIGINT System	670
6498.293	RW-23-1 Radar Warning Equipment	647
6499.253	APECS II ESM/ECM System	676
6500.253	NATO Sea Gnat	652
6501.181	Mygale Air Defence System	41
6502.181	Mace Command and Control System	55
6503.181	Mardis Air Defence Network	56
6504.393	GT-1 Chaff/IR Flare Dispenser	815
6505.393	Medusa Jammer	815
6506.393	TMV202 (DR4000A) ESM Suite	819
6507.393	SRS-25 Airborne Receiver	823
6508.393	LWS-20 Laser Warning System	823
6509.353	SPS-2000 Airborne Self-Protection System	824
6510.353	SPS-65 Helicopter Self-Protection System	823
6511.393	AQ-861 Noise and Deception Jamming System	831
6512.393	ERIJAMMER A100	831
6513.393	ERIJAMMER 200	831
6514.393	Marconi Laser Warning System	834
6515.393	RACJAM AIR	836
6516.393	AN/ALQ-186(V) Jammer	852
6517.393	AN/ALQ-187 Countermeasures System	852
6518.393	GENE-X Expendable Cartridge	841
6519.393	AN/APR-47 Wild Weasel CIS	861
6520.393	Airborne Expendable Decoys	866
6521.193	LRS 7 Fire Control System	348
6522.093	*HNV-1 Holographic Goggles*	*1987-88*
6523.093	Nokia Electro-Optical Equipment	350
6524.193	Irgo Hand-Held Thermal Imager	357
6525.093	CALIOPE Thermal Imager	355
6526.193	*Milos TV Observation System*	*1987-88*
6527.193	Eagle Eye Fire Control System	251
6528.193	SP-T-694 Stabilised Panoramic Sight	370
6529.193	*UA 1134 Individual Weapon Night Sight*	*1987-88*
6530.193	*Lunds Night Vision System*	*1987-88*
6531.193	TILAS Tank Laser Sight	371
6532.193	Artillery Sighting System	394
6533.193	SS170 Day/Night Sight	390
6534.193	SS550 Thermal Elbow	391
6535.193	SDFOV Thermal Imaging System	361
6536.153	Mithridat Tank Fire Control System	353
6537.153	OG 14 L2A Tank Fire Control System	78
6538.153	OG 14 L3 Tank Fire Control System	78
6539.193	FOG-M Missile Guidance System	405
6540.391	*BTT-1 Imp*	*1987-88*
6541.391	*ASR-Spectre*	*1987-88*
6542.391	*JTT-5 Voodoo*	*1987-88*
6543.391	*Chukar III Aerial Target*	*1987-88*
6544.391	*Soviet Union RPV Systems*	*1987-88*
6545.391	*TACIT Rainbow RPV Programme*	*1987-88*
6546.493	PIVAIR Submarine Periscope	681
6547.253	Medusa Optical Designation and Fire Control System	683
6548.293	SEID Infra-Red Tracker	678
6549.293	Murene Thermal Camera	678
6550.453	*AN/SSQ-531 DIFAR Sonobuoy*	*1987-88*
6551.353	AN/AQS-503 Helicopter Acoustic Processing System	545
6552.353	Barra Side Processor (AQA-801)	545

Entry number	Subject	Page No/ Last edition*
6553.453	Fiskars Hydroacoustic Surveillance System	546
6554.463	TSM 2930 Lamproie Linear Towed Array	553
6555.353	SS240 Omni Sonar	560
6556.393	SPDT-1A Passive Sonar	555
6557.393	*PSC 8400 Passive Sonar Surveillance System*	*1987-88*
6558.353	PMS 56 Pursuit Sonar	565
6559.353	TG-1 HF Sonar System	577
6560.493	AN/SSQ-58A Moored Sensor Buoy	597
6561.493	Model 4235 ROV Doppler Sonar	591
6562.453	AN/BSY-1(V) Integrated Combat System	584
6563.353	AB/SQQ-89 Integrated Combat System	588
6564.493	AN/SSQ-75 ERAP Sonobuoy	597
6565.331	PL-5B Air-to-Air Missile	692
6566.331	AAAM Long-Range Air-to-Air Missile	705
6567.393	Super Cyclope and Corsaire Infrared Scanners	872
6568.393	Aspic Infrared Detection System	872
6569.393	DASH (Display and Sight Helmet) System	875
6570.393	HALO Helicopter Avionics	875
6571.393	A6E/F HUD Sight	879
6572.353	MIG-21 HUDWAC System	879
6573.393	Multirole FLIR Turret	880
6574.393	Type 311/318 CO_2 Laser Rangefinder	887
6575.393	TIALD Designator Pod	885
6576.393	ZHUD Sight	882
6577.393	Night Owl Targetting Pod	900
6578.393	ES-250 Standoff Electro-Optical Reconnaissance System	902
6579.393	SLOS Stabilised Long-Range Observation System	873
6580.363	Eagle Eye III Electro-Optical Sensor	890
6581.221	*Chinese Missile Seekers*	*1987-88*
6582.181	Super Fledermaus Update	74
6583.163	Target Acquisition Aid System	358
6584.193	Capitole Fire Control System	352
6585.153	*TRS 2800 Weapon Locating Radar*	*1987-88*
6586.153	SA 90 Radar System	236
6587.493	*XBT/XSV Expendable Bathythermograph/Sound Velocimeter*	*1987-88*
6588.303	Rattler Weapon System	200
6589.463	Osborn Minesweeping System	532
6590.193	*Dimtis Thermal Imaging System*	*1987-88*
6591.193	LF 19 Laser Rangefinder	382
6592.093	IR26 Thermal Imaging Sensor	382
6593.393	*BASIL Infrared Linescanner*	*1987-88*
6594.281	Ferranti Integrated Combat Suite	433
6595.281	Ferranti 500 Series Command and Control Systems	434
6596.253	*NAUTIS-L Command and Control System*	*1987-88*
6601.331	PL-2 Air-to-Air Missile	692
6602.331	PL-7 Air-to-Air Missile	692
6603.331	PL-9 Air-to-Air Missile	693
6604.331	STINGER (AIM-92) Air-to-Air Missile	705
6605.331	XAAM-3 Air-to-Air Missile	697
6606.331	Grail SA-7 Air-to-Air Missile	701
6607.311	HY-4 Air-to-Surface Missile	707
6608.311	Modular Stand-Off Weapons	712
6609.311	Sidearm AGM-122 Anti-Radar Missile	723
6610.311	Advanced Cruise Missile AGM-129	721
6611.311	Tacit Rainbow AGM-136 Anti-Radar Missile	724
6701.181	Integrated Fire and Maneuver Control System	64
6702.161	CDC Ballistic Computer System	64
6703.161	CDC Control and Display Unit	65
6704.161	CDC Military Communications Calculator	65
6705.161	Mere Mortar Fire Control Computer	84
6706.181	TAD LRM Fire Control System	70
6707.181	Norwegian Army CCIS	84
6708.181	KMT 400 Multi-Purpose Military Terminal	85
6709.181	FELIS Anti-Aircraft Fire Control System	88
6710.181	Bofors Type U Fire Control System	91
6711.181	9KA 100 Coastal Artillery Fire Control System	92
6712.111	SAKR 80	118
6713.111	Indian Tactical Surface-to-Surface Missile	121
6714.111	CHETZ Tactical Surface-to-Surface Missile	121
6715.111	Brazilian Anti-Armour Missile	137
6716.111	Bofors 155mm Anti-Armour Ammunition	148
6717.111	Condor Tactical Battlefield Support Missile	114
6718.111	Indian Anti-Tank Missile	140
6719.111	Further Improvements to TOW (FITOW)	153
6720.131	HQ-2B Air Defence Missile	167
6721.131	Indian Air Defence Missiles	170
6722.131	Taiwan Air Defence Missiles	176
6723.121	HY-4 Coast Defence Missile System	160

Entry number	Subject	Page No/ Last edition*
6724.121	SSM-1 Coastal Defence Missile	161
6725.131	Fibre Optic Guided Missile (FOG-M)	191
6726.131	Saber Air Defence Anti-Armour Missile	187
6727.131	IDRA Air Defence Missile	174
6728.131	Mauser Arrow Twin-Gun Anti-Aircraft System	194
6729.131	Updating Kit for Bofors 40 mm L/60 Gun System	201
6730.131	Avenger Air Defence System	210
6731.131	MSA-3.1 Portable Anti-Aircraft Missile System	211
6732.193	ARES Helmet-Mounted Target Designator	358
6733.193	SADOT Fire Control System	364
6734.193	Yuval Long Range Surveillance System	363
6735.193	Reveal Video Enhancement System	363
6736.193	OG-C-102 Coaxial Telescope	366
6737.193	UAL 11636 Light Anti-Aircraft Laser Rangefinder	378
6738.193	Air Defence Alerting Device (ADAD)	392
6739.193	RADAMEC Electro-Optical Pod Tracking System (E-OPTS)	394
6740.193	Lightweight Target Acquisition/Surveillance System (LTASS)	399
6740.281	NA 25 Weapon Control System	415
6741.281	MSI-340/350 Fire Control System	426
6742.281	MSI-3100 Multiple Integration System	426
6743.281	Bazan SCPVZ Combat System	426
6744.281	880 Series Mine Countermeasures Systems	438
6745.481	Outfit DCB Submarine TDHS	432
6746.481	Outfit DCG Submarine TDHS	433
6747.281	'Duke' Class (Type 23) Frigate Command System	431
6748.121	SM-70 Barracuda Surface-to-Surface Anti-Ship Missile	160
6748.481	Successor Submarine Command System	435
6751.231	Support Defence Missile System	477
6752.131	MISAT Surface-to-Air Missile	169
6753.261	Thorn EMI Ship's Weapon Data Distribution	435
6754.281	Kanaris Fire Control System	412
6755.231	NATO Anti-Air Weapon Systems NAAWS	467
6756.203	Type 61 Twin 25 mm Gun Mounting	486
6757.203	BREDA Local Control Single 30 mm Naval Mounting	493
6758.203	Sea Vulcan 25 Weapon System	510
6759.203	Trinity Mounting	497
6760.241	CY-1 Anti-Submarine Missile	513
6761.241	Chinese MCM System	513
6762.441	Chinese Naval Mine Upgrading System	513
6763.441	Sterne 1 Minesweeping System	516
6764.341	MILAS Anti-Submarine Weapon System	519
6765.441	SGM-80 Sea Mine	520
6766.441	ATC-1 Torpedo Decoy	520
6767.441	Bandfish Expendable Torpedo Countermeasure	530
6768.441	British Aerospace Mine Modernisation	531
6769.441	ADC Mk2 Expendable Torpedo Countermeasure	538
6770.441	US Navy Torpedo Decoys	539
6771.441	Limpet Mine Disposal Equipment (LMDE)	532
6772.293	Director-Mounted Thermal Imaging System (DIMTIS)	687
6773.293	RADAMEC Naval Directors	688
6774.293	Ferranti FD 5000 Periscope Video Camera	689
6775.293	Kollmorgen Model 76 Periscope	690
6776.393	Optech Lidar	867
6777.393	Convertible Laser Designator POD (CLDP)	872
6778.393	Litening Infrared Targeting and Navigation Pod	876
6779.393	Ferranti Helmet Pointing System	885
6780.363	Pathfinder Navigation Pod	896
6781.353	Stabilised Thermal Imaging System (STIS)	893
6782.353	Helicopter Infrared Navigation System (HIRNS)	893
6783.363	IRVAT (Infra-red Video Automatic Tracker)	889
6784.353	Northrop Mast-Mounted Sight (MMS) Sensor Suite	890
6785.363	Laser Detector-Tracker/Strike Camera (LDT/SCAM)	902
6786.393	Airborne Electro-Optics and SDI	902
6787.353	Advanced Tactical Air Reconnaissance System (ATARS)	903
6788.363	Cruise Missile Advanced Guidance (CMAG)	903
6789.261	Nokia Coastal Defence and Surveillance System	904
6790.461	Krupp Atlas ISUS (Integrated Sensor Underwater System)	904
6791.293	Kollmorgen Type 8/15/18 Periscopes	904
6801.181	Angola Air Defence	904
6802.181	TLVS Medium Range Tactical Air Defence System	44
6803.181	CATRIN C^{3}I System	45
6804.181	Southern Africa Air Defence	904
6805.181	Canadian Air Defence	39
6806.181	Air Defense Iniative	59
6807.153	Type 378 Battlefield Radar	220
6808.153	TRS 22XX Long-Range 3-D Radar	229
6809.153	SB 14 IFF Interrogator	234
6810.153	TPN-720 Mobile Control and Landing System	243
6811.153	GAP Gate Early Warning Radar	249
6812.153	Japanese OTH Radar	248
6813.153	Tien Kung Radar	252
6814.153	Raids Detection Radar	305
6815.153	AN/MPS-39 Instrumentation Tracking Radar	284
6816.153	AN/PPX-3B Ground Interrogator	299
6817.153	MFU 402 Fire Control Radar	271
6818.153	ETAS Target Acquisition System	305
6819.153	Norwegian Acquisition Radar and Control System	249
6820.181	Norwegian Air Defence	48
6821.153	Flat Twin Radar	
6822.153	Pawn Shop Radar	258
6823.153	Rappist Fire Control Radar	230
6824.193	EL/S-8811/2 Tactical Intelligent System	315
6825.193	Japanese EW Equipment	318
6826.193	EN/TSD-500 Series COMINT Systems	319
6827.193	Soviet Land-Based EW Systems	319
6828.193	Corvus Elint Station	327
6829.153	Automated Computer Assisted EW System (ACEWS)	321
6830.153	Series 500 Mobile ECM Systems	320
6831.193	WJ-8990 Manpack Tactical Intelligence System (MANTIS)	336
6832.193	Responsive Surveillance Subsystem	336
6833.193	WJ-4810 Communications Jamming System	336
6834.193	WJ-3600 EW Alert and Collection System	336
6835.153	ADEWS (Air Defence EW System)	347
6836.193	AN/VLQ-5(V)1 Comfy Sword	346
6837.193	Seidal Laser Detector	319
6838.193	ESM1000 System	311
6839.193	Siemens HF Jammer	313
6840.193	Babylon HF Jammer	313
6841.193	Common Jammer Programme	343
6842.153	ASAS/ENSCE	347
6843.193	Alberich Laser/Radar Warning System	314
6844.193	RWR 1000 Hand-Held Radar Detector	332
6845.453	SCAN 522/545 Shallow Water Sonar	542
6846.453	AN/SQS-510 Digital Sonar System	542
6847.453	DUBM 42 Side-Looking Sonar	554
6848.453	Anaconda Tactas System	555
6849.453	Dolphin 1000 Series Submarine Sonars	574
6850.453	Getas Passive Surveillance Sonars	574
6851.253	GS 7110 Multichannel VLF/LF Receiver	574
6852.453	Type 2026 Sonar	568
6853.453	Type 2031I Sonar	568
6854.453	Type SP 2110 Sonar	572
6855.453	Towed Sonar Source Type 2058	578
6856.363	Type 605 Sonobuoy Receiver	574
6857.363	RASAT Sonar Target	569
6858.453	AS360 Sonar Systems	579
6859.453	AS370 Seabed Surveillance Sonar	579
6860.353	AS380 Dunking Sonar	579
6861.453	SSQ-981 Barra Sonobuoy	576
6862.453	SSQ-991 Sonar Transponder	576
6863.453	SSQ-937A Bathythermal Buoy	576
6864.453	DE1191 Sonar Modernisation	589
6865.453	Model 1110 Sonar System	591
6866.293	Coherent Countermeasures Processor	647
6866.453	AN/SSQ-101 HLA Sonobuoy	598
6867.453	AN/SSQ-102 Sonobuoy	598
6868.453	Type 2074 Sonar System	571
6869.453	Type 2074 Sonar	571
6870.453	Type 2057 Towed Array Sonar	571
6871.453	Plessey Arms Sonar	577
6872.453	AN/BQS-14A Submarine Sonar	594
6873.253	LN66 Radar Series	600
6874.253	MM/SPN-728 Search Radar	611

Entry number	Subject	Page No/ Last edition*
6875.253	Orion RTN-25X Naval Tracking Radar	609
6876.253	RA3-30S Naval Radar	610
6877.253	Sting Radar Fire Control System	612
6878.253	Trap Door Naval Radar	626
6879.253	Cross Sword Fire Control Radar	618
6880.253	Top Plate/Top Mesh Radars	625
6881.463	NS-9034 Submarine ESM/ELINT System	655
6882.253	Nettuno Naval EW System	658
6883.253	UAF-1 ESM System	669
6884.253	Scorpion ECM System	668
6885.293	SADIE ESM Processor	669
6887.353	Osprey Dipping Sonar	557
6887.453	Ferret Submarine EW System	671
6888.053	Direct Coupled Jamming/Surveillance System	675
6889.353	DAV Search and Alert Radar	777
6890.353	RDY Multi-Function Radar	773
6891.353	EL/M-2011 and EL/M-2032 Fire Control Radar	780
6892.353	Phalcon AEW Radar	780
6893.353	AM/APS-717 Search and Weather Radar	781
6894.353	SC-01 (SCIPIO) Multi-Mode Radar	782
6895.353	MM/APS-784 Search Radar	782
6896.393	SIT 434 IFF Interrogator	782
6897.353	AN/APQ-165 Airborne Radar	803
6898.353	AN/APQ-174 Multi-Function Radar	804
6899.353	Mitsubishi Airborne Radar	782
6900.353	Teledyne IFF Interrogators/Transponders	810
6901.393	KJ 8602 Radar Warning Receiver	816
6902.393	Barax Detector-Jammer	820
6903.393	DDM IR Detector	820
6904.393	MSPS EW System	818
6905.053	Miriade Radar Warning System	818
6906.393	Thomson-CSF Sigint Aircraft Receivers	819
6907.353	Maritime Patrol ESM	819
6908.353	Rohde & Schwarz COMINT System	821
6909.353	EL/L 8240 ECM System	822
6910.393	Fibre-Optic Laser Warning System	865
6911.393	AN/AAR-47 Missile Warning Set	838
7001.493	*PTC 100 Periscopic Television Camera*	*1973-74*
7002.253	PMS26 and PMS27 Surface Ship Sonars	564
7003.453	*Bathythermograph Slug*	*1970-71*
7004.453	*MO30 & MO31 Sound Velocity Meters*	*1969-70*
7005.193	*Direct Fire Weapon Effects Simulator*	*1974-75*
7006.191	*Surface Tactical & Blind Pilotage Trainer*	*1970-71*
7007.393	*UVR-700 TV Tracker System*	*1976*
7008.363	*AN/AYA-8 Data Analysis Programming Group*	*1970-71*
7009.393	*UVR-700 & 800 FPS TV Camera System*	*1973-74*
7010.393	*Night Window Low Light-Level TV*	*1973-74*
7011.353	*AN/APN-171(V) Radar Altimeter*	*1970-71*
7012.383	*A-7 Navaid/Weapon Delivery System*	*1970-71*
7014.353	*AN/APQ-97 Mapping Radar*	*1976*
7015.293	*CAINS (AN/ASN-92) Carrier Aircraft Inertial Navigation System*	*1970-71*
7016.363	*RAF Aircraft Digital Data Link*	*1970-71*
7017.353	*AN/APN-167 Radar Altimeter*	*1969-70*
7018.353	*HG 151 Drone Radar Altimeter*	*1970-71*
7019.193	*Solartron Naval Operations Trainers*	*1973-74*
7020.193	*Solartron Action Speed Tactical Teachers*	*1973-74*
7021.281	*Sea Hunter Fire Control System*	*1971-72*
7022.231	*Sea Indigo Shipborne Surface-to-Air Missile*	*1969-70*
7023.193	*DX43 Simulator*	*1970-71*
7024.193	*DX44 Simulator*	*1970-71*
7025.393	*DX47 Simulator*	*1970-71*
7026.393	*DX140 Simulator*	*1970-71*
7027.193	*DX143 Simulator*	*1970-71*
7028.193	*DX144 Simulator*	*1970-71*
7029.393	*DX145 Simulator*	*1970-71*
7030.393	*260F1 Gyro-Stabilised Sight*	*1979-80*
7031.393	*M296 Gyro-Stabilised Sight*	*1973-74*
7032.393	APX M334 Series Gyro-Stabilised Sights	867
7033.393	*M328 Gyro-Stabilised Sight*	*1972-73*
7034.221	*Sea Killer Mk 2 Surface-to-Surface*	*1969-70*
7035.163	*IRIS 35M Computer*	*1970-71*
7036.163	*L300 & L3000 Series Computers*	*1970-71*
7037.163	*Litton Computer Memory Products*	*1970-71*
7038.163	*TAC Digital Computer*	*1970-71*
7039.363	*AN/UYK-501 Computer*	*1970-71*
7040.163	*Alert Digital Computer*	*1970-71*
7041.163	*Myriad Computer*	*1970-71*
7501.441	*Acoustic Torpedoes L3, E14 and E15*	*1987-88*
7502.393	*Alkan Type 80 Training Bomb Rack*	*1972-73*
7503.393	*Alkan Type 500-OM Bomb Rack*	*1979-80*
7504.393	*Type 10B Sonobuoy Adapter*	*1981-82*
7505.393	*Alkan Cartridge Loader*	*1970-71*
7506.393	*Alkan Flare Launchers*	*1970-71*
7507.393	*Alkan Universal Adapter Type F1*	*1981-82*
7508.393	*Jaguar Weapon Pylons*	*1981-82*
7509.383	*7215E ECA for Helicopters*	*1970-71*
7510.393	*Matra LRC 100mm Rocket Launcher*	*1970-71*
7511.383	*7201B Automatic Control & Auto-Stabilisation System*	*1970-71*
7512.363	*AN/AYA-4 Navigation System*	*1970-71*
7513.393	*ML Ejector Release Units No 119 Series*	*1987-88*
7514.393	*Virgo 120 kg Fragmentation Bomb M/71*	*1987-88*

IAI Electronic Defense

Israel Aircraft Industries' **Electronics Division** can give you the edge on the battlefield. Its four plants lead the way to the future in electronic and opto-electronic warfare, smart ordnance, artificial intelligence in air, land, sea and space technology.

- **ELTA** – Airborne, Surface and Tactical Radars; Airborne, Shipborne and Ground EW and SIGINT Systems; Secure Communications Systems and Data links; Military Computers and Peripherals; Signal Processing and A1; Advanced Technologies.
- **MBT** – Artillery Location Systems, Naval Tactical Command and Control Systems, Simulators and Trainers, Barak Missile and Smart Ordnance.
- **TAMAM** – Stabilized payloads, optronics and navigation components and systems.
- **MLM** – C^3 Integration, Sensors and Image Processing.

Ben Gurion International Airport, Israel.
Tel: (3)971-3111.
Tlx: 371102, 371114 ISRAVIA IL.
Fax: (3)971-3131.

IAI Action in the Air

We at IAI can supply you with what you need. We've been supplying it to the Israel Defense Forces and other airforces for many years. We offer: design and production of new combat aircraft; upgrading and remanufacturing of combat/trainer aircraft like the Mirage, F4 and A4; multi-mission conversions of transport aircraft to electronic, air-refueling and other special mission configurations; ground and airborne radars and secure communications systems and the full range of ground services and equipment.

- **B3 Advanced Combat Technology Demonstrator.**
- **Kfir C7/F21A** multi-role platform.
- **Astra, Westwind, SeaScan, Arava** V.I.P./logistics/sea-recon. aircraft.
- **Phantom F-4** upgrading.
- **B707-C130** and other upgrades.
- **Simulators** and complete fixed and rotary-wing aircraft maintenance and service.
- Complete **Electronic Combat Systems; Weapon Delivery** and **Navigation Systems; Radars; EW** and **Optronics.**

Ben Gurion International Airport, Israel.
Tel: (3)971-3111.
Tlx: 371102, 371114 ISRAVIA IL.
Fax: (3)971-3131.

IAI and Action at Sea

We at IAI can supply you with what you need. We've been supplying it to the Israel Defense Forces and other modern and modernizing navies for many years. Our equipment can be integrated into your existing platforms to meet your mission requirements. IAI can supply you with what you need using sea, land, air and space technology.

- **AMDR** – Ship-borne Automatic Missile Detection Radar
- **Barak 1** Naval Point-Defense Missile System
- **ESM** and **Electronic Combat Systems**
- **Coastal Surveillance** and **Shipborne Radar** which can detect small sea targets
- **Reshet** Naval Tactical Command and Control System
- **Missile** and **Gun Fire Control Systems**
- **Civil** or **military transports** modified for surveillance, intelligence or command missions
- **Dvora** fast patrol gun boat or fast patrol boat equipped with combat-proven Gabriel SSMs.
- **LPH-292 Helicopter Landing Platform and Hangar**
- **Maintenance, training** and **logistics support** for ships, helicopters and missiles.

Ben Gurion International Airport, Israel.
Tel: (3)971-3111.
Tlx: 371102, 371114 ISRAVIA IL.
Fax: (3)971-3131.

IAI Action on the Ground

We at IAI can supply you with what you need. We've been supplying and integrating it into the Israel Defense Forces and other modern and modernizing armies for many years. IAI can supply what you need using ground, air, space and marine technology.

- **Gun stabilization system**
- **LANS** Land Area Navigation System
- **GONS** Gun Orientation Navigation System
- **EL/M 2121 Intelligence Battlefield Surveillance Radar**
- **Eagle-Eye** Air Defense Fire Control System
- **C³I, SIGINT** and **COMINT** systems for all tactical levels
- **TWMP** Mine-clearing attachment for engineering vehicles
- **G 360 20 mm Cannon**
- **20 mm Gun Mounts**
- **EL/M 2106 Point Defense Alert Radar**
- **SLOS** Stabilized Long-range Observation System for visual or video observation
- **NFS** North Finding System
- **PADS** Precision Azimuth Determination System
- **RAM** Multi-mission, multi-purpose armored vehicle
- **Secure Communication Systems**
- **Hydraulic platform** for TOW launch from APCs
- Specialized Ordnance

For further information on IAI and Action on the ground, please contact:

Ben Gurion International Airport, Israel.
Tel: (3)971-3111.
Tlx: 371102, 371114 ISRAVIA IL.
Fax: (3)971-3131.

Index

3 inch/50 single calibre gun mounting Mk 34 (USA) 507
3 inch/50 twin calibre gun mounting Mk 33 (USA) 508
3-in window dispensing rocket system (Corvus) (UK) 662
3RM series naval radars (Italy) 610
5 inch/38 calibre single gun mountings (USA) 506
5 inch/38 calibre twin gun mounting Mk 38 (USA) 506
5 inch/38 twin calibre gun mounting Mk 28 (USA) 505
5 inch/54 single calibre gun mounting Mk 39 (USA) 505
5 inch/54 single calibre gun mounting Mk 42 (USA) 506
5 inch/54 single calibre gun mounting Mk 42 (Lightweight Mod 9) 507
5 inch/54 single calibre lightweight automatic gun mounting Mk 45 (USA) 507
9AA 100 fire control system (Sweden) 90
9AA 200 fire control system (Trackfire) (Sweden) 90
9CSI 500 STINA (Sweden) 429
9CSI STINA (Sweden) 429
9EW 300 ECM systems (Sweden) 661
9EW 400 ECM systems (Sweden) 661
9EW countermeasures launcher (Sweden) 660
9FA 302 field artillery computer (Sweden) 89
9GR 600 naval radar (Sweden) 615
9KA 100 coastal artillery fire control system (Sweden) 92
9KA 400 Kastell coast fire control system (Sweden) 91
9 KA 500 Kardinal coastal fire control system (Sweden) 92
9 KR 400 coastal radar (Sweden) 251
9 LV + 50 ASW system (Sweden) 429
9 LV 100 fire control system (Sweden) 427
9 LV 200 fire control system (Sweden) 427
9 LV 200 Mk 2 tracking radar (Sweden) 615
9 LV 200 Mk 2 weapon control system (Sweden) 428
9 LV 200 Mk 3 tracking radar (Sweden) 615
9 LV 200 Mk 3 weapon control system (Sweden) 428
9 LV 300 naval weapon control C^3 system (Sweden) 428
9 LV 400 series C^3I system (Sweden) 429
10 Hz laser for anti-aircraft systems (Yugoslavia) 407
14.5mm twin gun mounting (China) 486
16 inch triple gun turret Mk 7 (USA) 505
20mm single naval gun S20 (Germany, FR) 488
20mm twin AA turret (Brazil) 485
20mm twin gun anti-aircraft system (Germany, FR) 194
25mm twin AA gun mountings (USSR) 502
27mm Mauser cannon (Germany, FR) 741
30mm close-in weapon system (USSR) 475
30mm twin AA gun mounting (USSR) 502
35mm OE/OTO twin mounting (Italy) 491
37mm single AA gun mounting 70-K (USSR) 502
37mm twin AA gun mounting (China) 487
37mm twin AA gun mountings (USSR) 502
40mm AA gun mountings (USA) 508
40mm L/70 single automatic gun mounting (Sweden) 496
40/60 single AA gun mounting (Sweden) 497
40mm single naval mounting Type 107 (Italy) 492
40mm single naval mounting Type 564 (Italy) 491
40mm twin naval mounting Type 64 (Italy) 491
40mm twin naval mounting Type 106 (Italy) 491
57mm aircraft rocket (China) 730
57mm single dual-purpose gun mounting (USSR) 501
57mm single Mk 2 naval gun (Sweden) 496
57mm quadruple AA gun mountings (USSR) 501
57mm twin AA gun mounting (China) 486
57mm twin AA gun mounting (France) 487
57mm twin AA gun mountings (USSR) 501
57mm twin automatic AA gun mounting (USSR) 501
74mm anti-armour grenade (France) 736
75mm single dual-purpose gun mounting (China) 486
76mm single dual-purpose gun mounting (USSR) 501
76mm twin dual-purpose gun mounting (USSR) 500
76mm/62 calibre gun mounting Mk 75 (USA) 508
76/62 compact single gun mounting (Italy) 489
76/62 OTO MMI single gun mounting (Italy) 489
76/62 super rapid single gun mounting (Italy) 489
84mm RCL Carl-Gustav (Sweden) 146
85mm single dual-purpose gun mounting (China) 486
85mm single dual-purpose gun mounting (USSR) 500
85mm twin AA gun mounting (USSR) 501
100kg aircraft bombs (China) 731
100mm compact gun mounting (France) 487
100mm single dual-purpose gun mounting (China) 486
100mm single dual-purpose gun mounting (automatic) (USSR) 500
100mm single dual-purpose gun mounting Creusot-Loire (China) 485
100mm single dual-purpose gun mounting Model 1968-11 (France) 487
100mm twin dual-purpose gun mounting (China) 485
100mm twin dual-purpose gun mounting (USSR) 500
107mm Type 63 multiple rocket launcher (China) 116
107mm Type 81 multiple rocket launcher (China) 117
122mm multiple rocket launcher (Egypt) 118
122mm rocket-launcher M-1972 (RM-70) (Czechoslovakia) 118
122mm rocket launcher (USSR) 502
122mm (40-round) SP MRS (China) 116
125kg retard close support bomb (France) 735
127/54 single gun mounting (Italy) 488
130mm Type 70 multiple rocket launcher (China) 115
130mm Type 75 multiple rocket launcher (Japan) 124
130mm (30-round) Type 82 multiple rocket system (China) 116
130mm single gun mounting (China) 485
130mm twin dual-purpose gun mounting (China) 485
130mm twin dual-purpose gun mounting (USSR) 500
130mm twin gun mounting (USSR) 499
140mm rocket launchers (USSR) 502
152mm triple gun mounting (USSR) 499
193M/193M Mod 1/minehunting sonar (UK) 566
273mm (4-round) Type 83 multiple rocket system (China) 115
290mm medium range rocket (MAR 290) (Israel) 121
400 series naval fire control radar (UK) 634
533mm torpedo (USSR) 527
660mm torpedo (USSR) 527
805 series naval tracking radars (UK) 633
880 series mine countermeasures systems (UK) 438
880 series IFF interrogators (UK) 274
890 series iff interrogators (UK) 275
921-A ESM equipment (China) 647
2000 series thermal imaging systems (USA) 900

A

A-6E/F HUD sight (UK) 879
A 184 submarine- or surface-launched torpedo (Italy) 521
A 244/S (Mod .0 and Mod .1) lightweight torpedo (Italy) 522
AA-2 (Atoll) air-to-air missile (USSR) 699
AA-3 (Anab) air-to-air missile (USSR) 699
AA-5 (Ash) air-to-air missile (USSR) 699
AA-6 (Acrid) air-to-air missile (USSR) 700
AA-7 (Apex) air-to-air missile (USSR) 700
AA-8 (Aphid) air-to-air missile (USSR) 700
AA-9 (Amos) air-to-air missile (USSR) 701
AA-10 (Alamo) air-to-air missile (USSR) 701
AA-11 (Archer) air-to-air missile (USSR) 701
AAAM long range air-to-air missile (USA) 705
AATCP Mistral helicopter missile system (France) 693
ABM-1B (Galosh) anti-ballistic missile missile (USSR) 17
ABO 2000 detector-jammer system (France) 820
Abu Dhabi
land-based air defence systems 34
AC 008 directional antenna (Germany, FR) 310
ACL-STRIM anti-tank rocket launcher (France) 138
Acoustic intruder detection system (UK) 577
Active electronic buoys (USA) 676
ACWAR (agile CW acquisition radar) (USA) 304
AD 130 sonar homing and DF receiver (UK) 563
ADAMS (Air Defence Missile System) (Israel) 173
ADATS (Air Defence Anti-Tank System) (International) 140, 171
ADAWS 2 action data automation weapons system (UK) 436
ADAWS 4 action data automation weapon system (UK) 436
ADAWS 5 action data automation weapon system (UK) 436
ADAWS 6 action data automation weapon system (UK) 437
ADAWS 10 action data automation weapon system (UK) 437
Aden 25mm cannon (UK) 753
Aden 30mm cannon (UK) 753
ADEWS (Air defence EW system) (USA) 347
ADIVA artillery division automation system (France) 70
ADLIPS (Automatic Data Link Plotting System) (Canada) 408
Adour II instrumentation tracking radar (France) 225
Advanced anti-tank weapon system – heavy (AAWS-H) (USA) 157
Advanced anti-tank weapon system-medium (AAWS-M) (USA) 153
Advanced missile surveillance radar (USA) 297
Advanced phased array radar (APAR) 808
Advanced sea mine (International) 519
Advanced tactical air reconnaissance system (ATARS) (USA) 903
AEG receivers and ancillary systems 308
AEGIS command and control system (USA) 448
AEGIS surface-to-air weapon system (USA) 478
AEL Type 6040 universal jammer (USA) 341
Aerea multi task pod (Italy) 743
Aerea twin gun pod 7.62 (Italy) 743
AF532 helicopter roof sight (UK) 883
AF533 helicopter roof sight (UK) 883
AF580 helicopter roof sight (UK) 883
Agave airborne radar (France) 776
AGIS combat information system (Germany, FR) 411
AGM-45 Shrike air-to-surface missile (USA) 722
AGM-65 Maverick air-to-surface missile system (USA) 722
AGM-69 SRAM short range attack missile (USA) 721
AGM-78 standard anti-radar missile (USA) 723
AGM-84 Harpoon air-to-surface anti-ship missile (USA) 724
AGM-86 ALCM air-launched cruise missile (USA) 721
AGM-88A HARM high-speed anti-radar missile (USA) 724
AGM-114 tactical air-to-surface missile Hellfire (USA) 726
AGM-122 Sidearm anti-radar missile (USA) 723
AGM-123A Skipper II guided missile (USA) 724
AGM-129 advanced cruise missile (USA) 721
AGM-130 air-to-surface missile (USA) 723
AGM-136 Tacit Rainbow anti-radar missile (USA) 723
Agrion airborne radar (France) 773
AIMS antenna system (USA) 645
Airborne electro-optics and SDI (USA) 902
Airborne expendable decoys (USA) 866
Airborne radar jamming system (USA) 865
Airborne radar tables 935
Airborne remote controlled ESM system (USA) 866
Aircraft bombs (SA) 746
Aircraft cannon (China) 731
Air defence alerting device (ADAD) (UK) 392
Air defence initiative (USA) 59
Air defence missiles (India) 170
Air defence missiles (Taiwan) 176
Air defence missiles tables 913
Air-launched ANS (International) 710
Air-launched anti-tank missiles (USSR) 718
Air-launch Exocet (AM 39) (France) 709
Air-to-air missiles tables 915
Air-to-surface missiles tables 916
Airship surveillance radar (USA) 813
ALARM air-to-surface missile (UK) 719
Albatros naval defence system (Italy) 469
Albatros weapon system (Italy) 419
Alberich laser/radar warning system (International) 314
Alkan chaff/infra-red countermeasures dispensers (France) 820
Alkan sonobuoy launchers (France) 555
ALT airborne laser tracker (USA) 892
Altair airborne radar (Italy) 781
AM/APS-717 search and weather radar (Italy) 781
AMC-11 mortar computer (China) 66
AMETS–artillery meteorological system (UK) 99
AMRAAM (AIM-120A) (International) 702
AN/AAQ-9 FLIR (USA) 898
AN/AAQ-10 FLIR (USA) 898
AN/AAQ-16 night vision system (USA) 897
AN/AAR-34 infra-red receiving set (USA) 838
AN/AAR-42 FLIR (USA) 897
AN/AAR-44 infra-red warning receiver (USA) 838
AN/AAR-45 FLIR pod (USA) 897
AN/AAR-47 missile warning set (USA) 838
AN/AAS-33 TRAM target recognition and attack multi-sensor (USA) 898
AN/AAS-35 Pave Penny laser tracker (USA) 898
AN/AAS-36 IRDS (USA) 899
AN/AAS-37 DRTS (USA) 899
AN/AAS-40 Seehawk FLIR (USA) 893
AN/AKT-22(V)4 telemetry data transmitting set (USA) 580
AN/ALE-29A countermeasures dispenser system (USA) 839
AN/ALE-29B countermeasures dispenser system (USA) 839
AN/ALE-36 countermeasures dispensing pod (USA) 839
AN/ALE-37 chaff dispenser (USA) 839
AN/ALE-38 bulk chaff dispenser (USA) 839
AN/ALE-39 ECM dispenser (USA) 839
AN/ALE-40 countermeasures dispenser system (USA) 839
AN/ALE-41 bulk chaff dispenser (USA) 839
AN/ALE-43 chaff cutter dispenser pod (USA) 840
AN/ALE-44 dispenser pod (USA) 841
AN/ALE-45 countermeasures dispenser system (USA) 841
AN/ALE-47 countermeasures dispensing system (USA) 841
AN/ALQ-78 countermeasures set (USA) 842
AN/ALQ-99 ECM equipment (USA) 842
AN/ALQ-101 ECM pods (USA) 843
AN/ALQ-108 jamming pod (USA) 843
AN/ALQ-117 jammer (USA) 843
AN/ALQ-119 ECM pod (USA) 844
AN/ALQ-122 ECM system (USA) 844
AN/ALQ-123 infra-red countermeasures (USA) 844
AN/ALQ-125 tactical electronic reconnaissance sensor (USA) 844
AN/ALQ-126 A/B ECM equipments (USA) 845
AN/ALQ-128 warning receiver (USA) 845
AN/ALQ-130 ECM equipment (USA) 845
AN/ALQ-131 ECM pod (USA) 845
AN/ALQ-131 receiver/processor (USA) 846
AN/ALQ-133 ELINT equipment (USA) 846
AN/ALQ-135 advanced jammer set (USA) 846
AN/ALQ-136 jammer (USA) 846
AN/ALQ-137 airborne jammer (USA) 846
AN/ALQ-142 ESM system (USA) 846
AN/ALQ-144 IR countermeasures set (USA) 847
AN/ALQ-147A IR countermeasures set (USA) 847
AN/ALQ-149 jammer (USA) 847
AN/ALQ-150 ESM receiver (USA) 848
AN/ALQ-151 airborne EW system (USA) 848
AN/ALQ-153 tail warning set (TWS) (USA) 848
AN/ALQ-155(V) EW power management system (USA) 848
AN/ALQ-156 missile detection system (USA) 848
AN/ALQ-157 infra-red countermeasures system (USA) 849
AN/ALQ-161 B-I RFS/ECM sub-system (USA) 849
AN/ALQ-162 radar jammer (USA) 849
AN/ALQ-164 ECM pod (USA) 850
AN/ALQ-165 airborne self protection jammer (USA) 850
AN/ALQ-171(V) tactical ECM system (USA) 850
AN/ALQ-172 noise deception jammer (USA) 851
AN/ALQ-176(V) ECM pod (USA) 851
AN/ALQ-178 ECM system (USA) 851
AN/ALQ-184 jamming pod (USA) 852
AN/ALQ-186(V) jammer (USA) 852
AN/ALQ-187 countermeasures system (USA) 852
AN/ALR-45 radar warning and control system (USA) 852
AN/ALR-45F radar warning and control systems (USA) 852
AN/ALR-46 radar warning system (USA) 853
AN/ALR-47 radar homing and warning system (USA) 853
AN/ALR-50 ECM equipment (USA) 853
AN/ALR-52 ECM receiver (USA) 853
AN/ALR-56 radar warning receiver (USA) 854
AN/ALR-60 communications intercept and analysis system (USA) 854
AN/ALR-62(V) radar warning system (USA) 854
AN/ALR-66 series radar warning receivers (USA) 854
AN/ALR-67(V) radar warning system (USA) 855
AN/ALR-68 advanced radar warning system (USA) 856
AN/ALR-69 radar warning receiver (USA) 856
AN/ALR-73 passive detection system (USA) 857
AN/ALR-74 radar warning system (USA) 857
AN/ALR-75(V) surveillance receiver (USA) 858
AN/ALR-76 ESM system (USA) 858
AN/ALR-77 ESM system (USA) 858
AN/ALR-80(V) radar warning receiver (USA) 858
AN/ALR-606(V)1 radar warning receiver (USA) 859
AN/ALR-606(VE) radar warning receiver (USA) 859
AN/ALT-28 noise jammer (USA) 859
AN/APG-63 airborne radar (USA) 793
AN/APG-65 airborne radar (USA) 794
AN/APG-66 F-16 radar (USA) 794
AN/APG-67 multimode airborne radar (USA) 795
AN/APG-68 airborne radar (USA) 796
AN/APG-69(V) multimode airborne radar (USA) 796
AN/APG-70 airborne radar (USA) 797

AN/APG-71 airborne radar (USA) ... 797
AN/APN-215 airborne radar (USA) ... 798
AN/APN-234 multi-mode radar (USA) ... 798
AN/APQ-99 radar (USA) ... 799
AN/APQ-113 radar (USA) ... 799
AN/APQ-114 radar (USA) ... 799
AN/APQ-120 fire control radar (USA) ... 799
AN/APQ-122(V) airborne radar (USA) ... 800
AN/APQ-126(V) airborne navigation and attack radar (USA) ... 800
AN-APQ-144 radar (USA) ... 800
AN/APQ-146 airborne radar (USA) ... 800
AN/APQ-148/156 multimode radar (USA) ... 801
AN/APQ-153 fire control radar (USA) ... 801
AN/APQ-157 fire control radar (USA) ... 802
AN/APQ-158 airborne radar (USA) ... 802
AN/APQ-159 airborne radar (USA) ... 802
AN/APQ-161 radar (USA) ... 803
AN/APQ-164 airborne radar (USA) ... 803
AN/APQ-165 airborne radar ... 803
AN/APQ-168 airborne radar (USA) ... 804
AN/APQ-171 airborne radar (USA) ... 804
AN/APQ-172 airborne radar (USA) ... 804
AN/APQ-174 multi-function radar ... 804
AN/APR-38 Wild Weasel CIS (USA) ... 859
AN/APR-39(V) radar warning system (USA) ... 860
AN/APR-39A(V) threat warning system (USA) ... 860
AN/APR-43 tactical radar warning receiver (USA) ... 860
AN/APR-44 radar warning receiver (USA) ... 861
AN/APR-47 Wild Weasel CIS (USA) ... 861
AN/APS-94F side-looking radar (USA) ... 804
AN/APS-115 airborne radar (USA) ... 804
AN/APS-116 airborne radar (USA) ... 804
AN/APS-124 airborne radar (USA) ... 805
AN/APS-125 airborne radar (USA) ... 808
AN/APS-127 airborne radar (USA) ... 805
AN/APS-128 Model D surveillance radar (USA) ... 805
AN/APS-130 airborne radar (USA) ... 806
AN/APS-131 side-looking radar (USA) ... 806
AN/APS-133 airborne radar (USA) ... 806
AN/APS-134(V) maritime surveillance radar (USA) ... 807
AN/APS-135 side-looking radar (USA) ... 807
AN/APS-137(V) airborne radar (USA) ... 807
AN/APS-138 airborne radar (USA) ... 808
AN/APS-139 airborne radar (USA) ... 808
AN/APS-145 AEW radar (USA) ... 808
AN/APS-503 airborne radar (International) ... 778
AN/APS-504(V) series airborne radar (International) ... 778
AN/APX-72 airborne IFF transponder (USA) ... 809
AN/APX-76 airborne IFF interrogator (USA) ... 809
AN/APX-83 airborne IFF interrogator (USA) ... 809
AN/APX-100(V) airborne IFF transponder (USA) ... 809
AN/APX-103 airborne interrogator (USA) ... 810
AN/AQS-13 helicopter sonar (USA) ... 592
AN/AQS-14 airborne minehunting sonar (USA) ... 592
AN/AQS-18 airborne sonar (USA) ... 592
AN/AQS-503 helicopter acoustic processing system (Canada) ... 545
AN/ARR-72 sonobuoy receiving set (USA) ... 581
AN/ARR-75 sonobuoy receiver (USA) ... 582
AN/ARR-78(V) advanced sonobuoy communication link (ASCL) (USA) ... 583
AN/ASA-64 submarine anomaly detector (Canada) ... 543
AN/ASA-65 compensation group adaptor (Canada) ... 543
AN/ASA-65(V) nine-term compensator (Canada) ... 543
AN/ASA-66 tactical data display (USA) ... 580
AN/ASA-82 tactical data display system (USA) ... 580
AN/ASB-19 ARBS angle-rate bombing set (USA) ... 896
AN/ASG-26A lead computing optical sight (USA) ... 888
AN/ASG-29 lead computing optical sight (USA) ... 889
AN/ASQ-81(V) airborne ASW magnetometer (USA) ... 598
AN/ASQ-145 low light-level TV system (USA) ... 890
AN/ASQ-153 Pave Spike laser target designator (USA) ... 895
AN/ASQ-191 communications /countermeasures set (USA) ... 862
AN/ASQ-502 submarine detecting set (MAD) (Canada) ... 541
AN/ASQ-504(V) MAD system (Canada) ... 541
AN/ASX-1 target identification system, electro-optical (TISEO) (USA) ... 889
AN/AVQ-26 Pave Tack (USA) ... 894
AN/AVQ-27 airborne laser target designator set (LTDS) (USA) ... 894
AN/AVR-2 laser warning receiver (USA) ... 861
AN/AWG-9 airborne weapon control system (USA) ... 810
AN/AWG-10 missile control system (USA) ... 811
AN/AWG-11 missile control system (USA) ... 811
AN/AWG-12 missile control system (USA) ... 811
AN/AXQ-14 weapon control equipment (USA) ... 901
AN/AYA-8B ASW data processing system (USA) ... 581
AN/BQQ-5 sonar (USA) ... 593
AN/BQQ-6 sonar (USA) ... 593
AN/BQR-15 sonar (USA) ... 593
AN/BQR-21 DIMUS sonar (USA) ... 593
AN/BQS-13 sonar (USA) ... 594
AN/BQS-14A submarine sonar (USA) ... 594
AN/BQS-15 submarine sonar (USA) ... 594
AN/BSY-1(V) integrated combat system (USA) ... 584
AN/FPN-62 precision approach radar (PAR) (USA) ... 277
AN/FPQ-6 precision instrumentation radar (USA) ... 278
AN/FPS-6 heightfinding radar (USA) ... 278
AN/FPS-7 3-D search and GCI radar (USA) ... 278
AN/FPS-8 search radar (USA) ... 279
AN/FPS-16 instrumentation radar (USA) ... 279
AN/FPS-20—AN/FPS-100 family of long-range search and GCI radars (USA) ... 279
AN/FPS-24 search radar (USA) ... 279
AN/FPS-49 early warning and tracking radar (USA) ... 280
AN/FPS-50 early warning radar (USA) ... 280
AN/FPS-85 long-range phased array radar (USA) ... 280
AN/FPS-88 surveillance radar (USA) ... 281
AN/FPS-89 heightfinding radar (USA) ... 281
AN/FPS-105(V) range instrumentation radar (CAPRI) (USA) ... 281
AN/FPS-115 Pave Paws radar (USA) ... 277
AN/FPS-117 air defence radar (USA) ... 282
AN/FPS-118 over-the-horizon radar (USA) ... 302
AN/GLQ-3B communications countermeasures set jammer (USA) ... 338
AN/GPN-22 precision approach radar (PAR) (USA) ... 282
AN/GVS-5 laser rangefinder (USA) ... 395
AN/GYG-501 milipac artillery computer (Canada) ... 65
AN/KAS-1 chemical warfare directional detector (USA) ... 676
AN/MLQ-34 TACJAM (USA) ... 340
AN/MLQ-T6 communications data link jammer (USA) ... 340
AN/MPQ-4 mortar locating radar (USA) ... 282
AN/MPQ-49 forward area alerting radar (USA) ... 287
AN/MPQ-53 guidance radar (USA) ... 282
AN/MPQ-54 forward area alerting radar (USA) ... 287
AN/MPQ-63 instrumentation ballistic radar (ARBAT) ... 292
AN/MPS-11 search radar (USA) ... 283
AN/MPS-11A search radar (USA) ... 283
AN/MPS-14 heightfinding radar (USA) ... 283
AN/MPS-36 instrumentation tracking radar (USA) ... 283
AN/MPS-39 instrumentation tracking radar (USA) ... 284
AN/MSQ-103 Teampack ESM system (USA) ... 341
AN/MSQ-117 IFF tracking system (USA) ... 299
AN/PAQ-1 laser target designator (USA) ... 402
AN/PAQ-3 modular universal laser equipment (MULE) (USA) ... 402
AN/PPS-5 combat surveillance radar (USA) ... 283
AN/PPS-11 battlefield surveillance radar (USA) ... 284
AN/PPS-12 battlefield surveillance radar (USA) ... 284
AN/PPS-15A infantry radar (USA) ... 284
AN/PPS-15B base security radar (USA) ... 285
AN/PPX-3B ground interrogator (USA) ... 299
AN/PQS-2A hand-held sonar (USA) ... 582
AN/PRD-11 man-portable df equipment (USA) ... 337
AN/PSC-2 digital communications terminal (USA) ... 106
AN/PVS-6 Melios hand-held laser rangefinder (USA) ... 403
AN/SAR-8 infra-red warning system (Canada) ... 647
AN/SLQ-17 naval electronic countermeasures suite (USA) ... 671
AN/SLQ-30 TRUMP (USA) ... 672
AN/SLQ-32 shipboard EW equipment (USA) ... 672
AN/SLQ-501 CANEWS ew systems (Canada) ... 646
AN/SLQ-503 (Ramses) shipborne ecm system (Canada) ... 646
AN/SPG-51 radar (USA) ... 635
AN/SPG-55 radar (USA) ... 636
AN/SPG-60 radar (USA) ... 636
AN/SPQ-9 radar (USA) ... 636
AN/SPQ-11 Cobra Judy radar (USA) ... 636
AN/SPS-6 radar (USA) ... 637
AN/SPS-10 radar (USA) ... 637
AN/SPS-30 radar (USA) ... 638
AN/SPS-37 radar (USA) ... 637
AN/SPS-39A (Frescan) radar (USA) ... 638
AN/SPS-40 radar (USA) ... 638
AN/SPS-43 radar (USA) ... 638
AN/SPS-48 air surveillance radar (USA) ... 638
AN/SPS-49 radar (USA) ... 639
AN/SPS-52 radar (USA) ... 640
AN/SPS-52A radar (USA) ... 640
AN/SPS-52B radar (USA) ... 640
AN/SPS-52C radar (USA) ... 639
AN/SPS-55 naval radar (USA) ... 640
AN/SPS-58 naval radar (USA) ... 640
AN/SPS-63 radar (USA) ... 641
AN/SPS-65 naval radar (USA) ... 640
AN/SPS-65(V) ER naval surveillance radar (USA) ... 641
AN/SPS-67 radar (USA) ... 641
AN/SPS-503 surveillance radar (Canada) ... 600
AN/SPY-1 multi-function array radar (USA) ... 641
AN/SQQ-14 minehunting sonar (USA) ... 587
AN/SQQ-30 mine-hunting sonar (USA) ... 587
AN/SQQ-32 advanced minehunting sonar system (USA) ... 587
AN/SQQ-89 integrated combat system (USA) ... 588
AN/SQQ-504 towing condition monitor (Canada) ... 546
AN/SQR-15 towed array sonar (USA) ... 584
AN/SQR-17 sonar signal processor (USA) ... 584
AN/SQR-18A tactical sonar towed array (TACTAS) (USA) ... 584
AN/SQR-19 tactical sonar towed array (USA) ... 585
AN/SQR-501 Canadian towed array sonar system (CANTASS) (Canada) ... 543
AN/SQS-35 variable depth sonar (USA) ... 586
AN/SQS-36 hull-mounted sonar (USA) ... 586
AN/SQS-38 hull-mounted sonar (USA) ... 586
AN/SQS-53/26 sonars (USA) ... 585
AN/SQS-56/DE1160 (USA) ... 586
AN/SQS-505 sonar (Canada) ... 542
AN/SQS-509 sonar (Canada) ... 542
AN/SQS-510 digital sonar system (Canada) ... 542
AN/SSQ-41B sonobuoy (USA) ... 595
AN/SSQ-47B sonobuoy (USA) ... 595
AN/SSQ-53B DIFAR sonobuoy (USA) ... 595
AN/SSQ-53B sonobuoy (USA) ... 596
AN/SSQ-53C Sparton Dwarf DIFAR sonobuoys (USA) ... 597
AN/SSQ-57A sonobuoy (USA) ... 596
AN/SSQ-58A moored sensor buoy (USA) ... 597
AN/SSQ-62B DICASS sonobuoy (USA) ... 596
AN/SSQ-71 ATAC buoy (USA) ... 597
AN/SSQ-72 outboard system (USA) ... 673
AN/SSQ-75 ERAP sonobuoy (USA) ... 597
AN/SSQ-77 (VLAD) sonobuoy (USA) ... 597
AN/SSQ-86(XN-1) down-link communication sonobuoy (USA) ... 598
AN/SSQ-101 HLA sonobuoy (USA) ... 598
AN/SSQ-102 sonobuoy (USA) ... 598
AN/SSQ-522 sonobuoy (Canada) ... 544
AN/SSQ-527B sonobuoy (Canada) ... 544
AN/SSQ-530 DIFAR sonobuoy (Canada) ... 544
AN/TAS-4 TOW night sight (USA) ... 403
AN/TAS-6 (AN/UAS-11) night observation device, long range (USA) ... 404
AN/TLQ-15 communications and countermeasures set (USA) ... 341
AN/TLQ-17A countermeasures set (USA) ... 342
AN/TPN-18A radar (USA) ... 285
AN/TPN-18B radar (USA) ... 285
AN/TPN-19 landing control central (USA) ... 285
AN/TPN-22 precision approach and landing radar (USA) ... 285
AN/TPN-24 airport surveillance radar (USA) ... 286
AN/TPN-25 precision approach radar (USA) ... 287
AN/TPQ-10 radar course directing central (USA) ... 287
AN/TPQ-18 precision instrumentation radar (USA) ... 278
AN/TPQ-32 forward area alerting radar (USA) ... 287
AN/TPQ-36 weapon locating radar (USA) ... 289
AN/TPQ-37 weapon locating radar (USA) ... 290
AN/TPQ-39(V) instrumentation radar (USA) ... 290
AN/TPS-21 pulse doppler battlefield radar (USA) ... 291
AN/TPS-32 3-D long-range surveillance radar (USA) ... 291
AN/TPS-33 pulse doppler battlefield radar (USA) ... 291
AN/TPS-34 transportable air defence radar (USA) ... 293
AN/TPS-43E tactical 3-D radar (USA) ... 293
AN/TPS-44 tactical surveillance radar (Alert Mk 11 A/O) (USA) ... 294
AN/TPS-59 3-D tactical radar (USA) ... 294
AN/TPS-63 tactical surveillance radar (USA) ... 295
AN/TPS-65 tactical surveillance radar (USA) ... 296
AN/TPS-70(V)-1 tactical radar (USA) ... 296
AN/TPX-42 PIDP IFF-ATCRBS equipment (USA) ... 298
AN/TPX-46(V) IFF interrogator (USA) ... 298
AN/TPX-54 IFF/SSR interrogator (USA) ... 298
AN/TSQ-73 air defence missile control and co-ordination system (USA) ... 62
AN/TSQ-112 TACELIS ESM system (USA) ... 343
AN/TSQ-114 intercept and df system (USA) ... 343
AN/TVQ-2 ground/vehicular laser locator designator (G/VLLD) (USA) ... 401
AN/ULQ-11 airborne jamming system (USA) ... 864
AN/ULQ-14 radar countermeasures (USA) ... 329
AN/ULQ-19 tactical communication jammer (USA) ... 329
AN/UPD-4 side-looking radar system (USA) ... 812
AN/UPD-8 side-looking reconnaissance radar (USA) ... 812
AN/UPS-3 tactical defence radar (USA) ... 301
AN/UPX-24(V) shipborne IFF system (USA) ... 645
AN/USQ-81(V) command control system (USA) ... 443
AN/UYS-503 sonobuoy processor (SBP) (Canada) ... 545
AN/VAS-3 driver's thermal viewer (USA) ... 402
AN/VLQ-5(V) 1 comfy sword (USA) ... 346
AN/VPS-2 ranging radar (USA) ... 299
AN/VSG-2 tank thermal sight (USA) ... 396
AN/VSG-X (TIS) thermal imaging system (USA) ... 401
AN/VVS-1 tank laser system (USA) ... 397
AN/VVS-2 driver's night vision viewer (USA) ... 397
AN/VVS-501 night driving/observation viewer (USA) ... 397
AN/WLQ-4 SIGINT system (USA) ... 670
AN/WLR-1H radar detection equipment (USA) ... 670
AN/WLR-8 tactical EW receiver system (USA) ... 671
AN/WLR-11 IFM receiver (USA) ... 671
AN/WQC-2A sonar communication set (USA) ... 582
Anaconda Tactas system (France) ... 555
Anaren shipborne tracking receiver (USA) ... 675
Anemone airborne radar (France) ... 776
Angola
land-based air defence systems ... 34
ANS anti-ship missile (International) ... 451
Antares heightfinding radar (France) ... 225
Anti-armour missile (Brazil) ... 137
Antilope V TC airborne radar (France) ... 746
Anti-submarine mortar system (Australia) ... 511
Anti-submarine rocket launchers (USSR) ... 527
Anti-submarine weapon system (France) ... 513
Apache/CWS 50 (International) ... 711
APECS II ESM/ECM system (USA) ... 676
Apollo airborne ECM system (UK) ... 834
APS series airborne torpedo control systems (Italy) ... 522
APX M334 series gyro-stabilised sights (France) ... 867
APX M334-04 gyro-stabilised sight (France) ... 868
APX M334-25 gyro-stabilised sight (France) ... 868
APX M363 fire control (France) ... 356
APX M520 panoramic sight (France) ... 356
APX M527 tank sight (France) ... 356
APX M550/TCV 80 laser rangefinder (France) ... 351
APX M586 tank FCS update (France) ... 67
AQ 31 jammer (Sweden) ... 830
AQ 800 jammer (Sweden) ... 830
AQ-861 noise and deception jamming system (Sweden) ... 831
AQ 900 airborne self-protection system (Sweden) ... 831
AQS 901 sonobuoy processing and display system (UK) ... 562
AQS-902 series acoustic processors (UK) ... 562
AQS-903 series acoustic processor systems (UK) ... 563
AQS-920 series acoustic processors (UK) ... 562
AWS-930 series acoustic processor systems (UK) ... 563
Aquila laser transmitter (USA) ... 891
Aquila mission payload sub-system (USA) ... 891
Aquila mobile radar (Chile) ... 218
Aquisition radar and control system (Norway) ... 249
AR-3D air defence radar (UK) ... 266
AR-15M mobile surveillance radar (UK) ... 267
AR-320 three-dimensional air defence radar (UK) ... 269
AR-325 commander air defence radar (UK) ... 269
AR 753 jammer set-on receiver (Sweden) ... 829
AR 765 radar warning receiver (Sweden) ... 829
AR 777 airborne microwave receiver (Sweden) ... 829
AR 830 airborne warning system (Sweden) ... 830
ARBB 33 naval jammer (France) ... 649
ARBR-17 ESM system (France) ... 649
Ares helmet-mounted target designator (France) ... 358
Argentina
air-to-surface missiles ... 707
aircraft armament ... 727
anti-tank/assault weapons ... 136
battlefield support weapons ... 113
land-based air defence systems ... 34
Argonaute sonar system (France) ... 548
Argonaute submarine sonar (France) ... 548
Argos-10 early warning radar (Italy) ... 245
Argos-45 tactical radar system (Italy) ... 246
Argus day/night sight (UK) ... 389
ARI 18240/1 ESM (USA) ... 862
ARI 23246/1 ECM pod (Skyshadow) (UK) ... 832
ARI 5983 aircraft transponder (UK) ... 789
Armat anti-radar missile (France) ... 707

Army command and control system programmes (Germany, FR) 72
ARR-81 COMINT receiver (USA) 862
ARR 901 sonar receiver (UK) 574
ARS field artillery computer system (Yugoslavia) 111
ARSR-3 surveillance radar (USA) 286
Artemis 30 twin 30mm light AA gun system (Greece) 195
Artillery delivered expendable jammer (AD/EXJAM) (USA) 346
Artillery fire control system (Italy) 79
Artillery rockets (USSR) 129
Artillery sighting system (UK) 394
Artois instrumentation tracking radar (France) 226
AS-13 air-to-surface missile (USSR) 718
AS.15 TT air-to-surface missile (France) 708
AS.30 laser air-to-surface missile (France) 708
AS80 artillery fire control system (South Africa) 86
AS-228 depth charge (Chile) 513
AS360 sonar systems (UK) 579
AS370 seabed surveillance sonar (UK) 579
AS380 dunking sonar (UK) 579
ASAR3-2 synthetic aperture radar (USA) 814
ASAS/ENSCE (USA) 347
ASAT anti-satellite weapon (USA) 33
ASM-1 air-to-surface missile (Type 80) (Japan) 714
ASMP air-to-surface missile (France) 709
ASN-902 tactical processing system (TATTIX) (UK) 563
Aspic infrared detection system (France) 872
Aspide multirole missile (Italy) 697
ASR-30 surveillance radar (USA) 286
ASR 360 airborne surveillance radar (UK) 790
ASRAAM programme (AIM-132) (International) 695
ASROC anti-submarine system (USA) 533
ASTAC tactical ELINT system (France) 818
ASTAR-2 air defence radar (USA) 297
ASTAR-3 air defence radar (USA) 297
ASTOR - Airborne Stand-off Radar (UK) 788
AT-2 (Swatter) anti-tank missile (USSR) 148
AT-3 (Sagger) anti-tank missile (USSR) 149
AT-4 (Spigot) anti-tank missile (USSR) 149
AT-5 (Spandrel) anti-tank missile (USSR) 150
AT-6 (Spiral) anti-tank missile (USSR) 150
AT-7 (Saxhorn) anti-tank missile (USSR) 150
AT-8 (Songster) anti-tank missile (USSR) 151
AT-101A triple band surveillance antenna (Israel) 608
AT-4910 comint/jamming system (Israel) 317
ATAC artillery mini computer (France) 68
ATAFCS equipment (USA) 891
ATAS active towed array sonar (UK) 573
ATC-1 torpedo decoy (Brazil) 520
ATIBA artillery fire control system (France) 69
Atila artillery automation system (France) 69
Atlas instrumentation tracking radar (France) 226
ATREOS laser tank fire control system (Italy) 79
Australia
electronic warfare equipment (sea) 646
ground radar 218
land-based air defence systems 37
sonar and underwater detection equipment 540
underwater warfare systems 511
Austria
land-based air defence systems 38
Automated computer assisted electronic warfare system (ACEWS) (UK) 321
Automatic countermeasures dispensing system (Israel) 653
Automatic missile detection radar (Israel) 607
AV-8 Harrier 25mm armament system (USA) 761
AV-8A head-up display and weapon aiming computing system (UK) 881
AV-8B head-up display (UK) 881
AV 62 commander's sight (UK) 380
AVC-101 artillery fire control computer (China) 66
Avenger air defence system (USA) 210
AVIBRAS artillery rockets (Brazil) 114
AVIBRAS EDT-FILA anti-aircraft fire control system (Brazil) 63
AWARE-3 radar warning receiver (International) 821
AWDATS–artillery weapon data transmission system (UK) 101
AWS-4 naval radar (UK) 628
AWS-5 series naval radar (UK) 628
AWS-6 series naval radar (UK) 629
AWS-7 naval radar (UK) 629
AWS-9 naval surveillance radar (UK) 629
AW-X self-propelled twin 35mm AA gun system (Japan) 199
AYB AL-SAQR portable anti-aircraft missile (Egypt) 212

B

B-52 airborne radar (USA) 814
B171-II IR aiming unit (Germany, FR) 359
B515/M/3 torpedo launching system (Italy) 522
Babylon hf jammer (Germany, FR) 313
Back Net GCI radar (USSR) 253
BADGE air defence system (Japan) 46
BAe ESM system (UK) 329
BAe improved TOW sight (UK) 887
BAe infra-red jammer (UK) 836
Ballistic Missile Early Warning System (BMEWS) (USA) 56
Band Stand naval radar (USSR) 617
Bandfish expendable torpedo countermeasure (UK) 530
Bantam anti-tank missile (BS 53) (Sweden) 147
BAP 100mm airfield denial and anti-runway weapon system (France) 736
Barak 1 point defence system (Israel) 468
Barax detector-jammer (France) 820
Barbican EW system (UK) 326
Barem self-protection jammer (France) 817
Barlock GCI radar (USSR) 253
Barr & Stroud submarine attack periscope (UK) 685
Barr & Stroud submarine optronics mast (UK) 686
Barr & Stroud submarine search periscope (UK) 685
Barr & Stroud tank laser sight (UK) 382
Barra side processor (AQA-801) (Canada) 545
Barricade naval decoy equipment (UK) 662
Basilisk jamming pod (France) 817
Bass Tilt fire control radar (USSR) 617
BAT-30 fire control system (Israel) 365
BAT 120 anti-armour weapon system (France) 736
BATES - battlefield artillery target engagement system (UK) 100
Bathythermograph data recorder RO-308/SSQ-36 (USA) 590
Battery Computer System (BCS) (USA) 104
Battlefield operations laser designator (BOLD) (UK) 385
Bavard LXT 462 adjustable practice bombs (France) 740
Bazan SCPV2 combat system (Spain) 426
BC81 (Re 83) (FARGO) artillery computer (International) 74
Beamtrap close defence rocket (Israel) 653
Béarn instrumentation tracking radar (France) 226
Belgium
army fire control systems 63
electro-optical equipment (land) 348
Belouga dispenser weapon (BLG 66) (France) 738
BI series sonobuoys (Italy) 558
Big Bar A GCI radar (USSR) 253
Big Bulge airborne radar (USSR) 784
Big Fred artillery/mortar locating radar (USSR) 253
Big Mesh GCI radar (USSR) 253
Big Net search radar (USSR) 618
Blast jamming system (Germany, FR) 309
Bloodhound Mk 2 surface-to-air weapon system (UK) 183
Blowpipe portable anti-aircraft missile system (UK) 214
Blue Fox radar (UK) 792
Blue Kestrel helicopter radar (UK) 793
Blue Thunder HADAS (Helmet Airborne Display and Sight) (Israel) 875
Blue Vixen radar (UK) 792
BM 400 multi-munition bomb (France) 735
BM 8025 passive night sight (Germany, FR) 361
BO series countermeasures dispensers 831
BODIL coast artillery FCS (Norway) 85
BOFI anti-aircraft fire control system (Sweden) 88
Bofors 40mm armour trinity (Sweden) 200
Bofors 40mm BOFI gun system (Sweden) 201
Bofors 80 kg aircraft parachute flare SUPER LEPUS (Sweden) 749
Bofors 135mm air-to-ground rocket system M70 (Sweden) 748
Bofors 155mm anti-armour ammunition (Taiwan) 148
Bofors AFV fire control system (Sweden) 90
Bofors-Breda 144 Mod 76, 40mm automatic gun SAK L/70 (Spain) 495
Bofors sea mines (Sweden) 525
Bofors Type U fire control system (Sweden) 91
Bomb retarding system (UK) 754
BPD 2 inch aircraft rockets (Italy) 744
BPD 122mm Falco air-to-ground rocket system (Italy) 744
BPD Medusa 81mm air-to-ground rocket system (Italy) 745
BPTSG modular FCS for AA guns (France) 66
Brazil
air-to-air missiles 692
aircraft armament 727
anti-tank/assault weapons 137
army fire control systems 63
battlefield support weapons 114
coastal defence weapons 160
ground radar 218
land-based air defence systems 38
naval guns and bombardment rockets 485
portable anti-aircraft guided missiles 211
BRD bomb and rocket dispensers (Italy) 744
Breda 81mm bombardment launcher (Italy) 494
Breda/Bofors naval mountings (Italy) 491
Breda compact twin 40 mm L/70 naval mounting type 70 (Italy) 490
Breda light chaff rocket launching system (Italy) 659
Breda lightweight 30mm AA field mounting (Sentinel) (Italy) 198
Breda local control single 30mm naval mounting (Italy) 493
Breda multi-calibre assault launcher (Italy) 494
Breda SCLAR multipurpose rocket launcher (Italy) 493
Breda single 30mm mounting (Italy) 493
Breda twin 30mm compact naval mounting (Italy) 492
Breda twin 40L70 AA field mounting (Italy) 198
Breda twin fast forty naval mounting (Italy) 490
Briefcase terminal (USA) 106
British Aerospace mine modernisation (UK) 531
Buck/Wegmann decoy system (Germany, FR) 651

C

C-101 supersonic anti-ship missile (China) 449
C-204/WJ-1140 analysis control unit (USA) 333
C 215 articulated telescopic sight (Italy) 365
C-601 anti-ship missile (China) 707
C-801 air-to-surface missile (China) 707
C801 coastal defence missile system (China) 160
C801 surface-to-surface anti-ship missile (China) 449
CAAIS 450 series fire control and action information system (UK) 431
CAC 101 field artillery computer (France) 68
CACS 1 (UK) 431
Cactus (SA) 175
Caiman noise/deception jamming pod 816
Caiquen airborne radar warning receiver (Chile) 815
CALIOPE thermal imager (France) 355
Calypso III submarine radar TRS 3100 (France) 606
Calypso IV submarine radar TRS 3110 (France) 606
CAM counter-mine charge (Italy) 521
Canada
airborne radar 772
aircraft armament 728
army fire control systems 64
electronic warfare equipment (sea) 646
electro-optical equipment (air) 867
land-based air defence 39
naval fire control systems 408
naval radar 600
sonar and underwater detection equipment 541
CANE/MAINS (UK) 438
Canopus (TRS906) lightweight electro-optical director (France) 678
Canopus naval weapon control system (France) 409
CAP underwater explosive charge (Italy) 521
Capitole fire control system (France) 352
Captor — encapsulated torpedo (USA) 537
Cardoen general-purpose bombs (Chile) 730
Cascade helicopter decoy system (UK) 834
Castor II fire control radars (France) 604
Castor thermal imaging sight (France) 354
CATRIN C31 system (Italy) 45
CATRIN communications system (Italy) 79
CB 130 cluster bomb (Chile) 729
CB 470 cluster bomb system (SA) 746
CB 500 cluster bomb (Chile) 729
CBAS-1 Albatros aircraft rocket (Argentina) 727
CCS-280 Command and Control Systems (DDH-280) Class (Canada) 408
CDC ballistic computer system (Canada) 64
CDC control and display unit (Canada) 65
CDC military communications calculator (Canada) 65
CDN 22 north seeker (France) 356
CE618 laser rangefinder (Germany, FR) 359
CE624 laser rangefinder/sight (Germany, FR) 359
CE626 laser transmitter (Germany, FR) 873
CE628 laser rangefinder (Germany, FR) 359
CE632 laser rangefinder (Germany, FR) 359
CEL 850/4 IFF interrogator (UK) 274
CEL 850/4 naval interrogator (UK) 627
CERES communications monitoring system (UK) 321
Chaff rocket (UK) 663
Chaparral surface-to-air weapon system (USA) 186
Cheetah artillery radar (China) 223
Chemring countermeasures (UK) 837
Chetz tactical surface-to-surface missile (Israel) 121
Chief EW combat system (USA) 345
Chile
aircraft armament 729
electronic warfare equipment (air) 815
ground radar 218
underwater warfare systems 513
China (People's Republic)
aircraft armament 730
air-to-air missiles 692
air-to-surface missiles 707
anti-tank/assault weapons 137
army fire control systems 66
battlefield support weapons 115
coastal defence weapons 160
electronic warfare equipment (air) 815
electronic warfare equipment (land) 306
electronic warfare equipment (sea) 647
electro-optical equipment (land) 349
ground radar 219
land-based air defence systems 40
mobile surface-to-air guided missile systems 167
naval guns and bombardment rockets 485
naval radar 600
portable anti-aircraft guided missiles 211
shipborne surface-to-air weapons 462
shipborne weapons 449
strategic weapon systems 2
underwater warfare systems 513
Ching Feng (Green Bee) surface-to-surface missile (Taiwan) 126
CH-TABO bomb dispersal pod (France) 739
CIDIS combat information display system (Netherlands) 421
Circe national secure iff systems (UK) 275
Circularly disposed antenna array (CDAA) (USA) 331
Class 122 frigate combat system (Germany, FR) 412
Classic (RGS 2740) (UK) 104
CMB8 mortar computer (Italy) 77
CMR-610A wideband DF receiver (USA) 337
CN2-500 series passive vehicle periscope (France) 353
C-NITE TOW missile sight (USA) 889
Coastal tracking radar (China) 222
Cobra 2000 anti-tank missile (Germany, FR) 138
Cobra Dane radar (USA) 276
COBRA weapon locating radar (International) 239
Coherent countermeasures processor (Canada) 647
COLDS laser detection system (International) 314
Combat artillery C^3 system (Israel) 74
Combat Grande air defence system (Spain) 49
Combat monitoring system (Germany, FR) 71
Combat vehicle sacs (USA) 346
COMKAFS (Outfit DCH) submarine combat system (UK) 433
Common jammer programme (USA) 343
Communications EW systems (CEWS) (USA) 339
Compact IFF/secondary radar (Spain) 250
Compact towed array sonar system (COMTASS) (UK) 571
Compass call ECM system (USA) 865
Computer sighting systems (CSS) (UK) 102
Condor commander's day/night sight (UK) 388
Condor tactical battlefield support missile (Argentina) 114
Co-news naval ESM system (Italy) 658
Contraves day/night remote video equipment (USA) 901
Control for Surface-Launched Torpedoes (CSLT) (UK) 435
Convertible laser designation pod (CLDP) (France) 872
Copperhead CLGP (M712) (USA) 158
Cormorant lightweight dipping sonar (UK) 568
Corvus ELINT station (UK) 327
Cossor monopulse secondary surveillance radar systems (UK) 275
COTAC APX M401 fire control system (France) 66
COTAC APX M581 fire control system (France) 66
Cotol optronic fire control system (France) 680
Cotrol optronic fire control system (France) 681
Covert strike radar (USA) 814
CR-2700 mobile ELINT system (Israel) 317
CR-2740 ELINT system (Israel) 316
CR-2800 ESM/ELINT system (Israel) 824
Cross Sword fire control radar (USSR) 618

Crotale low-altitude ground-to-air weapon system (TSE 5000) (France) ... 169
Crotale naval surface-to-air weapon system (TSE 5500) (France) ... 463
Crouzet Type 200 helicopter sight (France) ... 872
CRS 512 SSR antenna (UK) ... 276
Cruise missile advanced guidance (CMAG) (USA) ... 903
Cruise missile programme (USA) ... 27
CRV7 aircraft rocket ... 728
CS-AX-? point defence missile (China) ... 462
CSS-1 MRBM (China) ... 2
CSS-2 IRBM (China) ... 2
CSS-3 ICBM (China) ... 2
CSS-4 ICBM (China) ... 3
CSS-N-3 submarine-launched ballistic missile (China) ... 3
CSS-NX-4 submarine-launched ballistic missile (China) ... 4
Cutlass ESM/ECM equipment (UK) ... 667
CY-1 anti-submarine missile (China) ... 513
Cygnus ECM system (UK) ... 668
Cymbeline mortar locating radar (UK) ... 262
Cyrano IV attack radar (France) ... 776
Cyrano IVM attack radar (France) ... 776
Cyrano radar family (France) ... 776
Czechoslovakia
battlefield support weapons ... 118

D

D1800 DF and monitoring systems (USA) ... 344
DA.05 naval surveillance radar (Netherlands) ... 612
DA.08 naval surveillance radar (Netherlands) ... 613
Dagaie naval countermeasures equipment (France) ... 647
Daisy and Foresee systems (Netherlands) ... 421
Daldo target acquisition aid system (France) ... 358
Dardo close-in weapon system (Italy) ... 470
Dardo fire control system (Italy) ... 415
Dash (display and sight helmet system) (Israel) ... 875
DAV search and alert radar (France) ... 777
DAVID artillery battery computer (Israel) ... 75
DB 3141 jamming pod (low-band Remora) (France) ... 816
DB 3163 jamming pod (high-band Remora) (France) ... 816
DDM IR detector (France) ... 820
DE1164 sonar (USA) ... 588
DE1167 modernisation sonar (USA) ... 588
DE1191 modernisation sonar (USA) ... 589
Deadeye 5-inch guided projectile (USA) ... 461
DEFA 552A 30mm aircraft gun (France) ... 734
DEFA 553 30mm aircraft gun (France) ... 734
DEFA 554 30mm aircraft gun (France) ... 734
Denmark
electro-optical equipment (land) ... 349
electro-optical equipment (sea) ... 677
ground radar ... 225
naval radar ... 601
Depth compensated ring shell projector (Canada) ... 545
DES optical fire director (UK) ... 685
D-ESM shipborne receiver (USA) ... 675
DG naval fire control system (Israel) ... 414
Digital tactical system (DITACS) radar (USA) ... 806
Direct airfield attack combined munition (DAACM) ... 758
Direct coupled jamming/surveillance system (USA) ... 675
Director aiming sight (UK) ... 688
Director-mounted thermal imaging system (DIMTIS) (UK) 687
DLA anti-submarine fire control system (France) ... 410
DLT-D3 submarine torpedo fire control system (France) .. 410
Dog House ABM radar (USSR) ... 254
Dolphin 1000 series submarine sonars (UK) ... 574
Door gunner post (DGP) (Italy) ... 744
DR 810 Mk V muzzle velocimeter (USA) ... 304
DR2000 IFM ESM receiver (France) ... 650
DR4000 IFM ESM receiver (TMV 201) (France) ... 650
Dragon anti-tank/assault weapon (FGM-77A/FTM-77A) (USA) ... 154
Dragonfish mine (UK) ... 531
Dragonfly airborne vhf/uhf EW System (USA) ... 865
DRBC 32 naval radar (France) ... 601
DRBC 33 naval radar (France) ... 601
DRBI 10 3-D naval radar (France) ... 601
DRBI 23 3-D naval radar (France) ... 601
DRBJ 11 naval radar (France) ... 602
DRBR 51 naval radar (France) ... 602
DRBV 13 naval radar (France) ... 602
DRBV 20 naval radar (France) ... 602
DRBV 22 naval radar (France) ... 602
DRBV 23 naval surveillance radar (France) ... 603
Driver's periscope Type BM 8005 (Germany, FR) ... 361
Drum Tilt fire control radar (USSR) ... 618
DS 25 series naval gun mounting (UK) ... 504
DS 30 series naval gun mounting (UK) ... 504
DS-35 (SPIRTAS) target acquisition system (Israel) ... 682
DSBV 61A towed array sonar (France) ... 550
DSQS 21 series sonars (Germany, FR) ... 555
Dual band radar (USA) ... 296
DUAV-4 helicopter sonar (TSM 8251) (France) ... 549
DUBA 25 (TSM 2400) sonar (France) ... 551
DUBM 41B side-looking sonar (France) ... 553
DUBM 42 side-looking sonar (France) ... 554
DUBV-23D surface vessel sonar (France) ... 547
DUBV-43B/C variable depth sonar (France) ... 547
'Duke' class (Type 23) frigate command system (UK) ... 431
Durandal penetration bomb (France) ... 737
DUUA-2A active/passive sonar (France) ... 548
DUUA-2B active/passive sonar (France) ... 548

E

E-3 (AWACS) radar AN/APY-2 (USA) ... 798
E14 acoustic torpedo (France) ... 514
E15 acoustic torpedo (France) ... 514
E2030 night driverscope (South Africa) ... 376
E5010 laser rangefinder (South Africa) ... 376
Eagle coastal radar (China) ... 222
Eagle Eye III electro-optic sensor (USA) ... 890
Eagle Eye anti-aircraft fire control system (Israel) ... 74
Eagle fire control radar (Sweden) ... 251
ECA 38 (PAP 104) mine disposal system (France) ... 516
Eclipse chaff/IR decoy launcher (Chile) ... 815
ECP explosive cutter device (Italy) ... 523
EDO Model 610 long range hull-mounted scanning radar (USA) ... 589
EDO Model 700 series sonars (USA) ... 590
EDO Model 780 series sonar (USA) ... 590
EDO Model 900 underwater detection system (USA) ... 590
EDO Model 910 underwater detection system (USA) ... 590
EDO Model 1102/1105 submarine sonar (USA) ... 591
EFA radar (International) ... 779
EFCS-2 tank fire control system (Yugoslavia) ... 407
EFCS 600 (UK) ... 101
Egypt
battlefield support weapons ... 118
land-based air defence systems ... 35
portable anti-aircraft guided missiles ... 212
Elbit action information system (Israel) ... 413
Elbit fire control system (Israel) ... 76
Elbit MN-53 ESM system (Israel) ... 653
Elbit night vision head-up display (Israel) ... 874
Electronic warfare equipment tables ... 947
Eledone submarine sonar (France) ... 548
Elettronica EW equipment (Italy) ... 657
EL/K 7032 COMINT system (Israel) ... 821
EL/L-8202 ECM pod (Israel) ... 821
EL/L-8230 airborne ECM system (Israel) ... 822
EL/L-8231 ECM system (Israel) ... 822
EL/L 8240 ECM system (Israel) ... 822
EL/L-8300 airborne SIGINT system (Israel) ... 822
EL/L 8312A ELINT system (Israel) ... 822
EL/M-2001B radar (Israel) ... 779
EL/M-2011 fire control radar (Israel) ... 780
EL/M-2021 radar (Israel) ... 780
EL/M-2032 fire control radar (Israel) ... 780
EL/M-2035 multimode fire control radar (Israel) ... 780
EL/M 2121 battlefield surveillance radar (Israel) ... 240
EL/M-2207 naval radar Elta (Israel) ... 607
EL/M-2208 shipborne search radar (Israel) ... 608
EL/M-2215 surveillance radar (Israel) ... 241
EL/M-2216(V) coastal surveillance radar (Israel) ... 241
EL/M-2220 mobile surveillance radar (Israel) ... 241
EL/M-2221 naval fire control radar (Israel) ... 607
Elma ASW grenade system (Sweden) ... 524
ELOP head-up displays (Israel) ... 875
ELOP mini-laser tank rangefinder (Israel) ... 364
ELOP tank laser rangefinder (Israel) ... 364
EL/S-8811/2 tactical intelligent systems (Israel) ... 315
ELT/128 communications direction finder (Italy) ... 317
ELT/156 radar warning receiver (Italy) ... 825
ELT/211 naval ESM equipment (Italy) ... 655
ELT/263 airborne ESM equipment (Italy) ... 825
ELT/318 naval jammer (Italy) ... 656
ELT/521 naval deception jammer (Italy) ... 656
ELT/555 jammer pod (Italy) ... 826
ELT/562 deception jammer (Italy) ... 827
ELT Colibri helicopter ESM/ECM system (Italy) ... 827
ELT noise jammer pods (Italy) ... 826
Elta ground-based EW systems (Israel) ... 314
Elta naval EW systems (Israel) ... 653
Elta surface radars (Israel) ... 242
Eltro electro-optical equipment (Germany, FR) ... 360
Emerlec-30 twin 30mm mounting (USA) ... 508
EMPAR-European multifunction phased array radar (International) ... 606
EMVA Mk 10B muzzle velocity analyser (South Africa) ... 87
Enhanced radar warning system (USA) ... 859
ENOSA electro-optical systems (Spain) ... 376
EN/TSD-500 series COMINT systems (Spain) ... 319
EOS-400 fire director system (Sweden) ... 429
EP 1650 automatic VHF/UHF radio reconnaissance system (Germany, FR) ... 309
EPLO tactical data system (Sweden) ... 427
ERIJAMMER 39 (Sweden) ... 831
ERIJAMMER A 100 (Sweden) ... 831
ERIJAMMER 200 (Sweden) ... 831
Ericsson AEW radar (Sweden) ... 784
Ericsson anti-aircraft laser rangefinder (Sweden) ... 377
Ericsson artillery laser rangefinder (Sweden) ... 377
Ericsson coast artillery laser rangefinder (Sweden) ... 378
Ericsson command and control information system (CCIS) (Sweden) ... 427
Ericsson laser rangefinder (Sweden) ... 877
Ericsson laser tracker (Sweden) ... 378
Ericsson SLAR (Sweden) ... 783
Ericsson submarine action information and fire control system (Sweden) ... 430
Ericsson thermal imager (Sweden) ... 378
Eryx anti-tank missile (France) ... 138
ES-250 standoff electro-optical reconnaissance system (USA) ... 902
Escort ESM system (USA) ... 864
ESD 3300 IFF/SIF transponder (France) ... 778
ESD 3400 IFF/SIF transponder (France) ... 778
ESD LAD off-board decoy system (France) ... 650
ESM 500 series VHF/UHF receivers (Germany, FR) ... 311
ESM 1000 system (Germany, FR) ... 311
ESP scanning receiver (Germany, FR) ... 311
ESS-2 electronic surveillance system (Italy) ... 318
ETAS target acquisition system (USA) ... 305
Evade airborne decoy system (UK) ... 834
EW-1017 electronic surveillance system (USA) ... 862
EW equipment (Japan) ... 828
EW programme (Spain) ... 829
EWS-900 chaff launching equipment (Sweden) ... 660
EWS-905 ESM warning system (Sweden) ... 661
Excalibur coast defence missile system (UK) ... 165
Exocet coastal defence missile system (France) ... 161
EXPAL BME 330 cluster bomb (Spain) ... 747
EXPAL BRFA 330 bomb (Spain) ... 747
EXPAL BRP retarded bombs (Spain) ... 748
EXPAL general-purpose low drag bombs (Spain) ... 748
Eye Bowl naval radar (USSR) ... 618

F

F-5 E/F head-up display weapon aiming computer (HUDWAC) (UK) ... 878
F-16 A/B HUD sight (UK) ... 878
F-16 C/D HUD sight (UK) ... 878
F-16 LANTIRN holographic HUD sight (UK) ... 878
F-17 torpedo (France) ... 514
F100-L fire control predictor (UK) ... 439
F/A-18 FLIR (AN/AAS-38) (USA) ... 840,899
FACE–field artillery computer equipment (UK) ... 100
Fairchild Weston ECM payload packages (USA) ... 863
FALCO submarine locating system (Italy) ... 558
Falcon (AIM-4D) air-to-air missile (USA) ... 703
Falcon series surveillance radar (USA) ... 292
Falconer precision approach radar (UK) ... 263
Falke artillery computer system (Germany, FR) ... 71
FAMAE 3.5 kg anti-personnel bomb (Chile) ... 729
FAMAE general purpose low-drag bombs (Chile) ... 729
Fan Song B missile control radar (USSR) ... 254
Fan Song E naval radar (USSR) ... 619
Fan Song E missile control radar (USSR) ... 254
Fan Song missile control radars (USSR) ... 254
FARAD naval EW system (Italy) ... 657
Fast jam helicopter ECM system (Italy) ... 827
Fastar battlefield surveillance radar (UK) ... 273
Fathom sonar handling equipment (Canada) ... 545
Felis anti-aircraft fire control system (Spain) ... 88
Fenelon-DUUX-5 passive sonar and acoustic rangefinder (France) ... 549
Ferranti 500 series command and control systems (UK) ... 434
Ferranti FD 5000 periscope video camera (UK) ... 689
Ferranti FINAS (UK) ... 883
Ferranti fire control system (UK) ... 103
Ferranti helmet pointing system (UK) ... 386,885
Ferranti integrated combat suite (UK) ... 433
Ferranti laser ranger and marked target seeker (UK) ... 884
Ferranti laser warning receiver (UK) ... 836
Ferranti lightweight electro-optic multisensor (UK) ... 688
Ferranti mine countermeasures control system (MCCS) (UK) ... 439
Ferranti multi-role naval lasers (UK) ... 688
Ferret submarine EW system (USA) ... 671
FERRIMAR WB-500-F Avispa cluster bomb system (Chile) 730
Feu 80 electric multi-purpose bomb fuze (France) ... 735
FFV 30mm Aden gun pod (Sweden) ... 748
FFV Uni-pod 0127 (Sweden) ... 748
FG1 sea mine (Germany, FR) ... 519
Fibre-optic laser warning system (USA) ... 865
Fieldguard artillery fire direction system (Switzerland) ... 93
FIN 1064 (UK) ... 882
Finland
electro-optical equipment (land) ... 350
sonar and underwater equipment ... 546
Fire can fire control radar (USSR) ... 254
Fire wheel fire control radar (USSR) ... 255
FIROS 6 51mm field rocket system (Italy) ... 122
FIROS 25/30 122 mm field rocket system (Italy) ... 123
Fiskars hydroacoustic surveillance system (Finland) ... 546
FIST artillery system (Netherlands) ... 82
FL-1 (CSS-N-2) naval surface-to-surface missile (China) .. 449
FL-7 anti-ship missile (China) ... 450
FL 400 series EW suites (Germany, FR) ... 652
FL1800S naval ESM/ECM system (Germany, FR) ... 652
Flakpanzer 35mm AA tanks (Switzerland) ... 203
Flanking-beam array switching technique radar (FAST) ... 644
Flap Wheel fire control radar (USSR) ... 255
Flat Face target acquisition radar (USSR) ... 255
Flat Jack AWACS radar (USSR) ... 784
Flat Twin radar (USSR) ... 255
FLER-H tank fire control computer (Germany, FR) ... 73
FLEXAR weapon control system (USA) ... 643
FLORIDA air defence command control system (Switzerland) ... 50
FLP 10 tank fire control system (Germany, FR) ... 73
Flycatcher radar (Netherlands) ... 248
Flycatcher AA fire control system (Netherlands) ... 82
FMS series sonar systems (UK) ... 572
FOG-M missile guidance system (USA) ... 405
Folgore anti-tank weapon (Italy) ... 144
Forward area air defense system (USA) ... 206
Forward observer laser designator (FOLD) (UK) ... 385
Fourth generation intercontinental ballistic system (USSR) .. 8
Foxhunter airborne interceptor radar (UK) ... 786
Framing thermal imager (UK) ... 880
France
air-to-air missiles ... 693
air-to-surface missiles ... 707
airborne radar ... 772
aircraft armament ... 731
anti-tank/assault weapons ... 137
army fire control systems ... 66
battlefield support weapons ... 119
coastal defence weapons ... 161
electronic warfare equipment (air) ... 816
electronic warfare equipment (land) ... 306
electronic warfare equipment (sea) ... 647
electro-optical equipment (air) ... 867
electro-optical equipment (land) ... 350
electro-optical equipment (sea) ... 677
ground radar ... 225
land-based air defence systems ... 40
mobile guns and rocket systems ... 192
mobile surface-to-air guided missile systems ... 168
naval fire control systems ... 408
naval radar ... 601
portable anti-aircraft guided missiles ... 212
shipborne surface-to-air weapons ... 462
shipborne weapons ... 450
sonar and underwater detection equipment ... 546
strategic weapon systems ... 4
underwater warfare systems ... 513
FRAS-1 anti-submarine weapon (USSR) ... 526
Frog-7 battlefield support missile (USSR) ... 128
Front series radars (USSR) ... 619

Photo: The Bushmaster's lethal performance against one-inch armor plate.

G

G 733 SLUTT (UK) 564
G 738 towed decoy (UK) 564
G 740 SACACS (UK) 564
G 750 sonar (UK) 564
GA1 20 mm automatic cannon (SA) 746
GA 082 FSK analyser (Germany, FR) 313
GA 250/K antenna (Sweden) 252
GA 250/X antenna (Sweden) 252
Gabriel long range anti-ship missile (Israel) 453
Gabriel Mk III air-launched anti-ship missile (Israel) 712
Gabriel shipborne surface-to-surface missile (Israel) 452
Gage acquisition radars (USSR) 255
GapGate early warning radar (Poland) 249
GAU-8/A 30mm gun system (USA) 759
GAU-12/U Equaliser 25mm automatic gun (USA) 760
GAU-13/A 30mm lightweight gun (USA) 760
GBU-15 modular glide bomb (USA) 757
GE 592 transportable air defence radar (USA) 295
GEADGE (Germany, FR) 43
GEC Sensors CCD cameras (UK) 881
GEC Sensors FLIR systems (UK) 879
GEC Sensors V300 series CCTV equipment (UK) 687
GECAL 50 aircraft machine guns (USA) 766
GECAL 50 composite turret weapon system (USA) 766
GEMAG 25 AA gun system (USA) 209
Gemini naval fire control system (Netherlands) 420
GENE-X expendable cartridge (USA) 841
General Electric 7.62mm minigun (USA) 767
General Electric 20mm Vulcan gun system for AMX fighter (USA) 763
General Electric 25mm lightweight gun GE 225 (USA) 762
General Electric Air Defence Turrets (USA) 207
General Electric Vulcan 20mm aircraft gun and linkless feed system (USA) 762
Germany, FR
air-to-surface missiles 709
aircraft armament 741
anti-tank/assault weapons 138
army fire control systems 71
battlefield support weapons 120
electronic warfare equipment (air) 821
electronic warfare equipment (land) 308
electronic warfare equipment (sea) 651
electro-optical equipment (air) 873
electro-optical equipment (land) 359
ground radar 236
land-based air defence systems 42
mobile gun and rocket systems 193
naval fire control systems 411
naval guns and bombardment rockets 488
sonar and underwater detection equipment 555
underwater warfare systems 517
GETAS passive surveillance sonars (UK) 574
Gipsy submarine data handling and weapon control system (Netherlands) 423
Giraffe search radar (Sweden) 250
Goa shipborne surface-to-air missile (SA-N-1) (USSR) 472
Goalkeeper anti-missile close-in weapon (International) 467
Goalkeeper naval FCS (Netherlands) 422
Goblet shipborne surface-to-air missile (SA-N-3) (USSR) 472
GPU-2/A 20mm lightweight gun pod (USA) 760
GPU-5/A lightweight 30mm gun pod (USA) 761
Grail SA-7 air-to-air missile (USSR) 701
Greece
mobile gun and rocket systems 195
Green Baron pod (Sweden) 877
Grifo airborne radar (Italy) 781
Ground launched cruise missile (GLCM) (USA) 130
GS20 hostile fire locator Claribel radar (UK) 263
GS 7110 multi-channel vlf/lf receiver (UK) 574
GSA series gunsights (UK) 385
GSQ-250K remotely monitored sensor system (Korea, Republic) 81
GT-1 chaff/IR flare dispenser (China) 815
GTE computer controlled receiving systems (USA) 332
GTE EW antennas (USA) 329
Guardian naval EW system 666
Guardian star shipborne EW system (USA) 672
Guardrail airborne SIGINT system (AN/USD-9(V)2) (USA) 865
Guardsman coastal defence radar (UK) 268
Guardsman defence radar (UK) 268
Guideline shipborne surface-to-air missile (SA-N-2) (USA) 472
Guillemot dipping sonar (UK) 568
Gun Dish fire control radar (USSR) 255
Gun display system (Israel) 76
Gun King sighting system (Switzerland) 95
GUNZEN artillery fire data computer (UK) 98
Gyro-stabilised sighting system (GSSS) (Israel) 362, 682

H

H-930 series combat/weapon control systems (USA) 444
HADES tactical surface-to-surface missile (France) 119
HALO helicopter avionics (Israel) 875
Hand-emplaced expendable jammer (USA) 346
Hard 3-D search radar (Sweden) 251
Harpoon shipborne surface-to-surface missile (RGM-84A) (USA) 460
Hawk screech fire control radar (USSR) 619
Hawk surface-to-air missile (USA) 187
HBS 202 helicopter armament system (Germany, FR) 741
Head Light group (USSR) 619
Head Net A air surveillance radar (USSR) 620
Head Net B air surveillance radar (USSR) 620
Head Net C air surveillance radar (USSR) 620
Head-Up display for Harrier and Jaguar (UK) 881
Head-Up display for Sea Harrier (UK) 882
Heightfinding radars (USSR) 256
HELAS esm system (Germany, FR) 308
Helicopter gun mounting 19A (France) 733
Helicopter infra-red navigation system (HIRNS) (USA) 893
Helicopter laser designator (HELD) (UK) 886
Helicopter self screening launcher (Sweden) 877
Helios helicopter observation system (Sweden) 876
Hellfire modular missile (USA) 153
Helras airborne sonar (International) 556
Hen House ABM radar (USSR) 256
Hen series early warning radars (USSR) 256
Hera airborne radar (Sweden) 784
Hermes ESM system (UK) 833
HEROS army command, control and information system (Germany, FR) 71
HGS-55 7.62mm armament sub-system (USA) 768
HHI-8 hand-held thermal imager (UK) 387
High definition scanning sonar (UK) 573
High Lark A1 radar (USSR) 784
HIPAR (high power acquisition radar) (USA) 300
Hispano Oerlikon 20mm tactical air armament (Switzerland) 751
HN-5A portable anti-aircraft missile system (China) 211
HN-401R mobile surveillance radar (China) 221
Hofin hostile fire indicator (UK) 836
Honest John battlefield support missile (USA) 132
Honeywell AFV thermal sights (USA) 405
Honeywell FLIR systems (USA) 892
Honeywell MNS (USA) 539
Honeywell visually coupled systems (USA) 888
HOT air-to-surface missile (International) 711
HOT anti-tank missile (International) 140
HOWLS experimental airborne radar (USA) 811
HQ-2B air defence missile (China) 167
HQ-2J air defence missile (China) 167
HQ-61 surface-to-air missile system (China) 167
HRLR-04 AAC laser rangefinder (Israel) 365
HS 12 active/passive helicopter sonar (TSM 8252) (France) 550
HS-1000 series sonar (Canada) 541
HS 3125S ASW helicopter sonar (France) 550
Hsiung Feng anti-ship missile (Taiwan) 455
Hsiung Feng coastal defence missile (Taiwan) 163
HSS-1 harbour surveillance radar (Italy) 243
HSS-15 compact surveillance sonar (Netherlands) 559
HUD sight repeater (France) 871
Hughes Air Defence Radar (HADR) (USA) 300
Hughes ANVIS (USA) 902
Hughes head-up display (USA) 901
Hughes lightweight rocket launcher M260 (USA) 767
Hughes low altitude surveillance radar (USA) 288
HUMMEL vhf jammer (Germany, FR) 309
Hunting area denial system (HADES) (UK) 755
Hunting direction finding system (UK) 325
Hunting dog radar (China) 224
HWR-2 radar warning receiver (UK) 668
HY-2 coast defence missile system 160
HY-4 air-to-surface missile (China) 707
HY-4 coast defence missile system (China) 160

I

IBIS minehunting sonar systems (France) 554
ICM (USA) 157
Idra air defence missile (Italy) 174
IFCS-improved (tank) fire control system (UK) 102
IFF 2720 airborne transponder (UK) 789
IFF 2743 electronic control unit (UK) 789
IFF 2750 (Type 1013) naval transponder (UK) 627
IFF 3100 airborne transponder (UK) 789
IFF 3500 airborne interrogator (UK) 790
IFF 4700 series airborne transponders (UK) 790
IGS-1/IGS-3P mobile integrated ESM/ECM system (Italy) 318
Iguane airborne radar (France) 773
Ikara anti-submarine weapon system (Australia) 512
IMICS (Integrated mine-hunting combat system) (Italy) 418
India
anti-tank/assault weapons 140
battlefield support weapons 121
ground radar 239
land-based air defence systems 35
mobile surface-to-air guided missile systems 170
Indonesia
land-based air defence systems 35
Indra air defence radars (India) 239
Infra-red scanning seeker head (UK) 394
INS-3 naval ESM/ECM system (Italy) 659
Integrated electronic warfare system (USA) 863
Integrated fire and maneuver control system (Canada) 64
Integrated laser for main battle tanks (Sweden) 377
Integrated radar system for 35mm AA tank (Netherlands) 84
Integrated tank fire control system (USA) 105
Integrated tracking and fire control system for Wildcat (Netherlands) 83
Intelligent terminal (USA) 106
International
air-to-air missiles 695
air-to-surface missiles 710
airborne radar 778
anti-tank/assault weapons 140
army fire control systems 74
coastal defence weapons 161
electronic warfare equipment (air) 821
electronic warfare equipment (land) 314
electronic warfare equipment (sea) 652
electro-optical equipment (land) 361
ground radar 239
mobile surface-to-air guided missile systems 171
naval fire control systems 412
naval radar 606
shipborne surface-to-air weapons 466
shipborne weapons 451
sonar and underwater detection equipment 556
underwater warfare systems 519
IP64-MD64 sonars (Italy) 558
IPD70/S integrated sonar (Italy) 557
IPN tactical data display systems (Italy) 419
IR18 Mk II thermal scanner (UK) 381
IR26 thermal imaging sensor (UK) 382
IRCM self-defense system (USA) 864
IRCOM (UK) 687
IRGO hand-held thermal imager (France) 357
IRIS integrated radar imaging system (Canada) 772
IRS-10 secondary surveillance radar (Spain) 249
IRS-20MP monopulse secondary surveillance radar (Spain) 250
IRSCAN infra-red surveillance system (Netherlands) 684
IRSTS (airborne infra-red search and track system) (USA) 900
IRVAT (infra-red video automatic tracker) (USA) 889
ISC Cardion coastal radar system (USA) 303
ISCB-1 area denial cluster weapon (USA) 756
ISIS sigint systems (Italy) 317
ISIS weapon aiming system (UK) 883
Israel
air-to-air missiles 696
air-to-surface missiles 712
airborne radar 779
aircraft armament 742
anti-tank/assault weapons 144
army fire control systems 74
battlefield support weapons 121
electronic warfare equipment (air) 821
electronic warfare equipment (land) 314
electronic warfare equipment (sea) 652
electro-optical equipment (air) 873
electro-optical equipment (land) 362
electro-optical equipment (sea) 682
ground radar 240
land-based air defence systems 44
mobile gun and rocket systems 196
mobile surface-to-air guided missile systems 173
naval fire control systems 413
naval radar 607
shipborne surface-to-air weapons 468
shipborne weapons 452
underwater weapon systems 520
Italy
air-to-air missiles 697
air-to-surface missiles 713
airborne radar 781
aircraft armament 743
anti-tank/assault weapons 144
army fire control systems 77
battlefield support weapons 122
electronic warfare equipment (air) 825
electronic warfare equipment (land) 317
electronic warfare equipment (sea) 655
electro-optical equipment (air) 876
electro-optical equipment (land) 365
electro-optical equipment (sea) 682
ground radar 242
land-based air defence systems 45
mobile gun and rocket systems 197
mobile surface-to-air guided missile systems 173
naval fire control systems 414
naval guns and bombardment rockets 488
naval radar 608
shipborne surface-to-air weapons 469
sonar and underwater detection equipment 557
underwater warfare systems 520
ITATA airborne elint system (Chile) 815
ITT ANVIS (USA) 902

J

J-3400 mobile ECM system (USA) 330
J-3410 mobile ECM system (USA) 330
JAMCAT jamming unit (UK) 325
Jam Pac (V) ECM jammer system (USA) 863
Janet naval jammer (France) 650
JAN/MPQ-N1 (Type 92) mortar locating radar (Japan) 247
Japan
airborne radar 782
air-to-air missiles 697
air-to-surface missiles 714
anti-tank/assault weapons 145
battlefield support weapons 124
coastal defence weapons 161
electronic warfare equipment (air) 828
electronic warfare equipment (land) 318
electronic warfare equipment (sea) 660
ground radar 247
land-based air defence systems 46
mobile gun and rocket systems 199
mobile surface-to-air guided missile systems 174
naval radar 611
portable anti-aircraft guided missiles 213
sonar and underwater detection equipment 559
JAS-39 airborne radar (Sweden) 784
Javelin man-portable anti-aircraft missile (UK) 215
Jay Bird airborne radar (USSR) 785
Jezebel sonobuoy (UK) 574
Jindalee OTH radar (Australia) 218
JLG-43 heightfinding radar (China) 222
JLP-40 surveillance radar (China) 220
J/MPQ-P7 artillery locating radar (Japan) 247
Joint surveillance system (USA) 58
Joint surveillance and target attack radar system (USA) 814
Jordan
land-based air defence systems 35
JP233 low altitude airfield attack system (UK) 755
Jupiter naval air surveillance radar TRS 3010 (France) 605
JY-8 mobile tactical 3-D radar system (China) 219
JY-9 low-altitude search radar (China) 219
JY-I0RIPP radar information processing post (China) 220

K

KAFS submarine AIO and FCS (UK) 432
KAM-3D anti-tank missile (Type 64) (Japan) 145

KAM-9 (Type 79) battlefield missile (Japan) 145
Kanaris fire control system (International) 412
Kangaroo (AS-3) air-to-surface missile (USSR) 716
Karrawarra towed array sonar (Australia) 541
KATIE airborne radar warning system (UK) 835
Kelt (AS-5) air-to-surface missile (USSR) 715
Kent (AS-15) air-launched cruise missile (USSR) 718
Kenya
land-based air defence systems 35
Kestrel airborne ESM (UK) 835
Kingfish (AS-6) air-to-surface missile (USSR) 717
Kipper (AS-2) air-to-surface missile (USSR) 716
Kitchen (AS-4) air-to-surface missile (USSR) 717
Kite Screech fire control radar (USSR) 620
KJ 8602 radar warning receiver (China) 816
Klein MCM sonar systems (USA) 598
KMT 400 multi-purpose military terminal (Norway) 85
Knife Rest A early warning radar (USSR) 256
Knife Rest B early warning radar (USSR) 256
Knife Rest C early warning radar (USSR) 256
Kollmorgen electro-optical systems (USA) 399
Kollmorgen Model 76 periscope (USA) 690
Korea (Republic of)
army fire control systems 80
Kormoran air-to-surface anti-ship missile (Germany, FR) 709
Krupp Atlas-Elektronik/MCM systems (Germany, FR) 556
Krupp Atlas-Elektronik submarine sonars (Germany, FR) 556
Kukri air-to-air missile (SA) 698
Kung Feng III/IV 126mm multiple rocket system (Taiwan) 126
Kun-Wu anti-tank missile (Taiwan) 148
Kuwait
land-based air defence systems 35
KV-SK/20 dual-purpose 20mm naval mounting (Norway) 494

L

L1A2 artillery director (UK) 386
L3 acoustic torpedo (France) 515
L4 airborne acoustic torpedo (France) 515
L5 multi-purpose torpedo (France) 515
L-6/A-WJ-1140 microwave direction finding system (USA) 334
L20 laser rangefinder sight (UK) 393
L/46 single 120mm gun mounting (Sweden) 495
L50 conversion unit (UK) 394
L/70 Mk 1 57mm single gun mounting (Sweden) 495
LAADS Low Altitude Aircraft Detection System (USA) 288
LAAT sight (USA) 889
Lacroix airborne countermeasures ammunition (France) 820
Lacroix naval countermeasures (France) 651
LAMPS Mk III (USA) 532
Lance battlefield support missile (USA) 131
Lancelot tank FCS (Israel) 76
Lanco RC-06-100 (Spain) 747
Land-based communications EW systems (France) 306
Land-based EW systems (USSR) 319
Land-based radar tables 919
LANTIRN (USA) 895
LAR 160 160 mm light artillery rocket system (Israel) 122
LAR 200 local area radar (UK) 271
LARS 110mm artillery rocket launcher (Germany, FR) 120
Laser detector-tracker/strike camera (LDT/SCAM) (USA) 902
Laser irradiation detector (Yugoslavia) 347
LASR-2 radar (International) 778
Launching systems for lightweight torpedoes (UK) 436
LAW 80 light anti-armour weapon (UK) 151
LEMSTAR tank fire control system (Germany, FR) 72
LF 19 laser rangefinder (UK) 3822
LGA 389 artillery goniometer sight (France) 359
Libya
land-based air defence systems 35
Light torpedo (USSR) 527
Lightweight TACFIRE (USA) 107
Lightweight target acquisition/surveillance system (LTASS) (USA) 399
Limbo anti-submarine mortar system (AS Mk 10) (UK) 528
Limpet mine disposal equipment (LMDE) (UK) 532
LINDA C^{3}I artillery system (Sweden) 88
LIOD lightweight optronic fire control system (Netherlands) 683
LIROD 8 naval director (Netherlands) 420
Litening infra-red targeting and navigation pod (Israel) 876
Litton mast-mounted sight LRF/D (USA) 891
Long Talk recovery radar (USSR) 256
Long Track surveillance radar (USSR) 257
Look Two airborne radar (USSR) 785
Loral air inflatable bomb retarders - Ballutes (USA) 770
Loral infra-red countermeasures (USA) 863
Louxor air defence radar (France) 226
Low Blow tracking and missile control radar (USSR) 257
Low Cost Weapon Delivery System (LCWDS) (Israel) 873
LN66/HP radar system (Canada) 772
LN66 radar series (Canada) 600
LP3 laser rangefinder (Norway) 374
LP3 laser rangefinder (UK) 380
LP6 laser rangefinder (UK) 380
LP8 laser rangefinder (Norway) 374
LP100 laser rangefinder (Norway) 375
LP160 laser rangefinder (Norway) 374
LPD-20 search radar (Italy) 243
LQ-102A hand-emplaced expendable jammer (USA) 346
LR-5200 tactical ESM system (USA) 865
LRS 4 Mk II fire control system (Belgium) 348
LRS 5 fire control system (Belgium) 348
LRS 7 fire control system (Belgium) 348
LU250 EG-FT bomb (France) 740
Lundy chaff and flares countermeasures system (USA) 842
Lundy countermeasures (USA) 838
LV2 laser rangefinder (UK) 380
LV3 laser rangefinder (UK) 380
LV5 laser rangefinder (Norway) 374
LV350 series laser rangefinders (Norway) 375
LW.02 naval air surveillance radar (Netherlands) 613
LW.08 naval surveillance and weapon direction radar (Netherlands) 612
LWDR lightweight designator/rangefinder (USA) 396
LWS-20 laser warning system (Israel) 823
Lynx optical sight (France) 678

M

M1 laser rangefinder (USA) 400
M3-VDA 20mm AA gun system (France) 192
M5 aircraft armament system (USA) 769
M8 submarine torpedo fire control system (Netherlands) 423
M16 armament sub-system (USA) 770
M19A1 driver's night periscope (USA) 400
M20 series fire control systems (Netherlands) 420
M21 MAMEE helicopter armament system (USA) 765
M23 armament sub-system (USA) 770
M24 armament sub-system (USA) 770
M28A1 armament system (USA) 766
M32E1 Model 9880 night vision elbow (USA) 398
M40 series fire control systems (Netherlands) 423
M41 armament system (USA) 770
M60A3 laser tank fire control system (USA) 401
M61/M61A1 fire control sight (USA) 400
M-63 128mm rocket launcher (Yugoslavia) 135
M-130 countermeasures dispenser (USA) 841
M156 armament mount (USA) 764
M197 20mm three-barrel aircraft gun (USA) 765
M230 30mm chain gun weapon (USA) 769
M389 panoramic sight (France) 359
M391 episcopic sight (France) 353
M590 gyrostabilised sight (France) 359
M.621 20mm cannon pod (France) 734
M6043 fire control director sight (USA) 400
MA1122 tactical df processing and display unit (UK) 325
MAA-1 air-to-air missile (Brazil) 692
MACE command and control system (UK) 55
MAD Mk III magnetic anomaly detector (France) 555
Madis sighting and drive system (Italy) 367
MAGAIE lightweight antimissile decoy system (France) 648
Malafon surface-to-sub-surface missile (France) 513
Malaysia
land-based air defence systems 35
MAMBA artillery and mortar ballistic aid (UK) 99
MAMBA portable anti-tank weapon system (Germany, FR) 139
Man-portable surface-to-air missile (Japan) 213
Manta mine (Italy) 521
Manta submarine ESM system (UK) 665
MAPATS anti-tank weapon (Israel) 144
Marcal gun calibration system/Mogul (UK) 98
Marconi 800 series radars (land-based) (UK) 264
Marconi/BAe lightweight gun fire control system (Sapphire) (UK) 439
Marconi DFCS (UK) 103
Marconi gun control and stabilisation equipment (UK) 102
Marconi laser warning system (UK) 834
Marconi OTH radar (UK) 265
Marconi radar warning receivers (UK) 833
Marconi unattended expandable jammer (UK) 329
MARDIS air defence network (UK) 56
MAREC II maritime reconnaissance radar (UK) 790
Maritime patrol ESM (France) 819
Mk 1 No 1 cluster bomb (UK) 754
Mk 2 ADC expendable torpedo countermeasure (USA) 538
Mk 2 No 1 cluster bomb (UK) 754
Mk 2 submarine combat control system (USA) 446
Mk 3 No 1 cluster bomb (UK) 754
Mk 4 Mod O gun pod (USA) 764
Mk 4 No 1 cluster bomb (UK) 754
Mk 5 low light-level TV system (USA) 690
Mk 5 single 4.5 inch gun mounting (UK) 503
Mk 5 twin 40mm Bofors gun mounting (UK) 504
Mk 6 twin 3 inch gun mounting (USA) 503
Mk 6 twin 4.5 inch gun mounting (UK) 503
Mk 7 16-inch triple gun turret (USA) 505
Mk 7 single 40mm Bofors gun mounting (UK) 504
Mk 8 single 4.5 inch gun mounting (UK) 503
Mk 8 torpedo (UK) 528
Mk 9 single 40mm Bofors gun mounting (UK) 504
Mk 11 Mod 5 gun (USA) 764
Mk 15/16 Phalanx close-in weapon system (USA) 479
Mk 16 triple 6 inch/47 gun turret (USA) 505
Mk 19 twin 4 inch gun mounting (UK) 503
Mk 23 target acquisition system radar (TAS) (USA) 441, 644
Mk 24 Tigerfish torpedo (UK) 529
Mk 27 single 5 inch dual-purpose gun mounting (USA) 505
Mk 28 twin 5 inch/38 calibre gun mounting (USA) 505
Mk 32 torpedo (USA) 534
Mk 33 3 inch/50 twin calibre gun mounting (USA) 508
Mk 34 single 3 inch/50 calibre gun mounting (USA) 507
Mk 36 Super Rapid Bloom off-board countermeasures (USA) 674
Mk 37 torpedo Mod 0 (USA) 534
Mk 37 torpedo Mod 1 (USA) 535
Mk 37 torpedo Mod 2 (USA) 535
Mk 37 torpedo Mod 3 (USA) 534
Mk 38 5 inch/38 twin calibre gun mounting (USA) 506
Mk 39 single 5 inch/54 calibre gun mounting (USA) 505
Mk 41 vertical launch system (USA) 484
Mk 42 single 5 inch/54 calibre gun mounting (USA) 506
Mk 42 single 5 inch/54 calibre gun mounting (Lightweight Mod 9) (USA) 507
Mk 44 torpedo (USA) 535
Mk 45 single 5 inch/54 calibre lightweight automatic gun mounting (USA) 507
Mk 46 torpedo (USA) 535
Mk 48 torpedo (USA) 537
Mk 56 gun fire control system (USA) 441
Mk 50 advanced lightweight torpedo (ALWT) (USA) 536
Mk 73 gun and guided missile director (USA) 442
MK 75 76mm/62 calibre gun mount (USA) 508
Mk 82 (Aegis) gun and missile director (USA) 442
Mk 86 gun fire control system (USA) 442
Mk 92 fire control system (USA) 443
Mk 95 Mod 0 radar set (International) 606
Mark 105 (TAC) passive fire control system (USA) 445
Mk 113 fire control system (USA) 445
Mk 114 fire control system Asroc (USA) 445
Mk 116 fire control system (USA) 445
Marksman twin 35mm anti-aircraft turret (UK) 205
Marte anti-ship missile system (Italy) 713
Marte helicopter anti-ship missile system (Italy) 771
Martel AS.37/AJ.168 air-to-surface missile (International) 710
Martin Pescador air-to-surface missile (Argentina) 707
Masquerade airborne decoy system (UK) 834
Masurca surface-to-air missile system (France) 463
Matador (USA) 864
Matador fire control system (Israel) 365
Mathogo anti-tank missile (Argentina) 136
Matilda radar warning receiver system (UK) 664
Matra laser-guided bombs (LGB) (France) 739
Matra passive countermeasures (France) 820
Matra SFA bomb retarding system (France) 737
Matra Type 200 bomb retarding system (France) 737
Mauser Arrow twin gun anti-aircraft system (Germany, FR) 194
MBAR fire control radar (USA) 643
MCM system (China) 513
MCTNS man-portable common thermal night sights (USA) 403
MD100/S passive rangefinder set (Italy) 557
Medium surface-to-air missile (MSAM) system (USA) 190
Medusa jammer (Chile) 815
Medusa optical designation and fire control system (Italy) 683
MEL AEW radar (UK) 791
MEL IFM receivers (UK) 327
MEL modular EW programme (UK) 327
Mentor naval esm system (UK) 669
Mere mortar fire control computer (New Zealand) 84
Merlin anti-tank mortar bomb (UK) 152
Meroka 20mm anti-aircraft system (Spain) 199
Meroka naval CIWS (Spain) 471
Messenger secondary surveillance radar (UK) 276
MFU 402 fire control radar (UK) 271
MIA fire control computer and communication unit (Sweden) 89
MICA air-to-air missile (France) 693
MicroPUFFS passive ranging sonar (USA) 593
MIFASS (Marine Integrated Fire and Air Support System) (USA) 104
Mig-21 HUDWAC system (UK) 879
Milan anti-tank missile (International) 142
Milas anti-submarine weapon system (International) 519
Military radar (China) 219
Mine warfare programmes (USA) 538
Minigun pod type SUU 11B/A (USA) 767
Minilaser rangefinder (Netherlands) 371
MIN mine countermeasures system (Italy) 523
Minuteman II intercontinental ballistic missile (USA) 26
Minuteman III intercontinental ballistic missile (USA) 26
MIR-2 ESM equipment (UK) 835
Mira thermal imager (France) 354
Mirage 2000 integrated EW suite (ICMS) (France) 818
Mirage 2000 RDI radar (France) 772
Mirage 2000 RDM radar (France) 772
Miriade radar warning system (France) 818
MISAR MSR-810 sonobuoy (Italy) 558
MISAT surface-to-air missile (France) 169
Miscellaneous underwater detection equipment 944
Missile launcher systems (USA) 482
Mistral air defence missile (France) 212
MISTRAL passive meteorological system (UK) 97
Mithridat tank fire control system (France) 353
Mitsubishi airborne radar (Japan) 782
MM-38 Exocet surface-to-surface missile (France) 450
MM/APQ-706 radar (Italy) 781
MM/APS-705 radar (Italy) 781
MM/APS-707 radar (Italy) 781
MM/APS-784 search radar (Italy) 782
MM/BPS-704 submarine radar (Italy) 610
MM/SPN-703 navigation radar (Italy) 610
MM/SPN-728 search radar (Italy) 611
MM/SPQ-3 coastal radars (Italy) 242
MM/SPQ-701 naval radar (Italy) 611
MM/SPS-702 naval radar (Italy) 611
MM/SSN-714 minesweeper data processing system (Italy) 418
MM/SSN-715 navigation and tracking system (Italy) 418
MM/UPX-709 airborne IFF transponder (USA) 810
Mobile automatic reporting station (MARS-402) (Italy) 45
Mobile EW support system (MEWSS) (USA) 346
Model 17-C surveillance radar (China) 220
Model 100 naval DF set (Canada) 646
Model 200 TISS (USA) 890
Model 354 shipborne surface and low altitude warning radar (Canada) 600
Model 970 radar jamming system (China) 306
Model 971 colour imaging sonar (Canada) 543
Model 975 optronic director (USA) 689
Model 977 colour imaging sonar (Canada) 543
Model 985 optical director (USA) 690
Model 1110 sonar system (USA) 591
Model 2679 IFF interrogator (USA) 299
Model 4235 ROV doppler sonar (USA) 591
Model 9891 commander's periscope (USA) 398
Model 9892 driver's viewer (USA) 398
Model 9894 gunner's periscope (USA) 398
Model 9896 night vision elbow (USA) 398
Model 9897 laser rangefinder (USA) 398
Model 9898 gunner's periscope (USA) 399
Modular chaff/flare dispenser (USA) 842
Modular control equipment (USA) 61
Modular dispenser system (MDS) (Germany, FR) 741
Modular infra-red transmitter system (MIRTS) 838
Modular stand-off weapons (Germany, FR) 712
Monarch RPV EW system (UK) 834
Morcos mortar computing system (UK) 99

Morocco
land-based air defence systems ... 35
MORZEN 111 mortar computing system (UK) ... 99
MPI-5A miniature pulse analyser (USA) ... 332
MR-80 sea mine (Italy) ... 520
MRP sea mine (Italy) ... 520
MRS3/GWS22 fire control radars Types 903/904 (UK) ... 628
MS series ELINT/ESM systems (UK) ... 327
MSA-3. 1 portable anti-aircraft missile system ... 211
MS1-80S fire control system (Norway) ... 425
MSI-90U submarine and weapon control system (Norway) ... 425
MSI-340/350 fire control system (Norway) ... 426
MSI-3100 multiple integration system (Norway) ... 426
MSBS submarine launched ballistic missile family (France) ... 4
MSPS EW system (France) ... 818
MSS 1.1/MAF anti-tank missile (International) ... 140
MSTAR man-portable radar (UK) ... 273
MSZ2 day/night gunner peritelescope (Israel) ... 363
MT-18 laser rangefinder (Israel) ... 364
MT-19 laser rangefinder (Israel) ... 364
MTAS target acquisition system (USA) ... 305
Muff Cob fire control radar (USSR) ... 621
Mulloka sonar (Australia) ... 540
Multi-application sonar system (MAST) (Italy) ... 557
Multibeam multimode surveillance radar (USA) ... 811
Multi-dart helicopter anti-armour system (France) ... 732
Multi-function electronically scanned adaptive radar (MESAR) (UK) ... 635
Multi-function radar (MFR) (USA) ... 302
Multi-mode radar for F-5E (USA) ... 797
Multirole FLIR turret (UK) ... 880
Multiple Launch Rocket System (MLRS) (USA) ... 133
Multiple radar emitter system (USA) ... 346
Multi-role survivable radar (USA) ... 297
Murene lightweight torpedo (France) ... 515
Murene thermal camera (France) ... 678
MW-1 multi-purpose weapon (Germany, FR) ... 742
MW-5 fire control radar (China) ... 223
MW08 3-D surveillance radar (Sweden) ... 615
Mygale air defence system (France) ... 41

N

N-90 series laser (USA) ... 396
NA 18 fire control system (Italy) ... 414
NA 21 fire control system (Italy) ... 414
NA 25 weapon control system (Italy) ... 415
NA 30 series fire control system (Italy) ... 415
NADGE–NATO Air Defence Ground Environment ... 46
NADS Naval Antimissile Decoy System (France) ... 649
Naja electro-optical director (France) ... 679
Najir electro-optical director (France) ... 679
NATO anti-air weapon systems (NAAWS) (International) ... 467
NATO designations of Soviet systems and equipment tables ... 955
NATO identification system (NIS) (UK) ... 779
NATO Sea Gnat (International) ... 652
NATO Sea Sparrow surface-to-air missile system (International) ... 466
Nautis command and control systems (UK) ... 437
Naval mine upgrading system (China) ... 513
Naval ordnance (China) ... 485
Naval radar tables ... 930
Naval surface-to-air missiles table ... 912
Naval tactical data system (USA) ... 446
NAVKIS (Norway) ... 425
Navy torpedo decoys (USA) ... 539
ND-9010 naval ESM system (Israel) ... 655
Netherlands
army fire control systems ... 82
electronic warfare equipment (land) ... 318
electro-optical equipment (land) ... 370
electro-optical equipment (sea) ... 683
ground radar ... 248
land-based air defence systems ... 48
naval fire control systems ... 420
naval radar ... 612
sonar and underwater detection equipment ... 559
Nettuno naval EW system (Italy) ... 658
New Zealand
army fire control systems ... 84
Newsy naval electronic warfare system (France) ... 649
Newton naval EW equipment (Italy) ... 656
Night ATAC fire control system (USA) ... 900
Nigeria
land-based air defence systems ... 36
Nike Hercules (MIM-14B) air defence missile (USA) ... 190
Nimtan airborne night sight (UK) ... 887
Nite Owl targeting pod (USA) ... 900
North Atlantic Treaty Organisation (NATO)
land-based air defence systems ... 46
Nokia electro-optical equipment (Finland) ... 350
NOL series naval EW systems (Japan) ... 660
North warning system (DEW line) (USA) ... 57
Northrop focal plane array (USA) ... 891
Northrop mast-mounted sight (MMS) sensor suite (USA) ... 890
Northrop TCS (television camera set) (USA) ... 891
Norway
air-to-surface missiles ... 714
army fire control systems ... 84
coastal defence weapons ... 162
electronic warfare equipment (land) ... 318
electronic warfare equipment (sea) ... 660
electro-optical equipment (land) ... 374
electro-optical equipment (sea) ... 684
ground radar ... 249
land-based air defence systems ... 48
naval fire control systems ... 425
naval guns and bombardment rockets ... 494
shipborne weapons ... 453
sonar and underwater detection equipment ... 560
underwater warfare systems ... 523
NPG-880 3-D air defence radar (Japan) ... 248
NPM-510 mobile 3-D radar (Japan) ... 247
NR-A1-5A iff interrogator-decoder (France) ... 234
NRA1-6A iff interrogator-decoder (France ... 234
NRAI-7 airborne IFF transponder (France) ... 777
NRAI-9A IFF transponder (France) ... 777
NRAI-10A airborne IFF interrogator (France) ... 778
NRAI-11A IFF interrogator-decoder (France) ... 778
NRSI-3 IFF interrogator (France) ... 233
NS-9001 ELINT system (Israel) ... 654
NS-9003 naval ESM system (Israel) ... 654
NS-9005 ECM jammer (Israel) ... 655
NS-9009 naval EW system (Israel) ... 655
NS-9034 submarine ESM/ELINT system (Israel) ... 655
No 63 passive night sight (L3A1) (UK) ... 388
NVL53 day/night sight and rangefinder (UK) ... 381
Nysa C long-range radar (USSR) ... 257

O

OA-5154/ASQ automatic MAD compensation system (Canada) ... 544
OB-31 night-driving periscope (France) ... 354
OB-47 night sight (France) ... 352
Observer thermal imaging system (OTIS) (UK) ... 392
Odin artillery FCS (Norway) ... 85
OE-320/A DF system (USA) ... 344
Oeil Vert radar (France) ... 229
Oerlikon 25mm twin field AD gun (Diana) (Switzerland) ... 204
Oerlikon 30mm gun Type KCA (Switzerland) ... 751
Oerlikon 81mm aircraft rocket (Switzerland) ... 749
Oerlikon 81mm Type RWK014 rocket launcher (Switzerland) ... 125
Oerlikon Escorter twin 35mm SP air defence system (Switzerland) ... 203
Oerlikon rocket pods (Switzerland) ... 750
Oerlikon SNORA 81mm rocket (Switzerland) ... 750
Oerlikon twin field 35mm air defence guns type GDF (Switzerland) ... 202
OFRIS target tracker (Germany, FR) ... 360
OG 14 L2A tank fire control system (Italy) ... 78
OG 14 L2B tank fire control system (Italy) ... 78
OG 14 L3 (TURMS) laser tank fire control system (Italy) ... 78
OG 30A naval multi-sensor director (Italy) ... 416
OG 30B shipborne thermal imager (Italy) ... 683
OG C-102 coaxial telescope (Italy) ... 366
OG-PO7B periscope (Italy) ... 369
OG-P20 periscope (Italy) ... 369
OG-P101 periscope sight (Italy) ... 369
OGR7/2 optical director (Italy) ... 682
OGR7/2-A optical director (Italy) ... 682
Oldelft AIFV thermal sights (Netherlands) ... 373
Oman
land-based air defence systems ... 48
OMU-2 artillery laser rangefinder (Yugoslavia) ... 407
Opher terminal guidance kit for general purpose bombs (Israel) ... 742
OPS series shipborne search radar (Japan) ... 611
Optech Lidar (Canada) ... 867
OR-89/AA FLIR (USA) ... 897
OR-5008/AA FLIR (USA) ... 898
ORB-31 radar systems (France) ... 774
ORB 31 AS (France) ... 775
ORB 32 radar systems (France) ... 775
ORB 37-HL airborne radar (France) ... 775
Orchidee battlefield surveillance radar (France) ... 778
Orion RTN-10X naval tracking radar (Italy) ... 609
Orion RTN-20X naval radar (Italy) ... 609
Orion RTN-25X naval tracking radar (Italy) ... 609
Orion RTN-30X naval radar (Italy) ... 609
Osborn minesweeping system (UK) ... 532
Osprey day/night AFV sight (UK) ... 388
Osprey dipping sonar (International) ... 557
OTH radar (Japan) ... 248
OTH radar (USSR) ... 261
OTO 76/62 compact automatic gun (Spain) ... 495
Otomat anti-ship missile (France/Italy) (International) ... 451
Otomat coastal defence system (France/Italy) (International) ... 161
Otomatic 76mm 76/62 SPAAG (Italy) ... 197
Outfit DCB submarine TDHS (UK) ... 432
Outfit DCC submarine TDHS (UK) ... 432
Outfit DCG submarine TDHS (UK) ... 432
Over-the-horizon radar (China) ... 222
Over-the-Horizon (OTH) radar systems (USA) ... 57
Owl screech fire control radar (USSR) ... 619
OX-60/FPS-117 IFF interrogator (USA) ... 298

P

P0701 laser rangefinder (Italy) ... 683
P0702 airborne laser rangefinder (Italy) ... 876
P 56 series computing sight and gun drive (Italy) ... 366
P 186 telescopic sight (Italy) ... 368
P 192 night driving periscope (Italy) ... 368
P 204 periscope (Italy) ... 369
P 233 night vision periscope (Italy) ... 368
P 2072 (AN/SQQ-141(T) minehunting sonar (Italy) ... 558
P4674 thermal imager (Italy) ... 683
PA 010 doppler direction finder (Germany, FR) ... 311
PA 055 doppler direction finder (Germany, FR) ... 311
PA 510 mobile direction finding system (Germany, FR) ... 313
PAC 101 portable artillery computer (Italy) ... 80
Pacer Mk 2 muzzle velocity radar (UK) ... 273
PADS Position and Azimuth Determining System (UK) ... 100
Pagoda search radar (Italy) ... 244
Panda Mk 2 optical director (France) ... 679
PAR-80 precision approach radar (USA) ... 286
Passive infra-red surveillance equipment (UK) ... 383
Pat Hand fire control radar (USSR) ... 257
Pathfinder navigation pod (USA) ... 896
Patriot (MIM-104) tactical air defence system (USA) ... 189
Paveway laser-guided bombs (USA) ... 757
Pawn Shop radar (USSR) ... 258
Peacekeeper (MGM-118) ICBM (USA) ... 23
Peel Group fire control radar (USSR) ... 621
Pegaso fire control system (Italy) ... 416
Penguin Mk 2 anti-ship missile (Norway) ... 453
Penguin Mk 2 Mod 7 air-to-surface missile (Norway) ... 715
Penguin Mk 3 air-to-surface missile (Norway) ... 714
Penguin coastal defence system (Norway) ... 162
Pershing Ia battlefield support missile (USA) ... 131
Pershing II battlefield support missile (USA) ... 132
Phalcon AEW radar (Israel) ... 780
Philippines
land-based air defence systems ... 36
Phoenix (Aim-54) air-to-air missile (USA) ... 703
Phoenix ESM system (USA) ... 676
PHS-32 sonar (Netherlands) ... 559
PHS-36 series sonars (Netherlands) ... 559
Picador mobile 3-D radar (France) ... 231
Pinguin mine countermeasures vehicle (Germany, FR) ... 518
Pirana infra-red tracker (France) ... 677
Piranha add-on jammer (USA) ... 343
PIVADS (Product Improved Vulcan Air Defence System) (USA) ... 209
PIVAIR submarine periscope (France) ... 681
PJ-1 manually operated bomb (Chile) ... 729
PKD-8/KOL-2 sighting system for artillery (Sweden) ... 379
PL-2 air-to-air missile (China) ... 692
PL-5B air-to-air missile (China) ... 692
PL-7 air-to-air missile (China) ... 692
PL-9 air-to-air missile (China) ... 693
Plath direction finding systems (Germany, FR) ... 313
Plessey ARMS sonar (UK) ... 577
Plessey HF monitoring equipment (UK) ... 321
Plessey-Sippican XBT (UK) ... 576
Plessey-Sippican XSV (UK) ... 576
Plinth Net search radar (USSR) ... 622
PLRS (AN/TSQ-129) (USA) ... 108
Pluto low-coverage surveillance radar (Italy) ... 244
Pluton battlefield support missile (France) ... 119
PMS26 surface ship sonars (UK) ... 564
PMS27 surface ship sonars (UK) ... 564
PMS 56 pursuit sonar (UK) ... 565
PMS 75 speedscan (UK) ... 576
POET expendable jammer (USA) ... 841
Poland
ground radar ... 249
Pollux tracker radar TRS 3220 (France) ... 604
Pop Group fire control radar (USSR) ... 622
Pork Trough (SNAR-2) counter-battery radar (USSR) ... 258
Porpoise submarine ESM equipment (UK) ... 668
Poseidon C-3 (UGM-73A) fleet ballistic missile (USA) ... 27
Possum air defence surveillance radar (UK) ... 270
Pot Drum naval radar (USSR) ... 622
Pot Head naval radar (USSR) ... 622
PPE passive driving periscope (UK) ... 387
Project Barra (Australia) ... 540
Prophet radar warning receiver (UK) ... 836
Protean chaff launcher (UK) ... 665
PS-37/A airborne attack radar (Sweden) ... 783
PS-46/A multi-mode AI radar (Sweden) ... 783
PSM 33 mobile radar (India) ... 239
PSNR-1 battlefield radar (USSR) ... 261
Puff Ball airborne radar (USSR) ... 785
PV 017 azimuth analyser (Germany, FR) ... 313
PYK-80 field artillery computer (Korea, Republic of) ... 80
Python 3 air-to-air missile (Israel) ... 696

Q

QQS-4 surface ship sonar (Japan) ... 559
QRC 83-05 infra-red countermeasures system (USA) ... 837
QRC 84-02 infra-red countermeasures pod (USA) ... 837
QRC-259 receiving system (USA) ... 335
Quickfire artillery FCS (UK) ... 97
Quick fix IIB countermeasures system (USA) ... 864

R

R-76 naval fire control radar (USA) ... 643
R-76 fire control system (USA) ... 441
R.530 air-to-air missile (France) ... 694
R550 Magic air-to-air missile (France) ... 694
R-1651/ARA sonobuoy OTPI radio receiver (UK) ... 584
R-3030 surveillance receiver (USA) ... 332
R-3050 surveillance receiver (USA) ... 332
R-3080 surveillance receiver (USA) ... 332
R4000 superheterodyne receiver (USA) ... 344
RA-20S fire control radar (France) ... 230
RA1792 programmable receiver (UK) ... 321
RA1794 tactical EW receiver (UK) ... 321
RA1795 VHF/UHF receiver (UK) ... 322
RA1796 HF/VHF/UHF tactical EW receiver (UK) ... 322
RA3700 series modular hf receivers (UK) ... 322
RA6778-C receiver (USA) ... 337
RA6778-Q converter (USA) ... 337
RA-6793A hf receiver (USA) ... 338
RACAAS airborne radar (France) ... 773
Racal EW programmes (UK) ... 328
Racal helicopter-borne jamming system (USA) ... 864
Racal-Decca surveillance and navigation radars (UK) ... 627
Racal tactical direction finding system (UK) ... 322
RACEWS electronic warfare system (UK) ... 324
RACJAM Air (UK) ... 836
RACJAM-V responsive vhf jammers (UK) ... 323
Radamec electro-optical directors (UK) ... 394
Radamec electro-optical pod tracking system (E-OPTS) (UK) ... 394
Radamec EW antenna director mounts (UK) ... 670
Radamec lightweight missile launcher (UK) ... 663
Radamec naval directors (UK) ... 688

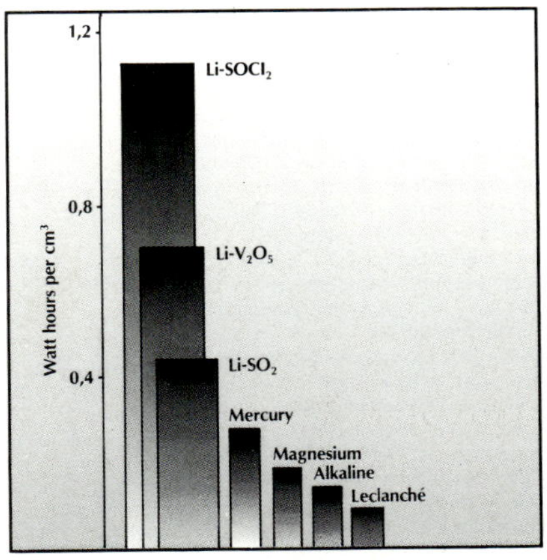

Radamec Series 2000 fire control and surveillance systems (UK) 688
Radar antenna pedestals (UK) 635
Radar elevator system (Norway) 249
RADIRS 70mm multiple rocket system (USA) 132
RADOP series naval fire control systems (France) 408
Rafael long range chaff rocket (Israel) 652
Rafael short range chaff rocket (Israel) 652
RAIDS detection radar (USA) 305
Rainbow laser rangefinder (Israel) 362
RAJ 101 ground radar jammer (Israel) 315
RAM (RIM-116A) system (USA) 480
Rampart (UK) 320
Ramsa low-altitude surveillance radar TRS 2140 (France) 226
RAN-3L naval radar (MM/SPS-768) (Italy) 608
RAN-10S naval radar (MM/SPS-774) (Italy) 608
RAN-11 L/X integrated naval radar (Italy) 609
RAN-12 L/X integrated naval radar (Italy) 609
RAN-30S naval radar (Italy) 610
RAN 30X surveillance radar (Italy) 610
Rank thermal imager (UK) 391
Raphael TH airborne surveillance radar (France) 774
RAPIDS naval ESM (UK) 653
Rapier 2000 surveillance and target acquisition radar (UK) 270
Rapier Blindfire tracking radar (UK) 272
Rapier mobile low-level anti-aircraft missile system (UK) 183
Rapier search and acquisition radar (UK) 270
Rappist fire control radar (France) 230
Rapport ECM systems (USA) 862
RASAT sonar target (UK) 569
Rasit E ground surveillance radar (France) 232
RAT-31S transportable 3-D radar (Italy) 245
RAT-31SL long range radar (Italy) 245
RATAC-S battlefield radar (Germany FR) 238
RATAC battlefield radar (International) 239
Rattler jammer (Israel) 316
Rattler weapon system (South Africa) 200
Raven day/night sight (UK) 388
Raytheon standard sonar transmitter (SST) (USA) 386
RB 08 coastal defence weapon (Sweden) 162
RBI2A (Rassur) ground surveillance radar (France) 233
RBOC 11 (USA) 674
RBS 15 anti-ship missile (Sweden) 715
RBS 15 coastal defence weapon (Sweden) 162
RBS 17 hellfire shore defence system (Sweden) 162
RBS 56 Bill anti-tank missile (Sweden) 147
RBS 70 anti-aircraft missile system (Sweden) 213
RBS 70 Rayrider missile system (Sweden) 471
RCA balloon-borne radar (USA) 813
RCM series ECM equipment (UK) 666
RDL series ESM equipment (UK) 667
RDR-1500 multi-mode search and surveillance radar (USA) 799
RDX airborne radar (France) 773
RDY multi-function radar (France) 773
Real-time velocimeter radar (RTVS) (USA) 282
Red Arrow 8 anti-tank missile (China) 137
Red Baron pod (Sweden) 877
Redeye portable anti-aircraft missile (USA) 216
Red Fox airborne radar (UK) 793
Relocatable over-the-horizon radar (USA) 303
REMBASS (Remotely Monitored Battlefield Sensor System) (USA) 109
Replica naval decoy (UK) 670
Reporter tactical early warning system (Netherlands) 83
Reshet naval command and control system (Israel) 413
Responsive surveillance subsystem (USA) 336
Reveal video enhancement system (Israel) 363
RG-1320A spectrum display unit (USA) 338
RG-1340 spectrum surveillance controller (USA) 338
RG-5202A VHF scanning receiver (USA) 337
RG-5540 VHF/UHF monitoring receiver (USA) 337
RG-5545A VHF/UHF receiver (USA) 337
RGS 2 optical sighting system (Sweden) 876
RG series spectrum monitors (USA) 339
Rheinmetall anti-armour ammunition developments (Germany, FR) 139
RIA gunsight (Sweden) 376
RJS3140 unattended expendable jammer (UK) 324
RLD-2 hand-held laser rangefinder (Yugoslavia) 406
RN Type 901 radar (UK) 629
RN Type 909 radar (UK) 629
RN Type 910 radar (UK) 630
RN Type 911 tracking radar (UK) 633
RN Type 912 radar (UK) 629
RN Type 965 radar (UK) 630
RN Type 967 radar (UK) 630
RN Type 968 radar (UK) 630
RN Type 975 radar (UK) 630
RN Type 992Q radar (UK) 631
RN Type 994 radar (UK) 631
RN Type 996 naval surveillance radar (UK) 631
RN Type 1022 naval radar (UK) 631
Rockeye II Mk 20 cluster bomb (USA) 756
Rockwell mast-mounted sight (USA) 891
Rodeo 2 fire control radar (France) 230
Rodeo naval air surveillance radar (France) 605
Rohde & Schwarz antenna systems (Germany, FR) 312
Rohde & Schwarz COMINT system (Germany, FR) 821
Rohde & Schwarz direction finding equipment (Germany, FR) 310
Rokit tank sight (USA) 401
Roland mobile anti-aircraft weapon system (International) 171
RQH-5(V) ESM/ELINT airborne systems (Italy) 828
RQN-1 radar intercept equipment (Italy) 658
RQN-5 naval ESM/ELINT system (Italy) 659
RSI tank laser rangefinder (Netherlands) 371
RS3153 radio surveillance equipment (UK) 323
RW-23-1 radar warning equipment (China) 647
RWR 1000 hand-held radar detector (USA) 332
RZ1001 thermal imaging equipment (Germany, FR) 360

S

S500 ESM system (USA) 344
5500 series primary airfield surveillance radar (UK) 265
S613 heightfinding radar (UK) 265
S700 meteorological station (South Africa) 87
S700 series mobile radar systems (UK) 265
S713 Martello 3-D surveillance radar (UK) 263
S723 Martello 3-D surveillance radar (UK) 264
S810 naval surveillance radar (UK) 632
S820 naval surveillance radar (UK) 632
S1810 shipborne and coastal defence radar (UK) 634
S1820 naval surveillance radar (UK) 634
S-2150 ESM systems (USA) 672
S-3000 surveillance system (USA) 675
SA-1 (Guild) surface-to-air missile (USSR) 176
SA-2 Challenger (USA) 444
SA-2 (Guideline) medium-range surface-to-air missile (USSR) 177
SA-3 (Goa) surface-to-air missile (USSR) 177
SA-4 (Ganef) surface-to-air missile (USSR) 177
SA-5 (Gammon) long-range air defence missile (USSR) 178
SA-6 (Gainful) surface-to-air missile (USSR) 179
SA-7 (Grail) man-portable anti-aircraft missile (USSR) 214
SA-8 (Gecko) surface-to-air missile (USSR) 179
SA-9 (Gaskin) surface-to-air missile (USSR) 180
SA-10 (Grumble) surface-to-air missile (USSR) 180
SA-11 (Gadfly) surface-to-air missile (USSR) 181
SA-12 (Gladiator) giant surface-to-air missile system (USSR) 181
SA-13 (Gopher) surface-to-air missile (USSR) 182
SA-14 (Gremlin) man-portable anti-aircraft missile (USSR) 214
SA 90 radar system (France) 236
SA-N-4 shipborne surface-to-air missile (USSR) 473
SA-N-5 surface-to-air missile (USSR) 473
SA-N-6 Grumble surface-to-air missile system (USSR) 473
SA-N-7 Gadfly surface-to-air missile system (USSR) 474
SA-N-9 naval surface-to-air missile (USSR) 475
Saab 04E air-to-surface missile (Sweden) 716
Saab 05A air-to-surface missile (Sweden) 716
Saab IRS-700 IR surveillance equipment (Sweden) 879
Saab JA37 Viggen head-up display (UK) 882
Saab RBS 15 anti-ship missile (Sweden) 454
Saab TVT-300 fire director system (Sweden) 685
SAAM naval air defence system (France) 462
SAAM surface-to-air missile system (France) 168
SABCA tank fire control systems (Belgium) 63
Saber air defence anti-armour missile (USA) 187
Sabre day/night vehicle sights (UK) 389
Sabre twin 30mm AA gun turret system (France) 192
SACU command and control (Israel) 75
SACTIS combat information system (Italy) 416
SADARM system (USA) 158
Sadie ESM processor (UK) 669
SADOT fire control system (Israel) 364
SADRAL naval air defence system (France) 465
SAGAIE naval countermeasures decoy system (France) 648
SAKR 80 (Egypt) 118
Samlet cruise missile (SSC-2B) (USSR) 163
SAMOS naval air defence system (France) 464
SAMP fragmentation bombs (France) 739
SAMP general-purpose bombs (France) 738
SAMP practice bombs (France) 738
SAMTOR thermal observation system (Netherlands) 371
SA-NX-? underwater-to-air missile (USSR) 475
SAPBA-1 127 mm multiple rocket system (Argentina) 114
SAPIENS EW system (France) 649
SARIE-Selective Automatic Radar Identification Equipment (UK) 666
SAT 3311 HEF signal analysis terminal (UK) 324
SATAN close-in weapon system (France) 464
SATIR action information system (Germany, FR) 412
SATT update package for anti-aircraft fire control system (Sweden) 92
Saudi Arabia
land-based air defence systems 36
Saviour laser warning system (UK) 328
Savoie instrumentation tracking radar (France) 227
SA-X-15 low-to-medium altitude surface-to-air missile system (USSR) 182
SB 14 IFF interrogator (France) 234
S-Band meteorological radar (Brazil) 218
SC-01 (Scipio) multi-mode radar (Italy) 782
Scan 522 shallow water sonar (Canada) 542
Scan 545 shallow water sonar (Canada) 542
Scan Fix airborne radar (USSR) 785
Scan Odd airborne radar (USSR) 785
Scarecrow RPV EW system (USA) 863
Sceptre ESM system (UK) 664
Schalmei countermeasures decoy system (Germany, FR) 651
Schermuly infra-red countermeasures (UK) 329
Scimitar jammer (UK) 664
SCLAR Mk 2 rocket launching system (Italy) 659
Scoop Pair surface target radars (USSR) 623
Score Board IFF radar (USSR) 258
Scorpion ECM system (UK) 668
Scorpion target illuminating radar (UK) 271
SCR series receiving and monitoring equipment (USA) 344
SDFOV thermal imaging system (International) 361
Sea Archer 1 FCS (UK) 440
Sea Archer 30 fire control system (UK) 440
Seacat shipborne surface-to-air missile (UK) 476
Sea Chaparral surface-to-air missile system (USA) 479
Sea Cobra fire control system 413
Sea Dart surface-to-air missile (UK) 476
Sea Eagle air-to-surface missile (UK) 720
Sea Eagle SL anti-ship missile (UK) 458
Sea Giraffe naval search radar (Sweden) 616
Seaguard CIWS (International) 468
Seaguard naval combat system (International) 412
Seaguard tracking module (TMX) (Switzerland) 617
Sea Lance anti-submarine warfare stand-off weapon (USA) 534
Seal/Seeschlange torpedoes (Germany, FR) 517
Sea mines (USSR) 527
Sea Owl passive identification device sensor turret (UK) 880
Sea Saviour EW system (UK) 669
Sea Searcher airborne radar ARI 5991 (UK) 791
Sea Serpent magnetic minesweeping system (UK) 532
Sea Skua helicopter air-to-surface missile (UK) 719
Sea Skua SL anti-ship missile system (UK) 458
Seaspray helicopter ASV radar (UK) 792
Sea Tiger (International) 413
Sea Tiger II surveillance radar TRS 3001 (France) 605
Searchwater surveillance radar (ARI 5980) (UK) 788
Sea Urchin mine (UK) 530
Sea Vulcan 20 20mm naval mount (USA) 509
Sea Vulcan 20p three-barrel 20mm pintle mount (USA) 509
Sea Vulcan 25 weapon system (USA) 510
Sea Vulcan 30 lightweight 30 mm naval gun mount (USA) 509
Sea Wolf surface-to-air missile (UK) 477
SECO electro-optical systems (USA) 406
SEDA artillery automation system (Italy) 77
SEDAB artillery automation system (Italy) 77
SEID infra-red tracker (France) 678
Seidal laser detector (Spain) 319
SEIMOR mortar computing system (Spain) 88
Selenia electronic surveillance system (Italy) 318
Selenia mobile GCA system (Italy) 246
SEN series combat systems (Italy) 416
SENIT naval tactical data handling systems (France) 410
Sensor fuzed weapon (SFW) (USA) 758
Sentry ESM system (UK) 320
SEPA naval fire control system (Italy) 417
SEPADS sonar prediction system (UK) 579
Sepal mobile cruise missile (SSC-1B) (USSR) 164
Series 52 precision approach radar (USA) 277
Series 320 3-D air defence radars (USA) 291
Series 500 mobile EW system (UK) 320
Series 2000 receiver system (USA) 339
Serval radar warning receiver (France) 817
Setter fire control radar (Italy) 781
SEWACO – sensor, weapon and command system (Netherlands) 421
SGM-80 sea mine (Germany, FR) 520
SGS-55 tank day/night sight (Yugoslavia) 406
SH-08 (Gazelle) anti-ballistic missile missile 18
Shafrir air-to-air missile (Israel) 696
Shahine air defence system (TSE 5100) (France) 170
Sherloc radar warning system (France) 817
Shield anti-ship missile decoy system (UK) 663
Shipboard electronic warfare suite (SEWS) (Israel) 655
Shipborne four-tube anti-submarine rocket launcher (Sweden) 524
Shipborne two-tube anti-submarine rocket launcher (Sweden) 525
Shorar search radar (Italy) 244
Short Horn airborne radar (USSR) 785
Siclamen IFF system (France) 233
Sidam 25 AA turret (Italy) 197
Side Net heightfinder (USSR) 258
Sidewinder (AIM-9) air-to-air missile family (USA) 704
Siemens artillery surveillance radar (Germany, FR) 237
Siemens Battery Co-ordination Post (BCP) (Germany, FR) 44
Siemens hf jammer (Germany, FR) 313
Siemens military radars (Germany, FR) 236
SIGINT aircraft (Israel) 825
Signaal 3-D multi-target tracking radar (Netherlands) 614
Signaal coastal radar (Netherlands) 249
Signaal STIR tracking radar (Netherlands) 612
SILLACS low-level air defence system (Germany, FR) 42
Simrad RL1 laser warning device (Norway) 318
SINBADS – Submarine Integrated Battle And Data System (Netherlands) 423
Singapore
land-based air defence systems 36
Sippican expendable ASW devices (USA) 594
Siren naval decoy system (UK) 670
Sirocco meteorological radar station (QR-MX-2) (France) 68
Siscea integrated airspace control system (Brazil) 38
SIT 421 (MM/UPX-709) transponder (Italy) 782
SIT 422 SHORAD interrogator (Italy) 782
SIT 431 (AN/TPX-54) IFF interrogator (Italy) 246
SIT 434 IFF interrogator (Italy) 782
Skin Head naval radar (USSR) 623
Skip Spin airborne radar (USSR) 785
Skorpioen anti-ship missile (South Africa) 454
Sky Flash air-to-air missile (UK) 701
Skyguard anti-aircraft fire control system (Switzerland) 93
Sky Guardian airborne radar warning receiver radar (UK) 834
Skymaster multi-role airborne radar (UK) 788
Skyranger airborne radar (UK) 787
Sky Sword air-to-air missile (Taiwan) 698
SL/ALQ-234 ECM pod (Italy) 828
SLAMMR side-looking radar (USA) 808
SLAM-Pampero 105 mm multiple rocket system (Argentina) 113
SLBM detection-system 474N (USA) 58
Slim Net surface warning radar (USSR) 623
SLOS — stabilised long-range reservation system (Israel) 873
SLR-600 series naval ESM equipment (USA) 673
SM 39 submarine-launched Exocet (Germany, FR) 517
SM-70 Barracuda surface-to-surface anti-ship missile 160
Small platform EW system (USA) 865
Smart Guard helicopter ESM system (Italy) 827
SMART naval surveillance radar (Netherlands) 614
SMT modular thermal imaging system (France) 356
SNAP-1 steerable hull antenna processor (USA) 345
SNEB 68 aircraft rockets (France) 731
Snoop class submarine surveillance radars (USSR) 624
SO-1 radar warning receiver (Yugoslavia) 866
Sonar 2048 (UK) 526
Sonar 2093 (UK) 571
Sonar 3000 PLUS series (UK) 569

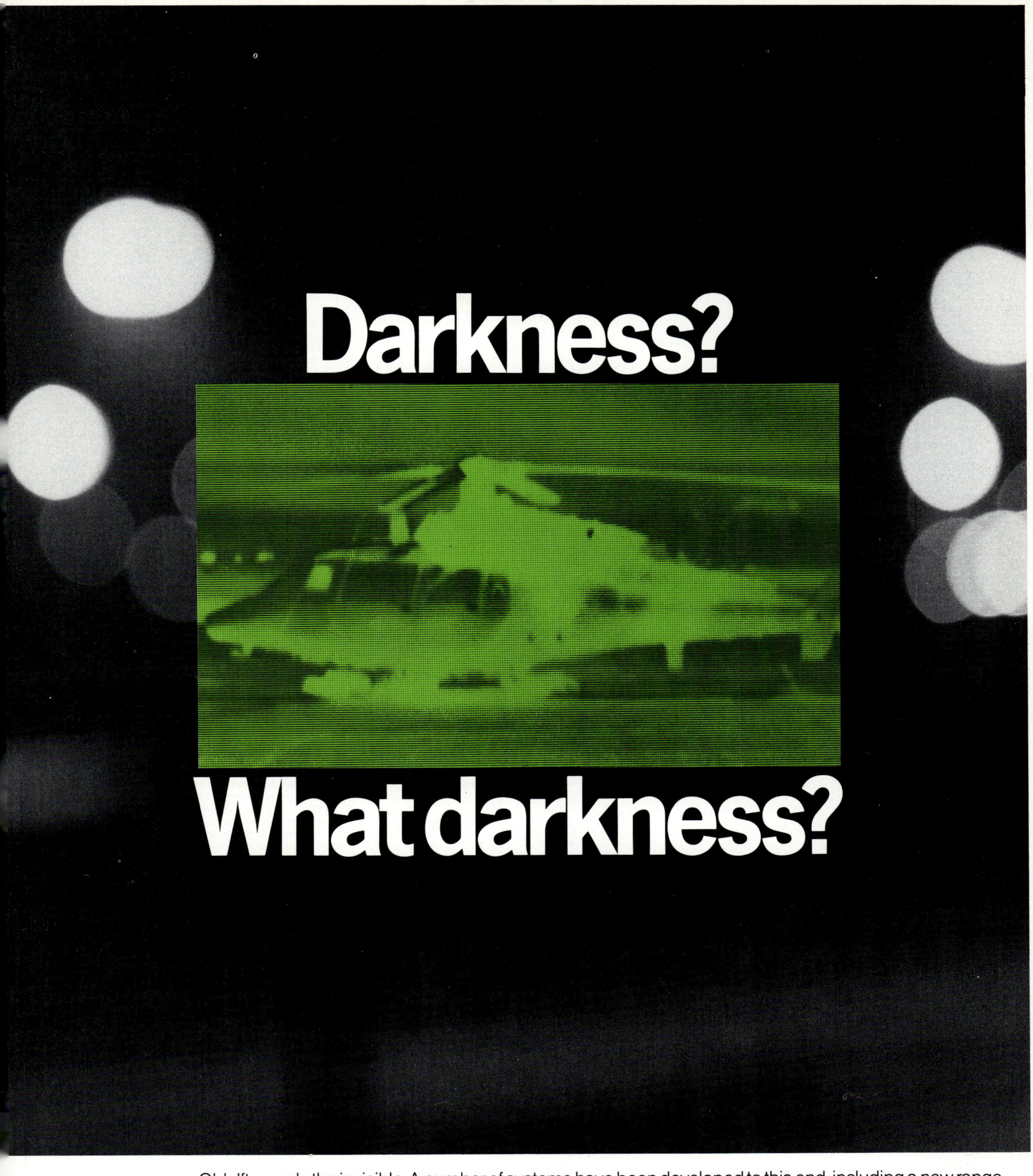

Darkness?

What darkness?

Oldelft reveals the invisible. A number of systems have been developed to this end, including a new range of highly sophisticated thermal imaging systems.

Oldelft day-and-night thermal observation and aiming sights are already successfully used for the Dutch army's AIFV'S. The systems provide high-contrast pictures of objects in smoke, haze, drizzle and complete darkness, through foliage and camouflage.

They are modular, highly compact and ideally suited for adaptation to different types of armoured vehicles, tanks, vessels and aircraft. For more information, call or write Oldelft. **Oldelft, the apogee of thermal imaging.**

Sonar equipment tables....940
SOPELEM SOPTAC 36 day/night fire control system (France)....67
SOPELEM submarine periscopes (France)....681
SOPTAC 11 fire control system (France)....67, 353
SOPTAC 18 fire control system (France)....353
SOPTAC 36 fire control system (France)....353
SOPTAL 29 fire control system (France)....352
Sound ray tracer (TSM 9310) (France)....551

South Africa
- aircraft armament....746
- air-to-air missiles....698
- army fire control systems....86
- battlefield support weapons....124
- electro-optical equipment (land)....376
- mobile gun and rocket systems....200
- mobile surface-to-air guided missile systems....175
- shipborne weapons....454

South Korea
- land-based air defence systems....36

SP2104 fast time acoustic processor (UK)....577
Spacetrack (USA)....59
Spada point defence missile system (Italy)....173

Spain
- aircraft armament....747
- army fire control systems....88
- battlefield support weapons....125
- electronic warfare equipment (air)....829
- electronic warfare equipment (land)....319
- electro-optical equipment (land)....376
- ground radar....249
- land-based air defence systems....49
- naval fire control systems....426
- mobile gun and rocket systems....199
- naval guns and bombardment rockets....495
- shipborne surface-to-air weapons....471

Sparrow (AIM-7) air-to-air missile (USA)....705
Sparton airborne sonar processing equipment (USA)....582
SPASYN visual target acquisition system (USA)....888
SPDT-1A passive sonar (France)....555
Spearfish torpedo (UK)....529
Spectra ELINT system (Netherlands)....318
Spectrum I and II submarine SEWACO system (Netherlands)....424
Sperry antenna systems equipment (USA)....331
Sphinx naval ESM (Netherlands)....665
SPI-04 sound ray path analyser (Netherlands)....560
Spider II air defence system (Israel)....196
Spin Scan airborne radar (USSR)....785
Spoon Rest A early warning radar (USSR)....259
SPQ-2D/F naval radar (Italy)....610
Sprat submarine ESM system (UK)....669
SPS series shipborne torpedo control systems (Italy)....522
SPS-20 airborne self protection system (Israel)....823
SPS-65 helicopter self-protection system (Israel)....823
SPS-200 airborne self protection systems (Israel)....824
SPS-2000 airborne self-protection system (Israel)....824
SP-T-694 stabilised panoramic sight (Italy)....370
Square Head naval radar (USSR)....624
Square Pair guidance radar (USSR)....258
Square Tie naval radar (USSR)....624
Squat Eye acquisition radar (USSR)....258
Squint Eye acquisition radar (USSR)....258
SR series surveillance receivers (USA)....339
SR-1A ESM receiver (Norway)....660
SR-200 ESM system (USA)....672
SR-2020 tactical ESM system (USA)....340
SR-2152 VHF/UHF receiver (USA)....340
SRL night vision/head-up display (USA)....901
SRS-5 reflecting sight (Sweden)....379
SRS-25 airborne receiver (Israel)....823
SRS-2500 VHF/UHF receiving system (USA)....340
SS-1 (Scud) battlefield support missile (USSR)....127
SS-4 (Sandal) MRBM (USSR)....6
SS-11 (Sego) ICBM (USSR)....7
SS 11 battlefield missile (France)....137
SS 12 small ship sonar (hull and VDS) (France)....549
SS-12A (Scaleboard A) battlefield support missile (USSR) 127
SS-12B (Scaleboard B) battlefield support missile (USSR) 128
SS-13 (Savage) ICBM (USSR)....7
SS-16 (Sinner) ICBM (USSR)....7
SS-17 (Spanker) strategic ICBM (USSR)....8
SS-18 (Satan) strategic ICBM (USSR)....8
SS-19 (Stilleto) strategic ICBM (USSR)....8
SS-20 (Saber) mobile IRBM (USSR)....9
SS-21 (Scarab) battlefield support missile (USSR)....128
SS-23 (Spider) battlefield support missile (USSR)....128
SS-24 (Scalpel) ICBM....10
SS-25 (Sickle) ICBM (USSR)....11
SS100/110 night sight (UK)....389
SS 105 scanning sonar (Norway)....560
SS120 night sight (UK)....389
SS130 driving periscope (UK)....390
SS141 commander periscope (UK)....390
SS142 commander periscope (UK)....390
SS162 commander periscope (UK)....390
SS170 day/night sight (UK)....390
SS200 fire control system (UK)....390
SS240 omni sonars (Norway)....560
SS304 scanning sonar (Norway)....561
SS500 armoured vehicle sight (UK)....391
SS550 thermal elbow (UK)....391
SSBS intermediate range ballistic missile Type S-3D/TN-61 (France)....4
SSBS intermediate range missile Type S-4 (SX) (France)....4
SSC-3 coast defence missile system (USSR)....164
SSC-4 coast defence missile system (USSR)....164
SSM-1 coastal defence missile (Japan)....161
SS-N-2 (Styx) shipborne surface-to-surface missile (USSR)....455
SS-N-3 (Shaddock) shipborne surface-to-surface missile (USSR)....456
SS-N-5 (Serb) submarine launched ballistic missile (USSR)....13

SS-N-6 (Sawfly) submarine launched ballistic missile (USSR) ... 14
SS-N-7 (Star Bright) submarine-launched cruise missile (USSR) ... 457
SS-N-7 submarine launched cruise missile (USSR) ... 484
SS-N-8 submarine launched ballistic missile (USSR) ... 14
SS-N-9 (Siren) shipborne and submarine launched surface-to-surface missile (USSR) ... 457
SS-N-12 (Sandbox) anti-ship missile (USSR) ... 457
SS-N-14 anti-submarine weapon (Silex) (USSR) ... 526
SS-N-15 (Starfish) anti-submarine weapon (USSR) ... 526
SS-N-16 (Stallion) anti-submarine weapon (USSR) ... 526
SS-N-17 (Snipe) submarine launched ballistic missile (USSR) ... 15
SS-N-18 (Stingray) submarine launched ballistic missile (USSR) ... 15
SS-N-19 (Shipwreck) anti-ship missile (USSR) ... 458
SS-N-20 (Sturgeon) submarine launched ballistic missile (USSR) ... 15
SS-N-22 (Sunburn) anti-ship missile (USSR) ... 458
SS-N-23 (Skiff) SLBM ... 16
SS-NX-21 cruise missile (USSR) ... 16
SS-NX-24 cruise missile (USSR) ... 16
SSQ-801 Barra sonobuoy (Australia) ... 540
SSQ-904A LOFAR sonobuoy (UK) ... 575
SSQ-905A LOFAR sonobuoy (UK) ... 575
SSQ-906 LOFAR sonobuoy (UK) ... 575
SSQ-907 LOFAR sonobuoy (UK) ... 575
SSQ-937 DEL miniature bathythermal buoy (UK) ... 575
SSQ-937A bathythermal buoy (UK) ... 576
SSQ-947B DEL Ranger sonobuoy (UK) ... 575
SSQ 954B DIFAR sonobuoy (UK) ... 575
SSQ-963B command active multi-beam sonobuoy (X17255) (UK) ... 575
SSQ-981 Barra sonobuoy (UK) ... 576
SSQ-991 sonar transponder (UK) ... 576
(X17255) (UK) ... 575
SST 4 special surface target torpedo (Germany, FR) ... 518
ST 802 naval tracking radar (UK) ... 632
ST850 Tigercat radar (UK) ... 272
ST1802 series fire control radars (UK) ... 633
Stabilised thermal imaging system (STIS) (USA) ... 893
Standard (RIM-66A and RIM-67A) surface-to-air missile (USA) ... 480
Startstreak high velocity missile system (UK) ... 216
Stentor battlefield radar (France) ... 234
Sterne 1 minesweeping system (France) ... 516
Stinger portable anti-aircraft missile (FIM-92A) (USA) ... 217
Sting radar fire control system (Netherlands) ... 612
Sting Ray torpedo (UK) ... 529
Stinger (AIM-92) air-to-air missile (USA) ... 705
Stockade-Seaflash decoy system (UK) ... 662
Stonefish mines (UK) ... 531
Storm rocket launcher system (France) ... 487
Straight Flush fire control radar (USSR) ... 258
STRAP programme (USA) ... 599
Strategic missiles tables ... 906
STRIDA air defence system (France) ... 41
STRIL air defence system (Sweden) ... 49
Strix anti-tank mortar bomb (Sweden) ... 146
Strut Curve search radar (USSR) ... 624
SUAWACS airborne AEW radar (USSR) ... 785
Subfar 100 submarine radar (Sweden) ... 616
Submarine integrated attack and surveillance sonar system (SIASS) (Netherlands) ... 559
Submarine integrated combat system (SICS) (Italy) .. 411, 417
Submarine periscope image intensifier (UK) ... 686
SUBROC (UUM-44A) anti-submarine missile (USA) ... 533
Successor submarine command system (UK) ... 435
Sun Visor fire control radar (USSR) ... 624
Super 530 air-to-air missile (France) ... 694
Super Corsare infrared scanner (France) ... 872
Super Cyclope infrared scanner (France) ... 872
Super-Fledermaus (Switzerland) ... 94
Super-Fledermaus update (Denmark) ... 74
Super Ikara anti-submarine weapon system (Australia) ... 512
Super Marec airborne radar (UK) ... 790
Super Searcher airborne radar (UK) ... 791
Support defence missile system (UK) ... 477
Surface and underwater target (SUT) torpedo (Germany, FR) ... 517
Surveillance thermal imager (UK) ... 393
SUVOA artillery fire control system (Yugoslavia) ... 407
SUVOA fire control system (Yugoslavia) ... 111
Sweden
air-to-surface missiles ... 715
airborne radar ... 783
aircraft armament ... 748
anti-tank/assault weapons ... 146
army fire control systems ... 88
coastal defence weapons ... 162
electronic warfare equipment (air) ... 829
electronic warfare equipment (sea) ... 660
electro-optical equipment (air) ... 876
electro-optical equipment (land) ... 376
electro-optical equipment (sea) ... 685
ground radar ... 250
land-based air defence systems ... 49
mobile gun and rocket systems ... 200
mobile surface-to-air guided missile systems ... 175
naval fire control systems ... 427
naval guns and bombardment rockets ... 495
naval radar ... 615
portable anti-aircraft guided missiles ... 213
shipborne surface-to-air weapons ... 471
shipborne weapons ... 454
underwater warfare systems ... 524
Swingfire anti-tank missile (UK) ... 151
Switzerland
aircraft armament ... 749
army fire control systems ... 93
battlefield support weapons ... 125
land-based air defence systems ... 50
mobile gun and rocket systems ... 202
naval guns and bombardment rockets ... 497
naval radar ... 617
SWORD decoy launcher (UK) ... 663
SWS 35 miniature towed array (UK) ... 577
Syrel ELINT pod (TMV 018) (France) ... 817
System 81 WDNS (Israel) ... 873
System 82 WDNS (Israel) ... 874

T

T178/278 goniometer (Italy) ... 370
TA20/RA20S twin 20mm AA weapon system (France) ... 192
T.ACCS (France) ... 40
TACFIRE tactical fire direction system (AN/GSG-10) (USA) ... 107
Tactical Air Defence System–TADS (USA) ... 59
Tactical air-to-surface missiles (USSR) ... 718
Tactical communications jammer system ... 342
Tactical communications surveillance system (USA) ... 342
Tactical Computer Terminal (TCT) AN/UYQ-30 (USA) ... 108
Tactical data handling projects (USA) ... 110
Tactical Electronic Reconnaissance (TEREC) Remote Terminal (TRT) (USA) ... 332
Tactical land-based surface-to-surface missiles tables ... 907
Tactical missiles (China) ... 117
Tactical missile system (USA) ... 156
Tactical shipborne surface-to-surface missiles tables ... 910
Tactical surface-to-surface missile (India) ... 121
TAD LRM fire control system (France) ... 70
Tadiran COMINT systems (Israel) ... 317
TADS/PNVS (USA) ... 892
Taiwan
air-to-air missiles ... 698
anti-tank/assault weapons ... 148
battlefield support weapons ... 126
coastal defence weapons ... 163
ground radar ... 252
land-based air defence systems ... 37
mobile surface-to-air guided missile systems ... 176
shipborne weapons ... 455
Tall King early warning radar (USSR) ... 260
TAL cluster bombs (Israel) ... 743
Tamam PADS (Israel) ... 363
Tango thermal imager (France) ... 869
Tank thermal sight (UK) ... 393
Tan-sam surface-to-air missile (Type 81) (Japan) ... 174
TAOC tactical air operations central (USA) ... 60
TAON TMY83 laser rangefinder/illuminator (France) ... 681
Target Acquisition Aid System (France) ... 358
Target designation sight (TDS) (UK) ... 685
Tartar surface-to-air missile (RIM-24) (USA) ... 481
TAS Tracking Adjunct System (Hawk) (USA) ... 405
TAV-38 laser rangefinder (France) ... 871
TAVITAC tactical data handling systems ... 409
TBA 100mm aircraft rockets (France) ... 732
TBA 100mm rocket launchers (France) ... 733
TCI0 thermal imager (Norway) ... 375, 684
TC-5100 series DF systems (USA) ... 344
TCM 30 naval air defence system (Israel) ... 469
TCM Mk 3 light AA system (Israel) ... 196
TCV 107 tank laser rangefinder (France) ... 350
TCV 115 airborne laser rangefinder (France) ... 867
TCV 186 laser rangefinder (France) ... 350
TCY 133 laser rangefinder/illuminator (France) ... 351
TCY 167 miniature laser rangefinder (France) ... 352
TD22 thermal imaging system (Israel) ... 362
Teledyne IFF interrogators/transponders ... 810
Terma radar and ballistic instrumentation (Denmark) ... 225
Terma scanter MIL navigation and surveillance radars (Denmark) ... 600
Terminal imaging radar system (USA) ... 304
Terne III surface to-sub-surface missile (Norway) ... 523
Terrier surface-to-air missile (RIM-2) (USA) ... 482
Teruel 140mm multiple rocket system (Spain) ... 125
TG-1 hf sonar system (UK) ... 577
Thailand
land-based air defence systems ... 37
Thermal cueing unit (UK) ... 879
Thermal imager vehicle sight (UK) ... 387
Thermal observation gunnery sight (TOGS) (UK) ... 382
THETIS submarine ESM system (Italy) ... 657
Thin Skin heightfinder radar (USSR) ... 260
Thomsea naval combat systems (France) ... 409
Thomson-CSF active expendable jammers (France) ... 818
Thomson-CSF ATLIS laser target designation pod (France) ... 869
Thomson-CSF canasta day/night aiming equipment (DIVT 13) (France) ... 357
Thomson-CSF helmet mounted sights (France) ... 872
Thomson-CSF laser warning receiver (France) ... 818
Thomson-CSF Series 121-196 attack sights (France) ... 870
Thomson-CSF SIGINT aircraft receivers (France) ... 819
Thomson-CSF Type 902 gunsight (France) ... 871
Thomson-CSF VE 110 head-up display (France) ... 869
Thomson-CSF Series VE 120/130 head-up displays (France) ... 870
Thomson-CSF VEH 3020 holographic head-up displays (France) ... 871
Thomson Sintra Sea Mines (France) ... 516
Thorn-EMI hand-held thermal imager (UK) ... 393
Thorn-EMI multi-role thermal imager (UK) ... 391
Thorn-EMI naval thermal imagers (UK) ... 689
Thorn-EMI ship's weapon data distribution (UK) ... 435
Thorn-EMI torpedo fire control equipment (UK) ... 435
Thorn-EMI underwater acoustic ranges, systems and targets (UK) ... 578
Threat adaptive countermeasures dispenser system (TADS) ... 840
TIALD designator pod (UK) ... 885
TICM II thermal imaging systems (UK) ... 386
Tien Kung radar (Taiwan) ... 252
Tigercat close-range air defence system (UK) ... 186
Tiger G (TRS 2105/TRS 2106) low altitude radars (France) 236
Tiger S (TRS 2100 series) radar (France) ... 235
TILAS tank laser sight (Netherlands) ... 371
Timnex-4CH automatic Elint/ESM system (Israel) ... 316
TIMS (USA) ... 404
TJN2 71 mixed day/night sight (France) ... 352
TK-640 laser rangefinder (USA) ... 396
TL10-T articulated telescope laser rangefinder (UK) ... 381
TLVS medium-range tactical air defence system ... 44
TM 17 (TM 12) artillery sight/laser rangefinder (France) ... 350
TM 18 (TPV 89) laser rangefinder (France) ... 351
TMLD low-level reporting and control system (Germany, FR) ... 43
TMLD series tank laser rangefinders (Yugoslavia) ... 406
TMV 026 airborne ESM system (France) ... 818
TMV 200 Elisa Elint receiver (France) ... 306
TMV 202 (DR 4000A) ESM suite (France) ... 819
TMV 630 laser rangefinder (France) ... 871
TMV 980A electronic multidisplay system (France) ... 870
TMY 113 laser rangefinder/illuminator (France) ... 681
TMY 156 laser rangefinder (France) ... 681
Token early warning and GCI radar (USSR) ... 260
Tomahawk CCD TV sensor (USA) ... 887
Tomahawk sea-launched cruise missile (BGM-109) (USA) 459
Top Dome radar group (USSR) ... 624
Top Mesh radar (USSR) ... 625
Top Pair radar group (USSR) ... 625
Top Plate radar (USSR) ... 625
Top Sail three-dimensional radar (USSR) ... 625
Top Steer radar group (USSR) ... 626
Top Trough surveillance radar (USSR) ... 626
Tornado head-up display (UK) ... 881
Tornado nose radar (USA) ... 797
Torpedoes tables ... 945

Tote missile sight (UK) 383
Totem naval fire control equipment (France) 680
TOW air-to-surface application (USA) 725
TOW anti-tank missile (BGM-71) (USA) 155
Towed arrays (UK) 577
TP1MC night periscope (Netherlands) 373
TPN-720 mobile control and landing system (Italy) 243
Trackstar low altitude air defence system (USA) 288
Trail blazer ROV (International) 519
Transponder reply receiver RRB (UK) 635
Transportable Watchman defence radar (UK) 267
Trap Door naval radar (USSR) 626
TRC series communications intelligence equipment (France) 306
Trident C-4 (UGM-93A) strategic weapon system (USA) 28
Trident fire control system Mk 98 (USA) 444
Trigat programme (International) 142
Trigat sighting system (International) 143
Trinity mounting (Sweden) 497
Triton II surveillance radar TRS 3030/TRS 3035 (France) 603
Triton G doppler surveillance radar TRS 3050 (France) 603
Triton S surveillance radar TRS 3033 (France) 603
Triton threat warning system (USA) 675
TRM/TRS C-band radar series (Germany, FR) 238
Troika MCM system (Germany, FR) 518
Tround international 12.7mm aircraft machine gun (USA) 759
TRS 22XX long-range 3-D radar (France) 229
TRS 2052 D-band early warning radar (France) 227
TRS 2053 D-band early warning radar (France) 227
TRS 2054 long-range surveillance radar family (France) 228
TRS 2056 D-band early warning radar (France) 227
TRS 2201 three-dimensional radar (France) 235
TRS 2206 Volex IV 3-D surveillance radar (France) 235
TRS 2215 3-D radar (France) 228
TRS 2230 3-D radars (France) 228
TRS 2300 SPARTIATE recovery radar (France) 231
TRS 2620 radars (Gerfaut) (France) 231
TRS 2630 Griffin shorads radar (France) 232
TRS 3405 coastal surface surveillance radar (France) 231
TRS 3410 mobile coastal surface surveillance radar (France) 232
TS-M thermal sight (Yugoslavia) 407
TSM 2021B minehunting sonar (France) 554
TSM 2022 minehunting sonar (France) 554
TSM 2314 Diodon sonar (France) 551
TSM 2630 Diodon Spherion sonar (France) 552
TSM 2640 Salmon sonar (France) 552
TSM 2820 Tarpon sonar (France) 552
TSM 2930 Lamproie linear towed array (France) 553
TSM 8020 sonobuoy (France) 555
TSM 8200-8202 airborne Sadang acoustic system (France) 546
TSM 8220 (Lamparo) lightweight airborne sonobuoy processing equipment (France) 547
TSM 8221 (Lamparo) lightweight airborne sonobuoy processing equipment (France) 547
TSM 8222 (Lamparo) lightweight airborne sonobuoy processing equipment (France) 547
Two Spot PAR (USSR) 260
Type 61 twin 25mm AA gun mounting (China) 486
Type 105 air-to-surface laser ranger (UK) 884
Type 106 laser target seeker (UK) 884
Type 116 laser rangefinder (UK) 884
Type 117 laser designator (UK) 885
Type 118 laser designator/ranger (UK) 886
Type 122 helicopter laser rangefinder (UK) 886
Type 162M sideways-looking sonar (UK) 565
Type 184M sonar (UK) 566
Type 195 helicopter sonar (UK) 565
Type 221 thermal surveillance system (UK) 886
Type 282 ranging radar (UK) 273
Type 306 laser designator/ranger (UK) 384
Type 311-A fire control radar (China) 224
Type 311 CO_2 laser devices (UK) 887
Type 312 laser beacon and target simulator (UK) 384
Type 318 CO_2 laser devices (UK) 887
Type 375 anti-submarine weapon system (Sweden) 524
Type 378 battlefield radar (China) 220
Type 386 ground surveillance radar (USA) 285
Type 404A coastal surveillance radar (China) 222
Type 405J mobile ECM equipment (UK) 320
Type 408-C long range air warning system (China) 221
Type 518 laser system (UK) 384
Type 520 laser rangefinder (UK) 384
Type 531 laser system (UK) 384
Type 571 surveillance radar (China) 221
Type 581 air warning radar (China) 221
Type 605 sonobuoy receiver (UK) 574
Type 771 low light TV camera (Denmark) 677
Type 772 sighting system (Denmark) 349
Type 791-A precision approach radar (China) 224
Type 864 airborne sonobuoy telemetry system (UK) 574
Type 865 airborne sonobuoy telemetry system (UK) 574
Type 866 airborne sonobuoy telemetry system (UK) 574
Type 1006 radar (UK) 632
Type 1007 RN radar (UK) 632
Type 2016 fleet escort sonar (UK) 566
Type 2020 submarine sonar (UK) 567
Type 2026 sonar (UK) 568
Type 2031Z towed array sonar (UK) 568
Type 2034 sonar (UK) 569
Type 2046 submarine sonar (UK) 569
Type 2050 sonar (UK) 570
Type 2051 submarine sonar (UK) 571
Type 2052 submarine sonar (UK) 570
Type 2057 towed array sonar (UK) 571
Type 2054 submarine sonar (UK) 571
Type 2058 towed sonar source (UK) 578
Type 2069 helicopter sonar (UK) 565
Type 2074 sonar system (UK) 571
Type 2075 sonar (UK) 571
Type 20311 sonar (UK) 568
Type A naval mounting for 20mm F2 gun (France) 488
Type A41A, 20mm naval gun mounting (Switzerland) 499
Type BF radar warning receiver (France) 816
Type GAM-BO1, 20mm naval gun (Switzerland) 499
Type GAM-CO1, 20mm naval gun (Switzerland) 499
Type GBM-AO1, 25mm naval gun (Switzerland) 498
Type GBM-BIZ, 25mm mount Sea Zenith (Switzerland) 498
Type GCM-A, 30mm twin naval AA gun (Switzerland) 497
Type GDM-A, 35mm twin AA gun (Switzerland) 497
Type MS7TS marine night sight (Netherlands) 684
Type S373 ECM equipment (UK) 326
Type SP 2110 sonar (UK) 572
Type TP1MS night periscope (Netherlands) 373
Type TP 43XO torpedo (Sweden) 525
Type TP 617 torpedo (Sweden) 525
Type TS7TS passive aiming sight (Netherlands) 370
Type UA 9053 thermal imaging camera (Netherlands) 684

U

UA 9092 tank thermal imager (Netherlands) 372
UA 9124/9126 day/night periscope system (Netherlands) 372
UA 9127 thermal imaging/daylight aiming and observation system (Netherlands) 373
UA 9630 series universal passive periscope (Netherlands) 372
UAF-1 ESM system (UK) 669
UAL 11636 light anti-aircraft laser rangefinder (Sweden) 378
UAP 13 Series airborne attack radars (Sweden) 783
UAR 1021 combined search and track pulse-doppler radar (Sweden) 251
Underwater combat systems (UCS) 257/280 (Canada) 408
Underwater ordnance locator Mk 24 Mod 0 (USA) 599

Union of Soviet Socialist Republics
air-to-air missiles 699
air-to-surface missiles 716
airborne radar 784
aircraft armament 752
anti-tank/assault weapons 148
army fire control systems 95
battlefield support weapons 127
coastal defence weapons 163
electronic warfare equipment (air) 832
electronic warfare equipment (land) 319
electronic warfare equipment (sea) 661
electro-optical equipment (land) 379
ground radar 252
land-based air defence systems 51
mobile gun and rocket systems 205
mobile surface-to-air guided missile systems 176
naval guns and bombardment rockets 499
naval radar 617
portable anti-aircraft guided missiles 214
shipborne surface-to-air weapons 472
shipborne weapons 455
sonar and underwater detection equipment 561
strategic weapon systems 6
underwater warfare systems 526

United Arab Emirates
land-based air defence systems 37

United Kingdom
air-to-air missiles 701
air-to-surface missiles 719
airborne radar 786
aircraft armament 753
anti-tank/assault weapons 151
army fire control systems 96
coastal defence weapons 165
electronic warfare equipment (air) 832
electronic warfare equipment (land) 320
electronic warfare equipment (sea) 662
electro-optical equipment (air) 878
electro-optical equipment (land) 380
electro-optical equipment (sea) 685
ground radar 262
land-based air defence systems 54
mobile gun and rocket systems 205
mobile surface-to-air guided missile systems 183
naval fire control systems 431
naval guns and bombardment rockets 503
naval radar 627
portable anti-aircraft guided missiles 214
shipborne surface-to-air weapons 476
shipborne weapons 458
sonar and underwater detection equipment 562
strategic weapon systems 20
underwater warfare systems 528

United States of America
air-to-air missiles 702
air-to-surface missiles 721
airborne radar 793
aircraft armament 756
anti-tank/assault weapons 153
battlefield support weapons 130
electronic warfare equipment (air) 837
electronic warfare equipment (land) 329
electronic warfare equipment (sea) 670
electro-optical equipment (air) 887
electro-optical equipment (land) 395
electro-optical equipment (sea) 689
ground radar 276
land-based air defence systems 56
mobile gun and rocket systems 206
mobile surface-to-air guided missile systems 186
naval fire control systems 441
naval guns and bombardment rockets 505
naval radar 635
portable anti-aircraft guided missiles 216
shipborne surface-to-air weapons 478
shipborne weapons 459
sonar and underwater detection equipment 579
strategic weapon systems 22
underwater warfare systems 532

Universal turret for helicopter gunships (USA) 765
Updating kit for Bofors 40mm L/60 gun system (Sweden) 201

V

V3G helicopter day/night sights (France) 868
V 200 vision block (Italy) 368
V3800 thermal imaging sensor (UK) 687
Valkiri 127mm artillery rocket system (South Africa) 124
VAMPIR surveillance equipment (France) 678
Vanguard 3-D radar (Netherlands) 248
Vanth optronic fire control system (Italy) 367
VAP 80mm multiple rocket launcher (Egypt) 118
VAQ-3 tank laser rangefinder (Italy) 366
VAQ-33 tank laser rangefinder (Italy) 366
Varan airborne radar (France) 773
Vassir 85 anti-aircraft optronic sight (France) 358
Vehicle-mounted RBS 70 anti-aircraft missile (Sweden) 175

Venezuela
land-based air defence systems 37

Vertical ballistic weapon/VBW airborne anti-tank weapon (Germany, FR) 742
VH 100 Series helicopter head-up displays (France) 871
Vicon 77 ESM pod (UK) 837
Vicon 78 airborne decoy dispenser (UK) 837
Vigilant anti-tank missile (UK) 152
Vigilant tactical surveillance radar (USA) 296
VIMOS vibration and noise monitoring system (UK) 573
Vinten tracking mount and tripods (UK) 395
Vinten Vicon 90 Series 101 FLIR pod (UK) 887
Vipere infra-red surveillance equipment (France) 357
Virtual Image Display (USA) 890
VISAA anti-aircraft sight (France) 355
Vison optical director (France) 680
Viviane/Venus HOT/helicopter night sight (France) 868
VM Series AFV equipment (UK) 393
VML23 vehicle laser rangefinder (UK) 381
VS 580 gyro-stabilised vehicle sight (France) 355
VSEL combat system data bus (UK) 434
VSTAR/VSTAR-PT air defence radar (USA) 301
VTG 120 light thermal imager (Italy) 366
VTS-800 video tracking equipment (Sweden) 378
Vulcan air defence system (USA) 208
Vulcan gun pods SUU 16/A (USA) 762
Vulcan gun pods SUU 23/A (USA) 762

W

W-120 shipboard fire control radar (USA) 642
W-160 shipboard fire control radar (USA) 642
W-1200 series fire control systems (USA) 447
W-2000 3-D surveillance radar (USA) 286
WAAM programme (USA) 758
Wallop infra-red decoy flares (UK) 837
Watchdog coastal defence systems (Netherlands) 48
Watchman air surveillance radar (UK) 267
Watchman coastal defence radar (UK) 268
Wavell ADP system (UK) 96
WDNS-391 weapon delivery systems (Israel) 873
Weasel ESM and ELINT collection system (UK) 328
Westinghouse balloon-borne radar (USA) 813
Westinghouse WX series fire control radars (USA) 812
Whiff fire control radar (USSR) 261
Wildcat twin 30mm AA weapon system (Germany, FR) 193
Winnin decoy system (Australia) 646
WJ-927 ESM equipment (USA) 333
WJ-945 receiving/DF system (USA) 333
WJ-1205B video digitiser unit (VDU) (USA) 333
WJ-1240 receiving system (USA) 334
WJ-1424 microwave receiver (TLR-31) (USA) 335
WJ-1440 microwave receiver system (USA) 335
WJ-1740 receiving system (USA) 865
WJ-1840 receiving system (USA) 335
WJ-1920 multiparameter distributed processing system (USA) 333
WJ-1988 amplitude monopulse DF processor (USA) 334
WJ-1996 ESM receiver (USA) 336
WJ-3600 EW alert and collection system (USA) 336
WJ-4810 communictions jamming system (USA) 336
WJ-8976 direction finding system (USA) 336
WJ-8990 manpack tactical intelligence system (MANTIS) (USA) 336
WM20 series fire control radar (Netherlands) 614
Working Bee-6 117mm artillery rocket system (Taiwan) 126
WSA-420 series fire control and action information systems (UK) 431

X

XAAM-3 air-to-air missile (Japan) 697
XJ/FPS-3 air defence radar (Japan) 248
XM27E1 7.62mm armament sub-system helicopter (USA) 768
XM28E1 armament sub-system (USA) 766
XM30 aircraft armament sub-system (USA) 766
XM35 armament sub-system (USA) 764
XM59 armament sub-system (USA) 770
XM76 anti-oscillation sighting system (USA) 887
XM94 armament sub-system (USA) 770
XM188E1 30mm three-barrel aircraft gun (USA) 765

Y

YMRL 32 128mm multiple rocket launcher M77 (Yugoslavia) 135
Yo-Yo missile control radar (USSR) 261

Yugoslavia
aircraft armament 770
army fire control systems 111
battlefield support weapons 135
electronic warfare equipment (air) 866
electronic warfare equipment (land) 347
electro-optical equipment (land) 406
land-based air defence systems 37

Yuval long range surveillance system (Israel) 363

Z

Z16 submarine-launched torpedo (France) 516
ZB 298 short range ground surveillance radar (UK) 272
Zeus airborne ECM system (UK) 833
ZHUD sight (UK) 882
ZSU-23-4 23mm SP AA vehicle (USSR) 205
Zulu millimetric surveillance radar (UK) 271
ZW series naval radars (Netherlands) 613

Printed and bound in Great Britain by Butler & Tanner Ltd, Frome and London